JANE'S
ALL THE WORLD'S AIRCRAFT
1981-82

aerospatiale

this is who we are:

one of the world's largest aerospace manufacturers

Not the first in the world, but already the first in Europe in terms of turnover (above 13 billion french francs in 1980) and diversity of products. We were incorporated in 1970 under the name of Société Nationale Industrielle Aérospatiale, as a result of the merger of Nord-Aviation, Sud-Aviation and Sereb. Some have been inclined to shorten our name, retaining only the initials S.N.I.A.S. ; with the same object in mind we prefer to be called commonly **aerospatiale** It is our signature.

top manpower

Men whose imagination and creativity have made our achievements possible. Close to 40,000 employees, specialists, engineers, chosen for their skill in particular fields.

ways and means

The most modern research, development and production facilities, where data processing plays a leading part:
Computer Aided Design (CAD), Computer Aided Manufacturing (CAM).
Behind these means, a policy: to come up with the best product at minimum cost.

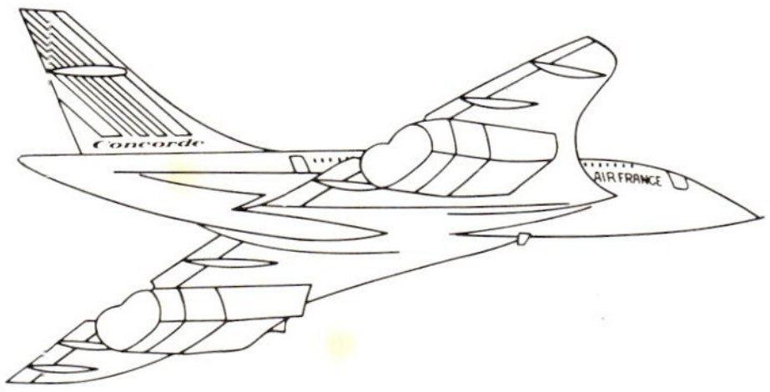

products

The most diversified range of aerospace products: airplanes, helicopters, tactical missiles, space and ballistic systems, combining top performance and low cost of operation in order to satisfy market requirements.

We are the largest manufacturer of helicopters and tactical missiles in Europe, and the French Nuclear Defence Forces are equipped with our strategic ballistic missiles.

Finally, in the framework of multinational industrial programs, we are responsible for the success of a number of famous products that are not signed only with our name: AIRBUS A 300/A 310 - ARIANE - CONCORDE - HOT - MILAN - ROLAND - INTELSAT V - METEOSAT...

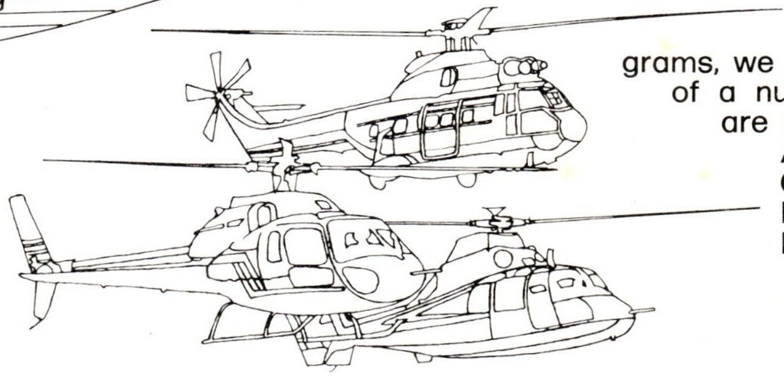

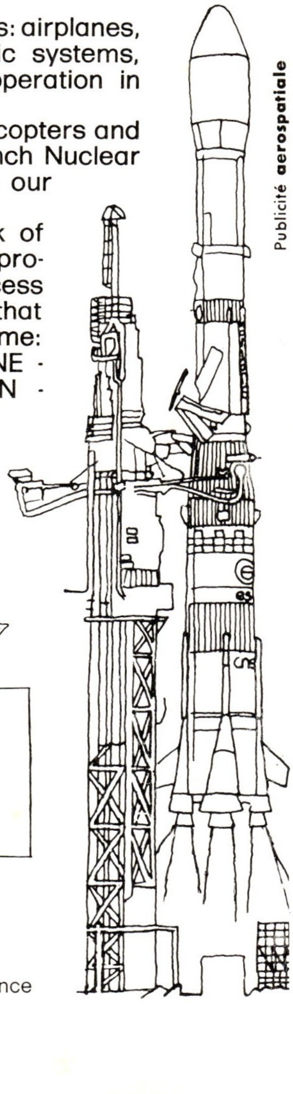

exports

We are exporting 80 % of our helicopter production, 70 % of our missiles, 90 % of our airplanes, as a mark of quality of our products - tailored to market requirements, and of our competitive sales policy.

Société Nationale Industrielle
aerospatiale
37, bd de Montmorency - 75781 Paris Cedex 16 - France

Alphabetical list of advertisers

A

Aeritalia SpA
Piazzale Tecchio 51, 80125 Naples, Italy .... [19]

Aermacchi SpA
Via Sanvito 80, 21100 Varese, Italy .... [66]

Aérospatiale
37 boulevard de Montmorency, 75781 Paris
Cedex 16, France .... [3]

Agusta
Via Caldera 21, 20100 Milan, Italy .... [56] & [57]

Airbus Industrie
avenue Lucien Servanty, BP 33,
31700 Blagnac, France .... [13]

Airline Publications
Noble Corner, Great West Road, Hounslow,
Middlesex TW5 0PA, England .... [49]

Alfa Romeo Avio SpA
Romigliano D'Arco, 80030 Naples, Italy .... [46]

Avco Lycoming Inc
550 South Main Street, Stratford,
Connecticut 06497, USA .... *front endpaper iv*

Aviaexport
32-34 Smolenskaja-Sennaja Square,
121200 Moscow, USSR ...... *facing inside back cover*

B

Barr & Stroud Ltd
Melrose House, 4-6 Savile Row,
London W1X 1AF, England .. *facing inside front cover*

Bell Helicopter Textron
PO Box 482, Fort Worth, Texas 76101, USA ...... [72]

Boeing Commercial Airplane Company
PO Box 3707, Seattle, Washington 98124, USA ... [70]

Bofors AB
Ordnance Division, Box 500,
S-691 80 Bofors, Sweden .... [21]

British Aerospace Public Limited Company
Brooklands Road, Waybridge, Surrey KT13 0SJ,
England .... [60]

C

Canadair Ltd
PO Box 6087, Station 'A', Montreal, Quebec H3C 3G9,
Canada .... [38]

Central Engineering International Company
2930 Anthony Lane, Minneapolis,
Minnesota 55418, USA .... [34]

John Curran Ltd
PO Box 72, Curran Road, Cardiff CF1 1TE,
Wales .... [5]

D

Difesa e Spazio SpA
Via Sicilia 162, 00187 Rome, Italy .... [51]

DMS
DMS Building, 100 Northfield Street,
Greenwich, Connecticut 06830, USA .... [44]

&

DMS (European Office)
31 Station Road, Henley-on-Thames,
Oxon RG9 1AT, England .... [44]

Dornier GmbH
PO Box 2160, D-8000 Munich 66,
Federal Republic of Germany .... [8] & [9]

E

Electronique Aérospatiale
BP 51, 93350 Le Bourget, France .... [50]

Elettronica SpA
Via Tiburtina Km 13.700, 00100 Rome,
Italy .... *inside front cover*

Elmer
Viale dell'Industria 4, 00040 Pomezia, Italy .... [27]

Eltro GmbH
Postfach 10 21 20, D-6900 Heidelberg 1,
Federal Republic of Germany .... [29]

Embraer International
12200 São José dos Campos, São Paulo, Brazil .. [52]

F

FIAR SpA
Via Montefeltro 8, 20156 Milan, Italy .... [28]

JOHN CURRAN

Aero·engine test facilities for civil and military aviation

J Mk 8 FACILITY
with TRISTAR RB 211

J Mk 9 FACILITY with TORNADO RB199

* Workshop Pre-Rigging
* Instant Clamping and Coupling
* Rapid Throughput of Engines
* Full All Round Access
* Minimum Maintenance
* Adaptable to other Engines within the Capacity Range

J Mk 6 FACILITY with HAWK-ADOUR 151

Mobile Test Facility with Industrial AVON

Engine Run-Up FACILITY with PHANTHOM-SPEY 201

JOHN CURRAN LTD ~ PO Box 72 ~ Curran Road ~ CARDIFF ~ Wales ~ UK ~ CF1 1TE   Tel: 0222 - 20641

FIAT Aviazione SpA
Casella Postale 1389, Via Nizza 312,
10127 Turin, Italy .... [33]

Flight Refuelling Ltd
Wimborne, Dorset BH21 2BJ, England ..... *bookmark*

Fokker BV
PO Box 7600, 1117 ZJ Schipol,
The Netherlands .... [40] & [41]

G

Garrett Corporation
PO Box 92248, Los Angeles,
California 90009, USA .... [74]

GIAT
10 place Georges-Clemenceau, 92211 Saint-Cloud,
France .... [53]

Goodyear Tyre & Rubber Company
Aviation Products Division (GB) Ltd,
Viscount Way, London (Heathrow Airport),
Hounslow, Middlesex, England .... [17]

Grumman Aerospace Corporation
Bethpage, Long Island, New York 11714, USA .... [47]

I

Italtel SpA
12 Piazza Zavattari, 20149 Milan, Italy .... [39]

L

Lucas Aerospace Ltd
Shirley, Solihull, West Midlands B90 2JJ,
England .... [62]

M

McDonnell Douglas Corporation
Box 14526, St. Louis, Missouri 63178,
USA .... [2]

Messerschmitt-Bölkow-Blohm GmbH
Postfach 80 11 09, D-8000 Munich,
Federal Republic of Germany .... [15]

Meteor SpA
Via Nomentana 146, 00162 Rome, Italy .... [34]

ML Aviation Company Ltd
White Waltham Aerodrome, Maidenhead,
Berkshire SL6 3JG, England .... [44]

Motoren-und Turbinen-Union München GmbH
PO Box 50 06 40, D-8000 Munich 50,
Federal Republic of Germany .... [23]

O

Oerlikon-Bührle Ltd
Machine Tool Works, Birchstrasse 155,
CH-8050 Zurich, Switzerland .... [25]

Omnipol
Foreign Trade Corporation, Nekázanka 11,
112 21 Praha 1, Czechoslovakia [68]

P

Panavia Aircraft GmbH
Arabellastrasse 16, D-8000 Munich 86,
Federal Republic of Germany .... [7]

PRB SA
Division Recticel, Damstraat 2,
B-9200 Wetteren, Belgium .... [32]

Précision Mecanique Labinal
5 avenue Newton, Parc d'Activités,
78390 Bois d'Arcy, France .... [32]

R

Racal Acoustics Ltd
Beresford Avenue, Wembley, Middlesex HA0 1RU,
England .... [64]

Rinaldo Piaggio SpA
Via Cibrario 4, 16154 Genoa, Italy .... [31]

S

Saab-Scania AB
Aerospace Division, S-581 88 Linköping,
Sweden .... [11]

SAFT
Storage Battery Division, 156 avenue de Metz,
93230 Romainville, France .... [30]

SAMM
224 quai de Stalingrad, 92130 Issy-les-Moulineaux,
France .... [42]

Sangamo Weston Inc
Data Systems Division, PO Box 3041,
Sarasota, Florida 33578, USA .... [42]

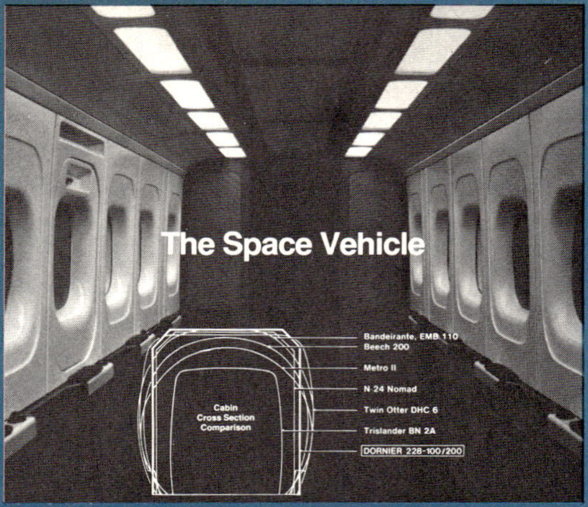

Dornier 228, the world's first production aircraft, equipped with DORNIER's low-drag/high lift New Technology Wing.

Dornier 228, setting standards of performance and economy, previously unknown to this class of aircraft.

Dornier 228 – the new fuel-saving technology, using 30 % less in comparison with competing aircraft.

Dornier 228 – the „space vehicle", offering the rectangular widebody cabin, free of any interfering structure over its full length.

Dornier 228 – the „hot and high" specialist, with outstanding STOL characteristics and exceptional safe-climb performance.

Dornier 228 – the optimum in commuter comfort, appealing to all passengers thanks to its first-class cabin.

Dornier 228 – the long-range specialist on a 1970 km (1065 nm) utility non stop run with maximum payload or on several commuter hops without refueling.

Dornier 228 – the only flexible and versatile answer to the specific and demanding requirements of utility and commuter operators in the 1980s.

Dornier 228. A technologically more advanced aircraft has yet to be invented.

Dornier 228. The Utility and Commuter Specialists.

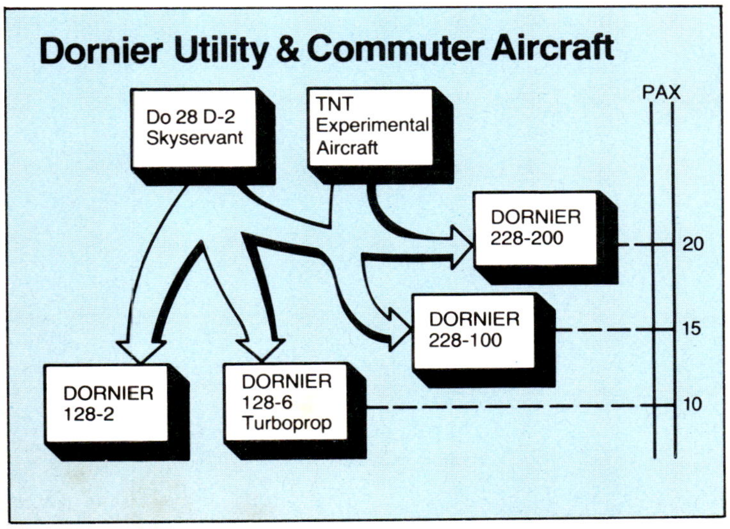

Technology Made in Germany

DORNIER

For further information, please contact
Dornier Aircraft Sales, Dornier GmbH,
D-8000 Munich 66, P. O. Box 2160
Federal Republic of Germany
Telephone (8153) 190, Telex 526450

Selenia SpA
Radar & Missile Systems Division,
Via Tiburtina Km 12, 400, 00131 Rome, Italy ...... [54]

SFENA
Aerodrome de Villacoublay, BP 59,
78140 Velizy Villacoublay Cedex, France .... [48]

Siai Marchetti SpA
Via Indipendenza 1, 21018 Sesto Calende-Varese,
Italy .... [43]

SNECMA
2 boulevard Victor, 75724 Paris Cedex 15,
France .... [45]

Standard Wire & Cable Company
2345-J Alaska Avenue, El Segundo,
California 90245, USA .... [37]

T

Thomson CSF/AVS
178 boulevard Gabriel Peri,
92240 Malakoff, France .... [55]

Thomson CSF/DSE
1 rue des Mathurins, BP 10,
92222 Bagneux, France .... *inside back cover*

Turbo-Union GmbH
Arabellastrasse 4, D-8000 Munich 81,
Federal Republic of Germany .... [35]

&

Turbo Union Ltd
PO Box 3, Filton, Bristol BS12 7QE, England ..... [35]

Viggen's aerodynamic configuration, a canard and delta-wing combination, is a design concept which will stay advanced for years to come.
The all-altitude pulse-doppler radar, the most advanced produced in Europe, ensures that Viggen reaches its target in any weather, independent of ground control systems.
Armament, long-range radar- and IR-homing missiles and a powerful 30 mm automatic cannon.
SAAB
Saab-Scania AB, Aerospace Division,
S-581 88 Linköping, Sweden
Europe's
most modern
all-weather fighter

Classified list of advertisers

AC motors
Aviaexport
Garrett Corporation
Lucas Aerospace
Thomson CSF

Accelerometers
Aeritalia
Aviaexport
SFENA
Thomson CSF

Accessories
Aviaexport
Garrett Corporation

Accumulators, cadmium nickel
Aviaexport
SAFT

Actuators, electric
Aviaexport
Garrett Corporation
Lucas Aerospace
Meteor
Precision Mecanique Labinel
SFENA
Thomson CSF

Aerials, aircraft
Aeritalia
British Aerospace Aircraft Group
Dornier
Elettronica
Siai Marchetti
Thomson CSF

Aerials, survey instrument
Elettronica
SFENA
Thomson CSF

Aero-auxiliary equipment
Garrett Corporation
Siai Marchetti

Aero-engine test plant
Avco Lycoming
Central Engineering
John Curran
FIAT
Siai Marchetti
SNECMA

Aero-engines
Avco Lycoming
Aviaexport
FIAT
Garrett Corporation
MTU
Omnipol
Rinaldo Piaggio
SNECMA
Turbo-Union

Aeronautical engineers and consultants
Aviaexport
Boeing Commercial Airplane Co
British Aerospace Aircraft Group
SFENA
Siai Marchetti

Aerosystems
Boeing Commercial Airplane Co
Meteor
Thomson CSF

Agricultural aircraft spray and dust systems and components
Aviaexport
Dornier
Siai Marchetti

Air compressors
Garrett Corporation

Air compressors (cabin) + (engine starting)
Garrett Corporation

Air compressors for engine starting
Lucas Aerospace
Meteor

Air-conditioning equipment
Aviaexport
Garrett Corporation
Lucas Aerospace
ML Aviation

Air-conditioning systems
Aviaexport
Garrett Corporation
Messerschmitt-Bölkow-Blohm

Air control equipment for cabins
Aviaexport
Garrett Corporation

Air cycle refrigeration packages
Aviaexport
Garrett Corporation

Air data computer systems
Aviaexport
Bofors
Garrett Corporation
SFENA

Air data computer, test set
SFENA
Thomson CSF

Air traffic control equipment
Aeritalia
Italtel
Thomson CSF

Airborne surveillance drone systems, recoverable, unmanned
Canadair
Lucas Aerospace

Aircraft, agricultural (dusters & sprayers)
Aviaexport
Dornier
Siai Marchetti

Aircraft, ambulance
Aérospatiale
Aviaexport
British Aerospace Aircraft Group
Dornier
Rinaldo Piaggio
Siai Marchetti

Aircraft, combat
British Aerospace Aircraft Group

Aircraft, commercial
Aeritalia
Aérospatiale
Airbus Industrie
Boeing Commercial Airplane Co
British Aerospace Aircraft Group
Canadair
Dornier
Fokker
Messerschmitt-Bölkow-Blohm
Rinaldo Piaggio
Saab-Scania
Siai Marchetti

Aircraft, executive
Aérospatiale
Aviaexport
Boeing Commercial Airplane Co
British Aerospace Aircraft Group
Canadair
Dornier
Fokker
Garrett Corporation
Rinaldo Piaggio

Aircraft, integrated data systems
Thomson CSF

Aircraft, military
Aeritalia
Aermacchi
Aérospatiale
Boeing Commercial Airplane Co
British Aerospace Aircraft Group
Canadair
Dornier
Fokker
GIAT
Messerschmitt-Bölkow-Blohm
Panavia
Rinaldo Piaggio
Saab-Scania
Siai Marchetti

Aircraft, naval
Aeritalia
Boeing Commercial Airplane Co
British Aerospace Aircraft Group
Fokker
Rinaldo Piaggio
Thomson CSF

Aircraft, private
Aérospatiale
Dornier
Fokker
Omnipol
Rinaldo Piaggio
Siai Marchetti

Aircraft, radio controlled
Flight Refuelling
Meteor

Aircraft, supersonic
Aeritalia
Aérospatiale
British Aerospace Aircraft Group
Panavia
Saab-Scania

Aircraft, training
Aeritalia
Aermacchi
Aérospatiale
Boeing Commercial Airplane Co
British Aerospace Aircraft Group
Canadair
Dornier
Messerschmitt-Bölkow-Blohm
Omnipol
Rinaldo Piaggio
Saab-Scania
Siai Marchetti

SAS
CRUZEIRO
garuda
PIA
KLM
TDA
TEA
LAKER
KUWAIT
VARIG
VASP
Wardair Canada
More and more tails to tell.
Reading left to right, top to bottom
South African Airways · Scandinavian Airlines System
Cruzeiro do Sul · Singapore Airlines · Garuda Indonesian Airways
Philippine Airlines · Korean Airlines · Air France
Pakistan International Airlines · Lufthansa · KLM Royal Dutch Airlines
Toa Domestic Airlines · Eastern Airlines · Air Inter · Hapag-Lloyd
Thai Airways International · Indian Airlines · Olympic Airways · Iran Air
Alitalia · Iberia · Swissair · Egyptair · Trans European Airways
Sabena · Air Afrique · Malaysian Airline System · Laker Airways
Trans Australia Airlines · Martinair · British Caledonian · Tunis Air
Austrian Airlines · Kuwait Airways Corporation · Varig Brazilian Airlines
VASP (Sao Paulo) · Saudi Arabian Airlines · Wardair Canada
Nigeria Airways · China Airlines
A310
A300
Airbus Industrie

Aircraft, transport
Aeritalia
Aérospatiale
Airbus Industrie
Boeing Commercial Airplane Co
British Aerospace Aircraft Group
Dornier
Fokker
Messerschmitt-Bölkow-Blohm
Rinaldo Piaggio

Aircraft V/STOL
Aviaexport
Boeing Commercial Airplane Co
British Aerospace Aircraft Group
Dornier
Messerschmitt-Bölkow-Blohm
Omnipol
Siai Marchetti

Aircraft canopies
Aeritalia
British Aerospace Aircraft Group
Goodyear Tyre & Rubber Co
Lucas Aerospace
Messerschmitt-Bölkow-Blohm

Aircraft development
Aeritalia
Aérospatiale
Boeing Commercial Airplane Co
British Aerospace Aircraft Group
Canadair
Fokker
Messerschmitt-Bölkow-Blohm

Aircraft escape systems
British Aerospace Aircraft Group
Garrett Corporation

Aircraft field operations & support
Boeing Commercial Airplane Co
British Aerospace Aircraft Group
Fokker

Aircraft floats
Dornier
Garrett Corporation

Aircraft freight handling equipment
Aviaexport
Dornier
Messerschmitt-Bölkow-Blohm

Aircraft integrated data systems
Eltro
Garrett Corporation
Thomson CSF

Aircraft mechanical handlers
Aviaexport
Dornier
ML Aviation

Aircraft modifications
Boeing Commercial Airplane Co
British Aerospace Aircraft Group
Flight Refuelling
Fokker
Messerschmitt-Bölkow-Blohm

Aircraft product support
Aermacchi
British Aerospace Aircraft Group
Fokker
Messerschmitt-Bölkow-Blohm
SFENA

Aircraft propellers
Aviaexport
Difesa e Spazio

Aircraft propeller governors
Rinaldo Piaggio

Aircraft wire & cable
Fokker
Standard Wire & Cable
Thomson CSF

Airfield lighting
Aviaexport
Omnipol
Standard Wire & Cable Co

Airline technical assistance
Boeing Commercial Airplane Co
British Aerospace Aircraft Group
Messerschmitt-Bölkow-Blohm
SFENA

Airport buildings and facilities, military and civil
British Aerospace Aircraft Group

Airport maintenance equipment
ML Aviation

Airspeed indicators
Aeritalia
Aviaexport
Dornier

Alternators
Aviaexport
Garrett Corporation
Lucas Aerospace
Precision Mecanique Labinal

Altimeters encoding
Aeritalia

Altitude control systems
Aviaexport
Eltro

Ammunition boosters
Difesa e Spazio
Meteor

Amplifiers
Thomson CSF

Antennas
Messerschmitt-Bölkow-Blohm
Meteor
Thomson CSF

Anti-skid systems
Aviaexport
Goodyear Tyre & Rubber Co
SNECMA
Thomson CSF

Armaments for aircraft
Aérospatiale
Bofors
Oerlikon-Bührle
Difesa e Spazio
GIAT
Messerschmitt-Bölkow-Blohm
ML Aviation

Automatic checkout systems
Aérospatiale
Aviaexport
Messerschmitt-Bölkow-Blohm
Selenia
Thomson CSF

Automatic digital data acquisition systems for engine testing
Central Engineering

Automatic parachute openers
Aviaexport

Automatic pilots
Aviaexport
Meteor
SFENA

Automatic voice alert devices
Racal Acoustics

Automatic voltage and current regulators
Aviaexport
EAS Electronique Aérospatiale
Lucas Aerospace
Precision Mecanique Labinal

Auxiliary power plant
Aviaexport
Garrett Corporation
Lucas Aerospace

Bars, stainless steel & heat resisting steel
Aviaexport

Batteries
Aviaexport
Meteor
SAFT

Batteries, aviation
Aviaexport
SAFT

Battery chargers
Aviaexport
EAS Electronique Aérospatiale
SAFT

Belts, safety
Aviaexport

Binoculars
Aeritalia

Blades, gas turbine
Avco Lycoming
Aviaexport
FIAT
SNECMA

Blades, rotor
Messerschmitt-Bölkow-Blohm

Bomb carriers
ML Aviation

Bombsights
Thomson CSF

Bonding jumpers bus bars
Standard Wire & Cable Co

Brake linings
Goodyear Tyre & Rubber Co

Brakes for aircraft
Aviaexport
Goodyear Tyre & Rubber Co
Israel Aircraft Industries
SNECMA

Cabin cooling (tropical airfield equipment)
Aviaexport
Garrett Corporation

Cabin pressure control system
Aviaexport
Garrett Corporation

BK117
KAWASAKI

Cabin pressurising test equipment
Aviaexport
Garrett Corporation

Cables, electric
Fokker
ML Aviation
Standard Wire & Cable

Cables, RF
Fokker
Standard Wire & Cable
Thomson CSF

Central air data computers
Garrett Corporation

Coatings, erosion resistant
Goodyear Tyre & Rubber Co
Lucas Aerospace

Combustion systems (gas turbine)
Lucas Aerospace

Communications control systems
Bofors
EAS Electronique Aérospatiale
Racal Acoustics
Thomson CSF

Components
Aviaexport
British Aerospace Aircraft Group
Dornier
Garrett Corporation
Thomson CSF

Computers
Dornier
Hollandse Signaalapparaten
Sangamo Weston
SFENA
Thomson CSF

Computers, aerodynamic analogue and digital
Dornier
Garrett Corporation
SFENA
Thomson CSF

Connectors/connector accessories
Aviaexport
Garrett Corporation
ML Aviation
Thomson CSF

Constant speed alternator drive units
Garrett Corporation
Lucas Aerospace

Constant speed drive test benches
Central Engineering

Control equipment for aircraft
Aviaexport
Dornier
Garrett Corporation
Thomson CSF

Controls, cockpit
Aviaexport

Controls, main engine fuel
Aviaexport
Lucas Aerospace
SNECMA
Thomson CSF

Cooling compressors
Aviaexport
Garrett Corporation

Cooling turbines
Garrett Corporation

Cryogenic turbines
Garrett Corporation

Cryogenic equipment
Messerschmitt-Bölkow-Blohm

Data processing equipment
Bofors
Central Engineering
Dornier
Garrett Corporation
Selenia
Thomson CSF

Data processing equipment for ATC
Dornier
Selenia
Thomson CSF

Data transmission equipment
Dornier
Elmer
Sangamo Weston
Thomson CSF

Dc generators
Lucas Aerospace
Precision Mecanique Labinal

Dc motors
Aviaexport
Garrett Corporation
Lucas Aerospace
Precision Mecanique Labinal
Thomson CSF

Defence contracts
British Aerospace Aircraft Group

De-icing equipment
EAS Electronique Aérospatiale
Garrett Corporation
Goodyear Tyre & Rubber Co
Lucas Aerospace
Precision Mecanique Labinal

Direction-finding equipment (triangulation)
Aviaexport
Thomson CSF

Drogue Guns
ML Aviation

Drones
Aérospatiale
Canadair
Dornier
Flight Refuelling
Meteor

Drop hammer work
British Aerospace Aircraft Group

Ejection seats
Aermacchi
SNECMA

Ejector release units
ML Aviation

Electric auxiliaries
Aviaexport
Garrett Corporation

Electric tractors
ML Aviation

Electrical equipment
Aviaexport
EAS Electronique Aérospatiale
Garrett Corporation
ML Aviation

Electrical plugs and sockets (waterproof)
Thomson CSF

Electrical wiring assemblies
Aviaexport
Fokker
ML Aviation
Standard Wire & Cable

Electrical wire cable cord of all types
Standard Wire & Cable

Electro-optical systems
Barr & Stroud
Eltro
Flight Refuelling
Saab-Scania
SFENA
Thomson CSF

Electronic equipment
Aeritalia
Aviaexport
Boeing Commercial Airplane Co
EAS Electronique Aérospatiale
Elettronica
Eltro
Fokker
Garrett Corporation
ML Aviation
Saab-Scania
SFENA
SNECMA
Thomson CSF

Electronic fuel control systems
Fokker
Lucas Aerospace
SNECMA
Thomson CSF

Electronics & guidance
Israel Aircraft Industries
SFENA
Thomson CSF

Engine compressor cleaning rigs
John Curran

Engine—design and manufacture
Alfa Romeo

Engine handling equipment
Central Engineering
John Curran

Engine parts fabrication
Avco Lycoming
FIAT
Lucas Aerospace
Rinaldo Piaggio

Engine starting equipment
Central Engineering
Lucas Aerospace
SAFT

Engine testing equipment
Avco Lycoming
Aviaexport
Central Engineering
John Curran
Garrett Corporation

Engines, aircraft
Alfa Romeo
Avco Lycoming
Aviaexport
FIAT
Garrett Corporation
MTU
Rinaldo Piaggio
SNECMA

Engines, auxiliary
Avco Lycoming
Aviaexport
Lucas Aerospace

Engines, V/STOL
Aviaexport
Garrett Corporation
MTU

Environmental control systems
Garrett Corporation
Lucas Aerospace
Selenia

Executive transport, twin fan jet
British Aerospace Aircraft Group
Canadair
Fokker

Feel simulator controls
Fokker
Garrett Corporation
Lucas Aerospace

Fibre optics
Barr & Stroud
ML Aviation
Thomson CSF

Filters, air
Aviaexport

Filters, electronic
Aviaexport
Barr & Stroud
EAS Electronique Aérospatiale
Thomson CSF

Filters, fuel & oil
Aviaexport
Flight Refuelling

Fire suppression systems
Eltro

Flight instrument test sets
EAS Electronique Aérospatiale
Garrett Corporation
SFENA
Thomson CSF

Flotation gear
Garrett Corporation

Flow gauges
Aviaexport

Forgings, Steel
SNECMA

Fuel control test benches
Central Engineering

Fuel flow proportioners
Flight Refuelling
Garrett Corporation
Lucas Aerospace

Fuel pump test benches
Central Engineering

Fuel pumps
Aviaexport
Garrett Corporation
Lucas Aerospace

Fuel systems protection
PRB

Fuel systems & refuelling equipment
Aviaexport
Flight Refuelling

Fuel tank pressurisation equipment
Flight Refuelling
Garrett Corporation

Furnishings & aircraft cabins
Aviaexport
Garrett Corporation
Rinaldo Piaggio

Gas turbine starting systems
Lucas Aerospace

Gas turbines
Avco Lycoming
Aviaexport
FIAT
Garrett Corporation
Lucas Aerospace
MTU
Rinaldo Piaggio
SNECMA

Gas turbines, equipment & accessories
Avco Lycoming
Aviaexport
Garrett Corporation
Lucas Aerospace

Gauges
Aviaexport
Flight Refuelling
Thomson CSF

Generator test benches
Central Engineering

Generators
Aviaexport
Garrett Corporation
Lucas Aerospace
Thomson CSF

Ground refuelling equipment
Flight Refuelling
Goodyear Tyre & Rubber Co

Ground support equipment
Aermacchi
Canadair
GIAT
ML Aviation
Sangamo Weston
Thomson CSF

Guidance control test set
Thomson CSF

Guided missile ground handling equipment
Eltro
Fokker
Garrett Corporation
Messerschmitt-Bölkow-Blohm
ML Aviation

Guided missiles
Aérospatiale
Bofors
Dornier
Messerschmitt-Bölkow-Blohm
Saab-Scania
Selenia
Thomson CSF

Gunnery training apparatus
Oerlikon-Bührle
Saab-Scania
SFENA
Thomson CSF

Hand sets
Racal Acoustics

Hangar test stands
John Curran

Headphones
Racal Acoustics

Heat exchangers
Fokker

Heat transfer systems
Fokker
Garrett Corporation

Heated windows
Barr & Stroud
Lucas Aerospace

Heated windscreen controllers
Lucas Aerospace

Helicopter deck restraint systems
ML Aviation

Helicopter parts & components
Agusta
Aviaexport
Bell Helicopter Textron
Dornier
Messerschmitt-Bölkow-Blohm

Helicopter searchlights
Garrett Corporation

Helicopter surface-to-air refuelling equipment
Flight Refuelling

Helicopter training & support
Bell Helicopter Textron
Messerschmitt-Bölkow-Blohm

Helicopters, ambulance
Aérospatiale
Agusta
Bell Helicopter Textron
Messerschmitt-Bölkow-Blohm

Helicopters, commercial-executive
Aérospatiale
Agusta
Aviaexport
Bell Helicopter Textron
Boeing Commercial Airplane Co
Dornier
Messerschmitt-Bölkow-Blohm

Helicopters, military-naval
Aérospatiale
Agusta
Bell Helicopter Textron
Boeing Commercial Airplane Co
Messerschmitt-Bölkow-Blohm

High pressure couplings
Flight Refuelling

High-speed research cameras
Thomson CSF

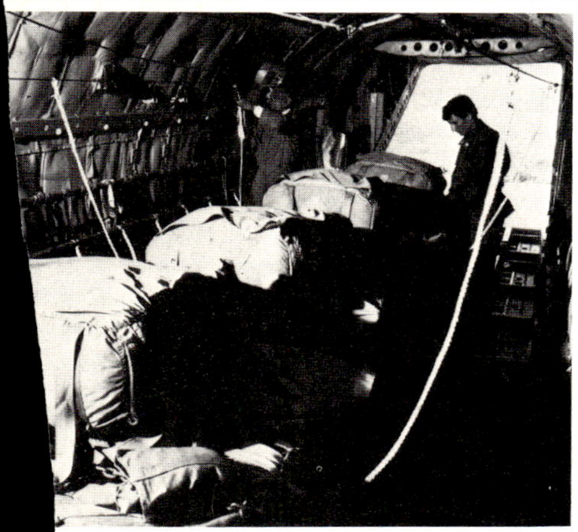

82

Hydraulic equipment
ML Aviation
Precision Mecanique Labinal
SAMM
SNECMA

Hydraulic pressure pumps
Lucas Aerospace
SAMM

Hydraulic test units, mobile & static
Aermacchi
Central Engineering
ML Aviation
SAMM

Inertial navigation systems
SFENA

Inertial navigation systems (strap-down technology)
Bofors

Inflatable structures
Garrett Corporation

Infra-red linescan
Eltro

Infra-red materials
Barr & Stroud
Selenia

Infra-red systems
Barr & Stroud
Elettronica
Eltro
Garrett Corporation
Meteor
Selenia
Thomson CSF

Instrument components (mechanical)
Thomson CSF

Instruments, aircraft
Aeritalia
EAS Electronique Aérospatiale

Instruments, electronic
Aeritalia
EAS Electronique Aérospatiale
SFENA
SNECMA
Thomson CSF

Instruments, navigation
Aeritalia
Aviaexport
EAS Electronique Aérospatiale
SFENA

Instruments, test equipment
Aeritalia
Aviaexport
Central Engineering
EAS Electronique Aérospatiale
Garrett Corporation
SFENA

Integrated total pneumatic systems
Garrett Corporation

Intercommunication equipment
Elmer

Jet engine parts
FIAT
Garrett Corporation
MTU
Rinaldo Piaggio
SNECMA

Jet engine test plant
Central Engineering
John Curran
FIAT
Garrett Corporation
SNECMA

Jet fuel starters
Garrett Corporation
Lucas Aerospace

Jet propulsion engines
Avco Lycoming
FIAT
Garrett Corporation
Lucas Aerospace
SNECMA

Jet trainer, military
Aermacchi
British Aerospace Aircraft Group
Canadair
Embraer

Jointing compound
Goodyear Tyre & Rubber Co

Lamps, cockpit
Aviaexport
Lucas Aerospace

Landing lamps
Aviaexport

Lasers
Barr & Stroud
Eltro
Garrett Corporation
Selenia
SFENA

Laser rangefinder
Barr & Stroud
Eltro
Selenia
Thomson CSF

Life saving equipment
Garrett Corporation
Saab-Scania

Light aircraft
Aérospatiale
Embraer

Lights, aircraft
Aviaexport
Precision Mecanique Labinal

Lights, landing
Aviaexport

Lights, navigation
Aviaexport
Lucas Aerospace

Linear actuator test benches
Central Engineering

Linear actuators
Garrett Corporation
Lucas Aerospace
SAMM

Linings, brakes
Goodyear Tyre & Rubber Company

Mach number transducers
Garrett Corporation

Machine tools
SNECMA

Machining
British Aerospace Aircraft Group

Marine engines
FIAT
Garrett Corporation

Market intelligence reports
DMS

Materials technology
Boeing Commercial Airplane Company
Fokker
Thomson CSF

Metal fittings
Aviaexport

Microphones
Aviaexport
Thomson CSF

Missile optics
Barr & Stroud
Bofors
Messerschmitt-Bölkow-Blohm
Thomson CSF

Missiles, guided
Aérospatiale
Bofors
Dornier
Messerschmitt-Bölkow-Blohm
Saab-Scania
Selenia
Thomson CSF

Motor generators
Aviaexport
Garrett Corporation
Lucas Aerospace

Motors, electric
Aviaexport
Garrett Corporation
Lucas Aerospace
Precision Mecanique Labinal
Thomson CSF

Motors, hydraulic
Aviaexport
Garrett Corporation
Lucas Aerospace

Navigation beacons
Thomson CSF

Night vision equipment
Barr & Stroud
Bofors
Eltro
Selenia
Thomson CSF

Non-destructive inspect equipment
Fokker

Oil valves
Flight Refuelling
Garrett Corporation
SAMM

Optical equipment
Barr & Stroud
Bofors

Optical gun sights
Barr & Stroud
Bofors
Eltro
GIAT
Saab-Scania
Thomson CSF

Overhaul & modification kits
Dornier
Fokker
Garret Corporation
Messerschmitt-Bölkow-Blohm
Siai Marchetti

Oxygen apparatus
Aviaexport

Oxygen breathing apparatus
Garrett Corporation

Parachute flares
Bofors
Oerlikon-Bührle

Parachutes
Aviaexport
Meteor

Parachutes, special purpose
Aviaexport

Parts for US built aircraft
British Aerospace Aircraft Group
Fokker
Garrett Corporation
Messerschmitt-Bölkow-Blohm

Patrol aircraft, maritime
Aérospatiale
British Aerospace Aircraft Group
Embraer
Fokker

Periscopes
Barr & Stroud
Bofors

Photographic interpretation
Bofors
Oerlikon-Bührle

Pilot handgrips
SAMM

Plastic fabrications
Aeritalia
Fokker
PRB

Plastic fabrications (reinforced with fibreglass)
Aeritalia
Boeing Commercial Airplane Co
Fokker
Lucas Aerospace

Plastic mouldings
British Aerospace Aircraft Group
Fokker
PRB

Pneumatic component test benches
Central Engineering

Pneumatic controls
Garrett Corporation
Lucas Aerospace

Power control for aircraft
Garrett Corporation
Lucas Aerospace

Power, solar cell panels & arrays
Fokker
Messerschmitt-Bölkow-Blohm

Precision gears
Barr & Stroud
FIAT

Precision potentiometers
Thomson CSF

Pressure control equipment
Garrett Corporation

Pressure radio transducers
Garrett Corporation

Pressure regulation valves, fluids & gases
Flight Refuelling
Garrett Corporation
Lucas Aerospace
SAMM

Pressure switches
Lucas Aerospace
SAMM

Pressure transducers
Garrett Corporation
SFENA
Thomson CSF

Private planes, civil & military
Aérospatiale
British Aerospace Aircraft Group
Embraer
Fokker

Propeller governors
Rinaldo Piaggio

Propeller test stands
John Curran

Propellers
Rinaldo Piaggio

Proposals for aircraft ground support operations
Boeing Commercial Airplane Co
British Aerospace Aircraft Group
Fokker
Rinaldo Piaggio
Siai Marchetti

Provisioning parts breakdown lists
British Aerospace Aircraft Group

Pumps, agricultural spray
Dornier

Pumps, air compressor
Garrett Corporation

Pumps, fuel and oil
Garrett Corporation
Lucas Aerospace
SAMM
Siai Marchetti

Pumps, hydraulic
Garrett Corporation
Lucas Aerospace
Precision Mecanique Labinal
SAMM
SNECMA
Thomson CSF

Radar for navigation, warning interception, fire control & airfield supervision
Aviaexport
Omnipol
Selenia
Thomson CSF

Radar reflectors
Fokker
Thomson CSF

Radar test set
Thomson CSF

Radar towers
John Curran
Thomson CSF

Radar turning gears & equipment
John Curran
Thomson CSF

Radio equipment
Aviaexport
Bofors
EAS Electronique Aérospatiale
Elmer
Italtel
Thomson CSF

Radio equipment, ground hf & airborne hf
Bofors
EAS Electronique Aérospatiale
Elmer
Italtel
Thomson CSF

Radio equipment, ground and airborne vhf
Bofors

Radio navigation equipment
EAS Electronique Aérospatiale
Elmer
Thomson CSF

Ramjet propulsion engines
Messerschmitt-Bölkow-Blohm

Rangefinders
Eltro
Selenia
Thomson CSF

Recirculating precision ball-screws
Lucas Aerospace

Reconnaissance equipment
Oerlikon-Bührle

Refrigeration compressors
Garrett Corporation

Refuelling equipment hose
Goodyear Tyre & Rubber Co

Repair & maintenance of aircraft
Aeritalia
Avco Lycoming
Boeing Commercial Airplane Co
British Aerospace Aircraft Group
Dornier
Fokker
Garrett Corporation
Messerschmitt-Bölkow-Blohm
Rinaldo Piaggio
Siai Marchetti

HEER

98

Repair & overhaul of aero-engines
Alfa Romeo
FIAT
Garrett Corporation
MTU
Rinaldo Piaggio
SAMM
SNECMA
Sochata SNECMA

Repair of aircraft instruments
Aviaexport
Fokker
Siai Marchetti

Rocket engine test plant
Difesa e Spazio
Messerschmitt-Bölkow-Blohm

Rocket propulsion
Bofors
Difesa e Spazio
Messerschmitt-Bölkow-Blohm
Oerlikon-Bührle

Rocket sounding
Dornier
Saab-Scania

Rotory actuator test benches
Central Engineering

Rotary actuators
Garrett Corporation
Lucas Aerospace
Precision Mecanique Labinal

RPVs
Canadair
Flight Refuelling
Messerschmitt-Bölkow-Blohm
Meteor
ML Aviation

Runway friction measuring equipment
John Curran
ML Aviation
Saab-Scania

Seals
Flight Refuelling

Seat belts
Aviaexport

Servo actuators
Garrett Corporation
Lucas Aerospace
Meteor
Precision Mecanique Labinal
SAMM
Thomson CSF

Sheet metal work
British Aerospace Aircraft Group
Lucas Aerospace

Simulators
Dornier
Messerschmitt-Bölkow-Blohm
Siai Marchetti
Thomson CSF

Simulators, combat
British Aerospace Aircraft Group

Space hardware recovery
Dornier
Fokker

Space launchers
Aérospatiale

Space satellites
Aeritalia
Aérospatiale
Dornier
Fokker
Messerschmitt-Bölkow-Blohm

Space systems
Aeritalia
Boeing Commercial Airplane Co
Difesa e Spazio
Dornier
Fokker
Messerschmitt-Bölkow-Blohm
Saab-Scania
Thomson CSF

Spacecraft
Aeritalia
Dornier
Fokker
Messerschmitt-Bölkow-Blohm

Spare parts for US built aircraft
Bestobell
British Aerospace Aircraft Group
Goodyear Tyre & Rubber Co
Siai Marchetti

Stability augmentation system
Dornier
SFENA

Stall warning systems
Fokker

Starter pods, airborne
Garrett Corporation

Starting systems, airborne
Garrett Corporation
Lucas Aerospace

Static inverters
EAS Electronique Aérospatiale
Lucas Aerospace
SFENA

Surveillance systems
Barr & Stroud
Canadair
Elettronica
Garrett Corporation
Meteor
Selenia
Thomson CSF

Switches
Lucas Aerospace
SAMM

Switches, miniature electrical
SAMM

Switchgear
Aviaexport
Lucas Aerospace

Tachometers
Aviaexport
Thomson CSF

Targets, aerial
Flight Refuelling

Target release & exchanger mechanisms
Flight Refuelling
Meteor

Target towing winches
Flight Refuelling
Garrett Corporation
Meteor

Technical publications
British Aerospace Aircraft Group
Dornier
Flight Refuelling
Fokker
Messerschmitt-Bölkow-Blohm
Rinaldo Piaggio
Siai Marchetti

Technical publications, special studies
British Aerospace Aircraft Group
Dornier
Fokker

Temperature control equipment
EAS Electronique Aérospatiale
Garrett Corporation

Test equipment
Aérospatiale
Aviaexport
British Aerospace Aircraft Group
EAS Electronique Aérospatiale
Fokker
Garrett Corporation
Messerschmitt-Bölkow-Blohm
ML Aviation
Racal Acoustics
Selenia
SFENA
Thomson CSF

Test equipment, radar, air data computer, fire control system, avionics etc
ML Aviation
Selenia
SFENA
Thomson CSF

Test equipment, airfield radio
Bofors
EAS Electronique Aérospatiale
Thomson CSF

Test equipment, metal bonding
Aviaexport
Fokker
Siai Marchetti

Test facilities
British Aerospace Aircraft Group

Thermal imaging systems
Barr & Stroud
Eltro
Thomson CSF

Thermo couple cables
Standard Wire & Cable Co

Tractors, electric
ML Aviation

Training devices
British Aerospace Aircraft Group
Oerlikon-Bührle
Thomson CSF

Transformer rectifier units
EAS Electronique Aérospatiale
Lucas Aerospace
Thomson CSF

Tubes, stainless steel
Aviaexport

Turbines, ram air
Garrett Corporation

Turbofan engines
FIAT
Garrett Corporation
MTU
SNECMA

Turnkey airport projects
British Aerospace Aircraft Group

Tyres for aircraft
Goodyear Tyre & Rubber Co

Undercarriage equipment
SNECMA

Undercarriage Gear, retractable
British Aerospace Aircraft Group
SNECMA

Valves
Flight Refuelling
Garrett Corporation
Siai Marchetti
Thomson CSF

Valves & miniature relays
Flight Refuelling
Thomson CSF

Valves, control hydraulic
Garrett Corporation
Goodyear Tyre & Rubber Co
SAMM
Siai Marchetti

Valves, electronic
Garrett Corporation
Thomson CFS

Valves, non-return fuel
Flight Refuelling
Garrett Corporation

Valves, non-return hydraulic
Flight Refuelling
Garrett Corporation
SAMM

Valves, relief hydraulic
Garrett Corporation
SAMM

Valves & miniature relays
Garrett Corporation

Vertical take-off aircraft
British Aerospace Aircraft Group
Dornier

Visibility measuring equipment
SNECMA

Voltage & current regulators
EAS Electronique Aérospatiale
Garrett Corporation
Lucas Aerospace
Precision Mecanique Labinal

Warning tone generators
Racal Acoustics

Water separators
Flight Refuelling
Garrett Corporation

Wheels for aircraft
Goodyear Tyre & Rubber Co
SNECMA

Wind tunnel testing plant
Boeing Commercial Airplane Co
British Aerospace Aircraft Group
John Curran
Dornier

Wire & cables all types
Fokker
Standard Wire & Cable Co
Thomson CSF

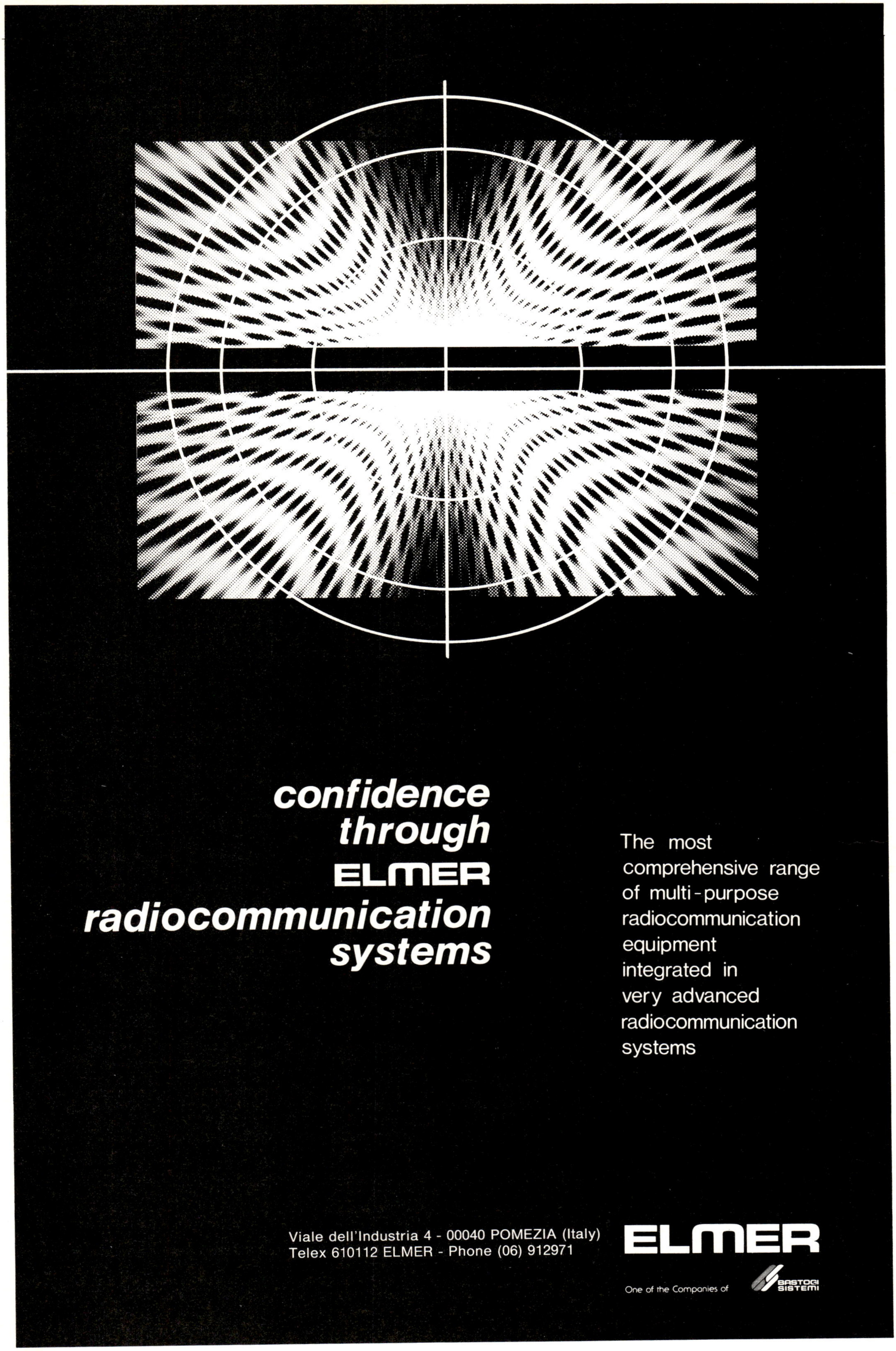
confidence through ELMER radiocommunication systems
The most comprehensive range of multi-purpose radiocommunication equipment integrated in very advanced radiocommunication systems
Viale dell'Industria 4 - 00040 POMEZIA (Italy)
Telex 610112 ELMER - Phone (06) 912971
ELMER
One of the Companies of
BASTOGI SISTEMI

ELTRO
know how and cooperation

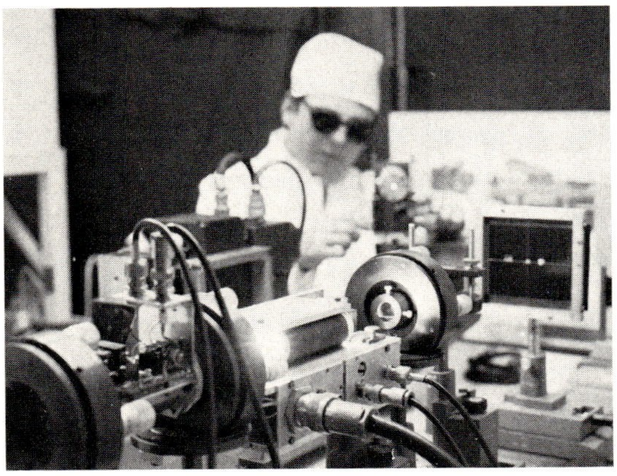

Years spent on research and development, the successful transfer of know-how and cooperation in both national and international military programs have accounted for Eltro's achievements in the field of optronics.

We specialize in:

- Solid-state, semiconductor and gas lasers for rangefinding, target illumination and target tracking
- Thermal detection units
- IR-missile guidance
- Image intensifier and converter units
- Fire suppression systems
- Mobile workshops for maintenance service

NC-automats, high-precision measuring and test installations, modern air-conditioned workshops and qualified personnel assure the quality standard called out for military equipment.

Eltro OPTRONIK

ELTRO GmbH, Gesellschaft für Strahlungstechnik
Postfach 102120, 6900 Heidelberg 1/W.-Germany

A real utility aircraft: **PIAGGIO-DL3**

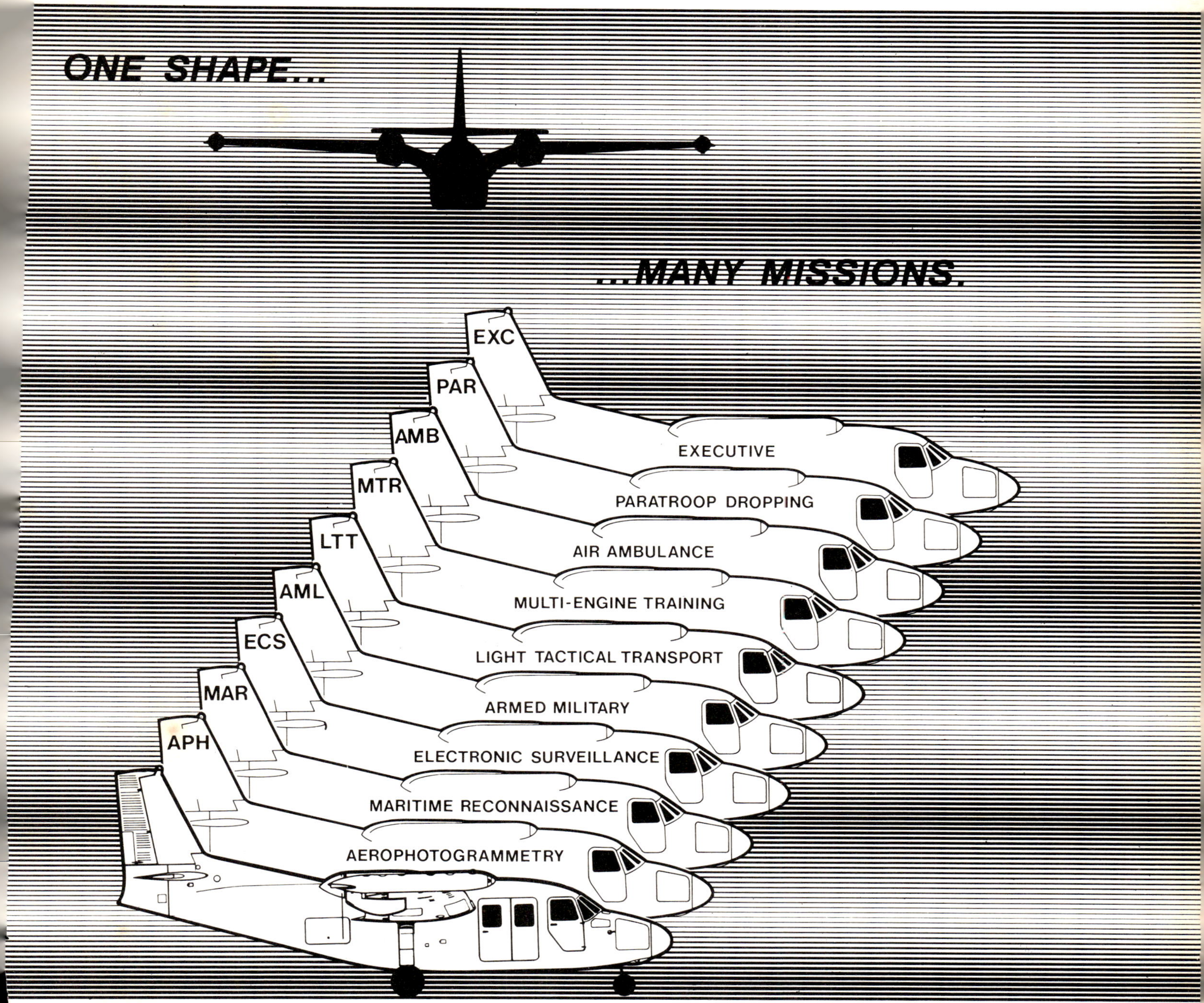

The shape makes the difference!

There are many advantages to this flexible and useful general purpose aircraft . . .

— Exceptional handling and stability
— Excellent visibility for operators and passengers
— Extra large passengers or cargo cabin
— Extraordinary loading ease with the main deck only 22 inches high
— Extra rugged, sturdy airframe
— Economical operating and maintenance costs
resulting from fuel efficent modular turboprop engines

. . . These and many other features makes of the Piaggio -DL3 a real utility aircraft.

RINALDO PIAGGIO

Via Cibrario n. 4 - 16154 GENOVA - Italy

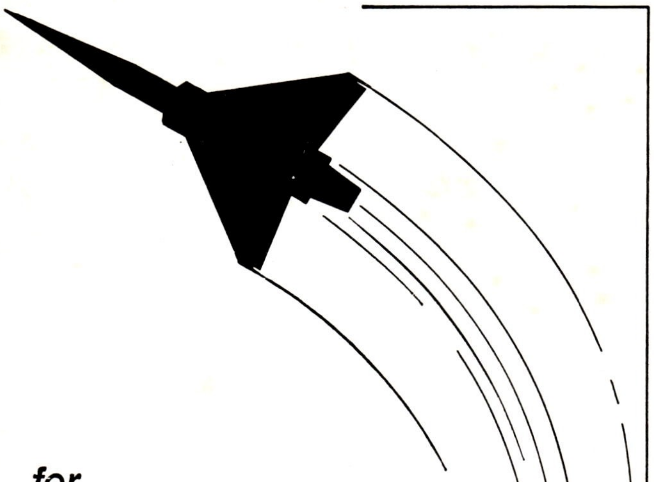

for the suppression of fuel systems explosions

be protected by

safom®

Reticulated Urethane Safety Foam
Produced to military specifications

When fitted into a fuel tank it will completely suppress the explosion which follows the ignition of a fuel/air mixture in a confined space.

Weight: 30 kg or 20 kg per m³.

Reduction of tank capacity: 3%

Can be fitted during construction or retrofitted on existing aircraft fuel-system by our specialised personnel.

S.A. PRB

Division RECTICEL
Damstraat, 2
B-9200 Wetteren/Belgium
Tel. 091/69.28.01
Telex. 11363 PRBWTN B

---

précision mécanique
labinal

division
aérospatiale
électronique

5, Avenue Newton B.P. 64
Parc d'Activités - 78390 BOIS-D'ARCY
Tél. (3) 043.82.12 - Télex : LABIN-A 698330

AIRCRAFT TYRE PRESSURE MONITORING SYSTEM

PM LABINAL's Design Office has developed an automatic monitoring system for tyre pressure.

The system operates at all times (parked, taxiing and in flight) and comprises :

- A rotating section (wheel) containing the pressure sensor and an electronic module (power supply to piezoelectric sensor, conversion of pressure reading into a frequency) ;
- A stationary section in the aircraft receiving and processing the pressure signals prior to transmitting them to a visual display system.

Connection between the revolving part and the fixed part is effected at the axle through a rotating transformer, so arranged as to provide power supply to the sensor in one direction, and in the opposite direction, to transmit the valve of pressure as a sinusoïdal signal with a frequency which is a function of the magnitude being measured.

The time taken to check each wheel is less than 40 ms. The weight of the system, covering 10 wheels, is less than 11 kg.

A permanent display feature is available on request; otherwise, automatic indication in the event of an abnormal reading. Indication is via a cathode tube (CRT) but other types of display units may be used.

Developed originally for the AIRBUS and already ordered by a number of airlines, the system can be adapted for other aircraft now in service.

1494

Sea and sky: a story begun in 1908.

1908: SA 8/75 8 cylinders - 50 HP. A new century begins.
Fiat brings the future into shape by building its first aircraft engine for use in military airships and Farman aeroplanes.

1970's: RB 199 Three shafts - over 7,000 kg thrust - a multirole engine for an all-weather aircraft. Built in collaboration with Rolls Royce and MTU.

1980's: PW 2037 18,000 kg thrust. The commercial engine for the new-generation airliners.
Fuel consumption 7-9% lower than the competition, and 30% fuel saving as compared to the engines currently in service.
Already chosen by Delta and American Airlines.

LM 2500 LM 500 Jointly built by Fiat and General Electric, the LM 2500 marine turbine is derived from the CF6, power plant of the DC10, Airbus, and Jumbo. The LM 500 stems from the TF 34 turbofan, which powers the S3A and the A 10.

Gearboxes for helicopters and engines

Fiat Aviazione specialises in the design and manufacture of the mechanical transmission assemblies . The CF6 engines mount the engine accessory drive gearbox made by Fiat Aviazione.

FIAT AVIAZIONE

AEROSYSTEM ELECTRONIC
AEROSYSTEM ELECTRONIC
AEROSYSTEM ELECTRONIC
ANDROMEDA
METEOR
METEOR
METEOR
METEOR
MULTIROLE REMOTELY PILOTED OR AUTOMATIC AIRCRAFT SYSTEM
AEROSYSTEM ELECTRONIC A.G.
International Division
Am Balsberg, 30
8302 KLOTEN-ZURICH (SWITZERLAND)
Phone: (01) 8141117
Telex: 55660 AERSY CH
METEOR S.p.A.
Via Nomentana, 146
00162 ROMA (ITALY)
Phone: (06) 830991
Telex: 680 136 METEOR I

Central Engineering . . .
We're the world's most experienced designer and builder of aircraft engine test facilities
Since 1960, Central Engineering has been serving the best names in the commercial and military aircraft industries with a full range of engine test facilities . . . from specialized test systems to complete turnkey installations. We think experience is important. Contact us for the full story.
CE
Central Engineering Company
CENTRAL ENGINEERING INTERNATIONAL CO.
2930 Anthony Lane, Minneapolis, MN 55418 U.S.A.
Phone (612) 781-6557
TELEX 290 376

The future's looking bright.

The power unit of Tornado, the West's outstanding multirole combat aircraft, is the result of resources pooled over the last decade by Britain's Rolls-Royce, Germany's MTU and Italy's Fiat Aviazione.

Technically speaking, the end-product, the TURBO-UNION RB 199, is a three shaft, reheated turbofan of short, rigid design with compact afterburner and integral thrust reverser. Its modular construction also means unprecedented ease of service, fault-diagnosis and repairs.

In effect, it has an exceptional thrust-to-weight ratio, ample power for combat manoeuvres and supersonic acceleration, and low fuel-consumption for long range cruising.

More than 2,000 units are being produced for the British, German and Italian air forces. Real evidence that with the RB 199 TURBO-UNION gives you today the power-plant of tomorrow.

TURBO-UNION RB 199
There's really no alternative.

AD 4/80 ER

Turbo-Union Ltd. **Head-Office:** P.O. Box 3, Filton, Bristol BS 12 7 QE, England. **Munich-Office:** Arabellastrasse 4, D-8000 München 81, West Germany

AIR FRANCE

Tarmac

Makers of world aerospace history

Today, it's the Challenger — the third-generation business jet — that's keeping us growing. And the world-famous CL-215 — the only aircraft specifically designed to fight forest fires. And the CL-89 Surveillance Drone System — among the very first such systems in operation around the world. And major airframe component sub-contracts, including the Boeing 767 aft fuselage.

Yesterday, it was the 24-hour-patrol Argus, the pioneering 'swingtail', and over 3,200 jets including 580 supersonics.

Tomorrow, it promises to be everything from our state-of-the-art CL-227 and CL-289 surveillance systems to a family of energy-efficient transport aircraft.

Come and talk to us. Our people are not only uniquely versatile, but long experienced in applied aerospace technology.

canadair

CANADAIR LIMITED • P.O. BOX 6087, STATION A, MONTREAL, QUÉBEC, CANADA • H3C 3G9 • (514) 744-1511

ITALTEL

SOCIETÀ ITALIANA
TELECOMUNICAZIONI s.p.a.
20149 Milan (Italy) - 12, Piazzale Zavattari
phone (+ 392) 4388.1

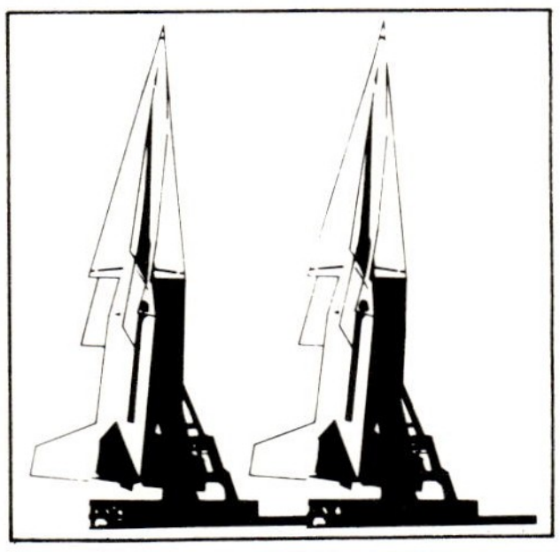

IFF SYSTEM BY ITALTEL

A complete line of equipment, for airborne, naval and ground applications.

TRANSPONDER SIT 421
(MM/UPX-709)

- All solid state
- AIMS and stanag 5017-ED3 fully compatible transponder
- Diversity operation for spherical coverage
- Specified MTBF 1,000 hours
- Continuous performance monitoring
- Mode 4 reply indication
- Front panel and remote test indication.

INTERROGATOR SIT 431
(MM/UPX-708)

- All solid state AIMS compatible interrogator, receiver crystal-controlled CODER/DECODER
- Specified MTBF 2,000 hours
- Built-in test
- Continuous performance monitoring
- Front panel and remote test indication

INTERROGATOR SIT 432
(AN/APX-104)

- All solid state AIMS 65 - 1000 and stanag 5017-ED3

DECODER GROUP
AN/UPA-59A(V)

- Designed to MIL - D - 28736B (EC) for use with Mark X A (SIF) and Mark XII interrogators
- Compatible with all AIMS modes - comply with characteristics of ATCRBS (Air Traffic Control Radar Beacon System)
- Designed for shipboard use

SP 8801/03

We don't
Fokker F28 Fellowship
CityHopper
PH-KFK
Fokker F27 Friendship

waste our energy.

Fly the fuel-efficient skies of Fokker.

Two totally different aircraft. Two totally different propulsion systems. But the same high standard of proven fuel efficiency. That's what makes the Fokker F28 and F27 the first choices of short-haul operators around the world.

The Fokker F28 is the only short-haul jet offering jet speed and comfort to passengers and cost-efficient operation to airlines. It is the only jet specifically designed to fly short sectors, where flights average 30-45 minutes with quick turn-around times, and do it profitably. With the lowest investment cost of any jet flying today, the F28 further economizes with the lowest fuel consumption of any jet airliner. Over a stage length of 250 nautical miles, the F28, with its two Rolls-Royce RB183 engines, burns up to 35% less fuel than the nearest competitor in operation.

The Fokker F27, with its wide-body look interior and pressurized cabin, is used by major airlines all over the world as feeder aircraft on their short-haul networks and to generate traffic on low density routes. Its remarkable energy efficiency is the result of exceptionally low drag qualities, low structure weight and efficient Rolls-Royce Dart 7 turboprop engines. These factors combine to assure fuel consumption that is less than any of the F27's competitors.

The F28 Fellowship and the F27 Friendship—setting new standards of fuel-efficiency and comfort for the airlines of the world.

Fokker

Fokker B.V.
P.O. Box 7600
1117 ZJ Schiphol
Netherlands
Phone: #31 20 5449111
Telex: 11526

Fokker Aircraft U.S.A., Inc.
2361 Jefferson Davis Hwy.
Arlington, Virginia
22202, U.S.A.
Phone: (703) 979-6400
Telex: 899462

Dutch. Dedicated. Dependable.

siaimarchetti
SF 260 TP
SF 260
siaimarchetti
siaimarchetti
SM 1019
siaimarchetti
S 211
siaimarchetti
SF 600 TP
siaimarchetti
SIAI

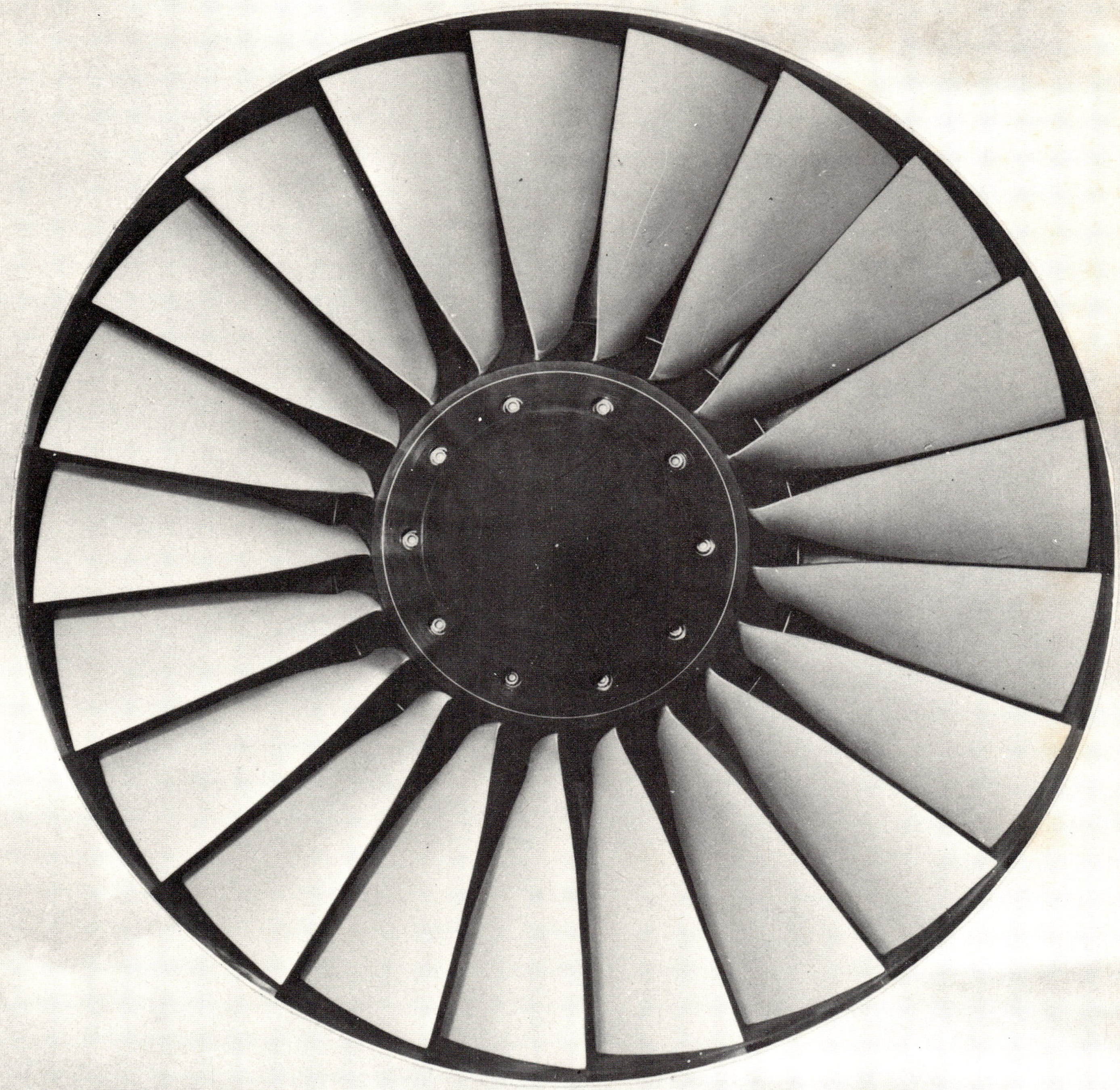

New books from Airline Publications

WORLD AIRLINE FLEETS 1982

by Günter G. Endres

Brand new edition of the most comprehensive and accurate fleet listing available. 600 pages of data processed information checked for accuracy by hundreds of sub-editors in their respective countries throughout the world. Over 150 photographs. Nearly 2000 airlines most with brief histories, logos and motifs as well as full fleet lists.

☐ £8.50 including postage/packing in the UK and Europe (publication November 20th).

BIZ-JET 1982

by Brian Gates

Latest completely updated and revised edition of the standard reference work listing all civil and executive jets throughout the world country by country in alpha, numerical sequence by registrations. Includes manufacturers serial numbers, owner/operator and previous identity and for the first time BIZ-JET production list of Falcon 10, 20, 50, Gulf Stream II, III, HS 125-700.

☐ £3.80 including postage/packing in the UK and Europe (publication November 20th).

AVIATION DATA CENTRE

BIZ-JET DATA UPDATE — MONTHLY — A completely updated and reprinted edition of Biz-Jet each month. A5 size, spiral bound. Published directly from our daily updated data processed programme. £25 per issue post free UK/Europe or £250 annual subscription for 12 editions including postage/packing in the UK and Europe. $600 including airmail USA, $700 Cdn including airmail Canada.

CIVIL AIRCRAFT REGISTRATIONS 1982

Current UK/Eire civil register listing aircraft registration letters, manufacturers serial number, owner/operator for all civil aircraft, light/executive/airliners registered in both countries plus fleet lists of those airlines likely to be seen in the UK. Civil Aircraft Register is pocket sized and completely updated and reprinted every 6 months.

☐ £1.50 + 25p postage/packing in the UK and Europe. (publication November 20th).

AVIATION DATA CENTRE

AIRLINE MARKET INTELLIGENCE INFORMATION

World Airline Fleets, Airliner Production List, Military Transport Fleets and Biz-Jet are now fully programmed onto our own Amtext 425 data processors. All information is now available on copy floppy disk or computer tape with a weekly updated hard copy print-out. Let us quote you for supplying and servicing your commercial aviation market intelligence requirement.

MILITARY TRANSPORT FLEETS 81/82

by Ricky D. Halliday

This new data processed publication lists for the first time the world military transport fleets and Government operated VIP aircraft. Each fleet includes the individual aircraft serial numbers and manufacturers serial numbers where known. Nearly 200 pages and 200 photographs.

☐ £6.00 including postage/packing in the UK and Europe.

EXECUTIVE DISPLAY MODELS

(Travel agent type display models — complete not kits)

1/200 Concorde 'British' 12" long
£12.95 + £1.00 postage/packing

1/200 Boeing 747 'British'
£29.95 + £2.50 postage/packing

1/200 Douglas DC-10 'Laker Skytrain'
£29.95 + 2.50 postage/packing

1/200 A300 Airbus 'Lufthansa' PIA, TDA each £29.95 + £2.50 postage/packing

1/100 DC-10 'British Caledonian'
£79.50 + £10.00 Securicor

1/100 Concorde 'British' 24" long
£45.00 + £7.50 Securicor

1/36 Spitfire 1
£55.75 + £5.00 Securicor

1/36 Mosquito
£86.00 + £10.00 Securicor

1/72 scale fibreglass models hand painted to order in the colour scheme of your choice from £175.00 per model.
-DC-6, Viscount, Britannia, Ambassador, Stratocruiser, F-27, F-28, Trident, 707, 727, C-130 Hercules, Belfast, Vanguard, Super Constellation, Convair Comet 4, Boeing 747, DC-10 plus many others. Also larger travel agent type models.

AERO MINI MODELS

Bought for cash or exchanged.

1/240 DC-3 metal kit with American Airlines and Air Atlantique transfers
£12.50 + £1.00 postage/packing

SEE ALL THESE MODELS AT OUR VHF SUPPLIES SHOWROOM NEAR HEATHROW

OTHER AIRLINE PUBLICATIONS TITLES

Boeing 747 (hard cover)	£ 9.75	☐
S E Caravelle (hard cover)	£ 9.75	☐
HP Herald (hard cover)	£ 8.75	☐
Convair 240-640 (hard cover)	£ 8.25	☐
DC-3 parts 1-12	£15.00	☐
DC-3 parts 13-24 (Subscription)	£18.00	☐
DC-4 Production List	£ 5.50	☐
DC-7 Production List	£ 2.50	☐
Constellation Production List	£ 4.95	☐
World Airline Fleets Monthly (Subscription-12 editions)	£15.00	☐
UK VHF Airband Guide — (2nd Edition)	£ 1.75	☐
1981 Airport Schedules	£ 2.75	☐
1981 Airline Handbook	£ 6.50	☐
Aviation Postcard Collector (Editions 1, 2, 3, and 4)	£ 4.00	☐
Jane's Civil Aircraft — in clour	£ 5.50	☐
Visitors Guide to Gatwick/Heathrow	£ 2.50	☐

SPECIAL OFFERS

World Airline Fleets 1981	£ 3.50	☐
Airliner Production Lists 1980/81	£ 3.50	☐
Herk Hero of the Skies	£ 3.50	☐
World Airline Fleets 1980	£ 2.50	☐
Air Transport Hulks	£ 3.00	☐
Bis-Jet 1981	£ 1.75	☐
Airport Schedules 1981	£ 1.50	☐
Prop-Jet 1981	£ 1.50	☐
Biz-Jet 1980	£ 1.25	☐
Prop-Jet 1980	£ 1.25	☐
C-130 Production List	£ 1.25	☐
Cambrian Airways	£ 1.25	☐

COLOUR SLIDES

Sealed pack of 10 airliner slides (including original and duplicates). £1.25 including postage/packing in the UK and Europe. Slide auction first Saturday in each month at 2 pm at VHF Supplies Showroom.

EXECUTIVE JET MODELS

Superb quality — hand painted to the colours of your private or company aircraft.

Gulfstream II, Falcon 20, Falcon 50, Jetstar, Learjet 25, Learjet 35, Citation, HS125, 300, 400, or 700, Boeing 727-100 — Please make cheques/postal orders payable to Executive Display Models.

Jane's Yearbooks

All the Worlds Aircraft 1981/82 (Dec) £47.50 ☐

AIRLINE PUBLICATIONS

(Dept J) Noble Corner,
Gt. West Road, Hounslow,
Middx. TW5 0PA
(Telephone: 01-572 0225)

MAIL ORDER — Tick the boxes of any of the titles above which you wish to order and return this page and your payment directly to AIRLINE PUBLICATIONS at the above address. Prices include postage and packing UK/Europe. Add 25% for airmail outside Europe — 50% for Australia/NZ/Far East. We no longer send by surface mail. (Write order on separate paper if you wish).

Please send me ________ books as ticked. I enclose payment of £ ________ including postage and packing (Add 25% or 50% as above if outside Europe).

Name ____________________

Address ____________________

____

____

or pay by credit card — Visa/Master Charge/American Express/Access.

Card type __________ Number __________ Expiry Date __________

If paying by Eurocheque you must put Cheque Card Number on reverse __________
Our Postal Giro Account No. is 357-0355.

00
YT-27
1300
YT-27
1301
01

The outstanding weapon of tomorrow...
available to day.
The DEFA 30 mm cannon 553 and 554:
• proven performances
• already operational on 17 types of aircraft worldwide.
DEFA 30 mm cannon 553 or 554: the effective calibre to counter modern air and surface targets.
GIAT
10, Place Georges Clemenceau
92211 St Cloud France
Tél. : 602.52.00
Télex : DTATSCLOU 260010 F

A 109
A 109 TOW
A 129
AB 206
AB 206 L1
AB 205
AB 412

JANE'S
ALL THE WORLD'S AIRCRAFT

FOUNDED IN 1909 BY FRED T. JANE

COMPILED AND EDITED BY
JOHN W R TAYLOR FRAeS, FRHistS, FSLAET
ASSISTANT EDITOR
KENNETH MUNSON, Associate RAeS, ARHistS

1981-82

JANE'S
LONDON • NEW YORK • SYDNEY

Published in the United Kingdom by
Jane's Publishing Company Limited, 238 City Road, London EC1V 2PU

ISBN 0 7106-0729-6

Published in the United States of America by
Jane's Publishing Incorporated, 730 Fifth Avenue, New York, NY 10019

ISBN 0 531-03975-7

146
A300
A300

CONTENTS

The Editor has been assisted in the compilation of this edition as follows:

Kenneth Munson	AIRCRAFT SECTION: ARGENTINA TO FINLAND, GREECE TO TURKEY, UNITED KINGDOM (pp 252-272); RPVs; SAILPLANES; MICROLIGHTS AND HANG GLIDERS
David Mondey	AIRCRAFT SECTION: WEST GERMANY, UNITED KINGDOM (pp 235-251), UNITED STATES OF AMERICA; BALLOONS; INDEX
Bill Gunston	GLOSSARY; AERO ENGINES
Michael Taylor	HOMEBUILTS; METRIC CONVERSIONS
Maurice Allward	SPACEFLIGHT
The Lord Ventry	AIRSHIPS

Alphabetical list of advertisers .... [6]
Classified list of advertisers .... [12]
Foreword .... [63]
Glossary .... [71]
Some first flights made during the period 1 July 1980 to 1 October 1981 ... [77]
Official Records .... [78]

Aircraft
Argentina 1
Australia 4
Belgium 8
Brazil 9
Canada 18
Chile 28
China (People's Republic) 28
Colombia 36
Czechoslovakia 37
Finland 42
France 43
Germany (Federal Republic) 76
Greece 86
India 86
Indonesia 90
International Programmes 91
Israel 112
Italy 118
Japan 141
Korea (Republic) 149
Libya 150
Mexico 150
Netherlands 150
New Zealand 155
Pakistan 157
Philippines 157
Poland 157
Portugal 168
Romania 168
South Africa 171
Spain 173
Sweden 176
Switzerland 178
Taiwan 183
Turkey 185
Union of Soviet Socialist Republics 186
United Kingdom 235
United States of America 273
Yugoslavia 471

Homebuilt Aircraft (including Racing Aircraft)
Australia 473
Austria 473
Brazil 473
Canada 474
Czechoslovakia 478
Finland 479
France 480
Germany (Federal Republic) 494
Italy 496
Japan 496
New Zealand 498
Poland 498
South Africa 498
Sweden 498
Switzerland 500
Union of Soviet Socialist Republics 500
United Kingdom 500
United States of America 506

Sailplanes
Australia 567
Austria 567
Brazil 567
Canada 568
China (People's Republic) 569
Czechoslovakia 570
Denmark 570
Finland 570
France 571
Germany (Federal Republic) 575
India 584
Italy 584
Japan 585
Poland 586
Romania 586
Spain 588
Switzerland 588
Union of Soviet Socialist Republics 588
United Kingdom 590
United States of America 590
Yugoslavia 594

Microlight Aircraft and Hang Gliders
Argentina 605
Australia 605
Belgium 606
Bulgaria 606
Canada 607
China (People's Republic) 608
Czechoslovakia 608
France 608
Germany (Federal Republic) 610
Japan 612
Poland 612
Switzerland 612
Union of Soviet Socialist Republics 613
United Kingdom 614
United States of America 616

Lighter-than-Air: Airships
Japan 635
United Kingdom 635
United States of America 637

Lighter-than-Air: Balloons
France 640
Germany (Federal Republic) 640
United Kingdom 640
United States of America 641

RPVs and Targets
Argentina 645
Australia 645
Canada 645
France 646
Germany (Federal Republic) 648
Indonesia 650
International Programmes 650
Israel 650
Italy 652
Japan 653
Saudi Arabia 654
Sweden 654
Union of Soviet Socialist Republics 654
United Kingdom 654
United States of America 658

Air-launched Missiles
Argentina 671
Brazil 671
France 671
Germany (Federal Republic) 672
International Programmes 674
Israel 674
Italy 674
Japan 676
Norway 676
South Africa 676
Sweden 676
Union of Soviet Socialist Republics 676
United Kingdom 680
United States of America 680

Spaceflight
Canada 689
China (People's Republic) 689
Czechoslovakia 690
France 690
Germany (Federal Republic) 691
India 692
International Programmes 693
Italy 694
Japan 695
Sweden 696
Union of Soviet Socialist Republics 696
United Kingdom 700
United States of America 701

Satellites and Spacecraft launched during 1980 710

Aero-Engines
Australia 713
Belgium 713
Canada 713
China (People's Republic) 717
Czechoslovakia 717
Finland 718
France 719
Germany (Federal Republic) 727
India 730
Indonesia 730
International Programmes 731
Israel 735
Italy 736
Japan 736
Poland 738
Romania 741
South Africa 742
Sweden 742
Union of Soviet Socialist Republics 743
United Kingdom 750
United States of America 758

Addenda 798

Indexes 814

FOREWORD

The deterrent is dead. For 36 years since the agony of Hiroshima and Nagasaki, a 'peace through fear' nuclear policy has inhibited every threatened outbreak of a third World War. Today, this policy and the East/West balance of military power that made it feasible are disintegrating. While leaders of the so-called super-powers plan the manufacture of new weapons such as neutron bombs, and no longer regard a limited nuclear war as unthinkable, the peoples of their satellite nations become increasingly restless. Poland rumbles because its leaders fail to feed its families adequately. The Socialist opposition to Britain's government talks of unilateral disarmament, withdrawal from NATO, and isolationist withdrawal from the European Economic Community if it is returned to power.

In October 1981, a quarter of a million people gathered peacefully in West Germany to emphasise that they wanted no additional Pershing strategic rockets in their land, with consequent risk of a pre-emptive attack from the East in any future period of international tension. Even in the homeland of the western super-power, the citizens of Utah made so clear their opposition to the shuttling of MX ICBMs around their State, between mostly-empty shelters, that the whole absurd proposal of multiple basing has been rejected in favour of concealing MX initially in second-hand Titan silos in other places.

Paradoxically, any weakening of the cohesion among groups of nations making up the NATO and Warsaw Pact alliances could lessen the chances of eventual, permanent, peaceful coexistence. The SALT I agreement would never have materialised if the USA and USSR had not been able to speak for the two most formidable military groupings in the world at the original strategic arms limitation talks. Their mistake was to follow with the ill-conceived draft of SALT II, which authorised an increase in the number of weapons it was supposed to limit, and left the junior partners with the feeling that they had been selfishly overlooked. Missiles pointing in their direction were unrestricted by the proposed treaty. Even worse, proposals to base new generations of strategic nuclear weapons within their territory seemed to indicate the super-powers' determination that lesser nuclear conflicts would be conducted far outside the US and Soviet borders.

Neither a growing acceptance by military leaders of the feasibility of limited nuclear warfare, nor a drift by the civil population towards anarchy, via protest, offers a path to survival. Unilateralists cannot point to any period of history when abandonment of all forms of defence deterred aggression. Certainly, Hiroshima was not saved in 1945 by the fact that Japan had no nuclear weapons. Four years earlier, America's isolationist policy had not enabled it to escape direct and costly involvement in the second World War.

Eventually, world military and political leaders may learn what common folk already know, that there is nothing to be gained from modern war, which nobody wins. Even if, for example, the Soviet Union succeeded in defeating America, the US prairies from which it now purchases much-needed wheat would be blackened and poisonous for a thousand years. Nor should the anti-war factions concentrate their attention exclusively on nuclear weapons. Revelation during the past year of British/American proposals to attack Germany with anthrax 'bombs' in the second World War reminds us that far more horrible forms of weapons are under constant study in hidden places. How, then, can the fragile peace of the past three decades be maintained, following loss of credibility of the nuclear deterrent policy?

First, it is important to understand why, at a time when the USA and USSR have invested huge sums in a combined total of 3,976 intercontinental and submarine-launched ballistic missiles, these weapons have lost their effectiveness as a deterrent. Everyday life supplies the answer: empires, commercial enterprises, even human gluttons, are likely to collapse if they grow too large. After Hiroshima, the ability to eliminate with a single warhead any potential target on Earth, including a capital city with all its people, seemed the ultimate deterrent. This was logical. One or two cities might be (and had been!) considered expendable to avert years of major war. But who, today, would initiate the launch of 3,976 ICBMs and SLBMs in a nuclear exchange that would, inevitably, erase life from vast areas in his own homeland, as well as in that of an enemy?

Except as a consequence of miscalculation by those who toy with ideas of a limited nuclear war, the wholesale use of ICBMs and SLBMs must remain unthinkable as a means of averting or winning a global war. Equally, at a time of conflict and civil strife everywhere from Iran to Namibia, El Salvador, Northern Ireland and Kampuchea, it will be many years before all peoples can be assured of their inalienable right to 'life, liberty and the pursuit of happiness'. The first essential, therefore, is to devise a deterrent to war (at every level) that will supersede effectively those 3,976 ICBMs and SLBMs.

To cope with a major confrontation, the answer must bear some relationship to what we have lived with since 1945—the threatened elimination of one or two specific targets, which need not be cities, as a demonstration of intent if aggression is not halted. Ten or 20 highly-accurate missiles, deployed invulnerably in hardened silos by each of the two super-powers, would be adequate for this. To render ineffective defensive countermeasures, including evacuation of known target areas, the missiles could be targeted against different places week by week. Supported by reconnaissance satellites and other forms of intelligence, they would also make it possible to monitor and deter nuclear proliferation into the control of less-responsible nations. Countries like Britain could dispense with all strategic missiles in the knowledge that, even now, they simply add to an existing NATO 'overkill' and could be used only for pointless revenge after destruction of the UK.

* * * * * *

Universal and complete abandonment of nuclear weapons is out of the question, as it would leave the door open to the kind of seemingly-endless war that was endured in 1914-18 and in Viet-Nam. The vital second component of a future deterrent is, therefore, a tri-service force capable of absorbing an attack by opposing forces, on any scale short of nuclear, for as long as it would take to stabilise the situation, militarily and politically. This is the area in which NATO is currently least capable of meeting its commitment, which could prove ominous for both East and West. Over-anxious reaction to a minor fracas might lead all too easily to premature use of nuclear weapons.

The Foreword to last year's *Jane's* included a brief description of how a serious assault on NATO's central front might be mounted. Since then, both the quality and quantity of Soviet weapons have continued to increase. During the single year of 1980, the 135 major Soviet military plants delivered a huge variety of weapons, including 3,000 tanks, 5,500 other armoured fighting vehicles, 150 self-propelled field guns, 30 supersonic bombers, 1,300 fighters and fighter-bombers, 350 transport aircraft, 750 helicopters, 200 new-generation ICBMs, 100 IRBMs, 700 submarine-launched cruise missiles, 175 SLBMs, 1,500 air-to-surface missiles, 50,000 surface-to-air missiles, 11 submarines and 11 major combat surface ships.

More than 80 per cent of the tanks were new, fast, heavily-armoured T-72s. The submarines included the first of the 25,000-ton Typhoon class, 170 m (557 ft) long, with tubes for 20 SS-NX-20 missiles, each capable of delivering 12 nuclear warheads over a range of nearly 4,500 nm; and the deep-diving titanium-hulled Alfa attack class with a submerged speed of over 40 knots. Sea trials of the 23,000-ton guided missile cruiser *Kirov*, the first Soviet nuclear-powered surface combatant, began in May 1980. Thirteen months later, an equally-impressive new giant of the air was demonstrated publicly for the first time, when a Mil Mi-26, the world's largest production helicopter, put in an appearance at the 1981 Paris Air Show.

A detailed description of the Mi-26 is included in the Soviet section of this edition of *Jane's*. Added to photographs and much-expanded information on other Soviet military and civil aircraft, it presents a fascinating picture of the current Soviet state of the art. Helicopter engineers who studied and discussed the Mi-26 at Le Bourget were immensely impressed by its unusual and efficient rotor head and gearbox, the large plastics rotor blades, and the TV-monitored in-flight-adjustable telescopic main landing gear legs which facilitate landing on a slope.

Good photographs of the Sukhoi Su-24 in this edition reveal for the first time features of the Soviet Union's counterpart to the F-111 and Tornado. The power plant of the bulged-fuselage export models of the variable-geometry Sukhoi 'Fitter' is identified for the first time, with an unexpectedly-higher thrust rating than that of

Soviet Air Force 'Fitters'. Never before illustrated in the Western press, at the time this Foreword is being written, are a version of the Ka-25 helicopter with a container for wire-guided torpedoes, and the new turbine-powered Ka-126.

The Soviet Union has nearly 900,000 scientists and engineers engaged full-time in research and development—50 per cent more than even the technology-conscious USA. *Jane's* makes clear the capability of those engaged in aerospace programmes, and only regrets that so much more has to remain cloaked in military security. References are made in this edition to new fighters, and a counterpart to the USAF's A-10 Thunderbolt II attack aircraft, which are known to have been in existence for some time but of which no photographs or accurate drawings have yet become available. The US Department of Defense has referred to a 'Modified Foxbat' which will be "the Soviet's first look-down/shoot-down fighter", armed with four new AA-X-9 missiles; a new airborne warning and control system (AWACS) which will replace the Tu-126 from the mid-1980s; and "evidence indicating that the Soviets are in the process of developing a new long-range bomber, and possibly a strategic cruise missile carrier".

Summarising its latest beliefs in a publication entitled *Soviet Military Power,* the DoD states:

> "During the 1970s, the Soviets have dramatically reduced the US lead in virtually every important basic technology. The United States is losing its lead in key technologies, including electro-optical sensors, guidance and navigation, hydro-acoustics, optics and propulsion . . . . The Soviet Union is estimated to have taken the lead in the development of directed energy weapons such as high-power lasers and possibly radio frequency devices. It has been interested in particle beam weapon concepts since the early 1950s. In many areas where the United States continues to lead the Soviets, their technology has achieved a level of adequacy with respect to present military requirements. (However,) the United States still leads by two to seven years in micro-electronics, computers and jet engines critical to the development of advanced weapon systems."

* * * * * *

Both America's Reagan administration and Britain's Conservative government were elected on programmes which included promised improvements in defence capability. Their ability to keep their word has been affected by world recession and economic problems. What *has* been done provides a strange contrast with the mighty efforts being made to build up the Soviet armed forces.

After a confident start, the US administration was compelled to announce a reduction of $1,300 million in defence expenditure over the next three years. If all the proposals are accepted, production of the KC-10 Extender flight refuelling tanker will be terminated, putting continuation of the entire DC-10 programme in jeopardy. The next procurement of A-10 Thunderbolts will be cut from 60 to 20. F-15 Eagle production will be reduced and stretched out. The B-52D fleet and Titan II ICBM force will be retired earlier than intended. The Anglo/American JP233 airfield attack weapon system will be dropped by the US, and cuts made in a whole range of other high-technology programmes, including those for KC-135 re-engining, the Martin Marietta LANTIRN infra-red nav/attack system for the F-16 and A-10, the Wasp anti-armour missile and the AMRAAM NATO-standardised air-to-air missile.

Nuclear force priority will be maintained, and in September 1981 came news which indicated a fresh and realistic approach to the problem of providing at affordable cost a fully-effective deterrent for the remaining years of our century.

Last year's Foreword to *Jane's* reflected the views of many who had watched anxiously the growing imbalance of East/West military power on which peace and stability depend. In its opening paragraphs it stated that "America needs urgently a modern aircraft like the B-1", the supersonic strategic bomber which former President Carter had cancelled in 1977. This was a requirement that *Jane's* had urged repeatedly since the B-1 cancellation, which had left the USAF with B-52s, none now less than nineteen years old. It was followed by the remark that "we have reached the time when America's Minuteman force is vulnerable to Soviet ICBM attack, making development and deployment of the MX missile system utterly essential."

It is good to record that the Reagan Strategic Program of September 1981 calls for development and deployment of 100 variants of the B-1; development and deployment of at least 100 MX missiles; and continued vigorous R & D effort on the so-called 'stealth' bomber for deployment in the 1990s. Fears that the new-production B-1B would be ponderous and toothless by comparison with the original version have proved groundless. T-O weight has increased sharply, to 216,360 kg (477,000 lb) with a full load of 30 air-launched cruise missiles, and maximum wing sweep will be about 59° 30′ now that Mach 2 performance is no longer required. But, compared with the original B-1, the all-important avionics of the 'B' will be greatly extended; and embodiment of some 'stealth' technology, added to reduced wing sweep and power plant installational changes, will reduce the radar signature.

Greater flexibility compared with a missile force, and vastly improved survivability compared with the B-52, will make the B-1B in every way a sharp spearhead of the US deterrent into the 21st century. The modest number on order makes sense. It is adequate to form a *real* deterrent, yet not wasteful in a period when all people of goodwill pray that future SALT agreements will achieve genuine, speedy and massive reductions in strategic nuclear forces.

Certainly, cruise missiles are better air-launched than ground-launched. The success rate attained so far in the ALCM development programme must be discouraging for anyone living in eastern England, between proposed GLCM launch areas and the coast. It is clearly preferable to designate wartime launch areas for all kinds of cruise missiles in places where the nuclear warheads would no longer threaten friend as well as foe.

By comparison with the Reagan Strategic Program, the much-revised UK defence programme seems all wrong, with the emphasis on false economies, an anachronistic desire to retain the trappings of a military super-power, and an apparent wish to earn the approval of a transatlantic senior partner. The motives could be right, but the consequences are indefensible.

Despite promises to improve UK defences, the Conservative government decided to put one of the Royal Navy's three new Invincible class anti-submarine aircraft carriers into mothballs within days of the launch of HMS *Ark Royal* by the Queen Mother, and to plan early retirement in 1983 for the newly-modified assault carrier *Hermes.* Earlier in the year, it had been announced that some RAF Vulcan, Canberra and Shackleton AEW.2 aircraft were to be retired ahead of schedule to reduce fuel and maintenance costs, which are higher for older types. Plans to form a third squadron of Lightning interceptors with mothballed aircraft were dropped; and work on the improved Skyflash Mk 2 air-to-air missile for the Tornado F.Mk 2 was terminated.

In July 1981 came the decree that Britain cannot afford a direct and early replacement for the Jaguar ground attack aircraft which fills a key role in NATO's front-line force in Europe. Yet, throughout all these parings of an already-slender defence effort, the proposed Trident missile submarine fleet cruised on, its estimated total cost rising from £5,000 million to £8,000 million, with a likely massive supplement if (as seems inevitable) the UK follows America in having improved D-5 missiles instead of the earlier model once specified.

With equal inevitability, in the Summer, came confirmation that Britain would enter into partnership with the USA on the AV-8B programme, rather than continue with BAe's Big Wing Harrier, designed specifically to meet RAF requirements. From a purely short-term financial viewpoint, the decision could not be faulted. The RAF and US Marine Corps have a combined initial requirement for 400 aircraft, the export potential is enormous, and no-one will dispute the advantage of even one step involving standardisation of equipment between NATO partners. Furthermore, the AV-8B is a superb new version of a wholly unique British concept, which brought fresh meaning to the term close support and was clearly the progenitor of whole generations of future designs. Here was the hidden barb, for by choosing the AV-8B the UK government was compelled to concede programme leadership to McDonnell Douglas of the USA, as well as 75 per cent of the airframe work on export orders. It will be interesting to see if British Aerospace ever regains prime rights to a great British technological breakthrough as it progresses to supersonic V/STOL designs in partnership with McDonnell Douglas.

* * * * * *

At this stage, it may be interesting to take a quick backward glance at the figures quoted earlier for Soviet combat aircraft production in 1980. The number of fighters and bombers built in that one year was more than double the entire first-line combat aircraft strength of the Royal Air Force. This is clearly an unfair comparison; but if, instead, we study the FY 1981 US budget procurement

(which set an expenditure record for any US peace or wartime defence budget), we find that the USAF was allowed a total of only 300 new aircraft, including 60 A-10s, 42 F-15s and 180 F-16s.

The FY 1982 budget inherited by the Reagan administration was even more depressing, with *total* provision for 96 F-16s, 30 F-15s and 4 TR-1s for the USAF. Little wonder that NATO remains so heavily outnumbered by the Warsaw Pact air forces in Europe.

Elsewhere throughout the world, the military aviation scene has changed little in the past year. Brazil and Italy have continued to develop jointly their AM-X fighter-bomber. Israel's Lavi strike fighter has also progressed steadily, and the Swedish industry now seems certain to beat off the challenge of foreign competition to provide a Viggen replacement for the 1990s. A much-expanded Chinese section in this edition reflects the growing capability of that nation's aerospace industry which, among other things, succeeded in launching a three-satellite payload for the first time in the late Summer and has flight tested an experimental four-engined airliner of its own design.

Libya does not appear to have succeeded yet in getting its embryo aircraft industry established, but the nations of South America are becoming more adventurous, following the spectacular achievements of EMBRAER. Peru is planning a new start, after a gap of more than 30 years, by assembling M.B 339s with assistance from Aermacchi. Chile is undertaking manufacture of the Pillan trainer, developed to its specific requirements by Piper, using Cherokee components, and has acquired a Falco kit from Sequoia to evaluate the potential of this 'Frati homebuilt' as a do-it-yourself aerobatic trainer. If the latest reports are true, these modest projects could be followed by licence production in Chile of the turbofan-powered CASA Aviojet.

International co-operation of this kind is increasing in an era when high costs and ever-advancing technology make it difficult for less-affluent or smaller countries to maintain a fully-competent design/manufacturing organisation. Thus, Indonesia has discontinued development of the indigenous Islander-like LAPAN XT-400 utility transport, and has linked its Nurtanio company with CASA of Spain in joint development of the far more ambitious 34/38-passenger Airtec CN-235.

Such programmes have caused the 'International' aircraft section of *Jane's* to expand from eight pages to 21 in just ten years. Even more significant is the high quality of the aircraft resulting from such collaboration—AM-X, Airbus, Alpha Jet, Concorde, BK 117 helicopter, Tornado, Saab-Fairchild 340, Jaguar, Orao and Transall. Technologically, the Concorde must be regarded as one of the major aviation triumphs of all time. The Tornado is at the threshold of what promises to be an outstanding military career. The Airbus family of transports has put its European manufacturers second only to Boeing as suppliers of economical, efficient airliners to operators in every part of the globe.

* * * * * *

Boeing itself continues to set production and operating records of such magnitude that they soar beyond comprehension. The 4,000th Boeing jet transport was delivered in July 1981, with well over 500 more on order. Those in service had flown a total of 44,709 million miles by 1 September, carrying 3,602 million passengers—equivalent to four out of every five people in the world. There could be few more dramatic examples of the extent to which air travel has expanded in the jet age, or of the contribution made by this single manufacturer.

Sadly, the achievements of commercial aviation are not always reflected in the financial results of the manufacturers and airlines. Reference was made earlier to the problems which McDonnell Douglas may face if KC-10 Extender procurement planned for the next two years does not materialise. The company was relying on these contracts to keep the DC-10 line busy in 1983, as there are not yet any firm commercial orders for the airliner which call for delivery in that year. This is frustrating, as market surveys suggest that the end of the recession will bring a strong demand for DC-10s during the decade from 1985, to the extent that Douglas had been planning two additional versions that would offer major fuel savings. The DC-10 Super 10 would be a Series 10 with winglets and PW2037 or RB.211-535F4 engines; the Super 30 would be a Series 30 stretched to seat 326 passengers in a mixed-class configuration, with CF6-80C1 or JT9D-7R4H engines and no winglets.

The future holds similar uncertainty for many major operators, hit by every kind of problem from effects of the recession to the lengthy strike by US air traffic controllers, high fuel costs and cut-throat low-fare competition. Pan American has had to ask its worldwide staff to accept a 10 per cent pay cut and a wage freeze throughout 1982. It had already sold its Intercontinental Hotels group to raise capital, and introduced fare cuts which added to the troubles of its domestic and long-haul competitors. British Airways sold its London terminal; made savage cuts in its route network; announced plans to reduce its 52,000 employees by 9,000 before June 1982; and offered for sale its entire all-cargo fleet of three 707-320Cs and a 747F, as well as a TriStar 500, two 707-320C passenger airliners and two undelivered 747-200Bs. Airline after airline has announced staggering financial losses—British Airways £145 million for the year ending 31 March; Pan Am $217 million in the first six months of 1981; Qantas between A$25 and A$30 million and Air New Zealand £14 million in 1980-81 . . . . .

In its mid-1981 analysis of airline results, IATA had no hesitation in referring to the previous year as financially the worst in airline history. During 1980, its members had taken delivery of 300 new aircraft, including 120 wide-bodied jets. This was the highest annual rate of fleet increase since 1968, in a year when passenger seat occupancy on international scheduled services slipped 2·4 points to 60·9 per cent. However, it was not all gloom. In the Far East and south-west Pacific area, traffic grew 21·8 per cent, and Latin American operations showed above-average rates of expansion. Safety statistics were also pleasing, with the lowest number of fatal accidents since 1976, equivalent to 0·13 per 100 million km flown.

The picture has been much the same for general aviation. In the Summer of this year, Beech suspended production of its Skipper two-seat trainer. Cessna decided to discontinue production of the Hawk XP, Skywagon 180, Ag Wagon and Model 310. Piper sold its complete existing inventory of Pawnee/Pawnee Brave agricultural aircraft and Super Cubs to WTA Inc of Lubbock, Texas, and terminated its factory-direct sales programme for these types. In France, a receiver was appointed at Avions Pierre Robin's Dijon works; but the company has six months in which to get its finances on a firm footing and its many friends will be hoping for better news soon.

Even for the largest and most renowned French independent aircraft manufacturer, the news was not all good in 1981. Following the election of President Mitterrand, the new Socialist administration took immediate steps to nationalise some of the major privately-owned manufacturers of military equipment, including Avions Marcel Dassault/Breguet Aviation and the military sectors of Matra and Thomson-Brandt. Those familiar with the years of chaos which followed nationalisation of the French aircraft industry in the mid-1930s will be grateful for the assurance that there is no intention of changing the identity, autonomy, executive appointments or programmes of these companies.

Together with Aérospatiale, Dassault/Breguet played a major role in increasing the turnover of the French aerospace industry from 20,500 million Francs in 1979 to more than 35,000 million in 1980, of which nearly 60 per cent came from exports. These exceeded imports by 13,700 million Francs, placing aerospace in second position among French exporting industries. Even more significant is that, in the Western world, the French aerospace industry now comes second to the United States in this respect—a remarkable success that owes much to the Airbus and CFM56 turbofan international programmes.

Britain's aerospace industry also reported record export receipts of more than £1,775 million in 1980, representing an increase of 39·8 per cent compared with the previous year, against inflation for the period of 15·1 per cent.

A glance through the list of aircraft which have made their first flights between July 1980 and October 1981 (on page [77]) will emphasise the strength of the advanced-technology base on which the aerospace industries of the world are building for the future. They include civil and military aircraft of every category, from the 211-passenger Boeing 767 and 71/93-passenger BAe 146 airliners, through commuter and business transports, supersonic combat aircraft and spyplanes to large and small helicopters, touring and aerobatic lightplanes, trainers, sailplanes, motor gliders and the comparatively new generation of microlights.

This last category of highly-personal aircraft has generated a surprising volume of comment, correspondence and controversy considering its diminutive, often primitive, vehicles. Is 'microlight' the correct generic term, or should we prefer ultralight, featherweight or one of the other printable alternatives? What particular features qualify a design as a microlight? . . . . and so on. With the expert assistance of Mrs Ann Welch and the FAI, the guidelines

were established. To everyone's surprise, by the time the microlights and hang gliders had been sorted into a fit state for the printer, they filled 29 pages—nearly as many as the main French aircraft section. If anyone considers this unjustified, he should remember the full title of our book and the fact that some 9,000 microlights are now flying in 15 or more countries worldwide, including 6,000 in the USA and 200 in the UK.

An aircraft which almost qualified as a microlight, but seemed more at home among the 'professional' types, is Paul MacCready's Solar Challenger, which made history on 7 July 1981 by completing the first flight from France to England (no less than 163 nm; 302 km; 188 miles from T-O to landing) under solar power. The design team led by this same remarkable American had earlier been responsible for the manpowered aircraft which made the first flight in figure-of-eight pattern around two pylons half a mile apart, and the first cross-Channel flight, by aircraft of this type.

The most spectacular single achievement of the past year, in a very different category, provides the subject for the frontispiece illustration in this edition of *Jane's*. The blast-off of Space Shuttle Orbiter OV-102 *Columbia* from Kennedy Space Center, Florida, took place on 12 April 1981, exactly 20 years after Yuri Gagarin became the first man to orbit the Earth. Never has any branch of science progressed so rapidly, or to such effect. On this occasion, *Columbia* encircled the Earth fairly uneventfully for just over two days, before making an immaculate unpowered touchdown at Edwards Air Force Base, California, watched on TV by millions of viewers worldwide. With its sister-craft, it will soon begin the space missions for which it was designed, to the benefit of everyone on Earth.

Also in this year of 1981, our TV screens have brought us close-up photographs of the rings, moons and swirling clouds of the planet Saturn, as seen from Voyager 2. Next port of call for this spacecraft is Uranus, in January 1986, after which it will journey on to Neptune by August 1989. The skill with which it can be manoeuvred through the vast distances of interplanetary space, and positioned precisely to take photographs of particular features such as tiny moons never seen from Earth, can only fill Earthbound engineers and airmen with wonder. Some folk have become more blasé, and there are suggestions that the cameras of Voyager 2 may remain switched off when it reaches the distant planets because NASA is short of funds. The world in which we live is sometimes as difficult to comprehend as are the mysteries of space.

* * * * * *

For a variety of reasons, this edition of *Jane's* has been particularly difficult to compile to the standards of up-to-the-minute accuracy essential to its highly-professional readers. Worldwide economic problems have delayed or prevented a go-ahead on programmes for many new aircraft, which explains the inclusion of a few important designs that might not qualify under our usual rule of coverage only when metal is being cut on the prototype. There was just time to delete some of the non-starters at the final proof stage. More difficult was to cope at that time with developments like the merging of MBB and VFW aircraft production centres and programmes in Germany, and the planned restructuring of almost the entire Italian industry.

Thanks to the enthusiasm and effort put into the job by many people, in countries all over the world, the work is now almost complete for 1980-81. Two things remain for the Editor to do. The first is to ask forgiveness if thoughts or suggestions expressed in this Foreword appear presumptuous. They were written in the spirit of the wise counsel of the late King Abdul Aziz of Saudi Arabia which was quoted in last year's edition: "Your friend is he who tells you the truth, not he who keeps telling you you're right." At *Jane's*, we depend on friends living in every nation that builds aircraft, to ensure that our information is as accurate, up-to-date and comprehensive as national security permits.

The second duty that the Editor must undertake before this final page goes to press is to thank them all. To do this sincerely is easy; to do so adequately is impossible. A representative of one US manufacturer apologised for the loss of a little time in meeting the deadline for his 19 pages of material, explaining that a senior engineer had spent dozens of man-hours checking carefully every individual fact and figure.

My warm thanks to such people, everywhere—and to the smaller band of friends and correspondents who provide so much valuable additional material: the editorial staffs of *Air Force Magazine* (Washington) and *Flug Revue* (West Germany), with whom we have worked so happily for many years through our *Jane's* Supplements; Delden Badcock in Australia; Ronaldo S. Olive and Roberto Pereira de Andrade in Brazil; Vico Rosaspina in Italy; Eiichiro Sekigawa in Japan; Antonio Camarasa and Javier Taibo in Spain; Roland Eichenberger and Dr Ulrich Haller in Switzerland; Norman Polmar and Tom DeFrank in the USA; Wolfgang Wagner in West Germany; Dipl Ing Andrzej Glass at the Instytut Lotnictwa in Poland; William Green and Gordon Swanborough of *Air International,* David Dorrell of *Air Pictorial,* and Alan Hall of *Aviation News* in the UK; Jay Miller of *Aerophile* (USA); and the editorial staffs of *Flight International* (UK), *Aviation Magazine* (France); *FLYGvapenNYTT* (Sweden); *Herkenning* and *de Vliegende Hollander* (Netherlands); *Skrzydlata Polska* (Poland); and *Letectvi + Kosmonautika* (Czechoslovakia).

Despite the ever-increasing costs of materials and postage, the familiar small band of highly-professional photographers has again met every demand for prints this year. Their names often appear in captions, but special thanks must go to Howard Levy, J. M. G. Gradidge, Brian M. Service, Denis Hughes, Gordon S. Williams, Don Berliner, Peter M. Bowers, Austin J. Brown, Don Dwiggins, Katsumi Hinata, Geoffrey P. Jones and Neil Macdougall. Our book would be incomplete and far less attractive without the contributions made by these friends, and by Dennis Punnett of Pilot Press and Michael A. Badrocke whose unique skills produce our three-view drawings.

Most names in the preceding paragraphs have appeared on this page of *Jane's* for ten or twenty years. Such continuity and experience provide the solid foundation on which each annual edition is built. At its heart is the editorial team, whose seven members have given a combined total of 118 years of service to this unique publication. The task of Assistant Editor Kenneth Munson was made more arduous this year by his agreement to accept responsibility for the new Microlights section. It would be quite impossible to attempt compilation of this book without his loyal and tireless contributions, and those of Maurice Allward, Bill Gunston, David Mondey, Michael Taylor and the Lord Ventry. Thanks—and well done!

October 1981 — J W R T

YOUR FORTUNE

747
757
767

GLOSSARY

AAM Air-to-air missile.
AC Alternating current.
ACLS (1) Automatic carrier landing system; (2) Air cushion landing system.
ADAC Avion de décollage et attérrissage court (STOL).
ADAV Avion de décollage et attérrissage vertical (VTOL).
ADC (1) US Air Force Aerospace Defense Command (no longer active); (2) air data computer.
ADF Automatic direction finding (equipment).
ADG Accessory-drive generator.
ADI Attitude/director indicator.
aeroplane (N America, airplane) Heavier-than-air aircraft with propulsion and a wing that does not rotate in order to generate lift.
AEW Airborne early warning.
AFB Air Force Base (USA).
AFCS Automatic flight control system.
afterburning Temporarily augmenting the thrust of a turbofan or turbojet by burning additional fuel in the jetpipe.
AGREE Advisory Group on Reliability in Electronic Equipment.
Ah Ampère-hours.
AHRS Attitude/heading reference system.
AIDS Airborne integrated data system.
aircraft All man-made vehicles for off-surface navigation within the atmosphere, including helicopters and balloons.
airstair Retractable stairway built into aircraft.
AM Amplitude modulation.
anhedral Downward slope of wing seen from front, in direction from root to tip.
AP Ammonium perchlorate.
APFD Autopilot flight director.
aphelion The point in a solar (Sun-centred) orbit furthest from the Sun.
apogee The point in an Earth-centred orbit furthest from the Earth.
approach noise Measured 1 nm from downwind end of runway with aircraft passing overhead at 112·6 m (370 ft).
APS Aircraft prepared for service; a fully equipped weight.
APU Auxiliary power unit (part of aircraft).
ARINC Aeronautical Radio Inc, US company whose electronic box sizes (racking sizes) are the international standard.
ASE Automatic stabilisation equipment.
ASI Airspeed indicator.
ASIR Airspeed indicator reading.
ASM Air-to-surface missile.
aspect ratio Measure of wing (or other aerofoil) slenderness seen in plan view, usually defined as the square of the span divided by area.
ASV (1) Air-to-surface-vessel; (2) Anti-surface-vessel.
ASW Anti-submarine warfare.
ATC Air traffic control.
ATR Airline transport radio, series of ARINC standard box sizes.
attack, angle of Angle at which airstream meets aerofoil (angle between mean chord and free-stream direction). Not to be confused with angle of incidence (which see).
augmented Boosted by afterburning.
autogyro Rotary-wing aircraft propelled by a propeller (or other thrusting device) and lifted by a freely running autorotating rotor.
AUW All-up weight (term meaning total weight of aircraft under defined conditions, or at specific time during flight). Not to be confused with MTOGW (which see).
avionics Aviation electronics, such as communications radio, radars, navigation systems and computers.
AWACS Airborne warning and control system (aircraft).

bar Non-SI unit of pressure adopted by this yearbook pending wider acceptance of Pa. 1 bar = 10^5 Pa, and ISA pressure at S/L is 1,013·2 mb, or just over 1 bar.
bare weight Undefined term meaning unequipped empty weight.
basic operating weight MTOGW minus payload (thus, including crew, fuel and oil, bar stocks, cutlery etc).
BCAR British Civil Airworthiness Requirements.
Beta mode Propeller or rotor operating regime in which pilot has direct control of pitch.
BFO Beat-frequency oscillator.
BITE Built-in test equipment.
bladder tank Fuel (or other fluid) tank of flexible material.
bleed air Hot high-pressure air extracted from gas-turbine engine compressor or combustor and taken through valves and pipes to perform useful work such as driving machinery or anti-icing by heating surfaces.
blown flap Flap across which bleed air is discharged at high (often supersonic) speed to prevent flow-breakaway.
BOW Basic operating weight.
BPR Bypass ratio.
BTU Non-SI unit of energy (British Thermal Unit) = 0·9478 J.
bus Busbar, main terminal in electrical system to which battery or generator power is supplied.
bypass ratio Airflow through fan duct (not passing through core) divided by airflow through core.

CAA Civil Aviation Administration (UK).
CAB Civil Aeronautics Board (USA).
CAB Pt 298 Sets the commercial standards for non-certificated carriers, mainly commuter airlines.
cabin altitude Height above S/L at which ambient pressure is same as inside cabin.
CAM Cockpit-angle measure (crew field of view).
CAN 5 Committee on Aircraft Noise (ICAO) rules for new designs of aircraft.
canards Foreplanes, fixed or controllable aerodynamic surfaces ahead of CG.
CAR Civil Airworthiness Regulations.
CAS Calibrated airspeed, ASI calibrated to allow for air compressibility according to ISA S/L.
CBR California bearing ratio, measure of ability of airfield surface (paved or not) to support aircraft.
CCV Control configured vehicle.
CEAM Centre d'Expériences Aériennes Militaires.
CEAT Centre d'Essais Aéronautiques de Toulouse.
CEP Circular error probability (50/50 chance of hit being inside or outside) in bombing, missile attack or gunnery.
CEV Centre d'Essais en Vol.
CFRP Carbonfibre-reinforced plastics.
CG Centre of gravity.
chaff Thin slivers of radar-reflective material cut to length appropriate to wavelengths of hostile radars and scattered in clouds to protect friendly aircraft.
chord Distance from leading-edge to trailing-edge measured parallel to longitudinal axis.
clean In-flight configuration with landing gear, flaps, slats etc retracted.
'clean' Without any optional external stores.
c/n Construction (or constructor's) number.
comint Communications intelligence
composite material Made of two constituents, such as filaments or short whiskers plus adhesive.
CONUS Continental USA (ie, excluding Hawaii, etc).
convertible Transport aircraft able to be equipped to carry passengers or cargo.
core Gas generator portion of turbofan comprising compressor(s), combustion chamber and turbine(s).
CRT Cathode-ray tube.
CSAS Command and stability augmentation system (part of AFCS).
CSD Constant-speed drive (output shaft speed held steady, no matter how input may vary).

daN Decanewtons (Newtons force × 10).
dB Decibel.
DC Direct current.
derated Engine restricted to power less than potential maximum (usually such engine is flat rated).
design weight Different authorities have different definitions; weight chosen as typical of mission but usually much less than MTOGW.
DF Direction finder, or direction finding.
DGAC Direction Générale à l'Aviation Civile.
dibber bomb Designed to cause maximum damage to concrete runways.
dihedral Upward slope of wing seen from front, in direction from root to tip.
DINS Digital inertial navigation system.
disposable load Sum of masses that can be loaded or unloaded, including payload, crew, usable fuel etc; MTOGW minus OWE.
DME Distance-measuring equipment; gives slant distance to a beacon directly ahead.
dog-tooth A step in the leading-edge of a plane resulting from an increase in chord. (See also saw-tooth.)
Doppler Short for Doppler radar—radar using fact that received frequency is a function of relative velocity between transmitter or reflecting surface and receiver.
double-slotted flap One having an auxiliary aerofoil ahead of main surface to increase maximum lift.
dP Maximum design differential pressure between pressurised cabin and ambient (outside atmosphere).
drone Pilotless aircraft, usually winged, following preset programme of manoeuvres.

EAA Experimental Aircraft Association (divided into local branches called Chapters).
EAS Equivalent airspeed, RAS minus correction for compressibility.
ECCM Electronic counter-countermeasures.
ECM Electronic countermeasures.
ehp Equivalent horsepower, measure of propulsive power of turboprop made up of shp plus addition due to residual thrust from jet.
ekW Equivalent kilowatts, SI measure of propulsive power of turboprop (see ehp).
elevon Wing trailing-edge control surface combining functions of aileron and elevator.
elint Electronics intelligence.
ELT Emergency locator transmitter, to help rescuers home on to a disabled or crashed aircraft.
EPA Environmental Protection Agency.
EPNdB Effective perceived noise decibel, SI unit of EPNL.
EPNL Effective perceived noise level, measure of noise effect on humans which takes account of sound intensity, frequency, character and duration, and response of human ear.
EPU Emergency power unit (part of aircraft, not used for propulsion).
ERP Effective radiated power.
ESA European Space Agency.
ESM (1) Electronic surveillance (or support) measures; (2) Electronic signal monitoring.
EVA Extra-vehicular activity, ie outside spacecraft.

FAA Federal Aviation Administration.
factored Multiplied by an agreed number to take account of extreme adverse conditions, errors, design deficiencies or other inaccuracies.
FAI Fédération Aéronautique Internationale.
fail-operational System which continues to function after any single fault has occurred.
fail-safe Structure or system which survives failure (in case of system, may no longer function normally).
FAR Federal Aviation Regulations.
FAR Pt 23 Defines the airworthiness of private and air-taxi aeroplanes of 5,670 kg (12,500 lb) MTOGW and below.
FAR Pt 25 Defines the airworthiness of public transport aeroplanes exceeding 5,670 kg (12,500 lb) MTOGW.
FBW Fly by wire (which see).
FDS Flight director system.
feathering Setting propeller or similar blades at pitch aligned with slipstream, to give resultant torque (not tending to turn shaft) and thus minimum drag.
FEBA Forward edge of battle area.
fence A chordwise projection on the surface of a wing, used to modify the distribution of pressure.
fenestron Helicopter tail rotor with many slender blades rotating in short duct.
ferry range Extreme safe range with zero payload.
FFAR Folding-fin (or free-flight) aircraft rocket.
FFVV Fédération Française de Vol à Voile (French gliding authority).
field length Measure of distance needed to land and/or take off; many different measures for particular purposes, each precisely defined.
flaperon Wing trailing-edge surface combining functions of flap and aileron.
flat-four Engine having four horizontally opposed cylinders; thus, flat-twin, flat-six etc.
flat rated Propulsion engine capable of giving full thrust or power for take-off up to high airfield height and/or high ambient temperature (thus, probably derated at S/L).
FLIR Forward-looking infra-red.
fly by wire Flight control system with electrical signalling (ie, without mechanical interconnection between cockpit flying controls and control surfaces).
FM Frequency modulation.
footprint A precisely delineated boundary on the surface, inside which the perceived noise of an aircraft exceeds a specified level during take-off and/or landing.
Fowler flap Moves initially aft to increase wing area and then also deflects down to increase drag.
free turbine Turbine mechanically independent of engine upstream, other than being connected by rotating bearings and the gas stream, and thus able to run at its own speed.
Frise aileron Most common manual aileron, with leading-edge that projects below wing to increase drag when aileron is raised.
FY Fiscal year (1 July to 30 June in US government affairs).

g Acceleration due to mean Earth gravity, ie of a body in free fall.
gallons Non-SI measure; 1 Imp gal (UK) = 4·546 litres, 1 US gal = 3·785 litres.
GCI Ground-controlled interception.
geostationary Of an Earth satellite, rotating with the Earth and thus always overhead the same point. Corresponds to altitude above Earth's surface of about 35,800 km (22,245 miles).
geostationary orbit An Earth-centred orbit at a height above the Earth's surface of about 35,800 km (22,245 miles) and lying approximately in the plane of the equator. A satellite in such an orbit travelling eastwards will remain over the same point, rotating precisely with the Earth.
geosynchronous See geostationary.

GfK Glassfibre-reinforced plastics (German).
glide ratio Of a sailplane, distance travelled along track divided by height lost in still air.
glove In a swing-wing aeroplane with pivots well out from the centreline it is geometrically impossible to have one-piece pivoted wings because at zero sweep the inner ends would overlap; the answer is fixed inner leading portions called gloves.
GPU Ground power unit (not part of aircraft).
GPWS Ground-proximity warning system.
green aircraft Aircraft flyable but lacking furnishing and customer's choice of avionics.
gross wing area See wing area.
GRP Glassfibre-reinforced plastics.
GS Glideslope, of ILS.
GSE Ground-support equipment (such as special test gear, steps and servicing platforms).
GTS Gas-turbine starter (ie starter is miniature gas turbine).
gunship Helicopter designed for battlefield attack, normally with slim body carrying pilot and weapon operator only.

h Hour(s).
hardened Protected as far as possible against nuclear explosion.
hardpoint Reinforced part of aircraft to which external load can be attached, eg weapon or tank pylon.
helicopter Rotary-wing aircraft both lifted and propelled by one or more power-driven rotors turning about substantially vertical axes.
HF High frequency.
'hot and high' Adverse combination of airfield height and high ambient temperature, which lengthens required TOD.
hovering ceiling Ceiling of helicopter (corresponding to air density at which maximum rate of climb is zero), either IGE or OGE.
HP High pressure.
hp Horsepower.
HSI Horizontal situation indicator.
HUD Head-up display (bright numbers and symbols projected on pilot's windscreen and focused on infinity so that pilot can simultaneously read display and look ahead).
HVAR High-velocity aircraft rocket.
Hz Hertz, cycles per second.

IAS Indicated airspeed, ASIR corrected for instrument error.
IATA International Air Transport Association.
ICAO International Civil Aviation Organization.
IFF Identification friend or foe.
IFR Instrument flight rules (ie, not VFR).
IGE In ground effect: helicopter performance with theoretical flat horizontal surface just below the wheels.
ILS Instrument landing system.
IMC Instrument meteorological conditions, basically IFR.
IMK Increased manoeuvrability kit.
INAS Integrated nav/attack system.
incidence Strictly, the angle at which the wing is set in relation to the fore/aft axis. Wrongly used to mean angle of attack (which see).
inertial navigation Measuring all accelerations imparted to a vehicle and, by integrating these with respect to time, calculating speed at every instant (in all three planes) and by integrating a second time calculating total change of position in relation to starting point.
INS Inertial navigation system.
integral construction Machined from solid instead of assembled from separate parts.
integral tank Fuel or other liquid tank formed by sealing part of structure.
intercom Wired telephone system for communication within aircraft.
inverter Electric or electronic device for inverting (reversing polarity of) alternate waves in AC power to produce DC.
IP Intermediate pressure.
IR Infra-red.
IRAN Inspect and repair as necessary.
IRLS Infra-red linescan (builds TV-type picture showing cool regions as dark and hot regions as light).
ISA International Standard Atmosphere.
ISIS (1 Boeing Vertol) Integral spar inspection system; (2 Ferranti) integrated strike and interception sight.
ITE Involute throat and exit (rocket nozzle).

J Joule(s). SI unit of energy.
JASDF Japan Air Self-Defence Force.
JATO Jet-assisted take-off (actually means rocket-assisted).
JCAB Japan Civil Airworthiness Board.
JDA Japan Defence Agency.
JGSDF Japan Ground Self-Defence Force.
JMSDF Japan Maritime Self-Defence Force.

Kevlar A filament-wound high-strength composite material.
km/h Kilometres per hour.
kN Kilonewtons (the Newton is the SI unit of force; 1 lbf = 4·448 N).
knot 1 nm per hour.
Krüger flap Hinges down and then forward from below the leading edge.
Küchemann tip Wing tip of curving planform intended to minimise drag at high subsonic speed.
kVA Kilovolt-ampères.
kW Kilowatt, SI measure of all forms of power (not just electrical).

LABS Low-altitude bombing system.
LARC Low-altitude ride control.
LBA Luftfahrtbundesamt (Federal German civil aviation authority).
lbf Pounds of thrust.
LCN Load classification number, measure of 'flotation' of aircraft landing gear linking aircraft weight, weight distribution, tyre numbers, pressures and disposition.
LED Light-emitting diode.
LITVC Liquid-injection thrust vector control.
LLTV Low-light TV (thus, LLLTV, low-light-level).
Load factor (1) percentage of max payload; (2) stress limit.
LOC Localiser.
localiser Element giving steering guidance in ILS.
loiter Flight for maximum endurance, such as supersonic fighter on patrol.
longerons Principal fore-and-aft structural members (eg, in fuselage).
Loran (Long Range Navigation) Family of hyperbolic navaids based on ground radio emissions.
lox Liquid oxygen.
LP Low pressure.
LRMTS Laser ranger and marked-target seeker.

m Metre(s), SI unit of length.
M or Mach number The ratio of the speed of a body to the speed of sound (1,116 ft; 340 m/sec in air at 15°C) under the same ambient conditions.
MAC US Air Force Military Airlift Command.
MAD Magnetic anomaly detector.
Madar Maintenance analysis, detection and recording.
Madge Microwave aircraft digital guidance equipment.
marker, marker beacon Ground beacon giving position guidance in ILS.
MASTACS Manoeuvrability augmentation system for tactical air combat simulation.
mb Millibars, bar $\times 10^{-3}$.
MBR Marker beacon receiver.
MEPU Monofuel emergency power unit.
MF Medium frequency.
mg Milligrammes, grammes $\times 10^{-3}$.
MLS Microwave landing system.
MLW Maximum landing weight.
mm Millimetres, metres $\times 10^{-3}$.
MMH Monomethyl hydrazine.
MMO Maximum permitted operating Mach number.
MNPS Minimum navigation performance specification.
monocoque Structure with strength in outer shell, devoid of internal bracing.
MPA Man-powered aircraft.
mph Miles per hour.
MRW Maximum ramp weight.
MTBF Mean time between failures.
MTOGW Maximum take-off gross weight (MRW minus taxi/run-up fuel).
MZFW Maximum zero-fuel weight.

NACA US National Advisory Committee for Aeronautics (now NASA).
Nadge NATO air defence ground environment.
NASA National Aeronautics and Space Administration.
NASC US Naval Air Systems Command (also several other aerospace meanings).
NATC US Naval Air Training Command or Test Center (also several other aerospace meanings).
NBAA US National Business Aircraft Association.
NH_4ClO_4 Ammonium perchlorate.
nm Nautical mile, 1·8532 km, 1·15152 miles.
NOAA US National Oceanic and Atmospheric Administration.
NOE Nap-of-the-Earth (low flying in military aircraft, using natural cover of hills, trees, etc).
NOGS Night observation gunship.
NOS Night observation surveillance.
Ns Newton-second (1 N thrust applied for 1 second).

OBS Omni-bearing selector.
OCU Operational Conversion Unit.
OGE Out of ground effect: helicopter hovering far above nearest surface.
Omega Long-range hyperbolic navaid.
OMI Omni-bearing magnetic indicator.
omni Generalised word meaning equal in all directions (as in omni-range, omni-flash beacon).
OTPI On-top position indicator (indicates overhead of submarine in ASW).
OUV Oskar-Ursinus-Vereinigung (West German chapter of EAA).
OWE Operating weight empty, MTOGW minus payload, usable fuel and oil and other consumables.

PA system Public-address.
pallet (1) for freight, rigid platform for handling by fork-lift or conveyor; (2) for missile, mounting and electronics box outside aircraft.
payload Disposable load generating revenue (passengers, cargo, mail and other paid items); in military aircraft loosely used to mean total load carried of weapons, cargo or other mission equipment.
PD radar Pulse-Doppler.
penaids Penetration aids, such as jammers, chaff or decoys to help aircraft fly safely through hostile airspace.
perigee The point in an Earth-centred orbit nearest to the Earth.
perihelion The point in a solar (Sun-centred) orbit closest to the Sun.
PFA Popular Flying Association (UK).
PHI Position and heading (or homing) indicator.
plane A lifting surface (eg, wing, tailplane).
pneumatic de-icing Covered with flexible surfaces alternately pumped up and deflated to throw off ice.
port Left side, looking forward.
power loading Aircraft weight (usually MTOGW) divided by total propulsive power or thrust at T-O.
pressure fuelling Fuelling via a leakproof connection through which fuel passes at high rate under pressure.
pressure ratio In gas-turbine engine, compressor delivery pressure divided by ambient pressure (in supersonic aircraft, divided by ram pressure downstream of inlet).
primary flight controls Those used to control trajectory of aircraft (thus, not trimmers, tabs, flaps, slats, airbrakes or lift dumpers etc).
pulse-Doppler Radar sending out pulses and measuring frequency-shift of returns from target(s).
pylon Structure linking aircraft to external load (engine nacelle, drop-tank, bomb etc). Also used in conventional sense in pylon racing.

radius In terms of performance, the distance an aircraft can fly from base and return without intermediate landing.
RAE Royal Aircraft Establishment.
RAI Registro Aeronautico Italiano.
ram pressure Increased pressure in forward-facing aircraft inlet, generated by converting (relative) kinetic energy to pressure.
ramp weight Maximum weight at start of flight (MTOGW plus taxi/run-up fuel).
range Too many definitions to list, but essentially the distance an aircraft can fly (or is permitted to fly) with specified load and usually whilst making allowance for specified additional manoeuvres (diversions, stand-off, go-around etc).
RAS Rectified airspeed, IAS corrected for position error.
raster Generation of large-area display, eg TV screen, by close-spaced horizontal lines scanned either alternately or in sequence.
RATT Radio teletype.
redundant Provided with spare capacity or data channels and thus made to survive failures.
refanned Gas-turbine engine fitted with new fan of higher BPR.
rigid rotor Helicopter rotor without articulating hinges (eg, flapping hinge, drag hinge) but with pitch variation.
RLD Rijksluchtvaartdienst, Netherlands civil aviation department.
RMI Radio magnetic indicator (compass).
R/Nav Area navigation, navaid covering whole of local area instead of just crowded airways.
Rotor-kite Rotary-wing aircraft with no internal power, lifted by a freely running autorotating rotor and towed by an external vehicle.
roving Multiple strands of fibre, as in a rope (but usually not twisted).
RPV Remotely piloted vehicle (pilot in other aircraft or on ground).
RSA Réseau du Sport de l'Air.
RVR Runway visual range.

s Second(s).
SAC US Air Force Strategic Air Command.
safe-life A term denoting that a component has proved by testing that it can be expected to continue to function safely for a precisely defined period before replacement.
salmon (French saumon) Streamlined fairings, usually at wingtip of sailplane, serving same function as endplate and acting also as tip-skid.
SAR (1) Search and rescue; (2) synthetic aperture radar.
SATS (1) Small airfield for tactical support; (2) Small Arms Target System.

Nearly every Free World airline flies Garrett flight systems today.

All the more reason why they'll want to fly them tomorrow, too.

For almost four decades, Garrett flight systems have been meeting the constantly-changing needs of airframe builders and airlines the world over. That's why we'll be ready with even more reliable, more cost-effective and more fuel-efficient systems needed for the advanced-technology jetliners that are on the way.

saw-tooth Same as dog-tooth.
second-source Production of identical item by second factory or company.
semi-active Homing on to radiation reflected from target illuminated by radar or laser energy beamed from elsewhere.
service ceiling Usually height equivalent to air density at which maximum attainable rate of climb is 100 ft/min.
servo A device which acts as a relay, usually augmenting the pilot's efforts to move a control surface or the like.
sfc Specific fuel consumption.
SGAC Secrétariat Général à l'Aviation Civile (now DGAC).
or other driven unit. Two-shaft engine has second shaft, rotating at different speed, surrounding the first (thus, HP surrounds inner LP or fan shaft).
Shoran Short range navigation (radio).
shp Shaft horsepower, measure of power transmitted via rotating shaft.
sideline noise EPNdB measure of aircraft landing and taking off, at point 0·25 nm (2- or 3-engined) or 0·35 nm (4-engined) from runway centreline.
SIF Selective identification facility.
sigint Signals intelligence.
signature Characteristic 'fingerprint' of all electromagnetic radiation (radar, IR etc).
single-shaft Gas turbine in which all compressors and turbines are on common shaft rotating together.
S/L Sea level.
SLAR Side-looking airborne radar.
snap-down Air-to-air interception of low-flying aircraft by AAM fired from fighter at a higher altitude.
soft target Not armoured or hardened.
specific fuel consumption Rate at which fuel is consumed divided by power or thrust developed, and thus a measure of engine efficiency. For jet engines (air-breathing, ie not rockets) unit is mg/Ns, milligrams per Newton-second; for shaft engines unit is μg/J, micrograms (millionths of a gram) per Joule (SI unit of work or energy).
specific impulse Measure of rocket engine efficiency; thrust divided by rate of fuel/oxidant consumption per second, the units for mass and force being the same so that the answer is expressed in seconds.
spool One complete axial compressor rotor; thus a two-shaft engine may have a fan plus an LP spool.
SSB Single-sideband (radio).
SSR Secondary surveillance radar.
st Static thrust.
stabiliser Fin (thus, horizontal stabiliser = tailplane).
stall strips Sharp-edged strips on wing leading-edge to induce stall at that point.
stalling speed TAS at which aircraft stalls at 1*g*, ie wing lift suddenly collapses.
standard day ISA temperature and pressure.
starboard Right side, looking forward.
static inverter Solid-state electric inverter (ie, not rotary machine).
stick-pusher Stall-protection device that forces pilot's control column forward as stalling angle of attack is neared.
stick-shaker Stall-warning device that noisily shakes pilot's control column as stalling angle of attack is neared.
STOL Short take-off and landing. (Several definitions, stipulating allowable horizontal distance to clear screen height of 35 or 50 ft or various SI measures).
store Object carried as part of payload on external attachment (eg bomb, drop-tank).
strobe light High-intensity flashing beacon.
substrate The underlying layer on which something (such as a solar cell or integrated circuit) is made.
supercritical wing Wing of relatively deep, flat-topped profile generating lift right across upper surface instead of concentrated close behind leading edge.
sweepback Backwards inclination of wing or other aerofoil, seen from above, measured relative to fuselage or other reference axis, usually measured at quarter-chord (25%) or at leading-edge.
synchronous See geostationary.
synchronous satellite Geostationary.

t Tonne, 1 Megagram, 1,000 kg.
tabbed flap Fitted with narrow-chord tab along entire trailing-edge which deflects to greater angle than main surface.
tabs Small auxiliary surfaces hinged to trailing-edge of control surfaces for purposes of trimming, reducing hinge moment (force needed to operate main surface) or in other way assisting pilot.
TAC US Air Force Tactical Air Command.
Tacan Tactical air navigation, simple military navaid using ground beacons.
taileron Left and right tailplanes used as primary control surfaces in both pitch and roll.
tailplane Main horizontal tail surface, originally fixed and carrying hinged elevator(s) but today often a single 'slab' serving as control surface.
TANS Tactical air navigation system (Doppler-based computer, control and display unit).
TAS True airspeed, EAS corrected for density (often very large factor) appropriate to aircraft height.
TBO Time between overhauls.
TFR Terrain-following radar (for low-level attack).
thickness Depth of wing or other aerofoil; maximum perpendicular distance between upper and lower surfaces.
T-O Take-off.
T-O noise EPNdB measure of aircraft taking off, at point directly under flight path 3·5 nm from brakes-release (regardless of altitude).
TOD Take-off distance.
TOGW Take-off gross weight (not necessarily MTOGW).
ton Imperial (long) ton = 1·016 t (Mg), US (short) ton = 0·9072 t.
track Distance between centres of contact areas of main landing wheels measured left/right across aircraft (with bogies, distance between centres of contact areas of each bogie).
transceiver Radio transmitter/receiver.
transfer orbit Orbit, or part of an orbit, linking two others at different heights around the same planetary body.
transponder Radio transmitter triggered automatically by a particular received signal.
TRU Transformer/rectifier unit.
TSO Technical Standard Order (FAA).
turbofan Gas-turbine jet engine generating most thrust by a large-diameter cowled fan, with small part added by jet from core.
turbojet Simplest form of gas turbine comprising compressor, combustion chamber, turbine and propulsive nozzle.
turboprop Gas turbine in which as much energy as possible is taken from gas jet and used to drive reduction gearbox and propeller.
turboshaft Gas turbine in which as much energy as possible is taken from gas jet and used to drive high-speed shaft (which in turn drives external load such as helicopter gearbox).
TVC Thrust vector control (rocket).
TWT Travelling-wave tube.
tyre sizes In simplest form, first figure is rim diameter (in or mm) and second is rim width (in or mm). In more correct three-unit form, first figure is outside diameter, second is max width and third is wheel diameter.

UHF Ultra-high frequency.
unfactored Performance level expected of average pilot, in average aircraft, without additional safety factors.
usable fuel Total mass of fuel consumable in flight, usually 95-98 per cent of system capacity.

variable-geometry Capable of grossly changing shape in flight, especially by varying sweep of wings.
VD Maximum permitted diving speed.
vernier Small thruster, usually a rocket, for final precise adjustment of a vehicle's trajectory and velocity.
VFR Visual flight rules.
VHF Very high frequency.
VLF Very low frequency.
VMO Maximum permitted operating flight speed (IAS, EAS or CAS must be specified).
VNE Never-exceed speed (aerodynamic or structural limit).
VOR VHF omni-directional range, ground navaid usable only when flying along predetermined airways.
VTOL Vertical take-off and landing.

washout Inbuilt wing twist reducing angle of incidence towards the tip.
wheelbase Minimum distance from nosewheel or tailwheel (centre of contact area) to line joining main wheels (centres of contact areas).
wing area Total projected area of clean wing (no flaps, slats etc) including all control surfaces and area of fuselage bounded by leading- and trailing-edges projected to centreline (inapplicable to slender-delta aircraft with extremely large leading-edge sweep angle). Sometimes called gross wing area; net area excludes projected areas of fuselage, nacelles, etc.
wing loading Aircraft weight (usually MTOGW) divided by wing area.
winglet Small auxiliary aerofoil, usually sharply upturned and often swept back, at tip of wing.
wire guidance Guidance of missile or RPV by signals transmitted through fine wire(s) linking it with operator.

zero-fuel weight MTOGW minus usable fuel and other consumables, in most aircraft imposing severest stress on wing.
zero/zero seat Ejection seat designed for use even at zero speed on ground.
ZFW Zero-fuel weight.
μg Microgrammes, grammes × 10^{-6}.

Jane's Aerospace Dictionary by Bill Gunston (Jane's Publishing Company, £15) provides a complete listing of current aerospace expressions with definitions.

FIRST FLIGHTS

Some first flights made during the period 1 July 1980 to 1 October 1981

July 1980

1 Quickie Aircraft Corporation Quickie Q2 (USA)
2 Maggi MG3-15L Condor sailplane (I-CXOS) (Italy)
11 Borg Rose Parakeet replica (N80RG) (USA)
12 Aérospatiale Epsilon, second prototype (France)
12 McDonnell Douglas KC-10 Extender (79-0433) (USA)
14 Sikorsky YSH-60B Seahawk, fifth prototype (161173) (USA)
16 BAe Nimrod AEW. Mk 3, first development aircraft (XZ286) (UK)
16 Wallis WA-122/R-R (G-BGGW) (UK)
18 Panavia Tornado F. Mk 2, second prototype (ZA267) (International)
21 Caproni Vizzola C22J (I-CAVJ) (Italy)
23 Aérospatiale AS 366G/HH-65A Dolphin (France)

August 1980

2 Pilatus Britten-Norman BN-2T Turbine Islander (G-BPBN) (UK)
7 MacCready Gossamer Penguin (USA)
11 Gates Learjet 50 series, first production (USA)
14 Lockheed C-5A Galaxy, first re-winged aircraft (USA)
16 EMBRAER EMB-312/YT-27 (1300) (Brazil)
16 Fournier RF6B-120 (F-WANF, later F-GANF) (France)
19 Boeing Vertol Model 234 Commercial Chinook (USA)

September 1980

3 Boeing B-52G, first aircraft retrofitted with OAS (offensive avionics system) (USA)
25 Cameron D-38 hot-air airship (G-BGEP) (UK)
30 Monnett Monex (N82MX) (USA)

October 1980

3 Boeing 747/JT9D-7R4D, engine testbed for Boeing 767 (USA)
9 Hoffmann H-36 Dimona motor glider (D-KDIM) (German Federal Republic)
10 Aérospatiale AS 332L Super Puma (F-WZJN) (France)
11 Dassault-Breguet Mirage 2000 B (B 01) (France)
16 Pottier P.100TS (F-WYJC) (France)
23 Striplin Lone Ranger microlight (USA)
24 Mancro (Fairchild) C-123T (56-4357) (USA)
29 General Dynamics F-16/79 (converted F-16B, 75-0752) (USA)
30 Panavia Tornado, first IDS aircraft for Luftwaffe (43 + 12) (International)
31 Aérospatiale Epsilon, first prototype, after modifications to tail etc (01) (France)

November 1980

6 MacCready Solar Challenger (first flight, but powered by battery) (USA)
8 Nash Petrel (G-AXSF) (UK)
14 Socata TB 20 Trinidad (F-WDBA) (France)
18 Panavia Tornado F. Mk 2, third prototype (ZA283) (International)
19 Ikarusflug Bodensee Windspiel 2 hang glider (German Federal Republic)
20 MBB F-104G/CCV, first flight in E1 (marginally stable) configuration (98 + 36) (German Federal Republic)
20 MacCready Solar Challenger (first solar-powered flight) (USA)

December 1980

2 Gyroflug Speed-Canard (D-EEEX) (German Federal Republic)
8 Robin R 3140 (France)
10 Airbus A300, first flight guided entirely by Cat.III digital autopilot (International)
12 Dassault-Breguet/Dornier Alpha Jet A1, first flight as TST testbed (98 + 33) (German Federal Republic)
13 Sikorsky CH-53E Super Stallion, first production aircraft (159876) (USA)
16 Glasflügel 402 sailplane (D-2611) (German Federal Republic)
17 Rochelt Solair I solar-powered microlight (German Federal Republic)
18 Boeing E-3A Sentry, first for NATO (minus rotodome) (79-0442) (USA)
19 General Dynamics F-16/101 (converted F-16A, 75-0745) (USA)
31 SZD-51 Junior sailplane (Poland)

January 1981

1 Lear Fan Model 2100 (N626BL) (USA)
8 McDonnell Douglas DC-9 Super 82 (USA)
8 McDonnell Douglas DC-10 Series 15 (N19B) (USA)
23 BAe Nimrod AEW. Mk 3, second development aircraft (XZ287) (UK)
26 Gulfstream American Hustler 500 (N501GA) (USA)

February 1981

5 ENSMA FS-25S Cuervo sailplane (F-WRFM) (France)
6 Sikorsky YEH-60B (USA)
11 Socata MS 892 Rallye (SACMA 1-150 engine) (France)
19 SOKO SL-40 Liska motor glider (001) (Yugoslavia)
19 Marconi Avionics Machan RPV (UK)
21 Schempp-Hirth Nimbus 3 sailplane (D-2111) (German Federal Republic)
26 Aerotec A-132 Tangará (1000) (Brazil)
28 Valentin Taifun 17E motor glider (D-KONO) (German Federal Republic)

March 1981

6 MBB/Kawasaki BK 117, first pre-production (D-HBKB) (International)
6 Fournier RF-10 motor glider (F-WARG) (France)
17 Gulfstream American Gulfstream II-B (N711SC) (USA)
21 Dornier MTC II RPV (German Federal Republic)
28 Dornier 228-100 (D-IFNS) (German Federal Republic)
30 Fairchild Swearingen Merlin IIIC, first production (USA)
30 Fairchild Swearingen Merlin IVC, first production (USA)

April 1981

7 Milomei M2 sailplane (D-2502) (German Federal Republic)
8 SIAI-Marchetti SF.600TP Canguro (I-CANG) (Italy)
9 Transall C-160, first aircraft of second series (International)
10 SIAI-Marchetti S.211 (I-SITF) (Italy)
12 ThunderColt Balloons A.S.80 hot-air airship (UK)
12 Space Shuttle Orbiter OV-102 *Columbia* (USA)
15 Dassault-Breguet Gardian (France)

May 1981

8 Dassault-Breguet Atlantic NG (01) (France)
9 Dornier 228-200 (D-ICDO) (German Federal Republic)
9 Boeing 747/RB.211-535C, engine testbed for Boeing 757 (USA)
15 CNPSL-Warszawa PZL-106B Kruk (SP-PKW) (Poland)
15 Slingsby T.67A (G-BIOW) (UK)
16 PIK-21 Super Sytky (OH-XTM) (Finland)
21 MBB BO 105 'Giraffe', with Ophelia mast-mounted sensor package (D-HABV) (German Federal Republic)
21 Mitsubishi Diamond I, first production (Japan/USA)
22 Gulfstream American Peregrine 600 (N600GA) (USA)
27 Transall C-160, second aircraft of second series (International)

June 1981

1 Shorts 360 (G-ROOM) (UK)
17 Panavia Tornado F. Mk 2, third prototype, first flight fitted with Foxhunter radar (ZA283) (International)
26 Grumman (General Dynamics) EF-111A, first production (66-049) (USA)

July 1981

8 Schleicher ASW 22 sailplane (D-7122) (German Federal Republic)
14 Grumman F-14 Tomcat, development aircraft with General Electric F101DFE engines (157986) (USA)
17 Piper T-1040 (USA)
24 Rockwell International HIMAT RPV, second prototype (USA)
31 HAL HPT-32 (improved version), third aircraft to fly (India)
31 SEPECAT Jaguar, testbed aircraft with Ferranti FIN 1064 inertial navigation system (XX-108) (International/UK)

August 1981

1 Lockheed TR-1A (80-1066) (USA)
13 Wiweko WEL-1 (R1-X) (Indonesia)
15 McDonnell Douglas DC-8 Series 71 (Srs 61 re-engined with CFM56s) (N8093U) (USA)
19 CNPSL-Warszawa PZL-106AS Kruk (SP-PBD) (Poland)
26 McDonnell Douglas F-15J Eagle, first Japanese-assembled (USA/Japan)
31 McDonnell Douglas DC-10 Series 10 fitted with winglets (N68048) (USA)

September 1981

3 BAe 146 Series 100 (G-SSSH) (UK)
10 Fairchild/Ames scale research aircraft for NGT (USA)
24 Sikorsky YEH-60A (USA)
25 Panavia Tornado, first Italian production (International)
26 Boeing 767 (N767BA) (USA)
28 Airship Industries Skyship 500 (G-BIHN) (UK)

Prototype Boeing 767 landing at Paine Field, Everett, Washington, after its first flight on 26 September 1981

OFFICIAL RECORDS

Corrected to October 1981

ABSOLUTE WORLD RECORDS

Seven records are classed as Absolute World Records for aeroplanes by the Fédération Aéronautique Internationale, as follows:

Distance in a straight line (USA)
Major Clyde P. Evely, USAF, in a Boeing B-52H Stratofortress, on 10-11 January 1962, from Okinawa to Madrid, Spain. 10,890·27 nm (20,168·78 km; 12,532·3 miles).

Distance in a closed circuit (USA)
Captain William M. Stevenson, USAF, in a Boeing B-52H Stratofortress, on 6-7 June 1962. Seymour Johnson AFB-Bermuda-Sondrestrom (Greenland)-Anchorage (Alaska)-March AFB-Key West-Seymour Johnson AFB. 9,851·54 nm (18,245·05 km; 11,337 miles).

Height (USSR)
Alexander Fedotov in an E-266M (MiG-25) on 31 August 1977. 37,650 m (123,523 ft).

Height in sustained horizontal flight (USA)
Captain Robert C. Helt and Major Larry A. Elliott (USAF) in a Lockheed SR-71A on 28 July 1976 at Beale AFB, California. 25,929·031 m (85,069 ft).

Height, after launch from a 'mother-plane' (USA)
Major R. White, USAF, in the North American X-15A-3 on 17 July 1962, at Edwards AFB, California. 95,935·99 m (314,750 ft).

Speed in a straight line (USA)
Captain Eldon W. Joersz and Major George T. Morgan Jr (USAF) in a Lockheed SR-71A on 28 July 1976 over a 15/25 km course at Beale AFB, California. 1,905·81 knots (3,529·56 km/h; 2,193·17 mph).

Speed in a closed circuit (USA)
Major Adolphus H. Bledsoe Jr and Major John T. Fuller (USAF) in a Lockheed SR-71A on 27 July 1976, over a 1,000 km closed circuit from Beale AFB, California. 1,818·154 knots (3,367·221 km/h; 2,092·294 mph).

Seven records are classed as Absolute World Records for manned spacecraft by the Fédération Aéronautique Internationale, as follows:

Endurance in Earth orbit (USSR)
Vladimir Lyakhov and Valery Ryumin on board Soyuz 32/Salyut 6/Soyuz 34, from 25 February to 19 August 1979. 175 days 0 h 35 min 36 s.

Awaiting confirmation is a subsequent record of 185 days set by Soyuz 35/Salyut 6/Soyuz 37 cosmonauts Leonid Popov and Valery Ryumin.

Altitude (USA)
F. Borman, J. A. Lovell and W. Anders in Apollo 8, on 21-27 December 1968. 203,925 nm (377,668·9 km; 234,673 miles).

Greatest mass lifted to altitude (USA)
F. Borman, J. A. Lovell and W. Anders in Apollo 8, on 21-27 December 1968. 127,980 kg (282,147 lb).

Distance in Earth orbit (USSR)
Vladimir Lyakhov and Valery Ryumin on board Soyuz 32/Salyut 6/Soyuz 34, from 25 February to 19 August 1979. 62,857,205 nm (116,411,557 km; 72,334,783 miles).

Awaiting confirmation is the greater distance covered by Soyuz 35/Salyut 6/Soyuz 37 cosmonauts Leonid Popov and Valery Ryumin.

Extravehicular duration (USA)
Eugene A. Cernan, from the Apollo 17 lunar module *Challenger*, on 12, 13 and 14 December 1972, during mission of 7-19 December 1972. 21 h 31 min 44 s.

Number of astronauts remaining simultaneously outside spacecraft (USSR)
A. Eliseiev and E. Khrounov, from Soyuz 4 and 5, for 37 min on 14-18 January 1969. Two astronauts.

Accumulated time in spaceflight (USSR)
Valery Ryumin, on board Soyuz 25, Soyuz 32, Salyut 6, Soyuz 34, Soyuz 35 and Soyuz 37, a total of 361 days 21 h 31 min 55 s.

WORLD CLASS RECORDS

Following are details of some of the more important world class records confirmed by the FAI:

CLASS C, GROUP I (Aeroplanes with piston engines)

Distance in a straight line (USA)
Cdr Thomas D. Davies, USN, and crew of three in a Lockheed P2V-1 Neptune, on 29 September-1 October 1946, from Perth, Western Australia, to Columbus, Ohio, USA. 9,763·49 nm (18,081·99 km; 11,235·6 miles).

Distance in a closed circuit (USA)
James R. Bede in the Bede BD-2, on 7-10 November 1969, between Columbus, Ohio, and Toledo, Ohio, USA. 7,797·66 nm (14,441·26 km; 8,973·38 miles).

Height (Italy)
Mario Pezzi, in a Caproni Ca 161*bis*, on 22 October 1938. 17,083 m (56,046 ft).

Speed in a straight line (USA)
Steve Hinton in a modified North American P-51D Mustang, with 2,834 kW (3,800 hp) Rolls-Royce Griffon engine, on 14 August 1979, over 3 km course at Tonopah, Nevada. 433·660 knots (803·138 km/h; 499·047 mph).

Last illustrated in *Jane's* more than ten years ago, the little Moscow Aviation Institute Kwant has set a series of new records in Class C1b for piston-engined aircraft weighing 500-1,000 kg. Piloted by V. I. Loitchikov, it averaged 205 knots (379·7 km/h; 235·9 mph) over a 15/25 km course on 4 September 1979. On 4 June 1981, piloted by Victor Zabolotsky, it set a sustained height record of 6,550 m (21,490 ft), and climbed to 6,000 m in 16 min 0·6 s. The Kwant is powered by a 268 kW (360 hp) Vedeneev M-14 II radial engine *(Tass)*

CLASS C, GROUP II (Aeroplanes with turboprop engines)

Distance in a straight line (USA)
Lt Col E. L. Allison and crew in a Lockheed HC-130H Hercules, on 20 February 1972. 7,587·99 nm (14,052·95 km; 8,732·098 miles).

Distance in a closed circuit (USA)
Cdr Philip R. Hite and crew in a Lockheed RP-3D Orion, on 4 November 1972. 5,455·46 nm (10,103·51 km; 6,278·03 miles).

Height (USA)
Donald R. Wilson in an LTV Electrosystems L450F, on 27 March 1972, at Majors Field, Greenville, Texas. 15,549 m (51,014 ft).

Speed in a straight line (USA)
Cdr Donald H. Lilienthal and crew in a Lockheed P-3C Orion, over a 15/25 km course on 27 January 1971. 435·26 knots (806·10 km/h; 500·89 mph).

Speed in a closed circuit (USSR)
Ivan Sukhomlin and crew in a Tupolev Tu-114, on 9 April 1960, carrying a 25,000 kg payload over a 5,000 km circuit. 473·66 knots (877·212 km/h; 545·07 mph).

CLASS C, GROUP III (Aeroplanes with jet engines)

Distance in a straight line, distance in a closed circuit, height, speed in straight line and speed in 1,000 km closed circuit:
See Absolute World Records.

Speed over a 3 km course at restricted altitude (USA)
Darryl Greenamyer in the modified Red Baron F-104RB Starfighter, on 24 October 1977, at Mud Lake, Tonopah, Nevada. 858·77 knots (1,590·45 km/h; 988·26 mph).

Speed in a 100 km closed circuit (USSR)
Alexander Fedotov in a Mikoyan E-266 (MiG-25), on 8 April 1973. 1,406·641 knots (2,605·1 km/h; 1,618·734 mph).

Speed in a 500 km closed circuit (USSR)
M. Komarov in a Mikoyan E-266 (MiG-25), on 5 October 1967, near Moscow. 1,609·88 knots (2,981·5 km/h; 1,852·62 mph).

Speed around the World (USA)
Walter H. Mullikin and crew of four, in a Boeing 747SP of Pan American, on 1-3 May 1976, from New York City, via Delhi and Tokyo, back to New York, in 1 day 22 h 0 min 50 s. 436·95 knots (809·24 km/h; 502·84 mph).

Greatest mass lifted to a height of 2,000 m (USA)
William J. Allsopp and crew in a Boeing 747-236B with Rolls-Royce RB.211 engines, at Le Moore NAS, California, on 1 November 1976. 381,108·25 kg (840,200 lb).

CLASS C.2, ALL GROUPS (Seaplanes)

Distance in a straight line (UK)
Capt D. C. T. Bennett and First Officer I. Harvey, in the Short-Mayo *Mercury*, on 6-8 October 1938, from Dundee, Scotland, to the Orange River, South Africa. 5,211·66 nm (9,652 km; 5,997·5 miles).

Height (USSR)
Georgi Buryanov and crew of two in a Beriev M-10, on 9 September 1961, over the Sea of Azov. 14,962 m (49,088 ft).

Speed in a straight line (USSR)
Nikolai Andrievsky and crew of two in a Beriev M-10, on 7 August 1961, at Joukovski-Petrovskœ, over a 15/25 km course. 492·44 knots (912 km/h; 566·69 mph).

CLASS D, GROUP I (Single-seat sailplanes)

Distance in a straight line (Germany, Federal Republic)
Hans W. Grosse in a Schleicher ASW 12, on 25 April 1972. 788·77 nm (1,460·8 km; 907·70 miles).

Height (USA)
Paul F. Bickle, in a Schweizer SGS 1-23E, on 25 February 1961, at Mojave-Lancaster, California. 14,102 m (46,266 ft).

CLASS D, GROUP II (Two-seat sailplanes)

Distance in a straight line (Australia)
Ingo Renner and Hilmer Geissler in a Caproni Vizzola Calif A-21, on 27 January 1975, from Bendigo Aerodrome to Langley Station, Australia. 523·97 nm (970·4 km; 602·98 miles).

Height (USA)
L. E. Edgar and H. E. Klieforth in a Pratt-Read sailplane, on 19 March 1952, at Bishop, California. 13,489 m (44,256 ft).

CLASS E.1 (Helicopters)

Distance in a straight line (USA)
R. G. Ferry in a Hughes OH-6A, on 6-7 April 1966, 1,923·08 nm (3,561·55 km; 2,213 miles).

Height (France)
Jean Boulet in an Aérospatiale SA 315B Lama on 21 June 1972. 12,442 m (40,820 ft).

Speed in a straight line (USSR)
Gourguen Karapetyan in a Mil A-10 (Mi-24), on 21 September 1978, over a 15/25 km course near Moscow. 198·9 knots (368·4 km/h; 228·9 mph).

Speed in a 100 km closed circuit (USSR)
Boris Galitsky and crew of five in a Mil Mi-6, on 26 August 1964, near Moscow. 183·67 knots (340·15 km/h; 211·36 mph).

CLASS E.2 (Convertiplanes)

Height (USSR)
D. Efremov and crew of two, in the Kamov Ka-22 Vintokryl, on 24 November 1961 at Bykovo. 2,588 m (8,491 ft).

Speed in a straight line (USSR)
D. Efremov and crew of five, in the Kamov Ka-22 Vintokryl, on 7 October 1961, at Joukovski-Petrovskœ, over a 15/25 km course. 192·39 knots (356·3 km/h; 221·4 mph).

Speed in a 100 km closed circuit (New Zealand)
Sqd Ldr W. R. Gellatly and J. G. P. Morton, in the Fairey Rotodyne, on 5 January 1959, White Waltham-Wickham-Radley Bottom-Kintbury-White Waltham. 165·89 knots (307·22 km/h; 190·90 mph).

CLASS E.3 (Autogyros)

Height (UK)
Wing Cdr K. H. Wallis, in a Wallis WA-116/Mc, on 11 May 1968. 4,639 m (15,220 ft).

Distance in a straight line (UK)
Wing Cdr K. H. Wallis, in a Wallis WA-116/F, from Lydd Airport, Kent, to Wick, Scotland, on 28 September 1975. 472·092 nm (874·315 km; 543·274 miles).

Distance in a closed circuit (UK)
Wing Cdr K. H. Wallis, in a Wallis WA-116/F on 13 July 1974. 361·91 nm (670·26 km; 416·48 miles).

Speed in a straight line (UK)
Wing Cdr K. H. Wallis, in a Wallis WA-116/Mc, over a 3 km course, on 12 May 1969. 96·589 knots (179 km/h; 111·225 mph).

AIRCRAFT

ARGENTINA

AERO BOERO

AERO BOERO SRL

HEAD OFFICE: Hipólito Irigoyen 505, 2421 Morteros, Córdoba
Telephone: Morteros 409 and 2121
DIRECTORS: Cesar E. Boero and Hector C. Boero

Aero Boero's activities were seriously affected in 1979 by a tornado and floods, which caused severe damage to a batch of production aircraft, and also to the prototype AB 260 Ag agricultural monoplane described in the 1979-80 *Jane's*. As a result, the Aero Boero 180 series (see 1980-81 *Jane's*) is no longer being built, but manufacture of two versions of the lower-powered AB 150 is continuing against firm orders. A new factory, to replace the one damaged in 1979, was nearing completion in mid-1981.

AERO BOERO 180

The Aero Boero 180 was produced in the following versions:

AB 180. Original three-seat version; described in 1972-73 *Jane's*.

AB 180 RV. Later standard version, flown for first time in October 1972. Extended-span all-metal wings, increased fuel capacity, recontoured fuselage and sweptback vertical tail surfaces. Total of 45 (including 180 RVR, which see) built by Spring 1978. Detailed description in 1980-81 *Jane's*.

AB 180 RVR. Glider-towing version of AB 180 RV, to which it is generally similar except for a transparent cockpit roof panel and provision of a towing hook.

AB 180 Ag. Agricultural version, certificated in Restricted category. Wings have 3° 30′ incidence at root, 2° at tip. Normal accommodation for pilot only in agricultural role. Flush-fitting underfuselage pod containing agricultural chemical. Spraybars fitted along rear bar of V strut and horizontally below wings. Electrically-operated rotary atomisers (two each side) fitted to rear bar of V strut. Ten built by Spring 1978, at which time a further six were on order.

AB 180 SP. Biplane version of AB 180 Ag, with short-span lower wings added (approx 6 m; 19 ft 8¼ in). Improved T-O and landing performance, wider speed range, and enhanced payload/range capability. Tankage for approx 330 litres (72·5 Imp gallons) of liquid chemical in lower wings, instead of underfuselage pack of AB 180 Ag. Undergoing redesign in 1981, to have a 170 litre (37·5 Imp gallon) chemical tank beneath each lower wingtip, with dispersal by an engine-driven pump.

AB 180 SP biplane version of the Aero Boero 180

Aero Boero 150 Ag agricultural aircraft (Avco Lycoming O-320 engine)

AERO BOERO 150

Essentially a lower-powered version of the AB 180, the Aero Boero 150 is certificated in two versions:

AB 150 RV. Four built by Spring 1978, at which time five more were on order. Certificated in Normal category. Forty being built for January 1983 World Gliding Championships in Argentina.

AB 150 Ag. Certificated in Restricted category for use as agricultural aircraft. Glassfibre non-corrosive underfuselage tank, with capacity of 270 litres (59·4 Imp gallons) of liquid chemical. Sales of about 10 anticipated during 1981.

The following description applies to the AB 150 RV:

TYPE: Three-seat light aircraft.

WINGS: Strut-braced high-wing monoplane. Streamline-section V bracing strut each side. Wing section NACA 23012. Dihedral 1° 45′. Incidence 3° at root, 1° at tip. Light alloy structure, including skins. Ailerons and flaps of aluminium alloy construction.

FUSELAGE: Welded steel tube structure (SAE 4130), covered with Ceconite.

TAIL UNIT: Wire-braced welded steel tube structure, covered with Ceconite. Sweptback vertical surfaces. Ground-adjustable tab on rudder.

LANDING GEAR: Non-retractable tailwheel type, with shock-absorption by helicoidal springs inside fuselage. Main wheels carried on faired-in V struts and half-axles. Main wheels and tyres size 6·00-6, pressure 1·65 bars (24 lb/sq in). Hydraulic disc brakes on main units. Tailwheel steerable and fully castoring.

POWER PLANT: One 112 kW (150 hp) Avco Lycoming O-320-A2B flat-four engine, driving a Sensenich 74-DM6-0-54 fixed-pitch propeller. Two wing fuel tanks, total capacity 134 litres (29·5 Imp gallons).

ACCOMMODATION: Normal accommodation for pilot and two passengers in enclosed cabin. Baggage compartment on port side, aft of cabin.

AVIONICS AND EQUIPMENT: One 40A alternator and one 12V battery. VHF radio standard. Provision for night or blind-flying instrumentation at customer's option.

DIMENSIONS, EXTERNAL:

Wing span	10·72 m (35 ft 2 in)
Wing chord (constant)	1·61 m (5 ft 3½ in)
Wing aspect ratio	7·05
Length overall	7·273 m (23 ft 10¼ in)
Height overall	2·10 m (6 ft 10½ in)
Wheel track	2·05 m (6 ft 8¾ in)
Wheelbase	5·10 m (16 ft 8¾ in)

AREAS:

Wings, gross	16·47 m^2 (177·3 sq ft)
Ailerons (total)	1·84 m^2 (19·81 sq ft)
Flaps (total)	1·94 m^2 (20·88 sq ft)
Fin	0·93 m^2 (10·01 sq ft)
Rudder, incl tab	0·41 m^2 (4·41 sq ft)
Tailplane	1·40 m^2 (15·07 sq ft)
Elevators	0·97 m^2 (10·44 sq ft)

PERFORMANCE (at max T-O weight):

Never-exceed speed:	
RV	132 knots (245 km/h; 152 mph)
Ag	123 knots (228 km/h; 141 mph)
Max level speed:	
Ag	119 knots (220 km/h; 137 mph)
Max cruising speed:	
RV	114 knots (211 km/h; 131 mph)
Econ cruising speed:	
Ag	82 knots (152 km/h; 94·5 mph)
Stalling speed, flaps down:	
RV	42 knots (77 km/h; 48 mph)
Max rate of climb at S/L: RV	270 m (885 ft)/min
T-O run: RV	100 m (328 ft)
T-O to 15 m (50 ft): RV	220 m (722 ft)

AMC

AREA DE MATERIAL CÓRDOBA, FUERZA AÉREA ARGENTINA (Fábrica Militar de Aviones)

Avenida Fuerza Aérea Argentina Km 5½, 5103 Guarnicion Aérea Córdoba
Telephone: 45011/15
Telex: 51965 AMCOR AR

DIRECTOR: Brigadier Guillermo Héctor Marotta
PRODUCTION DIRECTOR: Brigadier Oscar José Julia
VICE-DIRECTOR, AIRCRAFT AND ARMAMENT RESEARCH AND DEVELOPMENT: Vice-Comodoro Ricardo José Angel Valenzuela

The original Fábrica Militar de Aviones (Military Aircraft Factory) was founded in 1927 as a central organisation for aeronautical research and production in Argentina. Its name was changed to Instituto Aerotécnico in 1943 and then to Industrias Aeronáuticas y Mecánicas del Estado (IAME) in 1952. In 1957 it became a State enterprise under the title of Dirección Nacional de Fabricaciones e Investigaciónes Aeronáuticas (DINFIA), but reverted to its original title in 1968. It is now a component of the Area de Material Córdoba division of the Argentinian Air Force.

FMA comprises two large divisions. The Instituto de Investigaciónes Aeronáuticas y Espacial (IIAE) is responsible for the design of aircraft, and the design, manufacture and testing of rockets, sounding equipment and other equipment. The Fábrica Militar de Aviones itself controls the aircraft manufacturing facilities (Grupo Fabricación) situated in Córdoba, as well as the Centro de Ensayos en Vuelo (Flight Test Centre), to which all aircraft produced in Argentina are sent for certification tests. The laboratories, factories and other aeronautical division buildings occupy a total covered area of 261,970 m^2 (2,819,845 sq ft); the Area de Material Córdoba employs 4,835 persons, of whom 2,303 are in the Grupo Fabricación.

In current production is the IA 58 Pucará, a nationally designed counter-insurgency aircraft. Other versions of this aircraft, and the IA 63 primary and advanced jet

IA 58A Pucará combat aircraft of the Argentinian Air Force

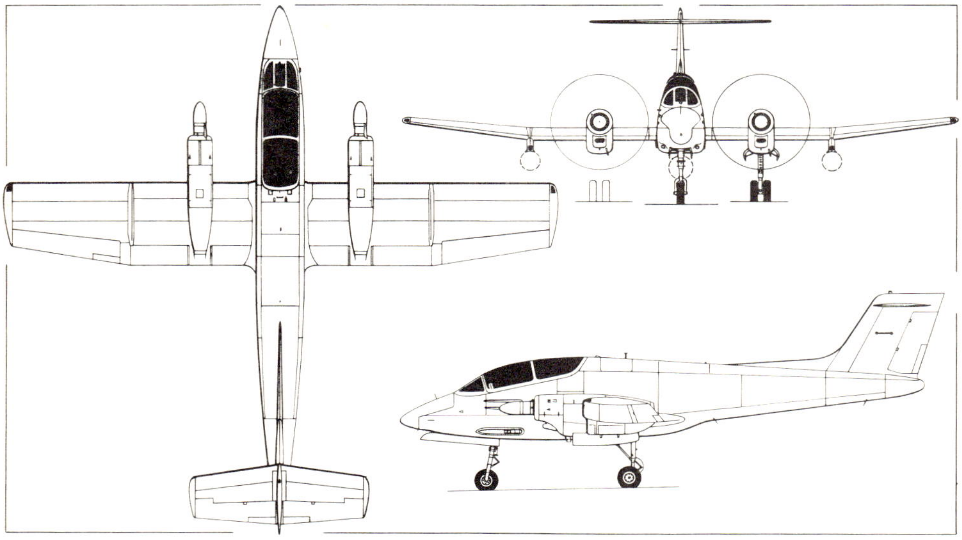

FMA IA 58B Pucará Bravo twin-turboprop counter-insurgency aircraft *(Pilot Press)*

trainer, are under development.

The Grupo Fabricación also produced Cessna single-engined aircraft under licence, under agreements announced in October 1965 and subsequently. These aircraft were repurchased by Cessna for sale through its distributors and dealers in Latin America or sold directly by FMA to Argentinian government agencies. Details of this production were given in the 1980-81 *Jane's*. Cessna production by FMA ended in October 1980, the Model A182 being the only type produced during the final period.

IA 58A PUCARÁ

Design of this twin-turboprop counter-insurgency aircraft, to meet an Argentinian Air Force requirement, began in August 1966. Known originally as the Delfin (Dolphin), it was later renamed Pucará after a type of stone fortress built by the early South American Indians. An unpowered aerodynamic prototype, which first flew on 26 December 1967, was described in the 1968-69 *Jane's*. The first powered prototype, designated A-X2, flew for the first time on 20 August 1969 with 674 kW (904 ehp) Garrett TPE331-U-303 turboprop engines, and was described in the 1971-72 *Jane's*. It was later redesignated AX-01.

A second prototype, designated AX-02, flew for the first time on 6 September 1970, powered by 761 kW (1,022 ehp) Turboméca Astazou XVI G turboprops. This power plant was adopted as standard for the production version, for which the prototype was the similarly-powered AX-03, first flown in mid-1973; the AX-01 also was re-engined with Astazou XVI Gs.

An initial order for 30 IA 58A Pucarás, subsequently increased to 60, was placed by the Argentinian Air Force, and the first of these (A-501) flew for the first time on 8 November 1974. Approx 60 Pucarás had been delivered by mid-1981, with production then at the rate of two per month. These are in service with Argentinian Air Force units, including the II Escuadron de Exploración y Ataque at Reconquista air base, with which the Pucará was deployed operationally in late 1976 against terrorist groups in north-western Argentina. Six IA 58As have been sold to the Uruguayan Air Force; deliveries of these were due to begin in May 1981.

To establish a second source of power plants for the Pucará, a prototype has been re-engined with Garrett TPE331-11-601W turboprop engines, each driving a Dowty R316/4-82F/G propeller. Flight tests in this form began in 1980, and were continuing in early 1981. Details of this version, which is designated **IA 66**, are given separately.

An **IA 58B Pucará Bravo** version, currently under development, is described separately.

The following description applies to the basic IA 58A, production of which was at the rate of approx two per month in early 1981:

Type: Twin-turboprop counter-insurgency aircraft.

Wings: Cantilever low-wing monoplane. Wing section NACA 64_2A215 at root, NACA 64_1A212 at tip. Dihedral 7° on outer panels. Incidence 2°. No sweepback. Conventional semi-monocoque fail-safe structure of duralumin. Frise-type fabric-covered duralumin ailerons, and all-dural slotted trailing-edge flaps, actuated by pushrods. No slats. Balance tab in starboard aileron, electrically-operated trim tab in port aileron.

Fuselage: Conventional semi-monocoque fail-safe structure of duralumin frames and stringers.

Tail Unit: Cantilever semi-monocoque structure of duralumin. Fixed-incidence tailplane and elevators mounted near top of fin. Rudder and elevators actuated by pushrods, and each fitted with inset trim tab.

Landing Gear: Retractable tricycle type. All units retract forward hydraulically, steerable nose unit into fuselage, main units into engine nacelles. Shock-absorbers of Kronprinz Ring-Feder type. Single wheel on nose unit, twin wheels on main units, all with Dunlop tubeless Type III tyres size 7·50-10. Tyre pressures: 2·82 bars (41 lb/sq in) on main units, 2·41 bars (35 lb/sq in) on nose unit. Dunlop hydraulic disc brakes on main units. No anti-skid units.

Power Plant: Two 761 kW (1,022 ehp) Turboméca Astazou XVI G turboprop engines, each driving a Hamilton Standard 23LF/1015-0 three-blade metal propeller with spinner. Fuel in two fuselage tanks (total 800 litres; 176 Imp gallons) and one 230 litre (50·5 Imp gallon) self-sealing tank in each wing, giving overall internal capacity of 1,260 litres (277 Imp gallons). Refuelling point on top of fuselage aft of cockpit. Fuel system includes provision for up to 30 s of inverted flight. A long-range auxiliary tank, capacity 1,200 litres (264 Imp gallons), can be attached to the fuselage centreline pylon, and a 330 litre (72·5 Imp gallon) auxiliary tank on each underwing pylon. Max internal and external fuel capacity 3,120 litres (686 Imp gallons). Oil capacity 11·75 litres (2·6 Imp gallons).

Accommodation: Pilot and co-pilot in tandem on Martin-Baker AP06A zero-zero ejection seats beneath transparent moulded canopy which is hinged at rear and opens upward. Rear seat elevated 25 cm (9·8 in) above front seat. Bulletproof windscreen, with wiper. Dual controls standard.

Systems: Hydraulic system, pressure 207 bars (3,000 lb/sq in), supplied by two engine-driven pumps, actuates landing gear, flaps and wheel brakes. Electrical system includes two 28V 300A starter/generators for DC power and three 500/750VA static inverters for 115/200V AC power. One 24V 36Ah SAFT Voltabloc 4006 battery. No APU. Liquid oxygen bottle.

Avionics and Equipment: Blind-flying instrumentation standard. Avionics include Bendix DFA-73A-1 ADF, Bendix RTA-42A VHF com, Bendix RNA-2bc VHF nav, Northern N-420 HF/SSB com, amplifier and audio-selector system with AS-A-31 panel. Optional avionics include weather radar, IFF, and VHF/FM tactical communications system. Landing/taxying light in leading-edge of each underwing pylon.

Armament and Operational Equipment: Two 20 mm Hispano HS-2804 cannon, each with 270 rds, in underside of forward fuselage; and four 7·62 mm FN-Browning machine-guns, each with 900 rds, in sides of fuselage abreast of cockpit. Aero 7A-1 pylon on centreline beneath fuselage, capacity 1,000 kg (2,205 lb). Aero 20A-1 pylon, capacity 500 kg (1,102 lb), beneath each wing outboard of engine nacelle. Total external stores load 1,620 kg (3,571 lb), including gun and rocket pods, bombs, or auxiliary fuel tanks. Matra 83-4-3 reflector gunsight and AN/AWE-1 programmer.

Dimensions, external:	
Wing span	14·50 m (47 ft 6¾ in)
Wing chord at root	2·24 m (7 ft 4¼ in)
Wing chord at tip	1·60 m (5 ft 3 in)
Wing aspect ratio	6·95
Length overall	14·25 m (46 ft 9 in)
Length of fuselage	13·32 m (43 ft 8½ in)
Fuselage: Max width	1·24 m (4 ft 0¾ in)
Height overall	5·36 m (17 ft 7 in)
Tailplane span	4·70 m (15 ft 5 in)
Wheel track (c/l of shock-absorbers)	4·20 m (13 ft 9¼ in)
Wheelbase	3·48 m (11 ft 5 in)
Propeller diameter	2·59 m (8 ft 6 in)
Dimensions, internal:	
Cabin: Length	2·85 m (9 ft 4¼ in)
Max width	0·81 m (2 ft 8 in)
Max height	1·25 m (4 ft 1¼ in)
Floor area	2·90 m^2 (31·2 sq ft)
Volume	2·74 m^3 (96·8 cu ft)

The IA 66, a converted IA 58A Pucará with Garrett TPE331 turboprop engines

Areas:
Wings, gross 30·30 m^2 (326·1 sq ft)
Ailerons (total) 3·29 m^2 (35·41 sq ft)
Trailing-edge flaps (total) 3·58 m^2 (38·53 sq ft)
Fin 3·465 m^2 (37·30 sq ft)
Rudder, incl tab 1·565 m^2 (16·84 sq ft)
Tailplane 4·60 m^2 (49·51 sq ft)
Elevators, incl tabs 2·612 m^2 (28·11 sq ft)

Weights and Loadings:
Weight empty, equipped 4,037 kg (8,900 lb)
Max T-O weight 6,800 kg (14,991 lb)
Max landing weight 5,806 kg (12,800 lb)
Max wing loading 224·4 kg/m^2 (46 lb/sq ft)
Max power loading 4·46 kg/kW (7·3 lb/ehp)

Performance (at max T-O weight except where indicated):
Never-exceed speed 404 knots (750 km/h; 466 mph)
Max level speed at 3,000 m (9,845 ft) 270 knots (500 km/h; 310 mph)
Max cruising speed at 6,000 m (19,685 ft) 259 knots (480 km/h; 298 mph)
Econ cruising speed 232 knots (430 km/h; 267 mph)
Stalling speed, flaps and landing gear up 68 knots (125 km/h; 78 mph)
Stalling speed, flaps and landing gear down, at 4,790 kg (10,560 lb) gross weight 77·5 knots (142·5 km/h; 89 mph)
Max rate of climb at S/L 1,080 m (3,543 ft)/min
Service ceiling 10,000 m (32,810 ft)
Service ceiling, one engine out 6,000 m (19,685 ft)
T-O run 300 m (985 ft)
T-O to 15 m (50 ft) 705 m (2,313 ft)
Landing from 15 m (50 ft) at 5,100 kg (11,243 lb) gross weight 603 m (1,978 ft)
Landing run at 5,100 kg (11,243 lb) gross weight 200 m (656 ft)
Range with max fuel at 5,000 m (16,400 ft) 1,641 nm (3,042 km; 1,890 miles)
g limits +6; −3

IA 58B PUCARÁ BRAVO

Design of this improved version of the Pucará started in September 1977, and construction of a prototype began 12 months later. This aircraft (AX-05) flew for the first time on 15 May 1979, and made its public debut at the Paris Air Show in May/June of that year. A total of 40 production Bravos had been ordered by early 1980, and manufacture of these was scheduled to begin in 1981.

The airframe and power plant remain essentially the same as for the IA 58A, except for deepening of the forward fuselage to accommodate a heavier nose armament. Other improvements include a new avionics installation.

Type, Wings, Fuselage, Tail Unit and Landing Gear: As for IA 58A, except that forward fuselage is deeper and tyre pressure (all units) is increased to 3·10 bars (45 lb/sq in).

Power Plant: As for IA 58A. Internal fuel tank capacity differs slightly (782 litres; 172 Imp gallons in fuselage tanks, 478 litres; 105 Imp gallons in wing tanks), but overall internal capacity of 1,260 litres (277 Imp gallons) remains the same. Provision for one underfuselage and two underwing drop-tanks, as on IA 58A.

Accommodation: As for IA 58A.

Systems: Hydraulic system and liquid oxygen bottle as in IA 58A. No APU, pneumatic or wing/tail de-icing systems. Electrical system includes two 28V 175A starter/generators for DC power; two 250VA static inverters for 115/26V AC power, and one giving 1,000/200V AC for windscreen demisting/de-icing. One 24V 36Ah SAFT Voltabloc battery.

Avionics and Equipment: Standard avionics include Collins 718US HF/AM and dual Bendix VHF/AM radio; dual Bendix VOR/ILS, dual ADF, DME, ATC and IFF/SSB. Optional avionics include Bendix VHF/FM with homing, Omega, and weather radar. Blind-flying instrumentation standard.

Armament: Two 30 mm DEFA 553 cannon, each with 140 rds, in underside of forward fuselage; and four 7·62 mm FN-Browning machine-guns, each with 900 rds, in fuselage sides abreast of cockpit. One underfuselage and two underwing hardpoints for up to twelve 125 kg bombs; one Dassault 30 mm gun pod, plus four Alkan 530 rocket launchers or two drop-tanks; or six Alkan 530 launchers; or three drop-tanks.

Dimensions, external and internal: As IA 58A except:
Length of fuselage 13·68 m (44 ft 10½ in)
Fuselage: Max depth 2·10 m (6 ft 10¾ in)
Wheelbase 3·885 m (12 ft 9 in)

Areas: As IA 58A

Weights and Loadings:
Weight empty, equipped 4,030 kg (8,884 lb)
Max external stores load 1,686 kg (3,717 lb)
Max fuel load: internal 1,005 kg (2,215 lb)
external 1,250 kg (2,755 lb)

The deepened front fuselage of the IA 58B Pucará Bravo is shown clearly in this view

Full-scale mockup of FMA IA 63 tandem two-seat trainer

Max T-O weight 6,800 kg (14,991 lb)
Max zero-fuel weight 4,546 kg (10,022 lb)
Max landing weight 5,600 kg (12,345 lb)
Max wing loading 224·4 kg/m^2 (46 lb/sq ft)
Max power loading 4·46 kg/kW (7·3 lb/ehp)

Performance (at max T-O weight except where indicated):
Never-exceed speed 404 knots (750 km/h; 466 mph)
Max level speed at 1,000 m (3,280 ft) 252 knots (467 km/h; 290 mph)
Max cruising speed at 1,000 m (3,280 ft) 237 knots (440 km/h; 273 mph)
Stalling speed, flaps up 107 knots (198 km/h; 123 mph)
Stalling speed, flaps down, power reduced 86 knots (159 km/h; 99 mph)
Max rate of climb at S/L 1,080 m (3,543 ft)/min
Climbing speed at S/L, one engine out 182 knots (338 km/h; 210 mph)
Service ceiling 7,400 m (24,275 ft)
Service ceiling, one engine out 2,300 m (7,550 ft)
T-O run 710 m (2,330 ft)
T-O to 15 m (50 ft) 1,040 m (3,412 ft)
T-O to 15 m (50 ft) at 5,800 kg (12,787 lb) gross weight 750 m (2,460 ft)
Landing from 15 m (50 ft) at max landing weight 605 m (1,985 ft)
Landing run at max landing weight 470 m (1,542 ft)
Range with max internal fuel 728 nm (1,350 km; 839 miles)
Range with max internal and external fuel 1,620 nm (3,000 km; 1,865 miles)

IA 66

This designation applies to the version of the Pucará fitted with 746 kW (1,000 shp) Garrett TPE331-11-601W turboprop engines and Dowty R-316/4-82F/G four-blade propellers. A prototype (AX-06), converted from a standard production IA 58A, began flight testing in late 1980, and had completed more than 50 hours' flying with TPE331 engines by mid-1981. The nacelles and engine mounting for this installation were designed and installed by Volpar Inc (see US section). Two further examples of the IA 66, for evaluation by the Argentinian Air Force, were due to be completed in the Spring of 1981.

FMA IA 63

Design of this tandem two-seat basic and advanced jet trainer, which began in 1979, has been finalised, and construction of the first of four prototypes began in March 1981. First flight is planned for October 1983, and entry into service for 1985. Two static test airframes are also being built.

Technical and general assistance is being provided by Dornier GmbH during the development stage, which is due to end in late 1982.

General appearance of the IA 63 can be seen in the accompanying illustration. The shoulder-mounted non-swept wings are of supercritical section, with thickness/chord ratios of 14·5% at root and 12·5% at tip, and have two-segment large-area flaps inboard of the ailerons. There is a door-type airbrake on each side of the rear fuselage. The wide-track hydraulically-retractable tricycle landing gear has emergency free-fall extension. Power plant will consist of a single 15·57 kN (3,500 lb st) Garrett TFE731-2-2N turbofan engine in the first three prototypes; the fourth will have a Pratt & Whitney Aircraft of Canada JT15D-5 turbofan for comparison purposes. Pupil and instructor sit on Martin-Baker Mk 8 lightweight ejection seats beneath individual cockpit canopies, the rear (instructor's) seat being elevated in the now-customary fashion. Accommodation for both occupants is pressurised and air-conditioned. Underwing attachments for drop-tanks and/or armament are intended to be optional, and the aircraft will be capable of up to 20 s of inverted flight. Dual controls are standard.

Dimensions, external:
Wing span 9·686 m (31 ft 9¼ in)
Wing area, gross 15·63 m^2 (168·2 sq ft)
Length overall 10·928 m (35 ft 10¼ in)
Height overall 4·28 m (14 ft 0½ in)

Weight and Loading:
T-O weight, 'clean' 3,490 kg (7,694 lb)
Wing loading, 'clean' 223·25 kg/m^2 (45·75 lb/sq ft)

Performance (estimated, at 'clean' T-O weight):
Max limiting Mach number Mach 0·8
Max level speed at 9,000 m (29,525 ft) Mach 0·75
Stalling speed, flaps and landing gear down 83 knots (154 km/h; 96 mph)
Service ceiling (50% fuel) 14,000 m (45,925 ft)
T-O to 15 m (50 ft):
ISA + 15°C 640 m (2,100 ft)
ISA + 30°C 1,110 m (3,645 ft)
Landing from 15 m (50 ft), AUW of 3,330 kg (7,341 lb):
ISA + 30°C 880 m (2,890 ft)
Ferry range, with auxiliary fuel, at 12,000 m (39,370 ft) 1,350 nm (2,500 km; 1,550 miles)
g limits: 'clean' +6·0; −3·0
at max T-O weight +4·5; −2·0

CHINCUL

CHINCUL S.A.C.A.I.F.I.

Head Office: 25 de Mayo 489, 6° Piso, Buenos Aires
Telephone: 32 5671/5
Telex: 22706 MACUB AR
Works: Calle Mendoza entre 6 y 7 (Casilla de Correo 80) San Juan
President: Juan José Beraza
Vice-President: José María Beraza
Executive Director: Aquiles Luis Uriarte
Works Director: Héctor Carlos Delgado
Chief of Engineering Dept: Fernando Leon Reisin

Prototype Chincul military trainer developed from the Piper Arrow

Chief of Production Dept: Angel F. Videla
Chief of Quality Control and Flight Test Dept: Jaime A. Ferreyra

This company, a wholly-owned subsidiary of La Macarena SA, Piper's Argentinian distributor, concluded an agreement with Piper Aircraft Corporation in November 1971 for manufacture of a range of Piper products in Argentina. The plan called for a progression through manufacturing phases of increasing complexity, and some Chincul products (such as the Pawnee) are now more than 60% manufactured in Argentina.

The aircraft plant at San Juan has a covered area of 16,500 m² (177,600 sq ft), including a maintenance and overhaul workshop, and a work force of 320 people.

Chincul's production programme involves most Piper aircraft except the Warrior, Turbo Dakota, Seminole (both models), Turbo Saratoga, and all Aerostars. Production of Piper types began in December 1972, and more than 500 aircraft had been delivered by May 1981.

All Piper kits delivered to Chincul are for Phase 3 completion, involving the assembly and riveting of wings and control surfaces, manufacture of interiors, upholstery, electrical harness, and other systems installation. Batteries, upholstery, fabrics, tyres, engine instruments, fire extinguishers and glassfibre components are of Argentinian manufacture. Chincul also has an agreement with King Radio Corporation, under which it assembles KMA 24 audio, KN 53 nav, KI 208 VOR/LOC converter indicators, KI 209 VOR/LOC/glideslope converter indicators, KR 87 ADF and KY 196 VHF sets.

The company has developed a fully-aerobatic two-seat military trainer based on the Piper Cherokee Arrow.

CHINCUL (PIPER) CHEROKEE ARROW TRAINER

In January 1978 Chincul began flight testing the prototype (LV-X67) of a military training aircraft which it had developed from the Piper Cherokee Arrow four-seat light aircraft. The principal modifications from the standard Arrow are a more powerful engine, a two-seat cockpit with new canopy, revised internal equipment, and provision for a built-in machine-gun and underwing weapons for armament training. Although the basic airframe of the Arrow is retained, it has been entirely restressed to permit certification in the fully-aerobatic category.

Following certification, production of the Arrow Trainer was expected to begin in the second half of 1981.

Type: Two-seat military trainer.

Wings, Fuselage, Tail Unit and Landing Gear: Generally similar to those of Piper Arrow (see US section), but with fuselage modified in cabin area. Entire airframe restressed for aerobatic flying.

Power Plant: One 194 kW (260 hp) Avco Lycoming AEIO-540 series flat-six engine, driving a two-blade propeller with spinner. Fuel tank in each wing leading-edge, combined capacity 272 litres (71·8 US gallons). Oil capacity 11·5 litres (3 US gallons).

Accommodation: Seats for instructor and pupil side by side under rearward-sliding framed canopy. Dual controls standard.

Avionics and Equipment: Two VHF and one HF com; two VOR; one ILS; one DME; two ADF; audio selector panel; oxygen system.

Armament: Provision for one 7·62 mm machine-gun in lower front fuselage; underwing pylons for bombs and rockets.

Dimensions, external:	
Wing span	10·67 m (35 ft 0 in)
Wing chord (constant portion inboard of ailerons)	1·60 m (5 ft 3 in)
Wing chord at tip	1·07 m (3 ft 6¼ in)
Length overall	7·25 m (23 ft 9½ in)
Height overall	2·23 m (7 ft 3½ in)
Tailplane span	3·92 m (12 ft 10½ in)
Wheel track	3·05 m (10 ft 0 in)
Wheelbase	2·40 m (7 ft 10½ in)
Propeller diameter	2·03 m (6 ft 8 in)
Area:	
Wings, gross	15·79 m² (170·0 sq ft)
Weights and Loadings:	
Weight empty	785 kg (1,730 lb)
Max T-O weight	1,315 kg (2,900 lb)
Max wing loading	83·25 kg/m² (17·06 lb/sq ft)
Max power loading	6·78 kg/kW (11·15 lb/hp)
Performance (at max T-O weight):	
Max level speed	169 knots (314 km/h; 195 mph)
Max cruising speed (75% power)	156 knots (290 km/h; 180 mph)
Speed for optimum climb	87 knots (161 km/h; 100 mph)
Stalling speed, flaps and landing gear down	54 knots (100 km/h; 62 mph)
Max rate of climb at S/L	238 m (780 ft)/min
Service ceiling	3,962 m (13,000 ft)
T-O run	302 m (990 ft)
Landing run	227 m (744 ft)
Range (75% power, optimum mixture, at optimum altitude)	729 nm (1,352 km; 840 miles)

RACA
REPRESENTACIONES AERO COMERCIALES ARGENTINAS SA

Head Office: Lavalle 715, 5° Piso, 1047 Buenos Aires
Telephone: 393 7334 and 392 9488
Telex: 22844 RACA AR
Works: Aeródromo San Fernando, Provincia de Buenos Aires
President: J. R. Fernández Racca
Programme Director: H. Guerra

This company is the representative or dealer in Argentina for the Canadair CL-215 and CASA C-212 Aviocar aircraft, and the MBB BO 105 and Hughes Model 500 helicopters. Under a licence agreement concluded in December 1972 RACA is undertaking, with Argentinian government approval (granted in mid-1973), the progressive local manufacture of a minimum of 120 Hughes Model 500s from knock-down components. Those in current production are known locally as RACA-Hughes 500Ds, and are identical to the Hughes-built 500D described in the US section.

The programme is covering, in three phases, a period of eight years, to supply military and civil customers in Argentina and neighbouring countries. In anticipation of this programme, RACA expanded its workshop facilities at San Fernando aerodrome to a covered area of 4,600 m² (49,514 sq ft).

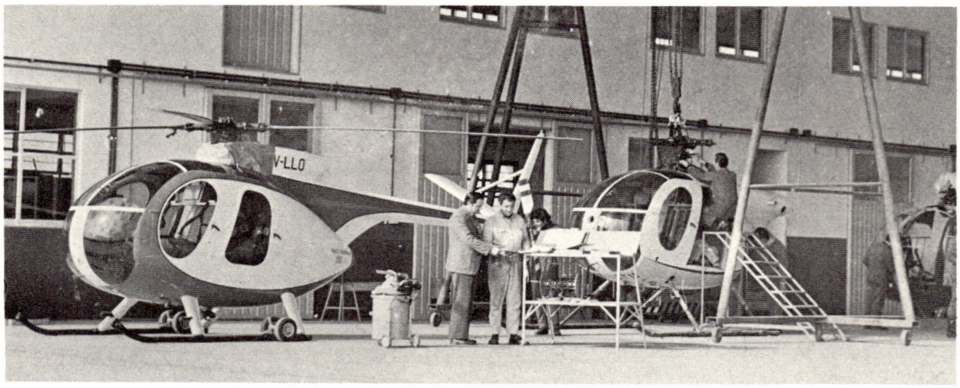

Hughes Model 500D helicopters under construction by RACA

RRA
RONCHETTI, RAZZETTI AVIACIÓN SA

Aeropuerto Internacional Rosario, Casilla Correo 7, 2132 Funes, Provincia de Santa Fé
Telephone: Rosario 58251 or Funes 93276
President: Julio E. Razzetti
Head of Design and Development Group (RRAFAGA): Ing Norberto S. Cobelo
Head of Production: Julio Di Giuseppe

RRAFAGA J-1 MARTIN FIERRO

The J-1 Martin Fierro single-seat agricultural aircraft was designed in Argentina by a team led by Ing Norberto S. Cobelo. Design began in September 1972, and the first of two prototypes made its first flight on 18 December 1975.

Five more J-1s were under construction during 1978, including one for structural testing. Certification by the Argentinian Air Force was anticipated during that year, and plans were then being made for series production. No later news than this has been received. A description and illustration of the J-1 can be found in the 1980-81 *Jane's*.

AUSTRALIA

CAC
COMMONWEALTH AIRCRAFT CORPORATION LIMITED

Head Office and Works: 304 Lorimer Street, Port Melbourne, Victoria 3207
Telephone: (03) 647 6111
Telex: AA 30721

Chairman of Directors: N. F. Stevens
General Manager: D. J. Dalziel

Managers:
D. W. Burton (Aircraft Division)
G. K. Doleman (Gas Turbine Division)
D. R. Rees (Market Development)
R. H. R. Relf (Publicity)

Commonwealth Aircraft Corporation Pty Ltd was formed in 1936. On 18 August 1975 it became a public company, changing its name to Commonwealth Aircraft Corporation Limited.

CAC is a major supplier of equipment and services to the Royal Australian Air Force, and holds Dept of Transport approval for civil aviation activities. The company maintains a capability for initial design and in support of manufacturing and overhaul activities.

Major current programmes include life-of-type extension for RAAF Aermacchi M.B.326H training aircraft; modification of RAAF Lockheed P-3C aircraft to Australian long-range maritime patrol (LRMP) requirements; repair and overhaul, including life-of-type extension, of RAAF Mirage major airframe components; repair and overhaul of RAAF Atar 9C, Viper and Avon engines; manufacture of components and assemblies for GAF Nomad airframes; design and manufacture of in-flight catering equipment for Australian airlines; manufacture of passenger doors for the Lockheed L-1011 TriStar and escape slides for the McDonnell Douglas DC-10.

Ground support equipment is supplied through Static General Engineering Pty Ltd, a CAC subsidiary.

GAF

GOVERNMENT AIRCRAFT FACTORIES

HEADQUARTERS: Fishermen's Bend, Private Bag No. 4, Post Office, Port Melbourne, Victoria 3207
Telephone: 64 0661
Telex: AA 34397
WORKS: Avalon Airfield, Beach Road, Lara, Victoria 3212
Telephone: Lara 82 1202
GENERAL MANAGER: M. C. Morrison
MARKETING MANAGER: E. A. Morris
PUBLIC RELATIONS OFFICER: C. Collinson

The Government Aircraft Factories are units of the Defence Production facilities owned by the Australian government and operated by the Department of Industry and Commerce. They have a work force of approximately 2,500 persons. Their functions include the design, development, manufacture, assembly, maintenance and modification of aircraft, target drones and guided weapons. At Avalon airfield, subassembly of components, final assembly, modification, repair and test-flying of jet and other aircraft are undertaken.

Current activity includes development and production of the Nomad twin-turboprop STOL aircraft, the Ikara anti-submarine missile, and the Jindivik target drone; and the design of a new trainer aircraft for the Royal Australian Air Force.

The GAF are producing elevators for the Boeing 727 and wing flaps for the Fokker F28 Fellowship.

N22B Surveymaster geophysical survey version of the GAF Nomad

GOVERNMENT AIRCRAFT FACTORIES NOMAD

The N2 prototype of the Nomad twin-turboprop light utility aircraft was flown for the first time on 23 July 1971, and received an Australian Department of Transport type certificate in August 1972. Type certificates for the N22 and N24 initial production versions, described in previous editions of *Jane's*, were issued in May 1975 and October 1977 respectively.

The N22B and N24A current versions were type certificated in August 1975 and May 1978 respectively, and in December 1978 both of these versions were awarded certification under US FAR Pt 135 Appendix A (Transport category). Type certificates have also been issued in various European, Asian, South American and Pacific countries.

Current versions are as follows:

N22B. Short-fuselage civil version, intended primarily as a STOL utility aircraft for short/medium-range transportation of up to 13 passengers and/or cargo. Also used for aerial ambulance, geophysical and geographical survey duties, particularly in areas with minimal maintenance and landing field facilities. Transparent acrylic flight deck roof panels, as in Missionmaster, are available optionally. Total of 27 delivered by early 1981, including 19 Commuter Liners, two Medicmasters, three Surveymasters, one Cargomaster, one Commuter/Survey aircraft and one Cargo/Survey.

Floatmaster. Flight trials were concluded in Minneapolis, USA, in late 1978 of an N22B equipped with twin Wipline 9812 floats manufactured by Wipaire Inc of Minnesota. These trials, which had started on 30 May 1978, showed exceptionally good surface and in-flight stability. FAA certification for the floatplane was awarded in September 1979. Flight trials of an amphibious version began on 5 April 1980, and certification was received in the same year. The Wipline 9500 amphibious gear uses the same large floats, which are of all-aluminium construction with bonded chines, keels and gunwales, and have a double afterbody which gives the aircraft an additional 3½° of nose-up attitude before the rear of the float enters the water during rotation. The amphibious version has a single size 6·00-6 Cessna wheel and tyre at the front of each float, and twin size 8·00-6 Cessna wheels and tyres at the rear. All wheels are retractable.

Missionmaster. Short-fuselage military version, in service for forward area support and surveillance, and as a light military transport for personnel and equipment. Four underwing pylons, load-bearing drop doors in the cabin floor, self-sealing fuel cells, transparent acrylic flight deck roof panels, military avionics, and special equipment according to role. Ordered by Australian Army (11), Philippine Air Force (12) and Papua New Guinea Defence Force (7).

N24A. Civil version with lengthened fuselage, offering seating for up to 17 passengers and a considerably improved payload capability. Design includes the insertion of a 1·14 m (3 ft 9 in) plug in the cabin, and increased forward baggage capacity. The basic aircraft includes a full commuter interior and IFR avionics. Total of 23 delivered by early 1981, including 13 Commuter Liners, seven Medicmasters and three Cargomasters.

Searchmaster B. Basic coastal patrol version, equipped with Bendix RDR 1400 search radar, with 46 cm (18 in) forward-looking flat-plate scanner in nose radome. Ordered by Indonesian Navy (16), Papua New Guinea Defence Force and Airline of the Marshall Islands (4), Northern Airlines (3), Bush Pilots Airways (1) and Executive Air Services (1). Normal crew of four.

Searchmaster L. More sophisticated coastal patrol version, operated by Indonesian Navy (2), Northern Airlines (2), Executive Air Services (2) and Reprographics of Australia (1). Litton APS-504(V) 2 search radar and a 100 cm (40 in) flat-plate phased array scanner rotating 360° in undernose 'guppy' radome. Search capacity significantly increased; choice of Doppler, Omega or INS long-range navigation system also ground-stabilises radar display. Both B and L versions have long-range fuel tanks, provide excellent field of view for the crew, and have a spacious cabin. Options include a floor hatch, bubble windows and underwing pylons. Normal crew of five.

By the beginning of 1981, Nomad deliveries (all versions) totalled 104, of which 79 were variants of the short-fuselage version and 25 of the long-fuselage version.

The following description applies generally to all versions, except where a specific model is indicated:

TYPE: Twin-turboprop STOL utility aircraft.

WINGS: Braced high-wing monoplane. Basic NACA 23018 wing section, modified to incorporate increased nose radius and camber. Dihedral 1° from roots. Incidence 2°. No sweepback. Two-spar fail-safe torsion-box structure of riveted light alloy. Full-span double-slotted trailing-edge flaps. All-metal ailerons, which droop with the flaps and transfer their motion progressively to slot-lip ailerons as the flaps extend, resulting in full-span flap. Controls actuated manually by cables and push-rods. Pneumatic de-icing of leading-edges optional. Small stub wings at cabin floor level support the main landing gear fairings from which a single strut on each side braces the main wing.

GAF N22B Floatmaster, with amphibious landing gear

FUSELAGE: Conventional semi-monocoque riveted light alloy structure of stringers and frames.

TAIL UNIT: Cantilever all-metal structure. One-piece all-moving tailplane, with inset trim and anti-balance tab. Tailplane and rudder actuated manually by cables. Trim tab in rudder. Pneumatic de-icing of leading-edges optional. Ventral fin on Floatmaster.

LANDING GEAR: Retractable tricycle type, with electrical retraction by means of single actuator in the fuselage. GAF oleo-pneumatic shock-absorbers. Single rearward-retracting steerable nosewheel, tyre size 8·00-6, pressure 3·17 bars (46 lb/sq in) on N22B, 3·03 bars (44 lb/sq in) on N24A. Twin wheels on each main unit, tyre size 8·00-6, pressure 2·34 bars (34 lb/sq in) on N22B, 2·69 bars (39 lb/sq in) on N24A. Main wheels retract forward into streamline fairings at outer ends of stub wings. Dual hydraulically-operated single-disc brakes on main units. No anti-skid units. Wipaire Wipline 9812 twin-float gear on Floatmaster. Amphibious Wipline 9500 gear also available.

POWER PLANT: Two 313 kW (420 shp) Allison 250-B17C turboprop engines, each driving a Hartzell three-blade constant-speed fully-feathering reversible-pitch metal propeller. Standard fuel capacity 1,018 litres (224 Imp gallons) plus 20 litres (4·4 Imp gallons) unusable in flexible bag tanks. Provision for internal auxiliary tanks for ferry purposes. An additional fuel capacity of 335 litres (73·7 Imp gallons) is provided by two optional integral tanks, one in each wingtip. Gravity refuelling via overwing point above each pair of tanks. Oil capacity 8·5 litres (1·9 Imp gallons) per engine.

ACCOMMODATION (N22B): Flight deck accommodates a crew of two on side-by-side seats, but certification covers single-pilot operation in countries where this applies. Access to flight deck by forward-opening door on each side. Main cabin has individual seats for up to 12 passengers, at 74 cm (29 in) pitch, with continuous seat tracks and readily-removable seats which allow rapid rearrangement of the cabin to suit alternative loads. Access to main cabin via double doors on port side, with single emergency exit on starboard side. Baggage compartments in nose (with door on each side) and optionally in rear of fuselage (with internal and external access). Whole interior, including flight deck, is heated and ventilated.

ACCOMMODATION (N24A): Flight deck accommodation and access as for N22B. Lengthened main cabin, with similar internal provision to N22B for up to 16 passengers, and access via double port-side doors as in N22B. Enlarged nose baggage compartment. Rear baggage compartment of same capacity as N22B. Ventilation and heating system with individual adjustable outlets.

SYSTEMS: No air-conditioning, hydraulic or pneumatic systems normally, but air-conditioning and pneumatic airframe de-icing are available optionally. Electrical system comprises a 28V 150A DC starter/generator on each engine, and a 22Ah battery with AC inverters. Other optional systems include oxygen demand system for crew and continuous-flow system for passengers; electrical de-icing for propellers, cabin floor hatch and underwing pylon racks.

AVIONICS AND EQUIPMENT: Provision is made for a wide range of nav/com and radar equipment to meet specific customer requirements. Full IFR avionics, co-pilot's instrumentation and commuter interior in standard N24A. Wide range of customer options available for all models.

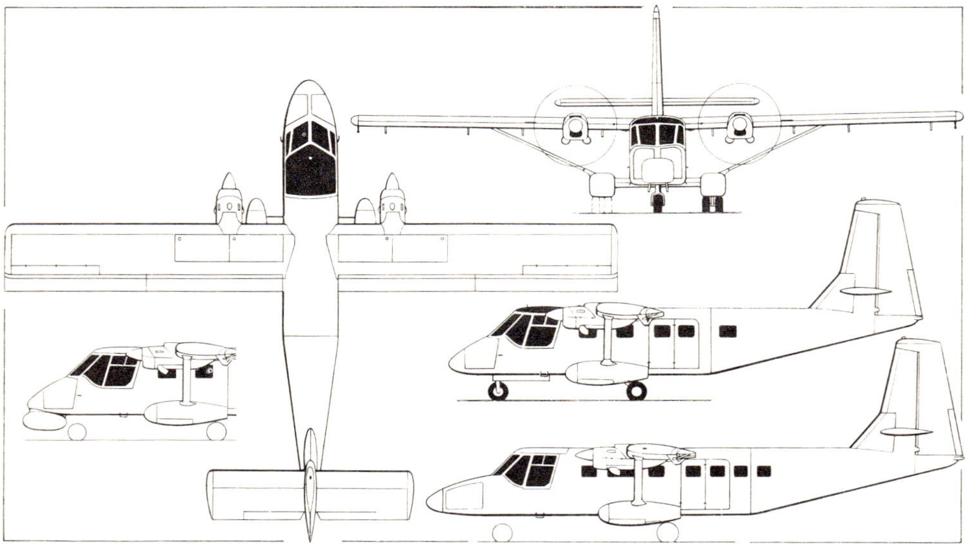

Government Aircraft Factories N22B Nomad, with additional side view (bottom) of N24A and scrap view of forward fuselage of Searchmaster L *(Pilot Press)*

N24A long-fuselage Commuter Liner version of the GAF Nomad, seating up to 17 passengers

Armament and Operational Equipment (Missionmaster/Searchmaster): The military variants have provision for four underwing hardpoints capable of accepting up to 227 kg (500 lb) loads, including gun and rocket pods. The nose bay can be utilised to accommodate surveillance and night vision aid equipment. Removable seat armour and self-sealing fuel tanks can be fitted for added protection. Other equipment as detailed under model listings.

Dimensions, external:
Wing span 16·52 m (54 ft 2·3 in)
Wing chord (constant) 1·81 m (5 ft 11¼ in)
Wing aspect ratio 9·11
Length overall: N22B 12·56 m (41 ft 2·4 in)
N24A 14·36 m (47 ft 1¼ in)
Fuselage: Max depth 1·85 m (6 ft 1 in)
Height overall 5·52 m (18 ft 1½ in)
Tailplane span 5·39 m (17 ft 8·4 in)
Wheel track 2·90 m (9 ft 6 in)
Wheelbase: N22B 3·73 m (12 ft 3 in)
N24A 4·45 m (14 ft 7 in)
Length of floats: Floatmaster 8·15 m (26 ft 9 in)
Distance between c/l of floats:
Floatmaster 3·28 m (10 ft 9 in)
Propeller diameter 2·29 m (7 ft 6 in)
Propeller ground clearance 1·22 m (4 ft 0 in)
Distance between propeller centres
5·22 m (17 ft 1½ in)
Crew doors (each): Height 0·86 m (2 ft 10 in)
Width 0·69 m (2 ft 3 in)
Passenger double doors (port):
Height 1·32 m (4 ft 4 in)
Width 1·22 m (4 ft 0 in)
Height to sill 0·89 m (2 ft 11 in)
Emergency exit (stbd): Height 0·58 m (1 ft 11 in)
Width 0·63 m (2 ft 1 in)

Dimensions, internal:
Cabin, excl flight deck and rear baggage compartment:
Length: N22B 5·33 m (17 ft 6 in)
N24A 6·50 m (21 ft 4 in)
Max width 1·30 m (4 ft 3 in)
Max height 1·58 m (5 ft 2·4 in)
Floor area: N22B 7·06 m² (76·0 sq ft)
N24A 8·55 m² (92·0 sq ft)
Volume: N22B 10·34 m³ (365·0 cu ft)
N24A 13·22 m³ (467·0 cu ft)
Baggage compartment volume (nose):
N22B 0·76 m³ (27·0 cu ft)
N24A 1·13 m³ (40·0 cu ft)
Baggage compartment volume (optional in N22B, rear):
N22B, N24A 0·85 m³ (30·0 cu ft)

Areas:
Wings, gross 30·10 m² (324·0 sq ft)
Ailerons (total net) 2·55 m² (27·4 sq ft)
Trailing-edge flaps (total net) 9·81 m² (105·6 sq ft)
Fin 3·63 m² (39·1 sq ft)
Rudder, incl tab 2·89 m² (31·1 sq ft)
Tailplane, incl tabs 7·25 m² (78·0 sq ft)

Weights and Loadings (landplane versions):
Manufacturer's basic weight empty:
N22B 2,092 kg (4,613 lb)
N24A 2,377 kg (5,241 lb)
Typical operating weight empty:
N22B 2,150 kg (4,741 lb)
N24A 2,377 kg (5,241 lb)
Max fuel load (usable), N22B and N24A:
standard 803 kg (1,770 lb)
extended range 1,066 kg (2,350 lb)
Max T-O and landing weight:
N22B 3,855 kg (8,500 lb)
Searchmaster L 4,127 kg (9,100 lb)
N24A 4,173 kg (9,200 lb)
Max wing loading: N22B 127·9 kg/m² (26·2 lb/sq ft)
N24A 141·5 kg/m² (29·0 lb/sq ft)
Max power loading:
N22B 6·16 kg/kW (10·12 lb/shp)
N24A 6·67 kg/kW (10·95 lb/shp)

Weights (Floatmaster. F: floatplane; A: amphibian):
Weight empty: F 2,290 kg (5,050 lb)
A 2,522 kg (5,560 lb)
Max T-O weight: F 3,855 kg (8,500 lb)
A 3,764 kg (8,300 lb)

Performance (landplane versions, at max T-O weight, ISA at S/L, except where indicated otherwise):
Normal cruising speed:
N22B, N24A 168 knots (311 km/h; 193 mph)
Stalling speed, power off, flaps up, at AUW of 3,402 kg (7,500 lb):
N22B, N24A 65 knots (121 km/h; 75 mph)
Stalling speed, power off, flaps down, at AUW of 3,402 kg (7,500 lb):
N22B, N24A 47 knots (88 km/h; 54·5 mph)
Max rate of climb at S/L, both engines, T-O rating for 5 min: N22B 445 m (1,460 ft)/min
N24A 390 m (1,280 ft)/min
N22B (ISA+20°C) 396 m (1,300 ft)/min
N24A (ISA+20°C) 325 m (1,066 ft)/min
Rate of climb at S/L, one engine out, max continuous rating: N22B 73 m (240 ft)/min
N24A 67 m (220 ft)/min
N22B (ISA+20°C) 52 m (170 ft)/min
N24A (ISA+20°C) 49 m (160 ft)/min
Service ceiling, both engines, climbing at 30·5 m (100 ft)/min, max cruise rating:
N22B at 3,630 kg (8,000 lb) AUW
6,400 m (21,000 ft)
N24A at 4,082 kg (9,000 lb) AUW
6,100 m (20,000 ft)
Min ground turning radius:
N22B, N24A 11·66 m (38 ft 3 in)
Runway LCN at max T-O weight:
N22B 2·3
N24A 2·5
T-O run:
N22B (FAR 23) 223 m (730 ft)
N24A (FAR 23) 296 m (970 ft)
N22B (STOL) 183 m (600 ft)
N22B (FAR 23), ISA+20°C 296 m (970 ft)
N24A (FAR 23), ISA+20°C 366 m (1,200 ft)
N22B (STOL), ISA+20°C 213 m (700 ft)
T-O to 15 m (50 ft):
N22B (FAR 23) 360 m (1,180 ft)
N24A (FAR 23) 521 m (1,710 ft)
N22B (STOL) 320 m (1,050 ft)
N22B (FAR 23), ISA+20°C 463 m (1,520 ft)
N24A (FAR 23), ISA+20°C 610 m (2,000 ft)
Landing from 15 m (50 ft), AUW of 3,630 kg (8,000 lb):
N22B (FAR 23) 408 m (1,340 ft)
N22B (STOL) 216 m (710 ft)
N22B (FAR 23), ISA+20°C 353 m (1,160 ft)
Landing from 15 m (50 ft), AUW of 4,082 kg (9,000 lb):
N24A (FAR 23) 408 m (1,340 ft)
N24A (FAR 23), ISA+20°C 439 m (1,440 ft)
Landing run, AUW of 3,630 kg (8,000 lb):
N22B (FAR 23) 212 m (695 ft)
N22B (STOL) 76 m (250 ft)
N22B (FAR 23), ISA+20°C 204 m (670 ft)
Landing run, AUW of 4,082 kg (9,000 lb):
N24A (FAR 23) 238 m (780 ft)
N24A (FAR 23), ISA+20°C 256 m (840 ft)
Max range at 90% power, standard fuel, reserves for 45 min hold:
N22B, N24A at S/L 580 nm (1,074 km; 668 miles)
N22B, N24A at 3,050 m (10,000 ft)
730 nm (1,352 km; 840 miles)

Performance (Floatmaster. F: floatplane; A: amphibian):
Max cruising speed at 1,525 m (5,000 ft):
F 150 knots (278 km/h; 173 mph)
A 147 knots (272 km/h; 169 mph)
Climbing speed, one engine out:
F, A 87 knots (161 km/h; 100 mph)
Min control speed: F, A 63 knots (117 km/h; 73 mph)
Stalling speed: F, A 53 knots (99 km/h; 61·5 mph)
T-O to 15 m (50 ft), FAR 135 standard
701 m (2,300 ft)
Accelerate/stop distance, FAR 135 standard
747 m (2,450 ft)

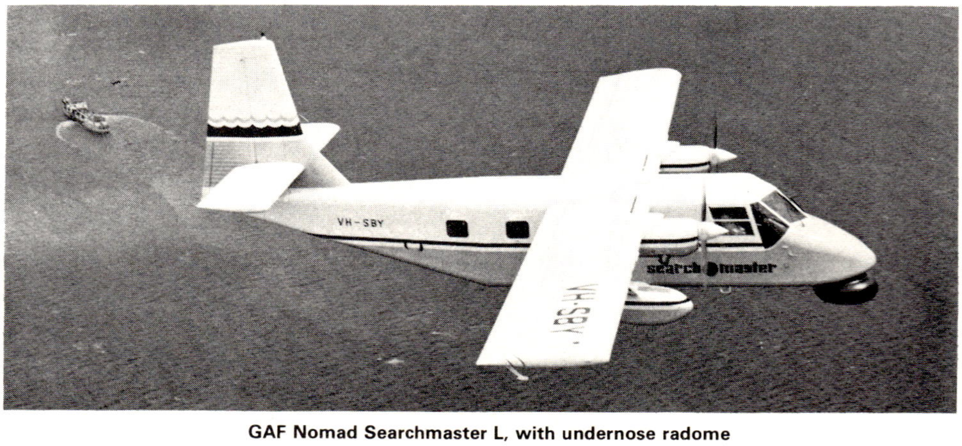

GAF Nomad Searchmaster L, with undernose radome

HAWKER DE HAVILLAND

HAWKER DE HAVILLAND AUSTRALIA PTY LTD (Member Company of Hawker Siddeley Group)

HEAD OFFICE: PO Box 30, Bankstown, NSW 2200
Telephone: 772 8111
Telex: 20719
CHAIRMAN OF DIRECTORS: R. Kingsford-Smith
MANAGING DIRECTOR: B. S. Price

Hawker de Havilland is primarily a manufacturing, maintenance and overhaul organisation, concentrating on the defence, aviation and marine fields in Australia and overseas, with centres at Sydney, Perth and Salisbury, and at Los Angeles in the USA.

Major activities include the manufacture of components for the GAF Nomad aircraft, and sole source production of selected airframe and engine components for Airbus Industrie, Boeing, Boeing Vertol, Lockheed, McDonnell Douglas, Pratt & Whitney, and Westland. Other aerospace products have included infra-red target mini-RPVs (see RPVs and Targets section). Hawker de Havilland's main activity continues to be the provision of overhaul, modification and repair services to the Australian defence forces and a wide variety of other Australian and regional customers.

MINTY

E. R. MINTY

29 Benning Avenue, Turramurra, Sydney, NSW 2074

MINTY SKYHOOK

Mr Ted Minty began preliminary design of this single-seat lightweight autogyro in 1975, and construction of the prototype was started in early 1978. The completed aircraft was exhibited for the first time at the Schoefields Air Show, in October 1978. Mr Minty's selection of a twin tail unit for this autogyro, which he has named Skyhook, was influenced by Mr Martin Hollmann in the USA. (Details of the Hollmann HA-2M Sportster appear in the Homebuilts section, and this aircraft has a tail unit of similar configuration.) The use of twin tails, which are mounted within the wash of the pusher propeller, provides good directional stability and power-off manoeuvrability. Furthermore, the tails are connected at their base by a horizontal stabiliser which is intended to counteract the pitching moment or 'porpoising' experienced with some autogyro designs. The power plant installation includes ultra-lightweight glassfibre cylinder head cooling cowls, designed by Mr Minty, which allow air to be drawn through inlet ducts and over the cylinder heads, the airflow being induced by the rotation of the pusher propeller.

During the past year the tailplane has been slightly reduced in span and area, and the rudder area substantially increased to improve power-off control. It was hoped to begin production of the Skyhook in late 1981 or early 1982.

Minty Skyhook single-seat autogyro, intended for commercial production

TYPE: Single-seat lightweight autogyro.

ROTOR SYSTEM: Two-blade rotor of Haller Rotor-Hawk design, with light alloy blades attached directly to a fully-adjustable hub bar. Rotor mast attached to engine mounting frame, and reinforced by two stainless steel cables which are attached to the rotor head so that, in the event of a failure of the mast structure, the rotor blades, mast and head cannot be lost.

FUSELAGE: Single keel to which are attached the landing gear, fuselage structure, seat, engine mounting frame, rotor mast and tail unit. The majority of structural attachments are clamped, rather than bolted or riveted, to reduce to a minimum the number of potential fracture locations in the airframe. Glassfibre fuselage shell weighs only 5·4 kg (12 lb). During the Summer months the aircraft can be flown if desired without the removable Plexiglas windscreen enclosure.

TAIL UNIT: Twin fins and rudders, the base of the fins united by a horizontal stabiliser attached to the keel. The upper ends of the fins are braced by double V triangular frames of light alloy and chrome molybdenum tube. Fins and rudders each constructed with an Airrx foam core, covered by two layers of continuous woven glassfibre.

LANDING GEAR: Non-retractable tricycle type with single wheel on each unit. Single small tailwheel at aft end of keel. Fully sprung steerable nosewheel. Main wheels are 5 in go-kart wheels, each with a 5·00-5 tyre. Provision for disc brakes.

POWER PLANT: One modified Volkswagen motorcar engine of 1,835 cc capacity, mounted on triangular frame structures supported from the keel, and driving a handmade two-blade silver ash pusher propeller. Fuel tank, incorporated in the base of the pilot's seat, has a capacity of 40·5 litres (8·9 Imp gallons).

ACCOMMODATION: Pilot only, in enclosed cockpit. Aircraft can be flown without Plexiglas windscreen enclosure if desired. Adjustable vents in fuselage nose provide ventilation when the aircraft is flown with the cockpit enclosed.

EQUIPMENT: Standard equipment includes cylinder head high temperature and low fuel warning lights.

DIMENSION, EXTERNAL:
Propeller diameter 1·32 m (4 ft 4 in)

WEIGHTS:
Weight empty 161 kg (354 lb)
Max T-O weight 271 kg (597 lb)

PERFORMANCE:
Max level speed more than 87 knots (161 km/h; 100 mph)
Range with max fuel 121 nm (225 km; 140 miles)

TRANSAVIA

TRANSAVIA DIVISION, TRANSFIELD (NSW) PTY LTD

73 Station Road, Seven Hills, NSW 2147
Telephone: 624 4244
Telex: 21396 and 70300
CHAIRMAN: F. Belgiorno-Nettis, CBE
DIRECTOR: C. Salteri
GENERAL MANAGER: J. Corby

Transavia, formed in 1964, is a division of Transfield (NSW) Pty Ltd, one of Australia's largest construction companies.

TRANSAVIA PL-12 AIRTRUK

The Airtruk, designed by Mr Luigi Pellarini, was originally type-certificated on 10 February 1966, for spreading fertiliser and for seeding. Swath width is up to 32 m (35 yd) and of unusual uniformity. A liquid-spraying conversion, developed in 1968, is capable of covering a 30·2 m (33 yd) swath. This version has an engine-driven spraypump and a liquid chemical capacity of 818 litres (180 Imp gallons). The PL-12's unconventional layout keeps the tails clear of chemicals, and also permits rapid loading by a vehicle which approaches the aircraft between the tails.

The three-seat prototype Airtruk flew for the first time on 22 April 1965. It was followed in December 1970 by a prototype of the **PL-12-U**, a multi-purpose cargo/passenger/ambulance/aerial survey version which received certification in February 1971. Deliveries of the PL-12 began in December 1966, and of the PL-12-U in 1971.

Production of both versions was continuing in 1981, and by January of that year had reached a total (all series) of 116, including ten assembled by Flight Engineers Ltd in New Zealand and five PL-12-Us. A developed version, the T-300 Skyfarmer, is also in production; this is described separately.

Transavia T-300 Skyfarmer, showing separated-tails configuration and upper-wing fences

The following description applies to both the PL-12 and PL-12-U, except where a particular version is indicated:

TYPE: Single-engined agricultural (PL-12) or multi-purpose (PL-12-U) aircraft.

WINGS: Strut-braced sesquiplane. Wing section NACA 23012. Dihedral 1° 30′ on upper wings. Incidence (upper wings) 3° 30′, stub-wings 4°. Conventional all-metal structure, covered with Alclad sheet. All-metal trailing-edge flaps and ailerons, covered with ribbed Alclad sheet, and operated manually. Upper-wing fence on each side of each tailboom to ensure full aileron control, even below stalling speeds. Small stub-wings below fuselage, braced to cabin by a single strut and to upper wings by a V strut on each side.

FUSELAGE: Pod-shaped structure, of 4130 welded steel tube construction with 2024 Alclad covering and glassfibre tailcone.

TAIL UNIT: Twin units, each comprising a fin, rudder and separate T tailplane and elevator, and each carried on a cantilever tubular Alclad boom extending from the upper wings. Small bumper fairing underneath each fin. Manually-operated control surfaces. Adjustable tab in each elevator. Fixed tab on starboard rudder.

LANDING GEAR: Non-retractable tricycle type, each of the three wheels being carried on a pivoted trailing leg. Oleo-pneumatic shock-absorbers in all units. All wheels and tyres same size, 8·00-6. Nosewheel tyre pressure 1·38 bars (20 lb/sq in); main-wheel tyre pressure 2·21 bars (32 lb/sq in). Cleveland hydraulic disc brakes with parking lock.

POWER PLANT: One 224 kW (300 hp) Rolls-Royce Continental IO-520-D flat-six engine, driving a McCauley D2A34C58/90AT-2 two-blade constant-speed metal propeller with spinner. Two upper-wing fuel tanks, total capacity 189 litres (41·5 Imp gallons). Optional long-range installation of second tank in each upper mainplane, increasing total capacity to 379 litres (83·4 Imp gallons). Refuelling point above each upper wing. Oil capacity 11·4 litres (2·5 Imp gallons).

Accommodation (PL-12): Single-seat cockpit, with door on starboard side. Two-seat cabin aft of chemical hopper/tank for carriage of ground crew, with door at rear of lower deck. Accommodation heated and ventilated.

Accommodation (PL-12-U): Single-seat cockpit as in PL-12. By removing the central hopper or tank, passenger cabin is enlarged to seat one passenger on upper deck (back to back with pilot's seat) and four more passengers on lower deck. Doors on upper deck (starboard side) and lower deck (port side). Lower-deck cabin is heated.

System: 12V electrical system standard.

Avionics and Equipment: Optional avionics for PL-12-U include VHF (available optionally for PL-12), HF, ADF, artificial horizon and directional gyro. Standard 907 kg (2,000 lb) capacity hopper aft of cockpit, with twin nozzles for dry chemical dispersal and seeding. Optional Powermist spray system (up to 454 litres; 120 US gallons; 100 Imp gallons/min) for liquid chemical (max capacity 816 litres; 216 US gallons; 180 Imp gallons). Hopper can be filled in 15-20 s and contents jettisoned in 3·2-4·5 s. Other optional equipment includes spray nozzles; additional wing fuel tanks; radio; lighting system, including iodine quartz lamps, for night spraying; windscreen demister.

Dimensions, external:
Upper wing span 11·98 m (39 ft 3½ in)
Upper wing chord (constant portion) 1·76 m (5 ft 9¼ in)
Upper wing chord at tip 1·27 m (4 ft 2 in)
Stub-wing span 4·93 m (16 ft 2 in)
Length overall 6·35 m (20 ft 10 in)
Length of fuselage 4·19 m (13 ft 9 in)
Height overall 2·79 m (9 ft 2 in)
Fuselage: Max width 0·97 m (3 ft 2 in)
Tailplane span (each) 2·13 m (7 ft 0 in)
Distance between tailplanes 3·48 m (11 ft 5 in)
Wheel track 2·44 m (8 ft 0 in)
Wheelbase 1·64 m (5 ft 4½ in)
Propeller diameter 2·23 m (7 ft 4 in)
Min propeller ground clearance 0·30 m (1 ft 0 in)
Passenger door (PL-12, rear):
Height 0·97 m (3 ft 2 in)
Passenger doors (PL-12-U, stbd upper and port lower, each): Height 0·91 m (3 ft 0 in)

Dimensions, internal (PL-12):
Rear passenger cabin: Length 1·83 m (6 ft 0 in)
Max width 0·97 m (3 ft 2 in)
Max height 2·03 m (6 ft 8 in)
Floor area 0·37 m² (4 sq ft)
Volume: Passenger cabin 0·85 m³ (30 cu ft)
Chemical hopper 1·02 m³ (36 cu ft)

Dimensions, internal (PL-12-U):
Passenger cabin: Length 2·74 m (9 ft 0 in)
Max width 0·97 m (3 ft 2 in)
Max height 2·11 m (6 ft 11 in)
Floor area 1·67 m² (18 sq ft)
Volume 2·10 m³ (74 cu ft)

Areas:
Wings, gross 23·78 m² (256·0 sq ft)
Ailerons, total 1·67 m² (18·0 sq ft)
Trailing-edge flaps, total 1·67 m² (18·0 sq ft)
Fins, total 1·30 m² (14·0 sq ft)
Rudders, total 0·56 m² (6·0 sq ft)
Tailplanes, total 2·60 m² (28·0 sq ft)
Elevators, total, incl tabs 1·30 m² (14·0 sq ft)

Weights and Loadings:
Weight empty: PL-12 839 kg (1,850 lb)
PL-12-U 830 kg (1,830 lb)
Max T-O weight:
PL-12 (agricultural category) 1,855 kg (4,090 lb)
PL-12-U (normal category) 1,723 kg (3,800 lb)
Max landing weight (both) 1,723 kg (3,800 lb)
Max wing loading: PL-12 79 kg/m² (16·2 lb/sq ft)
PL-12-U 73 kg/m² (15·0 lb/sq ft)
Max power loading: PL-12 8·28 kg/kW (13·7 lb/hp)
PL-12-U 7·69 kg/kW (12·7 lb/hp)

Performance (at max T-O weight except where indicated):
Never-exceed speed:
PL-12, PL-12-U 155 knots (286 km/h; 178 mph)
Max level speed at S/L, ISA:
PL-12 103 knots (192 km/h; 119 mph)
PL-12-U 112 knots (208 km/h; 129 mph)
Max cruising speed (75% power) at S/L, ISA:
PL-12 95 knots (175 km/h; 109 mph)
PL-12-U 102 knots (188 km/h; 117 mph)
Stalling speed, flaps up:
PL-12 55 knots (103 km/h; 64 mph)
PL-12-U 52 knots (97 km/h; 60 mph)
Stalling speed, flaps down:
PL-12 52 knots (97 km/h; 60 mph)
PL-12-U 50 knots (94 km/h; 58 mph)
Max rate of climb at S/L: PL-12 183 m (600 ft)/min
PL-12-U 244 m (800 ft)/min
Service ceiling (both versions) 3,200 m (10,500 ft)
*T-O run: PL-12 334 m (1,095 ft)
PL-12-U 274 m (900 ft)
*T-O to 15 m (50 ft): PL-12 564 m (1,850 ft)
PL-12-U 457 m (1,500 ft)
Landing run (both versions, at max landing weight) 183 m (600 ft)
Normal range with standard fuel 286 nm (531 km; 330 miles)
Ferry range, standard fuel 330 nm (611 km; 380 miles)
DCA Australia technique

TRANSAVIA T-300 SKYFARMER

A prototype of the Skyfarmer was flown for the first time in July 1978, and nine had been completed by January 1981.

The Skyfarmer is a development of the PL-12, powered by a 224 kW (300 hp) Avco Lycoming IO-540-K1A5 flat-six engine driving a 2·13 m (7 ft 0 in) diameter Hartzell HC-C3YR-1RF/F8468-2R three-blade constant-speed propeller. It also incorporates some of the improved features of the T-320 (1979-80 *Jane's*). The description of the PL-12 Airtruk applies also to the Skyfarmer, except for the power plant and the following details:

Dimensions and Areas:
Propeller diameter 2·13 m (7 ft 0 in)
Wing area, gross 23·48 m² (252·7 sq ft)

Weights and Loadings:
Weight empty 907 kg (2,000 lb)
Max T-O weight (agricultural category) 1,855 kg (4,090 lb)
Max landing weight 1,723 kg (3,800 lb)
Max wing loading 78·6 kg/m² (16·1 lb/sq ft)
Max power loading 8·28 kg/kW (13·6 lb/hp)

Performance (at max T-O weight, ISA at S/L, except where indicated):
Never-exceed speed 155 knots (286 km/h; 178 mph)
*Max light-weight level speed 111 knots (206 km/h; 128 mph)
Max cruising speed (75% power) 103 knots (190 km/h; 118 mph)
*Max light-weight cruising speed (75% power) 106 knots (196 km/h; 122 mph)
Stalling speed, flaps up 55 knots (102 km/h; 63·5 mph)
Stalling speed, flaps down 52 knots (96·5 km/h; 60 mph)
*Light-weight stalling speed, flaps up, power off 45 knots (83·5 km/h; 52 mph)
*Light-weight stalling speed, flaps down, power off 42 knots (78 km/h; 48·5 mph)
Max rate of climb at S/L 183 m (600 ft)/min
*Max light-weight rate of climb at S/L 457 m (1,500 ft)/min
Service ceiling 3,810 m (12,500 ft)
*Light-weight service ceiling 6,890 m (22,600 ft)
T-O run 275 m (902 ft)
*Light-weight T-O run 78 m (255 ft)
T-O to 15 m (50 ft) 475 m (1,560 ft)
*Light-weight landing run 82 m (270 ft)
**Weight of empty aircraft plus pilot and 50% standard fuel*

TRANSAVIA PL-12/T-300A SKYFARMER

Developed during 1980, this updated version of the Airtruk was undergoing certification testing in early 1981; the first production example was displayed at the Paris Air Show in June 1981.

Although substantially similar to the T-300 Skyfarmer, the T-300A model has a much larger and more comfortable cabin for the pilot, with adjustable rudder pedals and ample space for extra optional avionics; electrically-operated flaps; and a calibrated hydraulic pressure gauge, connected to the hydraulic shock-absorber system, which warns the pilot of inadvertent overloading.

BELGIUM

SABCA
SOCIÉTÉ ANONYME BELGE DE CONSTRUCTIONS AÉRONAUTIQUES

General Management: Chaussée de Haecht 1470, B-1130 Brussels
Telephone: Brussels (02) 216 80 10
Telex: SABUSH 21 237
Chairman of the Board: J. Groothaert
Director, General Manager: P. G. Willekens
Works:
Haren-Brussels: Chaussée de Haecht 1470, B-1130 Brussels
Telephone: Brussels (02) 216 80 10
Telex: SABUSH 21 237
Gosselies: Aéroport de Gosselies-Charleroi, B-6200 Gosselies
Telephone: Charleroi (071) 35 01 70
Telex: SABGO 51 251

Founded in 1920, SABCA is the oldest aerospace company in Belgium. Before the second World War, it produced more than 600 military and civil aircraft of 19 different types, some of its own design and some manufactured under licence. Already, at that time, SABCA had participated in various European aircraft programmes.

SABCA's current activities are carried out at two manufacturing plants: Haren-Brussels, where 70 per cent of its activities are concentrated, and Gosselies-Charleroi.

At Haren, the company's main effort is directed towards production of aerospace structures and hydraulic systems. In these areas SABCA is manufacturing main frame structures such as wings and nose sections, and other structural components and equipment, for the General Dynamics F-16; Dassault-Breguet/Dornier Alpha Jet; Dassault Mirage F1, Mirage III and Mirage 5; Fokker F27 and F28; Aérospatiale SA 330 Puma; Spacelab; and Ariane launchers. Servo controls are produced for the F-16 and the Ariane launchers.

At Gosselies, SABCA is assembling and testing the 174 General Dynamics F-16s for Belgium and Denmark. Deliveries have been made regularly since the handover of the first F-16 in January 1979. SABCA is also continuing to maintain and overhaul Mirage 5 and F-104G aircraft for the Belgian Air Force, and is engaged in overhaul and repair of other military aeroplanes and helicopters for the Belgian and foreign armed forces.

SABCA's Electronic Division is manufacturing IFF, SATT and Doppler equipment, and a variety of aircraft electronic ground equipment. It undertakes work on electronic equipment under F-104G overhaul contracts, and is currently integrating ECM equipment in Mirage 5 aircraft. This division produces a laser tank fire control system for the Belgian, Canadian and Australian Armies and other armed forces, for new and updated main battle tanks.

SABCA is a member of various European industrial consortia; Dassault-Breguet and Fokker have parity holdings in the company.

In 1981 the Haren and Gosselies works occupied a total area of approx 72,000 m² (775,000 sq ft) and between them employed 2,175 people.

SONACA
SOCIÉTÉ NATIONALE DE CONSTRUCTION AÉROSPATIALE

Head Office, Works and Airport: Parc Industriel, Route Nationale Cinq, B-6200 Gosselies
Telephone: Charleroi (071) 35 01 90
Telex: 51241

Directors:
M. Claisse (Managing)
J. Lodewijckx (Financial and Personnel)
P. Wacquez (Manufacturing)
R. Bouniton (Commercial)
E. Barthelemy (Research and Development)
J. Maitre (Quality Control)

SONACA SA, formerly Fairey SA, was formed on 1 May 1978 with an initial capital of BF 96 million, of which 50% was funded by the Belgian government. It currently has a capital of BF 260 million.

Based at Gosselies, SONACA participates in civil and military aviation manufacturing programmes, co-producing the General Dynamics F-16 combat aircraft and components for the Airbus A310. It also supplies parts for the Dassault-Breguet Atlantic, Mirage III, Mirage 5 and Mirage F1, and the Dassault-Breguet/Dornier Alpha Jet, as well as carrying out repairs and maintenance in the civil and military fields.

SONACA has a licence to manufacture and sell one of the world's lightest passenger seats. It also designed and sells aircraft galley polycarbonate standard units.

The company's work force in 1981 was 1,800 persons. At the beginning of that year an additional 2,850 m² (30,677 sq ft) of factory space was nearing completion, virtually doubling the previous area of the facility. An additional plant of 3,200 m² (34,444 sq ft), is under construction for the manufacture of composite materials.

BRAZIL

AEROTEC
AEROTEC S/A INDUSTRIA AERONÁUTICA

HEAD OFFICE AND WORKS: Caixa Postal 286, 12200 São José dos Campos, São Paulo State
Telephone: (0123) 21 8011 and 21 8877
GENERAL AND INDUSTRIAL DIRECTOR: Eng Carlos Gonçalves
COMMERCIAL AND ADMINISTRATIVE DIRECTOR: Almir Medeiros

This company was formed in 1962. It designed and built the Uirapuru light aircraft, which, under the military designation T-23, was ordered by the Brazilian, Bolivian and Paraguayan air forces and for civil flying clubs.

Aerotec builds wings for the EMBRAER Ipanema agricultural aircraft (which see). The company is also participating in EMBRAER's general aviation aircraft construction programme, by building fuselages for the EMB-720C/D Minuano, EMB-721C/D Sertanejo and EMB-810C Seneca II.

In 1979 Aerotec employed 110 persons and its premises occupied approx 7,000 m² (75,350 sq ft) of covered space.

AEROTEC A-132 TANGARÁ

Brazilian Air Force designation: YT-17

Development of this aircraft, known originally as the Uirapuru II, was started by Aerotec in the late 1970s. After being placed temporarily in abeyance, the programme was resumed in 1980, and the first prototype (serial-numbered 1000) made its initial flight on 26 February 1981. A second has been completed for static and fatigue testing, and it was hoped to obtain clearance for series production within six months.

Named after an Amazonian songbird, the Tangará is intended to replace the earlier T-23 Uirapuru (Aerotec A-122A) as the Brazilian Air Force's standard primary trainer, and orders for up to 100 are anticipated. Entry into service should take place in late 1982 or early 1983. The T-17 designation was chosen to avoid giving the new primary trainer a 'later' one than the EMBRAER T-27 advanced trainer. (Brazilian Air Force trainer designations began with a projected design, the T-18.)

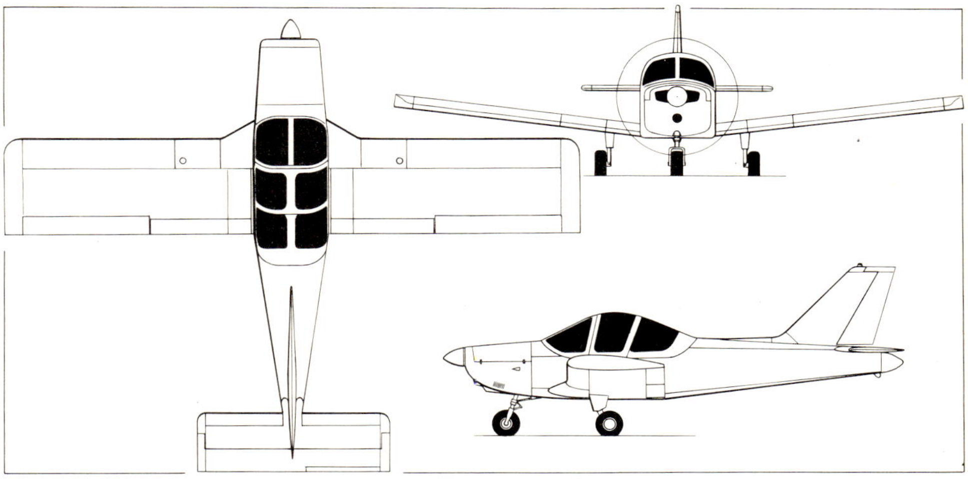

Aerotec YT-17 Tangará (Avco Lycoming O-320-B2B engine) *(Pilot Press)*

The Tangará is based substantially upon the Uirapuru, and utilises the same power plant. Principal changes concern the wings, which have a slight increase in span and dihedral; the fuselage, which is of simplified construction and lacks the underfuselage fin/strake of the T-23; and the cockpit, which has an improved layout and a redesigned canopy with enhanced field of view.

TYPE: Two-seat primary trainer.

WINGS: Cantilever low-wing monoplane. Wing section NACA 43013. Dihedral 7° from roots. Light alloy structure, with all-metal ailerons and trailing-edge split flaps. No tabs.

FUSELAGE: All-metal semi-monocoque structure, of aluminium (steel in critical areas).

TAIL UNIT: Cantilever metal structure, with sweptback vertical and non-swept horizontal surfaces. Ground-adjustable tab on rudder; trim tab in starboard half of one-piece balanced elevator.

First flying prototype of the Aerotec A-132 (YT-17) Tangará primary trainer

LANDING GEAR: Non-retractable tricycle type, with steerable nosewheel. Shock-absorption on all units. Small fairings on main-wheel legs. Disc brakes on main units.

POWER PLANT: One 119 kW (160 hp) Avco Lycoming O-320-B2B flat-four engine, driving a Sensenich M-74 two-blade fixed-pitch metal propeller.

ACCOMMODATION: Two fully-adjustable seats side by side under rearward-sliding framed canopy. Dual controls standard. Baggage space aft of seats.

AVIONICS AND EQUIPMENT: VHF transceiver and intercom standard; VOR and ADF optional.

DIMENSIONS, EXTERNAL:	
Wing span	9·00 m (29 ft 6⅓ in)
Wing area, gross	13·77 m² (148·2 sq ft)
Wing aspect ratio	5·88
Length overall	7·00 m (22 ft 11½ in)
Height overall	2·70 m (8 ft 10¼ in)
Wheel track	2·35 m (7 ft 8½ in)
Propeller diameter	1·87 m (6 ft 1½ in)
WEIGHTS AND LOADINGS:	
Weight empty, equipped	560 kg (1,234 lb)
Max T-O weight	860 kg (1,896 lb)
Max wing loading	62·45 kg/m² (12·79 lb/sq ft)
Max power loading	7·21 kg/kW (11·85 lb/hp)
PERFORMANCE (at max T-O weight):	
Max level speed at S/L	128 knots (238 km/h; 148 mph)
Cruising speed:	
75% power	105 knots (195 km/h; 121 mph)
60% power	94 knots (174 km/h; 108 mph)
Stalling speed:	
flaps up	56·5 knots (104 km/h; 65 mph)
flaps down	39 knots (72 km/h; 45 mph)
Max rate of climb at S/L	276 m (905 ft)/min
Service ceiling	4,500 m (14,760 ft)
T-O to 15 m (50 ft)	200 m (656 ft)
Landing from 15 m (50 ft)	160 m (525 ft)
Endurance, 30 min reserves	4 h 18 min

CTA
CENTRO TÉCNICO AEROESPACIAL

HEADQUARTERS: São José dos Campos, São Paulo State

The Centro Técnico Aeroespacial (Aerospace Technical Centre) is a Ministry of Aeronautics establishment for training aeronautical and aerospace personnel and for conducting aeronautical and aerospace research and development. It is composed of four institutes: the Instituto Tecnológico de Aeronáutica (ITA); the Instituto de Pesquisas e Desenvolvimento (IPD); the Instituto de Atividades Espaciais (IAE); and the Instituto de Fomento e Coordenação Industrial (IFI).

The **ITA** is a college of engineering for aeronautical and aerospace personnel. The **IPD** conducts aeronautical research and development. The **IAE** conducts space research and development. The **IFI** is responsible for the fostering and co-ordination of the Brazilian aerospace industry, and for the certification of civil aircraft and approval of other aeronautical products.

EMBRAER
EMPRESA BRASILEIRA DE AERONÁUTICA SA

HEAD OFFICE AND WORKS: Av Brig Faria Lima 2170, Caixa Postal 343, 12200 São José dos Campos, SP
Telephone: (123) 21 5400
Telex: (391) 1133589 EBAE BR or (391) 1133917 EBAE BR
RIO OFFICE: Aeroporto Santos-Dumont, Sobreloja, Salão de Embarque No. 2, Rio de Janeiro, RJ
Telephone: (21) 222 8981
CHAIRMAN AND CHIEF EXECUTIVE: Ozires Silva
COMMERCIAL DIRECTOR: Ozilio Carlos da Silva
TECHNICAL DIRECTOR: Guido Fontegalante Pessotti
PRODUCTION DIRECTOR: Antonio Garcia da Silveira
FINANCIAL DIRECTOR: Alberto Franco Faria Marcondes
INDUSTRIAL RELATIONS DIRECTOR: Renato José da Silva
PUBLIC RELATIONS: Mário Leme Galvão
PRESS RELATIONS: Fernando Laux

EMBRAER was created on 19 August 1969, and came into operation on 2 January 1970 to promote the development of the Brazilian aircraft industry. It now has an authorised capital of Cr $4,840 million, of which Cr $3,615 million had been subscribed by December 1980. The Brazilian government owns 51% of the voting shares, 90% of the subscribed capital being held by private shareholders. EMBRAER had a work force in May 1981 of 5,929 persons and a factory area of 130,022 m² (1,399,540 sq ft). By the end of 1980 EMBRAER had built a total of 2,483 aircraft.

Since August 1974, EMBRAER has had a comprehensive co-operative agreement with Piper Aircraft Corporation involving the manufacture in Brazil of the Seneca II and Navajo Chieftain twin-engined aircraft and five models of four- and six-passenger single-engined types.

Since 1976, EMBRAER has manufactured in Brazil components for the Northrop F-5E Tiger II combat aircraft.

EMBRAER has in current production the EMB-110 Bandeirante, the EMB-111 maritime surveillance version of the Bandeirante, the pressurised EMB-121 Xingu twin-turboprop transport aircraft, and the EMB-326GB Xavante licence-built version of the Italian Aermacchi M.B.326GC jet trainer and ground attack aircraft. Types under development include the EMB-120 Brasilia commuter transport, EMB-121B Xingu II executive aircraft, the EMB-312 military trainer, and, in partnership with Aeritalia and Aermacchi of Italy, the AM-X tactical fighter. Manufacture of the EMB-201A Ipanema agricultural aircraft, and various EMBRAER-built versions of Piper single- and twin-engined light aircraft, is now the responsibility of Neiva (which see), which became a subsidiary of EMBRAER in March 1980.

EMBRAER EMB-110 BANDEIRANTE (PIONEER)

The Bandeirante twin-turboprop light transport was developed to a Brazilian Ministry of Aeronautics specification calling for a general-purpose aircraft capable of carrying out transport, navigation training and aeromedical evacuation missions.

The first YC-95 prototype (2130) flew for the first time on 26 October 1968, followed by the second (2131) on 19 October 1969, and the basically similar third aircraft

(PP-ZCN) on 26 June 1970. These prototypes, designated EMB-100, were described in the 1970-71 *Jane's*.

The first production EMB-110 Bandeirante (C-95/2133) flew for the first time on 9 August 1972. Following the completion of testing to FAR Pt 23, the EMB-110 was certificated by the Aerospace Technical Centre of the Ministry of Aeronautics, and the first three Bandeirantes were delivered to the Brazilian Air Force on 9 February 1973.

By 1 January 1981 a total of 309 Bandeirantes of various models had been sold, to more than 70 operators in 23 countries worldwide. The 300th Bandeirante was delivered on 8 November 1980.

Models no longer in regular production include the following (seating capacity in parentheses): EMB-110 (2+8), EMB-110/C-95 (2+12), EMB-110A/EC-95 (2+6), EMB-110B/R-95 (2+5), EMB-110B1 (2+14), EMB-110C (2+15), EMB-110E(J) (2+7/8), EMB-110K1/C-95A (2+1,650 kg; 3,637 lb cargo), EMB-110P (2+18) and EMB-110S1 (2+2). Details of these versions, which can still be produced to special order, can be found in the 1980-81 and previous editions of *Jane's*. Principal current models in production are the EMB-110P1 and P2, and, from 1981, the P1/41 and P2/41.

EMB-110P1. Quick-change version of EMB-110P2, for passenger and cargo operations.

EMB-110P1K. Revised version of EMB-110K1 (C-95A) all-cargo military version, described in 1980-81 *Jane's*. Available to special order only.

EMB-110P2. Third-level commuter transport version, carrying up to 21 passengers. First flown on 3 May 1977. Detailed description applies to this version.

EMB-110P1/41 and EMB-110P2/41. New versions of P1 and P2, available from 1981, certificated under SFAR Pt 41 for a max T-O weight of 5,896 kg (13,000 lb). Power plant and dimensions unchanged. Available also as retrofit to existing P1 and P2 versions. First delivery of a P1/41 was in the Spring of 1981, to Provincetown-Boston Airlines of the USA.

EMB-110P3. Pressurised version with T tail, seating up to 19 passengers; powered by 875 kW (1,173 shp) PT6A-65 turboprop engines. In design stage.

EMB-111. Maritime surveillance version, described separately.

The following description, except where noted, applies to the standard production EMB-110P2:

TYPE: Twin-turboprop general-purpose transport.

WINGS: Cantilever low-wing monoplane. Wing section NACA 23016 (modified) at root, NACA 23012 (modified) at tip. Sweepback 19′ 48″ at quarter-chord. Dihedral 7° at 28% chord. Incidence 3°. All-metal two-spar structure, of 2024-T3 and -T4 aluminium alloy, with detachable glassfibre wingtips. Glassfibre wing/fuselage fairing. All-metal statically-balanced Frise-type ailerons and double-slotted flaps. Trim tab in port aileron. De-icing system optional.

FUSELAGE: All-metal semi-monocoque structure of 2024-T3 aluminium alloy. Two upward-hinged doors, one on each side of nose, provide access to avionics. Enlarged rear cargo door on P1K.

TAIL UNIT: Cantilever all-metal structure, with sweptback vertical surfaces. Glassfibre dorsal fin. Ventral fin. Trim tabs in rudder and port elevator. Tab in starboard elevator linked to flaps, to offset pitching moment during flap extension. De-icing system optional.

LANDING GEAR: Hydraulically-retractable tricycle type, with single wheel and ERAM oleo-pneumatic (nitrogen) shock-absorber on each unit. Main-wheel tyre size 670 × 270-12 (10 ply rating), pressure 5·86-6·20 bars (85-90 lb/sq in). Steerable, forward-retracting nosewheel unit has tyre size 6·50-8, pressure 4·27-4·69 bars (62-68 lb/sq in).

POWER PLANT: Two 559 kW (750 shp) Pratt & Whitney Aircraft of Canada PT6A-34 turboprop engines, each driving a Hartzell HC-B3TN-3C/T10178H-8R constant-speed three-blade metal propeller with autofeathering and full reverse-pitch capability. Four integral fuel tanks in wings, with total capacity of 1,720 litres (378 Imp gallons). Oil capacity 8·7 litres (1·9 Imp gallons). Gravity refuelling point on top of each wing. Optional de-icing system for engine air inlets and propellers.

ACCOMMODATION: Pilot and co-pilot side by side on flight deck. Seats for up to 21 passengers in main cabin of P2, at 74 cm (29 in) pitch. P1 has quick-change cabin seating up to 18 persons. Crew/passenger door at front and passenger/baggage door at rear, both on port side; emergency exit over wing on each side, and opposite crew/passenger door on starboard side. Crew/passenger door can also be used as emergency exit. Cabin floor stressed for uniformly distributed loads of up to 488 kg/m² (100 lb/sq ft). Baggage compartment at rear of cabin, with total capacity of 2·0 m³ (70·6 cu ft). Flush-type toilet in compartment at rear of cabin. Toilet/lavatory standard. Windscreen de-icing optional.

SYSTEMS: Air-cycle-type air-conditioning system with cooling capacity of 25,000 BTU/h and engine bleed heating. Hydraulic system, pressure 207 bars (3,000 lb/sq in), for landing gear actuation, dual independent braking systems, nosewheel steering and parking brake. Electrical system utilises two starter/generators, giving 200A continuously or 300A for one minute, and one 24V 34Ah nickel-cadmium battery with two 250VA static inverters to supply 115/26V 400Hz AC power. External power receptacle on port side of forward fuselage. Oxygen system for crew and passengers (standard in P2, optional in P1), using oxygen cylinder in rear of fuselage with capacity of 3·3 m³ (115 cu ft) at 128 bars (1,850 lb/sq in) pressure.

AVIONICS AND EQUIPMENT: Collins Proline, Collins Microline and King Silver Crown II avionics packages available. *Proline package* includes two VHF-20 com transceivers with 37R-2 antennae; VIR-30A VHF nav VOR/ILS/marker beacon receiver with 837B-1, 37P-5 and 37X-2 antennae; second VIR-30A with 837B-1 antennae only; 313N-2 VHF nav/com control panels; PN-101 pictorial navigation system with 331A-3G pilot's HSI; 331H-3G co-pilot's VOR/ILS indicator; ADF-60 ADF receiver with 614L-12 control panel, Bendix LPA 73C-1 loop antenna and EMBRAER sense antenna; 332C-10 RMI with NAV 1/ADF 1 on single needle and NAV 2/ADF 2 on double needle; two 387C-4 audio control panels with two 356F-3 speaker amplifiers and eight speakers (flight deck and cabin); PA system; pilot's interphone; and radio master switch. Options to Proline package include second ADF-60 with 614L-13 control panel, Bendix LPA 73C-1 loop antenna and EMBRAER sense antenna (exchange 614L-12 control panel); second 332C-10 RMI; DME-40 with 339F-12A indicator, NAV 1/HOLD/NAV 2 switching and 237Z-1 antenna (exchange both 313N-2 VHF nav/com control panels for 313N-2D); TDR-90 transponder with 613L-3-20 single control panel and 237Z-1 antenna; second TDR-90 with 613L-3-50 dual control panel and 237Z-1 antenna (exchange 613L-3-20 single control panel); second PN-101 with 331A-3G co-pilot's HSI (exchange co-pilot's AIM-200-DC-FM directional gyro and 331H-3G VOR/ILS indicator); and ALT-50 radar altimeter with 339H-4 indicator and two 437X-1 antennae. *Microline package* includes two VHF-251 com transceivers with 37R-2 antennae; VIR-351 VHF nav receiver with Dorne & Margolin DMH 21-1 nav adapter plus Collins GLS-350 glideslope receiver, 837B-1 nav antenna and 37P-5 dual glideslope antenna; second VIR-351 with GLS-350 glideslope receiver only (all VHF com and nav systems include three PWC-150 converters with power transfer unit); ADF-650A ADF, including RCR-650A receiver and ANT-650A antenna; Aeronetics 7137 RMI with Aeronetics 7100 pilot's dual RMI converter; MKR-350 pilot's marker beacon receiver with 37X-2 antenna; PN-101 pilot's pictorial navigation system with 331A-3G HSI; IND-351A co-pilot's VOR/ILS indicator (coupled to NAV 2); two 387C-4 audio control panels with two 356F-3 speaker amplifiers and eight speakers (flight deck and cabin); PA system; pilot's interphone; and radio master switch. Options to Microline package include ANS-351 R/Nav (only if aircraft is equipped also with DME-451); second ADF-650A, with RCR-

EMBRAER EMB-110P1 Bandeirante, with side views (right, top to bottom) of B1, S1, P1K and P1 *(Pilot Press)*

The 300th Bandeirante, an EMB-110P2 commuter transport version, was delivered to Talair of Papua-New Guinea

650A and ANT-650A; one or two DME-451, each with TCR-451 transceiver, IND-451 indicator and ANT-451 antenna; one or two TDR-950 transponders, each with 237Z-1 antenna; second Aeronetics 7137 RMI for co-pilot, with one or two ADF adapters as appropriate; second PN-101 with 331A-3G co-pilot's HSI (substitutes IND-351A and AIM-200-DC-FM directional gyro); second MKR-350 with 37X-2 antenna; King KI 207 pilot's VOR/ILS indicator repeater, coupled to NAV 2; second KI 207 for co-pilot, coupled to NAV 1; ALT-50A radio altimeter with 339H-4 indicator and two 437X-1 antennae; AP-106 autopilot with pilot's FD-112V flight director (substitutes 331A-3G HSI and RAI-303 artificial horizon on pilot's side); manual electric trim for pitch control, with command for both pilots (Collins 334D-6 trim servo, as used in autopilot); and NAV 1/NAV 2 transfer switch. *Silver Crown II package* includes two KY 196 VHF com transceivers with Collins 37R-2 antennae; two KN 53 VHF nav receivers, one with Dorne & Margolin DMH 21-1 nav antenna coupler, Collins 837B-1 VOR/ILS antenna and 37P-5 dual glideslope antenna; KR 87 ADF receiver with KA 44 antenna; KNI 582 RMI with NAV 1/ADF 1 on single needle and NAV 2/ADF 2 on double needle; KNR 633 RMI converter; KR 21 pilot's marker beacon receiver with Collins 37X-2 antenna; KCS 55A compass system with KI 525A pilot's HSI; KN 72 nav converter; KI 204 co-pilot's VOR/ILS indicator, coupled to NAV 2; two Collins 387C-4 audio control panels with two 356F-3 speaker amplifiers and eight speakers (flight deck and cabin); PA system; pilot's interphone; and radio master switch. Options to Silver Crown II package include one or two KNS 81 R/Nav (only if aircraft is equipped also with KN 63 DME system); second KR 87 with KA 44 antenna; one or two KN 63 DME systems, each with KDI 572 indicator and KA 60 antenna; one or two KT 76A transponders, each with KA 60 antenna; second KNI 582 system; second KCS 55A with KI 525A co-pilot's HSI (substitutes directional gyro and KI 204); KN 72 nav converter (required when ordering second KCS 55A without second KNS 81); KI 204 pilot's VOR/ILS indicator, coupled to NAV 2; second KI 204 for co-pilot, coupled to NAV 1; second KR 21 for co-pilot, with Collins 37X-2 antenna; KRA 405 radio altimeter, with KNI 415 indicator and two KA 54 antennae; and two KY 196E wide-bandwidth VHF com transceivers (exchange for both KY 196). *In addition to these packages,* other avionics options include Collins HF-200 SSB 20-channel HF transceiver; Bendix M4-D autopilot; King KA 52 autopilot adapter (necessary with King package); Collins AVR-101 cockpit voice recorder (mandatory for French certification); Dorne & Margolin DMELT-6 emergency locator transmitter system (mandatory for US, Canadian and French certification); encoding altimeter (IDC, Smiths or Jaeger); and weather radar (Bendix RDR-1200 or RDR-130).

DIMENSIONS, EXTERNAL:

Wing span	15·32 m (50 ft 3¼ in)
Wing chord at root	2·45 m (8 ft 0½ in)
Wing chord at tip	1·35 m (4 ft 5 in)
Wing aspect ratio	8·10
Length overall	15·10 m (49 ft 6½ in)
Length of fuselage	14·59 m (47 ft 10½ in)
Height overall	4·92 m (16 ft 1¾ in)
Fuselage: Max width	1·72 m (5 ft 7¾ in)
Tailplane span	7·54 m (24 ft 9 in)
Propeller diameter	2·36 m (7 ft 9 in)
Distance between propeller centres	4·80 m (15 ft 9 in)
Propeller ground clearance	0·276 m (10¾ in)
Wheel track	4·94 m (16 ft 2½ in)
Wheelbase	5·01 m (16 ft 5¼ in)
Passenger door (rear, port):	
Height	1·35 m (4 ft 5¼ in)
Width	0·85 m (2 ft 9½ in)
Crew/passenger door (fwd, port):	
Height	1·42 m (4 ft 8 in)
Width	0·63 m (2 ft 1 in)
Passenger emergency exits (two, each):	
Height	0·80 m (2 ft 7½ in)
Width	0·63 m (2 ft 1 in)
Crew emergency exit: Height	0·80 m (2 ft 7½ in)
Width	0·55 m (1 ft 9¾ in)
Cargo door (P1K, rear, port):	
Height	1·42 m (4 ft 8 in)
Width	1·80 m (5 ft 11 in)

DIMENSIONS, INTERNAL:

Cabin: Max length	9·53 m (31 ft 3¼ in)
Width	1·60 m (5 ft 3 in)
Height	1·60 m (5 ft 3 in)
Floor area	12·00 m² (129·2 sq ft)
Volume	20·4 m³ (720·4 cu ft)

AREAS:

Wings, gross	29·10 m² (313·23 sq ft)
Ailerons (total)	2·16 m² (23·25 sq ft)
Flaps (total)	4·90 m² (52·74 sq ft)
Fin, excl dorsal fin	3·81 m² (41·01 sq ft)
Dorsal fin	0·82 m² (8·83 sq ft)
Ventral fin	0·80 m² (8·61 sq ft)
Rudder, incl tab	1·69 m² (18·19 sq ft)
Tailplane	5·51 m² (59·31 sq ft)
Elevators, incl tabs	4·31 m² (46·39 sq ft)

WEIGHTS AND LOADINGS (A: P2; B: P1/41 and P2/41; C: P1 in passenger configuration):

Weight empty, equipped: A	3,516 kg (7,751 lb)
B, commercial	3,624 kg (7,990 lb)
B, cargo	3,386 kg (7,465 lb)
Max payload: A	1,681 kg (3,706 lb)
B, commercial	1,561 kg (3,443 lb)
B, cargo	1,800 kg (3,970 lb)
C	1,633 kg (3,600 lb)
Max T-O weight: A	5,670 kg (12,500 lb)
B	5,900 kg (13,007 lb)
Max ramp weight: B	5,930 kg (13,073 lb)
Max landing weight: A	5,450 kg (12,015 lb)
B	5,700 kg (12,566 lb)
Max zero-fuel weight: A, B	5,450 kg (12,015 lb)
Max wing loading: A	195·52 kg/m² (40·04 lb/sq ft)
B	202·61 kg/m² (41·50 lb/sq ft)
Max power loading: A	5·07 kg/kW (8·33 lb/shp)
B	5·27 kg/kW (8·67 lb/shp)

PERFORMANCE (at max T-O weight, ISA, except where indicated. A: P2; B: P1/41 and P2/41; C: P1 in passenger configuration):

Max level speed at 2,440 m (8,000 ft):	
A	248 knots (460 km/h; 286 mph)
Max cruising speed at 3,050 m (10,000 ft):	
A	225 knots (417 km/h; 259 mph)
B	224 knots (415 km/h; 258 mph)
Econ cruising speed at 3,050 m (10,000 ft):	
A	176 knots (326 km/h; 203 mph)
B	178 knots (330 km/h; 205 mph)
Stalling speed at max landing weight:	
A	71 knots (132 km/h; 82 mph) CAS
Max rate of climb at S/L:	
A	545 m (1,788 ft)/min
B	500 m (1,640 ft)/min
Rate of climb at S/L, one engine out:	
A	136 m (446 ft)/min
B	112 m (370 ft)/min
Time to 3,050 m (10,000 ft): A	6 min
Time to 4,575 m (15,000 ft): A	10 min
Service ceiling:	
A at AUW of 5,300 kg (11,684 lb)	7,350 m (24,100 ft)
B at max T-O weight	6,550 m (21,500 ft)
Service ceiling, one engine out:	
A at AUW of 5,300 kg (11,684 lb)	3,780 m (12,400 ft)
B at max T-O weight	3,050 m (10,000 ft)
T-O run: A	430 m (1,410 ft)
T-O to 15 m (50 ft): A	675 m (2,215 ft)
B	815 m (2,674 ft)
Landing from 15 m (50 ft): A	850 m (2,790 ft)
B	820 m (2,690 ft)

EMB-110S1 remote sensor geophysical survey version of the EMBRAER Bandeirante

Brazilian Air Force C-95A Bandeirante (EMB-110P1K), with a typical cargo load

Artist's impression of the T-tailed pressurised EMB-110P3 Bandeirante

EMB-111 patrol version of the Bandeirante, developed by EMBRAER *(Pilot Press)*

EMBRAER EMB-111 maritime surveillance aircraft, with searchlight and underwing rockets

Landing run: A 565 m (1,854 ft)
B 580 m (1,903 ft)
Range at 3,050 m (10,000 ft), 45 min reserves:
A with max fuel 1,025 nm (1,900 km; 1,180 miles)
B with max fuel 1,003 nm (1,858 km; 1,155 miles)
A with 1,440 kg (3,175 lb) standard payload 268 nm (497 km; 309 miles)
B, commercial, with standard payload 345 nm (639 km; 397 miles)
B, cargo, with standard payload 238 nm (441 km; 274 miles)
C with 1,440 kg (3,175 lb) standard payload 225 nm (417 km; 259 miles)

EMBRAER EMB-111

Brazilian Air Force designation: P-95

This land-based maritime surveillance aircraft, based on the EMB-110 Bandeirante, was designed to meet specifications issued by the Comando Costeiro, the Brazilian Air Force's Coastal Command, which ordered 12. The main external differences in this version are the large nose radome, housing search radar, and the addition of wingtip fuel tanks.

The first EMB-111 (2262) flew for the first time on 15 August 1977; Brazilian Air Force aircraft, the first three of which were delivered on 11 April 1978, serve with the 1° Esquadrão of the 7° Grupo de Aviação, located at Salvador AFB, Bahia. The 12th aircraft was due to be delivered in June 1981.

Six have been delivered to the Chilean Navy. These aircraft have some mission equipment changes, including full de-icing system, and passive ECM antennae under the nose and at the tail. One EMB-111 was delivered in June 1981 to the Gabon Air Force, which has a second on order.

Type: Twin-turboprop maritime surveillance aircraft.

Wings: As EMB-110P2, but with reinforced leading-edges and fitted with tip-tanks.

Fuselage: Similar to EMB-110P2, but with large nose radome.

Tail Unit and Landing Gear: As EMB-110P2.

Power Plant: Two 559 kW (750 shp) Pratt & Whitney Aircraft of Canada PT6A-34 turboprop engines, each driving a Hartzell three-blade reversible-pitch propeller with spinner. Four integral fuel tanks in wings (total capacity 1,914 litres; 421 Imp gallons), and two permanent wingtip tanks (total capacity 636 litres; 140 Imp gallons). Overall total fuel capacity 2,550 litres (561 Imp gallons), of which 2,454 litres (540 Imp gallons) are usable. Oil capacity 8·7 litres (1·9 Imp gallons).

Accommodation: Pilot and co-pilot side by side on flight deck. Main cabin can accommodate search radar/radio operator, ECM operator, one or two observers, and a second radar or ECM operator. Port-side door at rear, for crew and cargo, opens inwards and can be used for airdrop of paratroops and survival equipment. Galley and toilet in main cabin.

Avionics and Equipment: One Collins 618T-3B or 718U-5 AM/SSB/CW transceiver, two Collins 618M-3 VHF transceivers, one Collins AN/ARC-159 UHF transceiver, Collins Audio interphone, two Sperry C-14 gyromagnetic compasses, two Bendix DFA-74A or Collins DF-206 ADF receivers, two Collins VIR-30A or VIR-31A VOR/ILS/marker beacon receivers, one Collins AN/APX-92 IFF transponder, one Collins DF-301E VHF/DF or UHF/DF, one Collins DME-40 DME, one Bendix ALA-51 or Collins ALT-50 radio altimeter, one Litton LN-33 inertial navigation system, Thomson-CSF passive ECM, and one AIL AN/APS-128 (SPAR-1) sea patrol radar. Optional avionics include Bendix M4-D autopilot and ONS-25 Omega, single or dual Sperry STARS IVB or IVC flight directors, entertainment radio, tape deck, PA systems and complete de-icing and anti-icing system.

Armament and Operational Equipment: Four underwing pylons for eight 5 in HVAR air-to-surface rockets (two per pylon), or four launchers each with seven 2·75 in FFAR rockets; or three stores pylons, plus a leading-edge-mounted 50 million candlepower searchlight on the starboard wing. For target marking, six Brazilian-built Mk 6 smoke grenades are carried, as well as a Motorola SST-121 transponder. Flares of 200,000 candlepower also available for illumination of targets at night. Ventral chute for smoke marker buoys, high-intensity flares or chaff dispensing. Provision for 1·4 kW loud-hailer system.

Dimensions, external:
Wing span over tip-tanks 15·95 m (52 ft 4 in)
Wing mean chord (aerodynamic) 1·95 m (6 ft 4¾ in)
Wing area, gross 29·10 m^2 (313·2 sq ft)
Wing aspect ratio 8·07
Length overall 14·91 m (48 ft 11 in)
Length of fuselage 14·38 m (47 ft 2¼ in)
Height overall 4·91 m (16 ft 1¼ in)
Wheel track 4·94 m (16 ft 2½ in)
Wheelbase 4·26 m (13 ft 11¾ in)
Propeller diameter 2·36 m (7 ft 9 in)
Propeller ground clearance 0·28 m (11 in)
Main door (rear, port):
Height 1·35 m (4 ft 5¼ in)
Width 0·80 m (2 ft 7½ in)
Height to sill 1·26 m (4 ft 1½ in)
Emergency exit (overwing, stbd):
Height 0·80 m (2 ft 7½ in)
Width 0·63 m (2 ft 0¾ in)

Dimensions, internal:
Cabin: Max length 8·65 m (28 ft 4½ in)
Width 1·60 m (5 ft 3 in)
Height 1·60 m (5 ft 3 in)
Floor area 11·60 m^2 (124·9 sq ft)
Areas: As EMB-110P2

Weights and Loadings:
Weight empty, equipped 3,760 kg (8,289 lb)
Max T-O weight 7,000 kg (15,432 lb)
Max ramp weight 7,030 kg (15,498 lb)
Max zero-fuel weight 5,150 kg (11,354 lb)
Max landing weight 5,450 kg (12,015 lb)
Max wing loading 241·38 kg/m^2 (49·44 lb/sq ft)
Max power loading 6·26 kg/kW (10·29 lb/shp)

Performance (at max T-O weight, ISA+15°C, except where indicated):
Max cruising speed at 3,050 m (10,000 ft) 208 knots (385 km/h; 239 mph)
Econ cruising speed at 3,050 m (10,000 ft) 190 knots (352 km/h; 218 mph)
Stalling speed at max landing weight 71 knots (132 km/h; 82 mph) CAS
Max rate of climb at S/L 362 m (1,190 ft)/min
Time to 3,050 m (10,000 ft) 13 min
Time to 4,575 m (15,000 ft) 28 min
Service ceiling, at AUW of 5,300 kg (11,684 lb) 7,770 m (25,500 ft)
Service ceiling, one engine out, at AUW of 5,300 kg (11,684 lb) 4,025 m (13,200 ft)
T-O run 940 m (3,084 ft)
T-O to 15 m (50 ft) 1,590 m (5,217 ft)
Landing from 15 m (50 ft) 725 m (2,379 ft)
Landing run 405 m (1,329 ft)
Min ground turning radius 12·80 m (42 ft 0 in)
Range at 3,050 m (10,000 ft), max fuel, 45 min reserves 1,590 nm (2,945 km; 1,830 miles)

EMBRAER EMB-120 BRASILIA

The EMB-120 Brasilia is a twin-turboprop passenger and cargo transport, design of which was started in September 1979. The first of two prototypes is due to fly in July 1983. Certification, scheduled for 1984, will be to FAR Pt 25. A mockup of the passenger version was displayed for the first time in Rio de Janeiro on 25 April 1980; general appearance can be seen in the accompanying illustration. Options on approx 115 Brasilias were held by May 1981, from 23 operators in eight countries (eight in Brazil, five in the USA, three each in Finland and France, and one each in Australia, Colombia, Mexico and the UK).

Military versions, including maritime patrol and ECM models, are expected to be developed.

Type: Twin-turboprop general-purpose transport.

Wings: Cantilever low-wing monoplane. Wing section NACA 23018 (modified) at root, NACA 23012 (modified) at tip. Dihedral 6° 30′ from roots. Incidence 2°. Sweepback 0° at 66 per cent chord. Three-spar fail-safe structure between engine nacelles (across fuselage), two-spar on outer panels, forming a continuous tip-to-tip torsion box. Machined and bonded extrusions of 2024 aluminium alloy on spar caps. Chemically milled skin. Double-slotted semi-Fowler all-metal trailing-edge flaps (three segments each side), driven by hydraulic actuator. All-metal Frise-type ailerons, each with inset trim and balance tab. Pneumatic-boot de-icing of leading-edges.

Fuselage: Conventional metal semi-monocoque fail-safe structure, of circular cross-section. Machined windscreen frames and wing attachment fittings. Chemically milled skin of aluminium sheet. Structure pressurised, except for nosecone and tailcone.

Tail Unit: Cantilever T tail, of two-spar all-metal construction with chemically milled skin. Fixed-incidence swept tailplane, with horn-balanced elevators. Swept-back fin and double-hinged rudder. Trim and anti-servo tabs in elevators. Pneumatic-boot de-icing of leading-edges.

Landing Gear: Retractable tricycle type, with twin wheels and oleo-pneumatic shock-absorber on each unit. Hydraulic actuation; all units retract forward (main units into engine nacelles). Hydraulically-powered nosewheel steering. Tyre sizes 24 × 7·25 in (main), 18 × 5·5 in (nose); tyre pressure 5·17 bars (75 lb/sq in) on main units, 5·86 bars (85 lb/sq in) on nose unit. Multiple-disc brakes, actuated hydraulically by means of a proportional anti-skid system. Autobrake optional.

Power Plant: Two Pratt & Whitney Aircraft of Canada PW115 turboprop engines, each flat rated to 1,118·5 kW (1,500 shp) and driving a Hamilton Standard 14RF-1 four-blade constant-speed reversible-pitch fully-feathering double-actuated metal and glassfibre propeller. Fuel in two-cell 1,674 litre (368 Imp gallon) integral tank in each wing; total capacity 3,348 litres (736·5 Imp gallons). Single-point pressure refuelling, plus gravity point in upper surface of each wing. Oil capacity 16 litres (3·5 Imp gallons).

Accommodation: Pilot and co-pilot on flight deck, with dual controls; cabin attendant in passenger version.

Main cabin accommodates 30 passengers in three-abreast seating at 79 cm (31 in) pitch, with overhead lockable baggage racks, in pressurised and air-conditioned environment. Pressurised baggage compartment aft of passenger cabin, with door on port side. Provisions for wardrobe, galley and toilet. Downward-opening main passenger door, with airstairs, forward of wing on port side. Type II emergency exit on starboard side at rear. Overwing Type III emergency exit on each side. Large cargo door at rear on port side. Also available with all-cargo interior, with reduced rear baggage compartment volume; executive or military transport interior; or in mixed-traffic version with 24 or 26 passengers (toilet omitted in latter case), and 900 kg (1,984 lb) of cargo in enlarged rear baggage compartment.

Systems: Air-conditioning and pressurisation system (max differential 0·48 bars; 7 lb/sq in), with dual packs of recirculating equipment. Duplicated hydraulic system (pressure 207 bars; 3,000 lb/sq in), powered by dual pumps (one on each engine), for landing gear, flap and brake actuation, and nosewheel steering. Emergency standby electric pumps on each system, plus emergency handpump, for landing gear extension. Rudder booster actuated pneumatically by engine bleed air. Main electrical power supplied by two 28V 400A DC starter/generators; two 28V 100A DC auxiliary generators and 24V 40Ah nickel-cadmium battery for secondary power. Main and standby 600VA solid-state static inverters for 26/115V AC power at 400 Hz. Single high-pressure (124 bars; 1,800 lb/sq in) oxygen cylinder for crew; individual chemical system for passengers. Pneumatic de-icing for wing and tail leading-edges; electrical heaters to de-ice windscreens, propellers and pitot tubes; bleed air de-icing of engine air intakes. The starboard engine, with propeller stopped, can serve as an APU to provide bleed air for air-conditioning and electrical power on the ground. Halon 1301 engine fire extinguishing system.

Avionics and Equipment: Blind-flying instrumentation standard. Provisions for full Category II operation, depending upon avionics and equipment selection by individual customers. Special navigation systems (INS, Omega, satellite) also available optionally. Standard avionics package includes two VHF com, two VHF nav (VOR/ILS/marker beacon receiver), two audio and interphone, ADF, ATC transponder, DME, weather radar, two RMI, two course indicators, ELT, two gyromagnetic compasses and two attitude indicators. Options include third VHF com, HF com, passenger PA and radio/tape deck, second ADF, second ATC transponder, second DME, one or two radio altimeters, GPWS, one or two R/Nav, one or two Omega nav, altitude alerter, voice recorder, flight recorder, third gyromagnetic compass, third attitude indicator, attitude and course monitor, and one or two air data computers.

Dimensions, external:
- Wing span 19·78 m (64 ft 10¾ in)
- Wing chord at root 2·50 m (8 ft 2½ in)
- Wing chord at tip 1·40 m (4 ft 7 in)
- Wing aspect ratio 10·29
- Length overall 19·725 m (64 ft 8½ in)
- Length of fuselage 18·448 m (60 ft 6¼ in)
- Fuselage: Max diameter 2·28 m (7 ft 5¾ in)
- Height overall 6·30 m (20 ft 8 in)
- Tailplane span 6·94 m (22 ft 9¼ in)
- Wheel track (c/l of shock-struts) 6·07 m (19 ft 11 in)
- Wheelbase 6·81 m (22 ft 4 in)
- Propeller diameter 3·20 m (10 ft 6 in)
- Propeller ground clearance (min) 0·49 m (1 ft 7¼ in)
- Passenger door (fwd, port):
  - Height 1·70 m (5 ft 7 in)
  - Width 0·80 m (2 ft 7½ in)
  - Height to sill 1·50 m (4 ft 11 in)
- Emergency exit (rear, stbd):
  - Height 1·12 m (3 ft 8 in)
  - Width 0·51 m (1 ft 8 in)
  - Height to sill 1·50 m (4 ft 11 in)
- Cargo door (rear, port):
  - Height 1·30 m (4 ft 3¼ in)
  - Width 1·30 m (4 ft 3¼ in)
  - Height to sill 1·50 m (4 ft 11 in)
- Emergency exits (overwing, each):
  - Height 0·91 m (3 ft 0 in)
  - Width 0·51 m (1 ft 8 in)
- Emergency exits (flight deck side windows, each):
  - Min height 0·48 m (1 ft 7 in)
  - Min width 0·51 m (1 ft 8 in)

Dimensions, internal:
- Cabin, excl flight deck and baggage compartment:
  - Length 9·26 m (30 ft 4½ in)
  - Max width 2·15 m (7 ft 0½ in)
  - Max height 1·78 m (5 ft 10 in)
  - Floor area 15·00 m² (161·5 sq ft)
- Rear baggage compartment (30-passenger version):
  - Length 2·50 m (8 ft 2½ in)
  - Floor area 3·15 m² (33·9 sq ft)
  - Volume 6·40 m³ (226 cu ft)
- Rear baggage compartment volume:
  - all-cargo version 3·00 m³ (106 cu ft)
  - passenger/cargo version 11·00 m³ (388·5 cu ft)
- Max available cabin volume (all-cargo version) 29·90 m³ (1,056 cu ft)
- Cabin, incl flight deck and baggage compartment:
  - Total volume approx 33·0 m³ (1,165 cu ft)

Areas:
- Wings, gross 38·025 m² (409·30 sq ft)
- Ailerons (total) 3·18 m² (34·23 sq ft)
- Trailing-edge flaps (total) 6·514 m² (70·12 sq ft)
- Fin 6·90 m² (74·27 sq ft)
- Rudder 2·59 m² (27·88 sq ft)
- Tailplane 10·00 m² (107·64 sq ft)
- Elevators, incl tabs 3·90 m² (41·98 sq ft)

Weights and Loadings:
- Basic operating weight empty 5,576 kg (12,295 lb)
- Max payload (30 passengers) 3,024 kg (6,666 lb)
- Max payload (all-cargo) 3,178 kg (7,006 lb)
- Max fuel 2,370 kg (5,225 lb)
- Max T-O and landing weight 9,600 kg (21,165 lb)
- Max ramp weight 9,680 kg (21,340 lb)
- Max zero-fuel weight 8,600 kg (18,960 lb)
- Max wing loading 252 kg/m² (51·6 lb/sq ft)
- Max power loading 3·83 kg/kW (7·05 lb/shp)

Performance (estimated, at max T-O weight, ISA):
- Never-exceed speed 270 knots (500 km/h; 310 mph) EAS
- Max level and max cruising speed at 6,400 m (21,000 ft) 294 knots (545 km/h; 338 mph)
- Econ cruising speed 250 knots (463 km/h; 288 mph)
- Stalling speed, flaps up, power off 105 knots (195 km/h; 121 mph) EAS
- Stalling speed, flaps down, power off 83 knots (154 km/h; 96 mph) EAS
- Max rate of climb at S/L 808 m (2,650 ft)/min
- Rate of climb at S/L, one engine out 217 m (710 ft)/min
- Service ceiling 9,750 m (32,000 ft)
- Service ceiling, one engine out 5,485 m (18,000 ft)
- Balanced T-O field length at S/L, and 0·6 weighted landing field length at S/L 1,200 m (3,937 ft)
- Min ground turning radius 15·50 m (50 ft 10¼ in)
- Range with 30 passengers, no optional fittings, reserves for 100 nm (185 km; 115 mile) alternate plus 45 min 545 nm (1,010 km; 628 miles)
- Range with max fuel, reserves as above 1,580 nm (2,928 km; 1,819 miles)

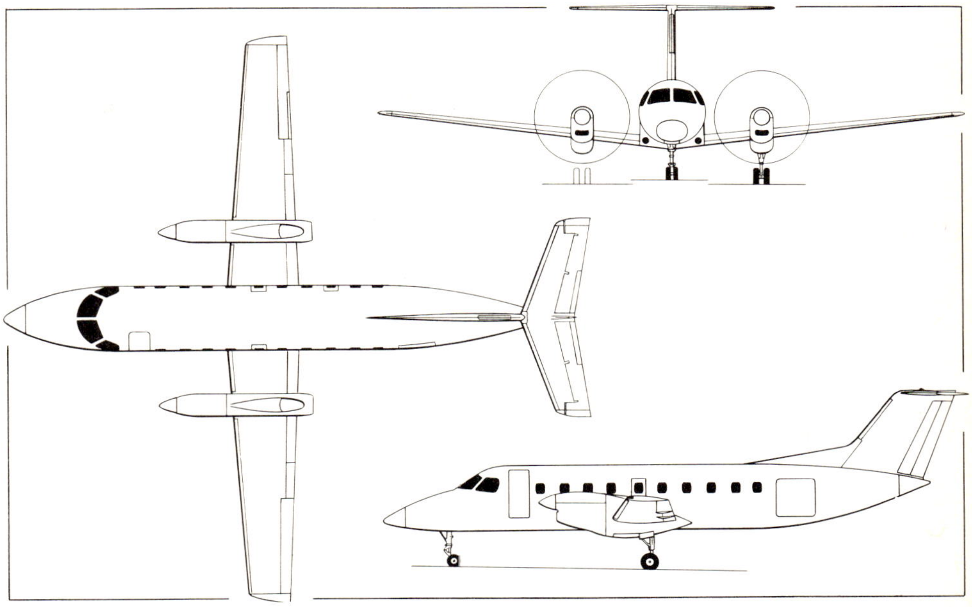

EMBRAER EMB-120 Brasilia twin-turboprop transport *(Pilot Press)*

EMBRAER EMB-121 Xingu twin-turboprop executive transport aircraft

EMBRAER EMB-121 XINGU

Brazilian Air Force designation: VU-9

The prototype Xingu (PP-ZXI) flew for the first time on 10 October 1976. A second airframe was built for static testing, and a third is being used for fatigue testing. First production Xingu (PP-ZCT) was flown on 20 May 1977, and Brazilian CTA certification was awarded in May 1979. British CAA certification was granted in July of the same year. Xingu production totalled 47 by April 1981. Six VU-9s have been delivered to the Grupo de Transporte Especial (Special Transport Group) of the Brazilian Air Force and are in service with the 6° Esquadrão de Transporte Aéreo (6th Air Transport Squadron) at Brasilia. Other customers include CSE in the UK, Sabena, and Brazilian civil operators. The French Air Force and Navy have ordered 41 Xingus (25 and 16 respectively), for aircrew training and liaison duties.

Type: Twin-turboprop general-purpose transport and advanced training aircraft.

Wings: Cantilever low-wing monoplane. Utilises same wing as EMB-110P2 (which see), but with reduced span and modified tips. Leading-edge pneumatic de-icing boots optional.

Fuselage: All-metal semi-monocoque fail-safe structure of aluminium alloy, with circular cross-section, designed for a max operating pressure differential of 0·414 bars (6·0 lb/sq in).

Tail Unit: Cantilever metal T tail, with glassfibre dorsal fin. Pneumatic de-icing boots optional for fin and tailplane leading-edges. Trim and balance tabs in rudder and each elevator. Ventral fin.

Landing Gear: Hydraulically-retractable tricycle type, with oleo-pneumatic shock-absorber on each unit. Single main wheels, with tyres size 670 × 210-12 (10 ply rating), pressure 5·52-5·86 bars (80-85 lb/sq in). Steerable, forward-retracting twin nosewheels, with tyres size 16 × 44 (six ply rating), pressure 5·52-5·86 bars (80-85 lb/sq in).

Power Plant: Two 507 kW (680 shp) Pratt & Whitney Aircraft of Canada PT6A-28 turboprop engines, each driving a Hartzell HC-B3TN-3D/T10178HB-8R three-blade constant-speed metal propeller with autofeathering and full reverse-pitch capability. Four integral fuel tanks in wings, with total capacity of 1,666 litres (366 Imp gallons). Gravity refuelling point on top of each wing. Oil capacity 8·3 litres (1·8 Imp gallons).

Accommodation: Two seats side by side on flight deck. Cabin seats up to nine passengers. Downward-hinged door on port side, aft of wing, with built-in airstairs.

Emergency exit over wing on starboard side. Baggage compartments in nose (with external access) and at rear of cabin (with internal access). Toilet/lavatory and galley standard.

SYSTEMS: Air-cycle type air-conditioning system, max capacity 20,000 BTU/h for cooling and 40,000 BTU/h for heating. Pressurisation system (max differential 0·41 bars; 6 lb/sq in) maintains S/L cabin environment up to 4,270 m (14,000 ft) and 2,440 m (8,000 ft) environment up to 8,335 m (27,350 ft). Primary electrical system is 28V DC, supplied by two 9 kW starter/generators and a 40Ah alkaline battery with a temperature sensor. Two 125VA or 600VA static inverters provide 115/26V single-phase AC power at 400Hz. Electrical anti-icing of windscreen, engine air intakes and propellers; pneumatic-boot de-icing of wing and tail unit leading-edges.

AVIONICS: Standard avionics include dual RCA AVC-110A VHF com transceivers, dual RCA AVN-220A VOR/ILS/marker beacon receivers, dual Collins ADF-60A ADF, one VOR/ILS OBS indicator (co-pilot), dual RCA AVA-310 audio panels, one Sperry C-14 gyromagnetic compass with dual Sperry RD-44 course indicator (pilot), one AIM-200 DC FM directional gyro (co-pilot), one Sperry GH-14-330 gyro horizon (pilot), one AIM-500 DC FM gyro horizon (co-pilot), one Collins TDR-90 transponder with Smiths 01-200-105 encoding altimeter, one Collins DME-40 DME, two Telex TEL-66C microphones, two pairs of Telex A1210 earphones, dual Flite-Tronics PC-15 BC(D) static converters, and one SunAir ASB-100A HF transceiver. Optional avionics are available in three standard packages, with dual Collins DF-206 ADF, SunAir ASB-100A HF/AM/SSB, dual RCA AVA-310 and Sperry SPZ-200 autopilot common to all three. In addition, package No. 1 offers dual RCA AVC-110A VHF com, one RCA AVN-220A VHF nav and one Sperry C-14 gyromagnetic compass; package No. 2 offers dual RCA AVC-110A VHF com, dual RCA AVN-220A VHF nav, dual Sperry C-14 gyromagnetic compasses, and Garrett Rescu/88L ELT; package No. 3 offers dual Collins VHF-20A VHF com, dual Collins VIR-30A VHF nav, dual Sperry C-14 gyromagnetic compasses, Garrett Rescu/88L ELT, Collins ALT-50 radio altimeter, and dual Sperry STARS IVC flight directors for pilot and co-pilot. Bendix RDR-1200 or RCA Primus 40 weather radar, cabin music and PA systems are optional extras on all three packages; Collins TDR-90 transponder on packages 2 and 3; Collins DME-40 on package 3; and pilot's Sperry STARS IVB or IVC on package 2.

EQUIPMENT: Standard equipment comprises maximum permissible airspeed indicator, eight-day clock, chronometer, cabin rate of climb indicator, cabin altitude and differential pressure indicator, annunciator panel, heated stall warning system, dual heated pitot tubes and heated static ports, external power sockets, wing ice light, dual landing lights, dual taxi lights, dual anti-collision strobe lights, dual map lights, cabin dome lights, instrument lighting system, low profile glareshield, and hand-type cabin fire extinguishers. Optional equipment includes three-light strobe system, fire extinguishing system, de-icing and anti-icing system, toilet, and a range of galley equipment.

DIMENSIONS, EXTERNAL:
Wing span 14·45 m (47 ft 5 in)
Wing chord at fuselage c/l 2·47 m (8 ft 1¼ in)
Wing chord at root 2·33 m (7 ft 7¾ in)
Wing chord at structural tip 1·49 m (4 ft 10¾ in)
Wing aspect ratio 7·18
Length overall 12·25 m (40 ft 2¼ in)
Length of fuselage 11·16 m (36 ft 7½ in)
Fuselage: Max width 1·86 m (6 ft 1¼ in)
Height overall 4·74 m (15 ft 6½ in)
Tailplane span 5·58 m (18 ft 3¾ in)
Wheel track 5·24 m (17 ft 2¼ in)
Wheelbase 2·88 m (9 ft 5½ in)
Propeller diameter 2·36 m (7 ft 9 in)
Distance between propeller centres 5·10 m (16 ft 8¾ in)
Passenger door (rear, port):
Height 1·32 m (4 ft 4 in)
Width 0·63 m (2 ft 0¾ in)
Emergency exit (one, overwing):
Height 0·96 m (3 ft 1¾ in)
Width 0·51 m (1 ft 8 in)

DIMENSIONS, INTERNAL:
Cabin, incl flight deck:
Max length 5·18 m (17 ft 0 in)
Max width 1·74 m (5 ft 8½ in)
Max height 1·52 m (4 ft 11¾ in)
Baggage compartment volume:
nose 0·30 m³ (10·6 cu ft)
rear 0·71 m³ (25·1 cu ft)

AREAS:
Wings, gross 27·50 m² (296·0 sq ft)
Ailerons (total) 1·84 m² (19·81 sq ft)
Trailing-edge flaps (total) 4·90 m² (52·74 sq ft)
Vertical tail surfaces (total, excl dorsal fin) 4·00 m² (43·06 sq ft)
Rudder, incl tab 1·30 m² (13·99 sq ft)
Dorsal fin 0·54 m² (5·81 sq ft)
Ventral fin 0·94 m² (10·12 sq ft)
Horizontal tail surfaces (total) 5·84 m² (62·86 sq ft)
Elevators, incl tabs 2·17 m² (23·36 sq ft)

WEIGHTS AND LOADINGS:
Weight empty, equipped 3,500 kg (7,716 lb)
Max payload 1,477 kg (3,256 lb)
Max T-O weight 5,670 kg (12,500 lb)
Max zero-fuel weight 4,660 kg (10,273 lb)
Max landing weight 5,340 kg (11,772 lb)
Max cabin floor loading 488 kg/m² (100 lb/sq ft)
Max wing loading 206·2 kg/m² (42·2 lb/sq ft)
Max power loading 5·59 kg/kW (9·19 lb/shp)

PERFORMANCE (at max T-O weight, ISA, except where indicated):
Max cruising speed at 3,350 m (11,000 ft) 243 knots (450 km/h; 280 mph)
Econ cruising speed at 6,100 m (20,000 ft) 203 knots (376 km/h; 234 mph)
Stalling speed at max T-O weight, flaps up 96 knots (178 km/h; 111 mph)
Stalling speed at max landing weight, flaps down 74 knots (137·5 km/h; 85·5 mph)
Max rate of climb at S/L 426 m (1,400 ft)/min
Service ceiling 7,925 m (26,000 ft)
Service ceiling, one engine out 3,960 m (13,000 ft)
T-O to 15 m (50 ft) 865 m (2,840 ft)
Landing from 15 m (50 ft) at max landing weight 850 m (2,790 ft)
Range with 900 kg (1,985 lb) payload at 6,100 m (20,000 ft), 45 min reserves 900 nm (1,666 km; 1,035 miles)
Range with max fuel and 610 kg (1,345 lb) payload at 6,100 m (20,000 ft), ISA, 45 min reserves 1,270 nm (2,352 km; 1,461 miles)

EMBRAER EMB-121B XINGU II

This 'stretched' fuselage version of the current Xingu, with PT6A-42 engines, is under development for delivery from 1983. It was flying by mid-1981 and will be produced in parallel with the existing version.

WINGS: Similar to those of EMB-121, but with larger tip fairings, and integral leading-edge tanks outboard of engine nacelles to increase fuel capacity. Trim and balance tabs in ailerons. Pneumatic-boot de-icing of leading-edges.

FUSELAGE: Generally similar to that of EMB-121, but cabin lengthened by 0·89 m (2 ft 11 in). Fail-safe structure except for wing carry-through members, which are safe-life. Chemically milled skin. Pressurised, except for nosecone and tailcone.

TAIL UNIT: Generally similar to EMB-121, but with larger-area fin and tailplane, and smaller elevators.

LANDING GEAR: Similar to that of EMB-121, but with size 40 × 110 mm nosewheel tyres and Goodrich multiple-disc brakes.

POWER PLANT: Two 634 kW (850 shp) Pratt & Whitney Aircraft of Canada PT6A-42 turboprop engines, each driving a Dowty R314 constant-speed reversible-pitch propeller. Integral fuel tank in each wing leading-edge, outboard of nacelle, combined usable capacity 2,170 litres (477 Imp gallons). Gravity refuelling point in upper surface of each wing. Pitot-type engine intakes, with inertial separator to prevent solid particle ingestion. Electrical de-icing of engine air intakes; NACA flush-type intakes for oil coolers.

ACCOMMODATION: Normal flight crew of two; to be certificated also for single-pilot operation. Basic layout for eight or nine passengers, with seats at 79 cm (31 in) pitch, plus toilet. Alternative club layout with seven seats, two folding tables, galley and toilet. Baggage compartments in nose (unpressurised) and at rear of cabin (pressurised). Passenger door aft of wing on port side; overwing emergency exit on starboard side. Nose baggage compartment has door on each side. Entire accommodation pressurised and air-conditioned.

SYSTEMS: Hamilton Standard air-conditioning system; pressurisation system max differential 0·41 bars (6 lb/sq in). Single hydraulic system (pressure 207 bars; 3,000 lb/sq in) for landing gear extension/retraction, brakes (incl parking brake) and nosewheel steering. Emergency gear extension by manual pump, brake actuation by pressure accumulators. Primary electrical system is 28V DC, supplied by two 9kW starter/generators and a 40Ah nickel-cadmium battery, for flap actuation, propeller and engine intake anti-icing, autopilot and trim servos, and windscreen anti-icing. Pneumatic-boot de-icing of wing and tail unit leading-edges. High-pressure oxygen bottle for crew, individual chemical system for passengers.

AVIONICS AND EQUIPMENT: Basic single King Silver Crown II avionics system, plus Bendix RDR-1200 weather radar and Sperry SPZ-200 autopilot. Blind-flying instrumentation standard.

DIMENSIONS, EXTERNAL: As for EMB-121 except:
Wing span 14·83 m (48 ft 7¾ in)
Wing aspect ratio (geometric) 7·88
Length overall 13·44 m (44 ft 1¼ in)
Length of fuselage 12·05 m (39 ft 6½ in)
Tailplane span 6·05 m (19 ft 10¼ in)
Wheelbase 3·19 m (10 ft 5½ in)
Passenger door (rear, port):
Height 1·32 m (4 ft 4 in)
Width 0·63 m (2 ft 0¾ in)
Height to sill 1·20 m (3 ft 11¼ in)
Baggage doors (fwd, port and stbd) and service door (rear, stbd): Height 0·47 m (1 ft 6½ in)
Width 0·90 m (2 ft 11½ in)
Height to sill 1·50 m (4 ft 11 in)
Emergency exit (overwing, stbd):
Height 0·96 m (3 ft 1¾ in)
Width 0·51 m (1 ft 8 in)

DIMENSIONS, INTERNAL:
Cabin, excl flight deck: Length 4·50 m (14 ft 9¼ in)
Max width 1·70 m (5 ft 7 in)
Max height 1·52 m (5 ft 0 in)
Floor area 5·30 m² (57·05 sq ft)
Volume 9·00 m³ (317·8 cu ft)
Baggage compartment volume:
nose 0·34 m³ (12·00 cu ft)
rear of cabin 0·80 m³ (28·25 cu ft)

AREAS: As EMB-121 except:
Wings, gross 27·90 m² (300·3 sq ft)
Fin (incl dorsal and ventral fins) 5·48 m² (58·99 sq ft)
Tailplane 5·07 m² 54·57 sq ft)
Elevators, incl tabs 1·83 m² 19·70 sq ft)

WEIGHTS AND LOADINGS:
Basic operating weight empty 3,850 kg (8,487 lb)
Max payload 950 kg (2,094 lb)
Max fuel 1,690 kg (3,725 lb)
Max T-O weight 6,140 kg (13,536 lb)
Max ramp weight 6,210 kg (13,690 lb)
Max zero-fuel weight 4,800 kg (10,582 lb)
Max landing weight 5,700 kg (12,566 lb)
Max wing loading 220 kg/m² (45·1 lb/sq ft)
Max power loading 4·85 kg/kW (7·97 lb/shp)

PERFORMANCE (estimated, at max T-O weight, ISA):
Never-exceed speed 253 knots (468 km/h; 291 mph) EAS
Max cruising speed 272 knots (504 km/h; 313 mph)
Econ cruising speed 232 knots (430 km/h; 267 mph)
Stalling speed, flaps up, power off 96 knots (178 km/h; 111 mph) EAS
Stalling speed, flaps down, power off 76 knots (141 km/h; 88 mph) EAS
Max rate of climb at S/L 731 m (2,400 ft)/min
Rate of climb at S/L, one engine out 176 m (580 ft)/min
Service ceiling 8,840 m (29,000 ft)
Service ceiling, one engine out 4,360 m (14,300 ft)

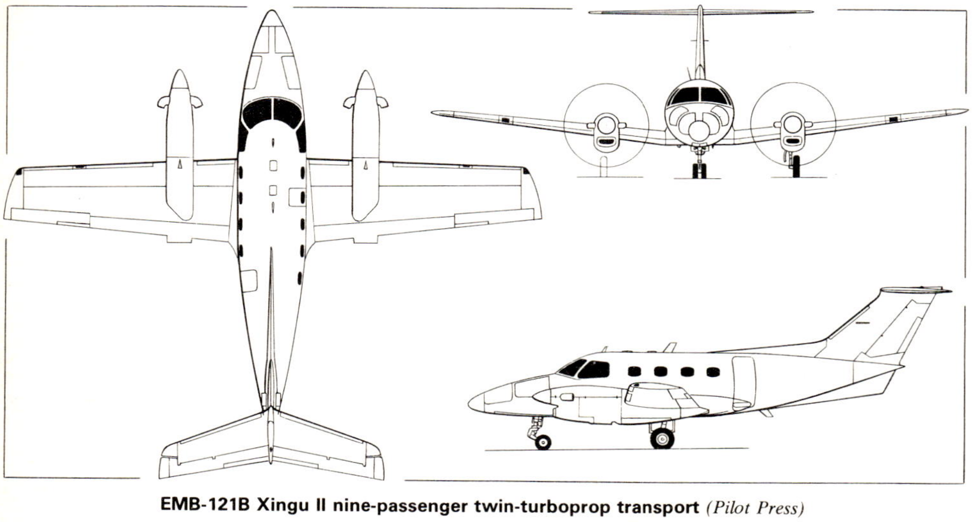

EMB-121B Xingu II nine-passenger twin-turboprop transport *(Pilot Press)*

T-O run 480 m (1,575 ft)
T-O to 15 m (50 ft) 820 m (2,690 ft)
Accelerate/stop distance 1,200 m (3,937 ft)
Landing from 15 m (50 ft) 960 m (3,150 ft)
Landing run 730 m (2,395 ft)
Min ground turning radius 11·20 m (36 ft 9 in)
Range, cruising at 7,620 m (25,000 ft), 45 min reserves:
with 900 kg (1,984 lb) payload
1,200 nm (2,224 km; 1,382 miles)
with max fuel and 460 kg (1,014 lb) payload
1,670 nm (3,095 km; 1,923 miles)

EMBRAER (AERMACCHI) EMB-326GB XAVANTE

Brazilian Air Force designation: AT-26

Under an agreement signed in May 1970, EMBRAER is assembling under licence the Aermacchi M.B.326GB jet trainer/ground attack aircraft for the Brazilian Air Force, by whom the type is known as the AT-26 Xavante, the name of a Brazilian Indian tribe.

The initial order for 112 aircraft was subsequently increased to an overall total of 182, of which production is due to end in 1982. The first Brazilian-assembled Xavante made its first flight on 3 September 1971, and the first two aircraft were handed over to the Brazilian Air Force a few days later. Six Xavantes have been delivered to the Togolese Air Force and nine to the Paraguayan Air Force.

Those of the Brazilian Air Force are in service with the 3° Esquadrão Misto de Reconhecimento e Ataque (3rd Mixed Reconnaissance and Attack Squadron) at Santa Cruz AFB, Rio de Janeiro; the 4° Esquadrão Misto de Reconhecimento e Ataque at Santa Maria AFB, Rio Grande do Sul State; the 4° Grupo de Aviação (4th Aviation Group) at Fortaleza, Ceará State; the CATRE (Centro de Aplicações Táticas e Recompletamento de Equipagens: Tactical and Aircrew Training Centre) at Natal, Rio Grande do Norte State; the 10° Grupo de Aviação at Cumbica AFB, São Paulo State; and other units.

A detailed description of the standard M.B. 326GB can be found in the Italian section of the 1980-81 *Jane's*. Its production by Aermacchi has ended; the following shortened description applies to the Brazilian version:

TYPE: Two-seat basic and advanced trainer, and light ground attack aircraft.

POWER PLANT: One Rolls-Royce Viper 20 Mk 540 turbojet engine, rated at 15·17 kN (3,410 lb st). Fuel in flexible rubber main tank in fuselage, capacity 782 litres (172 Imp gallons), and two 305 litre (67 Imp gallon) non-jettisonable wingtip tanks. Total standard fuel capacity 1,392 litres (306 Imp gallons). Provision for two 332 litre (73 Imp gallon) jettisonable underwing tanks, to give total capacity of 2,056 litres (452 Imp gallons). Single-point pressure refuelling receptacle under fuselage. Fuel dump valves permit quick emptying of tip-tanks.

AVIONICS AND EQUIPMENT: Two Collins Type 618M-2B 360-channel VHF transceivers, Collins CIA-102A interphone system, Bendix DFA 73A-1 ADF, and a complete VOR/ILS system using a Collins 51RV-1 VOR/LOC/glideslope receiver, Collins 51Z-4 marker beacon receiver and AN/APX-72. Provision for IFF transponder.

ARMAMENT: Six underwing points for bombs, gun pods or other stores. Typical loads include six 250 lb bombs; two 500 lb bombs; two 500 lb bombs and two twin 7·62 mm gun pods; four 250 lb bombs and two twin 7·62 mm gun pods; two twin 7·62 mm gun pods and two underwing drop-tanks; two twin 7·62 mm gun pods and four LM-70/7 rocket pods (each with seven SBAT 70 mm folding-fin air-to-ground projectiles); two twin 7·62 mm gun pods and two LM-37/36 rocket pods (each with thirty-six SBAT 37 mm air-to-ground rockets); six LM-70/7 rocket pods; or two LM-70/19 rocket pods (each with nineteen SBAT 70 mm air-to-ground rockets); or photographic reconnaissance pods. All armament loads are designed and manufactured in Brazil.

DIMENSIONS, EXTERNAL:
Wing span over tip-tanks 10·854 m (35 ft 7¼ in)
Wing area, gross 19·30 m² (207·7 sq ft)
Length overall 10·673 m (35 ft 0¼ in)
Height overall 3·72 m (12 ft 2 in)

WEIGHTS AND LOADINGS (A: Trainer; B: Attack):
Basic operating weight, excl crew:
A 2,685 kg (5,920 lb)
*B 2,558 kg (5,640 lb)
Max T-O weight (full internal fuel, wingtip and underwing tanks): A 4,577 kg (10,090 lb)
B, no armament 4,447 kg (9,805 lb)
B, with 769 kg (1,695 lb) armament
5,216 kg (11,500 lb)
Max T-O weight (max armament):
*B, with fuel in fuselage tank only and 1,962 kg (4,325 lb) armament 5,216 kg (11,500 lb)
Max wing loading 269·5 kg/m² (55·2 lb/sq ft)
Max power loading 343·8 kg/kN (3·37 lb/lb st)
* *Without tip-tanks and aft ejection seat*

PERFORMANCE: (A: Trainer at typical weight of 3,937 kg; 8,680 lb, representing max T-O weight without underwing tanks; B: Attack version at combat weight of 4,763 kg; 10,500 lb; C: Attack version at max T-O weight):
Never-exceed speed:
A Mach 0·82 (469 knots; 871 km/h; 541 mph EAS)
B Mach 0·75 (419 knots; 778 km/h; 483 mph EAS)
Max level speed: A 468 knots (867 km/h; 539 mph)
Max cruising speed:
A 430 knots (797 km/h; 495 mph)
Max rate of climb at S/L: A 1,844 m (6,050 ft)/min
B 1,082 m (3,550 ft)/min
C 945 m (3,100 ft)/min
Time to 3,050 m (10,000 ft): B 3 min 10 s
C 4 min 0 s
Time to 6,100 m (20,000 ft): A 4 min 10 s
B 8 min 0 s
C 9 min 20 s
Time to 9,150 m (30,000 ft): A 7 min 40 s
B 15 min 0 s
C 18 min 40 s
Time to 12,200 m (40,000 ft): A 13 min 5 s
Service ceiling: A 14,325 m (47,000 ft)
B 11,900 m (39,000 ft)
T-O run, ISA: A 412 m (1,350 ft)
B 640 m (2,100 ft)
C 845 m (2,770 ft)
T-O to 15 m (50 ft), ISA: A 555 m (1,820 ft)
B 866 m (2,840 ft)
C 1,140 m (3,740 ft)
Landing from 15 m (50 ft), ISA:
A at landing weight of 3,175 kg (7,000 lb)
631 m (2,070 ft)
B at landing weight of 4,195 kg (9,250 lb)
802 m (2,630 ft)
Range (A, with 113 litres; 25 Imp gallons reserve):
fuselage and tip-tanks
998 nm (1,850 km; 1,150 miles)
fuselage, tip and underwing tanks
1,320 nm (2,445 km; 1,520 miles)
Combat radius (C):
max fuel, 769 kg (1,695 lb) armament, 90 kg (200 lb) fuel reserve, out at 6,100 m (20,000 ft), return at 7,620 m (25,000 ft)
350 nm (648 km; 403 miles)
fuselage tank only, 1,814 kg (4,000 lb) armament, 90 kg (200 lb) fuel reserve, cruise at 3,050 m (10,000 ft), five minutes over target
69 nm (130 km; 80 miles)
max fuel, 771 kg (1,700 lb) armament, 90 kg (200 lb) fuel reserve, cruise at 3,050 m (10,000 ft), 1 h 50 min patrol at 150 m (500 ft) over target
49·5 nm (92 km; 57 miles)

EMBRAER AT-26 Xavante (Aermacchi M.B.326GB) of the Brazilian Air Force

EMBRAER EMB-312

Brazilian Air Force designation: T-27

Design of the EMB-312, by a team under the leadership of Ing Joseph Kovacs, began in January 1978 as part of a programme to develop a new basic trainer for the Brazilian Air Force. On 6 December that year a contract was received from the Departamento de Pesquisas e Desenvolvimento (Department of Research and Development) of the Brazilian Ministry of Aeronautics, for two flying prototypes plus two other airframes for static and fatigue testing.

Characteristics of the EMB-312 include high manoeuvrability, short take-off and landing, the ability to operate from unprepared runways, and a high degree of stability. In addition to meeting the requirements of FAR Pt 23 Appendix A, the aircraft will meet MIL and CAA Section K specifications. Its construction embodies such modern techniques as integral numerical control machining, chemical milling, and metal-to-metal bonding.

The EMB-312 is designated T-27 by the Brazilian Air Force, which has ordered 118 with a further 50 on option. These are scheduled to be delivered from October 1982, initially to replace the Cessna T-37C. Sixty are scheduled to be in Brazilian Air Force service by February 1984. The first prototype (Brazilian Air Force serial number 1300) made its initial flight on 16 August 1980, and the second (1301) on 13 December 1980. The prototypes had accumulated more than 190 h of flying by early May 1981.

TYPE: Tandem two-seat basic trainer.

WINGS: Cantilever low-wing monoplane. Wing section NACA 63_2A-415 at root, NACA 63A-212 at tip. Dihedral 5° 30′ at 30% chord. Incidence 1° 25′. Geometric twist 2° 13′. Sweepback 0° 43′ 26″ at quarter-chord. Aluminium alloy two-spar torsion-box structure of 2024T-3511 extrusions and 2024T-3 sheet. Single-slotted trailing-edge flaps of 2024T-3, supported on 4130 steel tracks. Frise-type constant-chord balanced ailerons. Electro-mechanically actuated spring compensation in each aileron.

FUSELAGE: Conventional semi-monocoque structure of 2024T-3 aluminium alloy.

TAIL UNIT: Cantilever all-metal structure, of similar construction to wings. Non-swept fin, with dorsal fin, and horn-balanced rudder. Non-swept fixed-incidence tailplane and balanced elevators. Electro-mechanically actuated spring compensation in port elevator.

LANDING GEAR: Hydraulically retractable tricycle type, with single wheel and Piper oleo-pneumatic shock-absorber on each unit. Shimmy damper on nose unit. Rearward-retracting nose unit; main units retract inward into wings. Cleveland 40-130 main wheels, 40-76B nosewheel. Tyre sizes 6·50-10 (Type III, 8-ply

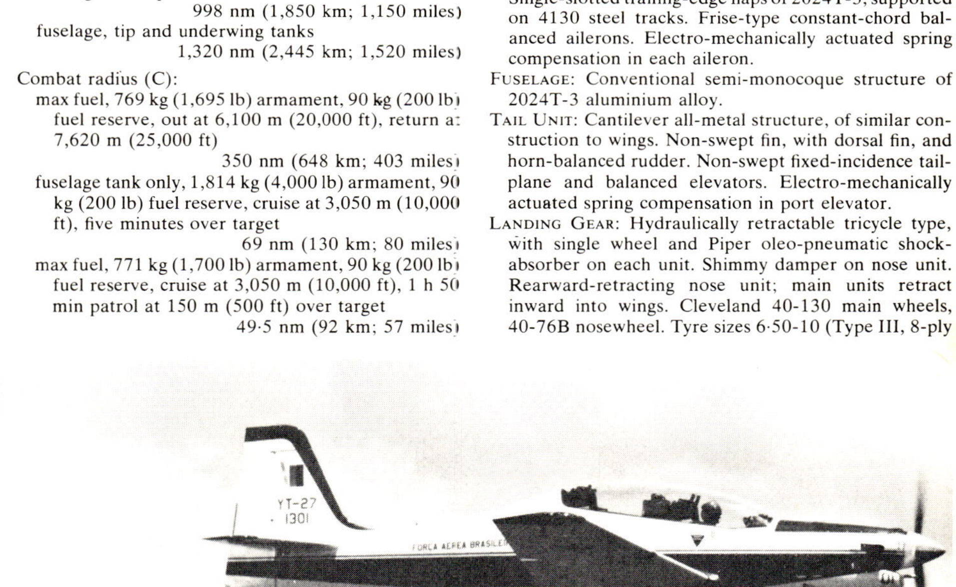

First and second YT-27 prototypes of the EMBRAER EMB-312 tandem two-seat trainer

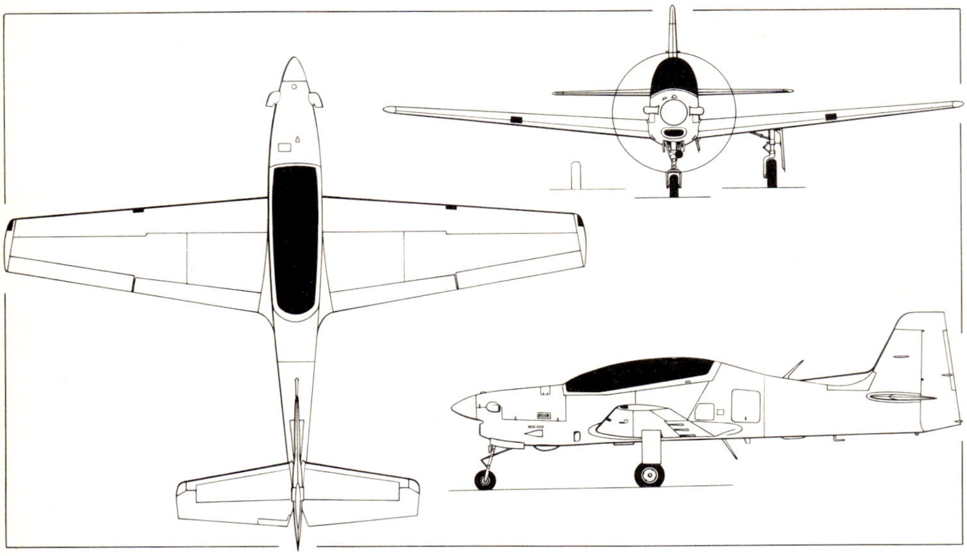

EMBRAER EMB-312 basic trainer for the Brazilian Air Force, in production form *(Pilot Press)*

rating) on main wheels, 5·00-5 (Type III, 6-ply rating) on nosewheel. Cleveland 30-95A brakes.

Power Plant: One 559 kW (750 shp) Pratt & Whitney Aircraft of Canada PT6A-25C turboprop engine, driving a Hartzell HC-B3TN-3C/T10178H-8R three-blade constant-speed propeller with spinner. Two integral fuel tanks in each wing, total capacity 710 litres (156 Imp gallons). Fuel tanks lined with anti-detonation plastics foam. Gravity refuelling point in each wing upper surface. Fuel system allows for up to 35 s of inverted flight.

Accommodation: Instructor and pupil in tandem, on Martin-Baker BR8LC ejection seats, in air-conditioned cockpit. One-piece fully-transparent vacuum-formed canopy, opening sideways to starboard. Rear seat elevated. Dual controls standard. Baggage compartment in rear fuselage, with access via door on port side.

Systems: Freon cycle air-conditioning system, with engine-driven compressor. Single hydraulic system, pressure 131 bars (1,900 lb/sq in), for landing gear extension and retraction. No pneumatic system. 28V DC electrical power provided by a 6kW starter/generator, 26Ah battery and, for AC power at 115V 400Hz, a 250VA inverter. Oxygen system for occupants conforms to MIL-C-5887 and is supplied by six MS 21227 D2 type cylinders (total capacity approx 1,200 litres; 264 Imp gallons) at a pressure of 31 bars (450 lb/sq in).

Avionics and Equipment: Standard avionics include two Collins VHF-20A VHF transceivers; two Collins 387C-4 audio systems, one EMBRAER radio transferring system; one Collins VIR-31A VOR/ILS/marker beacon receiver; one Collins TRD-90 ATC transponder; one Collins DME-40 DME; one Collins PN-101 gyromagnetic compass; and one Collins ADF-60A ADF. Landing light in each wing leading-edge, taxying lights on nosewheel unit.

Armament: Two hardpoints under each wing, each stressed for a max load of 150 kg (330 lb). Max external stores load 560 kg (1,234 lb). Typical loads, on Aermacchi MA-4A pylons, include two 0·50 in MS10-21/22-10A machine-gun pods, each with 350 rds; four 25 lb Mk 76 practice bombs; four 250 lb Mk 81 general-purpose bombs; or four LM-37/7A or LM-70/7 launchers, each with seven rockets (Avibras SBAT-37 and SBAT-70 respectively). Fixed reflex-type gunsight for use with machine-gun pods.

Dimensions, external:
Wing span 11·14 m (36 ft 6½ in)
Wing chord at root 2·30 m (7 ft 6½ in)
Wing chord at tip 1·07 m (3 ft 6¼ in)
Wing aspect ratio 6·4
Length overall 9·86 m (32 ft 4¼ in)
Length of fuselage (excl rudder) 8·53 m (27 ft 11¾ in)
Fuselage: Max width 1·00 m (3 ft 3¼ in)
Max depth 1·55 m (5 ft 1 in)
Height overall (static) 3·40 m (11 ft 1¾ in)
Tailplane span 4·66 m (15 ft 3½ in)
Wheel track 3·76 m (12 ft 4 in)
Wheelbase 3·16 m (10 ft 4½ in)
Propeller diameter 2·36 m (7 ft 9 in)
Propeller ground clearance (static) 0·25 m (9¾ in)
Baggage compartment door:
Height 0·60 m (1 ft 11½ in)
Width 0·54 m (1 ft 9¼ in)
Height to sill 1·25 m (4 ft 1¼ in)

Dimensions, internal:
Cockpits: Combined length 2·90 m (9 ft 6¼ in)
Max height 1·55 m (5 ft 1 in)
Max width 0·85 m (2 ft 9½ in)
Baggage compartment volume 0·17 m³ (6·0 cu ft)

Areas:
Wings, gross 19·40 m² (208·82 sq ft)
Ailerons (total) 1·97 m² (21·20 sq ft)
Trailing-edge flaps (total) 2·58 m² (27·77 sq ft)
Fin 2·08 m² (22·39 sq ft)
Rudder 1·46 m² (15·72 sq ft)
Tailplane 4·57 m² (49·19 sq ft)
Elevators, incl tab 2·00 m² (21·53 sq ft)

Weights and Loadings:
Basic weight empty 1,582 kg (3,487 lb)
Max fuel load (usable) 546 kg (1,203 lb)
Max T-O and landing weight 2,350 kg (5,180 lb)
Max ramp weight 2,368 kg (5,220 lb)
Max zero-fuel weight 1,900 kg (4,188 lb)
Max wing loading 121·1 kg/m² (24·82 lb/sq ft)
Max power loading 4·2 kg/kW (6·9 lb/shp)

Performance (estimated, at max T-O weight):
Never-exceed speed
292 knots (541 km/h; 336 mph) EAS
Max level speed at 4,115 m (13,500 ft)
247 knots (458 km/h; 284 mph)
Max cruising speed at 4,570 m (15,000 ft)
236 knots (437 km/h; 272 mph)
Econ cruising speed at 3,050 m (10,000 ft)
185 knots (343 km/h; 213 mph)
Stalling speed, flaps up, power off
72 knots (134 km/h; 83 mph) EAS
Stalling speed, flaps down, power off
65 knots (121 km/h; 75 mph) EAS
Max rate of climb at S/L 649 m (2,130 ft)/min
Service ceiling 9,935 m (32,600 ft)
T-O run 290 m (951 ft)
T-O to 15 m (50 ft) 510 m (1,673 ft)
Landing from 15 m (50 ft) 505 m (1,657 ft)
Landing run 240 m (787 ft)
Range at 6,100 m (20,000 ft) with max fuel, 30 min reserves 1,106 nm (2,050 km; 1,274 miles)
Endurance with max fuel 6 h
g limits (Aerobatic) +6·0; −3·0

EMBRAER EMB-201A IPANEMA

The original version of this agricultural aircraft was designed and developed to specifications laid down by the Brazilian Ministry of Agriculture. Design was started in May 1969, and the EMB-200 prototype (PP-ZIP) made its first flight on 30 July 1970. A type certificate was granted on 14 December 1971. Ipanema is the name of a famous beach in Rio, and also of a farm which is the headquarters of the Agricultural Air School of the Ministry of Agriculture, where the EMB-200 was flight tested.

Details of the earlier EMB-200/200A (73 built), EMB-201 (200 built) and EMB-201R (three built) can be found in the 1977-78 and previous editions of *Jane's*. The current production version, first flown on 10 March 1977, is the EMB-201A, of which 219 had been sold by January 1981. This incorporates several modifications requested by operators as a result of field experience, including new wing profile, wingtips and other aerodynamic improvements, improved systems, and revised cockpit layout.

Manufacture of the EMB-201A was being transferred to EMBRAER's Neiva subsidiary during the second half of 1981.

Type: Single-seat agricultural aircraft.

Wings: Cantilever low-wing monoplane. Wing section NACA 23015 (modified), with cambered leading-edges. Dihedral 7° from roots. Incidence 3°. All-metal single-spar structure of 2024 aluminium alloy with all-metal Frise-type ailerons outboard and all-metal slotted flaps on trailing-edge, and all-detachable cambered leading-edges. No tabs. Cambered wingtips standard.

Fuselage: Rectangular-section all-metal safe-life structure, of welded 4130 steel tube with removable skin panels of 2024 aluminium alloy. Structure is specially treated against chemical corrosion.

Tail Unit: Cantilever two-spar all-metal structure of 2024 aluminium alloy. Slight sweepback on fin and rudder. Fixed-incidence tailplane. Trim tab in starboard elevator.

Landing Gear: Non-retractable main and tail wheels, with oleo shock-absorbers on main units. Tailwheel has tapered spring shock-absorber. Main wheels and tyres size 8·50-10. Tailwheel diameter 250 mm (10 in). Tyre pressures: main, 2·07-2·41 bars (30-35 lb/sq in); tailwheel, 3·79 bars (55 lb/sq in). Hydraulic disc brakes on main wheels.

Power Plant: One 224 kW (300 hp) Avco Lycoming IO-540-K1J5D flat-six engine, driving a Hartzell two-blade constant-speed metal propeller with spinner. Integral fuel tanks in each wing leading-edge, with total capacity of 292 litres (64·2 Imp gallons). Refuelling point on top of each tank. Oil capacity 12 litres (2·6 Imp gallons).

Accommodation: Single horizontally/vertically-adjustable seat in fully-enclosed cabin with bottom-hinged window/door on each side. Ventilation system in cabin. Inertial type shoulder harness standard.

System: 28V DC electrical system supplied by a 24Ah BB639/U battery and a Bosch K.1 28V 35A alternator. Power receptacle for external battery (AN-2552-3A type) on port side of forward fuselage.

Avionics and Equipment: Standard VFR avionics include 720-channel Collins VHF-251S transceiver and Collins RCR-650 ADF transceiver. Hopper for agricultural chemicals has capacity of 680 litres (149·5 Imp gallons) liquid or 750 kg (1,653 lb) dry. Dusting system below centre of fuselage. Spraybooms and Micronair atomisers aft of and above wing trailing-edges.

Dimensions, external:
Wing span 11·69 m (38 ft 4¼ in)
Wing chord (constant) 1·71 m (5 ft 7½ in)
Wing aspect ratio 6·85
Length overall (tail up) 7·43 m (24 ft 4½ in)
Height overall (tail down) 2·22 m (7 ft 3½ in)
Fuselage: Max width 0·93 m (3 ft 0½ in)
Tailplane span 3·66 m (12 ft 0 in)
Wheel track 2·20 m (7 ft 2½ in)
Wheelbase 5·20 m (17 ft 7¼ in)
Propeller diameter 2·13 m (7 ft 0 in)

Dimensions, internal:
Cockpit: Max length 1·20 m (3 ft 11¼ in)
Max width 0·85 m (2 ft 9½ in)
Max height 1·34 m (4 ft 4¾ in)

Areas:
Wings, gross 19·94 m² (214·63 sq ft)
Ailerons (total) 1·60 m² (17·22 sq ft)
Trailing-edge flaps (total) 2·30 m² (24·76 sq ft)
Fin 0·58 m² (6·24 sq ft)
Rudder 0·63 m² (6·78 sq ft)
Tailplane 3·17 m² (34·12 sq ft)
Elevators (total, incl tab) 1·50 m² (16·15 sq ft)

Weights and Loadings (N: Normal; R: Restricted category):
Weight empty: N, R 1,011 kg (2,229 lb)
Max payload: N, R 750 kg (1,653 lb)
Max T-O and landing weight: N 1,550 kg (3,417 lb)
R 1,800 kg (3,968 lb)

EMB-201A Ipanema agricultural aircraft (Avco Lycoming IO-540-K1J5D engine)

Max wing loading: N 77·75 kg/m² (15·92 lb/sq ft)
R 90·29 kg/m² (18·49 lb/sq ft)
Max power loading: N 6·92 kg/kW (11·39 lb/hp)
R 8·03 kg/kW (13·23 lb/hp)

PERFORMANCE (at max T-O weight, 'clean' configuration, ISA):
Never-exceed speed:
N 165 knots (305 km/h; 190 mph)
R 104 knots (193 km/h; 120 mph)
Max level speed at S/L:
N 124 knots (230 km/h; 143 mph)
R 121 knots (225 km/h; 140 mph)
Max cruising speed (75% power) at 1,830 m (6,000 ft):
N 115 knots (212 km/h; 132 mph)
R 110 knots (204 km/h; 127 mph)
Stalling speed, power off (N):
flaps up 56 knots (103 km/h; 64 mph)
8° flap 54 knots (100 km/h; 62 mph)
30° flap 49·5 knots (92 km/h; 57 mph)
Stalling speed, power off (R):
flaps up 59·5 knots (109·5 km/h; 68 mph)
8° flap 57·5 knots (106·5 km/h; 66 mph)
30° flap 53 knots (98·5 km/h; 61 mph)
Max rate of climb at S/L, 8° flap:
N 283 m (930 ft)/min
R 201 m (660 ft)/min
Service ceiling, 8° flap: R 3,470 m (11,385 ft)
T-O run at S/L, 8° flap, asphalt runway:
N 282 m (925 ft)
R 368 m (1,208 ft)
T-O to 15 m (50 ft), conditions as above:
N 438 m (1,437 ft)
R 707 m (2,320 ft)
Landing from 15 m (50 ft) at S/L, 30° flap, asphalt runway: N 489 m (1,605 ft)
R 507 m (1,664 ft)
Landing run, conditions as above: N 150 m (492 ft)
R 168 m (551 ft)
Range at 1,830 m (6,000 ft), no reserves:
N 506 nm (938 km; 583 miles)
R 474 nm (878 km; 545 miles)

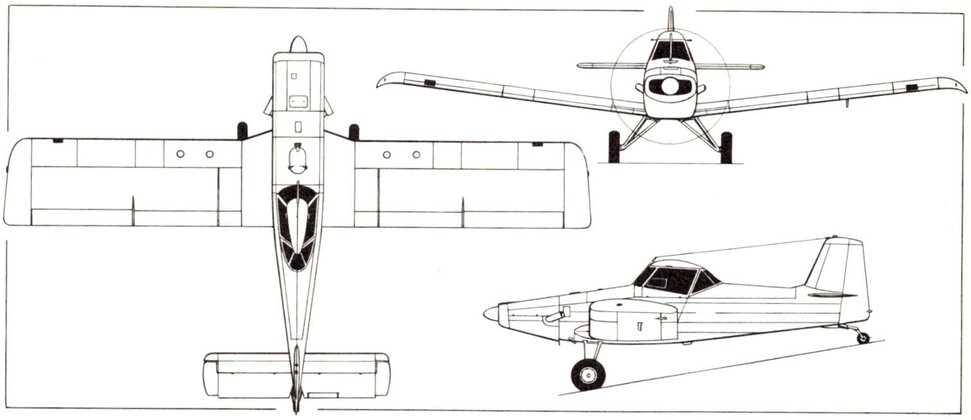

EMBRAER EMB-201A Ipanema single-seat agricultural aircraft *(Pilot Press)*

EMBRAER-PIPER LIGHT AIRCRAFT PROGRAMME

Detailed descriptions of the Piper aircraft built by EMBRAER and Neiva can be found in the US section, except for the EMB-710 Carioca. Manufacture of the EMB-710, 711, 712, 720 and 721 is now undertaken by EMBRAER's subsidiary, Neiva.

EMB-710C Carioca. Piper PA-28-235 Cherokee Pathfinder; named after inhabitants of Rio de Janeiro. Total of 264 licence-built by EMBRAER, which continued production in Brazil after manufacture by Piper had ended. Brazilian production ended in 1979. Replaced from August 1980 by EMB-710D.

EMB-710D Carioca. Piper PA-28-236 Dakota, with tapered outer wings. Replaced EMB-710C in production from August 1980. Total of three sold by January 1981.

EMB-711C Corisco. Piper PA-28R-200 Cherokee Arrow II. Total of 251 sold. Replaced in 1980 by EMB-711T and 711ST.

EMB-711T Corisco. Piper PA-28RT-201 Arrow II Total of 32 sold by January 1981.

EMB-711ST Corisco. Piper PA-28RT-201T turbocharged version of Arrow II. Total of 52 sold by January 1981.

EMB-712 Tupi. Piper PA-28-181 Cherokee Archer II. Production, under subcontract by Neiva, began in October 1979. Total of 78 sold.

EMB-720C Minuano. Piper PA-32-300 Cherokee SIX. Total of 126 sold. Replaced by EMB-720D.

EMB-720D Minuano. Piper PA-32-301 Cherokee SIX. First example completed in August 1980. Total of 17 sold by January 1981.

EMB-721C Sertanejo. Piper PA-32R-300 Lance. Total of 150 sold. Replaced by EMB-721D.

EMB-721D Sertanejo. Piper PA-32R-301 Lance. First example completed in June 1980. Total of 15 sold by January 1981.

EMB-810C Seneca II. Piper PA-34-200T Seneca II. Total of 394 sold by January 1981, including 21 (designated **U-7**) and 11 with Robertson STOL kit **(U-7A)** for Brazilian Air Force. Production to be transferred to Neiva (which see) in second half of 1982.

EMB-820C Navajo. Piper PA-31-350 Navajo Chieftain. First flight of Brazilian-built example 31 December 1975. Total of 124 sold by January 1981.

HELIBRAS
HELICÓPTEROS DO BRASIL SA

HEADQUARTERS: Rua Projetada Um 200, Distrito Industrial, Caixa Postal 184, 37500 Itajubá, Estado de Minas Gerais
Telephone: (035) 622 1366 and 622 2455
SUPERINTENDENT DIRECTOR: Nivaldo Alves da Silva
COMMERCIAL DIRECTOR: Alfredo Mendes de Macedo
PRODUCTION DIRECTOR: Jean Raquin

Formation of this company was announced in October 1977. Owned jointly by Aérospatiale of France and the State of Minas Gerais (45% each) and Aerofoto (10%), it is engaged in a 10-year programme involving the assembly (graduating to local manufacture) of 30 Aérospatiale SA 315B Lama and 200 AS 350B Ecureuil helicopters, in a new factory at Itajubá in Minas Gerais State. The first assembly hall to be completed, which is 5,500 m² (59,200 sq ft) in area, was officially inaugurated on 28 March 1980. The complete facility extends over an area of nearly 210,000 m² (2,260,420 sq ft).

The first Lamas were assembled in Brazil during the second half of 1979; the first Brazilian-assembled Ecureuils were also delivered in 1979. Brazilian names of the two helicopters are **Gavião** and **Esquilo** respectively. By March 1981, Helibras had sold six Esquilos to the Brazilian Navy, 15 to civil customers, and seven Gaviãos for export, of which six were for the Bolivian Air Force.

In March 1981 Helibras was beginning licence assembly of six Aérospatiale SA 330 Puma helicopters; assembly of the AS 332 Super Puma is under consideration.

Left: Helibras Gavião (Aérospatiale Lama); right: Helibras Esquilo (Aérospatiale Ecureuil), both for civil customers in Brazil

NEIVA
INDÚSTRIA AERONÁUTICA NEIVA S/A (subsidiary of EMBRAER)

HEAD OFFICE AND WORKS: Rua Nossa Senhora de Fátima 360, Vila Antártica, Botucatu, SP, Caixa Postal 10, Código 18600
Telephone: (0149) 22 1010
Telex: 0142 423 SOAN BR

ENGINEERING DEPARTMENT: Estrada Velha Rio-São Paulo 2176, São José dos Campos, SP, Caixa Postal 247, Código 12200
Telephone: (0123) 21 6333
Telex: 011 33673 SOAN BR

DIRECTORS:
Ozires Silva (President)
Antonio Garcia da Silveira

GENERAL MANAGER: Marcos Baptista dos Santos Jr

Neiva became a wholly-owned subsidiary of EMBRAER on 10 March 1980.

The company participates in EMBRAER's general aviation programme, being responsible for all production of the EMB-710 (PA-28-236) Carioca, EMB-711 (PA-28RT-201/201T) Corisco, EMB-712 (PA-28-181) Tupi, EMB-720 (PA-32-301) Minuano and EMB-721 (PA-32R-301) Sertanejo. It will assume production of the EMB-810 (PA-34-200T) Seneca II in the second half of 1982.

Neiva also participates in EMBRAER's agricultural aircraft programme, building at Botucatu the fuselage structure of the EMB-201A Ipanema. EMBRAER was to transfer the entire Ipanema assembly line to Botucatu in the second half of 1981.

NEIVA N-621 UNIVERSAL

Brazilian Air Force designation: T-25 and T-25A

The Universal flew for the first time on 29 April 1966. The first production T-25 was flown on 7 April 1971, and 140 were delivered to the Brazilian Air Force between Autumn 1971 and late 1979. Ten T-25s were supplied to the Chilean Army; these were transferred subsequently to the Chilean Air Force.

The T-25 is no longer in production. A description and illustration can be found in the 1980-81 *Jane's*.

NEIVA N-622 UNIVERSAL II

Brazilian Air Force designation: YT-25B

A YT-25B prototype of the Universal II was ordered by the Brazilian Air Force. This aircraft made its first flight on 22 October 1978, and in early 1980 had completed nearly 100 hours' flying.

The Universal II programme was abandoned in June 1980. Details, and illustrations of the prototype, can be found in the 1980-81 *Jane's*.

SÃO CARLOS

SÃO CARLOS ENGINEERING SCHOOL (IPAI AERONAUTICAL DIVISION), UNIVERSITY OF SÃO PAULO

São Carlos, São Paulo State

DIRECTORS:
Dr Eng Romeu Corsini
Eng Sylvio de Oliveira

Under the direction of Dr Corsini, formerly of the IPT, the IPAI (Aeronautical Division of the São Carlos Engineering School) provides facilities for training aircraft engineers, developing new aircraft and carrying out associated test programmes. It is adjacent to a 2,000 m (6,560 ft) paved runway, hangarage and other airport facilities, and a large dam from which waterborne aircraft can operate. Its most recent ventures are described below.

IPAI-26 TUCA

The Tuca light aircraft was designed by Prof Dawilson Lucato, and built in two-seat prototype form by a group of students of the São Carlos Engineering School. Some features are similar to those of the SP-18 Onça built by Dr Eng Romeu Corsini in the mid-1960s, which is now known as the IPAI-27 and is still flown in São Carlos as a flying testbed. Flight testing of the Tuca began in 1979 and was continuing in 1981.

TYPE: Two-seat fully-aerobatic light aircraft.

WINGS: Strut-braced high-wing monoplane. Wing section NACA 23012.

FUSELAGE: Welded steel tube structure, with easily-removable metal and glassfibre skin panels.

TAIL UNIT: Cantilever T tail, with sweptback vertical surfaces.

LANDING GEAR: Non-retractable tricycle type, with single wheel on each unit. Oleo-sprung nose unit; cantilever main-wheel legs of flexible reinforced glassfibre.

POWER PLANT: One 80·5/86 kW (108/115 hp) Avco Lycoming O-235-C1 flat-four engine, driving a two-blade propeller with spinner. Other engines expected to be tested during trials programme: structure is capable of accepting engines of up to 134 kW (180 hp) without modification, in which form the Tuca could be used as a utility four-seater, with fuel for up to four hours' endurance.

ACCOMMODATION (prototype): Seats for two persons side by side in fully enclosed, extensively glazed and reinforced cabin. Dual controls standard.

DIMENSIONS, EXTERNAL:

Wing span	11·00 m (36 ft 1 in)
Wing mean aerodynamic chord	1·58 m (5 ft 2¼ in)
Wing area, gross	16·99 m² (182·88 sq ft)
Wing aspect ratio	7·12
Length overall	6·82 m (22 ft 4½ in)

WEIGHT AND LOADING (prototype):

Max T-O weight	700 kg (1,543 lb)
Max wing loading	40·61 kg/m² (8·32 lb/sq ft)

PERFORMANCE (Normal category):

Max level speed at S/L	102 knots (190 km/h; 118 mph)
Cruising speed at S/L	100 knots (185 km/h; 115 mph)
Stalling speed at S/L	41·5 knots (76 km/h; 47·5 mph)
Service ceiling	3,600 m (11,810 ft)
T-O and landing run at S/L	250 m (820 ft)
Endurance	4 h 0 min

IPAI-27 JIPE VOADOR (FLYING JEEP)

This is the current designation of the four-seat Corsini-designed SP-18 Onça (Jaguar), which is now fitted with a 134 kW (180 hp) Avco Lycoming flat-four engine and a specially-designed cooling fan. Construction is of steel tube (fuselage) and wood (wings and tail), with covering of metal, fabric, plastics and glassfibre.

DIMENSIONS, EXTERNAL:

Wing span	12·00 m (39 ft 4½ in)
Wing area, gross	18·00 m² (193·75 sq ft)
Length overall	7·50 m (24 ft 7¼ in)
Height overall	2·50 m (8 ft 2½ in)

WEIGHTS AND LOADING:

Max T-O weight	900 kg (1,984 lb)
Max wing loading	50 kg/m² (10·24 lb/sq ft)

PERFORMANCE (at max T-O weight):

Stalling speed	30 knots (55 km/h; 34·5 mph)

IPAI-28 SUPER SURUBIM

First flown in 1980, the Super Surubim was designed by Eng Sylvio de Oliveira and completed at São Carlos City Flying Club. Design parameters were calculated with the assistance of the São Carlos Engineering School; assistance was also provided by EMBRAER, the Oldi aircraft instruments factory, the Conal aircraft maintenance company, Neiva, and several aero club pilots. Certification was expected during 1980.

TYPE: Tandem two-seat aerobatic trainer; *g* limits ±8·0.

WINGS: Cantilever low-wing monoplane, with highly tapered wings. Wing section NACA 23016·7 at root, NACA 23012 at tip. Dihedral from roots. All-metal two-spar structure, with rear auxiliary spar, with skins of 2024-T3 Alclad aluminium alloy, chemically treated against corrosion. Wings built in one piece and attached to underside of fuselage by bolts. Trailing-edge split flaps between ailerons and fuselage; ailerons have 100% balancing. No tabs. 'Letterbox' fixed slots near outer leading-edges, forward of ailerons.

FUSELAGE: Conventional all-metal semi-monocoque structure of frames and longerons.

TAIL UNIT: Cantilever metal structure, of similar construction to wings. Trim tab in rudder and starboard elevator. Rudder and elevators have 100% balancing.

LANDING GEAR: Non-retractable type, with single main wheels and tailwheel. Main units have strut bracing, leg and wheel fairings, 190 mm diameter wheels from a Neiva T-25 Universal, and are fitted with Oldi brakes. Main-wheel tyre size 7·00-6, pressure 2·76 bars (40 lb/sq in). Small tailwheel, carried on leaf spring.

POWER PLANT: One 146 kW (196 hp) Ranger 6-440-C4 six-cylinder inverted in-line aircooled engine, driving a two-blade propeller with spinner. Fuel in two wing tanks, combined capacity 140 litres (30·75 Imp gallons).

ACCOMMODATION: Seats for two persons in tandem. Rearward-sliding framed canopy, jettisonable in flight.

DIMENSIONS, EXTERNAL:

Wing span	9·10 m (29 ft 10¼ in)
Wing area, gross	10·12 m² (108·9 sq ft)
Wing aspect ratio	8·18
Length overall	8·10 m (26 ft 7 in)
Height overall	2·12 m (6 ft 11½ in)
Wheel track	2·40 m (7 ft 10½ in)
Propeller diameter	2·20 m (7 ft 2½ in)

WEIGHTS AND LOADINGS:

Weight empty	650 kg (1,433 lb)
Max T-O weight	950 kg (2,094 lb)
Max wing loading	93·8 kg/m² (19·21 lb/sq ft)
Max power loading	6·5 kg/kW (10·68 lb/hp)

PERFORMANCE (estimated, at max T-O weight):

Never-exceed speed	215·5 knots (400 km/h; 248 mph)
Max level speed	146 knots (270 km/h; 168 mph)
Max cruising speed	129 knots (240 km/h; 149 mph)
Stalling speed, flaps up	59·5 knots (110 km/h; 68·5 mph)
Stalling speed, flaps down	54·5 knots (100 km/h; 62·5 mph)
Max rate of climb at S/L	320 m (1,050 ft)/min

IPAI-29 TIRA PROSA

The IPAI-29 is a very lightweight single-seat aircraft, powered by a locally-produced modified Volkswagen engine designed to use sugar-cane alcohol fuel. Of wooden construction, it has a twin-boom layout, with the 48·5 kW (65 hp) engine driving a pusher propeller. Design is by Prof Dawilson Lucato. First flight was expected in 1981.

IPAI-30

The IPAI-30, developed by Prof Dawilson Lucato, is the intended production version of the IPAI-26 Tuca, and is also powered by an 86 kW (115 hp) Avco Lycoming O-235-C1 flat-four engine. Applications include basic, aerobatic and navigation training, observation, search and rescue, and touring. Improvements include a new-design wing, of supercritical section and incorporating integral flaps. Max T-O weight is reduced to 580 kg (1,278 lb); cruising speed is estimated at 113 knots (210 km/h; 130 mph).

IPAI-26 Tuca light cabin monoplane

São Carlos IPAI-28 Super Surubim tandem two-seat light aircraft in uncompleted state
(José Fernandez)

CANADA

BOEING

BOEING OF CANADA LTD (Vertol Division)

Arnprior Airport, Box 338, Arnprior, Ontario
Telephone: (613) 233 7414 and 623 4215

BOEING VERTOL/CAF SARCUP PROGRAMME

CAF designations: CH-113 Labrador and CH-113A Voyageur

In March 1980 the Canadian government awarded a $20 million contract jointly to Boeing Vertol (see US section) and Boeing of Canada Ltd for the second phase of the Canadian Armed Forces' programme known as SARCUP (Search And Rescue Capability Upgrade Project).

The programme involves the upgrading of six CH-113 and five CH-113A helicopters to a single, improved maritime search and rescue standard by mid-1984. The first phase, started in 1978, involved improvements to the communications and navigation avionics, and the installation of high-powered searchlights. In the second phase each aircraft's flight deck will be modified, as far as possible, to achieve a common configuration. An APU, weather radar and external rescue system will be added to the CH-113s; the CH-113As will be upgraded by adding two 1,893 litre (500 US gallon) auxiliary fuel tanks, a water dam, external rescue system and weather radar.

These improvements will extend the aircraft's range, and enable them to fly search and rescue missions in adverse weather conditions by day and by night. Production of the improvement kits, aircraft modification and flight testing are being undertaken by Boeing of Canada; Boeing Vertol in Philadelphia provides engineering, training and other support for the SARCUP programme.

One of the CH-113 helicopters being modified by Boeing of Canada in the SARCUP programme

CANADAIR

CANADAIR LIMITED

Head Office and Works: Cartierville Airport, 1800 Laurentien Boulevard, St Laurent H4R 1K2, Quebec
Postal Address: PO Box 6087, Station 'A', Montreal, Quebec H3C 3G9
Telephone: (514) 744 1511
Telex: 05-826747

Chairman of the Board and of the Executive Committee: Guy J. Desmarais
President and Chief Executive Officer:
Frederick R. Kearns
Executive Vice-Presidents:
Harry Halton (Challenger)
Robert D. Richmond (Operations)
Senior Vice-Presidents:
Peter J. Aird (Finance)
Frank M. Francis (Materiel)
Jacques E. Ouellet (Resources)
James B. Taylor
Andreas Throner (Operations)
Robert A. Wohl (Administration and Legal)
Vice-Presidents:
Vincent Ambrico (Manufacturing)
Justin Battle (Public Relations)
Leonard B. Box (Engineering)
Brig Gen John C. Henry (Flight Operations)
James R. Humphreys (Core Engineering)
James R. Ross (Challenger Engineering)
Jacques A. Sincennes (CL-215 and Boeing 767)
Roger G. Tétreault (Materiel)
Harvie D. Walford (Corporate Planning)
Director, Product Engineering: Dr Roy Swanson

Canadair Inc, 274 Riverside Avenue, Westport, Connecticut 06880, USA (US sales subsidiary)
Telephone: (203) 226 1581
Telex: 0096-4282

President: James B. Taylor

Canadair Limited, formerly the Canadian subsidiary of General Dynamics Corporation and owned since 1976 by the Canadian government, has been engaged in the development and manufacture of military and commercial aircraft since 1944. It has also been employed in the research, design, development and production of missile components, pilotless surveillance systems and a variety of non-aerospace products. Canadair has three plants in the St Laurent complex at Cartierville Airport, and a fourth at Dorval International Airport, Montreal. Total covered floor space is 260,127 m² (2,800,000 sq ft), and total work force 6,500.

The CL-600 Challenger twin-turbofan executive transport entered production during 1978; manufacture of the fourth series of CL-215 tanker/utility amphibians (aircraft No. 66 onwards) began in 1980; production of the CL-89 and development of the CL-289 surveillance systems (see RPVs and Targets section) continue. Major subcontracts concern structural components for US Navy and JMSDF P-3C Orions and the CP-140 Aurora Canadian version of this aircraft (see under Lockheed in US section), various components for the McDonnell Douglas F-15, and aft fuselage sections for the Boeing 747SP and 767. Production of aircraft spares, and the modification, repair and overhaul of aircraft, are also included in the current work programme.

CANADAIR CL-600 and CL-601 CHALLENGER

In April 1976, Canadair acquired from the late Mr William P. Lear Sr the worldwide exclusive rights to design, manufacture, market and support the latter's LearStar 600; this concept envisaged an aircraft using an advanced-technology wing and two high bypass ratio turbofan engines. The programme was launched on 29 October 1976. In March 1977, major design changes were announced and the aircraft became known as the Canadair Challenger.

The first of three pre-production Challengers (C-GCGR-X, c/n 1001) made its first flight on 8 November 1978. The second flew on 17 March and the third on 14 July 1979. First flight by a production Challenger was made on 21 September 1979, and this joined the programme in October 1979.

The Challenger received Canadian DoT type approval on 11 August 1980, and FAA certification on 7 November of that year. By 17 February 1981 a total of 13 had been delivered. Firm orders and options had by that time reached 187, of which 130 were for the standard **CL-600** Challenger with Avco Lycoming engines; others were for the **CL-601**, a variation of the standard Challenger with General Electric engines which is due to fly for the first time in April 1982. Customers include the Canadian DoT (one).

The following description applies to the basic CL-600 passenger version, except where indicated:

Type: Twin-turbofan business, cargo and commuter transport.

Wings: Cantilever low-wing monoplane, built in one piece. Advanced-technology wing section. Thickness/chord ratio 14% at root, 12% at leading-edge sweep break and 10% at tip. Dihedral 2° 20′. Incidence 3° at root. Sweepback at quarter-chord 25°. Two-spar structure, primarily of aluminium alloy; spars covered with skin-stringer panels to form rigid torsion box. Two-section double-slotted trailing-edge flaps. Hydraulically-powered aluminium plain ailerons. Inboard spoilers for descent control and ground lift dumping. No tabs. Thermal anti-icing of leading-edges by engine bleed air.

Fuselage: Aluminium alloy damage-tolerant semi-monocoque pressurised structure of circular cross-section, with clad frames, stringers, and chemically-milled skins.

Tail Unit: Cantilever multi-spar aluminium alloy T tail, with swept vertical and horizontal surfaces. All control surfaces powered. Tailplane incidence adjusted by electric trim motor. No tabs.

Landing Gear: Hydraulically-retractable tricycle type, with twin wheels and Dowty Rotol oleo-pneumatic shock-absorbers on each unit. Main wheels retract inward into wing centre-section, nose unit forward. Nose unit steerable and self-centering. Main wheels have Goodyear 26 × 6·65 tyres, pressure 11·38 bars (165 lb/sq in); nosewheels have Goodyear 18 × 4·4 tyres, pressure 8·27 bars (120 lb/sq in). Goodyear hydraulically-operated multiple-disc carbon brakes with fully-modulated anti-skid system.

Power Plant: Two 33·36 kN (7,500 lb st) Avco Lycoming ALF 502L turbofan engines (38·48 kN; 8,650 lb st General Electric CF34-1A in CL-601), one pylon-mounted on each side of rear fuselage, fitted with

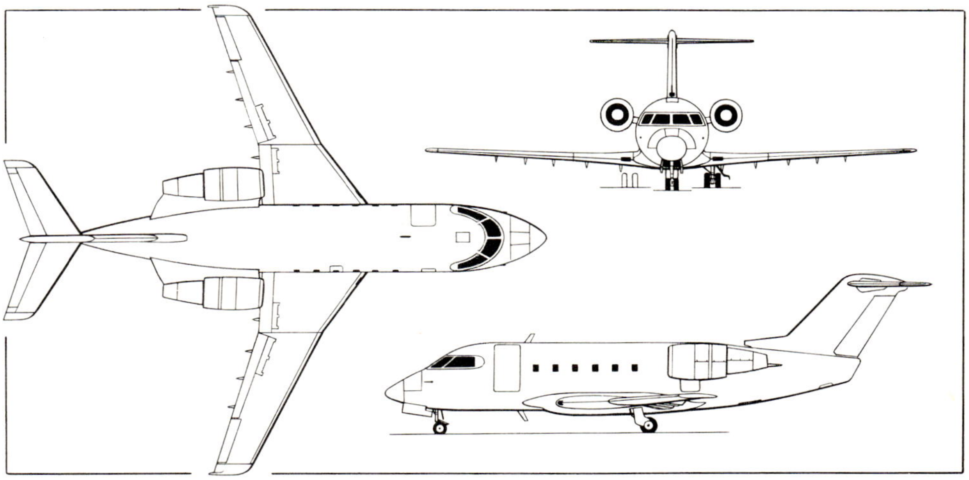

Canadair CL-601 Challenger (two General Electric CF34-1A turbofan engines) *(Pilot Press)*

Canadair CL-600 Challenger business, cargo and commuter transport (two Avco Lycoming ALF 502L turbofan engines)

Canadair CL-600 Challenger operated by the Canadian Department of Transport

cascade-type fan-air thrust reversers. Integral fuel tank in centre-section (capacity 2,839 litres; 624 Imp gallons; 750 US gallons) and one in each wing (each 2,744 litres; 604 Imp gallons; 725 US gallons). Total fuel capacity 8,327 litres (1,832 Imp gallons; 2,200 US gallons). Pressure and gravity fuelling and defuelling. Oil capacity 13·6 litres (3 Imp gallons).

Accommodation: Pilot and co-pilot side by side on flight deck with dual controls. Blind-flying instrumentation standard. Typical accommodation includes three pairs of facing seats at front of cabin, with tables between; a further three pairs at rear, with inward-facing three-place settee on each side (total 18 seats). Wardrobe at front on port side. Provision for crew locker, buffet, bar, storage cabinet(s) or second wardrobe, depending on number and layout of seats. Toilet and/or baggage area aft of cabin. Upward-opening door on port side, forward of wing. (Downward-opening door to be fitted from c/n 24 onwards.) Overwing emergency exit on starboard side. Entire accommodation heated, ventilated and air-conditioned.

Systems: Sundstrand pressurisation and Garrett air-conditioning systems, max pressure differential 0·64 bars (9·3 lb/sq in). Three independent hydraulic systems, each of 207 bars (3,000 lb/sq in). No. 1 system powers flight controls (via servo-actuators positioned by cables and pushrods); No. 2 system for flight controls and brakes; No. 3 system for flight controls, landing gear extension/retraction, brakes and nosewheel steering. Nos. 1 and 2 systems each powered by an engine-driven pump, supplemented by an AC electrical pump; No. 3 system by two AC pumps. Two 300kVA engine-driven generators supply primary three-phase AC electrical power at 115/200V 400Hz. Three transformer-rectifiers to convert AC power to 28V DC; one nickel-cadmium battery. Alternative primary power provided by APU and an air-driven generator, the latter being deployed automatically if the engine-driven generators and APU are inoperative. Stall warning system, with stick shakers and stick pusher. Garrett gas turbine APU (to be certificated for in-flight operation) for engine start, ground air-conditioning and other services. Electrical anti-icing of windscreen, flight deck side windows and pitot heads; Sundstrand bleed air anti-icing of wings, tailplane, engine intake cowls and guide vanes. Gaseous oxygen system, pressure 124 bars (1,800 lb/sq in). Continuous-element fire detectors in each engine nacelle and APU; two-shot extinguishing system for engines, single-shot system for APU.

Avionics: Standard avionics include dual Collins VHF-20A com, dual Collins VIR-30A VOR/ILS/marker beacon receiver, dual Sperry SPZ 600 flight directors, SPZ 600 autopilot with dual servos, dual TDR-90 transponders, dual DME (with HSI display), one ALT-55B radio altimeter, dual C-14 compasses, standby compass, one ADF-60 ADF, dual intercom system, comparator warning system, RCA Primus 400 weather radar, HF and other antennae. Provision, at customer's option, for HF com, third VHF com, second ADF, VLF nav, INS, GPWS, ELT, flight data recorder and cockpit voice recorder.

Dimensions, external:
Wing span 18·85 m (61 ft 10 in)
Wing chord at fuselage c/l 4·89 m (16 ft 0½ in)
Wing chord at tip 1·27 m (4 ft 1·9 in)
Wing area (basic) 41·81 m² (450 sq ft)
Wing aspect ratio 8·5
Length overall 20·85 m (68 ft 5 in)
Fuselage: Max diameter 2·69 m (8 ft 10 in)
Height overall 6·30 m (20 ft 8 in)
Tailplane span 6·20 m (20 ft 4 in)
Wheel track (c/l of shock-struts) 3·18 m (10 ft 5 in)
Wheelbase 7·99 m (26 ft 2½ in)
Passenger door (port, fwd): Height 1·78 m (5 ft 10 in)
 Width 0·91 m (3 ft 0 in)
 Height to sill 1·61 m (5 ft 3½ in)
Baggage door (port, aft): Height 0·84 m (2 ft 9 in)
 Width 0·71 m (2 ft 4 in)
 Height to sill 1·61 m (5 ft 3½ in)
Overwing emergency exit (stbd):
 Height 0·91 m (3 ft 0 in)
 Width 0·51 m (1 ft 8 in)

Dimensions, internal:
Cabin: Length, incl galley, toilet and baggage area, excl flight deck 8·61 m (28 ft 3 in)
 Max width 2·49 m (8 ft 2 in)
 Width at floor level 2·18 m (7 ft 2 in)
 Max height 1·85 m (6 ft 1 in)
 Floor area 18·81 m² (202·5 sq ft)
 Volume 32·6 m³ (1,150 cu ft)

Weights:
Manufacturer's weight empty:
 600 8,437 kg (18,600 lb)
 601 8,931 kg (19,690 lb)
Typical operating weight empty:
 600 10,285 kg (22,675 lb)
 601 10,852 kg (23,925 lb)
Max fuel:
 600, 601 7,586 kg (16,725 lb)
Max payload:
 600 3,400 kg (7,500 lb)
Payload with max fuel:
 600, 601 454 kg (1,000 lb)
Max T-O weight:
 600 18,325 kg (40,400 lb)
 601 18,824 kg (41,500 lb)
Max ramp weight:
 600 18,393 kg (40,550 lb)
 601 18,892 kg (41,650 lb)
Max landing weight:
 600, 601 16,329 kg (36,000 lb)
Max zero-fuel weight:
 600 12,474 kg (27,500 lb)
 601 12,700 kg (28,000 lb)

Performance (CL-600, at max T-O weight):
Max cruising speed Mach 0·84 (480 knots; 889 km/h; 553 mph)
Long-range cruising speed Mach 0·70 (402 knots; 745 km/h; 463 mph)
Time to 13,715 m (45,000 ft) 20 min
Max certificated operating altitude 13,715 m (45,000 ft)
Service ceiling, one engine out 7,315 m (24,000 ft)
Balanced T-O field length (ISA at S/L) 1,984 m (6,510 ft)
Landing distance 1,356 m (4,450 ft)
Min ground turning radius 14·63 m (48 ft 0 in)
Range with NBAA IFR reserves (200 nm; 370 km; 230 mile alternate):
 with standard fuel 3,400 nm (6,300 km; 3,915 miles)
 with additional fuel 3,600 nm (6,671 km; 4,145 miles)
Design *g* limit +2·7

Performance (CL-601, estimated, at max T-O weight):
As for CL-600 except:
Time to 13,715 m (45,000 ft) 21 min
Balanced T-O field length (ISA at S/L) 1,707 m (5,600 ft)
Range with NBAA IFR reserves (200 nm; 370 km; 230 mile alternate):
 with standard fuel 3,600 nm (6,671 km; 4,145 miles)

Operational Noise Levels (CL-600, FAR Pt 36, estimated):
 T-O 81·6 EPNdB
 Sideline 89·3 EPNdB
 Approach 91·2 EPNdB

Operational Noise Levels (CL-601, FAR Pt 36, estimated):
 T-O 82 EPNdB
 Sideline 87 EPNdB
 Approach 91 EPNdB

CANADAIR CL-610 CHALLENGER E

It was announced by Canadair in August 1981 that development of this version of the Challenger has been postponed.

CANADAIR CL-215

The Canadair CL-215 is a twin-engined amphibian, intended primarily for firefighting but adaptable to a wide variety of other duties. It is designed for simplicity of operation and maintenance, and can operate from small airstrips, lakes, ocean bays etc.

The CL-215 made its first flight on 23 October 1967, and its first water take-off on 2 May 1968. Canadian DoT certification in the Utility and Restricted categories was obtained in March 1969, followed by FAA certification in the Restricted category in May of the same year.

As detailed in previous editions of *Jane's,* sales have been made to the provinces of Quebec (15) and Manitoba (2); the Securité Civile of France (15); the Greek government (11); the Spanish government (17, equipped for SAR but capable of firefighting and other roles); the Royal Thai Navy (2 for coastal patrol, transport and firefighting); CVG Ferrominera Orinoco CA of Venezuela (2 dual-purpose aircraft for firefighting/passenger transport); and the government of Yugoslavia (4).

Production of the 65 aircraft built to date was completed in three series. Details of the first and second series can be found in the 1977-78 *Jane's,* and of the third series in the 1979-80 edition. The fourth series (c/n 66 onward) is being produced at the rate of one per month, with deliveries due to begin in the last quarter of 1981. By early 1981, five had been ordered, including the four for Yugoslavia already mentioned.

The CL-215 firefighting installation consists of two internal tanks, two retractable probes and two drop doors, plus the associated operating systems. It attacks fires in the following ways:

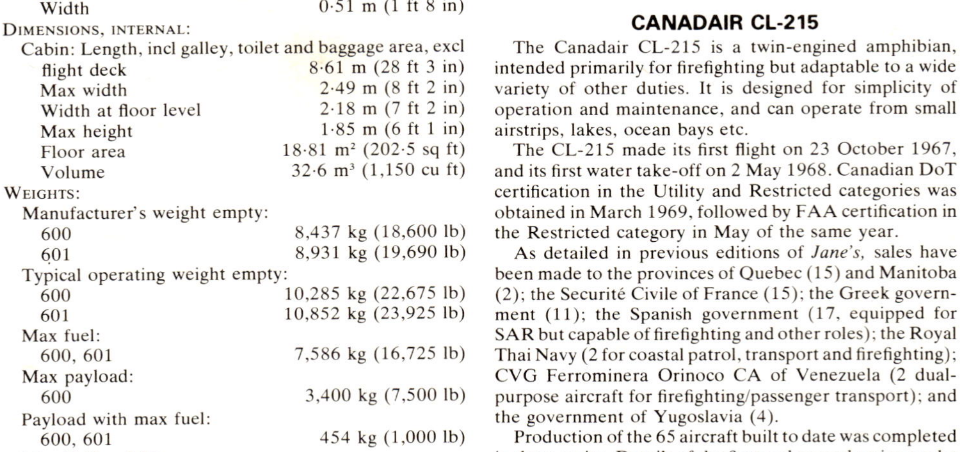

Search and rescue version of the Canadair CL-215 twin-engined amphibian, with nose radome *(Pilot Press)*

Canadair CL-215 utility amphibian (two Pratt & Whitney R-2800 radial engines) operated by the government of Quebec

(a) with water or chemical retardants ground-loaded at airports; or

(b) with fresh or salt water scooped from a suitable body of water as the aircraft skims across the surface.

The aircraft carries a maximum water or retardant load of 5,455 litres (1,200 Imp gallons). The tanks can be ground-filled in 90 s, or scoop-filled in 10 s while the aircraft planes at 60 knots (111 km/h; 69 mph). Pickup distance in still air, from 15 m (50 ft) above the surface during landing to 15 m (50 ft) above the surface during climb-out, is 1,220 m (4,000 ft).

On a number of occasions single CL-215s have made over 100 drops totalling more than 545,520 litres (120,000 Imp gallons) in one day. Full loads have been scooped from the Mediterranean in wave heights of up to 2 m (6 ft). In June 1978 a Manitoba government CL-215 made 160 drops (totalling 872,824 litres; 192,000 Imp gallons) on fires in one day.

A lightweight integrated liquid spray system has been developed, and the first example installed and tested on the company demonstration aircraft. Production units are available for installation on existing or new aircraft. Uses include the application of oil dispersants and pesticides.

Tests conducted at Canadair have shown that the CL-215 can be used to extinguish oil fires by airdropping a suitable foaming agent.

Type: Twin-engined multi-purpose amphibian.

Wings: Cantilever high-wing monoplane. No dihedral. All-metal one-piece fail-safe structure, with front and rear spars at 16% and 49% chord. Spars of conventional construction, with extruded caps and webs stiffened by vertical members. Aluminium alloy skin, with riveted spanwise extruded stringers, is supported at 762 mm (30 in) pitch by interspar ribs. Leading-edge consists of aluminium alloy skin attached to pressed nose-ribs and spanwise stringers. Hydraulically-operated all-metal single-slotted flaps, supported by four external hinges on interspar ribs on each wing. Trim tab and geared tab in port aileron, rudder/aileron interconnect tab in starboard aileron. Detachable glassfibre wingtips.

Fuselage: All-metal single-step flying-boat hull of conventional fail-safe construction.

Tail Unit: Cantilever all-metal fail-safe structure with horizontal surfaces mounted midway up fin. Structure of aluminium alloy sheet, honeycomb panels, extrusions and fittings. Elevators and rudder fitted with dynamic balance, trim tab (port elevator only), spring tabs and geared tabs. Provision for de-icing of leading-edges.

Landing Gear: Hydraulically-retractable tricycle type. Fully-castoring, self-centering twin-wheel nose unit retracts rearward into hull and is fully enclosed by doors. Main gear support structures retract into wells in sides of hull. A plate mounted on each main gear assembly encloses bottom of wheel well. Main-wheel tyre pressure 5·31 bars (77 lb/sq in); nosewheel tyre pressure 6·55 bars (95 lb/sq in). Hydraulic disc brakes. Non-retractable stabilising floats are each carried on a pylon cantilevered from wing box structure, with breakaway provision.

Power Plant: Two 1,566 kW (2,100 hp) Pratt & Whitney R-2800-CA3 eighteen-cylinder radial engines, each driving a Hamilton Standard Hydromatic constant-speed fully-feathering three-blade propeller, with 43E60 hub and type 6903 blades. Two fuel tanks, each of eight flexible cells, in wing spar box, with total usable capacity of 5,910 litres (1,300 Imp gallons). Gravity refuelling through two points above each tank. Oil in two tanks, with total capacity of 272·75 litres (60 Imp gallons), aft of engine firewalls.

Accommodation (water bomber version): Crew of two side by side on flight deck. Dual controls standard. Two 2,673 litre (588 Imp gallon) water tanks in main fuselage compartment, with retractable pickup probe in each side of hull bottom. Water-drop door in each side of hull bottom. Flush-type doors on port side of fuselage forward and aft of wings. Emergency exit on starboard side aft of wing trailing-edge. Emergency hatch above starboard cockpit. Mooring hatch on top of hull nose below flight deck windows. Side-facing canvas folding seats for eight people are located in the forward cabin area.

Accommodation (other roles): When configured for patrol and search and rescue missions, the aircraft has additional stations for a flight engineer, navigator and two observers. The navigator's station, located immediately behind the flight deck, includes a search and radar display. The observers' stations are located in the rear fuselage, and have sliding seats which can be positioned alongside blister windows. A toilet is provided in the rear of the cabin and a galley is installed. Additional seats and/or stretchers are available. In the passenger transport configuration, up to 26 forward-facing seats can be fitted in a fully-furnished interior with toilet and galley. Utility passenger interiors provide foldup side-facing canvas seats. Cargo tiedown fittings are provided for the carriage of loads up to 3,630 kg (8,000 lb). Provision exists for the installation of extra cabin windows, to a maximum of 14.

Systems: Hydraulic system, pressure 207 bars (3,000 lb/sq in), utilises two engine-driven pumps to actuate landing gear, flaps, water-drop doors and pickup probes, and wheel brakes. Electric pump in system provides power for emergency actuation of landing gear and brakes and closure of water doors. Electrical system includes two 250VA 115V 400Hz single-phase inverters, two 28V 200A DC generators, one 34Ah lead-acid battery and one aircooled petrol-engine-driven 28V 200A generator GPU. In the SAR version, two 800VA inverters are installed.

Avionics and Equipment: Standard installation includes dual VHF and VHF/FM com, dual VOR/ILS receivers, ADF, marker beacon receiver and ATC transponder. Optional avionics include HF, DME, radio altimeter, second ADF, and ELT. The SAR version includes a search radar, DME and radio altimeter as standard.

Dimensions, external:

Wing span	28·60 m (93 ft 10 in)
Wing chord (constant)	3·54 m (11 ft 7½ in)
Wing aspect ratio	8·15
Length overall	19·82 m (65 ft 0½ in)
Beam	2·59 m (8 ft 6 in)
Length/beam ratio	7·5
Height overall (on land)	8·92 m (29 ft 3 in)
Tailplane span	10·97 m (36 ft 0 in)
Wheel track	5·28 m (17 ft 4 in)
Wheelbase	7·23 m (23 ft 8½ in)
Propeller diameter	4·34 m (14 ft 3 in)
Forward door: Height	1·37 m (4 ft 6 in)
Width	1·03 m (3 ft 4 in)
Rear door: Height	1·12 m (3 ft 8 in)
Width	1·03 m (3 ft 4 in)
Water-drop door: Length	1·60 m (5 ft 3 in)
Width	0·81 m (2 ft 8 in)
Emergency exit: Height	0·91 m (3 ft 0 in)
Width	0·51 m (1 ft 8 in)

Dimensions, internal:

Cabin, excl flight deck: Length	9·38 m (30 ft 9½ in)
Max width	2·39 m (7 ft 10 in)
Max height	1·90 m (6 ft 3 in)
Floor area	19·69 m^2 (212 sq ft)
Volume	35·03 m^3 (1,237 cu ft)

Areas:

Wings, gross	100·33 m^2 (1,080 sq ft)
Ailerons (total)	8·05 m^2 (86·6 sq ft)
Flaps (total)	22·39 m^2 (241 sq ft)
Vertical tail surfaces (total)	17·23 m^2 (185·5 sq ft)
Rudder, incl tabs	6·02 m^2 (64·75 sq ft)
Horizontal tail surfaces (total)	28·43 m^2 (306 sq ft)
Elevators, incl tabs	7·88 m^2 (84·8 sq ft)

Weights and Loadings:

Manufacturer's weight empty	12,160 kg (26,810 lb)
Typical operating weight empty	12,672 kg (27,938 lb)
Max payload: Water bomber	5,443 kg (12,000 lb)
Utility version	3,838 kg (8,462 lb)
Max T-O weight (land)	19,731 kg (43,500 lb)
Max T-O weight (water)	17,100 kg (37,700 lb)
Max zero-fuel weight	19,275 kg (42,500 lb)
Max landing weight: on land	15,603 kg (34,400 lb)
on water	16,780 kg (37,000 lb)
Cabin floor loading	732 kg/m^2 (150 lb/sq ft)
Max wing loading	196·66 kg/m^2 (40·3 lb/sq ft)
Max power loading	6·23 kg/kW (10·36 lb/hp)

Performance:

Cruising speed (max recommended power) at AUW of 18,595 kg (41,000 lb) at 3,050 m (10,000 ft) 157 knots (291 km/h; 181 mph)

Stalling speed, 15° flap, AUW of 19,731 kg (43,500 lb) 78·5 knots (145 km/h; 90 mph)

Stalling speed, 25° flap, AUW of 15,603 kg (34,400 lb), power off 66 knots (123 km/h; 76 mph)

Max rate of climb at S/L at AUW of 19,731 kg (43,500 lb) at max continuous power 305 m (1,000 ft)/min

Rate of climb at S/L, one engine out, at AUW of 17,100 kg (37,700 lb) at T-O power 75 m (245 ft)/min

T-O to 15 m (50 ft):

from land at AUW of 19,731 kg (43,500 lb) 811 m (2,660 ft)

from water at AUW of 17,100 kg (37,700 lb) 800 m (2,625 ft)

Landing from 15 m (50 ft):

on land at AUW of 15,603 kg (34,400 lb) 732 m (2,400 ft)

on water at AUW of 16,780 kg (37,000 lb) 835 m (2,740 ft)

Range with 1,587 kg (3,500 lb) payload:

at max cruise power 925 nm (1,714 km; 1,065 miles)

at long-range cruise power 1,130 nm (2,094 km; 1,301 miles)

DE HAVILLAND CANADA

THE DE HAVILLAND AIRCRAFT OF CANADA LTD

Head Office and Works: Garrett Boulevard, Downsview, Ontario M3K 1Y5

Telephone: (416) 633 7310

Telex: 06-22128

Chairman: D. N. Kendall

President: John W. Sandford

Vice-Presidents:

D. B. Annan (Senior Vice-President)

M. C. W. Davy (Engineering)

W. J. Easdale (Personnel and Industrial Relations)

W. T. Heaslip (Special Projects)

F. A. Johnson (Customer Support)

S. B. Kerr (Finance)

R. G. McCall (Operations)

J. A. Timmins (Marketing and Sales)

Director, Product Engineering: John Thompson

Manager, Public Relations: Scott Langdon

The de Havilland Aircraft of Canada Ltd was established in early 1928 as a subsidiary of The de Havilland Aircraft Co Ltd, and became subsequently a member of the Hawker Siddeley Group. On 26 June 1974 ownership was transferred to the Canadian government, which planned to operate the company only until responsible Canadian investors were found to purchase and operate de Havilland.

In January 1981 facilities covered a total area of 113,246 m^2 (1,218,980 sq ft), comprising a 77,023 m^2 (829,070 sq ft) main plant on the southern border of Downsview airport, 30,455 m^2 (327,815 sq ft) of leased space on the northern boundary of the airport, and an additional 4,988 m^2 (53,690 sq ft) of leased storage and warehousing space. A further 780 m^2 (8,400 sq ft) of office space for the Marketing and Sales Engineering depart-

de Havilland Canada DHC-5D Buffalo STOL transport aircraft for the Sultan of Oman's Air Force *(Austin J. Brown)*

ments is also included. To handle the increased production rates, an expansion programme at the main plant was begun in 1979. This consists of a 3,326 m^2 (35,800 sq ft) extension for small parts manufacturing, plus a 7,153 m^2 (77,000 sq ft) high-bay aircraft assembly and preparation area. Approx 4,700 people were employed at the company's Downsview headquarters in early 1981.

DHC-5D BUFFALO

The Buffalo, first flown on 9 April 1964, is a developed version of the DHC-4 Caribou with an enlarged fuselage and two General Electric T64 turboprop engines. Development costs were shared equally by the US Army, the Canadian government and de Havilland Canada.

Details have appeared in earlier editions of *Jane's* of the DHC-5 (four for US Army evaluation, now designated C-8A), DHC-5A (15 for the Canadian Armed Forces, designated CC-115), and DHC-5A/C-115 (24 for the Brazilian Air Force and 16 for the Peruvian Air Force). The DHC-5B and C were proposed versions with CT64-P4C and Rolls-Royce Dart RDa.12 engines respectively.

The current production version is the DHC-5D, with a higher gross weight and improved performance. The first production DHC-5D set up, in February 1976, the following time-to-height records, subsequently confirmed by the FAI: 2 min 12·75 s to 3,000 m; 4 min 27·5 s to 6,000 m and 8 min 3·5 s to 9,000 m. These times qualify both in the C1 class for turboprop-powered aircraft of unlimited weight category, and Class C1h for those in the 12,000-16,000 kg weight category.

Deliveries of the DHC-5D began in early 1976 and a total of 41 had been sold by January 1981. Part of these orders are being fulfilled from a new batch of 24 aircraft, and production was continuing at the rate of one per month during 1981. Customers for the DHC-5D include the air forces of Cameroun (two), Ecuador (three, including one for the Army), Kenya (six), Mauritania (two), Mexico (one, plus one for Mexican Navy), Togo (two), Zaïre (three), Zambia (seven), Oman (one), Sudan (four), Tanzania (four) and the United Arab Emirates (four).

A civil version of the Buffalo, known as the **DHC-5E Transporter**, is also available; this is described separately.

The following description applies to the military DHC-5D:

Type: Twin-turboprop STOL utility transport.

Wings: Cantilever high-wing monoplane. Wing section NACA 64_3A417·5 (mod) at root, NACA 63_2A615 (mod) at tip. Dihedral 0° inboard of nacelles, 5° outboard. Incidence 2° 30′. Sweepback at quarter-chord 1° 40′. Conventional fail-safe two-spar box structure of high-strength aluminium alloys. Full-span double-slotted aluminium alloy flaps, outboard sections functioning as ailerons. Aluminium alloy slot-lip spoilers, forward of inboard flaps, are actuated by Jarry Hydraulics unit. Spoilers coupled to manually-operated ailerons for lateral control, uncoupled for symmetrical ground operation. Electrically-actuated trim tab in starboard aileron. Geared tab in each aileron. Rudder/aileron interconnect tab in port aileron. Outer wing leading-edges fitted with electrically-controlled flush pneumatic rubber de-icing boots.

Fuselage: Fail-safe non-pressurised structure of high-strength aluminium alloy. Longitudinal keel members support cargo floor.

Tail Unit: Cantilever structure of high-strength aluminium alloy, with fixed-incidence T tailplane. Elevator aerodynamically and mass balanced. Fore and trailing serially-hinged rudders are powered by tandem jacks operated by two independent hydraulic systems manufactured by Jarry Hydraulics. Trim tab in port half of elevator, spring tab in starboard half. Electrically-controlled flush pneumatic rubber de-icing boot on tailplane leading-edge.

Landing Gear: Retractable tricycle type, with twin wheels on each unit. Hydraulic actuation, nose unit retracting aft, main units forward. Jarry Hydraulics oleo-pneumatic shock-absorbers. Goodrich main wheels and tyres, size 37 × 15-12, pressure 3·10 bars (45 lb/sq in) for STOL assault role, 4·14 bars (60 lb/sq in) as STOL transport. Goodrich nosewheels and tyres, size 8·9 × 12·5, pressure 3·17 bars (46 lb/sq in). Goodrich multi-disc anti-skid brakes.

Power Plant: Two General Electric CT64-820-4 turboprop engines, each flat rated to 2,336 kW (3,133 shp) and driving a Hamilton Standard 63E60-25 three-blade constant-speed reversible-pitch fully-feathering metal propeller with Beta control. Propellers have integral hydraulic systems and electrical de-icing of spinners, cuffs and blades. Fuel in one integral tank in each inner wing, capacity 4,841 litres (1,065 Imp gallons; 1,279 US gallons), and ten interconnected rubber bag tanks in each outer wing, capacity 3,137 litres (690 Imp gallons; 829 US gallons). Total fuel capacity 7,978 litres (1,755 Imp gallons; 2,108 US gallons). Refuelling points above wings and in starboard side of fuselage for pressure refuelling. Total oil capacity 45·5 litres (10 Imp gallons; 12 US gallons).

Accommodation: Crew of three, comprising pilot, co-pilot and crew chief. Main cabin can accommodate roll-up troop seats or folding forward-facing seats for 34-41 troops or 28 combat-equipped troops, or 24 stretchers and six seats. Provision for toilet in forward part of cabin. Door on each side at rear of cabin. Loading height with rear cargo loading door up and ramp down 2·90 m (9 ft 6 in). 508 mm (20 in) grid of tiedown points, with additional tiedowns at sides of cabin.

Systems: Garrett bleed air cabin heating and cooling system. Two independent hydraulic systems, each of 207 bars (3,000 lb/sq in), actuate landing gear, flaps, spoilers, rudders, brakes, nosewheel steering, winch and APU starting. 3·45 bar (50 lb/sq in) pneumatic system for engine starting, de-icing and environmental control. Two Lucas Aerospace engine-driven variable-frequency 3-phase 20kVA AC generators with 28V DC and 400Hz conversion subsystems. Solar T-62T-40-5 gas turbine APU in port engine nacelle provides electric (10kVA generator), hydraulic and pneumatic power for environmental control, hydraulic operation of cargo winch, electrical systems, and other utility functions.

Avionics and Equipment: Full IFR instrumentation and weather radar standard, as are 34 troop seats, cargo buffer rail, cargo winch, roller conveyors, parachute anchor cables and retrieval system, pendulum release system, Brooks and Perkins palletised loading system with integral side rail restraint, tiedown straps and chains, and crew oxygen system. Optional items include seven forward-facing troop seats, airframe de-icing, toilet, cabin oxygen system, and self-sealing fuel cells.

Dimensions, external:

Wing span	29·26 m (96 ft 0 in)
Wing chord at root	3·59 m (11 ft 9¼ in)
Wing chord at tip	1·19 m (5 ft 11 in)
Wing aspect ratio	9·75
Length overall	24·08 m (79 ft 0 in)
*Height overall	8·73 m (28 ft 8 in)
Tailplane span	9·75 m (32 ft 0 in)
Wheel track	9·29 m (30 ft 6 in)
Wheelbase	8·48 m (27 ft 10 in)
Propeller diameter	4·42 m (14 ft 6 in)
Propeller/fuselage clearance	0·97 m (3 ft 2½ in)
Propeller ground clearance	0·97 m (3 ft 2½ in)
Cabin doors (each side):	
Height	1·68 m (5 ft 6 in)
Width	0·84 m (2 ft 9 in)
*Height to sill	1·17 m (3 ft 10 in)
Emergency exits (each side, below wing leading-edge):	
Height	1·02 m (3 ft 4 in)
Width	0·66 m (2 ft 2 in)
*Height to sill	approx 1·52 m (5 ft 0 in)
Rear cargo loading door and ramp:	
Height	6·33 m (20 ft 9 in)
Width	2·34 m (7 ft 8 in)
*Height to ramp hinge	1·17 m (3 ft 10 in)

**will vary with aircraft configuration and loading conditions*

Dimensions, internal:

Cabin, excl flight deck:	
Length, cargo floor	9·58 m (31 ft 5 in)
Width at floor	2·36 m (7 ft 9 in)
Max width	2·67 m (8 ft 9 in)
Max height (aft of wings)	2·08 m (6 ft 10 in)
Height forward of rear spar	1·98 m (6 ft 6 in)
Floor area	22·48 m^2 (242 sq ft)
Volume (rectangular)	44·74 m^3 (1,580 cu ft)

Areas:

Wings, gross	87·8 m^2 (945 sq ft)
Ailerons (total)	3·62 m^2 (39 sq ft)
Trailing-edge flaps (total, incl ailerons)	26·01 m^2 (280 sq ft)
Spoilers (total)	2·34 m^2 (25·2 sq ft)
Fin	8·55 m^2 (92 sq ft)
Rudder	5·57 m^2 (60 sq ft)
Tailplane	14·07 m^2 (151·5 sq ft)
Elevator, incl tabs	7·57 m^2 (81·5 sq ft)

Weights and Loadings (A: STOL assault mission from unprepared airfield; B: STOL transport mission, firm smooth airfield surface):

Operational weight empty (incl 3 crew and 680 kg; 1,500 lb allowance for options and avionics):	
A, B	11,412 kg (25,160 lb)
Max payload: A	5,370 kg (11,840 lb)
B	8,164 kg (18,000 lb)
Max normal fuel: A, B	6,212 kg (13,696 lb)
Max unit load for airdrop:	
A, B	2,721 kg (6,000 lb)
Manoeuvring limit load factor: A	3·0
B	2·5
Max T-O weight: A	18,597 kg (41,000 lb)
B	22,316 kg (49,200 lb)
Max landing weight: A	17,735 kg (39,100 lb)
B	21,273 kg (46,900 lb)
Max zero-fuel weight: A	16,782 kg (37,000 lb)
B	19,731 kg (43,500 lb)
Max wing loading: A	211·8 kg/m^2 (43·4 lb/sq ft)
B	254·4 kg/m^2 (52·1 lb/sq ft)
Max uniform cabin floor loading:	
A, B	976 kg/m^2 (200 lb/sq ft)
Max power loading: A	3·98 kg/kW (6·54 lb/shp)
B	4·78 kg/kW (7·85 lb/shp)

Performance (at max T-O weight except where indicated. A: STOL assault mission from unprepared airfield; B: STOL transport mission from firm smooth airfield surface):

Max cruising speed at 3,050 m (10,000 ft):	
***A	252 knots (467 km/h; 290 mph)
*B	227 knots (420 km/h; 261 mph)
Stalling speed, 40° flap:	
A at 17,690 kg (39,000 lb) AUW	67 knots (124 km/h; 77 mph)
B at 21,273 kg (46,900 lb) AUW	73 knots (135 km/h; 84 mph)

DHC-5E Transporter civil passenger/cargo version of the DHC-5D Buffalo *(Brian M. Service)*

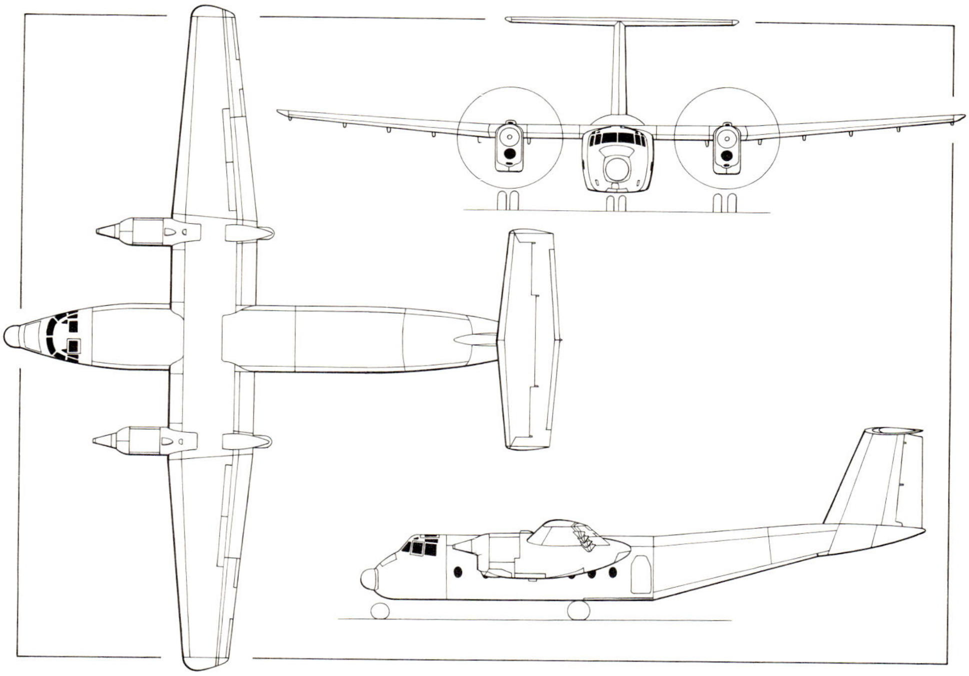

de Havilland Canada DHC-5E Transporter twin-turboprop STOL utility transport *(Michael A. Badrocke)*

Max rate of climb at S/L, normal rated power:
- A — 710 m (2,330 ft)/min
- B — 555 m (1,820 ft)/min

Rate of climb at S/L, one engine out:
- A, max power — 205 m (675 ft)/min
- B, military power — 116 m (380 ft)/min

†Service ceiling, normal rated power:
- A — 9,450 m (31,000 ft)
- B — 8,380 m (27,500 ft)

Service ceiling, one engine out:
- A, military power — 5,575 m (18,300 ft)
- *B, max power — 3,810 m (12,500 ft)

STOL T-O run:
- **A — 289 m (950 ft)
- B — 701 m (2,300 ft)

STOL T-O to 15 m (50 ft), mid-CG:
- **A — 381 m (1,250 ft)
- B — 876 m (2,875 ft)

STOL landing from 15 m (50 ft):
- **A — 346 m (1,135 ft)
- B — 613 m (2,010 ft)

STOL landing run:
- **A — 168 m (550 ft)
- B — 259 m (850 ft)

Range at 3,050 m (10,000 ft):
- A, max payload — 225 nm (416 km; 259 miles)
- B, max payload — 600 nm (1,112 km; 691 miles)
- A, B, zero payload — 1,770 nm (3,280 km; 2,038 miles)

† *Recommended max operating altitude of 7,620 m (25,000 ft)*
* *at 21,200 kg (46,737 lb) AUW*
** *with 5,443 kg (12,000 lb) payload*
****at 17,667 kg (38,950 lb) AUW*

DHC-5E TRANSPORTER

The Transporter is a civil version of the DHC-5D Buffalo which will be certificated to CAR 4b and SR 442b standards for commercial operation. A demonstration aircraft (C-GTLW) was exhibited at the Paris Air Show in June 1979. Airframe, engines, propellers and APU are the same as those of the DHC-5D; some systems are modified, and military installations are replaced by a civil passenger interior which can be transformed quickly for freight-carrying.

The Transporter is available with a 44-seat utility interior as standard, with seats folding into the side walls to convert for freight loading. Optional configurations include a VIP/executive version, seating up to 19 persons in two cabins; and a de luxe 'hard wall' interior with increased soundproofing and air ventilation system.

Civil certification by the Canadian DoT was granted in March 1981 at a max T-O weight of 18,597 kg (41,000 lb). Operated to civil regulations, the Transporter will require a 1,106 m (3,630 ft) field length; but it retains the capability to operate from 457 m (1,500 ft) strips, provided a clear stopway is available at each end. Subsequently, it is intended to obtain certification at a max T-O weight of 22,316 kg (49,200 lb), permitting carriage of a max payload of 8,165 kg (18,000 lb). At this higher weight the Transporter would require a 1,463 m (4,800 ft) field length, but would retain the capability to operate from 876 m (2,875 ft) strips.

Type: Twin-turboprop general-purpose transport aircraft.

Wings, Fuselage, Tail Unit, Landing Gear: As for DHC-5D Buffalo.

Power Plant: As for DHC-5D Buffalo. Fuel in one integral tank in each inner wing, capacity 4,841 litres (1,065 Imp gallons; 1,279 US gallons), and one tank in each outer wing, capacity 3,134 litres (689 Imp gallons; 828 US gallons). Total fuel capacity 15,950 litres (3,508 Imp gallons; 4,214 US gallons). Oil capacity as for DHC-5D.

Accommodation: Pilot and co-pilot on flight deck; cabin attendant. Standard utility layout for 44 passengers in four-abreast seating at 76 cm (30 in) pitch. Alternative VIP/executive and quick-change cargo/passenger layouts available optionally. Toilet and baggage compartment optional. Access to passenger cabin via airstair door at rear on port side, and side-hinged door at rear on starboard side. Push-out emergency door beneath wing on each side; crew escape hatch in flight deck roof. Entire accommodation heated, ventilated and air-conditioned.

Systems: Nose-mounted refrigeration unit supplies cooled bleed air at 11·3 m³ (400 cu ft)/min. Two independent hydraulic systems, as in DHC-5D. Additional, entirely independent system for rudder power control. Emergency hydraulic system powered by handpump. Pneumatic system confined to use of compressed air cylinders for emergency lowering of nose gear and emergency operation of main-wheel brakes. Electrical power supplied by two AC generators, each with separate distribution system. DC system has 28V nickel-cadmium battery for APU starting, and to power fuel system, engine controls, and fire detection/extinguishing systems. 110V AC available for propeller and spinner de-icing, windscreen and miscellaneous heating, fuel booster pump and landing lamps. All AC, 400Hz and DC systems are duplicated for safety. Fixed-diluter oxygen systems: demand type for flight crew, portable equipment with mask and eye protection for occupants of main cabin. De-icing available for propellers, spinners, air intakes, windscreen, and windscreen wipers/washers. Anti-icing of wings and tailplane available optionally. Solar T-62T-40-5 gas turbine APU provides electric and hydraulic power for ground operation, bleed air for flight deck and passenger cabin heating and cooling, and main engine starting.

Avionics and Equipment: Four-position audio system, comprising two VHF-20B VHF com; one 718U-5 HF com; one TDR-90 transponder; and one C-3000 emergency locator transmitter. Other avionics include Primus 40 radar; TCN-40 Tacan; and dual VIR-30AGM VHF nav, ADF-60 ADF, and C-14 compasses.

Dimensions, external: As DHC-5D Buffalo

Dimensions, internal: As DHC-5D Buffalo except:
- Cabin, excl flight deck: Volume — 43·6 m³ (1,540 cu ft)
- Freight hold (rear cargo-loading door and ramp): Volume — 7·7 m³ (272 cu ft)

Areas: As DHC-5D Buffalo

Weights and Loadings:
- Operating weight empty — 11,430 kg (25,200 lb)
- Max useful load (fuel and payload) — 7,167 kg (15,800 lb)
- Max fuel load — 6,212 kg (13,696 lb)
- Max T-O weight — 18,597 kg (41,000 lb)
- Max zero-fuel and landing weight — 18,144 kg (40,000 lb)
- Max wing loading — 211·8 kg/m² (43·4 lb/sq ft)
- Max power loading — 3·98 kg/kW (6·54 lb/shp)

Performance (estimated, at max T-O weight except where indicated):
- Max cruising speed at 3,050 m (10,000 ft) — 250 knots (463 km/h; 287 mph)
- Econ cruising speed at 3,050 m (10,000 ft) — 183 knots (339 km/h; 210 mph)
- Stalling speed, 25° flap — 71 knots (132 km/h; 82 mph)
- Max rate of climb at S/L — 710 m (2,330 ft)/min
- Rate of climb at S/L, one engine out — 183 m (600 ft)/min
- Recommended operating altitude — 7,620 m (25,000 ft)
- Service ceiling — 9,450 m (31,000 ft)
- Service ceiling, one engine out — 4,875 m (16,000 ft)
- Runway LCN (10 in pavement) — 7·5
- * STOL T-O run — 290 m (950 ft)
- * STOL T-O to 15 m (50 ft) — 381 m (1,250 ft)
- T-O field length (SR 422B), S/L, ISA — 860 m (2,820 ft)
- Landing field length at max landing weight (SR 422B) — 853 m (2,800 ft)
- * STOL landing from 15 m (50 ft) — 346 m (1,135 ft)
- * STOL landing run — 168 m (550 ft)
- Range with max fuel, long-range cruising speed at 3,050 m (10,000 ft), 45 min VFR reserves at 1,525 m (5,000 ft) — 1,720 nm (3,185 km; 1,980 miles)
- Range with 6,713 kg (14,800 lb) fuel and payload, ISA, at econ cruising speed with anti-icing off — 100 nm (185 km; 115 miles)

* *With 5,443 kg (12,000 lb) payload*

Twin Otter 300 for the People's Republic of China, equipped for geophysical survey

DHC-6 TWIN OTTER SERIES 300

CAF designation: CC-138
US Army designation: UV-18A
USAF designation: UV-18B

Announced in 1964, the Twin Otter is a STOL transport powered by two Pratt & Whitney Aircraft of Canada PT6A series turboprop engines. Design began in January 1964, and construction of five aircraft began in November of the same year. The first of these (CF-DHC-X), powered by two 432 kW (579 ehp) PT6A-6 engines, flew on 20 May 1965.

The fourth and subsequent aircraft of the initial Series 100 version were fitted with PT6A-20 engines, and the first delivery of a production aircraft, to the Ontario Department of Lands and Forests, was made in July 1966, shortly after the Twin Otter received FAA Type Approval. All Series are certificated to FAR 23 Pt 135.

By 1 February 1981, more than 720 Twin Otters had been sold in more than 80 countries, and operating hours totalled more than 5 million. Production in mid-1981 was at the rate of five aircraft per month. The 700th aircraft was delivered to Widerøe's Flyveselskap Air Service of Norway, bringing that operator's total fleet of DHC-6s to 18.

Military operators of Twin Otters include the Argentinian Air Force (six) and Army (two); Chilean Air Force (five); Ecuadorean Air Force (three); Ethiopian Army Air Corps (three); Jamaica Defence Force (one); Royal Norwegian Air Force (four); Peruvian Air Force (twelve); the Canadian Armed Forces (eight CC-138 for SAR and utility duties); the US Army (two); and the US Air Force Academy (two).

The Twin Otter is available as a maritime surveillance aircraft. One has been modified for Greenlandair Charter for use on ice patrol/maritime surveillance duties. Changes include operation at a higher MTOGW (6,350 kg; 14,000 lb), provision of a Litton LASR-2 search radar in a chin radome, paradrop doors, four observers' stations with bubble windows, additional fuel tanks in the cabin with a max capacity of 567 kg (1,250 lb) each, a VLF navigation system, and provision for photographic equipment.

The Twin Otter is used as a photo survey aircraft in Switzerland and the Sudan. Three such aircraft have also been delivered to the People's Republic of China. A fourth aircraft, especially modified for geophysical surveying, was delivered in April 1979.

Four production versions of the Twin Otter have been built, of which the Series 100 (115 built), Series 200 (115 built) and Series 300S (six built) were described in the 1967-68, 1970-71 and 1976-77 *Jane's* respectively. The current production version is:

Series 300. Deliveries began in the Spring of 1969 with the 231st Twin Otter off the line. Available, with short nose, as floatplane. Ten of the 12 aircraft supplied to Peru were fitted with floats, for operation by Grupo Aéreo No. 42 of the Peruvian Air Force, based at Iquitos.

Type: Twin-turboprop STOL transport.

Wings: Strut-braced high-wing monoplane. Wing section NACA 6A series mean line; NACA 0016 (modified) thickness distribution. Dihedral 3°. No sweepback. All-metal safe-life structure, each wing being attached to the fuselage by two bolts at the front and rear spar fitting and braced by a single streamline section strut on each side. Light alloy riveted construction is used throughout except for the upper skin panels, which have spanwise corrugated stiffeners bonded to them. All-metal double-slotted full-span trailing-edge flaps. Spoilers fitted to Series 300S aircraft only. All-metal ailerons which also droop for use as flaps. Electrically-actuated tab in port aileron; geared trim tabs in both port and starboard ailerons. Optional pneumatic-boot de-icing equipment.

Fuselage: Conventional semi-monocoque safe-life structure, built in three sections. Primary structure of frames, stringers and skin of aluminium alloy. Windscreen and cabin windows of acrylic plastics. Cabin floor is of low-density aluminium-faced sandwich construction and is designed to accommodate distributed loads of up to 976 kg/m² (200 lb/sq ft).

Tail Unit: Cantilever all-metal structure of high-strength aluminium alloys. Fin and fixed-incidence tailplane are bolted to rear fuselage. Manually-operated trim tabs in rudder and elevators. A geared tab is fitted to the rudder to lighten control forces, and a tab fitted to the starboard elevator is linked to the flaps to control longitudinal trim during flap retraction and extension. Optional pneumatic-boot de-icing of tailplane leading-edge.

Landing Gear: Non-retractable tricycle type, with single wheel on each unit. Fully-steerable nosewheel. Urethane compression-block shock-absorption on main units. Oleo-pneumatic nosewheel shock-absorber. Goodyear main-wheel tyres size 11·00-12, pressure 2·62 bars (38 lb/sq in). Goodyear nosewheel tyre size 8·90-12·50, pressure 2·28 bars (33 lb/sq in). Goodrich independent, hydraulically-operated disc brakes on main wheels. Alternatively, high-flotation wheels and tyres, for operation in soft-field conditions, are available at customer's option, size 15·0-12·0 for nosewheel and main wheels. Provision for alternative wheel/ski landing gear. Twin-float gear available for short-nose Srs 300, with added wing fences and small auxiliary fins.

Power Plant: Two 486 kW (652 ehp) Pratt & Whitney Aircraft of Canada PT6A-27 turboprop engines, each driving a Hartzell HC-B3TN-3DY three-blade reversible-pitch fully-feathering metal propeller with Beta control (zero-pitch propeller on floatplane). Two underfloor fuel tanks (eight cells), total capacity of 1,446 litres (318 Imp gallons). Refuelling point for each tank on port side of fuselage. Oil capacity 9·1 litres (2 Imp gallons) per engine. Optional electrical de-icing system for propellers and air intakes.

Accommodation: Side-by-side seats for one or two pilots on flight deck, access to which is by a forward-opening car-type door on each side or via the passenger cabin. Dual controls standard. Windscreen demisting and defrosting standard. Cabin divided by bulkhead into main passenger or freight compartment and baggage compartment. Seats for up to 20 passengers in main cabin. Standard interior is 20-seat commuter layout, with Douglas track, carpets, double windows, individual air vents and reading lights, and airstair door. Optional layouts include 18- or 19-seat commuter versions, 13/20-passenger utility version with foldaway seats and double cargo doors with ladder, and 11-passenger layout in Series 300S. Access to cabin by door on each side of rear fuselage; airstair door on the port side. Optional double door for cargo on port side instead of airstair door. Compartments in nose and aft of main cabin, each with upward-hinged door on port side, for 136 kg (300 lb) and 227 kg (500 lb) of baggage respectively; rear baggage hold accessible from cabin in emergency. Emergency exits near front of cabin on each side. Heating of flight deck and passenger cabin by engine bleed air; ventilation via a ram-air intake on the port side of the fuselage nose. Oxygen system for crew and passengers optional. Executive, survey or ambulance interiors can be fitted at customer's option. Tiedown cargo rings are installed as standard for the freighter role.

Systems: Hydraulic system, pressure 103·5 bars (1,500 lb/sq in), for flaps, brakes, nosewheel steering and (where fitted) ski retraction mechanism. A handpump in the crew compartment provides emergency pressure for standby or ground operation if the electric pump is inoperative. Accumulators smooth the system pressure pulses and provide pressure for parking and emergency braking. Optional low-pressure pneumatic system (1·24 bars; 18 lb/sq in) for operation of autopilot or wing and tail de-icing boots, if fitted. Primary electrical system is 28V DC, with one 200A starter/generator on each engine. One 40Ah 20-cell nickel-cadmium battery (optionally a 36Ah lead-acid battery) for emergency power and engine starting. Separate 3·6Ah battery supplies independent power for engine starting relays and ignition. 250VA main and standby static inverters provide 400Hz AC power for instruments and avionics. External DC receptacle aft of port side cabin door permits operation of complete system on the ground.

Avionics and Equipment: Blind-flying instrumentation standard. Navigation and communications equipment, including weather radar, to customer's specification.

Dimensions, external:

Wing span	19·81 m (65 ft 0 in)
Wing chord (constant)	1·98 m (6 ft 6 in)
Wing aspect ratio	10
Length overall: landplane	15·77 m (51 ft 9 in)
seaplane	15·09 m (49 ft 6 in)
*Height overall: landplane	5·94 m (19 ft 6 in)
seaplane (from waterline)	6·04 m (19 ft 10 in)
Tailplane span	6·30 m (20 ft 8 in)
Wheel track (landplane)	3·71 m (12 ft 2 in)
Wheelbase (landplane)	4·53 m (14 ft 10½ in)
Length of floats (seaplane)	9·65 m (31 ft 8 in)
Width over floats (seaplane)	5·18 m (17 ft 0 in)
Seaplane track (c/l of floats)	4·06 m (13 ft 4 in)
Propeller diameter	2·59 m (8 ft 6 in)
Passenger door (port side): Height	1·27 m (4 ft 2 in)
Width	0·76 m (2 ft 6 in)
*Height to sill	1·32 m (4 ft 4 in)
Passenger door (starboard side):	
Height	1·15 m (3 ft 9½ in)
Width	0·77 m (2 ft 6¼ in)
*Height to sill	1·32 m (4 ft 4 in)
Baggage compartment door (nose):	
Mean height	0·69 m (2 ft 3¼ in)
Width	0·76 m (2 ft 5¾ in)
*Height to sill	1·32 m (4 ft 4 in)
Baggage compartment door (port, rear):	
Max height	0·97 m (3 ft 2 in)
Width	0·65 m (2 ft 1½ in)
Cargo double door (port, rear):	
Height	1·27 m (4 ft 2 in)
Width	1·42 m (4 ft 8 in)
*Height to sill	1·32 m (4 ft 4 in)

**will vary with aircraft configuration and loading conditions*

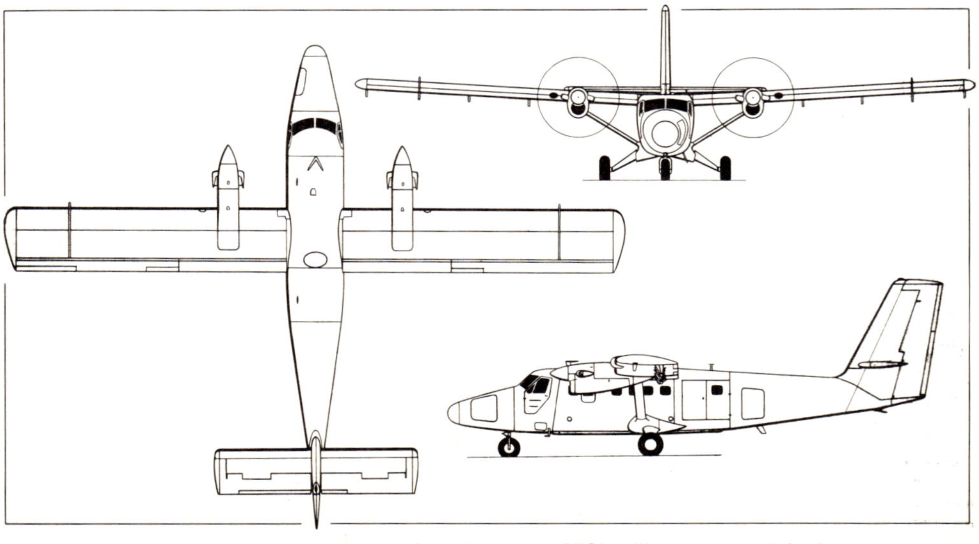

de Havilland Canada DHC-6 Twin Otter Series 300 STOL utility transport *(Pilot Press)*

DIMENSIONS, INTERNAL:
Cabin, excl flight deck, galley and baggage compartment: Length 5·64 m (18 ft 6 in)
Max width 1·61 m (5 ft 3¼ in)
Max height 1·50 m (4 ft 11 in)
Floor area 7·45 m² (80·2 sq ft)
Volume 10·87 m³ (384 cu ft)
Baggage compartment (nose):
Volume 1·08 m³ (38 cu ft)
Baggage compartment (rear):
Length 1·88 m (6 ft 2 in)
Volume 2·49 m³ (88 cu ft)

AREAS:
Wings, gross 39·02 m² (420 sq ft)
Ailerons (total) 3·08 m² (33·2 sq ft)
Trailing-edge flaps (total) 10·42 m² (112·2 sq ft)
Fin 4·46 m² (48·0 sq ft)
Rudder, incl tabs 3·16 m² (34·0 sq ft)
Tailplane 9·29 m² (100·0 sq ft)
Elevator, incl tabs 3·25 m² (35·0 sq ft)

WEIGHTS:
Typical operating weight (20-seat commuter, incl 2 crew and 59 kg; 130 lb of avionics) 3,363 kg (7,415 lb)
Max payload for 100 nm (185 km; 115 miles) 1,941 kg (4,280 lb)
Max T-O weight 5,670 kg (12,500 lb)
Max landing weight:
wheels and skis 5,579 kg (12,300 lb)
floats 5,670 kg (12,500 lb)

PERFORMANCE (at max T-O weight, ISA):
Max cruising speed at 3,050 m (10,000 ft) 182 knots (338 km/h; 210 mph)
Stalling speed, flaps up 74 knots (137·5 km/h; 85·5 mph) EAS
Stalling speed, flaps down 58 knots (108 km/h; 67 mph) EAS
Max rate of climb at S/L 488 m (1,600 ft)/min
Rate of climb at S/L, one engine out 104 m (340 ft)/min
Service ceiling 8,140 m (26,700 ft)
Service ceiling, one engine out 3,530 m (11,600 ft)
T-O run: STOL 213 m (700 ft)
CAR Pt 3 262 m (860 ft)
T-O to 15 m (50 ft): STOL 366 m (1,200 ft)
CAR Pt 3 457 m (1,500 ft)
Landing from 15 m (50 ft): STOL 320 m (1,050 ft)
CAR Pt 3 591 m (1,940 ft)
Landing run: STOL 157 m (515 ft)
CAR Pt 3 290 m (950 ft)
Range at long-range cruising speed with 1,134 kg (2,500 lb) payload 700 nm (1,297 km; 806 miles)
Range at long-range cruising speed with 862 kg (1,900 lb) payload and wing tanks 920 nm (1,704 km; 1,059 miles)

DHC-7 DASH 7

CAF designation: CC-132

The Dash 7 'quiet STOL' airliner project was begun by de Havilland Canada in late 1972, following a worldwide market survey of short-haul transport requirements.

Two pre-production aircraft were built, the first of these (C-GNBX-X) flying on 27 March 1975 and the second (C-GNCA-X) on 26 June 1975. A third airframe was built for structural testing and a fourth for fatigue testing. The first production Dash 7 (C-GQIW, c/n 3), flew on 30 May 1977 and, together with the two pre-production aircraft, participated in the flight test programme.

Certification by the Canadian Department of Transport to FAR 25 was received on 2 May 1977; STOL performance is approved under conventional FAR 25 and FAR 121 regulations. In addition, certification has been given for 7° 30′ glideslope and 10·7 m (35 ft) landing reference height adopted by the FAA for STOL aircraft.

Standard DHC-6 Twin Otter Series 300 in the insignia of Aerolineas Centrales de Colombia

The first Dash 7 to enter service was c/n 4, with Rocky Mountain Airways (USA), on 3 February 1978. By February 1980 a total of 18 Dash 7s were in service with Spantax (Spain), the Canadian Armed Forces, Air Wisconsin (USA), Ransome (USA) and Air Pacific (USA). The all-cargo **DHC-7 Series 101** was then in service with Emirates Air Service (Abu Dhabi), Gronlandsfly (Greenland), the Canadian Armed Forces (replacing the Convair CC-109 Cosmopolitan transports based in West Germany), and Alyemda (South Yemen). The 50th Dash 7 entered service in June 1981.

Orders and options for the Dash 7 by March 1981 had reached a total of 125. Most recent customers were Widerøe (Norway), Golden West (USA), Time Air (Canada), Brymon (UK), ANHSA (Honduras), Henson (USA), Yemen Airways, Tyrolean (Europe), Hawaiian Airlines, Rio Airways (USA), Ethiopian Airlines, Loganair (UK), CIS (USA), Air Niugini (Papua New Guinea), Maersk Air (Denmark), and Mobil Oil. The Dash 7 production rate remained at two per month throughout 1980.

Details of a proposed maritime reconnaissance variant, the **DHC-7R Ranger,** can be found in the 1980-81 *Jane's*. The following description applies to the standard production DHC-7:

TYPE: Four-engined short/medium-range quiet STOL transport.

WINGS: Cantilever high-wing monoplane, with 4° 30′ dihedral from centre-section. Wing section NACA 63A418 (modified) at root, NACA 63A415 (modified) at tip. Incidence 3° at root. Conventional all-metal two-spar bonded skin/stringer structure. Double-slotted flaps, extending over approx 80% of trailing-edge, are actuated mechanically for take-off, by irreversible screwjacks, and hydraulically for landing. Two inboard ground spoilers/lift dumpers and two outboard air spoilers in each upper surface, forward of flaps, also actuated hydraulically. Outboard sections can be operated symmetrically, or differentially in combination with the cable-operated ailerons. Trim tab in starboard aileron; servo-tab in each aileron. Pneumatic-boot de-icing of leading-edges outboard of the inner nacelles.

FUSELAGE: Conventional all-metal stressed-skin pressurised structure, of bonded skin/stringer construction. Basically circular cross-section, with flattened profile under floor level.

TAIL UNIT: Cantilever all-metal T tail, with large dorsal fin. Fixed-incidence tailplane, and one-piece cable-operated horn-balanced elevator with trim tabs. Two-piece vertically-split rudder, actuated hydraulically. Pneumatic-boot de-icing of leading-edges of tailplane and of elevator horns.

LANDING GEAR: Menasco retractable tricycle type, with twin wheels on all units. Oleo-pneumatic shock-absorbers. Hydraulic retraction, main units forward into inboard engine nacelles, steerable nose unit rearward into fuselage. Main-wheel tyres size 30 × 9·00-15, pressure 7·38 bars (107 lb/sq in); nosewheel tyres size 6·50-10, pressure 5·31 bars (77 lb/sq in). Larger, low-pressure tyres optional, with pressures of 4·83 bars (70 lb/sq in) on main units, 4·76 bars (69 lb/sq in) on nose unit. Anti-skid hydraulic braking system for all units. Small retractable tailskid under rear fuselage.

POWER PLANT: Four 835 kW (1,120 shp) Pratt & Whitney Aircraft of Canada PT6A-50 turboprop engines, each driving a Hamilton Standard 24PF-305 constant-speed fully-feathering reversible-pitch four-blade glassfibre propeller, with Beta control, of slow-turning type (1,210 rpm) to reduce noise level. Fuel in two integral tanks in each wing, total capacity 5,602 litres (1,232 Imp gallons). Single pressure refuelling/defuelling point on underside of rear fuselage, aft of pressure dome. Pneumatic de-icing of engine air intakes; electrical de-icing for propellers. Oil capacity 23 litres (5 Imp gallons).

ACCOMMODATION: Flight crew of two, plus one or two cabin attendants. Dual controls standard. Seats for 50 passengers at 81 cm (32 in) pitch, in pairs on each side of centre aisle, with generous provision for underseat carry-on baggage. Outward-opening airstair door at rear on port side. Emergency exits on each side at front of cabin and on starboard side at rear. Baggage compartment in rear fuselage (capacity 998 kg; 2,200 lb), with external access on starboard side and internal access from cabin. Galley, coat rack and toilet at rear of cabin. Optional arrangements include movable bulkhead for mixed freight/passenger loads with large forward freight door on port side. Up to five standard pallets can be accommodated in an all-cargo role. Quick-change cargo handling system available optionally. Entire accommodation pressurised and air-conditioned.

SYSTEMS: Cabin pressure differential 0·294 bars (4·26 lb/sq in). Two air-cycle systems, driven by engine bleed air, for cabin air-conditioning. Two independent hydraulic systems, each of 207 bars (3,000 lb/sq in). No. 1 system actuates flaps, rudder, wing spoilers and main-wheel brakes; No. 2 system actuates landing gear, nosewheel and backup main-wheel brakes, parking brakes, nosewheel steering, rudder and outboard wing spoilers. Primary DC power provided by four Phoenix 28V 250A 7·5kW starter/generators. 115/200V three-phase AC power at 400Hz from four 10kVA Lucas brushless generators for propeller and windscreen de-icing and standby fuel pumps. Lucas static inverters supply constant-frequency 400Hz loads, including engine instrumentation and navigational systems. Nickel-cadmium batteries for engine starting. APU, for cabin air-conditioning and electrics, and engine starting, available optionally.

AVIONICS AND EQUIPMENT: Standard avionics include crew interphone system; cabin PA system; flight data recorder; flight compartment voice recorder; emergency locator transmitter; two independent VHF communications systems; two independent VHF (VOR/ILS) radio navigation systems; one LF (ADF) radio navigation system; one ATC transponder; one DME; one RCA Primus 40 weather radar; one marker beacon receiver; Sperry SPZ-700 autopilot/flight director system, incorporating Z-500 flight computer and ADC-200 central air data computer; Sperry STARS ADI and HSI; Sperry AA-215 radio altimeter; and two Sperry C-14 slaved gyro compasses and VG-14 vertical gyros. Provision for variety of optional avionics to customer's requirements. Standard options include Collins 618M-3 com transceiver, Collins 51RV-4D nav receiver, Collins 51Z-4 glideslope/marker beacon receiver, and Collins 621A-6 transponder.

DIMENSIONS, EXTERNAL:
Wing span 28·35 m (93 ft 0 in)
Wing chord at root 3·81 m (12 ft 6 in)
Wing chord at tip 1·68 m (5 ft 6 in)
Wing mean aerodynamic chord 2·99 m (9 ft 9¾ in)
Wing aspect ratio 10

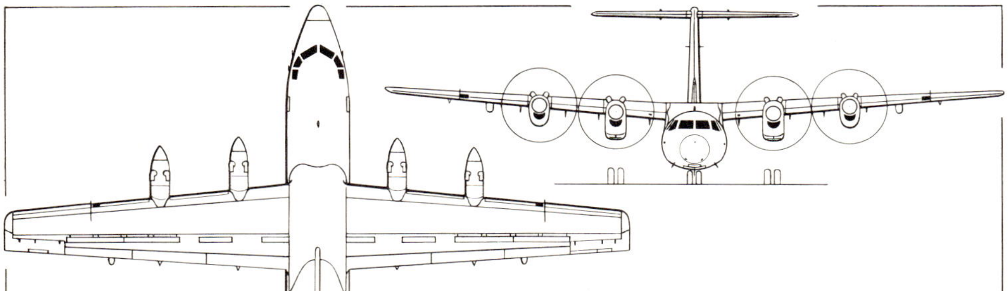

de Havilland Canada DHC-7 Dash 7 four-turboprop STOL transport *(Pilot Press)*

DHC-7 Dash 7 four-turboprop quiet STOL transport, in the insignia of Yemen Airways Corporation

Length overall 24·58 m (80 ft 7·7 in)
*Height overall 7·98 m (26 ft 2 in)
Tailplane span 9·45 m (31 ft 0 in)
Fuselage: Max diameter 2·79 m (9 ft 2 in)
Wheel track 7·16 m (23 ft 6 in)
Wheelbase 8·38 m (27 ft 6 in)
Propeller diameter 3·43 m (11 ft 3 in)
*Propeller ground clearance (inboard engines) 1·60 m (5 ft 3 in)
Min propeller/fuselage clearance 0·75 m (2 ft 5·4 in)
Passenger door (rear, port):
Height 1·75 m (5 ft 9 in)
Width 0·76 m (2 ft 6 in)
*Height to sill 1·09 m (3 ft 7 in)
Emergency exit doors (fwd, each):
Height 0·91 m (3 ft 0 in)
Width 0·51 m (1 ft 8 in)
*Height to sill 1·55 m (5 ft 1 in)
Emergency exit door (rear, stbd):
Height 1·35 m (4 ft 5 in)
Width 0·61 m (2 ft 0 in)
*Height to sill 1·09 m (3 ft 7 in)
Baggage hold door (rear, stbd):
Height 1·02 m (3 ft 4 in)
Width 0·84 m (2 ft 9 in)
*Height to sill 1·47 m (4 ft 10 in)
Cargo door (fwd, port, optional):
Height 1·78 m (5 ft 10 in)
Width 2·31 m (7 ft 7 in)
*Height to sill approx 1·22 m (4 ft 0 in)
will vary with aircraft configuration and loading conditions

Dimensions, internal:
Cabin, excl flight deck: Length 12·04 m (39 ft 6 in)
Max width 2·59 m (8 ft 6 in)
Floor width 2·13 m (7 ft 0 in)
Max height 1·94 m (6 ft 4½ in)
Height under wing 1·85 m (6 ft 1 in)
Volume 54·1 m³ (1,910 cu ft)
Baggage compartment (rear fuselage):
Max length 2·30 m (7 ft 6½ in)
Volume 6·8 m³ (240 cu ft)

Areas:
Wings, gross 79·90 m² (860·0 sq ft)
Ailerons (total) 2·16 m² (23·22 sq ft)
Trailing-edge flaps (total) 27·33 m² (294·20 sq ft)
Spoilers (total) 3·63 m² (39·04 sq ft)
Vertical tail surfaces (total, excl dorsal fin) 15·79 m² (170·0 sq ft)
Horizontal tail surfaces (total) 20·16 m² (217·0 sq ft)

Weights and Loadings:
Basic weight empty (standard 50-passenger layout) 12,247 kg (27,000 lb)
Operating weight empty 12,542 kg (27,650 lb)
Max payload (50 passengers or cargo) 5,148 kg (11,350 lb)
Max fuel (standard tanks) 4,563 kg (10,060 lb)
Max T-O weight 19,958 kg (44,000 lb)
Max zero-fuel weight 17,690 kg (39,000 lb)
Max landing weight 19,050 kg (42,000 lb)
Max cabin floor loading 366·2 kg/m² (75 lb/sq ft)
Max wing loading 249·8 kg/m² (51·17 lb/sq ft)
Max power loading 5·98 kg/kW (9·82 lb/shp)

Performance (at max T-O weight, FAR Pt 25, at S/L, ISA, except where indicated):
Max cruising speed at 2,440 m (8,000 ft) at AUW of 18,597 kg (41,000 lb) 231 knots (428 km/h; 266 mph)
Max cruising speed at 4,570 m (15,000 ft) at AUW of 18,597 kg (41,000 lb) 227 knots (420 km/h; 261 mph)
En-route rate of climb, flaps and landing gear up:
4 engines, max climb power 372 m (1,220 ft)/min
3 engines, max continuous power 220 m (720 ft)/min
Service ceiling at AUW of 18,597 kg (41,000 lb):
4 engines, max climb power 6,400 m (21,000 ft)
3 engines, max continuous power 3,855 m (12,650 ft)
T-O field length, 25° flap 689 m (2,260 ft)
T-O field length at 3,050 m (10,000 ft), 15° flap 1,829 m (6,000 ft)
Landing field length at max landing weight, 45° flap 594 m (1,950 ft)
Landing field length at 3,050 m (10,000 ft) at 18,915 kg (41,700 lb) landing weight, 45° flap 823 m (2,700 ft)
Min ground turning radius 8·84 m (29 ft 0 in)
Runway LCN with 32 × 11·50-15 low-pressure tyres, rigid, 30 in relative stiffness 16·2
Range at 3,960 m (13,000 ft) with 50 passengers and baggage, at long-range cruising speed, IFR reserves 730 nm (1,352 km; 840 miles)
Max range at 4,570 m (15,000 ft) with standard fuel and 3,040 kg (6,700 lb) payload, long-range cruising speed 1,160 nm (2,148 km; 1,335 miles)

Operational Noise Levels (FAR Pt 36 at S/L, ISA + 10°C, confirmed):
T-O 80·5 EPNdB
Approach on 3° glideslope 91·4 EPNdB
Sideline 82·8 EPNdB

DHC-8 DASH 8

The DHC-8 is being developed by de Havilland Canada to fill the growing demand for a quiet, fuel-efficient short-haul transport in the 30/40-seat category, and is scheduled to make its first flight in mid-1983.

Sized to accommodate 32 to 36 passengers, the Dash 8 fits in between the company's 19-passenger Twin Otter and 50-passenger Dash 7, and is designed to meet the requirements of FAR Pt 25. Certification by the Canadian DoT will be to FAR Pts 25 and 36, and SFAR Pt 27. Worldwide applications include scheduled airline service on routes of up to 485 nm (805 km; 500 miles); convertible passenger/cargo operations in developing countries overseas; or comparable corporate and military transport roles.

Initial brief details of the Dash 8 were released on 2 April 1980, concurrently with the news that norOntair, the air service sector of the Ontario Northland Transportation Commission, had ordered two of these aircraft. By June 1981, de Havilland Canada had received options totalling 115 Dash 8s, from 31 operators in 10 countries.

The first DHC-8 for norOntair is due for delivery in September 1984, and to enter service later that year.

Type: Twin-turboprop quiet short-range transport.

Wings: Cantilever high-wing monoplane, with constant-chord centre-section and tapered outer panels. Dihedral 2° 30′ on outer panels. Single-slotted Fowler-type flaps inboard and outboard of engine nacelles. Hydraulically actuated roll control spoilers/lift dumpers forward of each outer flap segment; independent ground spoiler/lift dumper inboard and outboard of each engine nacelle. Mechanically actuated balanced ailerons, with inset tabs.

Fuselage: Conventional semi-monocoque pressurised structure, of circular cross-section.

Tail Unit: Cantilever T tailplane; full-span balanced elevator, with tabs. Sweptback fin, with large dorsal fin, and two-segment serially-hinged rudder. Tailplane, elevator and rudder built by Canadian Aircraft Products.

Landing Gear: Retractable tricycle type, by Dowty Equipment of Canada Ltd, with twin wheels on each unit. Nose unit retracts forward, main units rearward into engine nacelles.

Power Plant: Two 1,342 kW (1,800 shp) Pratt & Whitney Aircraft of Canada PW120 turboprop engines, each driving a Hamilton Standard 14SF-1 four-blade constant-speed fully-feathering propeller with reverse pitch.

Accommodation: Crew of two on flight deck, plus one attendant in cabin. Dual controls standard, though aircraft will be certificated for single-pilot operation. Standard layout provides four-abreast seating in main cabin for 36 passengers at 79 cm (31 in) pitch, plus buffet, toilet and large rear baggage compartment. Alternative 32-passenger, mixed passenger/cargo or corporate layouts available at customer's option. Passenger door on port side at front; large port-side door aft of wing for cargo loading. Emergency exit each side, in line with wing leading-edge, and opposite passenger door on starboard side. Entire accommodation pressurised and air-conditioned.

Systems: Hamilton Standard air-conditioning system. Cabin max pressurisation differential 0·38 bars (5·5 lb/sq in).

Avionics: Sperry DFZ-800 automatic flight control system.

Dimensions, external:
Wing span 25·60 m (84 ft 0 in)

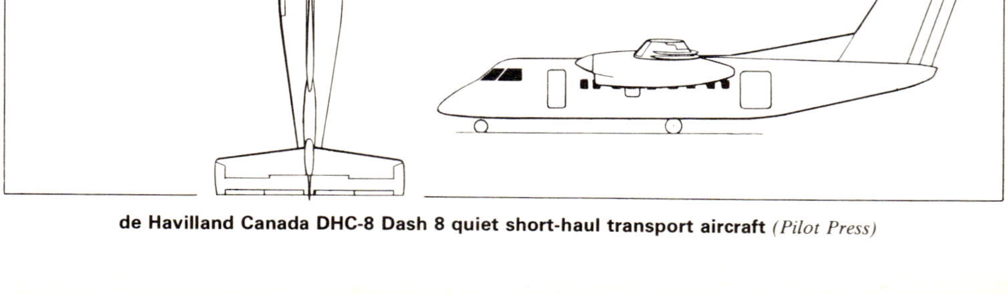

de Havilland Canada DHC-8 Dash 8 quiet short-haul transport aircraft *(Pilot Press)*

Wing area, gross	54·35 m² (585·0 sq ft)
Wing aspect ratio	12·06
Length overall	22·25 m (73 ft 0 in)
Height overall	7·62 m (25 ft 0 in)
Tailplane span	7·92 m (26 ft 0 in)
Wheel track (c/l of shock-struts)	7·87 m (25 ft 10 in)
Wheelbase	7·90 m (25 ft 11 in)
Propeller diameter	3·96 m (13 ft 0 in)
Propeller ground clearance	0·94 m (3 ft 1 in)
Propeller/fuselage clearance	0·76 m (2 ft 6 in)
Passenger door (fwd, port):	
Height to sill	1·09 m (3 ft 7 in)
Cargo door (rear, port):	
Height	1·52 m (5 ft 0 in)
Width	1·27 m (4 ft 2 in)
Height to sill	1·09 m (3 ft 7 in)
DIMENSIONS, INTERNAL:	
Cabin: Length	9·19 m (30 ft 2 in)
Max width	2·49 m (8 ft 2 in)
Max height	1·88 m (6 ft 2 in)
Cargo compartment volume	8·5 m³ (300 cu ft)
WEIGHTS AND LOADING:	
Max fuel	2,676 kg (5,900 lb)
Max payload	3,549 kg (7,824 lb)
Max T-O weight	13,834 kg (30,500 lb)
Max landing weight	13,607 kg (30,000 lb)
Max wing loading	254·5 kg/m² (52·1 lb/sq ft)
PERFORMANCE (estimated, at max T-O weight):	
Max cruising speed at 4,575 m (15,000 ft), ISA	270 knots (500 km/h; 310 mph)
Max operating altitude, standard aircraft	7,620 m (25,000 ft)
FAR Pt 25 T-O field length at S/L, ISA	826 m (2,710 ft)
FAR Pt 25 landing field length at S/L	908 m (2,980 ft)
Range with IFR reserves:	
normal	570 nm (1,056 km; 656 miles)
max	1,100 nm (2,038 km; 1,266 miles)

HAWKER SIDDELEY

HAWKER SIDDELEY CANADA INC (Member Company of Hawker Siddeley Group)

HEAD OFFICE: 7 King Street East, Toronto, Ontario M5C 1A3
Telephone: (416) 362 2941
Telex: 06 217711

CHAIRMAN: Sir Arnold Hall, FRS
VICE-CHAIRMEN:
J. F. Howard, QC
F. H. Wood
PRESIDENT AND CHIEF EXECUTIVE OFFICER: R. F. Tanner
DIRECTOR OF CORPORATE AFFAIRS: J. F. A. Painter

Known as A. V. Roe Canada Ltd until 1962, and Hawker Siddeley Canada Ltd until 1 July 1980, this company controls operating units and subsidiaries in Canada, the UK and the USA employing about 7,600 people.

The company's chief aviation unit is Orenda Division (see Aero-engines section).

McKINNON

McKINNON-VIKING ENTERPRISES

HEAD OFFICE AND WORKS: Hangar No. 2, Victoria International Airport (PO Box 2004), Sidney, British Columbia V8L 3S3
Telephone: (604) 656 7227
PRESIDENT, OWNER AND MANAGER: N. Christensen

McKinnon-Viking, formed in 1978, superseded the former McKinnon Enterprises Inc, included previously in the US section. McKinnon (formerly McKinnon-Hickman Company) entered the aircraft conversion field in 1953 when it began to undertake the conversion of Grumman Widgeon twin-engined light amphibians into executive aircraft. The success of this programme led to development and manufacture of a four-engined conversion of the larger Grumman Goose amphibian, known as the McKinnon G-21 Goose, details of which can be found in the 1966-67 *Jane's*. It was superseded by the turboprop-powered G-21C and D Turbo-Goose, described in the 1976-77 *Jane's*.

Details of more recent conversions (G-21G Turbo-Goose, turboprop Goose conversion, and Super Widgeon) can be found in the 1980-81 *Jane's*.

No news has been received from McKinnon-Viking since its re-formation in 1978.

NWI

NORTHWEST INDUSTRIES LIMITED (A subsidiary of CAE Industries Ltd)

Edmonton Municipal and International Airports, PO Box 517, Edmonton, Alberta T5J 2K5
Telephone: (403) 455 3161
Telex: 037 2681
PRESIDENT: L. H. Prokop
VICE-PRESIDENT, OPERATIONS: F. A. Maybee
DIRECTORS:
C. H. Fraser (Finance)
T. Luykenaar (Industrial Relations and Facilities)
R. H. Sharples (Quality Control and Engineering)
SECRETARY: C. H. Fraser

Employing some 500 people, NWI is one of Canada's largest and most experienced aircraft maintenance, repair, overhaul and modification centres for military and commercial aircraft, including the Lockheed C-130 Hercules, Dassault Falcon, Lockheed T-33 and Canadair CL-41 jet trainers, and the CF-104 Starfighter. The company has permanent mobile repair parties at CFB Cold Lake and CFB Edmonton in support of the CF-5, C-130 Hercules and CF-104 aircraft of the Canadian Armed Forces. NWI operates western Canada's largest airframe manufacturing facility, and since 1969 has produced major airframe components and air-conditioning ducting for the Lockheed L-1011 TriStar, including the L-1011-500.

The company is actively involved also in aircraft instrument and accessory overhaul, aircraft electronics, avionics design and interface, structural and systems engineering, non-destructive testing, and technical publications.

Canadian Armed Forces Lockheed CF-104 undergoing a run-up check after overhaul by NWI

By February 1981, NWI had completed approx 50 per cent of a service life extension programme (SLEP) for the Canadian Armed Forces' CF-104 aircraft, under a Canadian government contract valued at more than 27 million dollars. This programme involves the embodiment of extensive fuselage modifications, sophisticated non-destructive testing, wiring replacements, and the major overhaul of all aircraft systems, including in-depth visual inspections and repair. Extending into 1982, the programme employs more than 200 of the company's 550 personnel, and will ensure the continued active role of the CF-104 ground strike support aircraft in its NATO and Canadian defence assignments until its replacement by the McDonnell Douglas CF-18 Hornet.

NWI airlifts two CF-104s per round trip to Edmonton from West Germany, utilising a Lockheed L-100-30 Commercial Hercules transport aircraft owned and operated by Pacific Western Airlines under subcontract to NWI.

ROBIN

AVIONS ROBIN CANADA

Lachute, Quebec
SALES MANAGER: Jacques Lacrivain

This company was established by Avions Pierre Robin of France (which see) to assemble in Canada and market certain models of the current Robin range. The government of Quebec has a 48 per cent holding in the company, and the Canadian government contributed 25 per cent of the cost of the buildings and machinery.

The three models chosen for initial production in Canada are the Robin R 1180, R 2112 and R 2160. Assembly is from French-built subassemblies and components, with US engines, propellers, tyres, instruments and radios being added on the Canadian production line.

First example of the R 2160 was completed in 1980, and this type is now in production. Canadian certification was obtained in 1980, and FAA certification was expected in late 1980 or early 1981.

TRIDENT

TRIDENT AIRCRAFT LIMITED

PO Box 2428, Sidney, British Columbia V8L 3Y4

TRIDENT TR-1 TRIGULL

The Trigull six-seat light amphibian flew for the first time on 5 August 1973. Certification from the Canadian Dept of Transport and the FAA was received on 28 October and 16 December 1976 respectively.

Production of the Trigull, which was described and illustrated in the 1980-81 *Jane's*, was dependent upon adequate funding, but despite some government financial assistance the company found it impossible to launch production, and ceased trading in late 1980.

ZENAIR

ZENAIR LTD

HEAD OFFICE: 236 Richmond Street, Richmond Hill, Ontario L4C 3Y8
Telephone: (416) 884 9044
WORKS: 25 King Road, Nobleton, Ontario
Telephone: (416) 859 4556
PRESIDENT AND DESIGNER: Christophe Heintz

This company was formed by M Christophe Heintz, formerly designer with Avions Pierre Robin of France, to market plans, materials, parts and complete kits of his

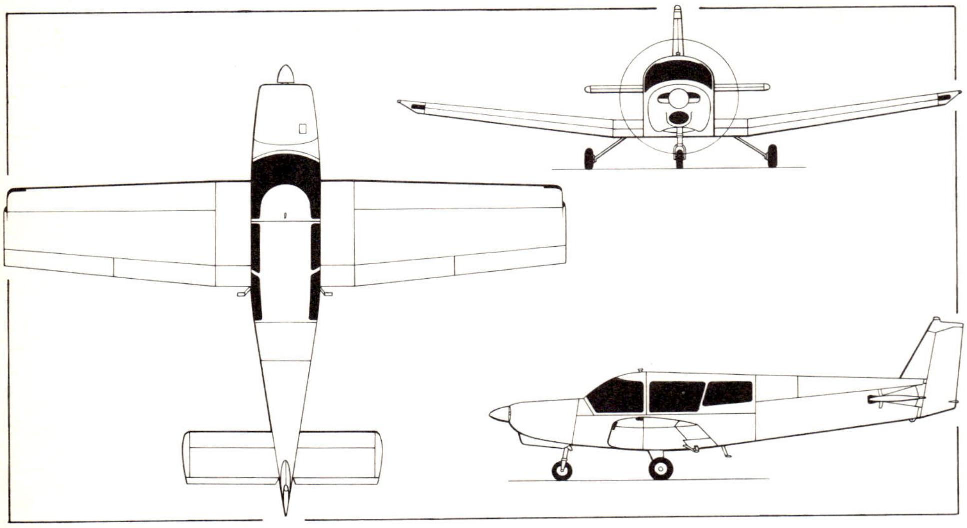

Heintz Zénith-CH 400 four-seat light aircraft (Avco Lycoming O-360-A engine) *(Michael A. Badrocke)*

single-seat, two-seat and three-seat Zénith light aircraft (see Homebuilt Aircraft section).

Additionally, Zenair is building prototypes of the four-seat Zénith-CH 400, intended for eventual factory production.

HEINTZ ZÉNITH-CH 400

Design began in 1976 of a new four-seat version of the Zénith which retains the basic configuration of the earlier aircraft but in scaled-up form. Construction of three prototypes at Zenair's plant began in September 1977, and the first of these was expected to fly for the first time in the Summer of 1981. The CH 400 was expected to become available in 1982 as a factory-built aircraft.

TYPE: Four-seat light aircraft.

WINGS: Cantilever low-wing monoplane. Wing section NACA 33015. Dihedral 6° 30′ on tapered outer panels. Incidence 3° 30′. Single-spar aluminium alloy structure. Frise-type ailerons and slotted flaps of aluminium alloy. No tabs.

FUSELAGE: Conventional aluminium alloy stressed-skin structure of basically rectangular section, with rounded top-decking.

TAIL UNIT: Cantilever one-piece all-moving tailplane, with automatic and controllable anti-servo trim tab, and one-piece all-moving rudder (no fin). Single-spar structures, with ribs and skins of aluminium.

LANDING GEAR: Non-retractable tricycle type, with steel spring main-gear legs. All three Cleveland wheels and tyres size 6·00-6, pressure 2·21 bars (32 lb/sq in). Hydraulically-actuated disc brakes on main gear. Steerable nosewheel. Wheel fairings optional.

POWER PLANT: One 134 kW (180 hp) Avco Lycoming O-360-A flat-four engine, driving a Sensenich M76EMM-2-66 metal fixed-pitch propeller. Fuel tanks in wings, standard capacity 210 litres (46 Imp gallons). Refuelling points in top of wings. Fuselage tank of 80 litres (17·5 Imp gallons) capacity optional. Oil capacity 9 litres (2 Imp gallons).

ACCOMMODATION: Pilot and passenger on individual front seats, with rear bench seat for two further passengers. Forward-sliding Plexiglas canopy. Baggage compartment, with two separate doors, capacity 60 kg (132 lb).

SYSTEM: 12V alternator for electrical system.

AVIONICS: As required. Full IFR potential.

DIMENSIONS, EXTERNAL:

Wing span	9·20 m (30 ft 2¼ in)
Wing chord at root	1·82 m (5 ft 11¾ in)
Wing chord at tip	1·38 m (4 ft 6¼ in)
Wing aspect ratio	5·6
Length overall	7·80 m (25 ft 7 in)
Fuselage: Max width	1·12 m (3 ft 8 in)
Height overall	2·70 m (8 ft 10¼ in)
Tailplane span	3·00 m (9 ft 10 in)
Wheel track	3·00 m (9 ft 10 in)
Wheelbase	1·60 m (5 ft 3 in)
Propeller diameter	1·88 m (6 ft 2 in)
Propeller ground clearance	0·27 m (10¾ in)

AREAS:

Wings, gross	15·20 m² (163·6 sq ft)
Ailerons (total)	0·94 m² (10·12 sq ft)
Flaps (total)	1·62 m² (17·44 sq ft)
Rudder	1·15 m² (12·38 sq ft)
Tailplane (incl tabs)	2·70 m² (29·06 sq ft)

WEIGHTS AND LOADINGS:

Weight empty, equipped	638 kg (1,406 lb)
Max T-O weight	1,170 kg (2,580 lb)
Max wing loading	77 kg/m² (15·77 lb/sq ft)
Max power loading	8·73 kg/kW (14·33 lb/hp)

PERFORMANCE (estimated):

Never-exceed speed	188 knots (350 km/h; 217 mph)
Max level speed	146 knots (270 km/h; 168 mph)
Max cruising speed	135 knots (250 km/h; 155 mph)
Econ cruising speed	124 knots (230 km/h; 143 mph)
Stalling speed, flaps up	56 knots (103 km/h; 64 mph)
Stalling speed, flaps down	48·5 knots (90 km/h; 56 mph)
Max rate of climb at S/L	274 m (900 ft)/min
Service ceiling	4,875 m (16,000 ft)
T-O and landing run	244 m (800 ft)
T-O to, and landing from, 15 m (50 ft)	457 m (1,500 ft)
Range with standard fuel and max payload	647 nm (1,200 km; 745 miles)
Range with max fuel	863 nm (1,600 km; 994 miles)
Endurance with standard fuel and max payload	4 h 45 min
Endurance with max fuel	6 h 30 min

CHILE

FAC

FUERZA AÉREA CHILENA (Chilean Air Force)

HEADQUARTERS: Av Bernardo O'Higgins 1170, 10° Piso, CC 1152, Correo 21, Santiago

COMMANDER-IN-CHIEF: Gen Fernando Matthei Aubel

WORKS: Ala de Mantenimiento, Fuerza Aérea Chilena, El Bosque Air Base

As recorded under the Piper entry in the US section, the Maintenance Wing of the Chilean Air Force has begun the licence assembly in Chile of **Piper PA-28 Dakotas**, of which 30 had been completed by mid-1981. Beginning in 1982 it will undertake the assembly for the FAC of up to 100 Pillan two-seat basic/intermediate trainers, of which two prototypes have been built by Piper derived from the Cherokee series.

In July 1981 Sequoia Aircraft Corporation (see US Homebuilts section) announced delivery to the FAC of an assembly kit for one **F.8L Falco** two-seat aerobatic trainer, and was preparing for anticipated production of this aircraft in Chile, both for the FAC and for civilian flying clubs.

CHINA
(PEOPLE'S REPUBLIC)

The Chinese language is gradually undergoing a process of simplification, from the pictograms in which the language has been written for some 3,000 years to a new Latinised or Westernised form of spelling known as Pinyin. Under this system, the transliterated spellings with which Western readers have been familiar for years have, in many cases, undergone some change. For example, Peking is now written as Beijing, which conforms more closely with the Chinese pronunciation. Place names in this section are shown first in the current Pinyin spelling, followed where applicable by the more familiar 'old' spelling. Similarly, the designations of Chinese aircraft are wherever possible given first in their Chinese form, with the Westernised equivalent in parentheses. Both Chinese and Westernised forms may be found in Chinese publications.

STATE AIRCRAFT FACTORIES

WORKS: Shenyang, Liaoning Province; Xian (Sian), Shaanxi (Shensi) Province; Harbin, Heilongjiang (Heilunkiang) Province; Shanghai, Zhejiang (Chekiang) Province; Beijing (Peking), Hebei (Hopei) Province; Nanzhang (Nanchang), Jiangxi (Kiangsi) Province; and elsewhere

Longest-established of the Chinese national aircraft factories is the works at Shenyang, which had its origin in the Shenyang (Mukden) plant of the Manshu Aeroplane Manufacturing Company, one of several aircraft and aero-engine manufacturing facilities established in Manchukuo (Manchuria) by the Japanese invaders in 1938. After the Communist regime became responsible for the whole of mainland China in 1949 the Manchurian factories were re-established and re-equipped with Soviet assistance. Today Shenyang and Harbin (also known as Binjiang) are the main centres of Chinese aircraft and aero-engine production, under the jurisdiction of the Third Ministry of Machine Building. There are design and development centres at Shenyang, Beijing (Peking), Harbin and elsewhere.

In the middle and late 1950s the Shenyang factory produced in large numbers under licence several aircraft types, the first of these being the Yak-18, for which a licence agreement was signed in November 1952. In October 1954 this was followed by a licence for the Antonov An-2 biplane, the Mil Mi-4 helicopter, and their ASh-62 and ASh-82 engines.

First jet aircraft to be built in China were the single-seat MiG-15bis and two-seat MiG-15UTI. These do not have Chinese designations. They were followed by the MiG-17F, which has the Chinese designation **J-5** (or, in Westernised form, F-5); and MiG-17PF (J-5 Jia or J-5A, Westernised as F-5A). The MiG fighters and their Klimov VK-1 engines were produced at Shenyang, deliveries of both the MiG-15UTI and J-5 beginning in late 1956; by mid-1959 these were almost totally of Chinese manufacture. A 'UTI' tandem two-seat conversion of the J-5 was also produced by the Chinese: this, too, is a standard advanced trainer type. Several hundred MiG-15s (mostly UTIs) remain in service, and over 1,000 J-5/5As still equip more than 20 air regiments in air-to-air and fighter-bomber roles. A small number of J-5s also continues in service with the Aviation of the People's Navy. J-5s were exported to Albania (30), Kampuchea and North Viet-Nam. These types were followed by Chinese versions of the MiG-19 (J-6), licence production of the Yak-18A (CJ-5), and design and manufacture of the CJ-6 trainer.

Aircraft built at the Xian works include the Soviet Tupolev Tu-16 bomber (Chinese designation H-6), Chinese versions of the MiG-21 (J-7), and the Yinan No. 1 utility aircraft. Harbin's products have included the Czechoslovak Super Aero 45 two/three-passenger twin-engined utility aircraft, the Heilongjiang No. 1 agricultural/utility aircraft resembling the Soviet Yak-12 and, currently, the Soviet Ilyushin Il-28 jet bomber (Chinese H-5), Mil Mi-4 helicopter (Chinese Z-5), and the Y-11 agricultural/utility light twin. Another Yak-12 development, known as the Red Banner No. 1, was produced at Beijing, together with two light piston-engined transports known as the Capital No. 1 and Beijing (Peking) No. 1. Most of the older types were illustrated in *Jane's* between 1959 and 1964.

The present capability of China's aircraft industry was first revealed openly by study of the J-6 single-seat day fighters supplied to Pakistan, generally similar to the Soviet MiG-19SF. An assessment of the J-6 by a Western observer described the general standard of workmanship of the airframe as very good. At low altitudes this fighter was said to outmanoeuvre any type of combat aircraft then in service in Asia except the F-86, and to outclimb the MiG-21 and F-104 Starfighter. The potential of the Pakistani J-6s was much enhanced by supplementing their standard cannon armament with two Sidewinder missiles.

The indications are that output of older fighters is now diminishing, with increasing emphasis being placed on the development of new aircraft making use of China's increasing technological capability. The Q-5 'Fantan' twin-jet fighter-bomber is in production and service, and prototypes have been built of an all-weather interceptor and of a four-jet transport aircraft, both of indigenous design.

The Chairman of the US Joint Chiefs of Staff, in his FY 1979 Military Posture statement, reported that the Chinese Air Force of the People's Liberation Army then had an operational home defence fighter force of about 4,000 MiG-15s, -17s, -19s and -21s, and a tactical air force of more than 1,000 MiG-15s, Tupolev Tu-2s, Ilyushin Il-28s and 'Fantan-As'. The Chinese Air Force currently deploys a medium bomber force of about 95 Tu-16s, which are nuclear-capable. More than 100 MRBMs and IRBMs are also deployed in a mobile strategic role, supplemented by a few limited-range ICBMs.

Chinese two-seat training versions of the Shenyang J-5 (MiG-17)

An FT-6 two-seat fighter trainer of the Egyptian Air Force *(Denis Hughes)*

SHENYANG J-5 TRAINER (FT-5)

The two-seat advanced training version of the J-5 (Chinese-built MiG-17), shown in an accompanying illustration, represents a unique Chinese development of the basic MiG-17 fighter. An instructor's cockpit, with dual controls and raised seat, is installed to the rear of the standard pilot's cockpit. The canopy over the front seat is sideways-hinged, to starboard. The rear seat is enclosed by a rearward-sliding canopy. Other changes by comparison with the single-seat fighter include use of a non-afterburning Klimov VK-1A turbojet engine, rated at 26·47 kN (5,950 lb st); reduction of the armament to a single 23 mm NR-23 gun, carried in a removable belly pack, with the barrel to the starboard side of the nose-wheel doors; and enclosure of the ranging radar by a small lip at the top of the nose air intake.

This version of the J-5 is the standard advanced trainer of the Chinese air forces, to which pupil pilots graduate after basic training on the CJ-6.

Dimensions, external:
- Wing span: 9·628 m (31 ft 7 in)
- Length overall: 11·50 m (37 ft 9 in)
- Height overall: 3·80 m (12 ft 5¾ in)
- Wheel track: 3·85 m (12 ft 7½ in)

Weight:
- Normal T-O weight: 5,401 kg (11,907 lb)

Performance:
- Max level speed at 5,000 m (16,400 ft): Mach 0·932 (565 knots; 1,048 km/h; 651 mph)
- Normal operating speed: 418 knots (775 km/h; 482 mph)
- Rate of climb at S/L: 1,620 m (5,315 ft)/min
- Service ceiling: 14,300 m (46,900 ft)
- Max endurance at 13,700 m (45,000 ft) with two 400 litre (88 Imp gallon) drop-tanks: 2 h 38 min

SHENYANG (MIKOYAN) J-6 (F-6)

Chinese name: Type 6 Jianjiji (Type 6 Fighter)
NATO reporting names: Farmer-C (MiG-19SF) and Farmer-D (MiG-19PF)

The J-6 (F-6) is basically a MiG-19 fighter built in China. Its original design was initiated by the Mikoyan bureau in the USSR, where the I-350 (M) or I-360 prototype, with non-afterburning Mikulin AM-5 engines, flew for the first time on or about 18 September 1953.

The initial production **MiG-19** day fighter, known to NATO as 'Farmer-A', was powered by two AM-5F (Forsirovanny: increased power) turbojets, each rated at 22·06 kN (4,960 lb st) dry and 29·81 kN (6,702 lb st) with afterburning. This version, which had a maximum level speed of Mach 1·1 at altitude, began to enter service with the Soviet air defence force in early 1955. It was later redesignated **MiG-19F** after being refitted with R-9BF engines and other improvements developed for subsequent versions.

The latter included a ventral airbrake, dorsal spine fairing and enlarged dorsal fin, but the first major airframe improvement was the introduction of an all-moving tailplane to replace the original tailplane/elevators configuration, the latter having proved ineffective. This brought a change of designation to **MiG-19S** (Stabilisator: tailplane), and in this form the day fighter received the NATO reporting name 'Farmer-C'. Other changes introduced in the MiG-19S included Tumansky (Mikulin) R-9B engines (rated at 25·50 kN; 5,732 lb st dry and 32·36 kN; 7,275 lb st with afterburning); built-in armament of three 30 mm NR-30 cannon (instead of the single 37 mm N-37 and two wing-mounted 23 mm NR-23 guns in the original MiG-19); and an attachment under each wing for a bomb or air-to-surface rocket. Normal T-O weight of this version was 7,400 kg (16,314 lb). When fitted with R-9BF engines, with a modified afterburner, it has a 200 kg (441 lb) greater T-O weight and is designated **MiG-19SF.**

Meanwhile, in about 1957, a version with limited all-weather capability was put into production as the **MiG-19P** (Perekhvatchik: interceptor), and was allocated the NATO reporting name 'Farmer-B'. Powered by R-9BF engines, it had a small Izumrud (Emerald) radar scanner inside its engine air intake and a ranging unit in the intake top lip. An armament of two wing-mounted NR-23 cannon was carried. The later **MiG-19PF** differed in having NR-30 wing guns and two 212 mm ARS-212 underwing rockets; the **MiG-19PM** (Modifikatsirovanny: modified) differed from the PF in having four first-generation radar-homing missiles (NATO 'Alkali') instead of guns. The PF and PM were both given the NATO reporting name 'Farmer-D'.

Other versions produced in the USSR included the **MiG-19R** (Razvedchik: reconnaissance), with cameras in the bottom of the front fuselage, and two wing-mounted NR-30 cannon; and the **MiG-19UTI** tandem two-seat operational trainer.

In the Soviet Union the MiG-19 was phased out of production by the end of the 'fifties, but a licence agreement for its manufacture in China was signed in January 1958. Many MiG-19s had been delivered to China in knocked-down form before the deterioration of Moscow-Beijing relations. The designation **J-6** was given to the Chinese version of the MiG-19S fighter, which first flew in December 1961 and from mid-1962 became standard equipment in the Chinese Air Force of the People's Liberation Army.

Production of the J-6 was stepped up from about 1966, and several thousand have since been built, including counterparts of the Soviet MiG-19PF/PM and MiG-19SF versions. China has also developed a number of variants. One of these is a tactical reconnaissance model, similar to the MiG-19R, with a camera pack in the lower forward fuselage. The J-6 from Xinjiang (Sinkiang) air base in Fujian (Fukien) Province, whose pilot defected to Taiwan in July 1977, was one of the original Soviet-supplied MiG-19s converted to this configuration. It was equipped with a vertical/oblique camera installation in the lower forward fuselage, from which the underfuselage 30 mm cannon had been deleted. This aircraft was reported to belong to the 1st Reconnaissance Air Unit; the two wing-root cannon showed no traces of having been fired. Some of the limited all-weather models have a different centre-body radar housing of longer, slim-conical form. Another variant in service is the **FT-6**, a trainer version with a tandem two-seat cockpit installation (see accompanying illustration) generally similar to that of the Soviet-built MiG-19UTI. The Q-5 'Fantan-A' strike fighter, derived from the J-6, is described separately.

Immediately after the Indo-Pakistan war of September 1965, China offered J-6s to Pakistan. Forty were supplied initially and the first PAF squadron was operational within a year. Subsequent deliveries brought to 135 the total of J-6s acquired by Pakistan, and by 1981 these equipped nine PAF air defence and ground-attack squadrons. PAF J-6s have been fitted with rails for a Sidewinder missile under each wing, and are currently being equipped to carry an underbelly auxiliary fuel tank. By the Spring of 1974 the Air Wing of the Tanzanian People's Defence Force had received sufficient J-6s for a single squadron; 40 have been delivered to Egypt, including some two-seat FT-6s, and well over 100 others to Albania, Bangladesh, Kampuchea and Viet-Nam.

In the Chinese Air Force of the PLA, the J-6 equips more than 40 air regiments, each regiment having three or four squadrons. Roles for which the aircraft are employed include air-to-air interception, battlefield interdiction, close support, counter air and tactical reconnaissance. The J-6 also serves in small numbers with the Aviation of the People's Navy.

In early 1980, US aerospace industry visitors to China were told that Shenyang had the capacity to build 30 J-6s per month; but it appeared to be involved primarily on overhaul programmes at that time. It is believed that production now continues at a rate sufficient only to provide aircraft for domestic replacements and to fulfil possible export orders.

The following description is based on known details of the basic MiG-19SF, modified where possible to apply specifically to the Chinese J-6:

Type: Single-seat day fighter, attack and tactical reconnaissance aircraft.

Wings: Cantilever all-metal mid-wing monoplane. Wing section TsAGI S-12S at root, SR-7S at tip. Thickness/chord ratio 8·74% (root), 8·00% (tip). Anhedral 4° 30′. Sweepback at quarter-chord 55°. Three-spar structure, with auxiliary spar, ribs, and stressed skin of 1·5-2·0 mm sheet duralumin. Main spar is of 30 HGNSA; auxiliary spar provides mountings for wing guns and main landing gear members. Entire trailing-edge of each wing formed by aerodynamically balanced aileron (outboard) and large Fowler-type TsAGI flap, both hydraulically powered. Compressed-air

J-6 single-seat day fighter supplied by China to the Egyptian Air Force *(Denis Hughes)*

Chinese-built MiG-19PF (J-6) all-weather fighter

Note the unusual slim pointed radome fitted to the J-6 (MiG-19PF) at rear

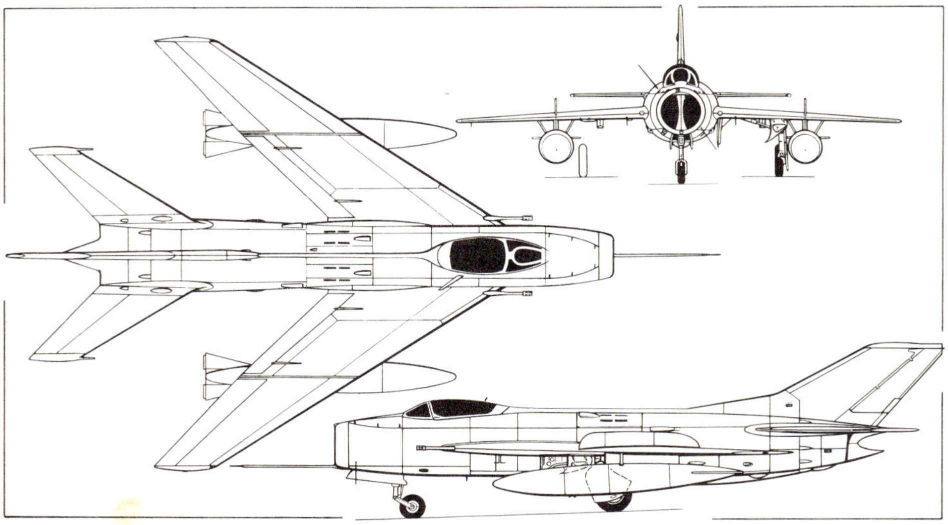

Shenyang J-6 (Chinese-built MiG-19SF) single-seat day fighter *(Pilot Press)*

emergency extension system for flaps. Trim tab in port aileron. Large full-chord boundary layer fence, 32 mm (1¼ in) deep, above each wing at mid-span to enhance aileron effectiveness. Plate-type spoiler beneath each wing, forward of aileron, to improve lateral control. Spoiler actuation is coupled with that of aileron, and takes place only when aileron is deflected downward.

Fuselage: Conventional all-metal semi-monocoque structure of circular section, with divided air intake in nose and side-by-side twin orifices at rear. Top and bottom 'pen-nib' fairings aft of nozzles. Entire rear fuselage detaches at wing trailing-edge for engine servicing. Forward-hinged door-type airbrake, operated hydraulically, on each side of fuselage aft of wing trailing-edge. Forward-hinged perforated door-type airbrake under centre of fuselage. Shallow ventral strake under rear of fuselage. Upward-hinged pitot boom mounted on lower lip of nose intake.

Tail Unit: Conventional all-metal structure, of similar construction to wings, with stressed skin of 1·2 mm duralumin. Hydraulically-actuated one-piece horizontal surfaces, with electrical emergency actuation in the event of hydraulic failure. Anti-flutter weight projecting forward from each tailplane tip. Stick-to-tailplane gearing, via electro-mechanical linkage, reduces required stick forces during high-*g* manoeuvres. Sweepback on fin leading-edge 57° 37′. Mass-balanced rudder, with electrically-actuated trim tab. Large dorsal fin between fin and dorsal spine enclosing actuating rods for tail control surfaces. Fin-tip incorporates antenna for tail warning radar.

Landing Gear: Wide-track retractable tricycle type, with single wheel on each unit. Hydraulic actuation, nosewheel retracting forward, main units inward into wing roots. Pneumatic emergency extension system. All units of levered-suspension type, with oleo-pneumatic shock-absorbers. Nose unit is steerable, self-centering, and fitted with hydraulic shimmy damper. Main units have KT-37 wheels, and 660 × 220 mm tyres, pressure 8·83 bars (128 lb/sq in). Nosewheel tyre size 500 × 180; pressure 5·88 bars (85·3 lb/sq in). Pneumatically-operated brakes on main wheels, with pneumatic emergency backup. Pneumatically-deployed brake parachute housed in bottom of rear fuselage above ventral strake. Small tail bumper.

Power Plant: Two Shenyang-built WP-6 developments of Tumansky (Mikulin bureau) R-9BF turbojet, each rated at 25·50 kN (5,732 lb st) dry and 31·88 kN (7,167 lb st) with afterburning. Hydraulically-actuated nozzles. Two main fuel tanks in tandem between cockpit and engines, and two smaller tanks under forward end of engine tailpipes, total capacity 2,155 litres (474 Imp gallons). Provision for two 800 litre (176 Imp gallon) or 1,520 litre (334 Imp gallon) underwing drop-tanks, raising max total fuel capacity to 3,755 litres (826 Imp gallons) and 5,195 litres (1,143 Imp gallons) respectively; provision on Pakistan Air Force J-6s for underfuselage tank.

Accommodation: Pilot only, on Martin-Baker PKD10 zero-zero ejection seat, under rearward-sliding blister canopy. In emergency, canopy is jettisoned by an explosive charge at the lock, after which it is carried away by the slipstream. Fluid anti-icing system for windscreen. Cockpit pressurised, heated and air-conditioned.

Systems: Cockpit pressurised by air-conditioning system mounted in top of fuselage aft of cockpit, using compressor bleed air. Constant temperature maintained by adjustable electric thermostat. Two independent hydraulic systems. Main system, powered by pump on starboard engine, actuates landing gear retraction and extension, flaps, spoilers, airbrakes and afterburner nozzle mechanism. System for tailplane and aileron boosters is powered by a pump on the port engine, and can also be supplied by the main system should the booster system fail. Electrical system powered by two DC starter/generators, supplemented by a battery, providing 27V DC, and 115V 400Hz and 36V 400Hz AC.

Avionics and Equipment: Standard avionics include VHF transceiver, blind-flying equipment, radio compass, radio altimeter, tail-warning system, navigation lights,

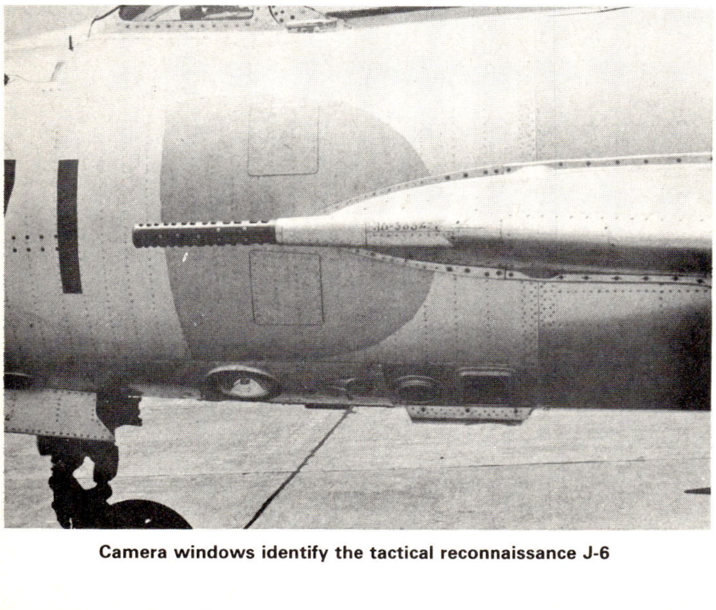

Camera windows identify the tactical reconnaissance J-6

This view of the underside of the Q-5 shows the fuselage weapon bay and external stores racks *(Camera Press)*

Production line of Q-5 'Fantan' tactical strike aircraft for the Chinese Air Force *(Liu Zhibin)*

taxying light on nosewheel leg, and landing light in bottom of front fuselage.

ARMAMENT: Installed armament of two or three 30 mm NR-30 belt-fed cannon, one in each wing root (deleted on FT-6) and (not on MiG-19PF) one under starboard side of nose. Aircraft supplied to Pakistan have an attachment under each wing for a Harbin-built Sidewinder air-to-air missile, outboard of drop-tank. More usual is the provision of one or two attachments inboard of each tank. Packs of eight air-to-air rockets can be carried on these inboard points, or on the drop-tank attachments. Alternative underwing loads can include four air-to-air guided weapons, two 250 kg (or 500 lb) bombs, or single rockets of up to 212 mm calibre. Optical gunsight. Gun camera in top lip of air intake of MiG-19SF; Izumrud airborne interception radar in centre of nose intake of MiG-19PF, with ranging unit in top lip of intake.

DIMENSIONS, EXTERNAL:
Wing span 9·20 m (30 ft 2¼ in)
Wing chord at root 3·73 m (12 ft 2¾ in)
Wing chord at tip 1·278 m (4 ft 2¼ in)
Wing chord, mean 3·02 m (9 ft 10¾ in)
Wing aspect ratio 3·24
Length overall (MiG-19SF):
incl nose probe 14·90 m (48 ft 10½ in)
excl nose probe 12·60 m (41 ft 4 in)
Length overall (FT-6):
excl nose probe approx 13·44 m (44 ft 1 in)
Length of fuselage 11·82 m (38 ft 9½ in)
Fuselage: Max diameter 1·45 m (4 ft 9 in)
Height overall 3·88 m (12 ft 8¾ in)
Tailplane span 5·00 m (16 ft 4¾ in)
Wheel track 4·15 m (13 ft 7½ in)

AREAS:
Wings, gross 25·00 m^2 (269 sq ft)
Ailerons (total) 1·56 m^2 (16·79 sq ft)
Trailing-edge flaps (total) 3·44 m^2 (37·03 sq ft)
Airbrakes (three, total) 1·50 m^2 (16·15 sq ft)
Ventral strake 0·614 m^2 (6·61 sq ft)
Fin, incl dorsal fin 4·62 m^2 (49·73 sq ft)
Rudder, incl tab 1·90 m^2 (20·45 sq ft)
Tailplane 4·62 m^2 (49·73 sq ft)

WEIGHTS AND LOADINGS:
Weight empty, nominal 5,760 kg (12,700 lb)
Normal T-O weight 7,600 kg (16,755 lb)
Max T-O weight 8,700 kg (19,180 lb)
Max wing loading 348 kg/m^2 (71·28 lb/sq ft)
Max power loading 136·4 kg/kN (1·34 lb/lb st)

PERFORMANCE:
Max level speed at 10,000 m (32,800 ft)
783 knots (1,452 km/h; 902 mph)
Cruising speed 512 knots (950 km/h; 590 mph)
Stalling speed, flaps up
189 knots (350 km/h; 218 mph)
Landing speed 127 knots (235 km/h; 146 mph)
Max rate of climb at S/L 6,900 m (22,635 ft)/min
Time to service ceiling 8 min 12 s
Service ceiling 17,900 m (58,725 ft)
Absolute ceiling 19,870 m (65,190 ft)
T-O run, with afterburning 515 m (1,690 ft)
T-O run, with underwing tanks, no afterburning
900 m (2,953 ft)
T-O to 25 m (82 ft), with afterburning
1,525 m (5,000 ft)
T-O to 25 m (82 ft), with underwing tanks, no afterburning 1,880 m (6,170 ft)
Landing from 25 m (82 ft), with brake-chute
1,700 m (5,580 ft)
Landing from 25 m (82 ft), without brake-chute
1,980 m (6,495 ft)
Landing run, with brake-chute 600 m (1,970 ft)
Landing run, without brake-chute 890 m (2,920 ft)
Combat radius with 800 litre external tanks
370 nm (685 km; 426 miles)
Normal range at 14,000 m (46,000 ft)
750 nm (1,390 km; 863 miles)
Max range with 1,520 litre external tanks
1,187 nm (2,200 km; 1,366 miles)
Max endurance at 14,000 m (46,000 ft) 2 h 38 min

NANZHANG Q-5 (A-5)

Chinese name: Type 5 Qianjiji (Type 5 Attack aircraft)
NATO reporting name: Fantan-A

This twin-engined fighter-bomber, derived from the J-6/MiG-19 produced in China, was referred to formerly, but incorrectly, by the Westernised designations F-9 and F-6bis. Its correct designation was indicated by Chinese officials in 1980 when discussing the aircraft with visiting members of the US aerospace industry. According to Chinese spokesmen, the design was then "about ten years old".

The airframe of the Q-5 is based substantially upon that of the J-6, but with overall dimensions increased to a modest extent. The wings are basically unchanged and retain the four external attachment points and large boundary layer fences of the J-6. More extensive changes had to be made to the centre and front fuselage. To make room for the internal weapons bay considered essential in an attack aircraft, it was decided to transfer various items of equipment from the centre fuselage to a 'solid' ogival nose, and to provide lateral air intakes for the twin engines, abreast of the single-seat cockpit; the shape of these increases the area-rule 'waisting' of the central portion of the fuselage. The cockpit canopy differs from that on the J-6 in being hinged at the rear to open upwards, with a much deeper spine fairing behind it. The centre and rear fuselage sections retain the various louvres and external airscoops of the J-6, supporting the belief that the latter's power plant of two side-by-side Shenyang-built Tumansky (Mikulin bureau) R-9BF turbojet engines (Chinese designation WP-6) remains basically unchanged.

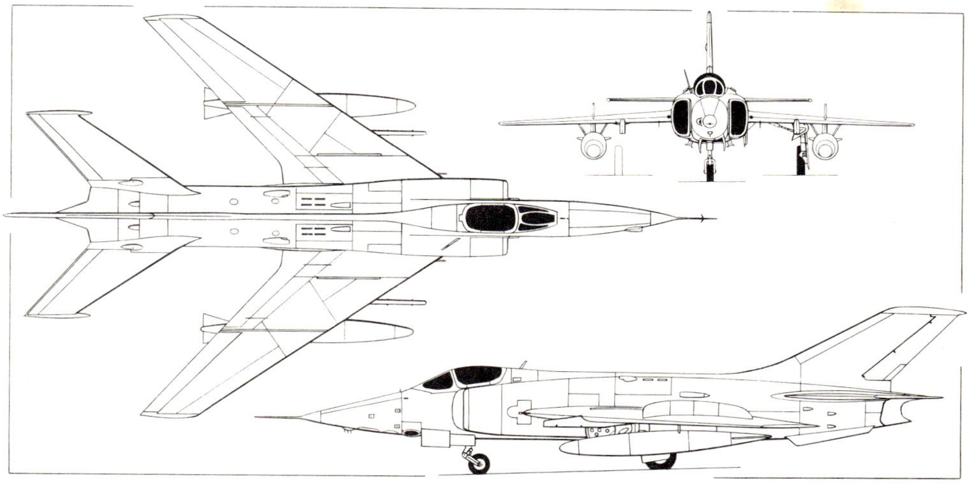

Q-5 'Fantan-A' single-seat twin-jet combat aircraft *(Pilot Press)*

J-7 Chinese-built counterparts of the MiG-21F ('Fishbed-C') *(Zhou Yi)*

One report has suggested that the engines in the Q-5 are of the R-9B-811 type, having a rating of 25·50 kN (5,732 lb st) dry and 36·77 kN (8,267 lb st) with afterburning. The taller main fin has a smaller dorsal fin than the J-6, and the strake fairing below the tail is shorter. Horizontal tail surfaces, including anti-flutter weights at their tips, appear to be similar to those of the J-6. As on the J-6, the nosewheel retracts forward, the main units inward into the wings, and there is a brake parachute housed in the tail-cone.

The Q-5 retains the two wing-mounted 30 mm cannon of the J-6, these now occupying the revised wing-root position outboard of the engine air intake trunks. Underwing stores load normally comprises two 800 litre (176 Imp gallon) drop-tanks on the outboard pylons and two seven-round pods of 57 mm air-to-surface rockets (S-5 type or similar) or two 250 kg bombs on the inboard pylons. The internal fuselage weapons bay, capable of housing four 250 kg bombs, is located aft of the airbrake. External stores racks are mounted to each side of this bay, for two more 250 kg bombs.

Details of the Q-5's avionics are necessarily speculative, but one report has suggested that equipment includes items based on, or similar to, the SRO-2 IFF (NATO 'Odd Rods'), ARK-5 radio compass, RV-UM low-altitude radio altimeter, MRP-48P marker beacon receiver, and RSIU-4 VHF com radio. There is a gun camera on the starboard side of the nose and a large landing light in the bottom of the front fuselage on the port side.

According to one report, at least 210 'Fantan-As' were in service with the Chinese Air Force by 1979, serving with tactical strike fighter squadrons of the People's Liberation Army. A 1980 report claimed that the aircraft had been built in "relatively large numbers". It is known to serve also in an air defence role with the Aviation of the People's Navy; and the Pakistan Air Force was reportedly receiving about 65 to re-equip three F-6 squadrons during 1980-81.

DIMENSIONS, EXTERNAL (estimated):
Wing span — 10·20 m (33 ft 5 in)
Length overall — 15·25 m (50 ft 0 in)

WEIGHTS (estimated):
Weight empty — 6,200 kg (13,670 lb)
Max T-O weight — 10,700 kg (23,590 lb)
Max landing weight — 9,200 kg (20,280 lb)

PERFORMANCE (estimated):
Max level speed at high altitude
Mach 1·35 (775 knots; 1,435 km/h; 890 mph)
Max level speed at low altitude
Mach 0·95 (625 knots; 1,160 km/h; 722 mph)
Max rate of climb at S/L — 6,000 m (19,685 ft)/min
Service ceiling — 16,000 m (52,500 ft)
T-O run — 620 m (2,035 ft)
T-O to 15 m (50 ft) — 920 m (3,020 ft)
Landing run — 900 m (2,950 ft)
Combat radius with four 250 kg bombs and two underwing drop-tanks:
lo-lo-lo — 200 nm (370 km; 230 miles)
hi-lo-hi — 350 nm (650 km; 400 miles)
Ferry range with two underwing drop-tanks
1,000 nm (1,850 km; 1,150 miles)

XIAN (MIKOYAN) J-7 (F-7)

Chinese name: Type 7 Jianjiji (Type 7 Fighter)
NATO reporting name: Fishbed-C

Design of this Chinese copy of the Mikoyan MiG-21 fighter was based on that of a number of Soviet-built MiG-21Fs ('Fishbed-Cs') delivered to China prior to the political break in 1960. The difficult task of copying the airframe, the R-11 afterburning turbojet (built at Shenyang) and equipment was completed so quickly that the J-7 made its first flight in December 1964 and began to enter service with the Chinese Air Force of the PLA in 1965.

In early 1975 General George S. Brown, USAF, told the Senate Armed Services Committee that China "has produced a number of MiG-21s, but for reasons which are not yet fully clear . . . production was suspended and only a small number of PRC (People's Republic of China) produced MiG-21s are operational with the PRC Air Force. The balance of the operational MiG-21s were Soviet-provided some years ago".

Chinese production of the MiG-21 was reported to have ended in 1966 after between 60 and 80 had been completed, but has been resumed more recently, perhaps with engine, cockpit and avionics changes. The J-6 is now the major type of fighter in production in China. A reconnaissance version is also said to have been developed. The MiG-21/J-7 is currently believed to equip two air regiments of the People's Liberation Army; it was reported in 1980 that deliveries to Egypt were expected to begin before the end of that year.

The MiG-21 is fully described and illustrated in the USSR section of this edition.

SHENYANG J-8 (F-8)

Chinese name: Type 8 Jianjiji (Type 8 Fighter)

The J-8 is of Chinese design, and incorporates technology gleaned from the Soviet MiG-23 variable-geometry combat aircraft, one or more examples of which were received from Egypt in 1976.A delegation from the US Department of Defense, headed by Under-Secretary for Research and Engineering William J. Perry and including US aerospace industry officials, examined the prototype during a visit to the Shenyang works in September 1980. It is described as a Mach 2 aircraft, powered by a Chinese-built Tumansky R-11 turbojet. This power plant is described as being less efficient than that of the MiG-23, giving rise to the suggestion that it may be replaced in production by the Chinese-built Rolls-Royce Spey turbofan; the nose has a ram-air intake, similar to that of the MiG-21F/J-7, with provision for a conical centrebody radome. Members of the delegation also reported that development was under way of a radar-guided air-to-air missile for carriage by the J-8 and that air testing had begun of a 30 mm anti-armour gun. The J-8 was still undergoing development in 1981.

J-12 (F-12)

Agency reports originating in Hong Kong in late 1977 referred to a new fighter allegedly designated F-12, said to be under development in China. Subsequent reports, which should be regarded as very provisional, suggest that it is intended to have a maximum speed of up to Mach 2·4 and to be able to operate from 600 m (1,970 ft) airstrips. Gross weight is thought to be in the order of 20,000 kg (44,100 lb), the power plant probably being a pair of Xian-built Rolls-Royce RB.168-25R Spey Mk 202M afterburning turbofan engines (each 54·5 kN; 12,250 lb st dry, 91·25 kN; 20,515 lb st with afterburning). The first Chinese-built Spey engine completed a 150 h acceptance test at Rolls-Royce Derby in the Spring of 1980.

HARBIN (ILYUSHIN) H-5 (B-5)

Chinese name: Type 5 Huongzhaji (Type 5 Bomber)
NATO reporting names: Beagle (Il-28) and Mascot (Il-28U)

Believed to equip about a dozen air regiments of the People's Liberation Army, with about 100 more in service with the Aviation of the People's Navy, the H-5 is the Chinese-built equivalent of the Soviet Ilyushin Il-28 three-seat tactical light bomber.

According to the US FY 1979 Military Posture statement, 400 Il-28s were then operational in China; the Military Posture statement for FY 1981 added that "a few Il-28 'Beagle' medium-range bombers may be configured for nuclear weapon delivery. Both the 'Badger' and the 'Beagle' are still in production. The limited range of the 'Beagle' makes it only marginally suitable for long-range operations". The latter statement was the first recent official indication that production of the Il-28 is still undertaken in China.

Designed in the USSR by the Ilyushin bureau, the Il-28 was developed to meet a 1946 requirement for which Sukhoi Su-10 and various Tupolev prototypes were also built. The clear leader among the competing designs was the Il-28, which was given a high development priority following a personal order from the Soviet leader, Josef Stalin, in the Spring of 1947.

Three Il-28 prototypes were completed, the first of these making its initial flight on 8 August 1948, powered by two 22·3 kN (5,004 lb st) RD-45 centrifugal-flow turbojet engines, a Soviet derivative of the Rolls-Royce Nene. State acceptance trials were completed in the Spring of 1949, and, again under direct order from Stalin, 25 Il-28s were available in time to take part in the 1950 May Day flypast. Most or all of these are believed to have been pre-production aircraft, possibly powered by RD-45FA engines, an improved version of the RD-45 rated at 26·9 kN (6,040 lb st).

V-VS (Soviet tactical aviation) units began working up with the Il-28 in the late Summer of 1950, although major deliveries did not begin until the following year. From then until about 1960, several thousand Il-28s were built, in a number of Soviet factories and also for a time in Czechoslovakia, where the aircraft was known by the Czechoslovak designation B-228. Production aircraft have VK-1A engines, an improved development of the RD-45 produced by the Klimov design bureau.

About 500 Il-28s were supplied to other Communist and Socialist states, including Afghanistan, Algeria, Bulgaria, Cuba, Czechoslovakia, Egypt, Finland, Germany (Democratic Republic), Hungary, Indonesia, Korea (People's Republic), Nigeria, Poland, Romania, Somalia, South Yemen, Syria, North Viet-Nam and Yemen Arab Republic. A similar number was supplied to the People's Republic of China, where the Il-28 entered licence production after the political break with the Soviet Union. Some Chinese-built Il-28s were exported to Albania. A dual-control version is also produced at Harbin.

Five main versions of the Il-28 received Soviet type designations, as follows:

Il-28. Standard three-seat tactical light bomber. Detailed description applies to this version except where indicated.

Il-28U (Uchebny: instruction). Two-seat operational and pilot training version, which appeared in 1951. NATO reporting name 'Mascot'. Armament and ventral ground-mapping radar fairing deleted; 'solid' nose; second, 'stepped' cockpit (with full dual controls) ahead of and below pilot's cockpit. Two or three supplied to each operational Il-28 unit.

Il-20. 'Demilitarised' version, with armament and some military equipment deleted, used by Aeroflot in the mid-1950s for the high-speed transportation of newspaper matrices. Cargo loaded via navigator's hatch in forward fuselage.

Il-28R (Razvedchik: reconnaissance). Three-seat tactical reconnaissance version. Wingtip auxiliary fuel tanks standard; weapons bay occupied by alternative packs containing cameras or electronic sensors. Some examples fitted with second radome under centre of fuselage.

Il-28T (Torpedonosets: torpedo carrier). Three-seat torpedo-bomber version for AV-MF (Soviet Naval Aviation), contemporary with Il-28R. Modified avionics; one large or two smaller torpedoes, or mines or depth charges, in weapons bay.

In addition to the above, many Il-28s when obsolescent in their original roles were converted for target towing (with hook attachment under tailcone), meteorological reconnaissance, and other second-line duties.

The description which follows applies primarily to the standard Il-28 bomber, but is modified where possible to apply to other models, including the Chinese-built H-5:

TYPE: Three-seat tactical light bomber.

WINGS: Cantilever shoulder-wing monoplane, with non-swept leading-edges and tapered trailing-edges. TsAGI SR-5S wing section, with max thickness/chord ratio of 12%. Incidence 0° 38′. Dihedral 3° from roots. Two-spar torsion-box structure, comprising integral centre-section (carrying straight through fuselage) and outer panels; built in upper and lower halves and riveted together. Construction mainly of D16-T duralumin, with skins varying from 2 to 4 mm in thickness. Hydraulically-actuated trailing-edge slotted flaps, inboard and outboard of each engine nacelle, with settings of 0°, 20° and 50°. Plain ailerons, which deflect 15° up and 20° down. Trim tab in each aileron. Hot-air de-icing of leading-edges.

FUSELAGE: Conventional all-metal semi-monocoque structure, of circular cross-section. Construction mainly of D16-T duralumin, with flush-riveted skins varying in thickness from 0·8 to 2·0 mm. Glazed nose, with optically-flat bomb-aiming panel. 'Solid' fairing aft of pilot's cockpit is of ATIM-X and ANZM magnesium alloys and incorporates a dielectric panel. Single ventral radome standard, forward of weapons bay; some aircraft have two such radomes, others none. Compartment in rear of fuselage for radio, batteries, air-conditioning and other equipment.

TAIL UNIT: All-swept cruciform structure, of similar construction to wings. Fin, on root platform built integrally with fuselage, has leading-edge sweep of 45°. Fixed-incidence tailplane has 33° sweepback on leading-edges and 7° dihedral. Trim tabs in rudder and each elevator.

Ilyushin Il-28 twin-jet light bomber (built in China as the H-5) in the insignia of the Egyptian Air Force *(Denis Hughes)*

Hot-air de-icing of fin and tailplane leading-edges; de-icing air vents in fin and tailplane tips.

LANDING GEAR: Hydraulically-actuated retractable tricycle type, with pneumatic emergency extension. Oleo-pneumatic shock-absorber on each unit. Twin-wheel nose unit, with shimmy damper, retracts rearward into fuselage. Single-wheel main units retract forward and upward into engine nacelles, the legs rotating through 90° during retraction to enable the wheels to lie flat behind doors which form a bulge under the nacelle. Main-wheel tyres size 1,150 × 355 mm, pressure 6·86 bars (99·5 lb/sq in); nosewheel tyres size 600 × 180 mm, pressure 4·41 bars (64 lb/sq in).

POWER PLANT: Two Klimov VK-1A non-afterburning turbojet engines, each rated at 26·5 kN (5,952 lb st) and mounted in an underwing pod. Fuel in five flexible fuselage tanks (three forward and two aft of weapons bay), integral wing tanks, and (standard on Il-28R, optional on other models) wingtip auxiliary tanks. Total fuel capacity, including tip-tanks, 7,908 litres (1,740 Imp gallons). Refuelling points in fuselage (four), wings and each tip-tank. Bifurcated intakes, each with central 'bullet' fairing to facilitate distribution of airflow. Provision for assisted take-off using JATO rocket under fuselage on each side.

ACCOMMODATION: Flight crew of three (instructor and pupil only in Il-28U), all in pressurised and air-conditioned accommodation. Pilot on ejection seat in single 'fighter' type cockpit, under jettisonable canopy which opens sideways to starboard. Navigator/bomb-aimer, also on ejection seat, occupies a position forward, below and to starboard of pilot, access to which is via an upward-opening jettisonable hatch above the nose and offset to starboard. (In Il-28U, roof hatch of forward cockpit hinges sideways to starboard.) Access to radio operator/rear gunner's position is via a power-operated downward-opening hatch in underside of rear fuselage, which also serves as escape hatch for this member of the crew. Dual controls in Il-28U.

SYSTEMS: Pressurisation system (max differential 0·4 bars; 5·8 lb/sq in) maintains accommodation at a 2,500 m (8,200 ft) environment at altitudes up to 5,000 m (16,400 ft) and at 4,250 m (13,940 ft) up to 12,000 m (39,370 ft). MWP hydraulic system, pressure 110 bars (1,595 lb/sq in), actuates flaps and landing gear via an AK-150 engine-driven compressor. Pneumatic system, max pressure 150 bars (2,175 lb/sq in), with operating pressures of 55 bars (800 lb/sq in) and 25 bars (362 lb/sq in), for emergency landing gear extension, sealing of weapons bay doors and crew hatches, and gun charging. Electrical system incorporates a 9 kW GSR-9000 starter/generator and two 28V 30Ah 12-A-30 batteries. PSR-1500-15 APU, rated at 14·7 kN (3,306 lb st) for 15 s, provides power for engine start via ST-2-48 starter motor in each air intake.

AVIONICS AND EQUIPMENT: HF and UHF radio (HF antenna on fairing aft of pilot's cockpit); radio compass; radio altimeter; IFF; PSB-N ground-mapping radar in underfuselage fairing forward of weapons bay; tail warning radar in fairing beneath tailcone; landing light in nosewheel leg door.

ARMAMENT AND OPERATIONAL EQUIPMENT: Two fixed, forward-firing Nudelman-Richter 23 mm NR-23 cannon (each with 100 rds) in lower forward fuselage, one each side of nosewheel bay; associated gyro gunsight in pilot's cockpit. Two similar guns, each with 225 rds, on Il-K6 ball-type movable mounting in tail turret. Internal weapons bay in mid-fuselage, with normal and max capacities of 1,000 kg (2,205 lb) and 3,000 kg (6,614 lb) respectively. Typical loads may include four 500 kg or eight 250 kg bombs or (Il-28T) one large or two smaller torpedoes, mines or depth charges. Some H-5s may be configured for nuclear weapon delivery. FAB-3000 primary bombsight is a modification of the US Norden M-9 of the second World War. Provision in standard Il-28 for single AFA 33/20, 33/75-50 or 33/100 vertical camera, installed beneath rearmost forward-fuselage fuel tank. Il-28R can carry from three to five cameras in the weapons bay, plus 12 to 18 flares or photoflash bombs.

Final assembly line of H-6 strategic bombers at Xian *(Liu Zhibin)*

DIMENSIONS, EXTERNAL:
- Wing span (excl tip-tanks) 21·45 m (70 ft 4½ in)
- Wing chord, mean 2·955 m (9 ft 8½ in)
- Wing area, gross 60·80 m² (654·45 sq ft)
- Wing aspect ratio 7·55
- Length of fuselage (excl tail guns) 17·65 m (57 ft 11 in)
- Fuselage: Max diameter 1·80 m (5 ft 10¾ in)
- Distance between c/l of engine nacelles 6·80 m (22 ft 3¾ in)
- Height overall 6·70 m (21 ft 11¾ in)
- Tailplane span 7·10 m (23 ft 3½ in)
- Wheel track 7·40 m (24 ft 3½ in)
- Wheelbase approx 8·10 m (26 ft 7 in)

WEIGHTS AND LOADINGS:
- Weight empty, equipped 12,890 kg (28,417 lb)
- Fuel load: normal 3,800 kg (8,377 lb)
  - max (incl 200 kg; 441 lb in tip-tanks) 6,600 kg (14,550 lb)
- Internal weapon load: normal 1,000 kg (2,205 lb)
  - max 3,000 kg (6,614 lb)
- Normal T-O weight 18,400 kg (40,565 lb)
- Max T-O weight 21,200 kg (46,738 lb)
- Wing loading: at normal T-O weight approx 303 kg/m² (62 lb/sq ft)
  - at max T-O weight approx 349 kg/m² (71·5 lb/sq ft)
- Power loading: at normal T-O weight approx 347·5 kg/kN (3·4 lb/lb st)
  - at max T-O weight approx 400 kg/kN (3·9 lb/lb st)

PERFORMANCE (at normal T-O weight except where indicated):
- Max level speed:
  - at S/L 432 knots (800 km/h; 497 mph)
  - at 1,750 m (5,740 ft) 473 knots (876 km/h; 544 mph)
  - at 4,500 m (14,760 ft) 487 knots (902 km/h; 560 mph)
  - at 8,000 m (26,250 ft) 473 knots (876 km/h; 544 mph)
  - at 10,000 m (32,800 ft) 461 knots (855 km/h; 531 mph)
  - at 12,000 m (39,370 ft) 434 knots (805 km/h; 500 mph)
- Typical cruising speed 415 knots (770 km/h; 478 mph)
- Unstick speed:
  - at normal T-O weight 119 knots (220 km/h; 137 mph)
  - at max T-O weight 126 knots (234 km/h; 145 mph)
- Touchdown speed 100 knots (185 km/h; 115 mph)
- Rate of climb:
  - max, at S/L 900 m (2,952 ft)/min
  - at 5,000 m (16,400 ft) 630 m (2,067 ft)/min
  - at 8,000 m (26,250 ft) 420 m (1,378 ft)/min
  - at 12,000 m (39,370 ft) 72 m (236 ft)/min
- Time to 5,000 m (16,400 ft) 6 min 30 s
- Time to 10,000 m (32,800 ft) 18 min 0 s
- Service ceiling 12,300 m (40,350 ft)
- T-O run: at normal T-O weight 875 m (2,870 ft)
  - at max T-O weight 1,150 m (3,773 ft)
- Landing run at landing weight of 14,690 kg (32,385 lb) 1,170 m (3,838 ft)
- Range with max fuel, at max T-O weight:
  - at 410 knots (760 km/h; 472 mph) at 1,000 m (3,280 ft) 612 nm (1,135 km; 705 miles)
  - at 415 knots (770 km/h; 478 mph) at 10,000 m (32,800 ft) 1,176 nm (2,180 km; 1,355 miles)
  - at 232 knots (430 km/h; 267 mph) at 10,000 m (32,800 ft) 1,295 nm (2,400 km; 1,490 miles)

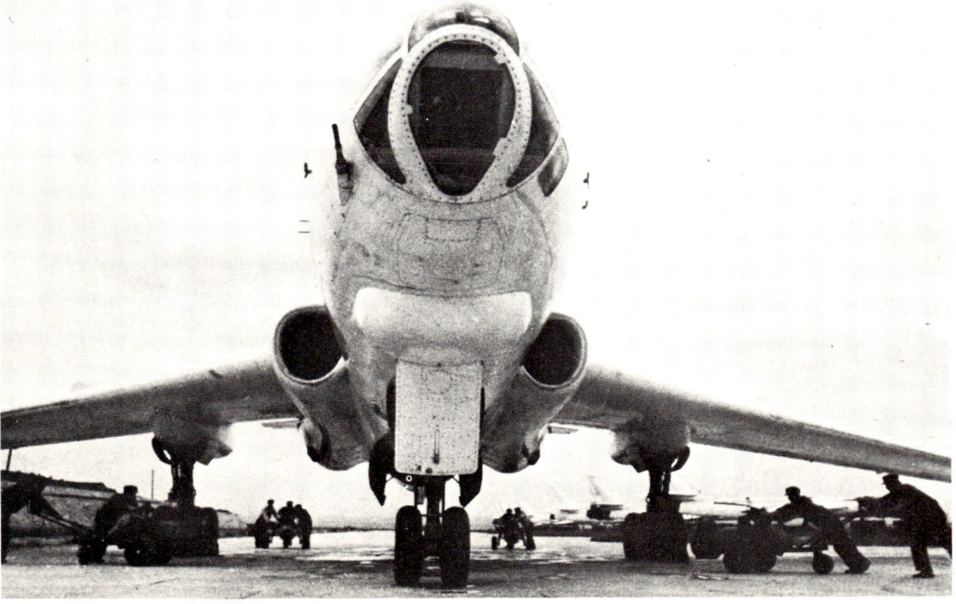

H-6 strategic bomber, based on the Tupolev Tu-16

XIAN (TUPOLEV) H-6 (B-6)

Chinese name: Type 6 Huongzhaji (Type 6 Bomber)
NATO reporting name: Badger

First steps to assemble the Tupolev Tu-16 bomber under licence in China were taken in 1958, but work was suspended in 1960 after the political break with the USSR. A production programme was reinstated some two years later, and the formidable task was undertaken of copying the design without Soviet assistance. Deliveries of the Chinese-built H-6 version did not begin until about 1968. Six of the 12 nuclear devices tested at Lop Nur (Lop Nor) up to 1971 were airdropped from Tu-16/H-6s, but production of this aircraft has been relatively slow. By the early 1970s, about 60 had been completed.

The US FY 1979 Military Posture statement suggested that "More than 80 of these Soviet-designed bombers are now operational. Although no radical changes to the 'Badger' basic design are expected, limited production of ECM, reconnaissance or tanker variants could be initiated at any time. An air-to-surface missile carrier still remains a

Nose of Harbin-built Z-5 helicopter, re-engined with PT6T-6 Turbo Twin Pac power plant

Harbin Y-5 equipped for agricultural spraying duty

possibility in the future." The FY 1981 statement confirmed that production of the Tu-16 in China was then continuing. China is also supplying spares for the Tu-16 bombers of the Egyptian Air Force.

HARBIN (MIL) Z-5 (H-5)

Chinese name: Type 5 Zhishengji
(Type 5 Vertical take-off aircraft)
NATO reporting name: Hound

The Z-5 is the Chinese-built military version of the Mil Mi-4 general-purpose helicopter, some 300-350 of which are estimated to be in service with the Air Force of the People's Liberation Army. A further 50 or so serve with the Aviation of the People's Navy for anti-submarine and search and rescue duties.

A civil version of the Mi-4/Z-5 has the Chinese name **Xuanfeng** (Syuan Fen) (Whirlwind). Combined military and civil production is believed to total about 1,000.

At least two other types of helicopter have been developed in China; one of these is an Mi-4/Z-5 converted to utilise a Pratt & Whitney Aircraft of Canada PT6T-6 Turbo Twin Pac turboshaft power plant and first flown in 1979 (see accompanying illustration). Approximately 600 Mi-4/Z-5s, in military and civil use, are available for similar conversion.

Z-6 (H-6)

At Shanghai in early 1980 the Chinese displayed a component said to belong to a helicopter having the Chinese designation H-6. Positive identification of this type is still awaited; the designation may refer to the PT6T-engined Z-5 mentioned in the preceding entry.

PRC (AÉROSPATIALE) SA 365N DAUPHIN 2

A licence agreement was signed on 2 July 1980 between Aérospatiale of France and the Chinese government for the former's Dauphin 2 twin-turboshaft helicopter (which see) to be manufactured in China. The agreement is for an initial batch of 50, most of which will be used for offshore oil exploration work.

HARBIN (BELL) MODELS 212 and 412

Eight Bell Model 212 helicopters were purchased by China in 1979, and photographs of these being assembled after arrival at Tianjin (Tientsin) Airport appeared in the Western press in early 1980. In the Spring of 1980, negotiations were under way with a view to a four-year programme in which the Harbin factory would assemble, and eventually manufacture, a further 50 Bell helicopters. Plans envisaged the assembly of 20 Model 212s from knocked-down components, after which Bell would provide assistance at Harbin during the manufacture of 30 Model 412s. Similar negotiations with Pratt & Whitney Aircraft of Canada, for China to produce the helicopters' PT6 turboshaft engines, were in hand at the same time.

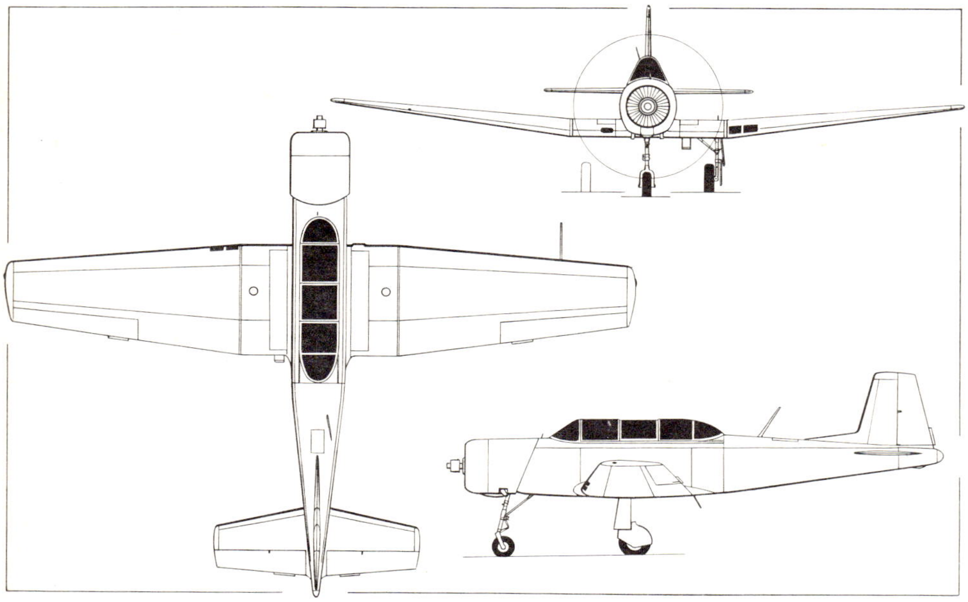

Shenyang CJ-6 tandem two-seat basic training aircraft *(Pilot Press)*

Shenyang CJ-6 basic trainer (Chinese-built Housai-6 engine)

SHENYANG (YAKOVLEV) CJ-5 (BT-5)

This designation refers to the Chinese licence-built version of the Soviet Yakovlev Yak-18A basic trainer. It is no longer in production in China; for further details, see the 1980-81 edition of *Jane's*.

SHENYANG CJ-6 (BT-6)

Chinese name: Type 6 Chuji Jiaolianji
(Type 6 Basic Trainer)

Despite a close similarity to the Yak-18A/BT-5 (see 1980-81 *Jane's*), the CJ-6 is an essentially indigenous design of the Chinese aerospace industry, and was the first such design to enter large-scale production.

Developed to replace the CJ-5 (BT-5) in the basic training role, the two-seat CJ-6 is powered by a 213 kW (285 hp) Housai-6 nine-cylinder aircooled radial engine, based on the Ivchenko AI-14RF and driving a J9-G1 two-blade propeller. It is reported to have been in production since 1961, with more than 2,000 delivered, including exports to Bangladesh, Korea, Viet-Nam and Zambia. As shown in the accompanying illustrations, features include a tricycle landing gear and large, square-cut vertical tail surfaces.

DIMENSIONS, EXTERNAL:	
Wing span	10·70 m (35 ft 1¼ in)
Length overall	8·40 m (27 ft 6¾ in)
Height overall	3·30 m (10 ft 10 in)
WEIGHTS:	
Weight empty	1,095 kg (2,415 lb)
Max fuel	110 kg (243 lb)
Max T-O weight	1,400 kg (3,088 lb)
PERFORMANCE:	
Max level speed	155 knots (286 km/h; 178 mph)
Landing speed	62 knots (115 km/h; 71·5 mph)
Max rate of climb at S/L	380 m (1,248 ft)/min
Service ceiling	5,080 m (16,680 ft)
T-O run	280 m (920 ft)
Landing run	350 m (1,150 ft)
Endurance	3 h 36 min

HARBIN (ANTONOV) Y-5 (C-5)

Chinese name: Type 5 Yunshuji
(Type 5 Transport aircraft)
NATO reporting name: Colt

The Antonov An-2 general-purpose biplane was supplied to, and built under licence in, China in considerable numbers (more than 1,000). It has been used extensively in a civil capacity for agricultural and general transport work, and several hundred still serve with the Chinese Air Force.

Chinese-produced examples have included a number of variants differing from the standard Soviet- or Polish-built production models, and some of them have reportedly been fitted with turboprop power plants.

SHANGHAI (?) Y-7 (C-7)

According to an unconfirmed press report, the Chinese designation Y-7 applies to a 50-passenger twin-turboprop transport aircraft which has been undergoing testing for about three years.

It is possible that this may be a Chinese-developed version of the Antonov An-24: Chinese sources have stated that local versions of both this and the An-12 four-turboprop military transport are under construction in the country.

SHANGHAI Y-10 (C-10)

Chinese name: Type 10 Yunshuji
(Type 10 Transport aircraft)

The first jet airliner of Chinese design and manufacture was quoted as "being readied for flight test" in mid-1980, and is believed to have flown for the first time in September 1980. This prototype is the second aircraft, the No.1 airframe having been used at Xian for structural testing. Components for a third aircraft were in assembly at the time of the first flight.

Although of similar configuration to the Boeing 707, the Y-10 is a smaller aircraft, seating approx 120 passengers in a five-abreast layout. Initial design began in about 1970, some two years before the first 707s were delivered to China, and was undertaken to demonstrate the Chinese industry's capability to develop an aircraft of this type. The prototype is powered by four Pratt & Whitney JT3D-7 turbojet engines, already available in China as 707 spares. The engine pods, which incorporate thrust reversers, and the underwing pylons, are of Chinese design; some parts of the aircraft are said to be of British origin. The Y-10 has an all-moving tailplane, and the landing gear is of welded steel tube construction.

Other details of the Y-10, according to US and British aerospace representatives who visited China in 1980, include a wing span of 42·00 m (137 ft 9½ in), length of 43·00 m (141 ft 1 in), height of 13·00 m (42 ft 7¾ in), payload of 17,000 kg (37,500 lb), max T-O weight of 108,000 kg (238,100 lb), cruising speed of Mach 0·88, and range of nearly 4,340 nm (8,045 km; 5,000 miles).

HARBIN Y-11 (C-11)

Chinese name: Type 11 Yunshuji (Type 11 Transport aircraft)

First details of this Chinese designed and developed utility aircraft were given on 26 August 1977, in the French/Chinese publication *Chine Nouvelle,* which stated that it was in production as a replacement for the Y-5 (Chinese-built Antonov An-2). First flight is believed to have taken place in 1975.

The Y-11 is intended for use primarily in agricultural and forestry applications, and in June/July 1977 underwent operational trials for crop-dusting and spraying missions. A small pre-production series of about 15 aircraft was built, for use in further top-dressing and pest control operations over 16,000 hectares (40,000 acres) of farmland and forest in 1977-78. Other applications include geophysical survey and short-haul transportation.

In overall size and general configuration the Y-11 resembles the Australian GAF Nomad. Its engines are fitted with louvred intakes to control cooling, and are probably related to the Soviet Vedeneev-developed Ivchenko AI-14RF.

It was reported in 1980 that China was negotiating for licence manufacture of Pratt & Whitney Aircraft of Canada PT6A-110 turboprop engines (354 kW; 475 shp) to power a new version of the Y-11, and an aircraft powered by two of four imported examples of this engine was flying by early 1981. Referred to by the Westernised designation **C-11T**, it was described as a 16-passenger (or 1,500 kg; 3,300 lb cargo) transport, suitable for utility, commuter and agricultural operations.

The following description applies to the standard piston-engined version:

TYPE: Twin-engined agricultural and general-purpose aircraft.

WINGS: Braced high-wing monoplane, with constant chord from root to tip. Wing section NACA 4412. No dihedral. Two-spar structure with aluminium alloy skin, bonded between spars, riveted elsewhere. All-metal drooping ailerons and electrically-actuated fabric-covered two-section double-slotted flaps along full span of trailing-edges. All-metal leading-edge automatic slats from nacelle to tip of each wing, with smaller inboard fixed slat on each side between nacelle and fuselage. Tab in each aileron. Small stub-wings at cabin floor level support the main landing gear units; bracing strut from each stub-wing out to approx mid-span.

FUSELAGE: Conventional semi-monocoque all-metal structure of basically rectangular cross-section, swept upward at rear.

TAIL UNIT: Cantilever non-swept metal structure, with low-set tailplane and small dorsal fin. Fabric-covered horn-balanced rudder and elevators. Inset tab in rudder and port elevator.

LANDING GEAR: Non-retractable tricycle type, with oleo-pneumatic shock-absorber in each unit. Twin-wheel main units, attached to underside of stub-wings. Single steerable nosewheel. Main-wheel tyres size 500 × 150 mm, pressure 2·90-3·45 bars (42-50 lb/sq in). Nosewheel tyre size 400 × 150 mm, pressure 2·90 bars (42 lb/sq in). Small bumper under tailcone. Pneumatic brakes.

POWER PLANT: Two 213 kW (285 hp) Housai 6-A nine-cylinder radial aircooled engines, each driving a two-blade variable-pitch propeller, underslung from wings and fitted with louvred intakes in front of cylinders to control cooling. Two metal fuel tanks between spars of each outer wing, with smaller tank in each engine nacelle. Total fuel capacity 530 litres (116 Imp gallons).

ACCOMMODATION: Crew of two on flight deck, with separate forward-opening door(s) for access. Dual controls. Cabin accommodates seven passengers normally (with removable folding jump-seat for an eighth passenger), or equivalent cargo. Cargo/passenger double door on port side of fuselage, in line with wing trailing-edge. Underside of rear fuselage, aft of this door, probably lets down to act as a loading ramp for bulky cargo.

Y-11 utility aircraft under construction at Harbin

SYSTEMS: Pneumatic system for engine starting and wheel brakes, supplied by pump driven by each engine and backup air cylinder. Electrical system includes 1·5 kW generator on each engine and 30Ah storage battery.

AVIONICS AND EQUIPMENT: Radio; operational equipment according to mission. Agricultural version has hopper with capacity of 855 kg (1,885 lb) or 975 litres (214 Imp gallons), and six rotary atomisers for spraying. Electromagnetic survey version has magnetometer boom projecting 1·30 m (4 ft 3¼ in) beyond the tail and a camera installation in the cabin floor.

DIMENSIONS, EXTERNAL:
- Wing span: 17·00 m (55 ft 9¼ in)
- Wing aspect ratio: 8·5
- Length overall: 12·017 m (39 ft 5⅛ in)
- Height overall: 4·64 m (15 ft 2¾ in)
- Fuselage depth: 1·73 m (5 ft 8¼ in)
- Fuselage width: 1·47 m (4 ft 10 in)
- Wheel track (c/l of shock-absorbers): approx 3·00 m (9 ft 10 in)
- Wheelbase: approx 3·70 m (12 ft 1¾ in)
- Propeller diameter: 2·40 m (7 ft 10½ in)

DIMENSIONS, INTERNAL:
- Cabin: Length: 3·58 m (11 ft 9 in)
  - Max width: 1·37 m (4 ft 6 in)
  - Max height: 1·53 m (5 ft 0¼ in)

AREA:
- Wings, gross: 34·00 m² (365·97 sq ft)

WEIGHTS:
- Weight empty: 2,050 kg (4,519 lb)
- Max fuel load: 390 kg (860 lb)
- Max payload: 870 kg (1,918 lb)
- Normal T-O weight: 3,250 kg (7,165 lb)
- Max T-O weight: 3,500 kg (7,715 lb)

PERFORMANCE (at max T-O weight):
- Max level speed: 119 knots (220 km/h; 137 mph)
- Cruising speed, 57% power: 89 knots (165 km/h; 102 mph)
- Stalling speed, flaps up: 57 knots (105 km/h; 65 mph)
- Max rate of climb at S/L: 252 m (825 ft)/min
- Service ceiling: 4,200 m (13,775 ft)
- T-O and landing run: 150 m (492 ft)
- Range with max fuel, 45 min reserves: 431 nm (800 km; 497 miles)
- Range with max payload and 250 kg (551 lb) fuel, 45 min reserves: 215 nm (400 km; 248 miles)

Y-11 (C-11) twin-engined general-purpose aircraft, designed and built in China

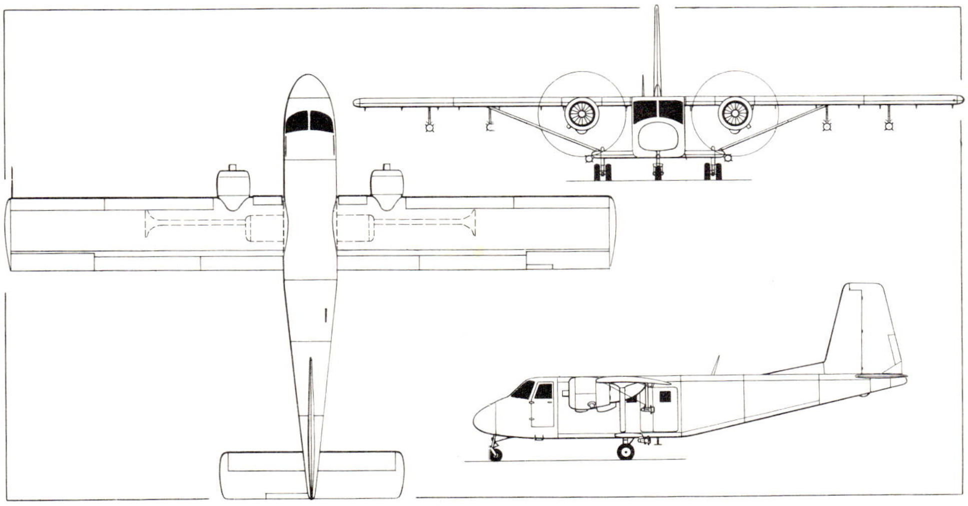

Harbin Y-11 (C-11) twin-engined agricultural and utility aircraft *(Pilot Press)*

COLOMBIA

AGRICOPTEROS

AGRICOPTEROS LTDA

Apartado Aéreo 1789, Cali

President: Eng Maximo Tedesco

This company, which undertakes crop-spraying operations in Colombia, is assembling kits of a modified agricultural version of the Aerosport Scamp (see under USA in the Homebuilt Aircraft section).

AGRICOPTEROS (AEROSPORT) SCAMP MODEL B

Following a visit to the USA in 1976 the President of Agricopteros Ltda ordered two kits of the Aerosport Scamp homebuilt biplane, with a view to the possible adaptation of this aircraft for agricultural duties. In collaboration with Aerosport, he introduced a number of modifications into the new Scamp B version, and the first example made its initial flight on 27 May 1977.

In this basic aircraft, the major changes included a slight increase in wing span, the provision of ailerons on the lower as well as the upper wings, and a different power plant. No agricultural equipment was installed at that time; this was designed subsequently by Sr Tedesco, and consists primarily of an underfuselage glassfibre chemical tank, wind-driven pressure pump, spraybars, and nozzles. The entire installation weighs only 17 kg (38 lb). A T-type hydraulic valve system forces the ultra-low-volume liquid chemical into the spraybars at a pressure of 5·52 bars (80 lb/sq in), and it is discharged through the nozzles at 4·14 bars (60 lb/sq in). The tank is fitted with a quick-release trapdoor, and the entire chemical load can be jettisoned in 2 s in an emergency. If the aircraft is required for cross-country journeys, the tank can be used to carry auxiliary fuel instead of chemical.

The Scamp B entered production by Agricopteros by the assembly of kits supplied by Aerosport. No recent news of the programme has been received.

Type: Single-seat Restricted category (export) agricultural light aircraft.

Wings: Braced biplane structure, with V-type interplane strut each side. Flying and landing wires of streamline section. Single 5 × 12·5 cm (2 × 5 in) extruded section of 6063-T3 light alloy tubing forms a pylon to support the centre-section of the upper wing. Wing section NACA 23012. Dihedral 3° on lower wings only. All-metal light alloy tubular front and main spars. Light alloy plain ailerons, with piano hinge at upper surface, on upper and lower wings. Lower ailerons actuated by slaved push/pull tube connected to upper pair. No flaps or tabs.

Fuselage: All-metal light alloy semi-monocoque structure.

Tail Unit: Braced T tail of light alloy construction. Single bracing strut each side. Fixed-incidence tailplane. Ground-adjustable trim tab on rudder.

Landing Gear: Non-retractable tricycle type. Cantilever spring main-gear struts of light alloy. Wheel fairing optional for each unit.

Power Plant: One 74·5 kW (100 hp) Revmaster 2,100 cc modified Volkswagen engine, driving an Aerial 56-38 two-blade fixed-pitch wooden propeller. Fuel tank in fuselage nose, aft of firewall, capacity 30·5 litres (6·7 Imp gallons). Refuelling point on fuselage upper surface, forward of windscreen.

Accommodation: Single seat in open cockpit.

Equipment: Underfuselage tank for ultra-low-volume chemical or auxiliary fuel, capacity 60 litres (13·2 Imp gallons).

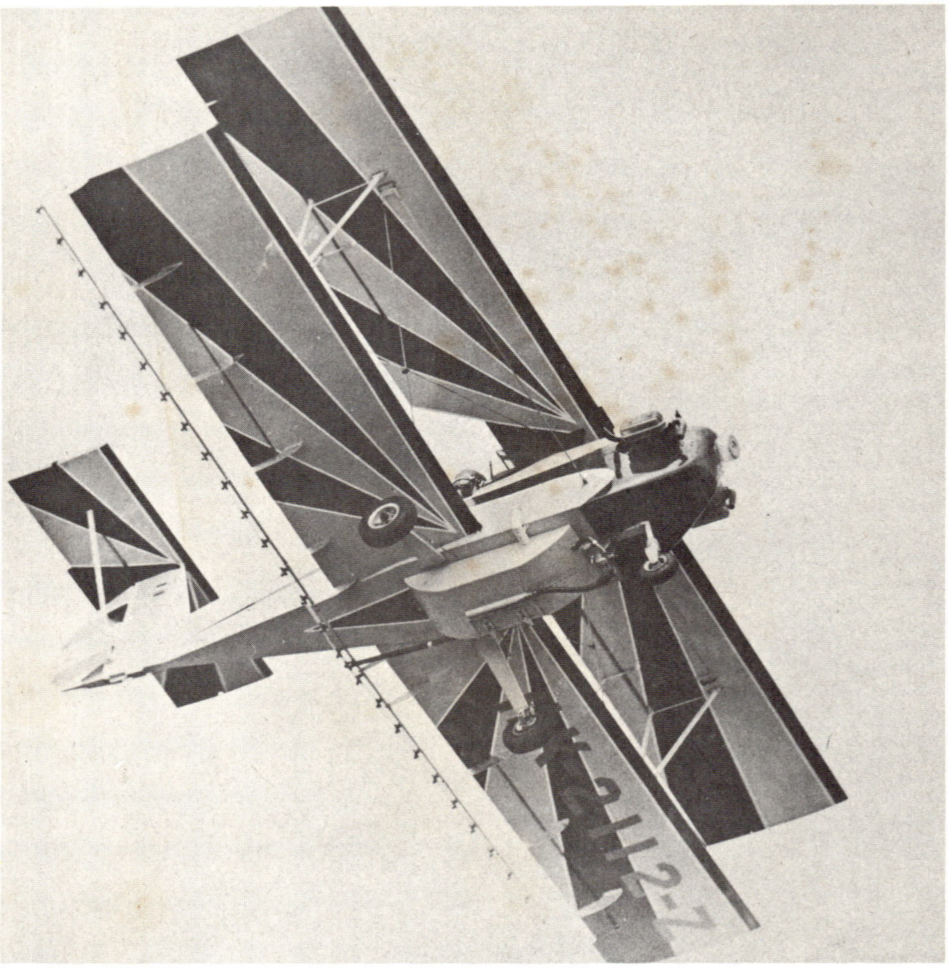

Agricopteros-built Aerosport Scamp B, with belly tank and spraybars

Dimensions, external:

Wing span	5·94 m (19 ft 6 in)
Wing chord, constant	0·91 m (3 ft 0 in)
Wing area, gross	10·82 m² (116·5 sq ft)
Length overall	4·37 m (14 ft 4 in)
Height overall	1·73 m (5 ft 8 in)
Tailplane span	1·98 m (6 ft 6 in)
Wheel track	1·52 m (5 ft 0 in)
Wheelbase	1·22 m (4 ft 0 in)
Propeller diameter	1·42 m (4 ft 8 in)

Weights:

Weight empty	259 kg (572 lb)
Normal T-O weight	360 kg (795 lb)
Max T-O weight with chemical	428 kg (945 lb)

Performance:

Never-exceed speed	100 knots (185 km/h; 115 mph)
Max level speed	82 knots (153 km/h; 95 mph)
Cruising speed	76 knots (140 km/h; 87 mph)
Max manoeuvring speed	72 knots (134 km/h; 83 mph)
Stalling speed	43·5 knots (81 km/h; 50 mph)
Service ceiling: tested	2,590 m (8,500 ft)
estimated	3,810 m (12,500 ft)
T-O run	122 m (400 ft)
Landing run	152 m (500 ft)
Range at cruising speed	130 nm (241 km; 150 miles)
Max range with auxiliary fuel in underfuselage tank	477 nm (885 km; 550 miles)
Max endurance with auxiliary fuel in underfuselage tank	6 h 45 min
Swath width	8·5 m (28 ft)

AVIONES DE COLOMBIA

AVIONES DE COLOMBIA SA

Head Office: Calle 26 No. 4A-45 Piso 8, Bogotá

Telephone: (282) 9648, 9668 and 9728

Telex: 45 220

Works: Aeropuerto Guaymaral, Apartado Aéreo 6876, Bogotá

Telephone: (254) 1515 and 8831

Sales Manager: Rafael Urdaneta

This company, established in the 1950s and known formerly as Urdaneta y Galvez Ltda, has been a South American distributor for Cessna aircraft since 1961. In 1969 it began assembling and partly building selected Cessna types under licence. Sixty-five aircraft were so produced in 1973, and a further 93 in 1974, some 40 per cent of these being Cessna Model A188B Ag Wagons (see US section). Production in 1974 was limited to wings, tail units and seats for the Cessna range, but welding and other techniques were learnt, and the company is now qualified to manufacture complete airframes. Facilities include 464·5 m² (5,000 sq ft) of office space in Bogotá, and 13,935 m² (150,000 sq ft) at Guaymaral, the general aviation airport for Bogotá. Service facilities include complete engine, propeller and avionics workshops. A new 1,350 m² (14,530 sq ft) paint shop was being added in 1980.

By 31 January 1980 the company had a work force of 250 persons, and had assembled a total of 668 Cessna aircraft including 50 Model 150s, 48 Model 172s, six Model 177RGs and 70 Model 182s. Of the types continuing in production, Aviones de Colombia had completed at that date 19 Model 152s, 31 Model 172XPs, four Model 182RGs, 24 Model 185s, 241 Model 188s, 117 Model 206s, 30 Model 210s, seven Model 310s, nine Model 337s, five Model 340s and seven Model 402s. No later details have been received for publication. About 20 per cent of total production is exported to Bolivia, Ecuador and Peru.

First prototype of the Aviones de Colombia/Cessna AgTrainer, a modified Cessna 188 Ag Truck

AVIONES DE COLOMBIA/CESSNA AGTRAINER

Illustrated in an accompanying photograph, the AgTrainer is modified by Aviones de Colombia from the Cessna Model 188 Ag Truck. The cabin has been widened to accommodate two persons side by side, increasing the empty weight by approx 91 kg (200 lb). Flight characteristics remain unchanged. Two prototypes have flown (the first of them on 16 September 1976), and by 1 January 1979 these two aircraft had accumulated nearly 2,000 flying hours. The first prototype has been operated by Aeroandes, a local crop-spraying flying school.

It was hoped that production of the AgTrainer would begin in 1980.

CZECHOSLOVAKIA

Central direction of the Czechoslovak aircraft industry is by a body known as the Generlní Reditelstvi Aero—Ceskoslovenské Letecke Podniky (Trust Aero—Czechoslovak Aeronautical Works, Prague-Letnany), whose General Manager is Josef Skarohlid.

About 29,000 people are employed by the Czechoslovak aircraft industry. Principal factories concerned with aircraft manufacture are the Aero Vodochody National Corporation, Let National Corporation and Zlin Aircraft-Moravan National Corporation, whose current products appear under the appropriate headings in this section. Other Czechoslovak factories engaged in the production of aero-engines and sailplanes are listed in the relevant sections of this edition.

Sales of all aircraft products outside Czechoslovakia are handled by the Omnipol Foreign Trade Corporation, whose address is given below.

OMNIPOL
FOREIGN TRADE CORPORATION

Nekázanka 11, 112 21 Prague 1
Telephone: 268261/8 and 263071/7
Telex: 121299 and 121297
General Manager: Ing Tomás Marecek, GE
Sales Manager: Ing Ludvik Skocdopole
Publicity Manager: Jirí Matula

This concern handles the sales of products of the Czechoslovak aircraft industry outside Czechoslovakia and furnishes all information requested by customers with regard to export goods.

AERO
AERO VODOCHODY NÁRODNÍ PODNIK (Aero Vodochody National Corporation)

Vodochody, p. Odelená Voda, near Prague
Managing Director: Jirí Chmelícek
Vice-Directors:
Ing Josef Sedlácek (Technical)
Jan Spára (Production)
Ing Otakar Stella (Sales)
Ing Václav Klouda (Works Economy)
Chief Designer: Dipl Ing Jan Vlcek
Project Engineer, L-39: Ing Vlastimil Havelka
Chief Pilot: Antonin Saller

This factory perpetuates the name of one of the three founder companies of the Czechoslovak aircraft industry, which began activities shortly after the first World War with the manufacture of Austrian Phönix fighters. Subsequent well-known products included the A 11 military general-purpose biplane and its derivatives, and licence manufacture of the French Bloch 200 twin-engined bomber. The present works was established on 1 July 1953, since when it has seven times received the Red Banner award of the Ministry of Engineering and UVOS, as well as many other awards including those of Exemplary Exporting Corporation and the Order of Labour.

Aero's major product from 1963-74 was the L-29 Delfin jet basic and advanced trainer, of which approx 3,600 were built. A full description of this can be found in the 1974-75 edition of *Jane's*. It has been superseded in production by the L-39, a description of which follows:

Aero L-39C version of the Albatros, in Czechoslovak insignia

AERO L-39 ALBATROS

The L-39 basic and advanced jet trainer was developed in the Aero works at Vodochody by a team led by the chief designer, Dipl Ing Jan Vlcek, working in close cooperation with the USSR. Two prototype airframes were built initially, of which the first was used for structural testing. The first flight, on 4 November 1968, was made by the second aircraft. By the end of 1970, five flying prototypes and two for ground testing had been completed. Slightly larger and longer air intake trunks were fitted after preliminary flight tests.

A pre-production batch of ten aircraft began to join the flight test programme in 1971, and series production started in late 1972, following official selection of the L-39 to succeed the L-29 Delfin (1974-75 *Jane's*) as the standard jet trainer for the air forces of the Soviet Union, Czechoslovakia and the German Democratic Republic. Service acceptance trials, in Czechoslovakia and the USSR, took place in 1973, and by the Spring of 1974 the L-39 had begun to enter service with the Czechoslovak Air Force.

The L-39 forms part of a comprehensive training system which includes a specially designed pilot training flight simulator (TL-39), a pilot ejection ground training simulator (NK-TL-29/39), and vehicle-mounted mobile automatic test equipment (AKZ-KL-39). The aircraft is capable of operation from grass strips (with a bearing strength of 6 kg/cm²; 85 lb/sq in) at up to 4,600 kg (10,141 lb) T-O weight, or from unprepared runways.

By May 1977, when the L-39 made its first appearance in the West, at the Paris Air Show, some 400-500 were in service with several air forces. Production had totalled almost 1,000 by early 1981. The Albatros is used in Czechoslovakia for all pilot training, including that of helicopter pilots. On average, pupils solo after approx 12 hours' dual instruction on the L-39.

Official Czechoslovak designations for the different L-39 versions are as follows:

L-39C (C for Cvicný: training). Basic and advanced flying training version, to which the detailed description chiefly applies. In service with the air forces of Afghanistan, Czechoslovakia, Germany (Democratic Republic) and USSR. In production.

L-39Z0 (Z for Zbrojní: armed). Jet trainer with four underwing weapon stations and reinforced wings. Export customers include the air forces of Iraq and Libya. In production.

L-39Z (Z for Zbrojní: armed). Weapon system training, ground attack and reconnaissance version, with underfuselage gun and four underwing weapon stations; reinforced wings and landing gear. In production.

The following description applies to the current production L-39C trainer version, except where indicated:

Type: Two-seat basic and advanced jet trainer (L-39C, L-39Z0 and L-39Z); L-39Z also has light attack and reconnaissance capability.

Wings: Cantilever low-wing monoplane, with 2° 30′ dihedral from roots. Wing section NACA 64A012 mod. 5. Incidence 2°. Sweepback 6° 26′ on leading-edges, 1° 45′ at quarter-chord. One-piece all-metal stressed-skin structure, with main spar and auxiliary spar; four-point attachment to fuselage. All-metal double-slotted trailing-edge flaps, operated by push/pull rods actuated by a single hydraulic jack. Flaps retract automatically when airspeed reaches 167 knots (310 km/h; 193 mph). Small fence above and below each trailing-edge between flap and aileron. Electrically-operated servo tab in each aileron; port tab, used also for trim, is operated by electromechanical actuator. Flaps deflect 25° for take-off, 44° for landing; ailerons deflect 16° up or down; airbrakes deflect 55° downward. Non-jettisonable wingtip fuel tanks, incorporating landing/taxying lights.

Fuselage: Metal semi-monocoque structure, built in two portions. Front portion consists of three sections, the first of which is a laminated glassfibre nosecone housing avionics, antennae, battery, compressed air and oxygen bottles and the nose landing gear. Next comes the pressurised compartment for the crew. The third section incorporates the fuel tanks, air intakes and the engine bay. The rear fuselage, carrying the tail unit, is attached by five bolts and can be removed quickly to provide access for engine installation and removal. Two airbrakes side by side under fuselage, just forward of wing leading-edge, actuated by single hydraulic jack; these are lowered automatically as airspeed nears a maximum of Mach 0·8.

Tail Unit: Conventional all-metal cantilever structure, with sweepback on vertical surfaces. Variable-incidence tailplane. Control surfaces actuated by pushrods. Electrically-operated trim tab in each elevator; servo tab in rudder. Elevators deflect 30° up, 20° down; rudder 30° to right and left.

Landing Gear: Retractable tricycle type, with single wheel and oleo-pneumatic shock-absorber on each unit. Gear is designed for a touchdown sink rate of 3·4 m (11·15 ft)/s at AUW of 4,600 kg (10,141 lb). Retraction/extension is operated hydraulically, with electrical control. All wheel-well doors close automatically after wheels are lowered, to prevent ingress of dirt and debris. Main wheels retract inward into wings (with automatic braking during retraction), nosewheel forward into fuselage. K24 main wheels, fitted with Barum tubeless tyres size 610 × 215 mm (610 × 185 mm on early production aircraft), pressure 5·88 bars (85·34 lb/sq in). K25 castoring and self-centering nosewheel, fitted with Barum tubeless tyre size 450 × 165 mm (430 × 150 mm on early production aircraft), pressure 3·92 bars (56·89 lb/sq in). Hydraulic disc brakes and anti-skid units on main wheels; shimmy damper on nosewheel leg. Landing gear of L-39Z reinforced to cater for higher operating weights.

Power Plant: One 16·87 kN (3,792 lb st) Ivchenko AI-25 TL turbofan engine mounted in rear fuselage, with semi-circular lateral air intake, fitted with splitter plate, on each side of fuselage above wing centre-section. Fuel

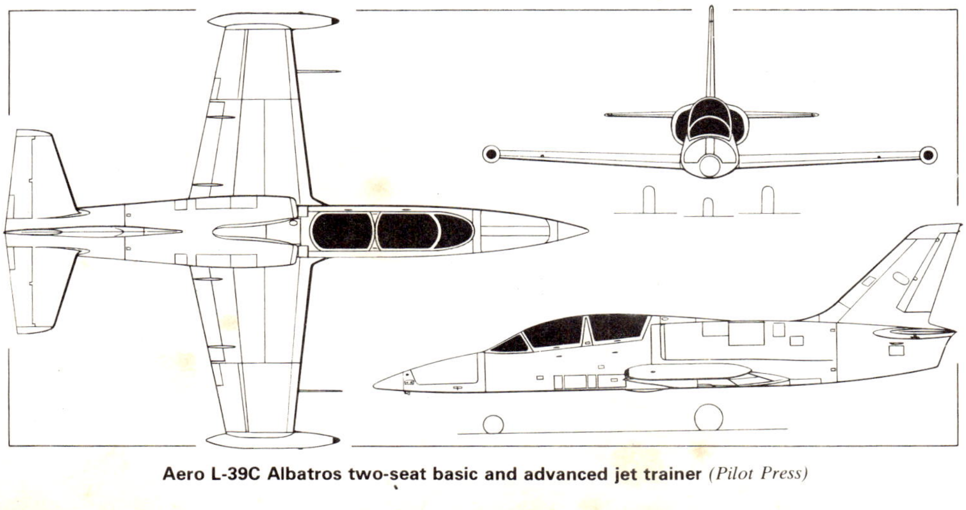

Aero L-39C Albatros two-seat basic and advanced jet trainer *(Pilot Press)*

Close-up of underfuselage gun installation on Aero L-39Z

in five rubber bag-type main tanks aft of cockpit, with combined capacity of 1,055 litres (232 Imp gallons), and two 100 litre (22 Imp gallon) non-jettisonable wingtip tanks. Total internal fuel capacity 1,255 litres (276 Imp gallons). Gravity refuelling points on top of fuselage and on each tip-tank. Provision for two 150 or 350 litre (33 or 77 Imp gallon) underwing drop-tanks on inboard underwing pylons, increasing total overall fuel capacity to a maximum of 1,955 litres (430 Imp gallons). Fuel system permits up to 20 s of inverted flight.

Accommodation: Crew of two in tandem, on VS-1-BRI rocket-assisted ejection seats, operable at zero height and at speeds down to 81 knots (150 km/h; 94 mph), beneath individual transparent canopies which hinge sideways to starboard and are jettisonable. One-piece windscreen hinges forward to provide access to front instrument panel. Internal transparency between front and rear cockpits. Dual controls standard.

Systems: Cabin pressurised (max differential 0·23 bars; 3·34 lb/sq in) and air-conditioned, using engine bleed air and cooling unit. Air-conditioning system provides automatic temperature control from 10° to 25°C at ambient air temperatures from −55°C to +45°C. Main and standby interconnected hydraulic systems, the main system having a variable-flow pump with an operating pressure of 147 bars (2,133 lb/sq in) for actuation of landing gear, flaps, airbrakes, ram-air turbine and wheel brakes. Emergency system, for all of above except airbrakes, incorporates three accumulators. Pneumatic canopy seals supplied by one 2 litre compressed air bottle in nose (pressure 147 bars; 2,133 lb/sq in). Electrical system (28·5V DC) is powered by a 9kW VG 7500 Ja engine-driven generator. If primary generator fails, a V 910 ram-air turbine is extended automatically into the airstream and generates up to 3kW of emergency power for essential services. 12V 28Ah SAM 28 lead-acid battery for standby power and for APU starting. Two 800VA static inverters (the first for radio equipment, ice warning lights, engine vibration measurement and air-conditioning, the second for navigation and landing systems, IFF and air-to-air missiles) provide 115V single-phase AC power at 400Hz. A second circuit incorporates a 500VA rotary inverter and 40VA static inverter to provide 36V three-phase AC power, also at 400Hz. Saphir 5 APU and SV-25 turbine for engine starting. Air intakes and windscreen anti-iced by engine bleed air; normally, anti-icing is sensor-activated automatically, but a manual standby system is also provided. Six-bottle oxygen system for crew, pressure 147 bars (2,133 lb/sq in).

Avionics and Equipment: Standard avionics include R-832 M two-band radio (VHF 118-140MHz, UHF 220-389MHz); SPU-9 crew intercom; RKL-41 ADF (150-1,800kHz); RV-5 radio altimeter; MRP-56 P/S marker beacon receiver; SRO-2 IFF; and RSBN-5S navigation and landing system. VOR/ILS system available at customer's option. Landing and taxying light in forward end of each tip-tank.

Armament (L-39Z): Underfuselage pod, below front cockpit, housing a single 23 mm Soviet GSh-23 two-barrelled cannon; ammunition for this gun (max 150 rds) is housed in fuselage, above gun pod. Gun/rocket firing and weapon release controls, including electrically-controlled ASP-3 NMU-39 Z gyroscopic gunsight and FKP-2-2 gun camera, in front cockpit only. Four underwing hardpoints, the inboard pair each stressed for loads of up to 500 kg (1,102 lb) and the outer pair for loads of up to 250 kg (551 lb) each; max external underwing stores load 1,100 kg (2,425 lb). Non-jettisonable pylons, each comprising a D3-57D stores rack. Typical underwing stores can include various combinations of bombs up to 500 kg in size; four UB-16-57 M pods each containing sixteen S-5 57 mm air-to-surface rockets; infra-red air-to-air missiles (outer pylons only); a five-camera day reconnaissance pod (port inboard pylon only); or (on inboard stations only) two 150 or 350 litre (33 or 77 Imp gallon) drop-tanks.

Dimensions, external:
Wing span 9·46 m (31 ft 0½ in)
Wing chord (mean) 2·15 m (7 ft 0½ in)
Wing aspect ratio (geometric) 4·4
Length overall 12·13 m (39 ft 9½ in)
Height overall 4·77 m (15 ft 7¾ in)
Tailplane span 4·40 m (14 ft 5 in)
Wheel track 2·44 m (8 ft 0 in)
Wheelbase 4·39 m (14 ft 4¾ in)

Areas:
Wings, gross 18·80 m² (202·36 sq ft)
Ailerons (total) 1·23 m² (13·26 sq ft)
Trailing-edge flaps (total) 2·68 m² (28·89 sq ft)
Airbrakes (total) 0·50 m² (5·38 sq ft)
Vertical tail surfaces (total) 3·51 m² (37·78 sq ft)
Tailplane 3·93 m² (42·30 sq ft)
Elevators, incl tabs 1·14 m² (12·27 sq ft)

Weights and Loadings (A1: two-seat, aircraft 'clean'; A2: two-seat, gun only; F1: single-seat, gun plus one camera pod and one 350 litre drop-tank; G1: single-seat, gun only; G2: single-seat, gun plus two 150 litre drop-tanks; G3: single-seat, gun plus two 350 litre drop-tanks; P1: single-seat, gun plus four 250 kg bombs; P2: single-seat, gun plus two 500 kg bombs; P3: single-seat, gun plus six 100 kg bombs; P4: single-seat, gun plus four 100 kg bombs; P5: single-seat, gun plus eight 50 kg bombs; P6: single-seat, gun plus two 250 kg bombs and two 150 litre drop-tanks; P7: single-seat, gun plus two 100 kg bombs and two 350 litre drop-tanks; PR1: single-seat, gun plus two rocket launchers and two 250 kg bombs; R1: single-seat, gun plus four rocket launchers; R2: single-seat, gun plus two rocket launchers and two 150 litre drop-tanks; R3: single-seat, gun plus two rocket launchers and two 350 litre drop-tanks; R4: single-seat, gun plus two missiles and two 350 litre drop-tanks; T1: two-seat, gun plus two 100 kg bombs; T2: two-seat, gun plus two rocket launchers; T3: two-seat, gun plus two missiles):
Weight empty: single-seat 3,490 kg (7,694 lb)
two-seat 3,565 kg (7,859 lb)
Crew: single-seat 80 kg (176 lb)
two-seat 160 kg (352 lb)
Fuel load: A1, A2, P1, P2 824 kg (1,816 lb)
F1 1,252 kg (2,760 lb)
G1, P3, P4, P5, PR1, R1, T1, T2, T3 980 kg (2,160 lb)
G2, P6, R2 1,214 kg (2,676 lb)
G3, P7, R3, R4 1,524 kg (3,360 lb)
Max underwing weapon load 1,100 kg (2,425 lb)
Mission T-O weight: A1 4,549 kg (10,028 lb)
A2 4,682 kg (10,322 lb)
F1 5,138 kg (11,327 lb)
G1 4,720 kg (10,405 lb)
G2 5,063 kg (11,162 lb)
G3 5,396 kg (11,896 lb)
P1, P3 5,602 kg (12,350 lb)
P2 5,596 kg (12,337 lb)
P4 5,306 kg (11,697 lb)
P5 5,458 kg (12,032 lb)
P6 5,539 kg (12,211 lb)
P7 5,646 kg (12,447 lb)
PR1 5,510 kg (12,147 lb)
R1 5,262 kg (11,600 lb)
R2 5,291 kg (11,664 lb)
R3 5,624 kg (12,398 lb)
R4 5,588 kg (12,319 lb)
T1 5,131 kg (11,312 lb)
T2 5,109 kg (11,263 lb)
T3 5,029 kg (11,087 lb)
Max landing weight: L-39C 4,300 kg (9,480 lb)
L-39Z 4,600 kg (10,141 lb)
Wing loading 'clean' 243 kg/m² (49·77 lb/sq ft)
Power loading 'clean' 270·9 kg/kN (2·65 lb/lb st)

General Performance (A: at 'clean' AUW of 4,570 kg; 10,075 lb with tip-tanks empty; B: at AUW of 5,270 kg; 11,618 lb with full internal fuel and four underwing rocket pods; C: at AUW of 4,300 kg; 9,480 lb, except where indicated):
Max limiting Mach number (Vne) above 1,300 m (4,265 ft) 0·85
Max permitted diving speed (Vd)
491 knots (910 km/h; 565 mph)
Max level speed at S/L:
C 378 knots (700 km/h; 435 mph)
Max level speed at altitude:
A at 6,000 m (19,685 ft)
421 knots (780 km/h; 485 mph)
B at 6,000 m (19,685 ft)
340 knots (630 km/h; 391 mph)
C at 5,000 m (16,400 ft)
405 knots (750 km/h; 466 mph)
Cruising speed at 5,000 m (16,400 ft):
C 367 knots (680 km/h; 423 mph)
Stalling speed:
C, flaps up 97 knots (180 km/h; 112 mph)
C, 25° flap 90 knots (165 km/h; 103 mph)
C, 44° flap 84 knots (155 km/h; 97 mph)
Touchdown speed:
A at 4,500 kg (9,920 lb) AUW
94·5 knots (175 km/h; 109 mph)
B at 4,600 kg (10,141 lb) AUW
98 knots (182 km/h; 113 mph)
Max rate of climb at S/L: A 1,320 m (4,330 ft)/min
B 960 m (3,150 ft)/min
Optimum climbing speed:
C 210 knots (390 km/h; 242 mph)
Service ceiling: A 11,500 m (37,730 ft)
B 9,000 m (29,525 ft)
T-O run, 25° flap, ISA:
at 4,300 kg (9,480 lb) AUW, concrete
480 m (1,575 ft)
at 4,300 kg (9,480 lb) AUW, grass 630 m (2,067 ft)
at 4,570 kg (10,075 lb) AUW 500 m (1,640 ft)
at 5,270 kg (11,618 lb) AUW 800 m (2,625 ft)
T-O to 25 m (82 ft):
at 4,300 kg (9,480 lb) AUW 630 m (2,067 ft)
Landing from 25 m (82 ft):
at 4,300 kg (9,480 lb) AUW 1,120 m (3,675 ft)
Landing run, 44° flap, ISA:
at 4,100 kg (9,039 lb) AUW 620 m (2,034 ft)
at 4,300 kg (9,480 lb) AUW 690 m (2,264 ft)
g limits:
at 4,200 kg (9,259 lb) AUW +8; −4
at 4,400 kg (9,700 lb) AUW +7·5; −3·75
at 4,600 kg (10,141 lb) AUW +7; −3·5
at 5,500 kg (12,125 lb) AUW +5·2; −2·6

Typical Mission Performance (at mission weights as defined under 'Weights and Loadings' heading):
Max permissible Mach No: A1, A2 0·80
G3, P7, R1 0·75
Max level speed at S/L:
A1 378 knots (700 km/h; 435 mph)
A2 375 knots (695 km/h; 432 mph)
G3 351 knots (650 km/h; 404 mph)
P7 329 knots (610 km/h; 379 mph)
R1 318 knots (590 km/h; 366 mph)
Max level speed at 3,000 m (9,845 ft):
A1 399 knots (740 km/h; 460 mph)
A2 396 knots (735 km/h; 457 mph)
G3 372 knots (690 km/h; 429 mph)
P7 340 knots (630 km/h; 391 mph)
R1 332 knots (615 km/h; 382 mph)
Max level speed at 5,000 m (16,400 ft):
A1 405 knots (750 km/h; 466 mph)
A2 402 knots (745 km/h; 463 mph)
G3 372 knots (690 km/h; 429 mph)
P7 340 knots (630 km/h; 391 mph)
R1 326 knots (605 km/h; 376 mph)
Max rate of climb at S/L:
A1 1,230 m (4,035 ft)/min
A2 1,140 m (3,740 ft)/min
G3 960 m (3,150 ft)/min
P7 780 m (2,560 ft)/min
R1 870 m (2,854 ft)/min
Max rate of climb at 5,000 m (16,400 ft):
A1 690 m (2,264 ft)/min
A2 600 m (1,968 ft)/min
G3 450 m (1,476 ft)/min
P7 270 m (886 ft)/min
R1 360 m (1,181 ft)/min
Time to 5,000 m (16,400 ft): A1 5 min 30 s
A2 6 min 12 s
G3 8 min 30 s
P7 11 min 0 s
R1 9 min 30 s
Time to 8,000 m (26,250 ft): A1 11 min 0 s
A2 13 min 18 s
G3 20 min 0 s
Service ceiling: A1 11,000 m (36,100 ft)
A2 10,200 m (33,460 ft)
G3 9,000 m (29,525 ft)
P7 7,500 m (24,600 ft)
R1 8,400 m (27,560 ft)
T-O run (concrete): A1 530 m (1,739 ft)
A2 660 m (2,165 ft)
G3 950 m (3,117 ft)
P7 1,050 m (3,445 ft)
R1 880 m (2,887 ft)
Landing run (concrete): A1, A2 660 m (2,165 ft)
P7, R1 750 m (2,460 ft)
Radius of action, lo-lo-lo at 500 m (1,640 ft), last 27 nm (50 km; 31 miles) at max rpm for 3 min, 5% reserves:
G3 286 nm (530 km; 329 miles)
P7 270 nm (500 km; 310 miles)
R1 148 nm (275 km; 170 miles)
Radius of action, hi-lo-hi at 5,000/500/5,000 m (16,400/1,640/16,400 ft), conditions as above:
G3 410 nm (760 km; 472 miles)
P7 323 nm (600 km; 373 miles)
R1 167 nm (310 km; 192 miles)
Max range, 5% reserves:
A1 593 nm (1,100 km; 683 miles)
A2 566 nm (1,050 km; 652 miles)
G3 944 nm (1,750 km; 1,087 miles)
P7 658 nm (1,220 km; 758 miles)
R1 453 nm (840 km; 522 miles)
Max endurance at econ speed, 5% reserves:
A1, R1 2 h 0 min
A2 1 h 54 min
G3 3 h 36 min
P7 3 h 0 min

LET

LET NÁRODNÍ PODNIK (Let National Corporation)

Uherské Hradiste-Kunovice
Telephone: Uherské Hradiste 5121
Telex: 060180 and 060181
MANAGING DIRECTOR: Ing Stanislav Boura
CHIEF DESIGNER: Ing Ladislav Smrcek
CHIEF PILOT: Frantisek Srnec

The Let plant at Kunovice was established in 1950, its early activities including licence production of the Soviet Yak-11 piston-engined trainer under the Czechoslovak designation C-11. It contributed to the production of the Aero 45, was responsible for the L 200 Morava twin-engined air taxi and Z-37 Cmelák agricultural aircraft, and the L 13 Blanik sailplane; it is currently responsible for development and manufacture of the L-410 twin-turboprop light transport aircraft.

The factory also produces apparatus and equipment for radar and computer technology.

LET L-410UVP TURBOLET

Design of the L-410 twin-turboprop light transport was started in 1966, by a team led by Ing Ladislav Smrcek. The first prototype (OK-YKE), powered by Pratt & Whitney Aircraft of Canada PT6A-27 turboprop engines, flew for the first time on 16 April 1969. Three additional PT6A-engined prototypes were completed subsequently; the second of these was later test-flown with Hartzell four-blade propellers in a successful demonstration of reduced aircraft vibration and cabin noise levels.

Details of the earlier L-410A, L-410AF and L-410M versions can be found in the 1980-81 and earlier editions of *Jane's*. The current version is:

L-410UVP. Standard production version from beginning of 1979; first of three prototypes flown on 1 November 1977. Changes include increased wing span and area; fuselage lengthened by 0·47 m (1 ft 6½ in) compared with L-410M; enlarged vertical tail surfaces; dihedral tailplane; improved cockpit systems and additions to standard instrumentation; introduction of spoilers, automatic bank control flaps, automatic propeller feathering, and anti-skid system for main landing gear units; fabric-covered elevators and rudder; and later-model M 601 engines and V 508 propellers. Basic version is for passenger transportation, but cabin can be converted easily to all-cargo, aeromedical, parachutist or firefighting configuration; aircraft can also be equipped for aerial photography or calibration of ground navigation aids. The L-410UVP can operate from grass, sand and gravel strips as well as from paved runways, and in snow and ice conditions. It is being manufactured in large numbers, and following its certification in 1980 will become standard Aeroflot equipment on Soviet internal feederline services, eventually in company with the PZL Mielec (Antonov) An-28. Stringent Aeroflot requirements included the ability to operate in temperatures ranging from −50°C to +45°C; systems were required to be survivable in temperatures as low as −60°C.

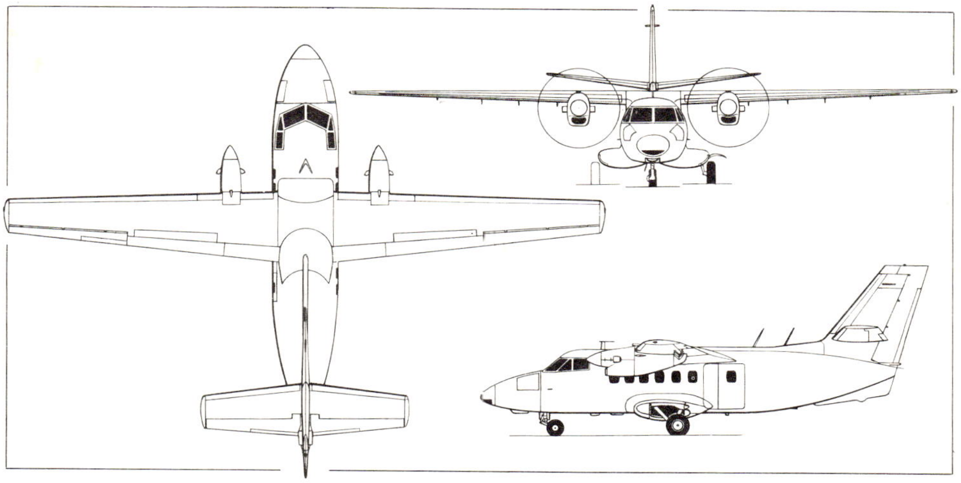

Let L-410UVP Turbolet twin-turboprop 15-passenger light transport *(Pilot Press)*

Production of the L-410UVP is planned to reach about 100 a year, most of them for Aeroflot. By 20 January 1980, 45 production L-410UVPs had been completed. Three specially-equipped aircraft have been ordered by the East German airline Interflug for air photography and environmental studies.

TYPE: Twin-turboprop general-purpose light transport.

WINGS: Cantilever high-wing monoplane. Wing section NACA 63A418 at root, NACA 63A412 at tip. Dihedral 1° 45′. Incidence 2° at root, −0° 30′ at tip. No sweepback at front spar. Conventional all-metal two-spar torsion-box structure, attached to fuselage by four-point mountings. Chemically machined skin with longitudinal reinforcement. Hydraulically actuated double-slotted metal flaps, with both slots variable. Spoiler forward of each flap. All-metal ailerons, forward of which are 'pop-up' bank control surfaces that come into operation automatically during single-engine operation and decrease the lift on the side of the running engine. Kléber-Colombes pneumatic de-icing of leading-edges.

FUSELAGE: Conventional all-metal semi-monocoque spot welded and riveted structure, built in three main portions.

TAIL UNIT: Conventional cantilever structure, of all-metal construction except for elevators and rudder, which are fabric-covered. Vertical tail surfaces swept back 35°; shallow dorsal fin and deeper ventral fin. One-piece tailplane, with 7° dihedral from roots, mounted part-way up fin. Balance tab in rudder and each elevator. Kléber-Colombes pneumatic de-icing of leading-edges.

LANDING GEAR: Retractable tricycle type, with single wheel on each unit. Hydraulic retraction, nosewheel forward, main wheels inward to lie flat in fairing on each side of fuselage. Technometra Radotin oleo-pneumatic shock-absorbers. Non-braking nosewheel, with servo-assisted steering, fitted with 548 × 221 mm (9·00-6) tubeless tyre, pressure 2·74 bars (39·8 lb/sq in). Nose-wheel is also steerable by rudder pedals. Main wheels fitted with 718 × 306 mm (12·50-10) tubeless tyres, pressure 3·14 bars (45·5 lb/sq in). All wheels manufactured by Moravan Otrokovice, tyres by Rudy Rijen, Gottwaldow. Moravan Otrokovice hydraulic disc brakes, parking brake and anti-skid units on main wheels. Metal ski landing gear, with plastics under-surface, optional.

POWER PLANT: Two 544 kW (730 ehp) Walter M 601 B turboprop engines, each driving an Avia V 508 B three-blade reversible-pitch fully-feathering metal propeller. At higher ambient temperatures, engine power can be increased to 590 kW (790 ehp) for short periods by water injection into compressor. De-icing for propeller blades (electrical) and lower intakes; anti-icing flaps inside each nacelle. Eight bag-type fuel tanks in wings, total capacity 1,290 litres (284 Imp gallons).Total oil capacity (incl oil in cooler) 22 litres (4·8 Imp gallons). Water tank capacity (for injection into compressor) 11 litres (2·4 Imp gallons).

ACCOMMODATION: Crew of one or two on flight deck. Dual controls standard. Standard accommodation in main cabin for 15 passengers, with pairs of adjustable seats on starboard side of aisle and single seats opposite, all at 76 cm (30 in) pitch. Baggage compartment (at rear, accessible from cabin), toilet and wardrobe standard in this version. Cabin heated by engine bleed air. Alternative layouts include all-cargo; ambulance, accommodating

L-410UVP Turbolet general-purpose transport aircraft, in service with Aeroflot, at Nizhneudinsk Airport *(Tass)*

Let L-410UVP camouflaged for military service *(Letectvi + Kosmonautika)*

six stretchers, five sitting patients and a medical attendant; accommodation for 14 parachutists and a despatcher/instructor; firefighting configuration, carrying 12 firefighters and a pilot/observer. All-cargo version has protective floor covering, crash nets on each side of cabin, and tiedown provisions; floor is at truck-bed height. Aircraft can also be equipped for aerial photography or for calibration of ground navigation aids. Double upward-opening doors aft on port side, with stowable steps; right hand door serves as passenger entrance and exit. Both doors open for cargo loading, and can be removed for paratroop training missions. Rearward-opening door, at front on starboard side, serves as emergency exit.

Systems: No APU, air-conditioning or pressurisation systems. Duplicated hydraulic systems, No. 1 system actuating landing gear, flaps, spoilers, automatic pitch trim surfaces, main-wheel brakes, nosewheel steering and windscreen wipers. No. 2 system for emergency landing gear extension, flap actuation and parking brake. Electrical system includes AC power from three three-phase 36V 400Hz rotary inverters and two single-phase 115V 400Hz inverters, guaranteeing against a loss of power for essential instruments; DC power from two 5·6kW generators and two 25Ah batteries.

Avionics and Equipment: Standard instrumentation provides for flight in IMC conditions, with all basic instruments duplicated and three artificial horizons. Communications include two VHF with a range of 65 nm (120 km; 75 miles) at 1,000 m (3,280 ft) altitude, and crew intercom. Standard navigation instruments include artificial horizons (three); barometric altimeters, airspeed indicators, rate of climb indicators, turn indicators, RMIs, gyro-compasses, ILS, and ARK-15M ADF with range of 97 nm (180 km; 112 miles) at 1,000 m (3,280 ft) altitude (two of each); and radio altimeter with ground proximity warning, ASI with stall warning, magnetic compass, GMK-1GE VOR, and ILS with marker beacon receiver (one of each). Cockpit, instrument and passenger cabin lights, navigation lights, three landing lights in nose (each with two levels of light intensity), crew and cabin fire extinguishers, windscreen wipers, and alcohol spray for windscreen and wiper de-icing, are also standard. Flight data recorder, cockpit voice recorder, SSR repeater and encoding altimeter, transponder, and electrically heated flight deck windows, are all optional.

Dimensions, external:
- Wing span: 19·478 m (63 ft 10¾ in)
- Wing chord at root: 2·534 m (8 ft 3¾ in)
- Length overall: 14·467 m (47 ft 5½ in)
- Fuselage: Max width: 2·08 m (6 ft 10 in)
  - Max depth: 2·10 m (6 ft 10¾ in)
- Height overall: 5·829 m (19 ft 1½ in)
- Tailplane span: 6·736 m (22 ft 1¼ in)
- Wheel track: 3·65 m (11 ft 11½ in)
- Wheelbase: 3·666 m (12 ft 0¼ in)
- Propeller diameter: 2·50 m (8 ft 2½ in)
- Distance between propeller centres: 4·816 m (15 ft 9½ in)
- Passenger/cargo door (port, aft):
  - Height: 1·46 m (4 ft 9½ in)
  - Width overall: 1·25 m (4 ft 1¼ in)
  - Width (passenger door only): 0·80 m (2 ft 7½ in)
  - Height to sill: 0·79 m (2 ft 7 in)
- Emergency exit door (stbd, fwd):
  - Height: 0·97 m (3 ft 2¼ in)
  - Width: 0·66 m (2 ft 2 in)
  - Height to sill: 0·80 m (2 ft 7½ in)

Dimensions, internal:
- Cabin, exl flight deck: Length: 6·34 m (20 ft 9½ in)
  - Max width: 1·95 m (6 ft 4¾ in)
  - Max height: 1·658 m (5 ft 5¼ in)
  - Aisle width at 0·4 m (1 ft 3¾ in) above cabin floor: 0·34 m (1 ft 1½ in)
  - Floor area: 9·69 m^2 (104·3 sq ft)
  - Volume: 17·86 m^3 (630·7 cu ft)
- Baggage compartment volume (rear): 0·77 m^3 (27·2 cu ft)

Areas:
- Wings, gross: 35·18 m^2 (378·67 sq ft)
- Ailerons (total): 2·89 m^2 (31·11 sq ft)
- Automatic bank control flaps (total): 0·49 m^2 (5·27 sq ft)
- Trailing-edge flaps (total): 5·92 m^2 (63·72 sq ft)
- Spoilers (total): 0·87 m^2 (9·36 sq ft)
- Fin: 4·49 m^2 (48·33 sq ft)
- Rudder, incl tab: 2·81 m^2 (30·25 sq ft)
- Tailplane: 6·41 m^2 (69·00 sq ft)
- Elevators, incl tabs: 3·15 m^2 (33·91 sq ft)

Weights:
- Basic empty weight: 3,725 kg (8,212 lb)
- Max fuel: 1,000 kg (2,205 lb)
- Max payload: 1,310 kg (2,888 lb)
- Max T-O weight: 5,700 kg (12,566 lb)
- Max landing weight: 5,500 kg (12,125 lb)
- Max zero-fuel weight: 5,170 kg (11,398 lb)

Performance (at max T-O weight, ISA, except where indicated):
- Never-exceed speed: 194 knots (360 km/h; 224 mph) EAS
- Max cruising speed: 197 knots (365 km/h; 227 mph)
- Econ cruising speed: 162 knots (300 km/h; 186 mph)
- Stalling speed, flaps up: 78 knots (144 km/h; 90 mph) EAS
- Stalling speed, flaps down, at max landing weight: 61 knots (112 km/h; 70 mph) EAS
- Max rate of climb at S/L: 468 m (1,535 ft)/min
- Rate of climb at S/L, one engine out: 90 m (295 ft)/min
- Max operating altitude: 6,000 m (19,700 ft)
- Service ceiling, one engine out: 2,850 m (9,350 ft)
- T-O run: 400 m (1,312 ft)
- T-O to 10·5 m (35 ft): 520 m (1,706 ft)
- Landing from 9 m (30 ft) at max landing weight: 810 m (2,657 ft)
- Landing run at max landing weight: 328 m (1,000 ft)
- Range at 3,000 m (9,850 ft) with max fuel and 850 kg (1,874 lb) payload, 30 min reserves: 561 nm (1,040 km; 646 miles)
- Range with max payload and 505 kg (1,113 lb) fuel, 30 min reserves: 248 nm (460 km; 285 miles)

VZLU
VYZKUMNY A ZKUSEBNÍ LETECKY USTAV (Aeronautical Research and Test Institute)

Beranovych 130, Prague 9-Letnany 19905
Telephone: Prague 827041 and 826541
Telex: Prague 1493
Managing Director: Ing Josef Kurz
Public Relations: Jirí Kucera

This Institute, whose title is self-explanatory, was founded in 1922 and undertakes a range of activities corresponding broadly to those carried out by the RAE in Britain. Details of its principal facilities appeared in the 1970-71 and 1972-73 *Jane's*. It is a member of the Czechoslovak aircraft manufacturing group, under the general management of Aero (which see).

ZLIN
MORAVAN NÁRODNÍ PODNIK (Zlin Aircraft Moravan National Corporation)

Otrokovice 76581
Telephone: Gottwaldov 92 2041-44
Telex: Gottwaldov 067 334
Managing Director: Frantisek Klapil
Vice-Directors:
Ing Jan Bartoň (Technical)
Jan Munclinger (Production)
Frantisek Muzny (Sales)
Ing Adolf Dolezal (Works Economy)
Chief Designer: Ing Jirí Navrátil
Chief Pilot: Zdenek Polásek

The Moravan works, responsible for production of the famous range of Zlin aerobatic and light touring aircraft, was formed originally on 8 July 1935 as Zlinská Letecká Akciová Spolecnost (Zlin Aviation Joint Stock Co) in Zlin, although manufacture of Zlin aircraft was actually started two years earlier by the Masarykova Letecká Liga (Masaryk League of Aviation). The factory was renamed Moravan after the second World War. At present, in addition to production of the Zlin 142 and Zlin 50 L, Moravan is building items of aircraft equipment.

Zlin 142 two-seat aerobatic, training and touring aircraft

ZLIN 42 M

The initial version of the Zlin 42 entered production in 1971 and was described in the 1973-74 *Jane's*.

The **Zlin 42 M** entered production in 1974, and by early 1981 total Zlin 42 production (all versions) was in excess of 235. A number of early Zlin 42s have during overhaul been fitted retrospectively with Avia V 503 A two-blade metal propellers of the type fitted to the Z 42 M. The aircraft so modified are then redesignated **Zlin 42 MU**.

A detailed description of the Z 42 M can be found in the 1980-81 *Jane's*. An improved version, the Zlin 142, is now in production, as described in the entry which follows.

ZLIN 142

The Zlin 142 is intended for basic and advanced flying training, aerobatic flying and the training of aerobatic pilots, glider towing, and (when equipped with appropriate instrumentation) for night and IFR flying training. It is a progressive development of the Zlin 42 M (see 1980-81 *Jane's*), from which it differs mainly in having a more powerful engine, improved cockpit design, increased useful load and higher max T-O weights. Design began in the Winter of 1977-78, and the prototype (OK-078) flew for the first time on 29 December 1978. In 1980 it received FAR Pt 23 certification in the Aerobatic, Utility and Normal categories, and production began during that year. A total of 30 had been completed by early 1981.

Type: Two-seat fully-aerobatic (A), light training (U) and touring (N) aircraft.

Wings: Cantilever low-wing monoplane. Wing section NACA $63_2$416·5. Dihedral 6° from roots. Sweep-forward 4° 20′ at quarter-chord. All-metal structure with single main spar and auxiliary spar; skins (fluted on control surfaces) of aluminium-plated duralumin sheet. All-metal slotted ailerons and flaps all have same dimensions. Mass-balanced flaps and ailerons, operated mechanically by control rods. Ground-adjustable tab on each aileron.

Fuselage: Engine cowlings of sheet metal. Centre-fuselage of welded steel tube truss construction, covered with laminated glassfibre panels. Rear fuselage is all-metal semi-monocoque structure.

Tail Unit: Cantilever all-metal structure with skins (fluted on control surfaces) of duralumin sheet. Control surfaces have partial mass and aerodynamic balance. Trim tabs on elevator and rudder. Rudder actuated by control cables, elevator by control rods.

Landing Gear: Non-retractable tricycle type, with nosewheel offset to port. Oleo-pneumatic nosewheel shock-absorber. Main wheels carried on flat spring steel legs. Nosewheel steering by means of rudder pedals. Main wheels and Barum tyres size 420 × 150, pressure 1·90 bars (27·6 lb/sq in); nosewheel and Barum tyre size 350 × 135, pressure 2·50 bars (36·3 lb/sq in). Hydraulic disc brakes on main wheels can be operated from either seat. Parking brake standard.

Power Plant: One 156·5 kW (210 hp) Avia M 337 AK inverted six-cylinder aircooled in-line engine, with compressor and low-pressure injection pump, driving a two-blade Avia V 500 A constant-speed metal propeller. Fuel tanks in each wing leading-edge, with combined capacity of 125 litres (27·5 Imp gallons). Normal category version has auxiliary 50 litre (11 Imp gallon) tank at each wingtip, increasing total fuel capacity to 225 litres (49·5 Imp gallons). Fuel and oil systems permit inverted flying for up to 3 min. Oil capacity 12 litres (2·6 Imp gallons).

Accommodation: Individual side-by-side seats for two persons, the main pilot's seat being to port. Both seats are adjustable and permit the use of back-type parachutes. Baggage space aft of seats. Cabin and windscreen heating and ventilation standard. Forward-sliding cockpit canopy. Dual controls standard.

Systems: Electrical system includes a 600W 27V engine-driven generator and 24V 25Ah Teledyne battery. External power source can be used for starting the engine.

Avionics and Equipment: VHF radio with IC (Mesit LUN 3524.20) and IFR instrumentation optional. Standard equipment includes cockpit, instrument and cabin lights; navigation lights; landing and taxying lights; and anti-collision light. Towing gear, for gliders of up to 500 kg (1,102 lb) weight, optional.

Dimensions, external:

Wing span	9·16 m (30 ft 0½ in)
Wing chord (constant over most of span)	1·42 m (4 ft 8 in)
Length overall	7·33 m (24 ft 0½ in)
Height overall	2·75 m (9 ft 0¼ in)
Elevator span	2·904 m (9 ft 6⅓ in)
Wheel track	2·33 m (7 ft 7¾ in)
Wheelbase	1·66 m (5 ft 5¼ in)
Propeller diameter	2·00 m (6 ft 6¾ in)
Propeller ground clearance	0·40 m (1 ft 3¾ in)

Dimensions, internal:

Cabin: Length	1·80 m (5 ft 10¾ in)
Max width	1·12 m (3 ft 8 in)
Max height	1·20 m (3 ft 11¼ in)
Baggage space	0·2 m^3 (7·1 cu ft)

Areas:

Wings, gross	13·15 m^2 (141·5 sq ft)
Ailerons (total)	1·408 m^2 (15·16 sq ft)
Trailing-edge flaps (total)	1·408 m^2 (15·16 sq ft)
Fin	0·54 m^2 (5·81 sq ft)
Rudder, incl tab	0·81 m^2 (8·72 sq ft)
Tailplane	1·23 m^2 (13·24 sq ft)
Elevator, incl tabs	1·36 m^2 (14·64 sq ft)

Weights and Loadings (A: Aerobatic; U: Utility; N: Normal category):

Basic weight empty (all versions)	730 kg (1,609 lb)
Max T-O weight: A	970 kg (2,138 lb)
U	1,020 kg (2,248 lb)
N	1,090 kg (2,403 lb)
Max landing weight: A	970 kg (2,138 lb)
U	1,020 kg (2,248 lb)
N	1,050 kg (2,315 lb)
Max wing loading: A	73·76 kg/m^2 (15·11 lb/sq ft)
U	77·57 kg/m^2 (15·89 lb/sq ft)
N	82·89 kg/m^2 (16·98 lb/sq ft)
Max power loading: A	6·19 kg/kW (10·17 lb/hp)
U	6·51 kg/kW (10·69 lb/hp)
N	6·96 kg/kW (11·43 lb/hp)

Performance (at max T-O weight):

Never-exceed speed (all versions)	179 knots (333 km/h; 206 mph) IAS
Max level speed at 500 m (1,640 ft):	
A, U	125 knots (231 km/h; 143 mph)
N	122 knots (227 km/h; 141 mph)
Max cruising speed at 500 m (1,640 ft):	
A, U	106 knots (197 km/h; 122 mph)
N	102 knots (190 km/h; 118 mph)
Econ cruising speed at 500 m (1,640 ft):	
A	97 knots (180 km/h; 112 mph)

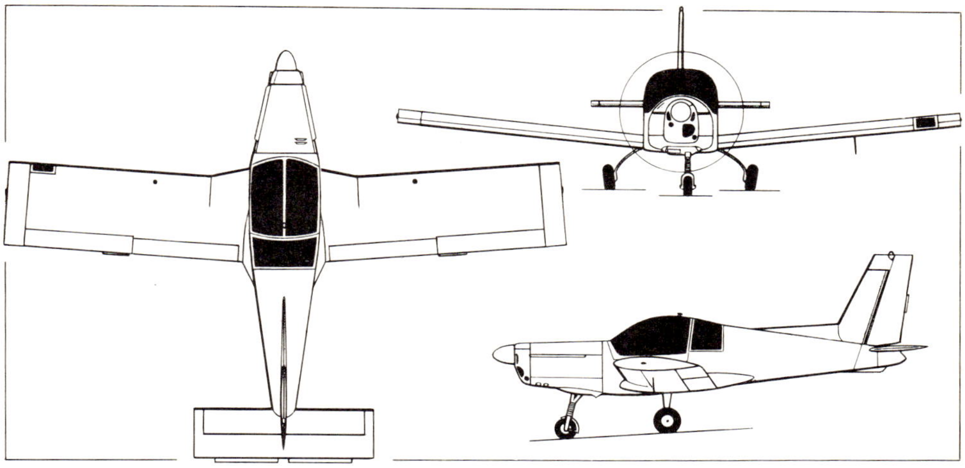

Zlin 142 two-seat light aircraft, developed from the Z 42 M *(Pilot Press)*

Stalling speed, flaps up:	
A	56 knots (103 km/h; 64 mph) IAS
U	58 knots (107 km/h; 67 mph) IAS
N	60 knots (110 km/h; 69 mph) IAS
Stalling speed, T-O flap setting:	
A	54 knots (99 km/h; 62 mph) IAS
U	55·5 knots (102 km/h; 63·5 mph) IAS
N	57 knots (105 km/h; 66 mph) IAS
Stalling speed, flaps down:	
A	48 knots (88 km/h; 55 mph) IAS
U	49·5 knots (91 km/h; 57 mph) IAS
N	51·5 knots (95 km/h; 59·5 mph) IAS
Max rate of climb at S/L, ISA:	
A	330 m (1,082 ft)/min
U	306 m (1,004 ft)/min
N	264 m (866 ft)/min
Service ceiling: A	5,000 m (16,400 ft)
U	4,700 m (15,425 ft)
N	4,300 m (14,100 ft)
T-O run: A	220 m (722 ft)
T-O to 15 m (50 ft): A	440 m (1,444 ft)
U	475 m (1,560 ft)
N	540 m (1,772 ft)
Landing from 15 m (50 ft): A	400 m (1,313 ft)
U	425 m (1,395 ft)
N	460 m (1,510 ft)
Landing run: A	190 m (624 ft)
Range at max cruising speed:	
A, U	283 nm (525 km; 326 miles)
N	513 nm (950 km; 590 miles)
Max range: N	566 nm (1,050 km; 652 miles)
g limits: A	+6·0; −3·5
U	+5·0; −3·0
N	+3·8; −1·5

ZLIN Z 50 L

The Z 50 L is a fully-aerobatic single-seat competition light aircraft, the design of which began in 1973. Construction of a prototype was started in 1975, and this aircraft flew for the first time on 18 July 1975. Two additional prototypes and 25 production aircraft (including some for export) had been completed by 1 January 1981.

Thirteen Z 50 Ls took part in the 1978 World Aerobatic Championships at Hosin, Czechoslovakia, gaining 1st and 3rd places in the men's individual competitions and 1st place in the team event.

Type: Single-seat aerobatic aircraft.

Wings: Cantilever low-wing monoplane. Wing section NACA 0018 at root, NACA 0012 at tip. Dihedral 1° 7′ 24″. All-metal structure, with single continuous main spar, rear auxiliary spar, and aluminium-clad duralumin skin. All-metal mass-balanced ailerons, actuated by pushrods, occupy most of each trailing-edge. Ground-adjustable tab on port outer aileron; automatic trim tab on each inboard aileron. No flaps. Provision for fitting wingtip fuel tanks for cross-country flights.

Fuselage: All-metal semi-monocoque stressed-skin structure, with duralumin skin.

Tail Unit: Conventional metal structure. Braced tailplane and fin duralumin-covered, balanced elevators and rudder fabric-covered. One mechanically-adjustable tab and one automatic trim tab on elevators; automatic trim tab on rudder. Elevators actuated by pushrods, rudder by cables.

Landing Gear: Non-retractable tailwheel type. Main wheels carried on flat-spring titanium cantilever legs. Mechanical main-wheel brakes actuated by rudder pedals. Fully-castoring tailwheel, with flat-spring shock-absorption, has automatic locking device to maintain aircraft on a straight track during taxying, take-off and landing. Main wheels size 350 × 135 mm, pressure 2·5 bars (36 lb/sq in); tailwheel size 200 × 80 mm, pressure 1·0 bar (14·5 lb/sq in). Main-wheel fairings optional.

Power Plant: One 194 kW (260 hp) Avco Lycoming AEIO-540-D4B5 flat-six engine, without reduction gear, driving a Hoffmann HO-V123K-F/200AH three-blade constant-speed variable-pitch wooden propeller with spinner. Single main fuel tank in fuselage, aft of firewall, capacity 60 litres (13·2 Imp gallons). Auxiliary 50 litre (11 Imp gallon) tank can be attached to each wingtip for cross-country flights only. Fuel and oil systems designed for full aerobatic manoeuvres, including inverted flight. Oil capacity 12 litres (2·64 Imp gallons).

Accommodation: Single seat under fully-transparent sideways-opening (to starboard) bubble canopy, which can be jettisoned in an emergency. Seat and backrest are adjustable, and permit the use of a back-type parachute. Cockpit ventilated by sliding panel in canopy.

System: Electrical system includes an alternator as main power source and a Varley battery. External power socket in fuselage side for engine starting.

Avionics and Equipment: VHF radio optional.

Dimensions, external:

Wing span	8·58 m (28 ft 1¾ in)
Wing span over tip-tanks	9·03 m (29 ft 7½ in)
Wing chord at root	1·73 m (5 ft 8¼ in)
Wing chord at tip	1·21 m (3 ft 11¾ in)

Zlin Z 50 L aerobatic aircraft (Avco Lycoming AEIO-540-D4B5 engine) *(Letectvi + Kosmonautika)*

Wing mean aerodynamic chord 1·4853 m (4 ft 10½ in)
Wing aspect ratio 5·88
Length overall (tail up) 6·62 m (21 ft 8¾ in)
Height over tail (static) 1·985 m (6 ft 6¼ in)
Elevator span 3·44 m (11 ft 3½ in)
Wheel track 1·90 m (6 ft 2¾ in)
Wheelbase 5·05 m (16 ft 7 in)
Propeller diameter 2·00 m (6 ft 6¾ in)
Propeller ground clearance (tail up) 0·31 m (1 ft 0¼ in)

Areas:
Wings, gross 12·50 m² (134·55 sq ft)
Ailerons (total) 2·80 m² (30·14 sq ft)
Fin 0·59 m² (6·35 sq ft)
Rudder, incl tab 0·81 m² (8·72 sq ft)
Tailplane 1·66 m² (17·87 sq ft)
Elevators (total, incl tabs) 1·20 m² (12·92 sq ft)

Weights and Loadings (Aerobatic category):
Weight empty, equipped 570 kg (1,256 lb)
Max T-O weight 720 kg (1,587 lb)
Max wing loading 57·6 kg/m² (11·80 lb/sq ft)
Max power loading 3·72 kg/kW (6·11 lb/hp)

Performance (at max Aerobatic T-O weight):
Never-exceed speed 181·5 knots (337 km/h; 209 mph) CAS
Max level speed at S/L, ISA 156 knots (290 km/h; 180 mph)
Max cruising speed at 1,000 m (3,280 ft), ISA 147 knots (273 km/h; 169 mph)
Econ cruising speed at 1,000 m (3,280 ft), ISA 134 knots (248 km/h; 154 mph)
Stalling speed 53 knots (98 km/h; 61 mph) CAS
Max rate of climb at S/L 720 m (2,360 ft)/min
Service ceiling 7,000 m (23,000 ft)
T-O run 170 m (558 ft)
T-O to 15 m (50 ft) 290 m (952 ft)
Landing from 15 m (50 ft) 530 m (1,740 ft)
Landing run 250 m (820 ft)
Range with max fuel (incl wingtip tanks) 345 nm (640 km; 397 miles)
g limits: Aerobatic +9; −6

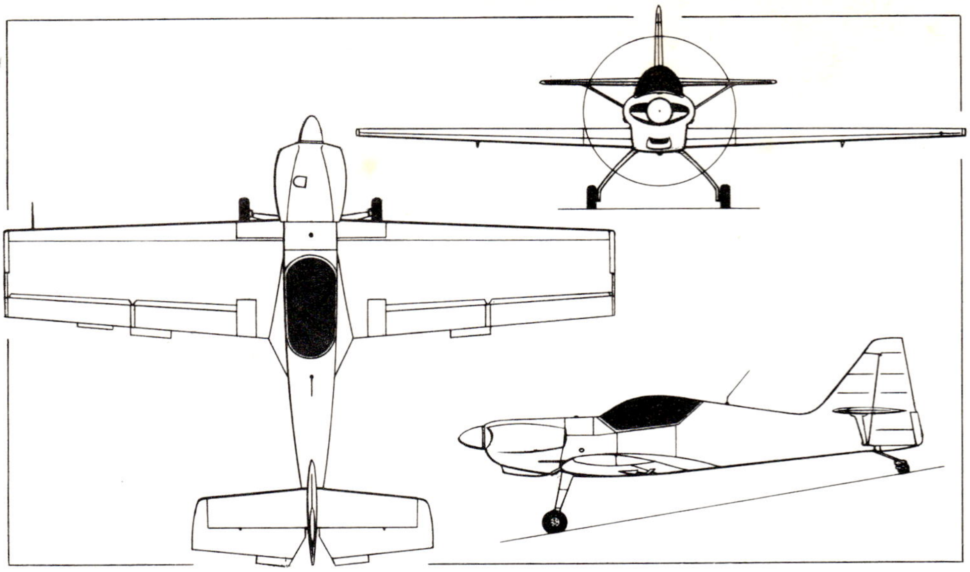

Zlin Z 50 L single-seat aerobatic aircraft *(Pilot Press)*

FINLAND

VALMET
VALMET OY KUOREVESI WORKS

Office and Works: 35600 Halli
Telephone: (358) 42 82810
Telex: 28269 Valku SF

Corporate Vice-President, Defence Equipment Group: Heikki Mäntylä
General Manager, Kuorevesi Works: Juhani Mäkinen

Valmet Oy Kuorevesi Works is affiliated to Valmet Oy, a State-owned company consisting of several metal-working factories. It continues the traditions of Ilmailuvoimien Lentokonetehdas, established in 1921, and was formerly a part of the Valmet Oy Tampere factory group, from which it was separated in 1974. It is now an independent factory directly responsible to Valmet's Head Office in Helsinki, and is currently the largest aircraft industry establishment in Finland. Since 1922, Valmet Oy Kuorevesi Works and its predecessors have built 30 different types of aircraft, of which 18 have been of Finnish design. Valmet Oy was responsible for assembly of the 12 Saab 35XS Drakens ordered by Finland in 1970.

Current activities of Valmet Oy Kuorevesi Works include the overhaul and repair of military and civil aircraft, piston engines and instruments. The factory has a covered area of approximately 14,000 m² (150,695 sq ft). Valmet Oy Linnavuori Works, at Siuro, is concerned primarily with the overhaul and repair of aircraft jet engines.

The Kuorevesi and Linnavuori Works are also participating in the manufacture and assembly of 46 of the 50 BAe Hawk Mk 51 jet trainers and their Adour Mk 851 jet engines, purchased by the Finnish Air Force from the UK in early 1978. The four UK-built Hawks were delivered in 1980-81. The first Valmet-assembled Hawk was handed over on 20 February 1981; deliveries are scheduled to continue until 1985. For these, Valmet manufactures the wing flaps, airbrake, tailplane and fin.

The latest aircraft of Finnish design to be built by Valmet are the L-70 Miltrainer piston-engined trainer and the PIK-23 Towmaster two-seat light aircraft.

First production Valmet L-70 Vinka two-seat trainer for the Finnish Air Force

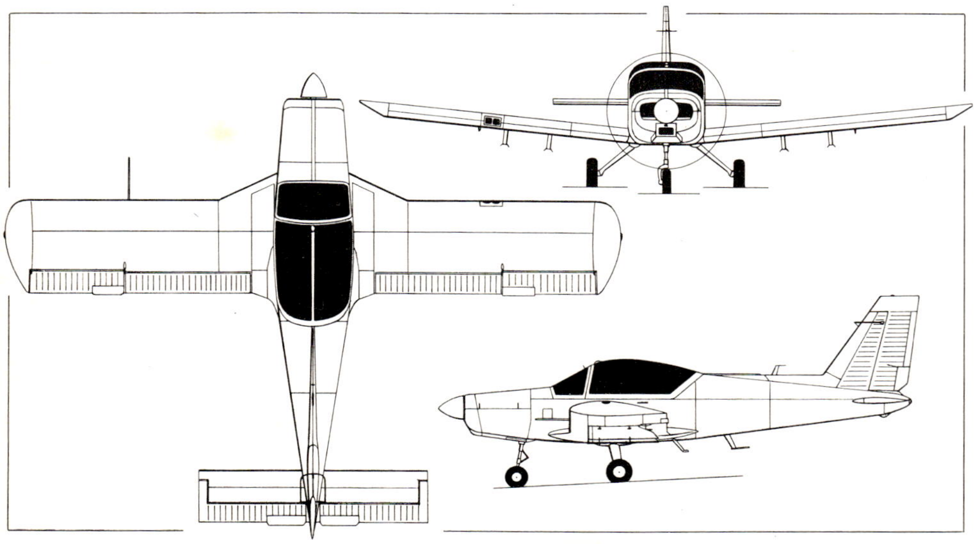

Valmet L-70 Miltrainer two/four-seat training and touring aircraft *(Pilot Press)*

VALMET L-70 MILTRAINER

Finnish Air Force name: Vinka

In late 1970 an aeronautical research and development group was established in Finland, its first major task being to study a Finnish Air Force requirement for a basic training aircraft to replace the Saab 91 Safir. It was decided to produce an entirely Finnish design to fulfil this need, and a development contract was placed with Valmet by the Finnish Air Force on 23 March 1973. The aircraft, which was originally designated Leko-70, an abbreviation of 'Lentokone', the Finnish word for 'aeroplane', is named Vinka (a cold Arctic wind) by the Finnish Air Force.

The prototype made its first flight at Kuorevesi on 1 July 1975. A second prototype was used for static and fatigue testing; a full-size cockpit mockup and components for a third aircraft were also completed.

The Finnish Air Force ordered 30 Vinkas on 28 January 1977. Manufacture of these began on 15 January 1977, and the first production Vinka made its initial flight on 29 December 1979. Deliveries were due to begin in February 1980, but were delayed by a fire in the assembly hangar. The first two Vinkas were handed over to the Finnish Air Force on 7 October 1980, and five were in service with the Air Academy by mid-1981. Production for the Finnish Air Force is due to end in late 1982.

The Miltrainer is designed for Aerobatic or Utility use as a two-seater. In civil use, in Normal category, it is capable of seating up to four persons, depending upon the amount of baggage carried, and fulfils the requirements of FAR Pt 23 in all three categories. Those for the Finnish

Air Force fulfil additional military strength and other requirements.

TYPE: Two-seat training or two/four-seat touring aircraft.

WINGS: Cantilever low-wing monoplane. Wing section NACA 63_2A615 (modified). Dihedral 6° from roots. Incidence 2°. Fail-safe structure of main spar, auxiliary spar, ribs and stringers, of constant chord except for forward-swept wing-root leading-edges, and attached to fuselage by steel fittings. Riveted aluminium alloy skin. Electrically-operated slotted flaps, and mass-balanced slotted ailerons, on trailing-edges, all of aluminium alloy riveted construction. Ailerons actuated by stainless steel control cables. Flaps and ailerons have fluted skins. Spring tab in each aileron.

FUSELAGE: Conventional aluminium alloy semi-monocoque fail-safe structure of frames and longerons, with riveted skin. Welded steel tube engine mount and wing carry-through structure; stainless steel firewall. Cockpit floor panels of bonded sandwich.

TAIL UNIT: Cantilever aluminium alloy structure, with fluted and riveted skin. Slight sweepback on vertical surfaces; shallow dorsal fin. Elevators and rudder aerodynamically and mass balanced, and actuated by stainless steel control cables. Geared trim tabs in rudder and each elevator.

LANDING GEAR: Non-retractable tricycle type. Cantilever main legs. Automotive Products oleo-pneumatic shock-absorbers on all units. Cleveland 40-75S main wheels, with Goodyear 6·00-6 six-ply tyres, pressure 1·86 bars (27 lb/sq in); Goodyear 395-32926 nose-wheel, with Goodyear 5·00-5 four-ply tyre, pressure 2·07 bars (30 lb/sq in). Cleveland 30-52K disc brakes. Provision for fitting Finncraft skis.

POWER PLANT: One 149 kW (200 hp) Avco Lycoming AEIO-360-A1B6 flat-four engine, driving a Hartzell HC-C2YK-4F/FC 7666A-2 two-blade constant-speed propeller with spinner. Christen-801 inverted fuel and oil systems. Semi-integral bonded sandwich fuel tank in each wing root ahead of main spar; total capacity 170 litres (37·4 Imp gallons). Overwing fuelling point above each tank. Oil capacity 7·5 litres (1·65 Imp gallons).

ACCOMMODATION: Side-by-side seats for instructor and pupil in trainer version, with integral longitudinal central console which serves also to reinforce fuselage floor. Dual controls standard, but instructor's or pupil's control column can be removed if desired. Windscreen and one-piece rearward-sliding fully-transparent jettisonable canopy, with steel tube turnover frame. Canopy can be locked in partially open position if required. Provision for two more seats at rear, which can be removed to make room for additional baggage. Up to 280 kg (617 lb) of baggage or freight can be carried internally, or externally if flown as a single-seater. As ambulance, can accommodate one stretcher patient and medical attendant in addition to pilot. Cockpit heated and ventilated.

SYSTEM: 28V 70A DC electrical system, with brushless generator and 25Ah nickel-cadmium battery. No hydraulic or pneumatic systems.

AVIONICS AND EQUIPMENT: Two VHF, one ADF, one VOR/ILS, one ATC transponder, one RMI, and one gyromagnetic compass. Blind-flying instrumentation standard. Landing/taxying lights in starboard wing leading-edge. Equipment for secondary roles may include glider or target towing hook; one long-focus or four short-focus vertical cameras (provision for 35 × 40 cm; 13·8 × 15·75 in aperture, with hinged doors, in floor of rear cockpit); four underwing pylons (total capacity 300 kg; 661 lb: see following paragraph) for stores; and dispersal equipment for agricultural missions.

ARMAMENT AND OPERATIONAL EQUIPMENT: Four underwing attachments, the inner pair each stressed for 150 kg (330·5 lb) and the outer pair for 100 kg (220 lb) each; max external load 300 kg (661 lb). As single-seater, can carry four 50 kg bombs; two 100 kg bombs and two flare pods; four pods each with nineteen 37 mm or six 68 mm rockets; two pods each with twin 7·62 or 5·56 mm machine-guns and a total of 2,000 rds of ammunition; two such gun pods and two flare pods; two pods each with single 12·7 mm machine-gun and a total of 300 rds; or two reconnaissance or photographic pods. As two-seater, typical loads can include four or eight anti-tank missiles, depending upon type and size; one TV pod (with transmitter) and one searchlight pod; three 10-person life rafts and a searchlight pod; or three 6-person emergency rescue packs and a searchlight pod.

DIMENSIONS, EXTERNAL:

Wing span	9·85 m (32 ft 3¾ in)
Wing chord (constant over most of span)	1·53 m (5 ft 0¼ in)
Wing aspect ratio	6
Length overall	7·50 m (24 ft 7¼ in)
Height overall	3·31 m (10 ft 10¼ in)
Tailplane span	3·60 m (11 ft 9¾ in)
Wheel track	2·30 m (7 ft 6½ in)
Wheelbase	1·61 m (5 ft 3½ in)
Propeller diameter	1·88 m (6 ft 2 in)
Propeller ground clearance	0·25 m (9¾ in)

Valmet PIK-23 Towmaster side-by-side two-seat light aircraft

AREAS:

Wings, gross	14·00 m² (150·70 sq ft)
Ailerons (total)	1·412 m² (15·20 sq ft)
Trailing-edge flaps (total)	1·90 m² (20·45 sq ft)
Fin	0·87 m² (9·36 sq ft)
Rudder, incl tab	0·79 m² (8·50 sq ft)
Tailplane	2·01 m² (21·64 sq ft)
Elevators, incl tabs	1·01 m² (10·87 sq ft)

WEIGHTS AND LOADINGS:

Operating weight empty, equipped	767 kg (1,691 lb)
Max payload with full fuel	380 kg (838 lb)
Max T-O weight: Aerobatic	1,040 kg (2,293 lb)
Normal	1,250 kg (2,756 lb)
Max wing loading	89·3 kg/m² (18·53 lb/sq ft)
Max power loading	8·38 kg/kW (13·78 lb/hp)

PERFORMANCE (at max Aerobatic T-O weight):

Never-exceed speed	193 knots (360 km/h; 223 mph)
Max level speed at S/L	129 knots (240 km/h; 149 mph)
Cruising speed (75% power) at 1,525 m (5,000 ft)	120 knots (222 km/h; 138 mph)
Stalling speed, flaps up, power off	53 knots (98 km/h; 61 mph)
Stalling speed, flaps down, power off	46 knots (85 km/h; 53 mph)
Max rate of climb at S/L	342 m (1,120 ft)/min
Service ceiling	5,000 m (16,400 ft)
T-O run	180 m (590 ft)
T-O to 15 m (50 ft)	260 m (855 ft)
Landing from 15 m (50 ft)	300 m (985 ft)
Landing run	150 m (490 ft)
Min ground turning radius	8·00 m (26 ft 3 in)
Range with max fuel	547 nm (1,015 km; 630 miles)
Range with max payload, no reserves	464 nm (860 km; 534 miles)
Endurance at S/L (65% power)	4 h 48 min
g limits: Aerobatic	+6·00; −3·00
Utility	+4·40; −2·02
Normal	+3·80; −1·80

VALMET PIK-23 TOWMASTER

Designed by the students of Helsinki University Flying Club, the PIK-23 (known originally as the Suhino) is a side-by-side two-seat sporting, training and glider-towing aircraft developed from the PIK-19 Muhinu (1975-76 *Jane's*). Two prototypes are being completed by Valmet, the first of which (OH-TOW) made its debut at the 1981 Paris Air Show and was scheduled to make its first flight in August of that year.

Many components of the PIK-23 are of composite construction. These are manufactured by PIK; final assembly and flight testing are undertaken by Valmet. Production aircraft are expected to be available from late 1982/early 1983. It is also planned to market the Towmaster in kit form for amateur construction.

TYPE: Two-seat towing, training and club aircraft.

WINGS: Cantilever low-wing monoplane, constructed in two parts. Wing section NACA 63_2 415 (modified). Constant chord. Dihedral from roots. Carbonfibre main spar. Glassfibre/epoxy/PVC foam sandwich structure.

FUSELAGE: Glassfibre/epoxy/PVC foam sandwich structure.

TAIL UNIT: Cantilever structure, with sweptback vertical and non-swept horizontal surfaces. Construction similar to that of wings. Balanced rudder and one-piece elevator. Tab in starboard half of elevator.

LANDING GEAR: Non-retractable tricycle type, with CFRP main-wheel legs and hydraulic brakes.

POWER PLANT: One 134 kW (180 hp) Avco Lycoming O-360-A4M flat-four engine, driving a Hoffmann two-blade fixed-pitch propeller with spinner. Fuel tank in each wing, combined capacity 180 litres (39·5 Imp gallons). Overwing gravity refuelling point for each tank.

ACCOMMODATION: Two glassfibre seats side by side under rearward-sliding bubble canopy. Dual controls standard. Baggage compartment aft of seats.

AVIONICS AND EQUIPMENT: VFR avionics and towing hook standard. Panel has provision for full IFR equipment.

DIMENSIONS, EXTERNAL:

Wing span	10·00 m (32 ft 9¾ in)
Wing chord, constant	1·40 m (4 ft 7 in)
Wing aspect ratio	7·14
Length overall	7·10 m (23 ft 3½ in)
Height overall	2·85 m (9 ft 4¼ in)

AREAS:

Wings, gross	14·00 m² (150·69 sq ft)
Ailerons (total)	1·42 m² (15·28 sq ft)
Trailing-edge flaps (total)	1·46 m² (15·71 sq ft)
Fin	0·68 m² (7·32 sq ft)
Rudder	0·63 m² (6·78 sq ft)
Tailplane	1·76 m² (18·94 sq ft)
Elevator, incl tab	1·13 m² (12·16 sq ft)

WEIGHTS AND LOADINGS:

Weight empty, equipped	560 kg (1,234 lb)
Baggage (max)	40 kg (88 lb)
Max T-O and landing weight: Utility	794 kg (1,750 lb)
Normal	870 kg (1,918 lb)
Max wing loading: Utility	56·7 kg/m² (11·6 lb/sq ft)
Normal	62·1 kg/m² (12·7 lb/sq ft)
Max power loading: Utility	5·9 kg/kW (9·7 lb/hp)
Normal	6·5 kg/kW (10·6 lb/hp)

PERFORMANCE (estimated, at max T-O weight, with propeller optimised for climb unless otherwise stated):

Never-exceed speed	151 knots (281 km/h; 174 mph)
Max level speed at S/L, propeller at cruise setting	135 knots (250 km/h; 155 mph)
Cruising speed (75% power) at S/L, propeller at cruise setting	121 knots (225 km/h; 140 mph)
Stalling speed: flaps up	47 knots (87 km/h; 54·5 mph)
flaps down	43 knots (79 km/h; 49·5 mph)
Rate of climb at S/L (75% power) with single-seat sailplane	228 m (748 ft)/min
T-O to 15 m (50 ft)	320 m (1,050 ft)
Landing run	150 m (492 ft)
Range with max fuel	502 nm (930 km; 578 miles)
Endurance	5 h

FRANCE

AÉROSPATIALE
SOCIÉTÉ NATIONALE INDUSTRIELLE AÉROSPATIALE

HEAD OFFICE: 37 boulevard de Montmorency, 75781 Paris Cédex 16

Telephone: 524 43 21

Telex: AISPA X 620059 F

BOARD OF DIRECTORS

CHAIRMAN: Jacques Mitterrand, Général d'Armée Aérienne (CPN)

REPRESENTATIVES OF THE SHAREHOLDERS:

René Ravaud, Ingénieur Général de l'Armement (President and Director General of SNECMA)

Roger Martin (President d'Honneur of the Saint-Gobain company)

Banque Nationale de Paris, represented by René Thomas (Deputy Director General)
SOGEPA, represented by Général Maurice Saint Cricq (President)

Representatives of the State:
Pierre Dario (Controller General, representing the Secretary of State for Transport)
Jacques Villiers (representing the Minister of State for Transport)
Mme Isabelle Cheyvialle (representing the Minister for Economic Affairs)

Representatives of the Employees:
Paul Bienfait (representing executives)
Jean Masse and Vincent Valente (representing workmen, office staff, technicians and supervisors)

GENERAL MANAGEMENT

President and Director General: Jacques Mitterrand
Director General Delegate: Roger Chevalier
Deputy General Manager, Economics and Finance: Yves Barbé
Deputy General Manager, Industrial and Technical Matters: Joseph Millara
General Secretary: Roger Courot
Director of the President's Office: Jean-Charles Poggi
Press and Information Adviser to the President: Jean E. Lagrange
Director of Public Relations: Claude Loiseau
Industrial Director: Roger Berthier
Central Financial Director: Michel Euvrard
Central Technical Director: Jacques Balazard
Director (Industrial Relations): André Escoulin
Director (Quality Assurance): Georges Sertour
Deputy Director of Industrial Relations (Director, Headquarters Establishment): Didier Godechot

AIRCRAFT DIVISION

Managing Director: André Etesse
Assistant Managing Director: René Dor
Assistant Managing Director, Programmes: Georges Roche
Airbus Programme Director: Alain Bruneau
Military Adviser: Gen (Retd) Pierre Delachenal
Technical Director: Pierre Lecomte
Financial Director: Joseph Carpentier
Commercial Director: Henri Paul Puel
Flight Test Director: Henri Perrier
Director, Support Services: Pierre Schaffner
Director (Quality Assurance): Pierre Lagarde
Works and Facilities:
Toulouse. Director: Jean Pierson
Nantes-Bouguenais. Director: Jean-Louis Fache
Saint-Nazaire. Director: Jean Renon
Méaulte. Director: Jean-Paul Chandez

HELICOPTER DIVISION

Managing Director: François Legrand
Technical Director: Georges Petit
Director of Engineering: René Mouille
Director of Commercial Programme: André Renaud
Sales Director: Jean-Claude Rebuffel
Director (Product Support): Marc Fourcade
Flight Test Director: Jean-Marie Besse
Director (Quality Assurance): André Breton
Works and Facilities:
Marignane. Director: Fernand Carayon
La Courneuve. Director: Lucien Fournier

TACTICAL MISSILES DIVISION

Managing Director: Michel Allier
Assistant Managing Director: Philippe Girard
Sales Director: Pierre Frojet
Technical Director: Jean Guillot
Director of Design: Yves de Rougemont
Director of Economic Affairs: Jacques Pottier
Works and Facilities:
Châtillon. Director: Jean-Claude Renaut
Bourges. Director: Georges Barroy

SPACE AND BALLISTIC SYSTEMS DIVISION

Managing Director: Pierre Usunier
Assistant Managing Director—Director of Systems Subdivision: Gérard Payelle
Commercial Director: Charles Bigot
Works and Facilities:
Aquitaine. Director: Jean-Rémy Hugues
Les Mureaux. Director: Pierre Madon
Cannes. Director: Gérard Chauvallon

SUBSIDIARIES

Société Girondine d'Entretien et de Réparation de Matériel Aéronautique (SOGERMA)
Société de Construction d'Avions de Tourisme et d'Affaires (Socata)
Société d'Exploitation et de Constructions Aéronautiques (SECA)
Electronique Aérospatiale (EAS)
Société Charentaise d'Equipements Aéronautiques (SOCEA)
Aérospatiale Inc (USA)
Aérospatiale Aircraft Corporation (USA)
Aérospatiale Helicopter Corporation (USA)
Aérospatiale Assistencia Tecnica do Brasil Ltda (Brazil)

The Société Nationale Industrielle Aérospatiale was formed on 1 January 1970, by decision of the French government, as a result of the merger of the former Sud-Aviation, Nord-Aviation and SEREB companies. It is the biggest aerospace company in the Common Market countries on the Continent of Europe, with a registered capital of 447,400,000 francs, facilities extending over a total area of 9,790,000 m² (105,380,000 sq ft), of which 1,778,000 m² (19,138,000 sq ft) are covered, and a staff (including subsidiary companies) of 38,173 persons at the beginning of 1980.

In the aircraft field, products include the Epsilon primary/basic trainer; the short/medium-range large-capacity A300/A310 European Airbus, in co-operation with Deutsche Airbus GmbH, British Aerospace, Fokker and CASA; the Transall heavy multi-purpose transport, in co-operation with MBB; and the ATR 42 regional transport, in co-operation with Aeritalia.

Aérospatiale produces a range of light piston-engined aircraft through its subsidiary, Socata (which see). It is also a major manufacturer of helicopters, guided missiles, spacecraft and research rockets.

Helicopter activities, concentrated at Marignane, involve the development and production of a wide range of turbine-powered types, described under this entry. Agreements concluded with Westland in the UK covered joint development and production of the Puma and Gazelle, and the Westland-designed Lynx, after all three types had been chosen to equip the French and British armed forces.

By the beginning of 1980, the Helicopter Division and its US subsidiary, Aérospatiale Helicopter Corporation, had sold a total of 5,950 production helicopters to 497 customers in 94 countries. Orders for 639 helicopters were received in 1980, about 85 per cent for export, and 345 helicopters were delivered.

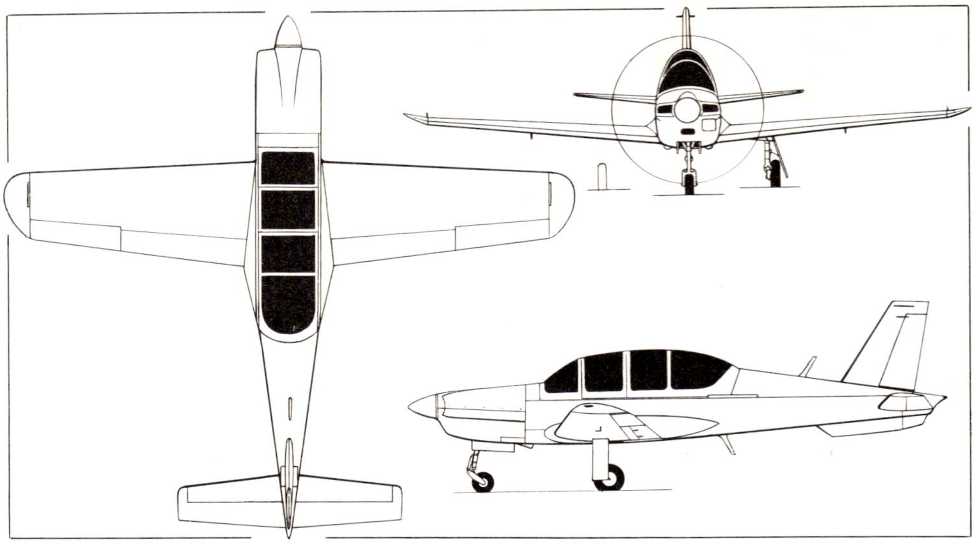
Aérospatiale Epsilon tandem two-seat primary/basic trainer *(Pilot Press)*

AÉROSPATIALE EPSILON

First details of this tandem two-seat primary/basic trainer were released at the Farnborough Air Show in September 1978. Purpose of the project was to meet a French Air Force requirement for a propeller-driven aircraft for use in the initial stages of a more cost-effective pilot training scheme than that currently operated.

A development contract from the Air Force, for two prototypes and two ground test airframes, was announced by Aérospatiale at the Paris Air Show in June 1979. The first prototype flew for the first time on 22 December that year, followed by the second prototype on 12 July 1980. By the beginning of February 1981, these aircraft had logged a combined 160 h of test flying, and discussions were under way concerning a planned production order for about 150 aircraft, to be delivered over a six-year period. An initial production contract for 30 aircraft was reported in June 1981.

The Epsilon programme is handled by the Aircraft Division of Aérospatiale, as prime contractor responsible for the entire programme. Design and manufacture are subcontracted to Socata, the company's light aircraft subsidiary at Tarbes.

The following description applies to the export version of the Epsilon:

Type: Two-seat military primary/basic trainer.

Wings: Cantilever low-wing monoplane. Wing section RA 1643 at root, RA 1243 at tip. Thickness/chord ratio 16% at root, 12% at tip. Dihedral 5°. Incidence 2°. All-metal light alloy structure, with single main spar and rear auxiliary spar, built in two panels attached directly to sides of fuselage. Press-formed ribs and heavy-gauge skin without stringers. Electrically-actuated single-slotted flaps. Light alloy ailerons, with spring tabs.

Fuselage: Light alloy semi-monocoque structure of four longerons, frames and heavy-gauge skin, without stringers.

Tail Unit: Cantilever single-spar light alloy structure. Fixed surfaces metal-covered; elevators and rudder covered with polyester fabric. Fixed-incidence tailplane. Balanced elevators and rudder, with controllable trim tabs.

Landing Gear: Hydraulically-retractable tricycle type, with single wheel on each unit. Inward-retracting main units and rearward-retracting nosewheel. Independent hydraulic single-disc brake on each main wheel.

Power Plant: One 224 kW (300 hp) Avco Lycoming IO-540-L1B5D flat-six engine, driving a Hartzell HC-C2YR-4-F/FC 8475-6 two-blade constant-speed metal propeller, with spinner. Fuel in two wing leading-edge tanks, with total capacity of 210 litres (46 Imp gallons). Refuelling points on wing upper surface.

Accommodation: Two seats in tandem, with rear seat raised by 70 mm (2·7 in). Rudder pedals are mechanically adjustable fore and aft. Sideways-hinged Plexiglas canopy, with emergency jettison system. Baggage compartment aft of cabin.

Systems: Hydraulic systems for actuating landing gear and brakes. 28V electrical system includes engine-driven alternator, battery for engine starting and emergency use. Cabin heated and ventilated. Windscreen demister.

Avionics and Equipment: Radio com and blind-flying instrumentation standard. Second prototype has VHF, UHF, automatic VOR and a transponder, with Tacan to be added later.

Dimensions, external:

Wing span	7·92 m (25 ft 11¾ in)
Wing chord at root	1·46 m (4 ft 9½ in)
Wing chord at tip	0·92 m (3 ft 0¼ in)
Length overall	7·59 m (24 ft 10¾ in)
Height overall	2·66 m (8 ft 8¾ in)
Tailplane span	3·20 m (10 ft 6 in)
Wheel track	2·30 m (7 ft 6½ in)
Wheelbase	1·80 m (5 ft 10¾ in)
Propeller diameter	1·98 m (6 ft 6 in)
Propeller ground clearance	0·25 m (10 in)

Areas:

Wings, gross	9·60 m² (103·3 sq ft)

Second prototype of the Aérospatiale Epsilon (Avco Lycoming IO-540-L1B5D engine)

Fin 1·02 m² (10·98 sq ft)
Tailplane 2·00 m² (21·53 sq ft)

WEIGHTS AND LOADINGS:
Weight empty, equipped 878 kg (1,936 lb)
Max T-O and landing weight (aerobatic) 1,200 kg (2,645 lb)
Max wing loading (aerobatic) 125 kg/m² (25·6 lb/sq ft)
Max power loading (aerobatic) 5·36 kg/kW (8·82 lb/hp)

PERFORMANCE (at max aerobatic T-O weight):
Never-exceed speed 280 knots (518 km/h; 322 mph)
Max level speed at S/L 200 knots (370 km/h; 230 mph)
Max cruising speed (75% power) at 1,830 m (6,000 ft) 184 knots (340 km/h; 212 mph)
Stalling speed, flaps down, power off 62 knots (115 km/h; 72 mph)
Max rate of climb at S/L 516 m (1,690 ft)/min
Service ceiling 6,100 m (20,000 ft)
T-O run 350 m (1,150 ft)
T-O to 15 m (50 ft) 550 m (1,805 ft)
Landing from 15 m (50 ft) 575 m (1,890 ft)
Landing run 250 m (820 ft)
Endurance (75% power) 3 h 45 min
g limits +7; −3·5

AÉROSPATIALE/MBB TRANSALL

Details of the Transall programme can be found in the International section of this edition.

AÉROSPATIALE/AERITALIA ATR 42

Details of the ATR 42 programme can be found in the International section of this edition.

AÉROSPATIALE SA 315B LAMA

Indian Army name: Cheetah

Design of the SA 315B Lama began in late 1968, initially to meet a requirement of the Indian armed forces, and a prototype was flown for the first time on 17 March 1969. French certification was granted on 30 September 1970 and FAA Type Approval on 25 February 1972.

The Lama combines features of the Alouette II and III, having the airframe (with some reinforcement) of the former and the dynamic components, including the Artouste power plant and rotor system, of the SA 316 Alouette III.

During demonstration flights in the Himalayas in 1969 a Lama, carrying a crew of two and 140 kg (308 lb) of fuel, made the highest landings and take-offs ever recorded, at a height of 7,500 m (24,600 ft).

On 21 June 1972, a Lama set a helicopter absolute height record of 12,442 m (40,820 ft). The pilot was Jean Boulet, holder of the previous record in an SE 3150 Alouette.

The production Lama is capable of transporting an external load of 1,135 kg (2,500 lb) at an altitude of more than 2,500 m (8,200 ft). In an agricultural role, it can be fitted with spraybars and an underbelly tank of 1,135 litres (250 Imp gallons; 300 US gallons) capacity, developed jointly by Aérospatiale Helicopter Corporation and Simplex Manufacturing Company. The tank is equipped with an electrical emergency dump system.

An alternative agricultural installation, using two side-mounted glassfibre tanks of Simplex manufacture, was shown for the first time at the 1976 US National Agriculture Aviation Association convention. Up to 1,000 kg (2,200 lb) of liquid chemicals can be carried in these tanks, which weigh 132 kg (290 lb) empty. A high-performance electric pump dispenses up to 455 litres (100 Imp gallons; 120 US gallons)/min at 2·06 bars (30 lb/sq in) boom pressure.

A total of 285 Lamas had been delivered by 1 January 1981. In addition to manufacture by Aérospatiale, the SA 315B is produced under licence by HAL for the Indian Army, under the name Cheetah; and is assembled by Helibras in Brazil under the name Gavião.

TYPE: Turbine-driven general-purpose helicopter.
ROTOR SYSTEM AND ROTOR DRIVE:
As for Alouette III.
FUSELAGE: Glazed cabin has light metal frame. Centre and rear of fuselage have a triangulated steel tube framework.
LANDING GEAR: Skid type, with removable wheels for ground manoeuvring. Pneumatic floats for normal operation from water, and emergency flotation gear, inflatable in the air, are available.
POWER PLANT: One 649 kW (870 shp) Turboméca Artouste IIIB turboshaft engine, derated to 410 kW (550 shp). Fuel tank in fuselage centre-section, with capacity of 575 litres (126·5 Imp gallons), of which 573 litres (126 Imp gallons) are usable.
ACCOMMODATION: Glazed cabin seats pilot and passenger side by side in front and three passengers behind. Provision for external sling for loads of up to 1,135 kg (2,500 lb). Can be equipped for rescue (hoist capacity 160 kg; 352 lb), liaison, observation, training, agricultural, photographic and other duties. As an ambulance, can accommodate two stretchers and a medical attendant.

DIMENSIONS, EXTERNAL:
Main rotor diameter 11·02 m (36 ft 1¾ in)
Tail rotor diameter 1·91 m (6 ft 3¼ in)
Main rotor blade chord (constant) 0·35 m (13·8 in)
Length overall, both rotors turning 12·92 m (42 ft 4¾ in)
Length of fuselage 10·24 m (33 ft 7¼ in)
Height overall 3·09 m (10 ft 1¾ in)
Skid track 2·38 m (7 ft 9¾ in)

WEIGHTS:
Weight empty 1,021 kg (2,251 lb)
Normal max T-O weight 1,950 kg (4,300 lb)
Max T-O weight with externally-slung cargo 2,300 kg (5,070 lb)

PERFORMANCE (A, at AUW of 1,950 kg; 4,300 lb. B, at AUW of 2,300 kg; 5,070 lb, with slung load):
Never-exceed speed at S/L:
A 113 knots (210 km/h; 130 mph)
Max cruising speed:
A 103 knots (192 km/h; 119 mph)
B 65 knots (120 km/h; 75 mph)
Max rate of climb at S/L: A 330 m (1,080 ft)/min
B 234 m (768 ft)/min
Service ceiling: A 5,400 m (17,715 ft)
B 3,000 m (9,840 ft)
Hovering ceiling IGE: A 5,050 m (16,565 ft)
B 2,950 m (9,675 ft)
Hovering ceiling OGE: A 4,600 m (15,090 ft)
B 1,550 m (5,085 ft)
Range with max fuel: A 278 nm (515 km; 320 miles)

AÉROSPATIALE SA 316B ALOUETTE III

The Alouette III helicopter was developed from the Alouette II, with larger cabin, greater power, improved equipment and higher performance. The prototype flew for the first time on 28 February 1959, and a total of 1,419 Alouette IIIs had been delivered by 1 January 1981.

Those delivered up to the end of 1969 were designated **SE 3160.** The subsequent Artouste-engined **SA 316B** has strengthened main and rear rotor transmissions, higher AUW and increased payload. It flew for the first time on 27 June 1968; first deliveries were made in 1970, and this version received FAA Type Approval on 25 March 1971. The **SA 319B,** with Astazou engine, is described separately, but is included in the total sales figures above.

The sale of Alouette IIIs to India, Romania and Switzerland included licence agreements for manufacture of the aircraft in those countries. Quantities involved were 250 in India, 180 in Romania and 60 in Switzerland.

In 1977, an SA 316B operated by Trans North Turbo Air for the Canadian Park Service evacuated a mountain climber, suffering from pulmonary edema and frostbite, from a point at the 4,235 m (13,900 ft) level on the east ridge of Mount Logan. This is thought to be the greatest height at which a hoist rescue has been effected.

TYPE: Turbine-driven general-purpose helicopter.
ROTOR SYSTEM: Three-blade main and anti-torque rotors. All-metal main rotor blades, of constant chord, on articulated hinges, with hydraulic drag-hinge dampers. Rotor brake standard.
ROTOR DRIVE: Ma n rotor driven through planetary gearbox, with freewheel for autorotation. Take-off drive for tail rotor at lower end of main gearbox, from where a torque shaft runs to a small gearbox which supports the tail rotor and houses the pitch-change mechanism. Cyclic and collective pitch controls are powered.
FUSELAGE: Welded steel tube centre-section, carrying the cabin at the front and a semi-monocoque tailboom.
TAIL UNIT: Cantilever all-metal fixed tailplane, with twin endplate fins, mounted on tailboom.
LANDING GEAR: Non-retractable tricycle type, manufactured by Messier-Hispano-Bugatti. Nosewheel is fully-castoring. Provision for pontoon landing gear.
POWER PLANT: One 649 kW (870 shp) Turboméca Artouste IIIB turboshaft engine, derated to 425 kW (570 shp). Fuel in single tank in fuselage centre-section, with capacity of 575 litres (126·5 Imp gallons), of which 573 litres (126 Imp gallons) are usable.
ACCOMMODATION: Normal accommodation for pilot and six persons, with three seats in front and a four-person folding seat at the rear of the cabin. Two baggage holds in centre-section, on each side of the welded structure and enclosed by the centre-section fairings. Provision for carrying two stretchers athwartships at rear of cabin, and two other persons, in addition to pilot. All passenger seats removable to enable aircraft to be used for freight-carrying. Provision for external sling for loads of up to 750 kg (1,650 lb). One forward-opening door on each side, immediately in front of two rearward-sliding doors. Dual controls and cabin heating optional.
OPERATIONAL EQUIPMENT (military version): In the assault role, the Alouette III can be equipped with a wide range of weapons. A 7·62 mm AA52 machine-gun (with 1,000 rds) can be mounted athwartships on a tripod behind the pilot's seat, firing to starboard, either through a small window in the sliding door or through the open doorway with the door locked open. The rear seat is removed to allow the gun mounting to be installed. In this configuration, max accommodation is for pilot, co-pilot, gunner and one passenger, although normally only the pilot and gunner would be carried. Alternatively, a 20 mm MG 151/20 cannon (with 480 rds) can be carried on an open turret-type mounting on the port side of the cabin. For this installation all seats except that of the pilot are removed, as is the port side cabin door, and the crew consists of pilot and gunner. Instead of these guns, the Alouette III can be equipped with four AS.11 or two AS.12 wire-guided missiles on external jettisonable launching rails, with an APX-Bézu 260 gyro-stabilised sight, or 68 mm rocket pods.

DIMENSIONS, EXTERNAL:
Diameter of main rotor 11·02 m (36 ft 1¾ in)
Main rotor blade chord (each) 0·35 m (13·8 in)
Diameter of tail rotor 1·91 m (6 ft 3¼ in)
Length overall, rotors turning 12·84 m (42 ft 1½ in)
Length of fuselage, tail rotor turning 10·17 m (33 ft 4½ in)
Width overall, blades folded 2·60 m (8 ft 6¼ in)
Height to top of rotor head 3·00 m (9 ft 10 in)
Wheel track 2·60 m (8 ft 6¼ in)

WEIGHTS:
Weight empty 1,143 kg (2,520 lb)
Max T-O weight 2,200 kg (4,850 lb)

PERFORMANCE (standard version, at max T-O weight):
Never-exceed speed at S/L 113 knots (210 km/h; 130 mph)
Max cruising speed at S/L 100 knots (185 km/h; 115 mph)
Max rate of climb at S/L 260 m (850 ft)/min
Service ceiling 3,200 m (10,500 ft)
Hovering ceiling IGE 2,850 m (9,350 ft)
Hovering ceiling OGE 1,500 m (4,920 ft)
Range with max fuel at S/L 267 nm (495 km; 307 miles)
Range at optimum altitude 290 nm (540 km; 335 miles)

AÉROSPATIALE SA 319B ALOUETTE III ASTAZOU

Indian military name: Chetak

The SA 319B Alouette III Astazou is a direct development of the SA 316B, from which it differs principally in having an Astazou XIV turboshaft engine (649 kW; 870 shp, derated to 447 kW; 600 shp) with increased thermal efficiency and a 25 per cent reduction in fuel consumption.

A prototype SA 319 was completed in 1967. The production total is included in the figures given under the SA 316B entry. In addition to versions comparable with those of the SA 316B, a specially-equipped naval version is available, as follows:

OPERATIONAL EQUIPMENT (naval version): The Alouette III can fulfil a variety of shipborne roles; features common to all naval configurations include a quick-mooring harpoon to ensure instant and automatic mooring on landing and before take-off, a nosewheel locking device, and folding main rotor blades. For detecting and destroying small surface craft such as torpedo-boats, it can be equipped with a SFENA three-axis stabilisation system, OMERA ORB 31 radar, APX-Bézu 260 gyro-stabilised sight and two AS.12 wire-guided missiles. For the ASW role, it can carry two Mk 44 homing torpedoes beneath the fuselage, or one torpedo and MAD (magnetic anomaly detection) gear in a streamlined

Aérospatiale SA 315B Lama (Turboméca Artouste IIIB turboshaft engine)

Aérospatiale SA 319B Alouette III Astazou operated by the French Gendarmerie

container which is towed behind the helicopter on a 50 m (150 ft) cable. The aircraft can be used for air/sea rescue when the cabin floor is protected by an anti-corrosion covering to prevent sea water from reaching vital components. Rescue hoist (capacity 225 kg; 500 lb) mounted on port side of fuselage.

WEIGHTS (standard SA 319B):
- Weight empty 1,146 kg (2,527 lb)
- Max T-O weight 2,250 kg (4,960 lb)

PERFORMANCE (standard SA 319B, at max T-O weight):
- Max level speed at S/L 118 knots (220 km/h; 136 mph)
- Max cruising speed at S/L 106 knots (197 km/h; 122 mph)
- Max rate of climb at S/L 270 m (885 ft)/min
- Hovering ceiling IGE 3,100 m (10,170 ft)
- Hovering ceiling OGE 1,700 m (5,575 ft)
- Range with 6 passengers (80 kg; 176 lb each), T-O at S/L 325 nm (605 km; 375 miles)

AÉROSPATIALE SA 321 SUPER FRELON

The Super Frelon is a three-engined multi-purpose helicopter derived from the smaller SA 3200 Frelon (see 1961-62 *Jane's*).

Under a technical co-operation contract, Sikorsky Aircraft, USA, provided assistance in the development of the Super Frelon, in particular with the detail specifications, design, construction and testing of the main and tail rotor systems. Under a further agreement, the main gearcase and transmission box are produced in Italy by Fiat.

The first prototype of the Super Frelon (originally designated SA 3210-01) flew on 7 December 1962, powered by three 985 kW (1,320 shp) Turmo IIIC$_2$ engines, and represented the troop transport version. In July 1963 this aircraft set up several international helicopter records (since beaten), including a speed of 184 knots (341 km/h; 212 mph) over a 3 km course, and a speed of 189·115 knots (350·47 km/h; 217·77 mph) over a 15/25 km course.

The second prototype, flown on 28 May 1963, was representative of the naval version, with stabilising floats on the main landing gear supports. Four pre-production aircraft followed, and the French government ordered an initial production series of 17, designated SA 321G, in October 1965. By 1 January 1981, a total of 99 Super Frelons had been sold and 95 delivered.

Passenger and utility versions of the Super Frelon are available, and the main differences between the current versions are summarised as follows:

SA 321F. Commercial airliner, designed to carry 34-37 passengers in a standard of comfort comparable to that of fixed-wing airliners, over 94 nm (175 km; 108 mile) stage lengths at a cruising speed of 124 knots (230 km/h; 143 mph), with 20 min reserve fuel. The prototype was designed in accordance with US FAR Pt 29 regulations and flew for the first time on 7 April 1967. Type certification was granted by the SGAC on 27 June 1968 and by the FAA on 29 August 1968.

SA 321G. Anti-submarine helicopter. First version of the SA 321 to enter production. The first SA 321G flew on 30 November 1965 and deliveries began in early 1966. Twenty-four built. In service with Flottille 32F of Aéronavale, which was commissioned at Lanvéoc-Poulmic on 5 May 1970. Duties of this squadron include patrols in support of *Redoutable* class nuclear submarines entering and leaving their base on the Île Longue. The SA 321G can also be operated from the French helicopter carrier *Jeanne d'Arc*. Under a contract placed in late 1979, the original ORB 31 Héraclès I radar fitted to these aircraft is being replaced with ORB 32 Héraclès II of doubled performance in terms of power (80 kW) and scan (360°).

SA 321H. Version for air force and army service, without stabilising floats or external fairings on each side of lower fuselage. Turmo IIIE$_6$ engines instead of Turmo IIIC$_6$ in other versions. No de-icing equipment fitted.

SA 321Ja. Utility and public transport version, intended to fulfil the main roles of personnel and cargo transport. Designed to carry a maximum of 27 passengers. External loads of up to 5,000 kg (11,023 lb) can be suspended from the cargo sling and carried 27 nm (50 km; 31 miles), the aircraft returning to base without load. An internal payload of 4,000 kg (8,818 lb) can be carried over 100 nm (185 km; 115 miles) at 124 knots (230 km/h; 143 mph) with 20 min fuel reserves. The SA 321J prototype flew for the first time on 6 July 1967. A French certificate of airworthiness was granted in December 1971.

The following description applies generally to current production models of the Super Frelon, except where specific variants are indicated:

TYPE: Three-engined heavy-duty helicopter.

ROTOR SYSTEM: Six-blade main rotor and five-blade anti-torque tail rotor. Main rotor head consists basically of two six-armed star-plates carrying the drag and flapping hinges for each blade. The root of each blade carries a fitting for pitch control and each blade has an individual hydraulic damper to govern movement in the drag plane. Each main blade is 8·60 m (28 ft 2½ in) long, with constant chord and NACA 0012 section. All-metal construction, with D-section main spar forming leading-edge. Tail rotor of similar construction to main rotor, with blades 1·60 m (5 ft 3 in) long. Rearward folding of all six main rotor blades of SA 321G is accomplished automatically by hydraulic jacks, simultaneously with automatic folding of the tail rotor pylon.

ROTOR DRIVE: The driveshaft from the rear engine is geared directly to the shaft from the port forward engine. The two forward engines have a common reduction gear from which an output shaft drives the main rotor shaft through helical gearing. There are two reduction gear stages on the main rotor shaft. The tail rotor shaft is driven by gearing from the shaft linking the rear and port forward engines and incorporates two-stage reduction. The rotor can be stopped within 40 s by a boosted disc-type rotor brake fitted to this shaft. Main rotor rpm 207 and 212. Tail rotor rpm 990.

FUSELAGE: Boat-hull fuselage of conventional metal semi-monocoque construction, with watertight compartments inside planing bottom. On the SA 321G, there is a small stabilising float attached to the rear landing gear support structure on each side. The tail section of the SA 321G folds for stowage. Small fixed stabiliser on starboard side of tail rotor pylon on all versions. The SA 321F does not have stabilising floats, but large external fairings on each side of the centre-fuselage serve a similar purpose and also act as baggage containers.

LANDING GEAR: Non-retractable tricycle type, by Messier-Hispano-Bugatti. Twin wheels on each unit. Oleo-pneumatic shock-absorbers can be shortened on the SA 321G to reduce height of aircraft for stowage. Magnesium alloy wheels, all of same size. Tyre pressure 6·9 bars (100 lb/sq in). Optionally, low-pressure (3·45 bars; 50 lb/sq in) tyres may be fitted. Hydraulic disc brakes on main wheels. Nosewheel unit is steerable and self-centering.

POWER PLANT: Three 1,170 kW (1,570 shp) Turboméca Turmo IIIC$_6$ turboshaft engines (IIIE$_6$ in SA 321H); two mounted side by side forward of main rotor shaft and one aft of rotor shaft. Fuel in flexible tanks under floor of centre fuselage, with total standard capacity of 3,975 litres (874 Imp gallons) in SA 321G/H and 3,900 litres (858 Imp gallons) in SA 321Ja. Optional auxiliary fuel tankage comprises two 500 litre (110 Imp gallon) external tanks on all models, two 500 litre (110 Imp gallon) internal tanks in the SA 321G, and three 666 litre (146·5 Imp gallon) internal tanks in the SA 321H/Ja.

ACCOMMODATION (military versions): Crew of two on flight deck, with dual controls and advanced all-weather equipment. Equipment in the SA 321G, which carries a flight crew of five, includes a tactical table and a variety of devices for anti-submarine detection and attack, towing, minesweeping and other duties. This version also has provision for carrying 27 passengers. SA 321H transport accommodates 27-30 troops, 5,000 kg (11,023 lb) of internal or external cargo, or 15 stretchers and two medical attendants. Rescue hoist of 275 kg (606 lb) capacity. Main cabin is ventilated and soundproofed. Sliding door on starboard side of front fuselage. Rear loading ramp is actuated hydraulically and can be opened in flight.

ACCOMMODATION (SA 321F): Airliner seats for up to 37 passengers (34 if toilets are installed) in three-abreast rows with centre aisle. Alternative layouts for 8, 14 or 23 passengers, with toilets, or 11, 17 or 26 passengers

Aérospatiale SA 321G Super Frelon of Aéronavale, equipped for development launches of Exocet anti-ship missiles

without toilets, the remainder of the cabin space being blanked off by movable partitions and used for the carriage of freight; with these configurations, unused seats are folded against the cabin wall. All seats and interior furnishings are designed for quick removal when the helicopter is to be used for all-freight services. To cater for operations over marshland or water, the hull and lateral cargo compartments are sealed sufficiently to permit an occasional landing on water.

ACCOMMODATION (SA 321Ja): Seating for up to 27 passengers in the personnel transport role. As a cargo transport, external loads of up to 5,000 kg (11,023 lb) can be suspended from the cargo sling. Loading of internal cargo (up to 5,000 kg; 11,023 lb) is effected via the rear ramp-doors, with the assistance of a Tirefor hand winch.

OPERATIONAL EQUIPMENT: The ASW SA 321G operates normally in tactical formations of three or four aircraft, each helicopter carrying the full range of detection, tracking and attack equipment, including a self-contained navigation system associated with a Doppler radar, a 360° radar with transponder and display console, and dipping sonar. Four homing torpedoes can be carried in pairs on each side of the main cabin. Both the SA 321G and H can be fitted with an anti-surface-vessel weapon system, consisting of two Exocet missiles and launch installation, associated with an Omera-Segid Héraclès ORB 31D or ORB 32 radar for target designation.

DIMENSIONS, EXTERNAL:

Diameter of main rotor	18·90 m (62 ft 0 in)
Main rotor blade chord (each)	0·54 m (1 ft 9¼ in)
Diameter of tail rotor	4·00 m (13 ft 1½ in)
Tail rotor blade chord (each)	0·30 m (11¾ in)
Length overall, rotors turning	23·03 m (75 ft 6⅝ in)
Length of fuselage, tail rotor turning	20·08 m (65 ft 10¾ in)
Length of fuselage	19·40 m (63 ft 7¾ in)
Length overall: SA 321G, blades and tail folded	17·07 m (56 ft 0 in)
Width overall: SA 321F, incl baggage containers	5·04 m (16 ft 6⅜ in)
SA 321G, blades and tail folded	5·20 m (17 ft 0¾ in)
Width of fuselage	2·24 m (7 ft 4¼ in)
Height over tail rotor (normal)	6·76 m (22 ft 2¼ in)
Height overall: SA 321G, blades and tail folded	4·94 m (16 ft 2½ in)
Wheel track	4·30 m (14 ft 1 in)
Wheelbase	6·56 m (21 ft 6¼ in)
Cabin door: Height	1·55 m (5 ft 1 in)
Width	1·20 m (3 ft 11¼ in)
Rear loading ramp: Length	1·90 m (6 ft 2¾ in)
Width	1·90 m (6 ft 2¾ in)

DIMENSIONS, INTERNAL:

Cabin: Length: SA 321F	9·67 m (31 ft 9 in)
SA 321G and Ja	7·00 m (22 ft 11½ in)
Width: SA 321F	1·96 m (6 ft 5 in)
SA 321G and Ja, at floor	1·90 m (6 ft 2¾ in)
Height: SA 321F	1·80 m (5 ft 11 in)
SA 321G and Ja	1·83 m (6 ft 0 in)
Usable volume: SA 321G and Ja	25·3 m³ (893 cu ft)

WEIGHTS:

Weight empty, standard aircraft:	
SA 321G	6,863 kg (15,130 lb)
SA 321H	6,702 kg (14,775 lb)
SA 321Ja	6,868 kg (15,141 lb)
Max T-O weight	13,000 kg (28,660 lb)

PERFORMANCE (at max T-O weight):

Never-exceed speed at S/L	148 knots (275 km/h; 171 mph)
Cruising speed at S/L	134 knots (248 km/h; 154 mph)
Cruising speed at S/L, one engine out	113 knots (210 km/h; 130 mph)
Max rate of climb at S/L	300 m (985 ft)/min
Rate of climb at S/L, one engine out	146 m (479 ft)/min
Service ceiling	3,100 m (10,170 ft)
Service ceiling, one engine out	1,200 m (3,940 ft)
Hovering ceiling IGE	1,950 m (6,400 ft)
Normal range at S/L	440 nm (815 km; 506 miles)
Normal range at S/L, one engine out	496 nm (920 km; 572 miles)
Range at S/L with 3,500 kg (7,716 lb) payload	549 nm (1,020 km; 633 miles)
Endurance in ASW role	4 h

AÉROSPATIALE SA 330 PUMA

The twin-engined SA 330 Puma was developed initially to meet a French Army requirement for a medium-sized *hélicoptère de manoeuvre,* able to operate by day or night in all weathers and all climates. In 1967, the SA 330 was selected for the RAF Tactical Transport Programme, and was included in the joint production agreement between Aérospatiale and Westland in the UK.

The first of two SA 330 prototypes flew on 15 April 1965, and the last of six pre-production models on 30 July 1968, followed in September 1968 by the first production aircraft.

Details of six early versions of the Puma can be found in the 1976-77 *Jane's.* Current production versions are as follows:

SA 330J/L. Civil (J) and military (L) versions introduced in 1976 with main rotor blades of composite materials. Increased max T-O weight.

On 25 April 1978, the SA 330J became the first helicopter outside the Soviet Union to be certificated for all-weather operations, including flight in icing conditions. Equipment for this comprises thermal de-icing of the main rotor blades; thermal anti-icing of the tail rotor blades; special lengthened air intakes to ensure normal air supply into engines regardless of ambient conditions, including protection against sand and sea spray; and installation of weather radar.

A total of 646 Pumas had been delivered by 1 January 1981. Nurtanio of Indonesia (which see) is engaged in the assembly under licence of both Puma and Super Puma helicopters.

TYPE: Medium-sized transport helicopter.

ROTOR SYSTEM: Four-blade main rotor, with a fully-articulated hub and integral rotor brake. The blade cuffs, equipped with horns, are connected by link-rods to the swashplate, which is actuated by three hydraulic twin-cylinder servo-control units. Each of the moulded blades is made up of a glassfibre roving spar, a composite glassfibre and carbonfibre fabric skin, with Moltoprene/honeycomb filler. The leading-edge is covered with a stainless steel protective section. Attachment of each blade to its sleeve by two quick-disconnect pins enables blades to be folded back quickly by manual methods. The five-blade tail rotor has flapping hinges only, and is located on the starboard side of the tailboom. Optional blade de-icing system, with heating mat protected by titanium shielding on leading-edge of each main and tail rotor blade.

ROTOR DRIVE: Mechanical shaft and gear drive. Main gearbox, mounted on top of cabin behind engines, has two separate inputs from the engines and five reduction stages. The first stage drives, from each engine, an intermediate shaft directly driving the alternator and the ventilation fan, and indirectly driving the two hydraulic pumps. At the second stage the action of the two units becomes synchronised on a single main driveshaft by means of freewheeling spur gears. If one or both engines are stopped, this enables the drive gears to be rotated by the remaining turbine or the autorotating rotor, thus maintaining drive to the ancillary systems when the engines are stopped. Drive to the tail rotor is via shafting and an intermediate angle gearbox, terminating at a right-angle tail rotor gearbox. Turbine output 23,000 rpm, main rotor shaft 265 rpm. Tail rotor shaft 1,278 rpm. The hydraulically-controlled rotor brake, installed on the main gearbox, permits stopping of the rotor 15 s after engine shutdown.

Aérospatiale SA 330J Puma (two Turboméca Turmo IVC turboshaft engines) *(Richard Killen)*

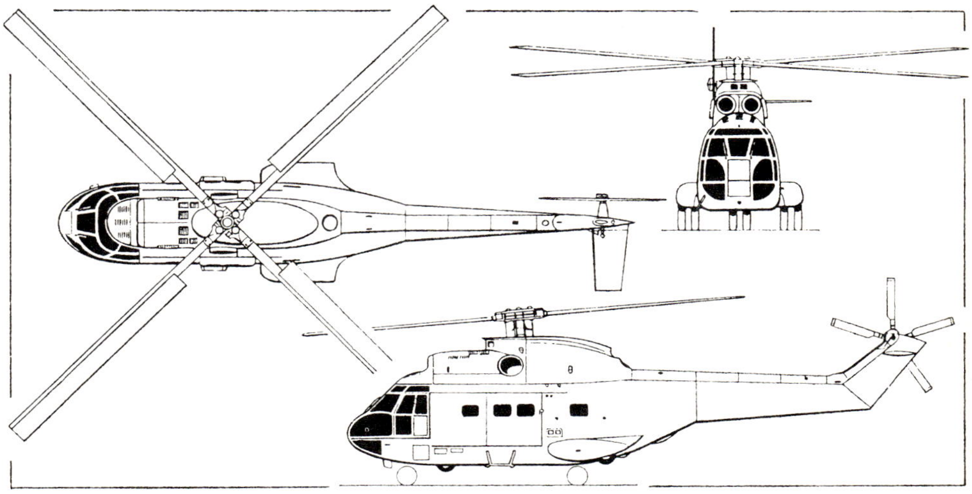
Aérospatiale SA 330 Puma transport helicopter *(Pilot Press)*

FUSELAGE: Conventional all-metal semi-monocoque structure. Local use of titanium alloy under engine installation, which is outside the main fuselage shell. Monocoque tailboom supports the tail rotor on the starboard side and a horizontal stabiliser on the port side.

LANDING GEAR: Messier-Hispano-Bugatti semi-retractable tricycle type, with twin wheels on each unit. Main units retract upward hydraulically into fairings on sides of fuselage; self-centering nose unit retracts rearward. When landing gear is down, the nosewheel jack is extended and the main-wheel jacks are telescoped. Dual-chamber oleo-pneumatic shock-absorbers. All tyres same size (7·00-6), of Dunlop or Kléber-Colombes tubeless type, pressure 6·0 bars (85 lb/sq in) on all units. Hydraulic differential disc brakes, controlled by foot pedals. Lever-operated parking brake. Emergency pop-out flotation units can be mounted on rear landing gear fairings and forward fuselage.

POWER PLANT: Two Turboméca Turmo IVC turboshaft engines, each with max rating of 1,175 kW (1,575 shp) and fitted with intake anti-icing. Engines are mounted side by side above cabin forward of the main rotor assembly and separated by a firewall. They are coupled to the main rotor transmission box, with shaft drive to the tail rotor, and form a completely independent system from the fuel tanks up to the main gearbox inputs. Fuel in four flexible tanks and one auxiliary tank beneath cargo compartment floor, with total capacity of 1,544 litres (339·5 Imp gallons). Provision for additional 1,900 litres (418 Imp gallons) in four auxiliary ferry tanks installed in cabin. External auxiliary tanks (two, each 350 litres; 77 Imp gallons capacity) are available. For long-range missions (mainly offshore) one or two special internal tanks (each 215 litres; 47·25 Imp gallons) can be fitted in the cabin. Each engine is supplied normally by a pair of interconnected primary tanks, the lower halves of which have self-sealing walls for protection against small-calibre projectiles. Refuelling point on starboard side of main cabin. Oil capacity 22 litres (4·8 Imp gallons) for engines, 25·5 litres (5·6 Imp gallons) for transmission.

ACCOMMODATION: Crew of one or two side by side on anti-crash seats on flight deck, with jump-seat for third crew member if required. Door on each side of flight deck on current versions. Internal doorway connects flight deck to cabin, with folding seat in doorway for an extra crew member or cargo supervisor. Dual controls standard. Accommodation in main cabin for 16 individually-equipped troops, six stretchers and six seated patients, or equivalent freight. The number of troops can be increased to 20 in the high-density version. Strengthened floor for cargo-carrying, with lashing points. Jettisonable sliding door on each side of main cabin; or port-side door with built-in steps and starboard-side double door in VIP or airline configurations. Removable panel on underside of fuselage, at rear of main cabin, permits longer loads to be accommodated and also serves as emergency exit on SA 330L version. Removable door with integral steps for access to baggage racks on SA 330J version. A hatch in the floor below the centreline of the main rotor is provided for carrying loads of up to 3,200 kg (7,055 lb) on an internally-mounted cargo sling. A fixed or retractable rescue hoist (capacity 275 kg; 606 lb) can be mounted externally on the starboard side of the fuselage. The cabin can be equipped in 8/9/12-seat VIP, 17-seat commuter or 20-seat high-density layouts, with baggage compartment and/or toilet facilities in rear of cabin. Cabin and flight deck are heated, ventilated and soundproofed. Demisting, de-icing, washers and wipers for pilots' windscreens.

Systems: Two independent hydraulic systems, each 172 bars (2,500 lb/sq in), supplied by self-regulating pumps driven by the main gearbox. Each system supplies one set of servo unit chambers, the left-hand system supplying in addition the autopilot, landing gear, rotor brake and wheel brakes. Freewheels in main gearbox ensure that both systems remain in operation, for supplying the servo-controls, if the engines are stopped in flight. Other hydraulically-actuated systems can be operated on the ground from the main gearbox, or by external power through the ground power receptacle. There is also an independent auxiliary system, fed through a handpump, which can be used in an emergency to lower the landing gear and pressurise the accumulator for the parking brake on the ground. Three-phase 200V AC electrical power supplied by two 15kVA 400Hz alternators, driven by the port side intermediate shaft from the main gearbox and available on the ground under the same conditions as the hydraulic ancillary systems. 28·5V 10kW DC power provided from the AC system by two transformer-rectifiers. Main aircraft battery used for self-starting and emergency power in flight. For the latter purpose, an emergency 400VA inverter can supply the essential navigation equipment from the battery, permitting at least 20 min continued flight in the event of a main power failure. De-icing of engines and engine air intakes by warm air bled from compressor. Anti-snow shield for Winter operations.

Avionics and Equipment: Optional communications equipment includes VHF, UHF, tactical HF and HF/SSB radio installations and intercom system. Navigational equipment includes radio compass, radio altimeter, VLF Omega, Decca navigator and flight log, Doppler, and VOR/ILS with glidepath. Autopilot, with provision for coupling to self-contained navigation and microwave landing systems. Full IFR instrumentation available optionally. The search and rescue version has nose-mounted Bendix RDR 1400 or RCA Primus 40 or 50 search radar, Doppler, and Decca self-contained navigation system, including navigation computer, polar indicator, roller-map display, hover indicator, route mileage indicator and ground speed and drift indicator.

Armament (optional): A wide range of armament can be carried, including side-firing 20 mm cannon, axial-firing 7·62 mm machine-guns, missiles and rockets.

Dimensions, external:	
Diameter of main rotor	15·00 m (49 ft 2½ in)
Diameter of tail rotor	3·04 m (9 ft 11½ in)
Distance between rotor centres	9·20 m (30 ft 2¼ in)
Blade chord, main rotor	0·60 m (1 ft 11½ in)
Ground clearance of tail rotor	2·00 m (6 ft 6¾ in)
Length overall	18·15 m (59 ft 6½ in)
Length of fuselage	14·06 m (46 ft 1½ in)
Length, blades folded	14·80 m (48 ft 6¾ in)
Width, blades folded	3·50 m (11 ft 5¾ in)
Height overall	5·14 m (16 ft 10½ in)
Height to top of rotor hub	4·38 m (14 ft 4½ in)
Width over wheel fairings	3·00 m (9 ft 10 in)
Wheel track	2·38 m (7 ft 10¾ in)
Wheelbase	4·045 m (13 ft 3 in)
Passenger cabin doors, each:	
Height	1·35 m (4 ft 5 in)
Width	1·35 m (4 ft 5 in)
Height to sill	1·00 m (3 ft 3¼ in)
Floor hatch, rear of cabin:	
Length	0·98 m (3 ft 2¾ in)
Width	0·70 m (2 ft 3½ in)

Dimensions, internal:	
Cabin: Length	6·05 m (19 ft 10 in)
Max width	1·80 m (5 ft 10¾ in)
Max height	1·55 m (5 ft 1 in)
Floor area	7·80 m² (84 sq ft)
Usable volume	11·40 m³ (403 cu ft)

Areas:	
Main rotor blades (each)	4·00 m² (43 sq ft)
Tail rotor blades (each)	0·28 m² (3·01 sq ft)
Main rotor disc	177·0 m² (1,905 sq ft)
Tail rotor disc	7·30 m² (78·6 sq ft)
Horizontal stabiliser	1·34 m² (14·4 sq ft)

Weights:	
Weight empty, standard aircraft:	
SA 330J	3,766 kg (8,303 lb)
SA 330L	3,615 kg (7,970 lb)
Max T-O and landing weight	7,400 kg (16,315 lb)*

*Certification for T-O weight of 7,500 kg (16,535 lb), *for cargo-sling mission only,* is anticipated

Performance (SA 330J/L: A at 6,000 kg; 13,230 lb AUW, B at 7,400 kg; 16,315 lb AUW):	
Never-exceed speed:	
A	158 knots (294 km/h; 182 mph)
B	142 knots (263 km/h; 163 mph)
Max cruising speed:	
A	146 knots (271 km/h; 168 mph)
B	139 knots (258 km/h; 160 mph)
Max rate of climb at S/L:	
A	552 m (1,810 ft)/min
B	366 m (1,200 ft)/min
Service ceiling (30 m; 100 ft/min rate of climb):	
A	6,000 m (19,680 ft)
B	4,800 m (15,750 ft)
Hovering ceiling IGE:	
A, ISA	4,400 m (14,435 ft)
A, ISA +20°C	3,700 m (12,135 ft)
B, ISA	2,300 m (7,545 ft)
B, ISA +20°C	1,600 m (5,250 ft)
Hovering ceiling OGE:	
A, ISA	4,250 m (13,940 ft)
A, ISA +20°C	3,600 m (11,810 ft)
B, ISA	1,700 m (5,575 ft)
B, ISA +20°C	1,050 m (3,445 ft)
Max range at normal cruising speed, no reserves:	
A	309 nm (572 km; 355 miles)
B	297 nm (550 km; 341 miles)

AÉROSPATIALE AS 332 SUPER PUMA

Design of this derivative of the SA 330 Puma was started in 1974, and the programme received a formal go-ahead from the French government in June 1975. As a first stage, Aérospatiale retrofitted a Puma airframe with two Turboméca Makila turboshaft engines and an uprated transmission. This experimental helicopter, designated AS 331, flew for the first time on 5 September 1977. It was followed, on 13 September 1978, by the first flight of the prototype AS 332 Super Puma (F-WZJA), embodying more extensive changes to provide increased payload and performance, simplified maintenance, reduced cabin noise level, reduced vulnerability to hostile fire in combat areas, and better crew survivability in a crash. Original plans to fit a 'fenestron' ducted tail rotor were dropped after evaluation of a 'fenestron' on the SA 330Z testbed (see 1976-77 *Jane's*, page 812) indicated no worthwhile performance gains.

Aérospatiale AS 332C Super Puma twin-turbine multi-role helicopter

Externally evident airframe changes compared with the SA 330 Puma include a lengthened nose; increased wheelbase and wheel track; a new landing gear with a single wheel on each of the main units, which offer an optional 'kneeling' capability to reduce overall dimensions for shipboard stowage; and an added ventral fin. The main and tail rotor blades have a new and more efficient profile.

There are five versions of the Super Puma:

AS 332B. Military version. Standard fuselage, seating 20 troops.

AS 332C. Civil version. Standard fuselage, seating 17 passengers.

AS 332F. Naval version, with folding tail rotor pylon, deck landing assist device and anti-corrosion treatment. Suitable for search and rescue, ASW and anti-ship roles.

AS 332L. As 332C, but with cabin lengthened by 0·76 m (2 ft 6 in) to provide four more seats and two additional windows. Fuel capacity increased. Fourth production Super Puma (F-WZJN), to this standard, flew for the first time on 10 October 1980.

AS 332M. As 332B, but with cabin lengthened by 0·76 m (2 ft 6 in) to provide four more seats and two additional windows. Fuel capacity increased.

All variants will offer all-weather flight capability, including operation in forecast icing conditions. Authorised flight envelope for the AS 332C in mid-1981 was from −30° to +50°C (ISA +35°C).

DGAC certification of the AS 332C Super Puma, for VFR flight and IFR category A and B operation, was received on 24 April 1981. Orders for nearly 80 Super Pumas had been received by the beginning of May 1981, including 35 AS 332Ls for Bristow Helicopters, whose specially-equipped aircraft will be known as **Super Tigers**. Deliveries were scheduled to begin in mid-1981 (AS 332B/C) or 1982 (AS 332L/M). To facilitate integration of the aircraft into existing fleets of Pumas (notably in French Army service), Aérospatiale hopes to launch a programme under which SA 330 Pumas will be retrofitted with the engines, some dynamic components and the composite-blade tail rotor of the Super Puma.

Type: Twin-turbine multi-role helicopter.

Rotor System: Four-blade main rotor, with a fully-articulated hub and integral rotor brake. Each drag hinge is fitted with an elastomeric frequency adaptor. The blade cuffs, equipped with horns, are connected by link-rods to the swashplate, which is actuated by three hydraulic twin-cylinder servo-control units. Each of the moulded blades is made up of a glassfibre roving spar and a composite glassfibre and carbonfibre fabric skin, with Moltoprene filler. The leading-edge is covered with a titanium protective section. The tips are swept. Attachment of each blade to its sleeve by means of two quick-disconnect pins enables the blades to be folded back quickly by manual methods. The five-blade tail rotor has flapping hinges only, and is located on the starboard side of the tailboom. The rotors may be equipped with a de-icing system similar to that certificated for the Puma.

Rotor Drive: Mechanical shaft and gear drive. Modular main gearbox is fitted with two torquemeters and has two separate lubrication circuits. It is mounted on top of the cabin behind the engines, has two separate inputs from the engines and five reduction stages. The first stage drives, from each engine, an intermediate shaft directly driving the alternator and indirectly driving the two hydraulic pumps, with a further shaft drive to the ventilation fan. At the second stage the action of the two units becomes synchronised on a single main driveshaft by means of freewheeling spur gears. If one or both engines are stopped, this enables the drive gears to be rotated by the remaining turbine or the autorotating rotor, thus maintaining drive to the ancillary systems when the engines are stopped. Drive to the tail rotor is via shafting and an intermediate angle gearbox, terminating at a right-angle tail rotor gearbox. Turbine output 23,840 rpm, main rotor shaft 265 rpm. Tail rotor shaft 1,278 rpm. The hydraulically-controlled rotor brake, installed on the main gearbox, permits stopping of the rotor 15 s after engine shutdown.

Fuselage: Conventional all-metal semi-monocoque structure, embodying anti-crash features. Local use of titanium alloy under engine installation, which is outside the main fuselage shell. Monocoque tailboom supports the tail rotor on the starboard side and a horizontal stabiliser with fixed leading-edge slat on the port side. Optional folding tailboom for aircraft that will serve on ships such as frigates.

Landing Gear: Retractable tricycle type, of Messier-Hispano-Bugatti high energy-absorbing design. All units retract rearward hydraulically, main wheels into fairings on sides of fuselage. Dual-chamber oleo-pneumatic shock-absorbers. Optional 'kneeling' capability for main units. Twin-wheel self-centering nose unit, tyre size 466 × 176, pressure 6.0 bars (85 lb/sq in). Single wheel on each main unit with tyre size 615 × 225-10, pressure 6.0 bars (85 lb/sq in). Hydraulic differential disc brakes, controlled by foot pedals. Lever-operated parking brake. Emergency pop-out flotation units can be mounted on main landing gear fairings and forward fuselage.

Power Plant: Two Turboméca Makila IA turboshaft engines, each with max contingency rating of 1,327 kW (1,780 shp) and max continuous rating of 1,145 kW (1,535 shp). Air intakes protected by a grille against ingestion of ice, snow and foreign objects; but multi-purpose intake is necessary for flight into sandy areas. AS 332B/C have five flexible fuel tanks under cabin floor, with total capacity of 1,560 litres (343 Imp gallons). AS 332L/M have a basic fuel system of six flexible tanks with total capacity of 2,060 litres (453 Imp gallons). Provision for additional 1,900 litres (418 Imp gallons) in four auxiliary ferry tanks installed in cabin. Two external auxiliary tanks are available, with total capacity of 700 litres (154 Imp gallons). For long-range missions (mainly offshore), special internal auxiliary and external tanks can be fitted to raise the total fuel capacity to 2,590 litres (570 Imp gallons) in AS 332C, and 3,090 litres (680 Imp gallons) in AS 332L. This auxiliary tank fits in cargo-sling well beneath cabin floor and is quickly removable to permit use of sling. Refuelling point on starboard side of cabin. Fuel system is designed to avoid fuel leakage following a crash, with flexible fuel lines and interconnections between tanks, self-sealing valves and automatic fuel pump shutdown in a crash. Options include a fuel dumping system, pressure refuelling, and crash-resistant or self-sealing tanks.

Accommodation: One pilot (VFR) or two pilots side by side (IFR) on flight deck, with jump-seat for third crew member or paratroop dispatcher. Provision for composite light alloy/Kevlar armour for crew protection on military models. Door on each side of flight deck and internal doorway connecting flight deck to cabin. Dual controls, co-pilot instrumentation and anti-crash flight deck floor. AS 332B/C accommodate in main cabin up to 20 troops in normal seating, 16 troops in anti-crash seats, six stretchers and six seated casualties/attendants, 17 civilian passengers (18/20 in 'offshore' role), or eight, nine or twelve VIP passengers in special interiors with toilet and galley. AS 332L/M accommodate in main cabin up to 24 troops in normal seating, 18 troops in anti-crash seats, nine stretchers and three seated casualties/attendants, 21 civilian passengers (22/24 in 'offshore' role), or nine, ten or 15 VIP passengers with toilet and galley. Strengthened floor for cargo carrying, with lashing points. Jettisonable sliding door on each side of main cabin; or port-side door with built-in steps and starboard-side double door in VIP or airline configurations. Removable panel on underside of fuselage, at rear of main cabin, permits longer loads to be accommodated, and also serves as emergency exit. Removable door with integral steps for access to baggage racks optional. A hatch in the floor below the centreline of the main rotor is provided for carrying loads of up to 4,500 kg (9,920 lb) on an internally-mounted cargo sling. A fixed or retractable rescue hoist (capacity 275 kg; 606 lb) can be mounted externally on the starboard side of the fuselage. Cabin and flight deck are heated, ventilated and soundproofed. Demisting, de-icing, washers and wipers for pilots' windscreens.

Systems: Two independent hydraulic systems, supplied by self-regulating pumps driven by the main gearbox. Each system supplies one set of servo unit chambers, the left-hand system supplying in addition the autopilot, landing gear, rotor brake and wheel brakes. Freewheels in main gearbox ensure that both systems remain in operation, for supplying the servo-controls, if the engines are stopped in flight. Other hydraulically-actuated systems can be operated on the ground from the main gearbox (when a special disconnect system is installed to permit running of port engine with rotors stationary), or by external power through the ground power receptacle. There is also an independent auxiliary system, fed through a handpump, which can be used in an emergency to lower the landing gear. Three-phase 200V AC electrical power supplied by two 20kVA 400Hz alternators, driven by the port side intermediate shaft from the main gearbox and available on the ground under the same conditions as the hydraulic ancillary systems. 28·5V DC power provided from the AC system by two transformer-rectifiers. Main aircraft battery used for self-starting and emergency power in flight.

Avionics and Equipment: Optional communications equipment includes VHF, UHF, tactical HF and HF/SSB radio installations and intercom system. Navigational equipment includes radio compass, radio altimeter, VLF Omega, Decca navigator and flight log, Doppler, and VOR/ILS with glidepath. SFIM 155 autopilot, with provision for coupling to self-contained navigation and microwave landing systems. Full IFR instrumentation available optionally. 'Offshore' models have nose-mounted radar. The search and rescue version has nose-mounted Bendix RDR 1400 or RCA Primus 40 or 50 search radar, Doppler, and Crouzet Nadir or Decca self-contained navigation system, including navigation computer, polar indicator, roller-map display, hover indicator, route mileage indicator and ground speed and drift indicator. For naval ASW and ASV missions, aircraft can be fitted with nose-mounted OMERA type ORB 32 ASD 360° radar, linked to a tactical table.

Large additional window at the front of the cabin identifies the AS 332L version of the Super Puma

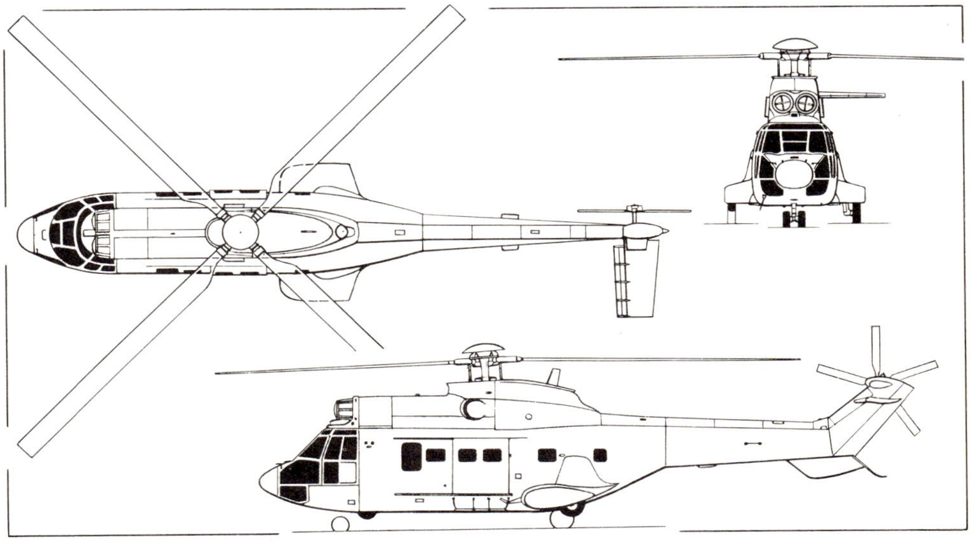
Aérospatiale AS 332L Super Puma, with lengthened cabin *(Pilot Press)*

Armament and Operational Equipment (optional): Typical alternatives for army/air force missions are one 20 mm gun, two 7·62 mm machine-guns, or two pods each containing twenty-two 68 mm rockets or nineteen 2·75 in rockets. Armament and equipment for naval missions includes two Exocet missiles, or two torpedoes and sonar, or MAD and sonobuoys.

Dimensions, external:

Diameter of main rotor	15·08 m (49 ft 5¾ in)
Diameter of tail rotor	3·04 m (9 ft 11½ in)
Blade chord, main rotor	0·60 m (1 ft 11½ in)
Length overall	18·48 m (60 ft 7½ in)
Length of fuselage, excl tail rotor:	
AS 332B/C	14·76 m (48 ft 5 in)
AS 332C 'Offshore'	15·49 m (50 ft 10 in)
AS 332L/M	15·52 m (50 ft 11 in)
AS 332L 'Offshore'	16·25 m (53 ft 3¾ in)
Length, blades folded:	
AS 332B/C	15·60 m (51 ft 2 in)
AS 332L/M	16·36 m (53 ft 8 in)
Length, blades and tail pylon folded:	
AS 332B/C	12·64 m (41 ft 5½ in)
AS 332F	12·84 m (42 ft 1½ in)
AS 332L/M	13·40 m (43 ft 11½ in)
Width, blades folded:	
AS 332B/C/L/M	3·79 m (12 ft 5¼ in)
AS 332F	4·04 m (13 ft 3 in)
Height overall	4·92 m (16 ft 1¾ in)
Height, blades and tail pylon folded:	
AS 332F	4·80 m (15 ft 9 in)
Height to top of rotor hub	4·48 m (14 ft 8½ in)
Width over wheel fairings	3·79 m (12 ft 5¼ in)
Wheel track	3·00 m (9 ft 10 in)
Wheelbase	4·49 m (14 ft 8¾ in)
Passenger cabin doors, each:	
Height	1·35 m (4 ft 5 in)
Width	1·35 m (4 ft 5 in)
Floor hatch, rear of cabin:	
Length	0·98 m (3 ft 2¾ in)
Width	0·70 m (2 ft 3½ in)

Dimensions, internal:

Cabin: Length: AS 332B/C	6·05 m (19 ft 10½ in)
AS 332L/M	6·81 m (22 ft 4 in)
Max width	1·80 m (5 ft 11 in)
Max height	1·55 m (5 ft 1 in)
Floor area: AS 332B/C	7·80 m² (84 sq ft)
Usable volume: AS 332B/C	11·40 m³ (403 cu ft)
AS 332L/M	13·30 m³ (469·5 cu ft)

Weights:

Weight empty:	
AS 332B/C, basic	4,120 kg (9,080 lb)
AS 332F	4,370 kg (9,635 lb)
AS 332L/M, basic	4,265 kg (9,402 lb)
Payload for offshore mission, excl fuel reserves:	
AS 332C, VFR	2,834 kg (6,248 lb)
AS 332C, IFR	2,734 kg (6,027 lb)
AS 332L, VFR	2,727 kg (6,012 lb)
AS 332L, IFR	2,627 kg (5,791 lb)
Max T-O weight:	
All versions, normal	8,200 kg (18,080 lb)
AS 332F, alternative	8,700 kg (19,180 lb)
All versions, with sling load	9,000 kg (19,840 lb)

Performance (all versions at AUW of 8,200 kg; 18,080 lb, except where noted):

Never-exceed speed	160 knots (296 km/h; 184 mph)
Max cruising speed at S/L:	
AS 332B/C/L/M	151 knots (280 km/h; 173 mph)
AS 332F	150 knots (278 km/h; 172 mph)
Econ cruising speed at S/L	140 knots (260 km/h; 161 mph)
Max rate of climb at S/L	551 m (1,810 ft)/min
Service ceiling	4,750 m (15,580 ft)
Hovering ceiling IGE	3,000 m (9,840 ft)
Hovering ceiling OGE	2,300 m (7,545 ft)
Range at S/L, no reserves:	
AS 332B/C, standard tanks	348 nm (644 km; 400 miles)
AS 332F/L/M, standard tanks	458 nm (848 km; 527 miles)

AÉROSPATIALE SA 341/342 GAZELLE

The first prototype of this five-seat lightweight helicopter (designated SA 340) made its first flight on 7 April 1967, powered by an Astazou III engine. It was followed by a second prototype on 12 April 1968 and then by four pre-production SA 341 Gazelles.

The first production SA 341 Gazelle flew for the first time on 6 August 1971, with a longer cabin than its predecessors, enlarged tail unit, additional door on the starboard side at rear (optional on production aircraft) and uprated Astazou IIIA engine.

Details of early versions of the Gazelle can be found in the 1979-80 *Jane's*. Versions being marketed by Aérospatiale in 1981 were as follows:

SA 341G. Civil version, with Astazou IIIA engine. Certificated by SGAC on 7 June 1972 and by the FAA on 18

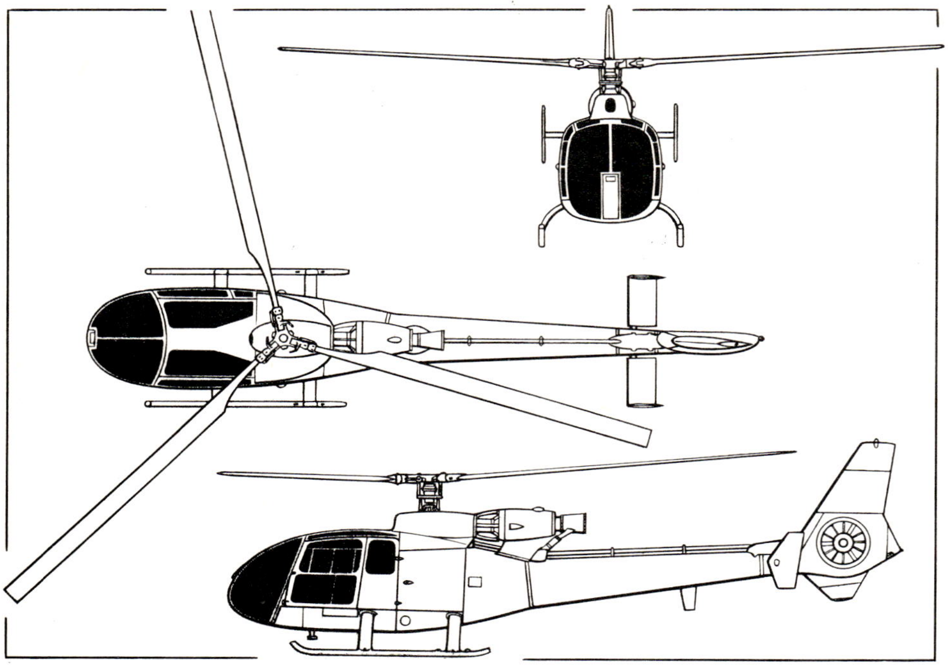

Aérospatiale SA 341 Gazelle five-seat light utility helicopter *(Pilot Press)*

September 1972. In January 1975, it was announced that the SA 341G had become the first helicopter in the world authorised to be flown by a single pilot under IFR Cat I conditions. It is now certificated for IFR Cat II operation, with a ceiling of 30 m (100 ft) and 365 m (1,200 ft) forward visibility. Equipment fitted to the aircraft which qualified for this FAA certification comprised a Sperry flight director coupled to SFENA servo-dampers. A variant, known as the **Stretched Gazelle,** has the rear portion of the cabin modified to provide an additional 20 cm (8 in) of legroom for the rear-seat passengers.

SA 342J. Similar to SA 342L, for commercial operators. Higher max T-O weight. Improved 'fenestron' tail rotor. Certificated by DGAC on 27 April 1976. Deliveries began in 1977.

SA 342L. Military counterpart of SA 342J, with improved 'fenestron' tail rotor. Astazou XIVH turboshaft engine.

SA 342M. For ALAT (French Army Light Aviation Corps). Differs from SA 342L in having an ALAT instrument panel and 640 kW (859 shp) Astazou XIVM turboshaft with automatic startup and high-energy ignition. Optional equipment specified by ALAT includes SFIM PA 85G autopilot, Crouzet Nadir self-contained navigation system, Decca 80 Doppler, infra-red homing, exhaust deflector, and night flying equipment. Order for first increment of planned total of about 120 announced in December 1978, each armed with four Hot missiles and gyro-stabilised sight for anti-tank warfare. Delivery to ALAT began on 9 June 1980.

A two-stretcher ambulance configuration has received FAA Standard Type Certification. No major modification is necessary to convert the aircraft to carry two patients longitudinally on the port side of the cabin, one above the other, leaving room for the pilot and a medical attendant in tandem on the starboard side. The dual spineboard arrangement weighs 27 kg (60 lb) and stows into the baggage compartment when not in use.

Under an Anglo-French agreement signed in 1967, Gazelles are produced jointly with Westland Helicopters Ltd, and are also built under licence in Yugoslavia. A total of 955 had been sold by Spring 1981.

Three Class E1c records were set by the SA 341-01 at Istres on 13 and 14 May 1971 and were unbeaten by mid-1981. These were: 167·28 knots (310·00 km/h; 192·62 mph) in a straight line over a 3 km course; 168·36 knots (312·00 km/h; 193·87 mph) in a straight line over a 15/25 km course; and 159·72 knots (296·00 km/h; 183·93 mph) over a 100 km closed circuit.

The following details apply to the SA 341, except where indicated:

TYPE: Five-seat light utility helicopter.

ROTOR SYSTEM: Three-blade semi-articulated main rotor and 13-blade shrouded-fan anti-torque tail rotor (known as a 'fenestron' or 'fan-in-fin'). Rotor head and rotor mast form a single unit. The main rotor blades are of NACA 0012 section, attached to NAT hub by flapping hinges. There are no drag hinges. Each blade has a single leading-edge spar of plastics material reinforced with glassfibre, a laminated glass-fabric skin and honeycomb filler. Tail rotor blades are of die-forged light alloy, with articulation for pitch change only. Main rotor blades can be folded manually for stowage. Rotor brake standard.

ROTOR DRIVE: Main reduction gearbox forward of engine, which is mounted above the rear part of the cabin. Intermediate gearbox beneath engine, rear gearbox supporting the tail rotor. Main rotor/engine rpm ratio 378·3 : 6,179. Tail rotor/engine rpm ratio 5,774 : 6,179.

FUSELAGE: Cockpit structure is based on a welded light alloy frame which carries the windows and doors. This is mounted on a conventional semi-monocoque lower structure consisting of two longitudinal box sections connected by frames and bulkheads. Central section, which encloses the baggage hold and fuel tank and supports the main reduction gearbox, is constructed of light alloy honeycomb sandwich panels. Rear section, which supports the engine and tailboom, is of similar construction. Honeycomb sandwich panels are also used for the cabin floors and transmission platform. Tailboom is of conventional sheet metal construction, as are the horizontal tail surfaces and the tail fin.

TAIL UNIT: Small horizontal stabiliser on tailboom, ahead of tail rotor fin.

LANDING GEAR: Steel tube skid type. Wheel can be fitted at rear of each skid for ground handling. Provision for alternative float or ski landing gear.

POWER PLANT: One Turboméca Astazou IIIA turboshaft engine, installed above fuselage aft of cabin and delivering 440 kW (590 shp) for take-off (max continuous rating also 440 kW; 590 shp). Main fuel tank in fuselage, usable capacity 445 litres (98 Imp gallons). Provision for 90 litre (19·8 Imp gallon) auxiliary tank beneath baggage compartment and/or 200 litre (44 Imp gallon) ferry tank inside rear cabin. Total possible usable fuel capacity 735 litres (161 Imp gallons). Refuelling point on starboard side of cabin. Oil capacity 13 litres (2·8 Imp gallons) for engine, 3·5 litres (0·77 Imp gallons) for gearbox.

ACCOMMODATION: Crew of one or two on side-by-side seats in front of cabin, with bench seat to the rear for a further three persons. The bench seat can be folded into floor wells to leave a completely flat cargo floor. Access to baggage compartment via rear cabin bulkhead, or via optional door on starboard side. Cargo tiedown points in cabin floor. Forward-opening car-type door on each side of cabin, immediately behind which are rearward-opening auxiliary cargo loading doors. Baggage compartment at rear of cabin. Ventilation standard. Dual controls optional.

SYSTEMS: Hydraulic system, pressure 40 bars (570 lb/sq in), serves three pitch change jacks for main rotor head and one for tail rotor. 28V DC electrical system supplied by 4kW engine-driven generator and 40Ah battery. Optional 26V AC system, supplied by 0·5kVA alternator at 115/200V 400Hz.

AVIONICS AND EQUIPMENT: Optional communications equipment includes UHF, VHF, HF, intercom systems and homing aids. Optional navigation equipment includes radio compass, radio altimeter and VOR. Blind-flying instrumentation optional. A variety of operational equipment can be fitted, according to role, including a 700 kg (1,540 lb) cargo sling, 135 kg (300 lb) rescue hoist, one or two stretchers (internally), or photographic and survey equipment.

ARMAMENT (SA 342): Military loads can include two pods of Matra or Brandt 2·75 in or 68 mm rockets, four AS.11 or two AS.12 wire-guided missiles with APX-Bézu 334 gyro-stabilised sight, four or six Hot wire-guided missiles with APX 397 gyro-stabilised sight, two forward-firing 7·62 mm machine-guns, reconnaissance flares or smoke markers.

DIMENSIONS, EXTERNAL:

Diameter of main rotor	10·50 m (34 ft 5½ in)
Diameter of tail rotor	0·695 m (2 ft 3⅜ in)
Distance between rotor centres	5·85 m (19 ft 2¼ in)
Main rotor blade chord (constant)	0·30 m (11·8 in)
Length overall	11·97 m (39 ft 3 5/16 in)
Length of fuselage	9·53 m (31 ft 3 3/16 in)
Width, rotors folded	2·015 m (6 ft 7 5/16 in)
Height to top of rotor hub	2·72 m (8 ft 11⅛ in)
Height overall	3·18 m (10 ft 5¼ in)
Skid track	2·015 m (6 ft 7 5/16 in)
Main cabin doors, each:	
Height	1·05 m (3 ft 4 9/16 in)
Width	1·00 m (3 ft 3¼ in)
Height to sill	0·63 m (2 ft 0¾ in)
Auxiliary cabin doors, each:	
Height	1·05 m (3 ft 4 9/16 in)
Width	0·48 m (1 ft 6¾ in)
Height to sill	0·63 m (2 ft 0¾ in)

DIMENSIONS, INTERNAL:

Cabin: Length	2·20 m (7 ft 2 9/16 in)
Max width	1·32 m (4 ft 4 in)
Max height	1·21 m (3 ft 11⅝ in)
Floor area	1·50 m^2 (16·1 sq ft)
Volume	1·80 m^3 (63·7 cu ft)
Baggage hold volume	0·45 m^3 (15·9 cu ft)

AREAS:

Main rotor blades, each	1·57 m^2 (16·9 sq ft)
Tail rotor blades, each	0·007 m^2 (0·075 sq ft)
Main rotor disc	86·5 m^2 (931 sq ft)
Tail rotor disc	0·37 m^2 (3·98 sq ft)
Fin	0·45 m^2 (4·84 sq ft)
Tailplane	1·80 m^2 (19·4 sq ft)

WEIGHTS AND LOADING:

Weight empty: 341G	917 kg (2,022 lb)
342L	975 kg (2,150 lb)
Max T-O and landing weight:	
341G	1,800 kg (3,970 lb)
342J/L	1,900 kg (4,190 lb)
Max disc loading: 341G	19·5 kg/m^2 (4 lb/sq ft)

PERFORMANCE (at max T-O weight):

Never-exceed speed at S/L:	
341, 342	167 knots (310 km/h; 193 mph)
Max cruising speed at S/L:	
341, 342	142 knots (264 km/h; 164 mph)

Aérospatiale SA 342M Gazelle, armed with four Hot missile launchers

Econ cruising speed at S/L:
341 — 126 knots (233 km/h; 144 mph)
342 — 128 knots (238 km/h; 148 mph)
Max rate of climb at S/L: 341 — 540 m (1,770 ft)/min
342 — 510 m (1,670 ft)/min
Service ceiling: 341 — 5,000 m (16,400 ft)
342 — 4,300 m (14,100 ft)
Hovering ceiling IGE: 341 — 2,850 m (9,350 ft)
342 — 3,650 m (11,970 ft)
Hovering ceiling OGE: 341 — 2,000 m (6,560 ft)
342 — 2,875 m (9,430 ft)
Range at S/L with max fuel:
341 — 361 nm (670 km; 416 miles)
Range with max fuel, econ cruising speed, no reserves:
342 — 407 nm (755 km; 469 miles)
Range with pilot and 500 kg (1,102 lb) payload:
341 — 193·5 nm (360 km; 223 miles)

AÉROSPATIALE AS 350 ECUREUIL/ASTAR

Intended as a successor to the Alouette, the AS 350 Ecureuil (Squirrel) was designed with an emphasis on low operating and maintenance costs, and low noise and vibration levels. It embodies Aérospatiale's Starflex type of main rotor hub, made of glassfibre, with elastomeric spherical stops and oleo-elastic frequency matchers.

The decision to build prototypes of the Ecureuil was taken in April 1973. The first of these (F-WVKH) flew on 27 June 1974, powered by an Avco Lycoming LTS 101 turboshaft engine. It was followed on 14 February 1975 by a second prototype (F-WVKI) with a Turboméca Arriel turboshaft.

The Avco Lycoming-powered version is marketed only in North America, as the **Astar**, and the Arriel-powered **AS 350B Ecureuil** throughout the rest of the world. French certification of the AS 350B was obtained on 27 October 1977 and deliveries began in March 1978. FAA certification of the original **AS 350C** Astar was obtained on 21 December 1977 and the first production delivery was made in April 1978. The AS 350C was superseded in 1978 by the **AS 350D** with a more powerful version of the Avco Lycoming engine. By Spring 1981 a total of 680 Ecureuils and Astars had been sold. Production by Aérospatiale was intended to build up to 7 Ecureuils and 16 Astars per month by February 1980. Ecureuils are also produced under licence by Helibras of Brazil, with the name Esquilo.

Details of a specially-equipped air ambulance version of the Astar were announced in Autumn 1980. Designed to meet American Hospital Association specifications, this accommodates two stretchers, one above the other, across the rear of the cabin, with a bubble door on the starboard side. Stretchers are staggered to facilitate access by the attendant, whose rearward-facing seat is on the port side, next to the pilot. Options include a quick-disconnect oxygen system, dual intravenous system, lower stretcher lighting and attendant's swivelling seat.

TYPE: Five/six-seat light general-purpose helicopter.

ROTOR SYSTEM: Three-blade main rotor, with Starflex glassfibre hub in which the three conventional hinges for each blade are replaced by a single balljoint of rubber/steel sandwich construction, requiring no maintenance. Glassfibre blades, with stainless steel leading-edge sheath, produced by an entirely mechanised process. Symmetrical blade section. Two-blade tail rotor; each blade comprises a sheet metal skin around a glassfibre spar, the flexibility of which obviates the need for hinges.

ROTOR DRIVE: Simplified transmission, with single epicyclic main gear train. By comparison with Alouette II, number of gear wheels is reduced from 22 to 9 and number of bearings from 23 to 9. Tail rotor driveshaft coupling on engine.

FUSELAGE: Basic structure of light alloy pressings, with skin mainly of thermoformed plastics, including baggage compartment doors.

TAIL UNIT: Horizontal stabiliser, of inverted aerofoil section, mid-mounted on tailboom. Sweptback fin, in two sections above and below tailboom.

LANDING GEAR: Steel tube skid type. Emergency flotation gear optional.

POWER PLANT: One 478 kW (641 shp) Turboméca Arriel (AS 350B) or 459 kW (616 shp) Avco Lycoming LTS 101-600A.2 (AS 350D) turboshaft engine, mounted above fuselage to rear of cabin. Plastics fuel tanks with total capacity of 530 litres (116·5 Imp gallons).

ACCOMMODATION: Two individual bucket seats at front of cabin and two two-place bench seats are standard. In the alternative layout the two benches are replaced by three armchair seats. Optional ambulance layout described in detail in introductory notes. Large forward-hinged door on each side. Optional sliding door at rear of cabin on port side. Baggage compartment aft of cabin, with full-width upward-hinged door on starboard side. Top of baggage compartment reinforced to provide platform on each side for inspecting and servicing rotor head.

SYSTEMS: Hydraulic system includes four single-body servo units, operating at 40 bars (570 lb/sq in) pressure, and accumulators to protect against a hydraulic power supply failure. Electrical system includes a 4·5kW engine-driven starter/generator, a 16Ah 24V nickel-cadmium battery and a ground power receptacle connected to the busbar which distributes power to the electrical equipment. Cabin air-conditioning system optional.

Aérospatiale AS 350B Ecureuil (Turboméca Arriel turboshaft engine)

AVIONICS AND EQUIPMENT: Optional com/nav radio equipment includes VHF/AM, ICS, VOR/LOC/glide-slope, marker beacon indicator, radio compass, HF/SSB, transponder and DME. IFR instrumentation optional. Available equipment includes a SFIM PA 85E, Sperry HelCis or Collins APS-841H autopilot, 750 kg (1,650 lb) cargo sling, a 900 kg (1,984 lb) sling associated with a max T-O weight of 2,100 kg (4,630 lb), a 135 kg (297 lb) electrical hoist, a TV camera for aerial filming, and a 735 litre (161 Imp gallon) Simplex agricultural spray tank and boom system.

DIMENSIONS, EXTERNAL:
Diameter of main rotor — 10·69 m (35 ft 0¾ in)
Main rotor blade chord — 300 mm (11·8 in)
Diameter of tail rotor — 1·86 m (6 ft 1¼ in)
Length overall — 12·99 m (42 ft 7½ in)
Length of fuselage — 10·91 m (35 ft 9½ in)
Width of fuselage — 1·80 m (5 ft 10¾ in)
Height overall — 3·15 m (10 ft 4 in)
Skid track — 2·10 m (6 ft 10¾ in)
Cabin doors (standard, each):
Height — 1·15 m (3 ft 9¼ in)
Width — 1·10 m (3 ft 7¼ in)

DIMENSIONS, INTERNAL:
Cabin: Length — 2·42 m (7 ft 11¼ in)
Width at rear — 1·65 m (5 ft 5 in)
Height — 1·35 m (4 ft 5 in)
Baggage compartment volume — 1·00 m^3 (35·31 cu ft)

WEIGHTS:
Weight empty: 350B — 1,065 kg (2,348 lb)
350D — 1,070 kg (2,359 lb)
Max T-O weight: normal — 1,950 kg (4,300 lb)
with max sling load — 2,100 kg (4,630 lb)

PERFORMANCE (at T-O weight of 1,950 kg; 4,300 lb):
Never-exceed speed below 500 m (1,640 ft):
350B, 350D — 147 knots (272 km/h; 169 mph)
Max cruising speed:
350B — 125 knots (232 km/h; 144 mph)
350D — 124 knots (230 km/h; 143 mph)
Max rate of climb at S/L:
350B, 350D — 480 m (1,575 ft)/min
Service ceiling: 350B, 350D — 4,575 m (15,000 ft)
Hovering ceiling IGE: 350B — 2,950 m (9,675 ft)
350D — 2,500 m (8,200 ft)
Hovering ceiling OGE: 350B — 2,250 m (7,380 ft)
350D — 1,800 m (5,900 ft)
Range with max fuel at S/L, no reserves:
350B — 378 nm (700 km; 435 miles)
350D — 410 nm (760 km; 472 miles)

AÉROSPATIALE AS 355 ECUREUIL 2/TWINSTAR

The AS 355 represents a twin-engined addition to the AS 350 Ecureuil/Astar family. Many components, such as the main rotor mast and head, tail rotor hub, servo units, cabin and landing gear, are identical to those of the AS 350. Major changes apply to the main rotor blades, power plant, transmission, fuel system and fuselage structure.

Development of the basic **AS 355E** began in mid-1978. Construction of the first of two prototypes (F-WZLA) started at the beginning of 1979, and this aircraft flew for the first time on 28 September 1979, followed by the second prototype on 14 November. Each was fitted with rotor blades identical to those of the AS 350, and the helicopter entered production in this form. The first production AS 355E flew on 11 March 1980, and this version received French DGAC certification for day and night VFR operation on 24 October 1980. FAA certification for VFR flight followed on 19 January 1981, and deliveries of the AS 355E were planned to begin in April 1981. On the 14th of that month DGAC certification was received for day and night VFR operation of the improved **AS 355F** version of the helicopter, with modifications that include main rotor blades of increased chord and OA 209 section, twin-body servo-command units, and two electrical generators, permitting an increase in max T-O weight and payload. DGAC certification for IFR operation of both versions is expected in early 1982, in which year deliveries of the AS 355F will begin. Operators of the AS 355E will be offered retrofit kits to convert their aircraft to 'F' standard if they so wish.

The AS 355E/F is intended primarily for the civil market, in particular for use by companies working in the oil industry. By 1 May 1981, sales totalled more than 400, and production was scheduled to reach 14 aircraft a month by the beginning of 1982. The version for the North American market is known as the **Twinstar**; aircraft marketed elsewhere are named **Ecureuil 2**.

TYPE: Twin-turbine light general-purpose helicopter.

ROTOR SYSTEM: As for AS 350B, except that main rotor blades of AS 355F have increased chord and OA 209 section.

ROTOR DRIVE: Single main gearbox, made up of three modules (coupling gearbox with freewheel, angle gearing with spiral-bevel gears, and epicyclic gear train including five oscillating planetary gears). Power take-offs for the accessories and tail rotor.

FUSELAGE: Light alloy centre-fuselage structure, with deep-drawn sheet-metal forms of simple geometric design. Cabin skin of thermoformed plastics. Tapered tailboom of light alloy sheet wrapped and riveted around deep-drawn sheet metal cylindrical frames.

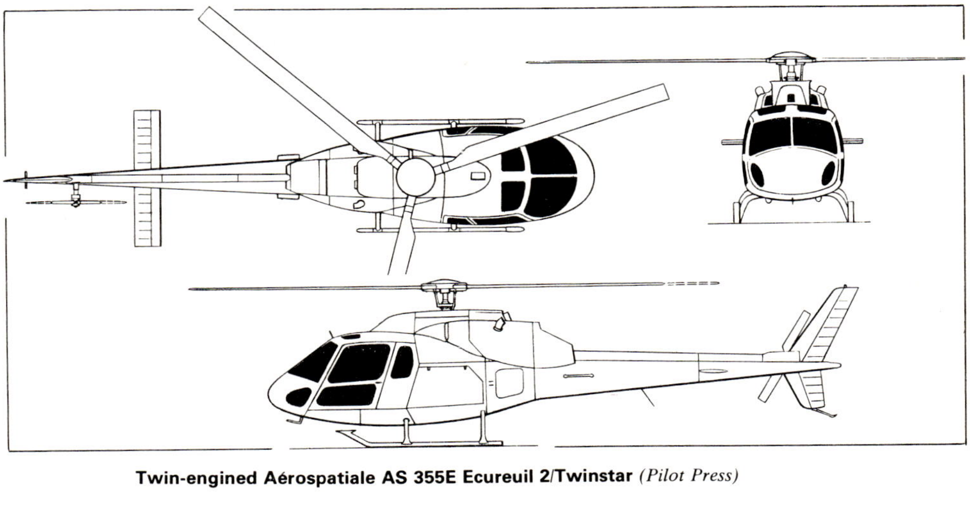

Twin-engined Aérospatiale AS 355E Ecureuil 2/Twinstar *(Pilot Press)*

Aérospatiale AS 355E Ecureuil 2/Twinstar twin-turbine five/six-seat light helicopter

Tail Unit and Landing Gear: As for AS 350B.

Power Plant: Two Allison 250-C20F turboshaft engines, each rated at 317 kW (425 shp) for take-off, mounted above fuselage to rear of cabin. Two structural fuel tanks, with total usable capacity of 720 litres (158 Imp gallons), in body structure.

Accommodation: As for AS 350B, except sliding doors are optional on both sides, and there are three baggage holds with external doors.

Systems, Avionics and Equipment: As for AS 350B, except that twin-body servo-command units and a second electrical generator are standard on AS 355F. Options include a second VHF/AM, radio altimeter, autopilot and casualty installations.

Dimensions, external and internal:
As for AS 350B, except:
Main rotor blade chord: 355F 350 mm (13·8 in)

Weights:
Weight empty: 355E 1,230 kg (2,712 lb)
355F 1,260 kg (2,778 lb)
Max sling load: 355F 1,045 kg (2,300 lb)
Max T-O weight: 355E 2,100 kg (4,630 lb)
355F, internal load 2,300 kg (5,070 lb)
355F, max sling load 2,400 kg (5,291 lb)

Performance (at max T-O weight, ISA):
Never-exceed speed (structural limitation):
355E/F 147 knots (272 km/h; 169 mph)
Max cruising speed at S/L:
355E 129 knots (240 km/h; 149 mph)
355F 127 knots (235 km/h; 146 mph)
Econ cruising speed at S/L:
355E 119 knots (220 km/h; 137 mph)
355F 113 knots (209 km/h; 130 mph)
Cruising speed at max power, one engine out:
355E 97 knots (180 km/h; 112 mph)
355F 94 knots (175 km/h; 108 mph)
Max rate of climb at S/L:
355E 522 m (1,710 ft)/min
355F 492 m (1,615 ft)/min
Service ceiling: 355E 4,000 m (13,125 ft)
355F 4,500 m (14,760 ft)
Service ceiling, one engine out:
355E 2,450 m (8,030 ft)
355F 1,825 m (6,000 ft)
Hovering ceiling IGE: 355E 2,200 m (7,215 ft)
355F 3,100 m (10,170 ft)
Hovering ceiling OGE: 355E 1,500 m (4,920 ft)
355F 2,400 m (7,875 ft)
Range with max fuel at S/L, no reserves:
355E 459 nm (850 km; 528 miles)
355F 407 nm (755 km; 469 miles)
Range with six occupants, each 80 kg (176 lb):
355E 200 nm (370 km; 230 miles)
Endurance with max fuel, no reserves:
355F 4 h 45 min

AÉROSPATIALE SA 360C DAUPHIN

The SA 360 Dauphin was developed, with the twin-engined SA 365 variant (described separately), as a replacement for the Alouette III. The first of two SA 360 prototypes (F-WSQL) flew for the first time on 2 June 1972, powered by a 730 kW (980 shp) Turboméca Astazou XVI turboshaft engine. After 180 flights, it was re-engined with an Astazou XVIIIA turboshaft and modified in certain respects, including the addition of small weights to the rotor blades, to eliminate ground resonance and reduce vibration to an unprecedented level, even at high speed. The aircraft flew for the first time in its modified form on 4 May 1973, having been joined by the second prototype (F-WSQX) on 29 January 1973.

DGAC certification of the basic SA 360C was awarded on 18 December 1975, followed by FAA certification on 31 March 1976. By Spring 1981, Aérospatiale had received orders for a total of 30 SA 360C and 370 SA 365 helicopters, including a first increment of 23 AS 366Gs, out of 90 required by the US Coast Guard.

Three helicopter speed records in Class E1d (1,750 to 3,000 kg weight) were set at Istres by the first prototype of the SA 360 on 15, 16 and 17 May 1973, piloted by Roland Coffignot. Carrying a payload equivalent to eight persons and fuel for one hour's flying, the SA 360 achieved, successively, 161·4 knots (299 km/h; 185·8 mph) over a 100 km closed circuit; 168·4 knots (312 km/h; 193·9 mph) over a 3 km course; and 163·5 knots (303 km/h; 188·3 mph) over a 15 km course. These records had not been beaten by mid-1981.

Type: Turbine-powered general-purpose helicopter.

Rotor System: Four-blade semi-articulated main rotor and 13-blade shrouded-fan anti-torque tail rotor (known as a 'fenestron' or 'fan-in-fin'). Main rotor blades are of symmetrical NACA 0012 section, with a theoretical twist of 8° and constant chord, and are attached to the NAT hub via flapping hinges. There are no drag hinges. Each blade has a single leading-edge spar of polyester plastics, extending back to about 30% chord at top and bottom. The outer skin is of glassfibre, with an inner skin of carbonfibre, and the entire blade is filled with Nomex honeycomb. The leading-edge is formed by a layer of Vulkollan plastics with an outer protective shield of thin-gauge stainless steel. Tail rotor blades are of die-forged light alloy, with articulation for pitch change only. Main rotor blades can be folded manually for stowage. Rotor brake and main rotor blade de-icing optional.

Rotor Drive: Main reduction gearbox forward of engine, which is mounted above the fuselage to the rear of the cabin. Output shaft enters main transmission box above the driveshaft to the tail rotor. Self-lubricating bearings. Main rotor rpm: 348 normal; 393 in autorotation. Tail rotor rpm: 4,700.

Fuselage: Conventional all-metal assembly of cabin and semi-monocoque tailboom. Cabin built on a strong box structure embodying two transverse frames and the cabin floor.

Tail Unit: Horizontal stabiliser mid-set on tailboom, forward of shrouded tail rotor, with endplate fins. Tailboom terminates in large fin of unsymmetrical section, housing the tail rotor. The section of this fin is such that in cruising flight it counters the torque of the main rotor; the tail rotor is thus required to provide only yaw control, with minimal variation of pitch, requiring only small power intake.

Landing Gear: Steel tube skid type, or ERAM non-retractable tailwheel-type landing gear, with single wheel on each unit. Main legs of wheel gear embody hydraulic shock-absorbers. Tailwheel carried on anti-shimmy leg which can be locked manually in central position. Dunlop main-wheel tyres size 355 × 150-4, pressure 5 bars (73 lb/sq in). Dunlop tailwheel tyre size 260 × 80-4, pressure 5 bars (73 lb/sq in). Disc brakes on main wheels. Wheel fairings standard. Provision for emergency flotation gear, and skis.

Power Plant: One Turboméca Astazou XVIIIA turboshaft engine, delivering 783 kW (1,050 shp) for take-off. Four Kléber-Colombes bag-type fuel tanks under cabin floor, total normal capacity 640 litres (141 Imp gallons). Provision for ferry tank of 475 litres (104 Imp gallons) capacity on the cabin floor and one auxiliary tank of 215 litres (47 Imp gallons) capacity at the back of the cabin.

Accommodation: Standard 10-seat version has seats for pilot (to starboard) and co-pilot or passenger in front, and two rows of four seats to the rear. Interior of the cabin is clear except for a vertical duct, housing the flying control rods, positioned centrally aft of the centre row of seats. Two large forward-hinged doors on each side. Optional rearward-sliding doors. Compartment for hand baggage or coats aft of rear row of seats. Separate main baggage compartment aft of cabin, with door on starboard side. Alternative 14-seat layout has an extra row of four seats attached to the cabin rear wall, and no space for hand baggage or coats. Ambulance version carries four stretcher patients, a medical attendant and two crew. Mixed-traffic version carries six persons at front of cabin, with 2·50 m³ (88·3 cu ft) of cargo space to the rear. The floor in this area will support a loading of 610 kg/m² (125 lb/sq ft). Executive versions are available with VIP interiors for four or six passengers. Cabin is heated and ventilated.

Systems: Two sets of three single-body servo controls mounted in series on the flight control channel, each supplied by an independent hydraulic system: one main set at the main rotor head, one auxiliary set capable of integrating an autopilot, and one tail rotor servo unit. Electrical system includes a 4·5kW engine-driven starter/generator, a 0·5kVA 115/200V 400Hz generator, a 23Ah nickel-cadmium battery and a ground power receptacle.

Avionics and Equipment: Optional equipment includes IFR instrumentation, an autopilot, 1,300 kg (2,865 lb) capacity cargo sling, 272 kg (600 lb) capacity rescue hoist, air-conditioning system, VHF/AM, ICS, VOR/LOC/glideslope, marker beacon indicator, HF/SSB, transponder and DME.

Dimensions, external:
Diameter of main rotor 11·50 m (37 ft 8¾ in)
Main rotor blade chord (constant)
0·35 m (1 ft 1¾ in)
Diameter of tail rotor 0·90 m (2 ft 11⁷⁄₁₆ in)
Length overall 13·20 m (43 ft 3½ in)
Length of fuselage 10·98 m (36 ft 0 in)
Height overall 3·50 m (11 ft 6 in)
Stabiliser span 3·15 m (10 ft 4 in)
Wheel track 1·95 m (6 ft 4¾ in)
Wheelbase 7·23 m (23 ft 8¾ in)
Cabin doors (fwd, each): Height 1·16 m (3 ft 9½ in)
Width 1·14 m (3 ft 9 in)
Cabin doors (aft, each): Height 1·16 m (3 ft 9½ in)
Width 0·87 m (2 ft 10¼ in)
Freight compartment door: Height 0·56 m (1 ft 10 in)
Width 0·75 m (2 ft 5½ in)

Dimensions, internal:
Cabin: Usable length 2·30 m (7 ft 6½ in)
Height at front 1·40 m (4 ft 7 in)
Height at rear 1·06 m (3 ft 5¾ in)
Width at front 1·92 m (6 ft 3½ in)
Width at rear 1·60 m (5 ft 3 in)

Aérospatiale SA 360C Dauphin 10-seat general-purpose helicopter

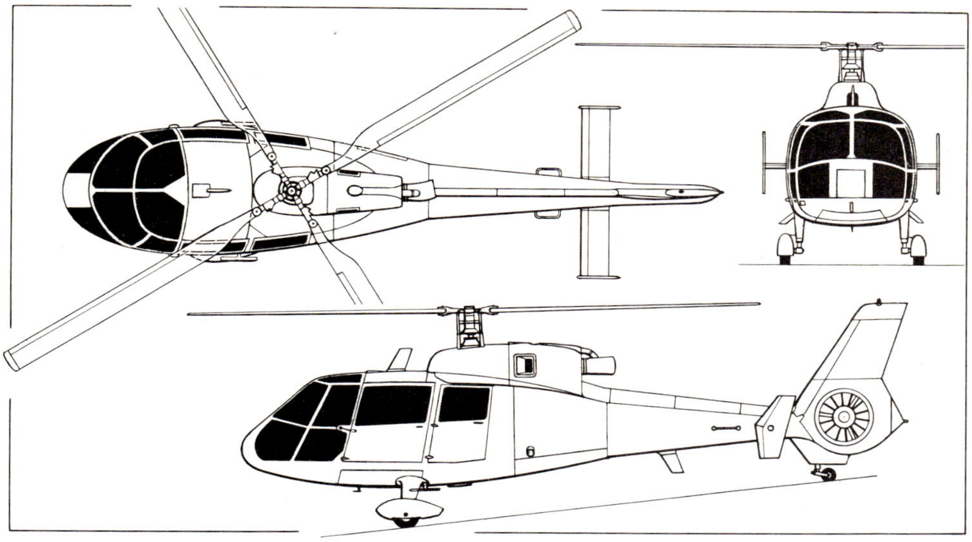

Aérospatiale SA 360C Dauphin (Turboméca Astazou XVIIIA turboshaft engine) with wheel-type landing gear *(Pilot Press)*

Floor area 4·20 m² (45·20 sq ft)
Volume 5·0 m³ (176 cu ft)
Baggage compartment volume 1·00 m³ (35·31 cu ft)

WEIGHTS:
Basic operating weight 1,637 kg (3,609 lb)
Max payload: internal 1,363 kg (3,005 lb)
slung 1,300 kg (2,865 lb)
Max T-O weight 3,000 kg (6,614 lb)

PERFORMANCE (at AUW of 2,600 kg; 5,732 lb):
Never-exceed speed at S/L 170 knots (315 km/h; 196 mph)
Cruising speed at S/L 148 knots (275 km/h; 171 mph)
Max rate of climb at S/L 540 m (1,770 ft)/min
Hovering ceiling IGE 3,850 m (12,630 ft)
Range at S/L with max fuel 367 nm (680 km; 423 miles)
Endurance at S/L 4 h

AÉROSPATIALE SA 365C DAUPHIN 2

Announced in early 1973, the SA 365C is a twin-engined version of the SA 360, powered by Turboméca Arriel turboshaft engines. The prototype (F-WVKE) flew for the first time on 24 January 1975. DGAC certification for VFR flight was received in July 1978, and for IFR flight in December 1978, followed in each case by FAA and CAA certification. Delivery of production SA 365Cs began in December 1978, initially to Offshore Helicopters Inc of the USA. A total of 37 had been delivered by 1 January 1980.

The SA 365C is designed for single-pilot IFR flight. It differs from the SA 360C in the following details:

ROTOR SYSTEM: Main rotor blades are attached to Starflex glassfibre hub, as described under entry for AS 350 Ecureuil.

ROTOR DRIVE: The installation of free-turbine engines has eliminated the need for a clutch in the output drive from each engine into the main gearbox.

FUSELAGE: The profile of the firewall between the two turboshaft engines is extended rearward in the form of a curved fairing which blends into the dorsal spine fairing over the tail rotor driveshaft.

TAIL UNIT: Horizontal stabiliser has inverted-camber aerofoil section, and the fixed vertical tail fins are offset to produce a lateral component which enhances the anti-torque function of the unsymmetrical surfaces. These features increase the efficiency of the 'fenestron' tail rotor, notably during hover.

POWER PLANT: Two Turboméca Arriel free-turbine turboshaft engines, each rated at 507 kW (680 shp), mounted side by side above the fuselage, aft of the main rotor driveshaft. Four separate bag-type fuel tanks, filling full width of fuselage under cabin floor. Two refuelling points aft of rear cabin door on port side.

DIMENSIONS, EXTERNAL:
Diameter of main rotor 11·68 m (38 ft 4 in)
Length overall 13·32 m (43 ft 8½ in)

WEIGHTS:
Weight empty 1,876 kg (4,136 lb)
Max T-O weight 3,400 kg (7,495 lb)

PERFORMANCE (at AUW of 3,000 kg; 6,613 lb):
Never-exceed speed 170 knots (315 km/h; 196 mph)
Cruising speed 137 knots (255 km/h; 158 mph)
Max rate of climb at S/L 600 m (1,970 ft)/min
Service ceiling (60 m; 200 ft/min climb) 6,000 m (19,680 ft)
Hovering ceiling IGE 3,570 m (11,710 ft)
Hovering ceiling OGE 2,840 m (9,315 ft)
Max range at econ cruising speed, no reserves 251 nm (465 km; 289 miles)

OPERATING LIMITS:
Max pressure altitude 4,570 m (15,000 ft)
Max temperature +40°C
Min temperature −40°C

AÉROSPATIALE AS 365N DAUPHIN 2

Although the AS 365 N resembles closely the earlier SA 365C, about 90 per cent of its components are different. Only 25 per cent of the airframe is of conventional construction. Composites such as glassfibre-Nomex, glassfibre-Kevlar and glassfibre-Rohacell are used for 20 per cent of the structure. Carbonfibre is used for the spars, skins and tapered tips of each main rotor blade, in the main rotor hub, and for the horizontal stabiliser. Some 35 per cent of the fuselage is made of light alloy-Nomex sandwich.

The prototype AS 365N (F-WZJD) flew for the first time on 31 March 1979 and was exhibited at the Paris Air Show in June of that year. A second prototype followed a few months later. The first production model (F-WZJJ) introduced further changes to the rotor mast fairings, engine cowlings, crew doors, transmission and main rotor blades, as well as larger tail surfaces. On 6 February 1980, it established a record for the round trip between Issy-les-Moulineaux, Paris, and Battersea Heliport, London, at an average speed of 158·89 knots (294·26 km/h; 182·84 mph). Its T-O weight of 3,800 kg (8,377 lb) included ten occupants and 30 min fuel reserves. The Paris-London leg was covered in 1 h 7 min 48 s, the return journey in 1 h 11 min 8 s. No landing at Battersea was necessary, as the total distance of 367 nm (680 km; 422 miles) was within the aircraft's range with full normal payload.

Two days later, the same aircraft, carrying six persons, set new records between Issy and Battersea, with a landing in London. Time for the Paris-London flight was 1 h 3 min 30 s at an average speed of 173·82 knots (321·91 km/h; 200·03 mph); the return flight took 1 h 12 min 9 s, at an average speed of 151·75 knots (281·05 km/h; 174·64 mph).

French civil certification for VFR operation by day and night was received on 10 April 1981, at which time further testing for single/two-pilot IFR operation was under way. The basic AS 365N will be built in China, as well as France, following signature on 2 July 1980 of a licence agreement covering a first batch of 50 aircraft. Orders total more than 320 helicopters for civil and military use, including AS 366Gs for the US Coast Guard and AS 365F/AS 15TTs with special equipment for attacking surface ships. These two variants are described separately.

Under development in the Spring of 1981 was a special aeromedical version of the AS 365N. This will carry a flight crew of two and will be available in two forms. An 'intensive care' layout will be arranged to carry two patients, one on each side of the cabin on a standard NATO stretcher, with space between for the doctor's seat and medical equipment. One of the stretchers can be replaced by seats for two patients, if required. The alternative 'ambulance' configuration provides space for four stretchers, one above the other on each side of the cabin, plus room for the doctor; or a single pair of stretchers, with room for four seated persons on the other side, and a doctor. Stretchers will be loaded through nose doors, with 180° opening, on both models. Those in the ambulance layout will be fixed to the sides of the cabin, and the patients will be carried to them on special mattresses.

The following structural description refers to the standard AS 365N, but is generally applicable to all versions:

Aérospatiale SA 365C Dauphin 2 with twin Turboméca Arriel turboshaft engines *(The Age,* **Melbourne***)*

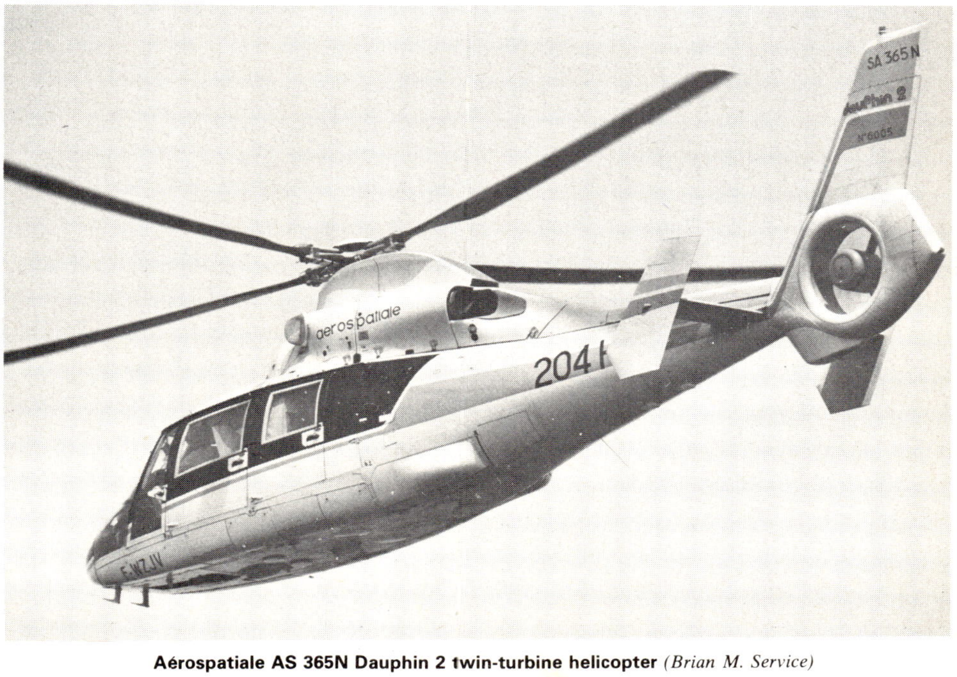

Aérospatiale AS 365N Dauphin 2 twin-turbine helicopter *(Brian M. Service)*

TYPE: Twin-turbine military and commercial general-purpose helicopter.

ROTOR SYSTEM: Four-blade main rotor. Blades attached by quick-disconnect pins to Starflex glassfibre/carbonfibre hub, in which the three conventional hinges for each blade are replaced by a single balljoint of rubber/steel sandwich construction, requiring no maintenance. Blades of new OA 2 section, developed in collaboration with Onera: varying from OA 212 (12%) at root to OA 207 (7%) at tip, with 10° negative twist from root to tip. Each blade comprises two Z-section carbonfibre spars and carbonfibre skin, a solid glassfibre-resin leading-edge covered with a stainless steel sheath, and Nomex honeycomb filling. Leading-edge of carbonfibre tip is swept back at 45°. Ground-adjustable tab on trailing-edge of each blade towards tip. Blade chord extended outboard of tab to align with tab trailing-edge. Rotor brake standard. Thirteen-blade 'fenestron' type of metal ducted-fan anti-torque tail rotor.

ROTOR DRIVE: Mechanical shaft and gear drive. Transmission shaft from each engine extends forward, through freewheel, to helical and epicyclic reduction stages of main gearbox. Shaft to 'fenestron' driven off bottom of main rotor shaft. Main rotor rpm 349. 'Fenestron' rpm 4,706.

FUSELAGE: Semi-monocoque structure. Bottom structure and framework of front fuselage, primary machined frames fore and aft of the main gearbox platform and at the rear of the centre fuselage, floors under main gearbox and engines, cabin doors, 'fenestron' and fin are all of light alloy (AU4G). Nose and power plant fairings and fin tip of glassfibre-Nomex sandwich. Centre and rear fuselage assemblies, flight deck floor, roof, walls and bottom skins of fuel tanks of light alloy-Nomex sandwich.

TAIL UNIT: Horizontal stabiliser mid-set on rear fuselage, forward of 'fenestron'; swept endplate fins offset 10° to port. Construction of carbonfibre and Nomex-Rohacell sandwich.

LANDING GEAR: Hydraulically-retractable tricycle type. Twin-wheel steerable and self-centering nose unit retracts rearward. Single wheel on each rearward-retracting main unit, fully enclosed by doors of Kevlar-Nomex sandwich when retracted. All three units embody oleo-pneumatic shock-absorber. Tyre pressure 7 bars (101 lb/sq in) for main wheels, 4 bars (58 lb/sq in) for nosewheels. Hydraulic disc brakes.

POWER PLANT: Two Turboméca Arriel IC free-turbine turboshaft engines, each rated at 530 kW (710 shp), mounted side by side aft of main rotor driveshaft, with stainless steel firewall between them. Standard fuel in four tanks under cabin floor and a fifth tank in the bottom of the centre-fuselage; total capacity 1,140 litres (250 Imp gallons). Provision for auxiliary tank in baggage compartment, with capacity of 180 litres (39·5 Imp gallons); or ferry tank in place of rear seats in cabin, capacity 475 litres (104·5 Imp gallons). Refuelling point above landing gear door on port side. Oil capacity 14 litres (3 Imp gallons).

ACCOMMODATION: Standard accommodation for pilot and co-pilot or passenger in front, and two rows of four seats to rear. High-density seating for one pilot and 13 passengers. VIP configurations for four to six persons in addition to pilot. Three forward-opening doors on each side. Freight hold aft of cabin rear bulkhead, with door on starboard side. Cabin heated and ventilated.

SYSTEMS: SEMCA air-conditioning system optional. Duplicated hydraulic system. Electrical system includes two 4·5kW starter/generators, one 17Ah 24V battery and two 250VA 115V 400Hz inverters. Provision for de-icing system.

AVIONICS AND EQUIPMENT: Optional avionics include VHF and HF com/nav, VOR, ILS, ADF, transponder, DME, radar and self-contained nav system. Optional equipment includes a SFIM 85 duplex autopilot, a 1,700 kg (3,750 lb) capacity cargo sling, and 275 kg (605 lb) capacity hoist with 80 m (260 ft) cable length.

Mockup of AS 365M anti-tank helicopter at 1981 Paris Air Show *(Brian M. Service)*

DIMENSIONS, EXTERNAL:	
Diameter of main rotor	11·93 m (39 ft 1¾ in)
Diameter of 'fenestron'	0·90 m (2 ft 11 7/16 in)
Blade chord, main rotor, basic	0·385 m (1 ft 3¼ in)
Blade chord, main rotor, outboard of tab	0·405 m (1 ft 4 in)
Length overall	13·46 m (44 ft 2 in)
Length of fuselage	11·44 m (37 ft 6½ in)
Width, rotor blades folded	3·21 m (10 ft 6½ in)
Height to top of rotor hub	3·47 m (11 ft 4¾ in)
Height overall (tip of fin)	4·01 m (13 ft 2 in)
Wheel track	1·90 m (6 ft 2¾ in)
Wheelbase	3·61 m (11 ft 10¼ in)
Main cabin door (fwd, each side):	
Height	1·16 m (3 ft 9½ in)
Width	1·14 m (3 ft 9 in)
Main cabin door (rear, each side):	
Height	1·16 m (3 ft 9½ in)
Width	0·87 m (2 ft 10¼ in)
Baggage compartment door (stbd):	
Height	0·51 m (1 ft 8 in)
Width	0·73 m (2 ft 4¾ in)
DIMENSIONS, INTERNAL:	
Cabin: Length	2·30 m (7 ft 6½ in)
Max width	2·03 m (6 ft 8 in)
Max height	1·40 m (4 ft 7 in)
Floor area	4·20 m^2 (45·20 sq ft)
Volume	5·00 m^3 (176 cu ft)
Baggage compartment volume	2·20 m^3 (77·7 cu ft)
WEIGHTS:	
Weight empty, equipped	1,945 kg (4,288 lb)
Max T-O weight	3,850 kg (8,488 lb)
PERFORMANCE (at max T-O weight):	
Never-exceed speed at S/L	165 knots (306 km/h; 190 mph)
Max cruising speed at S/L	150 knots (278 km/h; 172 mph)
Econ cruising speed at S/L	135 knots (250 km/h; 155 mph)
Max rate of climb at S/L	390 m (1,280 ft)/min
Service ceiling	3,750 m (12,300 ft)
Hovering ceiling IGE	1,005 m (3,300 ft)
Hovering ceiling OGE	950 m (3,115 ft)
Max range with standard fuel	475 nm (880 km; 546 miles)

AÉROSPATIALE AS 365F/AS 15TT DAUPHIN 2

On 13 October 1980, the government of Saudi Arabia placed in France orders for military equipment valued at 14,400 million francs. Known as the Sawari contract, it included the supply of 24 AS 365F Dauphin 2 helicopters. The first four of these will be equipped with an Omera ORB 32 radar for search and rescue duties. The remaining 20 will be anti-ship helicopters, equipped with Thomson-CSF Agrion 15 radar and Aérospatiale AS 15TT all-weather air-to-surface missiles, for operation from both shore bases and frigates. Deliveries are scheduled to begin in 1983, and will include 200 AS 15TT missiles.

A full-scale mockup of the anti-ship AS 365F/AS 15TT was displayed for the first time at the 7th Naval Equipment Exhibition at Le Bourget at the end of October 1980. Generally similar to the AS 365N, it carries the Agrion 15 radar on a roll-stabilised pivot-mounting under its nose, to ensure a 360° field of sweep, and a total of four AS 15TT missiles, in pairs on an outrigger on each side of the fuselage. Agrion 15 is derived from the Iguane radar fitted to the Atlantic NG maritime patrol aircraft, and possesses a track-while-scan capability that enables it to detect threats over long ranges while tracking ten targets simultaneously. In addition to locating and attacking hostile warships, the AS 365F/AS 15TT can be utilised for coastal surveillance and ship escort duties, and to provide over-the-horizon target designation for long-range anti-ship missiles launched from ship or shore. An anti-submarine version is available, initially with MAD and homing torpedoes, but with provision for sonar.

Powered, like the AS 365N, by two Turboméca Arriel IC turboshaft engines, the AS 365F carries a normal crew of two and has a max T-O weight of 3,900 kg (8,598 lb). Endurance is 2 h 45 min when armed with four missiles, equivalent to a radius of action of nearly 125 nm (230 km; 143 miles). Alternatively, with two missiles, endurance can be increased to 3 h 45 min and radius of action to 165 nm (305 km; 190 miles). Range of the AS 15TT missile is greater than 8 nm (15 km; 9·3 miles).

Addition of the radar and missiles makes the following changes to the overall dimensions of the helicopter:

DIMENSIONS, EXTERNAL:	
Length of fuselage	12·15 m (39 ft 10½ in)
Width over missiles	4·20 m (13 ft 9½ in)

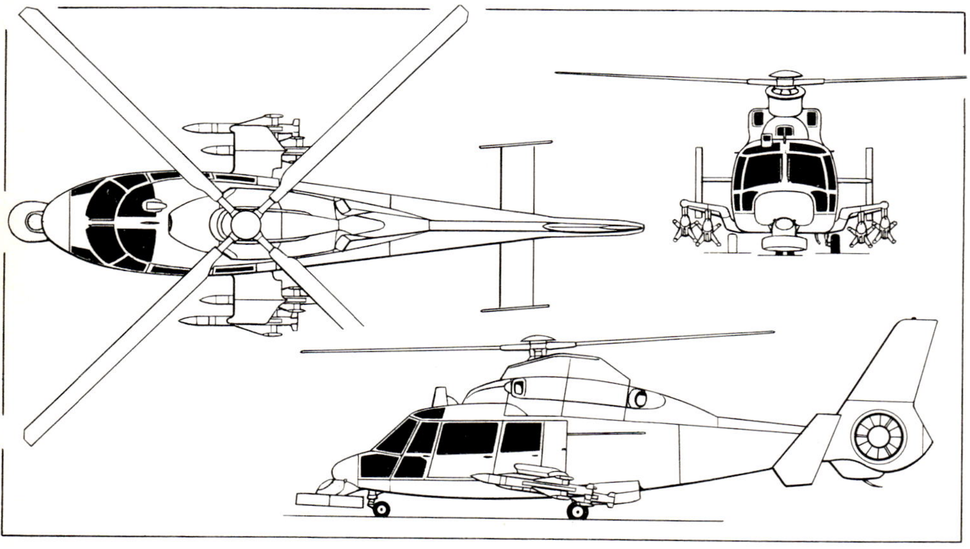

Aérospatiale AS 365F/AS 15TT anti-shipping helicopter *(Pilot Press)*

AÉROSPATIALE AS 365M DAUPHIN 2

This military version of the AS 365N is under development, with a variety of armament, for several potential customers and was first displayed in mockup form at the 1981 Paris Air Show. It could provide a high-speed assault transport for 10 to 12 troops. The fuselage-side outriggers could each carry a pack of 22 SNEB 68 mm rockets, or a four-round pack of Hot anti-tank missile, with associated stabilised sight. The mockup was shown with a Venus night sight system in a nose mounting. This spherical sensor houses a SFIM gyro-stabilised platform, TRT Hector infra-red camera to detect and identify the target, and a SAT deviation measuring instrument to assist missile guidance. It was fitted successfully to the SA 361H Dauphin for missile firing trials, as described in the 1979-80 *Jane's*.

AÉROSPATIALE AS 366 DAUPHIN 2
US Coast Guard designation: HH-65A Dolphin

At the 1979 Paris Air Show, Aérospatiale announced that it had won with this aircraft the competition for a helicopter to perform SRR (Short Range Recovery) duties from 18 shore bases, and from icebreakers and cutters, of the US Coast Guard. The initial requirement is for 90 **AS 366Gs**, basically similar to the AS 365N but with engines and equipment of US manufacture accounting for about 60 per cent of the total cost of each aircraft. The first order, for 23, was received in 1979.

The AS 366G (known to the Coast Guard as the HH-65A Dolphin) is powered by two Avco Lycoming LTS 101-750A-1 turboshafts, each rated at 507 kW (680 shp), and normally carries a crew of three (pilot, co-pilot and aircrewman/hoist operator). Rockwell Collins is prime contractor for the advanced communications, navigation and all-weather search equipment. The communications package includes dual UHF/VHF transceivers and single UHF/FM and HF systems, plus a data link for automatic transmission of data, such as aircraft position, flight path, ground speed, wind and fuel state, to ship or shore base. Under development is a nose-mounted Northrop See Hawk forward-looking infra-red sensor to aid rescue operations in bad weather, darkness or high seas. Important design features include the passive-failure characteristics of the Dolphin's automatic flight control system, and an omnidirectional airspeed system able to provide information while the aircraft is hovering. Inflatable flotation bags would enable occupants to evacuate the aircraft safely after a forced landing in sea state 5, and would keep the helicopter afloat for subsequent salvage.

The first Aérospatiale HH-65A Dolphin for the US Coast Guard

The first AS 366G flew for the first time at Marignane on 23 July 1980. It was later shipped to Aérospatiale Helicopter Corporation in Texas for installation of avionics, and flight testing for FAA certification. Deliveries to the Coast Guard are scheduled to extend from the late Spring of 1982 to 1986.

A civil counterpart, the **AS 366N**, is planned for the North American market.

WEIGHTS:
- Weight empty, incl mission equipment 2,530 kg (5,577 lb)
- Max T-O weight 3,900 kg (8,600 lb)

PERFORMANCE (at max T-O weight):
- Never-exceed speed 175 knots (324 km/h; 201 mph)
- Max cruising speed 145 knots (268 km/h; 167 mph)
- Econ cruising speed 128 knots (237 km/h; 147 mph)
- Hovering ceiling IGE 2,290 m (7,510 ft)
- Hovering ceiling OGE 1,627 m (5,340 ft)
- SRR range 166 nm (307 km; 191 miles)
- Range with max passenger load 216 nm (400 km; 248 miles)
- Max range, one engine out 327 nm (605 km; 376 miles)
- Range with max fuel 420 nm (778 km; 483 miles)
- Endurance with max fuel 4 h 11 min

C.A.A.R.P. — *See 'Mudry'*

DASSAULT-BREGUET
AVIONS MARCEL DASSAULT/BREGUET AVIATION

HEAD OFFICE: 33 rue du Professeur Victor Pauchet, 92420 Vaucresson
POSTAL ADDRESS: BP 32, 92420 Vaucresson
Telephone: 741 79 21
Telex: AMADAS 203944 F
PRESS INFORMATION OFFICE: 46 avenue Kléber, 75116 Paris
Telephone: 727 61 19
WORKS: 92214 Saint-Cloud, 77000 Melun-Villaroche, 95100 Argenteuil, 92100 Boulogne/Seine, 78140 Vélizy-Villacoublay, 33610 Martignas, 33700 Bordeaux-Mérignac, 91120 Bretigny, 33630 Cazaux, 31770 Toulouse-Colomiers, 64600 Biarritz-Anglet, 64200 Biarritz-Parme, 13800 Istres, 74370 Argonay, 93350 Le Bourget, 59113 Lille-Seclin, 86000 Poitiers
FOUNDER: Marcel Dassault
PRESIDENT AND CHIEF EXECUTIVE: B. C. Vallières
GENERAL MANAGER, INTERNATIONAL AFFAIRS: H. de l'Estoile
DEPUTY GENERAL MANAGER: X. D'Iribarne
DEPUTY GENERAL MANAGER: J. Estebe
SECRETARY GENERAL: C. Edelstenne
GENERAL TECHNICAL MANAGER: J. Cabrière
EXPORT TECHNICAL MANAGER: Y. Thiriet
MILITARY AIRCRAFT SALES MANAGER: F. Serralta
MILITARY AIRCRAFT SALES MANAGER: P. E. Jaillard
CIVIL AIRCRAFT SALES MANAGER: B. Latreille
PRODUCTION MANAGER: J. C. Veber
FLIGHT TEST MANAGER: J. F. Cazaubiel
PRESS INFORMATION MANAGER: A. Segura

Avions Marcel Dassault/Breguet Aviation resulted from the merger on 14 December 1971 of the Avions Marcel Dassault and Breguet Aviation companies. On 1 January 1979 it announced that, in accordance with an option taken in 1976, it had transferred 21 per cent of its stock to the French government.

Dassault-Breguet is engaged in the development and production of military and civil aircraft, guided missiles and servo control equipment. Series production of its aircraft is undertaken under a widespread subcontracting programme, with final assembly and flight testing handled by the company. Its 18 separate works and facilities covered 635,400 m² (6,839,450 sq ft), with a total of 16,265 employees in mid-1981.

Dassault-Breguet has established close links with the industries of other countries. The original programme for the Atlantic maritime patrol aircraft (now undergoing major new development) associated manufacturers in Belgium, France, West Germany, Italy and the Netherlands under the overall responsibility of their respective governments. In the same way the British and French governments are associated in the SEPECAT concern, formed to control the Dassault-Breguet/BAe Jaguar programme; and the West German and French governments are associated in the Dassault-Breguet/Dornier Alpha Jet programme. Purchase of Mirage fighters by Belgium and Spain led to Belgian and Spanish participation in Mirage III/5 and Mirage F1 production.

In 1980, Dassault-Breguet built a total of 310 aircraft, made up of 18 Mirage III/5/50s, 60 Mirage F1s, 18 Jaguars, 128 Alpha Jets, 20 Super Etendards, 2 Mirage 2000 prototypes, 22 Falcon 10s, 16 Falcon 20s, 1 Falcon 200 prototype and 25 Falcon 50s. In addition, its Biarritz factory delivered 13 fuselages for the Fokker F27. Orders for 101 aircraft were received during 1980.

DASSAULT-BREGUET MIRAGE III

The Mirage III was designed initially as a Mach 2 high-altitude all-weather interceptor, capable of performing ground support missions and requiring only small airstrips. Developed versions include a two-seat trainer, long-range fighter-bomber and reconnaissance aircraft. A total of more than 1,400 Mirage III/5/50s of all types had been ordered and 1,336 delivered for service in 21 countries by 1 March 1981, including licence production abroad. Current contracts will maintain production, at a current rate of two per month, until at least the end of 1984.

The experimental prototype flew for the first time on 17 November 1956, powered by a SNECMA Atar 101G turbojet with afterburner (44·1 kN; 9,900 lb st). Versions currently available include the following:

Mirage III-D. Two-seat version, built initially in Australia for the RAAF. Similar, French-built models ordered by 12 countries, including six more for Australia.

Dassault-Breguet Mirage III-E fighter-bomber/intruder of the French Air Force

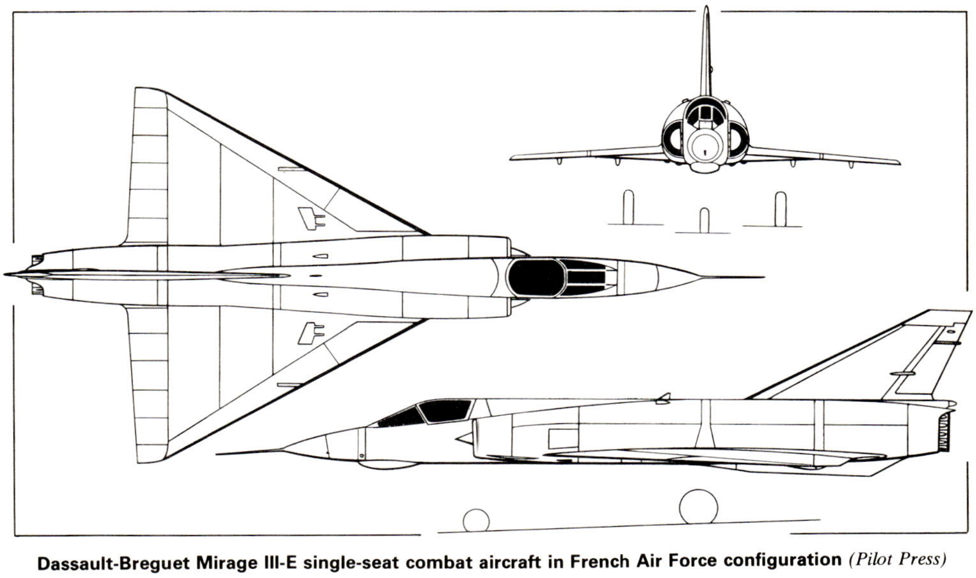

Dassault-Breguet Mirage III-E single-seat combat aircraft in French Air Force configuration *(Pilot Press)*

Dassault-Breguet Mirage III-RD reconnaissance aircraft of the Pakistan Air Force

Tandem seating under one-piece canopy; radar deleted, but fitted with radio beacon equipment. Intended primarily as a trainer, but suitable for strike sorties, carrying air-to-surface armament. Total of 180 Mirage III-B/III-D/5 two-seaters sold to 20 countries.

Mirage III-D2Z. For South Africa. Generally similar to III-D but with SNECMA Atar 9K-50 turbojet. Delivered 1974-75.

Mirage III-E. Long-range fighter-bomber/intruder version, of which 523 have been built for 13 air forces. First of three prototypes flew on 5 April 1961, and the first delivery of a production III-E was made in January 1964. Thirty III-Es of the 4e Escadre of the French Air Force, equipping two squadrons at Luxeuil, are carriers of the 15 kT AN 52 tactical nuclear weapon.

Mirage III-R. Reconnaissance version of III-E. Set of five OMERA type 31 cameras, in place of radar in nose, can be focused in four different arrangements for very low altitude, medium altitude, high altitude and night reconnaissance missions. Self-contained navigation system. Provision for air-to-surface armament. Two prototypes, of which the first flew in November 1961. Total of 159 production models ordered, including Mirage 5-Rs, for nine air forces.

Mirage III-R2Z. For South Africa. Generally similar to III-R but with SNECMA Atar 9K-50 turbojet. Delivered 1974-75.

Mirage III-RD. Similar to III-R but with improved Doppler navigation system in fairing under front fuselage, gyro gunsight and nose pack containing, typically, EMI side-looking airborne radar and OMERA 40 and 33 cameras. Provision for carrying SAT Cyclope infra-red tracking equipment in ventral fairing, and two 1,700 litre (374 Imp gallon) underwing auxiliary fuel tanks. Twenty built for French Air Force.

The following description refers to the Mirage III-E, but is generally applicable to all versions:

TYPE: Single-seat fighter-bomber/intruder aircraft.

WINGS: Cantilever low-wing monoplane of delta planform, with conical camber. Thickness/chord ratio 4·5% to 3·5%. Anhedral 1°. No incidence. Sweepback on leading-edge 60° 34′. All-metal torsion-box structure; stressed skin of machined panels with integral stiffeners. Elevons hydraulically powered by Dassault twin-cylinder actuators with artificial feel. Airbrakes, comprising small panels hinged to upper and lower wing surfaces, near leading-edge.

FUSELAGE: All-metal structure, 'waisted' in accordance with the area rule.

TAIL UNIT: Cantilever fin and hydraulically-actuated powered rudder only. Dassault twin-cylinder actuators with artificial feel.

LANDING GEAR: Retractable tricycle type, with single wheel on each unit. Hydraulic retraction, nosewheel rearward, main units inward. Messier-Hispano-Bugatti shock-absorbers and disc brakes. Main-wheel tyre pressure 5·9-9·8 bars (85·5-142 lb/sq in). Braking parachute.

POWER PLANT: One SNECMA Atar 9C turbojet engine (60·8 kN; 13,670 lb st with afterburning), fitted with an overspeed system which is engaged automatically from Mach 1·4 and permits a thrust increase of approx 8 per cent in the high supersonic speed range. Optional and jettisonable SEPR 844 single-chamber rocket motor (14·7 kN; 3,300 lb st) or interchangeable fuel tank. Movable half-cone centrebody in each air intake. Total internal fuel capacity 3,330 litres (733 Imp gallons) when rocket motor is not fitted. Provision for this to be augmented by two 625, 1,100, 1,300 or 1,700 litre (137, 242, 285 or 374 Imp gallon) underwing drop-tanks; 500 litre (110 Imp gallon) non-jettisonable supersonic tanks; JL-100 jettisonable tanks each housing both 280 litres (61 Imp gallons) fuel and air-to-surface rockets; Bidon Cyclope jettisonable tanks each housing 1,100 litres (242 Imp gallons) fuel and electronic equipment; or Bidon Homing jettisonable tanks housing 850 litres (187 Imp gallons) fuel and electronic equipment.

ACCOMMODATION: Single seat under rearward-hinged canopy. Hispano-built Martin-Baker Type RM4 zero-altitude ejection seat.

SYSTEMS: Two separate air-conditioning systems for cockpit and avionics. Two independent hydraulic systems, pressure 207 bars (3,000 lb/sq in), for flying controls, landing gear and brakes. Power for DC electrical system from 24V 40Ah batteries and a 26·5V 9kW generator. AC electrical system power provided by one 200V 400Hz transformer and one 200V 400Hz 9kVA alternator.

AVIONICS AND EQUIPMENT: Duplicated UHF, Tacan, Doppler, CSF Cyrano II fire-control radar in nose, navigation computer, bombing computer, automatic gunsight.

The Mirage III-E has a normal magnetic detector mounted in the fin, and a central gyro and other avionics to provide accurate and stabilised heading information. The pilot's equipment determines at any instant the geographical co-ordinates of the aircraft and compares them with the co-ordinates of the target, the differences between the two being presented to the pilot as a 'course to steer' and 'distance to run'. Associated with this facility is a rotative magazine in the cockpit in which it is possible to insert up to twelve plastics punch-cards. Each card represents the co-ordinates of a geographical position. Therefore it is possible before take-off at point A to select point B on the rotating magazine. During take-off, ie after reaching 150 knots (278 km/h; 173 mph), the computer will switch on and the heading and distance to point B will be presented to the pilot. When overhead point B (assuming a pure navigational sortie) he can either select point A or the next turning point, or if required this sequence can continue until a maximum of twelve pre-set turning points have been used. Another facility available in the computer is known as the 'additional base'. Assuming that between points A and B the pilot receives instructions by radio to go to point C (and that there is no punch-card in the magazine for point C) the pilot can, by means of setting knobs, wind on the bearing and distance of point C from point B; then, when he selects the switch 'additional base', the heading to steer and distance to run to point C will be indicated.

Marconi Doppler equipment provides the ground speed and drift information for the above, while Tacan is presented as a 'bearing and distance' on the navigation indicator located on the starboard side of the instrument panel.

The Cyrano II installation in the aircraft's nose provides orthodox air-to-air interception radar, and has the additional mode available of control from the ground. In the latter case the pilot simply obeys his gunsight instructions, and radio silence is maintained. Cyrano II also functions in an air-to-ground role for high-level navigation, presenting a radar picture of the ground; for low-level navigation, presenting the obstacles above a preselected altitude; for blind descent, presenting obstacles that intercept the descent path; for anti-collision, presenting the obstacles that can be avoided by applying a 0·1g pull-up; and for distance measuring, by presenting in the sight the oblique aircraft-to-ground distance.

Allied to the Cyrano II installation is the CSF 97 sighting system, of illuminated points, dots, bars and figures, giving air-to-air facility for cannon and missiles, air-to-ground facility for dive-bombing or LABS, and navigation facility for horizon and heading.

ARMAMENT: Ground attack armament consists normally of two 30 mm DEFA cannon in fuselage, each with 125 rounds of ammunition, and two 1,000 lb bombs, or an AS.30 air-to-surface missile under the fuselage and 1,000 lb bombs under the wings. Alternative underwing stores include JL-100 pods, each with 18 rockets, and jettisonable underwing fuel tanks. For interception duties, one Matra R.530 air-to-air missile can be carried under fuselage, with optional guns and two Sidewinder missiles.

DIMENSIONS, EXTERNAL:

Wing span	8·22 m (26 ft 11½ in)
Wing aspect ratio	1·94
Length overall: III-E	15·03 m (49 ft 3½ in)
III-R	15·50 m (50 ft 10¼ in)
Height overall	4·50 m (14 ft 9 in)
Wheel track	3·15 m (10 ft 4 in)
Wheelbase: III-E	4·87 m (15 ft 11¾ in)

AREAS:

Wings, gross	34·85 m² (375 sq ft)
Vertical tail surfaces (total)	4·5 m² (48·4 sq ft)

WEIGHTS AND LOADING:

Weight empty: III-E	7,050 kg (15,540 lb)
III-R	6,600 kg (14,550 lb)
T-O weight 'clean': III-E	9,600 kg (21,165 lb)
Max T-O weight: III-E, R	13,700 kg (30,200 lb)
Max wing loading: III-E, R	393·1 kg/m² (80·53 lb/sq ft)

PERFORMANCE (Mirage III-E, in 'clean' condition with guns installed, except where indicated):

Max level speed at 12,000 m (39,375 ft)	Mach 2·2 (1,268 knots; 2,350 km/h; 1,460 mph)
Max level speed at S/L	750 knots (1,390 km/h; 863 mph)
Cruising speed at 11,000 m (36,000 ft)	Mach 0·9
Approach speed	183 knots (340 km/h; 211 mph)
Landing speed	157 knots (290 km/h; 180 mph)
Time to 11,000 m (36,000 ft), Mach 0·9	3 min
Time to 15,000 m (49,200 ft), Mach 1·8	6 min 50 s
Service ceiling at Mach 1·8	17,000 m (55,775 ft)
Ceiling, using rocket motor	23,000 m (75,450 ft)
T-O run, according to mission (up to max T-O weight)	700-1,600 m (2,295-5,250 ft)
Landing run, using brake parachute	700 m (2,295 ft)
Combat radius, ground attack	647 nm (1,200 km; 745 miles)

DASSAULT-BREGUET MIRAGE 5

The Mirage 5 is a ground attack aircraft using the same airframe and engine as the Mirage III-E. The basic VFR version has simplified avionics, 500 litres (110 Imp gallons) greater fuel capacity than the III-E and considerably extended stores carrying capability. It combines the full Mach 2+ capability of the Mirage III, and its ability to operate from semi-prepared airfields, with simpler maintenance. In ground attack configuration, up to 4,000 kg (8,820 lb) of weapons and 1,000 litres (220 Imp gallons) of fuel can be carried externally on seven wing and fuselage attachment points. The Mirage 5 can also be flown as an interceptor, with two Magic or Sidewinder air-to-air missiles and 4,700 litres (1,034 Imp gallons) of external fuel. At customer's option, any degree of IFR/all-weather operation can be provided for, with reduced fuel or weapons load. The Mirage 5 was flown for the first time on 19 May 1967.

Up to 1 February 1980, a total of 480 Mirage 5s had been ordered for eleven air forces, including Mirage 5-R reconnaissance variants and two-seat Mirage 5-Ds.

The structural description of the Mirage III-E is generally applicable to the Mirage 5, with the following exceptions:

ARMAMENT: Seven attachment points for external loads, with multiple launchers permitting a max load of more

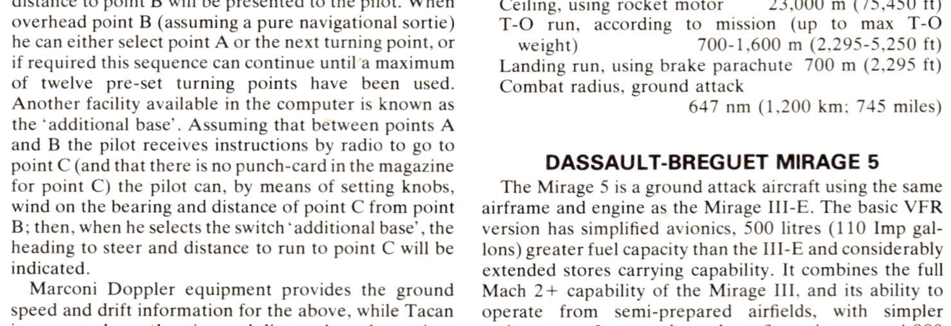

Dassault-Breguet Mirage 5 single-seat ground attack aircraft *(Pilot Press)*

Dassault-Breguet Mirage 5 single-seat ground attack aircraft (SNECMA Atar 9C turbojet engine)

than 4 tonnes. Ground attack armament consists normally of two 30 mm DEFA cannon in fuselage, each with 125 rounds of ammunition, and two 1,000 lb bombs or an AS.30 air-to-surface missile under the fuselage and 1,000 lb bombs under the wings. Alternative underwing stores include tank/bomb carriers, each with 500 litres (110 Imp gallons) of fuel and four 500 lb or two 1,000 lb bombs, and JL-100 pods, each with eighteen 68 mm rockets and 250 litres (55 Imp gallons) of fuel. For interception duties, two Sidewinder missiles can be carried under the wings.

EQUIPMENT: Current versions can have an inertial nav system and nav/attack system, with head-up display and either Agave multi-purpose radar or an air-to-surface laser rangefinder and Aïda II radar.

DIMENSIONS, EXTERNAL: As III-E, except:
- Length overall — 15·55 m (51 ft 0¼ in)

WEIGHTS AND LOADING: As III-E, except:
- Weight empty — 6,600 kg (14,550 lb)

PERFORMANCE (in 'clean' condition, with guns installed, except where indicated): As III-E, plus:
- Combat radius with 907 kg (2,000 lb) bomb load:
  - hi-lo-hi — 700 nm (1,300 km; 808 miles)
  - lo-lo-lo — 350 nm (650 km; 404 miles)
- Ferry range with three external tanks — 2,158 nm (4,000 km; 2,485 miles)

DASSAULT-BREGUET MIRAGE 50

First displayed in representative form at the 1975 Paris Air Show, the Mirage 50 retains the basic airframe of the Mirage III/5 series, but is powered by the higher-rated SNECMA Atar 9K-50 turbojet, as fitted in the Mirage F1-Cs of the French Air Force and Mirage III-R2Zs of the South African Air Force. This gives 70·6 kN (15,873 lb st) with afterburning, representing a 16 per cent thrust increase compared with standard Mirage III/5s.

The prototype Mirage 50 flew for the first time on 15 April 1979. First customer is the air force of Chile.

The Mirage 50 is a multi-mission fighter, suitable for air superiority duties with guns and dogfight missiles, air patrol and supersonic interception, and ground attack combined with self-defence capability. It can carry the full range of operational stores, armament and equipment developed for the Mirage III/5 series, plus Agave or Cyrano IVM multi-function radar, with Matra Magic or 530 air-to-air missiles respectively, an inertial or Doppler nav/attack system, and head-up display. It is available in reconnaissance configuration. A two-seat training version is also available. Improvements compared with other delta-wing Mirages include a 15-20 per cent shorter take-off run, improved armament/fuel load, higher rate of climb, faster acceleration and better manoeuvrability. Maximum internal fuel capacity is 3,475 litres (764 Imp gallons). Underwing and underfuselage tanks can increase total capacity to 4,700 litres (1,034 Imp gallons).

DIMENSIONS, EXTERNAL:
- Wing span — 8·22 m (27 ft 0 in)
- Length overall — 15·56 m (51 ft 0½ in)
- Height overall — 4·50 m (14 ft 9 in)

AREA:
- Wings, gross — 35·00 m² (376·7 sq ft)

WEIGHTS:
- Weight empty, equipped — 7,150 kg (15,765 lb)
- T-O weight, 'clean' — 9,900 kg (21,825 lb)
- Max T-O weight — 13,700 kg (30,200 lb)

PERFORMANCE:
- Max level speed at altitude — Mach 2·2 (750 knots; 1,390 km/h; 863 mph IAS)
- Max rate of climb at S/L — 11,100 m (36,400 ft)/min
- Time to 12,200 m (40,000 ft): 'clean' — 4 min 24 s
  - with 2 Magic missiles — 5 min 24 s
- Service ceiling at Mach 2 — 18,000 m (59,055 ft)
- T-O run with 2 Magic missiles — 915 m (3,000 ft)
- T-O run at max T-O weight — 1,830 m (6,000 ft)
- Combat radius at low altitude with two 400 kg bombs — 340 nm (630 km; 391 miles)

DASSAULT-BREGUET MIRAGE F1

Details of the early history of the Mirage F1 can be found in the 1977-78 *Jane's*. The prototype flew for the first time on 23 December 1966 and was followed by three pre-series aircraft.

The primary role of the Mirage F1 is that of all-weather interception at any altitude, and the **F1-C** production version, to which the detailed description applies, utilises weapon systems similar to those of the Mirage III-E. It is equally suitable for attack missions, carrying a variety of external loads beneath the wings and fuselage. A ground attack version designated **F1-A** is also in production, with much of the more costly avionic equipment deleted and the space so vacated occupied by an additional fuel tank. Further versions include the **F1-B** two-seat trainer, the first of which made its first flight on 26 May 1976; the **F1-E**, similar to the F1-C but with more comprehensive nav/attack system; and the **F1-R** reconnaissance variant.

Twenty-five F1-Cs of the French Air Force have been modified to **F1-C-200** standard by permanent installation of a flight refuelling probe for long-range reinforcement capability. Four of these aircraft, which serve with the 5e Escadre at Orange, made a six-hour nonstop flight from Solenzarra (Corsica) to Djibouti on 29 January 1980. In-flight refuelling was provided by a C-135F tanker. Point-to-point distance flown was about 2,700 nm (5,000 km; 3,100 miles).

By 1 March 1981, a total of 622 Mirage F1s had been ordered, comprising 225 for the French Air Force and 397 for service with the air forces of Ecuador (F1-B and C), Greece (F1-C), Iraq (F1-B and E), Jordan (F1-B, C and E), Kuwait (F1-B and C), Libya (F1-A, B and E), Morocco (F1-C and E), South Africa (F1-A and C) and Spain (F1-B, C and E). The first production F1 flew on 15 February 1973 and was delivered officially to the French Air Force on 14 March 1973. The first unit to receive the F1 was the 30e Escadre at Reims, which became operational in early 1974. It was followed by the 5e Escadre at Orange and the 12e Escadre at Cambrai. Each Escadre comprised two interceptor squadrons in 1980, but a third squadron was formed within the 12e Escadre in 1981, together with one squadron in the 10e Escadre at Creil. The first three of 20 two-seat F1-Bs on order entered service with the French Air Force in June 1980; each aircraft is equipped with the same Cyrano IV radar, weapon system and air-to-air missiles as the F1-C.

In February 1979, it was announced that the French Air Force had decided to purchase about 30 F1-Rs, to replace Mirage III-R/RD aircraft equipping two squadrons of the 33e Escadre de Reconnaissance, at Strasbourg. These aircraft will carry an OMERA 40 panoramic camera and an OMERA 35 camera internally, together with an infra-red sensor and an OMERA 360 sight recorder. Further electromagnetic or optical sensors will be carried in an underbelly pod. Other new equipment will include a SAGEM Uliss 47 inertial navigation system and EMD navigation radar. The first F1-R squadron is expected to become operational in 1983.

By the beginning of 1981 a total of 392 Mirage F1s had been completed, with production continuing at the rate of seven per month. Current orders will sustain production until at least the end of 1984.

The Mirage F1 is produced by Dassault-Breguet in co-operation with the Belgian companies SABCA, in

Dassault-Breguet Mirage 50 fighter (SNECMA Atar 9K-50 turbojet engine)

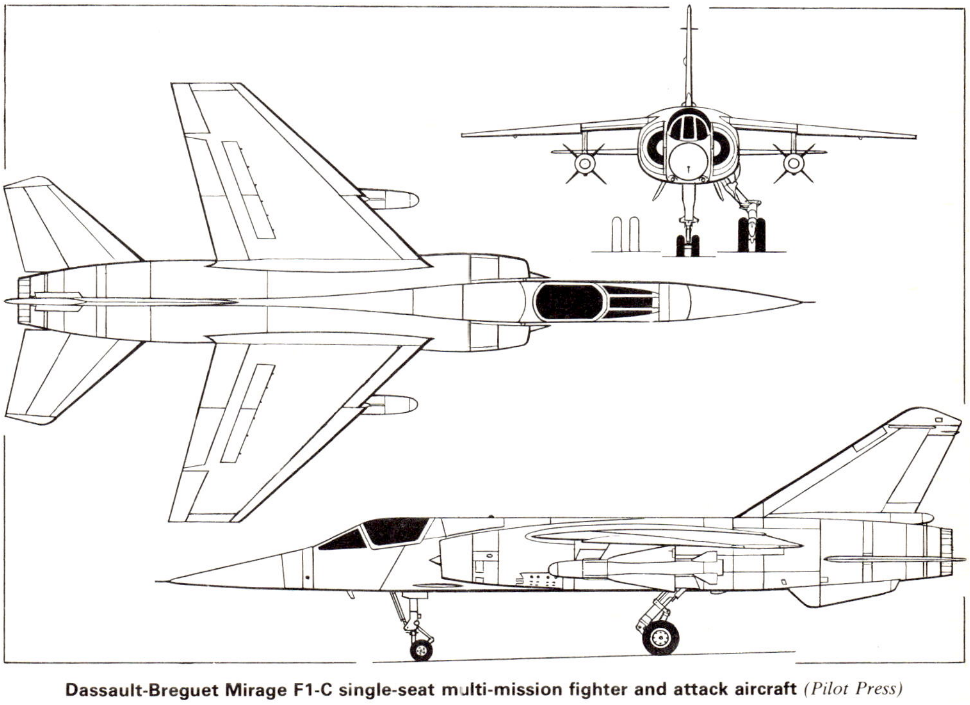

Dassault-Breguet Mirage F1-C single-seat multi-mission fighter and attack aircraft *(Pilot Press)*

Dassault-Breguet Mirage F1 in advanced ground attack configuration

Dassault-Breguet Mirage F1-B two-seat combat trainer

which Dassault-Breguet has a parity interest, and Sonaca, which is building rear fuselage sections for all Mirage F1s ordered. Dassault-Breguet also has a technical and industrial co-operation agreement with the Armaments Development and Production Corporation of South Africa Ltd, whereby the latter company has rights to build the Mirage F1 under licence.

The following description applies to the F1-C production version for the French Air Force:

TYPE: Single-seat multi-mission fighter and attack aircraft.

WINGS: Cantilever shoulder-wing monoplane. Anhedral from roots. Sweepback 47° 30′ on leading-edges, with extended chord on approximately the outer two-thirds of each wing. All-metal two-spar torsion-box structure, making extensive use of mechanically or chemically milled components. Trailing-edge control surfaces of honeycomb sandwich construction. Entire leading-edge can be drooped hydraulically (manually for T-O and landing, automatic in combat). Two differentially-operating double-slotted flaps and one aileron on each trailing-edge, actuated hydraulically by servo controls. Ailerons are compensated by trim devices incorporated in linkage. Two spoilers on each wing, ahead of flaps.

FUSELAGE: Conventional all-metal semi-monocoque structure. Primary frames are milled mechanically, secondary frames and fuel tank panels chemically. Electrical spot-welding for secondary stringers and sealed panels, remainder titanium flush-riveted or bolted and sealed. Titanium alloy also used for landing gear trunnions, engine firewall and certain other major structures. High-tensile steel wing attachment points. Nosecone over radar, and antennae fairings on fin, are of plastics. Large hydraulically-actuated door-type airbrake in forward underside of each intake trunk.

TAIL UNIT: Cantilever all-metal two-spar structure, with sweepback on all surfaces. All-moving tailplane mid-set on fuselage, and actuated hydraulically by electrical or manual control. Tailplane trailing-edge panels are of honeycomb sandwich construction. Auxiliary fin beneath each side of rear fuselage.

LANDING GEAR: Retractable tricycle type, by Messier-Hispano-Bugatti. Hydraulic retraction, nose unit rearward, main units upward into rear of intake trunk fairings. Twin wheels on each unit. Nose unit steerable and self-centering. Oleo-pneumatic shock-absorbers. Main-wheel tyre pressure 9-11 bars (130-160 lb/sq in), permitting operation from semi-prepared airfields. Messier-Hispano-Bugatti brakes and anti-skid units. Brake parachute in bullet fairing at base of rudder.

POWER PLANT: One SNECMA Atar 9K-50 turbojet engine, rated at 70·6 kN (15,873 lb st) with afterburning. Movable semi-conical centrebody in each intake. Fuel in integral tanks in wings and fuselage, on each side of intake trunks, able to be pressure-refuelled completely in about 6 min. Provision for three jettisonable auxiliary fuel tanks (each 1,200 litres; 264 Imp gallons) to be carried under fuselage and on inboard wing pylons. Non-retractable flight refuelling probe on starboard side of nose optional.

ACCOMMODATION: Single SEMMB (Martin-Baker FIRM4) ejection seat for pilot, under rearward-hinged canopy. Cockpit is air-conditioned, and is heated by warm air bled from engine which also heats the radar compartment and certain equipment compartments. Intertechnique liquid oxygen system for pilot.

SYSTEMS: Two independent hydraulic systems, for landing gear retraction, flaps and flying controls, supplied by pumps similar to those fitted in Mirage III. Electrical system includes two Auxilec 15kVA variable-speed alternators, either of which can supply all functional and operational requirements. Emergency and standby power provided by SAFT Voltabloc 40Ah nickel-cadmium battery and EMD static converter. DC power provided by transformer-rectifiers operating in conjunction with battery. Liquid oxygen system standard.

AVIONICS AND EQUIPMENT: Thomson-CSF Cyrano IV fire-control radar in nose. This permits all-sector interception at any altitude and incorporates a system to eliminate 'fixed' echoes when following low-flying aircraft. Two UHF transceivers (one UHF/VHF), Socrat 6200 VOR/ILS with Socrat 5600 marker beacon receiver, LMT Tacan, LMT NR-AI-4-A IFF, remote-setting interception system, three-axis generator, central air data computer, Bézu Sphere with ILS indicator, Crouzet Type 63 navigation indicator and SFENA 505 autopilot. CSF head-up display, with magnifying lens, provides all necessary data for flying and fire control. SAGEM Uliss 47 inertial navigation system optional. Equipment for attack role can include Doppler radar and bombing computer, navigation computer, position indicator, laser rangefinder and terrain-avoidance radar.

ARMAMENT AND OPERATIONAL EQUIPMENT: Standard fixed armament of two 30 mm DEFA 553 cannon, with 125 rds/gun, mounted in lower front fuselage. Two Alkan universal stores attachment pylons under each wing and one under centre-fuselage, plus provision for carrying one air-to-air missile at each wingtip. Max external combat load 4,000 kg (8,820 lb). Externally-mounted weapons for interception role include Matra R.530 or Super 530 radar homing or infra-red homing air-to-air missiles on underfuselage and inboard wing pylons, and/or a Sidewinder or Matra 550 Magic infra-red homing air-to-air missile at each wingtip station. For ground attack duties, typical loads may include one AS.37 Martel anti-radar missile or AS.30 air-to-surface missile, eight 450 kg bombs, four launchers each containing 18 air-to-ground rockets, or six 600 litre (132 Imp gallon) napalm tanks. Other possible external loads include three 1,200 litre (264 Imp gallon) auxiliary fuel tanks, or two photoflash containers and a reconnaissance pod incorporating an SAT Cyclope infra-red system and EMI side-looking radar.

DIMENSIONS, EXTERNAL:

Wing span	8·40 m (27 ft 6¾ in)
Length overall	15·00 m (49 ft 2½ in)
Height overall	4·50 m (14 ft 9 in)
Wheel track	2·50 m (8 ft 2½ in)
Wheelbase	5·00 m (16 ft 4¾ in)

AREA:

Wings, gross	25·00 m² (269·1 sq ft)

WEIGHTS AND LOADING:

Weight empty	7,400 kg (16,314 lb)
T-O weight, 'clean'	10,900 kg (24,030 lb)
Max T-O weight	15,200 kg (33,510 lb)
Max wing loading	608 kg/m² (124·5 lb/sq ft)

PERFORMANCE:

Max level speed (high altitude)	Mach 2·2
Max level speed (low altitude)	Mach 1·2
Approach speed	141 knots (260 km/h; 162 mph)
Landing speed	124 knots (230 km/h; 143 mph)
Max rate of climb at S/L (with afterburning)	12,780 m (41,930 ft)/min
Max rate of climb at high altitude (with afterburning)	14,580 m (47,835 ft)/min
Service ceiling	20,000 m (65,600 ft)
Stabilised supersonic ceiling	18,500 m (60,700 ft)
T-O run (AUW of 11,500 kg; 25,355 lb)	450 m (1,475 ft)
T-O run (typical interception mission)	640 m (2,100 ft)
Landing run (AUW of 8,500 kg; 18,740 lb)	500 m (1,640 ft)
Landing run (typical interception mission)	610 m (2,000 ft)
Endurance	3 h 45 min

DASSAULT-BREGUET MIRAGE 2000

Following cancellation of the ACF (Avion de Combat Futur) programme, described briefly in the 1975-76 *Jane's*, the Mirage 2000 was selected on 18 December 1975 as the primary combat aircraft of the French Air Force from the mid-eighties. Under French government contract, it is being developed initially as an interceptor and air superiority fighter, powered by a single SNECMA M53 turbofan engine. The Mirage 2000 will be equally suitable for reconnaissance, close support, and low-altitude attack missions in areas to the rear of a battlefield.

Reversion to a Mirage III/5 type of delta-wing design, without horizontal tail surfaces, caused some surprise after Dassault's choice of a tailed sweptwing configuration for the later Mirage F1 and ACF. It resulted from considerable study of the requirements of a smaller and less ambitious aircraft than the ACF. Research left no doubt that a delta wing embodying the latest aerodynamic concepts

Dassault-Breguet Mirage F1-C of the Picardie squadron, newly-formed in the 12e Escadre of the French Air Force at Cambrai (*Air Portraits*)

offers an excellent compromise between structural simplicity, light weight, high speed characteristics and the demands of rapid acceleration, high rate of climb and manoeuvrability for an aeroplane of relatively modest size and installed power. In particular, a delta layout offers low drag over a wide range of angles of attack in flight, while providing the largest practicable wing area, with attendant benefits in terms of tight turning capability and high service ceiling.

Former shortcomings, such as higher landing speed than a comparable sweptwing type, are overcome by the addition of automatic leading-edge flaps which, used in conjunction with the elevons, constitute a variable-camber wing. At the same time, the adoption of fly-by-wire control for the wing surfaces and rudder, with artificial stability ensured by a central computer, permits acceptance of a far-aft centre of gravity. This makes possible a much reduced landing speed for the Mirage 2000, and improves its manoeuvrability in combat.

Having tested successfully a carbonfibre rudder on a Mirage III, and boron horizontal tail surfaces on a Mirage F1 throughout the flight regime to Mach 2·2, Dassault-Breguet decided to utilise both materials in the Mirage 2000, achieving weight saving of 15-20 per cent in the components so constructed.

Five prototypes have been built, of which four were funded by the French Air Force and one by the manufacturers. The first of these made its first flight, at Istres, on 10 March 1978, only 27 months after programme launch in December 1975. The second flew on 18 September 1978, the third on 26 April 1979 and the fourth on 12 May 1980. The fifth, a **Mirage 2000B** two-seat trainer version, flew on 11 October 1980 and, like its four predecessors, achieved supersonic speed (between Mach 1·3 and 1·5) during its first flight. By 1 June 1981, the five prototypes had amassed a total of nearly 1,000 hours in 1,042 flights by more than 40 pilots, including about 15 from foreign air forces. On the basis of structural testing, the airframe has been approved for a load factor of +9*g* and rate of roll of 270°/s in subsonic and supersonic flight, clean or with four air-to-air missiles.

Delivery of production Mirage 2000s is expected to begin in 1983. The manufacturers' prototype will be used to develop equipment and other changes proposed for future variants and for export models of the Mirage 2000. Further airframes have been built for static and fatigue testing.

Initial production contracts, in 1979-82, were expected to finance 127 single-seat and two-seat Mirage 2000s in air defence configuration, with an eventual requirement for 200 aircraft in this role; the first four were ordered in the 1979 budget, 22 were requested in 1980 and 22 more in 1981. Dassault believes that a further 200 Mirage 2000s will be required for reconnaissance and strike duties. A single basic type would then make up a high proportion of the French Air Force's planned first-line strength of 450 combat aircraft by the second half of the 1980s.

Because of development delays, the first 50 production Mirage 2000s will be delivered from 1983 with RDM multi-mission radar (as offered to export customers) instead of the RDI pulse-Doppler interception radar specified by the French Air Force. RDI systems, able to detect a 5 m² (54 sq ft) target at a range of 54 nm (100 km; 62 miles), will be fitted from 1985.

Two prototypes of the **Mirage 2000N** two-seat low-altitude penetration version are expected to fly in 1983. Strengthened for flight at a typical 600 knots (1,110 km/h; 690 mph) at 60 m (200 ft) above the terrain, this version will have EMD Antilope V terrain-following radar, two SAGEM inertial platforms, improved TRT AHV12 radio altimeter, Thomson-CSF colour CRT, an OMERA vertical camera, and special ECM. Production deliveries of this model, with provision for carrying ASMP nuclear missiles, are scheduled to start in 1986.

The following description applies to the single-seat air defence version of the Mirage 2000:

TYPE: Single-seat interceptor and air superiority fighter.

WINGS: Cantilever multi-spar low-wing monoplane of delta planform, with cambered profile. Leading-edge sweepback 58°. Large-radius root fairings. Full-span automatic leading-edge flaps operate in conjunction

Prototype Dassault-Breguet Mirage 2000 carrying two Magic air-to-air missiles, eight 250 kg bombs and two 1,700 litre drop-tanks

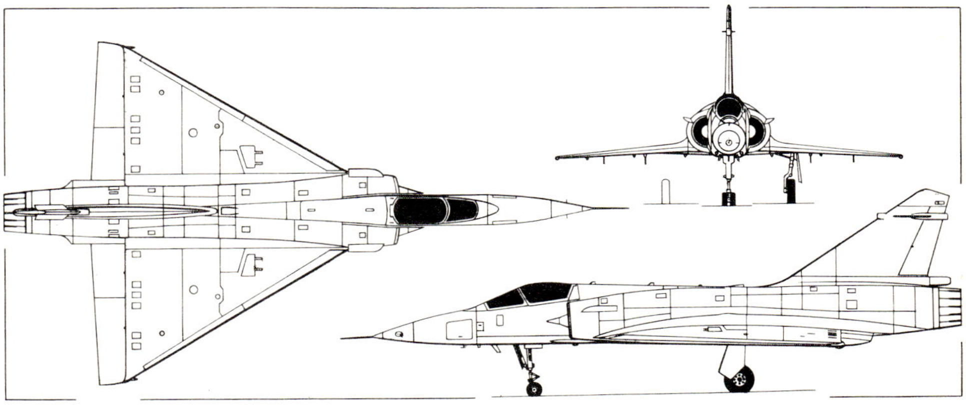

Dassault-Breguet Mirage 2000 (one SNECMA M53 afterburning turbofan engine) *(Pilot Press)*

with two-section elevons which form entire trailing-edge of each wing, to provide variable camber in combat and during landing approach. Leading-edge flaps are retracted during all phases of acceleration and low-altitude cruise, to reduce drag. Elevons have carbonfibre skin, with AG5 light alloy honeycomb core. Fly-by-wire control system for elevons and flaps, with surfaces actuated by hydraulic servo-units. No tabs. Retractable airbrake above and below each wing.

FUSELAGE: Conventional semi-monocoque structure, 'waisted' in accordance with area rule; of conventional all-metal construction except for carbonfibre/light alloy honeycomb panel over avionics compartment, immediately aft of canopy. Small fixed strake, with marked dihedral, near leading-edge of each air intake trunk.

TAIL UNIT: Cantilever fin and inset rudder only; latter actuated by fly-by-wire control system via hydraulic servo-units. Much of fin skin and all rudder skin of boron/epoxy/carbon composites with honeycomb core of Nomex (fin) or light alloy (rudder). Sweepback on fin leading-edge 45°. No tab.

LANDING GEAR: Retractable tricycle type by Messier-Hispano-Bugatti, with twin nosewheels, and single wheel on each main unit. Hydraulic retraction, nosewheels rearward, main units inward. Oleo-pneumatic shock-absorbers. Electro-hydraulic nosewheel steering, through 45° to each side. Manual disconnect permits nosewheel unit to castor through 360° for ground towing. Light alloy wheels and tubeless tyres, size 360 × 135-6 on nosewheels, 750 × 230-15 on main wheels. Messier-Hispano-Bugatti hydraulically-actuated graphite composite disc brakes on main wheels, with anti-skid units. Runway arrester gear standard. Brake-chute in canister above jet nozzle.

POWER PLANT: One '10,000 kg class' SNECMA M53 afterburning turbofan engine. The 83·4 kN (18,740 lb st) M53-2 fitted for early prototype testing was replaced in 1980 by the 88·3 kN (19,840 lb st) M53-5, which will also power initial production aircraft. Under development for subsequent use is the M53-P2, rated at 95·1 kN (21,385 lb st). Movable half-cone centrebody in each air intake. Internal fuel capacity 3,800 litres (835 Imp gallons). (Fuel capacity of 2000B reduced by less than 100 litres; 22 Imp gallons.) Provision for a jettisonable fuel tank, capacity 1,700 litres (374 Imp gallons), under each wing. Flight refuelling probe forward of cockpit on starboard side.

ACCOMMODATION: Pilot only, on Martin-Baker F10Q zero-zero ejection seat, under transparent canopy, in air-conditioned and pressurised cockpit.

SYSTEMS: Two independent hydraulic systems, pressure 280 bars (4,000 lb/sq in), to actuate flying control servo-units, landing gear and brakes. Electrical system includes two Auxilec 20110 aircooled 20kVA 400Hz constant-frequency alternators, two Bronzavia DC transformers, a SAFT 40Ah battery and ATEI static inverter.

AVIONICS AND EQUIPMENT: Pulse-Doppler RDI radar, developed and produced by Thomson-CSF in collaboration with Electronique Marcel Dassault, with operating range of 54 nm (100 km; 62 miles), capability of detecting targets at all altitudes, and good ECCM characteristics (interchangeable with Thomson-CSF RDM multi-role radar. Strike version will have EMD/Thomson-CSF Antilope 5 ground-scan radar). SAGEM Uliss 52 inertial platform. EMD central digital computer. Thomson-CSF VE-130 head-up and VMC-180 head-down displays. SFENA 605 autopilot. Thomson-CSF/EMD ECM, including passive radar warning, LMT Deltac Tacan and IFF, Socrat VOR/ILS/marker beacon receiver, TRT radio altimeter, TRT or EAS UHF and V/UHF com, Thomson-CSF laser designator and marked target seeker.

ARMAMENT: Two 30 mm DEFA 554 cannon, with 125 rds per gun. Nine attachments for external stores, five under fuselage and two under each wing. Typical interception weapons comprise two Matra Super 530 missiles (inboard) and two Matra 550 Magic missiles (outboard)

The two-seat Dassault-Breguet Mirage 2000B, first flown on 11 October 1980 *(J. M. G. Gradidge)*

under wings. Alternatively, a single Super 530 can be carried under the fuselage instead of two underwing, or each of the four underwing hardpoints can carry a Magic. (Strike version will carry more than 6,000 kg; 13,225 lb of external stores, including 18 250 kg bombs or Durandal penetration bombs; three 1,000 kg bombs; four 18-round packs of 68 mm rockets; two packs of 100 mm rockets; seven Beluga cluster bombs; two cannon pods; one reconnaissance pod; three AS.30 Laser air-to-surface missiles; or three Exocet anti-ship missiles.)

Dimensions, external:	
Wing span	9·00 m (29 ft 6 in)
Length overall: 2000	14·35 m (47 ft 1 in)
2000B	14·55 m (47 ft 9 in)
Wheel track	3·40 m (11 ft 1¾ in)
Wheelbase	5·00 m (16 ft 4¾ in)
Area:	
Wings, gross	41 m² (441 sq ft)
Weights:	
Weight empty	7,400 kg (16,315 lb)
Max T-O weight	16,500 kg (36,375 lb)
Performance:	
Max level speed	over Mach 2·2
Max continuous speed	Mach 2·2 (800 knots IAS)
Max speed at low altitude without afterburning, carrying eight 250 kg bombs and two Magic missiles	over 600 knots (1,110 km/h; 690 mph)
Approach speed	140 knots (260 km/h; 162 mph)
Min speed in stable flight	90 knots (167 km/h; 104 mph)
Rate of climb at S/L	more than 18,000 m (59,050 ft)/min
Time to 15,250 m (50,000 ft) and Mach 2	4 min
Time from brake release to intercept target flying at Mach 3 at 24,400 m (80,000 ft)	less than 5 min
Service ceiling	20,000 m (65,600 ft)
Range with four 250 kg bombs	more than 800 nm (1,480 km; 920 miles)
Range with two 1,700 litre drop-tanks	more than 972 nm (1,800 km; 1,118 miles)

DASSAULT-BREGUET SUPER MIRAGE 4000

When the French Air Force abandoned development of the ACF (Avion de Combat Futur) programme, in favour of the single-engined Mirage 2000, M Marcel Dassault announced, in December 1975, that Dassault-Breguet would develop at the French industry's own expense a twin-turbofan scale-up of the Mirage 2000, intended primarily for interception and low-altitude penetration attacks on targets a considerable distance from its base. Potential export customers were assured that the new aircraft would offer overall performance superior to that of any aircraft in its class known to be in production or under development.

A mockup of the new type, now designated Super Mirage 4000 (originally Super Mirage Delta), was unveiled in December 1977. The prototype achieved a speed of Mach 1·2 during its first flight on 9 March 1979, Mach 1·6 on its second flight three days later, and Mach 2·2 during its sixth flight on 11 April, when an initial spin analysis was also made and it was flown at angles of attack up to 25°.

Its general configuration is shown in the accompanying illustrations. Dimensions, weights, performance, and details of armament are generally classified; but installation of two engines of the type fitted in the single-engined Mirage 2000 will give the Super Mirage 4000 a power:weight ratio well above 1:1 in an interceptor role. It was said to have taken off at a loaded weight of about 20,000 kg (44,000 lb) for early flight tests.

The Super Mirage 4000 has computer-derived aerodynamics, with a rearward CG made possible by a fly-by-wire active control system. Other features include foreplanes, a blister-type cockpit canopy giving a 360° field of view, a very large nose radome, and extensive use of boron and carbonfibre composites for structures such as the fin, rudder, elevons, fuselage access panels and foreplanes.

The following details should be regarded as provisional:

Type: Single-seat multi-role combat aircraft (two-seat version under study).

Wings: Cantilever mid-wing monoplane of delta planform, with computer-derived aerodynamics. Large-radius root fairings. Two-section elevons form entire trailing-edge of each wing. Full-span automatic leading-edge flaps provide variable camber in combat. Fly-by-wire active control system for elevons and flaps.

Fuselage: Conventional semi-monocoque structure, 'waisted' in accordance with area rule. Door-type airbrake in each intake trunk above wing-root leading-edge.

Tail Unit and Foreplanes: Cantilever fin and inset rudder; latter actuated by fly-by-wire control system. Fin is made of carbon composite and contains fuel tankage. Variable-incidence sweptback foreplane near lip of each engine air intake duct.

Landing Gear: Retractable tricycle type, of Messier-Hispano-Bugatti design, with twin nosewheels and single wheel on each main unit. Hydraulic retraction, nosewheels forward, main units inward. Oleo-pneumatic shock-absorbers. Electro-hydraulic nosewheel steering. Aluminium alloy main wheels, with tubeless tyres and steel disc brakes on prototype; graphite composite brakes planned on production aircraft.

Power Plant: Two '10,000 kg class' SNECMA M53 afterburning turbofan engines side by side in rear fuselage. Movable half-cone centrebody in each air intake. Provision for a large jettisonable fuel tank under each wing and under fuselage. Fuel tankage in fin helps to give total capacity about three times that of the Mirage 2000.

Accommodation: Pilot only, on Martin-Baker F10R zero-zero ejection seat, under sideways-opening (to starboard) transparent canopy; 360° field of view.

Systems: Messier-Hispano-Bugatti hydraulic system, pressure 280 bars (4,000 lb/sq in), powered by four advanced pumps and using lightweight titanium pipelines. Two Auxilec electrical generators. Turboméca Palouste gas turbine APU, in compartment aft of pilot's seat, for engine starting.

Avionics and Equipment: Provision for a radar of 80 cm (31·5 in) diameter in nose, to provide search range of up to 65-70 nm (120-130 km; 75-80 miles). Prototype is fitted initially with the same RDM multi-function Doppler radar as the Mirage 2000, to be replaced later by pulse-Doppler RDI radar. Digital autopilot, multimode displays, SAGEM Uliss 52 inertial navigation system, Crouzet Type 80 air data computer, Thomson-CSF VE-130 head-up display and digital automated weapon delivery system.

Armament: Provision for two 30 mm DEFA guns in bottom of air intake trunks and underwing rails for Matra 550 Magic air-to-air missiles, plus a wide range of air-to-air and air-to-surface weapons. Total of eleven hardpoints for external stores.

Dimensions, external:	
Wing span	12·00 m (39 ft 4½ in)
Length overall	18·70 m (61 ft 4¼ in)
Wheel track	4·36 m (14 ft 3½ in)
Wheelbase	6·90 m (22 ft 7½ in)
Area:	
Wings, gross	73·00 m² (786 sq ft)

Dassault-Breguet Super Mirage 4000 (two SNECMA M53 afterburning turbofan engines)

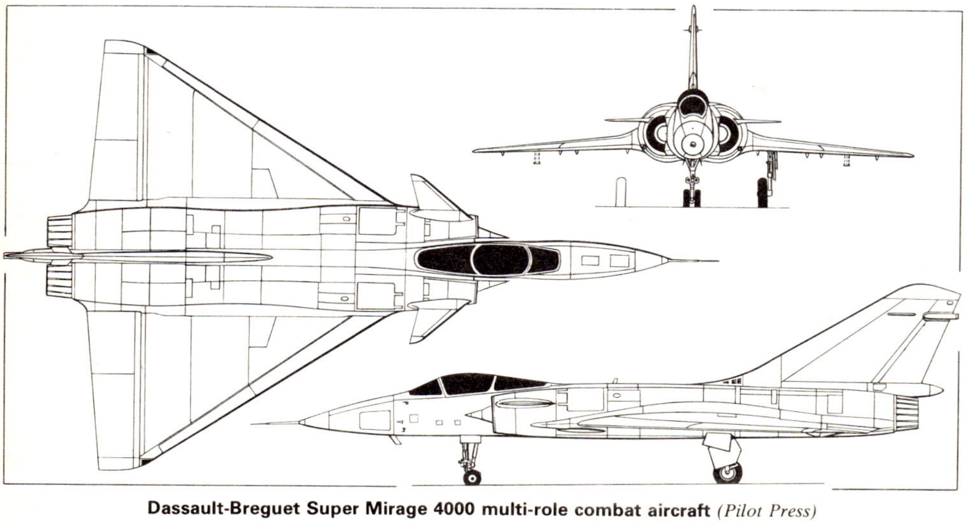

Dassault-Breguet Super Mirage 4000 multi-role combat aircraft *(Pilot Press)*

DASSAULT-BREGUET SUPER ETENDARD

Dassault-Breguet is producing for the French Navy an updated version of its Etendard IV-M carrier-based fighter which has served with the Navy's operational squadrons since 1962, and was last described in the 1965-66 *Jane's*. The airframe and equipment of the new version, known as the Super Etendard, were expected to be 90 per cent common with those of the Etendard IV-M, except for the nav/attack system. In fact, the installation of a more powerful turbojet engine and equipment of enhanced capability, together with the adoption of improved aerodynamic features and modern manufacturing techniques, has made the Super Etendard 90 per cent new.

The Super Etendard is a transonic single-seat strike fighter, for low and medium altitude operations from ships in the class of the French Navy's *Clémenceau* and *Foch*. Its equipment includes a highly sophisticated and accurate nav/attack integrated avionics system. Inherent long range is increased by flight refuelling capability, and it is able to operate as a tanker for other aircraft.

The Atar 8K-50 turbojet engine is a non-afterburning version of the Atar 9K-50 used in the Mirage F1 multimission fighter and attack aircraft. It has a lower specific fuel consumption than the Atar 8 fitted in the Etendard IV-M. The thrust increase of about ten per cent, combined with a new wing leading-edge and redesigned flaps, allows a significant increase in gross weight for catapulting and, hence, permits increased fuel load and armament.

Two prototypes were produced by conversion of standard IV-M airframes. The first of these flew for the first time on 28 October 1974. Its programme included engine development, followed in 1978 by tests of the Super Etendard's external load-carrying capability and firing trials of the Exocet AM39 air-to-surface anti-shipping missile.

The second prototype, which flew for the first time on 25 March 1975, was used for tests of the Super Etendard's navigation system and bombing capabilities. Its subsequent tasks included shipboard operation under open-sea conditions in waters other than the Mediterranean, where all early trials took place.

It was intended originally to build 100 production aircraft, but the number was reduced to 71 in order to conform with budget limitations. The first aircraft flew on 24 November 1977. Deliveries began on 28 June 1978, when the third production aircraft was accepted officially by the French Navy, and Dassault-Breguet had delivered 51 Super Etendards to this service by 1 May 1981, with production continuing at the rate of two aircraft per month. The first export order, for 14, was placed by the Argentine Navy in 1979.

It was announced in 1980 that the French Navy was studying the possibility of developing a reconnaissance version of the Super Etendard.

Type: Single-seat transonic carrier-based strike fighter.

Wings: Cantilever mid-wing monoplane. Thickness/chord ratio varies from 6% at root to 5% at tip. Anhedral 1°. Sweepback at quarter-chord 45°. All-metal two-spar torsion-box structure; stressed skin of machined panels with integral stiffeners. Tips fold upward for carrier stowage. Inset ailerons, hydraulically-powered by Dassault irreversible dual circuits with artificial feel. Spoiler on top surface of each wing, ahead of special double-slotted flap with second slot in form of an integral 'gutter'. Flap travel increased by comparison with Etendard IV-M. Hydraulically-powered drooping leading-edges, with extended chord on outer panels.

Fuselage: All-metal semi-monocoque structure, 'waisted' in accordance with area rule. Perforated airbrake under each side of centre-fuselage.

Tail Unit: Cantilever all-metal structure, with tailplane mid-set on fin. All surfaces swept. All-moving tailplane (with electrically-powered pitch trim) and rudder are powered in same way as ailerons.

Landing Gear: Retractable tricycle type, with single wheel on each unit, manufactured by Messier-Hispano-Bugatti. Nosewheel retracts rearward, main units inward into wings and fuselage. Messier-Hispano-Bugatti oleo-pneumatic shock-absorbers and disc brakes. Main-wheel tyres size 30 × 7·7-16; nose-wheel tyre size 490 × 155-9. Brake-chute in fairing at junction of fin and tailplane trailing-edges.

Power Plant: One SNECMA Atar 8K-50 non-afterburning turbojet, rated at 49 kN (11,025 lb st). Fuel in integral tanks in wings and rubber tanks in fuselage, with total capacity of 3,200 litres (704 Imp gallons; 845 US gallons). Provision for an external tank of 1,100 litres (242 Imp gallons; 290 US gallons) under each wing, and a 600 litre (132 Imp gallon; 158 US gallon) centreline tank or flight refuelling 'buddy' pack under the fuselage. Retractable flight refuelling probe in front of windscreen.

Accommodation: Pilot only, on Hispano-built Martin-Baker SEMMB CM4A lightweight ejection seat in pressurised and air-conditioned cockpit. Extensively armoured.

Systems: Duplicated hydraulic circuits for flying controls, landing gear, brakes and airbrakes, and wing leading-edge droop.

Avionics and Equipment: SAGEM-Kearfott ETNA inertial navigation and attack system; Thomson-CSF/EMD Agave lightweight search/track/designation/telemetry/navigation radar; Thomson-CSF VE-120 head-up display; Crouzet Type 97 navigation display, armament control panel and selector box, and Type 66 air data computer; TRT radio altimeter; SFIM three-axis attitude indicator; LMT micro-Tacan and IFF, and SOCRAT VOR.

Armament: Two DEFA 30 mm guns, each with 125 rds, in bottom of engine air intake trunks. Underfuselage attachments for two 250 kg bombs. Four underwing attachments for 400 kg bombs, Magic air-to-air missiles or rocket pods. Optionally, one Exocet AM39 air-to-surface missile under starboard wing, and one external fuel tank under port wing.

Dimensions, external:

Wing span	9·60 m (31 ft 6 in)
Width, wings folded	7·80 m (25 ft 7 in)
Wing aspect ratio	3·23
Length overall	14·31 m (46 ft 11½ in)
Height overall	3·86 m (12 ft 8 in)
Wheel track	3·50 m (11 ft 6 in)
Wheelbase	4·80 m (15 ft 9 in)

Area:

Wings, gross	28·4 m² (305·7 sq ft)

Weights:

Weight empty	6,450 kg (14,220 lb)
Max fuel, incl two 1,100 litre underwing tanks	4,800 kg (10,580 lb)
Max weapon load, internal fuel only	2,100 kg (4,630 lb)
Mission T-O weight	9,200-11,500 kg (20,280-25,350 lb)

Performance:

Max level speed at height	approx Mach 1
Max level speed at low altitude	650 knots (1,204 km/h; 748 mph)
Approach speed for shipboard landing at AUW of 7,800 kg (17,200 lb)	135 knots (250 km/h; 155 mph)
Service ceiling	13,700 m (45,000 ft)
Radius of action, with AM39 missile	350 nm (650 km; 403 miles)

DASSAULT-BREGUET/DORNIER ALPHA JET

Details of the Alpha Jet programme can be found in the International section of this edition.

DASSAULT-BREGUET/BAe JAGUAR

Details of the Jaguar programme can be found under 'SEPECAT' in the International section of this edition.

BREGUET ALIZÉ

The French Navy is modernising 28 of its remaining 34 Alizé carrier-based anti-submarine aircraft, to extend their effectiveness until the 1990s. The work includes installation of new Thomson-CSF Iguane retractable

Dassault-Breguet Super Etendard carrier-based fighter armed with Exocet AM39 anti-ship missile

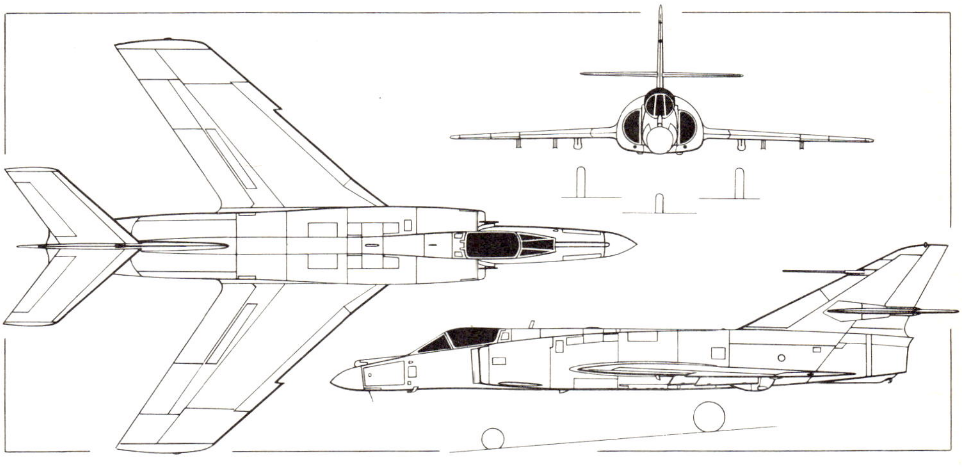

Dassault-Breguet Super Etendard naval fighter (SNECMA Atar 8K-50 turbojet engine) *(Pilot Press)*

radar (as fitted to the Atlantic ANG), a new navigational system and ECM equipment. The programme is scheduled for completion in 1983.

Since the Alizés entered service in 1959, they have accumulated a total of about 250,000 flying hours with the French Navy. The type was last described in the 1963-64 *Jane's*.

DASSAULT-BREGUET ATLANTIC ANG

The Atlantic ANG (Atlantic Nouvelle Génération) is a twin-turboprop maritime patrol aircraft derived directly from the earlier Atlantic that was produced in 1964-74 for operation by the armed services of France (40, of which 3 were sold subsequently to Pakistan), the German Federal Republic (20, including 5 special-purpose ECM aircraft), Italy (18) and the Netherlands (9). Design definition of the new version was initiated by the French government in July 1977, with the aim of providing a replacement for the first generation Atlantic and the Neptune during the period from 1985 to 1990. This led to launch of the development phase of the ANG programme in September 1978. Initial requirement is for 42 aircraft for the French Navy.

Two prototypes have been produced by modification of first generation Atlantic airframes. Work started in January 1979, and the first prototype flew for the first time in its new form on 8 May 1981. The second is expected to fly in March 1982, by which time series production will have begun, to permit completion of the first production Atlantic ANG in early 1986. The work is to be shared by the same European SECBAT (Société d'Etudes et de Construction du Breguet Atlantic) consortium that was responsible for the earlier programme, with possible modification of the work-split to reflect varying national interests in the ANG aircraft. Companies involved, under Dassault-Breguet direction, are SABCA and Sonaca of Belgium, Fokker of the Netherlands, MBB and Dornier of Federal Germany, Aeritalia of Italy and Aérospatiale of France. The Tyne engines are being produced by SNECMA of France, Rolls-Royce of the UK and FN of

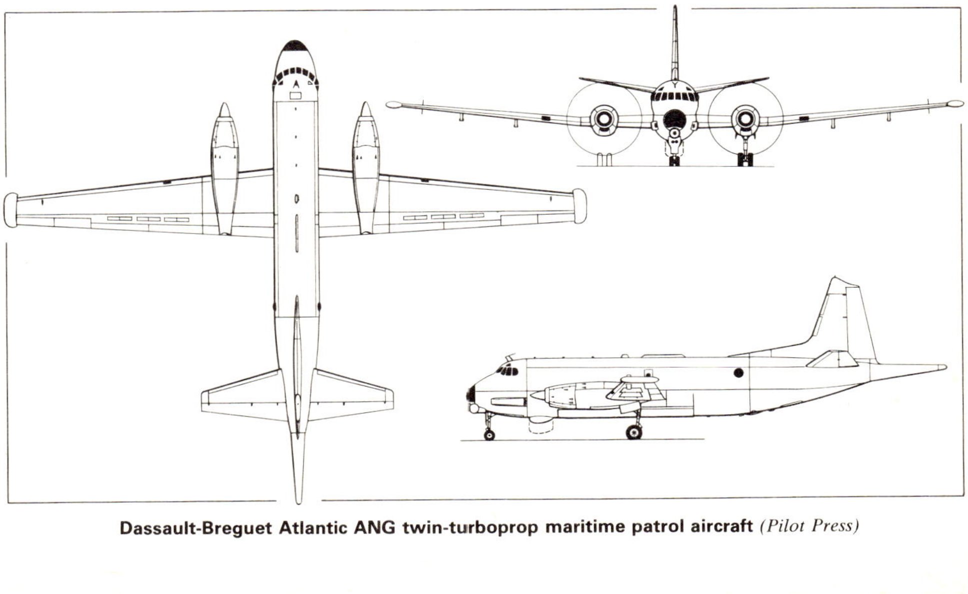

Dassault-Breguet Atlantic ANG twin-turboprop maritime patrol aircraft *(Pilot Press)*

First prototype of the Dassault-Breguet Atlantic ANG maritime patrol aircraft *(Brian M. Service)*

Belgium; and propellers by Ratier of France and British Aerospace.

Structural changes by comparison with the first generation Atlantic include use of a refined bonding technique, improved anti-corrosion protection, better sealing between skin panels, and design improvements offering longer fatigue life and more economical maintenance. These are intended to ensure increased serviceability, with 75 per cent of squadron aircraft permanently available for operations; readiness to take off within 20 minutes of an order to go; and an aircraft life of more than 25 years, or at least 15,000 flying hours.

The basic mission performance requirements envisaged for the ANG are quite similar to those of the Atlantic now in service: a high cruising speed to the operational area, quick descent from cruising altitude to patrol height, lengthy patrol endurance at low altitude, and a high degree of manoeuvrability at sea level. It is able to carry a wide variety of weapons and equipment for finding and attacking both submarines and surface targets in all weathers. Like the original Atlantic, the ANG is able to perform minelaying, logistic support, and passenger and freight transport missions. It could be adapted for advanced AEW and flight refuelling duties, and is suitable for civilian tasks such as air/sea rescue and patrol of offshore fishing and oil interests.

Type: Twin-turboprop maritime patrol aircraft.

Wings: Cantilever mid-wing monoplane, with streamlined ESM pods on tips. Wing section NACA 64 series. Dihedral 6° on outer panels only. Incidence 3°. Tapered planform, with 9° sweepback on leading-edge. All-metal three-spar fail-safe structure, with bonded light alloy honeycomb skin panels on torsion box and on main landing gear doors. Two conventional all-metal ailerons on each wing, actuated by SAMM twin-cylinder jacks. All-metal slotted flaps, with bonded light alloy honeycomb filling, in three segments on each wing, over 75 per cent of span. Three hinged spoilers on upper surface of each outer wing, forward of flaps. Metal airbrake above and below each wing. No trim tabs. Air Equipment/Kléber-Colombes pneumatic de-icing system on leading-edges.

Fuselage: All-metal 'double-bubble' fail-safe structure, with bonded honeycomb sandwich skin on pressurised central section of upper lobe, upward-sliding weapons bay doors and nosewheel door. Larger air intake and duct for air-conditioning system on each side of nose.

Tail Unit: Cantilever all-metal structure, with bonded honeycomb sandwich skin panels on torsion boxes. Slightly bulged housing for ESM antennae at top of fin leading-edge. Fixed-incidence tailplane, with dihedral. Control surfaces operated through SAMM twin-cylinder jacks. No trim tabs. Air Equipement/Kléber-Colombes pneumatic de-icing system on leading-edges.

Landing Gear: Retractable tricycle type, supplied by Messier-Hispano-Bugatti, with twin wheels on each unit. Hydraulic retraction, nosewheels rearward, main units forward into engine nacelles. Kléber-Colombes or Dunlop tyres; size 39 × 13-20 on main wheels, pressure 12 bars (170 lb/sq in), 26 × 8-13 on nosewheels, pressure 6·5 bars (94 lb/sq in). New Messier-Hispano-Bugatti disc brakes with higher braking energy, and Modistop anti-skid units.

Power Plant: Two 4,638 kW (6,220 ehp) Rolls-Royce Tyne RTy.20 Mk 21 turboprop engines, each driving a four-blade Ratier/British Aerospace constant-speed metal propeller type PD 249/476/3 on prototypes. For production aircraft Ratier-Figeac has proposed a lighter-weight propeller of increased diameter, with blades of composite construction. Four pressure-refuelled integral fuel tanks in wings, with total capacity of 23,000 litres (5,059 Imp gallons). Updated gauging system. Oil capacity 100 litres (22 Imp gallons).

Accommodation: Normal flight crew of 12, comprising observer in nose; pilot, co-pilot and flight engineer on flight deck; a radio-navigator, ESM-ECM-MAD operator, radar-IFF operator, tactical co-ordinator and two acoustic sensor operators at stations on the starboard side of the tactical compartment; and two observers in beam positions at the rear. Provision for carrying relief crew, or 12 other personnel. Rest compartment, with eight seats, in centre-fuselage, forward of crew room with tables and seats, galley, toilet and wardrobe. Primary access via extending airstair door in bottom of rear fuselage. Emergency exits above and below flight deck and on each side of fuselage, above wing trailing-edge.

Systems: Air-conditioning system supplied by two compressors driven by gearboxes. Heat exchangers and bootstrap system for cabin temperature control. Duplicated hydraulic system, to operate flying controls, landing gear, flaps, weapons bay doors and retractable radome. Three basic electrical systems: variable-frequency three-phase 115/208V AC system, with two 60/80kVA Auxilec alternators and modernised control and protection equipment; fixed-frequency three-phase 115/208V 400Hz AC system, with four 15kVA Auxilec Auxivar generators, two on each engine; 28V DC system, with four 6kW transformer-rectifiers supplied from the variable-frequency AC system, and one 40Ah battery. One 60kVA emergency AC generator, driven at constant speed by APU. Individual oxygen bottles for emergency use. Electrical anti-icing for engine air intake lips, propeller blades and spinners. Turboméca/ABG/SEMCA Astadyne gas turbine APU for engine starting, emergency electrical supply, and air-conditioning on ground.

Armament, Avionics and Operational Equipment: Main weapons bay in unpressurised lower fuselage can accommodate all NATO standard bombs, depth charges, eight homing torpedoes or two air-to-surface missiles (typical load comprises three torpedoes and one AM39 Exocet missile). Four underwing attachments for up to 3,500 kg (7,716 lb) of stores, including rockets, air-to-surface missiles or containers. Up to 78 sonobuoys in compartment aft of weapons bay, where whole of upper and lower fuselage provides storage for sonobuoys and marker flares. SAT/TRT forward-looking infra-red sensor in turret under nose. Thomson-CSF Iguane retractable radar immediately forward of weapons bay, with LMT IFF interrogator and SECRE decoder. Agiflite and OMERA cameras in port side of nose and in bottom of rear fuselage. Crouzet-manufactured MAD in lengthened tail sting. Thomson-CSF Arar 13 radar detector for ESM. Other equipment includes HF com, Tacan and DME by Thomson-CSF, VHF/AM com by Socrat, VOR/ILS by EAS, TRT radio altimeter, Collins MF radio compass, ADF, HSI and autopilot/flight director by SFENA, dual SAGEM Uliss 53 inertial navigation systems, SAGEM teleprinters, Crouzet geographical display and air data computer. UHF and VHF/FM suppliers not yet designated. Thomson-CSF Sadang system for processing active and passive acoustic detection data. Optional equipment includes EMD/Decca Doppler and Crouzet Omega.

Dimensions, external:
- Wing span, incl wingtip pods 37·30 m (122 ft 4½ in)
- Wing aspect ratio 11·63
- Length overall 32·62 m (107 ft 0¼ in)
- Height overall 11·35 m (37 ft 3 in)
- Fuselage: Max depth 4·00 m (13 ft 1½ in)
- Tailplane span 12·31 m (40 ft 4½ in)
- Wheel track 9·00 m (29 ft 6¼ in)
- Wheelbase 9·45 m (31 ft 0 in)
- Propeller diameter: prototypes 4·88 m (16 ft 0 in)
  - proposed for series aircraft 5·20 m (17 ft 0¾ in)
- Distance between propeller centres 9·00 m (29 ft 6¼ in)
- Main weapons bay: Length 9·00 m (29 ft 6¼ in)
  - Width 2·10 m (6 ft 10¾ in)

Dimensions, internal:
- Cabin, incl rest compartment, galley, toilet, aft observers' stations: Length 18·50 m (60 ft 8½ in)
  - Max width 3·60 m (11 ft 9½ in)
  - Max height 2·00 m (6 ft 6¾ in)
  - Floor area 155 m² (1,668 sq ft)
  - Volume 92 m³ (3,250 cu ft)

Areas:
- Wings, gross 120·34 m² (1,295·3 sq ft)
- Ailerons (total) 5·26 m² (56·62 sq ft)
- Flaps (total) 26·80 m² (288·48 sq ft)
- Spoilers (total) 1·66 m² (17·87 sq ft)
- Vertical tail surfaces (total) 16·64 m² (179·11 sq ft)
- Rudder 5·96 m² (64·15 sq ft)
- Horizontal tail surfaces (total) 33·00 m² (355·21 sq ft)
- Elevators 8·30 m² (89·34 sq ft)

Weights and Loadings:
- Weight empty, equipped, standard mission 25,300 kg (55,775 lb)
- Military load: ASW mission 2,200 kg (4,850 lb)
  - ASV mission 3,000 kg (6,610 lb)
- Max fuel 18,500 kg (40,785 lb)
- Mission T-O weight: ASW 43,900 kg (96,780 lb)
  - ASV 45,400 kg (100,090 lb)
- Max overload T-O weight 46,200 kg (101,850 lb)
- Max zero-fuel weight 29,000 kg (63,935 lb)
- Normal design landing weight 36,000 kg (79,365 lb)
- Max landing weight, emergency 46,000 kg (101,400 lb)
- Max wing loading 385 kg/m² (78·96 lb/sq ft)
- Max power loading 5·07 kg/kW (8·34 lb/ehp)

Performance (estimated, with metal propellers, at T-O weight of 45,000 kg; 99,200 lb, except where indicated):
- Never-exceed speed Mach 0·7
- Max level speed at optimum height 355 knots (657 km/h; 408 mph)
- Max level speed at S/L 320 knots (592 km/h; 368 mph)
- Max cruising speed at 7,600 m (25,000 ft) 300 knots (555 km/h; 345 mph)
- Normal patrol speed, S/L to 1,525 m (5,000 ft) 170 knots (315 km/h; 195 mph)
- Stalling speed, flaps down 90 knots (167 km/h; 104 mph)
- Max rate of climb at S/L, AUW of 30,000 kg (66,140 lb) 884 m (2,900 ft)/min
- Max rate of climb at S/L, AUW of 40,000 kg (88,185 lb) 610 m (2,000 ft)/min
- Rate of climb at S/L, one engine out, AUW of 30,000 kg (66,140 lb) 365 m (1,200 ft)/min
- Rate of climb at S/L, one engine out, AUW of 40,000 kg (88,185 lb) 213 m (700 ft)/min
- Service ceiling 9,150 m (30,000 ft)
- Runway LCN at max T-O weight 60

T-O to 10·5 m (35 ft)	1,620 m (5,315 ft)
T-O to 10·5 m (35 ft), one engine out	2,240 m (7,350 ft)
Landing from 15 m (50 ft)	1,500 m (4,922 ft)
170 knot turning radius at AUW of 40,000 kg (88,185 lb) at:	
30° bank	1,500 m (4,925 ft)
45° bank	600 m (1,970 ft)
60° bank	500 m (1,640 ft)

Typical mission profiles, with fuel reserves of 5% total fuel, 5% of fuel consumed and 20 min hold-off:

Anti-ship mission: T-O with max fuel and one AM39 missile; fly 1,800 nm (3,333 km; 2,071 miles) to target area; descend for two-hour search and attack at 90 m (300 ft); return to base

Anti-submarine mission: T-O at 44,300 kg (97,665 lb) AUW with 15,225 kg (33,565 lb) of fuel, four Mk 46 torpedoes, 78 sonobuoys, and a full load of markers and flares; cruise to search area at 290 knots (537 km/h; 333 mph) at 7,600 m (25,000 ft); descend for 8 h patrol at 600 nm (1,110 km; 690 miles) from base, or 5 h patrol at 1,000 nm (1,850 km; 1,150 miles) from base; return to base at 9,145 m (30,000 ft). Total mission time 12 h 31 min

Ferry range with max fuel	4,400 nm (8,150 km; 5,065 miles)
Max endurance	18 h

DASSAULT-BREGUET MYSTÈRE-FALCON 20 SERIES F

Details of the early history of this twin-turbofan light transport can be found in the 1977-78 *Jane's*.

Dassault-Breguet builds the wings and Aérospatiale the fuselages and tail units of production aircraft, which are marketed in the USA under the name Falcon and elsewhere as Mystère 20s. The first production aircraft flew on 1 January 1965, and by 1 May 1981 sales of Mystère-Falcon 20s had reached 476, of which 438 had been delivered, more than 60 per cent of them to the Business Jets Division of Pan American and its successor, Falcon Jet Corporation, formed jointly by Dassault-Breguet and Pan American in 1973.

Basic current production version is the **Mystère-Falcon 20 Series F,** which introduced high-lift devices to improve T-O and landing performance, more powerful engines than earlier Falcons and increased wing fuel tank capacity. The prototype was displayed at the Paris Air Show in June 1969 and deliveries began in July 1970. During 1970 the Series F became the first aircraft to receive type approval under FAR Pt 36 anti-noise regulations. This approval was subsequently extended to other versions.

The **Mystère-Falcon 20 Series G** and **200** are described separately.

All versions of the Mystère-Falcon 20 can be modified as follows for specific duties:

Calibration: Two aircraft ordered by the SGAC, one by the Spanish government and one by the Indonesian government are used for the calibration of radio navigation aids. The equipment includes a removable console, thus retaining the full passenger-carrying capability of the aircraft. The Indonesian aircraft is fitted with an infra-red tracking system supplied by SAT.

Airline crew training: Since 20 September 1966, several Mystère-Falcon 20s have been used by Air France to train pilots for its jet airliners, with up to five aircraft being used simultaneously. Japan Air Lines also bought three of this version.

Cross-country: Similar to basic aircraft, but with low-pressure tyres for soft-field operation at the same take-off and landing weights. Described in 1968-69 *Jane's*.

Quick-change and cargo: A quick-change kit, consisting of an assembly of nets and supports, keeps the centre aisle free and allows direct access to nine freight compartments. Total usable volume of these compartments is 6·65 m³ (235 cu ft), and transformation from executive configuration to cargo configuration, or vice versa, takes less than one hour. A cargo version of the Falcon is also available and is described separately. For both versions an increase of the maximum zero-fuel weight from 8,900 kg (19,600 lb) to 9,980 kg (22,000 lb) allows an increased payload of up to 3,000 kg (6,615 lb).

Target towing: A Mystère-Falcon 20 is used by the French Air Force for target towing missions, with one Secapem target under each wing.

Aerial photography: The French Institut Géographique National has a Mystère-Falcon 20 fitted with two cameras (Zeiss RMK 610 mm focal length, and Wild RC8, RC9 or RC10) and an intervalometer. This enables the aircraft to be used for high-altitude photography and photogrammetry duties.

Systems trainer: Two aircraft fitted with Mirage III-E combat radar and navigation systems are in service with the French Air Force for training its Mirage pilots. This version, known as the **Falcon ST,** has been sold also to Libya.

TYPE: Twin-turbofan executive transport.

WINGS: Cantilever low-wing monoplane. Thickness/chord ratio varies from 10·5 to 8%. Dihedral 2°. Incidence 1° 30′. Sweepback at quarter-chord 30°. All-metal (copper-bearing alloys) fail-safe torsion-box structure with machined stressed skin. Ailerons are each operated by Dassault twin-body actuators, from dual hydraulic systems, and have artificial feel. Non-slotted slats inboard of fence, and slotted slats outboard, on each wing. Hydraulically-actuated spoilers forward of the hydraulically-actuated two-section single-slotted flaps. Leading-edges anti-iced by engine bleed air.

Dassault-Breguet Mystère-Falcon 20 of the Spanish Air Force, by which it is designated TM.11 *(Antonio Camarasa)*

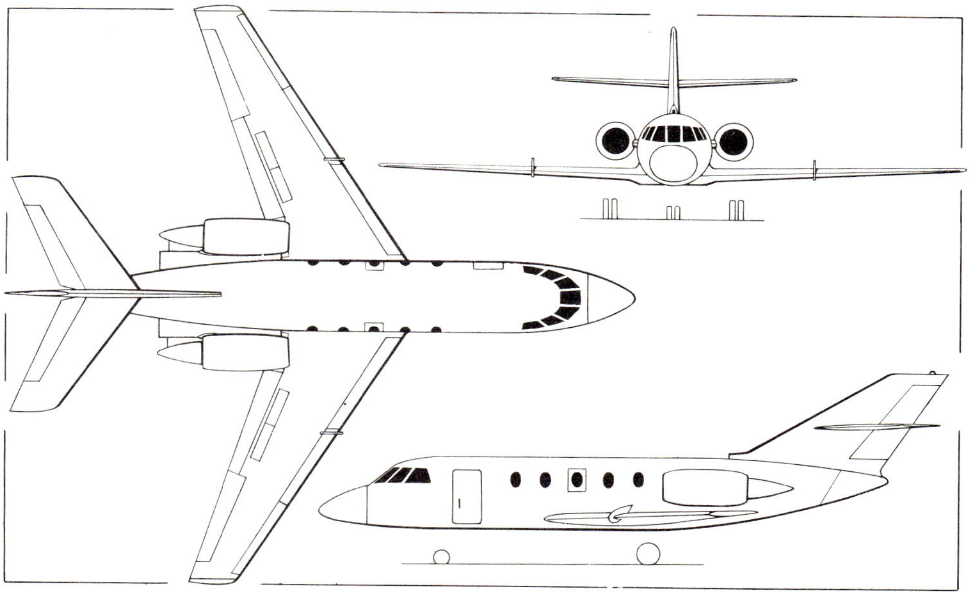
Dassault-Breguet Mystère-Falcon 200 described on page 64 *(Pilot Press)*

FUSELAGE: All-metal semi-monocoque structure of circular cross-section, built on fail-safe principles.

TAIL UNIT: Cantilever all-metal structure, with electrically-controlled variable-incidence tailplane mounted halfway up fin. Elevators and rudder each actuated by twin hydraulic servos. No trim tabs.

LANDING GEAR: Retractable tricycle type, by Messier-Hispano-Bugatti, with twin wheels on all three units. Hydraulic retraction, main units inward, nosewheels forward. Oleo-pneumatic shock-absorbers. Goodyear disc brakes and anti-skid units. Normal tyre pressure 9·15 bars (133 lb/sq in) on all units. Low-pressure gear (4·5 bars; 65 lb/sq in) available optionally. Steerable and self-centering nosewheels. Braking parachute standard.

POWER PLANT: Two General Electric CF700-2D-2 turbofan engines (each 20 kN; 4,500 lb st), pod-mounted on sides of rear fuselage. Fuel in two integral tanks in wings and two auxiliary tanks aft of rear pressure bulkhead in fuselage, with total capacity of 5,180 litres (1,139 Imp gallons; 1,368 US gallons). Separate fuel system for each engine, with provision for cross-feeding. Single-point pressure refuelling. Emergency refuelling by gravity.

ACCOMMODATION: Crew of two on flight deck, with full dual controls and airline-type instrumentation. Normal seating for eight or ten passengers in individual reclining chairs, with tables between forward pairs of seats and a central 'trench' aisle, or 12-14 passengers at reduced pitch without tables. Toilet at rear. Baggage space and wardrobe on starboard side, immediately aft of flight deck opposite door, and at rear of cabin. Buffet with ice-box, food and liquid storage at front of cabin on port side. Downward-opening door has built-in steps.

SYSTEMS: Duplicated air-conditioning and pressurisation system, supplied with air bled from both engines. Pressure differential 0·57 bars (8·3 lb/sq in). Two independent hydraulic systems, pressure 207 bars (3,000 lb/sq in), with twin engine-driven pumps and emergency electric pump, actuate primary flying controls, flaps, landing gear, wheel brakes, spoilers and nosewheel steering. 28V DC electrical system with a 9kW 28V DC starter/generator on each engine, one 1500VA and two 750VA 400Hz 118/208V inverters and two 40Ah batteries. Automatic emergency oxygen system. 9kW Microturbo Saphir II APU optional.

AVIONICS AND EQUIPMENT: Standard equipment includes duplicated VHF and VOR/glideslope, single ADF and DME, marker beacon receiver, ATC transponder, cockpit audio and duplicated blind-flying instrumentation. Optional equipment includes integrated flight instrument system, weather radar, HF communications radio, autopilot, second ADF and DME, and cabin address system.

DIMENSIONS, EXTERNAL:	
Wing span	16·30 m (53 ft 6 in)
Wing chord (mean)	2·85 m (9 ft 4 in)
Wing aspect ratio	6·4
Length overall	17·15 m (56 ft 3 in)
Length of fuselage	15·55 m (51 ft 0 in)
Height overall	5·32 m (17 ft 5 in)
Tailplane span	6·74 m (22 ft 1 in)
Wheel track	3·69 m (12 ft 1¼ in)
Wheelbase	5·74 m (18 ft 10 in)
Passenger door: Height	1·52 m (5 ft 0 in)
Width	0·80 m (2 ft 7½ in)
Height to sill	1·09 m (3 ft 7 in)
Emergency exits (each side, over wing):	
Height	0·66 m (2 ft 2 in)
Width	0·48 m (1 ft 7 in)
DIMENSIONS, INTERNAL:	
Cabin, incl fwd baggage space and rear toilet:	
Length	7·08 m (23 ft 2¾ in)
Max width	1·87 m (6 ft 1¾ in)
Max height	1·72 m (5 ft 8 in)
Volume	20·0 m³ (700 cu ft)
Baggage compartment (fwd)	0·70 m³ (24·7 cu ft)
Baggage compartment (aft)	0·37 m³ (13·1 cu ft)
AREAS:	
Wings, gross	41·00 m² (440 sq ft)
Horizontal tail surfaces (total)	11·30 m² (121·6 sq ft)
Vertical tail surfaces (total)	7·60 m² (81·8 sq ft)
WEIGHTS:	
Weight empty, equipped	7,530 kg (16,600 lb)
Payload with max fuel	1,180 kg (2,600 lb)
Max fuel	4,160 kg (9,170 lb)
Max T-O and ramp weight	13,000 kg (28,660 lb)
Max zero-fuel weight	8,900 kg (19,600 lb)
	or 9,980 kg (22,000 lb)
Typical landing weight	8,930 kg (19,685 lb)
PERFORMANCE:	
Never-exceed speed at S/L	350 knots (650 km/h; 404 mph) IAS
Never-exceed speed at 7,000 m (23,000 ft)	390 knots (725 km/h; 450 mph) IAS

Max cruise Mach No. at 10,050 m (33,000 ft) 0·78
Max cruising speed at 7,620 m (25,000 ft) at AUW of 9,071 kg (20,000 lb)
465 knots (862 km/h; 536 mph)
Econ cruising speed at 12,200 m (40,000 ft)
405 knots (750 km/h; 466 mph)
Stalling speed 82 knots (152 km/h; 95 mph)
Absolute ceiling 12,800 m (42,000 ft)
Service ceiling, one engine out, at AUW of 8,500 kg (18,700 lb) 7,480 m (24,500 ft)
T-O to 10·7 m (35 ft) at AUW of 12,580 kg (27,735 lb) (full tanks, 8 passengers and baggage)
915 m (3,000 ft)
FAR 25 balanced T-O field length, AUW as above
1,450 m (4,750 ft)
FAR 121 landing field length (8 passengers, 940 kg; 2,070 lb reserves) 1,000 m (3,280 ft)
Landing from 15 m (50 ft) 590 m (1,930 ft)
Range with max fuel and 8 passengers at econ cruising speed, with reserves for 45 min cruise
1,780 nm (3,300 km; 2,050 miles)

DASSAULT-BREGUET MYSTÈRE-FALCON 20 SERIES G

US Coast Guard designation: HU-25A Guardian

It was announced in the Autumn of 1976 that a tender by Falcon Jet Corporation, distributor and support centre for Falcons in the USA, had proved the lowest bid to meet a US Coast Guard requirement for a medium-range surveillance aircraft known by the project designation HX-XX. This was confirmed on 5 January 1977, when William T. Coleman Jr, then US Secretary of Transportation, authorised the Coast Guard to award a contract for 41 aircraft to Falcon Jet Corporation. Known as **HU-25A Guardians**, these are planned to have 44 per cent French content and 56 per cent US content.

A Falcon 20F (F-WATF) modified to have Garrett ATF 3 turbofans, as specified for the Guardian, flew for the first time, in France, on 28 November 1977. It was followed on 4 August 1978 by the first flight, in the USA, of another Falcon 20F, powered by General Electric CF700 engines but fitted out to US Coast Guard specifications. The Series G received DGAC certification on 21 June 1981. Deliveries to the US Coast Guard were expected to begin in the Autumn of 1981, at the rate of about one per month.

On 8 June 1979, the French Navy ordered five Mystère-Falcons for maritime duties under the name Gardian. These are variants of the new civil Mystère-Falcon 200 (which see), with an additional structural fuel tank in the fuselage. The Coast Guard HU-25As are now expected to be the only Falcon 20Gs built.

The basic airframe of the HU-25A is little changed from that of the Mystère-Falcon 20F. The most significant new features are as follows:

Airframe: Fuselage is modified to embody a drop hatch, and one search window on each side. Four hardpoints under fuselage: two for 500 kg (1,100 lb) loads, two for 200 kg (440 lb) loads. Four underwing hardpoints: two for 660 kg (1,455 lb) loads, two for 230 kg (507 lb) loads.

Power Plant: Two Garrett ATF 3-6-2C turbofan engines (each 24·65 kN; 5,538 lb st), meeting current and proposed FAR Pt 36 noise standards. Entire engine open to borescope inspection. Fuel tankage, total capacity 5,770 litres (1,269 Imp gallons; 1,524 US gallons), divided into two identical halves, one for each engine with cross-feed capability. Wing feeder tanks pressurised with bleed air, so that fuel will continue to flow to engines with all pumps turned off. Provision for auxiliary fuel tank in rear of cabin. Single-point refuelling in about nine minutes. Fuel heaters and bacterial protection standard.

Accommodation: Normal crew of five to seven. Typical complement will comprise two pilots, one surveillance system operator (SSO) at a console on the starboard side at the rear of the cabin, two search crew members at side windows. A three-seat sofa is provided for passengers, on the port side. A drop-hatch for stores, with floor-mounted roller conveyor, is located towards the front of the cabin. Galley and retractable toilet on port side. Provision for carrying four stretchers.

Systems: Pressurisation and air-conditioning by engine bleed air; max pressure differential 0.585 bars (8·5 lb/sq in). Two independent hydraulic systems, with twin engine-driven pumps; electric standby pump to power primary flight control system in emergency. All primary flight controls utilise dual hydraulic actuators, artificial feel, electric trim and manual backup. Each half of the dual actuator is fed by one of the hydraulic systems; failure of either system will not affect handling, as each actuator has sufficient power for full control deflection. DC electrical system, with two 9kW engine-driven starter/generators, two nickel-cadmium batteries and two 1,000VA static inverters. Ground power receptacle. One 20kVA alternator driven by hydraulic motor, plus one 4kVA alternator driven by APU. Wings and nacelles anti-iced by engine bleed air, permitting flight under maximum icing conditions with one engine out.

Avionics: Basic avionics package includes dual HF, VHF-AM, IFF, single VHF-FM and UHF. Nav equipment includes inertial sensor system, Omega, dual VOR/ILS/marker beacon receivers, DME, ADF, radio altimeters, area navigation system and single Tacan. Sensors include maritime search and weather radar and optional SLAR, infra-red and ultra-violet scanners, FLIR, aerial reconnaissance camera and steerable TV camera with laser illumination invisible to human eyes. Equipment ordered for the HU-25As includes six Aerojet Electro-Systems Aireye pollution control systems, combining a TV camera, still camera, infra-red/ultra-violet line scanner and radar to detect and record on videotape oil spills and infringement of fishing laws and treaties up to 43 nm (80 km; 50 miles) to each side of the aircraft's flight path by day or night, through clouds and in virtually any kind of weather.

Dimensions, external:
Wing span 16·30 m (53 ft 6 in)
Wing aspect ratio 7·02
Length overall 17·15 m (56 ft 3 in)
Height overall 5·32 m (17 ft 5 in)
Tailplane span 6·74 m (22 ft 1 in)

Area:
Wings, gross 41·80 m² (450 sq ft)

Weights:
Weight empty 8,620 kg (19,000 lb)
Operating weight empty, with 5 crew and complete avionics package 9,475 kg (20,890 lb)
Max fuel 4,820 kg (10,625 lb)
Max T-O weight 15,200 kg (33,510 lb)
Max landing weight 13,100 kg (28,880 lb)
Max zero-fuel weight 10,500 kg (23,150 lb)

Performance (at max T-O weight, except where indicated):
Max cruising speed at 12,200 m (40,000 ft)
Mach 0·8 (461 knots; 855 km/h; 531 mph)
Econ cruising speed at 12,500 m (41,000 ft) Mach 0·72
Min manoeuvring speed at low altitude
150 knots (278 km/h; 173 mph)
Initial cruising height 12,500 m (41,000 ft)
T-O run 1,235 m (4,050 ft)
FAR 25 landing run at typical landing weight
625 m (2,050 ft)
Range with 6 crew, reserves of 5% total fuel plus 30 min at S/L 2,250 nm (4,170 km; 2,590 miles)

DASSAULT-BREGUET MYSTÈRE-FALCON 200 and GARDIAN

This new version of the Mystère-Falcon 20 was announced as the Series H at the 1979 Paris Air Show, and was ordered into production for commercial sale in September 1980. As well as representing the latest variant of the 8/14-seat executive transport, it is the basic aircraft for maritime surveillance, under the name Gardian/Guardian; five have been ordered by the French Navy, to replace P-2H Neptunes, and others by the Japan Maritime Safety Agency.

The Falcon 200 retains almost unchanged the airframe, varied configurations and operational flexibility of the Mystère-Falcon 20 Series F, but is powered by two Garrett turbofans of the type used in the Series G/HU-25A and is fitted with the new rear fuselage structural fuel tank developed for the Mystère-Falcon 50. This combination of engines with improved sfc and increased fuel capacity enables the Falcon 200 to offer a 42 per cent longer range than the Series F with eight passengers and 45 min fuel reserves, without any sacrifice of cabin volume.

Many other improvements are embodied in the civil Falcon 200, based on experience with the Falcon 50. Dual hydraulic systems are supplied separately by two pumps, with a new emergency system. Two increased-power electrical circuits are completely segregated. The ABG/SEMCA air-conditioning system operates on the ground, with one engine idling, without use of the APU. A Solar T62-T40 APU is available optionally. Avionics options include a Collins APS 80 three-axis autopilot, INS, Omega, and HF com. Other options include automatic stabilisation, Mach trim, yaw damper and flap asymmetry limiter. Avionics can be installed in the nose to avoid a radio bay on the flight deck. An additional baggage compartment, capacity 0·75 m³ (26·5 cu ft), is situated behind the new fuselage structural fuel tank.

To hasten development, Dassault-Breguet flew for the first time on 24 April 1979 an aircraft known as the **Mystère-Falcon 20FH** (F-WZAH), embodying the new fuselage fuel tank and associated system changes, but retaining the General Electric CF700 engines of the Series F. The same aircraft was then re-engined with Garrett ATF 3 turbofans and resumed flying on 30 April 1980 as a fully-representative Falcon 200 prototype. DGAC certification was received on 21 June 1981

The first of the five Gardians ordered for the French Navy flew for the first time on 15 April 1981. Equipment of this version includes Thomson-CSF Varan radar, able to detect small objects in a rough sea, and a special Crouzet navigation system for display of tactical and geographical data. Large items such as liferafts can be dropped in flight from a 1·0 m × 0·6 m (3 ft 3¼ in × 1 ft 11½ in) floor hatch. The front cabin window on each side has been much enlarged for use by an observer. Four underwing hardpoints can carry a variety of loads, including sensors, target towing equipment, weapons and countermeasures.

The description of the Mystère-Falcon 20F applies also to the Falcon 200, except as follows:

Power Plant: Two Garrett ATF 3-6-1C turbofan engines (each 22·45 kN; 5,050 lb st). Fuel in two integral tanks in wings and large integral tank in rear fuselage, with total capacity of 6,000 litres (1,320 Imp gallons; 1,585 US gallons).

First of five Dassault-Breguet Gardian maritime surveillance aircraft for the French Navy

Dassault-Breguet Mystère-Falcon 200 prototype (Garrett ATF 3 turbofan engines)

ACCOMMODATION (Gardian): Pilot and co-pilot side-by-side on flight deck; swivelling centre seat to rear, from which camera can be operated and items airdropped through floor hatch. Seats for observers by large windows at front of cabin. SAR containers retained by straps on starboard side of centre-fuselage, with inward-facing seats opposite. Navigator's table and radar operator's console at rear of cabin on starboard side, with track-mounted side-facing seats.

AVIONICS AND EQUIPMENT (Gardian): Integrated nav/com package includes VHF, VOR, ADF, DME, ATC, Tacan, IFF, HF, VHF-FM and UHF equipment. Collins APS 80 autopilot, Crouzet Omega, inertial navigation system and Thomson-CSF Varan radar. Teletype for navigator. OMERA 35 camera, forward of cabin door, can photograph vertically or obliquely at 55°, and is coupled with an electronic flash for night photography located under the rear of the port engine mounting. Launcher for Lofar-Difar-Dicass acoustic sonobuoys and markers. Provision for carrying underwing an infra-red sensor or two Secapem targets, each with 1,500 m (4,900 ft) towing cable. Garrett or Microturbo Saphir APU, driving a 20kVA generator.

WEIGHTS:

Weight empty, equipped	8,150 kg (17,970 lb)
Operating weight empty	8,332 kg (18,370 lb)
Max fuel	4,818 kg (10,620 lb)
Max payload	1,875 kg (4,130 lb)
Payload with max fuel	752 kg (1,658 lb)
Max T-O weight	13,900 kg (30,650 lb)
Max landing weight	12,500 kg (27,560 lb)
Max zero-fuel weight	10,200 kg (22,500 lb)

PERFORMANCE (estimated):

Max cruise Mach No. at 10,050 m (33,000 ft) 0·80
Max cruising speed 460 knots (850 km/h; 528 mph)
Balanced T-O field length with 8 passengers and max fuel 1,600 m (5,250 ft)
Landing field length with 8 passengers and reserves 1,100 m (3,610 ft)
Max range at Mach 0·71 at 12,000 m (39,375 ft), with 8 passengers, 45 min reserves 2,370 nm (4,390 km; 2,730 miles)
Range at Mach 0·78 at 9,150 m (30,000 ft), with 8 passengers and max fuel 1,727 nm (3,200 km; 1,988 miles)

DASSAULT-BREGUET FALCON CARGO JET

Under contract from Pan American Business Jets, Little Rock Airmotive converted a Falcon 20 into a specialised cargo aircraft. Known as the Falcon Cargo Jet, the prototype flew for the first time on 28 March 1972. By the Summer of the same year, Federal Express Corporation, of Little Rock, had three similar aircraft in service and expanded its fleet subsequently to a total of 32 Falcon D Cargo Jets.

The cargo conversion can be applied to any Mystère-Falcon 20 and is offered on the current Series F aircraft.

Basic feature of the conversion is replacement of the standard cabin door by a hydraulically-actuated cargo door 1·88 m wide by 1·44 m high (6 ft 2 in × 4 ft 9 in), forward of the wing on the port side. This door opens upward, with its sill at cabin floor level. The flooring itself is new and offers a completely flat area 7·01 m long by 1·62 m wide (23 ft × 5 ft 4 in). Made of aluminium honeycomb, it can sustain loadings of up to 488 kg/m² (100 lb/sq ft) and affords a vast number of alternative tiedown points for retainer nets and pallets. Floor-mounted rollers are optional.

The Falcon Cargo Jet's Category II solid-state avionics standard includes dual com/nav, dual flight directors, autopilot and weather radar. Specifically, in the case of the Federal Express fleet, the fit comprises RCA com/nav with DME, dual Collins FD-108 flight directors, RCA AVQ-21 radar, Collins AP-105 autopilot, Teledyne angle-of-attack system, Collins ADF and RCA transponder. Also installed on these aircraft is a Fairchild integral electronic weight and balance system, which indicates as a cockpit readout whether or not cargo weight and distribution are within the legal limits. Standard safety provisions include a quick-release cargo restraint system able to withstand 9*g*.

Dimensions, weights and performance are largely unchanged by this conversion scheme. Nominal empty weight is 6,963 kg (15,350 lb) and max zero-fuel weight is increased to 9,980 kg (22,000 lb). Range varies from 1,215 nm (2,250 km; 1,400 miles) with max payload to 1,736 nm (3,215 km; 2,000 miles) with max fuel and a 2,040 kg (4,500 lb) payload. Usable cabin volume is 14·15 m³ (500 cu ft).

DASSAULT-BREGUET MYSTÈRE-FALCON 10

The Mystère-Falcon 10 is basically a scaled-down version of the Mystère-Falcon 20, with similar wing high-lift devices to those of the Mystère-Falcon 20F and powered by two small turbofans in the 14·6 kN (3,300 lb st) class. It is designed to fail-safe principles and to comply with US FAR 25 transport category requirements.

A prototype (F-WFAL), with General Electric CJ610 turbojets, made its first flight on 1 December 1970. Flight testing was resumed on 7 May 1971 following modifications to the angles of wing incidence and dihedral and an increase in wing sweepback. The modified aircraft set a 1,000 km closed-circuit speed record of 502·05 knots (930·4 km/h; 578·13 mph) in FAI Class C1f on 1 June 1971.

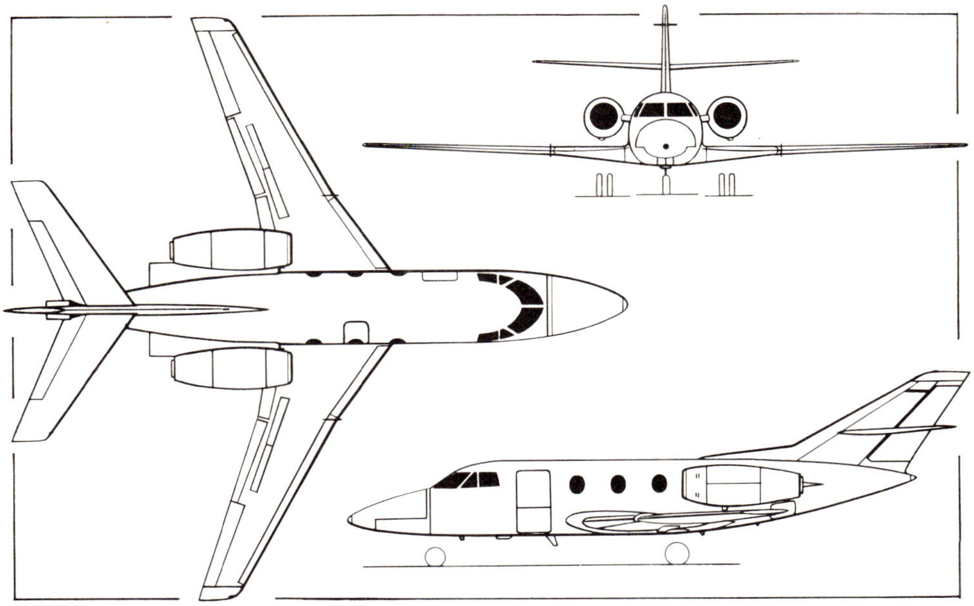

Dassault-Breguet Mystère-Falcon 10 four/seven-passenger executive transport *(Pilot Press)*

Dassault-Breguet Mystère-Falcon 10 taking off from an unpaved airfield

A second prototype (F-WTAL), with Garrett TFE731-2 engines, flew for the first time on 15 October 1971, followed by a third aircraft on 14 October 1972, with similar engines. This third Mystère-Falcon 10 set a 2,000 km closed-circuit speed record of 494·83 knots (917·02 km/h; 569·809 mph) in Class C1f on 29 May 1973.

The first production Mystère-Falcon 10 with TFE731-2 engines flew on 30 April 1973. French certification was granted on 11 September 1973, followed by FAA certification nine days later, allowing deliveries of production aircraft to begin on 1 November. A total of 179 had been delivered by 1 May, 1981, at which time 217 Mystère-Falcon 10s had been ordered. They include three **Mystère-Falcon 10MER** aircraft for the French Navy, which uses them as intruders for interception training of Super Etendard pilots, as conventional instrument and night flying trainers, for calibration of shipboard radars and for medical evacuation, as well as for communications duties. The Navy has an option on two more.

In mid-1981, rate of production at the Istres assembly plant was two per month. Fuselages are provided by the Potez works at Aire-sur-Adour, which assembles components built by SOGERMA, SOCEA and Socata. Wings come from CASA of Spain; tail units and nose assemblies from IAM of Italy; and many other components such as tail fins, doors and emergency exits from Latécoère's Toulouse works.

Like the Mystère-Falcon 20, the Mystère-Falcon 10 can be equipped for liaison, executive transport, navigation/attack system training, aerial photography, radio navigation aid calibration and ambulance duties.

As part of a continuing research programme into the use of composites for structural components, Aérospatiale hopes to flight test on a Falcon 10, in 1982, a complete V10F upper wing skin panel of resin-impregnated carbonfibre.

TYPE: Twin-turbofan executive transport.

WINGS: Cantilever low-wing monoplane with increased sweepback on inboard leading-edges. All-metal torsion-box structure, with leading-edge slats and double-slotted trailing-edge flaps and plain ailerons. Two-section spoilers above each wing, forward of flaps.

FUSELAGE: All-metal semi-monocoque structure, designed to fail-safe principles.

TAIL UNIT: Cantilever all-metal structure, similar to that of Falcon 20.

LANDING GEAR: Retractable tricycle type, manufactured by Messier-Hispano-Bugatti, with twin wheels on main gear, single wheel on nose gear. Hydraulic retraction, main units inward, nosewheel forward. Oleo-pneumatic shock-absorbers. Low-pressure tyres for soft-field operation.

POWER PLANT: Two Garrett TFE731-2 turbofan engines (each 14·4 kN; 3,230 lb st), pod-mounted on sides of rear fuselage. Fuel in two integral tanks in wings and two feeder tanks aft of rear bulkhead, with total capacity of 3,340 litres (735 Imp gallons; 882 US gallons). Separate fuel system for each engine, with provision for cross-feeding. Pressure refuelling system.

ACCOMMODATION: Crew of two on flight deck, with dual controls and airline-type instrumentation. Provision for third crew member on a jump-seat. Normal seating for four passengers (two individual seats and a three-seat sofa) or for seven passengers, with two individual seats added. Each pair of single seats is separated by a table. Coat compartment on starboard side, immediately aft of flight deck opposite door; rear baggage compartment behind sofa. Galley on left of entrance. Optional front toilet compartment. Downward-opening door with built-in steps.

SYSTEMS: Duplicated air-conditioning and pressurisation systems supplied with air bled from both engines. Pressure differential 0·61 bars (8·8 lb/sq in). Two independent hydraulic systems, each of 207 bars (3,000 lb/sq in) pressure and with twin engine-driven pumps and emergency electric pump, to actuate primary flight controls, flaps, landing gear, wheel brakes, spoilers, yaw damper and nosewheel steering. 28V DC electrical system with a 9kW DC starter/generator on each engine, three 750VA 400Hz 115V inverters and two 23Ah batteries. Automatic emergency oxygen system.

AVIONICS AND EQUIPMENT: Standard avionics include duplicated VHF and VOR/glideslope, single ADF, marker beacon receiver, ATC transponder, autopilot, intercom system and duplicated blind-flying instrumentation. Optional avionics include duplicated DME and flight director, second ADF, weather radar and radio altimeter.

DIMENSIONS, EXTERNAL:

Wing span	13·08 m (42 ft 11 in)
Wing chord (mean)	2·046 m (6 ft 8½ in)
Wing aspect ratio	7·1
Length overall	13·86 m (45 ft 5¾ in)
Length of fuselage	12·47 m (40 ft 11 in)

Height overall	4·61 m (15 ft 1½ in)
Tailplane span	5·82 m (19 ft 1 in)
Wheel track	2·86 m (9 ft 5 in)
Wheelbase	5·38 m (17 ft 8 in)
Passenger door: Height	1·47 m (4 ft 10 in)
Width	0·80 m (2 ft 7 in)
Height to sill	0·884 m (2 ft 10¾ in)
Emergency exit (stbd side, over wing):	
Height	0·914 m (3 ft 0 in)
Width	0·508 m (1 ft 8 in)
DIMENSIONS, INTERNAL:	
Cabin, excl flight deck: Length	5·00 m (16 ft 5 in)
Max width	1·46 m (4 ft 9 in)
Max height	1·49 m (4 ft 10½ in)
Volume	7·50 m³ (264·6 cu ft)
Baggage compartment volume:	
front (wardrobe)	0·35 m³ (12·35 cu ft)
rear	0·70 m³ (24·7 cu ft)
AREAS:	
Wings, gross	24·1 m² (259 sq ft)
Horizontal tail surfaces (total)	6·75 m² (72·65 sq ft)
Vertical tail surfaces (total)	4·54 m² (48·87 sq ft)
WEIGHTS:	
Weight empty, equipped	4,880 kg (10,760 lb)
Max payload	1,090 kg (2,400 lb)
Max fuel	2,680 kg (5,910 lb)
Max T-O weight	8,500 kg (18,740 lb)
Max zero-fuel weight	6,150 kg (13,560 lb)
Max landing weight	8,000 kg (17,640 lb)
PERFORMANCE:	
Never-exceed speed at S/L	350 knots (648 km/h; 402 mph)
Max operating Mach No.	0·87
Max cruise Mach No. at 10,670 m (35,000 ft)	0·84
Max cruising speed at 7,620 m (25,000 ft)	492 knots (912 km/h; 566 mph)
Approach speed	100 knots (185 km/h; 115 mph)
FAR 25 balanced T-O field length with four passengers and fuel for a 1,000 nm (1,850 km; 1,150 mile) stage, 45 min reserves	930 m (3,050 ft)
FAR 25 balanced T-O field length, with four passengers and max fuel	1,242 m (4,075 ft)
FAR 121 landing field length, with four passengers and 45 min reserves	1,052 m (3,450 ft)
Range with four passengers and 45 min reserves	1,920 nm (3,560 km; 2,210 miles)

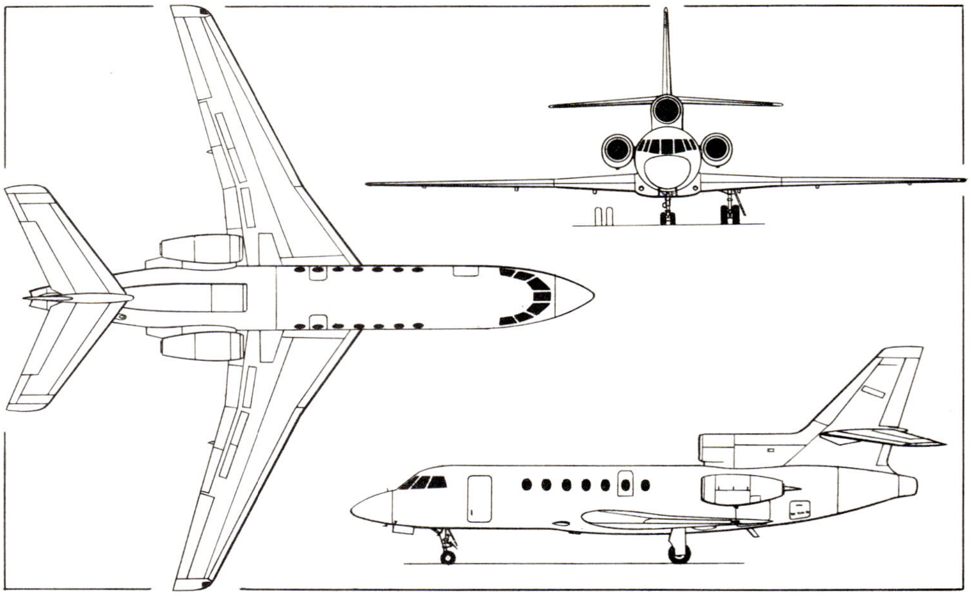
Dassault-Breguet Mystère-Falcon 50 extended-range three-turbofan executive transport *(Pilot Press)*

DASSAULT-BREGUET MYSTÈRE-FALCON 50

The Mystère-Falcon 50 is a three-turbofan executive transport derived from the Mystère-Falcon 20. It has an entirely new wing of supercritical section, adapted to flight at high Mach numbers and embodying efficient high-lift devices, and complies with FAR Pt 36 noise requirements. Standard accommodation is for a crew of two and eight passengers; but in 1980 an ambulance version became available, with the interior laid out for three stretchers (or two stretchers and heavy medical equipment) and two doctors.

The original prototype (F-WAMD) flew for the first time on 7 November 1976, followed by a second prototype (F-WINR) on 16 February 1978 and the third (and sole pre-production) aircraft on 13 June 1978. DGAC certification was received on 27 February 1979, followed by FAA type approval on 8 March. Falcon 50 No. 4, flown on 2 March 1979, was the first built on Dassault-Breguet's Mérignac assembly line and became Falcon Jet's US demonstrator. Orders totalled 173 on 1 June 1981, by which date 52 had been delivered. The first delivery was made in July 1979, when Falcon 50 No. 3 was handed over to a customer who had owned a Falcon 20 for 12 years. In early 1980, the fifth aircraft was delivered to GLAM (Groupe de Liaisons Aériennes Ministérielles) for priority use by the President of the Republic. Another was supplied to the King of Morocco. Production was at the rate of three per month in mid-1981.

Fuselages for the Mystère-Falcon 50 are produced at Aérospatiale's Saint-Nazaire works, wings at the Colomiers plant of Dassault-Breguet, tail units by Aérospatiale at Méaulte, and cowlings by Hurel-Dubois at Vélizy-Villacoublay.

TYPE: Three-turbofan executive transport.

WINGS: Cantilever low-wing monoplane, with compound leading-edge sweepback and supercritical section. Each wing is attached to the central box structure by multiple bolts and forms an integral fuel tank. Full-span leading-edge slats, of which the outboard sections are slotted. Double-slotted trailing-edge flaps and ailerons. Three-section two-position airbrakes on top surface of each wing.

FUSELAGE: All-metal semi-monocoque structure of circular cross-section, with aft baggage compartment included in pressure-cell.

TAIL UNIT: Cantilever all-metal structure, with horizontal surfaces mounted partway up fin. Tailplane incidence adjustable by screwjack, driven by two electric motors controlled by 'normal' and 'emergency' controls located respectively on the control wheels and pedestal.

LANDING GEAR: Retractable tricycle type by Messier-Hispano-Bugatti, with twin wheels on each unit. Hydraulic retraction, main units inward, nosewheels forward. Four-disc brakes designed for 400 landings with normal-energy braking.

POWER PLANT: Three Garrett TFE731-3 turbofan engines, each rated at 16·5 kN (3,700 lb st) for take-off. Two engines pod-mounted on sides of rear fuselage, the third attached by two top mounts. Thrust reverser on centre engine. Fuel in wing and fuselage tanks, with total capacity of 8,765 litres (1,928 Imp gallons; 2,315 US gallons). Single-point pressure fuelling.

ACCOMMODATION: Crew of two side by side on flight deck, with full dual controls and airline-type instrumentation. Third seat to rear of co-pilot. Normal accommodation for eight passengers in individual reclining seats, with first two rows facing each other and separated by stowable tables. Lavatory with water supply and flushing toilet, and pressurised baggage compartment, at rear. Wardrobe and galley in forward part of cabin. Downward-opening airstair door. Optional 12-passenger high-density layout; or two pairs of reclining seats with five-passenger lounge at rear. All passenger seats removable for freight carrying. An ambulance version is available.

SYSTEMS: Air-conditioning system utilises bleed air from all three engines. Max pressure differential 0·61 bars (8·8 lb/sq in). Pressurisation maintains a max cabin altitude of 2,440 m (8,000 ft) to a flight altitude of 13,700 m (45,000 ft). Two independent hydraulic systems, pressure 207 bars (3,000 lb/sq in), with three engine-driven pumps and one emergency electric pump, actuate primary flying controls, flaps, slats, landing gear, wheel brakes, airbrakes and nosewheel steering. 28V DC electrical system, with a 9kW 28V DC starter-generator on each engine and two 23Ah batteries. Automatic emergency oxygen system. Optional 9kW Garrett APU.

AVIONICS: Standard items include Omega, inertial navigation system, duplicated VHF and VOR, ADF, DME, ATC and HF, radio altimeter and weather radar. Basic aircraft includes Collins AP-580 autopilot and two FCS-80 flight directors.

DIMENSIONS, EXTERNAL:	
Wing span	18·86 m (61 ft 10½ in)
Wing chord (mean)	2·84 m (9 ft 3¾ in)
Wing aspect ratio	7·6
Length overall	18·50 m (60 ft 8½ in)
Length of fuselage	17·66 m (57 ft 11 in)
Height overall	6·97 m (22 ft 10½ in)
Tailplane span	7·74 m (25 ft 4¾ in)
Wheel track	3·98 m (13 ft 0¾ in)
Wheelbase	7·24 m (23 ft 9 in)
Passenger door: Height	1·52 m (4 ft 11¾ in)
Width	0·80 m (2 ft 7½ in)
Height to sill	1·30 m (4 ft 3¼ in)
Emergency exits (each side, over wing):	
Height	0·92 m (3 ft 0¼ in)
Width	0·51 m (1 ft 8 in)
DIMENSIONS, INTERNAL:	
Cabin, incl forward baggage space and rear toilet:	
Length	7·15 m (23 ft 5½ in)
Max width	1·86 m (6 ft 1¼ in)
Max height	1·79 m (5 ft 10½ in)
Volume	18·3 m³ (646·2 cu ft)
Baggage compartment (aft)	2·34 m³ (82·6 cu ft)
AREAS:	
Wings, gross	46·83 m² (504·1 sq ft)
Horizontal tail surfaces (total)	13·35 m² (143·7 sq ft)
Vertical tail surfaces (total)	9·82 m² (105·7 sq ft)
WEIGHTS:	
Weight empty, equipped	9,000 kg (19,840 lb)
Max payload	1,720 kg (3,792 lb)
Max fuel	7,040 kg (15,520 lb)
Max T-O and ramp weight	17,600 kg (38,800 lb)
Max zero-fuel weight	11,000 kg (24,250 lb)
Max landing weight	16,200 kg (35,715 lb)

Dassault-Breguet Mystère-Falcon 50 (three Garrett TFE731-3 turbofan engines)

PERFORMANCE:
Max operating Mach No. 0·86
Max operating speed at S/L
350 knots (648 km/h; 402 mph) IAS
Max operating speed at 7,225 m (23,700 ft)
370 knots (685 km/h; 425 mph) IAS
Max cruising speed
Mach 0·82 or 475 knots (880 km/h; 546 mph)
Service ceiling 13,800 m (45,300 ft)
FAR 25 balanced field length with 8 passengers and fuel for 3,400 nm (6,295 km; 3,910 miles)
1,340 m (4,400 ft)
FAR 121 landing distance with 8 passengers and FAR 121 reserves 1,050 m (3,450 ft)
Range at Mach 0·75 with 8 passengers and FAR 121 reserves 3,400 nm (6,300 km; 3,910 miles)

FOURNIER
FOURNIER AVIATION

OFFICE AND WORKS: Aérodrome d'Athée-Nitray, 37270 Montlouis
Telephone: (47) 50 68 30
Telex: 696 254
MANAGING DIRECTOR: René Caillet
SALES MANAGER: Patrice Caillet
CONSULTANT: René Fournier

Fournier Aviation was formed in 1978, bringing together the assets and activities of the former Avions Fournier SA and the Fournier Design Office, in close collaboration with M René Fournier.

While completing the manufacture of a total of 45 RF-6B two-seat aerobatic aircraft, each powered by a 74·5 kW (100 hp) Rolls-Royce Continental O-200-A engine, it designed a more powerful version with an 88 kW (118 hp) Avco Lycoming O-235-L2A engine. The prototype (F-GANF) of this new model, known as the RF6B-120 and produced by modification of the 44th airframe off the assembly line, flew for the first time on 16 August 1980. It received DGAC certification on 7 November 1980, and a licence to manufacture this version in series has been granted to Slingsby Engineering Ltd in the UK (which see).

Fournier Aviation will specialise in the production of motor gliders designed by M René Fournier (see Sailplanes section), and will distribute in France and in French-speaking nations overseas aircraft built by Slingsby Engineering.

FOURNIER RF-6B

Full details of the basic RF-6B with O-200-A engine can be found in the 1980-81 *Jane's*. The new RF6B-120 is being manufactured under licence, as the T.67, by Slingsby Engineering Ltd. A full description can be found under the Slingsby entry in the UK section.

HELICOP-JET

Héliport de Paris, 4 avenue de la Porte de Sèvres, 75015 Paris
Telephone: 554 69 13
PROPRIETOR: Charles Déchaux

HELICOP-JET

A full-scale mockup of this 'cold-jet' tip-driven light helicopter was first exhibited at the Paris Air Show in June 1969. Construction of a pre-production prototype was started by Établissements Charles Déchaux in 1970, and this aircraft made its first flight in December 1976 at Issy-les-Moulineaux. It logged about 60 flying hours, powered by a 186 kW (250 hp) Turboméca Palouste air generator, with which it could lift only the pilot.

The definitive prototype, with an Astazou engine, was displayed at the Paris Air Show in June 1979, and was expected to fly in October 1979. There has been no news of such a flight, or of further progress with planned production of the Helicop-Jet, of which an illustrated description can be found in the 1979-80 *Jane's*.

ISSOIRE — *see Sailplanes section*

MICROTURBO
MICROTURBO SA

21 avenue Clément Ader, 31770 Colomiers
Telephone: (61) 78 54 44
Telex: 521804 MIJET
MJ 200 PROGRAMME MANAGER: J. Grangette

MICROTURBO MICROJET 200 B

To promote utilisation of the small gas turbine engines that it manufactures, Microturbo SA initiated the design and development of an aircraft known as the Microjet 200 (MJ 200), of which a wooden prototype (F-WZJF) was flown for the first time by M Jacques Grangette on 24 June 1980. Aim of the programme is to offer economies in military pilot training by use of very small high-performance jet aircraft with comparatively low initial and operating costs.

The Microjet 200 is a side-by-side two-seater, of simple configuration, powered by two Microturbo TRS 18-046 turbojets, each rated at 1·08 kN (243 lb st). Since the prototype first flew the tail unit has been given additional sweepback, the canopy has been made more rounded, the fuselage lengthened, and the landing gear improved. It is being followed by three pre-production models of the same design, known as the Microjet 200 B, with more powerful engines and with wings, tail unit and control surfaces of glassfibre/epoxy and a metal fuselage. The following description applies to the Microjet 200 B, of which manufacture started in December 1980:

TYPE: Two-seat lightweight training aircraft.

WINGS: Cantilever low-wing monoplane of tapered planform. Wing section RA163C3. Thickness/chord ratio 16%. Dihedral 4° constant from roots. Incidence 3°. Sweepback 0° at 30% chord. Wings, Frise ailerons and electrically-operated single-slotted trailing-edge flaps all of glassfibre/epoxy. Small airbrake of glassfibre/epoxy forward of outer end of flap on upper surface of each wing. Ailerons embody adjustable artificial feel.

FUSELAGE: Conventional light alloy semi-monocoque structure. NACA type flush engine air intake on each side of fuselage aft of cockpit; exhaust through lateral jetpipes forward of tail unit.

TAIL UNIT: Cantilever V type, comprising fixed surfaces and elevators of glassfibre/epoxy. Sweepback 26° at 50% chord. Included angle 110°. Controllable tab at root end of each elevator.

LANDING GEAR: Retractable tricycle type, with single wheel on each unit. Electrical retraction, nosewheel forward, main units inward into fuselage. All wheels fully enclosed by doors when retracted. Microjet oleo-pneumatic shock-absorber in all three units. Nosewheel offset to starboard. Goodyear wheels, tyres and two-disc brakes.

POWER PLANT: Two Microturbo TRS 22 turbojet engines, each rated at 1·25 kN (281 lb st). Fuel in two interconnected tanks, with total capacity of 400 litres (88 Imp gallons). Two refuelling points in sides of fuselage. Total oil capacity 1·6 litres (0·35 Imp gallon).

ACCOMMODATION: Pilot and instructor in side-by-side seats, under one-piece rearward-hinged transparent canopy. Starboard seat staggered 55 cm (1 ft 9¾ in) aft of port seat. Cockpit heated and ventilated but not pressurised.

SYSTEMS: Two engine-driven generators drive electrical motors for actuation of flaps and landing gear. One nickel-cadmium storage battery. Gaseous oxygen supply for two crew for two hours.

AVIONICS AND EQUIPMENT: Blind-flying instrumentation standard. Avionics to customer's requirements.

DIMENSIONS, EXTERNAL:
Wing span 7·56 m (24 ft 9¾ in)
Wing chord at root 0·85 m (2 ft 9½ in)
Wing aspect ratio 8
Length overall 6·55 m (21 ft 6 in)
Length of fuselage 6·37 m (20 ft 10¾ in)
Width of fuselage 1·10 m (3 ft 7¼ in)
Height overall 2·27 m (7 ft 5½ in)
Tailplane span 3·07 m (10 ft 1 in)
Wheel track 1·87 m (6 ft 1¾ in)
Wheelbase 2·10 m (6 ft 10¾ in)

Prototype Microturbo Microjet 200 two-seat lightweight training aircraft *(J. M. G. Gradidge)*

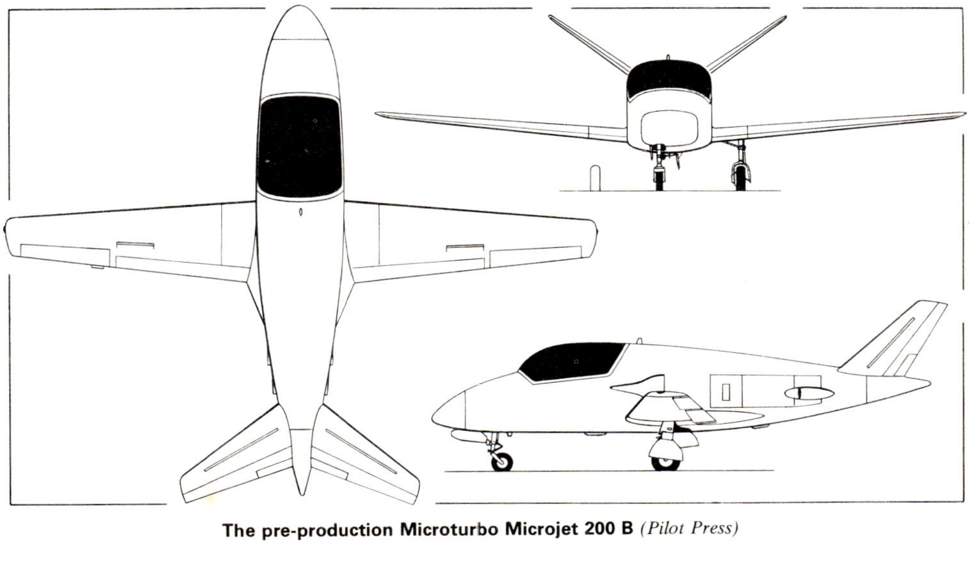
The pre-production Microturbo Microjet 200 B *(Pilot Press)*

Areas:

Wings, gross	6·10 m² (65·66 sq ft)
Ailerons (total)	0·446 m² (4·80 sq ft)
Trailing-edge flaps (total)	0·69 m² (7·43 sq ft)
Tail surfaces (total)	2·50 m² (26·91 sq ft)

Weights and Loadings:

Weight empty	650 kg (1,433 lb)
Max fuel	312 kg (687 lb)
Max T-O weight	1,150 kg (2,535 lb)
Max ramp weight	1,160 kg (2,557 lb)
Max zero-fuel weight	825 kg (1,818 lb)
Max landing weight	1,130 kg (2,491 lb)
Max wing loading	188 kg/m² (38·6 lb/sq ft)
Max power loading	460 kg/kN (4·51 lb/lb st)

Performance (estimated, at max T-O weight):

Never-exceed speed	300 knots (555 km/h; 345 mph)
Max level speed and max cruising speed at 5,500 m (18,000 ft)	250 knots (463 km/h; 287 mph)
Econ cruising speed	210 knots (389 km/h; 241 mph)
Stalling speed, flaps down, engines idling	64 knots (119 km/h; 74 mph)
Max rate of climb at S/L	640 m (2,100 ft)/min
Rate of climb at S/L, one engine out	213 m (700 ft)/min
Service ceiling	9,150 m (30,000 ft)
Service ceiling, one engine out	3,050 m (10,000 ft)
T-O run	560 m (1,838 ft)
T-O to 15 m (50 ft)	680 m (2,231 ft)
Landing from 15 m (50 ft)	510 m (1,674 ft)
Landing run	390 m (1,280 ft)
Range with max fuel, 20 min hold	556 nm (1,030 km; 640 miles)

MUDRY

AVIONS MUDRY et CIE

Aérodrome de Bernay, BP 47, 27300 Bernay
Telephone: (32) 43 47 34
Director: Auguste Mudry

M Auguste Mudry established this company in the works of the former Société Aéronautique Normande at Bernay, and operated it in parallel with his other aircraft manufacturing company, C.A.A.R.P. of Beynes (see 1977-78 *Jane's*). All activities of C.A.A.R.P. were subsequently combined with those of Avions A. Mudry, at Bernay, where the CAP 10 B, CAP 20 L and CAP 21 aerobatic light aircraft are in production. Under development are two new light aircraft designated CAP X and Mini-Cap.

MUDRY CAP 10 B

Developed from the Piel Emeraude two-seat light aircraft (see Homebuilts section), via the prototype C.P. 100 aerobatic version built by C.A.A.R.P., the CAP 10 is intended for use as a training, touring or aerobatic aeroplane. The prototype was flown for the first time in August 1968, and certification of the CAP 10 was granted on 4 September 1970. Construction is to French AIR 2052 (CAR 3) Category A standards for aerobatic flying, with load factors of +6*g* and −4·5*g*.

A total of 130 CAP 10s had been built by January 1981, including 36 ordered initially for the French Air Force and six for the French Navy. Some are in service with the Air Force's Equipe de Voltige Aérienne (EVA) at Salon-de-Provence, others with its basic flying training schools at Clermont-Ferrand-Aulnat and Cognac, and the Navy's pilot selection centre at St Raphaël. Fourteen more of the current CAP 10 B version were ordered for the French Air Force in 1979. Two were sold, for export to Brazil, during the Paris Air Show in June 1981.

Type: Two-seat aerobatic light aircraft.

Wings: Cantilever low-wing monoplane. Wing section NACA 23012. Dihedral 5° from roots. Incidence 0°. No sweepback. All-spruce single-spar torsion-box structure, with trellis ribs, rear auxiliary spar and okoumé plywood covering. Inner section of each wing is rectangular in plan, outer section semi-elliptical. Wooden trailing-edge plain flaps and slotted ailerons.

Fuselage: Conventional spruce girder structure, built in two halves and joined by three main frames. Of basically rectangular section with rounded top-decking. Polyester fabric covering. Forward section also has an inner plywood skin for added strength. Engine cowling panels of non-inflammable laminated plastics.

Tail Unit: Conventional cantilever structure. All-wood single-spar fin, integral with fuselage, and tailplane. All surfaces plywood-covered except rudder, which is covered with both plywood and polyester fabric. Tailplane incidence adjustable on ground. Trim tab in each elevator. Automatic rudder trim.

Landing Gear: Non-retractable tailwheel type. Mainwheel legs of light alloy, with ERAM type 9 270 C oleo-pneumatic shock-absorbers. Single wheel on each main unit, tyre size 380 × 150. Solid tailwheel tyre, size 6 × 200. Tailwheel is steerable by rudder linkage but can be disengaged for ground manoeuvring. Hydraulically-actuated main-wheel disc brakes (controllable from port seat) and parking brake. Streamline fairings on main wheels and legs.

Power Plant: One 134 kW (180 hp) Avco Lycoming IO-360-B2F flat-four engine, driving a Hoffmann two-blade fixed-pitch wooden propeller. Standard fuel tank aft of engine fireproof bulkhead, capacity 72 litres (16 Imp gallons). Optional auxiliary tank, capacity 75 litres (16·5 Imp gallons), beneath baggage compartment. Fuel and oil systems modified to permit periods of inverted flying.

Accommodation: Side-by-side adjustable seats for two persons, with provision for back parachutes, under rearward-sliding and jettisonable moulded transparent canopy. Special aerobatic shoulder harness standard. Space for 20 kg (44 lb) of baggage aft of seats in training and touring models.

Systems: Electrical system includes Delco-Rémy 40A engine-driven alternator and SAFT 12V DC nickel-cadmium battery.

Avionics and Equipment: CSF 262 12-channel VHF radio and *g* meter fitted. Optional equipment includes starboard brake pedals; Narco, Jolliet or Badin VHF; Narco VOR; radio compass; IFR instrumentation; navigation and landing lights; and heated pitot.

Dimensions, external:

Wing span	8·06 m (26 ft 5¼ in)
Wing aspect ratio	5·96
Length overall	7·16 m (23 ft 6 in)
Height overall	2·55 m (8 ft 4½ in)
Tailplane span	2·90 m (9 ft 6 in)
Wheel track	2·06 m (6 ft 9 in)

Dimension, internal:

Cabin: Max width	1·054 m (3 ft 5½ in)

Areas:

Wings, gross	10·85 m² (116·79 sq ft)
Ailerons (total)	0·79 m² (8·50 sq ft)
Vertical tail surfaces (total)	1·32 m² (14·25 sq ft)
Horizontal tail surfaces (total)	1·86 m² (20·0 sq ft)

Weights (A: Aerobatic, U: Utility):

Weight empty, equipped: A, U	540 kg (1,190 lb)
Fuel load: A	54 kg (119 lb)
U	108 kg (238 lb)
Max T-O weight: A	760 kg (1,675 lb)
U	830 kg (1,829 lb)

Performance (at max T-O weight):

Never-exceed speed	183 knots (340 km/h; 211 mph)
Max level speed at S/L	146 knots (270 km/h; 168 mph)
Max cruising speed (75% power)	135 knots (250 km/h; 155 mph)
Stalling speed, flaps up	52 knots (95 km/h; 59·5 mph)
Stalling speed, flaps down	44 knots (80 km/h; 50 mph)
Max rate of climb at S/L	over 360 m (1,180 ft)/min
Service ceiling	5,000 m (16,400 ft)
Range with max fuel	647 nm (1,200 km; 745 miles)

MUDRY CAP 20L

First flown on 15 January 1976, the CAP 20L ('léger') is a lightweight development of the earlier CAP 20. It is intended as both a comparatively inexpensive high-performance type for private individuals to fly in competitive aerobatics, and as an aircraft superior to any previous version of the CAP 20 for the Armée de l'Air and for international competition against the best aerobatic types of foreign design.

The prototype (F-WVKY) flew as a **CAP 20L-180**, with 134 kW (180 hp) Avco Lycoming AEIO-360 engine. Standard production model was the **CAP 20LS-200**, with 149 kW (200 hp) Avco Lycoming engine and constant-speed propeller, the first example of which flew for the first time on 6 November 1976. A total of 12 had been delivered by mid-1980 when the decision was taken to terminate production in favour of the new CAP 21.

The 100th aircraft of the CAP 10/20 series, delivered on 4 March 1978, was a CAP 20LS-200. The following data apply to this version:

Type: Single-seat aerobatic light aircraft.

Wings: Cantilever low-wing monoplane. All-wood single-spar wings, similar in construction to those of CAP 10 but with only ailerons on trailing-edge.

Fuselage: Conventional all-wood structure, of basically triangular section with rounded top-decking. Wooden covering, except for laminated plastics engine cowling.

Tail Unit: Cantilever all-wood structure. Trim tab in rudder and each elevator.

Landing Gear: Non-retractable tailwheel type. Oleo-pneumatic shock-absorbers. Streamline fairings on main wheels and legs. Disc brakes.

CAP 10 two-seat aerobatic light aircraft of the French Air Force

Mudry CAP 20LS-200 single-seat aerobatic aircraft

POWER PLANT: One 149 kW (200 hp) Avco Lycoming AIO-360-B1B flat-four engine, driving a Hartzell two-blade constant-speed metal propeller. Fuel tank aft of cockpit, capacity 80 litres (17·5 Imp gallons), with system modified to permit periods of inverted flight.

ACCOMMODATION: Single seat under transparent moulded canopy which opens sideways to starboard and is jettisonable. Special aerobatic shoulder harness.

DIMENSIONS, EXTERNAL:

Wing span	7·57 m (24 ft 10 in)
Length overall	6·46 m (21 ft 2½ in)
Height overall	1·52 m (5 ft 0 in)

AREA:

Wings, gross	10·47 m² (112·7 sq ft)

WEIGHTS:

Weight empty	500 kg (1,102 lb)
Max T-O weight (Aerobatic)	650 kg (1,433 lb)

PERFORMANCE:

Max cruising speed (75% power)	162 knots (300 km/h; 186 mph)
Stalling speed	46 knots (85 km/h; 53 mph)
Max rate of climb at S/L	840 m (2,755 ft)/min
Endurance with max fuel	2 h
g limits	+8; −6

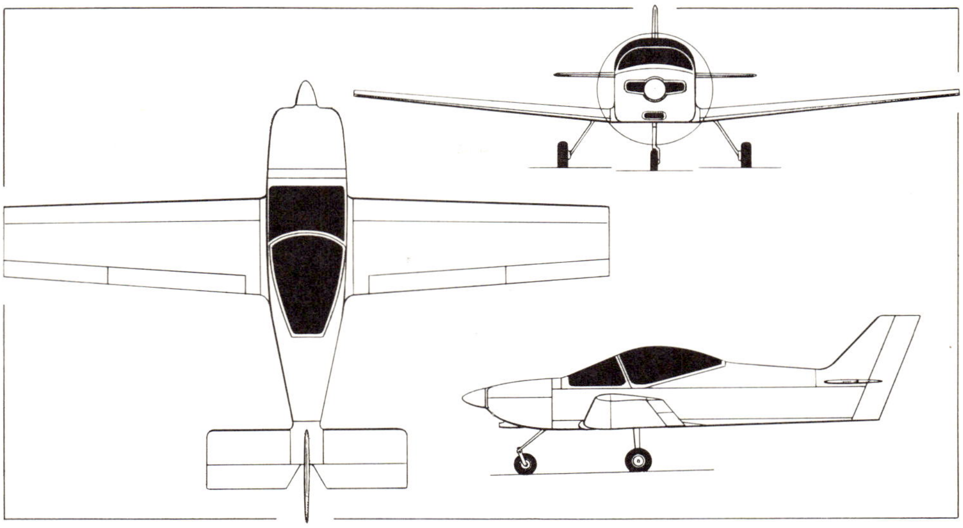

Mudry CAP X two-seat trainer (Mudry-Buchoux MB-4-80 engine) *(Pilot Press)*

MUDRY CAP 21

The prototype of this new single-seat aerobatic competition aircraft (F-WZCH) was displayed at the 1979 Paris Air Show. It flew for the first time on 23 June 1980, and 10 pre-production CAP 21s were under construction at the beginning of 1981. Initial orders for four aircraft had been received, from customers in Belgium, Brazil, France and Italy. The CAP 21 will be marketed in both ready-to-fly and kit form.

The CAP 21 retains the fuselage and tail unit of the CAP 20LS-200, but has cantilever main landing gear legs of glassfibre, and an entirely new wing of V16 F section, with a different planform and built by a new production method. This wing has improved the rate of roll to 180°/s at 135 knots (250 km/h; 155 mph) by comparison with the CAP 20L's 130°/s, and facilitates the execution of snap manoeuvres. It has a thickness/chord ratio of 16%, dihedral of 1° 30′ and no twist. Aileron stick forces are reduced by an automatic tab inserted in the trailing-edge of each aileron. Other features of the CAP 21 include a max fuel capacity of 75 litres (16·5 Imp gallons), a 15 litre (3·3 Imp gallon) gravity tank for inverted flying, and a glassfibre seat for the pilot. The 149 kW (200 hp) Avco Lycoming AEIO-360-A1B engine of the prototype drives a two-blade Hartzell variable-pitch propeller, but a fixed-pitch propeller is available optionally.

DIMENSIONS, EXTERNAL:

Wing span	8·08 m (26 ft 6 in)
Wing aspect ratio	6·95
Length overall	6·46 m (21 ft 2½ in)
Height overall	1·52 m (5 ft 0 in)

AREA:

Wings, gross	9·2 m² (99·0 sq ft)

WEIGHTS:

Weight empty	490 kg (1,080 lb)
Max T-O weight, Aerobatic	600 kg (1,323 lb)

PERFORMANCE:

Never-exceed speed	205 knots (380 km/h; 236 mph)
Max cruising speed	172 knots (320 km/h; 199 mph)
Stalling speed	46 knots (85 km/h; 53 mph)
Endurance with max fuel	2 h
g limits	+8; −6

Prototype CAP 21 (Avco Lycoming AEIO-360-A1B engine) *(J. M. G. Gradidge)*

MUDRY CAP X

In early 1981, Avions Mudry announced its intention to develop a side-by-side two-seat training aircraft powered by its new 59 kW (80 hp) Mudry-Buchoux MB-4-80 flat-four engine. It is a low-wing monoplane, with OAAG 03 wing section, designed for low initial cost, minimum maintenance requirements and a fuel consumption of 17-18 litres (3·75-4 Imp gallons) per hour. Composite materials are used in its construction. Range with max fuel is expected to be 450 nm (830 km; 515 miles) at 102 knots (190 km/h; 118 mph).

The partially-completed prototype was exhibited at the 1981 Paris Air Show. First flight was scheduled for Autumn 1981.

DIMENSIONS, EXTERNAL:

Wing span	8·00 m (26 ft 3 in)
Length overall	5·90 m (19 ft 4½ in)

WEIGHT:

Max T-O weight	520 kg (1,146 lb)

MUDRY MINI-CAP

The Mini-Cap is planned as an 'economy' single-seat aerobatic trainer, having the same engine and many components in common with the CAP X. The prototype is scheduled to fly in 1982.

REIMS AVIATION
REIMS AVIATION SA

OFFICE AND WORKS: Reims-Prunay Airport, BP 2745, 51062 Reims Cédex
Telephone: (26) 06 96 55
Telex: REMAVIA 830754
PARIS OFFICE: 18 quai Alphonse le Gallo, 92100 Boulogne-Billancourt
Telephone: 604 81 36

CHAIRMAN: Pierre Clostermann
CHIEF OPERATING OFFICER AND PRODUCTION MANAGER: Jean Pichon
FINANCIAL DIRECTOR: Jean Luc Varga
ADMINISTRATIVE DIRECTOR: Armand Blang
PUBLIC RELATIONS: Frédéric Amanou
CHIEF PILOT: Michel Jacquet

Under an agreement signed on 16 February 1960, the Cessna Aircraft Company of Wichita, Kansas, USA, acquired a 49 per cent holding in this company, which was then known as Société Nouvelle des Avions Max Holste.

Reims Aviation has the right to manufacture under licence Cessna designs for sale in Europe, Africa and Asia. By 1 January 1981 it had assembled a total of 1,833 Cessna F 150 and F 152, and 375 FRA 150 and FA 152 two-seat aircraft; 2,079 F 172 and 669 FR 172 four-seat aircraft; and 61 FTB 337 5/6-seat STOL aircraft, plus a large number of other Cessna models which are no longer assembled in France. Deliveries totalled 322 aircraft in the 1980 calendar year.

Reims Aviation is a subcontractor to Dassault-Breguet in the Mystère-Falcon 10, 20, 20H and 50, Mirage F1 and Mirage III programmes; and a subcontractor to Aérospatiale in the Airbus programme and for miscellaneous parts. It had 550 employees on 1 January 1981. Its offices and factory at Reims-Prunay Airport have an area of 23,000 m² (260,500 sq ft).

Reims/Cessna FA 152 Aérobat at Reims-Prunay Airport

REIMS/CESSNA F 152 and FA 152 AÉROBAT

Standard and aerobatic versions of the Cessna 152 assembled under licence by Reims Aviation are designated F 152 and FA 152 Aérobat respectively. The first FA 152 was flown on 25 April 1977, followed four days later by the first F 152. Production was at the rate of 80 F 152s and five FA 152s a year in early 1981.

Details of the current Cessna 152 series can be found in the US section. They apply also to aircraft assembled by Reims Aviation, except that conical camber wingtips are standard on the F 152 and FA 152, which have the following empty weights:

Reims/Cessna FR 172K Hawk XP II (Continental IO-360-K engine)

Reims Aviation FTB 337 G, equipped with remote sensing systems in underbelly pack, including Super Cyclope infra-red sensor and Hasselblad electric cameras

WEIGHTS: As Cessna 152 and Aerobat, except:

Weight empty: F 152	501 kg (1,105 lb)
F 152 II	519 kg (1,144 lb)
FA 152	517 kg (1,140 lb)

REIMS/CESSNA F 172 SKYHAWK/100 and FR 172K HAWK XP

Cessna Skyhawk and Skyhawk II aircraft assembled under licence by Reims Aviation are designated **F 172 Skyhawk/100** and **Skyhawk/100 II** respectively.

Until 1971, Reims Aviation retained a 145 hp Rolls-Royce Continental engine in the F 172, of which the first example was flown on 4 January 1963, with 805 delivered by the end of 1971. The current F 172 has a 119 kW (160 hp) Avco Lycoming O-320-D2J. Production in early 1981 was at the rate of 100 aircraft per year.

In early 1967 Reims Aviation flew the prototype of a more powerful version of the F 172, which it named the Reims Rocket. Details of this aircraft, which the company produced exclusively for worldwide sale, can be found in the 1976-77 *Jane's*. Production ended after 590 Rockets had been completed, and Reims Aviation is now assembling standard Cessna R172 Hawk XP and Hawk XP II aircraft as the **FR 172K Hawk XP** and **Hawk XP II** respectively. The prototype flew for the first time on 21 September 1976, and current rate of production is 10 per year.

Full details of the current Cessna Skyhawk and Hawk XP series can be found in the US section. They apply also to aircraft assembled by Reims Aviation, which have the following empty weights:

WEIGHT EMPTY: As Cessna models, except:

Skyhawk/100	640 kg (1,411 lb)
Skyhawk/100 II	653 kg (1,440 lb)
Hawk XP	704 kg (1,552 lb)
Hawk XP/II	717 kg (1,581 lb)
Skyhawk/100 floatplane	721 kg (1,590 lb)
Skyhawk/100 II floatplane	729 kg (1,607 lb)
Hawk XP floatplane	823 kg (1,815 lb)
Hawk XP/II floatplane	831 kg (1,832 lb)

REIMS/CESSNA F 182 and FR 182 RG

The first Reims-assembled **F 182** flew for the first time on 2 June 1975. Production has now ended.

REIMS AVIATION FTB 337 G

The airframe of this five/six-seat push-and-pull light twin is basically similar to that of the Cessna Model 337 Skymaster described in 1980-81 *Jane's*), but embodies STOL (ADAC) modifications, comprising high-lift trailing-edge flaps, and is fitted with two 168 kW (225 hp) Continental TSIO-360-D turbocharged engines. The FTB 337 G is not pressurised but can be equipped for maritime or overland patrol duties, sea or land rescue, or other specialised tasks by day and night, with four underwing pylons for containers of food and medicine, dinghies and locator beacons, radar, or equipment to detect illegal oil jettison and slicks at sea, or forest fires, including a SAT Super Cyclope infra-red sensor. The rear of the cabin can be cleared to carry cargo or two stretchers. The aircraft can also be equipped for navigation and IFR training.

An FTB 337 G was specially equipped with underwing containers and a belly pack and searchlight for anti-pollution operations over the English Channel in 1977.

By 1 January 1981 Reims Aviation had delivered 61 FTB 337 Gs, and expected to build at least two more during 1981.

DIMENSIONS, EXTERNAL: As Cessna 337, except:

Wing span	12·10 m (39 ft 8½ in)
Height overall	2·84 m (9 ft 4 in)

AREA:

Wings, gross	18·81 m² (202·5 sq ft)

WEIGHTS AND LOADING:

Weight empty	1,454 kg (3,206 lb)
Max T-O weight	2,100 kg (4,630 lb)
Max wing loading	113 kg/m² (23·2 lb/sq ft)

PERFORMANCE (at max T-O weight):

Max level speed at S/L	205 knots (380 km/h; 236 mph)
Cruising speed (75% power): at 3,000 m (10,000 ft)	186 knots (344 km/h; 214 mph)
at 6,000 m (20,000 ft)	200 knots (370 km/h; 230 mph)
Stalling speed, power reduced, wheels and flaps up	67 knots (124 km/h; 77 mph)
Max rate of climb at S/L	375 m (1,230 ft)/min
Rate of climb at S/L, one engine out	100 m (328 ft)/min
Rate of climb at 3,000 m (10,000 ft)	346 m (1,135 ft)/min
Service ceiling	7,300 m (23,950 ft)
Service ceiling, one engine out	6,000 m (20,000 ft)
STOL T-O to 15 m (50 ft)	245 m (804 ft)
STOL landing from 15 m (50 ft)	260 m (853 ft)
Max range, no reserves: 75% power at 6,000 m (20,000 ft)	955 nm (1,770 km; 1,100 miles)
econ power at 3,000 m (10,000 ft)	1,085 nm (2,012 km; 1,250 miles)
econ power at 6,000 m (20,000 ft)	1,150 nm (2,132 km; 1,325 miles)
Max endurance at 120 knots (222 km/h; 138 mph), no reserves	approx 5 h

ROBIN
AVIONS PIERRE ROBIN

HEAD OFFICE AND WORKS: Aérodrome de Dijon Val-Suzon Darois, 21121 Fontaine-les-Dijon Cédex
Telephone: (80) 31 61 01
Telex: 350 818 Robin F
PRESIDENT DIRECTOR GENERAL: Pierre Robin
GENERAL MANAGER: Michel Brandt
COMMERCIAL MANAGER: Thérèse Robin
SALES SUPERVISOR: Jacques Bernardin
PRODUCTION DIRECTOR: Philippe Estassy
TECHNICAL DIRECTOR: Daniel Muller

This company was formed in October 1957 as Centre Est Aéronautique to design, manufacture and sell touring aircraft. In 1969 the name of the company was changed to Avions Pierre Robin. Details of its subsidiary, Avions Pierre Robin Inc, can be found in the Canadian section.

Since 1973, Avions Pierre Robin has manufactured the DR 400 series of wooden light aircraft, all of which represent highly-refined developments of the company's earlier Jodel designs and were first flown in prototype form in 1972. They are described in detail, together with the company's current range of all-metal light aircraft, some with retractable landing gear.

Robin is also engaged on extensive subcontracting, including the manufacture of components for the Pilatus Turbo-Porter and Turbo-Trainer, Alpha Jet and Aérospatiale helicopters.

The company's works currently cover an area of about 11,000 m² (118,400 sq ft) and it employs 160 people.

ROBIN DR 400/120 DAUPHIN 80

The prototype of this DR 400 series lightplane flew for the first time on 15 May 1972 and received DGAC certification on the 10th of that month, followed by CAA certification in December 1972. The original version had a 93 kW (125 hp) engine and was manufactured as the DR 400/125 Petit Prince. It was superseded in 1975 by the DR 400/120 Petit Prince, with 88 kW (118 hp) engine, as described in the 1979-80 *Jane's*. The current version has a fine-pitch propeller and new instrument panel, and entered production in 1979 as the Dauphin 80, to which the following details apply:

TYPE: Three/four-seat light training and touring aircraft.

WINGS: Cantilever low-wing monoplane. Wing section NACA 23013·5 (modified). Centre-section has constant chord and no dihedral; outer wings have a dihedral of 14°. All-wood one-piece structure, with single box-spar. Leading-edge plywood-covered; Dacron covering overall. Wooden ailerons, covered with Dacron. Aluminium alloy flaps. Ailerons and flaps interchangeable port and starboard. Manually-operated airbrake under spar outboard of landing gear on each side. Picketing ring under each wingtip.

FUSELAGE: Wooden semi-monocoque structure of basic rectangular section, plywood-covered.

TAIL UNIT: Cantilever all-wood structure, covered with Dacron. Sweptback fin and rudder. All-moving one-piece horizontal surface, with tab.

LANDING GEAR: Non-retractable tricycle type, with oleo-pneumatic shock-absorbers and Manu hydraulically-

Robin DR 400/120 Dauphin 80 three/four-seat light aircraft

actuated drum brakes. All three wheels and tyres are size 380 × 150, pressure 1·57 bars (22·8 lb/sq in) on nose unit, 1·77 bars (25·6 lb/sq in) on main units. Nosewheel steerable via rudder bar. Fairings over all three legs and wheels. Tailskid with damper. Parking brake.

Power Plant: One 83·5 kW (112 hp) Avco Lycoming O-235-L2A flat-four engine, driving a Sensenich 72 CKS 6-0-56 two-blade fixed-pitch metal propeller, or Hoffmann two-blade fixed-pitch wooden propeller. Fuel tank in fuselage, usable capacity 100 litres (22 Imp gallons); optional 50 litre (11 Imp gallon) auxiliary tank. Oil capacity 5·7 litres (1·25 Imp gallons).

Accommodation: Enclosed cabin, with seats for three or four persons, in pairs, up to a max weight of 154 kg (340 lb) on front pair and 136 kg (300 lb), including baggage, at rear. Access via forward-sliding jettisonable transparent canopy. Dual controls standard. Cabin heated and ventilated. Baggage compartment with internal access.

Systems and Equipment: Standard equipment includes a 12V 50A alternator, 12V 32Ah battery, push-button starter, audible stall warning, and windscreen de-icing. Radio, blind-flying equipment, and navigation, landing and anti-collision lights to customer's requirements.

Dimensions, external:
Wing span 8·72 m (28 ft 7¼ in)
Wing chord, centre-section (constant) 1·71 m (5 ft 7½ in)
Wing chord at tip 0·90 m (3 ft 0 in)
Wing aspect ratio 5·6
Length overall 6·96 m (22 ft 10 in)
Height overall 2·23 m (7 ft 3¾ in)
Tailplane span 3·20 m (10 ft 6 in)
Wheel track 2·60 m (8 ft 6¼ in)
Wheelbase 5·20 m (17 ft 0¾ in)
Propeller diameter 1·78 m (5 ft 10 in)

Dimensions, internal:
Cabin: Length 1·62 m (5 ft 3¾ in)
Max width 1·10 m (3 ft 7¼ in)
Max height 1·23 m (4 ft 0½ in)
Baggage space, volume 0·39 m³ (13·75 cu ft)

Areas:
Wings, gross 13·60 m² (146·39 sq ft)
Ailerons, total 1·15 m² (12·38 sq ft)
Flaps, total 0·70 m² (7·53 sq ft)
Fin 0·61 m² (6·57 sq ft)
Rudder 0·63 m² (6·78 sq ft)
Horizontal tail surfaces, total 2·88 m² (31·00 sq ft)

Weights and Loadings:
Weight empty, equipped 530 kg (1,169 lb)
Max baggage 40 kg (88 lb)
Max T-O and landing weight 900 kg (1,984 lb)
Max wing loading 66·2 kg/m² (13·56 lb/sq ft)
Max power loading 10·23 kg/kW (16·8 lb/hp)

Performance (at max T-O weight):
Never-exceed speed 166 knots (308 km/h; 191 mph)
Max level speed at S/L 130 knots (241 km/h; 150 mph)
Max cruising speed at 2,250 m (7,400 ft) 116 knots (215 km/h; 133 mph)
Econ cruising speed at 3,000 m (9,800 ft) 105 knots (195 km/h; 121 mph)
Stalling speed, flaps down 45 knots (82 km/h; 51 mph)
Stalling speed, flaps up 51 knots (94 km/h; 59 mph)
Max rate of climb at S/L 183 m (600 ft)/min
Service ceiling 3,650 m (12,000 ft)
T-O run 235 m (771 ft)
T-O to 15 m (50 ft) 535 m (1,755 ft)
Landing from 15 m (50 ft) 460 m (1,510 ft)
Range with standard fuel at max cruising speed, no reserves 464 nm (860 km; 534 miles)

ROBIN DR 400/160 MAJOR 80

The first DR 400/160 flew on 29 June 1972. It was awarded DGAC certification on 6 September 1972, and CAA certification in December of the same year, and was manufactured as the Chevalier (see 1979-80 *Jane's*). The current version, with wing-root fuel tanks, a baggage hold door, a propeller of finer pitch and a new instrument panel based on experience with the Aiglon, has been in production since 1980 as the Major 80.

Type: Four-seat light aircraft.

Wings, Fuselage, Tail Unit, Landing Gear: Generally as for DR 400/120, but with external baggage door aft of cabin, in top of fuselage on port side.

Power Plant: One 119 kW (160 hp) Avco Lycoming O-320-D flat-four engine, driving a Sensenich two-blade metal fixed-pitch propeller. Fuel tank in fuselage, capacity 110 litres (24 Imp gallons), and two tanks in wing-root leading-edges, giving total capacity of 190 litres (41·75 Imp gallons), of which 182 litres (40 Imp gallons) are usable. Provision for auxiliary tank, raising total capacity to 240 litres (52·75 Imp gallons). Oil capacity 7·55 litres (1·66 Imp gallons).

Accommodation: Seating for four persons, on two side-by-side adjustable front seats (max load 154 kg; 340 lb total) and rear bench seat (max load 154 kg; 340 lb total). Forward-sliding transparent canopy gives access to all seats. Up to 40 kg (88 lb) of baggage can be stowed aft of rear seats when four occupants are carried.

Systems and Equipment: As for DR 400/120.

Dimensions and Areas: As for DR 400/120, except:
Propeller diameter 1·83 m (6 ft 0 in)
Baggage door: Height 0·47 m (1 ft 6½ in)
Width 0·55 m (1 ft 9½ in)
Wing area 14·20 m² (152·8 sq ft)

Weights and Loadings:
Weight empty, equipped 570 kg (1,257 lb)
Max T-O and landing weight 1,050 kg (2,315 lb)
Max wing loading 74·2 kg/m² (15·20 lb/sq ft)
Max power loading 8·82 kg/kW (14·47 lb/hp)

Performance (at max T-O weight):
Never-exceed speed 166 knots (308 km/h; 191 mph)
Max level speed at S/L 146 knots (271 km/h; 168 mph)
Max cruising speed (75% power) at 2,440 m (8,000 ft) 132 knots (245 km/h; 152 mph)
Econ cruising speed (65% power) at 3,200 m (10,500 ft) 130 knots (241 km/h; 150 mph)
Stalling speed, flaps up 56 knots (103 km/h; 64 mph)
Stalling speed, flaps down 50 knots (93 km/h; 58 mph)
Max rate of climb at S/L 255 m (836 ft)/min
Service ceiling 4,115 m (13,500 ft)
T-O run 300 m (985 ft)
T-O to 15 m (50 ft) 500 m (1,640 ft)
Landing from 15 m (50 ft) 545 m (1,788 ft)
Landing run 250 m (820 ft)
Range with standard fuel at econ cruising speed, 45 min reserves 693 nm (1,285 km; 798 miles)

ROBIN DR 400/180 RÉGENT

First flown on 27 March 1972, this most powerful, four/five-seat member of the wooden DR 400 series received DGAC certification on 10 May 1972, and CAA certification in December 1972.

The DR 400/180 is generally similar to the DR 400/160 Major 80, except in the following details:

Power Plant: One 134 kW (180 hp) Avco Lycoming O-360-A flat-four engine. Fuel tankage as for DR 400/160.

Accommodation, Systems and Equipment: Basically as for DR 400/160, but optional seating for three persons on rear bench seat. Baggage capacity 55 kg (121 lb).

Dimensions and Areas: As for DR 400/160, except:
Propeller diameter 1·93 m (6 ft 4 in)

Weights and Loadings:
Weight empty, equipped 600 kg (1,322 lb)
Max T-O and landing weight 1,100 kg (2,425 lb)
Max wing loading 77·7 kg/m² (15·91 lb/sq ft)
Max power loading 8·21 kg/kW (13·47 lb/hp)

Performance (at max T-O weight):
Never-exceed speed 166 knots (308 km/h; 191 mph)
Max level speed at S/L 150 knots (278 km/h; 173 mph)
Max cruising speed (75% power) at 2,440 m (8,000 ft) 144 knots (267 km/h; 166 mph)
Econ cruising speed (60% power) at 3,660 m (12,000 ft) 134 knots (249 km/h; 155 mph)
Stalling speed, flaps up 56·5 knots (105 km/h; 65 mph)
Stalling speed, flaps down 51·5 knots (95 km/h; 59 mph)
Max rate of climb at S/L 252 m (825 ft)/min
Service ceiling 4,720 m (15,475 ft)
T-O run 315 m (1,035 ft)
T-O to 15 m (50 ft) 610 m (2,000 ft)
Landing from 15 m (50 ft) 530 m (1,740 ft)
Landing run 249 m (817 ft)
Range with standard fuel at 65% power, no reserves 783 nm (1,450 km; 900 miles)

ROBIN DR 400/180R REMORQUEUR

The DR 400/180R is a member of the DR 400 range designed for use as a glider-towing aircraft, although it can also be flown as a normal four-seat tourer. The prototype first flew on 6 November 1972 and received DGAC certification on the 28th of that month. Details are generally the same as for the DR 400/180 Régent, except for the following items:

Fuselage: No external baggage door. The baggage compartment is covered with transparent Plexiglas as an extension of the canopy, allowing optimum rearward vision.

Power Plant: One 134 kW (180 hp) Avco Lycoming O-360-A flat-four engine, driving (for glider-towing) a Sensenich 76 EM 8S5 058 or Hoffmann HO-27-HM-180/138 two-blade propeller. For touring operation a Sensenich 76 EM 8S5 064 propeller of the same diameter is fitted. Fuel capacity as for DR 400/120.

Dimensions and Areas: As for DR 400/120, except:
Propeller diameter 1·83 m (6 ft 0 in)

Weights and Loadings:
Weight empty, equipped 560 kg (1,234 lb)
Max T-O and landing weight 1,000 kg (2,205 lb)
Max wing loading 73·5 kg/m² (15·05 lb/sq ft)
Max power loading 7·46 kg/kW (12·25 lb/hp)

Performance (glider tug, at max T-O weight):
Never-exceed speed 166 knots (308 km/h; 191 mph)
Max level speed at S/L (70% power) 124 knots (230 km/h; 143 mph)
Max cruising speed at 2,440 m (8,000 ft) 124 knots (230 km/h; 143 mph)
Econ cruising speed (56% power) at 3,660 m (12,000 ft) 122 knots (226 km/h; 140 mph)
Stalling speed, flaps up 54 knots (99 km/h; 62 mph)
Stalling speed, flaps down 47 knots (87 km/h; 54 mph)
Max rate of climb at S/L towing Bijave sailplane 210 m (690 ft)/min
Service ceiling 6,000 m (19,685 ft)
T-O run 205 m (673 ft)
T-O to 15 m (50 ft) 400 m (1,313 ft)
Landing from 15 m (50 ft) 470 m (1,542 ft)
Landing run 220 m (722 ft)
Range at econ cruising speed, max fuel, no reserves 444 nm (825 km; 512 miles)

Robin DR 400/180 Régent (Avco Lycoming O-360-A engine)

Robin DR 400/180R Remorqueur glider-towing aircraft

Robin R 1180 Aiglon four-seat light aircraft (Avco Lycoming O-360 engine)

Robin R 2112 Alpha, basic current model of the R 2000 series

ROBIN R 1180 AIGLON

The R 1180 Aiglon is an all-metal four-seat light aircraft of conventional low-wing monoplane layout. It has non-retractable landing gear, and is powered by a 134 kW (180 hp) Avco Lycoming O-360-A3AD engine which drives a two-blade Sensenich fixed-pitch propeller. Wing section is NACA 23015. The prototype (F-WVKU) flew for the first time in late 1976; the first production Aiglon (F-GBAM), with a larger, tinted canopy and other changes, received DGAC certification on 19 September 1978. Eighteen have been ordered for use at French civil aviation training schools. These introduced a number of refinements, including a redesigned instrument panel, a control column adjustable for height, increased fuel capacity of 258 litres (56·75 Imp gallons) in integral wing tanks, and improved soundproofing.

DIMENSIONS, EXTERNAL:
- Wing span: 9·08 m (29 ft 9½ in)
- Wing aspect ratio: 5·46
- Length overall: 7·26 m (23 ft 9¾ in)
- Height overall: 2·38 m (7 ft 9¾ in)
- Propeller diameter: 1·93 m (6 ft 4 in)

DIMENSIONS, INTERNAL:
- Cabin: Length: 2·00 m (6 ft 6¾ in)
  - Width: 1·12 m (3 ft 8 in)
  - Height: 1·22 m (4 ft 0 in)

AREA:
- Wings, gross: 15·1 m² (162·5 sq ft)

WEIGHTS:
- Weight empty: 650 kg (1,433 lb)
- Max T-O weight: 1,150 kg (2,535 lb)

PERFORMANCE (at max T-O weight):
- Max level speed at S/L: 135 knots (251 km/h; 156 mph)
- Max cruising speed (75% power) at height: 133 knots (247 km/h; 153 mph)
- Cruising speed (65% power) at height: 127 knots (235 km/h; 146 mph)
- Stalling speed, flaps up: 58 knots (107 km/h; 66·5 mph)
- Stalling speed, flaps down: 50 knots (92 km/h; 57 mph)
- Max rate of climb at S/L: 258 m (846 ft)/min
- Service ceiling: 5,030 m (16,500 ft)
- T-O to 15 m (50 ft): 970 m (3,185 ft)
- Landing from 15 m (50 ft): 460 m (1,510 ft)
- Range with max fuel at 65% power, with reserves: 876 nm (1,624 km; 1,009 miles)

ROBIN R 2000 ALPHA SERIES

First of a new generation of light aircraft designated R for Robin, the R 2160 prototype (F-WZAC) flew for the first time in September 1976 and was announced officially in the following month. Further variants of the R 2000 series have since been developed. Those currently available differ as follows:

R 2112 Alpha. Two-seat aerobatic light aircraft with an 83·5 kW (112 hp) Avco Lycoming engine. Second example built was demonstrated at the 1979 Paris Air Show.

R 2160 Alpha Sport. Generally similar to R 2112, but with a 119 kW (160 hp) Avco Lycoming O-320-D engine. Prototype of the R 2000 series flew originally in this form. Seventeen ordered by the Service de la Formation Aéronautique in Spring 1978.

The R 2000 series were derived from the Robin HR 200, which they superseded, and embody many components of that aircraft, including the basic fuselage and tail fin. The wing is completely new, of increased chord and different section; the rudder is larger and there is a long, shallow ventral fin to improve spinning characteristics. All R 2000 variants are being certificated in the Aerobatic category, and are being fully tested for spin recovery, including inverted spin. They will be manufactured in Canada, by Avions Pierre Robin Inc, as well as in France.

The following details refer specifically to the R 2160 Alpha Sport, but are generally applicable to the R 2112:

TYPE: Two-seat aerobatic light aircraft.

WINGS: Cantilever low-wing monoplane. Wing section NACA 23015. Dihedral 6° 20′ from roots. Incidence 3°. No sweep. Conventional single-spar aluminium alloy structure. Entire trailing-edge of each wing comprises an inboard slotted flap and a fully balanced slotted aileron, both of aluminium alloy construction.

FUSELAGE: Conventional aluminium alloy semi-monocoque structure.

TAIL UNIT: Cantilever structure of aluminium alloy construction. All-moving single-spar horizontal surfaces, with anti-tab on trailing-edge each side. Rudder is horn-balanced, with no tab.

LANDING GEAR: Non-retractable tricycle type, with oleo-pneumatic shock-absorbers. All three tyres size 380 × 150. Cleveland disc brakes on main wheels. Three wheel and leg fairings standard.

POWER PLANT: One 119 kW (160 hp) Avco Lycoming O-320-D flat-four engine, driving a Sensenich two-blade fixed-pitch propeller. Fuel tank in fuselage, capacity 120 litres (26·5 Imp gallons). Refuelling point at side of fuselage.

ACCOMMODATION: Two seats side by side under large transparent canopy, the front half of which slides forward for access. Seats are adjustable.

SYSTEM: 12V electrical system.

AVIONICS AND EQUIPMENT: Blind-flying instruments, radio, VOR, ADF and ILS available to customer's requirements.

DIMENSIONS, EXTERNAL:
- Wing span: 8·33 m (27 ft 4 in)
- Wing chord (constant): 1·554 m (5 ft 1¼ in)
- Wing aspect ratio: 5·34
- Length overall: 7·10 m (23 ft 3½ in)
- Height overall: 2·135 m (7 ft 0 in)
- Tailplane span: 3·04 m (9 ft 11¾ in)
- Wheel track: 2·91 m (9 ft 6½ in)
- Wheelbase: 1·434 m (4 ft 8½ in)
- Propeller diameter: 1·88 m (6 ft 2 in)

DIMENSION, INTERNAL:
- Cabin: Width: 1·06 m (3 ft 5¾ in)

AREA:
- Wings, gross: 13·00 m² (140 sq ft)

WEIGHTS:
- Weight empty: R 2112: 535 kg (1,180 lb)
  - R 2160: 550 kg (1,213 lb)
- Max T-O weight:
  - R 2112 and R 2160: 800 kg (1,764 lb)

PERFORMANCE (at max T-O weight):
- Never-exceed speed:
  - R 2112 and R 2160: 180 knots (333 km/h; 207 mph)
- Max level speed at S/L:
  - R 2112: 112 knots (208 km/h; 129 mph)
  - R 2160: 138 knots (257 km/h; 160 mph)
- Max cruising speed (75% power) at 2,285 m (7,500 ft):
  - R 2112: 114 knots (212 km/h; 131 mph)
  - R 2160: 130 knots (242 km/h; 150 mph)
- Cruising speed (65% power) at 3,350 m (11,000 ft):
  - R 2112: 94 knots (175 km/h; 108 mph)
  - R 2160: 126 knots (234 km/h; 145 mph)
- Stalling speed, flaps down:
  - R 2112: 45 knots (83 km/h; 52 mph)
  - R 2160: 46 knots (85 km/h; 53 mph)
- Max rate of climb at S/L:
  - R 2112: 222 m (728 ft)/min
  - R 2160: 312 m (1,025 ft)/min
- Service ceiling:
  - R 2112: 3,660 m (12,000 ft)
  - R 2160: 4,575 m (15,000 ft)
- Range with max fuel at 65% power:
  - R 2112: 631 nm (1,170 km; 727 miles)
  - R 2160: 430 nm (796 km; 495 miles)

ROBIN R 3000 SERIES

Development of this new series of all-metal light aircraft began in 1978, to replace types in current production. All models will embody common components, and design emphasis has been placed on reduced manufacturing time. Versions currently projected are as follows:

R 3100. Two-seat trainer, powered by an Avco Lycoming O-235-L engine rated at 74·5 kW (100 hp). Non-retractable landing gear. Max T-O weight 800 kg (1,763 lb).

R 3120. Four-seat (2+2) light aircraft, with Avco Lycoming O-235 engine rated at 89 kW (120 hp). Endurance of 5 h with pilot, three passengers and no baggage. Non-retractable landing gear. Max T-O weight 900 kg (1,984 lb).

R 3140. Full four-seater, with 106 kW (143 hp) Avco Lycoming O-320 engine. This was the first version to fly, on 8 December 1980, with certification and series production scheduled for late 1982.

R 3160L. Glider-towing version. Basically as R 3140, but with 119 kW (160 hp) Avco Lycoming O-360 engine.

R 3160 GT. One of three top-of-the-range GT models, basically as R 3140 but with longer and slightly higher cabin. Powered by 119 kW (160 hp) Avco Lycoming O-360 engine. Max T-O weight 1,100 kg (2,425 lb).

Robin R 2160 Alpha Sport two-seat aerobatic aircraft (Avco Lycoming O-320-D engine) *(J. M. G. Gradidge)*

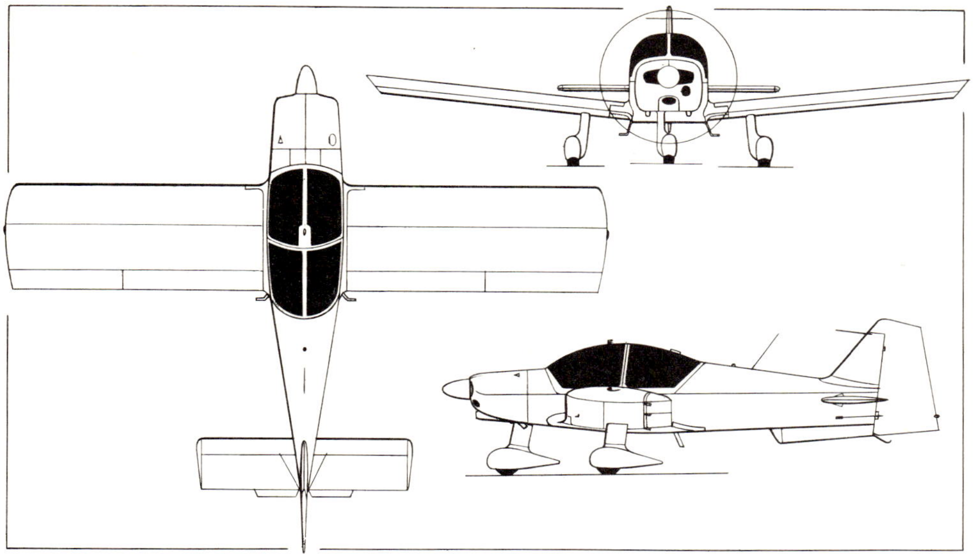

Robin R 2160 side-by-side two-seat training aircraft *(Pilot Press)*

R 3180 GT. As R 3160 GT, but with 134 kW (180 hp) TO-360 engine, constant-speed propeller, and max T-O weight of 1,150 kg (2,535 lb).

R 3180 GTR. As R 3180 GT, but with retractable landing gear. Max T-O weight 1,200 kg (2,645 lb).

The following details apply to the R 3140:

Type: Four-seat all-metal light aircraft.

Wings: Cantilever low-wing monoplane. Wing section NACA 23015. Dihedral 6° from roots. Incidence 3°. No sweep. Conventional single-spar aluminium alloy structure, of constant chord. Entire trailing-edge of each wing comprises an inboard slotted flap and slotted aileron, both of aluminium alloy construction, over 25% of chord.

Fuselage: Conventional aluminium alloy semi-monocoque structure.

Tail Unit: Cantilever T tail of aluminium alloy construction. Pitch trim tab on elevator trailing-edge.

Landing Gear: Non-retractable tricycle type, with steerable nosewheel. Oleo-pneumatic shock-absorbers. Main-wheel tyres size 380 × 150. Nosewheel tyre size 5·00-5. Cleveland disc brakes. Streamline fairings on all three wheels.

Power Plant: One 106 kW (143 hp) Avco Lycoming O-320-D2A flat-four engine, driving a Sensenich two-blade fixed-pitch propeller. Two fuel tanks in wing leading-edges, with total capacity of 240 litres (52·75 Imp gallons). Oil capacity 7·5 litres (1·6 Imp gallons).

Accommodation: Four seats in pairs in enclosed cabin. Forward-sliding transparent canopy.

System: Electrical system includes 12V 50A alternator and 12V 32Ah battery.

Avionics and Equipment: Prototype has VHF, VOR, DME, ADF, transponder, gyro horizon and Century II autopilot.

Dimensions, external:
Wing span 9·46 m (31 ft 0½ in)
Wing chord, constant 1·55 m (5 ft 1 in)

Dimension, internal:
Cabin: Max width 1·14 m (3 ft 8¾ in)

Area:
Wings, gross 14·46 m² (155·65 sq ft)

Weights:
Weight empty 576 kg (1,270 lb)
Max T-O and landing weight 1,050 kg (2,315 lb)

Performance (estimated, at max T-O weight):
Never-exceed speed 171 knots (318 km/h; 197 mph)
Max cruising speed (75% power) at 2,600 m (8,500 ft) 133 knots (246 km/h; 153 mph)
Econ cruising speed (65% power) at 3,350 m (11,000 ft) 126 knots (234 km/h; 145 mph)
Stalling speed, flaps down 47 knots (86 km/h; 54 mph)
Max rate of climb at S/L 260 m (850 ft)/min
Service ceiling 4,420 m (14,500 ft)
Range with max fuel, 45 min reserves 1,000 nm (1,850 km; 1,150 miles)

Robin R 3140, prototype for the new R 3000 series *(Air Portraits)*

SOCATA

SOCIÉTÉ DE CONSTRUCTION D'AVIONS DE TOURISME ET D'AFFAIRES (Subsidiary of Aérospatiale)

Head Office, Works and After-Sales Service: Aérodrome de Tarbes-Ossun-Lourdes, BP 38, 65001 Tarbes Cédex
Telephone: (62) 93 97 30
Telex: SOCATA 520828 F
Sales: 12 rue Pasteur, 92150 Surèsnes
Telephone: (1) 506 37 60
Telex: AIRSPA 620 059 F
Flying School: Aérodrome de Toussus-le-Noble, BP 2, 78350 Buc
Telephone: 956 21 00
Telex: SOCAERO 697836 F
President and Director General: Pierre Gautier
Commercial Manager: J. C. Godiveau

This company was formed in 1966, as a subsidiary of Aérospatiale (then Sud-Aviation), to be responsible for producing all of the group's piston-engined light aircraft. As well as those described in this entry, Socata is responsible for producing the Aérospatiale Epsilon military primary/basic trainer, described under the Aérospatiale heading in this section.

Socata also produces components for the A300 Airbus, Super Étendard, Alpha Jet, Mystère-Falcon 10, 20 and 50 business aircraft, and Puma, Dauphin and Ecureuil helicopters. It is responsible for overhaul and repair of MS 760 Paris light jet aircraft.

Socata's works cover an area of 54,000 m² (581,250 sq ft), and employ a total of about 890 people.

SOCATA RALLYE SERIES

The Rallye had its origin in a competition organised by the SFACT in 1958 and was developed originally by the old-established Morane-Saulnier company. The prototype (67 kW; 90 hp MS 880A) Rallye-Club flew on 10 June 1959, and the initial production versions were the MS 880B Rallye-Club and the MS 885 Super Rallye. FAA certification was obtained on 21 November 1961.

Aircraft built up to 1972 had rounded wingtips. Those in the 67-112 kW (90-150 hp) range had a 'light' structure and wing area of 12·28 m² (132·2 sq ft); those in the 112-175 kW (150-235 hp) range had a 'heavy' structure and wing area of 12·38 m² (133·3 sq ft), the additional area resulting from use of modified ailerons. In 1972, rectangular wingtips became standard, increasing the wing area of all models as noted in this entry. Use of 'light' and 'heavy' airframes for different versions of the Rallye series continues.

The 3,000th Rallye built in France was delivered to the Aéro-Club Renault on 26 May 1977. During 1980, a total of 43 Rallyes was sold and 67 delivered by Socata. Versions of the Rallye are also built under licence in Poland, by the PZL-Warszawa organisation (which see).

Since 1979, the seven aircraft constituting the current Rallye series have had names, instead of the former designations, as follows:

Galopin (formerly Rallye 110 ST). This version may be operated either as a two-seater cleared for spinning or as a three/four-seater with spins prohibited. It has a 'light' airframe and an 82 kW (110 hp) Avco Lycoming O-235-L2A engine, driving a two-blade McCauley fixed-pitch propeller.

Garnement (formerly Rallye 160 ST). Similar to Galopin, but with 115 kW (155 hp) Avco Lycoming O-320-D2A engine. Developed as a high-performance aircraft for flying clubs and for use in countries where high temperatures or high altitude might impose limitations on lower-powered aircraft.

Galérien (formerly Rallye 180 T). Specialised glider and banner towing version, developed to meet West German noise regulations. 'Light' airframe and 134 kW (180 hp) Avco Lycoming O-360-A3A engine, driving a two-blade fixed-pitch propeller. Certificated in September 1977.

Gaillard (formerly Rallye 180 GT). Four-seat version with 'heavy' airframe and 134 kW (180 hp) Avco Lycoming O-360-A3A engine, driving a fixed-pitch or constant-speed propeller. Strengthened structure, permitting increased AUW; larger rudder and ailerons; fillets of increased size between wing trailing-edges and fuselage; longer nosewheel leg to give increased propeller clearance; enlarged dorsal fin; modified cockpit canopy; and a baggage compartment. Streamline wheel fairings optional. Prototype flew for first time on 7 December 1964. French certification received on 27 April 1965; FAA type approval on 23 June 1971. Total of 742 built by 1 January 1979; 100 ordered by the SFACT for duty as glider tugs at French gliding centres. A variant with a TV camera pod mounted under the starboard wing was demonstrated in 1978. Known as ATAL, this equipment allows the aircraft

Aircraft of the basic Rallye series are similar externally. This light Rallye airframe is used for the Galopin, Garnement and Galérien. The other variants have a heavy airframe

to be used for aerial surveillance of highway traffic, transmitting images to ground stations over a radius of about 27 nm (50 km; 31 miles). The camera is steerable by the pilot, who also sees the images on a cockpit monitor.

Gabier (formerly Rallye 235 GT). High-performance version, with a 'heavy' airframe, a 175 kW (235 hp) Avco Lycoming O-540-B4B5 engine, driving a Hartzell HC-C2YK-1/8468-6 two-blade constant-speed metal propeller, and accommodation for four persons. Construction of prototype began in January 1975. First flight was made on 1 April 1975. Total of 103 built by 1 January 1979. A variant (formerly Rallye 235 A) is available with dual control columns instead of control wheels. Another variant (formerly Rallye 235 C) has a tailwheel type landing gear instead of the standard tricycle type.

Gaucho (formerly Rallye 235 CA). Agricultural version of Gabier with tailwheel landing gear. Described separately.

Guerrier (formerly Rallye 235 G). Military version. Described separately.

The Gabier and Gaillard are authorised for use as ambulance aircraft carrying a pilot, one stretcher patient and medical attendant. They can also be used for glider towing, and some 400 Rallyes are employed in this role, including more than 250 in France. Agricultural spraygear can be fitted and tests have been conducted with various models on ski landing gear.

The following details apply to all French-built versions listed:

Type: Two/four-seat light monoplane.

Wings: Cantilever low-wing monoplane. Wing section NACA 63A416 (modified). Dihedral 7°. Incidence 4°. All-metal single-spar structure. Wide-chord slotted ailerons. Full-span automatic slats. Long-span slotted flaps. Ailerons and flaps have corrugated metal skin. Ground-adjustable aileron tabs. No anti-icing equipment.

Fuselage: All-metal semi-monocoque structure.

Tail Unit: Cantilever all-metal structure with corrugated skin on the mass-balanced control surfaces. Fixed-incidence tailplane. One automatic tab and one controllable tab on elevator. One controllable tab on rudder.

Landing Gear: Non-retractable tricycle type on all but Gaucho and non-standard Gabier. Oleo-pneumatic shock-adsorbers. Castoring nosewheel. Standard Gabier has Cleveland main wheels with tyres size 6·00-6, pressure 1·8 bars (26·1 lb/sq in); nosewheel tyre size 5·00-4, pressure 1·4 bars (20·3 lb/sq in). Cleveland hydraulic disc brakes. Provision for fitting skis or floats.

Power Plant: One flat-four or flat-six engine (details under entries for individual models), driving a two-blade fixed-pitch or constant-speed metal propeller. Fuel in two metal tanks in wings, with total capacity of 105 litres (23 Imp gallons) in Galopin, 184 litres (40·5 Imp gallons) in Garnement and Galérien, 235 litres (52 Imp gallons) in Gaillard, and 282 litres (62 Imp gallons) in Gabier. Refuelling points above wings. Oil capacity 6 litres (1·3 Imp gallons) in Galopin, 8 litres (1·75 Imp gallons) in Garnement, Galérien and Gaillard, 12 litres (2·6 Imp gallons) in Gabier.

Accommodation: Two seats side by side at front, and rear bench seat for one/two persons in Galopin, under large rearward-sliding canopy. Other versions are full four-seaters. Dual control columns on Galopin, Garnement, Galérien, Gaucho, Guerrier and both non-standard variants of Gabier. Dual control wheels on Gaillard and standard Gabier. Individual adjustable front seats and baggage space aft of rear seats (accessible internally) on the Gaillard, Gabier and the Guerrier. Heating and ventilation standard.

Systems: Electrical system includes 12V 18A battery and 12V 40A alternator in Galopin; 12V 32A battery and 12V 40A alternator in Garnement, Galérien and Gaillard; 12V 32A battery and 12V 55A alternator in Gabier.

Avionics and Equipment: The instrument panel is fitted with an anti-glare visor, and is designed to take full radio-navigation equipment to customer's requirements.

Dimensions, external (A, Galopin; B, Garnement; C, Galérien; D, Gaillard; E, Gabier):
Wing span 9·74 m (31 ft 11 in)
Wing chord (constant) 1·30 m (4 ft 3 in)
Wing aspect ratio 7·57
Length overall: A, B, C, D 7·24 m (23 ft 9 in)
E 7·25 m (23 ft 9½ in)
Height overall 2·80 m (9 ft 2¼ in)
Tailplane span 3·67 m (12 ft 0½ in)
Wheel track 2·01 m (6 ft 6½ in)
Wheelbase 1·71 m (5 ft 7¼ in)

Dimensions, internal:
Cabin: Length: A, B, C 1·80 m (5 ft 11 in)
D, E 2·25 m (7 ft 4 in)
Width 1·13 m (3 ft 8½ in)

Areas:
Wings, gross: A, B, C 12·66 m² (136·3 sq ft)
D, E 12·76 m² (137·3 sq ft)
Trailing-edge flaps (total) 2·40 m² (25·83 sq ft)
Vertical tail surfaces (total) 1·74 m² (18·73 sq ft)
Horizontal tail surfaces (total) 3·48 m² (37·50 sq ft)

Weights and Loadings:
Weight empty: A 520 kg (1,145 lb)
B 540 kg (1,191 lb)
C 545 kg (1,200 lb)
D 570 kg (1,257 lb)
E 694 kg (1,530 lb)
Max T-O and landing weight: A 770 kg (1,697 lb)
B 870 kg (1,918 lb)
C 950 kg (2,094 lb)
D 1,050 kg (2,315 lb)
Max T-O weight: E 1,200 kg (2,645 lb)
Max landing weight: E 1,140 kg (2,513 lb)
Max wing loading: E 95 kg/m² (19·45 lb/sq ft)
Max power loading: E 6·69 kg/kW (10·98 lb/hp)

Performance (at max T-O weight):
Max level speed at S/L:
A 110 knots (204 km/h; 126 mph)
B 119 knots (220 km/h; 137 mph)
C 122 knots (226 km/h; 140 mph)
D 130 knots (240 km/h; 150 mph)
E 148 knots (275 km/h; 171 mph)
Max cruising speed (75% power):
A 104 knots (192 km/h; 119 mph)
B 115 knots (213 km/h; 132 mph)
C 113 knots (209 km/h; 130 mph)
D 122 knots (226 km/h; 140 mph)
E 132 knots (245 km/h; 152 mph)
Econ cruising speed (65% power):
A 95 knots (176 km/h; 109 mph)
B 109 knots (202 km/h; 125 mph)
C 111 knots (205 km/h; 127 mph)
D 115 knots (213 km/h; 132 mph)
E 125 knots (231 km/h; 144 mph)
Stalling speed, flaps down:
A 41 knots (75 km/h; 47 mph)
B 45 knots (83 km/h; 52 mph)
C 46 knots (85 km/h; 53 mph)
D 50 knots (92 km/h; 57·5 mph)
E 54 knots (100 km/h; 62 mph)
Max rate of climb at S/L: A 192 m (630 ft)/min
B 246 m (805 ft)/min
C 265 m (870 ft)/min
D 231 m (760 ft)/min
E 300 m (984 ft)/min
Service ceiling: A 3,200 m (10,500 ft)
B 3,950 m (12,950 ft)
C 4,500 m (14,750 ft)
D 3,600 m (11,800 ft)
E 4,500 m (14,760 ft)
T-O run: A, B 214 m (700 ft)
C 240 m (790 ft)
D 250 m (820 ft)
E 360 m (1,180 ft)
T-O to 15 m (50 ft): A 418 m (1,370 ft)
B 365 m (1,200 ft)
C 408 m (1,340 ft)
D 445 m (1,460 ft)
E 500 m (1,640 ft)
Landing from 15 m (50 ft): A 268 m (880 ft)
B 335 m (1,100 ft)
C 342 m (1,120 ft)
D 360 m (1,180 ft)
E 397 m (1,300 ft)
Landing run: A 113 m (370 ft)
B 170 m (560 ft)
C 146 m (480 ft)
D 150 m (490 ft)
E 165 m (540 ft)
Range with max fuel (allowances for T-O, climb and descent, 45 min reserves):
A 400 nm (740 km; 460 miles)
B 520 nm (963 km; 598 miles)
C 460 nm (852 km; 529 miles)
D 600 nm (1,110 km; 690 miles)
E 590 nm (1,090 km; 679 miles)

Socata Guerrier in ground support configuration, with two F2 rocket packs and bombs on underwing attachments

SOCATA R 235 GUERRIER

This military aircraft (known originally as the Rallye 235 G) is generally similar to the high-performance Gabier four-seat light aircraft, with a 175 kW (235 hp) Avco Lycoming O-540-B4B5 engine, but has four Alkan 663 underwing stores pylons which enable it to be used for a variety of armed and support missions. The pylons are attached under each wing between ribs 8 and 9, and ribs 15 and 16, and are connected to a weapon selection box installed centrally on the radio panel in the cockpit.

Stores that can be carried on these pylons include Matra F2 rocket launchers, each containing six 68 mm rockets; Type AA 52 pods, each containing two 7·62 mm machine-guns with 500 rds/gun, and large enough to retain all spent cartridge cases and links after firing; 50 kg operational or practice bombs; rescue packs for airdropping over water, desert, jungle or polar regions; flares for use during operational or rescue missions by night; a surveillance pack containing a TV camera and transmitter to send images to a ground station. The camera is fitted with a zoom lens, and can scan to 45° on each side of the aircraft, with a vertical scan of 110°. The pilot has a control box (normal and zoom), and a monitor on which to check precisely the images the camera is viewing. All underwing loads can be jettisoned in an emergency.

The cockpit of the Guerrier contains two side-by-side seats, with dual controls, enabling the aircraft to be used for both basic and operational training, as well as combat missions. A rear bench seat can be installed to permit the carriage of two passengers and a quantity of baggage or freight. It can carry a stretcher patient on casevac missions. Structure of the aircraft is basically unchanged, except for some reinforcement, notably to the wings in the vicinity of the weapon pylons.

Performance:
Range/endurance:
Armed reconnaissance with 2 gun pods at 70% power, 30 min fuel reserves
5 h or 556 nm (1,030 km; 640 miles)
Armed reconnaissance with 4 rocket launchers at 70% power, 30 min fuel reserves
2 h 40 min or 286 nm (530 km; 329 miles)
Ground support with 2 gun pods at 75% power at 915 m (3,000 ft), 30 min fuel reserves, 10 min over target 243 nm (450 km; 280 miles)
Ground support with 4 rocket launchers at 75% power at 915 m (3,000 ft), 30 min fuel reserves, 10 min over target 130 nm (240 km; 149 miles)
Ground support with 2 rocket launchers and 2 gun pods at 75% power at 915 m (3,000 ft), 15 min fuel reserves, 10 min over target
43 nm (80 km; 50 miles)
Unarmed reconnaissance with TV pod at 70% power, 30 min fuel reserves
545 nm (1,010 km; 627 miles)

SOCATA RALLYE GAUCHO

The high-lift and safety characteristics inherent in the basic Rallye series well suit it for agricultural operations. It has been possible to purchase standard models adapted for spraying/dusting for some years. The Gaucho represents a more specialised development, with a tailwheel-type landing gear instead of the normal tricycle type; a faired-in rear cabin to house a 580 litre (127·5 Imp gallon) chemical tank; reinforced structure, with anti-corrosive treatment on metal surfaces; and a propeller specially designed for heavy duty at low speeds. The airframe is based on that of the Gabier, and a generally similar 175 kW (235 hp) Avco Lycoming O-540-B2B5 engine is fitted.

The cockpit is equipped normally with a single seat, on the port side, with entrance via an upward-opening canopy/door hinged on the centreline. The forward portion of the chemical tank projects into the starboard side of the cockpit, alongside the pilot. If desired, it can be removed, and replaced by a cover plate and second seat. The aircraft can then be used as a dual-control agricultural pilot trainer, with tank capacity reduced to 500 litres (110 Imp gallons).

Between the seasons for agricultural flying, the Gaucho's cockpit canopy, cabin fairing, chemical tank, spraybars and other specialised equipment can be removed and replaced by conventional Gabier seats and sliding canopy, converting the aircraft into a four-seat touring aircraft or light freighter.

A variety of dispersal equipment is available for the Gaucho, including four Micronair units, two above each wing trailing-edge; a Sorensen underfuselage pump with a capacity of 2 to 40 litres/hectare (1 to 22 Imp gallons/acre) and spraybars with 24 or 32 nozzles; or a Transland spreader for solids. Steel wire-cutters are fitted on the front of each main landing gear leg and on the windscreen centreline, with a steel cable from the latter to the tip of the fin.

The Gaucho was certificated in May 1978.

DIMENSIONS: As for Gabier, except:
- Length overall 7·26 m (23 ft 10 in)
- Height overall 2·31 m (7 ft 7 in)

WEIGHTS:
- Weight empty 694 kg (1,530 lb)
- Max chemical payload 425 kg (935 lb)
- Max T-O weight 1,350 kg (2,980 lb)

PERFORMANCE (at max T-O weight):
- Max cruising speed at S/L 119 knots (220 km/h; 137 mph)
- Econ cruising speed (65% power) 115 knots (213 km/h; 132 mph)
- Spraying speed 70 knots (130 km/h; 80 mph)
- Stalling speed, flaps down 54 knots (100 km/h; 62 mph)
- Max rate of climb at S/L 300 m (984 ft)/min
- T-O run 365 m (1,200 ft)
- T-O to 15 m (50 ft) 495 m (1,620 ft)
- Landing from 15 m (50 ft) 435 m (1,430 ft)
- Landing run 315 m (1,030 ft)
- Range, with allowances for T-O, climb and descent, 45 min reserves 485 nm (900 km; 558 miles)

This view of the Socata Gaucho shows the wing spraybars

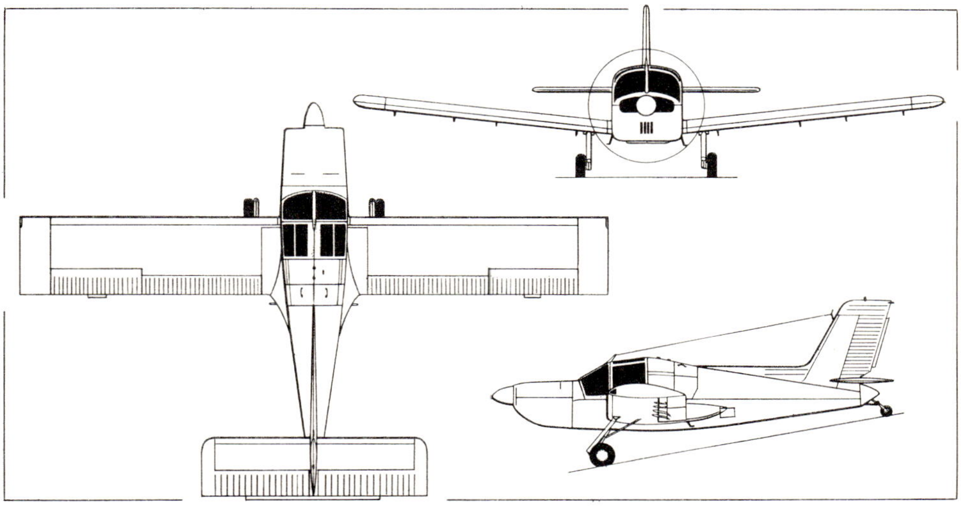

Socata Gaucho, with hopper aft of cabin and tailwheel landing gear *(Pilot Press)*

SOCATA RALLYE TB 9 TAMPICO AND TB 10 TOBAGO

The prototype for this new series of all-metal light aircraft was the original TB 10 (F-WZJP), of which design was initiated by Socata's Research and Development Department in February 1975. Construction began in February 1976, and it made a 25 min first flight at Tarbes on 23 February 1977, powered by a 119 kW (160 hp) Avco Lycoming O-320-D2A engine. The second prototype of the TB 10 was fitted with a 134 kW (180 hp) Lycoming engine. A further airframe underwent static tests at the CEAT, Toulouse.

Production versions are as follows:

TB 9 Tampico. Four-seater, with 119 kW (160 hp) Avco Lycoming O-320-D2A engine, Sensenich fixed-pitch propeller, fuel capacity of 155 litres (34 Imp gallons), and non-retractable landing gear. Hartzell constant-speed propeller optional.

TB 10 Tobago. Four/five-seater, with 134 kW (180 hp) engine and non-retractable landing gear.

The more powerful TB 20 Trinidad, with retractable landing gear, is described separately.

Following certification of the TB 10 by the DGAC, on 26 April 1979, the TB 9 and TB 10 entered production. By February 1981, sales totalled 58 TB 9s and 167 TB 10s, with production at the rate of 15 aircraft per month.

Manufacture of these aircraft is simplified by comparison with the Rallye, by the elimination of certain features such as wing leading-edge slats. The following description applies to the Tobago:

TYPE: Four/five-seat all-metal light aircraft.

WINGS: Cantilever low-wing monoplane. Wing section RA 163 C3. Thickness/chord ratio 16%. Dihedral 4° 30′ from roots. No incidence at root. No sweep. Conventional light alloy single-spar structure of constant chord, with glassfibre tips. Balanced ailerons and electrically-actuated slotted flaps, of light alloy. Ground-adjustable tabs.

FUSELAGE: Light alloy semi-monocoque structure. Shallow strake under each side of fuselage immediately aft of wing root fillet. Glassfibre engine cowlings.

TAIL UNIT: Cantilever all-metal type, with sweptback vertical surfaces and constant-chord all-moving horizontal surfaces mounted at extreme tail, aft of rudder. Ground-adjustable tab at top of rudder. Anti-tab in horizontal surfaces.

LANDING GEAR: Non-retractable tricycle type, with steerable nosewheel. Oleo-pneumatic shock-absorber in all three units. Glassfibre wheel fairings on all three units. Hydraulic disc brakes. Parking brake.

Socata Tobago four/five-seat light aircraft (Avco Lycoming O-360-A1AD engine) *(Geoffrey P. Jones)*

POWER PLANT: One 134 kW (180 hp) Avco Lycoming O-360-A1AD flat-four engine, driving a Hartzell two-blade constant-speed propeller with spinner. Two integral fuel tanks in wing leading-edges; total capacity 208 litres (45·75 Imp gallons). Oil capacity 7·5 litres (1·6 Imp gallons).

ACCOMMODATION: Four/five seats in enclosed cabin. Sharply-inclined low-drag windscreen. Access via upward-hinged window/doors of glassfibre. Baggage compartment aft of cabin, with external door on port side. Cabin carpeted, soundproofed, heated and ventilated. Windscreen defrosting standard.

SYSTEMS: Electrical system includes 12V 60A alternator and 12V 32A battery, landing and navigation lights, four individual cabin lights and instrument panel lighting. Hydraulic system for wheel brakes only.

AVIONICS AND EQUIPMENT: Avionics to customer's specification. Standard equipment includes seat belts for all seats, armrests, glove box, map pockets, antiglare visors, stall warning indicator, tiedown fittings and towbar.

DIMENSIONS, EXTERNAL (Tampico and Tobago):
- Wing span 9·76 m (32 ft 0¼ in)
- Wing chord (constant) 1·22 m (4 ft 0 in)
- Wing aspect ratio 8
- Length overall 7·64 m (25 ft 0¾ in)
- Height overall 3·20 m (10 ft 6 in)
- Tailplane span 3·20 m (10 ft 6 in)
- Wheelbase 1·95 m (6 ft 5 in)
- Propeller diameter 1·88 m (6 ft 2 in)
- Propeller ground clearance 0·10 m (4 in)
- Cabin doors (each): Width 0·89 m (2 ft 11 in)
  - Height 0·76 m (2 ft 6 in)
- Baggage door: Width 0·63 m (2 ft 1 in)
  - Height 0·43 m (1 ft 5 in)

DIMENSIONS, INTERNAL (Tampico and Tobago):
- Cabin: Length, firewall to rear bulkhead 2·54 m (8 ft 4 in)
  - Length, panel to rear bulkhead 2·00 m (6 ft 6¾ in)
  - Max width, at rear seats 1·27 m (4 ft 2 in)
  - Max width, at front seats 1·15 m (3 ft 9¼ in)
  - Max height, floor to roof 1·12 m (3 ft 8 in)

AREAS:
- Wings, gross 11·9 m² (128·1 sq ft)
- Ailerons (total) 0·91 m² (9·80 sq ft)
- Trailing-edge flaps (total) 3·72 m² (40·04 sq ft)
- Horizontal tail surfaces (total) 2·56 m² (27·56 sq ft)

WEIGHTS (A, Tampico; B, Tobago):
- Weight empty: A 650 kg (1,433 lb)
  - B 670 kg (1,477 lb)

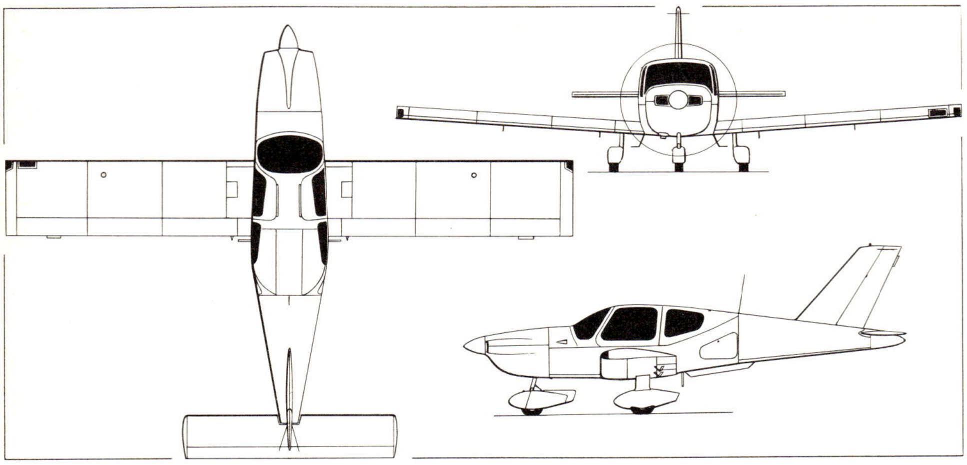

Socata Tobago four/five-seat all-metal light aircraft *(Pilot Press)*

Max T-O weight: A 1,060 kg (2,340 lb)
B 1,150 kg (2,530 lb)

PERFORMANCE (at max T-O weight):
Max level speed: A 125 knots (231 km/h; 144 mph)
B 133 knots (246 km/h; 153 mph)
Max cruising speed (75% power):
A 121 knots (224 km/h; 139 mph)
B 127 knots (235 km/h; 146 mph)
Econ cruising speed (65% power):
A 115 knots (213 km/h; 132 mph)
B 119 knots (220 km/h; 137 mph)
Stalling speed, flaps up:
A 59 knots (110 km/h; 68 mph)
B 60 knots (111 km/h; 69 mph)
Stalling speed, flaps down:
A 50 knots (92 km/h; 57·5 mph)
B 52 knots (97 km/h; 60 mph)
Max rate of climb at S/L: A 214 m (700 ft)/min
B 240 m (790 ft)/min
Service ceiling: A 4,270 m (14,000 ft)
B 3,960 m (13,000 ft)
T-O run: A 380 m (1,250 ft)
B 360 m (1,180 ft)
T-O to 15 m (50 ft): A 570 m (1,870 ft)
B 540 m (1,770 ft)
Landing from 15 m (50 ft): A 380 m (1,250 ft)
B 420 m (1,380 ft)
Landing run: A 183 m (600 ft)
B 190 m (620 ft)
Range with max fuel, allowances for T-O, climb and descent, 45 min reserves:
A 460 nm (852 km; 529 miles)
B 580 nm (1,074 km; 667 miles)

Socata TB 20 Trinidad with retractable landing gear

SOCATA TB 20 TRINIDAD

The TB 20 Trinidad is a four/five-seat light aircraft, basically similar to the TB 10 Tobago (which see) but with a more powerful engine and retractable landing gear. The prototype (F-WDBA) flew for the first time, at Tarbes, on 14 November 1980. Certification was expected by September 1981, enabling deliveries to begin in 1982.

The description of the Tobago applies also to the Trinidad, except as follows:

WINGS: Dihedral 6° 30′ from roots. Flap preselector standard.

TAIL UNIT: Span and chord of horizontal tail surfaces increased. Mechanical rudder trim standard. Optional electrically-actuated elevator trim tab.

LANDING GEAR: Hydraulically-retractable tricycle type, with single wheel on each unit. Free-fall emergency extension. Steerable nosewheel retracts rearward. Main units retract inward into fuselage. Hydraulic disc brakes. Parking brake.

POWER PLANT: One 186 kW (250 hp) Avco Lycoming IO-540-C4D5D flat-six engine, driving a Hartzell HC-C2YK-1BF/F8477-4 two-blade metal propeller. (175 kW; 235 hp engine optional.) Fuel tanks in wings; total usable capacity 320 litres (70 Imp gallons). Oil capacity 12 litres (2·5 Imp gallons).

SYSTEMS: Self-contained electro-hydraulic system for landing gear actuation.

DIMENSIONS:
Generally as for Tobago, except:
Tailplane span 3·64 m (11 ft 11¼ in)

AREAS:
Generally as for Tobago, except:
Horizontal tail surfaces (total) 3·06 m² (32·94 sq ft)

WEIGHTS:
Weight empty 762 kg (1,680 lb)
Max baggage 45 kg (100 lb)
Max T-O weight 1,335 kg (2,943 lb)
Max ramp weight 1,341 kg (2,956 lb)

PERFORMANCE (at max T-O weight):
Max level speed 168 knots (312 km/h; 193 mph)
Max cruising speed (75% power) at 2,600 m (8,500 ft)
164 knots (303 km/h; 188 mph)
Econ cruising speed (65% power) at 3,660 m (12,000 ft)
160 knots (296 km/h; 184 mph)
Stalling speed, flaps up 66 knots (121 km/h; 76 mph)
Stalling speed, flaps down
54 knots (100 km/h; 63 mph)
Max rate of climb at S/L 384 m (1,260 ft)/min
Service ceiling 6,100 m (20,000 ft)
T-O run 340 m (1,115 ft)
T-O to 15 m (50 ft) 540 m (1,772 ft)
Landing from 15 m (50 ft) 450 m (1,476 ft)
Landing run 230 m (755 ft)
Range with max fuel, allowances for T-O, climb and descent, 45 min reserves:
at 75% power
908-934 nm (1,682-1,730 km; 1,045-1,075 miles)
at 65% power
969-1,012 nm (1,795-1,875 km; 1,115-1,165 miles)

GERMANY
(FEDERAL REPUBLIC)

AIRCONCEPT
AIRCONCEPT FLUGZEUG UND GERÄTEBAU GmbH und Co KG

Flughafen, an der Niersbrucke, 4050 Mönchengladbach 1
Telephone: 02161 661489
Telex: 08 52 168

DIRECTOR: Alois Neikes

The 1977-78 *Jane's* contained an illustrated entry on an ultra-lightweight single-seat sporting aircraft known as the VoWi 10, designed and built by the late Ing Helmut Wilden. This aircraft (D-EGWI) consisted basically of a wire-braced all-moving wing and V tail unit, linked by a light alloy tube 'backbone', and with an exposed seat for the pilot carried on a tubular structure beneath the wing. It flew for the first time on 16 April 1975, powered by two 6 kW (8 hp) Stihl-Baumsägen single-cylinder two-stroke aircooled engines.

Subsequently, the VoWi 10 was almost completely redesigned to make it suitable for series production, and Airconcept was formed to manufacture and market the aircraft in its developed form.

No recent information has been received from Airconcept. Details of the VoWi 10 can be found in the 1980-81 *Jane's*.

DEUTSCHE AIRBUS
DEUTSCHE AIRBUS GmbH

8000 München 81, Arabellastrasse 30, Postfach 810260
Telephone: (089) 92 63 1
Telex: 5215149

CHAIRMAN OF THE SUPERVISORY BOARD: Dr Franz-Josef Strauss

MANAGEMENT:
Dipl Kfm Rolf Siebert
Dipl-Ing Klaus Hamann

PUBLIC RELATIONS: Jochen H. Eichen

This company is the West German partner in the consortium formed for development and production of the A300 and A310 high-capacity transport aircraft described under the Airbus heading in the International section. It is a wholly-owned subsidiary of MBB (which see).

DORNIER
DORNIER GmbH

HEAD OFFICE: Postfach 1420, 7990 Friedrichshafen/Bodensee
Telephone: Immenstaad (07545) 81+
Telex: 0734372

WORKS:
Research and Development: 7759 Immenstaad/Bodensee (near Friedrichshafen)
Production: Postfach 2160, Trimburgstrasse, 8000 München 66

AIRFIELD AND FLIGHT TEST CENTRE: 8031 Oberpfaffenhofen, near München

BONN OFFICE: Allianzplatz, 5300 Bonn

BOARD OF DIRECTORS:
Prof Dipl-Ing Claudius Dornier Jr (Chairman)
Dr Ing Bernhard Schmidt (Deputy Chairman)
Dipl-Ing Dr jur Karl-Wilhelm Schäfer (Deputy Chairman)
Rainer Hainich
Werner Kresin
Dr Ing Fritz Mader

PUBLIC RELATIONS: Gerhard Patt
Postfach 2160, 8000 München 66
Telephone: 089 87 15 480
Telex: 52 35 43

Dornier GmbH, formerly Dornier-Metallbauten, was formed in 1922 by the late Professor Claude Dornier as the successor to the 'Do' division of the former Zeppelin Werke, Lindau, GmbH. It has been operated in the form of a Gesellschaft mit beschränkter Haftung since 22 December 1972. Of more than 8,500 employees in the Dornier group, approximately 43% are production staff,

32% in research and development, and 25% engaged in technical and logistic support.

Member companies, in addition to Dornier GmbH, include Dornier-Reparaturwerft GmbH, at Oberpfaffenhofen (aircraft servicing and maintenance), Dornier System GmbH of Friedrichshafen (spaceflight, new technologies, electronics, management consultancy and contract research) and Lindauer Dornier GmbH of Lindau, which produces machinery for the textile industry and for the manufacture of plastics foils.

The Dornier 128-2 Skyservant twin-engined STOL transport and utility aircraft continues in production and is available also in turboprop form as the Dornier 128-6 Turbo-Skyservant. Dornier is supplementing these types with a larger twin-turboprop transport designated Dornier 228, of which two versions began development in prototype form and entered production during 1981. These embody advanced technology features, including the TNT wing, which is continuing its test programme on a specially converted Skyservant airframe.

Dornier is also active in development of various types of RPV and tethered rotor platform (see RPVs and Targets section).

The company continues in partnership with the Dassault-Breguet group in France to develop and produce the Alpha Jet training/light attack aircraft, described in the International section. Under Federal German government contract, Dornier has designed and constructed a transonic supercritical wing, to explore the aerodynamic and mechanical properties of such a wing for use by subsonic combat aircraft. Following initial wind tunnel tests by Dornier, and in the large wind tunnel by ONERA, in France, flight testing of the wing on an Alpha Jet began on 12 December 1980.

Dornier is also involved in licence production of components for the McDonnell Douglas Phantom fighter; and is collaborating with Pilatus of Switzerland (which see) in marketing the latter company's Turbo-Trainer. In another important programme, Dornier is responsible for integrating the operational avionics in the 18 Boeing E-3A Sentry AWACS aircraft being acquired between 1982 and 1985 by NATO for use in Europe. The first of these arrived at Oberpfaffenhofen on 19 March 1981.

Dornier is responsible also for the technical and logistic servicing of the Breguet Br 1150 Atlantic aircraft operated by the Dutch and Federal German Navies. Under a contract awarded by the German Federal Procurement Agency, Dornier is prime contractor for the modernisation of the Federal German Navy's Atlantics. This programme, known as Atlantic KWS (Kampfwertsteigerung: enhanced combat capability) covers the sonobuoy launch, ESM, navigation, radar and sonar systems, and magnetic tape data recorder.

Dornier was one of the companies within the West German aerospace industry which was carrying out independent studies that it was hoped would lead to definition and, ultimately, design of the TKF 90 tactical fighter project. This project has now been terminated on the grounds of unacceptable launch and development costs, and the Luftwaffe is evaluating existing aircraft as potential replacements for its F-4F Phantoms.

DORNIER Do 28 D-2 SKYSERVANT

Under contract to the Federal Ministry of Defence, Dornier re-engined one of the Luftwaffe's Do 28 D-2s with Avco Lycoming TIGO-540 turbocharged engines. This aircraft, redesignated **Do 28 D-2T**, flew for the first time in its new form in March 1980. All Do 28 D-2s in service with the Federal German armed forces may be upgraded to this standard.

Details of the standard Do 28 D-2 can be found in the 1979-80 *Jane's*. It has been superseded in production by the Dornier 128-2.

DORNIER 128-2 SKYSERVANT

This 10-seat utility aircraft became available in 1980, replacing the Do 28 D-2 (1979-80 *Jane's*), on which it is based. A version fitted with undernose radar, for maritime patrol duties, was first exhibited at the 1980 Hanover Air Show. A version with turbocharged engines, designated **Dornier 128-7**, is also available.

The following details apply specifically to the standard 128-2:

TYPE: Twin-engined STOL transport and utility aircraft.

WINGS: Cantilever high-wing monoplane. Wing section NACA 23018 (modified), with nose slot in the outer half of each wing. Dihedral 1° 30′. Incidence 4°. All-metal box-spar structure. Double-slotted ailerons and flaps have metal structure, partly Eonnex-covered. Balance tabs in ailerons. Pneumatic de-icing optional.

FUSELAGE: Conventional all-metal stressed-skin structure.

TAIL UNIT: Cantilever all-metal structure, with rudder and horizontal surfaces partly Eonnex-covered. All-moving horizontal surface, with combined anti-balance and trim tab. Trim tab in rudder. Pneumatic de-icing optional.

LANDING GEAR: Non-retractable tailwheel type. Dornier oleo-pneumatic shock-absorbers on main units, glassfibre sprung tailwheel unit. Main-wheel tyres size 8·50-10, pressure 3·38 bars (49 lb/sq in). Twin-contact tailwheel tyre size 5·50-4, pressure 2·76 bars (40 lb/sq in). Double-disc hydraulic brakes. Fairings on main legs and wheels standard.

Dornier 128-2 Skyservant equipped with auxiliary fuel tanks on wing pylons

POWER PLANT: Two 283 kW (380 hp) Avco Lycoming IGSO-540-A1E flat-six engines, mounted on stub-wings and each driving a Hartzell Type HC-B3W30-2B/W10151 B-8R three-blade constant-speed and fully-feathering propeller. Fuel tanks in engine nacelles, with total usable capacity of 893 litres (196·5 Imp gallons). Refuelling points above nacelles. Provision for two underwing auxiliary fuel tanks with combined capacity of 474 litres (104 Imp gallons). Total capacity of separate oil tanks, 33 litres (7·25 Imp gallons).

ACCOMMODATION: Pilot and either co-pilot or passenger side by side on flight deck. Dual controls standard. Main cabin equipped normally to carry eight passengers in pairs, with centre aisle, or five stretchers and five folding seats, all layouts including toilet and/or baggage compartment and/or darkroom for aerial survey missions aft of cabin. Alternatively, cabin can be stripped for cargo-carrying. Door on each side of flight deck. Emergency exit on starboard side of cabin. Combined two-section passenger and freight door on port side of cabin, at rear.

AVIONICS AND EQUIPMENT: Provision for com/nav antennae installation, and avionics, to customer's requirements. Standard equipment includes directional slaved gyro, cabin heating, dual brake system, 100A alternators, and provisions for optional de-icing system.

DIMENSIONS, EXTERNAL:

Wing span	15·55 m (51 ft 0¼ in)
Wing chord (constant)	1·90 m (6 ft 2¾ in)
Wing aspect ratio	8·33
Length overall (in flying attitude)	11·41 m (37 ft 5¼ in)
Height overall (static)	3·90 m (12 ft 9½ in)
Tailplane span	6·61 m (21 ft 8¼ in)
Wheel track	3·52 m (11 ft 6 in)
Wheelbase	8·63 m (28 ft 3¾ in)
Propeller diameter	2·36 m (7 ft 9 in)
Passenger door (port, rear):	
Height	1·34 m (4 ft 4¾ in)
Width	0·64 m (2 ft 1¼ in)
Height to sill	0·60 m (1 ft 11½ in)
Freight door (port, rear):	
Height	1·34 m (4 ft 4¾ in)
Width, incl passenger door	1·28 m (4 ft 2½ in)

DIMENSIONS, INTERNAL:

Cabin, excl flight deck and rear baggage compartment:	
Max length	3·97 m (13 ft 0½ in)
Max width	1·37 m (4 ft 6 in)
Max height	1·61 m (5 ft 3½ in)
Floor area	5·30 m² (57·05 sq ft)
Volume	8·00 m³ (282·5 cu ft)

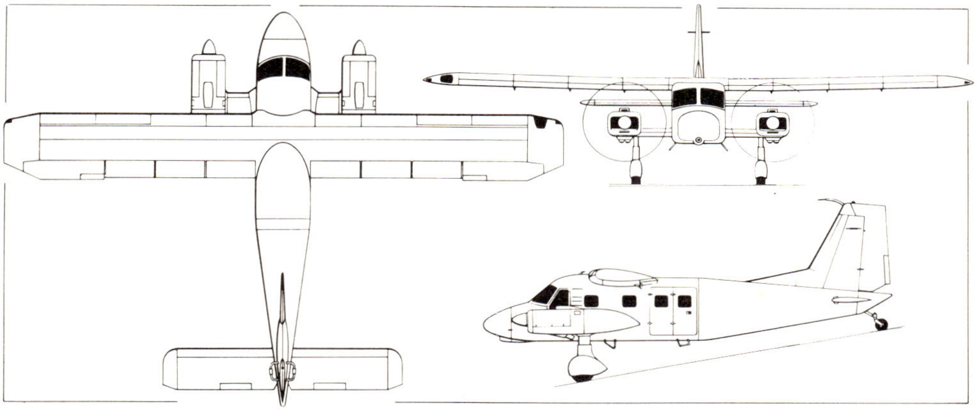

Dornier 128-2 Skyservant STOL utility light transport aircraft *(Pilot Press)*

Dornier 128-2 with undernose radome for maritime surveillance duties *(Brian M. Service)*

AREAS:
Wings, gross 29·00 m² (312·2 sq ft)
Ailerons (total) 2·64 m² (28·4 sq ft)
Trailing-edge flaps (total) 4·80 m² (51·6 sq ft)
Fin, incl dorsal fin 3·65 m² (39·3 sq ft)
Rudder, incl tab 1·40 m² (15·1 sq ft)
Tailplane, incl tab 7·65 m² (82·3 sq ft)

WEIGHTS AND LOADINGS:
Weight empty, standard 2,328 kg (5,132 lb)
Max payload 805 kg (1,775 lb)
Max T-O weight 3,842 kg (8,470 lb)
Max T-O weight with optional external fuel tanks 4,015 kg (8,852 lb)
Max ramp weight 4,035 kg (8,895 lb)
Max landing weight 3,650 kg (8,050 lb)
Max wing loading 132 kg/m² (27·1 lb/sq ft)
Max power loading 6·79 kg/kW (11·14 lb/hp)

PERFORMANCE (at max T-O weight):
Max level speed at 3,050 m (10,000 ft) 175 knots (325 km/h; 202 mph)
Max cruising speed at 3,050 m (10,000 ft) 164 knots (304 km/h; 189 mph)
Max cruising speed at S/L 147 knots (273 km/h; 170 mph)
Econ cruising speed, 50% power at 3,050 m (10,000 ft) 114 knots (211 km/h; 131 mph)
Stalling speed, power off, flaps down 56·5 knots (104 km/h; 65 mph)
Min control speed, power on, flaps down 35 knots (65 km/h; 40 mph)
Max rate of climb at S/L 312 m (1,025 ft)/min
Rate of climb at S/L, one engine out 24 m (80 ft)/min
Service ceiling 7,680 m (25,200 ft)
Service ceiling, one engine out 2,620 m (8,595 ft)
T-O run 280 m (920 ft)
T-O to 15 m (50 ft) 560 m (1,835 ft)
Landing from 15 m (50 ft) 390 m (1,280 ft)
Landing run 228 m (748 ft)
Range at max cruising speed 346 nm (642 km; 399 miles)
Range with max fuel 1,551 nm (2,875 km; 1,786 miles)

DORNIER 128-6 TURBO-SKYSERVANT

Dornier flew for the first time on 9 April 1978 the Do 28 D-5X prototype of a turboprop-powered version of the Skyservant (D-IBUF). In this prototype the piston engines of the Do 28 D-2 were replaced by two 447 kW (600 shp) Avco Lycoming LTP 101-600-1A turboprop engines, derated to 298 kW (400 shp) and each driving a three-blade constant-speed metal propeller. Two Pratt & Whitney Aircraft of Canada PT6A-110 free-turbine turboprops, each derated to 298 kW (400 shp), are standard on production Dornier 128-6 Turbo-Skyservants, each driving a Hartzell Type B3TN-3D T10282B-9·5 three-blade constant-speed metal propeller. The first flight of the prototype with this power plant installation (128-6X) was made on 4 March 1980.

Except for reinforced landing gear, to permit a higher landing weight, the airframe of the Dornier 128-6 is generally similar to that described for the 128-2, and can also be made available in maritime patrol form with a large undernose radome.

Dornier TNT research aircraft to evaluate the company's new technology wing

Certification by the LBA was granted in March 1981; the first production Dornier 128-6, an aircraft for Lesotho Airways, was delivered in July 1981.

WEIGHTS AND LOADINGS (PT6A-110 engines):
Weight empty 2,184 kg (4,815 lb)
Operating weight empty 2,540 kg (5,600 lb)
Max payload 1,273 kg (2,806 lb)
Max T-O weight 4,300 kg (9,480 lb)
Max landing weight 4,100 kg (9,039 lb)
Max wing loading 148 kg/m² (30·4 lb/sq ft)
Max power loading 7·21 kg/kW (11·85 lb/shp)

PERFORMANCE (at max T-O weight except where indicated; PT6A-110 engines):
Max cruising speed at 3,050 m (10,000 ft) 183 knots (340 km/h; 211 mph)
Cruising speed, 75% power at 3,050 m (10,000 ft) 165 knots (305 km/h; 190 mph)
Cruising speed, max range power at 3,050 m (10,000 ft) 140 knots (259 km/h; 161 mph)
Cruising speed, 50% power at 3,050 m (10,000 ft) 138 knots (256 km/h; 159 mph)
Landing speed 67·5 knots (125 km/h; 78 mph)
Max rate of climb at S/L 384 m (1,260 ft)/min
Rate of climb at S/L, one engine out 54 m (177 ft)/min
Service ceiling 8,580 m (28,150 ft)
T-O to 15 m (50 ft) 554 m (1,820 ft)
Landing from 15 m (50 ft) 435 m (1,430 ft)
Range with max fuel at max cruising speed 607 nm (1,125 km; 699 miles)
Range with 805 kg (1,774 lb) payload, no reserves 788 nm (1,460 km; 907 miles)
Range with max fuel at max range power 894 nm (1,658 km; 1,030 miles)

DORNIER NEW TECHNOLOGY WING (TNT) TESTBED AIRCRAFT

The advanced technology wing (*Tragflügel Neuer Technologie*: TNT) has been developed by Dornier with support from the Federal Ministry of Research and Technology. The first phase of the TNT programme, including design, aerodynamic and structural studies, began in mid-1975 and was completed at the end of 1976. Phase II covered engineering, construction of the wing, and its installation on the testbed aircraft, a Do 28 D-2 Skyservant. This was modified by extending the centre-fuselage 1·36 m (4 ft 5½ in) and installing a longer nose; fitting standard Alpha Jet landing gear in a non-retractable form; and fitting two Garrett TPE331-5-252D turboprop engines, one on each wing. The first flight of this research aircraft (D-IFNT) was made on 14 June 1979, beginning a full year of testing under Phase III of the programme.

By June 1981 the aircraft had made more than 200 test flights, confirming the performance expectations of the new wing. Testing took place in 1980 of seven of the most modern types of Hartzell and Dowty propeller, featuring supercritical and extra-wide-chord blades, and three- or four-blade types with Kevlar blades. Additional propellers, and other advanced technologies, are the subject of a continuing test programme. These include the investigation of alternative spoiler systems for roll control, thus making the entire wing trailing-edge available for full-span Fowler flaps to enhance low-speed performance and permit a major reduction of wing area.

The first production application of the TNT wing is in the Dornier 228 commuter transport (which see). The choice of an optimum propeller design for the 228 was also a result of the TNT test programme.

The description of the Dornier 128-2 applies also to the TNT testbed aircraft, except as follows:

WINGS: Cantilever high-wing monoplane. Special Dornier wing section. Thickness/chord ratio reduces from 16% over the full centre-section to 13% at the tip. Dihedral 0° on centre-section; 4° on outer panels. Incidence at root 2° 12′. Twist on outer panels −3°. Composite structure, with spars, panels and ribs of torsion box integrally milled from light alloy; leading-edge boxes of conventional sheet metal construction; trailing-edge boxes and sharply-swept wingtips of glassfibre sandwich. Electrically-operated single-slotted Fowler type trailing-edge flaps. Adjustable slotted ailerons, which can be drooped up to 10°, have Flettner tabs to offload control forces, and trim tabs.

FUSELAGE AND TAIL UNIT: Generally similar to 128-2, but fuselage extended in length.

LANDING GEAR: Non-retractable tricycle type, consisting of standard production Alpha Jet landing gear units.

POWER PLANT: Two 533 kW (715 shp) Garrett TPE 331-5-252D turboprop engines, mounted on the wings and each driving a Hartzell type HC-B4TN-5CL/L 10282+4 four-blade constant-speed metal propeller. Integral fuel tanks in wing centre-section, outside fuselage area, with combined capacity of 2,000 kg (4,409 lb).

SYSTEMS: Generally as for 128-2, but the electrical system has been modified to meet the requirements of the turboprop engines and their new location.

AVIONICS: Have been selected to satisfy the test assignment and meet IFR requirements.

DIMENSIONS, EXTERNAL:
Wing span 16·97 m (55 ft 8 in)
Wing chord (mean aerodynamic) 2·04 m (6 ft 8¼ in)
Wing aspect ratio 9
Length overall (excl nose probe) 13·35 m (43 ft 9½ in)
Height overall 4·96 m (16 ft 3¼ in)
Wheel track 3·02 m (9 ft 10¾ in)
Wheelbase 4·19 m (13 ft 9 in)
Propeller diameter 2·73 m (8 ft 11½ in)

AREA:
Wings, gross 32·00 m² (344·5 sq ft)

WEIGHTS AND LOADINGS:
Weight empty 3,010 kg (6,636 lb)
Max T-O weight 4,500 kg (9,921 lb)
Max wing loading 140·6 kg/m² (28·80 lb/sq ft)
Max power loading 4·22 kg/kW (6·94 lb/shp)

PERFORMANCE (estimated, at max T-O weight except where indicated):
Never-exceed speed 231 knots (428 km/h; 266 mph)
Cruising speed 185 knots (342 km/h; 213 mph)
Max rate of climb at S/L, AUW of 4,000 kg (8,818 lb) 942 m (3,091 ft)/min
Rate of climb, one engine out, conditions as above 348 m (1,142 ft)/min
T-O to 15 m (50 ft) at AUW of 4,000 kg (8,818 lb) 250 m (820 ft)

DORNIER 228

Under this general designation, Dornier is marketing two new types of increased-capacity utility and commuter aircraft, as follows:

Dornier 228-100. Basic version, with accommodation for 15 passengers in standard airline-type seats at 76 cm (30 in) pitch. Suitable for a wide range of other duties, including freight or mixed cargo/passenger transport,

Dornier 128-6 Turbo-Skyservant (two Avco Lycoming LTP 101-600-1A turboprop engines)

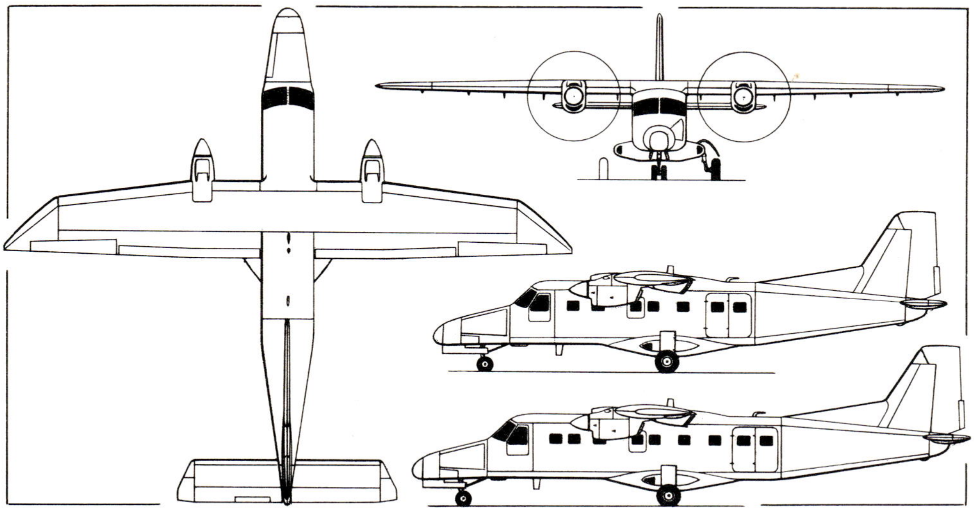

Dornier 228-100 light transport, with additional side view (bottom) of 228-200 *(Pilot Press)*

executive travel, air taxi service, maritime surveillance (with undernose search radar), airways calibration, training, ambulance or search and rescue operations, and paramilitary missions. Main fuselage segments, cabin door and cockpit equipment standardised with those of Dornier 128. Fitted with the Dornier new technology (TNT) wing, and Garrett turboprop power plant.

Dornier 228-200. Lengthened fuselage, providing accommodation for 19 passengers at 76 cm (30 in) seat pitch and a larger rear baggage compartment, but otherwise generally similar to 228-100.

The design of these aircraft has been formulated to comply with US certification regulations. One prototype of each version has been built; the first of these, the Dornier 228-100 (D-IFNS) made its first flight on 28 March 1981, and the 228-200 (D-ICDO) on 9 May 1981. A static test airframe of the 228-200 has also been completed.

By early June 1981 firm orders had been received for 21 Dornier 228s from Australia (five -200s), Kenya (one), Nigeria (nine), Norway (two -100s), Singapore (two -200s) and South Africa (two). Orders and options totalled 78 by 1 July 1981. The first delivery, of a 228-100 for Norving Flyservice of Norway, was due to be made in December 1981.

Type: Twin-turboprop light transport.

Wings: Cantilever high-wing monoplane, comprising two-spar rectangular centre-section and two tapered outer panels ending in raked tips. Dornier Do A-5 supercritical wing section. Wing leading-edge and raked wingtips of glassfibre/Kevlar composites. Fowler-type single-slotted trailing-edge flaps and ailerons of carbonfibre composites. Ailerons can be drooped symmetrically to augment trailing-edge flaps, and are operated differentially to serve as conventional ailerons. Remainder of wing of light alloy construction.

Fuselage: Conventional stressed-skin unpressurised structure of light alloy.

Tail Unit: Cantilever all-metal structure, with rudder and horizontal surfaces partly Eonnex-covered. All-moving tailplane, with combined anti-balance and trim tab. Trim tab in rudder and port elevator.

Landing Gear: Retractable tricycle type, with single wheel on each unit. Main units retract forward and inward into fairings built on to the lower fuselage. Nosewheel retracts forward. Low-pressure tyres optional.

Power Plant: Two 533 kW (715 shp) Garrett TPE331-5 turboprop engines, each driving a Hartzell four-blade constant-speed fully-feathering reversible-pitch metal propeller. Alternative engines which may become available at a later date, to meet differing customer requirements, include the Garrett TPE331-10, and Pratt & Whitney Aircraft of Canada PT6A-41 and PT6A-135. Primary wing box forms an integral fuel tank.

Accommodation: Crew of two, and 15 or 19 passengers as described under model listings. Individual seats down each side of the cabin with a central aisle. Combined two-section passenger and freight door, with integral steps, on port side of cabin at rear. Baggage compartment at rear of cabin accessible from cabin. Additional baggage space in fuselage nose. Modular units for rapid changes of role.

Dimensions, external:
- Wing span: 16·97 m (55 ft 8 in)
- Length overall: 228-100: 15·03 m (49 ft 3¾ in)
  - 228-200: 16·55 m (54 ft 3½ in)
- Height overall: 4·86 m (15 ft 11½ in)
- Wheel track: 3·30 m (10 ft 10 in)
- Wheelbase: 228-100: 5·50 m (18 ft 0½ in)
  - 228-200: 6·30 m (20 ft 8 in)
- Propeller diameter: 2·73 m (8 ft 11½ in)
- Passenger door (port, rear):
  - Height: 1·34 m (4 ft 4¾ in)
  - Width: 0·64 m (2 ft 1¼ in)
  - Height to sill: 0·60 m (1 ft 11½ in)
- Freight door (port, rear):
  - Height: 1·34 m (4 ft 4¾ in)
  - Width, incl passenger door: 1·28 m (4 ft 2½ in)

Dimensions, internal:
- Cabin, excl flight deck and rear baggage compartment:
  - Length: 228-100: 6·33 m (20 ft 9¼ in)
  - 228-200: 7·18 m (23 ft 6¾ in)
  - Max width: 1·346 m (4 ft 5 in)
  - Max height: 1·55 m (5 ft 1 in)
  - Floor area: 228-100: 8·50 m^2 (91·49 sq ft)
  - 228-200: 10·55 m^2 (113·56 sq ft)
  - Volume: 228-100: 13·00 m^3 (459·1 cu ft)
  - 228-200: 14·70 m^3 (519·1 cu ft)
- Rear baggage compartment volume:
  - 228-100: 0·90 m^3 (31·8 cu ft)
  - 228-200: 2·28 m^3 (80·5 cu ft)
- Nose baggage compartment volume: 0·89 m^3 (31·4 cu ft)

Area:
- Wings, gross: 32·00 m^2 (344·4 sq ft)

Weights (estimated):
- Weight empty: 228-100: 2,798 kg (6,168 lb)
  - 228-200: 2,908 kg (6,411 lb)
- Max passenger payload: 228-100: 2,100 kg (4,630 lb)
  - 228-200: 1,960 kg (4,321 lb)
- Max payload (freighter):
  - 228-100: 2,207 kg (4,865 lb)
  - 228-200: 2,057 kg (4,535 lb)
- Max zero-fuel weight (both): 5,400 kg (11,905 lb)
- Max T-O weight (both): 5,700 kg (12,566 lb)

Performance (at max T-O weight, S/L, ISA, except where indicated):
- Max cruising speed at 3,050 m (10,000 ft): 233 knots (432 km/h; 268 mph)
- Econ cruising speed at 3,050 m (10,000 ft): 179 knots (332 km/h; 206 mph)
- Max rate of climb at S/L: 624 m (2,050 ft)/min
- Rate of climb at S/L, one engine out: 162 m (531 ft)/min
- Service ceiling, 30·5 m (100 ft)/min rate of climb: 9,020 m (29,600 ft)
- Service ceiling, one engine out, 30·5 m (100 ft)/min rate of climb: 4,265 m (14,000 ft)
- T-O run: 415 m (1,362 ft)
- T-O to 15 m (50 ft): 525 m (1,725 ft)
- Range at max cruising speed:
  - 228-100: 933 nm (1,730 km; 1,075 miles)
  - 228-200: 555 nm (1,030 km; 640 miles)
- Range at 3,050 m (10,000 ft) with max passenger payload, at econ cruising speed, no reserves:
  - 228-100: 1,063 nm (1,970 km; 1,224 miles)
  - 228-200: 621 nm (1,150 km; 715 miles)

Dornier's two new twin-turboprop commuter transports: the 228-100 (above) and 228-200 (below)

EQUATOR
EQUATOR AIRCRAFT GESELLSCHAFT FÜR FLUGZEUGBAU mbH ULM

Head Office: Benzstrasse 15, D-7904 Erbach
Telephone: (073) 04 6116 and 05 6116
President: Günther Pöschel

This company was known until 1974 as Pöschel Aircraft GmbH. Its President designed and built the prototype of a five/six-seat light STOL aircraft named the P-300 Equator. This flew for the first time on 8 November 1970, powered by a 216 kW (290 hp) Avco Lycoming IO-540 flat-six engine, and was last described in the 1972-73 *Jane's*. Simultaneously with flight testing the P-300, a turboprop-powered version of the same design was built as the P-400 Turbo-Equator STOL amphibian (see Addenda to 1977-78 *Jane's*). Following an incident caused by engine failure when the P-400 was taking off for its eighth test flight, the configuration has been changed

Prototype of the current Equator series

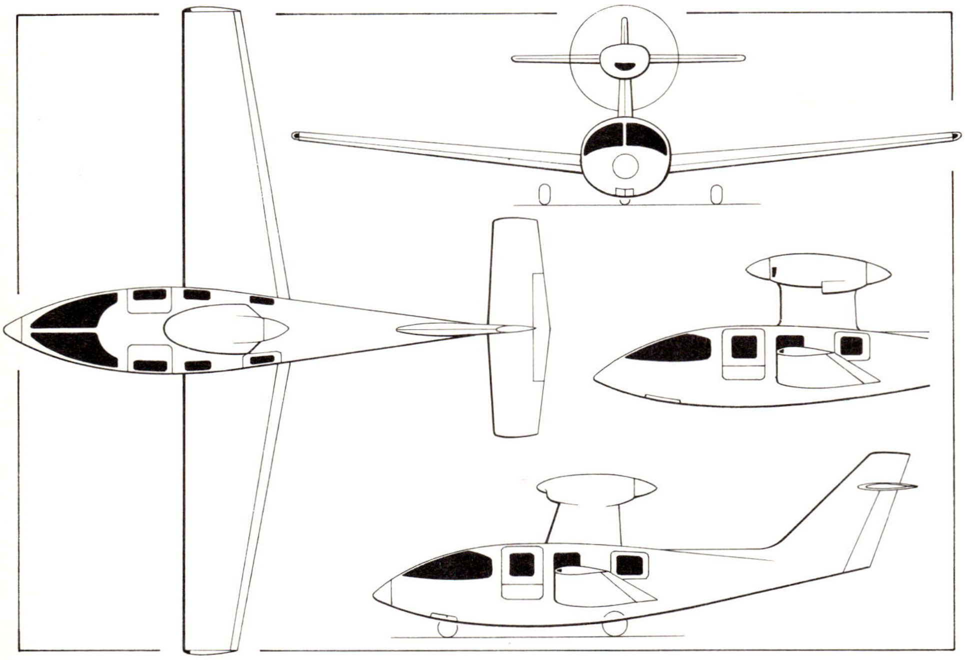

P-450 Equator eight/ten-seat light transport, with scrap view of the P-420 Twin Equator *(Michael A. Badrocke)*

extensively, and a common basic airframe is being developed for a family of different models, as follows:

EQUATOR, TWIN EQUATOR and TURBO EQUATOR

This series of STOL aircraft utilises a common airframe, related to that of the earlier P-300 and P-400 but with the wing lowered from high-wing to mid-wing position, swept vertical tail surfaces, and a pylon-mounted power plant, above the cabin instead of being carried on the tail unit. Differences between the seven projected models are as follows:

P-300 Equator F. Similar to P-350, except for non-retractable tricycle landing gear with streamline wheel and main-leg fairings. TIO-540 engine rated at 231 kW (310 hp).

P-350 Equator P. Basic retractable-gear version with a 261 kW (350 shp) Avco Lycoming TIO-540-J2BD flat-six engine, driving a three-blade constant-speed and reversible-pitch metal tractor propeller. Prototype had fuselage and wings manufactured by the Glasflügel and Glaser-Dirks sailplane companies respectively.

P-400 Equator SA. Generally similar to P-350, but with a 298 kW (400 hp) Avco Lycoming IO-720-B1BD flat-eight engine and 'pusher' propeller.

P-420 Turbo Equator. Generally similar to P-350, but with an Allison 250-B17C turboprop engine of 313 kW (420 shp), driving a pusher propeller.

P-420 Twin Equator SAP. Generally similar to P-350, but with 'push and pull' tandem pair of 157 kW (210 hp) Avco Lycoming TO-360-C1A6D turbocharged flat-four engines, driving three-blade constant-speed fully-feathering and reversible-pitch propellers.

P-450 Equator. Generally similar to P-350, but with a 336 kW (450 hp) Avco Lycoming TIGO-541-D1B flat-six engine.

P-550 Turbo Equator SP. Generally similar to P-350, but with a Pratt & Whitney Aircraft of Canada PT6A-27 turboprop engine, flat rated to 410 kW (550 shp) and driving a three-blade constant-speed and reversible-pitch metal 'pusher' propeller.

Although bearing the designation P-300, the prototype of the current series (D-EULM) has a pusher propeller and retractable landing gear.

All production models are intended normally to be completed as light transports for eight or ten persons, including one or two pilots, and with retractable tricycle landing gear. Non-retractable gear can be fitted to any version. Non-STOL versions are available, with simplified flaps; these have increased useful load, fuel capacity and range, but longer T-O and landing runs and higher stalling speeds than those quoted. The fuselage can be adapted for amphibious operation, with a retractable step, water rudder and bilge pump. Other major options include cabin pressurisation, with a loss-of-pressure warning system, emergency oxygen, a cabin altimeter and cabin rate of climb indicator.

The interior can be converted for a wide variety of alternative roles, including freight transport, search and rescue, aerial photography and survey. The following description applies to the basic P-350 STOL version:

TYPE: Eight/ten-seat cabin monoplane.

WINGS: Cantilever mid-wing monoplane. Wing section Wortmann FX-78-P-160/200. Dihedral 6°. Incidence 2°. Forward sweep at quarter-chord 2°. Laminar-flow single-spar wing of fail-safe construction, utilising a fibre epoxy honeycomb sandwich structure. Drooping ailerons and double-slotted trailing-edge flaps of similar construction. Electrical de-icing of wing leading-edges optional. 'Float-wing' design eliminates need for sponsons or wingtip floats when operating from water.

FUSELAGE: Conventional fail-safe semi-monocoque structure, using construction materials similar to those of wing.

TAIL UNIT: Cantilever structure, with all-moving tailplane mounted approximately two-thirds up the fin. Constructed of the same materials as wing and fuselage. Electrically-operated trim tab in tailplane and rudder. Electrical de-icing of tailplane and fin leading-edges optional.

LANDING GEAR: Hydraulically-retractable tricycle type, with single wheel on each unit. Main units retract outward. Oleo-pneumatic shock-absorbers. Main-wheel tyres size 6·50-8. Steerable nosewheel with tyre size 6·00-6. Goodyear brakes. Parking brake.

POWER PLANT: One or two engines as detailed in model listings. Fuel capacity 1,227·5 litres (270 Imp gallons), in integral tanks in wings.

ACCOMMODATION: Standard accommodation for a pilot and seven passengers, or pilot and co-pilot with six passengers. Rear baggage and bar compartment may be used optionally to accommodate 9th and 10th seats, or toilet. Door on each side at forward end of cabin. Dual controls standard. Shoulder harness for two front seats standard. Baggage space, with internal and external access, at rear of cabin. Accommodation heated and ventilated.

SYSTEMS: 24V electrical system. Vacuum system for blind-flying instrumentation. Air-conditioning, oxygen system, pressurisation system, and electrical de-icing system with 100A alternator, optional.

AVIONICS AND EQUIPMENT: Standard avionics include dual nav/com with VOR/LOC and VOR/ILS, marker beacon receiver, ADF, transponder, and audio panel. Standard equipment includes blind-flying instrumentation, clock, exhaust gas temperature gauge, outside air temperature gauge, map compartment, sun visors, heated pitot, stall warning device, cabin fresh air vents, dome lights, beacon lights, landing light, navigation lights, taxi light, static wicks and tiedown rings. Optional avionics include autopilot, DME, R/Nav, radar altimeter and weather radar. Optional equipment includes cabin altimeter and rate of climb indicator; tinted windows; amphibian conversion with retractable step, water rudder and bilge pump; fixed landing gear, oversize wheels, ski gear; and cargo doors.

DIMENSIONS, EXTERNAL:

Wing span	12·20 m (40 ft 0¼ in)
Length overall	10·00 m (32 ft 9¾ in)
Height overall	3·66 m (12 ft 0 in)

AREA:

Wings, gross	18·00 m² (193·75 sq ft)

WEIGHTS AND LOADING (estimated, A: P-300; B: P-350; C: P-400; D: P-420 Turbo; E: P-420 Twin; F: P-450; G: P-550):

Weight empty: A	1,070 kg (2,359 lb)
B	1,100 kg (2,425 lb)
C	1,123 kg (2,475 lb)
D	943 kg (2,080 lb)
E	1,242 kg (2,738 lb)
F	1,162 kg (2,561 lb)
G	989 kg (2,180 lb)
Max T-O weight: A	1,900 kg (4,190 lb)
* all other versions	2,000 kg (4,410 lb)
Max wing loading, all versions	111 kg/m² (22·7 lb/sq ft)

**Can be increased to 2,520 kg (5,555 lb), subject to conditions of CAR 8, for increased range*

PERFORMANCE (estimated, at 2,000 kg; 4,410 lb max T-O weight, ISA, except A at 1,900 kg; 4,190 lb max T-O weight):

Max cruising speed, retractable landing gear:	
A at 5,550 m (18,200 ft)	250 knots (463 km/h; 288 mph)
B at 6,705 m (22,000 ft)	264 knots (489 km/h; 304 mph)
C at 1,525 m (5,000 ft)	230 knots (426 km/h; 265 mph)
D at 1,525 m (5,000 ft)	235 knots (436 km/h; 271 mph)
E at 5,975 m (19,600 ft)	270 knots (500 km/h; 311 mph)
F at 5,485 m (18,000 ft)	280 knots (519 km/h; 322 mph)
G at 7,620 m (25,000 ft)	300 knots (556 km/h; 345 mph)
Max cruising speed, non-retractable landing gear:	
A	210 knots (389 km/h; 242 mph)
B	220 knots (408 km/h; 253 mph)
C	195 knots (361 km/h; 225 mph)
D	200 knots (371 km/h; 230 mph)
E	230 knots (426 km/h; 265 mph)
F	240 knots (445 km/h; 276 mph)
G	255 knots (473 km/h; 294 mph)
Max cruising speed, amphibious version:	
A	230 knots (426 km/h; 265 mph)
B	242 knots (448 km/h; 279 mph)
C	212 knots (393 km/h; 244 mph)
D	217 knots (402 km/h; 250 mph)
E	250 knots (463 km/h; 288 mph)
F	260 knots (482 km/h; 299 mph)
G	282 knots (523 km/h; 325 mph)
Stalling speed, standard wing:	
A	62 knots (115 km/h; 71·5 mph)
all other versions	64 knots (119 km/h; 74 mph)
Stalling speed, STOL wing:	
A	51 knots (95 km/h; 59 mph)
all other versions	52 knots (96·5 km/h; 60 mph)
Max rate of climb at S/L: A	450 m (1,475 ft)/min
B	515 m (1,690 ft)/min
C	587 m (1,925 ft)/min
D	600 m (1,970 ft)/min
E	619 m (2,030 ft)/min
F	671 m (2,200 ft)/min
G	797 m (2,615 ft)/min

Service ceiling: A	10,170 m (33,360 ft)
B	10,695 m (35,080 ft)
C	6,400 m (21,000 ft)
D	10,795 m (35,410 ft)
E	10,095 m (33,115 ft)
F	11,990 m (39,345 ft)
G	11,690 m (38,360 ft)
T-O to 15 m (50 ft), standard wing:	
A	658 m (2,160 ft)
B	634 m (2,080 ft)
C	524 m (1,720 ft)
D	533 m (1,750 ft)
E	579 m (1,898 ft)
F	451 m (1,480 ft)
G	355 m (1,165 ft)
T-O to 15 m (50 ft), STOL wing:	
A	453 m (1,485 ft)
B	433 m (1,420 ft)
C, E	390 m (1,280 ft)
D	386 m (1,265 ft)
F	335 m (1,100 ft)
G	265 m (870 ft)
Landing from 15 m (50 ft), standard wing:	
A	799 m (2,620 ft)
all other versions	823 m (2,700 ft)
Landing from 15 m (50 ft), STOL wing:	
A	544 m (1,785 ft)
all other versions	562 m (1,845 ft)
Range with max fuel at max cruising speed, retractable landing gear:	
A	2,345 nm (4,345 km; 2,700 miles)
B, D	1,737 nm (3,219 km; 2,000 miles)
C	1,954 nm (3,621 km; 2,250 miles)
E	1,823 nm (3,379 km; 2,100 miles)
F	1,650 nm (3,058 km; 1,900 miles)
G	2,127 nm (3,943 km; 2,450 miles)
Range with max fuel at econ cruising speed, retractable landing gear:	
A	4,759 nm (8,819 km; 5,480 miles)
B	4,550 nm (8,433 km; 5,240 miles)
C	3,882 nm (7,194 km; 4,470 miles)
D	4,029 nm (7,467 km; 4,640 miles)
E	3,673 nm (6,807 km; 4,230 miles)
F	3,925 nm (7,274 km; 4,520 miles)
G	3,890 nm (7,210 km; 4,480 miles)

FUS
FLUGZEUG-UNION-SÜD GmbH (subsidiary of MBB)

HEAD OFFICE: PO Box 83 17 20, 8000 München 83
Telephone: (089) 60 00 3017, 60 00 4965
Telex: 5 29 132 fus d
LIGHT AIRCRAFT DEPARTMENT: Otto-Hahn-Strasse 4, 8012 Ottobrunn (Riemerling)
GENERAL MANAGER: Josef Fuchshuber

MBB FLAMINGO-TRAINER

Flugzeug-Union-Süd, a subsidiary of Messerschmitt-Bölkow-Blohm, is marketing a trainer derived from the MBB 223 Flamingo, last described under the CASA heading in the Spanish section of the 1975-76 *Jane's*. It is available in three versions:

Flamingo-Trainer A1. Standard version, with 149 kW (200 hp) Avco Lycoming IO-360 engine and Hartzell HC-E2YR-1BF/F8467-7R two-blade constant-speed metal propeller.

Flamingo-Trainer K1. Aerobatic version, with 149 kW (200 hp) Avco Lycoming AIO-360 engine, and propeller as for A1.

Flamingo-Trainer T1. Turbocharged version, with 156 kW (210 hp) Avco Lycoming TO-360-C1A6D turbocharged engine and two-blade propeller as for A1 or, optionally, with a Hoffmann HO-Y123K/180 R+8 three-blade constant-speed composite propeller.

A prototype of the Flamingo-Trainer (D-EFWC) flew for the first time on 25 April 1979. By comparison with the original Flamingo, it benefits from the improvements in construction, equipment and materials which have been developed during the 16 years which separate these aircraft. It is being marketed as a multi-role trainer for both civil and military use. In early 1981 sales negotiations were being conducted with ten potential customers in nine countries.

TYPE: Two/three-seat light trainer.

WINGS: Cantilever low-wing monoplane. Wing section NACA 64_2-A215. Dihedral 3°. Incidence 3°. No sweepback. All-metal constant-chord two-spar structure. Main spars pass through fuselage side walls and are bolted together on aircraft centreline. Rear spars are attached to fuselage sides. Wingtip fairings of GRP are easily removable. Frise-type ailerons of metal construction. Trim tab in starboard aileron. Electrically-actuated trailing-edge plain flaps of similar construction to ailerons.

FUSELAGE: Conventional semi-monocoque structure of light alloy with riveted skins.

TAIL UNIT: Cantilever all-metal structure with fixed-incidence tailplane. Shallow dorsal fin from canopy to fin. Horn-balanced elevators and rudder. Trim tab in rudder, and in port elevator.

LANDING GEAR: Non-retractable tricycle type. Shock-absorption of all units by rubber in compression, assisted by hydraulic dampers. Steerable and self-centering nosewheel. All three wheels have tyres size 6·00-6, pressure 2·15 bars (31 lb/sq in). Hydraulic disc brakes. Parking brake.

FUS MBB Flamingo-Trainer T1, a two/three-seat multi-role trainer

POWER PLANT: One Avco Lycoming flat-four engine as detailed in model listings, driving a two-blade, or optional three-blade, propeller with spinner. Integral fuel tank in each wing, combined capacity 100 litres (22 Imp gallons) in Utility version, 170 litres (37·4 Imp gallons) in Normal category version. Refuelling point on upper surface of each wing.

ACCOMMODATION: Two seats side by side, with bench seat aft for additional trainee, beneath aft-sliding transparent canopy. Rear bench seat removable. Removal of starboard front seat permits the carriage of a stretcher or bulky cargo. Canopy is jettisonable, and incorporates a foul-weather window in the port side. Dual controls standard; starboard control column easily removed. Accommodation is air-conditioned.

SYSTEMS: Hydraulic system for brakes only. Electrical system includes a 28V engine-driven generator and two 24V 25Ah storage batteries.

AVIONICS AND EQUIPMENT: Avionics to customer's specific requirements. Standard equipment includes blind-flying instrumentation on port side, heated pitot, stall warning device, instrument lights; anti-collision, landing, navigation and taxi lights; and external power socket. Blind-flying instrumentation for starboard side is optional.

DIMENSIONS, EXTERNAL:

Wing span	8·28 m (27 ft 2 in)
Wing chord, constant	1·40 m (4 ft 7 in)
Wing aspect ratio	5·96
Length overall	7·60 m (24 ft 11¼ in)
Height overall	2·70 m (8 ft 10¼ in)
Wheel track	2·70 m (8 ft 10¼ in)
Wheelbase	1·60 m (5 ft 3 in)
Propeller diameter: two-blade	1·95 m (6 ft 4¾ in)
optional three-blade	1·88 m (6 ft 2 in)

AREAS:

Wings, gross	11·50 m² (123·8 sq ft)
Ailerons (total)	0·77 m² (8·24 sq ft)
Trailing-edge flaps (total)	0·94 m² (10·12 sq ft)
Horizontal tail surfaces (total)	2·45 m² (26·37 sq ft)
Vertical tail surfaces (total)	1·84 m² (19·81 sq ft)

WEIGHTS AND LOADINGS (T1, A: Normal; B: Utility category):

Weight empty: A, B	700 kg (1,543 lb)
Max fuel weight: A	122·4 kg (270 lb)
B	72 kg (159 lb)
Max payload: A	227·6 kg (502 lb)
B	208 kg (458 lb)
Max T-O weight: A	1,050 kg (2,315 lb)
B	980 kg (2,160 lb)
Max wing loading: A	91·3 kg/m² (18·70 lb/sq ft)
B	85·2 kg/m² (17·45 lb/sq ft)
Max power loading: A	6·69 kg/kW (11·02 lb/hp)
B	6·24 kg/kW (10·29 lb/hp)

PERFORMANCE (T1, at max Normal T-O weight):

Max level speed at 4,570 m (15,000 ft)	150 knots (278 km/h; 173 mph)
Max cruising speed, 75% power at 3,050 m (10,000 ft)	130 knots (241 km/h; 150 mph)
Stalling speed, flaps up	65 knots (120 km/h; 75 mph) IAS
Stalling speed, flaps down	56 knots (102 km/h; 64 mph) IAS
Max rate of climb at S/L	323 m (1,060 ft)/min
Certificated ceiling	6,690 m (22,000 ft)
T-O to 15 m (50 ft)	365 m (1,200 ft)
Landing from 15 m (50 ft)	375 m (1,230 ft)
Range with max fuel, 65% power at 4,570 m (15,000 ft)	482 nm (893 km; 555 miles)
g limits, aerobatic	+6; −4

GYROFLUG
GYROFLUG INGENIEURGESELLSCHAFT mbH

Gebelsbergstrasse 95, 7000 Stuttgart 1
Telephone: (07 11) 609853 and (07257) 3927
Telex: 0723268 AFI D
MANAGEMENT:
Dipl-Ing Wolfgang Schiller
Jörg Elzenbeck

Gyroflug was formed in July 1978 to manufacture and market, as a ready-to-fly production aircraft, a developed version of the two-seat VariEze lightweight sporting aircraft designed by Burt Rutan for amateur construction (see Rutan entry in US Homebuilt section). The company's executives had earlier built and flown, in April 1977, the first VariEze completed in Europe (D-EEEZ).

Gyroflug's development of the VariEze is known as the Speed Canard. It was hoped to begin delivery of production aircraft in late 1981.

GYROFLUG SPEED CANARD

As well as being slightly larger overall than the Rutan VariEze, on which it is based, the Speed Canard differs in a number of important details. Its 20° swept wing retains the NASA winglets that have proved so successful on the original design, but utilises a new aerofoil designed by Prof Eppler which is claimed to reduce drag by 30 per cent. Instead of retaining the glassfibre/foam core structure of the VariEze, which is well suited to amateur construction, the Speed Canard is made of glassfibre and carbonfibre reinforced plastics laid up in female moulds, giving a lighter structure optimised for series production. The forward fuselage embodies a flush canopy similar to that of the latest sailplanes, with the rear seat raised to ensure a good field of view for both occupants, and there is limited baggage space aft of the rear seat. The cabin is heated and ventilated. Tricycle landing gear has fixed main units of glassfibre with the size 5·00-5 wheels enclosed in speed fairings; the nosewheel is electrically retractable. The wings are detachable for transport and storage.

A more powerful engine is standard, in the form of an 86 kW (115 hp) Avco Lycoming O-235-M1 flat-four engine, driving a Hoffmann HO-V72G/LD170U-10 two-blade constant-speed pusher propeller with spinner. Integral fuel tank in wing centre-section with capacity of 160 litres (35·2 Imp gallons). Gyroflug is studying the possibility of fitting two Klöckner-Humboldt-Deutz KHD 117 lightweight turbojets in a later version of the Speed Canard, giving an estimated max level speed of 237 knots (440 km/h; 273 mph).

Construction of the moulds in which to build components of the piston-engined prototype had been completed by the end of 1979, and the first pre-production prototype (D-EEEX) made its first flight on 2 December 1980. A second prototype was due to fly during the Summer of 1981, with the first production examples scheduled for delivery in late 1981.

DIMENSIONS, EXTERNAL:

Wing span	7·60 m (24 ft 11¼ in)
Foreplane span	3·60 m (11 ft 9¾ in)
Foreplane chord (constant)	0·34 m (1 ft 1⅜ in)
Length overall	5·20 m (17 ft 0¾ in)
Length of fuselage	4·40 m (14 ft 5¼ in)
Max width of fuselage	0·74 m (2 ft 5 in)
Height overall	1·90 m (6 ft 2¾ in)

Prototype Gyroflug Speed Canard, at 1981 Paris Air Show *(Air Portraits)*

AREAS:
- Wings, gross: 7·40 m² (79·65 sq ft)
- Foreplane, gross: 1·20 m² (12·91 sq ft)

WEIGHTS AND LOADINGS (estimated):
- Weight empty: 380 kg (837 lb)
- Max fuel: 130 kg (287 lb)
- Max T-O weight: 680 kg (1,499 lb)
- Max wing loading: 78·9 kg/m² (16·17 lb/sq ft)
- Max power loading: 7·93 kg/kW (13·03 lb/hp)

PERFORMANCE (at max T-O weight):
- Never-exceed speed: 202 knots (375 km/h; 233 mph)
- Max level speed: 183 knots (340 km/h; 211 mph)
- Max cruising speed at S/L, 75% power: 162 knots (300 km/h; 186 mph)
- Cruising speed at S/L, 50% power: 132 knots (245 km/h; 152 mph)
- Stalling speed: 57 knots (105 km/h; 66 mph)
- Max rate of climb at S/L: 390 m (1,280 ft)/min
- T-O to 15 m (50 ft): 360 m (1,180 ft)
- Landing from 15 m (50 ft): 400 m (1,312 ft)
- Range with max fuel (45% power): 1,133 nm (2,100 km; 1,305 miles)
- g limits: ±4·4

PERFORMANCE (at 500 kg; 1,102 lb AUW):
- Stalling speed: 49 knots (90 km/h; 56 mph)
- Rate of climb at S/L: 600 m (1,968 ft)/min
- T-O run: 220 m (722 ft)

MBB
MESSERSCHMITT-BÖLKOW-BLOHM GmbH

HEAD OFFICE: Ottobrunn bei München, 8000 München 80, Postfach 80 11 09
Telephone: (089) 6000 1
Telex: 5287 0 mbb d
PRESIDENT AND GENERAL MANAGER:
Prof Dipl-Ing Gero Madelung
EXECUTIVE VICE-PRESIDENT AND DEPUTY GENERAL MANAGER: Sepp Hort
EXECUTIVE VICE-PRESIDENTS:
Dr Johannes Broschwitz
Fritz Killguss
Ernst-Georg Pantel
Dipl-Ing Johann Schäffler
CHAIRMAN OF THE SUPERVISORY BOARD: Max Streibl
PUBLIC RELATIONS: Eduard Roth

In May 1969 the former Messerschmitt-Bölkow GmbH and Hamburger Flugzeugbau GmbH (see 1968-69 *Jane's*) merged to form a new group known as Messerschmitt-Bölkow-Blohm GmbH. Then, in 1980, the Federal German government expressed its wish that MBB and VFW (which see) should merge, to strengthen the capabilities and competitive position of the two companies, and to help improve the structure of the German aerospace industry. Following approval by the shareholders of both companies, MBB acquired all shares of VFW on 1 January 1981. Shareholders in the company are Fides GmbH (25·7%), the State of Hamburg (18·23%), Bayerische Landesanstalt für Aufbaufinanzierung (16·5%), ABM GmbH (13·56%), VFW Verwaltungsgesellschaft (10·0%), the State of Bavaria (7·02%), the Messerschmitt Foundation (6·75%), Dr-Ing h.c. Ludwig Bölkow (1·57%), and the Blohm family (0·67%). The resulting combine employed approximately 39,000 people in Spring 1981.

Integration of the two companies (including VFW's wholly-owned subsidiary, RFB) is taking place progressively, and in April 1981 their activities were reorganised into six product-orientated divisions, as detailed hereafter. Of these, two are new divisions: the Transport and Commercial Aircraft Division at Hamburg, and the Marine and Special Technologies Division at Bremen. The Space Division combines MBB's former Space Division at Ottobrunn with VFW's ERNO subsidiary, though for the time being ERNO (see Spaceflight section) is continuing as a legally independent operation. The former RFB company has become a part of the MBB Military Aircraft Division, and Deutsche Airbus GmbH (which see) has become a wholly-owned subsidiary of MBB.

Military Aircraft Division

Ottobrunn bei München, 8000 München 80, Postfach 801160
Telephone: (089) 6000 5916
Telex: 5287 910 mbb d
WORKS: Ottobrunn (MBB), Augsburg (MBB), Manching (MBB) and Mönchengladbach (RFB)
DIRECTOR: Oskar Friedrich

Principal activities of the Military Aircraft Division involve the Panavia Tornado (see International section), for which MBB is the German prime contractor; weapon system leadership for F-104G Starfighters of the Luftwaffe; modifications to improve the combat capability of Luftwaffe F-4F and RF-4E Phantoms; development of control configured vehicle (CCV) technology, using a modified F-104G aircraft; continuing development of the RFB Fantrainer; and participation in the European manufacturing programme for the General Dynamics F-16 combat aircraft. It also manufactures major assemblies for the BO 105 helicopter, Airbus A300 and A310, and Transall C-160 Series 2.

In the Tornado programme, MBB is building centre-fuselages for all production aircraft, and is responsible for final assembly and flight testing of the 324 aircraft required by the Luftwaffe and Marineflieger. Improvements are currently being made to the optical and electronic reconnaissance equipment in the Luftwaffe's RF-4E Phantoms, which are also being armed to enable them to be used also in a fighter-bomber role. An enhancement programme for the F-4F tactical fighter is also under way.

In 1981 the company was engaged in studies to provide a new-generation tactical fighter aircraft for the Luftwaffe and other European air forces in the 1990s. In addition, it is engaged in research and development programmes concerned with carbonfibre technology, and the creation of more simple, more reliable aircraft subsystems.

Helicopter and Transport Systems Division

Ottobrunn bei München, 8000 München 80, Postfach 801140
Telephone: (089) 6000 2945
Telex: 5287 740 mbb d
WORKS: Ottobrunn (MBB), Donauwörth (MBB), Laupheim (MBB), Speyer (VFW), Kassel (VFW) and Lemwerder (VFW)
DIRECTOR: Kurt Pfleiderer

This Division is concerned primarily with the manufacturing programmes for the MBB BO 105 helicopter, including its VBH and PAH-1 military versions, and the MBB/Kawasaki BK 117 (described in the International section). Main production centre for the BO 105 is MBB's Donauwörth factory. The Division also now assumes the former VFW responsibility for overhaul and repair of Alouette, Sikorsky CH-53G and Westland Sea King Mk 41 helicopters in service with the West German armed forces.

Under study, with the company designation BO 125, is a third-generation transport helicopter (TH 3) for service in the 1990s. International collaboration on this programme is planned.

Transport and Commercial Aircraft Division

2103 Hamburg 95, Postfach 950109
Telephone: (040) 747 1
Telex: 217 684 mbbh d
WORKS: Hamburg-Finkenwerder (MBB), Stade (MBB), Bremen (VFW), Einswarden (VFW), Lemwerder (VFW) and Varel (VFW)
DIRECTOR: Dipl-Ing Johann Schäffler

Throughout MBB/VFW, some 9,900 employees—about 25 per cent of the total work force— are engaged in the manufacturing programme for the Airbus A300 and A310 wide-bodied transport aircraft. Most of them are in this new Division, which is based primarily upon MBB's Hamburg facility but includes also the VFW factories engaged in Airbus manufacture and conversion.

In the Spring of 1981, a Lufthansa Airbus A300 began operating with a newly-installed rudder made of carbonfibre reinforced plastics by MBB. Largest CFRP aircraft component made in Europe up to that time, it will be followed by an A300 fin box primary structure, of which MBB expected to complete design by the end of 1981. Its weight will be 20 per cent less than that of the present conventional structure, which weighs almost 1,200 kg (2,650 lb). VFW was responsible for development of the C4 convertible cargo version of the Airbus A300, and has also converted more than 30 A300B-100s into A300B4-203s with increased max T-O weight and extended range.

This MBB Division is also the German partner in the current (Series 2) production programme for the Transall C-160 military transport (see International section), and participates in the production programmes for the F27 and F28 described under the Fokker entry in the Netherlands section. Other work includes overhaul and repair of Transall C-160 (original series), Lockheed JetStar and other transport aircraft.

Dynamics Division

Ottobrunn bei München, 8000 München 80, Postfach 801149
Telephone: (089) 6000 2206
Telex: 5287 0 mbb d
WORKS: Ottobrunn (MBB), Nabern/Teck (MBB) and Schrobenhausen (MBB)
DIRECTOR: Günther Kuhlo

Dynamics Division is engaged in a number of missile programmes including those for the MBB Kormoran and Euromissile Hot described in the Air-Launched Missiles section. Others include Armbrust, Cobra/Mamba, Milan, MLRS and Roland.

Since 1966, MBB has been developing a new type of conventional weapon system, designated **MW-1**, which will be carried by the Tornado and can equip other types such as the F-4F Phantom. It consists of a dispenser carried under the aircraft's belly, from which a large number of small munitions can be ejected pyrotechnically. For use against tanks, the munitions may be shaped-charge bombs which detonate on impact or mines that are set off when the tanks run over them. They can be of types designed to destroy airfield runways, splinter mines with fuses triggered by aircraft movements, or a type designed to destroy aircraft inside shelters.

The Division is also involved extensively in RPV development and production, as described under the MBB and VFW entries in the appropriate section of this edition.

Space Division

2800 Bremen 1, Postfach 107845
Telephone: (0421) 5382305, 5382500 and 5394349
WORKS: Ottobrunn (MBB) and Bremen (VFW)
DIRECTOR: Dipl-Ing Johannes Schubert

Space programmes in which MBB and ERNO are or have recently been involved include Exosat, TV-Sat (TDF-1), Spacelab, SPAS (Space Shuttle Satellites), the Galileo Jupiter probe, and the Ariane launch vehicle. Details of most of these can be found in the Spaceflight section of this edition.

Marine and Special Technologies Division

2800 Bremen 1
WORKS: Ottobrunn (MBB), Donauwörth (MBB) and Bremen (VFW)
DIRECTOR: Dr jur Carl-Peter Fichtmüller

This new Division was formed in April 1981 to continue the maritime programmes of VFW, and to initiate new activities. The former include the Pinguin B3 remotely controlled submarine drone and the MiJ 331 B minesweeper.

The Division is also responsible for diversified MBB activities which include various surface transport systems, solar power stations, solar and wind energy systems, medical technology, mechanical handling systems, data acquisition systems, and simulation techniques.

MBB F-104 CCV

MBB's Military Aircraft Division has for several years been undertaking, under contract to the Federal German Ministry of Defence, a programme to develop and demonstrate an advanced control-configured vehicle (CCV) flight control system. A Luftwaffe F-104G Starfighter (98 + 36, formerly 23 + 91) is being used as the test aircraft in this programme, the results of which are expected to have particular relevance to the Luftwaffe's requirement for an advanced tactical combat aircraft for service in the 1990s.

The CCV test aircraft is equipped with a digital fly-by-wire flight control system which incorporates modular, self-monitoring, triple-redundant series actuators for the primary control surfaces. Those for the ailerons are located in large fairings on the fuselage sides, on the upper surface of the engine air intake trunks; actuators for the rudder and all-moving tailplane are located in the fin, and all utilise the aircraft's standard power jacks. All other elements of the FBW system, including the four 16-bit 16K onboard computers, are quadruplexed to provide parallel redundancy. Aircraft control column and rudder pedals, instead of fulfilling their conventional functions, are used instead to initiate normal acceleration, rate of roll, and changes in angle of sideslip. The necessary control inputs are generated by the computers, each of which is supplied with information via a strapdown inertial platform and skewed vane-type airflow direction sensor. The computers are interconnected to provide access to each other's memory, and between them manage the aircraft's pre-flight checks, stabilisation and control, autopilot, inertial navigation, air data calculations, and systems redundancy management.

The main purpose of the CCV programme is to investigate the use of an artificial stability system in an unstable aircraft, and the performance improvements expected from this concept, although the current investigation is of aircraft behaviour rather than of specific performance gains. The system is required to be able to continue a mission after a single failure, and to permit a safe return to base after any additional failure. Safety problems are also being studied as they arise, and as a precautionary measure the test aircraft retains a conventional mechanical backup system of flight control.

Work on the F-104 CCV programme began in December 1974, and installation of the FBW flight control system was completed in May 1977. The first stage of flight testing, with the aircraft fully stable (approx 20 per cent positive stability) and otherwise unmodified, began in December of that year and was completed in October 1979. This is being followed by four further stages (B2, E1, E2, and E3) in which the aircraft is made progressively less stable, culminating in the final stage with a shift in CG of up to 20 per cent of the wing mean chord aft of the aerodynamic centre. In the first of these stages (B2), the test aircraft was rendered moderately stable by adding 600 kg (1,323 lb) of lead shot ballast aft of the CG, beneath the jetpipe. Conversion to B2 configuration, and flight trials, which included flutter and ballast jettison tests, took place in October/November 1979. In this configuration a modified fuel transfer system was also adopted, in which the normal (tip-tanks first) sequence of fuel usage was reversed. With internal tanks depleted, the CG is displaced to the most rearward position; before landing the CG is moved forward, to 22 per cent of the mean aerodynamic chord, by transferring fuel to the fuselage tanks from the wingtip tanks.

By December 1979 the testbed aircraft had been further modified to E1 (marginally stable) configuration. This was achieved by deleting the aft ballast, pylon-mounting a canard surface (actually the tailplane of another F-104) on top of the forward fuselage just aft of the cockpit with a fixed angle of incidence of −4°, and adding forward ballast.

Lockheed F-104G modified by MBB as part of its five-year CCV research programme

Installation of the canard has the effect of relocating the aircraft's aerodynamic centre further forward by some 35 per cent of the wing mean aerodynamic chord. The purpose of the forward ballast is to restore it to a position just aft of the CG for the E1 phase; when aft ballast is added in the two final stages, the aerodynamic centre will revert to a position forward of the CG. The canard, which supports approx 20 per cent of the weight of the aircraft in flight, becomes lift-free at the same angle of attack as the wings. It stalls at an angle of 11°, restoring positive stability. To prevent the onset of wing stall (which would occur at 16° wing angle of attack), artificial stall warning is given by a stick-shaker when the wing angle reaches 12°. The F-104 CCV made a successful 46 min first flight in the E1 configuration on 20 November 1980.

In the two remaining stages of the programme, the F-104 CCV is rendered moderately unstable in stage E2, and highly unstable (20 per cent negative stability) in stage E3, by the reintroduction of aft ballast (400 kg; 882 lb and 600 kg; 1,323 lb respectively). Stage E2 flight testing began in May 1981. The reversed fuel transfer system will again be used in the E3 trials, to act as a ballast trim. All flight testing is being carried out within a flight envelope limited to speeds of Mach 1·3 and 650 knots (1,205 km/h; 748 mph) IAS. The overall flight test programme calls for a total of 120 flights, of which 76 were with the aircraft in partly de-stabilised form (B1 and B2 configurations).

RFB FANTRAINER 400/600

This tandem two-seat training aircraft was first projected in 1970, at which time a model was exhibited at the Hanover Air Show. It utilises a ducted fan propulsion system which, in the original concept, comprised a Dowty variable-pitch fan, integral with the rear fuselage and driven by two 112 kW (150 hp) Wankel two-disc rotary engines.

In March 1975 the Federal German Defence Ministry awarded RFB a contract to develop and build two Fantrainer prototypes, for evaluation as potential replacements for the Piaggio P.149D primary trainers now used by the Luftwaffe. They were designed to conform to US FAR Pt 23 specifications in the Aerobatic and Utility categories, and it was suggested that pupils might be able to make direct transition from the Fantrainer to the Alpha Jet.

The first prototype, designated **AWI-2**, flew for the first time on 27 October 1977, powered by two 112 kW (150 hp) Wankel engines driving a variable-pitch fan through a reduction gearbox, with airbrakes in the fan shroud. The **ATI-2** second prototype (D-EATI) flew for the first time on 31 May 1978 and had one Allison 250-C20B turboshaft engine. It logged 62 flying hours before being lost on 7 September 1978.

In 1978, the first prototype (D-EATJ) was brought up to proposed production configuration for the Luftwaffe, as the **Fantrainer 400**, with Allison 250-C20B turboshaft engine. It had logged 252 flying hours by June 1980, including 55 hours with its original Wankel engines. It was then fitted with an Allison 250-C30 turboshaft engine, and redesignated **Fantrainer 600.** Except for the changed power plant, the Fantrainer 400 and 600 are virtually identical. Flight testing and demonstration of the prototype with the higher-powered Allison engine had brought its total flight hours to 382 by 1 January 1981. In accumulating these hours, more than 1,100 landings were made. A decision on whether to initiate production was anticipated during 1981.

On the basis of the Fantrainer propulsion system, an uprated twin-engined project has been proposed to the US Air Force by Vought Corporation as a contender to meet the USAF's requirement for a T-37 replacement. Known as the **V-539 Eaglet**, it would be a side-by-side two-seater powered by two turboshaft engines; alternative power plants which have been proposed are the 559 kW (750 shp) Pratt & Whitney Aircraft of Canada PT6A-34, or a developed version of the Allison 250-C30 rated at 533 kW (715 shp). The V-539 is entered as a joint Vought/RFB team design: if selected, MBB would have a 50 per cent share in production of the aircraft structure.

The description which follows applies to the Fantrainer 400 and 600:

TYPE: Two-seat basic and IFR training aircraft; *g* limits +6/−3.

WINGS: Cantilever mid-wing monoplane. Wing section Eppler 502. Thickness/chord ratio 15·7%. Dihedral 2° 30′. No incidence. Sweepforward 2° 30′ at quarter-chord. Constructed mainly of glassfibre and plastics tube sandwich. Frise-type ailerons and electrically-actuated Fowler-type trailing-edge flaps. No tabs.

FUSELAGE: The load-carrying fail-safe structure of the forward and centre-fuselage is of light alloy, with non-load-bearing glassfibre skin, sections of which are removable for servicing purposes. Cruciform metal rear fuselage is connected to the centre-fuselage at three points. The integral fan duct is free of structural loads. Large airbrake on each side of fan duct, operation of which causes no lift or stability changes.

TAIL UNIT: All-metal T tail of light alloy, with conventional rudder and elevators. Servo tab in trailing-edge of each elevator. Trim tab in rudder.

LANDING GEAR: Retractable tricycle type, with single wheel on each unit. Hydraulic actuation, with manual emergency extension. All units retract into fuselage, nosewheel rearward, main units inward and upward into wing roots. Steel tube legs, acting as torsional/bending springs. Cleveland main wheels size 15 × 6·00-6, tyre pressure 4·7 bars (68 lb/sq in). Goodyear nosewheel size 5·00-5, tyre pressure 3·45 bars (50 lb/sq in). Cleveland wheel brakes.

RFB Fantrainer 600 two-seat basic and IFR training aircraft *(Jay Miller)*

Power Plant: One turboshaft engine, driving a Dowty Rotol five-blade constant-speed ducted fan. Fantrainer 400 has a 313 kW (420 shp) Allison 250-C20B, Fantrainer 600 a 485 kW (650 shp) Allison 250-C30, derated to 447 kW (600 shp). Engine air intakes under wing leading-edges. Four integral fuel tanks in wings, with combined capacity of 430 litres (94 Imp gallons). Refuelling points on wing upper surface. Fantrainer 600 is able to carry 300 kg (661 lb) of auxiliary fuel externally. Oil capacity 16 litres (3·5 Imp gallons).

Accommodation: Two seats in tandem cockpit, meeting US MIL specifications in terms of dimensions and layout. Seats and rudder pedals adjustable. Provision for back parachutes. Fighter-type side consoles. Canopy over each seat hinges sideways (to stbd) independently. Accommodation heated and ventilated.

Systems: Electrical system includes a starter/generator and battery. Hydraulic system for operation of landing gear and airbrakes.

Dimensions, external:
Wing span 9·70 m (31 ft 10 in)
Wing chord at root 1·89 m (6 ft 2½ in)
Wing chord at tip 1·02 m (3 ft 4 in)
Wing aspect ratio 6·6
Length overall, incl nose probe 9·00 m (29 ft 6¼ in)
Height overall 2·90 m (9 ft 6 in)
Tailplane span 3·29 m (10 ft 9½ in)
Wheel track 1·94 m (6 ft 4¼ in)
Wheelbase 3·80 m (12 ft 5½ in)
Fan diameter 1·20 m (3 ft 11¼ in)

Areas:
Wings, gross 13·9 m² (149·6 sq ft)
Ailerons (total) 1·19 m² (12·81 sq ft)
Trailing-edge flaps (total) 1·40 m² (15·07 sq ft)
Rudder, incl tab 2·20 m² (23·68 sq ft)
Tailplane 2·90 m² (31·22 sq ft)
Elevators, incl tab 0·77 m² (8·29 sq ft)

Weights and Loadings (A: Aerobatic; B: Normal category):
Weight empty:
400: A, B 939 kg (2,070 lb)
600: A, B 1,060 kg (2,337 lb)
Fuel weight (internal):
400: A 145 kg (320 lb)
B 340 kg (750 lb)
600: A 190 kg (419 lb)
B, internal fuel only 340 kg (750 lb)
B, with drop-tanks 640 kg (1,411 lb)
Max T-O weight:
400: A 1,350 kg (2,976 lb)
B 1,580 kg (3,483 lb)
600: A 1,450 kg (3,196 lb)
B 2,300 kg (5,070 lb)
Max landing weight:
600: A 1,450 kg (3,196 lb)
B 2,000 kg (4,409 lb)
Max wing loading:
400 114 kg/m² (23·35 lb/sq ft)
600 165·5 kg/m² (33·90 lb/sq ft)
Max power loading:
400 5·05 kg/kW (8·29 lb/shp)
600 5·15 kg/kW (8·45 lb/shp)

Performance (at max aerobatic T-O weight):
Max level speed:
400 at 3,050 m (10,000 ft) 200 knots (370 km/h; 230 mph)
600 at 5,485 m (18,000 ft) 232 knots (430 km/h; 267 mph)
Cruising speed at 3,050 m (10,000 ft):
400 175 knots (325 km/h; 202 mph)
600 200 knots (370 km/h; 230 mph)
Approach speed:
400 72 knots (133 km/h; 83 mph)
600 75 knots (140 km/h; 87 mph)
Max rate of climb at S/L:
400 612 m (2,000 ft)/min
600 960 m (3,145 ft)/min
Service ceiling: 400 6,100 m (20,000 ft)
600 7,620 m (25,000 ft)
T-O run: 400 230 m (755 ft)
600 200 m (656 ft)
Landing run: 400 250 m (820 ft)
600 270 m (886 ft)
Range with max internal fuel, optimum cruising speed at 3,050 m (10,000 ft), no reserves:
400 950 nm (1,760 km; 1,093 miles)
600 750 nm (1,390 km; 863 miles)
Endurance, conditions as above:
400 6 h 18 min
600 4 h 48 min

MBB/KAWASAKI BK 117

Following signature of an agreement on 25 February 1977, MBB is developing and building in conjunction with Kawasaki of Japan an 8/10-seat multi-purpose helicopter known as the BK 117. A description of this aircraft can be found in the International section.

MBB BO 105

Design of this light utility helicopter was started in July 1962 and construction of prototypes began in 1964, under Federal German government contract. The first BO 105 prototype was fitted with an existing conventional rotor and two Allison 250-C18 turboshaft engines; subsequent aircraft have had a rotor system based on a rigid titanium hub, with feathering hinges only, and hingeless flexible glassfibre blades. From the Spring of 1970 'droop-snoot' rotor blades of MBB design have been standard.

Details of subsequent prototypes, early production helicopters and special variants can be found in previous editions of *Jane's*. Current models are as follows:

BO 105 CB. Standard production version since 1975, with two Allison 250-C20B engines, operable in air temperatures ranging from −45° to +54°C, and strengthened rotor gearing. LBA certification received in November 1976.

BO 105 CBS. Version with increased seating or cargo capacity in a 0·25 m (10 in) longer fuselage. Available in five-seat executive or six-seat high-density configurations. Identified by small additional window aft of rear door on each side.

BO 105 D. As supplied to the UK, with modified equipment. The BO 105 D has been approved since 1973 by the CAA for commercial single-pilot IFR operation, even in controlled airways.

BO 105 LS. Variant due for certification by early 1982. Described separately.

BO 105 M (VBH). Liaison and light observation helicopter for the Federal German Army, with strengthened transmission gearing, reinforced rotor components, a tail rotor with improved thrust and performance, a rupture-proof fuel system, and a landing gear able to absorb higher energy levels. Production of 227 approved by the Federal government, to replace Alouette IIs. Deliveries began in 1980.

BO 105 P (PAH-1). Anti-tank version, with same airframe improvements as BO 105 M, outriggers to carry six Euromissile Hot missiles, a stabilised sight above the co-pilot's position, and Singer AN/ASN-128 Doppler navigation system. The German Federal government has given approval for the procurement of 212 examples for the Federal German Army. Deliveries began on 4 December 1980 and will be completed in mid-1983; 31 had been delivered by the beginning of 1981. First PAH-1 unit is Heeresfliegerregiment 16 at Celle. Three anti-tank regiments, attached to three Army corps, will each operate two squadrons of 28 helicopters, in four flights. A further squadron of 21 PAH-1s will be attached to No. 6 Panzergrenadier Division, in Schleswig-Holstein, for special duties. Empty equipped weight of the PAH-1, including pilot and weapons operator, is 1,458 kg (3,215 lb). Max T-O weight is 2,400 kg (5,291 lb). At T-O power and 25°C the PAH-1 has a forward rate of climb at 1,000 m (3,280 ft) of 420 m (1,378 ft)/min and can hover OGE at 1,000 m (3,280 ft). At max continuous power it has a max level speed of 113 knots (210 km/h; 130 mph) and a service ceiling (at 15°C) of 3,000 m (9,845 ft). Endurance, including 20 min fuel reserves, with 456 kg (1,005 lb) of fuel, is 1 h 30 min.

In the Philippines and Indonesia, NAM and Nurtanio respectively (which see) are participating in licence assembly programmes for the BO 105. A contract for 60 BO 105s was signed by the Spanish Ministry of Defence in late June 1979. These aircraft are being assembled in Spain, by CASA, and are operated by Spanish Army aviation units for communications, observation and anti-tank missions. Approx 10 had been delivered by the beginning of 1981.

BO 105 Ophelia (Optique Platforme Hélicoptère Allemand). Version sponsored by Federal Ministry of Research and Technology, with two-axis gyro-stabilised mast-mounted visual aids for day and night observation. Spherical package, which weighs approx 115 kg (253·5 lb), contains TRT Calipso FLIR camera, SFIM TV camera and SFIM laser rangefinder; VDO computer symbol generator in cabin. Sensors being tested in conjunction with VDO head-up display, Thomson-CSF colour or VDO monochromatic head-down display, and Ferranti HMS/D helmet-mounted sight and display. Flight testing on a BO 105 CB prototype (D-HABV) began in May 1981.

Orders for all versions of the BO 105 exceeded 1,000 by early 1981. About 500 had been delivered.

The description which follows applies to the BO 105 CB production version:

Type: Five-seat light helicopter.

Rotor System: Four-blade main rotor, comprising rigid titanium hub and GRP blades, with titanium anti-erosion strip forming leading-edge of each blade. MBB-designed 'droop-snoot' blades of NACA 23012 asymmetrical section, having a specially-designed trailing-edge giving improved control in pitching moment. Flexible tension/torsion blade retention, to take up centrifugal forces. Roller bearings for pitch change. Main rotor blade folding optional. Two-blade semi-rigid tail rotor; blades of GRP, with titanium anti-erosion strip on leading-edge. Main rotor rpm 424. Tail rotor rpm 2,220.

Rotor Drive: Main transmission utilises two bevel gear input stages with freewheeling clutches and a spur collector gear stage. Planetary reduction gear; three auxiliary drives for accessories. Main transmission rated for twin-engine input of 257 kW (345 shp) per engine, or a single engine input of 283 kW (380 shp). Tail rotor

MBB BO 105 CBS, the 'stretched' five/six-seat version of the BO 105, here fitted with nose radar

gearbox on fin. Main rotor/engine rpm ratio 1 : 14·2. Tail rotor/engine rpm ratio 1 : 2·7.

FUSELAGE: Conventional light alloy semi-monocoque structure of pod and boom type. Glassfibre-reinforced cowling over power plant. Titanium sheet engine deck.

TAIL UNIT: Horizontal stabiliser of conventional light alloy construction with small endplate fins.

LANDING GEAR: Skid type, with cross-tubes designed for energy absorption by plastic deformation in the event of a heavy landing. Inflatable emergency floats can be attached to skids.

POWER PLANT: Two 313 kW (420 shp) Allison 250-C20B turboshaft engines, each with a max continuous rating of 298 kW (400 shp). Bladder-type fuel tanks under cabin floor, capacity 580 litres (127·5 Imp gallons), of which 570 litres (125·3 Imp gallons) are usable. Fuelling point on port side of cabin. Provision for fitting auxiliary tanks in freight compartment. Oil capacity: engine 9 litres (1·98 Imp gallons), gearbox 7 kg (15·4 lb).

ACCOMMODATION: Pilot and co-pilot or passenger on individual front seats, with safety belts. Optional dual controls. Bench seat at rear for three persons, removable for cargo and stretcher carrying. Entire rear fuselage aft of seats and under power plant available as freight and baggage space, with access through two clamshell doors at rear. Two standard stretchers can be accommodated in ambulance role. One forward-opening hinged and jettisonable door and one sliding door on each side of cabin. Heating system optional.

SYSTEMS: Tandem fully-redundant hydraulic system for powered main rotor controls. Electrical system powered by two 150A 28V DC starter/generators and a 24V 22Ah nickel-cadmium battery; external power socket.

AVIONICS AND EQUIPMENT: Standard equipment includes clock, outside air temperature gauge, battery overheat warning lights, heated pitot, tiedown rings in cargo compartment, cabin and cargo compartment dome lights, position lights and collision warning lights. A wide range of optional avionics and equipment is available, including Doppler navigation, search radar, pilot and co-pilot adjustable shoulder harness, rear seat assembly, dual controls, heating system, windscreen wiper, rescue winch, landing light, searchlight, externally-mounted loudspeaker, fuel dump valve, external load hook, settling protectors, snow skids, rotor brake, and main rotor blade folding. A completely equipped ambulance version is available.

ARMAMENT (military versions): Provision for a variety of alternative military loads, including six Hot anti-tank missiles and associated stabilised sight.

DIMENSIONS, EXTERNAL:

Diameter of main rotor	9·84 m (32 ft 3·4 in)
Diameter of tail rotor	1·90 m (6 ft 2¾ in)
Main rotor blade chord	0·27 m (10·63 in)
Tail rotor blade chord	0·18 m (7·09 in)
Distance between rotor centres	5·95 m (19 ft 6¼ in)
Length, incl main and tail rotors	11·86 m (38 ft 11 in)
Length, excl main rotor:	
CB	8·56 m (28 ft 1 in)
CBS	8·81 m (28 ft 11 in)
Height to top of main rotor head	3·00 m (9 ft 10 in)
Width over skids, unladen	2·53 m (8 ft 3½ in)
Width over skids, laden	2·58 m (8 ft 5½ in)
Rear loading doors: Height	0·64 m (2 ft 1 in)
Width	1·40 m (4 ft 7 in)

DIMENSIONS, INTERNAL:

Cabin, incl cargo compartment:	
Length	4·30 m (14 ft 1 in)
Max width	1·40 m (4 ft 7 in)
Max height	1·25 m (4 ft 1 in)
Volume	4·80 m^3 (169 cu ft)
Cargo compartment: Length	1·85 m (6 ft 0¾ in)
Max width	1·20 m (3 ft 11¼ in)
Max height	0·57 m (1 ft 10½ in)
Floor area	2·20 m^2 (23·7 sq ft)
Volume	1·50 m^3 (53 cu ft)

WEIGHTS AND LOADINGS:

Weight empty, basic	1,189 kg (2,622 lb)
Weight empty, operating	1,279 kg (2,820 lb)
Standard fuel	460 kg (1,014 lb)
Fuel, incl auxiliary tanks	780 kg (1,720 lb)
Max T-O weight	2,400 kg (5,291 lb)
Normal disc loading	26·5 kg/m^2 (5·43 lb/sq ft)
Max disc loading	30·5 kg/m^2 (6·25 lb/sq ft)

PERFORMANCE (at T-O weight of 2,300 kg: 5,070 lb):

Never-exceed speed at S/L	145 knots (270 km/h; 167 mph)
Max cruising speed at S/L	132 knots (245 km/h; 152 mph)
Normal cruising speed	125 knots (232 km/h; 144 mph)
Max rate of climb at S/L	540 m (1,773 ft)/min
Vertical rate of climb at S/L	258 m (847 ft)/min
Max operating height	5,180 m (17,000 ft)
Single-engine service ceiling	1,280 m (4,200 ft)
Hovering ceiling IGE	2,900 m (9,515 ft)
Hovering ceiling OGE	1,980 m (6,500 ft)
Range with standard fuel and max payload, no reserves:	
at S/L	310 nm (575 km; 357 miles)
at 1,525 m (5,000 ft)	355 nm (656 km; 408 miles)
Ferry range with auxiliary tanks, no reserves:	
at S/L	540 nm (1,000 km; 621 miles)
at 1,525 m (5,000 ft)	600 nm (1,112 km; 691 miles)

MBB BO 105 P (PAH-1) fitted with six Hot anti-tank missiles and SFIM stabilised target acquisition and tracking sight *(Brian M. Service)*

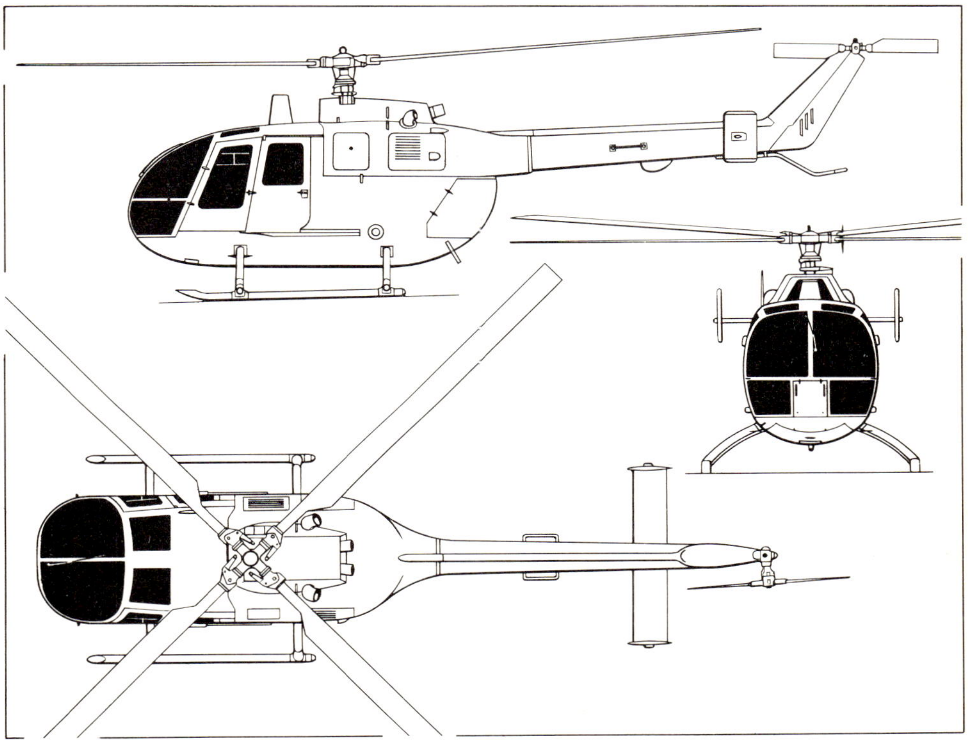

BO 105 CB five-seat light helicopter (two Allison 250-C20B turboshaft engines) *(Pilot Press)*

MBB BO 105 LS

This version of the BO 105 has more powerful engines and an uprated transmission, permitting operation at a higher gross weight with external load. It was first flown in late March 1979. The description of the BO 105 CB applies also to the BO 105 LS, except as follows:

ROTOR DRIVE: Main transmission, type ZF-FS 112, is rated for a twin-engine input of 294 kW (395 shp) per engine, or a single-engine input of 368 kW (493 shp) continuous, or 404 kW (542 shp) for 2·5 min.

POWER PLANT: Two Allison 250-C28C turboshaft engines, each rated at 410 kW (550 shp) for 2·5 min, and with a 5 min take-off or max continuous power rating of 373 kW (500 shp). Fuel system generally as for BO 105 CB, but max standard capacity 380 kg (838 lb).

SYSTEMS: As for BO 105 CB, except 24V battery is of 25Ah capacity.

WEIGHTS:

Weight empty	1,250 kg (2,756 lb)
Max T-O weight, internal load	2,400 kg (5,291 lb)
Max T-O weight, external load	2,600 kg (5,732 lb)

PERFORMANCE (at max T-O weight):

Never-exceed speed at S/L	145 knots (270 km/h; 167 mph)
Max cruising speed at S/L	136 knots (252 km/h; 157 mph)
Max rate of climb at S/L	600 m (1,970 ft)/min
Max operating height	6,100 m (20,000 ft)
Service ceiling, one engine out	2,880 m (9,450 ft)
Hovering ceiling IGE	4,000 m (13,120 ft)
Hovering ceiling OGE	3,440 m (11,290 ft)
Range at S/L, standard fuel, max internal payload, no reserves	248 nm (460 km; 286 miles)

MYLIUS

LEICHTFLUGZEUGE-ENTWICKLUNGEN DIPL ING HERMANN MYLIUS

Kuckucksweg 6, 8011 Brunnthal

In addition to heading the light aircraft technical development activities of MBB, Dipl Ing Hermann Mylius develops and builds sporting aircraft privately. In July 1971 he began the development of a single-seat version of the MHK 101, intended for competitive aerobatics. Construction of a prototype (D-EMYS) was started in December 1971, and this flew for the first time on 7 July 1973. Known as the MY 102 Tornado, it has achieved considerable success in competition. Details of this aircraft and projected developments can be found in the 1980-81 *Jane's*.

RFB
RHEIN-FLUGZEUGBAU GmbH

This company, founded in 1956, acquired in 1976 100% of the stock of Sportavia-Pützer, which was then integrated into the RFB organisation. RFB was itself a wholly-owned subsidiary of VFW GmbH, and with VFW has now become part of MBB (which see).

RFB specialised in the development and manufacture of airframe structural components, with particular reference to wings and fuselages made entirely of glassfibre-reinforced resins. Recent programmes included series and individual production of aircraft components and assemblies made of light alloy, steel and glassfibre-reinforced resin, for aircraft in quantity production by other West German companies, as well as spare parts and ground equipment.

Under contract to the Federal German government, RFB was, prior to the merger, servicing military aircraft, and was providing target-towing flights and other services with special aircraft. It operated a factory-certificated service centre for all types of Piper aircraft and for the Mitsubishi MU-2 utility transport aircraft. General servicing of other types of all-metal aircraft was also undertaken.

In the aircraft propulsion field, RFB engaged for some years in the development of specialised applications for ducted propellers, leading to the Fantrainer military multi-purpose training aircraft (now described under the MBB entry).

RFB RS 180 SPORTSMAN

M René Fournier began design of the Sportsman in December 1970. The first of two prototypes made its first flight on 1 March 1973, powered by a 93 kW (125 hp) Avco Lycoming O-235-F2A engine.

The second prototype, a completely new four-seat design developed by Sportavia, had the designation RF6C, and this flew for the first time on 28 April 1976. It differed from the first prototype in having a more powerful engine, increased cabin volume, and a GRP outer skin, and entered production in late 1976. The change of designation to RS 180 took effect in early 1978, simultaneously with raising of the tailplane from its original position at the top of the fuselage.

At the end of 1980 Sportavia became integrated into the RFB company. No details are known of the number of RS 180s which were built, but in early 1981 production had been suspended. A description and illustration of the Sportsman can be found under the Sportavia-Pützer entry in the 1980-81 *Jane's*.

SPORTAVIA *(see RFB)*

VFW
VEREINIGTE FLUGTECHNISCHE WERKE GmbH

Vereinigte Flugtechnische Werke GmbH (VFW) was formed at the end of 1963 by a merger of the two Bremen-based aircraft companies of Focke-Wulf GmbH and 'Weser' Flugzeugbau GmbH. They were joined in 1964 by Ernst Heinkel Flugzeugbau GmbH. Between 1970–1980 it was an equal partner in Zentralgesellschaft VFW-Fokker GmbH. Details of the history and products of these former companies can be found in earlier editions of *Jane's*.

All VFW shares were acquired by MBB on 1 January 1981, in accordance with a Federal German government decision that the two companies should amalgamate. The former shareholders of VFW GmbH are now represented by VFW Verwaltungsgesellschaft, which has a 10% shareholding in MBB. The activities of the former VFW facilities are now described under the MBB heading in this section, and those of ERNO in the Spaceflight section.

GREECE

HAI
HELLENIC AEROSPACE INDUSTRY

Athens Office: 2-4 Mesogion Avenue, PO Box 3110, Ampelokipi, Athens 610
Telephone: 779 9678/9
Telex: 21-9528 HAI GR

Works: PO Box 23, Schimatari, Viotia
Telephone: (0262) 30080/1
Telex: 27-2106 HAI GR

Chairman of the Board: J. Stratos
General Manager: H. E. Mack
Marketing Director: J. W. Eagle Jr
Technical Director: R. F. Bennett
Administrative Director: D. Georgiou
Management Services Director: E. Contogeorgis
Public Relations Manager: Spiros Karayiannis

The establishment of an aerospace industry in Greece became effective in November 1975. Its main purposes are to provide manufacturing and repair facilities in the country, to execute both civil and military work, and to achieve independence in this field from foreign companies. The Hellenic Aerospace Industry (HAI) venture was formed following the signing of contracts between the Greek government and a group of leading international aerospace and construction firms. The total investment will exceed $320 million. The company is owned by the Greek government and is headed by a Greek nine-member board of directors.

Contracts were signed with the Lockheed, Westinghouse, General Electric and Austin companies to assist in the buildup and initial stages of the operation. Construction began on 4 February 1977, and the HAI facility was opened officially on 18 December 1979 by the Greek Prime Minister, Mr C. Karamanlis. It is the most modern of its type in the Mediterranean area, and represents one of the largest single construction programmes ever undertaken in Greece. The complex consists of an aircraft depot, engine depot, electronics depot and accessory depot, in addition to the usual flight line support, utility and ancillary buildings.

The aircraft depot, with high bay and low bay hangars, covers 27,500 m² (296,007 sq ft) of space. The commercial hangar section has a 22 m (72 ft) clearance for aircraft such as the Boeing 747 and Airbus A300. Its activities include the manufacture of selected structural components for the Aeritalia G222 twin-turboprop transport aircraft, and it has a similar agreement with Dassault-Breguet of France. The Egyptian Air Force has selected HAI to overhaul and repair its C-130 Hercules transport aircraft.

The 16,000 m² (172,222 sq ft) engine depot, with overhaul and test shops, has capability for more than 20 types of engine, and has a US Air Force contract for the overhaul of J79 turbojet engines. The computerised engine test facility has a present capacity for engines of up to 133·45 kN (30,000 lb st); it is planned to increase this capacity to cover all types of modern high-power engines. The accessory and shop depot, which covers 15,500 m² (166,840 sq ft) of space, includes 33 shops and facilities, led by the engine and electronic departments. Among these are workshops for sheet-metal fabrication and repair, machine tooling, heat treatment and welding, plastics and woodworking, pneumatics, instruments and electrical work.

The electronics division covers 9,500 m² (102,257 sq ft) and contains 18 shops for servicing airborne and ground-to-air electronics equipment and instrumentation, accelerometers and inertial guidance platforms. Current activities include the overhaul of NATO Sidewinder air-to-air missiles.

HAI also offers technical and management training covering a wide variety of aeronautical skills related to military and civil aircraft maintenance. HAI is located in Tanagra, approx 60 km (37 miles) north of Athens; it is served by a 3,660 m (12,000 ft) runway, and is adjacent to main highways and international rail/sea links. More than 2,500 people were employed at the facility in early 1981; within two years it is anticipated that this figure will rise to more than 3,000.

INDIA

HAL
HINDUSTAN AERONAUTICS LIMITED

Indian Express Building, Vidhana Veedhi, PO Box 5150, Bangalore 560 001
Telephone: 76901 (8 lines)
Telex: 845-266 HAL IN

Chairman: Gp Capt B. K. Kapur (Retd)

Directors:
- Gp Capt B. K. Kapur (Retd)
- Gp Capt S. C. Keshu
- Raj Mahindra
- L. K. Joshi
- G. Narasimhan
- K. K. Kirtikar
- B. M. Prabhu
- Dr D. N. Prasad
- Dr Raja Ramanna
- Dr S. R. Valluri
- A. H. Mehta
- P. V. Desai
- Air Marshal C. V. Gole

General Managers:
- Bangalore Complex:
  - K. N. R. Swamy (Aircraft Division)
  - S. K. Khanna (Engine Division)
  - Gp Capt Willie Raj (Overhaul Division)
  - N. Nagaraja Rao (Foundry and Forge Division)
  - S. K. Ohrie (Helicopter Division)
- MiG Complex:
  - B. S. Balooja (Nasik Division)
  - C. V. Vijayaraghavan (Koraput Division)
- Accessories Complex:
  - Gp Capt R. S. Sivaswamy (Retd) (Hyderabad Division)
  - H. K. Singh (Lucknow Division)

Chief Test Pilot: Wg Cdr M. W. Tilak (Retd)

Chief of Marketing: Air Cdre R. J. M. Upot (Retd)

Hindustan Aeronautics Limited (HAL) was formed on 1 October 1964, amalgamating the former Hindustan Aircraft Ltd (formed 1940) and Aeronautics India Ltd (formed 1963), and has 10 Divisions, five at Bangalore and one each at Nasik, Koraput, Hyderabad, Kanpur and Lucknow, plus a Design and Development Complex. The total work force is about 40,000 people. A new factory, to manufacture inertial navigation systems, is being established. The company, whose principal customer is the Indian Air Force, is currently manufacturing and overhauling many types of aircraft, helicopters, and their related aero-engines, avionics, instruments and accessories. It is also manufacturing components for satellites on behalf of the Indian Space Research Organisation (ISRO).

The Bangalore Complex is engaged in the manufacture of military aircraft and aero-engines, both under licence and of indigenous design. This Complex has a large organisation undertaking repair and overhaul of airframes, engines, and allied instruments and accessories. It has also a self-contained foundry and forge.

Kanpur Division has been engaged mainly in the manufacture of different versions of the British Aerospace HS 748 under licence. It was responsible for series production of the Basant agricultural aircraft, designed at the Bangalore design bureau.

Nasik, Koraput and Hyderabad Divisions are manufacturing the airframe, engine and avionics respectively of the Soviet MiG-21 fighter with the collaboration of the USSR.

Lucknow Division, formed in late 1969, is producing aircraft accessories under licence from various manufacturers in the UK, France and the USSR, including wheels and brakes, ejection seats, instruments, fuel accessories, air-conditioning and pressurisation systems. The Division has successfully developed a number of electrical and hydraulic accessories which have entered production. It is planned to begin manufacturing gyros for use in airborne systems.

In addition to its manufacturing programmes, HAL is pursuing design and development activities relating to aircraft, helicopters, avionics and accessories. The design bureau at Bangalore is engaged in development of the HPT-32 piston-engined basic trainer, an armed version of the HJT-16 Kiran, and a trainer version of the Ajeet. As noted in the International section, HAL is to be responsible for assembly and/or licence construction of approximately three-quarters of the SEPECAT Jaguar International combat aircraft ordered by the Indian government. The helicopter design department at Bangalore is designing an Advanced Light Helicopter (ALH) with the assistance of Aérospatiale of France. The avionics design department at Hyderabad is pursuing the design and development of airborne electronic equipment, and the design department at Lucknow is developing aircraft instruments and accessories.

BANGALORE COMPLEX

Bangalore Complex 560 017 (Karnataka State)
Telephone: 565201

The Bangalore Complex of HAL consists essentially of the former Hindustan Aircraft Limited, the activities of which, since its formation in 1940, were described in previous editions of *Jane's*. The Complex is subdivided into an Aircraft Division, Helicopter Division, Engine Division, Overhaul Division, Foundry and Forge Division, and Design and Development Complex.

Bangalore Complex is engaged in producing aircraft designed and developed by the HAL design bureau, and also in manufacturing various aircraft and aero-engines under licence. The Engine Division's activities are described in the appropriate section of this edition.

Licence production of Aérospatiale SA 315B Lama and SA 316B Alouette III helicopters, undertaken initially by the Aircraft Division, is undertaken by the Helicopter Division, officially inaugurated on 19 July 1974.

The Overhaul Division of Bangalore Complex repairs and overhauls HAL HF-24 and HT-2 and Hawker Siddeley (de Havilland) Dove/Devon aircraft, DHC-4 Caribou, Fairchild C-119 Packet transports and English Electric Canberra bombers, and various piston engines. Jet engines are overhauled at the Engine Division. The branch factory at Barrackpore near Calcutta continues to concentrate on repair and overhaul of C-47/DC-3s of the Indian Air Force and non-scheduled operators.

HAL (SEPECAT) JAGUAR INTERNATIONAL

The Bangalore Complex is responsible for the assembly and/or licence construction of approximately three-quarters of the SEPECAT Jaguar International combat aircraft (see International section) ordered for the Indian Air Force. Current orders are for 40 aircraft, with components for a further 45. The first UK-built airframe components for final assembly in India were delivered to HAL in early May 1981, and the first Indian-assembled Jaguars are due to enter Indian Air Force service in 1982.

HAL HJT-16 Mk I KIRAN (RAY OF LIGHT)

In December 1959, the government of India approved the design and development by HAL of a side-by-side two-seat jet basic trainer designated HJT-16 Mk I, powered by a Rolls-Royce Viper 11 turbojet.

Detailed design work on the HJT-16 Mk I began in April 1961 under the leadership of Dr V. M. Ghatage. The first prototype flew for the first time on 4 September 1964. It was followed by a second aircraft in August 1965.

A total of 24 pre-production HJT-16 Mk Is were delivered to the Indian Air Force, the initial delivery (of six aircraft) being made in March 1968. The Mk IA is now in series production. By 1 January 1981 a total of 180 Mk I/IAs had been produced, to meet the requirements of both the Indian Air Force and Indian Navy. Production was reportedly being slowed down in 1981 in favour of the armed **Kiran Mk II**, which is described separately.
rately.

The Kiran is suitable for use in armament training or light attack roles, and the 119th and subsequent Kirans, designated **Mk IA**, are fitted with a hardpoint beneath each wing capable of carrying weapons or a drop-tank.

HAL HJT-16 Mk IA Kiran two-seat jet basic trainer

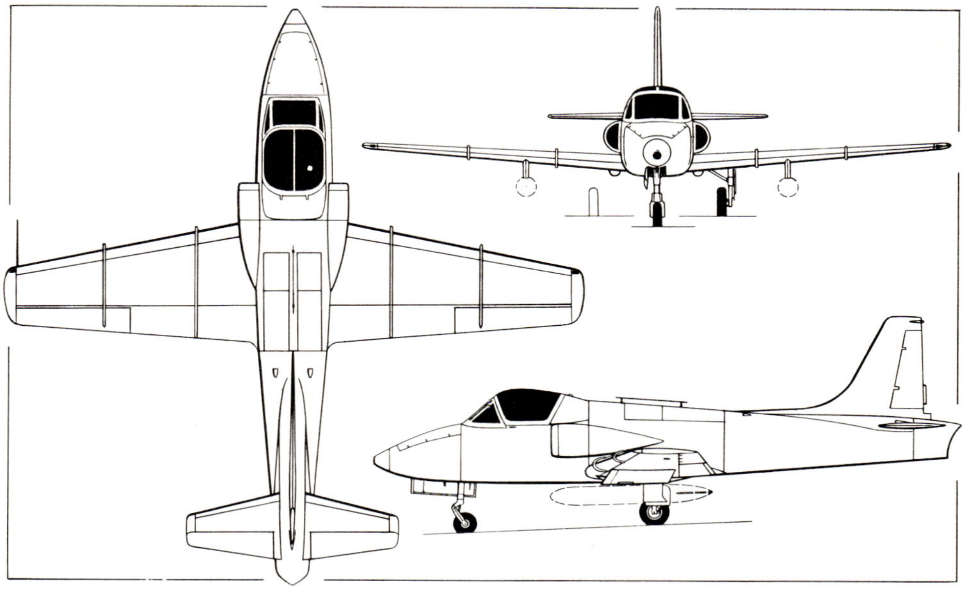

HAL HJT-16 Mk IA Kiran side-by-side two-seat basic trainer and light attack aircraft *(Pilot Press)*

Type: Two-seat jet basic trainer.

Wings: Cantilever low-wing monoplane. Wing section NACA 23015 at root, NACA 23012 at tip. Dihedral 4° from roots. Incidence 0° 30′ at root. Conventional all-metal three-spar structure. Frise-type differential ailerons. Hydraulically-actuated trailing-edge split flaps. Two full-chord boundary layer fences on upper surface of each wing.

Fuselage: All-metal semi-monocoque structure of light alloy. Hydraulically-actuated door-type airbrake under centre of fuselage.

Tail Unit: Cantilever all-metal structure. Electrically-operated variable-incidence tailplane. Ground-adjustable tab on rudder.

Landing Gear: Retractable tricycle type, of HAL manufacture. Hydraulic actuation. Main units retract inward into fuselage; self-centering twin-contact non-steerable nosewheel retracts forward. Oleo-pneumatic shock-absorbers. Main-wheel tyres size 19 × 6·25-9, pressure 6·21 bars (90 lb/sq in). Nosewheel tyre size 15·4 × 4-6, pressure 4·83 bars (70 lb/sq in). Hydraulic brakes, without cooling.

Power Plant: One 11·12 kN (2,500 lb st) Rolls-Royce Viper 11 turbojet engine. Internal fuel in main saddle tanks in fuselage (two 209 litre; 46 Imp gallon), wing centre-section collector tank (282 litres; 62 Imp gallons) and outboard wing tanks (two 218 litre; 48 Imp gallon), giving total capacity of 1,137 litres (250 Imp gallons). Provision for two underwing tanks with total capacity of 454 litres (100 Imp gallons). System permits 30 s of inverted flight.

Accommodation: Crew of two side by side in air-conditioned and pressurised cockpit, on Martin-Baker H4HA zero-altitude fully-automatic ejection seats. Clamshell-type canopy. Dual controls and duplicated blind-flying instruments.

Systems: Air-conditioning system has max pressure differential of 0·12 bars (1·75 lb/sq in). Dowty hydraulic system for landing gear, flaps and airbrake, pressure 207 bars (3,000 lb/sq in). Accumulator for manual emergency system. Electrical system is of 28V DC single-wire earth return type, with two 24V 25Ah batteries. Normalair pressure-demand oxygen system.

Avionics and Equipment: STR 9X/M 10-channel VHF transceiver, AX-3 single-channel VHF standby set and Marconi Avionics DFA-73 ADF manufactured by BEL-India. Landing light in nose.

First prototype HAL Kiran Mk II armament training and counter-insurgency aircraft

Armament: Hardpoint beneath each wing of 119th and subsequent aircraft, each capable of carrying a 500 lb bomb, an HAL pod containing two 7·62 mm FN machine-guns, a pod containing seven 68 mm SNEB rockets, or a 227 litre (50 Imp gallon) drop-tank.

Dimensions, external:
- Wing span: 10·70 m (35 ft 1¼ in)
- Wing chord at root: 2·35 m (7 ft 8½ in)
- Wing chord at tip: 1·02 m (3 ft 4 in)
- Wing aspect ratio: 6
- Length overall: 10·60 m (34 ft 9 in)
- Height overall: 3·635 m (11 ft 11 in)
- Tailplane span: 3·90 m (12 ft 9½ in)
- Wheel track: 2·42 m (7 ft 11 in)
- Wheelbase: 3·50 m (11 ft 6 in)

Areas:
- Wings, gross: 19·00 m² (204·5 sq ft)
- Ailerons (total): 1·55 m² (16·68 sq ft)
- Flaps (total): 2·34 m² (25·19 sq ft)
- Vertical tail surfaces (total): 2·10 m² (22·60 sq ft)
- Rudder, incl tab: 0·714 m² (7·69 sq ft)
- Horizontal tail surfaces (total): 3·72 m² (40·04 sq ft)
- Elevators: 1·14 m² (12·27 sq ft)

Weights and Loadings:
- Weight empty: 2,560 kg (5,644 lb)
- Normal T-O weight: 3,600 kg (7,936 lb)
- Max T-O weight (with two 50 Imp gallon drop-tanks): 4,100 kg (9,039 lb)
- Max wing loading: 190 kg/m² (38·9 lb/sq ft)
- Thrust/weight ratio: 0·315

Performance (at normal T-O weight):
- Max level speed at S/L: 375 knots (695 km/h; 432 mph)
- Max level speed at 9,150 m (30,000 ft): 371 knots (688 km/h; 427 mph)
- Max cruising speed: 175 knots (324 km/h; 201 mph)
- Stalling speed, flaps and landing gear up: 81 knots (151 km/h; 94 mph)
- Stalling speed, flaps and landing gear down: 71 knots (132 km/h; 82 mph)
- Ceiling: 9,150 m (30,000 ft)
- Time to 9,150 m (30,000 ft): 20 min
- Min ground turning radius: 5·50 m (18 ft 0½ in)
- T-O run: 442 m (1,450 ft)
- Endurance on internal fuel at 230 knots (426 km/h; 265 mph) at 9,150 m (30,000 ft): 1 h 45 min

HAL KIRAN Mk II

This version of the Kiran, suitable for armament training or counter-insurgency duties, is being developed by the HAL design bureau at Bangalore. Principal differences include improved weapon-carrying capability, a more powerful engine, updated avionics and an improved hydraulic system. The engine is a derated version of the Orpheus 701 turbojet, rated at 15·1 kN (3,400 lb st), which gives the Kiran Mk II improved maximum speed, climb and manoeuvrability. The range, with two underwing drop-tanks fitted as standard, remains the same as for the Mk I. The first prototype made its initial flight on 30 July 1976. A second prototype was flown in February 1979, and others are reported to have been completed subsequently.

Development flying has been completed, and an Indian Air Force order for 24 Kiran Mk IIs was announced in the Summer of 1981. Deliveries are due to begin in early 1983.

Armament: Two 7·62 mm machine-guns in nose, with 250 rds/gun. Two Ferranti F 195R ISIS gunsights, with common control unit. Two hardpoints beneath each wing, each capable of carrying a 227 litre (50 Imp gallon) drop-tank, a 500 lb high-explosive bomb, a Type 122·1 68 mm rocket pod or a 25 lb practice bomb carrier.

HAL AJEET (INVINCIBLE)

The Hawker Siddeley Gnat light fighter and fighter-bomber was built under licence by HAL between 1962 and 1974, as described in the 1974-75 and 1975-76 editions of *Jane's*.

The design bureau completed in 1974 the design of a Mk II version of the Gnat known as the Ajeet, with improved performance characteristics and equipment, including improved communications and navigation systems; more reliable longitudinal control; and increased combat capability. The last-named characteristic is achieved by a redesigned fuel system, dispensing with the underwing drop-tanks in favour of integral wing tanks, so permitting additional underwing armament to be carried.

The last two Gnat Mk I aircraft were converted as prototypes for the Ajeet; the first of these was flown on 5 March 1975, and the second on 5 November 1975.

First flight of a production Ajeet was made on 30 September 1976. By January 1981 a total of 50 Ajeets had been completed, of 89 on order. Production in 1980-81 was reported to be at the rate of approx 20 per year. Ajeets are in service with Nos. 9 and 18 Squadrons of the Indian Air Force, and two further squadrons are to be formed.

Type: Single-seat lightweight interceptor and ground attack aircraft.

Wings: Cantilever shoulder-wing monoplane. Sweptback wings, of RAE 102 section. Thickness/chord ratio 8%. Anhedral 5°. Sweepback 40° at quarter-chord. One-piece wing of two-spar thick-skin light alloy construction, fitting into recess in top of fuselage and secured by bolts at four points. Inboard ailerons, powered by hydraulic actuators, droop 22° to serve as flaps when the landing gear is lowered.

Fuselage: Light alloy semi-monocoque structure of pressed frames and extruded stringers.

Tail Unit: Cantilever all-metal structure. Three-spar sweptback fin, integral with fuselage. One-piece three-spar variable-incidence tailplane, operated hydraulically by modified Hobson PFC 1003 actuator. Rear portions of tailplane can be unlocked to perform as elevators; or locked to provide the functions of an all-moving tailplane. Ground-adjustable tab on rudder.

Landing Gear: Retractable tricycle type, all units retracting rearward hydraulically into fuselage. Dowty Rotol oleo-pneumatic shock-absorber struts. Wheel well fairings attached to individual landing gear units serve as airbrakes when landing gear is partly lowered, the relative movements of the airbrakes being so adjusted that no change of trim occurs at any speed. Dunlop mainwheel tyres size 20 × 5·25, pressure 9·3 bars (135 lb/sq in); twin nosewheel tyres, size 17 × 3·25, pressure 5·65 bars (82 lb/sq in). Hydraulically operated brakes and Maxaret anti-skid units on main wheels. Braking parachute in fairing at base of fin.

Power Plant: One Rolls-Royce Orpheus 701-01 non-afterburning turbojet engine, rated at 20 kN (4,500 lb st). Compressed-air starting. Air intakes in sides of fuselage. Seven crashproof flexible tanks and two metal tanks in fuselage, and two 250 litre (55 Imp gallon) integral wing tanks. Total internal fuel capacity 1,350 litres (297 Imp gallons). Fuel supplied to engine by electrically-driven booster pump in one of the tanks. Provision for two 150 litre (33 Imp gallon) underwing drop-tanks.

Accommodation: Pilot only, on Martin-Baker GF4 zero-height/90 knot (167 km/h; 104 mph) lightweight ejection seat. Pressurised, heated and air-conditioned cockpit, with jettisonable canopy which is hinged at rear and opens upward.

Systems: Normalair air-conditioning and pressurisation system, max differential of 0·24 bars (3·5 lb/sq in) at 12,800 m (42,000 ft). Oxygen system with demand-type regulator. Dowty hydraulic system of 207 bars (3,000 lb/sq in), with Abex pump, for aileron, landing gear, main-wheel brake and tailplane actuation. 28V DC electrical system, with 3·5kW Rotax generator and two 12V 25Ah Varley batteries. Oxygen system for pilot.

Avionics and Equipment: Bendix TA/RA-22 VHF transceiver (initially; V/UHF later) and BEL AX-3 standby VHF set; Bendix DFA-73 ADF; IFF Mk 10 (BAT) transponder. Ferranti navigation system.

Armament: Two 30 mm Aden Mk 4 cannon in air intake fairings, one on each side of fuselage, with 90 rds/gun. Ferranti F 195R/3 ISIS weapons sight. Vinten G90 gun camera. Four underwing hardpoints able to carry two 500 lb bombs (inner pylons), four Arrow Type 122 pods each containing eighteen 68 mm rockets, or two 150 litre (33 Imp gallon) drop-tanks (outer pylons).

Dimensions, external:

Wing span	6·73 m (22 ft 1 in)
Wing chord at c/l	2·58 m (8 ft 5·6 in)
Wing chord at tip	1·17 m (3 ft 10 in)
Wing area, gross	14·65 m² (157·7 sq ft)
Wing aspect ratio	3·575
Length overall	9·04 m (29 ft 8 in)
Height overall	2·46 m (8 ft 1 in)
Tailplane span	2·84 m (9 ft 4 in)
Wheel track	1·55 m (5 ft 1 in)
Wheelbase	2·36 m (7 ft 9 in)

Weights and Loading:

Basic weight empty	2,307 kg (5,086 lb)
T-O weight 'clean'	3,539 kg (7,803 lb)
Max T-O weight	4,170 kg (9,195 lb)
Normal landing weight	2,767 kg (6,100 lb)
Max wing loading	284·6 kg/m² (58·3 lb/sq ft)

Performance (in configurations indicated; A: ISA; B: ISA + 15°C; C: ISA + 30°C):

Max Mach No. at 12,000 m (39,375 ft) at 'clean' T-O weight: A 0·96
B 0·953
C 0·948

Max level speed at S/L at 'clean' T-O weight:
A 595 knots (1,102 km/h; 685 mph)
B 612 knots (1,134 km/h; 705 mph)
C 622 knots (1,152 km/h; 716 mph)

Time to 12,000 m (39,375 ft) from brakes off, at 'clean' T-O weight: A 6 min 2 s
B 7 min 43 s
C 9 min 33 s

Service ceiling: A, B, C 13,720 m (45,000 ft)

Turning performance at 450 knots (834 km/h; 518 mph) IAS at S/L: A 5·30g
B 5·28g
C 5·00g

T-O run at S/L, zero wind, at T-O weight of 4,136 kg (9,118 lb) with two Arrow rocket pods and two 33 Imp gallon drop-tanks: A 1,034 m (3,390 ft)
B 1,180 m (3,870 ft)
C 1,376 m (4,515 ft)

Landing run 'clean' at S/L, zero wind, at normal landing weight, no brake 'chute: A 951 m (3,120 ft)
B 997 m (3,270 ft)
C 1,047 m (3,435 ft)

Landing run 'clean' at S/L, zero wind, at normal landing weight, with brake 'chute: A 658 m (2,160 ft)
B 695 m (2,280 ft)
C 725 m (2,379 ft)

Combat radius (A, B and C), low level ground attack mission:
with two 500 lb bombs on inboard stations 110 nm (204 km; 127 miles)
with two Arrow rocket pods inboard and two 33 Imp gallon drop-tanks outboard 140 nm (259 km; 161 miles)
with four Arrow rocket pods 104 nm (193 km; 120 miles)

Prototype of the Ajeet lightweight combat aircraft, developed by HAL from the Hawker Siddeley Gnat

HAL Ajeet lightweight interceptor/ground attack aircraft, with additional side view (bottom) of trainer version *(Pilot Press)*

HAL AJEET TRAINER

This tandem two-seat trainer version of the Ajeet, which retains the four underwing hardpoints and full combat capability of the single-seater, is evolved in a similar manner to that in which the T.Mk 1 trainer was evolved from the original Gnat fighter. To accommodate the second cockpit, two of the fuselage fuel tanks are deleted, although this can be offset by deleting the Aden cannon to make room for an additional 273 litres (60 Imp gallons) of fuel internally; in this event, provision is made for carrying a 7·62 mm gun pod on each of the inboard wing pylons. Sections 0·70 m (2 ft 3½ in) long are inserted in the fuselage fore and aft of the wings, increasing the overall length of the Ajeet Trainer to 10·44 m (34 ft 3 in).

The Ajeet Trainer is powered by a single Orpheus 701 turbojet engine, and retains the main hydraulic system and powered flying controls of the single-seater. Normal and emergency operation of the landing gear is also similar, but with duplicated controls and a mechanical override facility in the rear cockpit. All instruments, including those for blind flying, are duplicated in the rear cockpit, which will be illuminated, air-conditioned and pressurised similarly to the front cockpit. A gunsight is fitted in the front cockpit only. Both occupants are provided with Martin-Baker GF4 ejection seats. Avionics and equipment include a multi-channel VHF transceiver (to be replaced eventually by a V/UHF set), ADF and IFF Mk 10.

A prototype is under construction, and was expected to undergo evaluation by the end of 1981.

HAL HPT-32

Currently under development for the Indian Air Force, the HPT-32 is a fully-aerobatic piston-engined basic trainer, with side-by-side seats for instructor and pupil and a third seat at the rear. A four-seat version is under consideration for the liaison role. The trainer can be used for a wide range of ab initio training, including instrument, navigation, night flying and formation flying; for armed patrol; for observation, liaison or sport flying; or for weapon training, light strike duties, supply dropping, search and rescue, reconnaissance, or glider or target towing. The airframe, which is of all-metal construction, is

First prototype HAL HPT-32 two/three-seat basic training aircraft

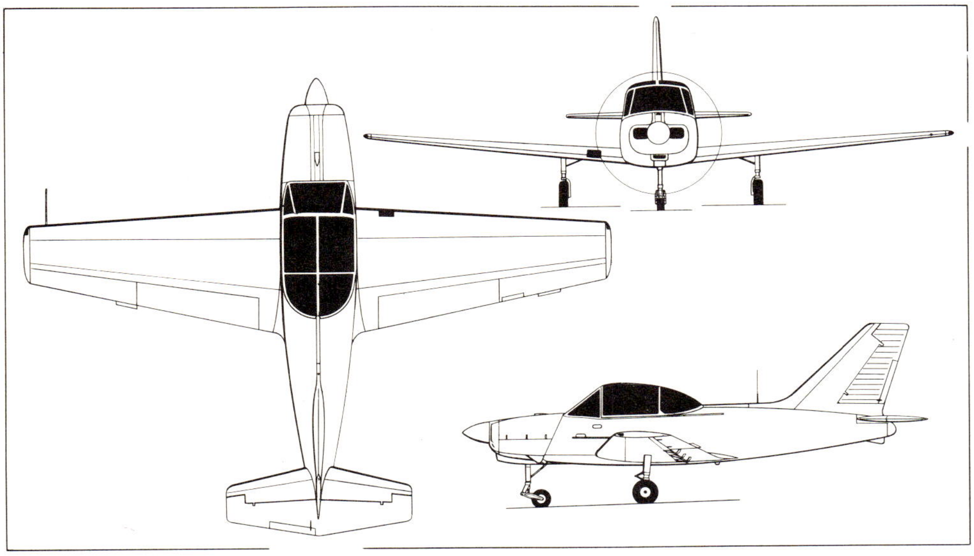

HAL HPT-32 two/three-seat basic trainer in production form *(Pilot Press)*

designed to FAR Pt 23, and is expected to have a fatigue life of 6,500 h.

The first prototype (X2157) made its first flight on 6 January 1977. The second was flown on 12 March 1979; a third, flown on 31 July 1981, represents an improved version, substantially lighter in weight and with aerodynamic refinements. Indian Air Force orders are expected to cover an eventual total of 160 HPT-32s to this latest standard. An initial batch of 24, for delivery from 1983, was ordered in the Summer of 1981. These will be built by HAL's Kanpur Division.

Type: Two/three-seat basic trainer or four-seat liaison aircraft.

Wings: Cantilever low-wing monoplane of all-metal construction. Dihedral 5° from roots. Incidence 2° 30′ at root. Ground-adjustable tab on each aileron.

Fuselage: All-metal semi-monocoque structure.

Tail Unit: Cantilever all-metal structure, with sweptback vertical surfaces. One-piece elevator. Trim tabs in rudder and port half of elevator.

Landing Gear: Non-retractable tricycle type on prototype and ab initio version. Main wheels size 6·00-6·5, nosewheel 5·00-5. Dunlop tyres on all wheels, pressure 2·41-2·76 bars (35-40 lb/sq in) on main units, 3·10 bars (45 lb/sq in) on nose unit..

Power Plant: Prototype powered by 194 kW (260 hp) Avco Lycoming AEIO-540-D4B5 flat-six engine, driving a Hartzell two-blade constant-speed metal propeller with spinner. Two integral wing fuel tanks, with total capacity of 227 litres (50 Imp gallons); provision for 136·5 litre (30 Imp gallon) tank in place of rear seat. For production aircraft, an indigenous engine of similar power is under development.

Accommodation: Side-by-side seats for two persons in front, with provision for one or two seats plus baggage space at rear, under rearward-sliding jettisonable framed canopy. Front two seats adjustable in height by 127 mm (5 in); rear seat(s) not adjustable. Full dual controls, and adjustable rudder pedals, for instructor and pupil.

Armament and Equipment: Four underwing attachments for armament or other stores, up to a total of 255 kg (562 lb). VHF radio.

Dimensions, external:

Wing span	9·50 m (31 ft 2 in)
Wing chord at root	2·24 m (7 ft 4¼ in)
Wing chord at tip	0·92 m (3 ft 0¼ in)
Wing aspect ratio	6·013
Length overall	7·715 m (25 ft 3¾ in)
Height overall	3·27 m (10 ft 8¾ in)
Wheel track	3·45 m (11 ft 4 in)
Wheelbase	2·10 m (6 ft 10¾ in)
Propeller diameter	2·032 m (6 ft 8 in)
Propeller ground clearance (static)	0·241 m (9½ in)

Areas:

Wings, gross	15·01 m² (161·6 sq ft)
Ailerons (total)	1·04 m² (11·19 sq ft)
Trailing-edge flaps (total)	1·83 m² (19·70 sq ft)
Fin	2·316 m² (24·93 sq ft)
Rudder, incl tab	1·078 m² (11·60 sq ft)
Tailplane	3·02 m² (32·50 sq ft)
Elevator, incl tab	1·35 m² (14·53 sq ft)

Weights and Loadings:

Weight empty	926 kg (2,041 lb)
Normal T-O weight	1,250 kg (2,756 lb)
Max T-O weight	1,375 kg (3,031 lb)
Max wing loading	91·6 kg/m² (18·76 lb/sq ft)
Max power loading	7·09 kg/kW (11·66 lb/hp)

Performance (at Normal T-O weight, ISA):

Max level speed at S/L	117 knots (217 km/h; 135 mph)
Stalling speed, flaps up	59 knots (110 km/h; 68 mph)
Stalling speed, flaps down	55 knots (102 km/h; 63 mph)
Max rate of climb at S/L	335 m (1,100 ft)/min
Service ceiling	4,750 m (15,580 ft)
T-O run	205 m (673 ft)
Range at 1,525 m (5,000 ft) with 50 Imp gallons fuel	378 nm (700 km; 435 miles)
Endurance at 1,525 m (5,000 ft) with 50 Imp gallons fuel	4 h 0 min
g limits	+6·0; −3·0

HAL (AÉROSPATIALE) SA 315B LAMA

Indian name: Cheetah

The Bangalore Complex's Helicopter Division is building the French Aérospatiale SA 315B Lama five-seat general-purpose helicopter (which see) under licence in India, where it is known as the Cheetah.

Initial production was from French-built components. The first Indian-assembled Cheetah was test-flown on 6 October 1972, and a total of 133 had been delivered by early 1981. Production of aircraft manufactured from raw materials began in 1976-77.

HAL (AÉROSPATIALE) SA 316B ALOUETTE III

Indian name: Chetak

The Bangalore Complex's Helicopter Division is building the French Aérospatiale SA 316B Alouette III under a licence granted in June 1962. The first Indian-assembled Alouette III (Indian name Chetak) was flown for the first time on 11 June 1965.

By early 1981 a total of 251 Alouette IIIs had been manufactured by HAL, with production continuing at the rate of about four per month. A few Alouette IIIs were presented to the Royal Nepal Army in 1974, and the company now markets both military and civil models. It also supplies Indian-built components for French-built Alouette IIIs.

An armed version of the Chetak is being developed by HAL for the Indian Air Force and Navy, carrying four air-to-surface missiles on laterally-mounted booms. Target identification and fire control are via a monocular periscopic sight on the cabin roof. Successful firing trials have been carried out.

HAL LIGHT HELICOPTER

The Design Bureau of HAL has under development a twin-engined advanced light helicopter (ALH). Two versions are being developed: a standard version for Indian Air Force/Army use, and a variant for the Indian Navy. The former will have a capability for combat missions, communications duties, armed reconnaissance and surveillance, casualty evacuation, crew rescue, external cargo carrying and training. The naval version will be able to perform anti-submarine search and strike, air to surface vessel search and strike, search and rescue, reconnaissance, casualty evacuation, and vertical replenishment duties at sea.

It has been reported that the ALH may be produced in collaboration with a European helicopter manufacturer. The total Indian armed forces requirement for the ALH has been quoted as 300-400 aircraft.

HAL Cheetah, Indian-assembled version of the Aérospatiale SA 315B Lama helicopter

HAL-assembled Aérospatiale Alouette III (Chetak) helicopter in Indian military insignia

KANPUR DIVISION

Chakeri, Kanpur
Telephone: HAL PABX 62471-4
Telex: HAL KP 243

When the decision was taken to build the Hawker Siddeley (now British Aerospace) 748 twin-turboprop transport in India, as a replacement for the C-47s of the Indian Air Force, four hangars at Kanpur were taken over, on 23 January 1960, as the IAF Aircraft Manufacturing Depot. The Depot was incorporated in Aeronautics (India) Ltd in June 1964 and subsequently became the Kanpur Division of Hindustan Aeronautics Ltd.

HAL (ANTONOV) An-32

NATO reporting name: Cline

India's Prime Minister, Mrs Indira Gandhi, confirmed in late 1980 that the An-32 had been selected to replace the Douglas C-47 and Fairchild C-119 transport aircraft in Indian Air Force service.

Reports indicate that assembly of 45 An-32s for the IAF will be undertaken under licence by HAL, following the delivery of 50 Soviet-built examples.

HAL (BAe (HS)) 748

The first Indian-built 748 flew on 1 November 1961, followed by the second on 13 March 1963. The first four Indian 748s were Srs 1 aircraft, utilising components imported from the UK.

The first Indian-built Srs 2 flew for the first time on 28 January 1964, and 17 were delivered to Indian Airlines between 1967 and 1972. Since then three more have been modified for a photographic survey role and delivered to the Indian Air Force. Three others are being delivered to the Indian Directorate General of Civil Aviation (for a calibration role) and the National Remote Sensing Agency.

HAL has also produced the aircraft in several versions for the Indian Air Force. These include 10 VIP executive transports, 10 as navigation trainers and 18 as trainers for pilots of multi-engined aircraft.

A prototype BAe (HS) 748(M) military freighter, developed by Kanpur Division, flew for the first time on 16 February 1972 and successfully completed flight trials with the Indian Air Force. An order for 10 more BAe (HS) 748s was announced in June 1975. These aircraft are built to the military freighter standard. A further 10 BAe (HS) 748(M)s were ordered in 1978, extending production until 1983. Production rate in 1980 was approx four per year.

HAL HA-31 Mk II BASANT (SPRING)

HAL HA-31 Mk II Basant single-seat agricultural and utility aircraft

Design of this agricultural aircraft began at the Bangalore design bureau in mid-1968. The prototype, designated HA-31 Mk I, was powered by a 186 kW (250 hp) Rolls-Royce Continental engine, and was described in the 1971-72 *Jane's.* The aircraft was subsequently completely redesigned as the HA-31 Mk II, with a 298 kW (400 hp) engine, and a prototype of this version flew for the first time on 30 March 1972. A second, pre-production prototype flew in September 1972, and certification was obtained in March 1974. A pre-production batch of 20 Basants was built by the design bureau, and the first eight of these were handed over to the Indian Ministry of Food and Agriculture on 21 June 1974. Responsibility for series manufacture of the Basant was then assigned to the Kanpur Division.

The original intention was to manufacture 100 production aircraft, but the programme was reportedly terminated in 1980 after only 40 had been completed. A full description can be found in the 1980-81 *Jane's*.

Type: Single-seat agricultural and utility aircraft.

Power Plant: One 298 kW (400 hp) Avco Lycoming IO-720-C1B flat-eight engine, driving a Hartzell three-blade constant-speed metal propeller. Total fuel capacity 318 litres (70 Imp gallons).

Accommodation: Single seat in fully-enclosed cockpit. Forward-hinged door on starboard side; emergency door on port side. Cockpit heated and ventilated.

Avionics and Equipment: VHF radio optional. Glassfibre hopper, installed between engine firewall and front wall of cockpit enclosure, has 0·93 m³ (33 cu ft) capacity and can carry up to 605 kg (1,333 lb) of pesticide for Normal category operation and up to 907 kg (2,000 lb) in Restricted category.

Dimensions, external:
Wing span 12·00 m (39 ft 4½ in)
Wing area, gross 23·34 m² (251·23 sq ft)
Length overall 9·00 m (29 ft 6¼ in)
Height overall 2·55 m (8 ft 4½ in)
Wheel track 2·70 m (8 ft 10¼ in)
Wheelbase 6·00 m (19 ft 8¼ in)
Propeller diameter 2·13 m (7 ft 0 in)

Weights and Loadings:
Weight empty 1,200 kg (2,645 lb)
Basic operating weight 1,954 kg (4,300 lb)
Max T-O weight 2,270 kg (5,000 lb)
Max wing loading 97·25 kg/m² (19·92 lb/sq ft)
Max power loading 7·62 kg/kW (12·6 lb/hp)

Performance (at basic operating weight):
Never-exceed speed 164 knots (305 km/h; 189 mph)
Max level speed at S/L 121 knots (225 km/h; 140 mph)
Max cruising speed at 2,440 m (8,000 ft) 100 knots (185 km/h; 115 mph)
Econ cruising speed 87 knots (161 km/h; 100 mph)
Stalling speed, flaps up 52 knots (96 km/h; 60 mph)
Stalling speed, flaps down 49 knots (91 km/h; 57 mph)
Max rate of climb at S/L 228 m (750 ft)/min
Service ceiling 3,800 m (12,500 ft)
Min ground turning radius 7·00 m (22 ft 11 in)
T-O run 214 m (700 ft)
T-O to 15 m (50 ft) 365 m (1,200 ft)
Landing from 15 m (50 ft) 305 m (1,000 ft)
Landing run 183 m (600 ft)
Range with max fuel, no payload 348 nm (645 km; 400 miles)
Endurance with max payload, 30 min reserves 1 h

MiG COMPLEX

The MiG Complex was originally formed with the Nasik, Koraput and Hyderabad Divisions of HAL, which, under an agreement concluded in 1962, built respectively the airframes, power plants and avionics equipment of MiG-21 fighters under licence from the USSR.

As a result of the reorganisation of HAL, the Hyderabad Division, with Lucknow Division, now constitutes the Accessories Complex. Nasik had a work force of approx 8,000 in 1980, of whom about 800 were technical and management staff.

HAL (MIKOYAN) MiG-21

Indian Air Force designations: Type 66-400, 66-600, 74, 76, 77 and 96

NATO reporting name: Fishbed

Several versions of the MiG-21 have been supplied to or, since 1966, manufactured in India. These have included the MiG-21F (IAF designation Type 74) and MiG-21PF (Type 76) of which details can be found in the 1976-77 *Jane's;* and the MiG-21FL (Type 77) and MiG-21U (Types 66-400 and 66-600), as described in the 1977-78 *Jane's.* Other versions include:

MiG-21M. Production version from 1972, with R-11F2S-300 engine. First aircraft, assembled from imported components, handed over to IAF on 14 February 1973. Manufacture from locally produced materials began in 1975. IAF designation **Type 96**. Fifty Soviet-built MiG-21PFMAs, to supplement Indian production, were placed in service with Nos. 7 and 108 Squadrons; No. 26 Squadron is among those also reportedly equipped with MiG-21M. Production scheduled to phase out in 1981 in favour of MiG-21 bis.

MiG-21MF ('Fishbed-J'). Improved production version, having more powerful Tumansky R-13-300 turbojet engine and increased fuel.

MiG-21 bis ('Fishbed-N'). Deliveries of Soviet-built examples of this version, initially for service with No. 21 Squadron of IAF, were reported in 1977. Indian production of the MiG-21 bis, and its Tumansky R-25 engine, were reported to have begun by early 1980, and this was to supersede the MiG-21M as the major production version from 1981. To reduce cost and engineering problems, modified MiG-21M wings are reportedly used on the Indian MiG-21 bis.

Indian production of MiG-21s was reported to be at the rate of about 30 per year in 1980. It is at present expected to end in 1983, in favour of the MiG-23.

A full description of the MiG-21 appears in the USSR section of this edition.

HAL (MIKOYAN) MiG-23

NATO reporting name: Flogger

The Indian Air Force has begun to receive 15 two-seat MiG-23UM ('Flogger-C') combat trainers and 80 single-seat MiG-23BNs ('Flogger-F') from the USSR. Deliveries began in December 1980, and by mid-1981 the MiG-23BN was in service (replacing HAL HF-24 Maruts) with Nos. 10 and 230 Squadrons. A third Marut squadron is due to re-equip with MiG-23BNs in 1982. In the following year, licence production of up to 200 of this version is expected to begin at HAL.

INDONESIA

LAPAN

LEMBAGA PENERBANGAN DAN ANTARIKSA NASIONAL (National Aeronautics and Space Institute)

Headquarters: Jalan Pemuda Persil No. 1, PO Box 3048, Jakarta
Telephone: (021) 482802 and 485125
Telex: 45675 LAPAN IA
Chairman: Air Vice-Marshal Dr Sunaryo
Vice-Chairman: Prof Wiranto Arismunandar

Pusat Teknologi Dirgantara (Aerospace Technology Center)

Rumpin Airfield, Bogor, West Java

Head of Aerospace Technology Center: Dr Harijono Djojodihardjo

Manager, Aerospace Technology Development: Ir Jaidun Kromodihardjo

Established in 1963 under the original chairmanship of Air Vice-Marshal J. Salatun, LAPAN had 525 personnel at the beginning of 1979, accommodated within four centres: the Aerospace Technology Center at Rumpin Airfield, near Bogor; the Aerospace Study Center and the Space Applications Center, both at Jakarta; and the Atmospheric and Space Research Center at Bandung.

LAPAN XT-400

In the same class as the Britten-Norman Islander, the XT-400 twin-engined light STOL transport differs in having an upswept rear fuselage with clamshell rear-loading doors. Designed under the leadership of Dipl-Ing Suharto, it is of all-metal construction, and in its initial form was intended to accommodate one pilot and up to seven passengers. Power plant comprised two 186·5 kW (250 hp) Avco Lycoming IO-540-C flat-six engines.

Construction of a prototype was started in FY 1977, but was suspended in 1981. A full description, and a three-view drawing of the XT-400, can be found in the 1980-81 *Jane's*.

Nurtanio-built NC-212 Srs 200 in the insignia of PLP (Pendidikan Latihan Penerbangan), the Indonesian Civil Training Centre at Curug, West Java

Nurtanio's own NBO 105 executive transport helicopter (PK-XBF)

SA 330 Puma built by Nurtanio under Aérospatiale licence

PT NURTANIO

PT INDUSTRI PESAWAT TERBANG NURTANIO (Nurtanio Aircraft Industries Ltd)

Lanuma Husein Sastranegara (Husein Sastranegara Air Force Base), Jalan Pajajaran 154, Bandung
Telephone: Bandung 613662, 613840, 613835 and 613836
Telex: 28295 NUR BD
President Director: Prof Dr-Ing B. J. Habibie
Secretary to President Director: Tatang Endan
Directors:
Mr Suwondo (General Affairs)
Ir Yuwono (Production)
Ir Harsono Pusponegoro (Technology)
Managers:
Ir Ernis Burhan (General Workshop Division)
Ir Sutadi Suparlan (Fixed Wing Division)
Ir Djermani (Rotary Wing Division)
Ir Paramayuda (Aircraft Service Division)
Asst for Commercial Affairs: Ir Paramayuda
Asst for Finance: Dr Bambang Ekoyono
Asst for Programme Co-ordination: Ir Rahadi Ramelan
Asst for Security Co-ordination: Tatang Endan
General Inspector: Dr S. Parlin Napitupulu
Chief of Public Relations: Suripto Sugondo

This company was officially inaugurated in August 1976 when, as a result of a government order dated 5 April 1976, the former Lipnur (Lembaga Industri Penerbangan Nurtanio: see 1977-78 *Jane's*) combined its aircraft industry activities with those of the Pertamina oil company. It is named in honour of the late Air Marshal Nurtanio Pringgoadisuryo, who was largely responsible for establishing an aircraft industry in Indonesia.

Construction of a new 36 hectare (89 acre) administration and manufacturing facility at Husein Sastranegara AFB has been under way since the Spring of 1980, and is due to be completed in 1986. Nurtanio had a work force of approx 5,000 persons in 1981.

CASA/NURTANIO (AIRTEC) CN-235

Nurtanio and CASA (see Spanish section) are developing jointly a 34/38-passenger transport aircraft of their own design, known as the CN-235. Details of this aircraft can be found under the Airtec heading in the International section.

NURTANIO (CASA) NC-212 AVIOCAR

The C-212 Aviocar twin-turboprop transport aircraft has been manufactured in Indonesia since 1976, under licence from CASA of Spain (which see). Indonesian-built Aviocars have the designation NC-212.

By 1 April 1981 Nurtanio had delivered 29 NC-212-5 Series 100 Aviocars; and has now completed production of this version. By the same date it had also completed 12 examples of the Srs 200, of which two had been delivered, and had orders for a further 51 of this version, bringing the total orders for Indonesian-built Aviocars to 92. Principal customers to date include the Indonesian Air Force (two Srs 100 and 14 Srs 200), Merpati Nusantara Airlines (six Srs 100 and ten Srs 200), Pelita Air Service (four Srs 100 and 12 Srs 200), and Mandala Airlines (ten Srs 200). Export orders have been received from Thailand, Bangladesh and Pakistan.

NURTANIO (MBB) NBO 105

The BO 105 helicopter has been manufactured in Indonesia since 1976, under licence from MBB of the Federal Republic of Germany (which see). Indonesian designation is NBO 105.

By 1 April 1981 a total of 52 had been completed, of which 37 had been delivered; a further 74 were then on order. Major customers are the Indonesian Army, Navy and Air Force (16 each), Pelita Air Service (36), and the Indonesian Forestry Department (12).

NURTANIO (AÉROSPATIALE) SA 330 PUMA and AS 332 SUPER PUMA

Nurtanio has rights to manufacture under licence the Aérospatiale SA 330 Puma and AS 332 Super Puma helicopters. Knocked-down components for three SA 330s were delivered in mid-December 1980, and assembly of these began in early January 1981. The first was scheduled to make its initial flight by the end of March, with deliveries beginning in April 1981.

The initial batch of helicopters, covered by an agreement signed on 13 May 1980, comprises eight Pumas for the Indonesian Air Force and seven Super Pumas for Pelita Air Service. The first Super Puma is due to be completed by early 1982.

INTERNATIONAL PROGRAMMES

AERITALIA/AERMACCHI/EMBRAER

Participating Companies:
Aeritalia (Combat Aircraft Group), Piazzale Vincenzo Tecchio 51 (Casella Postale 3065), 80125, Naples, Italy
Telephone: (081) 619522, 619721, 619845, 619149 and 619703
Telex: N 710370 AERIT
Aermacchi SpA, Via Sanvito 80 (Casella Postale 246), 21100 Varese, Italy
Telephone: (0332) 283100
Telex: 380070 AERMAC I
Empresa Brasileira de Aeronáutica SA, Av Brig Faria Lima 2170, Caixa Postal 343, 12200 São José dos Campos, SP, Brazil
Telephone: (123) 21 5400
Telex: (391) 1133589 EBAE BR or (391) 1133917 EBAE BR

AM-X

Development of this subsonic single-seat and single-engined combat aircraft was initiated by Aeritalia to meet the requirements of the Italian Air Force. Aermacchi was nominated associate contractor in the programme.

In the Italian Air Force the AM-X is intended to take over duties performed currently by four types of aircraft: the G91, which will be phased out of its close air support role by 1985/86; the G91Y interdictor, also due for phase-out by 1985/86; and the F-104G and F-104S versions of the Starfighter, scheduled for replacement in the strike role by 1986/87 and 1990 respectively.

The close air support and interdiction tasks will be undertaken fully by the AM-X; counter-air duties will be shared with the longer-range Tornado. A total of 187 aircraft is needed to meet these re-equipment plans, with deliveries beginning ideally in 1986, to ensure initial operational capability by the Spring of 1987.

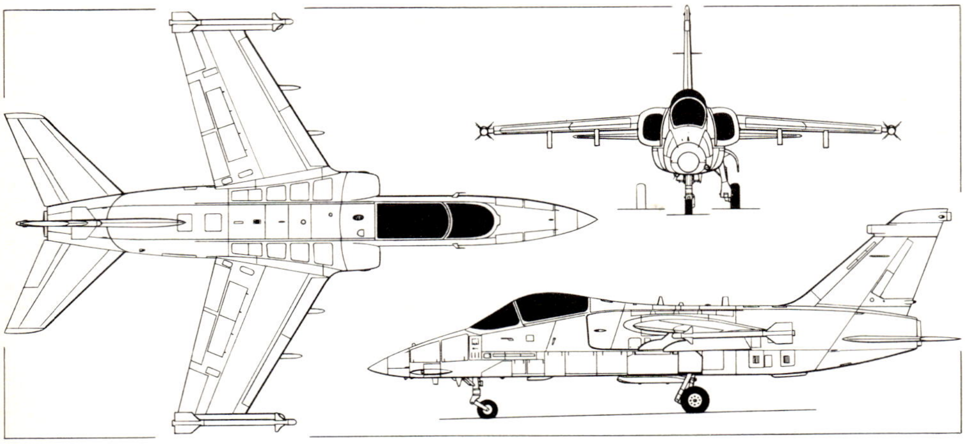

Aeritalia/Aermacchi/EMBRAER AM-X, under development for the Air Forces of Italy and Brazil *(Pilot Press)*

Work on the AM-X project was started by Aeritalia in 1977. By early 1979 the definition phase, due to last for 18 months, had been initiated under Italian Air Force contract. Six prototypes have been ordered by the Italian Air Force. Major programme milestones include a first flight scheduled to take place in 1983.

In October 1978 the Italian Air Force officially selected as the AM-X power plant a non-afterburning version of the Rolls-Royce Spey Mk 807 turbofan engine, currently rated at approx 49·1 kN (11,030 lb st). This engine was undergoing detail definition for completion by the end of 1981.

In June 1980 EMBRAER of Brazil joined Aeritalia and Aermacchi as a partner in the AM-X programme, in order to meet a Brazilian Air Force requirement for about 140 generally similar aircraft. Differences in the Brazilian version will be restricted mainly to the avionics and weapon systems.

Basic design features include very good field performance, good gust response and high manoeuvrability rather than extreme subsonic speeds at low level. This implies use of advanced high-lift devices of a purely aerodynamic type, with valuable crossfeed from Aeritalia's current Tornado work, but not a 'swing-wing'.

Primary roles of the AM-X were laid down as close interdiction over a combat radius of 180 nm (333 km; 205 miles), anti-ship operations, reconnaissance and close air support. Secondary roles are foreseen as counter-air missions, notably against enemy forward airfields, and air defence against low-level intruders. Emphasis is being placed on the ability to carry a wide variety of external weapons, on four underwing pylons, and flexibility in the avionics package; a typical combat load would comprise 1,360 kg (3,000 lb) of external ordnance, two wingtip-mounted infra-red air-to-air missiles for self-defence, one internally-mounted 20 mm rotary-barrel gun, and internal ECM equipment. The possibility of developing an advanced training version as a G91T replacement is being borne in mind.

DIMENSIONS, EXTERNAL:

Wing span over missiles	10·00 m (32 ft 9¾ in)
Wing span (excl wingtip missiles and rails)	8·88 m (29 ft 1½ in)
Wing area, gross	21·00 m² (226·0 sq ft)
Wing aspect ratio	3·75
Length overall	13·57 m (44 ft 6¼ in)
Height overall	4·58 m (15 ft 0¼ in)
Wheel track	2·15 m (7 ft 0¾ in)
Wheelbase	approx 4·75 m (15 ft 7 in)

WEIGHTS:

Operational weight empty	6,500 kg (14,330 lb)
Max external stores load	3,800 kg (8,377 lb)
Max T-O weight	12,000 kg (26,455 lb)

AÉROSPATIALE/AERITALIA

PARTICIPATING COMPANIES:

Aérospatiale (Aircraft Division), 37 boulevard de Montmorency, 75781 Paris Cédex 16, France
Telephone: 524 43 21
Telex: AISPA X 620059 F

Aeritalia (Transport Aircraft Group), Piazzale Vincenzo Tecchio 51 (Casella Postale 3065), 80125 Naples, Italy
Telephone: (081) 619522, 619721, 619845, 619149 and 619703
Telex: N 710370 AERIT

ATR 42

Implementing their agreement of July 1980 to study the joint development of a regional transport aircraft, Aérospatiale of France and Aeritalia of Italy have completed the definition of this aircraft, which has the designation ATR 42 (ATR corresponding to the initial letters of the French and Italian words for 'regional transport aircraft', and 42 to the seating capacity of the basic aircraft at 81cm; 32 in pitch).

The ATR 42 combines many features of the former Aérospatiale AS 35 and Aeritalia AIT 230 projects, but incorporates a number of refinements suggested by potential airline customers during a worldwide market survey conducted by teams from the two manufacturers during the fourth quarter of 1980. Further discussions with key airlines were held during the first half of 1981, when Aérospatiale and Aeritalia were also seeking approval from their respective governments to launch the ATR 42 programme, in order to make possible a first flight in mid-1984 and initial deliveries to airlines in June 1985. The launch decision was expected during the third quarter of 1981.

The ATR 42 is a high-wing twin-turboprop aircraft embodying a number of advanced-technology features in the fields of aerodynamics, structures and equipment. Design is to FAR Pt 25 and to European Joint Airworthiness Requirement JAR 25 for the certification of transport aircraft. Choice of the Pratt & Whitney Aircraft of Canada PW100/2 turboprop as the aircraft's power plant was announced on 8 June 1981. The General Electric CT7-9 turboprop may be offered as an alternative power plant.

Basic design targets are good economy, a high level of field performance, and a wide-body standard of comfort. The fully-pressurised cabin has a width greater than that of any present or projected aircraft in this category, and will provide four-abreast seating for up to 46 passengers at 76 cm (30 in) pitch, with a galley, toilet and two baggage compartments. Development possibilities include an initial 'stretch' (ATR XX) to carry 54-58 passengers, a quick-change passenger/cargo version (ATR 42QC), an all-freight version (ATR 42F), and commercial or military transport versions with rear-loading capability.

Main characteristics of the basic passenger version are as follows:

TYPE: Twin-turboprop regional transport aircraft.

WINGS, FUSELAGE AND TAIL UNIT: General configuration as shown in accompanying three-view drawing. Dihedral 2° 30′ on outer wing panels. Wings have double-slotted trailing-edge flaps, with fixed slat. All primary flight controls operated manually by cables and push/pull rods, with no servos. Trim tabs (in each aileron, each elevator and rudder) are actuated electrically.

LANDING GEAR: Retractable tricycle type, with twin wheels on each unit. Nose unit retracts forward, main units inward into large fairing on underside of fuselage. Disc brakes and anti-skid units on main gear.

POWER PLANT: Two Pratt & Whitney Aircraft of Canada PW100/2 turboprop engines, each flat rated to 1,342 kW (1,800 shp) and driving a Hamilton Standard 14SF four-blade propeller. Fuel in two integral tanks formed by wing spar box, total capacity 5,700 litres (1,254 Imp gallons). Single-point pressure refuelling, plus gravity refuelling points.

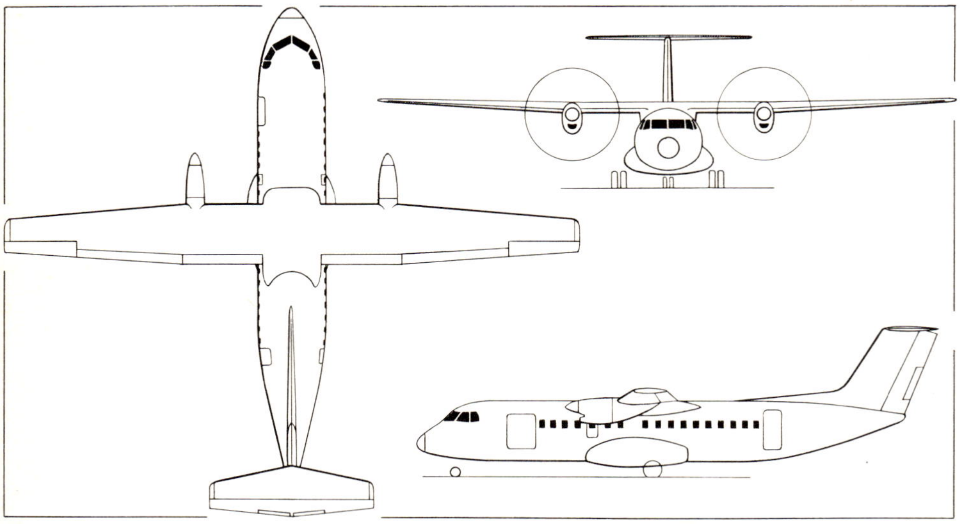

Aérospatiale/Aeritalia ATR 42 twin-turboprop regional transport aircraft *(Pilot Press)*

ACCOMMODATION: Seating for 42 passengers at 81 cm (32 in) pitch, or 46 passengers at 76 cm (30 in) pitch, in four-abreast layout with central aisle. Passenger door, with integral steps, at rear of cabin on port side. Main baggage/cargo compartment between flight deck and passenger cabin, with separate loading door on port side. Rear baggage compartment, toilet, galley and wardrobe aft of passenger cabin, with service door on starboard side. Additional baggage space provided by overhead bins and underseat stowage. Entire accommodation pressurised and air-conditioned. Quick-change passenger/cargo version (42 passengers or five LD3 containers) fitted with ball transfer plates aft, roller tracks, and anti-crash net at front of cabin.

SYSTEMS: Independent air-conditioning and pressurisation systems, utilising engine bleed air. Pressurisation system (max differential 0·41 bars; 6·0 lb/sq in) provides cabin altitude of approx 2,000 m (6,560 ft) at flight altitude of up to 7,620 m (25,000 ft), and a sea level cabin environment at flight levels up to approx 4,100 m (13,500 ft). Hydraulic system includes two independent groups, each including an electrically-driven pump and an accumulator, and actuates wing flaps, landing gear, wheel brakes and nosewheel steering. Pneumatic system for de-icing of outer wing leading-edges, tailplane, and engine air intakes. Main electrical system is 28V DC, supplied by two 9kW engine-driven starter/generators and two nickel-cadmium batteries, with two inverters for 115/26V single-phase AC supply. A 115/200V three-phase supply from two 20kVA engine-driven generators is used for anti-icing of windscreen, flight deck side windows, stall warning and airspeed indicator pitots, and for de-icing of propeller blades and spinners.

AVIONICS: Standard avionics package includes two VHF, two VOR/ILS, ADF, ATC transponder, cockpit voice recorder, and PA system. Optional avionics include HF radio, autopilot, flight director, radio altimeter, GPWS and weather radar.

DIMENSIONS, EXTERNAL:

Wing span	24·572 m (80 ft 7½ in)
Wing area, gross	54·00 m² (581·25 sq ft)
Wing aspect ratio	11·2
Length overall	22·70 m (74 ft 5¾ in)
Height overall	6·45 m (21 ft 2 in)
Wheel track	4·10 m (13 ft 5½ in)
Wheelbase	8·262 m (27 ft 1¼ in)
Propeller diameter	3·96 m (13 ft 0 in)
Passenger door: Height	1·75 m (5 ft 9 in)
Width	0.75 m (2 ft 5½ in)
Height to sill	1·265 m (4 ft 1¾ in)
Forward cargo/baggage door:	
Height	1·53 m (5 ft 0¼ in)
Width	1·275 m (4 ft 2¼ in)
Height to sill	1·265 m (4 ft 1¾ in)

DIMENSIONS, INTERNAL:

Cabin: Length	13·849 m (45 ft 5¼ in)
Max width	2·57 m (8 ft 5¼ in)
Max width at floor	2·19 m (7 ft 2¼ in)
Max height	1·90 m (6 ft 2¾ in)

Baggage volume:
front compartment 6·00 m³ (212 cu ft)
rear compartment 2·70 m³ (95·3 cu ft)
overhead bins 1·90 m³ (67·1 cu ft)

Weights and Loading:
Operational weight empty 9,062 kg (19,978 lb)
Max fuel load 4,500 kg (9,920 lb)
Max payload 4,720 kg (10,405 lb)
Max T-O weight 14,510 kg (31,990 lb)
Max ramp weight 14,540 kg (32,055 lb)
Max zero-fuel weight 13,780 kg (30,380 lb)
Max landing weight 14,220 kg (31,350 lb)
Max wing loading 268·7 kg/m² (55·0 lb/sq ft)

Performance (estimated, to FAR Pt 25, incl Amendment 42, PW100/2 engines, ISA, except where indicated):
Max cruising speed at 6,100 m (20,000 ft)
275 knots (510 km/h; 317 mph)
Cruise ceiling (91·5 m; 300 ft/min rate of climb)
7,620 m (25,000 ft)
Service ceiling, one engine out, at 97 per cent of max T-O weight, ISA + 10°C 3,960 m (13,000 ft)
T-O balanced field length at max T-O weight:
at S/L 990 m (3,250 ft)
at 915 m (3,000 ft) (ISA + 10°C) 1,190 m (3,905 ft)
Landing field length at S/L, at max landing weight
830 m (2,725 ft)
Block time for 200 nm (370 km; 230 mile) stage length at max cruising speed at 6,100 m (20,000 ft), IFR reserves 54 min

AIRBUS

AIRBUS INDUSTRIE

Head Office: Avenue Lucien Servanty, BP No. 33, 31700 Blagnac, France
Telephone: (61) 71 11 11
Telex: AI TO 530526 F
Paris Office: 12bis avenue Bosquet, 75007 Paris, France
Telephone: 551 40 95

Airframe Prime Contractors:
Aérospatiale, 37 boulevard de Montmorency, 75781 Paris Cédex 16, France
Deutsche Airbus GmbH, 8000 München 81, Arabellastrasse 30, Postfach 810260, Federal Republic of Germany
British Aerospace PLC Aircraft Group, Richmond Road, Kingston upon Thames, Surrey KT2 5QS, England

Chairman of Supervisory Board: Dr Franz-Josef Strauss
President and Chief Executive: Bernard Lathière
Executive Vice-President and General Manager: Roger Béteille
Senior Vice-Presidents:
J. Roeder (Technical)
G. Warde (Commercial)
F. Kracht (Production)
G. Ville (Finance and Administration)
B. Ziegler (Flight and Support)
Programme Managers:
J. Thomas (A300)
J. Plénier (A310)
D. Brown (Future Projects)

Airbus Industrie was set up in December 1970 as a Groupement d'Intérêt Economique to manage the development, manufacture, marketing and support of a twin-engined large-capacity short/medium-range transport aircraft known as the A300. This management now extends to the A310 and other types included in this entry. Airbus Industrie is responsible for all work on these programmes by the partner companies, made up of Aérospatiale of France, which has a 37·9% interest in the programme, Deutsche Airbus (MBB) of West Germany (37·9%), British Aerospace PLC (20%), and CASA of Spain (4·2%). Fokker (Netherlands) is an associate in the A300 and A310 programmes; Belairbus (Belgium) in the programme for the A310; and SOKO (Yugoslavia) in that for the A300 and/or A310. Some of the Deutsche Airbus work on the A300/A310 is subcontracted to the Italian aerospace industry.

AIRBUS A300

The Airbus A300 is a wide-bodied aircraft with underwing pods for two turbofan engines. The early history of the programme has appeared in previous editions of *Jane's*.

Construction of the first A300, a B1, began in September 1969. This aircraft (F-WUAB, later F-OCAZ) made its first flight on 28 October 1972, and was followed by the second B1 (F-WUAC) on 5 February 1973. The B1 was described in detail in the 1971-72 *Jane's*. Initial certification covered automatic approach and landing in Category II weather conditions. Certification for Category IIIA automatic approach and landing was granted on 30 September 1974.

On 10 December 1980 the first flight was made by an A300 equipped with a SFENA/Smiths/Bodenseewerk Category III digital all-weather automatic landing system, and this is now available as a customer option. The first airline to purchase A300s having this system was Garuda, to which deliveries are to begin in 1982.

The A300 is offered currently with two General Electric CF6-50C or C2 turbofans, or two Pratt & Whitney JT9D-59As or Bs. All CF6-50 engines are available with a weight and fuel saving 'short' nozzle. The first A300 with JT9D engines, a B2 for SAS, made its first flight on 28 April 1979. The CF6-80C1, JT9D-7R4H and Rolls-Royce RB.211-524D4 are also available as customer options.

The following versions are available:

A300B2-100 (formerly A300B2). Initial production version. Third and fourth aircraft are to this configuration, flying for the first time on 28 June and 20 November 1973 respectively. Aircraft No. 4 subsequently delivered to Air Inter on 22 January 1977 as F-BUAE. Type certificated by DGAC and LBA on 15 March 1974; entered service, with Air France, between Paris and London on 30 May 1974. Designated **B2-101** with CF6-50C engines.

A300B2-200 (formerly A300B2K). Basically as B2-100 series, but fitted with wing-root leading-edge Krueger flaps developed originally for A300B4. With same wheels and brakes also as B4, is suitable for operation from 'hot and high' airports. First aircraft of this type, No. 32 (ZS-SDA), first flew on 30 July 1976. First delivery, of ZS-SDB to South African Airways, made on 23 November 1976. Designation 'dash' numbers, according to power plant, are **B2-201** (CF6-50C) and **B2-220** (JT9D-59A).

A300B2-300. Basically as B2-200, but increased zero-fuel and landing weights for increased payload and multi-stop flexibility. First delivery, to SAS, made on 15 January 1980.

A300B2-600. Advanced version of B2-200. Described separately.

A300B4-100 (formerly A300B4 Stage I and Stage II). Basic longer-range version. Developed from original A300B2, with same external dimensions and volumetric payload capacity, but with increased design weights and fuel capacity, and Krueger flaps at wing-root leading-edges to improve T-O performance. Aircraft No. 9 (F-WLGA) was the first built to this configuration, and first flew on 26 December 1974. French and West German certification granted on 26 March 1975, and FAA Type Approval on 30 June 1976. First delivery, of D-AMAX to Germanair, made in May 1975; this aircraft entered service on 1 June 1975. Designation 'dash' numbers, according to power plant, are **B4-101** (CF6-50C), **B4-103** (CF6-50C2) and **B4-120** (JT9D-59A).

A300B4-200 (formerly A300B4 Stage III). Announced on 19 January 1978, concurrently with order for three by Air France, to which first example was delivered on 27 April 1979. Compared with -100 series, has reinforced wings and fuselage, strengthened landing gear, and size 49 × 17-20 main-wheel tyres. This allows certification at higher T-O weight, enabling full payload of passengers and cargo to be carried on stages of up to 1,954 nm (3,621 km; 2,250 miles) with standard airline reserves. Optional additional fuel tank available with this version: installed in rear cargo hold in place of two LD3 containers, it extends the range with full complement of passengers to 3,100 nm (5,740 km; 3,565 miles). Designation 'dash' numbers, according to power plant, are **B4-201** (CF6-50C), **B4-203** (CF6-50C2), **B4-220** (JT9D-59A), and **B4-221** (JT9D-59B).

A300B4-600. Advanced version of B4-200. Described separately.

A300C4 Freighter conversion of B4; described separately.

A310. Developed version, with shorter fuselage and redesigned wings; described separately.

A320. Advanced-technology single-aisle developed version; described separately.

Airbus A300B2-200 twin-turbofan transport in the insignia of Toa Domestic Airlines of Japan

By 1 August 1981, orders and options totalling 316 A300s (including C4s) had been received, as follows:

	Orders	*Options*
Air Afrique (B4-200)	3	—
Air France (B2-100, B4-100/200)	23	12
Air Inter (France) (B2-100)	8	—
Alitalia (B4-200)	8	3
China Airlines (B4-200)	4	—
Cruzeiro do Sul (Brazil) (B4-203)	2	—
Eastern Air Lines (USA) (B2-200, B4-100/600)	34	26
Egyptair (B4-203)	5	2
Garuda (Indonesia) (B4-220)	9	3
Hapag-Lloyd (West Germany) (B4-100, C4-200)	7	—
Iberia (B4-120)	6	—
Indian Airlines (B2-100)	8	3
Iran Air (B2-200)	6	3
Korean Air Lines (B4-100)	8	—
Kuwait Airways (C4-600)	3	—
*Laker Airways (UK) (B4-203)	10	—
Lufthansa (West Germany) (B2-100, B4-100)	11	—
Malaysian Airline System (B4-200)	4	—
Olympic Airways (Greece) (B4-100)	8	2
Pakistan International (B4-200)	4	6
Philippine Air Lines (B4-200)	5	—
SAS (Scandinavia) (B2-320)	4	4
Saudi Arabian Airlines (B4-600)	11	—
Singapore Airlines (B4-200)	6	6
South African Airways (B2-200, B4-200)	7	1
Thai International (B4-100/200/600)	12	2
Toa Domestic Airlines (Japan) (B2-200)	9	—
Trans Australia Airlines (B4-203)	5	2
Trans European Airways (Belgium) (B1)	1	—
Tunis Air (B4-200)	1	1
Varig (Brazil) (B4-203)	2	—
VASP (Brazil) (B2-203)	3	—
Undisclosed	—	3

**option for last four to be A310s*

A total of 146 A300s (46 B1/B2s and 100 B4s) had been delivered by 1 August 1981; the 100th was an Alitalia B4-200.

Aérospatiale is responsible for manufacturing the entire nose section (including the flight deck), lower centre fuselage and engine pylons, and for final assembly. Deutsche Airbus is responsible for manufacturing the forward fuselage, between the flight deck and wing box, the upper centre fuselage, the rear fuselage and the vertical tail surfaces. British Aerospace has design responsibility for the wings, builds the wing fixed structures, and is working in collaboration with Fokker, which is building the wing-tips and wing moving surfaces. CASA manufactures the horizontal tail surfaces, the two (port and starboard) forward passenger doors and landing gear doors.

Large A300 sections are flown from their places of manufacture in Europe to the final assembly line in Toulouse on board two Super Guppy outsize cargo aircraft, acquired by Airbus Industrie in November 1971 and September 1973 respectively. After assembly, painting in customers' colour scheme is carried out at Toulouse. Aircraft are then flown to Hamburg for installation of interior furnishings and equipment before returning to Toulouse for final customer acceptance. Two additional Super Guppies are being built for this and similar work, as described in a later entry.

General Electric engines are assembled under licence by SNECMA; some components are also licence-built by SNECMA (France) and MTU (West Germany). The whole power plant assembly is virtually identical to that of the McDonnell Douglas DC-10-30, and nacelles are supplied by McDonnell Douglas. Nacelles for Pratt & Whitney engines are manufactured by Rohr (West Germany) and are interchangeable with those of JT9D-59-powered DC-10-40 aircraft and JT9D-70-powered Boeing 747s.

In 1980 Airbus Industrie received 50 firm orders and 36

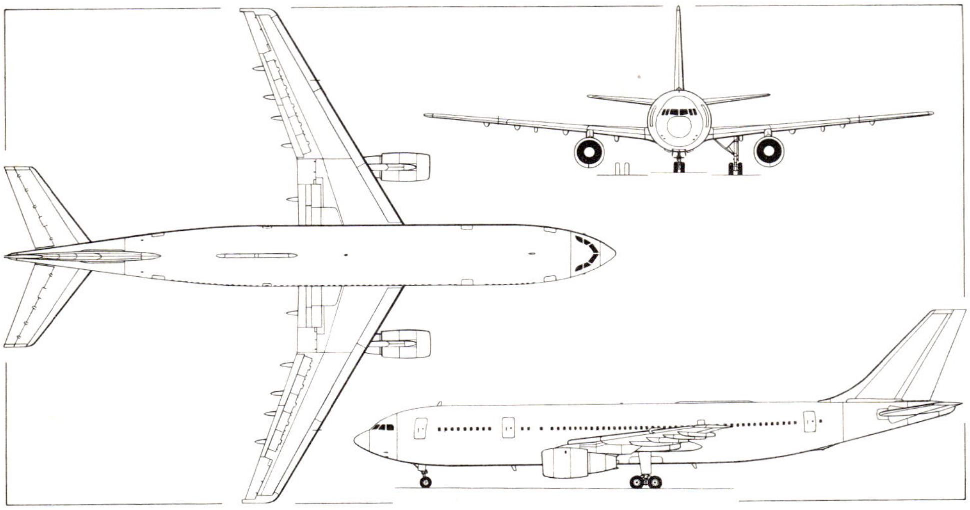

Airbus A300B4 wide-bodied short/medium-range transport (CF6 engines) *(Pilot Press)*

options for the A300/310, including orders from eight new customers. The latter included Saudia, the launch customer for the A300B4-600. Production was at the rate of four per month in 1981, and is planned to reach eight per month in 1984 and ten per month in 1985.

Type: Large-capacity wide-bodied short/medium-range transport.

Wings: Cantilever mid-wing monoplane. Thickness/chord ratio 10·5%. Sweepback 28° at quarter-chord. Primary two-spar box-type structure, integral with fuselage and incorporating fail-safe principles, built of high-strength aluminium alloy. Third spar across inboard sections. Machined skin with open-sectioned stringers. Each wing has three-section leading-edge slats (no slat cutout over the engine pylon), and three Fowler-type double-slotted flaps on trailing-edge; a Krueger flap on the leading-edge wing root (B2-200/B4); an all-speed aileron between inboard flap and outer pair; and a low-speed aileron outboard of the outer pair of flaps. Lift dump facility by combination of three spoilers (outboard) and two airbrakes (inboard) on each wing, forward of outer pair of flaps, plus two additional airbrakes forward of inboard flap. CFRP spoilers, developed by MBB/VFW, being evaluated on 12 aircraft in scheduled service. The flaps extend over 84% of each half-span, and increase the wing chord by 25% when fully extended. The datum of the all-speed aileron is deflected downward by up to 10° with flap operation to maintain trailing-edge continuity with deflected flaps. Drive mechanisms for flaps and slats are similar to one another, each powered by twin motors driving ball screwjacks on each surface with built-in protection against asymmetric operation. Two slat positions for take-off and landing. Pre-selection of the airbrake/lift dump lever allows automatic extension of the lift dumpers on touchdown. All flight controls are powered by triplex hydraulic servo-jacks, with no manual reversion. Anti-icing of wing leading-edges, outboard of engine pods, is by hot air bled from engines.

Fuselage: Conventional semi-monocoque structure of circular cross-section, with frames and open Z-section stringers. Built mainly of high-strength aluminium alloy, with steel or titanium for some major components. Skin panels integrally machined in areas of high stress. Honeycomb panels or restricted glassfibre laminates for secondary structures.

Tail Unit: Cantilever all-metal structure, with sweepback on all surfaces. Variable-incidence tailplane and separately-controlled elevators. Tailplane powered by two motors driving a fail-safe ball screwjack. No anti-icing of leading-edges.

Landing Gear: Hydraulically-retractable tricycle type, of Messier-Hispano-Bugatti design, with Messier-Hispano-Bugatti/Liebherr/Dowty shock-absorbers and wheels standard. Twin-wheel nose unit retracts forward, main units inward into fuselage. Free-fall extension. Each four-wheel main unit comprises two tandem-mounted bogies, interchangeable left with right. Standard bogie size is 927 × 1,397 mm (36½ × 55 in); wider bogie of 978 × 1,397 mm (38½ × 55 in) is available on B4-100/200. Main-wheel tyre sizes and pressures as follows: B2, 46 × 16-20 (standard bogie) at 12·4 bars (180 lb/sq in); B2-200, 49 × 17-20 (standard bogie) at 11·4 bars (165 lb/sq in); B4-100, 46 × 16-20 (standard), 49 × 17-20 (standard) or 49 × 17-20 (wide bogie), with respective pressures of 14·2, 11·9 and 10·6 bars (206, 172 and 154 lb/sq in); B4-200, 49 × 17-20 (standard) or 49 × 17-20 (wide bogie), with respective pressures of 12·4 and 11·1 bars (180 and 161 lb/sq in). Nosewheel tyres size 40 × 14-16 on all these B2/B4 models, with pressures of 8·6 bars (125 lb/sq in) on B2 variants and 9·0 bars (131 lb/sq in) on B4 variants. Steering angles 65°/95°. Messier-Hispano-Bugatti/Liebherr/Dowty hydraulic disc brakes standard on all main wheels. Duplex anti-skid units fitted, with a third standby hydraulic supply for wheel brakes. Bendix or Goodrich wheels and brakes available optionally.

Power Plant: Underwing location of the power plant enables the A300 to use any advanced technology turbofan engine in the 222·5 kN (50,000 lb thrust) class. It is currently being offered with the following engines:
Two 227 kN (51,000 lb st) General Electric CF6-50C turbofans (B2-101/201 and B4-101/201);
Two 233·5 kN (52,500 lb st) General Electric CF6-50C2 (B4-103/203); or
Two 236 kN (53,000 lb st) Pratt & Whitney JT9D-59A (B2-220 and B4-120/220) or -59B (B4-221).
These engines are installed in pods interchangeable with those of the McDonnell Douglas DC-10 Series 30 and 40. For details of other engines available, see introductory copy.
Fuel in two integral tanks in each wing, with total usable capacity of 43,000 litres (9,460 Imp gallons) in B2. Fifth integral tank in wing centre-section of B4, increasing total usable capacity to 59,700 litres (13,133 Imp gallons). For B4-200 series, an optional self-contained fuselage fuel tank is available, capacity 6,000 litres (1,320 Imp gallons). This unit fits into the rear cargo hold, where it takes the place of two standard LD3 containers. Two standard refuelling points beneath each wing, outboard of engine pylons.

Accommodation (B2 and B4): Crew of three on flight deck, plus two observer's seats. Seating for between 220 and 320 passengers in main cabin in six, seven, eight or nine-abreast layout with 79/86 cm (31/34 in) seat pitch. Typical economy class layout has 269 seats, eight abreast with two aisles, at 86 cm (34 in) seat pitch. This layout includes one galley and one toilet forward, two galleys in mid-cabin, and one galley and four toilets aft, with provision for a second toilet forward and an additional galley aft. Up to 336 passengers can be carried at 76 cm (30 in) seat pitch in single-class high-density layout. Closed hatracks on each side, forming baggage lockers (max individual capacity 0·062 m^3; 2·19 cu ft). Provision for central double-sided rack. Two outward parallel-opening plug-type passenger doors ahead of wing leading-edge on each side, and one on each side at rear. Underfloor baggage/cargo holds fore and aft of wings, with doors on starboard side. The forward hold will accommodate four 2·23 × 3·17 m (88 × 125 in) pallets or twelve LD3 or IATA A1 containers on aircraft with the original door with a width of 2·44 m (8 ft 0 in). When the wider (2·69 m; 8 ft 10 in) door is fitted, pallet size can be increased to 2·43 × 3·17 m (96 × 125 in). The rear hold will accommodate eight LD3 containers each of 4·25 m^3 (150 cu ft) capacity. Additional bulk loading of freight provided for in an extreme rear compartment with usable volume of 16·0 m^3 (565 cu ft). The latter compartment can be used for the transport of livestock. Entire accommodation is pressurised, including freight, baggage and avionics compartments.

Systems: Air for air-conditioning system can be provided from engines, the APU or a high pressure ground source. Supply is controlled by separate and parallel bootstrap-type units, each of which includes a flow limiting unit, cooler unit, water separator and temperature control unit. In addition, air from each engine passes through a pressure control pre-cooler unit. Distribution in flight deck and three cabin areas, with independent regulation. Two independent automatic systems, with manual override, control the cabin altitude, its rate of change and the differential pressure. Cabin pressure differential for normal operations is 0·57 bars (8·25 lb/sq in). Hydraulic system comprises three fully-independent circuits, operating simultaneously. Fluid used is a fire-resistant phosphate-ester type, working at a pressure of 207 bars (3,000 lb/sq in). The three circuits provide triplex power for primary flying controls; if any circuit fails, full control of the aircraft is retained without any necessity for action by the crew. All three circuits supply the all-speed and low-speed ailerons, rudder and elevator; 'blue' circuit additionally supplies tail trim, spoilers, slats and rudder variable-gear unit; 'green' circuit additionally supplies airbrakes, spoilers, slats, elevator artificial feel units, flaps, steering, wheel brakes and normal landing gear requirements; 'yellow' circuit additionally supplies tail trim, airbrakes, lift dumpers, rudder variable-gear unit, elevator artificial feel units, flaps, wheelbrakes and steering. Each circuit normally powered by engine-driven self-regulating pumps, one on each engine for the green circuit and one each for the blue and yellow circuits. Dowty Rotol ram-air turbine-driven pump provides standby hydraulic power should both engines become inoperative. Main electrical power is supplied by two Westinghouse three-phase constant-frequency AC generators mounted on the engines. A third identical generator, driven by the APU, can supply power in flight, to replace a failed engine-driven generator, and on the ground. Supply frequency is 400Hz and voltage is 115/200V. Any one generator can supply sufficient power to operate all equipment and systems necessary for take-off and landing. A conventional generator CSD system is installed, the two units being mounted on opposite sides of the engine gearbox with the CSD driving an aircooled generator at a constant 8,000 rpm. Each generator is rated at 90kVA, with overload ratings of 135kVA for 5 min and 180kVA for 5 s. The APU generator is driven at constant speed through a gearbox. Three unregulated transformer-rectifier units (TRUs) supply 28V DC power. Three 24V 25Ah nickel-cadmium batteries are used for APU starting and fuel control, engine starter control, standby lights and, by

Airbus A300B4-203 wide-bodied commercial transport aircraft, in the 'Skytrain' livery of Laker Airways

selection, emergency busbar. This busbar and a 115V 400Hz static inverter provide standby power in flight if normal power is unavailable. This system is separated completely from the main system. Hot air protection for engine intakes and slat sections on the wings outboard of the engine. Garrett TSCP 700-5 or (from c/n 246) GTCP 331-250 APU in tailcone, exhausting upward. The installation incorporates APU noise attenuation. Fire protection system is self-contained, and firewall panels protect main structure from an APU fire. The APU can be operated on the ground, in flight up to 10,675 m (35,000 ft), and in icing conditions. Relights are possible up to 7,620 m (25,000 ft). Aircraft is completely independent of ground power sources, since all major services can be operated by the APU.

Avionics and Equipment: Standard communications avionics include two VHF sets and one Selcal system, plus interphone and passenger address systems. An accident recorder and voice recorder are also installed. Standard navigation avionics include two VOR, two ILS, two radio altimeters, marker beacon receiver, two ADF, two DME, two ATC transponders and a weather radar. Sperry digital air data computer standard. Most other avionics are to customer's requirements, only those related to the blind landing system (ILS and radio altimeter) being selected and supplied by the manufacturer. Additional optional avionics include one or two HF, third VHF, second marker beacon receiver, second radar, navigation computer and pictorial display. Pilot and co-pilot each have an integrated instrument system combining heading and attitude (three SAGEM MGC 10/ARINC 569 are standard in B2, but they can be replaced by MGC 30/ARINC 571 Mk 1 inertial sensors, which are modular with the MGC 10); SFENA autopilot/flight director system; and radio information. The SFENA/Smiths/Bodenseewerk automatic flight control system includes a comprehensive range of en-route facilities such as VOR coupling, heading select, height acquire, turbulence, rate of descent (if required) and control wheel steering, in addition to the normal height, speed, pitch-and-roll attitude and heading locks. An optional speed reference system with built-in windshear protection is available. Dual automatic landing system provides coupled approach and automatic landing facilities suitable for Category IIIA operation. The system is designed to allow extension to Category IIIB automatic landing capability. An automatic braking system is available.

Dimensions, external (B2-100/200/300, B4-100/200):
Wing span 44·84 m (147 ft 1 in)
Wing aspect ratio 7·73
Length overall 53·62 m (175 ft 11 in)
Length of fuselage 52·03 m (170 ft 8½ in)
Fuselage max diameter 5·64 m (18 ft 6 in)
Height overall 16·53 m (54 ft 2¾ in)
Tailplane span 16·94 m (55 ft 7 in)
Wheel track 9·60 m (31 ft 6 in)
Wheelbase (c/l of shock-absorbers) 18·60 m (61 ft 0 in)
Passengers doors (each):
Height 1·93 m (6 ft 4 in)
Width 1·07 m (3 ft 6 in)
Height to sill: fwd 4·60 m (15 ft 1 in)
centre 4·80 m (15 ft 9 in)
rear 5·50 m (18 ft 0½ in)
Emergency exits (each): Height 1·60 m (5 ft 3 in)
Width 0·61 m (2 ft 0 in)
Height to sill 4·87 m (15 ft 10 in)
Underfloor cargo door (fwd):
Height 1·71 m (5 ft 7½ in)
Width 2·44 m (8 ft 0 in)
or 2·69 m (8 ft 10 in)
Height to sill 2·56 m (8 ft 4¾ in)
Underfloor cargo door (rear):
Height 1·71 m (5 ft 7½ in)
Width 1·81 m (5 ft 11¼ in)
Height to sill 2·96 m (9 ft 8½ in)
Underfloor cargo door (extreme rear):
Height 0·95 m (3 ft 1 in)
Width 0·95 m (3 ft 1 in)
Height to sill 3·30 m (10 ft 10 in)

Dimensions, internal (B2-100/200/300, B4-100/200):
Cabin, excl flight deck:
Length 39·15 m (128 ft 6 in)
Max width 5·35 m (17 ft 7 in)
Max height 2·54 m (8 ft 4 in)
Underfloor cargo hold:
Length: fwd 10·60 m (34 ft 9¼ in)
rear 6·89 m (22 ft 7¼ in)
extreme rear 3·10 m (10 ft 2 in)
Max height 1·76 m (5 ft 9 in)
Max width 4·20 m (13 ft 9¼ in)
Underfloor cargo hold volume:
fwd 75·1 m^3 (2,652 cu ft)
rear 46·8 m^3 (1,652 cu ft)
extreme rear 16·0 m^3 (565 cu ft)
Max total volume for bulk loading
140·0 m^3 (4,944 cu ft)

Areas (B2, B4):
Wings, gross 260·0 m^2 (2,798·6 sq ft)
Vertical tail surfaces (total) 45·2 m^2 (486·5 sq ft)
Horizontal tail surfaces (total) 69·5 m^2 (748·1 sq ft)

Weights and Loadings:
Manufacturer's weight empty:
B2-100/120 77,062 kg (169,892 lb)
B2-200/220 77,427 kg (170,697 lb)
B4-100/120 79,070 kg (174,319 lb)
B4-200/220 79,833 kg (176,000 lb)
Typical operating weight empty:
B2-100/120, B2-200 85,910 kg (189,400 lb)
B2-220 87,500 kg (192,905 lb)
B4-100 (basic) 88,100 kg (194,227 lb)
B4-120 (basic) 89,300 kg (196,873 lb)
B4-200 (optional) 88,500 kg (195,109 lb)
B4-220 (optional) 89,700 kg (197,755 lb)
Max payload (structural):
B2-100/120 34,590 kg (76,258 lb)
B2-200 34,600 kg (76,280 lb)
B2-220 33,000 kg (72,752 lb)
B4-100 (basic) 35,900 kg (79,146 lb)
B4-120 (basic) 34,700 kg (76,500 lb)
B4-200 (optional) 35,200 kg (77,602 lb)
B4-220 (optional) 34,300 kg (75,618 lb)
Max usable fuel:
B2-200/220 (basic) 34,000 kg (74,957 lb)
B4-100/120 (basic) 47,500 kg (104,720 lb)
Max T-O weight:
B2-100 137,000 kg (302,030 lb)
or 142,000 kg (313,055 lb)
B2-200 142,000 kg (313,055 lb)
B4-100 150,000 kg (330,690 lb)
or 153,000 kg (337,305 lb)
or 157,500 kg (347,230 lb)
B4-200 165,000 kg (363,760 lb)
Max ramp weight:
B2-200 142,900 kg (315,040 lb)
B4-200 165,900 kg (365,745 lb)
Max landing weight:
B2-100 127,500-134,000 kg (281,090-295,420 lb)
B2-200 130,000-134,000 kg (286,600-295,420 lb)
B4-100 133,000-136,000 kg (293,215-299,825 lb)
B4-200 134,000 kg (295,420 lb)
Max zero-fuel weight:
B2-100
116,500-124,000 kg (256,835-273,375 lb)
B2-200 120,500 kg (265,655 lb)
or 124,000 kg (273,375 lb)
B4-100
122,000-126,000 kg (268,960-277,780 lb)
B4-200 124,000 kg (273,375 lb)
Max wing loading:
B2-200 546 kg/m^2 (111·8 lb/sq ft)
B4-100 606 kg/m^2 (124·1 lb/sq ft)
B4-200 635 kg/m^2 (130·0 lb/sq ft)
Max power loading:
B2-201 (CF6-50C) 312·8 kg/kN (3·07 lb/lb st)
B2-220 (JT9D-59A) 300·8 kg/kN (2·95 lb/lb st)
B4-101 (CF6-50C) 346·9 kg/kN (3·40 lb/lb st)
B4-103 (CF6-50C2) 337·3 kg/kN (3·31 lb/lb st)
B4-203 (CF6-50C2) 353·3 kg/kN (3·46 lb/lb st)
B4-220 (JT9D-59A) 349·6 kg/kN (3·43 lb/lb st)
B4-221 (JT9D-59B) 340·2 kg/kN (3·34 lb/lb st)

Performance (at max T-O weight except where indicated):
Max operating speed (Vmo): B2 from S/L to 8,475 m (27,800 ft), B4 from S/L to 7,740 m (25,400 ft)
345 knots; 639 km/h; 397 mph) CAS
Max operating Mach number (Mmo):
B2 above 8,475 m (27,800 ft) 0·86
B4 above 7,740 m (25,400 ft) 0·82
Max cruising speed at 7,620 m (25,000 ft):
B2, B4 492 knots (911 km/h; 567 mph)
Typical high-speed cruise at 9,145 m (30,000 ft):
B2, B4 495 knots (917 km/h; 570 mph)
Typical long-range cruising speed at 9,450 m (31,000 ft): B2, B4 457 knots (847 km/h; 526 mph)
Approach speed at typical weight:
B2 131 knots (243 km/h; 151 mph) CAS
B4 128 knots (237 km/h; 147 mph) CAS
Max operating altitude: B2 10,675 m (35,000 ft)
Min ground turning radius (wingtips)
33·51 m (109 ft 11¼ in)
Runway LCN at max T-O weight (A: flexible pavement of 51 cm; 20 in thickness, B: rigid pavement of 76 cm; 30 in radius of relative stiffness):
B2-100: A 75
B 68
B2-200: A 73
B 67
B4-100 (46 × 16-20 tyres): A 86
B 79
B4-100 (49 × 17-20 tyres): A 78
B 70
B4-100 (49 × 17-20 tyres, wide bogie):
A 70
B 62
B4-200 (49 × 17-20 tyres): A 85
B 74
B4-200 (49 × 17-20 tyres, wide bogie):
A 77
B 67
T-O field length (S/L, ISA + 15°C):
B2 1,951 m (6,400 ft)
B4-100/120 2,750 m (9,020 ft)
B4-200/220 3,000 m (9,845 ft)
Landing field length at typical weight:
B2 1,630 m (5,350 ft)
B4-100/120 1,660 m (5,445 ft)
Range with 269 passengers and baggage:
B2-200/220 1,800 nm (3,334 km; 2,074 miles)
B4-100/120 2,600 nm (4,818 km; 2,994 miles)
B4-200/220 2,750 nm (5,095 km; 3,165 miles)
Range with max fuel:
B2-200/220 2,300 nm (4,261 km; 2,648 miles)
B4-100/120 3,200 nm (5,930 km; 3,685 miles)

Operational Noise Levels (FAR Pt 36):
T-O: B2 90 EPNdB
B4-100/120 92 EPNdB
B4-200/220 94 EPNdB
Approach: B2, B4 101 EPNdB
Sideline: B2, B4 95 EPNdB

AIRBUS A300B2-600 and A300B4-600

These advanced versions of the B2-200 and B4-200 incorporate a number of improvements, including increased passenger and freight capacity. Modifications include use of the A310 rear fuselage, which is shorter but volumetrically more efficient than that of the A300, with a 0·52 m (1 ft 9 in) extension of the parallel section of the fuselage to restore tail moment arm. Passenger capacity is thus increased by two seat rows for an increase in overall length equivalent to only one frame pitch.

An extensive weight reduction programme, including simplified systems and the use of composite materials for some secondary structural components, allows greater payload capacity to be offered with very little change in empty weight. Performance improvements, offering better payload/range capability and greater fuel economy, result from a comprehensive 'drag clean-up' programme carried out on an A300 development aircraft (c/n 3).

Power plants currently offered for the -600 are the General Electric CF6-80C1 (249 kN; 56,000 lb st), Pratt & Whitney JT9D-7R4H (also 249 kN; 56,000 lb st), and Rolls-Royce RB.211-524D (235·75 kN; 53,000 lb st).

Definition of the -600 version was completed in 1980 and the first order, from Saudi Arabian Airlines for the B4-600, was received in December of that year. Saudia's B4-600s will be powered by JT9D-7R4H1 engines.

Accommodation: Typical seating for 267 passengers in mixed-class layout, with first class seats at 100 cm (39/40 in) and tourist class seats at 86 cm (34 in) pitch.

Dimensions, external and internal: As for B2-100/200/300 and B4-100/200 except:
Length overall 54·08 m (177 ft 5 in)
Length of fuselage 53·30 m (174 ft 10½ in)
Height overall 16·62 m (54 ft 6¼ in)
Tailplane span 16·26 m (53 ft 4 in)
Underfloor cargo door (fwd):
Height 1·71 m (5 ft 7½ in)
Width 2·69 m (8 ft 10 in)
Cabin, excl flight deck:
Length 40·21 m (131 ft 11 in)
Underfloor cargo hold length:
rear 7·95 m (26 ft 1 in)
extreme rear 3·40 m (11 ft 2 in)
Underfloor cargo hold volume:
rear 55·0 m^3 (1,942 cu ft)
extreme rear 17·3 m^3 (611 cu ft)

Weights and Loadings:
Max T-O weight:
B2-600 142,000 kg (313,055 lb)
B4-600 165,000 kg (363,760 lb)
Max ramp weight:
B2-600 142,900 kg (315,040 lb)
B4-600 165,900 kg (365,745 lb)
Max landing weight:
B2-600 134,000 kg (295,420 lb)
B4-600 138,000 kg (304,240 lb)
Max zero-fuel weight:
B2-600 124,000 kg (273,375 lb)
B4-600 130,000 kg (286,600 lb)

AIRBUS A300C4/F4

The A300C4 is a convertible freighter version of the A300B4, available with the same range of power plant options. Main differences from the B4 are a large upper-deck cargo door, a reinforced cabin floor, a smoke detection system in the main cabin, and an interior trim adaptable to the freighter role. The upper-deck cargo door is on the opposite side to that of the underfloor holds, enabling loading or unloading to be carried out simultaneously at all positions.

In the freight mode the loading system, consisting of ball mats, roller tracks and electrical drive units, is fitted to the existing seat rails, and permits the carriage of up to thirteen 2·24 × 3·17 m (88 × 125 in) pallets, twelve 2·44 × 3·17 m (96 × 125 in) pallets, or eight of the former plus five of the latter, in basic configurations. Total upper-deck volume thus varies between 173 m^3 (6,100 cu ft) and 179 m^3 (6,315 cu ft). A 9*g* barrier net is installed in the front of the cabin. Total cargo-carrying capability of the A300C4 is approx 41,000 kg (90,390 lb). Modification kits covering the forward fuselage section, in which the upper-deck

Airbus A300C4-200 freighter for Hapag-Lloyd of West Germany, converted by VFW

cargo door is installed, are built in Bremen by MBB.

For customers requiring it, the A300C4 has the capability for conversion to passenger or mixed passenger/cargo configuration. Typical options include accommodation (in mainly eight-abreast seating) for up to 315 passengers on the upper deck; or 145 passengers (seven/eight-abreast) plus six 88 × 125 in pallets; or 75 passengers plus nine 88 × 125 in pallets; or fourteen 88 × 125 in or 96 × 125 in pallets; or twenty 86 × 125 in pallets.

The first A300C4 (c/n 83), for Hapag-Lloyd of West Germany, made its initial flight in mid-1979; after fitting of the upper-deck cargo door, it was delivered in January 1980.

Airbus Industrie has also defined an **A300F4** freighter version, with a maximum payload of 46,000 kg (101,410 lb). This is similar to the C4, but with all passenger provisions removed completely and the cabin windows replaced by metal blanking plates. It can carry up to twenty 88 × 125 in pallets on the upper deck, or eight 96 × 96 × 120 in ISO containers plus five 96 × 125 in pallets. Loadable volume of the upper deck of the A300F4 is 294·5 m³ (10,400 cu ft).

The description of the A300B4 applies generally also to the convertible C4, except as outlined above and detailed below:

Dimensions, external: As A300B4, plus:
Upper-deck cargo door (fwd, port):
Height (projected) 2·57 m (8 ft 5¼ in)
Width 3·58 m (11 ft 9 in)

Dimensions, internal: As A300B4, except:
Cabin upper deck usable for cargo:
Length 31·78 m (104 ft 3 in)
Min height 2·16 m (7 ft 1 in)
Max height 2·44 m (8 ft 0 in)
Volume 173–179 m³ (6,100–6,315 cu ft)
Max total volume for bulk loading (upper and lower decks) 286 m³ (10,087 cu ft)

Weights:
Manufacturer's weight empty (basic):
passenger mode 81,000 kg (178,575 lb)
freight mode 81,900 kg (180,560 lb)
pure freighter 78,300 kg (172,620 lb)
Manufacturer's weight empty (optional):
passenger mode 81,300 kg (179,235 lb)
freight mode 82,200 kg (181,220 lb)
pure freighter 78,600 kg (173,285 lb)
Operating weight empty (basic):
*passenger mode 89,000 kg (196,210 lb)
freight mode 82,900 kg (182,765 lb)
pure freighter 79,000 kg (174,165 lb)
Operating weight empty (optional):
*passenger mode 89,200 kg (196,650 lb)
freight mode 83,200 kg (183,425 lb)
pure freighter 79,400 kg (175,045 lb)
Max payload (structural) (basic):
passenger mode 35,000 kg (77,160 lb)
freight mode 41,100 kg (90,610 lb)
pure freighter 45,000 kg (99,210 lb)
Max payload (structural) (optional):
passenger mode 36,800 kg (81,130 lb)
freight mode 42,800 kg (94,360 lb)
pure freighter 46,600 kg (102,735 lb)
Max T-O weight:
basic 157,500 kg (347,230 lb)
optional 165,000 kg (363,765 lb)
Max landing weight:
basic 134,000 kg (295,420 lb)
optional 136,000 kg (299,830 lb)
Max zero-fuel weight:
basic 124,000 kg (273,375 lb)
optional 126,000 kg (277,780 lb)

**incl weight of underfloor cargo hold containers and pallets*

Performance (estimated):
Range with max (structural) payload, allowances for ground manoeuvring, 30 min hold at 460 m (1,500 ft), no diversion:
basic 2,100 nm (3,890 km; 1,825 miles)
optional 2,500 nm (4,635 km; 2,170 miles)
Range with max fuel, no payload, allowances as above:
passenger mode 3,900 nm (7,230 km; 3,390 miles)
freight mode 4,100 nm (7,600 km; 3,560 miles)
pure freighter 4,200 nm (7,785 km; 3,650 miles)

AIRBUS A310

Known originally as the A300B10, the A310 short-fuselage development of the A300 was launched in July 1978. Compared with the A300, the cabin is shorter by 11 frames and the overall fuselage by 13 frames. The cabin thus normally seats from 210–234 passengers, although the aircraft will be certificated to carry up to 255 persons. The A310 retains the same fuselage cross-section as the A300, thus being able to carry standard LD3 containers two abreast, and/or standard pallets installed crosswise.

The A310 also has new, advanced-technology wings, of reduced span and area, designed by the British Aerospace Aircraft Group; new and smaller horizontal tail surfaces; new 'multi-role' pylons able to support all types of engine offered; and landing gear modified to cater for these changes in size and weight.

Manufacturing breakdown of the A310 differs in some respects from that of the A300. Aérospatiale builds the nose section (including flight deck), lower centre-fuselage and wing box, engine pylons, and CFRP airbrakes. MBB is responsible for the forward passenger cabin, upper centre-fuselage, rear fuselage, fin and rudder, flaps and flap tracks, and CFRP spoilers and lift dumpers. BAe Chester produces the wing fixed structures. CASA's contribution includes the horizontal tail surfaces, nose-gear and main-wheel doors, and forward passenger doors. Fokker manufactures the CFRP main landing gear leg doors, wingtips, all-speed ailerons, and CFRP flap track fairings. The wing leading-edge slats and wing/fuselage fairings are produced by Belairbus. Landing gear is by Messier-Hispano-Bugatti.

Like the A300, the A310 can be powered by General Electric, Pratt & Whitney or Rolls-Royce turbofan engines. The customers so far identified (up to mid-1981) have selected for their aircraft the General Electric CF6-80A1 or the Pratt & Whitney JT9D-7R4D1; and the aircraft is to be certificated initially with these two types of engine. Studies to power the A310 with the 222·4 kN (50,000 lb st) Rolls-Royce RB.211-524D4 are under way.

At present, the A310 is offered in one basic version, the A310-200, for short/medium-haul operations, including transcontinental routes. However, it is to be certificated and offered at different weight options, to match different airline needs, and further developments will offer increases in payload/range capability.

Ground testing of the A310 was expected to begin in November/December 1981. First flight is scheduled for late March 1982, with entry into service (by Swissair and Lufthansa) following approximately one year later. The following versions have been announced:

A310-200. Basic designation of passenger version. Model designations, according to power plant, include **A310-202** (CF6-80A1), **A310-203** (CF6-80A3), **A310-221** (JT9D-7R4D1), **A310-240** (RB.211-524B4) and **A310-241** (RB.211-524D4).

A310C-200. Convertible version.

A310F-200. Freighter version.

By 1 September 1981, orders and options totalling 153 had been received, as follows:

	Orders	*Options*
Air France (A310-202)	5	10
Austrian Airlines (A310-221)	2	2
British Caledonian Airways (A310-202)	3	3
KLM (Netherlands) (A310-202)	10	10
Kuwait Airways (A310-221)	8	—
Lufthansa (West Germany) (A310-202)	25	25
Martinair (Netherlands) (A310C-202)	3	1
Nigeria Airways	4	4
Sabena (Belgium) (A310-221)	3	3
Swissair (A310-221)	10	10
Wardair International (Canada)	6	6

Type: Large-capacity wide-bodied short/medium-range transport.

Wings: Cantilever mid-wing monoplane. Thickness/chord ratio 15·2% at root, 11·8% at 'break' in trailing-edge, and 10·8% at tip. Dihedral at trailing-edge 11° 8′ (inboard) and 4° 3′ (outboard). Incidence 5° 3′ at root. Sweepback 28° at quarter-chord. Wing box is two-spar multi-rib structure of high-strength light alloy, with top and bottom load-carrying skins. Three-section leading-edge slats on each wing over almost full span, with no cutout over engine pylon; Krueger flap between inboard slat and wing root. Simple Fowler-type trailing-edge flap on outboard section of each wing; vaned Fowler-type flap inboard. All-speed aileron between flaps on each wing. Two airbrakes between root and engine, two airbrakes outboard of engine, and three spoilers outboard of outer airbrakes, on each wing; all 14 surfaces are used also as lift dumpers. Outer slat leading-edges de-iced by engine bleed air.

Fuselage: Generally similar to A300B4, except for reduced length (see introductory copy), resulting in deletion of two passenger doors. Redesigned rear fuselage, between parallel section and tailcone.

Tail Unit: Vertical surfaces as for A300; horizontal surfaces similar to A300, but with slightly reduced dimensions and area.

Landing Gear: Hydraulically-retractable tricycle type. Twin-wheel steerable nose unit (steering angle 65°/95°) as for A300. New main gear by Messier-Hispano-Bugatti, each bogie comprising two tandem-mounted twin-wheel units. Retraction as for A300. Under-tail bumper beneath rear fuselage, to protect structure against excessive nose-up attitude during T-O and landing. Standard tyre sizes: main, 46 × 16-20, pressure 11·2 bars (163 lb/sq in); nose, 40 × 14-16, pressure 9·0 bars (131 lb/sq in). Two options for low-pressure tyres on main units: (1) size 49 × 17-20, pressure 9·8 bars (143 lb/sq in); (2) size 49 × 19-20, pressure 8·9 bars (129 lb/sq in). Messier-Hispano-Bugatti brakes and anti-skid units standard; Bendix type optional.

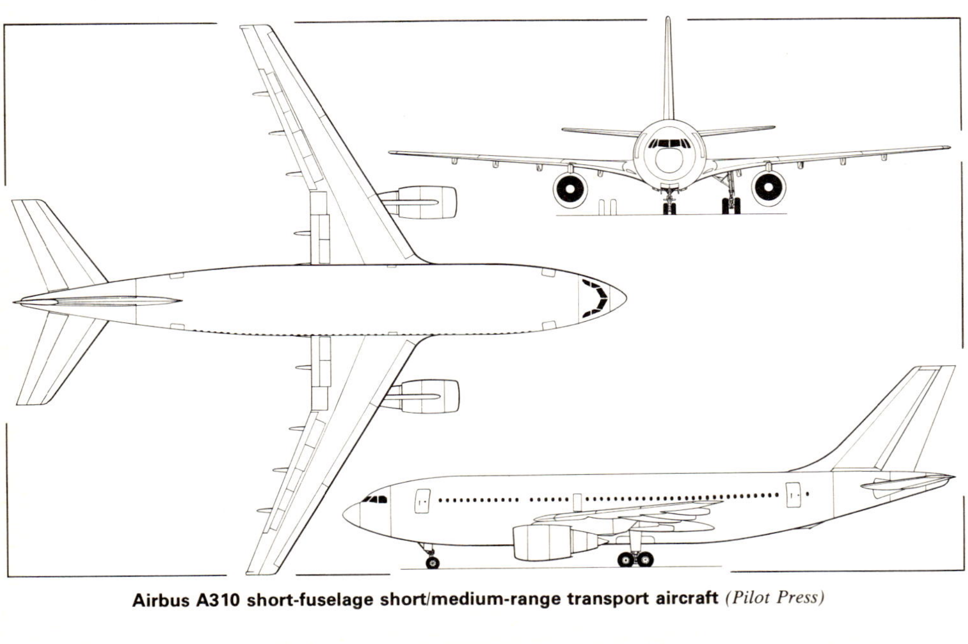

Airbus A310 short-fuselage short/medium-range transport aircraft *(Pilot Press)*

Power Plant: Currently ordered or available with the following turbofan engines:

Two 213·5 kN (48,000 lb st) General Electric CF6-80A1 (A310-202) or Pratt & Whitney JT9D-7R4D1 (A310-221);

Two 222·4 kN (50,000 lb st) General Electric CF6-80A3 (A310-203) or Rolls-Royce RB.211-524B4 (A310-240); or

Two 235·75 kN (53,000 lb st) Rolls-Royce RB.211-524D4 (A310-241).

Usable total fuel capacity approx 55,000 litres (12,098 Imp gallons), comprising 3,938 litres (866 Imp gallons) in each outer-wing tank; 13,937 litres (3,066 Imp gallons) in each inboard wing tank; and 19,250 litres (4,234 Imp gallons) in centre-section tank. Two standard refuelling points, one beneath each wing outboard of engines.

Accommodation: Crew of two or three in forward-facing crew cockpit (FFCC), in which all crew members face forward at all times. Provision for fourth crew member. Standard cabin arranged for 195-255 seats in six/seven/eight-abreast layout, at pitch of 96·5-103 cm (38-40 in) first class, 76-79 or 86 cm (30-31 or 34 in) economy class. Nine-abreast seating at 76 cm (30 in) pitch for 282 passengers in high-density configuration. Standard layout has galley and toilet at forward end of cabin, plus larger galley and four toilets at aft end. Depending upon customer requirements, a second toilet and galley can be added forward, and an additional galley aft. Toilets and galleys can be located at the forward end at the class divider position. Overhead baggage stowage as for A300, rising to average of 0·07 m³ (2·33 cu ft) per passenger with optional central stowage in 212-seat layout. Four passenger doors only, one forward and one aft on each side. Type I emergency exit over wing on each side. Underfloor baggage/cargo holds fore and aft of wings, each with door on starboard side. Forward hold will accommodate eight LD3 containers or three 2·24 × 3·17 m (88 × 125 in) standard pallets; the enlarged cargo door, optional on the A300, is standard on the A310, enabling 2·44 × 3·17 m (96 × 125 in) pallets to be loaded. Rear hold will accommodate six LD3 containers, with an optional seventh LD3 position which reduces the volume of the additional compartment aft of rear hold for bulk loading.

Systems: Garrett GTCP 331-250 APU. Air-conditioning system, powered by compressed air from engines, APU, or a ground supply unit, comprises two separate packs; air is distributed to flight deck, three separate cabin zones, electrical and electronic equipment, avionics bay and bulk cargo compartment. Ventilation of forward cargo compartments optional. Pressurisation system has a max normal differential of 0·57 bars (8·25 lb/sq in). Air supply for wing ice protection, engine starting and thrust reverser system is bled from various stages of the engine compressors, or supplied by the APU or a ground supply unit. Hydraulic system (three fully-independent circuits operating at 207 bars; 3,000 lb/sq in under normal conditions) is similar to that of A300, with some modifications. Electrical system, similar to that of A300, consists of a three-phase 115/200V 400Hz constant-frequency AC system and a 28V DC system. Two 90kVA engine-driven brushless generators for normal single-channel operation, with automatic transfer of busbars in the event of a generator failure. Each has an overload rating of 135kVA for 5 min and 180kVA for 5 s. A third (identical) AC generator, directly driven at constant speed by the APU, can be used during ground operations, and also in flight to compensate for the loss of one or both engine-driven generators. Any one generator can provide sufficient power to operate all equipment and systems necessary for an indefinite period of safe flight. DC power is generated via three 150A transformer-rectifiers. Three nickel-cadmium batteries are supplied. Flight crew oxygen system fed from rechargeable pressure bottle of 2,166 litres (76·5 cu ft) capacity. Standard options are a second 76·5 cu ft bottle, a 3,256 litre (115 cu ft) bottle, and an external filling connection. Emergency oxygen sets for passengers and cabin attendants. Anti-icing of outer wing leading-edge slats and engine air intakes by hot air bled from engines; and of pitot probes, static ports and plates, and sensors, by electrical heating.

Avionics and Equipment: Basic standard flight deck displays include flight guidance, navigation, configuration and engine management/monitoring information presented by electro-mechanical indicators; warning information presented by conventional warning lights and on a master warning CRT; system information presented on a system CRT. Electronic flight instrument system, provided as an option, comprises a CRT primary flight display, replacing the ADI and radio altimeter, and a CRT navigation display replacing the HSI and weather radar. The latter will display data from the flight management system (map mode, flight data display). A flight data recorder is also installed. Head-up display is optional. The basic aircraft is fitted with an AIDS (airborne integrated data system) providing a basic 80-parameter (40 mandatory plus 40 additional) system with an option to extend the system to 160 parameters. Standard com system includes HF radio (ARINC 719) (full provision); two VHF transceivers and space provision for a third (all to ARINC 716); Selcal system (ARINC 714); passenger address system (the amplifier conforming to ARINC 715); audio systems, comprising service interphone, audio integrating and flight interphone systems, and a ground crew call circuit. Digital navigation system, to ARINC 429 and ARINC 600, includes ADF (ARINC 712); two radio altimeters (ARINC 707); two DME (ARINC 709); two ATC transponders (ARINC 718); two VOR, one including marker beacon receiver (ARINC 711); weather radar (ARINC 708); two ILS (ARINC 710); and three AHRS (ARINC 705). Options include a second ADF; two or three IRS (ARINC 704); Omega system; and a second weather radar. The digital automatic flight control system (AFCS), in its basic definition, comprises a single flight control computer (FCC) for automatic flight control (to ARINC 701); a single thrust control computer (TCC) for speed and thrust control (to ARINC 703); and a duplicated flight augmentation computer (to ARINC 701). The flight management system (to ARINC 702) comprises a computer unit and control display unit. The FCC, functioning as autopilot, flight director and speed reference system, has the following basic modes: pitch hold, heading/roll altitude hold, altitude hold, altitude acquire, level change, vertical speed select and hold, heading select, VOR, heading, take-off and go-around. The installation of a second FCC will provide Cat. III autolands. The TCC provides the following functions: permanent computation of N1 or EPR limits, autothrottle functions, throttle pusher with windshear protection, speed and angle of attack protection, and a test function. A delayed flap approach mode is available as an option.

Dimensions, external:
Wing span 43·90 m (144 ft 0¼ in)
Wing chord: at root 8·38 m (27 ft 6 in)
at tip 2·18 m (7 ft 1¾ in)
Wing aspect ratio 8·8
Length overall 46·67 m (153 ft 1½ in)
Length of fuselage 45·13 m (148 ft 0¾ in)
Fuselage: Max diameter 5·64 m (18 ft 6 in)
Height overall 15·81 m (51 ft 10½ in)
Tailplane span 16·26 m (53 ft 4¼ in)
Wheel track 9·60 m (31 ft 6 in)
Wheelbase (c/l of shock-absorbers) 15·21 m (49 ft 10¾ in)
Passenger door (fwd, port):
Height 1·93 m (6 ft 4 in)
Width 1·07 m (3 ft 6 in)
Height to sill at OWE 4·54 m (14 ft 10¾ in)
Passenger door (aft, port):
Height 1·93 m (6 ft 4 in)
Width 1·07 m (3 ft 6 in)
Height to sill at OWE 4·85 m (15 ft 11 in)
Servicing doors (fwd and aft, stbd):
As corresponding passenger doors
Upper-deck cargo door (A310C/F): As A300C4/F4
Emergency exits (overwing, port and stbd, each):
Height 1·39 m (4 ft 6¾ in)
Width 0·67 m (2 ft 2½ in)
Underfloor cargo door (fwd):
Height 1·71 m (5 ft 7½ in)
Width 2·69 m (8 ft 10 in)
Height to sill at OWE 2·611 m (8 ft 6¾ in)
Underfloor cargo door (rear):
Height 1·71 m (5 ft 7½ in)
Width 1·81 m (5 ft 11¼ in)
Height to sill at OWE 2·72 m (8 ft 11 in)
Underfloor cargo door (aft bulk hold):
Height 0·95 m (3 ft 1½ in)
Width 0·95 m (3 ft 1½ in)
Height to sill at OWE 2·751 m (9 ft 0¼ in)

Dimensions, internal:
Cabin, excl flight deck:
Length 33·24 m (109 ft 0¾ in)
Max width 5·29 m (17 ft 4¼ in)
Max height 2·33 m (7 ft 7¾ in)
Fwd cargo hold: Length 7·63 m (25 ft 0½ in)
Max width 4·18 m (13 ft 8½ in)
Height 1·71 m (5 ft 7¼ in)
Volume 50·3 m³ (1,776·3 cu ft)
Aft cargo hold: Length 5·07 m (16 ft 7½ in)
Max width 4·17 m (13 ft 8¼ in)
Height 1·67 m (5 ft 5¾ in)
Volume 34·5 m³ (1,218·4 cu ft)
Aft bulk hold: Volume 17·3 m³ (610·9 cu ft)
Total overall volume 102·1 m³ (3,605·6 cu ft)

Areas:
Wings, gross 219 m² (2,357·3 sq ft)
Vertical tail surfaces (total) 45·2 m² (486·5 sq ft)
Horizontal tail surfaces (total) 64·0 m² (688·9 sq ft)

Weights:
Typical operating weight empty:
202 76,616 kg (168,910 lb)
221 76,895 kg (169,525 lb)
Max payload: 202 31,884 kg (70,292 lb)
221 31,605 kg (69,677 lb)
A310C 36,800 kg (81,130 lb)
A310F 39,400 kg (86,860 lb)
Max usable fuel 43,000 kg (94,798 lb)
Max T-O weight 132,000 kg (291,010 lb)
or 138,600 kg (305,560 lb)
Max ramp weight 132,900 kg (292,995 lb)
or 139,500 kg (307,540 lb)
Max landing weight 118,500 kg (261,250 lb)
or 121,500 kg (267,860 lb)
Max zero-fuel weight 108,500 kg (239,200 lb)
or 111,500 kg (245,810 lb)

Performance (A310-200, estimated, with 237 passengers except where indicated):
Max operating speed: all versions
Mach 0·84 (360 knots; 667 km/h; 414 mph CAS)
Typical high-speed cruise at 9,145 m (30,000 ft)
Mach 0·82 (483 knots; 895 km/h; 556 mph)
Typical long-range cruise at 11,275 m (37,000 ft)
Mach 0·78 (447 knots; 828 km/h; 515 mph)
Approach speed at max landing weight
133 knots (246 km/h; 153 mph)
Approach speed at typical landing weight
125 knots (232 km/h; 144 mph)
T-O field length at max T-O weight, S/L, ISA:
202 1,845 m (6,053 ft)
221 1,815 m (5,955 ft)
Landing field length (S/L airport):
at max landing weight 1,665 m (5,463 ft)
at typical landing weight 1,475 m (4,840 ft)
Runway LCN at max T-O weight (A: 20 in flexible pavement, B: 30 in radius rigid pavement):
A, standard main-wheel tyres 64
A, option 1 main-wheel tyres 61
A, option 2 main-wheel tyres 55
B, standard main-wheel tyres 59
B, option 1 main-wheel tyres 55
B, option 2 main-wheel tyres 50
Range, with allowances for ground manoeuvring and approach manoeuvring, reserves for overshoot at destination, 5% of stage fuel, 200 nm (370 km; 230 mile) diversion, and 45 min hold at 1,525 m (5,000 ft):
with max passenger payload:
202 2,530 nm (4,688 km; 2,913 miles)
221 2,470 nm (4,577 km; 2,844 miles)

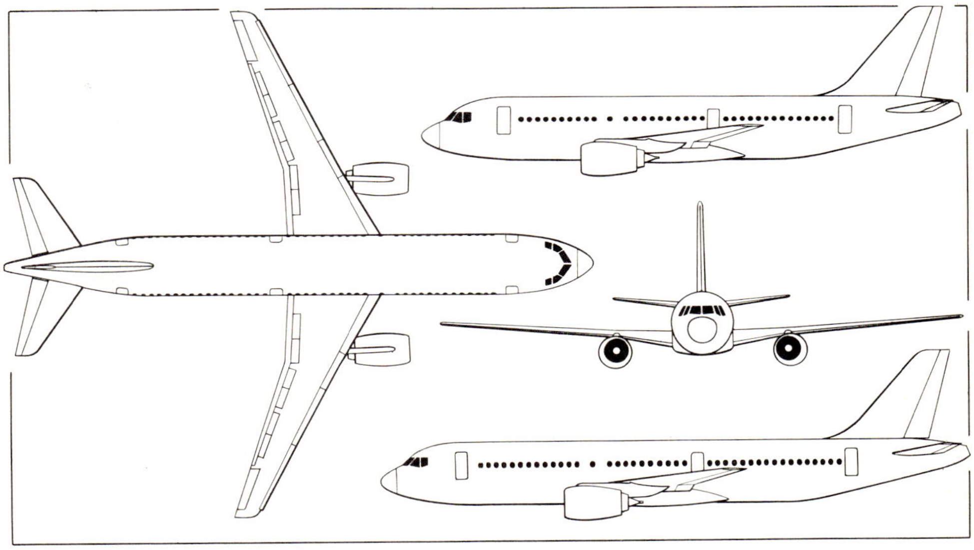

Airbus A320-200 twin-turbofan transport, with additional side view (top) of A320-100 *(Pilot Press)*

Super Guppy outsize cargo aircraft, used to transport major components of the Airbus A300 and A310 to the final assembly centre at Toulouse

with max structural payload:

202, 221	1,460 nm (2,705 km; 1,681 miles)

with max fuel:

202	3,510 nm (6,504 km; 4,041 miles)
221	3,490 nm (6,467 km; 4,018 miles)

AIRBUS INDUSTRIE A320

The A320, an advanced-technology single-aisle aircraft for short to medium ranges, is intended to offer considerably lower fuel consumption and seat-mile costs than any other aircraft existing or on offer in this category. Known formerly as the SA, it has a seating capacity of up to 154 passengers in a normal **A320-100** mixed-class configuration, and up to 172 passengers in the **A320-200** series. It will be powered by two underwing-mounted engines in the 111·2 kN (25,000 lb st) category, available from any of three manufacturers: the CFM International (General Electric/SNECMA) CFM56-2000, Pratt & Whitney PW2025, and Rolls-Royce/Japan Aero Engines RJ500-35B2. Its initial range of up to 2,000 nm (3,700 km; 2,300 miles), with full airline allowances, will be extended later to 2,500 nm (4,630 km; 2,875 miles).

Compared with existing narrow-body aircraft, the fuselage cross-section is significantly increased, being 3·95 m (12 ft 11½ in) wide. This permits the use of wider (1·575 m; 5 ft 2 in) triple seats, to provide higher standards of passenger comfort, and five-abreast business class seating provides a standard equal to that offered as first class on major competitive aircraft. Overhead stowage room of around 0·06 m³ (2·0 cu ft) per seat is superior to that available on existing aircraft of similar capacity, and provides ample carry-on baggage space. Improved seat design and optimised positioning of the seat rails makes optimum use of the additional room provided under the seats. The fuselage cross-section has a double-bubble form and is 4·14 m (13 ft 7 in) deep, providing increased freight hold volume and working height, and the ability to carry containers derived from the standard interline LD3 type. As the base is the same as that of the LD3, all existing wide-body aircraft and ground handling equipment can accept these new containers, without modification.

Wing design incorporates the latest advances in technology, and experience from the A310 wing. The A320 will be certificated for two-crew operation and for Category IIIB all-weather landings. It will offer significant commonality with other Airbus Industrie aircraft, where this is cost-effective, particularly in respect of the flight deck and equipment. The wide use of composite materials, stemming from A300 and A310 experience, will allow a lower structural weight for a given payload.

A launch decision was anticipated by the end of 1981. If given, this would permit customer deliveries to begin in early 1986. Air France signed an agreement to purchase 25 A320s, with options on 25 more (16 A320-100 and 34 A320-200), on 6 June 1981.

DIMENSIONS, EXTERNAL:

Wing span	34·57 m (113 ft 5 in)
Fuselage length: A320-100	36·04 m (118 ft 3 in)
A320-200	39·24 m (128 ft 9 in)
Height overall	11·85 m (38 ft 10½ in)

WEIGHTS:

Operating weight empty:	
A320-100	38,478 kg (84,830 lb)
A320-200	40,835 kg (90,026 lb)
Max weight-limited payload:	
A320-100	16,422 kg (36,204 lb)
A320-200	19,565 kg (43,133 lb)
Max T-O weight:	
A320-100	66,000 kg (145,505 lb)
A320-200	71,900 kg (158,510 lb)
Max landing weight:	
A320-100	60,700 kg (133,820 lb)
A320-200	65,700 kg (144,840 lb)
Max zero-fuel weight:	
A320-100	54,900 kg (121,030 lb)
A320-200	60,400 kg (133,160 lb)

AIRBUS FUTURE DEVELOPMENTS

In 1981, Airbus Industrie was studying the following possible future developments of the A300/310/320 family:

TA9 (formerly A300B9). 'Stretched' version of A300-600, with a fuselage approx 8·70 m (28 ft 6½ in) longer to provide a 25 per cent increase in passenger seating (approx 400 in a high-density layout) and increased cargo capacity. New and bigger wing. Optimised for short/medium-range operation. Eight or nine-abreast seating, with twin aisles, in same fuselage cross-section as A300. Advanced forward-facing crew cockpit (FFCC) with CRT displays, fully digital systems, flight management computer, electrical signalling for some control surfaces, quieter APU. Extensive use of composites and improved metallic materials. Two variants proposed: **TA9-100** short-range version (1,500 nm; 2,780 km; 1,727 mile range with full passenger payload), and **TA9-200** medium range version (3,200 nm; 5,930 km; 3,685 mile range with full passenger payload). Power plants would be high-thrust versions of current CF6-80, JT9D-7R4 and RB.211-524 engines.

TA11 (formerly A300B11). Long-range twin-aisle aircraft, using structural components of A300 and A310. Same fuselage cross-section as A300/A310/TA9, with seating for 200-230 passengers. Wing similar to that of TA9. Powered by four turbofan engines in 133·45 kN (30,000 lb st) class. Advanced flight deck with CRT displays, digital avionics, electrical signalling, and active controls. Extensive use of composites and improved metallic materials. Proposed in two versions, both using the same basic airframe: **TA11-100** basic version (5,500 nm; 10,190 km; 6,333 mile range with full passenger payload), and **TA11-200** developed version (6,500 nm; 12,046 km; 7,485 mile range with full passenger payload).

TA12. Combining the wings/engines of the TA9 with the A310/TA11 fuselage. Applicable to long-range non-trans-oceanic routes with stage lengths of up to 4,500 nm (8,340 km; 5,180 miles), and for operation from 'hot and high' airfields.

AIRBUS INDUSTRIE SUPER GUPPY

To cater for the currently increasing A300 production rate, Airbus Industrie is to acquire two more Super Guppy cargo transports in Spring 1982 and Spring 1983 respectively. Conversion of basic Stratocruiser airframes will be undertaken under subcontract to Airbus Industrie by UTA Industries at Le Bourget, France, in conjunction with Aero Spacelines of Santa Barbara, California, from whom Airbus Industrie has now acquired the production rights and technical drawings. Operation of its Super Guppies is subcontracted by Airbus Industrie to Aéromaritime, a charter subsidiary of UTA. A description of the Super Guppy can be found under the Aero Spacelines entry in the US section of the 1973-74 *Jane's*.

AIRTEC

AIRCRAFT TECHNOLOGY INDUSTRIES

HEAD OFFICE: Rey Francisco 4, Apartado 193, Madrid 8, Spain

PRESIDENT: Prof Dr-Ing B. J. Habibie

VICE-PRESIDENT: Dr Carlos Marin Ridruejo

PARTICIPATING COMPANIES:

Construcciones Aeronauticas SA, Rey Francisco 4, Apartado 193, Madrid 8, Spain
Telephone: 247 25 00
Telex: 27418

PT Industri Pesawat Terbang Nurtanio, Lanuma Husein Sastranegara (Husein Sastranegara Air Force Base), Jalan Pajajaran 154, Bandung, Indonesia
Telephone: Bandung 613662 and 613835
Telex: 28295 NUR BD

Airtec is a joint company formed by CASA of Spain and PT Nurtanio of Indonesia to develop a 34/38-passenger twin-turboprop transport aircraft. Design and production work will be shared 50-50 between the two companies. CASA will build the wing centre-section, forward and centre fuselage and horizontal tail surfaces; the outer wings, rear fuselage and vertical tail surfaces will be manufactured by Nurtanio.

AIRTEC (CASA/NURTANIO) CN-235

Preliminary design of the CN-235 was initiated in January 1980, and detail design work began a year later. Two prototypes are to be built, one in each country, and simultaneous first flights are planned for October 1983. Deliveries are intended to begin in the second half of 1984, following certification under FAR Pt 25. By mid-1981 firm orders had been received from Merpati Nusantara Airlines (14), Pelita (10), Deraya Air Taxi (10), and Iberia/Aviaco/Transeuropa (20).

TYPE: Twin-turboprop commuter and utility transport aircraft.

WINGS: Cantilever high-wing monoplane. Constant-chord centre-section; tapered outer panels, with raked tips. Trim tab in each aileron.

FUSELAGE: Conventional fail-safe, pressurised semi-monocoque structure, of flattened circular cross-section, upswept at rear.

TAIL UNIT: Conventional all-metal sweptback fin, large dorsal fin and non-swept tailplane. Elevators and rudder of composite materials. Trim tabs in rudder and each elevator.

LANDING GEAR: Retractable tricycle type. Main units each comprise two wheels in tandem, retracting into fairing on side of fuselage. Single steerable nosewheel.

POWER PLANT: Two General Electric CT7-7 turboprop engines, each flat rated to 1,268 kW (1,700 shp) for take-off and driving an advanced-technology four-blade constant-speed propeller with full feathering and reverse-pitch capability. Refuelling point in starboard main landing gear fairing. Propeller braking permits engine to be used as an on-ground APU.

ACCOMMODATION: Crew of two on flight deck, plus cabin attendant. Standard accommodation in commuter version for 34 or 38 passengers, in four-abreast seating, at 81 cm (32 in) and 76 cm (30 in) pitch respectively. Toilet, wardrobe, galley and overhead lockers standard. Baggage compartment at rear of cabin. Can also be equipped as mixed passenger/cargo combi, or for all-cargo configuration, in latter version able to carry standard LD2, LD3 and LD6 containers and to accept pallets up to 2·24 m (88 in) wide. Main passenger door, with integral stairs, aft of wing on port side. Emergency exits facing this door on starboard side, and on port side at front of cabin. Opposite forward emergency exit is a service door, which serves also as passenger door in combi version. Ramp/door in underside of upswept rear fuselage, for loading of bulky cargo. Accommodation fully air-conditioned, and pressurised to 6,100 m (20,000 ft) flight cruise level.

DIMENSIONS, EXTERNAL:

Wing span	25·80 m (84 ft 7¾ in)
Wing chord: at root	3·00 m (9 ft 10 in)
at tip	1·20 m (3 ft 11¼ in)
Wing aspect ratio	10·2
Length overall	21·30 m (69 ft 10½ in)
Fuselage: Max diameter	2·90 m (9 ft 6 in)
Height overall	7·90 m (25 ft 11 in)
Wheel track	3·50 m (11 ft 5¾ in)
Passenger door: Width	0·70 m (2 ft 3½ in)

DIMENSIONS, INTERNAL:

Cabin, excl flight deck:	
Length	9·35 m (30 ft 8 in)
Max width	2·70 m (8 ft 10½ in)
Width at floor	2·36 m (7 ft 9 in)
Max height	1·90 m (6 ft 2¾ in)
Volume	42·0 m³ (1,483·2 cu ft)
Baggage compartment volume	7·0 m³ (247·2 cu ft)

AREAS:

Wings, gross	60·00 m² (645·8 sq ft)
Vertical tail surfaces (total)	11·11 m² (119·6 sq ft)
Horizontal tail surfaces (total)	20·00 m² (215·3 sq ft)

WEIGHTS (estimated):

Max fuel load	4,000 kg (8,818 lb)
Nominal payload	4,000 kg (8,818 lb)
Max payload (cargo)	4,500 kg (9,921 lb)
Max T-O weight	13,000 kg (28,660 lb)
Max landing weight	12,800 kg (28,220 lb)
Max zero-fuel weight	11,800 kg (26,015 lb)

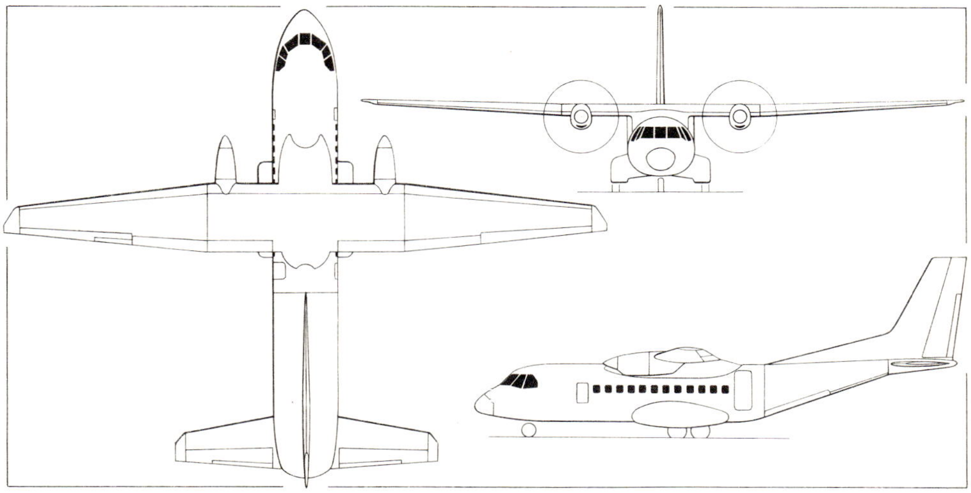

CASA/Nurtanio (Airtec) CN-235 twin-turboprop commuter transport *(Pilot Press)*

PERFORMANCE (estimated, at max T-O weight, ISA):

Max cruising speed	250 knots (463 km/h; 288 mph)
Max rate of climb at S/L	660 m (2,165 ft)/min
Rate of climb at S/L, one engine out	200 m (655 ft)/min
Service ceiling (30·5 m; 100 ft/min rate of climb):	
two engines	8,840 m (29,000 ft)
one engine out	3,960 m (13,000 ft)
T-O to 10·7 m (35 ft) at S/L	900 m (2,953 ft)
Landing from 15 m (50 ft) at S/L	600 m (1,968 ft)
Range with max payload at 6,100 m (20,000 ft), no reserves	900 nm (1,668 km; 1,036 miles)

ALPHA JET

AIRFRAME PRIME CONTRACTORS:

Avions Marcel Dassault/Breguet Aviation, 27 rue du Professeur Pauchet, BP 32, 92420 Vaucresson, France
Telephone: 741 7921
Telex: AMADAS 203944 F

Dornier GmbH, Postfach 1420, 7990 Friedrichshafen, Federal Republic of Germany
Telephone: (07545) 82307
Telex: 0734372

On 22 July 1969 the French and Federal German governments announced a joint requirement for a new subsonic basic/advanced training and light attack aircraft to enter service with their armed forces in the 1970s. Each government had a potential requirement for about 200 such aircraft, to replace Magister, Lockheed T-33A and Mystère IV-A trainers, and Fiat G91 attack aircraft, then in service.

On 24 July 1970, it was announced that the Alpha Jet had been selected to meet the requirement. The programme received joint French-West German government approval in late 1972; approval to proceed with the production phase was announced on 26 March 1975.

DASSAULT-BREGUET/DORNIER ALPHA JET

Dassault-Breguet and Dornier are jointly producing the Alpha Jet, with Dassault-Breguet as main contractor and Dornier as industrial collaborator, the total work load being shared primarily between the two groups.

All production Alpha Jets have identical structure, power plant, landing gear and standard equipment; there are assembly lines in France, West Germany, Egypt and Belgium. The outer wings, tail unit, rear fuselage, landing gear doors and cold-flow exhaust are manufactured in West Germany; the forward and centre fuselage (with integrated wing centre-section) are manufactured in France. Fuselage nosecones and wing flaps are manufactured in Belgium by SABCA. The power plant prime contractors are Turboméca and SNECMA in France, and MTU and KHD in West Germany; and, for the landing gear, Messier-Hispano-Bugatti in France and Liebherr Aero Technik in West Germany.

Four flying prototypes were built, plus two airframes for static and fatigue testing. The 01 made its first flight, at Istres, on 26 October 1973; all four had flown by the end of 1974. Details of the prototypes can be found in the 1978-79 and earlier editions of *Jane's*.

There are three production versions, as follows:

Alpha Jet E (for École). Trainer version, ordered for the air forces of France (200), Abu Dhabi (6), Belgium (33), Cameroun (6), Egypt (22), Ivory Coast (6), Morocco (24), Nigeria (12), Qatar (6) and Togo (5). Those for Nigeria are from German assembly line, all others from French production. First production aircraft (E1 for French Air Force) flown on 4 November 1977; deliveries, beginning with E2, began in the Summer of 1978. Now equips Groupement-École 314 at Tours (65 aircraft) and the Patrouille de France (12 aircraft) at Salon de Provence; 14 due for delivery to the Centre d' Entraînement au Vol Sans Visibilité in 1981, followed by 30 in 1981-82 to the 8e Escadre de Transformation at Cazaux to replace Mystère IV-As in the weapons training role. Aircraft for Belgian Air Force, delivered in 1978-80, serve with Nos. 2 and 7 Squadrons.

Alpha Jet A (for Appui). Close support version, ordered for Federal German Luftwaffe (175). First flown (A1) on 12 April 1978; deliveries began in mid-March 1979. Three Jagdbombergeschwader (fighter-bomber groups) being equipped: JaboG 49 at Fürstenfeldbruck (from 20 March 1980), JaboG 43 at Oldenburg (from January 1981), and JaboG 41 at Husum (from June 1981), replacing Fiat G91Rs in the close support and reconnaissance roles. Each of these units is to have 51 aircraft. Eighteen others are assigned to the Luftwaffe base at Beja in Portugal for weapons training; deliveries of these began in 1980. The remaining four aircraft will be used for continuing flight test and evaluation work, including A1, which is currently undergoing a flying programme fitted with an experimental TST (Transsonik Tragflügel: transonic wing) of supercritical section, developed by Dornier under a Federal Ministry of Defence contract. First flown on 12 December 1980, the TST Alpha Jet (98 + 33) has a wing extended forward at the root as far as the rear frame of the front cockpit. A large leading-edge radius increases the thickness/chord ratio by 20%. Upper-surface curvature is relatively flat near the leading-edge, and more pronounced towards the trailing-edge. Other changes include leading- and trailing-edge manoeuvring flaps, and a slightly modified 'sawtooth' position. The 100th Alpha Jet A was delivered in March 1981, and 160 were scheduled to be delivered by the end of that year.

Alpha Jet MS2. Improved close support version, developed by Dassault-Breguet. Equipped with new nav/attack system which includes a Sagem Uliss 81 inertial platform, Thomson-CSF VE 110C head-up display, Thomson-CSF TMV 630 laser rangefinder and TRT AHV 9 radio altimeter. Ordered by Egypt (8) and one other air force by mid-1981.

By Summer 1981 more than 300 of the 503 Alpha Jets then on order had been delivered. The 100th aircraft was delivered on 22 February 1980, the 200th on 1 December 1980, and the 300th in July 1981. Production is continuing at the rate of 12 per month.

Dassault-Breguet and Dornier teamed with Lockheed-California to enter the Alpha Jet in the US Navy's VTX-TS advanced trainer competition; if selected,

French Air Force Alpha Jet Es of the École de Chasse at Tours, a unit of Groupement-École 314 'Christian Martel'

it would be built under licence in the USA by Lockheed. Alpha Jet A58 undertook a four-week demonstration tour of the USA in September/October 1980 in connection with this competition.

The following description applies to the standard production versions, as indicated:

Type: Tandem two-seat basic, low-altitude and advanced jet trainer (E) and close support and battlefield reconnaissance aircraft (A).

Wings: Cantilever shoulder-wing monoplane, with 6° anhedral from roots. Thickness/chord ratio 10·2% at root, 8·6% at tip. Sweepback 28° at quarter-chord. All-metal numerically- or chemically-milled structure, consisting of two main wing panels bolted to a centre frame. Extended chord on outer wings. Hydraulically actuated Fowler-type slotted flaps on each trailing-edge. Ailerons actuated by double-body irreversible hydraulic servo, with trimmable artificial feel system.

Fuselage: All-metal semi-monocoque structure, numerically or chemically milled, of basically oval cross-section. Built in three sections: nose (including cockpit), centre-section (including engine air intake trunks and main landing gear housings) and rear (including engine mounts and tail assembly). Narrow strake on each side of nose of Alpha Jet E. Pointed nose, with pitot probe, on A version. Electrically controlled, hydraulically actuated airbrake on each side of rear upper fuselage, of carbonfibre-reinforced epoxy resin.

Tail Unit: Cantilever type, of similar construction to wings, with 45° sweepback on fin leading-edge and 30° on tailplane leading-edge. Dorsal spine fairing between cockpit and fin. All-flying tailplane, with trimmable and IAS-controlled artificial feel system. Glassfibre fin tip and tailplane tips. Double-body irreversible hydraulic servo-actuated rudder, with trimmable artificial feel system. Yaw damper on close support version.

Landing Gear: Forward-retracting tricycle type, of Messier-Hispano-Bugatti/Liebherr design. All units retract hydraulically, main units into underside of engine air intake trunks. Single wheel and low-pressure tyre (approx 4 bars; 58 lb/sq in at normal T-O weight) on each unit. Tyre sizes 615 × 255-10 on main units, 380 × 150-4 on nose unit. Steel disc brakes and anti-skid units on main gear (Minispad on A version, Ministop on E). Emergency braking system. Hydraulic nosewheel steering and arrester hook on close support version. Nosewheel offset to starboard to permit ground firing from gun pod.

Power Plant: Two SNECMA/Turboméca Larzac 04-C5 turbofan engines, each rated at 13·24 kN (2,976 lb st), mounted on sides of fuselage. Splitter plate in front of each intake. Fuel in two integral tanks in outer wings, one in centre-section and three fuselage tanks. Internal fuel capacity 1,900 litres (418 Imp gallons). Provision for 310 litre (68·2 Imp gallon) capacity drop-tank on each outer wing pylon. Pressure refuelling standard for all tanks, including drop-tanks. Gravity system for fuselage tanks and drop-tanks. Pressure refuelling point near starboard engine air intake. Fuel system incorporates provision for inverted flying.

Accommodation: Two persons in tandem, in pressurised cockpit under individual upward-opening canopies. Dual controls standard. Rear seat (for instructor in trainer versions) is elevated. French trainer versions fitted with Martin-Baker AJRM4 ejection seats, operable (including ejection through canopy) at zero height and speeds down to 90 knots (167 km/h; 104 mph). Martin-Baker B10N zero-zero seats in aircraft for Belgium, E10N in those for Egypt and Q10N in those for Qatar. Aircraft for West Germany fitted with licence-built (by MBB) Stencel S-III-S3AJ zero-zero ejection seats. Baggage compartment in tailcone, with access door on starboard side.

Systems: Cockpit air-conditioning and demisting system. Cabin pressure differential 0·30 bars (4·3 lb/sq in). Two independent and redundant hydraulic systems, each 207 bars (3,000 lb/sq in), with engine-driven pumps (emergency electric pump on one circuit), for actuating control surfaces, landing gear, brakes, flaps, airbrakes, and (when fitted) nosewheel steering. Pneumatic system, for cockpit pressurisation and air-conditioning, occupants' pressure suits and fuel tank pressurisation, is supplied by compressed air from engines. Main electrical power supplied by two 28V 9kW starter/generators, one on each engine. Circuit includes a 36Ah nickel-cadmium battery for self-starting and two static inverters for supplying 115V AC current at 400Hz to auxiliary systems. External ground DC power receptacle in port engine air intake trunk. Hydraulic and electrical systems can be sustained by either engine in the event of the other engine becoming inoperative. Liquid-film anti-icing system; de-icing by electrical heater mats. Oxygen mask for each occupant, supplied by liquid oxygen converter of 10 litres (2·2 Imp gallons) capacity. Emergency gaseous oxygen bottle for each occupant.

Avionics and Equipment: Large avionics bay in rear fuselage, containing most of the radio and navigation equipment. Standard avionics, according to version, include VHF and UHF transceivers, IFF/SIF, VOR/ILS/marker beacon receiver, Tacan, gyro platform and intercom. Landing light on starboard mainwheel leg, taxying light on port leg. Trainer version has SFIM 550 gyro platform, Thomson-CSF 920 weapon-aiming computer, LMT micro-Tacan, EAS Socrat 6200NRAX-7 VOR/ILS/marker beacon receiver, TEAM com radio and intercom, and LMT IFF/SIF. Close support version has Kaiser/VDO KM 808 head-up display, TRT AHV 6-18 radar altimeter, Lear Siegler LSI 6000 E attitude and heading reference system, and Litef LDN Doppler navigation system with LR-1416 navigation computer, Litef ABE control unit and Teledyne Ryan speed sensor, Elettronica (Italy) ECM, SEL Mitac/Setac Tacan with ILS, Siemens STR 700 IFF/SIF, Becker VCS 220 intercom, Rohde und Schwarz XT 3011 com radio, and EAS IMT 565 BDHI. See model listing for nav/attack avionics in MS2 attack version.

Armament and Operational Equipment: Alpha Jet users have, to date, defined 35 different weapon configurations for training and tactical air support missions. For armament training and close support, the Alpha Jet can be equipped with an underfuselage detachable pod containing a 30 mm DEFA (E) or 27 mm Mauser (A) cannon with 150 rds; or an under-fuselage pylon for one 250 kg bomb, one 400 kg modular bomb, or a target towing system. Provision also for two hardpoints under each wing, with non-jettisonable adaptor pylons. On these can be carried M155 launchers for eighteen 68 mm rockets; HE or retarded bombs of 50, 125, 250 or 400 kg; 625 lb cluster dispensers; 690 or 825 lb special-purpose tanks; combined external multistore adaptor (CEM-1) for six to eighteen rockets and four practice bombs, or one 500 lb bomb, or six penetration bombs, or grenades or other stores; practice launchers for bombs or rockets; Dassault-Breguet CC-420 30 mm gun pods; or two 310 litre (68·2 Imp gallon) drop-tanks (outer pylons only). Provision also for carrying air-to-air or air-to-surface missiles such as Magic or Maverick, target demonstration devices or reconnaissance pod. Total load for all five stations 2,500 kg (5,510 lb). At customer's option, the standard pylons can be replaced by CEM-1 (combined external multistore) carriers, developed by Dassault-Breguet, which permit the simultaneous carriage of mixed fuel/bomb/rocket loads underwing. A special version of the CEM-1 allows carriage of a reconnaissance pod containing a Super Cyclope infra-red system, four cameras, and a decoy launcher. Fire control system for air-to-air or air-to-ground firing, dive bombing and low-level bombing. Firing by trainee pilot (in front seat) is governed by a safety interlock system controlled by the instructor, which energises the forward station trigger circuit and illuminates a fire clearance indicator in the trainee's cockpit. Thomson-CSF Type 902 sight and gun camera in French version; Kaiser/VDO KM 808 sight, head-up display and gun camera in West German attack version; Thomson-CSF VE 110C head-up display and gun camera in MS2 attack version.

Dimensions, external:

Wing span	9·11 m (29 ft 10¾ in)
Wing aspect ratio	4·8

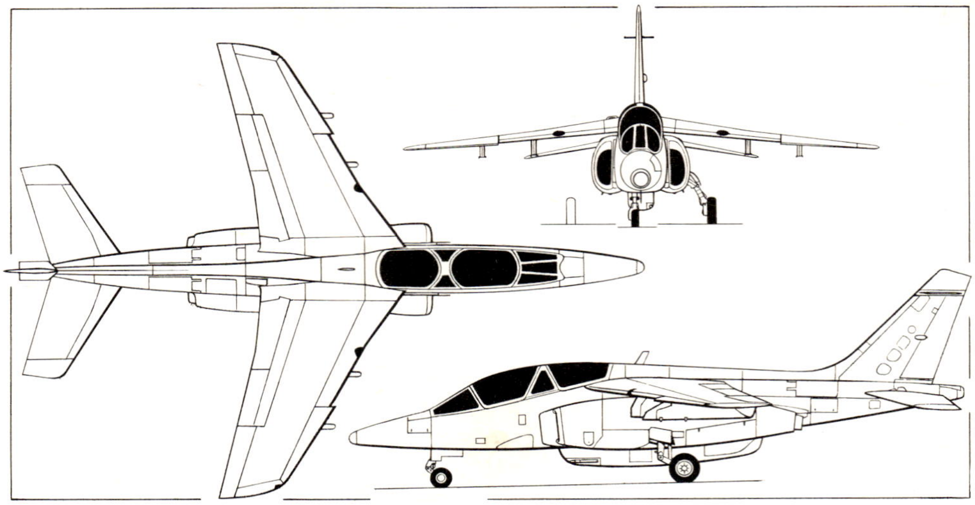

Alpha Jet E basic and advanced trainer, with underfuselage gun pod *(Pilot Press)*

Export Alpha Jet Es for the air forces of Togo (foreground), Ivory Coast (centre) and Qatar (rear)

Alpha Jet A1 testbed with Dornier's experimental TST supercritical wing

Length overall: trainer	12·29 m (40 ft 3¾ in)
close support version, incl probe	13·23 m (43 ft 5 in)
Height overall (at normal T-O weight)	4·19 m (13 ft 9 in)
Tailplane span	4·33 m (14 ft 2½ in)
Wheel track	2·71 m (8 ft 10¾ in)
Wheelbase	4·72 m (15 ft 5¾ in)

Areas:

Wings, gross	17·50 m² (188·4 sq ft)
Ailerons (total)	1·04 m² (11·19 sq ft)
Trailing-edge flaps (total)	2·86 m² (30·78 sq ft)
Airbrakes (total)	0·74 m² (7·97 sq ft)
Fin	2·97 m² (31·97 sq ft)
Rudder	0·62 m² (6·67 sq ft)
Horizontal tail surfaces (total)	3·94 m² (42·41 sq ft)

Weights:

Weight empty, equipped:	
trainer	3,345 kg (7,374 lb)
close support version	3,515 kg (7,749 lb)
Max fuel load:	
internal	1,415 kg (3,120 lb)
external	500 kg (1,102 lb)
Max external load	2,500 kg (5,510 lb)
Normal T-O weight:	
trainer, 'clean'	5,000 kg (11,023 lb)
Max T-O weight:	
with external stores	7,500 kg (16,535 lb)

Performance (at normal 'clean' T-O weight, except where indicated):

Max level speed at 10,000 m (32,800 ft)	Mach 0·85
Max level speed at S/L	540 knots (1,000 km/h; 621 mph)
Max speed for flap and landing gear extension	200 knots (370 km/h; 230 mph)
Approach speed	110 knots (204 km/h; 127 mph)
Stalling speed, flaps and landing gear up	116 knots (216 km/h; 134 mph)
Stalling speed, flaps and landing gear down	90 knots (167 km/h; 104 mph)
Landing speed at normal landing weight	92 knots (170 km/h; 106 mph)
Max rate of climb at S/L	3,420 m (11,220 ft)/min
Rate of climb at S/L, one engine out, at 4,782 kg (10,542 lb) AUW, in landing configuration	330 m (1,085 ft)/min
Time to 9,145 m (30,000 ft)	less than 7 min
Service ceiling	14,630 m (48,000 ft)
T-O run	410 m (1,345 ft)
Landing run	610 m (2,000 ft)
Low altitude radius of action (trainer):	
'clean', max internal fuel	291 nm (540 km; 335 miles)
with external tanks	361 nm (670 km; 416 miles)
High altitude radius of action (trainer), reserves of 15% internal fuel:	
'clean', max internal fuel	664 nm (1,230 km; 764 miles)
with external tanks	782 nm (1,450 km; 901 miles)
Lo-lo-lo mission radius (close support version), incl 5 min combat at max continuous thrust:	
with belly gun pod and underwing weapons	229 nm (425 km; 264 miles)
with belly gun pod, underwing weapons and external tanks	329 nm (610 km; 379 miles)
Hi-lo-hi mission radius (close support version), incl 5 min combat at max continuous thrust, reserves for two GCA landings:	
with belly gun pod and underwing weapons	364 nm (675 km; 419 miles)
with belly gun pod, underwing weapons and external tanks	591 nm (1,095 km; 680 miles)
Ferry range (internal fuel and two 310 litre external tanks)	1,586 nm (2,940 km; 1,827 miles)
Endurance (internal fuel only):	
low altitude	more than 2 h 30 min
high altitude	more than 3 h 30 min
g limits (ultimate)	+12; −6·4

CONCORDE

CONCORDE SUPERSONIC TRANSPORT

Airframe Prime Contractors:

British Aerospace PLC Aircraft Group, Richmond Road, Kingston upon Thames, Surrey KT2 5QS, England

Aérospatiale, 37 boulevard de Montmorency, 75781 Paris Cédex 16, France

CONCORDE

Full details of the Concorde programme have been given in the 1980-81 and earlier editions of *Jane's*. The final delivery, of aircraft No. 215 to Air France, was made on 23 October 1980.

EHI

EH INDUSTRIES LIMITED

Granville House, 132-135 Sloane Street, London SW1X 9BB

Directors:

The Rt Hon Lord Aldington, PC, KCMG, CBE, DSO (Chairman)

Cav del lavoro Conte Corrado Agusta (alternate Chairman)

B. D. Blackwell, MA, BSc(Eng), FEng, FIMechE, FRAeS, FBIM

Dott Ing P. Fascione

Executive Director:

D. A. P. Saunders-Davies

Participating Companies:

Costruzioni Aeronautiche Giovanni Agusta SpA, 21017 Cascina Costa di Samarate, Gallarate, Italy
Telephone: (0331) 220478
Telex: 332569

Westland Helicopters Ltd, Yeovil, Somerset BA20 2YB, England
Telephone: Yeovil (0935) 5222
Telex: 46277

This company was formed in June 1980 by Westland Helicopters and Agusta to undertake the joint development, production and marketing of an SKR (Sea King Replacement) helicopter, for which the Royal Navy and Italian Navy both have a requirement. Such a programme was initiated by Westland in the UK in 1977, leading to the WG 34 helicopter described under that company's heading in the 1979-80 *Jane's*. The design is now being adapted, under the designation EH 101, to meet the detailed requirements of both navies, and for other civil, military and naval roles. British and Italian government approval for the nine-month project definition phase was given on 12 June 1981. The programme is being handled on behalf of both governments by the British Ministry of Defence. Technical responsibility rests with Westland Helicopters and Agusta, each of which has a 50% interest in EHI. Subcontractors include Fiat Aviazione, which will develop the main transmission system.

EH INDUSTRIES EH 101

Development of this helicopter began, initially under British Ministry of Defence (Navy) contract, as a replacement for the Royal Navy's Sea King HAS. Mk 2 shipboard anti-submarine helicopters.

In the Spring of 1977 the MoD(N) completed a series of feasibility studies to examine how an SKR (Sea King Replacement) would operate, and what sensors and performance standards it would require.

Westland's WG 34 design, marginally smaller than the Sea King but with substantially more payload capability, was selected by the MoD (Navy) for development in the late Summer of 1978. The Italian Navy has a requirement broadly similar to that of the Royal Navy, and in 1980 Westland and Agusta decided to join forces to develop a joint design, the EH 101, to meet the requirements of both services and for other military and civil applications.

The EH 101 will have three engines, and will incorporate composite materials, plus the latest available electronics and data handling systems. Present plans are for the completion of several prototypes, with a first flight due in the mid-1980s and the start of production deliveries in the late 1980s. Westland's partially-completed WG 34A prototype is being analysed as a part of the development study programme.

Type: Multi-purpose helicopter.

Airframe: For general appearance, see accompanying illustration. Landing gear is fully retractable, main units retracting into fairings on fuselage sides.

Power Plant: Three 1,193 kW (1,600 shp) General Electric T700-GE-401 turboshaft engines are proposed for prototypes. No decision yet made regarding production power plant, but the Rolls-Royce Turboméca RTM 321 remains an option.

Accommodation: Crew of three normally in ASW version (pilot, observer, and acoustics systems operator); provision for co-pilot if required. Up to 31 troops or 30 passengers in military or civil transport versions.

Armament and Operational Equipment (ASW version): Bendix 360° search radar under development, for production in Italy by FIAR. Fully enclosed weapons bay, capable of accommodating homing torpedoes or other weapons. No details of individual weapons yet available.

Dimensions, external:

Diameter of main rotor	18·29 m (60 ft 0 in)
Diameter of tail rotor	4·05 m (13 ft 3½ in)

Dimensions, internal:

Cabin: Length	6·50 m (21 ft 4 in)
Width	2·50 m (8 ft 2½ in)
Height	1·91 m (6 ft 3 in)

Weights:

Basic weight empty	6,827 kg (15,050 lb)
Disposable load	6,000 kg (13,227 lb)
Max T-O weight	13,000 kg (28,660 lb)
	or 13,600 kg (29,980 lb)

Performance (estimated, at max T-O weight of 13,000 kg; 28,660 lb, T700 engines):

Max level speed at S/L:	
ISA	164 knots (304 km/h; 189 mph)
ISA + 20°C	158 knots (293 km/h; 182 mph)
Hovering ceiling OGE:	
ISA	1,465 m (4,800 ft)
ISA + 20°C	855 m (2,800 ft)
Still-air range, no reserves	1,050 nm (1,945 km; 1,210 miles)
Endurance:	
at S/L, 3 engines	7 h 12 min
at 915 m (3,000 ft), 2 engines	8 h 36 min

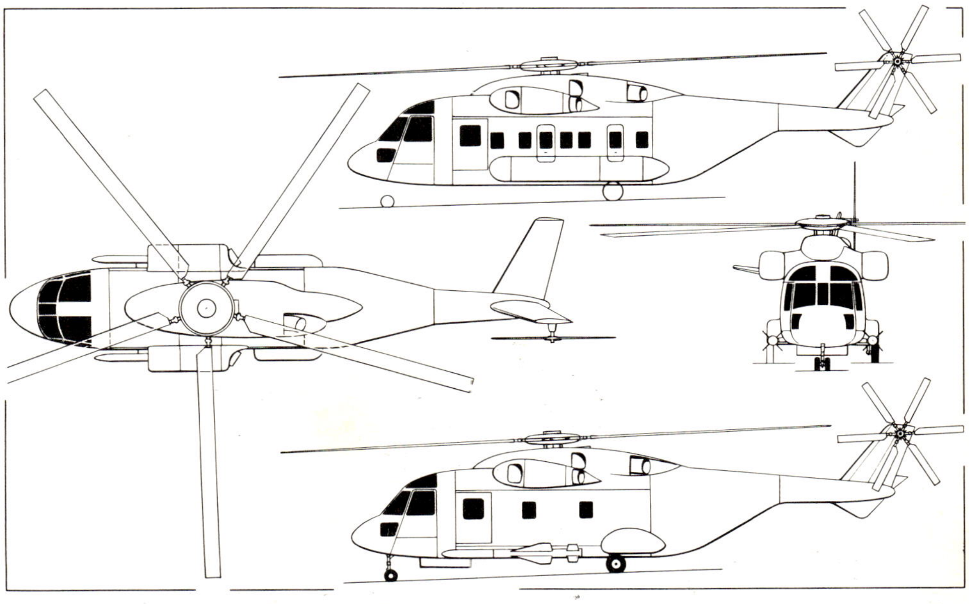

EH Industries EH 101 anti-submarine helicopter, with additional side view (top) of possible commercial version *(Pilot Press, provisional)*

MBB/AÉROSPATIALE

Participating Companies:

Messerschmitt-Bölkow-Blohm GmbH, Helicopter and Transport Systems Division, Ottobrunn bei München, 8000 München 80, Postfach 801140, Federal Republic of Germany

Telephone: (089) 6000 2734

Aérospatiale, 37 boulevard de Montmorency, 75781 Paris Cédex 16, France

Telephone: 524 43 21

MBB PAH-2/AÉROSPATIALE HAC

MBB and Aérospatiale were designated by their respective governments to design, develop and produce an anti-tank helicopter that would meet the requirements of the French and Federal German armies for service in the second half of the 1980s. This programme is known in West Germany as the PAH-2 (Panzerabwehr Hubschrauber 2) and in France as the HAC (Helicoptère Anti-Char). Executive authority for the programme is the Bundesamt für Wehrtechnik und Beschaffung (German federal defence technology and procurement agency). MBB, as programme leader, was made responsible for the rotor system, fuselage centre-section, equipment, controls, hydraulic and fuel systems, engine installation, electrical and automatic flight control systems, and overall design integration. Aérospatiale's responsibilities include the main and tail rotor gearboxes, forward fuselage and cockpit, tailboom, control surfaces, landing gear, integration of avionics and weapon systems.

The programme entered a 1½-year joint definition phase during 1980. The Federal German Army has a requirement for 212 PAH-2s; the French forces' requirement is for 120. Armament will include eight improved Hot anti-tank missiles.

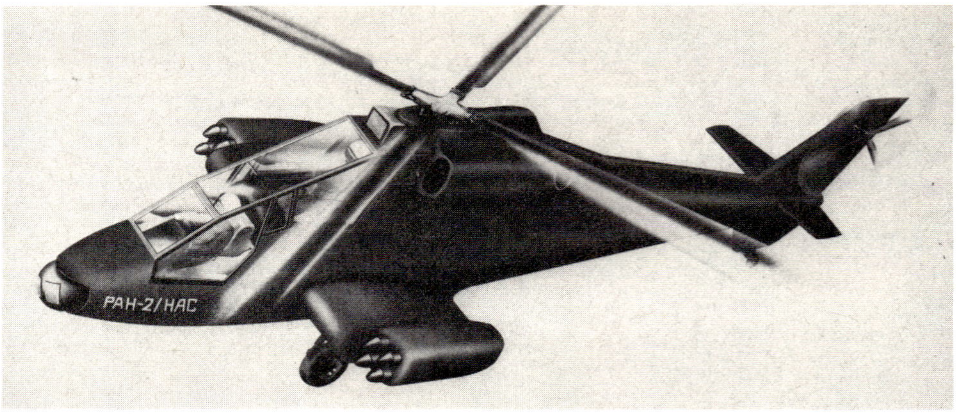

Artist's impression (provisional) of the MBB/Aérospatiale PAH-2/HAC anti-tank helicopter

MBB/KAWASAKI

Airframe Prime Contractors:

Messerschmitt-Bölkow-Blohm GmbH, Helicopter and Transport Systems Division, Ottobrunn bei München, 8000 München 80, Postfach 801140, Federal Republic of Germany

Telephone: (089) 6000 2734

Kawasaki Heavy Industries Ltd, World Trade Center Building, 4-1 Hamamatsu-cho 2-chome, Minato-ku, Tokyo, Japan

Telephone: Tokyo (03) 435 2971

MBB/KAWASAKI BK 117

Following nearly two years of negotiations, an agreement was signed on 25 February 1977 between MBB and Kawasaki to develop jointly a multi-purpose helicopter known as the BK 117. This superseded two earlier, separate projects known as the MBB BO 107 and the Kawasaki KH-7.

Both civil and military applications are foreseen, and the BK 117 has many components and accessories interchangeable with those of the MBB BO 105. Its rotor head is identical to that of the BO 105, from which aircraft the hydraulic boost system is also adapted. The transmission is based on that developed by Kawasaki for its earlier KH-7 design. The two-blade tail rotor is mounted on the central fin, forward of which is a horizontal stabiliser carrying twin endplate fins.

Development costs of the BK 117 programme are being shared equally between the two companies, with support for MBB from the West German government. MBB is responsible for production of the main and tail rotor systems, tailboom and tail unit, skid landing gear, hydraulic system, cabin floor, engine firewall and cowlings, power-amplified controls and systems integration; Kawasaki is responsible for the fuselage, transmission and electrical systems, and some items of equipment.

Four prototypes have been built, of which the P1 first prototype is being used at Gifu in Japan as a ground test vehicle for endurance trials, and the fourth aircraft (P4) for static and fatigue testing. Initial flight testing has been undertaken by the second and third prototypes, which flew for the first time, in West Germany and Japan respectively, on 13 June 1979 (D-HBKA) and 10 August 1979 (JQ-0003). An S-01 pre-production aircraft (D-HBKB) was completed in West Germany, and made its first flight on 6 March 1981. This aircraft, together with the third prototype in Japan (now modified and known as the P5), is undertaking the certification flying programme. The S-01 does not have the ventral fin fitted to the two prototypes.

By February 1981 more than 130 BK 117s had been ordered, and initiation of the first 100 production aircraft had been authorised. West German, Japanese and FAA certification for the initial VFR version was expected in late 1981, with IFR certification to follow; certification will be to FAR Pt 29, Categories A and B. There will be two production centres, at Munich and Gifu, and deliveries of production aircraft are expected to begin in early 1982.

Type: Twin-turbine multi-purpose helicopter.

Rotor System: Four-blade 'System Bölkow' rigid main rotor; head identical to that of BO 105; main rotor blades similar to those of BO 105, but larger. Two-blade tail rotor. Main rotor has a titanium head, to which are attached hingeless, fail-safe GRP blades of NACA 23012/23010 (modified) section with a stainless steel anti-erosion strip on each leading-edge. Provision for folding two blades of main rotor. Main rotor rpm: 383. Two-blade semi-rigid (teetering) tail rotor, mounted on port side of vertical fin and rotating clockwise when viewed from that side. Blades are of GRP, with high impact resistance and MBB-S102E performance/noise-optimised section. Tail rotor rpm: 2,169.

Rotor Drive: Each engine has separate drive input into Kawasaki KB 03 main transmission (see 'Power Plant' paragraph for transmission ratings), via single bevel gear and collector. Auxiliary drives for accessories. Dual redundant lubrication system.

Fuselage: Of typical pod-shaped configuration, comprising flight deck, cabin, cargo compartment and engine deck. Structure, designed to fulfil requirements of FAR Pt 29, is generally similar to that of BO 105, main components being of semi-monocoque riveted aluminium construction with single-curvature sheets and bonded aluminium sandwich panels. Secondary components are compound-curvature shells with sandwich panels and Kevlar skins. Floor extends throughout cockpit, cabin and cargo compartment at same level. Engine deck forms roof of cargo compartment and, adjacent to engine bays, is of titanium to serve as a firewall.

Tail Unit: Semi-monocoque tailboom, of tapered conical section, attached integrally to engine deck at forward end. Rear end, which is detachable, carries main fin/tail rotor support, and horizontal stabiliser with smaller, endplate fins. General design similar to that of BO 105, except for shape of outer fins.

Landing Gear: Non-retractable tubular skid type, of aluminium construction, similar to that of BO 105. Skids are detachable from cross-tubes. Ground handling wheels standard. Emergency flotation gear, settling protectors and snow skids available optionally.

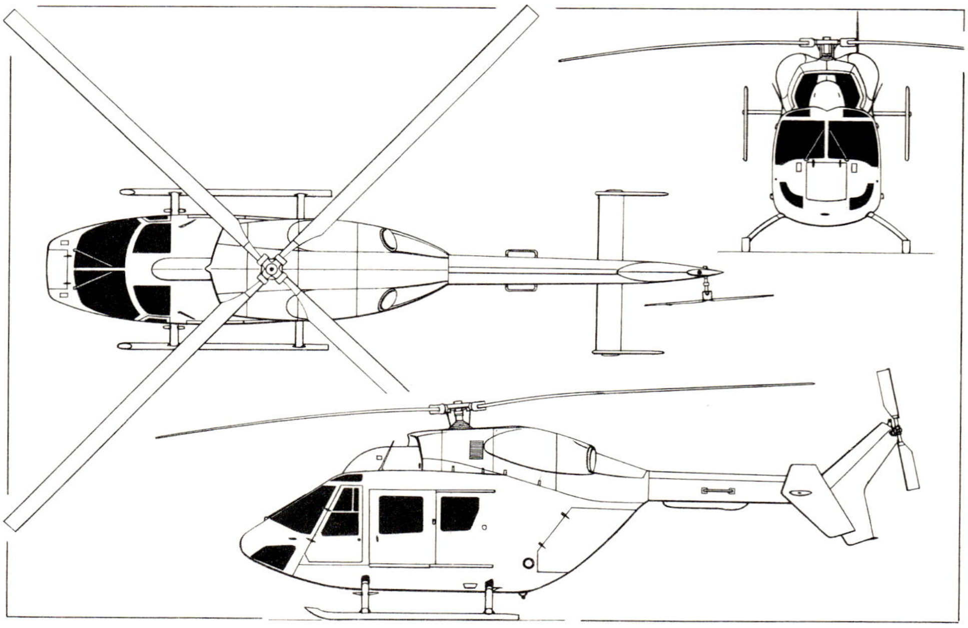

MBB/Kawasaki BK 117 twin-turboshaft multi-purpose helicopter, prototype with ventral fin *(Pilot Press)*

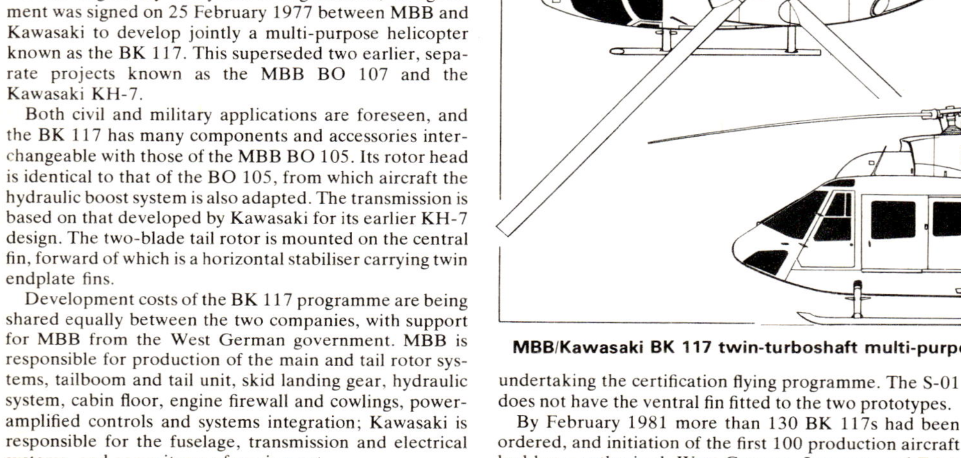

Pre-production MBB/Kawasaki BK 117, flown for the first time in March 1981

POWER PLANT: Two Avco Lycoming LTS 101-650B-1 turboshaft engines, each flat rated at 316 kW (424 shp) for take-off and max continuous operation. Transmission rated at 633 kW (849 shp) for twin-engine take-off and max continuous operation; and, for single-engine operation, at 442 kW (592 shp) for 2½ min, 405 kW (543 shp) for 30 min, and 368 kW (493 shp) max continuous. Fuel in four flexible bladder-type tanks (forward and aft main tanks, with two supply tanks between), in compartments under cabin floor. Two independent fuel feed systems, each able to supply both engines. Total standard fuel capacity 605 litres (133 Imp gallons). Provision for two 200 litre (44 Imp gallon) auxiliary tanks, raising total capacity to 1,005 litres (221 Imp gallons).

ACCOMMODATION: Pilot and up to six (executive version) or seven passengers (standard or offshore IFR-equipped versions). High-density layouts available for nine or eleven passengers in addition to pilot. Provision for two-pilot operation. Jettisonable forward-hinged door on each side of flight deck, with openable window on pilot's side. Jettisonable rearward-sliding passenger door on each side of cabin, lockable in open position. Fixed steps on each side. Two hinged, clamshell doors at rear of cabin, providing access to cargo compartment. Rear cabin window on each side. Aircraft can be equipped, according to mission, for offshore, medical evacuation (pilot, four stretchers and two attendants), firefighting, search and rescue, law enforcement, cargo transport or other operations. Cabin floor hatch optional.

SYSTEMS: Ram-air and electrical ventilation system. Fully redundant tandem hydraulic boost system (one operating and one standby) for flight controls. Main DC electrical power from two 150A 30V starter/generators (one on each engine) and a 24V 22Ah nickel-cadmium battery. AC power can be provided optionally by two independent inverters. Emergency busbar provides direct battery power to essential services in event of a double generator failure. External DC power receptacle.

AVIONICS AND EQUIPMENT: Basic aircraft has instrumentation for single-pilot VFR operation, including airspeed indicator, electrically-heated pitot tube, encoding altimeter, rate of climb indicator, 10 cm (4 in) attitude indicator with turn and slip indication, directional gyro, RMI, and magnetic compass. Dual controls and dual VFR instrumentation available optionally. Com/nav and other avionics available to customer's requirements, including VHF and HF transceivers; nav, RNav, ADF and VLF/Omega systems; radar altimeter; encoding altimeter; DME; ATC transponder; multi-mode radar; IFR instrumentation packages; and stability augmentation system. Standard basic equipment includes annunciator panel, master caution light, rotor rpm/engine fail warning control unit, fuel quantity indicator and low level sensor, outside air temperature indicator, clock, engine and transmission oil pressure and temperature indicators, dual exhaust temperature indicators, dual torque indicators, triple tachometer, NI tachometer, mast moment indicator, instrument panel lights, cockpit/cabin/cargo compartment dome lights, utility lights, emergency exit lights, position lights, anti-collision warning lights, retractable landing light, portable flashlight, ground handling wheels, pilot's windscreen wiper, floor covering, interior panelling and sound insulation, ashtrays, map/document case, tiedown rings in cabin and cargo compartment, engine compartment fire warning indicator, engine fire extinguishing system, portable fire extinguisher, first aid kit, and single colour exterior paint scheme. Optional equipment includes high-density seating arrangement, bleed air heating system, fuel dump valve, two long-range fuel tanks, emergency flotation gear, snow skids, main rotor blade folding kit, non-retractable landing light, co-pilot's windscreen wiper, stretcher installation, external cargo hook, multi-mode radar, rescue hoist, SX 16 remotely controlled searchlight, external loudspeaker, and sand filter. Special optional equipment planned to include special mission kits for rescue, law enforcement and VIP transport.

DIMENSIONS, EXTERNAL:
Main rotor diameter 11·00 m (36 ft 1 in)
Tail rotor diameter 1·90 m (6 ft 2¾ in)
Main rotor blade chord 0·31 m (1 ft 0¼ in)
Tail rotor blade chord 0·18 m (7·1 in)
Length overall, main and tail rotors turning 13·00 m (42 ft 8 in)
Length of fuselage, tail rotor blade vertical 9·88 m (32 ft 5 in)
Fuselage: Max width 1·58 m (5 ft 2¼ in)
Height overall, main and tail rotors turning 3·84 m (12 ft 7¼ in)
Height to top of main rotor hub 3·30 m (10 ft 10 in)
Height to top of main fin 3·30 m (10 ft 10 in)
Tailplane span (over endplate fins) 2·75 m (9 ft 0¼ in)
Tail rotor ground clearance 1·93 m (6 ft 4 in)
Width over skids 2·50 m (8 ft 2½ in)
Cabin floor hatch (optional):
Length 0·50 m (1 ft 7¾ in)
Width 0·40 m (1 ft 3¾ in)

DIMENSIONS, INTERNAL:
Passenger cabin: Length 2·02 m (6 ft 7½ in)
Max width 1·43 m (4 ft 8¼ in)
Max height 1·29 m (4 ft 2¾ in)
Volume 3·22 m³ (113·7 cu ft)
Cargo compartment: Length 1·10 m (3 ft 7¼ in)
Max width 1·23 m (4 ft 0½ in)
Max height 1·23 m (4 ft 0½ in)
Volume 1·34 m³ (47·3 cu ft)

AREAS:
Main rotor disc 95·03 m² (1,022·9 sq ft)
Tail rotor disc 2·84 m² (30·6 sq ft)

WEIGHTS:
Weight empty, equipped 1,520 kg (3,351 lb)
Fuel: standard 470 kg (1,036 lb)
incl auxiliary tanks 780 kg (1,720 lb)
Max T-O weight:
internal payload 2,800 kg (6,173 lb)
with externally-slung load 3,000 kg (6,614 lb)

PERFORMANCE (at max T-O weight, ISA):
Never-exceed speed at S/L 148 knots (275 km/h; 171 mph)
Max cruising speed at S/L 142 knots (264 km/h; 164 mph)
Econ cruising speed at S/L 126 knots (234 km/h; 145 mph)
Max forward rate of climb at S/L 600 m (1,968 ft)/min
Max vertical rate of climb at S/L 420 m (1,378 ft)/min
Service ceiling 4,500 m (14,760 ft)
Hovering ceiling IGE 4,020 m (13,175 ft)
Hovering ceiling OGE 3,150 m (10,335 ft)
Service ceiling, one engine out, 46 m (150 ft)/min climb reserve 3,000 m (9,845 ft)
Range at S/L with pilot and 7 passengers, standard fuel, no reserves 294 nm (545 km; 338 miles)
Ferry range at S/L with two 200 litre auxiliary tanks, no reserves 491 nm (910 km; 565 miles)
Endurance, conditions as above 3 h 0 min

MCDONNELL DOUGLAS/FOKKER

PARTICIPATING COMPANIES:
McDonnell Douglas Corporation, Box 516, St Louis, Missouri 63166, USA
Telephone: (314) 232 0232
Telex: 44 857
Fokker BV, PO Box 1065, 1000 BB Amsterdam, Netherlands
Telephone: (020) 5449111
Telex: 12227 Foa NL

SUPERVISORY BOARD:
John C. Brizendine (McDonnell Douglas)
D. Krook (Fokker)
Robert C. Little (McDonnell Douglas)
F. Nel (Fokker)

PROGRAMME MANAGER: Ray E. Bates (McDonnell Douglas)
DEPUTY PROGRAMME MANAGER: J. Cornelis (Fokker)

MCDONNELL DOUGLAS/FOKKER MDF-100

The MDF-100 is a proposed new commercial transport aircraft, to seat approximately 150 passengers and be ready for entry into service in about 1986. On 4 May 1981 McDonnell Douglas and Fokker announced a memorandum of understanding to form a co-equal joint venture to study its design, production and marketing; while exploring the market, the two companies will proceed with preliminary design of the aircraft. Other companies are to be invited to join the MDF-100 programme on a risk-sharing basis.

No details of the MDF-100 had been released at the time of closing for press, but it is expected to be a twin-turbojet or turbofan medium-range aircraft, in which will be reflected much of the design effort already expended in two earlier, separate programmes, the McDonnell Douglas ATMR (or DC-XX) and Fokker F29.

Model of McDonnell Douglas/Fokker MDF-100 in its original form. A T tail is now proposed (see Addenda) *(Air Portraits)*

The joint venture programme office is headed by a McDonnell Douglas executive as programme manager, with an executive from Fokker as his deputy.

PANAVIA

PANAVIA AIRCRAFT GmbH

HEAD OFFICE: 8 München 86, Postfach 860629, Arabellastrasse 16, Federal Republic of Germany
Telephone: (089) 92171
Telex: 05 29 825

DIRECTORS:
Prof Gero Madelung (MBB) (Chairman)
Sir Frederick Page (BAe) (Deputy Chairman)
Dott R. Bonifacio (Aeritalia) (Deputy Chairman)
A. H. C. Greenwood (BAe)
A. F. Atkin (BAe)
I. R. Yates (BAe)
Dott C. Innocenti (Aeritalia)
G. Sarzotti (Aeritalia)
Dott R. Mautino (Aeritalia)
F. Forster-Steinberg (MBB)
O. Friedrich (MBB)
H. Plückthun (MBB)

MANAGING DIRECTOR: Hans-Joachim Klapperich
DEPUTY MANAGING DIRECTOR: Dr I. A. M. Hall

FUNCTIONAL DIRECTORS:
J. L. Dell (Flight Operations)
V. von Tein (Systems Engineering, Munich)
Dr I. A. M. Hall (Programme Management)
B. O. Heath (Systems Engineering, Warton)
Dott P. Ricci (Systems Engineering, Turin)
Prof Dr-Ing R. Riccius (Marketing)
R. Sanitz (Product Support)
Dott R. Sassi (Production)
J. A. Thornber (Procurement)

PUBLICITY MANAGER: F. Oelwein

Panavia was formed on 26 March 1969 to design, develop and produce an all-weather multi-role combat aircraft (MRCA) for the air forces of the United Kingdom, the Federal Republic of Germany and Italy, and the Federal German Navy. The name Tornado for this aircraft was adopted officially in March 1976. This programme is one of the largest European industrial programmes ever undertaken. The three component companies of Panavia are British Aerospace PLC (42·5% participation), MBB (42·5%) and Aeritalia (15%).

The Federal German, British and Italian governments set up a joint organisation known as NAMMO (NATO MRCA Management and production Organisation). This has its executive agency NAMMA (NATO MRCA Management Agency) in the same building as Panavia, in Munich.

On 29 July 1976 the three governments signed a Memorandum of Understanding for the production of 809 Tornados (805 new, and four of the pre-production aircraft), enabling the three partner countries to embark upon the production programme.

PANAVIA TORNADO

RAF designation (IDS version): Tornado GR. Mk 1

The Tornado is a twin-engined two-seat supersonic aircraft capable of fulfilling the agreed operational requirements of its three sponsoring countries. The use of a variable-geometry wing, and avionics which enable the aircraft to fly 'blind' in all weathers, day and night, at very low level, with automatic terrain-following, give it the necessary flexibility to achieve all-weather penetration.

The aircraft is intended to fulfil six major requirements, some of which are shared by more than one of the partners. These are:

(a) Close air support/battlefield interdiction
(b) Interdiction/counter air strike
(c) Air superiority
(d) Interception/air defence
(e) Naval strike
(f) Reconnaissance

The 809 aircraft to be produced for the participating nations, which will include four of the six pre-series aircraft brought up to production standard, will comprise 644 of the IDS (Interdictor Strike) version, and 165 examples of the ADV (Air Defence Variant) for the RAF. A total of 671 Tornados will be operational aircraft, and 138 will be dual-control trainers with full operational capability.

The RAF is to have 385 Tornados, of which 220 will be of the interdictor/strike version and 165 of the air defence variant. These are due to become operational with Strike Command in 1982 and will, in the first instance, replace the Vulcans and Buccaneers of Nos. 9, 12, 15, 16, 35, 44, 50, 101 and 617 Squadrons in the overland strike and reconnaissance roles. Later, the air defence version will succeed the Phantom and Lightning; and finally the Tornado will replace the Buccaneer for maritime strike tasks. Some two-thirds of the RAF's front-line aircraft will eventually be Tornados.

The Luftwaffe is to receive 212 Tornados, to replace the Lockheed F-104G in the battlefield interdiction, counter air and close air support roles. Four wings (Jabos 31, 32, 33 and 34) and one training squadron are to be equipped, starting in 1982. The 112 for Marinefliegergruppen 1 and 2 of the Federal German Navy will be equipped for strike missions against sea and coastal targets, and for reconnaissance. MFG 1 is due to begin converting to the Tornado in early 1982.

The Italian Air Force will use 54 of its 100 Tornados to replace F-104G and G91R aircraft of the 20°, 102°, 154° and 186° Gruppi in the air superiority, ground attack and reconnaissance roles. Of the remainder, 34 will be kept in reserve and 12 will be equipped as dual-control trainers.

Structural design of the Tornado was completed in August 1972. Nine flying prototypes were built—four in the UK, three in West Germany and two in Italy. Static tests with airframe No. 10 began at Warton in the Spring of 1974. The 01 first prototype (D-9591), assembled by MBB, made its first flight at Manching, West Germany, on 14 August 1974. Details of all nine prototypes can be found in the 1978-79 and earlier editions of *Jane's*.

Of the six pre-series Tornados which followed, all had flown by 1979: three in West Germany, two in the UK and one in Italy. Details of these aircraft can be found in the 1980-81 and earlier editions.

By 1 January 1981, prototype, pre-series and production Tornados had accumulated a total of 5,000 hours' flying in 4,350 flights. During test flights the Tornado has been flown at indicated airspeeds of up to 800 knots (1,480 km/h; 920 mph) at comparatively low levels, and has exceeded Mach 2 at high altitude.

With the major part (95%) of the flight envelope already cleared, delivery to service test centres began with pre-series aircraft No. 12 to the Aeroplane and Armament Experimental Establishment at Boscombe Down on 3 February 1978, followed by No. 11 to the official test centre at Manching, West Germany, and No. 14 to Pratica di Mare, Italy.

Four production contracts have so far been placed for the Tornado. The first, for 40 aircraft, covered 20 Tornado GR. Mk 1 interdictor/strike versions for the RAF; three F. Mk 2 prototypes for the RAF (see separate description); and 17 Tornados for the Federal German Luftwaffe and Marineflieger. The second contract, for 110 Tornados, provides 55 aircraft for the RAF, 40 for Germany, and 15 for the Aeronautica Militare Italiana. The third contract is for 164 aircraft (68 each for the UK and West Germany and 28 for Italy). The fourth contract, covering a further 162 aircraft, includes the first production batch of 18 ADV Tornados, plus 53 GR. Mk 1s, and was signed in 1981, bringing the total then on order to 476.

Panavia Tornado GR. Mk 1 with eight underfuselage 1,000 lb bombs, and drop-tanks and ECM pods underwing

The first production Tornado to be completed was BT 001 (British Trainer 001), which made its initial flight on 10 July 1979. The first production West German Tornado (GT 001) flew for the first time on 27 July 1979; the first Italian production aircraft was scheduled to fly in Autumn 1981. Initial production aircraft have RB.199-34R-04 Mk 101 engines.

By the end of 1980, fifteen production Tornados had been delivered to the Tri-national Tornado Training Establishment (TTTE) at RAF Cottesmore. This unit, to which the first two Tornados (BT 002 and BS 002) were delivered on 1 July 1980, was formally opened on 29 January 1981. When training reaches its peak, there will be nearly 50 Tornados based at Cottesmore. UK weapons training will be performed from RAF Honington.

By 30 April 1981 all 40 first-batch aircraft had flown, and deliveries to the services concerned were completed during the first three months of the year. At that time the first 20 of the second batch were in the final assembly stage.

Nominal max weapons load of the IDS Tornado is 8,165 kg (18,000 lb) on three twin hardpoints in tandem under the fuselage, two inboard points in tandem underwing, and two single outboard points underwing. A new missile known as Sea Eagle, a turbojet-powered active-radar sea-skimming development of the BAe/Matra Martel, is being developed for the RAF's GR. Mk 1, which is also expected to have the capability to carry Pave Spike pods for laser weapons guidance. Primary armament of the Federal German Navy Tornados will be four MBB Kormoran anti-shipping missiles. Italy's Tornados are expected to be equipped with the Selenia Aspide 1A air-to-air missile.

Initial bomb-aiming and loft-bombing trials were completed during 1980. Bomb-aiming resulted in all four bombs being 'target hits' or 'close', ie within a few yards of a target some 3·05 m (10 ft) in diameter. During loft-bombing trials, four out of the seven bombs 'tossed' some 5 to 6·4 km (3 to 4 miles) at a 3·05 m (10 ft) target landed within 9·1 m (30 ft) of the aiming mark.

The following details apply to the basic IDS production version:

Type: Twin-engined multi-purpose military aircraft.

Wings: Cantilever shoulder-wing monoplane. All-metal wings, of variable geometry, the outer panels having a leading-edge sweep of 25° in the fully forward position and 68° when fully swept. Fixed inboard portions have a leading-edge sweep of 60°. Wing carry-through box is of electron-beam-welded titanium alloy; majority of remaining wing structure is of aluminium alloy, with integrally stiffened skin. There is a Krueger flap on the leading-edge of each wing glove box. The wings each pivot hydraulically, on Teflon-plated bearings, from a point in the centre-section just outboard of the fuselage. The root of the outer wing mates with the pivot pin through attachment members made of titanium alloy and fixed to the upper and lower light alloy panels of the outer wing box, and a so-called 'round rib', also of titanium alloy, transmitting the normal aerodynamic force. Sweep actuators are of the ballscrew type, with hydraulic motor drive. In the event of wing sweep failure, the aircraft can land safely with the wings fully swept. High-lift devices on the outer wings include full-span leading-edge slats (three sections on each side), full-span double-slotted fixed-vane trailing-edge flaps (four sections each side), and spoilers (two on upper surface on each side). Spoilers give augmented roll control at unswept and intermediate wing positions at low speed, and also act as lift dumpers after touchdown. All flying control surfaces actuated by electrically-controlled tandem hydraulic jacks. No ailerons. Entire outer wings, including control surfaces, are Italian-built, Aeritalia having prime responsibility for final assembly and production, assisted by Aermacchi, Aeronavali Venezia, Piaggio, Saca and SIAI-Marchetti as subcontractors. Microtecnica (Italy) is prime subcontractor for the wing sweep system.

Fuselage: Conventional all-metal semi-monocoque structure, mainly of aluminium alloy, built in three main sections. MBB in West Germany is prime contractor for the centre fuselage section, including the engine air intake trunks and wing centre-section box and pivot mechanism. This task includes responsibility for the surface interface between the movable wing and the fixed portion, to ensure both a smooth and slender external contour and proper sealing against aerodynamic pressure over a range of wing sweep positions. The present design uses fibre-reinforced plastics in these areas, and an elastic seal between the outer wings and the fuselage sides. Responsibility for the front fuselage, including both cockpits, and for the rear fuselage, including the engine installation, is undertaken by BAe (Warton). Radar-transparent nose-cone by AEG-Telefunken, assisted by Aeritalia and BAe, hinges sideways to starboard. Door-type airbrake on each side at top of rear fuselage.

Tail Unit: Cantilever all-metal structure, consisting of single sweptback two-spar fin and rudder, and low-set all-moving horizontal surfaces ('tailerons') which operate together for pitch control and differentially for roll control, assisted by use of the wing spoilers when the wings are not fully swept. Rudder and tailerons actuated by electrically-controlled tandem hydraulic jacks. Passive ECM antenna fairing near top of fin. Ram-air intake for heat exchanger at base of fin. Entire tail unit is the responsibility of BAe.

Landing Gear: Hydraulically-retractable tricycle type, with forward-retracting twin-wheel steerable nose unit. Single-wheel main units retract forward and upward into centre section of fuselage. Development and manufacture of the complete landing gear and associated hydraulics is headed by Dowty Rotol (UK). Dunlop aluminium alloy wheels, brakes and low-pressure tyres (to permit operation from soft, semi-prepared surfaces) and Goodyear anti-skid units. Main-wheel tyres size 30 × 11·50-14·5, Type VIII (20 ply); nosewheel tyres size 18 × 5·5, Type VIII (12 ply). Runway arrester hook beneath rear of fuselage.

Power Plant: Two Turbo-Union RB.199-34R-04 turbofan engines, each rated at approx 40·0 kN (9,000 lb st) dry and 71·2 kN (16,000 lb st) with afterburning, fitted with bucket-type thrust reversers and installed in rear fuselage with downward-opening doors for servicing and engine change. Four large 'blow-out' doors in top of each trunk, above the double-wedge variable-ramp intake. Dowty Boulton Paul air intake ramp actuators and afterburner control system. All integral fuel in multi-cell Uniroyal self-sealing integral fuselage tanks and/or wing box tanks, all fitted with press-in fuel sampling and water drain plugs, and all refuelled from a single-point NATO connector. Detachable and retractable in-flight refuelling probe can be mounted on starboard side of fuselage, adjacent to cockpit. System also designed to accept a buddy-to-buddy refuelling pack. Provision for drop-tanks of various sizes up to 1,500 litres (330 Imp gallons) to be carried beneath outer wings. Dowty Fuel Systems/Lucas/Microtecnica afterburning fuel control system. AEG-Telefunken intake de-icing system.

Accommodation: Crew of two on tandem Martin-Baker Mk 10A zero-zero ejection seats under Kopperschmidt/AIT one-piece canopy, which is hinged at rear and opens upward. Flat centre windscreen panel and curved side panels, built by Lucas Aerospace, incorporate Sierracote electrically-conductive heating film for de-icing and internal demisting. Seats provide safe escape at zero altitude and at speeds from zero up to 630 knots (1,166 km/h; 725 mph) IAS.

Systems: Nordmicro/BAe/Microtecnica air intake control system, and Dowty Boulton Paul/Liebherr Aerotechnik engine intake ramp control actuators. Two separate independent hydraulic systems, one driven by each engine, provide fully duplicated power for wing sweep, flaps, slats, spoilers, airbrakes, landing gear, tailerons and rudder. Main system includes Vickers pump, Dowty accumulators and Teves power pack. Fairey Hydraulics system for actuation of spoilers, rudder and taileron control. Provision for reversion to single-engine drive of both systems, via a mechanical cross-connection between the two engine auxiliary gearboxes, in the event of a single engine failure. In the event of a double engine flameout, an emergency pump in No. 1 system has sufficient duration for re-entry into the engine cold relight boundary. Flying control circuits are protected from loss of fluid due to leaks in other circuits by isolating valves which shut off the utility circuits if the reservoir contents drop below a predetermined safety limit level. Duplicated AC and DC electrical power is provided by two alternators, each driven by its respective engine auxiliary gearbox, to two separate main AC busbars and one essential AC busbar, and through two fan-cooled transformer-rectifier units (TRUs) to two main DC busbars. Lucas/Siemens 40/60kVA 200V 400Hz three-phase constant-frequency AC generating system. Either generator can cope with the full demand of the electrical systems in the event of a single generator failure. If both TRUs fail, an onboard Varta battery supplies the essential DC busbar. In the event of a total loss of power a one-shot battery is used to power the hydraulic pump, independently from the battery used to start the APU under normal conditions. Normalair-Garrett precooler and cold-air unit, Marston Excelsior intercooler and Teddington temperature control system. Normalair-Garrett/Draegerwerk/OMI demand-type oxygen system, using a lox converter. KHD accessory drive gearboxes and Rotax/Lucas/Siemens integrated drive generator. Marconi flow-metering system. Eichweber fuel gauging system and Flight Refuelling flexible couplings. Graviner fire detection and extinguishing systems. Rotax contactors. Smiths engine speed and temperature indicators.

Avionics and Equipment: Communications equipment includes Plessey PTR 1721 (UK and Italy) or Rohde und Schwarz (West Germany) UHF/VHF transceiver; AEG-Telefunken UHF/DF (UK and West Germany only); Chelton UHF homer aerial; SIT/Siemens emergency UHF with Rohde und Schwarz switch; BAe HF/SSB aerial tuning unit; Rohde und Schwarz (UK and West Germany) or Montedel (Italy) HF/SSB radio; Ultra communications control system; Marconi Avionics central suppression unit; Epsylon voice recorder; and Chelton communications and landing system aerials.

Primary self-contained nav/attack system includes Texas Instruments multi-mode forward-looking radar; Ferranti FIN 1010 three-axis digital inertial navigation system (DINS) and combined radar display; Decca Type 72 Doppler radar system, with Kalman filtering of the Doppler and inertial inputs for extreme navigational accuracy; Microtecnica air data computer; Litef Spirit 3 16-bit central digital computer; Aeritalia radio/radar altimeter; Smiths/Teldix/OMI electronic head-up display with Davall camera; Ferranti nose-mounted laser ranger and marked target receiver; Marconi Avionics

Panavia Tornado IDS multi-role combat aircraft *(Pilot Press)*

TV tabular display produced in partnership with AEG and Selenia; Astronautics (USA) bearing distance heading indicator and contour map display. Defensive equipment includes Siemens (West Germany) or Cossor SSR-3100 (UK) IFF transponder; Elettronica warning radar; and MSDS/Plessey/Decca Sky Shadow passive ECM system.

Flight control system includes a Marconi Avionics/Bodenseewerk triplex command stability augmentation system (CSAS), incorporating fly-by-wire and autostabilisation; Marconi Avionics/Aeritalia autopilot and flight director (APFD), using two self-monitoring digital computers; Marconi Avionics triplex transducer unit (TTU), with analogue computing and sensor channels; Marconi Avionics terrain-following E-scope (TFE), produced in partnership with Selenia; Fairey/Marconi Avionics quadruplex electro-hydraulic actuator; and Microtecnica air data set. The APFD provides preselected attitude, heading or barometric height hold, heading and track acquisition, and Mach number or airspeed hold with autothrottle. Flight director operates in parallel with, and can be used as backup for, the autopilot, as a duplex digital system with an extensive range of modes. Automatic approach, terrain-following and radio height-holding modes are also available. Other instrumentation includes Smiths horizontal situation indicator, vertical speed indicator and standby altimeter; AEG-Telefunken ADF; Lital standby attitude and heading reference system; SEL (with Setac) or (in UK aircraft) Marconi Avionics AD2770 (without Setac) Tacan; Cossor CILS 75 ILS; Bodenseewerk attitude direction indicator; Marconi Avionics central suppression unit (CSU); and Dornier System flight data recorder.

Overall responsibility for the avionics rests with Panavia, with EASAMS (UK) as the avionics prime contractor, and ESG (Germany) and SIA (Italy) as subcontractors. The avionics systems, while standardised as far as possible, retain the flexibility necessary to perform the various roles required. They provide accurate low- and high-level navigation; precision visual attack on ground targets in blind and poor weather conditions; air-to-ground and air-to-air attack with a wide variety of weapons; manually controlled and automatic attack; and comprehensive onboard checkout and mission data recording; with minimisation of ground support facilities at bases and the front line.

Armament: Fixed armament comprises two 27 mm IWKA-Mauser cannon, one in each side of the lower forward fuselage. Other armament varies according to version, with emphasis on the ability to carry a wide range of advanced non-nuclear weapons on three underfuselage attachments and up to four swivelling hardpoints beneath the outer wings. A Marconi Avionics/Selenia stores management system is fitted, and Sandall Mace 355 and 762 mm (14 and 30 in) ejector release units are standard. Initial weapon systems evaluation included trials of a modified Raytheon Sparrow missile, fitted with a British warhead and fuse. The battlefield interdiction version is capable of carrying defensive 'streuwaffen' (scatter weapons) such as the MBB MW-1 munitions dispenser, and of carrying weapons to suit 'hard' or 'soft' targets. The naval and interdictor strike versions have provision for carrying additional, externally-mounted fuel tanks. For German Navy and Italian Air Force Tornados, MBB has developed (first flight 14 April 1981) a multi-sensor reconnaissance pod to be carried on the centreline pylon. Among the weapons already specified for, or suitable for carriage by, the IDS Tornado are the Sidewinder air-to-air missile; JP 233 Paveway laser-guided bomb, AS.30, Maverick, GBU-15, Sea Eagle and Kormoran air-to-surface missiles; napalm; BL-755 600 lb cluster bombs; MW-1 munitions dispenser; Mk 83 or other 1,000 lb bombs; 'smart' or retarded bombs; Lepus flare bombs; LR-25 rocket launchers; and active or passive ECM pods.

Dimensions, external:
- Wing span: fully spread 13·90 m (45 ft 7¼ in)
- fully swept 8·60 m (28 ft 2½ in)
- Length overall 16·70 m (54 ft 9½ in)
- Height overall 5·70 m (18 ft 8½ in)
- Tailplane span 6·80 m (22 ft 3½ in)
- Wheel track 3·10 m (10 ft 2 in)
- Wheelbase 6·20 m (20 ft 4 in)

Weights:
- Weight empty, equipped approx 14,000 kg (30,865 lb)
- Max weapon load carried more than 7,257 kg (16,000 lb)
- Max T-O weight:
  - 'clean', full internal fuel 20,411 kg (45,000 lb)
  - with external stores 26,490 kg (58,400 lb)

Performance (prototype/pre-series aircraft up to mid-1980, with development engines):
- Max level speed at altitude, 'clean' Mach 2·2 (1,261 knots; 2,337 km/h; 1,452 mph)
- Max level speed with external stores Mach 0·92 (600 knots; 1,112 km/h; 691 mph)
- Max level speed at low altitude above 800 knots (1,480 km/h; 920 mph) IAS
- Landing speed approx 115 knots (213 km/h; 132 mph)
- Time to 9,145 m (30,000 ft) from brakes off less than 2 min
- Terrain following down to 61 m (200 ft)
- T-O and landing run approx 366 m (1,200 ft)
- Max normal acceleration 7·5*g*
- Max 360° rapid-roll clearance with full lateral control 4·0*g*
- Radius of action with heavy weapons load, hi-lo-lo-hi 750 nm (1,390 km; 863 miles)
- Ferry range approx 2,100 nm (3,890 km; 2,420 miles)
- *g* limit +7·5

PANAVIA TORNADO ADV

RAF designation: Tornado F. Mk 2

Full-scale development of the Tornado ADV (air defence variant) was authorised by the British government on 4 March 1976. This version is being developed specifically for the RAF, which will include 165 of this long-range interceptor model, designated F. Mk 2, in its total procurement of 385 Tornados, to re-equip two Lightning squadrons and seven squadrons of Phantoms.

Most of the 165 ADV Tornados will be based in the United Kingdom, both for the air defence of the UK and to protect the northern and western approaches of NATO. The F. Mk 2 will also fulfil the RAF's commitments to provide long-range air defence of Britain's maritime forces, over a wide UK defence region extending from the Atlantic approaches to the Baltic and from Iceland to the English Channel; and to contribute towards air defence in the Central Region of Europe. It will be able to loiter on patrol for several hours, using in-flight refuelling when necessary, and to detect, identify and destroy enemy aircraft approaching at supersonic speeds at high, medium or low altitudes, using its snap-up or snap-down missiles. Its fire control system will be able to engage multiple targets in rapid succession; its weapon systems will be highly resistant to enemy ECM; and it will be able to operate from damaged airfields by virtue of its good short-field performance. Supersonic acceleration is better than that of the IDS version. A genuine long-range autonomous capability will enable it to operate more than 350 nm (645 km; 400 miles) from its base at night, in bad weather, in heavy ECM conditions, against multiple targets at low level.

Armament of the Tornado F. Mk 2 consists of a single 27 mm built-in IWKA-Mauser cannon in the starboard

Third prototype Tornado F. Mk 2 air defence variant for the RAF, in light grey finish with low-visibility markings

side of the lower forward fuselage, four BAe Dynamics Sky Flash medium-range air-to-air missiles semi-recessed under the centre-fuselage, and two NWC AIM-9L Sidewinder short-range infra-red air-to-air missiles on the inboard wing stations.

These weapons will be operated in conjunction with a new all-British track-while-scan pulse-Doppler airborne interception radar named Foxhunter, designed and developed by Marconi Avionics, with Ferranti as subcontractor for the transmitter and aerial scanning mechanism. Foxhunter will enable the Tornado F. Mk 2 to detect targets more than 100 nm (185 km; 115 miles) away, and to track several targets simultaneously; the Sky Flash missiles, each fitted with an MSDS monopulse seeker head, will be able to engage targets at high altitude or at levels below 75 m (250 ft), in the face of heavy ECM, and at stand-off ranges of more than 22 nm (40 km; 25 miles). A new release system, designed specially for Sky Flash, permits the missile to be fired over the Tornado's full flight envelope. Furthermore, the missile is highly capable of tracking targets in a ground clutter environment, and of discriminating between closely-spaced targets. An EMI Electronics active fusing system allows these benefits to be realised fully in snap-down attacks against targets flying at very low level.

The Tornado F. Mk 2 carries a two-man crew, the rear (navigator's) cockpit being equipped with a Ferranti FH 31A AC-driven 3 in horizon gyro which, in addition to providing an attitude display for the navigator, feeds pitch and roll signals to other avionics systems in the aircraft in certain modes. A pilot's head-up display, ESM (electronic surveillance measures) and ECCM are also standard, as are an extremely advanced modular radar homing and warning receiver being developed by Marconi Space and Defence Systems, and a low light level TV visual augmentation system, developed by Marconi Avionics, for identifying enemy aircraft. Other avionics include a Cossor IFF 3500 interrogator, and an ECM-resistant data link system which is interoperable with all other NATO systems. Its comprehensive avionics enable the Tornado ADV to contribute significantly to the transfer of vital information over the entire tactical area; it can, if necessary, partially fulfil the roles of both AEW and ground-based radar.

Although possessing some 80 per cent commonality with the IDS (interdictor/strike) version, the Tornado F. Mk 2 differs in several important respects, and the initial production contract for 40 Tornados included provision for three prototypes of the fighter version. The first of these (ZA254) made its first flight on 27 October 1979. This aircraft, not fitted with the Foxhunter radar at the time of its first flight, is acting as the aerodynamic and missile release test aircraft. The second prototype (ZA 267) flew on 18 July 1980, and the third (ZA283) on 18 November 1980. The first two aircraft are being used for avionics and weapon system testing and integration. The third prototype is concerned primarily with evaluating the AI radar, the first flight with this installed being made on 17 June 1981. By early August 1980 the A 01 first prototype had extended the flight envelope clearance to an IAS of 800 knots (1,480 km/h; 920 mph), a milestone passed by the IDS Tornado in 1979, and had exceeded Mach 2 at high altitude.

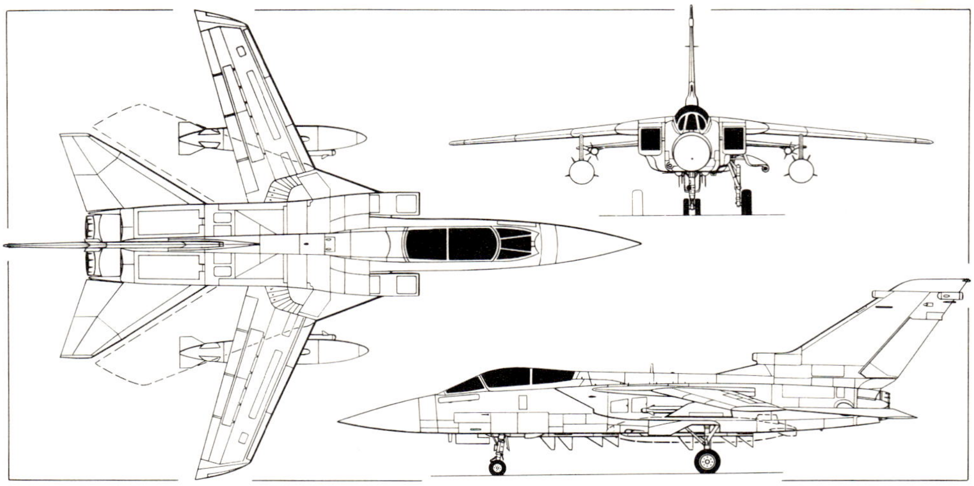

Panavia Tornado F. Mk 2, armed with Sky Flash and Sidewinder missiles *(Pilot Press)*

The contract for the first production batch of F. Mk 2s was signed in 1981. These are now under construction, and the first of them is due to fly in 1983; IOC (initial operational capability) is scheduled to be achieved by late 1984. The production F. Mk 2 will be powered by an uprated version of the RB.199-34R-04 engine at present standard for the IDS Tornado.

Two main airframe modifications distinguish the ADV Tornado from the IDS version. The principal one is a 1·36 m (4 ft 5½ in) increase in fuselage length, created by the longer nose radome and the need for a small 'stretch' aft of the cockpit to allow the four Sky Flash missiles to be carried in two tandem pairs. The other external difference is that the fixed inboard portions of the wings are extended forward at the leading-edges (sweep angle 68° instead of 60°), to give additional chord. Extension of the fuselage provides additional space for avionics and for an extra 909 litres (200 Imp gallons) of internal fuel. A fully-retractable flight refuelling probe is mounted in the port side of the nose, and drop-tanks can be carried on the inner underwing stations.

Dimensions, external: As for the IDS version, except:
- Length overall — 18·06 m (59 ft 3 in)

Performance (estimated):
- Normal touchdown speed — 115 knots (213 km/h; 132 mph)
- T-O run with normal weapon and fuel load — 762 m (2,500 ft)
- Landing run, with thrust reversal — 366 m (1,200 ft)
- Endurance (combat air patrol) at 260-347 nm (483-644 km; 300-400 miles) from base, incl time for interception and 10 min combat — more than 2 h

SAAB/FAIRCHILD

Airframe Prime Contractors:

Saab-Scania Aktiebolag, S-581 88 Linköping, Sweden
Telephone: 46 13 12 90 20
Telex: 50040 SAABLGS

Fairchild Industries Inc, 20301 Century Boulevard, Germantown, Maryland 20767, USA
Telephone: (301) 428 6000

Board of Management:
Dr Knut Hagrup (Sweden) (Chairman)
Edward G. Uhl (Fairchild)
Sten Gustafsson (Saab-Scania)
George S. Attridge (Fairchild)
James J. Foody (Fairchild)
Dr Tore Gullstrand (Saab-Scania)
C. B. Winqvist (Sweden)

Chief Executive, Saab-Fairchild 340:
Alan R. Buley

Marketing, Saab-Fairchild 340:
Saab-Fairchild HB, S-581 88 Linköping, Sweden (worldwide except USA, Canada and Mexico)
Telephone: 46 13 12 9700
Fairchild Swearingen Corporation, PO Box 32486, San Antonio, Texas 78284, USA (USA, Canada and Mexico)
Telephone: (512) 824 9421

Saab-Scania and Fairchild Industries announced in January 1980 that the two companies had signed an agreement jointly to develop, produce and market a new transport aircraft. A 25,000 m² (269,100 sq ft) factory is under construction at Linköping for final assembly of the aircraft.

SAAB-FAIRCHILD 340

This transport aircraft, the first collaborative venture of its kind between members of the European and US aerospace industries, is being developed jointly by Saab-Scania and Fairchild Industries for entry into service in 1984. First flight is scheduled to take place in late 1982. Certification, scheduled for early 1984, will be to both FAR Pt 25 and JAR Pt 25. Orders and paid options for more than 100 aircraft had been received by August 1981.

Design emphasis is being placed on simplicity of systems, operation and maintenance, with quick turnarounds made possible by a number of built-in features which will make the aircraft independent of ground handling equipment. The airliner is designed specially for short-haul, low-density routes and will have two new-generation turboprop engines offering low fuel consumption, low operating costs and low operating noise levels. It is also expected to appeal to corporate aviation and other non-airline markets.

Fairchild Republic will build the wings, tail unit and engine nacelles; Saab-Scania will be responsible for fuselage construction, final assembly, flight testing and certification.

Type: Twin-turboprop transport aircraft.

Wings: Cantilever low-wing monoplane. Wing sections MS(1)-0316 at root, MS(1)-0312 at tip, with thickness/chord ratios of 16% and 12% respectively. Dihedral 7° from roots. Incidence 2° at root. Sweepback 3° 36′ at quarter-chord. Tapered two-spar wings embodying fail-safe principles. Stringers and skins of 2024/7075 aluminium alloy. Hydraulically-actuated single-slotted trailing-edge flaps, with aluminium alloy spars and honeycomb skins. Electrically-operated geared/trim tab in each aileron. Pneumatic-boot de-icing of leading-edges.

FUSELAGE: Conventional fail-safe/safe-life semi-monocoque structure, of circular cross-section. Built in three sections: nose (incl flight deck), passenger compartment, and tail section incorporating baggage compartment.

TAIL UNIT: Cantilever structure, with sweptback vertical and non-swept horizontal surfaces with marked dihedral. Fin integral with fuselage. Geared/trim tab in each elevator and in rudder. Construction similar to that of wings, except for bonded glassfibre honeycomb skins on rudder and elevators.

LANDING GEAR: Retractable tricycle type, of AP Precision Hydraulics design and manufacture, with twin Goodyear wheels and oleo-pneumatic shock-absorber on each unit. Hydraulic actuation. All units retract forward, main units into engine nacelles. Hydraulically steerable nose unit (60° to left and right), with shimmy damper. Main-wheel tyres size 24 × 7·7-10, pressure 6·89 bars (100 lb/sq in); nosewheel tyres size 18 × 6·0-6, pressure 3·79 bars (55 lb/sq in). Independent Goodyear hydraulic disc brakes on main units, with anti-skid control. Steerable nosewheels.

POWER PLANT: Two General Electric CT7-5A turboprop engines, each rated at 1,249 kW (1,675 shp) and driving a Dowty Rotol four-blade variable-pitch propeller with spinner and composite blades. Fuel in two integral tanks in wings, combined capacity 3,331 litres (733 Imp gallons; 880 US gallons). Single-point pressure refuelling inlet in starboard outer wing panel. Overwing gravity refuelling point in each wing.

ACCOMMODATION: Crew of two on flight deck; seat for attendant at front of passenger cabin on port side. Main cabin accommodates up to 34 passengers, in 10 rows of three, with aisle, and a final row of four. Seat pitch 76 cm (30 in). Movable bulkhead aft of last row of seats. Toilet and wardrobe on starboard side at forward end of cabin; provision for optional galley. Sixteen-seat corporate/executive version is planned. Passenger door at front of cabin on port side, with separate airstair. Type II emergency exit opposite this on starboard side; Type III emergency exit over wing on each side. Baggage space under each passenger seat; overhead storage bins optional. Main baggage/cargo compartment aft of passenger cabin, with large door on port side. Entire accommodation pressurised, including baggage compartment.

SYSTEMS: Hamilton Standard environmental control system (max pressure differential 0·48 bars; 7 lb/sq in) maintains a S/L cabin environment up to an altitude of 3,660 m (12,000 ft) and a 1,525 m (5,000 ft) environment up to the max cruising altitude of 7,620 m (25,000 ft). Hydraulic system, operating at 138-207 bars (2,000-3,000 lb/sq in), for actuation of landing gear, wheel and propeller braking, nosewheel steering, anti-skid control, and wing flaps. Electrical power supplied by two 28V 400A DC engine-driven starter/generators each connected to a separate busbar. Two main solid-state inverters, and a third as standby, provide 115V AC power at 400Hz. Two 36-40Ah nickel-cadmium batteries for ground power and engine starting in temperatures down to −18°C. Pneumatic-boot de-icing of wing, fin and tailplane leading-edges, using engine bleed air. Flight deck windows have electrical anti-icing and electrically-driven windscreen wipers. Electrical anti-icing is provided also for engine air intakes, propellers and pitot heads. Plug-in connections for oxygen masks. Kidde engine fire detection system.

AVIONICS AND EQUIPMENT: Standard avionics include two VHF com, two VOR/LOC receivers, glideslope receiver, marker beacon receiver, ADF, DME, GPWS, slaved compass system, radio altimeter, weather radar, ATC transponder, interphone, PA system, flight data recorder and cockpit voice recorder. Provision for additional optional flight control systems, including autopilot and flight director, dual ADF/DME and dual HF.

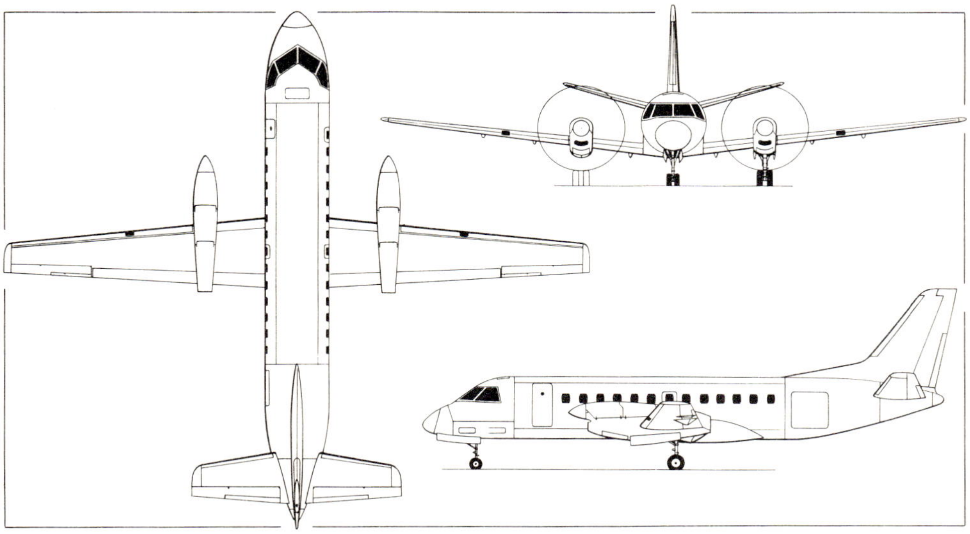

Saab-Fairchild 340 airliner, for service in 1984 *(Pilot Press)*

DIMENSIONS, EXTERNAL:
- Wing span: 21·44 m (70 ft 4 in)
- Wing chord: at root: 2·837 m (9 ft 3·7 in)
  - at tip: 1·0645 m (3 ft 5·9 in)
- Wing aspect ratio: 11
- Length overall: 19·72 m (64 ft 8·4 in)
- Fuselage: Max diameter: 2·31 m (7 ft 7 in)
- Height overall: 6·87 m (22 ft 6½ in)
- Wheel track: 6·71 m (22 ft 0 in)
- Wheelbase: 7·15 m (23 ft 5½ in)
- Propeller diameter: 3·20 m (10 ft 6 in)
- Propeller ground clearance: 0·58 m (1 ft 11 in)
- Distance between propeller centres: 6·71 m (22 ft 0 in)
- Passenger door: Height: 1·60 m (5 ft 3 in)
  - Width: 0·69 m (2 ft 3 in)
  - Height to sill: 1·63 m (5 ft 4 in)
- Cargo door: Height: 1·29 m (4 ft 3 in)
  - Width: 1·35 m (4 ft 5 in)
  - Height to sill: 1·63 m (5 ft 4 in)
- Emergency exit (fwd, stbd):
  - Height: 1·32 m (4 ft 4 in)
  - Width: 0·51 m (1 ft 8 in)
- Emergency exits (overwing, each):
  - Height: 0·91 m (3 ft 0 in)
  - Width: 0·51 m (1 ft 8 in)

DIMENSIONS, INTERNAL:
- Cabin, excl flight deck, incl toilet and galley:
  - Length: 10·57 m (34 ft 8 in)
  - Max width: 2·16 m (7 ft 1 in)
  - Width at floor: 1·70 m (5 ft 7 in)
  - Max height: 1·83 m (6 ft 0 in)
- Baggage/cargo compartment volume: 6·4 m³ (225·0 cu ft)

*AREAS:
- Wings, gross: 41·81 m² (450·0 sq ft)
- Ailerons (total): 2·00 m² (21·53 sq ft)
- Trailing-edge flaps (total): 8·25 m² (88·80 sq ft)
- Fin, incl dorsal fin: 10·34 m² (111·30 sq ft)
- Rudder, incl tab: 2·97 m² (31·97 sq ft)
- Tailplane: 12·70 m² (136·70 sq ft)
- Elevators (total, incl tabs): 3·51 m² (37·78 sq ft)

*WEIGHTS AND LOADINGS:
- Max fuel load: 2,676 kg (5,900 lb)
- Max T-O and landing weight: 11,340 kg (25,000 lb)
- Max zero-fuel weight: 9,979 kg (22,000 lb)
- Max wing loading: 271·2 kg/m² (55·5 lb/sq ft)
- Max power loading: 4·75 kg/kW (7·81 lb/shp)

*PERFORMANCE (at max T-O weight, ISA, estimated):
- Max operating speed (VMO): 250 knots (463 km/h; 288 mph) EAS
- Max operating Mach No. (MMO): 0·5
- Max cruising speed at 4,570 m (15,000 ft): 274 knots (508 km/h; 315 mph)
- Econ cruising speed at 7,620 m (25,000 ft): 232 knots (430 km/h; 267 mph)
- Stalling speed: 0° flap: 100 knots (185 km/h; 116 mph)
  - T-O flap setting: 88 knots (164 km/h; 102 mph)
  - Approach flap setting: 84 knots (156 km/h; 97 mph)
  - Landing flap setting: 78 knots (145 km/h; 90 mph)
- Max rate of climb at S/L: 579 m (1,900 ft)/min
- Rate of climb at S/L, one engine out: 213 m (700 ft)/min
- Service ceiling: 7,620 m (25,000 ft)
- Service ceiling, one engine out: 3,780 m (12,400 ft)
- FAR Pt 25 required T-O field length:
  - at S/L, ISA: 1,082 m (3,550 ft)
  - at S/L, ISA +15°C: 1,280 m (4,200 ft)
  - at 1,525 m (5,000 ft), ISA: 1,402 m (4,600 ft)
  - at 1,525 m (5,000 ft), ISA +15°C: 1,707 m (5,600 ft)
- Landing field length at 10,205 kg (22,500 lb) weight:
  - at S/L: 1,036 m (3,400 ft)
  - at 1,525 m (5,000 ft): 1,158 m (3,800 ft)
- Min ground turning radius: 14·07 m (46 ft 2 in)
- Runway LCN: flexible pavement: 8
  - rigid pavement: 10
- Range, allowances for 130 nm (240 km; 150 mile) diversion and 45 min hold:
  - with max payload: 970 nm (1,795 km; 1,115 miles)
  - with max fuel: 1,770 nm (3,280 km; 2,040 miles)

** Revised data in Addenda*

OPERATIONAL NOISE LEVELS (FAR Pt 36, estimated):
- T-O: 87 EPNdB
- Sideline: 91 EPNdB
- Approach: 94 EPNdB

SEPECAT

SOCIÉTÉ EUROPÉENNE DE PRODUCTION DE L'AVION E.C.A.T.

AIRFRAME COMPANIES:

British Aerospace PLC Aircraft Group, Richmond Road, Kingston upon Thames, Surrey KT2 5QS, England
Telephone: 01 546 7741

Avions Marcel Dassault/Breguet Aviation, BP 32, 92420 Vaucresson, France
Telephone: 970 38 50

DIRECTORS:
I. R. Yates (alternate President)
P. E. Jaillard (alternate President)
Sir Frederick Page
J. Bonnet
C. Edelstenne
M. Berjon
R. H. Evans
A. Constantine

PUBLIC RELATIONS:
G. B. Hill (BAe)
C. P. Raffin (Dassault-Breguet)

This Anglo-French company was formed in May 1966 by Breguet Aviation and British Aircraft Corporation, to design and produce the Jaguar supersonic strike fighter/trainer. The Jaguar project was initiated by the Defence Ministries of Britain and France on 17 May 1965. The governments of the two countries appointed an official Jaguar Management Committee to look after their interests. SEPECAT is the complementary industrial organisation.

In 1980, BAe's Warton Division took over responsibility for all future Jaguar development.

SEPECAT JAGUAR

The Jaguar, which was developed from the Breguet Br121 project, was designed by Breguet and BAe to meet a common requirement of the French and British air forces laid down in early 1965. This requirement called for a dual-role aircraft, to be used as an advanced and operational trainer and a tactical support aircraft of light weight and high performance, to enter French service in 1972 and with the RAF in 1973.

The following versions of the Jaguar have been built:

Jaguar A. French single-seat tactical support version. Prototypes (A-03 and A-04) first flown on 23 March and 27 May 1969. Total of 160 ordered, of which 150 had been delivered by 31 March 1981. France's final 30 Jaguar As are being fitted with a Martin Marietta/Thomson-CSF target TV acquisition and laser designation pod, named ATLIS II, and can carry two AS.30 air-to-surface missiles, as detailed under the appropriate 'Avionics and Operational Equipment' paragraph.

The first operational Armée de l'Air Jaguar unit (Esc. 1/7 'Provence') was formed at St Dizier in eastern France on 19 June 1973. The French Air Force has nine Jaguar A squadrons: three each with the 7th and 11th Escadres at St Dizier and Toul-Rosières, one with the 3rd Escadre at Nancy, and two with the 92nd Escadre. In December 1977 and May 1978, Jaguars of the Armée de l'Air were the first of their type to enter combat, when they were used to attack guerrilla forces in Mauritania.

Jaguar B (RAF designation: Jaguar T.Mk 2). British two-seat operational training version. Prototype B-08 (XW566) first flown on 30 August 1971. Total of 37 built initially; one more ordered in 1980. First T.Mk 2 delivered to RAF was XX137.

Jaguar E. French two-seat advanced training version. Prototypes (E-01 and E-02) first flown on 8 September

AS.30 Laser missile being fired from a French Air Force Jaguar A equipped with an underfuselage ATLIS II pod

1968 and 11 February 1969. Total of 40 built. First production Jaguar, designated E-1, flew for the first time on 2 November 1971, and deliveries to the CEAM at Air Base 118, Mont de Marsan, began in May 1972. The first unit to equip with this version was Esc. 1/7 at St Dizier.

Jaguar S (RAF designation: Jaguar GR. Mk 1). British single-seat tactical support version, basically similar to A but with an advanced inertial navigation and weapon-aiming system (NAVWASS) controlled by a digital computer. Prototypes first flown on 12 October 1969 and 12 June 1970. Total of 165 built. The first production GR. Mk 1 (XX108) flew on 11 October 1972, and this version now equips eight RAF front-line squadrons: Nos. 2 (AC), 14, 17, 20 and 31 with the Second Allied Tactical Air Force in West Germany and Nos. 6, 41 and 54 at Coltishall in the UK. Nos. 2 (AC) and 41 Squadrons are reconnaissance units. Since 1978, RAF Jaguars have been refitted with uprated Adour Mk 104 engines, equivalent to the Mk 804 which powers the Jaguar International (see following paragraph). This programme was due for completion in 1981.

Jaguar International. Export version, first flown (G27-266) on 19 August 1976. Differs little from Jaguar S or B except in having more powerful Adour turbofan engines (initially the Adour Mk 804, rated at 23·40 kN; 5,260 lb st dry and 38·25 kN; 8,600 lb st with afterburning), which give improved combat performance with substantially enhanced manoeuvrability and acceleration in the low-level speed range. The next standard of engine, the Mk 811, is now available, and uprating to this standard (see under 'Power Plant' for details) is possible during overhaul of the Mk 804 engines. Flight trials with the Mk 811 began on 14 August 1978; since then the aircraft has frequently demonstrated its ability to take off and land on grass strips.

Other customer options include overwing pylons compatible with Matra R.550 Magic or similar dogfight missiles; a multi-purpose radar such as the Thomson-CSF Agave, with which flight trials were completed in March 1977; up to four anti-shipping weapons such as Harpoon, Exocet and Kormoran on the underwing and underfuselage hardpoints; and night sensors such as low light level TV.

Initial orders were placed by the Sultan of Oman's Air Force (12) and Ecuadorean Air Force (12), each order including two two-seaters. Deliveries to Ecuador were made in January-November 1977, and to Oman between March 1977 and July 1978. SOAF aircraft are fitted with a Marconi Avionics 920ATC NAVWASS computer and have provision for carrying overwing Magic missiles. A further 12 Jaguar Internationals, powered by Adour Mk 811 engines, were ordered by Oman in 1980.

Under a 1979 agreement, an initial batch of 40 complete Jaguar Internationals is being purchased from Britain by the Indian government; the agreement then provides for a further 45 to be assembled in India from British-built components, leading eventually to full manufacture of a third batch of 55 aircraft under licence by Hindustan Aeronautics Ltd, Bangalore. The first components for Indian assembly left BAe Warton on 5 May 1981. Meanwhile, the Indian Air Force has on loan 18 RAF Jaguars, the first two of which were handed over on 19 July 1979. Two more arrived in that country on 14 October 1979, and the Indian Air Force's first Jaguar squadron (No. 14) had become operational by the Summer of 1980.

The Jaguar International is also available in a maritime strike version, similarly powered by Adour Mk 804 (current standard) or Mk 811 (optional) turbofan engines. This also would have the Agave nose radar, and ability to carry Harpoon, Exocet or Kormoran missiles, as shown in an accompanying illustration. The capacity to carry overwing Magic missiles is retained.

A total of 202 Jaguars have been built for the RAF and 200 are being built for the French air force. The first formal production contract, placed in the Autumn of 1969, covered 50 Jaguars for France; the second was for 30 for the RAF. Subsequent contracts brought total Anglo-French orders to 402 by early 1976; one additional Jaguar B was ordered in 1980. Of this total of 403, SEPECAT had delivered 392 by 1 March 1981.

Dassault-Breguet factories at Toulouse and Biarritz are responsible for the front and centre fuselage. BAe has responsibility for the rear fuselage, air intakes, wings and tail unit. There are final assembly lines for complete aircraft in both Britain and France.

The Jaguar is fully power-controlled in all three axes and is automatically stabilised as a weapons platform by gyros which sense disturbances and feed appropriate correcting data through a computer to the power control assemblies, in addition to the human pilot manoeuvre demands. The power controls are all of duplex tandem arrangement, with both mechanical and electrical servo-valves of the established Fairey platen design. Air-to-air combat capability can be enhanced by the inclusion of roll/yaw dampers, to increase lateral stability, and by increasing the slat and flap angles.

Under contract to BAe, Dowty Boulton Paul has designed and manufactured the primary flying controls for a fly-by-wire Jaguar demonstration programme.

TYPE: Single-seat tactical support aircraft (Jaguar A, S and International) and two-seat operational or advanced trainer (Jaguar B and E).

WINGS: Cantilever shoulder-wing monoplane. Anhedral 3°. Sweepback 40° at quarter-chord. All-metal two-spar torsion-box structure, the skin of which is machined from solid aluminium alloy, with integral stiffeners. Entire wing unit is British-built. Main portion built as single unit, with three-point attachment to each side of fuselage. Outer panels fitted with slat which also gives effect of extended-chord leading-edge. No conventional ailerons. Lateral control by two-section spoilers, forward of outer flap on each wing, in association (at low speeds) with differential tailplane. Hydraulically-operated (by screwjack) full-span double-slotted trailing-edge flaps. Fairey Hydraulics powered flying controls. Leading-edge slats can be used in combat.

FUSELAGE: All-metal structure, mainly aluminium, built in three main units and making use of sandwich panels and, around the cockpit(s), honeycomb panels. Local use of titanium alloy in engine bay area. The forward and centre fuselage, up to and including the main undercarriage bays, and including cockpit(s), main systems installations, forward fuel tanks and landing gear, are of French construction. The air intakes, and entire fuselage aft of the main-wheel bays, including engine installation, rear fuel tanks and complete tail assembly, are British-built. Two door-type airbrakes under rear fuselage, immediately aft of each main-wheel well. Structure and systems, aft of cockpit(s), are identical for single-seat and two-seat versions.

TAIL UNIT: Cantilever all-metal two-spar structure, covered with aluminium alloy sandwich panels. Rudder and outer panels and trailing-edge of tailplane have honeycomb core. Sweepback at quarter-chord 40° on horizontal, 43° on vertical surfaces. All-moving slab-type tailplane, with 10° of anhedral, the two halves of which can operate differentially to supplement the spoilers. No separate elevators. Fairey Hydraulics powered flying controls. Ventral fins beneath the rear fuselage. Entire tail unit is British-built.

LANDING GEAR: Messier-Hispano-Bugatti retractable tricycle type, all units having Dunlop wheels and low-pressure tyres for rough-field operation. Hydraulic retraction, with oleo-pneumatic shock-absorbers. Forward-retracting main units each have twin wheels, tyre size 615 × 225-10, tyre pressure 5·8 bars (84 lb/sq in). Wheels pivot during retraction to stow horizontally in bottom of fuselage. Single rearward-retracting nose-wheel, with tyre size 550 × 250-6 and pressure of 3·9 bars (57 lb/sq in). Twin landing taxying lights in nose-wheel door. Dunlop hydraulic brakes. Anti-skid units and arrester hook standard. Irvin brake parachute of 5·5 m (18 ft 0½ in) diameter housed in fuselage tailcone.

POWER PLANT: Two Rolls-Royce Turboméca Adour Mk 102 turbofan engines initially (each rated at 22·75 kN; 5,115 lb st dry and 32·5 kN; 7,305 lb st with afterburning). French Jaguars still have this power plant. RAF Jaguars now nearly all have Adour Mk 104, rated at 23·7 kN (5,320 lb st) dry and 35·75 kN (8,040 lb st) with afterburning. Jaguar International offered initially with Adour Mk 804 (same ratings as Mk 104); now available with Adour Mk 811, rated at 24·6 kN (5,520 lb st) dry and 37·4 kN (8,400 lb st) with afterburning. Aircraft built in India, and the second 12 for Oman, will have Adour Mk 811 engines. Lateral-type fixed-geometry air intake on each side of fuselage aft of cockpit. Fuel in six tanks, one in each wing and four in fuselage. Total internal fuel capacity 4,200 litres (924 Imp gallons). Armour protection is provided for critical fuel system components. In the basic tactical sortie the loss of fuel from one tank at the halfway point would not prevent the aircraft from regaining its base. Provision for carrying three auxiliary drop-tanks, each of 1,200 litres (264 Imp gallons) capacity, on fuselage and inboard wing pylons. Jaguar A and S equipped for in-flight refuelling, with a retractable probe forward of the cockpit on the starboard side.

ACCOMMODATION (Jaguar B and E): Crew of two in tandem on Martin-Baker Mk 9B zero-zero ejection seats (Jaguar B) or JRM4 seats (Jaguar E) giving zero-altitude ejection at speeds down to 90 knots (167 km/h; 104 mph). Individual rearward-hinged canopies. Rear seat is 38 cm (15 in) higher than front seat. Windscreen bullet-proof against 7·5 mm rifle fire.

ACCOMMODATION (Jaguar A and S): Enclosed cockpit for pilot, with rearward-hinged canopy and Martin-Baker Mk 9B (Jaguar S), EB9 (Ecuador), OB9 (Oman) or JRM4 (Jaguar A) ejection seat as in two-seaters. Bulletproof windscreen, as in two-seat versions.

SYSTEMS: Air-conditioning and pressurisation systems maintain automatically, throughout the flight envelope, comfortable operating conditions for the crew, and also control the temperature in certain equipment bays. Two independent hydraulic systems, powered by two Vickers engine-driven pumps. Hydraulic pressure 207 bars (3,000 lb/sq in). First system (port engine) supplies one channel of each actuator for the flying controls, the hydraulic motors which actuate the flaps and slats, the landing gear retraction and extension system, the brakes and anti-skid units. The second system supplies the other half of each flying control actuator, two further

JI004, the fourth (and second single-seat) Jaguar International delivered to the Indian Air Force

hydraulic motors actuating the slats and flaps, the air-brake and landing gear emergency extension jacks, nosewheel steering system and the wheel brakes. In addition to the duplicated hydraulic power systems, there is an emergency hydraulic power transfer unit. Electrical power provided by two 15kVA AC generators, either of which can sustain functional and operational equipment without load-shedding. DC power provided by two 4kW transformer-rectifiers. Emergency power for essential instruments provided by 15Ah battery and static inverter. De-icing, rain clearance and demisting standard. Liquid oxygen system installed, which also pressurises the pilot's anti-*g* suit.

Avionics and Operational Equipment (French versions): Avionics in Jaguar E include VHF/UHF radio, VOR/ILS and IFF; Tacan with Crouzet Type 90 navigation indicator; SFIM 153-6 twin-gyro inertial platform with two SFIM 810 all-attitude roll and pitch spherical indicators; SFIM 511 directional compass; Jaeger ELDIA air data system with Jaeger altitude indicator; CSF RL 50Pj incidence probe with angle of attack indicator; CSF 121 fire control sighting unit with weapon selector and adaptor for sighting head camera. Except for the use of a SFIM 250-1 twin-gyro platform, and the addition of a vector adder to the navigation indicator, this equipment is repeated in the Jaguar A, which has in addition an EMD Decca RDN 72 Doppler radar, Crouzet Type 90 navigation computer with target selector, CFTH passive radar warning (ECM) detector, CSF 31 weapon aiming computer, a Dassault fire control computer for Martel anti-radar missiles and a CSF laser rangefinder. Provision for the addition to these basic installations of such other items as terrain-following radar or sighting equipment for low light level targets. About six Jaguar As in each French squadron have an Omera 40 panoramic camera installed under the nosecone to give 180° horizon-to-horizon coverage. The final 30 Jaguar As for France carry an ATLIS II (Automatic Tracking Laser Illumination System) pod, developed by Martin Marietta and Thomson-CSF, on the centreline pylon. The pod contains a laser designator, and a wide-angle TV camera with its field of view centred along the line of the laser beam. The second part of the system is a modular laser guidance unit known as Ariel, which is implanted in the nosecone of the aircraft's rockets, missiles or bombs. The system is such that one aircraft can illuminate a target with its ATLIS pod while another releases its laser-guided weapons. The missile chosen by the French Air Force to operate with the ATLIS system is the Aérospatiale AS.30; Ariel can also be fitted to the Thomson-Brandt 100 mm rocket. A development of Ariel, known as TMV 630 Eblis, can be fitted to free-fall laser-guided bombs, and can be adapted to fit other 250, 400 or 1,000 kg bombs.

Avionics and Operational Equipment (British versions): Basic avionics of both the Jaguar B and S are similar. They include a Smiths-built Honeywell radio altimeter, slip indicator, E2B standby compass and autostabilising system, Plessey PTR 377 VHF/UHF radio and Marconi Avionics HF radio; Cossor CILS 75 ILS (VOR/ILS in Jaguar International for Oman, being replaced by Magnavox system in 1981; planned to retrofit RAF Jaguars similarly in due course); IFF; Tacan; Marconi Avionics digital/inertial navigation and weapon aiming subsystem (NAVWASS) with MCS 920M digital computer, E3R three-gyro inertial platform, inertial velocity sensor, navigation control unit and projected-map display; Marconi Avionics air data computer; Smiths electronic head-up display; Smiths FS6 horizontal situation indicator; Sperry C2J gyro amplifier master unit, compass controller and magnetic detector; Plessey weapon control system. Jaguar S fitted with Ferranti Type 105 laser rangefinder and Type 106 marked target seeker in

At the 1981 Paris Air Show, this Jaguar was displayed in maritime strike configuration, with Agave nose radar and underwing armament of Harpoon and Sidewinder missiles *(Brian M. Service)*

modified nose. Ferranti FIN 1064 digital inertial navigation and weapon aiming equipment (first test-flown in a Jaguar on 31 July 1981) ordered to replace major elements of NAVWASS equipment in RAF Jaguars. The BAe-designed flush-fitting reconnaissance pod for RAF Jaguars, carried on the fuselage centreline, has optical cameras for horizon-to-horizon coverage and BAe Dynamics (HS) 401 infra-red linescan (IRLS) for additional daylight, poor weather and night capability. Cameras are installed in two rotatable drums within the pod, the forward drum (for low/medium altitude missions) containing two Vinten F95 Mk 10 oblique and one F95 Mk 7 forward-looking cameras. The rear drum can carry alternative modules: two F95 Mk 10 cameras for low-level sorties, or a single F126 for medium altitude reconnaissance. The IRLS package is installed in the rear end of the pod, adjacent to a data conversion unit linked to the onboard NAVWASS digital computer.

Avionics and Operational Equipment (Jaguar International): Differ according to individual customer requirements; details of these are generally still classified, but it has been announced that the first 40 or so for India will have a Smiths head-up display similar to that in RAF Jaguars. Indian-built Jaguars will have a new head-up display and weapon-aiming system (HUDWAS) similar to that in the Sea Harrier. Indian Jaguars also have a Ferranti COMED 2045 combined map and electronic display.

Armament (Jaguar A, S and International): Two 30 mm cannon (DEFA 553 in Jaguar A, Aden in Jaguar S) in lower fuselage aft of cockpit. One stores attachment point on fuselage centreline and two under each wing. Centreline and inboard wing points can each carry up to 1,134 kg (2,500 lb) of weapons, and the outboard underwing points up to 567 kg (1,250 lb) each. Maximum external stores load, including overwing loads on Jaguar International, is 4,763 kg (10,500 lb). Jaguar As in service can carry the AN 52 tactical nuclear weapon. Typical alternative loads include one Martel AS.37 anti-radar missile and two 1,200 litre (264 Imp gallon) drop-tanks; eight 1,000 lb bombs; various combinations of free-fall and retarded bombs, Hunting BL755 or Beluga cluster bombs, Matra R.550 Magic missiles and air-to-surface rockets, including the 68 mm SNEB rocket; a reconnaissance-camera pack; or two drop-tanks. Jaguar International has full weapon capability of Jaguar S, plus the ability to carry two Matra Magic or Sidewinder air-to-air missiles on overwing pylons.

Armament (Jaguar B and E): Two 30 mm DEFA 553 cannon in Jaguar E; Jaguar B has single 30 mm Aden cannon on port side. The two-seat versions have similar weapons capability to the tactical models, and can be employed for operational missions as required.

Dimensions, external:
- Wing span: 8·69 m (28 ft 6 in)
- Wing chord at root: 3·58 m (11 ft 9 in)
- Wing chord at tip: 1·13 m (3 ft 8½ in)
- Wing aspect ratio: 3·12
- Length overall, incl probe:
  - A and S: 16·83 m (55 ft 2½ in)
  - B and E: 17·53 m (57 ft 6¼ in)
- Length overall, excl probe:
  - A and S: 15·52 m (50 ft 11 in)
  - B and E: 16·42 m (53 ft 10½ in)
- Height overall: 4·89 m (16 ft 0½ in)
- Tailplane span: 4·53 m (14 ft 10¼ in)
- Wheel track: 2·41 m (7 ft 11 in)
- Wheelbase: 5·69 m (18 ft 8 in)

Areas:
- Wings, gross: 24·18 m² (260·27 sq ft)
- Leading-edge slats (total): 1·05 m² (11·30 sq ft)
- Trailing-edge flaps (total): 4·12 m² (44·35 sq ft)
- Spoilers (total): 0·90 m² (9·67 sq ft)
- Vertical tail surfaces (total): 3·90 m² (42·00 sq ft)
- Horizontal tail surfaces (total): 7·80 m² (83·96 sq ft)

Weights and Loadings:
- Typical weight empty: 7,000 kg (15,432 lb)
- Normal T-O weight (single-seater, with full internal fuel and ammunition for built-in cannon): 10,954 kg (24,149 lb)
- Max T-O weight, with external stores: 15,700 kg (34,612 lb)
- Max wing loading: 649·3 kg/m² (133 lb/sq ft)
- Max power loading:
  - Adour Mk 102: 241·5 kg/kN (2·37 lb/lb st)
  - Adour Mks 104, 804: 219·6 kg/kN (2·15 lb/lb st)
  - Adour Mk 811: 209·9 kg/kN (2·06 lb/lb st)

Performance:
- Max level speed at S/L: Mach 1·1 (729 knots; 1,350 km/h; 840 mph)
- Max level speed at 11,000 m (36,000 ft): Mach 1·6 (917 knots; 1,699 km/h; 1,056 mph)
- Landing speed: 115 knots (213 km/h; 132 mph)
- T-O run: 'clean': 565 m (1,855 ft)
  - with four 1,000 lb bombs: 880 m (2,890 ft)
  - with eight 1,000 lb bombs: 1,250 m (4,100 ft)
- T-O to 15 m (50 ft) with typical tactical load: 940 m (3,085 ft)
- Landing from 15 m (50 ft) with typical tactical load: 785 m (2,575 ft)
- Landing run:
  - normal weight, with brake-chute: 470 m (1,540 ft)
  - normal weight, without brake-chute: 680 m (2,230 ft)
  - overload weight, with brake-chute: 670 m (2,200 ft)
- Typical attack radius, internal fuel only:
  - hi-lo-hi: 460 nm (852 km; 530 miles)
  - lo-lo-lo: 290 nm (537 km; 334 miles)
- Typical attack radius with external fuel:
  - hi-lo-hi: 760 nm (1,408 km; 875 miles)
  - lo-lo-lo: 495 nm (917 km; 570 miles)
- Ferry range with external fuel: 1,902 nm (3,524 km; 2,190 miles)
- *g* limits: +8·6; +12 (ultimate)

SEPECAT JAGUAR DEVELOPMENT

In addition to earlier developments listed in the previous entry, BAe is considering other means of developing the power plant and Jaguar airframe for possible future versions.

The Adour engine is capable of considerable further growth, the greatest potential being represented currently by the 'Dash 63', which has 38 per cent more T-O thrust with afterburning than the Mk 102, and 65 per cent more

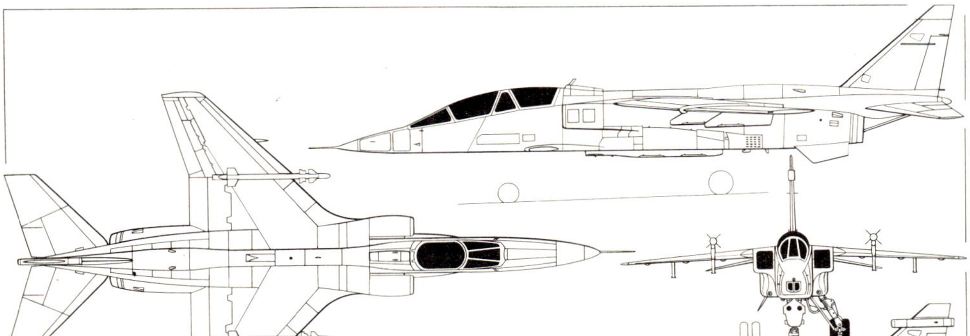

SEPECAT Jaguar International single-seat strike aircraft, with additional side view (top) of two-seat Jaguar B and scrap view of version with Agave nose radar and Ferranti 105 laser *(Pilot Press)*

in the combat regime. Various Adour models, with dash numbers between 58 (now the Adour Mk 811) and 63, are being studied.

BAe is also studying designs for a new wing which could either be fitted to a new aircraft with the minimum of design change, or be retrofitted to existing Jaguars.

Development contracts have been placed for an advanced fly-by-wire installation, and for a carbonfibre wing, for the Jaguar. The Marconi Avionics/Dowty quadruplex FBW flight control system, which has all-electric signalling and no manual backup, is being installed by BAe in an RAF Jaguar GR. Mk 1 (XX765), and was due to be test-flown in the Autumn of 1981. After an initial flight in stable configuration, the test aircraft is to have wing leading-edge strakes added to destabilise it in the pitch axis, so becoming a control configured vehicle (CCV) for the remainder of its test programme.

SOKO/CNIAR

PARTICIPANTS:

SOKO, Mostar, Yugoslavia

Centrul National al Industriei Aeronautice Române, Bucharest, Romania

SOKO/CNIAR ORAO (EAGLE)/IAR-93

This twin-jet ground attack fighter is in production to meet a joint requirement of the air forces of Romania and Yugoslavia. In the latter country it is known as the Orao (Eagle); in Romania it is known as the IAR-93. The joint programme is known as 'Yurom' (from *Yu*goslavia-*Rom*ania).

The Orao/IAR-93 was designed jointly by Yugoslav engineers from the Vazduhoplovno Tehnicki Institut in Zarkovo, near Belgrade, and by Romanian engineers from the Institutul de Mecanica Fluidelor si Constructii Aerospatiale in Bucharest. Manufacture of single-seat prototypes began simultaneously in the two countries, and a first flight in each country was made on 31 October 1974. On 15 April 1975 the Yugoslav prototype, bearing serial number 25001, was demonstrated publicly during the Victory Day parade at Batajnica military airfield near Belgrade. A tandem two-seat dual-control prototype was then built and flown in each country, after which Yugoslavia and Romania each completed a number of pre-production aircraft. A formation of three IAR-93s was flown publicly on 23 August 1979, during the Liberation Day celebrations in Bucharest.

More than 33 factories in Yugoslavia are involved in the Orao manufacturing programme, led by SOKO, which has prime responsibility for final assembly and flight testing. In Romania, CNIAR is co-ordinating the factories involved and is responsible for delivering completed aircraft to the Ministry of National Defence from the IAR-93 final assembly plant at Craiova.

It is anticipated that 200 or more of these aircraft will be built for each country, including a proportion of two-seat dual-control operational trainers. The initial production batch is estimated to be nearly 40, including two-seaters.

TYPE: Single-seat close support aircraft and interceptor.

WINGS: Cantilever shoulder-wing monoplane, of low aspect ratio. Anhedral 3° 30′ from roots. Sweepback approx 43° on leading-edges. Leading-edge slats. Two small boundary layer fences on upper surface of each wing. Wide-chord semi-Fowler-type trailing-edge flaps. Trim tab in starboard aileron. Ailerons controlled by Dowty servo-actuators with autostabiliser system.

FUSELAGE: All-metal semi-monocoque structure. Door-type perforated airbrake under each side of fuselage, forward of main-wheel bays. Dorsal spine fairing houses systems circuits and flight control rods. 'Pen-nib' fairing above exhaust nozzles. Rear fuselage detachable to facilitate access to engines. Space provision in nose for ranging radar.

TAIL UNIT: Cantilever metal structure, with sweepback on all surfaces. Low-set all-moving tailplane. Small dorsal fin. Tailplane and rudder controlled by Dowty servo-actuators with autostabiliser system. Auxiliary ventral fin on each side beneath rear fuselage.

LANDING GEAR: Retractable tricycle type, with single-wheel steerable nose unit and twin-wheel main units. Manufactured by PPT factories (for Yugoslav aircraft) and CNIAR factories (Romanian aircraft). All units have two-stage oleo-pneumatic shock-absorbers. Hydraulic actuation, all units retracting forward into fuselage. Braking parachute in bullet fairing at base of rudder.

POWER PLANT: Two 17·8 kN (4,000 lb st) Rolls-Royce Viper Mk 632-41 non-afterburning turbojet engines, mounted side by side in rear fuselage, with lateral air intakes. Internal fuel capacity approx 3,200 litres (704 Imp gallons). Production aircraft have Rolls-Royce-developed afterburners, increasing power of each engine to approx 22·3 kN (5,008 lb st); afterburners are produced under licence in Yugoslavia and Romania.

ACCOMMODATION: Pilot only, on Martin-Baker YU10J or RU10J zero-zero ejection seat beneath rear-hinged, upward-opening canopy; or crew of two in tandem under elongated canopy in operational training version. Dual controls in training version.

SYSTEMS AND EQUIPMENT: Autostabiliser system for control surface servo-actuators. Standard equipment varies according to operator and mission/performance requirements. Landing lights under nose, forward of nosewheel bay, and on nosewheel leg. Ram-air scoop aft of cockpit on each side; additional airscoops on top of fuselage, aft of canopy and at front of dorsal fin, and below rear fuselage.

ARMAMENT: Two 23 mm cannon in lower front fuselage, aft of nosewheel bay; one underfuselage and four underwing stations for external stores. Max external load approx 2,500 kg (5,510 lb).

DIMENSIONS, EXTERNAL:

Wing span	9·63 m (31 ft 7⅛ in)
Wing area, gross	26·00 m² (279·86 sq ft)
Wing aspect ratio	3·57
Length overall:	
incl probe	14·88 m (48 ft 9⅞ in)
excl probe	13·99 m (45 ft 10¾ in)
Height overall	4·45 m (14 ft 7¼ in)
Wheel track (c/l of shock-struts)	2·50 m (8 ft 2½ in)
Wheelbase	5·35 m (17 ft 6⅝ in)

WEIGHTS AND LOADINGS (A: prototypes, B: production version):

Weight empty, equipped: A	6,100 kg (13,448 lb)
B	5,700 kg (12,566 lb)
Internal fuel load: A	2,600 kg (5,732 lb)
B	2,700 kg (5,952 lb)
T-O weight 'clean': A	8,880 kg (19,577 lb)
B	8,600 kg (18,959 lb)
Max T-O weight with external stores:	
A, B	10,500 kg (23,148 lb)
Wing loading:	
A at T-O weight 'clean'	341 kg/m² (69·8 lb/sq ft)
B at T-O weight 'clean'	330 kg/m² (67·6 lb/sq ft)
A and B at max T-O weight	404 kg/m² (82·7 lb/sq ft)
Power loading:	
A at T-O weight 'clean'	250 kg/kN (2·45 lb/lb st)
A at max T-O weight	295 kg/kN (2·89 lb/lb st)
B at T-O weight 'clean'	192 kg/kN (1·88 lb/lb st)
B at max T-O weight	235 kg/kN (2·30 lb/lb st)

PERFORMANCE (A: without afterburning; B: production version with afterburning):

Max level speed at low level:	
A	556 knots (1,030 km/h; 640 mph)
B	610 knots (1,130 km/h; 702 mph)
Max level speed at high altitude:	
A	529 knots (980 km/h; 609 mph)
B	577 knots (1,070 km/h; 665 mph)
Landing speed:	
A, B	130 knots (240 km/h; 149 mph)
Max rate of climb at S/L at 'clean' T-O weight:	
A	2,310 m (7,578 ft)/min
B	4,536 m (14,882 ft)/min
Time to 12,000 m (39,370 ft): A	10 min 15 s
Time to 13,000 m (42,650 ft): B	5 min 30 s
Service ceiling: A	12,000 m (39,370 ft)
B	13,000 m (42,650 ft)
T-O run:	
A at 8,500 kg (18,740 lb) AUW	925 m (3,035 ft)
B at max T-O weight	1,000 m (3,280 ft)
Landing run:	
A at 8,500 kg (18,740 lb) AUW	900 m (2,953 ft)
B at max T-O weight	1,100 m (3,609 ft)
Combat radius (A and B) with 2,000 kg (4,410 lb) external stores:	
lo-lo-lo	162 nm (300 km; 186 miles)
hi-lo-hi	194 nm (360 km; 224 miles)
g limit: A, B	+7·0

Single-seat prototype of the Orao tactical fighter (two Rolls-Royce Viper turbojet engines)

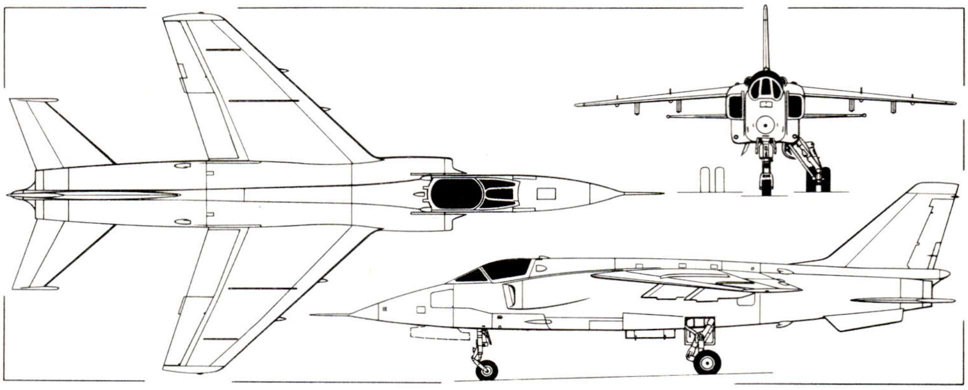
Orao/IAR-93 single-seat tactical fighter, developed jointly by Romania and Yugoslavia *(Pilot Press)*

Yugoslav-built two-seat Orao (nearest camera), with single-seater

TRANSALL
ARBEITSGEMEINSCHAFT TRANSALL

AIRFRAME COMPANIES:

Aérospatiale, 37 boulevard de Montmorency, 75781 Paris Cédex 16, France
Telephone: 524 43 21
Telex: 620059 F

MBB, PO Box 950109, 2103 Hamburg 95, Federal Republic of Germany
Telephone: (040) 747 5164
Telex: 217684

PROGRAMME DIRECTOR:
Jacques Hablot (Aérospatiale)

The Transall (Transporter Allianz) group was formed in January 1959 by MBB, Aérospatiale and VFW, to undertake joint development and production of the C-160 twin-turboprop military transport for the French and German air forces. Others were built for the air forces of South Africa and Turkey. Initial production, of the C-160 D (90), C-160 F (50), C-160 T (20) and C-160 Z (9), was shared between the three participating companies and ended in 1972, as described in earlier editions of *Jane's*. Production of a second series was authorised in 1977 to meet an additional French order and requests from other countries.

TRANSALL C-160 (Second Series)

The Transall C-160 was developed originally to meet the specific requirements of the Federal German and French governments for a military transport aircraft capable of carrying troops, casualties, freight, supplies and vehicles, and of operating from semi-prepared surfaces.

An industrial agreement was signed on 29 October 1976, sharing the work on new production between Aérospatiale (50%) and the two German companies (50%), with a single final assembly line at Toulouse. Aérospatiale builds the wings, wing/fuselage fairings, fuselage doors, emergency exits and engine nacelles. The fuselage, main landing gear fairings and all tail surfaces are built by MBB. The engines, as before, are manufactured jointly by Rolls-Royce, SNECMA, MTU and FN-Herstal. Components are airlifted to Toulouse by Super Guppy transport for final assembly and flight testing.

The main improvements in this second production series are updated avionics, and extended range resulting from a reinforced wing with an optional additional fuel tank in the centre-section.

In July 1977 the French government gave its approval to the launching of the second production series, primarily to satisfy a requirement of the French Air Force, which subsequently placed an initial order for 25. Ten of these are being fitted at the outset with in-flight refuelling equipment (hose-reel and drogue type) in the port main landing gear fairing to permit their operation as tankers; five others will incorporate provisions for this equipment and be capable of rapid adaptation to the tanker role if needed. All 25 will have a 4·00 m (13 ft 2 in) receiver boom mounted above and behind the flight deck. In addition to the French order, three second-series Transalls have been ordered by the Indonesian government.

Rollout of the first aircraft of the new series took place at Toulouse on 2 December 1980, and the first flight was made on 9 April 1981. Deliveries to the French Air Force were due to begin in October 1981, and to Indonesia in early 1982.

New capabilities for the Transall, with corresponding conversion kits, are being offered for maritime surveillance and attack (**C-160S** and **C-160ASF**, described separately), electronic warfare, airborne command, VIP transport, firefighting and anti-pollution missions (with a 12-ton liquid-dropping capability) and for service as an airborne surgical unit.

TYPE: Twin-engined turboprop transport.

WINGS: Cantilever high-wing monoplane. Dihedral on outer wings 3° 26′. All-metal two-spar structure designed on fail-safe principles. Wing in three sections, comprising a centre-section, which carries the engines, and two outer panels. All-metal ailerons and hydraulically-operated double-slotted flaps. Hydraulically-operated airbrakes (inboard, above and below wings) and spoilers (outboard) forward of flaps on each wing. Electrical de-icing of leading-edges.

FUSELAGE: Aluminium alloy (2024-T3) semi-monocoque structure of circular basic section, flattened at the bottom, and designed on fail-safe principles. Underside of upswept rear fuselage lowers to form loading ramp for vehicles.

TAIL UNIT: Cantilever aluminium alloy (2024-T3) structure. Large dorsal fin.

LANDING GEAR: Retractable tricycle type, built by Messier-Hispano-Bugatti and Liebherr-Aerotechnik. Hydraulic retraction and hydraulic/pneumatic shock-absorption. Each main unit comprises two pairs of wheels in tandem and is mounted inside a fairing on the side of the fuselage. Wheels can be raised to lower the fuselage for loading. Steerable twin-wheel nose unit. Main-wheel tyres size 15·00 × 16; nosewheel tyres size 12·5 × 16. Tyre pressure 3·79 bars (55 lb/sq in) on main units, 3·14 bars (45·5 lb/sq in) on nose unit. Messier-Hispano-Bugatti brakes.

POWER PLANT: Two 4,549 kW (6,100 ehp) Rolls-Royce Tyne RTy.20 Mk 22 turboprop engines, each driving a Ratier Forest-built BAe Dynamics 4/8000/6 four-blade constant-speed fully-feathering reversible-pitch propeller. Single-point pressure refuelling; gravity refuelling available optionally. Fuel in four integral wing tanks with total capacity of 19,050 litres (4,190 Imp gallons). Additional wing centre-section tank optional, capacity 9,000 litres (1,980 Imp gallons). Provision for in-flight refuelling. Water-methanol usable capacity 318·5 litres (70 Imp gallons). Oil capacity (total) 68·4 litres (15 Imp gallons).

ACCOMMODATION: Pressurised accommodation for crew of three, comprising pilot, co-pilot and flight engineer. Typical payloads include 93 troops or 61-88 fully-equipped paratroops; 62 stretchers and four attendants; armoured vehicles, tanks and tractors not exceeding max permissible payload weight. Flight deck and cargo compartment air-conditioned and pressurised in flight and on the ground. Power-assisted controls. Paratroop door on each side immediately aft of the landing gear fairings; hydraulically-operated rear loading ramp. The floor and all doors are at truckbed height. The floor is provided with lashing points of 5,000 kg (11,023 lb) capacity, arranged in a 51 cm (20 in) grid, and 12,000 kg (26,455 lb) capacity on the sidewalls, and is stressed to carry large military vehicles. Loads which cannot be driven in can be taken on board rapidly by an automatic translation and stowing system. Individual loads of up to 8,000 kg (17,637 lb) can be airdropped, including drops at low altitude (3-9 m; 10-30 ft) or during touch-and-go.

SYSTEMS: Normalair pressurisation and air-conditioning system, differential 0·302-0·322 bars (4·38-4·67 lb/sq in). Two separate primary hydraulic systems, pressure 175 bars (2,500 lb/sq in), for flying controls, loading ramp, landing gear, wheel brakes, flaps, spoilers, airbrakes, nosewheel steering and other auxiliaries. Two more systems, pressure 175 bars (2,538 lb/sq in), for emergency and ground services, as well as a handpump driven emergency system. AC electrical system includes two 60kVA 380-580Hz generators, one 60kVA 400Hz generator and two 9kVA 400Hz generators. 28V DC system and 40Ah batteries. Garrett GTCP-85-160A APU in forward part of port main undercarriage fairing.

AVIONICS AND EQUIPMENT: Socrat TRAP-138 VHF; TRT TRAP-139 UHF; LMT 3527C or Collins 628T-1 HF; TEAM AS-1227B PA system; TEAM TF-AP 14 intercom; EAS RNA-720 VOR/ILS; Collins NRAN-19 or DF 206 ADF; LMT DM-820 or Collins 860E-5 (without micro-Tacan) DME; LMT-3560 or Collins 621A-6A ATC transponder; Omera ORB-37 weather radar; TRT AHV-6 radio altimeter; EAS RM-671 or Collins 51Z-4 marker beacon receiver; Jaeger 60571 or Jaeger 64111 encoding altimeter; SFIM CADV automatic flight control system, incorporating two vertical gyros, two gyro-compasses, PA-51 autopilot and DV-86 flight director; and EMD RDN-72 Doppler navigation radar. Export version available with LMT micro-Tacan; Collins DF-301E UHF/DF (with UHF com system); Sercel Crouzet type Equinox Omega; and TRT APS-500 GPWS.

DIMENSIONS, EXTERNAL:	
Wing span	40·00 m (131 ft 3 in)
Wing chord at root	4·84 m (15 ft 10½ in)
Wing chord at tip	2·428 m (7 ft 11½ in)
Wing chord (mean)	4·176 m (13 ft 8½ in)
Wing aspect ratio	10
Length overall	32·40 m (106 ft 3½ in)
Fuselage: Max diameter	4·30 m (14 ft 1¼ in)
Height overall	11·65 m (38 ft 2¾ in)
Tailplane span	14·50 m (47 ft 7 in)
Wheel track	5·10 m (16 ft 9 in)
Wheelbase	10·48 m (34 ft 4½ in)
Propeller diameter	6·50 m (21 ft 4 in)
Distance between propeller centres	10·90 m (35 ft 9¼ in)
Crew door (fwd, port): Height	1·22 m (4 ft 0 in)
Width	0·62 m (2 ft 0½ in)
Paratroop door (each side): Height	1·90 m (6 ft 2½ in)
Width	0·90 m (3 ft 0 in)
Rear loading ramp: Length	3·70 m (12 ft 1½ in)
Width	3·15 m (10 ft 3½ in)
Emergency exits:	
Main hold, fwd, stbd side (one): Height	0·88 m (2 ft 10½ in)
Width	0·54 m (1 ft 9¼ in)
Flight deck roof (one); roof of main hold, fwd (one); and two in roof of main hold at rear (one each side of dorsal fin): Height	0·54 m (1 ft 9¼ in)
Width	0·64 m (2 ft 1¼ in)
DIMENSIONS, INTERNAL:	
Cabin, excl flight deck and ramp: Length	13·51 m (44 ft 4 in)
Max width	3·15 m (10 ft 3½ in)
Max height	2·98 m (9 ft 8½ in)
Floor area	42·6 m^2 (458·5 sq ft)
Volume	115·0 m^3 (4,061 cu ft)
Cabin, incl ramp: Length	17·21 m (56 ft 6 in)
Floor area	54·25 m^2 (584 sq ft)
Volume	139·9 m^3 (4,940 cu ft)
AREAS:	
Wings, gross	160·00 m^2 (1,722 sq ft)
Ailerons (total)	6·88 m^2 (74·06 sq ft)
Trailing-edge flaps (total, extended)	34·54 m^2 (371·8 sq ft)
Spoilers (total)	0·80 m^2 (8·61 sq ft)
Fin: excl dorsal fin	29·50 m^2 (317·5 sq ft)
incl dorsal fin	36·00 m^2 (387·5 sq ft)
Rudder	10·20 m^2 (109·8 sq ft)
Tailplane	43·80 m^2 (471·5 sq ft)
Elevators	10·30 m^2 (110·9 sq ft)

First new-production Transall C-160 for the French Air Force, with in-flight refuelling receiver boom

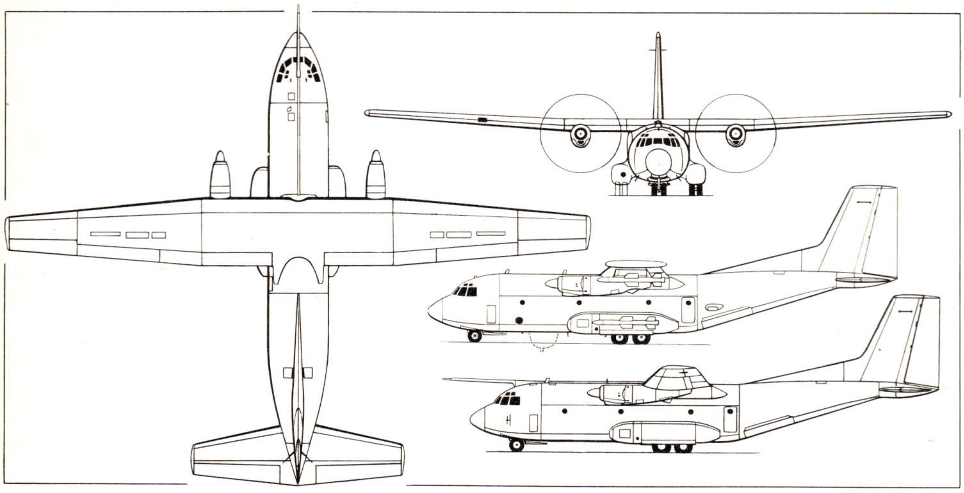

Transall C-160 transport, with additional side view (upper) of C-160ASF maritime version *(Pilot Press)*

Weights and Loadings:
- Min operating weight empty 28,000 kg (61,730 lb)
- Typical operating weight empty 29,000 kg (63,935 lb)
- Max payload 16,000 kg (35,275 lb)
- Max T-O weight 51,000 kg (112,435 lb)
- Max zero-fuel weight 45,000 kg (99,210 lb)
- Max landing weight 47,000 kg (103,615 lb)
- Max wing loading 319 kg/m² (65·34 lb/sq ft)
- Max power loading 5·61 kg/kW (9·22 lb/ehp)

Performance (at max T-O weight except where indicated):
- Never-exceed speed:
  - at 4,875-9,145 m (16,000-30,000 ft) Mach 0·64
  - below 4,875 m (16,000 ft) 320 knots (593 km/h; 368 mph)
- Max level speed at 4,875 m (16,000 ft) 277 knots (513 km/h; 319 mph)
- Stalling speed, flaps down 95 knots (177 km/h; 110 mph)
- Max rate of climb at S/L 396 m (1,300 ft)/min
- Rate of climb at S/L, one engine out 91 m (300 ft)/min
- Service ceiling at 45,000 kg (99,208 lb) AUW 8,535 m (28,000 ft)
- Service ceiling, one engine out at 45,000 kg (99,208 lb) AUW 3,050 m (10,000 ft)
- T-O run, 20° flap 730 m (2,395 ft)
- T-O to 10·5 m (35 ft), 20° flap 990 m (3,248 ft)
- Landing from 15 m (50 ft), 40° flap, at max landing weight without propeller reversal 869 m (2,850 ft)
- Landing run, normal 550 m (1,800 ft)
- Min ground turning radius 28·60 m (93 ft 10 in)
- Range, reserve of 5% initial fuel, allowance for 30 min hold at S/L, OWE of 29,000 kg (63,934 lb):
  - with 8,000 kg (17,640 lb) payload 2,750 nm (5,095 km; 3,166 miles)
  - with 16,000 kg (35,275 lb) payload 1,000 nm (1,853 km; 1,151 miles)
- Max ferry range with centre-section wing tank 4,780 nm (8,858 km; 5,504 miles)

TRANSALL C-160 (MARITIME VERSIONS)

By means of modular conversion kits, existing or new-production Transalls can be adapted to a maritime role, and in early 1981 two versions were proposed:

C-160S. Surveillance version, available from late 1983/early 1984, with search radar in nose, observers' bubble windows, cameras, cartridge launcher and universal launcher.

C-160ASF. Anti-surface attack version, with nose radar replaced by a retractable Omera ORB-32 ventral radar, or a Thomson-CSF Varan TMV-118B (or Varan derivative), for 360° scan. Attachment points underwing and on landing gear fairings for air-to-surface missiles, or for containers for ECM or survival equipment. Weapons include mines; D/F and elint antennae in wingtip fairings.

The following details apply to the C-160S:

Airframe and Power Plant: Generally similar to standard C-160, except as described under model listings.

Accommodation: Flight crew of three (pilot, co-pilot and flight engineer). Basic four-man tactical crew (two radar navigators and two observers), plus additional members as required or as relief crew. Observers' bubble windows located at floor level on port side, and in forward emergency exit on starboard side.

Avionics and Equipment: Two Socrat 4600 VHF com, single TRT ERA-8250 UHF com, Collins 628-T1 HF com, loudspeaker and interphone. SFIM-51 autopilot; SFENA flight director system; two EAS RNA-720 VOR/ILS; Collins 51Z-4 marker beacon receiver; Collins-Socrat DF-206 ADF; twin SFM VG-75 control gyros and amplifier; two SFIM CG-512-5 heading data generators; LMT-3592D micro-Tacan; two Sperry C6E RMIs; Omera ORB-32 or Thomson-CSF Varan TMV-118B multi-mode radar, with 224° scan; EMD RDN-72 Doppler; Crouzet-Nadir navigation computer; Collins 621A-6 ATC transponder; Jaeger 64-11 encoding altimeter; TRT AHV-6 radio altimeter; and SFENA 7054 emergency horizon.

Operational Equipment: Two Omera 35 cameras, mounted in lobe on each side of rear fuselage, linked to the navigation computer and capable of continuous or intermittent (manual) operation. Sixteen or 32 flares in port main landing gear fairing, with observer-actuated Alkan ejectors. Three alternative sea rescue systems: (1) a cylindrical container accommodating a pneumatic raft, float line, extractor parachute and stabilising parachute. This container can also be used to parachute 60 kg (132 lb) of supplies to ships and ground forces; (2) SAR system, jettisonable via paratroop doors; (3) 12·7 cm (5 in) diameter universal launch tube, on port side at rear of cabin, for dropping markers (luminous floats or radio markers) without need to open fuselage doors.

Weights:
- Operating weight empty 29,830 kg (65,763 lb)
- Max payload 16,000 kg (35,275 lb)
- Max T-O weight 51,000 kg (112,435 lb)

Performance (at max T-O weight):
- Optimum patrol speed 170 knots (315 km/h; 196 mph)
- Endurance at optimum patrol speed, at 500 nm (926 km; 576 miles) from base 14 h
- Range at optimum patrol speed 1,500 nm (2,780 km; 1,727 miles)
- Ferry range with crew of 4 and max payload, standard reserves 1,000 nm (1,853 km; 1,151 miles)

ISRAEL

IAI
ISRAEL AIRCRAFT INDUSTRIES LTD

Head Office and Works: Ben Gurion International Airport, Lydda (Lod)
Telephone: 03 973111
Telex: Isravia 31114
President: G. Gidor
Executive Vice-President: A. Ostrinski
Vice-Presidents:
- M. Keret (Aircraft and Airborne Systems Marketing)
- N. Rosen (Naval Systems and Missiles Marketing)
- A. Ezroni (General Manager, Bedek Aviation Division)
- M. Blumkin (General Manager, Engineering Division)
- Y. Ben-Bassat (General Manager, Manufacturing Division)
- Y. Giladi (General Manager, Combined Technologies Division)
- M. Na'aman (General Manager, Electronics Division)
- M. Dvir (Research and Development)
- D. Mozes (Finance)
- Mrs. H. Ron (General Counsel)
- Y. Zinger (Comptroller)

Commercial Director, Arava: Y. Avidan
Commercial Director, Westwind: S. Samach
Press Manager: Shai Tadmor

This company was established in 1953 as Bedek Aircraft Company. The change of name, to Israel Aircraft Industries, was made on 1 April 1967.

IAI employs more than 22,000 people in all its facilities, which occupy a total covered floor area of 450,000 m² (4,843,750 sq ft). It is licensed by, among others, the Israel Civil Aviation Administration, US Federal Aviation Administration, British Civil Aviation Authority and the Israeli Air Force as an approved repair station and maintenance organisation.

Israel Aircraft Industries Ltd is composed of several divisions, plants and subsidiary companies. These underwent a major reorganisation in the latter part of 1977, and are now disposed in five divisions, as follows:

Bedek Aviation Division, incorporating Turbochrome, is an internationally approved multi-faceted single-site civil and military aircraft service centre. Present programmes include the turnaround inspection, overhaul, repair, retrofitting, outfitting and testing of 30 types of aircraft, including the Boeing 707/720, 727, 737, and McDonnell Douglas DC-8; 28 types of civil and military engines, including the JT3D, JT4A, JT8D and JT9D; and 8,000 types of components, accessories and systems. Offshore workload includes the supply of total technical support to several international operators. The division holds warranty and/or approved service centre approvals from many of the world's leading component manufacturers. Bedek has refurbished and resold numerous Boeing 707/720s, often after conversion from passenger to cargo configuration. The procedure for the structural modification was developed jointly by Boeing and Bedek Aviation.

Bedek Aviation has a total floor area of more than 74,322 m² (800,000 sq ft). Its workforce includes more than 3,500 technicians, engineers and supporting personnel.

The **Aircraft Manufacturing Division** produces the Kfir fighter, the Arava STOL transport, the turbofan-powered Westwind business aircraft, the Sea Scan maritime surveillance version of the Westwind, and the Scout mini-RPV. In addition, it is engaged in the manufacture of a vast variety of spares and assemblies for aircraft and jet engines, to meet Israeli Air Force requirements. As a subcontractor to many US and European aircraft manufacturers, the Division produces major aircraft structures, flight control surfaces, cargo loading systems and spares.

The **Engineering Division**, the largest establishment of its kind in Israel, employs some 1,800 technical, scientific and other skilled personnel. It is responsible for engineering research, design, development and testing of aerospace systems. It provides engineering support in system analysis, aerodynamics, materials and processing, landing and control systems, and in structural, flight and environmental testing. The Division performed modification and production support for the manufacture of the Magister jet trainer for the IAF; and major structural conversions of the Boeing Model 377, for military applications such as swing-tail freighter and hose-refuelling tanker. The Division designed and developed the Arava STOL transport aircraft and developed both the 1123 and 1124 Westwind.

The Division is or has recently been engaged in a variety of programmes which include development of a supercritical wing for the Westwind, a fly-by-wire system for flight testing in the Kfir, and research into materials, structures and electronic countermeasures. It is now developing a new single-seat strike fighter known as the Lavi (Young Lion), intended to replace eventually the Israeli Air Force's A-4 Skyhawks and F-4 Phantoms.

The **Electronics Division** incorporates Elta Electronics Industries, MBT Weapons Systems, Tamam Precision Instruments, MLM Systems Engineering and Integration, and Magal Detection and Alert Systems. It specialises in the design, development and production of sophisticated electronic equipment such as airborne, ground and shipborne communications and radars, transceivers and navigational aids, general communications equipment, automatic test systems, and such electronic medical devices as cardiac resuscitation instruments. MBT participated in the development of the Division's Gabriel shipborne surface-to-surface missile system, among others, as well as of an electronic warning fence and an audible bomb release altimeter. Tamam manufactures and assembles high-precision electromechanical components and servosystems for such mechanisms as aerosystems, torque motors and gyroscopes.

Ramta Structures and Systems, SHL Servo-Hydraulics, Orlite Engineering, MATA Helicopters, Golan Industries, and PML Precision Mechanisms make up the **Combined Technologies Division.** This designs, develops and manufactures hydraulic and fuel system components, hydraulic flight control servo-systems, landing gears and brake systems; produces air-actuated chucks, miniature gears, clutches and brakes; is a custom-moulder of reinforced plastics, producing parts for aircraft, cabs and trucks, concrete casting moulds and sheet products; man-

IAI Kfir-C2 multi-mission combat aircraft, with typical weapon loads

ufactures ground support equipment, stainless steel tanks, the Dabur and Dvora patrol boats and the RBY armoured vehicle; manufactures high precision metal products for the aircraft and military industries; and produces electronic assemblies and subassemblies for aircraft.

Through its **Airborne Systems Marketing Group,** IAI is able to offer a number of services to foreign customers, based on the considerable capability of its five main divisions. Among these are Combat Aircraft Upgrading, a retrofit package that can include improved systems, engines, avionics, design configuration and structures. Another is the OFT (operational flight trainer), a simulator for procedure and cockpit training; weapon delivery and navigation training; and ILS, VOR and GCA approach and landing.

IAI LAVI (YOUNG LION)

Israeli government approval has been given for IAI to develop a new lightweight single-seat air defence and ground attack fighter for the Israeli Air Force. Now in the detail design stage, with a first flight planned for 1984, the Lavi will have a CCV configuration, with a delta main wing and canard surfaces, incorporating proven state-of-the-art technology. Approx 20 per cent of the structure, by weight, will be built of composite materials. Power plant will be a single Pratt & Whitney PW1120 afterburning turbojet engine in the 89 kN (20,000 lb st) class. Equipment will include Israeli-developed fly-by-wire control, fire control radar, ECM, and weapons delivery and navigation systems. The four prototypes will include one two-seater.

The Lavi is planned to enter service in 1986, replacing first the Israeli Air Force's A-4 Skyhawks, then the F-4 Phantom II, and eventually the Kfir-C2.

IAI KFIR (LION CUB)

Following manufacture of the Nesher fighter, powered by an Atar turbojet (see 1977-78 *Jane's*), IAI developed a more extensively modified and further improved version of the same airframe, powered by a General Electric J79 afterburning turbojet engine. One of these engines was installed in a much-modified Israeli Air Force Mirage III-B, with which it was first test-flown in September 1971, and production of the J79-engined Mirage began in 1972. A prototype of the Kfir was flown in 1973, before the outbreak of the October 'Yom Kippur' war, and details of this further-modified version were made public officially for the first time on 14 April 1975, when two of the new aircraft were displayed at Ben Gurion Airport.

The Kfir utilises a basic airframe similar to that of the Dassault Mirage 5, the main changes being a shorter but larger-diameter rear fuselage, to accommodate the J79 engine; an enlarged and flattened undersurface to the forward portion of the fuselage; introduction of four small fuselage airscoops, plus a larger dorsal airscoop in place of the triangular dorsal fin, to provide cooling air for the afterburner; and a strengthened landing gear, with longer-stroke oleos. Several internal changes have also been made, including a redesigned cockpit layout, addition of a considerable amount of Israeli-built avionics equipment, and increased revised fuel tankage compared with the Mirage 5. Intended for both air defence and ground attack roles, the Kfir retains the standard Mirage fixed armament of two 30 mm DEFA cannon, and can carry a variety of external weapons including the Rafael Shafrir 2 air-to-air and Luz-1 air-to-surface missiles. It has demonstrated stall-free gun firing throughout the flight envelope. Two squadrons of the Israeli Air Force were equipped with this initial Kfir-C1 version.

On 20 July 1976, at the Israeli Air Force base at Hatzerim, in the Negev, the first public demonstration took place of a modified version known as the **Kfir-C2**, which by that time was already in Israeli Air Force service, having entered production in 1974. The Kfir-C2 has a number of changes from the C1, the most significant of which is the addition of non-retractable, sweptback foreplanes just aft of the engine air intakes; a small strake on each side of the extreme nose; and extended wing leading-edges, created by increasing the chord on approximately the outer 40 per cent of each wing. The foreplanes can be detached for missions not requiring high manoeuvrability.

The Kfir-C2 is the principal production version, both for the Israeli Air Force and for export. The modifications, reportedly being retrofitted to Kfir-C1s in service, were designed to improve the aircraft's dogfighting manoeuvrability at the lower end of the speed range and to enhance take-off and landing performance. It is claimed that, in particular, they give a better sustained turning performance, with improved lateral, longitudinal and directional control; contribute to a very low gust response at all operational altitudes, especially at very low level; offer improved handling qualities at all angles of attack, high g loadings, and low speeds; reduce take-off and landing distances, and landing speeds; and permit a more stable (and, if required, a steeper) approach, with a flatter angle of approach and touchdown. Later versions of the C2 have Elta EL/M-2001B nose radar in an extended nose, increasing the overall length by 0·80 m (2 ft 7½ in).

A two-seat version, known as the **Kfir-TC2**, was flown for the first time in February 1981, and is now in production and service. Overall dimensions, power plant and performance are similar to those of the single-seat version, except for the insertion of a 0·84 m (2 ft 9 in) plug in the forward fuselage to accommodate a second cockpit in tandem. The nose is drooped in order to maintain a good field of view from both seats, and the second cockpit accommodates additional systems not present in the single-seat version.

Approximately 200 Kfirs, of all versions, are thought to have been built by the Spring of 1981, with production continuing.

The following description applies to the Kfir-C2:

TYPE: Single-seat interceptor, long range patrol fighter and ground attack aircraft.

WINGS: Cantilever low-wing monoplane of delta planform, with conical camber. Thickness/chord ratio 4·5% to 3·5%. Anhedral 1°. Incidence 1°. Sweepback on leading-edges 60° 35′. All-metal torsion-box structure, with stressed skin of machined panels with integral stiffeners. Two-section elevons on each trailing-edge, with smaller elevator/trim flap inboard of inner elevon. Elevons powered by hydraulic jacks; trim flaps are servo-assisted. Small, hinged plate-type airbrake above and below each wing, near leading-edge. Extended chord on outer leading-edges. Small leading-edge fence on some aircraft, at approx one-third span.

FOREPLANES: Detachable sweptback canard surface above and forward of each wing, near top lip of engine air intake.

FUSELAGE: All-metal semi-monocoque structure, 'waisted' in accordance with area rule. Cross-section of forward fuselage has a wider and flatter undersurface than that of Mirage 5. Nosecone built of locally-developed composite materials, with a small horizontal strake or 'body fence' on each side near the tip. UHF antenna under front of fuselage, forward of nosewheel door. Enlarged-diameter rear fuselage, compared with Mirage 5, with approx 0·61 m (2 ft) shorter tailpipe. Ventral fairing under rear fuselage.

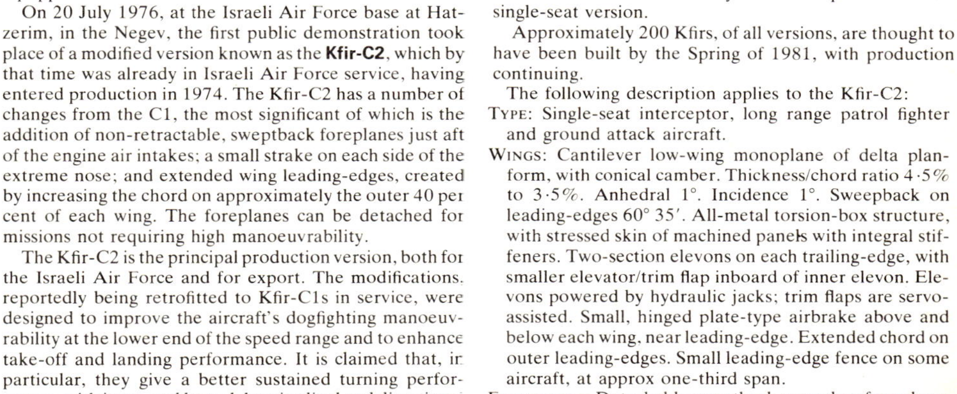

IAI Kfir-TC2 two-seater, with drooped and lengthened nose

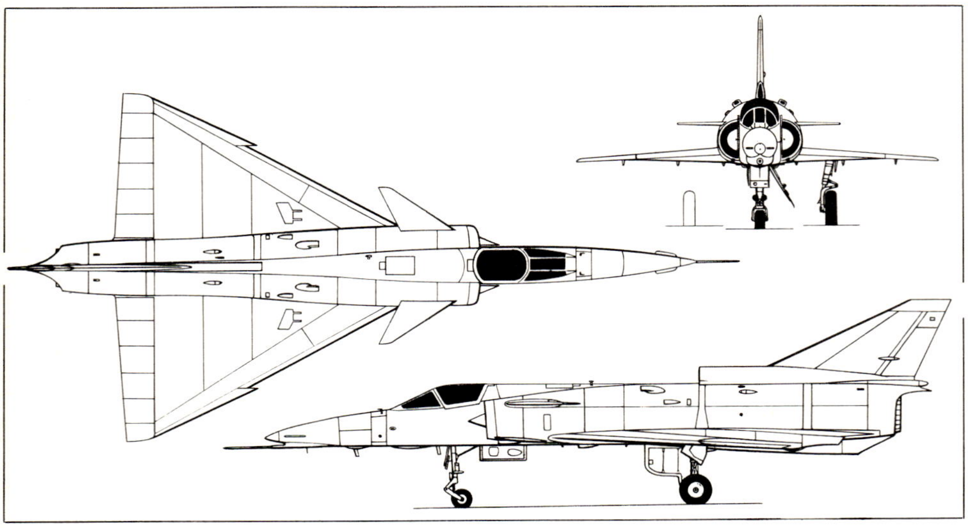

IAI Kfir-C2 (General Electric J79-J1E afterburning turbojet engine) *(Pilot Press)*

Tail Unit: Cantilever all-metal fin; rudder powered by hydraulic jack, with servo-assisted trim. UHF antenna in tip of fin. Triangular-section dorsal airscoop forward of fin, to provide cold air for afterburner cooling. No horizontal tail surfaces.

Landing Gear: Retractable tricycle type, with single SHL wheel on each unit. Electrically operated hydraulic actuation, nose unit retracting rearward, main units inward into fuselage. Longer-stroke oleos than on Mirage 5, and all units strengthened to permit higher operating weights. Low-pressure tubeless tyres on all units. Main-gear leg fairings shorter than on Mirage; inner portion of each main-leg door is integral with fuselage-mounted wheel door. Steerable nosewheel, with anti-shimmy damper. Oleo-pneumatic shock-absorbers, SHL hydraulic disc brakes and anti-skid units. Braking parachute in bullet fairing below rudder.

Power Plant: One General Electric J79-J1E turbojet engine (modified GE-17), with variable-area nozzle, rated at 52·9 kN (11,890 lb st) dry and 79·62 kN (17,900 lb st) with afterburning. Air intakes enlarged, compared with Mirage 5, to allow for higher mass flow. Adjustable half-cone centrebody in each air intake. Internal fuel in five fuselage and four integral wing tanks. Total internal capacity 3,243 litres (713 Imp gallons). There is a refuelling point on top of the fuselage, above the forward upper tank. In addition, there are wet-points for the carriage of one or two drop-tanks beneath each wing, and one under the fuselage; these tanks may be of 500, 600, 1,300 or 1,700 litres (110, 132, 286 or 374 Imp gallons) capacity. Max external fuel load 3,900 litres (858 Imp gallons).

Accommodation: Pilot only, on Martin-Baker IL10P zero-zero ejection seat, under rearward-hinged upward-opening canopy. Cockpit pressurised, heated and air-conditioned. A two-seat version is under development.

Systems: Two separate environmental control systems (ECS), one (using engine bleed air) for cockpit heating, pressurisation and air-conditioning, and one for avionics compartments. Two independent hydraulic systems, probably of 207 bars (3,000 lb/sq in) pressure. No. 1 system actuates flying control surfaces and landing gear; No. 2 actuates flying controls, airbrakes, landing gear, wheel brakes and utilities. Fully redundant primary AC electrical system, with two 15kVA (115V 400Hz) alternators, each driven by a CSD (constant-speed drive) unit, and a 750VA Oram static inverter connected for split-bus non-synchronised operation. DC system includes two Elta 200A 28V transformer-rectifiers and a 24V 40Ah nickel-cadmium battery. External AC and DC power receptacles. Oxygen system for pilot.

Avionics and Equipment: MBT twin-computer flight control system (ASW-41 control augmentation and ASW-42 stability augmentation systems), with Elta integrated memory unit (IMU), Tamam two-axis gyro and standby compass, autopilot, MBT radar altimeter, angle of attack transmitter and indicator, and accelerometer indicator. Elbit S-8600 multi-mode navigation (Singer-Kearfott licence) and Rafael Mahat weapon delivery system or IAI WDNS-141 weapon delivery and navigation system, Tamam central air data computer, Tacan, Elta EL/M-2001B or EL/M-2021 X-band air-to-air and air-to-surface pulse-Doppler target acquisition and tracking radar, IFF/SIF and fire control, Israel Electro-optics head-up display and automatic gunsight. Two Elta AN/ARC-51 UHF transceivers. Twin landing lights on nosewheel leg; anti-collision light in fin leading-edge.

Armament: Fixed armament of one IAI-built 30 mm DEFA 552 cannon in underside of each engine air intake (140 rds/gun). Seven hardpoints (three under fuselage and two under each wing) for up to 4,295 kg (9,468 lb) of external stores. For interception duties, one Rafael Shafrir 2 infra-red homing air-to-air missile can be carried under each outer wing. Ground attack version can carry two 1,000 lb bombs, four 500 lb bombs, or a Rafael Luz-1 or similar air-to-surface missile under the fuselage, and two 1,000 lb or six 500 lb bombs (conventional or 'concrete dibber' type) under the wings. Alternative external stores may include IMI rocket pods; napalm; Shrike, Maverick or Hobos air-to-surface missiles; ECM pods; or drop-tanks.

Dimensions, external:

Wing span	8·22 m (26 ft 11½ in)
Wing chord at root	8·04 m (26 ft 4½ in)
Wing aspect ratio	1·94
Foreplane span	3·73 m (12 ft 3 in)
Length overall, incl probe	15·65 m (51 ft 4¼ in)
Height overall	4·55 m (14 ft 11¼ in)
Wheel track	3·20 m (10 ft 6 in)
Wheelbase	4·87 m (15 ft 11¾ in)

Areas:

Wings, gross	34·8 m² (374·6 sq ft)
Foreplanes (total)	1·66 m² (17·87 sq ft)

Weights and Loadings:

Weight empty (interceptor, estimated)	7,285 kg (16,060 lb)
Max internal fuel	2,572 kg (5,670 lb)
Max external fuel	3,075 kg (6,780 lb)
Typical combat weight:	
interceptor, 50% internal fuel, two Shafrir missiles	9,390 kg (20,700 lb)
interceptor, two 500 litre drop-tanks, two Shafrir missiles	11,603 kg (25,580 lb)
combat air patrol, three 1,300 litre drop-tanks, two Shafrir missiles	14,270 kg (31,460 lb)
ground attack, two 1,300 litre drop-tanks, seven 500 lb bombs, two Shafrir missiles	14,670 kg (32,340 lb)
Max combat T-O weight	14,700 kg (32,408 lb)
Wing loading at 9,390 kg (20,700 lb) combat weight	270 kg/m² (55·3 lb/sq ft)
Thrust/weight ratio at 9,390 kg (20,700 lb) combat weight	0·87

Performance:

Max speed above 11,000 m (36,100 ft) over Mach 2·3	(1,317 knots; 2,440 km/h; 1,516 mph)
Max sustained level speed at height, 'clean'	Mach 2·0
Max level speed at S/L, 'clean'	750 knots (1,389 km/h; 863 mph)
Max rate of climb at S/L	14,000 m (45,950 ft)/min
Time to 15,250 m (50,000 ft), full internal fuel, two Shafrir missiles	5 min 10 s
Height attainable in zoom climb	22,860 m (75,000 ft)
Stabilised ceiling (combat configuration)	17,680 m (58,000 ft)
T-O run at max T-O weight	1,450 m (4,750 ft)
Landing from 15 m (50 ft) at 11,566 kg (25,500 lb) landing weight	1,555 m (5,100 ft)
Landing run at 11,566 kg (25,500 lb) landing weight	1,280 m (4,200 ft)
Combat radius, 20 min fuel reserves:	
interceptor, two 500 litre drop-tanks, two Shafrir missiles	187 nm (346 km; 215 miles)
combat air patrol, three 1,300 litre drop-tanks, two Shafrir missiles	377 nm (699 km; 434 miles)
ground attack, hi-lo-hi, seven 500 lb bombs, two Shafrir missiles, two 1,300 litre drop-tanks	415 nm (768 km; 477 miles)
g limit	+7

IAI 101, 102 and 201 ARAVA

Design of the Arava light STOL transport started in 1966, and construction of a prototype began towards the end of that year. This airframe was used for structural testing; it was followed by two flying prototypes, of which the first (4X-IAI) made its initial flight on 27 November 1969 and the second (4X-IAA) on 8 May 1971.

The Arava was first certificated as a civil aircraft, by the FAA in April 1972. This version, designated IAI 101, did not go into production, but formed the basis for the initial production IAI 102 (civil) and IAI 201 (military) transport versions.

The following models of the Arava are currently available:

IAI 101B. Updated version, with PT6A-36 engines and accommodation for 18 passengers or more than 1,814 kg (4,000 lb) of cargo with full fuel load. Improved cabin interior, and enhanced performance at higher ambient temperatures. Certificated by FAA under SFAR Pt 41 on 17 November 1980. Intended primarily for US commuter market; first customer was Key West Airlines; Airspur of Los Angeles ordered 10 of cargo version (marketed in USA as **Cargo Commuterliner**) in June 1981.

IAI 102. Initial production civil transport version, based on IAI 101; certificated by Israel Civil Aviation Administration in April 1976. Accommodation for 20 passengers in airline-standard four-abreast configuration, with toilet. Available also in a VIP configuration for up to 12 passengers, as an all-cargo transport, as a medical clinic for flying doctor services, and in versions for mapping, mining research, rainmaking and bridge construction, as flying laboratories for agriculture and health ministries, and for supplying oil prospecting units. Total of ten sold by Spring 1981, including six in Argentina and two in Africa. One of the Argentinian Aravas (LV-MRR) has been adapted in that country by CATA to a water bomber configuration, with a 209 kg (461 lb) self-transportable water tank installation loaded into the cabin via the clamshell rear doors. The installation can deliver approx 2·72 tonnes (3 US tons) of water in 4 s.

IAI 201. Military transport version, based on IAI 101. Prototype (4X-IAB) first flew on 7 March 1972. Standard equipment enables a wide variety of missions to be undertaken. Total of 60 sold by Spring 1981, including some equipped for maritime surveillance duties, fitted with either an AD-9 modification to extend the range and detection capability of the standard search/weather radar, or a more advanced detection system. Available also in electronic warfare configuration, with various pallet-mounted elint and ESM packages, ventral 'dustbin' radome, rearward-facing scanners mounted on the clamshell doors, 60 kVA APU for additional electrical power generation, and numerous blade and whip-type antennae above and below fuselage, on top of tailbooms, and elsewhere. This version is reportedly in service with the Israeli Air Force. Customers for the IAI 201 include the Israeli Air Force (14), Bolivian Air Force (5), Colombian Air Force (3), Ecuadorean Army (6) and Navy (3), Guatemalan Air Force (10), Honduran Air Force (3), Mexican Air

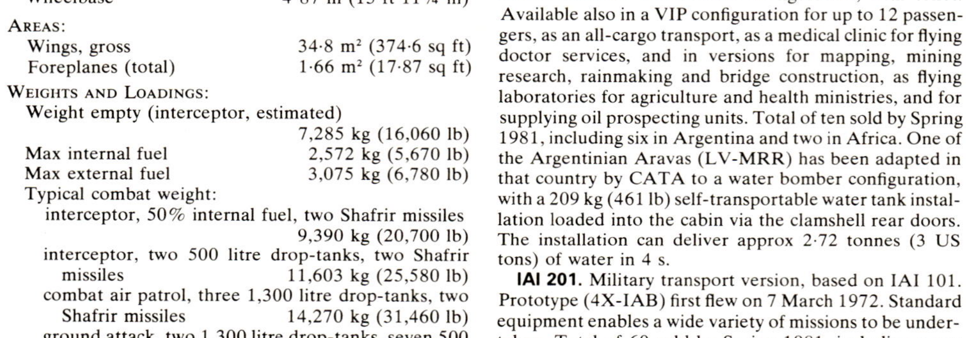

IAI 101B cargo and commuter version of the Arava twin-turboprop transport aircraft

Force (10), Nicaraguan Air Force (2) and Salvadorean Air Force (4).

IAI 202. Modified version, described in 1979-80 and earlier editions of *Jane's*. Differs principally in being longer, and in having a fully 'wet' wing with wingtip winglets, and a boundary layer fence just inboard of each tip. Powered by 559 kW (750 shp) PT6A-36 engines; single-point pressure refuelling system. The winglet modification (but not the increased fuel capacity) is available as a retrofit modification of existing Aravas. None sold or retrofitted by Spring 1981.

The following description applies to the IAI 201, except where indicated otherwise:

TYPE: Twin-turboprop STOL light military transport.

WINGS: Braced high-wing monoplane, with single streamline-section strut each side. Wing section NACA 63(215)A 417. Dihedral 1° 30'. Incidence 0° 27'. No sweepback. Light alloy two-spar torsion-box structure. Frise-type light alloy ailerons. Electrically-operated double-slotted light alloy flaps. Scoop-type light alloy spoilers, for lateral control, above wing at 71% chord. Electrically-actuated trim tab in port aileron.

FUSELAGE: Conventional semi-monocoque light alloy structure of stringers, frames and single-skin panels.

TAIL UNIT: Cantilever light alloy structure, with twin fins and rudders, carried on twin booms extending rearward from engine nacelles. Fixed-incidence tailplane. Geared tab and electrically-actuated trim tab in elevator and geared trim tab in each rudder. Tailbooms are built by IAI Combined Technologies Division.

LANDING GEAR: Non-retractable tricycle type, of Servo-Hydraulics Lod manufacture, with single main wheels and steerable nosewheel. Main wheels carried on twin struts, incorporating oleo-pneumatic shock-absorbers. Main wheels size 11·00-12, tyre pressure 3·31 bars (48 lb/sq in); nosewheel size 9·00-6, tyre pressure 2·90 bars (42 lb/sq in). Disc brakes on main units.

POWER PLANT: Two 559 kW (750 shp) Pratt & Whitney Aircraft of Canada PT6A-34 turboprop engines, each driving a Hartzell HC-B3TN three-blade hydraulically-actuated fully-feathering reversible-pitch metal propeller. (PT6A-36 engines of same T-O rating in IAI 101B and IAI 202.) Electrical de-icing of propellers optional. Two integral fuel tanks in each wing, with total usable capacity of 1,663 litres (366 Imp gallons). Four overwing refuelling points. Optional pressure refuelling point in fuselage/strut fairing. Two cabin-mounted tanks, each of 1,022 litres (225 Imp gallons), are available optionally for self-ferry flights.

ACCOMMODATION: Crew of one or two on flight deck, with door on starboard side. IAI 102 has airline-type seating for up to 20 passengers (18 in IAI 101B), plus toilet. IAI 201 has similar seating for 24 fully-equipped troops or 17 paratroops and a dispatcher. Outward-opening door at rear of cabin, opposite which, at floor level, is an emergency exit/baggage door on the starboard side. Rear doors are built by IAI Combined Technologies Division. Aft section of fuselage is hinged to swing sideways through more than 90° to provide unrestricted access to main cabin. Alternative interior configurations available for ambulance role (12 stretchers and two sitting patients or medical attendants); as all-freight transport carrying (typically) a Jeep-mounted recoilless rifle and its four-man crew; or as a maritime patrol aircraft fitted with search radar and other special equipment.

SYSTEMS: Hydraulic system (pressure 172 bars; 2,500 lb/sq in) for brakes and nosewheel steering only. Electrical system includes two 28V 170A DC engine-driven starter/generators, a 28V 40Ah nickel-cadmium battery, and two 250VA 115/26V 400Hz static inverters.

AVIONICS AND EQUIPMENT: Blind-flying instrumentation standard. Optional avionics include VHF, VOR/ILS, ADF, marker beacon receiver and PA system.

ARMAMENT: Optional 0·50 in Browning machine-gun pack on each side of fuselage, above a pylon for a pod containing six 82 mm rockets. Provision for aft-firing machine-gun. Librascope gunsight.

DIMENSIONS, EXTERNAL:
Wing span 20·96 m (68 ft 9 in)
Wing chord (constant) 2·09 m (6 ft 10½ in)
Wing aspect ratio 10
Length overall 13·03 m (42 ft 9 in)
Length of fuselage pod 9·33 m (30 ft 7 in)
Diameter of fuselage 2·50 m (8 ft 2 in)
Height overall 5·21 m (17 ft 1 in)
Tailplane span (c/l of tailbooms) 5·21 m (17 ft 1 in)
Wheel track 4·01 m (13 ft 2 in)
Wheelbase 4·62 m (15 ft 2 in)
Propeller diameter 2·59 m (8 ft 6 in)
Propeller ground clearance 1·75 m (5 ft 9 in)
Crew door (fwd, stbd): Height 0·93 m (3 ft 0½ in)
Width 0·48 m (1 ft 7 in)
Passenger door (rear, port):
Height 1·57 m (5 ft 2 in)
Width 0·62 m (2 ft 0½ in)
Airdrop opening, tailcone removed:
Height 1·75 m (5 ft 9 in)
Width 2·33 m (7 ft 8 in)

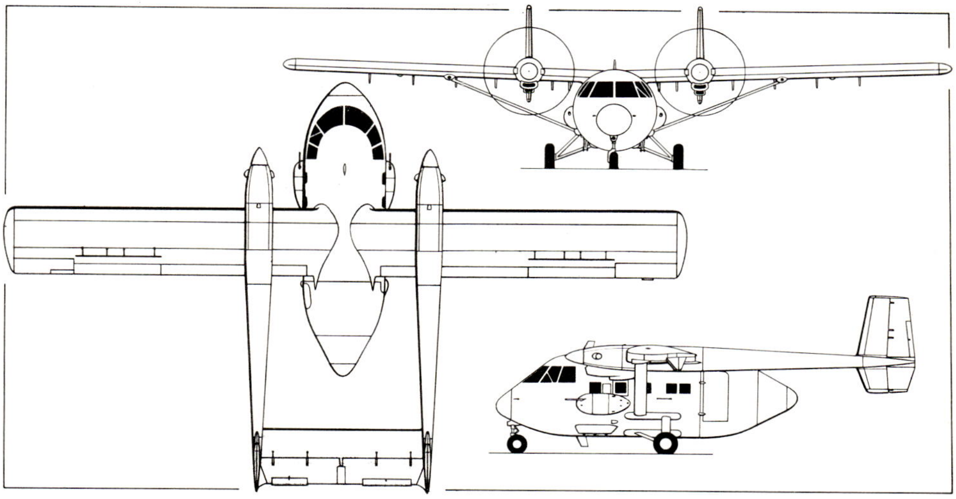

IAI 201 Arava twin-turboprop STOL light military transport *(Pilot Press)*

IAI 201 Arava military transport of the Israeli Air Force

Emergency/baggage door (rear, stbd):
Height 1·12 m (3 ft 8 in)
Width 0·61 m (2 ft 0 in)
Emergency window exits (each):
Height 0·66 m (2 ft 2 in)
Width 0·48 m (1 ft 7 in)

DIMENSIONS, INTERNAL:
Cabin, excl flight deck and hinged tailcone:
Length 3·87 m (12 ft 8 in)
Max width 2·33 m (7 ft 8 in)
Max height 1·75 m (5 ft 9 in)
Floor area 7·16 m² (77 sq ft)
Volume 12·7 m³ (449·2 cu ft)
Baggage compartment volume 2·60 m³ (91·8 cu ft)
Tailcone volume 3·20 m³ (113 cu ft)

AREAS:
Wings, gross 43·68 m² (470·2 sq ft)
Ailerons (total) 1·75 m² (18·84 sq ft)
Trailing-edge flaps (total) 8·80 m² (94·72 sq ft)
Spoilers (total) 0·85 m² (9·2 sq ft)
Fins (total) 4·86 m² (52·31 sq ft)
Rudders (total, incl tabs) 3·44 m² (37·03 sq ft)
Tailplane 9·36 m² (100·75 sq ft)
Elevator, incl tabs 2·79 m² (30·03 sq ft)

WEIGHTS AND LOADINGS:
Basic operating weight 3,999 kg (8,816 lb)
Max payload 2,351 kg (5,184 lb)
Max T-O and landing weight 6,803 kg (15,000 lb)
Max zero-fuel weight 6,350 kg (14,000 lb)
Max wing loading 153·5 kg/m² (31·44 lb/sq ft)
Max power loading 6·08 kg/kW (10·00 lbs/hp)

PERFORMANCE (at max T-O weight):
Never-exceed speed 215 knots (397 km/h; 247 mph)
Max level speed at 3,050 m (10,000 ft)
176 knots (326 km/h; 203 mph)
Max cruising speed at 3,050 m (10,000 ft)
172 knots (319 km/h; 198 mph)
Econ cruising speed at 3,050 m (10,000 ft)
168 knots (311 km/h; 193 mph)
Stalling speed, 0° flap 75 knots (140 km/h; 87 mph)
Stalling speed, 54° flap
62 knots (115 km/h; 71·5 mph)
Max rate of climb at S/L 393 m (1,290 ft)/min
Rate of climb at S/L, one engine out
55 m (180 ft)/min
Service ceiling 7,620 m (25,000 ft)
Service ceiling, one engine out 2,375 m (7,800 ft)
STOL T-O run 293 m (960 ft)
STOL T-O to 15 m (50 ft) 463 m (1,520 ft)
STOL landing from 15 m (50 ft) 469 m (1,540 ft)
STOL landing run 250 m (820 ft)
Range with max payload, 45 min reserves
151 nm (280 km; 174 miles)
Range with max fuel, 45 min reserves
705 nm (1,306 km; 812 miles)

IAI 1124/WESTWIND I

Israeli Navy designation: 1124 Sea Scan

The Westwind had its origins in the Jet Commander designed in the USA by Mr Ted Smith and flown for the first time on 27 January 1963. Production was transferred in 1968 to Israel Aircraft Industries, which has continued to develop and market successively improved versions.

A total of 186 early-model Jet Commander/Commodore Jet/1123 Westwind executive aircraft, with General Electric CJ610 turbojet engines, were built by Aero Commander in the USA (150) and IAI (36). Details of these have appeared in previous editions of *Jane's*. Aircraft from c/n 187 onwards have Garrett TFE731 turbofan engines and are designated as follows:

1124 Westwind. Initial turbofan-powered production version, announced in September 1974 and introduced in the following year. Prototypes were two converted 1123 Westwinds, the first of which made its first flight on 21 July 1975. FAA certification received in Spring 1976, with deliveries beginning shortly afterwards. Described in detail in 1978-79 *Jane's*. Total of 53 built (c/n 187-239), of which c/n 239 later became prototype for 1124A Westwind 2.

1124 Westwind I. Current basic turbofan-powered production version, announced in September 1978 and introduced from c/n 240 onwards. Improved version of 1124 Westwind, differing chiefly in having an optional 317 kg (700 lb) increase in fuel load, installed in a removable tank in the forward baggage compartment; an increase of approx 5 per cent in cabin useful volume, achieved by relocation of some avionics and by lowering the floor in the toilet compartment; RCA Primus 400 colour weather radar as standard; and improved fuel and environmental control systems. In production. Recent orders include one from Rhein-Flugzeugbau for four aircraft, equipped for

IAI 1124 Westwind I (two Garrett TFE731-3-1G turbofan engines)

IAI 1124N Sea Scan maritime reconnaissance version of the Westwind

target simulation on behalf of the West German armed forces; these were due for delivery in 1981.

1124N Sea Scan. Maritime version, announced in 1976. Prototype was converted from an aircraft (4X-CJA, c/n 154) which had served previously as prototype for the 1124 Westwind. Three specially-equipped 1123N Sea Scans, delivered to the Israeli Navy for coastal patrol, tactical support and anti-terrorist duties, have since been brought up to 1124N standard and are equipped with thrust reversers, single-point pressure refuelling, anti-corrosion protection, fuselage-side stores pylons, bubble windows, Litton APS-504(V)2 360° search radar, Global GNS-500A VLF/Omega navigation system, operators' consoles, galley, and toilet. A low-altitude search range of 1,379 nm (2,555 km; 1,588 miles), and search endurance of more than 6 h 30 min, enables the Sea Scan to cover a search area of 82,740 nm² (268,056 km²; 103,496 sq miles) along a 60 nm (111 km; 69 mile) search band at a height of 915 m (3,000 ft). Increased search range and endurance to 2,500 nm (4,633 km; 2,878 miles) and over 8 h can be attained at altitudes up to 13,720 m (45,000 ft). Available for specific operational requirements, with equipment to customer's specification.

Preliminary design and evaluation studies of a second-generation Sea Scan have been completed by IAI. In addition to its routine anti-terrorist low-level maritime patrol functions, the new Sea Scan multi-mission maritime patrol aircraft (MPA) can be deployed for ASW, signal intelligence (sigint) and anti-shipping air-to-surface missile attack operations. In the ASW role, search, detection, tracking, identification and attack are carried out using high-performance maritime search radar, ESM, sonobuoys, onboard signal analysis, colour multi-purpose displays (MPDs), trailing MAD, long-range gyro-stabilised sighting system (GSSS), and torpedos. Search, localisation and attack at 100 nm (185 km; 115 miles) from base can be performed for approx 5 h, enabling a landing back at base with 45 min reserve fuel. Replacing torpedos with Gabriel Mk III air-to-surface missiles, and removing some specific ASW mission equipment (sonobuoys, MAD etc), allows anti-shipping missile attacks to be made from a standoff range of 32 nm (60 km; 37 miles) at distances greater than 1,000 nm (1,853 km; 1,151 miles) from base. Comint, elint and IDF equipment installed in the aircraft permits long-range high-altitude sigint operations with an endurance of more than 8 h.

1124A Westwind 2. Developed version of Westwind I for improved range and economy of operation, announced in September 1979. Prototype (4X-CMK, c/n 239) flown for first time on 24 April 1979. Certificated by Israeli CAA on 11 December 1979 and by FAA on 17 April 1980. New modified 'Sigma' wing of IAI section, NASA-type winglets above tip-tanks, flat (instead of 'trenched') cabin floor, increased seated headroom, airline-type flushing toilet, relocated overhead passenger service units, and other improvements. First delivery (of the prototype, to Helicol of Colombia) made on 16 May 1980. In production. Total of 48 ordered by Spring 1981, of which 13 had been delivered. Described separately.

1125 Astra. Proposed version, under development for deliveries beginning in 1984, combining new 'Sigma 2' wing (mounted beneath cabin floor) with 10-passenger fuselage of Westwind 2 and TFE731 engines. Described separately.

Deliveries of turbofan-powered Westwinds (all versions) totalled approx 130 by early 1981. Westwind I/2 production was scheduled to be increased from four to five per month in the Autumn of 1981. The following description applies to the 1124 Westwind I:

TYPE: Twin-turbofan business transport.

WINGS: Cantilever mid-wing monoplane. Wing section NACA 64A212. Dihedral 2°. Incidence 1° at root, −1° at tip. Sweepback 4° 37′ at quarter-chord. Aluminium alloy flush-riveted two-spar fail-safe structure. Manually-operated all-metal ailerons. Electrically-operated all-metal double-slotted Fowler-type trailing-edge flaps. Drooped and cambered glassfibre-covered leading-edges. Electrically-operated trim tab in port aileron. Hydraulically-actuated speed brake and two lift dumpers above each wing, forward of flap. All skins chemically milled and fully sealed. All primary control surfaces, including aileron tab, are fully mass-balanced. Goodyear pneumatic de-icing boots standard. Permanently-attached wingtip fuel tanks.

FUSELAGE: All-metal semi-monocoque flush-riveted structure of aluminium alloy and steel sheet, with chemically milled skins. Built in two main sections and joined at aft pressure bulkhead. Forward section, except for nosecone, is fully pressurised and fail-safe.

TAIL UNIT: Cantilever all-metal structure, with 28° sweepback at tailplane quarter-chord and 35° sweepback at fin quarter-chord. Variable-incidence tailplane, actuated electrically. Manually-operated statically-balanced elevators and rudder. Electrically-operated trim tab in rudder. Goodyear pneumatic de-icing boots on tailplane leading-edges.

LANDING GEAR: Hydraulically-retractable tricycle type, main wheels retracting outward into wings, twin nosewheels rearward. No doors over main wheels when retracted. Oleo-pneumatic shock-absorbers. Single wheels on main units, pressure 10·69 bars (155 lb/sq in). Nose unit steerable and self-centering. Nosewheel tyre pressure 3·45 bars (50 lb/sq in). Goodyear multiple-disc brakes, with Hydro-Aire fully-modulated anti-skid system having automatic computer/sensor to prevent wheel lock and maintain brake effectiveness. Parking brake fitted.

POWER PLANT: Two 16·46 kN (3,700 lb st) Garrett TFE731-3-1G turbofan engines, with Grumman thrust reversers, pod-mounted on each side of rear fuselage. 85% of wing area forms an integral fuel tank, and additional fuel is carried separately in wingtip tanks and single rear fuselage tank. Total usable capacity 4,920 litres (1,082 Imp gallons; 1,300 US gallons), including wingtip tanks. Increased-weight option permits additional 317 kg (700 lb) of fuel (397 litres; 87 Imp gallons; 105 US gallons) to be carried in a removable tank in forward baggage compartment. Single-point pressure refuelling.

ACCOMMODATION: Standard seating for two pilots and up to 10 passengers in pressurised cabin. Interior layout to customer's requirements, with galley and toilet standard. Two separate heated compartments for up to 476 kg (1,050 lb) of baggage. Passenger door at front on port side; emergency exit on each side, forward of wing. Entire accommodation heated, ventilated and air-conditioned.

SYSTEMS: Garrett three-spool freewheeling turbine air-conditioning system: pressurisation differential 0·61 bars (8·8 lb/sq in) normal, 0·62 bars (9·0 lb/sq in) maximum. Primary hydraulic system, pressure 138 bars (2,000 lb/sq in), operates through two engine-driven pumps to actuate landing gear, wheel brakes, nosewheel steering, speed brakes, lift dumpers and thrust reversers. Electrically-operated emergency system, pressure 69 bars (1,000 lb/sq in), for brakes only. Pneumatic system, using engine bleed air, for wing and tailplane de-icing boots only. DC electrical system with two 350A 28V engine-driven starter/generators and two 28V 37Ah long-life nickel-cadmium batteries. One main bus for each generator, connected to the central battery bus. Two 1,000VA solid-state static inverters provide 115V AC power at 400Hz, each being independently capable of supplying the entire AC load if required. Engine air intakes anti-iced by engine bleed air. Oxygen system supplied by pressurised cylinder of 1·36 m³ (48 cu ft) capacity. Electrically-heated windscreen, pitot system and angle of attack sensor. Engine fire extinguishing system. No APU.

AVIONICS AND EQUIPMENT: Full dual IFR instrumentation standard, including Collins dual VHF-20A com, dual VIR-30A nav, dual DME-40 and ADF-60A. Other avionics include Collins NCS-31A radar navigation and control system (Global Navigation GNS-500A VLF on Sea Scan), RCA Primus 400 weather radar, and dual Sperry C-14 compass system. Collins FCS-105 flight control system (FD-109Z flight director and AP-105 autopilot). Canadian Marconi CMA-734 Omega navigation system approved for use in US and North Atlantic airspace.

DIMENSIONS, EXTERNAL:

Wing span: incl tip-tanks	13·65 m (44 ft 9½ in)
excl tip-tanks	13·16 m (43 ft 2 in)
Wing chord: at root	3·20 m (10 ft 6 in)
at tip	1·07 m (3 ft 6 in)
Wing aspect ratio	6·51
Length overall	15·93 m (52 ft 3 in)
Fuselage: Max width	1·57 m (5 ft 2 in)
Max depth	1·83 m (6 ft 0 in)
Height overall	4·81 m (15 ft 9½ in)

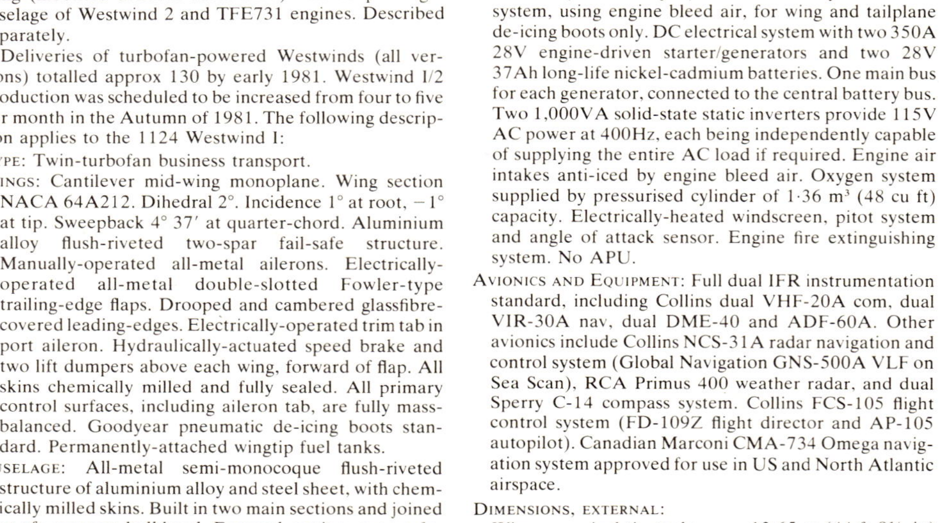

IAI 1124 Westwind twin-turbofan light executive transport *(Pilot Press)*

Tailplane span 6·40 m (21 ft 0 in)
Wheel track 3·35 m (11 ft 0 in)
Wheelbase 7·79 m (25 ft 6¾ in)
Passenger door: Height 1·32 m (4 ft 4 in)
Width 0·61 m (2 ft 0 in)
Height to sill 0·51 m (1 ft 8 in)
Baggage compartment door (main):
Height 0·61 m (2 ft 0 in)
Width 0·56 m (1 ft 10 in)
Height to sill 0·91 m (3 ft 0 in)
Baggage compartment door (rear):
Height 0·38 m (1 ft 3 in)
Width 0·51 m (1 ft 8 in)
Height to sill 1·27 m (4 ft 2 in)
Emergency exits (each):
Height 0·66 m (2 ft 2 in)
Width 0·51 m (1 ft 8 in)

Dimensions, internal:
Cabin, excl flight deck: Length 4·72 m (15 ft 6 in)
Max width 1·45 m (4 ft 9 in)
Max height 1·50 m (4 ft 11 in)
Floor area 6·52 m² (70·2 sq ft)
Volume 9·83 m³ (347 cu ft)
Baggage compartments: fwd (main) 1·13 m³ 40 cu ft)
rear 0·40 m³ (14 cu ft)
cabin 0·25 m³ (9 cu ft)

Areas:
Wings, gross 28·64 m² (308·26 sq ft)
Ailerons (total) 1·43 m² (15·40 sq ft)
Trailing-edge flaps (total) 3·86 m² (41·58 sq ft)
Fin 3·52 m² (37·94 sq ft)
Rudder, incl tab 0·99 m² (10·66 sq ft)
Tailplane 4·87 m² (52·42 sq ft)
Elevators 1·64 m² (17·66 sq ft)

Weights and Loadings:
Weight empty, equipped 5,578 kg (12,300 lb)
Typical basic operating weight (incl two pilots and service load) 5,760 kg (12,700 lb)
Max payload 1,496 kg (3,300 lb) or 1,542 kg (3,400 lb)
Max T-O weight 10,365 kg (22,850 lb) or 10,660 kg (23,500 lb)
Max ramp weight 10,430 kg (23,000 lb) or 10,725 kg (23,650 lb)
Max landing weight 8,620 kg (19,000 lb)
Max zero-fuel weight 7,255 kg (16,000 lb) or 7,485 kg (16,500 lb)
Max cabin floor loading 976 kg/m² (200 lb/sq ft)
Max wing loading 361·9 kg/m² (74·13 lb/sq ft) or 372·2 kg/m² (76·23 lb/sq ft)
Max power loading 314 kg/kN (3·09 lb/lb st) or 324 kg/kN (3·18 lb/lb st)

Performance (at max T-O weight of 10,365 kg; 22,850 lb, except where stated):
Max level speed, S/L to 5,900 m (19,400 ft) 471 knots (872 km/h; 542 mph)
Max operating speed, S/L to 5,900 m (19,400 ft) 360 knots (666 km/h; 414 mph) IAS
Max operating Mach No. from 5,900 m (19,400 ft) to 13,725 m (45,000 ft) Mach 0·765
Econ cruising speed at 12,500 m (41,000 ft) 400 knots (741 km/h; 460 mph)
Stalling speed, flaps and landing gear down, at max landing weight 99 knots (183 km/h; 114 mph) CAS
Max rate of climb at S/L 1,524 m (5,000 ft)/min
Max operating height 13,725 m (45,000 ft)
FAA T-O balanced field length 1,495 m (4,900 ft)
T-O balanced field length at 8,165 kg (18,000 lb) AUW 945 m (3,100 ft)
Landing distance from 15 m (50 ft) at max landing weight, with thrust reversal 625 m (2,050 ft)
Landing distance from 15 m (50 ft) at 6,350 kg (14,000 lb) AUW, with thrust reversal 518 m (1,700 ft)
Range with 7 passengers and baggage, IFR reserves more than 2,150 nm (3,983 km; 2,475 miles)
Max range with 2 passengers and baggage, 45 min reserves more than 2,600 nm (4,815 km; 2,993 miles)
Range with long-range fuel tank, 5 passengers and baggage, IFR reserves, at T-O weight of 10,660 kg (23,500 lb) 2,400 nm (4,446 km; 2,763 miles)
Range with long-range fuel tank, 2 passengers and baggage, 45 min reserves, at T-O weight of 10,660 kg (23,500 lb) 2,900 nm (5,373 km; 3,339 miles)

Operational Noise Levels (FAR Pt 36 at max T-O weight):
T-O 84·2 EPNdB
Approach 93·0 EPNdB
Sideline 88·4 EPNdB

IAI 1124A/WESTWIND 2

Wings: Modified Sigma 1 aerofoil section, developed by IAI. Dihedral 2° at 50 per cent chord. Incidence 1° at root. No drooped and cambered leading-edges. NASA-type winglet on upper surface of each tip-tank.

Fuselage and Tail Unit: As Westwind I.

Landing Gear: Similar to Westwind I, but with Goodyear main wheels (tyre size 16 × 4·4, pressure 9·86 bars; 143 lb/sq in) and nosewheels (tyre size 24 × 9·50-10·5, pressure 3·79 bars; 55 lb/sq in).

Power Plant: Engines as for Westwind I. Increased fuel capacity, with 2,089 litres (460 Imp gallons; 552 US gallons) in each main wing tank, 428 litres (94 Imp gallons; 113 US gallons) in each wingtip tank, and 379 litres (83 Imp gallons; 100 US gallons) in rear fuselage auxiliary tank; total usable capacity 5,413 litres (1,191 Imp gallons; 1,430 US gallons). Single-point pressure refuelling on starboard side of fuselage; gravity points in each wing upper surface, each tip-tank, and for fuselage auxiliary tank. Oil capacity 5·7 litres (1·25 Imp gallons; 1·5 US gallons) per engine.

New 'Sigma' wings and winglets identify IAI's Westwind 2

Accommodation: Standard seating for pilot, co-pilot and seven passengers, or up to a maximum of ten passengers, in pressurised and air-conditioned cabin. Elliptical cabin section increases seated headroom and allows a flat rather than 'trenched' cabin floor, an airline-type flushing toilet, and improved placing of the overhead passenger service units. Standard passenger layout comprises six individual tracked and swivelling seats, with two tables, plus a one-person divan. Fully-enclosed toilet compartment at rear of cabin on starboard side. Plug-type door, at front on port side, provides access to both cabin and flight deck. Pressurised baggage compartment in rear of cabin, adjacent to toilet; two heated but unpressurised main baggage compartments in rear of fuselage, each with separate external access on port side.

Avionics and Equipment: Standard avionics and equipment (all Collins except where indicated) include dual VHF-20A VHF com, dual VIR-30A VHF nav, IAI nav switching system, FCS-80 flight control system, FDS-85 flight director, APS-80 autopilot, ADS-80 air data system, FMS VLF/Omega, DME-40 DME, ADF-60A ADF, dual RMI-36 RMIs, dual TDR-90 transponders, ALT-50A radio altimeter, ALI-80A encoding altimeter (pilot), Kollsman B4420 digital altimeter (co-pilot), MSI-80C Mach/airspeed indicator (pilot), IDC Mach/airspeed indicator (co-pilot), VNI-80A vertical nav indicator (pilot), Teledyne SLZ-9706-DGLE vertical nav indicator (co-pilot), PRE-80A preselector/alerter, dual 346B-3 audio systems, RNS-300 radar navigation system, WXR-300 weather radar, Teledyne SLZ-9618-5 angle of attack system, dual Sperry C-14 compasses, HSI-84 co-pilot's HSI, Sperry GH-14B co-pilot's attitude gyro, J.E.T. A1-804 standby attitude gyro, and Davco 811-B digital clock. Landing light in nose of each wingtip tank. Optional avionics include dual VHF-20B (instead of VHF-20A) and single VHF-251 VHF com; HF-718U-5 and HF-220 HF com; Litton LTN-211, Collins LRN-85 or Global GNS-500A-2 VLF/Omega nav; second FDS-85, second ADC-80J for co-pilot's FDS-85, and comparator warning annunciator system; second ADF-60A, second DME-40, FPA-80 Flight Profile ADV, Collins TAI-80A SAT/TAS indicator, DRI-55 digital radio altimeter, ALT-55B (instead of ALT-50A) radio altimeter; Fairchild 5424-501 flight data recorder; Fairchild A-100 cockpit voice recorder; Kollsman ALT B4515 co-pilot's encoding altimeter (instead of B4420); Davco 811-B co-pilot's digital clock, Hobbs hour meter, Dorne and Margolin ELT-6 emergency locator transmitter, Devore Tel-Tail lights, Wulfsberg Flitefone III system, and ICD cabin display.

Dimensions, external: As Westwind I except:
Wing chord at tip 1·17 m (3 ft 10·232 in)

Dimensions, internal: As Westwind I except:
Cabin: Length, incl flight deck and toilet 6·08 m (19 ft 11¼ in)
Length, excl flight deck 4·74 m (15 ft 6½ in)

Areas: As Westwind I except:
Trailing-edge flaps (total) 3·85 m² (41·40 sq ft)
Speed brakes/lift dumpers (total) 1·37 m² (14·80 sq ft)
Fin 3·02 m² (32·52 sq ft)
Rudder, incl tab 1·02 m² (11·00 sq ft)
Tailplane 4·86 m² (52·28 sq ft)

Weights and Loadings:
Basic operating weight 6,010 kg (13,250 lb)
Max payload 1,474 kg (3,250 lb)
Max fuel load 4,345 kg (9,580 lb)
Max T-O weight 10,660 kg (23,500 lb)
Max ramp weight 10,725 kg (23,650 lb)
Max zero-fuel weight 7,485 kg (16,500 lb)
Max landing weight 8,620 kg (19,000 lb)
Max wing loading 372·0 kg/m² (76·23 lb/sq ft)
Max power loading 332·6 kg/kN (3·26 lb/lb st)

Performance (at max T-O weight except where indicated):
Never-exceed, max level and max cruising speed at 8,840 m (29,000 ft) 469 knots (868 km/h; 539 mph)
Econ cruising speed between 11,890 and 12,500 m (39,000-41,000 ft) 390 knots (723 km/h; 449 mph)
Stalling speed at max landing weight, flaps down, engines idling 99 knots (184 km/h; 114 mph) CAS
Max rate of climb at S/L 1,524 m (5,000 ft)/min
Rate of climb at S/L, one engine out 250 m (820 ft)/min
Max certificated ceiling 13,725 m (45,000 ft)
Service ceiling, one engine out:
at 9,072 kg (20,000 lb) gross weight 6,400 m (21,000 ft)
at 7,030 kg (15,500 lb) gross weight 9,450 m (31,000 ft)
Min ground turning radius 14·50 m (47 ft 7 in)
T-O run 1,218 m (3,995 ft)
T-O balanced field length 1,600 m (5,250 ft)
Landing from 15 m (50 ft) at max landing weight 747 m (2,450 ft)
Landing run at max landing weight 534 m (1,750 ft)
Range, NBAA VFR reserves:
with max payload (10 passengers) 2,390 nm (4,430 km; 2,750 miles)
with max fuel and 4 passengers 2,905 nm (5,385 km; 3,345 miles)

Operational Noise Levels (FAR Pt 36 at max T-O weight):
T-O 85·1 EPNdB
Approach 92·8 EPNdB
Sideline 88·5 EPNdB

IAI 1125 ASTRA

Known originally as the 1125 Westwind, the Astra is a more fuel-efficient, environmentally-acceptable development of the Westwind series, featuring also an improved standard of passenger comfort. First flight is expected to take place in the second half of 1983, with customer deliveries beginning in mid-1985. Certification will be to FAR Pts 25 and 36.

The major difference from earlier Westwind models is to be found in the wings, which have a new-design aerofoil section, are sweptback, and are mounted low on the fuselage. Whereas the Westwind's mid-mounted wings pass through the rear of the passenger cabin, those of the Astra pass beneath the cabin floor, so avoiding any interruption of the available internal space. This relocation of the wings results in a somewhat deeper fuselage profile, allowing 0·25 m (8 in) more cabin headroom than in the Westwind 2. The cabin is nearly 0·61 m (2 ft) longer than in the Westwind, but otherwise the fuselage is little changed structurally.

Design of the Sigma 2 wing section, a progressive improvement by IAI of the Sigma 1 (modified NASA 64A-212) section employed in the Westwind 2, has been computer-assisted and is intended to provide more efficient high-subsonic cruising flight over long ranges, with reduced operating costs. Construction of the Astra will make a wider use than the Westwind of composite materials, notably for the control surfaces.

The following description should be regarded as provisional:

Type: Twin-turbofan business transport.

Wings: Cantilever low-wing monoplane, with moderate sweepback (approx 30°) on leading-edges and outboard

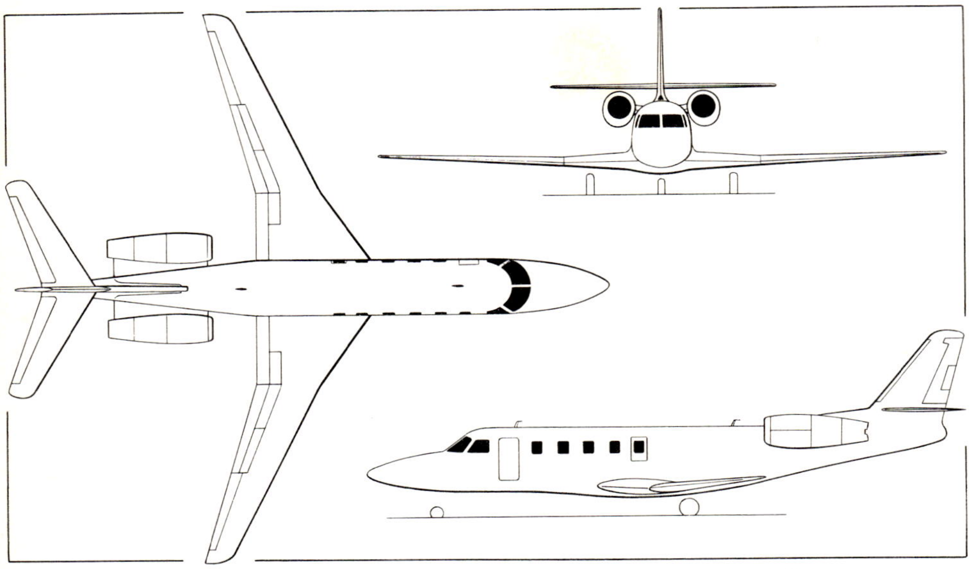

IAI 1125 Astra business transport (two Garrett TFE731-100 turbofan engines) *(Pilot Press)*

trailing-edges. Thin, high-efficiency Sigma 2 aerofoil section, of IAI design. Automatic leading-edge slats. Full-span trailing-edge flaps and ailerons. Spoilers forward of flaps. No winglets or tip-tanks.

FUSELAGE: Generally similar structurally to that of Westwind 2 except in wing carry-through area. Constant cross-section throughout passenger cabin.

TAIL UNIT: Generally similar to that of Westwind 2.

LANDING GEAR: Retractable tricycle type, of SHL (Servo Hydraulics) design.

POWER PLANT: Two approx 16·2 kN (3,650 lb st) Garrett TFE731-100 turbofan engines, with Grumman thrust reversers, pod-mounted in Grumman nacelle on each side of rear fuselage.

ACCOMMODATION: Crew of two on flight deck. Standard accommodation in pressurised cabin for seven persons on individual swivelling seats, with galley, coatrack and toilet at rear. Maximum accommodation for 10 persons. Baggage compartment aft of passenger cabin, with external access. Cabin soundproofing improved compared with Westwind 2.

SYSTEMS: Include Garrett environmental control system.

AVIONICS AND EQUIPMENT: Cathode ray tube (CRT) displays on flight deck. Area navigation system for operation in Category II weather conditions. Avionics contractor to be selected from Bendix, Collins, RCA and Sperry.

DIMENSIONS, EXTERNAL:

Wing span	15·80 m (51 ft 10 in)
Wing area, gross	29·40 m² (316·6 sq ft)
Wing aspect ratio	8·5
Length overall	16·60 m (54 ft 5½ in)

DIMENSIONS, INTERNAL:

Cabin, excl flight deck: Length	5·23 m (17 ft 2 in)
Max width	1·45 m (4 ft 9 in)
Max height	1·70 m (5 ft 7 in)
Baggage compartment volume	1·56 m³ (55 cu ft)

WEIGHTS:

Weight empty	5,900 kg (13,005 lb)
Max fuel load	4,300 kg (9,480 lb)
Max T-O weight	10,600 kg (23,370 lb)

PERFORMANCE (estimated):

Cruising speed	Mach 0·80
Balanced field length at max T-O weight	1,372 m (4,500 ft)
Cruising range, IFR reserves	more than 2,500 nm (4,635 km; 2,880 miles)
Range with 4 passengers and baggage, econ cruising speed, NBAA VFR reserves	3,000 nm (5,560 km; 3,455 miles)

ITALY

AERITALIA

AERITALIA—SOCIETÀ AEROSPAZIALE ITALIANA p.a.

HEAD OFFICE: Piazzale Vincenzo Tecchio 51 (Casella Postale 3065), 80125 Naples
Telephone: (081) 619522, 619721, 619845, 619149, 619703
Telex: N. 710370 (AERIT)
OFFICE OF THE CHAIRMAN, AND BRANCH OFFICE (COMMERCIAL): Via Panama 52, 00198 Rome
Telephone: (06) 841441
Telex: N. 611395 (AERIT)
HONORARY PRESIDENT: Dott Egidio Ortona
PRESIDENT AND CHAIRMAN OF THE BOARD:
Ing Renato Bonifacio
VICE-PRESIDENT: Ing Luigi D'Agostini
GENERAL MANAGER: Ing Fausto Cereti
JOINT GENERAL MANAGER: Ing Giovanni Sarzotti
BOARD OF DIRECTORS:
Ing Fausto Cereti
Ing Giulio Ciampolini
Ing Luigi D'Agostini
Ing Corrado Innocenti
Ing Francesco La Via
Dott Roberto Lenci
Dott Giorgio Massone
Amb Egidio Ortona
Dott Francesco Palma
Ing Amilcare Porro
Ing Beppe Sacchi
Ing Franco Schepis
EXECUTIVE DIRECTORS:
Ing Stefano Abbà (Space and Alternative Energy Group)
Ing Roberto Mannu (Commercial)
Ing Riccardo Mautino (Strategic Development)
Dott Massimo Rizzo (General Secretary)
PUBLIC RELATIONS AND PRESS DIRECTOR: Cesare Falessi

Aeritalia is a joint stock company which was formed on 12 November 1969 by an equal shareholding of Fiat and IRI-Finmeccanica, to combine Fiat's aerospace activities (except those which concerned aero-engines) with those of Aerfer and Salmoiraghi of the Finmeccanica group. The company became fully operational under the new title on 1 January 1972. On 28 September 1976 IRI-Finmeccanica purchased the Aeritalia stock owned by Fiat, thus acquiring complete control of the company's stock capital. Aeritalia has a total work force of approx 12,000.

Aeritalia's organisation is based upon a centralised general management and four operational groups: Combat Aircraft Group, Transport Aircraft Group, Equipment Group, and Space and Alternative Energy Group. The production centres are located in Turin (Corso Marche, Caselle Nord and Caselle Sud), Milan (Nerviano) and Naples (Pomigliano d'Arco, Casoria and Capodichino).

Following an inter-company agreement on 2 June 1978, Aeritalia and Aeronautica Macchi (which see) are continuing, with EMBRAER of Brazil, the definition phase of the AM-X strike fighter, under an Italian Defence Ministry contract. Details of the AM-X programme can be found in the International section, and in the Addenda. In August 1978 Aeritalia and Boeing of the USA signed a formal agreement under which Aeritalia is responsible for designing and manufacturing components for the Boeing 767 airframe, including wing spoilers, outboard ailerons and trailing-edge flaps, and the fin and rudder. The first set of these items was delivered to Boeing by March 1981 for final assembly.

Aeritalia is also a 50-50 partner with Aérospatiale of France in the ATR 42 regional transport aircraft programme, described in the International section.

In August 1981 it was reported that Aeritalia was to take over Partenavia (which see) in the proposed reorganisation of the Italian aerospace industry.

COMBAT AIRCRAFT GROUP

HEADQUARTERS AND TURIN WORKS: Corso Marche 41, 10146 Turin
Telephone: (011) 33321
Telex: N. 221076 (AERITOR)
CASELLE WORKS: Turin Airport, 10100 Turin
Telephone: (011) 991362
Telex: 210086

The Turin area factories are engaged in the design of the AM-X combat aircraft; design and manufacture of structural components, final assembly and flight testing of the Panavia Tornado (see International section); and repair, overhaul and maintenance of test equipment. Other activities include the repair, overhaul and maintenance of F-104G, F-104S and TF-104G aircraft.

TRANSPORT AIRCRAFT GROUP

HEADQUARTERS AND NAPLES AREA WORKS: 80038 Pomigliano d'Arco, Naples
Telephone: (081) 84511
Telex: N. 710082 and 710522 (AERITPOM)
CAPODICHINO WORKS: Via del Riposo alla Doganella, Aeroporto di Capodichino, 80144 Naples
Telephone: (081) 444166
Telex: N. 710356

Aeritalia's principal activities in the Naples area comprise construction of the complete series of fuselage structural panels for the McDonnell Douglas DC-9; fuselage upper panels and the vertical tail surfaces for the DC-10 commercial airliner; engine support pylons for the Boeing 747; numerous components for the Boeing 767 (see earlier paragraph in this introduction); outboard ailerons, trailing-edge flaps, spoilers and fins, and manufacture, assembly and flight testing, of the Aeritalia G222. Other activities include the repair, overhaul and modification of aircraft of various nations, including Italy and the United States, and the repair and maintenance of Breguet 1150 Atlantic, and G91R, G91T and G91Y aircraft.

EQUIPMENT GROUP

CASELLE WORKS: 10072 Caselle-Turin
Telephone: (011) 991362
Telex: N. 210411 and 210095

Since its creation within the Fiat Aircraft Division, the activities of this Group have been extended to include research and experimentation with equipment installed in new prototypes, and the series production of electronic equipment and systems, both of original design and under licence. The Group has also taken an important part in Italian and international space programmes. Avionics include complex electronics systems and specific equipment for aerospace applications. Design work includes aerospace systems and new systems for other civil purposes. Activities include the manufacture, repair and overhaul of sophisticated electronics and space equipment, for aircraft and space components of Aeritalia's own production and for the international market.

NERVIANO WORKS: Viale Europa, 20014 Nerviano, Milan
Telephone: (0331) 587330
Telex: 330675

The Nerviano works manufactures a wide range of avionics equipment and instrumentation, including systems and instruments for aeronautical, missile and space applications. It is scheduled to become a part of Agusta SpA in the planned reorganisation of the Italian aerospace industry.

SPACE AND ALTERNATIVE ENERGY GROUP

See Spaceflight section.

AERITALIA (LOCKHEED) F-104S

Details of Lockheed production of earlier models of the F-104 Starfighter can be found in the 1972-73 and previous editions of *Jane's*.

The first of two Lockheed-built F-104S prototypes flew during December 1966. Aeritalia then built 246 production examples under licence, including 40 for the Turkish Air Force, before ending production in March 1979. Deliveries to the Italian Air Force began in the Spring of 1969, and these aircraft are in service with the 4°, 5°, 6°, 9°, 36°, 51° and 53° Wings.

In 1980 Aeritalia began work on a feasibility study aimed at increasing substantially the weapon system of the F-104S, and also the aircraft's operational capability in the interception and interdiction/strike roles. Go-ahead for this programme was expected during 1981. If approved, it will offer:

(1) a high-precision navigation system capable of pin-

point navigation and target acquisition in all weathers and at low altitudes;

(2) high-precision weapon delivery at low level, in all conditions of weather and visibility;

(3) lookdown capability against low-flying targets; and

(4) advanced defensive electronic aids and air-to-air self-defence capability.

The following description applies to the standard production F-104S without these modifications. A full structural description can be found in the 1980-81 *Jane's*.

TYPE: Single-seat multi-purpose combat aircraft.

POWER PLANT: One General Electric J79-GE-19 turbojet engine, rated at 52·8 kN (11,870 lb st) dry and 79·62 kN (17,900 lb st) with afterburning. Electrical de-icing elements fitted to air intakes. Internal fuel in five bag-type fuselage tanks with total standard capacity of 3,392 litres (896 US gallons). Provision for external fuel in two 740 litre (195 US gallon) pylon tanks and two 645 litre (170 US gallon) wingtip tanks. Pressure refuelling of all internal and external tanks through single point on upper port fuselage just forward of air intake duct. Gravity fuelling point for internal tanks aft of pressure refuelling point, with individual gravity fuelling of external tanks. In-flight refuelling can be provided through Lockheed-designed probe-drogue system. Probe, mounted below port sill of cockpit, is removable but when installed is non-retractable. Oil capacity 15 litres (4 US gallons).

ACCOMMODATION: Pressurised and air-conditioned cockpit well forward of wings. Canopy hinged to starboard for access. Martin-Baker IQ7A zero-zero ejection seat.

AVIONICS AND EQUIPMENT: Integrated avionics system in which various communications and navigation components may be installed as a series of interconnecting but self-sustaining units which may be varied to provide for different specific missions. Equipment includes autopilot with 'stick steering', which includes modes for pre-selecting and holding altitude, speed, heading and constant rate of turn; multi-purpose R21G/H radar for air-to-air interception, ground and contour mapping, and terrain avoidance modes of operation; fixed-reticle gunsight; bombing computer; air data computer; dead reckoning navigation device; Tacan radio air navigation system; provision for data link-time division set and AN/ARC-552 UHF radio; Litton LN-3-2A lightweight fully-automatic inertial navigation system; Sperry C-2G compass system; AN/APN-198 radar altimeter; AIC-18 intercom; and AN/APX-46 IFF/SIF. Provision for fitting a camera pod under the fuselage for reconnaissance duties. (Avionics and equipment of Italian Air Force F-104S to be upgraded: details still classified at time of closing for press.)

ARMAMENT: Nine external attachment points, at wingtips, under wings and under fuselage, for bombs, rocket pods, auxiliary fuel tanks and air-to-air missiles. Normal primary armament consists of two AIM-7 Sparrow air-to-air missiles under wings and/or two Sidewinders under fuselage and either a Sidewinder or 645 litre (170 US gallon) fuel tank on each wingtip. Alternatively, an M61 20 mm multi-barrel rotary cannon can be fitted in the port underside of the fuselage instead of the AIM-7 missile control package. Max external weapon load 3,402 kg (7,500 lb).

DIMENSIONS, EXTERNAL:

Wing span without tip-tanks	6·68 m (21 ft 11 in)
Wing chord (mean)	2·91 m (9 ft 6·6 in)
Wing area, gross	18·22 m² (196·1 sq ft)
Wing aspect ratio	2·45
Length overall	16·69 m (54 ft 9 in)
Length of fuselage	15·62 m (51 ft 3 in)
Height overall	4·11 m (13 ft 6 in)
Tailplane span	3·63 m (11 ft 11 in)
Wheel track	2·74 m (9 ft 0 in)
Wheelbase	4·59 m (15 ft 0½ in)

WEIGHTS AND LOADINGS:

Weight empty	6,760 kg (14,900 lb)
Max internal fuel load	2,641 kg (5,824 lb)
Max internal and external fuel load	5,153 kg (11,362 lb)
T-O weight ('clean')	9,840 kg (21,690 lb)
Max T-O weight	14,060 kg (31,000 lb)
Max zero-fuel weight ('clean')	6,806 kg (15,006 lb)
Max zero-fuel weight (fighter-bomber)	7,148 kg (15,760 lb)
Max wing loading	772 kg/m² (158 lb/sq ft)
Max power loading, with afterburning	176·6 kg/kN (1·73 lb/lb st)

PERFORMANCE (at 9,840 kg; 21,690 lb AUW except where indicated):

Never-exceed speed	Mach 2·2
Max level speed at 11,000 m (36,000 ft)	Mach 2·2 (1,259 knots; 2,330 km/h; 1,450 mph)
Max level speed at S/L	Mach 1·2 (790 knots; 1,464 km/h; 910 mph)
Max cruising speed at 11,000 m (36,000 ft)	530 knots (981 km/h; 610 mph)
Econ cruising speed	Mach 0·85
T-O speed at S/L, interceptor with two AIM-7 missiles	189 knots (350 km/h; 217 mph)
Typical landing speed at S/L	159 knots (295 km/h; 183 mph)
Max rate of climb at S/L	16,765 m (55,000 ft)/min
Service ceiling	17,680 m (58,000 ft)
Zoom altitude	more than 27,400 m (90,000 ft)
Time to accelerate from Mach 0·92 to Mach 2·0	2 min
Time to climb to 10,670 m (35,000 ft)	1 min 20 s
Time to climb to 17,070 m (56,000 ft)	2 min 40 s
T-O run at S/L, interceptor with two AIM-7 missiles	823 m (2,700 ft)
Typical landing run at S/L	762 m (2,500 ft)
Radius with max fuel	673 nm (1,247 km; 775 miles)
Ferry range (excl flight refuelling)	1,576 nm (2,920 km; 1,815 miles)

Aeritalia-built Lockheed F-104S combat aircraft in the insignia of the Italian Air Force

AERITALIA G222

The G222 was originally conceived in four separate configurations, three of which were halted at the research project stage. Two prototypes were built of the military transport version, of which the first (MM582) made its initial flight on 18 July 1970 and the second (MM583) on 22 July 1971. The first prototype was handed over to the Italian Air Force on 21 December 1971 for operational evaluation. One airframe was completed for static and fatigue testing.

The Italian Air Force has ordered 44 G222s, the first of which flew on 23 December 1975. Deliveries began in late 1976, and 30 had been delivered by the beginning of May 1981. These are in service with the 46th Brigade at Pisa, and are operated in the primary roles of troop, paratroop or cargo transport, or for aeromedical duties. The G222 can operate from semi-prepared airstrips and in all weathers.

The Argentinian Army has three G222s, and in November 1976 one was delivered to the Dubai government, which has an option on a second. The Somali Aeronautical Corps has ordered four, of which two had been delivered by May 1981.

Several major Italian airframe companies share in the construction programme, including Aermacchi (outer wings); Piaggio (wing centre-section); SIAI-Marchetti (tail unit); CIRSEA (landing gear); and IAM (miscellaneous airframe components). Fuselages are built by Aeritalia's Transport Aircraft Group, in the Pomigliano d'Arco Works near Naples; final assembly takes place at Capodichino Airport, Naples.

Design of the G222 makes it suitable for adaptation to such other roles as maritime patrol, VIP transport, navigation training, electronic warfare, and for such civil applications as firefighting, crop-spraying, aerial photogrammetry, and radio calibration. Firefighting tests with the second prototype (I-MAXB), designated **G222 SAMA** (Sistema Aeronautico Modulare Antincendio) and equipped with a dispersal system designed by Food Machinery Corporation, were completed successfully in 1976. Electronic warfare, navaid calibration (**G222RM**) and maritime surveillance versions are described separately, as is the G222T version powered by Rolls-Royce Tyne turboprop engines.

The following description applies to the standard G222 military transport version currently in production, except where indicated:

TYPE: Twin-turboprop general-purpose transport aircraft.

WINGS: Cantilever high-wing monoplane, with thickness/chord ratio of 15%. Dihedral 2° 30′ on outer panels. Aluminium alloy three-spar fail-safe box structure, built in three portions. One-piece constant-chord centre-section fits into recess in top of fuselage and is secured by bolts at six main points. Outer panels tapered on leading- and trailing-edges. Upper-surface skins are of 7075-T6 alloy, lower-surface skins of 2024-T3 alloy. All control surfaces have bonded metal skins with metal honeycomb core. Double-slotted flaps extend over 60% of trailing-edge. Two-section spoilers ahead of each outboard flap segment, used also as lift dumpers on landing. Spoilers and flaps fully powered by tandem hydraulic actuators. Manually-operated ailerons, each with inset servo tab. Pneumatically-inflated de-icing boots on leading-edges, using engine bleed air.

FUSELAGE: Pressurised fail-safe structure of aluminium alloy stressed-skin construction and circular cross-section. Easily removable stiffened floor panels.

TAIL UNIT: Cantilever safe-life structure of aluminium alloy, with sweptback three-spar fin and non-swept two-spar variable-incidence tailplane. Pneumatically-inflated de-icing boots on fin and tailplane leading-edges, using engine bleed air. Rudder and elevators of metal honeycomb construction. Two tabs in each elevator; no rudder tabs. Rudder fully powered by tandem hydraulic actuators; elevators operated manually.

LANDING GEAR: Hydraulically-retractable tricycle type, suitable for use from prepared runways or grass fields. Messier-Hispano-Bugatti design, built under licence by CIRSEA (Nardi-Magnaghi). Steerable twin-wheel nose unit retracts forward. Main units, each consisting of two single wheels in tandem, retract into fairings on sides of fuselage. Oleo-pneumatic shock-absorbers. Gear can be lowered by gravity in emergency, the nose unit being aided by aerodynamic action and the main units by the shock-absorbers, which remain compressed in the retracted position. Oleo pressure in shock-absorbers is adjustable to permit variation in height and attitude of the cabin floor from the ground. Low-pressure tubeless tyres on all units, size 37·91 × 12·35 in (Type III) on main wheels, 27·56 × 10·51 in (Type III) on nosewheels. Tyre pressures 4·41 bars (64 lb/sq in) on main units, 3·92 bars (56·88 lb/sq in) on nose unit. Hydraulic multi-disc brakes. No anti-skid units.

POWER PLANT: Two 2,535 kW (3,400 shp) Fiat-built General Electric T64-GE-P4D turboprop engines, each driving a Hamilton Standard 63E60 three-blade variable-pitch propeller with spinner. Fuel in integral tanks: two in the outer wings, combined capacity 6,800 litres (1,495 Imp gallons), and two centre-section tanks, combined capacity 5,200 litres (1,143 Imp gallons), with cross-feed provision to either engine. Total overall fuel capacity 12,000 litres (2,638 Imp gallons). Single pressure refuelling point on starboard side of fuselage. Overwing gravity refuelling point above each tank.

ACCOMMODATION: Normal crew of three (two pilots and radio operator/flight engineer) on flight deck. Provision

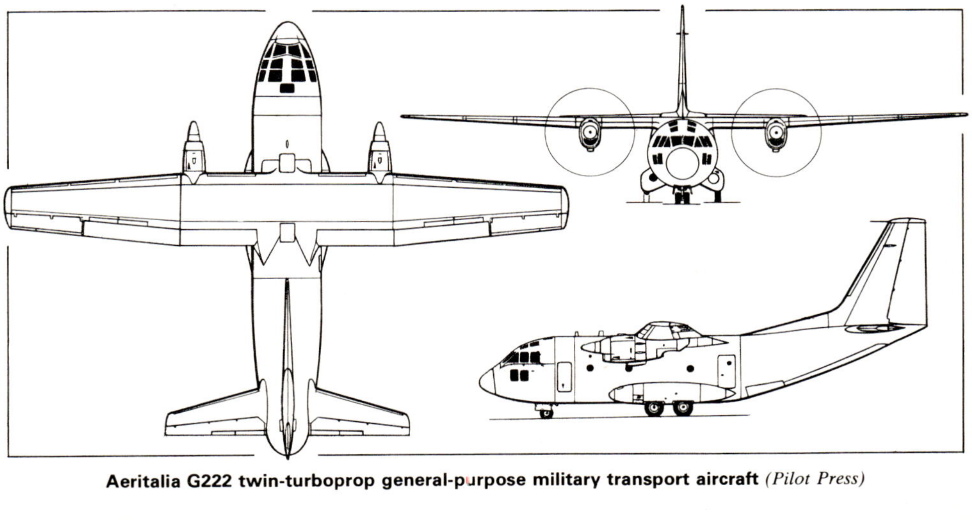

Aeritalia G222 twin-turboprop general-purpose military transport aircraft *(Pilot Press)*

Aeritalia G222 twin-turboprop military transport aircraft in the insignia of the Italian Air Force

for fourth crew member or jumpmaster when required. Standard troop transport version has 32 foldaway sidewall seats and 21 stowable seats for 53 fully-equipped troops, and carries also two 20-man life rafts stowed in the wing/fuselage fairing and a single 9-man life raft in the cargo compartment. Paratroop transport version can carry up to 42 fully-equipped paratroops, and is fitted with the 32 sidewall seats and life rafts as in the troop transport version, plus 10 stowable seats, door jump platforms and static lines. Cargo transport version can accept standard pallets of up to 2·24 m (7 ft 4 in) wide, and can carry up to 9,000 kg (19,841 lb) of freight. Provision is made for 135 cargo tiedown points and a 1,500 kg (3,306 lb) capacity cargo hoist. Typical Italian military equipment loads can include two CL-52 light trucks; one CL-52 with a 105 mm L4 howitzer or one-ton trailer; Fiat AR-59 Campagnola reconnaissance vehicle with 106 mm recoilless gun or 250 kg (550 lb) trailer; or five standard A-22 freight containers. In the aeromedical role the G222 can accommodate 36 stretchers, two sitting patients and four medical attendants. A second toilet can be installed, and provision can be made to increase the water supply and to install electrical points and hooks for medical treatment bottles. In this version, the cabin oxygen system is available to all stretcher positions. Crew access door is forward of cabin on port side. Passenger doors, at front and rear of main cabin on starboard side and at rear on port side, can be used also as emergency exits. Two emergency hatches in cabin roof, forward and aft of wing carry-through structure. Hydraulically operated rear loading ramp and upward-opening door in underside of upswept rear fuselage, which can be opened in flight for airdrop operations. In cargo version, five loads of up to 1,000 kg (2,205 lb) each can be airdropped from rear opening, or a single load of up to 5,000 kg (11,023 lb). Paratroop jumps can be made either from this opening or from the rear side doors. Windscreens and quarter-light panels are de-iced and demisted electrically. Wipers and screen wash for both windscreens. Entire accommodation pressurised and air-conditioned.

Systems: Pressurisation system maintains a cabin differential of 0·41 bars (5·97 lb/sq in), giving a 1,200 m (3,940 ft) environment at altitudes up to 6,000 m (19,685 ft). Air-conditioning system uses engine bleed air during flight; on ground, it is fed by compressor bleed air from APU to provide cabin heating to a minimum of 18°C. Two independent hydraulic systems, each of 207 bars (3,000 lb/sq in) pressure. No. 1 system actuates flaps, spoilers, rudder and wheel brakes; No. 2 system actuates flaps, spoilers, rudder, wheel brakes, nosewheel steering, landing gear extension and retraction, rear ramp/door and windscreen wipers. Auxiliary hydraulic system, fed by APU-powered pump, can take over from No. 2 system in flight, if both main systems fail, to operate essential services. In addition, a standby handpump is provided for emergency use to lower the landing gear and, on the ground, to operate the ramp/door and parking brakes. Three 45kVA alternators, one driven by each engine through constant-speed drive units and one by the APU, provide 115/200V three-phase AC electrical power at 400Hz. 28V DC power is supplied from the main AC buses via two transformer-rectifiers, with 24V 34Ah nickel-cadmium battery and static inverter for standby and emergency power. External AC power socket. Garrett 113·3 kW (152 hp) APU, installed in starboard main landing gear fairing, provides power for engine starting, hydraulic pump and alternator actuation, air-conditioning on ground, and all hydraulic and electrical systems necessary for loading and unloading on ground. Electrical de-icing of spinners and propeller leading-edges. Engine intakes anti-iced by electrical/hot air system. Liquid oxygen system for crew and passengers (with cabin wall outlets); this system can be replaced by a gaseous oxygen system if required. Emergency oxygen system available for all occupants in the event of a pressurisation failure.

Avionics and Equipment: Standard communications equipment includes 3,500-channel UHF, two 1,360-channel VHF/AM, 920-channel VHF/FM, 28,000-channel HF/SSB, crew intercom and PA system. Navigation equipment includes Omega system, with TAS computer, autopilot, flight director, two compasses, and two vertical gyros; and an integrated ground-based system incorporating two VOR, marker beacon receiver, two ILS, ADF, two Tacan or DME, and horizontal situation indicator. Other avionics include Meteo weather radar, with secondary terrain-mapping mode; radar altimeter; and IFF/ATC transponder including altitude reporting. Provision for head-up display. Landing light on nosewheel leg.

Equipment (G222 SAMA): Modular palletised firefighting pack can be installed in under two hours without any modification to the basic transport aircraft. The module consists of a 6,300 litre (1,385 Imp gallon) tank and four pressurised air containers to activate the pneumatic actuators and discharge the retardant/fertiliser through two nozzles. Length of area covered averages 300 m (985 ft).

Dimensions, external:
Wing span 28·70 m (94 ft 2 in)
Wing chord at root 3·40 m (11 ft 1¾ in)
Wing chord at tip 1·685 m (5 ft 6¼ in)
Wing aspect ratio 9·15
Length overall 22·70 m (74 ft 5½ in)
Height overall 9·80 m (32 ft 1¾ in)
Fuselage: Max diameter 3·55 m (11 ft 7¾ in)
Tailplane span 12·40 m (40 ft 8¼ in)
Wheel track 3·668 m (12 ft 0½ in)
Wheelbase (to c/l of main units) 6·23 m (20 ft 5¼ in)
Propeller diameter 4·42 m (14 ft 6 in)
Distance between propeller centres 9·50 m (31 ft 2 in)
Rear-loading ramp/door: Width 2·45 m (8 ft 0½ in)
Height 2·25 m (7 ft 4½ in)

Dimensions, internal:
Main cabin: Length 8·58 m (28 ft 1¾ in)
Width 2·45 m (8 ft 0½ in)
Height 2·25 m (7 ft 4½ in)
Volume 58·0 m³ (2,048 cu ft)

Areas:
Wings, gross 82·00 m² (882·6 sq ft)
Ailerons (total) 3·65 m² (39·29 sq ft)
Trailing-edge flaps (total) 18·40 m² (198·06 sq ft)
Spoilers (total) 1·65 m² (17·76 sq ft)
Fin (incl dorsal fin) 12·19 m² (131·21 sq ft)
Rudder 7·02 m² (75·56 sq ft)
Tailplane 19·09 m² (205·48 sq ft)
Elevators (total) 4·61 m² (49·62 sq ft)

Weights and Loadings (standard transport):
Weight empty 14,590 kg (32,165 lb)
Weight empty, equipped 15,400 kg (33,950 lb)
Operating weight empty 15,700 kg (34,610 lb)
Max payload 9,000 kg (19,840 lb)
Max fuel load 9,400 kg (20,725 lb)
Max T-O weight 28,000 kg (61,730 lb)
Max landing weight 26,500 kg (58,420 lb)
Max zero-fuel weight 24,400 kg (53,790 lb)
Max cargo floor loading 750 kg/m² (155 lb/sq ft)
Max wing loading 341·5 kg/m² (69·9 lb/sq ft)
Max power loading 5·52 kg/kW (9·1 lb/shp)

Weights (G222 SAMA):
Operating weight empty 15,700 kg (34,614 lb)
SAMA equipment module 2,200 kg (4,850 lb)
Retardant 6,800 kg (14,990 lb)
Fuel 3,330 kg (7,340 lb)
Max T-O weight 28,000 kg (61,730 lb)

Performance (standard G222 transport, at max T-O weight except where indicated):
Max level speed at 4,575 m (15,000 ft) 291 knots (540 km/h; 336 mph)
Long-range cruising speed at 6,000 m (19,685 ft) 237 knots (439 km/h; 273 mph)
Airdrop speed (paratroops or cargo) 110-140 knots (204-259 km/h; 127-161 mph) IAS
Stalling speed, flaps and landing gear down 84 knots (155 km/h; 96·5 mph)
Time to 4,500 m (14,750 ft) 8 min 35 s
Max rate of climb at S/L 520 m (1,705 ft)/min
Rate of climb at S/L, one engine out 125 m (410 ft)/min
Service ceiling 7,620 m (25,000 ft)
Service ceiling, one engine out 5,000 m (16,400 ft)
T-O run 662 m (2,172 ft)
T-O to 15 m (50 ft) 1,000 m (3,280 ft)
Landing from 15 m (50 ft) 775 m (2,543 ft)
Landing run at max landing weight 545 m (1,788 ft)
Accelerate/stop distance 1,200 m (3,937 ft)
Min ground turning radius 20·80 m (68 ft 3 in)
Range with max payload, optimum cruising speed and height 740 nm (1,371 km; 852 miles)
Range with 53 troops 1,300 nm (2,409 km; 1,497 miles)
Range with 36 stretchers and 4 medical attendants 1,349 nm (2,500 km; 1,553 miles)
Ferry range with max fuel 2,500 nm (4,633 km; 2,879 miles)
g limit +2·5

Performance (G222 SAMA at max T-O weight):
As standard transport except:
Drop speed (T-O configuration) 120 knots (222 km/h; 138 mph)
Optimum height above ground during drop 50-100 m (165-330 ft)
Range with max retardant load 540 nm (1,000 km; 622 miles)

AERITALIA G222 (ELECTRONIC WARFARE, NAVAID CALIBRATION and MARITIME SURVEILLANCE VERSIONS)

Aeritalia has developed an electronic warfare version of the G222 and is developing flight inspection (radio/radar calibration) and maritime patrol, search and rescue versions. Dimensions, weights and performance of all three are similar to those of the standard troop transport.

Carrying a pilot, co-pilot and up to 10 systems operators, the electronic warfare version (first flown in prototype form on 9 March 1978) has a modified cabin fitted with racks and consoles for detection, signal processing and data recording equipment, and an electrical system providing up to 40kW of power for its operation. Externally, it is distinguishable by a small 'thimble' radome beneath the nose and a larger 'doughnut' radome on top of the tail-fin. Other versions, for such duties as airborne warning and control, elint, ECM/ESM, drone launching, and navigation training, are being studied.

The flight inspection version, of which the Italian Air Force has ordered six, is externally similar to the standard troop transport. Known as the **G222RM**, it is equipped for flights below 3,050 m (10,000 ft) to calibrate airport flight paths and radio assistance, enabling it to check VOR, ILS, DME, Tacan, PAR, NDB, marker beacons, and air traffic control systems, in addition to VHF and UHF radio transmissions. Onboard equipment includes separate receivers and displays, a central computer to collect inertial navigation data (updated continually by DME), and data on the state of the radio aid(s) being calibrated. Only one equipment operator is necessary, in addition to the two-man flight crew, and ample space remains in the rear of the hold to carry a Jeep-type vehicle for ground-based operations. This version has an optional secondary capability to perform survey missions, at altitudes between 6,100 and 7,620 m (20,000 and 25,000 ft), for multiple control of flight path assistance, using SAFI screened flight path equipment.

AERITALIA G222T

This Tyne-engined version of the G222 flew for the first time on 13 May 1980, and deliveries of 20 to the Libyan Arab Air Force began in February 1981. It differs from the standard version in the following respects:

Power Plant: Two 3,624 kW (4,860 shp) Rolls-Royce Tyne RTy.20 Mk 801 turboprop engines, each driving a BAeDG 4/7000/6 four-blade variable-pitch propeller. Fuel capacity 12,000 litres (2,638 Imp gallons), as in basic version.

Systems: In flight, air-conditioning system uses air supplied by a centrifugal compressor; on ground it is fed by APU. Alternators (three) each have 60kVA capacity. No electrical de-icing of spinners.

Dimensions, external:
Propeller diameter 4·88 m (16 ft 0 in)

Weights:
Operating weight empty 18,000 kg (39,685 lb)
Max T-O weight 29,000 kg (63,935 lb)
Max landing weight 27,200 kg (59,965 lb)

Performance (at max T-O weight except where indicated):
Max level speed at 9,145 m (30,000 ft) 310 knots (574 km/h; 357 mph)
Long-range cruising speed at 9,145 m (30,000 ft) 300 knots (556 km/h; 345 mph)
Time to 4,570 m (15,000 ft) 6 min 48 s
Max rate of climb at S/L, one engine out 305 m (1,000 ft)/min
Service ceiling, one engine out 4,725 m (15,500 ft)
T-O run 649 m (2,130 ft)
T-O to 15 m (50 ft) 1,006 m (3,300 ft)
Landing from 15 m (50 ft) 655 m (2,150 ft)
Landing run at max landing weight 378 m (1,240 ft)
Range with max payload, optimum cruising speed and height 1,020 nm (1,890 km; 1,174 miles)
Range with 53 troops 1,000 nm (1,853 km; 1,151 miles)
Ferry range with max fuel 2,750 nm (5,096 km; 3,166 miles)

AERITALIA AP 68B

This turbine-engined version of the Partenavia P.68R (which see), known originally as the P.68 Turbo, was developed initially by Partenavia. The improved power/weight ratio resulting from the change to turboprop engines is claimed to give the AP 68B the lowest operating costs of any aircraft in its class. Multi-role versatility includes the ability to carry a wide variety of external stores on four underwing hardpoints, to cover a broad spectrum of possible missions. These include air taxi, liaison, executive or cargo transport, training, ground support, reconnaissance, coastal patrol and aerial survey.

The AP 68TP prototype (NC 6001; I-PAIT) flew for the first time on 11 September 1978, at Naples, piloted by Comm Lionello Bellio, an Aeritalia test pilot. Certification of this aircraft, which has a retractable landing gear, was expected by mid-1981.

The basic civil version, with non-retractable landing gear, is designated **AP 68B Series 100**. The first production AP 68 was due to fly in late 1981, with deliveries to customers beginning in early 1982.

The following description applies to the AP 68B Series 100:

Wings, Fuselage and Tail Unit: As P.68R, except for lengthening of fuselage by 0·44 m (1 ft 5¼ in).

Landing Gear: Non-retractable tricycle type, with single wheel on each unit. Available at customer's option with retractable tricycle gear.

Power Plant: Two Allison 250-B17C turboprop engines, each flat rated to 246 kW (330 shp) for T-O and max continuous operation. Hartzell HC-B3TF-7A/T10173B-11R three-blade constant-speed fully-feathering reversible-pitch metal propellers with spinners. Fuel in two main and two auxiliary tanks in wings, total capacity 772 litres (170 Imp gallons), of which 757 litres (166·5 Imp gallons) are usable. Oil capacity 8·05 litres (1·75 Imp gallons) per engine.

Prototype of the Tyne-engined Aeritalia G222T

Aeritalia AP 68TP seven-seat prototype for the eight-passenger AP 68B

Accommodation: Seating for eight or nine persons, including pilot. Access to cabin via pilot's door and passenger/cargo door on starboard side, plus additional passenger door on port side. Dual controls, cabin heating, ventilation and soundproofing standard.

Systems: Primary electrical power supplied by two 150A 28V DC starter/generators. In the event of primary electrical failure, power is supplied by a 28V nickel-cadmium battery (self-sufficient for engine starting), and a 115V inverter for AC power. Control unit to protect electrical system from under- or over-voltage and reverse currents.

Avionics: To customer's requirements. Typical installations may include dual VHF/AM com, HF, dual NAV/ILS, ADF, DME, ATC transponder, marker beacon receiver, weather radar, autopilot, audio control and gyro compass in civil version; or VHF/FM com with homing, UHF, Tacan and IFF/ATC in military version.

Armament and Operational Equipment (military version): Four underwing hardpoints, each of 182 kg (400 lb) capacity, with standard NATO MA-4A racks. Typical loads may include four SUU-11B/A 7·62 mm Minigun pods, four LAU-32B/A rocket launchers (each containing seven rockets), four 400 lb bombs, flare dispensers, air-to-surface missiles, supply containers, or auxiliary fuel tanks.

Dimensions, external:
Wing span 12·00 m (39 ft 4½ in)
Wing chord, constant 1·55 m (5 ft 1 in)
Wing area, gross 18·60 m² (200·2 sq ft)
Wing aspect ratio 7·742
Length overall 9·70 m (31 ft 10 in)
Height overall 3·65 m (11 ft 11¾ in)
Wheel track 2·115 m (6 ft 11¼ in)
Wheelbase 3·613 m (11 ft 10¼ in)
Propeller diameter 2·29 m (7 ft 6 in)

Weights and Loadings:
Weight empty 1,472 kg (3,245 lb)
Operating weight empty 1,549 kg (3,415 lb)
Max usable fuel 608 kg (1,340 lb)
Max payload 726 kg (1,600 lb)
Max T-O weight 2,600 kg (5,732 lb)
Max ramp weight 2,625 kg (5,787 lb)
Max zero-fuel weight 2,303 kg (5,077 lb)
Max landing weight 2,470 kg (5,445 lb)
Max wing loading 139·8 kg/m² (28·6 lb/sq ft)
Max power loading 5·28 kg/kW (8·68 lb/shp)

Performance (estimated, at max T-O weight, ISA, except where indicated):
Max operating speed (Vmo) 200 knots (370 km/h; 230 mph) EAS
Max cruising speed at 4,570 m (15,000 ft), average cruise weight of 2,200 kg (4,850 lb) 208 knots (385 km/h; 240 mph)
Econ cruising speed at 5,475 m (18,000 ft), average cruise weight of 2,200 kg (4,850 lb) 165 knots (306 km/h; 190 mph)
Max manoeuvring speed 145 knots (269 km/h; 167 mph) EAS
Min control speed 80 knots (148 km/h; 92·5 mph) EAS
Stalling speed, power off:
'clean' 73 knots (135 km/h; 84 mph) CAS
T-O configuration 68 knots (126 km/h; 78·5 mph) CAS
Landing configuration 67 knots (124 km/h; 77·5 mph) CAS
Max rate of climb at S/L 620 m (2,025 ft)/min
Rate of climb at S/L, one engine out 210 m (690 ft)/min
Service ceiling (max operating altitude) 7,620 m (25,000 ft)
Service ceiling, one engine out, 30·5 m (100 ft)/min rate of climb 4,570 m (15,000 ft)
T-O to 15 m (50 ft): at S/L 488 m (1,600 ft)
at 1,525 m (5,000 ft) 631 m (2,070 ft)
Landing from 15 m (50 ft), at max landing weight, 36° flap: at S/L 497 m (1,630 ft)
at 1,525 m (5,000 ft) 558 m (1,830 ft)
Range at 3,050 m (10,000 ft) with max payload and 281 kg (619 lb) fuel:
at 204 knots (378 km/h; 235 mph) 311 nm (576 km; 358 miles)
at 158 knots (293 km/h; 182 mph) 437 nm (810 km; 503 miles)
Range at 3,050 m (10,000 ft) with max fuel and 468 kg (1,032 lb) payload:
at 205 knots (380 km/h; 236 mph) 643 nm (1,191 km; 740 miles)
at 157 knots (291 km/h; 181 mph) 908 nm (1,682 km; 1,045 miles)

AERITALIA P.78

Originally projected by Partenavia as a derivative of the P.68 Victor, the P.78 has undergone a number of major design revisions to emerge as an entirely new aircraft. Of high-wing configuration, with retractable landing gear, it has an unpressurised fuselage (with pressurisation as an option) and is powered by two Alfa Romeo/Rolls-Royce AR.318 turboprop engines, each flat rated to 388 kW (520 shp) and driving a three-blade Dowty Rotol propeller. The cabin accommodates up to 14 passengers, with a flight crew of two. A large door at the rear makes possible the use of the cabin for alternative loads such as light cargo and stretchers.

Dimensions, external:
Wing span 15·00 m (49 ft 2½ in)
Wing area, gross 27·00 m² (290·6 sq ft)
Wing aspect ratio 8·33

Weights:
Basic weight empty 2,500 kg (5,511 lb)
Max T-O weight 4,500 kg (9,921 lb)
Max landing weight 4,280 kg (9,435 lb)

PERFORMANCE (estimated, at max T-O weight, ISA):
Max level speed 223 knots (413 km/h; 257 mph)
Max cruising speed 207 knots (383 km/h; 238 mph)
Econ cruising speed 189 knots (350 km/h; 217 mph)
Best climbing speed 120 knots (222 km/h; 138 mph)
Stalling speed, flaps down 68 knots (126 km/h; 79 mph)
Max rate of climb at S/L 533 m (1,750 ft)/min
Rate of climb at S/L, one engine out 122 m (400 ft)/min
Service ceiling 9,145 m (30,000 ft)
Service ceiling, one engine out 4,875 m (16,000 ft)
T-O run 448 m (1,470 ft)
T-O to 15 m (50 ft) 747 m (2,450 ft)
Landing from 15 m (50 ft) 640 m (2,100 ft)
Landing run 381 m (1,250 ft)
Accelerate/stop distance 700 m (2,300 ft)
Range with max fuel, econ cruising speed at 6,100 m (20,000 ft) 1,000 nm (1,853 km; 1,151 miles)

AERONAUTICA MACCHI
AERONAUTICA MACCHI SpA

HEAD OFFICE: Corso Vittorio Emanuele 15, 20122 Milan
Telephone: (02) 792696 and 702020

OFFICES: Via Sanvito Silvestro 80, CP 246, 21100 Varese
Telephone: (0332) 254111
Telex: 380070 AERMAC I

PRESIDENT: Dott Fabrizio Foresio

VICE-PRESIDENT AND MANAGING DIRECTOR: Gen Ing Mario Matacotta

FINANCIAL DIRECTOR: Dott Gino Bavastro

COMMERCIAL DIRECTOR: Dott Ing Giovanni Cattaneo

INDUSTRIAL DEVELOPMENT DIRECTOR: Dott Ing Gianluigi della Torre

The original Macchi company was founded in 1912 in Varese, and its first aeroplane was built in 1913. On 1 January 1981 the Aeronautica Macchi group reorganised its structure, transforming itself into a holding company and transferring all of its operating activities to a newly-formed, wholly-owned company known as Aermacchi SpA. The group includes, besides Aermacchi SpA, other subsidiary companies operating in the fields of airframe and equipment manufacturing, precision machining and EDP. Aeronautica Macchi, the holding company, co-ordinates the corporate strategies in finance, commercial and industrial investments.

AERMACCHI
AERMACCHI SpA (Subsidiary of Aeronautica Macchi SpA)

HEAD OFFICE, OFFICES AND WORKS: Via Sanvito Silvestro 80, CP 246, 21100 Varese
Telephone: (0332) 254111
Telex: 380070 AERMAC I
PRESIDENT: Dott Fabrizio Foresio
MANAGING DIRECTORS:
Dott Ing Ermanno Bazzocchi
Gen Ing Mario Matacotta
GENERAL MANAGER: Dott Ing Giulio Cesare Valdonio
MARKETING MANAGER: Dott Ing Ferruccio Tommasi
COMMERCIAL DIRECTOR: Dott Ing Giovanni Cattaneo
TECHNICAL MANAGER: Dott Ing Alberto Notari
MARKETING AND PUBLIC RELATIONS: Isabella Stifani

Aermacchi is the aircraft manufacturing company of the Aeronautica Macchi group. In addition to its former factory area of 36,900 m² (397,200 sq ft), it has a more recently completed plant at Venegono Airfield, with a covered area of about 35,000 m² (376,750 sq ft). Total work force in 1981 was more than 2,200.

The M.B.339A two-seat trainer is in series production for the Italian Air Force and for export. A single-seat version, the M.B.339K Veltro 2, is also in production. Other activities include the manufacture of major assemblies for the Aeritalia G222 transport aircraft and Panavia Tornado; Aermacchi is also active in the field of aerospace ground equipment, with a complete line of hydraulic, electric and pneumatic ground carts for servicing civil and military aircraft. Aermacchi is teamed with Aeritalia and EMBRAER in developing the AM-X combat aircraft (see International section) for the Italian and Brazilian air forces.

AERMACCHI M.B. 326

The M.B. 326 jet trainer flew for the first time on 10 December 1957, powered by a Rolls-Royce Viper 8 turbojet engine. The more powerful Viper 11 powers six production versions built for the air forces of Italy (M.B. 326 and 326E), Tunisia (M.B. 326B), Ghana (M.B. 326F), Australia (M.B. 326H) and South Africa (M.B. 326M/Impala Mk 1), and one version built for Alitalia (M.B. 326D).

Versions with more powerful Viper engines, armament changes and other modifications included the M.B.326GB, for the Argentinian Navy and the air forces of Zaïre and Zambia; the M.B.326GC built in Brazil for the air forces of Brazil, Paraguay and Togo; the single-seat M.B.326K for the air forces of Dubai, Ghana, South Africa, Tunisia and Zaïre; and the M.B.326L for the air forces of Dubai and Tunisia.

Except for the two-seat M.B. 326L, Italian production of the M.B.326 has now ended; descriptions of the most recent versions can be found in the 1980-81 and earlier editions of *Jane's*. Manufacture of the M.B.326GC, as the EMBRAER AT-26 Xavante, continues during 1981-82 in Brazil, and a shortened description of this version can be found in that section of this edition. A modified version of the M.B.326K, with a Viper 540 (instead of Viper 632) engine, is produced in South Africa by Atlas (which see) as the Impala Mk 2.

AERMACCHI M.B. 339A

The M.B. 339A is a tandem two-seat trainer/ground attack aircraft, powered by a Fiat/Piaggio-built Rolls-Royce Viper 632 engine.

The first of two M.B. 339X flying prototypes (MM588) was flown for the first time on 12 August 1976. The second aircraft (MM589), which made its first flight on 20 May 1977, was built to pre-production standard; the third airframe was used for static and fatigue testing. The first production aircraft made its initial flight on 20 July 1978, and the first of 100 M.B.339As for the Italian Air Force were handed over for pre-service trials on 8 August 1979. Deliveries have since begun to the 14th Radio Aids Survey Squadron (February 1981) and to the Scuola di Volo at Lecce Galatina in southern Italy (April 1981).

Deliveries have been completed of ten M.B.339As ordered in 1980 by the Argentinian Navy. Delivery of 14 ordered by the Peruvian Air Force was due to be completed by the end of 1981.

TYPE: Two-seat basic and advanced trainer and ground attack aircraft.

AIRFRAME: Structural design criteria based on MIL-A-8860A; 8*g* limit load factor in 'clean' configuration. Cockpit designed for 40,000 pressurisation cycles. Service life requirement 10,000 flying hours and 20,000 landings in the training role. Entire structure specially treated to prevent corrosion.

WINGS: Cantilever low/mid-wing monoplane. Wing section NACA 64A-114 (mod) at centreline, NACA 64A-212 (mod) at tip. Leading-edge swept back 11° 18′. Sweepback at quarter-chord 8° 29′. All-metal stressed-skin structure, with single main spar and auxiliary rear spar, built in two portions and bolted to fuselage. Skin stiffened by spanwise stringers, closely spaced ribs, and false ribs. Wingtip tanks permanently attached. Single fence on each wing at approx two-thirds span. Servo-powered ailerons embody 'Irving'-type aerodynamic balance provisions, and are statically balanced along their entire span. Balance tabs facilitate reversion to manual operation in the event of servo failure. Hydraulically actuated single-slotted flaps, operated by push/pull rods.

Aermacchi M.B. 339A basic/advanced jet trainer with underwing weapons and drop-tanks

FUSELAGE: All-metal semi-monocoque structure, built in two main portions: forward (nose to engine mounting bulkhead), and rear (engine bulkhead to tailcone). Forward portion built of C-section frames, four C-section spars, longitudinal L-section stringers, and skin panels. Rear section manufactured entirely from aluminium alloy except for firewall and most of tailcone, which are of stainless steel; four-bolt attachment to forward fuselage to facilitate access to engine. Hydraulically actuated, electrically controlled airbrake under centre of fuselage, just forward of CG.

TAIL UNIT: Cantilever all-metal structure, of similar construction to wings. Slightly sweptback vertical surfaces. Rudder and elevators are statically balanced, each having an electrically actuated dual-purpose balance and trim tab. Two auxiliary fins under rear fuselage.

LANDING GEAR: Hydraulically-retractable tricycle type, with oleo-pneumatic shock-absorbers; suitable for operation from semi-prepared runways. Nosewheel retracts forward, main units outward into wings. Steerable nosewheel, fitted with shimmy damper. Low-pressure main-wheel tubeless tyres size 545 × 175-10 (12 ply rating); nosewheel tubeless tyre size 380 × 150-4 (6 ply rating). Emergency extension system. Hydraulic disc brakes with anti-skid system.

POWER PLANT: One Rolls-Royce Viper Mk 632-43 turbojet engine, rated at 17·8 kN (4,000 lb st). Engines built in Italy under Rolls-Royce/Fiat licence; final assembly by Piaggio. Fuel in two-cell rubber fuselage tank, capacity 781 litres (172 Imp gallons), and two integral wingtip tanks, combined capacity 632 litres (139 Imp gallons). Total internal capacity 1,413 litres (311 Imp gallons) usable. Single-point pressure refuelling receptacle in port side of fuselage, below wing trailing-edge. Gravity refuelling points on top of fuselage and each tip-tank. Provision for two drop-tanks, each of 325 litres (71·5 Imp gallons) capacity, on centre underwing stations. Anti-icing system for engine air intakes.

ACCOMMODATION: Crew of two in tandem, on Martin-Baker IT10F zero-zero ejection seats in pressurised cockpit. Rear seat elevated 32·5 cm (12¾ in). Rearview mirror for each occupant. Two-piece moulded transparent jettisonable canopy, opening sideways to starboard.

SYSTEMS: Hydraulic system, pressure 172·5 bars (2,500 lb/sq in), for actuation of flaps, aileron servos, airbrake, landing gear, wheel brakes and nosewheel steering. Backup system for wheel brakes and emergency extension of landing gear. Main electrical DC power from one 28V 9kW engine-driven starter/generator and one 28V 6kW secondary generator. Two 24V 22Ah nickel-cadmium batteries for engine starting. Fixed-frequency 115/26V AC power from two 600VA single-phase static inverters. External power receptacle. Pressurised cockpit, with max differential of 0·24 bars (3·5 lb/sq in). Bootstrap-type air-conditioning system, which also provides air for windscreen and canopy demisting. Low-pressure demand-type oxygen system, operating at 28 bars (400 lb/sq in).

AVIONICS AND EQUIPMENT: Typical avionics installation includes Perkin-Elmer/Magnavox AN/ARC-150(V) UHF or Perkin-Elmer/Magnavox SRT-194B VHF primary com transceiver; Collins 618M-3A VHF/AM or equivalent ARINC 566A, or Collins AN/ARC-186(V) VHF/AM and FM secondary com transceiver; Collins IA-210 interphone; Collins AN/ARN-118(V)1 Tacan or Collins 860E-5 DME nav system; Fiar/Bendix AN/APX-100(V) IFF; Collins 51RV-4D VOR/ILS, including localiser and glideslope receivers; Collins

MKI-3 marker beacon receiver; Collins ADF-60A ADF; or (M.B.339A) Marconi Avionics AD-620C computerised area and dead reckoning navigation system. Standard instrumentation includes ARU-2B/A attitude director indicator, AQU-6/A HSI, Sperry AS-339 attitude and heading reference system, AG-5 standby attitude indicator, and flight director system. Retractable landing light beneath port wing; taxying light on nosewheel leg.

Armament and Operational Equipment: Up to 1,815 kg (4,000 lb) of external stores can be carried on six underwing hardpoints, the inner four of which are stressed for loads of up to 454 kg (1,000 lb) each and the outer two for up to 340 kg (750 lb) each. Provisions are made, on the two inner stations, for the installation of two Macchi gun pods, each containing either a 30 mm DEFA 553 cannon with 120 rds, or a 12·7 mm AN/M-3 machine-gun with 350 rds. Other typical loads can include two Matra 550 Magic or AIM-9 Sidewinder air-to-air missiles on the two outer stations; four 1,000 lb or six 750 lb bombs; six SUU-11A/A 7·62 mm Minigun pods with 1,500 rds/pod; six Matra 155 launchers, each for eighteen 68 mm rockets; six Matra F-2 practice launchers, each for six 68 mm rockets; six LAU-68/A or LAU-32G launchers, each for seven 2·75 in rockets; six Aerea AL-25-50 or AL-18-50 launchers, each with twenty-five or eighteen 50 mm rockets respectively; six Aerea AL-12-80 launchers, each with twelve 81 mm rockets; four LAU-10/A launchers, each with four 5 in Zuni rockets; four Thomson-Brandt 100-4 launchers, each with four 100 mm Thomson-Brandt rockets; six Aerea BRD bomb/rocket dispensers; six Aermacchi 11B29-003 bomb/flare dispensers; six Thomson-Brandt 14-3-M2 adaptors, each with six 100 mm anti-runway bombs or 120 mm tactical support bombs; or two 325 litre (71·5 Imp gallon) drop-tanks; or a photographic pod with four 70 mm Vinten cameras; or a single underwing Elettronica ECM pod, combined with a rear-fuselage flare/chaff dispenser, onboard RHAW receiver and indicators, and ECM antennae in the nose and rear of the port wingtip tank. Provision for Aeritalia 8.105.924 fixed reflector sight, Saab RGS 2 gunsight or Thomson-CSF RD-21 self-contained gyroscopic sight; a gunsight can also be installed in rear cockpit, to enable instructor to evaluate manoeuvres performed by student pilot. All gunsights can be equipped with fully automatic Teledyne TSC 116-2 gun camera. Head-up display system under study. Provision for towing type A-6B (1·83 × 9·14 m; 6 × 30 ft) aerial banner target; tow attachment point on inner surface of ventral airbrake.

Dimensions, external:

Wing span over tip-tanks	10·858 m (35 ft 7½ in)
Wing aspect ratio	5·26
Length overall	10·972 m (36 ft 0 in)
Height overall	3·600 m (11 ft 9¾ in)
Tailplane span	4·164 m (13 ft 8 in)
Wheel track	2·483 m (8 ft 1¾ in)
Wheelbase	4·369 m (14 ft 4 in)

Areas:

Wings, gross	19·30 m² (207·74 sq ft)
Ailerons (total)	1·328 m² (14·29 sq ft)
Trailing-edge flaps (total)	2·21 m² (23·79 sq ft)
Airbrake	0·68 m² (7·32 sq ft)
Fin	2·370 m² (25·51 sq ft)
Rudder, incl tab	0·610 m² (6·57 sq ft)
Tailplane	3·380 m² (36·38 sq ft)
Elevators (total, incl tabs)	0·979 m² (10·54 sq ft)

Weights and Loading:

Weight empty, equipped	3,125 kg (6,889 lb)
Basic operating weight empty	3,136 kg (6,913 lb)
Fuel load (internal, usable)	1,100 kg (2,425 lb)
T-O weight, 'clean'	4,400 kg (9,700 lb)

Typical T-O weights with armament indicated:

A: four Mk 82 bombs and two drop-tanks	5,895 kg (13,000 lb)
B: six Mk 82 bombs	5,895 kg (13,000 lb)
C: two Macchi 30 mm gun pods, two LR-25-0 rocket launchers and two drop-tanks	5,808 kg (12,805 lb)
D: four LR-25-0 launchers and two drop-tanks	5,642 kg (12,440 lb)
E: six LR-25-0 launchers	5,323 kg (11,735 lb)
Max T-O weight with external stores	5,895 kg (13,000 lb)
Wing loading (50 per cent fuel)	205 kg/m² (42·00 lb/sq ft)

Performance (at 'clean' T-O weight, ISA, except where stated):

EAS limit/Mach limit	Mach 0·82 (500 knots; 926 km/h; 575 mph)
Max level speed at S/L	485 knots (898 km/h; 558 mph)
Max level speed at 9,150 m (30,000 ft)	Mach 0·77 (441 knots; 817 km/h; 508 mph)
Max speed for landing gear extension	175 knots (324 km/h; 201 mph) IAS
T-O speed	100 knots (185 km/h; 115 mph)
Approach speed over 15 m (50 ft) obstacle	98 knots (182 km/h; 113 mph) IAS
Landing speed	89 knots (165 km/h; 102·5 mph) IAS
Stalling speed	80 knots (148·5 km/h; 92·5 mph)
Max rate of climb at S/L	2,010 m (6,595 ft)/min
Time to 9,150 m (30,000 ft)	7 min 6 s
Service ceiling (30·5 m; 100 ft/min rate of climb)	14,630 m (48,000 ft)
Min ground turning radius	8·45 m (27 ft 8¾ in)
T-O run at S/L: 'clean' T-O weight	465 m (1,525 ft)
max T-O weight	915 m (3,000 ft)
Landing run at S/L, ISA	415 m (1,362 ft)
Max ferry range with two underwing drop-tanks, 10% reserves	1,140 nm (2,110 km; 1,310 miles)
Max endurance at 7,620 m (25,000 ft) with two underwing drop-tanks, 10% reserves	3 h 45 min
g limits	+8·0; −4·0

Performance (armed configuration, at T-O weights given above):

Radius of action, hi-lo-hi (no run-in or run-out):

A	320 nm (593 km; 368 miles)
B	212 nm (393 km; 244 miles)
C	275 nm (510 km; 317 miles)
D	305 nm (565 km; 351 miles)
E	165 nm (306 km; 190 miles)

Radius of action, lo-lo-lo (no run-in or run-out):

A	200 nm (371 km; 230 miles)
B	146 nm (271 km; 168 miles)
C	190 nm (352 km; 219 miles)
D	193 nm (358 km; 222 miles)
E	123 nm (228 km; 142 miles)

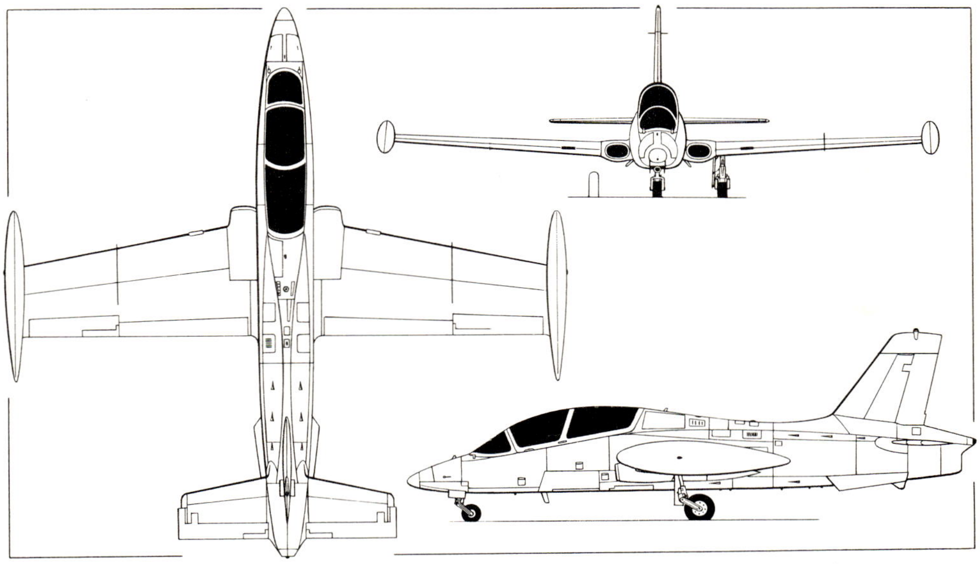

Aermacchi M.B. 339A two-seat jet trainer and light attack aircraft *(Pilot Press)*

Prototype Aermacchi M.B.339K Veltro 2 light close air support aircraft, developed from the two-seat M.B.339A

AERMACCHI M.B. 339K VELTRO 2 (GREYHOUND)

The Veltro 2, which perpetuates the name of the Macchi M.C. 205V fighter of the second World War, is a single-seat development of the M.B. 339A, optimised for the roles of light close air support and operational training. It is based on experience gained in the production and operation of the M.B. 326K. The M.B. 339K private-venture prototype (I-BITE) flew for the first time on 30 May 1980 and from October 1980 to May 1981 was engaged in weapon firing and compatibility trials. Construction of an initial production batch began in March 1981.

The airframe and power plant of the M.B. 339A are retained, except for a new forward fuselage with redesigned single-seat cockpit and internally-mounted cannon armament; the other major changes concern the avionics and equipment relevant to the different roles performed by the Veltro 2. The following description applies to the basic M.B. 339K; operational capability can, at customer's option, be extended by adopting such additional features as a head-up display, cockpit TV display, ECM, and other improved avionics.

Type: Single-seat ground attack aircraft and operational trainer.

Airframe: Structural design criteria based on MIL-A-8860 series of specifications; +8/−4*g* limit load factor in 'clean' configuration. With a fatigue spectrum based on the ground attack role only, service life will be more than 5,000 flying hours with 10,000 landings. Entire structure is specially treated to prevent corrosion.

Wings: As M.B.339A.

Fuselage: As M.B.339A, except for modified forward section.

Tail Unit: As M.B.339A.

Landing Gear: Similar to M.B.339A.

Power Plant: As M.B.339A. Fuel in one fuselage tank, consisting of three rubber cells with a total capacity of 1,030 litres (226·5 Imp gallons), and two integral wing-tip tanks with a combined capacity of 630 litres (138·5 Imp gallons). Latter can be increased to 1,000 litres (220 Imp gallons) by the use of enlarged tip-tanks with a constant circular section. Total usable internal fuel capacity 1,660 litres (365 Imp gallons) standard, 2,030 litres (446·5 Imp gallons) with circular-section tip-tanks.

Accommodation: Pilot only, on Martin-Baker IT10F zero-zero ejection seat in pressurised cockpit. Rearview mirror standard. One-piece moulded transparent jettisonable canopy, opening sideways to starboard.

Systems: As M.B.339A.

Avionics and Equipment: Typical avionics installation includes Collins AN/ARC-159(V) UHF (or Perkin-Elmer SRT-194B VHF) primary com transceiver; Collins 618M-3A or equivalent ARINC 566A VHF/AM (or Collins AN/ARC-186(V) VHF/AM and FM) secondary com transceiver; Collins IA-210 interphone; Collins AN/ARN-118(V)1 Tacan (or Collins 860E-5 or equivalent ARINC 568 DME) nav system; Fiar/Bendix AN/APX-100(V) IFF; Collins 51RV-4B or equivalent ARINC 547A-4 VOR/ILS, including localiser and glideslope receivers; Collins MKI-3 marker beacon receiver; Collins DF-206 or equivalent ARINC 570 ADF; Marconi Avionics AD-620C computerised radio and dead reckoning area navigation system. Standard instrumentation includes Astronautics ARU-2B/A attitude director indicator; AQU-6/A horizontal situation indicator and flight director system; Sperry P-140 attitude and heading reference system; and Microtecnica AG-5 standby attitude indicator. Retractable landing light beneath port wing; taxying light on nosewheel leg.

Armament and Operational Equipment: Two 30 mm DEFA cannon, with 125 rds/gun, mounted internally in lower forward fuselage, with external fairings. Firing rate 1,200 rds/min. Loads on six underwing hardpoints generally as for M.B.339A. Has also been flight tested

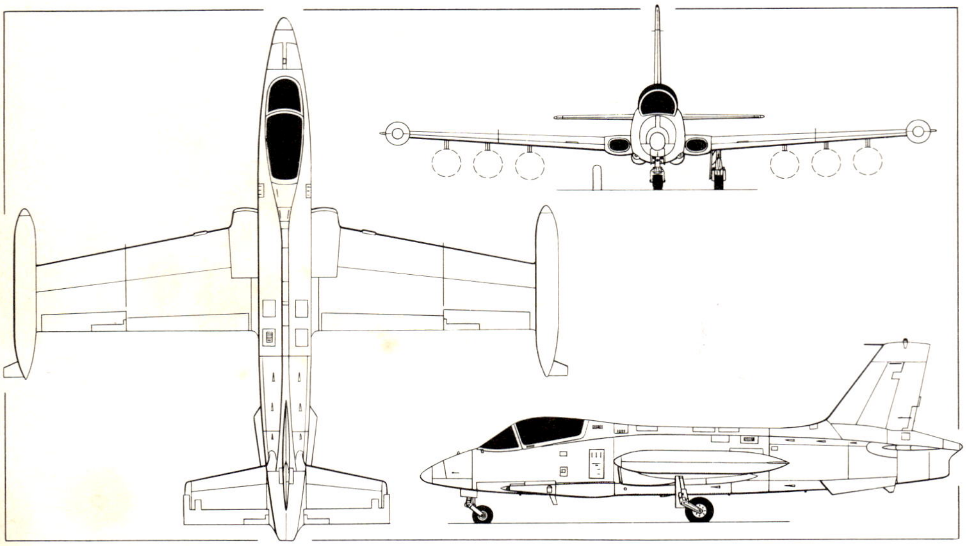

Aermacchi M.B.339K Veltro 2 light ground attack aircraft *(Pilot Press)*

successfully with four free-fall and retarded Mk 82 bombs of 500 lb; Matra 500 lb bombs; Expal BRP 250 lb bombs; 120 mm light attack bombs; Thomson-Brandt 100 mm anti-runway bombs; and Thomson-Brandt 100 mm rockets. Saab-Scania RGS 2 gunsight, with gyro lead computer; gunsight can be equipped with a fully-automatic Teledyne TSC 116-2 gun camera. Provision for towing type A-6B (1·83 × 9·14 m; 6 × 30 ft) aerial banner target.

DIMENSIONS, EXTERNAL: As for M.B.339A except:
- Wing span over tip-tanks:
  - standard tanks 10·858 m (35 ft 7½ in)
  - circular-section tanks 11·045 m (36 ft 2¾ in)
- Length overall 10·792 m (35 ft 5 in)
- Height overall 3·90 m (12 ft 9½ in)

AREAS: As for M.B.339A

WEIGHTS:
- Weight empty, equipped 3,174 kg (6,997 lb)
- Fuel load (internal, usable, with circular-section tip-tanks) 1,582 kg (3,487 lb)
- T-O weight 'clean', incl ammunition for internal guns 4,978 kg (10,974 lb)
- Max T-O weight with external stores 6,150 kg (13,558 lb)

PERFORMANCE (with full gun ammunition load):
- Max level speed at S/L 485 knots (899 km/h; 558 mph) IAS
- Landing speed 89 knots (165 km/h; 102 mph) IAS
- Max rate of climb at S/L 2,286 m (7,500 ft)/min
- Service ceiling 13,565 m (44,500 ft)
- T-O run 'clean' 575 m (1,886 ft)
- Landing run 410 m (1,345 ft)
- Combat radius with two 30 mm cannon (125 rds/gun) and four 500 lb Mk 82 bombs (total military load 1,088 kg; 2,400 lb):
  - lo-lo-lo 203 nm (376 km; 234 miles)
  - hi-lo-hi 350 nm (648 km; 403 miles)

AGUSTA
AGUSTA SpA

LEGAL ADDRESS: 21017 Cascina Costa
Telephone: (0331) 229 111
Telex: 332569 AGUCA I

OFFICES:
Via Caldera 21, 20153 Milan
Telephone: (02) 452 5051
Telex: 333280 AGUMI I
Via Abruzzi 11, 00187 Rome
Telephone: (06) 475 6551
Telex: 614398 AGURO I

MANAGEMENT:
PRESIDENT AND CHAIRMAN OF THE BOARD:
Cav del lavoro Conte Corrado Agusta
EXECUTIVE VICE-PRESIDENT:
Dott Ing Pietro Fascione
VICE-PRESIDENT:
Dott Ermenegildo Marelli
ORGANISATION AND DEVELOPMENT:
Dott Ing Arnaldo Antichi

Formed originally in 1977, the Agusta Group (see 1980-81 *Jane's*) completely reorganised its structure from 1 January 1981 under a new holding company known as Agusta SpA. This has three operational divisions, as detailed below. Agusta is reportedly to acquire the Breda-Nardi and Caproni Vizzola companies (which see) in the reorganisation of the Italian aerospace industry proposed by the Italian government.

DIVISIONE ELICOTTERI (Helicopter Division)
MANAGING DIRECTOR: Dott Ing Giorgio Brazzelli

Costruzioni Aeronautiche Giovanni Agusta SpA
See following entry

Elicotteri Meridionali SpA
See following entry

Agusta Aviation Corporation
One West Loop South, Suite 600, Houston, Texas 77027, and Philadelphia, Pennsylvania, USA (marketing and support in North and Central America)

EH Industries Ltd (50% holding)
See International section

DIVISIONE AEROPLANI (Aircraft Division)
MANAGING DIRECTOR: Dott Ing C. Camposampiero

SIAI-Marchetti SpA
See following entry

Industria Aeronautica Meridionale SpA
Via Provinciale San Vito 221, 72100 Brindisi (aircraft co-production, and overhaul of aircraft and helicopters)

DIVISIONE ATTIVITÁ VARIE (Other Products and Support Division)

Fonderie e Officine Meccaniche di Benevento SpA
Corso Garibaldi 8, 82100 Benevento (production of lightweight alloys and components)

Agusta International
Avenue Louise 523, Brussels, Belgium (aviation support services)

Meccanica Verghera SpA
V. le Adriatico 50, 21010 Verghera (product support for MV Agusta motorcycles)

Società Italiana Sistemi di Addestramento
c/o EMP, 04011 Aprilia (maintenance, overhaul and training systems for aircraft and helicopters, in equal partnership with Aeritalia and Elettronica)

COSTRUZIONI AERONAUTICHE GIOVANNI AGUSTA SpA

LEGAL ADDRESS, MAIN OFFICES AND WORKS:
21017 Cascina Costa di Samarate, Gallarate
Telephone: (0331) 229 111
Telex: 332569 AGUCA I
COMMERCIAL OFFICES:
Via Caldera 21, 20153 Milan
Telephone: (02) 452 5051
Telex: 333280 AGUMI I
PRESIDENT AND CHAIRMAN OF THE BOARD:
Cav del lavoro Conte Corrado Agusta
VICE-PRESIDENT:
Dott Ing Pietro Fascione
MANAGING DIRECTOR AND GENERAL MANAGER:
Dott Ing Giorgio Brazzelli
INTERNATIONAL MARKETING:
Dott Enrico Guerra
ENGINEERING: Dott Ing Luigi Passini
CUSTOMER SERVICES: Ilario Cernich
PUBLIC RELATIONS: Francesco Lami

This company was established in 1907 by Giovanni Agusta and built many experimental and production aircraft before the second World War.

In 1952 Agusta acquired a licence to manufacture the Bell Model 47 helicopter and the first Agusta-built Model 47G made its initial flight on 22 May 1954.

In addition to the A 109A and A 129 of its own design, Agusta is producing under licence the Bell Models 205, 206 and 212. It has collaborated with Bell Helicopter Textron (see US section) in developing the Model 412 version of the Bell 212, fitted with a new four-blade rotor system. Under licence from Sikorsky, production of SH-3D helicopters began in 1967, and manufacture of the HH-3F (S-61R) started in 1974. Agusta also produces under licence the Boeing Vertol CH-47C Chinook helicopter, and is collaborating with Westland Helicopters of the UK in developing the EH 101 three-turboshaft helicopter (see under EHI in the International section).

AGUSTA A 109A

The basic version of the Agusta A 109A high-speed, high-performance twin-engined helicopter accommodates a pilot and seven passengers, and has a large baggage compartment in the rear of the fuselage. Alternatively, the A 109A can be adapted for freight-carrying, as an ambulance, or for search and rescue and other duties. Military and naval versions, and the improved A 109A Mk II, are described separately.

The first of three A 109 flying prototypes flew for the first time on 4 August 1971. RAI and FAA certification for VFR operation was announced on 1 June 1975, and certification for IFR single-pilot operation was obtained on 20 January 1977. Certification has since been granted in Canada, France, West Germany, the Philippines, Sweden, Switzerland and the UK. Deliveries of production aircraft, designated A 109A, started in early 1976 and had reached a total of about 120 by the beginning of 1981.

Agusta has successfully carried out trials of a four-unit flotation gear for the A 109A, manufactured by Garrett Air Cruisers of New Jersey.

The following description applies to the standard A 109A, production of which was at the rate of six to seven per month in mid-1981.

TYPE: Twin-engined general-purpose helicopter.

ROTOR SYSTEM: Fully-articulated four-blade single main rotor and port-side two-blade semi-rigid delta-hinged tail rotor. Main rotor blades have a 'droop-snoot' aerofoil section, with thickness/chord ratios of 11·3% at root and 6% at tip, and are attached to hub by tension/torsion straps. They are of aluminium alloy bonded construction, with a honeycomb core, have swept tips, stainless steel tip caps and leading-edge strips, and are protected against corrosion. A manual blade-folding capability and rotor brake are optional. Tail rotor blades are of aluminium alloy, bonded at the trailing-edge, with a honeycomb core and stainless steel leading-edge strip.

ROTOR DRIVE: Main transmission assembly housed in fairing above passenger cabin, driving main rotor through a coupling gearbox and 90° main reduction gearbox. Take-off drive from coupling gearbox drives tail rotor via an output shaft and tail rotor gearbox. Transmission

Agusta A 109A general-purpose helicopter (two Allison 250-C20B turboshaft engines)

ratings 516 kW (692 shp) for take-off and max continuous twin-engined operation, with max contingency rating of 598 kW (802 shp) for 6 s. Ratings for single-engined operation are 298 kW (400 shp) for take-off, 287 kW (385 shp) max continuous, and 334 kW (448 shp) max contingency for 10 s. Main rotor/engine rpm ratio 1:15·62; tail rotor/engine rpm ratio 1:2·80.

Fuselage and Tail Unit: Pod and boom type, of aluminium alloy construction, built in four main sections: nose, cockpit, passenger cabin and tailboom. Sweptback vertical fins (above and below fuselage), and non-swept elevator, mounted on rear of tailboom. Elevator linked to collective pitch control.

Landing Gear: Retractable tricycle type, with oleo-pneumatic shock-absorber on each unit. Single main wheels and castoring (45° each side of centre) and self-centering nosewheel. Hydraulic retraction, nosewheel forward, main wheels upward into fuselage. Hydraulic emergency extension and locking. Disc brakes on main wheels. All tyres are of tubeless type, and of same size (360 × 135-6) and pressure (5·9 bars; 85 lb/sq in). Tailskid under ventral fin. Emergency pop-out flotation gear and fixed snow skis optional.

Power Plant: Two Allison 250-C20B turboshaft engines (each 313 kW; 420 shp for T-O, 298 kW; 400 shp max continuous power, 276 kW; 370 shp max cruise power, derated to 258 kW; 346 shp for twin-engine operation), mounted side by side in upper rear fuselage and separated from passenger cabin and from each other by firewalls. Two fuel tanks in lower rear fuselage, combined capacity 560 litres (123 Imp gallons), of which 550 litres (121 Imp gallons) are usable. Refuelling point in each side of fuselage, near top of each tank. Oil capacity 7·7 litres (1·7 Imp gallons) for each engine and 12 litres (2·6 Imp gallons) for transmission. Provision for internal auxiliary tanks containing 138 or 198 kg (304 or 436 lb) of fuel.

Accommodation: Crew of one or two on flight deck, with pilot seated on right. Dual controls optional. Main cabin seats up to six passengers on three forward- or rearward-facing seats in centre, plus three forward-facing seats at rear. A seventh passenger can be carried in lieu of second crew member. Four/five-seat VIP layout available, with refreshment and music centre. Forward-opening crew door and passenger door on each side. Large space at rear of cabin for up to 150 kg (331 lb) of baggage, with access via forward-opening door on port side. Centre row of seats removable to permit use as freight transport. Ambulance version can accommodate two stretchers, one above the other, and two medical attendants, in addition to the pilot. External freight can be transported on a centre-of-gravity hook. Sliding doors can be installed for rescue missions.

Systems: Two identical independent Magnaghi hydraulic systems, pressure 103·5 bars (1,500 lb/sq in), supply dual flight servo-controls and provide emergency power in the event of engine failure. Magnaghi utility hydraulic system, pressure 69 bars (1,000 lb/sq in), for actuation of landing gear, wheel and rotor braking, nosewheel locking, and emergency backup. 28V DC electrical system, using two 30V 150A engine-driven starter/generators, and one 24V 13Ah nickel-cadmium battery (22Ah heavy-duty battery on IFR version). Single-phase AC power at 400Hz supplied by two 115/26V 250VA solid-state static inverters. Third inverter as emergency backup on IFR version. External power receptacle. Engine anti-icing system, using engine bleed air.

Avionics: Standard instrumentation, plus Collins avionics for VFR or IFR operation, to customer's requirements, including VHF-20A VHF/AM com (dual in IFR version), AG-06 intercom, VIR-31A VOR/ILS with VOR/LOC, glideslope and marker beacon receiver, TDR-90 ATC transponder, ADF-60A ADF and DME-40 DME. Optional avionics include AA-215 radio altimeter with LGCS, Helcis II flight director and autotrim, AFCS, pilot's navigation instruments, co-pilot's flight and navigation instruments, standby attitude indicator, two- or three-axis autopilot, weather radar and area nav.

Equipment: Depending upon mission, may include internal cargo platform, external cargo sling, externally-mounted rescue hoist, first aid kit, stretchers, container for up to 9,800 litres (2,155 Imp gallons) of fire retardant, or equipment for exploration, thermal mapping, survey, or powerline control duties.

Dimensions, external:
Diameter of main rotor — 11·00 m (36 ft 1 in)
Diameter of tail rotor — 2·03 m (6 ft 8 in)
Length overall, rotors turning — 13·05 m (42 ft 9¾ in)
Length of fuselage — 10·706 m (35 ft 1½ in)
Fuselage: Max width — 1·42 m (4 ft 8 in)
Height over tail fin — 3·30 m (10 ft 10 in)
Elevator span — 2·88 m (9 ft 5½ in)
Width over main wheels — 2·45 m (8 ft 0½ in)
Wheelbase — 3·535 m (11 ft 7¼ in)
Passenger doors (each): Height — 1·06 m (3 ft 5¾ in)
Width — 1·15 m (3 ft 9¼ in)
Height to sill — 0·65 m (2 ft 1½ in)

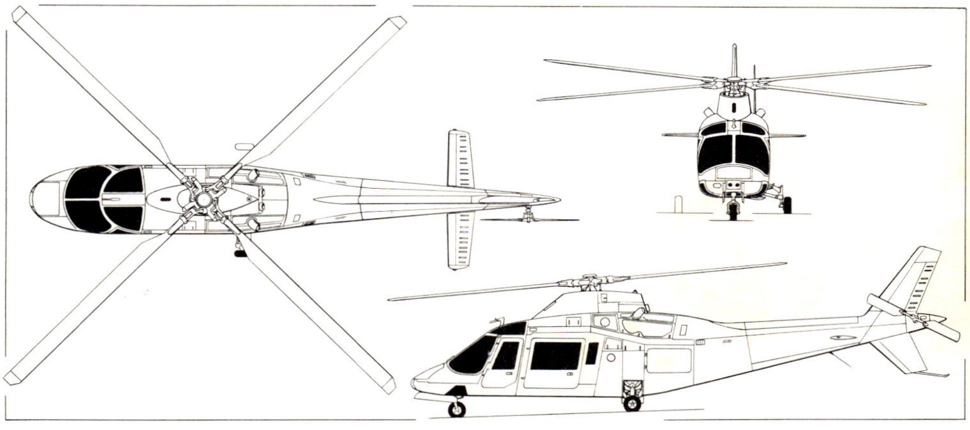

Agusta A 109A twin-engined general-purpose helicopter *(Pilot Press)*

Agusta A 109A in anti-tank configuration, armed with four TOW missiles and undernose telescopic sight

Baggage door (port, rear):
Height — 0·51 m (1 ft 8 in)
Width — 1·00 m (3 ft 3¼ in)

Dimensions, internal:
Cabin, excl flight deck: Length — 1·63 m (5 ft 4¼ in)
Max width — 1·32 m (4 ft 4 in)
Max height — 1·28 m (4 ft 2½ in)
Volume — 2·82 m³ (100 cu ft)
Baggage compartment volume — 0·52 m³ (18·4 cu ft)

Areas:
Main rotor blades (each) — 1·84 m² (19·8 sq ft)
Tail rotor blades (each) — 0·203 m² (2·185 sq ft)
Main rotor disc — 95·03 m² (1,022·9 sq ft)
Tail rotor disc — 3·24 m² (34·87 sq ft)

Weights and Loadings:
Basic weight empty:
standard (pilot and 7 passengers) — 1,415 kg (3,120 lb)
offshore oil support (IFR) — 1,604 kg (3,536 lb)
ambulance (IFR) — 1,657 kg (3,653 lb)
firefighting — 1,596 kg (3,518 lb)
Max external slung load — 907 kg (2,000 lb)
Max baggage — 150 kg (331 lb)
Typical T-O weight:
offshore oil support (IFR) — 2,596 kg (5,723 lb)
ambulance (IFR) — 2,409 kg (5,311 lb)
Max normal T-O weight — 2,450 kg (5,400 lb)
Max certificated T-O weight — 2,600 kg (5,732 lb)
Max disc loading — 27·4 kg/m² (5·60 lb/sq ft)
Max power loading — 4·15 kg/kW (6·82 lb/shp)

Performance (S/L, ISA, except where indicated. A: AUW of 2,250 kg; 4,960 lb, B: AUW of 2,450 kg; 5,400 lb, C: AUW of 2,600 kg; 5,732 lb):
Never-exceed speed:
A, B — 168 knots (311 km/h; 193 mph)
C — 160 knots (296 km/h; 184 mph)
Max cruising speed:
A — 151 knots (280 km/h; 174 mph)
B — 144 knots (267 km/h; 166 mph)
C — 142 knots (263 km/h; 163 mph)
Econ cruising speed:
A — 126 knots (233 km/h; 145 mph)
B — 125 knots (232 km/h; 144 mph)
C — 124 knots (230 km/h; 143 mph)
Max rate of climb at S/L: A — 567 m (1,860 ft)/min
B — 494 m (1,620 ft)/min
C — 451 m (1,480 ft)/min
Rate of climb at S/L, one engine out:
A — 152 m (500 ft)/min
B — 104 m (340 ft)/min
C — 67 m (220 ft)/min
Service ceiling, 30·5 m (100 ft)/min rate of climb, at max continuous power:
A — 5,485 m (18,000 ft)
B — 4,970 m (16,300 ft)
C — 4,600 m (15,100 ft)
Service ceiling, one engine out, 30·5 m (100 ft)/min rate of climb, at max continuous power:
A — 2,440 m (8,000 ft)
B — 1,615 m (5,300 ft)
C — 945 m (3,100 ft)
Hovering ceiling IGE: A — 3,750 m (12,300 ft)
B — 2,985 m (9,800 ft)
C — 2,410 m (7,900 ft)
Hovering ceiling IGE, ISA+20°C:
A — 2,955 m (9,700 ft)
B — 2,135 m (7,000 ft)
C — 1,400 m (4,600 ft)
Hovering ceiling OGE: A — 2,835 m (9,300 ft)
B — 2,040 m (6,700 ft)
Hovering ceiling OGE, ISA+20°C:
A — 2,040 m (6,700 ft)
B — 1,220 m (4,000 ft)
Range with max fuel, no reserves:
A — 315 nm (583 km; 363 miles)
B — 305 nm (565 km; 351 miles)
C — 296 nm (548 km; 341 miles)
Endurance with max fuel, no reserves:
A — 3 h 43 min
B — 3 h 30 min
C — 3 h 15 min

AGUSTA A 109A (MILITARY, NAVAL and POLICE VERSIONS)

Several non-commercial versions of the A 109A have been or are being developed by Agusta. In general, their configuration, structure and power plant are similar to those of the standard civil production versions, although certain versions can be made available, if required, with non-retractable landing gear. Features of some or all military and naval versions include, as standard, dual controls and instrumentation; rotor brake; tail rotor control magnetic brake; sliding doors; environmental control system; emergency flotation gear; armoured seats; crash-worthy fuel tanks; heavy-duty battery; particle separator; external cargo hook; multi-purpose universal supports for external stores; rescue hoist; high-load cargo floor; and infra-red suppression system. The naval versions, specially configured for shipboard compatibility, can be equipped with four-axis ASE, radar altimeter, internal auxiliary fuel tanks, non-retractable landing gear, search radar, anchorage points for deck lashings, and an automatic navigation system.

The principal military, naval and other non-commercial versions available or under development are as follows:

Aerial scout. For forward area combat reconnaissance, and command and direction of attack helicopter team. Secondary capability for support of covert operations; artillery observation and adjustment; radio relay; and emergency rescue of combat aircrew. Can be armed with a 7·62 mm flexible machine-gun, with stabilised sight, plus two XM157 launchers each for seven 2·75 in rockets. Normal crew of three.

Agusta A 109A in electronic warfare configuration
(Air Portraits)

Light attack against tanks and other hard-point targets such as air defence weapons, vehicles and bunkers for which it can be armed with TOW or Hot missile system. Has been extensively and successfully demonstrated (and evaluated by Italian Army Aviation) with Hughes M65 TOW system incorporating undernose telescopic sight unit (TSU), plus four or eight Hughes BGM-71A TOW missiles. Normal crew of two.

Light attack against soft-point targets such as automatic weapons and/or troop formations. Various combinations of armament include a pintle-mounted MG3 7·62 mm machine-gun in each doorway, with 600 rds/gun; a flexible, remotely controlled externally-mounted 7·62 mm gun with 1,000 rds; twin trainable, remotely controlled externally-mounted MG3s, with a total of 2,500 rds; two external pods each containing one or two 7·62 mm machine-guns or one 12·7 mm gun, with varying ammunition capacities; or two gun pods and two launchers each for six 68 mm, seven 70, 75 or 81 mm, or fourteen, eighteen or twenty-eight 50 mm rockets. Normal crew of two.

Command and control. For target designation and direction of helicopter attack force. Can be armed with combination of rockets and flexible machine-guns, as described in preceding paragraph.

Utility. With accommodation or equipment for up to seven troops (transport role); two stretcher patients and two medical attendants in addition to pilot (as ambulance); externally-mounted electrically-operated 150 kg (331 lb) capacity rescue hoist above rear door on starboard side; or underfuselage hook for 907 kg (2,000 lb) slung load. In ambulance version, cabin sidewalls are extended outwards, to enable stretchers to be installed across width of cabin, and modified port-side loading doors are fitted.

ESM/ECM. Electronic warfare version, for military and naval use. Available in two basic forms: with passive ESM equipment only in cabin, plus weapon systems if required; and with passive ESM plus modularised active ECM (jamming), plus any required weapons. Passive ECM include radar warning and locating equipment, interferometer, and an electromagnetic emission analyser. Provision for chaff dispenser to be mounted on tailboom.

Naval. Primary naval missions are anti-submarine classification and attack, anti-surface vessel, electronic warfare, stand-off missile guidance, and reconnaissance. Secondary capabilities for search and rescue, troop transportation, ambulance, flying crane, coastguard patrol, and inter-ship liaison duties. Configurations for electronic warfare and utility roles are generally similar to those described in preceding 'Utility' and 'ESM/ECM' paragraphs. There is standard accommodation for a two- or three-man crew, and complete instrumentation for day and night sea operation in all weathers; ASE is supplied by the standard duplicated hydraulic systems, and MAD by the self-contained third system. Electrical system capacity is increased to cater for higher power demand; a four-axis cross-country autopilot system and emergency flotation gear are optional.

For the ASW role, specialised equipment includes one or two homing torpedoes and six marine markers. Detection of the submarine can be carried out either by the parent ship (in which case the A 109A is acting as a weapon carrier system) or by the helicopter's onboard retractable classification and localisation equipment (MAD). For the ASV role the naval A 109A carries a high-performance long-range search radar with high discrimination in rough sea conditions. The surface attack is performed with AS.12 or AM-10 air-to-surface wire-guided missiles. For the TG-2 (stand-off missile guidance) mission, the helicopter is equipped with a special system to control and guide a ship-launched Otomat missile. For armed patrol, the naval A 109A is equipped with a search radar and armament to customer's requirements. The coastguard patrol configuration includes a search radar, a low light level TV camera and a special installation for external high efficiency loudspeakers.

Police and other patrol duties. For police patrol (including armed patrol) and surveillance, coastal patrol, pollution patrol, overland and oversea search and rescue, forestry patrol and firefighting, and similar utility missions. Principal equipment for SAR versions includes search radar, rescue hoist, stretcher/first aid kits, radar altimeter, skis or emergency flotation gear, ASE, and flare/smoke grenades. For aerial patrol it can include external loudspeakers, low light level TV, pollution monitoring equipment, system for spraying chemical retardants, and other items depending upon requirements of mission.

Typical Weights (A: armed scout, 2 machine-gun pods and 2 flexible machine-guns; B: anti-tank with 4 TOW; C: anti-tank with 6 Hot; D: attack/suppression with 2 machine-gun pods and 14 × 2·75 in rockets; E: attack/area target destruction, with armament as for D; F: transport for 7 fully-equipped troops; G: ambulance with 2 stretchers and 2 medical attendants; H: electronic warfare with ESM and ECM; J: ASW with 1 torpedo; K: ASW with 2 torpedoes; L: ASV with 2 air-to-surface missiles; M: naval stand-off missile guidance with TG-2 system; N: naval armed patrol; P: naval electronic warfare with passive ESM only; Q: armed civil patrol; R: unarmed patrol/surveillance; S: sea search and rescue):

Weight empty, equipped: J, K	1,551 kg (3,419 lb)
F	1,560 kg (3,439 lb)
H	1,627 kg (3,587 lb)
G	1,630 kg (3,594 lb)
P	1,672 kg (3,686 lb)
L	1,677 kg (3,697 lb)
Q, R, S	1,683 kg (3,710 lb)
M	1,697 kg (3,741 lb)
N	1,710 kg (3,770 lb)
A	1,721 kg (3,794 lb)
D	1,743 kg (3,843 lb)
E	1,765 kg (3,891 lb)
C	1,806 kg (3,982 lb)
B	1,889 kg (4,164 lb)
Crew: F, G (1)	80 kg (176 lb)
B, C, D, E, H, J, K, L, N, P, Q, R (2)	160 kg (353 lb)
A, M, S (3)	240 kg (529 lb)
Armament/mission equipment/payload:	
N	15 kg (33 lb)
M	75 kg (165 lb)
B	96 kg (212 lb)
S	156 kg (344 lb)
R	165 kg (364 lb)
P	167 kg (368 lb)
A	178 kg (392 lb)
C	192 kg (423 lb)
L	257 kg (567 lb)
H	270 kg (595 lb)
D	275 kg (606 lb)
Q	318 kg (701 lb)
G	320 kg (705 lb)
J	345 kg (761 lb)
E	352 kg (776 lb)
K	595 kg (1,312 lb)
F	630 kg (1,389 lb)
Fuel: K	294 kg (648 lb)
E	323 kg (712 lb)
F	330 kg (728 lb)
A, B, C, D, G, H	415 kg (915 lb)
J, L	416 kg (917 lb)
Q	439 kg (968 lb)
S	521 kg (1,149 lb)
M, N, P	545 kg (1,202 lb)
R	560 kg (1,234 lb)
T-O weight: N	2,430 kg (5,358 lb)
G	2,445 kg (5,390 lb)
H, J	2,472 kg (5,450 lb)
L	2,510 kg (5,534 lb)
P	2,544 kg (5,609 lb)
A	2,554 kg (5,630 lb)
M	2,557 kg (5,637 lb)
B	2,560 kg (5,644 lb)
R	2,568 kg (5,661 lb)
C	2,573 kg (5,673 lb)
D	2,593 kg (5,717 lb)
E, F, K, Q, S (= max T-O weight)	2,600 kg (5,732 lb)

Performance: Generally similar to that of civil models

AGUSTA A 109A Mk II

This updated version of the A 109A, of which deliveries were expected to begin in September 1981, has a number of changes resulting from improved design, better manufacturing techniques, and operating experience with the initial model. All operating parameters are improved as a result of an increase in transmission rating to 552 kW (740 shp) max continuous (two engines) and 313 kW (420 shp) for single-engined operation. A new tail rotor driveshaft with two-point suspension increases power transmission and virtually eliminates vibration; tail rotor blade life and reliability are increased by the use of new Nomex core material and improved bonding techniques. New self-damping engine mounts provide better load path distribution and eliminate transmitted vibration and noise. New integral-design oil coolers and blowers permit operation in higher ambient temperatures (−50° to +50°C). Structural redesign of the tailboom combines increased fatigue resistance with improved maintainability. A new, higher-pressure hydraulic system (107 bars; 1,550 lb/sq in), with less plumbing and lower weight, improves system performance at extreme low temperatures. Avionics and instrument layout are improved. Additional access panels, and a removable floor in the baggage compartment, make inspection and maintenance easier.

Typical Weights (military Mk II):

Basic weight empty	1,418 kg (3,126 lb)
Weight empty, equipped:	
utility	1,560 kg (3,439 lb)
ESM/ECM	1,627 kg (3,587 lb)
ambulance	1,630 kg (3,594 lb)
scout, atttack, air defence	1,650 kg (3,638 lb)
anti-tank	1,790 kg (3,946 lb)
Crew: utility, ambulance (1)	80 kg (176 lb)
attack, anti-tank, air defence, ESM/ECM (2)	160 kg (353 lb)
scout (3)	240 kg (529 lb)
Armament/equipment/payload:	
ambulance (1 medical attendant)	80 kg (176 lb)
air defence (8 missiles)	150 kg (331 lb)
anti-tank (8 missiles)	196 kg (432 lb)
ESM/ECM (radar warning, deception jammer, noise jammer, ESM equipment)	270 kg (595 lb)
scout (2 podded 12·7 mm and 2 pintle-mounted 7·62 mm machine guns)	287 kg (633 lb)
attack (2 podded 12·7 mm machine-guns and 14 rockets in pods)	344 kg (758 lb)
utility (7 equipped troops)	630 kg (1,389 lb)
Fuel: utility	330 kg (728 lb)
scout	423 kg (932 lb)
attack	446 kg (983 lb)
anti-tank	454 kg (1,001 lb)
air defence, ambulance	540 kg (1,190 lb)
ESM/ECM	543 kg (1,197 lb)
T-O weight: ambulance	2,330 kg (5,136 lb)
air defence	2,500 kg (5,512 lb)
scout, attack, anti-tank, ESM/ECM, utility (= max T-O weight)	2,600 kg (5,732 lb)

Performance (S/L, ISA, except where indicated. A: AUW of 2,250 kg; 4,960 lb, B: AUW of 2,450 kg; 5,400 lb, C: AUW of 2,600 kg; 5,732 lb):

Never-exceed speed:	
A, B, C	168 knots (311 km/h; 193 mph)
Max cruising speed: A	155 knots (287 km/h; 178 mph)
B	150 knots (278 km/h; 173 mph)
C	147 knots (272 km/h; 169 mph)
Econ cruising speed: A, B, C	As A 109A
Max rate of climb at S/L: A	640 m (2,100 ft)/min
B	555 m (1,820 ft)/min
C	503 m (1,650 ft)/min
Rate of climb at S/L, one engine out:	
A	158 m (520 ft)/min
B	108 m (355 ft)/min
C	78 m (255 ft)/min
Service ceiling, 30·5 m (100 ft)/min rate of climb, at max continuous power: A	5,180 m (17,000 ft)
B	4,815 m (15,800 ft)
C	4,570 m (15,000 ft)
Hovering ceiling IGE: A	3,750 m (12,300 ft)
B	2,985 m (9,800 ft)
C	2,410 m (7,900 ft)
Hovering ceiling OGE: A	2,880 m (9,450 ft)
B	2,075 m (6,800 ft)
C	1,495 m (4,900 ft)
Range with max fuel, no reserves:	
A	320 nm (593 km; 368 miles)
B	310 nm (574 km; 357 miles)
C	300 nm (556 km; 345 miles)
Endurance with max fuel, no reserves:	
A, B, C	As A 109A

AGUSTA A 129

Preliminary design of this light military helicopter began in 1978, since when it has undergone a number of configuration changes. It entered the final design stage in 1980, and is due to make its first flight in September 1983.

Initially, the A 129 is intended primarily for specialised attack against armoured targets with anti-tank or area suppression weapons, and will have full night/bad weather combat capability. It is also suitable, in an export version, for the advanced scouting role.

Agusta is building four flying prototypes, including one funded by the company, plus a ground test aircraft. Italian government approval has been given for an initial production batch of 67 A 129s, of which 60 will equip two Italian Army Aviation operational squadrons, the other seven being used for training. Subject to a production go-ahead, scheduled to follow the completion of flight testing in mid-1984, the A 129 should enter service by the end of 1985. A requirement exists for an additional 30 aircraft, plus reserves, to equip a third operational squadron.

On 28 May 1981 Agusta proposed the submission to the US Army in 1984 of two A 129 prototypes configured to the latter's Advanced Scout Helicopter (ASH) requirements. These A-129/ASH aircraft would have a Martin Marietta day and night mast-mounted sight (MMS) incorporating a laser rangefinder, laser target designator, laser spot tracker, daytime TV and FLIR sighting; Martin Marietta pilot's night vision sensor (PNVS) for nap-of-the-earth operation at night and in bad weather; and provision to carry multipurpose lightweight missiles (MLMs) for air-to-air and defence suppression use.

The following description applies to the A 129 prototypes currently under construction:

TYPE: Light anti-tank, attack and advanced scout helicopter.

ROTOR SYSTEM: Fully-articulated four-blade main rotor and two-blade tail rotor, each with elastomeric bearings. Main rotor blades, which have a very low vibration level, each consist of a glassfibre spar, Nomex honeycomb leading- and trailing-edge, and skin of composite materials. They have a ballistic tolerance against hits from 12·7 mm ammunition. Hub has a swashplate of glassfibre composites; all mechanical linkages and moving parts are housed inside the rotor mast. There are no lubricated bearings in the rotor head. Transmission rating 895 kW (1,200 shp) (two engines), 626 kW (840 shp) for single-engined operation. Tail rotor blades are also of composite materials.

WINGS: Cantilever mid-mounted stub-wings, aft of rear cockpit in plane of main rotor mast.

FUSELAGE: Conventional semi-monocoque structure. Forward fuselage of metal construction; nosecone and tailboom of composite materials. Small and narrow frontal area. Anti-roll bar in forward fuselage for crew protection; armour protection for vital areas of power plant. Airframe has a ballistic tolerance against 12·7 mm armour-piercing ammunition, and meets the crashworthiness standards of MIL-STD-1290.

TAIL UNIT: Sweptback main fin, with tail rotor mounted near top on port side. Small underfin, serving also as mount for tailwheel. Tailplane mid-mounted on tailboom in line with fin leading-edges. All tail surfaces built of composite materials.

LANDING GEAR: Non-retractable tailwheel type, of Magnaghi/Messier-Hispano-Bugatti design, with single wheel on each unit.

POWER PLANT: Two Rolls-Royce Gem-2-3 turboshaft engines, each with a max continuous rating of 602·5 kW (808 shp) for normal twin-engined operation; a max contingency rating (one engine out) of 716·5 kW (961 shp) for 2½ min; and an emergency rating (one engine out, S/L, ISA) of 772 kW (1,035 shp) for 20 s. Production aircraft will have engines licence-built in Italy by Piaggio. Two separate fuel systems, with cross-feed capability, self-sealing and crash-resistant tanks, self-sealing lines, and digital fuel feed control. Low noise levels and low infra-red signature.

ACCOMMODATION: Pilot and co-pilot/gunner in tandem-mounted separate cockpits, each cockpit having a low-glint canopy with Explosive Technology blow-out side panels for emergency exit. Elevated rear (pilot's) cockpit. Energy-absorbing armoured seats.

SYSTEMS, AVIONICS AND EQUIPMENT: All main functions of aircraft are handled and monitored by a Harris Corporation digital integrated multiplex system which controls com, nav, engine, fire control, power distribution and utility systems. Full automatic stabilisation equipment standard. Avionics include advanced com/nav equipment, and both active and passive self-protection systems (Elettronica/E-Systems radar warning receiver, Tracor chaff/flare dispenser, ITT radar jammer, Sanders infra-red jammer, Perkin-Elmer radar and laser warning receiver, and engine infra-red suppressor). Automatic fire extinguishing system.

ARMAMENT: Four underwing attachments, inner pair stressed for loads of up to 300 kg (661 lb) each, outer pair (at wingtips) for up to 200 kg (441 lb) each). Initial armament of up to eight Hughes M65 TOW wire-guided anti-tank missiles (two, three or four in pod suspended from each wingtip station); with these can be carried, on the inboard stations, either two 12·7 mm machine-gun pods or two launchers each for seven 2·75 in air-to-surface rockets. For general attack missions, rocket launchers can be carried on all four stations. Telescopic sight unit for TOW missiles, fitted with laser rangefinder and FLIR for target acquisition and designation. Alternatively, will be able to carry eight Hellfire anti-tank missiles (four beneath each wingtip), plus Martin Marietta mast-mounted sight, pilot's night vision equipment, and integrated helmet and display sight system. Other underwing loads can include two 12·7 mm machine-gun pods plus two nineteen-tube launchers for 2·75 in rockets; or two nineteen-tube plus two seven-tube launchers; or grenade launchers. Provision for auxiliary fuel tanks to be carried on inboard underwing stations.

DIMENSIONS, EXTERNAL:

Diameter of main rotor	11·90 m (39 ft 0½ in)
Diameter of tail rotor	2·24 m (7 ft 4¼ in)
Wing span	3·20 m (10 ft 6 in)
Width over TOW pods	3·60 m (11 ft 9¾ in)
Length overall, both rotors turning	14·29 m (46 ft 10½ in)
Length of fuselage	12·275 m (40 ft 3¼ in)
Fuselage: Max width	0·95 m (3 ft 1½ in)
Height over tail fin, tail rotor horizontal	2·65 m (8 ft 8¼ in)
Height over tail, tail rotor turning	3·315 m (10 ft 10½ in)
Height to top of rotor hub	3·35 m (11 ft 0 in)
Tailplane span	3·00 m (9 ft 10 in)
Wheel track	2·20 m (7 ft 3½ in)
Wheelbase	6·955 m (22 ft 9¾ in)

Full-size mockup of the Agusta A 129 anti-tank helicopter

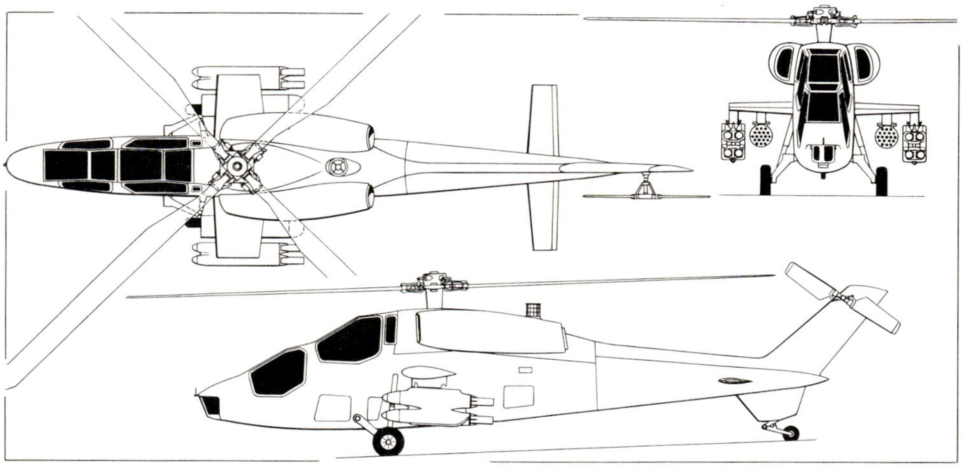

Agusta A 129 light anti-tank, attack and advanced scout helicopter *(Pilot Press)*

AREAS:

Main rotor disc	111·2 m² (1,196·95 sq ft)
Tail rotor disc	3·94 m² (42·42 sq ft)

WEIGHTS (estimated):

Weight empty, equipped	2,529 kg (5,575 lb)
Max internal fuel load	650 kg (1,433 lb)
Max mission T-O weight	3,655 kg (8,058 lb)

PERFORMANCE (estimated):

At max mission T-O weight, at 2,000 m (6,560 ft), ISA+20°C, except where indicated, the A 129 is designed to exceed the following performance requirements:

Max level speed at S/L	145 knots (270 km/h; 168 mph)
Cruising speed	135 knots (250 km/h; 155 mph)
Max rate of climb at S/L	600 m (1,968 ft)/min
Hovering ceiling IGE	2,920 m (9,580 ft)
Hovering ceiling OGE	2,000 m (6,560 ft)
Endurance (anti-tank mission), no reserves	2 h 30 min
Max range, standard tanks, no reserves	312 nm (580 km; 360 miles)
Max endurance, standard tanks, no reserves	3 h 48 min

AGUSTA-BELL 205

The Agusta-Bell 205 is a multi-purpose utility helicopter, corresponding to the UH-1D/UH-1H versions described under the Bell Helicopter Textron heading in the US section. It is fitted with IFR and night flying instruments, and for normal operation only one pilot is needed. Power plant is a 1,044 kW (1,400 shp) Avco Lycoming T53-L-13B turboshaft engine, flat rated to 820 kW (1,100 shp) for take-off.

The AB 205 is in service with the Italian armed forces and has been ordered by many other countries.

WEIGHTS: As Bell Model 205 except:

Weight empty (standard)	2,177 kg (4,800 lb)

PERFORMANCE (at max T-O weight, ISA):

Max level speed at S/L	120 knots (222 km/h; 138 mph)
Cruising speed	110 knots (204 km/h; 127 mph)
Max rate of climb at S/L	512 m (1,680 ft)/min
Service ceiling	4,570 m (15,000 ft)

AGUSTA-BELL 206B JETRANGER III

The JetRanger has been manufactured under licence from Bell since the end of 1967; deliveries began in 1972 of the Agusta-Bell 206B JetRanger II, and of the JetRanger III at the end of 1978. A description of the JetRanger III appears under the Bell Helicopter Textron entry in the US section.

AGUSTA-BELL 206L-1 LONGRANGER II

This aircraft is described under the Bell Helicopter Textron entry in the US section. Licence production by Agusta began in 1976.

AGUSTA-BELL 212

The Agusta-Bell 212 is a twin-engined utility transport helicopter particularly suited to military or civilian passenger transport duties. Its general configuration is similar to that of the Bell Model 212 Twin Two-Twelve, described in the US section.

Recent customers for the AB 212 have included the Spanish Army, which received four in early 1981.

The extensively-modified AB 212ASW naval version produced by Agusta is described separately.

DIMENSIONS, EXTERNAL: As Bell Model 212 except:

Diameter of main rotor	14·63 m (48 ft 0 in)
Length overall, rotors turning	17·40 m (57 ft 1 in)

WEIGHTS: As Bell Model 212 except:

Weight empty (standard)	2,630 kg (5,800 lb)

PERFORMANCE (at AUW of 4,536 kg; 10,000 lb, ISA):

Cruising speed at S/L	110 knots (204 km/h; 127 mph)
Max rate of climb at S/L	567 m (1,860 ft)/min
Service ceiling	5,180 m (17,000 ft)
Hovering ceiling IGE	3,960 m (13,000 ft)
Hovering ceiling OGE	3,050 m (10,000 ft)

Max range at 1,525 m (5,000 ft) with standard fuel, no reserves:

on two engines	267 nm (494 km; 307 miles)
on one engine	318 nm (589 km; 366 miles)

AGUSTA-BELL 212ASW

The AB 212ASW is an extensively modified version of the AB 212, intended primarily for anti-submarine search

Agusta-Bell 212ASW anti-submarine and anti-surface-vessel helicopter

and attack missions, and for attacks on surface vessels, but suitable also for search and rescue and utility roles. It benefits from considerable naval operational experience gained with the single-engined AB 204AS, and because of its similarity in size to the 204AS can operate from the same small ship decks.

Apart from some local strengthening and the provision of deck-mooring equipment, the airframe structure remains essentially similar to that of the commercial Model 212 and military UH-1N, described under the Bell Helicopter Textron entry in the US section. Main differences from the Agusta-Bell 212 are as follows:

TYPE: Twin-engined anti-submarine and anti-surface-vessel helicopter.

POWER PLANT: One Pratt & Whitney Aircraft of Canada PT6T-6 Turbo Twin Pac, rated at 1,398 kW (1,875 shp). Protection against salt water corrosion. Provision for one internal or two external auxiliary fuel tanks.

ACCOMMODATION: Normal crew of three or four. Volume of cabin is 6·1 m^3 (215 cu ft), with floor area of 5·0 m^2 (54 sq ft). With sonar installed, volume is reduced to 5·1 m^3 (180 cu ft). Naval 212 can accommodate two pilots and seven passengers; or two pilots, four stretcher patients and attendant. Single sliding door, with jettisonable emergency exit panel, on each side.

SYSTEMS: Standard duplicated hydraulic systems for flight controls, as in AB 212. The hydraulic system operates the automatic flight control system. Self-contained hydraulic system for operation of sonar, rescue hoist and other utilities. Electrical system capacity increased to cater for higher power demand (28V DC, and three-phase 200/115V or single-phase 26V AC at 400Hz); the two standard generators are integrated with a 20kVA alternator.

AVIONICS AND EQUIPMENT: Complete instrumentation for day and night sea operation in all weathers. Avionics installed are AN/ARC-159 UHF transceiver, Collins SSB/DSB 718 U-5 HF transceiver, and Agusta AG-03-M intercom, for communications; Marconi Avionics AD 370B ADF, Hoffman AN/ARN-91 Tacan and Collins AN/ARA-50 homing UHF, for navigation assistance; Aeritalia (Honeywell) AN/APN-171 radar altimeter, Canadian Marconi AN/APN-208(V)2 Doppler radar, Canadian Marconi CMA-708B/ASW navigation computer, and automatic flight control system with General Electric SR-3 gyro platform, Agusta ASE-531A automatic stabilisation equipment and Agusta AATH-547A automatic approach to hover, for automatic navigation; Siemens AN/APX-77 IFF/SIF transponder; SMA/APS series search radar and Motorola SST-119X radar transponder; and Bendix AN/AQS-13B sonar for ASW search.

ARMAMENT AND OPERATIONAL EQUIPMENT: Weapons may consist of two homing torpedoes, depth charges or two air-to-surface missiles. Rescue hoist, capacity 270 kg (600 lb), standard. Provisions for auxiliary installations such as a 2,270 kg (5,000 lb) capacity cargo sling, inflatable emergency pontoons, internal and external auxiliary fuel tanks, according to mission.

ASW MISSION: The basic sensor system employed for the ASW search and attack mission is a low-frequency variable-depth sonar, with a max operating depth of 137 m (450 ft). The automatic navigation system permits the positioning of the helicopter over any desired 'dip' point of a complex search pattern. The position of the helicopter, computed by the automatic navigation system, is integrated with sonar target information in the radar tactical display where both the surface and the underwater tactical situations can be continuously monitored. Additional navigation and tactical information is provided by accurate UHF direction-finding equipment, from an A/A mode-capable Tacan and from a radar transponder. The automatic flight control system (AFCS) integrates the basic automatic stabilisation equipment with signal output from the radar altimeter, the Doppler radar, sonar cable angle signals and outputs from the dry cable transducer. The effectiveness of this system results in hands-off flight from cruise condition to sonar hover in all weathers and under rough sea conditions. A specially designed cockpit display shows the pilots all flight parameters for each phase of the ASW operation. The attack mission is carried out with two homing torpedoes, or with depth charges.

ASV MISSION: For this mission the AB 212ASW carries a high-performance long-range search radar, with a very efficient scanner design and installation possessing high discrimination in rough sea conditions. Provisions have also been made to permit incorporation of future radar system developments. The automatic navigation systems and the search radar are integrated to permit a continuously updated picture of the tactical situation. Provisions are also incorporated for the installation of the most advanced ECM systems. The surface attack is performed with air-to-surface wire-guided missiles. In operation, the co-pilot aims and 'flies' the missiles to the target through an XM-58 gyro-stabilised sight system.

STAND-OFF MISSILE GUIDANCE MISSION: In this mission the AB 212ASW, with special equipment, can provide mid-course passive guidance for the ship-launched Otomat 2 surface-to-surface missile. Equipment includes an SMA/APS series search radar and a TG-2 real-time target data transmission system for guidance of the missile.

DIMENSIONS, EXTERNAL: As AB 212, except:
- Max width:
  - with torpedoes 3·95 m (12 ft 11½ in)
  - with missiles 4·17 m (13 ft 8¼ in)

WEIGHTS (A: ASW mission with Mk 46 torpedoes; B: ASV mission with AS.12 missiles; C: search and rescue mission; all at S/L, ISA):
- Weight empty, equipped:
  - A, B, C 3,420 kg (7,540 lb)
- Crew of three: A, B, C 240 kg (529 lb)
- Mission equipment:
  - A (two Mk 46 torpedoes) 490 kg (1,080 lb)
  - B (AS.12 installation and XM-58 sight) 180 kg (396 lb)
  - C (rescue hoist) 40 kg (89 lb)
- Full fuel (normal tanks) 1,021 kg (2,250 lb)
- Auxiliary external tanks 32 kg (70 lb)
- Auxiliary fuel 356 kg (785 lb)
- Mission T-O weight: A 5,070 kg (11,176 lb)
  - B 4,973 kg (10,961 lb)
  - C 4,937 kg (10,883 lb)

PERFORMANCE (at max T-O weight, except where indicated, ISA):
- Never-exceed speed 130 knots (240 km/h; 150 mph)
- Max level speed at S/L 106 knots (196 km/h; 122 mph)
- Max cruising speed with armament 100 knots (185 km/h; 115 mph)
- Max rate of climb at S/L: A 396 m (1,300 ft)/min
- Rate of climb at S/L, one engine out:
  - A 61 m (200 ft)/min
- Hovering ceiling IGE: A 3,200 m (10,500 ft)
- Hovering ceiling OGE:
  - A at AUW of 4,763 kg (10,500 lb) 396 m (1,300 ft)
- Search endurance (A) with 50% at 90 knots (167 km/h; 103·5 mph) cruise and 50% hovering OGE, 10% reserve fuel 3 h 12 min
- Search range (B) with 10% reserve fuel 332 nm (615 km; 382 miles)
- Endurance (B), no reserves 4 h 7 min
- Endurance (C) at 90 knots (167 km/h; 103·5 mph) search speed 5 h 4 min
- Max range with auxiliary tanks, 100 knots (185 km/h; 115 mph) cruise at S/L, 15% reserves 360 nm (667 km; 414 miles)
- Max endurance with auxiliary tanks, no reserves 5 h 0 min

AGUSTA-BELL 412

Series production by Agusta of the Bell Model 412 (see US section) was due to begin in late 1981 or early 1982. Civil, military and naval versions will be available.

The Agusta military version will be capable of performing medevac, tactical support, logistics transport, SAR and patrol duties, and of being used effectively against armoured vehicles. Equipment will include a secure voice communications system, and a self-contained navigation system permitting nap-of-the-earth flying at night or in low-visibility conditions by day. Armament system development is aimed primarily at the installation of a large-calibre cannon and high-efficiency rocket launchers, coupled with the use of a target acquisition sight. A prototype military AB 412 was due to be completed by the end of 1981, and systems qualification by the Summer of 1982.

Proposed weapons include a 20 mm cannon, 19-tube launchers for 2·75 in rockets, integral 12·7 mm gun/rocket pods, and four air-to-air or defence suppression missiles. Jeeps, self-propelled guns, mortars and other field equipment can be transported on the external cargo sling. The cabin can accommodate 14 combat-equipped troops, or six stretcher patients and two medical attendants. For SAR duties, an externally-mounted 272 kg (600 lb) capacity rescue hoist can be fitted.

AGUSTA-SIKORSKY SH-3D and SH-3H

During 1967, Agusta began the construction under licence of Sikorsky SH-3D anti-submarine helicopters for the Italian Navy. Deliveries began in 1969. Additional orders have since been placed, both for the Italian armed forces and for other navies, in various configurations including ASW, VIP transport and rescue. The VIP transport version, designated **SH-3D/TS** (Trasporto Speciale), serves with the 31° Stormo of the Italian Air Force and with some foreign air forces. Current production aircraft are to SH-3H standard.

Apart from some local strengthening and an improved horizontal tail surface, the Agusta-built airframe remains essentially similar to that of the SH-3D/H described under the Sikorsky heading in the US section of this edition of *Jane's*. The Agusta SH-3H is capable of operation in the roles of anti-submarine search, classification and strike; anti-surface-vessel (ASV); anti-surface-missile defence (ASMD); electronic warfare (EW); tactical troop lift; search and rescue (SAR); vertical replenishment; and casualty evacuation.

POWER PLANT: Two 1,118 kW (1,500 shp) General Electric T58-GE-100 turboshaft engines, mounted side by side above the cabin. An optional anti-ice/sand shield can be provided. Fuel in underfloor bag tanks with a total capacity of 3,180 litres (840 US gallons). Internal auxiliary fuel tank may be fitted for long-range ferry purposes. Pressure and gravity refuelling points.

ACCOMMODATION: Crew of four in ASW role; accommodation for up to 31 paratroops in troop lift role, 15 stretchers and a medical attendant in casualty evacuation configuration, and up to 25 survivors in SAR role.

SYSTEMS: Three main hydraulic systems. Primary and auxiliary systems operate main rotor control. Utility system for landing gear, winches and blade folding, pressure 207 bars (3,000 lb/sq in). Electrical system includes two 20kVA 200V three-phase 400Hz engine-driven generators, a 26V single-phase AC supply fed from the aircraft's 22Ah nickel-cadmium battery through an inverter, and DC power provided as a secondary system from two 200A transformer-rectifier units.

OPERATIONAL EQUIPMENT (ASW/ASV role): As equipped for this role the Agusta SH-3H is a fully integrated all-weather weapon system, capable of operating independently of surface vessels, and has the following equipment and weapons to achieve this task: low-frequency 360° depth sonar; Doppler radar and ASW automatic navigation system; SMA/APS series radar with one or two transceivers, with ventral radome for 360° coverage; radio altimeter; AFCS; marine markers and smoke floats; four homing torpedoes or four depth charges. The AFCS provides three-axis stabilisation in pilot-controlled manoeuvres, attitude hold, heading hold and height hold in cruising flight; controlled transition manoeuvres to and from hover; automatic height control and plan position control in the hover; and trim facility. According to the threat, the Agusta SH-3H can be equipped with medium-range (four AS.12 air-to-surface wire-guided) missiles or long-range (two Sea Killer Mk 2 or Exocet AM-39/Harpoon type) missiles. The Sistel Sea Killer Mk 2 is an all-weather day and night anti-ship missile with a range of 13·5 nm (25 km; 15·5 miles); guidance: sea skimming in elevation, radar in azimuth. The SMA/APS series radar has been specially designed to operate in a dense electronic emission environment and has a special interface to draw out target data to feed the computer for the long-range missiles. Provisions are also incorporated for the installation of the most advanced EW systems.

OPERATIONAL EQUIPMENT (Search and rescue and transport roles): Search radar, and variable-speed hydraulic rescue hoist of 272 kg (600 lb) capacity mounted above starboard side cargo door.

WEIGHTS:
- Internal load capacity (cargo) 2,720 kg (6,000 lb)
- Max external load capacity (with low-response sling) 3,630 kg (8,000 lb)

Agusta-Sikorsky SH-3H ASW/SAR helicopter in the insignia of the Peruvian Navy

Max T-O weight	9,525 kg (21,000 lb)
PERFORMANCE (at max T-O weight):	
Never-exceed speed	144 knots (267 km/h; 165 mph)
Typical cruising speed	120 knots (222 km/h; 138 mph)
Max rate of climb at S/L	670 m (2,200 ft)/min
Service ceiling	3,720 m (12,200 ft)
Hovering ceiling IGE	2,500 m (8,200 ft)
Hovering ceiling OGE	1,130 m (3,700 ft)
Range with 31 troops	314 nm (582 km; 362 miles)
Range with max standard fuel	630 nm (1,166 km; 725 miles)

AGUSTA-SIKORSKY HH-3F (S-61R)

Agusta began production of this multi-purpose search and rescue helicopter in 1974. Twenty were built initially, 12 for SAR duties with the Italian Air Force and others for foreign operators. Deliveries started in 1976.

Details of the HH-3F can be found under the Sikorsky heading in the US section of this edition.

Agusta-Sikorsky HH-3F (S-61R) search and rescue helicopter of the Italian Air Force

ELICOTTERI MERIDIONALI SpA

WORKS: Via Giovanni Agusta 1, Frosinone
Telephone: (0775) 82801
Telex: 860026 ELMEFI
PRESIDENT: Cav del lavoro Conte Corrado Agusta
VICE-PRESIDENT: Dott Ing Pietro Fascione
MANAGING DIRECTOR: Dott Ing Giorgio Brazzelli
GENERAL MANAGER: Dott Ing Pietro Tana

This company was formed with assistance from Agusta and began to operate in October 1967. Initially, its activities consisted of overhauling helicopters of the Italian armed forces and other organisations, and the manufacture of helicopter components and sub-assemblies. In 1968 Meridionali acquired rights to the co-production, marketing and servicing of the Boeing Vertol CH-47C Chinook transport helicopter for customers in Italy and several foreign countries. Italian production of the CH-47C airframe is undertaken by SIAI-Marchetti.

Meridionali, whose works occupy a total area of more than 300,000 m² (3,229,170 sq ft), participates in the manufacturing programmes for the Agusta A 109A, Agusta-Bell 205, 206B and 212, and Agusta-Sikorsky SH-3H helicopters. It is also distributor in Italy for the Allison 250 series of turboshaft engines, and has complete facilities for overhaul, repair and field assistance.

Italian-built Boeing Vertol CH-47C helicopter in the insignia of Italian Army Aviation

EM (BOEING VERTOL) CH-47C

Italian manufacture of the CH-47C began in the Spring of 1970, initially to meet an order for Italian Army Aviation. Total sales had reached nearly 190 by the Spring of 1981, customers including the armed forces of Egypt (15), Greece (6), Iran (95), Libya (20), Morocco (6) and Tanzania (2).

Agreement has been reached for EM to produce a modernised version of the Chinook, generally similar to the Boeing Vertol CH-47D model (which see). A ground test vehicle should begin tests in early 1983, and the improved version will become available both as a new-production aircraft and as a retrofit for existing CH-47Cs.

SIAI-MARCHETTI SpA

MANAGEMENT: Via Indipendenza 2, 21018 Sesto Calende (Varese)
Telephone: (0331) 924421
Telex: 332601 SIAIAV I
AERODROME AND MAIN WORKS: Vergiate (Varese)
OTHER WORKS: Sesto Calende (Varese), Malpensa and Borgomanero
ROME OFFICE: Via Barberini 36, 00187 Rome
Telephone: (06) 482811
PRESIDENT: Dott Ermenegildo Marelli
MANAGING DIRECTOR: Dott Ing C. Camposampiero
ADVERTISING MANAGER: Alberto Menozzi

Founded in 1915, the SIAI-Marchetti company was known originally as Savoia-Marchetti, producing a wide range of military and civil landplanes and flying-boats. Its current products include civil and military light aircraft of its own design or development. Since the 1970s it has been engaged in the co-production with Agusta of Boeing Vertol CH-47C, Bell 204/205/212/412, and Sikorsky SH-3D/H and HH-3F helicopters. On 1 January 1981 it combined with Industria Aeronautica Meridionale, under a single management and with SIAI-Marchetti as the leading company, to form the Aircraft Division of Agusta SpA.

SIAI-Marchetti is engaged in the overhaul and repair of various types of large aircraft (notably the C-130 Hercules) serving with the Italian Air Force. It also participates in national or multi-national programmes for the Aeritalia G222, Panavia Tornado, Airbus A310, Boeing 767, Dassault-Breguet Falcon 10 and Lockheed TriStar.

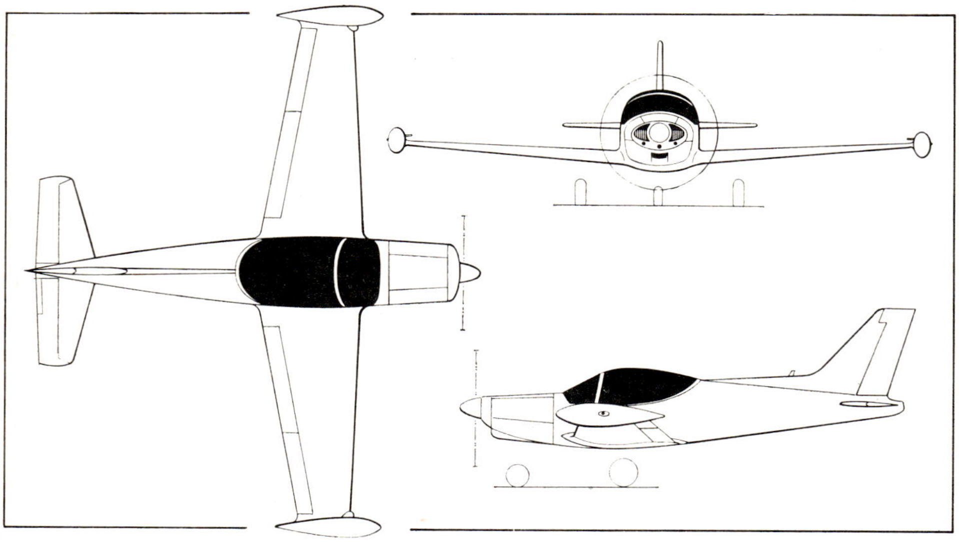

SIAI-Marchetti SF.260M military trainer (Avco Lycoming O-540-E4A5 engine) *(Roy J. Grainge)*

SIAI-Marchetti SF.260WS Warrior in the insignia of the Somali Aeronautical Corps

The company's works at Vergiate, Sesto Calende, Malpensa and Borgomanero total 1,345,000 m² (14,477,450 sq ft) in area, of which 116,000 m² (1,248,600 sq ft) are covered, and employ nearly 4,000 people.

SIAI-MARCHETTI SF.260M and SF.260W WARRIOR

The prototype for the SF.260 series, known as the F.250, was designed by Dott Ing Stelio Frati and built by Aviamilano. Flown for the first time on 15 July 1964, it was powered by a 186·5 kW (250 hp) Avco Lycoming engine and was certificated for aerobatic flying. A description appeared in the 1965-66 *Jane's*.

The version developed initially, for civil production, was manufactured, at first under licence from Aviamilano, by SIAI-Marchetti, and is designated SF.260. It received FAA type approval on 1 April 1966. Subsequently SIAI-Marchetti became the official holder of the type certificate and of all manufacturing rights in the SF.260; it continued to develop the civil version, described separately, and has also developed three basic versions for military use. Current military models are as follows:

SF.260M. Two/three-seat military trainer, developed from the initial civil SF.260A and flown for the first time on 10 October 1970. Introduced a number of important structural and aerodynamic improvements, many of which were subsequently applied to the later civil versions. Meets the necessary requirements for basic flying training; instrument flying; aerobatics, including deliberate spinning and recovery; night flying; navigation flying; and formation flying.

Production to date of SF.260Ms has included orders from the Italian Air Force (25 SF.260AMI), Belgian Air Force (36 SF.260M), Bolivian Air Force (6), Burmese Air Force (10 SF.260MB), Ecuadorean Air Force (12 SF.260ME), Libyan Arab Air Force (240 SF.260ML, of which 160 were intended to be assembled in that country for various Arab air forces), Moroccan Air Force (2 SF.260MM), Philippine Air Force (32 SF.260MP), Singapore Air Force (22 SF.260MS), Royal Thai Air Force (12 SF.260MT), Tunisian Air Force (6), Zaïre Air Force (23 SF.260MC) and Zambian Air Force (8 SF.260MZ).

SF.260W Warrior. Trainer/tactical support version of SF.260M, first flown (I-SJAV) in May 1972. Two or four underwing pylons, for up to 300 kg (661 lb) of external stores, and cockpit stores selection panel. Able to undertake a wide variety of roles, including low-level strike; forward air control; forward air support; armed reconnaissance; and liaison. Also meets same requirements as SF.260M for use as a trainer. One aircraft (described in 1980-81 and earlier editions of *Jane's*) completed as **SF.260SW Sea Warrior** surveillance/SAR/supply version.

Customers to date for the SF.260W include the Dubai Air Wing (1 SF.260WD), Irish Air Corps (10 SF.260WE), Philippine Air Force (16 SF.260WP), Somali Aeronautical Corps (12 SF.260WS), Tunisian Air Force (12 SF.260WT), Union of Burma Air Force (9 SF.260WB), Comoros (8 SF.260WC) and Zimbabwe Air Force (14).

SF.260TP. Turboprop-powered development of SF.260M/W. Described separately.

By mid-1981 more than 700 civil and military SF.260s of all models had been completed, of which most were for export. Total sales were then in excess of 800.

The following description applies primarily to the SF.260M, but is generally applicable also to the SF.260W Warrior unless otherwise stated.

Type: Two/three-seat fully-aerobatic military light aircraft.

Wings: Cantilever low-wing monoplane. Wing section NACA 64_1-212 (modified) at root, NACA 64_1-210 (modified) at tip. Dihedral 6° 20′ from roots. Incidence 2° 45′ at root, 0° at tip. No sweepback. All-metal light alloy safe-life structure, with single main spar and auxiliary rear spar, built in two portions bolted together at centreline and attached to fuselage by six bolts. Pressformed ribs, with dimpled stiffening holes. Skin, which is butt-joined and flush-riveted, stiffened by stringers between main and rear spars. Differentially-operating Frise-type light alloy mass-balanced ailerons, and electrically-actuated light alloy single-slotted flaps. Flaps operated by torque tube and mechanical linkage, ailerons by pushrods and cables. Servo tab in each aileron.

Fuselage: Semi-monocoque safe-life structure of frames, stringers and flush-riveted skin, exclusively of light alloy except for welded steel tube engine mounting, glassfibre front panel of engine cowling, stainless steel firewall and detachable glassfibre tailcone.

Tail Unit: Cantilever light alloy safe-life structure, with sweptback vertical surfaces, fixed-incidence tailplane and one-piece elevator. Two-spar fin and one-piece tailplane, bolted to fuselage; single-spar elevator, statically and aerodynamically balanced, and balanced rudder. Military models have reinforced tail unit/fuselage joints compared with SF.260C. Rudder and elevator operated by cables. Controllable trim tab in starboard half of elevator; ground-adjustable tab on rudder.

Landing Gear: Electrically-retractable tricycle type, with manual emergency actuation. Inward-retracting main gear, of trailing-arm type, and rearward-retracting nose unit, each embodying Magnaghi oleo-pneumatic shock-absorber (type 2/22028 in main units). Each welded steel tube main leg is hinged to the main and rear spars. Nose unit is of leg-and-fork type, with co-axial shock-absorber and torque strut. Cleveland P/N 3080A main wheels, with size 6·00-6 tube and tyre (6-ply rating), pressure 2·45 bars (35·5 lb/sq in). Cleveland P/N 40-77A nosewheel, with size 5·00-5 tube and tyre (6-ply rating), pressure 1·96 bars (28·4 lb/sq in). Cleveland P/N 3000-500 independent hydraulic single-disc brake and parking brake on each main wheel. Nosewheel steering (20° to left or right) is operated directly by the rudder pedals, to which it is linked by pushrods. Up-lock secures main gear in retracted position during flight; anti-retraction system prevents main gear from retracting whenever strut is compressed by weight of aircraft. Compared with SF.260C, the military models have a reinforced nosewheel drag brace attachment and landing gear retraction supports; increased use of light alloy forgings, instead of welded steel, in certain landing gear components; and an improved retraction locking mechanism. On all models, the mooring point beneath the rear fuselage acts as a tail bumper.

Power Plant: One 194 kW (260 hp) Avco Lycoming O-540-E4A5 flat-six engine, driving a Hartzell HC-C2YK-1BF/8477-8R two-blade constant-speed metal propeller with spinner. Fuel in two light alloy tanks in wings, capacity of each 49·5 litres (10·9 Imp gallons); and two permanent wingtip tanks, capacity of each 72 litres (15·85 Imp gallons). Total internal fuel capacity 243 litres (53·5 Imp gallons), of which 235 litres (51·7 Imp gallons) are usable. Individual refuelling point on top of each tank. In addition, SF.260W may be fitted with two 80 litre (17·5 Imp gallon) auxiliary tanks on underwing pylons. Oil capacity (all models) 11·4 litres (2·5 Imp gallons).

Accommodation (SF.260M; W similar): Side-by-side front seats (for instructor and pupil in SF.260M), with third seat centrally at rear. Front seats are individually adjustable fore and aft, and have forward-folding backs and provision for back-type parachute packs. Dual controls standard. All three seats equipped with lap belts and shoulder harnesses. Baggage compartment aft of rear seat. One-piece fully-transparent rearward-sliding Plexiglas canopy, upper portion of which is tinted. Emergency canopy release handle for each front seat occupant. Steel tube windscreen frame for protection in the event of an overturn. Cabin carpeted, air-conditioned, heated and ventilated; walls thermally insulated and soundproofed by a glassfibre lining. Slots at base of windscreen admit air for windscreen defrosting.

Systems (SF.260M; other models generally similar): Hydraulic equipment for main-wheel brakes only. No pneumatic system. 24V DC electrical system of single-conductor negative earth type, including 70A Prestolite engine-mounted alternator/rectifier and 24V 24Ah Varley battery, for engine starting, flap and landing gear actuation, fuel booster pumps, electronics and lighting. Sealed battery compartment in rear of fuselage on port side. Connection of an external power source automatically disconnects the battery. Heating system for carburettor air intake. Emergency electrical system for extending landing gear if normal electrical actuation fails; provision for mechanical extension in the event of total electrical failure. Cabin heating, and windscreen de-icing and demisting, by heat exchanger using engine exhaust air. Additional manually controlled warm-air outlets for general cabin heating. Oxygen system optional.

Avionics and Equipment (SF.260M; W generally similar): Basic instrumentation and military equipment to customer's requirements. Blind-flying instrumentation and communications equipment optional: typical selection includes dual Collins 20B VHF com; Collins VIR-31A VHF nav; Collins ADF-60A ADF; Collins TDR-90 ATC transponder; Collins PN-101 compass; ID-90-000 RMI; and Gemelli AG04-1 intercom. Landing light in nose, below spinner. Instrument panel can be slid rearward to provide access to rear of instruments.

Armament (SF.260W): Two or four underwing hardpoints, able to carry external stores on NATO standard pylons up to a maximum of 300 kg (661 lb) when flown as a single-seater. Typical alternative loads can include one or two SIAI gun pods, each with one or two 7·62 mm FN machine-guns and 500 rds; two Aerea AL-8-70 launchers each with eight 2·75 in rockets; two LAU-32 launchers each with seven 2·75 in rockets; two Aerea AL-18-50 launchers each with eighteen 2 in rockets; two Aerea AL-8-68 launchers each with eight 68 mm rockets; two Aerea AL-6-80 launchers each with six 81 mm rockets; two LUU-2/B parachute flares; two SAMP EU 32 125 kg general purpose bombs or EU 13 120 kg fragmentation bombs; two SAMP EU 70 50 kg general purpose bombs; Mk 76 11 kg practice bombs; two cartridge throwers for 70 mm multi-purpose cartridges, F 725 flares, or F 130 smoke cartridges; one or two photo-reconnaissance pods with two 70 mm automatic cameras; two supply containers; or two 80 litre (17·5 Imp gallon) auxiliary fuel tanks.

Dimensions, external:

Wing span over tip-tanks	8·35 m (27 ft 4¾ in)
Wing chord at root	1·60 m (5 ft 3 in)
Wing mean aerodynamic chord	1·325 m (4 ft 4¼ in)

Wing chord at tip 0·784 m (2 ft 6⅞ in)
Wing aspect ratio (excl tip-tanks) 6·33
Wing taper ratio 2·24
Length overall 7·10 m (23 ft 3½ in)
Fuselage: Max width 1·10 m (3 ft 7¼ in)
Max depth 1·042 m (3 ft 5 in)
Height overall 2·41 m (7 ft 11 in)
Elevator span 3·01 m (9 ft 10½ in)
Wheel track 2·274 m (7 ft 5½ in)
Wheelbase 1·66 m (5 ft 5¼ in)
Propeller diameter 1·93 m (6 ft 4 in)
Propeller ground clearance 0·32 m (1 ft 0½ in)

DIMENSIONS, INTERNAL:
Cabin: Length 1·66 m (5 ft 5¼ in)
Max width 1·00 m (3 ft 3¼ in)
Height (seat cushion to canopy) 0·98 m (3 ft 2½ in)
Volume 1·50 m³ (53 cu ft)
Baggage compartment volume 0·18 m³ (6·36 cu ft)

AREAS:
Wings, gross 10·10 m² (108·70 sq ft)
Ailerons (total, incl tabs) 0·762 m² (8·20 sq ft)
Trailing-edge flaps (total) 1·18 m² (12·70 sq ft)
Fin 0·76 m² (8·18 sq ft)
Dorsal fin 0·16 m² (1·72 sq ft)
Rudder, incl tab 0·60 m² (6·46 sq ft)
Tailplane 1·46 m² (15·70 sq ft)
Elevator, incl tab 0·96 m² (10·30 sq ft)

WEIGHTS AND LOADINGS:
Manufacturer's basic weight empty:
M 755 kg (1,664 lb)
W 770 kg (1,697 lb)
Weight empty, equipped: M 815 kg (1,797 lb)
W 830 kg (1,830 lb)
Fuel:
in-wing and wingtip tanks (all versions) 169 kg (372·5 lb)
underwing tanks (W only) 114 kg (251·5 lb)
Typical mission weights:
M, trainer ('clean') 1,140 kg (2,513 lb)
W, two 47 kg (103·5 lb) machine-gun pods and full internal fuel 1,163 kg (2,564 lb)
W, one Alkan 500B cartridge thrower, one two-camera reconnaissance pod and full internal fuel 1,182 kg (2,605 lb)
W, trainer with 94 kg (207 lb) external stores 1,249 kg (2,753 lb)
W, self-ferry with two 83 litre (18·25 Imp gallon) underwing tanks 1,285 kg (2,833 lb)
W, two 125 kg bombs and 150 kg (331 lb) internal fuel 1,300 kg (2,866 lb)
W, two AL-8-70 rocket launchers and 160 kg (353 lb) internal fuel 1,300 kg (2,866 lb)
Max T-O weight:
M, Aerobatic 1,100 kg (2,425 lb)
M, Utility 1,200 kg (2,645 lb)
W, max permitted 1,300 kg (2,866 lb)
Max wing loading: M 119 kg/m² (24·4 lb/sq ft)
W 129 kg/m² (26·4 lb/sq ft)
Max power loading:
M 6·19 kg/kW (10·17 lb/hp)
W 6·70 kg/kW (11·01 lb/hp)

PERFORMANCE (M at AUW of 1,200 kg; 2,645 lb, W at 1,300 kg; 2,866 lb, unless stated otherwise):
Never-exceed speed:
M 235 knots (436 km/h; 271 mph)
Max level speed at S/L:
M 180 knots (333 km/h; 207 mph)
W 165 knots (305 km/h; 190 mph)
Max cruising speed (75% power) at 1,500 m (4,925 ft):
M 162 knots (300 km/h; 186 mph)
W 152 knots (281 km/h; 175 mph)
Stalling speed, flaps and landing gear up:
M 74 knots (137 km/h; 85·5 mph)
W 88 knots (163 km/h; 101·5 mph)
Stalling speed, flaps and landing gear down:
M 68 knots (126 km/h; 78·5 mph)
W 72 knots (134 km/h; 83 mph)
Max rate of climb at S/L: M 457 m (1,500 ft)/min
W 381 m (1,250 ft)/min
Time to 1,500 m (4,925 ft): M 4 min 0 s
W 6 min 20 s
Time to 2,300 m (7,550 ft): M 6 min 50 s
W 10 min 20 s
Time to 3,000 m (9,850 ft): M 10 min 0 s
W 18 min 40 s
Service ceiling: M 4,665 m (15,300 ft)
W 4,480 m (14,700 ft)
T-O run at S/L: M 384 m (1,260 ft)
T-O to 15 m (50 ft) at S/L: M 606 m (1,988 ft)
W 825 m (2,707 ft)
Landing from 15 m (50 ft) at S/L:
M 539 m (1,768 ft)
W 645 m (2,116 ft)
Landing run at S/L: M 345 m (1,132 ft)
Operational radius:
W, 6 h 25 min single-seat armed patrol mission at 1,163 kg (2,564 lb) AUW, incl 5 h 35 min over operating area, 20 kg (44 lb) fuel reserves 50 nm (92 km; 57 miles)

SIAI-Marchetti SF.260TP (Allison 250-B17C turboprop engine)

W, 3 h 38 min single-seat strike mission, incl two 5 min loiters over separate en-route target areas, 20 kg (44 lb) fuel reserves 250 nm (463 km; 287 miles)
W, 4 h 54 min single-seat strike mission, incl 5 min over target area, 20 kg (44 lb) fuel reserves 300 nm (556 km; 345 miles)
W, 4 h 30 min single-seat photo-reconnaissance mission at 1,182 kg (2,605 lb) AUW, incl three 1 h loiters over separate en-route operating areas, 20 kg (44 lb) fuel reserves 150 nm (278 km; 172 miles)
W, 6 h 3 min two-seat self-ferry mission with two 83 litre (18·25 Imp gallon) underwing tanks, at 1,285 kg (2,833 lb) AUW, 30 kg (66 lb) fuel reserves 926 nm (1,716 km; 1,066 miles)
Range with max fuel:
M (two-seat) 890 nm (1,650 km; 1,025 miles)
g limits (M):
at max Aerobatic T-O weight +6·0; —3·0
at max Utility T-O weight without external load +4·4; —2·2

SIAI-MARCHETTI SF.260TP

First flown in July 1980, the SF.260TP is a turboprop-powered development of the SF.260M/W, the airframe remaining virtually unchanged aft of the firewall except for substitution of an inset rudder trim tab and provision of an automatic fuel feed system. It is to be made available in trainer and Warrior configurations (and, presumably, in civil form) from early 1982, subject to certification during the second half of 1981, and will be offered both as a conversion kit for existing SF.260s and as a new-production aircraft.

AIRFRAME: As SF.260M/W, except for increased overall length and provision of trim tab in rudder.

POWER PLANT: One Allison 250-B17C turboprop engine, flat rated to 261 kW (350 shp) and driving a Hartzell HC-B3TF-7A/T10173-25R three-blade constant-speed fully-feathering and reversible-pitch propeller with spinner. Fuel capacity as for SF.260M/W; automatic fuel feed system.

ACCOMMODATION, SYSTEMS, AVIONICS AND EQUIPMENT: Generally as for SF.260M/W.

DIMENSIONS, EXTERNAL AND INTERNAL, AND AREAS: As for SF.260M/W except:
Length overall 7·40 m (24 ft 3¼ in)

WEIGHTS AND LOADINGS:
Weight empty, equipped 795 kg (1,752 lb)
Max T-O weight: Trainer 1,200 kg (2,645 lb)
Warrior 1,300 kg (2,866 lb)
Max power loading:
Trainer 4·60 kg/kW (7·56 lb/shp)
Warrior 4·98 kg/kW (8·19 lb/shp)

PERFORMANCE (at AUW of 1,200 kg; 2,645 lb, ISA):
Max level speed at S/L 206 knots (382 km/h; 237 mph)
Max cruising speed at 3,050 m (10,000 ft) 200 knots (370 km/h; 230 mph)
Stalling speed at S/L, flaps down 68 knots (126 km/h; 79 mph)
Max rate of climb at S/L 661 m (2,170 ft)/min
Service ceiling 8,535 m (28,000 ft)
T-O run 298 m (978 ft)
Landing run, without reverse pitch 307 m (1,007 ft)
Range at 4,570 m (15,000 ft) with max fuel, 30 min reserves 512 nm (949 km; 589 miles)

SIAI-MARCHETTI SF.260C

As indicated in a preceding entry, the SF.260 was developed from the F.250 prototype originally as a civil aircraft. Details of the first **(SF.260A)** and second **(SF.260B)** civil series can be found in the 1980-81 and earlier editions of *Jane's*.

Approx 30-40 civil SF.260s, including SF.260As and SF.260Bs, had been built by the Spring of 1978.

The current civil version, certificated by the RAI and FAA on 23 October and 30 December 1976 respectively, is designated **SF.260C**. Four were delivered to Alitalia in 1980 for pilot training, and 17 are operated for training by the Zimbabwe Air Force. The following description applies to the SF.260C:

TYPE: Three-seat fully-aerobatic light aircraft.

WINGS: As SF.260M/W except dihedral 5°.

FUSELAGE: As SF.260M/W.

TAIL UNIT: As SF.260M/W, but without extra reinforcement.

LANDING GEAR: See description for SF.260M/W.

POWER PLANT: As for SF.260M/W, but driving a Hartzell HC-C2YK-1BF/8477-8R two-blade constant-speed metal propeller with spinner. Fuel capacity 49·5 litres (10·9 Imp gallons) in each wing tank and 72 litres (15·8 Imp gallons) in each wingtip tank; total internal capacity 243 litres (53·4 Imp gallons). Oil capacity as for military versions.

ACCOMMODATION: Three seats in enclosed cockpit, two side by side in front, one at rear. Two children with a combined weight not exceeding 113 kg (250 lb) may occupy rear seat. One-piece fully-transparent rearward-sliding Plexiglas canopy, with rubber-cord canopy release. Baggage compartment, capacity 40 kg (88 lb), aft of rear seat. Cabin heated, ventilated and soundproofed with glassfibre.

SYSTEMS, AVIONICS AND EQUIPMENT: Generally similar to SF.260M.

DIMENSIONS, EXTERNAL: As SF.260M/W

DIMENSIONS, INTERNAL: As SF.260M/W

AREAS: As SF.260M/W

SIAI-Marchetti SF.260C three-seat light aircraft (Avco Lycoming O-540 flat-six engine)

WEIGHTS AND LOADINGS:
Weight empty, equipped 755 kg (1,664 lb)
Max T-O weight: Aerobatic 1,000 kg (2,205 lb)
Utility 1,102 kg (2,430 lb)
Max wing loading 109 kg/m^2 (22·4 lb/sq ft)
Max power loading 5·68 kg/kW (9·33 lb/hp)

PERFORMANCE (at max Utility T-O weight):
Max level speed at S/L 187 knots (347 km/h; 215 mph)
Max cruising speed (75% power) at 3,050 m (10,000 ft) 178 knots (330 km/h; 205 mph)
Stalling speed, flaps and landing gear down 60 knots (111 km/h; 70 mph)
Max rate of climb at S/L 546 m (1,791 ft)/min
Service ceiling 5,790 m (19,000 ft)
T-O run 560 m (1,837 ft)
Landing from 15 m (50 ft) at S/L 490 m (1,608 ft)
Landing run at S/L 345 m (1,132 ft)
Range with max fuel (two persons) 805 nm (1,490 km; 925 miles)
g limits: Aerobatic +6·0; −3·0
Utility +4·4; −2·2

SIAI-MARCHETTI S.205/20R and S.208A

Full details can be found in the 1972-73 *Jane's* of the S.205 light aircraft, and in the 1974-75 and 1976-77 editions of the S.208 series.

SIAI-Marchetti has offered, since early 1977, further-improved versions of these aircraft, which have the same basic airframe. Improvements applicable to both types include a new cockpit and cabin layout, with better cabin soundproofing; modified flaps and landing gear; and new heating and ventilation system.

Differences between these versions are as follows:

S.205/20R. Four-seater, with 149 kW (200 hp) Avco Lycoming IO-360 engine. Selected as second-level trainer by Italian Aero Club, to which deliveries (of an order for 140 designated **S.205AC**) began in 1977. Production now believed to have ended; details in 1980-81 *Jane's*.

S.208A. Five-seater with 194 kW (260 hp) Avco Lycoming O-540 engine. Also available in cargo, ambulance or agricultural versions.

The following details apply to the S.208A:

POWER PLANT: One 194 kW (260 hp) Avco Lycoming O-540-E4A5 flat-six engine, driving a Hartzell HC-C2YK two-blade constant-speed propeller. Fuel in two wing tanks, total capacity 215 litres (47·3 Imp gallons), and two wingtip tanks, total capacity 115 litres (25·3 Imp gallons). Overall total fuel capacity 330 litres (72·6 Imp gallons).

ACCOMMODATION: Seats for five persons in fully-enclosed cabin. Baggage space aft of rear seats. Forward-opening car-type door on starboard side of cabin. Additional small window at rear of cabin on each side, and door each side.

DIMENSIONS, EXTERNAL:
Wing span over tip-tanks 11·24 m (36 ft 10½ in)
Wing area, gross 16·09 m^2 (173·2 sq ft)
Length overall 8·10 m (26 ft 7 in)
Height overall 2·89 m (9 ft 5¾ in)

DIMENSIONS, INTERNAL:
Cabin: Max length 2·80 m (9 ft 2¼ in)
Max width 1·14 m (3 ft 9 in)
Max height 1·32 m (4 ft 4 in)
Volume 3·00 m^3 (106 cu ft)

WEIGHTS AND LOADINGS:
Weight empty, equipped 835 kg (1,841 lb)
Max T-O weight 1,505 kg (3,318 lb)
Max wing loading 87·0 kg/m^2 (17·82 lb/sq ft)
Max power loading 7·76 kg/kW (12·7 lb/hp)

PERFORMANCE (at max T-O weight):
Max level speed at S/L 173 knots (320 km/h; 199 mph)
Max cruising speed 162 knots (300 km/h; 186 mph)
Stalling speed, flaps down 50 knots (92 km/h; 57·5 mph)
Max rate of climb at S/L 300 m (984 ft)/min
Service ceiling 5,400 m (17,725 ft)
T-O run 302 m (991 ft)
Landing run 171 m (561 ft)
Max range 971 nm (1,800 km; 1,118 miles)

SIAI-MARCHETTI SF.600 CANGURO (KANGAROO)

The prototype F.600 Canguro (I-CANG), built by General Avia and powered by 261 kW (350 hp) Avco Lycoming TIO-540-J flat-six engines, made its first flight on 30 December 1978. This aircraft was described under the General Avia heading in the 1979-80 *Jane's*.

SIAI-Marchetti began flight testing the Canguro in 1979, and is responsible for certification and production. The basic aircraft is offered with non-retractable landing gear and either Avco Lycoming piston engines (as the **SF.600**) or Allison turboprops (as the **SF.600TP**). The prototype, refitted with 313 kW (420 shp) Allison 250-B17C turboprop engines, made its first flight in this form on 8 April 1981. Retractable-gear, swing-tail, maritime surveillance and other versions have been projected.

The following description applies to both the SF.600 and SF.600TP, except where a specific version is indicated:

TYPE: Twin-engined freight, ambulance and general utility transport.

WINGS: Cantilever high-wing monoplane. Wing section NASA GAW-1, with 17% thickness/chord ratio. Dihedral 2°. Incidence (constant) 1° 30′. All-metal riveted structure in light alloy, with stressed skin. Centre-section has main spar and two auxiliary spars; outboard of engines, wings have two spars. All-metal ailerons and electrically-operated double-slotted flaps. Trim tab in starboard aileron.

FUSELAGE: All-metal semi-monocoque structure, with stressed skin.

TAIL UNIT: Cantilever all-metal stressed-skin structure. Trim tabs in rudder and each elevator. Small dorsal fin.

LANDING GEAR: Non-retractable tricycle type, with oleo-pneumatic shock-absorber in each unit. Twin-wheel main units, mounted on small stub-wings attached to fuselage floor; single steerable nose unit. All five wheels and tyres same size, 7·00-6, tyre pressure 2·48 bars (36 lb/sq in). Retractable-gear version also available.

POWER PLANT: Two 261 kW (350 hp) Avco Lycoming TIO-540-J flat-six engines in SF.600, each driving a Hartzell three-blade fully-feathering constant-speed propeller; SF.600TP has two 313 kW (420 shp) Allison 250-B17C turboprop engines. Fuel in four wing tanks, total capacity 1,100 litres (242 Imp gallons). Provision for underwing tanks, total capacity 600 litres (132 Imp gallons), on turboprop version. Oil capacity 11·4 litres (2·5 Imp gallons).

ACCOMMODATION: Pilot and co-pilot or passenger on flight deck. Cabin accommodates up to nine passengers (plus toilet in VIP version) or 12 paratroops, or four stretcher patients and two medical attendants, or freight. Baggage compartment at rear of cabin in standard passenger version; in centre of cabin, opposite toilet, in VIP version. Forward door on port side for crew; wider, sliding door at rear on port side for freight loading. Cargo version can accept three 1·30 × 1·15 × 1·07 m (51 × 45 × 42 in) containers, two of size 2·20 × 1·15 × 1·07 m (87 × 45 × 42 in), or a single 4·50 × 1·15 × 1·07 m (177 × 45 × 42 in) container. Swing-tail rear-loading version under development.

EQUIPMENT: Both models can have pylons for the carriage of external stores, two under outer wings and two under main-wheel stub-wings at junction with fuselage. Each pylon is stressed for a 300 kg (661 lb) load, subject to a max external stores load of 900 kg (1,983 lb). The underwing points on the SF.600TP are 'wet', permitting the carriage of auxiliary fuel tanks. Cabin floor rollers (two longitudinal rows) and balls in cargo version. Twelve inward-facing seats in troop transport/parachutist version. Can be equipped for target towing, with floor-mounted winch, 2,000 m (6,560 ft) of cable, electrical power unit (100A/28V DC), and miss-distance indicator system; with undertail hook for towing one or more gliders; with one or two photogrammetric automatic cameras (plus additional avionics at customer's option); with equipment for in-flight inspection and calibration of ground radio/navigation aids; or with appropriate sensors for elint, sigint or other ECM missions. Other specialised applications include agricultural duties (two underwing chemical tanks or single underfuselage tank, plus bubble-type and additional lower windows in pilot's door, windscreen and nose-gear wire cutters, ceiling-mounted airscoop, and anti-corrosion paint finish). The maritime surveillance version under development can be equipped with nose-mounted Bendix RDR-1400 search and navigation radar, underfuselage side/down-looking surveillance radar, belly-mounted panoramic camera and forward-looking oblique camera; FLIR or low light level TV camera under fuselage or in an underwing pod, Omega-VLF area navigation system, advanced compass system, periscopic sextant, bubble side windows for observers, and a searchlight. External stores can include auxiliary fuel tanks, weapons and survival gear.

DIMENSIONS, EXTERNAL:
Wing span 15·00 m (49 ft 2½ in)
Wing chord (constant) 1·60 m (5 ft 3 in)
Wing area, gross 24·00 m^2 (258·3 sq ft)
Wing aspect ratio 9·1
Length overall 12·15 m (39 ft 10½ in)
Height overall 4·60 m (15 ft 1 in)
Tailplane span 5·06 m (16 ft 7¼ in)
Wheel track 2·40 m (7 ft 10½ in)
Wheelbase 4·88 m (16 ft 0 in)
Propeller diameter 2·03 m (6 ft 8 in)
Crew door (fwd, port):
Height 1·14 m (3 ft 9 in)
Width 0·86 m (2 ft 10 in)
Height to sill 0·90 m (2 ft 11½ in)
Cargo door (rear, port):
Height 1·13 m (3 ft 8½ in)
Width 1·49 m (4 ft 10¾ in)
Height to sill 0·90 m (2 ft 11½ in)

DIMENSIONS, INTERNAL:
Cabin, excl flight deck:
Length 5·05 m (16 ft 6¾ in)
Width 1·23 m (4 ft 0½ in)
Height 1·27 m (4 ft 2 in)
Floor area 5·57 m^2 (60 sq ft)
Volume 7·80 m^3 (275 cu ft)

WEIGHTS AND LOADING (A: SF.600; B: SF.600TP):
Weight empty, equipped:
A (cargo) 1,950 kg (4,299 lb)
B (cargo) 1,800 kg (3,968 lb)
Operating weight empty: A 2,090 kg (4,607 lb)
B 1,947 kg (4,292 lb)
Max payload (cargo) 1,050 kg (2,315 lb)
Max external load: A, B 900 kg (1,983 lb)
Max T-O weight: A 3,200 kg (7,054 lb)
B, internal fuel 3,300 kg (7,275 lb)
B, internal and external fuel 3,700 kg (8,157 lb)
Max cargo floor loading 400 kg/m^2 (81·93 lb/sq ft)

PERFORMANCE (estimated, at max T-O weight. A: SF.600; B: SF.600TP with internal fuel only, except where indicated):
Max level speed at 3,050 m (10,000 ft):
A 166 knots (307 km/h; 191 mph)
B 161 knots (298 km/h; 185 mph)
Cruising speed (75% power) at 3,050 m (10,000 ft):
A, B 154 knots (285 km/h; 177 mph)
Stalling speed, flaps down:
A 58 knots (107 km/h; 66·5 mph)
B 59 knots (109 km/h; 68 mph)
Max rate of climb at S/L: A 381 m (1,250 ft)/min
B 426 m (1,400 ft)/min
Rate of climb at S/L, one engine out:
A 84 m (275 ft)/min
B 90 m (295 ft)/min
Service ceiling: A 7,100 m (23,300 ft)
B 6,700 m (21,980 ft)
Service ceiling, one engine out:
A 3,200 m (10,500 ft)
B 2,400 m (7,875 ft)
T-O run: A 240 m (788 ft)
B 220 m (722 ft)
T-O to 15 m (50 ft): A 395 m (1,296 ft)
B 350 m (1,150 ft)
Landing from 15 m (50 ft): A 340 m (1,115 ft)
B 320 m (1,050 ft)
Landing run: A 200 m (656 ft)
B 210 m (690 ft)
Range with max fuel:
A (IFR reserves) 1,295 nm (2,400 km; 1,490 miles)
B (internal fuel) 853 nm (1,580 km; 981 miles)
B (internal and external fuel, IFR reserves) 1,079 nm (2,000 km; 1,242 miles)
g limits (Normal category) +3·8; −1·52

Prototype Canguro utility transport in SF.600TP turboprop-engined configuration

SIAI-MARCHETTI SM.1019E

The SM.1019 light STOL observation, light ground attack and utility aircraft flew for the first time on 24 May 1969, with an Allison 250-B15C engine, and was granted Normal and Utility category certification by the RAI on 25 October 1969.

A second prototype (I-SJAR), which first flew on 18 February 1971, was designated SM.1019A. It had an improved fuel system, two doors and two instrument panels, and received RAI civil certification in the Normal and Utility categories.

Production began in 1974 of more than 100 SM.1019EIs for the Aviazione Leggera dell'Esercito (ALE, or Italian Army Light Aviation), powered by the 298 kW (400 shp) Allison 250-B17B engine. All of these have been delivered.

In 1981 SIAI-Marchetti was still describing the SM.1019 as a current production aircraft, apparently now available also with a 313 kW (420 shp) Allison 250-B17C engine. At least two aircraft (I-AMET and I-JUMA) were being operated at that time as company demonstrators.

The following description applies to the SM.1019E with 298 kW (400 shp) engine, except where indicated:

Type: Two-seat STOL light aircraft.

Wings: High-wing monoplane, braced by single strut on each side. Wing section NACA 2412. Dihedral 2° 8′. Incidence 1° 30′. Washout 3°. Conventional two-spar aluminium alloy structure, with detachable tapered outer panels. Metal Frise-type ailerons and electrically-actuated trailing-edge slotted flaps. Electrically-actuated trim tab in starboard aileron. Tiedown point at each wingtip.

Fuselage: Conventional all-metal stressed-skin structure.

Tail Unit: Conventional cantilever all-metal structure, with horizontal surfaces mounted on top of fuselage. Dorsal fin. Fixed-incidence tailplane. Elevators and rudder horn-balanced. Electrically-actuated or manually-operated mechanically-actuated trim tab in starboard elevator; servo tab in port elevator. Ground-adjustable tab on rudder.

Landing Gear: Non-retractable tailwheel type, with cantilever leaf-type spring steel main-wheel legs. Goodyear 511960 main wheels, with low-pressure tyres, size 7·00-6, pressure 2·07 bars (30 lb/sq in); Scott 3200A steerable tailwheel, with size 8-3·00 tyre, pressure 2·41 bars (35 lb/sq in). Goodyear independent hydraulic multiple-disc brakes on main wheels, controllable from either seat. Parking brake. Combined wheel/ski gear, with hydraulic retraction and extension of skis, is optional.

Power Plant: One 298 kW (400 shp) Allison 250-B17B turboprop engine, derated to 238·5 kW (320 shp) and driving a Hartzell HC-B3TF-7/T10173-11R three-blade constant-speed reversible-pitch metal propeller with automatic feathering. Fuel in two aluminium alloy tanks in each wing, each of 80 litres (17·5 Imp gallons) capacity; total capacity 320 litres (70 Imp gallons). Refuelling point for each tank on top of wings. Provision for two 80 litre (17·5 Imp gallon) auxiliary underwing tanks. Oil capacity 8 litres (1·75 Imp gallons).

Accommodation: Pilot and co-pilot or observer/systems operator seated in tandem in fully-enclosed and extensively-glazed cabin. Two forward-hinged doors on starboard side. Cabin heated, by engine bleed air, and ventilated by ram air. Dual controls standard.

Systems: 24V DC electrical power provided by 30V 150A Lear Siegler P/N230320020 engine-driven starter/generator and 24V 22Ah nickel-cadmium battery, with static inverter to provide 115V AC power at 400Hz. Transformer supplies slaved gyro with 26V AC at 400Hz. External ground power receptacle. Windscreen defrosting and engine compressor inlet anti-icing (both by engine bleed air) are standard. Oxygen system optional.

Avionics and Equipment: Choice of UHF, VHF/AM, VHF/FM or HF communication systems. VLF/Omega navigation. ADF; VOR/ILS; IFF; high-performance intercom and compass system. Provision for specialised avionics (radar warning, Tacan and microwave landing system) to customer's requirements. Twin taxying and landing lights in port outer wing leading-edge. Anti-collision light on top of rudder.

Armament and Operational Equipment (SM.1019EI): Two hardpoints beneath each wing, stressed for loads of 160 kg (353 lb) inboard, 60 kg (132 lb) outboard; max external stores load 320 kg (706 lb). Standard NATO stores racks. Typical underwing loads can include two or four 7·62 mm gun pods; two launchers each with six or eight 68 mm, seven or eight 2·75 in, or eighteen 2 in rockets; two dispensers for 74 mm cartridges; four LUU-2/B parachute flares; two or four bombs (up to 150 kg size); two photo-reconnaissance pods; four survival packs; and two 80 litre (17·5 Imp gallon) drop-tanks.

Dimensions, external:
Wing span 10·972 m (36 ft 0 in)
Wing chord at root 1·63 m (5 ft 4¼ in)
Wing chord at tip 1·09 m (3 ft 7 in)
Wing aspect ratio 7·44
Length overall (tail up) 8·52 m (27 ft 11½ in)
Height overall (tail down) 2·86 m (9 ft 4½ in)

SIAI-Marchetti SM.1019E demonstration aircraft fitted with wheel/ski landing gear

Elevator span 3·42 m (11 ft 2¾ in)
Wheel track 2·29 m (7 ft 6¼ in)
Wheelbase 6·23 m (20 ft 5¼ in)
Propeller diameter 2·29 m (7 ft 6 in)
Propeller ground clearance 0·23 m (9 in)
Cabin doors, each: Height 1·06 m (3 ft 5¾ in)
Width 0·60 m (1 ft 11½ in)
Baggage door: Height 0·47 m (1 ft 6½ in)
Width 0·53 m (1 ft 9 in)
Height to sill 0·62 m (2 ft 0¼ in)

Dimensions, internal:
Cabin: Max length 2·00 m (6 ft 6¾ in)
Max width 0·63 m (2 ft 0¾ in)
Max height 1·25 m (4 ft 1¼ in)
Volume 1·10 m³ (38·8 cu ft)
Baggage compartment volume 0·1 m³ (3·5 cu ft)

Areas:
Wings, gross 16·16 m² (173·95 sq ft)
Ailerons (total) 1·70 m² (18·30 sq ft)
Trailing-edge flaps (total) 1·96 m² (21·10 sq ft)
Fin 0·957 m² (10·30 sq ft)
Rudder 1·295 m² (13·94 sq ft)
Tailplane 1·896 m² (20·41 sq ft)
Elevators (total) 1·584 m² (17·05 sq ft)

Weights and Loadings (without external stores):
Basic weight empty 755 kg (1,664 lb)
Weight empty, equipped:
'clean' 812 kg (1,790 lb)
with external stores 834 kg (1,838 lb)
T-O weight:
SM.1019EI (training) 1,300 kg (2,866 lb)
Max T-O weight:
SM.1019EI (helicopter escort, reconnaissance) 1,450 kg (3,196 lb)
Wing loading at 1,300 kg (2,866 lb) AUW 80·4 kg/m² (16·5 lb/sq ft)
Max wing loading 89·7 kg/m² (18·4 lb/sq ft)
Power loading at 1,300 kg (2,866 lb) AUW 4·36 kg/kW (7·2 lb/shp)
Max power loading 4·87 kg/kW (8·0 lb/shp)

Performance (A: Utility, AUW of 1,300 kg; 2,866 lb. B: helicopter escort, AUW of 1,450 kg; 3,196 lb. C: reconnaissance, AUW of 1,450 kg; 3,196 lb; D: 420 shp engine and AUW of 1,450 kg; 3,196 lb, except where indicated):
Never-exceed speed 169 knots (313 km/h; 194 mph)
Max level speed: D 159 knots (295 km/h; 183 mph)
Max cruising speed at S/L:
A 152 knots (281 km/h; 175 mph)
B, C 147 knots (272 km/h; 169 mph)
Max cruising speed at 1,525 m (5,000 ft):
A 159 knots (295 km/h; 183 mph)
B, C 154 knots (285 km/h; 177 mph)
D 142 knots (263 km/h; 163 mph)
Cruising speed (75% power) at 1,525 m (5,000 ft):
A 142 knots (263 km/h; 163 mph)
B, C 134 knots (248 km/h; 154 mph)
Stalling speed, flaps up:
A 57 knots (106 km/h; 66 mph)
B, C 68 knots (126 km/h; 78·5 mph)
Stalling speed, flaps down:
A, D 51 knots (94·5 km/h; 59 mph)
B, C 62 knots (115 km/h; 71·5 mph)
Max rate of climb at S/L: A, D 499 m (1,640 ft)/min
B, C 439 m (1,440 ft)/min
Service ceiling: A, B, C, D 7,620 m (25,000 ft)
T-O run at S/L: A 85 m (280 ft)
B, C 122 m (400 ft)
T-O to 15 m (50 ft) at S/L: A 195 m (640 ft)
B, C 280 m (920 ft)
Landing from 15 m (50 ft) at S/L:
A 220 m (722 ft)
B, C 281 m (922 ft)
Landing run at S/L: A 91·5 m (300 ft)
B, C 135 m (443 ft)
Typical operational radius:
B, with two rocket launchers and two gun pods 60 nm (111 km; 69 miles)
Max range at S/L: A 499 nm (925 km; 575 miles)
B 421 nm (780 km; 485 miles)
C 610 nm (1,130 km; 702 miles)
Max range at 3,000 m (9,845 ft):
A 588 nm (1,090 km; 677 miles)
B 505 nm (935 km; 581 miles)
C 750 nm (1,390 km; 864 miles)
Max range with two external tanks, AUW of 1,400 kg (3,086 lb): C, at 610 m (2,000 ft) 623 nm (1,154 km; 717 miles)
C, at 2,745 m (9,000 ft) 730 nm (1,352 km; 840 miles)
Max endurance at S/L: A 5 h 45 min
B 5 h 0 min
C 7 h 20 min
Max endurance at 3,000 m (9,845 ft): A 6 h 40 min
B 6 h 5 min
C, with auxiliary fuel tanks 8 h 45 min
g limits:
AUW of 1,300 kg (2,866 lb) +4·4; −1·76
AUW of 1,450 kg (3,196 lb) +3·8; −1·52

SIAI-MARCHETTI S.211

This lightweight, low-cost basic trainer and light attack aircraft was first revealed in the form of a model at the Paris Air Show in May/June 1977. Two flying prototypes have since been built, and the first of these (I-SITF) made its initial flight on 10 April 1981. The second (I-SIJF) was due to fly approx three months later; a static test airframe has also been completed.

Preparations for initial production were under way in mid-1981, with deliveries scheduled to begin in the second half of 1982.

Type: Two-seat basic trainer and light attack aircraft.

Wings: Cantilever shoulder-wing monoplane, with supercritical section evolved by computer with the assistance of the US universities of New York and Kansas. Thickness/chord ratio 15% at root, 13% at tip. Incidence 2° 13′ at root, −1° 17′ at tip. Anhedral 2° from roots. Sweepback 15° 30′ at quarter-chord. Two-spar metal torsion box structure, forming integral fuel tank; attached to fuselage by four bolts. Upper and lower skins each formed by two one-piece panels joined along centreline and to the spars. Ailerons and large-area Fowler-type flaps on trailing-edges. Trim tab in each aileron.

Fuselage: Conventional metal and glassfibre semi-monocoque structure. Hydraulically actuated airbrake under centre-fuselage. Equipment bay in nose.

Tail Unit: Cantilever metal structure. Sweptback fin and horn-balanced rudder; sweptback leading-edge on tailplane. No tabs.

Landing Gear: Hydraulically-retractable tricycle type, of Messier-Hispano-Bugatti/Magnaghi design. All units retract forward into fuselage (main units turning through 90° to lie flat in undersides of engine air intake trunks). Main wheels size 6·50-8; nosewheel size 5·00-5. Designed for sink rate of 4 m (13 ft)/s. Wheel brakes

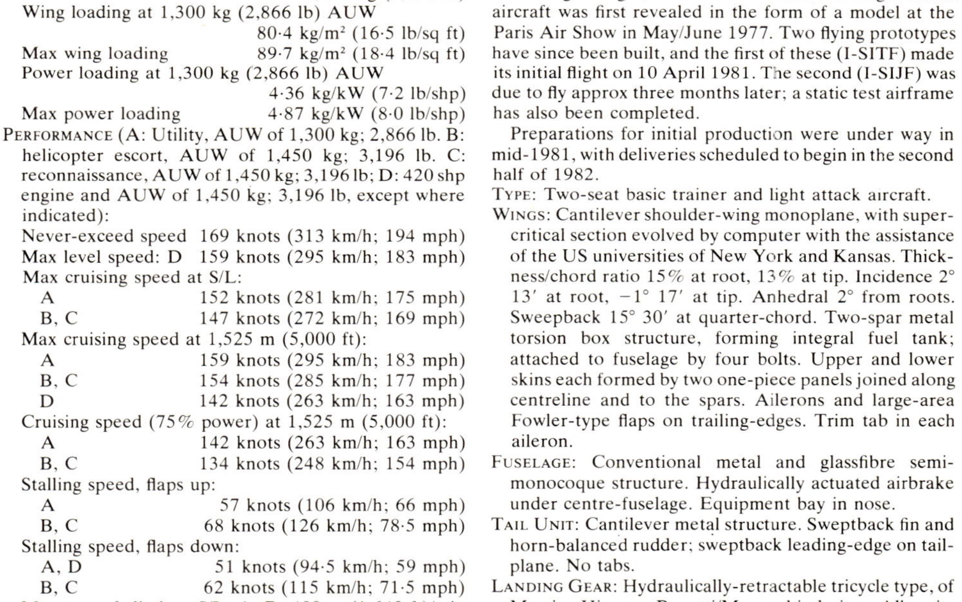

Prototype of the SIAI-Marchetti S.211 two-seat training and light attack aircraft

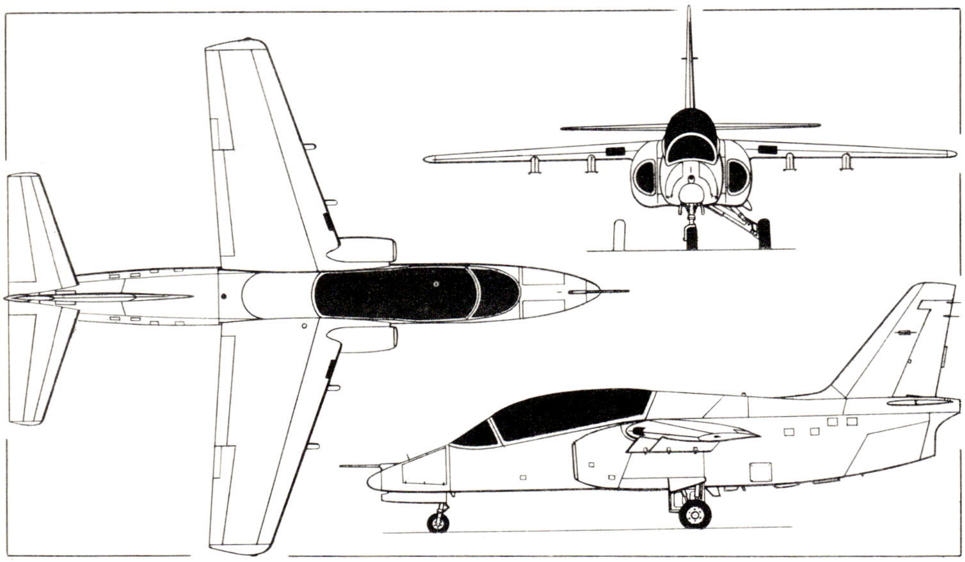

SIAI-Marchetti S.211 basic trainer and light attack aircraft *(Pilot Press)*

actuated hydraulically, independently of main hydraulic system. Provision for emergency free-fall extension.

POWER PLANT: One 11·12 kN (2,500 lb st) Pratt & Whitney Aircraft of Canada JT15D-4M non-afterburning turbofan engine mounted in rear of fuselage; lateral intake each side of fuselage. Fuel in 600 litre (132 Imp gallon) integral wing tank, 150 litre (33 Imp gallon) fuselage main tank, and fuselage reservoir tank; total usable capacity 832 litres (183 Imp gallons). Single gravity refuelling point in top surface of starboard wing. Provision for two 350 litre (77 Imp gallon) drop-tanks on inboard underwing stores points. Oil capacity 10 kg (22 lb).

ACCOMMODATION: Seats for two persons in tandem in pressurised and air-conditioned cockpit under one-piece canopy opening sideways to starboard: pupil in front, instructor on rear seat elevated 28 cm (11 in). Martin-Baker Mk 8 ejection seats for both occupants, capable of operation at all altitudes and at speeds between 60-400 knots (111-741 km/h; 69-461 mph), including ejection through canopy; capable of conversion in field to Martin-Baker IT10LA zero-zero seats by addition of rocket assistance.

SYSTEMS: Environmental control system for cockpit pressurisation and air-conditioning, using engine bleed air. Max pressure differential 0·29 bars (4·2 lb/sq in). Hydraulic system, pressure 207 bars (3,000 lb/sq in), for landing gear and airbrake actuation, and independent actuation of wheel brakes. Primary electrical system is 28V DC, using an engine-driven starter/generator; nickel-cadmium battery; two static inverters supply AC power for instruments and avionics. External power receptacle in port side of lower fuselage aft of wing. Demand-type main oxygen system, at 124 bars (1,800 lb/sq in) pressure, sufficient to supply both occupants for 4 hours, plus bottles for emergency oxygen supply.

AVIONICS AND EQUIPMENT: To customer's requirements. Communications system has, as standard, dual U/VHF and one HF/SSB, all with dual control. BAe Dynamics AN16/D suppressed HF antenna. Choice of ADF, Tacan or VOR/ILS nav. IFF, flight director and radio altimeter standard. Provision for Doppler radar, nose-mounted attack radar, head-up display, radar warning system and ECM. Landing and taxying lights in inboard wing leading-edges.

ARMAMENT: Four underwing hardpoints, stressed for loads of up to 300 kg (660 lb) inboard, 150 kg (330 lb) outboard; max external load 600 kg (1,320 lb). Typical loads can include four single- or twin-gun 7·62 mm machine-gun pods, four 12·7 mm gun pods, or (inboard only) two 20 mm gun pods; four AL-18-50 (18 × 50 mm), Matra F2 (6 × 68 mm), LAU-32 (7 × 2·75 in), or AL-6-80 (6 × 81 mm) rocket launchers, or (inboard only) two Matra 155 (18 × 68 mm) or SNORA RWK-020 (12 × 81 mm) launchers; two Sidewinder or Magic air-to-air, or two Maverick air-to-surface, missiles on the inboard points; four bombs or practice bombs of up to 150 kg size, or (inboard only) two bombs or napalm containers of up to 300 kg; four 74 mm cartridge throwers; or (inboard only) two photo-reconnaissance pods each with four cameras and infra-red linescan; or (inboard only) two 350 litre (77 Imp gallon) auxiliary fuel tanks.

DIMENSIONS, EXTERNAL:
- Wing span: 8·00 m (26 ft 3 in)
- Wing chord: at root: 2·151 m (7 ft 0¾ in)
- at tip: 1·00 m (3 ft 3¼ in)
- Wing chord (mean aerodynamic): 1·646 m (5 ft 4¾ in)
- Wing aspect ratio: 5·08
- Length overall: 9·28 m (30 ft 5½ in)
- Height overall: 3·73 m (12 ft 2¾ in)
- Tailplane span: 3·96 m (13 ft 0 in)
- Wheel track: 2·29 m (7 ft 6 in)
- Wheelbase: 4·02 m (13 ft 2¼ in)

AREAS:
- Wings, gross: 12·60 m² (135·63 sq ft)
- Airbrake: 0·42 m² (4·52 sq ft)
- Vertical tail surfaces (total): 1·936 m² (20·84 sq ft)
- Horizontal tail surfaces (total): 3·378 m² (36·36 sq ft)

WEIGHTS:
- Weight empty, equipped: 1,445 kg (3,185 lb)
- Max T-O weight: trainer, 'clean': 2,300 kg (5,070 lb)
- armed version: 2,800 kg (6,173 lb)

PERFORMANCE (estimated, at T-O weight of 2,300 kg; 5,070 lb except where indicated):
- Never-exceed speed: Mach 0·80 (400 knots; 740 km/h; 460 mph EAS)
- Max level speed at 7,620 m (25,000 ft): 390 knots (723 km/h; 449 mph)
- Max cruising speed at 7,620 m (25,000 ft): 380 knots (704 km/h; 438 mph)
- Stalling speed, flaps down: 71 knots (131 km/h; 82 mph)
- Max rate of climb at S/L: 1,508 m (4,950 ft)/min
- Service ceiling: 12,800 m (42,000 ft)
- T-O and landing run (S/L, ISA): less than 305 m (1,000 ft)
- T-O to 15 m (50 ft): 347 m (1,140 ft)
- Landing from 15 m (50 ft): 649 m (2,130 ft)
- Min air turning radius at S/L: less than 305 m (1,000 ft)
- Typical attack radius with four rocket launchers, AUW of 2,800 kg (6,173 lb):
  - hi-lo-hi, out and back at 265 knots (491 km/h; 305 mph) at 9,145 m (30,000 ft), 3 h mission (incl 5 min over target), 60 kg (132 lb) of fuel remaining: 350 nm (649 km; 403 miles)
  - lo-lo-lo, out and back at 300 knots (556 km/h; 345 mph) at less than 305 m (1,000 ft), 1 h 12 min mission (incl 5 min over target), 60 kg (132 lb) of fuel remaining: 130 nm (241 km; 150 miles)
- Range at 9,145 m (30,000 ft) with max fuel, 30 min reserves: 1,030 nm (1,908 km; 1,186 miles)
- Ferry range (AUW of 2,800 kg; 6,173 lb, max internal and external fuel) at 290 knots (537 km/h; 334 mph) at 9,145 m (30,000 ft), 150 kg (330 lb) of fuel remaining: 1,453 nm (2,692 km; 1,673 miles)
- Endurance, 30 min reserves: 4 h
- Sustained *g* limit at 4,575 m (15,000 ft): 3·45
- Design *g* limits: +6; −3

BREDANARDI

BREDANARDI COSTRUZIONI AERONAUTICHE SpA

OFFICE AND WORKS: Casella Postale 108, 63039 San Benedetto del Tronto (Ascoli Piceno)
Telephone: (0735) 67246/7/8/9
Telex: 560165 BRENAR I
MARKETING OFFICES:
Aeroporto Forlanini, 20090 Milan
Telephone: (02) 7560241
Telex: 312666 NARDI I
Via XXIV Maggio 46, 00187 Rome
Telephone: (06) 6797360 and 6796834
Telex: 614111 BRENAR I

PRESIDENT: Dott Eng Ermanno Raffetto
MANAGING DIRECTOR: Dott Elto Nardi
MARKETING DIRECTOR: Dott Elio Nardi

BredaNardi is a partnership of Nardi and INSUD, a state-owned financial holding, and was established in 1971.

At Monteprandone (AP) BredaNardi produces, under licence from Hughes Helicopters of the USA, the NH-300C, NH-500D and NH-500MD light helicopters, including TOW and ASW versions of the military NH-500MD. Descriptions of these aircraft can be found under the Hughes heading in the US section of this edition.

Some 20 NH-300Cs were delivered in 1980 to commercial and agricultural operators in Europe. The Fire Patrol Department of the Italian Ministry of Agriculture increased its fleet of NH-500Ds by acquiring three in 1980; it was due to receive three more in 1981, and was expected to order a further six. The Italian Customs, which has for some time operated a fleet of NH-500MDs, received seven of these helicopters during 1980 and was to receive a further nine in 1981. Twenty more NH-500MDs were expected to be ordered in the Spring of 1981. Also expected during 1981 was a contract for 44 NH-500MDs from the Italian Air Force, which has selected the type for basic and advanced training.

Under Italian government proposals for reorganising the Italian aerospace industry, announced in August 1981, BredaNardi is reportedly to be acquired by Agusta SpA.

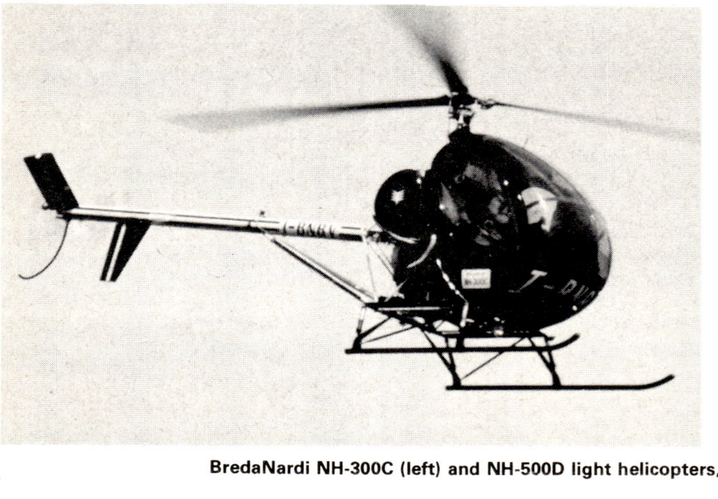

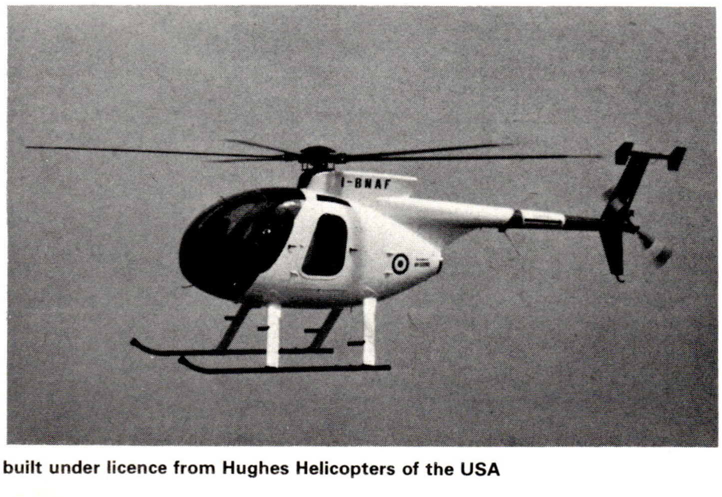

BredaNardi NH-300C (left) and NH-500D light helicopters, built under licence from Hughes Helicopters of the USA

CAPRONI VIZZOLA

CAPRONI VIZZOLA COSTRUZIONI AERONAUTICHE SpA

Via Montecchio 2, 21010 Vizzola Ticino (Va)
Telephone: (0331) 230 826
Telex: 332554 CAVIZ I
PRESIDENT: Dott Giovanni Caproni di Taliedo
COMMERCIAL DIRECTOR: Dott Livio Sonzio

In 1981 a 50 per cent share in the C22J programme was acquired by Agusta, which is reportedly to take over Caproni Vizzola under the proposed aerospace industry reorganisation outlined by the Italian government.

CAPRONI VIZZOLA C22J

The C22J is a two-seat lightweight training aircraft, developed by Caproni Vizzola as a private venture. Its configuration bears a close resemblance to that of the company's A-21SJ Calif jet-powered sailplane (which see); construction is largely of metal, with the forward fuselage skin, some fairings and other unstressed areas of glassfibre. A prototype (I-CAVJ) made its first flight on 21 July 1980.

Intended primarily for student pilot screening, basic and proficiency training, the C22J is also suitable for ECM evaluation, ground and air navaid calibration, ecological survey and high-speed liaison. It can be converted easily for photographic survey duties, or for use as an RPV.

Plans have been made by Agusta (SIAI-Marchetti) to develop a **C22R** version of the aircraft for tactical intelligence-gathering, reconnaissance and forward air control roles. The C22R would be capable of carrying substantially more sensor equipment than an RPV configured for similar roles, and is claimed to have extremely low infra-red and electromagnetic signatures. It would also be less susceptible to ECM than an RPV. The piloted C22R could be launched (with rocket motor boost) from a mobile platform, and recovered by an arrester system.

TYPE: Two-seat basic training aircraft.

WINGS: Cantilever shoulder-wing monoplane. Constant-chord wings, of Wortmann FX-67K-170 section. Dihedral 2°. Incidence 1° 9′. No sweepback. Single-spar structure, built as two panels and joined on centreline, with aluminium alloy skin and extruded leading-edge. Electrically-actuated trailing-edge plain flaps can be set in any position throughout their full range of movement. Flaps operate in conjunction with aluminium alloy airbrakes/spoilers, of which there is one in the upper surface of each wing, forward of the flap. Airbrakes are opened manually, but move with the flaps to provide balanced control. Aerodynamically balanced ailerons operate differentially and are drooped to provide additional flap area. All movable surfaces are of extruded aluminium alloy, and are operated by Aviac actuators. No tabs.

FUSELAGE: Tadpole-shaped structure, designed as a laminar lifting body. Primary load-bearing keel and wing spar pick-up cross-structure is of light alloy, inside moulded glassfibre shell. NACA-type flush engine air inlet of glassfibre in top of fuselage, aft of cockpits. Tailboom is of light alloy.

TAIL UNIT: Cantilever T tail, tailplane being of light alloy stressed-skin construction. Full-span balanced elevator is a chemically milled extrusion, and has spring trim, actuated by an Industria electrical system located in the fin. All-metal two-spar stressed-skin fin, bolted to tailboom. All control surfaces operated by push/pull rods. No tabs. Rudder pedals adjustable in flight.

LANDING GEAR: Retractable tricycle type, actuated electrically with manual backup. All units retract forward into fuselage. Cantilever sprung main legs, of glassfibre epoxy; rubber-in-torsion shock-absorption on nose unit. Cleveland main wheels, with Goodyear size 5·00-5 tyres (6 ply rating), pressure 3·0 bars (43·5 lb/sq in). Tost nosewheel, with Tost tyre size 260 × 85 mm (4 ply rating), pressure 2·5 bars (36·3 lb/sq in). Cleveland independent hydraulic disc brakes on main wheels. Steerable nosewheel, linked to rudder pedals. Safety lock for up and down positions. Electrical warning system.

POWER PLANT: Two Microturbo TRS 18-046 turbojet engines, each rated at 1·0 kN (220 lb st), mounted side by side in fuselage aft of cockpits. Production version may alternatively be fitted with KHD T 317 turbojets. Integral fuel tank in each wing leading-edge, combined capacity 290 litres (64 Imp gallons). Fuel system incorporates fuselage collector tank which permits inverted flight. Refuelling point at each wingtip. Provision for two 112 litre (24·5 Imp gallon) underwing drop-tanks. Oil capacity 0·8 litres (1·4 Imp pints).

ACCOMMODATION: Seats for two persons side by side under jettisonable canopy which is hinged at rear and opens upward. Seats are semi-supine. Dual controls on production version. Single instrument panel and centre console, eliminating need for dual instruments and avionics. Cockpit heated, ventilated and demisted.

SYSTEMS: Hydraulic system for main-wheel brakes only. No pneumatic system. Electrical system is 28V DC, incorporating two Microturbo 800W or KHD 1·2kW starter/generators and a 24V 18Ah lead-acid battery. Cockpit ventilation and demisting by heat exchangers on jetpipes. Demand-type low-pressure oxygen system, capacity 8·5 litres (0·3 cu ft), for each occupant.

AVIONICS AND EQUIPMENT: Avionics bay in top of fuselage, aft of cockpits. Collins Microline radio, Pro Line flight director, navigation, landing and anti-collision lights, standard.

ARMAMENT: Provision for two or four standard NATO underwing pylons, for a wide range of stores (max external load 200 kg; 440 lb) for gunnery/weapon training, photographic reconnaissance and target-towing missions. Typical loads include one auxiliary fuel tank and one three-camera pod; two auxiliary fuel tanks; two 7·62 mm gun pods and 500 rds of ammunition; two Simpres AL-18-50 pods with eighteen 2 in rockets; four SAMP EU70 50 kg general-purpose bombs; four Mk 70 11 kg or M38-A2 50 kg practice bombs; or two Dornier DATS 1 50 kg towed targets.

Prototype Caproni Vizzola C22J (two Microturbo TRS 18 turbojet engines)

DIMENSIONS, EXTERNAL:

Wing span	10·00 m (32 ft 9¾ in)
Wing chord (constant)	0·90 m (2 ft 11½ in)
Wing aspect ratio	11·42
Length overall	6·188 m (20 ft 3½ in)
Fuselage: Max width	1·228 m (4 ft 0½ in)
Height overall	1·88 m (6 ft 2 in)
Tailplane span	2·66 m (8 ft 8¾ in)
Wheel track	1·81 m (5 ft 11¼ in)
Wheelbase	1·81 m (5 ft 11¼ in)

AREAS:

Wings, gross	8·75 m^2 (94·18 sq ft)
Ailerons (total)	0·718 m^2 (7·73 sq ft)
Trailing-edge flaps (total)	0·824 m^2 (8·87 sq ft)
Airbrakes/spoilers (total)	0·572 m^2 (6·16 sq ft)
Fin	0·808 m^2 (8·70 sq ft)
Rudder	0·225 m^2 (2·42 sq ft)
Tailplane	1·40 m^2 (15·07 sq ft)
Elevator	0·338 m^2 (3·64 sq ft)

WEIGHTS AND LOADING:

Weight empty	720 kg (1,587 lb)
Max fuel load	245 kg (540 lb)
Max T-O and landing weight	1,135 kg (2,502 lb)
Max wing loading	129·7 kg/m^2 (26·56 lb/sq ft)

PERFORMANCE (at max 'clean' T-O weight):

Never-exceed speed	300 knots (556 km/h; 345 mph) EAS
Max cruising speed at S/L	260 knots (482 km/h; 299 mph)
Max design manoeuvring speed	234 knots (433 km/h; 269 mph) EAS
Max design speed with airbrakes fully deployed	178 knots (330 km/h; 205 mph) EAS
Econ cruising speed at 3,050 m (10,000 ft)	175 knots (324 km/h; 202 mph)
Max landing gear extension speed	128 knots (238 km/h; 147 mph) EAS
Stalling speed, flaps down, power off	65 knots (121 km/h; 75 mph) EAS
Max rate of climb at S/L	552 m (1,810 ft)/min
Rate of climb at S/L, one engine out	177 m (580 ft)/min
Time to climb to 5,000 m (16,400 ft)	12 min
Service ceiling	7,620 m (25,000 ft)
Service ceiling, one engine out	6,500 m (21,325 ft)
T-O run at S/L, ISA, zero wind	350 m (1,150 ft)
T-O to 15 m (50 ft), conditions as above	550 m (1,805 ft)
T-O to 15 m (50 ft) at 1,500 m (4,920 ft), ISA, zero wind	800 m (2,625 ft)
Max range on internal fuel	400 nm (741 km; 461 miles)
Max endurance	2 h
g limits	+6·0; −3·0

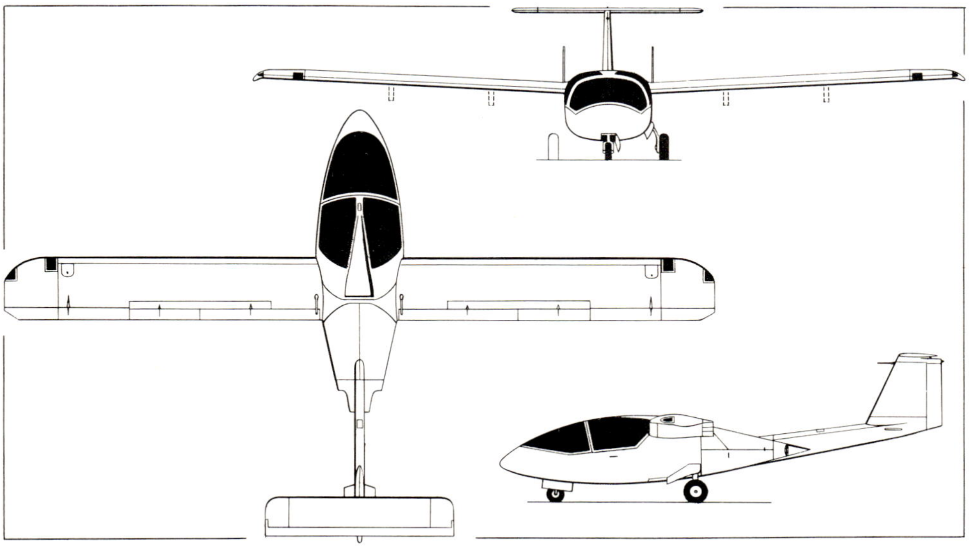

Caproni Vizzola C22J two-seat basic training aircraft *(Pilot Press)*

EM — *See Agusta*

GENERAL AVIA
GENERAL AVIA COSTRUZIONI AERONAUTICHE Srl

ADDRESS: Via Trieste 22-24, 20096 Pioltello, Milan
Telephone: (02) 9046774
TECHNICAL DIRECTOR: Dott Ing Stelio Frati
SECRETARY-TREASURER: Lamberto Frati
PRODUCTION: Enrico Fanoli
PUBLIC RELATIONS: Carla Bielli

Dott Ing Stelio Frati is well known for the many successful light aircraft which, as a freelance designer, he has developed since 1950. These have been built in prototype and production series by several Italian manufacturers, and have included the Procaer F15 Picchio and the F.250, now manufactured by SIAI-Marchetti as the SF.260.

General Avia was established by Dott Ing Frati in early 1970, primarily to develop prototypes of his own design for production by other companies. These have included the F15F, a derivative of the Procaer F15E Picchio which is to be placed in production by Procaer; and the F.20 Pegaso. General Avia also developed and built the prototype of the Canguro transport aircraft described under the SIAI-Marchetti heading in this section.

GENERAL AVIA F.20 PEGASO/CONDOR

Design of the F.20 Pegaso six-seat light business twin was started in January 1970. Two prototypes were built by General Avia, these making their first flights on 21 October 1971 and 11 August 1972, as described in previous editions of *Jane's*. Following type certification by the RAI (19 November 1974) and FAA (14 May 1975), the first series-built example flew on 17 December 1979.

The production version introduces a number of improvements compared with the prototypes. These include better cabin heating and soundproofing, and the use of three-blade propellers to improve certain aspects of performance.

The Pegaso is also offered in a four-seat military version, the F.20 **Condor**, suitable for weapon training, long-range maritime surveillance, and search and rescue. This has turboprop engines and can be fitted with three stores pylons under each wing, the centre one on each side being capable of carrying a 300 litre (66 Imp gallon) auxiliary fuel tank. A prototype of the Condor is under construction.

The following description applies to the civil Pegaso, except where indicated:

TYPE: Twin-engined six-seat light executive transport aircraft.

WINGS: Cantilever low-wing monoplane. Wing section NACA 65_2-415. Dihedral 5° from roots. Incidence 1° 45′. All-metal single-spar structure in light alloy, with flush-riveted stressed skin. Differentially-operated all-metal ailerons and electrically actuated double-slotted metal trailing-edge flaps. No tabs. Anti-icing optional.

FUSELAGE: All-metal semi-monocoque structure, with flush-riveted aluminium alloy skin.

TAIL UNIT: Cantilever all-metal structure with flush-riveted skin. Fixed-incidence tailplane. Trim tabs in rudder and each elevator. Anti-icing optional.

LANDING GEAR: Retractable tricycle type, with single wheel on each unit. Nosewheel steerable 18° to left and right. Electrical actuation, with manual backup. Oleo-pneumatic shock-absorbers. Main wheels and tyres size 7·00-6, pressure 3·24 bars (47 lb/sq in); nosewheel and tyre size 6·00-6, pressure 1·86 bars (27 lb/sq in). Cleveland brakes.

POWER PLANT (Pegaso): Two 223·5 kW (300 hp) Continental IO-520-K flat-six engines, each driving a Hartzell PHC-C3YF-2UF three-blade constant-speed fully-feathering propeller. Fuel in two wing tanks, each of 124 litres (27·25 Imp gallons), and two wingtip tanks each of 126 litres (27·7 Imp gallons) capacity. Total capacity 500 litres (110 Imp gallons). Oil capacity 22·8 litres (5 Imp gallons). Electrical propeller de-icing available optionally.

POWER PLANT (Condor): Two 298 kW (400 shp) Allison 250-B17B turboprop engines, each driving a Hartzell HC-B3TF-7A/T10173-21R three-blade propeller. Total internal fuel capacity 1,200 litres (264 Imp gallons); provision for a further 600 litres (132 Imp gallons) to be carried in two underwing auxiliary tanks. Oil capacity 16 litres (3·5 Imp gallons).

ACCOMMODATION: Normal seating, in fully enclosed cabin, for six persons (four in Condor), including pilot. Space for up to 104 kg (230 lb) of baggage in four compartments in forward and rear fuselage and each engine nacelle. Access to cabin via large door on each side. Cabin heated, ventilated, and soundproofed with glass-wool insulation.

AVIONICS AND EQUIPMENT: IFR instrumentation and dual controls standard; other installations to customer's requirements.

DIMENSIONS, EXTERNAL:

Wing span over tip-tanks	10·34 m (33 ft 11 in)
Wing chord at root	1·65 m (5 ft 5 in)
Wing chord at tip	1·50 m (4 ft 11 in)
Wing aspect ratio	6·67
Length overall	8·22 m (26 ft 11½ in)
Height overall	3·50 m (11 ft 5¾ in)
Tailplane span	4·80 m (15 ft 9 in)
Wheel track	3·50 m (11 ft 5¾ in)
Wheelbase	2·40 m (7 ft 10½ in)
Propeller diameter	1·98 m (6 ft 6 in)
Min propeller ground clearance	0·20 m (8 in)
Distance between propeller centres	3·42 m (11 ft 2½ in)

DIMENSIONS, INTERNAL:

Cabin: Max length	3·66 m (12 ft 0 in)
Max width	1·17 m (3 ft 10 in)
Max height	1·13 m (3 ft 8½ in)
Baggage compartments (four): Total volume	0·70 m³ (25 cu ft)

AREAS:

Wings, gross	16·02 m² (172·4 sq ft)
Ailerons (total)	1·42 m² (15·28 sq ft)
Trailing-edge flaps (total)	1·59 m² (17·13 sq ft)
Fin	1·24 m² (13·35 sq ft)
Rudder, incl tab	0·79 m² (8·50 sq ft)
Tailplane	2·88 m² (31·00 sq ft)
Elevators, incl tabs	2·28 m² (24·54 sq ft)

WEIGHTS AND LOADINGS (A: Pegaso; B: Condor without underwing tanks; C: Condor with underwing tanks):

Weight empty, equipped: A	1,510 kg (3,330 lb)
B	1,400 kg (3,086 lb)
C	1,500 kg (3,307 lb)
Max payload: A	514 kg (1,134 lb)
Max T-O and landing weight: A	2,250 kg (4,960 lb)
B	2,200 kg (4,850 lb)
C	2,500 kg (5,511 lb)
Max wing loading: A	145·0 kg/m² (29·7 lb/sq ft)
B	137·3 kg/m² (28·12 lb/sq ft)
C	156·0 kg/m² (31·95 lb/sq ft)
Max power loading: A	5·30 kg/kW (8·71 lb/hp)
B	3·69 kg/kW (6·06 lb/shp)
C	4·19 kg/kW (6·89 lb/shp)

PERFORMANCE (at max T-O weight except where indicated):

Never-exceed speed: A	232 knots (431 km/h; 267 mph)
Max level speed at S/L: A	216 knots (400 km/h; 249 mph)
B	248 knots (460 km/h; 286 mph)
C	227 knots (420 km/h; 261 mph)
Max cruising speed (75% power) at 2,440 m (8,000 ft): A	203 knots (377 km/h; 234 mph)
B	234 knots (435 km/h; 270 mph)
C	205 knots (380 km/h; 236 mph)
Econ cruising speed (60% power) at 3,660 m (12,000 ft): A, C	183 knots (340 km/h; 211 mph)
B	210 knots (390 km/h; 242 mph)
Stalling speed, flaps down: A	66 knots (121 km/h; 76 mph)
B at 2,000 kg (4,409 lb) AUW	62 knots (115 km/h; 71·5 mph)
C at 2,000 kg (4,409 lb) AUW	68 knots (125 km/h; 78 mph)
Max rate of climb at S/L: A	550 m (1,805 ft)/min
B	900 m (2,950 ft)/min
C	720 m (2,360 ft)/min
Rate of climb at S/L, one engine out: A	120 m (394 ft)/min
Service ceiling: A	6,250 m (20,500 ft)
B	8,500 m (27,900 ft)
C	7,000 m (22,975 ft)
Service ceiling, one engine out: A	2,290 m (7,515 ft)
B	4,800 m (15,750 ft)
C	4,000 m (13,125 ft)
T-O run: A	230 m (755 ft)
B	180 m (590 ft)
C	220 m (720 ft)
T-O to 15 m (50 ft): A	320 m (1,050 ft)
Landing from 15 m (50 ft): A	380 m (1,250 ft)
Landing run: A, B	260 m (853 ft)
C	300 m (984 ft)
Range: A with max fuel	955 nm (1,770 km; 1,100 miles)
A with max payload	547 nm (1,013 km; 630 miles)
B at 75% power	540 nm (1,000 km; 621 miles)
B at 50% power	593 nm (1,100 km; 684 miles)
C at 75% power	1,187 nm (2,200 km; 1,367 miles)
C at 50% power	1,349 nm (2,500 km; 1,553 miles)
Max endurance: B	5 h
C	10 h

Condor four-seat turboprop-powered military version of the General Avia F.20 Pegaso *(Pilot Press)*

First production example of the General Avia F.20 Pegaso six-seat light business twin

PARTENAVIA

PARTENAVIA COSTRUZIONI AERONAUTICHE SpA

HEAD OFFICE AND WORKS: Via Cava, CP 2179, 80026 Casoria (Naples)
Telephone: (081) 7596311 (PBX)
Telex: 720199 PARTNA I
PRESIDENT: Prof Ing Luigi Pascale
DIRECTORS:
Franco Capanna
Dott Ing Giulio Ciampolini
Ing Beppe Sacchi
Ing Giovanni Sarzotti
PRODUCTION DIRECTOR: Ing Nino Pascale

This company was founded in 1957 by Prof Ing Luigi Pascale and his brother, Ing Nino Pascale, and has since built a series of light aircraft designed by Prof Ing Pascale.

Since 1974 the company has occupied a 12,000 m^2 (129,165 sq ft) facility on Capodichino Airport, Naples, where it is now concentrating on production and development of the P.68C Victor twin-engined seven-seat light aircraft and its derivatives, and on the P.66C Charlie.

Details of the AP 68B and P.78 aircraft, originated by Partenavia, can now be found under the entry for Aeritalia, which is reportedly to take over Partenavia in a forthcoming reorganisation of the Italian aerospace industry.

Partenavia P.66C Charlie two/four-seat basic training aircraft in Italian Aero Club insignia

PARTENAVIA P.66C-160 CHARLIE

This two/four-seat basic training aircraft is an improved version of the P.64B/P.66B Oscar series (see 1975-76 *Jane's*), and is certificated in the FAR Pt 23 Utility category, with clearance for all positive-*g* aerobatic manoeuvres and six-turn spins. The prototype P.66C, with a 112 kW (150 hp) Avco Lycoming O-360-A1A engine, flew in early January 1976, and was certificated in Spring 1976.

The P.66C was chosen by the Aero Club d'Italia as the standard basic trainer for aero clubs throughout Italy, and an order for 70 P.66C Charlies was placed, with an option to increase this order to 250. Deliveries began at the end of 1977, and totalled 96 by the end of 1980, with production continuing.

TYPE: Two/four-seat light monoplane.

WINGS: Braced high-wing monoplane with single streamline-section bracing strut each side. Wing section NACA 63 series. Thickness/chord ratio 15%. Dihedral 1° 30′. Incidence at root 1° 40′. No sweepback. Stressed-skin single-spar torsion-box structure of aluminium alloy, with one-piece GRP moulded leading-edges. Ailerons and electrically-operated slotted trailing-edge flaps of similar construction to wings.

FUSELAGE: Forward portion, to rear of cabin, has a welded steel tube basic structure to which are attached light alloy skin panels. Rear fuselage is of conventional light alloy stressed-skin construction.

TAIL UNIT: Cantilever stressed-skin metal structure with sweptback vertical surfaces. All-moving tailplane in two symmetrical halves joined by steel cross-tube. Anti-balance tab in trailing-edge of tailplane, over 80% of span.

LANDING GEAR: Non-retractable tricycle type, with steerable nosewheel. Cantilever spring steel main legs. Oleo nosewheel shock-absorber. Cleveland main wheels type 40-28, with Pirelli tyres size 6·00-6. Goodyear nosewheel tyre size 5·00-5. Cleveland type 30-18 hydraulic disc brakes.

POWER PLANT: One 119 kW (160 hp) Avco Lycoming O-320-H2AD flat-four engine, driving a Hoffmann HO 23-183.150 two-blade fixed-pitch propeller. Two integral fuel tanks in wing roots, total usable capacity 162 litres (35·5 Imp gallons). Refuelling points above wings. Oil capacity 7·5 litres (1·7 Imp gallons).

ACCOMMODATION: Enclosed cabin seating two or four persons in pairs; front seats are of the adjustable sliding type. Two forward-hinged doors, one by each front seat. Baggage space aft of rear seats, with separate door on starboard side. Dual controls, heating, ventilation and soundproofing standard.

AVIONICS AND EQUIPMENT: Optional items include full IFR instrumentation, Grimes rotating beacon, VHF radio, VOR and ADF.

DIMENSIONS, EXTERNAL:	
Wing span	9·986 m (32 ft 9¼ in)
Wing chord (constant)	1·36 m (4 ft 5½ in)
Wing aspect ratio	7·45
Length overall	7·24 m (23 ft 9 in)
Height overall	2·77 m (9 ft 1 in)
Tailplane span	3·10 m (10 ft 2 in)
Wheel track	2·10 m (6 ft 10½ in)
Wheelbase	1·63 m (5 ft 4¼ in)
Propeller diameter	1·88 m (6 ft 2 in)
DIMENSIONS, INTERNAL:	
Cabin: Max width	1·06 m (3 ft 5¾ in)
Max height	1·20 m (3 ft 11¼ in)
AREAS:	
Wings, gross	13·40 m^2 (144·2 sq ft)
Ailerons (total)	1·29 m^2 (13·88 sq ft)
Trailing-edge flaps (total)	1·71 m^2 (18·40 sq ft)
Fin	0·73 m^2 (7·86 sq ft)
Rudder	0·45 m^2 (4·84 sq ft)
Tailplane, incl tab	2·17 m^2 (23·36 sq ft)
WEIGHTS AND LOADINGS:	
Weight empty	600 kg (1,322 lb)
Max T-O and landing weight	990 kg (2,183 lb)
Max wing loading	73·88 kg/m^2 (15·13 lb/sq ft)
Max power loading	8·30 kg/kW (13·63 lb/hp)
PERFORMANCE (at max T-O weight, ISA):	
Max level speed at S/L	130 knots (241 km/h; 150 mph)
Max cruising speed (75% power) at 1,980 m (6,500 ft)	118 knots (218 km/h; 135 mph)
Econ cruising speed (65% power) at 2,745 m (9,000 ft)	111 knots (206 km/h; 128 mph)
Stalling speed at S/L:	
flaps up	53 knots (98·5 km/h; 61·5 mph)
flaps down (T-O)	50 knots (93 km/h; 58 mph)
flaps down (landing)	44 knots (82 km/h; 51 mph)
Max rate of climb at S/L	289 m (950 ft)/min
Service ceiling	4,570 m (15,000 ft)
T-O run at S/L	245 m (805 ft)
T-O to 15 m (50 ft) at S/L	420 m (1,378 ft)
Landing from 15 m (50 ft) at S/L	350 m (1,148 ft)
Landing run at S/L	150 m (492 ft)
Range at 1,980 m (6,500 ft), 75% power, with reserves	461 nm (854 km; 531 miles)
Range at 2,745 m (9,000 ft), 65% power, with reserves	526 nm (975 km; 605 miles)
Endurance at 1,980 m (6,500 ft), 75% power, with reserves	3 h 54 min
Endurance at 2,745 m (9,000 ft), 65% power, with reserves	4 h 44 min

Partenavia P.68C current production version of the Victor six/seven-seat light aircraft

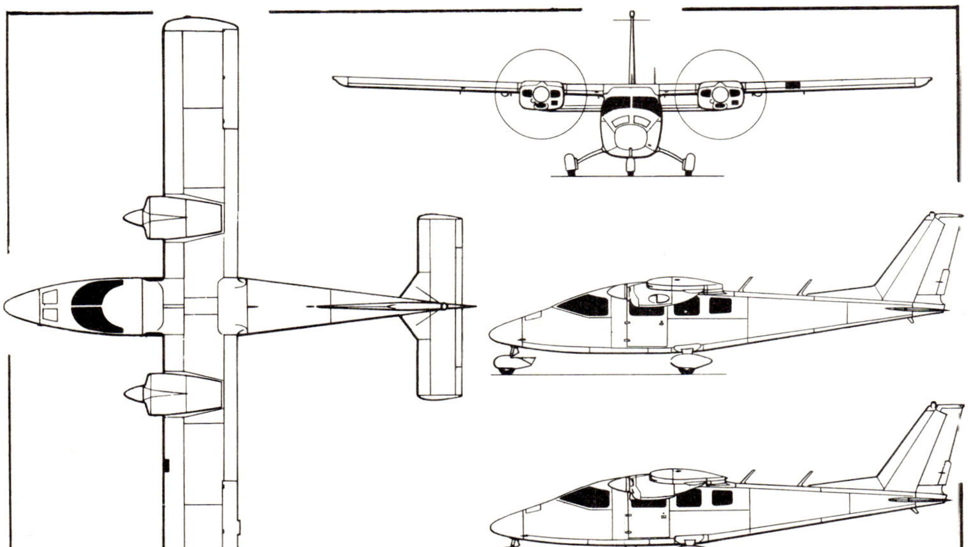
Partenavia P.68C Victor, with additional side view of P.68C-TC (centre) *(Pilot Press)*

PARTENAVIA VICTOR

The original P.68, designed by Prof Ing Luigi Pascale in 1968, was described in the 1975-76 *Jane's*. From it was developed the P.68B Victor twin-engined light transport, which entered production in Partenavia's factory at Naples Airport in the Spring of 1974.

The following versions have been built:

P.68B. Initial production version, described in 1979-80 *Jane's*. Max T-O weight increased to 1,990 kg (4,387 lb). Production phased out in late 1979 in favour of P.68C.

P.68C. Improved version of P.68B, with lengthened nose, increased fuel capacity, and several internal changes. Detailed description applies primarily to this current production version.

P.68C-R. Retractable landing gear version of P.68C. Not intended for production.

P.68C-TC. Similar to P.68C, but with 156·5 kW (210 hp) Avco Lycoming TO-360-C1A6D turbocharged engines. Certificated in June 1980; in production.

P.68 floatplane/amphibian. One P.68B, modified for the installation of both amphibious and non-amphibious floats. All known details in 1979-80 *Jane's*.

P.68C-TC turbocharged version of the Victor light transport aircraft

P.68 Observer. Special observation version; described separately.

P.68R. Retractable landing gear version of P.68B, first flown in December 1976. Certificated, but not intended for production. Described in 1978-79 *Jane's*.

AP.68B. Turboprop-powered version of P.68R, now being developed by Aeritalia. Described under Aeritalia entry.

By December 1980, approx 225 Victors had been delivered, the majority of them for export to operators in Australia, Austria, Belgium, Chile, Congo, Denmark, Finland, France, Gabon, Germany (Federal Republic), Israel, Kenya, the Middle East, Morocco, Netherlands, Papua New Guinea, Norway, South Africa, Sweden, Switzerland, Tanzania, the UK, USA and Venezuela.

Type: Six/seven-seat light transport and trainer.

Wings: Cantilever high-wing monoplane. Wing section NACA 63_3-515. Dihedral 1°. Incidence 1° 30′. No sweepback. Stressed-skin single-spar torsion-box structure of aluminium alloy. All-metal ailerons and electrically-operated single-slotted trailing-edge flaps. Hoerner GRP wingtips. No tabs.

Fuselage: Conventional all-metal semi-monocoque structure of frames and longerons, with four main longerons and stressed-skin covering. Fuselage/wing intersection mainly of GRP.

Tail Unit: Cantilever stressed-skin metal structure. All-moving tailplane, in two symmetrical halves joined by steel cross-tube and of constant chord except for increase at leading-edge roots. Balance tab in tailplane trailing-edge, over 80 per cent of span. Sweptback fin and rudder, with small dorsal fin. Trim tab in rudder.

Landing Gear: Non-retractable tricycle type, with steerable nosewheel. Cantilever spring steel main legs. Oleo-pneumatic shock-absorber on nosewheel. Cleveland main wheels, type 40-96, with Pirelli eight-ply tyres size 6·00-6. Goodyear six-ply nosewheel tyre, size 5·00-5. Cleveland type 30-61 hydraulic disc brakes. Parking brake. Streamline wheel fairings standard.

Power Plant (P.68C): Two 149 kW (200 hp) Avco Lycoming IO-360-A1B6 flat-four engines, each driving a Hartzell HC-C2YK-2C/C-7666A-4 two-blade constant-speed fully-feathering propeller with spinner. Integral fuel tank in each wing, total capacity 538 litres (118 Imp gallons), of which 520 litres (114 Imp gallons) are usable. Refuelling point above each wing. Oil capacity 15 litres (3·3 Imp gallons).

Accommodation: Seating for seven persons in cabin, including pilot, in two rows of two seats and a rear bench seat for three persons. A 'club' seating arrangement is available optionally, having the two middle seats facing rearward with a folding table between them and the bench seat. Front seats are of the adjustable sliding type. Access to all seats via large car-type door on port side of cabin. Up to 181 kg (400 lb) of baggage can be carried in compartment aft of rear bench seat. Access to baggage compartment from inside cabin, or via large forward-hinged door on starboard side, which serves also as emergency exit. Two stretchers or other loads can be carried when all passenger seats are removed. Dual controls, cabin heating, ventilation and soundproofing standard.

Systems: Electrical power supplied by two 24V 70A alternators and a 24V 17Ah battery. No hydraulic system. Goodrich pneumatic de-icing system optional.

Avionics and Equipment (P.68C): Wide range of Collins Micro Line, Edo-Aire Mitchell Century III Autopilot, or King Silver Crown avionics, to customer's requirements. Provision for SunAir ASB 100 HF radio. Standard equipment includes airspeed indicator, gyro horizon, directional gyro, two cylinder head temperature gauges, clock, exhaust gas temperature indicator, outside air temperature gauge, rate of climb indicator, sensitive altimeter, electrical turn rate indicator, inertia-reel shoulder harness for pilot and co-pilot, stall warning system, four upholstered seats with back pockets, and one bench seat with folding back (with safety belts on all seats), cabin fire extinguisher, six individual fresh-air outlets and six floor warm-air vents, windscreen defrosters, cabin soundproofing, annunciator panel warning lights, two map lights, individual reading lights, individual instrument panel floodlights with rheostat, anti-collision strobe light, two landing/taxying lights, navigation lights, anti-static kit, external power receptacle, oil coolers with thermostatic control, quick-drain fuel and oil valves, and towbar. Optional equipment includes Janitrol 45,000 BTU combustion heater, wing and tail pneumatic de-icing system, electrothermal propeller de-icing system, 0·46 × 0·58 m (18 × 23 in) floor panel for photogrammetric camera, including periscope sight hatch, second airspeed indicator, second gyro horizon, chronometer, second altimeter, vertically adjustable pilot's and co-pilot's seats, alcohol windscreen de-icing, heated stall warning indicator, all-leather interior, forced ventilation blower, ice light and second oil cooler.

Dimensions, external:

Wing span	12·00 m (39 ft 4½ in)
Wing chord (constant)	1·55 m (5 ft 1 in)
Wing aspect ratio	7·75
Length overall	9·55 m (31 ft 4 in)
Height overall	3·40 m (11 ft 1¾ in)
Tailplane span	3·90 m (12 ft 9½ in)
Wheel track	2·40 m (7 ft 10½ in)
Wheelbase	3·50 m (11 ft 5¾ in)
Propeller diameter	1·88 m (6 ft 2 in)
Baggage door, stbd: Height	0·80 m (2 ft 7½ in)
Width	0·80 m (2 ft 7½ in)

Dimensions, internal:

Cabin: Length	3·58 m (11 ft 9 in)
Max width	1·16 m (3 ft 9½ in)
Max height	1·20 m (3 ft 11¼ in)
Baggage space	0·56 m^3 (20 cu ft)

Areas:

Wings, gross	18·60 m^2 (200·2 sq ft)
Ailerons (total)	1·79 m^2 (19·27 sq ft)
Trailing-edge flaps (total)	2·37 m^2 (25·51 sq ft)
Fin	1·59 m^2 (17·11 sq ft)
Rudder, incl tab	0·44 m^2 (4·74 sq ft)
Tailplane, incl tab	4·41 m^2 (47·47 sq ft)

Weights and Loadings:

Weight empty: C	1,230 kg (2,711 lb)
C-TC	1,300 kg (2,866 lb)
Max T-O weight: C, C-TC	1,990 kg (4,387 lb)
Max landing weight: C, C-TC	1,890 kg (4,166 lb)
Max wing loading: C, C-TC	107 kg/m^2 (21·9 lb/sq ft)
Max power loading: C	6·68 kg/kW (10·97 lb/hp)
C-TC	6·36 kg/kW (10·45 lb/hp)

Performance (at max T-O weight):

Max level speed: C at S/L	174 knots (322 km/h; 200 mph)
C-TC at 5,335 m (17,500 ft)	195 knots (361 km/h; 224 mph)
Max cruising speed (75% power): C at 2,290 m (7,500 ft)	166 knots (307 km/h; 191 mph)
C-TC at 6,100 m (20,000 ft)	183 knots (339 km/h; 211 mph)
C-TC at 3,660 m (12,000 ft)	172 knots (318 km/h; 198 mph)
Cruising speed (65% power): C at 3,350 m (11,000 ft)	161 knots (298 km/h; 185 mph)
C-TC at 3,050 m (10,000 ft)	158 knots (293 km/h; 182 mph)
Cruising speed (55% power): C at 3,660 m (12,000 ft)	150 knots (278 km/h; 173 mph)
C-TC at 3,050 m (10,000 ft)	147 knots (272 km/h; 169 mph)
Stalling speed, flaps up: C, C-TC	65 knots (120 km/h; 75 mph)
Stalling speed, flaps down: C, C-TC	57·5 knots (106 km/h; 66 mph)
Max rate of climb at S/L: C	457 m (1,500 ft)/min
C-TC	472 m (1,550 ft)/min
Rate of climb at S/L, one engine out: C	82 m (270 ft)/min
C-TC	88 m (290 ft)/min
Service ceiling: C	5,850 m (19,200 ft)
C-TC	7,620 m (25,000 ft)
Service ceiling, one engine out: C	2,100 m (6,900 ft)
C-TC	4,420 m (14,500 ft)
T-O run: C, C-TC	230 m (755 ft)
T-O to 15 m (50 ft): C	396 m (1,300 ft)
C-TC	385 m (1,263 ft)
Landing from 15 m (50 ft): C, C-TC	488 m (1,600 ft)
Landing run: C, C-TC	215 m (705 ft)
Accelerate/stop distance: C	473 m (1,550 ft)
C-TC	510 m (1,673 ft)
Optimum cruising range (C), 45 min reserves: 75% power at 2,290 m (7,500 ft)	1,050 nm (1,945 km; 1,209 miles)
65% power at 3,350 m (11,000 ft)	1,140 nm (2,112 km; 1,312 miles)
55% power at 3,660 m (12,000 ft)	1,210 nm (2,242 km; 1,393 miles)
Optimum cruising range (C-TC) at 3,660 m (12,000 ft), 45 min reserves: 75% power	775 nm (1,436 km; 892 miles)
65% power	940 nm (1,742 km; 1,082 miles)
55% power	1,020 nm (1,890 km; 1,175 miles)
Range with max fuel (C-TC): 65% power at 6,400 m (21,000 ft)	1,100 nm (2,037 km; 1,266 miles)

PARTENAVIA P.68 OBSERVER

Developed in collaboration with Sportavia-Pützer, Partenavia's West German distributor, the Observer has a forward and downward view for the crew equal to that of a helicopter. The Plexiglas nose, cockpit and associated structure were designed by Sportavia-Pützer; the prototype (D-GERD) was constructed at that company's Dahlemer-Binz factory, and first flew on 20 February 1976.

With its good low-speed handling characteristics, the Observer is considered to be capable of performing many roles allocated normally to helicopters. It is intended particularly for patrol and observation operations. The prototype was evaluated in mid-1978 by the Rhineland-Palatinate police department.

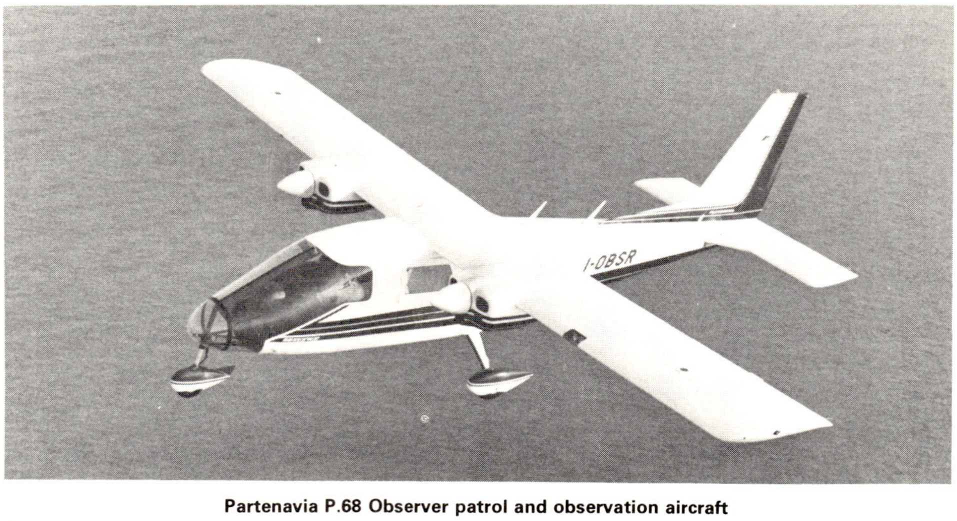

Partenavia P.68 Observer patrol and observation aircraft

The first Partenavia-built Observer was flown in the Spring of 1980, and certification was obtained in June of that year. Improvements have been made to the flight deck and instrument panel. The Observer was introduced on to the P.68 production line in July 1980, and by mid-1981 three had been delivered, to customers in Belgium, West Germany and Italy. Additional orders were then being negotiated.

Dimensions, external: As P.68C except:
Length overall 9·35 m (30 ft 8 in)

Weights and Loadings:
Weight empty 1,280 kg (2,822 lb)
Max T-O weight 1,960 kg (4,321 lb)
Max wing loading 105·3 kg/m² (21·58 lb/sq ft)
Max power loading 6·58 kg/kW (10·80 lb/hp)

Performance (at max T-O weight):
Max level speed at S/L 174 knots (322 km/h; 200 mph)
Cruising speed:
75% power at 2,290 m (7,500 ft) 165 knots (306 km/h; 190 mph)
65% power at 3,350 m (11,000 ft) 160 knots (296 km/h; 184 mph)
55% power at 3,660 m (12,000 ft) 149 knots (276 km/h; 171 mph)
Stalling speed, flaps up 64 knots (118 km/h; 74 mph)
Stalling speed, flaps down 56 knots (101 km/h; 64 mph)
Max rate of climb at S/L 488 m (1,600 ft)/min
Rate of climb at S/L, one engine out 98 m (320 ft)/min
Service ceiling 6,100 m (20,000 ft)
Service ceiling, one engine out 2,375 m (7,800 ft)
T-O run 229 m (750 ft)
T-O to 15 m (50 ft) 387 m (1,270 ft)
Landing from 15 m (50 ft) 479 m (1,570 ft)
Landing run 210 m (690 ft)
Accelerate/stop distance 473 m (1,550 ft)
Optimum cruising range, 45 min reserves:
75% power at 2,290 m (7,500 ft) 1,015 nm (1,880 km; 1,168 miles)
65% power at 3,350 m (11,000 ft) 1,108 nm (2,053 km; 1,275 miles)
55% power at 3,660 m (12,000 ft) 1,180 nm (2,186 km; 1,358 miles)

PIAGGIO

INDUSTRIE AERONAUTICHE E MECCANICHE RINALDO PIAGGIO SpA

Head Office: Via Cibrario 4, 16154 Genoa
Telephone: 600831
Telex: 270695 AERPIA I
Branch Office: Via A. Gramsci 34, Rome
Works: Genoa (Aircraft Division); Finale Ligure (Aero-Engine Division)
Chairman and Managing Director: Dott Rinaldo Piaggio
Director General: Ing Umberto Barnato
Director of International Programmes: Ing B. Mori
Technical Consultant: Ing Giovanni P. Casiraghi
Technical Director: Ing Alessandro Mazzoni
Marketing Director: Commander G. B. Pizzinato

The original Piaggio company began the construction of aeroplanes in its Genoa-Sestri plant in 1916, and later in the Finale Ligure works. The present company was formed on 29 February 1964, and has since operated as an independent concern. It employs about 1,400 people and has a total covered works area (Genoa-Sestri and Finale-Ligure) of approx 100,000 m² (1,076,390 sq ft). In addition to building aircraft of its own design, Piaggio is producing components for the Aeritalia G222, Panavia Tornado, Boeing 767 and McDonnell Douglas DC-10.

The company is organised into two production Divisions: the activities of the Aero-Engine Division are described in the appropriate section of this edition.

R. Piaggio P.166-DL3 twin-turboprop light transport

R. PIAGGIO P.166-DL3

The P.166 has been produced in several basic versions, of which the original piston-engined P.166 was described in the 1963-64 *Jane's*; the P.166M, P.166B Portofino and P.166C in the 1971-72 *Jane's*; the P.166S in the 1974-75 *Jane's*; and the P.166-DL2 in the 1978-79 *Jane's*. Current versions are as follows:

P.166-DL3. This turboprop version flew for the first time on 3 July 1976, and received FAA and RAI certification in 1978. It can be configured and equipped for a wide variety of duties, including executive transport (EXC); transport and dropping of up to ten paratroops (PAR); air ambulance for two stretchers and two medical attendants (AMB); multi-engine aircrew training (MTR); light tactical transport (LTT); armed military counter-insurgency, field support, and search and rescue (AML); maritime reconnaissance (MAR); environmental control and geophysical survey (ECS); aerophotogrammetry (APH); and aerial firefighting (AFF).

Piaggio is currently building a batch of 22 P.166-DL3s. Sales include two of the MTR version to Alitalia, each of which is equipped to accommodate two trainees, and four other DL3s to the Somali Aeronautical Corps.

P.166-DL3-MAR. Medium-range maritime surveillance version. Applications include coastguard, anti-smuggling, maritime traffic, fishery and pollution control, and search and rescue. Equipped with integrated search/detection/identification/plotting and reporting system, plus four underwing pylons for external weapons or other stores. Capable of operating from unprepared strips, day or night, in all weathers.

Type: Twin-turboprop light transport (DL3) and maritime surveillance aircraft (DL3-MAR).

Wings: Shoulder gull-wing cantilever monoplane. NACA 230 wing section. Dihedral 21° 30′ on inner portion, 2° 30′ on outer wings. Incidence 2° 43′ at root. Sweepback 7° 30′ at quarter-chord. Aluminium alloy flush-riveted torsion-box structure, with single main spar and auxiliary rear spar. All-metal slotted ailerons, with geared and trim tab in starboard aileron. All-metal hydraulically-actuated slotted flaps. Rubber-boot leading-edge de-icing optional.

Fuselage: Aluminium alloy flush-riveted semi-monocoque structure of frames and L-section extruded stringers: no longerons.

Tail Unit: Cantilever aluminium alloy structure, with flush-riveted smooth skin on fixed surfaces and beaded skin on control surfaces. Rudder and elevators statically and dynamically balanced. Geared and trim tabs in elevators; trim tab in rudder. Rubber-boot leading-edge de-icing of fin and tailplane optional.

Landing Gear: Retractable tricycle type. Magnaghi oleo-pneumatic shock-absorbers on all units. Hydraulic actuation. Nosewheel retracts rearward, main units upward. Goodyear 24 × 7·7 main wheels with size 8·50-10 tyres, pressure 3·79 bars (55 lb/sq in). Goodyear steerable and self-centring nosewheel with size 6·00-6 tyre, pressure 2·90 bars (42 lb/sq in). Goodyear or Magnaghi hydraulic brakes.

Power Plant: Two Avco Lycoming LTP 101-600 turboprop engines, each developing 446·5 kW (599 shp) at take-off, and each driving a Hartzell HC-B3TN-3DL/LT10282-9·5 three-blade constant-speed fully-feathering metal pusher propeller. Available optionally with 503 kW (675 shp) LTP 101-700A1A engines. Fuel in two 212 litre (46·5 Imp gallon) outer-wing main tanks and two 323 litre (71 Imp gallon) wingtip tanks; total internal fuel capacity 1,070 litres (235 Imp gallons). Gravity refuelling points in each main tank and tip-tank. Provision for two 177 or 284 litre (39 or 62·5 Imp gallon) underwing drop-tanks. Air intakes and propeller blades de-iced by engine exhaust.

Accommodation: Crew of two on raised flight deck, with dual controls. Aft of flight deck, accommodation consists of a passenger cabin, utility compartment and baggage compartment. Access to flight deck is via passenger/cargo double door on port side, forward of wing, or via individual crew door on each side of flight deck. External access to baggage compartment via port-side door aft of wing. Passenger cabin extends from rear of flight deck to bulkhead at wing main spar; fitting of passenger-carrying, cargo or other interiors is facilitated by two continuous rails on cabin floor, permitting considerable flexibility in standard or customised interior layouts. Standard seating in DL3 is for eight passengers, with individual lighting, ventilation and (optionally) oxygen controls. Flight deck can be separated from passenger cabin by a screen. Door in bulkhead at rear of cabin provides access to utility compartment, in which can be fitted a toilet, bar, or mission equipment for certain roles. Entire accommodation is heated, ventilated and soundproofed. Emergency exit forward of wing on starboard side. Windscreen hot-air demisting standard. Windscreen wipers, washers and methanol spray de-icing optional.

Systems: Hydraulic system, pressure 127 bars (1,840 lb/sq in), for landing gear, flap and brake actuation, nosewheel steering and lock, and (on APH version) actuation of ventral doors. Handpump for emergency extension of landing gear. Standard electrical system is 28V DC, supplied by two engine-driven starter/generators and a nickel-cadmium battery. External power receptacle. Static or rotary inverters, to supply AC power for avionics and instruments, available optionally. Oxygen system, to customers' requirements, optional.

Avionics: Standard avionics packages available to individual customer's requirements: minimum recommended package includes two VHF com, two VHF nav (VOR/ILS), ADF, ATC transponder, compass system and intercom. Optional avionics include radar, autopilot, navigation system and synthesiser-type HF radio.

Equipment: According to mission configuration. Quickly interchangeable individual seats of various types, bench seat, divan or stretchers for EXC, PAR, AMB, MTR and LTT versions; strengthened floor in LTT. Four underwing pylons standard on AML, for ordnance, supply containers and auxiliary fuel tanks. Four pylons and integrated search/detection/identification/plotting

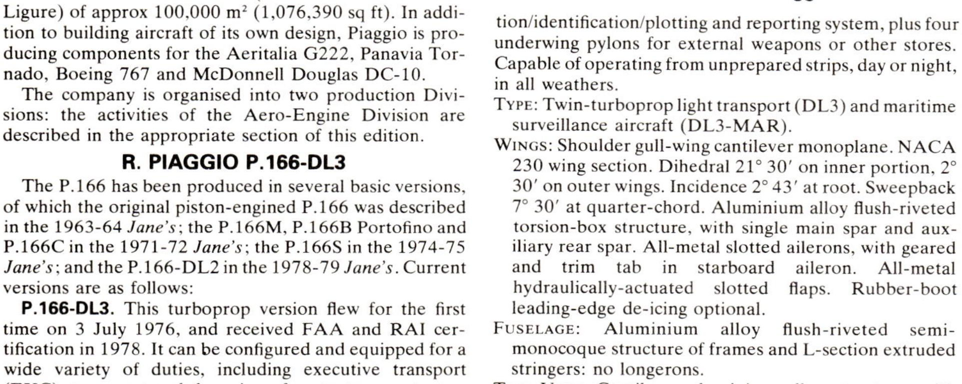

P.166-DL3 version of the Piaggio P.166, with two Avco Lycoming LTP 101 turboprop engines *(Pilot Press)*

and reporting system on MAR (see separate paragraph following). Magnetometer, multiscanner, multiple-head camera and associated equipment in ECS version. Two cameras, associated equipment, and ventral sliding door in APH, with option for four underwing pylons. Internal removable water/extinguisher container and rapid charge/discharge system for AFF.

Avionics and Equipment (DL3-MAR): According to customer and mission requirements. Typical packages may include 180° scan search radar, two Vinten aerial cameras, Doppler or other suitable navigation system, and operator's console; or 360° search radar, low light level TV camera with monitor and recorder, Doppler/Omega navigation system and tactical co-ordinator's console.

Armament (P.166-DL3-AML): Four underwing pylons for ordnance or other stores. Typical loads include two 7·5 mm machine-gun pods, with 1,000 rds/pod; four rocket launchers, each with six 68 mm, eighteen 37 mm or thirty-six 37 mm SNEB rockets; four bombs of up to 500 lb each; four clusters, each of four 36 lb bombs; two Mk 44 torpedoes; four SAR supply containers; or two jettisonable auxiliary fuel tanks.

Dimensions, external:
Wing span: without tip-tanks 13·51 m (44 ft 4 in)
with tip-tanks 14·69 m (48 ft 2½ in)
Wing chord at root 2·40 m (7 ft 10½ in)
Wing chord at tip 1·15 m (3 ft 9¼ in)
Wing aspect ratio 7·3
Length overall 11·90 m (39 ft 3 in)
Height overall 5·00 m (16 ft 5 in)
Tailplane span 5·10 m (16 ft 9 in)
Wheel track 2·66 m (8 ft 9 in)
Wheelbase 4·71 m (15 ft 5½ in)
Propeller diameter 2·36 m (7 ft 9 in)
Cabin door: Height 1·38 m (4 ft 6 in)
Width 1·28 m (4 ft 2 in)

Dimensions, internal:
Cabin, incl flight deck: Length 3·20 m (10 ft 6 in)
Max width 1·57 m (5 ft 2 in)
Max height 1·76 m (5 ft 9 in)
Floor area 5·14 m² (55·3 sq ft)
Volume 6·63 m³ (234·1 cu ft)
Utility compartment: Length 0·65 m (2 ft 1½ in)
Max width 1·52 m (5 ft 0 in)
Max height 1·70 m (5 ft 7 in)
Volume 2·27 m³ (80·2 cu ft)
Baggage compartment volume 1·80 m³ (63·6 cu ft)

Areas:
Wings, gross 26·56 m² (285·9 sq ft)
Ailerons (total) 1·95 m² (21·00 sq ft)
Trailing-edge flaps (total) 2·38 m² (25·60 sq ft)
Fin 1·62 m² (17·44 sq ft)
Rudder, incl tab 1·23 m² (13·24 sq ft)
Tailplane 3·50 m² (37·67 sq ft)
Elevators, incl tabs 1·29 m² (13·88 sq ft)

Weights and Loadings:
Weight empty, equipped: DL3 2,600 kg (5,732 lb)
DL3-MAR 2,860 kg (6,306 lb)
Operational weight empty (DL3-MAR) 3,206 kg (7,069 lb)
Max payload (DL3) 1,188 kg (2,619 lb)
Max T-O weight (both) 4,300 kg (9,480 lb)
Max zero-fuel weight (DL3) 3,800 kg (8,377 lb)
Max landing weight (DL3) 3,800 kg (8,377 lb)
Max wing loading 162 kg/m² (33·2 lb/sq ft)
Max power loading 4·91 kg/kW (8·07 lb/shp)

Performance (at AUW of 3,600 kg; 7,936 lb except where indicated. A: LTP 101-600 engines; B: LTP 101-700A1A):
Max level speed at 3,050 m (10,000 ft):
A 215 knots (400 km/h; 248 mph)
B 224 knots (415 km/h; 258 mph)
Max cruising speed at 3,050 m (10,000 ft):
A, B 200 knots (370 km/h; 230 mph)
Econ cruising speed at 3,050 m (10,000 ft):
A, B 162 knots (300 km/h; 186 mph)
Stalling speed, flaps and landing gear up:
A, B 78·5 knots (145 km/h; 90 mph) CAS
Stalling speed, flaps and landing gear down:
A, B 69 knots (128 km/h; 80 mph) CAS
Max rate of climb at S/L, AUW of 3,855 kg (8,500 lb):
A 710 m (2,330 ft)/min
B 783 m (2,570 ft)/min
Rate of climb at S/L, one engine out, AUW of 3,855 kg (8,500 lb): A 189 m (620 ft)/min
B 226 m (740 ft)/min
Service ceiling, AUW of 3,855 kg (8,500 lb):
A 8,840 m (29,000 ft)
B 9,300 m (30,500 ft)
Service ceiling, one engine out, AUW of 3,855 kg (8,500 lb): A 3,960 m (13,000 ft)
B 4,875 m (16,000 ft)
T-O run at max T-O weight: A 494 m (1,620 ft)
B 475 m (1,560 ft)
T-O to 15 m (50 ft) at max T-O weight:
A 640 m (2,100 ft)
B 610 m (2,000 ft)
Landing from 15 m (50 ft), at AUW of 3,400 kg (7,500 lb), with propeller reversal: A, B 415 m (1,360 ft)
Range at 3,050 m (10,000 ft), 1,424 litres (313·25 Imp gallons) fuel, 30 min reserves:
A, B at max cruising speed 975 nm (1,806 km; 1,123 miles)
A, B at econ cruising speed 1,100 nm (2,038 km; 1,266 miles)
Mission profile, DL3-MAR, estimated, with crew of three, cruising to search area at 170 knots; 315 km/h; 196 mph at 3,050 m; 10,000 ft, maintaining search at 140 knots; 259 km/h; 161 mph at 1,220 m; 4,000 ft, 30 min reserves:
At 40 nm (74 km; 46 miles) from base:
time on station 6 h 0 min
total mission time 6 h 30 min
At 200 nm (370 km; 230 miles) from base:
time on station 4 h 0 min
total mission time 6 h 20 min
At 360 nm (667 km; 415 miles) from base:
time on station 2 h 0 min
total mission time 6 h 15 min
At 440 nm (815 km; 506 miles) from base:
time on station 1 h 0 min
total mission time 6 h 0 min

PROCAER
PROGETTI COSTRUZIONI AERONAUTICHE Srl

Viale Gramsci 2, 20091 Bresso (Milan)
Telephone: 6105742
President: Dott Ing Rico Neeff

Work by this company in recent years has concentrated on various versions of the F15 four-seat light aircraft. The latest of these are the all-metal F15E Picchio, and a developed version initiated by General Avia (which see) and known as the F15F. Both are part of a licence package offered by Procaer.

First production example of the Procaer F15E Picchio four-seat light aircraft

PROCAER F15E PICCHIO

The F15E prototype (I-PROM) flew for the first time on 21 December 1968. The aircraft has been certificated by both the RAI and FAA, in November 1970 and July 1971 respectively. A second prototype, embodying modifications, was flight tested in 1976 and the first production F15E (I-PROD) was certificated on 19 November 1977.

The following description applies to the first prototype:

Type: Four-seat light aircraft.

Wings: Cantilever low-wing monoplane. NACA 64-215/64-210 wing sections. Dihedral 6° from roots. Incidence 4°. One-piece metal structure with single main spar, rear spar carrying aileron and flap hinges, and short front spar to carry landing gear loads. All-metal Frise ailerons and electrically actuated Fowler flaps.

Fuselage: All-metal semi-monocoque structure.

Tail Unit: Cantilever all-metal structure. Trim tab on rudder and in starboard elevator.

Landing Gear: Retractable tricycle type. Electrical or mechanical actuation. Oleo-pneumatic shock-absorbers. Inward-retracting main wheels, size 6·00-6. Steerable, rearward-retracting nosewheel, size 5·00-5. Hydraulic disc brakes.

Power Plant: One 224 kW (300 hp) Continental IO-520-F flat-six engine, driving a Hartzell HC-C2YF-1B/8475-6 two-blade constant-speed metal propeller. Fuel in two wing tanks and two wingtip tanks, total capacity 318 litres (70 Imp gallons). Oil capacity 11·5 litres (2·5 Imp gallons).

Accommodation: Four persons in pairs in enclosed cabin. Forward-opening door on each side. Space for 45 kg (100 lb) of baggage behind rear bench seat. Dual controls. Cabin soundproofed, heated and ventilated.

Systems and Equipment: Two 12V 35Ah batteries, connected in series, provide power for landing gear and flap actuation. 600W engine-driven generator. Blind-flying instruments and radio optional.

Dimensions, external:
Wing span over tip-tanks 9·90 m (32 ft 5¾ in)
Wing chord at root 1·72 m (5 ft 8 in)
Wing chord at tip 0·85 m (2 ft 9 in)
Wing aspect ratio 7·37
Length overall 7·50 m (24 ft 7¼ in)
Height overall 2·80 m (9 ft 2½ in)
Tailplane span 3·55 m (11 ft 8 in)
Wheel track 2·78 m (9 ft 1¼ in)
Wheelbase 1·73 m (5 ft 8½ in)
Propeller diameter 1·98 m (6 ft 6 in)

Dimensions, internal:
Cabin: Length 2·75 m (9 ft 0 in)
Max width 1·20 m (3 ft 11¼ in)
Max height 1·35 m (4 ft 5¼ in)

Areas:
Wings, gross 13·30 m² (143·2 sq ft)
Ailerons (total) 1·19 m² (12·81 sq ft)
Trailing-edge flaps (total) 1·72 m² (18·50 sq ft)
Fin 0·88 m² (9·50 sq ft)
Rudder, incl tab 0·49 m² (5·27 sq ft)
Tailplane 1·67 m² (17·97 sq ft)
Elevators, incl tab 1·27 m² (13·67 sq ft)

Weights and Loadings:
Weight empty 842 kg (1,856 lb)
Weight empty, equipped 861 kg (1,900 lb)
Max T-O weight: Normal 1,360 kg (3,000 lb)
Utility 1,225 kg (2,700 lb)
Max wing loading:
Normal 101·8 kg/m² (20·85 lb/sq ft)
Utility 92·0 kg/m² (18·85 lb/sq ft)
Max power loading:
Normal 6·42 kg/kW (10·55 lb/hp)
Utility 5·75 kg/kW (9·45 lb/hp)

Performance (N: at max Normal T-O weight; U: at max Utility T-O weight):
Never-exceed speed:
N 191 knots (354 km/h; 220 mph)
U 200 knots (370 km/h; 230 mph)

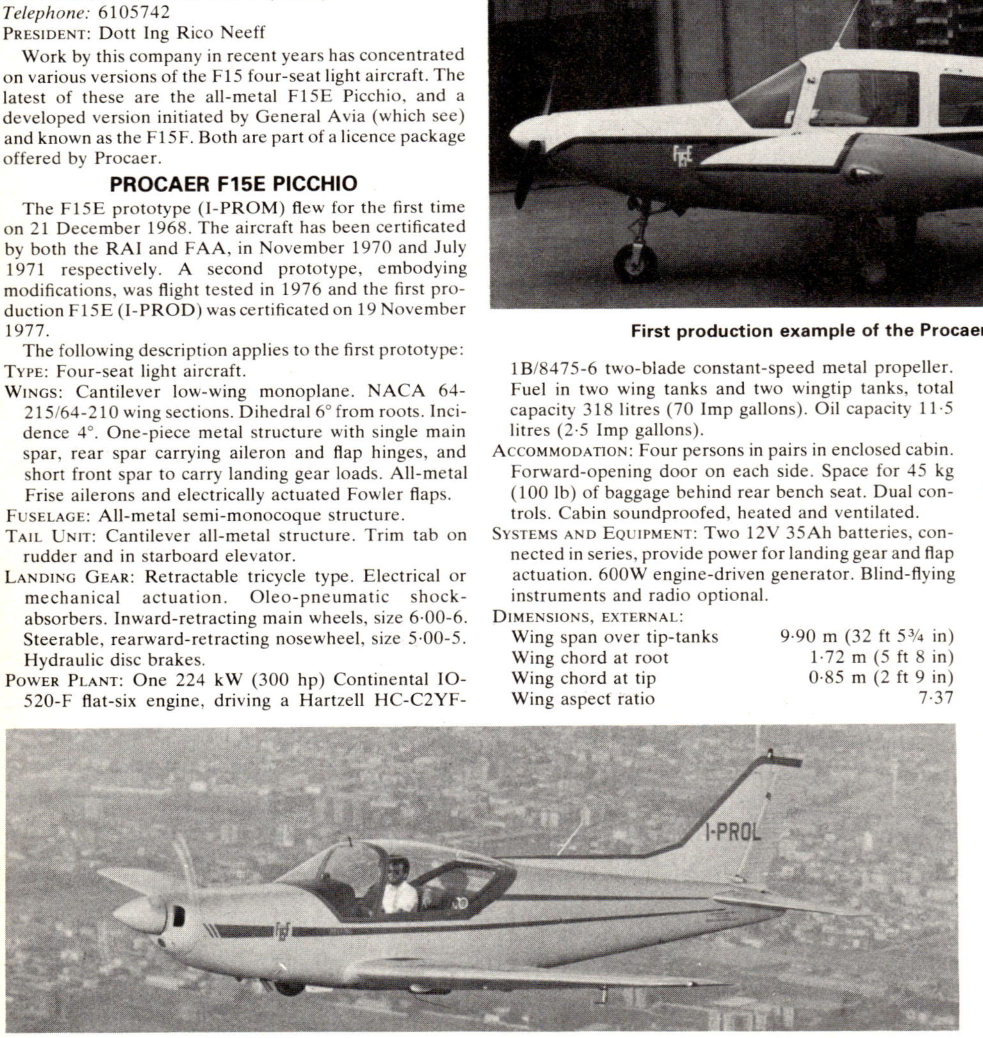

Prototype of the F15F, with modified canopy, lower-powered engine and no tip-tanks

Max level speed:
N 173 knots (320 km/h; 199 mph)
U 174 knots (322 km/h; 200 mph)
Max cruising speed:
N, U 165 knots (306 km/h; 190 mph)
Econ cruising speed:
N, U 144 knots (267 km/h; 166 mph)
Stalling speed, flaps and landing gear up:
N 68 knots (127 km/h; 78·5 mph)
U 63 knots (118 km/h; 73 mph)
Stalling speed, flaps and landing gear down:
N 60 knots (111 km/h; 69 mph)
U 56 knots (103 km/h; 64 mph)
Service ceiling: N 5,300 m (17,390 ft)
T-O run: N 360 m (1,181 ft)
T-O to 15 m (50 ft): N 560 m (1,837 ft)
Landing from 15 m (50 ft): N 625 m (2,050 ft)
Landing run: N 375 m (1,230 ft)
Max range (N), no reserves:
at max cruising speed 647 nm (1,200 km; 746 miles)
at econ cruising speed 863 nm (1,600 km; 994 miles)
Max endurance (N):
at max cruising speed 5 h 0 min
at econ cruising speed 6 h 0 min

PROCAER/GENERAL AVIA F15F

The F15F is a derivative of the F15E Picchio, designed by Dott Ing Frati of General Avia. The prototype (I-PROL), completed by General Avia, made its first flight on 20 October 1977. The certification programme was still under way in early 1981, at which time discussions were taking place with parties interested in the possibility of licence manufacture.

Principal differences from the F15E are as follows:

Power Plant: One 149 kW (200 hp) Avco Lycoming IO-360-A1B1 flat-four engine, driving a two-blade constant-speed metal propeller. Wingtip fuel tanks deleted.

Accommodation: Fully-transparent moulded canopy instead of enclosed cabin.

Dimensions, external:
Length overall 7·75 m (25 ft 5¼ in)

Weights and Loadings: As F15E (Utility category) except:
Weight empty 765 kg (1,686 lb)
Max power loading 8·22 kg/kW (13·45 lb/hp)

Performance (at max T-O weight):
Max level speed 164 knots (305 km/h; 189 mph)
Cruising speed 151 knots (280 km/h; 174 mph)
Stalling speed 55·5 knots (102 km/h; 63·5 mph)
Max rate of climb at S/L 270 m (885 ft)/min
Service ceiling 5,200 m (17,050 ft)

SIAI-MARCHETTI — *See Agusta*

JAPAN

FUJI
FUJI HEAVY INDUSTRIES LTD (Fuji Jukogyo Kabushiki Kaisha)

Head Office: Subaru Building, 7-2, 1-chome, Nishi-shinjuku, Shinjuku-ku, Tokyo
Telephone: Tokyo (03) 347 2505
Telex: 0 232 2268
Aircraft Factory (Utsunomiya Manufacturing Division): Utsunomiya City, Tochigi Prefecture
Telephone: Utsunomiya (0286) 58 1111
Chairman of the Board: Eiichi Ohara
President: Sadamichi Sasaki
Senior Executive Vice-President: Nobuhiro Sakata
Executive Managing Directors:
Sukemitsu Irie
Shoji Nagashima
Managing Directors:
Iwao Shibuya
Kiyoyuki Kawabata
Yoshishige Suzuki
Hiroshi Yamamoto
Kiyoshi Ogawa
Aircraft Division Management:
Atsushi Kasai (General Manager)
Yasuo Kaneta (Deputy General Manager)
Kiichi Nomura (Asst General Manager, Commercial Marketing)
Takaaki Hosaka (Asst General Manager, Military Marketing)
Manager of Aircraft Sales Department: Kenshi Miura
Superintendent of Utsunomiya Aircraft Factory:
Yasumasa Honda

Fuji Heavy Industries Ltd was established on 15 July 1953. It is a successor to the Nakajima aircraft company, which was established in 1914 and built 29,925 aircraft up to the end of the second World War.

The present Utsunomiya Manufacturing Division (Aircraft and Rolling Stock Factories) occupies a site of 512,070 m² (5,511,870 sq ft) including a floor area of 161,532 m² (1,738,710 sq ft) and in 1981 employed 3,273 people.

Under licence from Cessna, Fuji produced 22 L-19E Bird Dog observation aircraft for the Japan Ground Self-Defence Force. Under licence from Beech, it built also the Beechcraft Mentor, and several modified versions of the Mentor designated LM-1 Nikko, LM-2, KM and KM-2. Five additional KM-2s were ordered for the JMSDF in FY 1979; earlier details can be found in previous editions of *Jane's*. The modified KM-2B, combining features of the KM-2 and the Beechcraft T-34A, has been ordered by the JASDF as the T-3 primary trainer. An army version of the KM-2, designated TL-1, has been chosen by the JGSDF as an LM-1/2 replacement.

Fuji is also producing the Bell Model 204/205 series of helicopters. The first 204B covered by the agreement arrived in Japan in kit form in May 1962 for assembly.

Fuji is building wing main assemblies for the Lockheed P-3C Orion maritime patrol aircraft being manufactured under licence in Japan for the JMSDF, and main landing gear doors and some titanium airframe parts for Japanese-built McDonnell Douglas F-15J fighters delivered to the JASDF. As a part of Japan's YX civil transport aircraft programme, Fuji is responsible, under subcontract through the CTDC (Civil Transport Development Corporation), for building wing/fuselage body fairings and main landing gear doors for the Boeing 767 jet transport.

FUJI FA-200 AERO SUBARU

Fuji began detail design of this light aircraft in 1964 and the prototype flew for the first time on 12 August 1965. It went into production as the FA-200-160, FA-200-180 and FA-200-180AO, which differ as follows:

FA-200-160. Basic four-seat light aircraft, with 119 kW (160 hp) Avco Lycoming engine. Received JCAB and FAA Normal category type certificate as a four-seater, Utility category as a three-seater and Aerobatic category as a two-seater.

FA-200-180. Developed version with 134 kW (180 hp) Avco Lycoming engine.

FA-200-180AO. Version with 134 kW (180 hp) Avco Lycoming engine and fixed-pitch propeller

Production of the FA-200 began in March 1968, and 296 had been completed by 1 February 1981, of which more than 170 were for export.

Production of the FA-200 was suspended in early 1980. A full description of all versions can be found in the 1979-80 *Jane's*; the following shortened description applies to the basic FA-200-160:

Power Plant: One 119 kW (160 hp) Avco Lycoming O-320-D2A flat-four engine, driving a McCauley 1C 160/FGM 7656 two-blade fixed-pitch metal propeller. Fuel in two integral tanks in inner wings with total capacity of 204·5 litres (45 Imp gallons).

Accommodation: Four seats in pairs in enclosed cabin. Individual adjustable front seats, with dual controls. Optional shoulder harness on each of the four seats. Large rearward-sliding canopy, which can be opened in flight. Two tinted roof windows optional. Cabin heating and ventilation, and windscreen defrosting, standard. Baggage compartment in rear of fuselage, capacity 80 kg (176 lb). Baggage shelf aft of rear seats, capacity 20 kg (44 lb).

Dimensions, external:
Wing span 9·42 m (30 ft 11 in)
Wing chord (constant) 1·525 m (5 ft 0 in)
Wing area, gross 14·0 m² (150·7 sq ft)
Wing aspect ratio 6·34
Length overall 8·17 m (26 ft 9½ in)
Height overall 2·59 m (8 ft 6 in)
Tailplane span 3·47 m (11 ft 4½ in)
Wheel track 2·63 m (8 ft 7½ in)
Wheelbase 1·75 m (5 ft 8¾ in)
Propeller diameter 1·93 m (6 ft 4 in)

Weights and Loadings (N: Normal; U: Utility; A: Aerobatic category):
Weight empty 620 kg (1,366 lb)
Max T-O weight: N 1,060 kg (2,335 lb)
U 970 kg (2,138 lb)
A 880 kg (1,940 lb)
Max wing loading: N 75·7 kg/m² (15·5 lb/sq ft)
Max power loading: N 8·91 kg/kW (14·59 lb/hp)

Performance (N: Normal category; A: Aerobatic category, at max T-O weight):
Max level speed at S/L:
N 120 knots (222 km/h; 138 mph)
A 122 knots (225 km/h; 140 mph)
Max cruising speed (75% power):
N at 1,525 m (5,000 ft) 106 knots (196 km/h; 122 mph)
A at 2,290 m (7,500 ft) 114 knots (211 km/h; 131 mph)
Econ cruising speed (55% power):
N at 1,525 m (5,000 ft) 89 knots (164 km/h; 102 mph)
A at 2,290 m (7,500 ft) 96 knots (177 km/h; 110 mph)
Stalling speed, flaps down:
N 49 knots (90 km/h; 56 mph)
A 45·5 knots (84 km/h; 52·5 mph)
Max rate of climb at S/L: N 207 m (680 ft)/min
A 302 m (991 ft)/min
Service ceiling: N 3,480 m (11,400 ft)
A 4,725 m (15,500 ft)
T-O run: A 160 m (525 ft)
T-O to 15 m (50 ft): N 465 m (1,530 ft)

Fuji FA-200-180 Aero Subaru four-seat light aircraft in inverted flight

One of the Japanese-built prototypes of the Fuji FA-300/Model 710

Fuji T-3 (KM-2B) two-seat primary trainer of the JASDF (Avco Lycoming IGSO-480-A1F6 engine)

A 310 m (1,020 ft)
Landing from 15 m (50 ft): N 340 m (1,115 ft)
A 315 m (1,033 ft)
Landing run: A 115 m (377 ft)
Range with max fuel (55% power at 2,290 m; 7,500 ft, no reserves): N 655 nm (1,215 km; 755 miles)
A 816 nm (1,512 km; 940 miles)

FUJI FA-300/MODEL 710

This version of the FA-300 family (335 kW; 450 hp engines) was built under a collaboration agreement between Fuji and Rockwell International (see 1980-81 *Jane's*). The first flight was made on 22 December 1976. Two prototypes were flown in Japan, and after about 580 h of flight testing by these two aircraft JCAB certification was awarded on 26 February 1979. Two other Model 710s were used in a programme to obtain FAA certification under FAR Pts 21-29, and this was granted on 29 May 1980. No production has yet been undertaken.

FUJI KM-2B

JASDF designation: T-3

The KM-2B is a modification of the original KM-2 development of the Beechcraft T-34A Mentor, described in the 1969-70 *Jane's,* combining the airframe and power plant of the KM-2 with the two-seat cockpit installation of the T-34A. The first KM-2B (JA3725) was flown for the first time on 26 September 1974, and received JCAB category A certification on 28 November 1974.

In 1975 the JASDF selected the KM-2B to replace the T-34A in the primary trainer role, and 50 were ordered with the designation T-3. The first of these made its first flight on 17 January 1978, and 44 had been delivered to the JASDF by February 1981.

Type: Two-seat primary trainer.

Wings: Cantilever low-wing monoplane. Wing section NACA 23016·5 at root, NACA 23012 at tip. Dihedral 6° from roots. Incidence 4° at root. 3° geometric twist. No sweep at quarter-chord. All-metal tapered two-spar structure with stressed skin. Aluminium alloy single-slotted flaps and plain ailerons. Servo tab in each aileron; port tab controllable for trim.

Fuselage: Conventional all-metal semi-monocoque structure.

Tail Unit: Cantilever all-metal structure. Fixed-incidence tailplane, with elevators. Controllable tab in each elevator; anti-servo tab in rudder.

Landing Gear: Electrically retractable tricycle type, with single wheel and oleo-pneumatic shock-absorber on each unit. Nosewheel retracts rearward, main wheels inward into wings. Goodyear wheels and Type III tyres on all units: size 6·50-8 (6 ply), pressure 2·34 bars (34 lb/sq in) on main units; size 5·00-5 (4 ply), pressure 2·76 bars (40 lb/sq in) on nose unit. Goodyear automatically adjustable single-disc brakes on main units.

Power Plant: One 254 kW (340 hp) Avco Lycoming IGSO-480-A1F6 flat-six engine, driving a Hartzell HC-A3V20-1F/V9333N-3 three-blade constant-speed propeller with spinner. One metal and one bladder-type fuel tank in each wing, total capacity 265 litres (70 US gallons; 58·3 Imp gallons). Refuelling points on top of wings. Oil capacity 11·35 litres (3 US gallons; 2·5 Imp gallons).

Accommodation: Crew of two in tandem, on adjustable seats in heated and ventilated cabin. Dual controls standard. Framed canopy, with rearward-sliding section over each seat. Space for 13·6 kg (30 lb) of baggage aft of rear seat.

Systems and Avionics: Blind-flying instrumentation standard. Electrical system includes 28V 100A DC generator and two 160VA static inverters. Standard avionics include ICS, VHF, Tacan, and ATC transponder with Mode C.

Dimensions, external:
Wing span 10·004 m (32 ft 10 in)
Wing chord at root 2·13 m (7 ft 0 in)
Wing chord at tip 1·07 m (3 ft 6 in)
Wing aspect ratio 6·1
Length overall 8·036 m (26 ft 4¼ in)
Height overall 3·023 m (9 ft 11 in)
Tailplane span 3·712 m (12 ft 2¼ in)
Wheel track 2·924 m (9 ft 7 in)
Wheelbase 2·266 m (7 ft 5¼ in)
Propeller diameter 2·286 m (7 ft 6 in)
Propeller ground clearance 0·25 m (9¾ in)

Dimensions, internal:
Cabin: Max width approx 0·90 m (2 ft 11½ in)
Max height approx 1·30 m (4 ft 3¼ in)

Areas:
Wings, gross 16·50 m² (177·6 sq ft)
Ailerons (total) 1·07 m² (11·52 sq ft)
Trailing-edge flaps (total) 1·98 m² (21·31 sq ft)
Fin 0·97 m² (10·44 sq ft)
Rudder, incl tab 0·61 m² (6·57 sq ft)
Tailplane 3·46 m² (37·24 sq ft)
Elevators, incl tab 1·39 m² (14·96 sq ft)

Weights and Loadings:
Weight empty 1,120 kg (2,469 lb)
Normal operational T-O weight 1,510 kg (3,329 lb)
Max T-O weight 1,542 kg (3,400 lb)
Max wing loading 93·5 kg/m² (19·15 lb/sq ft)
Max power loading 6·07 kg/kW (10·00 lb/hp)

Performance (at normal operational T-O weight):
Never-exceed speed 223 knots (413 km/h; 257 mph) EAS
Max level speed at 2,440 m (8,000 ft) 198 knots (367 km/h; 228 mph)
Max cruising speed at 2,440 m (8,000 ft) 177 knots (328 km/h; 204 mph)
Econ cruising speed at 2,440 m (8,000 ft) 137 knots (254 km/h; 158 mph)
Stalling speed, flaps up 66 knots (123 km/h; 76 mph)
Stalling speed, flaps down 54 knots (100 km/h; 62 mph)
Max rate of climb at S/L 463 m (1,520 ft)/min
Service ceiling 8,170 m (26,800 ft)
T-O run 265 m (870 ft)
T-O to 15 m (50 ft) 503 m (1,650 ft)
Landing from 15 m (50 ft) 436 m (1,430 ft)
Landing run 238 m (780 ft)
Range with max fuel 520 nm (965 km; 600 miles)

FUJI KM-2

JMSDF designation: KM-2
JGSDF designation: TL-1

Under the new designation TL-1, the Japan Ground Self-Defence Force has selected the Fuji KM-2 as a replacement for its LM-1s and LM-2s. The first order, for two aircraft, was placed in FY 1980.

The TL-1 is almost identical to the KM-2 for the JMSDF (of which production has ended), except for minor changes of instrumentation and the installation of ARC-3 VHF radio, JARN-A-2 ADF, and JAN/APX-68 IFF.

FUJI-BELL 204B/204B-2 and 205A-1/UH-1H

Fuji is manufacturing Bell Model 204B and UH-1H helicopters under sub-licence from Mitsui and Co Ltd, Bell's Japanese licensee. By March 1981 a total of 49 Fuji-Bell 204B/204B-2s had been produced for civilian operators.

Following the delivery of 90 UH-1Bs (military version of the 204B) to the Japan Ground Self-Defence Force by early 1973, Fuji production continued with the UH-1H version, of which the first example flew for the first time on 17 July 1973; 67 had been ordered by March 1981.

The Fuji-Bell 204B is identical with that built by Bell Helicopter Textron (see US section). It is powered by an 820 kW (1,100 shp) Kawasaki-built Avco Lycoming T53-K-11A turboshaft engine. The Fuji-Bell UH-1H has the same airframe and dynamic components as the Bell-built UH-1H, but has a tractor-type tail rotor and is powered by a 1,044 kW (1,400 shp) Kawasaki-built Avco Lycoming T53-K-13B turboshaft engine.

In October 1973 Fuji developed a higher-powered **204B-2** version of the 204B. Powered by a 1,044 kW (1,400 shp) Avco Lycoming T53-K-13B turboshaft engine, it has the same basic airframe and dynamic components as the 204B, but has a tractor-type tail rotor. The first example of this version was delivered in early 1974.

The following details apply to the standard Fuji-Bell 204B/204B-2/UH-1H:

Dimensions, external:
Diameter of main rotor 14·63 m (48 ft 0 in)
Diameter of tail rotor 2·59 m (8 ft 6 in)
Length overall, tail rotor turning:
204B/B-2 13·61 m (44 ft 7¾ in)
UH-1H 13·67 m (44 ft 10 in)
Length of fuselage: 204B/B-2 12·31 m (40 ft 4¾ in)
UH-1H 12·37 m (40 ft 7 in)
Height overall, tail rotor turning 4·42 m (14 ft 6 in)

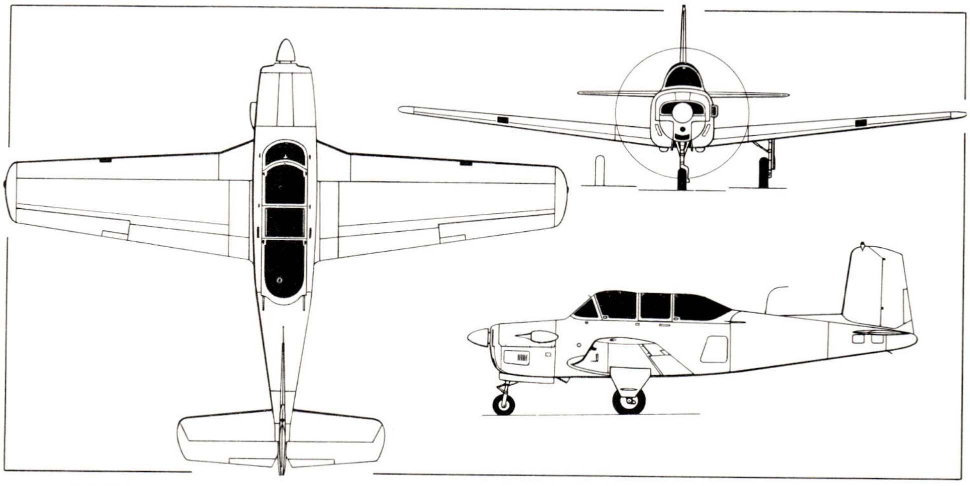

Fuji T-3 two-seat primary trainer, developed from the KM-2 and Beechcraft T-34A *(Pilot Press)*

Height to top of rotor hub:
- 204B/B-2: 3·77 m (12 ft 4½ in)
- UH-1H: 3·98 m (13 ft 0¾ in)

Max width over landing skids:
- 204B/B-2: 2·64 m (8 ft 8 in)
- UH-1H: 2·60 m (8 ft 6½ in)

Tailplane span: 2·84 m (9 ft 4 in)

WEIGHTS AND LOADINGS:

Weight empty: 204B/B-2: 2,177 kg (4,800 lb)
- UH-1H: 2,390 kg (5,270 lb)

Max T-O weight:
- 204B/B-2, internal load: 3,855 kg (8,500 lb)
- 204B/B-2, external load: 4,309 kg (9,500 lb)
- UH-1H: 4,309 kg (9,500 lb)

Max disc loading: 25·6 kg/m² (5·25 lb/sq ft)

Max power loading:
- 204B: 5·25 kg/kW (8·64 lb/shp)
- 204B-2, UH-1H: 4·13 kg/kW (6·78 lb/shp)

PERFORMANCE (at max T-O weight):

Max level and cruising speed: 110 knots (204 km/h; 127 mph)

Max rate of climb at S/L:
- 204B: 463 m (1,520 ft)/min
- 204B-2: 588 m (1,930 ft)/min
- UH-1H: 488 m (1,600 ft)/min

Service ceiling: 204B: 4,480 m (14,700 ft)
- 204B-2: 5,790 m (19,000 ft)
- UH-1H: 3,840 m (12,600 ft)

Hovering ceiling IGE: 204B: 2,985 m (9,800 ft)
- 204B-2: 4,635 m (15,200 ft)
- UH-1H: 4,145 m (13,600 ft)

Hovering ceiling OGE: 204B: 1,310 m (4,300 ft)
- 204B-2: 3,200 m (10,500 ft)
- UH-1H: 335 m (1,100 ft)

Range at S/L: 204B: 206 nm (381 km; 237 miles)
- 204B-2: 207 nm (383 km; 238 miles)
- UH-1H: 252 nm (467 km; 290 miles)

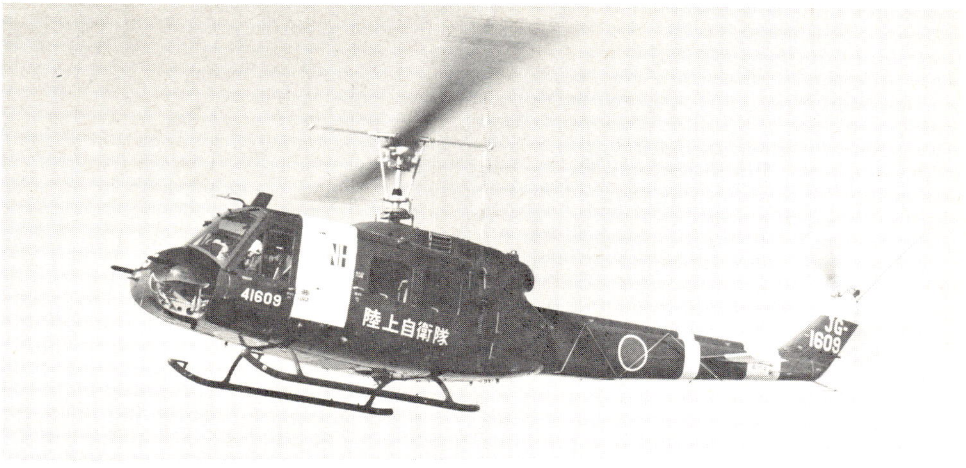

Fuji-Bell UH-1H medium helicopter of the JGSDF (Avco Lycoming T53-K-13B turboshaft engine)

KAWASAKI

KAWASAKI JUKOGYO KABUSHIKI KAISHA (Kawasaki Heavy Industries Ltd)

HEAD OFFICE: 2-1-18 Nakamachi-Dori, Chuo-ku, Kobe
TOKYO AND AIRCRAFT GROUP OFFICE: World Trade Center Building, 4-1, Hamamatsu-cho 2-chome, Minato-ku, Tokyo
Telephone: Tokyo (03) 435 2971
Telex: 242-4371 KAWAJU J

CHAIRMAN: Kiyoshi Yotsumoto
PRESIDENT: Zenji Umeda
EXECUTIVE VICE-PRESIDENTS:
Tsuneo Ando
Toraichi Imai
Kenko Hasegawa
Renzo Nihei
WORKS: Gifu
MANAGING DIRECTOR, AND SENIOR GENERAL MANAGER, AIRCRAFT GROUP: Teruaki Yamada
DIRECTOR, AND ASST SENIOR GENERAL MANAGER, AIRCRAFT GROUP, AND GENERAL MANAGER, AIRCRAFT MANUFACTURING DIVISION: Masahiko Iwata

With effect from 1 April 1969, Kawasaki Aircraft Co Ltd was amalgamated with the Kawasaki Dockyard Co Ltd and the Kawasaki Rolling Stock Mfg Co Ltd, to form Kawasaki Heavy Industries Ltd. The Aircraft Division of the former Kawasaki Aircraft Co Ltd, which employs some 3,500 people, continues its activities as the Aircraft Group of this company. Kawasaki has a 32·08% holding in Nippi (which see).

In addition to extensive overhaul work, Kawasaki has built many US aircraft under licence since 1953, including 48 Lockheed P2V-7 (P-2H) Neptune anti-submarine aircraft and 239 Bell Model 47 helicopters, plus another 211 Model KH-4 helicopters developed from the Bell 47 by its own design staff. From the Neptune it developed the P-2J anti-submarine aircraft, of which it delivered one prototype and 82 production examples (see 1978-79 *Jane's*).

In February 1977 Kawasaki was awarded a JDA contract to produce an experimental variable-stability aircraft (VSA) by converting a P2V-7 of the JMSDF. This aircraft has large direct side-force control panels on each wing, a fly-by-wire control system, direct-lift control flaps, and a miniature computer.

Design studies have been undertaken for a new intermediate jet trainer, provisionally designated KA-850, to meet the Japan Defence Agency's MTX requirement for a successor to the Lockheed T-33A, of which the company built 210 under licence during 1956-59. The KA-850 is also intended as a possible replacement for the Fuji T-1.

Kawasaki is developing, jointly with MBB of West Germany, the BK 117 twin-engined multi-purpose helicopter described in the International section.

Kawasaki is prime contractor for the JASDF's C-1 transport; and for licence production of the Lockheed P-3C/Update II Orion, 45 of which are to be purchased by the JMSDF during the 1980s. Eighteen of these have so far been ordered, of which the first three (US-built) P-3Cs were handed over to the JMSDF in April 1981. The next four are being assembled by Kawasaki from knocked-down assemblies, with deliveries due to begin to Atsugi Air Base in May 1982. Kawasaki is responsible for building the centre-fuselages, and for final assembly and flight testing. Participants in the programme include Fuji, Mitsubishi, Nippi and Shin Meiwa for the airframe, and IHI for the engines. Kawasaki is also a subcontractor for rear fuselages, wings and tail units of the McDonnell Douglas F-15J Eagles being licence-built in Japan by Mitsubishi (which see). As a part of Japan's YX civil transport aircraft programme, Kawasaki is subcontracted, through CTDC, to build forward and mid fuselage sections, and wing ribs, for the Boeing 767 jet transport. In late 1979 the company was nominated by the JASDF as prime contractor for overhaul of the eight Grumman E-2C Hawkeye AEW aircraft which are to be introduced into service from 1982.

Kawasaki has exclusive rights to manufacture and sell the twin-engined Boeing Vertol 107 Model II helicopter and its own KV-107/IIA development of it. The Hughes Model 369 (500, 500C and 500D) light observation helicopter is also being manufactured by Kawasaki under a licence agreement concluded in October 1967. By 1 March 1981 a total of 126 KV-107 helicopters had been delivered to customers in Japan and other countries including Saudi Arabia, Sweden, Thailand and the USA; and approx 200 Hughes 500s to government and commercial operators in Japan.

Kawasaki is engaged in missile development and production; its aero-engine activities are described in the appropriate section of this edition.

KAWASAKI UP-2J

A full description of the P-2J can be found in the 1978-79 *Jane's*.

By March 1981, three P-2Js had been converted to **UP-2J** configuration with equipment for target towing and ECM operations. It is planned to convert a fourth in FY 1982.

KAWASAKI P2V-7 (P-2H) VSA VARIABLE STABILITY AIRCRAFT

Kawasaki modified the 39th production P2V-7 (P-2H) Neptune ASW aircraft to experimental configuration as a Variable Stability Aircraft (VSA) for training duty at the JMSDF test pilots' school.

The aircraft made its first flight at Gifu on 23 December 1977, and was described in the 1979-80 *Jane's*.

Transfer of the VSA system to a P-2J is scheduled to be accomplished by March 1984.

KAWASAKI C-1

The C-1 medium-sized troop and freight transport was designed to meet the JASDF's requirement for a replacement for its former fleet of Curtiss C-46s. Preliminary design, and the construction of two XC-1 flying prototypes (plus one airframe for static tests), was undertaken by NAMC. The first flying prototype, assembled at Kawasaki's Gifu factory, made its first flight on 12 November 1970, followed by the second on 16 January 1971. These were handed over to the JDA on 24 February and 20 March 1971 respectively. Further development and evaluation tests by the JDA were completed in March 1973.

Two pre-production aircraft had been delivered by the end of February 1974. Airframe fatigue testing by the JDA was completed in November 1974.

Eleven production C-1s were delivered between December 1974 and the end of March 1976. An additional 13 C-1s were ordered during FY 1975; these had all been delivered by the end of February 1978, including two long-range aircraft with an additional 4,732 litre (1,250 US gallon) wing centre-section fuel tank. Two more long-range C-1s were ordered during FY 1977, and one in FY 1979; production was expected to end in 1981 with the delivery of the 31st aircraft. All C-1s were to be camouflaged by 1981.

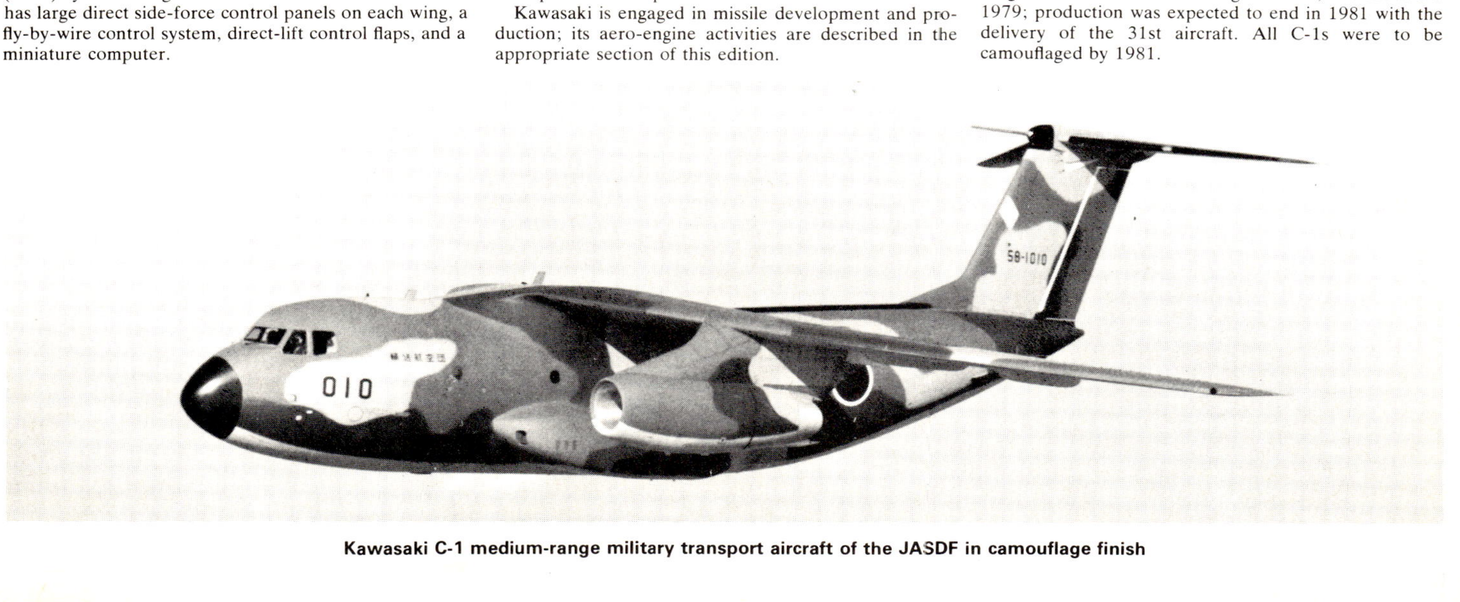

Kawasaki C-1 medium-range military transport aircraft of the JASDF in camouflage finish

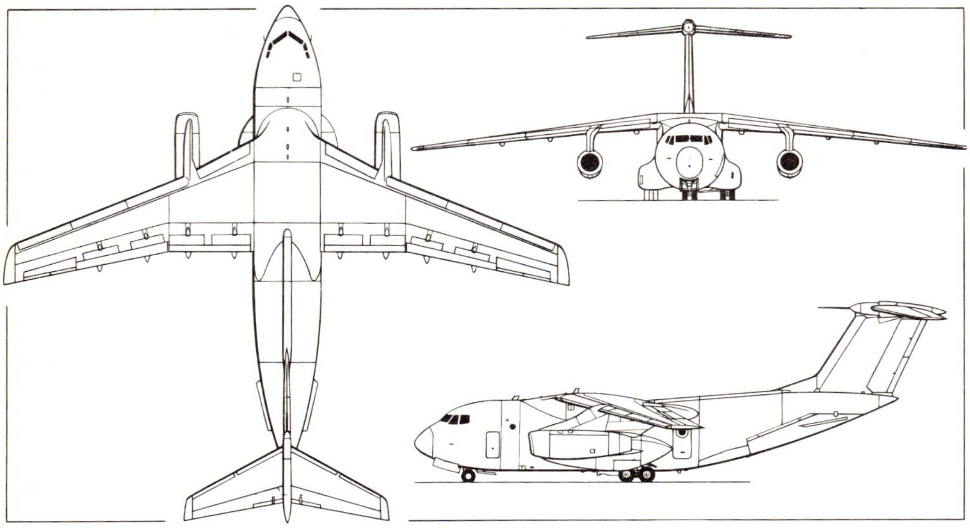

Kawasaki C-1 twin-turbofan medium-range military transport *(Pilot Press)*

The National Aerospace Laboratory (see NAL entry) is building an experimental STOL aircraft utilising the airframe of a C-1. The Technical Research and Development Institute of the Japan Defence Agency is retrofitting one JASDF C-1 as a flying testbed for the XF-3 lightweight turbofan engine. This work is expected to be completed in 1982.

Prime contractor in the C-1 programme is Kawasaki, which builds the front fuselage and wing centre-section and undertakes final assembly and flight testing. Major subcontractors are Fuji (outer wing panels); Mitsubishi (centre and aft fuselage sections and tail surfaces); Nihon Hikoki (Nippi) (flaps, slats, spoilers, ailerons, engine pylons and pods); and Shin Meiwa (cargo loading equipment).

Type: Twin-turbofan medium-range transport.

Wings: Cantilever high-wing monoplane. Wings have 20° sweepback at quarter-chord, with slightly increased leading-edge sweep inboard of the engine pylons. Thickness/chord ratio 12% at root, 11% at tip. Anhedral 5° 30′ from centre-section. Conventional two-spar fail-safe structure of aluminium alloy, including control surfaces. Two quadruple-slotted flaps on each trailing-edge. Forward of these, on each wing, are three flight spoilers and a ground spoiler. Drooping leading-edge slats, in four sections, on each wing. Manually-operated aileron outboard of each outer flap. Trim tab in port aileron. Thermal anti-icing of leading-edges, using engine bleed air.

Fuselage: Conventional semi-monocoque fail-safe structure of aluminium alloy, with a circular cross-section.

Tail Unit: Aluminium alloy cantilever T tail, with sweepback on all surfaces (30° at fin quarter-chord, 25° at tailplane quarter-chord). Tailplane has 5° anhedral. Variable-incidence tailplane, with elevators. Balance tab in each elevator and anti-balance tab in rudder. Elevators and rudder are each operated by two independent hydraulic actuator systems; the elevators can be operated manually in an emergency. Thermal de-icing of tailplane, using electric heater mat.

Landing Gear: Hydraulically-retractable tricycle type, of Sumitomo design. Each main unit has two pairs of wheels in tandem, retracting forward into fairings built on to the sides of the fuselage. Forward-retracting nose unit has twin wheels. Main-wheel tyres size 35 × 10·7-16, nosewheel tyres size 28 × 9-12; tyre pressure (all units) 6·21 bars (90 lb/sq in). Oleo shock-absorbers. Kayaba wheels with Dunlop tyres, which on main units have pressure of 5·17 bars (75 lb/sq in). Kayaba hydraulic brakes (two-rotor on first 10 production aircraft, three-rotor from 11th aircraft onward); Sumisei anti-skid units.

Power Plant: Two 64·5 kN (14,500 lb st) Mitsubishi (Pratt & Whitney) JT8D-M-9 turbofan engines, installed in pylon-mounted underwing pods and fitted with thrust reversers. Four integral wing fuel tanks with total capacity of 15,200 litres (3,344 Imp gallons). Single pressure refuelling point for all tanks, plus overwing gravity refuelling point for each tank.

Accommodation: Crew of five, comprising pilot, co-pilot, navigator, flight engineer and load supervisor. Escape hatch in flight deck roof on starboard side. Flight deck and main cabin pressurised and air-conditioned. Standard complements are as follows: troops (max) 60, paratroops (max) 45, stretchers 36 plus attendants. As a cargo carrier, loads can include a 2½ ton truck, a 105 mm howitzer, two ¾ ton trucks or three Jeeps. Up to three preloaded freight pallets, 2·24 m (7 ft 4 in) wide and 2·74 m (9 ft 0 in) long, can be carried. Floor is stressed for loads of up to 7 kg/cm^2 (100 lb/sq in). Access to flight deck via downward-opening door, with built-in stairs, on port side of forward fuselage. Paratroop door on each side of fuselage, aft of wing trailing-edge. For airdropping, the rear-loading ramp/door can be opened in flight to the full cabin cross-section.

Systems: Pressurisation and air-conditioning systems utilise engine bleed air. APU in front section of starboard landing gear fairing supplies electrical power on ground and in the air, and bleed air on ground. Three independent hydraulic systems, each 207 bars (3,000 lb/sq in). No. 1 system actuates flight controls; No. 2 actuates flight controls, high-lift devices, landing gear, nosewheel steering and brakes; No. 3 system is used for cargo door operation, and provides emergency backup for brakes and high-lift devices. Electrical power supplied by three 40kVA AC generators, two engine-driven and aircooled and one driven by the APU. 28V DC power is obtained from AC source through a transformer-rectifier. One 24V 30Ah nickel-cadmium battery for emergency DC power.

Avionics and Equipment: Standard avionics include autopilot, Doppler radar, radio altimeter, HF, VHF and UHF radio, ADF, UHF/DF, marker beacon, VOR/ILS, Tacan, SIF, dual compass system and flight director system. Optional avionics include Loran and weather radar.

Dimensions, external:	
Wing span	30·60 m (100 ft 4¾ in)
Wing chord at root	6·30 m (20 ft 8 in)
Wing chord at tip	2·00 m (6 ft 6¾ in)
Wing aspect ratio	7·8
Length overall	29·00 m (95 ft 1¾ in)
Length of fuselage	26·50 m (86 ft 11¼ in)
Height overall	9·99 m (32 ft 9¼ in)
Tailplane span	11·30 m (37 ft 1 in)
Wheel track	4·40 m (14 ft 5¼ in)
Wheelbase	9·33 m (30 ft 7¼ in)
Rear-loading ramp/door:	
Length	2·67 m (8 ft 9¼ in)
Width	2·70 m (8 ft 10¼ in)
Height to sill	1·25 m (4 ft 1¼ in)
Dimensions, internal:	
Cabin: Max length	10·80 m (35 ft 5¼ in)
Max width	3·60 m (11 ft 9¾ in)
Max height	2·55 m (8 ft 4½ in)
Floor area	28·6 m^2 (308 sq ft)
Volume (excl ramp area)	73·8 m^3 (2,606 cu ft)
Areas:	
Wings, gross	120·5 m^2 (1,297 sq ft)
Ailerons (total)	3·4 m^2 (36·6 sq ft)
Trailing-edge flaps (total)	22·9 m^2 (246·5 sq ft)
Spoilers (total)	8·9 m^2 (95·8 sq ft)
Fin	15·8 m^2 (170·1 sq ft)
Rudder, incl tabs	6·4 m^2 (68·9 sq ft)
Tailplane	18·3 m^2 (197·0 sq ft)
Elevators, incl tabs	6·5 m^2 (70·0 sq ft)
Weights and Loadings:	
Weight empty	23,320 kg (51,410 lb)
Weight empty, equipped	24,300 kg (53,572 lb)
Normal payload	7,900 kg (17,416 lb)
Max payload	11,900 kg (26,235 lb)
Normal T-O weight	38,700 kg (85,320 lb)
Max T-O weight	45,000 kg (99,210 lb)
Max wing loading	373·4 kg/m^2 (76·48 lb/sq ft)
Max power loading	350 kg/kN (3·42 lb/lb st)

Performance (at normal T-O weight except where indicated):

Max level speed at 7,620 m (25,000 ft) at 35,450 kg (78,150 lb) AUW	435 knots (806 km/h; 501 mph)
Econ cruising speed at 10,670 m (35,000 ft) at 35,450 kg (78,150 lb) AUW	354 knots (657 km/h; 408 mph)
Max rate of climb at S/L	1,065 m (3,500 ft)/min
Service ceiling	11,580 m (38,000 ft)
Service ceiling, one engine out	5,485 m (18,000 ft)
T-O run	640 m (2,100 ft)
T-O to 15 m (50 ft)	910 m (3,000 ft)
Landing from 15 m (50 ft) at 36,860 kg (81,260 lb) weight	823 m (2,700 ft)
Landing run at 36,860 kg (81,260 lb) weight	455 m (1,500 ft)
Range with max fuel and 2,200 kg (4,850 lb) payload	1,810 nm (3,353 km; 2,084 miles)
Range with 7,900 kg (17,416 lb) payload	700 nm (1,300 km; 807 miles)

KAWASAKI KA-850

Under the company designation KA-850, Kawasaki is developing an intermediate jet trainer to meet a JASDF requirement for a replacement for its Lockheed T-33As and Fuji T-1s. The KA-850 has a tandem two-seat cockpit arrangement, and will be powered by two approx 15·7 kN (3,530 lb st) F3 turbofan engines, the production version of the XF3 currently under development by the Technical Research and Development Institute of the Japan Defence Agency. It will have an additional capability for light attack, carrying underwing 250 kg bombs.

The KA-850 is one of three designs submitted for JASDF consideration, the others being from Fuji (FT-20) and Mitsubishi. The Japan Defence Agency selected the KA-850 for development in 1981, with a prototype first flight scheduled for FY 1985.

For further details see Addenda.

Specification:	
Wing area, gross	20·4 m^2 (219·6 sq ft)
Max T-O weight	approx 7,300 kg (16,094 lb)
Econ cruising speed at 7,620 m (25,000 ft)	Mach 0·75 (450 knots; 834 km/h; 518 mph)
Max range with 454 litre (120 US gallon; 100 Imp gallon) external tanks	900 nm (1,670 km; 1,035 miles)

KAWASAKI (BOEING VERTOL) KV-107/II and KV-107/IIA

Swedish Navy designation: HKP 4C

Kawasaki has exclusive rights to manufacture and sell the Boeing Vertol 107 Model II helicopter. The first KV-107 to be produced by Kawasaki under this licence agreement flew for the first time in May 1962.

In 1965, Kawasaki obtained world-wide sales rights in the KV-107 from The Boeing Company's Vertol Division. In November 1965, it was awarded a type certificate for the KV-107 by the FAA.

Kawasaki KV-107/IIA-4 of the JGSDF with external fuel tanks *(Katsumi Hinata)*

An improved model, the **KV-107/IIA,** is available in any of the KV-107/II forms, powered by two 1,044 kW (1,400 shp) General Electric CT58-140-1 or Ishikawajima-Harima CT58-IHI-140-1 turboshaft engines (max continuous rating 932 kW; 1,250 shp), which give improved performance during VTOL and in 'hot and high' conditions. Fuel capacity 1,324 litres; 350 US gallons (standard), 3,785 litres; 1,000 US gallons (max). A prototype (JA9509) was converted from a standard KV-107/II-2 and first flown on 3 April 1968. Type approval granted by JCAB on 26 September 1968 and by FAA on 15 January 1969.

The following versions of the KV-107/II and IIA have been announced:

KV-107/II-1. Basic utility helicopter. None yet built.

KV-107/II-2. Basic airline helicopter. Eleven built. Some now operated by Columbia Helicopters Inc of USA (five) and AirLift Inc of Japan (two). Prototype (JA9509) upgraded as IIA-2 prototype, currently used as a company test aircraft. One **KV-107/IIA-2** was delivered to AirLift Inc in February 1981.

KV-107/II-3. Mine countermeasures (MCM) helicopter for JMSDF with extended-range fuel tanks, towing hook and cargo sling. Nine ordered, all of which had been delivered by February 1975, including seven of the **KV-107/IIA-3** version with uprated power. All of these are fitted with minesweeping and retrieval equipment and serve with the 111th Air Wing of the JMSDF.

KV-107/II-4. Tactical cargo/troop transport for JGSDF, with foldable seats for 26 troops or 15 casualty litters. Strengthened floor for carrying heavy vehicles. Orders for 60 placed, of which 59 had been delivered by December 1980, including one equipped as a VIP transport. Sixtieth aircraft due for delivery by October 1981. The latest 18 aircraft are of the **KV-107/IIA-4** version with uprated power plant, and four of them are fitted with extended-range fuel tanks.

KV-107/II-5. Long-range search and rescue helicopter for JASDF. Orders for 32 placed, of which 29 had been delivered by December 1980, including 16 uprated **KV-107/IIA-5s**. Last three aircraft of the 32 due for delivery by end of FY 1981. Approval for four more anticipated. Extended-range fuel tank each side of fuselage, making total capacity 3,785 litres (1,000 US gallons). Extensive nav/com equipment, four searchlights, domed observation window and rescue hoist. Eight aircraft for Swedish Navy and three for JASDF are equipped with a Kawasaki/Boeing automatic flight control system, enabling them to cruise at preselected altitude and speed, descend at a programmed rate and distance, and come to hover at a preselected altitude. Also provided are automatic climb-out to the cruise mode; standard distance approach; a turns coupler to a preselected heading; altitude sensing, with dual radar altimeters, for added safety in IFR operations; and a vernier flight control to permit critical positioning during rescue hoist operations. The Swedish aircraft, supplied in 1972-74, have Kawasaki-built airframes and rotor assemblies but were fitted in Sweden with Rolls-Royce Gnome H.1200 turboshaft engines and a Decca navigation system. Details of other equipment fitted to Swedish aircraft as listed in 1977-78 *Jane's*.

KV-107/II-6. De luxe transport version. None yet built.

KV-107/II-7. De luxe VIP transport with 6-11 seats. One built: currently owned by Columbia Helicopters Inc of USA.

KV-107/IIA-17. Long-range passenger and cargo transport version for Tokyo Metropolitan Police Department; one delivered in February 1973. Cabin divided into two compartments: front section with 12 passenger seats, rear section capable of accommodating 2,268 kg (5,000 lb) of cargo, six stretcher patients or 12 passengers.

KV-107/IIA-SM-1. Firefighting version for Saudi Arabian government. Four delivered by early 1979. All can be fitted with specially designed equipment for various forms of firefighting (foam agent, chemical powder, water, and co-ordinated firefighting). Additional order for three anticipated.

KV-107/IIA-SM-2. Rescue and aeromedical version for Saudi Arabian government, with external rescue hoist, medical equipment, stretcher kit and other rescue gear; 303 litre (80 US gallon) additional fuel tank can be mounted on each side of fuselage. Two delivered by November 1978; additional order for two anticipated. For self-ferry flights, both SM-1 and SM-2 can be fitted with internally-mounted 1,893 litre (500 US gallon) auxiliary fuel tank.

The description which follows applies to the commercial KV-107/II-2, except where shown:

TYPE: Twin-engined transport helicopter.

ROTOR SYSTEM: Two three-blade rotors in tandem, rotating in opposite directions. Each blade is made up of a steel D spar to which is bonded a trailing-edge box constructed of aluminium ribs and glassfibre or aluminium skin.

ROTOR DRIVE: Power is transmitted from each engine through individually-overrunning clutches into the aft transmission, which combines the engine outputs, thereby providing a single power output to the interconnecting shaft which enables both rotors to be driven by either engine.

FUSELAGE: Basically square-section semi-monocoque structure built primarily of high-strength bare and Alclad aluminium alloy. Transverse bulkheads and built-up frames support transmission, power plant and landing gear. Loading ramp forms undersurface of upswept rear fuselage on utility and military models. Baggage container replaces ramp on airliner version. Fuselage is sealed to permit operation from water.

LANDING GEAR: Non-retractable tricycle type, with twin wheels on all three units. Oleo-pneumatic shock-absorbers. Tubeless tyres, size 18 × 5·5, pressure 10·34 bars (150 lb/sq in), on all wheels. Disc brakes. Wheel/ski gear optional.

POWER PLANT (KV-107/II): Two 932 kW (1,250 shp) General Electric CT58-110-1 or Ishikawajima-Harima CT58-IHI-110-1 turboshaft engines, mounted side by side at base of rear rotor pylon. Alternatively, two Rolls-Royce Gnome H.1200 turboshafts (in HKP 4C). Fuel tanks in sponsons, capacity 1,324 litres (350 US gallons). KV-107/IIA has more powerful CT58 engines and provision for increased fuel capacity (see introductory copy).

ACCOMMODATION: Standard accommodation for two pilots, stewardess and 25 passengers in airliner version. Seats in eight rows, in pairs on port side and single seats on starboard side (two pairs at rear of cabin) with central aisle. Airliner fitted with parcel rack and a roll-out baggage container, with capacity of approximately 680 kg (1,500 lb), located in underside of rear fuselage. Ramp of utility model is power-operated on the ground or in flight and can be removed or left open to permit carriage of extra-long cargo.

AVIONICS AND EQUIPMENT: Standard avionics include stability augmentation system (SAS) and automatic speed trim system (AST). Optional avionics include automatic stabilisation equipment (ASE); automatic flight control system (AFCS); Doppler radar; radio altimeter; HF, VHF and UHF radio; ADF; VOR/ILS; Tacan; compass system and attitude director indicator system; and intercom system.

DIMENSIONS, EXTERNAL:
- Rotor diameter (each) — 15·24 m (50 ft 0 in)
- Length overall, blades turning — 25·40 m (83 ft 4 in)
- Length of fuselage — 13·59 m (44 ft 7 in)
- Height to top of rear rotor hub — 5·09 m (16 ft 8½ in)
- Wheel track — 3·92 m (12 ft 10½ in)
- Wheelbase — 7·57 m (24 ft 10 in)
- Passenger door (fwd):
  - Height — 1·60 m (5 ft 3 in)
  - Width — 0·91 m (3 ft 0 in)

DIMENSIONS, INTERNAL:
- Cabin, excl flight deck:
  - Length — 7·37 m (24 ft 2 in)
  - Normal width — 1·83 m (6 ft 0 in)
  - Max height — 1·83 m (6 ft 0 in)
  - Floor area — 13·47 m^2 (145 sq ft)
  - Volume (usable) — 24·5 m^3 (865 cu ft)

AREAS:
- Rotor blades (each) — 3·48 m^2 (37·50 sq ft)
- Rotor discs (total) — 364·6 m^2 (3,925 sq ft)

WEIGHTS AND LOADINGS:
- Weight empty, equipped:
  - II-2 — 4,868 kg (10,732 lb)
  - IIA-1 — 4,589 kg (10,118 lb)
  - IIA-2 — 5,250 kg (11,576 lb)
- Max T-O and landing weight — 8,618 kg (19,000 lb) *or* 9,706 kg (21,400 lb)
- Max disc loading — 23·6 kg/m^2 (4·84 lb/sq ft)
- Max power loading — 4·62 kg/kW (7·6 lb/shp)

PERFORMANCE (A: KV-107/II-2 at 8,618 kg; 19,000 lb AUW. B: KV-107/IIA at same AUW):
- Never-exceed speed:
  - A, B — 146 knots (270 km/h; 168 mph)
- Max speed at S/L, normal rated power:
  - A — 136 knots (253 km/h; 157 mph)
  - B — 137 knots (254 km/h; 158 mph)
- Cruising speed at 1,525 m (5,000 ft):
  - A, B — 130 knots (241 km/h; 150 mph)
- Max rate of climb at S/L:
  - A, normal rated power — 463 m (1,520 ft)/min
  - B — 625 m (2,050 ft)/min
- Max vertical rate of climb at S/L:
  - B — 381 m (1,250 ft)/min
- Service ceiling:
  - A, normal rated power — 4,570 m (15,000 ft)
  - B — 5,180 m (17,000 ft)
- Service ceiling, one engine out:
  - A, military power, yaw, 248 rpm — 107 m (350 ft)
  - B — 1,740 m (5,700 ft)
- Hovering ceiling IGE: A — 2,895 m (9,500 ft)
  - B — 3,565 m (11,700 ft)
- Hovering ceiling OGE: A — 1,890 m (6,200 ft)
  - B — 2,680 m (8,800 ft)
- Min landing area (A, B): Length — 38 m (126 ft)
  - Width — 23 m (75 ft)
- T-O to 15 m (50 ft): B — 131 m (430 ft)
- Landing from 15 m (50 ft), one engine out:
  - B — 84 m (275 ft)
- Range:
  - A with 3,000 kg (6,600 lb) payload, 10% reserves — 94 nm (175 km; 109 miles)
  - B with standard fuel — 192 nm (357 km; 222 miles)
  - B with max fuel — 592 nm (1,097 km; 682 miles)

Kawasaki KV-107/IIA-SM-1 firefighting version, with water bucket on cargo sling

KAWASAKI (HUGHES) 369/MODEL 500 and 500C

JGSDF and JMSDF designation: OH-6J

A total of 50 Model 369HS helicopters, including 19 Model 500Cs, had been delivered to civil operators by the end of 1980. Japanese production of these models has now ceased.

KAWASAKI (HUGHES) 369D/MODEL 500D

JGSDF designation: OH-6D

The first Model 369D (500D) built by Kawasaki under licence from Hughes Helicopters was flown for the first time on 2 December 1977; JCAB Normal category certification was awarded on 20 April 1978. Six Model 500Ds had been delivered to civil operators in Japan by March 1981. The JGSDF ordered 32 as OH-6Ds, 22 of which have been delivered; it plans to order a further eight OH-6Ds in FY 1982.

Kawasaki (Hughes) Model 369D helicopter

MITSUBISHI

MITSUBISHI JUKOGYO KABUSHIKI KAISHA (Mitsubishi Heavy Industries Ltd)

HEAD OFFICE: 5-1, Marunouchi 2-chome, Chiyoda-ku, Tokyo 100
Telephone: Tokyo (03) 212 3111
Telex: J22282 and J22443
NAGOYA AIRCRAFT WORKS: 10, Oye-cho, Minato-ku, Nagoya 455
CHAIRMAN OF BOARD OF DIRECTORS: Gakuji Moriya
PRESIDENT: Masao Kanamori
EXECUTIVE VICE-PRESIDENTS:
Kaname Taniguchi
Soichiro Suenaga
Masao Suzuki
Toshimasa Mitsui
MANAGING DIRECTOR AND GENERAL MANAGER OF AIRCRAFT AND SPECIAL VEHICLE HEADQUARTERS: Kenji Ikeda
DIRECTOR AND DEPUTY GENERAL MANAGER OF AIRCRAFT AND SPECIAL VEHICLE HEADQUARTERS: Takashi Tamaki
ASST GENERAL MANAGER OF AIRCRAFT AND SPECIAL VEHICLE HEADQUARTERS: Yoshiaki Kato
GENERAL MANAGER, AIRCRAFT AND SPECIAL VEHICLE ADMINISTRATION DEPARTMENT: Yoshiaki Kato
GENERAL MANAGER, AIRCRAFT DEPARTMENT:
Toshiro Murai
GENERAL MANAGER, AIRCRAFT EQUIPMENT DEPARTMENT:
Naruo Taketa
GENERAL MANAGER, SPACE SYSTEM DEPARTMENT:
Masahiko Hamada
GENERAL MANAGER, SPECIAL VEHICLE DEPARTMENT:
Tetsuya Matsumura
MANAGER OF MU-2 ADMINISTRATION SECTION:
Yukiya Naramoto
GENERAL MANAGER, NAGOYA AIRCRAFT WORKS:
Yoshio Sasaki

Mitsubishi began the production of aircraft in the present Oye plant of its Nagoya Engineering Works in 1921, and manufactured a total of 18,000 aircraft of approximately 100 different types during the 24 years prior to the end of the second World War in 1945. The company was also one of the leading aero-engine manufacturers in Japan, and produced a total of 52,000 engines in the 1,000-2,500 hp range. The conclusion of the Peace Treaty in 1952 enabled the aircraft industry in Japan to recommence, and in December of that year the company constructed its present Komaki South plant. This factory, together with Mitsubishi's Oye, Daiko and Komaki North plants, was later consolidated as Nagoya Aircraft Works, with a combined floor area of 285,500 m² (3,073,100 sq ft).

In co-operation with Kawasaki as subcontractor, Mitsubishi was the JDA's prime contractor in producing F-4EJ Phantom tactical fighters for the JASDF, under licence from McDonnell Douglas Corporation. The first two F-4EJs, built by McDonnell Douglas, were delivered and ferried to Japan in July 1971. The next eight were assembled from knock-down subassemblies, the first of these making its first flight on 12 May 1972. The remaining 130, of which the first was delivered to the JASDF in September 1974, were built entirely in Japan. Production by Mitsubishi ended on 20 May 1981.

The JASDF is to acquire 100 McDonnell Douglas F-15 Eagles, of which 88 will be single-seat F-15Js and 12 will be two-seat F-15DJs. Two of the F-15Js and all 12 F-15DJs will be US-built aircraft, supplied under FMS (Foreign Military Sales); of these, the two Js and four of the DJs had been delivered by the Spring of 1981 to the Air Proving Wing of the JASDF at Gifu. The remaining 86 single-seaters will be completed in Japan, eight from US-supplied knocked-down assemblies and 78 built entirely in Japan. Mitsubishi was selected in April 1978 as prime contractor for the Japanese assembly and production programme; it will build the forward and centre-fuselages, and be responsible for final assembly and flight testing. The first 'knock-down' F-15J was flown on 26 August 1981 and was to be delivered in the following December. Funds were provided in the FY 1978 and FY 1980 budgets for the first 47 F-15Js (17 and 30), and for ten of the F-15DJs (six and four respectively). Participants in the Japanese production programme include Fuji, Kawasaki, Nippi and Shin Meiwa for the airframe, and IHI for the engines.

Mitsubishi holds licence agreements to manufacture the Sikorsky S-61A, S-61A-1, S-61B (HSS-2) and S-61B-1 (HSS-2A and -2B) helicopters. By 31 March 1981, Mitsubishi had delivered three S-61As and one S-61A-1 (for the JMSDF, for use in support of the Japanese Antarctic Expedition) and had delivered five S-61A-1s (of seven ordered) to the JMSDF for rescue duties. It had also delivered 91 of an order for 107 S-61Bs (HSS-2s) and S-61B-1s (HSS-2As and -2Bs) to the JMSDF for anti-submarine duties. From 1981, three S-61A-1s for rescue duties, and 16 S-61B-1s for anti-submarine duties with the JMSDF, are expected to be produced.

Mitsubishi is producing the MU-2 twin-turboprop utility transport; is prime contractor for the T-2 supersonic trainer and F-1 close-support combat aircraft for the JASDF, with Fuji, Nippi and Shin Meiwa as principal subcontractors; and is a subcontractor in the production programme for the Kawasaki C-1 (which see). It is producing forward and rear fuselages for the Lockheed P-3C Orions ordered by the JMSDF, under subcontract to Kawasaki (which see). As a part of Japan's YX civil transport aircraft programme, Mitsubishi is subcontracted, through the CTDC, to build aft passenger cabin sections of the Boeing 767 jet transport. Part of this work is, in turn, subcontracted by Mitsubishi to Nippi and Shin Meiwa.

Mitsubishi's aero-engine activities are described in the appropriate section of this edition.

MITSUBISHI MU-2

The MU-2 is a twin-turboprop STOL multi-purpose transport, the basic design of which was begun in 1960. Prototype construction began in 1962 and the first aircraft was flown on 14 September 1963. By 31 March 1981, total orders for the MU-2 (all versions) had reached 688, including 636 for export and 52 for Japanese customers. Fifteen versions have been announced, of which details have appeared in the 1965-66 and subsequent editions of *Jane's*. The two current versions are the Marquise and Solitaire, of which 103 and 52 respectively had been built by 31 March 1981. These aircraft are described under the Mitsubishi Aircraft International entry in the US section of this edition.

MITSUBISHI MU-300

The MU-300 is marketed worldwide (except in Japan), under the name **Diamond I,** by Mitsubishi Aircraft International in the USA. A full description of the aircraft can be found under that company's entry in this edition.

MITSUBISHI T-2

The T-2, the first supersonic aircraft developed by the Japanese aircraft industry, is a twin-engined two-seat jet trainer designed to meet the requirements of the JASDF.

Mitsubishi was selected as prime contractor for the development programme in September 1967. Preliminary and detailed design, under the leadership of Dr Kenji Ikeda, were followed by the completion of two XT-2 flying prototypes plus a static test airframe. The first XT-2 (19-5101) flew for the first time on 20 July 1971, and the second (29-5102) on 2 December 1971. These two aircraft were delivered to the JASDF in December 1971 and March 1972 respectively for further flight testing. Two additional development aircraft made their first flights on 28 April and 20 July 1972; the flight test programme was completed in March 1974. A fatigue test airframe was delivered in January 1975.

Production orders have been placed for 81 T-2s (31 **T-2** advanced trainers, 48 **T-2A** combat trainers, and two as prototypes for the F-1 close-support fighter version, described separately). Sixty-six of the T-2/2As had been delivered by 31 March 1981, to the 4th Air Wing at Matsushima. The current schedule calls for the 15 remaining T-2/2As to be delivered by August 1983. In 1982, the JASDF's Blue Impulse aerobatic team is expected to receive six T-2As in place of its F-86F Sabres.

Mitsubishi, as prime contractor, is responsible for fuselage construction, final assembly and flight testing of production aircraft. Major programme subcontractors are Fuji (wings and tail unit), Nippi (pylons and launchers) and Shin Meiwa (drop-tanks).

Under contract to the Technical Research and Development Institute, Mitsubishi is to convert one T-2 as a CCV (control configured vehicle). Initially, this is to have horizontal and vertical canard surfaces, manoeuvring flaps and slats, and a fly-by-wire control system. First flight is expected to take place in the Autumn of 1982.

The following description applies to the standard production T-2/2A:

TYPE: Two-seat supersonic jet trainer.

WINGS: Cantilever all-metal shoulder-wing monoplane. Wing section NACA 65 series (modified). Thickness/chord ratio 4·66%. Anhedral 9° from roots. Sweepback on leading-edges 68° at root, 42° 29′ inboard of outer extended-chord panels and 36° on outer panels; basic sweepback at quarter-chord 35° 47′. Multi-spar torsion box machined from tapered thick panels and constructed mainly of 7075 and 7079 aluminium alloy. Electrically-actuated aluminium honeycomb leading-edge flaps, the outer portions of which have extended chord. Electrically-actuated all-metal single-slotted flaps, with aluminium honeycomb trailing-edges over 70% of each half-span. No conventional ailerons. Lateral control by hydraulically-actuated all-metal two-section slotted spoilers ahead of flaps.

FUSELAGE: Conventional all-metal semi-monocoque structure, mainly of 7075 and 7079 aluminium alloy. Approx 10% of structure, by weight, is of titanium, mostly around engine bays. Two hydraulically-actuated door-type airbrakes under rear fuselage, aft of main-wheel bays.

TAIL UNIT: Cantilever all-metal structure. One-piece hydraulically-actuated all-moving swept tailplane, with 15° anhedral. Inner leading-edges of titanium; outer leading-edges of aluminium. Trailing-edges of aluminium honeycomb construction. Small ventral fin under each side of fuselage at rear. Hydraulically-actuated rudder.

LANDING GEAR: Hydraulically-retractable tricycle type, with pneumatic backup for emergency extension. Main units retract forward into fuselage, nose unit rearward. Single wheel on each unit. Nosewheel steerable through 72°. Oleo-pneumatic shock-absorbers. Hydraulic brakes and Hydro-Aire anti-skid units. Runway arrester hook beneath rear fuselage. Brake parachute in tailcone.

POWER PLANT: Two Rolls-Royce Turboméca Adour turbofan engines, each rated at 20·95 kN (4,710 lb st) dry and 31·45 kN (7,070 lb st) with afterburning, mounted side by side in centre of fuselage. (Engines licence-built by Ishikawajima-Harima, under designation TF40-IHI-801A.) Fixed-geometry air intake, with auxiliary 'blow-in' intake doors, on each side of fuselage aft of rear cockpit. Fuel in seven fuselage tanks with total capacity of 3,823 litres (841 Imp gallons;

Mitsubishi-built F-4EJ Phantom II fighter of the Japan Air Self-Defence Force *(Katsumi Hinata)*

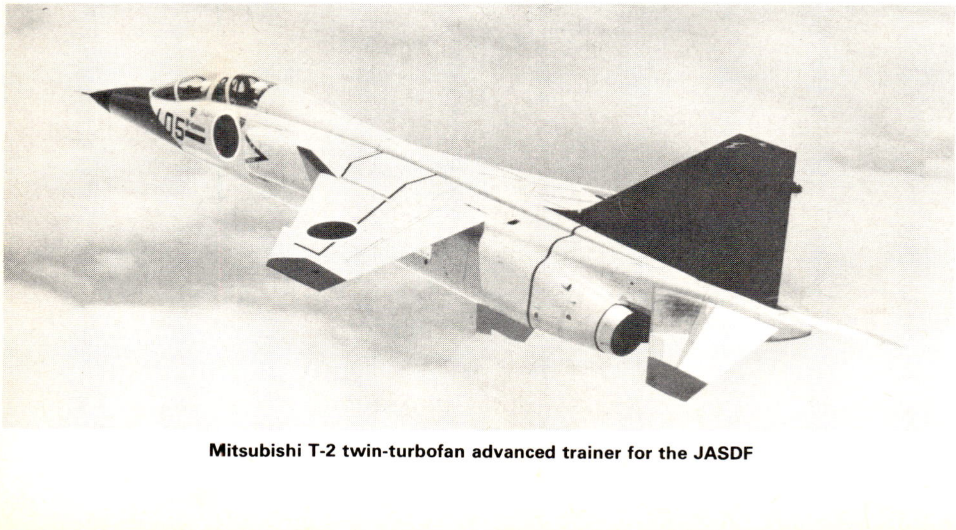
Mitsubishi T-2 twin-turbofan advanced trainer for the JASDF

1,010 US gallons). Pressure refuelling point in starboard side of fuselage, forward of main-wheel bay. Provision for carrying up to three 833 litre (183 Imp gallon; 220 US gallon) drop-tanks under wings and fuselage.

ACCOMMODATION: Crew of two in tandem on Daiseru-built Weber ES-7J zero-zero ejection seats in pressurised and air-conditioned cockpits, separated by windscreen. Rear seat elevated 0·28 m (11 in) above front seat. Individual manually-operated rearward-hinged jettisonable canopies. Liquid oxygen equipment.

SYSTEMS: Cockpit air-conditioning system. Two independent hydraulic systems, each 207 bars (3,000 lb/sq in), for flight controls, landing gear and utilities. Pneumatic bottle for landing gear emergency extension. Primary electrical power from two 12/15kVA AC generators.

AVIONICS AND EQUIPMENT: Mitsubishi Electric J/ARC-51 UHF. Nippon Electric J/ARN-53 Tacan and Toyo Communication J/APX-101 SIF/IFF. Mitsubishi Electric search and ranging radar in nose, with Mitsubishi Electric (Thompson-CSF) J/AWG-11 head-up display in cockpit. Lear 5010BL attitude and heading reference system.

ARMAMENT (combat trainer version): One Vulcan JM-61A1 multi-barrel 20 mm cannon in lower fuselage, aft of cockpit on port side. Attachment point on underfuselage centreline and two under each wing for drop-tanks or other stores. Wingtip attachments for air-to-air missiles.

DIMENSIONS, EXTERNAL:

Wing span	7·88 m (25 ft 10¼ in)
Wing chord at root	4·172 m (13 ft 8¼ in)
Wing chord at tip	1·133 m (3 ft 8½ in)
Wing aspect ratio	3
Wing taper ratio	3·7
Length overall, incl probe	17·84 m (58 ft 6¼ in)
Length of fuselage	17·31 m (56 ft 9½ in)
Height overall	4·38 m (14 ft 4¼ in)
Wheel track	2·82 m (9 ft 3 in)
Wheelbase	5·72 m (18 ft 9 in)

AREAS:

Wings, gross	21·18 m² (228·0 sq ft)
Airbrakes (total)	0·952 m² (10·25 sq ft)
Vertical tail surfaces (total, excl ventral fins)	5·00 m² (53·82 sq ft)
Horizontal tail surfaces (total)	6·70 m² (72·12 sq ft)

WEIGHTS:

Operational weight empty	6,302 kg (13,893 lb)
Max T-O weight, 'clean'	9,805 kg (21,616 lb)

PERFORMANCE (at max 'clean' T-O weight except where indicated):

Max level speed at 11,000 m (36,000 ft)	Mach 1·6
Max rate of climb at S/L	10,670 m (35,000 ft)/min
Service ceiling	15,240 m (50,000 ft)
Required field length	1,525 m (5,000 ft)
Max ferry range with external tanks	1,400 nm (2,593 km; 1,610 miles)

MITSUBISHI F-1

Following the JASDF's decision to develop a single-seat close air support fighter from the T-2 supersonic trainer, design began in 1972. During the development period the fighter was provisionally designated FS-T2-Kai, as described in previous editions of *Jane's*.

The second and third production T-2 trainers (59-5106 and 59-5107) were converted as prototypes, in which form they made their first flights on 7 and 3 June 1975 respectively. These aircraft retained the rear cockpit and canopy of the T-2, but this area was occupied by the fire control system and test equipment instead of a second occupant. Externally, they differed from the T-2 by the presence of a tubular fairing at the top of the fin, housing a passive warning radar antenna.

These prototypes were delivered to the JASDF Air Proving Wing at Gifu in July and August 1975, and after a year of flight test and evaluation the aircraft was type approved in November 1976 and officially designated F-1.

Production orders had been placed by March 1981 for 67 F-1s, of an anticipated total order for about 70. The first production F-1 (70-8201) made its first flight on 16 June 1977, and was delivered to the JASDF on 26 September 1977. Fifty-seven F-1s had been delivered, to the 3rd Squadron of the 3rd Air Wing at Misawa and the 8th Air Wing at Tsuiki, by 31 March 1981.

Mitsubishi F-1 single-seat close-support fighters of the JASDF

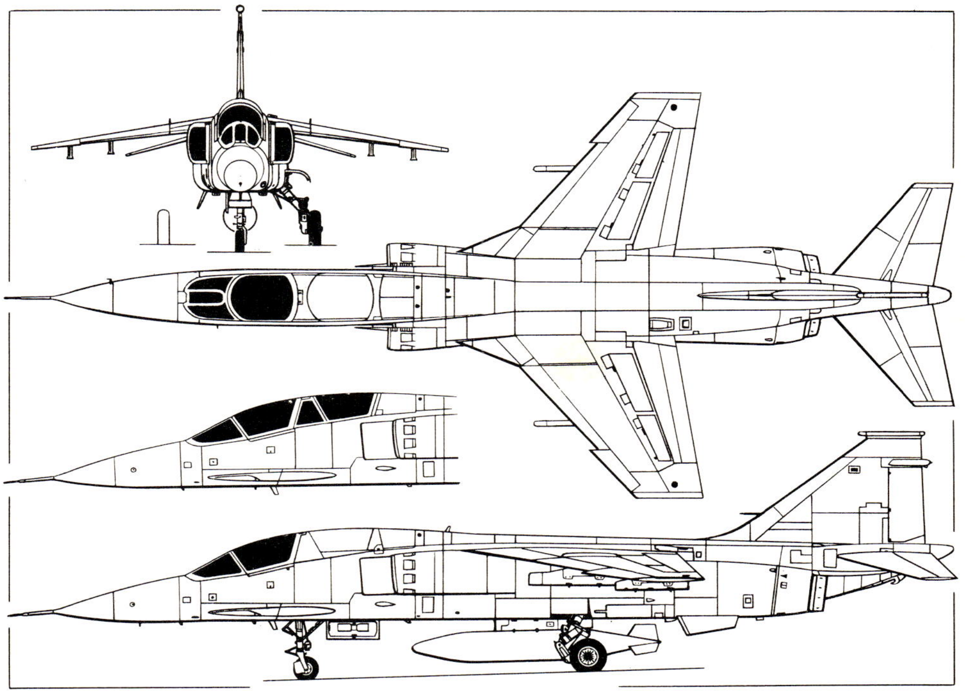

Mitsubishi F-1 (two Rolls-Royce Turboméca Adour turbofan engines), with additional scrap view of two-seat nose of T-2 trainer *(Pilot Press)*

TYPE: Single-seat close-support fighter.

AIRFRAME, POWER PLANT AND SYSTEMS: Generally similar to T-2, but with the rear cockpit area modified as avionics compartment for bombing computer, inertial navigation system and radar warning system. Up to three 821 litre (180 Imp gallon; 217 US gallon) auxiliary fuel tanks can be carried beneath the wings and fuselage.

ACCOMMODATION: Generally similar to T-2, but without rear seat and with 'solid' fairing in place of second canopy.

AVIONICS AND EQUIPMENT: Dual UHF; Tacan; IFF/SIF; Mitsubishi Electric nose-mounted air-to-air and air-to-ground radar, with Mitsubishi Electric (Thomson-CSF) J/AWG-12 head-up display; Ferranti 6TNJ-F inertial navigation system; radio altimeter; air data computer; Mitsubishi Electric J/AWG-12 fire control system and bombing computer (replacing original FCS from January 1982, for compatibility with ASM-1 missile); strike camera system; radar homing and warning system; attitude and heading reference system.

ARMAMENT: Single JM61 multi-barrel 20 mm cannon. One underfuselage and four underwing hardpoints, as in T-2, with detachable multiple ejector racks. Bombs of 250, 500 or 750 lb can be carried on all five external stations, up to a maximum weight of 2,721 kg (twelve 500 lb bombs). The four underwing stations can each be used for rocket pods such as the JLAU-3A (with nineteen 70 mm), RL-7 (seven 70 mm) and RL-4 (four 125 mm). Primary weapon will be the Mitsubishi ASM-1 air-to-surface missile, of which two can be carried on underwing stations. For air-to-air combat the F-1 can carry up to four AIM-9 Sidewinder missiles, one at each wingtip and one on each of the outboard underwing hardpoints. For long-range missions, the F-1 can carry up to three auxiliary fuel tanks (see 'Power Plant' paragraph).

DIMENSIONS AND AREAS:
As for T-2

WEIGHTS:

Operational weight empty	6,358 kg (14,017 lb)
Max T-O weight	13,674 kg (30,146 lb)

PERFORMANCE (at max 'clean' T-O weight except where indicated):

Generally similar to T-2 'clean', except for:

Time to 11,000 m (36,000 ft)	2 min
T-O run	1,280 m (4,200 ft)
Combat radius with four Sidewinders, internal fuel only, incl reserves	150 nm (278 km; 173 miles)
Combat radius (lo-lo-lo-hi) with eight 500 lb bombs and external tanks	190 nm (351 km; 218 miles)
Combat radius (hi-lo-hi) with two ASM-1s and one 821 litre drop-tank, incl reserves	300 nm (556 km; 346 miles)

NAL

NATIONAL AEROSPACE LABORATORY

1880 Jindaiji-machi, Chofu City, Tokyo
Telephone: Musashino (0422) 47 5911
DIRECTOR-GENERAL: Dr Toshio Kawasaki
DEPUTY DIRECTOR-GENERAL: Dr Shun Takeda
DIRECTOR OF V/STOL AIRCRAFT RESEARCH GROUP: Dr Norio Inumaru

The National Aerospace Laboratory (NAL) is a government establishment responsible for research and development in the field of aeronautical and space sciences. Since 1962 it has extended its activity in the field of V/STOL techniques.

NAL QSTOL RESEARCH AIRCRAFT

The NAL was allotted Y3,455 million in the FY 1981 budget for continuation of an 11-year programme to develop a large experimental Quiet STOL transport aircraft. This will be based upon the airframe of the Kawasaki C-1 tactical transport (which see), with the following modifications: replacement of the two Pratt & Whitney JT8D engines by four 48 kN (10,800 lb st) MITI/NAL FJR-710-600S high bypass ratio turbofan engines, installed above and far ahead of the wing leading-edges in nacelles with upper surface blowing, as on the Boeing YC-14 prototype STOL transport (see 1978-79 *Jane's*); installation of wing leading-edge and aileron BLC systems; replacement of the existing inboard flaps by USB flaps; structural strengthening of the fuselage and landing gear; and installation of a digital stability and control augmentation system.

The basic and detailed design work is expected to take four years; aircraft modification began in 1979; first flight is due in 1983. Total cost of the development programme is estimated at Y23,000 million, including flight testing. Data obtained from this programme will, it is hoped, enable NAL to develop, in co-operation with the Japanese aerospace industry, a commercial STOL transport aircraft able to operate from 800 m (2,625 ft) runways with 150 passengers, and to cruise at around Mach 0·62. Tentative plans call for development of this to be completed in 1990.

DIMENSIONS, EXTERNAL:
Wing span 30·60 m (100 ft 4¾ in)
Length overall, excl nose probe 29·00 m (95 ft 2 in)
Height overall 10·157 m (33 ft 4 in)
Tailplane span 11·30 m (37 ft 1 in)
Wheel track (c/l of shock-struts) 4·40 m (14 ft 5¼ in)
Wheelbase (c/l of shock-struts) 9·33 m (30 ft 7¼ in)
WEIGHTS (estimated):
Weight empty 30,430 kg (67,086 lb)
Max T-O weight 38,700 kg (85,320 lb)
PERFORMANCE (estimated):
Cruising speed Mach 0·62
Landing speed 72 knots (133 km/h; 83 mph)
T-O to 10·7 m (35 ft) 547 m (1,795 ft)
Landing from 10·7 m (35 ft) 488 m (1,600 ft)
Max range 1,400 nm (2,590 km; 1,610 miles)

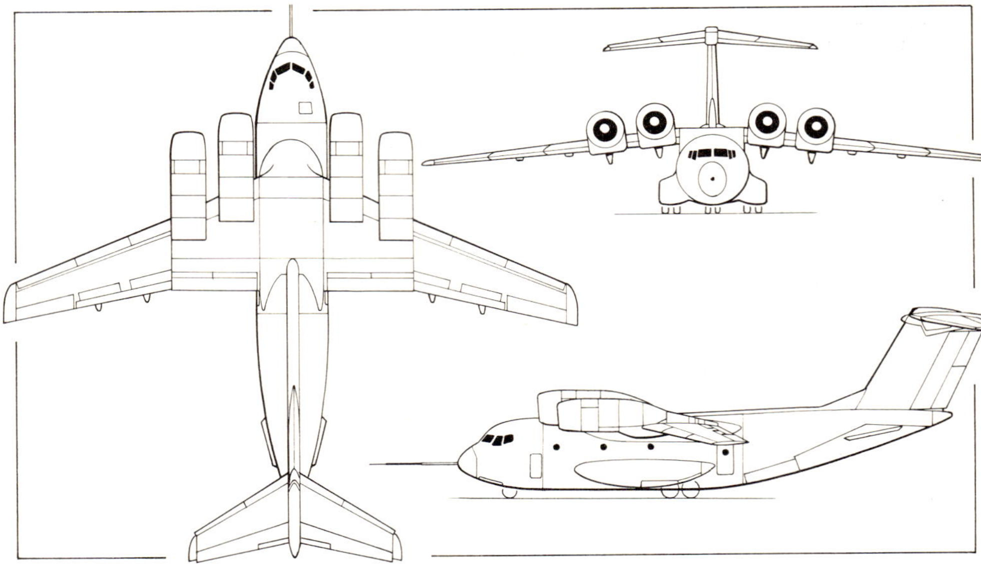

NAL QSTOL research aircraft, based on a Kawasaki C-1 airframe *(Michael A. Badrocke)*

NIPPI

NIHON HIKOKI KABUSHIKI KAISHA (Japan Aircraft Manufacturing Co Ltd)

HEAD OFFICE AND SUGITA WORKS: No. 3175 Showa-machi, Kanazawa-ku, Yokohama 236
Telephone: Yokohama (045) 771 1251
Telex: (3822) 267 Nippi J
OTHER WORKS: Atsugi
PRESIDENT: Masao Nagahisa
PUBLIC RELATIONS MANAGER: Taketoshi Kitamura

Nippi's Sugita plant, to which the head office was transferred in early 1971, has a floor area of 49,144 m^2 (528,800 sq ft) and employs 913 persons. The Atsugi plant, which employs 744 persons, has a floor area of 36,334 m^2 (390,950 sq ft). Kawasaki has a 32·08 per cent holding in Nippi.

The Atsugi plant is engaged chiefly in the overhaul, repair and maintenance of various types of aircraft and helicopters, including those of the Japan Defence Agency and Maritime Safety Agency, and carrier-based aircraft of the US Navy. The Sugita plant manufactures in-spar ribs for the Boeing 767; components and assemblies for the Kawasaki C-1, Mitsubishi T-2, F-1, F-15J (pylons and launchers), P-3C (engine nacelles), and Shin Meiwa US-1; airframe and dynamic components for the Kawasaki KV-107; dynamic components for the Fuji-Bell UH-1H and Kawasaki-Hughes OH-6D; body structures for Japanese satellites; tail units for Japanese-built rocket vehicles; and targets for the Japan Defence Agency.

The Nippi Pilatus B4 sailplane is described in the Sailplanes section.

SHIN MEIWA

SHIN MEIWA INDUSTRY Co Ltd

TOKYO OFFICE: c/o Shin Ohtemachi Building, 5th Floor, 2-1, 2-chome, Ohtemachi, Chiyoda-ku, Tokyo 100
Telephone: Tokyo (03) 279 3531
Telex: 222 2431 SMICAIR TOK
HEAD OFFICE: 1-5-25, Kosone-Cho, Nishinomiya-Shi, Hyogo-Ken
Telephone: Nishinomiya (0798) 47 0331
Telex: 5644493
WORKS (AIRCRAFT DIVISION): Konan and Itami
PRESIDENT: Yoshio Yagi
VICE-PRESIDENT: Itaru Morita
EXECUTIVE MANAGING DIRECTOR AND GENERAL MANAGER, AIRCRAFT DIVISION: Hajime Kawanishi
DIRECTOR AND ASST GENERAL MANAGERS:
Susumu Ishimoto
Kozo Kishikawa (Konan Works Manager)
Shigemi Matsui (Tokyo Office)
DIRECTOR, AND CHIEF OF ENGINEERING, HEAD OFFICE: Yukio Koya
SENIOR TECHNICAL CONSULTANT: Dr Shizuo Kikuhara
TECHNICAL CONSULTANT: Dr Koichi Tokuda
SALES MANAGER AND PUBLIC RELATIONS:
Motohiro Matsushita (Tokyo Office)

The former Kawanishi Aircraft Company became Shin Meiwa in 1949 and established itself as a major overhaul centre for Japanese and US military and commercial aircraft.

Shin Meiwa's principal current activities concern the series production of the US-1 medium-range STOL search and rescue amphibian for the JMSDF, and overhaul work on flying-boats and amphibians.

Shin Meiwa is engaged in the manufacture of components for other aircraft, including underwing drop-tanks for the Mitsubishi T-2 supersonic jet trainer and Japanese-built McDonnell Douglas F-15J Eagle jet fighters; the cargo loading system for the Kawasaki C-1 transport aircraft; nose and tail cones, ailerons and trailing-edge flaps for Japanese licence-built examples of the Lockheed P-3C. Shin Meiwa is also taking part in co-production of the Boeing 767, under subcontract to Mitsubishi.

SHIN MEIWA SS-2

JMSDF designation: PS-1

Shin Meiwa was awarded a contract in January 1966 to develop a new anti-submarine flying-boat for the Japan Maritime Self-Defence Force. Company designation for the basic flying-boat is SS-2; in ASW configuration this has the JMSDF designation PS-1.

The first PS-1 prototype (5801) flew for the first time on 5 October 1967, and the second on 14 June 1968. These aircraft were delivered to the 51st Flight Test Squadron of the JMSDF at Iwakuni. JDA type approval was granted in Autumn 1970.

In addition to the two prototypes, Shin Meiwa delivered 23 production PS-1s. Nineteen of these remain in service with the 31st Air Group of the JMSDF for ASW duties, most of them with No. 31 Squadron at Iwakuni.

No further procurement of the PS-1 is intended, it being considered that the JMSDF's additional requirements for ASW aircraft are met by its acquisition of Lockheed P-3C Orions. A full description of the PS-1 can be found in the 1980-81 *Jane's*.

SHIN MEIWA SS-2A

JMSDF designation: US-1

The US-1 (manufacturer's designation SS-2A) is an amphibious adaptation of the PS-1 (SS-2) flying-boat, configured for search and rescue duties with the JMSDF.

Design of the US-1 began in June 1970; the first example (9071) made its first flight, following a waterborne take-off, on 16 October 1974, and its first flight from a land base on 3 December 1974. The first US-1 was delivered on 5 March 1975, and six had been delivered by February 1981, of eight currently on order. From the seventh aircraft, due for delivery in March 1982, the US-1 will be fitted with T64-IHI-10J turboprops, offering a 10 per cent increase in power available for take-off. It is planned to retrofit the first six aircraft with these engines in due course. The US-1 is at present in service with No. 71 SAR Squadron of the JMSDF, based at Iwakuni.

To make possible very low landing and take-off speeds, the US-1 has both a boundary layer control system and extensive flaps for propeller slipstream deflection. Control and stability in low-speed flight are enhanced by 'blowing' the rudder, flaps and elevators, and by use of an automatic flight control system.

TYPE: Four-turboprop STOL air/sea rescue amphibian.
WINGS: Cantilever high-wing monoplane. Conventional all-metal two-spar structure with rectangular centre-section and tapered outer panels. High-lift devices include outboard leading-edge slats extending over nearly 17 per cent of the span and large outer and inner

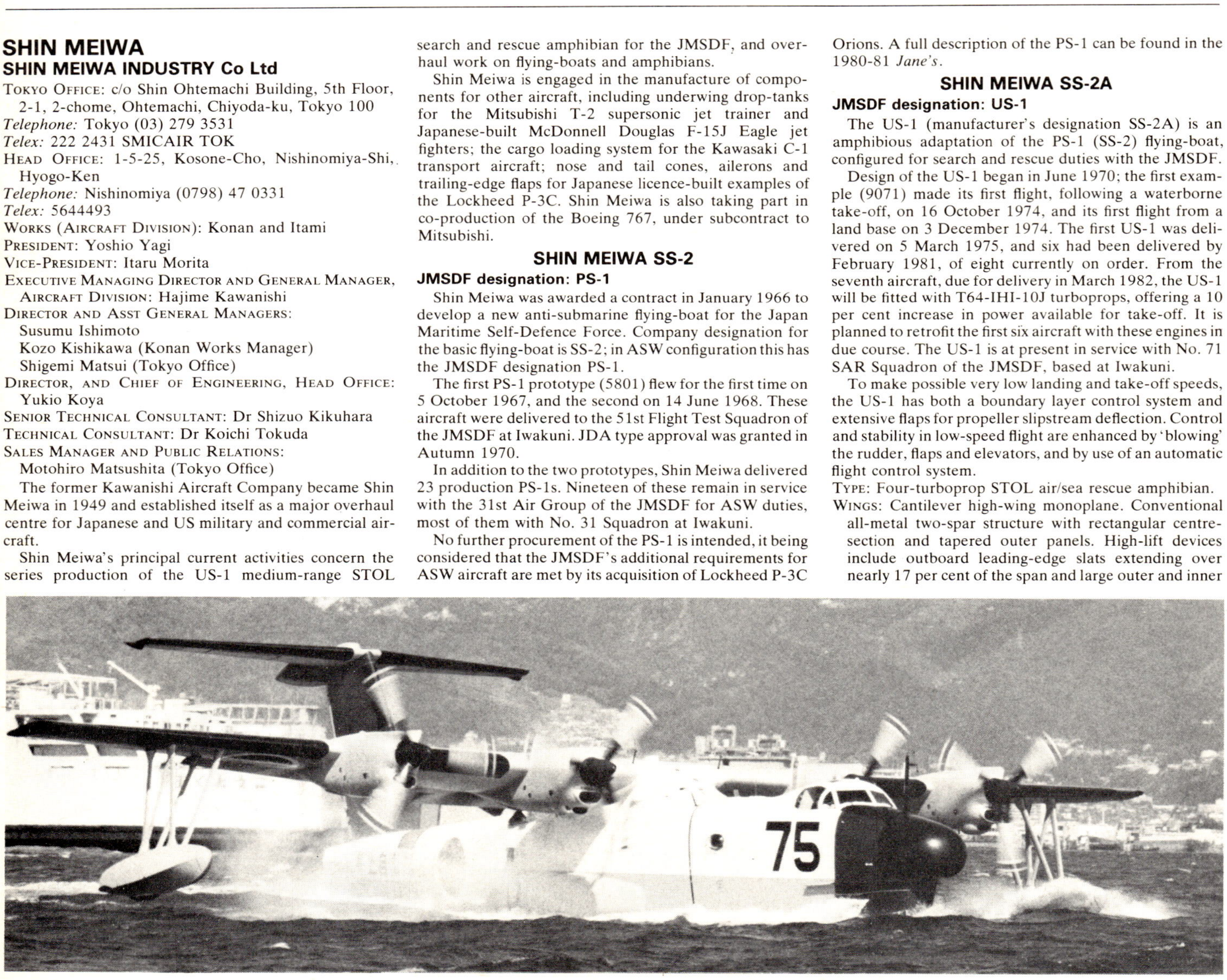

Shin Meiwa US-1 air/sea rescue amphibian (four Ishikawajima/General Electric T64-IHI-10 turboprop engines)

blown trailing-edge flaps extending 60° and 80° respectively. Two spoilers in front of outer flap on each wing. Powered ailerons. Leading-edge de-icing boots.

FUSELAGE: All-metal semi-monocoque hull, with high length/beam ratio. V-shaped single-step planing bottom, with curved spray suppression strakes along sides of nose and spray suppressor slots in fuselage undersides aft of inboard propeller line. Double-deck interior.

TAIL UNIT: Cantilever all-metal T tail. Large dorsal fin. Tailplane has slats and de-icing boots on leading-edge. Blown rudder and elevators. Tab in each elevator.

LANDING GEAR: Flying-boat hull, plus hydraulically-retractable Sumitomo tricycle landing gear with twin wheels on all units. Steerable nose unit. Oleo-pneumatic shock-absorbers. Main units, which retract rearward into fairings on hull sides, have size 40 × 14-22 (Type VII) tyres, pressure 7·79 bars (113 lb/sq in). Nosewheel tyres size 25 × 6·75-18 (Type VII), pressure 20·69 bars (300 lb/sq in). Three-rotor hydraulic disc brakes. No anti-skid units.

POWER PLANT: Four 2,282 kW (3,060 ehp) Ishikawajima-built General Electric T64-IHI-10 (more powerful T64-IHI-10Js from c/n 7) turboprop engines, each driving a Sumitomo-built Hamilton Standard 63E60-19 three-blade constant-speed reversible-pitch propeller. Additionally, one 932 kW (1,250 shp) Ishikawajima-built General Electric T58-IHI-10-M1 gas turbine is housed in the upper centre portion of the fuselage to provide power for boundary layer control system on rudder, flaps and elevators. Fuel in five wing tanks, with total usable capacity of 10,851 litres (2,387 Imp gallons) and two rear-fuselage tanks (11,649 litres; 2,563 Imp gallons); total capacity 22,500 litres (4,950 Imp gallons). Pressure refuelling point on port side, near bow hatch. Oil capacity 152 litres (33·4 Imp gallons). The US-1 can be refuelled on open sea, either from a surface vessel or from another US-1, fitted with detachable at-sea refuelling equipment.

ACCOMMODATION: Search and rescue version has accommodation for crew of nine and 20 seated survivors or 12 stretchers, one auxiliary seat and two observers' seats. Sliding rescue door on port side of fuselage, aft of wing. Transport version can seat up to 69 passengers in mainly four-abreast seating with centre aisle; rear portion of cabin convertible to cargo compartment.

SYSTEMS: Cabin air-conditioning system. Two independent hydraulic systems, each 207 bars (3,000 lb/sq in). No. 1 system actuates ailerons, outboard flaps, spoilers, elevators, rudder and control surface 'feel'; No. 2 system actuates ailerons, inboard and outboard flaps, wing leading-edge slats, elevators, rudder, landing gear extension/retraction and lock/unlock, nosewheel steering, main-wheel brakes and windscreen wipers. Emergency system, also of 207 bars (3,000 lb/sq in), driven by 24V DC motor, for actuation of inboard flaps, landing gear extension/retraction and lock/unlock, and main-wheel brakes. Oxygen system for all crew and stretcher stations. Garrett GTCP85-131J APU provides power for starting main engines and shaft power for 40kVA emergency AC generator. BLC system includes a C-2 compressor, driven by T58-IHI-10-M1 gas turbine, which delivers compressed air at 14 kg (30·9 lb)/s and pressure of 1·86 bars (27 lb/sq in) for ducting to inner and outer flaps, rudder and elevators. Electrical system includes 115/200V three-phase 400Hz constant-frequency AC and three transformer-rectifiers to provide 28V DC. Two 40kVA AC generators, driven by Nos. 2 and 3 main engines. Emergency 40kVA AC generator driven by APU. 24V emergency DC power from two 34Ah nickel-cadmium batteries. Anti-icing, air-conditioning, fire detection and extinguishing systems standard.

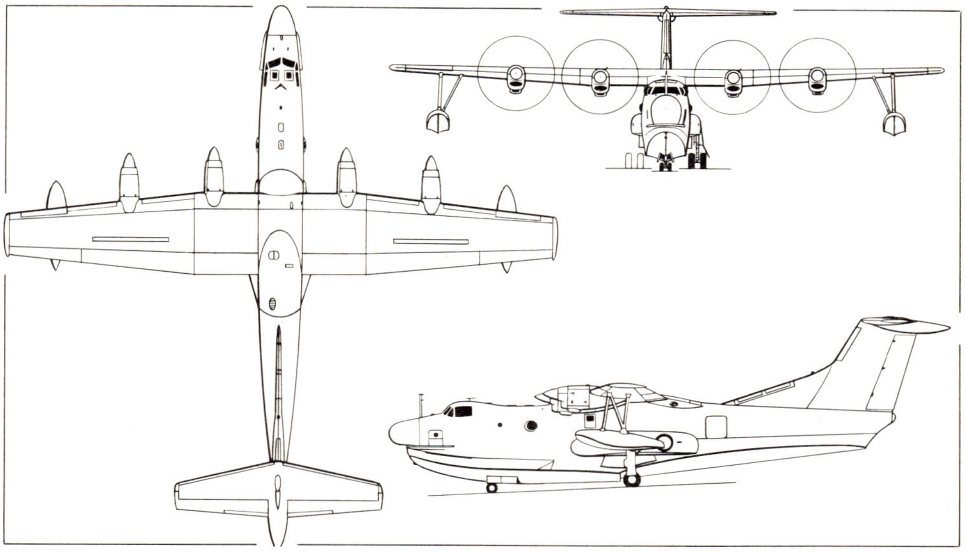

Shin Meiwa US-1 search and rescue amphibian, developed from the PS-1 *(Pilot Press)*

AVIONICS AND EQUIPMENT: HIC-3 interphone, HRC-107 HF, N-CU-58/HRC antenna coupler, HGC-102 teletypewriter, HRC-106 radio, HRC-110 radio, HRN-101 ADF, AN/ARA-50 UHF/DF, HRN-105 Tacan, HRN-104 Loran, HRA-4 Loran signal processor, HRN-106 ILS marker beacon receiver, AN/APN-171 (N2) radar altimeter, HPN-101B wave height meter, AN/APN-187C Doppler radar, AN/AYK-2 navigation computer, A/A24G-9 TAS transmitter, N-PT-3 dead reckoning plotting board, N-OA-35/HSA tactical plotter group, AN/APS-80N search radar, AN/APA-125N indicator group, AN/APX-68N IFF transponder, RRC-15 emergency transmitter and N-ID-66/HRN BDHI.

OPERATIONAL EQUIPMENT: Marker launcher, 10 marine markers, 6 green markers, 2 droppable message cylinders, 10 float lights, pyrotechnic pistol, parachute flares, 2 flare storage boxes, binoculars, 2 rescue equipment kits, 2 droppable life-raft containers, rescue equipment launcher, lifeline pistol, lifeline, 3 lifebuoys, loudspeaker, hoist unit, rescue platform, lifeboat with outboard motor, camera, and 12 stretchers. Stretchers can be replaced by troop seats.

DIMENSIONS, EXTERNAL:

Wing span	33·15 m (108 ft 9 in)
Wing chord at root	5·00 m (16 ft 4¾ in)
Wing chord at tip	2·39 m (7 ft 10 in)
Wing aspect ratio	8
Length overall	33·46 m (109 ft 9¼ in)
Height overall	9·82 m (32 ft 2¾ in)
Tailplane span	12·36 m (40 ft 8½ in)
Wheel track	3·56 m (11 ft 8¼ in)
Wheelbase	8·33 m (27 ft 4 in)
Propeller diameter	4·42 m (14 ft 6 in)
Rescue hatch, (port side, rear fuselage):	
Height	1·58 m (5 ft 2¼ in)
Width	1·46 m (4 ft 9½ in)

AREAS:

Wings, gross	135·8 m^2 (1,462 sq ft)
Ailerons (total)	6·40 m^2 (68·9 sq ft)
Inner flaps (total)	9·40 m^2 (101·18 sq ft)
Outer flaps (total)	14·20 m^2 (152·85 sq ft)
Leading-edge slats (total)	6·01 m^2 (64·7 sq ft)
Spoilers (total)	2·10 m^2 (22·60 sq ft)
Fin	17·56 m^2 (189 sq ft)
Dorsal fin	6·32 m^2 (68·03 sq ft)
Rudder	7·01 m^2 (75·5 sq ft)
Tailplane	23·05 m^2 (248 sq ft)
Elevators, incl tab	8·78 m^2 (94·5 sq ft)

WEIGHTS AND LOADINGS (search and rescue):

Manufacturer's weight empty	23,300 kg (51,367 lb)
Weight empty, equipped	25,500 kg (56,218 lb)
Usable fuel: JP-4	17,518 kg (38,620 lb)
JP-5	18,397 kg (40,560 lb)
Max oversea operating weight	36,000 kg (79,365 lb)
Max T-O weight from water	43,000 kg (94,800 lb)
Max T-O weight from land	45,000 kg (99,200 lb)
Max wing loading	331·4 kg/m^2 (67·9 lb/sq ft)
Max power loading	4·93 kg/kW (8·11 lb/ehp)

PERFORMANCE (search and rescue, at max T-O weight from land, except where indicated):

Max level speed 260 knots (481 km/h; 299 mph)
Max level speed at 3,050 m (10,000 ft), AUW of 36,000 kg (79,365 lb) 268 knots (496 km/h; 308 mph)
Cruising speed at 3,050 m (10,000 ft) 230 knots (426 km/h; 265 mph)
Max rate of climb at S/L 460 m (1,510 ft)/min
Max rate of climb at S/L, AUW of 36,000 kg (79,365 lb) 725 m (2,380 ft)/min
Service ceiling 6,520 m (21,400 ft)
Service ceiling, AUW of 36,000 kg (79,365 lb) 8,230 m (27,000 ft)
T-O to 15 m (50 ft) from land, 30° flap, BLC on 620 m (2,035 ft)
T-O to 15 m (50 ft) from water, AUW of 43,000 kg (94,800 lb), 40° flap, BLC on 600 m (1,970 ft)
Landing from 15 m (50 ft) on land, AUW of 36,000 kg (79,365 lb), 50° flap, BLC on, with reverse pitch 810 m (2,655 ft)
Landing from 15 m (50 ft) on water, AUW of 43,000 kg (94,800 lb), 60° flap, BLC on 290 m (950 ft)
Min ground turning radius:
self-powered 21·20 m (69 ft 6¾ in)
towed 18·80 m (61 ft 8¼ in)
Runway LCN requirement at AUW of 43,000 kg (94,800 lb) 42
Radius of search operation, incl 2·3 h search 900 nm (1,665 km; 1,035 miles)
Max range at 230 knots (426 km/h; 265 mph) at 3,050 m (10,000 ft) 2,270 nm (4,207 km; 2,614 miles)

SHOWA

SHOWA HIKOKI KOGYO KABUSHIKI KAISHA (Showa Aircraft Industry Co Ltd)

HEAD OFFICE AND SALES OFFICE: Mitsui Building, No. 3, 3-chome, Nihonbashi-Muromachi, Chuo-ku, Tokyo
Telephone: Tokyo (03) 270 1451 and 279 1451
Telex: 2223766 SHOWA J
WORKS: No. 600 Tanaka-machi, Akishima-shi, Tokyo
NARITA SERVICE CENTRE: 20 Araizumi, Naritasi, Chiba
PRESIDENT: Kakuro Tsukamoto

Showa was the first Japanese aircraft manufacturing company to resume post-war operations when it undertook the overhaul and repair of US Air Force aircraft.

The company's present activities comprise mainly the manufacture of wingtip floats, tail fin, partition and other doors and hatches for the Shin Meiwa US-1; and the supply of aluminium and non-metal honeycomb and honeycomb sandwich panels for aircraft floors and airframe construction. Showa also manufactures a variety of airborne equipment.

KOREA

(REPUBLIC)

KAL

KOREAN AIR LINES

KAL Building, CPO Box 864, Seoul
Telephone: 771 66 and 771 67
Telex: KALHO K27526
PRESIDENT: C. H. Cho
SENIOR VICE-PRESIDENT, SPECIAL PROJECT DEVELOPMENT: W. B. Lee

Following delivery by Hughes Helicopters of the USA of 34 Model 500MD Defender light helicopters to the Republic of Korea Air Force, Korean Air Lines is currently assembling both Defenders and Model 500D commercial helicopters from US-built components, and will proceed gradually to local manufacture. By early 1980, KAL had completed more than 75 Model 500MD Standard Scouts and 25 Model 500MD/TOW Defenders, in addition to a number of Model 500D commercial helicopters, for sale to customers in Korea. Descriptions of the Defender and Model 500D can be found under the Hughes entry in the US section of this edition.

KAL is also to assemble under licence 68 Northrop F-5E Tiger IIs and F-5F combat trainers (reported totals are 36 and 32 respectively), ordered for the Republic of Korea Air Force. Deliveries of completed aircraft are due to begin in late 1982.

LIBYA

It was reported in 1978 that a new aircraft factory was under construction in Libya, approx 216 nm (400 km; 248 miles) from Tripoli, with assistance from SIAI-Marchetti of Italy. Main purpose of the new factory was the licence assembly of 160 **SIAI-Marchetti SF.260s**, of an overall total of 240 ordered by Libya for its own and other Arab air forces. Libyan production was expected to begin in 1980, but it has since been reported that this programme has been abandoned.

MEXICO

AAMSA
AERONAUTICA AGRICOLA MEXICANA SA
(Subsidiary of Industrias Unidas SA)

171 Oriente No. 398, Colonia Aragon, Apartado 14783, Mexico 14, DF
Telephone: 760 60 00
Telex: 17-74359
ADMINISTRATIVE MANAGER: Armando Soto Ortega

As the result of an agreement between Rockwell International Corporation of the USA (which see) and Industrias Unidas SA of Mexico, this company was formed in 1971 to take over from the former's Commercial Products Group the manufacture of Aero Commander Quail Commander and Sparrow Commander agricultural aircraft. AAMSA purchased the type design, tooling and all production materials for the Sparrow and Quail Commander agricultural aircraft, in order to build them at a new manufacturing complex in Pasteje, Mexico.

Up to the end of 1975, only a dozen of these aircraft had been completed in Mexico, and the Sparrow programme was terminated. The Quail Commander, now known as the AAMSA A9B-M Quail, supersedes an earlier version designated A9 with a 175 kW (235 hp) Avco Lycoming O-540-B2B5 engine and lower operating weights.

AAMSA A9B-M QUAIL

The A9B-M current production version of the Quail agricultural aircraft has a 795 litre (210 US gallon) hopper and low operating cost. The entire primary structure is coated with Copon, an epoxy resin catalyst paint resistant to all known agricultural chemicals.

Component sets of the Quail are sent by AAMSA to its subsidiary, Aircraft Parts and Development Corporation of Laredo, Texas, which holds the FAA type certificate and is responsible for marketing the aircraft.

Compared with the A9B (described in the 1979-80 *Jane's*), the A9B-M has a number of significant modifications. The wings have increased dihedral, increased aileron chord and area, reduced flap area, increased trim tab area, and a root fairing at the junction with the fuselage. The cockpit and canopy have been redesigned, and the rear fuselage cut down, to afford a view to the rear, the resulting loss of fuselage keel area being offset by a slightly larger dorsal fin. A balanced rudder is now fitted. The air filter is relocated, and a stainless steel cowl flap is added on the underside of the engine cowling.

A total of 103 AAMSA-built Quails (all versions) had been built by the beginning of 1981. An output of 60 during 1981 was planned.

TYPE: Single-seat agricultural monoplane.

WINGS: Braced low-wing monoplane. Modified Clark Y wing section. Incidence 0° 20′. Composite structure with spruce spars, metal-covered leading-edge and Eonnex fabric covering on remainder of wing. Multiple steel tube overwing bracing struts on each side. Hoerner wingtips. Fabric-covered wooden ailerons. Flaps and drooping ailerons.

AAMSA A9B-M Quail single-seat agricultural monoplane, in spraying configuration

FUSELAGE: Steel tube structure with Eonnex fabric covering. Removable side panels.

TAIL UNIT: Wire-braced steel tube structure with Eonnex fabric covering. Trim tab in starboard elevator.

LANDING GEAR: Non-retractable tailwheel type. CallAir spring shock-absorbers. Cleveland main wheels, with Goodyear tyres, size 8·50-6 (6-ply). Scott 203 mm (8 in) steerable tailwheel. Cleveland toe-actuated brakes. Parking brake. Wire-cutters on main legs.

POWER PLANT: One 224 kW (300 hp) Avco Lycoming IO-540-K1A5 flat-six engine, driving a McCauley Type 1A200-DFA9050 two-blade fixed-pitch metal propeller. Two-position adjustable-pitch McCauley Type 2D34CT-84HF two-blade metal propeller optional. Fuel tank in each wing, combined capacity 151 litres (40 US gallons). Oil capacity 11 litres (3 US gallons).

ACCOMMODATION: Single seat in enclosed cockpit aft of hopper. Side doors. Steel tube overturn structure. Cabin heater standard.

SYSTEMS: Electrical system includes 12V 50A starter/generator and 30Ah battery.

AVIONICS AND EQUIPMENT: Radio and other avionics to customer's requirements. Standard equipment includes 0·64 m³ (22·5 cu ft) glassfibre/polyester hopper, capacity 795 litres (175 Imp gallons; 210 US gallons); wing tiedown rings; landing gear and windscreen wire-cutters; canopy-to-fin deflector cable; and electrically-operated auxiliary fuel pumps. Optional equipment includes Transland Boommaster or Micronair spray system, Transland Swathmaster dry spreader system, lights for night flying, 0·17 m³ (6·0 cu ft) hopper extension increasing capacity to 795 litres (210 US gallons) or 725 kg (1,600 lb), invert emulsion spray system or low volume spray system, bottom-loading equipment, rotating beacon, and 30 × 13-6 main wheels.

DIMENSIONS, EXTERNAL:

Wing span	10·59 m (34 ft 9 in)
Wing chord, constant	1·59 m (5 ft 2¾ in)
Wing area, gross	16·90 m² (182 sq ft)
Wing aspect ratio	6·63
Length overall (tail up)	7·32 m (24 ft 0 in)
Height overall	2·34 m (7 ft 8 in)
Tailplane span	3·20 m (10 ft 6 in)
Wheel track	2·08 m (6 ft 10 in)
Wheelbase	5·21 m (17 ft 1 in)
Propeller diameter: standard	2·29 m (7 ft 6 in)
optional	2·13 m (7 ft 0 in)

WEIGHTS AND LOADINGS:

Weight empty	817 kg (1,800 lb)
Max payload	725 kg (1,600 lb)
Max T-O weight	1,725 kg (3,800 lb)
Max wing loading	102 kg/m² (20·88 lb/sq ft)
Max power loading	7·70 kg/kW (12·67 lb/hp)

PERFORMANCE (at max T-O weight, except where indicated):

Max level speed at S/L	104 knots (193 km/h; 120 mph)
Max cruising speed (75% power) at 1,360 kg (3,000 lb) AUW	100 knots (185 km/h; 115 mph)
Normal operating speed	78-87 knots (145-161 km/h; 90-100 mph)
Stalling speed	52 knots (97 km/h; 60 mph)
Stalling speed as usually landed	35 knots (65 km/h; 40 mph)
Max rate of climb at S/L	259 m (850 ft)/min
Service ceiling	4,875 m (16,000 ft)
T-O run	328 m (1,000 ft)
Landing run at normal landing weight	136 m (447 ft)
Range at 50% power	260 nm (483 km; 300 miles)

ANAHUAC
FABRICA DE AVIONES ANAHUAC SA

ADDRESS: Calzada Adolfo López Mateos 478, Aeropuerto Internacional, Mexico 9, DF
Telephone: 558 27 57
PRESIDENT, FOUNDER AND GENERAL ADMINISTRATOR: Dr Alejandro Elizondo

This company was formed to initiate in Mexico the development of aircraft suited to the particular needs of agricultural aviation in that country, taking its name from the former Aztec valley where Mexico City is now situated.

ANAHUAC TAURO 350 (BULL)

Design of the Tauro single-seat agricultural aircraft was begun in January 1967, and the prototype first flew on 3 December 1968. By the end of 1971 Anahuac had built seven production aircraft, designated Tauro 300; for a description and illustration see the 1974-75 *Jane's*.

This initial version was followed by the improved Tauro 350, of which three had been delivered by early 1978. A further six were scheduled for completion by the Spring of 1978, but no later news than this has been received from the company.

A description and illustration of the Tauro 350 can be found in the 1980-81 *Jane's*.

NETHERLANDS

FOKKER
FOKKER BV

HEAD OFFICE AND MAIN FACTORY: PO Box 1065, 1000 BB Amsterdam
Telephone: (020) 5449111
Telex: 12227 Foa NL
OTHER FACTORIES AND COMPANIES:
Fokker Drechtsteden plant, with plants at Papendrecht, Dordrecht and Hoogeveen; Fokker Ypenburg plant at Ypenburg Air Base, near the Hague; and Fokker Woensdrecht plant, at Woensdrecht Air Base, near Bergen op Zoom
Trading Company Avio-Diepen BV
SUPERVISORY BOARD:
Mr drs H. Langman (Chairman)
ir A. Meijer (Vice-Chairman)
dr J. H. Greidanus
T. V. Jones
T. A. McDougall
H. G. Buiter
Prof Dr W. H. J. Reynaerts
ir J. W. Hillege
ir L. J. van Ameyden (Adviser)
BOARD OF MANAGEMENT:
F. Swarttouw (Chairman)
J. Cornelis (New Aircraft Projects)
J. Donders (Personnel and Social Affairs)
D. Krook (Marketing and Sales)
F. Nel (Technology)
HEAD OF DEPT OF CORPORATE RELATIONS: C. H. Verweij

MEDIA RELATIONS:
G. W. Knook (Manager)
L. J. N. Steijn

Fokker forms the entire aircraft industry in the Netherlands, with six plants, in which about 8,900 people are employed. Earlier collaborative ventures, other than those with VFW, included participation in the manufacturing programmes for the Gloster Meteor, Hawker Hunter and Lockheed F-104G, with final assembly lines at Schiphol; and for the Breguet Atlantic and Canadair (Northrop) CF-5/NF-5. Fokker has an important share in the European manufacturing programme for the General Dynamics F-16 fighter, being responsible for centre-fuselages and wing moving surfaces for 671 aircraft (for the Dutch final assembly line for 196 F-16s for the Netherlands and Norway; and the remainder for the Belgian assembly line, and for General Dynamics, for aircraft to be delivered to the USAF). In addition, Fokker is producing for the F-16 programme 435 main landing gear doors, 161 tailplanes, 304 rudders and 302 fin leading-edges. Delivery to the RNethAF of F-16s assembled by Fokker began in June 1979, following the first flight at Schiphol by a Dutch-assembled F-16 (J-259) on 3 May 1979. The first Dutch-assembled F-16 for the Royal Norwegian Air Force flew for the first time on 12 December 1979 and was handed over on 15 January 1980. The Norwegian F-16s are distinguished by a lengthened structure at the base of the rudder, housing a brake-chute. On 19 March 1981 the 100th European-built F-16, a Fokker-built F-16B, was delivered to the RNethAF.

Some 5,300 people are employed at the Schiphol-Oost works, Amsterdam, which accommodates the company headquarters and administration together with the main F27, F28 and F-16 assembly lines and test flying facilities. Production is continuing of the F27 and F28, each in various versions, and wing moving surfaces and other components are being produced for the Airbus A300 and A310. Also at Schiphol are the design offices, research department, numerically-controlled milling department, metal bonding department, electronics division, space division and scientific and administrative computer facilities.

Fokker has devoted considerable effort towards defining the optimum characteristics of a new 130/150-seat short/medium-haul twin-turbofan transport that would be marketed as the F29. In mid-1981 it announced that it was collaborating with McDonnell Douglas of the USA in a joint programme for an aircraft in this category, known as the MDF-100. A reference to this can be found in the International section.

The Drechtsteden plant, formed by the integrated production facilities at Dordrecht, Papendrecht and Hoogeveen, employs some 1,800 people. Most of these are engaged on detail production and component assembly for the General Dynamics F-16, Fokker F27 and F28 and Airbus A300/A310; other work includes the manufacture of antennae and other specialised products.

At Ypenburg the installation of F-16 centre-fuselages, and maintenance, overhaul, repair and modification work on a wide variety of military and civil aircraft, is carried out by a work force of 900 people.

Woensdrecht, which has a current work force of 850, has been building outer wings and struts for the Shorts 330 since 1974.

Reinforced plastics components for the Friendship, Fellowship, Airbus A300/A310 and Shorts 330, and radomes and fairings for the Westland Lynx helicopter, are manufactured at Ypenburg.

F-16 final assembly line at Fokker's Schiphol (Amsterdam) factory

At Woensdrecht the ELMO division produces electrical and electronic systems and wire harnesses.

Hoogeveen Division, a facility of the Drechtsteden plant, is engaged in the manufacture of parts for the aerospace industry, radar and telecommunications and other industries. Quantity production of aluminium shelters is also undertaken in this factory.

A new production facility at Hoogeveen, to become operational in 1982, is being set up to produce carbonfibre composite tail components for the F-16.

FOKKER F27 FRIENDSHIP

The first of two F27 prototypes made its first flight on 24 November 1955, and was designed to accommodate 28 passengers in a 22·3 m (73 ft) long fuselage. The second, which flew on 29 January 1957, was representative of Series 100 production aircraft, with Dart 511 engines and 32 seats in a 23·1 m (76 ft) fuselage. Two further airframes were built for static and fatigue testing.

The F27 has been in series production for many years, both by Fokker and, for a period, by Fairchild Industries in the United States. Deliveries by Fokker began in November 1958. US production of the F-27 and FH-227 totalled 205; details have appeared in previous editions of *Jane's*.

Features of current production aircraft include a low-noise interior; engine 'hush-kits'; and a Sperry SPZ-600 AFCS comprising a fail-passive autopilot and a flight director with Cat. II option.

The Mks 100 (85 built), 300 (13 built), 500M and 700 are no longer available. Details of these versions can be found in earlier editions of *Jane's*.

Fokker is standardising currently on the Mks 200, 400M, 500, 600 and Maritime, but any of the following versions of the F27 are available to order:

F27 Mk 200. Basic airliner or executive model with Dart RDa.7 Mk 536-7R turboprops. First flight 20 September 1959.

F27 Mk 400 Combiplane. Available on request only. Details in 1978-79 *Jane's*.

F27 Mk 400M. Military version, with accommodation for 46 parachute troops, 6,025 kg (13,283 lb) of freight or 24 stretchers and 9 attendants. Large cargo door and enlarged parachuting door on each side. First flight 24 April 1965. Four Iranian Air Force Mk 400Ms were modified by Fokker in 1977 for target towing duties.

F27 Mk 400M cartographic version. Aerial survey version with two super-wide-angle cameras, remotely controlled from central navigation station, and navigation sight. Inertial navigation system, with digital readout at navigation station and recorded on each picture. Photography through optical glass window panes. Electrically-operated window doors. First flight 24 August 1973.

F27 Mk 500. Similar to F27 Mk 200, but with lengthened fuselage and large cargo door. The 15 aircraft for the French Ministère des Postes et Télécommunications (Air France) have special large doors on both sides. First flight 15 November 1967.

F27 Mk 600. Similar to Mk 200, but with a large cargo door. Does not have the reinforced and watertight flooring of the Combiplane. Can be fitted with quick-change interior, featuring roller tracks and palletised seats and/or cargo pallets. First flight 28 November 1968.

Fokker F27 Mk 500 Friendship in the insignia of Air New Zealand

F27 Maritime. Maritime patrol version; described separately.

By 1 August 1981, total sales by Fokker had reached 530, bringing overall Dutch/US sales to 735. Sales of currently-available models are as follows:

Mk 200 (113 sold. Orders as listed in 1973-74 *Jane's*, except 1 only for Swissair/Balair; plus 1 for Icelandic government, previously recorded as Mk 700; plus 1 corporate.)

Mks 400/600 (217 sold, incl 10 corporate). Orders as listed in 1974-75 *Jane's*, except 5 instead of 6 for Ghana Air Force, plus the following:

AeroPeru (Mk 600)	2
Air Ivoire (Mk 600)	2
Air Niger (Mk 600)	1
Air Tanzania (Mk 600)	3
Algerian government (for Air Algérie)	8
Aramco	2
Argentinian government	1
Bolivian government	6
Burma Airways (Mk 600)	3
Caltex Pacific Indonesia	1
Indonesian government (Mk 400M)	8
Iranian Coppermines	1
Iranian government (Mk 400M/600)	2
Ivory Coast government (Mk 400)	1
Lauda Air (Mk 600)	1
Libyan Arab Airlines (Mk 600)	10
Oasis Oil Co of Libya	2
Occidental of Libya	1
Pelita/Pertamina (Indonesia)	3
Senegal government	6
Somali Airlines (Mk 600)	2
Swift Aire Lines (USA) (Mk 600)	3
TAAG/Angola Airlines	1
Uganda Airlines	2
Undisclosed	2

Mk 500 (91 sold, incl 6 corporate); orders as listed in 1974-75 and 1977-78 *Jane's*, plus:

Air New Zealand	6
Air Zaïre	4
Aramco	1
KLM (NLM CityHopper)	3
Malaysian Airline System	2
Mississippi Valley Airlines	4
Sonangol (Angola)	1
Undisclosed	8

F27 Maritime (11 sold)

Angola	1
Netherlands	2
Peru	2
Philippines	3
Spain	3

Outer wings for the F27 are manufactured in Belgium by SABCA; mid- and aft fuselage sections in France by Dassault-Breguet; dorsal fins, flaps and ailerons in West Germany by MBB.

Any of the current airliner models can be fitted, at customer's option, with a Dowty Rotol very-rough-field landing gear having two-stage oleos with a 100 mm (4 in) increase in stroke, giving increased overall height and propeller ground clearance. Rough-field gear versions are currently in operation with Air Tanzania, Aramco, Burma Airways and Somali Airlines.

Type: Twin-turboprop medium-range airliner.

Wings: Cantilever high-wing monoplane. Wing section NACA 64-421 at root, 64-415 at tip. Dihedral 2° 30′. Incidence 3° 30′. All-metal riveted and metal-bonded two-spar stressed-skin structure, consisting of centre-section and two detachable outer sections. Detachable honeycomb-core sandwich leading-edges with Kléber-Colombes rubber-boot de-icers. GRP trailing-edges. Mechanically-operated single-slotted flaps, divided by engine nacelles. Electrically-operated trim tab in each aileron.

Fuselage: All-metal stressed-skin structure, built to fail-safe principles, with cylindrical portions metal bonded and conical parts riveted. Fuselage is pressurised between rear bulkhead of nosewheel compartment and circular pressure bulkhead aft of the baggage compartment. Length of pressurised section 16·16 m (53 ft 0 in), except for Mk 500 in which the pressurised section is 17·66 m (57 ft 11 in) long. The slightly flattened fuselage bottom is reinforced by underfloor members.

Tail Unit: Cantilever all-metal stressed-skin structure. Fin and tailplane, as well as leading-edges of surfaces, are detachable. Trim tab in each elevator. Kléber-Colombes pneumatic-boot anti-icing.

Landing Gear: Retractable tricycle type. Pneumatic retraction. Dowty oleo-pneumatic shock-absorbers; Dunlop wheels, tyres and brakes. Twin-wheel main units retract backward into engine nacelles. Single-wheel steerable nose unit retracts forward into non-pressurised nosecone. Main-wheel tyre pressure 5·62 bars (81·5 lb/sq in), nosewheel tyre pressure 3·87 bars (56 lb/sq in). Pneumatic brakes on main wheels, with Dunlop Maxaret automatic anti-skid system. Provision on all currently-available models for Dowty Rotol very-rough-field landing gear in which, at 19,730 kg (43,500 lb) AUW, the total stroke in the main gear is lengthened from 305 mm (12 in) to 406 mm (16 in), increasing the aircraft's static height and propeller ground clearance by 76 mm (3 in). Low-pressure main-wheel tyres are fitted, pressure 4·2 bars (61 lb/sq in) below 18,143 kg (40,000 lb) AUW and 4·57 bars (66 lb/sq in) at higher operating weights. Nose unit is of levered-suspension type, with tyre pressure of 3·87 bars (56 lb/sq in).

Power Plant (all current versions): Two Rolls-Royce Dart Mk 536-7R (RDa.7 rating) turboprop engines, each developing 1,596 kW (2,140 shp) plus 2·34 kN (525 lb st) for take-off. Four-blade Dowty Rotol constant-speed propellers. Integral fuel tanks in outer wings, capacity 5,136 litres (1,130 Imp gallons). Optionally, wing bag tanks for an additional 2,289 litres (503·5 Imp gallons) may be fitted. Overwing fuelling, but pressure refuelling optional. Provision for carrying two 950 litre (209 Imp gallon) external fuel tanks under wings. Methyl-bromide fire-extinguishing system with flame detectors.

Accommodation (Mks 200 and 600): Flight compartment seats two pilots side by side, with folding seat for third crew member if required. Main cabin has standard four-abreast seating for 44 passengers at 78/84 cm (31/33 in) pitch; alternative arrangements allow this number to be increased to 48 in Mk 200. Passenger door at rear of cabin, on port side, with toilet opposite. Standard cargo door at front of Mk 200 on port side; large cargo door in same position on Mk 600, with sill at truck-bed level. Cargo holds forward and aft of main cabin, size dependent on interior arrangement.

Accommodation (executive and VIP versions): Can be furnished to customer's specification, but a basic layout is available. In this, the cabin is divided into three sections: a conference room with six seats, a rest room with settee and divan, and a lounge with four seats. Toilet, galley, wardrobe, baggage space and seat for attendant in forward fuselage. Second toilet and baggage space at rear.

Accommodation (Mk 400M): Folding canvas seats, with safety harnesses, along cabin sides for up to 46 paratroops. Toilet and provision for medical supply box or pantry unit at rear. Ambulance version can accommodate 24 USAF-type stretchers, in eight tiers of three, with seats at front and rear for up to nine medical attendants or sitting casualties. All-cargo version fitted with skid strips, tiedown fittings, protection plates and hinged hatracks. Dispatch door on each side of fuselage at rear for dropping supplies and personnel.

Accommodation (Mk 500): Main cabin has standard seating for 52 passengers four abreast at 89·5 cm (35·25 in) seat pitch; alternative layouts enable up to 60 passengers to be carried at 72 cm (28·5 in) pitch.

Systems: Garrett pressurisation and air-conditioning system utilises two Rootes-type engine-driven blowers. Choke heating and air-to-air heat exchanger; optional bootstrap cooling system. Pressure differential 0·29 bars (4·16 lb/sq in) in Mks 500 and 600; 0·38 bars (5·5 lb/sq in) in Mk 200. No hydraulic system. Dunlop pneumatic system, pressure 235 bars (3,400 lb/sq in), for landing gear retraction, nosewheel steering and brakes. Emergency pneumatic circuits for landing gear extension and brakes. Bendix primary 28V electrical system supplied by two 375A 28V DC engine-driven generators. Secondary system supplied via two 115V 400Hz AC constant-frequency inverters. Variable-frequency AC power supply, from 120/208V 15kVA engine-driven alternators, for anti-icing and heating. Two 24V 40Ah nickel-cadmium batteries. 1·12 m³ (39·4 cu ft) oxygen system for pilots.

Avionics and Equipment: VHF and HF transceivers, VHF navigation system (including glideslope), ADF, ILS, marker beacon receiver, dual gyrosyn compass system, Fairchild flight data recorder, intercom system, Bendix weather radar and Smiths autopilot. Marquette windscreen wipers.

Dimensions, external:

Wing span	29·00 m (95 ft 2 in)
Wing chord at root	3·45 m (11 ft 4 in)
Wing chord at tip	1·40 m (4 ft 7 in)
Wing aspect ratio	12
Length overall: except Mk 500	23·56 m (77 ft 3½ in)
Mk 500	25·06 m (82 ft 2½ in)
Fuselage: Max width	2·70 m (8 ft 10¼ in)
Max height	2·79 m (9 ft 1¾ in)
Height overall, standard landing gear:	
except Mk 500	8·50 m (27 ft 11 in)
Mk 500	8·71 m (28 ft 7¼ in)
Height overall, rough-field landing gear:	
except Mk 500	8·59 m (28 ft 2 in)
Tailplane span	9·75 m (32 ft 0 in)
Wheel track (c/l shock-absorbers)	7·20 m (23 ft 7½ in)
Wheelbase: except Mk 500	8·74 m (28 ft 8 in)
Mk 500	9·74 m (31 ft 11¼ in)
Propeller diameter	3·50 m (11 ft 6 in)
Propeller ground clearance:	
standard landing gear:	
except Mk 500	0·94 m (3 ft 1 in)
Mk 500	0·99 m (3 ft 3 in)
rough-field landing gear:	
except Mk 500	1·02 m (3 ft 4¼ in)
Passenger door (rear, port):	
Height	1·65 m (5 ft 5 in)
Width	0·73 m (2 ft 4¾ in)
Height to sill: except Mk 500	1·22 m (4 ft 0 in)
Mk 500	1·39 m (4 ft 6¾ in)
Service/emergency door (rear, stbd):	
Height	1·12 m (3 ft 8 in)
Width	0·74 m (2 ft 5 in)
Height to sill	0·99 m (3 ft 3 in)
Standard cargo door (Mk 200 only):	
Height	1·19 m (3 ft 11 in)
Width	1·04 m (3 ft 5 in)
Height to sill	0·99 m (3 ft 3 in)
Large cargo door (Mks 500 and 600):	
Height	1·78 m (5 ft 10 in)
Width	2·32 m (7 ft 7½ in)
Height to sill: except Mk 500	0·99 m (3 ft 3 in)
Mk 500	1·03 m (3 ft 4½ in)
Dispatch doors (Mk 400M only, rear, port and stbd, each): Height	1·65 m (5 ft 5 in)
Width	1·19 m (3 ft 11 in)
Height to sill	1·22 m (4 ft 0 in)

Dimensions, internal:

Cabin, excl flight deck:	
Length: except Mk 500	14·46 m (47 ft 5 in)
Mk 500	15·96 m (52 ft 4 in)
Max width	2·49 m (8 ft 2 in)
Max height	1·93 m (6 ft 4 in)
Floor area (excl toilet):	
except Mk 500	26·0 m² (280 sq ft)
Mk 500	30·2 m² (325 sq ft)
Volume (excl toilet):	
except Mk 500	56·0 m³ (1,978 cu ft)
Mk 500	65·5 m³ (2,313 cu ft)
Freight hold (fwd), max: Mk 200	4·78 m³ (169 cu ft)
Mks 500, 600	5·58 m³ (197 cu ft)
Freight hold (rear), max:	
all versions	2·83 m³ (100 cu ft)

Areas:

Wings, gross	70·00 m² (753·5 sq ft)
Ailerons (total)	3·51 m² (37·80 sq ft)
Trailing-edge flaps (total)	12·72 m² (136·90 sq ft)
Vertical tail surfaces (total)	14·20 m² (153 sq ft)
Horizontal tail surfaces (total)	16·00 m² (172 sq ft)

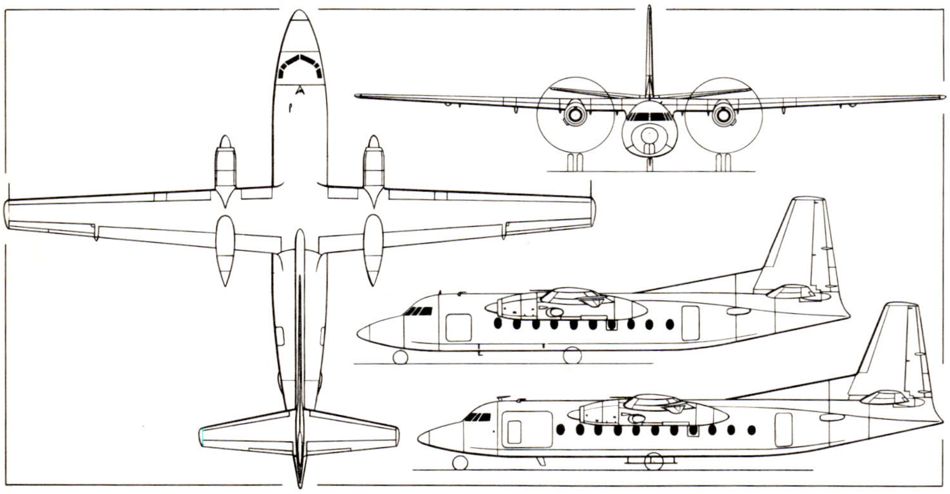

Fokker F27 Friendship Mk 200, with additional side view (bottom) of Friendship Mk 500
(Pilot Press)

Weights and Loadings:
Manufacturer's weight empty:
Mk 200, 44 seats 11,578 kg (25,525 lb)
Mk 400M 11,213 kg (24,720 lb)
Mk 500, 52 seats 12,243 kg (26,992 lb)
Mk 600, 44 seats 11,714 kg (25,825 lb)
Operating weight empty:
Mk 200, 44 seats 12,011 kg (26,480 lb)
Mk 400M, all-cargo 11,479 kg (25,307 lb)
Mk 400M, medical evacuation 11,902 kg (26,240 lb)
Mk 400M, paratrooper 11,655 kg (25,696 lb)
Mk 500, 52 seats 12,684 kg (27,964 lb)
Mk 600, 44 seats 12,148 kg (26,781 lb)
Max payload (weight limited):
Mk 200, 44 seats 5,906 kg (13,020 lb)
Mk 400M, all-cargo 6,438 kg (14,193 lb)
Mk 400M, medical evacuation 6,015 kg (13,260 lb)
Mk 400M, paratrooper 6,261 kg (13,804 lb)
Mk 500, 52 seats 5,233 kg (11,536 lb)
Mk 600, 44 seats 5,769 kg (12,719 lb)
Fuel load:
standard (all versions) 4,123 kg (9,090 lb)
with optional wing bag tanks 5,978 kg (13,180 lb)
Max T-O weight: all versions 20,412 kg (45,000 lb)
Max landing weight:
Mks 200, 400M and 600 18,597 kg (41,000 lb)
Mk 500 19,050 kg (42,000 lb)
optional, all versions 19,731 kg (43,500 lb)
Max zero-fuel weight (all versions) 17,917 kg (39,500 lb)
Max wing loading (all versions) 291·6 kg/m² (59·75 lb/sq ft)
Max power loading (all versions) 6·39 kg/kW (10·5 lb/shp)

Performance (at weights indicated):
Normal cruising speed at 6,100 m (20,000 ft) and AUW of 17,237 kg (38,000 lb):
all versions 259 knots (480 km/h; 298 mph)
Rate of climb at S/L, AUW of 18,143 kg (40,000 lb):
all civil versions 451 m (1,480 ft)/min
military versions 494 m (1,620 ft)/min
Service ceiling at AUW of 17,237 kg (38,000 lb):
all civil versions 8,990 m (29,500 ft)
military versions 9,145 m (30,000 ft)
Service ceiling, one engine out, at AUW of 17,237 kg (38,000 lb):
all civil versions 3,565 m (11,700 ft)
military versions 4,055 m (13,300 ft)
Runway LCN at max T-O weight, standard landing gear: rigid pavement 18
flexible pavement 15
Required T-O field length (ICAO-PAMC) at AUW of 18,143 kg (40,000 lb), all civil versions:
S/L, ISA 988 m (3,240 ft)
S/L, ISA +15°C 1,088 m (3,570 ft)
915 m (3,000 ft), ISA 1,210 m (3,970 ft)
Required T-O field length (military) at AUW of 18,143 kg (40,000 lb), military versions:
S/L, ISA 704 m (2,310 ft)
S/L, ISA +15°C 765 m (2,510 ft)
915 m (3,000 ft), ISA 838 m (2,750 ft)
Required landing field length (ICAO-PAMC) at AUW of 16,329 kg (36,000 lb), all civil versions:
S/L 1,003 m (3,290 ft)
1,525 m (5,000 ft) 1,076 m (3,530 ft)
Required landing field length (military) at AUW of 17,010 kg (37,500 lb), military versions:
S/L 579 m (1,900 ft)
915 m (3,000 ft) 622 m (2,040 ft)
Range (ISA, zero wind conditions) with FAR 121.645 reserves for alternate, 30 min hold at 3,050 m (10,000 ft) and 10% flight fuel:
Mks 200 and 600, 44 passengers 1,020 nm (1,926 km; 1,197 miles)
Mk 500, 52 passengers 935 nm (1,741 km; 1,082 miles)
Military transport range (ISA, zero wind conditions) at max T-O weight, reserves for 30 min hold at S/L and 5% initial fuel:
Mk 400M, all-cargo, max standard fuel 1,195 nm (2,213 km; 1,375 miles)
Mk 400M, all-cargo, max possible fuel 2,370 nm (4,389 km; 2,727 miles)
Military combat radius, conditions as above:
Mk 400M, all-cargo, max standard fuel 625 nm (1,158 km; 719 miles)
Mk 400M, all-cargo, max possible fuel 1,230 nm (2,278 km; 1,416 miles)
Max endurance at 6,100 m (20,000 ft):
Mk 400M, max standard fuel 7 h 25 min
Mk 400M, max possible fuel 12 h 47 min

Operational Noise Levels (FAR Pt 36):
T-O 90·6 EPNdB
Approach: Mk 500 98·9 EPNdB
Mk 600 100·3 EPNdB
Sideline 92·2 EPNdB

Fokker F27 Maritime, one of three ordered by the Philippine Air Force

FOKKER F27 MARITIME

The F27 Maritime is a medium-range maritime patrol version of the Friendship, designed to meet the requirements of various coastal agencies throughout the world which require a cost-effective surveillance aircraft for coastal patrol, fishery protection, search and rescue, and similar offshore duties. The basic design was defined in July 1975, and shortly afterwards Fokker began converting an ex-airline F27 to serve as a prototype/demonstration aircraft (PH-FCX). This prototype made its first flight in February 1976.

A total of 11 F27 Maritimes had been sold by mid-1981, as listed under the preceding main F27 entry.

The duties of the F27 Maritime include patrol of fishery areas and coastal shipping lanes, surveillance of offshore oil industry operations, search and rescue, environmental control and similar duties. It is operated by a crew of up to six persons, and the standard fuel capacity of 9,326 litres (2,051 Imp gallons; 2,463 US gallons) in centre-wing bag tanks and wing pylon tanks gives the aircraft an endurance of 10-12 h, or a range of up to 2,700 nm (5,000 km; 3,107 miles), depending on the mission to be flown.

Type: Twin-turboprop maritime patrol aircraft.

Wings, Fuselage and Tail Unit: As described for F27, except that airframe is heavily treated with anti-corrosive measures; in tail unit, only the port elevator has a trim tab; and 'teardrop' windows are fitted to flight deck.

Landing Gear: As described for F27, but with tyre pressures of 5·52 bars (80 lb/sq in) on main units and 3·80 bars (55 lb/sq in) on nose unit. With long-stroke main gear fitted, pressure in the low-pressure main-wheel tyres is 4·50 bars (65 lb/sq in), and in tyre on the levered-suspension nose unit is 3·80 bars (55 lb/sq in).

Power Plant: Two Rolls-Royce Dart Mk 536-7R (RDa.7 rating) turboprop engines, each developing 1,730 kW (2,320 shp) for T-O. Four-blade Dowty Rotol propellers. Integral fuel tanks in outer wings, total capacity 5,140 litres (1,130 Imp gallons; 1,357 US gallons). Overwing (gravity) and pressure refuelling. Additional centre-wing tank of 2,310 litres (508 Imp gallons; 610 US gallons) capacity, and two 938 litre (206·5 Imp gallon; 248 US gallon) tanks on underwing pylons, giving overall total fuel capacity of 9,326 litres (2,051 Imp gallons; 2,463 US gallons). Methyl bromide fire extinguishing system, with flame detectors. Water-methanol tank in each engine nacelle, combined capacity 303 litres (67 Imp gallons; 80 US gallons).

Accommodation: Flight compartment seats two pilots side by side, with folding seat for third crew member if required. Main cabin fitted out as tactical compartment (for a minimum of two operators), containing advanced avionics, galley, toilet and crew rest area. Bubble windows for observers are provided aft in main cabin. Rear cabin door is openable in flight. Standard cargo door at front on port side, with sill at truckbed height. Cargo holds forward and aft of main cabin.

Systems: Generally as described for F27, except that bootstrap cooling system is standard; cabin pressure differential is 0·38 bars (5·5 lb/sq in); secondary electrical system has a third 115V 400Hz AC constant-frequency inverter; and oxygen system includes individual supply for each tactical crew member.

Avionics and Equipment: Com/nav equipment comprises Collins 618T-3 HF transceiver, two Collins 618M-3 VHF transceivers, Collins AN/ARC-159 UHF transceiver, interphone, Litton LTN-72 inertial navigation system, IDC air data computer, dual Sperry C-9 gyro compasses, Collins DF-206 radio compass, Collins 51Z-4 marker beacon receiver, Honeywell AN/APN-198 radar altimeter, Collins DF-301E VHF/UHF direction finder, two Collins 51RV-4 VOR/ILS receivers, two Collins HSIs, Smiths SEP-2E/M autopilot, Collins 621A-6A ATC transponder. Operational equipment includes Litton AN/APS-504(V)-2 search radar, with 360° coverage, mounted in ventral radome.

Dimensions: As for F27 Mks 200/400/600, except:
Height overall 8·70 m (28 ft 6½ in)
Cabin volume (excl flight deck) 60·5 m³ (2,136 cu ft)

Weights and Loadings:
Manufacturer's weight empty 12,519 kg (27,600 lb)
Operating weight empty 13,314 kg (29,352 lb)
Normal max T-O weight 20,410 kg (45,000 lb)
Operational necessity weight 21,320 kg (47,500 lb)
Max landing weight 18,600 kg (41,000 lb)
Max zero-fuel weight 17,900 kg (39,500 lb)
Max wing loading 291·5 kg/m² (59·7 lb/sq ft)
Max power loading 6·39 kg/kW (10·5 lb/shp)

Performance (at normal max T-O weight except where indicated):
Never-exceed speed, AUW of 17,237 kg (38,000 lb), S/L to 6,100 m (20,000 ft) 259 knots (480 km/h; 298 mph) CAS
Normal cruising speed at 6,100 m (20,000 ft), AUW of 17,237 kg (38,000 lb) 250 knots (463 km/h; 287 mph)
Normal operating speed at 6,100 m (20,000 ft), AUW of 17,237 kg (38,000 lb) 227 knots (420 km/h; 261 mph) CAS
Patrol speed at 457 m (1,500 ft) 150-180 knots (277-333 km/h; 172-207 mph)
Stalling speed, flaps up 96 knots (178 km/h; 111 mph) CAS
Max rate of climb at S/L, ISA, AUW of 18,143 kg (40,000 lb) 442 m (1,450 ft)/min
Service ceiling 8,990 m (29,500 ft)
Service ceiling, one engine out 3,565 m (11,700 ft)
Runway LCN (42% tyre deflection) at 15,875 kg (35,000 lb) AUW:
rigid pavement, L = 76·2 cm (30 in) 10·4
flexible pavement, h = 25·4 cm (10 in) 11·4
flexible pavement, h = 12·7 cm (5 in) 9·0
Runway LCN (42% tyre deflection) at 20,410 kg (45,000 lb) AUW:
rigid pavement, L = 76·2 cm (30 in) 16·0
flexible pavement, h = 25·4 cm (10 in) 14·8
flexible pavement, h = 12·7 cm (5 in) 12·0
Runway CBR, unpaved soil, h = 25·4 cm (10 in), 3,000 passes:
AUW of 15,875 kg (35,000 lb) 6·2%
AUW of 20,410 kg (45,000 lb) 7·8%
T-O run at S/L: ISA 975 m (3,200 ft)
ISA + 20°C 1,080 m (3,545 ft)
Landing distance (unfactored, ISA at S/L):
AUW of 19,731 kg (43,500 lb) 610 m (2,000 ft)
AUW of 13,607 kg (30,000 lb) 530 m (1,740 ft)
Transport range at 6,100 m (20,000 ft) with 4,536 kg (10,000 lb) payload, 30 min loiter and 5% reserves 1,000 nm (1,850 km; 1,150 miles)

FOKKER F28 FELLOWSHIP

The F28 Fellowship twin-turbofan short/medium-haul transport was developed in collaboration with other European aircraft manufacturers and with pre-financing from the Netherlands Agency for Aerospace Programmes and through a loan guaranteed by the Dutch government.

Production is undertaken by Fokker in association with MBB in Germany and Short Bros in the UK. Fokker is responsible for the front fuselage, to a point just aft of the flight deck, the centre fuselage and wing-root fairings. MBB builds the cylindrical fuselage section between the wing leading-edge and flight deck, the fuselage from the wing trailing-edge to the rear pressure bulkhead, the rear fuselage and tail unit, and the engine nacelles and support stubs. Shorts are responsible for wings and other components, including main-wheel and nosewheel doors.

First flight of the first prototype F28 (PH-JHG) was made on 9 May 1967, and the second prototype, PH-WEV, flew on 3 August 1967. The third F28 (PH-MOL) flew for the first time on 20 October 1967 and was brought up to production standard in the early Summer of 1968.

The Dutch RLD granted a C of A to the F28 on 24 February 1969, and the first delivery (of the fourth aircraft, to LTU) was made on the same day. The aircraft received FAA type approval on 24 March 1969, German certification on 30 March 1969 and British CAA type approval in June 1979. RLD certification for operation from unpaved runways was granted in mid-1972. The Mk 1000 was granted FAA-approved noise certification on 31 December 1971. The 100th Fellowship, a Mk 1000 for the Peruvian Air Force, was delivered on 30 March 1976.

Fokker F28 Mk 4000 Fellowship short/medium-range transport, in the insignia of Altair Airlines of Philadelphia

A total of 182 Fellowships had been ordered by 9 June 1981, as follows:

Mk 1000/1000C (97 ordered, incl 7 Mk 1000C; orders as listed in 1975-76 *Jane's,* plus 5 Mk 1000 for Garuda)

Mk 2000 (10 ordered; orders as listed in 1975-76 *Jane's*)

Mk 3000

Argentinian Air Force	3
Cimber Air (Denmark)	2
Garuda (Indonesia)	7
Ghana Air Force	1
Royal Swazi National Airways	1
Tanzanian People's Defence Force Air Wing	1
Undisclosed	1

Mk 4000

Aerolineas Argentinas	1
Air Anglia (UK)	2
Altair Airlines (USA)	10
Bangladesh Biman	2
Burma Airways	1
East-West Airlines (Australia)	2
Empire Airlines	6
Garuda (Indonesia)	8
GATL/Air Ivoire	3
Iran Asseman Airlines-Pars Air	2
KLM-NLM	4
Linjeflyg (Sweden)	10
Nigeria Airways	2
Pelita Air Service (Indonesia)	3
Undisclosed	3

Seven versions have been announced. Details of the Mks 1000 and 2000 can be found in the 1976-77 *Jane's*; of the Mk 5000 in the 1975-76 edition; and of the Mk 6000 in the 1979-80 edition. The Mk 6600 project (1979-80 *Jane's*) was abandoned.

The two current production versions are as follows:

Mk 3000. Similar to Mk 4000, but with short fuselage seating up to 65 passengers. Available also in 15-passenger VIP or executive layout, with range of up to 2,200 nm (4,074 km; 2,533 miles). Two of those ordered by Garuda and two for the Argentinian Air Force have the optional large cargo door.

Mk 4000. High-density long-fuselage version, first flown on 20 October 1976, to seat up to 85 passengers at 74 cm (29 in) pitch. Two additional overwing emergency exits (making a total of four).

All current production Mk 3000s and Mk 4000s have uprated RB.183 Mk 555-15P engines, higher design weights, to enable more fuel to be carried, a strengthened wing of increased span, an engine noise reduction kit, new-look cabin decor and improved flight deck layout. From mid-1981 all F28 engines have been fitted with a new ten-lobe internal mixer, which reduces the aircraft's flyover noise footprint by approx 50 per cent. In addition, the new device permits a slight increase in take-off thrust, increased maximum attainable cruising flight levels, and a weight saving of 63 kg (140 lb). More importantly, use of the new internal mixer can reduce F28 block fuel consumption by 5·5 per cent on a sector of 300 nm (555 km; 345 miles), and by 6·1 per cent on an 800 nm (1,480 km; 920 mile) sector.

The following details apply to both the Mk 3000 and Mk 4000, except where a specific model is indicated:

Type: Twin-turbofan short/medium-range airliner.

Wings: Cantilever low/mid-wing monoplane. Wing section NACA 0000-X 40Y series with camber varying along span. Thickness/chord ratio up to 14% on inner panels, 10% at tip. Dihedral 2° 30′. Sweepback at quarter-chord 16°. Single-cell two-spar light alloy torsion-box structure, comprising centre-section, integral with fuselage, and two outer panels. Fail-safe construction. Lower skin made of three planks. Taper-rolled top skin. Forged ribs in centre-section, built-up ribs in outer panels. Double-skin leading-edge with ducts for hot-air de-icing. Irreversible hydraulically-operated ailerons. Emergency manual operation of ailerons, through tabs. Hydraulically-operated Fowler double-slotted flaps over 70% of each half-span with electrical emergency extension. Five-section hydraulically-operated lift dumpers in front of flaps on each wing. Trim tab in each aileron.

Fuselage: Circular-section semi-monocoque light alloy fail-safe structure, made up of skin panels with Redux-bonded Z-stringers. Bonded doubler plates at door and window cutouts. Quickly-detachable sandwich (metal/end-grain balsa) floor panels. Hydraulically-operated petal airbrakes form aft end of fuselage.

Tail Unit: Cantilever light alloy structure, with hydraulically-actuated variable-incidence T tailplane. Electrical emergency actuation of tailplane. Hydraulically-boosted elevators. Hydraulically-operated rudder with duplicated actuators and emergency manual operation. Honeycomb sandwich skin panels used extensively, in conjunction with multiple spars. Double-skin leading-edges for hot-air de-icing.

Landing Gear: Retractable tricycle type of Dowty Rotol manufacture, with twin wheels on each unit. Hydraulic retraction, nosewheels forward, main units inward into fuselage. Oleo-pneumatic shock-absorbers. Goodyear wheels, tyres and electronically-controlled braking system. Steerable nosewheel. Main-wheel tyres size 40 × 14, 16-ply rating, pressure 5·17 bars (75 lb/sq in).

Power Plant: Two Rolls-Royce RB.183-2 Mk 555-15P turbofan engines with blade-cooling (each 44 kN; 9,900 lb st, flat rated to 29·7°C), mounted in pod on each side of rear fuselage and fitted with acoustic intake liner and internal mixer. No water injection or thrust reversers. Thermal anti-icing for air intakes. Integral fuel tank in each outer wing panel with total usable capacity of 9,740 litres (2,143 Imp gallons). Optional seven bladder-type tank units in wing centre-section with total usable capacity of 3,300 litres (726 Imp gallons). Single refuelling point under starboard wing, near root.

Accommodation: Crew of two side by side on flight deck, with jump-seat for third crew member. Electrically-heated windscreen. Pantry/baggage space immediately aft of flight deck on starboard side, followed by entrance lobby with hydraulically-operated airstair door on port side, service and emergency door on starboard side, and seat for cabin attendant. Additional emergency door on each side of main cabin, over wing (two each side on Mk 4000). Main cabin layout of Mk 3000 can be varied to accommodate 55, 60 or 65 passengers five abreast at 94, 81/84 or 79 cm (37, 32/33 or 31 in) seat pitch respectively. In Mk 4000, layout can accommodate up to 85 passengers at 74 cm (29 in) pitch. Aft of cabin are a wardrobe (port), baggage compartment (port) and toilet compartment (starboard). Underfloor cargo compartments fore and aft of wing, with single door on starboard side of forward hold, with one door on rear hold of each version.

Systems: Garrett air-conditioning system, using engine bleed air. Max pressure differential 0·51 bars (7·45 lb/sq in). Two independent hydraulic systems, pressure 207 bars (3,000 lb/sq in). Primary system for flight controls, landing gear, nosewheel steering and brakes; secondary system for duplication of certain essential flight controls. Westinghouse all-AC electrical system utilises two 20kVA engine-driven generators to supply three-phase constant-frequency 115/200V 400Hz power. One 20Ah battery for starting APU and for emergency power. Garrett GTCP 36-4A APU, mounted aft of rear pressure bulkhead, for engine starting, ground air-conditioning and ground electrical power, and to drive a third AC generator for standby use on essential services in flight.

Avionics and Equipment: Standard avionics include Collins VHF transceivers, Collins VHF navigation system (with glideslope), DME, marker beacon receiver, RCA weather radar, ADF, ATC transponder, dual compass system, interphone and public address systems, Smiths SEP6 autopilot, Collins FD 108 flight director, flight

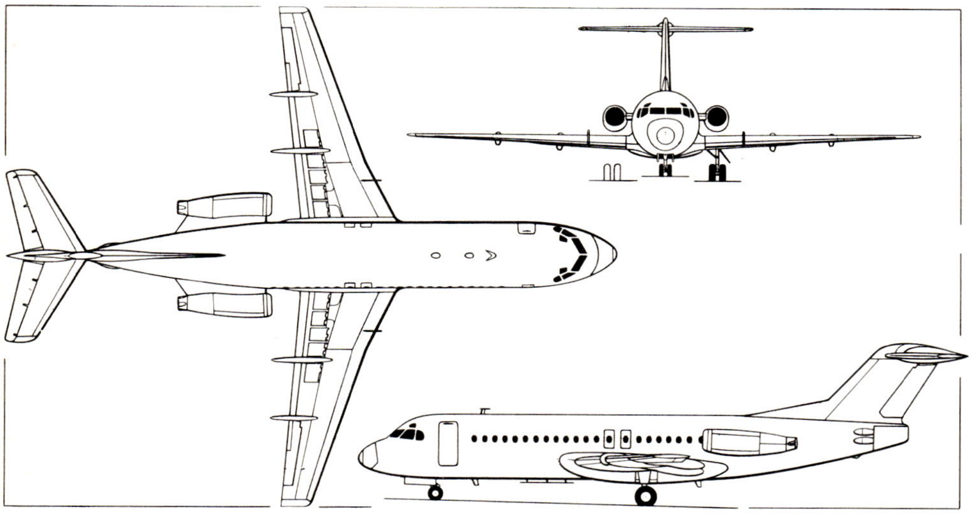

Fokker F28 Mk 4000 Fellowship high-density short/medium-range transport aircraft *(Pilot Press)*

guidance caution system, Fairchild flight data recorder and Fairchild voice recorder. Menasco powered flight controls. Thermal bleed air system for wing leading-edges, tailplane leading-edge and engine air intakes. Optional equipment to customer's requirements, including equipment for operation in Cat. II weather minima.

DIMENSIONS, EXTERNAL:
Wing span 25·07 m (82 ft 3 in)
Wing chord at root 4·80 m (15 ft 9 in)
Length overall: 3000 27·40 m (89 ft 10¾ in)
4000 29·61 m (97 ft 1¾ in)
Length of fuselage: 3000 24·55 m (80 ft 6½ in)
4000 26·76 m (87 ft 9½ in)
Fuselage: Max width 3·30 m (10 ft 10 in)
Height overall 8·47 m (27 ft 9½ in)
Tailplane span 8·64 m (28 ft 4¼ in)
Wheel track (c/l of shock-absorbers)
5·04 m (16 ft 6½ in)
Wheelbase: 3000 8·90 m (29 ft 2½ in)
4000 10·35 m (33 ft 11½ in)
Passenger door (fwd, port):
Height 1·93 m (6 ft 4 in)
Width 0·86 m (2 ft 10 in)
Service/emergency door (fwd, stbd):
Height 1·27 m (4 ft 2 in)
Width 0·61 m (2 ft 0 in)
Emergency exits (centre, each):
Height 0·91 m (3 ft 0 in)
Width 0·51 m (1 ft 8 in)
Freight hold doors (each):
Height (fwd, each) 0·90 m (2 ft 11½ in)
Height (aft) 0·80 m (2 ft 7½ in)
Width (fwd, each) 0·95 m (3 ft 1½ in)
Width (aft) 0·89 m (2 ft 11 in)
Height to sill (fwd, each) 1·47 m (4 ft 10 in)
Height to sill (aft) 1·59 m (5 ft 2½ in)
Baggage door (rear, port, optional):
Height 0·60 m (1 ft 11½ in)
Width 0·51 m (1 ft 8 in)
Cargo door (fwd, port, optional):
Height 1·87 m (6 ft 1¾ in)
Width 2·49 m (8 ft 2 in)
Height to sill 2·24 m (7 ft 4¼ in)

DIMENSIONS, INTERNAL:
Cabin, excl flight deck:
Length: 3000 13·10 m (43 ft 0 in)
4000 15·31 m (50 ft 3 in)
Max length of seating area:
3000 10·74 m (35 ft 2¾ in)
4000 12·95 m (42 ft 6¾ in)
Max width 3·10 m (10 ft 2 in)
Max height 2·02 m (6 ft 7¼ in)
Floor area: 3000 38·4 m² (413·3 sq ft)
4000 44·8 m² (482·2 sq ft)
Volume: 3000 71·5 m³ (2,525 cu ft)
4000 83·0 m³ (2,931 cu ft)
Freight hold (underfloor, fwd):
3000 6·90 m³ (245 cu ft)
4000 8·70 m³ (308 cu ft)
Freight hold (underfloor, rear):
3000 3·80 m³ (135 cu ft)
4000 4·84 m³ (171 cu ft)
Baggage hold (aft of cabin), max:
3000, 4000 2·30 m³ (81·22 cu ft)

AREAS:
Wings, gross 79·00 m² (850 sq ft)
Ailerons (total) 2·67 m² (28·74 sq ft)
Trailing-edge flaps (total) 14·00 m² (150·7 sq ft)
Fuselage airbrakes (total) 3·62 m² (38·97 sq ft)
Fin (incl dorsal fin) 12·30 m² (132·4 sq ft)
Rudder 2·30 m² (24·6 sq ft)
Tailplane 19·50 m² (209·9 sq ft)
Elevators (total) 3·84 m² (41·33 sq ft)

WEIGHTS AND LOADINGS:
Operating weight empty:
3000, 65 seats 16,781 kg (36,997 lb)
4000, 85 seats 17,546 kg (38,683 lb)
Max weight-limited payload:
3000 8,620 kg (19,003 lb)
4000 10,576 kg (23,317 lb)
Fuel load: standard 7,820 kg (17,240 lb)
with optional centre-section tanks
10,469 kg (23,080 lb)
Max T-O weight:
3000, 4000 33,110 kg (73,000 lb)
Max zero-fuel weight:
3000 25,400 kg (56,000 lb)
4000 28,122 kg (62,000 lb)
Max landing weight: 3000 29,030 kg (64,000 lb)
4000 31,524 kg (69,500 lb)
Max cabin floor loading:
all passenger versions 366 kg/m² (75 lb/sq ft)
Max wing loading:
3000, 4000 407 kg/m² (83·4 lb/sq ft)
Max power loading:
3000, 4000 367·5 kg/kN (3·6 lb/lb st)

PERFORMANCE (Mks 3000 and 4000 at AUW of 29,000 kg; 63,934 lb, ISA, except where indicated):
Never-exceed speed
Mach 0·83 (390 knots; 723 km/h; 449 mph) EAS
Max permissible operating speed
Mach 0·75 (330 knots; 611 km/h; 380 mph) EAS
Max cruising speed at 7,000 m (23,000 ft)
455 knots (843 km/h; 523 mph)
Econ cruising speed at 9,150 m (30,000 ft)
366 knots (678 km/h; 421 mph)
Max cruising altitude 10,675 m (35,000 ft)
Min ground turning radius:
3000 9·60 m (31 ft 6 in)
4000 10·90 m (35 ft 9 in)
Runway LCN at max T-O weight:
rigid pavement 25
flexible pavement 20
FAR T-O field length at max T-O weight:
S/L 1,585 m (5,200 ft)
S/L, ISA + 10°C 1,635 m (5,364 ft)
S/L, ISA + 15°C 1,710 m (5,610 ft)
610 m (2,000 ft), ISA 1,710 m (5,610 ft)
915 m (3,000 ft), ISA 1,820 m (5,970 ft)
FAR landing field length at max landing weight:
S/L 1,065 m (3,495 ft)
1,525 m (5,000 ft) 1,276 m (4,185 ft)
Range, high-speed schedule, FAR 121.645 reserves:
*3000, 65 passengers
1,480 nm (2,743 km; 1,704 miles)
4000, 85 passengers
1,025 nm (1,900 km; 1,180 miles)
Range, long-range schedule, FAR 121.645 reserves:
*3000, 65 passengers
1,710 nm (3,169 km; 1,969 miles)
4000, 85 passengers
1,125 nm (2,085 km; 1,295 miles)
With wing centre-section tanks

OPERATIONAL NOISE LEVELS (ICAO Annex 16):
T-O 86·3 EPNdB
Approach 94·0 EPNdB
Sideline 99·9 EPNdB

NEW ZEALAND

AEROSPACE
NEW ZEALAND AEROSPACE INDUSTRIES LIMITED

HEAD OFFICE AND WORKS: Hamilton Airport, R.D.2, Hamilton
Telephone: Hamilton 436 144 and 436 069
Telex: NZAIL 21242
GENERAL MANAGER: P. W. Goldsbro'
CHIEF DESIGNER: P. W. C. Monk
MARKETING MANAGER: H. J. I. Baxter
ENGINEERING MANAGER: L. C. Burrows
COMPANY SECRETARY: J. D. Linch

Aero Engine Services Ltd and Air Parts (NZ) Ltd (see 1972-73 *Jane's*) amalgamated on 1 April 1973 to form New Zealand Aerospace Industries Ltd. This company has a share capital of $NZ 1·6 million, half of which is held by Air New Zealand (representing government support), one-third by New Zealand's largest agricultural operator, and the remainder by shareholders of the two constituent companies.

The company produces the piston-engined Fletcher FU-24-954 and turboprop-powered Cresco agricultural aircraft, in a 5,574 m² (60,000 sq ft) main facility located at Hamilton Airport.

AEROSPACE AIRTRAINER CT4A

In 1981, New Zealand Aerospace Industries began remanufacturing, to RAAF specifications, 14 Airtrainer CT4A all-metal light piston-engined trainers. The aircraft had been held in store since their export to Rhodesia was embargoed.

A full description of the Airtrainer can be found in the 1976-77 *Jane's*.

AEROSPACE FLETCHER FU-24-954

The FU-24 was developed by the Sargent-Fletcher Company of El Monte, California, USA, initially for agricultural top-dressing work in New Zealand. The prototype flew in July 1954, followed by the first production aircraft five months later. Type certification was granted on 22 July 1955. All manufacturing and sales rights for the FU-24 were acquired by Air Parts (NZ) Ltd in 1964, and production was undertaken subsequently in that company's factory at Hamilton Airport, New Zealand.

The initial production series of 100 was delivered to New Zealand operators for top-dressing work. By June 1979, a total of 272 completed Fletcher aircraft had been produced, including 56 for export to Australia, Bangladesh, Dubai, Iraq, Pakistan, Thailand, Uruguay and the USA.

Aerospace Fletcher FU-24-954, fitted with an Easton hopper box under the fuselage and showing the now-standard large port-side door

Up to early 1979 all foreign deliveries had been effected by solo delivery flights, but export then began of PKD and CKD (parts knock-down and component knock-down) aircraft to Frontier-Aerospace Inc, a subsidiary in Long Beach, California, USA. The FU-24-954 is known in the USA by the name **TaskMaster**; a military utility version, known as the **Pegasus I**, has been developed by Frontier-Aerospace (which see).

The current model, to which the following description applies, is the FU-24-954. This has a large cargo door as standard, the earlier small door (see 1979-80 *Jane's*) having been discontinued.

TYPE: Agricultural and general-purpose aircraft.

WINGS: Cantilever low-wing monoplane. Constant-chord non-swept wings of NACA 4415 section. Dihedral (outer panels) 8°. Incidence 2°. Light alloy two-spar structure. Plain horn-balanced ailerons and single-slotted flaps, all of light alloy construction. Single row of vortex generators forward of each aileron.

FUSELAGE: All-metal semi-monocoque structure. Cockpit area stressed for 25*g* impact.

TAIL UNIT: Cantilever light alloy structure. All-movable horizontal tail with full-span anti-servo and trim tab. Ground-adjustable tab on rudder.

LANDING GEAR: Non-retractable tricycle type, with steerable nosewheel. Fletcher air-oil shock-absorber struts. Cleveland wheels and hydraulic disc brakes on main units. Goodyear tyres, size 8·50-6 (6-ply), pressure range 0·76-2·07 bars (11-30 lb/sq in).

POWER PLANT: One 298 kW (400 hp) Avco Lycoming IO-720-A1A or A1B flat-eight engine, driving a Hartzell HC-C3YR-1R/847SR three-blade constant-speed variable-pitch metal propeller with spinner. Fuel tanks

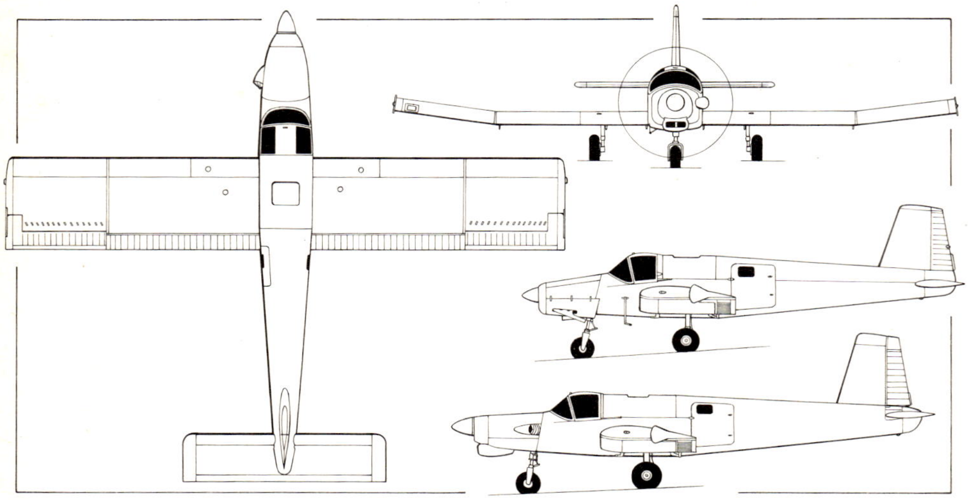

Aerospace Cresco 600 agricultural aircraft, with additional side view (centre) of Fletcher FU-24-954
(Michael A. Badrocke)

in wing leading-edges; total usable capacity 254 litres 67 US gallons) normal, 481 litres (127 US gallons) with optional long-range tanks.

Accommodation (Agricultural models): Enclosed cockpit for pilot and one passenger on side-by-side seats under rearward-sliding canopy. Cockpit reinforced for overturn/crash protection. Large port-side cargo door, optional on early production aircraft, is now standard. Optional features include additional cargo floor area and dual controls.

Accommodation (Utility models): Enclosed cabin for pilot and up to seven passengers or equivalent freight. Dual controls optional. Rearward-sliding hood over front two seats. Large passenger/cargo door on port side.

Agricultural Equipment: Glassfibre hopper aft of cockpit, capacity 1,045 litres (230 Imp gallons; 276 US gallons) of liquid, 1,066 kg (2,350 lb) of dry chemicals. Hopper outlets for spreading of solids (fertiliser, dry ice, poison bait, etc). Transland Swathmaster for top-dressing, seeding and high-volume spraying. Transland Boommaster for liquid spraying with booms, nozzles, fan-driven pump, etc, for low- and high-volume spraying. Micronair spraying equipment with electrically- or fan-driven pump, varied control systems, side-loading valve for liquids, and special adaptor plate for interchangeability of equipment.

Optional Equipment (all models): Full blind-flying instrumentation with ADF, VHF, VOR and DME. Full dual controls; dual main wheels and brakes, wheel and leg fairings; long-range fuel tanks; cabin heating and air-conditioning systems; metric instrumentation.

Dimensions, external:
- Wing span: 12·81 m (42 ft 0 in)
- Wing chord (constant): 2·13 m (7 ft 0 in)
- Wing aspect ratio: 6
- Length overall: 9·70 m (31 ft 10 in)
- Height to top of fin: 2·84 m (9 ft 4 in)
- Fuselage: Max width: 1·22 m (4 ft 0 in)
  - Max depth: 1·52 m (5 ft 0 in)
- Tailplane span: 4·22 m (13 ft 10 in)
- Wheel track: 3·71 m (12 ft 2 in)
- Wheelbase: 2·28 m (7 ft 6 in)
- Propeller diameter: 2·18 m (7 ft 2 in)
- Propeller ground clearance: 0·33 m (12·8 in)
- Passenger/cargo door (port, rear):
  - Height: 0·97 m (3 ft 2 in)
  - Width: 0·94 m (3 ft 1 in)

Dimensions, internal:
- Cabin: Length: 3·18 m (10 ft 5 in)
  - Max width: 1·22 m (4 ft 0 in)
  - Max height: 1·27 m (4 ft 2 in)
  - Floor area: 3·87 m² (41·7 sq ft)
  - Volume aft of hopper: 3·37 m³ (119·0 cu ft)
- Hopper volume: 1·22 m³ (43·0 cu ft)

Areas:
- Wings, gross: 27·31 m² (294 sq ft)
- Ailerons (total): 1·82 m² (19·6 sq ft)
- Trailing-edge flaps (total): 3·16 m² (34·0 sq ft)
- Fin: 1·26 m² (13·6 sq ft)
- Rudder: 0·64 m² (6·9 sq ft)
- Tailplane: 4·00 m² (43·1 sq ft)
- Tailplane tab: 0·45 m² (4·9 sq ft)

Weights and Loadings:
- Weight empty, equipped: 1,188 kg (2,620 lb)
- Max payload (agricultural): 1,052 kg (2,320 lb)
- Normal max T-O weight: 2,204 kg (4,860 lb)
- Max agricultural T-O weight: 2,463 kg (5,430 lb)
- Cabin floor loading: 1,885 kg/m² (386 lb/sq ft)
- Normal wing loading: 80·6 kg/m² (16·5 lb/sq ft)
- Normal power loading: 7·40 kg/kW (12·15 lb/hp)

Performance (at Normal max T-O weight):
- Never-exceed speed: 143 knots (265 km/h; 165 mph)
- Max level speed at S/L: 126 knots (233 km/h; 145 mph)
- Max cruising speed (75% power): 113 knots (209 km/h; 130 mph)
- Operating speed for spraying (75% power): 90-115 knots (167-212 km/h; 104-132 mph)
- Stalling speed:
  - flaps up: 55 knots (102 km/h; 63·5 mph)
  - flaps down: 49 knots (91 km/h; 57 mph)
- Max rate of climb at S/L: 280 m (920 ft)/min
- Service ceiling: 4,875 m (16,000 ft)
- T-O run: 244 m (800 ft)
- T-O to 15 m (50 ft): 372 m (1,220 ft)
- Landing from 15 m (50 ft): 390 m (1,280 ft)
- Landing run: 207 m (680 ft)
- Swath width (agricultural models):
  - oily: 23 m (75 ft)
  - aqueous: 21·3-24·4 m (70-80 ft)
  - dust: 7·6-15·2 m (25-50 ft)
- Range with max normal fuel, 45 min reserves: 383 nm (709 km; 441 miles)

AEROSPACE CRESCO 600

Design of this turboprop development of the FU-24 began in 1977. Construction of a prototype started in the following year, and this aircraft (ZK-LTP) first flew on 28 February 1979. Powered by an Avco Lycoming LTP 101-600 turboprop engine, and having a cabin approx 0·61 m (2 ft) longer, the Cresco 600 has many components interchangeable with the FU-24-954. Use of the LTP 101 engine, together with some structural refinement, permits a reduction in empty weight and a substantial increase in agricultural max T-O weight. The name Cresco is Latin for 'I grow'.

Aerospace Cresco 600 first prototype during an early test flight

Despite the loss of the prototype in a flying accident, certification of the Cresco 600 was well under way in 1980, and the first production aircraft had flown by the Summer of that year. Five Cresco 600s had been ordered by the beginning of 1980.

A military utility version known as the **Pegasus II** is being offered by Frontier-Aerospace Inc (see US section), with either a 462 kW (620 shp) Pratt & Whitney Aircraft of Canada PT6A-28 turboprop engine or the standard Avco Lycoming LTP 101. Applications include cargo and personnel transport, patrol and reconnaissance, medevac, defoliation, parachute and paramedic operations, aerial mapping and surveillance, tactical ground support and light attack. The civil version of the Cresco is known in the USA as the **AgMaster**.

The description of the FU-24-954 applies also to the Cresco 600, except in the following respects:

Type: Turboprop-powered agricultural and general-purpose aircraft.

Wings: As FU-24-954, constructed mainly of 2014, 2024 and 6061 light alloys. Ground-adjustable tab in each aileron.

Fuselage: Similar to FU-24-954, with slight increase in length.

Tail Unit: As FU-24-954, but larger and constructed mainly of 2024, 6061 and L65 light alloys.

Landing Gear: As FU-24-954 but with NZ Aerospace Industries air-oil shock-absorber struts; tyres size 8·50-6 on nosewheel and size 8·50-10 on main wheels.

Power Plant: One 447 kW (600 shp) Avco Lycoming LTP 101-600A-1A turboprop engine, driving a Hartzell HC-B3TN-3D/T10282 H three-blade constant-speed metal propeller with spinner. Four fuel tanks in wing centre-section, total capacity 545·5 litres (120 Imp gallons). Two refuelling points in upper surface of each wing. Oil capacity 5·5 litres (1·2 Imp gallons). Variable-geometry inertia separating intake under nose.

Accommodation: Pilot only, or crew of two side by side, under rearward-sliding bulged cockpit hood. Tinted windscreen and canopy side panels standard. Dual controls available optionally. Large forward-hinged door, with window, aft of wing on port side. Generous cargo space immediately aft of hopper. Cockpit ventilated; heating system optional.

Systems: No air-conditioning, pressurisation, hydraulic, pneumatic or oxygen systems. Electrical system powered by 24V 150A Auxilec starter/generator and two 24V 25Ah lead-acid batteries.

Avionics: Range of Narco or Becker avionics available, including VHF, VOR, ADF and transponder. Stall warning system standard.

Agricultural Equipment: Generally similar to FU-24-954, except for substantially larger hopper, increasing capacity to 1,777 litres (391 Imp gallons) of liquid or 1,860 kg (4,100 lb) of dry chemical. Range of dispersal systems available to customer's requirements, from ultra-high-volume solids dispersal to ultra-low-volume spray.

Dimensions, external: As FU-24-954 except:
- Length overall: 11·06 m (36 ft 3¼ in)
- Length of fuselage: 10·74 m (35 ft 2¾ in)
- Height overall: 3·42 m (11 ft 2½ in)
- Tailplane span: 4·61 m (15 ft 1½ in)
- Wheelbase: 2·77 m (9 ft 1¼ in)
- Propeller diameter: 2·59 m (8 ft 6 in)
- Propeller ground clearance (static): 0·38 m (1 ft 3 in)
- Cargo door (port): Height: 0·94 m (3 ft 1 in)
  - Width: 0·94 m (3 ft 1 in)
  - Height to sill: 0·91 m (3 ft 0 in)

Dimensions, internal:
- Cargo compartment volume (aft of hopper): 3·40 m³ (120·0 cu ft)
- Hopper volume: 1·77 m³ (62·5 cu ft)

Areas:
- Wings, gross: 27·31 m² (294·0 sq ft)
- Ailerons (total): 2·08 m² (22·4 sq ft)
- Trailing-edge flaps (total): 3·06 m² (32·9 sq ft)
- Fin: 1·53 m² (16·5 sq ft)
- Rudder, incl tab: 0·63 m² (6·8 sq ft)
- Tailplane, incl tab: 5·08 m² (54·7 sq ft)

Weights and Loadings:
- Weight empty, equipped: 1,161 kg (2,560 lb)
- Max disposable load (Agricultural, incl fuel): 1,929 kg (4,254 lb)
- Max fuel load: 435 kg (960 lb)
- Max T-O weight: Normal: 2,925 kg (6,450 lb)
  - Agricultural: 3,175 kg (7,000 lb)
- Max landing weight: 2,925 kg (6,450 lb)
- Max wing loading:
  - Normal: 107·07 kg/m² (21·94 lb/sq ft)
  - Agricultural: 116·19 kg/m² (23·81 lb/sq ft)
- Max power loading:
  - Normal: 6·54 kg/kW (10·75 lb/shp)

PERFORMANCE (at max Normal T-O weight, ISA, except where indicated):
Never-exceed speed 177 knots (328 km/h; 204 mph)
Max level speed at S/L 148 knots (274 km/h; 170 mph)
Max cruising speed (75% power) 127 knots (235 km/h; 146 mph)
Stalling speed at 2,767 kg (6,100 lb) AUW, flaps down, power off 52 knots (96·5 km/h; 60 mph)
Max rate of climb at S/L 290 m (950 ft)/min
Service ceiling 6,400 m (21,000 ft)
T-O to 15 m (50 ft) 436 m (1,430 ft)
Landing from 15 m (50 ft) 500 m (1,640 ft)
Range with max fuel, no reserves 460 nm (852 km; 529 miles)

PAKISTAN

PAC
PAKISTAN AERONAUTICAL COMPLEX

HEAD OFFICE: Director General of Aeronautical Projects, 273 Abid Majid Road, Rawalpindi Cantt
Telephone: 62688
WORKS: Aircraft Manufacturing Factory (P-751), Pakistan Aeronautical Complex, Kamra, District Attock
Telex: 5601

Established with considerable assistance from the People's Republic of China, the Pakistan Aeronautical Complex is responsible for overhaul and maintenance of aircraft of the Pakistan armed forces, including the Shenyang J-6 (Chinese-built MiG-19) and Dassault Mirage III/5. It is also assembling under licence the Saab Safari/Supporter multi-purpose light aircraft, with an increasing proportion of components manufactured at Kamra.

PAC MASHSHAQ

The Mashshaq (Urdu for 'proficient') is the local name for the Saab Safari/Supporter assembled under licence in Pakistan. The first example was completed in 1976 in an Army workshop at Risalpur, from which 87 of the aircraft were delivered before assembly of the next batch of 30 was transferred to Kamra in 1981. The Mashshaq is in service with both the Pakistan Air Force and Army Aviation, in a variety of roles which include basic flying training, forward air control, liaison, and other Army duties.

The Safari/Supporter, last described in detail in the 1978-79 *Jane's*, is a two/three-seat light aircraft, with dual controls, and is powered by a 149 kW (200 hp) Avco Lycoming IO-360-A1B6 flat-four engine, driving a Hartzell two-blade constant-speed propeller.

PAC Mashshaq, Pakistan-assembled version of the Saab Safari/Supporter

PHILIPPINES

PADC
PHILIPPINE AEROSPACE DEVELOPMENT CORPORATION

PADC Building, Domestic Terminal Avenue, Pasay, Metro Manila
Telephone: 839081 to 839088
Telex: 63883 PADC PN and 2114 AER PH
PRESIDENT: Roberto H. Lim
EXECUTIVE VICE-PRESIDENT: Pedro Q. Molina
TREASURER: Mila Abarro
CONTROLLER: Eloisa Valerio
DIRECTOR, MANAGEMENT SERVICES: Mario B. Antonio
BOARD OF DIRECTORS:
José Dans
Juan Ponce Enrile
Vicente Paterno
Geronimo Velasco
Panfilo Domingo
Roberto H. Lim
Placido Mapa Jr
Roman Cruz Jr

SUBSIDIARIES:
National Aero Manufacturing Corporation (NAM) (aircraft manufacture and assembly)
Philippine Aero Transport Inc (PATI) (air cargo operations)
Philippine Helicopter Services Inc (PHSI) (maintenance and overhaul of BO 105 helicopters)
Phillippine Resource Helicopters Inc (PRHI) (helicopter support and services for oil drilling operations)

PADC is a government corporation established in 1973 by President Ferdinand E. Marcos to promote the development of an aerospace industry in the Philippines.

NAM
NATIONAL AERO MANUFACTURING CORPORATION (A subsidiary of Philippine Aerospace Development Corporation)

PO Box 2023, Makati Commercial Centre; or Manila International Airport, Pasay City
WORKS: PADC Hangar No. 3, MIA Road, Pasay City; PADC Hangar No. 4, Sangley Point, Cavite City
PRESIDENT: Roberto H. Lim

PADC (MBB) BO 105

In 1974, PADC began an assembly and manufacturing programme for the BO 105C helicopter, under a licence agreement with MBB of West Germany. By mid-1979 a total of 38 had been assembled from knocked-down kits.

A new programme for six additional helicopters for export, and another six for PADC's own marketing needs, was then being negotiated; no further details of this programme have been received from the company, but the Philippine Navy is reported to have ordered an additional nine BO 105s.

NAM assembles/fabricates 45 per cent of the necessary components, importing the rest and exporting some components to MBB, including cargo and battery doors, oil drip pans, main relay boxes, overhead panels, nose doors, tailbooms, windscreen wipers and duct halves.

PADC (PILATUS BRITTEN-NORMAN) ISLANDER

Under a 1974 agreement between PADC and Britten-Norman, six Islanders were delivered to PADC from the UK. Fourteen unpainted aircraft, delivered to PADC without cabin trim, furnishings and avionics, were completed in 1975.

In 1976 NAM completed the assembly of 35 Islanders from completely knocked-down Britten-Norman kits. This phase included manufacture by NAM of ailerons, flaps, elevators, tailplanes, vertical fins, rudders, left centre boxes, fuselages and GRP parts.

The fourth phase, involving the assembly of 60 aircraft (of which 25 were to have been repurchased by Britten-Norman), was suspended when Pilatus acquired Britten-Norman in 1978, but was resumed later.

PADC/NSDB FIXED-WING AIRCRAFT PROTOTYPE

This joint venture between PADC and the Philippine government's National Science Development Board (NSDB) is for an all-metal, braced high-wing monoplane, accommodating four persons including the pilot. It is intended to carry passengers or cargo, and to be easily convertible into an agricultural crop dusting or seeding aircraft.

The detail design phase began in early 1976, but prototype construction had not begun by mid-1979, and no news of this project has been given since that time.

A description and three-view drawing can be found in the 1980-81 *Jane's*.

PAF
PHILIPPINE AIR FORCE

Headquarters PAF, Nichols Air Base, Pasay City

PAF (AJI) T-610 CALI (SUPER PINTO)

The Super Pinto, last described in the 1972-73 edition of *Jane's*, is a light strike version of the Temco TT-1 Pinto trainer, powered by a 12·7 kN (2,850 lb st) General Electric CJ610-4 turbojet engine and equipped with six underwing hardpoints for external stores.

The prototype, and all design and production rights, were acquired by the PAF with a view to possible manufacture in the Philippines (see 1979-80 *Jane's*), but this programme has now been abandoned.

POLAND

PZL
ZJEDNOCZENIE PRZEMYSLU LOTNICZEGO I SILNIKOWEGO PZL (Aircraft and Engine Industry Union)

HEAD OFFICE: ul. Miodowa 5, 00251 Warszawa
Telephone: Warszawa 279985
Telex: 814281
MANAGING DIRECTOR: Kazimierz Brejnak, ME

The Polish aircraft industry remains under the central management of the Zjednoczenie Przemyslu Lotniczego i Silnikowego PZL (Aircraft and Engine Industry Union).

The manufacture of aircraft in Poland began in 1910. In 1928 an industrial syndicate was established, grouping the existing aircraft factories into the Panstwowe Zaklady Lotnicze (State Aviation Works) to produce aircraft to meet domestic and export needs.

More than half a century of tradition in design and

manufacture resulted in the production of several tens of thousands of aircraft, helicopters and gliders of various types, as well as of aero-engines and equipment.

The PZL Aircraft and Engine Industry Union currently comprises 24 factories, scientific and development units, technical and commercial organisations, which between them employ about 100,000 qualified workers. The work of the Union has a broad base which includes research, design, development, manufacture, foreign trade, agricultural aviation services, and technical support for its own products which are operated by other countries.

Production by the Polish aviation industry relies substantially on aircraft, engines and equipment of its own design, as well as on co-operation and co-production with leading foreign aircraft manufacturers in both the East and the West. These programmes currently include the PZL-104 Wilga, PZL-106 Kruk, M-18 Dromader and M-15 Belphegor low/medium/high-capacity agricultural aircraft; component assembly (deliveries of which began in March 1979) for the Soviet Il-86 wide-bodied transport aircraft; the TS-11 Iskra, PZL-110 Koliber and M-20 Mewa light multi-purpose, training and sporting aircraft; local-service passenger transports such as the Soviet-designed An-2 and An-28; the Mi-2, Sokół and Kania low/medium-capacity helicopters for agricultural, medical and transport operations; sailplanes, including the Jantar series and Puchacz; piston, turbojet and turboprop engines; plus aircraft military equipment, propellers, and ground equipment for agricultural aircraft and helicopters, under the control of the PZL Union.

The export sales of all Polish aviation products are handled by Pezetel Foreign Trade Enterprise of the Aviation Industry, under the control of the Aircraft and Engine Industry Union.

Pezetel Foreign Trade Enterprise of the Aviation Industry

PO Box 6, 00-991 Warszawa
Telephone: 10 80 01
Telex: 813430
Managing Director: Jerzy Krezlewicz, MSc
Manager of Aviation Department: Kazimierz Niepsuj

IL

INSTYTUT LOTNICTWA (Aviation Institute)

Al. Krakowska 110/114, 02-256 Warszawa-Okecie
Telephone: Warszawa 460993
Telex: 813537
Managing Director: Dr Ing Andrzej Wierzba
Chief Consultant for Scientific and Technical Co-operation: Dipl Ing Jerzy Grzegorzewski

The Instytut Lotnictwa was founded in 1926. It belongs to the PZL Polish Aviation Works group, under the general management of the PZL Aircraft and Engine Industry Union. The IL is responsible for the control of all research and development work in the Polish aircraft industry. It conducts scientific research, including the investigation of problems associated with low-speed and high-speed aerodynamics, static and fatigue tests, development and testing of aero-engines, flight instruments and other equipment, flight tests, and materials technology. It is also responsible for the construction of experimental aircraft and aero-engines.

CNPSL-PZL WARSZAWA

CENTRUM NAUKOWO-PRODUKCYJNE SAMOLOTOW LEKKICH PZL WARSZAWA (Light Aircraft Science and Production Centre, PZL Warsaw)

Head Office and Works: Al. Krakowska 110/114, 02-256 Warszawa-Okecie
Telephone: Warszawa 461173 and 460031
Telex: 814649
Managers:
Jozef Lipinski, Eng MSc (General Manager)
Andrzej Jaworowski, Eng (Technical)
Jacek Makles, Eng MSc (Sales)
Public Relations: Jerzy Pasterski, Eng MSc

The Okecie factory, founded in 1928, is responsible for light aircraft development and production, and for the design and manufacture of associated agricultural equipment for its own aircraft and for those built at other factories in the Aircraft and Engine Industry Union. Formerly known as WSK-PZL-Okecie, it adopted its present title on 1 July 1976.

PZL-104 WILGA (THRUSH)

The PZL-104 Wilga is a light general-purpose aircraft intended for a wide variety of general aviation and flying club duties. The prototype Wilga 1 flew for the first time on 24 April 1962. This aircraft, the Wilga 2, C and 3 prototypes, the 3A and 3S production versions and other early models, were described in the 1968-69 *Jane's*.

In 1967 the basic design was further modified, with improved cabin comfort, redesigned landing gear and glassfibre tailwheel leg. Production began in 1968 as the Wilga 35 (first flight 28 July 1967) with a 194 kW (260 hp) AI-14R engine, and Wilga 32 (first flown 12 September 1967) with a 171·5 kW (230 hp) Continental O-470-K, -L or -R engine and shorter landing gear. Both received a Polish type certificate on 31 March 1969; the Wilga 32 was described in the 1974-75 *Jane's*, and its Indonesian-built modified version, the Lipnur Gelatik, in the 1975-76 edition. Details of the experimental Wilga 40 and 43 were given in the 1972-73 *Jane's*.

Major versions of the Wilga 35 (see 1979-80 *Jane's*) are the Wilga 35A (Aeroclub), 35P (Passenger/liaison), and 35R (agricultural version, described separately). Examples of the Wilga 35A have been sold to customers in Austria, Belgium, Bulgaria, Czechoslovakia, Egypt, Germany (Democratic Republic), Germany (Federal Republic), Hungary, Romania, Spain, Switzerland, the UK, the USA, the USSR, Venezuela and Yugoslavia.

The Wilga 35H floatplane prototype (SP-WBK) was flown for the first time on 31 October 1979. Fitted with twin CAP Model 3000E floats and a large dorsal fin, it successfully completed initial flight and water tests in Canada later that year.

PZL-104 Wilga 80 floatplane in service in Canada *(Neil A. Macdougall)*

The generally-similar **Wilga 80**, which conforms to FAR Pt 23, flew for the first time on 30 May 1979 and received Canadian government certification in March 1980. Fourteen Wilga 80s were delivered to Canada in that year, including several floatplanes.

By 1 January 1981 a total of 660 Wilgas, of all versions, had been built in Poland. The 500th Wilga was completed on 22 November 1979. In early 1981 both the Wilga 35A and Wilga 80 were in production, but manufacture of the 35A was not expected to continue for much longer.

The following description applies to the Wilga 35A and 80, except where a specific version is indicated:

Type: Single-engined general-purpose monoplane.

Wings: Cantilever high-wing monoplane. Wing section NACA 2415. Dihedral 1°. All-metal single-spar structure, with leading-edge torsion box and beaded metal skin. Each wing attached to fuselage by three bolts, two at spar and one at forward fitting. All-metal aerodynamically and mass-balanced slotted ailerons, with beaded metal skin. Ailerons can be drooped to supplement flaps during landing. Manually-operated all-metal slotted flaps with beaded metal skin. Fixed metal slat on the leading-edge along the full span of the wing and over the fuselage. Tab on starboard aileron.

Fuselage: All-metal semi-monocoque structure in two portions, riveted together. Forward section incorporates main wing spar carry-through structure. Rear section is in the form of a tailcone. Beaded metal skin. Floor in cabin is of metal sandwich construction, with a paper honeycomb core, covered with foam rubber.

Tail Unit: Braced all-metal structure, with sweptback vertical surfaces. Stressed-skin single-spar tailplane attached to fuselage by a single centre fitting and supported by a single aluminium alloy strut on each side. Stressed-skin two-spar fin structure of semi-monocoque construction. Rudder and one-piece elevator are aerodynamically horn-balanced and mass-balanced. Trim tab at centre of elevator trailing-edge.

Landing Gear: Non-retractable tailwheel type. Semi-cantilever main legs, of rocker type, have oleo-pneumatic shock-absorbers. Low-pressure tyres size 500 × 200 mm on main wheels. Hydraulic brakes. Steerable tailwheel, tyre size 255 × 110 mm, carried on rocker frame with oleo-pneumatic shock-absorber. Metal ski landing gear, and CAP Model 3000E twin-float landing gear, optional.

Power Plant: One 194 kW (260 hp) PZL AI-14RA nine-cylinder supercharged radial aircooled engine, driving a PZL US-122000 two-blade constant-speed wooden propeller. Two removable fuel tanks in each wing, with total capacity of 195 litres (43 Imp gallons). Refuelling point on each side of fuselage, at junction with wing. Oil capacity 16 litres (3·5 Imp gallons).

Accommodation: Passenger version accommodates pilot and three passengers, in pairs, with adjustable front seats. Baggage compartment aft of seats, capacity 35 kg (77 lb). Upward-opening door on each side of cabin, jettisonable in emergency. In the parachute training version the starboard door is removed and replaced by two tubular uprights with a central connecting strap, and the starboard front seat is rearward-facing. Jumps are facilitated by a step on the starboard side and by a parachute hitch. A controllable towing hook can be attached to the tail landing gear permitting the Wilga, in this role, to tow a single glider of up to 650 kg (1,433 lb) weight or two or three gliders with a total combined weight of 1,125 kg (2,480 lb).

Systems: Hydraulic system pressure 39 bars (570 lb/sq in). Engine starting is effected pneumatically by a built-in system of 7 litres (427 cu in) capacity with a pressure of 49 bars (710 lb/sq in). Electrical system powered by DC generator and 24V 10Ah battery.

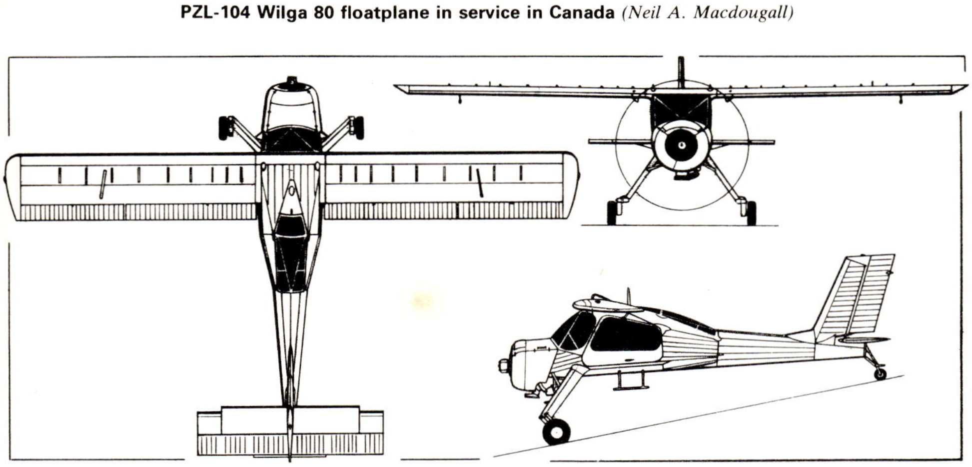

PZL-104 Wilga 35A general-purpose monoplane *(Pilot Press)*

Agricultural version of the Wilga 35, with underfuselage hopper and spraybooms

AVIONICS AND EQUIPMENT: Standard equipment includes VHF transceiver and blind-flying instrumentation. Optional avionics include RS-6102 (of Polish design), R-860 II, R860 IIM, King KY 195 or Bendix radio; and ARL-1601 VHF, ARK-9, King KR 85 or Bendix AV-200 ADF.

DIMENSIONS, EXTERNAL:

Wing span	11·12 m (36 ft 5¾ in)
Wing chord (constant)	1·40 m (4 ft 7¼ in)
Wing aspect ratio	7·95
Length overall	8·10 m (26 ft 6¾ in)
Height overall	2·94 m (9 ft 7¾ in)
Tailplane span	3·70 m (12 ft 1¾ in)
Wheel track	2·85 m (9 ft 4¼ in)
Wheelbase	6·70 m (21 ft 11¾ in)
Propeller diameter	2·65 m (8 ft 8 in)
Passenger doors (each): Height	1·00 m (3 ft 3¼ in)
Width	1·50 m (4 ft 11 in)

DIMENSIONS, INTERNAL:

Cabin: Length	2·20 m (7 ft 2½ in)
Max width	1·20 m (3 ft 10 in)
Max height	1·50 m (4 ft 11 in)
Floor area	2·20 m² (23·8 sq ft)
Volume	2·40 m³ (85 cu ft)
Baggage compartment	0·50 m³ (17·5 cu ft)

AREAS:

Wings, gross	15·50 m² (166·8 sq ft)
Ailerons (total)	1·57 m² (16·90 sq ft)
Trailing-edge flaps (total)	1·97 m² (21·20 sq ft)
Fin	0·97 m² (10·44 sq ft)
Rudder	0·92 m² (9·90 sq ft)
Tailplane	3·16 m² (34·01 sq ft)
Elevator, incl tab	1·92 m² (20·67 sq ft)

WEIGHTS AND LOADINGS (Wilga 35A):

Weight empty, equipped	900 kg (1,984 lb)
Max T-O and landing weight	1,300 kg (2,866 lb)
Max wing loading	83·9 kg/m² (17·18 lb/sq ft)
Max power loading	6·70 kg/kW (11·02 lb/hp)

PERFORMANCE (Wilga 35A at max T-O weight):

Never-exceed speed	150 knots (279 km/h; 173 mph)
Max level speed	108 knots (201 km/h; 125 mph)
Max cruising speed	104 knots (193 km/h; 120 mph)
Econ cruising speed	69·5 knots (128 km/h; 80 mph)
Stalling speed, power on	37 knots (68 km/h; 42·5 mph)
Max rate of climb at S/L	300 m (985 ft)/min
Service ceiling	4,000 m (13,125 ft)
T-O run	80 m (260 ft)
T-O to 15 m (50 ft)	186 m (610 ft)
Landing from 15 m (50 ft)	230 m (755 ft)
Landing run	95 m (310 ft)
Range with max fuel, 30 min reserves	366 nm (680 km; 422 miles)

PERFORMANCE AS GLIDER TUG (Wilga 35A):

Rate of climb at S/L: with 1 glider	234 m (770 ft)/min
with 2 gliders	204 m (670 ft)/min
with 3 gliders	132 m (435 ft)/min
Service ceiling: with 1 glider	4,000 m (13,125 ft)
with 2 gliders	3,900 m (12,800 ft)
with 3 gliders	2,500 m (8,200 ft)
Time to reach service ceiling:	
with 1 glider	37 min 12 s
with 2 gliders	43 min 12 s
with 3 gliders	36 min 36 s
Range:	
with 1 glider	344 nm (637 km; 395 miles)
with 2 gliders	272 nm (504 km; 313 miles)
with 3 gliders	227 nm (420 km; 260 miles)

PZL-104 WILGA 35R (AGRICULTURAL VERSION)

First flown on 13 February 1978, this agricultural version of the Wilga 35 takes advantage of the STOL capabilities of this aeroplane, its steep angles of climb and descent, and its ability to make tight turns, particularly over small fields where approach is a problem. Flight testing continued in 1979, and since early 1980 sets of equipment for further conversions have been under construction.

The agricultural equipment can be installed with any type of landing gear except floats, and is made entirely of materials resistant to chemical corrosion. Maximum chemical load, in addition to fuel for 1 hour's flying, is 270 kg (595 lb), and the limited carrying capacity qualifies the aircraft for ultra-low-volume spraying missions. With dual controls, this version can be used to give practical training to future agricultural pilots under genuine field conditions.

AGRICULTURAL EQUIPMENT: Glassfibre hopper for chemicals, attached beneath fuselage between main landing gear units; centrifugal pump, attached to port landing gear leg and driven by a six-blade fan; control valves; filter; sprayboums, supported well beneath wings on V struts and fitted with four Micronair atomisers or 40 pressurised spray outlets; fluid pressure gauges; Micronair flow meter. Pump brake actuated pneumatically; cut-off and release valves, and jettison system, operated manually.

PZL-106A KRUK (RAVEN)

The PZL-106 was designed in early 1972 by a team led by Andrzej Frydrychewicz. The first prototype (SP-PAS) flew for the first time on 17 April 1973, powered by a 298 kW (400 hp) Avco Lycoming IO-720 engine. It was followed in October of that year by a second Avco Lycoming-engined prototype (SP-PBG) and, from October 1974, by four prototypes fitted with the 441 kW (592 hp) PZL-3S radial engine that powers PZL-106A production aircraft. The latter also have a low-mounted tailplane instead of the earlier T tail, and a greater chemical load, in a larger hopper. Manufacture of some 600 aircraft for the member countries of the CMEA (Council for Mutual Economic Aid) is anticipated.

Production of the PZL-106A began in 1976; the first export delivery, to Hungary, was made in 1977, and 128 had been built by the beginning of 1981, of which 49 were delivered to the German Democratic Republic, two to Hungary, one to Cuba and two to Czechoslovakia. A tropical version of the standard PZL-106A, with uncowled engine, is available.

A version known as the **PZL-106AR**, first flown on 15 November 1978, is under development; this is powered by a geared PZL-3SR engine (gear ratio 0·7), driving a 3·10 m (10 ft 2 in) diameter PZL US-133000 propeller, and has an improved performance.

A **PZL-106AT Turbo-Kruk** version was exhibited at a Polish air show at Okecie in November 1979. This aircraft (SP-WUK) was fitted with a 567 kW (760 shp) Pratt & Whitney Aircraft of Canada PT6A-34AG turboprop engine, driving a 2·59 m (8 ft 6 in) diameter Hartzell HC-B3TN-3B/TIO282 three-blade constant-speed fully-feathering reversible-pitch metal propeller with spinner. First flight was made in July 1981.

The following description applies to the standard production PZL-106A, except where indicated:

TYPE: Single-engined agricultural aircraft.

WINGS: Braced low-wing monoplane with upward-cambered tips. Clark Y wing section throughout span, except at tips. Thickness/chord ratio 11·7%. Dihedral 4° from roots. Incidence 6° 6′. Sweepback 4° at quarter-chord. All-metal two-spar structure, of constant chord throughout most of span. Metal and polyester fabric covering. Glassfibre wingtips. Full-span six-section fixed leading-edge slats of glassfibre sandwich construction, with foam core. Three-section slotted ailerons of duralumin, with polyester fabric covering. No flaps or tabs. Streamline-section V bracing struts, with jury struts.

FUSELAGE: Welded steel tube structure, covered with quickly-removable panels of light alloy and GRP. Steel tube structure can be pressure-tested for crack detection.

TAIL UNIT: Conventional duralumin structure, with single bracing strut each side. Fixed surfaces metal-covered; rudder and mass-balanced elevators polyester fabric-covered. Trim tab in port elevator, automatic tab in starboard elevator; automatic tab on rudder.

LANDING GEAR: Non-retractable tailwheel type, with oleo-pneumatic shock-absorber in each unit. Main wheels, with low-pressure tyres (size 800 mm × 260 mm), each carried on side V and half-axle. Main-wheel tyre pressure 2·0 bars (29 lb/sq in). Pneumatically-operated hydraulic disc brakes on main wheels. Parking brake. Steerable tailwheel, with tubeless tyre size 350 mm × 135 mm, pressure 2·5 bars (36·25 lb/sq in).

POWER PLANT: One 441 kW (592 hp) PZL-3S seven-cylinder radial aircooled supercharged engine, driving a PZL US-132000/A four-blade constant-speed metal propeller. Fuel in two wing tanks, total capacity 300 litres (66 Imp gallons). Gravity refuelling point on each wing; semi-pressure refuelling point on starboard side of fuselage. Oil capacity 54 litres (11·9 Imp gallons) max, 7 litres (1·54 Imp gallons) min. Air filter fitted.

ACCOMMODATION: Pilot in enclosed, ventilated and heated cockpit. Second (mechanic's) rearward-facing seat to rear. Combined window/door on each side of cabin. Cockpit area strengthened to resist 40*g* impact.

SYSTEMS: Pneumatic system, rated at 49 bars (710 lb/sq in), for brakes and agricultural equipment. Electrical power, from 27·5V DC alternator and battery, for

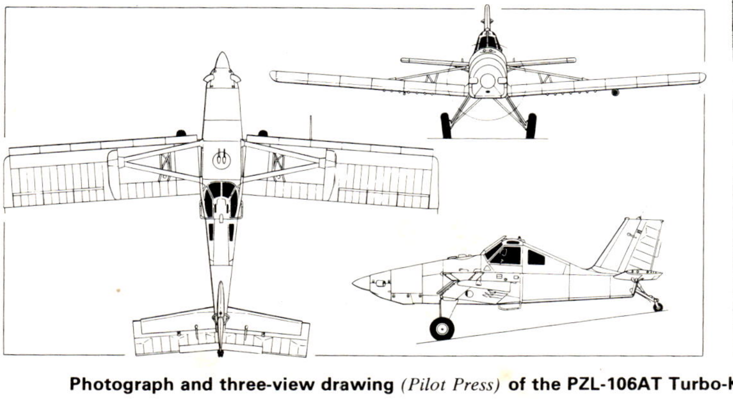

Photograph and three-view drawing *(Pilot Press)* **of the PZL-106AT Turbo-Kruk (Pratt & Whitney Aircraft of Canada PT6A-34AG turboprop engine)**

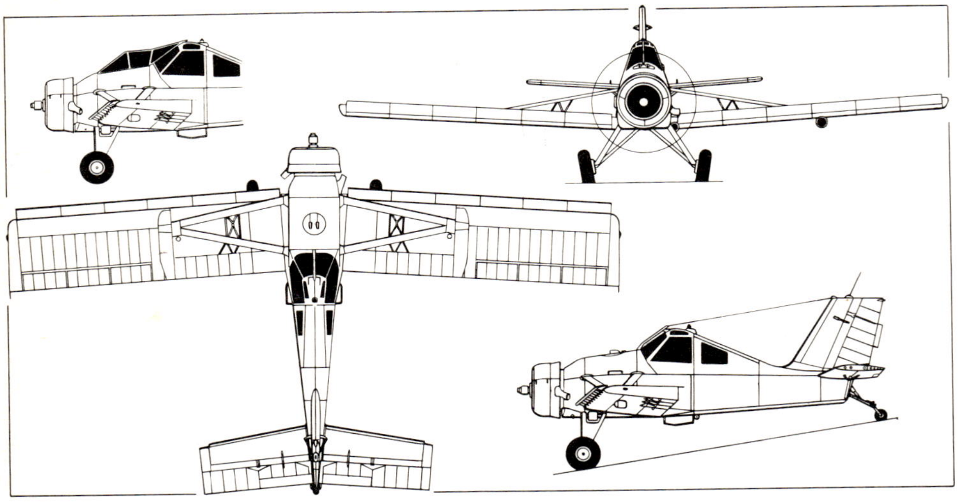

PZL-106A Kruk (PZL-3S engine), with scrap view of the two-seat version *(Pilot Press)*

PZL-106AR Kruk, with PZL-3SR geared engine *(A. Glass)*

engine starting, pneumatic system control, aircraft lights, instruments, VHF transceiver and semi-pressure refuelling.

Equipment: VHF transceiver. Easily removable non-corrosive (GRP) hopper/tank, forward of cockpit, can carry 1,000 kg (2,205 lb) of dry or liquid chemical, and has a maximum capacity of 1,400 litres (308 Imp gallons). The hopper has a quick-dump system that can release 1,000 kg of chemical in less than 5 s. A pneumatically operated intake for the loading of dry chemicals is optional. Distribution system for liquid chemical is powered by a fan-driven centrifugal pump. A precise and reliable dispersal system, with positive on/off action for dry chemicals, gives effective swath widths of 10-35 m (33-115 ft). Hopper can be replaced easily by a special container with instructor's seat, controls, basic instruments and windscreen, to convert any version of the Kruk into a two-seat trainer.

Dimensions, external (A, PZL-106A; B, PZL-106AR):

Wing span	14·80 m (48 ft 6½ in)
Wing chord (constant)	1·90 m (6 ft 2¾ in)
Wing aspect ratio	7·8
Length overall: A	9·10 m (29 ft 10¼ in)
B	9·16 m (30 ft 0¾ in)
Height overall	3·32 m (10 ft 10¾ in)
Tailplane span	5·77 m (18 ft 11¼ in)
Wheel track	3·10 m (10 ft 2¼ in)
Wheelbase	7·41 m (24 ft 3¾ in)
Propeller diameter: A	2·62 m (8 ft 7 in)
B	3·10 m (10 ft 2 in)
Propeller ground clearance (tail up):	
A	0·36 m (1 ft 2¼ in)
B	0·12 m (4¾ in)
Crew doors (each): Height	0·91 m (2 ft 11¾ in)
Width	1·06 m (3 ft 5¾ in)
Baggage door: Height	0·70 m (2 ft 3½ in)
Width	0·60 m (1 ft 11¾ in)

Dimensions, internal:

Cabin: Length	1·37 m (4 ft 6 in)
Max width	1·25 m (4 ft 1¼ in)
Max height	1·30 m (4 ft 3¼ in)
Floor area	1·12 m² (12·05 sq ft)
Baggage compartment:	
Length	1·40 m (4 ft 7 in)
Width	1·00 m (3 ft 3¼ in)
Depth	0·60 m (1 ft 11¾ in)

Areas:

Wings, gross	28·40 m² (305·7 sq ft)
Ailerons (total)	2·46 m² (26·50 sq ft)
Fin	1·26 m² (13·56 sq ft)
Rudder, incl tab	1·62 m² (17·44 sq ft)
Tailplane	3·34 m² (35·95 sq ft)
Elevators, incl tab	4·22 m² (45·42 sq ft)

Weights and Loadings (A, PZL-106A; B, PZL-106AR):

Weight empty, equipped: A	1,610 kg (3,550 lb)
B	1,685 kg (3,715 lb)
Max chemical payload	1,000 kg (2,205 lb)
Normal T-O weight: A	2,800 kg (6,173 lb)
B	3,000 kg (6,614 lb)
Max T-O and landing weight: A, B	3,000 kg (6,614 lb)
*Max ramp weight: A, B	3,000 kg (6,614 lb)
Max wing loading	105·6 kg/m² (21·63 lb/sq ft)
Max power loading	6·7 kg/kW (11·02 lb/hp)

* *Aircraft stressed for 3,000 kg (6,614 lb) T-O weight in Normal Category, BCAR Section K*

Performance (A, PZL-106A; B, PZL-106AR; at Normal T-O weight):

Never-exceed speed	140 knots (260 km/h; 161 mph)
Max level speed at S/L	114 knots (211 km/h; 131 mph)
Max cruising speed at S/L	97 knots (180 km/h; 112 mph)
Operating speed with 1,000 kg (2,205 lb) of chemical	65-86·5 knots (120-160 km/h; 74·5-99·5 mph)
Stalling speed at S/L	50 knots (92 km/h; 57·5 mph)
Max rate of climb at S/L with 1,000 kg (2,205 lb) of chemical: A	240 m (787 ft)/min
B	270 m (885 ft)/min
Service ceiling: A	4,000 m (13,125 ft)
B	4,200 m (13,780 ft)
*T-O run: A	220 m (722 ft)
B	240 m (788 ft)
*T-O to 15 m (50 ft): A	480 m (1,575 ft)
B	390 m (1,280 ft)
*Landing from 15 m (50 ft): A	410 m (1,345 ft)
B	370 m (1,215 ft)
*Landing run: A, B	210 m (689 ft)
Min ground turning radius	10·00 m (32 ft 10 in)
Range with max fuel: A	215 nm (400 km; 248 miles)
B	367 nm (680 km; 422 miles)

* *with agricultural equipment*

PZL-110 KOLIBER (HUMMING-BIRD)

Under this designation, CNPSL-PZL Warszawa is producing under licence a two/three-seat version of the Socata Rallye 100 ST, the lowest-powered model in the Rallye light aircraft family, and one which is no longer in production in France. The first PZL-110, modified to receive an 86·5 kW (116 hp) PZL-Franklin engine, made its initial flight on 18 April 1978.

The first production PZL-110 was flown on 8 May 1979, and by the beginning of 1980 a total of 10 Series I and 20 Series II aircraft had been completed. A Polish type certificate was awarded on 24 August 1979. A version with a 164 kW (220 hp) PZL-built Franklin six-cylinder engine is under development.

A description of the standard Rallye 100 ST can be found under the Socata heading in the French section of the 1978-79 *Jane's*. The PZL-110 version differs in the following details:

Power Plant: One 86·5 kW (116 hp) PZL-built Franklin 4A-235-B31 flat-four engine, driving a US 135 two-blade ground-adjustable propeller. Fuel system and capacity as for Rallye 100 ST. Oil capacity 6·2 litres (1·4 Imp gallons).

Systems and Equipment: 12V electrical system, with alternator and 18Ah battery. Equipment optional for Series I and standard for Series II includes VHF transceiver, ADF, electrically powered gyro attitude indicator, turn and bank indicator, and directional gyro.

Dimensions, external: As Rallye 100 ST except:

Length overall	7·20 m (23 ft 7½ in)
Propeller diameter	1·78 m (5 ft 10 in)

Areas: As Rallye 100 ST except:

Wings, gross	12·66 m² (136·3 sq ft)

Weights and Loading:

Weight empty, equipped	516 kg (1,137 lb)
Max design T-O and landing weight (Normal category)	850 kg (1,874 lb)
Max T-O weight (Utility category)	770 kg (1,700 lb)
Max wing loading	67·1 kg/m² (13·74 lb/sq ft)

Performance (at 770 kg; 1,700 lb Utility max T-O weight except where indicated):

Never-exceed speed	145 knots (270 km/h; 167 mph)
Max level speed at S/L	105 knots (195 km/h; 121 mph)
Max cruising speed at S/L	92 knots (170 km/h; 106 mph)
Stalling speed, flaps up	43·5 knots (80 km/h; 50 mph)
Stalling speed, flaps down	40 knots (74 km/h; 46 mph)
Max rate of climb at S/L at AUW of 830 kg (1,830 lb)	171 m (560 ft)/min
Service ceiling	3,500 m (11,480 ft)
T-O run at S/L	155 m (509 ft)
T-O to 15 m (50 ft) at S/L	380 m (1,247 ft)
Landing from 15 m (50 ft) at S/L	275 m (902 ft)
Landing run at S/L	115 m (377 ft)
Range at 500 m (1,640 ft) with max fuel, no reserves	394 nm (730 km; 453 miles)

PZL-110 Koliber, Polish-built version of the Socata Rallye 100 ST

WSK-PZL MIELEC

WYTWORNIA SPRZETU KOMUNIKACYJNEGO-PZL MIELEC (Transport Equipment Manufacturing Centre, Mielec)

Head Office and Works: ul. Ludowego Wojska Polskiego 3, 39-301 Mielec
Telephone: Mielec 70
Telex: 83293
General Manager: Dipl Ing Tadeusz Ryczaj

Largest and best-equipped aircraft factory in Poland, the WSK factory at Mielec was founded in 1938. It was engaged mainly in licence production of MiG-15/17 single-seat jet fighters for several years. These aircraft carry the Polish designation of LiM, meaning Licence MiG, and included the LiM-1 (MiG-15), SBLiM-1 (MiG-15UTI), LiM-2 (MiG-15bis) and LiM-5 (MiG-17). Polish-developed versions of the LiM-5 were designated LiM-5M and LiM-6bis. Licence production of MiG fighters ceased in Poland in about 1959.

Following a reduction in orders for combat aircraft in 1955, Mielec began production in 1956 of 240 TS-8 Bies basic trainers, described in the 1962-63 *Jane's*. Four years later, the Soviet-designed An-2 general utility biplane went into production at Mielec. In parallel production with the An-2 at the present time is the TS-11 Iskra jet trainer and light attack aircraft. In 1977 Mielec began to manufacture components, including fins, tailplanes, engine pylons, ailerons, and wing slats and flaps, for the Ilyushin Il-86 Soviet wide-bodied transport, and in 1978 it was announced that Mielec would assume responsibility for series production of the Soviet Antonov An-28 twin-turboprop light general-purpose aircraft.

There is a design office at the factory for development of original aircraft. Its most recent designs are the M-15 agricultural aircraft, M-17 light aircraft, and M-18 Dromader agricultural aircraft, of which the M-15 and M-18 are now in production.

PZL MIELEC (ANTONOV) An-2

NATO reporting name: Colt

The prototype of the An-2, designed to a specification of the Ministry of Agriculture and Forestry of the USSR, made its first flight on 31 August 1947. In 1948 the aircraft went into production in the USSR as the An-2, with a 746 kW (1,000 hp) ASh-62 engine.

By 1960, more than 5,000 An-2s had been built in the Soviet Union for service with the Soviet armed forces, Aeroflot and other civilian organisations; the various Soviet-built versions have been fully described in previous editions of *Jane's*. Many were exported, to all of the Socialist States, and to Greece, Afghanistan, Mali, Nepal, India and Cuba, and licence rights were granted to China, where the first locally-produced An-2 was completed in December 1957.

Since 1960, apart from a small Soviet-built quantity of a developed version known as the An-2M (see 1971-72 *Jane's*), the continued production of the An-2 has been the responsibility of the PZL factory at Mielec, the original licence arrangement providing for two basic versions: the An-2T transport and An-2R agricultural version. The first 10 Polish-built An-2s were completed in 1960 (first example flown on 23 October 1960). Mielec has since built considerable numbers of this aircraft for domestic use and for export to the USSR, Bulgaria, Czechoslovakia, Egypt, France, the German Democratic Republic, Hungary, North Korea, Mongolia, the Netherlands, Romania, Sudan, Tunisia and Yugoslavia. The 5,000th Polish-built An-2 was delivered, to the USSR, on 3 February 1973. Since beginning An-2 production, Mielec has made numerous improvements to the airframe of the An-2R, resulting in an increase in TBO from 900 hr in 1961 to 1,500 hr in 1970 and 2,000 hr in 1973.

By the beginning of 1981 about 9,000 An-2s (all versions) had been built at Mielec, including more than 5,500 of the An-2R agricultural version. More than 90 per cent of these were for export, including about 7,500 to the USSR. Orders in 1980 included 22 for Czechoslovakia and five for Venezuela.

The Polish-built versions have different designations from those built in the USSR. These are as follows:

An-2LW. Twin-float version of An-2T, similar to Soviet-built An-2V and originally given Polish designation An-2M. Built in small numbers only.

An-2P. Passenger version, with seating for 12 adult passengers and two children. Compared with Soviet-built An-2P (see 1971-72 *Jane's*) has improvements in passenger cabin layout and comfort, better soundproofing, a new propeller and spinner, and weight-saving instrumentation and equipment. Entered production in 1968.

An-2PK. Five-seat executive version, having two seats on starboard side and three on port side, with foldaway table between pairs of seats on each side.

An-2P-Photo. Photogrammetry version.

An-2PR. Television relay version.

An-2R. Agricultural version, with 1,350 kg (2,976 lb) capacity glassfibre-reinforced epoxy-resin hopper or 1,400 litre (308 Imp gallon) capacity tank for dry or liquid chemicals. Similar to Soviet-built An-2S. Agricultural equipment being modernised.

PZL Mielec (Antonov) An-2P passenger transport biplane

An-2S. Ambulance version, equipped to carry six stretcher patients and their medical attendants.

An-2T. Basic general-purpose version, with accommodation for 12 passengers and baggage or 1,500 kg (3,306 lb) of cargo.

An-2TD. Paratroop transport and training version with six tip-up seats along each side of cabin. Granted French type certificate No. IM-55 (for import licence) in 1972.

An-2TP. Cargo/passenger version, similar to AN-2TD with six tip-up seats along each side of cabin.

An-2 Geofiz. Geophysical survey version, developed for the State Prospecting Company in Warsaw.

The following details apply to the PZL Mielec An-2P:

Type: Single-engined general-purpose biplane.

Wings: Unequal-span single-bay biplane. Wing section RPS 14% (constant). Dihedral, both wings, approx 2° 48′. All-metal two-spar structure, fabric-covered aft of front spar. I-type interplane struts. Differential ailerons and full-span automatic leading-edge slots on upper wings, slotted trailing-edge flaps on both upper and lower wings. Flaps operated electrically, ailerons mechanically by cables and push/pull rods. Electrically-operated trim tab in port aileron.

Fuselage: All-metal stressed-skin semi-monocoque structure of circular section forward of cabin, rectangular in the cabin section and oval in the tail section.

Tail Unit: Braced metal structure. Fin integral with rear fuselage. Fabric-covered tailplane. Elevators and rudder operated mechanically by cables and push/pull rods. Electrically-operated trim tab in rudder and port elevator.

Landing Gear: Non-retractable split-axle type, with long-stroke oleo shock-absorbers. Main-wheel tyres size 800 × 260 mm, pressure 2·25 bars (32·7 lb/sq in). Pneumatic shoe brakes on main units. Fully-castoring and self-centering PZL-Krosno tailwheel with electro-pneumatic lock. For rough-field operation the oleo-pneumatic shock-absorbers can be charged from a compressed air cylinder installed in the rear fuselage. Interchangeable ski landing gear available optionally.

Power Plant: One 746 kW (1,000 hp) Shvetsov ASz-621R nine-cylinder radial aircooled engine, driving an AW-2 four-blade variable-pitch metal propeller. Six fuel tanks in upper wings, with total capacity of 1,200 litres (264 Imp gallons). Oil capacity 120 litres (26·4 Imp gallons).

Accommodation: Crew of two on flight deck, with access via passenger cabin. Standard accommodation for 12 passengers, in four rows of three with centre aisle. Two foldable seats for children in aisle between first and second rows, and infant's cradle at front of cabin on starboard side. Toilet at rear of cabin on starboard side. Overhead racks for up to 160 kg (352 lb) of baggage, with space for coats and additional 40 kg (88 lb) of baggage between rear pair of seats and toilet. Emergency exit on starboard side at rear. Walls of cabin are lined with glass-wool mats and inner facing of plywood to reduce internal noise level. Cabin floor is carpeted. Cabin heating and starboard windscreen de-icing by engine bleed air; port and centre windscreens are electrically de-iced. Cabin ventilation by ram-air intakes on underside of top wings. Air-conditioning system in An-2R.

Systems: Compressed air cylinder, of 8 litres (490 cu in) capacity, for pneumatic charging of shock-absorbers and operation of tailwheel lock at 49 bars (711 lb/sq in) pressure and operation of main-wheel brakes at 9·80 bars (142 lb/sq in). Contents of cylinder are maintained by AK-50 M engine-driven compressor, with AD-50 automatic relief device to prevent overpressure. DC electrical system is supplied with basic 27V power (and 36V or 115V where required) by an engine-driven generator and a storage battery. CO_2 fire extinguishing system with automatic fire detector.

Avionics and Equipment: Dual controls and blind-flying instrumentation standard. R-842 short wave and R-860 ultra short wave lightweight radio transceivers, RW-UM radio altimeter, ARK-9 radio compass, MRP-56P marker, GB-1 gyro compass, GPK-48 gyroscopic direction indicator and SPU-7 intercom.

Dimensions, external:

Wing span: upper	18·18 m (59 ft 8½ in)
lower	14·24 m (46 ft 8½ in)
Wing chord (constant): upper	2·40 m (7 ft 10½ in)
lower	2·00 m (6 ft 6¾ in)
Wing aspect ratio: upper	7·57
lower	7·12
Wing gap	2·17 m (7 ft 1½ in)
Length overall: tail up	12·74 m (41 ft 9½ in)
tail down	12·40 m (40 ft 8¼ in)
Height overall: tail up	6·10 m (20 ft 0 in)
tail down	4·00 m (13 ft 1½ in)
Tailplane span	7·20 m (23 ft 7½ in)
Wheel track	3·45 m (11 ft 3¾ in)
Wheelbase	3·36 m (11 ft 0¼ in)
Propeller diameter	3·60 m (11 ft 9¾ in)
Propeller ground clearance	0·70 m (2 ft 3½ in)
Emergency exit (stbd, rear):	
Height	0·65 m (2 ft 1½ in)
Width	0·51 m (1 ft 8 in)

Areas:

Wings, gross: upper	43·6 m² (469 sq ft)
lower	28·0 m² (301 sq ft)
Ailerons (total)	5·90 m² (63·5 sq ft)
Trailing-edge flaps (total)	9·60 m² (103 sq ft)
Fin	5·85 m² (62·97 sq ft)
Rudder, incl tab	2·65 m² (28·52 sq ft)
Tailplane	12·28 m² (132·18 sq ft)
Elevators (total, incl tab)	4·72 m² (50·81 sq ft)

Weights and Loadings:

Weight empty	3,450 kg (7,605 lb)
Max T-O weight	5,500 kg (12,125 lb)
Max landing weight	5,250 kg (11,574 lb)
Max wing loading	76·82 kg/m² (15·7 lb/sq ft)
Max power loading	7·38 kg/kW (12·13 lb/hp)

Performance (at AUW of 5,250 kg; 11,574 lb):

Max level speed at 1,750 m (5,740 ft)	139 knots (258 km/h; 160 mph)
Econ cruising speed	100 knots (185 km/h; 115 mph)
Min flying speed	49 knots (90 km/h; 56 mph)
T-O speed	43 knots (80 km/h; 50 mph)
Landing speed	46 knots (85 km/h; 53 mph)
Max rate of climb at S/L	210 m (689 ft)/min
Service ceiling	4,400 m (14,425 ft)
Time to 4,400 m (14,425 ft)	30 min
T-O run: hard runway	150 m (492 ft)
grass	170 m (558 ft)
T-O to 10·7 m (35 ft): hard runway	300 m (984 ft)
grass	320 m (1,050 ft)
Landing run: hard runway	170 m (558 ft)
grass	185 m (607 ft)
Range at 1,000 m (3,280 ft) with 500 kg (1,102 lb) payload	485 nm (900 km; 560 miles)

PZL MIELEC (ANTONOV) An-28

NATO reporting name: Cash

Oleg Antonov first referred to planned production of an enlarged turboprop version of the piston-engined An-14 light general-purpose aircraft in the early 1960s, but until the Spring of 1972 there was no proof that such an aircraft had been built. Photographs of the prototype (CCCP-1968) were then published in the Polish press. It had flown for the first time in the USSR in September 1969, powered by two 604 kW (810 shp) Isotov TVD-850 turboprop engines, and was described in this form in the Soviet section of the 1974-75 and previous editions of *Jane's*.

Initially, the new aircraft was designated An-14M. Its official flight testing was completed in 1972, and during 1973 it was allocated the production designation An-28. The first pre-production An-28 (CCCP-19723) retained the original TVD-850 engines, but flight trials suggested that field performance and climb, in particular, could be improved by fitting more powerful engines. Thus, in April 1975, the same development aircraft (re-registered CCCP-19753) flew for the first time with 716 kW (960 shp) Glushenkov TVD-10 turboprop engines, which were specified also for production An-28s. It won a subsequent competitive evaluation against the Be-30, in which the emphasis was placed by the evaluators on concept rather than detail design.

Following Polish-Soviet talks in February 1978, it was announced that series production of the An-28 was to be entrusted to PZL Mielec. Mr Antonov stated at the Paris Air Show in June 1979 that the aircraft was at the final stages of testing for certification prior to production, which he then expected to begin in 1980-81. The second pre-production aircraft (originally CCCP-19754, now

Antonov An-28 light general-purpose transport (Glushenkov TVD-10B turboprop engines) *(Tass)*

Antonov An-28 light transport, to be produced in Poland by WSK-PZL Mielec *(Pilot Press)*

CCCP-48105) was displayed at the Show. A total of 1,700 development flights had been completed by that time, including tests with ski landing gear. Subsequently, in September 1979, the Moscow representative of Pezetel stated that the first Polish-built An-28s would appear in 1982, with large-scale production beginning in 1983. Export of 1,200 to the Soviet Union by 1990 is envisaged.

In general configuration the An-28 differs from the piston-engined An-14 mainly in having a much-enlarged fuselage to carry up to 20 passengers or equivalent alternative payloads. The original An-14M had a retractable landing gear, with small fairings on the sides of the fuselage into which the main units retracted. Subsequently it was decided that retraction was unnecessary for flights over short distances at low speeds, and the pre-production An-28s have fixed gear. The shape of the vertical tail surfaces was also changed as a result of early flight testing.

The Antonov design bureau developed the An-28 for service on Aeroflot's shortest routes, particularly those operated by An-2 biplanes into places which are relatively inaccessible to other types of fixed-wing aircraft. The turboprop engines make possible full-payload operation under high-temperature conditions and in mountainous regions; and the An-28 is described as being suitable for carrying passengers, cargo and mail, for scientific expeditions, geological surveying, forest fire patrol, firefighting, air ambulance or rescue operations, and parachute training. In agricultural form it can carry an 800 kg (1,764 lb) chemical payload for dusting and spraying operations.

Mr Antonov has stated that Aeroflot pilots will begin their flying careers on the An-28, which will not stall, even with the control column held in the extreme rearward position, because of the action of its automatic slots. If an engine fails, the upper-surface spoiler forward of the aileron on the opposite wing is opened automatically; as a result, the wing bearing the 'dead' engine drops only 12° in 5 s instead of the 30° that it would drop through loss of lift without the action of the Antonov-patented spoiler. The fixed tailplane slot, also patented, improves handling during a high angle of attack climb-out. Under icing conditions, if the normal anti-icing system fails, ice collects on the slat rather than the tailplane, to retain controllability.

Type: Twin-turboprop light general-purpose aircraft.

Wings: Braced high-wing monoplane, with single streamline-section bracing strut each side. Conventional two-spar structure. Full-span automatic leading-edge slots. Entire trailing-edges hinged, the single-slotted ailerons being designed to droop with the large double-slotted flaps. No tabs. Spoiler forward of each aileron. Short stub-wing extends from each side of the lower fuselage, carrying the main landing gear units, and providing lower attachments for the wing bracing struts. Anti-icing of wing leading-edges by engine bleed air.

Fuselage: Conventional all-metal semi-monocoque structure, longer, wider and deeper than that of the piston-engined An-14. Underside of rear fuselage upswept and made up of clamshell doors.

Tail Unit: Cantilever all-metal structure. Twin fins and rudders mounted vertically on an inverted-aerofoil tailplane that lacks the dihedral of that on the An-14. Fixed leading-edge slat under full span of tailplane leading-edge. Anti-icing of leading-edges by engine bleed air. Twin tabs in each rudder.

Landing Gear: Non-retractable tricycle type, with single wheel on each unit. Wide-tread balloon tyres, size 720 × 320, pressure 3·5 bars (51 lb/sq in), on main units. Steerable and self-centering nosewheel, with size 595 × 185 tyre. Brakes on main wheels. Provision for skis or floats.

Power Plant: Two 715 kW (960 shp) Polish-built flat rated Glushenkov TVD-10B turboprop engines (Polish designation PZL-10W), each driving an AW-24AN three-blade reversible-pitch metal propeller. Two 310 litre (68 Imp gallon) centre-wing and two 670 litre (147 Imp gallon) outer-wing integral fuel tanks; total fuel capacity 1,960 litres (430 Imp gallons). Oil capacity 30 litres (6·5 Imp gallons) per engine. Electrical anti-icing of propellers and engine air intakes.

Accommodation: Crew of one or two on flight deck, which has bulged side windows and windscreen electrical anti-icing. Crew door forward of cabin on port side. Cabin of passenger version contains 15 seats in five rows at 72 cm (28 in) pitch, or up to 20 seats in high-density configuration, with double units on starboard side of aisle. Seats fold back against walls when aircraft is operated as a freighter or in mixed passenger/cargo role. Provision for baggage and toilet compartments and wardrobe space. Electrically-actuated ramp-door under upswept rear fuselage can slide forward under cabin to facilitate direct loading from trucks on to cabin floor. Overhead winch on rails, capacity 500 kg (1,102 lb), for handling cargo. Emergency exit at rear on starboard side. Six/seven-passenger executive version has four folding tables, which can be joined together in pairs to give working tops measuring 160 × 55 cm (63 × 21·5 in). Ambulance version accommodates six stretchers, five seated patients, a medical attendant and medical equipment. Can also be equipped to carry six parachutists and a dispatcher.

Avionics and Equipment: Flight and navigation equipment includes blind-flying instrumentation; R-80W UHF, Karat short-wave, RSB-5 medium-wave, Landysz-5 and R-851 radios; Wint-2 navigation computer; DISS Maszt-FK and ARK-U2 'special installations'; ADF; emergency locator transmitter; DWS-8 airspeed indicator; AR-C7 turn indicator; MS-61 drift recorder; three-axis autopilot; and dual AK-59P astrocompasses. Landing light in nose. Current level of equipment is intended to permit operation in ICAO Category II conditions, with extension later to Category III.

Dimensions, external:	
Wing span	22·06 m (72 ft 4½ in)
Wing chord at root	2·20 m (7 ft 2½ in)
Wing aspect ratio	12
Length overall	12·98 m (42 ft 7 in)
Height overall	4·60 m (15 ft 1 in)
Tailplane span	5·20 m (17 ft 0¾ in)
Wheel track	3·41 m (11 ft 2 ¼ in)
Wheelbase	4·35 m (14 ft 3¼ in)
Propeller diameter	2·80 m (9 ft 2¼ in)
Dimensions, internal:	
Cabin, excl flight deck: Length	5·26 m (17 ft 3 in)
Max width	1·66 m (5 ft 5 in)
Max height	1·70 m (5 ft 7 in)
Floor area	8·73 m² (93·97 sq ft)
Volume	14·84 m³ (524 cu ft)
Area:	
Wings, gross	40·28 m² (433·6 sq ft)
Weights and Loadings:	
Weight empty (approx)	3,500 kg (7,716 lb)
Normal payload	1,550 kg (3,415 lb)
Max payload	1,700 kg (3,750 lb)
Normal T-O weight	5,800 kg (12,785 lb)
Max T-O weight	6,100 kg (13,450 lb)
Max wing loading	151·4 kg/m² (31·03 lb/sq ft)
Max power loading	4·22 kg/kW (6·93 lb/shp)

Performance (A: at normal T-O weight; B: at max T-O weight):

Max cruising speed:	
A, B	188 knots (350 km/h; 217 mph)
Econ cruising speed:	
A, B	162 knots (300 km/h; 186 mph)
Stalling speed, flaps up:	
A	70 knots (130 km/h; 81 mph)
B	73 knots (135 km/h; 84 mph)
Stalling speed, flaps down:	
A	65 knots (120 km/h; 75 mph)
B	67·5 knots (125 km/h; 78 mph)
Max rate of climb at S/L: A	750 m (2,460 ft)/min
B	708 m (2,320 ft)/min
Rate of climb at S/L, one engine out:	
A	192 m (630 ft)/min
B	174 m (570 ft)/min
T-O run: A	180 m (590 ft)
B	210 m (690 ft)
T-O to 15 m (50 ft): A	330 m (1,085 ft)
B	360 m (1,180 ft)
Landing from 15 m (50 ft): A	287 m (942 ft)
B	305 m (1,000 ft)
Landing run: A	150 m (492 ft)
B	170 m (558 ft)
Range at econ cruising speed at 3,000 m (9,850 ft), 30 min reserves:	
A, 15 passengers	356 nm (660 km; 410 miles)
A, 18 passengers	202 nm (375 km; 233 miles)
B, 18 passengers	372 nm (690 km; 428 miles)
B, 20 passengers	275 nm (510 km; 317 miles)
A, max fuel	702 nm (1,300 km; 807 miles)
B, max fuel	696 nm (1,290 km; 801 miles)

PZL MIELEC TS-11 ISKRA (SPARK)

Designed in 1957 under the supervision of Docent Ing T. Soltyk, the TS-11 Iskra two-seat jet trainer was produced as a replacement for the piston-engined TS-8 Bies.

Single-seat Iskra-Bis C reconnaissance version of the PZL Mielec TS-11

Four prototypes were built during 1958-59, the first of these being used for static testing. First flight, on 5 February 1960, was made by the second aircraft, followed later in the same year by the third and fourth prototypes. Type approval was received in mid-1961, and quantity production began at Mielec in 1963. The formal handing over of the first Iskra to the Polish Air Force took place in March 1963, and the aircraft entered service in 1964. Iskras of the Polish Air Force logged a total of 100,000 flying hours by 1978. Another 30,000 flying hours had been accumulated in India.

Early production aircraft were powered by a 7·66 kN (1,720 lb st) HO-10 Polish-designed axial flow turbojet engine. In April 1964, flight testing began using the intended power plant, the more powerful SO-1; from the latter half of the 1960s, production Iskras were powered either by the SO-1 or by the modified but similarly rated SO-3.

About 500 Iskras had been built by mid-1979, in the following versions:

Iskra-Bis A. Basic two-seat version, for primary and advanced training. Two underwing hardpoints for external weapons.

Iskra-Bis B. Two-seat primary and advanced trainer, with four underwing attachments for missiles or other weapons. Prototype designated Iskra 100.

Iskra-Bis C. Single-seat reconnaissance version. Prototype, designated Iskra 200, first flew in June 1972. Increased fuel capacity for greater range; camera in fuselage floor, aft of cockpit.

Iskra-Bis D. Similar to Bis B, but able to carry a wider selection of external weapons. Prototype also designated Iskra 200. Export version supplied to Indian Air Force (50) in 1976. In production.

Iskra-Bis DF. Two-seat combat and reconnaissance trainer, with increased armament capability of Bis D, plus provision for three cameras: one in each air intake fairing and one in fuselage floor beneath rear cockpit. In production.

The following description applies to the Iskra-Bis B, D and DF, except where indicated:

TYPE: Fully aerobatic two-seat jet primary and advanced trainer (A, B and D) and single-seat reconnaissance aircraft (C and DF).

WINGS: Cantilever mid-wing monoplane. Wing section NACA 64209 at root, NACA 64009 at tip. Sweepback at quarter-chord 7°. Marked dihedral. All-metal torsion box structure with two steel main spars and duralumin stressed skin. Hydraulically servo-assisted, aerodynamically-balanced ailerons. Hydraulically-actuated two-section double-slotted flaps and airbrakes (max deflection 87°). One boundary layer fence on each wing. Anti-flutter weight fairing projecting from each wing near tip.

FUSELAGE: All-metal semi-monocoque structure of pod and boom type.

TAIL UNIT: Cantilever all-metal structure. Two-spar fin, integral with fuselage. Variable-incidence two-spar tailplane, actuated electrically. Mass- and aerodynamically-balanced elevators and rudder. Anti-flutter weight fairing projecting from each half of tailplane at tip. Ground-adjustable tab on rudder; fixed balance tab in port elevator.

LANDING GEAR: Retractable tricycle type with single wheel on each unit. Nosewheel retracts forward, main wheels inward into wing-root air intake trunks. Hydraulic actuation, with pneumatic emergency extension. Main wheels size 600 × 180, tyre pressure 5·38 bars (78 lb/sq in). Nosewheel size 400 × 150, tyre pressure 3·45 bars (50 lb/sq in). Oleo-pneumatic shock-absorbers. Disc brakes on main wheels. Castoring and self-centering nosewheel, with shimmy damper.

POWER PLANT: One SO-3 turbojet, rated at 9·81 kN (2,205 lb st), or (Iskra-Bis DF) higher-powered SO-3W, mounted in fuselage aft of cockpit section, with nozzle under tailboom. Fuel in two 315 litre (69 Imp gallon) integral wing tanks, one rubber 500 litre (110 Imp gallon) fuselage main tank (700 litre; 154 Imp gallon in single-seaters) and one rubber 70 litre (15·5 Imp gallon) fuselage collector tank. Total fuel capacity 1,200 litres (263·5 Imp gallons) in two-seaters, 1,400 litres (308 Imp gallons) in single-seaters. Fuel system permits up to 40 s of inverted flight.

ACCOMMODATION: Crew of one, or two in tandem, on lightweight ejection seat(s), under a one-piece hydraulically-actuated rearward-hinged upward-opening jettisonable canopy. Cockpit pressurised and air-conditioned. Rear seat of trainers slightly raised.

SYSTEMS: Hydraulic system, pressure 138 bars (2,000 lb/sq in), for actuation of ailerons, flaps, airbrakes, landing gear, canopy, and main-wheel brakes. Pneumatic system, pressure 118 bars (1,710 lb/sq in), for cockpit pressurisation, anti-icing and gun charging. Emergency pneumatic system for landing gear extension, flaps and emergency braking. Electrical power provided by 28·5V AW-30 generator and 24V 28Ah battery, for engine starting, instruments, lights and armament control system; 115V converter for AC power. Air-conditioning, oxygen, ethyl alcohol anti-icing and CO_2 fire extinguishing systems standard.

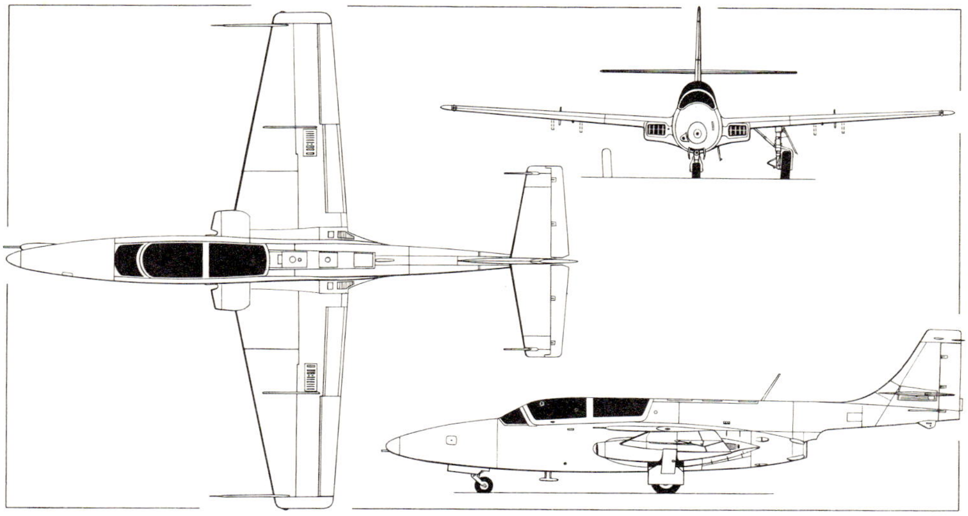

PZL Mielec Iskra-Bis D two-seat trainer, with four underwing pylons *(Pilot Press)*

Standard two-seat Iskra-Bis D trainer

AVIONICS AND EQUIPMENT: Trainers have complete dual controls and instrumentation, including blind-flying panels. Standard avionics also include R-800, R-802G or R-802W VHF com; ARK-9 or ARL-1601 radio compass; RW-UM radio altimeter; MRP-56P marker beacon receiver; SPU-2P crew intercom in two-seaters; and IFF.

ARMAMENT: Forward-firing 23 mm cannon in nose on starboard side, with S-13 gun camera. Two or four attachments for a variety of underwing stores, including bombs of up to 100 kg (220 lb), eight-barrel rocket pods and 7·62 mm gun pods.

DIMENSIONS, EXTERNAL:
Wing span 10·06 m (33 ft 0 in)
Wing chord at root 2·254 m (7 ft 4¾ in)
Wing chord at tip 1·162 m (3 ft 9¾ in)
Wing aspect ratio 5·71
Length overall 11·15 m (36 ft 7 in)
Height overall 3·50 m (11 ft 5½ in)
Tailplane span 3·84 m (12 ft 7¼ in)
Wheel track 3·47 m (11 ft 4½ in)
Wheelbase 3·44 m (11 ft 3½ in)

AREAS:
Wings, gross 17·50 m² (188·37 sq ft)
Ailerons (total) 1·48 m² (15·93 sq ft)
Trailing-edge flaps (total) 1·74 m² (18·73 sq ft)
Fin 1·55 m² (16·68 sq ft)
Rudder (incl tab) 0·70 m² (7·53 sq ft)
Tailplane 2·38 m² (25·62 sq ft)
Elevators (incl tab) 1·16 m² (12·48 sq ft)

WEIGHTS AND LOADINGS:
Weight empty (trainer versions):
B 2,494 kg (5,498 lb)
D/DF 2,560 kg (5,644 lb)
Normal T-O weight with 570 litres (125·5 Imp gallons) internal fuel: B 3,184 kg (7,019 lb)
D/DF 3,243 kg (7,150 lb)
Normal T-O weight with 1,200 litres (263·5 Imp gallons) internal fuel: B 3,704 kg (8,166 lb)
D/DF 3,734 kg (8,232 lb)
T-O weight (reconnaissance version):
DF 3,787 kg (8,349 lb)
Max T-O weight with full external armament:
B 3,810 kg (8,400 lb)
D/DF 3,840 kg (8,465 lb)
Max wing loading:
D/DF 219 kg/m² (44·85 lb/sq ft)
Max power loading:
D (SO-3) 387·4 kg/kN (3·8 lb/lb st)
DF (SO-3W) 355·9 kg/kN (3·5 lb/lb st)

PERFORMANCE (at normal T-O weight with full internal fuel, except where indicated):
Never-exceed speed
Mach 0·8 (404 knots; 750 km/h; 466 mph)
Max level speed at 5,000 m (16,400 ft):
B, D 388 knots (720 km/h; 447 mph)
DF 415 knots (770 km/h; 478 mph)
Normal cruising speed:
B, D/DF 324 knots (600 km/h; 373 mph)
Unstick speed:
B, D/DF 102 knots (190 km/h; 118 mph)
Landing speed:
B, D/DF 92 knots (170 km/h; 106 mph)
Stalling speed, power off, flaps down:
B, D/DF 81 knots (150 km/h; 93·5 mph)
Max rate of climb at S/L:
B, D 840 m (2,755 ft)/min
DF 1,164 m (3,820 ft)/min
Time to 5,000 m (16,400 ft): DF 5 min 18 s
Time to 6,000 m (19,685 ft): B, D 9 min 36 s
Time to 7,000 m (22,975 ft): B, D 13 min 36 s
Time to 11,000 m (36,000 ft): DF 26 min 0 s
Service ceiling: B 11,140 m (36,550 ft)
D/DF 11,000 m (36,000 ft)
T-O run: B 725 m (2,380 ft)
D 750 m (2,460 ft)
DF 650 m (2,135 ft)
T-O to 15 m (50 ft), flaps down:
B 1,100 m (3,609 ft)
D 1,190 m (3,904 ft)
DF 1,090 m (3,575 ft)
Landing from 15 m (50 ft), flaps down:
B 1,000 m (3,280 ft)
D/DF 1,110 m (3,642 ft)
Landing run: B 660 m (2,165 ft)
D/DF 700 m (2,300 ft)
Range at 7,000 m (22,975 ft):
B with 570 litres fuel 235 nm (435 km; 270 miles)
D/DF with 570 litres fuel 243 nm (450 km; 280 miles)
B with 1,200 litres fuel 626 nm (1,160 km; 720 miles)
g limits (ultimate): B, D/DF +8·0; −4·0

PZL MIELEC M-15 BELPHEGOR

As recorded in earlier editions of *Jane's*, the Polish and Soviet governments discussed in 1971 the development of a new, large agricultural aircraft known as the M-15, together with associated agricultural and ground support equipment. The Soviet government indicated a requirement for about 3,000 such aircraft, and on 2 December 1971 signed an agreement with the Polish government for large-scale production of the M-15.

Initial design of the M-15 was undertaken by a design bureau at Mielec under Soviet chief consulting engineer R. A. Ismailov and Polish designer K. Gocyla, and staffed by Polish and Soviet specialists. The agricultural equipment for the aircraft was developed jointly by the Instytut Lotnictwa at Warsaw (which see) and the Soviet Research Institute of Special and Utility Aviation at Krasnodar.

A prototype, designated LLP-M15, was flown on 30 May 1973; the first fully-representative M-15 prototype made its first flight on 9 January 1974. On 2 April 1975 five pre-series M-15s were sent to the USSR for evaluation; Soviet acceptance tests were completed successfully in 1976 and were followed by the award of a provisional Soviet GosAvia certificate of airworthiness. Completion

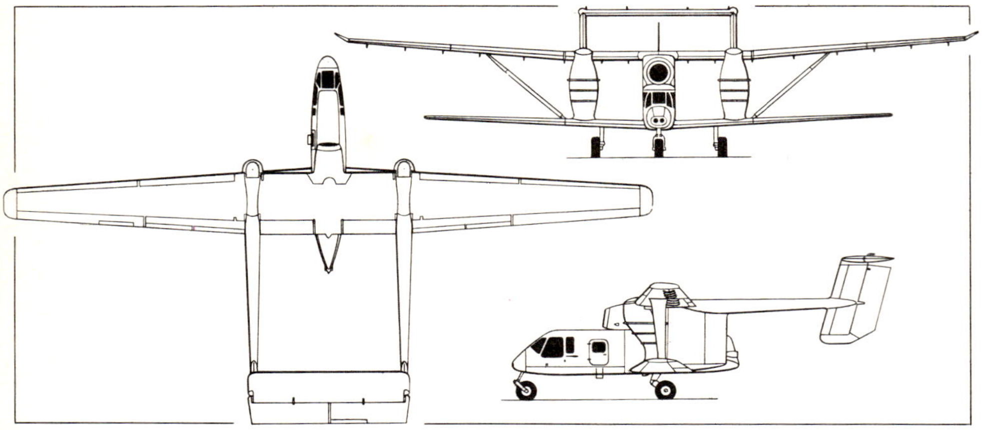

PZL Mielec M-15 turbofan-engined agricultural aircraft *(Pilot Press)*

PZL Mielec M-15 Belphegor agricultural biplane (Ivchenko AI-25 turbofan engine)

of the M-15 development programme, and the granting of a full Soviet certificate, was marked by a ceremony at the Mielec factory on 4 April 1979. The name Belphegor was registered for the M-15 at about the same time.

The prototype of a tandem two-seat training version, with large cabin side windows, was completed in 1980, and an improved standard model is under development for future production. Firefighting and cargo-carrying versions are expected to be developed.

The following description refers to the agricultural production version:

TYPE: Three-seat agricultural aircraft.

WINGS: Biplane wings, of mainly metal construction and unequal span, built chiefly of aluminium and steel alloys and glassfibre laminates. Wing section P-II-14. The upper wing has a constant-chord centre-section and tapered outer panels; dihedral at tips is increased on current aircraft. The centre-section is faired to the top of the engine pod. The shorter-span lower wings, which house the agricultural dispersal pipes, are of generally similar planform and are joined to the fuselage nacelle at floor level. The entire trailing-edge of the upper wing is hinged, and is made up of hydraulically-operated double-slotted flaps and single-slotted ailerons. There are automatically-operated slats on the leading-edge. In line with each tailboom, and occupying the full depth of the gap between the upper and lower wings, is a narrow streamlined hopper for agricultural chemical, and there is a single outward-sloping bracing strut outboard of each hopper fairing. Trim tab in port aileron.

FUSELAGE: Central semi-monocoque nacelle, of narrow rectangular section, built of same materials as wings.

TAIL UNIT: Cantilever metal/glassfibre structure, consisting of twin sweptback two-spar endplate fins and rudders, bridged by a high-mounted tailplane and full-span elevator, supported on two slender tailbooms located at approx one-quarter span on the upper wing. Trim tabs in starboard half of elevator and starboard rudder.

LANDING GEAR: Non-retractable tricycle type, with single wheel and oleo-pneumatic shock-absorber on each unit. Main wheels and tyres size 720 × 320 mm, tyre pressure 3·0 bars (43 lb/sq in); nosewheel and tyre size 700 × 250 mm, tyre pressure 2·5 bars (36 lb/sq in). Nosewheel steerable hydraulically, 50° to left or right. Hydraulic disc brakes on main wheels.

POWER PLANT: One 14·7 kN (3,306 lb st) Ivchenko AI-25 turbofan engine, mounted in a pod on top of the fuselage. Eight rubber fuel tanks in upper wing, total capacity 1,460 litres (321 Imp gallons).

ACCOMMODATION: Seat for pilot in fully-enclosed cockpit in extreme nose of fuselage. Two seats in cabin, to rear of pilot's compartment, for carrying ground staff during ferry flights. Cockpit air-conditioning by engine compressor bleed air.

SYSTEMS: Hydraulic system, pressure 150 bars (2,175 lb/sq in), for actuation of nosewheel steering, brakes, flaps, and dispersal systems (including emergency jettison). Freon-type fire extinguishing system.

AVIONICS AND EQUIPMENT: Full flight and navigation instrumentation, including stall-warning indicator.

Two-seat training version of the M-15 Belphegor, with larger side windows *(A. Glass)*

VFR radio/navigation equipment optional. The two between-wings hoppers have a combined capacity for 2,900 litres (638 Imp gallons) of liquid or 2,200 kg (4,850 lb) of dry (powdered or granulated) chemical. Ivchenko AI-9 APU, normally removed from aircraft during agricultural operations, provides power for engine starting, ground refuelling and filling of hoppers with liquid chemical. Compressed air for chemical dispersal is bled from the AI-25 engine. Twin landing lights in nose.

DIMENSIONS, EXTERNAL:

Wing span: upper	22·33 m (73 ft 3 in)
lower	16·428 m (53 ft 10¾ in)
Wing mean aerodynamic chord (upper)	1·84 m (6 ft 0½ in)
Length overall	13·135 m (43 ft 1 in)
Height overall	5·339 m (17 ft 6¼ in)
Height of fuselage above ground	1·008 m (3 ft 3½ in)
Wheel track	4·319 m (14 ft 2 in)
Wheelbase	4·879 m (16 ft 0 in)
Distance between c/l of hoppers	5·20 m (17 ft 0¾ in)

AREAS:

Wings (total)	67·90 m² (730·9 sq ft)
Ailerons (total)	9·03 m² (97·20 sq ft)
Trailing-edge flaps (total)	4·99 m² (53·71 sq ft)
Fins (total)	5·53 m² (59·52 sq ft)
Rudders (total)	4·00 m² (43·06 sq ft)
Tailplane	5·92 m² (63·72 sq ft)
Elevators (total)	4·08 m² (43·92 sq ft)

WEIGHTS AND LOADINGS:

Weight empty: dusting	3,230 kg (7,120 lb)
spraying	3,270 kg (7,210 lb)
with atomisers	3,275 kg (7,220 lb)
Max hopper load	2,200 kg (4,850 lb)
Max T-O weight	5,750 kg (12,675 lb)
Max landing weight	4,000 kg (8,815 lb)
Max wing loading	85·19 kg/m² (17·44 lb/sq ft)
Max power loading	391·19 kg/kN (3·83 lb/lb st)

PERFORMANCE (at max T-O weight):

Max cruising speed	108 knots (200 km/h; 124 mph)
Normal operating speed	86-95 knots (160-175 km/h; 99-109 mph)
Stalling speed, flaps up	60·5 knots (112 km/h; 70 mph)
Stalling speed, flaps down	48 knots (89 km/h; 55·5 mph)
Max rate of climb at S/L	390 m (1,280 ft)/min
T-O run: on grass	380 m (1,247 ft)
on concrete	260 m (853 ft)
Landing run: on grass	190 m (624 ft)
on concrete	330 m (1,083 ft)
Max range at 3,000 m (9,850 ft)	259 nm (480 km; 298 miles)
Swath width: dusting	65 m (213 ft)
spraying	40 m (131 ft)
with atomisers	70 m (230 ft)
Application rate:	
dusting	10-50 kg (22-110 lb)/s
spraying	0-25 litres (0-5·5 Imp gallons)/s
with atomisers	0·2-6·5 litres (0·05-1·4 Imp gallons)/s

PZL MIELEC M-18 DROMADER (DROMEDARY)

Although superficially similar to the CNPSL-PZL Warszawa PZL-106A Kruk, the M-18 Dromader is an entirely different and much larger agricultural aircraft, and is designed to meet the requirements of FAR Pt 23. Particular attention was paid in the design to pilot safety, and all parts of the structure exposed to contact with chemicals are treated with polyurethane or epoxy enamels, or manufactured from stainless steel.

The M-18 was first flown on 27 August 1976, and a second flying prototype followed on 2 October 1976; a third was completed for static tests. The M-18 made its public debut at the Paris Air Show in May/June 1977. The prototypes were followed by seven pre-series aircraft, of which two were used for static and fatigue testing. The remainder were employed for operating trials, two of them spraying and dusting cotton in Egypt during the Summer of 1978, prior to the award of a Polish type certificate on 27 September 1978. Other pre-series aircraft were exhibited and demonstrated in 1978 in Hungary, Iraq, Italy, the USA and Yugoslavia. A total of 50 production Dromaders had been built by the beginning of 1981, with production then planned to rise to a rate of eight per month. Orders during 1980 included 15 for Hungary and two for Cuba.

Future programmes are planned to include lighter-weight wings, and the development of new agricultural and other equipment, to decrease the reliance on imported items. A firefighting version of the Dromader was flown for the first time on 11 November 1978. Six were delivered to a customer in Canada in 1980, following Canadian certification of this version on 10 March of that year.

The following description refers to the standard agricultural version:

TYPE: Single-seat agricultural aircraft; *g* limits +3·4/−1·4 (FAR 23); +2·8 (CAM 8).

WINGS: Cantilever all-metal low-wing monoplane, of constant chord, with 2° 30′ dihedral on centre-section and 5° on outer panels. Wing sections NACA 4416 at root, NACA 4412 at tip. Incidence 3°. Single steel-capped duralumin spar. All-metal two-section trailing-edge slotted flaps, actuated hydraulically. All-metal slotted ailerons, mass and aerodynamically balanced, actuated by pushrods. No tabs.

FUSELAGE: All-metal structure. Main frame, of helium-arc welded chrome-molybdenum steel tube, oiled internally against corrosion. Duralumin side panels, detachable for airframe inspection and cleaning. Fixed stainless steel bottom covering.

TAIL UNIT: All-metal structure, with braced tailplane. Corrugated skin on vertical surfaces. Aerodynamically and mass balanced rudder and elevators. Elevator actuated by pushrods; rudder by cables. Trim tab on each elevator.

LANDING GEAR: Non-retractable tailwheel type. Oleo-pneumatic shock-absorber in each unit. Main units have tyres size 720 × 320 mm, and are fitted with hydraulic disc brakes, parking brake and wire cutters. Fully-castoring tailwheel, lockable for take-off and landing, with size 318 × 114 mm tyre.

POWER PLANT: One 736 kW (987 hp) PZL Kalisz ASz-621R nine-cylinder radial aircooled supercharged engine, driving a PZL four-blade constant-speed aluminium propeller. Integral fuel tank in each outer wing panel, combined usable capacity 400 litres (88 Imp gallons; 105·7 US gallons). Gravity-feed header tank in fuselage.

ACCOMMODATION: Single adjustable seat in fully enclosed, sealed and ventilated cockpit which is stressed to withstand 40*g* impact. Glassfibre cockpit roof and rear fairing. Adjustable shoulder-type safety harness. Adjustable rudder pedals. Baggage compartment aft of seat. Quick-opening door on each side; port door jettisonable.

SYSTEMS: Hydraulic system, pressure 98-137 bars (1,421-1,987 lb/sq in), for flap actuation, disc brakes and dispersal system. Electrical system powered by 28·5V 100A generator, with 24V 25Ah nickel-cadmium battery and overvoltage protection relay.

AVIONICS AND EQUIPMENT: King KX 170B communications transceiver, KI 201C navigation receiver, VOR-OBS indicator, landing lights, taxi light and night working light optional. Navigation lights, cockpit light, instrument panel lights, and two rotating beacons standard. Built-in jacking and tiedown points in wings and aft fuselage; towing lugs on main landing gear. Cockpit fire extinguisher and first aid kit.

AGRICULTURAL EQUIPMENT: Glassfibre epoxy hopper, with stainless steel tube bracing, forward of cockpit; capacity 2,500 litres (550 Imp gallons; 660 US gallons) of liquid or 1,500 kg (3,306 lb) of dry chemical. Deflector cable from cabin roof to fin. Transland gatebox, control valve and strainer, Root pump, and 48/96-nozzle spraybooms for spraying; Transland gatebox, control valve and high output spreader for dusting with dry chemical; or eight AV 3000 atomisers for fine spraying. Aircraft can also be fitted with Rockwell International water bombing installation for fire suppression.

DIMENSIONS, EXTERNAL:

Wing span	17·70 m (58 ft 0¾ in)
Wing chord (constant)	2·286 m (7 ft 6 in)
Wing aspect ratio	7·83
Length overall	9·47 m (31 ft 1 in)
Height overall	3·70 m (12 ft 1¾ in)
Tailplane span	5·00 m (16 ft 4¾ in)
Wheel track	3·58 m (11 ft 9 in)
Propeller diameter	3·30 m (10 ft 10 in)
Propeller ground clearance (tail up)	0·23 m (9 in)

AREAS:

Wings, gross	40·00 m² (430·5 sq ft)
Ailerons and flaps (total)	9·53 m² (102·6 sq ft)
Vertical tail surfaces (total)	2·65 m² (28·5 sq ft)
Horizontal tail surfaces (total)	6·50 m² (70·0 sq ft)

WEIGHTS AND LOADINGS:

Weight empty	2,550 kg (5,622 lb)
Payload: FAR 23	1,050-1,350 kg (2,315-2,976 lb)
CAM 8	1,550-1,850 kg (3,417-4,078 lb)
Max T-O weight: FAR 23	4,200 kg (9,259 lb)
CAM 8	4,700 kg (10,362 lb)
Max landing weight	4,200 kg (9,259 lb)
Max wing loading (FAR 23)	105·0 kg/m² (21·51 lb/sq ft)
Max power loading (FAR 23)	5·63 kg/kW (9·26 lb/hp)

PERFORMANCE (at 4,200 kg; 9,259 lb max T-O weight, ISA. A: without agricultural equipment; B: with spreader equipment):

Never-exceed speed:	
A	151 knots (280 km/h; 174 mph)
Max level speed: A	138 knots (256 km/h; 159 mph)
B	128 knots (237 km/h; 147 mph)
Cruising speed: A	110 knots (205 km/h; 127 mph)
B	102 knots (190 km/h; 118 mph)
Operating speed:	
B	92-102·5 knots (170-190 km/h; 106-118 mph)
Stalling speed, power off, flaps up:	
A, B	69 knots (127 km/h; 79 mph)
Stalling speed, power off, flaps down:	
A, B	59 knots (109 km/h; 68 mph)
Max rate of climb at S/L: A	348 m (1,141 ft)/min
Service ceiling: A	6,500 m (21,325 ft)
T-O run: A	275 m (903 ft)
Landing run: A, B	330 m (1,085 ft)
Max range, no reserves:	
A	280 nm (520 km; 323 miles)

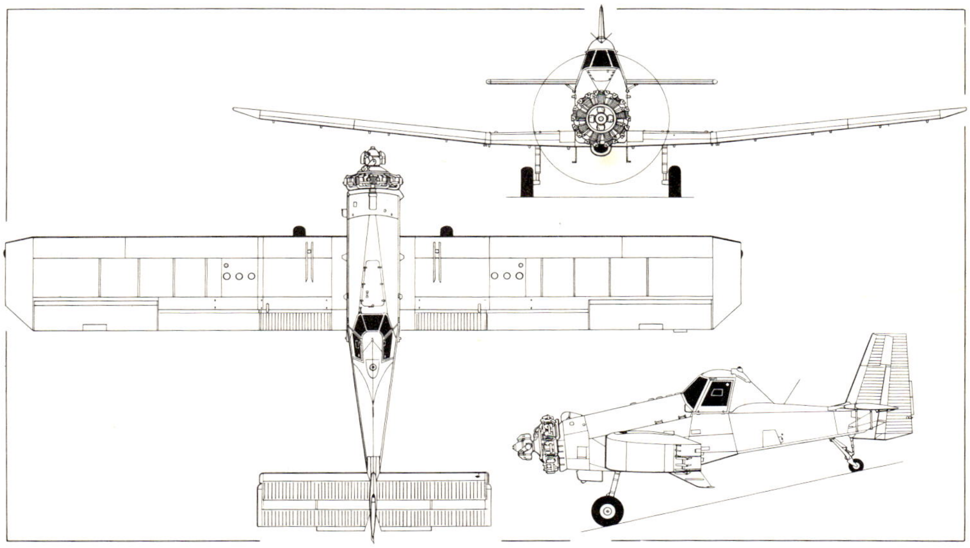

Three-view drawing *(Pilot Press)* and photograph of the PZL Mielec M-18 Dromader agricultural aircraft

PZL Mielec M-20 Mewa, a version of the Piper Seneca with PZL-F engines

PZL MIELEC M-20 MEWA (GULL)

It was announced in early 1977 that Pezetel had signed an agreement with Piper Aircraft Corporation of the USA whereby the Polish aircraft industry will assemble, manufacture and distribute in Eastern Europe the Piper Seneca twin-engined business aircraft.

This work, which will be undertaken by PZL Mielec, will begin with the assembly of imported kits, and will progress eventually to full licence manufacture. The aircraft (production of 400 or more has been authorised by the US government) will be powered by 164 kW (220 hp) PZL-F (Franklin) 6A-350C-1L/R engines, and will be known as the M-20 Mewa. The Polish prototype made its initial flight on 23 July 1979; two others were flying by mid-1980. Production version will be equivalent to the Seneca III.

WSK-PZL SWIDNIK

WYTWORNIA SPRZETU KOMUNIKACYJNEGO Im. ZYGMUNTA PULAWSKIEGO-PZL SWIDNIK (Zygmunt Pulawski Transport Equipment Manufacturing Centre, Swidnik)

HEAD OFFICE AND WORKS: 21-045 Swidnik k/Lublina
Telephone: Lublin 12061 and 12071
Telex: 84212 and 84302
GENERAL MANAGER: Dipl Ing Jan Czogala

The factory at Swidnik was built in 1951-52 and was engaged initially in manufacturing components for the LiM-1 (MiG-15) jet fighter.

In 1955, when the manufacture of combat aircraft was drastically reduced in Poland, the WSK at Swidnik began licence production of the Soviet-designed Mi-1 helicopter, some 1,700 of which were built under the designation SM-1. A small design office was formed subsequently at the factory to work on variants and developments of the basic SM-1 design and on original projects.

In September 1957, the Swidnik works was named after the famous pre-war PZL designer Zygmunt Pulawski, and currently employs about 10,000 people. Production is concentrated at present on various versions of the Soviet-designed Mil Mi-2 turbine-powered helicopter.

Swidnik, together with other PZL factories at Mielec and Kalisz, is manufacturing components for the Soviet Ilyushin Il-86 wide-bodied airliner.

PZL SWIDNIK (MIL) Mi-2

NATO reporting name: Hoplite

The Mil Mi-2, first flown in September 1961, was designed in the USSR by the Mikhail L. Mil bureau. It retains the basic configuration of Mil's earlier Mi-1 helicopter, but instead of a single piston engine has two Isotov turboshaft engines mounted side by side above the cabin.

Development of the Mi-2 prototype continued in the USSR until the helicopter had completed its initial State trials programme of flying. Then, in accordance with an agreement signed in January 1964, further development, production and marketing of the Mi-2 were assigned exclusively to the Polish aircraft industry, which flew its own first example of the Mi-2 on 4 November 1965.

Production at Swidnik began in 1965, and this factory has since built more than 3,000 in 24 different civil and military versions; the majority of these have been exported, including more than 2,000 to the USSR. Among the operators of the Mi-2 are the air forces of Czechoslovakia, Poland, Romania and the USSR, and civil operators in European and various developing countries. Production in 1979 was at a rate of approx 300 a year, and orders are reportedly sufficient to ensure continuation of production until 1985.

In the 1970s Swidnik initiated a development programme to improve and modernise the original design. The first modernised version was powered by uprated engines of 335 kW (450 shp) each and was essentially similar to the basic Mi-2. It flew for the first time in 1974. On another version, the metal stabiliser, tail rotor blades and main rotor blades were replaced with similar components made of plastics, intended to simplify production and improve performance. The new rotor blades were designed, manufactured and tested at Swidnik.

There are several versions of the basic Mi-2, as follows:

(a) Convertible passenger/cargo transport;
(b) Passengers-only version, for six or eight passengers;
(c) Ambulance version (Mi-2R);
(d) Agricultural version, for conventional or ultra-low-volume spraying, known in Poland as the **Bazant** (Pheasant). In service in Bulgaria, Czechoslovakia, Denmark, Egypt, Finland, Iran, Iraq, Libya, Poland, Sudan, Sweden and USSR;
(e) Search and rescue version, with electrically-operated external hoist;
(f) Freighter version, with external cargo sling;
(g) Pilot training version, designed by WSK Swidnik;
(h) Photogrammetric version, produced in small numbers only;
(i) Television version (for transmission from the air), produced in small numbers only;
(j) Version with 260 kg (573 lb) capacity hoist.

Agricultural version of the PZL Swidnik (Mil) Mi-2 twin-turbine helicopter

The following details apply specifically to the basic Mi-2, except where indicated:

Type: Twin-turbine general-purpose light helicopter.

Rotor System: Three-blade main rotor fitted with hydraulic blade vibration dampers. All-metal blades of NACA 230-13M section. Flapping, drag and pitch hinges on each blade. Main rotor blades and those of two-blade tail rotor each consist of an extruded duralumin spar with bonded honeycomb trailing-edge pockets. Anti-flutter weights on leading-edges, balancing plates on trailing-edges. Hydraulic boosters for longitudinal, lateral and collective pitch controls. Coil spring counterbalance mechanism in main and tail rotor systems. Pitch-change centrifugal loads on tail rotor carried by ribbon-type steel torsion elements. Rotors do not fold. Electrical blade de-icing system for main and tail rotors. Rotor brake fitted.

Rotor Drive: Main rotor shaft driven via gearbox on each engine; three-stage WR-2 main gearbox, intermediate gearbox and tail rotor gearbox. Main rotor/engine rpm ratio 1 : 24·6; tail rotor/engine rpm ratio 1 : 4·16. Main gearbox provides drive for auxiliary systems and take-off for rotor brake. Freewheel units permit disengagement of a failed engine and also autorotation.

Fuselage: Conventional semi-monocoque structure of pod and boom type, made up of three main assemblies: the nose (including cockpit), central section and tail-boom. Construction is of sheet duralumin, bonded and spot-welded or riveted to longerons and frames. Main load-bearing joints are of steel alloy.

Tail Unit: Variable-incidence horizontal stabiliser controlled by collective-pitch lever.

Landing Gear: Non-retractable tricycle type, plus tail-skid. Twin-wheel nose unit. Single wheel on each main unit. Oleo-pneumatic shock-absorbers in all units, including tailskid. Main shock-absorbers designed to cope with both normal operating loads and possible ground resonance. Main-wheel tyres size 600 × 180, pressure 4·41 bars (64 lb/sq in). Nosewheel tyres size 300 × 125, pressure 3·45 bars (50 lb/sq in). Pneumatic brakes on main wheels. Metal ski landing gear optional.

Power Plant: Two 298 or 335 kW (400 or 450 shp) Polish-built Isotov GTD-350P turboshaft engines, mounted side by side above cabin. Fuel in single rubber tank, capacity 600 litres (131 Imp gallons), under cabin floor. Provision for carrying a 238 litre (52·4 Imp gallon) external tank on each side of cabin. Refuelling point in starboard side of fuselage. Oil capacity 25 litres (5·4 Imp gallons). Engine air intake de-icing by engine bleed air.

Accommodation: Normal accommodation for one pilot on flight deck (port side). Seats for up to eight passengers in air-conditioned cabin, there being back-to-back bench seats for three persons each, with two optional extra starboard side seats at the rear, one behind the other. All seats are removable for carrying up to 700 kg (1,543 lb) of internal freight. Access to cabin via forward-hinged doors on each side at front of cabin and aft on port side. Pilot's sliding window jettisonable in emergency. Ambulance version has accommodation for four stretchers and a medical attendant or for two stretchers and two sitting casualties. Side-by-side seats and dual controls in pilot training version. Cabin heating, ventilation and air-conditioning standard. Electrical de-icing of windscreen.

Systems: Cabin heating, by engine bleed air, and ventilation; heat exchangers warm atmospheric air for ventilation system during cold weather. Hydraulic system, pressure 59-78·6 bars (855-1,140 lb/sq in), for cyclic and collective pitch control boosters. Pneumatic system, pressure 49 bars (710 lb/sq in), for main-wheel brakes. AC electrical system, with two STG-3 3kW engine-driven starter/generators and 208V 16kVA three-phase alternator. 24V DC system, with two 28Ah lead-acid batteries.

Avionics and Equipment: Standard equipment includes two transceivers (medium and short wave), gyro compass, radio compass, radio altimeter, intercom system and blind-flying panel. Electrically-operated wiper for pilot's windscreen. Fire extinguishing system, for engine bays and main gearbox compartment, is generally similar to, but simpler than, the freon system fitted to the Soviet Mil Mi-8, and can be actuated automatically or manually.

Operational Equipment: Bazant agricultural version carries a hopper on each side of the fuselage (total capacity 1,000 litres; 220 Imp gallons of liquid or 750 kg; 1,650 lb of dry chemical) and either a spraybar to the rear of the cabin on each side or a distributor for dry chemicals under each hopper. Swath width covered by the spraying version is 40-45 m (130-150 ft). As a search and rescue aircraft, an electric hoist, capacity 120 kg (264 lb), is fitted. In the freight role an underfuselage hook can be fitted for suspended loads of up to 800 kg (1,763 lb).

Armament: Some Mi-2s of the Polish Air Force are equipped with rocket pods or air-to-surface missiles mounted on pylons on each side of the cabin.

Dimensions, external:
- Diameter of main rotor: 14·56 m (47 ft 9¼ in)
- Main rotor blade chord (constant, each): 0·40 m (1 ft 3¾ in)
- Diameter of tail rotor: 2·70 m (8 ft 10¼ in)
- Length overall, rotors turning: 17·42 m (57 ft 2 in)
- Length of fuselage: 11·94 m (39 ft 2 in)
- Height to top of rotor hub: 3·75 m (12 ft 3½ in)
- Stabiliser span: 1·85 m (6 ft 0¾ in)
- Wheel track: 3·05 m (10 ft 0 in)
- Wheelbase: 2·71 m (8 ft 10¾ in)
- Tail rotor ground clearance: 1·59 m (5 ft 2¾ in)
- Cabin door (port, rear):
  - Height: 1·065 m (3 ft 5¾ in)
  - Width: 1·115 m (3 ft 8 in)
- Cabin door (stbd, front):
  - Height: 1·11 m (3 ft 7¾ in)
  - Width: 0·75 m (2 ft 5½ in)
- Cabin door (port, front):
  - Height: approx 1·40 m (4 ft 7 in)
  - Width: approx 1·20 m (3 ft 11¼ in)

Dimensions, internal:
- Cabin:
  - Length, incl flight deck: 4·07 m (13 ft 4¼ in)
  - Length, excl flight deck: 2·27 m (7 ft 5½ in)
  - Mean width: 1·20 m (3 ft 11¼ in)
  - Mean height: 1·40 m (4 ft 7 in)

Areas:
- Main rotor blades (each): 2·40 m² (25·83 sq ft)
- Tail rotor blades (each): 0·22 m² (2·37 sq ft)
- Main rotor disc: 166·4 m² (1,791·11 sq ft)
- Tail rotor disc: 5·73 m² (61·68 sq ft)
- Horizontal stabiliser: 0·70 m² (7·53 sq ft)

Weights and Loading:
- Weight empty, equipped:
  - passenger version: 2,402 kg (5,295 lb)
  - cargo version: 2,372 kg (5,229 lb)
  - ambulance version: 2,410 kg (5,313 lb)
  - Bazant: 2,372 kg (5,229 lb)
- Basic operating weight empty:
  - single-pilot versions: 2,365 kg (5,213 lb)
  - dual-control version: 2,424 kg (5,344 lb)
- Max payload, excl pilot, oil and fuel: 800 kg (1,763 lb)
- Normal T-O weight (and max T-O weight of Bazant): 3,550 kg (7,826 lb)
- Max T-O weight: 3,700 kg (8,157 lb)
- Max disc loading: 22·4 kg/m² (4·6 lb/sq ft)

Performance (at 3,550 kg; 7,826 lb T-O weight):
- Never-exceed speed at 500 m (1,640 ft):
  - Bazant: 84 knots (155 km/h; 96·5 mph)
  - other versions: 113 knots (210 km/h; 130·5 mph)
- Max cruising speed at 500 m (1,640 ft):
  - Bazant (without agricultural equipment): 102 knots (190 km/h; 118 mph)
  - other versions: 108 knots (200 km/h; 124 mph)
- Max level speed with agricultural equipment (Bazant): 84 knots (155 km/h; 96 mph)
- Econ cruising speed for max range at 500 m (1,640 ft): 102 knots (190 km/h; 118 mph)
- Econ cruising speed for max endurance at 500 m (1,640 ft): 54 knots (100 km/h; 62 mph)
- Max rate of climb at S/L: 270 m (885 ft)/min
- Time to 1,000 m (3,280 ft): 5 min 30 s
- Time to 4,000 m (13,125 ft): 26 min 0 s
- Service ceiling: 4,000 m (13,125 ft)
- Hovering ceiling IGE: approx 2,000 m (6,560 ft)
- Hovering ceiling OGE: approx 1,000 m (3,280 ft)
- Min landing area: 30 × 30 m (100 × 100 ft)

Armed version of the PZL Swidnik Mi-2, with four pylon-mounted air-to-surface missiles

- Range at 500 m (1,640 ft):
  - max payload, 5% fuel reserves: 91 nm (170 km; 105 miles)
  - max internal fuel, no reserves: 237 nm (440 km; 273 miles)
  - max internal and auxiliary fuel, 30 min reserves: 313 nm (580 km; 360 miles)
  - max internal and auxiliary fuel, no reserves: 430 nm (797 km; 495 miles)
- Endurance at 500 m (1,640 ft), no reserves:
  - max internal fuel: 2 h 45 min
  - max internal and auxiliary fuel: 5 h 0 min
- Endurance (Bazant), 5% reserves:
  - spraying: 40 min
  - dusting: 50 min

PZL SWIDNIK KANIA/KITTY HAWK and TAURUS

In collaboration with the Detroit Diesel Allison Division of General Motors Corporation in the USA, PZL Swidnik is developing a modified export version of the Mi-2 light helicopter, known as the Kania or Kitty Hawk, powered by two Allison 250-C20B turboshaft engines. In addition to a standard passenger version, the Kania is intended for cargo, agricultural, ambulance and other roles, similar to those performed by the Mi-2. Two examples were scheduled for completion in 1979, and the first of these (SP-PSA) made its initial flight on 3 June that year. At least two more were due for completion by the end of 1980.

In the USA, Spitfire Helicopters (which see) has rights to market a modified version of the Kania under the name **Taurus**. External differences in this version include a modified engine air intake cowling, with single orifice, revised nose contours, and a ventral tail fin. The power plant of the Taurus comprises two of the more powerful Allison 250-C28 turboshaft engines.

The following description applies to the standard Kania:

Type: Twin-turboshaft general-purpose light helicopter.

Rotor System: Three-blade fully articulated main rotor and two-blade seesaw tail rotor. Laminated plastics blades on both rotors. Three hydraulic boosters for longitudinal, lateral and collective pitch control of main rotor. Rotor brake fitted. Electrical anti-icing of rotor blades optional.

Rotor Drive: Transmission driven via main rotor, intermediate and tail rotor gearboxes, each with oil sight gauge. Oil temperature gauge, oil cooling system, pressure indicator and tachometer for main gearbox. Anti-friction bearings on tail rotor shaft.

Fuselage and Tail Unit: Conventional semi-monocoque fuselage, with circular-section monocoque tailboom. Laminated horizontal stabiliser at end of tailboom. Hoist and cargo sling attachment points standard.

Landing Gear: Non-retractable tricycle type, plus laminated tailskid. Twin-wheel castoring nose unit; single wheel on each main unit. Pneumatic brakes on main wheels.

Power Plant: Two Allison 250-C20B turboshaft engines, mounted side by side above cabin; each rated at 313 kW (420 shp) for T-O and 30 min power, 298 kW (400 shp) max continuous, and 276 kW (370 shp) for max cruise. Standard usable fuel capacity of 600 litres (131 Imp gallons), with provision for additional 480 litres usable (105·5 Imp gallons) in optional auxiliary tanks.

Accommodation: Pilot (port side), and co-pilot or passenger, on adjustable and removable front seats, each fitted with safety belt. Dual controls optional. Accommodation for up to eight more persons, on two three-person bench seats and a single or double seat at rear of cabin, all with safety belts. Seats removable for carriage of 800 kg (1,763 lb) of cargo, stretchers, agricultural or other equipment. Access to cabin via jettisonable door on each side at front (port door of sliding type) and larger passenger/cargo door at rear on port side. Pilot's windscreen wiper standard. Cabin heating or air-conditioning, and electrical anti-icing of pilot's windscreen, optional.

PZL Swidnik Kania twin-turboshaft light helicopter (left), and mockup (right) of the Taurus modified version for the North American market *(A. Glass and Brian M. Service)*

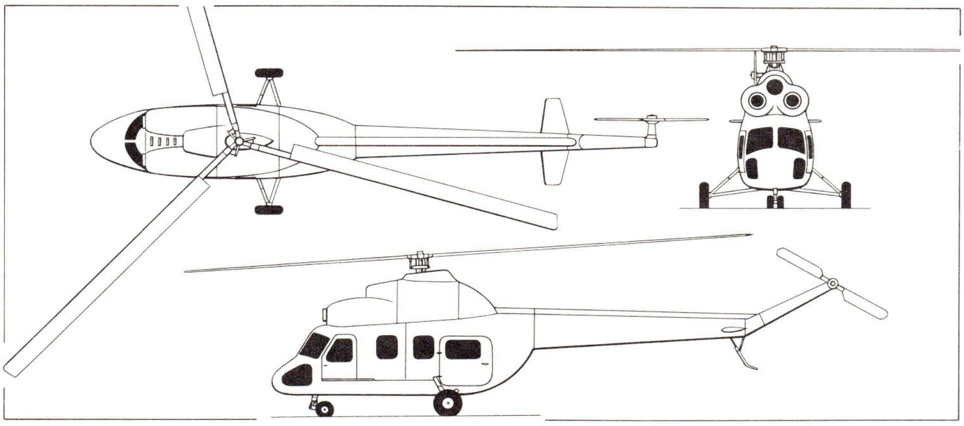

PZL Swidnik Kania (two Allison 250-C20B turboshaft engines) *(Pilot Press)*

SYSTEMS: Electrical system includes two 28V 150A DC starter/generators, two 115V 250A 400Hz static inverters, two 26V/115V 55VA 400Hz transformers, and a 25Ah nickel-cadmium battery; 16kVA alternator for anti-icing system optional. External power receptacle. Dual fire detection and extinguishing systems for engines; single system for transmission.

AVIONICS AND EQUIPMENT: King KX 175BE com/nav, KR 85 digital ADF, KT 76 transponder, and VFR instrumentation, all standard. Optional avionics include King KWX 50 digital weather radar, KRA 10 radar altimeter and KN 65A-03 DME. Dual instrument lighting systems, pilot's cabin extension light, cabin dome light, three hold dome lights, three position lights, adjustable landing light, and anti-collision light, are all standard, as are cargo and stretcher tiedown rings. Optional equipment includes dual controls, co-pilot's windscreen wiper, and auxiliary fuel tanks. Cabin carpet, basic or executive passenger seats, heating and air-conditioning systems, and sand filters, are also optional.

OPERATIONAL EQUIPMENT: According to mission, the Kania can be equipped with a 1,000 kg (2,205 lb) capacity stabilised cargo sling; 120 kg (265 lb) capacity hoist; stretchers and casualty care equipment; agricultural spraying or dusting gear.

DIMENSIONS, EXTERNAL:
Diameter of main rotor 14·56 m (47 ft 9¼ in)
Diameter of tail rotor 2·70 m (8 ft 10¼ in)
Length overall, rotors turning 17·41 m (57 ft 1½ in)
Length of fuselage 11·95 m (39 ft 2½ in)
Fuselage: Max width 1·60 m (5 ft 3 in)
Height to top of rotor hub 3·75 m (12 ft 3½ in)
Stabiliser span 1·84 m (6 ft 0½ in)
Wheel track 3·25 m (10 ft 8 in)
Wheelbase 2·71 m (8 ft 10¾ in)
Tail rotor ground clearance 1·80 m (5 ft 10¾ in)

DIMENSIONS, INTERNAL:
Cabin: Length, incl flight deck 4·07 m (13 ft 4¼ in)
Max width 1·50 m (4 ft 11 in)
Height at front 1·30 m (4 ft 3¼ in)
Mean height at rear 1·51 m (4 ft 11½ in)
Max height at rear 1·62 m (5 ft 3¾ in)
Floor area 5·68 m² (61·1 sq ft)
Volume 7·766 m³ (274·25 cu ft)

AREAS:
Main rotor disc 166·50 m² (1,792·2 sq ft)
Tail rotor disc 5·725 m² (61·6 sq ft)

WEIGHTS:
Weight empty, standard 2,140 kg (4,717 lb)
Normal T-O weight 3,350 kg (7,385 lb)
Max T-O weight 3,550 kg (7,826 lb)

PERFORMANCE (estimated, for 'clean' aircraft at S/L, ISA, zero wind. A: at 3,150 kg; 6,944 lb gross weight; B: at normal T-O weight; C: at max T-O weight):
Max cruising speed:
A, B, C 113 knots (210 km/h; 130 mph)
Econ cruising speed:
A, B 105 knots (195 km/h; 121 mph)
C 104·5 knots (194 km/h; 120·5 mph)
Rate of climb: A 504 m (1,653 ft)/min
B 450 m (1,476 ft)/min
C 402 m (1,319 ft)/min
Rate of climb, one engine out:
A 96 m (315 ft)/min
B 66 m (216 ft)/min
C 42 m (138 ft)/min
Service ceiling (30·5 m; 100 ft/min climb rate):
A, B above 4,000 m (13,125 ft)
C 4,000 m (13,125 ft)
Hovering ceiling IGE at T-O power:
A 2,560 m (8,400 ft)
B 1,940 m (6,365 ft)
C 1,320 m (4,330 ft)
Hovering ceiling OGE at T-O power:
A 1,860 m (6,100 ft)
B 1,240 m (4,070 ft)
C 480 m (1,575 ft)
Max range at econ cruising speed, standard fuel, no reserves: A 282 nm (522 km; 324 miles)
B 275 nm (510 km; 317 miles)
C 268 nm (497 km; 309 miles)
Max endurance, conditions as above:
A 3 h 41 min
B 3 h 27 min
C 3 h 15 min

PZL SWIDNIK W-3 SOKÓL (FALCON)

Development of this all-new Polish helicopter took place in 1978-79, and the first flight of the prototype was made on 16 November 1979. The design team was headed by Stanislav Kaminski. Larger than the Mi-2/Kania, the Sokól accommodates a crew of two, and 12 passengers or a maximum 1,500 kg (3,306 lb) of internal cargo. Power plant consists of two Polish-built Glushenkov TVD-10 turboshaft engines. Agricultural and ambulance versions are also planned.

TYPE: Twin-turboshaft medium-weight multi-purpose helicopter.

ROTOR SYSTEM: Four-blade fully-articulated main rotor and three-blade tail rotor. Blades of both rotors constructed of laminated glassfibre impregnated with epoxy resin. Main rotor blades have tapered tips. Tail rotor has titanium hub. Three hydraulic boosters for longitudinal, lateral and collective pitch control of main rotor. Blade anti-icing by electrically heated elements.

ROTOR DRIVE: Transmission driven via main rotor, intermediate, and tail rotor gearboxes. Tail rotor driveshaft of duralumin tube with splined couplings.

FUSELAGE AND TAIL UNIT: Conventional light alloy semi-monocoque structure, with circular-section monocoque tailboom. Fin integral with tailboom structure. Horizontal stabiliser, under end of tailboom, has a single continuous spar, is built up of laminated glassfibre impregnated with epoxy resin, and is interconnected with the main rotor pitch control system.

LANDING GEAR: Non-retractable tricycle type, plus tail-skid beneath tailboom. Twin-wheel nose unit; single wheel on each main unit. Main-wheel tyres size 500 × 250 mm; nosewheel tyres size 400 × 150 mm. Pneumatic brakes on main wheels. Float or ski installations optional.

POWER PLANT: Two PZL-10W (Polish-built Glushenkov TVD-10) turboshaft engines, each with take-off rating of 649 kW (870 shp), and emergency rating of 746 kW (1,000 shp). Engines and main rotor gearbox are mounted to a bed frame, eliminating any drive misalignment due to deformations of the fuselage structure.

PZL Swidnik W-3 Sokól twin-turboshaft helicopter

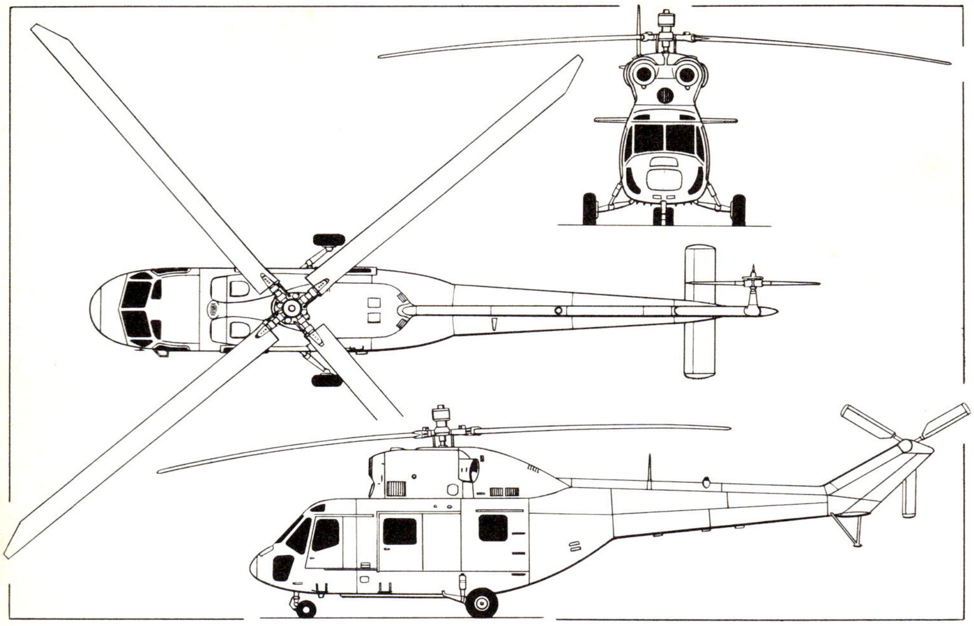

PZL Swidnik W-3 Sokól (two PZL-10W turboshaft engines) *(Pilot Press)*

It is intended to investigate the potential of reducing noise and vibration by attaching this frame to the fuselage by elastomeric pads. Bladder-type fuel tanks beneath the cabin floor, with combined capacity of 1,100 litres (242 Imp gallons).

Accommodation: Pilot (port side), and co-pilot or flight engineer, side by side on flight deck. Dual controls and dual flight instrumentation optional. Accommodation for 12 passengers in main cabin. Seats removable for carriage of internal cargo. Ambulance version will permit the loading of four stretcher cases and a medical attendant. Baggage space at rear of cabin. Door to flight deck on each side; large sliding door for passenger and/or cargo loading on port side at forward end of cabin; second sliding door at rear of cabin on starboard side. Design of flight deck permits use of optically flat windscreens, improving view and enabling wipers to sweep a large area. Accommodation heated and ventilated.

Systems: Two independent hydraulic systems for controlling main and tail rotors, unlocking collective pitch control lever, and feeding dumper of directional steering system. Pneumatic system for main-wheel brakes only. Electrical system providing both AC and DC power. Automatic power control system linking power plant and rotor pitch for optimum performance. Fire protection system. Air-conditioning and oxygen systems optional.

Avionics and Equipment: Standard nav/com avionics which permit adverse weather operation by day or night. Two-axis autopilot standard. Stability augmentation warning system. Cargo version equipped with external hook and onboard hoist.

Dimensions, external:	
Main rotor diameter	15·70 m (51 ft 6 in)
Tail rotor diameter	3·00 m (9 ft 10 in)
Length overall, rotors turning	18·85 m (61 ft 10¼ in)
Length of fuselage	13·65 m (44 ft 9½ in)
Height overall	4·15 m (13 ft 7½ in)
Wheel track	3·25 m (10 ft 8 in)
Passenger/cargo doors (each):	
Height	1·20 m (3 ft 11¼ in)
Width	1·25 m (4 ft 1¼ in)
Dimensions, internal:	
Cabin: Length	3·20 m (10 ft 6 in)
Max width	1·56 m (5 ft 1½ in)
Max height	1·40 m (4 ft 7 in)
Weights:	
Basic operating weight empty	2,490 kg (5,489 lb)
Max payload	1,500 kg (3,307 lb)
Normal T-O weight	5,810 kg (12,808 lb)
Max T-O weight	6,000 kg (13,227 lb)
Performance (estimated):	
Max level speed	140 knots (260 km/h; 161 mph)
Cruising speed	119 knots (220 km/h; 137 mph)
Max rate of climb at S/L	564 m (1,850 ft)/min
Max vertical rate of climb at S/L	120 m (394 ft)/min
Service ceiling	5,000 m (16,400 ft)
Hovering ceiling OGE	1,850 m (6,070 ft)
Range: standard fuel	324 nm (600 km; 373 miles)
auxiliary fuel	593 nm (1,100 km; 683 miles)

PZL SWIDNIK SMALL HELICOPTER

Work has begun at Swidnik on a new 4/5-seat helicopter.

PORTUGAL

OGMA
OFICINAS GERAIS DE MATERIAL AERONÁUTICO (General Aeronautical Material Workshops)

Works: 2615 Alverca
Telephone: 2580 786; 2581 803, 882, 923 and 979
Telex: 14479 OGMA P
Director: Maj Gen Eng Rui do Carmo da Conceição Espadinha
Asst Director: Col Eng António Pedro da Silva Gonçalves

OGMA was founded in 1918 and has been in continuous operation since then. It is the Air Force department responsible for maintenance and repair, at depot level, of all aircraft, avionics, engines, structures, ground communications and radar equipment of the Portuguese Air Force. OGMA has a total covered area of 110,000 m² (1,184,030 sq ft), and a work force of approx 2,500 people.

A new aero-engine repair shop was scheduled to go into operation during the second half of 1981, to improve major overhaul capacity on different types of military and commercial turbojets and turbofans (up to 160 kN; 36,000 lb st), and turboprop and turboshaft engines of up to 4,475 kW (6,000 shp).

Under a contract signed in 1959, OGMA also undertakes IRAN, refurbishing and rehabilitation, periodic inspection and emergency maintenance and crash repair of US Air Force and US Navy aircraft.

Under contract to Aérospatiale of France, OGMA is manufacturing structures for the SA 315B Lama and SA 318 Alouette II helicopters, and small structural components for the SA 330 Puma. Other component manufacture is undertaken for Socata, Omera and Turboméca. For Turboméca, OGMA is also overhauling Artouste III engines.

ROMANIA

CNIAR
CENTRUL NATIONAL AL INDUSTRIEI AERONAUTICE ROMÂNE (National Centre of the Romanian Aeronautical Industry)

Headquarters: 133 Calea Victoriei, Sector 1, 71102 Bucharest
Telephone: 50 27 14
Telex: 11648 AEROM
General Director: Eng Teodor Zanfirescu
Commercial Director: Vasile Racovitan

Romania has had a tradition of aviation since the earliest days of flying, dating from the first monoplane built in France in early 1906 by the Romanian engineer Traian Vuia, the original monoplane of Aurel Vlaicu which, in Bucharest on 17 June 1910, became the first nationally-designed aeroplane to be flown in Romania, and the aeroplanes designed and built in France and Britain by Henri Coanda in 1910-14.

Since that time the Romanian aircraft industry (IAR) has produced some 90 different types of landplane, including helicopters (of which 80 were Romanian-designed), and about 40 different types of sailplane. Many other achievements in the fields of theoretical and experimental aerodynamics have been made by teams of Romanian engineers, led by Prof Elie Carafoli, Prof Ion Stroiescu, Prof Ion Grosu, Dipl Eng Radu Manicatide, Dipl Eng Iosif Silimon and others, using latterly the Canadian-licensed trisonic wind tunnel at INCREST in Bucharest.

The foundations for the present industry were laid at Brasov in 1926, and by 1939 the most important centre of the Romanian aircraft industry was the IAR (Industria Aeronautica Româna) at Brasov, with 8,000 employees. Other factories included Astra-Arad, SET and ICAR. The IAR factory, destroyed by bombing in 1944, was rebuilt after the war for the manufacture of tractors. It resumed its aeronautical activities with the IAR-811 training aircraft, flown for the first time in March 1949, and the IAR-813, -814 and -817 utility, agricultural and ambulance aircraft. Known until 1959 as URMV-3 (aircraft component repair factory 3), the Brasov factory during that period produced more than 200 aircraft of different types and more than 20 types of sailplane. Two other repair factories were set up at Medias (ARMV-1) and Bucharest (ARMV-2).

In 1959 the URMV-3 was divided into two factories. The Brasov team was renamed ICA (Intreprinderea de Constructii Aeronautice) Brasov-Ghimbav, and up to 1968 produced more than 100 sailplanes designed by Dipl Eng Iosif Silimon, who died in early 1981. The second team, relocated in Bucharest and combined with ARMV-2, became known as IRMA (Intreprinderea de Reparat Material Aeronautic), under the leadership of Dipl Eng Radu Manicatide. Here, up to 1968, were built more than 140 ambulance, training, agricultural and other aircraft.

The industry was again reorganised in 1968, and its activities are now undertaken, within the Ministry of Machine Building Industry, by the CNIAR, which combines the activities of the former CIAR and GAB (see 1979-80 *Jane's*). Major production activities of the CNIAR are at present carried out in five factories. The two main aircraft factories, details of whose products follow, are the ICA at Brasov and IAv Bucharest. The latter is currently being enlarged to provide a total floor area of 300,000 m² (3,229,170 sq ft). A third factory, IAv Bacau, builds the Soviet Yak-52 under licence and manufactures various components for the IAR-823 and IAR-316B; the SOKO/CNIAR Orao/IAR-93 (see International section) is manufactured at IAv Craiova. Viper 632-41 engines for the IAR-93 and Turmo IV CA engines for the IAR-330 (Puma) helicopter, are built by the Turbomecanica Enterprise in Bucharest. Two new factories are currently being built: one at Bucharest for avionics and airborne equipment, and the other at Baneasa for forgings and castings.

Exports and imports of aircraft and aero-engines are the responsibility of the CNA (see next entry), formed in 1979; avionics and electronic equipment sales are dealt with by Electronum and Electroexportimport. Aeronautical research and development are undertaken by INCREST (formerly IMFCA), the Aerospace Research and Design Institute at Bucharest. INCREST also designs and manufactures aerospace equipment, including anti-skid brakes, engine stands, fuel monitoring equipment, intercoms and various aviation raw materials. The flight test centre is the CIIAR (Centrul de Incercari in zbor) at Craiova.

The Baneasa area of Bucharest is being developed as the headquarters and main centre for CNIAR activities, and for the laboratories and design offices of INCREST.

Romanian-built IAR-93 prototype strike fighter (two Rolls-Royce Viper Mk 632 turbojet engines)

CNA

CENTRUL NATIONAL AERONAUTIC (Intreprindere de Comert Exterior) (National Centre for Aeronautical Foreign Trade)

Bulevardul Dacia 13, POB 22-149, R-70185 Bucharest

Telephone: 12 08 78
Telex: 10660 CNAER
General Director: Eng C. Zgavirdici
Deputy Director: Mircea Costescu
Publicity: George Popescu

IAv BACAU

INTREPRINDEREA DE AVIOANE BACAU (Bacau Aircraft Enterprise)

Bacau

This factory manufactures hydraulic, pneumatic, air-conditioning, fuel system and landing gear components for the ICA IAR-823 light aircraft and IAR-316B helicopter, and for the IAR-93 strike fighter. It is also building under licence the Soviet Yakovlev Yak-52 light aircraft.

IAv BACAU (YAKOVLEV) Yak-52

Announced in late 1978, the Yak-52 is a tandem-cockpit variant of the Yak-50, with unchanged span and length, but with a semi-retractable tricycle landing gear. Although aesthetically unattractive, this last feature is intended to reduce damage in a wheels-up landing.

The Yak-52 is expected to replace the Yak-18, which has been the standard ab initio trainer for Soviet pilots since the mid-1940s. Alexander Yakovlev has believed for many years that aeroplanes to be flown by young people should be designed by members of the Komsomol youth brigades and light aircraft enthusiasts, under experienced leadership. The enthusiasm engendered by this policy led to first flight of the prototype Yak-52 less than a year after its design was started. Flight testing was then undertaken by pilots qualified as Soviet Masters of Sport, as well as professional test pilots. Production has been entrusted to the Romanian aircraft industry, under the Comecon (Council for Mutual Economic Assistance) programme.

Manufacture began at Bacau in 1978, and in the Summer of 1979 a Yak-52 fuselage was installed in a structural test rig at INCREST in Bucharest. By September of that year, four complete airframes and two fuselages were on the final assembly line at Bacau, the aircraft at the head of the line being described as the third 'pre-production' example. The aircraft is now in series production; it does not have an IAR designation number.

Type: Tandem two-seat piston-engined primary trainer.

Wings: Cantilever low-wing monoplane of single-spar stressed-skin all-metal construction. Each wing comprises a single straight-tapered panel, attached directly to the side of the fuselage. Fabric-covered slotted ailerons. Light alloy trailing-edge split flaps. Ground-adjustable tab on each aileron.

Fuselage: Conventional light alloy semi-monocoque structure.

Tail Unit: Cantilever light alloy structure. Fixed surfaces metal covered; control surfaces fabric covered. Horn-balanced rudder, with ground-adjustable tab. Mass-balanced elevators. Controllable tab in port elevator.

Landing Gear: Semi-retractable tricycle type, with single wheel on each unit. Pneumatic retraction, nosewheel rearward, main units forward. All three wheels remain fully exposed to airflow, against the undersurface of the fuselage and wings respectively, to offer greater safety in the event of a wheels-up emergency landing. Oleo-pneumatic shock-absorbers. Main-wheel tyre size 500 × 150; nosewheel tyre size 400 × 150. Pneumatic brakes. Skis can be fitted in place of wheels for Winter operations, permissible at temperatures down to −42°C.

Power Plant: One 269 kW (360 hp) Vedeneev M-14P nine-cylinder aircooled radial engine, driving a two-blade variable-pitch propeller type V-530TA-D35, without spinner. Louvres in front of cowling to regulate cooling. Two-part cowling, split on horizontal centre-line. Two fuel tanks, in wing roots forward of spar, each with capacity of 65 litres (14·25 Imp gallons). Additional tank of 5·5 litres (1·25 Imp gallons) capacity supplies engine during inverted flight. Oil capacity 20 litres (4·5 Imp gallons).

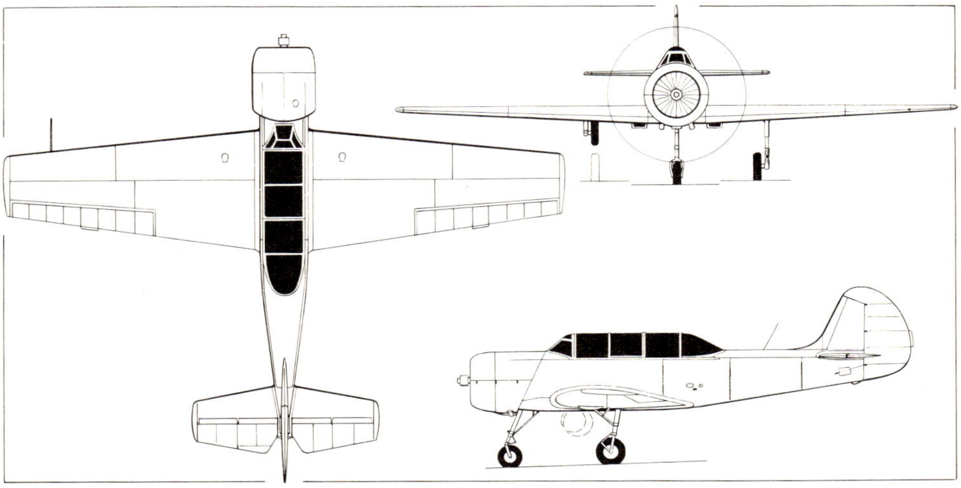

Yakovlev Yak-52 tandem two-seat primary trainer *(Pilot Press)*

Accommodation: Tandem seats for pupil and instructor (at rear) under long 'glasshouse' canopy, with separate rearward-sliding hood over each seat. Seats and dual flying controls are adjustable. Sides of cockpit have a soft synthetic lining. Heating and ventilation standard.

Systems: No hydraulic system. Independent main and emergency pneumatic systems, for flap actuation, landing gear retraction, engine starting, and wheel brake control. Pneumatic systems supplied by two compressed air bottles, mounted behind rear seat and recharged in flight by an AK-50T compressor. GSR-3000M engine-driven generator supplies 27V electrical system. Battery in port wing.

Avionics and Equipment: Dual engine and flying instruments. Equipment includes GMK-1A gyro-compass, ARK-15M automatic radio compass, Landysh-5 VHF com and SPU-9 intercom.

Dimensions, external:	
Wing span	9·50 m (31 ft 2 in)
Length overall	7·676 m (25 ft 2¼ in)
Height overall	2·95 m (9 ft 8¼ in)
Propeller diameter	2·40 m (7 ft 10½ in)
Area:	
Wings, gross	15·00 m² (161·5 sq ft)
Weights and Loadings:	
Basic operating weight	1,000 kg (2,205 lb)
Max T-O weight	1,290 kg (2,844 lb)
Max wing loading	86·0 kg/m² (17·61 lb/sq ft)
Max power loading	4·80 kg/kW (7·90 lb/hp)
Performance:	
Never-exceed speed	194 knots (360 km/h; 223 mph)
Max level speed	154 knots (285 km/h; 177 mph)
Landing speed	60 knots (110 km/h; 69 mph)
Max rate of climb at S/L	600 m (1,970 ft)/min
Service ceiling	6,000 m (19,685 ft)
T-O run	170 m (558 ft)
Landing run	200 m (656 ft)
Range with max fuel	297 nm (550 km; 341 miles)
Endurance with max fuel	2 h 50 min
g limits	+7; −5

Yakovlev Yak-52 two-seat primary trainer, built in Romania by IAv Bacau *(Air Portraits)*

IAv BUCURESTI
INTREPRINDEREA DE AVIOANE BUCURESTI (Bucharest Aircraft Enterprise)

Baneasa Airport, Bd Ficusului 44, Bucharest

IAv Bucuresti's predecessor, IRMA, was formed in 1959 from part of the former URMV-3 at Brasov (see 1979-80 and earlier editions of *Jane's*). The factory is currently responsible for manufacture of the BAe One-Eleven (components and complete aircraft) and the Pilatus Britten-Norman Islander. It specialises in the development and manufacture of commercial and agricultural aircraft; in the repair and overhaul of various large and small aircraft; is agent and repair centre for Avco Lycoming engines; and manufactures aircraft equipment.

IAv BUCURESTI (BAe) ONE-ELEVEN (ROMBAC 1-11)

IAv Bucuresti is the Romanian contractor for the licence manufacture of BAe One-Eleven Series 475 and 500 twin-turbofan transports (Romanian designations Rombac 1-11 Series 495 and 550 respectively). A corresponding programme provides for Romanian manufacture of the Rolls-Royce Spey engines.

Initially, three complete aircraft are being delivered from British Aerospace's Hurn (Bournemouth) factory. The first of these was handed over on 22 February 1980. The transfer of industrial technology will take place in seven stages, permitting manufacture of the first 22 aircraft by 1985, after which production of complete aircraft will continue both to cover domestic requirements and for export. Romania will eventually become the sole production source for the One-Eleven. The first fuselage for a Romanian-assembled One-Eleven (a Srs 500) was delivered by Super Guppy transport aircraft in January 1980; two more fuselages were delivered by the end of that year. First flight by a Romanian-assembled aircraft is scheduled for Spring 1982.

As part of the agreement BAe is also supplying three new One-Elevens to Tarom, the Romanian state airline. The first of these, a Srs 525, was delivered in mid-January 1981; a Srs 487 freighter was delivered in July 1981 and a second Srs 525 is due for delivery in early 1982.

BAe One-Eleven fuselage for airlift to Romania by Super Guppy

Pilatus Britten-Norman BN-2A-27 Islander built in Romania

IAv BUCURESTI (PILATUS BRITTEN-NORMAN) ISLANDER

The Pilatus Britten-Norman Islander (see UK section) has been manufactured under licence in Romania, originally by IRMA, for several years. The first Romanian-built example flew for the first time at Baneasa Airport, Bucharest, on 4 August 1969, and the initial commitment to build 215 Islanders was completed in 1976. A total of 617 had been completed by June 1981, at which time production was continuing at the rate of about 50 Islanders per year.

IAv CRAIOVA
INTREPRINDEREA DE AVIOANE CRAIOVA (Craiova Aircraft Enterprise)

Craiova

This new factory is responsible for manufacture of the IAR-93 ground attack aircraft and operational trainer built in collaboration with Yugoslavia (see SOKO/CNIAR entry in the International section).

ICA
INTREPRINDEREA DE CONSTRUCTII AERONAUTICE (Aeronautical Construction Enterprise)

Ghimbav, near Brasov

ICA, created in 1968, continues the work that was begun in 1926 by IAR-Brasov and was then undertaken in 1950-59 as URMV-3 Brasov. Today, it manufactures the Romanian-designed IAR-823, the Alouette III and Puma helicopters under licence from Aérospatiale of France (as the IAR-316B and IAR-330 respectively), and the IS-28/29 series of Romanian sailplanes and motor gliders. It also produces aircraft components and equipment.

ICA IAR-823

Design of the IAR-823 two/five-seat training and touring light aircraft was started at IMFCA in May 1970, by a team led by Dipl Ing Radu Manicatide. Construction of a prototype began at ICA-Brasov in the Autumn of 1971, and this aircraft made its first flight in July 1973. The first production aircraft flew in 1974, and more than 70 had been delivered to the Romanian Air Force and Romanian flying clubs by the beginning of 1981.

As a two-seater, the IAR-823 is fully aerobatic and is intended for training duties. With a rear bench seat for up to three more persons it is suitable as an executive, taxi or touring aircraft. Provision is made for two underwing pylons for the carriage of drop-tanks or practice weapons.

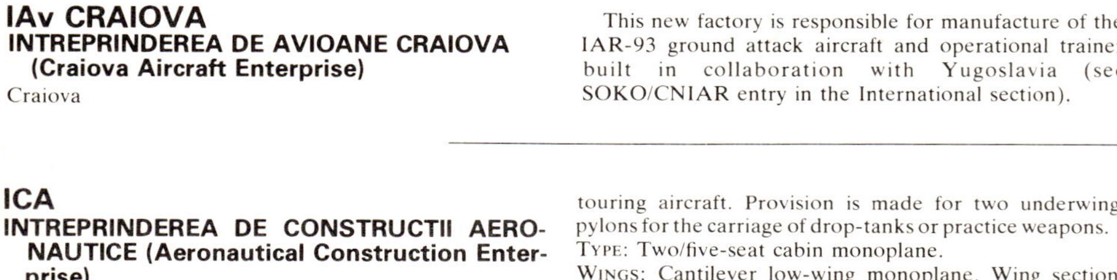

IAR-823 two/five-seat light aircraft (Avco Lycoming IO-540 engine) *(Peter R. March)*

TYPE: Two/five-seat cabin monoplane.

WINGS: Cantilever low-wing monoplane. Wing section NACA 23012 (modified). Dihedral 7° from roots. Incidence 3° at root, 1° at tip. Conventional all-metal structure, with single main spar and rear auxiliary spar; three-point attachment to fuselage. Riveted spars, ribs and skin of corrosion-proof aluminium alloy. Leading-edges riveted, and sealed to ribs and main spar to form main torsion box and integral fuel tanks. Electrically-actuated fabric-covered metal single-slotted flaps and fabric-covered Frise-type slotted metal ailerons. Ground-adjustable tab.

FUSELAGE: All-metal semi-monocoque structure. Glassfibre engine cowling.

TAIL UNIT: Cantilever metal structure. Two-spar duralumin-covered fin and tailplane; fabric-covered duralumin horn-balanced rudder and elevators. Electrically-actuated automatic trim tabs in elevators; controllable tab in rudder.

LANDING GEAR: Retractable tricycle type, with steerable nosewheel. Electrical retraction, main units inward, nose unit rearward. Emergency manual actuation. Oleo-pneumatic shock-absorbers. Main-wheel tyres size 6·00-6, pressure 2·93 bars (42·5 lb/sq in). Nosewheel tyre size 355 × 150 mm. Independent hydraulic main-wheel brakes, pedal-controlled from left front seat. Shimmy damper on nose unit. No wheel doors.

POWER PLANT: One 216 kW (290 hp) Avco Lycoming IO-540-G1D5 flat-six engine, driving a Hartzell two-blade variable-pitch constant-speed metal propeller. Fuel in four integral wing tanks, total capacity 360 litres (79 Imp gallons). Provision for two 70 litre (15·4 Imp gallon) drop-tanks on underwing pylons.

ACCOMMODATION: Fully-enclosed cabin, seating two persons side by side on individual adjustable front seats, with removable bench seat at rear for up to three more people. Dual controls standard in training version, optional in other versions. Upward-hinged door (optionally jettisonable) on each side of cabin, which is soundproofed, heated and ventilated. Compartment at rear of cabin for up to 40 kg (88 lb) of baggage. Equipment and layout can be varied for use as air taxi, executive or freight transport, ambulance, liaison or photographic aircraft.

SYSTEMS AND AVIONICS: Electrical system, including 50A alternator and 24V 30Ah battery, for engine starting, elevator tab and landing gear actuation, radio communications, landing and navigation lights, and cabin and instrument lighting. Standard avionics include VFR instrumentation and Bendix RT 221-AE transceiver. Optional equipment, according to mission, includes blind-flying instrumentation and, in civil transport version, marker beacon receiver, nav/com radio, VOR/ILS, ADF and autopilot.

DIMENSIONS, EXTERNAL:

Wing span	10·00 m (32 ft 9¾ in)
Wing chord at c/l	2·00 m (6 ft 6¾ in)
Wing chord at tip	1·00 m (3 ft 3¼ in)
Wing aspect ratio	6·66
Length overall	8·24 m (27 ft 0¼ in)
Height overall	2·52 m (8 ft 3¼ in)
Wheel track	2·48 m (8 ft 1¾ in)
Wheelbase	1·86 m (6 ft 1¼ in)
Propeller diameter	2·23 m (7 ft 4 in)

AREAS:
- Wings, gross — 15·00 m^2 (161·5 sq ft)
- Ailerons (total) — 1·20 m^2 (12·92 sq ft)
- Trailing-edge flaps (total) — 1·78 m^2 (19·16 sq ft)
- Horizontal tail surfaces (total) — 3·30 m^2 (35·52 sq ft)
- Vertical tail surfaces (total) — 1·50 m^2 (16·15 sq ft)

WEIGHTS AND LOADINGS (A: Aerobatic; U: Utility category):
- Weight empty: A — 900 kg (1,984 lb)
- U — 910 kg (2,006 lb)
- Max T-O weight: A — 1,190 kg (2,623 lb)
- U — 1,380 kg (3,042 lb)
- Max permissible weight for special missions — 1,500 kg (3,307 lb)
- Max normal wing loading:
  - A — 79·0 kg/m^2 (16·2 lb/sq ft)
  - U — 92·0 kg/m^2 (18·8 lb/sq ft)
- Max normal power loading:
  - A — 5·51 kg/kW (9·15 lb/hp)

PERFORMANCE (at 1,400 kg; 3,086 lb AUW except where indicated):
- Never-exceed speed — 215 knots (400 km/h; 248 mph)
- Max level speed at S/L — 167 knots (310 km/h; 192·5 mph)
- Max cruising speed (75% power) at 1,750 m (5,750 ft) — 162 knots (300 km/h; 186 mph)
- Econ cruising speed (60% power) at 3,050 m (10,000 ft) — 156 knots (290 km/h; 180 mph)
- Landing speed — 56·5 knots (105 km/h; 65 mph)
- Stalling speed, flaps up — 62·5 knots (115 km/h; 71·5 mph)
- Stalling speed, flaps down, power off — 49 knots (90 km/h; 56 mph)
- Max rate of climb at S/L — 450 m (1,475 ft)/min
- Time to 1,000 m (3,280 ft) — 2 min 20 s
- Service ceiling — 5,600 m (18,375 ft)
- T-O run — 160 m (525 ft)
- T-O to 15 m (50 ft) — 310 m (1,017 ft)
- Landing run — 200 m (656 ft)
- Range, according to mission and payload, 1 h reserves — 431-970 nm (800-1,800 km; 497-1,118 miles)
- Endurance, according to mission and payload — 3-6 h
- *g* limits (at 1,190 kg; 2,623 lb AUW) — +6; −3

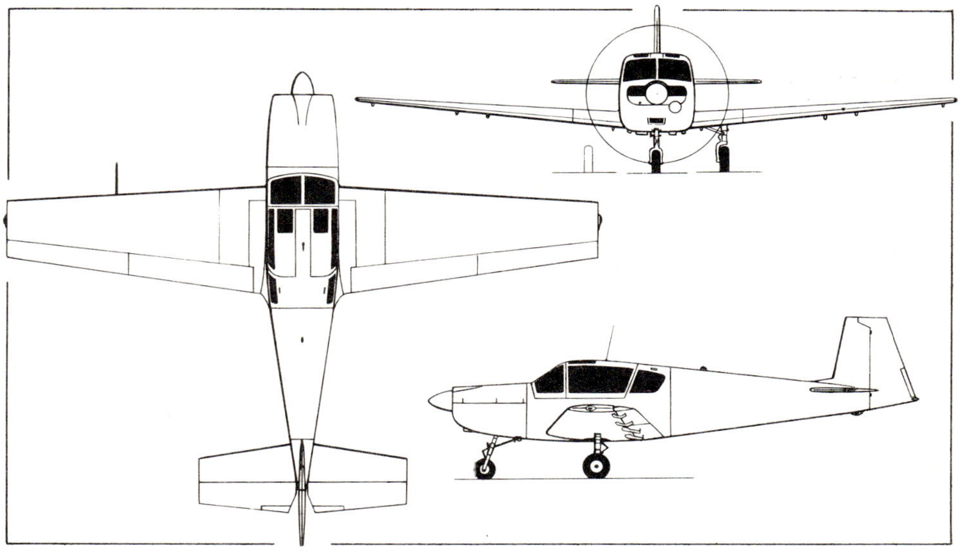

ICA IAR-823 light touring and training aircraft *(Pilot Press)*

ICA IAR-827A

Design of the basic IAR-827 began in the early part of 1973. The objective was to develop an agricultural aircraft with an airframe life of 4,000 h (equivalent to 22,000 flights) and able to carry a useful load equivalent to 2 kg (4·4 lb) for every horsepower. Particular attention was paid to minimising corrosion problems.

Static testing was completed before the first flight, which took place in July 1976. Early flight testing revealed the need for a more powerful engine, and the original 298 kW (400 hp) Avco Lycoming IO-720-DA1B flat-eight engine was replaced by a 447 kW (600 hp) PZL-3S radial engine in the first prototype. Certification testing with the first and second prototypes began in 1977, and was followed by the manufacture, at IAv Bucuresti, of five pre-series aircraft.

A description of the Lycoming-engined IAR-827 prototype can be found in the 1979-80 *Jane's*. Certification of the IAR-827A, as the radial-engined production version is known, was received in 1979, and the first production IAR-827A was displayed at the Aero Technica '80 exhibition in Bucharest in June 1980.

Series manufacture of the PZL-engined IAR-27A, which was to begin in 1981, is the responsibility of ICA Brasov. The following description applies to this version:

TYPE: Single/two-seat agricultural aircraft.

WINGS: Cantilever low-wing monoplane. Wing section NACA 23015. Constant-chord safe-life structure, with 6° dihedral from roots. Welded chrome-molybdenum steel tube centre-section with duralumin skin. All-metal single-spar outer panels. Plain slotted ailerons. Electrically-operated all-metal single-slotted Fowler-type flaps. Automatic trim tab in each aileron.

FUSELAGE: Forward structure of welded chrome-molybdenum steel tube, with duralumin and glassfibre skin panels attached by quick-release fastenings. Rear fuselage is a light alloy monocoque.

TAIL UNIT: Cantilever all-metal structure, with slightly-sweptback fin and rudder. Balanced elevators. Automatic trim tab in port elevator, ground-adjustable tab on rudder, and manually-controlled tab in starboard elevator.

LANDING GEAR: Non-retractable tailwheel type. Main units comprise 140 mm stroke hydraulic shock-struts and side Vs, and are fitted with Dunlop wheels (tyre size 615 × 225) and brakes. Gear is of Romanian design and is designed to withstand a vertical velocity of 6 m (19·5 ft)/s or a free drop of 1·2 m (3 ft 11¼ in).

POWER PLANT: One 447 kW (600 hp) PZL-3S seven-cylinder radial aircooled engine, driving a PZL four-blade constant-speed propeller. Fuel tank in each wing leading-edge, each of 180 litres (39·6 Imp gallons) capacity.

ACCOMMODATION: Side-by-side seats for pilot and mechanic in fully-enclosed cockpit, with window/door on each side. Provision for dual controls and emergency door jettison. Seat height and rudder pedals adjustable. Crash pylon in fairing aft of seats. Heated and ventilated cockpit is sealed and slightly pressurised to exclude dust.

EQUIPMENT: All-glassfibre hopper in forward fuselage, with a volume of 1·23 m^3 (43·44 cu ft). Hopper is stressed for loads of up to 1,000 kg (2,205 lb), but normal max load is 800 kg (1,763 lb) of dry or 1,200 litres (264 Imp gallons) of liquid chemical.

DIMENSIONS, EXTERNAL:
- Wing span — 14·00 m (45 ft 11¼ in)
- Wing aspect ratio — 6·67
- Length overall — 8·80 m (28 ft 10½ in)
- Height overall — 2·60 m (8 ft 6½ in)
- Tailplane span — 4·90 m (16 ft 1 in)
- Wheel track — 3·42 m (11 ft 2¾ in)
- Wheelbase — 6·20 m (20 ft 4 in)
- Propeller diameter — 2·62 m (8 ft 7 in)

AREA:
- Wings, gross — 29·40 m^2 (316·46 sq ft)

WEIGHTS:
- Weight empty, with agricultural equipment — 1,660 kg (3,660 lb)
- Max T-O weight — 2,800 kg (6,173 lb)

PERFORMANCE (with agricultural equipment, at max T-O weight):
- Max level speed — 113 knots (210 km/h; 130 mph)
- Cruising speed — 105 knots (195 km/h; 121 mph)
- Operating speed range — 73-97 knots (135-180 km/h; 84-112 mph)
- Stalling speed, 10° flap, power off — 59·5 knots (110 km/h; 68·5 mph)
- Max rate of climb at S/L — 270 m (885 ft)/min
- Service ceiling — 3,500 m (11,475 ft)
- T-O to 15 m (50 ft) — 440 m (1,445 ft)
- Max endurance — 2 h 30 min

IAR-827A single/two-seat agricultural aircraft, in production form with Polish radial engine *(Brian M. Service)*

ICA (AÉROSPATIALE) IAR-316B ALOUETTE III

ICA and Aérospatiale concluded an agreement in 1971 for manufacture in Romania of SA 316B Alouette III helicopters. Production of these continues, and had totalled approx 175 by early 1981. Production is currently at the approximate rate of 20-30 per year. Romanian-built components are also supplied for incorporation in French-built Alouette IIIs.

ICA (AÉROSPATIALE) IAR-330 PUMA

An agreement for licence production of the Aérospatiale SA 330 Puma in Romania was concluded in 1977, an initial quantity of 100 being involved. Approx 90 of these had been delivered by early 1981, most of them (and most Alouette IIIs) reportedly to the Romanian Air Force. Production is currently at the approximate rate of 20-30 per year.

SOUTH AFRICA

AERONICS

AERONICS (PTY) LTD

PO Box 391090, Bramley 2018

Telephone: (011) 786 9600

DIRECTOR: J. M. Cohoe

Aeronics (Pty) Ltd has production rights to produce exclusively in South Africa the **Sequoia Models 300 and 301** light aircraft, following an agreement signed in November 1978 with Sequoia Aircraft Corporation of Richmond, Virginia, USA. The start of production is dependent upon the arrangement of suitable financing.

A detailed description of the Sequoia 300/301 can be found in the Homebuilts section.

ATLAS

ATLAS AIRCRAFT CORPORATION OF SOUTH AFRICA (PTY) LIMITED

Head Office and Works: PO Box 11, Atlas Road, Kempton Park 1620, Transvaal
Telephone: 973 0111
Telex: 87965
Manager: G. W. Ward
Manager (Commercial): A. J. Bester

Atlas built the Impala Mk 1 (M.B. 326M) jet trainer under licence from Aermacchi. Its present programmes include production of the Impala Mk 2 and of the Atlas C4M STOL light transport.

Atlas is manufacturing under licence components for Dassault Mirage F1-AZ and -CZ multi-purpose combat aircraft currently in service with Nos. 1 and 3 Squadrons respectively of the South African Air Force. It also undertakes maintenance and overhaul of SAAF aircraft.

ATLAS IMPALA Mk 2

The Impala Mk 2 is based on the single-seat Aermacchi M.B. 326K. It was developed by Atlas as an advanced trainer for the SAAF and is in production. Four M.B. 326Ks are reported to have been delivered initially by Aermacchi. No details of the Impala Mk 2 have been released officially, except that it has a Rolls-Royce Viper Mk 540 turbojet engine; a description of the Aermacchi M.B. 326K can be found in the Italian section of the 1980-81 and earlier editions of *Jane's*. Impala Mks 1 and 2 are reported to equip Nos. 4, 5, 6, 7 and 8 Squadrons of the SAAF's Active Citizen Force, in addition to the Flying Training School at Langebaanweg.

ATLAS C4M

South African Air Force name: Kudu

The C4M is a six/eight-seat light transport aircraft, developed by Atlas, which can be converted rapidly from the passenger to the freight role, and vice versa, and can operate from unprepared surfaces. The prototype (ZS-IZF) flew for the first time on 16 February 1974, and certification to FAR Pt 23 was granted on 16 June 1975. The third C4M (ZS-IZG) flew shortly afterwards.

A military prototype flew for the first time on 18 June 1975, and was later handed over to the South African Air Force (SAAF serial number 961) for evaluation. A military C4M bearing the SAAF serial number 978, and belonging to No. 41 Squadron, was displayed at the Air Africa air show in 1977, indicating that the C4M is in production. No confirmation of its civil or military status has been given by Atlas, but the Kudu military version is believed also to form part of the equipment of No. 42 Squadron of the SAAF. It is reported that the SAAF has ordered 40.

Type: Single-engined cabin monoplane.

Wings: High-wing monoplane, with single elliptical-section bracing strut each side. Wing section NACA 23016 at root, NACA 4412 (modified) at tip. Dihedral 3°. Incidence 4° at root, 0° 27′ at tip. All-metal D-spar torsion-box structure, bolted to fuselage. Electrically-operated Fowler flaps, of all-metal two-spar construction, interchangeable right with left. All-metal piano-hinged ailerons, with inset electrically-operated trim tab (port) and balance tab (starboard). Vortex generators forward of ailerons. Glassfibre wingtips.

Fuselage: All-metal stressed-skin structure, comprising a sturdy floor, bulkheads and stringers, of basically rectangular section. No longerons. Titanium alloy firewall. Glassfibre tailcone.

Tail Unit: Cantilever all-metal structure. Electrically-operated variable-incidence tailplane, with manual standby. Aerodynamically and mass balanced elevators and rudder. Balance tab in port elevator. Servo tab in each elevator. Vortex generators forward of elevators.

Landing Gear: Non-retractable tailwheel type. Main units each have an independent cantilever leg, and are connected to oleo-pneumatic shock-absorbers mounted below cabin floor level in small underfuselage blister fairings. Main-wheel tube-type tyres size 7·00-8, 6-ply rating, pressure 2·55 bars (37 lb/sq in). Single-disc hydraulic brakes. Steerable tailwheel, mounted on an oleo-pneumatic levered-suspension strut, has a size 5·00-4 6-ply tubed-type tyre, pressure 2·45 bars (35·5 lb/sq in).

Power Plant: One 254 kW (340 hp) Piaggio-built Avco Lycoming GSO-480-B1B3 flat-six engine, driving a Hartzell HC-B3R20-4/R10160-1 three-blade constant-speed variable-pitch metal propeller with spinner. Three removable and partially self-sealing bag-type fuel tanks in each wing, total capacity 432 litres (95 Imp gallons). Overwing refuelling points. Oil capacity 11·3 litres (2·5 Imp gallons).

Accommodation: Pilot and co-pilot side by side at front, with four individual seats in pairs at rear, or two bench seats each for three persons. Baggage compartment aft of seats. Passenger seats can be removed to provide space for up to 560 kg (1,235 lb) of cargo. Heating, ventilation and windscreen demisting standard. Forward-hinged door on each side for pilot and co-pilot. Main cabin door is on port side, and is in two sections: forward-opening front section for passengers, rearward-opening rear section for loading cargo. A sliding door at the rear of the cabin on the starboard side may be opened in flight for parachuting and is available for emergency exit. A trap-door is provided in the cabin floor for aerial survey or supply-dropping purposes.

Systems: Hydraulic system for main-wheel brakes only. 28V DC electrical system supplied by an engine-driven generator and a 24V 15Ah battery. Static inverter provides 115/26V AC power for instruments and radio.

Avionics and Equipment: To customer's specification, but probably including two VHF transceivers, intercom, HF radio and ADF. Blind-flying instrumentation standard.

Dimensions, external:
Wing span — 13·075 m (42 ft 10¾ in)
Wing chord at root — 1·727 m (5 ft 8 in)
Wing chord at tip — 1·168 m (3 ft 10 in)
Wing aspect ratio — 8·07
Length overall (tail down) — 9·31 m (30 ft 6½ in)
Height overall (tail down) — 3·66 m (12 ft 0 in)
Elevator span — 4·79 m (15 ft 8½ in)
Wheel track (full fuel, no load) — 2·935 m (9 ft 7½ in)
Wheelbase — 6·625 m (21 ft 9 in)
Propeller diameter — 2·54 m (8 ft 4 in)
Propeller ground clearance — 0·233 m (9¼ in)
Crew doors (each): Height — 1·16 m (3 ft 9¾ in)
Mean width — 0·53 m (1 ft 9 in)
Double doors (port): Height — 1·05 m (3 ft 5¼ in)
Width — 1·42 m (4 ft 8 in)
Sliding door (stbd): Height — 1·04 m (3 ft 5 in)
Width — 0·74 m (2 ft 5 in)
Trap-door (floor): Length — 0·70 m (2 ft 3½ in)
Width — 0·50 m (1 ft 7¾ in)

Dimensions, internal:
Cabin: Length, incl flight deck — 3·20 m (10 ft 6 in)
Max width — 1·14 m (3 ft 9 in)
Height at front — 1·31 m (4 ft 3½ in)
Height at rear — 1·14 m (3 ft 9 in)

Areas:
Wings, gross — 20·971 m² (225·7 sq ft)
Ailerons (total) — 2·620 m² (28·19 sq ft)
Trailing-edge flaps (total) — 3·774 m² (40·62 sq ft)
Fin — 1·625 m² (17·49 sq ft)
Rudder — 0·945 m² (10·17 sq ft)
Tailplane — 3·430 m² (36·92 sq ft)
Elevators, incl tabs — 2·063 m² (22·21 sq ft)

Weights and Loadings:
Weight empty — 1,230 kg (2,711 lb)
Max payload — 560 kg (1,235 lb)
Max T-O and landing weight — 2,040 kg (4,497 lb)
Max wing loading — 97·3 kg/m² (19·93 lb/sq ft)
Max power loading — 8·03 kg/kW (13·23 lb/hp)

Performance (at max T-O weight):
Never-exceed speed — 165 knots (305 km/h; 190 mph) IAS
Max level speed at 2,440 m (8,000 ft) — 140 knots (259 km/h; 161 mph)
Max cruising speed at 3,050 m (10,000 ft) — 126 knots (233 km/h; 145 mph)
Econ cruising speed at 3,050 m (10,000 ft) — 108 knots (200 km/h; 124 mph)
Stalling speed, flaps up, power on — 65 knots (121 km/h; 75 mph) IAS
Stalling speed, flaps down, power on — 48 knots (89 km/h; 55·5 mph) IAS
Max rate of climb at S/L — 244 m (800 ft)/min
Service ceiling — 4,270 m (14,000 ft)
T-O run — 215 m (705 ft)
T-O to 15 m (50 ft) — 370 m (1,214 ft)
Landing from 15 m (50 ft) — 260 m (853 ft)
Landing run — 140 m (460 ft)
Range with max fuel, 45 min reserves — 700 nm (1,297 km; 806 miles)
Range with 400 kg (882 lb) payload, 45 min reserves — 400 nm (740 km; 460 miles)
Endurance with max fuel, no reserves — 8 h

Atlas Impala Mk 2 single-seat light strike aircraft, based on the Aermacchi M.B. 326K

Atlas C4M six/eight-seat light transport (Avco Lycoming GSO-480-B1B3 engine)

CSIR

COUNCIL FOR SCIENTIFIC AND INDUSTRIAL RESEARCH (National Institute for Aeronautics and Systems Technology)

PO Box 395, Pretoria 0001
Telephone: (012) 86 9211
Telex: 3-461 SA

Head of Aeronautics Department:
Dr C. G. van Niekerk, FRAeS

Current activities of the CSIR include research into lifting rotors, airframe fatigue, synthetic materials, separation of underwing stores, aircraft and missile stability, flutter, atmospheric turbulence, and aircraft noise problems.

The prototype SARA II (South African Research Autogyro) made its first free flight in Pretoria during November 1972; a description of this can be found in the 1975-76 *Jane's*, and of the modified SARA 3 in the 1979-80 edition.

SPAIN

AISA
AERONAUTICA INDUSTRIAL SA

HEAD OFFICE, WORKS AND AIRFIELD: Cuatro Vientos (Carretera del Aeroclub Carabanchel Alto), Apartado 984, Madrid 25
Telephone: 208 75 40 and 208 96 40
Telex: 23593 E Madrid
PRESIDENT: Gonzalo Suárez
MANAGING DIRECTOR: José A. Delgado
DESIGN MANAGER: Juán del Campo
WORKS MANAGER: Angel Romero

This company has been engaged since 1923 in the manufacture, repair and maintenance of fixed-wing aircraft and helicopters. Its design office has been responsible for several liaison, training and sporting aircraft for the Spanish Air Force and aeroclub flying schools, including the I-11, I-11B, AVD-12 and I-115. The Cuatro Vientos factory has a covered area of 8,000 m² (86,110 sq ft) and employs some 140 persons.

AISA is engaged in IRAN work and general overhaul of several types of US aircraft, in particular the North American T-6, Beechcraft B55, C90 and F33 Bonanza, and Piper PA-23 aircraft operated by the Spanish Air Force and the National School of Aeronautics. It is also engaged in the repair and overhaul of Bell 47, 204, 205 and 206, and Boeing Vertol CH-47 helicopters, and their dynamic components, for the Spanish Army, Spanish Air Force and civilian operators. As a subcontractor to Messier-Hispano-Bugatti, it is producing landing gear shock-absorbers and hydraulic actuators for the Dassault-Breguet Mirage and Falcon series, Dassault-Breguet/Dornier Alpha Jet, and other European aviation programmes. Under subcontract to CASA, it produces structural components for the C-212 Aviocar, the C-101 Aviojet, and the MBB BO 105 helicopter.

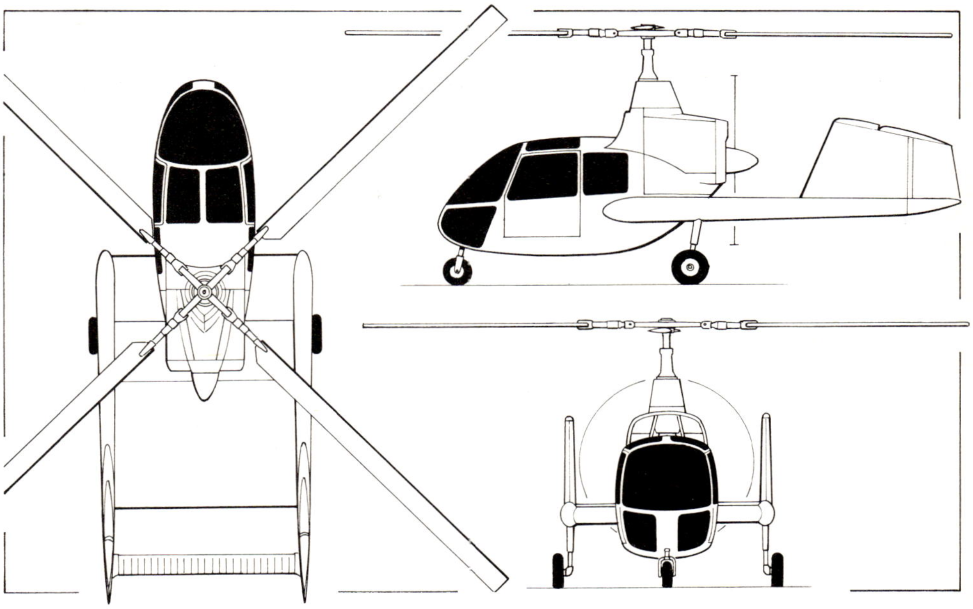

AISA GN autogyro (Avco Lycoming IO-540-K1A5 engine) *(Michael A. Badrocke)*

AISA GN

AISA, which in 1927 built some of the earliest Cierva Autogiros, is currently developing this four-seat autogyro. Construction of two prototypes (one for ground testing) began in 1979, and most structural static tests had been completed by January 1980. First flight was scheduled for the Summer of 1981.

TYPE: Four-seat light autogyro, with jump take-off capability.

ROTOR SYSTEM: Fully-articulated four-blade rotor, with all-metal blades of type fitted to Bell 47G. Hydraulic blade dampers. Rotor brake fitted. Rotor rpm 1,380-1,680 in normal forward flight (autorotation), increasable to 1,884 rpm for jump take-off. Conventional collective and cyclic pitch controls.

ROTOR DRIVE: By power take-off from engine, via two-stage reduction gear, for spin-up on ground. Transmission can be disconnected manually, or will occur automatically when correct rpm and blade pitch angle for T-O are reached.

WINGS: Short-span all-metal wings, of NACA 0024 section, enclosed at each tip by tailboom. No tabs.

FUSELAGE: Pod-shaped semi-monocoque structure of aluminium alloy, with moulded Plexiglas transparencies.

TAIL UNIT: Twin tailbooms, with enclosed tailplane, cable-actuated elevator, and twin fins and rudders, all of metal construction. No tabs.

LANDING GEAR: Non-retractable tricycle type, with single wheel and oleo-pneumatic shock-absorber on each unit. Steerable nosewheel. Independent disc brakes on main wheels. Tailskid under each tailboom.

POWER PLANT: One 224 kW (300 hp) Avco Lycoming IO-540-K1A5 flat-six engine, with direct drive to a Hartzell two-blade constant-speed metal pusher propeller with spinner. Engine also provides power for rotor spin-up on ground. Four bag-type fuel tanks in wings. Overwing gravity refuelling points.

ACCOMMODATION: Seats for four persons in fully-enclosed cabin. Front two seats are independently three-way adjustable; bench seat for two persons at rear. In prototype, all four seats have provision for seat-type parachutes; production version will have space for baggage aft of rear seat. Forward-opening car-type door on each side (jettisonable on prototype). Full dual controls, except for single centrally-located collective pitch control. Cabin ventilation standard.

SYSTEMS: 24V battery for engine starting; provision for ground power receptacle. Fire extinguishing system.

DIMENSIONS, EXTERNAL:
- Rotor diameter — 11·76 m (38 ft 7 in)
- Rotor blade chord (each) — 0·28 m (11 in)
- Rotor disc area — 108·62 m² (1,169·2 sq ft)
- Wing span (incl tailbooms) — 2·60 m (8 ft 6¼ in)
- Wing chord (constant) — 0·85 m (2 ft 9½ in)
- Length overall, excl rotor — 6·52 m (21 ft 4¾ in)
- Height overall — 3·32 m (10 ft 10¾ in)
- Max width (over main wheels) — 2·83 m (9 ft 3½ in)
- Wheel track (c/l of main wheels) — 2·67 m (8 ft 9 in)
- Wheelbase — 2·89 m (9 ft 5¾ in)
- Propeller diameter — 2·14 m (7 ft 0 in)
- Doors (each): Height — 1·085 m (3 ft 6¾ in)
- Max width — 0·93 m (3 ft 0½ in)

DIMENSIONS, INTERNAL:
- Cabin: Max width — 1·14 m (3 ft 9 in)
- Max height — 1·24 m (4 ft 1 in)

WEIGHTS:
- Weight empty — 978 kg (2,156 lb)
- Max payload — 308 kg (679 lb)
- Max fuel — 120 kg (265 lb)
- Max T-O weight — 1,400 kg (3,086 lb)

PERFORMANCE (estimated, at max T-O weight):
- Max level speed at S/L — 129 knots (240 km/h; 149 mph)
- Max cruising speed at S/L — 114 knots (212 km/h; 132 mph)
- Min speed, power on — 29 knots (54 km/h; 33·5 mph)
- Max rate of climb at S/L — 360 m (1,180 ft)/min
- Min sinking speed, power off — 3·73 m (12·24 ft)/s
- Landing run (zero wind) — 0-5 m (0-16·5 ft)
- Range: pilot and 3 passengers — 378 nm (700 km; 435 miles)
- pilot and 2 passengers — 540 nm (1,000 km; 621 miles)
- Endurance: pilot and 3 passengers — 4 h
- pilot and 2 passengers — 6 h
- g limits — +3·5/−1·0

AISA I-122 and I-124

These two designs are, respectively, for a two-seat trainer and a four-seat light aircraft, the I-124 being essentially similar to the I-122 except for an enlarged cockpit. The I-122 is under consideration, together with the CASA C-102, to fulfil a Spanish Air Force requirement.

CASA
CONSTRUCCIONES AERONAUTICAS SA

HEAD OFFICE: Rey Francisco 4, Apartado 193, Madrid 8
Telephone: 247 25 00
Telex: 27418
WORKS: Getafe, Seville, San Pablo, Cádiz, Madrid and Ajalvir
PRESIDENT AND CHAIRMAN OF THE BOARD: Enrique de Guzman
DEPUTY PRESIDENT: Carlos Marin
VICE-PRESIDENT: Eugenio Aguirre
DIRECTOR-GENERAL: Restituto Estirado
DEPUTY DIRECTOR-GENERAL: Pablo Palomar
EXECUTIVE DIRECTOR, SALES AND MARKETING: Pablo de Bergia
DIRECTORS:
Alberto Elvira (Programmes)
Juan Alonso (Sales)
Carlos Grandal (Marketing)
Jesus Salas (Product Support)
PUBLIC RELATIONS AND PRESS MANAGER: Francisco Flores

This company was formed in March 1923 for the primary purpose of producing metal aircraft for the Spanish Air Force. It began by building under licence the Breguet XIX and has since manufactured many other aircraft of foreign design, the most recent being the Northrop F-5 fighter. It is assembling the 60 MBB BO 105C helicopters ordered for the Spanish Army in July 1979.

CASA's own Project Office has designed several aircraft under contract to the Spanish Air Ministry, including the C-212 Aviocar transport and the C-101 Aviojet jet trainer, both of which are currently in production. It also undertakes design and development work for foreign companies and, for example, collaborated in the design of the MBB HFB 320 Hansa executive transport; MBB in turn co-operated in the design of the Aviocar. In order to promote sales in the Far East, CASA established a C-212 assembly line in Indonesia, as well as full after-sales support in that area (see PT Nurtanio entry in Indonesian section). As described in the International section, it is also collaborating with Nurtanio to develop the Airtec CN-235, a new 34/38-passenger transport aircraft.

Under contract to Dassault-Breguet (which see), CASA is responsible for manufacturing outer wings for the Falcon 10 light business aircraft and centre-fuselages for the Mirage F1 combat aircraft. It is a full member (4·2%) of Airbus Industrie (see International section), and manufactures the horizontal tail surfaces, landing gear doors and forward passenger doors for the Airbus A300 wide-bodied transport aircraft. It also manufactures glassfibre honeycomb components, including underwing fillets for McDonnell Douglas DC-10 and upper-rudder segments for Boeing 727 jet transports, outboard flaps for the Boeing 757, and wing units for the Otomat air-launched missile.

CASA undertakes maintenance and modernisation work for the Spanish Air Force and for the US Air Force in Europe. Its principal current activities of this kind concern overhaul and maintenance of McDonnell Douglas F-4 combat aircraft and Bell 47G, 204 and 205 and Sikorsky H-19 helicopters.

In 1972 the former Hispano Aviación SA (see 1972-73 *Jane's*) was merged with CASA, the latter company taking over all of Hispano's offices and other facilities, aircraft production programmes and personnel. In June 1973 ENMASA (Empresa Nacional de Motores de Aviación) was merged into CASA, and now constitutes the CASA División de Motores (see Aero-Engines section). CASA has six factories, employing about 8,300 people in early 1981. Including Hispano production, the company had by then manufactured more than 3,500 aircraft and overhauled approx 6,250. CASA has a total covered area in the region of 200,000 m² (2,152,780 sq ft); majority shareholder in the company is the INI (Instituto Nacional de Industria).

CASA C-212-5 SERIES 100 AVIOCAR

The C-212 Aviocar twin-turboprop light utility STOL transport was developed by CASA to fulfil a variety of military or civil roles, but primarily to replace the mixed fleet of Junkers Ju 52/3m (T.2), Douglas DC-3 (T.3) and CASA-207 Azor (T.7) transport aircraft formerly in service with the Spanish Air Force. The first of two prototypes flew for the first time on 26 March 1971.

In addition to the prototypes and eight pre-production Aviocars, 154 were completed to C-212-5 Series 100 initial production standard (125 by CASA and 29 by PT Nurtanio in Indonesia), as recorded in the 1980-81 edition of *Jane's*. This total included six VIP transports, ten for photographic survey, and six for navigation training. Military sales included CASA-built examples for the air forces of Jordan (four), Nicaragua (five), Portugal (24)

CASA C-212 Series 200 Aviocar twin-turboprop transport aircraft in the insignia of Allegheny Commuter Airlines

and Spain (71); plus two Nurtanio-built Srs 100s for the Indonesian Air Force and four for the Royal Thai Air Force.

CASA C-212 SERIES 200 AVIOCAR

The current standard version of the Aviocar since 1979, the Series 200 is an improved version of the original C-212-5 (1979-80 *Jane's*) with more powerful TPE331-10 engines and increased max T-O weight. Aircraft Nos. 138 and 139 served as prototypes for this version, making their first flights on 30 April and 20 June 1978 respectively.

Designed to meet the requirements of FAR Pt 25, the Aviocar can be operated under FAR Pt 121 and Pt 135 conditions, and is well within the noise requirements of FAR Pt 36.

By mid-1981 total sales of the Aviocar (all versions) had reached 264, of which approx 190 had been delivered by CASA and Nurtanio. Sales of 85 Srs 200s by CASA included nine to the Spanish Air Force, one to the Spanish Navy, three to the Uruguayan Army and two to the Venezuelan Navy. Nurtanio orders for the Srs 200 included 14 for the Indonesian Air Force.

ASW/maritime patrol and elint/ECM versions of the Srs 200 are under development, and are described separately. The following description applies to the standard version:

Type: Twin-turboprop STOL utility transport.

Wings: Cantilever high-wing monoplane. Wing section NACA 653-218. Incidence 2° 30′. No dihedral or sweepback. All-metal light alloy fail-safe structure. All-metal ailerons and double-slotted trailing-edge flaps. Trim tab in port aileron. Rubber-boot de-icing of leading-edges.

Fuselage: Semi-monocoque fail-safe structure of light alloy construction.

Tail Unit: Cantilever two-spar all-metal structure, with dorsal fairing forward of fin. Tailplane mid-mounted on rear of fuselage. Trim tab in rudder and each elevator. Rubber-boot de-icing of leading-edges.

Landing Gear: Non-retractable tricycle type, with single main wheels and single steerable nosewheel. CASA oleo-pneumatic shock-absorbers. Goodyear wheels and tyres, main units size 11·00-12 Type III (10-ply rating), nose unit size 24-7·7 Type VII (8-ply rating). Tyre pressure 3·72 bars (54 lb/sq in) on main units, 3·52 bars (51 lb/sq in) on nose unit. Goodyear hydraulic disc brakes on main wheels.

Power Plant: Two Garrett TPE331-10-501C turboprop engines, each flat-rated to 671 kW (900 shp) and driving a Hartzell HC-B4NM-5AL four-blade constant-speed fully-feathering reversible-pitch propeller (Beta mode on ground). Electrical anti-icing of propeller blades. Fuel in four integral outer-wing tanks, with total capacity of 2,040 litres (449 Imp gallons), of which 2,000 litres (440 Imp gallons) are usable. Oil capacity 6 litres (1·32 Imp gallons) per engine.

Accommodation: Crew of two on flight deck. For troop transport role, main cabin can be fitted with 21 inward-facing seats along cabin walls, plus three forward-facing seats, to accommodate 23 paratroops with an instructor/jumpmaster; or seats for 24 fully-equipped troops. As an ambulance, cabin would normally be equipped to carry 12 stretcher patients and up to four medical attendants. As a freighter, Srs 200 can carry up to 2,770 kg (6,107 lb) of cargo in main cabin, including two LD1, LD727/DC-8 or LD3 pallets, or light vehicles. Photographic version is equipped with two Wild RC-10 vertical cameras and a darkroom. Navigation training version accommodation consists of individual desks for an instructor and five pupils, in two rows, fitted with appropriate instrument installations. Civil passenger transport version has standard seating for up to 26 persons in mainly four-abreast layout at 72 cm (28·5 in) pitch, with provision for quick change to all-cargo or mixed passenger/cargo interior. High-density layout can accommodate 28 passengers. Toilet and 400 kg (882 lb) capacity baggage compartment standard. VIP transport version can be furnished to customer's requirements. Access to main cabin is via two doors on port side, one aft of (and providing access to) flight deck and one aft of wing trailing-edge. In addition, a two-section underfuselage loading ramp/door aft of main cabin is openable in flight for discharge of paratroops or cargo, and is fitted with external wheels, to allow door to remain open during ground manoeuvring. Emergency exit aft of wing trailing-edge on starboard side. Interior of rear-loading door can be used for additional baggage stowage.

Systems: Unpressurised cabin. Hydraulic system, pressure 138 bars (2,000 lb/sq in), operates main-wheel brakes, flaps, nosewheel steering and ventral cargo door. Electrical system is supplied by two 9kW starter/generators, three batteries and three static converters.

Avionics and Equipment: Standard avionics include King or Collins VHF com, VHF nav, interphone, ADF, DME, ATC transponder, radio altimeter, weather radar, flight director and PA system. Blind-flying instrumentation standard. Optional avionics include Tacan, SIF/IFF, and a second ADF.

Dimensions, external:
Wing span 19·00 m (62 ft 4 in)
Wing chord at root 2·49 m (8 ft 2 in)
Wing chord at tip 1·50 m (4 ft 11 in)
Wing aspect ratio 9
Length overall 15·16 m (49 ft 9 in)
Height overall 6·30 m (20 ft 8 in)
Tailplane span 8·40 m (27 ft 6¾ in)
Wheel track 3·10 m (10 ft 2 in)
Wheelbase 5·55 m (18 ft 2½ in)
Propeller diameter 2·79 m (9 ft 2 in)
Distance between propeller centres 5·30 m (17 ft 4¾ in)
Passenger door (port, rear):
Max height 1·58 m (5 ft 2¼ in)
Max width 0·70 m (2 ft 3½ in)
Crew and servicing door (port, fwd):
Max height 1·10 m (3 ft 7¼ in)
Max width 0·57 m (1 ft 10⅝ in)
Rear-loading door: Max length 3·66 m (12 ft 0 in)
Max width 1·70 m (5 ft 7 in)
Max height 1·80 m (5 ft 11 in)

Dimensions, internal:
Cabin (excl flight deck and rear-loading door):
Length 6·50 m (21 ft 4 in)
Max width 2·10 m (6 ft 10¾ in)
Max height 1·80 m (5 ft 11 in)
Floor area 10·50 m² (113·0 sq ft)
Volume 22·00 m³ (776·9 cu ft)
Cabin: volume incl flight deck and rear-loading door 26·00 m³ (918·2 cu ft)

Areas:
Wings, gross 40·0 m² (430·56 sq ft)
Ailerons (total) 2·45 m² (26·37 sq ft)
Trailing-edge flaps (total) 7·38 m² (79·44 sq ft)
Fin 4·25 m² (45·75 sq ft)
Rudder, incl tab 2·02 m² (31·74 sq ft)
Tailplane 12·57 m² (135·30 sq ft)
Elevators, incl tabs 4·36 m² (46·93 sq ft)

Weights and Loadings:
Manufacturer's weight empty 3,780 kg (8,333 lb)
Weight empty, equipped (cargo) 4,115 kg (9,072 lb)
Max payload (cargo) 2,770 kg (6,106 lb)
Max fuel 1,600 kg (3,527 lb)
Max T-O weight 7,450 kg (16,424 lb)
Max ramp weight 7,500 kg (16,534 lb)
Max landing weight 7,350 kg (16,204 lb)
Max zero-fuel weight 7,050 kg (15,542 lb)
Max cabin floor loading 732 kg/m² (149·9 lb/sq ft)
Max wing loading 186·2 kg/m² (38·1 lb/sq ft)
Max power loading 5·55 kg/kW (9·12 lb/shp)

Performance (at max T-O weight, ISA):
Max operating speed (Vmo) 202 knots (374 km/h; 232 mph) IAS
Max cruising speed at 3,050 m (10,000 ft) 197 knots (365 km/h; 227 mph)
Normal cruising speed at 3,050 m (10,000 ft) 187 knots (346 km/h; 215 mph)
Stalling speed, T-O configuration 78 knots (145 km/h; 90 mph) EAS
Max rate of climb at S/L 516 m (1,700 ft)/min
Rate of climb at S/L, one engine out 114 m (374 ft)/min
Service ceiling 8,535 m (28,000 ft)
Service ceiling, one engine out 3,500 m (11,475 ft)
T-O run 440 m (1,445 ft)
T-O distance (FAR Pt 25, unfactored) 610 m (2,000 ft)
T-O to 15 m (50 ft) 630 m (2,065 ft)
Landing from 15 m (50 ft) 505 m (1,655 ft)
Landing distance (FAR Pt 25, unfactored) 595 m (1,950 ft)
Landing run 200 m (656 ft)
Range at max cruising speed, no reserves:
with max payload 220 nm (408 km; 253 miles)
with max fuel 950 nm (1,760 km; 1,094 miles)

CASA C-212 SERIES 200 AVIOCAR (ASW and MARITIME PATROL VERSION)

For service with the Spanish Air Force, and for certain foreign countries, CASA has developed a version of the C-212 Srs 200 equipped for anti-submarine and maritime patrol duties. A total of 15 had been ordered by mid-1981 (nine by the Spanish Air Force for SAR duties, three by the Spanish Ministry of Finance, two by the Venezuelan Navy and one by the Uruguayan Navy).

As can be seen from the accompanying illustration, the principal external differences from the transport version are the addition of a nose radome and the appearance of various antennae on the fuselage and tail-fin. Underwing pylons are provided for the carriage of torpedoes, rocket pods and other weapons.

Type: Twin-turboprop ASW and maritime patrol aircraft.

Airframe: Generally similar to standard C-212 Srs 200 except for addition of nose radome and various external antennae.

Power Plant: As for C-212 Srs 200.

Accommodation (ASW version): Pilot and co-pilot on flight deck, with OTPI and additional central console for radar repeater; control for radio-navigation, Doppler, DME, ADF, UHF/DF, Omega and VOR/ILS; weapons delivery controls; and intervalometer for rockets.

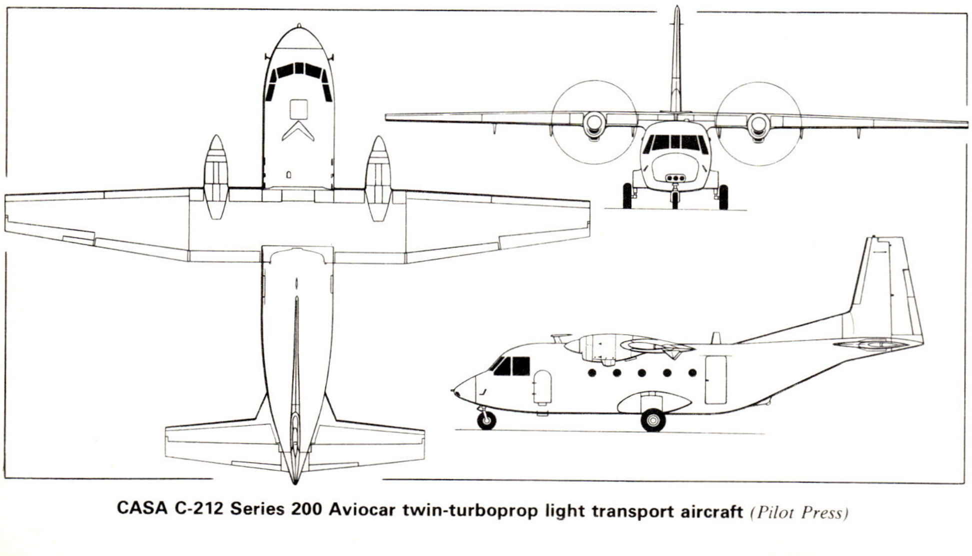

CASA C-212 Series 200 Aviocar twin-turboprop light transport aircraft *(Pilot Press)*

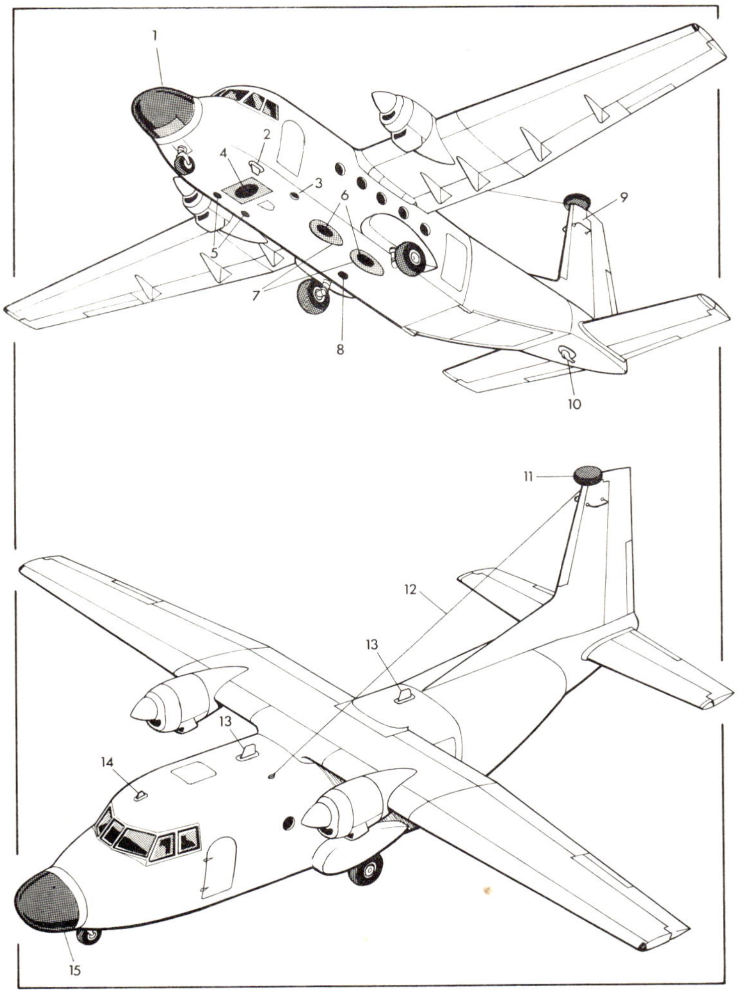

1 Search radar
2 IFF
3 Sonobuoys
4 Doppler
5 Radar altimeter
6 ADF loop
7 ADF sense antenna
8 Marker beacon receiver
9 VHF nav, VOR/LOC
10 Omega
11 Electronic warfare
12 HF com
13 UHF/VHF com
14 DME
15 Glideslope

Maritime patrol version of CASA C-212 Srs 200 *(Michael A. Badrocke)*

Avionics rack on port side, aft of pilot, for com/nav equipment; second rack on starboard side, aft of co-pilot, contains avionics for mission equipment (radar, sonobuoys, MAD and ESM). Immediately aft of the latter rack, along the starboard side of the cabin, are three control consoles for the mission crew members. The first console has the radar control and display, ESM control and display, and ICS control. The second has the tactical display and control, MAD recorder and control, and ICS. The rearmost of the three incorporates ICS, sonobuoy receiver control unit, acoustic control panel; and HCU and ADU units.

Accommodation (maritime patrol version): Pilot and co-pilot on flight deck, with central console for radar repeater; control for radio-navigation, Doppler, DME, ADF, UHF/DF, Omega, VOR/ILS and searchlight. Avionics rack on port side, aft of pilot, for com/nav and radar equipment. On starboard side of cabin is a console for the radar operator that incorporates radar PPI and ICS controls. Posts for two observers are located at the rear of the cabin.

Avionics: Communications equipment includes two HF and two VHF transceivers, single UHF, and interphone. Navigation equipment includes automatic flight control system, flight director, VOR/ILS (including VOR/LOC), glideslope and marker beacon receiver, DME, two ADF, UHF/DF, radar altimeter, Doppler radar, VLF/Omega, autopilot and compass.

Operational Equipment (ASW version): Search radar with 360° scan, electronic support measures (ESM), sonobuoy processing system (SPS), OTPI, MAD, tactical processing system (TPS), IFF/SIF transponder, sonobuoy and smoke marker launcher, torpedoes, rockets and other weapons.

Operational Equipment (maritime patrol version): Search radar with 240° scan, searchlight, smoke markers and camera.

Weights (ASW version):
Max T-O weight 8,400 kg (18,519 lb)
Max landing weight 7,000 kg (15,432 lb)

Performance (at max T-O weight, ISA):
Max cruising speed at 3,050 m (10,000 ft) 190 knots (353 km/h; 219 mph)
Loiter speed at 457 m (1,500 ft) 105 knots (195 km/h; 121 mph)
Service ceiling 7,315 m (24,000 ft)

CASA C-212 SERIES 200 AVIOCAR (ELINT/ECM VERSION)

A version of the Srs 200 Aviocar for electronic intelligence and electronic countermeasures duties was under development in early 1981, at which time four had been ordered by the air force of the United Arab Emirates (Abu Dhabi).

The elint/ECM version carries equipment for automatic signal interception, classification and identification in dense signal environments, data from which enable a map to be drawn plotting the position and characteristics of hostile radars. Emitters for the jamming part of the mission are also carried.

CASA C-101 AVIOJET

Spanish Air Force designation: E.25

On 16 September 1975, CASA and the Spanish Ministerio del Aire signed a contract for a new basic and advanced military jet trainer aircraft, the C-101. The contract covered design, development, and the construction of four prototype aircraft for flight test and two for static and fatigue testing. First flights were made on 27 June and 30 September 1977, and 26 January and 17 April 1978. All four were handed over to the Spanish Air Force, for service trials, in 1978.

To minimise cost and maintenance, the C-101 is built on modular lines, with ample space within the airframe for equipment for any training mission likely to be required in the 1980s. The C-101 is fully aerobatic, and is able to carry out such additional duties as ground attack, reconnaissance, escort, weapons training, electronic countermeasures (ECM), and photographic missions. The Spanish Air Force trainer version has the manufacturer's designation **C-101EB**; an armed version for export, with TFE731-3 engine, is designated **C-101BB**.

The C-101 was designed with the co-operation of MBB (West Germany) and Northrop (USA), the latter company providing assistance with the inlet design and that of the 'Norcasa' wing section. Manufacture of production aircraft is carried out entirely by CASA except for the nosewheel unit, which is produced in the UK by Dowty Rotol. Wings and main landing gear units are built at Getafe and fuselages at Seville.

Production of the first 10 aircraft started at the beginning of 1978, and the first of these flew for the first time on 8 November 1979. A total of 88 has been ordered by the Spanish Air Force, which has an overall requirement for about 120. Eight C-101BBs have been ordered by an undisclosed customer (reportedly the Chilean Air Force). The first four C-101EB trainers were delivered on 17 March 1980, and by February 1981 a total of 25 had been delivered to the Spanish Air Force Academy (No. 793 Squadron).

Type: Tandem two-seat basic and advanced trainer and light tactical aircraft.

Wings: Cantilever low-wing monoplane. Wing section Norcasa 15, thickness/chord ratio 15%. Dihedral 5°. Incidence 1°. Sweepback at quarter-chord 1° 53'. All-metal (aluminium alloy) three-spar fail-safe stressed-skin structure, with six-bolt attachment to fuselage. Plain ailerons and slotted trailing-edge flaps, of glassfibre/honeycomb sandwich construction. Flap track guides of titanium. Ailerons actuated hydraulically, with manual backup. Ground-adjustable tab on port aileron.

Fuselage: All-metal semi-monocoque fail-safe structure. Hydraulically-operated aluminium honeycomb airbrake under centre of fuselage.

Tail Unit: Cantilever all-metal structure, with electrically actuated variable-incidence tailplane. Aluminium honeycomb rudder and elevators, actuated manually. Electrically-actuated trim tab in rudder.

Landing Gear: Hydraulically-retractable tricycle type, with single wheel and oleo-pneumatic shock-absorber on each unit. Forward-retracting Dowty Rotol nose unit, with non-steerable nosewheel size 450 × 140-8. Inward-retracting main units, wheel size 615 × 225-10, with disc brakes.

Power Plant: One 15·6 kN (3,500 lb st) Garrett TFE731-2-2J non-afterburning turbofan engine in C-101EB; C-101BB has a 16·5 kN (3,700 lb st) TFE731-3-1J turbofan. Lateral intake on each side of fuselage, abreast of second cockpit. Fuel in one 1,115 litre (245 Imp gallon) fuselage bag tank, one 550 litre (121 Imp gallon) integral tank in wing centre-section, and two outer-wing integral tanks each of 335 litres (74 Imp gallons). Total usable internal fuel capacity 2,335 litres (514 Imp gallons). Pressure refuelling point in port air intake; gravity fuelling point above each tank. No provision for drop-tanks. Oil capacity 8·5 litres (1·8 Imp gallons).

Accommodation: Crew of two in tandem, on Martin-Baker E10C zero-zero ejection seats, under individual canopies which open sideways to starboard and are separated by internal windscreen. Rear seat elevated 0·325 m (1 ft 0¾ in). Cockpit pressurised and air-conditioned by engine bleed air. Dual controls standard.

Systems: Hamilton Standard three-wheel bootstrap-type air-conditioning and pressurisation system, differential 0·28 bars (4·1 lb/sq in). Single hydraulic system, pressure 207 bars (3,000 lb/sq in), for landing gear, ailerons, flaps, airbrake and wheel brakes. Backup system comprising compressed nitrogen bottle for landing gear extension and accumulator for braking power. Pneumatic system for air-conditioning, pressurisation and canopy seal. Electrical system includes 28V 8·4 kW DC starter/generator, two 700VA inverters for AC power, and two 22Ah nickel-cadmium batteries. High-pressure gaseous oxygen system.

Avionics and Equipment: Spanish Air Force C-101EB equipped with Magnavox RT-1168/ARC-164 UHF; Wilcox AN/ARC-134 VHF; Bendix AN/ARN-127 VOR/ILS/marker; Collins AN/ARN-118 Tacan; Teledyne Electronic RT-1063B/APX-101 IFF/SIF; Sperry SPI-402 flight director system, including Tarsyn vertical and directional gyro package, dual HZ-444 attitude director indicators, RD-500A horizontal situation indicators with remote course selection, RH-405 radio magnetic indicators and 807A com transceivers. Wide range of alternative avionics available for export version(s).

CASA C-101EB (E.25) Aviojet trainer in Spanish Air Force service

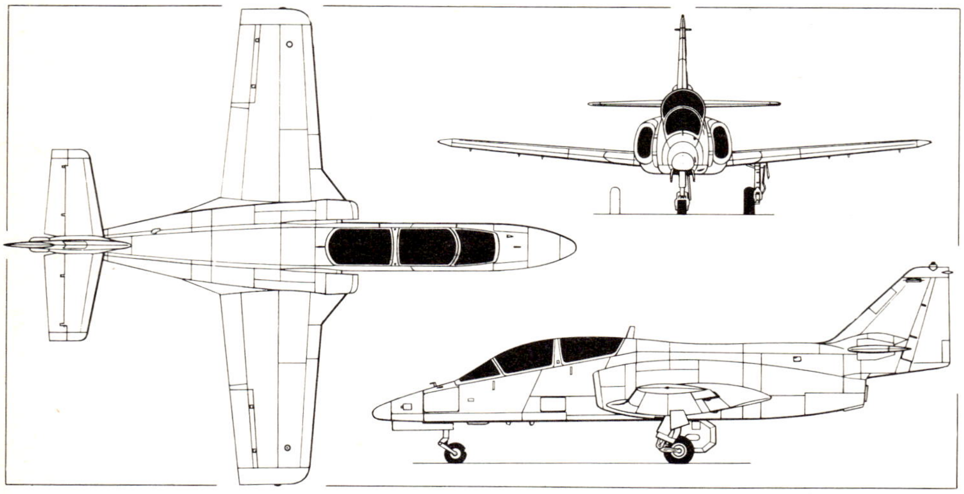

CASA C-101 Aviojet basic/advanced training and light strike aircraft *(Pilot Press)*

ARMAMENT AND OPERATIONAL EQUIPMENT: Large bay below rear cockpit suitable for quick-change packages, including 30 mm DEFA cannon pod, a twin 12·7 mm M3 machine-gun pod, reconnaissance camera, ECM package or laser designator. Six underwing hardpoints, capacities 500 kg (1,102 lb) inboard, 375 kg (827 lb) centre and 250 kg (551 lb) outboard. Typical armament may include one 30 mm cannon or two 12·7 mm guns in the fuselage; and four LAU-10 pods of 5 in rockets, six LAU-68 pods of 2·75 in rockets, six 250 kg BR-250 bombs, four 375 kg BR-375 bombs, two 500 kg BR-500 bombs, four BLU-27 napalm canisters or six SUU-25 flare pods under the wings.

DIMENSIONS, EXTERNAL:

Wing span	10·60 m (34 ft 9⅜ in)
Wing chord at c/l	2·36 m (7 ft 9 in)
Wing chord at tip	1·41 m (4 ft 7½ in)
Wing aspect ratio	5·6
Length overall	12·25 m (40 ft 2¼ in)
Height overall	4·25 m (13 ft 11¼ in)
Tailplane span	4·32 m (14 ft 2 in)
Wheel track	2·83 m (9 ft 3½ in)
Wheelbase	4·87 m (15 ft 11¾ in)

AREAS:

Wings, gross	20·00 m² (215·3 sq ft)
Ailerons (total)	1·60 m² (17·22 sq ft)
Trailing-edge flaps (total)	2·50 m² (26·91 sq ft)
Fin	1·90 m² (20·45 sq ft)
Rudder	1·30 m² (13·99 sq ft)
Tailplane	3·44 m² (37·03 sq ft)
Elevators	1·00 m² (10·76 sq ft)

WEIGHTS AND LOADINGS:

Weight empty, equipped	3,350 kg (7,385 lb)
Max internal fuel weight	1,850 kg (4,078 lb)
T-O weight: trainer, 'clean'	4,850 kg (10,692 lb)
ground attack	5,600 kg (12,345 lb)
Max landing weight:	
3·66 m (12 ft)/s sink rate	4,700 kg (10,361 lb)
3·05 m (10 ft)/s sink rate	5,400 kg (11,905 lb)
Wing loading: trainer	240 kg/m² (49·16 lb/sq ft)
ground attack	280 kg/m² (57·35 lb/sq ft)
Power loading: trainer	308 kg/kN (3·02 lb/lb st)
ground attack	359 kg/kN (3·53 lb/lb st)

PERFORMANCE (C-101BB at 4,400 kg; 9,700 lb AUW):

Max limiting Mach No.	Mach 0·80
Max level speed at S/L	373 knots (691 km/h; 430 mph)
Max level speed at 7,620 m (25,000 ft)	Mach 0·71 (428 knots; 793 km/h; 493 mph)
Econ cruising speed at 9,145 m (30,000 ft)	Mach 0·60 (354 knots; 656 km/h; 407 mph)
Stalling speed, flaps up	99 knots (183 km/h; 114 mph)
Stalling speed, flaps down	88 knots (164 km/h; 102 mph)
Max rate of climb at S/L	1,152 m (3,780 ft)/min
Time to 7,620 m (25,000 ft)	8 min 40 s
Service ceiling	12,200 m (40,000 ft)
T-O run	630 m (2,065 ft)
T-O to 15 m (50 ft)	850 m (2,790 ft)
Landing from 15 m (50 ft)	900 m (2,955 ft)
Landing run	530 m (1,740 ft)

Typical combat radius:
- interdiction (lo-lo-hi) with four 250 kg bombs and 30 mm gun, 3 min over target, 30 min reserves — 205 nm (380 km; 236 miles)
- close air support (lo-lo-hi) with four 19 × 2·75 in rocket launchers and 30 mm gun, 50 min loiter over battle area, 8 min over target, 30 min reserves — 150 nm (278 km; 173 miles)
- ECM, 3 h 15 min loiter over target, 30 min reserves — 330 nm (611 km; 380 miles)
- photo-reconnaissance (hi-lo-lo), 30 min reserves — 520 nm (964 km; 599 miles)

Armed patrol with 30 mm gun or two 12·7 mm guns, no underwing stores, 45 min reserves:
3 h at 205 knots (380 km/h; 236 mph) at S/L, with 100 nm (185 km; 115 mile) transit distance from base to patrol area

Ferry range, 30 min reserves — 1,950 nm (3,613 km; 2,245 miles)

Typical endurance, training missions:
two 1 h 10 min general handling missions, incl aerobatics, with 20 min reserves after second mission

g limits:

at 4,800 kg (10,582 lb) AUW	+7·5/−3·9
at 5,600 kg (12,345 lb) AUW	+5·5/−2·4

CASA C-102

Under this designation CASA has completed preliminary design of a side-by-side two-seat primary training aircraft with a tricycle landing gear and a 149-186·5 kW (200-250 hp) engine.

SWEDEN

JAS
INDUSTRI GRUPPEN JAS

JAS

In June 1980 the Swedish government approved the allocation of SKr 200 million to finance project definition and initial development during 1980-82 of a Viggen replacement to enter service from about 1990. Tentatively known as JAS (Jakt/Attack/Spaning: fighter/attack/reconnaissance), the new multi-role combat aircraft would replace, successively, the AJ/SH/SF/JA 37 versions of the Viggen. A similar financial commitment is being made by Industri Gruppen JAS, a Swedish aerospace industry group formed in 1980 by Saab-Scania, Volvo Flygmotor, L. M. Ericsson, SRA Communications and FFV (Förenade Fabriksverken).

On 3 June 1981 the group submitted to the Swedish Defence Materiel Administration (FMV) its detailed proposals for an aircraft to meet the JAS requirement. The airframe has the Saab project design number 2105, and would be powered by a single General Electric F404J afterburning turbofan engine in the 80 kN (18,000 lb thrust) class developed and produced in collaboration with Volvo Flygmotor. Saab, which would be airframe prime contractor, has announced technology exchange agreements with MBB of West Germany and Rockwell International of the USA. Of similar aerodynamic configuration to the Viggen (see accompanying illustration), the Saab 2105 JAS structure would use some 30 per cent of composite materials, permitting the max T-O weight to be kept down to approx 8,000 kg (17,635 lb). An air-to-air missile can be carried at each wingtip, and there are four underwing hardpoints for other external stores. Armament includes heavy anti-shipping missiles, and reconnaissance equipment can be carried externally. Other features include an L. M. Ericsson multi-mode pulse-Doppler radar, Ericsson FLIR camera (to be flight tested during 1982), and Ericsson/SRA/Datasaab SDS 80 onboard computer. The cockpit head-up display will use diffraction optics (holography), and three other cockpit displays by SRA will (a) replace all conventional instruments (although four conventional dials will be installed as back-ups); (b) give a computer-generated map showing the area immediately surrounding the aircraft; (c) indicate to the pilot targets acquired by the radar and FLIR.

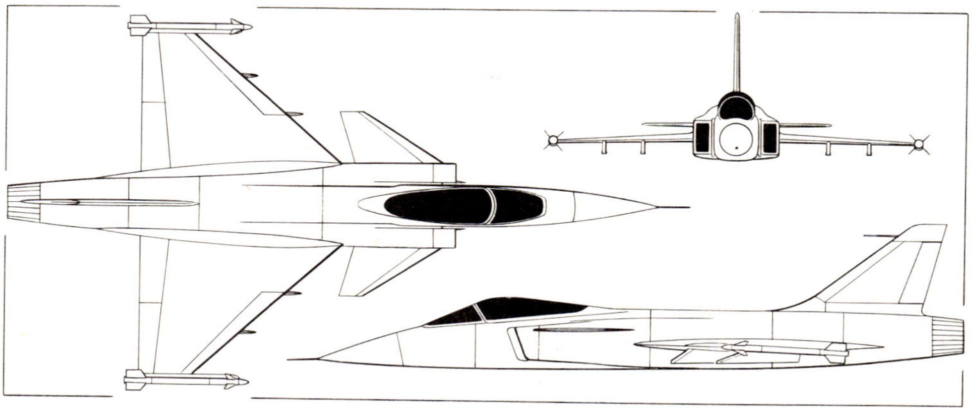

Saab 2105 single-seat fighter, proposed to meet the Swedish Air Force's JAS requirement *(Pilot Press)*

The FMV has evaluated the Swedish industry proposals against those involving aircraft produced in other countries, and has recommended to the Swedish Parliament the adoption of the Saab 2105 design. Government approval was anticipated as part of the new five-year defence plan due to begin in 1982.

SAAB-SCANIA
SAAB-SCANIA AKTIEBOLAG

HEAD OFFICE: S-581 88 Linköping
Telephone: 46 13 11 54 00
Telex: 50040 SAABLGS
PRESIDENT: Sten Gustafsson

Aerospace Division
Telephone: 46 13 12 90 20
HEAD OF DIVISION: Dr Tore Gullstrand
HEAD OF MILITARY AIRCRAFT SECTOR: H. Schröder
HEAD OF CIVIL AIRCRAFT SECTOR: R. Ljungkvist
HEAD OF MISSILE AND ELECTRONICS SECTOR: I. K. Olsson
INFORMATION: Hans G. Andersson

The original Svenska Aeroplan AB was founded at Trollhättan in 1937 for the production of military aircraft. In 1939 this company was amalgamated with the Aircraft Division (ASJA) of the Svenska Järnvägsverkstäderna rolling stock factory in Linköping. Following this merger, Saab moved its head office and engineering departments to Linköping, which is now the main aerospace factory. The company's name was changed to Saab Aktiebolag in May 1965.

Post-war expansions include a bombproof underground factory in Linköping, as well as important new production and engineering facilities in Linköping, Jönköping, Trollhättan and Gothenburg. A new 25,000 m² (269,100 sq ft) factory, scheduled for completion by 1 July 1982, is under construction at Linköping for production of the Saab-Fairchild 340 airliner.

During 1968 Saab merged with Scania-Vabis, to

Saab JA 37 version of the Viggen, in the markings of F13 Wing, Swedish Air Force

strengthen the two companies' position in automotive product development, production and export. Malmö Flygindustri (MFI) was acquired in the same year.

Saab-Scania has nearly 40,000 employees, organised in four operating divisions. Of these, about 6,000 are employed by the Aerospace Division, including 5,500 at Linköping.

Saab-Scania's current aerospace activities include production of the Saab 37 Viggen supersonic multi-purpose STOL combat aircraft and development of the Saab-Fairchild 340 airliner. Since 1949 Saab-Scania has delivered more than 2,000 military jet aircraft to the air forces of four nations. It has also delivered more than 1,500 piston-engined aircraft to military and civil customers around the world. Since 1962, Saab-Scania has had a dealership for Hughes helicopters in Scandinavia and Finland. Since 1978, it has manufactured inboard wing flaps and vanes for the McDonnell Douglas DC-9 Super 80, and is currently also manufacturing tailplanes, elevators, rudders, ailerons and spoilers for the British Aerospace 146 four-turbofan feederliner.

Saab-Scania has greatly expanded its activities in the electronics field in recent years. Current production items include computer systems, autopilots, fire control and bombing systems for piloted aircraft, and electronics for guided missiles. A major production programme is the airborne computer for the Saab 37. The Saab RGS2 lead-computing optical sighting system has been selected for the Royal Netherlands Air Force Northrop NF-5A, the Italian Air Force Aermacchi M.B.339A, and the British Aerospace Hawks for the Finnish Air Force. Space-borne computers, optronic fire control systems and field artillery computer systems are also under development and in production.

Saab-Scania's guided missile activities are now conducted by Saab-Bofors Missile Corporation, formed jointly with AB Bofors. Initial task of this company is to develop a new anti-ship weapon, the RBS 15.

Saab-Scania is a member of the MESH space technology consortium responsible for the TD-1A solar research satellite, the OTS satellite, and the MAROTS maritime communications satellite. For the French space organisation CNES, Saab-Scania delivered the OBC-AR compact guidance and control computer for the Ariane launch vehicle. In September 1980, Saab-Scania was appointed main contractor for the Swedish Viking research satellite, scheduled for launch by Ariane in early 1984.

SAAB-FAIRCHILD 340

Details of the Saab-Fairchild 340 programme can be found in the International section of this edition.

SAAB 37 VIGGEN (THUNDERBOLT)

Swedish Air Force designations: AJ 37, JA 37, SF 37, SH 37 and SK 37

The Saab 37 Viggen multi-mission combat aircraft is the major component in the System 37 manned weapon system for the Swedish Air Force. System 37 comprises the Viggen airframe and its power plant, airborne equipment, armament, ammunition and photographic equipment; ground servicing equipment, including test equipment; and training equipment, including simulators.

The Viggen is a basic platform which was readily adaptable to fulfil the four primary roles of attack, interception, reconnaissance and training. Its STOL characteristics enable it to operate from narrow runways of about 500 m (1,640 ft) length.

The first of seven prototypes flew for the first time on 8 February 1967, and by April 1969 all six single-seat prototypes were flying. The seventh Viggen was the prototype for the two-seat SK 37 operational trainer.

The following versions have been announced:

AJ 37. Single-seat all-weather attack version, with secondary interceptor capability. Initial production version, of which deliveries began on 21 June 1971. Equips two squadrons of F6 Wing at Karlsborg, two squadrons of F7, and one squadron of F15 at Söderhamn. Production completed; details in 1980-81 and earlier editions of *Jane's*.

JA 37. Single-seat interceptor, with Volvo Flygmotor RM8B engine. Improved performance, and secondary capability for attack missions. Four elevon hydraulic actuators under each wing, instead of three as on other versions, and a modified, taller tail-fin similar to that of the SK 37. In production: total of 149 ordered, to re-equip eight Draken fighter squadrons of the Swedish Air Force in 1979-85. First flight by a production JA 37 (serial No. 301) was made on 4 November 1977. Deliveries, to a squadron of F13 Wing at Norrköping, began in 1979. Second squadron is with F17 Wing at Ronneby. Third squadron will form part of F21 Wing at Luleå.

SF 37. Single-seat all-weather overland armed photographic reconnaissance version to replace the S 35E Draken. Modified nose, containing cameras and other equipment. Deliveries, to F13 at Norrköping, began in April 1977. Also in service with two mixed (SF 37/SH 37) squadrons: one with F21 at Luleå and one with F17 at Kallinge. Production completed; details in 1980-81 and earlier editions of *Jane's*.

SH 37. Single-seat all-weather maritime reconnaissance version, to replace the S 32C version of the Lansen. Can also be used for attack missions. In service with F13 at Norrköping (one squadron); and in two mixed (SF 37/SH 37) squadrons: one each with F17 at Kallinge and F21 at Luleå. Production completed; details in 1980-81 and earlier editions of *Jane's*.

SK 37. Tandem two-seat dual-control training version; rear cockpit fitted with bulged hood and twin periscopes. Modified, taller tail-fin of increased area. Capable of secondary attack role, with full range of attack armament as in AJ 37. Prototype first flown on 2 July 1970. First production SK 37 delivered in June 1972. In service with conversion unit of F15 at Söderhamn. Production completed; details in 1980-81 and earlier editions of *Jane's*.

Saab 37X. Proposed export version, essentially similar to JA 37.

Production of the AJ 37, SF/SH 37 and SK 37 versions totalled 180. More than 200 of the 329 Viggens on order had been delivered by mid-1981, then equipping 10 of the planned total of 17 squadrons.

The following description applies to the JA 37:

TYPE: Single-seat all-weather multi-purpose combat aircraft.

WINGS: Tandem arrangement of delta foreplane, with trailing-edge flaps, and a rear-mounted delta main wing with two-section hydraulically-actuated powered elevons on each trailing-edge, which can be operated differentially or in unison. Main wing has compound sweep on leading-edge. Outer sections have extended leading-edge. Extensive use of metal-bonded honeycomb panels for wing control surfaces, foreplane flaps and main landing gear doors.

FUSELAGE: Conventional all-metal semi-monocoque structure, using light metal forgings and heat-resistant plastics bonding. Local use of titanium for engine firewall and other selected areas. Four plate-type airbrakes, one on each side and two below fuselage. Metal-bonded honeycomb construction is used to a large extent. Quick-release handle permits nosecone to be pulled forward on tracks to give access to radar compartment.

TAIL UNIT: Vertical surfaces only, comprising main fin and powered rudder, supplemented by a small ventral fin. Rudder of metal-bonded honeycomb construction. The main fin can be folded downward to port.

LANDING GEAR: Retractable tricycle type of Saab origin, built by Motala Verkstad and designed for a max rate of sink of 300 m (985 ft)/min. Power-steerable twin-wheel nose unit retracts forward. Each main unit has two main wheels in tandem and retracts inward into main wing and fuselage. Main oleos shorten during retraction. Nosewheel tyres size 18 × 5·5, pressure 10·7 bars (155 lb/sq in). Main-wheel tyres size 26 × 6·6, pressure 14·8 bars (215 lb/sq in). Goodyear wheels and brakes. Dunlop anti-skid system.

POWER PLANT: One Volvo Flygmotor RM8B (supersonic development of the Pratt & Whitney JT8D-22) turbofan engine, fitted with a Swedish-developed afterburner and thrust reverser. This engine is rated at 72·1 kN (16,203 lb st) dry and 125 kN (28,108 lb st) with afterburning. Thrust reverser doors are actuated automatically by the compression of the oleo as the nose gear strikes the runway, the thrust being deflected forward

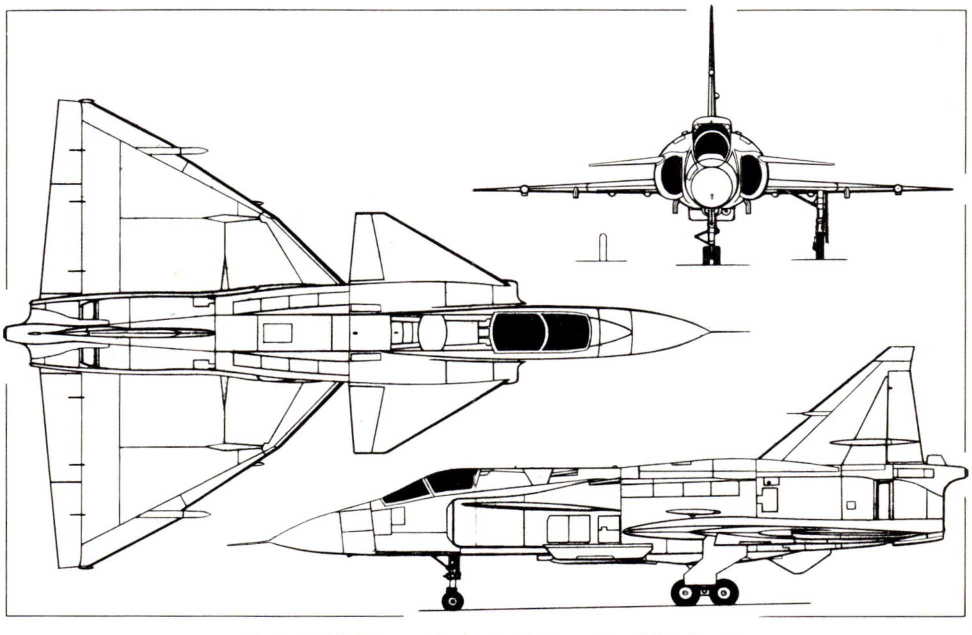

Saab JA 37 Viggen single-seat interceptor *(Pilot Press)*

via three annular slots in the ejector wall. The ejector is normally kept open at subsonic speeds to reduce fuselage base drag; at supersonic speeds, with the intake closed, the ejector serves as a supersonic nozzle. Fuel is contained in one tank in each wing, a saddle tank over the engine, one tank in each side of the fuselage, and one aft of the cockpit. Electrically-powered pumps deliver fuel to the engine from the central fuselage tank, which is kept filled continuously from the peripheral tanks. Pressure refuelling point beneath starboard wing. Provision for jettisonable external auxiliary tank on underfuselage centreline pylon.

Accommodation: Pilot only, on Saab-Scania fully-adjustable rocket-assisted ejection seat beneath rearward-hinged clamshell canopy. Cockpit pressurisation, heating and air-conditioning by engine bleed air, via Delaney Gallay heat exchangers, cooling turbines and water separator. Birdproof windscreen.

Systems: Two independent hydraulic systems, each of 207 bars (3,000 lb/sq in) pressure, each with engine-driven pump; auxiliary electrically-operated standby pump for emergency use. Three-phase AC electrical system supplies 210/115V 400Hz power via a Westinghouse 75kVA liquid-cooled brushless generator, which also provides 28V DC power via 24V nickel-cadmium batteries and rectifier. Emergency standby power from 6kVA turbogenerator, which is extended automatically into the airstream in the event of a power failure. External power receptacle on port side of fuselage. Graviner fire detection system.

Avionics and Flight Equipment: Altogether, about 50 avionics units, with a total weight of approx 600 kg (1,323 lb), are installed in the Viggen. Flight equipment includes an automatic speed control system, a Smiths electronic head-up display, AGA aircraft attitude instruments and radio, Garrett air data computer and instruments, L.M. Ericsson radar, Honeywell radar altimeter, Decca Doppler Type 72 navigation equipment, SATT radar warning system, Svenska Radio radar display system and electronic countermeasures, and AIL Tactical Instrument Landing System (TILS), a microwave scanning beam landing guidance system. Most avionics equipment is connected to the central digital computer, which is programmed to check out and monitor these systems both on the ground and during flight. Ram-air intake on underfuselage centreline, for cooling avionics compartment.

Armament and Operational Equipment: Permanent underbelly pack, offset to port side of centreline, containing one 30 mm Oerlikon KCA long-range cannon with a muzzle velocity of 1,050 m (3,445 ft)/s, a rate of fire of 1,350 rds/min, and a projectile weight of 0·36 kg (0·79 lb). Improved fire control equipment. This gun installation permits retention of the three underfuselage stores attachment points, in addition to the four underwing hardpoints. Advanced target search and acquisition system, based on a high-performance long-range L.M. Ericsson UAP-1023 X-band pulse-Doppler radar which is unaffected by variations of weather and altitude. This radar is not disturbed by ground clutter, and is highly resistant to ECM. Singer-Kearfott SKC-2037 central digital computer and Garrett LD-5 digital air data computer. Singer-Kearfott KT-70L inertial measuring equipment. Honeywell/Saab-Scania SA07 digital automatic flight control system. Armament can include a total of six BAe Dynamics Sky Flash (Swedish designation RB71) and Sidewinder (RB24) air-to-air missiles. For air-to-surface attack, a total of twenty-four 135 mm rockets can be carried in four pods.

Dimensions, external:	
Main wing span	10·60 m (34 ft 9¼ in)
Main wing aspect ratio	2·45
Foreplane span	5·45 m (17 ft 10½ in)
Length overall (incl probe)	16·40 m (53 ft 9¾ in)
Length of fuselage	15·58 m (51 ft 1½ in)
Height overall	5·90 m (19 ft 4¼ in)
Height overall, main fin folded	4·00 m (13 ft 1½ in)
Wheel track	4·76 m (15 ft 7½ in)
Wheelbase (c/l of shock-absorbers)	5·69 m (18 ft 8 in)
Areas:	
Main wings, gross	46·00 m² (495·1 sq ft)
Foreplanes, outside fuselage	6·20 m² (66·74 sq ft)
Weights (approx):	
T-O weight: 'clean'	15,000 kg (33,070 lb)
with normal armament	17,000 kg (37,478 lb)
Performance:	
Max level speed: at high altitude	above Mach 2
at 100 m (300 ft)	Mach 1·2
Approach speed:	approx 119 knots (220 km/h; 137 mph)
Time to 10,000 m (32,800 ft) from brakes off, with afterburning	less than 1 min 40 s
T-O run	approx 400 m (1,310 ft)
Landing run	approx 500 m (1,640 ft)
Required landing field length:	
conventional landing	1,000 m (3,280 ft)
no-flare landing	500 m (1,640 ft)
Tactical radius with external armament:	
hi-lo-hi	over 540 nm (1,000 km; 620 miles)
lo-lo-lo	over 270 nm (500 km; 310 miles)

SAAB SAFARI

The prototype Safari (formerly Saab-MFI 15) flew for the first time on 11 July 1969 with a 119 kW (160 hp) engine. After being re-engined with a 149 kW (200 hp) Avco Lycoming IO-360-A1B6, it resumed flying on 26 February 1971. The production version with this engine was described and illustrated in the 1978-79 *Jane's*.

Production of the Safari by Saab has ended. In February 1981 the final 16 production aircraft were ordered by the Royal Norwegian Air Force, for primary training duties.

One prototype/demonstrator (SE-FIT) was completed of a turbocharged version, the **Safari TS**, with a 157 kW (210 hp) Continental engine. This aircraft was described and illustrated in the 1980-81 *Jane's*.

Licence assembly continues of the non-turbocharged version by the Pakistan Aeronautical Complex (which see).

SAAB 114

The journal of the Royal Swedish Aero Club has reported that Saab-Scania is designing a two-seat turboprop-powered military training aircraft under this company designation. No details are available.

SWITZERLAND

ALR
ARBEITSGRUPPE FÜR LUFT- UND RAUMFAHRT (Aerospace Task Group)

Postfach 63, CH-8050 Zürich
Telephone: (01) 311 58 14
Telex: 56970 FEYER
Programme Director: Dr Georges Bridel

ALR PIRANHA

This group of Swiss scientists started work on the design of the Piranha in September 1977, in an attempt to develop a new-generation lightweight supersonic combat aircraft at a cost that could be afforded by the world's less affluent air forces, so enabling them to embody both quality and quantity of new aircraft in their re-equipment programmes.

Primary missions of the Piranha are seen as: clear-weather, low/medium-altitude air defence; low-level FEBA strike, mainly under VFR conditions; reconnaissance; electronic warfare; and operational training. A limited amount of active ECM and reconnaissance equipment is carried as standard, but all-weather interception and strike systems are not. This avoids the weight and performance penalties that such systems would inevitably attract, and keeps the pilot's workload to a minimum. STOL capability was considered essential, to permit operation from dispersed airfields with runways no more than 1,000 m (3,280 ft) long.

Principal design features include a short-coupled canard configuration, with all-moving foreplanes; shoulder-mounted wings, to provide an optimum stores attachment layout; fly-by-wire flight control; and a centrally-located high-performance gun. Other inherent advantages claimed for the Piranha are low procurement and operating costs, small radar and infra-red signatures, and transonic speed capability.

Details have been released of four basic proposed single-seat versions:

Piranha 2C. Transonic ground attack version, with single Rolls-Royce Turboméca Adour Mk 811 turbofan engine, rated at 24·6 kN (5,520 lb st) dry and 37·4 kN (8,400 lb st) with afterburning. No radar.

Piranha 2D. Combined ground attack/air superiority version of 2C, with more powerful RT.172-63 Adour turbofan rated at 29·2 kN (6,570 lb st) dry and 44·9 kN (10,100 lb st) with afterburning. General appearance shown in accompanying side-view drawing.

Piranha 4. Twin-engined ground attack/air superiority version, with two 15·5 kN (3,485 lb st) Turboméca-SNECMA Larzac 05 turbofan engines (each 25·0 kN; 5,620 lb st with afterburning) in shorter and wider fuselage. General appearance shown in accompanying three-view drawing.

Piranha 5. Similar to Piranha 4, but powered by two Garrett/Volvo Flygmotor TFE1042-7 turbofans (each 18·4 kN; 4,135 lb st dry, 30·2 kN; 6,800 lb st with afterburning).

Each of the above models could be produced in tandem two-seat form for ECM and training purposes.

Preliminary design, and selection of the major systems, has been completed for the Piranha 2D and Piranha 4, and negotiations with possible development partners were under way in the Spring of 1981. At that time, the Piranha 4 was considered to be the principal version.

The following description applies to all currently-projected versions, except where specifically indicated otherwise:

Type: Proposed multi-purpose lightweight combat aircraft.

Wings: Tandem arrangement, utilising shoulder-mounted, all-moving foreplanes, each with powered elevator on trailing-edge for pitch control. Immediately aft of foreplanes, and mounted slightly lower on fuselage, are low aspect ratio sweptback main wings, each with single flap/aileron ('flaperon') on trailing-edge. Main wings have compound sweep on leading-edges, which have extended chord outboard on outer panels. Thickness/chord ratio of main wings decreases from 5·5% at root to 4% at tip. Main wings have 4° of anhedral from roots, and are set at incidence of 1° 24′. Sweepback at quarter-chord 45° on outer panels, 31° on inner panels. Multi-spar main wings, which form torsion box passing through fuselage, are of light alloy construction with machined skins. Flaperons are of honeycomb sandwich construction. Fly-by-wire control of foreplanes, elevators and flaperons. No tabs or slats.

Fuselage: Conventional metal semi-monocoque structure. Door-type airbrake in upper surface of rear fuselage on each side of fin.

Tail Unit: Vertical surfaces only, comprising sweptback fin (with small dorsal fin) and fly-by-wire powered rudder.

Landing Gear: Retractable type, of Dowty (Piranha 2) or Messier-Hispano-Bugatti design (Piranha 4 and 5), with single wheel on each unit; all units retract forward into fuselage. Nose unit offset from centreline to provide clear field of fire for centrally-mounted gun. Oleo-pneumatic shock-absorbers. Dowty gear for Piranha 2

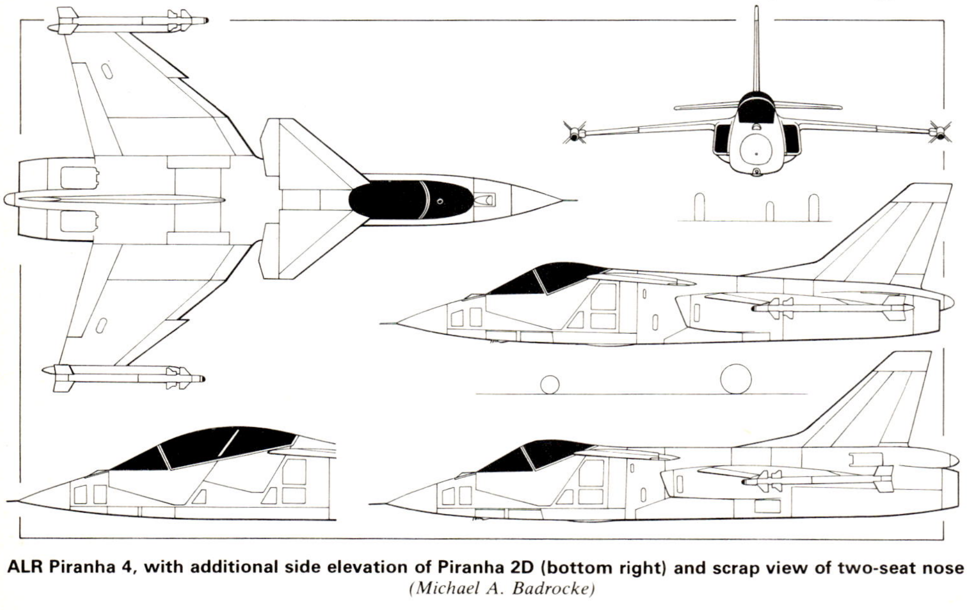

ALR Piranha 4, with additional side elevation of Piranha 2D (bottom right) and scrap view of two-seat nose
(Michael A. Badrocke)

has size 350 × 157 mm nosewheel tyre and 560 × 215 mm main-wheel tyres, all at nominal pressure of 8·6 bars (125 lb/sq in). Anti-skid units for all versions. Irvin braking parachute (9·5 m^2; 102·3 sq ft or 13·0 m^2; 140·0 sq ft) in bullet fairing on top of rear fuselage.

Power Plant: One or two turbofan engines (for details see under model listings), installed in rear fuselage. Lateral intake on each side of fuselage, with auxiliary inlet doors, spring-loaded for low-speed and high-angle-of-attack manoeuvres. Fuel in one main fuselage tank and two integral wing tanks. Provision for one external auxiliary tank on underfuselage station and one on each inboard underwing station. No in-flight refuelling capability.

Accommodation: Single Martin-Baker Mk 10L ejection seat under rearward-sliding canopy; or, on ECM and operational training versions, tandem seats under sideways-opening canopy.

Systems: BAe Dynamics cockpit pressurisation system, with pre-cooler and two-wheel bootstrap cold-air unit. Duplex hydraulic system (pressure 207 bars; 3,000 lb/sq in). Two generators, each of 12kVA minimum rating, for AC electrical power. Electrical triplexed fly-by-wire system, with four actuators powering all control surfaces. Oxygen system designed for pilot only. No pneumatic or de-icing systems.

Avionics and Equipment (Piranha 2D, 4 and 5): Ferranti LINAS nav/attack system with lightweight inertial platform; laser rangefinder; Smiths head-up display; Ferranti Comed multi-purpose head-down display; Thomson-CSF Agave or advanced (lookdown) lightweight search/track/designation/telemetry/navigation radar; internally-mounted passive ECM, chaff and flares.

Armament: One high-performance cannon (27 mm Mauser BK 27 or 30 mm Oerlikon KCA), mounted in underside of fuselage, on aircraft centreline, beneath cockpit. One underfuselage, four underwing and two wingtip attachment points for air-to-surface weapons such as Maverick, Beluga and Durandal; or air-to-air missiles such as AIM-9L Sidewinder and Matra Magic. Provision for active ECM and reconnaissance equipment to be carried in underfuselage pod.

Dimensions, external (Piranha 4):
Wing span (excl missile rails) 6·49 m (21 ft 3½ in)
Wing aspect ratio 2·63
Foreplane span 3·14 m (10 ft 3½ in)
Length overall 10·50 m (34 ft 5½ in)
Height overall 4·12 m (13 ft 6¼ in)

Areas:
Wings, gross (reference area) 16·00 m^2 (172·22 sq ft)
Flaperons (total) 1·28 m^2 (13·78 sq ft)
Foreplanes, gross (reference area) 4·57 m^2 (49·19 sq ft)
Elevators (total) 0·73 m^2 (7·86 sq ft)
Fin 3·33 m^2 (35·84 sq ft)
Rudder 0·56 m^2 (6·03 sq ft)

Weights and Loadings (Piranha 4*):
Max external weapons/fuel load approx 2,000 kg (4,410 lb)
Max T-O weight (air superiority configuration, with gun ammunition and two air-to-air missiles) 6,200 kg (13,668 lb)
Wing loading at T-O (air superiority configuration) 388 kg/m^2 (79·5 lb/sq ft)
Thrust/weight ratio (50% internal fuel) greater than 1·0

2D slightly lighter, 5 heavier

Performance (Piranha 4, estimated, at max T-O weight; other versions not significantly different):
Never-exceed speed Mach 1·9
Max level speed at 11,000 m (36,100 ft) Mach 1·8 (1,033 knots; 1,915 km/h; 1,190 mph)
Max rate of climb at S/L more than 13,800 m (45,275 ft)/min
Service ceiling 16,000 m (52,500 ft)
T-O run at S/L, ISA (air superiority configuration) less than 500 m (1,640 ft)
Landing run at S/L, ISA (minimum fuel, using brake 'chute) less than 500 m (1,640 ft)
Range with typical combat load, incl 5 min combat (lo-lo-lo) more than 189 nm (350 km; 217 miles)

FWA

FLUGZEUGWERKE ALTENRHEIN AG

Head Office and Works: CH-9423 Altenrhein
Telephone: (071) 43 01 01
Telex: 77 230 arag ch
President, Board of Management: Dr C. Caroni
Vice-President: Dr L. Caroni
Director: Dipl Ing H. Eisenring
Chief Engineer: Dipl Ing P. Spalinger
Marketing: R. Boehm

This company had its origin in AG für Dornier Flugzeuge, which was formed as the Swiss branch of the West German Dornier company. It became subsequently an entirely Swiss company named Flug- und Fahrzeugwerke AG Altenrhein. In 1980 the aircraft division was renamed Flugzeugwerke Altenrhein AG (FWA).

FWA's current activities include production of the AS 202 Bravo light aircraft, and the overhaul, modification and servicing of military and civil aircraft.

The company has about 900 employees, approximately one-quarter of whom are engaged in its aviation activities.

FWA AS 202 BRAVO

Following an agreement concluded with SIAI-Marchetti of Italy, FWA is engaged in production and development of the AS 202 Bravo light trainer and sporting aircraft.

The first Bravo to fly was a Swiss-assembled AS 202/15 prototype (HB-HEA), which flew for the first time on 7 March 1969. The Italian-built second prototype flew on 7 May 1969. The third aircraft (HB-HEC) made its first flight on 16 June 1969, and the first production aircraft on 22 December 1971.

Three versions are available, as follows:

AS 202/15. Two/three-seat initial production version, with 112 kW (150 hp) Avco Lycoming O-320-E2A engine. Optional third seat. Swiss certification granted on 15 August 1972; FAA certification awarded on 16 November 1973. Deliveries totalled 32 by January 1981.

AS 202/18A. Two/three-seat aerobatic version with a 134 kW (180 hp) Avco Lycoming engine, Hartzell constant-speed propeller and inverted oil system. First example (HB-HEY) flew for the first time on 22 August 1974. Swiss certification granted on 12 December 1975; FAA certification awarded on 17 December 1976. In production: 94 ordered by January 1981.

AS 202/26A. First flown in 1978. Two/three-seat training and aerobatic version with a 194 kW (260 hp) Avco Lycoming engine and systems for unlimited inverted flying. None ordered up to 1 March 1981.

Aircraft currently in production incorporate a number of modifications, including riveted wing skins, rubber fuel tanks, and a glassfibre engine cowling. About 150 Bravos had been sold by mid-1981, in Switzerland and to foreign customers including Royal Air Maroc, the Royal Flight of Oman, the Uganda Central Flying School, and the air forces of Indonesia (20 AS 202/18A), Iraq (48 AS 202/18A), Jordan, and Morocco (14).

The following description applies to all versions of the Bravo, except where a specific model is indicated:

Type: Two/three-seat light aircraft.

Wings: Cantilever low-wing monoplane. Wing section NACA $63_2$618 (modified) at centreline, $63_2$415 at tip. Thickness/chord ratio 17·63% at root, 15% at tip. Dihedral 5° 43′ from roots. Incidence 3°. Sweepback at quarter-chord 0° 40′. Conventional aluminium single-spar fail-safe structure, with honeycomb laminate skin. Aluminium single-slotted flaps and single-slotted ailerons. Ground-adjustable tab on each aileron.

Fuselage: Conventional aluminium semi-monocoque fail-safe structure, with several glassfibre fairings.

FWA AS 202/18A Bravo for the Indonesian Air Force

Tail Unit: Cantilever aluminium single-spar structure with sweptback vertical surfaces. Rudder mass-balanced, with provision for anti-collision beacon. Fixed-incidence tailplane. Two-piece elevator with full-span trim tab on starboard half. (AS 202/26A has electrically-operated tailplane trim; elevator tab is retained for anti-balance purposes only.) Ground-adjustable tab on rudder.

Landing Gear: Non-retractable tricycle type, with steerable nosewheel. Rubber-cushioned shock-absorber struts of SIAI-Marchetti design. Main-wheel tyres size 6·00-6; nosewheel tyre size 5·00-5. Tyre pressure (all units) 2·41 bars (35 lb/sq in). Independent hydraulically-operated disc brake on each main wheel.

Power Plant (AS 202/15): One 112 kW (150 hp) Avco Lycoming O-320-E2A flat-four engine, driving a McCauley 1C172 MGM two-blade fixed-pitch metal propeller with spinner. Two wing leading-edge fuel tanks with total capacity of 170 litres (37·4 Imp gallons). Refuelling point above each wing. Oil capacity 7·6 litres (1·6 Imp gallons). Additional exhaust muffler available optionally.

Power Plant (AS 202/18A): One 134 kW (180 hp) Avco Lycoming AEIO-360-B1F flat-four engine, driving a Hartzell HC-C2YK-1BF/F7666A-2 two-blade constant-speed propeller with spinner. Hoffmann three-blade propeller optional. Fuel capacity as for AS 202/15; starboard tank has additional flexible fuel intake for aerobatics. Christen 801 fully-aerobatic oil system, capacity 7·6 litres (1·6 Imp gallons).

Power Plant (AS 202/26A): One 194 kW (260 hp) Avco Lycoming fuel-injection engine, driving a Hartzell two-blade constant-speed propeller. Fuel capacity 174 litres (38·3 Imp gallons). Fuel and oil systems permit unrestricted inverted flight.

Accommodation: Seats for two persons side by side, in Aerobatic versions, under rearward-sliding jettisonable transparent canopy. Space at rear in Utility versions for a third seat or 100 kg (220 lb) of baggage. Dual controls, cabin ventilation and heating standard.

Systems: Hydraulic system for brake actuation. One 12V 60A engine-driven alternator and one 25Ah battery provide electrical power for engine starting, lighting, instruments, communications and navigation installations, and (AS 202/26A only) tailplane trim. 28V electrical system optional.

Avionics and Equipment: Provision for VHF radio, VOR, ADF, Nav-O-Matic 200A autopilot, blind-flying instrumentation or other special equipment at customer's option. Clutch-and-release mechanism for glider towing optional.

Dimensions, external:
Wing span 9·75 m (31 ft 11¾ in)
Wing chord at root 1·88 m (6 ft 2 in)
Wing chord at tip 1·16 m (3 ft 9½ in)
Wing aspect ratio 6·51
Length overall 7·50 m (24 ft 7¼ in)
Length of fuselage 7·15 m (23 ft 5½ in)

AS 202 Bravo, engine testbed for AS 32T with Allison 250-B17C turboprop power plant

Height overall 2·81 m (9 ft 2¾ in)
Tailplane span 3·67 m (12 ft 0½ in)
Wheel track 2·25 m (7 ft 4½ in)
Wheelbase 1·78 m (5 ft 10 in)
Propeller diameter 1·88 m (6 ft 2 in)
Propeller ground clearance 0·31 m (1 ft 0¼ in)

Dimensions, internal:
Cabin: Max length 2·15 m (7 ft 0½ in)
Max width 1·02 m (3 ft 4¼ in)
Max height 1·10 m (3 ft 7¼ in)
Floor area 2·15 m² (23·14 sq ft)

Areas:
Wings, gross 13·86 m² (149·2 sq ft)
Ailerons (total) 1·09 m² (11·7 sq ft)
Trailing-edge flaps (total) 1·49 m² (16·04 sq ft)
Fin 0·45 m² (4·84 sq ft)
Rudder, incl tab 0·94 m² (10·12 sq ft)
Tailplane 1 ` m² (20·24 sq ft)
Elevators, incl tab 0·76 m² (8·18 sq ft)

Weights and Loadings:
Weight empty, equipped: 15 630 kg (1,388 lb)
18A 700 kg (1,543 lb)
26A 793 kg (1,748 lb)
Max useful load (incl fuel):
15, Aerobatic 175 kg (386 lb)
15, Utility 270 kg (595 lb)
18A, Aerobatic 172 kg (379 lb)
18A, Utility 258 kg (568 lb)
26A, Aerobatic 188 kg (414 lb)
26A, Utility 292 kg (643 lb)
Max T-O and landing weight:
15, Aerobatic 885 kg (1,951 lb)
15, Utility 999 kg (2,202 lb)
18A, Aerobatic 950 kg (2,094 lb)
18A, Utility 1,050 kg (2,315 lb)
26A, Aerobatic 1,075 kg (2,370 lb)
26A, Utility 1,200 kg (2,645 lb)
Max wing loading:
15, Utility 72·1 kg/m² (14·8 lb/sq ft)
18A, Utility 75·8 kg/m² (15·52 lb/sq ft)
26A, Utility 86·6 kg/m² (17·75 lb/sq ft)
Max power loading:
15, Utility 8·92 kg/kW (14·68 lb/hp)
18A, Utility 7·84 kg/kW (12·86 lb/hp)
26A, Utility 6·18 kg/kW (10·17 lb/hp)

Performance (Utility category, at max T-O weight):
Never-exceed speed:
15, 18A 173 knots (320 km/h; 199 mph)
26A 208 knots (385 km/h; 240 mph)
Max level speed at S/L:
15 114 knots (211 km/h; 131 mph)
18A 130 knots (241 km/h; 150 mph)
Max cruising speed (75% power) at 2,440 m (8,000 ft):
15 114 knots (211 km/h; 131 mph)
18A 122 knots (226 km/h; 141 mph)
26A 138 knots (256 km/h; 159 mph)
Econ cruising speed at 3,050 m (10,000 ft):
15 (66% power) 109·5 knots (203 km/h; 126 mph)
18A (55% power) 109·5 knots (203 km/h; 126 mph)
Stalling speed, flaps up, engine idling:
15 59·5 knots (110 km/h; 68·5 mph)
18A 62 knots (114 km/h; 71 mph)
26A 63·5 knots (117 km/h; 73 mph)
Stalling speed, flaps down, engine idling:
15 48·5 knots (89 km/h; 55·5 mph)
18A 49 knots (90 km/h; 56 mph)
26A 53 knots (98 km/h; 61 mph)
Max rate of climb at S/L: 15 193 m (633 ft)/min
18A 276 m (905 ft)/min
26A 360 m (1,181 ft)/min
Service ceiling: 15 4,265 m (14,000 ft)
18A 5,490 m (18,000 ft)
26A 5,670 m (18,600 ft)
T-O run at S/L: 15 235 m (771 ft)
18A 210 m (689 ft)
26A 185 m (607 ft)
T-O to 15 m (50 ft) at S/L: 15 475 m (1,558 ft)
18A 400 m (1,312 ft)
26A 340 m (1,115 ft)
Landing from 15 m (50 ft): 15 415 m (1,362 ft)
18A, 26A 465 m (1,525 ft)
Landing run: 15 130 m (427 ft)
18A, 26A 210 m (690 ft)
Range with max fuel, no reserves:
15 480 nm (890 km; 553 miles)
18A 521 nm (965 km; 600 miles)
26A 459 nm (850 km; 528 miles)
Max endurance: 18A 5 h 30 min
26A 4 h 54 min

FWA AS 32T TURBO TRAINER

Preliminary data for the proposed AS 32T were released during the Paris Air Show in June 1979; detail design began in October of that year. The accompanying photograph illustrates the general configuration. The wings and tailplane are almost identical with those of the AS 202 Bravo. Under this programme, an Allison 250-B17C engine, as specified for the AS 32T, has been installed experimentally in a standard Bravo (HB-HEC), which began flight testing in August 1980.

Model of the FWA AS 32T Turbo Trainer

Type: Tandem two-seat basic and advanced trainer.

Wings: Identical geometrically to those of Bravo, except where it is necessary to modify them because of the narrower fuselage and retractable landing gear. Electrically-operated trim tab in port aileron.

Fuselage: Uses similar constructional techniques and materials to Bravo, but crew are seated in tandem.

Tail Unit: As AS 202/26A version of Bravo, except for electrically-operated trim tab in rudder.

Landing Gear: Retractable tricycle type, with single wheel and oleo-pneumatic shock-absorber on each unit. Nose unit retracts rearward, main units inward into wings. Tyre sizes 6·00-6 (main) and 5·00-5 (nose).

Power Plant: One 313 kW (420 shp) Allison 250-B17C turboprop engine, flat rated to 239 kW (320 shp) or 268 kW (360 shp) for military training, and driving a Hartzell HC-B3TF-2 three-blade constant-speed propeller. Fuel in one 130 litre (29 Imp gallon) centre-section tank and an 85 litre (18·5 Imp gallon) tank in each outer wing; total capacity 300 litres (66 Imp gallons). Overwing refuelling points. Oil capacity 5·7 litres (1·25 Imp gallons; 1·5 US gallons).

Accommodation: Instructor and pupil in tandem under fully-transparent rearward-sliding canopy. Dual controls standard.

Systems: No hydraulic or pneumatic systems; 24V electrical system; oxygen system.

Avionics and Equipment: Instrumentation and avionics in production version to customer's requirements.

Dimensions, external:
Wing span 9·68 m (31 ft 9 in)
Wing aspect ratio 6·67
Length overall 9·31 m (30 ft 6½ in)
Height overall 3·39 m (11 ft 1½ in)
Tailplane span 3·67 m (12 ft 0½ in)
Wheel track 2·60 m (8 ft 6½ in)
Propeller diameter 2·18 m (7 ft 2 in)

Areas:
Wings, gross 14·056 m² (151·3 sq ft)
Ailerons (total) 1·203 m² (12·95 sq ft)
Trailing-edge flaps (total) 1·472 m² (15·84 sq ft)
Fin 0·45 m² (4·84 sq ft)
Rudder, incl tab 1·027 m² (11·05 sq ft)
Tailplane 1·88 m² (20·24 sq ft)
Elevators, incl tab 0·76 m² (8·18 sq ft)

Weights and Loadings:
Weight empty, equipped 838 kg (1,847 lb)
Max internal fuel 240 kg (529 lb)
Max T-O weight:
trainer, 'clean' 1,300 kg (2,866 lb)
with external stores 1,900 kg (4,189 lb)
Max landing weight 1,805 kg (3,979 lb)
Wing loading at AUW of 1,900 kg (4,189 lb)
139 kg/m² (28·47 lb/sq ft)
Power loading at AUW of 1,900 kg (4,189 lb)
6·07 kg/kW (9·97 lb/shp)

Performance (estimated, at AUW of 1,200 kg; 2,645 lb, ISA, zero wind):
Never-exceed speed
270 knots (500 km/h; 311 mph) EAS
Cruising speed at 3,000 m (9,845 ft):
80% power 190 knots (352 km/h; 219 mph)
55% power 170 knots (315 km/h; 196 mph)
Rate of climb at 500 m (1,640 ft)
534 m (1,750 ft)/min
Service ceiling 10,050 m (33,000 ft)
T-O run at 500 m (1,640 ft), zero wind
165 m (540 ft)
T-O to 15 m (50 ft) 330 m (1,080 ft)
Landing from 15 m (50 ft) 460 m (1,510 ft)
Landing run 210 m (690 ft)
Range with max internal fuel, 55% power, no reserves 777 nm (1,440 km; 895 miles)
Max endurance at 3,000 m (9,845 ft), 55% power, no reserves 4 h 30 min

PILATUS

PILATUS FLUGZEUGWERKE AG

Head Office and Works: CH-6370 Stans, near Lucerne
Telephone: (041) 63 11 33
Telex: 78 329
General Manager: H. Uehlinger
Managers:
Dr A. Canal (Vice-President, Sales)
P. Ebner (Vice-President, Administration)
W. Gubler (Vice-President, Production)
K. Bergius (Research and Development)
W. Volkart (Product Support)
Public Relations: Ulrich Wenger

Pilatus Flugzeugwerke AG was formed in December 1939; details of its early history can be found in previous editions of *Jane's*. It is part of the Oerlikon-Bührle Group.

Current Pilatus products are the Turbo-Porter single-engined utility transport and the PC-7 Turbo-Trainer.

On 24 January 1979 Pilatus purchased the assets of Britten-Norman (Bembridge) Ltd of the UK, and the latter company is operated under the name Pilatus Britten-Norman Ltd (which see) as a subsidiary of Pilatus Aircraft Ltd.

PILATUS PC-6 TURBO-PORTER

US Army designation: UV-20A Chiricahua

The Pilatus PC-6 is a single-engined multi-purpose utility aircraft, with STOL characteristics permitting operation from unprepared strips under harsh environmental and terrain conditions. The aircraft can be converted rapidly from a pure freighter to a passenger transport, and can be adapted for a great number of different missions, including supply dropping, search and rescue, ambulance, aerial survey and photography, parachuting, crop spraying, water bombing, rainmaking and glider or target towing as well as operation from soft ground, snow, glacier or water, and long-range operations.

The first of five PC-6 piston-engined prototypes made its first flight on 4 May 1959, and 20 pre-series PC-6s, with 253·5 kW (340 hp) Avco Lycoming engines, had been delivered by the Summer of 1961. Subsequent versions have included the piston-engined PC-6 and PC-6/350 Porters; the PC-6/A, A1, A2, B and C2-H2 Turbo-Porters, with various turboprop power plants. Descriptions of all these can be found in the 1974-75 and earlier editions of *Jane's*.

Swiss-built piston-engined variants have the name Porter, and turboprop-powered variants are known as Turbo-Porters. In the USA, where the PC-6 was manufactured by Fairchild, it is known simply as the Porter, irrespective of the type of power plant fitted.

The current production version is the **PC-6/B2-H2 Turbo-Porter,** certificated on 30 June 1970 and powered by a PT6A-27 turboprop engine. Other versions can be made available on request.

By mid-1981, more than 430 PC-6 aircraft, of all models, had been delivered (including US licence manufacture), and were operating in more than 50 countries. Production was then continuing at a rate of one to two per month. Military operators include the air forces of Angola, Argentina, Australia, Austria, Bolivia, Burma, Chad, Ecuador, Oman, Peru, Sudan, Switzerland, Thailand and the US Army. In 1980-81 Pilatus converted the Swiss Air Force's 12 piston-engined Porters to Turbo-Porter standard.

Pilatus markets a Q-STOL (Quiet STOL) conversion kit for the B1 and B2 Turbo-Porters fitted with PT6A-20 or -27 turbine engines. This includes a system whereby propeller speed can be altered independently of the engine power setting, and is claimed to reduce the noise level by more than 10 dB for T-O and 20 dB for landing.

The structural description which follows is applicable to the current B2-H2 version. Details of the agricultural Turbo-Porter are given separately.

Type: Single-engined STOL utility transport.

Wings: Braced high-wing monoplane, with single streamline-section bracing strut each side. Wing section NACA 64-514 (constant). Dihedral 1°. Incidence 2°. Single-spar all-metal structure. Entire trailing-edge hinged, inner sections consisting of electrically-operated all-metal double-slotted flaps and outer sections of all-metal single-slotted ailerons. No airbrakes or de-icing equipment. Trim tabs and/or Flettner tabs on ailerons optional; ground-adjustable tabs are mandatory if these are not fitted.

Fuselage: All-metal semi-monocoque structure.

Tail Unit: Cantilever all-metal structure. Variable-incidence tailplane. Flettner tabs on elevator.

Landing Gear: Non-retractable tailwheel type. Oleo shock-absorbers of Pilatus design in all units. Steerable/lockable tailwheel. Goodyear Type II main wheels and GA 284 tyres size 24 × 7 or 7·50 × 10 (pressure 2·21 bars; 32 lb/sq in); oversize Goodyear Type III wheels and tyres optional, size 11·0 × 12, pressure 0·88 bars (12·8 lb/sq in). Goodyear tailwheel with size 5·00-4 tyre. Goodyear disc brakes. Pilatus wheel/ski gear or Edo 58-4580 or 679-4930 floats optional.

Power Plant (PC-6/B2-H2): One 507 kW (680 shp) Pratt & Whitney Aircraft of Canada PT6A-27 turboprop engine (flat rated to 410 kW; 550 shp at S/L), driving a Hartzell HC-B3TN-3D/T-10178 fully-feathering reversible-pitch propeller with Beta mode control. Standard fuel in integral wing tanks, capacity 480 litres (127 US gallons; 105·5 Imp gallons) normal, 644 litres (170 US gallons; 142 Imp gallons) maximum. Two underwing auxiliary tanks, each of 190 litres (50 US gallons; 42 Imp gallons), available optionally.

Accommodation: Cabin has pilot's seat forward on port side, with one passenger seat alongside, and is normally fitted with six quickly-removable seats, in pairs, to the rear of these for additional passengers. Up to 11 persons, including the pilot, can be carried in 2-3-3-3 high-density layout; or eight parachutists, who can be dropped from heights up to 5,000 m (16,400 ft). Floor is level, flush with door sill, and is provided with seat rails. Forward-opening door beside each front seat. Large rearward-sliding door on starboard side of main cabin. Double doors, without central pillar, on port side. Hatch in floor 0·58 × 0·90 m (1 ft 10¾ in × 2 ft 11½ in), openable from inside cabin, for aerial camera or for supply dropping. Hatch in cabin rear wall 0·50 × 0·80 m (1 ft 7 in × 2 ft 7 in) permits stowage of six passenger seats or accommodation of freight items up to 5·0 m (16 ft 5 in) in length. Walls lined with lightweight sound-proofing and heat-insulation material. Adjustable heating and ventilation systems provided. Dual controls optional.

Systems: Cabin heated by engine bleed air. Scott 8500 oxygen system optional. 200A 30V starter/generator and 24V 34Ah nickel-cadmium battery.

Equipment: Generally to customer's requirements, but can include stretchers for ambulance role, agricultural equipment (see separate description) or a 1,330 litre (292·5 Imp gallon) water tank in cabin, with quick-release system, for firefighting role.

PC-6/B1 Turbo-Porter agricultural aircraft fitted with 46-nozzle sprayboom

Dimensions, external:
- Wing span: 15·13 m (49 ft 8 in)
- Wing span over navigation lights: 15·20 m (49 ft 10½ in)
- Wing chord (constant): 1·90 m (6 ft 3 in)
- Wing aspect ratio: 7·95
- Length overall: 10·90 m (35 ft 9 in)
- Height overall (tail down): 3·20 m (10 ft 6 in)
- Elevator span: 5·12 m (16 ft 9½ in)
- Wheel track: 3·00 m (9 ft 10 in)
- Wheelbase: 7·87 m (25 ft 10 in)
- Propeller diameter: 2·56 m (8 ft 5 in)
- Cabin double door (port) and sliding door (starboard):
  - Height: 1·04 m (3 ft 5 in)
  - Width: 1·58 m (5 ft 2¼ in)

Dimensions, internal:
- Cabin, from back of pilot's seat to rear wall:
  - Length: 2·30 m (7 ft 6½ in)
  - Max width: 1·16 m (3 ft 9½ in)
  - Max height (at front): 1·28 m (4 ft 2½ in)
  - Height at rear wall: 1·18 m (3 ft 10½ in)
  - Floor area: 2·67 m² (28·6 sq ft)
  - Volume: 3·28 m³ (107 cu ft)

Areas:
- Wings, gross: 28·80 m² (310 sq ft)
- Ailerons (total): 3·83 m² (41·2 sq ft)
- Flaps (total): 3·76 m² (40·5 sq ft)
- Fin: 1·70 m² (18·3 sq ft)
- Rudder, incl tab: 0·96 m² (10·3 sq ft)
- Tailplane: 3·78 m² (40·7 sq ft)
- Elevator, incl tab: 2·24 m² (24·1 sq ft)

Weights and Loadings:
- Weight empty, equipped: 1,218 kg (2,685 lb)
- Max T-O and landing weight:
  - Normal (CAR 3): 2,200 kg (4,850 lb)
  - Restricted (CAR 8): 2,770 kg (6,100 lb)
- Max cabin floor loading: 488 kg/m² (100 lb/sq ft)
- Max wing loading (Normal): 76·4 kg/m² (15·65 lb/sq ft)
- Max power loading (Normal): 5·37 kg/kW (8·82 lb/shp)

Performance (at max T-O weight, Normal category):
- Never-exceed speed: 151 knots (280 km/h; 174 mph) IAS
- Max cruising speed at 3,050 m (10,000 ft): 140 knots (259 km/h; 161 mph)
- Econ cruising speed at 3,050 m (10,000 ft): 129 knots (240 km/h; 150 mph)
- Max manoeuvring speed: 106 knots (196 km/h; 122 mph)
- Max speed with flaps extended: 82 knots (152 km/h; 94·5 mph)
- Stalling speed, power off, flaps up: 50 knots (93·5 km/h; 58 mph)
- Stalling speed, power off, flaps down: 44 knots (82 km/h; 51 mph)
- Normal rate of climb at S/L: 387 m (1,270 ft)/min
- Service ceiling: 8,535 m (28,000 ft)
- T-O run at S/L: 110 m (360 ft)
- T-O to 15 m (50 ft) at S/L: 235 m (771 ft)
- Landing from 15 m (50 ft) at S/L: 218 m (715 ft)
- Landing run at S/L: 80 m (262 ft)
- Max range, no reserves:
  - internal fuel only: 560 nm (1,036 km; 644 miles)
  - with external fuel: 875 nm (1,620 km; 1,007 miles)
- Endurance, no reserves: internal fuel only: 4 h 20 min
  - with external fuel: 6 h 45 min
- *g* limits: +3·72; −1·50

Pilatus PC-6/B2-H2 Turbo-Porter (P&WC PT6A-27 turboprop engine)

PILATUS PC-6 TURBO-PORTER (AGRICULTURAL VERSIONS)

The Turbo-Porter can, if required, be equipped for agricultural duties, the necessary equipment being easily removable when not required, to permit the use of the aircraft for other work. Approx 40 Turbo-Porters have been completed in agricultural configuration: these are in service in Indonesia, Sudan, Switzerland, Thailand and Venezuela.

For liquid spraying, a stainless steel tank (capacity 1,330 litres; 292·5 Imp gallons; 351·4 US gallons) is installed behind the two front seats, and 46- or 62-nozzle sprayborns are fitted beneath the wings. In this configuration the aircraft can cover a swath width of 45 m (148 ft). An ultra-low-volume system, using four to six atomisers or two to four Micronairs, is also available, permitting increase in swath width up to 400 m (1,310 ft).

For dusting with granulated materials, the lower part of the standard tank can be replaced by a discharge and dispersal door permitting coverage of a swath width of up to 20 m (66 ft). A Transland spreader can be fitted for dust application (swath up to 30 m; 100 ft). Effective swath width of these versions is 13-40 m (43-131 ft), the optimum being approx 20 m (66 ft).

Both versions are fitted with small doors in the fuselage sides, giving access to the tank/hopper for servicing, removal or replenishment, and two single seats or a bench seat for three persons can be installed aft of the tank. Optional items include an engine air intake screen and a loading door for chemical in the top of the fuselage.

Avionics and Equipment: Optional avionics include Decca Mk 8A navigator, Decca Hi-Fix radio, Decca Doppler 72 radar, gyrosyn CL-11 compass and SR 54A radio altimeter.

Weights (L: liquid spray system; D: dry chemicals system):
- Weight empty: L, D: 1,215 kg (2,678 lb)
- Agricultural installation: L: 133 kg (293 lb)
  - D: 105 kg (231 lb)
- Chemical: L: 1,132 kg (2,497 lb)
  - D: 1,160 kg (2,559 lb)
- Fuel, oil and pilot: L, D: 286 kg (630 lb)
- Max T-O and landing weight:
  - L, D: 2,770 kg (6,100 lb)

Performance (liquid spray version, PT6A-27 engine, at max T-O weight):
- Never-exceed speed: 120 knots (222 km/h; 138 mph)
- Operating speed: approx 90 knots (167 km/h; 104 mph)
- Operating height: 6-8 m (20-26 ft)
- Stalling speed, power off, flaps down: 49 knots (91 km/h; 57 mph)
- T-O run: 180 m (590 ft)
- T-O to 15 m (50 ft): 390 m (1,280 ft)
- Landing from 15 m (50 ft): 345 m (1,132 ft)
- Landing run: 130 m (427 ft)
- Spraying duration with full spray tank: 6 min

PILATUS PC-7 TURBO-TRAINER

Swiss Air Force designation: PC-7/CH

Pilatus announced in the Spring of 1975 details of the PC-7 Turbo-Trainer fully aerobatic two-seat training aircraft, fitted with a 410 kW (550 shp) Pratt & Whitney Aircraft of Canada PT6A-25A turboprop engine.

The PC-7 can be used for basic, transition and aerobatic training, and, with suitable equipment installed, for IFR and tactical training. It meets the requirements of FAR Pt 23 (Aerobatic and Utility categories) and is also designed to comply with a selected group of US Air Force military specifications (Trainer category). As a single-seater, it is flown from the front seat. Six underwing hardpoints are standard, the inner pair stressed for 250 kg (551 lb) loads, the centre pair for 160 kg (353 lb) each and the outer pair for 110 kg (242·5 lb) each. Max external load comprises 1,040 kg (2,293 lb) of underwing stores.

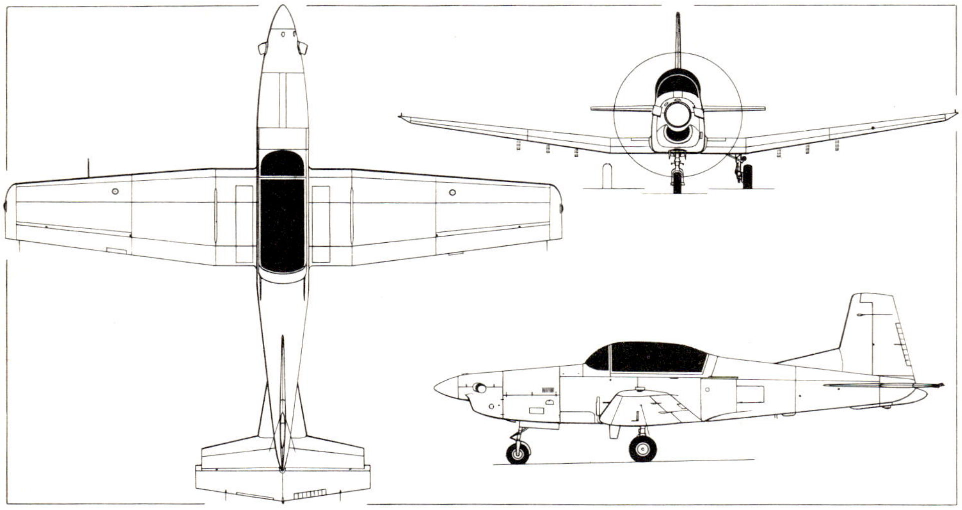

Pilatus PC-7 Turbo-Trainer (Pratt & Whitney Aircraft of Canada PT6A-25A turboprop engine) *(Pilot Press)*

Pilatus PC-7/CH Turbo-Trainer in Swiss Air Force insignia

The first production PC-7 was flown on 18 August 1978. Certification under FAR Pt 23 was awarded on 5 December that year, and deliveries began immediately afterwards. By 4 June 1981 firm orders had been received for 248 aircraft, of which 112 had been delivered. Customers include the air forces of Abu Dhabi (14), Bolivia (12), Burma (16), Chile (10 for Navy), Guatemala (12), Iraq (52), Mexico (55), Switzerland (40), and one undisclosed African country (12). Five of the customers have placed repeat orders. Production capacity is absorbed until 1983, and output was being increased from slightly more than five per month in mid-1981 to six per month in 1982.

Type: Single-engined single/two-seat training aircraft.

Wings: Cantilever low-wing monoplane. Wing section NACA 64_2A series at root, NACA 64_1A series at tip. Thickness/chord ratio 15% at root, 12% at tip. Dihedral 7° on outer panels. Sweepback 1° at quarter-chord. One-piece all-metal single-spar structure, with auxiliary spar, ribs and stringer-reinforced skin. Constant-chord centre-section and tapered outer panels. Aluminium alloy (2022 or 2024) skin, reinforced by stringers. Some fairings of GRP. Plain mass-balanced ailerons; split trailing-edge flaps, extending under fuselage. Flaps actuated electrically, ailerons mechanically by push-rods. Trim tab in port aileron; ground-adjustable tab on starboard aileron.

Fuselage: All-metal semi-monocoque structure, with stringers, bulkheads and aluminium alloy skin. Some fairings of GRP.

Tail Unit: Cantilever all-metal structure, of similar construction to wings. Trim tabs in rudder and starboard elevator. All control surfaces mass-balanced and cable-operated.

Landing Gear: Retractable tricycle type, of Dornier design and manufacture. Electrical actuation, with emergency manual extension. Main wheels retract inward, nosewheel rearward. Oleo-pneumatic shock-absorbers in all units. Castoring nosewheel, with shimmy dampers. Goodrich main wheels and tyres, size 6·50-8, pressure 4·83 bars (70 lb/sq in). Goodrich hydraulic disc brakes on main wheels. Parking brake.

Power Plant: One 485 kW (650 shp) Pratt & Whitney Aircraft of Canada PT6A-25A turboprop engine, flat rated to 410 kW (550 shp at S/L), driving a Hartzell HC-B3TN-2 three-blade constant-speed fully-feathering propeller with spinner. Fuel in integral tanks in outer wings, total usable capacity 474 litres (104 Imp gallons). Fuel system permits up to 30 s of inverted flight. Provision for underwing drop-tanks.

Accommodation: Adjustable seats for two persons in tandem, beneath rearward-sliding jettisonable Plexiglas canopy. Dual controls standard. Cockpits ventilated and heated by engine bleed air, which can also be used for windscreen de-icing. Space for 25 kg (55 lb) of baggage aft of seats, with external access.

Systems: Hydraulic system for main-wheel brakes only. No pneumatic system. 28V DC operational (24V nominal) electrical system, incorporating 30V 200A Lear Siegler starter/generator and 24V 34Ah or 40Ah nickel-cadmium battery. Ground power receptacle fitted.

Avionics and Equipment: Rosemount Srs 92 angle of attack indicator standard. Optional equipment includes IFR training shield, to screen rear cockpit, radio, oxygen and air-conditioning. Six underwing hardpoints, stressed for loads of 250 kg (551 lb) (inner), 160 kg (352 lb) (centre) and 110 kg (242 lb) (outer). Centre hardpoints are 'wet' for carriage of auxiliary fuel tanks. Max external stores load 1,040 kg (2,290 lb).

Dimensions, external:

Wing span	10·40 m (34 ft 1½ in)
Wing mean aerodynamic chord	1·64 m (5 ft 4½ in)
Wing mean geometric chord	1·59 m (5 ft 2·6 in)
Wing aspect ratio	6·52
Length overall	9·775 m (32 ft 0¾ in)
Height overall	3·21 m (10 ft 6½ in)
Propeller diameter	2·36 m (7 ft 9 in)

Area:

Wings, gross	16·60 m² (178·7 sq ft)

Weights and Loadings:

Weight empty, equipped	1,330 kg (2,932 lb)
Max T-O weight:	
Aerobatic	1,900 kg (4,188 lb)
Utility	2,700 kg (5,952 lb)
Max landing weight:	
Aerobatic	1,900 kg (4,188 lb)
Utility (with underwing loads)	2,565 kg (5,655 lb)
Max wing loading:	
Aerobatic	114·5 kg/m² (23·44 lb/sq ft)
Utility	162·7 kg/m² (33·31 lb/sq ft)
Max power loading:	
Aerobatic	4·63 kg/kW (7·61 lb/shp)
Utility	6·59 kg/kW (10·82 lb/shp)

Performance (at max T-O weight, ISA. A: Aerobatic category, B: Utility category with underwing loads):

Never-exceed speed:	
A, B	270 knots (500 km/h; 310 mph) EAS
Max operating speed:	
A, B	270 knots (500 km/h; 310 mph) EAS
Max cruising speed at 6,100 m (20,000 ft):	
A	222 knots (412 km/h; 256 mph)
B	196 knots (364 km/h; 226 mph)
Econ cruising speed at 6,100 m (20,000 ft):	
A	171 knots (317 km/h; 197 mph)
B	165 knots (305 km/h; 190 mph)
Manoeuvring speed:	
A	175 knots (325 km/h; 202 mph) EAS
B	189 knots (350 km/h; 218 mph) EAS
Max speed with flaps and landing gear down:	
A, B	135 knots (250 km/h; 155 mph) EAS
Stalling speed, flaps up, power off:	
A	71 knots (131 km/h; 81·5 mph) EAS
B	85 knots (158 km/h; 98 mph) EAS
Stalling speed, flaps down, power off:	
A	64 knots (119 km/h; 74 mph) EAS
B	74 knots (137·5 km/h; 85·5 mph) EAS
Normal rate of climb at S/L:	
A	610 m (2,000 ft)/min
B	364 m (1,195 ft)/min
Time to 5,000 m (16,400 ft): A	10 min
B	27 min
Service ceiling: A	9,755 m (32,000 ft)
B	6,705 m (22,000 ft)
T-O run at S/L: A	248 m (814 ft)
B	780 m (2,560 ft)
T-O to 15 m (50 ft) at S/L: A	365 m (1,198 ft)
B	1,080 m (3,543 ft)
Landing from 15 m (50 ft) at S/L:	
A	494 m (1,620 ft)
B	800 m (2,625 ft)
Landing run at S/L: A	336 m (1,100 ft)
B	505 m (1,655 ft)
Max range at cruise power at 5,000 m (16,400 ft), 5% plus 20 min reserves:	
A, B	810 nm (1,500 km; 932 miles)
Endurance, with reserves:	
A, at max speed	3 h 0 min
A, for max range	4 h 22 min
B, at max speed	2 h 36 min
B, for max range	3 h 54 min
g limits: A	+ 6·0; − 3·0
B	+ 4·5; − 2·25

SWISS FEDERAL AIRCRAFT FACTORY (F+W)

EIDGENÖSSISCHES FLUGZEUGWERK—FABRIQUE FÉDÉRALE D'AVIONS—FABBRICA FEDERALE D'AEROPLANI

Head Office and Works: CH-6032 Emmen
Telephone: (041) 59 41 11
Telex: 7 84 80 fwe ch
Director: Lucien Othenin-Girard
Deputy Director: Dr Peter Burkhardt
Chief Designers:
Hansjoerg Kobelt (Electronics)
André Jordi (Structures)
Head of Research Department: Heini Kamber
Head of Production: Hanspeter Arnold

The F+W is the Swiss government's official aircraft establishment for research, development, production, maintenance and modification of military aircraft. It employs about 650 people in its works at Emmen, near Lucerne, which cover 35,300 m² (380,000 sq ft). Included are four wind tunnels for speeds of up to Mach 4·5, test cells for piston and turbojet engines with or without afterburners, and a modern data acquisition and processing system.

The F+W is made up of six technical departments. Research Department operates the wind tunnel facilities and performs scientific research and development in the fields of aerodynamics, thermodynamics, systems analysis and flight mechanics. Engineering Department is responsible for the design and development of aircraft, aircraft subsystems and components, and the development of space hardware. Electronics Department undertakes maintenance and modification of the avionics of the Swiss Air Force's aircraft, and provides support for flight evaluations. Quality Assurance Department assures maintenance of high quality and adherence to standards in the production and maintenance of the aircraft and space hardware. The M & P Department maintains up-to-date awareness of new materials and processes, and develops its own process techniques to meet particular requirements. Thus, for example, fabrication of titanium parts and Inconel pressure vessels is undertaken at the F+W. Production Department is competent to undertake the fabrication of aircraft subassemblies and components, as well as final assembly and checkout of complete aircraft.

The Factory is engaged in the assembly programme of Northrop F-5E/F Tiger IIs for the Swiss Air Force, which began in 1976 and involves final assembly of an initial total of 72 aircraft, of which 66 had been completed by 1 January 1981. A second procurement order, for 32 F-5Es and six F-5Fs, received Swiss government approval in mid-1981. In addition to final assembly, work under this second contract will include the manufacture and subassembly of some components.

Beginning in 1980, the F+W is general manager of a Swiss industry programme for the production of Rapier surface-to-air missiles under licence from the Dynamics Group of British Aerospace.

Since 1977, when purchase of the Dragon anti-tank weapon system was authorised by the Swiss government, the F+W has been responsible for licence manufacture of the tracker, its field test equipment, and the appropriate training equipment. The pilot quantities of these items had been completed by the beginning of 1980.

The F+W conducts wind tunnel programmes for foreign aircraft manufacturers, ground transportation developers and users, and for the building industry. It also offers proprietary products to potential customers, including an electronic acoustic warntone generator, water separators for aircraft conditioning systems, strain-gauge force measuring scales, and POHWARO hot water rockets. Details of these rockets can be found in the 1977-78 *Jane's*.

Northrop F-5E Tiger II for the Swiss Air Force, assembled by F+W

TAIWAN

AIDC/CAF
AERO INDUSTRY DEVELOPMENT CENTER—CHINESE AIR FORCE

PO Box 8676-1, Taichung, Taiwan 400
Telephone: Taichung (042) 523051 and 523052
Telex: 51140
Other Works: Kang Shan
Director: Lieutenant General Y. C. Lee
Deputy Directors:
Dr Hsichun M. Hua (Engineering and Research)
Major General E. Y. Chu (Manufacturing)

The Aero Industry Development Center was established on 1 March 1969 as a successor to the Bureau of Aircraft Industry (BAI), which was established in 1946 in Nanking and moved to Taiwan in 1948. It employs approx 2,000 people.

In October 1968 the Aeronautical Research Laboratory, then a branch of BAI, constructed the first Chinese-built PL-1A (see 1970-71 *Jane's*), a slightly modified version of the US Pazmany PL-1 which flew for the first time on 26 October 1968. After further modifications, 55 PL-1B Chienshou production models were built by AIDC in 1970-74. The PL-1B was described and illustrated in the 1975-76 *Jane's*.

Between 1969 and 1976, the AIDC produced in Taiwan 118 Bell UH-1H (Bell Model 205) helicopters under licence for the Chinese Nationalist Army.

The AIDC is currently engaged in licence building 212 Northrop F-5E Tiger II tactical fighter aircraft (see US section) and 36 two-seat F-5Fs for the Chinese Nationalist Air Force. The first Chinese-built F-5E (CAF name **Chung Cheng**) was rolled out on 30 October 1974, and more than 160 had been delivered by the beginning of 1980. No recent details have been received.

The AIDC designed and is producing the T-CH-1 tandem two-seat turboprop trainer for the Chinese Nationalist Air Force. The Avco Lycoming T53 engines for the T-CH-1 are licence-built at Kang Shan. Under development are two other indigenous designs, the XC-2 twin-turboprop transport and a tandem two-seat twin-engined military trainer designated XAT-3.

AIDC/CAF-built Chung Cheng (Northrop F-5F) for the Chinese Nationalist Air Force

AIDC T-CH-1

This aircraft is a tandem two-seat trainer, the design of which was started by AIDC in November 1970. Two prototypes were ordered, designated XT-CH-1A and XT-CH-1B; construction began in January 1972.

The first aircraft was flown for the first time on 23 November 1973; it was described and illustrated in the 1974-75 *Jane's*. The second prototype flew for the first time on 27 November 1974.

The following description applies to the T-CH-1, of which production began in May 1976. Fifty have been ordered for the Chinese Nationalist Air Force, of which more than 35 had been delivered by early 1980. It is believed that production was due to end in 1981.

Type: Turboprop-powered trainer and light ground attack aircraft.

Wings: Cantilever low-wing monoplane. Wing section NACA 64-2A215 (constant). Dihedral 8° from roots. Incidence 2°. No sweepback. Conventional aluminium alloy stressed-skin structure, with aluminium alloy ailerons and slotted trailing-edge flaps. Link-balance type trim tab in each aileron.

Fuselage: Conventional semi-monocoque structure of aluminium alloy.

Tail Unit: Cantilever aluminium alloy structure, with fixed-incidence tailplane. Dorsal fin. Link-balance type trim tabs in rudder and each elevator.

Landing Gear: Retractable tricycle type. Hydraulic retraction, main wheels inward into wings, nosewheel rearward. Telescopic shock-absorbers. Goodyear brakes. Small tail bumper under rear fuselage.

Power Plant: One 1,082 kW (1,451 ehp) Taiwan-built Avco Lycoming T53-L-701 turboprop engine, driving a Hamilton Standard 53C51-27 three-blade metal propeller with spinner. Fuel in two tanks in each wing and one in fuselage, with total capacity of 963 litres (212 Imp gallons). Oil capacity 30 litres (6·6 Imp gallons).

Production line-up of AIDC/CAF T-CH-1 basic trainers for the Chinese Nationalist Air Force

ACCOMMODATION: Crew of two in tandem. Rearward-sliding fully-transparent canopy over each cockpit. Cockpits heated and ventilated.

SYSTEMS: Midland-Ross Corporation heating and ventilating system. 115V 300A system provides AC electrical power at 250VA 400Hz. 28V DC system includes 24V 34Ah battery. Oxygen bottle with volume of 3·5 litres (2,100 cu in).

AVIONICS AND EQUIPMENT: Collins AN/ARC-51BX UHF radio and Collins AN/ARN-83 ADF.

DIMENSIONS, EXTERNAL:

Wing span	12·19 m (40 ft 0 in)
Wing chord at root	2·44 m (8 ft 0 in)
Wing chord at tip	1·52 m (5 ft 0 in)
Wing aspect ratio	6
Length overall	10·26 m (33 ft 8 in)
Height overall	3·66 m (12 ft 0 in)
Tailplane span	5·56 m (18 ft 3 in)
Wheel track	3·86 m (12 ft 8 in)
Wheelbase	2·39 m (7 ft 10 in)
Propeller diameter	3·05 m (10 ft 0 in)
Propeller ground clearance	0·74 m (2 ft 5 in)

AREAS:

Wings, gross	25·18 m² (271·0 sq ft)
Ailerons (total)	2·42 m² (26·0 sq ft)
Flaps (total)	5·02 m² (54·0 sq ft)
Fin	1·67 m² (18·0 sq ft)
Rudder, incl tab	1·11 m² (12·0 sq ft)
Elevators, incl tabs	1·81 m² (19·5 sq ft)

WEIGHTS AND LOADINGS:

Weight empty	2,608 kg (5,750 lb)
T-O weight: 'clean'	3,402 kg (7,500 lb)
max	5,057 kg (11,150 lb)
Max wing loading	200·9 kg/m² (41·14 lb/sq ft)
Max power loading	4·68 kg/kW (7·68 lb/ehp)

PERFORMANCE (at AUW of 3,447 kg; 7,600 lb):

Never-exceed speed	370 knots (685 km/h; 426 mph)
Max level speed at 4,570 m (15,000 ft)	320 knots (592 km/h; 368 mph)
Max cruising speed at 4,570 m (15,000 ft)	220 knots (407 km/h; 253 mph)
Econ cruising speed at 4,570 m (15,000 ft)	170 knots (315 km/h; 196 mph)
Stalling speed	50 knots (93 km/h; 58 mph)
Max rate of climb at S/L	1,036 m (3,400 ft)/min
Service ceiling	9,755 m (32,000 ft)
T-O run	146 m (480 ft)
T-O to 15 m (50 ft)	244 m (800 ft)
Landing from 15 m (50 ft)	381 m (1,250 ft)
Landing run	183 m (600 ft)
Range with max fuel	1,085 nm (2,010 km; 1,250 miles)

AIDC XAT-3

Up to July 1981 no details of the XAT-3 had been received from Taiwan, although the prototype is reported to have been completed in August 1980 and to have been due to make its first flight by the end of that year. Assistance in its design has been given by Northrop Corporation of the USA, and the XAT-3 is said to resemble both the Northrop/Vought contender in the US Navy's VTX-TS competition and the CASA C-101 Aviojet, for which Northrop also provided design assistance.

Powered by two Garrett TFE731 turbofan engines, the XAT-3 is a tandem two-seat aircraft with a reported gross weight of about 5,000 kg (11,025 lb) and max level speed of Mach 0·85. The Chinese Nationalist Air Force is said to have a requirement for about 60.

AIDC XC-2

The basic design of the XC-2 twin-turboprop transport, which was started in January 1973, incorporates features of common interest to military and civil operators, including quick-change capability and the ability to operate from short fields and unprepared surfaces. The XC-2 can carry up to 38 passengers or 3,855 kg (8,500 lb) of cargo.

The prototype (serial number 68-5001) was rolled out on 31 October 1978, and made its first flight on 26 February 1979. No recent details have been received.

TYPE: Twin-turboprop transport aircraft.

WINGS: Cantilever high-wing monoplane. Wing section NACA 65_3-218. Incidence 4°. No dihedral or sweepback at quarter-chord. Light alloy three-spar fail-safe structure, in three sections: constant-chord centre-section and tapered outer panels. All-metal manually-operated ailerons and hydraulically-actuated Fowler-type trailing-edge flaps. Servo tab in each aileron.

Prototype AIDC/CAF XC-2 twin-turboprop transport taking off for its first flight

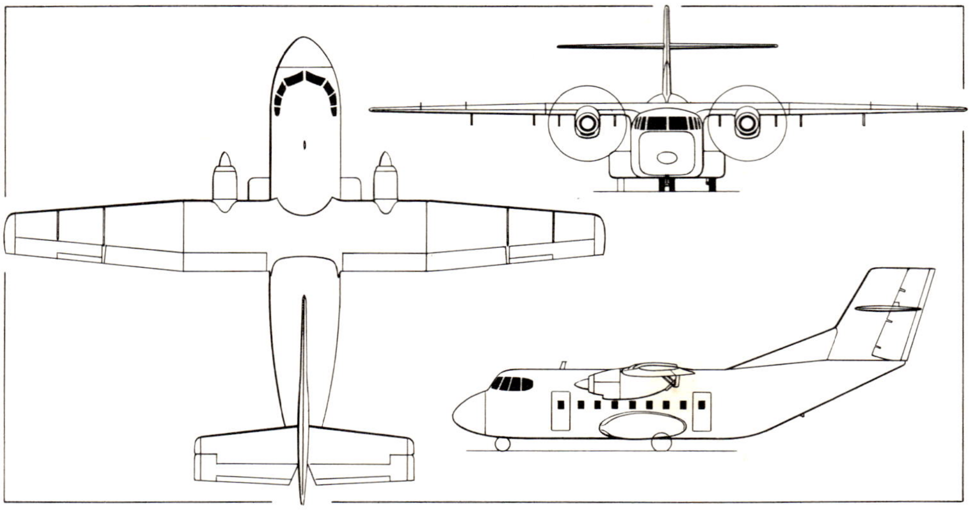

AIDC XC-2 twin-turboprop utility transport aircraft *(Pilot Press)*

FUSELAGE: Conventional all-metal semi-monocoque fail-safe structure, of basically rectangular section, upswept at rear to provide clearance for rear loading. Cabin pressurisation optional.

TAIL UNIT: Cantilever aluminium alloy three-spar structure, with sweptback fin and rudder and non-swept horizontal surfaces. Dorsal fin. Horizontal surfaces mounted halfway up fin. Trim and balance tab in rudder and each elevator.

LANDING GEAR: Retractable tricycle type, with hydraulically-steerable twin-wheel nose unit. Single-wheel main units retract into fairings on sides of fuselage.

POWER PLANT: Two 1,082 kW (1,451 ehp) Avco Lycoming T53-L-701A turboprop engines, each driving a Hamilton Standard 53C51-27 three-blade variable-pitch metal propeller with spinner. Fuel in rubber tanks in wings, with combined standard capacity of 3,028 litres (666 Imp gallons).

ACCOMMODATION: Crew of three (pilot, co-pilot and flight engineer) on flight deck. Standard seating in main cabin for 38 passengers, four abreast at 79 cm (31 in) pitch. Interior layout has quick-change capability to passenger/cargo or all-cargo configuration. Access to main cabin via forward and rear doors on port side; single door on starboard side; and a two-section loading ramp/door in underside of rear fuselage, aft of main cabin, which is openable in flight for airdrop operations. Provision for toilet, galley and baggage compartment in passenger version.

SYSTEMS: Anti-icing and cabin heating systems standard. Hydraulic system, pressure 207 bars (3,000 lb/sq in), for flaps, landing gear and nosewheel steering. 28V DC primary electrical system, with 300A starter/generator on each engine. Two nickel-cadmium batteries for engine starting and emergency power.

AVIONICS AND EQUIPMENT: Standard avionics include UHF and VHF com; and ADF, Tacan and transponder navigation equipment. Optional avionics include VOR/ILS and HF.

DIMENSIONS, EXTERNAL:

Wing span	24·90 m (81 ft 8·4 in)
Wing chord (centre-section, constant)	3·05 m (10 ft 0 in)
Wing aspect ratio	9·5
Length overall	20·10 m (65 ft 11·3 in)
Height overall	7·72 m (25 ft 3·8 in)
Tailplane span	9·12 m (29 ft 10·9 in)
Wheel track	3·86 m (12 ft 7·8 in)
Wheelbase	7·10 m (23 ft 3·5 in)
Propeller diameter	3·05 m (10 ft 0 in)
Propeller ground clearance	0·90 m (2 ft 11·5 in)

DIMENSIONS, INTERNAL:

Cabin, excl flight deck:

Length	8·095 m (26 ft 6·7 in)
Width	2·57 m (8 ft 5 in)
Height	2·23 m (7 ft 3·7 in)
Floor area	20·85 m² (224·4 sq ft)
Volume	45·45 m³ (1,605·0 cu ft)

AREAS:

Wings, gross	65·40 m² (704·00 sq ft)
Ailerons (total)	4·24 m² (45·63 sq ft)
Trailing-edge flaps (total)	11·69 m² (125·80 sq ft)
Fin (incl dorsal fin)	9·35 m² (100·64 sq ft)
Rudder (incl tabs)	4·75 m² (51·12 sq ft)
Tailplane	10·41 m² (112·05 sq ft)
Elevators (incl tabs)	8·97 m² (96·55 sq ft)

WEIGHTS AND LOADINGS:

Weight empty	7,031 kg (15,500 lb)
Max payload	3,855 kg (8,500 lb)
Max T-O weight	12,474 kg (27,500 lb)
Max landing weight	12,247 kg (27,000 lb)
Max zero-fuel weight	11,254 kg (24,810 lb)
Max wing loading	190·7 kg/m² (39·06 lb/sq ft)
Max power loading	5·77 kg/kW (9·48 lb/ehp)

PERFORMANCE (estimated, at max T-O weight):

Never-exceed speed	250 knots (463 km/h; 287 mph)
Max level speed at S/L	212 knots (392 km/h; 244 mph)
Max cruising speed at 3,050 m (10,000 ft)	200 knots (370 km/h; 230 mph)
Econ cruising speed at 3,050 m (10,000 ft)	180 knots (333 km/h; 207 mph)
Stalling speed, flaps down	78 knots (145 km/h; 90 mph)
Max rate of climb at S/L	457 m (1,500 ft)/min
Service ceiling	8,015 m (26,300 ft)
Service ceiling, one engine out	2,740 m (9,000 ft)
T-O run	625 m (2,050 ft)
T-O to 15 m (50 ft)	778 m (2,550 ft)
Landing from 15 m (50 ft)	826 m (2,710 ft)
Landing run	582 m (1,910 ft)
Range with max payload, reserves for 87 nm (161 km; 100 mile) alternate and 45 min hold	259 nm (480 km; 298 miles)
Range with max fuel, 45 min reserves	897 nm (1,661 km; 1,032 miles)

TURKEY

THvK (KIBM)

TÜRK HAVA KUVVETLERI, HAVA IKMAL BAKIM MERKEZI KOMÜTANGLICI KAYSERI (Turkish Air Force, Kayseri Air Supply and Maintenance Centre Command)

Kayseri, Turkey
Telephone: (351) 155 56, extn 301
COMMANDER, KIBM: Major General Ahmet Cesur
TECHNICAL DIRECTOR, AND HEAD OF MAVI ISIK PROJECT GROUP: Col Ergün Belirgen, MSAeE
SECRETARY GENERAL, MAVI ISIK PROJECT GROUP: Capt Durmus Salar, MSAeE

The Kayseri Air Supply and Maintenance Centre Command was formed by the Turkish Ministry of Defence in 1932 as the Kayseri Aircraft Factory. Between 1933 and 1940, in addition to aircraft maintenance work, it produced Gotha Go 145, Curtiss Hawk and PZL P-24 aircraft under licence from their German, US and Polish manufacturers. In 1950, after Turkey became a member of NATO, its name was changed to Air Supply Centre and its function was restricted to the overhaul of aircraft and aero-engines. In 1970, to reflect an increasing emphasis on maintenance work, it was renamed as the Kayseri Supply and Maintenance Centre General Directorate; the present title was adopted in 1975.

Currently, the KIBM overhauls all of the Turkish Air Force's propeller-driven aircraft, their engines and accessories, as well as the anti-submarine aircraft of the Turkish Navy and their engines. The Centre also manufactures aircraft spare parts and various types of parachute. It is responsible for development of the Mavi Isik nationally-designed series of aircraft, of which a description follows. Another agricultural aircraft, of all-metal construction and with a more powerful engine, is being constructed; this was expected to fly in September 1981.

KIBM MAVI ISIK (BLUE LIGHT)

The Mavi Isik programme, to design and develop an aircraft for primary training, agricultural and surveillance duties, was launched in April 1978, and construction of a prototype began in the following month. This aircraft, designated **Mavi Isik 78-XA,** flew for the first time in February 1979, and made its public debut at Kayseri on 25 May of that year after completing ground and flight testing. It was awarded a C of A by the Turkish Civil Aviation General Directorate and registered to the Turkish Air Force on 19 June 1979.

Development of the agricultural version, designated **Mavi Isik B,** has continued under the direction of the Turkish Air Force. The prototype, modified to this configuration, resumed flight testing in February 1980 and was displayed publicly in Ankara on 19 June of that year.

The following description applies to the Mavi Isik B:

TYPE: Single-seat agricultural aircraft.

WINGS: Low-wing monoplane, braced to fuselage by overwing inverted-V struts and jury struts on each side. Wing section NACA 4412 (modified). Dihedral 7° from roots. Incidence 1° 30'. Constant-chord structure, with aluminium leading-edge, spars and ribs. Dacron covering, with Duraclad fire-resistant plastic finish overall. Ailerons and trailing-edge flaps are of similar construction. No tabs.

FUSELAGE: Welded SAE 4130 steel tube structure, with Dacron covering and Duraclad plastic finish overall.

TAIL UNIT: Conventional cruciform structure of welded SAE 4130 steel tube, with fabric covering. Wire-braced tailplane. No tabs.

LANDING GEAR: Non-retractable tailwheel type, with main wheels carried on side Vs. Castoring tailwheel. Oleo-pneumatic shock-absorption. Tyre sizes 6·50 × 8 in on main wheels, 2·50 × 8 in on tailwheel.

POWER PLANT: One 157 kW (210 hp) Continental IO-360-D flat-six engine, driving a McCauley two-blade constant-speed propeller. Single aluminium fuel tank aft of firewall, capacity 80 litres (17·5 Imp gallons), of which 75 litres (16·5 Imp gallons) are usable. Oil capacity 9·5 litres (2 Imp gallons).

ACCOMMODATION: Pilot only, in heated cockpit. Combined window/door on each side, hinged at bottom to open outward and downward.

Turkish Air Force Mavi Isik 78-XA in its original form

Turkish Air Force Mavi Isik B prototype agricultural aircraft

SYSTEMS: Hydraulic system; 24V 35Ah battery.

DIMENSIONS, EXTERNAL:

Wing span	10·77 m (35 ft 4 in)
Wing chord, constant	1·60 m (5 ft 3 in)
Wing aspect ratio	6·73
Length of fuselage	7·48 m (24 ft 6½ in)
Height overall	2·21 m (7 ft 3 in)
Tailplane span	2·86 m (9 ft 4½ in)
Wheel track	2·08 m (6 ft 10 in)
Propeller diameter	1·93 m (6 ft 4 in)

AREAS:

Wings, gross	17·23 m² (185·46 sq ft)
Ailerons (total)	1·75 m² (18·84 sq ft)
Trailing-edge flaps (total)	0·80 m² (8·61 sq ft)
Fin	0·37 m² (3·98 sq ft)
Rudder	0·66 m² (7·10 sq ft)
Tailplane	1·18 m² (12·70 sq ft)
Elevators (total)	1·32 m² (14·21 sq ft)

WEIGHTS AND LOADINGS:

Weight empty, equipped	741 kg (1,633 lb)
Max payload, incl pilot	450 kg (992 lb)
Max fuel load	60 kg (132 lb)
Max T-O and landing weight	1,270 kg (2,800 lb)
Max wing loading	73·71 kg/m² (15·10 lb/sq ft)
Max power loading	8·11 kg/kW (13·32 lb/hp)

PERFORMANCE (at max T-O weight):

Max level speed at S/L	98 knots (182 km/h; 113 mph)
Max cruising speed (75% power)	92 knots (171 km/h; 106 mph)
Stalling speed, flaps down	44·5 knots (82 km/h; 51 mph)
Max rate of climb at S/L	202 m (662 ft)/min
T-O run	256 m (840 ft)
T-O to 15 m (50 ft)	435 m (1,427 ft)
Landing run	285 m (935 ft)

TUSAS

TÜRK UÇAK SANAYII ANONIM ORTAKLIGI (Turkish Aircraft Industries Inc)

HEAD OFFICE: Atatürk Bulvari 227, Kavaklidere, Ankara
Telephone: 27 81 25
Telex: 42127 TUS TR
BOARD:
Saim Dilek (President and General Manager)
Hulki Yanat, MSAeEng (representing Ministry of Industry and Technology)
Gen Cemal Tural (Retd); former Chief of Staff, Turkish Armed Forces (representing Ministry of Finance)
Gen Etem Ayan (Retd); former Commander of Turkish Air Force (representing Air Force Foundation)
Gen Cengiz Sakaryali (Retd) (representing Ministry of Defence)
Yildirim Özdamar (Asst General Manager, Financial)

TUSAS is financed jointly by the Turkish government (55%) and the Turkish Air Force Foundation, a non-profit-making organisation (45%). It was officially established with effect from 11 July 1973 with an initial capital of 1,500 million Turkish liras. Its head office is in Ankara, and an area of 5 million m² (1·93 sq miles) at Murted, near Ankara, is provided for an aircraft manufacturing facility and auxiliary buildings.

In early 1981 negotiations were under way for licence assembly by TUSAS of the Northrop F-5E Tiger II combat aircraft, to begin by the end of that year.

UNION OF SOVIET SOCIALIST REPUBLICS

Antonov An-12BP ('Cub-A') four-turboprop general-purpose military transport aircraft *(Pilot Press)*

Electronic intelligence version of the Antonov An-12 known to NATO as 'Cub-B' *(Swedish Air Force)*

ANTONOV

GENERAL DESIGNER IN CHARGE OF BUREAU:
Oleg Konstantinovich Antonov

After establishing his reputation with a series of successful glider and sailplane designs, Oleg K. Antonov became one of the Soviet Union's leading designers of transport aircraft, particularly those types intended for short-field operation.

In addition to the current products described in this section, his design bureau is studying the requirements for a large transport to replace the An-22.

ANTONOV An-2

NATO reporting name: Colt

Following manufacture of the An-2M specialised agricultural version of this large single-engined biplane, in the mid-sixties, production of the An-2 came to an end in the Soviet Union. Details of the various versions that were built can be found in the 1971-72 *Jane's*.

Several versions of the An-2 continue in production under licence in Poland (see WSK-PZL Mielec entry). Others were built at Harbin, in China, under the Chinese designation Y-5 (C-5).

ANTONOV An-3

It was reported in the Spring of 1972 that the Antonov design bureau was engaged on design studies for a turboprop development of the An-2 biplane (see WSK-PZL Mielec in Polish section). Designated An-3, the aircraft was intended specifically for agricultural duties and was then expected to compete with the Polish turbofan-engined WSK-Mielec M-15 as the next-generation agricultural aircraft for use throughout the countries of eastern Europe and the Soviet Union.

At the 1979 Paris Air Show, Mr Antonov emphasised his continued interest in agricultural aircraft, and confirmed that a prototype of the An-3 (CCCP-30576) had been produced by converting an An-2 to have a 701 kW (940 shp) Glushenkov TVD-10B turboprop engine, driving a slow-turning large-diameter propeller optimised for an aircraft operating speed of 75-97 knots (140-180 km/h; 87-112 mph). He added that an important feature was the ability of the turboprop to ensure adequate cockpit air-conditioning, with clean air at 18-20°C, during operation in ambient temperatures of 40-45°C.

A photograph of the An-3 prototype, under test, was released to the East European press in 1979. This showed that it differs externally from the standard An-2 only in having a longer and slimmer nose, housing the TVD-10B turboprop. A chemical payload of 1,800 kg (3,968 lb) can be carried, representing an increase of nearly 40 per cent over that of the An-2; the cost of spraying each hectare of land is reduced by 25 to 30 per cent.

The only performance figure yet released is a rate of climb of 240 m (785 ft)/min after take-off with a full load. See Addenda for illustration.

ANTONOV An-12

NATO reporting name: Cub

About 850 An-12 freighters were built for military and civil use before production ended in 1973. Versions in service with the Soviet air forces, and identified by NATO reporting names, are as follows:

Cub-A. Standard Soviet military paratroop and freight transport for many years. At peak strength, 'Cub-As' could carry two full army divisions, totalling 14,000 men and their equipment, over a radius of 651 nm (1,207 km; 750 miles). In 1981, about 560 'Cub-As', designated **An-12BP** in the USSR, continued in service with the VTA (Military Transport Aviation) force, although replacement with Ilyushin Il-76s started in 1974. Others are operated by the air forces of Afghanistan, Algeria, Bangladesh, China, Egypt, India, Iraq, Malagasy Republic, Poland, Syria and Yugoslavia. Current Soviet Air Force An-12s have a larger undernose radome than that originally fitted. Civil An-12s serve with Aeroflot, Polish Airlines (LOT), Balkan Air, Air Guinée and Iraqi Airways.

Cub-B. Conversion of 'Cub-A' for electronic intelligence (elint) duties. Example photographed off Sweden had four additional blister fairings under forward- and centre-fuselage, plus other antennae. Few produced.

Cub-C. About 30 used by Soviet Air Force and Navy for ECM duties. Ogival 'solid' fuselage tailcone, housing electronic equipment, instead of gun position. Additional electronic pods faired into forward fuselage and ventral surfaces. Glazed nose and undernose radar of other versions retained.

'Cub-A' has a tail gunner's position. In the refined commercial version, first demonstrated at the 1965 Paris Air Show, the turret is removed and replaced by a streamline fairing.

The following abbreviated details apply to the standard military An-12BP transport. A full description can be found in the 1979-80 *Jane's*.

TYPE: Four-engined cargo transport.

POWER PLANT: Four 2,983 kW (4,000 ehp) Ivchenko AI-20K turboprops, driving AV-68 four-blade reversible-pitch propellers. All fuel in 22 bag-type tanks in wings, total normal capacity 13,900 litres (3,058 Imp gallons). Max capacity 18,100 litres (3,981 Imp gallons).

ACCOMMODATION: Pilot and co-pilot side by side on flight deck. Engineer's station on starboard side, behind co-pilot. Radio operator in well behind pilot, facing outward. Navigator in glazed nose compartment. Rear gunner in tail turret. Crew door on port side forward of wing. Access to freight hold via ramp-door at rear, under upswept rear fuselage. Ramp-door is divided into two longitudinal halves, which can be hinged upward inside cabin to provide access for direct loading of freight from trucks. Undersurface of fuselage aft of ramp is formed by door which hinges upward into fuselage to facilitate loading and unloading. Equipped to carry 100 paratroops, all of whom can be despatched in under one minute, with ramp-doors folded upward.

ARMAMENT: Two 23 mm NR-23 guns in tail turret.

DIMENSIONS, EXTERNAL:	
Wing span	38·00 m (124 ft 8 in)
Wing chord (mean)	3·452 m (11 ft 4 in)
Wing area, gross	121·70 m² (1,310 sq ft)
Wing aspect ratio	11·85
Length overall	33·10 m (108 ft 7¼ in)
Height overall	10·53 m (34 ft 6½ in)
Tailplane span	12·20 m (40 ft 0¼ in)
Wheel track	5·42 m (17 ft 9½ in)
Wheelbase	10·82 m (35 ft 6 in)
Propeller diameter	4·50 m (14 ft 9 in)
Rear loading hatch: Length	7·70 m (25 ft 3 in)
Width	2·95 m (9 ft 8 in)
DIMENSIONS, INTERNAL:	
Cargo hold: Length	13·50 m (44 ft 3½ in)
Max width	3·50 m (11 ft 5¾ in)
Max height	2·60 m (8 ft 6¼ in)
Volume	97·2 m³ (3,432·6 cu ft)
WEIGHTS:	
Weight empty	28,000 kg (61,730 lb)
Max payload	20,000 kg (44,090 lb)
Normal T-O weight	55,100 kg (121,475 lb)
Max T-O weight	61,000 kg (134,480 lb)
PERFORMANCE:	
Max level speed	419 knots (777 km/h; 482 mph)
Max cruising speed	361 knots (670 km/h; 416 mph)
Min flying speed	88 knots (163 km/h; 101 mph)
Landing speed	108 knots (200 km/h; 124 mph)

ECM version of the Antonov An-12 ('Cub-C'), operated temporarily in Egyptian Air Force insignia

Antonov An-22 Antheus long-range heavy transport aircraft (four Kuznetsov NK-12MA turboprop engines) *(Tass)*

Max rate of climb at S/L	600 m (1,970 ft)/min
Service ceiling	10,200 m (33,500 ft)
T-O run	700 m (2,300 ft)
Landing run	500 m (1,640 ft)
Range with max payload	1,942 nm (3,600 km; 2,236 miles)
Range with max fuel	3,075 nm (5,700 km; 3,540 miles)

ANTONOV An-22 ANTHEUS

NATO reporting name: Cock

The prototype of this very large transport aircraft flew for the first time on 27 February 1965. Production aircraft were delivered to the Soviet Air Force and to Aeroflot, which uses the An-22 mainly in underdeveloped areas of the northern USSR, Siberia and the Far East. Deliveries were completed during 1974, after about 50 aircraft had been delivered. In 1981 they remained the only Soviet transports capable of airlifting the T-62 tank.

A full list of the 27 FAI-confirmed records for payload-to-height, and speed with payload, held currently by the An-22 can be found in the 1979-80 *Jane's*.

TYPE: Long-range heavy turboprop transport.

WINGS: Cantilever high-wing monoplane. Marked anhedral on outer panels. All-metal structure, appearing to have three main spars which attach to three strong fuselage ring-frames. Double-slotted trailing-edge flaps. Tab in each aileron.

FUSELAGE: All-metal semi-monocoque structure, with upswept rear fuselage containing loading-ramp/door for direct loading. Retractable jacks support rear fuselage at point where rear loading ramp is hinged.

TAIL UNIT: Cantilever all-metal structure. Twin fins and rudders (each in two sections, above and below tailplane) mounted outboard of mid-span. Bullet fairing forward of each fin tip. Tabs in each elevator and in each of the four rudder sections.

LANDING GEAR: Retractable tricycle type, designed to permit off-runway operation. Steerable twin-wheel nose unit. Each main gear consists of three twin-wheel levered-suspension units in tandem, each unit mounted at the bottom of one of the fuselage ring frames that also picks up a wing spar. Main units retract upward into fairings built on to sides of fuselage. Tyre pressure adjustable in flight or on ground to suit airfield surface.

POWER PLANT: Four 11,186 kW (15,000 shp) Kuznetsov NK-12MA turboprop engines, each driving a pair of four-blade contra-rotating propellers.

ACCOMMODATION: Crew of five or six. Navigator's station in nose. Cabin for 28-29 passengers aft of flight deck, separated from main cabin by bulkhead containing two doors. Uninterrupted main cabin, with reinforced titanium floor, tiedown fittings and rear loading ramp. When ramp lowers, a large door which forms the underside of the rear fuselage retracts upward inside fuselage to permit easy loading of tall vehicles. Rails in roof of cabin for four travelling gantries continue rearward on underside of this door. Two winches, used in conjunction with the gantries, each have a capacity of 2,500 kg (5,500 lb). Door in each landing gear fairing, forward of wheels, for crew and passengers.

AVIONICS AND EQUIPMENT: Pressurisation equipment and APU in forward part of starboard landing gear fairing. Two radars, in nose 'thimble' and undernose fairings.

DIMENSIONS, EXTERNAL:	
Wing span	64·40 m (211 ft 4 in)
Length overall	approx 57·92 m (190 ft 0 in)
Height overall	12·53 m (41 ft 1½ in)
Propeller diameter	6·20 m (20 ft 4 in)
DIMENSIONS, INTERNAL:	
Main cabin: Length	33·0 m (108 ft 3 in)
Max width	4·4 m (14 ft 5 in)
Max height	4·4 m (14 ft 5 in)
AREA:	
Wings, gross	345 m² (3,713 sq ft)
WEIGHTS:	
Weight empty, equipped	114,000 kg (251,325 lb)
Max payload	80,000 kg (176,350 lb)
Max fuel	43,000 kg (94,800 lb)
Max T-O weight	250,000 kg (551,160 lb)
PERFORMANCE:	
Max level speed	399 knots (740 km/h; 460 mph)
T-O run	1,300 m (4,260 ft)
Landing run	800 m (2,620 ft)
Range with max fuel and 45,000 kg (99,200 lb) payload	5,905 nm (10,950 km; 6,800 miles)
Range with max payload	2,692 nm (5,000 km; 3,100 miles)

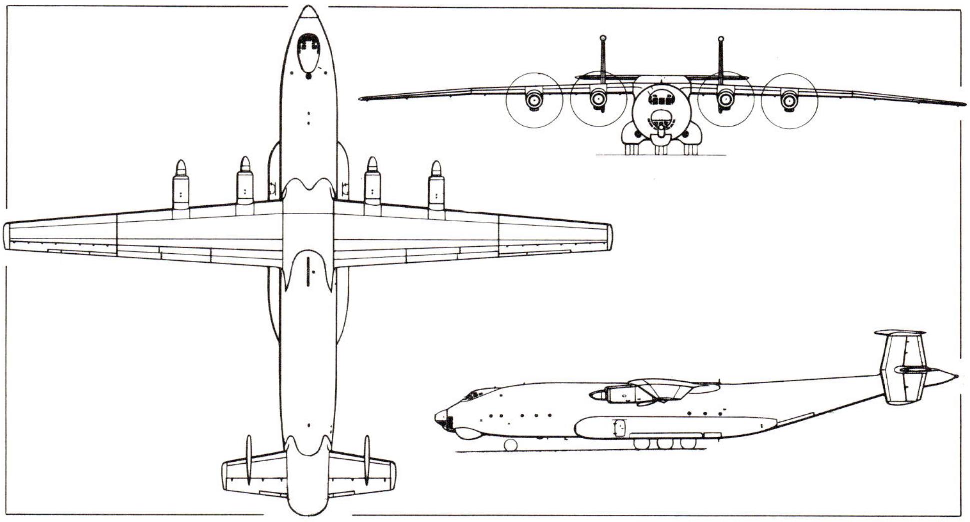

Antonov An-22 Antheus long-range heavy transport aircraft *(Pilot Press)*

ANTONOV An-24

NATO reporting name: Coke

Development of this twin-turboprop transport was started in 1958, to replace piston-engined types on Aeroflot's internal feederline routes. The first prototype flew in April 1960 and the An-24 entered service on Aeroflot's routes from Moscow to Voronezh and Saratov in September 1963. When production ended in 1978, a total of about 1,100 had been delivered. Aeroflot received several hundred; the last off the assembly line, delivered to the Romanian airline Tarom, was described as the 750th aircraft of the An-24/26 series to be exported.

A list of 15 airlines known to have operated An-24s can be found in the 1976-77 *Jane's*. The An-24 was also supplied for military service, usually in small numbers, with the air forces of the USSR, Afghanistan, Algeria, Bangladesh, Bulgaria, China, Congo, Cuba, Czechoslovakia, Egypt, East Germany, Hungary, Iraq, North Korea, Laos, Mali, Mongolia, Poland, Romania, the Somali Republic, Sudan, Syria, Viet-Nam, North Yemen, and South Yemen.

A full description of the An-24 can be found in the 1979-80 *Jane's*.

ANTONOV An-26

NATO reporting name: Curl

First displayed in public at the 1969 Paris Air Show, the An-26 is generally similar to the earlier An-24RT specialised freighter, with an auxiliary turbojet in the starboard engine nacelle, but has more powerful AI-24T turboprop engines and a completely redesigned rear fuselage of the 'beaver-tail' type. This embodies Oleg Antonov's special type of loading ramp, which forms the underside of the rear fuselage when retracted, in the usual way, but can be slid forward under the rear of the cabin to facilitate direct loading, or when the cargo is to be air-dropped.

There are two versions:

An-26. Original version, with electrically/manually-operated conveyor built-in flush with cabin floor to facilitate movement of freight.

An-26B. Improved version, announced in 1981. Equipped to carry three standard freight pallets, each 2·438 m (8 ft) long, 1·46 m (4 ft 9½ in) wide and 1·60 m (5 ft 3 in) high, with a total weight of 5,500 kg (12,125 lb). Rollgangs on floor, mechanism to move the pallets, and moorings enable two men to load and unload all three pallets in 30 min. Rollgangs can be stowed against sides of cabin when not required.

Although intended primarily for cargo-carrying, the An-26 can be adapted easily for passenger-carrying, ambulance or paratroop transport duties. Production was continuing in 1981.

The Air Wing of the Bangladesh Defence Force has two An-26s, as well as one An-24. Other An-26s serve with the Afghan, Angolan, Benin, Bulgarian, Chinese, Cuban, Czechoslovak, East German, Hungarian, Iraqi, Lao, Malagasy, Mozambique, Peruvian, Polish, Romanian, Somali, Syrian, Yugoslav and Soviet Air Forces. Aeroflot has more than 100, the Angolan freight carrier CTA has one, and three are operated by Tarom of Romania.

TYPE: Twin-turboprop pressurised short-haul transport.

WINGS: Cantilever high-wing monoplane, with 2° anhedral on outer panels. Incidence 3°. Sweepback at quarter-chord on outer panels 6° 50′. All-metal two-spar structure, built in five sections: centre-section, two inner wings and two detachable outer wings. Wing skin is attached by electrical spot-welding. Mass-balanced servo-compensated ailerons, with large trim tabs of glassfibre construction. Hydraulically-operated Fowler flaps along entire wing trailing-edges inboard of unpowered ailerons; single-slotted flaps on centre-section, double-slotted outboard of nacelles. Servo tab and electrically-operated trim tab in each aileron. Thermal de-icing system.

FUSELAGE: All-metal semi-monocoque structure in front, centre and rear portions, of bonded/welded construction. Skin on lower portion of fuselage is made of 'bimetal' (duralumin-titanium) sheet for protection during operations from unpaved airfields.

TAIL UNIT: Cantilever all-metal structure, with ventral fin strake on each side of rear ramp. 9° dihedral on tailplane. All controls manually operated. Electrically-operated trim tab in each elevator. Electrically-operated combined trim/servo tab in rudder. All leading-edges incorporate thermal de-icing system.

LANDING GEAR: Hydraulically-retractable tricycle type, with twin wheels on each unit. Emergency extension by gravity. All units retract forward. Shock-absorbers of oleo-nitrogen type on main units; nitrogen-pneumatic type on nose unit. Main-wheel tyres size 1,050 × 400, pressure 5·86 bars (85 lb/sq in). Nosewheel tyres size

700 × 250, pressure 3·93 bars (57 lb/sq in). Main wheels are fitted with hydraulic disc brakes and anti-skid units. Nosewheels can be steered hydraulically through 45° each side while taxying and are controllable through ±10° during take-off and landing.

Power Plant: Two 2,103 kW (2,820 ehp) Ivchenko AI-24VT turboprop engines, each driving a four-blade constant-speed fully-feathering propeller. Electrical de-icing system for propeller blades and hubs; hot air system for engine air intakes. One 7·85 kN (1,765 lb st) RU 19A-300 auxiliary turbojet in starboard nacelle for use, as required, at take-off, during climb and in level flight, and for self-contained starting of main engines. Fuel load 5,500 kg (12,125 lb), contained in integral tanks in inner wings and ten bag-type tanks in centre-section. Pressure refuelling socket in starboard engine nacelle. Gravity fuelling point above each tank area. Carbon dioxide inert gas system to create fireproof condition inside fuel tanks.

Accommodation: Basic crew of five (pilot, co-pilot, radio operator, flight engineer and navigator), with station at rear of cabin on starboard side for loading supervisor or load dispatcher. Electrical de-icing system for windscreens. Toilet on port side aft of flight deck; crew door, small galley and oxygen bottle stowage on starboard side. Emergency escape hatch in door immediately aft of flight deck. Large downward-hinged rear ramp/door, which can also slide forward under fuselage for direct loading on to cabin floor or for air-dropping of freight. Electrically-powered mobile winch, capacity 2,000 kg (4,409 lb), hoists crates through rear entrance and runs on a rail in the cabin ceiling to position payload in cabin. Electrically- and manually-operated conveyor, capacity 4,500 kg (9,920 lb), built-in flush with cabin floor of original An-26, facilitates loading and airdropping of freight. An-26B has removable rollgangs, mechanism for moving pallets inside hold, and moorings (see introductory notes). Both versions can accommodate a variety of motor vehicles, including GAZ-69 and UAZ-469 military vehicles, or cargo items up to 1·50 m (59 in) high by 2·10 m (82·6 in) wide. Height of rear edge of cargo door surround above the cabin floor is 1·50 m (4 ft 11 in). Cabin is pressurised and air-conditioned, and is fitted with a row of tip-up seats along each wall to accommodate a total of 38 to 40 persons. Conversion to troop transport role, or to an ambulance for 24 stretcher patients and a medical attendant, takes 20 to 30 min in the field.

Systems: Air-conditioning system uses hot air tapped from the 10th compressor stage of each engine, with a heat exchanger and turbocooler in each nacelle. Cabin pressure differential 0·29 bars (4·27 lb/sq in). Main and emergency hydraulic systems, pressure 151·7 bars (2,200 lb/sq in), for landing gear retraction, nosewheel steering, flaps, brakes, windscreen wipers, propeller feathering and operation of cargo ramp and emergency escape doors. Handpump to operate doors only and build up pressure in main system. Electrical system includes two 27V DC starter/generators on engines, a standby generator on the auxiliary turbojet, and three storage batteries for emergency use. Two engine-driven alternators provide 115V 400Hz single-phase AC supply, with standby inverter. Basic source of 36V 400Hz three-phase AC supply is two inverters, with standby transformer. Permanent oxygen system for pilot, installed equipment for other crew members and three portable bottles for personnel in cargo hold.

Avionics and Equipment: Standard com/nav avionics comprise two VHF transceivers, HF, intercom, two ADF, radio altimeter, glidepath receiver, glideslope receiver, marker beacon receiver, weather/navigation radar, directional gyro and flight recorder. Optional equipment includes a flight director system, astrocompass and autopilot. Standard operational equipment includes parachute static line attachments and retraction devices, tiedowns, jack to support ramp sill, flight deck curtains, sun visors and windscreen wipers. Optional items include a navigator's observation blister on port side of flight deck, OPB-1R sight for pinpoint dropping of freight, medical equipment, and liquid heating system.

Dimensions, external:
Wing span 29·20 m (95 ft 9½ in)
Wing aspect ratio 11·7
Length overall 23·80 m (78 ft 1 in)
Height overall 8·575 m (28 ft 1½ in)
Width of fuselage 2·90 m (9 ft 6 in)
Depth of fuselage 2·50 m (8 ft 2½ in)
Tailplane span 9·973 m (32 ft 8¾ in)
Wheel track (c/l shock-struts) 7·90 m (25 ft 11 in)
Wheelbase 7·651 m (25 ft 1¼ in)
Propeller diameter 3·90 m (12 ft 9½ in)
Propeller ground clearance 1·227 m (4 ft 0¼ in)
Crew door (stbd, front): Height 1·40 m (4 ft 7 in)
Width 0·60 m (1 ft 11¾ in)
Height to sill 1·47 m (4 ft 9¾ in)
Loading hatch (rear): Length 3·40 m (11 ft 1¾ in)
Width at front 2·40 m (7 ft 10½ in)
Width at rear 2·00 m (6 ft 6¾ in)
Height to sill 1·47 m (4 ft 9¾ in)
Height to top edge of hatchway 3·014 m (9 ft 10¾ in)
Emergency exit (in floor at front):
Length 1·02 m (3 ft 4¼ in)
Width 0·70 m (2 ft 3½ in)
Emergency exit (top):
Diameter 0·65 m (2 ft 1½ in)
Emergency exits (one each side of hold):
Height 0·60 m (1 ft 11¾ in)
Width 0·50 m (1 ft 7½ in)

Dimensions, internal:
Cargo hold: Length of floor 11·50 m (37 ft 8¾ in)
Width of floor 2·40 m (7 ft 10½ in)
Max height 1·91 m (6 ft 3 in)

Areas:
Wings, gross 74·98 m² (807·1 sq ft)
Horizontal tail surfaces (total) 19·83 m² (213·45 sq ft)
Vertical tail surfaces (total, incl dorsal fin) 15·85 m² (170·61 sq ft)

Weights:
Weight empty 15,020 kg (33,113 lb)
Normal payload 4,500 kg (9,920 lb)
Max payload 5,500 kg (12,125 lb)
Normal T-O and landing weight 23,000 kg (50,706 lb)
Max T-O and landing weight 24,000 kg (52,911 lb)

Performance (at normal T-O weight):
Cruising speed at 6,000 m (19,675 ft) 237 knots (440 km/h; 273 mph)
T-O speed 108 knots (200 km/h; 124 mph) CAS
Landing speed 102 knots (190 km/h; 118 mph) CAS
Max rate of climb at S/L ~480 m (1,575 ft)/min
Service ceiling 7,500 m (24,600 ft)
T-O run, on concrete 780 m (2,559 ft)
T-O to 15 m (50 ft) 1,240 m (4,068 ft)
Landing from 15 m (50 ft) 1,740 m (5,709 ft)
Landing run, on concrete 730 m (2,395 ft)
Min ground turning radius 22·3 m (73 ft 2 in)
Range with max payload, no reserves 594 nm (1,100 km; 683 miles)
Range with max fuel, no reserves 1,376 nm (2,550 km; 1,584 miles)

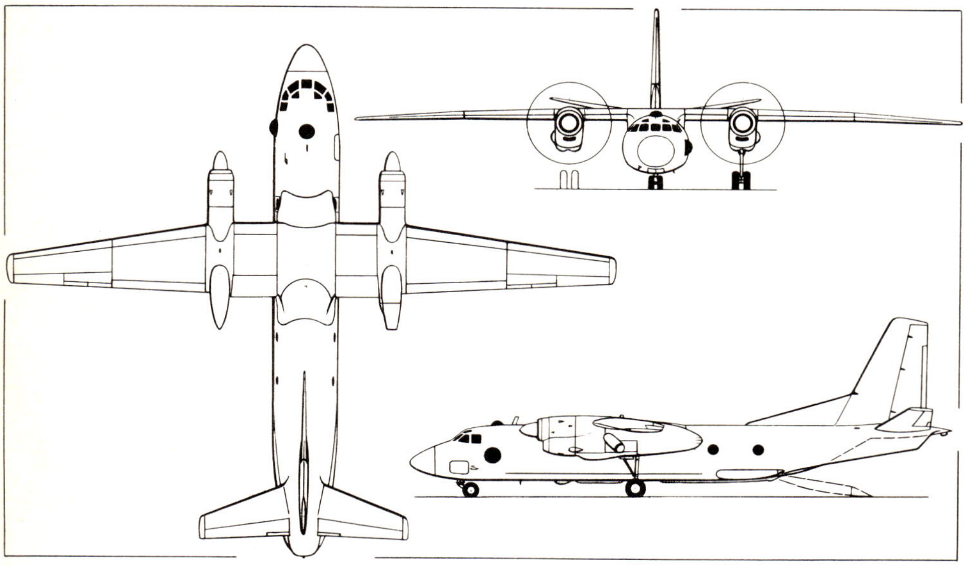

Antonov An-26 twin-turboprop short-haul transport *(Pilot Press)*

ANTONOV An-28

NATO reporting name: Cash

Responsibility for An-28 production has been allocated to the WSK-PZL Mielec works in Poland (see Polish section).

ANTONOV An-30

NATO reporting name: Clank

Described as the first specialised aerial survey aeroplane produced in the Soviet Union, the An-30 is evolved from the An-24RT and An-26 twin-turboprop transports, to which it is generally similar. The major modifications are made to the nose, which is extensively glazed to give the navigator a wide field of view, and to the flight deck, which is raised to improve the pilots' view and increase the size of the navigator's compartment. There are fewer windows in the main cabin, which contains a darkroom and film storage cupboard, as well as survey cameras and a control desk. Other amenities include a toilet, buffet and crew rest

One of five Antonov An-26 transports which passed through Gatwick Airport, en route to Cuba, in December 1978 *(Austin J. Brown)*

area with armchairs and couches. All accommodation is pressurised and air-conditioned.

Photography can be automatic or semi-automatic if required, but two photographer/surveyors are normally carried, in addition to a flight crew of five (pilot, co-pilot, flight engineer, radio operator and navigator).

For the primary task of air photography for map-making, the An-30 is equipped with large survey cameras. These are mounted in the cabin above glazed apertures, of which there are five, each covered by a door. A crew photographer uncovers the apertures, as required, by remote control from his desk in the aircraft.

Standard equipment includes radio topographic distance measuring equipment and a radio altimeter, with recording units. The pre-programmed flight path of the aircraft over the area to be photographed is fed into an onboard computer, controlled from the navigator's station, which maintains the correct speed, altitude and direction of flight throughout the mission. The cartographic An-30 has an AFA-41/7·5 wide-angle camera, in a TAU-M gyro-stabilised mounting, over No. 1 aperture; and an A54/50-FK long focal-length camera over No. 3 aperture, each mounted vertically. Two further A54/50-FK cameras take oblique photographs at 28° to the vertical, port and starboard, through Nos. 4 and 5 apertures, and an SU-5 lightmeter is positioned over No. 2 window. One photogrammetric version has a vertically mounted AFA-41/7·5 camera in a TAU-M mount and an AFA-41/10 or AFA-41/20 camera in a fixed vertical mount. Another has the same installation without the gyroscopic mounting. A fourth variation offers an AFA-41/7·5 and an AFA-42/20, both in fixed mountings.

If required, the cameras can be replaced by other kinds of survey equipment, such as those used for mineral prospecting or for microwave radiometer survey, which measures the heat emission of land and ocean to obtain data on ocean surface characteristics, sea and lake ice, snow cover, flooding, seasonal vegetation changes, and soil types.

The power plant comprises two 2,103 kW (2,820 ehp) Ivchenko AI-24VT turboprop engines, with water injection, each driving an AV-72T four-blade constant-speed fully-feathering and reversible-pitch propeller. Main engines are supplemented by a 7·85 kN (1,765 lb st) RU 19A-300 auxiliary turbojet in the rear of the starboard engine nacelle. The latter is used for engine starting, and for take-off, climb and cruise power in the event of failure of the primary power plant. Max fuel capacity is 6,200 litres (1,364 Imp gallons).

Conversion into a transport aircraft is provided for, with cover plates to place over the camera apertures. The standard An-24 cabin door, on the port side of the fuselage at the rear, is retained, together with the standard forward freight compartment door on the starboard side and the load hoisting/conveying system.

The prototype An-30 flew for the first time in 1974. Operators include the Romanian Air Force, which has at least three.

Dimensions, external:

Wing span	29·20 m (95 ft 9½ in)
Wing aspect ratio	11·4
Length overall	24·26 m (79 ft 7 in)
Height overall	8·32 m (27 ft 3½ in)
Tailplane span	9·09 m (29 ft 10 in)
Fuselage, nominal diameter	2·90 m (9 ft 6¼ in)
Wheel track (c/l of oleos)	7·90 m (25 ft 11 in)
Wheelbase	7·65 m (25 ft 1¼ in)
Propeller diameter	3·90 m (12 ft 9½ in)
Propeller ground clearance	1·20 m (3 ft 11¼ in)

Areas:

Wings, gross	74·98 m² (807·1 sq ft)
Horizontal tail surfaces (total)	17·23 m² (185·46 sq ft)
Vertical tail surfaces (total, incl dorsal fin)	15·85 m² (170·61 sq ft)

Antonov An-30 aerial survey development of the An-24 twin-turboprop transport *(M. D. West)*

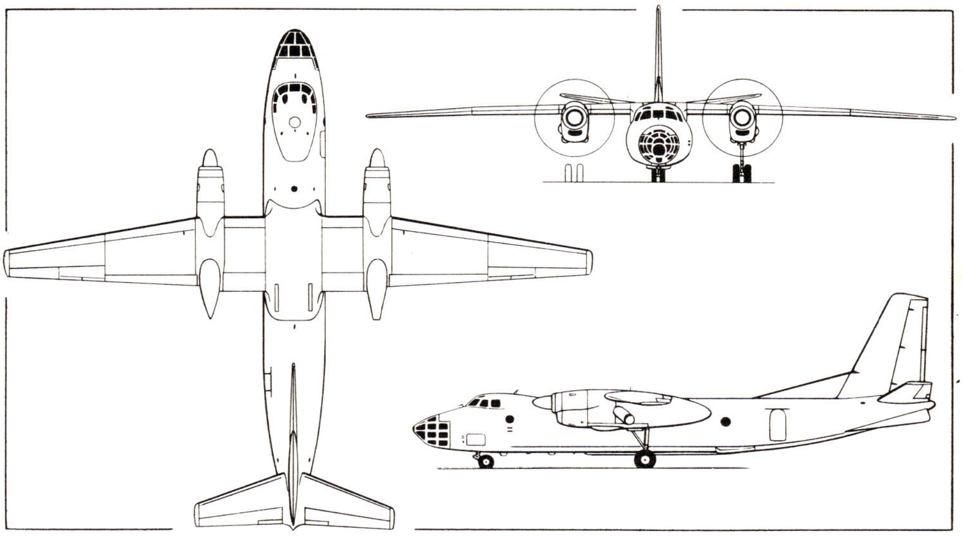

Antonov An-30 (two AI-24VT turboprops and RU 19A-300 auxiliary turbojet) *(Pilot Press)*

Weights:

Basic operating weight	15,590 kg (34,370 lb)
Weight of aerial photography equipment	650 kg (1,433 lb)
Max T-O and landing weight	23,000 kg (50,705 lb)

Performance:

Max level speed	291 knots (540 km/h; 335 mph)
Cruising speed at 6,000 m (19,685 ft)	232 knots (430 km/h; 267 mph)
Landing speed	95 knots (175 km/h; 109 mph)
Service ceiling with APU operating	8,300 m (27,230 ft)
Service ceiling without APU	7,300 m (23,950 ft)
T-O run on concrete	710 m (2,330 ft)
Landing run on concrete	670 m (2,198 ft)
Range with max fuel, no reserves	1,420 nm (2,630 km; 1,634 miles)

ANTONOV An-32

NATO reporting name: Cline

This pressurised short/medium-range transport, of which first details were released in May 1977, is a developed version of the An-26, intended for operation in high-temperature or high-altitude environments. Except for having much enlarged ventral fins and a full-span slotted tailplane, the airframe is generally similar to that of the An-26.

The An-32 is powered by two 3,862 kW (5,180 ehp) Ivchenko AI-20M turboprop engines, each driving a four-blade propeller. The large increase in power is intended specifically to improve take-off performance, service ceiling and payload. Thus, the An-32 is able to operate from airfields 4,000-4,500 m (13,125-14,750 ft) above sea level in an ambient temperature of ISA + 25°C, and can transport 3 tonnes of freight over a 594 nm (1,100 km; 683 mile) stage length, with fuel reserves. The overwing location of the engines requires nacelles of considerable depth, as the main landing gear units continue to retract into the underwing portions.

A rear loading hatch and forward-sliding ramp/door, similar to those of the An-26, are retained, as well as a hoist, capacity 2,000 kg (4,409 lb), to facilitate handling of the maximum payload of 6 tonnes of freight. The air-conditioned and pressurised accommodation enables alternative payloads to include 39 passengers or 30 parachutists, on a row of tip-up seats along each cabin wall, or 24 stretcher patients and a medical attendant; the normal crew of five comprises pilot, co-pilot, navigator, radio operator and flight engineer.

Low-pressure tyres (of the same sizes as those on the An-26) permit operation from unpaved strips; and the high position of the engines reduces the possibility of stone or debris ingestion. A TG-16M APU, housed in the rear of the starboard landing gear fairing, helps to make the An-32 independent of ground servicing equipment by providing onboard engine starting capability at airfields up to 4,500 m (14,750 ft) above sea level.

Antonov An-32, a version of the An-26 transport with a major increase in engine power for take-off from 'hot and high' airfields

Only a single prototype of the An-32 had been built by mid-1980. It was exhibited at the 1977 Paris Air Show; at the 1979 Show, Mr Antonov commented that production would be undertaken only if sufficient orders were received in advance to justify such a move. The first order was foreshadowed in December 1980, when India's Prime Minister, Mrs Indira Gandhi, stated in answer to a parliamentary question that An-32s would replace C-47s and C-119s in the Indian Air Force. It has been reported that as many as 95 aircraft could be involved, with assembly centred at the Kanpur plant of Hindustan Aeronautics, following delivery of an initial quantity from the Soviet Union.

DIMENSIONS, EXTERNAL AND INTERNAL:
As for An-26, except:
Tailplane span 10·01 m (32 ft 10 in)
Propeller diameter 4·70 m (15 ft 5 in)

WEIGHT:
Max T-O weight 26,000 kg (57,320 lb)

PERFORMANCE:
Normal cruising speed 275 knots (510 km/h; 317 mph)
Optimum cruising height 8,000 m (26,250 ft)
Service ceiling 9,500 m (31,150 ft)
Service ceiling, one engine out 5,000 m (16,400 ft)
T-O run 500 m (1,640 ft)
Range with max payload, 45 min reserves 432 nm (800 km; 497 miles)
Range with max fuel, 45 min reserves 1,188 nm (2,200 km; 1,367 miles)

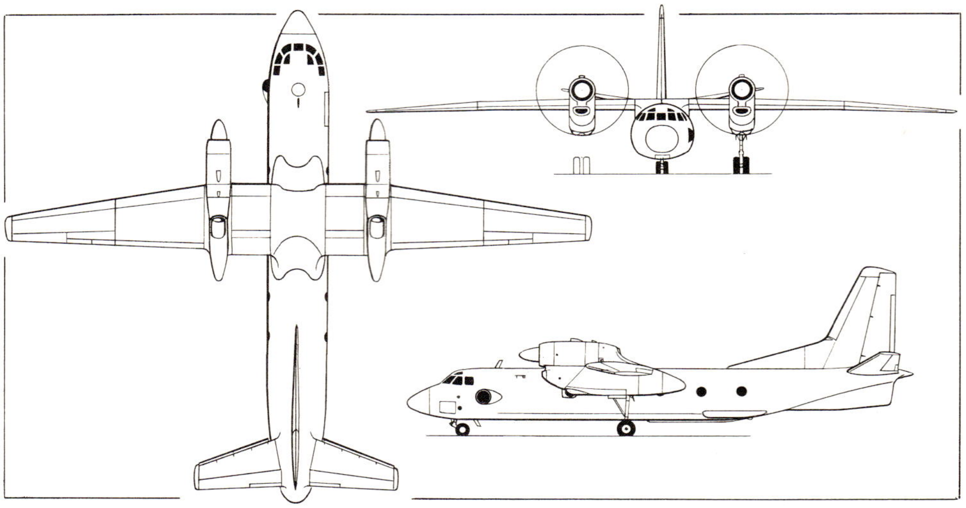

Antonov An-32 short/medium-range transport (two Ivchenko AI-20M turboprop engines) *(Pilot Press)*

ANTONOV An-72

NATO reporting name: Coaler

Two prototypes of this twin-turbofan STOL transport were built, plus a third airframe for static testing. First photographs of one of these aircraft (CCCP-19774) were released by the Soviet Tass news agency shortly after the first flight of an An-72 on 22 December 1977. By the time of the 1979 Paris Air Show the two An-72s had logged a total of just over 1,000 h in about 300 flights, and were described by Mr Antonov as "progressing faster than the An-28", which was itself to enter production at Mielec in Poland.

Design features of the An-72 make it uniquely suited to its primary role, as a STOL replacement for the turboprop An-26, with the emphasis on freight carrying. Its low-pressure tyres and multi-wheel landing gear enable it to operate from unprepared airfields, or from surfaces covered with ice or snow; and the high-set engines avoid problems caused by foreign object ingestion. The exhaust efflux is ejected over the upper surface of the wing and down over the very large multi-slotted flaps. By taking advantage of the so-called 'Coanda effect', which causes the airflow to 'attach to' the extended flaps, a considerable increase in lift can be achieved.

Aircraft exhibited to date in the West have been fitted with deflector doors on each side at the rear of each engine nacelle, to 'spread' the exhaust flow for maximum effectiveness during take-off and landing. Mr Antonov is not convinced that so small a transport has much to gain from such complications, which may be omitted from production An-72s.

To ensure optimum versatility and efficiency, the rear loading doors have undergone considerable refinement since 1977. The original prototype (see 1978-79 *Jane's*) had two large outward-canted ventral fins, one on each side at the rear of its ramp/door, aft of which the fuselage tailcone was made up of two petal airbrakes. By comparison, the An-72 (CCCP-83966) shown in Paris in 1979 had a flattened 'beaver-tail' aft of two outward-hinged clamshell doors and a simple downward-hinged ramp/door (see 1980-81 *Jane's*). Mr Antonov suggested that this arrangement might be superseded on production aircraft by a special ramp/door of the kind fitted to the An-26, which can be hinged down conventionally to allow wheeled or tracked vehicles to be driven into the hold or, alternatively, can slide forward under the fuselage to permit direct loading from a truck. Publicity material distributed at the 1981 Paris Air Show described and illustrated this sliding-ramp arrangement as standard (see accompanying illustration). It also stated that each main landing gear leg will carry twin wheels on production aircraft, compared with single wheels on early aircraft.

In addition to civil applications, the military potential of a transport able to utilise natural landing areas in undeveloped countries, or even small fields in Europe, is obvious. In particular, the An-72 might be an ideal aircraft with which to support operations by the new generation of V/STOL combat aircraft that will follow the Yak-36MP.

Particular care was taken to ensure easy handling of the An-72 in the air, and the designer commented that the aircraft had proved outstanding in this respect. Its Doppler-based automatic navigation system, linked to an onboard computer, is preprogrammed before take-off on a push-button panel to the right of the large cockpit map display. Failure warning panels above the windscreen display red lights for critical failures, yellow lights for non-critical failures, to minimise the time that needs to be spent on monitoring instruments and equipment.

The An-72 exhibited at the 1981 Paris Air Show (CCCP-19795) is believed to be from a pre-production series. The following details apply to the production version, as described in current publicity brochures from Aviaexport:

TYPE: Twin-turbofan light STOL transport.

WINGS: Cantilever high-wing monoplane, with 17° sweepback on leading-edges and straight trailing-edges. Multi-spar structure. Short constant-chord centre-section, without dihedral or anhedral, mounted above fuselage to avoid encroaching on internal space. Approx 10° anhedral on outer panels. Wing upper-surface blowing requires engines to be mounted above and forward of wings, to exhaust over upper surface. Aft of nacelles, wing skin, spoilers and flaps made of titanium. Hydraulically-actuated full-span leading-edge flaps outboard of nacelles, embodying thermal anti-icing. Wide-span trailing-edge flaps, double-slotted inboard in exhaust efflux, triple-slotted on outer panels. Normal T-O flap setting 25-30°; max deflection 60°. Five-section spoilers forward of flaps on each side; some sections opened automatically on touchdown by sensors actuated by weight on main landing gear. Conventional ailerons outboard of flaps, with tab in port aileron.

FUSELAGE: Conventional all-metal semi-monocoque structure of circular cross-section. Underside of upswept rear fuselage formed by ramp/door, which can hinge downward conventionally or slide forward under fuselage to permit direct loading of hold from truck. One-piece 'beaver-tail' bottom section (replacing original clamshell doors) drops and moves backward slightly to release ramp.

TAIL UNIT: Cantilever all-metal T tail, with wide-chord sweptback vertical surfaces and variable-incidence tailplane. Double-hinged rudder, with tab in lower portion of two-section aft panel. During normal flight only the lower rear segment is used. Both rear segments are used during low-speed flight. The forward segment is actuated automatically to offset thrust asymmetry. Tailplane leading-edge sweep slightly greater than that of wings, with straight trailing-edge on horn-balanced and aerodynamically-balanced elevators. Two tabs in each elevator. Leading-edges of fin and tailplane are de-iced. Tapered fairing forward of fin/tailplane junction. Two outward-canted ventral fins near extreme tip of fuselage (not on pre-series aircraft, but shown on latest Aviaexport drawing).

LANDING GEAR: Hydraulically-retractable tricycle type. Twin wheels on rearward-retracting steerable nose unit. Each main unit comprises two trailing-arm legs in tandem, each with twin wheels, retracting inward through 90° so that wheels lie horizontally in bottom of large fairings, outside fuselage pressure cell. Oleo-pneumatic shock-absorber in each unit. Low-pressure tyres, size 720 × 310 on nosewheels, 1050 × 400 on main wheels. Hydraulic disc brakes. Telescopic strut is hinged downward, from rear of each side fairing, to support fuselage during direct loading of hold with ramp/door under fuselage.

Second prototype of the Antonov An-72 twin-turbofan STOL transport *(Brian M. Service)*

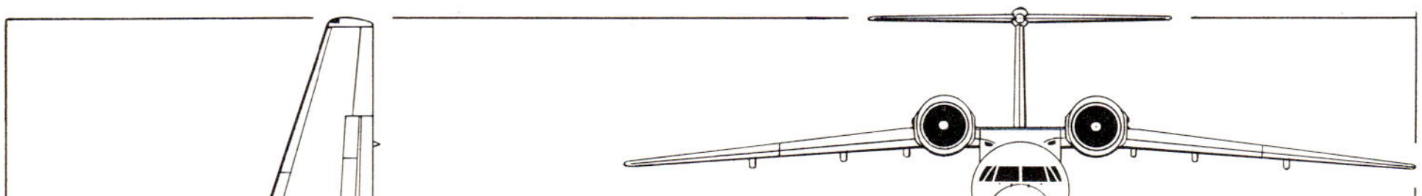

Second prototype of the Antonov An-72 (two Lotarev D-36 turbofan engines), with scrap view of rear fuselage of first prototype *(Pilot Press)*

Power Plant: Two Lotarev D-36 high bypass ratio turbofan engines, each rated at 63·74 kN (14,330 lb st). Integral fuel tanks between spars of outer wings. Thrust reversers standard.

Accommodation: Pilot and co-pilot/navigator side by side on very roomy flight deck, with heated windows. Two windscreen wipers. Flight engineer's seat, at rear on starboard side, slides forward on tracks to position between and slightly aft of pilots, to give access to controls on central console. Main cabin designed primarily for freight, but with folding seats for 32 passengers along side walls and provision for carrying 24 casualties and attendant in ambulance configuration. Flight deck and cabin pressurised and air-conditioned. Large downward-hinged and forward-sliding rear ramp/door for loading trucks and tracked vehicles, and for direct loading of hold from trucks. Mobile winch, capacity 2,500 kg (5,510 lb), assists loading of containers up to 1·90 × 2·44 × 1·46 m (6 ft 3 in × 8 ft × 4 ft 9½ in) in size, pallets 1·90 × 2·42 × 1·46 m (6 ft 3 in × 7 ft 11 in × 4 ft 9½ in) in size, and other bulky items. Cargo straps and nets are stowed in lockers on each side of hold when not in use. Provision for building roller conveyors into floor. Main crew and passenger door at front of cabin on port side. Small emergency exit and servicing door at rear of cabin on starboard side.

Systems: Air-conditioning system provides comfortable environment to altitude of 10,000 m (32,800 ft), with independent temperature control in flight deck and main cabin areas. It can be used to refrigerate main cabin when perishable goods are carried. Hydraulic system for actuating landing gear, flaps, ailerons, variable-incidence tailplane and ramp. Electrical system powers auxiliary systems, flight deck equipment, lighting and mobile hoist. Thermal de-icing system for leading-edges of wings and tail unit, engine air intakes and cockpit windows.

Avionics and Equipment: Large radome over navigation/weather radar in nose. Doppler-based automatic navigation system, with map display on flight deck.

Dimensions, external:
- Wing span 25·83 m (84 ft 9 in)
- Wing aspect ratio 7·4

This illustration from a 1981 brochure shows the rear ramp of the An-72 slid forward under the fuselage for direct loading. The landing gear retains the original single-wheel main units

- Length overall 26·576 m (87 ft 2¼ in)
- Height overall 8·235 m (27 ft 0¼ in)

Dimensions, internal:
- Main cabin: Length 9·00 m (29 ft 6¼ in)
  - Width at floor level 2·10 m (6 ft 10¾ in)
  - Height 2·20 m (7 ft 2½ in)

Area:
- Wings, gross approx 90 m² (969 sq ft)

Weights:
- Max payload: Normal 10,000 kg (22,045 lb)
  - STOL 3,500 kg (7,715 lb)
- Max T-O weight:
  - from 1,000 m (3,280 ft) runway 26,500 kg (58,420 lb)
  - from 1,200 m (3,935 ft) runway 30,500 kg (67,240 lb)
  - from 1,500 m (4,925 ft) runway 33,000 kg (72,750 lb)

Performance:
- Max level speed 410 knots (760 km/h; 472 mph)
- Max cruising speed 388 knots (720 km/h; 447 mph)
- T-O speed with light load 81 knots (150 km/h; 94 mph)
- T-O speed with heavier load 97 knots (180 km/h; 112 mph)
- Landing speed 89 knots (165 km/h; 103 mph)
- Service ceiling 11,000 m (36,100 ft)
- Normal operating height 8,000-10,000 m (26,250-32,800 ft)
- T-O run 470 m (1,542 ft)
- T-O run, one engine out 1,200 m (3,940 ft)
- Range with max payload, 30 min reserves 540 nm (1,000 km; 620 miles)
- Range with max fuel, 30 min reserves 2,050 nm (3,800 km; 2,360 miles)

Second prototype of the An-72, with modified rear fuselage and other changes, as shown at the 1979 Paris Air Show *(Tass)*

BERIEV

Georgi Mikhailovich Beriev, whose death at the age of 77 was reported in July 1979, was a graduate of the Leningrad Polytechnic Institute. He took up seaplane design in 1928 and became subsequently the best-known Soviet designer of water-based aircraft. He was appointed chief designer of the seaplane group at the TsKB (Central Design Bureau of Aviatrust) in 1930, and his first complete design, the single-engined MBR-2 flying-boat, was flown for the first time two years later, entering production in 1934. Other pre-war designs included the KOR-1 twin-float shipboard reconnaissance seaplane and the KOR-2 flying-boat, later redesignated Be-2 and Be-4 respectively.

In 1945 the Beriev bureau at Taganrog became the centre for all Soviet seaplane development, and the piston-engined Be-6 (first flown in 1947) was a standard military flying-boat during the 1950s. It was described in the 1959-60 *Jane's*. Only limited production was undertaken of the Be-8, also flown in 1947; and Beriev's next major flying-boat was the sweptwing twin-jet Be-10, based on the Be-R-1 prototype of 1949. A more recent maritime aircraft of Beriev design is the M-12 (Be-12) twin-turboprop amphibian, which is in standard service with Soviet Naval Air Force units as a successor to the Be-6.

BERIEV M-12 (Be-12) TCHAIKA (SEAGULL)

NATO reporting name: Mail

This twin-turboprop medium-range maritime reconnaissance amphibian was displayed for the first time in the 1961 Aviation Day flypast at Tushino Airport, Moscow. Subsequently, during the period 23-27 October 1964, it established six officially-recognised international height records in Class C3 Group II. Data submitted in respect of these records revealed that the designation of the aircraft was M-12 and the power plant two 4,000 shp Ivchenko AI-20D turboprop engines. The aircraft was also, clearly, able to lift a payload of around 10 tons under record conditions.

The records set by the M-12 in 1964 were altitude of 12,185 m (39,977 ft) without payload, altitude of 11,366 m (37,290 ft) with payloads of 1,000 kg and 2,000 kg, altitude of 10,685 m (35,055 ft) with 5,000 kg payload, altitude of 9,352 m (30,682 ft) with 10,000 kg payload, and maximum payload of 10,100 kg (22,266 lb) lifted to a height of 2,000 m (6,560 ft).

These 1964 records have never been bettered. Subsequent record attempts have ensured that the M-12 retains all 21 FAI records listed in Class C3 Group II for turboprop-powered amphibians, and all 19 current records in Class C2 Group II, for turboprop flying-boats. A list of these records can be found in the 1979-80 *Jane's*, except for two confirmed more recently, as follows: a C2/II sustained altitude record of 9,493 m (31,145 ft) set by E Kolkov on 2 November 1978; and a height record of 9,407 m (30,863 ft) with 2,000 kg payload set by N. Chkourko on 6 November 1978.

When three M-12s took part in the 1967 air display at Domodedovo, the commentator said that the unit to which they belonged was "one of those serving where the country's military air force began", implying that the aircraft were then in operational service. M-12s have since been identified in standard service at Soviet Northern and Black Sea Fleet air bases and were operational for a period from bases in Egypt, in Egyptian insignia. About 100 are believed to have been built, of which about 80 remain in service.

Type: Twin-turboprop maritime patrol amphibian.

Wings: Cantilever high-wing monoplane of sharply-cranked configuration to raise propellers clear of water. Unswept constant-chord centre-section; tapered outer panels. Two tabs in each aileron. Trailing-edge flaps in two sections on each wing, from aileron to centre-section (passing under engine) and on centre-section.

Fuselage: Single-step all-metal semi-monocoque hull of high length-to-beam ratio. Two long strakes, one above the other, on each side of front fuselage to prevent spray from enveloping the propellers at take-off.

Tail Unit: Two endplate fins and horn-balanced rudders at tips of 'dihedral' tailplane. Trim tab in each elevator and rudder.

Landing Gear: Retractable tailwheel type, comprising single-wheel main units which retract upward through 180° to lie flush within sides of hull, and a rearward-retracting tailwheel. Except for top of each main wheel, all units are fully enclosed by doors when retracted. Non-retractable wingtip floats.

Power Plant: Two Ivchenko AI-20D turboprop engines, each rated at 3,124 kW (4,190 ehp) and driving a four-blade propeller. Cowlings open downward in halves, permitting their use as servicing platforms.

Accommodation: Pilot and co-pilot side by side on flight deck. Glazed navigation and observation station in nose. Astrodome type of observation station in top of rear fuselage. Side hatches in rear fuselage permit loading while afloat.

Avionics and Equipment: Radome above nose glazing. MAD (magnetic anomaly detection) 'sting' extends rearward from tail. APU exhausts through aperture in port side of rear fuselage.

Armament: Internal weapons bay in bottom of hull aft of step. One large and one smaller external stores pylon under each outer wing panel.

Dimensions, external:

Wing span	29·71 m (97 ft 5¾ in)
Length overall	30·17 m (99 ft 0 in)
Height overall	7·00 m (22 ft 11½ in)
Propeller diameter	4·85 m (16 ft 0 in)

Area:

Wings, gross	105 m^2 (1,130 sq ft)

Weight:

Max T-O weight	29,450 kg (64,925 lb)

Performance:

Max level speed	328 knots (608 km/h; 378 mph)
Normal operating speed	172 knots (320 km/h; 199 mph)
Rate of climb at S/L	912 m (2,990 ft)/min
Service ceiling	11,280 m (37,000 ft)
Max range	2,158 nm (4,000 km; 2,485 miles)

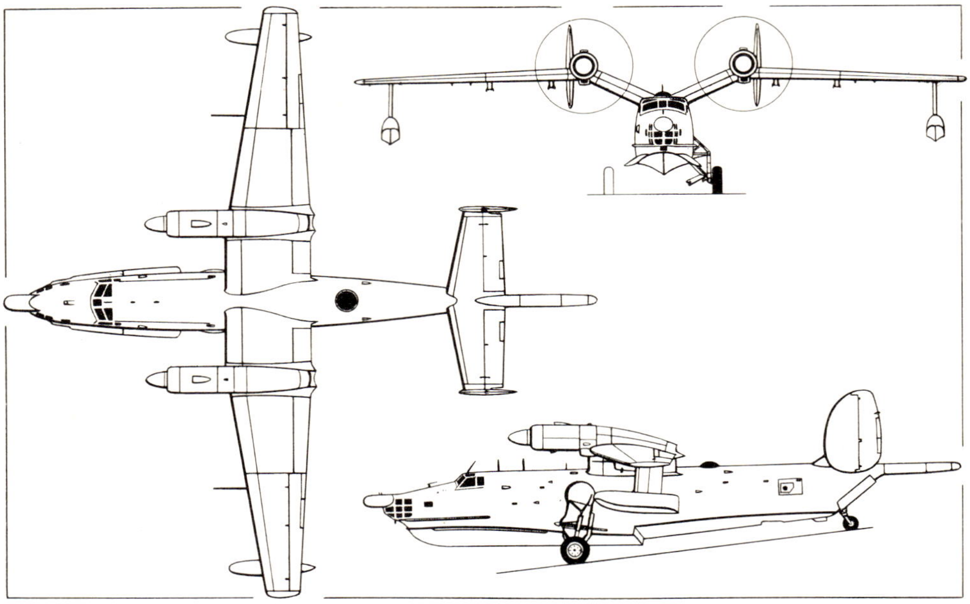

Beriev M-12 (Be-12) Tchaika twin-turboprop maritime reconnaissance amphibian *(Pilot Press)*

Beriev M-12 (Be-12) Tchaika maritime patrol amphibian flying-boat of the Soviet Naval Air Force *(Swedish Air Force)*

ILYUSHIN

Design Bureau Headquarters: Moscow Central Airport, Khodinka, Moscow

General Designer: G. Novozhilov

This design bureau is named after its former leader, Sergei Vladimirovich Ilyushin, who died on 9 February 1977, at the age of 83. Ilyushin began his aviation career as a member of the ground staff at an airport. One of the aircraft he designed in the 'thirties made a record flight from Moscow to North America, but he is best remembered for his Il-2 and Il-10 Shturmovik close support aircraft of the second World War. He was awarded the Order of Lenin and a third Hammer and Sickle gold medal on 29 March 1974, in recognition of his service in the development of aviation technology and the Soviet aircraft industry.

Aircraft designed by Ilyushin and currently in service include the Il-14 piston-engined light transport and four-turboprop Il-18 transport, of which details have been given in earlier editions of *Jane's*, and the Il-28 twin-jet bomber, still in production in China (see pages 32-33). More recent types from the Ilyushin bureau include an anti-submarine variant of the Il-18, designated Il-38; a four-turbofan rear-engined airliner known as the Il-62; the Il-76 turbofan-engined heavy freighter, AWACS and flight refuelling tanker aircraft; and a large wide-bodied transport designated Il-86.

ILYUSHIN Il-18

NATO reporting name: Coot

The Il-18 prototype flew for the first time on 4 July 1957 and production models entered service with Aeroflot in 1959. In the following 20 years they logged 12 million flying hours with Aeroflot, carrying 235 million passengers on 5·5 million flights. Production is believed to have exceeded 700 aircraft, of which more than 100 were exported for use by commercial airlines. Known civilian operators were listed in the 1975-76 *Jane's*. Military operators of Il-18s include the air forces of Afghanistan, Algeria, Bulgaria, China, Czechoslovakia, German Democratic Republic, Guinea, North Korea, Romania, the Soviet Union, Syria, Viet-Nam and Yugoslavia, mostly in small numbers.

Aeroflot is having some of its former passenger-carrying Il-18s converted into **freighters** by Factory 402 at Moscow/Bykovo. Modifications include installation of a freight door 3·50 m (11 ft 6 in) wide in the rear fuselage and a strengthened cabin floor.

This veteran Ilyushin Il-14 (NATO reporting name 'Crate') is equipped for ECM/elint duties with the Soviet Air Forces

An anti-submarine derivative, the **Il-38** (NATO reporting name 'May'), is in service and is described separately. Another military variant of the Il-18, seen for the first time in 1978, is the ECM or elint aircraft known to NATO as **Coot-A** and shown in an accompanying illustration. In this case, the airframe appears to be basically unchanged by comparison with the transport. It carries under its fuselage a container about 10·25 m long and 1·15 m deep (33 ft 7½ in × 3 ft 9 in), which is assumed to house side-looking radar. There is a further container, about 4·4 m long and 0·88 m deep (14 ft 5 in × 2 ft 10½ in) on each side of the forward fuselage, containing a door over a camera or other sensor. Numerous other antennae and blisters can be seen, about eight of them on the undersurface of the centre and rear fuselage, with two large plates projecting above the forward fuselage. Further variants of the Il-18, adapted for military support tasks, will probably appear as the airliners are replaced by jets.

A detailed description of the commercial airliner versions of the Il-18 can be found in the 1979-80 and earlier editions of *Jane's*. The following abbreviated details of the Il-18D are retained as an indication of likely features of the military 'Coot-A':

Wings: Cantilever low-wing monoplane. Mean thickness/chord ratio 14%. All-metal structure. Three spars in centre-section, two in outer wings. All-metal ailerons are mass-balanced and aerodynamically-compensated, and fitted with spring tabs. Manually-operated flying controls. Hydraulically-actuated double-slotted flaps. Electro-thermal de-icing.

Fuselage: Circular-section all-metal monocoque structure. The structure is of the fail-safe type, and appears to employ rip stop doublers around window cutouts, door frames and the more-heavily loaded skin panels.

Tail Unit: Cantilever all-metal structure. Trim tabs in rudder and elevators. Additional spring tab in rudder. Manually-operated flying controls. Electro-thermal de-icing.

Landing Gear: Retractable tricycle type. Hydraulic actuation. Four-wheel bogie main units, with 930 mm × 305 mm tyres and hydraulic brakes. Steerable (45° each way) twin nosewheel unit, with 700 mm × 250 mm

ECM or electronic intelligence (elint) version of the Ilyushin Il-18 (NATO 'Coot-A') *(Royal Air Force)*

tyres. Tyre pressures: main 7·86 bars (114 lb/sq in), nose 5·86 bars (85 lb/sq in). Hydraulic brakes and nosewheel steering. Pneumatic emergency braking.

Power Plant: Four 3,169 kW (4,250 ehp) Ivchenko AI-20M turboprop engines, each driving an AV-681 four-blade reversible-pitch propeller. Ten flexible bag-type fuel tanks in inboard panel of each wing and integral tank in outboard panel, with a total capacity of 23,700 litres (5,213 Imp gallons). Some Il-18 airliners have additional bag tanks in centre-section, giving a total capacity of 30,000 litres (6,600 Imp gallons). Pressure fuelling through four international standard connections in inner nacelles. Provision for overwing fuelling. Oil capacity 58·5 litres (12·85 Imp gallons) per engine.

Accommodation: Airliner was designed for a crew of five, comprising two pilots, navigator, radio operator and flight engineer. Flight deck is separated from remainder of fuselage by a pressure bulkhead to reduce the hazards following a sudden decompression of either.

Systems: Cabin pressurised to max differential of 0·49 bars (7·1 lb/sq in). Electrical system includes eight 12kW DC generators and 28·5V single-phase AC inverters. Hydraulic system, pressure 207 bars (3,000 lb/sq in), for landing gear actuation, nosewheel steering, brakes and flaps.

Dimensions, external:

Wing span	37·4 m (122 ft 8½ in)
Wing chord at root	5·61 m (18 ft 5 in)
Wing chord at tip	1·87 m (6 ft 2 in)
Wing aspect ratio	10
Length overall	35·9 m (117 ft 9 in)
Height overall	10·17 m (33 ft 4 in)
Tailplane span	11·8 m (38 ft 8½ in)
Wheel track	9·0 m (29 ft 6 in)
Wheelbase	12·78 m (41 ft 10 in)
Propeller diameter	4·50 m (14 ft 9 in)
Cabin doors (each): Height	1·40 m (4 ft 7 in)
Width	0·76 m (2 ft 6 in)
Height to sill	2·90 m (9 ft 6 in)

Dimensions, internal:

Flight deck: Volume	9·36 m³ (330 cu ft)
Cabin, excl flight deck:	
Length	approx 24·0 m (79 ft 0 in)
Max width	3·23 m (10 ft 7 in)
Max height	2·00 m (6 ft 6¾ in)
Volume	238 m³ (8,405 cu ft)

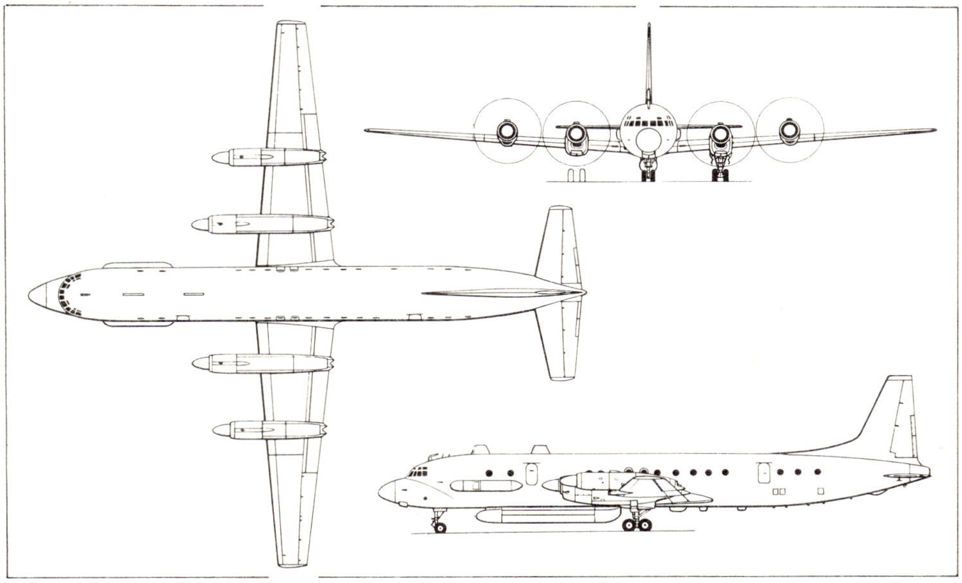

'Coot-A' elint development of the Ilyushin Il-18 airliner *(Pilot Press)*

Areas:

Wings, gross	140 m² (1,507 sq ft)
Ailerons (total)	9·11 m² (98·05 sq ft)
Trailing-edge flaps (total)	27·15 m² (292·2 sq ft)
Vertical tail surfaces (total)	17·93 m² (193·0 sq ft)
Rudder	6·83 m² (73·52 sq ft)
Horizontal tail surfaces (total)	27·79 m² (299·13 sq ft)
Elevators (total)	11·80 m² (127·0 sq ft)

Weights (Il-18D airliner):

Max payload	13,500 kg (29,750 lb)
Max T-O weight	64,000 kg (141,100 lb)

Performance (Il-18D airliner, at max T-O weight):

Max cruising speed	364 knots (675 km/h; 419 mph)
Econ cruising speed	337 knots (625 km/h; 388 mph)
Operating height	8,000-10,000 m (26,250-32,800 ft)
T-O run	1,300 m (4,265 ft)
Landing run	850 m (2,790 ft)
Range with max fuel, 1 h reserves	3,508 nm (6,500 km; 4,040 miles)

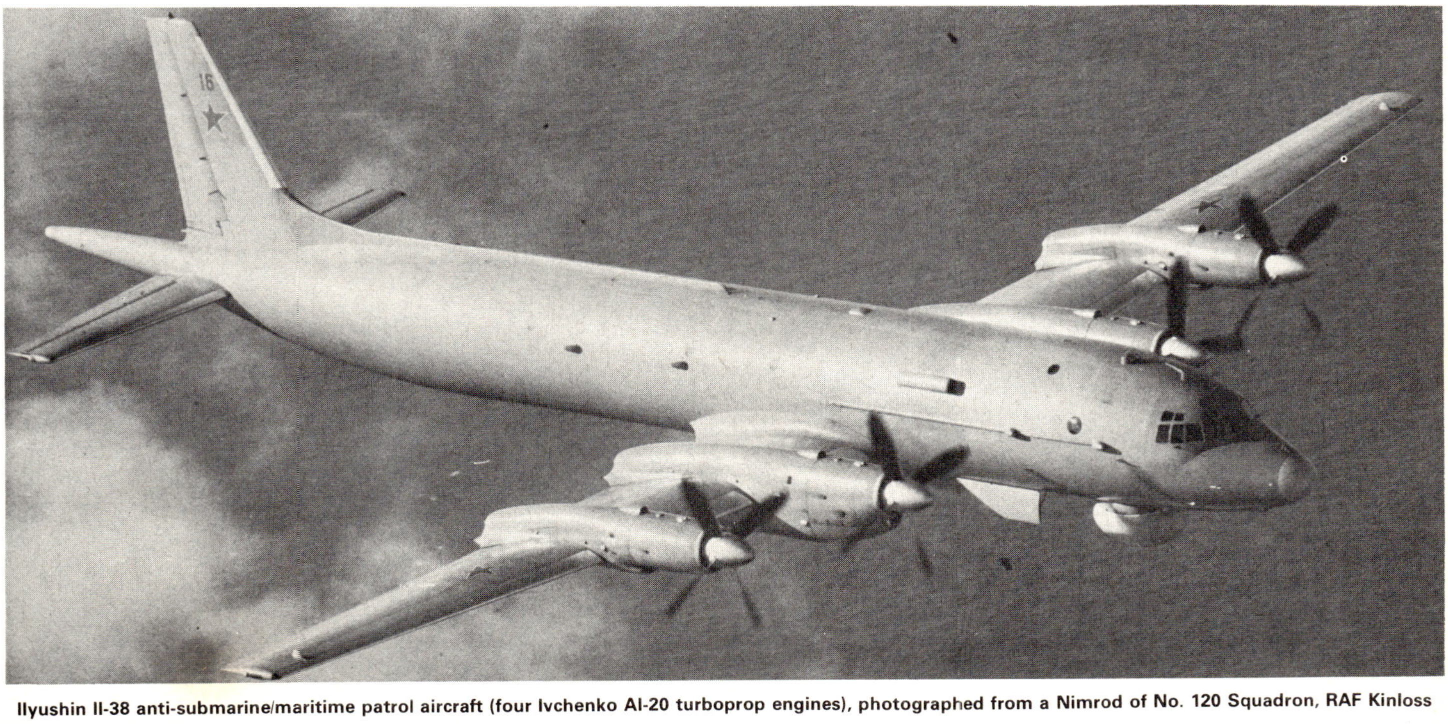

Ilyushin Il-38 anti-submarine/maritime patrol aircraft (four Ivchenko AI-20 turboprop engines), photographed from a Nimrod of No. 120 Squadron, RAF Kinloss

Range with max payload, 1 h reserves
1,997 nm (3,700 km; 2,300 miles)

ILYUSHIN Il-38

NATO reporting name: May

This anti-submarine/maritime patrol development of the Il-18 airliner represents a conversion similar to that by which the US Navy's P-3 Orion was developed from the Lockheed Electra transport. It has a lengthened fuselage fitted with an undernose radome similar in shape to that of the Ka-25 ASW helicopter but housing a different radar, an MAD tail 'sting', other specialised electronic equipment, and a weapon-carrying capability.

The main cabin of the Il-38 has few windows. The complete wing assembly is much further forward than on the Il-18, to cater for the effect on the CG position of internal equipment and stores. A crew of 12 is reported to be carried.

The Il-38 is a standard shore-based maritime patrol aircraft of the Soviet Naval Air Force, operating widely over the Atlantic and Mediterranean. In the latter area, some aircraft carried Egyptian Air Force insignia for a period, but are believed to have been manned by Soviet aircrew, operating from North African bases such as Matru, near Cairo. More recently, Il-38s of the Soviet Navy have operated over the Indian Ocean from an airfield in the People's Democratic Republic of Yemen.

In 1975, the Indian Navy ordered an initial batch of three refurbished ex-Soviet Navy Il-38s, of which delivery began in 1977, to equip INAS 315 at Dabolim, Goa. About 60 are believed to be operational with Soviet naval units.

Dimensions, external:
Wing span 37·4 m (122 ft 8½ in)
Length overall 39·6 m (129 ft 10 in)
Height overall 10·17 m (33 ft 4 in)
Performance (estimated):
Max cruising speed at 8,230 m (27,000 ft)
347 knots (645 km/h; 400 mph)
Max range 3,910 nm (7,240 km; 4,500 miles)

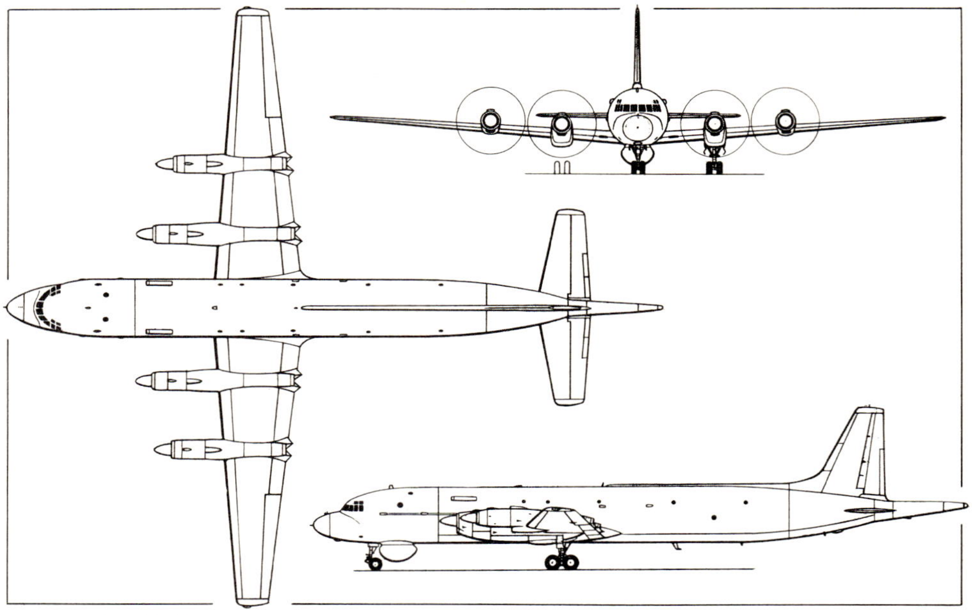

Ilyushin Il-38 anti-submarine/maritime patrol derivative of the Il-18 airliner *(Pilot Press)*

ILYUSHIN Il-62

NATO reporting name: Classic

Announced on 24 September 1962, when the first prototype (CCCP-06156) was inspected by the late Premier Krushchev, the standard Il-62 is a long-range airliner, with four Kuznetsov turbofan engines mounted in horizontal pairs on each side of the rear fuselage. It accommodates up to 186 passengers and was designed to fly on ranges equivalent to Moscow-New York (about 4,155 nm; 7,700 km; 4,800 miles) with more than 150 passengers and reserve fuel.

The Kuznetsov engines were not ready in time for the first flight of the first prototype, which took place in January 1963 with four 73·55 kN (16,535 lb st) Lyulka AL-7 engines installed. This aircraft was followed by a second prototype and three pre-production aircraft. Series production then started at Kazan, and Aeroflot introduced the Il-62 on to its Moscow-Montreal service on 15 September 1967, as a replacement for the Tu-114.

The Il-62 inaugurated Aeroflot's Moscow-New York service in July 1968, and has been used subsequently on many other routes, including Moscow-Paris and Moscow-Tokyo. Production is reported to have totalled at least 165 (including Il-62M/MKs), with export sales to CSA Czechoslovakian Airlines (8), Interflug of East Germany (6), Polish Airlines LOT (8), CAAC of China (5), Tarom of Romania (5), Cubana (1) and the Czechoslovakian government for VIP operation (1).

The Il-62's automatic flight control system is capable of taking over from a height of 200 m (660 ft) after take-off to a similar height during the landing approach. It can maintain a predetermined speed during climb and descent, and a selected cruising height, and can follow automatically a programmed track under command of the navigation computer.

The Il-62 is designed for an airframe service life of 25,000-30,000 flying hours, including 7,000-8,000 take-offs and landings.

A re-engined version, designated Il-62M, was flown for the first time in 1971. Details of this aircraft, and the further developed Il-62MK, are given separately, after the following description of the standard Il-62:

Type: Four-turbofan long-range airliner.

Wings: Cantilever low-wing monoplane. Sweepback 32° 30′ at quarter-chord. Extended-chord leading-edge on outer two-thirds of each wing. All-metal structure, with four spars inboard, two at tip. Removable leading-edge. Each wing fitted with three-section manually-operated ailerons, electrically-actuated slotted flaps and two hydraulically-operated spoiler sections forward of flaps. Trim tab and spring-loaded servo tab in each centre aileron, spring-loaded servo tab in each inner aileron. Hot-air anti-icing of leading-edges.

Fuselage: Conventional all-metal semi-monocoque structure. Frames are duralumin stampings and pressings. Integrally pressed skin panels at highly-stressed areas. Floors are sandwich panels with foam plastics filler. Nosecone hinges upward for access to radar.

Tail Unit: Cantilever all-metal structure, with electrically-actuated variable-incidence T tailplane. All surfaces sweptback. Manually-operated rudder, fitted with yaw damper, trim tab and spring servo tab. Manually-operated elevators have two automatic trim tabs and two manual trim tabs. Hot-air leading-edge anti-icing system.

Landing Gear: Hydraulically-retractable tricycle type. Forward-retracting twin-wheel steerable nose unit. Emergency extension by gravity. Oleo-nitrogen shock-absorber in each unit. Each main unit carries a four-wheel bogie and retracts inward into wing-roots. Main-wheel tyres size 1450 × 450, pressure 9·31 bars (135 lb/sq in). Nosewheel tyres size 930 × 305, pressure 7·86 bars (114 lb/sq in). Hydraulic disc brake and inertia-type electric anti-skid unit on each main wheel, supplemented by large tail parachute. Parking brakes. Hydraulic twin-wheel strut is extended downward to support rear fuselage during loading and unloading.

Power Plant: Four Kuznetsov NK-8-4 turbofan engines, each rated at 103 kN (23,150 lb st), mounted in horizontal pairs on each side of rear fuselage. Thrust reverser on each outboard engine. Hot-air anti-icing system for engine intakes. Automatically-controlled fuel system, with seven integral tanks, three in wing centre-section, two in each outer panel. Each engine has its own independent fuel system, with cross-feed. Total fuel capacity 100,000 litres (21,998 Imp gallons). Four standard international underwing pressure refuelling sockets. Eight gravity refuelling sockets. Total oil capacity 204 litres (45 Imp gallons).

Accommodation: Crew of five (two pilots, navigator, radio operator and flight engineer) on flight deck. Provision for two supernumerary pilot/navigators. Basic two-cabin layout, and galley, toilet and wardrobe facilities, are unchanged in the three main versions, only the width and pitch of the seats being varied. In the 186-passenger version, there are 72 seats in the forward cabin and 114 in the rear cabin, all six-abreast and all at a seat pitch of 86 cm (34 in). In the 168-seat configuration, increased pitch reduces capacity to 66 in the forward cabin and 102 in the rear cabin. The 114-passenger version has 45 seats in the forward cabin and 69 in the rear cabin, all five-abreast, except for four-abreast rear row by door. A first class/de luxe version for 85 passengers is available, with 45 seats in forward cabin and 40 four-abreast sleeperette chairs with footrests in rear cabin. Passenger doors forward of front cabin and between cabins on port side. Total of five toilets, opposite forward door, between cabins (starboard) and aft of rear cabin (both sides). Electrically-powered galley/pantry amidships and wardrobes in each version. Pressurised baggage and freight compartments under cabin floor, forward and aft of wing. Unpressurised baggage/cargo compartment at extreme rear of fuselage. All compartments have tiedown fittings and rails in floor, and removable nets to restrain cargo.

Systems: Air-conditioning and pressurisation system maintains sea level conditions up to 7,000 m (23,000 ft) and gives equivalent of 2,100 m (6,900 ft) at 13,000 m (42,600 ft). Pressure differential 0·62 bars (9·0 lb/sq in). Hydraulic system, pressure 207 bars (3,000 lb/sq in), for landing gear retraction, nosewheel steering, brakes, spoilers and windscreen wipers. Emergency hydraulic system, powered by electric motor, for nosewheel steering, main-wheel extension and spoiler control. Three-phase 200/115V AC electrical supply from four 40kVA engine-driven generators (optional 27V DC system with eight 18kW engine-driven generators). Four transformer-rectifiers and four batteries for DC supply. Electrical windscreen de-icing. Type TA-6 APU in tail-cone.

Avionics: Standard avionics include two-channel autopilot, navigation computer, air data system, HF and UHF radio, VOR/ILS, RMI, Doppler, radio altimeter and weather radar. Polyot automatic flight control system optional.

Dimensions, external:
Wing span 43·20 m (141 ft 9 in)
Length overall 53·12 m (174 ft 3½ in)
Length of fuselage 49·00 m (160 ft 9 in)
Height overall 12·35 m (40 ft 6¼ in)
Tailplane span 12·23 m (40 ft 1½ in)
Fuselage height 4·10 m (13 ft 5½ in)
Fuselage width 3·75 m (12 ft 3½ in)
Wheel track 6·80 m (22 ft 3½ in)
Wheelbase 24·49 m (80 ft 4½ in)
Passenger doors (each): Height 1·83 m (6 ft 0 in)
Width 0·86 m (2 ft 9¾ in)
Height to sill 3·55 m (11 ft 8 in)
Emergency exit (galley service) door:
Height 1·38 m (4 ft 6¼ in)
Width 0·61 m (2 ft 0 in)
Emergency exits (overwing):
Height 0·91 m (2 ft 11¾ in)
Width 0·51 m (1 ft 8 in)
Front cargo hold door: Height 1·31 m (4 ft 3½ in)
Width 1·26 m (4 ft 1½ in)
Height to sill 1·90 m (6 ft 3 in)
Second cargo hold door: Height 1·00 m (3 ft 3¼ in)
Width 1·26 m (4 ft 1½ in)
Height to sill 1·90 m (6 ft 3 in)
Third cargo hold door: Height 0·70 m (2 ft 3½ in)
Width 0·70 m (2 ft 3½ in)
Height to sill 2·26 m (7 ft 5 in)
Rear cargo hold door: Height 1·15 m (3 ft 9 in)
Width 1·07 m (3 ft 6 in)
Height to sill 3·68 m (12 ft 0¾ in)
Dimensions, internal:
Cabin: Max height 2·12 m (6 ft 11½ in)
Max width 3·49 m (11 ft 5¼ in)
Volume 163 m^3 (5,756 cu ft)
Total volume of pressure cell 396 m^3 (13,985 cu ft)
Cargo hold volume: Front 22·7 m^3 (801 cu ft)
Second 12·6 m^3 (445 cu ft)
Third 6·9 m^3 (243 cu ft)
Rear 5·8 m^3 (205 cu ft)
Areas:
Wings, gross 279·55 m^2 (3,009 sq ft)
Ailerons (total) 16·25 m^2 (174·9 sq ft)
Spoilers (total) 9·54 m^2 (102·7 sq ft)
Flaps (total) 43·48 m^2 (468·0 sq ft)
Horizontal tail surfaces (total) 40·00 m^2 (430·5 sq ft)
Vertical tail surfaces (total) 35·60 m^2 (383·2 sq ft)
Weights and Loading:
Weight empty 66,400 kg (146,390 lb)
Operating weight, empty 69,400 kg (153,000 lb)
Max payload 23,000 kg (50,700 lb)
Max fuel 83,325 kg (183,700 lb)
Max ramp weight 167,000 kg (368,000 lb)
Max T-O weight 162,000 kg (357,150 lb)
Max landing weight 105,000 kg (231,500 lb)
Max zero-fuel weight 93,500 kg (206,130 lb)
Max wing loading 572 kg/m^2 (117·2 lb/sq ft)
Performance (at max T-O weight):
Normal cruising speed
442-486 knots (820-900 km/h; 510-560 mph)
Normal cruising height
10,000-12,000 m (33,000-39,400 ft)
Landing speed
119-129 knots (220-240 km/h; 137-149 mph)
Max rate of climb at S/L 1,080 m (3,540 ft)/min

Ilyushin Il-62M long-range airliner (four Soloviev D-30KU turbofan engines) of Cubana *(Harold G. Martin)*

FAR T-O field length:
- ISA at S/L — 3,250 m (10,660 ft)
- ISA+20°C at S/L — 3,915 m (12,840 ft)

FAR landing field length:
- ISA at S/L — 2,800 m (9,185 ft)
- ISA+20°C at S/L — 2,950 m (9,680 ft)

Range with max payload, 66,700 kg (147,050 lb) fuel, 1 h fuel reserves 3,612 nm (6,700 km; 4,160 miles)

Range with 80,000 kg (176,370 lb) fuel and 10,000 kg (22,045 lb) payload, 1 h fuel reserves 4,963 nm (9,200 km; 5,715 miles)

ILYUSHIN Il-62M/MK

First displayed publicly at the 1971 Paris Air Show, the **Il-62M** is a developed version of the Il-62, with no dimensional changes to the airframe. It is fitted with more powerful turbofans, of a different type, with clamshell thrust reversers on the outboard engine of each pair, offering a lower approach speed and improved airflow over the rear of the nacelles. An additional fuel tank is installed in the tail-fin, contributing (with the improved specific fuel consumption of the engines) to the longer range of this version.

Revised layout of the flight deck equipment, and new navigation and radio communications equipment, are features of the Il-62M. Control wheels of new design allow the pilots a better field of view, and the aircraft's automatic flight control system permits automatic landings in ICAO Category II conditions, with extension to Category III conditions envisaged later. The wing spoilers of this version can be utilised differentially to enhance roll control.

Additional emergency and rescue equipment is installed on the Il-62M. Unlike the Il-62, it has a containerised baggage and freight system, with mechanised loading and unloading.

The Il-62M exhibited in Paris in 1971 and 1973 was the prototype (CCCP-86673). Production models entered service on Aeroflot's Moscow-Havana route in 1974 and took over progressively all of the airline's very-long-distance services. Six have been acquired by Cubana de Aviacion and three by the East German Interflug.

A variant announced in 1978 is the **Il-62MK**, still dimensionally unchanged and with the same power plant as the Il-62M, but with strengthened wings. Max T-O weight is increased to 167,000 kg (368,170 lb) and max landing weight to 110,000 kg (242,500 lb), permitting the carriage of up to 195 passengers. To ensure adequate cabin service with so many passengers, the interior has been redesigned to permit the more efficient use of service trolleys. Range with max fuel and 10,000 kg (22,045 lb) payload is 5,180 nm (9,600 km; 5,965 miles). Max payload is 25,000 kg (55,115 lb).

The basic structural description of the Il-62 applies also to the Il-62M. The main innovations are as follows:

Power Plant: Four Soloviev D-30KU turbofan engines, each rated at 107·9 kN (24,250 lb st), mounted in horizontal pairs on each side of rear fuselage. Clamshell-type thrust reverser on each outboard engine. Remainder of power plant installation basically as for Il-62, but additional fuel tank in tail-fin, giving total capacity of 105,300 litres (23,162 Imp gallons).

Accommodation: Alternative configurations for up to 174 economy class, 168 tourist class or 140 mixed class passengers. In the basic tourist class version there are two toilets opposite the forward door, on the starboard side, aft of the flight deck. The forward cabin contains 66 seats, all six-abreast in threes with centre aisle. Galley/pantry, coat stowage and toilet amidships. Rear cabin contains 102 seats, six-abreast in threes with centre aisle. Two toilets and wardrobe to rear of this cabin. Doors as on Il-62. Two emergency exits on each side, over wing. Forward underfloor baggage and freight hold accommodates nine containers, each weighing approximately 45 kg (100 lb) empty and with a capacity of 600 kg (1,322 lb) and 1·6 m³ (56·5 cu ft). Rear hold accommodates five similar containers. Two compartments for non-containerised cargo. Total baggage and freight capacity 48 m³ (1,695 cu ft)

Systems and Equipment: See introductory notes.

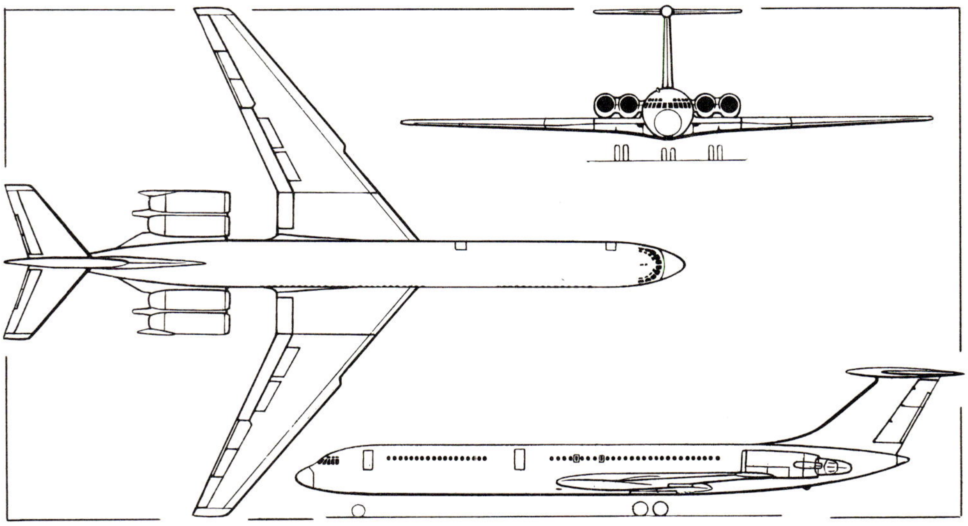

Ilyushin Il-62M long-range four-turbofan transport *(Pilot Press)*

Avionics: Duplicated SAU-1T automatic flight control system provides for automatic control from a height of 200 m (660 ft) after take-off to a height of 30 m (100 ft) on the approach to land. DISS-013 Doppler indicator and NV-PB-1 navigation computer. TKS-P course sensing system. TsGV-10P vertical master gyros. SVS-PN-15 air data system. Kurs-MP-2 radio navigation system, utilising VOR, ILS or SP-50 beacons. GROZA radar. SD-67 DME. ARK-15 ADF. RV-5 radio altimeter. SO-70 IFF transponder. MIKRON 2-24MHz HF radio. LANDASH 118-135MHz VHF radio. VESHANIE public address and in-flight entertainment system.

Dimensions and Areas:
- Same as for Il-62

Weights (Il-62M):
- Max payload — 23,000 kg (50,700 lb)
- Max T-O weight — 165,000 kg (363,760 lb)
- Max landing weight — 105,000 kg (231,500 lb)
- Max zero-fuel weight — 94,600 kg (208,550 lb)

Performance (Il-62M, at max T-O weight):
- Normal cruising speed — 442-486 knots (820-900 km/h; 509-560 mph)
- Normal cruising height — 10,000-12,000 m (33,000-39,400 ft)
- Balanced T-O distance (ISA, S/L) 3,300 m (10,830 ft)
- Landing run (ISA, S/L) — 2,500 m (8,200 ft)
- Range with max payload, with 5,100 kg (11,240 lb) fuel reserves — 4,210 nm (7,800 km; 4,846 miles)
- Range with 10,000 kg (22,045 lb) payload, with reserves — 5,400 nm (10,000 km; 6,215 miles)

ILYUSHIN Il-76

NATO reporting name: Candid

Towards the end of the 1960s, the Ilyushin design bureau, under the leadership of Mr G. V. Novozhilov, began design of a heavy transport aircraft for operation over long distances in the Siberian regions of the USSR. Nominal task for the aircraft was to transport 40 tonnes of freight for a distance of 2,700 nm (5,000 km; 3,100 miles) in less than six hours, as a replacement for the turboprop An-12. It had to be capable of operation from short unprepared airstrips, in the most difficult weather conditions experienced in Siberia, the north of the Soviet Union and the Far East, while being much simpler to service and flying much faster than the An-12.

The prototype of the new transport, known as the Il-76 (CCCP-86712), flew for the first time on 25 March 1971 and made its public debut at the 29th Salon de l'Aéronautique et de l'Espace in Paris in May 1971. Test flying continued until 1975, when the basic **Il-76T** entered series production. Subsequent operation in the most difficult weather and ground conditions of Central and Eastern Siberia revealed operating costs more than 25 per cent lower per tonne/km than for the An-12. This suggested that the Il-76T would be competitive with river transport, even during Summer months.

It was clear from the start that the Il-76 had considerable potential as a military transport. Evaluation by the Soviet Air Force had reached an advanced stage by 1974, when an official film depicted Il-76s with twin-gun rear turrets in use as vehicles for Soviet airborne troops. About 140 are believed to have been delivered subsequently to first-line squadrons of the Transport Aviation force, as An-12 replacements. Other customers for the military version are reported to include the air forces of Iraq, Czechoslovakia and Poland.

A version of the Il-76 was evaluated as a flight refuelling tanker for the 'Backfire' supersonic strategic bombers of the Soviet Air Force and Naval Air Fleet, as a successor to modified Myasishchev M-4 ('Bison') aircraft, and is expected to enter service in this role at an early date. Others are being equipped for an AWACS role, with a rotating 'saucer' radome, lengthened fuselage forward of the wing, and flight refuelling probe. At least 50 are expected to be operational in this form by the mid-80s.

It was announced in 1978 that, following extensive testing of the commercial Il-76T in Siberia, the type was to be introduced into service on the Moscow-Japan route of Aeroflot. By the beginning of 1980, this airline was reported to have about 40 Il-76Ts, and at least six of the basic military models are operated in airline insignia by Iraqi Airways, with the guns removed from the rear turret. Other operators include Libyan Arab Airlines, which has three Il-76Ts and two **IL-76M**s with unspecified modifications; and Syrianair, with one Il-76T and one Il-76M.

In July 1975, the Il-76 set a total of 25 officially recognised records for speed and altitude with payload. Piloted by Yakov I. Vernikov, on 4 July, it raised to 70,121 kg (154,590 lb) the record for greatest payload carried to a height of 2,000 m. The same flight recorded an altitude of 11,875 m (38,960 ft) with payloads of 60,000 kg, 65,000 kg and 70,000 kg. Also on 4 July, Alexander Turumine averaged 462·283 knots (856·697 km/h; 532·327 mph) around a 2,000 km circuit in an Il-76 carrying a payload of 60,000 kg, qualifying for additional records with 35,000 kg, 40,000 kg, 45,000 kg, 50,000 kg and 55,000 kg. On 7 July, Turumine averaged 462·801 knots (857·657 km/h; 532·923 mph) around 1,000 km, claiming nine records for payloads from 30,000 kg to 70,000 kg. The same pilot averaged 440·305 knots (815·968 km/h; 507·019 mph) around a 5,000 km circuit on 10 July, claiming records with 15,000 kg, 20,000 kg, 25,000 kg, 30,000 kg, 35,000 kg and 40,000 kg payloads. In the same year, a group of

Ilyushin Il-76T freight transport (four Soloviev D-30KP turbofan engines) in Aeroflot service *(Anton Wettstein)*

Soviet parachutists raised the world record for a high-altitude jump to 15,386 m (50,479 ft) from an Il-76.

The following description applies to the production Il-76T:

TYPE: Four-turbofan medium/long-range freight transport.

WINGS: Cantilever monoplane, mounted above fuselage to leave interior unobstructed, and with constant anhedral from junction with centre-section on each side. Sweepback 25° at quarter-chord. All-metal five-piece structure, comprising centre-section, two inner panels carrying engines, and two outer panels. Leading-edge sweepback constant. Trailing-edge sweep increases outboard of joint between each inner and outer panel. Multi-spar fail-safe construction. Centre-section integral with fuselage. Mass-balanced ailerons, with balance/trim tabs. Two-section triple-slotted flaps over approx 75 per cent of each semi-span, from wing root to inboard edge of aileron. Upper-surface spoilers forward of flaps in 16 segments, four on each inner and outer wing panel. Ten-segment leading-edge slats over almost entire span, two on each inner panel, three on each outer panel.

FUSELAGE: All-metal semi-monocoque fail-safe structure of basically circular section. Underside of upswept rear fuselage made up of two outward-hinged clamshell doors, upward-hinged panel between these doors, and downward-hinged loading ramp.

TAIL UNIT: Cantilever all-metal structure, with variable-incidence T tailplane. All surfaces sweptback. All control surfaces aerodynamically balanced. Tabs in rudder and each elevator.

LANDING GEAR: Hydraulically-retractable tricycle type, designed for operation from prepared and unprepared runways. Nose unit made up of two pairs of wheels, side by side, with central oleo. Main gear on each side is made up of two units in tandem, each unit with four wheels on a single axle. Low-pressure tyres size 1,300 × 480 on main wheels, 1,100 × 330 on nosewheels. Nosewheels retract forward. Main units retract inward into two large ventral fairings under fuselage, with an additional large fairing on each side of lower fuselage over actuating gear. During retraction main-wheel axles rotate around leg, so that wheels stow with axles parallel to fuselage axis (ie: wheels remain vertical but at 90° to direction of flight). All doors on wheel wells close when gear is down, to prevent fouling of legs by snow, ice, mud, etc. Oleo-pneumatic shock-absorbers. Tyre pressure can be varied in flight from 2·5 to 5 bars (36-73 lb/sq in) to suit different landing strip conditions. Hydraulic brakes on main wheels.

POWER PLANT: Four Soloviev D-30KP turbofan engines, each rated at 117·7 kN (26,455 lb st), in individual underwing pods. Each pod is carried on a large forwardly-inclined pylon and is fitted with a clamshell thrust reverser. Integral fuel tanks between spars of inner and outer wing panels. Total fuel capacity reported to be 81,830 litres (18,000 Imp gallons).

ACCOMMODATION: Crew of seven, including two freight handlers. Conventional side-by-side seating for pilot and co-pilot on spacious flight deck. Station for navigator below flight deck in glazed nose. Forward-hinged door on each side of fuselage forward of wing. Two windows on each side of hold serve as emergency exits. Hold has reinforced floor of titanium alloys, with folding roller-type conveyors, and is loaded via rear ramp. Entire accommodation is pressurised, and advanced mechanical handling systems are provided for containerised and other freight, which can include standard ISO containers, each 12 m (39 ft 4½ in) long, building machinery, heavy crawlers and mobile cranes. Typical loads include six containers measuring either 2·991 × 2·438 × 2·438 m (9 ft 9¾ in × 8 ft × 8 ft) or 2·991 × 2·438 × 1·900 m (9 ft 9¾ in × 8 ft × 6 ft 2¾ in) and with loaded weights of 5,670 kg (12,500 lb) or 5,000 kg (11,025 lb) respectively; or twelve containers measuring 1·456 × 2·438 × 1·900 m (4 ft 9¼ in × 8 ft × 6 ft 2¾ in) and each weighing 2,500 kg (5,511 lb) loaded; or six pallets measuring 2·991 × 2·438 m (9 ft 9¾ in × 8 ft) and each weighing 5,670 kg (12,500 lb); or twelve pallets measuring 1·456 × 2·438 m (4 ft 9¼ in × 8 ft) and each weighing 2,500 kg (5,511 lb). Quick configuration changes can be made by the use of modules, each able to accommodate 30 passengers in four-abreast seating, litter patients and medical attendants, or cargo. Three such modules can be carried, each approx 6·10 m (20 ft) long, 2·4 m (8 ft) wide and 2·4 m (8 ft) high. They are loaded through the rear doors by means of two overhead travelling cranes, and are secured to the cabin floor with cargo restraints. Cranes can utilise two hoists, each with capacity of 3,000 kg (6,615 lb), or four hoists, each with capacity of 2,500 kg (5,510 lb). Ramp can be used as additional hoist, with capacity of up to 30,000 kg (66,140 lb) to facilitate loading of large vehicles and those with caterpillar tracks. Wipers on pilot's windows.

SYSTEMS: Hydraulic system includes servo motors and motors to drive the flaps, slats, landing gear and its doors, ramp, rear fuselage clamshell doors and load hoists. Flying control boosters are supplied by electric pumps and are independent of the central hydraulic supply. Manual control is possible after booster failure. Electrical system includes engine-driven generators, auxiliary generators driven by an APU, DC converters and batteries. It powers the pumps for the flying control system boosters, radio and avionics, and lighting systems.

AVIONICS AND EQUIPMENT: Full equipment for all-weather operation by day and night, including a computer for automatic flight control and automatic landing approach. Large meteorological and ground-mapping radar in undernose radome. APU in port side landing gear fairing for engine starting and to supply all aircraft systems on ground, making aircraft independent of ground facilities.

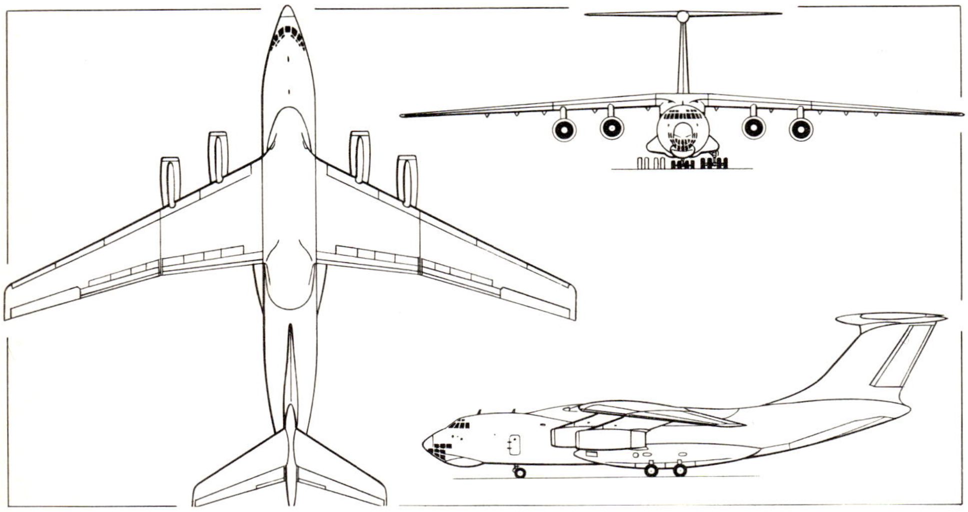

Ilyushin Il-76T four-turbofan heavy freight-carrying transport *(Pilot Press)*

Ilyushin Il-76 at Vienna Airport, in Aeroflot markings but retaining rear gun turret of military version, which can carry up to 140 troops as an alternative to freight *(Prof Johannes Zopp)*

DIMENSIONS, EXTERNAL:
- Wing span: 50·50 m (165 ft 8 in)
- Wing aspect ratio: 8·5
- Length overall: 46·59 m (152 ft 10½ in)
- Height overall: 14·76 m (48 ft 5 in)
- Rear loading aperture:
  - Width: 3·40 m (11 ft 1¾ in)
  - Height: 3·45 m (11 ft 4 in)

DIMENSIONS, INTERNAL:
- Cabin: Length, excl ramp: 20·00 m (65 ft 7½ in)
  - Length, incl ramp: 24·50 m (80 ft 4½ in)
  - Width: 3·40 m (11 ft 1¾ in)
  - Height: 3·46 m (11 ft 4¼ in)
  - Volume: 235·3 m³ (8,310 cu ft)

AREA:
- Wings, gross: 300·0 m² (3,229·2 sq ft)

WEIGHTS AND LOADINGS:
- Max payload: 40,000 kg (88,185 lb)
- Max T-O weight: 170,000 kg (374,785 lb)
- Permissible floor loading: 1,450-3,100 kg/m² (297-635 lb/sq ft)
- Permissible axle load (vehicles): 7,500-11,000 kg (16,535-24,250 lb)
- Max wing loading: 566·7 kg/m² (116·05 lb/sq ft)
- Max power loading: 361 kg/kN (3·54 lb/lb st)

PERFORMANCE:
- Max level speed: 459 knots (850 km/h; 528 mph)
- Cruising speed: 405-432 knots (750-800 km/h; 466-497 mph)
- T-O speed: 114 knots (210 km/h; 131 mph)
- Approach and landing speed: 119-130 knots (220-240 km/h; 137-149 mph)
- Normal cruising height: 9,000-12,000 m (29,500-39,370 ft)
- Absolute ceiling: approx 15,500 m (50,850 ft)
- T-O run: 850 m (2,790 ft)
- Landing run: 450 m (1,475 ft)
- Nominal range with max payload: 2,700 nm (5,000 km; 3,100 miles)
- Max range, with reserves: 3,617 nm (6,700 km; 4,163 miles)

ILYUSHIN Il-86

NATO reporting name: Camber

First indication that this aircraft was under development was given at the 1971 Paris Air Show. Mr Genrikh Novozhilov, successor to the late Sergei Ilyushin as chief of the Ilyushin design bureau, told visitors that a new wide-bodied transport known as the Il-86 was then in the early project design stage.

No final decision on the configuration, or number of engines, had been taken at that time; but in the Spring of 1972 a model of one projected configuration was displayed publicly in Moscow. This design was similar in layout to the Il-62, with four rear-mounted turbofan engines and a T tail, but was intended to be much larger, with a two-deck fuselage. It was described and illustrated in the 1972-73 *Jane's*.

Simultaneously with the display of this original model, it became known that the Il-86 had been chosen for development, after a competition in which it was matched against proposals from the Antonov and Tupolev design teams.

By the end of 1972, it was evident that the design of the Il-86 had evolved along different lines to those suggested by the model displayed six months earlier. In particular the engines had been repositioned into four underwing pods, permitting the tailplane to be lowered on to the rear fuselage. The first of two prototypes (CCCP-86000), in this form, made a first flight of about 40 min after taking off in 1,700 m (5,575 ft) from an 1,820 m (5,970 ft) runway at the old Moscow Central Airport of Khodinka, where the Ilyushin Bureau has its headquarters, to the official flight test centre on 22 December 1976, piloted by Hero of the Soviet Union A. Kuznetsov.

The Novosti official press agency announced in September 1977 that more than 100 test flights had been made and that basic testing was expected to be completed by 7 November 1977, the 60th anniversary of the October Revolution. The Soviet Civil Air Minister, Mr Boris Bugayev, said that the Il-86 would enter service with Aeroflot in time to carry visitors from Prague, Sofia and Berlin to the 1980 Olympic Games in Moscow. This was not achieved, although aircraft CCCP-86002, which flew for the first time at Voronezh on 24 October 1977, was described as the first production Il-86, and the first flight over an Aeroflot route, from Moscow to Mineralnye Vody, was made in September 1978, three months ahead of schedule. US reports suggest that the aim was to complete 1,200 h of flight testing before deliveries began in late 1979. Scheduled services began on 26 December 1980, operating three times each week between Moscow's Vnukovo Airport and Tashkent. Services between Moscow and Mineralnye Vody were being operated with Il-86s by Spring 1981, at which time Aeroflot planned to introduce the type on to 27 other major domestic routes, including those to Simferopol and Sochi. Delivery of four Il-86s to the Polish airline LOT was expected to begin in late 1981.

For production aircraft, ailerons, wing flaps and slats, pylons to carry the engine pods, and vertical and horizontal fixed tail surfaces are being manufactured in Poland. Final assembly is centred at Voronezh. The airframe is believed to be designed for 40,000 flying hours or 20,000 landings.

A long-range version of the Il-86, with more powerful engines, is reported to be under development, with a calculated range of 3,240 nm (6,000 km; 3,725 miles) with max fuel. The following description applies to the initial production model:

TYPE: Four-turbofan wide-bodied passenger transport.

WINGS: Cantilever monoplane of all-metal construction (made at Voronezh). Wings mounted in low-mid position on fuselage. Dihedral from roots. Sweepback 35° at quarter-chord. Large double-slotted trailing-edge flaps, in two sections along entire span of each wing inboard of aileron. Multi-section spoilers in top surface, forward of all four flap sections. Full-span leading-edge slats, with small cutaway to clear each inboard engine pylon. Shallow fence on top surface in line with each pylon.

FUSELAGE: Conventional semi-monocoque light alloy structure of circular cross-section. Floors of main and lower decks of honeycomb and carbonfibre reinforced plastics.

TAIL UNIT: Conventional sweptback cantilever structure, with tailplane dihedral. Each control surface in two sections.

LANDING GEAR: Retractable four-unit type. Forward-retracting steerable twin-wheel nose unit, and three four-wheel bogie main units. Two of the latter retract inward into the wing-root fairings; the third unit is mounted centrally under the fuselage, slightly forward of the others, and retracts forward. (Main landing gear made at Kuibyshev.) Main-wheel tyres size 1300 × 480; nosewheel tyres size 1120 × 450.

POWER PLANT: Four Kuznetsov NK-86 turbofan engines, each rated at 127·5 kN (28,660 lb st), mounted on pylons forward of wing leading-edges. Engines fitted with combined thrust reversers/noise attenuators. Fuel capacity 70,000-80,000 litres (15,400-17,600 Imp gallons). APU in tailcone.

ACCOMMODATION: Standard flight crew comprises two pilots and a flight engineer, with provision for a navigator if required. Flight engineer's seat normally faces to starboard, aft of co-pilot, but can pivot to central forward-facing position to enable the engineer to operate the throttles. Upper deck, on which all seats are located, is divided into three separate cabins by wardrobes, a serving area connected by elevator to the lower-deck galley, and cabin staff accommodation, with a total of eight toilets at front (2) and rear (6) of the aircraft. Cabins feature unusually large windows, indirect lighting in walls and in ceiling panels, and enclosed baggage lockers at top of side walls. Preponderance of metal and natural fibre materials rather than plastics throughout cabins to enhance safety in an emergency. Up to 350 passengers in basic nine-abreast seating throughout, with two aisles, each 55 cm (21·6 in) wide. Suggested mixed class alternative layout provides for 28 passengers six-abreast in the front cabin, and 206 passengers eight-abreast in the other two cabins. Passengers enter via three airstair-type doors (made in Kharkov), which hinge down from the port side of the lower deck. One of these doors is forward of the wing; the others are aft of the wing. Four further doors at upper-deck level on each side, for emergency use (using dual inflatable escape slides) and for use at airports where the utilisation of high-level boarding steps or bridges is preferred. Coats and hand baggage are stowed on the lower deck before passengers climb one of three fixed staircases to the main deck. Cargo holds on the lower deck are designed to accommodate heavy or

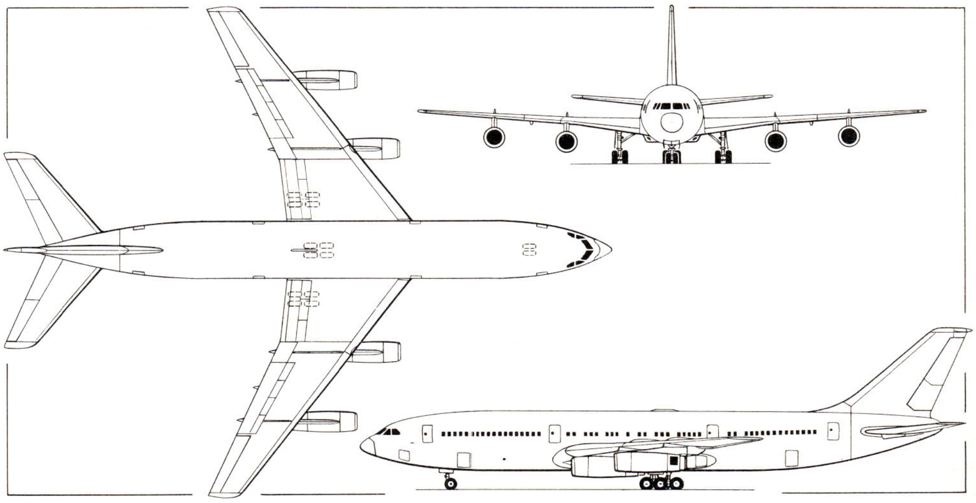

Ilyushin Il-86 four-turbofan wide-bodied passenger transport *(Pilot Press)*

Ilyushin Il-86 wide-bodied transport (four Kuznetsov NK-86 turbofans) in Aeroflot markings

registered baggage and freight in 8 standard LD3 containers, or 16 LD3 containers if some of the carry-on baggage racks are omitted. Access is via upward-hinged doors forward of the starboard wing-root leading-edge and at the side of the rear hold. Containers can be loaded and unloaded by means of a self-propelled truck with built-in roller conveyor. Films can be shown in flight, and there is a choice of 12 tape-recorded audio programmes. A bar-buffet can be provided on the lower deck in place of the baggage and freight accommodation in the forward vestibule.

Systems: Four completely self-contained hydraulic systems, each operated by one of the engines. All hot pipelines of air-conditioning system, and all fuel supply lines, outside pressure cell. New-type smoke detection system, with sensors in baggage, freight and equipment stowage areas. New-type pulse-generating de-icing system consuming 500 times less energy than a conventional hot air or electrical system.

Avionics: All avionics equipment located within pressurised part of fuselage. Flight control and nav systems provide for automatic climb to the selected height, control of the rate of climb and automatic descent, and permit automatic landing in ICAO Cat IIIA conditions. Pre-programmable Doppler nav system with readout display screen on flight deck, on which microfilmed maps can be projected. Position of aircraft is indicated by cursor, driven by system computer. Nav system is updated automatically by inputs from VOR or VOR/DME radio beacons.

Dimensions, external:
Wing span 48·06 m (157 ft 8¼ in)
Length overall 59·54 m (195 ft 4 in)
Diameter of fuselage 6·08 m (19 ft 11½ in)
Height overall 15·81 m (51 ft 10½ in)
Tailplane span 20·57 m (67 ft 6 in)
Wheel track (c/l of outer shock-struts) 11·15 m (36 ft 7 in)
Wheelbase 21·34 m (70 ft 0 in)

Dimensions, internal:
Main cabins: Height 2·61 m (8 ft 7 in)
Max width approx 5·70 m (18 ft 8½ in)

Area:
Wings, gross 320 m² (3,444 sq ft)

Weights:
Max payload 42,000 kg (92,600 lb)
Max fuel 86,000 kg (189,600 lb)
Max T-O weight (dependent on size and type of runway) 190,000-206,000 kg (418,875-454,150 lb)
Max landing weight 175,000 kg (385,800 lb)

Performance (estimated):
Normal cruising speed at 9,000-11,000 m (30,000-36,000 ft) 485-512 knots (900-950 km/h; 560-590 mph)
Approach speed 130-141 knots (240-260 km/h; 149-162 mph)
Field length for take-off and landing 2,300-2,600 m (7,550-8,530 ft)
Range with 40,000 kg (88,185 lb) payload 1,944 nm (3,600 km; 2,235 miles)
Range with max fuel 2,480 nm (4,600 km; 2,858 miles)

KAMOV

Chief of Design Bureau: S. Mikhéev

Nikolai I. Kamov, who died on 24 November 1973, aged 71, had been a leading designer of rotating-wing aircraft since the late 1920s and, with N. K. Skrzhinskii, was responsible for the first successful Soviet rotorcraft, the KaSkr-I, in 1929. He became well known internationally when he designed a series of one-man lightweight helicopters of the 'flying motorcycle' type in the late 1940s.

The Ka-15 and Ka-18 helicopters, developed by Kamov and his design team, under chief engineer Vladimir Barshevskii, were both put into large-scale production and service. Details of them can be found in the 1962-63 and 1963-64 editions of *Jane's* respectively. Later Kamov types include the following:

KAMOV Ka-25

NATO reporting name: Hormone

The prototype of this military helicopter was first shown in public in the Soviet Aviation Day flypast over Tushino Airport, Moscow, in July 1961. It was allocated the NATO code name 'Harp', but this was changed to 'Hormone' for the production versions, of which about 460 were built in 1966-75. Together with the newer Mil helicopters known to NATO as 'Haze-A', these replaced piston-engined Mi-4s in the Soviet Navy's ship and shore based force of around 250 helicopters. Five ex-Soviet Navy Ka-25s equip INAS 333 of the Indian Navy, for deployment on three Krivak class destroyers. Nine are operated on coastal anti-submarine duties by the Syrian Arab Air Force and others by Yugoslavia.

As well as serving as an anti-submarine and missile guidance aircraft, the Ka-25 fulfils a variety of other military roles. Only two versions may be identified at present by NATO reporting names:

Hormone-A. Basic ship-based anti-submarine version, operated from cruisers of 'Kresta' and 'Kara' classes, carrier/cruisers *Kiev* and *Minsk,* each of which carries 19 'Hormone-As' and 3 'Bs', and helicopter carrier/cruisers *Moskva* and *Leningrad,* each of which accommodates about 18 Ka-25s. Search radar in undernose radome, diameter 1·25 m (4 ft 1 in), towed magnetic anomaly detector, dipping sonar housed in compartment at rear of cabin, and electro-optical sensor. A major shortcoming is said to be lack of night and all-weather sonar-dipping capability.

Hormone-B. Special electronics variant, able to acquire targets for ship-launched missiles. Larger undernose radome with more spherical undersurface. Cylindrical radome under rear of cabin. Data link equipment.

Other versions of which photographs have appeared in the press include a utility model, generally similar to 'Hormone-A' but with unnecessary operational equipment and weapons removed. This version sometimes has a yagi aerial mounted on the nose; it has been photographed in non-operational red and white paint finish. Naval 'Hormones' have been seen carrying an external weapons pod housing long wire-guided torpedoes.

First photograph of a Kamov Ka-25 naval helicopter with underbelly container for wire-guided torpedoes

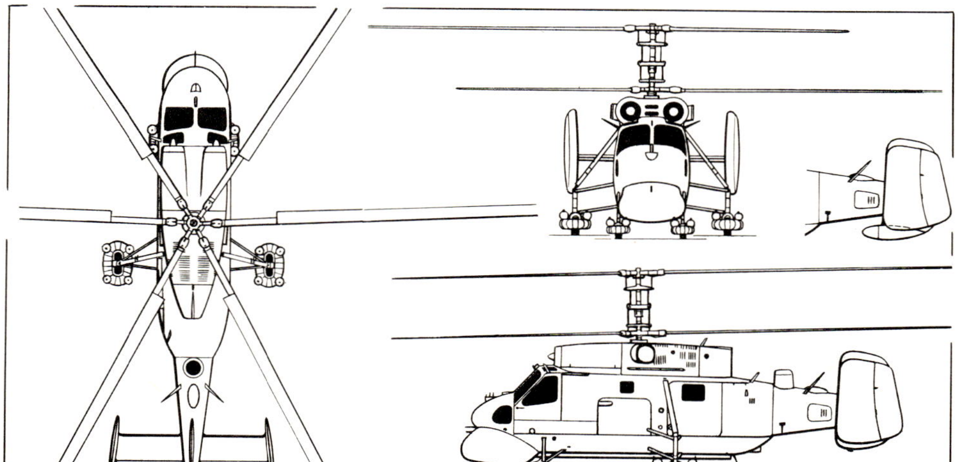

'Hormone-A' anti-submarine version of the Kamov Ka-25 helicopter. Scrap view shows optional blister fairing at base of central tail-fin *(Pilot Press)*

Type: Twin-turbine anti-submarine, missile support and general-purpose helicopter.

Rotor System: Two three-blade coaxial contra-rotating rotors. Automatic blade-folding.

Fuselage: Conventional all-metal semi-monocoque structure of pod and boom type.

Tail Unit: Cantilever all-metal structure, with central fin, ventral fin and twin endplate fins and rudders which are toed inward.

Landing Gear: Four-wheel type. Oleo-pneumatic shock-absorbers. Nosewheels are smaller than main wheels and are of castoring type. Each wheel can be enclosed in an inflatable pontoon surmounted by inflation bottles to provide flotation in event of an emergency alighting on water. Rear legs are pivoted to retract upward about their wishbone supports, so that the wheels can be moved to a position where they offer least interference to signals from the nose radar.

Power Plant: Two 671 kW (900 shp) Glushenkov GTD-3 turboshaft engines, mounted side by side above cabin, forward of rotor driveshaft, on early aircraft. Later aircraft believed to have two 738 kW (990 shp) GTD-3BM turboshafts. Provision for carrying external fuel tank on each side of cabin.

Accommodation: Pilot and co-pilot side by side on flight deck, with rearward-sliding door on each side. Entry to main cabin is via a rearward-sliding door to rear of main landing gear on port side. Cabin accommodates two or three systems operators in ASW role, but is large enough to contain 12 folding seats for passengers.

Avionics and Equipment: Equipment available for all versions includes autopilot, navigational system, radio compass, radio communications installations, and lighting system for all-weather operation by day or night. Dipping sonar housed in compartment at rear of main cabin, immediately forward of tailboom, and search radar under nose of anti-submarine version, which carries also a towed magnetic anomaly detector. Some aircraft have a blister fairing over equipment mounted at the base of the centre tail-fin; others have a cylindrical housing, with a transparent top, above the central point of the tailboom (see illustration), with a shallow blister fairing to the rear of this. Doors under the fuselage of some aircraft enclose a weapons bay for ASW torpedoes, nuclear depth charges and other stores. Two so-called 'air-to-surface missiles' carried on outriggers on each side of cabin of prototype during its Tushino appearance were dummies, but there is reason to believe that Ka-25s are being armed with newly-developed 'fire-and-forget' air-to-surface guided missiles.

'Hormone-B' target acquisition variant of ship-based Ka-25, with enlarged undernose radome and cylindrical radome under rear of cabin, but without belly weapon-bay doors *(US Navy)*

DIMENSIONS, EXTERNAL:
- Diameter of rotors (each): 15·74 m (51 ft 8 in)
- Length of fuselage: 9·75 m (32 ft 0 in)
- Height to top of rotor head: 5·37 m (17 ft 7½ in)
- Width over tail-fins: 3·76 m (12 ft 4 in)
- Wheel track: front: 1·41 m (4 ft 7½ in)
  - rear: 3·52 m (11 ft 6½ in)
- Cabin door: Height: 1·10 m (3 ft 7¼ in)
  - Width: 1·20 m (3 ft 11¼ in)

DIMENSIONS, INTERNAL:
- Cabin, excl flight deck:
  - Length: 3·95 m (12 ft 11½ in)
  - Max width: 1·50 m (4 ft 11 in)
  - Max height: 1·25 m (4 ft 1¼ in)

WEIGHTS (approx):
- Weight empty: 4,765 kg (10,500 lb)
- Max T-O weight: 7,500 kg (16,500 lb)

PERFORMANCE:
- Max level speed: 113 knots (209 km/h; 130 mph)
- Normal cruising speed: 104 knots (193 km/h; 120 mph)
- Service ceiling: 3,500 m (11,500 ft)
- Range with standard fuel, with reserves: 217 nm (400 km; 250 miles)
- Range with external tanks, with reserves: 351 nm (650 km; 405 miles)

KAMOV Ka-25K

The Ka-25K flying-crane helicopter, based on the military Ka-25, was shown publicly for the first time at the 1967 Paris Air Show. It did not enter production and service. Details can be found in the 1979-80 and earlier editions of *Jane's*.

KAMOV Ka-26

NATO reporting name: Hoodlum

First details of this twin-engined light helicopter were announced in January 1964, and the prototype flew for the first time in the following year. Kamov described it as an ideal helicopter for agriculture, possessing all the virtues of the Ka-15 (which was used in about a dozen countries) but able to lift three times as much chemical payload, and the Ka-26 entered large-scale service as an agricultural aircraft in the Soviet Union in 1970, being used primarily over orchards and vineyards. It is also used widely on Aeroflot's air ambulance services and is suitable for many other applications, including cargo and passenger transport, forest firefighting, mineral prospecting, pipeline construction and laying transmission lines.

The usual Kamov contra-rotating coaxial three-blade rotor system is retained, with hydraulic dampers fitted to each rotor head and the rotor shafts inclined forward at 6° to the vertical. The blades, made of glass-textolyte (plastics) materials, weigh only 25 kg (55 lb) each and are completely interchangeable. They, and the cabin windscreen, are equipped with an anti-icing system, activated automatically by a radioisotope ice warning device and utilising an alcohol glycerine mixture.

A powered control system is standard. The jacks are actuated by a single hydraulic system, with manual override in case of system failure.

The fully-enclosed cabin, with a door on each side, is fitted out normally for operation by a single pilot, but a second seat and dual controls are optional. The cabin is warmed and demisted by air from a combustion heater, which also heats the passenger compartment when fitted.

The tailplane, with twin fins and rudders toed inward at 15°, is carried on two plastics tailbooms. Short high-mounted stub-wings carry the two podded 242·5 kW (325 hp) M-14V-26 aircooled radial piston engines, designed by I. M. Vedeneev, and the main units of the non-retractable four-wheel landing gear. Each engine is cooled by a fan in the front of its nacelle, which absorbs about 18·6 kW (25 hp) from the engine output. Dust filters are fitted in the air delivery ducts, to protect the engines, each of which is connected to the rotor transmission by a shaft and two flexible couplings. Both rotors can be driven by either engine if the other fails; disengagement of the failed engine is automatic, and an autorotative landing can be made if both engines fail.

All four landing gear units embody oleo-pneumatic shock-absorbers. The forward wheels are of the castoring type and are not fitted with brakes. The rear wheels are fitted with pneumatically-operated brakes. Tyre size is 595 × 185 on the main wheels, 300 × 125 on the forward wheels.

The space aft of the cabin, between the main landing gear units and under the rotor transmission, is able to accommodate a variety of interchangeable payloads. For agricultural work the chemical hopper (capacity 900 kg; 1,985 lb) and dust-spreader or spraybars are fitted in this position, on the aircraft's centre of gravity. This equipment is quickly removable and can be replaced by a cargo/passenger pod accommodating six persons, with provision for a seventh passenger beside the pilot. Alternatively, the Ka-26 can be operated with either an open platform for hauling freight or a hook for slinging bulky loads at the end of a cable or in a cargo net.

A version for geophysical survey has an electromagnetic pulse generator in the cabin and is encircled by a huge 'hoop' antenna. It carries on the port side of the fuselage a mounting for the receiver 'bird' which is towed at the end of a cable, beneath the helicopter, when in use. The receiver is lowered by an electric winch and the cable is cut by automatic shears if its traction should exceed the authorised limit.

An aerial survey model is available with an AFA-31-MA camera mounted in the cabin. This aircraft can photograph 5 km² (2 sq miles) per hour at a scale of 1 : 10,000.

As an air ambulance, the Ka-26 can carry two stretcher patients, two seated casualties and a medical attendant. A winch, with a capacity of up to 150 kg (330 lb), enables it to be used for rescue duties.

When operating as an agricultural sprayer, the Ka-26 discharges its chemical payload at 1·5-12 litres/s (0·33-2·65 Imp gallons/s). The rate of discharge in a dusting role is 1·5-12 kg/s (3·3-26·5 lb/s). Up to 120 hectares (296 acres) can be sprayed during each flying hour at the rate of 50 kg/ha (44·5 lb/acre). As a duster, 140 ha (346 acres) can be treated at the same discharge rate. 50 ha (123 acres) can be topdressed with chemical fertilisers each flying hour, at a rate of 100 kg/ha (89 lb/acre).

To protect the pilot against toxic chemicals in the agricultural role, the cabin is lightly pressurised by a blower and air filter system which ensures that the cabin air is always clean. The flying and navigation equipment are adequate for all-weather operation, by day and night. VHF and HF radio are fitted, together with a radio compass and radio altimeter.

Because of its small size and manoeuvrability, the Ka-26 can be operated from platforms on small ships such as whalers and icebreakers, and a Soviet fishing boat operating in the North Atlantic in early 1970 carried a Ka-26 for fish-spotting duties. This aircraft was equipped with inflated pontoons to permit alighting on the water.

In mid-1969, a Ka-26 was tested in Siberia and the north-west USSR in a forest protection version able to deliver six firemen and their equipment speedily to the site of a forest fire. In the Spring of 1972, Ka-26s joined Mi-1, Mi-2 and Mi-4 helicopters in operations to clear ice from Soviet rivers, by landing demolition teams on thick ice-floes and destroying thinner ice-fields from the air.

More than 600 Ka-26s had been produced by 1977. They are in civilian service in 15 countries, including Bulgaria, East Germany, West Germany, Hungary, Japan, Romania and Sweden, as well as in the USSR. Military operators include the air forces of Hungary and Sri Lanka. Total flying time exceeded 500,000 hours by 1981.

DIMENSIONS, EXTERNAL:
- Diameter of rotors (each): 13·00 m (42 ft 8 in)
- Vertical separation between rotors: 1·17 m (3 ft 10 in)
- Length of fuselage: 7·75 m (25 ft 5 in)
- Height overall: 4·05 m (13 ft 3½ in)
- Width over engine pods: 3·64 m (11 ft 11½ in)
- Width over agricultural spraybars: 11·20 m (36 ft 9 in)
- Tailplane span: 4·60 m (15 ft 1 in)
- Wheel track: Main wheels: 2·42 m (7 ft 11½ in)
  - Nosewheels: 0·90 m (2 ft 11½ in)
- Wheelbase: 3·48 m (11 ft 5 in)
- Passenger pod door: Height: 1·40 m (4 ft 7 in)
  - Width: 1·25 m (4 ft 1¼ in)

DIMENSIONS, INTERNAL:
- Passenger pod:
  - Length, floor level: 1·83 m (6 ft 0 in)
  - Width, floor level: 1·25 m (4 ft 1¼ in)
  - Headroom: 1·40 m (4 ft 7 in)

WEIGHTS:
- Operating weight, empty:
  - Stripped: 1,950 kg (4,300 lb)
  - Cargo/platform: 2,085 kg (4,597 lb)
  - Cargo/hook: 2,050 kg (4,519 lb)
  - Passenger: 2,100 kg (4,630 lb)
  - Agricultural: 2,216 kg (4,885 lb)
- Fuel weight: Transport: 360 kg (794 lb)
  - Other versions: 100 kg (220 lb)
- Payload: Transport: 900 kg (1,985 lb)
  - Agricultural duster: 1,065 kg (2,348 lb)
  - Agricultural sprayer: 900 kg (1,985 lb)
  - With cargo platform: 1,065 kg (2,348 lb)
  - Flying crane: 1,100 kg (2,425 lb)
- Normal T-O weight: Transport: 3,076 kg (6,780 lb)
  - Agricultural: 2,980 kg (6,570 lb)
- Max T-O weight: all versions: 3,250 kg (7,165 lb)

PERFORMANCE (at max T-O weight):
- Max level speed: 91 knots (170 km/h; 105 mph)
- Max cruising speed: 81 knots (150 km/h; 93 mph)

Kamov Ka-26 used by Asahi Helicopter Co of Japan, mainly for crop-spraying in rice fields

Econ cruising speed
49-59 knots (90-110 km/h; 56-68 mph)
Agricultural operating speed range
16-62 knots (30-115 km/h; 19-71 mph)
Service ceiling 3,000 m (9,840 ft)
Service ceiling, one engine out 500 m (1,640 ft)
Hovering ceiling IGE at AUW of 3,000 kg (6,615 lb)
1,300 m (4,265 ft)
Hovering ceiling OGE at AUW of 3,000 kg (6,615 lb)
800 m (2,625 ft)
Range with 7 passengers, 30 min fuel reserves
215 nm (400 km; 248 miles)
Max range with auxiliary tanks
647 nm (1,200 km; 745 miles)
Endurance at econ cruising speed 3 h 42 min

KAMOV Ka-32

It was announced in early 1981 that a new helicopter, designated Ka-32, was to be put on display in the permanent Exhibition of Achievements of the National Economy (VDNKL) in Moscow. Primary applications for the aircraft were said to be reconnaissance, search and rescue, by day and night in all weathers, from ships that would include the atomic-powered icebreakers *Lenin, Sibir, Arktika* and *Rossiya*.

The Ka-32 had not entered service at the time of the announcement, and had not yet appeared in the Exhibition by mid-1981. All that is known is that it can lift up to 5,000 kg (11,000 lb) as an external slung load, and has a range of 100 nm (185 km; 115 miles) with such a load.

Kamov Ka-126, a turbine-powered development of the Ka-26

KAMOV Ka-126

A photograph of this helicopter was published in a book entitled *Soviet Aviation on the World Market* presented to *Jane's* by Aviaexport at the 1981 Paris Air Show. The Ka-126 appears to be identical to the Ka-26 except for the two very small closely-cowled shaft-turbine engines which replace the latter's M-14V-26 radial piston engines.

The fact that such a helicopter was under development was announced by Mr S. V. Mikhéev, Chief Designer of the Kamov Bureau, in 1979.

MiG

General Designer in Charge of Bureau:
Rostislav A. Belyakov

Colonel-General Artem I. Mikoyan, who died on 9 December 1970 at the age of 65, was head of the design bureau responsible for the MiG series of fighter aircraft. With Mikhail I. Gurevich, a mathematician, he collaborated in the design of the first really-modern Soviet jet-fighter, the MiG-15, which began to appear in squadron service in numbers in 1949.

The MiG-17, a progressive development of the MiG-15, appeared in Soviet squadrons in 1953 or 1954, and was followed into service by the supersonic MiG-19, which appeared in 1955 and has been manufactured also in large numbers in China (which see).

All available details of aircraft designed by the Mikoyan bureau which are currently in production or known to be under development follow:

One of the squadron of 12 MiG-21s (a 'Fishbed-C') flown by USAF pilots for air-to-air combat training at an air base in the western USA

MIKOYAN MiG-21

NATO reporting names: Fishbed and Mongol

The MiG-21 air superiority fighter was developed on the basis of experience of jet-to-jet combat between MiG-15s and US aircraft during the war in Korea. The emphasis was placed on good transonic and supersonic handling, high rate of climb, small size and light weight, using a turbojet engine of medium power, in contrast with the heavier and much more powerful Sukhoi Su-7 and Su-9 fighters that were developed simultaneously. The first versions of the MiG-21 were, therefore, day fighters of limited range, with comparatively light armament and limited avionics. Subsequent development of the type has been aimed primarily at improvements in range, weapons and all-weather capability, and the MiG-21 has become the most widely-used fighter in the world, as well as forming for many years the backbone of Soviet tactical air power.

The E-5 aerodynamic prototype of the MiG-21 flew for the first time on 16 June 1956, and made its public debut during the flypast in the Soviet Aviation Day display at Tushino Airport, Moscow, on 24 June. The initial production version (NATO 'Fishbed-A') was built in only limited numbers, with a Tumansky R-11 turbojet engine rated at 38·25 kN (8,600 lb st) dry and 50 kN (11,240 lb st) with afterburning, and with an armament of two 30 mm NR-30 cannon. Meanwhile, the Soviet Union had been developing a small infra-red homing air-to-air missile, designated K-13 (NATO 'Atoll') and generally similar to the US AIM-9B Sidewinder 1A. Underwing pylons for two K-13s were fitted on the MiG-21F, the suffix 'F' standing for *Forsirovanny* (boosted) and indicating that this model also had a slightly more powerful turbojet. To save weight and provide room for avionics associated with the missiles, the port NR-30 cannon was removed and its blast-tube fairing on the lower fuselage was blanked off. Further details of this and subsequent operational versions of the MiG-21 are as follows:

MiG-21F (Fishbed-C). First major production version, built also in Czechoslovakia. Short-range clear-weather fighter, with radar ranging equipment and a Tumansky R-11 turbojet rated at 42·25 kN (9,500 lb st) dry and 56·4 kN (12,676 lb st) with afterburning (designation of engine given in Soviet press statements as TDR Mk R37F). Two underwing pylons for UV-16-57 pods, each containing sixteen 57 mm rockets, or K-13 air-to-air missiles, and one NR-30 cannon in starboard side of fuselage (one each side on early aircraft and on the ten supplied to India). Internal fuel capacity of 2,340 litres (515 Imp gallons), plus underfuselage pylon for external fuel tank of 490 litres (108 Imp gallons) capacity. Small nose air intake of approximately 69 cm (27 in) diameter, with small movable three-shock centrebody housing the radar ranging equipment. Undernose pitot boom, which folds upward on the ground to reduce risk of ground personnel walking into it. Transparent blister cockpit canopy which hinges upward about base of integral flat bulletproof windscreen. Transparent rearview panel (not on aircraft built in Czechoslovakia) aft of canopy at front of shallow dorsal spine fairing. Large blade antenna at rear of this panel, with small secondary antenna midway along spine. Fowler-type flap between fuselage and aileron on each trailing-edge, with fairing plate under wing at outer extremity. Small forward-hinged airbrake under fuselage, forward of ventral fin; two further forward-hinged airbrakes, on each side of underfuselage in line with wing-root leading-edges, integral with part of cannon fairings. Brake-parachute housed inside small door on port underside of rear fuselage, with cable attachment under rear part of ventral fin. Semi-encapsulated escape system, in which canopy is ejected with seat, forming shield to protect pilot from slipstream, until the seat has been slowed by its drogue chute. Leading-edge of fin extended forward on all but early aircraft, to increase chord.

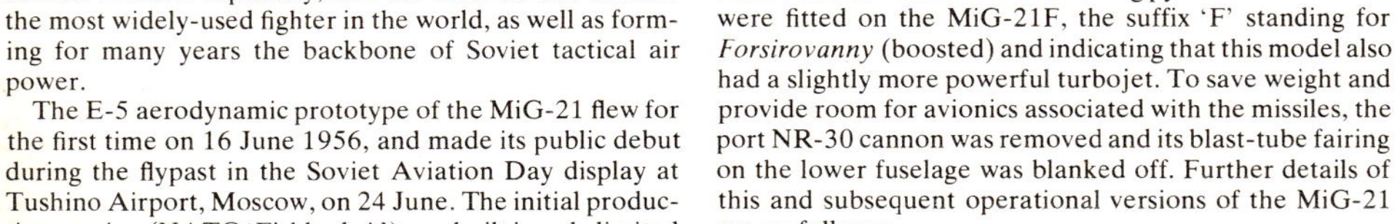

'Fishbed-H' reconnaissance version of MiG-21, with underbelly sensor pack, underwing fuel tanks and wingtip ECM *(Flug Revue)*

MiG-21PF (Fishbed-D). Basic model of second series of operational versions with forward fuselage of less-tapered form. Intake enlarged to diameter of approximately 91 cm (36 in) and housing larger centrebody for R1L search/track radar (NATO 'Spin Scan A') to enhance all-weather capability (designation suffix letter 'P', standing for *Perekhvatchik*, is applied to aircraft adapted for all-weather interception from an earlier designed role). Remainder of airframe generally similar to that of MiG-21F, but pitot boom repositioned above air intake; cannon armament and fairings deleted, permitting simplified design for forward airbrakes; larger main wheels and tyres, requiring enlarged blister fairing on each side of fuselage, over wing, to accommodate wheel in retracted position; dorsal spine fairing widened and deepened aft of canopy, to reduce drag and house additional fuel tankage, and rear-view panel deleted; primary blade antenna repositioned to mid-spine and secondary antenna deleted. Uprated R-11 turbojet, giving 58·4 kN (13,120 lb st) with afterburning. Internal fuel capacity increased to 2,850 litres (627 Imp gallons) in seven fuselage tanks. Late production aircraft have attachments for a rocket-assisted take-off unit (RATOG) aft of each main landing gear bay, and provision for a flap-blowing system known as *Sduva Pogranichnovo Sloya* (SPS), which reduces the normal landing speed by some 22 knots (40 km/h; 25 mph). Flaps are larger than original Fowler type, do not move aft, and lack outboard fairing plates. Prototype shown at Tushino in 1961 had dummy metal centrebody. Production aircraft in service with many air forces.

Fishbed-E. Basically similar to 'Fishbed-C' but with broad-chord vertical tail surfaces. Brake-parachute repositioned into acorn fairing, made up of clamshell doors, at base of rudder, above jet nozzle. Provision for GP-9 underbelly pack, housing GSh-23 twin-barrel 23 mm gun, in place of centreline pylon, with associated predictor sight and electrical ranging system. Identified in 1964.

MiG-21FL. Export version of late-model MiG-21PF series, with broad-chord vertical tail surfaces and brake-parachute housing at base of rudder but no provision for SPS or RATOG. About 200 were initially assembled and later built under licence in India by Hindustan Aeronautics Ltd (which see), with the IAF designation Type 77. R-11-300 turbojet rated at 38·25 kN (8,598 lb st) dry and 60·8 kN (13,668 lb) with afterburning. Suffix letter 'L' *(Lokator)* indicates the installation of Type R2L ('Spin Scan B') search/track radar. Can be fitted with GP-9 underbelly gun pack. Identified in 1966.

MiG-21PFS or MiG-21PF(SPS). Similar to 'Fishbed-D', but with SPS as standard production installation.

MiG-21PFM (Fishbed-F). Successor to interim MiG-21PFS, embodying all the improvements introduced progressively on the PF and PFS, the suffix letter 'M' indicating an exportable version of an existing design. Leading-edge of fin extended forward a further 45 cm (18 in). Small dorsal fin fillet eliminated. Additional refinements, including sideways-hinged (to starboard) canopy and conventional windscreen quarter-lights; simple ejection seat instead of semi-encapsulated type; and large dielectric portion at tip of tail-fin. R2L radar reported to have lock-on range of under 7 nm (13 km; 8 miles) and to be ineffective at heights below about 915 m (3,000 ft) because of ground 'clutter'. Max permissible speed at low altitude is reported to be 593 knots (1,100 km/h; 683 mph). Built also in Czechoslovakia.

MiG-21PFMA (Fishbed-J). Multi-role version. Basically similar to MiG-21PFM but with deeper dorsal fairing containing fuel tankage above fuselage, giving straight line from top of canopy to fin. Improved radar (NATO 'Jay Bird'). Pitot tube remains above air intake but is offset to starboard. Provision for GP-9 underbelly gun pack as alternative to centreline fuel tank. Four underwing pylons, instead of former two, for a variety of ground attack weapons and stores, as alternative or supplementary to two or four air-to-air missiles. Latter can include radar-homing 'Advanced Atoll' as well as infra-red K-13A 'Atoll'. Able to carry two underwing tanks in addition to standard underbelly tank, offsetting reduced internal fuel capacity of 2,600 litres (572 Imp gallons). Zero-zero ejection seat. Small boat-shape fairing with angle-of-attack indicator on port side of nose. Later production aircraft can have the GSh-23 gun installed inside the fuselage, with a shallow underbelly fairing for the twin barrels and splayed cartridge-ejection chutes to clear each side of centreline store.

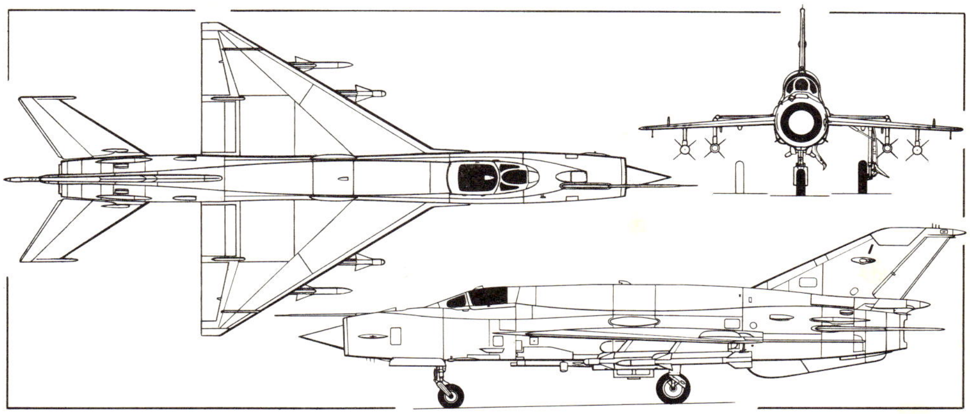

Mikoyan MiG-21SMT ('Fishbed-K') single-seat multi-role fighter *(Pilot Press)*

MiG-21bis fighter in service with the Finnish Air Force

MiG-21R. Generally similar to MiG-21PFMA, but belly gun replaced by a pack of three reconnaissance cameras mounted on a side-hinged (to starboard) door which protrudes from underfuselage, immediately aft of nosewheel leg. First identified in service with Egyptian Air Force in 1980.

MiG-21M. Generally similar to MiG-21PFMA with internal GSh-23 gun pack. R-11F2S-300 engine. Superseded MiG-21FL on Hindustan Aeronautics production line in India, with IAF designation Type 96. First Indian-built MiG-21M was handed over officially to IAF on 14 February 1973.

MiG-21R (Fishbed-H). Tactical reconnaissance version, basically similar to MiG-21PFMA. Equipment includes an external pod for forward-facing or oblique cameras, infra-red sensors or ECM devices, and fuel, on fuselage centreline pylon. Suppressed antenna at mid-fuselage and optional ECM equipment in wingtip fairings.

MiG-21MF (Fishbed-J). Generally similar to MiG-21PFMA but re-engined with a Tumansky R-13-300 turbojet, lighter in weight and with higher performance ratings. Small rearview mirror above cockpit canopy. Debris deflector beneath each suction relief door forward of wing root. Entered service with Soviet Air Force in 1970.

MiG-21RF (Fishbed-H). Tactical reconnaissance version of MiG-21MF. Equipment as for MiG-21R.

MiG-21SMT (Fishbed-K). Similar to MiG-21MF, except for having deep dorsal spine extended rearward as far as brake-parachute housing, to provide maximum possible fuel tankage and optimum aerodynamic form. Able to carry ECM equipment in small removable wingtip pods. Deliveries to Warsaw Pact air forces reported to have begun in 1971. Like the MiG-21PFMA and MiG-21MF, this version can carry K-13A 'Atoll' infra-red missiles and/or radar-homing 'Advanced Atolls'.

MiG-21bis (Fishbed-L). Third-generation multi-role air combat/ground attack version, with updated avionics and generally improved construction standards. Wider and deeper dorsal fairing. Max fuel capacity of seven internal self-sealing tanks 2,900 litres (638 Imp gallons).

MiG-21bis (Fishbed-N). Advanced version of 'Fishbed-L' with Tumansky R-25 turbojet engine, rated at 73·6 kN (16,535 lb st) with afterburning. Further improvement of avionics indicated by 'bow and arrow' antenna under nose. Radar detection range believed to be 16 nm (30 km; 18·5 miles). Standard equipment in Soviet Air Force for several years. Empty weight reported to be 5,200 kg (11,465 lb), normal T-O weight 7,960 kg (17,550 lb). Rate of climb at AUW of 6,800 kg (15,000 lb), with 50% fuel and two 'Atoll' missiles, is 17,700 m (58,000 ft)/min. Now carries two radar-homing 'Atolls' outboard, two 'Aphids' inboard.

MiG-21U (Mongol). Two-seat training versions. Initial version, 'Mongol-A', is generally similar to the MiG-21F but has two cockpits in tandem with sideways-hinged (to starboard) double canopy, larger main wheels and tyres of MiG-21PF, one-piece forward airbrake, and pitot boom repositioned above intake. Cannon armament is deleted. Later models, 'Mongol-B', have the broader-chord vertical tail surfaces and under-rudder brake-parachute housing of the later operational variants, with a deeper dorsal spine and no dorsal fin fillet.

MiG-21US (Mongol-B). Similar to later MiG-21U but with provision for SPS flap-blowing, and retractable periscope for instructor in rear seat.

MiG-21UM (Mongol-B). Two-seat trainer counterpart of MiG-21MF with R-13 turbojet and four underwing stores pylons.

Alternative designations, allocated by the Soviet authorities to MiG-21s used to set up FAI-recognised international records, are as follows:

E-33. This designation has been applied to MiG-21U 'Mongol' trainers used to establish women's records. Those confirmed by the FAI include an altitude of 24,336 m (79,842 ft) set up by Natalya Prokhanova on 22 May 1965, which has not been beaten.

E-66. Aircraft of basic MiG-21F series, used by Col Georgi Mossolov to set up a world absolute speed record (since beaten) of 1,288·6 knots (2,388 km/h; 1,484 mph) over a 15/25 km course on 31 October 1959. Engine is a 58·35 kN (13,120 lb st) Type TDR Mk R37F.

E-66A. Variant of E-66 used by Mossolov to set a world height record (since beaten) of 34,714 m (113,892 ft) on 28 April 1961, from an aerodrome near Moscow. Powered additionally by a 29·4 kN (6,615 lb st) GRD Mk U2 rocket engine in underbelly pack, exhausting between twin ventral fins. Other changes compared with then-standard operational model included a widened dorsal spine and repositioned blade antenna, as standardised for the MiG-21PF, and a blister fairing above the nose.

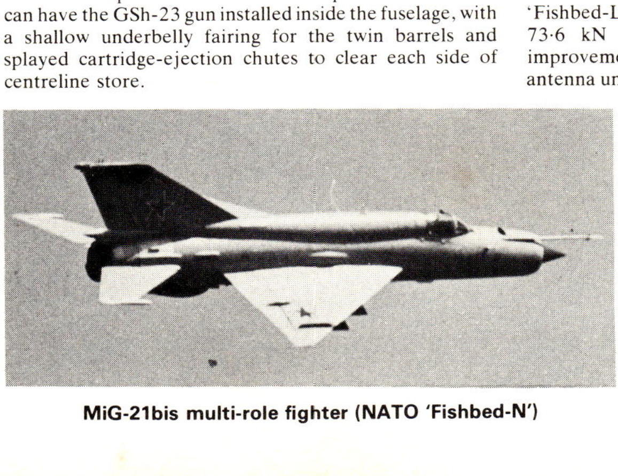

MiG-21bis multi-role fighter (NATO 'Fishbed-N')

MiG-21R of Egyptian Air Force, with belly reconnaissance pack and underwing 'Atoll' missiles
(Denis Hughes)

Close-up of wingtip pod, associated with ECM, on MiG-21 of Egyptian Air Force *(Denis Hughes)*

Two-seat training version of the MiG-21 ('Mongol-B') in Czechoslovak Air Force service *(Letectvi + Kosmonautika)*

E-66B. Used by Svetlana Savitskaya to set four women's time-to-height records on 15 November 1974, from an aerodrome near Moscow. Described as having one 68·7 kN (15,432 lb st) PDM engine (presumably afterburning turbojet) and two 22·6 kN (5,070 lb st) TTPDs (possibly assisted take-off rockets). Times recorded were 41·2 s to 3,000 m, 1 min 0·1 s to 6,000 m, 1 min 21 s to 9,000 m, and 1 min 59·3 s to 12,000 m.

E-76. Designation allocated to apparently-standard MiG-21PFs used by Soviet women pilots to establish international records. Those confirmed by the FAI and still unbeaten are for a speed of 485·78 knots (900·267 km/h; 559·40 mph) over a 2,000 km closed circuit by Yevgenia Martova on 11 October 1966; and a speed of 1,148·7 knots (2,128·7 km/h; 1,322·7 mph) over a 100 km closed circuit by the same pilot on 18 February 1967.

There is reason to believe that the similar designations E-74, E-77 and E-96 apply to versions of the export MiG-21F, MiG-21FL and MiG-21M respectively (see IAF type designations listed for MiG-21s built by HAL in India).

About 1,300 of the 5,000 aircraft estimated to equip Soviet tactical air forces are MiG-21s, including 250 of the reconnaissance versions ('Fishbed-H'). In addition, MiG-21s have been supplied to the Afghan, Algerian, Angolan, Bangladesh, Bulgarian, Chinese, Cuban, Czechoslovak, Egyptian, Ethiopian, Finnish, East German, Hungarian, Indian, Indonesian, Iraqi, North Korean, Laotian, Malagasy, Mozambique, Nigerian, Peruvian, Polish, Romanian, Somali, Sudanese, Syrian, Ugandan, Vietnamese, Yemen Arab Republic, People's Democratic Republic of Yemen, Yugoslav and Zambian air forces.

A version of the MiG-21 has been built at Xian, in China, under the designation J-7. This version is a standard fighter in the Chinese Air Force and has been exported to Albania, Egypt and Tanzania.

The Chinese industry is assisting with spares and overhaul services for the MiG-21s, MiG-17s and MiG-19s operated by Egypt. Twelve MiG-21s and four MiG-23s operated formerly by the Egyptian Air Force are now reportedly flown from an air base in the western USA, for realistic air-to-air combat training of USAF and allied pilots.

The following details refer to the MiG-21MF ('Fishbed-J'):

Type: Single-seat multi-role fighter.

Wings: Cantilever mid-wing monoplane of clipped-delta planform, with slight anhedral from roots. TsAGI section; thickness/chord ratio 5% at root, 4·2% at tip. No leading-edge camber. Sweepback approximately 57°. Small pointed fairing on each side of fuselage forward of wing-root leading-edge. Small boundary-layer fence above each wing near tip. All-metal construction. Inset ailerons, actuated hydraulically. Large 'blown' plain trailing-edge flaps, actuated hydraulically.

Fuselage: Circular-section all-metal semi-monocoque structure. Ram-air intake in nose, with three-position movable centrebody. Large dorsal spine fairing along top of fuselage from canopy to fin. Forward-hinged door-type airbrake on each side of underfuselage below wing leading-edge. A further forward-hinged airbrake under fuselage forward of ventral fin. All airbrakes actuated hydraulically. Blister fairings above and below wing on each side to accommodate main wheels when retracted.

Tail Unit: Cantilever all-metal structure, with all surfaces sharply swept. Conventional fin and hydraulically-powered rudder. Hydraulically-actuated one-piece all-moving horizontal surface, with two gearing ratios for use at varying combinations of altitude and airspeed. Tailplane trim switch on control column. No trim tabs. Single large ventral fin.

Landing Gear: Hydraulically-retractable tricycle type, with single wheel on each unit; all units housed in fuselage when retracted. Forward-retracting non-steerable nosewheel unit; inward-retracting main wheels which turn to stow vertically inside fuselage. Tyres on main wheels inflated to approximately 7·93 bars (115 lb/sq in), ruling out normal operation from grass runways. Pneumatic braking on all three wheels, supplied from compressed-air bottles. Steering by differential main-wheel braking. Wheel doors remain open when legs are extended. Brake parachute housed inside acorn fairing at base of rudder.

Power Plant: One Tumansky R-13-300 turbojet engine, rated at 50 kN (11,240 lb st) dry and 64·73 kN (14,550 lb st) with afterburning. Fuel tanks in fuselage, with total capacity of 2,600 litres (572 Imp gallons), of which approx 1,800 litres (396 Imp gallons) are usable within CG limits at low speed. Provision for carrying one finned external fuel tank, capacity 490 litres (108 Imp gallons), on underfuselage pylon and two similar drop-tanks on outboard underwing pylons. Two jettisonable solid-propellant JATO rockets can be fitted under rear fuselage, aft of wheel doors.

Accommodation: Pilot only, on ejection seat with spring-loaded arm at top which ensures that seat cannot be operated unless hood is closed. Canopy is sideways-hinged, to starboard, and is surmounted by a small rearview mirror. Flat bulletproof windscreen. Cabin air-conditioned. Armour plating forward and aft of cockpit.

Systems: Duplicated hydraulic system, supplied by engine-driven pump, with backup by battery-powered electric pump, and emergency electric tailplane trim and manual operation of flying controls. Autostabilisation in pitch and roll only.

Avionics and Equipment: Search and track radar (NATO 'Jay Bird') in intake centrebody, with search range of approx 15 nm (29 km; 18 miles). Other standard avionics include VOR/ADF and warning radar with an indicator marked in 45° sectors in front of and behind the aircraft. Gyro gunsight is reported to topple at 2·75*g*. Automatic ranging can be fed into gunsight. Full blind-flying instrumentation, with attitude and heading indicators driven by remote central gyro platform.

Armament: One twin-barrel 23 mm GSh-23 gun, with 200 rounds, in belly pack. Four underwing pylons for weapons or drop-tanks. Typical loads for interceptor role include two K-13A 'Atoll' air-to-air missiles on inner pylons and two radar-homing 'Advanced Atolls' or two UV-16-57 rocket packs (each sixteen 57 mm rockets) on outer pylons; four K-13As/'Advanced Atolls'; or two drop-tanks and two K-13As or 'Advanced Atolls'. Typical loads for ground attack role are four UV-16-57 rocket packs; two 500 kg and two 250 kg bombs; or four S-24 240 mm air-to-surface missiles.

Dimensions, external:
- Wing span: 7·15 m (23 ft 5½ in)
- Length, incl pitot boom: 15·76 m (51 ft 8½ in)
- Length, excl pitot boom and intake centrebody: 13·46 m (44 ft 2 in)
- Height overall: 4·10 m (13 ft 5½ in)
- Wheel track: 2·69 m (8 ft 10 in)

Area:
- Wings, gross: 23 m² (247 sq ft)

Weights:
- T-O weight:
  - with four K-13A missiles: 8,200 kg (18,078 lb)
  - with two K-13A missiles and two 108 Imp gallon drop-tanks: 8,950 kg (19,730 lb)
  - with two K-13As and three drop-tanks: 9,400 kg (20,725 lb)

Performance:
- Max level speed above 11,000 m (36,000 ft): Mach 2·1 (1,203 knots; 2,230 km/h; 1,385 mph)
- Max level speed at low altitude: Mach 1·06 (701 knots; 1,300 km/h; 807 mph)
- Landing speed: 146 knots (270 km/h; 168 mph)
- Design ceiling: 18,000 m (59,050 ft)
- Practical ceiling: about 15,250 m (50,000 ft)
- T-O run at normal AUW: 800 m (2,625 ft)
- Landing run: 550 m (1,805 ft)
- Range, internal fuel only: 593 nm (1,100 km; 683 miles)
- Ferry range, with three external tanks: 971 nm (1,800 km; 1,118 miles)

MIKOYAN MiG-23

NATO reporting names: Flogger-A, B, C, E, F and G

The prototype of this variable-geometry air combat fighter was first displayed in public during the 1967 Aviation Day flypast at Domodedovo Airport, Moscow. Initial deliveries of pre-series aircraft to the Soviet Air Force were made in 1970, but deployment of the MiG-23 in large numbers did not begin until 1973. Two Soviet fighter regiments, with a total of about 75 aircraft, were based in East Germany in 1973/74, and deliveries to the Soviet Air Force of all versions of the MiG-23 and the related MiG-27 were estimated to total about 2,000 aircraft by the Spring of 1981, when production was continuing at the rate of more than 600 a year. MiG-23s are flown by all of the Warsaw Pact air forces except that of Romania, and have been exported to at least nine other air forces.

US press reports suggest that four former Egyptian-operated MiG-23s are being flown by the US Air Force in the USA. At least one other is believed to have been presented to China.

There are seven versions of the MiG-23 of which details can be published in 1981:

MiG-23 (Flogger-A). Prototype shown at Domodedovo on 9 July 1967. Illustrated in 1973-74 and preceding editions of *Jane's*. It is now known that one or two development squadrons of 'Flogger-As' entered service with the Soviet Air Force. Experience with these dictated almost total redesign of the major production versions which followed.

MiG-23MF (Flogger-B). Single-seat air combat fighter which has now superseded the MiG-21 as the primary

'Flogger-E', the export version of 'Flogger-B', with smaller radome and other changes

Two views of the MiG-23MF 'Flogger-B' single-seat variable-geometry air combat fighter of the Soviet Air Force. Note the large splitter plates forward of variable-geometry intakes on this version, on the two-seat MiG-23U and on all export models of this family of fighters. Stripes on the stores under the wings of the aircraft in the lower picture imply that they are missile simulators

air-to-air tactical aircraft of the Soviet Air Forces, deployed in both forward areas and the interior of the USSR. Design changes compared with prototype include movement further rearward of all tail surfaces except ventral fin, giving much increased gap between wing and tailplane; a much larger dorsal fin; and the introduction of fixed inboard wing leading-edges. The US *Military Posture* statement for FY 1979 described 'Flogger-B' as "the first Soviet aircraft with a demonstrated ability to track and engage targets flying below its own altitude". Operated also by air forces of Czechoslovakia, East Germany and Hungary.

MiG-23U (Flogger-C). Tandem two-seat version suitable for both operational training and combat use. Individual canopy over each seat. Rear seat slightly higher than forward seat, with retractable periscopic sight for occupant. Dorsal fairing of increased depth aft of rear canopy. Otherwise identical to early MiG-23MF (with R-27 engine). In service with Soviet Air Force and air forces of other countries, including Cuba, Czechoslovakia, Egypt and Libya. Delivery of 15 to India began in December 1980.

MiG-23 (Flogger-E). Export version of 'Flogger-B'. Generally similar to Soviet Air Force version, but equipped to a lower standard. Smaller radar (NATO 'Jay Bird': search range 15 nm; 29 km; 18 miles, tracking range 10 nm; 19 km; 12 miles) in shorter nose radome. No undernose laser rangefinder or Doppler navigation equipment. Armed with 'Atoll' missiles and GSh-23 gun. In service in Algeria, Cuba, Iraq and Libya.

MiG-23BN (Flogger-F). Export counterpart of Soviet Air Force's MiG-27 ('Flogger-D') single-seat ground attack/interdictor version. Has the nose shape, raised seat, cockpit external armour plate and larger, low-pressure tyres of MiG-27, but retains the power plant, variable-geometry intakes and GSh-23 twin-barrel gun of the MiG-23MF interceptor. Operated by Algerian, Cuban, Egyptian, Ethiopian, Iraqi, Libyan, Syrian and Vietnamese Air Forces. Delivery of 70 to India began in December 1980.

MiG-23MF (Flogger-G). First identified when six aircraft from the air base of Kubinka made goodwill visits to Finland and France in the Summer of 1978. Although basically similar to 'Flogger-B', these aircraft had a much smaller dorsal fin. Absence of operational equipment, such as underwing pylons and infra-red and tracking pods, suggested that only a few aircraft had been modified to this standard for improved aerobatic capability as a display team. 'Flogger-G' has, however, been seen with an undernose sensor pod of new design, and is an operational variant.

MiG-23BN (Flogger-H). As 'Flogger-F' but with small avionics pod on each side at bottom of fuselage, immediately forward of nosewheel doors. Operated by Bulgarian, Czechoslovak and Polish Air Forces. Has also been used by Soviet Air Force.

Further versions can be expected, and it is likely that a seagoing variant of 'Flogger' will equip any future large aircraft carriers built for the Soviet Navy.

Early production aircraft were powered by a Tumansky R-27 turbofan engine, rated at 68·65 kN (15,430 lb st) dry and 100·0 kN (22,485 lb st) with afterburning. This power plant continues in use in the two-seat MiG-23U, but the current MiG-23MF and MiG-27 have a Tumansky R-29B turbojet.

The following description refers specifically to the single-seat MiG-23MF ('Flogger-B') as supplied to the Soviet Air Force:

Type: Single-seat variable-geometry air combat fighter.

Wings: Cantilever shoulder wing. Sweepback of main panels variable in flight or on the ground by manual control, reportedly at 16°, 45° or 72°. Fixed triangular inboard panels, with leading-edges swept at approximately 72°. Full-span trailing-edge single-slotted flaps, each in three sections, permitting independent actuation of outboard sections when wings are fully swept. Top-surface spoilers/lift dumpers forward of flaps, for differential operation in conjunction with horizontal tail surfaces, and for collective operation for improved runway adherence and braking after touchdown. Leading-edge flap on outboard two-thirds of each main (variable-geometry) panel.

Fuselage: Conventional semi-monocoque structure of basic circular section; flattened on each side of cockpit, forward of lateral air intake trunks which blend into circular shape of rear fuselage. Large flat boundary layer splitter plate forms inboard face of each intake. Two small rectangular 'blow-in' air intakes in each trunk, under inboard wing leading-edge. Perforations under rear fuselage, aft of main-wheel bays, are pressure-relief vents. Four door-type airbrakes, mounted two on each side of rear fuselage.

Tail Unit: All-moving horizontal surfaces, swept back at approximately 57° on leading-edge, operate both differentially and symmetrically to provide aileron and elevator function respectively. Conventional fin, swept back at approximately 65° on leading-edge, with inset rudder. Large dorsal fin. No tabs. Large ventral fin in two portions. Lower portion is hinged to fold to starboard when landing gear is extended, to increase ground clearance.

Landing Gear: Retractable tricycle type, with single wheel on each main unit and steerable twin-wheel nose unit. Main units retract inward into rear of air intake trunks. Main fairings to enclose these units are attached to legs. Small inboard fairing for each wheel bay hinged to fuselage belly. Nose unit, fitted with small mudguard, retracts rearward. Brake parachute housed in cylindrical fairing at base of rudder.

Power Plant: One Tumansky R-29B turbojet engine, rated currently at up to 122 kN (27,500 lb st) with max afterburning in aircraft for Soviet Air Force. Max internal fuel capacity 5,750 litres (1,265 Imp gallons). Variable-geometry air intakes and variable nozzle. Provision for carrying external fuel tank, capacity 800 litres (176 Imp gallons), on underfuselage centreline pylon.

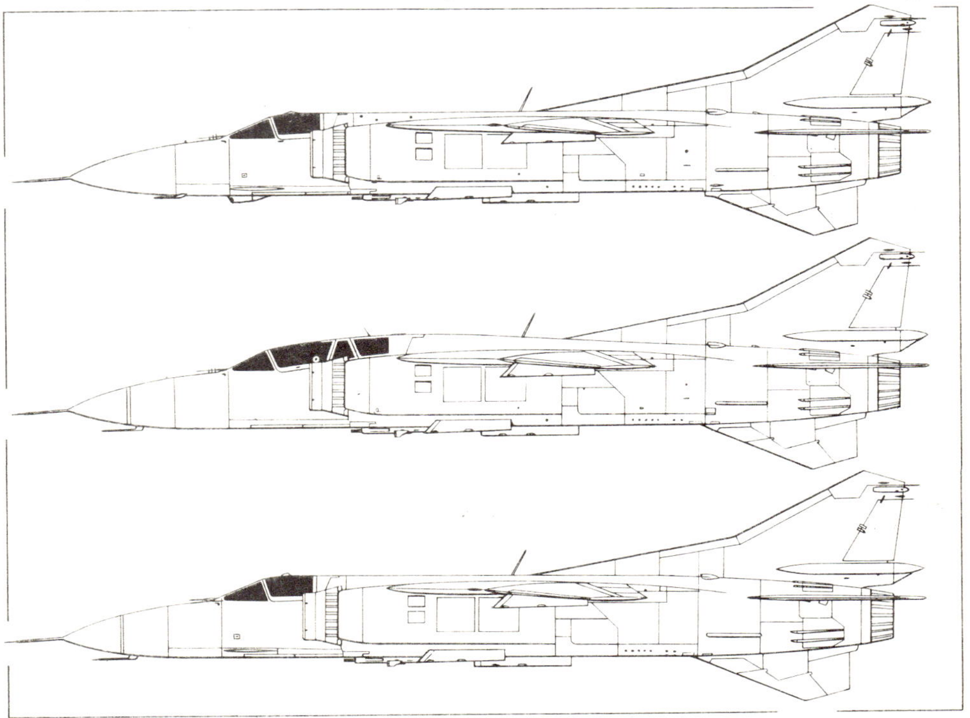

Top to bottom: Side views of the 'Flogger-B', 'Flogger-C' and 'Flogger-E' variants of the MiG-23 series *(Pilot Press)*

Accommodation: Single seat in air-conditioned cockpit, under small rearward-hinged canopy.

Avionics and Equipment: J-band radar dish (NATO 'High Lark': search range 46 nm; 85 km; 53 miles, tracking range 29 nm; 54 km; 34 miles) behind dielectric nosecone. ILS antenna (NATO 'Swift Rod') under radome; yaw sensor above fuselage forward of windscreen; angle of attack sensor on port side. Small cylindrical fairings forward of starboard underwing pylon and above rudder are believed to contain ECM equipment. Undernose laser rangefinder, Sirena 3 radar warning system, and Doppler equipment standard on Soviet Air Force version. Dr Robert C. Seamans, then US Secretary of the Air Force, stated his belief in early 1973 that the contemporary radar and missile systems were comparable with those of the US Air Force's F-4. Retractable landing light under nose, aft of radome.

Armament: One 23 mm GSh-23 twin-barrel gun in fuselage belly pack, with large flash eliminator around muzzles. One pylon under centre-fuselage, one under each engine air intake duct, and one under each fixed inboard wing panel, for rocket packs, air-to-air missiles (NATO 'Apex' and 'Aphid') or other external stores.

Dimensions, external (estimated):

Wing span: fully spread	14·25 m (46 ft 9 in)
fully swept	8·17 m (26 ft 9½ in)
Length overall	16·80 m (55 ft 1½ in)

Weight (estimated):

T-O weight	12,700-15,000 kg (28,000-33,050 lb)

Performance (estimated):

Max level speed at height with external stores	Mach 2·3
Max level speed at S/L	Mach 1·1
Service ceiling	18,600 m (61,000 ft)
T-O and landing run	900 m (2,950 ft)
Combat radius	485-650 nm (900-1,200 km; 560-745 miles)

The two-seat MiG-23U, identical to the early MiG-23MF except for second cockpit *(Letectvi + Kosmonautika)*

'Flogger-H' can be distinguished from 'Flogger-F' by the avionics pods near the nosewheel doors, and from the MiG-27 ('Flogger-D') by its variable-geometry intakes *(Letectvi + Kosmonautika)*

MIKOYAN MiG-27

NATO reporting name: Flogger-D

Although the single-seat ground attack aircraft known to NATO as 'Flogger-D' has many airframe features in common with the MiG-23, it differs in important respects and is designated MiG-27. Use of fixed air intakes and a fixed nozzle is consistent with the primary requirement of high subsonic speed at low altitude.

The forward portion of the fuselage is completely redesigned by comparison with the interceptor versions of the MiG-23. Instead of an ogival radome, 'Flogger-D' has a nose that is sharply-tapered in side elevation, with a small sloping window covering a laser rangefinder and marked target seeker, and additional armour on the flat sides of the cockpit. The seat and canopy are raised to improve the view from the cockpit. A six-barrel 23 mm Gatling-type underbelly gun replaces the GSh-23 of the interceptor, and there is a bomb rack under each side of the rear fuselage in addition to five pylons for external stores, which can include tactical nuclear weapons and, probably, the air-to-surface missile known to NATO as 'Kerry'. There is provision for carrying an external fuel tank for ferry flights under each outer wing, which must be kept in a fully-forward position when the tank is in place. Equipment includes an ECM antenna above the port glove pylon.

The somewhat similar aircraft known to NATO as 'Flogger-F and H' are members of the MiG-23 series, with variable-geometry intakes and a GSh-23 twin-barrel gun, although having the nose-shape, raised seat and larger, low-pressure tyres of 'Flogger-D'. Both versions have been operated by Soviet units; but the 'F' and 'H' are basically export counterparts of 'Flogger-D', with lower standards of equipment and performance, and are described under the MiG-23 entry.

The following data are estimated for the MiG-27 as operated by the Soviet and East German Air Forces:

Power Plant: Generally similar to MiG-23MF, but engine rated at 78·45 kN (17,635 lb st) dry and 112·8 kN (25,350 lb st) with max afterburning. Provision for three external tanks, each of 800 litres (176 Imp gallons) capacity, under fuselage and each outer wing.

Dimensions, external: As MiG-23, plus:

Wing aspect ratio (spread)	7·45
Tailplane span	5·75 m (18 ft 10¼ in)

Areas:

Wings, gross (spread)	27·26 m² (293·4 sq ft)
Horizontal tail surfaces	6·88 m² (74·06 sq ft)

Weights:

Max weapon load	3,500 kg (7,715 lb)
Max T-O weight, 'clean'	15,500 kg (34,170 lb)
Max T-O weight	20,100 kg (44,310 lb)

Performance (estimated):

Max level speed at height	Mach 1·6
Max level speed at S/L	Mach 0·95
T-O to 15 m (50 ft) at AUW of 15,700 kg (34,600 lb)	800 m (2,625 ft)
Combat radius, with underbelly fuel tank, four 500 kg bombs and two 'Atoll' missiles, lo-lo-lo	210 nm (390 km; 240 miles)
Max ferry range with three external tanks	1,350 nm (2,500 km; 1,550 miles)

MiG-23 with modified tail ('Flogger-G') photographed during a goodwill visit to France in 1978 *(Colair)*

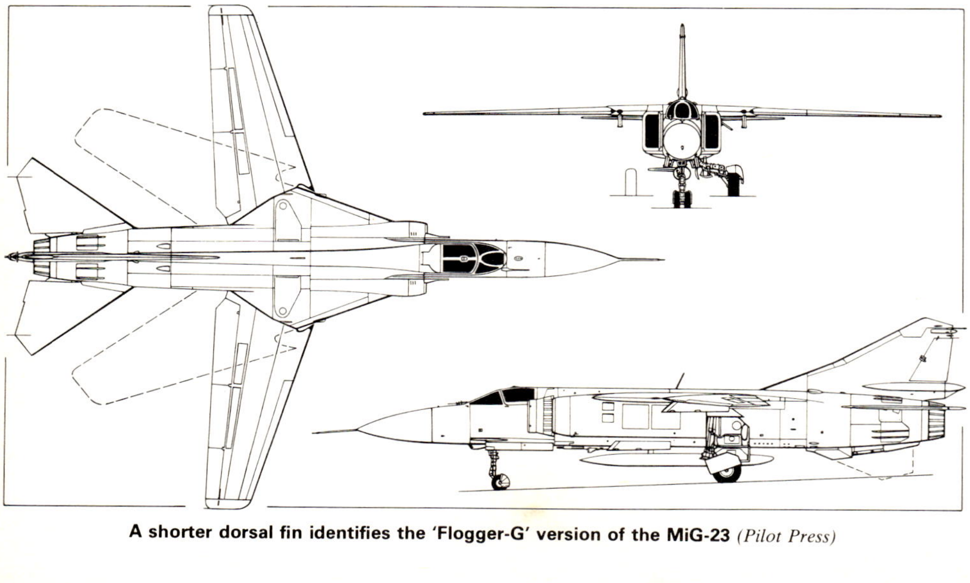

A shorter dorsal fin identifies the 'Flogger-G' version of the MiG-23 *(Pilot Press)*

MIKOYAN MiG-25 (E-266)

NATO reporting name: Foxbat

Development of the MiG-25 was initiated as a high-priority programme to counter the threat of the US Air Force's Mach 3 B-70 strategic bomber, for which North American Aviation Inc was chosen as prime contractor in December 1957. When the B-70 was cut back to a research project by President Kennedy, in March 1961, work on the MiG-25 continued, with increasing emphasis on the reconnaissance potential of the design.

First indication that the prototype had flown came with a Soviet claim, in April 1965, that a twin-engined aircraft designated E-266 had set a 1,000 km closed-circuit speed record of 1,251·9 knots (2,320 km/h; 1,441·5 mph), carrying a 2,000 kg payload. The attempt was made at a height of 21,000-22,000 m (69,000-72,200 ft) by Alexander Fedotov, who had earlier set a 100 km record in the Mikoyan E-166 research aircraft.

The same pilot set a payload-to-height record of 29,977 m (98,349 ft) with a 2,000 kg payload in the E-266 on 5 October 1967, after a rocket-assisted take-off. This qualified also for the record with a 1,000 kg payload. Photographs of the E-266 issued subsequently in the Soviet Union identified it as the twin-finned single-seat fighter of which four examples had taken part in the Domodedovo display in July 1967, and the designation MiG-25 was confirmed later. NATO had, meanwhile, allocated the reporting name 'Foxbat' to the type.

The aircraft's performance in level flight was demonstrated further on 5 October 1967, when M. Komarov set a speed record of 1,608·83 knots (2,981·5 km/h; 1,852·61 mph) over a 500 km closed circuit. On 27 October, P.

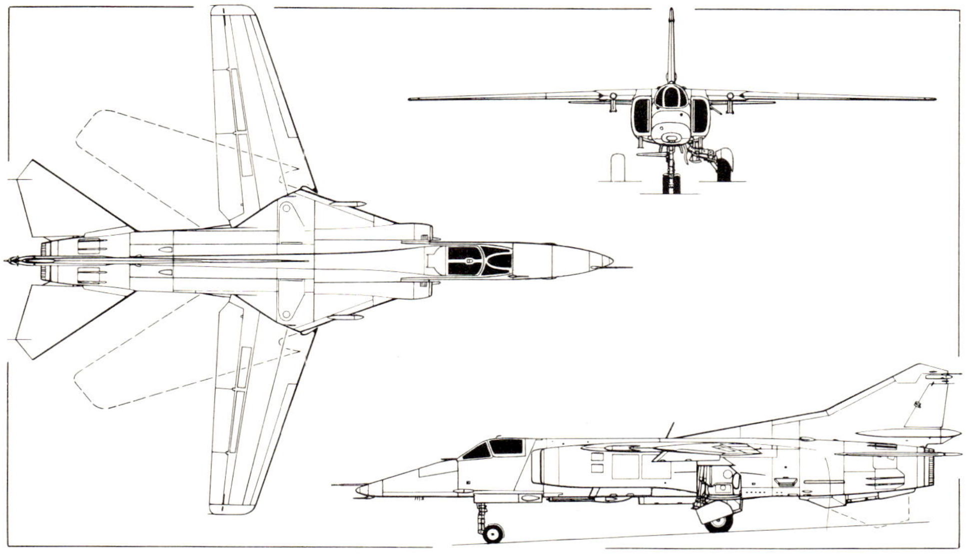

MiG-27 single-seat ground attack aircraft *(Pilot Press)*

MiG-27 ('Flogger-D') landing, with wings extended and ventral fin folded *(Flug Revue)*

Ostapenko raised the 1,000 km closed-circuit record to 1,576·00 knots (2,920·67 km/h; 1,814·81 mph) in an E-266, carrying a 2,000 kg payload and qualifying also for records with 1,000 kg payload and no payload. On 8 April 1973, Fedotov achieved a speed of 1,405·741 knots (2,605·1 km/h; 1,618·734 mph) over a 100 km closed circuit. Next, on 25 July 1973, he set an absolute height record by climbing to 36,240 m (118,898 ft) in an E-266.

Three time-to-height records were established by the E-266 on 4 June 1973, when Boris Orlov climbed to 20,000 m in 2 min 49·8 s, and P. Ostapenko climbed to 25,000 m in 3 min 12·6 s and 30,000 m in 4 min 3·86 s. All three records were beaten by the McDonnell Douglas F-15 *Streak Eagle* in January-February 1975; but two of them were recaptured by an E-266M (with uprated power plant) on 17 May 1975. Fedotov climbed to 25,000 m in 2 min 34·2 s and Ostapenko reached 30,000 m in 3 min 9·85 s. Fedotov also set a new record by climbing to 35,000 m in 4 min 11·7 s.

The current absolute height record was set by Fedotov on 31 August 1977, when he climbed to 37,650 m (123,524 ft) in an E-266M. He had, on 22 July, climbed to 37,080 m (121,654 ft) carrying a 2,000 kg payload, qualifying also for the record with 1,000 kg.

Four MiG-25 reconnaissance aircraft were deployed with Soviet Air Force units in Egypt in the Spring of 1971, having been airlifted to that country in An-22 transports. Between the Autumn of 1971 and the Spring of 1972, these aircraft were despatched in pairs from Cairo West airfield on at least four occasions to carry out high-speed reconnaissance missions off the Israeli coastline or down the full length of the Israeli-occupied Sinai Peninsula. Phantom interceptors sent up by the Israeli defence forces failed to make contact with the MiGs, which remained in Egypt until September 1975. In 1977, MiG-25 reconnaissance aircraft were based in the Soviet Union, at Brieg in Silesia, and in Syria. First foreign operators, in 1979, were the Algerian and Syrian Air Forces.

In early 1973 Dr Robert C. Seamans, then US Secretary of the Air Force, described the MiG-25 as "probably the best interceptor in production in the world today", and added "This Mach 3 aircraft performs both interceptor and reconnaissance missions, can operate at 24,400 m (80,000 ft), and has a highly capable avionics and missile system". In his FY 1975 US Defense Department Report, then-Secretary Schlesinger commented: "Should the Soviet Union develop and deploy an AWACS/'Foxbat' 'look-down, shoot-down' air defence system, we would have to counter it with new penetration devices and techniques such as the cruise missile, bomber defence missiles and improved ECM".

A first opportunity to study the MiG-25 interceptor outside the Soviet Union came when Lt Viktor Belenko defected in one from the Soviet air base of Sikharovka, 200 km (120 miles) from Vladivostok, to Hakodate airport, Japan, on 6 September 1976. Statements attributed to this pilot suggest that more than 400 MiG-25s had been built by that time, and that his particular aircraft left the production line less than three years earlier. Japanese and US military technicians who examined the aircraft reported that the airframe is constructed mainly of steel, with titanium only in places subjected to extreme heating such as the wing leading-edges. The inevitable weight penalty restricts the amount of equipment that can be carried. Belenko said that the aircraft took a considerable time to accelerate to high speeds, which were then difficult to maintain.

Examination of the aircraft is said to have shown that the fuselage weighs about 13,600 kg (30,000 lb) with the wings, tail surfaces and afterburners removed; the fire control system is bulky and lacking in advanced technology, with its very high power (600 kW) devoted to anti-jamming capability rather than range, and with vacuum tubes rather than solid-state circuitry throughout the avionics; the number of cockpit instruments was described as 50 per cent of those in F-4EJ Phantoms of the JASDF, with a smaller and less versatile weapon sight; and the Machmeter has a 'red-line' limit at Mach 2·8, which almost certainly represents a never-exceed speed when carrying missiles and pylons rather than the maximum speed of which the 'clean' aircraft is capable. Of particular interest is the aircraft's high-quality airborne computer which, in conjunction with a ground-based flight control system, enables the interceptor to be vectored automatically on to its target over long ranges.

There are now five variants of the MiG-25, as follows:

MiG-25 (Foxbat-A). Basic interceptor, with large radar (NATO 'Fox Fire') in nose and armed with four air-to-air missiles on underwing attachments. Slightly reduced wing leading-edge sweep towards tips. CW target illuminating radar in nose of wingtip anti-flutter bodies. In service also in Algeria, Libya (three squadrons) and Syria.

MiG-25R (Foxbat-B). Basic reconnaissance version, with five camera windows and various flush dielectric panels aft of very small dielectric nosecap for radar. Equipment believed to include Doppler navigation system and side-looking airborne radar (SLAR). No armament. Slightly reduced span. Wing leading-edge sweep constant from root to tip. In service also in Algeria, Libya (one squadron, including 'Foxbat-Ds') and Syria. Eight ordered to replace Canberras of No 106 Squadron, Indian Air Force, in 1982-83.

MiG-25U (Foxbat-C). Trainer, of which first photographs were published towards the end of 1975. Generally similar to operational versions, but with new nose, containing separate cockpit with individual canopy, forward of standard cockpit and at a lower level. No search radar or reconnaissance sensors in nose. The aircraft designated **E-133** in which Svetlana Savitskaya set a women's world speed record of 1,448·942 knots (2,683·44 km/h; 1,667·412 mph) on 22 June 1975 is believed to have been a MiG-25U. She has since set a women's sustained height record of 21,209·9 m (69,586 ft) in an E-133 on 31 August 1977, and a women's speed record of 1,331·70 knots (2,466·31 km/h; 1,532·49 mph) around a 500 km closed circuit on 21 October 1977.

MiG-25 ('Foxbat-A') interceptor, armed with four air-to-air missiles (NATO 'Acrid')

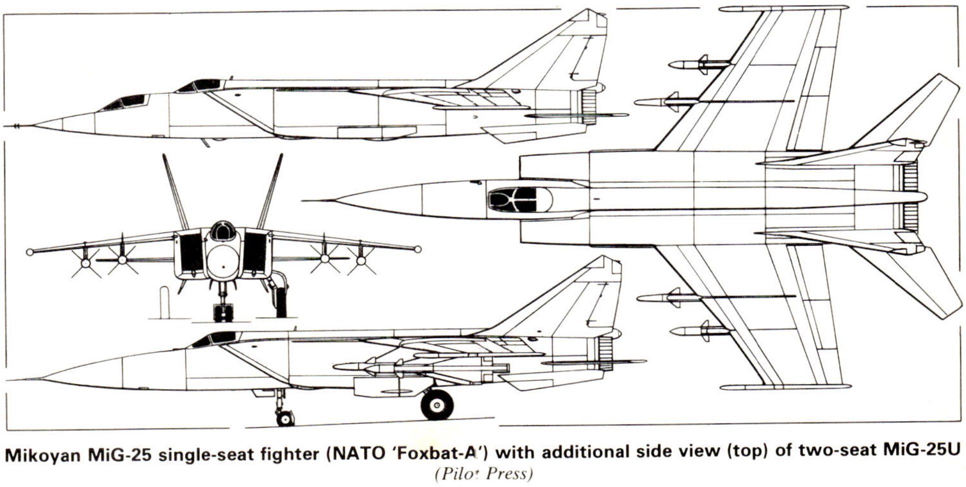

Mikoyan MiG-25 single-seat fighter (NATO 'Foxbat-A') with additional side view (top) of two-seat MiG-25U *(Pilot Press)*

MiG-25R (Foxbat-D). Generally similar to 'Foxbat-B', but with larger SLAR (side-looking airborne radar) dielectric panel, further aft on starboard side of nose, and no cameras. Operated by Soviet Air Force and, alongside 'Foxbat-Bs', in Libya.

E-266M. Designation of experimental aircraft with two 137·3 kN (30,865 lb st) engines which set three time-to-height records in 1975 and also holds the absolute height record. Lt Belenko, the Soviet pilot who defected to Japan in a MiG-25, referred to a developed interceptor version in which the airframe had been strengthened to permit supersonic flight near the ground; the engines had been uprated to give 137·3 kN (30,865 lb st) with afterburning; the avionics had been improved; and two fuselage attachments had been added to make possible the carriage of a total of six air-to-air missiles. Installation of a gun is also likely in the new operational model. This is known in the USA as **'Super Foxbat'** and is described as a tandem two-seater with an armament of radar-homing AA-X-9 air-to-air missiles and a radar that can display 20 targets and track four of them simultaneously.

In tests against simulated cruise missiles, a MiG-25 flying at 6,000 m (19,685 ft) has detected a target flying below 60 m (200 ft) at a range of 11 nm (20 km; 12·5 miles), fired an unarmed missile against it and achieved a theoretical 'kill'.

The following description applies to the MiG-25 ('Foxbat-A') interceptor:

Type: Single-seat interceptor.

Wings: Cantilever high-wing monoplane. Anhedral 4° from roots. Sweepback on leading-edge approx 40° inboard, 38° outboard of each outer missile attachment. Wing structure basically of steel, with titanium leading-edge. Upper surface fence in line with each inboard weapon attachment; shorter shallow fence in line with each outer attachment. Long anti-flutter body (max diameter 30 cm; 11·8 in) at each wingtip, housing avionics. Light alloy aileron at centre of each semi-span, with simple light alloy flap on inboard 37% of trailing-edge. No other movable wing surfaces.

Fuselage: Basic fuselage is quite slim, but is blended into the rectangular air intake trunks, which have wedge inlets. Inner walls of intakes are curved at top and do not run parallel with outer walls; a hinged panel forms the lower lip of each intake, enabling intake area to be varied electronically. Structure mainly of steel.

Tail Unit: Cantilever structure comprising twin outward-canted fins with inset rudders, and all-moving horizontal surfaces. All surfaces sweptback, without tabs. Main structures of steel, with titanium leading-edges and light alloy rear sections. Two outward-canted ventral fins. Large areas of each main and ventral fin form flush antennae.

Landing Gear: Retractable tricycle type. Single wheel, with high-pressure tyre of 1·20 m (47·25 in) diameter, on each forward-retracting main unit. Wheel stows vertically between air intake duct and outer skin of each trunk. Twin-wheel nose unit. Twin brake-chutes in fairing above and between jet nozzles.

Power Plant: Two Tumansky R-31 (R-266) single-shaft turbojet engines, each rated at 74·53 kN (16,755 lb st) dry, and 107·9 kN (24,250 lb st) with afterburning. Water-methanol injection standard. Fuel in two structural tanks in fuselage, between cockpit and engine bay, in saddle tanks around intake ducts, and in integral tank in each wing, filling almost the entire volume inboard of outer fence. Total fuel capacity approx 14,000 kg (30,865 lb) or 17,410 litres (3,830 Imp gallons).

Accommodation: Pilot only, on KM-1 zero-height, 80 knot (150 km/h; 93 mph) ejection seat similar to that fitted to some versions of MiG-21. Canopy hinged to open sideways, to starboard.

Avionics and Equipment: Main fire-control radar (NATO 'Fox Fire': range believed to be 45 nm; 85 km; 52 miles) in nose, forward of avionics compartment housing navigation radar. SRZO-2 IFF and SOD-57M ATC/SIF, with antennae in starboard fin tip. Sirena 3 360° radar warning system with receivers in centre of each wingtip anti-flutter body and starboard fin tip. Unidentified ECCM, decoys and jammers. RSB-70/RPS HF, RSIU-5 VHF, R-831 UHF communications equipment, SP-50 ILS, MRP-56P marker beacon receiver and ARK-15 radio compass. Retractable landing light under front of each intake trunk.

The reconnaissance version of the MiG-25 known to NATO as 'Foxbat-B'. Note the camera-carrying nose and dielectric panels *(Flug Revue)*

Close-up of tandem cockpits in nose of MiG-25U ('Foxbat-C') *(Flug Revue)*

Armament: Four air-to-air missiles on underwing attachments. These may comprise one infra-red and one radar homing example of the missile known to NATO as 'Acrid' under each wing. Alternatively, one 'Apex' and one 'Aphid' can be carried under each wing. Backup optical weapon sight.

Dimensions, external (estimated):
- Wing span: 'Foxbat-A' 13·95 m (45 ft 9 in)
- 'Foxbat-B' 13·40 m (44 ft 0 in)
- Wing aspect ratio: 'Foxbat-A' 3·4
- Length overall 23·82 m (78 ft 1¾ in)
- Length of fuselage 19·40 m (63 ft 7¾ in)
- Height overall 6·10 m (20 ft 0¼ in)

Area:
- Wings, gross: 'Foxbat-A' 56·83 m² (611·7 sq ft)

Weights (estimated):
- Basic operating weight, empty:
  - 'Foxbat-A' at least 20,000 kg (44,100 lb)
  - 'Foxbat-B' 19,600 kg (43,200 lb)
- Max T-O weight:
  - 'Foxbat-A' 36,200 kg (79,800 lb)
  - 'Foxbat-B' 33,400 kg (73,635 lb)

Performance (estimated):
- Max level speed at height:
  - 'Foxbat-B', 'clean' Mach 3·2
- Never-exceed combat speed: 'Foxbat-A', with four 'Acrid' missiles and 50% fuel Mach 2·8
- Max level speed at low altitude: 'Foxbat-A', with four 'Acrid' missiles and 50% fuel Mach 0·85
- Landing speed:
  - 'Foxbat-A' 146 knots (270 km/h; 168 mph)
- Max rate of climb at S/L:
  - 'Foxbat-A' 12,480 m (40,950 ft)/min
- Time to 11,000 m (36,000 ft) with afterburning:
  - 'Foxbat-A' 2 min 30 s
- Service ceiling: 'Foxbat-A' 24,400 m (80,000 ft)
  - 'Foxbat-B, D' 27,000 m (88,580 ft)
- T-O run: 'Foxbat-A' 1,380 m (4,525 ft)
- Landing run: 'Foxbat-A' 2,180 m (7,150 ft)
- Normal combat radius:
  - 'Foxbat-A' 610 nm (1,130 km; 700 miles)
  - 'Foxbat-B, D' 595 nm (1,100 km; 680 miles)
- Max combat radius, econ power:
  - 'Foxbat-A' 700 nm (1,300 km; 805 miles)

NEW MIKOYAN FIGHTER

Inevitably, there have been persistent reports in recent years concerning new fighter designs emanating from the Mikoyan bureau. The accompanying three-view drawing depicts an air superiority fighter in the class of the projected Northrop F-18L which is said to have been at the advanced flight testing stage in Spring 1979. US reports suggest that it is known in Washington as Ram L, having been identified initially on a photograph taken by reconnaissance satellite over Ramenskoye flight test centre in the USSR.

The drawing of this single-seat twin-jet fighter should be regarded as highly provisional. The aircraft's T-O weight is said to be in the 11,340 kg (25,000 lb) class, giving an optimum thrust to weight ratio of 1·2:1 in combat. Armament is said to include a 30 mm gun and up to eight air-to-air missiles, with look-down/shoot-down radar capability. Max level speed is estimated to be Mach 2·3 at height, Mach 1·2 at sea level, with a max combat radius of 500 nm (925 km; 575 miles).

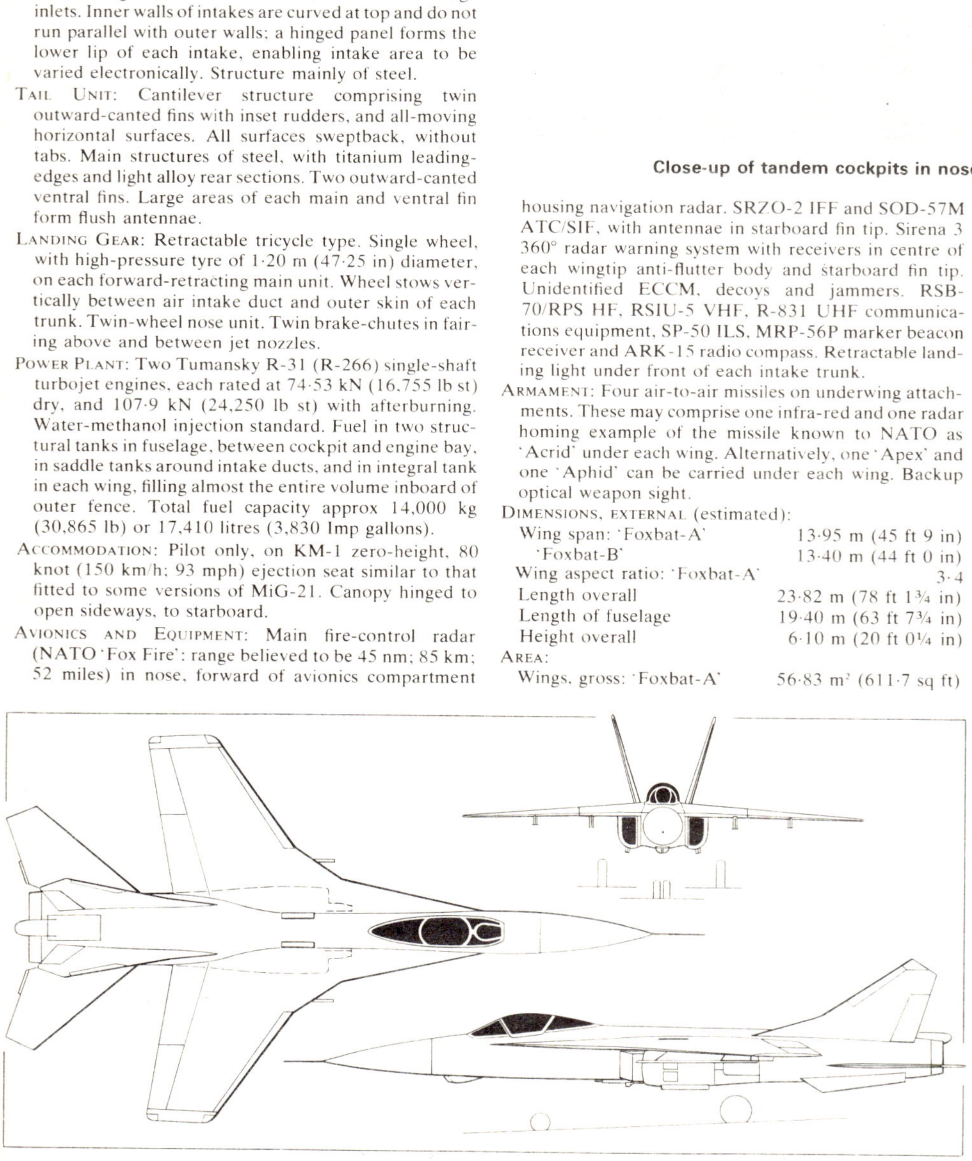

Provisional drawing of the single-seat twin-jet fighter said to have been identified by reconnaissance satellite and to be known in Washington as Ram L *(Pilot Press)*

MIL

General Designer in Charge of Bureau:
Marat N. Tishchenko

M. L. Mil was connected with Soviet gyroplane and helicopter development from at least 1930. His achievements were recognised by the award of the Order of Lenin on his 60th birthday in November 1969. He died on 31 January 1970.

His original Mi-l, which was designed in 1949, first flown in 1950 and introduced into squadron service in 1951, was the first helicopter to enter series production in the Soviet Union and was also produced in Poland. It was followed by the larger Mi-4 in a number of variants, and these types are still in service, in civil and military forms.

Subsequent products of the bureau that was headed by Mikhail Mil include the Mi-6, a large passenger and freight helicopter, the Mi-10 (V-10) and Mi-10K crane versions of the Mi-6, the smaller turbine-powered Mi-2 (V-2), Mi-8 (V-8) and Mi-17 passenger helicopters, the Mi-12 (V-12), which was the largest helicopter yet flown anywhere in the world, the Mi-14 coastal patrol development of the Mi-8, the Mi-24 military assault helicopter and the new heavy-lift Mi-26. Aviaexport has sold helicopters of Mil design in 40 countries.

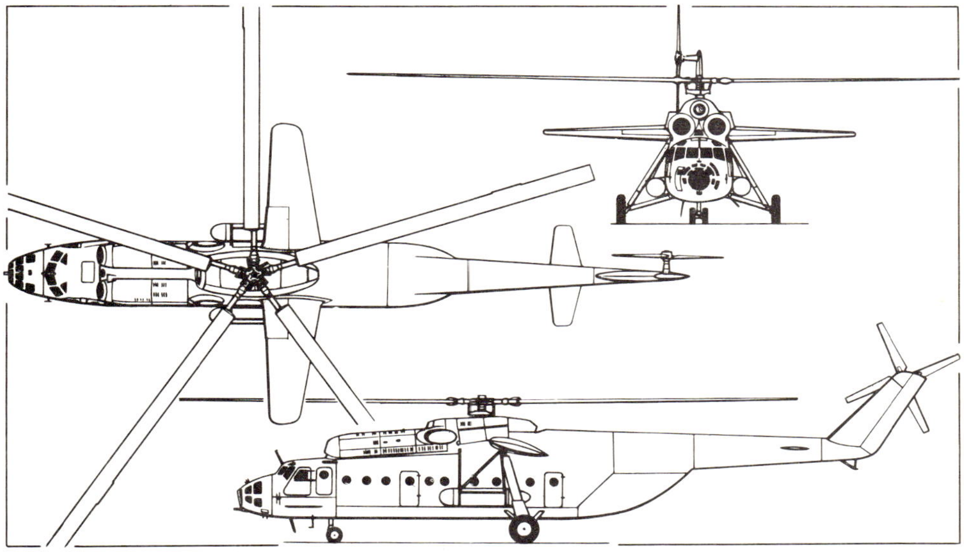

Military version of Mil Mi-6 heavy general-purpose helicopter *(Pilot Press)*

MIL Mi-2 (V-2)

Built exclusively in Poland and described under Polish aircraft industry entry for WSK-PZL Swidnik.

MIL Mi-4

NATO reporting name: Hound

Although Mi-4s have been largely replaced by turbine-powered helicopters in their original transport and anti-submarine roles, they continue in service with support units. A version first identified in 1977 is shown in an accompanying illustration. The multiple antennae projecting from the front and rear of the cabin, on each side, are communications jammers.

The designations given by NATO to military variants of the Mi-4 may now be listed, as follows:

Hound-A. Basic transport helicopter.

Hound-B. ASW version, with undernose radar, etc.

Hound-C. ECM version, shown in an accompanying illustration.

The Mil Mi-4 was last described in the 1971-72 *Jane's*. About 3,500 were built in the Soviet Union in 1952-69. Others were built at Harbin in China under the designation Z-5 (see Chinese section of this edition).

MIL Mi-6

NATO reporting name: Hook

First announced in the Autumn of 1957, the Mi-6 was then the largest helicopter flying anywhere in the world. From it were developed the Mi-10 and Mi-10K flying crane helicopters, and its dynamic components were used in duplicated form on the V-12 (Mi-12).

Layout of the Mi-6 is conventional. Clamshell rear loading doors and folding ramps facilitate the loading of bulky freight and vehicles. Freight can also be carried externally, suspended from a hook on the CG. When the aircraft is operated in this flying crane role, the small wings which normally offload the rotor in flight can be removed, permitting an increase in payload.

The stub-wings are deleted also from the firefighting version. First demonstrated at the 1967 Paris Air Show, this carries several tons of water in tanks inside its cabin and can either spray this slowly from nozzles or dump it through the hoist cutout in its belly.

In setting up 14 FAI-recognised records in Class E1, the Mi-6 has lifted payloads of up to 20,117 kg (44,350 lb). Records still standing in mid-1981 included a 100 km closed-circuit speed record of 183·54 knots (340·15 km/h; 211·36 mph), set up by Boris Galitsky on 26 August 1964. On 15 September 1962 the same pilot, and crew, in an Mi-6 had flown at 162·08 knots (300·377 km/h; 186·64 mph) over a 1,000 km circuit, setting the current records for speed with payload of 1,000 kg and payload of 2,000 kg. On 11 September 1962 Vasily Kolochenko and crew of four averaged 153·44 knots (284·354 km/h; 176·69 mph) over a 1,000 km closed circuit, with a payload of 5,000 kg.

Five Mi-6s are reported to have been built for development testing, followed by an initial pre-series of 30 and subsequent manufacture of some 800 for military and civil use. Six were supplied to the Indonesian Air Force; many others have been delivered to the Algerian, Bulgarian, Egyptian, Ethiopian, Iraqi, Peruvian, Syrian and Vietnamese air forces. The following details apply to the **Mi-6A** late production version:

Type: Heavy transport helicopter.

Rotor System: Five-blade main rotor and four-blade tail rotor. Main rotor blades each have a tapered steel tube spar, to which are bonded built-up metal aerofoil sections. Blades have coincident flapping and drag hinges and fixed tabs. Main rotor shaft inclined forward at 5° to vertical. Control via large welded swashplate. Hydraulically-actuated powered controls. Main rotor collective-pitch control interlocked with throttle controls. Main rotor blades incorporate electro-thermal anti-icing system. Tail rotor embodies a fluid-type anti-icing system.

Rotor Drive: Conventional transmission. Main reduction gearbox drives tail rotor, fan, AC generators and hydraulic pumps. Intermediate reduction gearbox fitted with special fan.

Fuselage: Conventional all-metal riveted semi-monocoque structure of pod and boom type.

Wings: Two small cantilever shoulder wings, mounted above main landing gear struts, offload rotor by providing some 20% of total lift in cruising flight. Removed when aircraft is operated as flying crane.

Tail Unit: Tail rotor support acts as vertical stabiliser. Variable-incidence horizontal stabiliser, near end of tailboom, for trim purposes.

Landing Gear: Non-retractable tricycle type, with castoring twin-wheel nose unit and single wheel on each main unit. Twin-chamber oleo-pneumatic (high-pressure and low-pressure) main landing gear shock-struts. High-pressure chambers interconnected through overflow system incorporating spring damper, to damp out oscillations at full landing gear loading and so eliminate ground resonance. Main wheels size 1,325 × 480 mm. Nosewheels size 720-310 mm. Brakes on main wheels. Small tail-bumper under end of tailboom.

Mil Mi-4 helicopter ('Hound-C') fitted with communications jamming equipment

Power Plant: Two 4,101 kW (5,500 shp) Soloviev D-25V (TV-2BM) turboshaft engines, mounted side by side above cabin, forward of main rotor shaft. Electro-thermal anti-icing system for air intakes. Eleven internal fuel tanks, with total capacity of 6,315 kg (13,922 lb), and two external tanks, on each side of cabin, with total capacity of 3,490 kg (7,695 lb). Provision for two additional ferry tanks inside cabin, with total capacity of 3,490 kg (7,695 lb). Automatic fuel-flow control system with manual override. Side panels of engine cowlings are opened and closed hydraulically and are used as platforms for inspection and maintenance of engines and rotor head.

Accommodation: Crew of five, consisting of two pilots, navigator, flight engineer and radio operator. Four jettisonable doors and overhead hatch on flight deck. Electro-thermal anti-icing system for glazing of flight deck and navigator's compartment. Equipped normally for cargo operation, with easily-removable tip-up seats

Mil Mi-6 heavy general-purpose helicopter, with nose gun, in service with the Egyptian Air Force *(Denis Hughes)*

along side walls. When these seats are supplemented by additional seats installed in centre of cabin, 65-90 passengers can be carried, with cargo or baggage in the aisles. As an air ambulance, 41 stretcher cases and two medical attendants on tip-up seats can be carried. One of attendant's stations is provided with intercom to flight deck, and provision is made for portable oxygen installations for the patients. Cabin floor is stressed for loadings of 2,000 kg/m^2 (410 lb/sq ft), with provision for cargo tiedown rings. Rear clamshell doors and ramps are operated hydraulically. Standard equipment includes an electric winch of 800 kg (1,765 lb) capacity and pulley block system. Central hatch in cabin floor for cargo sling system for bulky loads. Three jettisonable doors, fore and aft of main landing gear on port side and aft of landing gear on starboard side.

SYSTEMS: Main, standby and auxiliary hydraulic systems, each with separate pump mounted on main gearbox. Operating pressure 118-152 bars (1,705-2,205 lb/sq in). Main 27V DC electrical system, supplied by two 12kW starter/generators, with batteries for 30 min emergency supply. De-icing system and some radio equipment supplied by three-phase 360V 400Hz AC system, utilising two 90kVA generators. Trolley-mounted APU, consisting of 74·5 kW (100 hp) AI-8 gas turbine and 24kW generator, carried on board.

AVIONICS AND EQUIPMENT: Standard equipment includes VHF and HF communications radio, intercom, radio altimeter, radio compass, three-channel autopilot, marker beacon receiver, directional gyro and full all-weather instrumentation.

ARMAMENT: Some military Mi-6s are fitted with a 12·7 mm machine-gun in the fuselage nose.

DIMENSIONS, EXTERNAL:
Diameter of main rotor — 35·00 m (114 ft 10 in)
Diameter of tail rotor — 6·30 m (20 ft 8 in)
Distance between rotor centres — 21·09 m (69 ft 2½ in)
Length overall, rotors turning — 41·74 m (136 ft 11½ in)
Length of fuselage, excl nose gun and tail rotor — 33·18 m (108 ft 10½ in)
Height overall — 9·86 m (32 ft 4 in)
Wing span — 15·30 m (50 ft 2½ in)
Span of horizontal stabiliser — 5·04 m (16 ft 6½ in)
Wheel track — 7·50 m (24 ft 7¼ in)
Wheelbase — 9·09 m (29 ft 9¾ in)
Rear loading doors: Height — 2·70 m (8 ft 10¼ in)
Width — 2·65 m (8 ft 8¼ in)
Passenger doors:
Height: front door — 1·70 m (5 ft 7 in)
rear doors — 1·61 m (5 ft 3½ in)
Width — 0·80 m (2 ft 7½ in)
Sill height: front door — 1·40 m (4 ft 7¼ in)
rear doors — 1·30 m (4 ft 3¼ in)
Central hatch in floor — 1·44 m (4 ft 9 in) × 1·93 m (6 ft 4 in)

DIMENSIONS, INTERNAL:
Cabin: Length — 12·00 m (39 ft 4½ in)
Max width — 2·65 m (8 ft 8¼ in)
Max height: at front — 2·01 m (6 ft 7 in)
at rear — 2·50 m (8 ft 2½ in)
Cabin volume — 80 m^3 (2,825 cu ft)

WEIGHTS:
Weight empty — 27,240 kg (60,055 lb)
Max internal payload — 12,000 kg (26,450 lb)
Max slung cargo — 8,000 kg (17,637 lb)
Fuel load, internal — 6,315 kg (13,922 lb)
Fuel load, with external tanks — 9,805 kg (21,617 lb)
Max T-O weight with slung cargo at altitudes under 1,000 m (3,280 ft) — 38,400 kg (84,657 lb)
Normal T-O weight — 40,500 kg (89,285 lb)
Max T-O weight for VTO — 42,500 kg (93,700 lb)

PERFORMANCE (at max T-O weight):
Max level speed — 162 knots (300 km/h; 186 mph)
Max cruising speed — 135 knots (250 km/h; 155 mph)
Service ceiling — 4,500 m (14,750 ft)
Range with 8,000 kg (17,637 lb) payload — 334 nm (620 km; 385 miles)
Range with external tanks and 4,500 kg (9,920 lb) payload — 540 nm (1,000 km; 621 miles)
Max ferry range (tanks in cabin) — 781 nm (1,450 km; 900 miles)

Mil Mi-8 ('Hip-E') military helicopter. This differs from the commercial version in having circular cabin windows, an optional nose gun and weapon-carriers on outriggers

MIL Mi-8 (V-8)

NATO reporting name: Hip

This turbine-powered helicopter was shown in public for the first time during the 1961 Soviet Aviation Day display. Since then, more than 7,500 Mi-8s have been delivered for military and civil use, and production continues at a reported rate of about 750 a year. With Mi-24s, they form the standard equipment of Soviet tactical helicopter regiments, in a variety of forms, some carrying extremely heavy weapon loads. Military Mi-8s have also been supplied to the Afghan, Algerian, Angolan, Anguillan, Bangladesh, Bulgarian, Chinese, Cuban, Czechoslovak, Egyptian, Ethiopian, Finnish, East German, Guinea-Bissau, Hungarian, Indian, Iraqi, Kampuchean, North Korean, Laotian, Libyan, Malagasy, Mali, Mongolian, Mozambique, Pakistani, Peruvian, Polish, Romanian, Somali, Sudanese, Syrian, Ugandan, Vietnamese, North Yemen, South Yemen, Yugoslav and Zambian armed forces.

The commercial Mi-8, with larger, square windows in place of the circular cabin windows of the military version, is in service with Aeroflot for transport and air ambulance duties, and is operated by this airline in support of Soviet activities in the Antarctic. Standard Mi-8s are used there for ice patrol and reconnaissance, for rescue operations, and for carrying supplies and equipment to Vostok Station, near the South Pole.

Three international women's helicopter records for distance and speed in a 2,000 km closed circuit were still credited to the Mi-8 in mid-1981.

The original prototype (NATO **Hip-A**) had a single 2,013 kW (2,700 shp) Soloviev turboshaft engine and four-blade main rotor. When fitted with the five-blade rotor that became standard on subsequent aircraft, it was redesignated **Hip-B**. The second prototype, which flew for the first time on 17 September 1962, introduced the now-standard Isotov twin-turbine power plant and became **Hip-C** to NATO in both civil and military forms.

There are three civil versions, as follows:

Mi-8. Passenger version, with standard seating for 28-32 persons in main cabin.

Mi-8T. General utility version, equipped normally to carry internal or external freight, but able to accommodate 24 tip-up passenger seats along the cabin walls.

Mi-8 Salon. De luxe version. Main cabin is furnished normally for eleven passengers, with an eight-place couch facing inward on the port side, and two chairs and a swivelling seat on the starboard side. There is a table on the starboard side. An air-to-ground radio telephone and removable ventilation fans are standard equipment. Forward of the main cabin is a compartment for a hostess, with buffet and crew wardrobe. Aft of the main cabin are a toilet (port) and passenger wardrobe (starboard), to each side of the entrance. An alternative nine-passenger layout is available. The Mi-8 Salon has a max T-O weight of 10,400 kg (22,928 lb) and range of 205 nm (380 km; 236 miles) with 30 min fuel reserve. In other respects it is similar to the standard Mi-8.

Military versions are identified by the following NATO reporting names:

Hip-C. Basic assault transport. Twin-rack for stores on each side of cabin, able to carry total of 128 × 57 mm rockets in four packs, or other weapons.

Hip-D. Rectangular-section canisters on outer stores racks; added antennae. For electronic duties.

Hip-E. Described by US Department of Defense as world's most heavily armed helicopter. Standard equipment of Soviet tactical air forces. One flexibly-mounted 12·7 mm machine-gun in nose. Triple stores rack on each side of cabin, able to carry up to 192 rockets in six suspended packs, plus four 'Swatter' infra-red homing anti-tank missiles above racks.

Hip-F. Export counterpart of 'Hip-E'. Missile armament changed to six 'Saggers'.

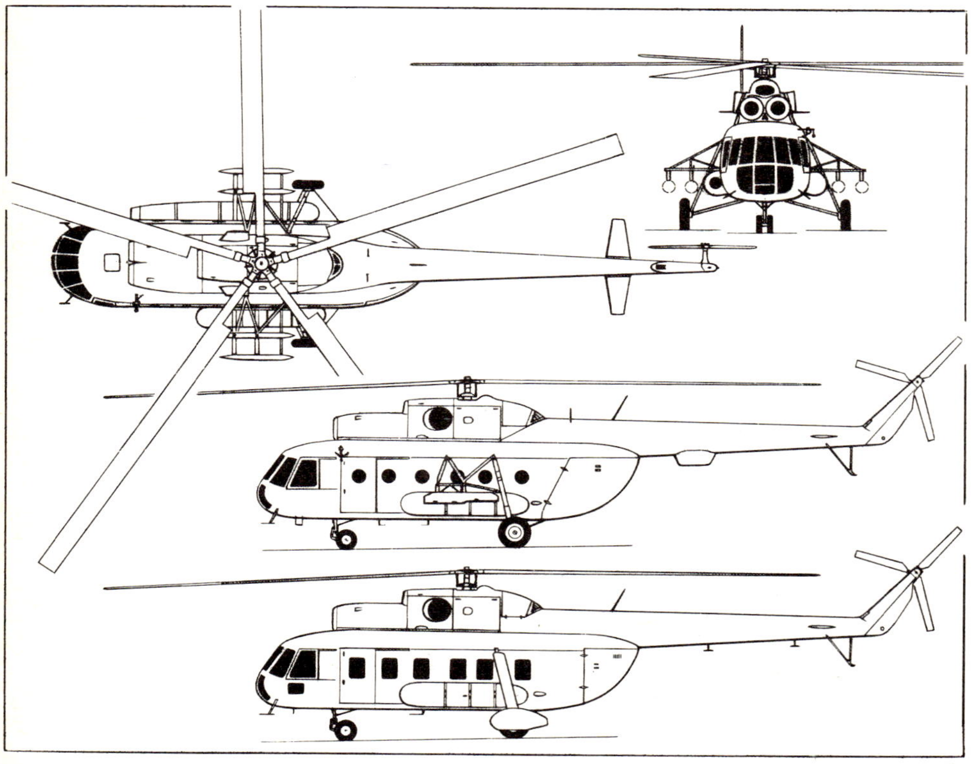

'Hip-C' military version of Mil Mi-8 twin-turbine helicopter, with additional side view (bottom) of commercial version *(Pilot Press)*

The Mi-17, with uprated power plant, is described separately.

TYPE: Twin-engined transport helicopter.

ROTOR SYSTEM: Five-blade main rotor and three-blade tail rotor. Transmission comprises a type VR-8 two-stage planetary main reduction gearbox giving main rotor shaft/engine rpm ratio of 0·016 : 1, intermediate and tail gearboxes, main rotor brake and drives off the main gearbox for the tail rotor, fan, AC generator, hydraulic pumps and tachometer generators. Main rotor shaft inclined forward at 4° 30′ from vertical. All-metal main rotor blades of basic NACA 230 section; solidity 0·0777. Each main blade is made up of an extruded light alloy spar carrying the blade root fitting, 21 trailing-edge pockets and the blade tip. Pockets are honeycomb-filled. Main rotor blades are fitted with balance tabs, embody a spar failure warning system, and are interchangeable. Their drag and flapping hinges are a few inches apart, and they are carried on a machined spider. Control system utilises irreversible hydraulic boosters. Main rotor collective-pitch control is interlocked to throttle controls. All-metal tail rotor blades, each made up of a spar and honeycomb-filled trailing-edge. Automatically-controlled electro-thermal de-icing system on all blades. In an emergency, the rotor blades of the Mi-8 and intermediate and tail gearboxes are interchangeable with those of the piston-engined Mi-4, although this prevents use of the de-icing system.

FUSELAGE: Conventional all-metal semi-monocoque structure of pod and boom type.

TAIL UNIT: Tail rotor support acts as small vertical stabiliser. Horizontal stabiliser near end of tailboom.

LANDING GEAR: Non-retractable tricycle type, with steerable twin-wheel nose unit, which is locked in flight, and single wheel on each main unit. All units embody oleo-pneumatic (gas) shock-absorbers. Main-wheel tyres size 865 × 280; nosewheel tyres size 595 × 185. Pneumatic brakes on main wheels. Pneumatic system can also recharge tyres in the field, using air stored in main landing gear struts. Optional main-wheel fairings.

POWER PLANT: Two 1,267 kW (1,700 shp) Isotov TV2-117A turboshaft engines. Main rotor speed governed automatically, with manual override. Single flexible internal fuel tank, capacity 445 litres (98 Imp gallons), and two external tanks, one each side of cabin, with capacity of 745 litres (164 Imp gallons) in the port tank and 680 litres (149·5 Imp gallons) in the starboard tank. Total standard fuel capacity 1,870 litres (411·5 Imp gallons). Provision for carrying one or two additional ferry tanks in cabin, raising max total capacity to 3,700 litres (814 Imp gallons). Fairing over starboard external tank houses optional cabin air-conditioning equipment at front. Engine cowling side panels form maintenance platforms when open, with access via hatch on flight deck. Engine air intake de-icing standard. Total oil capacity 60 kg (132 lb).

ACCOMMODATION: Two pilots side by side on flight deck, with provision for a flight engineer's station. Windscreen de-icing standard. Basic passenger version is furnished with 28 four-abreast track-mounted tip-up seats at a pitch of 72-75 cm (28-29·5 in), with a centre aisle 32 cm (12·5 in) wide, a wardrobe and baggage compartment; or 32 seats without wardrobe. Seats and bulkheads of basic version are quickly removable for cargo-carrying. Mi-8T has cargo tiedown rings in floor, a winch of 200 kg (440 lb) capacity and pulley block system to facilitate the loading of heavy freight, an external cargo sling system (capacity 3,000 kg; 6,614 lb), and 24 tip-up seats along the side walls of the cabin. All versions can be converted for air ambulance duties, with accommodation for 12 stretchers and a tip-up seat for a medical attendant. The large windows on each side of the flight deck slide rearward. The sliding, jettisonable main passenger door is at the front of the cabin on the port side. An electrically-operated rescue hoist (capacity 150 kg; 330 lb) can be installed at this doorway. The rear of the cabin is made up of large clamshell freight-loading doors, with a downward-hinged passenger airstair door inset centrally at the rear. Hook-on ramps are used for vehicle loading.

SYSTEMS: Standard heating system can be replaced by full air-conditioning system. Two independent hydraulic systems, each with own pump; operating pressure 44-64 bars (640-925 lb/sq in). DC electrical supply from two 27V 18kW starter/generators and six 28Ah storage batteries. AC supply for de-icing system and some radio equipment supplied by 208/115/36/7·5V 400Hz generator, with 36V three-phase standby system. Provision for oxygen system for crew and, in ambulance version, for patients. Freon fire-extinguishing system in power plant bays and service fuel tank compartments, actuated automatically or manually. Two portable fire extinguishers for use in cabin.

AVIONICS AND EQUIPMENT: Standard equipment includes a type R-842 HF transceiver with frequency range of 2 to 8 MHz and range of up to 540 nm (1,000 km; 620 miles), type R-860 VHF transceiver operating on 118 to 135·9 MHz over ranges of up to 54 nm (100 km; 62 miles), intercom, radio telephone, type ARK-9 automatic radio compass, type RV-3 radio altimeter with 'dangerous height' warning, and four-axis autopilot to give yaw, roll and pitch stabilisation under any flight conditions, stabilisation of altitude in level flight or hover, and stabilisation of pre-set flying speed, navigation equipment and instrumentation for all-weather flying by day and night, including two gyro horizons, two airspeed indicators, two main rotor speed indicators, turn indicator, two altimeters, two rate of climb indicators, magnetic compass, radio altimeter, radio compass, and astrocompass for Polar flying.

ARMAMENT: See individual model descriptions of military versions.

DIMENSIONS, EXTERNAL:
- Diameter of main rotor: 21·29 m (69 ft 10⅜ in)
- Diameter of tail rotor: 3·91 m (12 ft 9⅞ in)
- Distance between rotor centres: 12·65 m (41 ft 6 in)
- Length overall, rotors turning: 25·24 m (82 ft 9¾ in)
- Length of fuselage, excl tail rotor: 18·17 m (59 ft 7⅜ in)
- Width of fuselage: 2·50 m (8 ft 2½ in)
- Height overall: 5·65 m (18 ft 6½ in)
- Wheel track: 4·50 m (14 ft 9 in)
- Wheelbase: 4·26 m (13 ft 11¾ in)
- Fwd passenger door: Height: 1·41 m (4 ft 7¼ in)
  - Width: 0·82 m (2 ft 8¼ in)
- Rear passenger door: Height: 1·70 m (5 ft 7 in)
  - Width: 0·84 m (2 ft 9 in)
- Rear cargo door: Height: 1·82 m (5 ft 11½ in)
  - Width: 2·34 m (7 ft 8¼ in)

DIMENSIONS, INTERNAL:
- Passenger cabin: Length: 6·36 m (20 ft 10¼ in)
  - Width: 2·34 m (7 ft 8¼ in)
  - Height: 1·80 m (5 ft 10¾ in)
- Cargo hold (freighter):
  - Length at floor: 5·34 m (17 ft 6¼ in)
  - Width: 2·34 m (7 ft 8¼ in)
  - Height: 1·80 m (5 ft 10¾ in)
  - Volume: approx 23 m^3 (812 cu ft)

AREA:
- Main rotor disc: 355 m^2 (3,828 sq ft)

WEIGHTS:
- Weight empty:
  - passenger version: 6,799 kg (14,990 lb)
  - cargo version: 6,624 kg (14,603 lb)
- Max payload: internal: 4,000 kg (8,820 lb)
  - external: 3,000 kg (6,614 lb)
- Fuel, standard tanks: 1,450 kg (3,197 lb)
- Fuel, with 2 auxiliary tanks: 2,870 kg (6,327 lb)
- Normal T-O weight: 11,100 kg (24,470 lb)
- T-O weight with 28 passengers, each with 15 kg (33 lb) of baggage: 11,570 kg (25,508 lb)
- T-O weight with 2,500 kg (5,510 lb) of slung cargo: 11,428 kg (25,195 lb)
- Max T-O weight for VTO: 12,000 kg (26,455 lb)

PERFORMANCE:
- Max level speed at 1,000 m (3,280 ft):
  - Normal AUW: 140 knots (260 km/h; 161 mph)
- Max level speed at S/L:
  - Normal AUW: 135 knots (250 km/h; 155 mph)
  - Max AUW: 124 knots (230 km/h; 142 mph)
  - With 2,500 kg (5,510 lb) of slung cargo: 97 knots (180 km/h; 112 mph)
- Max cruising speed:
  - Normal AUW: 122 knots (225 km/h; 140 mph)
  - Max AUW: 97 knots (180 km/h; 112 mph)
- Service ceiling: 4,500 m (14,760 ft)
- Hovering ceiling IGE at normal AUW: 1,900 m (6,233 ft)
- Hovering ceiling OGE at normal AUW: 800 m (2,625 ft)
- Ranges:
  - Cargo version at 1,000 m (3,280 ft), with standard fuel, 5% reserves:
    - Normal AUW: 251 nm (465 km; 289 miles)
    - Max AUW: 240 nm (445 km; 276 miles)
  - With 28 passengers at 1,000 m (3,280 ft), with 20 min fuel reserves: 270 nm (500 km; 311 miles)
  - Ferry range of cargo version, with auxiliary fuel, 5% reserves: 647 nm (1,200 km; 745 miles)

MIL Mi-10 (V-10)

NATO reporting name: Harke

This flying crane development of the Mi-6 was demonstrated at the 1961 Soviet Aviation Day display at Tushino, having flown for the first time in the previous year. Above the line of the cabin windows the two helicopters are almost identical, but the depth of the fuselage is reduced considerably on the Mi-10, and the tailboom is deepened so that the flattened undersurface runs unbroken to the tail. The Mi-10 also lacks the fixed wings of the Mi-6.

Items which are interchangeable between the Mi-6 and Mi-10 include the power plant, transmission system and reduction gearboxes, swashplate assembly, main and tail rotors, control system and most items of equipment. The power of the Soloviev turboshaft engines remains constant up to 3,000 m (9,850 ft) and to an ambient air temperature of 40°C at sea level. The aircraft will maintain level flight on one engine. Full navigation equipment and an autopilot permit all-weather operation, by day and night.

The tall long-stroke quadricycle landing gear, with wheel track exceeding 6·0 m (19 ft 8 in) and clearance under the fuselage of 3·75 m (12 ft 3½ in) with the aircraft fully loaded, enables the Mi-10 to taxi over a load it is to carry and to accommodate loads as bulky as a pre-fabricated building.

Use can be made of interchangeable wheeled cargo platforms which are held in place by hydraulic grips controllable from either the cockpit or a remote panel. Using these grips without a platform, cargoes up to 20 m (65 ft 7 in) long, 10 m (32 ft 9½ in) wide and 3·1 m (10 ft 2 in) high can be lifted and secured in 1½ to 2 minutes. The cabin can accommodate additional freight or passengers.

A closed-circuit TV system, with cameras scanning forward from under the rear fuselage and downward through the sling hatch, is used to observe the payload and main landing gear, touchdown being by this reference. The TV system replaces the retractable undernose 'dustbin' fitted originally.

The following details refer to the standard Mi-10, which has been operated by both Aeroflot and the Soviet armed forces and is available for export. About 55 are believed to have been delivered by 1977, with production resuming at a modest rate after a six-year break. Some have been

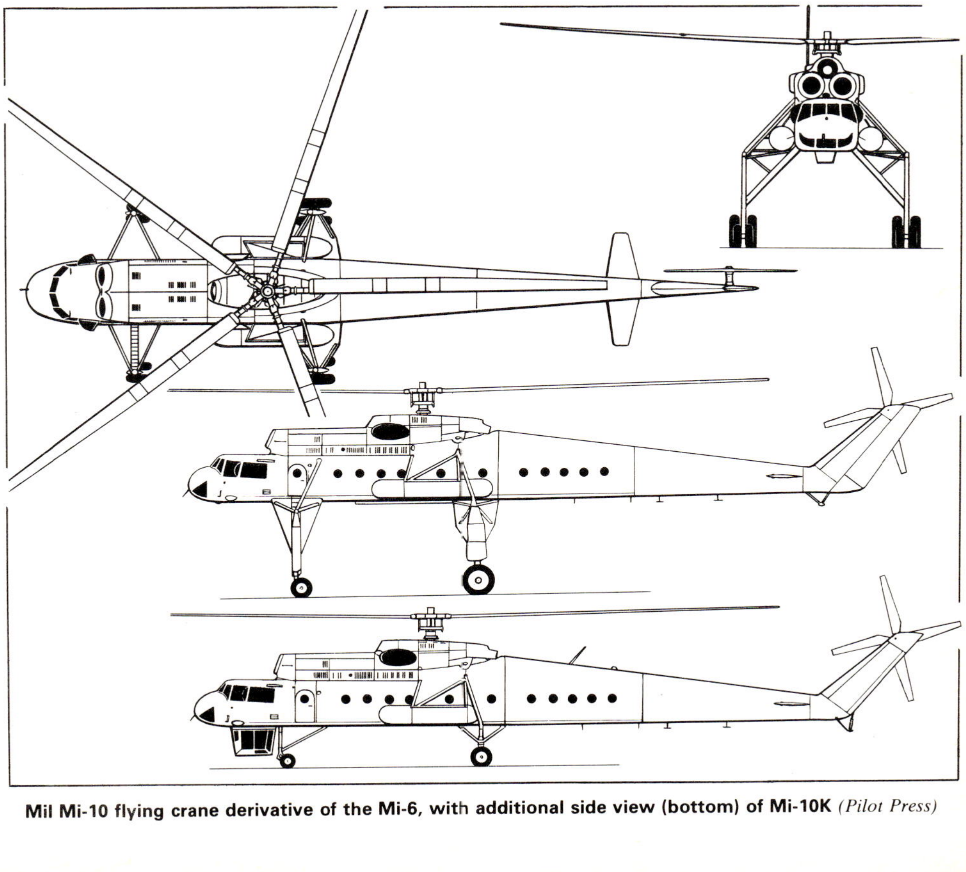

Mil Mi-10 flying crane derivative of the Mi-6, with additional side view (bottom) of Mi-10K *(Pilot Press)*

exported to Iraq and Pakistan.

Type: Heavy flying-crane helicopter.

Rotor System: Same as for Mi-6, except that main rotor shaft is inclined forward at an angle of only 0° 45′.

Fuselage: Conventional all-metal riveted semi-monocoque structure.

Tail Unit: Same as for Mi-6.

Landing Gear: Non-retractable quadricycle type, with twin wheels on each unit. All units fitted with oleo-pneumatic shock-absorbers. Telescopic main legs. Main wheels size 1,230 × 260 mm, each with brake. Levered-suspension castoring nose units. Nosewheels size 950 × 250 mm. All landing gear struts are faired. The port nose gear fairing incorporates steps to the crew door. Despite the height of the gear, the Mi-10 can make stable landing and take-off runs at speeds up to 54 knots (100 km/h; 62 mph).

Power Plant: Two 4,101 kW (5,500 shp) Soloviev D-25V turboshaft engines, mounted side by side above cabin, forward of main rotor drive-shaft. Single fuel tank in fuselage and two external tanks, on sides of cabin, with total capacity of 6,340 kg (13,975 lb). Provision for carrying two auxiliary tanks in cabin, to give total fuel capacity of 8,260 kg (18,210 lb). Engine cowling side panels (opened and closed hydraulically) can be used as maintenance platforms when open.

Accommodation: Two pilots and flight engineer accommodated on flight deck, which has bulged side windows to provide an improved downward view. Flight deck is heated and ventilated and has provision for oxygen equipment. Crew door is immediately aft of flight deck on port side. Main cabin can be used for freight and/or passengers, 28 tip-up seats being installed along the side walls. Freight is loaded into this cabin through a door on the starboard side, aft of the rear landing gear struts, with the aid of a boom and 200 kg (440 lb) capacity electric winch. In addition to the cargo platform described earlier, the Mi-10 has external sling gear as standard equipment. This can be used in conjunction with a winch controlled from a portable control panel inside the cabin. The winch can be also be used to raise loads of up to 500 kg (1,100 lb) while the aircraft is hovering on rescue and other duties, via a hatch in the cabin floor.

Systems, Avionics and Equipment: Generally as for Mi-6, including APU.

Dimensions, external:
- Diameter of main rotor: 35·00 m (114 ft 10 in)
- Diameter of tail rotor: 6·30 m (20 ft 8 in)
- Distance between rotor centres: 21·24 m (69 ft 8 in)
- Length overall, rotors turning: 41·89 m (137 ft 5½ in)
- Length of fuselage: 32·86 m (107 ft 9¾ in)
- Ground clearance under fuselage: 3·75 m (12 ft 3½ in)
- Height overall: 9·80 m (32 ft 2 in)
- Wheel track (c/l of shock-struts):
  - nosewheels: 6·01 m (19 ft 8¾ in)
  - main wheels: 6·92 m (22 ft 8½ in)
- Wheelbase: 8·29 m (27 ft 2½ in)
- Cargo platform: Length: 8·53 m (28 ft 0 in)
  - Width: 3·54 m (11 ft 7¼ in)
- Crew door: Height: 1·35 m (4 ft 5¼ in)
  - Width: 0·78 m (2 ft 6¾ in)
  - Height to sill: 3·91 m (12 ft 10¼ in)
- Freight loading door: Height: 1·56 m (5 ft 1½ in)
  - Width: 1·26 m (4 ft 1½ in)
  - Height to sill: 3·92 m (12 ft 10½ in)
- Cabin floor hatch: Diameter: 1·00 m (3 ft 3¼ in)

Dimensions, internal:
- Cabin: Length: 14·04 m (46 ft 0¾ in)
  - Width: 2·50 m (8 ft 2½ in)
  - Height: 1·68 m (5 ft 6 in)
  - Volume: approx 60 m³ (2,120 cu ft)

Weights:
- Weight empty: 27,300 kg (60,185 lb)
- Max payload on platform, incl platform: 15,000 kg (33,070 lb)
- Max slung payload: 8,000 kg (17,635 lb)
- T-O weight with slung cargo: 38,000 kg (83,775 lb)
- Max T-O weight: 43,700 kg (96,340 lb)

Performance:
- Max level speed at max T-O weight: 108 knots (200 km/h; 124 mph)
- Cruising speed at max T-O weight: 97 knots (180 km/h; 112 mph)
- Service ceiling (limited): 3,000 m (9,850 ft)
- Range with platform payload of 12,000 kg (26,455 lb): 135 nm (250 km; 155 miles)

MIL Mi-10K

First displayed publicly in Moscow on 26 March 1966, the Mi-10K is a development of the Mi-10 with a number of important design changes, most apparent of which are a reduction in the height of the landing gear and a more slender tail rotor support structure.

It can be operated by a crew of only two pilots. This is made possible by the provision of an additional cockpit gondola under the front fuselage, with full flying controls and a rearward-facing seat. By occupying this seat, one of the pilots can control the aircraft in hovering flight and, at the same time, have an unrestricted view of cargo loading, unloading and hoisting, which are also under his control.

Mil Mi-10K preparing to lift a 10-tonne sheet steel drum to the top of a tower at the Sinarski pipe works in Kamenski Uralski *(Tass)*

In the Mi-10K, the maximum slung payload was initially 11,000 kg (24,250 lb) and was expected to be increased further to 14,000 kg (30,865 lb) by using Soloviev D-25VF turboshaft engines, uprated to 6,500 shp each. Fuel capacity of the Mi-10K, in standard internal and external tanks, is 9,000 litres (1,980 Imp gallons). The rotor turns at 120 rpm.

Dimensions, external:
- Generally as for Mi-10, except:
- Height overall: 7·80 m (25 ft 7 in)
- Wheel track: 5·00 m (16 ft 4¾ in)
- Wheelbase: 8·74 m (28 ft 8 in)
- Door sill heights:
  - Crew door: 1·81 m (5 ft 11 in)
  - Freight door: 1·82 m (5 ft 11½ in)

Weights:
- Weight empty: 24,680 kg (54,410 lb)
- Max fuel load with ferry tanks in cabin: 8,670 kg (19,114 lb)
- Max T-O weight with slung cargo: 38,000 kg (83,775 lb)

Performance:
- Cruising speed, empty: 135 knots (250 km/h; 155 mph)
- Max cruising speed with slung load: 109 knots (202 km/h; 125 mph)
- Service ceiling: 3,000 m (9,850 ft)
- Ferry range with auxiliary fuel: 428 nm (795 km; 494 miles)

MIL Mi-14 (V-14)

NATO reporting name: Haze

An accompanying photograph shows the Mi-14 shore-based anti-submarine helicopter which is in standard service with the Soviet Navy as a replacement for the Mi-4. Clearly derived from the Mi-8, it is known to NATO as 'Haze'.

No details are available officially. Features evident in the photograph include a boat-hull planing bottom on the fuselage, and a sponson on each side at the rear, to provide a degree of amphibious capability; a large undernose radome; and a towed magnetic anomaly detection (MAD) bird stowed against the rear of the fuselage pod. The landing gear of 'Haze' is fully retractable.

Overall dimensions and dynamic components of 'Haze' are generally similar to those of the Mi-8. The power plant consists of two uprated Isotov TV3-117 turboshafts of the kind fitted also in the Mi-17 and Mi-24, and easily identified by their shorter nacelles (note intake position relative to cabin door).

It is assumed that 'Haze' is the 'float-equipped version of the Mi-8' which was known to be under test in the Soviet Union in early 1974, with the designation V-14. About 80 are believed to have been delivered, with production continuing at a rate of 25 per year. A few were reported to have been delivered to Bulgaria in 1979.

MIL Mi-17

First displayed in public at the 1981 Paris Air Show, the Mi-17 combines the airframe of the Mi-8 with the uprated power plant of the Mi-14. The example exhibited at Le Bourget (CCCP-17718) had flown from Moscow to Paris, in short stages, in company with the Mi-26. Production has already started.

The general description of the Mi-8 applies also to the Mi-17, except that the tail rotor is on the port side of the vertical stabiliser (as on Mi-14). Externally, the new power plant can be identified by the shorter nacelles, the air intakes extending forward only to the mid-point of the door on the port side at the front of the cabin. Also new is

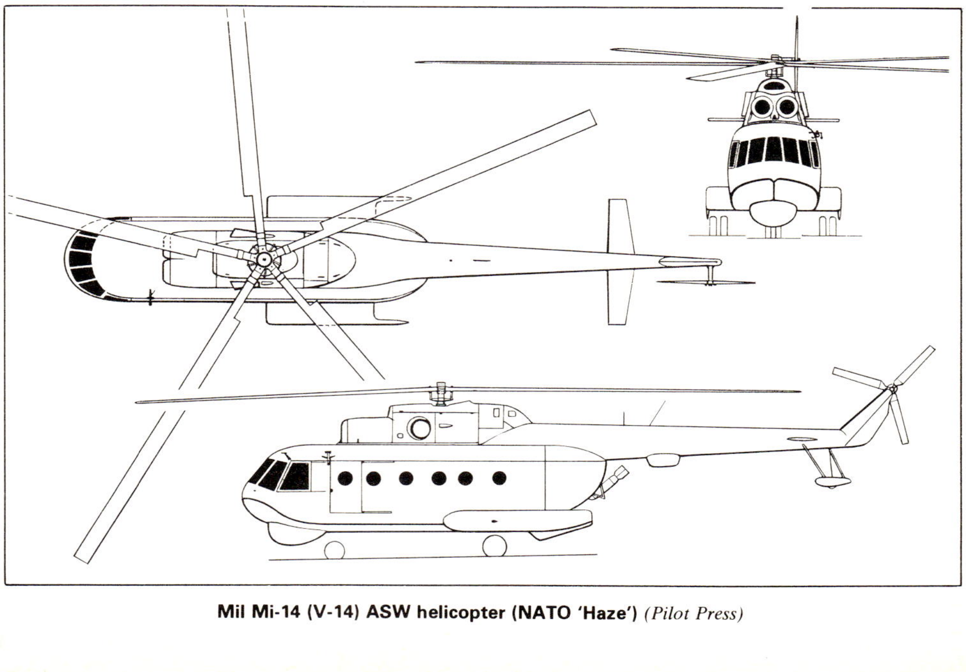

Mil Mi-14 (V-14) ASW helicopter (NATO 'Haze') *(Pilot Press)*

Mil Mi-14 twin-turboshaft helicopter, in service with the Soviet Navy for shore-based anti-submarine duties

the small orifice on each side forward of the jetpipe. Take-off rating of each of the two Isotov TV3-117MT turboshafts is 1,417 kW (1,900 shp), which offers a considerable improvement in performance compared with the Mi-8. Correct rotor speed is maintained automatically by a system which also synchronises the output of the two engines. Loss of power by one engine is offset automatically by increasing the output of the other. Should one engine stop, the output of the other is increased to a contingency rating of 1,640 kW (2,200 shp), enabling the flight to continue. An APU is carried to start the turboshafts pneumatically. If required, the engine air intakes can be fitted with deflectors to prevent the ingestion of sand, dust or foreign objects at unprepared landing sites.

Cabin configuration and payloads are unchanged by comparison with the Mi-8.

Dimensions, external and internal:
- As for Mi-8, except:
- Distance between rotor centres 12·661 m (41 ft 6½ in)
- Length overall, rotors turning 25·352 m (83 ft 2 in)
- Length of fuselage, excl tail rotor 18·424 m (60 ft 5⅜ in)
- Height to top of main rotor hub 4·755 m (15 ft 7¼ in)
- Wheel track 4·510 m (14 ft 9½ in)
- Wheelbase 4·281 m (14 ft 0½ in)

Weights:
- Weight empty, equipped 7,100 kg (15,653 lb)
- Normal T-O weight 11,100 kg (24,470 lb)
- Max T-O weight 13,000 kg (28,660 lb)

Performance (A, at normal T-O weight; B, at max T-O weight):
- Max level speed: B 135 knots (250 km/h; 155 mph)
- Max cruising speed: B 129 knots (240 km/h; 149 mph)
- Service ceiling: A 5,000 m (16,400 ft)
- B 3,600 m (11,800 ft)
- Hovering ceiling OGE: A 1,760 m (5,775 ft)
- Range with max standard fuel, 5% reserves:
  - A 267 nm (495 km; 307 miles)
  - B 251 nm (465 km; 289 miles)
- Range with auxiliary fuel:
  - A 513 nm (950 km; 590 miles)

MIL Mi-18

Aviaexport announced in 1981 that it expected soon to add a new helicopter designated Mi-18 to its range of aircraft available for export. No details were available in mid-1981.

MIL Mi-24 (and A-10)

NATO reporting name: Hind

This assault helicopter was known to exist for some two years before photographs became available to the technical press in early 1974. The two versions shown in those first photographs were each capable of carrying a squad of eight combat-equipped troops, and had attachments under their auxiliary wings for a variety of ordnance, to keep down the heads of enemy troops in the drop zone and to attack ground targets, including tanks. At least two units of approximate squadron strength were based in eastern Europe by the Spring of 1974. Subsequently, the Mi-24 has developed in two basic forms, one configured for optimum efficiency as an assault transport, while retaining heavy armament; the other as an advanced gunship, able to engage in air-to-air combat with other helicopters, and with secondary transport capability. Deliveries of all models are believed to exceed 1,000, with production of the gunship continuing at a rate of more than 15 a month. Well over 450 of the gunship versions were deployed in 1981 by the Soviet air forces, including full regiments of Mi-24s based at Parchim and Stendal, northwest and west of Berlin, near the border with West Germany. Other operators include the Warsaw Pact air forces of Bulgaria, Czechoslovakia, East Germany, Hungary and Poland. Export deliveries, mostly of the gunship version, have been made to Afghanistan (at least 30), Algeria (20, including 'Hind-As'), Iraq (41), Libya (26) and South Yemen (6). Many Mi-24s have been operated by Soviet forces in Afghanistan since December 1979.

The basic airframe, power plant and transmission system appear to be common to all versions of the Mi-24, with differences in armament, operational equipment and tail rotor location. In addition, the gunship has completely new crew accommodation forward of the engine inlets and above the fuselage floor. Five major variants of which details may be published are known by the following NATO reporting names:

Hind-A. Armed assault helicopter, with large enclosed flight deck for crew of four, comprising pilot, co-pilot, gunner/navigator and forward observer. Auxiliary wings, with considerable anhedral, each carry three weapon stations for heavy armament, supplemented by large-calibre machine-gun in nose. Anti-torque rotor, originally on starboard side of offset tail pylon, repositioned to port side on later and converted aircraft. Initial production Mi-24s were of this version.

Hind-B. Similar to 'Hind-A' except that auxiliary wings have neither anhedral nor dihedral, and carry only the two inboard weapon stations on each side. This version is believed to have preceded 'Hind-A' and was not built in large numbers.

Hind-C. Generally similar to late-model 'Hind-A' but without nose gun and undernose blister fairing, and no missile rails at wingtips.

Hind-D. Basically similar to late-model 'Hind-A', with tail rotor on port side, but with front fuselage completely redesigned for primary gunship role. Tandem stations for weapon operator (in nose) and pilot have individual canopies. Front canopy hinged to open sideways, to starboard; footstep under starboard side of fuselage for access to pilot's rearward-hinged door. Rear seat raised to give pilot an unobstructed forward view. Probe fitted forward

Mil Mi-17, a development of the Mi-8 with uprated engines *(Air Portraits)*

of top starboard corner of bulletproof windscreen at extreme nose may be part of low-airspeed sensing equipment, to indicate optimum conditions for minimum dispersion of 57 mm rockets. Under nose is a four-barrel Gatling-type 12·7 mm machine-gun in a turret with a wide range of movement in azimuth and elevation, providing air-to-air as well as air-to-surface capability. Undernose pack for sensors, possibly including radar and low-light-level TV. (Reports that forward-looking infra-red might be fitted were premature, as such equipment was not yet ready for service in the Soviet Union.) Wing armament of 'Hind-A' retained, but forward-looking (electro-optical?) sensor transferred from top of port inner pylon to wingtip. Many small antennae and blisters. Extended nosewheel leg to increase ground clearance of sensor pack; nosewheels semi-exposed when retracted.

Hind-E. As 'Hind-D', for Soviet armed forces, but with four laser-homing tube-launched anti-tank missiles (NATO 'Spiral') instead of 'Swatters', and structural hardening by substitution of steel and titanium for aluminium in critical components. According to Gen David Jones, Chairman of the US Joint Chiefs of Staff, more than 100 'Hind-Es' were deployed in Warsaw Pact forward areas in Europe in early 1981, with others in the Far East.

The helicopter known to the Soviet authorities as **A-10**, in which various FAI-approved records in class E1 have been set since 1975, is now known to be an Mi-24, with the 'Hind-A/C' type of front fuselage.

Pilot on all 1975 record flights was Galina Rastorgoueva, a test pilot and engineer. She was accompanied by Ludmila Polyanskaya, who is employed as navigator on Il-18 airliners of Aeroflot. On 18 July 1975, they averaged 180·480 knots (334·464 km/h; 207·826 mph) around a 100 km circuit, setting a new women's speed record. On 1 August, a speed of 178·624 knots (331·023 km/h; 205·688 mph) around 500 km set both general and women's records, as did an average speed of 179·500 knots (332·646 km/h; 206·697 mph) over 1,000 km on 13 August. Two women's time-to-height helicopter records followed, with a time of 2 min 33·5 s to 3,000 m on 8 August, and 7 min 43 s to 6,000 m on 26 August. At this period, the A-10 was stated to be powered by two 1,118·5 kW (1,500 shp) Isotov TV2-117A turboshaft engines, as fitted to the Mi-8.

Latest A-10 record confirmed by the FAI is a speed of 198·9 knots (368·4 km/h; 228·9 mph) set by Gourguen Karapetyan over a 15/25 km course near Moscow on 21 September 1978. This A-10 had two TV3-117 turboshaft engines, each rated at 1,640 kW (2,200 shp), which conforms with the known power plant of the production Mi-24.

Except where indicated, the following details apply to all current operational versions of the Mi-24:

ROTOR SYSTEM: Five-blade main rotor and three-blade tail rotor; latter on port side of offset tail fin. Main rotor blades, with glassfibre skin, on cast titanium head. Balance tab and electrical leading-edge de-icing on each blade.

FUSELAGE: Conventional all-metal semi-monocoque structure of pod and boom type. Forward portion, above shallow floor structure, differs with role.

AUXILIARY WINGS: Cantilever shoulder wings of tapered planform, with about 16° anhedral and 20° incidence. No movable surfaces.

TAIL UNIT: Swept fin, offset a few degrees, serves also as tail rotor pylon. Variable-incidence horizontal stabiliser at base of fin.

LANDING GEAR: Tricycle type, with rearward-retracting twin-wheel nose unit, and single-wheel main units with oleo-pneumatic shock-absorbers and low-pressure tyres. Main units retract rearward and inward into the aft end of the fuselage pod, turning through 90° to stow almost vertically, discwise to the longitudinal axis of the fuselage, under prominent blister fairings. Tubular tripod skid assembly protects tail rotor in a tail-down take-off or landing.

POWER PLANT: Two Isotov TV3-117 turboshaft engines, with max rating of 1,640 kW (2,200 shp), mounted side by side above the cabin, with their output shafts driving rearward to the main rotor shaft through a combining gearbox. Optional deflectors for dust and foreign objects forward of air intakes.

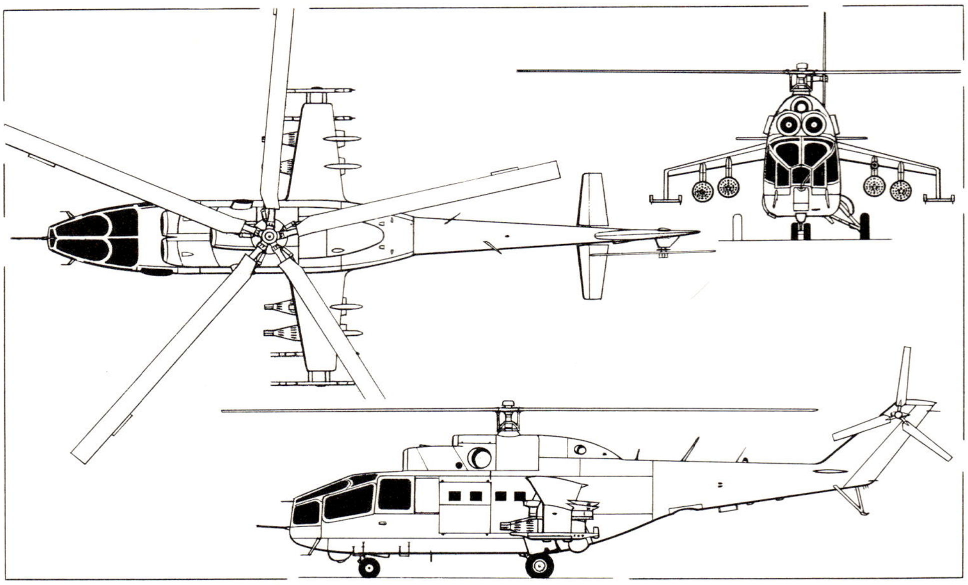

Mil Mi-24 assault helicopter, in the form known to NATO as 'Hind-A' *(Pilot Press)*

'Hind-A', first major production version of the Mil Mi-24 with original starboard-side tail rotor

ACCOMMODATION ('Hind-A'): Crew of four; eight fully-equipped troops in main cabin. Access to flight deck via large rearward-sliding blistered transparent panel which forms the aft flight deck window on the port side, and a large upward-hinged window forward of this. At front of passenger cabin on each side is a large door, divided horizontally into two sections which are hinged to open upward and downward respectively. Optically flat bulletproof glass windscreen, with wiper, for gunner. Armour protection for crew and some vital components.

SYSTEMS: Dual electrical system, with three generators. Stability augmentation system. Electro-thermal de-icing system for main and tail rotor blades.

AVIONICS: Include ADF navigation system with map display.

ARMAMENT ('Hind-A'): One 12·7 mm machine-gun in nose, probably slaved to undernose sighting system. Rails for four 'Swatter' anti-tank missiles under end-plate pylons at wingtips. Four underwing pylons for UB-32 rocket pods (each thirty-two S-5 type 57 mm rockets), special bombs, or other stores.

DIMENSIONS, EXTERNAL (estimated):

Diameter of main rotor	17·00 m (55 ft 9 in)
Diameter of tail rotor	3·90 m (12 ft 9½ in)
Length overall	17·00 m (55 ft 9 in)
Height overall	4·25 m (14 ft 0 in)

WEIGHTS (estimated):

Max external weapons	1,275 kg (2,800 lb)
Normal T-O weight	10,000 kg (22,000 lb)

MIL Mi-26

NATO reporting name: Halo

Design of the Mi-26 heavy-lift helicopter began in the early 1970s to meet the requirement for an aircraft of greater capability than the Mi-6 and Mi-10. Except for the four-engined twin-rotor Mi-12 (see 1977-78 *Jane's*), which did not progress beyond prototype testing, it is the heaviest helicopter yet flown anywhere in the world. Its rotor diameter is smaller than that of the Mi-6 and Mi-10, but this is offset by the fact that the Mi-26 is the first helicopter to operate successfully with an eight-blade main rotor.

The redesigned nose of the Mi-24 ('Hind-D') gunship is shown clearly in this view *(Letectvi + Kosmonautika)*

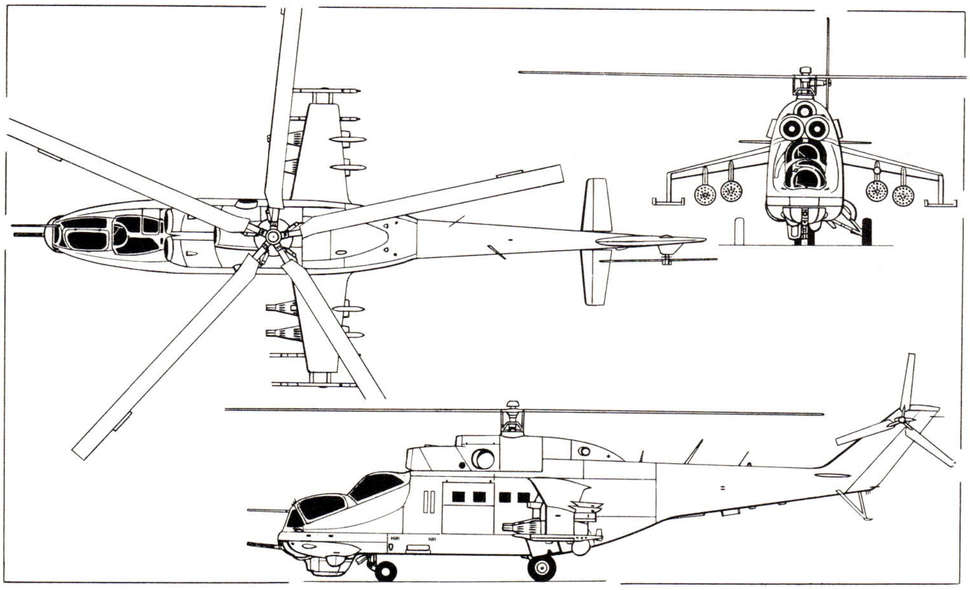

The gunship version of the Mil Mi-24 known to NATO as 'Hind-D' *(Pilot Press)*

It has obvious military applications, with a payload and cargo hold very similar in size to those of a Lockheed C-130 Hercules. To meet also Soviet Ministry of Civil Aviation requirements, for operation in Siberia and northern swamp and tundra areas of the USSR, emphasis had to be placed on reliability, especially when operating into unprepared landing sites. According to Mr Marat Tishchenko, General Designer in charge of the Mil bureau, this (plus, no doubt, the need to ensure torsional stiffness) explains why the main rotor blades have conventional steel spars.

Use of titanium for the rotor hub helped the Mil bureau to meet the official requirement of an empty weight only 50 per cent of the aircraft's maximum permissible take-off weight. A further contribution to weight-saving resulted from the decision to design the main gearbox in-house. The end product offers an impressive power to weight ratio, despite the need to absorb an unprecedented input from the two Lotarev D-136 turboshaft engines that power the Mi-26.

Representatives of the Mil bureau claimed at the 1981 Paris Air Show that the Mi-26 had already undergone two years of flight development, and that D-136 engines had amassed more than 13,000 hours of running on the test-bench and in the air. The Mi-26 (CCCP-06141) exhibited at the Show had been flown from Moscow to Le Bourget by Mil's chief test pilot, G. R. Karapetyan, via Smolensk, Vilnyus, Warsaw, Prague and Frankfurt. It was stated to be one of several prototypes or pre-production examples then flying. With development completed, production was said to be imminent.

Type: Twin-turboshaft heavy transport helicopter.

Rotor System: Eight-blade main rotor, with flapping and drag hinges; five-blade tail rotor, mounted on starboard side of tail fin. Each main rotor blade consists of a one-piece tubular steel spar and 26 carbonfibre aerofoil-shape full-chord pockets of varying form, with internal ribs and stiffeners. Blades attached to titanium hub of unconventional design. Hydraulically-powered cyclic and collective pitch controls actuated by small parallel jacks, with redundant autopilot and stability augmentation system inputs. Tail rotor blades made of carbonfibre. Leading-edge of main and tail rotor blades heated electrically for anti-icing.

Rotor Drive: Conventional transmission. Main gearbox type VR-26, rated at 14,914 kW (20,000 hp), is fan-cooled, with air intake above rear of engine cowlings.

Fuselage: Conventional all-metal riveted semi-monocoque structure of pod and boom type, with clam-shell rear loading doors and ramp. Flattened under-surface to tailboom.

Tail Unit: Sweptback vertical stabiliser, carrying tail rotor, is offset to port. Ground-adjustable variable-incidence horizontal stabiliser mounted on leading-edge of vertical stabiliser, a short distance above the tailboom.

Landing Gear: Non-retractable tricycle type, with twin wheels on each unit. Main-wheel tyres size 1,120 × 450. Retractable tailskid at end of tailboom to permit unrestricted approach to rear cargo doors. Length of main legs can be adjusted hydraulically to facilitate loading through rear doors and to permit landing on a slope.

Power Plant: Two 8,500 kW (11,400 shp) Lotarev D-136 free-turbine turboshaft engines, mounted side by side above cabin, forward of main rotor driveshaft. Air intakes designed to prevent foreign object ingestion, and provided with both electrical and bleed air anti-icing systems. System for synchronising the output of the engines and maintaining constant rotor rpm. If one engine fails, output of the other is increased to maximum automatically. Fuel in eight underfloor tanks, feeding into two supply tanks above engines, which permit gravity feed for a period in emergencies.

Accommodation: Crew of five, consisting of pilot (on port side) and co-pilot side by side, flight engineer behind pilot, navigator behind co-pilot on flight deck, and loadmaster in freight hold. About 20 tip-up seats along each side wall of hold. Four large blistered side windows

Loading vehicles on board a Mil Mi-26 *(Tass)*

Mil Mi-26 heavy-lift helicopter (two Lotarev D-136 turboshaft engines) photographed at 1981 Paris Air Show *(Brian M. Service)*

on flight deck. Forward pair swing open slightly outward and rearward. Downward-hinged doors, with integral airstairs, at front of hold on port side, and on each side of hold aft of main landing gear units. Hold is loaded via a downward-hinged lower door, with integral folding ramp, and two clamshell upper doors which form rear wall of hold when closed. Doors are opened and closed hydraulically, with backup handpump for emergency use. Two electric winches on overhead rails, each with capacity of 2,500 kg (5,511 lb), enable loads to be transported along cabin. Flight deck and hold fully air-conditioned.

SYSTEMS: Small size of jacks for rotor head controls implies use of hydraulic system of much higher pressure than usual in Soviet helicopters. APU supplies hydraulic, electrical and air-conditioning systems on ground.

AVIONICS AND EQUIPMENT: All items necessary for day and night operations in all weathers are standard, including weather radar in the hinged (to starboard) nosecone, Doppler, map display, HSI, and automatic hover system. APU under flight deck, with intake louvres (forming fuselage skin when closed) and exhaust on starboard side. Attachment for sling loads in bottom of centre-fuselage. Closed-circuit TV cameras to observe extendable main landing gear legs.

DIMENSIONS, EXTERNAL:
- Diameter of main rotor: 32·00 m (105 ft 0 in)
- Diameter of tail rotor: 7·61 m (24 ft 11½ in)
- Length of fuselage, excl tail rotor: 33·727 m (110 ft 8 in)
- Height to top of rotor head: 8·055 m (26 ft 5¼ in)
- Wheelbase: 8·95 m (29 ft 4½ in)

DIMENSIONS, INTERNAL:
- Freight hold:
  - Length, ramp trailed: 15·00 m (49 ft 2½ in)
  - Width: 3·20 m (10 ft 6 in)
  - Height: 3·15 m (10 ft 4 in)

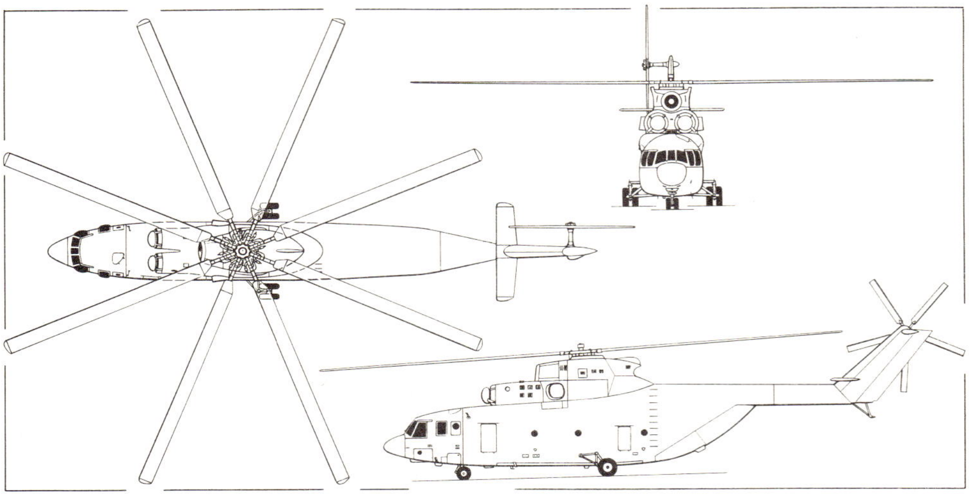

Mil Mi-26, first helicopter to operate successfully with an eight-blade main rotor *(Pilot Press)*

WEIGHTS:
- Weight empty: 28,200 kg (62,170 lb)
- Max payload, internal or external: 20,000 kg (44,090 lb)
- Normal T-O weight: 49,500 kg (109,125 lb)
- Max T-O weight: 56,000 kg (123,450 lb)

PERFORMANCE:
- Max level speed: 159 knots (295 km/h; 183 mph)
- Normal cruising speed: 137 knots (255 km/h; 158 mph)
- Service ceiling: 4,500 m (14,760 ft)
- Hovering ceiling OGE, ISA: 1,800 m (5,900 ft)
- Range with max internal fuel at max T-O weight, 5% reserves: 432 nm (800 km; 497 miles)

MIL NAVAL HELICOPTER

The US *Military Posture* statement for FY 1979 contained the remark: "Another new (Soviet) naval helicopter is projected in the mid-1980s for ASW and reconnaissance roles". No details were given.

MYASISHCHEV

This design bureau was formed in 1951, under the leadership of Professor Vladimir Mikhailovich Myasishchev, who died on 14 October 1978, at the age of 76. Although his work was little publicised, he had been responsible for 'translating into Russian' the drawings of the Douglas DC-3 transport, acquired from the USA under a licence agreement in 1936-38, and for the development of several important original types. First product of his own design bureau was the four-jet M-4 bomber (known in the West by the reporting name of 'Bison') which remains in service as a strategic bomber, maritime reconnaissance aircraft and flight refuelling tanker.

Myasishchev later designed a long-range four-jet heavy bomber known as the M-50 to replace the M-4. This was allocated the NATO code name of 'Bounder' and was described in the 1964-65 *Jane's*. Changing requirements limited 'Bounder' to a research role. Myasishchev spent the period from 1960 to 1967 as Director of TsAGI, before returning to design.

Myasishchev M-4 ('Bison-B') maritime reconnaissance aircraft *(Royal Air Force)*

MYASISHCHEV M-4

NATO reporting name: Bison

Three major production versions of this four-jet aircraft were identified by NATO code names, as follows:

Bison-A. The Soviet Union's first operational four-jet strategic bomber. Design work began in 1951. The last drawing was despatched to the works on 1 May 1952, and the prototype bomber was displayed initially over Moscow on 1 May 1954. Comparable with early versions of Boeing B-52 Stratofortress. Powered by four 85·3 kN (19,180 lb st) Mikulin AM-3D turbojets, buried in wing-roots. Defensive armament of ten 23 mm cannon in twin-gun turrets in tail, above fuselage fore and aft of wing and under fuselage fore and aft of bomb bays, believed necessary because of aircraft's limited operational ceiling. About 30 'Bison-As' are serving currently as flight refuelling tankers for the 43 remaining M-4 bombers and the Tu-95 'Bears' of the Soviet strategic bomber force, carrying a hose-reel unit in the bomb bay.

Bison-B. Maritime reconnaissance version, described in 1979-80 *Jane's*. Few remain.

Bison-C. Improved maritime reconnaissance version, described in 1979-80 *Jane's*. Few remain.

DIMENSIONS, EXTERNAL ('Bison-A'):
- Wing span: 50·48 m (165 ft 7½ in)
- Length overall: 47·20 m (154 ft 10 in)
- Tailplane span: 15·00 m (49 ft 2½ in)

WEIGHT ('Bison-A'):
- Max T-O weight: 158,750 kg (350,000 lb)

PERFORMANCE ('Bison-A', estimated):
- Max level speed at 11,000 m (36,000 ft): 485 knots (900 km/h; 560 mph)
- Service ceiling: 13,700 m (45,000 ft)
- Range at 450 knots (835 km/h; 520 mph) with 4,500 kg (10,000 lb) of bombs: 6,075 nm (11,250 km; 7,000 miles)

SUKHOI

CHIEF DESIGNER OF SUKHOI BUREAU: E. A. Ivanof

Pavel Osipovich Sukhoi, who headed this design bureau from 1938 until his death in September 1975, helped to design the ANT-25 and had a share in the construction of the 'Rodina' before the second World War; his Su-2 attack aeroplane was used in the war. He was also responsible for one of the jet aircraft in the 1947 Soviet Aviation Day display.

Nearly a decade later, on 24 June 1956, there appeared over Tushino new sweptwing and delta-wing fighters from Sukhoi's design team. Both aircraft subsequently entered squadron service with the Soviet Air Force, as the Su-7 and Su-9, and have been followed by other Sukhoi designs.

SUKHOI Su-7B

NATO reporting names: Fitter-A and Moujik

The prototype of this single-seat ground attack fighter, designated S-1, was flown for the first time by test pilot Kochetov in 1955, and was displayed in prototype form in the fly-past over Moscow on the 1956 Soviet Aviation Day. It was followed by a small pre-production series, which was allocated the official designation **Su-7,** as well as a second prototype, the S-2, embodying certain aerodynamic refinements. After evaluation of the Su-7s, the S-2 was ordered into series production in 1958 as the **Su-7B,** to which NATO gave the reporting name **Fitter-A.** This model appeared in formations of up to 21 aircraft at the 1961 Tushino display. It subsequently became the standard tactical fighter-bomber of the Soviet Air Force, with which about 200 continue in service. Others have been supplied to Afghanistan, Algeria, Czechoslovakia, Egypt, Hungary, India, Iraq, North Korea, Poland, Romania, Syria, Viet-Nam and South Yemen.

The fuselage and tail unit of the Su-7B were almost identical with those of the delta-wing Su-9. Early production models had the pitot boom mounted centrally above the air intake, but it is offset to starboard on later versions, the first of which was the **Su-7BKL**. This introduced the two slim duct fairings along the top of the centre-fuselage, which are seen also on the Su-11, and other detail changes. It was followed by the **Su-7BM**, with uprated engine; twin brake-chutes in a large container at the base of the rudder, instead of a single ribbon-type parachute attached under the rear fuselage; and larger blast panels on the sides of the front fuselage by the muzzles of the wing-root guns, implying the use of cannon with a higher muzzle velocity or rate of fire. A demand for improved capability in operation from short, unprepared fields led next to the **Su-7BMK**, with JATO attachments under the rear fuselage and a low-pressure nosewheel tyre, requiring blistered doors to enclose it when retracted.

A variant of the Su-7 seen first at Domodedovo in 1967 is the two-seat **Su-7U**, with the second cockpit in tandem, aft of the standard cockpit and with a slightly raised canopy. A prominent dorsal spine extends from the rear of the aft canopy to the base of the tail fin. The two-seater, which exists in **Su-7UM** and **Su-7UMK** versions corresponding to the single-seat 'M' and 'MK', is a standard operational trainer and has the NATO reporting name **Moujik**.

The following description applies to the standard Su-7BMK:

TYPE: Single-seat ground attack fighter.

WINGS: Cantilever mid-wing monoplane. No dihedral or anhedral. Sweepback 62° on leading-edges. Conventional all-metal construction. Wing-root chord is extended, giving a straight trailing-edge on inboard section of each wing. Very large area-increasing flaps over entire trailing-edge from root to inboard end of aileron on each wing. No slats or tabs. Two boundary-layer fences on each wing, at approx mid-span and immediately inboard of tip.

FUSELAGE: Conventional all-metal semi-monocoque structure of circular section. Two slim duct fairings along top of centre-fuselage. Two door-type airbrakes, at top and bottom, on each side of rear fuselage.

TAIL UNIT: Cantilever all-metal structure, with sweepback on all surfaces. All-moving horizontal surfaces, with anti-flutter bodies projecting forward from tips. Conventional rudder. No tabs.

LANDING GEAR: Retractable tricycle type, with single wheel on each unit. Nosewheel retracts forward, main units inward into wings. Twin brake-chutes in large container with clamshell doors, at base of rudder.

POWER PLANT: One Lyulka AL-7F-1 turbojet engine, rated at 68·65 kN (15,432 lb st) dry and 98·1 kN (22,046 lb st) with afterburning. Total internal fuel capacity 3,175 kg (7,000 lb). Provision for two external tanks, side by side under fuselage, with total capacity of 952 kg (2,100 lb). Two JATO solid-propellant rocket units can be attached under rear fuselage to shorten T-O run.

ACCOMMODATION: Pilot only, on rocket-powered ejection seat, under rearward-sliding blister canopy.

ARMAMENT: Two 30 mm NR-30 guns, each with 70 rounds, in wing-root leading-edges. Four underwing attachments for rocket pods or bombs (usually two 750 kg and two 500 kg), including nuclear weapons. When underbelly fuel tanks are fitted, max external weapon load is 1,000 kg (2,200 lb).

DIMENSIONS, EXTERNAL:

Wing span	8·93 m (29 ft 3½ in)
Length overall, incl probe	17·37 m (57 ft 0 in)
Height overall	4·57 m (15 ft 0 in)

WEIGHTS:

Weight empty	8,620 kg (19,000 lb)
Normal T-O weight	12,000 kg (26,450 lb)
Max T-O weight	13,500 kg (29,750 lb)

PERFORMANCE:

Max level speed at 11,000 m (36,000 ft):
'clean'
Mach 1·6 (917 knots; 1,700 km/h; 1,055 mph)
with external stores
Mach 1·2 (685 knots; 1,270 km/h; 788 mph)
Max level speed at S/L without afterburning
approx 460 knots (850 km/h; 530 mph)
Max rate of climb at S/L
approx 9,120 m (29,900 ft)/min
Service ceiling 15,150 m (49,700 ft)
Combat radius
172-260 nm (320-480 km; 200-300 miles)
Max range 780 nm (1,450 km; 900 miles)

Sukhoi Su-7BKL single-seat close support fighter of the Soviet Air Force

Su-7UMK two-seat training aircraft (NATO 'Moujik'), in service with the Egyptian Air Force *(Denis Hughes)*

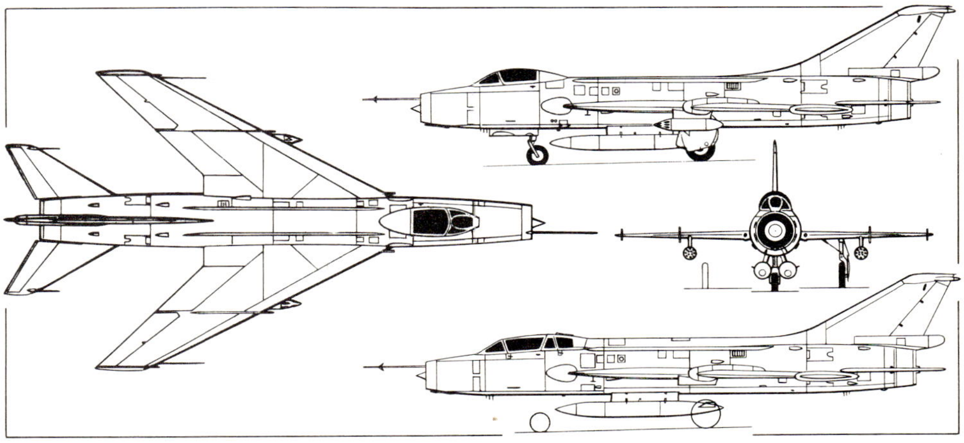

Sukhoi Su-7BMK single-seat close support fighter, with additional side view (bottom) of two-seat Su-7UM *(Pilot Press)*

SUKHOI Su-9 and Su-11

NATO reporting names: Fishpot and Maiden

First seen at Tushino during the 1956 Aviation Day Display, the prototype of these single-seat all-weather fighters (allocated the Sukhoi designation T-3 and the NATO reporting name 'Fishpot-A') had a small conical radome above its engine air intake. This was replaced by a centrebody air intake on the production version, which entered standard service in the Soviet Air Force in two forms as follows:

Su-9 (Fishpot-B). Initial version, with Sukhoi designation T-43, operational since 1959 and still in service. Powered by 88·25 kN (19,840 lb st) Lyulka AL-7F afterburning turbojet. Small-diameter air intake and centrebody housing R1L (NATO 'Spin Scan') S-band radar. Examples included in the Tushino display of 1961 carried four of the Soviet Air Force's then-standard radar-homing air-to-air missiles (NATO reporting name 'Alkali') on underwing attachments, plus two underfuselage fuel tanks side by side. No fixed armament.

The aircraft known as the T-431 in which test pilot Vladimir Ilyushin reached a record height of 28,852 m (94,659 ft) on 14 July 1959 was the first T-43.

Su-11 (Fishpot-C). First seen publicly at the Domodedovo Aviation Day display in 1967, the Su-11 is a much-improved development of the Su-9, with a Lyulka AL-7F-1 turbojet (63 kN; 14,200 lb st dry, 98·1 kN; 22,046 lb st with afterburning) and a standard armament of two underwing missiles (NATO 'Anab'), one with radar homing head and one with infra-red homing head. It also has a lengthened nose of less-tapered form than that of the Su-9, with an enlarged centrebody for the Uragan 5B (NATO 'Skip Spin') X-band radar, and two slim duct fairings along the top of the centre-fuselage, as on the Su-7BKL. The fuselage and tail unit of the two types are almost identical.

There is also a tandem two-seat training version known as the **Su-9U** (NATO reporting name **Maiden**), with a cockpit arrangement similar to that of the two-seat Su-7U ('Moujik'). When retired from operational service, some Su-9s were modified for use as radio-controlled pilotless targets for anti-aircraft missiles.

Although the Su-9 and Su-11 are generally similar in layout to their Mikoyan contemporary, the MiG-21, they are larger and heavier aircraft, with a much more powerful afterburning turbojet. They are less limited in all-weather

Sukhoi Su-9 all-weather fighter, with four underwing mountings for 'Alkali' missiles *(Tass)*

Sukhoi Su-11 single-seat fighter, armed with two of the missiles known to NATO as 'Anab' *(Novosti)*

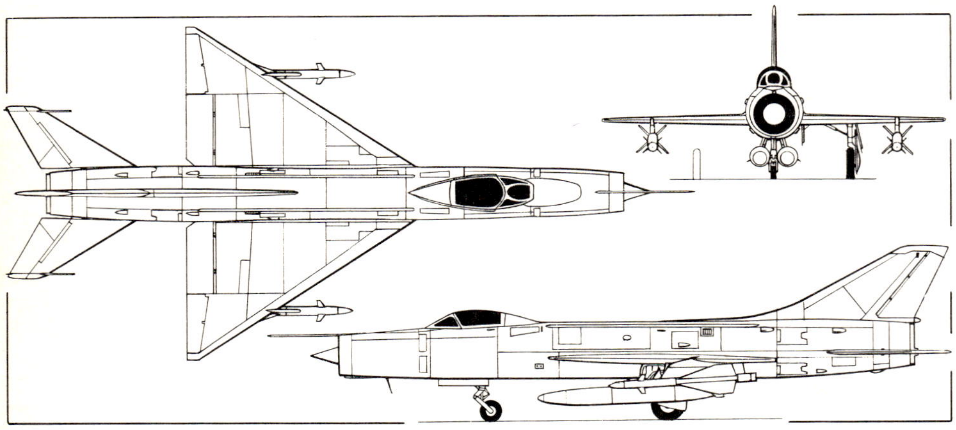

Sukhoi Su-11 single-seat all-weather interceptor *(Pilot Press)*

capability than early versions of the MiG-21. The Sukhoi and Mikoyan 'tailed deltas' were, therefore, regarded as complementary rather than competitive when ordered into production in the late 'fifties. In 1981, the Su-9 and Su-11 continued to form about 15 per cent of the Soviet home defence interceptor force of 2,600 aircraft.

The Su-9 and Su-11 can be distinguished from the MiG-21 by their cleaner airframe, and the absence of both a ventral stabilising fin and fairings on the fuselage forward of the wing-root leading-edges. The cockpit canopy of the single-seat versions is rearward-sliding, whereas that of the MiG-21 is hinged to open either forward about the base of the windscreen or sideways.

The tricycle landing gear of the Su-9 and Su-11 has a wide track, with a single wheel on each unit. The main units retract inward into the wings, the nosewheel forward. Control surfaces appear to be conventional, with a one-piece all-moving tailplane, carrying an anti-flutter body projection at each tip. There are four petal-type airbrakes, in pairs on each side of the rear fuselage.

DIMENSIONS, EXTERNAL (Su-11, estimated):
Wing span 8·43 m (27 ft 8 in)
Length overall, incl probe 17·0 m (56 ft 0 in)
Height overall 4·88 m (16 ft 0 in)

WEIGHT (Su-11, estimated):
Max T-O weight 13,600 kg (30,000 lb)

PERFORMANCE (Su-11, estimated):
Max level speed at 11,000 m (36,000 ft), 'clean'
Mach 1·8 (1,033 knots; 1,915 km/h; 1,190 mph)
Max level speed at 11,000 m (36,000 ft), with underbelly tanks and two 'Anab' missiles
Mach 1·2 (688 knots; 1,275 km/h; 790 mph)
Max level speed at 305 m (1,000 ft), 'clean'
Mach 0·95 (625 knots; 1,160 km/h; 720 mph)
Max rate of climb at S/L 8,200 m (27,000 ft)/min
Service ceiling 17,000 m (55,700 ft)

SUKHOI Su-15

NATO reporting name: Flagon

Ten examples of this twin-jet delta-wing fighter participated in the flying display at Domodedovo in July 1967. First to appear was a single black-painted machine, piloted by Vladimir Ilyushin, son of the famous designer and known to be a test pilot for Sukhoi. When a formation of nine similar aircraft appeared later, the identity of the design bureau responsible for them was confirmed by the obvious 'family likeness' to the Su-9 and Su-11 in the shape of the wings and tail unit.

It is now clear that this aircraft was developed to meet a Soviet Air Force requirement for a Mach 2·5 interceptor to follow the Su-11. It has served with the Soviet Air Force in several forms:

Flagon-A. Basic single-seater, with simple delta wings (estimated span 9·15 m; 30 ft), identical in form to those of the Su-11. Conical nose radome. Turbojets reported to be Tumansky R-11F2-300s, as used in MiG-21 series, each rated at 60·8 kN (13,668 lb st) with afterburning. Probably restricted to small initial quantity.

Flagon-C. Two-seat training version of 'Flagon-D', with probable combat capability. Individual rearward-hinged canopy over each seat. Periscope fitted above rear canopy for enhanced forward view.

Flagon-D. Generally similar to 'Flagon-A', but with longer-span wings of compound sweep, produced by reducing the sweepback at the tips via a very narrow unswept section. Conical radome. First major production version.

Flagon-E. Wings similar to those of 'Flagon-D'. New and more powerful propulsion system, increasing speed and range. Turbojets reported to be Tumansky R-13F-300s, as used in second-generation MiG-21MF, each rated at 64·73 kN (14,550 lb st). Uprated avionics. Major production version; operational since second half of 1973.

Flagon-F. Latest version in service. Can be identified by ogival nose radome. Generally similar to 'Flagon-E', but with uprated engines.

The number of Su-15s deployed with PVO-Strany appeared to have diminished to about 700 by 1981, as deliveries of the MiG-23MF and MiG-25 interceptors continued. All are based in the Soviet Union.

The following details apply to 'Flagon-F':

TYPE: Single-seat twin-jet all-weather interceptor.

WINGS: Cantilever mid-wing monoplane, basically similar to those of Su-11, but with new and extended outer panels. Sweepback approx 53° on inner wings, 37° on outer panels. No dihedral or anhedral. All-metal structure. Single boundary-layer fence above each wing at approx 70% span. Large area-increasing flap extends from inboard end of aileron to fuselage on each side.

FUSELAGE: Cockpit section is basically circular with large ogival dielectric nosecone. Centre-fuselage is faired into rectangular-section air intake ducts. Two door-type airbrakes at top and bottom on each side of rear fuselage, forward of tailplane.

TAIL UNIT: Cantilever all-metal structure, with sweepback on all surfaces. All-moving tailplane, with anhedral, mounted slightly below mid position and fitted with anti-flutter bodies near tips. Conventional rudder. No trim tabs.

LANDING GEAR: Tricycle type, with single wheel on each unit. Main wheels retract inward into wings and intake ducts; nosewheel retracts forward.

POWER PLANT: Two turbojets, with variable-area nozzles, mounted side by side in rear fuselage. These are reported to be Tumansky R-13F2-300s, each rated at 70·6 kN (15,875 lb st) with afterburning. Ram-air intakes, with variable ramps on splitter plates, embodying vertical slots for boundary layer control. Blow-in auxiliary inlets between main intake and wing leading-edge in side of each duct.

ACCOMMODATION: Single seat in enclosed cockpit, with rearward-sliding blister canopy. Rearview mirror above canopy of some aircraft.

ARMAMENT: Single pylon for external store under each wing, in line with boundary-layer fence. Normal armament comprises one radar homing and one infra-red homing air-to-air missile (NATO 'Anab'). Side-by-side pylons under centre-fuselage for further weapons or external fuel tanks.

AVIONICS AND EQUIPMENT: Large X-band radar (NATO 'Skip Spin') in nose, SOD-57M ATC/SIF nav system, SRO-2 (NATO 'Odd Rods') IFF, Sirena 3 radar warning system.

DIMENSIONS, EXTERNAL (estimated):
Wing span 10·53 m (34 ft 6 in)
Length overall 20·5 m (68 ft 0 in)

WEIGHT (estimated):
Max T-O weight 16,000 kg (35,275 lb)

PERFORMANCE (estimated):
Max level speed above 11,000 m (36,000 ft):
with external stores Mach 2·3
'clean' Mach 2·5
Time to 11,000 m (36,000 ft) 2 min 30 s
Service ceiling 20,000 m (65,600 ft)
Combat radius 390 nm (725 km; 450 miles)

SUKHOI Su-17, Su-20 and Su-22

NATO reporting names: Fitter-C, D, E, F, G, H and J

The variable-geometry Su-17, with more powerful engine and improved avionics, is in a completely different class from the veteran Su-7 ('Fitter-A'). The prototype, designated S-221 or Su-7IG (*Izmenyaemaya Geometriya*; variable-geometry), was an R & D aircraft shown at the Soviet Aviation Day display at Domodedovo Airport, Moscow, in July 1967, and was allocated the NATO reporting name 'Fitter-B'. Only some 4·0 m (13 ft) of each wing was pivoted, outboard of a very large fence, the remainder of the airframe being virtually identical with that of the Su-7. An attachment for an external store was

'Flagon-C' tandem two-seat combat trainer version of the Sukhoi Su-15

Servicing an Su-15 ('Flagon-E'), with rear fuselage removed for access to engines

'Flagon-F' version of the Su-15, armed with 'Anab' missiles *(Swedish Air Force)*

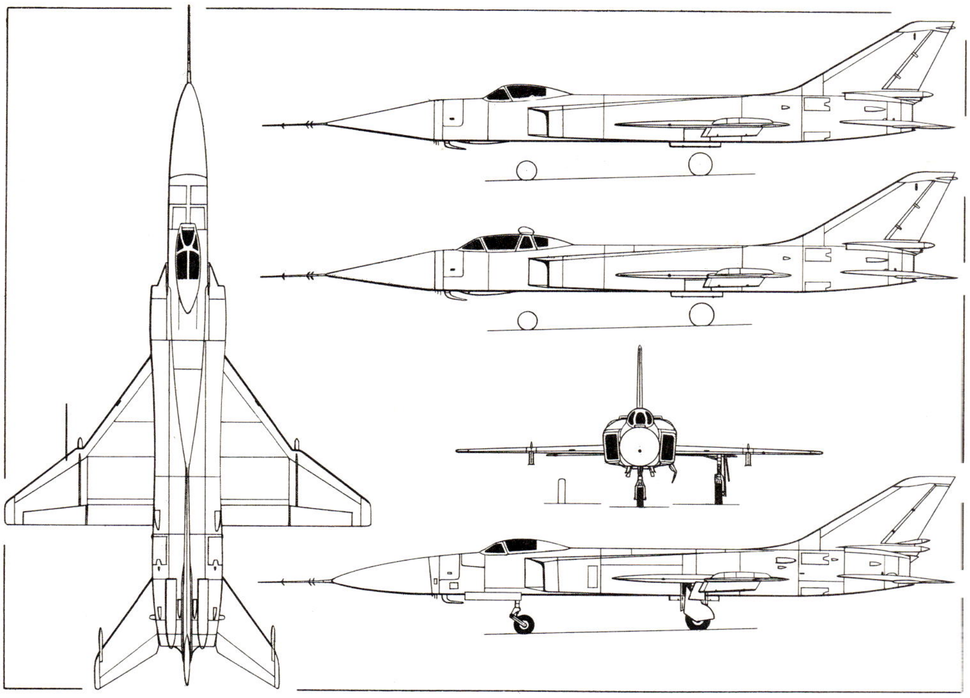

Sukhoi Su-15 ('Flagon-F'), with additional side views of two-seat 'Flagon-C' (centre) and 'Flagon-D' (top) *(Pilot Press)*

Fitter-E. Tandem two-seat trainer for Soviet Air Force. Generally similar to 'Fitter-C' but entire fuselage forward of wing drooped slightly to improve view from rear seat. Port wing-root gun deleted.

Fitter-F. Export counterpart of 'Fitter-D', with undernose radome. Single-seat. Gun in each wing-root. Tumansky R-29B turbojet, rated at 112·8 kN (25,350 lb st) with afterburning, in increased-diameter rear fuselage. Operators include Peruvian Air Force.

Fitter-G. Developed two-seater, with combat capability. Lyulka engine. Deepened dorsal spine fairing. Drooped front fuselage like 'Fitter-E'. Taller fin with straight top. Shallow ventral fin. Starboard gun only. Laser target seeker fitted.

Fitter-H. Improved single-seater for Soviet Air Force, basically as 'Fitter-C', with Lyulka engine. Wide and deep dorsal fairing aft of canopy, almost certainly providing additional fuel tankage. Taller fin of 'Fitter-G', with curved dorsal fin. Shallow ventral fin. Retains both wing-root guns. Small pylon for external store under wing centre-section on each side.

Fitter-J. Generally similar to 'Fitter-H' but with Tumansky engine. More angular dorsal fin.

In early 1980, the accompanying photograph of an unidentifiable tandem two-seat version was published in the Soviet press. This has the increased-diameter rear fuselage and fin shape of 'Fitter-F', and the front fuselage droop of 'Fitter-E'. The width and depth of the dorsal spine are increased aft of the rear canopy. Other features include a ventral fin, and a laser seeker in the intake centrebody like that of 'Fitter-D'.

The following description applies to the Su-17 ('Fitter-C'):

built into each wing fence, but the power plant appeared to be unchanged and there was no reason to expect 'Fitter-B' to form the basis of a production aircraft, in view of the modest improvement in overall performance offered by such minimal modification.

Discovery of at least one or two squadrons of 'improved Fitter-Bs' in service with the Soviet tactical air forces in 1972 came as a surprise, suggesting that even a small improvement in range and endurance by comparison with the Su-7 was considered worthwhile. Only after several years did the true measure of improvement become apparent. The combination of a more powerful engine and the variable-geometry wings permitted a doubled external load to be lifted from strips little more than half as long as those needed by the Su-7, and to be carried about 30 per cent further. Added to new avionics, this made the Su-17 so attractive that about 650 are deployed currently by Soviet tactical air forces, and 70 more by Soviet Naval Aviation units assigned to anti-shipping strike and amphibious support roles in the Baltic Sea area. Differences between the various versions identified to date are as follows:

Su-17 (Fitter-C). Basic single-seat attack aircraft for Soviet Air Force, with Lyulka AL-21F-3 turbojet, offering better specific fuel consumption than AL-7F-1 of Su-7. Curved dorsal fin between tail fin and dorsal spine fairing. No increase in fuselage diameter between wing and tailplane, in contrast to 'Fitter-F and J' export versions.

Su-17 (Fitter-D). Generally similar to 'Fitter-C', but forward fuselage lengthened by about 0·38 m (1 ft 3 in). Added undernose radome for terrain avoidance. Laser marked target seeker in intake centrebody.

Su-20 (Fitter-C). Export counterpart of Soviet basic 'Fitter-C', with reduced equipment standard. Supplied to Algeria, Czechoslovakia, Egypt, Iraq, Poland and Viet-Nam.

Su-22. Variant of Su-20 first delivered to Peru in 1977 (48 single-seat, 4 two-seat) and subsequently to India, Libya, Syria and North and South Yemen. Further reduced equipment standard, with Sirena 2 limited-coverage radar warning receiver, virtually no navigation aids, and IFF incompatible with Peru's SA-3 (NATO 'Goa') surface-to-air missiles. Weapons include 'Atoll' air-to-air missiles.

It has been known for some years that aircraft of the Su-17/20/22 series have been delivered with two different types of engine. As noted, the 'Fitter-C and D' operated by the Soviet Air Force have a rear fuselage of basically constant diameter between the wing and tailplane, and are known to be powered by a Lyulka AL-21F-3 turbojet. It was assumed that versions with a more bulged rear fuselage were export aircraft with a lower-rated engine. In fact, the Peruvian Air Force has stated that its Su-22s have a Tumansky R-29B turbojet, as fitted in the MiG-27, with a considerably higher rating than the AL-21F-3. It must be assumed that this is the standard power plant of all 'Fitters' with a heavily bulged rear fuselage.

It is not yet possible to relate to the Su-17, Su-20 and Su-22 type numbers the unclassified NATO reporting names allocated to later variants of the variable-geometry 'Fitter' which follow:

A bulged rear fuselage and angular dorsal fin distinguish this 'Fitter-J' of the Libyan Air Force from 'Fitter-H'. Note the underwing 'Atoll' air-to-air missile *(US Navy, via Angelo Romano)*

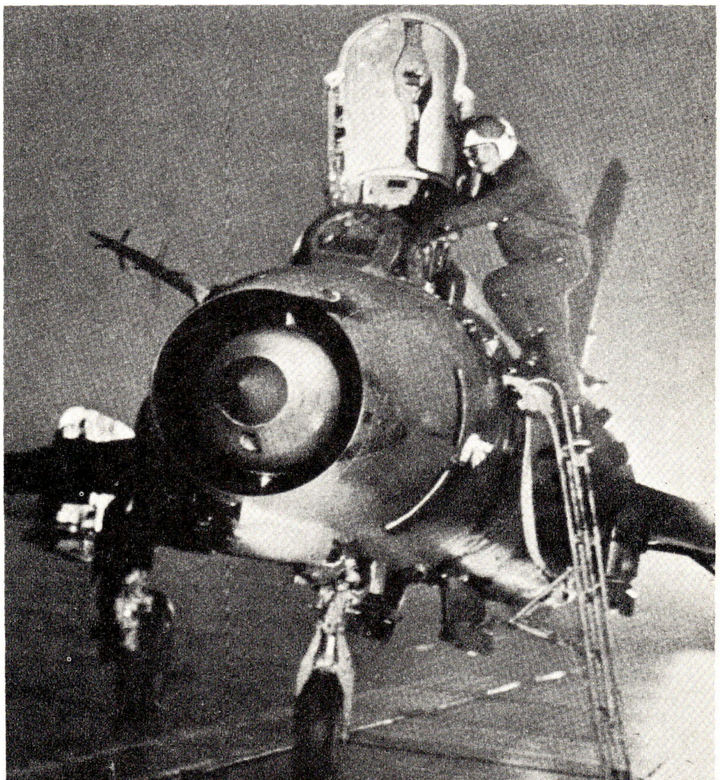

Sukhoi Su-17 ('Fitter-D') with undernose radar

Unidentified two-seat version of the Sukhoi Su-17/20/22 series

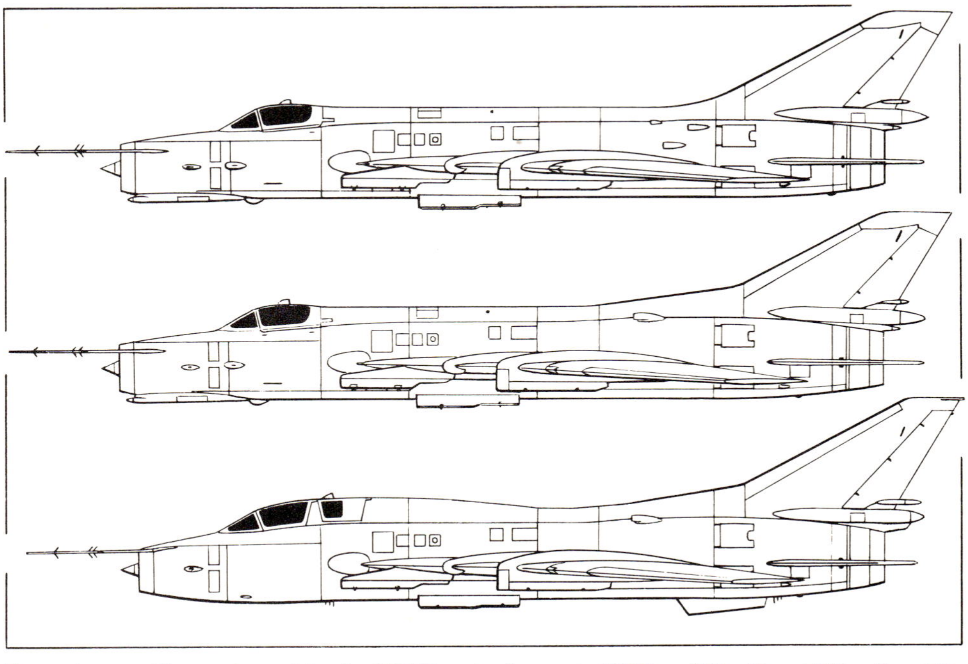

Top to bottom: The versions of the Su-17/20/22 series known to NATO as 'Fitter-D' and 'Fitter-F' and the so-far-unidentified two-seater respectively. Tumansky-powered versions can be identified by more bulged rear fuselage and single air intake by dorsal fin *(Pilot Press)*

Type: Single-seat ground attack fighter.

Wings: Cantilever mid-wing monoplane, with wide-span fixed centre-section and manually-actuated variable-geometry outer panels, with min sweep angle of 28° and max sweep angle of 62° approx. Centre-section appears to be generally similar to inner wings of Su-7, except for slight sweepback on trailing-edge of area-increasing centre-section flaps. Outboard of these flaps, centre-section trailing-edge is swept to align with trailing-edge of outer panels when they are fully swept. Full-span leading-edge slats on movable panels. Entire trailing-edge of each movable panel made up of a slotted flap, operable only when the wings are spread, and a slotted aileron operable at all times. Large main fence on each side, at junction of fixed and movable panels, is square-cut at front and incorporates attachments for external stores. Shorter fence above centre-section on each side, inboard of main fence.

Fuselage: Conventional all-metal semi-monocoque structure of circular section. Large dorsal spine fairing along top of fuselage, from canopy to fin. Ram-air intake in nose, with variable shockcone centrebody. Four door-type airbrakes, at top and bottom on each side of rear fuselage, forward of tailplane. Pitot on port side of nose; transducer to provide data for fire control computer on starboard side.

Tail Unit: Cantilever all-metal structure, with sweepback on all surfaces. All-moving horizontal surfaces, with anti-flutter body projecting forward on each side near tip. Conventional rudder. No tabs.

Landing Gear: Retractable tricycle type, with single wheel on each unit. Nosewheel retracts forward, requiring blistered door to enclose it. Main units retract inward into centre-section. Container for twin brake-chutes between the base of the rudder and the tailpipe.

Power Plant: One Lyulka AL-21F-3 turbojet engine, rated at 76·5 kN (17,200 lb st) dry and 110 kN (24,700 lb st) with afterburning. Fuel capacity increased to 4,550 litres (1,000 Imp gallons) by added tankage in dorsal spine fairing. Provision for carrying up to four 800 litre (176 Imp gallon) drop tanks on outboard wing pylons and under fuselage. When underfuselage tanks are carried, only the two inboard wing pylons may be used for ordnance, to a total weight of 1,000 kg (2,204 lb).

Sukhoi Su-17 ('Fitter-C') single-seat variable-geometry fighter with wings extended *(Flug Revue)*

Accommodation: Pilot only, on ejection seat, under rearward-hinged transparent canopy. Rearview mirror above canopy.

Armament: Two 30 mm NR-30 guns, each with 70 rds, in wing-root leading-edges. Total of eight weapon pylons (two tandem pairs under fuselage, one under each centre-section leading-edge, one under each main wing fence) for up to 4,000 kg (8,820 lb) of bombs, rocket pods and guided missiles such as the air-to-surface AS-7 (NATO 'Kerry').

Avionics and Equipment: SRD-5M (NATO 'High Fix') radar in intake centrebody; ASP-5ND fire control system; Sirena 3 radar homing and warning system providing 360° coverage, with antennae in slim cylindrical

A single-seat Su-17 ('Fitter-H') of the Soviet Air Force with (left) a two-seat training version ('Fitter-E')

housing above brake-chute container and in each centre-section leading-edge, between fences; SRO-2M IFF; SOD-57M ATC/SIF, with transponder housing beneath brake-chute container; RSIU-5/R-831 VHF/UHF and R5B-70 HF.

DIMENSIONS, EXTERNAL (estimated):

Wing span: fully spread	14·00 m (45 ft 11¼ in)
fully swept	10·60 m (34 ft 9½ in)
Wing aspect ratio: fully spread	4·9
fully swept	3·0
Length overall, incl probes	18·75 m (61 ft 6¼ in)
Fuselage length	15·40 m (50 ft 6¼ in)
Height overall	4·75 m (15 ft 7 in)

AREAS (estimated):

Wings, gross: fully spread	40·1 m² (431·6 sq ft)
fully swept	37·2 m² (400·4 sq ft)

WEIGHTS (estimated):

Weight empty	10,000 kg (22,046 lb)
Max internal fuel	3,700 kg (8,157 lb)
T-O weight, 'clean'	14,000 kg (30,865 lb)
Max T-O weight	17,700 kg (39,020 lb)

PERFORMANCE (estimated for 'clean' aircraft, 60% internal fuel, except where indicated):

Max level speed at height	Mach 2·17
Max level speed at S/L	Mach 1·05
Touchdown speed	143 knots (265 km/h; 165 mph)
Max rate of climb at S/L	13,800 m (45,275 ft)/min
Service ceiling	18,000 m (59,050 ft)
T-O run at AUW of 17,000 kg (37,478 lb)	620 m (2,035 ft)
T-O to 15 m (50 ft) at AUW of 17,000 kg (37,478 lb)	835 m (2,740 ft)
Landing run	600 m (1,970 ft)
Combat radius with 2,000 kg (4,409 lb) external stores:	
hi-lo-hi	340 nm (630 km; 391 miles)
lo-lo-lo	195 nm (360 km; 224 miles(

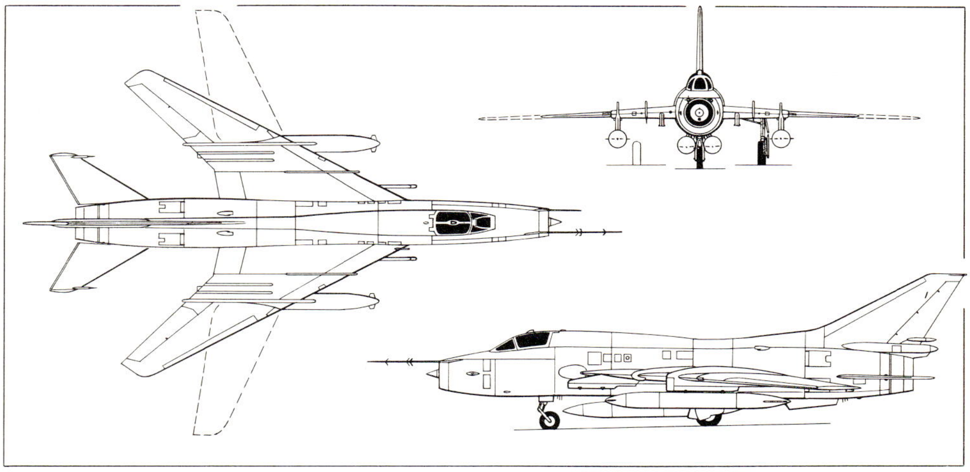

The 'Fitter-J' export single-seat version of the Sukhoi Su-20/22 series *(Pilot Press)*

Sukhoi Su-24 in landing configuration, with wings fully spread

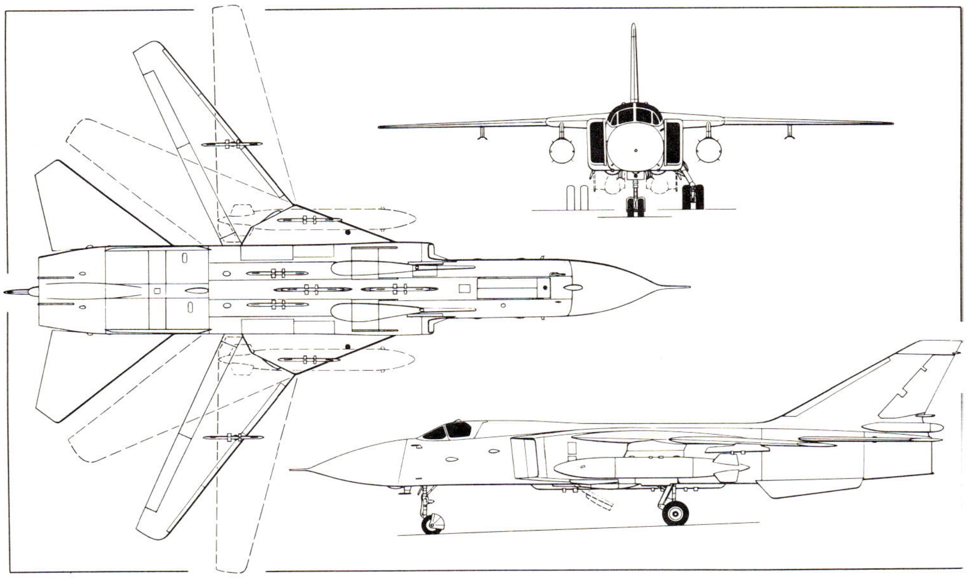

Sukhoi Su-24 variable-geometry attack aircraft *(Pilot Press, provisional)*

Su-24 with wings in intermediate position. Underbelly fairings and weapon racks are shown clearly

SUKHOI Su-24

NATO reporting name: Fencer

This variable-geometry attack aircraft was identified as a major new operational type by Admiral Thomas H. Moorer, then Chairman of the US Joint Chiefs of Staff, in early 1974, when he described it as "the first modern Soviet fighter to be developed specifically as a fighter-bomber for the ground attack mission". Initially, it was believed to be designated Su-19 in the Soviet Union, but the FY 1981 US Department of Defense Annual Report referred to it as the Su-24, and this designation is assumed to be correct for the production version, known to NATO as 'Fencer'.

Although smaller and lighter than its USAF counterpart, the F-111, it brings entirely new capability to Soviet Frontal Aviation. Lt Gen Donald R. Keith, US Army Deputy Chief of Staff for Research, Development and Acquisition, has said that 'Fencer' is credited with having terrain-avoidance radar, in addition to nav/attack radar, and "has the capability to deliver ordnance in all weather within 55 m (180 ft) of its target". The radar dish appears to have a diameter of at least 1·25 m (49 in), and is reported to be of the pulse-Doppler type. Equipment includes a laser rangefinder and marked target seeker.

'Fencer' entered squadron service in December 1974. In the Spring of 1981 at least 400 were serving with first-line squadrons in the European theatre, including two full regiments at Tukums in Latvia, near the Gulf of Riga, and at Chernyakhovsk, near Kaliningrad on the Soviet Baltic coast. There are two more at Starokonstantinov and Gorodok in the Ukraine, and a single regiment in the Soviet Far East. No 'Fencer' was allowed to fly outside the Soviet Union or its home waters until July 1979, when an Su-24 regiment was deployed briefly with the 16th Air Army, at Templin air base, north of Berlin in East Germany. Following this, three photographs of the aircraft were published in the Western aviation press in mid-1981; but the following description should be regarded as provisional:

TYPE: Two-seat variable-geometry attack aircraft.

WINGS: Cantilever shoulder-wing monoplane, each wing comprising a triangular fixed glove-box and three-position pivoted outer panel of all-metal construction. Slight anhedral from roots. Leading-edge sweepback on outer panels estimated at 16° fully-forward, and 68° fully-swept, with an intermediate sweep angle of 45°. Likely control surfaces include, on the outer panels, full-span leading-edge slats and two-section double-slotted trailing-edge flaps, of which the outer sections can operate independently when the wings are fully swept. Differential spoilers forward of flaps for roll control at low speeds and for use as lift dumpers on landing.

FUSELAGE: Conventional all-metal semi-monocoque structure of slab-sided rectangular section, with integral engine air intake trunks. Splitter plate and outer lip of each intake are inclined slightly downward. Variable intake ramps. Combined airbrake/main-wheel door under each side of centre-fuselage, curved to follow shape of underbelly fairing.

TAIL UNIT: Cantilever all-metal structure, comprising single sweptback fin with inset rudder, and high-mounted all-moving horizontal surfaces which operate together for pitch control and differentially for roll control, assisted by use of the wing spoilers when the wings are not fully swept. Two slightly-splayed ventral fins, one each side of fuselage undersurface.

LANDING GEAR: Retractable tricycle type, with twin wheels on each unit. Main units retract forward and inward into air intake duct fairings; nose unit retracts rearward. Trailing-link type of shock-absorbers in main units and low-pressure tyres for operation from semi-prepared fields. Mudguard on nosewheels.

POWER PLANT: Two unidentified afterburning engines side by side in rear fuselage. These could be Lyulka AL-21F turbojets, as fitted in Su-17, or Tumansky R-29B turbojets, as installed in the MiG-27. Internal fuel capacity, estimated at 13,000 litres (2,860 Imp gallons), can be supplemented by two very large external tanks, carried on glove pylons.

ACCOMMODATION: Crew of two (pilot and weapon systems officer) side by side on ejection seats.

ARMAMENT: Eight pylons under fuselage, each wing-root glove and outer wings for approx 8,000 kg (17,635 lb) of guided and unguided air-to-surface weapons. Two pivoting underwing pylons are the first of their kind observed on a Soviet aircraft. No internal weapons bay. One gun of unidentified type on port side of fuselage undersurface. Unidentified fairing on other side.

DIMENSIONS, EXTERNAL (estimated):

Wing span: spread	17·15 m (56 ft 3 in)
swept	9·53 m (31 ft 3 in)
Length overall	21·29 m (69 ft 10 in)
Height overall	5·50 m (18 ft 0 in)
Wheel track	4·25 m (14 ft 0 in)

WEIGHT (estimated):

Max T-O weight	39,500 kg (87,080 lb)

PERFORMANCE (estimated):

Max speed at height	above Mach 2
Service ceiling	17,500 m (57,400 ft)
Combat radius:	
lo-lo-lo	over 174 nm (322 km; 200 miles)
hi-lo-hi, with 2,000 kg (4,400 lb) weapons and two external tanks	970 nm (1,800 km; 1,115 miles)

This photograph shows clearly the unusual airbrakes under the fuselage of the Su-24

SUKHOI CLOSE-SUPPORT AIRCRAFT

Since 1978 there have been persistent reports concerning a Soviet counterpart to the USAF's A-10 Thunderbolt II single-seat close-support combat aircraft. A prototype is said to have been observed first by satellite at Ramenskoye experimental centre, leading to US designation of the type as Ram-J.

In late 1980, reports from the Pentagon suggested that, in general configuration, Ram-J is more like the Northrop A-9A (see 1972-73 *Jane's*) than the A-10. No reliable illustration is yet available, but it is said to be smaller than the A-10 and powered by two Tumansky R-13-300 non-afterburning turbojets, each rated at 50 kN (11,240 lb st). Normal T-O weight is given as 16,350 kg (36,050 lb), including 500 kg bombs, rocket pods or missiles on each of two underfuselage and eight underwing weapon pylons. Like the A-10, Ram-J is said to have also a heavy-calibre Gatling-type gun.

Ram-J is expected to be fully operational with Soviet tactical air forces by 1983-84.

TUPOLEV

CHIEF DESIGNER (Tu-144): Dr Alexei A. Tupolev
CHIEF DESIGNER (Tu-154): Dmitry Markov
DEPUTY CHIEF OF BUREAU: Andrei Kandolov

Andrei Tupolev, born in 1888, was a leading figure in the Central Aero-Hydrodynamic Institute (TsAGI) in Moscow from the time when it was founded, in 1929, until his death on 23 December 1972. He was for long the Soviet Union's outstanding designer, and the recent products of his design team range from turbofan civil transports to the first Soviet supersonic bomber to enter service and the first supersonic transport aircraft. Also in production in the Soviet Union are small amphibious aerosleighs of Tupolev design, powered by aircraft piston engines and capable of travelling over both water and snow. Current chief designers of the Tupolev bureau include Andrei Tupolev's son, Dr Alexei A. Tupolev.

TUPOLEV Tu-16

NATO reporting name: Badger

Known to NATO as 'Badger-G modified', this version of the Tu-16 carries 'Kingfish' missiles *(Swedish Air Force)*

The prototype of this bomber, which had the Tupolev design bureau designation Tu-88, was flown for the first time by N. Rybko in the Winter of 1952. The original strategic bomber version entered series production as the Tu-16 in 1953, and made its first major public appearance on 1 May 1954. Nearly half of the 2,000 that were built remain operational. About 300 are deployed with medium-range units of the Soviet strategic bomber force, as carriers of both nuclear and conventional weapons. They are supported by a few Tu-16 in-flight refuelling tankers, more than 90 of various versions equipped for ECM duties, and 15 equipped for reconnaissance. Naval units have about 275 Tu-16s for maritime attack, 70 tankers, and 40 reconnaissance and ECM models.

Early production Tu-16s had AM-3 turbojet engines. These were replaced in later aircraft by improved RD-3M (AM-3M) engines, which increased maximum speeds by up to 54 knots (100 km/h; 62 mph), and range with max fuel to 3,885 nm (7,200 km; 4,470 miles). Eleven versions of the Tu-16 have been identified by unclassified NATO reporting names. All except 'Badger-B' (see 1975-76 *Jane's*) remain in first-line service, as follows:

Badger-A. Basic strategic jet bomber, able to carry nuclear or conventional free-fall weapons. Glazed nose, with small undernose radome. Defensive armament of seven 23 mm cannon. Some equipped as flight refuelling tankers, using a unique wingtip-to-wingtip transfer technique first demonstrated publicly in 1956. Nine supplied

'Badger-H' version of the Tupolev Tu-16 twin-jet bomber, for stand-off or escort ECM duties

to Iraq. More than 80 operational with Chinese Air Force, and production continues in China under the designation Xian H-6.

Badger-C. Anti-shipping version, first seen at 1961 Soviet Aviation Day display. Large air-to-surface winged missile (NATO reporting name 'Kipper') carried under fuselage. Wide nose radome, in place of glazing and nose gun of 'Badger-A'.

Badger-D. Maritime/electronic reconnaissance version. Nose similar to that of 'Badger-C'. Enlarged undernose radome; three blister fairings in tandem under centre-fuselage.

Badger-E. Similar to 'Badger-A' but with cameras in bomb bay.

Badger-F. Basically similar to 'Badger-E' but with electronic intelligence pod on a pylon under each wing.

Badger-G. Similar to 'Badger-A' but with underwing pylons for two rocket-powered air-to-surface missiles (NATO reporting name 'Kelt'). One photographed by pilot of Japanese F-86F in December 1977 carried a new missile (NATO 'Kingfish') on port underwing pylon. Others seen subsequently with a 'Kingfish' under each wing (see Air-Launched Missiles section). Majority serve with anti-shipping squadrons of Soviet Naval Air Force. Some were included in the 25 Tu-16s supplied to Egypt as replacements for aircraft lost in the October 1973 war with Israel.

Badger-G modified. Specially equipped carrier for 'Kingfish' air-to-surface missiles, of which first photograph was released, by Swedish Air Force, in mid-1981. Large radome, presumably associated with missile operation, under centre-fuselage. Device mounted externally on glazed nose might help to ensure correct attitude of Tu-16 during missile launch.

Badger-H. Stand-off or escort ECM aircraft, with primary function of chaff dispensing. The chaff dispensers are probably located in the weapons bay area. Hatch aft of weapons bay. Two teardrop radomes, fore and aft of weapons bay. Two blade antennae aft of weapons bay.

Badger-J. Specialised ECM jamming aircraft, with at least some of the equipment located in a canoe-shape radome inside the weapons bay.

Badger-K. Electronic reconnaissance variant. Two teardrop radomes, inside and forward of weapons bay.

Maritime reconnaissance versions of 'Badger' make regular flights over units of the US Navy and other NATO naval forces at sea in the Atlantic, Pacific and elsewhere, and have been photographed while doing so. The aircraft often operate in pairs, with one 'Badger-F' accompanied by a different version. They also make electronic intelligence (elint) sorties around the coastlines of NATO and other non-Communist countries.

Type: Twin-jet medium bomber and maritime reconnaissance/attack aircraft.

Wings: Cantilever high mid-wing monoplane, with marked anhedral and with 35° of leading-edge sweep on outer panels; 42° sweep on inboard panels. Thickness/chord ratio 12½%. Two-spar light alloy structure, with two fences on each wing. Entire trailing-edge made up of slotted flaps (max deflection 35°) and mass-balanced ailerons, each with trim tab. Heavy engine nacelles form root fairings.

Fuselage: All-metal semi-monocoque structure of oval cross-section, made in five sections. The nose section houses the navigator's pressure cabin with double-glazed nose panels in a magnesium alloy frame, the pilots' pressure cabin, the forward gunner's cabin, and radar equipment. The second and fourth sections house the aircraft's fuel tanks, with the weapon compartment between them; the tail section contains a pressure cabin for the radio operator and rear gunner. Skin panels made of 3 mm light alloy sheet.

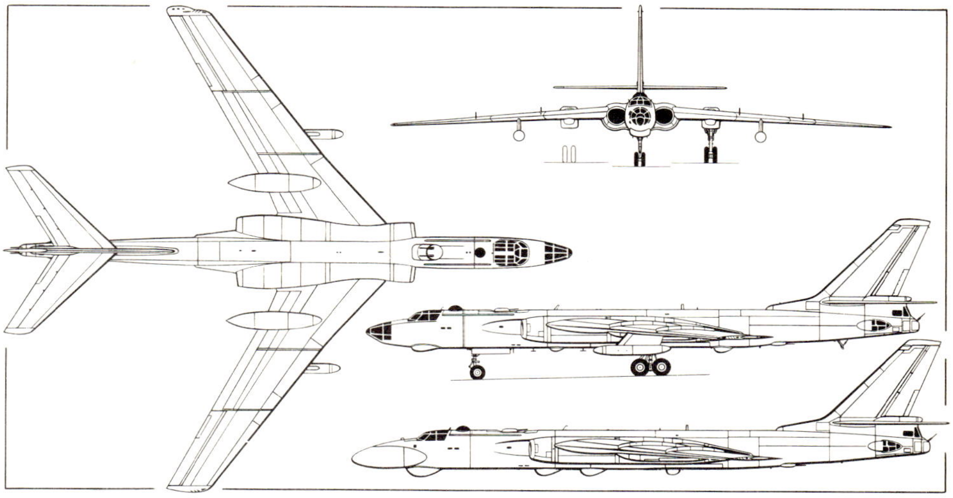

Tupolev Tu-16, in the form known to NATO as 'Badger-F', with additional side view (bottom) of 'Badger-D' *(Pilot Press)*

Tail Unit: Cantilever all-metal structure, with 42° leading-edge sweepback on all surfaces. Trim tabs in rudder and each elevator.

Landing Gear: Retractable tricycle type. Twin-wheel nose unit retracts rearward. Main four-wheel bogies retract into housings projecting beyond the wing trailing-edge.

Power Plant: Two Mikulin AM-3 turbojet engines, each rated at 85·8 kN (19,285 lb st) at sea level, in early Tu-16s. Later aircraft fitted with RD-3M (AM-3M) turbojets, each rated at 93·19 kN (20,950 lb st). Engines semi-recessed into sides of fuselage. Divided air intake ducts: main duct passes through wing torque box between spars; secondary duct passes under wing to feed into primary airflow in front of engine. Engines separated from wings and fuselage by firewalls. Jetpipes inclined outward 3° to shield fuselage from effects of exhaust gases. Fuel in wing and fuselage tanks, with total capacity of approx 45,450 litres (10,000 Imp gallons). Provision for underwing auxiliary fuel tanks and for flight refuelling. Tu-16 tankers trail hose from starboard wingtip; receiving equipment is in port wingtip extension.

Accommodation: Normal crew of six, with two pilots side by side on flight deck. Navigator, on seat with armoured sides and base, in glazed nose of all versions except 'Badger-C and D'. Manned tail position plus lateral observation blisters in rear fuselage under tailplane. Entry via two hatches in bottom of fuselage, in front and rear structural sections.

Armament: Forward dorsal and rear ventral barbettes each containing two 23 mm NR-23 cannon. Two further cannon in tail position controlled by an automatic gun-ranging radar set. Seventh, fixed, cannon on starboard side of nose of versions without nose radome. Bomb load of up to 9,000 kg (19,800 lb) delivered from weapons bay 6·5 m (21 ft) long in standard bomber, under control of navigator. Naval versions can carry air-to-surface winged stand-off missiles.

Avionics and Equipment: Radio and radar aids probably include HF and VHF R/T equipment, as well as IFF and a radio compass and radio altimeter. Other equipment differs according to role.

Dimensions, external ('Badger-A'):

Wing span	32·93 m (108 ft 0½ in)
Length overall	34·80 m (114 ft 2 in)
Height overall	10·80 m (35 ft 6 in)
Basic diameter of fuselage	2·50 m (8 ft 2½ in)
Tailplane span	11·75 m (38 ft 6½ in)
Wheel track	9·775 m (32 ft 0¾ in)

Area:

Wings, gross	164·65 m² (1,772·3 sq ft)

Weights (with AM-3 engines):

Weight empty, equipped	37,200 kg (82,000 lb)
Normal T-O weight	72,000 kg (158,730 lb)

Performance (estimated, with AM-3 engines, at max T-O weight):

Max level speed at 6,000 m (19,700 ft)	535 knots (992 km/h; 616 mph)
Service ceiling	12,300 m (40,350 ft)
Range with max bomb load	2,605 nm (4,800 km; 3,000 miles)
Range with max fuel	3,110 nm (5,760 km; 3,579 miles)

TUPOLEV Tu-95 and Tu-142

NATO reporting name: Bear

Documents issued in Washington concerning the SALT 2 negotiations, in 1979, revealed that the Soviet authorities use the designation **Tu-95** for the 113 'Bear-A/Bs' that form the mainstay of Dalnàya Aviatsiya's current strategic bomber force, but that the assorted 'Bears', totalling about 75, used by the Soviet Naval Air Force are known as **Tu-142s**. The Naval aircraft, being employed only for anti-submarine warfare, are not subject to SALT restrictions. Those operated from bases made available in Cuba and Angola are capable of covering the North and South Atlantic from the Mediterranean approaches westward to the US east coast, and southward to the Cape of Good Hope.

Long range and endurance are only two of the attributes that have kept the huge four-turboprop Tu-95s and Tu-142s in first-line service for a quarter of a century. Their high speed, exceeding that once considered possible for propeller-driven aircraft, eclipsed the contemporary four-jet Myasishchev M-4. Their size and payload potential enabled them to accommodate the largest air-to-surface missiles and radars yet carried by operational aircraft. Thus, the six versions identified by NATO reporting names all remain in service, as follows:

Bear-A. Basic strategic bomber, first flown in late Summer of 1954 and shown in Aviation Day display at Tushino in July 1955. Internal stowage for two nuclear or a variety of conventional free-fall weapons. Fitted with chin radar, and defensive armament comprising three pairs of 23 mm cannon in remotely-controlled dorsal and ventral barbettes and manned tail gun turret. Two glazed blisters on rear fuselage, under tailplane, are used for sighting by the gunner controlling all these weapons. The dorsal and ventral barbettes can also be controlled from a station aft of the flight deck.

Bear-B. First seen in 1961 Aviation Day flypast. As 'Bear-A' but able to carry a large air-to-surface aeroplane-type missile (NATO reporting name 'Kangaroo') under fuselage, with associated radar equipment (NATO 'Crown Drum') in wide undernose radome, replacing the original glazing. Some 'Bears' of Dalnaya Aviatsiya now carry 'Kitchen' air-to-surface missiles. Defensive armament retained. A few 'Bear-Bs' operate in maritime reconnaissance role, with flight refuelling nose-probe and, sometimes, a streamlined blister fairing on the starboard side of the rear fuselage. One was observed in 1978 with a pointed canister under each wing, presumably for air sampling.

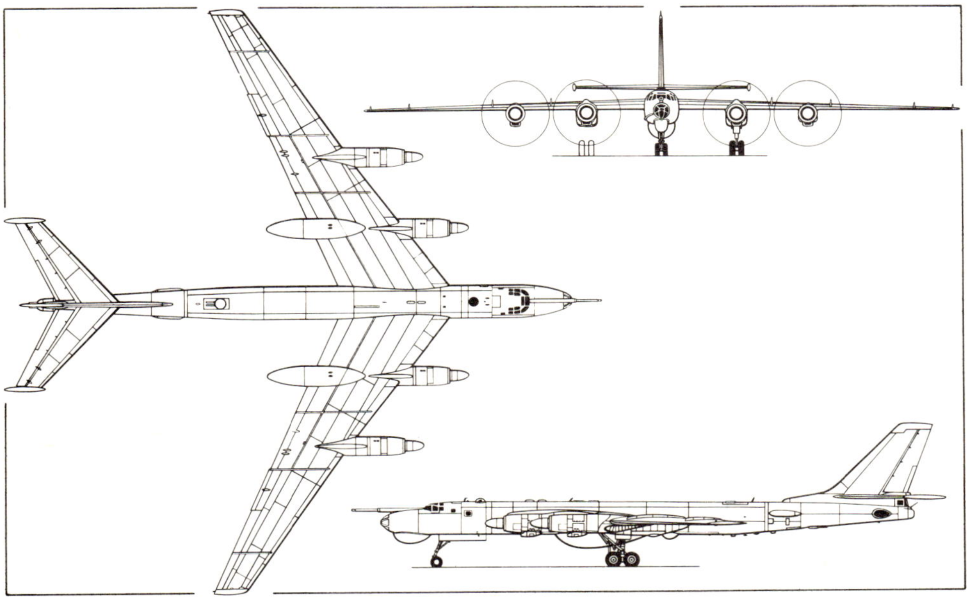

The version of the Tupolev Tu-142 known to NATO as 'Bear-D' *(Pilot Press)*

Tupolev Tu-142 ('Bear-D') photographed over the USS *Nimitz* **during NATO's Exercise Teamwork 80** *(US Navy)*

Bear-C. Maritime patrol version, first identified when it appeared in vicinity of NATO naval forces during Exercise Teamwork in September 1964. Generally similar to 'Bear-B' but with streamlined blister fairing on *both* sides of rear fuselage. Refuelling probe standard.

Bear-D. This version was first photographed extensively when several examples (together with Tu-16s) made low passes over the US Coast Guard icebreakers *Edisto* and *Eastwind* off Severnaya Zemlya, in the Soviet Arctic, in August 1967. These aircraft differed in detail, but each had a glazed nose, an undernose radar scanner, a large underbelly radome for X-band radar, a blister fairing on each side of the rear fuselage like 'Bear-C', a nose refuelling probe, and a variety of other blisters and antennae, including a streamlined fairing on each tailplane tip. The housing for I-band tail-warning radar above the tail turret is much larger than on previous versions. 'Bear-D' has an important function in support of operations involving surface-to-surface and air-to-surface missiles. It provides data on the location and nature of potential targets to missile launch crews on board ships and aircraft which are themselves too distant from the target to ensure precise missile aiming and guidance. About 40 serve with the Soviet Naval air fleet.

A 'Bear-D' photographed in the second half of 1978, after interception by US Navy F-4s, had in place of the normal tail turret and associated radome a faired tail housing special equipment (see accompanying illustration).

Bear-E. Maritime reconnaissance version, basically similar in configuration to 'Bear-A' but with refuelling probe and rear fuselage blister fairings as on 'Bear-C'. Six camera windows in bomb bay, in pairs in line with the wing flaps, sometimes with a seventh window to the rear on the starboard side.

Bear-F. First identified in 1973, this much-refined anti-submarine version has enlarged and lengthened fairings aft of its inboard engine nacelles, for purely aerodynamic reasons. The undernose radar of 'Bear-D' is missing on some aircraft; others have a radome in this position, but of considerably modified form. On both models the main underfuselage X-band radar housing is considerably further forward than on 'Bear-D' and smaller in size; the forward portion of the fuselage is longer; there are no large blister fairings under and on the sides of the rear fuselage; and the nosewheel doors are bulged prominently, suggesting the use of larger or low-pressure tyres. 'Bear-F' has two stores bays in its rear fuselage, one of them replacing the usual rear ventral gun turret and leaving the tail turret as the sole defensive gun position. About 40 of this version were operational in early 1981. One was photographed over the North Atlantic, by a USAF F-4 from Keflavik, Iceland, with an unidentified projection from the rear of its fin tip (see illustration).

Examples of all versions of the Tu-142 have made reconnaissance flights over units of the US Fleet at sea and have been photographed by US naval fighters while doing so. They are also encountered frequently over the North Sea by the RAF and Royal Navy, and off the US east coast during transits between Murmansk and Cuba, and elint missions from Cuba.

TYPE: Four-turboprop long-range bomber and maritime reconnaissance aircraft.

WINGS: Cantilever mid-wing monoplane. Sweepback 37° at quarter-chord on inner panels, 35° at quarter-chord on outer panels. All-metal three-spar structure. All-metal hydraulically-powered ailerons and Fowler flaps. Trim tabs in ailerons. Spoilers in top surface of wing forward of inboard end of ailerons. Three boundary layer fences on top surface of each wing. Thermal anti-icing system in leading-edges.

FUSELAGE: All-metal semi-monocoque structure of circular section, containing three pressurised compartments. Those forward and aft of the weapons bay are linked by a crawlway tunnel. The tail gunner's compartment is not accessible from the other compartments.

TAIL UNIT: Cantilever all-metal structure, with sweepback on all surfaces. Adjustable tailplane incidence. Hydraulically-powered rudder and elevators. Trim tabs in rudder and each elevator.

LANDING GEAR: Retractable tricycle type. Main units consist of four-wheel bogies, with tyres of approx 1·50 m (5 ft) diameter and hydraulic internal expanding brakes. Twin wheels on nose unit. All units retract rearward, main units into nacelles built on to wing trailing-edge. Retractable tail bumper consisting of two small wheels. Braking parachute may be used to reduce landing run.

POWER PLANT: Four Kuznetsov NK-12MV turboprop engines, each originally with max rating of approx 8,948 kW (12,000 ehp) but now uprated to 11,033 kW (14,795 ehp) and driving eight-blade contra-rotating reversible-pitch Type AV-60N propellers. Fuel in wing tanks, with normal capacity of 72,980 litres (16,540 Imp gallons).

ACCOMMODATION AND ARMAMENT: See notes applicable to individual versions and under 'Fuselage'.

OPERATIONAL EQUIPMENT ('Bear-D'): Large X-band radar in blister fairing under centre fuselage, for reconnaissance and to provide data on potential targets for anti-shipping aircraft or surface vessels. In latter mode, PPI presentation is data-linked to missile launch station. Four-PRF range J-band circular and sector scan bombing and navigation radar (NATO 'Short Horn'). I-band tail warning radar (NATO 'Bee Hind') in housing at base of rudder.

DIMENSIONS, EXTERNAL ('Bear-F', approx):

Wing span	51·10 m (167 ft 8 in)
Length overall	49·50 m (162 ft 5 in)
Height overall	12·12 m (39 ft 9 in)

WEIGHT ('Bear-F', estimated):

Max T-O weight	188,000 kg (414,470 lb)

PERFORMANCE ('Bear-A'):

Over-target speed at 12,500 m (41,000 ft) 435 knots (805 km/h; 500 mph)

Max range with 11,340 kg (25,000 lb) bomb load 6,775 nm (12,550 km; 7,800 miles)

TUPOLEV Tu-126

NATO reporting name: Moss

An officially-released Soviet documentary film, shown in the West in 1968, included sequences depicting a military version of the Tu-114 four-turboprop transport (see 1972-73 *Jane's*), carrying above its fuselage a rotating 'saucer' type early warning radar with a diameter of about 11 m (36 ft). This was a logical development, as the Tu-114 had a fuselage of larger diameter than the military Tu-95, and could accommodate more easily the extensive avionic equipment and crew of 12 required by what was soon confirmed as the Soviet air forces' first-generation AWACS (airborne warning and control system) aircraft, with the designation Tu-126. It proved to have also wings similar to those of the Tu-114, with extended-chord trailing-edge flaps, rather than the 'straight' trailing-edge of the Tu-95.

The general appearance of the Tu-126, which has the NATO reporting name 'Moss', is shown in the accompanying illustrations. It can be seen to have a flight refuelling nose-probe, ventral tail-fin and numerous additional antennae and blisters for electronic equipment. The power plant comprises four 11,033 kW (14,795 ehp) Kuznetsov NK-12MV turboprop engines.

The Tu-126 is intended to work in conjunction with advanced interceptors. After locating incoming low-level strike aircraft, the Tu-126 would ideally direct towards them fighters armed with 'snap-down' air-to-air missiles able to be fired from a cruising height of 6,100 m (20,000 ft) or higher. It has a further, obvious application in assisting strike aircraft to elude enemy interceptors picked up by its radar.

At least 10 Tu-126s are operational with the Soviet air

The faired tail, housing special equipment, on a 'Bear-D' photographed in 1978

'Bear-F' with unidentified projection from rear of fin tip, photographed in September 1980 *(USAF)*

The Tu-126 airborne warning and control system (AWACS) aircraft, known to NATO as 'Moss'

defence forces.They are said, by US defence experts, to have demonstrated some effectiveness in overwater exercises but to be ineffective over land.

DIMENSIONS, EXTERNAL:

Wing span	51·20 m (168 ft 0 in)
Wing aspect ratio	8·42
Length overall	55·20 m (181 ft 1 in)
Height overall	16·05 m (52 ft 8 in)
Wheel track	13·70 m (44 ft 11½ in)
Propeller diameter	5·60 m (18 ft 4½ in)

AREA:

Wings, gross	311·1 m² (3,349 sq ft)

WEIGHT (estimated):

Max T-O weight	170,000 kg (374,785 lb)

PERFORMANCE:

Max level speed	459 knots (850 km/h; 528 mph)
Normal operating speed	351 knots (650 km/h; 404 mph)
Max range without flight refuelling	6,775 nm (12,550 km; 7,800 miles)

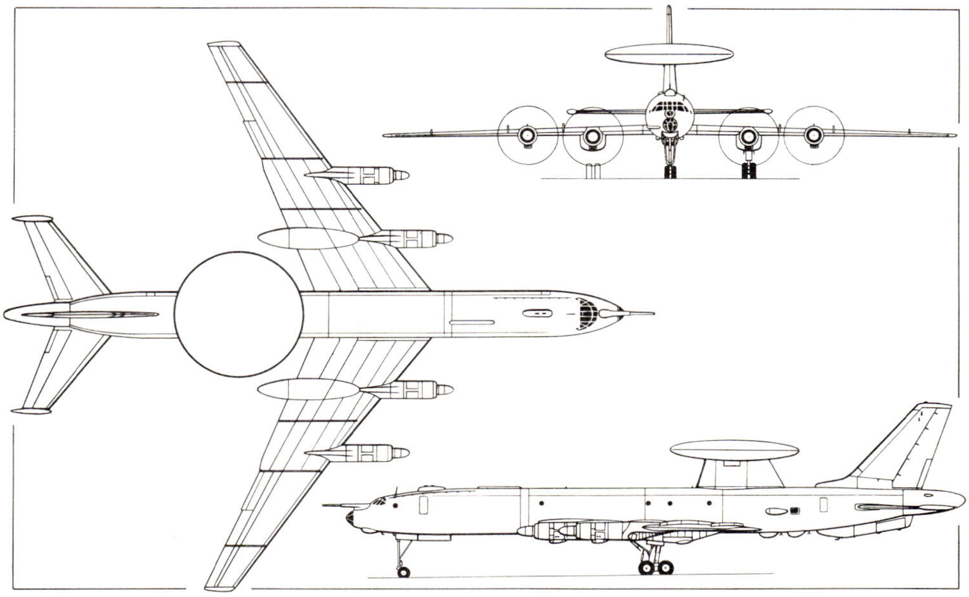

Tupolev Tu-126 (four Kuznetsov NK-12MV turboprops) *(Pilot Press)*

TUPOLEV Tu-22

NATO reporting name: Blinder

First shown publicly in the 1961 Aviation Day flypast over Moscow, the Tu-22 was the first operational Soviet supersonic bomber. Of the ten examples which took part in that display, only one carried visible weapons, in the form of an air-to-surface missile (NATO reporting name 'Kitchen'), some 11 m (36 ft) long, semi-submerged in the underside of its fuselage. This aircraft had also a wider nose radome.

A total of 22 Tu-22s took part in the 1967 display at Domodedovo. One was escorted by six MiG-21PFs, permitting a more accurate calculation of its overall dimensions than had previously been possible. Most carried 'Kitchen' missiles; all had a partially-retractable nose refuelling probe and the wide radome seen on the single missile-armed aircraft in 1961.

About 250 Tu-22s were built, in four versions, as follows:

Blinder-A. Basic reconnaissance bomber, with fuselage weapon bay for free-fall bombs. 'Blinder-A' entered only limited service, its max range being inadequate for the originally intended strategic role. Twelve supplied to Iraq.

Blinder-B. Generally similar to 'Blinder-A' but equipped to carry air-to-surface nuclear missile (NATO reporting name 'Kitchen') recessed in weapons bay. Larger radar and partially-retractable flight refuelling probe on nose. About 125 'Blinder-As' and 'Blinder-Bs' are believed to remain operational with the Soviet bomber force, plus 12 equipped for reconnaissance and one squadron serving with the Libyan Air Force.

Blinder-C. Maritime reconnaissance version, with six cameras in weapons bay and camera windows in weapons bay doors. Modifications to nosecone, dielectric panels, etc, suggest possible electronic intelligence role or equipment for electronic countermeasures (ECM) duties. About 60 delivered, of which 40 remain in service, for operation primarily over sea approaches to the Soviet Union, from bases in the Southern Ukraine and Estonia.

Blinder-D. Training version. Cockpit for instructor in raised position aft of standard flight deck, with stepped-up canopy. In service in the Soviet Union and Libya.

The following details apply to 'Blinder-A and B' but are generally applicable to all versions except as noted under model descriptions:

TYPE: Twin-jet supersonic bomber and maritime patrol aircraft.

WINGS: Cantilever mid-wing monoplane. Constant slight anhedral from roots. Sweepback approx 45° on leading-edge outboard of fence and 50° inboard of fence, increasing to acute sweep at roots. Conventional all-metal structure. Fully-powered two-section ailerons, with tab on each inboard section. Fowler flaps inboard and outboard of wheel pod on each wing trailing-edge.

FUSELAGE: All-metal semi-monocoque structure of circular section, with area-rule 'waisting' at wing-roots.

TAIL UNIT: Cantilever all-metal structure, with sweepback on all surfaces. Fully-powered slab-type horizontal surfaces at bottom of fuselage. Aerodynamically balanced rudder, with inset tab.

LANDING GEAR: Retractable tricycle type. Wide-track four-wheel bogie main units retract rearward into pods built on to wing trailing-edges. Oleo-pneumatic shock-absorbers. Main legs also designed to swing rearward for additional cushioning during taxying and landing on rough runways. Twin-wheel nose unit retracts rearward. Small retractable skid to protect rear fuselage in tail-down landing or take-off.

POWER PLANT: Two turbojet engines, each reportedly rated at 120·1 kN (27,000 lb st) with afterburning, mounted in pods above rear fuselage, on each side of tail fin. Lip of each intake is in the form of a ring which can be translated forward by jacks for take-off. Air entering

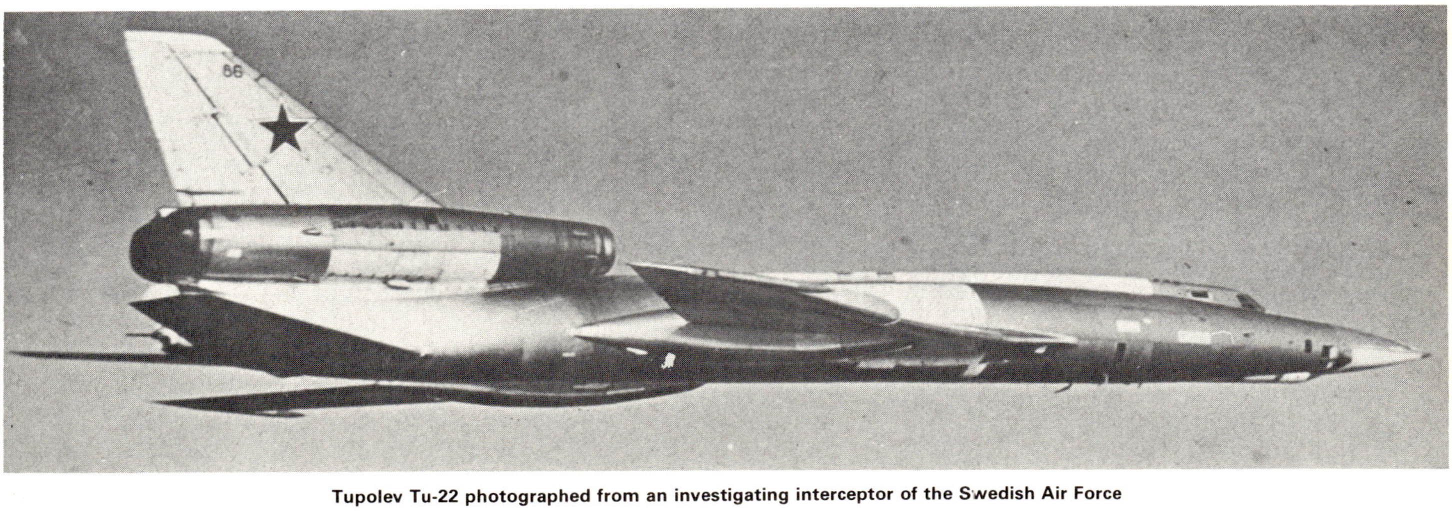

Tupolev Tu-22 photographed from an investigating interceptor of the Swedish Air Force

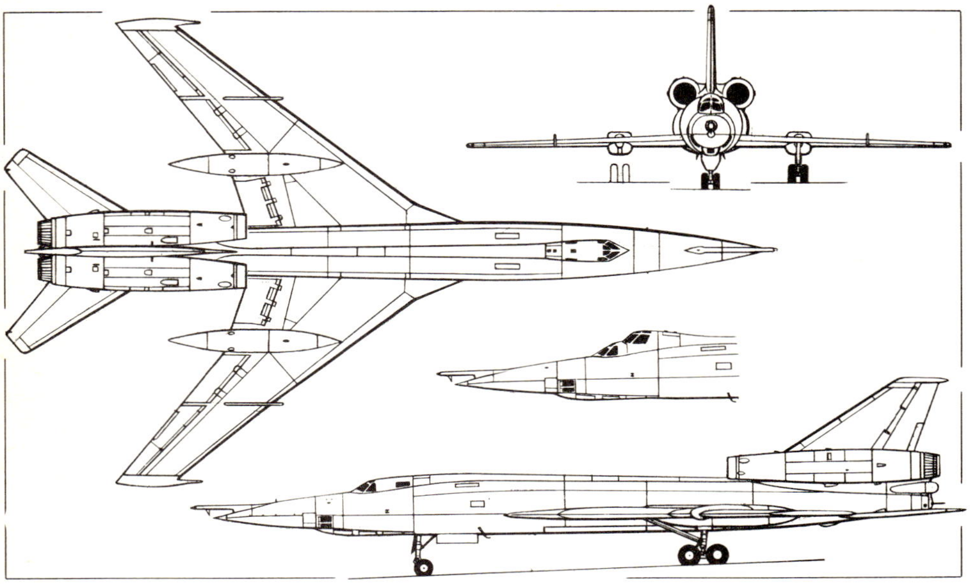

Tupolev Tu-22 twin-jet supersonic bomber ('Blinder-A') with additional view of nose of 'Blinder-D' training version *(Pilot Press)*

ram intake is then supplemented by air injected through annular slot between ring and main body of pod. Original nozzles had short fluted final section aft of short fixed section, with annular space between this and outer fairing. Current nozzles have longer-chord convergent-divergent nozzle inside the outer fairing, believed to have resulted in increased thrust and range. Semi-retractable flight refuelling probe on nose, with triangular guard underneath to prevent drogue damaging fuselage nosecone.

Accommodation: Crew of three in tandem. Row of windows in bottom of fuselage, aft of nose radome, at navigator/systems operator's station. Pilot has upward-ejection seat; other crew members have downward-ejection seats.

Armament and Operational Equipment: Weapons bay in centre-fuselage, with double-fold doors. Special doors with panels shaped to accommodate recessed 'Kitchen' missile on 'Blinder-B' version. Single 23 mm NS-23 gun in radar-directed tail turret, beneath 'Bee Hind' tail-warning radar antenna. Radar in nose (larger type in 'Blinder-B'). Chaff/flare countermeasures dispensers and bombing assessment cameras carried in rear of wheel pods of some aircraft.

Dimensions, external (estimated):

Wing span	27·70 m (90 ft 10½ in)
Length overall	40·53 m (132 ft 11½ in)
Height overall	10·67 m (35 ft 0 in)

Weight (estimated):

Max T-O weight	83,900 kg (185,000 lb)

Performance (estimated):

Max level speed at 12,200 m (40,000 ft)	
	Mach 1·4 (800 knots; 1,480 km/h; 920 mph)
Service ceiling	18,300 m (60,000 ft)
Max range	2,160 nm (4,000 km; 2,485 miles)

TUPOLEV Tu-22M (Tu-26)

NATO reporting name: Backfire

Official NATO sources first acknowledged the existence of a Soviet variable-geometry ('swing-wing') medium bomber in the Autumn of 1969. Such an aircraft was not unexpected, as the Tu-22 (NATO 'Blinder') was clearly incapable of fulfilling the long-range strategic bombing role for which it had been intended.

A prototype of the new bomber is said to have been observed in July 1970, on the ground near the Tupolev works at Kazan in Central Asia. Subsequent official statements confirmed the aircraft as a twin-engined design by the Tupolev bureau. At least two prototypes were built, and flight testing is believed to have started in 1971. Up to twelve pre-production models followed, for development testing, weapons trials and evaluation, by the beginning of 1973. The official designation was said to be **Tu-26**, but the Soviet Union referred to the type as **Tu-22M** in the SALT 2 treaty talks. This seems illogical if 'Blinder' is the Tu-22, as the suffix 'M' normally signifies a modification of an existing type, and is clearly inapplicable in the case of these very different aircraft. The NATO reporting name allocated to the aircraft is 'Backfire'.

When drawing up the basic parameters for the bomber, the Tupolev bureau is believed to have aimed at a maximum unrefuelled range of 4,775-5,200 nm (8,850-9,650 km; 5,500-6,000 miles) at high altitude. Unwillingness to depart from the Tupolev practice of retracting the main landing gear bogies into fairings on the wing trailing-edges limited the variable geometry to the outer wings, as on the Sukhoi Su-17 and Su-20. There is evidence to believe that the large size of these fairings, with the wheels stowed beneath the wing, caused excessive drag, so that 'Backfire's' range fell short of what had been planned. Redesign almost eliminated the fairings from later aircraft, after the main landing gear had been revised to retract inward into the fuselage. This accounts for the two versions of the Tu-22M/Tu-26 currently identified by NATO reporting names:

Backfire-A. Initial version, with large landing gear fairing pods on the wing trailing-edges. Believed to equip only one Dalnaya Aviatsiya squadron.

Backfire-B. Developed version, with landing gear fairing pods eliminated except for shallow underwing fairings, no longer protruding beyond the trailing-edge. Increased wing span. During the period of the abortive SALT 2 treaty negotiations, 'Backfire-Bs' were seen with the standard flight refuelling nose-probe removed, although the housing remained. This was assumed to stress Soviet assertions that the aircraft are intended for peripheral/theatre operations rather than long-range strategic use, and were therefore exempt from the restrictions that would have been imposed on intercontinental bombers by the treaty. External stores racks under the air intake trunks of aircraft shown in accompanying illustrations must impose speed limitations and may be fitted only for exercises or for short-range ground support missions.

Performance data given at the end of this entry conform with the latest estimates published openly by the Swedish Air Force and by other agencies. Some expert opinion continues to credit the 'Backfire' with a maximum speed in the Mach 2·25/2·5 bracket and a range adequate to cover virtually all of the continental USA with the aid of Arctic staging and flight refuelling, from bases in the USSR. US State Department document 12A on the SALT 2 agreement states: "Under certain flight conditions, the Backfire is assessed to have an intercontinental capability".

More than 150 'Backfire-Bs' are in service. Many of them have been allocated to medium-range bomber

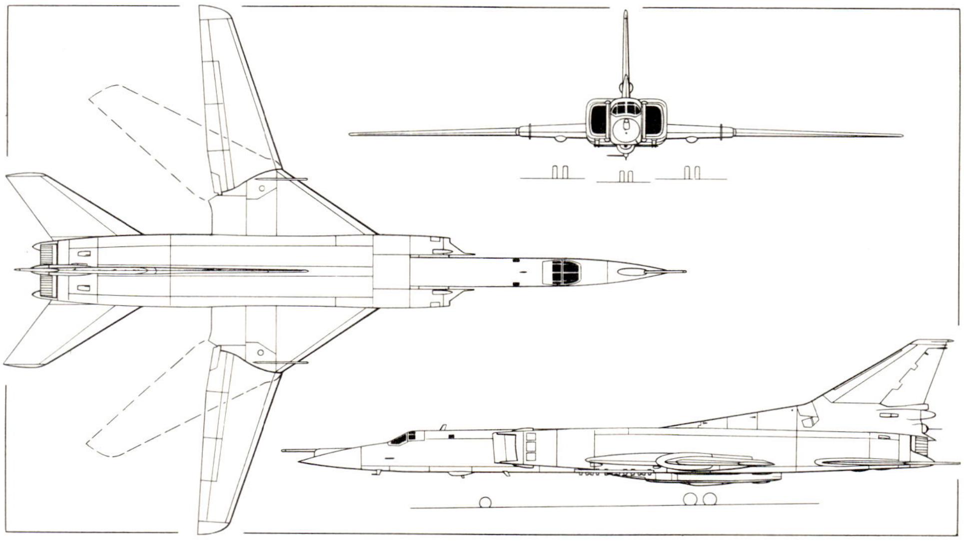

Tupolev Tu-22M/Tu-26 (NATO 'Backfire-B') bomber and maritime reconnaissance/attack aircraft *(Pilot Press)*

'Backfire-B' version of the Tupolev Tu-22M/Tu-26 photographed from an interceptor of the Swedish Air Force

squadrons of the Soviet Strategic Nuclear Forces opposing NATO in Europe and over the Atlantic; but at least 70 are deployed in a maritime role by Soviet Naval Aviation, and the FY 1979 Annual Report of the US Department of Defense stated: "There is increasing evidence that the Soviet bomber and cruise missile force may be overtaking their submarine force as a threat to our fleet and to our forces necessary for the resupply of Europe. They can concentrate aircraft, co-ordinate attacks with air, surface, or submarine-launched missiles, and use new technology to find our fleet units, jam our defences and screen their approach".

It is expected that the 'Backfire' strategic/maritime force will be built up to a total of 250-400 aircraft. Production was to be limited to a rate of 30 aircraft a year by the SALT 2 agreement, but is said to have been increased to 42 a year following non-ratification of the Treaty by the USA. One 'Backfire' unit is said to have been based at Komsomol'sk, about 430 nm (800 km; 500 miles) north of Vladivostok, in the Far East of the USSR, since Spring 1979.

TYPE: Twin-engined medium bomber and maritime reconnaissance/attack aircraft.

WINGS: Cantilever mid-wing monoplane, made up of a large-span fixed centre-section and two variable-geometry outer panels. No anhedral or dihedral, but wing section is so thin that considerable flexing of the outer panels takes place in flight. Leading-edge fence towards tip of centre-section on each side. Each outer wing panel is believed to be fitted with a full-span leading-edge slat, aileron, and slotted trailing-edge flaps aft of spoilers/lift dumpers. Wing sweep is believed to be variable from fully spread (20°) to fully swept (55°), rather than limited to one intermediate position as on the MiG-23.

FUSELAGE: Forward of wings, fuselage is basically circular with large ogival dielectric nosecone. Centre-fuselage is faired into rectangular-section air intake trunks, each fitted with a large splitter plate and assumed to embody complex variable-geometry ramps. There is no evidence to suggest external area-rule 'waisting' of these trunks.

TAIL UNIT: Cantilever structure, with sweepback on all surfaces. All-moving horizontal surfaces; conventional inset rudder.

LANDING GEAR: Retractable tricycle type, of which details remain largely speculative. Each main unit is assumed to carry a multi-wheel bogie, which pivots inward from the vestigial fairing under the centre-section into the bottom of the adjacent intake trunk.

POWER PLANT: Two turbofan engines with afterburners, mounted side by side in the rear fuselage. It is not yet possible to identify positively the type of engine fitted, but US sources have suggested the use of Kuznetsov turbofans similar to those installed in Tupolev's Tu-144 supersonic transport. This would be logical, as each engine is rated at 196·1 kN (44,090 lb st) with afterburning in the Tu-144. Uprated for military use, such engines would give an increase of at least 70% over the installed power in the Tu-22. A less-likely alternative is the turbojet evolved by the Koliesov bureau as a backup for the Tu-144, and which is said to be capable of supporting supersonic cruise without use of reheat. Fuel tankage is believed to include integral tanks in the entire fixed portion of the wings and much of the centre-fuselage above the weapon bay. A flight refuelling nose-probe can be fitted; after one observed refuelling, a 'Backfire' prototype is said to have remained airborne for a further 10 h.

ACCOMMODATION: Pilot and co-pilot side by side on flight deck. Other crew members further aft, as indicated by position of windows between flight deck and air intakes.

AVIONICS AND EQUIPMENT: Large bombing and navigation radar (NATO 'Down Beat') inside dielectric nosecone. Radar (NATO 'Fan Tail') for tail turret, above guns.

ARMAMENT: Aircraft observed to date have usually carried a primary armament of one of several different versions of 'Kitchen' air-to-surface missile semi-submerged in the underside of the centre-fuselage. Alternative weapon loads include up to 12,000 kg (26,450 lb) of conventional bombs. Aircraft photographed in recent years by the Royal Air Force and Swedish Air Force have also been fitted with multiple racks for lightweight stores under the air intake trunks. US reports have suggested that the Soviet Union is developing decoy missiles to assist penetration of advanced defence systems, in addition to very advanced ECM and ECCM. Twin 23 mm guns in radar-directed tail mounting.

DIMENSIONS, EXTERNAL:	
Wing span: fully spread	34·45 m (113 ft)
fully swept	26·21 m (86 ft)
Length overall	40·23 m (132 ft)
Height overall	10·06 m (33 ft)
WEIGHTS:	
Nominal weapon load	12,000 kg (26,450 lb)
Max T-O weight	122,500 kg (270,000 lb)
PERFORMANCE (estimated):	
*Max speed at high altitude	Mach 2·0
Max speed at low altitude	Mach 0·9
*Max unrefuelled combat range	4,350 nm (8,050 km; 5,000 miles)

See introductory copy

Tu-28P all-weather interceptor with 'Ash' infra-red missiles on inboard underwing pylons and 'Ash' radar homing missiles on outboard pylons

TUPOLEV Tu-28P/Tu-128

NATO reporting name: Fiddler

Largest fighter ever put into squadron service, this supersonic twin-jet interceptor was seen for the first time at Tushino in July 1961, with a large delta-wing air-to-air missile (NATO 'Ash') mounted under each wing. It is thought to have the service designation Tu-28P (US Department of Defense has used Tu-128); its NATO reporting name is 'Fiddler'.

The Tu-28P has a large ogival nose radome and carries a crew of two in tandem. The shoulder intakes for its two afterburning turbojet engines have half-cone shock-bodies, and the jetpipes are side by side in the bulged tail. Each engine is estimated to have a max rating of about 120·1 kN (27,000 lb st).

The sharply-swept wings are mid-set, with slight anhedral, and have considerably increased chord on the inboard panels, which have both increased sweep and a straight trailing-edge. The wide-track main landing gear units, comprising four-wheel bogies, retract into large fairings built on to the wing trailing-edges.

The tail unit is also sharply swept, and the two aircraft seen in 1961 were each fitted with two ventral fins. These were missing on the three Tu-28Ps which flew past at Domodedovo in July 1967, as was the large bulged fairing fitted under the fuselage in 1961.

The current armament is double that seen in 1961, each aircraft being equipped to carry two 'Ash' missiles under each wing, one usually of the radar homing type and the other of the infra-red homing type. This was confirmed as the standard armament of first-line service aircraft in a film released in 1969, showing units of the Soviet armed forces taking part in defence exercises.

About 130 'Fiddlers' are thought to be in service with the Soviet Union's PVO-Strany home defence fighter force.

DIMENSIONS, EXTERNAL (estimated):	
Wing span	20·00 m (65 ft 0 in)
Length overall	26·00 m (85 ft 0 in)
WEIGHT (estimated):	
Max T-O weight	45,000 kg (100,000 lb)
PERFORMANCE (estimated):	
Max speed at 11,000 m (36,000 ft)	Mach 1·75 (1,000 knots; 1,850 km/h; 1,150 mph)
Service ceiling	20,000 m (65,620 ft)
Range with max fuel	2,692 nm (4,989 km; 3,100 miles)

TUPOLEV Tu-104A (METEOROLOGICAL)

An accompanying illustration depicts a Tu-104A (CCCP-42454) that has been modified for operation by Gidrometsovcentr, a Soviet meteorological organisation, following retirement from airline service with Aeroflot. Its

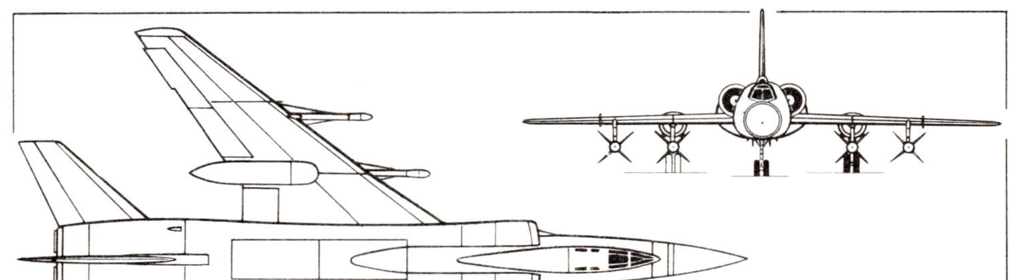

Tupolev Tu-28P taking off, with underwing armament of two 'Ash' missiles

Tupolev Tu-28P supersonic twin-jet all-weather interceptor *(Pilot Press)*

primary role is for surveillance of thunderstorm and hail-storm clouds, and for 'seeding' them with halogenide-filled rockets launched from its underwing racks.

The pointed nose of the aircraft houses a new weather radar. Specialised meteorological equipment is carried, together with personnel to operate it, within the former main passenger cabin and freight holds. Further meteorological equipment can be carried underwing, as an alternative to anti-storm-cell rockets.

Tupolev Tu-104A modified for meteorological operation by Gidrometsovcentr *(Oto Chudý)*

TUPOLEV Tu-134

NATO reporting name: Crusty

Known originally as the Tu-124A, this aircraft is a rear-engined twin-turbofan development of the Tu-124 (described in earlier editions of *Jane's*). It had completed more than 100 test flights when first details and photographs were released in mid-September 1964. The prototype was followed by five pre-production aircraft and the Tu-134 then went into series production at Kharkov. It entered international service on Aeroflot's Moscow-Stockholm route in September 1967, after a period on internal services, and was joined by the 'stretched' Tu-134A in the Autumn of 1970. The two versions differ as follows:

Tu-134. Initial version, with Soloviev D-30 turbofans, accommodating 64-72 passengers. Export orders included eleven for Interflug (East Germany), six for Balkan Bulgarian Airlines, five for LOT (Poland), six for Malev (Hungary) and three for Aviogenex (Yugoslavia). Described in detail in 1978-79 *Jane's*.

Tu-134A. Fuselage lengthened by 2·10 m (6 ft 10½ in) to accommodate 68-84 passengers and increase baggage space by 2·0 m³ (71 cu ft). Wider seats. Wings strengthened locally. Main landing gear units strengthened and fitted with Il-18 wheels and brakes. Thrust reversers on Soloviev D-30-II engines. New radio and navigation equipment to international standards. APU for self-contained engine starting, electrical power supply and air-conditioning on the ground. More than 200 Tu-134s and Tu-134As are believed to be in service with Aeroflot. Export orders for the Tu-134A have included 13 for CSA (Czechoslovakia), 20 for Interflug, three for Malev, seven for Balkan Bulgarian Airlines, seven for LOT, and six for Aviogenex. In some cases these replaced Tu-134s operated earlier.

The third aircraft delivered to Aviogenex differed from all Tu-134s seen previously in having the original glazed nose and undernose radome replaced by a more conventional conical nose radome. Subsequently this became optional on both the Tu-134 and the Tu-134A.

The following details apply to the Tu-134A:

TYPE: Twin-turbofan short/medium-range transport aircraft.

WINGS: Cantilever low-wing monoplane. Sweepback at quarter-chord 35°. Anhedral 1° 30′. Conventional light alloy two-spar structure, comprising centre-section, inner wings and detachable outer wings. Two-section aileron on each wing, operated manually through geared tabs, and fitted also with trim tabs. Electro-mechanically-actuated all-metal double-slotted flaps. Hydraulically-actuated spoilers. Hot-air de-icing system.

FUSELAGE: Conventional all-metal semi-monocoque structure of circular section. Electro-mechanically-actuated airbrake under fuselage, to steepen angle of approach.

TAIL UNIT: Cantilever all-metal structure, with variable-incidence T tailplane. Elevators operated manually through geared tabs. Rudder control is hydraulically powered, with yaw damper and Flettner/trim tab. Trim tab in each elevator. Fin leading-edge de-iced by hot air; tailplane leading-edge de-iced electrically.

LANDING GEAR: Retractable tricycle type. All units retract rearward. Main units consist of four-wheel bogies retracting into fairings built on to wing trailing-edge. Oleo-pneumatic shock-absorbers, supplemented by ability of legs to swing rearward to cushion taxying and landing on rough runways. Main wheels size 930 × 305, tyre pressure 5·86 bars (85 lb/sq in). Steerable twin nosewheels size 660 × 200, tyre pressure 6·38-6·90 bars (92·5-100 lb/sq in). Disc brakes and anti-skid units standard.

POWER PLANT: Two Soloviev D-30 Srs II turbofan engines, each rated at 66·7 kN (14,990 lb st), in pod on each side of rear fuselage, with thrust reversers, constant-speed drives and AC generators. Standard capacity of Tu-134A integral fuel tanks in inner wings is 16,600 litres (3,650 Imp gallons), with optional increase to 18,000 litres (3,960 Imp gallons). Single-point refuelling socket in starboard wing-root leading-edge. Gravity fuelling point above each tank. Hot-air de-icing system for nacelle intakes. Fire-warning and freon extinguishing system.

ACCOMMODATION: Flight crew of three, consisting of two pilots and a navigator, plus two stewardesses. All configurations have 28 seats in four-abreast rows in rear cabin. In standard configuration, front cabin seats 48 passengers, four-abreast at a seat pitch of 75 cm (29·5 in), with tables between front two rows. Alternative arrangements include an economy layout for a total of 84 passengers and a mixed-class layout for a total of 68, including 12 first class in a separate cabin at the front, with increased seat pitch. In each version there is a galley on the starboard side and baggage compartment on the port side immediately aft of the flight deck, two toilets at the rear and a large baggage and freight compartment aft, in line with the engines. Wardrobes between cabins. Reduced forward baggage space on 84-seater. Max loading on floor of freight compartment 400 kg/m² (82 lb/sq ft). The passenger door is on the port side, forward of the front cabin. There are two service doors, on the starboard side by the baggage compartments, and two emergency exits on each side over the wing. Crew cabin and canopy observation panel de-iced by electric heater and hot air.

SYSTEMS: Air-conditioning system, pressure differential 0·56 bars (8·10 lb/sq in), fed with bleed air from engine compressors. Hydraulic system operating pressure 207 bars (3,000 lb/sq in). Electrical system includes 27V DC supply from four 12kW starter/generators and two batteries, single-phase 115V 400Hz AC supply from two inverters and three-phase 36V 400Hz AC supply. APU available for main engine starting, electric power supply and operation of the air-conditioning system. Oxygen

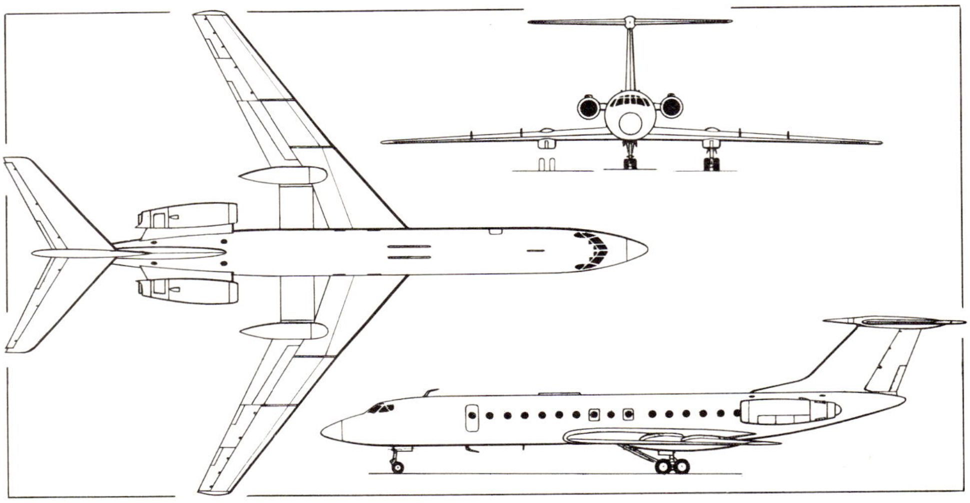
Tupolev Tu-134A twin-turbofan short/medium-range transport with conical nose radome *(Pilot Press)*

Tupolev Tu-134A medium-range transport (two Soloviev D-30 turbofan engines) of Aeroflot with glazed nose and undernose radar

Tupolev Tu-144 supersonic transport, in initial production form. This aircraft was displayed at the 1977 Paris Air Show *(Brian M. Service)*

available continuously for pilot, from 92 litre bottle, with 1 h supply for other crew members and portable supply for emergency use by passengers.

Avionics and Equipment: Typical installation includes autopilot, two ARK-15 radio compasses, Mikron HF communications radio, two UHF transceivers, RV-5 radio altimeter, two 'Course MP-2' VOR/ILS, two SO-70 transponders, ROZ-1 weather radar and DISS-013 Doppler.

Dimensions, external (Tu-134A):

Wing span	29·00 m (95 ft 1¾ in)
Wing chord at root	8·66 m (28 ft 5 in)
Wing chord at tip	1·92 m (6 ft 3½ in)
Wing aspect ratio	7·3
Length overall	37·05 m (121 ft 6½ in)
Length of fuselage	33·17 m (108 ft 10 in)
Fuselage max diameter	2·90 m (9 ft 6 in)
Height overall	9·14 m (30 ft 0 in)
Tailplane span	11·80 m (38 ft 8½ in)
Wheel track	9·45 m (31 ft 0 in)
Wheelbase	16·04 m (52 ft 7½ in)
Passenger door: Height	1·61 m (5 ft 3¼ in)
Width	0·70 m (2 ft 3½ in)
Height to sill	2·60 m (8 ft 6½ in)
Service door (fwd): Height	1·25 m (4 ft 1¼ in)
Width	0·75 m (2 ft 5½ in)
Rear door (stbd): Height	1·22 m (4 ft 0 in)
Width	0·90 m (2 ft 11½ in)
Emergency exits (each): Height	0·59 m (1 ft 11¼ in)
Width	0·60 m (1 ft 11½ in)

Dimensions, internal:

Cabin (portion containing seats only):	
Length	15·93 m (52 ft 3 in)
Width	2·71 m (8 ft 10½ in)
Height	1·96 m (6 ft 5 in)
Volume	68·0 m³ (2,400 cu ft)
Baggage compartment (fwd):	
Volume	4·0-6·0 m³ (141-212 cu ft)
Baggage compartment (aft):	
Height (mean)	1·75 m (5 ft 9 in)
Length (mean)	2·80 m (9 ft 2 in)
Width (mean)	1·75 m (5 ft 9 in)
Floor area	4·5 m² (48·4 sq ft)
Volume	8·50 m³ (300 cu ft)

Areas:

Wings, gross	127·3 m² (1,370·3 sq ft)
Ailerons (total)	9·68 m² (104·2 sq ft)
Trailing-edge flaps (total)	22·50 m² (242·2 sq ft)
Spoilers (total)	4·48 m² (48·2 sq ft)
Vertical tail surfaces (total)	21·25 m² (228·7 sq ft)
Horizontal tail surfaces (total)	30·68 m² (330·2 sq ft)
Elevators	6·42 m² (69·1 sq ft)

Weights (Tu-134A):

Operating weight empty	29,050 kg (64,045 lb)
Max fuel	14,400 kg (31,800 lb)
Max payload	8,200 kg (18,075 lb)
Max ramp weight	47,200 kg (104,000 lb)
Max T-O weight	47,000 kg (103,600 lb)
Max landing weight	43,000 kg (94,800 lb)

Performance (Tu-134A, at max T-O weight except where indicated):

Max cruising speed at AUW of 42,000 kg (92,600 lb) at 10,000 m (32,800 ft)
477 knots (885 km/h; 550 mph)

Normal cruising speed
405-458 knots (750-850 km/h; 466-528 mph)

Service ceiling at max T-O weight
11,900 m (39,000 ft)

T-O runway length (BCAR, at S/L)
2,400 m (7,875 ft)

Landing runway length at max landing weight (BCAR, at S/L)
2,200 m (7,218 ft)

Range at max AUW, cruising at 432 knots (800 km/h; 497 mph) at 11,000 m (36,100 ft):

with max payload
1,020 nm (1,890 km; 1,174 miles)

with payload of 5,000 kg (11,025 lb)
1,630 nm (3,020 km; 1,876 miles)

TUPOLEV Tu-144

NATO reporting name: Charger

A detailed history of the development of this supersonic transport aircraft can be found in the 1977-78 *Jane's*.

The first of the two prototypes of the Tu-144 (CCCP-68001) flew for the first time on 31 December 1968, this being the first flight by a supersonic airliner anywhere in the world. On 5 June 1969 the Tu-144 exceeded Mach 1 for the first time, at a height of 11,000 m (36,000 ft), half-an-hour after take-off. On 26 May 1970 it became the first commercial transport to exceed Mach 2, by flying at 1,160 knots (2,150 km/h; 1,335 mph) at a height of 16,300 m (53,475 ft) for several minutes. Highest speed reported subsequently was Mach 2·4, probably with the aircraft in its almost totally redesigned production form.

Supersonic proving flights by Aeroflot began on 26 December 1975, between Moscow's Domodedovo Airport and Alma-Ata, capital of Kazakhstan. Carrying a payload of freight and mail, Tu-144 No. CCCP-77106 took 1 h 59 min for the journey of about 1,760 nm (3,260 km; 2,025 miles), flying for most of the time at 1,187 knots (2,200 km/h; 1,367 mph) at a height of 16,000-18,000 m (52,500-59,000 ft). This aircraft was retired to the Soviet Air Force Museum at Monino, near Moscow, in early 1980.

Before passenger carrying started, 50 flights were made over the Alma-Ata route and between Moscow and Khabarovsk (first on 22 February 1977). Another eight flights were cancelled and one returned to base after take-off.

Scheduled passenger flights began on 1 November 1977, when aircraft CCCP-77109 flew from Moscow to Alma-Ata in just under two hours, with 80 of its 140 seats occupied. A public holiday and bad weather led to cancellation or postponement of five of the next six services; but reliability improved from 20 December and a total of 102 flights was made by 1 June 1978. However, the service was then terminated, following an accident to one of the Tu-144s on a non-commercial flight. There was no further news of route flying by the aircraft until the Tu-144D, with new engines, made a proving flight from Moscow to Khabarovsk in June 1979 (see separate entry on Tu-144D).

Thirteen Tu-144s were built, including prototypes.

Construction of the Tu-144 is mainly of VAD-23 light alloy, with extensive use of integrally-stiffened panels, produced by both chemical milling and machining from solid metal. Stainless steel and titanium are used for the leading-edges, elevons, rudder and undersurface of the rear fuselage, and the aircraft is stated to embody 10,000 parts made of plastics.

The wings have a double-delta planform, with a sweepback in the order of 76° on the inboard portions and 57° on the main panels. They are cambered over the full area, with a downward-curving trailing-edge like that of the Concorde. The structure is multi-spar, with large honeycomb panels. The powered control surfaces consist of four separate elevons on each wing and a two-section rudder, each operated by two separate actuators.

The fuselage blends with the low-set wings, giving a flat undersurface which contributes to fuselage lift and directional stability. There are doors forward of the passenger cabins and in the centre on the port side, and six emergency exits.

The 'moustache' foreplanes are pivoted from points near the top of the fuselage, immediately aft of the flight deck. Each is fitted with a double-slotted trailing-edge flap and a fixed leading-edge double-slat. The foreplanes retract rearward, protruding only a little externally but restricting to a narrow passage the space between flight deck and cabin. When extended, during take-off and landing only, they have anhedral but no sweep.

The twin-wheel steerable nose unit of the landing gear retracts forward into the fuselage. Each main eight-wheel bogie (two rows of four) retracts forward and up into one of the engine ducts, between the divided air-intake trunks. This requires the bogie first to pivot sideways through 90° about the base of the leg, before retraction. Nosewheel tyres are size 950 × 300. The main wheels are fitted with size 950 × 400 tyres and quadruple steel disc brakes. All wheel-bays are thermally insulated, and the nosewheel tyres are blown with cooling air after retraction, throughout cruising flight.

The first flight of the Tu-144 prototype was also the first time that the Kuznetsov NK-144 turbofan engine, selected for the original version of the aircraft, had been tested in the air. At that time the engine max ratings were 127·5 kN (28,660 lb st) without afterburning and 171·6 kN (38,580 lb) with full afterburning. On production aircraft the rating with full afterburning was raised to 196·1 kN (44,090 lb st). The engines were paired in two separate ducts; each intake trunk contained a central vertical wall, giving an individual flow of air to each engine. The intakes had fully-automatic movable ramps, with manual reversion, and with airflow dump doors midway from the inlet to the engines. Afterburning was normally maintained at 30 to 40 per cent of its maximum additional thrust throughout cruising flight. No thrust reversers were installed, but a twin brake-parachute was fitted solely for use on short runways.

Total fuel capacity was about 95,000 kg (209,440 lb), with a transfer tank in the fuselage tailcone to counterbalance CG movement in flight.

A flight crew of three is normally carried by both versions of the aircraft, consisting of two pilots and a flight engineer. The pilots have fully-adjustable armchair seats. During cruising flight, their windscreen is faired in by a retractable visor which has birdproof side windows and a 'solid' top. The entire nose can be drooped for improved view during take-off and landing.

The basic interior layout is for a total of 140 passengers in three cabins. The front cabin contains 11 seats for first class passengers, basically three-abreast, with tables between the front two rows. It is divided by a movable partition from the forward tourist class cabin, which contains six rows of five-abreast seats, with the three-seat units on the port side of the centre aisle. The rear tourist class cabin contains 15 rows of five-abreast seating at the front and six rows of four-abreast seating at the rear. Seat pitch is normally 102 cm (40 in) for first class and 87 cm (34·25 in) for tourist class; but alternative layouts are available.

Forward of the passenger accommodation there are toilet (starboard) and cloakroom compartments (port), with a bench seat for two cabin staff by the forward door. A second cloakroom, toilet and buffet kitchen are located between the two tourist class cabins, with two further toilets at the rear. Aft of these, in line with the engines, is a large compartment for containerised baggage and freight, which are loaded and unloaded semi-automatically through a large door on the starboard side of the hold, at the rear. There are no underfloor holds.

Little information is yet available on aircraft systems. The prototype had three independent hydraulic systems and two separate systems for pressurisation and air-conditioning. Preparation for flight, ground air-conditioning and engine starting can be performed independently of airport services. Advanced automatic flight control and navigation systems are standard, with the intention of progressing eventually to full automatic landing under all weather conditions. Six landing and taxi lights are mounted on the nosewheel leg.

Dimensions, external:

Wing span	28·80 m (94 ft 6 in)
Length overall	65·70 m (215 ft 6½ in)
Height, wheels up	12·85 m (42 ft 2 in)

Wheel track 6·05 m (19 ft 10¼ in)
Wheelbase 19·60 m (64 ft 3½ in)
DIMENSIONS, INTERNAL:
Cabin: Headroom 1·93 m (6 ft 4 in)
Baggage/cargo hold capacity 20 m^3 (706 cu ft)
AREA:
Wings, gross 438 m^2 (4,714·5 sq ft)
WEIGHTS:
Operating weight empty 85,000 kg (187,400 lb)
Max fuel 95,000 kg (209,440 lb)
Max payload (space limited) 14,000 kg (30,865 lb)
Max payload (structure limited) 15,000 kg (33,070 lb)
Max ramp weight 185,000 kg (407,850 lb)
Max T-O weight 180,000 kg (396,830 lb)
Max zero-fuel weight 100,000 kg (220,460 lb)
Max landing weight 110,000-120,000 kg (242,500-264,550 lb)
PERFORMANCE (nominal):
Max cruising speed
Mach 2·35 (1,350 knots; 2,500 km/h; 1,550 mph)
Normal cruising speed
Mach 1·98-2·2 (1,134-1,240 knots; 2,100-2,300 km/h; 1,305-1,430 mph)
Landing speed 151 knots (280 km/h; 174 mph)
Cruising height 16,000-18,000 m (52,500-59,000 ft)
Balanced field length at max T-O weight (approx):
ISA, S/L 3,000 m (9,845 ft)
ISA+15°C, S/L 3,200 m (10,500 ft)
Landing run 2,600 m (8,530 ft)
Max range with 140 passengers, at an average speed of Mach 1·9 (1,080 knots; 2,000 km/h; 1,243 mph) 3,500 nm (6,500 km; 4,030 miles)

TUPOLEV Tu-144D

NATO reporting name: Charger

First news of the development of this improved version of the Tu-144 was given after it had completed its first route proving flight from Moscow to Khabarovsk, in the Soviet Far East, on 23 June 1979. Time taken for the 3,340 nm (6,185 km; 3,840 mile) flight was 3 h 21 min. Passengers included the aircraft's designer, Dr Alexei A. Tupolev.

After this proving operation, Mr Ivan Razumovsky, Soviet Deputy Minister for Civil Aviation, said that the Tu-144D has new engines by comparison with the original Tu-144. Reported to be the Koliesov variable bypass engines first mentioned briefly in *Jane's* in 1973-74, these were said by the Minister to be 50 per cent more economical in operation, making possible a max range of 3,780 nm (7,000 km; 4,350 miles), and to meet international requirements in terms of noise emission.

No other details of the Tu-144D had become available by mid-1981. A Soviet Ministry of Aircraft Production official was reported at the beginning of the year to have said that, following tests which had assured the safety of the aircraft, it could soon be back in service. A further press report, in mid-1981, stated that work had started on modernising the existing Tu-144s, presumably by re-engining them, at the works in Voronezh. It suggested that the updated aircraft would operate the long-planned passenger service over the Moscow-Khabarovsk route.

TUPOLEV Tu-154

NATO reporting name: Careless

The three-engined Tu-154, announced in the Spring of 1966, was intended to replace the Tu-104, Il-18 and An-10 on medium/long stage lengths of up to 3,240 nm (6,000 km; 3,725 miles). It is able to operate from airfields with a class B surface, including packed earth and gravel. Normal flight can be maintained after shutdown of any one engine. Single-engine flight is possible at a lower altitude.

The first of six prototype and pre-production models flew for the first time on 4 October 1968. The seventh Tu-154 was delivered to Aeroflot for initial route proving and crew training in early 1971. Mail and cargo flights began in May. Initial passenger-carrying services were flown for a few days in the early Summer of 1971 between Moscow and Tbilisi. Regular services began on 9 February 1972, over the 700 nm (1,300 km; 800 mile) route between Moscow and Mineralnye Vody, in the North Caucasus. International services began with a proving flight between Moscow and Prague on 1 August 1972.

The following details apply to the basic Tu-154. This version was superseded in production successively by the Tu-154A and Tu-154B, which are described separately.

TYPE: Three-engined medium/long-range transport aircraft.

WINGS: Cantilever low-wing monoplane. Sweepback 35° at quarter-chord. Conventional all-metal three-spar fail-safe structure; centre spar extending to just outboard of inner edge of aileron on each wing. Five-section slat on outer 80% of each wing leading-edge. Triple-slotted flaps. Four-section spoilers forward of flaps on each wing. Outboard sections supplement ailerons for roll control. Section inboard of landing gear housing serves as airbrake and lift dumper; two middle sections can be used as airbrakes in flight. All control surfaces hydraulically actuated and of honeycomb construction. Hot-air de-icing of wing leading-edge. Slats are electrically heated.

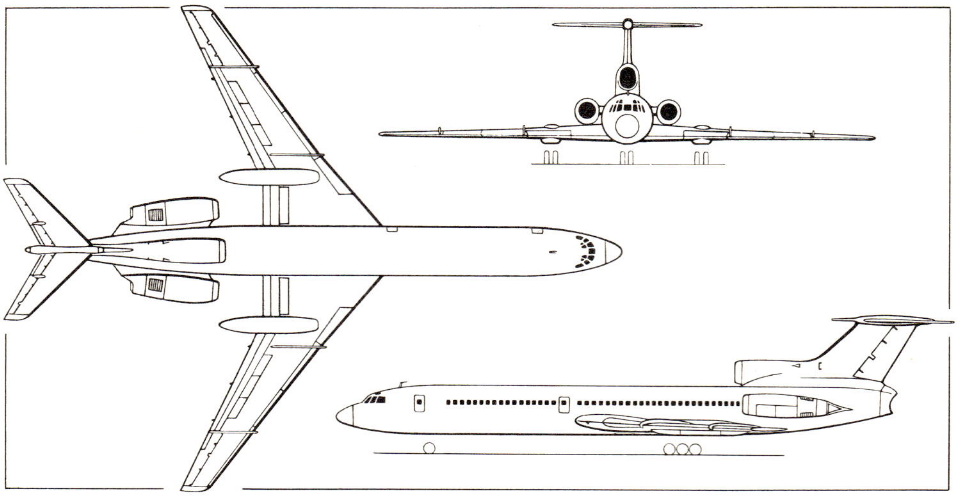

Tupolev Tu-154 medium/long-range three-turbofan transport aircraft *(Pilot Press)*

FUSELAGE: Conventional all-metal semi-monocoque fail-safe structure of circular section.

TAIL UNIT: Cantilever all-metal structure, with variable-incidence T tailplane. Rudder and elevator of honeycomb construction. Sweepback of 40° at quarter-chord on horizontal surfaces, 45° on leading-edge of vertical surfaces. Control surfaces hydraulically actuated by irreversible servo-controls. Leading-edges of fin and tailplane and engine air intake de-iced by hot air.

LANDING GEAR: Retractable tricycle type. Hydraulic actuation. Main units retract rearward into fairings on wing trailing-edge. Each consists of a bogie made up of three pairs of wheels, size 930 × 305, in tandem; tyre pressure 7·86 bars (114 lb/sq in). Steerable anti-shimmy twin-wheel nose unit has wheels size 800 × 225 and retracts rearward. Disc brakes and anti-skid units on main wheels.

POWER PLANT: Three Kuznetsov NK-8-2 turbofan engines, each rated at 93·2 kN (20,950 lb st), one on each side of rear fuselage and one inside extreme rear of fuselage. Two lateral engines fitted with upper and lower thrust-reversal grilles. Integral fuel tanks in wings; standard capacity 41,140 litres (9,050 Imp gallons). Max fuel capacity 46,825 litres (10,300 Imp gallons). Single-point refuelling standard.

ACCOMMODATION: Flight crew of two pilots and flight engineer; provision for navigator aft of pilot and folding seats for additional pilots or instructors. There are basic passenger versions for a total of 167, 158, 152, 146 and 128 passengers. Each has a toilet at the front (starboard), removable galley amidships and three toilets aft. Coat storage, folding seat and inflatable evacuation chute in each entrance lobby. Standard economy class version has 54 seats in six-abreast rows, with two tables between front rows, in forward cabin; and 104 seats in six-abreast rows (rear two rows four-abreast) in rear cabin at seat pitch of 75 cm (29·5 in). The 167-seat high-density version differs in having one further row of six seats in the forward cabin and reduced galley facilities. The tourist class versions carry 146 passengers at a seat pitch of 81 cm (32 in) or 152 at a pitch of 87 cm (34·25 in) with reduced galley facilities. The 128-seat version has only 24 first class seats, four-abreast at a pitch of 102 cm (40 in), in the forward cabin. There is also an all-cargo version. Passenger doors are forward of front cabin and between cabins on the port side, with emergency and service doors opposite. All four doors open outwards. Four emergency exits, two over wing on each side. Two pressurised baggage holds under main cabin floor, with two inward-opening doors. Normal provision for mechanised loading and unloading of baggage and freight in containers. Smaller unpressurised hold under rear cabin for carrying spare parts or special cargo such as radioactive isotopes.

SYSTEMS: Air-conditioning system pressure differential 0·62 bars (9·0 lb/sq in). Three independent hydraulic systems; working pressure 207 bars (3,000 lb/sq in). No. 1 system, powered by two pumps driven by centre engine and port engine, operates landing gear, brakes and all control surfaces. No. 2 system, powered by a pump driven by centre engine, actuates nosewheel steering, the second flying controls circuit and landing gear emergency extension. No. 3 system, powered by pump on starboard engine, actuates the third flying controls circuit and second landing gear emergency extension circuit. Three-phase 200/115V AC electrical system, supplied by three 40kVA alternators. 28V DC system. APU standard, driving 40kVA alternator and 12kW starter/generator.

AVIONICS AND EQUIPMENT: Automatic flight control system standard, including automatic navigation on pre-programmed route under control of navigational computer with en-route checks by ground radio beacons (including VOR, VOR/DME) or radar, and automatic approach by ILS to ICAO Category II standards (development to Category III standard in hand). Moving-map ground position indicator, HF and VHF radio, and radar standard. Safety equipment includes four inflatable life-rafts, each for 26 persons.

DIMENSIONS, EXTERNAL:
Wing span 37·55 m (123 ft 2½ in)
Length overall 47·90 m (157 ft 1¾ in)
Height overall 11·40 m (37 ft 4¾ in)
Diameter of fuselage 3·80 m (12 ft 5½ in)
Tailplane span 13·40 m (43 ft 11½ in)
Wheel track 11·50 m (37 ft 9 in)
Wheelbase 18·92 m (62 ft 1 in)
Passenger doors (each): Height 1·73 m (5 ft 7 in)
Width 0·80 m (2 ft 7½ in)
Height to sill 3·10 m (10 ft 2 in)
Servicing door: Height 1·28 m (4 ft 2½ in)
Width 0·61 m (2 ft 0 in)
Emergency door: Height 1·28 m (4 ft 2½ in)
Width 0·64 m (2 ft 1¼ in)
Emergency exits (each): Height 0·90 m (2 ft 11½ in)
Width 0·48 m (1 ft 7 in)
Main baggage hold doors (each):
Height 1·20 m (3 ft 11¼ in)
Width 1·35 m (4 ft 5 in)
Height to sill 1·80 m (5 ft 11 in)
Rear (unpressurised) hold:
Height 0·90 m (2 ft 11½ in)
Width 1·10 m (3 ft 7¼ in)
Height to sill 2·20 m (7 ft 2½ in)
DIMENSIONS, INTERNAL:
Cabin: Width 3·58 m (11 ft 9 in)
Height 2·02 m (6 ft 7½ in)
Volume 163·2 m^3 (5,763 cu ft)
Main baggage holds: front 21·5 m^3 (759 cu ft)
rear 16·5 m^3 (582 cu ft)
Rear underfloor hold 5·0 m^3 (176 cu ft)
AREAS:
Wings, gross 201·45 m^2 (2,169 sq ft)
Horizontal tail surfaces 40·55 m^2 (436·48 sq ft)
Vertical tail surfaces 31·72 m^2 (341·43 sq ft)
WEIGHTS:
Operating weight empty 43,500 kg (95,900 lb)
Normal payload 16,000 kg (35,275 lb)
Max payload 20,000 kg (44,090 lb)
Max fuel 33,150 kg (73,085 lb)
Max ramp weight 90,300 kg (199,077 lb)
Normal T-O weight 84,000 kg (185,188 lb)
Max T-O weight 90,000 kg (198,416 lb)
Normal landing weight 68,000 kg (149,915 lb)
Max landing weight 80,000 kg (176,370 lb)
Max zero-fuel weight 63,500 kg (139,994 lb)
PERFORMANCE (at max T-O weight, except where indicated):
Max level speed:
above 11,000 m (36,000 ft) Mach 0·90
at low altitudes 283 knots (525 km/h; 326 mph) IAS
Max cruising speed at 9,500 m (31,150 ft)
526 knots (975 km/h; 605 mph)
Best-cost cruising speed at 11,000-12,000 m (36,000-39,350 ft)
Mach 0·85 (486 knots; 900 km/h; 560 mph)
Long-range cruising speed at 11,000-12,000 m (36,000-39,350 ft)
Mach 0·80 (459 knots; 850 km/h; 528 mph)
Approach speed 127 knots (235 km/h; 146 mph)
Min ground turning radius 24·60 m (80 ft 8½ in)
T-O run at normal T-O weight, ISA 1,140 m (3,740 ft)
Balanced runway length at max T-O weight, FAR standard: ISA, S/L 2,100 m (6,890 ft)
ISA+20°C, S/L 2,420 m (7,940 ft)
Landing field length at max landing weight, FAR standard: ISA, S/L 2,060 m (6,758 ft)
ISA+20°C, S/L 2,217 m (7,273 ft)

Range at 11,000 m (36,000 ft) with standard fuel, reserves for 1 h and 6% of total fuel:
at 486 knots (900 km/h; 560 mph), with T-O weight of 84,000 kg and max payload (158 passengers, baggage and 5 tonnes of cargo and mail)
1,360 nm (2,520 km; 1,565 miles)
as above, T-O weight of 90,000 kg
1,867 nm (3,460 km; 2,150 miles)
at 459 knots (850 km/h; 528 mph), with T-O weight of 84,000 kg and max payload as above
1,510 nm (2,800 km; 1,740 miles)
as above, T-O weight of 90,000 kg
2,050 nm (3,800 km; 2,360 miles)
max range with 13,650 kg (30,100 lb) payload
2,850 nm (5,280 km; 3,280 miles)
Range at 11,000 m (36,000 ft) with optional centre-wing tanks, reserves as above:
with 9,000 kg (19,840 lb) payload (95 passengers)
3,453 nm (6,400 km; 3,977 miles)
with 6,700 kg (14,770 lb) payload (70 passengers)
3,723 nm (6,900 km; 4,287 miles)

TUPOLEV Tu-154A and Tu-154B

NATO reporting name: Careless

A developed version of the Tu-154, with the designation **Tu-154A**, was reported in early 1973, with the first flight scheduled for later that year. An article in the April 1975 issue of the Soviet magazine *Grazhdanskaya Aviatsiya* recorded that this aircraft had entered service with Aeroflot in April 1974 and that production Tu-154As were to be put into scheduled operation during 1975.

The Tu-154A is dimensionally unchanged by comparison with the original model, and is able to carry a normal payload of 152 passengers in Summer and 144 in Winter. Alternative configurations provide seats for 168 passengers on high-density routes, or 12 first class and 128 tourist class. Changes have centred mainly on the power plant, equipment and systems, to permit an increased gross weight, improve performance and reliability, and reduce servicing requirements.

The power plant consists of three Kuznetsov NK-8-2U turbofan engines, each uprated to 103 kN (23,150 lb st). Increased max take-off and landing weights allow extra fuel to be carried, raising the maximum capacity to 39,750 kg (87,630 lb). An additional tank, capacity 6,600 kg (14,550 lb), is mounted between the front and centre spars in the centre-section. It is intended primarily as a ballast tank for ferrying, and the fuel it contains can be pumped into the main system only on the ground. When the aircraft carries less than a full payload, this tank can be filled and its contents can be transferred to the main tanks at a destination airport, so reducing purchases of fuel outside the operator's home country. Other fuel system improvements have been made to the anti-icing fluid additive system; the centre-section tanks can be purged with CO_2 in the event of a forced landing with the wheels retracted.

The controls for the flaps, leading-edge slats and tailplane are interconnected, so that when the flaps are operated the tailplane is trimmed 3° down. An override switch caters for CG conditions which require a movement of more than 3°.

Additional emergency exits in the rear fuselage meet international requirements. The floor of the baggage holds has been strengthened to prevent damage by sharp-edged packages and baggage; and a smoke warning system has been introduced in the holds.

The electrical system has been modified by comparison with the Tu-154 and employs three alternators, on separate supply circuits, to provide 200/115V AC power. Two circuits supply all electrical services; the third supplies the electrical anti-icing system for the leading-edge slats. If one alternator fails, the remaining primary alternator can provide for all essential services, supplemented by the alternator on the APU. The duplicated DC electrical system embodies three rectifiers, of which one is for emergency use in the event of a failure of either of the others.

An ABSU automatic approach and landing system is fitted. This met ICAO Category I requirements initially, but was to be uprated to Category II later. Other equipment changes include the provision of duplicated radio compass, radio altimeter and DME; and the introduction of two-speed windscreen wipers and a system to indicate angle-of-bank limitations. An MSRP-64 flight recorder covers some 80 parameters, and a Mars-B voice recorder with open microphone is standard.

Servicing requirements and costs were reduced considerably on the Tu-154A, for which the servicing cycle is 300/900/1,800 h.

In 1977, production was switched to a further-improved version, designated **Tu-154B**, since refined as the **Tu-154B-2**. This retains the NK-8-2U turbofans of the Tu-154A, but is fitted with Thomson-CSF/SFIM automatic flight control and navigation equipment approved for ICAO Category II automatic landings. Max take-off and zero-fuel weights have been increased; and rearward extension of the usable cabin space enables 154-180 passengers to be carried despite the introduction of two more emergency exit doors, immediately forward of the engine nacelle on each side, making a total of six.

In the typical 169-seat high-density configuration, the passengers are seated mostly six-abreast with centre aisle at a pitch of 75 cm (29·5 in), with 68 in the forward cabin and 101 in the rear cabin. The forward door, on the port side between the flight deck and passenger accommodation, leads into a vestibule with a galley/pantry on the port side and a toilet opposite. There are three more toilets at the rear of the cabin, and each of the two doorways contains three seats for cabin staff. The basic all-tourist version differs in having 62 seats forward and 98 aft, with added coat stowage opposite the forward door and a further galley amidships. In the 154-seat mixed-class layout, the forward cabin is divided in two, with a separate lounge for up to 24 first class passengers at 96 cm; 38 in seat pitch.

Improvements have been made to the avionics, notably to simplify take-off and landing procedures. A different radar is fitted, and the fuel tank used as ballast on the Tu-154A can be used normally, as part of the standard fuel system of the Tu-154B/B-2.

Production of all versions of the Tu-154 exceeds 350. More than 300 are operated by Aeroflot, others by Balkan Bulgarian Airlines (13), Cubana (4), Malev (9) and Tarom (11).

WEIGHTS (A, Tu-154A; B, Tu-154B):
Basic operating weight: B 50,775 kg (111,940 lb)
Normal payload: A 16,000 kg (35,275 lb)
Max payload: A 18,000 kg (39,680 lb)
B 19,000 kg (41,887 lb)
Max fuel: B 39,750 kg (87,633 lb)
Max T-O weight: A 94,000 kg (207,235 lb)
B 98,000 kg (216,050 lb)
Max zero-fuel weight: B 71,000 kg (156,525 lb)
Max landing weight (normal):
A, B 78,000 kg (171,960 lb)
Max landing weight (emergency):
A 92,000-94,000 kg (202,825-207,235 lb)

PERFORMANCE (A, Tu-154A; B, Tu-154B, at max T-O weight, except where indicated):
Max level speed:
A 310 knots (575 km/h; 357 mph) IAS, except with less than 7,150 kg (15,763 lb) fuel at heights above 7,000 m (23,000 ft)
Normal cruising speed at up to 12,000 m (39,370 ft):
A Mach 0·85 (486 knots; 900 km/h; 560 mph)
B 486-513 knots (900-950 km/h; 560-590 mph)
Required runway length: B 2,200 m (7,218 ft)
Range with payload of 6,000 kg (13,225 lb), with reserves: B 2,700 nm (5,000 km; 3,105 miles)
Range with payload of 16,000 kg (35,275 lb):
A 1,725-1,780 nm (3,200-3,300 km; 1,985-2,050 miles)
Range with 164 passengers and baggage, with reserves:
B 1,890 nm (3,500 km; 2,175 miles)
Range with max payload, with reserves:
B 1,620 nm (3,000 km; 1,865 miles)

Tupolev Tu-154B-2 medium/long-range transport aircraft (three Kuznetsov NK-8-2U turbofan engines) at 1981 Paris Air Show *(Air Portraits)*

YAKOVLEV

GENERAL DESIGNER IN CHARGE OF BUREAU:
Alexander Sergeivich Yakovlev

Alexander Yakovlev is one of the most versatile Soviet designers, and products of his design bureau have ranged from transonic long-range fighters to the Yak-24 tandem-rotor helicopter, an operational VTOL carrier-based fighter and a variety of training and light general-purpose aircraft. Types in current production and service, or under development, are described hereafter.

YAKOVLEV Yak-18T

Details of this extensively-redesigned cabin version of the Yak-18 were given for the first time at the 1967 Paris Air Show, where an unregistered example was displayed statically. The first prototype flew for the first time in the Summer of that year, powered, like the Yak-18A and 18PM, with a 224 kW (300 hp) Ivchenko AI-14RF nine-cylinder radial engine, driving a two-blade variable-pitch propeller.

An initial evaluation programme of 450 test flights was completed by two prototypes during 1968-69. Together with experience gained during several months of operation at the Sasov flying school, this suggested that a number of improvements would be worthwhile. Most important of these was the installation of a more powerful M-14P radial engine of the type evolved by Vedeneev from the AI-14 and chosen also for the Kamov Ka-26 helicopter. In addition, improvements were made to the cabin layout and ventilation, and elevator effectiveness was reduced.

An aircraft embodying the modifications was approved by the Research Institute of the Soviet Ministry of Civil

Yakovlev Yak-18T basic trainer (Vedeneev M-14P engine)

Aviation. Four others logged 605 flying hours, including 1,591 take-offs and landings, over a five-month period of testing on every kind of airfield, under a wide variety of weather conditions. The Yak-18T was pronounced superior to earlier versions of the Yak-18 for basic training, and far more economical to operate than the An-2, which was also widely used.

Full production was ordered at Smolensk, and by 1974 it was possible to train the complete intake of 100 pupil pilots at Sasov on the new aircraft. Seventeen were then available for use at the school, of which two were being flown at double the normal rate of utilisation under a Research Institute programme.

As the standard basic trainer at Aeroflot flying schools, the Yak-18T is used for circuits, instrument training and navigation training, and as a flying classroom for an instructor and three pupils. Only one pupil accompanies the instructor on aerobatic flights.

Next to enter service, as a successor to the Yak-12, was the ambulance version, with light communications and forest fire patrol versions under consideration. A float-plane version is under development, and the Yak-18T will also operate eventually on skis.

Designer responsible for this variant of the Yak-18 was Mr Y. Yankievich.

TYPE: Four-seat multi-purpose light aircraft.

WINGS: Cantilever low-wing monoplane, in three sections: a constant-chord centre-section, integral with the fuselage, and two tapered outer panels. Wing section Clark YH, with thickness/chord ratio of 14·5% at root and 9·3% at tip. Dihedral on outer panels only. Two-spar light alloy construction. Light alloy covering on centre-section and on leading-edges of outer panels; inboard 25% of outer panels covered with light alloy, remainder with fabric. Slotted ailerons of light alloy construction, each hinged at three points and partly fabric-covered. Light alloy split flap across entire span of centre-section, actuated by two pneumatic servo-motors. Fixed step at port wing-root trailing-edge, with corrugated upper surface walkway to door on each side. Ailerons operated by pushrods. Ground-adjustable tab on each aileron.

FUSELAGE: Conventional light alloy semi-monocoque structure, of basically square section. Skin on rear fuselage spot welded to frames and stringers.

TAIL UNIT: Braced light alloy structure, with wire bracing above tailplane and wire and strut bracing below. All surfaces fabric-covered. Control surfaces operated by both pushrods and cables. Controllable trim tab in each elevator.

LANDING GEAR: Fully-retractable tricycle type, with single wheel on each unit. Pneumatic retraction, nosewheel rearward, main units inward into centre-section. No main-wheel doors. Oleo-nitrogen shock-absorbers. Castoring but non-steerable self-centering nosewheel with shimmy damper. All three tyres size 500 × 150. Differential pneumatic brakes on main wheels, with override button on instructor's control wheel.

POWER PLANT: One 269 kW (360 hp) Vedeneev M-14P nine-cylinder aircooled radial engine, driving a two-blade variable-pitch metal propeller, without spinner. Louvres in front of cowling to regulate cooling. Two-part cowling, split on horizontal centreline. Fuel tanks in wing roots.

ACCOMMODATION: Car-type cabin, seating four persons in pairs. Large forward-hinged door on each side, jettisonable in emergency. Provision for upholstered or parachute-type front seats. Bench-type rear seat removable for freight carrying. Ambulance configuration available, for pilot, stretcher patient and medical attendant. Large baggage compartment aft of rear seat, with external access on port side. Stretcher of ambulance version is loaded through this large baggage door. Cabin furnishings of non-inflammable synthetic materials. Dual control wheels. Glareshield above panel. Heating and ventilation standard.

SYSTEMS: Pneumatic system for actuating landing gear and flaps. Electrical system includes instrument panel red lighting, navigation and landing lights, and anti-collision beacon at top of fin.

AVIONICS AND EQUIPMENT: Standard equipment includes UHF radio, intercom, radio compass, radio altimeter and flight recorder.

DIMENSIONS, EXTERNAL:

Wing span	11·16 m (36 ft 7¼ in)
Length overall	8·35 m (27 ft 4¾ in)

AREA:

Wings, gross	18·75 m² (201·8 sq ft)

WEIGHTS AND LOADINGS (A, with instructor and one pupil; B, with instructor and three pupils):

Max payload: A	306 kg (675 lb)
B	436 kg (960 lb)
Max T-O weight: A	1,500 kg (3,307 lb)
B	1,650 kg (3,637 lb)
Max wing loading: A	80 kg/m² (16·4 lb/sq ft)
B	88 kg/m² (18·0 lb/sq ft)
Max power loading: A	5·59 kg/kW (11·0 lb/hp)
B	6·15 kg/kW (12·1 lb/hp)

PERFORMANCE (at max T-O weight: A, with instructor and one pupil; B, with instructor and three pupils):

Max level speed:	
A, B	159 knots (295 km/h; 183 mph)
Max cruising speed:	
B	135 knots (250 km/h; 155 mph)
Max rate of climb at S/L: B	300 m (985 ft)/min
Service ceiling: A, B	5,500 m (18,000 ft)
T-O run: A	330 m (1,085 ft)
B	400 m (1,315 ft)
Landing run: A	400 m (1,315 ft)
B	500 m (1,640 ft)
Range with max fuel, with reserves:	
A	350 nm (650 km; 403 miles)
B	485 nm (900 km; 560 miles)

Version of the Yak-28 known to NATO as 'Brewer-C' *(Flug Revue)*

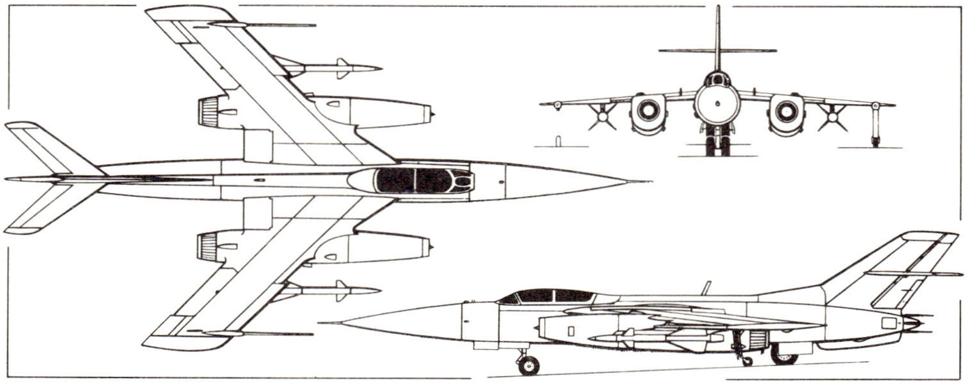

The long-nose version of the Yakovlev Yak-28P two-seat all-weather fighter ('Firebar') *(Pilot Press)*

YAKOVLEV Yak-28

NATO reporting names: Brewer, Firebar and Maestro

First seen in considerable numbers in the 1961 Soviet Aviation Day flypast were three successors to the Yak-25/27 series (see 1971-72 *Jane's*), described by the commentator as supersonic multi-purpose aircraft and identified subsequently by the designation Yak-28. These aircraft are shoulder-wing monoplanes, whereas all versions of the Yak-25, 26 and 27 were mid-wing. The Yak-28 series were, in fact, produced as entirely new designs, following only the general configuration of the earlier types.

The landing gear comprises two twin-wheel units in tandem, with the forward unit under the pilot's cockpit and the rear unit moved further aft than on the Yak-25/27, to a point immediately in front of the ventral fin. Wingtip balancer wheels are retained. The entire wing-root leading-edge is extended forward and the height of the fin and rudder increased. Tailplane sweep is also increased.

Several versions of the basic design have been identified, with the following NATO reporting names:

Brewer-A to C (Yak-28). Two-seat tactical attack versions. Single cockpit for pilot, with blister canopy, and glazed nose for navigator/bomb aimer. Corresponding to Yak-26 ('Mangrove') and produced to replace the Il-28 in the Soviet Air Force. Most examples have blister radome under fuselage just forward of wings. On some aircraft, long engine nacelles extend forward as far as the front of this radome. Others have shorter nacelles. Guns semi-submerged in each side of the fuselage on some aircraft; on starboard side only on others. Internal bomb bay between the underfuselage radome and the rear main landing gear unit. About 60 still active, in places of secondary importance.

Brewer-D. Reconnaissance version, with cameras in bomb bay. About 175 operational in 1981.

Brewer-E. First Soviet operational ECM escort aircraft, deployed in 1970. Underfuselage radome deleted. Active ECM pack built into bomb bay, from which it projects in form of a semi-cylindrical pack. Attachment under each outer wing, outboard of external fuel tank, for a rocket pod. About 20 in service in 1981.

Firebar. Tandem two-seat all-weather fighter derivative of Yak-28, corresponding to Yak-27. Nose radome. Internal weapons bay deleted. 'Anab' air-to-air missile under each wing instead of guns. Identified as **Yak-28P** (Perekhvatchik; interceptor) at 1967 Domodedovo display, the suffix 'P' indicating that the design had been *adapted* for the fighter role. Example shown in static park had a much longer dielectric nosecone than the standard operational 'Firebars' in the flying display and had two missile pylons under each wing, one for an 'Atoll' and one for an 'Anab'. This suggested that it was a weapons development aircraft. However, the lengthened nosecone has since been fitted retrospectively on many Yak-28Ps in squadron service. This does not indicate any increase in radar capability or aircraft performance.

Maestro (Yak-28U). Trainer version of 'Firebar'. Normal cockpit layout replaced by two individual single-seat cockpits in tandem, each with its own canopy. Front canopy sideways-hinged to starboard; rear canopy rearward-sliding.

About 300 Yak-28P 'Firebars' continue to form a significant component of the Soviet home defence interceptor force. The Yak-28 'Brewer' series is changing gradually from first-line attack to support roles, with the emphasis on ECM, reconnaissance and operational training.

The following details refer specifically to the Yak-28P, but are generally applicable to the other versions of the Yak-28:

TYPE: Two-seat transonic all-weather interceptor.

WINGS: Cantilever shoulder-wing monoplane of basically constant chord. Extended leading-edge on outer wings and also between fuselage and each engine nacelle. Outer extensions are drooped. Slotted flap, with unswept trailing-edge, between fuselage and each engine nacelle. Basic wing sweepback 45°. Anhedral from root. Single fence on upper surface of each wing, between fuselage and engine nacelle. Large trailing-edge flap and short aileron, with tab, outboard of nacelle on each wing. Balancer-wheel fairings, inset from wingtips, are extended forward as lead-filled wing balance weights.

FUSELAGE: All-metal semi-monocoque structure of basically circular section. Finely-tapered dielectric nosecone over radar scanner.

TAIL UNIT: Cantilever all-metal structure. Variable-incidence tailplane mounted midway up fin. All surfaces sweptback. Trim tab in rudder. Dorsal fin fairs into spine along top of fuselage. Shallow ventral stabilising fin.

LANDING GEAR: Two twin-wheel main units in tandem, retracting into fuselage. Front unit retracts forward, rear unit rearward. Small balancer wheel near each wingtip, retracting rearward under wing; fairing integral with leg.

POWER PLANT: Two afterburning turbojet engines, believed to be of same basic type as Tumansky R-11 fitted to some versions of MiG-21, with rating of 58·35 kN (13,120 lb st). Each fitted with centrebody shock-cone. A pointed slipper-type external fuel tank can be carried under the leading-edge of each wing, outboard of the engine nacelle.

ACCOMMODATION: Crew of two in tandem on ejection seats in pressurised cabin under long transparent blister canopy.

OPERATIONAL EQUIPMENT: Reported to include tail warning radar.

ARMAMENT: Pylon under each outer wing for 'Anab' air-to-air missile, with alternative infra-red or semi-active radar homing heads.

DIMENSIONS, EXTERNAL (estimated):

Wing span	12·95 m (42 ft 6 in)
Length overall: Yak-28	21·65 m (71 ft 0½ in)
Height overall	3·95 m (12 ft 11½ in)

WEIGHT (estimated):

Max T-O weight: Yak-28P	15,875 kg (35,000 lb)

PERFORMANCE (Yak-28P, estimated):

Max level speed at 10,670 m (35,000 ft)	Mach 1·1 (636 knots; 1,180 km/h; 733 mph)
Cruising speed	496 knots (920 km/h; 571 mph)
Service ceiling	16,750 m (55,000 ft)
Max combat radius	500 nm (925 km; 575 miles)
Max range	1,040-1,390 nm (1,930-2,575 km; 1,200-1,600 miles)

Yak-28P ('Firebar') fitted with original short radome and carrying two 'Anab' missiles *(Flug Revue)*

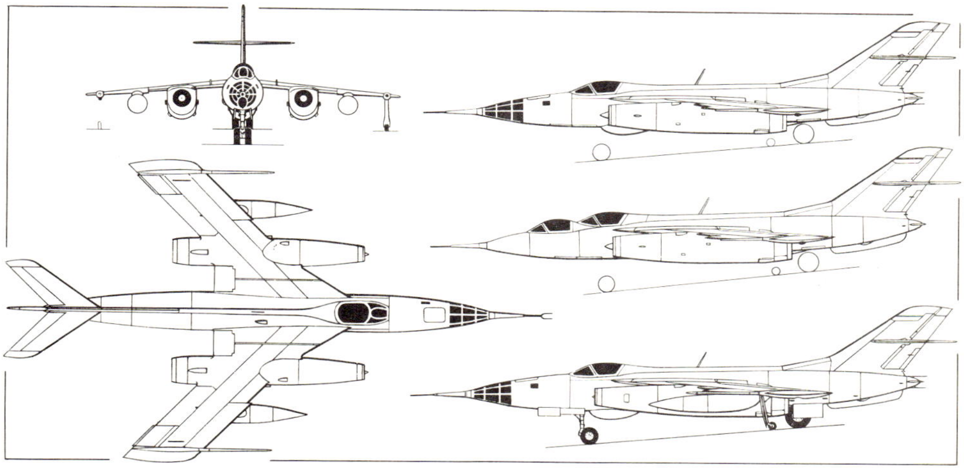

The version of the Yak-28 two-seat tactical attack aircraft known to NATO as 'Brewer-C', with additional side views of 'Brewer-B' (top) and the Yak-28U ('Maestro') tandem-cockpit trainer (centre) *(Pilot Press)*

YAKOVLEV Yak-36MP

NATO reporting name: Forger

This is the VTOL combat aircraft deployed by the Soviet Navy on the *Kiev* and *Minsk*, the first two of its 40,000 ton carrier/cruisers to put to sea. Two versions have been observed on the ships, as follows:

Forger-A. Basic single-seat combat aircraft. At least nine appear to be operational on each ship, in addition to about 15 Kamov Ka-25 anti-submarine and missile targeting helicopters. Primary operational roles are assumed to be reconnaissance, strikes against small ships, and fleet defence against shadowing, unarmed maritime reconnaissance aircraft.

Forger-B. Two-seat training version, of which one example was seen on the *Kiev* and another on the *Minsk*. A second cockpit is located forward of the normal cockpit, with the ejection seat at a lower level, under a continuous transparent canopy. To compensate for the longer nose, a 'plug' is inserted in the fuselage aft of the wing, lengthening the constant-section portion without requiring modification of the tapering rear fuselage assembly. In other respects this version appears to be identical to 'Forger-A', but has no ranging radar or weapon pylons.

The 1975-76 *Jane's* contained the remark that a strike/reconnaissance V/STOL aircraft was thought to have been developed by the Yakovlev bureau from the experimental VTOL aircraft known as 'Freehand', utilising a mixture of vectored thrust and direct jet-lift. This belief was confirmed when the *Kiev* entered the Mediterranean in July 1976 and subsequently operated its complement of 'Forgers' extensively during passage through that sea and the Atlantic en route to Murmansk. These aircraft, and those seen subsequently on the sister-ship *Minsk*, were assumed to be operated by a development squadron.

Observers of deck flying by 'Forger-As' report that the aircraft appear to be extremely stable during take-off and landing, which are performed with the vectored-thrust nozzles up to 10° forward of vertical. Take-offs are made vertically, with a smooth conversion about 5 to 6 m (15-20 ft) above the deck, initiated by lowering the aircraft's nose about 5° below the horizon and maintaining this attitude until the aircraft has accelerated to 30-40 knots (55-75 km/h; 35-46 mph). At this speed, a 5° nose-up attitude is assumed, and the accelerating transition is continued by vectoring aft the nozzles of the propulsion engine.

Landing procedure begins with a gradual descent from far astern, with the last 400 m (1,300 ft) flown essentially level, about 30 m (100 ft) above the water. The aircraft crosses the ship's stern with about a 5 knot (10 km/h; 6 mph) closure rate, 10-14 m (35-45 ft) above the flight deck, then flares gently to a hover and descends vertically. Landings are so precise that some form of control from the ship during take-off and approach has been suggested, perhaps in association with laser devices lining each side of the rear deck.

At no time was a STOL take-off observed, as practised by the Harrier/AV-8A combat aircraft of the Royal Air Force and US Marine Corps to increase their load-carrying capability. It is suggested that anything but direct vertical take-off might be difficult for the pilot of 'Forger-A', as take-off with forward speed over the deck would impose formidable stability and safety problems. The Soviet aircraft must also lack the Harrier's ability to increase its combat manoeuvrability by the use of thrust vectoring in forward flight (VIFF).

The following description applies to the single-seat 'Forger-A':

TYPE: Ship-based VTOL combat aircraft.

WINGS: Cantilever mid-wing monoplane, of very small area. Thickness/chord ratio estimated at 6% or less. Constant anhedral from roots. Sweepback on leading-edge approx 45°. Each wing comprises two all-metal panels of approx equal span, of which the outer panel folds vertically upward for stowage on board ship. Inboard panel has unswept trailing-edge, occupied entirely by a large single-slotted Fowler-type flap. Outer panel has a slightly sweptback trailing-edge, occupied almost entirely by an aileron with inset trim tab. No leading-edge flaps or slats. Jet reaction control valve with upper and lower slots in each wingtip.

FUSELAGE: Conventional all-metal semi-monocoque structure. Integral engine air intake ducts, with downward-inclined lips forward of rear edge of transparent cockpit canopy. Row of small blow-in auxiliary intake doors a short distance aft of each intake. Rearward-hinged door over lift-jets, immediately aft of canopy, with spring-loaded louvres. Position of corresponding side-hinged underfuselage doors conforms with forward tilt of lift engines. Yaw reaction control nozzle to each side of small tailcone. No reaction control system in nose.

TAIL UNIT: Conventional all-metal structure, with sweepback on all surfaces and considerable tailplane anhedral. Trim tab in rudder and each elevator. Air intake at front of long duct extending forward from base of fin, probably for air to cool avionics bay in rear fuselage.

LANDING GEAR: Retractable tricycle type. Single wheel on each unit, with legs of trailing-link type with oleo-pneumatic shock-absorption. Nose unit retracts rearward, main units forward into fuselage. Small bumper under upward-curving rear fuselage.

POWER PLANT: Primary power plant is believed to comprise a single large turbojet engine, mounted in the centre-fuselage and exhausting through a single pair of vectoring side-nozzles aft of the wings. No afterburner is fitted. Engine could be a version of the Lyulka AL-21, rated at approx 78 kN (17,500 lb st). Two lift-jet engines in tandem, immediately aft of cockpit, exhausting downward; believed to be of Koliesov design, each with thrust of 35·5 kN (8,000 lb). Fuel tanks in fuselage, probably forward and aft of main engine.

ACCOMMODATION: Pilot only, on ejection seat, under sideways-hinged (to starboard) transparent canopy.

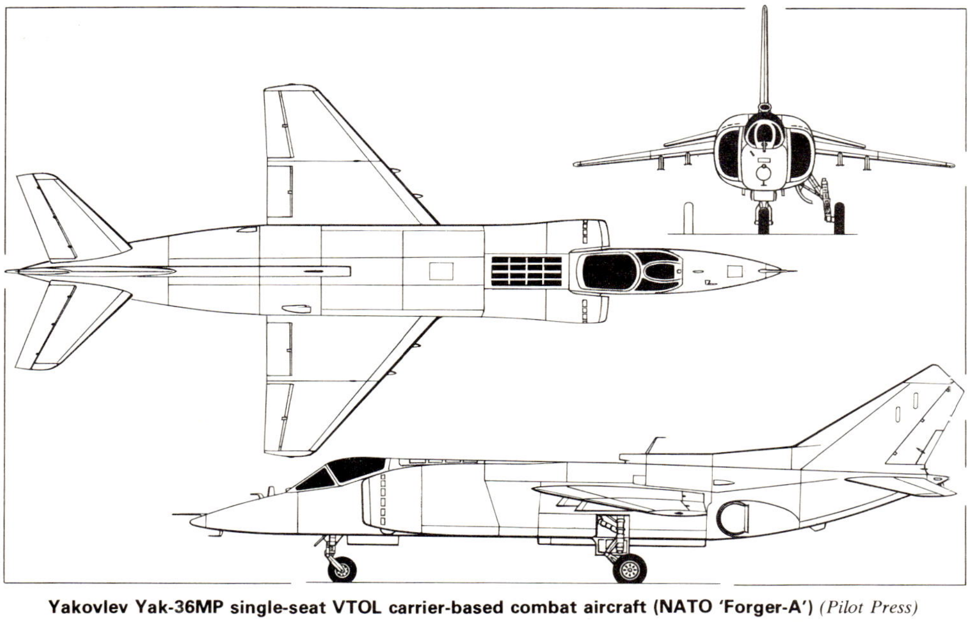

Yakovlev Yak-36MP single-seat VTOL carrier-based combat aircraft (NATO 'Forger-A') *(Pilot Press)*

Yakovlev Yak-36MP (NATO 'Forger-A') hovering over the flight deck of the *Kiev (Tass)*

The two-seat training version of the Yak-36 ('Forger-B')

Avionics: Ranging radar in nose. Probable fully-automatic control system for use during take-off and landing.

Armament: No installed armament. Two pylons under fixed panel of each wing for an estimated 1,360 kg (3,000 lb) of external stores, including gun pods, rocket packs, air-to-air missiles and auxiliary fuel tanks.

Dimensions, external (estimated):
- Wing span: 7·32 m (24 ft 0 in)
- Width, wings folded: 4·88 m (16 ft 0 in)
- Length overall: 'Forger-A': 15·25 m (50 ft 0 in)
  - 'Forger-B': 17·68 m (58 ft 0 in)
- Height overall: 4·37 m (14 ft 4 in)
- Tailplane span: 3·81 m (12 ft 6 in)
- Wheel track: 2·90 m (9 ft 6 in)
- Wheelbase: 5·50 m (18 ft 0 in)

Weights (estimated):
- Basic operating weight, incl pilot(s):
  - 'Forger-A': 7,485 kg (16,500 lb)
  - 'Forger-B': 8,390 kg (18,500 lb)
- Max T-O weight: 11,565 kg (25,500 lb)

Performance ('Forger-A', estimated, at max T-O weight):
- Max level speed at height: Mach 1·1 (630 knots; 1,170 km/h; 725 mph)
- Max level speed at S/L: Mach 0·8 (605 knots; 1,125 km/h; 700 mph)
- Max rate of climb at S/L: 4,500 m (14,750 ft)/min
- Service ceiling: 12,000 m (39,375 ft)
- Combat radius:
  - with air-to-air missiles and external tanks, 75 min on station: 100 nm (185 km; 115 miles)
  - with max weapons, lo-lo-lo: 130 nm (240 km; 150 miles)
  - with max weapons, hi-lo-hi: 200 nm (370 km; 230 miles)

YAKOVLEV Yak-40

NATO reporting name: Codling

This three-turbofan short-haul jet transport was designed to replace the Li-2 (Soviet-built DC-3) and to operate from Class 5 (grass) airfields. The prototype flew for the first time on 21 October 1966 and the Yak-40 made its first passenger flight in Aeroflot service on 30 September 1968.

By the Summer of 1976, more than 800 Yak-40s had been built. Most are in service with Aeroflot, some as air ambulances carrying patients to medical centres and to Black Sea convalescent centres. Others have been sold in Afghanistan, Bulgaria, Czechoslovakia, France, West Germany, Italy, Poland and Yugoslavia. Military operators include the Soviet, Bangladesh, Polish, Syrian and Yugoslav air forces. Production in the Soviet Union has ended. A company named ICX Aviation Inc announced plans to manufacture a re-engined version in the USA, as the X-Avia, but there has been no recent news of this programme.

All Soviet production Yak-40s are structurally similar, with AI-25 engines, and differ only in their standard of accommodation, usually for 27 or 32 airline passengers in a single class, 16 or 20 passengers in two-class layouts, or up to 11 passengers in an executive layout. All have clam-shell thrust reversers aft of the centre engine.

A freighter version is in service, with a cargo door, size approximately 1·50 m × 1·60 m (5 ft × 5 ft 2½ in), in the port side of the fuselage.

A full description of the Yak-40 can be found in the 1978-79 *Jane's*.

Power Plant: Three Ivchenko AI-25 turbofan engines, each rated at 14·7 kN (3,300 lb st). Fin and boundary layer splitter beneath and forward of intake for centre engine. Clamshell thrust reverser fitted to airframe aft of this engine. Hot-air anti-icing system for all three engine air intakes. Fire warning and extinguishing systems standard. Fuel in integral tanks between front auxiliary spar and main spar in each wing, from outboard of the fuselage to the inner end of the aileron, total capacity 3,910 litres (860 Imp gallons). Type AI-9 turbine APU mounted in rear of top engine intake fairing for engine starting. Provision for starting from ground compressed air supply.

Accommodation: Two pilots side by side on flight deck, on adjustable seats, with dual controls. Central jump seat at rear for third person. Automatically-actuated electrical windscreen de-icing system. Main cabin normally laid out for 27 passengers in three-abreast rows, with two-chair units on starboard side of aisle. Seat pitch 75·5 cm (29·7 in). Individual ventilator by each seat. Rack for hand baggage on starboard side of cabin ceiling. Cloakroom (port), buffet, baggage compartment and toilet (starboard) aft of main cabin. Seat for stewardess against rear face of partition separating cabin from rear compartments, on port side. Normal access via hydraulically-actuated ventral airstair door at rear. Service door on port side of cabin, at front. For high-density services, nine twin-seat units can be installed on the port side of the aisle and seven on the starboard side, giving a total of 32 seats. Pitch is unchanged, enabling the rear cabin partition to be moved forward and so giving a larger baggage compartment. The 18/20-seat mixed class version has two passenger cabins, separated by a partition. The forward cabin has two swivelling chairs, one each side of a table, on the port side; and an inward-facing four-place settee on the starboard side, with small cupboards fore and aft. Alternatively, two more swivelling seats and a table can replace the settee. The rear cabin contains 14 seats in basically four-abreast rows. A lobby between the flight deck and forward cabin provides access for the crew without passing through the main cabins, and contains a cloakroom (starboard) and seat for the stewardess. Compartments aft of the rear cabin are as in the standard 27-seat version. The executive version has a toilet and other facilities in a large compartment aft of the flight deck; a centre lounge furnished with a four-place settee, three armchairs, a writing desk and a sideboard; and rear cabin containing six seats, three-abreast. There is a further toilet aft, and the galley is equipped to special standards. A bar with two adjustable tables is built into the wall between the flight deck and lounge, on the port side. A freight-carrying version is in service.

Avionics and Equipment: Standard equipment includes full blind-flying instrumentation, two Landysh-5 VHF radio communications installations, an ARK-10

Yakovlev Yak-40 short-range transport (three Ivchenko AI-25 turbofan engines) in Aeroflot markings

automatic radio compass, KURS-MP-2 VOR/ILS system, RV-5 radio altimeter, Type SO 70 transponder, Grosa-40 weather radar, PRIVOD-ANE-1 flight director system, Kremenj 40E autopilot, AGD 1 artificial horizon and GMK-1GE gyro-compass, permitting automatic approach to ICAO Category II standards.

DIMENSIONS, EXTERNAL:
Wing span 25·00 m (82 ft 0¼ in)
Wing aspect ratio 9
Length overall 20·36 m (66 ft 9½ in)
Length of fuselage 17·00 m (55 ft 9 in)
Diameter of fuselage 2·40 m (7 ft 10½ in)
Height overall 6·50 m (21 ft 4 in)
Tailplane span 7·50 m (24 ft 7¼ in)
Wheel track 4·52 m (14 ft 10 in)
Wheelbase 7·47 m (24 ft 6 in)
Rear cabin door: Height 1·74 m (5 ft 8½ in)
Width 0·94 m (3 ft 1 in)
Service door: Height 1·20 m (3 ft 11¼ in)
Width 0·55 m (1 ft 9½ in)

DIMENSIONS, INTERNAL:
Cabin: Length 7·07 m (23 ft 2½ in)
Max width 2·15 m (7 ft 0¾ in)
Max height 1·85 m (6 ft 0¾ in)

AREA:
Wings, gross 70·00 m² (753·5 sq ft)

WEIGHTS AND LOADINGS (A: 27 seats, B: 32 seats, C: 16 seats, D: executive version):
Weight empty:
A 9,010-9,400 kg (19,865-20,725 lb)
B 9,400 kg (20,725 lb)
D 9,560-9,850 kg (21,075-21,715 lb)
Max payload: A 2,300 kg (5,070 lb)
B 2,720 kg (6,000 lb)
C 1,360 kg (3,000 lb)
D 990 kg (2,180 lb)
Max fuel weight:
A 2,125-4,000 kg (4,685-8,820 lb)
B 4,000 kg (8,820 lb)
D 3,000-4,000 kg (6,615-8,820 lb)
Normal T-O weight:
A 12,360-15,500 kg (27,250-34,170 lb)
B 15,500 kg (34,170 lb)
D 12,360-15,000 kg (27,250-33,070 lb)
Max T-O weight: A, B 16,000 kg (35,275 lb)
C 15,310 kg (33,750 lb)
D 15,400 kg (33,950 lb)
Max wing loading: A, B 230 kg/m² (47·1 lb/sq ft)

PERFORMANCE (corresponding to weights given above):
Max level speed at S/L
Mach 0·7 (324 knots; 600 km/h; 373 mph) IAS
Max cruising speed at 7,000 m (23,000 ft)
297 knots (550 km/h; 342 mph)
T-O speed: A, B 86 knots (160 km/h; 100 mph)
D 81-84 knots (150-156 km/h; 93-97 mph)
Max rate of climb at S/L 480 m (1,575 ft)/min
Normal T-O run: A, B 700 m (2,297 ft)
C 650 m (2,133 ft)
D 660 m (2,165 ft)
Normal landing run: A, C, D 320 m (1,050 ft)
B 360 m (1,182 ft)
Range with max payload at 254 knots (470 km/h; 292 mph) at 8,000 m (26,250 ft), with reserves:
A, C, D 971 nm (1,800 km; 1,118 miles)
B 782 nm (1,450 km; 900 miles)
Range with max fuel at 254 knots (470 km/h; 292 mph) at 8,000 m (26,250 ft), with reserves:
All versions 971 nm (1,800 km; 1,118 miles)
Max range at 254 knots (470 km/h; 292 mph) at 8,000 m (26,250 ft), no reserves:
All versions 1,080 nm (2,000 km; 1,240 miles)

YAKOVLEV Yak-42

NATO reporting name: Clobber

On the basis of experience with the Yak-40, the Yakovlev design bureau is developing for Aeroflot this larger civil airliner with a similar three-engined layout. According to Alexander Yakovlev, the basic design objectives were simple construction, reliability in operation, economy, and the ability to operate in remote areas with widely differing climatic conditions. Up to 2,000 aircraft in this category are needed, for use particularly on feederline services extending north and south from the main east-west trans-Siberian trunk routes.

Three prototypes of the Yak-42 were ordered initially. The first of these (CCCP-1974) flew for the first time on 7 March 1975, with a wing sweepback of 11°, and was furnished as a 100-passenger local-service version with carry-on baggage and coat stowage fore and aft of the cabin. The second prototype (CCCP-1975) had 23° of wing sweep, and cabin windows which extended further forward and rearward on each side, indicating that it was representative of the 120-seat version with three more rows of seats and no carry-on baggage areas. The third prototype (CCCP-1976; re-registered subsequently as CCCP-42303) differed from the second only in detail, having hot-air de-icing on the tail surfaces as well as the wings; fairing discs over the main wheels and longer leg fairings to improve airflow over the doorless main landing gear when retracted; and movement further forward of the overwing emergency exits.

It was made known that a decision on the degree of wing sweep to be standardised for production aircraft would be taken after simultaneous evaluation of the prototypes, in terms of high-speed cruise, economy and low-speed handling characteristics. The 23° wing showed itself superior, and aircraft No. CCCP-42303, exhibited at the 1977 Paris Salon, was generally typical of the first series of 200 production Yak-42s, which are being built at Smolensk to replace Tu-134s currently in Aeroflot service. Further changes introduced on production aircraft include the use of four-wheel main landing gear bogies instead of the

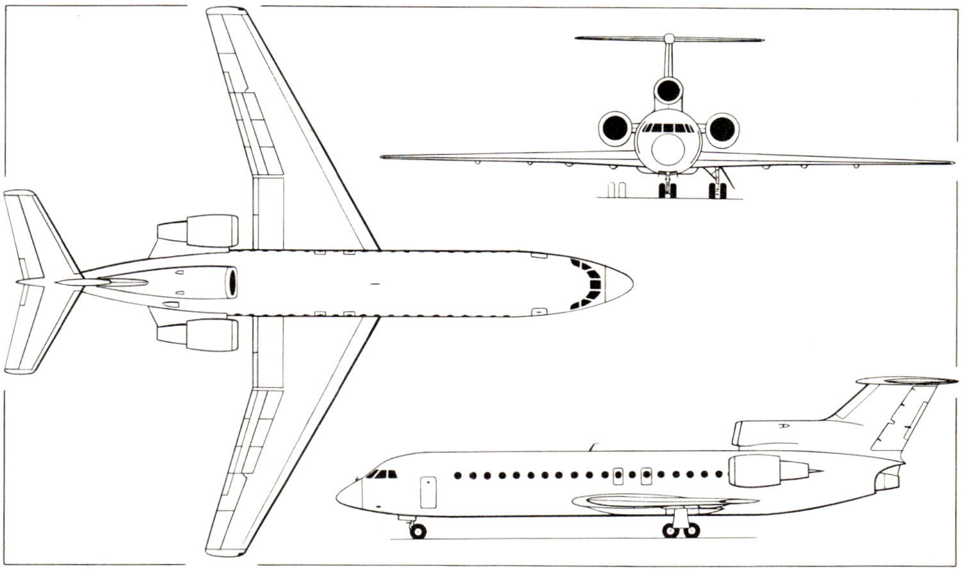

Yakovlev Yak-42 three-turbofan medium-range passenger transport *(Pilot Press)*

Yakovlev Yak-42 medium-range transport (three Lotarev D-36 turbofan engines) with new four-wheel main landing gear bogies

twin-wheel units fitted to the prototypes.

The Yak-42 entered scheduled passenger service with Aeroflot in late 1980, operating first over the Moscow-Krasnodar route. Ten aircraft had been flown by mid-1981, and it was hoped to complete 20 more by the end of the year. A first export order, for seven, had been placed by Aviogenex of Yugoslavia, with deliveries to begin in May 1982. By 1983, the Yakovlev bureau hopes to certificate a 'stretched' version of the aircraft, with 140 seats.

On 29 January 1981, a Yak-42 piloted by Valentin Mukhin set a record in FAI Class C1m (T-O weight 45,000-55,000 kg) by lifting a load of 20,186 kg (44,502 lb) to a height of 2,000 m (6,562 ft). In two subsequent flights, the aircraft climbed to 3,000 m in 2 min 37 s, 6,000 m in 5 min 11 s, and 9,000 m in 9 min 31 s to claim Class C1l (35,000-45,000 kg) records; and to 3,000 m in 3 min 6 s, 6,000 m in 6 min 27 s, and 9,000 m in 11 min 48 s to claim Class C1m records.

The Yak-42 is intended to use all three engines at cruise power during flight. It can, however, continue take-off after the failure of any one engine, and can maintain level cruising flight on a single engine.

Design is in accordance with the latest airworthiness standards of the Soviet civil authorities and US FAR 25 requirements. Special care has been taken during design to ensure that the D-36 engines conform with national and international limits on smoke and noise; and the Yak-42 is intended to operate in temperatures ranging from −50°C to +50°C. An APU is standard, for engine starting and ground services, making the aircraft independent of airport equipment. Airframe design life is 30,000 flying hours or 30,000 landings in 15 years. Engine life is 18,000 operating hours with two major overhauls.

TYPE: Three-turbofan medium-range passenger transport.

WINGS: Cantilever low-wing monoplane, consisting of a centre-section and two outer panels. No dihedral or anhedral. Sweepback 23° at quarter-chord. All-metal two-spar torsion-box structure. Two-section aileron on each wing, with servo tab on inner section and trim tab on outer section. Two-section single-slotted flaps on each wing. Multi-section spoilers forward of flaps. No leading-edge control surfaces. Ailerons and flaps actuated hydraulically.

FUSELAGE: All-metal riveted, bonded and welded semi-monocoque structure, of basic circular section, blending into an oval-section rear fuselage.

TAIL UNIT: Cantilever all-metal T-tail structure, with sweepback on all surfaces. One-piece tailplane; incidence variable from 4° upward to 8° downward. Trim tab in each elevator. Trim tab and spring servo tab in rudder. Control surfaces actuated hydraulically.

LANDING GEAR: Hydraulically-retractable tricycle heavy-duty type. Four-wheel bogie main units retract inward into flattened fuselage undersurface. Twin nosewheels retract forward. Hydraulic backup system for extension only. Emergency extension by gravity. Oleo-nitrogen shock-absorbers. Steerable nose unit of levered-suspension type. Low-pressure tyres; size 930 × 305 on nosewheels. Hydraulic disc brakes on main wheels. Nosewheel brakes to stop wheel rotation after take-off.

POWER PLANT: Three Lotarev D-36 three-shaft turbofan engines, each rated at 63·74 kN (14,330 lb st). Centre engine, mounted inside rear fuselage, has S-duct air intake. Outboard engines are mounted in pod on each side of rear fuselage. No thrust reversers. Integral fuel tanks between spars in wings. APU standard, for engine starting, and for power and air-conditioning supply on ground and, if necessary, in flight.

ACCOMMODATION: Crew of two side by side on flight deck. Single passenger cabin, with total of 120 seats in six-abreast rows, at pitch of 80 cm (31·5 in), with centre aisle, in high-density configuration. Alternative 100-passenger local-service configuration, with carry-on baggage and coat stowage compartments fore and aft of cabin. Main airstair door hinges down from under-surface of rear fuselage. Second door forward of cabin on port side, with integral airstairs. Service door opposite. Galley and crew coat stowage between flight deck and front vestibule. Passenger coat stowage and toilet between vestibule and cabin. Second coat stowage and toilet at rear of cabin. Two underfloor holds for cargo, mail and baggage in standard containers, loaded through a large door on the starboard side, forward of wing. Chain-drive handling system for containers built into cabin floor. Forward hold accommodates six containers, each with capacity of 2·2 m³ (77·7 cu ft); rear hold takes three similar containers. Provision for convertible passenger/cargo interior, with enlarged loading door on port side of front fuselage. Two emergency exits overwing on each side. All passenger and crew accommodation pressurised and air-conditioned, and furnished with non-inflammable materials.

AVIONICS AND EQUIPMENT: Flight and navigation equipment for operation by day and night under adverse

Yakovlev Yak-50 single-seat aerobatic and sporting aircraft

weather conditions, with landings on concrete or unpaved runways in ICAO Category II weather minima down to 40 m (131 ft) visibility at 300 m (985 ft). Type SAU-42 automatic flight control system and area navigation system standard.

DIMENSIONS, EXTERNAL:

Wing span	34·20 m (112 ft 2½ in)
Wing aspect ratio	7·8
Length overall	36·38 m (119 ft 4¼ in)
Fuselage diameter	3·80 m (12 ft 5½ in)
Height overall	9·80 m (32 ft 1¾ in)
Tailplane span	10·80 m (35 ft 5 in)
Wheel track	5·63 m (18 ft 5¾ in)
Wheelbase	14·78 m (48 ft 6 in)
Passenger door (fwd):	
Height	1·81 m (5 ft 11¼ in)
Width	0·83 m (2 ft 8½ in)
Passenger entrance (rear): Height	1·78 m (5 ft 10 in)
Width	0·81 m (2 ft 7¾ in)
Cargo door (convertible version):	
Height	2·025 m (6 ft 7¾ in)
Width	3·23 m (10 ft 7 in)
Baggage/cargo hold door: Height	1·35 m (4 ft 5 in)
Width	1·145 m (3 ft 9 in)
Height to sill	1·45 m (4 ft 9 in)

DIMENSIONS, INTERNAL:

Cabin: Length	19·89 m (65 ft 3 in)
Max width	3·60 m (11 ft 9¾ in)
Height	2·08 m (6 ft 9¾ in)
Forward baggage compartment volume (100-seater)	19·8 m³ (700 cu ft)
Rear baggage compartment volume (100-seater)	9·5 m³ (335 cu ft)

AREAS:

Wings, gross	150 m² (1,615 sq ft)
Horizontal tail surfaces (total)	27·60 m² (297·1 sq ft)
Vertical tail surfaces (total)	23·29 m² (250·7 sq ft)

WEIGHTS:

Weight empty	28,960 kg (63,845 lb)
Max payload	14,500 kg (32,000 lb)
Max fuel	18,500 kg (40,785 lb)
Max T-O weight	53,500 kg (117,950 lb)

PERFORMANCE:

Max cruising speed at 7,600 m (25,000 ft)	437 knots (810 km/h; 503 mph)
Econ cruising speed	405 knots (750 km/h; 466 mph)
Approach speed	114 knots (210 km/h; 131 mph)
Max cruising height	10,000 m (32,800 ft)
T-O field length	1,800 m (5,900 ft)
Landing from 15 m (50 ft)	1,100 m (3,610 ft)
Range at 415 knots (770 km/h; 478 mph) at 9,000 m (29,500 ft):	
with 6,500 kg (14,330 lb) payload	1,620 nm (3,000 km; 1,864 miles)
with 10,500 kg (23,150 lb) payload	1,080 nm (2,000 km; 1,242 miles)
with max payload	486 nm (900 km; 559 miles)

YAKOVLEV Yak-50

A Novosti Press Agency bulletin, dated 30 June 1975, stated that tests of a new Yakovlev sporting aircraft, designated Yak-50, had been carried out near Arsenyev in the Soviet Far East. Mr Nikolai Sazykin, director of the Progress Engineering Works in which all Yakovlev sporting aircraft are assembled, was quoted as saying that the Yak-50 was intended to participate in the 1976 world aerobatic championships.

No descriptive details, specification or illustrations of the new aircraft were released. Test pilot Anatoly Sergeyev stated only that it was more advanced than the familiar Yak-18 training and aerobatic monoplane, with a more powerful engine, better manoeuvrability, a speed of over 215 knots (400 km/h; 248 mph) in a dive, and the ability to perform all aerobatics with its landing gear retracted or extended.

When six Yak-50s participated in the 1976 world aerobatic championships at Kiev, their evolution from the Yak-18 was apparent, but with significant changes. Basic configuration is little different from that of the single-seat Yak-18PS, with tailwheel-type landing gear. This was deliberate, to keep the handling characteristics of the two types as similar as possible. However, overall dimensions are reduced; control surface hinge-lines have been moved to keep control forces light; and overall structural strength has been increased by switching entirely to metal covering. In particular, the fuselage is now semi-monocoque instead of steel tube with fabric covering to the rear of the cockpit. Designers responsible for these and other changes were Sergei Yakovlev (son of Alexander Yakovlev) and Yuri Yankievich.

The wings dispense with the Yak-18's centre-section and have 2° dihedral and 2° incidence, and retain an asymmetric section. To ensure a high power/weight ratio in such a large aircraft, the power plant is a 268 kW (360 hp) Vedeneev (Ivchenko) M-14P aircooled radial piston engine, driving a V-530TA-D35 two-blade variable-pitch propeller, instead of the 224 kW (300 hp) Ivchenko AI-14RF of the Yak-18PS. Main-wheel tyre size is 500 × 150, nosewheel tyre size 200 × 80. The main fuel tank, capacity 55 litres (12 Imp gallons), is aft of the engine firewall, the electrical system battery behind the pilot's seat. A Zyablik radio transceiver is standard.

Observers at the world championships at Kiev reported that the Yak-50s performed the all-important Aresti manoeuvres with smooth precision, their primary shortcoming being excessive directional stability. Yak-50s flown by V. Letsko and I. Egorov finished first and second in the men's competition. Others came fifth, seventh and ninth, to win the team prize. First five places in the women's championship were taken by Yak-50s.

DIMENSIONS, EXTERNAL:

Wing span	9·50 m (31 ft 2 in)
Length overall	7·676 m (25 ft 2¼ in)
Tailplane span	3·16 m (10 ft 4½ in)
Wheel track	2·00 m (6 ft 6¾ in)
Wheelbase	5·10 m (16 ft 8¾ in)
Propeller diameter	2·00 m (6 ft 6¾ in)

AREAS:

Wings, gross	15·00 m² (161·5 sq ft)
Ailerons (total)	1·95 m² (21·00 sq ft)
Vertical tail surfaces (total)	1·48 m² (15·93 sq ft)
Horizontal tail surfaces (total)	2·86 m² (30·78 sq ft)

WEIGHTS AND LOADINGS:

Weight empty, equipped	765 kg (1,686 lb)
Max T-O weight	900 kg (1,984 lb)
Max wing loading	60 kg/m² (12·29 lb/sq ft)
Max power loading	3·36 kg/kW (5·51 lb/hp)

PERFORMANCE:

Never-exceed speed	226 knots (420 km/h; 261 mph)
Max level speed	173 knots (320 km/h; 199 mph)
T-O speed	65 knots (120 km/h; 75 mph)
Rate of climb at S/L	960 m (3,150 ft)/min
Service ceiling	5,500 m (18,045 ft)
T-O run	200 m (657 ft)
Landing run	250 m (820 ft)
Max range at 1,000 m (3,280 ft), with 120 litres (26·4 Imp gallons) auxiliary fuel, reserve of 10 litres (2·2 Imp gallons)	267 nm (495 km; 307 miles)
Endurance at 500 m (1,640 ft) with 52 litres (11·4 Imp gallons) auxiliary fuel, reserve of 10 litres (2·2 Imp gallons)	48 min
g limits	+9; −6

YAKOVLEV Yak-52

Production of this tandem two-seat piston-engined primary trainer has been entrusted to the Romanian aircraft industry (which see), under the Comecon (Council for Mutual Economic Assistance) programme.

UNITED KINGDOM

ADA
AERONAUTICAL DEVELOPMENT ASSOCIATES LTD

14 Bishops Avenue, Elstree, Borehamwood, Hertfordshire WD6 3LZ
Telephone: 01 953 1250

DIRECTORS:
P. L. Cronbach, BSc (Eng), CEng, FRAeS
J. W. Allam, OBE, FRAeS

ADA APOLLO JETSTREAM

ADA is undertaking, on behalf of Apollo Airways Inc of Goleta, California, the modification of a number of Handley Page Jetstream Mk 1 light transport aircraft. Designed to meet the October 1979 requirements of SFAR 41, the purpose of the Apollo Jetstream is to exploit fully the potential of the basic Jetstream airframe in commuter operations under FAR Pt 135 in the western USA. Airframe alterations, which are minimal, are chiefly of a detail nature to meet requirements which differ slightly from those envisaged during the initial design. The major changes concern the power plant and air-conditioning system. The Turboméca Astazou XVI F3 turboprop engines installed by ADA are identical to the Astazou XVI F1s installed previously, but are flat rated to 626 kW (840 shp) each at 31°C at S/L. An improved air-conditioning distribution system, with Garrett cooling turbines, is also installed.

ADA's Apollo Jetstream modifications are cleared by the British CAA to US requirements under a bilateral agreement with the FAA. Flight testing, by ADA, is certificated by the CAA, this also qualifying as FAA type approval for US operation by Apollo Airways.

WEIGHTS AND LOADINGS:

Basic weight empty, equipped (18-seat commuter)	3,952 kg (8,714 lb)
Max fuel	1,370 kg (3,020 lb)
Max payload	1,717 kg (3,786 lb)
Max T-O weight	6,236 kg (13,750 lb)
Weight/altitude/temperature T-O limitations, ISA + 24°C:	
airfield at 305 m (1,000 ft)	6,236 kg (13,750 lb)
airfield at 915 m (3,000 ft)	6,087 kg (13,420 lb)
airfield at 1,375 m (4,500 ft)	5,969 kg (13,160 lb)
airfield at 1,830 m (6,000 ft)	5,670 kg (12,500 lb)
Max landing weight	6,123 kg (13,500 lb)
Max zero-fuel weight (limited by SFAR 41)	5,670 kg (12,500 lb)
Max wing loading	248·4 kg/m² (50·9 lb/sq ft)
Max power loading	4·99 kg/kW (8·2 lb/shp)

PERFORMANCE (at max T-O weight, ISA, except where indicated):

Never-exceed speed	300 knots (556 km/h; 345 mph)
Max level and cruising speed at 3,050 m (10,000 ft)	244 knots (452 km/h; 281 mph)
Cruising speed for max range, mean cruise weight, at 6,700 m (22,000 ft)	211 knots (391 km/h; 243 mph)
Stalling speed, flaps and landing gear down	80 knots (149 km/h; 93 mph) EAS
Max rate of climb at S/L, ISA + 15°C	610 m (2,000 ft)/min
Rate of climb at S/L, one engine out, ISA + 15°C	122 m (400 ft)/min
Service ceiling (certificated)	7,625 m (25,000 ft)
Service ceiling, one engine out	3,050 m (10,000 ft)
T-O run at S/L, ISA + 15°C	762 m (2,500 ft)
T-O to 15 m (50 ft) at S/L, ISA + 15°C	1,046 m (3,430 ft)
Landing from 15 m (50 ft) at S/L	959 m (3,144 ft)
Range, reserves for 45 min hold and 5% total fuel:	
with max payload (18 passengers)	323 nm (598 km; 372 miles)
with max fuel and 8 passengers	1,193 nm (2,210 km; 1,373 miles)

AIRCRAFT DESIGNS
AIRCRAFT DESIGNS (BEMBRIDGE) LTD

Bembridge Fort, Sandown, Isle of Wight
Telephone: 0983 406124
Telex: 86448 Micair G

CHAIRMAN AND CHIEF EXECUTIVE: R. C. Britten

DIRECTORS:
J. M. McMahon
N. J. R. James, MA (Cantab)
F. H. Mann

TECHNICAL DIRECTOR:
M. J. Brennan, BSc, CEng, FIMechE, FRAeS

CHIEF DESIGNER:
J. Allan, DipAe(Hull), CEng, FRAeS

Founded in 1978 to design and construct the Sheriff lightweight twin-engined aircraft, Aircraft Designs (Bembridge) Ltd has since expanded to become a consultant in all aspects of light aviation, gaining CAA approval in October 1979. A separate company, Sheriff Aerospace Ltd, has been formed, with Mr Denis Berryman, CEng, FRAeS, as Technical Director, to finance the Sheriff; but all design and marketing continue to be handled by Aircraft Designs (Bembridge) Ltd.

The initial Sheriff project, for an economical twin-engined trainer, developed subsequently into a lower/middle range four-seat touring or air taxi aircraft, while still retaining the originally conceived option of a twin-engined trainer, particularly suitable for low-cost military training.

Construction of a prototype has been subcontracted to Micronair (Aerial) Ltd, a company also situated at Bembridge Fort, and it is expected that this aircraft will fly for the first time in the second quarter of 1982.

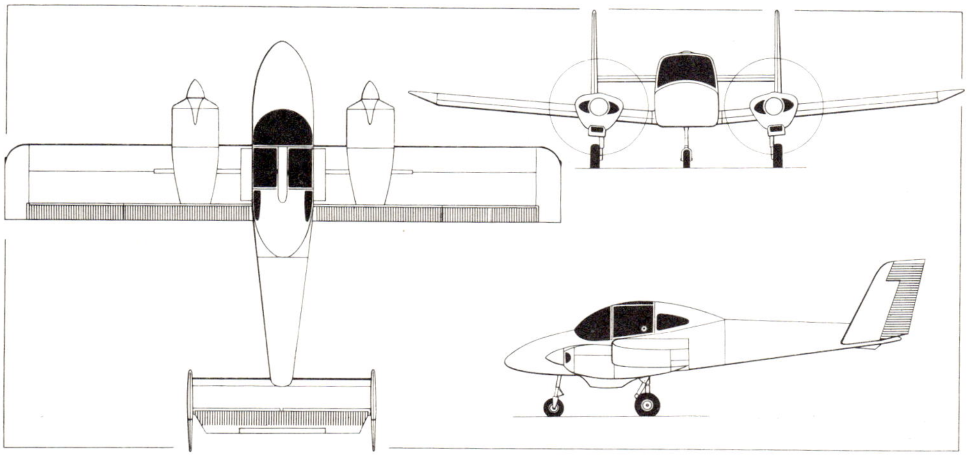

Aircraft Designs Britten Sheriff two/four-seat training and utility aircraft *(Pilot Press)*

BRITTEN SHERIFF

TYPE: Two/four-seat lightweight training and utility aircraft.

WINGS: Cantilever low-wing monoplane. Wing section NACA 23015. Conventional single-spar structure of 2024 light alloy. The main spar is a built-up I-section, and the basic structure includes light alloy ribs and a stressed leading-edge skin, part of which forms an integral fuel tank in each wing. Frise-type light alloy ailerons with corrugated skins. Plain light alloy trailing-edge flaps, actuated by a patented feature, the flap retracting step. No tabs.

FUSELAGE: Simple single-curvature rectangular box structure of 2024 light alloy. Four longerons and light alloy formers. Nosecone and fairings of glassfibre. Avionics/equipment bay in nose.

TAIL UNIT: Cantilever structure of 2024 light alloy, with twin endplate fins and rudders. Elevator and rudders have corrugated skins. Trim tab in centre of elevator.

LANDING GEAR: Retractable tricycle type with single wheel on each unit. Trailing-link main legs, with shock-absorption by air/oil struts. Nosewheel has oleo-pneumatic shock-strut and is fully castoring. Main wheels and tyres size 6·00-6. Nosewheel and tyre size 5·00-5.

POWER PLANT: Tourer has two 119 kW (160 hp) Avco Lycoming O-320-D1A flat-four engines, each flat rated to 110 kW (148 hp) and driving a two-blade constant-speed fully-feathering metal propeller with spinner. Integral wing fuel tanks with total capacity of 277 litres (61 Imp gallons). Refuelling point in each wing.

ACCOMMODATION: Tourer has four forward-facing seats. Trainer has two side-by-side seats, with dual controls as standard. Two 'gull-wing' doors provide access.

DIMENSIONS, EXTERNAL:

Wing span	10·06 m (33 ft 0 in)
Wing chord (constant)	1·40 m (4 ft 7 in)
Wing aspect ratio	7·26
Length overall	6·98 m (22 ft 11 in)
Propeller diameter	1·83 m (6 ft 0 in)

AREAS:

Wings, gross	13·94 m² (150·0 sq ft)
Vertical tail surfaces (total)	2·60 m² (28·0 sq ft)
Horizontal tail surfaces (total)	3·50 m² (37·7 sq ft)

WEIGHTS AND LOADINGS (estimated, four-seat tourer/air taxi):

Weight empty	855 kg (1,885 lb)
Max T-O weight	1,338 kg (2,950 lb)
Max wing loading	96·0 kg/m² (19·7 lb/sq ft)
Max power loading	6·08 kg/kW (9·95 lb/hp)

PERFORMANCE (estimated, engines 110 kW; 148 hp at 2,500 rpm at max T-O weight):

Max level speed	150 knots (278 km/h; 173 mph)
Cruising speed at 75% power	137 knots (254 km/h; 158 mph)
Cruising speed at 60% power	123 knots (229 km/h; 142 mph)
Max rate of climb at S/L	396 m (1,300 ft)/min
Rate of climb at S/L, one engine out	94 m (310 ft)/min
T-O to 15 m (50 ft)	456 m (1,495 ft)
Landing from 15 m (50 ft)	527 m (1,730 ft)
Range with max fuel, no reserves	580 nm (1,075 km; 668 miles)

BAe
BRITISH AEROSPACE PUBLIC LIMITED COMPANY

HEADQUARTERS: Brooklands Road, Weybridge, Surrey KT13 0SJ
Telephone: 0932 45522
Telex: 27111

REGISTERED OFFICE: 100 Pall Mall, London SW1Y 5HR
Telephone: 01 930 1020
Telex: 24353

BOARD OF DIRECTORS:
Sir Austin Pearce, CBE, PhD, FEng (Chairman)
A. H. C. Greenwood, CBE, JP, CEng, FRAeS (Deputy Chairman)
Sir Frederick Page, CBE, FRS, MA, FEng, FRAeS, FBIM (Chairman and Chief Executive, Aircraft Group)
Admiral Sir Raymond Lygo, KCB, FBIM, RN (Retd) (Chairman and Chief Executive, Dynamics Group)
B. E. Friend, FCA (Director of Finance)
E. G. Rubython, CBE (Deputy Chief Executive, Aircraft Group)
Air Chief Marshal Sir Peter Fletcher, KCB, OBE, DFC, AFC, RAF(Retd) (Corporate Strategy and Planning)
J. T. Stamper, MA, FEng, FRAeS (Technical Director)
L. W. Buck (Industrial Relations)
T. G. Kent, CBE, CEng, MIMechE, MRAeS (Deputy Chief Executive, Dynamics Group)

NON-EXECUTIVE DIRECTORS:
K. M. Bevins, CBE, TD
K. Durham, BSc
D. O. Gladwin, CBE
H. A. Hitchcock, DFC
Sir Jack Wellings, CBE

SECRETARY/LEGAL ADVISER: B. Cookson, LLB

CORPORATE PUBLIC RELATIONS MANAGER:
D. W. Bainbridge

British Aerospace was established by the Aircraft and Shipbuilding Industries Act 1977, as a result of which, on 29 April 1977, the ownership of British Aircraft Corporation (Holdings) Ltd, Hawker Siddeley Aviation Ltd, Hawker Siddeley Dynamics Ltd and Scottish Aviation Ltd was vested in the Corporation. Initially, these four companies continued to trade under their existing names. With effect from 1 January 1978, a new structure for British Aerospace was implemented, whereby the Corporation functioned through two operating groups, an Aircraft Group and a Dynamics Group (see Air-launched Missiles section). In mid-1981 these two groups had a home and overseas work force of some 78,500 people (60,000 in the Aircraft Group and 18,500 in the Dynamics Group).

On 23 July 1979 HM Government announced its intention to transfer the business of the Corporation to a company incorporated under the Companies Act, and to offer about half of that company's shares to the public, with employees being given a special opportunity to acquire shares. The passage through Parliament of the British Aerospace Act 1980 enabled all the property, rights, liabilities, and obligations of the Corporation to vest in British Aerospace Limited on 1 January 1981. On the following day Ordinary Shares were issued to a nominee of the Secretary of State, pursuant to Section 3 of the Act, and the company was re-registered as British Aerospace Public Limited Company under the Companies Act 1980. In February 1981 HM Government offered for sale up to 100 million Ordinary Shares, representing approximately half of the issued share capital of the company. This offer for sale was oversubscribed three and a half times, and the resulting ownership of British Aerospace has become: HM Government 48·43%, shareholders 48·43%, and employees 3·14%.

British Aerospace has the following overseas subsidiaries: British Aerospace Australia Ltd, British Aerospace Inc, and British Scandinavian Aviation AB; and the following UK subsidiaries: British Aerospace (Insurance) Ltd, British Aerospace (Insurance Brokers) Ltd, British Aerospace (Pension Fund Trustees) Ltd, British Aircraft Corporation (Pension Fund Trustees) Ltd and HSA/HSD (Pension Fund Trustees) Ltd.

Its associated companies are SEPECAT (formed in May 1966 by BAC and Breguet Aviation to control the development and production of the Jaguar tactical strike fighter and trainer), Panavia Aircraft GmbH (formed in March 1969 by BAC, MBB and Aeritalia to manage the development and production of the Tornado all-weather combat aircraft), Dulles International Services Inc (formed in 1976 by BAC (USA) Inc and Rolls-Royce Aero Engines of the USA to build and operate a facility at Washington's Dulles International Airport to supply customers in North America with spares and engineering support), Arab-British Dynamics Ltd (inaugurated in 1977 by BAC Guided Weapons Division and the Egyptian government to manufacture the Swingfire missile in Egypt), Frames Travel (Fylde) Ltd, and Remploy Services Ltd. British Aerospace is a 20% partner in the Airbus Industrie consortium, currently producing wings for the A300 and A310 wide-body civil transports. In January 1980 British Aerospace Dynamics Group announced, simultaneously with MBB and Aérospatiale's Division Engins Tactiques, the formation of a new European guided weapons company called Euromissile Dynamics Group (EMDG).

BRITISH AEROSPACE AIRCRAFT GROUP

HEADQUARTERS: Richmond Road, Kingston upon Thames, Surrey KT2 5QS
Telephone: 01 546 7741
Telex: 23726

GROUP BOARD:
- Sir Frederick Page, CBE, FRS, MA, FEng, FRAeS, FBIM (Chairman and Chief Executive)
- E. G. Rubython, CBE (Deputy Chief Executive)
- A. F. Atkin, CBE, BSc (Hons), DipAe (Hull), FEng, FIMechE, FRAeS (Managing Director, Marketing)
- J. L. Glasscock, BA, FCIS, JP (Managing Director, Military)
- J. L. Thorne, CBE (Managing Director, Civil)
- G. W. Carr, FCIS, MRAeS (Administration Director)
- P. Jefferson, CEng, MRAeS, MIMechE (Production Director)
- F. E. Roe, DIC, BSc, CEng, ACGI, FRAeS (Divisional Managing Director, Warton Division)
- R. H. Sawyer, FCA, FCMA, JDip, MA (Financial Director)
- A. F. Smith (Commercial Director)
- I. R. Yates, BEng, CEng, FRAeS, FIMechE (Director of Engineering and Project Assessment)
- N. V. Barber, BA, MSc (Divisional Managing Director, Manchester Division)
- C. M. Chandler, ACWA (Divisional Managing Director, Kingston-Brough Division)
- S. Gillibrand, MSc, CEng, FRAeS (Divisional Deputy Managing Director, Warton Division)
- M. J. Goldsmith, DIC, CEng, FRAeS (Divisional Managing Director, Hatfield-Chester Division)
- B. G. Thomas (Divisional Managing Director, Scottish Division)
- M. G. Wilde, OBE, BSc, DipAe, CEng, FRAeS (Divisional Managing Director, Weybridge-Bristol Division)

GROUP SECRETARY: J. A. Watson, ACIS

MILITARY ADVISER: Air Chief Marshal Sir Denis Smallwood, GBE, KCB, DSO, DFC (RAF Retd)

PUBLIC RELATIONS MANAGER: R. A. C. Gardner

PUBLIC RELATIONS OFFICER: D. M. Stroud

British Aerospace Aircraft Group was formed officially on 1 January 1978 by the reorganisation of the airframe interests of British Aircraft Corporation, Hawker Siddeley Aviation and Scottish Aviation. The Group is organised into six Divisions: Kingston/Brough; Manchester; Warton; Hatfield/Chester; Weybridge/Bristol; and Scottish.

Civil aircraft programmes include development and production of the British Aerospace 146 four-turbofan feederliner, the twin-turbofan HS 125 business aircraft and BAC One-Eleven transport, and the twin-turboprop HS 748 and Jetstream transports. Military aircraft programmes include manufacture of the Harrier and Sea Harrier V/STOL combat aircraft, Hawk ground attack/trainer and Bulldog basic and aerobatic trainer. Development and production of the Nimrod for anti-submarine warfare and airborne early warning duties continues. BAe also offers HS 748 variants for military transport and maritime patrol missions.

Major international collaborative programmes include the AV-8B development of the Harrier with McDonnell Douglas in the USA, the Tornado all-weather combat aircraft with MBB of Germany and Aeritalia of Italy, the Jaguar with Dassault-Breguet of France, and the One-Eleven with Romania. In 1978, British Aerospace became a full 20% risk-sharing partner in the Airbus Industrie consortium, currently producing wings for the A300 and A310 wide-body civil transport programmes.

In defence support, British Aerospace holds large contracts in the Middle East, notably in Saudi Arabia where a continuing programme of training and maintenance for the Royal Saudi Air Force involves more than 2,000 expatriate employees.

BRITISH AEROSPACE AIRCRAFT GROUP, HATFIELD/CHESTER DIVISION

Hatfield, Hertfordshire AL10 9TL
Telephone: 07072 62345
Telex: 22411

OTHER WORKS: Broughton, near Chester, Clwyd

DIVISIONAL BOARD:
- J. L. Thorne, CBE (Chairman)
- M. J. Goldsmith, DIC, CEng, FRAeS, (Managing Director)
- C. F. Bethwaite (Project Director, BAe 146)
- B. J. Champion, CA(SA) (Resources Director; and Finance Director, Chester)
- J. Humphreys, ACCA (Financial Director)
- J. A. Johnstone, OBE, CEng, FRAeS, FSLAET (Marketing Director, Civil)
- R. M. McKinlay (Director, Airbus)
- R. C. Meakins, MBIM (Administration Director; and Commercial Director, Chester)
- P. R. Owen, CEng, FRAeS (Technical Director)
- J. S. Paterson, CA (Commercial Director)
- A. E. Rowland (Director and General Manager, Chester)
- G. R. Wilkinson (Production Director)

SECRETARY: R. F. Kirkby

PUBLIC RELATIONS MANAGER: J. E. Scott, DSC, MRAeS

PUBLIC RELATIONS MANAGER, 146: Hugh Field

Main activities of this Division include development of the HS 125 and variants; design, development and production of A300 and A310 wings for Airbus Industrie; development and production of the British Aerospace 146, work on which is also taking place at other Group factories; responsibility for Trident aircraft in service.

BRITISH AEROSPACE AIRCRAFT GROUP, KINGSTON/BROUGH DIVISION

Richmond Road, Kingston upon Thames, Surrey KT2 5QS
Telephone: 01 546 7741
Telex: 23726

OTHER WORKS: Dunsfold Aerodrome, Godalming, Surrey; Hamble, Hampshire; Brough, North Humberside; Holme-on-Spalding Moor Aerodrome, Yorkshire; Bitteswell Aerodrome, Lutterworth, Leicestershire

DIVISIONAL BOARD:
- J. L. Glasscock, BA, FCIS, JP (Chairman)
- C. M. Chandler, ACWA (Managing Director)
- R. G. Adolphus, BSc, CEng, FRAeS (Production Director)
- A. C. Barber (Director and General Manager, Hamble)
- K. Essex-Crosby, DipAe, CEng, FRAeS (Director)
- J. W. Fozard, OBE, DCAe, BSc, CEng, FRAeS, FIMechE, FAIAA (Marketing Director)
- R. S. Hooper, DCAe, DAe, CEng, MIMechE, FRAeS (Technical Director)
- V. H. Lidstone (Commercial Director)
- L. W. Milsom (Divisional Resources Director; and General Manager, Brough)
- A. C. Spencer (Director)
- J. F. White, FCA, MBIM, MRAeS (Financial Director)
- A. H. Whitehouse, TEng, MRAeS (Administration Director; and General Manager, Bitteswell)

SECRETARY: T. A. Fletcher

PUBLIC RELATIONS MANAGER: A. D. F. Lewis

Main activities of this Division include design, development and production of the Harrier, Sea Harrier and Hawk; production of components for the BAe 146 and A300 Airbus; product support of the Buccaneer and Phantom; Phantom wing modifications; overhaul and conversion of Hunter, Shackleton, Vulcan and Buccaneer aircraft.

BRITISH AEROSPACE AIRCRAFT GROUP, MANCHESTER DIVISION

Greengate, Middleton, Manchester M24 1SA
Telephone: 061 681 2020
Telex: 667015
OTHER WORKS: Woodford Aerodrome, Woodford, Cheshire

DIVISIONAL BOARD:
- J. L. Glasscock, BA, FCIS, JP (Chairman)
- N. V. Barber, BA, MSc (Managing Director)
- H. T. Healy, ACMA (Financial Director)
- J. E. Perry, BSc, MSc, CEng, MIMechE, MIProdE (Production Director)
- J. B. Scott-Wilson, MA, FRAeS (Technical Director)
- S. C. Ward, BSc (Commercial Director)

SECRETARY: W. A. Hayhoe
PUBLIC RELATIONS MANAGER: J. R. Gray

Main activities of this Division include design, development and production of the HS 748, Nimrod MR. Mk 2 and AEW Mk 3, and production of A300 and A310 Airbus components.

BRITISH AEROSPACE AIRCRAFT GROUP, SCOTTISH DIVISION

Prestwick International Airport, Ayrshire KA9 2RW
Telephone: 0292 79888
Telex: 77432
DIVISIONAL BOARD:
- J. L. Thorne, CBE (Chairman)
- B. G. Thomas (Managing Director)
- W. Agnew (Production Director)
- G. J. Curran, ACMA (Financial Director)
- N. F. Harpur, MA, CEng, DIC, FRAeS (Technical Director)
- D. McConnell (Commercial Director)
- J. R. Woods (Administration Director)

SECRETARY: J. W. Connell

PUBLIC RELATIONS MANAGER: D. A. Dorman

Main activities of this Division include Bulldog and Jetstream production; production of fuselage sections for Lockheed C-130 Hercules and components for Lockheed TriStars; engine production and overhaul.

BRITISH AEROSPACE AIRCRAFT GROUP, WARTON DIVISION

Warton Aerodrome, Preston, Lancashire PR4 1AX
Telephone: 0772 633333
Telex: 67627

OTHER WORKS: Preston, Lancashire; Samlesbury Aerodrome, Blackburn, Lancashire

DIVISIONAL BOARD:
J. L. Glasscock, BA, FCIS, JP (Chairman)
F. E. Roe, DIC, BSc, CEng, ACGI, FRAeS (Managing Director)
S. Gillibrand, MSc, CEng, FRAeS (Deputy Managing Director)
A. H. Baxter (Production Director)
R. Dickson, MA (Cantab), CEng, FRAeS (Research Director)
R. H. Evans (Asst Managing Director, Commercial Director and Director-in-charge, India)
J. Glover, FCMA (Financial Director)
Dr I. A. M. Hall, MA, BSc, PhD, MSE (Technical Director)
B. O. Heath, OBE, BSc, DIC, CEng, MRAeS, Hon MIED (Advanced Engineering)
A. R. Keys, DFC, BSc (Marketing and product Support)
Air Chief Marshal Sir John Nicholls, KCB, CBE, DFC, AFC (Director, Saudi Arabia)
T. O. Williams, MA, CEng, FIMechE, MIEE (Administration Director)
F. G. Willox (Director of Projects)

SECRETARY: A. G. Leach

PUBLIC RELATIONS MANAGER: A. F. Johnston

Main activities of this Division include design, development and production with MBB and Aeritalia of the Panavia Tornado; joint development and production with Dassault-Breguet of the SEPECAT Jaguar; marketing of the Strikemaster; product support of the Jet Provost and Lightning; conversion programmes for the Canberra; provision of advanced defence support programmes.

BRITISH AEROSPACE AIRCRAFT GROUP, WEYBRIDGE/BRISTOL DIVISION

Brooklands Road, Weybridge, Surrey KT13 0SF
Telephone: 0932 45522
Telex: 27111

OTHER WORKS: Filton House, Filton, Bristol; Bournemouth (Hurn) Airport, Christchurch, Dorset

DIVISIONAL BOARD:
J. L. Thorne, CBE (Chairman)
M. G. Wilde, OBE, BSc, DipAe, CEng, FRAeS (Managing Director)
C. E. N. Arkell (Commercial Director)
K. Bentley, MA, CEng, MRAeS (Technical Director and General Manager, Weybridge)
J. A. Johnstone, OBE, CEng, FRAeS, FSLAET (Marketing Director, Civil)
D. F. McGregor, BSc (Hons), CEng, MRAeS (Production Director)
E. B. Trubshaw, CBE, MVO, FRAeS (Director and General Manager, Bristol)
D. Wynne, FCWA (Financial and Resources Director)

SECRETARY: L. F. Trueman, FCCA

DIVISIONAL PUBLIC RELATIONS MANAGER: N. A. Barfield

PUBLIC RELATIONS MANAGER, FILTON: H. Berry
PUBLIC RELATIONS OFFICER, HURN: D. C. Clarke

Main activities of this Division include production of the One-Eleven; support for the Concorde, One-Eleven, VC10, Vanguard, Viscount, Britannia and Bristol Freighter; major subcontract work; major component manufacture for the A310, BAe 146, Boeing 747, HS 748, Lockheed TriStar, Panavia Tornado, and SEPECAT Jaguar. Other major contracts include the Rombac 1-11 licence manufacturing programme, the conversion of five VC10 and four Super VC10 commercial transports to the air-to-air refuelling tanker (AART) configuration for the RAF, and maintenance of the F-111 for the USAF.

BAe JETSTREAM 31

The decision to proceed with development of this new version of the Jetstream was announced by British Aerospace on 5 December 1978. A production go-ahead was given in January 1981.

The Jetstream 31 will be certificated to BCAR Section D in the UK, and SFAR.41A in the USA. The development aircraft (G-JSSD), converted from a Handley Page-built Jetstream 1, flew for the first time on 28 March 1980.

The following versions will be available:

Commuter. Basic version, designed to carry 18/19 passengers and baggage. Able to operate three 110 nm (204 km; 127 mile) stage lengths, without refuelling, with maximum payload and full IFR reserves.

Corporate. Executive version, designed for eight to ten passengers, and able to carry eight passengers for 890 nm (1,649 km; 1,025 miles) with full IFR reserves.

Special. Intended for military communications, casualty evacuation, multi-engine training and cargo operations, and for specialist roles such as airfield calibration, resources survey and protection.

An initial batch of ten aircraft is under construction. Orders and options totalled nine by mid-August 1981, the firm orders being from Mall Airways of Albany, New York (two), Partnair of Norway (one), and Contactair of Stuttgart, West Germany (two). Deliveries are scheduled to begin in August 1982.

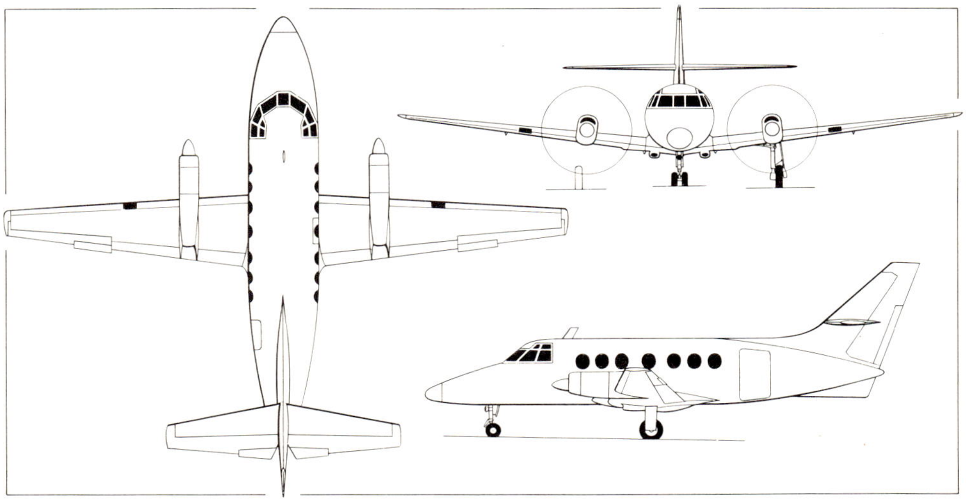

BAe Jetstream 31 twin-turboprop light commuter/executive transport *(Pilot Press)*

TYPE: Light commuter/executive transport.

WINGS: Cantilever low-wing monoplane. Wing section NACA 63A418 at root, NACA 63A412 at tip. Dihedral 7° from roots. Incidence 2° at root, 0° at tip. Sweepback 0° 34′ at quarter-chord. Aluminium alloy fail-safe structure. Aluminium alloy manually-operated Frise-type ailerons. Hydraulically-operated aluminium alloy double-slotted flaps. No slats or leading-edge flaps. Trim tab in each aileron. Goodrich pneumatic rubber-boot de-icing system for leading-edges.

FUSELAGE: Conventional aluminium alloy semi-monocoque fail-safe structure, with chemically-milled skin panels. Fully pressurised.

TAIL UNIT: Cantilever two-spar aluminium alloy structure. Fixed-incidence tailplane. Manually-operated control surfaces. Trim tabs in rudder and each elevator. Goodrich pneumatic rubber-boot de-icing system for leading-edges.

LANDING GEAR: Retractable tricycle type, with nosewheel steering. Hydraulic retraction, main wheels inward into wings, twin nosewheels forward. British Aerospace oleo-pneumatic shock-absorbers in all units. Dunlop wheels and tyres: main-wheel tyres size 28 × 9·00-12, pressure 3·93 bars (57 lb/sq in); nosewheel tyres size 6·00-6, pressure 2·34 bars (34 lb/sq in). No brake cooling. Anti-skid units.

POWER PLANT: Two 671 kW (900 shp) Garrett TPE331-10 turboprop engines, each driving a Dowty Rotol four-blade variable- and reversible-pitch fully-feathering metal propeller. Fuel in integral tank in each wing, total capacity 1,745 litres (384 Imp gallons; 461 US gallons). Refuelling point on top of each outer wing.

ACCOMMODATION: Two seats side by side on flight deck, with provision for dual controls, though aircraft can be approved (subject to local regulations) for single-pilot operation. Main cabin can be furnished in commuter layout for up to 18 passengers, or with executive interior for 8/10 passengers, but optional layouts are available. Downward-opening passenger door, with integral airstairs, at rear of cabin on port side. Emergency exit over wing on starboard side. Baggage compartment in rear of cabin, aft of main door. Entire accommodation pressurised, heated, ventilated and air-conditioned. Toilet, galley and bar optional.

SYSTEMS: Air-conditioning system with cabin pressurisation at max differential of 0·38 bars (5·5 lb/sq in), providing a 2,440 m (8,000 ft) cabin altitude at 7,620 m (25,000 ft). Single hydraulic system, pressure 138 bars (2,000 lb/sq in), with dual engine-driven pumps, for actuation of flaps, landing gear, brakes and nosewheel steering. Details of electrical and oxygen systems not finalised. APU optional.

BAe Jetstream 31 development aircraft (two Garrett TPE331-10 turboprop engines)

DIMENSIONS, EXTERNAL:

Wing span	15·85 m (52 ft 0 in)
Wing chord at root	2·19 m (7 ft 2½ in)
Wing chord at tip	0·80 m (2 ft 7¼ in)
Wing aspect ratio	10
Length overall	14·37 m (47 ft 1½ in)
Length of fuselage	13·40 m (43 ft 11½ in)
Height overall	5·32 m (17 ft 5½ in)
Fuselage: Max diameter	1·98 m (6 ft 6 in)
Tailplane span	6·60 m (21 ft 8 in)
Wheel track	5·94 m (19 ft 6 in)
Wheelbase	4·60 m (15 ft 1 in)
Propeller diameter	2·69 m (8 ft 10 in)
Passenger door: Height	1·42 m (4 ft 8 in)
Width	0·86 m (2 ft 10 in)
Emergency exit: Height	0·91 m (3 ft 0 in)
Width	0·56 m (1 ft 10 in)

DIMENSIONS, INTERNAL:

Cabin, excl flight deck: Length	7·32 m (24 ft 0 in)
Max width	1·83 m (6 ft 0 in)
Max height	1·80 m (5 ft 11 in)
Floor area	8·35 m² (90 sq ft)
Volume	16·92 m³ (598 cu ft)
Baggage compartment volume (according to layout)	1·94-2·53 m³ (68·5-89·5 cu ft)

AREAS:

Wings, gross	25·08 m² (270 sq ft)
Ailerons, aft of hinge line (total)	1·52 m² (16·4 sq ft)
Trailing-edge flaps (total)	3·25 m² (35·0 sq ft)
Vertical tail surfaces (total)	7·72 m² (83·1 sq ft)
Horizontal tail surfaces (total)	7·80 m² (84·0 sq ft)

WEIGHTS AND LOADINGS (estimated):

Manufacturer's weight empty	3,450 kg (7,606 lb)
Max T-O and landing weight	6,600 kg (14,550 lb)
Max ramp weight	6,650 kg (14,661 lb)
Max zero-fuel weight	6,000 kg (13,228 lb)

Max wing loading	263·1 kg/m² (53·89 lb/sq ft)
Max power loading	4·92 kg/kW (8·08 lb/shp)

PERFORMANCE (estimated, at max T-O weight):

Max cruising speed: max continuous power	263 knots (488 km/h; 303 mph)
Max cruise power	253 knots (469 km/h; 291 mph)
Stalling speed, flaps down	96 knots (179 km/h; 111 mph)
Max rate of climb at S/L	680 m (2,230 ft)/min
Rate of climb at S/L, one engine out	163 m (535 ft)/min
Service ceiling	9,630 m (31,600 ft)
Service ceiling, one engine out	4,665 m (15,300 ft)
T-O to 15 m (50 ft)	858 m (2,815 ft)
Landing from 15 m (50 ft)	818 m (2,684 ft)

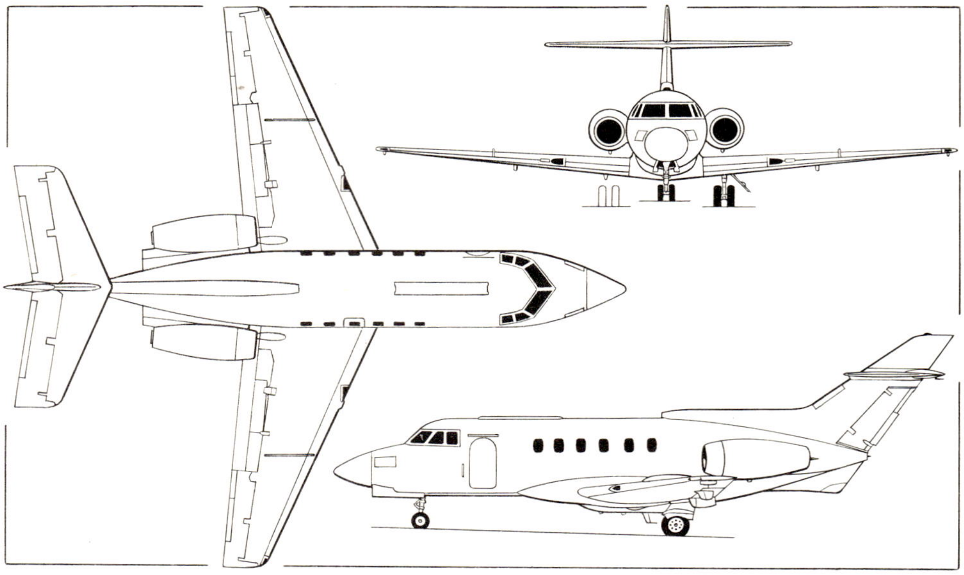

BAe HS 125 Series 700 twin-turbofan business transport *(Pilot Press)*

BAe HS 125 Series 700 twin-turbofan business transport (two Garrett TFE731 engines)

BAe HS 125 SERIES 700 and PROTECTOR

The BAe HS (formerly de Havilland) 125, produced by BAe's Hatfield/Chester Division, is a twin-engined business aircraft which is also suitable for use by armed forces in the communications role, as a troop carrier, as an ambulance aircraft, for airways inspection, and as an economical trainer for pilots, navigators and specialised radio and radar operators. It can operate from unpaved runways without modification.

The HS 125 was developed as a private venture, and the first of two prototypes flew for the first time on 13 August 1962. The Viper turbojet-engined Series 1/1A/1B/2/3/3A/3B/3A - R/3A - RA/3B - RA/400A/400B/600A/600B, which totalled 358, were superseded in 1976 by introduction of the Series 700, with Garrett TFE731 turbofan engines. The prototype (G-BFAN) was produced by conversion of a Series 600 airframe and flew for the first time on 28 June 1976. The Series 700 has now outsold any other model of the HS 125.

The use of turbofan engines gives an improved specific fuel consumption by comparison with that of the turbojets in earlier versions of the HS 125. The Series 700 also meets all existing and proposed international noise regulations. Turbofan conversions of existing turbojet-powered HS 125s are available, and more than 50 (mostly done by Garrett) had been ordered by mid-1981.

New-production Series 700s embody many other refinements in addition to the change of power plant. As in the case of earlier versions, the intended market is indicated by a suffix letter: the **Series 700A** is for the North American market, **Series 700B** for the rest of the world.

Improvements to the airframe, to reduce drag and enhance its appearance, include redesign of the wing keel skid, use of countersunk rivets instead of mushroom-head types in the flap bottom skin, replacement of the lower airbrake leading-edge castellations by internal castellations, replacement of the mushroom-head bolts and rivets in the inner tank doors and aileron trailing-edges by countersunk types, use of Harper radius countersunk rivets instead of mushroom-head rivets in the fuselage and tail unit, redesign of the ventral fin and adjacent fairings in glassfibre and enlargement of the area of the ventral fin to improve directional stability, deletion of the NACA cooling air intake introduced in the nose of the Series 600, and addition of fairings over the windscreen wiper blades and two ADF loop aerials.

New interior equipment and furnishings include the use of figured walnut veneer on cabin tables and toilet consoles, leather trim, provision of a Blaupunkt Bamberg combined radio/cassette stereo player and recorder, a luxury toilet compartment, digital cabin clock, slide-out portable bar box, full harness on sideways-facing seats, improved life-jacket stowage under seats, improved plug-in meal tray for divan occupants, and a new range of interior colour schemes.

A maritime surveillance version of the Series 700, known as the **Protector**, is also available. Equipped with specially-developed search radar, blister windows, cameras, and nav/com systems, it has a search endurance of more than 6 h.

The first flight of a production HS 125 Series 700 was made on 8 November 1976. UK certification was received on 7 April 1977. The first sale had been made before completion of the prototype, and by mid-August 1981 the sale of 172 Series 700s had been announced, including more than 120 for customers in North America. Military operators include the Irish Air Corps.

The following description applies specifically to the Series 700:

TYPE: Twin-turbofan business transport aircraft.

WINGS: Cantilever low-wing monoplane. Thickness/chord ratio 14% at root, 11% at tip. Dihedral 2°. Incidence 2° 6′ at root, −0° 24′ at tip. Sweepback 20° at quarter-chord. Wings built in one piece and dished to pass under fuselage, to which they are attached by four vertical links, a side link and a drag spigot. All-metal two-spar fail-safe structure, with partial centre spar of approx two-thirds span, sealed to form integral fuel tankage which is divided into two compartments by centreline rib. Skins are single-piece units on each of the upper and lower semi-spans. Detachable leading-edges. Fence on each upper surface at approx two-thirds span. Mass-balanced ailerons, operated manually by cable linkage. Trim tab and geared tab in port aileron, two geared tabs in starboard aileron. Aileron fences to improve lateral stability. Large, four-position double-slotted flaps, actuated hydraulically via a screwjack on each flap. Mechanically-operated hydraulic cutout prevents asymmetric operation of the flaps. Airbrakes above and below each wing, forming part of flap shrouds, provide lift-dumping facility during landing, and have interconnected controls to prevent asymmetric operation. TKS

BAe HS 748 Series 2B prototype which, after flying previously as the Coastguarder, has since been furnished to airline standard for DLT

liquid system, using porous stainless steel leading-edge panels, for de-icing or anti-icing.

Fuselage: All-metal semi-monocoque fail-safe structure, making extensive use of Redux bonding. Constant circular cross-section over much of its length.

Tail Unit: Cantilever all-metal structure, with fixed-incidence tailplane mounted on fin. Small fairings on tailplane undersurface to eliminate turbulence around elevator hinge cutouts. Triangular ventral fin, and extended dorsal fin. Control surfaces operated manually via cable linkage. Tabs in rudder and each elevator. TKS liquid de-icing or anti-icing of fin and tailplane leading-edges.

Landing Gear: Retractable tricycle type, with twin wheels on each unit. Hydraulic retraction of all units; nosewheels forward, main wheels inward into wings. Oleo-pneumatic shock-absorbers. Fully-castoring nose unit, steerable 45° to left or right. Dunlop main wheels and 10-ply tyres, size 23 × 7-12, pressure 8·75 bars (127 lb/sq in). Dunlop nosewheels and 6-ply tyres, size 18 × 4·25-10, pressure 5·51 bars (80 lb/sq in). Dunlop double-disc hydraulic brakes with Maxaret anti-skid units on all main wheels.

Power Plant: Two 16·46 kN (3,700 lb st) Garrett TFE731-3-1H turbofan engines, pod-mounted on sides of rear fuselage, in pods designed and manufactured by Grumman Aerospace. Engine intake anti-icing by engine bleed air. Integral fuel tanks in wings, with total capacity of 4,628 litres (1,018 Imp gallons). Single pressure refuelling point in lower starboard side of rear fuselage. Overwing refuelling point near each wingtip. Rear underfuselage tank of 509 litres (112 Imp gallons) capacity, with refuelling point on starboard side, and 232 litre (51 Imp gallon) dorsal fin tank, raising overall total capacity to 5,369 litres (1,181 Imp gallons; 1,418 US gallons) of which 5,323 litres (1,171 Imp gallons; 1,406 US gallons) are usable.

Accommodation: Crew of two on flight deck, which is fully soundproofed, insulated and air-conditioned. Dual controls standard. Seat provided for third crew member. Standard executive layout has seating for eight passengers, with forward baggage compartment, refreshment bar and coat compartment (forward) and toilet (aft). There are individual recessed lights and air louvres. Cabin styling offers the operator a choice of interchangeable furnishing units to suit individual requirements. The wide seats, which on the Srs 700A swivel through 180°, are adjustable fore and aft and sideways, have adjustable lumbar support, and can be reclined hydraulically up to 40°. Typical executive furnishing includes a couch for three persons, five individual seats, and individual foldaway wall tables. Alternative high-density layout is available, seating up to 14 passengers. Outward-opening door at front on port side, with integral airstairs. Emergency exit over wing on starboard side. Edge-heating for windscreen. Electrical windscreen anti-icing, with methanol spray backup.

Systems: Garrett air-conditioning and pressurisation system. Max cabin differential 0·58 bars (8·35 lb/sq in), maintaining S/L cabin pressure up to 6,550 m (21,500 ft). Oxygen system standard, with dropout masks for passengers. Hydraulic system, pressure 186-207 bars (2,700-3,000 lb/sq in), for operation of landing gear, main-wheel doors, flaps, spoilers, nosewheel steering, main-wheel brakes and anti-skid units. Two accumulators provide emergency hydraulic power for wheel brakes in case of a main system failure. Independent auxiliary system for lowering landing gear and flaps in the event of a main system failure. DC electrical system utilises two 12kW engine-driven starter/generators and two 24V 25Ah nickel-cadmium batteries. A 24V 3·5Ah battery provides separate power for igniter and starter control circuits. AC electrical system includes two 115V 2·5kVA 400Hz three-phase static inverters and one 250VA solid-state standby inverter for avionics, and one engine-driven 120V 4·4kVA frequency-wild alternator for windscreen anti-icing. Ground power receptacle on starboard side at rear of fuselage for 28V external DC supply. Garrett GTCP-30-92 auxiliary power unit is standard on Srs 700B. Engine ice protection system supplied by engine bleed air. Graviner triple FD Firewire fire warning system and two BCF engine fire extinguishers.

Avionics and Equipment: Standard avionics include dual Collins VHF-20A com transceivers, Collins VIR-30A VHF nav receivers with dual marker beacon indicators, dual Collins DF-206 ADF, dual Collins MC-103 compasses, Collins DME-40 DME, Marconi AD1540 audio control and passenger address system, RCA Primus 400 weather radar, Collins TDR-90 ATC transponder and Blaupunkt Bamberg stereo tape and AM/FM radio. Provisions for Collins 718U-5 HF com transceiver and second transponder and DME. Collins APS-80 autopilot and FDS-80 flight director system standard, to provide altitude hold, altitude pre-select, airspeed hold, Mach number hold, vertical speed hold, aircraft heading, VOR/LOC, ILS approach and pitch with electric trim.

Dimensions, external:
Wing span 14·33 m (47 ft 0 in)
Wing chord (mean) 2·29 m (7 ft 6¼ in)
Wing aspect ratio 6·25
Length overall 15·46 m (50 ft 8½ in)
Height overall 5·36 m (17 ft 7 in)
Fuselage: Max diameter 1·93 m (6 ft 4 in)
Tailplane span 6·10 m (20 ft 0 in)
Wheel track (c/l of shock-absorbers) 2·79 m (9 ft 2 in)
Wheelbase 6·34 m (20 ft 9½ in)
Passenger door (fwd, port):
Height 1·30 m (4 ft 3 in)
Width 0·69 m (2 ft 3 in)
Height to sill 1·07 m (3 ft 6 in)
Emergency exit (overwing, stbd):
Height 0·91 m (3 ft 0 in)
Width 0·51 m (1 ft 8 in)

Dimensions, internal:
Cabin (excl flight deck): Length 6·50 m (21 ft 4 in)
Max width 1·80 m (5 ft 11 in)
Max height 1·75 m (5 ft 9 in)
Floor area 5·11 m² (55·0 sq ft)
Volume 17·10 m³ (604·0 cu ft)
Baggage compartment 0·84 m³ (29·6 cu ft)

Areas:
Wings, gross 32·8 m² (353·0 sq ft)
Ailerons (total) 2·76 m² (29·76 sq ft)
Trailing-edge flaps (total) 5·21 m² (56·06 sq ft)
Fin, incl dorsal fin 5·31 m² (57·15 sq ft)
Ventral fin 0·61 m² (6·61 sq ft)
Horizontal tail surfaces (total) 9·29 m² (100 sq ft)

Weights and Loadings:
Weight empty 5,826 kg (12,845 lb)
Typical operating weight empty 6,270 kg (13,822 lb)
Max payload 1,010 kg (2,228 lb)
Max ramp and T-O weight 11,566 kg (25,500 lb)
Max zero-fuel weight 7,280 kg (16,050 lb)
Max landing weight 9,979 kg (22,000 lb)
Max wing loading 352·5 kg/m² (72·2 lb/sq ft)
Max power loading 351 kg/kN (3·45 lb/lb st)

Performance (at max T-O weight except where indicated):
Never-exceed speed Mach 0·85
Max level speed at S/L 320 knots (592 km/h; 368 mph) IAS
Max cruising speed at 8,380 m (27,500 ft) 436 knots (808 km/h; 502 mph)
Econ cruising speed at 11,275-12,500 m (37,000-41,000 ft) 390 knots (723 km/h; 449 mph)
Stalling speed, flaps down 83 knots (155 km/h; 96 mph) EAS
Service ceiling 12,500 m (41,000 ft)
T-O run 1,367 m (4,484 ft)
T-O to 10·7 m (35 ft), unfactored 1,448 m (4,750 ft)
T-O balanced field length 2,042 m (6,700 ft)
Landing from 15 m (50 ft) at landing weight of 7,167 kg (15,800 lb), unfactored 619 m (2,030 ft)
Landing run at landing weight of 6,804 kg (15,000 lb) 1,143 m (3,750 ft)
Range with max fuel and max payload, allowances for T-O, approach, landing, taxying and 45 min reserves 2,420 nm (4,482 km; 2,785 miles)

Operational Noise Levels (FAR Pt 36):
T-O 87·6 EPNdB
Approach 96·3 EPNdB
Sideline 88·9 EPNdB

BAe HS 748 SERIES 2B

Design of the HS 748 short/medium-range turboprop airliner started in January 1959. The first prototype flew on 24 June 1960, followed by a second on 10 April 1961. UK production of the Series 1 (18 built) and Series 2 (including two Andover CC.Mk 2s for The Queen's Flight and four for Air Support Command), described in previous editions of *Jane's*, has been completed. These models were followed by the Series 2A (see 1977-78 *Jane's*), which continues to be available to order.

In early 1979 the BAe HS 748 Series 2A was replaced as the standard version by a new **Series 2B** basic model, with improved 'hot and high' Dart 536-2 engines, a 1·22 m (4 ft) greater span wing with reduced drag, modified tail surfaces, and other refinements. Prototype for this version was G-BCDZ, flown earlier as the demonstration model of the Coastguarder. The Srs 2B is available in the same civil, military and Coastguarder configurations as the Srs 2A (see following paragraphs), and modification kits are available to existing Srs 2A operators. It is marketed in the USA under the name **Intercity 748**. First production Srs 2B (G-BGJV) made its initial flight on 22 June 1979, and nine had been ordered by 1 March 1981. First delivery of a Srs 2B was made to Air Madagascar in January 1980. Customers include the Colombian Air Force, which has ordered one (to be operated by Satena). On 4 March 1980 BAe, in conjunction with Rolls-Royce, began flight testing hush-kits for the Dart engines. Preliminary tests showed a minimum noise reduction of 5 EPNdB on approach, and useful attenuation levels achieved for ramp departure, taxi and take-off.

HS 748 Civil Transport. In addition to the basic transport configuration, the Series 2B is available optionally with a large rear freight door which has an opening of 2·67 m by 1·72 m (8 ft 9 in × 5 ft 7¾ in), together with a strengthened cabin floor capable of supporting an overall floor loading of 976 kg/m² (200 lb/sq ft).

HS 748 Military Transport. The military transport version has the large rear freight door and strengthened floor that are available for the civil transport and has, in addition, fixed fittings to undertake a wide range of military roles. Optional military overload take-off and landing weights give improved payload/range capabilities. A total of 52 military HS 748 Srs 2As had been exported by the Summer of 1981, of which 28 were fitted with the rear freight door and strengthened floor. These were for the air forces of Belgium (3), Brazil (12), Ecuador (5), three undisclosed air forces (6), the Royal Brunei Army (1), and the Nepal Royal Flight (1).

HS 748 Coastguarder. Variant for search and rescue and maritime surveillance roles. Described separately.

Sales of all Series (including 31 Andover C. Mk 1s for the RAF: see 1968-69 *Jane's*) totalled 355 by 1 August 1981, including more than 290 for export. Of these, over 335 had been delivered. Nine aircraft were sold with Dart RDa.8 engines, seven being supplied to Bundesanstalt für Flugsicherung (West Germany) with calibration equipment for radio navigational aids and two to the Royal Australian Navy with navigational and electronic training equipment. Six Andover C. Mk 1s were modified for flight checking and calibration duties, and redelivered to the RAF from 1977 to replace Argosy E. Mk 1 aircraft in this role. The modified Andovers are designated **E. Mk 3**.

In 1980 BAe's Manchester Division embarked upon a £500,000 design analysis of possible future developments of the HS 748, including a 60-passenger 'stretched' version that might be powered by General Electric T64 or other turboprop engines.

The HS 748 is the subject of a manufacturing agreement with the Indian government, and 89 aircraft (included in above overall totals) are being assembled from British-built components by Hindustan Aeronautics Ltd. Of these, 17 are for Indian Airlines and 72 for the Indian Air Force.

The following description applies to the current production HS 748 Series 2B, except where indicated:

Type: Twin-engined passenger or freight transport.

Wings: Cantilever low-wing monoplane. Wing section NACA 23018 at root, NACA 4412 at tip. Dihedral 7°. Incidence 3°. Sweepback 2° 54′ at quarter-chord. All-

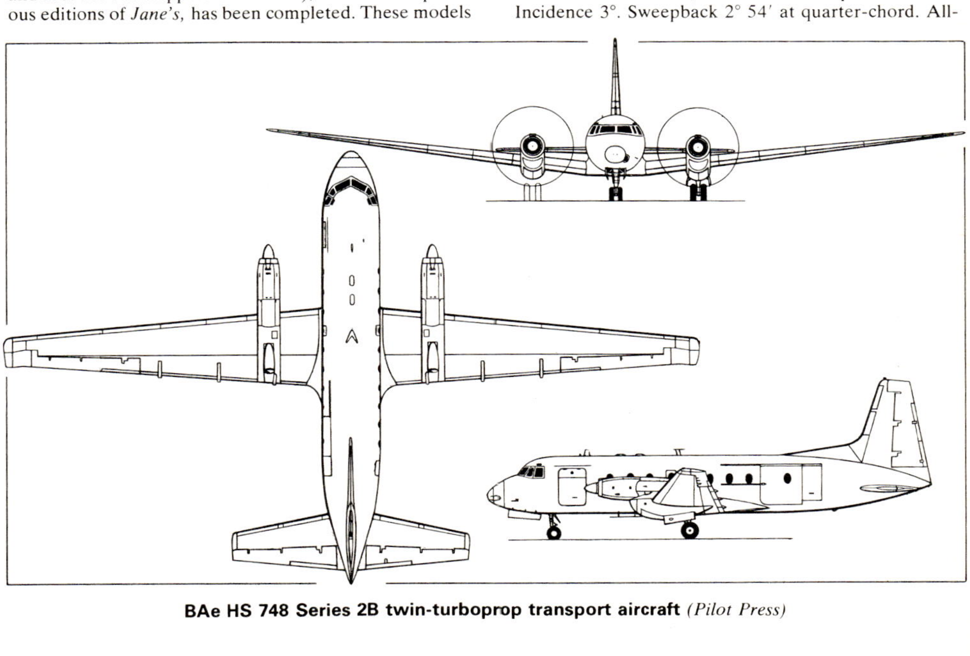

BAe HS 748 Series 2B twin-turboprop transport aircraft *(Pilot Press)*

BAe HS 748 Series 2A military transport aircraft with large rear freight door, in Belgian Air Force insignia

metal two-spar fail-safe structure. No cutouts in spars for engines or landing gear. All-metal set-back hinge, shielded horn-balance, manually-operated ailerons and electrically-actuated Fowler flaps. Geared tab in each aileron. Trim tab in starboard aileron. Pneumatic leading-edge de-icing boots.

Fuselage: All-metal semi-monocoque riveted fail-safe structure, of circular section.

Tail Unit: Cantilever all-metal structure. Fixed-incidence tailplane. Manually-operated controls. Trim tabs in elevators and rudder. Spring tab in rudder.

Landing Gear: Retractable tricycle type, with hydraulically-steerable nose unit. All wheels retract forward hydraulically. Main wheels retract into bottom of engine nacelles forward of front wing spar. Dowty Rotol shock-absorbers. Twin wheels, with Dunlop tyres, on all units. Main wheels size 32 × 10·75-14. Nosewheels size 25·65 × 8·5-10. Standard tyre pressures: main wheels 5·03 bars (73 lb/sq in); nosewheels 3·79 bars (55 lb/sq in). Minimum tyre pressures: main wheels 4·48 bars (65 lb/sq in); nosewheels 3·45 bars (50 lb/sq in). Dunlop disc brakes with Maxaret anti-skid units. No brake cooling.

Power Plant (Srs 2B): Two 1,700 kW (2,280 ehp) Rolls-Royce Dart RDa.7 Mk 536-2 turboprop engines, each driving a Dowty Rotol four-blade constant-speed fully-feathering propeller. Engine hush-kits available, externally evident by forward extension of nose cowl by approx 7·5 cm (3 in). Provision for automatic injection of water methanol into live engine in the event of an engine failure on take-off. Fuel in two integral wing tanks, with total capacity of 6,550 litres (1,440 Imp gallons). Tanks modified to provide increased wing bending relief. Underwing pressure refuelling and overwing gravity refuelling. Oil capacity 14·2 litres (25 Imp pints) per engine.

Accommodation (commercial): Crew of two on flight deck, and cabin attendant. Accommodation for 40-58 passengers in paired seats on each side of central gangway. Baggage compartment forward of cabin, with provision for steward's seat. Galley, toilet and baggage compartment aft of cabin. Forward baggage compartment and steward's seat can be replaced by freight hold with moving partition between hold and passenger cabin. Main passenger door, on port side at rear, with smaller door on starboard side to serve as baggage door and emergency exit. Crew and freight door on port side at front. Hydraulically-operated stairs.

Accommodation (military transport): Up to 58 troops in airline type seats. Provision for forward and aft baggage compartments and hydraulically-operated airstairs. In paratroop role up to 48 paratroops and dispatchers can be accommodated on sidewall folding seats with safety harness. Dropping by static line or free fall. For casualty evacuation up to 24 stretchers and nine nursing staff can be carried, with provision for medical supplies and equipment. For supply dropping a guided roller conveyor system allows twelve 340 kg (750 lb) or six 680 kg (1,500 lb) loads to be dropped within six seconds. Capacity for 5,886 kg (12,976 lb) freight. Large cargo door will accept items up to 1·42 m × 1·42 m × 3·66 m (4 ft 8 in × 4 ft 8 in × 12 ft) or small diameter pipes over 12 m (39 ft 4 in) in length. Onboard freight hoist and palletised freight system available. Quickly-removable VIP cabin available, and a variety of VIP layouts, with separate toilet, telephone and wide range of options.

Systems: Normalair automatic pressurisation and air-conditioning system, giving equivalent altitude of 2,440 m (8,000 ft) at 7,620 m (25,000 ft). Pressure differential 0·38 bars (5·5 lb/sq in). Hydraulic system, pressure 172 bars (2,500 lb/sq in), for landing gear retraction, nose-wheel steering, brakes and propeller brakes. No pneumatic system. One 9kW 28V DC generator and one 22kVA alternator on each engine. Two 1,800VA static inverters.

Avionics and Equipment: Collins or Bendix solid-state avionics. Blind-flying instrumentation and Bendix RDR colour weather radar (RDR-1300 in Intercity 748). Standard equipment by Sperry includes SPZ-500 multi-mode autopilot/flight director system, air data computer, SPI-402 10 cm (4 in) flight director instruments, VG-14A vertical gyro and C-14 Gyrosyn compass systems. Provision for flight data recorder. Marconi ARC 340 communications and homing system in Royal Brunei Army aircraft.

Dimensions, external:
Wing span: Srs 2A 30·02 m (98 ft 6 in)
Srs 2B 31·23 m (102 ft 5½ in)
Wing chord at root 3·49 m (11 ft 5¼ in)
Wing chord at tip: Srs 2A 1·34 m (4 ft 5 in)
Wing aspect ratio: Srs 2A 11·967
Srs 2B 12·668
Length overall 20·42 m (67 ft 0 in)
Fuselage: Max diameter 2·67 m (8 ft 9 in)
Height overall 7·57 m (24 ft 10 in)
Tailplane span 10·97 m (36 ft 0 in)
Wheel track 7·54 m (24 ft 9 in)
Wheelbase 6·30 m (20 ft 8 in)
Propeller diameter 3·66 m (12 ft 0 in)
Propeller ground clearance 0·61 m (2 ft 0 in)
Passenger door (port, rear):
Height 1·57 m (5 ft 2 in)
Width 0·76 m (2 ft 6 in)
Height to sill 1·84 m (6 ft 0½ in)
Freight and baggage door (fwd):
Height 1·37 m (4 ft 6 in)
Width 1·22 m (4 ft 0 in)
Height to sill 1·84 m (6 ft 0½ in)
Baggage door (rear, stbd):
Height 1·24 m (4 ft 1 in)
Width 0·64 m (2 ft 1 in)
Height to sill 1·84 m (6 ft 0½ in)
Optional freight door (rear, port):
Height 1·72 m (5 ft 7¾ in)
Width 2·67 m (8 ft 9 in)

Dimensions, internal:
Cabin, excl flight deck: Length 14·17 m (46 ft 6 in)
Max width 2·46 m (8 ft 1 in)
Max height 1·92 m (6 ft 3½ in)
Floor area 27·5 m^2 (296 sq ft)
Volume 56·35 m^3 (1,990 cu ft)
Max total freight holds 9·54 m^3 (337 cu ft)

Areas:
Wings, gross: Srs 2A 75·35 m^2 (810·75 sq ft)
Srs 2B 77·00 m^2 (828·87 sq ft)
Ailerons (total) 3·98 m^2 (42·90 sq ft)
Trailing-edge flaps (total) 14·83 m^2 (159·80 sq ft)
Fin 9·81 m^2 (105·64 sq ft)
Rudder, incl tabs 3·66 m^2 (39·36 sq ft)
Tailplane 17·55 m^2 (188·9 sq ft)
Elevators, incl tabs 5·03 m^2 (54·10 sq ft)

Weights and Loadings (A: standard Srs 2A; B: Srs 2A military transport; C: standard Srs 2B; D: Srs 2B military transport):
Basic operating weight, incl crew:
A 12,159 kg (26,806 lb)
B 11,577 kg (25,524 lb)
C 12,206 kg (26,910 lb)
D 11,671 kg (25,730 lb)
Max payload: A 5,304 kg (11,694 lb)
B 5,886 kg (12,976 lb)
B, optional overload 7,927 kg (17,476 lb)
C 5,257 kg (11,590 lb)
D 5,819 kg (12,829 lb)
D, optional overload 7,833 kg (17,270 lb)
Max T-O weight: A, B, C, D 21,092 kg (46,500 lb)
B, D, optional overload 23,133 kg (51,000 lb)
Max zero-fuel weight:
A, B, C, D 17,463 kg (38,500 lb)
B, D, optional overload 19,504 kg (43,000 lb)
Max landing weight: A, B, C, D 19,504 kg (43,000 lb)
B, D, optional overload 21,546 kg (47,500 lb)
Max wing loading: A 279·8 kg/m^2 (57·3 lb/sq ft)
C 273·9 kg/m^2 (56·1 lb/sq ft)
Max power loading: A, C 6·20 kg/kW (10·2 lb/ehp)

Performance (A: standard Srs 2A at max T-O weight unless otherwise indicated; B: military Srs 2A at normal max T-O weight with 20% fuel reserves; C: military Srs 2A at optional overload T-O weight with 20% fuel reserves; D: standard Srs 2B at max T-O weight unless otherwise indicated):
Cruising speed:
A, D at 17,236 kg (38,000 lb)
244 knots (452 km/h; 281 mph)
Max rate of climb at S/L:
A, D at 17,236 kg (38,000 lb) 433 m (1,420 ft)/min
Service ceiling: A, D 7,620 m (25,000 ft)
Min ground turning radius:
A, B, C, D 11·82 m (39 ft)
Runway LCN: A, D 9 to 18
T-O run: A (BCAR) 1,225 m (4,020 ft)
B 756 m (2,480 ft)
C 945 m (3,100 ft)
D (BCAR) 1,134 m (3,720 ft)
Balanced field length:
A (BCAR) 1,640 m (5,380 ft)
A (BCAR, 650 nm; 1,203 km; 748 mile sector, 40 passengers and reserves for 200 nm; 370 km; 230 miles plus 45 min hold) 892 m (2,925 ft)
D (BCAR) 1,393 m (4,570 ft)
D (BCAR, 840 nm; 1,557 km; 967 mile sector, 44 passengers and reserves for 200 nm; 370 km; 230 miles plus 45 min hold) 963 m (3,160 ft)
T-O to 15 m (50 ft): B 927 m (3,040 ft)
C 1,158 m (3,800 ft)
Landing field length:
A, D (BCAR) 1,036 m (3,400 ft)
Landing from 15 m (50 ft): B 567 m (1,860 ft)
C 625 m (2,050 ft)
Landing run: B 347 m (1,140 ft)
C 387 m (1,270 ft)
Radius of action:
B, supply drop mission with 12 × 340 kg (750 lb) containers 625 nm (1,158 km; 720 miles)
Range with max payload:
A, with reserves for 200 nm (370 km; 230 miles) plus 45 min hold 735 nm (1,361 km; 846 miles)
B 925 nm (1,714 km; 1,066 miles)
C 840 nm (1,556 km; 967 miles)
D, with reserves for 200 nm (370 km; 230 miles) plus 45 min hold 785 nm (1,455 km; 904 miles)
Range with max fuel:
A, with 3,662 kg (8,074 lb) payload, reserves for 200 nm (370 km; 230 miles) plus 45 min hold
1,340 nm (2,483 km; 1,543 miles)
B, with 4,321 kg (9,527 lb) payload
1,410 nm (2,613 km; 1,624 miles)
C, with 6,363 kg (14,027 lb) payload
1,280 nm (2,372 km; 1,474 miles)
D, with 3,660 kg (8,070 lb) payload, reserves for 200 nm (370 km; 230 miles) plus 45 min hold
1,420 nm (2,630 km; 1,635 miles)

Operational Noise Levels (FAR Pt 36):
T-O 92·5 EPNdB
Approach 103·8 EPNdB
Sideline 96·3 EPNdB

BAe HS 748 COASTGUARDER

The Coastguarder is a medium-range maritime patrol aircraft, based on the HS 748. Its development was initiated to meet the need for an aircraft in this category suitable for surface surveillance, fishery protection, pollution/contraband control, search and rescue, tactical surveillance and offshore oilfield patrol. A prototype/demonstration aircraft (G-BCDZ), constructed from

BAe HS 748 Coastguarder twin-turboprop maritime patrol aircraft

an ex-airline HS 748 Series 2A, flew for the first time on 18 February 1977. (This aircraft later served as the Series 2B prototype, and has since been delivered to an airline customer after refurbishing.) A later Coastguarder was outright winner of the Sea Search '81 competition held as part of the 1981 International Air Tattoo at Greenham Common, Berkshire.

The airframe of the Coastguarder is generally similar to that of the standard HS 748 civil and military transport; but there is crew accommodation for two pilots, two beam observers and a tactical navigator, to enable the Coastguarder to fulfil its primary roles. A 0·30 m (1 ft 0 in) diameter chute is mounted in the rear fuselage for the air launch of five-man rescue dinghies, and smoke or flame floats. Two optically flat, inward opening windows in the forward fuselage allow high definition photographs to be taken, with optional data annotation. The standard radio, radar and navigation equipment has been expanded to cover the normal naval radio frequencies, and to provide adequate navigation aids for long overwater flights.

The tactical navigator's station is situated midway down the cabin, on the starboard side, and is equipped with an MEL MAREC radar display and plotting board, Decca 72 Doppler, and a Decca 9447 TANS computer/display. The MAREC radar was chosen as standard on the basis of experience gained in previous ASV, ASW and SAR applications. It has an underfuselage antenna, a 0·43 m (1 ft 5 in) diameter main display and plotting board, with a 0·13 m (5 in) repeat display for the pilot. Used in conjunction with the Doppler, TANS computer and Marconi Omega VLF navigation system, the resulting tactical navigation system can, in addition to satisfying all normal search and navigation requirements, provide effective tactical plotting to control an exercise involving a group of friendly vessels and other radar targets, including aircraft. MAREC provides up to 210 nm (389 km; 242 miles) display range in all directions for the tactical navigator and up to 250 nm (460 km; 285 miles) for the pilot's repeater display. A choice of presentation scale between 1 and 30 nm (1·9-55·6 km; 1·2-34·5 miles) per inch allows enlargement of any selected part of the display.

To provide the additional range required for a maritime reconnaissance role, the fuel tankage has been increased to 10,047 litres (2,210 Imp gallons). The standard Coastguarder may be used for a number of maritime and other roles without any change to the basic configuration. Additional passengers can be accommodated by fitting seats to the standard rails which run the full length of the cabin. The optional rear freight door provides an airdropping capability, allowing the dispatch of large dinghies or supplies in an air/sea rescue role. As many as twelve 30-man dinghies can be transported and dropped for the rescue of a large number of aircraft/ship survivors. The Coastguarder can also be converted easily for cargo carrying, by removal of the tactical navigator's station and other equipment.

The description of the standard HS 748 Series 2B applies also to the Coastguarder, except as follows:

TYPE: Twin-turboprop maritime patrol aircraft.

POWER PLANT: Two 1,700 kW (2,280 ehp) Rolls-Royce Dart RDa.7 Mk 535-2 turboprop engines, each driving a Dowty Rotol four-blade metal constant-speed fully-feathering propeller. Fuel in integral wing tanks with a max combined capacity of 10,047 litres (2,210 Imp gallons). Underwing pressure refuelling and overwing gravity refuelling. Oil capacity 14·2 litres (25 Imp pints) per engine.

ACCOMMODATION: Standard Coastguarder layout has two pilots on flight deck; two beam observers seated at aft end of cabin, one each side, with domed windows; and tactical navigator approximately midway down cabin at tactical station on starboard side. Toilet on starboard side at aft end of cabin, with galley opposite. Main door on port side at rear of cabin; smaller door for emergency exit on starboard side. Crew door on port side at front of cabin. Large rear freight door optional. Four airline-type seats, forward of tactical navigator's station, on starboard side, serve as crew rest area.

AVIONICS AND EQUIPMENT: Avionics include Sylvania VHF/FM com transceiver, dual Collins 618M-3 VHF com transceivers, dual Collins 51RV-4B VHF nav receivers, Collins AN/ARC-159 UHF com transceiver, Collins 51Z-4 marker beacon receiver, dual Collins DF 206 ADF, Collins DF 301E UHF D/F, Collins 618T-3 HF transceiver, Collins 346D-1 address system, Ultra UA 60 interphone, Sperry RN 200 radio navigation display, Honeywell AN/APN-171 radar altimeter, MEL MAREC radar with 0·43 m (1 ft 5 in) main display and 0·13 m (5 in) pilot's repeat display, Marconi CMA 771 Omega VLF nav system, Decca 72 Doppler, and Decca 9447 TANS computer/display. Attitude stabilised antenna, size 0·91 m × 0·53 m (3 ft 0 in × 1 ft 9 in), in underfuselage radome provides 360° azimuth viewing, plus selected sector scan facilities. Provisions for optional ATC transponder, DME, and height encoding altimeter. Standard equipment includes a 0·30 m (1 ft 0 in) launch chute for five-man rescue dinghies and smoke or flame floats. Optional equipment includes large rear freight door, additional passenger seats, large dinghies, and other rescue equipment.

WEIGHTS AND LOADINGS:

*Typical operating weight empty	12,722 kg (28,048 lb)
Max T-O weight	21,092 kg (46,500 lb)
Max zero-fuel weight	17,463 kg (38,500 lb)
Max landing weight	19,504 kg (43,000 lb)
Max wing loading	279·8 kg/m² (57·3 lb/sq ft)
Max power loading	6·20 kg/kW (10·2 lb/ehp)

**depending upon customer requirements*

PERFORMANCE (at max T-O weight, unless stated otherwise):

Cruising speed at 4,570 m (15,000 ft) at AUW of 18,144 kg (40,000 lb)	233 knots (431 km/h; 268 mph)
Typical search speed at 610 m (2,000 ft)	140 knots (259 km/h; 161 mph)
Service ceiling	7,620 m (25,000 ft)
T-O to 15 m (50 ft), unfactored:	
S/L, ISA	951 m (3,120 ft)
S/L, ISA +20°C	1,045 m (3,430 ft)
Landing from 15 m (50 ft) at max landing weight, unfactored:	
S/L, ISA	570 m (1,870 ft)
S/L, ISA +20°C	608 m (1,995 ft)
Range with 4,536 kg (10,000 lb) payload, 20% fuel reserves, ISA	960 nm (1,779 km; 1,105 miles)
Range with max fuel, 20% reserves, ISA	2,300 nm (4,262 km; 2,648 miles)
Time on station at 200 nm (370 km; 230 mile) radius of action, at 3,050 m (10,000 ft), 20% fuel reserves, ISA	9 h 24 min

BAe 146

In August 1973, Hawker Siddeley announced that it was to produce with government support a four-turbofan quiet-operating transport aircraft known as the HS (now BAe) 146. Within a few months economic problems in the UK halted this programme, but research and design continued on a limited basis. With the absorption of Hawker Siddeley into British Aerospace in April 1977, BAe continued to provide limited funding to allow the manufacture of assembly jigs, systems test rigs, and continuing design and wind tunnel testing. On 10 July 1978, the British Aerospace Board's decision to give the 146 programme a full go-ahead was approved by the government, and production is being undertaken in several BAe factories, including Brough (fin), Filton (centre-fuselage), Manchester (rear fuselage), and Prestwick (engine pylons). Hatfield Division builds the forward fuselage and flight

Prototype BAe 146 Series 100 four-turbofan short-range transport, first flown on 3 September 1981

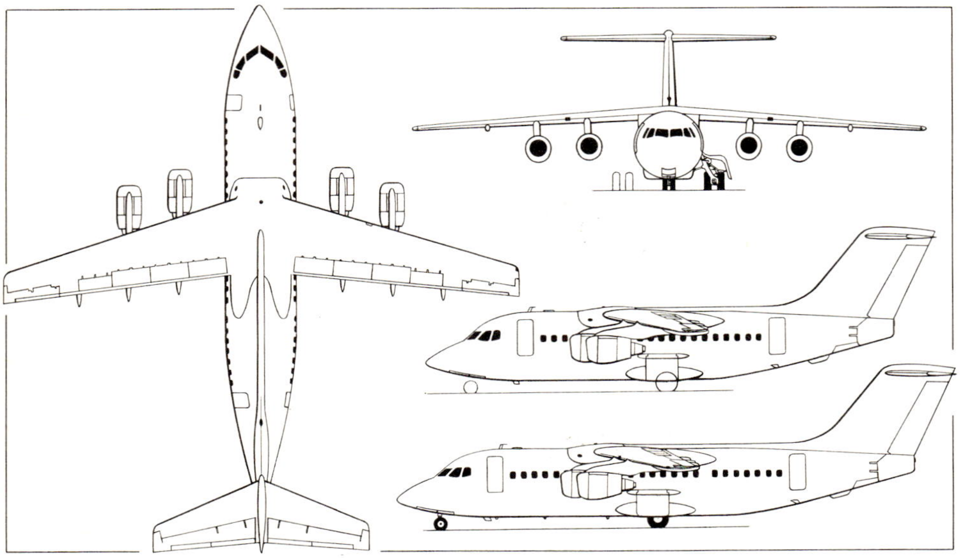

BAe 146 Series 200, with additional side view (centre right) of Series 100 *(Pilot Press)*

deck, and is responsible for final assembly and flight testing.

Following the production decision, risk-sharing agreements were signed with Avco Aerostructures (USA) for the manufacture of 20 sets of wing boxes; and with Saab-Scania (Sweden) for 20 sets of tailplanes and all movable control surfaces. Under an initial contract, Short Brothers (UK) are manufacturing 100 pods for the Avco Lycoming ALF 502 engines which power the 146. Planned initial production rate is three aircraft per month.

The basic aims of the BAe 146 are to provide a passenger seating standard comparable with present wide-bodied transports, combined with competitive operating costs, good airfield performance and low operating noise levels.

Two series are available:

Series 100. Designed to operate from short semi-prepared airstrips with minimal ground facilities, with a normal seating capacity of 71-93. A mixed passenger/freight version is planned.

Series 200. For operation from paved runways only, with seating capacity of 82-109 and greater range. Fuselage lengthened by five frame pitches (2·39 m; 7 ft 10 in). Increased maximum T-O weight and zero-fuel weight. Underfloor cargo volume increased by 35 per cent. Reduced max operating speed.

Rollout of the Series 100 prototype (G-SSSH) took place on 20 May 1981, and first flight was made on 3 September 1981. Full transport category CAA certification of the Series 100 is expected to be obtained in August 1982, with initial deliveries of production aircraft in September and October. The eighth aircraft will be the first Series 200, with first flight scheduled for 1982 and certification and first delivery in March 1983. Orders and options totalled 13 and 12 respectively by June 1981, as follows:

	Orders	*Options*
Air Wisconsin (Srs 200)	4	4
West Air Pacific Express (Srs 200)	6	8
Undisclosed (2 Srs 100/1 Srs 200)	3	–

Freight-carrying and military versions are proposed.

The following description applies to the BAe 146 Series 100, except where indicated:

Type: Four-turbofan short-range transport aircraft.

Wings: Cantilever high-wing monoplane. British Aerospace high-lift aerofoil section. Thickness/chord ratio 15·3% adjacent to fuselage, 12·2% at tip. Anhedral 3° at trailing-edge. Incidence 3° 6' at fuselage side, 0° at tip. Sweepback 15° at quarter-chord. All-metal fail-safe structure of light alloy with machined skins, integrally machined spars and ribs. Single-section hydraulically-actuated tabbed Fowler flaps of light alloy, spanning 78% of each wing trailing-edge, with Dowty Rotol actuators. Mechanically-actuated balanced ailerons, with hydraulically-operated power boost spoilers on upper surfaces. Trim and servo tab in each aileron. No leading-edge lift devices. Hot-air anti-icing of leading-edges.

Fuselage: All-metal fail-safe pressurised semi-monocoque structure. Flight deck and tailcone areas free of stringers. Remainder of structure has 'top hat' stringers bonded to skins above keel area. 'Z' section stringers 'wet' assembled with bonding agent and riveted to skin in keel area. Chemically-etched skins of light alloy. Petal-type airbrakes form tailcone when closed.

Tail Unit: Cantilever sweptback T tail, of all-metal construction. Chemically-etched light alloy skins bonded to 'top-hat' section stringers. Fixed-incidence tailplane. Manually-operated balanced elevators, each with trim and spring tab. Powered rudder. Hot-air anti-icing of tailplane leading-edges.

Landing Gear: Hydraulically-retractable tricycle type, of Dowty Rotol design, with twin Dunlop wheels on each unit. Main units retract inward into fairings on fuselage sides; steerable nose unit retracts forward. Oleo-pneumatic shock-absorbers with wheels mounted on trailing axle. Simple telescopic nosewheel strut. Main-wheel tyres size 12·50-16 Type III, pressure (Series 100) 7·59 bars (110 lb/sq in). Nosewheel tyres size 7·50-10 Type III, pressure (Series 100) 7·03 bars (102 lb/sq in). Low-pressure tyres optional. Dunlop multi-disc carbon brakes operated by duplicated hydraulic systems. Brake cooling and rudder pedal steering optional. Anti-skid units in both primary and secondary brake systems.

Power Plant: Four Avco Lycoming ALF 502R-3 turbofan engines, each rated at 29·8 kN (6,700 lb st), installed in pylon-mounted underwing pods. Fuel in two integral wing tanks and integral centre-section tank (the latter with a vented and drained sealing diaphrragm above passenger cabin), having a combined capacity of 11,547 litres (2,540 Imp gallons). Optional auxiliary tanks in wing-root fairings, with combined capacity of 1,363 litres (300 Imp gallons), giving total optional capacity of 12,910 litres (2,840 Imp gallons). Single-point pressure refuelling, with coupling situated in starboard wing outboard of outer engine.

Accommodation: Crew of two pilots on flight deck, and two or three cabin staff. Optional observer's seat. Series 100 has accommodation in main cabin for 71 passengers with five-abreast seating at 84 cm (33 in) pitch, and a maximum of 93 seats six-abreast at 74 cm (29 in) pitch. Series 200 will have maximum capacity for 109 passengers with six-abreast seating at 74 cm (29 in) pitch. Various alternative layouts for mixed passenger/freight configurations. All seating layouts have two toilets, one forward and one aft, and a forward galley as standard. One outward-opening passenger door forward and one aft on port side of cabin. Built-in airstairs optional. Servicing doors, one forward and one aft, on starboard side of cabin. Freight and baggage holds under cabin floor. All accommodation air-conditioned. Windscreen electrical anti-icing and demisting standard. Rain repellent system optional.

Systems: BAe Dynamics cabin air-conditioning and pressurisation system, using engine bleed air. Electro-pneumatic pressurisation control with discharge valves at fore and aft of cabin. Max differential 0·45 bars (6·5 lb/sq in). Hydraulic system, duplicated for essential services, for landing gear, flaps, rudder, roll and lift spoilers, airbrakes, nosewheel steering, brakes and auxiliary fuel pumps; pressure 207 bars (3,000 lb/sq in). Electrical system powered by two 40kVA integrated-drive alternators to feed 115/200V 3-phase 400Hz primary systems. 28V DC power supplied by transformer-rectifier in each channel. Hydraulically-powered emergency electrical power unit. Garrett GTCP 36-100 APU for ground air-conditioning and electrical power generation. High-pressure gaseous oxygen system, pressure 124 bars (1,800 lb/sq in).

Avionics: Smiths SEP 10 automatic flight control and flight guidance system incorporates a simplex Cat 1 autopilot with a flight director display and separate attitude reference for each pilot. Addition of extra equipment and wiring permits coupled approaches to Cat II minima. Standard ARINC interface with radio nav system allows choice of radio equipment. Basic avionics include dual VHF com, audio system, passenger address system, cockpit voice recorder, dual compass systems, dual ADIs with separate attitude reference driven by single computer, marker beacon receiver, weather radar, radio altimeter, ground proximity warning system, DME, ATC transponder, dual VHF nav and an ADF. Dowty-UEL flight deck warning system. Optional avionics include third VHF com, area navigation system, Selcal, tape reproducer, single or dual HF com, and second ADF, DME, transponder and radio altimeter.

Dimensions, external:
Wing span 26·34 m (86 ft 5 in)
Wing aspect ratio 8·97
Length overall: Series 100 26·16 m (85 ft 10 in)
Series 200 28·55 m (93 ft 8 in)
Height overall 8·61 m (28 ft 3 in)
Fuselage diameter 3·56 m (11 ft 8 in)
Tailplane span 11·09 m (36 ft 5 in)
Wheel track 4·72 m (15 ft 6 in)
Wheelbase: Series 100 10·43 m (34 ft 2¾ in)
Series 200 11·20 m (36 ft 9 in)
Passenger doors (port, fwd and rear):
Height 1·83 m (6 ft 0 in)
Width 0·85 m (2 ft 9½ in)
Height to sill 1·93 m (6 ft 4 in)
Servicing doors (stbd, fwd and rear):
Height 1·47 m (4 ft 10 in)
Width 0·85 m (2 ft 9½in)
Height to sill 1·93 m (6 ft 4 in)
Underfloor freight hold door (stbd, fwd):
Height 1·07 m (3 ft 6 in)
Width 1·22 m (4 ft 0 in)
Height to sill 0·86 m (2 ft 10 in)
Underfloor freight hold door (stbd, aft):
Height 1·07 m (3 ft 6 in)
Width 0·91 m (3 ft 0 in)
Height to sill 1·07 m (3 ft 6 in)

Dimensions, internal:
Cabin (excl flight deck, incl galley and toilets):
Length: Series 100 15·42 m (50 ft 7 in)
Series 200 17·81 m (58 ft 5 in)
Max width 3·38 m (11 ft 1 in)
Max height 2·02 m (6 ft 7½ in)
Floor area: Series 100 49·24 m^2 (530 sq ft)
Baggage/freight holds, underfloor:
Series 100 14·16 m^3 (500 cu ft)
Series 200 18·69 m^3 (660 cu ft)

Areas:
Wings, gross 77·30 m^2 (832 sq ft)
Ailerons (total) 3·53 m^2 (38 sq ft)
Trailing-edge flaps (total) 19·32 m^2 (208 sq ft)
Spoilers (total) 9·66 m^2 (104 sq ft)
Fin 11·61 m^2 (125 sq ft)
Rudder 9·20 m^2 (99 sq ft)
Tailplane 15·61 m^2 (168 sq ft)
Elevators, incl tabs 10·03 m^2 (108 sq ft)

Weights and Loadings (estimated):
Typical operating weight empty:
Series 100 20,670 kg (45,570 lb)
Series 200 21,405 kg (47,190 lb)
Max payload: Series 100 8,020 kg (17,680 lb)
Series 200 10,006 kg (22,060 lb)
Max T-O weight:
Series 100, standard 33,840 kg (74,600 lb)
Series 100, optional 36,628 kg (80,750 lb)
Series 200 40,030 kg (88,250 lb)
Max ramp weight: Series 100, standard 34,065 kg (75,100 lb)
Series 200 40,256 kg (88,750 lb)
Max zero-fuel weight:
Series 100 28,690 kg (63,250 lb)
Series 200 31,411 kg (69,250 lb)
Max landing weight:
Series 100 32,590 kg (71,850 lb)
Series 200 34,926 kg (77,000 lb)
Max wing loading:
Series 100, standard 437·5 kg/m^2 (89·66 lb/sq ft)
Series 200 517·6 kg/m^2 (106·07 lb/sq ft)
Max power loading:
Series 100, standard 283·9 kg/kN (2·78 lb/lb st)
Series 200 335·8 kg/kN (3·29 lb/lb st)

Performance (estimated, at max standard T-O weight, except where indicated):
Max operating speed:
Series 100
Mach 0·70 (310 knots; 574 km/h; 357 mph CAS)
Series 200
Mach 0·70 (300 knots; 555 km/h; 345 mph CAS)
Max cruising speed:
Series 100 and 200 at 7,925 m (26,000 ft)
419 knots (776 km/h; 482 mph)
Econ cruising speed at 9,145 m (30,000 ft):
Series 100 371 knots (687 km/h; 427 mph)
Series 200 379 knots (702 km/h; 436 mph)
Stalling speed, 30° flap:
Series 100 91 knots (169 km/h; 105 mph) EAS
Series 200 99 knots (183 km/h; 114 mph) EAS
Stalling speed, 33° flap, at max landing weight:
Series 100 87 knots (161 km/h; 100 mph) EAS
Series 200 90 knots (168 km/h; 104 mph) EAS
T-O to 10·7 m (35 ft), S/L, ISA:
Series 100 1,082 m (3,550 ft)
Series 200 1,585 m (5,200 ft)

FAR landing distance from 15 m (50 ft), S/L, ISA, at max landing weight:

Series 100	1,013 m (3,325 ft)
Series 200	1,061 m (3,480 ft)

Range with max fuel, incl 293 kg (645 lb) fuel for ground and airborne manoeuvres, plus fuel for 150 nm (278 km; 173 mile) diversion and 45 min hold at 1,525 m (5,000 ft):

Series 100	1,550 nm (2,872 km; 1,785 miles)
Series 200	1,480 nm (2,743 km; 1,704 miles)

Range with max payload, allowances as above:

Series 100	510 nm (945 km; 587 miles)
Series 200	1,080 nm (2,001 km; 1,243 miles)

Operational Noise Levels (FAR Pt 36-12, estimated):

T-O: Series 100	82·6 EPNdB
Series 200	87·8 EPNdB
Approach: Series 100 and 200	95·5 EPNdB
Sideline: Series 100 and 200	89 EPNdB

BAe 146M

This proposed military version of the BAe 146 is envisaged as a medium-lift aircraft suitable for the airlift of vehicles and palletised cargo; the delivery of heavy vehicles to forward airstrips, and casevac mission on the return flight; airdrop of paratroops and supplies, the latter by free-fall or parachute extraction technique; and troop or VIP transport, with mixed passenger/freight layouts optional.

Generally similar to the civil BAe 146, the 146M differs primarily in having a changed landing gear to simplify operation from unprepared surfaces at forward landing strips, and in redesign of the aft fuselage to provide rear ramp loading. Full details can be found in the 1980-81 *Jane's*.

BAe (BAC) ONE-ELEVEN

Details of the One-Eleven were announced on 9 May 1961, simultaneously with the news that British United Airways had ordered ten. Design and manufacture are shared between three Weybridge/Bristol Division factories, at Weybridge, Filton and Hurn.

Five commercial versions have been produced, and details of the Series 200 (56 built), 300 (nine built) and 400 (69 built) can be found in the 1974-75 *Jane's*. Two versions are currently in production, as follows:

Series 475. Combines standard fuselage and accommodation of Series 400 with wings and power plant of Series 500 and a modified landing gear system, using low-pressure tyres, to permit operation from secondary low-strength runways with poorer-grade surfaces. The prototype flew for the first time on 27 August 1970. First production Series 475 flew on 5 April 1971. Certification and first production delivery (to Faucett of Peru) in July 1971. The three Srs 475s supplied to the Sultan of Oman's Air Force have a quick-change passenger/cargo interior layout and a 3·05 × 1·85 m (10 ft 0 in × 6 ft 1 in) forward freight door. The first executive Series 475, for a Saudi Arabian customer, was delivered in May 1978. Nine Srs 475s sold by mid-1981.

Series 500. Derived from Series 300/400, this version incorporates a lengthened fuselage (2·54 m; 100 in fwd of wing, 1·57 m; 62 in aft) which accommodates 97-119 passengers, with a flight crew of two. Wingtip extensions increase span by 1·52 m (5 ft). Take-off performance improved by increased wing area and by installation of two Rolls-Royce Spey Mk 512 DW turbofans, each rated at 55·8 kN (12,550 lb st). Main landing gear strengthened and heavier wing planks used to cater for increased AUW. Prototype flew for first time on 30 June 1967. First production aircraft flew on 7 February 1968. ARB certification 15 August 1968. Deliveries to BEA (now British Airways) began on 29 August 1968. Total of 87 sold by mid-1981.

In addition, executive and freighter versions of the One-Eleven are available. More than 40 examples of the former are now in service, notably in the USA and with the Australian government, with interior and long-range tank conversions being made either by BAe or by specialist contractors. The One-Eleven freighter incorporates a 3·05 m by 1·85 m (10 ft 0 in by 6 ft 1 in) upward-opening hydraulically-powered loading door in the forward fuselage, together with a quickly-removable freight floor overlay and cargo handling system. First to be delivered, in November 1975, was one of three Series 475 freighters operated by the Sultan of Oman's Air Force.

In May 1977, BAe concluded an agreement with CNIAR of Romania (which see) for the Series 475 and Series 500 to be built under licence in Romania. Three complete aircraft (two **Srs 525/1s** and one **Srs 487** freighter) are being supplied from Hurn, of which the first two were delivered in January and July 1981. These will be followed, up to 1985, by a further 22 aircraft in component form. Thereafter, production of complete aircraft will take place in Romania. These will be known as the **Rombac 1-11 Series 550**. Orders for 230 One-Elevens (not including those to be built in Romania) had been received by 1 August 1981.

A One-Eleven 'hush kit', comprising an intake duct lining, a bypass duct lining, an acoustically-lined jetpipe, and a six-chute exhaust silencer, was flown for the first time, on the Srs 475 development aircraft G-ASYD, on 14 June 1974. It is designed to reduce the area within the 90 EPNdB noise contour by approximately 50 per cent, giving a noise footprint equivalent to that of a twin-turboprop aircraft. The first production aircraft to be fitted with 'hush kits' were five One-Eleven 500s for Tarom, the Romanian state airline, which were delivered between March and August 1977. In May 1979 Monarch Airlines in the UK became the first operator to order 'hush kits' for retrospective fitting to its fleet of One-Eleven Srs 500s.

The following description applies to the Series 475 and 500:

Type: Twin-turbofan short/medium-range transport.

Wings: Cantilever low-wing monoplane. Modified NACA cambered wing section. Thickness/chord ratio 12½% at root, 11% at tip. Dihedral 2°. Incidence 2° 30′. Sweepback 20° at quarter-chord. All-metal structure of copper-based aluminium alloy, built on fail-safe principles. Three-shear-web torsion box with integrally-machined skin/stringer panels. Ailerons of Redux-bonded light alloy honeycomb, manually operated through servo tabs. Port servo tab used for trimming. Light alloy Fowler flaps hydraulically operated through Hobson actuators. Light alloy spoiler/airbrakes on upper surface of wing, operated hydraulically through Dowty Boulton Paul actuators. Hydraulically-actuated lift dumpers, inboard of spoilers, are standard. Flaps on Series 475 have a glassfibre coating. Thermal de-icing of wing leading-edges with engine bleed air.

Fuselage: Conventional circular-section all-metal fail-safe structure with continuous frames and stringers. Skin made from copper-based aluminium alloy.

Tail Unit: Cantilever all-metal fail-safe structure, with variable-incidence T tailplane, controlled through duplicated Hobson hydraulic units. Fin integral with rear fuselage. Elevators and rudder actuated hydraulically through Dowty Boulton Paul tandem jacks. Leading-edges of fin and tailplane de-iced by engine bleed air.

Landing Gear: Retractable tricycle type, with twin wheels on each unit. Hydraulic retraction, nose unit forward, main units inward. Oleo-pneumatic shock-absorbers manufactured by BAC. Hydraulic nosewheel steering. Dunlop wheels, tubeless tyres and 5-plate heavy-duty hydraulic disc brakes. Hytrol Mk III anti-skid units. Main-wheel tyres size 40 × 12 on Srs 500, pressure 11·03 bars (160 lb/sq in). Dunlop 44 × 16 tyres on Srs 475, pressure 5·72 bars (83 lb/sq in). Nosewheel tyres size 24 × 7·25 on Srs 500, pressure 7·58 bars (110 lb/sq in). Dunlop 24 × 7·7 tyres on Srs 475, pressure 7·24 bars (105 lb/sq in). All tyre pressures are given for aircraft at mid-CG position and max taxi weight.

Power Plant: Two Rolls-Royce Spey Mk 512 DW turbofan engines, each rated at 55·8 kN (12,550 lb st), pod-mounted on sides of rear fuselage. Fuel in integral wing tanks of 10,160 litres (2,235 Imp gallons) and centre-section tank of 3,864 litres (850 Imp gallons) capacity; total fuel capacity 14,024 litres (3,085 Imp gallons). Optional 1,591 litre (350 Imp gallon) and 3,182 litre (700 Imp gallon) fuel tanks are available to increase total fuel capacity. Pressure refuelling point in fuselage forward of wing on starboard side. Provision for gravity refuelling. Oil capacity (total engine oil) 13·66 litres (3 Imp gallons) per engine.

Accommodation (Srs 475): Crew of two on flight deck and up to 89 passengers in main cabin. Single class or mixed class layout, with movable divider bulkhead to permit any first/tourist ratio. Typical mixed class layout has 16 first class (four abreast) and 49 tourist (five abreast) seats. Galley units normally at front on starboard side. Coat space available on port side aft of flight deck. Ventral entrance with hydraulically-operated airstair. Forward passenger door on port side incorporates

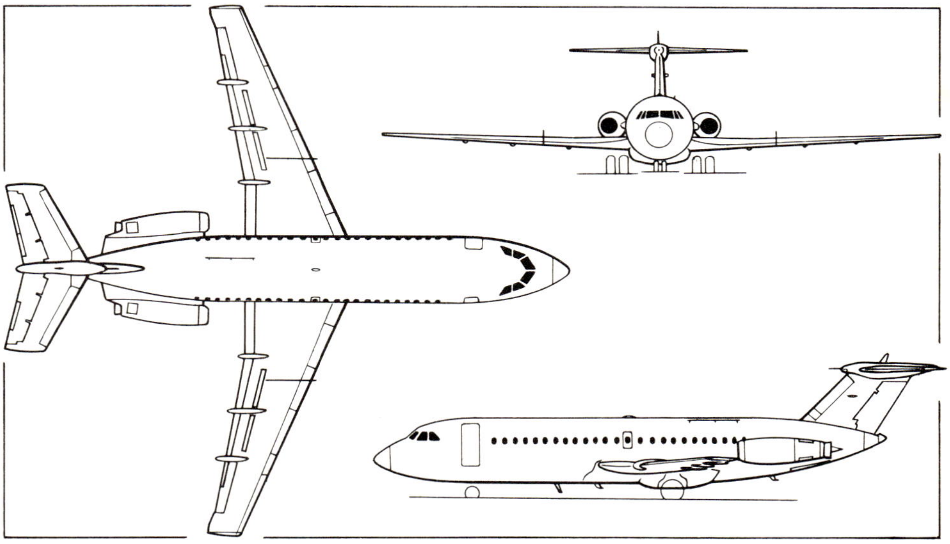

BAe (BAC) One-Eleven Series 475 twin-turbofan short/medium range airliner *(Pilot Press)*

The first of three new BAe (BAC) One-Elevens for Tarom. This Series 525/1 was delivered on 16 January 1981

optional power-operated airstair. Galley service door forward on starboard side. Two baggage and freight holds under floor, fore and aft of wings, with doors on starboard side. Upward-opening forward freight door available at customer's option. Entire accommodation air-conditioned.

Accommodation (Srs 500): Crew of two on flight deck and up to 119 passengers in main cabin. Two additional overwing emergency exits, making two on each side. One toilet on each side of cabin at rear. Otherwise generally similar to Srs 475.

Systems: Fully-duplicated air-conditioning and pressurisation systems with main components by Normalair-Garrett. Air bled from engine compressors through heat exchangers. Max pressure differential 0·52 bars (7·5 lb/sq in). Hydraulic system, pressure 207 bars (3,000 lb/sq in), operates flaps, spoilers, rudder, elevators, tailplane, landing gear, brakes, nosewheel steering, ventral and forward airstairs and windscreen wipers. No pneumatic system. Electrical system utilises two 30kVA Plessey/Westinghouse AC generators, driven by Plessey constant-speed drive and starter units, plus a similar generator mounted on the APU and shaft-driven. Garrett gas-turbine APU in tailcone to provide ground electrical power, air-conditioning and engine starting, also some system checkout capability. APU is run during take-off to eliminate performance penalty of bleeding engine air for cabin air-conditioning.

Avionics and Equipment: Communications and navigation avionics generally to customers' requirements. Typical installation includes dual VHF com to ARINC 546, dual VHF nav to ARINC 547A, including glideslope receivers, marker beacon receiver, flight/service interphone system, Marconi AD 370, Bendix DFA 73 or Collins DF 203 ADF, ATC transponder to ARINC 532D, Collins 860 E2 DME, Ekco E 190 or Bendix RDR 1E weather radar. Sperry C9 or CL11 compass systems and Collins FD 108 flight director system (dual) are also installed. Elliott 2000 Series autopilot system. Provision on the Srs 500 for additional equipment, including automatic throttle control, for low weather minima operation.

Dimensions, external:
Wing span 28·50 m (93 ft 6 in)
Wing chord at root 5·12 m (16 ft 9⅝ in)
Wing chord at tip 1·65 m (5 ft 5 in)
Wing aspect ratio 8·5
Length overall: Srs 475 28·50 m (93 ft 6 in)
Srs 500 32·61 m (107 ft 0 in)
Length of fuselage: Srs 475 25·55 m (83 ft 10 in)
Srs 500 29·67 m (97 ft 4 in)
Height overall 7·47 m (24 ft 6 in)
Tailplane span 8·99 m (29 ft 6 in)
Wheel track 4·34 m (14 ft 3 in)
Wheelbase: Srs 475 10·06 m (33 ft 0 in)
Srs 500 12·60 m (41 ft 4 in)
Passenger door (fwd, port):
Height 1·73 m (5 ft 8 in)
Width 0·81 m (2 ft 8 in)
Height to sill 2·13 m (7 ft 0 in)
Ventral entrance, bulkhead door:
Height 1·83 m (6 ft 0 in)
Width 0·66 m (2 ft 2 in)
Height to sill 2·13 m (7 ft 0 in)
Freight door (fwd, starboard):
Height (projected) 0·79 m (2 ft 7 in)
Width 0·91 m (3 ft 0 in)
Height to sill 1·09 m (3 ft 7 in)
Freight door (rear, starboard):
Height (projected) 0·66 m (2 ft 2 in)
Width 0·91 m (3 ft 0 in)
Height to sill 1·30 m (4 ft 3 in)
Freight door, main deck (optional, fwd, Srs 475 SOAF):
Height 1·85 m (6 ft 1 in)
Width 3·05 m (10 ft 0 in)
Galley service door (fwd, starboard):
Height (projected) 1·22 m (4 ft 0 in)
Width 0·69 m (2 ft 3 in)
Height to sill 2·13 m (7 ft 0 in)

Dimensions, internal (Srs 475):
Cabin, excl flight deck:
Length 17·32 m (56 ft 10 in)
Max width 3·15 m (10 ft 4 in)
Max height 1·98 m (6 ft 6 in)
Floor area approx 47·4 m² (510 sq ft)
Freight hold, fwd 10·02 m³ (354 cu ft)
Freight hold, rear 4·42 m³ (156 cu ft)

Dimensions, internal (Srs 500):
Cabin, excl flight deck:
Length 21·44 m (70 ft 4 in)
Total floor area approx 59·5 m² (640 sq ft)
Freight holds (total volume) 19·45 m³ (687 cu ft)

Areas (Srs 475, 500):
Wings, gross 95·78 m² (1,031 sq ft)
Ailerons (total) 2·86 m² (30·8 sq ft)
Flaps (total) 15·89 m² (171 sq ft)
Spoilers (total) 2·30 m² (24·8 sq ft)
Vertical tail surfaces (total) 10·91 m² (117·4 sq ft)
Rudder, incl tab 3·05 m² (32·8 sq ft)
Horizontal tail surfaces (total) 23·88 m² (257·0 sq ft)
Elevators, incl tab 6·54 m² (70·4 sq ft)

Weights and Loadings:
Operating weight empty, typical:
Srs 475 (89 seats) 23,286 kg (51,339 lb)
Srs 500 (109 seats) 24,386 kg (53,762 lb)
Max payload, typical:
Srs 475 10,733 kg (23,661 lb)
Srs 500 12,355 kg (27,238 lb)
Max T-O weight:
Srs 475 41,730-44,679* kg (92,000-98,500* lb)
Srs 500 45,200-47,400* kg (99,650-104,500* lb)
Max ramp weight:
Srs 475 41,957-44,906* kg (92,500-99,000* lb)
Srs 500 45,359-47,627* kg (100,000-105,000* lb)
Max landing weight:
Srs 475 38,102-39,463* kg (84,000-87,000* lb)
Srs 500 39,463 kg (87,000 lb)
Max zero-fuel weight:
Srs 475 33,112-34,019* kg (73,000-75,000* lb)
Srs 500 36,741 kg (81,000 lb)
Max wing loading:
Srs 475 466·3 kg/m² (95·5 lb/sq ft)
Srs 500 495·1 kg/m² (101·4 lb/sq ft)
Max power loading:
Srs 475 400·2 kg/kN (3·92 lb/lb st)
Srs 500 424·5 kg/kN (4·16 lb/lb st)
*optional

Performance (at standard max T-O weight):
Design diving speed (S/L)
410 knots (760 km/h; 472 mph) EAS
Max level and cruising speed at 6,400 m (21,000 ft)
470 knots (870 km/h; 541 mph)
Fuel econ cruising speed at 10,670 m (35,000 ft)
410 knots (760 km/h; 472 mph)
Stalling speed (landing flap setting, at standard max landing weight):
Srs 475 98 knots (182 km/h; 113 mph) EAS
Srs 500 100 knots (186 km/h; 115 mph) EAS
Rate of climb at S/L at 300 knots (555 km/h; 345 mph)
EAS: Srs 475 786 m (2,580 ft)/min
Srs 500 722 m (2,370 ft)/min
Max cruising height 10,670 m (35,000 ft)
Min ground turning radius (to outer wingtip):
Srs 475 15·70 m (51 ft 6 in)
Srs 500 17·07 m (56 ft 0 in)
Runway LCN, rigid pavement (l = 30):
Srs 475 32
Srs 500 53
T-O run at S/L, ISA: Srs 475 1,676 m (5,500 ft)
Srs 500 1,981 m (6,500 ft)
Balanced T-O to 10·7 m (35 ft) at S/L, ISA:
Srs 475 1,798 m (5,900 ft)
Srs 500 2,225 m (7,300 ft)
Landing distance (BCAR) at S/L, ISA, at standard max landing weight: Srs 475 1,440 m (4,725 ft)
Srs 500 1,455 m (4,775 ft)
Max still-air range, ISA, with reserves for 200 nm (370 km; 230 mile) diversion and 45 min hold:
Srs 475 2,063 nm (3,821 km; 2,374 miles)
Srs 500 1,922 nm (3,560 km; 2,212 miles)
Still-air range with typical capacity payload, ISA, reserves as above:
Srs 475 at 44,678 kg (98,500 lb)
1,627 nm (3,013 km; 1,872 miles)
Srs 500 at 47,400 kg (104,500 lb)
1,472 nm (2,726 km; 1,694 miles)
Srs 475 executive aircraft with additional 5,602 litres (1,232 Imp gallons) fuel and ten passengers
2,875 nm (5,325 km; 3,308 miles)

BAe (BAC/VICKERS) VC10 K. Mk 2/Mk 3 TANKERS

In early 1978 it was announced in Parliament by the Under-Secretary of State for Defence that the Royal Air Force had a requirement for additional flight refuelling tankers, and that it was intended to investigate the feasibility of converting civil VC10s for this role. By April 1978 Air Staff Requirement 406 had been formulated, and a contract for the design study awarded to British Aerospace. This study was completed before the middle of the year, proving that the aircraft could be converted effectively for such operations, and leading to the award to British Aerospace of a contract for the work in July 1978. Valued at that time at some £40 million, it covers the conversion of nine aircraft to tanker configuration, with delivery during 1983-84.

The nine VC10s which have been acquired to fulfil this programme comprise five of the 12 standard Model 1101s built during 1962-64 for service with British Overseas Airways Corporation, and four of the five Model 1154 Super VC10s that were delivered to East African Airways in the period 1966-70. RAF designations for the VC10 and Super VC10 tanker conversions will be VC10 K. Mk 2 and VC10 K. Mk 3 respectively. The modification of these commercial transports to a tanker configuration is complicated by the fact that the RAF's No. 10 Squadron already operates a fleet of 13 VC10 multi-mission transports under the designation VC10 C. Mk 1. These differ in several ways from the standard civil transports, and the need for these aircraft and the new tanker fleet to have generally similar configurations is desirable for both operating and engineering considerations.

The modification work is being carried out by British Aerospace at Filton, with one Model 1101 and one Model 1154 being worked upon initially to serve as prototype/development aircraft, and scheduled to fly in 1982.

A description of both the VC10 and Super VC10 can be found under the British Aircraft Corporation entry in the 1970-71 *Jane's*. Detailed below are the modifications required to make these commercial transports suitable for their new tanker role, as well as those needed to maintain an acceptable degree of commonality between these aircraft and the VC10 C. Mk 1s which are in service with the RAF's No. 10 Squadron.

To supplement the nine K. Mk 2/3 tankers, when the RAF's earlier Victor tankers have to be retired, the VC10 C. Mk 1 transports are to be modified for a secondary tanker role.

Type: Military flight refuelling aircraft.

Wings: As for VC10 and Super VC10, but pylons installed beneath the wings, immediately inboard of the ailerons, to carry Flight Refuelling Mk 32 refuelling pods. Aircraft can be flown without these pods being attached. Floodlights installed in each side of wing pylon fairings, and in the wing flap actuator fairings, to illuminate the aircraft for night operations.

Fuselage: As for VC10 and Super VC10, but with flight refuelling probe on nose, directly above weather radar radome, to conform with installation in C. Mk 1s. Remotely-controlled Flight Refuelling Mk 17B HDU (hose and drogue unit) installed in lower rear fuselage, involving the cutting of an aperture in the pressurised structure. This neccessitates new pressure bulkheads fore and aft of the cutout, new sidewalls, and a pressure roof over the HDU. Other items being incorporated in the underfuselage structure include a remotely-operated closed-circuit TV (CCTV) for examination of HDUs; and a floodlight in the rear fuselage to illuminate the engine nacelles.

Tail Unit and Landing Gear: As for VC10 and Super VC10.

Power Plant: Four 97 kN (21,800 lb st) Rolls-Royce Conway Mk 550B turbofan engines, interchangeable with the Conway 301 engines installed in the VC10 C. Mk 1s. Thrust reversers installed on outboard engines only, to conform with C. Mk 1s. Basic fuel capacity of the VC10 K. Mk 2 will be 81,486 litres

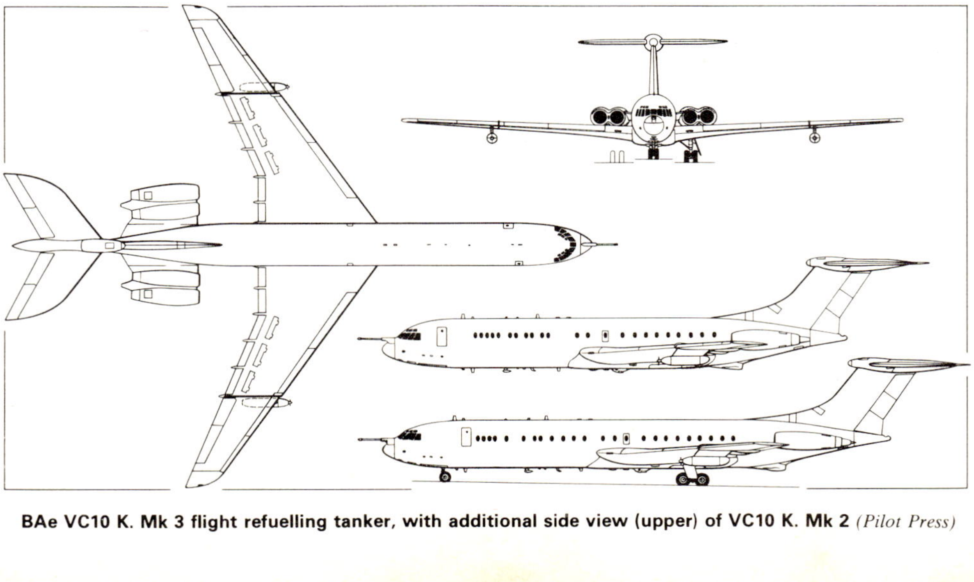

BAe VC10 K. Mk 3 flight refuelling tanker, with additional side view (upper) of VC10 K. Mk 2 *(Pilot Press)*

Artist's impression of one of the VC10 flight refuelling tankers for the RAF

(17,925 Imp gallons), the same as that of the standard VC10; that of the K. Mk 3 will be 88,032 litres (19,365 Imp gallons), the same as in the Super VC10. Additional fuel for flight refuelling operations accommodated in five cylindrical tanks installed within the fuselage. Each consists of a double-skinned metal container with an inner flexible bag tank, and each is mounted on two large beams and restrained from forward movement by a heavy A-frame in front of each tank. These supplementary tanks and the aircraft's basic fuel system are interconnected, and it is possible to transfer all but sufficient fuel for the tanker's mission requirement, or to take on board a similar volume via the nose-mounted refuelling probe. Installation of these fuselage tanks in the standard VC10s will require a section of the fuselage upper surface to be cut out and replaced. In the case of the Model 1154s which were built for East African Airways, these incorporate a large freight door forward of the wing, on the port side, which is large enough to accept these tanks. Following installation, the freight door will be sealed.

Accommodation: Primary flight crew of four, comprising pilot, co-pilot, navigator and flight engineer. Flight engineer's station, on starboard side of flight deck, equipped also for control of the air refuelling operation, has a cathode ray tube (CRT) display from the CCTV. At the forward end of the cabin, and separated from the tank bay by a bulkhead, limited rear-facing seating is provided for airlift of essential ground personnel when the tanker is deployed away from its home base: K. Mk 2 seats 18 persons, K. Mk 3 has seats for 17. Cabin windows and overhead baggage racks retained in this passenger area, but most windows will be blanked off in the tank bay, and all baggage racks removed. Access for crew and passengers through forward starboard door. Forward port door is to serve only for emergency exit by parachute. Remaining cabin doors and two of the four emergency exits to be sealed. Forward underfloor freight hold unchanged, and can be used to carry spares or accommodate refuelling pods during ferry flights.

Systems: Generally similar to those of C. Mk 1. A Rolls-Royce Turboméca Artouste Mk 520 APU is to be installed in the tailcone of both tanker versions to conform to the C. Mk 1s. This provides compressed air for engine starting, or essential electrical power when required on the ground.

Avionics and Equipment: Avionics will conform to those provided in VC10 C. Mk 1s, including dual VHF/UHF com, dual HF, ADF, IFF, Omega, Tacan and weather radar. A Smiths SFS6 flight system will be standard. Equipment includes Flight Refuelling Mk 32 pods and Mk 17B HDUs, one 10-man dinghy and one 26-man dinghy.

Dimensions, external:

Wing span	44·55 m (146 ft 2 in)
Length overall:	
K. Mk 2 (excl refuelling probe)	48·36 m (158 ft 8 in)
K. Mk 3 (excl refuelling probe)	52·32 m (171 ft 8 in)
Height overall	12·04 m (39 ft 6 in)
Tailplane span	13·36 m (43 ft 10 in)
Wheel track	6·53 m (21 ft 5 in)
Wheelbase: K. Mk 2	20·08 m (65 ft 10½ in)
K. Mk 3	21·98 m (72 ft 1½ in)

BAe SA-3-120 BULLDOG SERIES 120

First flight of the Beagle-built Bulldog prototype was made on 19 May 1969. A second prototype, completed by Scottish Aviation, was flown on 14 February 1971; it was later refurbished, issued with a Normal category C of A, and delivered to a private owner under the designation Model 104. The first production Bulldog flew on 22 June 1971 and received full ARB certification on 30 June 1971. The first 98 production Bulldogs were of the Series 100 version, described in the 1972-73 *Jane's*.

Production then continued with the Series 120, which was awarded full CAA certification on 12 February 1973. By mid-1981 orders for this series were as follows:

Model 121. For Royal Air Force, by whom it is designated **T. Mk 1.** First flight on 30 January 1973; 130 delivered in 1973-76. In service with No. 2 FTS at Leeming, the CFS at Leeming, and University Air Squadrons.

Model 122. For Ghana Air Force. Thirteen delivered, including seven **Mk 122A**.

Model 123. For Nigerian Air Force. Thirty-two delivered. Five more currently on order.

Model 124. One aircraft (G-ASAL) used as company demonstration aircraft.

Model 125. Built for Jordanian Royal Academy of Aeronautics. Total of 13 delivered in 1974-76. All transferred to Royal Jordanian Air Force in 1978. Further nine currently on order.

Model 126. For Lebanese Air Force. Six delivered.

Model 127. For Kenya. Nine delivered.

Model 128. For Royal Hong Kong Auxiliary Air Force. Two delivered.

Model 129. One civil aircraft delivered to Venezuela.

Model 130. Six delivered to Botswana Defence Force.

Production of the Bulldog will continue into 1982.

Type: Two/three-seat primary trainer.

Wings: Cantilever low-wing monoplane. Wing section NACA 63_2615. Dihedral 6° 30′. Incidence 1° 9′ at root. Conventional single-spar riveted stressed-skin structure of light alloy. Electrically-operated slotted trailing-edge flaps and slotted ailerons of similar construction. Ground-adjustable tab on starboard aileron.

Fuselage: Conventional light alloy stressed-skin semi-monocoque structure.

Tail Unit: Cantilever two-spar light alloy stressed-skin structure. Fixed-incidence tailplane. Full-span trim tab in starboard elevator. Manually-operated trim tab in rudder. Ventral fin.

Landing Gear: Non-retractable tricycle type, with single wheel on each unit. Steerable nosewheel with Automotive Products oleo-pneumatic shock-absorber and Goodyear wheel and tyre, size 5·00-5, pressure 2·76 bars (40 lb/sq in). Main units have Automotive Products oleo-pneumatic shock-absorbers and Goodyear wheels and tyres, size 6·00-6, pressure 2·07 bars (30 lb/sq in). Goodyear hydraulic disc brakes on main wheels. Optional ski landing gear.

Power Plant: One 149 kW (200 hp) Avco Lycoming IO-360-A1B6 flat-four engine, driving a Hartzell HC-C2YK-4F/FC7666A-2 two-blade constant-speed metal propeller with spinner. Avco Lycoming AEIO-360-A1B6 engine available optionally, permitting up to 20 s of inverted flight. Four removable metal fuel tanks, two in each wing, with total usable capacity of 145·5 litres (32 Imp gallons). Refuelling point on top of each wing. Oil capacity 7·6 litres (1·67 Imp gallons).

BAe Scottish Division Bulldog Model 121 trainer in service with the RAF

Accommodation: Enclosed cabin seating pilot and co-pilot or trainee side by side with dual controls, with space at rear for observer's seat or up to 100 kg (220 lb) of baggage. Rearward-sliding jettisonable transparent canopy. Cabin heated and ventilated.

Systems: Heat exchanger for cabin heating. Hydraulic system, pressure 40 bars (580 lb/sq in), for main-wheel brakes only. Vacuum-type pneumatic system available optionally. 24V DC power from engine-driven alternator and 24V 25Ah storage battery. No oxygen or de-icing systems.

Avionics and Equipment: VHF, UHF or HF com radio to individual customer's requirements; panel can accommodate dual VHF nav, DME, ATC transponder, ADF and other navaids. Blind-flying instrumentation standard. Glider towing attachment optional.

Armament and Operational Equipment: Standard aircraft is unarmed, but has provision for installation of four underwing hardpoints to which can be attached various loads including unguided or wire-guided air-to-surface weapons; 7·62 mm machine-gun pods; grenade launchers; practice or active bombs of up to 50 kg; markers; supply containers; leaflet dispensers; and rescue and survival equipment. Maximum underwing load 290 kg (640 lb).

Dimensions, external:	
Wing span	10·06 m (33 ft 0 in)
Wing chord at root	1·51 m (4 ft 11¼ in)
Wing chord at tip	0·86 m (2 ft 9¾ in)
Wing aspect ratio	8·4
Length overall	7·09 m (23 ft 3 in)
Height overall	2·28 m (7 ft 5¾ in)
Tailplane span	3·35 m (11 ft 0 in)
Wheel track	2·03 m (6 ft 8 in)
Wheelbase	1·40 m (4 ft 7 in)
Propeller diameter	1·88 m (6 ft 2 in)
Propeller ground clearance	0·26 m (10¼ in)
Dimensions, internal:	
Cabin: Length	2·11 m (6 ft 11 in)
Max width	1·14 m (3 ft 9 in)
Max height	1·02 m (3 ft 4 in)
Areas:	
Wings, gross	12·02 m^2 (129·4 sq ft)
Ailerons (total)	0·87 m^2 (9·4 sq ft)
Trailing-edge flaps (total)	1·30 m^2 (13·95 sq ft)
Vertical tail surfaces (total)	2·11 m^2 (22·72 sq ft)
Horizontal tail surfaces (total)	2·55 m^2 (27·50 sq ft)
Weights and Loadings:	
Weight empty, equipped	649 kg (1,430 lb)
Basic operating weight empty	669 kg (1,475 lb)
Max T-O weight:	
normal and semi-aerobatic	1,066 kg (2,350 lb)
fully aerobatic	1,015 kg (2,238 lb)
Max wing loading	88·6 kg/m^2 (18·15 lb/sq ft)
Max power loading	7·15 kg/kW (11·75 lb/hp)
Performance (at max T-O weight):	
Never-exceed speed (structural)	210 knots (389 km/h; 241 mph)
Max level speed at S/L	130 knots (241 km/h; 150 mph)
Max cruising speed at 1,220 m (4,000 ft)	120 knots (222 km/h; 138 mph)
Econ cruising speed at 1,220 m (4,000 ft)	105 knots (194 km/h; 121 mph)
Stalling speed, flaps down, power off	53 knots (98 km/h; 61 mph) EAS
Max rate of climb at S/L	315 m (1,034 ft)/min
Service ceiling	4,875 m (16,000 ft)
Min ground turning radius	9·75 m (32 ft 0 in)
T-O run	274 m (900 ft)
T-O to 15 m (50 ft)	427 m (1,400 ft)
Landing from 15 m (50 ft)	363 m (1,190 ft)
Landing run	153 m (500 ft)
Range with max fuel, 55% power, no reserves	540 nm (1,000 km; 621 miles)
Endurance with max fuel, conditions as above	5 h 0 min
g limits: semi-aerobatic	+4·4; −1·8
fully aerobatic	+6; −3

BAe (BAC 167) STRIKEMASTER

The BAe Strikemaster was developed from the BAC 145 series (see 1972-73 *Jane's*). It has the same airframe, but is powered by a Rolls-Royce Viper Mk 535 turbojet engine (15·2 kN; 3,410 lb st) and has four underwing hardpoints, enabling it to carry up to 1,360 kg (3,000 lb) of stores.

The first Strikemaster flew for the first time on 26 October 1967, and a total of 146 (including five BAC 145s) were ordered. By the end of 1978 deliveries had been completed, but ten aircraft are being assembled for stock in anticipation of further orders.

Details of the Strikemaster can be found in the 1978-79 *Jane's*.

BAe HAWK

RAF designation: Hawk T. Mk 1

After examining designs submitted by BAC and Hawker Siddeley to meet an RAF requirement for a basic and advanced jet trainer, the Ministry of Defence announced in October 1971 that the Hawker Siddeley

The first BAe Hawk Mk 51 for the Finnish Air Force, delivered on 16 December 1980

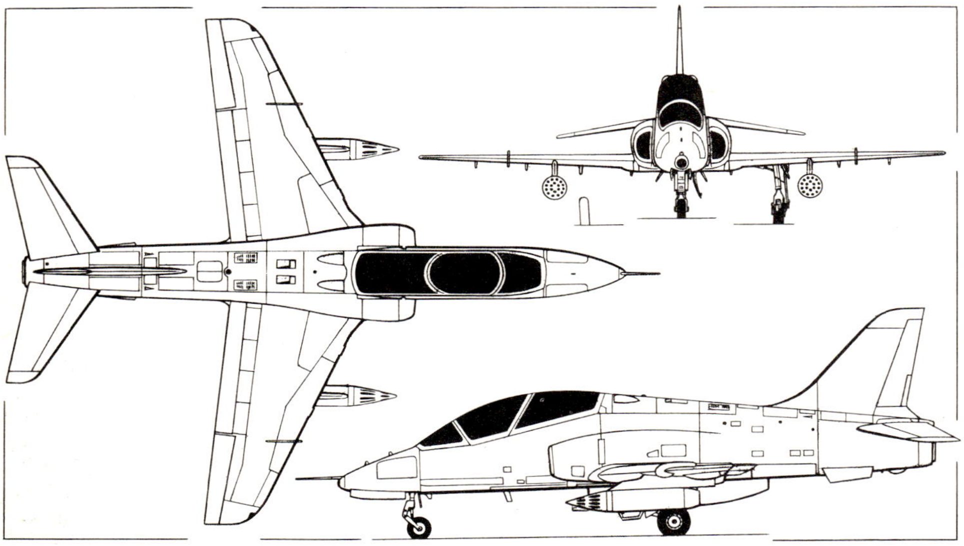

British Aerospace Hawk two-seat jet trainer/ground attack aircraft *(Pilot Press)*

1182 had been selected to meet this requirement. Selection of a non-afterburning version of the Rolls-Royce Turboméca Adour to power the aircraft was announced on 2 March 1972, and later in the same month the Ministry of Defence confirmed an initial order for 176 HS 1182s, which were given the RAF name of Hawk. These were to consist of one pre-production aircraft (XX154), which first flew on 21 August 1974, and 175 production Hawks.

There were no separate prototypes; instead, the first five production aircraft were allocated to the development programme. The first two production Hawks (XX162 and 163) were delivered to No. 4 Flying Training School at RAF Valley on 4 November 1976, and the 100th RAF Hawk was delivered on 27 March 1979. More than 150 had been delivered to the RAF by the beginning of 1981, now serving also with Nos. 1 and 2 Tactical Weapons Units at Brawdy, Wales, and Chivenor, Devon, and with the RAF's premier aerobatic team, the Red Arrows.

The Hawk is designed to be fully aerobatic (it is cleared for operation to $+8$ and $-4g$) and to have a fatigue life of 6,000 hours. It is replacing the Jet Provost, Gnat Trainer and Hunter in RAF service for advanced flying training, and for radio, navigation and weapons training. The basic design is capable of other operational roles, and studies of a number of variants have been made. One outcome of this is that 90 RAF Hawks will be modified to carry two AIM-9L Sidewinder air-to-air missiles (see accompanying illustration) to supplement home defence fighter squadrons in an emergency. (Of these, 72 will be allocated missiles under current plans.) Under a continuing development programme the more powerful Adour Mk 861 engine of 25·35 kN (5,700 lb st) has been installed in BAe's demonstrator aircraft, this providing some 8 per cent more thrust at take-off and, in conjunction with engine control refinements, more than 15 per cent increase at high speeds. Wing development is continuing, to improve both lift and stall characteristics.

The Finnish government has ordered 50 Hawk **Mk 51** trainers to replace its Fouga Magisters, and delivery of these began on 16 December 1980. Components for 46 of these, and final assembly, are being undertaken in Finland by Valmet (which see). British Aerospace announced on 5 April 1978 the receipt of a contract for the supply of eight **Mk 53** ground attack/trainer Hawks for the Indonesian Air Force; this order was increased to 12 in 1981. These, too, began to be delivered in 1980. Twelve Hawks have been ordered by Kenya, and eight by an undisclosed African country (reportedly Zimbabwe). British government approval has also been given for the sale of an undisclosed number (reportedly 24) to the United Arab Emirates Air Force (Abu Dhabi).

In January 1979, it was announced that the US Naval Air Development Center had awarded BAe a contract to study modification of the Hawk to meet the Navy's VTX/TS requirement for a replacement for the T-2C Buckeye and TA-4J Skyhawk. The modification, to make the Hawk capable of operation from aircraft carriers, would include a strengthened landing gear, including provision for nosewheel catapult launch; installation of an arrester hook; an avionics fit and cockpit display compatible with future US Navy operational aircraft; two fuselage side-mounted airbrakes instead of the current underfuselage airbrake; and use of carbonfibre composites for the nosecone, equipment bay access doors and nosewheel doors.

British Aerospace and McDonnell Douglas Corporation have finalised a partnership agreement under which BAe remains as prime contractor in the initial stages of VTX/TS bidding. Should the Hawk be selected for production for the US Navy, McDonnell Douglas would become prime contractor. The agreement also covers technical co-operation in making the Hawk suitable for carrier operations, and in the integration of studies and proposals on the training systems and support-in-service aspects of VTX/TS. An extensive US tour was flown by BAe's demonstrator Hawk (ZA 101) in the Summer of 1981, in connection with the VTX/TS programme.

The following description applies to the current standard production Hawk:

TYPE: Two-seat basic and advanced jet trainer, with capability for air defence and ground attack roles.

WINGS: Cantilever low-wing monoplane. Thickness/chord ratio 10·9% at root, 9% at tip. Dihedral 2°. Sweepback 26° on leading-edge, 21° 30′ at quarter-chord. One-piece wing, with six-bolt attachment to fuselage, employing a machined spars-and-skin torsion box, the greater part of which forms an integral fuel tank. Hydraulically-operated double-slotted flaps and ailerons, latter operated by Automotive tandem actuators.

FUSELAGE: Conventional all-metal structure of frames and stringers, cut out to accept the one-piece wing. Large airbrake under rear of fuselage, aft of wing.

TAIL UNIT: Cantilever all-metal structure, with sweepback on all surfaces. One-piece all-moving power-operated anhedral tailplane, with Automotive Products tandem hydraulic actuator. Manually-operated rudder, with electrically-actuated trim tab.

LANDING GEAR: Wide-track retractable tricycle type, with single wheel on each unit. Hydraulic actuation, using Automotive Products jacks. Main units retract inward into wing, ahead of front spar; nosewheel retracts forward. Main wheels and tyres size 6·50-10, pressure 9·86 bars (143 lb/sq in). Nosewheel and tyre size 4·4-16, pressure 8·27 bars (120 lb/sq in). Tail bumper fairing under rear fuselage. Anti-skid wheel brakes.Tail braking parachute optional (standard in Hawks for Indonesia).

POWER PLANT: One Rolls-Royce Turboméca Adour Mk 151 non-afterburning turbofan engine, rated at 23·13 kN (5,200 lb st) in RAF aircraft; initial export Hawks have an Adour Mk 851, rated at 23·75 kN (5,340 lb st). Adour Mk 861 (25·35 kN; 5,700 lb st) optional for export versions. Air intake on each side of fuselage, forward of wing leading-edge. Engine starting by integral gas turbine starter. Fuel in one fuselage bag tank of 868 litres (191 Imp gallons) capacity and integral wing tank of 836 litres (184 Imp gallons) capacity; total fuel capacity 1,704 litres (375 Imp gallons). Pressure refuelling point near front of port engine air intake trunk. Provision for carrying one 455 or 592 litre (100 or 130 Imp gallon) drop-tank on each inboard underwing pylon.

ACCOMMODATION: Crew of two in tandem under one-piece fully-transparent sideways-opening canopy. Fixed front windscreen and separate internal windscreen in front of rear cockpit. Rear seat elevated. Martin-Baker Mk 10B zero-zero rocket-assisted ejection seats, with MDC (miniature detonation cord) system to break canopy before seats eject. The MDC can also be operated from outside the cockpit in case of a ground emergency. Dual controls standard. Entire accommodation pressurised, heated and air-conditioned.

SYSTEMS: BAe Dynamics cockpit air-conditioning and pressurisation systems, using engine bleed air. Duplicated hydraulic systems, each 207 bars (3,000 lb/sq in), for actuation of control jacks, flaps, airbrake, landing gear and anti-skid wheel brakes. Compressed nitrogen accumulators provide emergency power for flaps and landing gear. Hydraulic accumulator for emergency operation of wheel brakes. No pneumatic system. DC electrical power from single brushless generator, with two static inverters to provide AC power and two batteries for standby power. Gaseous oxygen system for crew. Pop-up Dowty Rotol ram-air turbine in upper rear fuselage provides emergency power for flying controls in the event of an engine or No. 2 pump failure.

AVIONICS AND EQUIPMENT: The RAF standard of flight instrumentation includes Ferranti gyros and inverter, two Sperry Gyroscope RAI-4 4 in remote attitude indicators and a magnetic detector unit, and Louis Newmark compass system. Radio and navigation equipment includes Sylvania UHF and VHF, Cossor CAT.7000 Tacan, Cossor ILS with CILS.75/76 localiser/glideslope receiver and marker receiver, and IFF/SSR (Cossor 2720 Mk 10A IFF in aircraft for Finland).

ARMAMENT AND OPERATIONAL EQUIPMENT: Ferranti F.195 weapon sight and camera recorder in each cockpit. (Saab RGS2 sighting system in aircraft for Finland.) Trainer version has underfuselage centreline-mounted 30 mm Aden gun and ammunition pack, and two inboard underwing points each capable of carrying a nominal 454 kg (1,000 lb) stores load. Typical underwing armament training loads include two Matra 155 launchers, each with eighteen 2·75 in air-to-surface rockets, or two clusters of four practice bombs. Provision for two outboard underwing pylons, and a pylon in place of the ventral gun pack, also each capable of a 1,000 lb load (2,567 kg; 5,660 lb total external stores

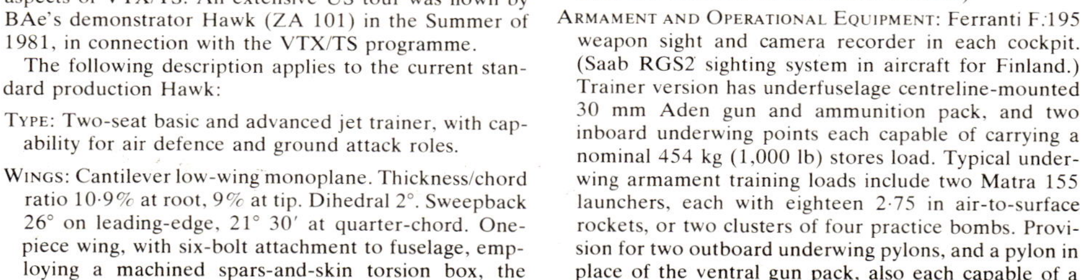

About 90 RAF Hawk T. Mk 1s are to be equipped to carry AIM-9L Sidewinder air-to-air missiles

load), for ground attack role. The Hawk has demonstrated its ability to carry a total external load of 3,084 kg (6,800 lb). In RAF training roles the normal max external load is about 680 kg (1,500 lb). AIM-9L Sidewinder air-to-air missiles will be carried by 72 RAF Hawks out of 90 which are to be modified for operational air defence role in an emergency.

DIMENSIONS, EXTERNAL:	
Wing span	9·39 m (30 ft 9¾ in)
Wing chord at root	2·65 m (8 ft 8¼ in)
Wing chord at tip	0·90 m (2 ft 11½ in)
Wing aspect ratio	5·284
Length overall, excl probe	11·17 m (36 ft 7¾ in)
Height overall	3·99 m (13 ft 1¼ in)
Tailplane span	4·39 m (14 ft 4¾ in)
Wheel track	3·34 m (10 ft 11½ in)
AREAS:	
Wings, gross	16·69 m² (179·6 sq ft)
Ailerons (total)	1·05 m² (11·30 sq ft)
Trailing-edge flaps (total)	2·50 m² (26·91 sq ft)
Airbrake	0·53 m² (5·70 sq ft)
Fin	2·51 m² (27·02 sq ft)
Rudder, incl tab	0·58 m² (6·24 sq ft)
Tailplane	4·33 m² (46·61 sq ft)
WEIGHTS:	
Weight empty	3,647 kg (8,040 lb)
T-O weight:	
trainer, 'clean'	5,035 kg (11,100 lb)
trainer, armed	5,572 kg (12,284 lb)
Max T-O weight	7,750 kg (17,085 lb)
Max landing weight	4,649 kg (10,250 lb)
PERFORMANCE:	
Max speed (in dive)	572 knots (1,060 km/h; 658 mph)
Max level speed	560 knots (1,038 km/h; 645 mph)
Max Mach number (in dive)	1·2
Max level speed Mach number	0·88
Max rate of climb at S/L	2,835 m (9,300 ft)/min
Time to 9,145 m (30,000 ft)	6 min 6 s
Service ceiling	15,240 m (50,000 ft)
T-O run	550 m (1,800 ft)
Landing run	488 m (1,600 ft)
Combat radius:	
with 2,540 kg (5,600 lb) weapon load	300 nm (556 km; 345 miles)
with 1,360 kg (3,000 lb) weapon load	560 nm (1,038 km; 645 miles)
Ferry range 'clean'	1,313 nm (2,433 km; 1,510 miles)
Ferry range with two 455 litre (100 Imp gallon) drop-tanks	1,669 nm (3,093 km; 1,922 miles)
Endurance	approx 4 h 0 min

BAe Harrier GR.Mk 3 of No. 1 Squadron, RAF, with special camouflage for Exercise Cold Winter, held in Norway in 1979 *(Rolls-Royce)*

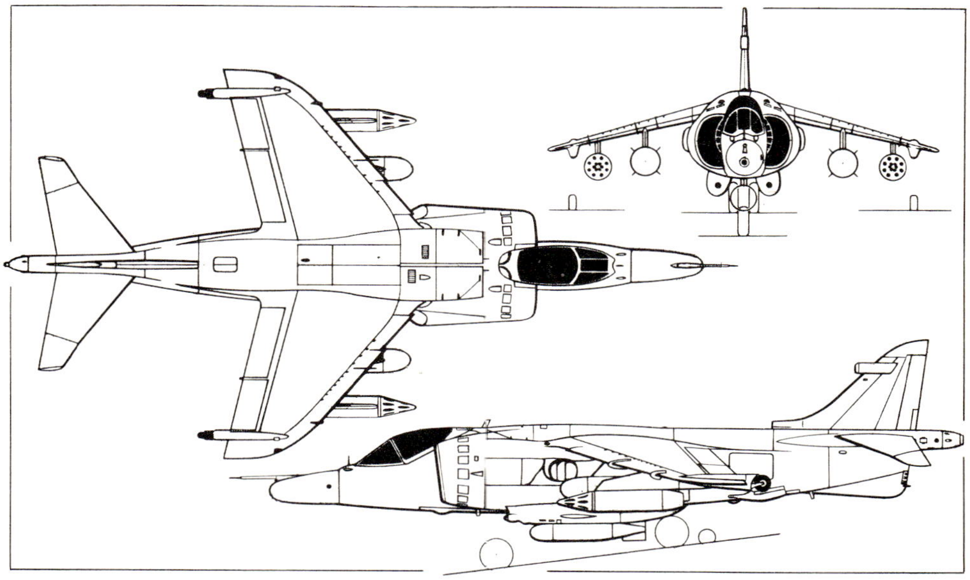

BAe Harrier GR.Mk 3 single-seat V/STOL close support and reconnaissance aircraft *(Pilot Press)*

BAe HARRIER

RAF designations: Harrier GR. Mk 3, and T. Mk 2A and 4

USMC designations: AV-8A (Mk 50) and TAV-8A (Mk 54)

Spanish Navy designation: Matador (AV-8S and TAV-8S)

The Harrier was the world's first operational fixed-wing V/STOL strike fighter. Developed from six years of operating experience with the P.1127/Kestrel series of aircraft (see 1968-69 *Jane's*), it is an integrated V/STOL weapon system, incorporating a Ferranti FE 541 inertial navigation and attack system and Smiths head-up display. The first of six single-seat prototypes (XV276) flew for the first time on 31 August 1966; the following versions have since been built:

Harrier GR. Mk 1, 1A and 3. Single-seat close-support and tactical reconnaissance versions for the RAF. First of initial series of 78 production GR. Mk 1s (XV738) flew on 28 December 1967. Entered service with the Harrier OCU, No. 233 Squadron, at RAF Wittering, on 1 April 1969. Delivered to No. 1 Squadron at Wittering and Nos. 3, 4 and 20 in West Germany. These aircraft were designated Harrier GR. Mk 1 when fitted initially with Pegasus 101 engines. When retrofitted subsequently with the Pegasus 102 they were redesignated GR.Mk 1A. Aircraft now in service have Pegasus 103 engines and are designated GR. Mk 3. A further 36 GR. Mk 3s, making a total for the RAF of 114 single-seat Harriers, were ordered subsequently. By 1 September 1980 a total of 110 (including two-seaters) had been delivered to the RAF.

A Harrier GR. Mk 1A, piloted by Sqn Ldr T. L. Lecky-Thompson, still holds two international time-to-height records after VTO, in Class H for jet-lift aircraft, set up on 5 January 1971. The aircraft reached 9,000 m (29,528 ft) in 1 min 44·7 s and 12,000 m (39,370 ft) in 2 min 22·7 s. The same RAF pilot also set up a Class H altitude record of 14,040 m (46,063 ft) in a Harrier GR. Mk 1A on 2 January 1971.

In early 1978 flight testing began of a Harrier GR. Mk 3 fitted with cushion augmentation devices (CADS) and wing leading-edge root extensions (LERX), both intended to improve lift capability. The CADS system consists of ventral strakes fitted to the underfuselage gun pods which, with the engine nozzles facing down for VTOL, trap the exhaust air from the engine as it rebounds from the ground; the air is prevented from re-entering the intakes by a laterally-positioned retractable surface fixed across the front of the two pods.

Harrier T. Mk 2, 2A, 4, 4A and 4RN. Two-seat versions, retaining the full combat capability of the single-seater in terms of equipment fit and weapon carriage. There is a large degree of commonality in structure and system components, ground support equipment and flight and ground crew training. Differences include a new, longer nose section forward of the wing leading-edge, with two cockpits in tandem; a tailcone approx 1·83 m (6 ft) longer than that of the single-seat model; and enlarged fin surfaces. The two-seat Harrier can be used operationally with the rear seat and compensating tail ballast removed, thus minimising the weight penalty over its single-seat counterpart. First development aircraft (XW174) flew on 24 April 1969, and the first of 21 production aircraft for the RAF (XW264) on 3 October 1969. The two-seater entered RAF service in July 1970. Four are on order for the Royal Navy.

The RAF Harrier T. Mk 2, like the GR. Mk 1, was powered originally by the Pegasus 101 engine. The designations T. Mk 2A and T. Mk 4 apply to aircraft retrofitted with, respectively, the Pegasus 102 and 103. Some have laser nose. Newly-built RAF two-seaters are designated T.Mk 4A; the Royal Navy version is T.Mk 4RN.

Harrier GR. Mk 5. Designation of 60 AV-8B aircraft to be ordered for the RAF. This version originated in an Advanced Harrier study completed in December 1973 by Hawker Siddeley, Rolls-Royce, McDonnell Douglas and Pratt & Whitney. This was funded jointly by the UK and US governments on behalf of the RAF, Royal Navy, US Marine Corps and US Navy, but in March 1975 the UK Secretary of State for Defence, Mr Roy Mason, stated that "there is not enough common ground on the Advanced Harrier for us to join in the programme with the US". The **AV-8B** advanced version of the Harrier has, therefore, been developed as a US programme since that time and is described under the McDonnell Douglas heading in this edition.

Adoption of the AV-8B for the RAF in 1981 has caused BAe's recent studies for a Big Wing Harrier (see 1980-81 *Jane's*) to be suspended.

Harrier Mk 50 (USMC designation AV-8A). Single-seat close-support and tactical reconnaissance version for the US Marine Corps, delivery of which began on 26 January 1971. Dimensionally as GR. Mk 3, but without laser ranger and marked target seeker, and with modifications to customer's specification, including provision for the carriage of Sidewinder missiles. Total of 102 ordered for US Marine Corps, plus eight Harrier **Mk 54s** (a two-seat version designated **TAV-8A**); all of those still in service now have Pegasus 103 engines.

The AV-8As equip three US Marine Corps combat squadrons: VMA 513, VMA 542 and VMA 231; and training squadron VMA(T) 203, at Cherry Point, North Carolina.

In the period from 1979 until FY 1984, the US Marine Corps is upgrading 61 AV-8As to **AV-8C** standard. This involves a service life extension programme (SLEP) of structural improvements aimed at extending the airframe fatigue life to up to 4,000 h; and a CILOP (conversion in lieu of procurement) programme under which the AV-8As are fitted with forward-looking passive radar warning equipment at the wingtips, tail warning radar in the tail 'bullet' fairing, improved UHF com radio, a flare/chaff dispenser in the rear-fuselage equipment bay, the LIDS (lift improvement devices: underfuselage strakes and forward flap) developed for the AV-8B, an onboard oxygen generating system, and KY 58 secure voice system. Conversion of the first few AV-8As to AV-8C standard was undertaken by McDonnell Douglas, from kits supplied by BAe. The remaining conversions will be carried out by the US Marine Corps at NAS Cherry Point, North Carolina.

Harrier Mk 52. One aircraft built as a demonstrator using BAe and equipment suppliers' private funding. It is similar to the Harrier T. Mk 4, and is fitted with a Pegasus 103 engine; in recognition of its status as the first civil-registered jet V/STOL aircraft in the UK, it was granted the civil registration G-VTOL. First flight was made on 16 September 1971, with a Pegasus 102 fitted initially.

Harrier Mk 55. Eleven AV-8As and two TAV-8As ordered, through USA, for the Spanish Navy, by whom they are known as **Matadors** and designated **AV-8S** and **TAV-8S** respectively. Ten delivered by September 1980. These equip the 8a Escuadrilla of the Spanish Navy at Rota, Cadiz. Like the AV-8As of the USMC, the Matador also has a Sidewinder capability.

Harrier T. Mk 60. Two-seat operational trainer version for Indian Navy; two ordered by mid-1980.

Sea Harrier. Version for Royal Navy and Indian Navy. Described separately.

The following details apply generally to the Harrier GR. Mk 3 and T. Mk 4, except where a specific version is indicated:

TYPE: V/STOL close support and reconnaissance aircraft.

WINGS: Cantilever shoulder-wing monoplane. Wing section of BAe (HS) design. Thickness/chord ratio 10% at root, 5% at tip. Anhedral 12°. Incidence 1° 45′. Sweepback at quarter-chord 34°. One-piece aluminium alloy three-spar safe-life structure with integrally-machined skins, manufactured by Brough factory of BAe, with

BAe Harrier T.Mk 4 two-seat combat trainer of the RAF's No. 4 Squadron, based at Gutersloh in Germany

six-point attachment to fuselage. Plain ailerons and flaps, of bonded aluminium alloy honeycomb construction. Ailerons irreversibly operated by Fairey tandem hydraulic jacks. Jet reaction control valve built into front of each outrigger wheel fairing. Entire wing unit removable to provide access to engine. For ferry missions, the normal 'combat' wingtips can be replaced by bolt-on extended tips to increase ferry range.

FUSELAGE: Conventional semi-monocoque safe-life structure of frames and stringers, mainly of aluminium alloy, but with titanium skins at rear and some titanium adjacent to engine and in other special areas. Access to power plant through top of fuselage, ahead of wing. Jet reaction control valves in nose and tailcone. Large forward-hinged airbrake under fuselage, aft of main-wheel well.

TAIL UNIT: One-piece variable-incidence tailplane, with 15° of anhedral, irreversibly operated by Fairey tandem hydraulic jack. Rudder and trailing-edge of tailplane are of bonded aluminium honeycomb construction. Rudder is operated manually. Trim tab in rudder. Ventral fin under rear fuselage. Fin tip carries suppressed VHF aerial.

LANDING GEAR: Retractable bicycle type of Dowty Rotol manufacture, permitting operation from rough unprepared surfaces of CBR as low as 3% to 5%. Hydraulic actuation, with nitrogen bottle for emergency extension of landing gear. Single steerable nosewheel retracts forward, twin coupled mainwheels rearward, into fuselage. Small outrigger units retract rearward into fairings slightly inboard of wingtips. Nosewheel leg is of levered-suspension Liquid Spring type. Dowty Rotol telescopic oleo-pneumatic main and outrigger gear. Dunlop wheels and tyres, size 26·00 × 8·75-11 (nose unit), 27·00 × 7·74-13 (main units) and 13·50 × 6·4 (outriggers). GR. Mk 3 tyre pressures 6·21 bars (90 lb/sq in) on nose and main units, 6·55 bars (95 lb/sq in) on outriggers. T. Mk 4 tyre pressures 6·90 bars (100 lb/sq in) on nose unit, 6·55 bars (95 lb/sq in) on main and outrigger units. Dunlop multi-disc brakes and Dunlop-Hytrol adaptive anti-skid system.

POWER PLANT: One Rolls-Royce Pegasus Mk 103 vectored-thrust turbofan engine (95·6 kN; 21,500 lb st), with four exhaust nozzles of the two-vane cascade type, rotatable through 98° from fully-aft position. Engine bleed air from HP compressor used for jet reaction control system and to power duplicated air motor for nozzle actuation. The low-drag intake cowls each have eight automatic suction relief doors aft of the leading-edge to improve intake efficiency by providing extra air for the engine at low forward or zero speeds. Fuel in five integral tanks in fuselage and two in wings, with total capacity of approx 2,865 litres (630 Imp gallons). This can be supplemented by two 455 litre (100 Imp gallon) jettisonable combat tanks or two 1,500 litre (330 Imp gallon) ferry tanks on the inboard wing pylons. Ground refuelling point in port rear nozzle fairing. Provision for in-flight refuelling probe above the port intake cowl.

ACCOMMODATION: Crew of one (Mk 3) or two (Mk 4) on Martin-Baker Mk 9D zero-zero rocket ejection seats which operate through the miniature detonating cord equipped canopy of the pressurised, heated and air-conditioned cockpit. AV-8As of the US Marine Corps retrofitted with Stencel SIIIS-3 ejection seats. Manually-operated canopy, rearward-sliding on single-seat, sideways-opening (to starboard) on two-seat versions. Birdproof windscreen, with hydraulically-actuated wiper. Windscreen de-icing.

SYSTEMS: Three-axis limited-authority autostabiliser for V/STOL flight. Pressurisation system of BAe design, with Normalair-Garrett and Marston major components; max pressure differential 0·24 bars (3·5 lb/sq in). Duplicated hydraulic systems, each of 207 bars (3,000 lb/sq in), actuate Fairey flying control and general services and include a retractable ram-air turbine inside top of rear fuselage, driving a small hydraulic pump for emergency power. AC electrical system with transformer-rectifiers to provide required DC supply. One 12kVA Lucas alternator. Two 28V 25Ah batteries, one of which energises a 24V motor to start Lucas gas-turbine starter/APU. This unit drives a 6kVA auxiliary alternator for ground readiness servicing and standby. Normalair-Garrett liquid oxygen system of 5 litres (1 Imp gallon) capacity. Bootstrap-type cooling unit for equipment bay, with intake at base of dorsal fin.

AVIONICS AND EQUIPMENT: Plessey U/VHF, Ultra standby UHF, Hoffman Tacan and Cossor IFF, Ferranti FE 541 inertial navigation and attack system (INAS), with Sperry C2G compass, Smiths electronic head-up display of flight information and Smiths air data computer. INAS can be aligned equally well at sea or on land. The weapon aiming computer provides a general solution for manual or automatic release of free-fall and retarded bombs, and for the aiming of rockets and guns, in dive and straight-pass attacks over a wide range of flight conditions and very considerable freedom of manoeuvre in elevation. Communication equipment ranges through VHF in the 100-156MHz band to UHF in the 220-400MHz band. Ferranti Type 106 laser ranger and marked target seeker (LRMTS) retrofitted to all RAF single-seat and some two-seat Harriers.

ARMAMENT AND OPERATIONAL EQUIPMENT: Optically-flat panel in nose, on port side, for F.95 oblique camera, which is carried as standard. A cockpit voice recorder with in-flight playback facility supplements the reconnaissance cameras, and facilitates rapid debriefing and mission evaluation. No built-in armament. Combat load is carried on four underwing and one underfuselage pylons, all with ML ejector release units. The inboard wing points and the fuselage point are stressed for loads of up to 910 kg (2,000 lb) each, and the outboard underwing pair for loads of up to 295 kg (650 lb) each; the two strake fairings under the fuselage can each be replaced by a 30 mm Aden gun pod and ammunition. The Harrier is cleared for operations with a maximum external load exceeding 2,270 kg (5,000 lb), and has flown with a weapon load of 3,630 kg (8,000 lb). It is able to carry 30 mm guns, bombs, rockets and flares of UK and US designs, and in addition to its fixed reconnaissance camera can also carry a five-camera reconnaissance pod on the underfuselage pylon. A typical combat load comprises a pair of 30 mm Aden gun pods, a 1,000 lb bomb on the underfuselage pylon, a 1,000 lb bomb on each of the inboard underwing pylons, and a Matra 155 launcher with 19 × 68 mm SNEB rockets on each outboard underwing pylon. A Sidewinder installation is provided in the AV-8A and Matador versions, to give the aircraft an effective air-to-air capability in conjunction with the two 30 mm Aden guns.

DIMENSIONS, EXTERNAL:

Wing span: combat	7·70 m (25 ft 3 in)
ferry	9·04 m (29 ft 8 in)
Wing chord at root	3·56 m (11 ft 8 in)
Wing chord at tip	1·26 m (4 ft 1½ in)
Wing aspect ratio: combat	3·175
ferry	4·08
Length overall: single-seat	13·89 m (45 ft 7 in)
single-seat (laser nose)	14·27 m (46 ft 10 in)
two-seat	17·00 m (55 ft 9½ in)
Height overall: single-seat	3·45 m (11 ft 4 in)
two-seat	4·17 m (13 ft 8 in)
Tailplane span	4·24 m (13 ft 11 in)
Outrigger wheel track	6·76 m (22 ft 2 in)
Wheelbase, nosewheel to main wheels	approx 3·45 m (11 ft 4 in)

AREAS:

Wings, gross: combat	18·68 m² (201·1 sq ft)
ferry	20·1 m² (216 sq ft)
Ailerons (total)	0·98 m² (10·5 sq ft)
Trailing-edge flaps (total)	1·29 m² (13·9 sq ft)
Fin (excl ventral fin): single-seat	2·40 m² (25·8 sq ft)
two-seat	3·57 m² (38·4 sq ft)
Rudder, incl tab	0·49 m² (5·3 sq ft)
Tailplane	4·41 m² (47·5 sq ft)

WEIGHTS AND LOADING:

Basic operating weight, empty, with crew:	
GR.Mk 3 and AV-8A	5,580 kg (12,300 lb)
T.Mk 4 (solo for combat)	5,896 kg (13,000 lb)
T.Mk 4 (dual)	6,237 kg (13,750 lb)
Internal fuel	2,295 kg (5,060 lb)
Max T-O weight (single-seat)	over 11,340 kg (25,000 lb)
Max wing loading (single-seat)	610 kg/m² (125 lb/sq ft)

PERFORMANCE:

Max speed at low altitude	over 640 knots (1,186 km/h; 737 mph) EAS
Max Mach number (in a dive)	1·3
Time to 12,200 m (40,000 ft) from vertical T-O	2 min 22·7 s
Ceiling	more than 15,240 m (50,000 ft)
Range with one in-flight refuelling	more than 3,000 nm (5,560 km; 3,455 miles)
Endurance with one in-flight refuelling	more than 7 h

BAe SEA HARRIER

RN designation: FRS. Mk 1
Indian Navy designation: FRS. Mk 51

On 15 May 1975, the British government announced its decision to proceed with full development of a maritime version of the Harrier, subsequently designated Sea Harrier **FRS. Mk 1**. The initial Royal Navy order is for 34 aircraft, of which 16 had been delivered by mid-1981. The Naval Intensive Flying Trials Unit for the Sea Harrier (No. 700A Squadron) was commissioned at RNAS Yeovilton on 19 September 1979. It has since had a change of title, becoming the shore-based No. 899 HQ squadron, with eight aircraft. Front-line units, each with five aircraft, will be Nos. 800, 801 and 802 Squadrons, of which the first two had formed by mid-1981; these squadrons will operate from the anti-submarine cruisers HMS *Invincible, Illustrious* and *Ark Royal*, and the assault carrier HMS *Hermes*. Six similar Sea Harriers, designated **FRS. Mk 51**, have been ordered by the Indian Navy. Four standard, non-navalised T. Mk 4RN two-seaters are being procured by the Royal Navy for land-based training, and two T. Mk 60s by the Indian Navy. Deliveries to India are due to begin in December 1982.

The first Sea Harrier to fly (XZ450) made its first flight on 20 August 1978, and the first for the Royal Navy (XZ451) was handed over on 18 June 1979. The first Sea Harrier ship trials were carried out on board HMS *Hermes* during November 1979.

Following proposals by Lt Cdr D. R. Taylor, RN, tests had been carried out successfully in 1977 with a 'ski-jump' launching ramp designed to boost the short take-off performance of vectored-thrust aircraft. This technique makes possible substantial benefits in Harrier operation both at sea and ashore, and will be a feature of all Royal Navy ships in which Sea Harriers are based. A ski-jump with a 20° angle enables the Harrier to take off at a 30 knot (55·5 km/h; 34·5 mph) lower speed than from a flat deck and at 60 knots (111 km/h; 69 mph) less than from a runway, so conserving fuel for the flight itself. It also permits well over 1,100 kg (2,500 lb) of extra fuel or weapons to be carried than would be possible in the same distance with a flat-deck take-off.

A 7° ski-jump T-O ramp is fitted to HMS *Invincible* and HMS *Illustrious*; that which was installed in HMS *Hermes* during 1980 is more steeply angled, at 12°. A similar 12° ramp is installed in HMS *Ark Royal*.

Major changes compared with the Harriers in service with the RAF, Spanish Navy and US Marine Corps comprise the elimination of magnesium components, introduction of a raised cockpit, revised operational avionics, and installation of multi-mode Ferranti radar in a redesigned nose that folds to port for carrier stowage. Known by the name Blue Fox, this radar has been under development since March 1973, when the Electronic Systems Department of Ferranti was awarded a study and preliminary development contract. It is a derivative of the frequency-agile Sea Spray radar fitted in the Lynx helicopter, but

BAe Sea Harrier FRS. Mk 1 armed with Sidewinder air-to-air missiles

embodies changes to suit its different role, with air-to-air intercept and air-to-surface modes of operation.

Three specially modified Hawker Hunter T. Mk 8s, redesignated T. Mk 8M, were fitted with nose-mounted Blue Fox radars. Intended initially to speed the development of this radar and a new nav/attack system, they will continue in use for radar training. Equipment of the Sea Harrier includes ECM in a fairing near the tip of the tail-fin, and underwing attachments for air-to-air missiles of the Sidewinder type.

The Royal Navy's Sea Harrier FRS. Mk 1 has a Rolls-Royce Pegasus 104 vectored-thrust turbofan engine, with the same rating as the Pegasus 103 fitted to current RAF Harriers. The two variants differ little in design, except that the Pegasus 104 incorporates additional anti-corrosion features and has the capability to generate more electrical power.

Estimated weights, loadings and detailed performance figures are not yet available for the Sea Harrier. It is expected that the FRS. Mk 1 will operate at approximately the same weights as the GR. Mk 3, and will be capable of lifting a full military load with a 152 m (500 ft) flat deck run into an overdeck wind of 30 knots (55·5 km/h; 34·5 mph).

The description of the GR. Mk 3 applies also to the FRS. Mk 1, except as follows:

TYPE: V/STOL fighter, reconnaissance and strike aircraft.

POWER PLANT: As GR. Mk 3, except one Rolls-Royce Pegasus Mk 104 vectored-thrust turbofan engine of 95·6 kN (21,500 lb st). Internal fuel capacity and external combat fuel capacity as for GR. Mk 3.

ACCOMMODATION: As GR. Mk 3, but with pilot raised 28 cm (11 in), on Martin-Baker Mk 10 zero-zero ejection seat.

SYSTEMS: As GR. Mk 3, except autopilot function on Fairey Hydraulics, giving throughput to aileron and tailplane power controls as well as to three-axis autostabs. Pressurisation system of BAe design with major components from Normalair-Garrett and Delaney Gallay. British Oxygen liquid oxygen system of 5 litres (1 Imp gallon) capacity in Royal Navy aircraft; those for Indian Navy have gaseous oxygen system. Lucas Mk 2 GTS/APU.

AVIONICS AND EQUIPMENT: Nose-mounted Ferranti Blue Fox multi-mode radar, with TV-raster daylight-viewing tube which conveys flight information, as well as radar data, to pilot. New and larger Smiths electronic head-up display and 20,000-word digital weapon aiming computer. Decca Doppler 72. Ferranti self-aligning attitude and heading reference platform and digital navigation computer. Radio navaids include UHF homing, Tacan with offset facility and I-band transponder. Radio com by multi-channel Plessey PTR 377 U/VHF, with VHF standby via D 403M transceiver. Passive electronic surveillance and warning of external radar illumination by receiver with forward and rear hemisphere antennae in fin and tailcone respectively.

ARMAMENT AND OPERATIONAL EQUIPMENT: As GR. Mk 3, except for addition of Sidewinder installation similar to that of AV-8A (Matra Magic instead of Sidewinder on Indian Navy aircraft), and provision for two air-to-surface missiles of Sea Eagle or Harpoon type.

DIMENSIONS, EXTERNAL: As GR. Mk 3 except:

Wing span	7·70 m (25 ft 3 in)
Length overall	14·50 m (47 ft 7 in)
Length overall, nose folded	12·73 m (41 ft 9 in)
Height overall	3·71 m (12 ft 2 in)

WEIGHT:

Max weapon load	3,630 kg (8,000 lb)

BAe HS Nimrod R. Mk 1 of No. 51 Squadron, RAF. Note the modified tailcone in place of the MAD boom and the revised contours of the port wing leading-edge pod *(Swedish Air Force)*

PERFORMANCE:

Max Mach No.	Mach 1·25
Max level airspeed	above 640 knots (1,185 km/h; 736 mph)
Typical cruising speed:	
High altitude, for well over 1 h on internal fuel	above Mach 0·8
Low altitude	350-450 knots (650-833 km/h; 404-518 mph), with rapid acceleration to 600 knots (1,110 km/h; 690 mph)
Time from alarm to 30 nm (55 km; 35 miles) combat area	under 6 min
High altitude intercept radius, with 3 min combat and reserves for VL	400 nm (750 km; 460 miles)
Strike radius	250 nm (463 km; 288 miles)

BAe HS NIMROD

The Nimrod was developed to replace the Shackleton maritime reconnaissance aircraft of RAF Strike Command, with which it is scheduled to serve until well into the 1990s. Design of the Nimrod, as the Hawker Siddeley 801, began in June 1964, and government authority to proceed was announced in June 1965.

Based substantially upon the airframe of the Hawker Siddeley (de Havilland) Comet 4C, the Nimrod is a new-production aircraft with a 1·98 m (6 ft 6 in) shorter, modified pressurised fuselage; an unpressurised, underslung pannier for operational equipment and weapons; and Rolls-Royce Spey turbofan engines (instead of the Avon turbojets of the Comet), with wider air intakes to allow for the greater mass flow. Other external changes include enlarged flight deck main windows and 'eyebrow' windows; ESM and MAD equipment, in glassfibre fairings on top of the fin and in the tailboom respectively; and a searchlight in the starboard wing external fuel tank. The search radar is housed in a glassfibre fairing which forms the nose of the unpressurised lower fuselage.

The Nimrod was designed to combine the advantages of high-altitude, fast transit speed with low wing loading and good low-speed manoeuvring capabilities when operating in its primary roles of anti-submarine warfare, surveillance and anti-shipping strike. When required, two of the four Spey engines can be shut down to extend endurance, and the aircraft can cruise and climb on only one engine. A wide range of weapons can be carried in the 14·78 m (48 ft 6 in) long bomb bay, and large numbers of sonobuoys and markers can be carried and released from the pressurised rear fuselage area.

In addition to its surveillance and ASW roles, the Nimrod can be used for day and night photography. As supplied originally to the RAF, these aircraft had a stand-off surface missile capability. This has since been deleted but could be reactivated if required. The Nimrod MR. Mk 1 can carry 16 additional personnel in the self-support role; the MR. Mk 2 can carry only 10 without the removal of equipment.

Two prototypes were built, both utilising existing Comet 4C airframes. The first of these (XV148), fitted with Spey engines, flew for the first time on 23 May 1967 and was used for aerodynamic testing. The second (XV147) retained its original Avon engines, was first flown on 31 July 1967, and was used for development of the nav/tac system and special maritime equipment.

The following versions have been produced:

Nimrod MR. Mk 1. Initial production version. First flown on 28 June 1968. In service with No. 42 Squadron at St Mawgan, Cornwall; and Nos. 120, 201 and 206 Squadrons at Kinloss, Scotland. Forty-six ordered, of which 43 delivered to Mk 1 standard; three others diverted as development aircraft for Nimrod AEW. Mk 3.

Nimrod R. Mk 1. Three aircraft (additional to the 46 MR. Mk 1s ordered for RAF Strike Command) delivered to No. 51 Squadron at Wyton, Huntingdonshire. These aircraft (XW664-666), are employed for electronic intelligence (elint) missions, and can be identified by the absence of an MAD tailboom.

Nimrod MR. Mk 2. Thirty-two aircraft of the RAF's Nimrod MR. Mk 1 fleet are being refitted with new communications equipment, and advanced tactical sensor, ESM and navigation systems, under a programme which began in 1975. Redelivery started on 23 August 1979 with XV236, the first completely refitted aircraft. After refit these aircraft are redesignated MR. Mk 2, and are repainted in a NATO-approved camouflage scheme. Equipment includes an advanced search radar, offering greater range and sensitivity coupled with a higher data processing rate; a new acoustic processing system, developed by Marconi Avionics Systems, which is compatible with a wide range of existing and projected sonobuoys, and an early warning support measures (EWSM) pod at each wingtip. Armament includes Stingray homing torpedoes.

Nimrod AEW. Mk 3. Airborne early warning version; described separately.

Ample space and power is available in the basic Nimrod design to accept additional or alternative sensors such as sideways-looking radar, forward-looking infra-red, infra-red linescan, low light level TV, digital processing of intercepted ESM signals, and other new developments.

The following description applies to the Nimrod MR. Mks 1 and 2:

TYPE: Four-turbofan maritime patrol aircraft.

WINGS: Cantilever low/mid-wing monoplane, of metal construction. Sweepback 20° at quarter-chord. All-metal two-spar structure, comprising a centre-section, two stub-wings and two outer panels. Extensive use of Redux metal-to-metal bonding. All-metal ailerons, operated through duplicated hydraulic and mechanical units. Trim tab in each aileron. Plain flaps outboard of engines, operated hydraulically. Hot-air anti-icing system.

FUSELAGE: All-metal semi-monocoque structure. The circular-section cabin space is fully pressurised. Below this is an unpressurised pannier housing the bomb bay, radome and additional space for operational equipment. Segments of this pannier are free to move relative to each other, so that structural loads in the weapons bay are not transmitted to the pressure-cell. A glassfibre nose radome and tailboom are provided.

TAIL UNIT: Cantilever all-metal structure. Rudder and elevators operated through duplicated hydraulic and mechanical units. A glassfibre pod on top of the fin houses ESM equipment. Trim tab in each elevator. Hot-air anti-icing system.

LANDING GEAR: Retractable tricycle type. Four-wheel tandem-bogie main units, with size 36 × 10-18 Dunlop tyres, pressure 12·76 bars (185 lb/sq in). Twin-wheel nose unit, with size 30 × 9-15 Dunlop tyres, pressure 6·21 bars (90 lb/sq in).

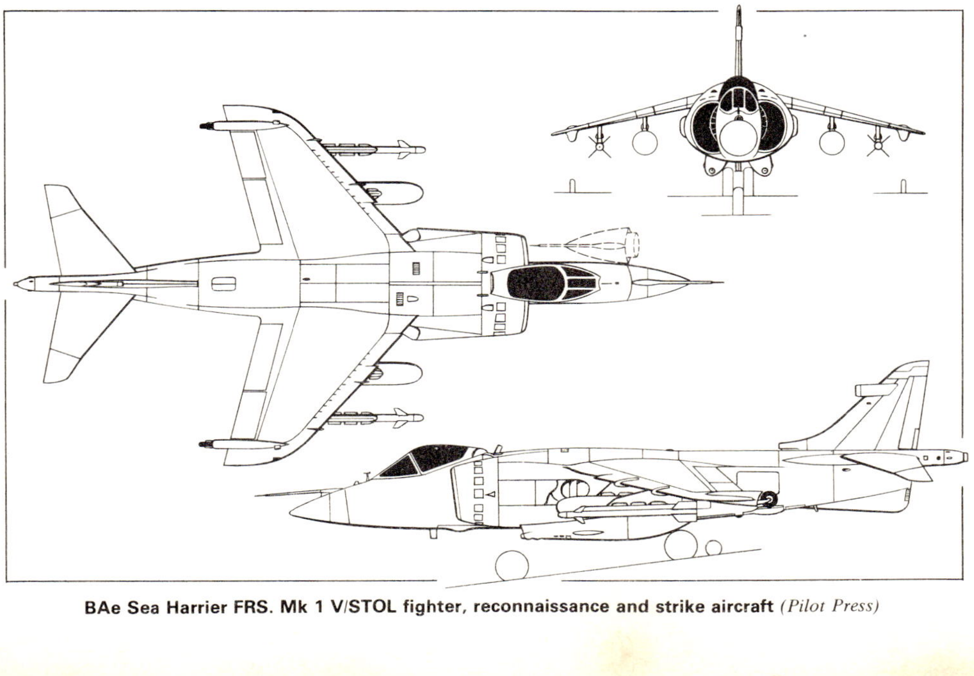

BAe Sea Harrier FRS. Mk 1 V/STOL fighter, reconnaissance and strike aircraft *(Pilot Press)*

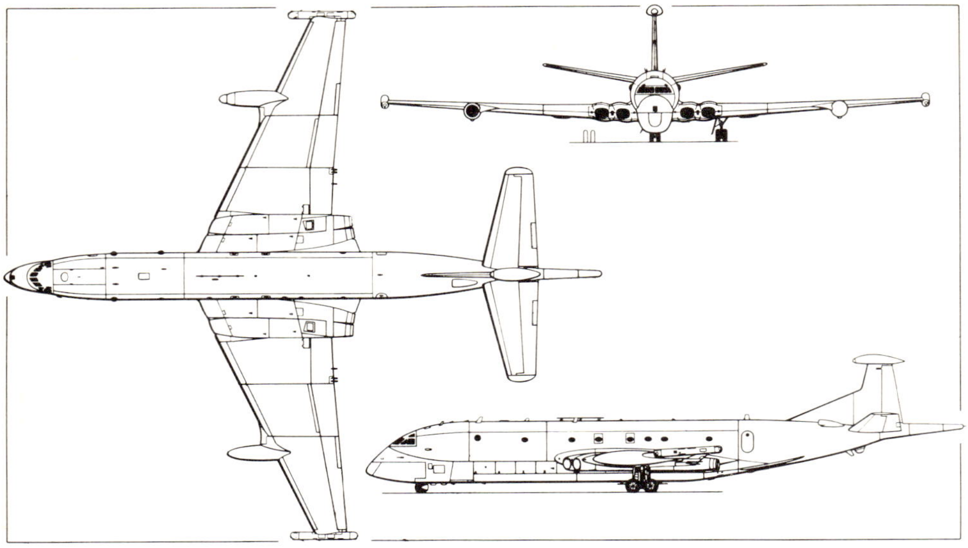

BAe HS Nimrod MR. Mk 2 four-turbofan maritime patrol aircraft *(Pilot Press)*

Power Plant: Four Rolls-Royce RB. 168-20 Spey Mk 250 turbofan engines, each rated at 54 kN (12,140 lb st). Reverse thrust fitted on two outer engines. Fuel in fuselage keel tanks, integral wing tanks, and permanent external tank on each wing leading-edge, with total capacity of 48,780 litres (10,730 Imp gallons). Provision for up to six removable tanks in weapons bay.

Accommodation: Normal crew of 12, comprising pilot, co-pilot, and flight engineer on flight deck; routine navigator, tactical navigator, radio operator, radar operator, two sonics systems operators, ESM/MAD operator, and two observers/stores loaders in main (pressurised) cabin, which is fitted out as a tactical compartment. In this compartment, from front to rear, are a toilet on the port side; stations for the two navigators (stbd), radio and radar operators (port), and sonics systems operators (stbd) in the forward section; ESM/MAD operator's station, galley, four-seat dining area, rest quarters and sonobuoy stowage in the middle section; and buoy and marker launch area in the rear section. Three hemispherical observation windows forward of wings (one port, two stbd), giving 180° field of view. Two normal doors, emergency door, and four overwing emergency exits. Weapons bay can be utilised for additional fuel tanks (see under 'Power Plant') or for the carriage of cargo. Provision is made for a trooping role, in which configuration 45 passengers can be accommodated if some rear-fuselage equipment is removed.

Systems: Air-conditioning by engine bleed air; Smith-Kollsman pressurisation system, with additional Normalair-Garrett conditioning pack on Mk 2 aircraft, max differential 0·603 bars (8·75 lb/sq in). Anti-icing and bomb-bay heating by engine bleed air. Lockheed hydraulic system, pressure 172 bars (2,500 lb/sq in), for duplicated flying control power units, landing gear shock-absorbers, steering and door jacks, weapons bay door jacks, camera aperture door jacks, and self-sealing couplings for water charging, ground test, engine bay and ancillary services. Lucas APU provides high-pressure air for engine starting. Electrical system utilises four 60kVA engine-driven alternators, with English Electric constant-speed drives, to provide 200V 400Hz three-phase AC supply. Secondary AC comes from two 115V three-phase static transformers, with duplicate 115/26V two-phase static transformers which also feed a 1kVA frequency changer providing a 115V 1,600Hz single-phase supply for radar equipment. Emergency supplies for flight instruments are provided by a 115V single-phase static inverter. DC supply is by four 28V transformer-rectifier units backed up by two nickel-cadmium batteries.

Avionics and Equipment (MR. Mk 1): Routine navigation by Decca Doppler Type 67M/Marconi E3 heading reference system, with reversionary heading from a Sperry GM7 duplicated gyro compass system, operating in conjunction with a Ferranti routine dynamic display. Tactical navigation, and stores selection and release, by Marconi nav/attack system utilising an 8K Marconi 920B digital computer. Tactical display station provides continually-updated information about aircraft position, with present and past track, sonobuoy positions, range circles from sonobuoys, ESM bearings, MAD marks, radar contacts and visual bearings. Course information can be displayed automatically to the pilots on the flight director system; alternatively, the computer can be coupled to the autopilot to allow the tactical navigator to direct the aircraft to a predicted target interception, weapon release point, or any other point on the tactical display. ASW equipment includes Sonics 1C sonar and a new long-range sonar system; EMI ASV-21D air-to-surface-vessel detection radar in nose; Thomson-CSF ESM (electronic support measures) equipment in pod on top of fin; and Emerson Electronics ASQ-10A MAD (magnetic anomaly detector) in extended tailboom. Strong Electric 70 million candlepower searchlight at front of starboard external wing fuel tank. Aeronautical and General Instruments F.126 and F.135 cameras for day and night photography respectively, the latter having Chicago Aero Industries electronic flash equipment. Smiths SFS.6 automatic flight control system, embodying SEP.6 three-axis autopilot, integrated with the navigation and tactical system. Twin Plessey PTR 175 UHF/VHF, and Marconi AD 470 HF, communications transceivers; twin Marconi AD 260 VOR/ILS; Hoffman ARN-72 Tacan; Decca Loran C/A; Marconi AD 360 ADF; Honeywell AN/APN-171(V) radar altimeter. Yaw damper and Mach trim standard.

Avionics and Equipment (MR. Mk 2): New and more flexible operational system, using three separate processors for tactical navigation, radar and acoustics. Marconi central tactical system, based on a 920 ATC computer with a greater storage capacity than that of MR. Mk 1, to provide improved computing and display facilities and, in conjunction with a Ferranti inertial navigation system, improved navigation capabilities. EMI Searchwater long-range air-to-surface-vessel radar, with its own data processing subsystem incorporating a Ferranti FM 1600D digital computer. This system presents a clutter-free picture, can detect and classify surface vessels, submarine snorts and periscopes at extreme ranges, can track several targets simultaneously, and is designed to operate in spite of countermeasures. AQS 901 acoustics processing and display system, based on twin Marconi 920 ATC computers, is compatible with a wide range of passive and active sonobuoys, either in existence or under development, including the Australian BARRA passive directional sonobuoy, the Canadian TANDEM, the US SSQ-41 and SSQ-53, and the new Ultra A-size X17255 command active multi-beam sonobuoys (CAMBS), with a performance similar to that of helicopter-mounted dipping sonars. Communications are being improved by the installation of twin Marconi AD 470 HF transceivers (instead of the original single AD 470), and a radio teletype and encryption system. EWSM equipment in two wingtip pods. Also being added, from 1981, is a new and possibly unique onboard crew training system developed by the Maritime Aircraft Systems Division of Marconi Avionics Ltd. Known as ACT-1 (Airborne Crew Trainer Mk 1), it consists of a single exercise control unit comprising a control and display panel with push-buttons, and a reel of magnetic tape containing the software programme, by means of which the AQS 901 processing and display system can operate in a training mode. Using the ACT-1, which physically resembles a TV game, one crew member can 'play' the part of a submarine, trying to outwit his colleagues operating the AQS 901 detection system. Although not a replacement for ground-based simulator training, the ACT-1 onboard system enables a Nimrod captain to train his crew in authentic operational conditions, without the expenditure of sonobuoys.

Armament (MR. Mk 1 and Mk 2): 14·78 m (48 ft 6 in) long weapons bay, with two pairs of doors, in unpressurised lower fuselage pannier, able to carry up to six lateral rows of ASW weapons, including up to nine torpedoes as well as bombs. (Capability of carrying depth charges and mines is not used by RAF.) Alternatively, to give greater range and endurance, up to six auxiliary fuel tanks can be fitted in the weapons bay, or a combination of fuel tanks and weapons can be carried. To ensure weapon serviceability, the weapons bay is heated when the ambient temperature falls below +5°C. Bay approx 9·14 m (30 ft) long in rear pressurised part of fuselage for storing and launching of active and passive sonobuoys and marine markers. Two rotary launchers, each capable of holding six size A

BAe HS Nimrod MR. Mk 2, without its wingtip EWSM pods

sonobuoys, are used when the cabin is unpressurised; two single-barrel launchers are used when the aircraft is pressurised. A hardpoint is provided beneath each wing, just outboard of the main-wheel doors, on which can be carried air-to-surface missiles, rocket or cannon pods, or mines, according to mission requirements. The missile capability has been deleted, but could be reactivated if required.

DIMENSIONS, EXTERNAL:
Wing span 35·00 m (114 ft 10 in)
Wing chord at root 9·00 m (29 ft 6 in)
Wing chord at tip 2·06 m (6 ft 9 in)
Wing aspect ratio 6·2
Length overall 38·63 m (126 ft 9 in)
Height overall 9·08 m (29 ft 8½ in)
Tailplane span 14·51 m (47 ft 7¼ in)
Wheel track 8·60 m (28 ft 2½ in)
Wheelbase 14·24 m (46 ft 8½ in)

DIMENSIONS, INTERNAL:
Cabin (incl flight deck, navigation and ordnance areas, galley and toilet): Length 26·82 m (88 ft 0 in)
Max width 2·95 m (9 ft 8 in)
Max height 2·08 m (6 ft 10 in)
Volume 124·14 m^3 (4,384 cu ft)

AREAS:
Wings, gross 197·0 m^2 (2,121 sq ft)
Ailerons (total) 5·63 m^2 (60·6 sq ft)
Trailing-edge flaps (total) 23·37 m^2 (251·6 sq ft)
Fin and rudder (above tailplane centreline) 10·96 m^2 (118 sq ft)
Dorsal fin 5·67 m^2 (61 sq ft)
Tailplane 40·41 m^2 (435 sq ft)
Elevators (incl tabs) 12·57 m^2 (135·3 sq ft)

WEIGHTS (MR. Mk 1):
Typical weight empty 39,000 kg (86,000 lb)
Max disposable load 6,120 kg (13,500 lb)
Fuel load:
standard tanks 38,940 kg (85,840 lb)
max with six auxiliary tanks in weapons bay 45,785 kg (100,940 lb)
Normal max T-O weight 80,510 kg (177,500 lb)
Max overload T-O weight 87,090 kg (192,000 lb)
Typical landing weight 54,430 kg (120,000 lb)

PERFORMANCE (MR. Mk 1):
Max operational necessity speed, ISA + 20°C 500 knots (926 km/h; 575 mph)
Max transit speed, ISA + 20°C 475 knots (880 km/h; 547 mph)
Econ transit speed, ISA + 20°C 425 knots (787 km/h; 490 mph)
Typical low-level patrol speed (two engines) 200 knots (370 km/h; 230 mph)
Operating height range S/L to 12,800 m (42,000 ft)
Min ground turning radius 27·1 m (89 ft 0 in)
Runway LCN at T-O weight of 82,550 kg (182,000 lb) 50
T-O run at 80,510 kg (177,500 lb) AUW, ISA at S/L 1,463 m (4,800 ft)
Unfactored landing distance at 54,430 kg (120,000 lb) landing weight, ISA at S/L 1,615 m (5,300 ft)
Typical ferry range 4,500-5,000 nm (8,340-9,265 km; 5,180-5,755 miles)
Typical endurance 12 h

BAe HS NIMROD AEW. Mk 3

This airborne early warning (AEW) version of the Nimrod was designed by Hawker Siddeley Aviation specifically for European defence. On 31 March 1977 the British Defence Secretary announced the government's intention to proceed with the procurement of 11 of the aircraft for the RAF, under the designation Nimrod AEW. Mk 3. Such a programme was made possible by the development by Marconi Avionics of a new radar system which, in addition to an essential maritime capability, satisfies also the air defence requirements of central Europe. Using this equipment, the Nimrod AEW. Mk 3 could provide, at long range and at low or high altitude, detection, tracking and classification of aircraft, missiles and ships; interceptor control; direction of strike aircraft; air defence; air traffic control; and search and rescue facilities. In doing so, it will be compatible with the USAF's Boeing E-3A Sentry AWACS, and with the E-3As that will be operated directly by NATO.

Designed specifically for installation in this modified version of the maritime reconnaissance Nimrod, the radar has made necessary some modifications to the nose and tail to permit installation of the newly developed and identically-shaped dual-frequency twisted cassegrain antennae in fore and aft positions. The aircraft's performance is affected only marginally by the structural changes, and a reduction in directional stability is compensated by a 0·91 m (3 ft 0 in) increase in fin height.

Mounting the scanners at the extremities of the airframe ensures good all-round coverage, and they do not suffer from airframe obscuration effects. Designed for very low sidelobe level, each sweeps through 180° in azimuth, the dual Cossor Jubilee Guardsman IFF 3500 interrogators using the same scanners to aid correlation of IFF and radar returns.

The associated radar is a pulsed Doppler system that, in addition to the detection of aircraft, has a ship surveillance capability. The rate at which pulses are transmitted can be varied to provide maximum detection in differing terrain conditions or sea states. The system has also highly sophisticated anti-jamming features to cope with the growing efficiency of electronic countermeasures.

The radar passes target plots in terms of range, azimuth, radial velocity and altitude to the advanced digital data handling system; this is based on an airborne computer that controls the flow of data from the scanners and correlates track information between the AEW aircraft and a surface control station. Six operator consoles are provided. Each has a tactical situation display, showing the tracks selected by the operator, and a tabular display for the selective presentation of detailed track and control information. Much of the data control is fully automatic; thus, association of radar, IFF and EWSM, track initiation, tracking and data storage requires no action from the operator. Control of the data handling system is achieved by rolling ball and functionally arranged keyboards, the operator interfacing with the system to carry out system control, track classification, fighter control and data link management.

High standards of communications and navigation are essential to complement the advanced radar and data handling system. For communications the AEW Nimrod carries tactical UHF transceivers, SIMOP HF transceivers, pilot's U/VHF, RATT, secure voice com, LF receiver and data links. Primary navigation avionics consist of dual inertial navigation systems (INS), plus a gyro magnetic compass, air data computer, twin VOR/ILS, ADF, Tacan, autopilot and a flight director. EWSM (early warning support measures) equipment is housed in the two pods at the wingtips. Other features of special significance for this role are the spacious cabin for avionics and crew, high transit speed and good low-speed characteristics.

The first of four development aircraft, a converted Comet 4C (XW626), made its first flight on 28 June 1977. This aircraft carried nose-mounted radar only. The other three development aircraft were diverted from the batch of eight extra Nimrod MR. Mk 1s ordered in 1972, and are being completed instead as AEW. Mk 3s. The first of these (XZ286), which made its initial flight on 16 July 1980, was the first aerodynamically representative AEW. Mk 3 airframe. The Nimrod AEW. Mk 3 is expected to enter RAF service, with No. 8 Squadron at RAF Waddington, Lincolnshire, in early 1982.

DIMENSIONS, EXTERNAL:
Wing span 35·08 m (115 ft 1 in)
Length overall 41·97 m (137 ft 8½ in)
Height overall 10·67 m (35 ft 0 in)

PERFORMANCE: Generally similar to Nimrod MR.Mk 2
Endurance in excess of 10 h

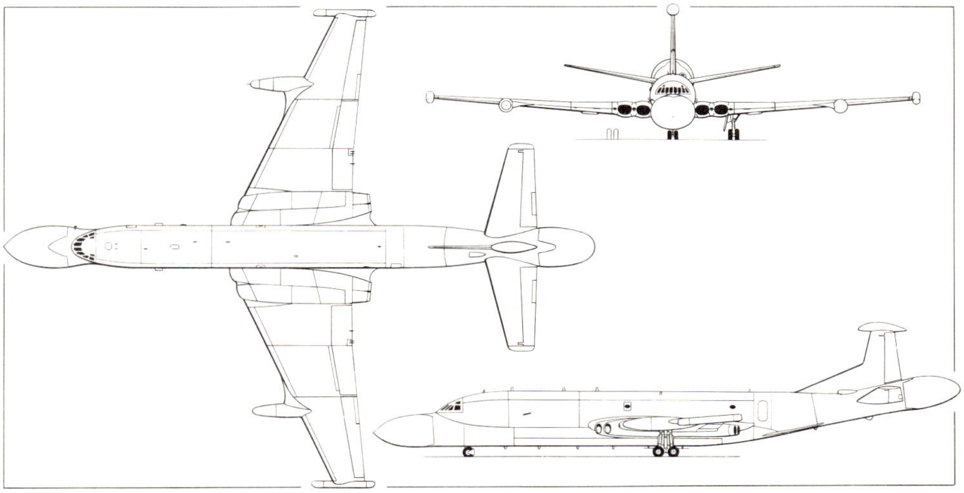

The AEW. Mk 3 version of the BAe HS Nimrod *(Pilot Press)*

The second fully-representative example (XZ287) of the BAe HS Nimrod AEW. Mk 3 airborne early warning aircraft

BGA
BRITISH GLIDING ASSOCIATION

Kimberley House, Vaughan Way, Leicester
Telephone: 0533 51051
CHIEF TECHNICAL OFFICER: R. B. Stratton

BGA SUPERMUNK

First flown on 20 August 1979, the Supermunk (G-BBNA) is a conversion of a British-built de Havilland Chipmunk Mk 22 to accept an Avco Lycoming flat-four engine in place of the original Gipsy Major in-line power plant. The conversion was started in April 1979 by officers of the BGA, and was undertaken in order to improve the performance and serviceability of the aircraft in the role of glider tug. By the Autumn of 1979, the aircraft had already satisfied expectations in regard to improvements in take-off performance and in turnround time during towing trials. Since completion and testing of G-BBNA, three other conversions have been carried out (G-AOSU, G-AOUO, and G-ATVF), and G-BCCX was being worked on in early 1981.

A description of the Chipmunk has appeared in previous editions of *Jane's*. The following details summarise the airframe, power plant and other changes embodied in the Supermunk version:

AIRFRAME: Basically as standard Chipmunk Mk 22 aft of firewall. Forward of firewall, engine is installed on a US-built welded teel tube dynafocal mounting, and is encased in a new two-piece cowling of fire-resistant polyester GRP. Not cleared for spinning or aerobatics; other limitations as for Chipmunk Mk 22.

POWER PLANT: One 134 kW (180 hp) Avco Lycoming O-360-A4A flat-four engine, driving a Sensenich two-blade metal propeller from a Piper Cherokee 180. Fuel system incorporates both mechanical and electrical pumps.

ACCOMMODATION: Seats for two persons in tandem, as in Chipmunk, under one-piece rearward-sliding framed canopy. Front cockpit instrumentation altered to accommodate different engine instruments. Original flight instruments retained in rear cockpit, but some engine instruments have been removed.

SYSTEMS: 12V (instead of 24V) electrical system includes generator, starter and single 35Ah battery. Vacuum pump to drive gyro instruments.

DIMENSIONS, EXTERNAL:

Wing span	10·46 m (34 ft 4 in)
Wing area, gross	15·98 m² (172 sq ft)
Length overall	7·84 m (25 ft 8½ in)
Height overall	2·13 m (7 ft 0 in)
Propeller diameter	1·93 m (6 ft 4 in)

WEIGHTS:

Weight empty	652 kg (1,438 lb)
Max T-O weight	952 kg (2,100 lb)

PERFORMANCE:
Cruising speed at 2,450 rpm
106 knots (196 km/h; 122 mph) IAS
Cruising speed at 2,300 rpm
97 knots (180 km/h; 112 mph) IAS
Max rate of climb at S/L (pilot only, plus max fuel)
360 m (1,180 ft)/min
Rate of climb as glider tug, depending on type and weight of glider 152-244 m (500-800 ft)/min

BGA Supermunk, a converted Chipmunk 22 with Avco Lycoming O-360 engine

CRANFIELD
CRANFIELD INSTITUTE OF TECHNOLOGY

College of Aeronautics, Cranfield Institute of Technology, Cranfield, Bedford MK43 0AL
Telephone: 0234 750911
Telex: 825072
PROFESSOR OF AIRCRAFT DESIGN: D. Howe, PhD, SM, DCAe, CEng, FIMechE, FRAeS

The College of Aeronautics at Cranfield Institute of Technology was responsible for the design, construction and development of the A1 Mk 1/Mk 2 aerobatic light-plane of which details can be found in the 1977-78 and 1979-80 *Jane's* respectively. The College also developed performance-improving wingtip sails for the Piper Pawnee agricultural aircraft, of which available details appeared in the 1980-81 *Jane's*.

Cranfield Institute of Technology, as a subcontractor to Marconi Avionics Ltd, designed and constructed the prototype airframe and launch dolly of Marconi's Machan remotely piloted research vehicle. The School of Electronic Systems at Cranfield was responsible for the design of Machan's flight control electronics. Details of this new research vehicle can be found under Marconi Avionics' entry in the RPVs and Targets section of this edition.

EDGLEY
EDGLEY AIRCRAFT LTD

31 Smith Street, Elsworth, Cambridgeshire CB3 8HY
Telephone: 09547 402
ALSO AT: Cranfield Institute of Technology, Cranfield, Bedford MK43 0AL
Telephone: 0234 750911 extn 479/294
DIRECTORS:
J. K. Edgley, BSc, DIC, CEng, MRAeS (Chairman)
F. M. Edgley
CONSULTANTS:
J. A. Eaton, BE, MSc, DIC, MAIAA
D. Kent

Edgley Aircraft Ltd was formed in 1974 to design, build and market an unorthodox observation aircraft, the EA7 Optica. Its designer, Mr John Edgley, was a director of Scenic Flying Ltd, which in 1972 projected and subsequently patented a slow-flying touring aircraft in which the pilot sat behind the two passengers. In 1974 Mr Edgley began the final aerodynamic design of what eventually became the Optica, and this underwent wind tunnel testing at the Imperial College of Science and Technology, London, where Mr Edgley was then a post-graduate student, in the Summer of 1975. The Optica's three-abreast seating configuration was introduced in 1977, and since Scenic Flying did not wish to be concerned directly in its production Edgley Aircraft Ltd was formed to continue its development. Mr David Kent, who was connected with the Tawney Owl light aircraft last described in the 1961-62 *Jane's*, joined Edgley Aircraft in 1976, assisting Mr Edgley in the final stages of design, stress analysis and construction of the Optica prototype, which began in August of that year in a London workshop. In mid-1978, with most of the airframe complete, the project was moved to the College of Aeronautics at Cranfield for final assembly.

EDGLEY EA7 OPTICA

First flown on 14 December 1979, the Optica (G-BGMW) is a most unusual design for a three-seat observation aircraft, particularly in the fields of pipeline and powerline inspection; forestry and coastal patrol; police traffic patrol; frontier patrol; aerial photography; film, TV and press reporting; and touring. The cabin configuration is based upon that of an insect's eye, to give the best possible all-round view from a fixed-wing aircraft. Power plant is a ducted propulsor unit, offering excellent quietness, both within the cabin and from the ground. A low wing loading, pre-set inboard flaps and a low stalling speed, facilitate continuous en-route flight at low speeds, and the generous flap area confers STOL capability from both hard and soft strips. Stability increases at low speed.

The Optica is intended for low volume production (up to 100 a year), using a minimum of sophisticated tooling. Since its public debut at Farnborough International 1980, and subsequently at the 1981 Paris Air Show, this aircraft has created worldwide interest. As a result it is intended to gain both CAA and FAA certification, and plans were under discussion in 1981 for production to start in the UK in early 1982.

TYPE: Three-seat slow-flying observation aircraft; stressed to BCAR Section K (non-aerobatic category) and FAR Pt 23 (Normal category).

WINGS: Cantilever mid-wing monoplane. Wing section NASA GAW-1, thickness/chord ratio 17%. Dihedral 3° on outer panels. Incidence 0°. Constant-chord single-spar non-swept wings of aluminium alloy stressed-skin construction. Wingtips (also fin/tailplane fillets, nosewheel mudguard and some power plant fairings) of glassfibre. Fowler trailing-edge flaps (29% of total wing chord) inboard and outboard of tailbooms. Electrically actuated outboard flaps can be set at angles up to 50° for landing; inboard flaps set permanently at 10°, giving the effect of a slotted wing, for continuous low-speed flying. Bottom-hinged, mass-balanced slotted ailerons outboard of outer flaps. No spoilers, airbrakes or tabs.

CABIN: 'Insect eye' shaped structure, built of aluminium alloy with Suntex vacuum formed acrylic windows. Cabin attached to fan shroud and rest of airframe by six stators of steel tube and aluminium alloy shear web construction. Steel tube and aluminium alloy nose beam supporting cabin floor. Horizontal window frame member just above floor level, designed to act in combination with forward-running nosewheel box to act as skid in case of nosewheel failure, is designed to withstand 9*g* impact. Depending upon number of occupants, the CG position can be adjusted by adding ballast weight; one possible means of accomplishing this is by installing a water tank in the nose, but the method to be adopted had not been finally decided up to early 1981.

TAIL UNIT: Twin-tailboom configuration, of aluminium alloy stressed-skin tubular construction. Tailboom pickup points at extremities of wing centre-section. Angular, inward-canted fins and balanced rudders. Fixed-incidence tailplane, with elevator, bridging space between tops of fins. Inset trim tab occupies port half of elevator trailing-edge; no rudder tabs.

LANDING GEAR: Non-retractable tricycle type, with steerable nosewheel offset to port. Modified Piper main-wheel legs, with rubber-in-compression shock-absorption. Nosewheel shock-absorption by bungee rubber in tension. Single wheel on each unit, tyre sizes 6·00-6 (main) and 5·00-5 (nose). Hydraulic disc brakes on main wheels. Retractable parking strut, operated from cabin, extends below fan shroud to keep aircraft level when empty on the ground, and to protect shroud in the event of main-wheel leg failure.

POWER PLANT: Ducted propulsor unit, with engine and fan forming a power pod separate from the main shroud. Pod is attached to fan shroud with four Lord rubber mountings, and supported by four stators of steel channel and aluminium alloy shear web construction, with steel tube engine bearers. Five-blade fixed-pitch birch

Edgley EA7 Optica three-seat slow-flying observation aircraft

fan, driven on the prototype by a 134 kW (180 hp) Avco Lycoming IO-360 flat-four engine (119 kW; 160 hp O-320 for early test flights) mounted in a duct downstream of the fan. Production aircraft are intended to have a 149 kW (200 hp) IO-360 or 156·5 kW (210 hp) turbocharged TIO-360 engine. Fuel tank of 91 litres (20 Imp gallons) in each wing leading-edge, immediately outboard of tailbooms and forward of wing spar. Tanks are of full wing section, but are designed not to be stressed by wing bending and torsion. Total fuel capacity 182 litres (40 Imp gallons). Refuelling point in upper surface of each wing. Oil capacity 7·6 litres (1·7 Imp gallons).

Accommodation: Cabin designed to seat up to three persons side by side, with either single- or two-pilot operation (left hand and centre setas). Dual controls optional. Fixed seats in prototype; adjustable seats optional. Baggage space aft of seats. Alternative accommodation for two stretchers, or one stretcher and one medical attendant, in addition to pilot. Single elliptical door on each side, hinged at front and opening forward. Cabin heated, by hot air from engine, and ventilated.

Systems: Hydraulics for main-wheel brakes only. Prototype has 12V electrical system (battery and alternator) for engine starting, and actuation of flaps and under-shroud strut; it has no lights. 24V system under consideration for production version.

Dimensions, external:

Wing span	12·16 m (39 ft 10¾ in)
Wing chord (basic, constant)	1·32 m (4 ft 4 in)
Wing chord (over 10° fixed flaps)	1·52 m (5 ft 0 in)
Wing aspect ratio	9·1
Length overall	8·16 m (26 ft 9¼ in)
Height over fan shroud (excl aerial)	1·92 m (6 ft 3½ in)
Diameter of fan shroud	1·68 m (5 ft 6¼ in)
Diameter of fan	1·20 m (3 ft 11¼ in)
Shroud ground clearance: at front	0·24 m (9½ in)
to aft leg mechanism	0·16 m (6¼ in)
Height over tailplane	2·31 m (7 ft 7 in)
Tailplane span: c/l of tailbooms	3·40 m (11 ft 1¾ in)
intersection fin chord	2·60 m (8 ft 6½ in)
Wheel track	3·40 m (11 ft 1¾ in)
Wheelbase	2·73 m (8 ft 11½ in)
Doors (each): Long axis	1·34 m (4 ft 4¾ in)
Short axis	0·96 m (3 ft 1¾ in)
Height to sill	0·50 m (1 ft 7¾ in)

Dimensions, internal:

Cabin: Length	2·45 m (8 ft 0½ in)
Max width (to door Perspex)	1·69 m (5 ft 6½ in)
Max height	1.35 m (4 ft 5¼ in)
Floor area	approx 0·72 m^2 (7·53 sq ft)

Areas:

Wings, gross	15·84 m^2 (170·5 sq ft)
Ailerons (total)	1·55 m^2 (16·68 sq ft)
Trailing-edge flaps: inboard (total)	0·61 m^2 (6·57 sq ft)
outboard (total)	1·49 m^2 (16·04 sq ft)
Fins (total)	2·59 m^2 (27·88 sq ft)
Rudders (total)	1·38 m^2 (14·85 sq ft)
Tailplane	1·62 m^2 (17·44 sq ft)
Elevator, incl tab	1·26 m^2 (13·56 sq ft)

Weights (A: prototype; B: production version with IO-360 engine):

Weight empty: A	895 kg (1,973 lb)
B	850 kg (1,875 lb)
Max T-O weight: A	1,220 kg (2,690 lb)
B	1,236 kg (2,725 lb)

Performance (IO-360 engine):

Never-exceed speed	142 knots (264 km/h; 164 mph)
Max level speed	109·5 knots (203 km/h; 126 mph)
Cruising speed (65% power)	94 knots (174 km/h; 108 mph)
Loiter speed	50 knots (92 km/h; 57 mph)
Stalling speed, flaps up	43 knots (80 km/h; 50 mph)
Stalling speed, 20° flap	40 knots (74 km/h; 46 mph)
Max rate of climb at S/L	219 m (720 ft)/min
Service ceiling	4,270 m (14,000 ft)
T-O to 15 m (50 ft)	200 m (655 ft)
Landing from 15 m (50 ft)	260 m (850 ft)
Range at 65% power cruising speed	640 nm (1,186 km; 737 miles)
Endurance: at loiter speed	10 h
at cruising speed	6 h 48 min

LEARFAN

LEARFAN LTD

62 Church Road, Newtownabbey, Co Antrim, Northern Ireland
Telephone: Whiteabbey (0231) 68862
and at Aldergrove Airport, near Belfast, Co Antrim, Northern Ireland

Managing Director and Chief Executive: Linden S. Blue
General Manager: William W. Surbey

LearAvia Corporation of the USA (which see) signed an agreement with the British government for sole-source production of the **LearFan 2100** twin-turboprop business aircraft at Newtownabbey and at the former RAF base at Aldergrove, near Belfast, Northern Ireland. To manage this programme, LearFan Ltd was established, with an initial investment of approx £3·5 million from each participant. LearAvia Corporation retains a 51% holding in the new company.

As noted in the US section, the first of four LearFan 2100 prototypes made its first flight on 1 January 1981; one of the other three prototypes under construction is intended for static testing. Deliveries are expected to begin shortly after certification in early 1983. The first 42 production aircraft are to be built in the USA; those from c/n 43 onward will be assembled in Northern Ireland, beginning in September 1983. In August 1981 it was reported that LearFan Ltd had some 400 employees, of whom about 50 had received training at the US plant in Reno. It is anticipated that by the time full production is under way, LearFan Ltd will have a work force of some 1,250 persons.

LOCKSPEISER

LOCKSPEISER AIRCRAFT LTD

Registered Office: Royal Chambers, High Street, Weston-super-Mare, Avon BS23
Telephone: Weston-super-Mare (0934) 29467
Managing Director: David Lockspeiser, MRAeS, CEng
Company Secretary: Christopher E. Bean, FCA

LOCKSPEISER LDA-01

Mr David Lockspeiser designed a utility aeroplane known as the LDA, or Land Development Aircraft, intended for operation as a passenger, freight or vehicle transport, as an agricultural, ambulance, survey or firefighting aircraft, or for other duties. In a military version, for use as a light troop transport or battlefield support aircraft, the initials stand for Light Defence Aircraft.

A 70% scale prototype, registered G-AVOR and known as the LDA-01, was flown for the first time by Mr Lockspeiser on 24 August 1971. A full description of this aircraft can be found in the 1979-80 and earlier editions of *Jane's*.

MARSHALL

MARSHALL OF CAMBRIDGE (ENGINEERING) LTD (Aircraft Division)

Head Office and Works: Airport Works, Cambridge CB5 8RX
Telephone: 0223 61133
Telex: 81208
Managing Director: Sir Arthur Marshall, OBE
Commercial Director: R. D. Horsbrough
Executive Director, Engineering: R. O. Gates
Sales Manager: Norman Sellars

The Aircraft Division of this company (known as Marshalls Flying School Ltd until 1962) has specialised for many years in the modification, overhaul and repair of military and commercial aircraft, including the design and installation of interior furnishing for executive transports and of avionics fits up to and including the complete outfitting of various aircraft for calibration and electronic countermeasures roles.

The company's design department is both CAA and MoD(PE) approved. As an approved service and repair centre for the Lockheed Hercules, Gulfstream American and Cessna Citation aircraft, Marshall of Cambridge also has FAA approval covering most types of American aircraft. The company's conversion, modification and overhaul facilities, which include some of the largest heated hangars in England, with workshop support to full aircraft factory standard, have enabled it to undertake numerous major programmes of work on Viscounts, Britannias, Comets, VC10s, Canberras and a large number of other civil and military aircraft. It has a separate hangar for specialised painting of the largest aircraft, and a sculpture milling shop for manufacture of major aircraft components.

In 1966, Marshall of Cambridge was appointed the designated centre for the Royal Air Force Hercules C. Mk 1 transport aircraft, and in 1973 the company completed the conversion of an RAF Hercules C. Mk 1 to W. Mk 2 configuration, as detailed in the 1979-80 and earlier editions of *Jane's*.

Since 1978 the company has become increasingly involved in work upon helicopters: much of this is concerned with the design and incorporation of updated avionics and stabilised/automatic flight systems, and modifications to permit these aircraft to operate on the UK CAA register. Other work on helicopters has included the design and incorporation of a rear escape hatch that can be actuated from the flight deck, plus the provision of an automatic dinghy release, and the capability to deploy an external sea anchor.

Shorts Belfast modified by Marshall for civil freight operation

MARSHALL (LOCKHEED) HERCULES CONVERSIONS

The **Hercules W. Mk 2** long-range meteorological aircraft was adapted by Marshall from a Hercules C. Mk 1; full details of the equipment and other changes involved were given in the 1974-75 *Jane's*.

It was announced in 1978 by the Ministry of Defence that a contract had been signed with Lockheed Corporation to modify 30 Hercules C. Mk 1s (C-130Ks) of the Royal Air Force to **Hercules C. Mk 3** configuration. The modification involves 'stretching' the fuselage by the insertion of a 2·54 m (8 ft 4 in) plug forward of the wings and a 2·03 m (6 ft 8 in) plug aft of the wings, thus producing the same fuselage dimensions and capacity as those of the L-100-30 Commercial Hercules (see US section). The first aircraft (XV223) was modified by Lockheed in Marietta, Georgia, in 1979; the remaining 29 are being modified by Marshall of Cambridge, with the last example scheduled for completion during 1983.

Examples of volumetric and load differences include:

Cabin volume, incl ramp: C. Mk 1	127·4 m^3 (4,500 cu ft)
C. Mk 3	171·5 m^3 (6,057 cu ft)
Palletised loads: C. Mk 1	5
C. Mk 3	7
Land-Rovers plus trailers: C. Mk 1	3 + 2
C. Mk 3	4 + 3

MARSHALL (SHORTS) BELFAST CONVERSION

In March 1980 the first of five Shorts Belfast ex-RAF military transports, purchased by HeavyLift Air Cargoes Ltd of Stansted Airport, Essex, entered civilian service. This is based at Stansted: the second, which was delivered later in the year, is based in Singapore. Intended for all-cargo use carrying unusual, outsize or heavy payloads, the Belfasts are being converted to HeavyLift requirements by Marshall of Cambridge. The programme involves an increase in minimum control speed of 10 knots (19 km/h; 12 mph); revision of the stall warning speeds; a reduction of max T-O weight to 104,325 kg (230,000 lb) from the normal military T-O weight of 104,780 kg (231,000 lb) and military operating standard T-O weight of 113,400 kg (250,000 lb); introduction of two-engine climb data to permit use of the fuel jettison system with 15° flap; changes to the water-methanol usage data for T-O; and a reassessment of aircraft performance standard.

In addition, in order to gain civil certification, changes include revision of the autopilot, avionics and electrical power installations; replacement of the liquid oxygen system by a gaseous system; introduction of a water-methanol contents indication system and a stall protection system; and other modifications to meet mandatory civil requirements. At the customer's request, the Belfasts are also being fitted with a dual Omega navigation system and an electrically driven auxiliary hydraulic pump.

Sikorsky S-61N modified for CAA register and repainted by Marshall

Dimensions, external:

Wing span	48·41 m (158 ft 10 in)
Length overall	41·58 m (136 ft 5 in)
Height overall	14·33 m (47 ft 0 in)
Height to ramp sill	1·63 m (5 ft 4 in)
Width of ramp	3·66 m (12 ft 0 in)
Height of rear loading entrance, ramp lowered	4·06 m (13 ft 4 in)
Height of rear loading entrance, ramp horizontal	3·12 m (10 ft 3 in)
Forward freight door: Height	1·98 m (6 ft 6 in)
Width	2·44 m (8 ft 0 in)

Dimensions, internal:

Cargo compartment:	
Length overall, incl ramp	25·70 m (84 ft 4 in)
Length overall, excl ramp	19·35 m (63 ft 6 in)
Max width	4·90 m (16 ft 1 in)
Max height	4·06 m (13 ft 4 in)
Max usable volume	321·4 m³ (11,350 cu ft)

Weights:

Typical operating weight empty	58,967 kg (130,000 lb)
Max payload	34,000 kg (75,000 lb)
Usable fuel	37,376 kg (82,400 lb)
Max T-O weight	104,325 kg (230,000 lb)
Max landing weight	97,520 kg (215,000 lb)
Max zero-fuel weight	92,986 kg (205,000 lb)

Performance (at max T-O weight, except where stated):

Average cruising speed	275 knots (510 km/h; 316 mph)
T-O runway length	2,500 m (8,200 ft)
Landing runway length at max landing weight	2,075 m (6,800 ft)
Range with max payload, standard reserves	approx 850 nm (1,575 km; 975 miles)
Range with 10,000 kg (22,000 lb) payload, standard reserves	approx 3,350 nm (6,200 km; 3,855 miles)

NASH

NASH AIRCRAFT LTD (a subsidiary of Kinetrol Ltd)

Registered Office: Trading Estate, Farnham, Surrey GU9 9NU
Telephone: 0252 723688
Telex: 858567
Directors:
A. R. B. Nash (Managing)
Roy G. Procter
R. C. Nash

Mr Alan Nash acquired a controlling interest in Procter Aircraft Associates Ltd in early 1978, as a result of which the company was renamed Nash Aircraft Ltd in 1980. Its principal current activity is the development and manufacture of the Petrel two-seat light aircraft.

NASH PETREL

This two-seat light aircraft is based upon the Procter Kittiwake I single-seat lightplane (see 1978-79 *Jane's*), with many components in common, but has increased wing area and is optimised for glider towing. It has a number of improvements and simplifications to the detail mechanical design, compared with Kittiwake I. Materials used throughout are L72 clad dural and S510 mild steel.

The company prototype (G-AXSF) was completed by Nash Aircraft Ltd and made its first flight on 8 November 1980. Since then, in initial glider-towing trials, it has towed Vega, Mini-Nimbus and Nimbus sailplanes to a height of 610 m (2,000 ft) in under 4 min. A pre-production batch of five Petrels was under construction in the summer of 1981.

Type: Two-seat light aircraft.

Wings: Cantilever low-wing monoplane. Wing section NACA 3415. Dihedral 5° on outer panels. No sweepback or washout. All-metal constant-chord structure, built in three sections; centre-section, integral with fuselage, to which outer panels are each attached with three bolts. Single main spar at 30% chord and lightweight auxiliary spar at 66% chord. Multiple ribs, with no spanwise stiffeners. All-metal NACA slotted flaps and ailerons on outer panels. Flaps are operated manually by pushrod and torque tube; ailerons are mass-balanced and operated by cables.

Fuselage: All-metal structure. Four longeron basic structure, with flat sides and bottom and single-curvature top-decking. Integral wing centre-section forms seat and main landing gear attachment structure.

Tail Unit: Cantilever all-metal structure. Fixed-incidence tailplane. Manually-operated tab in starboard elevator. Tab on rudder. Control surfaces mass-balanced and operated by cables.

Landing Gear: Non-retractable tricycle type. Nose unit is an oleo-pneumatic strut with Goodyear 5·00-6 wheel, and is steerable from the rudder pedals. Main gear is of cantilever spring type, with Goodyear 6·00-6 wheels and hydraulic disc brakes. Tyre pressure (all) 1·72 bars (25 lb/sq in).

Power Plant (prototype): One 119 kW (160 hp) Avco Lycoming O-320-D2A flat-four engine, driving a McCauley fixed-pitch two-blade metal propeller. Provision for alternative engines of 88-134 kW (118-180 hp). Removable fuel tanks in wing centre-section leading-edges, capacity 73 litres (16 Imp gallons).

Accommodation: Two persons side by side, on seats with individually adjustable backs. Baggage space aft of seats.

Equipment: Starter, generator and basic instrumentation. Radio, navigation and other equipment to customer's requirements.

Dimensions, external:

Wing span	9·14 m (30 ft 0 in)
Wing chord, constant	1·38 m (4 ft 6½ in)
Wing area, gross	12·5 m² (135·0 sq ft)
Wing aspect ratio	6·6
Length overall	6·30 m (20 ft 8 in)
Height overall	2·33 m (7 ft 8 in)
Tailplane span	2·79 m (9 ft 2 in)
Wheel track	2·24 m (7 ft 4 in)
Wheelbase	1·52 m (5 ft 0 in)

Weights:

Weight empty	515·5 kg (1,137 lb)
Max T-O weight	762 kg (1,680 lb)

Performance (at max T-O weight; estimated, based on measured performance of Procter Kittiwake I with 97 kW; 130 hp Rolls-Royce Continental O-240 engine):

Max level speed	113 knots (209 km/h; 130 mph)
Cruising speed	104 knots (193 km/h; 120 mph)
Max rate of climb at S/L	305 m (1,000 ft)/min

Company-owned prototype of the Nash Petrel two-seat light aircraft

NDN

NDN AIRCRAFT LTD

Isle of Wight Airport, Sandown, Isle of Wight
Telephone: (0983) 406421
Telex: 86501 NDNAIR G
Directors:
N. D. Norman, CBE, MRAeS (Managing Director)
K. A. F. Burke, ACA
A. J. Coombe, BSc (Technical)
B. E. Norman

Mr N. D. Norman founded NDN Aircraft Ltd in the early months of 1977 to develop and subsequently produce or arrange licence production of a new two-seat basic military trainer known as the NDN 1 Firecracker.

NDN 1 FIRECRACKER

The first prototype Firecracker (G-NDNI) made its initial flight on 26 May 1977, and a British type certificate in the Aerobatic category was issued by the CAA on 24 May 1979. The aircraft's design meets the requirements of FAR Pt 23-14.

The following description applies to the first prototype:

Type: Two-seat training aircraft.

Wings: Cantilever low-wing monoplane. Wing section NACA 23012 (modified). Dihedral 5° on outer wing panels only. Incidence 3°. Leading-edge of inner wing panels swept back approximately 20°. Light alloy structure, including aerodynamically and mass balanced ailerons and hydraulically-operated single-slotted trailing-edge flaps. Trim tab in starboard aileron; geared tab in each aileron. All controls manually operated.

Fuselage: Semi-monocoque stressed-skin structure of light alloy. Hydraulically-operated light alloy door-type airbrake in fuselage undersurface, at wing trailing-edge.

Tail Unit: Cantilever stressed-skin structure of light alloy. Fixed-incidence tailplane. Trim tab in starboard half of elevator and in rudder.

Landing Gear: Hydraulically retractable tricycle type, steerable nosewheel retracting aft, main units inward. Oleo-pneumatic shock-absorber and single wheel on each unit. Cleveland type 551-751 main wheels, with size 6·00-6 Goodyear tyres. Cleveland type 551-753 nosewheel, with size 5·00-5 Goodyear tyre. Cleveland type 551-705 hydraulically-operated disc brakes.

Power Plant: One 194 kW (260 hp) Avco Lycoming AEIO-540-B4D5 flat-six engine, driving a Hoffmann HO-V123K/200AH5 three-blade constant-speed propeller. Also offered with 224 kW (300 hp) Avco Lycoming AEIO-540-L flat-six engine. Four integral wing fuel tanks with a combined capacity of 427 litres (94 Imp gallons). Refuelling points on wing upper surface. Fuel and Christen oil systems permit inverted flight.

Prototype NDN 1 Firecracker two-seat military training aircraft with underwing SURA rockets

ACCOMMODATION: Two seats in tandem beneath sideways-opening (to starboard) transparent canopy. Canopy can be jettisoned in emergency. Rear seat is raised 10 cm (4 in) above level of forward seat for improved view. Baggage space (0·25 m³; 9 cu ft) aft of rear seat. Accommodation heated and ventilated.

SYSTEMS: Hydraulic system, pressure 103·5 bars (1,500 lb/sq in), supplied by electro-hydraulic pump with hand-operated emergency pump. Gas bottle for emergency nosewheel extension. Electrical system powered by a Prestolite engine-driven 24V 70A alternator. 12V 35Ah battery for aerobatics. Air-conditioning and oxygen systems optional.

AVIONICS AND EQUIPMENT: Typical avionics will include dual VHF com/nav, plus ILS, DME, ADF, transponder and marker beacon receiver. Blind-flying instrumentation is standard. Optional equipment includes target towing facilities, underwing weapon pylons, gunsight, survey camera and autopilot.

DIMENSIONS, EXTERNAL:
Wing span 7·92 m (26 ft 0 in)
Wing chord at root 1·83 m (6 ft 0 in)
Wing chord at tip 1·45 m (4 ft 9 in)
Wing aspect ratio 5·36
Length overall 7·70 m (25 ft 3 in)
Height overall 3·00 m (9 ft 10 in)
Tailplane span 2·90 m (9 ft 6 in)
Wheel track 3·05 m (10 ft 0 in)
Wheelbase 2·08 m (6 ft 10 in)
Propeller diameter 1·93 m (6 ft 4 in)

AREAS:
Wings, gross 11·71 m² (126 sq ft)
Ailerons (total) 1·23 m² (13·2 sq ft)
Trailing-edge flaps (total) 1·23 m² (13·2 sq ft)
Fin 0·60 m² (6·5 sq ft)
Rudder, incl tab 0·65 m² (7·0 sq ft)
Tailplane 1·28 m² (13·8 sq ft)
Elevator, incl tab 1·11 m² (12·0 sq ft)

WEIGHTS AND LOADINGS (260 hp engine):
Weight empty, equipped 875 kg (1,930 lb)
Max T-O weight, fully aerobatic 1,288 kg (2,840 lb)
Max landing weight 1,225 kg (2,700 lb)
Max wing loading 109·9 kg/m² (22·5 lb/sq ft)
Max power loading 6·64 kg/kW (10·9 lb/hp)

WEIGHTS AND LOADINGS (estimated, 300 hp engine):
Weight empty 902 kg (1,990 lb)
Max T-O weight, fully aerobatic 1,288 kg (2,840 lb)
Max landing weight 1,225 kg (2,700 lb)
Max wing loading 109·8 kg/m² (22·5 lb/sq ft)
Max power loading 5·76 kg/kW (9·47 lb/hp)

PERFORMANCE (with 260 hp engine, at max landing weight, ISA at S/L, zero wind, unless otherwise stated):
Never-exceed speed
238 knots (441 km/h; 274 mph) EAS
Max level speed at S/L
176 knots (326 km/h; 203 mph)
Max cruising speed, 75% power at 2,285 m (7,500 ft)
167 knots (309 km/h; 192 mph)
Stalling speed, flaps down
56·5 knots (105 km/h; 65 mph) EAS
Max rate of climb at S/L 442 m (1,450 ft)/min
Service ceiling 5,485 m (18,000 ft)
Absolute ceiling 6,100 m (20,000 ft)
T-O run 271 m (890 ft)
T-O to 15 m (50 ft) 360 m (1,180 ft)
Landing from 15 m (50 ft) 396 m (1,300 ft)
Landing run 259 m (850 ft)
Range with max fuel, no reserves
1,220 nm (2,260 km; 1,405 miles)

PERFORMANCE (estimated, with 300 hp engine, at max landing weight, ISA at S/L, zero wind, unless otherwise stated):
Never-exceed speed
247 knots (458 km/h; 285 mph) EAS
Max level speed at S/L
185 knots (343 km/h; 213 mph)
Max cruising speed, 75% power at 2,285 m (7,500 ft)
175 knots (323 km/h; 201 mph)
Stalling speed 56·5 knots (105 km/h; 65 mph) EAS
Max rate of climb at S/L 527 m (1,730 ft)/min
Service ceiling 6,700 m (22,000 ft)
Absolute ceiling 7,315 m (24,000 ft)
T-O and landing distances as 260 hp version
Range with max fuel, no reserves
1,116 nm (2,068 km; 1,285 miles)

NDN 5

NDN Aircraft is developing a turboprop version of the Firecracker with a 410 kW (550 shp) Pratt & Whitney Aircraft of Canada PT6A-25A engine. The prototype was expected to fly in late 1981 or early 1982. It is generally similar to the NDN 1, except for the power plant change. Brief specification details are as follows:

WEIGHTS (estimated):
Max T-O weight 1,338 kg (2,950 lb)
Max landing weight 1,270 kg (2,800 lb)

PERFORMANCE (estimated, at max T-O weight):
Max level speed 266 knots (493 km/h; 306 mph)
Max cruising speed 252 knots (467 km/h; 290 mph)
Max rate of climb at S/L 1,130 m (3,700 ft)/min
T-O to 15 m (50 ft) 335 m (1,100 ft)
Range with max fuel, no reserves
825 nm (1,529 km; 950 miles)

NDN 6 FIELDMASTER

NDN Aircraft has under development an agricultural aircraft known as the NDN 6 Fieldmaster, which is being financed jointly by the company and the UK National Research Development Corporation. Representing an entirely new approach to the design of agricultural aircraft, it has a titanium chemical hopper which is an integral part of the fuselage structure, its outer surface being contoured to serve as the skin of that fuselage section. The power plant is mounted on the front of this hopper, the aft fuselage to its rear, and the wings are built directly on to each side of the hopper's base. The cockpit, in the rear fuselage, is protected by a strong rollover structure, and is large enough to accommodate a second seat in tandem. Removable dual controls are standard, to simplify flight training and check-out procedures. The wing is fitted with full-span auxiliary aerofoil trailing-edge flaps, embodying a liquid spray dispersal system that discharges directly into the downwash of the flaps, and so ensures that the spray droplets achieve the best possible crop penetration.

The first flight of the Fieldmaster was expected to be made during the Autumn of 1981.

TYPE: Two-seat large-capacity agricultural aircraft.

WINGS: Low-wing monoplane, with an overwing streamline-section bracing strut each side. Conventional all-metal structure with full corrosion proofing. Electrically-actuated full-span auxiliary aerofoil trailing-edge flaps, which incorporate plumbing for 20 standard spray nozzles on each wing. Two narrow-chord ailerons on each wing, adjacent to wingtip and forward of trailing-edge flaps.

FUSELAGE: Forward fuselage comprises structural titanium hopper with capacity of 1,996 kg (4,400 lb) dry, or 2,642 litres (581 Imp gallons) of liquid chemicals, and incorporating large access door, vent system, inspection windows, and light. Rear fuselage, attached to the rear of this hopper, is of semi-monocoque light alloy construction, fully corrosion proofed, and with easy access for cleaning and maintenance.

TAIL UNIT: Braced conventional structure of light alloy. Trim tab in rudder and port elevator.

LANDING GEAR: Non-retractable tricycle type, with single wheel on each unit. Nosewheel has alternative steerable or castoring facility. Main units of levered-suspension type. Nosewheel tyre size 7·00 × 8; main wheels have tubed tyres, diameter 736 mm (29 in). Hydraulic disc brakes. Landing gear incorporates wire cutters.

POWER PLANT: One 559 kW (750 shp) Pratt & Whitney Aircraft of Canada PT6A-34AG turboprop engine, driving a Hartzell three-blade fully-feathering reversible-pitch metal propeller with spinner. Two integral fuel tanks in each wing, with combined capacity of 946 litres (208 Imp gallons). Engine air intake has a Centrisep filtration system.

ACCOMMODATION: Standard accommodation for pilot and trainee/passenger on fully-adjustable tandem seats in an enclosed cockpit, with roll-over protective structure. Dual controls standard, those for pupil easily removable. Crew safety helmets with headsets optional. Baggage space in fuselage. Birdproof windscreen standard; two-speed windscreen wiper and windscreen washer installation optional. Accommodation ventilated; heating system optional. Wirecutters forward of windscreen, and cable-deflecting wire from top of windscreen to tip of fin.

SYSTEMS: Electrical system includes 24V 300A starter/generator. Hydraulic system for brakes only. Central warning system standard, including engine fire warning.

AVIONICS AND EQUIPMENT: Intercom standard. Avionics available to customer requirements. Standard equipment includes an external power socket. Optional equipment includes blind-flying instrumentation and vacuum system; airframe and engine hour meter; instrument lighting, navigation lights, fin and wingtip strobe lights; two forward-looking and/or two retractable sideways-looking work lights, each 600,000 candlepower; automatic flagman installation; fire bombing dump door; adaptors to convert standard 40-nozzle to 80-nozzle spray system; Transland gatebox, high-volume spreader, quick-disconnect flange kit, and side loading system; and Micronair 8-unit installation, with flowmeter and rpm indicator.

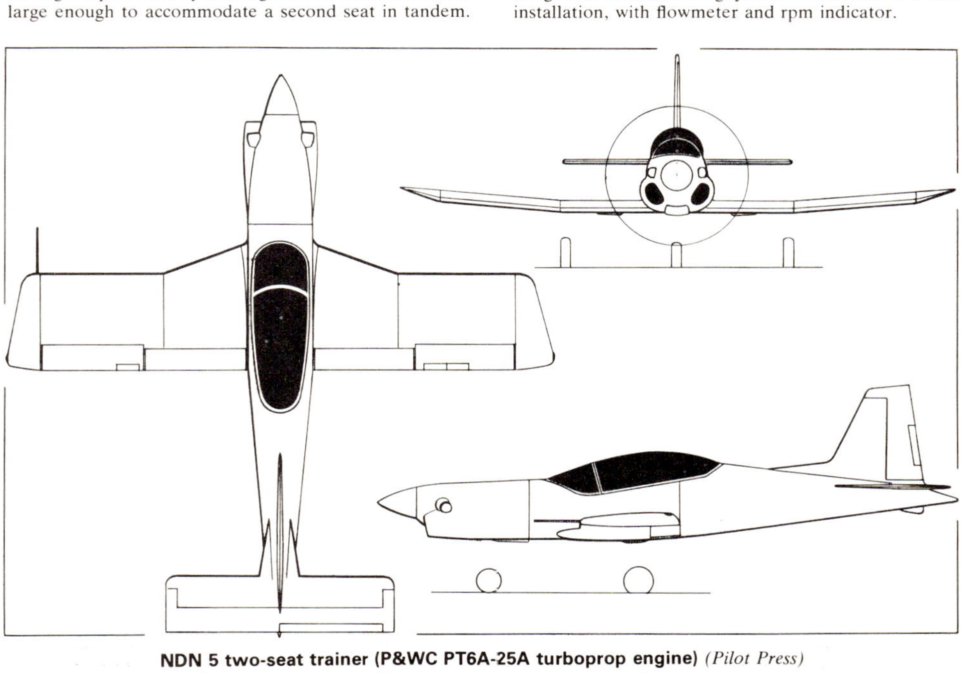

NDN 5 two-seat trainer (P&WC PT6A-25A turboprop engine) *(Pilot Press)*

Dimensions, external:
- Wing span 15·32 m (50 ft 3 in)
- Length overall 11·02 m (36 ft 2 in)
- Height overall 3·48 m (11 ft 5 in)
- Wheel track 5·08 m (16 ft 8 in)
- Wheelbase 3·35 m (11 ft 0 in)
- Propeller diameter 2·69 m (8 ft 10 in)

Area:
- Wings, gross 31·40 m² (338 sq ft)

Weights (estimated):
- Weight empty, equipped 1,588 kg (3,500 lb)
- Max payload 2,268 kg (5,000 lb)
- Max T-O weight 3,856 kg (8,500 lb)

Performance (estimated, at max T-O weight, except as indicated):
- Never-exceed speed 191 knots (354 km/h; 220 mph)
- Max level speed, 'clean' 163 knots (303 km/h; 188 mph)
- Cruising speed, 75% power 149 knots (275 km/h; 171 mph)
- Design manoeuvring speed 135 knots (249 km/h; 155 mph)
- Stalling speed, flaps down 59 knots (109 km/h; 68 mph)
- Stalling speed, flaps down, at 2,268 kg (5,000 lb) AUW 45 knots (84 km/h; 52 mph)
- Max rate of climb at S/L 366 m (1,200 ft)/min
- T-O run 177 m (580 ft)
- Landing run with propeller reversal, at normal operating weight 85 m (280 ft)
- Range with max fuel, no reserves 1,020 nm (1,889 km; 1,174 miles)

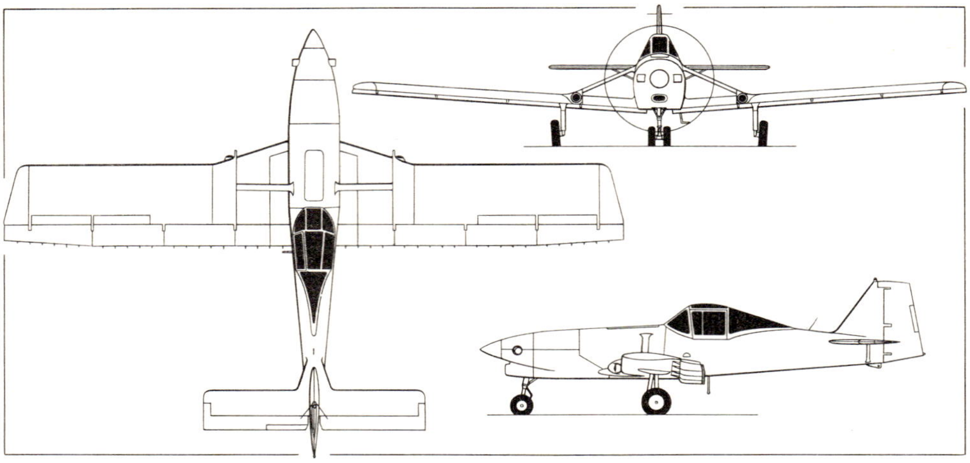

NDN 6 Fieldmaster agricultural aircraft. The landing gear is now intended to have a single wheel on each unit *(Pilot Press)*

PILATUS BRITTEN-NORMAN

PILATUS BRITTEN-NORMAN LTD (a subsidiary of Pilatus Aircraft Ltd)

Head Office: Bembridge Airport, Bembridge, Isle of Wight PO35 5PR
Telephone: 098387 2511/5
Telex: 86277

Directors:
- D. C. Kloeckner (Managing)
- P. Desai (Marketing)
- J. Keller (Production)
- D. A. Berryman (Technical)

Public Relations: Trevor Ward

The previous history of this company can be found in the 1978-79 and earlier editions of *Jane's*. It produces the Islander and Trislander transport aircraft; the Islander is manufactured also in Romania by IAv Bucuresti (which see), and approximately 200 had been delivered from that source by 1 March 1981. In addition to the Bembridge and Romanian production lines, Philippine Aerospace Development Corporation had an Islander production line in Manila, for the production of 115 aircraft under licence in four phases during the period 1974-1980. Only 55 aircraft were built, but assembly of 12 BN-2As is being continued in the Philippines by PADC's subsidiary, National Aero Manufacturing Corporation (which see).

By mid-1978, Britten-Norman had begun production of the latest series of the Islander, the BN-2B. Development of the Turbo Islander, Defender, Trislander and other versions of this widely used family of aircraft is continuing at Bembridge.

In September 1979 Pilatus Aircraft Ltd of Switzerland acquired all assets of Britten-Norman (Bembridge) Ltd, including the facilities on the Isle of Wight and the former Fairey SA Islander/Trislander production hardware at Gosselies in Belgium.

PILATUS BRITTEN-NORMAN BN-2B ISLANDER

The Islander is a modern replacement for aircraft in the class of the de Havilland Dragon Rapide, Detail design work began in April 1964 and construction of the prototype (G-ATCT) was started in September of the same year. It flew for the first time on 13 June 1965, powered by two 157 kW (210 hp) Rolls-Royce Continental IO-360-B engines and with wings of 13·72 m (45 ft) span. Subsequently, the prototype was re-engined with more powerful Avco Lycoming O-540 engines, with which it flew for the first time on 17 December 1965. The wing span was also increased by 1·22 m (4 ft) to bring the prototype to production standard.

The production prototype BN-2 Islander (G-ATWU) flew for the first time on 20 August 1966. The Islander received its domestic C of A on 10 August 1967 and an FAA Type Certificate on 19 December 1967.

Deliveries of Islanders began in August 1967, and various models have been delivered to operators in approx 120 countries, with the 1,000th example scheduled for delivery during June 1981.

Initial production aircraft were designated **BN-2**. Those built from 1 June 1969 until 1978 had the designation **BN-2A**, as described in the 1977-78 *Jane's*. The current standard piston-engined model is the **BN-2B** Islander, which has a higher max landing weight and improved interior design. Features include a new range of passenger seats and covers, more robust door locks, improved door seals, improved stainless steel sills, redesigned fresh air system to improve ventilation in hot and humid climates, smaller diameter propellers to decrease cabin noise, and redesigned flight deck and instrument panel.

On 2 August 1980 the prototype (G-BPBN) was flown of the **BN-2T Turbine Islander**, powered by two Allison 250-B17C turboprop engines each flat rated to 238·5 kW (320 shp), and this is described separately.

Military versions known as the **Defender** and **Maritime Defender**, and the three-engined civil **Trislander**, are also described separately.

The basic Islander is available with a choice of two piston-engine power plants, or turboprops, and either standard 14·94 m (49 ft 0 in) span wings or wingtip extensions having raked tips and containing auxiliary fuel tanks. A series of modification kits is available as standard or as an option for new production aircraft, and can also be supplied to operators in the field for retrospective fitting to existing aircraft. The version with 224 kW (300 hp) fuel-injection engines was introduced in 1970, deliveries beginning in November of that year. An extended nose, incorporating 0·62 m³ (22 cu ft) of additional baggage space, was introduced as an optional feature in 1972.

A Rajay turbocharging installation was developed in the United States by Jonas Aircraft, the New York based distributors for Britten-Norman aircraft. The Rajay installation is a bolt-on unit, for manual operation, which can be fitted on to standard 194 kW (260 hp) engines. The superchargers have the effect of increasing the single-engined ceiling to 3,810 m (12,500 ft) and twin-engined ceiling to 7,925 m (26,000 ft). Cruising speed is also increased, from 139 knots (257 km/h; 160 mph) at 2,135 m (7,000 ft) to 146 knots (270 km/h; 168 mph) at 3,050 m (10,000 ft).

The following description applies to the standard piston-engined landplane BN-2B, unless otherwise stated:

Type: Twin-engined feederline transport.

Wings: Cantilever high-wing monoplane. NACA 23012 constant wing section. No dihedral. Incidence 2°. No sweepback. Conventional riveted two-spar torsion-box structure in one piece, using L72 aluminium-clad aluminium alloys. Flared-up wingtips of Britten-Norman design. Raked-back extended wingtips optional. Slotted ailerons and single-slotted flaps of metal construction. Flaps operated electrically, ailerons by pushrods and cables. Ground-adjustable tab on starboard aileron. BTR-Goodrich pneumatic de-icing boots optional.

Fuselage: Conventional riveted four-longeron semi-monocoque structure of pressed frames and stringers and metal skin, using L72 aluminium-clad aluminium alloys. Optional 1·15 m (3 ft 9¼ in) nose extension for baggage stowage.

Tail Unit: Cantilever two-spar structure, with pressed ribs and metal skin, using L72 aluminium-clad aluminium alloys. Fixed-incidence tailplane and mass-balanced elevator. Rudder and elevator are actuated by pushrods and cables. Trim tabs in rudder and elevator. Pneumatic de-icing of tailplane and fin optional.

Landing Gear: Non-retractable tricycle type, with twin wheels on each main unit and single steerable nose-wheel. Cantilever main legs mounted aft of rear spar. All three legs fitted with Lockheed oleo-pneumatic shock-absorbers. All five wheels and tyres size 16 × 7-7, supplied by Goodyear. Tyre pressure: main 2·41 bars (35 lb/sq in); nose 2·00 bars (29 lb/sq in). Foot-operated aircooled Cleveland hydraulic brakes on main units. Parking brake. Wheel/ski gear available optionally.

Power Plant: Two Avco Lycoming flat-six engines, each driving a Hartzell HC-C2YK-2B or -2C two-blade constant-speed feathering metal propeller. Propeller synchronisers optional. Standard power plant is the 194 kW (260 hp) O-540-E4C5, but the 224 kW (300 hp) IO-540-K1B5 or 238·5 kW (320 shp) Allison 250-B17C turboprops can be fitted at customer's option. Optional Rajay turbocharging installation on 194 kW (260 hp) engines, to improve high-altitude performance. Integral fuel tank between spars in each wing, outboard of engine. Total fuel capacity (standard) 518 litres (114 Imp gallons; 137 US gallons). With auxiliary tanks in wingtip extensions, total capacity is increased to 741 litres (163 Imp gallons; 196 US gallons). Additional pylon-mounted underwing auxiliary tanks, each of 227 litres (50 Imp gallons; 60 US gallons) capacity, available optionally. Refuelling point in upper surface of wing above each internal tank. Total oil capacity 22·75 litres (5 Imp gallons).

Accommodation: Up to 10 persons, including pilot, on side-by-side front seats and four bench seats. No aisle. Seat backs fold forward. Access to all seats via three forward-opening doors, forward of wing and at rear of cabin on port side and forward of wing on starboard side. Baggage compartment at rear of cabin, with port-side loading door in standard versions. Exit in

Pilatus Britten-Norman BN-2B Islander with Micronair underwing spray unit

emergency by removing door windows. Special executive layouts available. Can be operated as freighter, carrying more than a ton of cargo; in this configuration the passenger seats can be stored in the rear baggage bay. In ambulance role, up to three stretchers and two attendants can be accommodated. Other layouts possible, including photographic and geophysical survey, parachutist transport or trainer (with accommodation for up to eight parachutists and a dispatcher), firefighting, public health spraying and crop spraying.

SYSTEMS: Southwind cabin heater standard. 45,000 BTU Stewart Warner combustion unit, with circulating fan, provides hot air for distribution at floor-level outlets and at windscreen demisting slots. Fresh air, boosted by propeller slipstream, is ducted to each seating position for on-ground ventilation. Electrical DC power, for instruments, lighting and radio, from two engine-driven 24V 50A self-rectifying alternators and a controller to main busbar and circuit-breaker assembly in nose bay. Emergency busbar with automatic changeover provides a secondary route for essential services. One 24V 17Ah heavy-duty lead-acid battery for independent operation. Ground power receptacle provided. Optional electrical de-icing of propellers and windscreen, and pneumatic de-icing of wing and tail unit leading-edges. Intercom system, including second headset, and passenger address system are standard. Oxygen system available optionally for all versions.

AVIONICS AND EQUIPMENT: Standard items include blind-flying instrumentation, autopilot, dual flying controls and brake system, and a wide range of VHF and HF communications and navigation equipment.

DIMENSIONS, EXTERNAL:
Wing span: standard 14·94 m (49 ft 0 in)
with extended tips 16·15 m (53 ft 0 in)
Wing chord (constant) 2·03 m (6 ft 8 in)
Wing aspect ratio: standard 7·4
with extended tips 7·95
Length overall: standard 10·86 m (35 ft 7¾ in)
optional nose extension 12·02 m (39 ft 5¼ in)
Fuselage: Max width 1·21 m (3 ft 11½ in)
Max depth 1·46 m (4 ft 9¾ in)
Height overall 4·18 m (13 ft 8¾ in)
Tailplane span 4·67 m (15 ft 4 in)
Wheel track (c/l of shock-absorbers) 3·61 m (11 ft 10 in)
Wheelbase: standard 3·99 m (13 ft 1¼ in)
optional nose extension 4·90 m (16 ft 0¾ in)
Propeller diameter 1·98 m (6 ft 6 in)
Cabin door (front, port):
Height 1·10 m (3 ft 7½ in)
Width: top 0·64 m (2 ft 1¼ in)
Height to sill 0·59 m (1 ft 11¼ in)
Cabin door (front, starboard):
Height 1·10 m (3 ft 7½ in)
Max width 0·86 m (2 ft 10 in)
Height to sill 0·57 m (1 ft 10½ in)
Cabin door (rear, port):
Height 1·09 m (3 ft 7 in)
Width: top 0·635 m (2 ft 1 in)
bottom 1·19 m (3 ft 11 in)
Height to sill 0·52 m (1 ft 8½ in)
Baggage door (rear, port):
Height 0·69 m (2 ft 3 in)

DIMENSIONS, INTERNAL:
Passenger cabin, aft of pilot's seat:
Length 3·05 m (10 ft 0 in)
Max width 1·09 m (3 ft 7 in)
Max height 1·27 m (4 ft 2 in)
Floor area 2·97 m² (32 sq ft)
Volume 3·68 m³ (130 cu ft)
Baggage space aft of passenger cabin:
standard 0·85 m³ (30 cu ft)
maximum 1·39 m³ (49 cu ft)
Nose baggage compartment (optional) 0·62 m³ (22 cu ft)
Freight capacity:
aft of pilot's seat, incl rear cabin baggage space 4·70 m³ (166 cu ft)
with four bench seats folded into rear cabin baggage space 3·68 m³ (130 cu ft)

AREAS:
Wings, gross: standard 30·19 m² (325·0 sq ft)
with extended tips 31·31 m² (337·0 sq ft)
Ailerons (total) 2·38 m² (25·6 sq ft)
Flaps (total) 3·62 m² (39·0 sq ft)
Fin 3·41 m² (36·64 sq ft)
Rudder, incl tab 1·60 m² (17·2 sq ft)
Tailplane 6·78 m² (73·0 sq ft)
Elevator, incl tabs 3·08 m² (33·16 sq ft)

WEIGHTS AND LOADINGS (A: standard wings, B: extended wings, C: 194 kW; 260 hp and D: 224 kW; 300 hp engines):
Weight empty, equipped (without avionics):
C 1,638 kg (3,612 lb)
D 1,695 kg (3,738 lb)
Max T-O and landing weight (A, B) 2,993 kg (6,600 lb)
Max zero-fuel weight (BCAR):
A/C, A/D 2,855 kg (6,300 lb)
B/C, B/D 2,810 kg (6,200 lb)

Pilatus Britten-Norman BN-2A Defender of the Botswana Defence Force

Max wing loading: A 99·1 kg/m² (20·3 lb/sq ft)
B 95·7 kg/m² (19·6 lb/sq ft)
Max floor loading, without cargo panels 586 kg/m² (120 lb/sq ft)
Max power loading: C 7·71 kg/kW (12·7 lb/hp)
D 6·68 kg/kW (11·0 lb/hp)

PERFORMANCE (C: at 2,855 kg; 6,300 lb AUW, ISA, 194 kW; 260 hp engines. D: at max T-O weight, ISA, 224 kW; 300 hp engines):
Never-exceed speed:
C, D (standard wings) 177 knots (327 km/h; 203 mph) IAS
C, D (extended wings) 184 knots (340 km/h; 211 mph) IAS
Max level speed at S/L:
C 148 knots (274 km/h; 170 mph)
D 151 knots (280 km/h; 173 mph)
Max cruising speed (75% power) at 2,135 m (7,000 ft):
C 139 knots (257 km/h; 160 mph)
D 142 knots (264 km/h; 164 mph)
Cruising speed (67% power) at 2,750 m (9,000 ft):
C 134 knots (248 km/h; 154 mph)
D 137 knots (254 km/h; 158 mph)
Cruising speed (59% power) at 3,660 m (12,000 ft):
C 130 knots (241 km/h; 150 mph)
D 132 knots (245 km/h; 152 mph)
Stalling speed, flaps up:
C 50 knots (92 km/h; 57 mph) IAS
Stalling speed, flaps down:
C, D 43 knots (79 km/h; 49 mph) IAS
Max rate of climb at S/L: C 290 m (950 ft)/min
D 344 m (1,130 ft)/min
Rate of climb at S/L, one engine out:
C 58·5 m (192 ft)/min
D 61 m (200 ft)/min
Absolute ceiling: C 4,938 m (16,200 ft)
D 5,974 m (19,600 ft)
Service ceiling: C 4,450 m (14,600 ft)
D 5,485 m (18,000 ft)
Service ceiling, one engine out:
C, standard wings 1,737 m (5,700 ft)
C, extended wings 1,981 m (6,500 ft)
D, standard wings 1,890 m (6,200 ft)
D, extended wings 2,133 m (7,000 ft)
Min ground turning radius 9·45 m (31 ft 0 in)
T-O run at S/L, zero wind, hard runway:
C 169 m (555 ft)
D 203 m (665 ft)
T-O run at 1,525 m (5,000 ft):
D 296 m (970 ft)
T-O to 15 m (50 ft) at S/L, zero wind, hard runway:
C 332 m (1,090 ft)
D 335 m (1,100 ft)
T-O to 15 m (50 ft) at 1,525 m (5,000 ft):
D 475 m (1,560 ft)
Landing from 15 m (50 ft) at S/L, zero wind, hard runway:
C 292 m (960 ft)
D 299 m (980 ft)
Landing from 15 m (50 ft) at 1,525 m (5,000 ft):
D 357 m (1,170 ft)
Landing run at 1,525 m (5,000 ft):
D 168 m (550 ft)
Landing run at S/L, zero wind, hard runway:
C 137 m (450 ft)
D 140 m (460 ft)
Range at 75% power at 2,135 m (7,000 ft):
C, standard wings 622 nm (1,153 km; 717 miles)
C, extended wings 903 nm (1,673 km; 1,040 miles)
D, standard wings 555 nm (1,028 km; 639 miles)
D, extended wings 816 nm (1,513 km; 940 miles)
Range at 67% power at 2,750 m (9,000 ft):
C, standard wings 713 nm (1,322 km; 822 miles)
C, extended wings 1,036 nm (1,920 km; 1,193 miles)
D, standard wings 577 nm (1,070 km; 665 miles)
D, extended wings 847 nm (1,569 km; 975 miles)
Range at 59% power at 3,660 m (12,000 ft):
C, standard wings 755 nm (1,400 km; 870 miles)
C, extended wings 1,096 nm (2,032 km; 1,263 miles)
D, standard wings 613 nm (1,136 km; 706 miles)
D, extended wings 905 nm (1,677 km; 1,042 miles)

PILATUS BRITTEN-NORMAN DEFENDER

The Defender is a variant of the civil Islander which can be adapted for a wide variety of government and military roles such as search and rescue, internal security, long-range patrol, forward air control, troop transport, logistic support and casualty evacuation. It is available with the same choices of wing configuration as the current civil versions and can be equipped with a wide range of highly sophisticated avionics, including nose-mounted weather radar, providing the aircraft with a marine search capability. Optional equipment includes four NATO standard underwing pylons for a variety of external stores, the inboard pair each carrying up to 317·5 kg (700 lb) and the outboard pair up to 204 kg (450 lb).

Typical underwing loads include twin 7·62 mm machine-guns in pod packs, 250 lb or 500 lb GP bombs, Matra rocket packs, SURA rocket clusters, wire-guided missiles, 5 in reconnaissance flares, anti-personnel grenades, smoke bombs, marker bombs and 227 litre (50 Imp gallon; 60 US gallon) drop-tanks.

Internal capacity for passengers, stretcher cases or cargo is the same as that of the civil Islander.

Britten-Norman Defenders/Islanders are in service with the Abu Dhabi Defence Force, Belgian Army, Botswana Defence Force, British Army Parachute Association, Ghana Air Force, Guyana Defence Force, Indian Navy, Jamaica Defence Force, Malagasy Air Force, Philippine Navy, Presidential Flight of the Mexican Air Force, Royal Hong Kong Auxiliary Air Force, Panamanian Air Force, Sultan of Oman's Air Force, Mauritania Islamic Defence Force, the Seychelles Ministry of Agriculture and Fisheries, the Malawi Army Air Wing and the Rwanda Air Force. Those operated by the air forces of Iraq, Israel and Qatar are military Islanders, and are not equipped to carry offensive weapons.

Pilatus Britten-Norman BN-2A Maritime Defender

The description given for the BN-2B Islander applies also to the Defender, except as follows:

Power Plant: Two 224 kW (300 hp) Avco Lycoming IO-540-K1B5 flat-six engines standard.

Avionics: Typical installation comprises King 360-channel VHF nav/com transceivers with VOR/LOC and VOR/ILS, ADF, marker beacon receiver, KT 76 transponder, SunAir ASB 500 HF com transceiver, Bendix radar and Collins autopilot.

Weights and Loadings:
Weight empty, equipped (excl avionics) 1,824 kg (4,020 lb)
Max T-O and landing weight 2,993 kg (6,600 lb)
Max wing loading 95·7 kg/m² (19·6 lb/sq ft)
Max power loading 6·68 kg/kW (11·0 lb/hp)

Performance (at max T-O weight, ISA. A: no stores on pylons; B: pylons loaded):
Max level speed:
A 151 knots (280 km/h; 174 mph)
B 144 knots (266 km/h; 166 mph)
Cruising speed, 67% power at 3,050 m (10,000 ft):
A 138 knots (255 km/h; 159 mph)
B 131 knots (242 km/h; 150 mph)
Cruising speed, 59% power at 610 m (2,000 ft):
A 123 knots (227 km/h; 141 mph)
B 116 knots (215 km/h; 133 mph)
Stalling speed, flaps down:
A, B 39 knots (73 km/h; 45 mph)
Max rate of climb at S/L: A 396 m (1,300 ft)/min
B 357 m (1,170 ft)/min
Service ceiling: A, B 5,180 m (17,000 ft)
Absolute ceiling: A, B 6,100 m (20,000 ft)
T-O to 15 m (50 ft): A, B 320 m (1,050 ft)
Landing from 15 m (50 ft): A, B 303 m (995 ft)
Range with max payload:
A 363 nm (672 km; 418 miles)
B 326 nm (603 km; 375 miles)
Range with standard fuel:
A 1,096 nm (2,027 km; 1,260 miles)
B 1,000 nm (1,850 km; 1,150 miles)
Max range with auxiliary fuel, no reserves, at full mission weight with max endurance power setting 1,497 nm (2,772 km; 1,723 miles)

PILATUS BRITTEN-NORMAN MARITIME DEFENDER

Generally similar to the Defender, the Maritime Defender differs by having a modified nose with a larger search radar, capable of detecting a 100 m² (1,076 sq ft) target in sea state 4-5 at a range of 36 nm (67 km; 41·5 miles). Scanning 60° on each side of the flight path, the radar provides a search width of 60 nm (111 km; 69 miles) at optimum altitude. The interior layout provides for pilot and co-pilot, a radar operator at a mid-cabin position on the starboard side, and two observers in the rear of the cabin, one aft of the radar operator, and one adjacent to a window on the port side.

Intended for coastal patrol, fishery and oil rig protection duties, as well as search and rescue support, the Maritime Defender is suitable for all-weather operation, by day or night, and carries the equipment necessary to fulfil such roles. This can include Collins PN 101 compass/HSI, horizon gyro (radar stabilisation), autopilot, Bendix RDR 1400 radar, Marconi CMA 734 Omega, Plessey PVI 712 radio altimeter, dual Collins VHF 20 com, dual Collins VIR 30M VHF nav/ILS, VHF marine band com, Collins ADF 60, Collins TDR 90 transponder, Collins DME 40, encoding altimeter, and SunAir ASB 500 SSB HF com. Specialised equipment includes a searchlight installation and hand-held camera; the four underwing pylons can be used to carry a loudspeaker pod, flares, parachute dinghy packs and a variety of weapons.

The description of the Defender applies also to the Maritime Defender, except that overall length is increased to 11·07 m (36 ft 3¾ in).

PILATUS BRITTEN-NORMAN BN-2T TURBINE ISLANDER

On 2 August 1980 the prototype (G-BPBN) was flown of the BN-2T Turbine Islander, powered by two Allison 250-B17C turboprop engines. These enable the BN-2T to use available low-cost jet fuel instead of scarce and costly Avgas, and offer a particularly low operating noise level. Improvements have also been made to the oil cooling system, engine cowling and propeller control.

British CAA certification of the Turbine Islander was received at the end of May 1981; FAA type approval was then expected to follow within a few months. Rollout of the first production aircraft was due in the Summer of 1981. The Turbine Islander is available in the same range of applications as the piston-engined Islander (which continues in production), including a military version known as the **Turbine Defender**.

The description of the BN-2B Islander applies also to the BN-2T, except as follows:

Type: Twin-turboprop feederline transport.

Wings: Generally as for BN-2B, except that the optional raked-back extended wingtips are not available.

Power Plant: Two 298 kW (400 shp) Allison 250-B17C turboprop engines, flat rated to 238·5 kW (320 shp), and each driving a Hartzell three-blade constant-speed fully-feathering metal propeller. Propeller synchronisers optional. Total standard fuel capacity 518 litres (114 Imp gallons). Pylon-mounted underwing auxiliary tanks, each of 227 litres (50 Imp gallons) capacity, available optionally. Total oil capacity 5·7 litres (1·25 Imp gallons).

Accommodation: Generally as for BN-2B, except in ambulance role can accommodate, in addition to the pilot, a single stretcher, one medical attendant, and five seated occupants; or two stretchers, one attendant, and three passengers; or three stretchers, two attendants, and one passenger. Other possible layouts include photographic and geophysical survey; parachutist transport or trainer (with accommodation for up to eight parachutists and a dispatcher): and pest control or other agricultural spraying. Maritime Turbine Islander/Defender versions available for fishery protection, coastguard patrol, pollution survey, search and rescue, and similar applications.

Avionics and Equipment: Standard avionics and equipment generally similar to BN-2B. Other equipment, according to mission, includes fixed tail 'stinger' or towed 'bird' magnetometer, spectrometer, or electromagnetic detection/analysis equipment (geophysical survey); one or two cameras, navigation sights, and appropriate avionics (photographic survey); 188·7 litre (41·5 Imp gallon) Micronair underwing spraypods complete with pump and rotary atomiser (pest control/agricultural spraying versions); radar, VLF Omega nav system, radar altimeter, marine band and VHF transceivers, dinghies, survival equipment, and special crew accommodation (maritime versions).

Prototype of Pilatus Britten-Norman BN-2T Turbine Islander

Dimensions, external:
As for BN-2B, except
Length overall: standard nose 10·87 m (35 ft 7¾ in)
weather radar nose 11·07 m (36 ft 3¾ in)
Propeller diameter 2·03 m (6 ft 8 in)

Weights (standard aircraft without auxiliary fuel):
Weight empty, equipped (incl pilot) 1,869 kg (4,120 lb)
Max usable fuel 395 kg (871 lb)
Payload with max fuel 730 kg (1,609 lb)
Max T-O and landing weight 2,994 kg (6,600 lb)

Performance (standard aircraft without auxiliary fuel, at max T-O weight, ISA, unless stated otherwise):
Max cruising speed at 3,050 m (10,000 ft) 171 knots (317 km/h; 197 mph)
Max cruising speed at S/L 156 knots (289 km/h; 180 mph)
Econ cruising speed at 3,050 m (10,000 ft) 141 knots (261 km/h; 162 mph)
Max rate of climb at S/L 335 m (1,100 ft)/min
Rate of climb at S/L, one engine out 69 m (225 ft)/min
Service ceiling over 6,100 m (20,000 ft)
Service ceiling, one engine out over 3,050 m (10,000 ft)
T-O to 15 m (50 ft) 396 m (1,300 ft)
Landing from 15 m (50 ft) 329 m (1,080 ft)
Range (IFR) with max fuel, reserves for 45 min hold plus 10% 334 nm (619 km; 384 miles)
Range (VFR) with max fuel, no reserves 447 nm (828 km; 514 miles)

PILATUS BRITTEN-NORMAN BN-2A Mk III TRISLANDER

In the Autumn of 1970 Britten-Norman introduced an enlarged development of the twin-engined Islander, having a third engine mounted at the rear and a lengthened fuselage seating up to 17 passengers.

The prototype Trislander was produced by converting the second prototype of the twin-engined Islander (G-ATWU), adding a 2·29 m (7 ft 6 in) length of parallel-section fuselage forward of the wing, reinforcing the rear fuselage and fitting a new main landing gear with larger wheels and tyres. The tail unit was modified to act as a mount for the third engine. This aircraft made its first flight on 11 September 1970, appearing at the SBAC Display at Farnborough later the same day. Production aircraft have additional fin area above the rear engine.

The prototype was later dismantled and its fuselage used for structural testing. By the end of 1970 construction had begun of three production aircraft by converting standard Islander airframes from the current production line. The first production Trislander (G-AYTU) was flown on 6 March 1971, and the first delivery (to Aurigny Air Services in the Channel Islands) was made on 29 June 1971.

ARB certification of the Trislander, granted on 14 May 1971, approved the aircraft for both VFR and IFR operation and for full public transport use with one pilot and 17 passengers. FAA certification followed on 4 August 1971, to FAR Pt 23 and to the latest air taxi requirements of SFAR Pt 23 and Appendix A of FAR Pt 135.

By mid-1981 orders had been received for more than 80 Trislanders. Of these more than 70 had been delivered, to customers in the UK, Africa, Australasia, USA, Canada, Indonesia and South America.

The following versions of the Trislander are currently available:

BN-2A Mk III-2. Standard version with extended nose containing baggage compartment.

BN-2A Mk III-3. As Mk III-2, with an autofeather system which feathers the propeller automatically should an engine fail on take-off.

BN-2A Mk III-4. As Mk III-3, plus a standby rocket engine to provide additional thrust should an engine fail on take-off.

Type: Three-engined feederline transport.

Wings: Cantilever high-wing monoplane. NACA 23012 constant wing section. No dihedral. Incidence 2°. No sweepback. Conventional riveted two-spar torsion-box structure in one piece, using aluminium-clad aluminium alloys. Increases in skin gauges and spar laminates compared with twin-engined versions. Structure is strictly 'safe-life', but exhibits several fail-safe features and principles. Flared-up wingtips of Britten-Norman design, with raked tips. Slotted ailerons and electrically-operated single-slotted permanently-drooped flaps of metal construction. Ground-adjustable tab in starboard aileron. BTR-Goodrich pneumatic de-icing boots optional.

Fuselage: Conventional riveted four-longeron semi-monocoque structure of pressed frames and stringers and metal skin, using L72 aluminium-clad aluminium alloys. Some reinforcement of fuselage aft of wing to support weight of rear engine. Structure is strictly 'safe-life', but has several fail-safe features and principles.

Tail Unit: Cantilever structure, using L72 aluminium-clad aluminium alloys, with low aspect ratio main fin which also acts as mount for the third engine. Fixed-incidence tailplane (with raked tips) and elevators are similar in construction to those of Islander. Trim tab in rudder. BTR-Goodrich pneumatic de-icing boots for tailplane optional.

LANDING GEAR: Non-retractable tricycle type, with twin-wheel main units and single steerable nosewheel. Cantilever main legs mounted aft of rear spar. All five wheels and tyres are Cleveland size 7·00-6. Tyre pressure 3·10 bars (45 lb/sq in) on main units, 2·00 bars (29 lb/sq in) on nose unit. Cleveland foot-operated air-cooled hydraulic disc brakes on main units. Parking brake. No anti-skid units. Fairings fitted to main gear extension tubes below the engine nacelle and above the shock-absorber attachment bolts.

POWER PLANT: Three 194 kW (260 hp) Avco Lycoming O-540-E4C5 flat-six engines (two mounted on wings and one on vertical tail), each driving a Hartzell HC-C2YK-2G/C8477-4 two-blade constant-speed fully-feathering metal propeller. Automatic feathering device available as an option. Standby rocket engine, mounted just below rear of tail-engine nacelle, is available as an option: weighing 21 kg (46·2 lb), this provides 1·56 kN (350 lb st) for 12 s. Fuel in two integral tanks between front and rear wing spars, outboard of the engine nacelles, and two tanks in wingtips. Total fuel capacity 746 litres (164 Imp gallons; 197 US gallons). Overwing refuelling point above each tank. Oil capacity 34 litres (7·5 Imp gallons; 9 US gallons).

ACCOMMODATION: Up to 18 persons, including pilot, in pairs on bench seats at approx 79 cm (31 in) pitch. Access to all seats provided by five doors, two on port side and three on starboard side. Baggage compartment at rear of cabin, with external baggage door on port side. Exit in emergency by removing window panels in front four passenger doors. Heating, ventilation and sound insulation standard. Ambulance or VIP interior layouts at customer's option. Dual controls standard.

SYSTEMS: One Southwind cabin heater fitted as standard. DC electrical system includes two 24V 50A self-rectifying alternators, supplying the instruments, lighting and radio, and a 24V 17Ah battery. No hydraulic or pneumatic systems, except for self-contained hydraulic brakes.

AVIONICS AND EQUIPMENT: Optional avionics include Collins autopilot, a wide range of Collins, Bendix, King or Narco VHF nav/com equipment, including ADF, DME, and Collins or SunAir HF transceiver. Optional equipment includes windscreen de-icing, second cabin heater, cargo tiedowns, anti-collision strobe beacons, emergency exit Beta lights, electric propeller de-icing and pneumatic airframe de-icing systems.

DIMENSIONS, EXTERNAL:

Wing span	16·15 m (53 ft 0 in)
Wing chord (constant)	2·03 m (6 ft 8 in)
Wing aspect ratio	7·95
Length overall	15·01 m (49 ft 3 in)
Fuselage: Max width	1·21 m (3 ft 11½ in)
Max depth	1·46 m (4 ft 9¾ in)
Height overall	4·32 m (14 ft 2 in)
Tailplane span	6·48 m (21 ft 3 in)
Wheel track (c/l of shock-absorbers)	3·35 m (11 ft 0 in)
Wheelbase	7·12 m (23 ft 4¼ in)
Propeller diameter	2·03 m (6 ft 8 in)
Propeller ground clearance	0·69 m (2 ft 3 in)
Distance between propeller centres (wing engines)	3·61 m (11 ft 10 in)
Passenger doors (stbd, fwd and centre):	
Height	1·10 m (3 ft 7½ in)
Max width	0·89 m (2 ft 10·9 in)
Height to sill	0·57 m (1 ft 10½ in)
Passenger doors (port, fwd and rear):	
Height	1·09 m (3 ft 7 in)
Max width	1·21 m (3 ft 11·9 in)
Height to sill	0·57 m (1 ft 10½ in)
Passenger door (stbd, rear):	
Height	1·09 m (3 ft 7 in)
Width	0·75 m (2 ft 5½ in)
Baggage compartment door (rear, port):	
Height	0·66 m (2 ft 1·95 in)
Width	0·44 m (1 ft 5·2 in)
Nose baggage compartment door (port, optional):	
Width	0·79 m (2 ft 7 in)

DIMENSIONS, INTERNAL:

Cabin: Length, excl flight deck but incl rear baggage compartment	8·24 m (27 ft 0½ in)
Max width	1·09 m (3 ft 7 in)
Max height	1·27 m (4 ft 2 in)
Floor area	7·85 m² (84·45 sq ft)
Volume	9·54 m³ (337 cu ft)
Rear baggage compartment volume	0·71 m³ (25·0 cu ft)
Nose baggage compartment volume (optional)	0·62 m³ (22·0 cu ft)

AREAS:

Wings, gross	31·31 m² (337·0 sq ft)
Ailerons (total)	2·38 m² (25·6 sq ft)
Trailing-edge flaps (total)	3·62 m² (39·0 sq ft)
Fin	5·83 m² (62·7 sq ft)
Rudder, incl tab	1·13 m² (12·2 sq ft)
Tailplane	8·36 m² (90·0 sq ft)
Elevators	2·42 m² (26·0 sq ft)

WEIGHTS AND LOADINGS:

Weight empty, equipped (without avionics)	2,650 kg (5,843 lb)
Max T-O and landing weight	4,536 kg (10,000 lb)
Max wing loading	144·8 kg/m² (29·67 lb/sq ft)
Max power loading	7·79 kg/kW (12·8 lb/hp)

PERFORMANCE (at max T-O weight, ISA):

Max level speed at S/L	156 knots (290 km/h; 180 mph)
Cruising speed (75% power) at 1,980 m (6,500 ft)	144 knots (267 km/h; 166 mph)
Cruising speed (67% power) at 2,470 m (9,000 ft)	138 knots (256 km/h; 159 mph)
Cruising speed (59% power) at 3,960 m (13,000 ft)	130 knots (241 km/h; 150 mph)
Max rate of climb at S/L	298 m (980 ft)/min
Rate of climb at S/L, one engine out	86 m (283 ft)/min
Absolute ceiling	4,450 m (14,600 ft)
Service ceiling	4,010 m (13,150 ft)
Service ceiling, one engine out	2,105 m (6,900 ft)
T-O run at S/L, zero wind, hard runway	393 m (1,290 ft)
T-O to 15 m (50 ft) at S/L, zero wind, hard runway	594 m (1,950 ft)
Landing from 15 m (50 ft) at S/L, zero wind, hard runway	440 m (1,445 ft)
Landing run at S/L, zero wind, hard runway	259 m (850 ft)
Max still-air range at 59% cruising power	868 nm (1,610 km; 1,000 miles)

Pilatus Britten-Norman BN-2A Mk III-2 Trislander of Inter-Island Airways, Seychelles

---

PRACTAVIA
PRACTAVIA LTD

HEAD OFFICE: Wycombe Air Park, Booker, near Marlow, Buckinghamshire
WORKS: Carlisle Airport, Carlisle, Cumbria
Telephone: 0228 73333

This company was formed initially to market plans and kits for amateur construction of the Sprite two-seat all-metal semi-aerobatic light aircraft, the design of which was initiated by *Pilot* magazine. First flight by one of these aircraft was made on 16 June 1976. Details of the homebuilt version can be found in the Homebuilt section of this edition.

In late 1980, Practavia released preliminary details of its intention to offer factory-built versions of the Sprite, with a choice of power plants and a modified fuel system. Manufacture of subassemblies for the first ten production aircraft was then under way at Practavia's Wycombe Air Park base; final assembly was scheduled to begin in March 1981 at Carlisle Airport, in a hangar used previously by CSE Aviation for the assembly of Piper Tomahawks. The first flight of a factory-built Sprite was expected before the end of 1981. At the time of the 1980 announcement, Practavia was engaged in negotiations with an undisclosed Arab nation to manufacture the Sprite under licence with kits supplied from the UK.

PRACTAVIA SPRITE

Compared with the homebuilt version, which has wingtip fuel tanks, the production Sprite has a wing of equivalent span without tip-tanks, and all fuel is carried in wing leading-edge tanks instead of in the fuselage and tip-tanks.

TYPE: Two-seat all-metal semi-aerobatic light aircraft.

WINGS: Cantilever low-wing monoplane. Wing section NACA 64315. Dihedral 6° on outer panels only. No incidence or sweepback. All-metal constant-chord structure of aluminium alloy, built in three portions. Outer panels are detachable for transportation and storage. Single main spar, with light rear spar, forming central torsion box. Skins and ribs of L72 alloy, extrusions of L65 alloy, and spar caps of L73 alloy. Single-slotted flaps and plain ailerons of L72 alloy extend over almost entire span of trailing-edge. No tabs.

FUSELAGE: All-metal semi-monocoque structure with no double curvature. Longerons of L65 aluminium alloy, skins and frames of L72 alloy. Sides and top curved to avoid drumming.

TAIL UNIT: Cantilever all-metal structure, with swept vertical surfaces, constructed of L72 alloy. Fixed-incidence tailplane. Trim tab in centre of elevator trailing edge, of one-third span; outer one-third on each side comprises anti-balance tab.

LANDING GEAR: Non-retractable tricycle type standard, although design of wing structure will allow for retractable gear as a future development. Shock-absorption by rubber in compression. Wheels and tyres size 5·00-5. Hydraulic disc brakes.

POWER PLANT: One 79 kW (106 hp) Avco Lycoming O-235 or 112 kW (150 hp) Avco Lycoming O-320 flat-four engine, driving a two-blade fixed-pitch propeller with spinner. Fuel contained in two wing leading-edge tanks, total capacity 136·4 litres (30 Imp gallons).

ACCOMMODATION: Two seats side by side in enclosed cockpit, with rearward-sliding transparent canopy. Space for baggage behind seats.

SYSTEMS: Hydraulic system for brakes only. 12V electrical system.

AVIONICS AND EQUIPMENT: Nav/com and blind-flying instrumentation standard.

DIMENSIONS, EXTERNAL:

Wing span	8·23 m (27 ft 0 in)
Wing chord, constant	1·22 m (4 ft 0 in)
Wing aspect ratio	6·75
Length overall	6·10 m (20 ft 0 in)
Width, outer wing panels removed	2·44 m (8 ft 0 in)
Height overall	2·51 m (8 ft 3 in)
Tailplane span	2·44 m (8 ft 0 in)
Wheel track	2·29 m (7 ft 6 in)
Wheelbase	1·40 m (4 ft 7 in)

DIMENSION, INTERNAL:

Cabin: Max width	1·17 m (3 ft 10 in)

AREA:

Wings, gross	10·03 m² (108 sq ft)

WEIGHTS AND LOADINGS (with 97 kW; 130 hp engine):

Weight empty	476 kg (1,050 lb)
Max T-O weight	748·5 kg (1,650 lb)
Max wing loading	74·7 kg/m² (15·3 lb/sq ft)
Max power loading	7·72 kg/kW (12·7 lb/hp)

PERFORMANCE (at max T-O weight, ISA, with 97 kW; 130 hp engine):

Never-exceed speed	250 knots (463 km/h; 288 mph)
Max level speed at 610 m (2,000 ft)	121 knots (224 km/h; 139 mph)
Cruising speed, 75% power, at 610 m (2,000 ft)	108 knots (200 km/h; 124 mph)
Stalling speed, flaps up	49 knots (91 km/h; 56·5 mph)
Stalling speed, 30° flap	43·5 knots (81 km/h; 50 mph)
Rate of climb at S/L, 75% power	219 m (720 ft)/min
T-O to 15 m (50 ft)	442 m (1,450 ft)
Landing from 15 m (50 ft)	427 m (1,400 ft)
Range with max fuel, incl reserves	543 nm (1,006 km; 625 miles)
Endurance with max fuel, incl reserves	5 h
g limits	+6·0/−3·0

SHORTS
SHORT BROTHERS LIMITED

HEAD OFFICE, WORKS AND AERODROME: PO Box 241, Airport Road, Belfast BT3 9DZ, Northern Ireland
Telephone: 0232 58444
Telex: 74688
OTHER FACTORIES: Newtownards, Castlereagh, Belfast (3)
LONDON OFFICE: Berkeley Square House, Berkeley Square, London W1X 5LB
CHAIRMAN: Sir George Leitch, KCB
MANAGING DIRECTOR: Sir Philip Foreman, CBE, DL
DIRECTORS:
Sir Sidney Bacon
F. F. H. Charlton

D. W. G. L. Haviland, CB
R. W. R. McNulty
A. F. C. Roberts, OBE
James Sim
Dr Llewellyn Smith, CBE
H. E. Trevan-Hawke
SECRETARY: R. Milnes
EXECUTIVE DIRECTORS:
B. Carlin (Manufacturing)
B. P. Laight, OBE (Engineering)
A. R. Manvell (Missiles)
R. W. R. McNulty (Finance and Administration)
A. F. C. Roberts, OBE (Aircraft)
K. W. Tyson (Commercial)
M. I. Wild
CHIEF TEST PILOT: L. L. Cumming
MANAGER, PUBLIC RELATIONS SERVICES: Derek S. Jones

Short Brothers were first established, as aeronauts, in 1898, and in 1901 they began the manufacture of balloons at Hove, Sussex. Works were later established in London, first at premises off the Tottenham Court Road and, in 1906, at Battersea. In March 1909 Shorts opened the United Kingdom's first purpose-built aircraft factory at Shellbeach, Isle of Sheppey. (The Short No. 1 had been ordered by F. T. McClean in January 1909.) The contract placed with Shorts by the Wright brothers in March 1909, for six Flyers, was the first aircraft production contract to be placed in the UK. The main aircraft works moved to Eastchurch in 1909-10, and to Rochester in 1913-14.

In June 1936 Short Brothers, in collaboration with Harland & Wolff Ltd, formed a new company known as Short & Harland Ltd to build aircraft in Belfast, and in 1947 activities were concentrated in Belfast under the name Short Bros & Harland Ltd. The name Short Brothers Ltd was re-adopted on 1 June 1977.

The British government now owns, directly or indirectly, 100% of the issued shareholding. In 1980 the company received the Queen's Award to Industry for the 11th time. It had 7,000 employees in February 1981.

The company's current products include the Shorts 330 30-seat and 360 36-seat commuter airliners, and the Skyvan turboprop STOL light transport, in use throughout the world for passenger, freight, survey, military and miscellaneous operations.

Internationally, Shorts is collaborating as risk-sharing partner with Fokker BV in production of the Fokker F28 Fellowship transport, with responsibility for the wings; and holds contracts to produce ailerons, spoilers, wingtips, landing gear doors, galley doors, environmental control system doors and tail unit rib assemblies for the Lockheed L-1011 TriStar, landing gear doors for the Boeing 747, and inboard trailing-edge flap assemblies for the Boeing 757. During 1967, Shorts began the design and manufacture of pods for Rolls-Royce jet engines, and is responsible for podding Rolls-Royce RB.211 turbofan engines for the TriStar. Deliveries of these direct to Lockheed at Palmdale began in the early Summer of 1970. The company is also podding RB.211 engines for the Boeing 747 and 757, and has produced pods for more than 600 RB.211s. A contract to produce pods for the Avco Lycoming ALF 502 turbofans of the British Aerospace 146 was announced in January 1979, and the first of these was delivered in January 1981. A contract from Pratt & Whitney, to design and build the first flight test pod for the PW2037 turbofan engine chosen to power some Boeing 757s, was announced in June 1981.

To cope with its involvement in the TriStar programme, Shorts installed some of Europe's most advanced facilities for the hot-forming of titanium and the manipulation of high-temperature creep-resistant alloys, and in 1981 announced the installation of a 110 ton autoclave with an internal working capacity of 12·8 × 4·57 m (42 × 15 ft) to increase the company's bonding facilities. It is conducting advanced research into jet-engine noise reduction and metal bonding. Shorts is also quality-approved subcontractor to many major US and UK aerospace companies. Conversely, production of the wings for the Shorts 330 is undertaken jointly by British Aerospace and Fokker, and production of the 330 landing gear by Menasco in Canada.

In addition to its activities in the field of piloted aircraft, Shorts is engaged in missile development and production, production of supersonic target drones, and production to MoD contract of the Skeet target drone.

The company's Flying Services Division operates maintenance units and airfields for various civil and military organisations, and flies and maintains aircraft and target drones for the Ministry of Defence. This includes operation of the target service, supply and recovery flight at the Woomera range in Australia.

SHORTS 330

The Shorts 330 (originally SD3-30) is a 30-passenger twin-turboprop transport aircraft designed primarily for commuter and regional air service operators whose current 18/20-seat aircraft require replacement by larger aircraft.

Design of the 330 is derived from that of the Skyvan STOL utility transport, and it retains many of the latter type's well-proven characteristics, including the large cabin cross-section. The same safe-life concept and design philosophy is employed in the structural components. The cabin, including the toilet and galley compartments, is 3·78 m (12 ft 5 in) longer than that of the Skyvan Srs 3.

Two prototypes and the first production aircraft were used for the development programme. The first prototype (G-BSBH) flew for the first time on 22 August 1974. Eight days earlier, the first order for the 330 was placed by Command Airways of Poughkeepsie, New York, for three aircraft. CAA certification to full Transport Category requirements was granted on 18 February 1976; this was followed on 18 June 1976 by US FAR Pt 25 and Pt 36 approval, and subsequently by approvals from the Canadian Dept of Transport and the West German LBA. The 330 conforms with CAB Pt 298 (US) and meets the noise requirements of FAR Pt 36 by a substantial margin. Initial deliveries began in June 1976; first to enter service, on 24 August 1976, was a Time Air 330.

Firm orders and options for nearly 90 Shorts 330s had been placed by 3 August 1981, the firm orders comprising:

Aeronaves del Centro (Venezuela)	3
Air North (Vermont)	2
Alidair (UK)	2
ALM Antillean Airlines (Curaçao)	2
Atlanta Express (Georgia)	2
Chautauqua Airlines (New York)	2
Command Airways (New York)	4
Crown Airways (Pennsylvania)	2
DLT (Germany)	7
Eastern Airways (UK)	1
Golden West Airlines (California)	5
Hawaiian Air	3
Henson Aviation (Maryland)	4
Jet Charter Airlines (Australia)	4
LAPA (Argentina)	2
Loganair (Scotland)	2
Metro Airlines (Texas)	5
Mississippi Valley Airlines (Wisconsin)	4
Olympic Airways (Greece)	2
Pennsylvania Commuter Airlines	1
Royale Airlines (Louisiana)	1
Suburban Airlines (Pennsylvania)	6
Time Air (Canada)	3

Of these, approx 60 had been delivered. They had carried more than four million passengers by 1 January 1981.

TYPE: Twin-turboprop transport aircraft.

WINGS: Braced high-wing monoplane, of all-metal safe-life construction, built in three sections. Wing sections NACA 63A series (modified). Thickness/chord ratio 18% at root, 14% on outer panels. Dihedral 3° on outer panels. Centre-section, integral with top of centre-fuselage, has taper on leading- and trailing-edges, and is a two-spar single-cell box structure of light alloy with conventional skin and stringers. The strut-braced outer panels, which are pin-jointed to the centre-section, are reinforced Skyvan constant-chord units, built of light alloy, and each consists of a two-cell box having wing skins made up of a smooth outer skin bonded to a corrugated inner skin. All-metal single-slotted ailerons. Geared trim tabs in ailerons. All-metal single-slotted flaps, each in three sections. Primary control surfaces are rod-actuated. Strengthened wing and struts of Shorts 360 to become available in 1982.

FUSELAGE: Light alloy structure, built in two main portions: nose (including flight deck, nosewheel bay and forward baggage compartment); and the centre (including main wing spar attachment frames and lower transverse beams which carry the main landing gear and associated fairings) and rear portion (including aft baggage compartment and tail unit attachment frames). The nose and rear underfuselage are of conventional skin/stringer design. The remainder is composed of a smooth outer skin bonded to a corrugated inner skin and stabilised by frames.

TAIL UNIT: Cantilever all-metal two-spar structure with twin fins and rudders, basically similar to that of the Skyvan. Fixed-incidence tailplane, with reinforced leading-edge. Full-span elevator, aerodynamically balanced by set-back hinges. Rudders each have an unshielded horn aerodynamic balance. Primary control

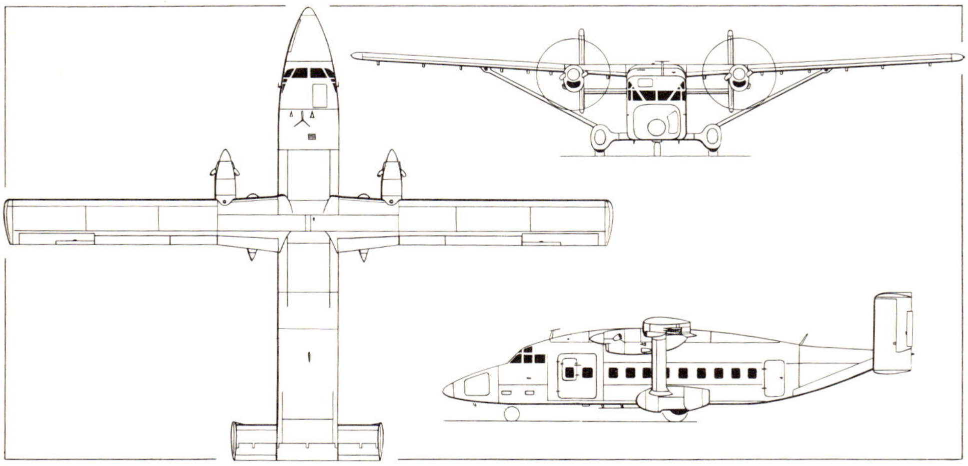

Shorts 330 twin-turboprop commuter and utility transport *(Pilot Press)*

Shorts 330 wide-bodied commuter airliner in the insignia of Aeronaves del Centro, Venezuela

surfaces are rod-actuated. Geared trim tabs in elevator and starboard rudder (port rudder, trim only).

Landing Gear: Menasco retractable tricycle type, with single wheel on each unit. Main units carried on short sponsons, into which the wheels retract hydraulically. Oleo-pneumatic shock-absorbers. Nosewheel is steerable. Normal tyre pressures: main units 5·38 bars (78 lb/sq in), nose unit 3·79 bars (55 lb/sq in). Special requirements for rough-field operation have been catered for in the design.

Power Plant: Two 862 kW (1,156 shp) (max continuous rating 761 kW: 1,020 shp) Pratt & Whitney Aircraft of Canada PT6A-45B turboprop engines, each driving a Hartzell five-blade low-speed propeller. Fuel in main tanks in wing centre-section/fuselage fairing, total usable capacity 2,182 litres (480 Imp gallons). Normal cross-feed provisions to allow for pump failure. Single-point pressure refuelling. Provision to increase total fuel capacity for special requirements.

Accommodation: Crew of two on flight deck, plus cabin attendant. Dual controls standard. Standard seating for 30 passengers, in ten rows of three at 76 cm (30 in) pitch, with wide aisle. Seat rails fitted to facilitate changes in configuration. Galley, toilet and cabin attendant's seat at rear. Large overhead baggage lockers. Entire accommodation soundproofed and air-conditioned. Baggage compartments in nose (1·27 m³; 45 cu ft) and to rear of cabin (2·83 m³; 100 cu ft), each with external access and capable of holding a combined total of 500 kg (1,100 lb) of baggage. Passenger door is at rear of cabin on port side. Passenger version has two emergency exits on the starboard side, two on the port side (including passenger door) and one in the flight deck roof. Mixed-traffic version has full access to these emergency exits. For mixed passenger/freight operation a partition divides the cabin into a rear passenger area (typically for 18 persons) and a forward cargo compartment, the latter being loaded through a large port-side door, capable of admitting ATA 'D' type containers. In all-cargo configuration the cabin can accommodate up to seven 'D' type containers, with ample space around them for additional freight. Cabin floor is flat throughout its length, and is designed to support loadings of 181 kg (400 lb) per foot run at 610·3 kg/m² (125 lb/sq ft). Locally-reinforced areas of higher strength are also provided. Seat rails can be used as cargo lashing points. Freight loading is facilitated by the low-level cabin floor.

Systems: Hydraulic system of 207 bars (3,000 lb/sq in), supplied by engine-driven pumps, operates landing gear, nosewheel steering, flaps and brakes (at lower pressure) and includes emergency accumulators. Main electrical system, for general services, is 28V DC and is of the split busbar type with cross-coupling for essential services. Lucas 28V 250A DC starter/generator for engine starting and aircraft services, with separate 1·5kW 200V AC output for windscreen anti-icing and demisting. Special AC sources of 115V and 26V available at 400Hz for certain instruments. Full air-conditioning system. Optional de-icing/anti-icing system for wing and tail unit leading-edges, engine intake ducts (inertial), and inlet lips and propellers (electric mats).

Avionics and Equipment: Passenger safety equipment standard. Wide range of radio and navigation equipment available to customer's requirements. Typical standard avionics comprise duplicated VHF communications and navigation systems, two glideslope/marker beacon receivers, two ILS repeaters, two radio magnetic indicators, one ADF, one transponder, one DME, PA system and weather radar. Flight data recorder and voice recorder available as standard options.

Dimensions, external:

Wing span	22·76 m (74 ft 8 in)
Wing chord (standard mean)	1·85 m (6 ft 0·7 in)
Length overall	17·69 m (58 ft 0½ in)
Height overall	4·95 m (16 ft 3 in)
Tailplane span	5·68 m (18 ft 7¾ in)
Wheel track	4·24 m (13 ft 11 in)
Wheelbase	6·15 m (20 ft 2 in)
Propeller diameter	2·82 m (9 ft 3 in)
Propeller ground clearance	1·83 m (6 ft 0 in)
Cabin floor: height above ground	0·94 m (3 ft 1 in)
Passenger door (port, rear):	
Height	1·57 m (5 ft 2 in)
Width	0·71 m (2 ft 4 in)
Forward cargo door (port):	
Height	1·68 m (5 ft 6 in)
Width	1·42 m (4 ft 8 in)

Dimensions, internal:

Cabin: Max length, incl toilet	9·47 m (31 ft 1 in)
Max width	1·93 m (6 ft 4 in)
Max height	1·93 m (6 ft 4 in)
Floor area	18·77 m² (202 sq ft)
Volume (all-cargo)	34·83 m³ (1,230 cu ft)
Baggage compartments volume (total usable)	4·11 m³ (145 cu ft)
Cabin overhead lockers (total)	1·13 m³ (40 cu ft)

Areas:

Wings, gross	42·1 m² (453·0 sq ft)
Ailerons (total, aft of hinges)	2·55 m² (27·5 sq ft)
Trailing-edge flaps (total)	7·74 m² (83·3 sq ft)
Fins (total)	8·65 m² (93·1 sq ft)
Rudders (total, aft of hinges)	2·24 m² (24·1 sq ft)
Tailplane	7·77 m² (83·6 sq ft)
Elevator (total, aft of hinges)	2·54 m² (27·3 sq ft)

Weights and Loadings:

Weight empty, equipped (incl crew of three): 330 for 30 passengers	6,690 kg (14,750 lb)
Fuel: standard tanks	1,741 kg (3,840 lb)
Max payload for normal max T-O weight: 30 passengers and baggage	2,653 kg (5,850 lb)
cargo	3,400 kg (7,500 lb)
Max T-O weight	10,250 kg (22,600 lb)
Max landing weight	10,115 kg (22,300 lb)
Max wing loading	243·6 kg/m² (49·89 lb/sq ft)
Max power loading	5·95 kg/kW (9·8 lb/shp)

Performance (at max T-O weight, ISA at S/L, except where indicated):

Max cruising speed at 3,050 m (10,000 ft), AUW of 9,525 kg (21,000 lb)	190 knots (352 km/h; 218 mph)
Econ cruising speed at 3,050 m (10,000 ft), AUW of 9,525 kg (21,000 lb)	160 knots (296 km/h; 184 mph)
Stalling speed, flaps and landing gear up	91 knots (168 km/h; 104 mph)
Stalling speed at max landing weight, flaps and landing gear down	74 knots (137 km/h; 85 mph)
Max rate of climb at S/L	353 m (1,160 ft)/min
Service ceiling, one engine out, AUW of 9,072 kg (20,000 lb)	3,500 m (11,500 ft)
T-O distance (FAR Pt 25 and BCAR Gp A): ISA	1,160 m (3,800 ft)
ISA + 15°C	1,295 m (4,250 ft)
Landing distance, AUW of 9,072 kg (20,000 lb): BCAR	1,143 m (3,750 ft)
FAR	1,030 m (3,380 ft)
Runway LCN at max T-O weight	10·4
Range with max passenger payload, cruising at 3,050 m (10,000 ft), no reserves	435 nm (805 km; 500 miles)
Range with max fuel, cruising at 3,050 m (10,000 ft), no reserves: passenger version, 1,830 kg (4,035 lb) payload	758 nm (1,403 km; 872 miles)
cargo version, 2,170 kg (4,785 lb) payload	758 nm (1,403 km; 872 miles)

Operational Noise Levels (FAR Pt 36):

Take-off	88·9 EPNdB
Sideline	84·1 EPNdB
Approach	92·8 EPNdB

SHORTS 330-200

First details of this new version of the Shorts 330 were given at the 1981 Paris Air Show. Principal differences are the adoption of more powerful PT6A-45R turboprop engines (893 kW; 1,198 shp instead of the 862 kW; 1,156 shp PT6A-45B engines in the current version); and a number of items as standard which are available only as options on the present Shorts 330. Certification, under FAR Pt 25, will be at the increased max T-O weight of 10,385 kg (22,900 lb).

Weights:

Operating weight empty	6,800 kg (15,000 lb)
Max T-O weight	10,385 kg (22,900 lb)
Max landing weight	10,250 kg (22,600 lb)

Performance (estimated, at max T-O weight except where indicated):

Max cruising speed at 3,050 m (10,000 ft), AUW of 9,525 kg (21,000 lb)	190 knots (352 km/h; 218 mph)
Econ cruising speed at 1,830 m (6,000 ft), AUW of 9,525 kg (21,000 lb)	157 knots (291 km/h; 181 mph)
Max rate of climb at S/L	360 m (1,180 ft)/min
T-O balanced field length, ISA at S/L	1,190 m (3,900 ft)
Landing field length	1,115 m (3,650 ft)
Range with max passenger payload	278 nm (515 km; 320 miles)
Range with max fuel	673 nm (1,247 km; 775 miles)

SHORTS 360

On 10 July 1980, Shorts released first details of this 'stretched' development of the Model 330, seating six more passengers in a lengthened fuselage and having strengthened outer wing panels and bracing struts, and a new tail unit, as well as more powerful and more fuel-efficient engines.

Designed specifically for short-haul airline operation, over typical commuter average stage lengths of about 104 nm (193 km; 120 miles), the Shorts 360 retains the basic configuration of the Model 330, the major external differences being the lengthened fuselage (a 0·91 m; 3 ft plug is inserted forward of the wings) and the introduction of a sweptback single fin and rudder instead of the latter's twin assembly. Power plant is a higher-rated version of the proven PT6A turboprop engine.

Pressurisation is considered unnecessary in view of the short stage lengths over which the aircraft will operate, and this enables the Shorts 360 to retain the same 'walk-about' headroom, square-section wide-bodied interior, seating comfort, air-conditioning and other amenities as its predecessor. The new rear fuselage/tail configuration is designed to reduce drag, improve fuel efficiency, and provide even greater baggage capacity. Considerable emphasis is placed on the 360's ability to provide more than 0·20 m³ (7 cu ft) of baggage space for each of its 36 passengers, a feature which Shorts claim is unique among today's commuter aircraft.

The prototype (G-ROOM) made its first flight on 1 June 1981, some six months ahead of schedule, powered initially by PT6A-45 engines. The intended PT6A-65 power plant will be installed later; it is planned to gain certification within about 12 months of the first flight, with delivery of production examples beginning immediately after.

The launching customer was Suburban Airlines of Reading, Pennsylvania, and by mid-June 1981 Shorts had received orders and options for 21 aircraft. Firm orders included:

Alidair (UK)	2
Chautauqua Airlines (New York)	2
Suburban Airlines (Pennsylvania)	4
Tavina (Colombia)	1

Type: Twin-turboprop commuter transport.

Wings: Generally as for Shorts 330, except that outer wing panels and bracing struts are strengthened for operation at higher gross weight. When in production, this wing will be standard also for the Shorts 330.

Fuselage: As Shorts 330, but lengthened by insertion of 0·91 m (3 ft) plug forward of wings. Rear fuselage modified to cater for redesigned tail unit.

Tail Unit: Cantilever all-metal two-spar structure, with single sweptback fin and rudder and constant-chord non-swept tailplane. Trim tabs in each elevator.

Landing Gear: Similar to Shorts 330, but with higher tyre pressures to cater for increased operating weights.

Power Plant: Two 965 kW (1,294 shp) Pratt & Whitney Aircraft of Canada PT6A-65R turboprop engines, each driving an advanced technology propeller with spinner. Fuel capacity as for Shorts 330.

Accommodation: Crew of two on flight deck, plus cabin attendant. Dual controls standard. Main cabin accommodation similar to Shorts 330, but seating a total of 36 passengers in 12 rows of three. Large overhead baggage lockers. Baggage compartments in nose and to rear of cabin, each with external access, giving equivalent of

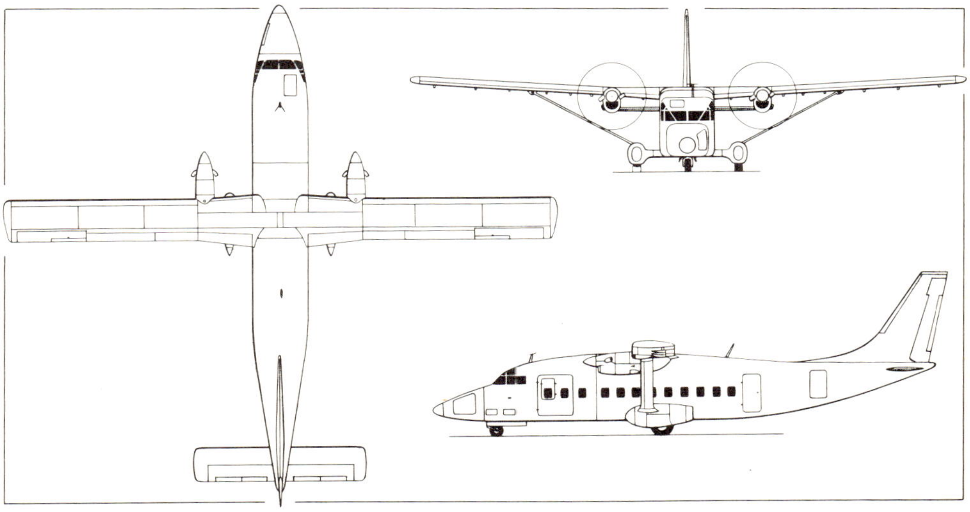

Shorts 360 commuter transport (two P&WC PT6A-65R turboprop engines) *(Pilot Press)*

almost 0·17 m³ (6 cu ft) of baggage space per passenger (0·20 m³; 7·2 cu ft per passenger if locker space is included).

SYSTEMS: Generally as for Shorts 330.

AVIONICS: Typical standard avionics comprise duplicated VHF com and nav systems (VOR/ILS and marker beacon receiver), two ATC transponders, one DME, one ADF, two gyromagnetic compass systems, two RMI, two ILS repeater indicators, flight director, HSI, weather radar, PA and stereo music systems, flight data recorder, voice recorder and GPWS.

DIMENSIONS, EXTERNAL: As for Shorts 330 except:

Length overall	21·49 m (70 ft 6 in)
Height overall	7·20 m (23 ft 7½ in)
Tailplane span	7·12 m (23 ft 4½ in)
Wheelbase	7·06 m (23 ft 2 in)
Passenger door (port, rear):	
Height	1·74 m (5 ft 8½ in)
Width	0·69 m (2 ft 3 in)
Height to sill	0·94 m (3 ft 1 in)
Forward cargo door (port):	
Height	1·68 m (5 ft 6 in)
Width	1·42 m (4 ft 8 in)
Height to sill	0·94 m (3 ft 1 in)

DIMENSIONS, INTERNAL:

Cabin: Length	11·02 m (36 ft 2 in)
Max width	1·93 m (6 ft 4 in)
Max height	1·93 m (6 ft 4 in)
Passenger compartment volume	41·06 m³ (1,450 cu ft)
Baggage compartment volume:	
forward	1·27 m³ (45 cu ft)
rear	4·81 m³ (170 cu ft)
lockers	1·27 m³ (45 cu ft)

AREAS: As for Shorts 330 except:

Vertical tail surfaces (total)	7·52 m² (81·0 sq ft)
Horizontal tail surfaces (total)	9·66 m² (104·0 sq ft)

WEIGHTS:

Typical operating weight empty	7,480 kg (16,490 lb)
Max payload (36 passengers)	3,102 kg (6,840 lb)
Max baggage load	635 kg (1,400 lb)
Max fuel load	1,741 kg (3,840 lb)
Max T-O weight	11,657 kg (25,700 lb)
Max ramp weight	11,702 kg (25,800 lb)
Max landing weight	11,521 kg (25,400 lb)

PERFORMANCE (estimated, at max T-O weight except where indicated):

Cruising speed at max recommended cruise power
211 knots (391 km/h; 243 mph)

T-O run at S/L: ISA	1,320 m (4,330 ft)
ISA + 20°C	1,440 m (4,720 ft)

Landing run at S/L at max landing weight, ISA
1,198 m (3,930 ft)

Runway LCN 12·8

Range at 3,050 m (10,000 ft), cruising at 211 knots (391 km/h; 243 mph), ISA, reserves for 87 nm (161 km; 100 mile) diversion and 45 min hold:
with max passenger payload
230 nm (425 km; 265 miles)
with max fuel 570 nm (1,055 km; 655 miles)

SHORTS SC.7 SKYVAN Series 3 and 3M

Design of the SC.7 Skyvan was started as a private venture in 1959, and construction of the first prototype began in 1960. This aircraft (G-ASCN) flew for the first time on 17 January 1963, with two 290 kW (390 hp) Continental GTSIO-520 piston engines, and completed its flight trials by mid-1963. It was then re-engined with 388 kW (520 shp) Astazou II turboprops and first flew in its new form on 2 October 1963. The change to Garrett TPE331 turboprops was made on the Skyvan Srs 3 in 1967.

Details of the Srs 1/1A/2/3A/Skyliner models have appeared in the 1980-81 and previous editions of *Jane's*. The two current versions are as follows:

Skyvan Srs 3. Current civil version, which superseded Srs 2 in 1968. First Srs 3 to fly was the second development aircraft, G-ASZI, which had been equipped originally with Astazous. The first flight with Garrett engines was made on 15 December 1967, and a second aircraft (G-ASZJ) re-engined with TPE331s flew on 20 January 1968. Total of 71 ordered by 1 March 1981, the most recent customers including Air Malawi, Gulf Air and the government of Saudi Arabia.

Skyvan Srs 3M. Military version of Srs 3, modified internally to accept optional equipment for typical military missions. Prototype (G-AXPT) flew for the first time in early 1970. Suitable for paratrooping and supply dropping, assault landing, troop transport, casualty evacuation, staff transport, and vehicle or ordnance transport.

A total of 54 had been ordered by March 1981, for armed services including the Argentinian Naval Prefectura (5), Austrian Air Force (2), Botswana Defence Force (2), Ecuador Army Air Force (1), Ghana Air Force (6), Indonesian Air Force (3), Lesotho Police (2), Malawi Police (1), Mauritanian Air Force (2), Royal Nepalese Army (2), No. 2 Squadron of the Sultan of Oman's Air Force (16), Panama National Guard (1), Singapore Air Force (6), Royal Thai Police (3) and Yemen Arab Republic Air Force (2). Three of the Singapore aircraft are equipped for search and rescue duties. Those of the Indonesian Air Force are equipped to civil standard and operate social services on behalf of the Ministry of the Interior.

Total orders for the Series 3 and 3M had reached 125 by 1 March 1981. In February 1970 the Skyvan became the first aircraft to be certificated under the British Air Registration Board's new Civil Airworthiness Requirements for STOL operations.

The following description applies to the standard current civil Srs 3 and military Srs 3M:

TYPE: Light civil or military STOL utility transport.

WINGS: Braced high-wing monoplane. Wing section NACA 63A series (modified). Thickness/chord ratio 14%. Dihedral 2° 2′. Incidence 2° 30′. Light alloy structure consisting of a two-cell box with wing skins made up of a uniform outer sheet bonded to a corrugated inner sheet. All-metal single-slotted ailerons. Geared tabs in port and starboard ailerons, with manual trim in starboard aileron. All-metal single-slotted flaps. Provision for sintered leading-edge de-icing system.

FUSELAGE: Light alloy structure. Nose and crew cabin section is of conventional skin/stringer design. Elsewhere, the fuselage structure consists of double-skin panels (flat outer sheets bonded to inner corrugated sheets), stabilised by frames.

TAIL UNIT: Cantilever all-metal two-spar structure, with twin fins and rudders. Fixed-incidence tailplane. Geared trim tabs in outer elevators and rudders. Provision for sintered leading-edge de-icing system.

LANDING GEAR: Non-retractable tricycle type. Single wheel on each unit. Steerable nosewheel. Main units carried on short sponsons. Electro-Hydraulics oleo-pneumatic shock-absorbers. Main-wheel tyres size 11·00-12, nosewheel tyre size 7·50-10. Tyre pressure (all units) 2·76 bars (40 lb/sq in). Hydraulically-operated disc brakes, with differential braking for steering. Low-pressure tyres available optionally.

POWER PLANT: Two 533 kW (715 shp) Garrett TPE331-201 turboprop engines, each driving a Hartzell HC-B3TN-5/T10282H three-blade variable-pitch propeller. Fuel in four tanks in pairs on top of fuselage between wing roots, each pair consisting of one tank of 182 litres

Shorts Skyvan STOL light transport in the livery of Hornbill Skyways, Malaysia

Prototype Shorts 360 twin-turboprop 36-passenger commuter transport

(40 Imp gallons) capacity and one of 484 litres (106·5 Imp gallons) capacity. Total fuel capacity of 1,332 litres (293 Imp gallons). Provision for increase in total fuel capacity to 1,773 litres (390 Imp gallons) by installing four specially-designed tanks in spaces between fuselage frames on each side, beneath main fuel tank. Oil capacity 7·73 litres (1·7 Imp gallons).

Accommodation: Crew of one, with provision for two. Accommodation (Srs 3) for up to 19 passengers, or 12 stretcher patients and attendants, or 2,085 kg (4,600 lb) of freight, vehicles or agricultural equipment. Srs 3M can accommodate 22 equipped troops; 16 paratroops and a dispatcher; 12 stretcher cases and two medical attendants; or 2,358 kg (5,200 lb) of freight. It carries its own lightweight vehicle loading ramps and has a one-piece door which leaves the fuselage threshold entirely clear of appendages. Executive version provides luxury accommodation and equipment for nine passengers. Full-width rear loading door, and forward door on each side of crew compartment. Rear door can be opened in flight to permit the parachuting of loads up to 1·37 m (4 ft 6 in) in height. Cockpit and cabin heated by engine bleed air mixed with fresh air from intake in nose. Cabin unpressurised. Some aircraft fitted with Rolamat cargo loading equipment.

Systems: Hydraulic system, pressure 172 bars (2,500 lb/sq in), operates flaps, wheel brakes and nosewheel steering. No pneumatic system. Electrical system utilises two busbars, operating independently, each connected to a 28V 125A DC starter/generator, a battery and a 115V 400Hz static inverter. General services are 28V DC; some radio and instruments 115V AC.

Avionics and Equipment: Radio optional. Typical installation for operations in Europe and USA consists of duplicated VHF, duplicated VOR/ILS, marker beacon receiver and ADF. Provision for HF, DME, transponder, Bendix M4D autopilot and weather radar. Blind-flying instrumentation standard.

Equipment (Srs 3M): Port-side blister window for an air dispatcher; two anchor cables for parachute static lines; a guard rail beneath the tail to prevent control surface fouling by the static lines; inward-facing paratroop seats with safety nets; parachute signal light; mounts for NATO-type stretchers; and roller conveyors for easy loading and paradropping of pallet-mounted supplies.

Dimensions, external:
Wing span 19·79 m (64 ft 11 in)
Wing chord (constant) 1·78 m (5 ft 10 in)
Wing aspect ratio 11
Length overall: without radome 12·21 m (40 ft 1 in)
with radome 12·60 m (41 ft 4 in)
Height overall 4·60 m (15 ft 1 in)
Tailplane span 5·28 m (17 ft 4 in)
Wheel track 4·21 m (13 ft 10 in)
Wheelbase 4·52 m (14 ft 10 in)
Propeller diameter 2·59 m (8 ft 6 in)
Propeller ground clearance 1·52 m (5 ft 0 in)
Crew and passenger doors (fwd, port and stbd):
Height 1·52 m (5 ft 0 in)
Width 0·51 m (1 ft 8 in)
Height to sill 1·14 m (3 ft 9 in)
Rear loading door: Height 1·98 m (6 ft 6 in)
Width 1·96 m (6 ft 5 in)
Height to sill 0·74 m (2 ft 5 in)

Dimensions, internal:
Cabin, excl flight deck: Length 5·67 m (18 ft 7 in)
Max width 1·98 m (6 ft 6 in)
Max height 1·98 m (6 ft 6 in)
Floor area 11·15 m² (120 sq ft)
Volume 22·09 m³ (780 cu ft)

Areas:
Wings, gross 34·65 m² (373 sq ft)
Ailerons (total) 3·00 m² (32·3 sq ft)
Trailing-edge flaps (total) 5·86 m² (63·1 sq ft)
Fins 7·62 m² (82·0 sq ft)
Rudders, incl tabs 2·41 m² (25·9 sq ft)
Tailplane 7·53 m² (81·0 sq ft)
Elevators, incl tabs 3·62 m² (39·0 sq ft)

Weights and Loadings (with 1,332 litres; 293 Imp gallons of fuel):
Basic operating weight: 3 3,331 kg (7,344 lb)
3M 3,356 kg (7,400 lb)
Typical operating weight as freighter:
3 3,447 kg (7,600 lb)
3M 3,456 kg (7,620 lb)
Typical operating weight with passengers or troops:
3 3,674 kg (8,100 lb)
3M 3,778 kg (8,330 lb)
Max payload for normal T-O weight:
3 2,086 kg (4,600 lb)
3M 2,358 kg (5,200 lb)
Max payload for overload T-O weight:
3M 2,721 kg (6,000 lb)
Max T-O weight: 3, normal 5,670 kg (12,500 lb)
3M, normal 6,214 kg (13,700 lb)
3M, overload 6,577 kg (14,500 lb)
Max landing weight: 3 5,670 kg (12,500 lb)
3M 6,123 kg (13,500 lb)
Max wing loading: 3 163·6 kg/m² (33·5 lb/sq ft)
3M 179·1 kg/m² (36·7 lb/sq ft)
Max power loading: 3 5·32 kg/kW (8·74 lb/shp)
3M 6·17 kg/kW (9·58 lb/shp)

Performance (at max T-O weight, with 1,332 litres; 293 Imp gallons of fuel):
Never-exceed speed 217 knots (402 km/h; 250 mph) EAS
Max cruising speed at 3,050 m (10,000 ft):
max cont power 176 knots (327 km/h; 203 mph)
cruise power 169 knots (314 km/h; 195 mph)
Econ cruising speed at 3,050 m (10,000 ft) 150 knots (278 km/h; 173 mph)
Stalling speed, flaps down:
3 60 knots (111 km/h; 69 mph) EAS
3M 62 knots (115 km/h; 71 mph) EAS
Max rate of climb at S/L: 3 500 m (1,640 ft)/min
3M 466 m (1,530 ft)/min
Service ceiling (30 m; 100 ft/min climb):
3 6,858 m (22,500 ft)
3M 6,705 m (22,000 ft)
Service ceiling, one engine out (15 m; 50 ft/min climb):
3 3,810 m (12,500 ft)
3M 2,895 m (9,500 ft)
Min ground turning radius 3·76 m (12 ft 4 in)
Runway LCN at AUW of 5,670 kg (12,500 lb):
standard tyres 3·5
low-pressure tyres 3·0
T-O run, STOL, unfactored: 3 213 m (700 ft)
3M 238 m (780 ft)
T-O run (normal): 3 (BCAR) 512 m (1,680 ft)
T-O to 10·7 m (35 ft), Transport Group A, ISA at S/L 1,020 m (3,350 ft)
T-O to 15 m (50 ft), STOL, unfactored:
3 320 m (1,050 ft)
3M 384 m (1,260 ft)
T-O to 15 m (50 ft):
3 (BCAR, normal) 610 m (2,000 ft)
3 (BCAR, STOL) 482 m (1,580 ft)
3 (FAR Pt 23) 488 m (1,600 ft)
Landing from 15 m (50 ft):
3 (BCAR, normal) 622 m (2,040 ft)
3 (BCAR, STOL) 567 m (1,860 ft)
3 (FAR Pt 23) 451 m (1,480 ft)
3M (STOL, unfactored) 425 m (1,395 ft)
Landing from 15 m (50 ft), Transport Group A, ISA at S/L 1,010 m (3,320 ft)
Landing from 9 m (30 ft):
3 (STOL, unfactored) 351 m (1,150 ft)
3 (BCAR, STOL) 500 m (1,640 ft)
Landing run: 3M (STOL, unfactored) 212 m (695 ft)
Range at long-range cruising speed, 45 min reserves:
3 600 nm (1,115 km; 694 miles)
3M 580 nm (1,075 km; 670 miles)
Range (typical freighter) at long-range cruising speed, 45 min reserves:
3 with 1,814 kg (4,000 lb) payload 162 nm (300 km; 187 miles)
3M with 2,268 kg (5,000 lb) payload 208 nm (386 km; 240 miles)

SLINGSBY

SLINGSBY ENGINEERING LTD (Aircraft Division)

Head Office and Works: Ings Lane, Kirkbymoorside, North Yorkshire YO6 6EZ
Telephone: 0751 31751
Telex: 57911
Officers: See Sailplanes section

SLINGSBY T67

The designation **T67A** applies to the Fournier RF6B-120 light aircraft, which is now built under licence by Slingsby. French production by Fournier, which totalled 45 RF6B-100s and one RF6B-120, has ended. The latter version flew for the first time on 14 August 1980 and received FAR Pt 23 certification on 7 November 1980. An initial batch of ten T67As is being built, of which the first (G-BIOW) flew for the first time on 15 May 1981.

Slingsby also plans to manufacture an all-GRP version of the aircraft as the **T67B**. This will be virtually identical externally to the T67A, with the same power plant, but is intended for certification, at the slightly higher gross weight of 816 kg (1,800 lb), in both the Utility and Aerobatic categories. The empty weight will be 530 kg (1,169 lb). First flight is planned for early 1982.

The following description applies to the T67A:

Type: Two-seat aerobatic, training and sporting aircraft.

Wings: Cantilever low-wing monoplane. Wing section NACA 23015 at root, NACA 23013 at tip. Dihedral 3° 30'. Incidence 3°. All-wood single-spar structure with plywood and Dacron covering. Frise-type ailerons of wooden construction, Dacron covered. No tabs. Plain trailing-edge flaps of wooden construction with Dacron covering.

Fuselage: All-wood oval structure, plywood covered.

Tail Unit: Cantilever structure of wood with Dacron covering. Fixed-incidence tailplane. Trim tab in port elevator.

Landing Gear: Non-retractable tricycle type. Oleo-pneumatic shock-absorber in each unit. Steerable nosewheel. Main-wheel tyres size 15 × 6·00-6, pressure 1·4 bars (20 lb/sq in). Nosewheel tyre size 4·00-4, pressure 2·5 bars (37 lb/sq in). Hydraulic disc brakes. GRP main-wheel fairings optional.

First Slingsby T67A, a licence-built version of the Fournier RF6B-120 two-seat light aircraft

Power Plant: One 88 kW (118 hp) Avco Lycoming O-235-L2A flat-four engine, driving a Hoffmann two-blade fixed-pitch composite propeller with spinner. Fuselage fuel tank, immediately aft of firewall, capacity 80 litres (17·6 Imp gallons). Refuelling point on fuselage upper surface, forward of windscreen. Oil capacity 4 litres (0·88 Imp gallons). Oil system permits short periods of inverted flight.

Accommodation: Two adjustable seats side by side under one-piece transparent canopy, which swings upward and aft for access to cockpit. Dual controls standard. Cockpit heated and ventilated. Baggage space aft of seats.

Systems: Hydraulic system for brakes only. Vacuum system optional, for blind-flying instrumentation when fitted. Electrical power supplied by 12V 60A engine-driven alternator and 12V 25Ah battery.

Avionics and Equipment: Range of Narco avionics available in basic package or a luxury option package offering full airline standard. Blind-flying instrumentation optional.

Dimensions, external:
Wing span 10·59 m (34 ft 9 in)
Wing chord at root 1·53 m (5 ft 0¼ in)
Wing chord at tip 0·83 m (2 ft 8¾ in)
Wing aspect ratio 8·88
Length overall 7·01 m (23 ft 0 in)
Height overall 2·51 m (8 ft 3 in)
Tailplane span 3·40 m (11 ft 1¾ in)
Wheel track 2·44 m (8 ft 0 in)
Wheelbase 1·495 m (4 ft 10¾ in)
Propeller diameter 1·78 m (5 ft 10 in)

Areas:
Wings, gross 12·63 m² (136·0 sq ft)
Ailerons (total) 1·24 m² (13·35 sq ft)
Trailing-edge flaps (total) 1·74 m² (18·73 sq ft)
Fin 0·80 m² (8·61 sq ft)
Rudder 0·81 m² (8·72 sq ft)
Tailplane 1·65 m² (17·76 sq ft)
Elevators (incl tab) 0·99 m² (10·66 sq ft)

Weights and Loadings:
Weight empty (basic) 518 kg (1,142 lb)
Max fuel 57·5 kg (126 lb)
Max baggage 30 kg (66 lb)
Max T-O weight: Aerobatic 720 kg (1,587 lb)
Utility 750 kg (1,653 lb)
Max wing loading 59·3 kg/m² (12·15 lb/sq ft)
Max power loading 8·52 kg/kW (14·0 lb/hp)

PERFORMANCE (at max T-O weight):

Never-exceed speed	138 knots (256 km/h; 159 mph)
Max level speed at S/L	113 knots (209 km/h; 130 mph)
Max cruising speed (75% power) at 2,440 m (8,000 ft)	108 knots (200 km/h; 124 mph)
Stalling speed, power off:	
flaps up	48 knots (89 km/h; 56 mph)
flaps down	39 knots (73 km/h; 45 mph)
Max rate of climb at S/L	247 m (810 ft)/min
Service ceiling	4,000 m (13,125 ft)
T-O run	200 m (656 ft)
T-O to 15 m (50 ft)	340 m (1,115 ft)
Landing from 15 m (50 ft)	450 m (1,477 ft)
Landing run	280 m (919 ft)
Max range, allowances for T-O and climb, and usable fuel	342 nm (633 km; 393 miles)
g limits	+6·0/−3·0

TO

SOLAR-POWERED AIRCRAFT DEVELOPMENTS

5A Lyndhurst Gardens, London NW3

This company was formed by Mr Freddie To, in order to build a solar-powered aircraft that he had devised. Design of the aircraft, known as the Solar One, was undertaken by Mr David Williams, and the first short hop was claimed on 19 December 1978. A flight of nearly three-quarters of a mile (1,200 m) was made by Mr Ken Stewart at Lasham Airfield, Hampshire, on 13 June 1979. The flight was made at a maximum height of about 25 m (80 ft) and at a maximum speed of about 35 knots (65 km/h; 40 mph). A second flight was made on the same day by Mr Bill Maidment. In this particular application, the electric motors which drive the propeller are powered by a battery that stores the energy received from 750 solar cells. Brief details of this aircraft can be found in the 1980-81 *Jane's*.

TO PHOENIX MAN-POWERED AIRCRAFT

Mr To has designed and built a unique man-powered aircraft of ultra-lightweight construction. Most of the structure comprises a wing of 31 m (102 ft) span, built up from Melinex and Mylar film, glued and sewn together, which can be inflated to a pressure of 0·03 bars (0·4 lb/sq in). This pressure is considered to be adequate to maintain the wing rigid in flight. A pilot's frame of carbonfibre, incorporating a seat, pedals, and a lightweight propeller, is mounted beneath the wing, and braced to it by 32 cords.

Three successful towed flights (the longest lasting approx 30 s) took place in August 1981, a height of 7·6 m (25 ft) being attained on the third flight. It was hoped to make a man-powered first flight within a few weeks. The aircraft has an empty weight of 50 kg (110 lb); its estimated flying speed is about 7 knots (13 km/h; 8 mph).

TRAGO MILLS

TRAGO MILLS LTD (Aircraft Division)

Treswithick Farm, Cardinham, Bodmin, Cornwall PL30 4BU

Telephone: Cardinham (020 882) 485

CHIEF DESIGNER: Sydney A. Holloway

Trago Mills sponsored and owns the MW2 Excalibur single-seat ultralight aircraft, designed by Mr Michael Whittaker and last described in the 1977-78 *Jane's*. More recently, the company has been building the prototype of Mr Sydney Holloway's SAH-1 two-seat light aircraft.

Trago Mills is CAA approved for light aircraft design, development, manufacture and testing, and has recently retained Mr Frank H. Robertson as design consultant. Mr Robertson, who will work on a part-time basis, has held engineering and senior design appointments with Vought Corporation in the USA, and with Miles, Saunders-Roe and Shorts in the UK.

TRAGO MILLS SAH-1

Design of the SAH-1 started in October 1977, and construction of a prototype began in January 1978. Both design and construction are to full CAA and FAR Pt 23 standards, and production is intended after certification has been obtained. It had been hoped to make the first flight in the Summer of 1981, but this has been delayed as the result of having to substitute an Avco Lycoming engine for the Rolls-Royce Continental intended originally as power plant of the SAH-1.

TYPE: Two-seat fully-aerobatic light aircraft.

WINGS: Cantilever low-wing monoplane. Wing section NACA 2413·6 (constant). Dihedral 5° from roots. Incidence 3° at root, 1° at tip. Tapered, non-swept aluminium alloy wings, with L65 spar booms and L72 sheet skins, stabilised with PVC foam. Trailing-edge single-slotted flaps and slotted ailerons of similar construction.

FUSELAGE: Aluminium alloy fail-safe stressed-skin structure, with radiused corners, incorporating centre-section spars.

TAIL UNIT: Aluminium alloy cantilever structure, stabilised with PVC foam. Constant-chord tailplane, attached to fuselage, with horn-balanced elevators; full-span trim tab in starboard elevator. Sweptback fin and horn-balanced rudder.

LANDING GEAR: Non-retractable tricycle type, with single Cleveland 5·00-5 wheel and tyre on each unit. Oleo-pneumatic shock-absorber in nosewheel leg; spring steel main legs. Tyre pressures 1·93 bars (28 lb/sq in) on nose unit, 2·07 bars (30 lb/sq in) on main units. Cleveland hydraulic brakes.

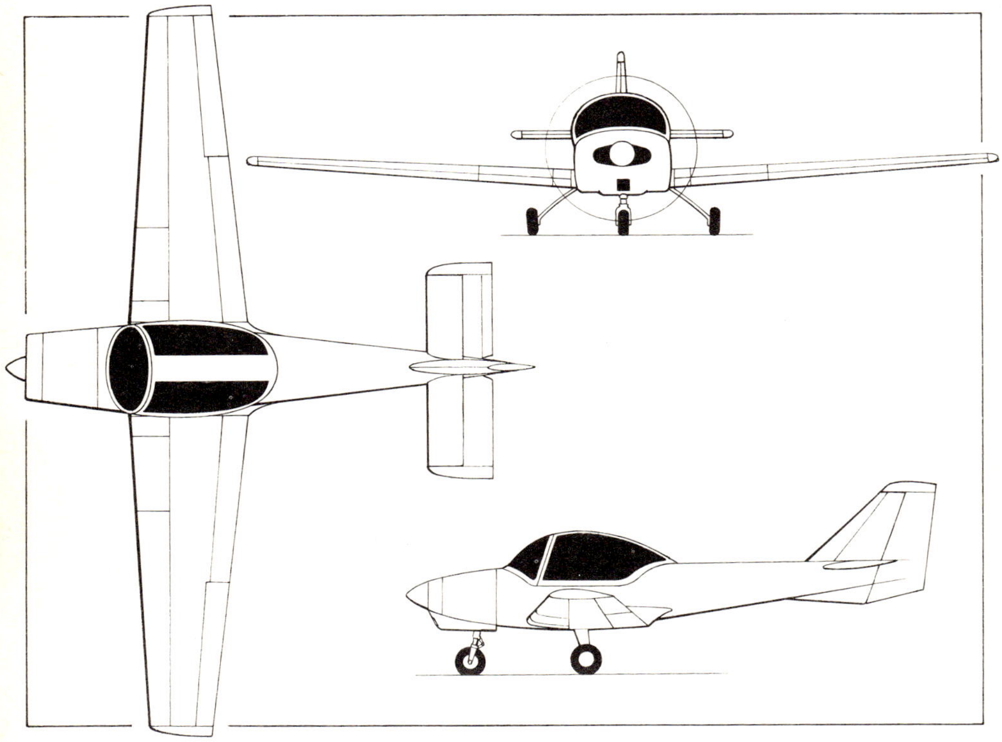

Trago Mills SAH-1 side-by-side two-seat light aircraft *(Michael A. Badrocke)*

POWER PLANT: One 88 kW (118 hp) Avco Lycoming O-235-L2A flat-four engine, driving a two-blade fixed-pitch propeller with spinner. Integral fuel tank in each leading-edge, total capacity 114 litres (25 Imp gallons). Refuelling point in upper surface of each wing. Oil capacity 5·7 litres (1·25 Imp gallons).

ACCOMMODATION: Two seats side by side under rearward-sliding bubble canopy. Baggage space aft of seats. Cockpit heated and ventilated.

SYSTEM: 60A engine-driven alternator.

AVIONICS AND EQUIPMENT: Blind-flying instrumentation standard. Radio to customer's specification.

DIMENSIONS, EXTERNAL:	
Wing span	9·36 m (30 ft 8·4 in)
Wing chord at root	1·515 m (4 ft 11⅔ in)
Wing chord at tip	0·81 m (2 ft 8 in)
Wing aspect ratio	7·5
Length overall	6·58 m (21 ft 7¼ in)
Height overall	2·38 m (7 ft 9·6 in)
Tailplane span	2·74 m (9 ft 0 in)
Wheel track	2·26 m (7 ft 4·8 in)
Wheelbase	1·46 m (4 ft 9·6 in)
Propeller diameter	1·68 m (5 ft 6 in)
Propeller ground clearance	0·25 m (10 in)
DIMENSIONS, INTERNAL:	
Cockpit: Length	1·52 m (5 ft 0 in)
Max width	1·19 m (3 ft 10·8 in)
Baggage space	0·4 m³ (14·0 cu ft)
AREAS:	
Wings, gross	11·15 m² (120·0 sq ft)
Ailerons (total)	0·89 m² (9·6 sq ft)
Trailing-edge flaps (total)	1·30 m² (14·0 sq ft)
Fin	0·96 m² (10·3 sq ft)
Rudder	0·63 m² (6·8 sq ft)
Tailplane	1·11 m² (12·0 sq ft)
Elevators, incl tab	0·93 m² (10·0 sq ft)
WEIGHTS AND LOADINGS:	
Weight empty, equipped	432 kg (951 lb)
Max fuel load	85 kg (188 lb)
Max T-O weight	725 kg (1,600 lb)
Max wing loading	65·05 kg/m² (13·33 lb/sq ft)
Max power loading	8·24 kg/kW (13·56 lb/hp)
PERFORMANCE (estimated, at max T-O weight):	
Never-exceed speed	202 knots (374 km/h; 232 mph) EAS
Max level speed at S/L	127 knots (235 km/h; 146 mph) EAS
Max cruising speed, 75% power at S/L	115 knots (213 km/h; 132 mph) EAS
Econ cruising speed, 50% power at S/L	95 knots (176 km/h; 109 mph) EAS
Stalling speed, flaps up	53 knots (98 km/h; 61 mph) EAS
Stalling speed, flaps down	46 knots (85 km/h; 53 mph) EAS
Max rate of climb at S/L	278 m (915 ft)/min
Service ceiling	5,410 m (17,750 ft)
T-O run	255 m (834 ft)
T-O to 15 m (50 ft)	376 m (1,234 ft)
Landing from 15 m (50 ft)	286 m (939 ft)
Max range, 83 knots (154 km/h; 95 mph) at 1,525 m (5,000 ft) with 18 litres (4 Imp gallons) reserves	665 nm (1,230 km; 765 miles)

WALLIS

WALLIS AUTOGYROS LTD

HEAD OFFICE: Reymerston Hall, Norfolk NR9 4QY

Telephone: 0362 850418

DIRECTORS:

Wg Cdr K. H. Wallis, CEng, FRAeS, FRSA, RAF (Retd)

P. M. Wallis

In 1908 the brothers H. S. and P. V. Wallis designed and began building the Wallbro monoplane, probably the first aeroplane to employ steel tubing for its primary structure; it made a number of hop flights two years later. A replica was flown in 1978, as recorded in the 1978-79 *Jane's*.

Wing Commander K. H. Wallis, the son of H. S. Wallis, became an amateur constructor of high-speed watercraft in the early 1930s, gained a pilot's licence in 1937, and was commissioned in the RAF in 1940. After two operational bombing tours he qualified as a specialist armament officer, and after the war converted, in his spare time, a small two-stroke engine with which he powered a Slingsby Petrel glider. He retired from the RAF in 1964 as a qualified jet pilot, and formed Wallis Autogyros Ltd to continue professionally an interest which he had pursued privately since 1958. Since then, in addition to contract work for major British companies and British government agencies, autogyros of his design have set many FAI-approved records, including those for altitude, distance, endurance, load-carrying, and speed.

The first Wallis single-seat ultralight autogyro (G-ARRT, flown for the first time in August 1961) introduced many patented features, including a rotor head with offset gimbal system to provide hands and feet off stability and to eliminate pitch-up and 'tuck-under' hazards; a high-speed flexible rotor spin-up shaft with positive disengagement during flight; an automatic system of controlling rotor

The photographic workhorse G-ASDY, a Wallis WA-116/F fitted with a Vinten 360 camera installation

The WA-116/F G-ATHM holds nine autogyro records and has a tenth awaiting confirmation

The Wallis WA-118/M, designed to fly at altitudes of more than 9,145 m (30,000 ft)

Rolls-Royce-engined Wallis WA-117/R-R, with Vinten 751 panoramic camera pack

drive on take-off which allows power to be applied until the last moment; centrifugal stops to control rotor blade teetering; and a novel safe starting arrangement.

Many other Wallis autogyros have since been completed.

It was announced in May 1981 that Wg Cdr Wallis, in conjunction with W. Vinten Ltd, was to develop a battlefield reconnaissance autogyro that would be available in both manned and remotely piloted configurations. Available details can be found in the RPVs and Targets section, and in the Addenda to this edition.

WALLIS WA-116 and WA-116-T

The WA-116 represents the original Wallis design, of which the prototype (G-ARRT) flew for the first time on 2 August 1961, powered by a 54 kW (72 hp) modified McCulloch 4318 piston engine. Four more WA-116s were built by Beagle and five by Wg Cdr Wallis, as described in the 1973-74 *Jane's*. The last of these was later dismantled for construction of the two-seat G-AXAS, a WA-116-T.

WA-116/Mc. The prototype (G-ARRT) currently holds the height record for autogyros in Classes E3 and E3a, at 4,639 m (15,220 ft), set on 11 May 1968 with its original 54 kW (72 hp) McCulloch engine; and the Class E3 and E3a records for speed in a straight line (96·589 knots; 179·000 km/h; 111·225 mph), set on 12 May 1969 after being re-engined with a 67 kW (90 hp) McCulloch. This aircraft, and the 'James Bond' WA-116/Mc (G-ARZB), continue to perform well. The latter has now appeared in more than 500 displays, as far afield as Australia and the USA.

WA-116/F. Designation applied following the refitting of G-ASDY in 1971 with a 44 kW (60 hp) Franklin 2A-120-A engine. It is currently fitted with a Franklin 2A-120-B engine, driving a specially-designed two-blade propeller, and has carried out a considerable amount of specialised aerial photographic work, as detailed in earlier editions of *Jane's*. This work is continuing, currently using a Vinten 360 camera with a 500-frame magazine of 70 mm film. The viewfinder, which is special to the fixed oblique beam installation, can be adjusted to suit the different focal lengths of lens. With this fixed mounting, it is easy for the pilot to hold a subject in the centre of the viewfinder for 360° coverage, literally 'flying the camera'. The pilot can thus concentrate on the photographic task, since rotor revolutions are maintained automatically, while airspeed is relatively unimportant. Changes in airspeed are easily sensed, and adjusted by power changes.

A particularly successful WA-116/F is G-ATHM which currently holds nine world rotorcraft records; a tenth was awaiting ratification in mid-1981. This aircraft underwent more extensive conversion than its predecessors, mainly to increase fuel capacity and pilot comfort to fit it for special long-range flights. It has a 50 litre (11 Imp gallon) internal fuel tank, and began test flying, for range, in April 1974. Fitted also with a 36 litre (8 Imp gallon) jettisonable long-range ventral tank, it set up on 13 July 1974 new Class E3 and E3a world records, for nonstop distance in a closed circuit, of 361·91 nm (670·26 km; 416·48 miles). Additionally, this flight set new 100 km (Class E3a only) and 500 km (Class E3 and E3a) closed-circuit speed records of 70·51 knots (130·67 km/h; 81·19 mph) and 68·07 knots (126·14 km/h; 78·38 mph) respectively. A 91 litre (20 Imp gallon) ventral tank was next fitted, and on 28 September 1975, with this tank containing some 70 litres (15·5 Imp gallons), Wg Cdr Wallis made a nonstop flight from Lydd, Kent, to Wick, Caithness. This flight set Class E3 and E3a records for nonstop distance in a straight line of 472·092 nm (874·315 km; 543·274 miles). In fact, the actual distance flown, to avoid airfield zones and other hazardous areas, was in the order of 521 nm (966 km; 600 miles), flown in 6 h 25 min at an average speed of approx 81 knots (150 km/h; 93 mph). The ventral tank was not jettisoned after being emptied, and G-ATHM landed with sufficient fuel remaining for a further 65 nm (121 km; 75 miles). Its time of 6 h 25 min was later confirmed by the FAI as a record for duration in both Class E3 and Class E3a. Thus, WA-116s now hold all the major autogyro world records for speed, height, range and endurance.

It was hoped during 1981 to fit G-ATHM with a prototype experimental British light aircraft engine, in collaboration with Weslake and Co of Rye, Sussex. This more economical and efficient engine was expected to enhance G-ATHM's performance, possibly leading to further record attempts. In early 1981 the cylinders of this engine, which are of special design to optimise combustion efficiency, were being bench tested on a Franklin 2A-120-B engine. Simultaneously, a new definitive WA-116 was being designed to utilise the totally new Weslake W65-118-02 engine.

WA-116-T/Mc. The McCulloch-engined WA-116-T/Mc, G-AXAS, is a tandem two-seat autogyro which flew for the first time on 3 April 1969. It remains the lightest two-seater in the Wallis range, and is expected to improve even further upon the 3·14 : 1 ratio of all-up weight to empty weight, which it established at RAE Bedford in 1971. It made more than 130 flights in a programme of multi-spectral experiments by Plessey Radar on behalf of the Home Office, as described in the 1979-80 *Jane's*. More recently it was operated on the volcanic island of Lanzarote, in the Canary Islands, during filming of the science-fiction motion picture *The Martian Chronicles*. In addition to its part in the film, as the vehicle carried on the 'Zeus 3' manned mission to Mars, the aircraft was used to carry a 35 mm Arriflex camera for some of the aerial photography.

DIMENSIONS (WA-116/F):

Rotor diameter	6·20 m (20 ft 4 in)
Length of fuselage	3·38 m (11 ft 1 in)
Height to top of rotor head	1·85 m (6 ft 1 in)
Wheel track	1·63 m (5 ft 4 in)
Wheelbase	1·07 m (3 ft 6 in)

WEIGHTS (WA-116/F):

Weight empty	143 kg (316 lb)
Max T-O weight	317·5 kg (700 lb)

PERFORMANCE (WA-116/F):

Max level speed	not fully explored
Cruising speed without long-range tank	87 knots (161 km/h; 100 mph)
Max rate of climb at S/L	305 m (1,000 ft)/min
Max range with long-range tank (estimated)	651 nm (1,207 km; 750 miles)

WALLIS WA-117/R-R

Started in 1964, the WA-117 combines the proven features of the WA-116 airframe with a fully-certificated engine, the 74 kW (100 hp) Rolls-Royce Continental O-200-B. An experimental test vehicle (G-ATCV) flew for the first time on 24 March 1965; this was later dismantled for the construction of a true WA-117 prototype (G-AVJV), which made its first flight on 28 May 1967. This aircraft took part in the Loch Ness investigations in 1970, and has since undertaken considerable evaluation of infra-red linescan and low light level TV systems. With a BAe Dynamics Type 213 Linescan and a photographic pack containing an electric-drive 70 mm wide-angle camera, it carried out experimental detection of leaks in main water pipelines buried 2·45 m (8 ft) underground. As shown in an accompanying photograph, it has also been fitted with a Vinten Type 751 panoramic camera, with which it recently made a photographic coverage of the Liverpool-Manchester railway on the occasion of the line's 150th anniversary celebrations. For that mission, the Vinten camera was used in the beam mode; it can also be used in a cross-track mode, as demonstrated during the 1979 Paris Air Show. With its special silencers and special four-blade 'quiet' propeller, G-AVJV is one of the quietest powered aircraft of any kind yet built.

WEIGHT:

Max T-O weight	approx 317·5 kg (700 lb)

PERFORMANCE:

Max level speed	104 knots (193 km/h; 120 mph)
Cruising speed	78 knots (145 km/h; 90 mph)
Max rate of climb	approx 305 m (1,000 ft)/min

WALLIS WA-118/M METEORITE

The WA-118 Meteorite flew for the first time on 6 May 1966, having at that time the registration G-ATPW. Its 89 kW (120 hp) supercharged Italian Meteor Alfa 1 engine was brought up to then-current standards during 1969-70.

The Wallis WA-121/Mc, which has already exceeded unofficially the world speed and altitude records for autogyros held by the WA-116/Mc prototype since 1968-69

Wg Cdr Wallis demonstrating the 'hands and feet off' stability of the WA-122/R-R at a preselected and trimmed airspeed

The aircraft was also fitted with a bubble canopy, reclining cockpit and other modifications, and was rebuilt as G-AVJW, making its first flight in this form on 9 August 1969. Intended for a long-term test programme, it was then completely redesigned and rebuilt with the latest features of the Wallis range. The aircraft has a partially stress-bearing GRP cockpit nacelle capable of being used as an open cockpit or, in inclement weather, of accepting an optional hardtop (not yet tested). The nacelle is being tested for a planned Franklin-powered autogyro of even better performance than the WA-116/F G-ATHM. The objective is to provide more space and comfort for the pilot, without sacrificing performance, and the cockpit will permit the internal stowage of oxygen equipment and an approved recording barograph. The WA-118/M is intended eventually to achieve altitudes in excess of 9,145 m (30,000 ft).

WALLIS WA-120/R-R

The WA-120 (G-AYVO) is powered by a 97 kW (130 hp) Rolls-Royce Continental O-240-A flat-four engine and cruises at a fuel consumption of 15·9 litres (3·5 Imp gallons)/h. Construction began under the original designation WA-117-S. It subsequently developed into more than a re-engined version of the WA-117, justifying the use of a new designation. It flew for the first time on 13 June 1971.

The WA-120 has a forward-sliding transparent cockpit canopy, and can be flown at speeds of up to 60 knots (111 km/h; 69 mph) with this canopy partly open.

During 1977 the WA-120/R-R was loaned to the Science Museum in London, where in mid-1981 it still formed part of an exhibition on the theme of 'Exploration', fitted with four Vinten 70 mm reconnaissance cameras in a multi-band pack. Still in flying order, the WA-120/R-R is expected to remain in this exhibition for a further two years.

WALLIS WA-121

The WA-121 is the smallest and lightest Wallis autogyro to date. Three versions are envisaged: a high-speed **WA-121/Mc** with a Wallis-McCulloch engine of about 74 kW (100 hp); a cross-country **WA-121/F** with a 44·5 kW (60 hp) Franklin 2A-120-B engine; and a high-altitude **WA-121/M Meteorite 2** with a supercharged 89 kW (120 hp) Meteor Alfa 1 radial two-stroke engine and transistorised ignition.

The prototype WA-121/Mc (G-BAHH) has a high-mounted tailplane and an open cockpit, and made its first flight on 28 December 1972. It has already exceeded unofficially the speed and altitude records held by the WA-116 prototype G-ARRT. It employs a number of improvements in control system design, resulting in greater stability at speed, better head resistance and greater pilot comfort. Special features in the rotor head suspension, originally incorporated in the WA-117 prototype G-AVJV, are incorporated also in the WA-120 and WA-121. Now fitted with a wider-track main landing gear, to standardise it with other autogyros in the Wallis range, the WA-121 is intended to undertake experimental flying using rotor blades designed for high speeds. The results so gained may be used in the design of a further autogyro, to be powered by the new Weslake engine referred to in the WA-116/F entry.

WALLIS WA-122/R-R

The basic layout of the open-frame, two-seat WA-116-T has proved very successful in providing a passenger with autogyro air experience. In addition, dual controls can be fitted when required, and other controls (eg, throttle and spin-up drive lever) can be reached by both occupants. However, the high noise level of the modified McCulloch engine would preclude the use of this aircraft for ab initio instruction, and reliability of this uncertificated engine could not be accepted for regular training use.

In consequence, a slightly larger two-seat aircraft (G-BGGW), known as the WA-122/R-R, has been built, and began flight testing on 16 July 1980. Powered by a 119 kW (160 hp) Rolls-Royce Continental O-240 flat-four engine, the WA-122/R-R retains the main characteristics of the WA-116-T, keeping the tandem layout but with more space between the seats. The rear seat is higher than the one in front. Rotor diameter is increased to approx 7·0 m (23 ft); the blades, and also the landing gear legs, can be quickly folded, for ease of transportation in a container.

WESTLAND

WESTLAND AIRCRAFT LTD

HEAD OFFICE, WORKS AND AIRFIELD: Yeovil, Somerset BA20 2YB

LONDON OFFICE: 4 Carlton Gardens, Pall Mall, London SW1

PRESIDENT: Sir Eric Mensforth, CB, DSc, FEng

CHAIRMAN: The Rt Hon Lord Aldington, PC, KCMG, CBE, DSO

DEPUTY CHAIRMAN: The Rt Hon Lord Aberconway

VICE-CHAIRMAN AND CHIEF EXECUTIVE: B. D. Blackwell, MA, BSc(Eng), FEng, FIMechE, FRAeS, FBIM

DIRECTORS:
Sir Alan Dalton
Sir Christopher Hartley, KCB, CBE, DFC, AFC, BA
Sir Ronald Melville, KCB
W. T. C. Miller, OBE, MA, CEng, MIMechE, MInstM
H. P. Stewart (Group Finance)
Sir John Treacher, KCB, FRAeS
S. W. Wiltshire

SECRETARY: J. R. Bayley, LLB

PUBLIC RELATIONS EXECUTIVE: John Teague, CEng, MRAeS, MIM

Westland Aircraft Ltd was formed in July 1935, to take over the aircraft branch of Petters Ltd, previously known as the Westland Aircraft Works, which had been engaged in aircraft design and construction since 1915.

Westland entered the helicopter industry in 1947 by acquiring the licence to build the Sikorsky S-51, of which it produced 133 under the name Westland Dragonfly. This technical association with Sikorsky has continued, and it was decided subsequently to concentrate on the design, development and construction of helicopters.

In 1959, Westland acquired Saunders-Roe Ltd. In 1960 it acquired the Helicopter Division of Bristol Aircraft Ltd and Fairey Aviation Ltd, and has since been the only major helicopter design and manufacturing organisation in the United Kingdom.

Since 1 October 1966, the company's helicopter business has been conducted through a wholly-owned company named Westland Helicopters Ltd.

Through the British Hovercraft Corporation Ltd, Westland is continuing development of the Hovercraft type of vehicle pioneered by Saunders-Roe.

One of Westland's subsidiary companies, Normalair-Garrett Ltd, specialises in the design, development and production of aircraft pressure control, air-conditioning, oxygen breathing and hydraulic systems. Most British pressurised aircraft, civil and military, use Normalair-Garrett equipment, as do the Panavia Tornado and many aircraft of foreign design. In addition, this company produces data loggers, trace readers and hydraulic equipment for aircraft flying controls.

WESTLAND HELICOPTERS LTD

HEAD OFFICE, WORKS AND AIRFIELD: Yeovil, Somerset BA20 2YB

Telephone: Yeovil (0935) 5222

Telex: 46277

DIRECTORS:
B. D. Blackwell, MA, BSc(Eng), FEng, FIMechE, FRAeS, FBIM (Chairman and Managing Director)
A. V. N. Reed, BSc(Tech), CEng, FRAeS, MBIM (Deputy Managing Director)
J. W. Bower, CEng, MIMechE, FBIM
A. R. B. Hobbs, FCCA
J. P. Jones, PhD, BSc(Eng), CEng, FRAeS
M. Jones, BSc(Econ), FCMA, FCCA, FBIM
V. A. B. Rogers, MSc, FEng, FRAeS, FIMechE

DIVISIONAL DIRECTORS:
D. K. Berrington, BA, CEng, MIMechE, MRAeS (Foreign Military Business)
I. P. Brennan, MA, Barrister (Civil Business)
C. J. Bunker, BSc, ACMA, FCCA, MBIM (Finance)

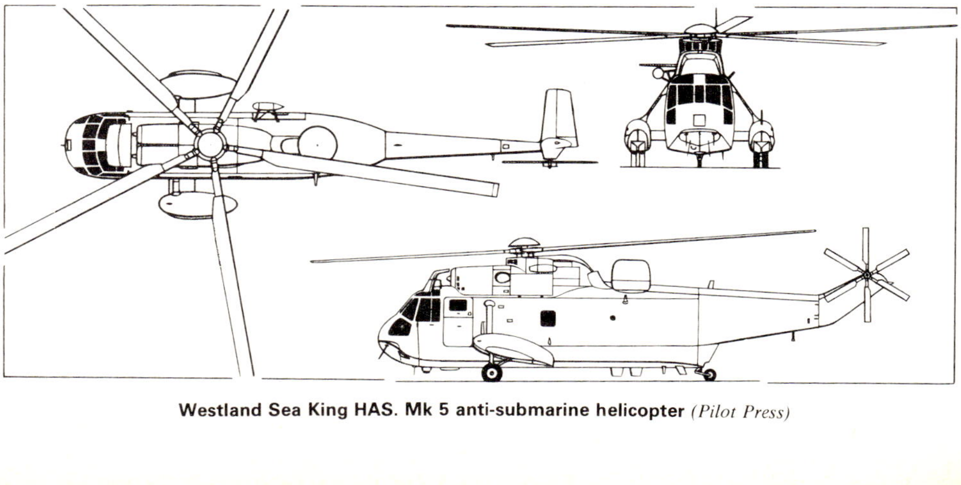

Westland Sea King HAS. Mk 5 anti-submarine helicopter *(Pilot Press)*

R. A. Doe, MSc, CEng, FRAeS (Engineering)
C. Hamshaw-Thomas, MA, CEng, MRAeS, MBIM (Marketing)
R. C. J. Hann (Business Systems)
M. V. Lowson, PhD, BSc, CEng, MRAeS, FIMA, FASA, MAIAA (Corporate Development)
G. Marsh (Materiel)
G. Marshall (Manufacture)
L. R. Moxam, MBE (Flight Operations)
M. A. Nedham, BSc(Eng), ACGI (Product Support Services)
D. Ridge, AIM, MRAeS (Quality)
J. W. Rosenthal, MIPS (Industrial Business)
F. P. Stanton (HM Government Business)
R. A. Wilson (Product Engineering)
PUBLIC RELATIONS MANAGER: Michael J. Farlam

Helicopters in current production at Yeovil are the Sea King, Commando, Army Lynx, Navy Lynx, WG.30 Lynx, and the Westland 30 for the civil market. At the Weston-super-Mare factory the Gazelle and Puma are in production, and Westland is actively involved with production of component sets for Pumas built by Aérospatiale.

In another international programme, Westland Helicopters and Agusta of Italy have formed jointly a company named EH Industries to develop and produce the EH 101 anti-submarine helicopter (see International section).

WESTLAND SEA KING

The Westland Sea King development programme stemmed from a licence agreement for the S-61 helicopter concluded originally with Sikorsky in 1959. This permitted Westland to utilise the basic airframe and rotor system of the Sikorsky SH-3D, of which a description can be found in the US section. Considerable changes were made in the power plant and in specialised equipment, initially to meet a Royal Navy requirement for an advanced anti-submarine helicopter with prolonged endurance. The Sea King can also undertake secondary roles, such as search and rescue, tactical troop transport, casualty evacuation, cargo carrying and long-range self-ferry. A land-based general-purpose version, the Commando, is described separately.

The fuselage is essentially similar to that of the basic Sikorsky aircraft, with a watertight hull which allows water landing in an emergency. The retractable main landing gear is housed in sponsons braced to the fuselage by fixed struts. To improve the lateral stability and flotation capability of the helicopter with the rotor stopped, inflatable buoyancy bags are fitted to the outside of each sponson.

The following versions of the Sea King had been announced up to mid-1981:

Sea King HAS. Mk 1. ASW version for Royal Navy, ordered in 1967. First production HAS. Mk 1 (XV642) flown for the first time on 7 May 1969. Total of 56 built, delivery of which was completed in May 1972. Served with Nos. 814, 819, 820, 824 and 826 Squadrons. Described in previous editions of *Jane's*. All Mk 1s have been uprated by Royal Navy to Mk 2 standard.

Sea King HAS. Mk 2. Uprated version for ASW and SAR duties with the Royal Navy. Twenty-one built; first flown (XZ570) on 18 June 1976. Equipment includes search radar and dunking sonar. To be upgraded to **Mk 5** standard (see below).

Sea King HAR. Mk 3. Uprated version for SAR duties with the Royal Air Force. Provision for flight crew of two pilots, air electronics/winch operator and loadmaster/winchman; up to six stretchers, or two stretchers and 11 seated survivors, or 19 persons. Nav system includes Decca TANS F computer, accepting inputs from Mk 19 Decca nav receiver and Type 71 Doppler. MEL radar. Sixteen ordered, to equip No. 202 Squadron at Lossiemouth, Scotland. First HAR. Mk 3 flew on 6 September 1977; deliveries completed in 1979.

Sea King HC. Mk 4. Utility version of Commando Mk 2 (which see) for Royal Navy.

Sea King Mk 4X. Two aircraft of this type, based on the Sea King Mk 4, are to be built for RAE Farnborough. The first is due for delivery in 1983.

Sea King HAS. Mk 5. Updated ASW and SAR version for the Royal Navy. Seventeen new-build aircraft ordered, of which the first two (ZA126/127) were handed over officially on 2 October 1980. Four delivered by 1 February 1981. The Sea King HAS. Mk 5 has a Decca 71 Doppler tactical air navigation system, passive sonobuoy dropping equipment, and associated Marconi LAPADS acoustic processing and display equipment, with provision for all-digital longer-range MEL Sea Searcher radar to be fitted later. The increased size of the rotating antenna has necessitated the Mk 5's larger dorsal radome.

Using this new equipment, the Sea King can pinpoint the position of an enemy submarine at far greater range than has been possible in the past, and attack it with torpedoes. In addition to monitoring signals from its own sonobuoys, the Sea King can handle information from buoys dropped by RAF Nimrod aircraft in a joint search. It can remain on station, up to 87 nm (160 km; 100 miles) from its parent ship, for long periods.

The Sea King HAS. Mk 5 carries a crew of four, with the dunking sonar operator also monitoring the LAPADS equipment at an additional crew station. To make room for the extra equipment, the cabin has been enlarged by moving the rear bulkhead nearly 1·83 m (6 ft 0 in) further into the tail. Improvements envisaged for the future include the carriage of more powerful torpedoes and further anti-submarine sensors.

Max T-O weight of the Sea King HAS. Mk 5 is 9,525 kg (21,000 lb), the same as that of the Mk 2. Royal Navy HAS. Mk 2s are to be upgraded to this standard.

Westland Sea King HAR. Mk 3 search and rescue helicopter of the Royal Air Force

Sea King Mk 41. Search and rescue version for Federal German Navy. First example (89 + 50) flown for the first time on 6 March 1972. Twenty-two ordered, of which production and delivery were completed in 1974. First unit to equip with these aircraft was MFG.5, based at Kiel-Holtenau.

Sea King Mk 42. ASW version for Indian Navy. Original order for six, which are in service with No. 330 Squadron. Delivery of a further six was completed in 1974, and these are in service with Nos. 330 and 336 Squadrons. Follow-on order announced in June 1977 for three uprated aircraft, designated **Mk 42A**, with hauldown capability for small-ship operation.

Sea King Mk 43. SAR version for Norwegian Air Force. Eleven delivered. In service with No. 330 Squadron at Bodo.

Sea King Mk 43A. Uprated SAR version for Norwegian Air Force; one ordered.

Sea King Mk 45. ASW version for Pakistan Navy. Six ordered, delivery of which was completed during 1975.

Sea King Mk 47. ASW version. Six ordered by Saudi Arabia on behalf of Egyptian Navy. Delivery completed in 1976.

Sea King Mk 48. SAR version for Belgian Air Force. Five ordered, including one aircraft with VIP interior capability. Delivery completed in November 1976. In service with No. 40 Squadron at Coxyde.

Sea King Mk 50. Version, developed from Mk 1, for No. 817 Squadron of Royal Australian Navy, which has ordered 12. First flight 30 June 1974. Production included offset manufacture in Australia to 30% of the contract value. Deliveries began in the Autumn of 1974. The Mk 50 was the first fully-uprated version of the Sea King to fly. It is capable of operation in the roles of anti-submarine search and strike, vertical replenishment, tactical troop lift, search and rescue, casualty evacuation, and self-ferry.

A total of 190 Sea King and 49 Commando aircraft had been ordered by February 1981. Deliveries by that time totalled 177 Sea Kings and 40 Commandos.

A description of the Sea King Mk 1, which is applicable also to most export versions except the Mks 47, 48 and 50, has appeared in previous editions of *Jane's*. Current production Sea Kings (Mk 2/3 standard) have uprated Gnome engines and transmission, a six-blade tail rotor, increased max T-O weight and other detail improvements; the following description applies to the current version. Most of the improvements are incorporated also in the Mk 50s built for Australia, and in export versions for India (Mk 42A), Norway (Mk 43A), Egypt (Mk 47) and Belgium (Mk 48).

POWER PLANT (all current versions): Two 1,238 kW (1,660 shp) (max contingency rating) Rolls-Royce Gnome H.1400-1 turboshaft engines, mounted side by side above cabin. Transmission rating 2,012 kW (2,700 shp). Fuel in underfloor bag tanks, total capacity (SAR versions) 3,636 litres (800 Imp gallons). Internal auxiliary tank may be fitted for long-range ferry purposes. Pressure refuelling point on starboard side, two gravity points on port side.

ACCOMMODATION: Crew of four in ASW role; accommodation for up to 22 survivors in SAR role. Two-section airstair door at front on port side, cargo door at rear on starboard side. Entire accommodation heated and ventilated. Cockpit doors and windows, and two windows each side of cabin, can be jettisoned in an emergency.

SYSTEMS: Three main hydraulic systems. Primary and auxiliary systems operate main rotor control. Utility system (207 bars; 3,000 lb/sq in) for main landing gear, sonar and rescue winches and blade folding. Pressure for windscreen wipers 86 bars (1,250 lb/sq in). Electrical system includes two 20kVA 200V three-phase 400Hz

Westland Sea King HC. Mk 4 utility version of the Commando for the Royal Navy

One of the first two Westland Sea King HAS. Mk 5 helicopters delivered to the Royal Navy on 2 October 1980

engine-driven generators, a 26V single-phase AC supply fed from the aircraft's 22Ah nickel-cadmium battery through an inverter, and DC power provided as a secondary system from two 200A transformer-rectifier units.

OPERATIONAL EQUIPMENT (ASW models): As equipped for this role, the Sea King is a fully-integrated all-weather hunter-killer weapon system, capable of operating independently of surface vessels, and has the following equipment and weapons to achieve this task: Plessey Type 195 dipping sonar, Bendix AN/AQS-13B dipping sonar (Mk 50), Marconi AD 580 Doppler navigation system, AW 391 search radar in dorsal radome, transponder beneath rear fuselage, Honeywell AN/APN-171 radar altimeter, Sperry GM7B Gyrosyn compass system, Louis Newmark Mk 31 automatic flight control system, two No. 4 marine markers, four No. 2 Mk 2 smoke floats, Ultra Electronics mini-sonobuoys, up to four Mk 46 homing torpedoes, or four Mk 11 depth charges or one Clevite simulator. Observer/navigator has tactical display on which sonar contacts are integrated with search radar and navigational information. Radio equipment comprises Plessey PTR 377 UHF/VHF and homer, Ultra D 403M standby UHF, Collins 618-T3 HF radio, Ultra UA 60M intercom, Telebrief system and IFF provisions. For secondary role a mounting is provided on the aft frame of the starboard door for a general-purpose machine-gun. The Mk 31 AFCS provides radio altitude displays for both pilots; artificial horizon displays; three-axis stabilisation in pilot-controlled manoeuvres; attitude hold, heading hold and height hold in cruising flight; controlled transition manoeuvres to and from the hover; automatic height control and plan position control in the hover; and an auxiliary trim facility.

OPERATIONAL EQUIPMENT (non-ASW models): A wide range of radio and navigation equipment may be installed, including VHF/UHF communications, VHF/UHF homing, radio compass, Doppler navigation system, radio altimeter, VOR/ILS, radar and transponder, of Collins, Plessey, Honeywell and Marconi manufacture. A Sperry compass system and a Louis Newmark automatic flight control system are also installed. Sea Kings equipped for search and rescue have in addition a Breeze BL 10300 variable-speed hydraulic rescue hoist of 272 kg (600 lb) capacity mounted above the starboard-side cargo door. Automatic main rotor blade folding and spreading is standard with this version, and for shipboard operation the tail pylon can also be folded. With search radar fitted, a total of 18 survivors and medical staff can be carried; this total can be increased to 22 if the search radar is omitted . In the casualty evacuation role, the Sea King can accommodate up to 9 stretchers and two medical attendants, or intermediate combinations of seats and stretchers; a typical layout might provide for 14 seats and two stretchers. In the troop transport role, the Sea King can accommodate 22 troops, with the majority of seats at 42 cm (16·5 in) pitch, and can carry this load over a range of 300 nm (555 km; 345 miles) under ISA sea level conditions. As a cargo transport, the aircraft has an internal capacity of 2,720 kg (6,000 lb) or a max external load capacity of 2,948 kg (6,500 lb) when a low-response sling is fitted.

DIMENSIONS, EXTERNAL:
- Diameter of main rotor — 18·90 m (62 ft 0 in)
- Diameter of tail rotor — 3·16 m (10 ft 4 in)
- Length overall (rotors turning) — 22·15 m (72 ft 8 in)
- Length of fuselage — 17·01 m (55 ft 9¾ in)
- Length overall (main rotor folded) — 17·42 m (57 ft 2 in)
- Length overall (rotors and tail folded) — 14·40 m (47 ft 3 in)
- Height overall (rotors turning) — 5·13 m (16 ft 10 in)
- Height overall (rotors spread and stationary) — 4·85 m (15 ft 11 in)
- Height to top of rotor hub — 4·72 m (15 ft 6 in)
- Width overall (rotors folded):
  - with flotation bags — 4·98 m (16 ft 4 in)
  - without flotation bags — 4·77 m (15 ft 8 in)
- Wheel track (c/l of shock-absorbers) — 3·96 m (13 ft 0 in)
- Cabin door (port): Height — 1·68 m (5 ft 6 in)
  - Width — 0·91 m (3 ft 0 in)
- Cargo door (stbd): Height — 1·52 m (5 ft 0 in)
  - Width — 1·73 m (5 ft 8 in)
  - Height to sill — 1·14 m (3 ft 9 in)

DIMENSIONS, INTERNAL:
- Cabin: Length: ASW — 5·87 m (19 ft 3 in)
  - SAR — 7·59 m (24 ft 11 in)
  - Max width — 1·98 m (6 ft 6 in)
  - Max height — 1·92 m (6 ft 3½ in)
  - Floor area (incl area occupied by radar, sonar etc):
    - ASW — 12·08 m² (130 sq ft)
    - SAR — 13·94 m² (150 sq ft)

WEIGHTS AND LOADINGS (A: anti-submarine, B: SAR, C: troop transport, D: casualty evacuation, E: internal cargo):
- Basic weight (depending on version) — approx 5,896 kg (13,000 lb)
- Weight, equipped: A — 6,201 kg (13,672 lb)
  - B, C — 5,613 kg (12,376 lb)
  - D — 5,797 kg (12,781 lb)
  - E — 5,558 kg (12,253 lb)
- Max T-O weight:
  - all current versions — 9,525 kg (21,000 lb)
- Max disc loading — 34·62 kg/m² (7·09 lb/sq ft)
- Max power loading — 4·73 kg/kW (7·77 lb/shp)

PERFORMANCE (at max T-O weight, all current versions):
- Cruising speed at S/L — 112 knots (208 km/h; 129 mph)
- Max rate of climb at S/L — 616 m (2,020 ft)/min
- Max vertical rate of climb at S/L — 119 m (390 ft)/min
- Service ceiling, one engine out — 1,220 m (4,000 ft)
- Hovering ceiling IGE — 1,525 m (5,000 ft)
- Hovering ceiling OGE — 975 m (3,200 ft)
- Range with max standard fuel — 664 nm (1,230 km; 764 miles)
- Ferry range with max standard and auxiliary fuel — 814 nm (1,507 km; 937 miles)

WESTLAND COMMANDO

First flown on 12 September 1973, the Commando is a tactical helicopter based on the Sea King.

The payload/range performance and endurance capabilities of the Sea King have been optimised in the design of the Commando, which is intended to operate with maximum efficiency in the primary roles of tactical troop transport, logistic support and cargo transport, and casualty evacuation. In addition, the Commando can operate effectively in the secondary roles of air-to-surface strike and search and rescue.

The following versions have been announced:

Commando Mk 1. Designation of first five Commandos, ordered on behalf of the Egyptian Air Force by the Saudi Arabian government. Minimally-modified version able to transport up to 21 troops. First two delivered to Egypt in January/February 1974.

Commando Mk 2. Major production version, to which detailed description applies. Flew for first time (G-17-12) on 16 January 1975. Saudi Arabian order included 17 Mk 2s and two VIP Mk 2Bs for the Egyptian Air Force. Four Mk 2s (three Mk 2As and one VIP Mk 2C) delivered to Qatar Emiri Air Force.

Sea King HC. Mk 4. Utility version of Commando Mk 2 for Royal Navy. Has folding main rotor blades and folding tail pylon of Sea King, but retains non-retractable wheeled landing gear of Commando. Designed to carry up to 27 fully-equipped troops in cabin, or 2,720 kg (6,000 lb) of cargo, and to operate in Arctic and tropical conditions. Max slung load 3,400 kg (7,500 lb). Equipped for parachuting and abseiling. Revised avionics, including Decca TANS with chart display and Decca 71 Doppler navigation system. For service with Nos. 845 and 846 (Naval Air Commando) Squadrons. First flight (by ZA290) 26 September 1979. Seventeen ordered, of which the first was handed over to the Royal Navy in November 1979. Eight delivered by 1 March 1981.

TYPE: Twin-turboshaft tactical military helicopter.

ROTOR SYSTEM: Five-blade single main rotor and six-blade tail rotor. Main rotor blades, of NACA 0012 section, attached to hub by multiple bolted joint. Blade construction consists of a light alloy extruded spar, with light alloy trailing-edge pockets. Tail rotor blades are of similar construction. Rotor brake fitted. Tail section folds for stowage; main rotor blades do not.

ROTOR DRIVE: Twin input four-stage reduction main gearbox, with single bevel intermediate and tail gearboxes. Main rotor/engine rpm ratio 93·43; tail rotor/engine rpm ratio: 15·26.

FUSELAGE: Light alloy stressed-skin structure, unpressurised. Sea King sponsons replaced by stub-wings.

TAIL UNIT: Similar to Sea King, with starboard-side half-tailplane at top of tail rotor pylon.

LANDING GEAR: Non-retractable tailwheel type, with twin-wheel main units. Oleo-pneumatic shock-absorbers. Main-wheel tyres size 6·50-10, tailwheel tyre size 6·00-6.

POWER PLANT: As for current versions of Sea King.

ACCOMMODATION: Crew of two on flight deck. Seats along cabin sides, and single jump seat, for up to 28 troops. Two-piece airstair door at front on port side, cargo door at rear on starboard side. Entire accommodation heated and ventilated. Cockpit doors and windows, and two windows each side of main cabin, are jettisonable in an emergency.

SYSTEMS: Primary and secondary hydraulic systems for flight controls. No pneumatic system. Electrical system includes two 20kVA alternators.

AVIONICS AND EQUIPMENT: Blind-flying instrumentation standard. Wide range of radio, radar and navigation equipment available to customer's requirements. Cargo sling and rescue hoist optional.

ARMAMENT: Wide range of guns, missiles, etc may be carried, to customer's requirements.

DIMENSIONS, EXTERNAL:
- Diameter of main rotor — 18·90 m (62 ft 0 in)
- Diameter of tail rotor — 3·16 m (10 ft 4 in)
- Distance between rotor centres — 11·10 m (36 ft 5 in)
- Main rotor blade chord — 0·46 m (1 ft 6¼ in)
- Length overall (rotors turning) — 22·15 m (72 ft 8 in)
- Length of fuselage — 17·02 m (55 ft 10 in)
- Height overall (rotors turning) — 5·13 m (16 ft 10 in)
- Height to top of rotor hub — 4·72 m (15 ft 6 in)
- Wheel track (c/l of shock-absorbers) — 3·96 m (13 ft 0 in)
- Wheelbase — 7·21 m (23 ft 8 in)

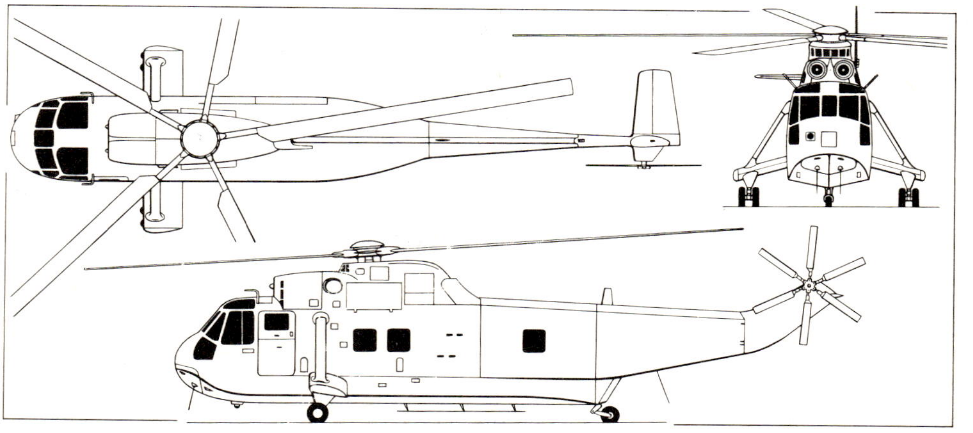

Westland Commando Mk 2 twin-turboshaft tactical military helicopter *(Pilot Press)*

Passenger door (fwd, port):
Height 1·68 m (5 ft 6 in)
Width 0·91 m (3 ft 0 in)
Cargo door (aft, stbd): Height 1·52 m (5 ft 0 in)
Width 1·73 m (5 ft 8 in)

DIMENSIONS, INTERNAL: As Sea King (SAR version)

AREAS:
Main rotor disc 280·5 m² (3,019 sq ft)
Tail rotor disc 7·79 m² (83·86 sq ft)
Main rotor blades (each) 4·14 m² (44·54 sq ft)
Tail rotor blades (each) 0·23 m² (2·46 sq ft)
Tailplane 1·80 m² (19·40 sq ft)

WEIGHTS:
Operating weight empty (troop transport, 2 crew) 5,700 kg (12,566 lb)
Max T-O weight 9,525 kg (21,000 lb)

PERFORMANCE (at max T-O weight): As given for Sea King, plus:
Range with max payload (28 troops), reserves for 30 min stand-off 240 nm (445 km; 276 miles)

WESTLAND LYNX

The Lynx is one of three types of aircraft (Puma, Gazelle and Lynx) covered by the Anglo-French helicopter agreement first proposed in February 1967 and confirmed on 2 April 1968. Westland has design leadership in the Lynx, which is intended to fulfil general-purpose, naval and civil transport roles.

The first of 13 Lynx prototypes (XW835) flew for the first time on 21 March 1971 and was followed by XW837, the third prototype (second Lynx to fly), on 28 September 1971. Details of subsequent development aircraft can be found in the 1975-76 *Jane's*.

A Lynx intensive flying trials unit, No. 700L Naval Air Squadron, was formed in September 1976 at RNAS Yeovilton, Somerset, as a joint Royal Navy/Royal Netherlands Navy operational evaluation unit. Deck handling tests, on board HMS *Birmingham* off Portland, were completed in February 1977. Another Lynx intensive flying trials unit was formed at the British Army Aviation headquarters at Middle Wallop, Hampshire, in mid-1977. The Army intensive flight trials were completed in December 1977.

The following versions have been announced:

Lynx AH. Mk 1. General-purpose and utility version for the British Army. Capable of operation on tactical troop transport, logistic support, armed escort of troop-carrying helicopters, anti-tank strike, search and rescue, casualty evacuation, reconnaissance and command post duties. Total of 114 ordered. First production aircraft (XZ170) flown on 11 February 1977; 84 delivered by 1 July 1981.

A Westland-owned aircraft (G-LYNX, first flown in May 1979), has demonstrated the helicopter's multi-role capability with a wide range of weapons which have included Hughes TOW and Euromissile Hot anti-tank guided missiles; SURA 80 mm, SNEB 68 mm and FZ 2·75 in rockets; twin 7·62 mm machine-gun pods and 20 mm automatic cannon; Matra Magic 550 air-to-air missiles; an AN/ALE-39 countermeasures dispenser with ECM chaff; and ECM warning equipment. It can also carry mine dispensers, or anti-tank teams armed with Milan missiles. Sixty Lynx AH. Mk 1 are to be equipped with TOW missiles, for service with BAOR in the anti-tank role. The first six of these were delivered in the Spring of 1981 to No. 654 Squadron of No. 4 Regiment, Army Aviation.

Lynx HAS. Mk 2. Version for Royal Navy, for advanced shipborne anti-submarine and other duties. Ferranti Sea Spray search and tracking radar in modified nose. Capable of operation on anti-submarine classification and strike, air to surface vessel search and strike, search and rescue, reconnaissance, troop transport, fire support, communication and fleet liaison, and vertical replenishment duties. Total of 60 ordered by 1 September 1980. First production aircraft (XZ229) flown on 10 February 1976; 59 delivered by 1 July 1981. First operational RN unit, No. 702 Squadron, formed on completion of Navy intensive flight trials in December 1977. Serving in 1980 with nine Ships' Flights (five in Type 21 frigates, two in 'Leander' class frigates, and two in Type 42 destroyers). May be fitted with additional submarine detection gear, possibly of the dipping sonar type.

Lynx Mk 2 (French Navy). Generally similar to British HAS. Mk 2 but with Alcatel dunking sonar, French radar and radio, and AS.12 wire-guided missiles. Delivery of initial batch of 26 began on 28 September 1978, and was completed during 1980.

Lynx HAS. Mk 3. Designation of 20 Royal Navy aircraft which have uprated power plant, comprising two 835 kW (1,120 shp) Rolls-Royce Gem 41-1 turboshafts.

Lynx Mk 4. Second batch of 14 aircraft ordered for French Navy in May 1980 with Gem 41-1 engines and uprated transmission to permit an increase in AUW to 4,763 kg (10,500 lb). Deliveries of this series had not started by 1 July 1981.

Lynx Mk 23. ASW version for Argentinian Navy, generally similar to HAS. Mk 2. Two delivered.

Lynx Mk 25. SAR version for Royal Netherlands Navy, with Gem 2 engines. In service with No. 7 Squadron; used also for communications and training. Six delivered; Dutch designation **UH-14A**.

Lynx Mk 27. Designation of initial ASW version for

British Army Westland Lynx during firing trials with TOW anti-tank missiles

Royal Netherlands Navy, generally similar to HAS. Mk 2, but equipped with Gem 41-1 engines and Alcatel dunking sonar. Ten delivered; Dutch designation **SH-14B**.

Lynx Mk 28. General-purpose military version, basically similar to AH. Mk 1, for State of Qatar Police. Three delivered, with uprated Gem 41-1 turboshafts.

Lynx Mk 80. ASW and maritime patrol version for Royal Danish Navy, generally similar to HAS. Mk 2 except for Gem 41-1 engines. Eight ordered, of which seven delivered by 1 July 1981.

Lynx Mk 81. ASW version for Royal Netherlands Navy, with Gem 41-1 uprated engines and equipped with MAD gear. Eight delivered; Dutch designation **SH-14C**.

Lynx Mk 86. SAR version for Royal Norwegian Air Force, generally similar to HAS. Mk 2 except for Gem 41-2 engines. Six ordered, of which one delivered by 1 July 1981.

Lynx Mk 87. For Argentinian Navy. Generally similar to Mk 23 but with Gem 41-2 engines and max AUW of 4,763 kg (10,500 lb). Two delivered.

Lynx Mk 88. ASW version for Federal German Navy. Generally similar to Mk 86, but with non-folding tail and with Bendix AN/AQS-18 sonar. Twelve ordered, of which one delivered by 1 July 1981.

Lynx Mk 89. ASW version for Brazilian Navy, generally similar to HAS Mk 2. Nine delivered.

In addition to the Westland demonstrator, a total of 306 Lynx had been ordered and 216 delivered by 1 July 1981. Production is shared in the ratio of 70 per cent by Westland to 30 per cent by Aérospatiale, and was at a rate of better than four per month in early 1981.

The following description applies to both the military general-purpose and naval versions with the Gem 2 power plant, except where otherwise indicated:

TYPE: Twin-engined multi-purpose helicopter.

ROTOR SYSTEM: Single four-blade semi-rigid main rotor and four-blade tail rotor. The main rotor blades, which are interchangeable, are of cambered aerofoil section and embody mass taper. Each blade consists of a two-piece, two-channel stainless steel D-shaped box-spar, to which is bonded a GRP rear skin stabilised by a Nomex plastics honeycomb core. Blade tips are of moulded GRP, with a stainless steel anti-erosion sheath forward of the 50% chord line. Each blade is attached to the main rotor hub by titanium root attachment plates and a flexible arm. The rotor hub and inboard portions of the flexible arms are built as a complete unit, in the form of a titanium monobloc forging. A two-pin jaw for blade attachment and manual blade folding is provided. Each tail rotor blade has a light alloy spar, machined integrally with the root attachment, which forms the nose portion of the aerofoil section and has a flush-fitting stainless steel sheath on the leading-edge. The rear section of each blade is of similar construction to that of the main rotor blades. Tail rotor blades are replaceable in opposing pairs, and each blade is attached to the hub by the outboard tiebar pin and a six-bolt root-end flanged joint. Main rotor blades of both versions can be folded, and tail rotor pylon of naval version can be folded and spread manually, to reduce overall length for stowage.

ROTOR DRIVE: Drives are taken from the front of the engines into the main gearbox, which is mounted above the cabin forward of the engines. This gearbox interconnects the two engines, with speed reduction in two stages. The first stage uses an involute-form spiral bevel pinion and gear. The second stage comprises a conformal pinion meshing with a gear fixed directly to the main rotor drive-shaft. In flight, the accessory gears, which are all at the front of the main gearbox, are driven by one of the two through shafts from the first-stage reduction gears. For system checking on the ground without the rotor turning, the accessories can be driven by the port engine. Freewheel units are mounted in each engine gearbox shaft, and also within the accessory drive chain of gears. Rotor head controls are actuated by three identical tandem servojacks, trunnion-mounted from the main rotor gearbox and powered by two independent hydraulic systems. Control system incorporates a simple stability augmentation system, which acts in a single channel to provide improved stability in pitch. Provision is made for in-flight blade tracking. Each engine embodies an independent control system which provides full-authority rotor speed governing, pilot control being limited to selection of the desired rotor speed range. In the event of an engine failure, this system will restore power up to single-engine maximum contingency rating. On the naval versions, the main rotor can provide negative thrust to increase stability on deck after touchdown. Tail rotor drive is taken from the main ring gear. A hydraulically-operated rotor brake is mounted on the main gearbox.

FUSELAGE AND TAIL UNIT: Conventional semi-monocoque pod and boom structure, mainly of light alloy, including a cantilever floor structure with unobstructed surface. Glassfibre components used for access panels, doors and fairings. The forward fuselage is free from bulkheads, giving an unrestricted field of view. Single large window in each of the main cabin sliding doors. Provision for internally-mounted defensive armament, and for universal flange mountings on each side of the exterior to carry weapons or other stores. Tailboom is a light alloy monocoque structure bearing the sweptback vertical fin/tail rotor pylon, which has a half-tailplane near the tip on the starboard side. Tailplane leading- and trailing-edges, and bullet fairing over tail rotor gearbox, are of glassfibre.

LANDING GEAR (general-purpose military version): Non-retractable tubular skid type. Provision for a pair of adjustable ground handling wheels on each skid. Flotation gear optional.

LANDING GEAR (naval versions): Non-retractable oleo-pneumatic tricycle type. Single-wheel main units,

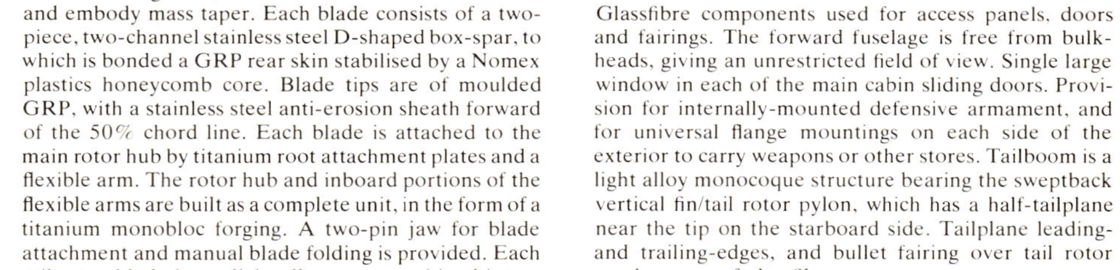

Westland Lynx AH. Mk 1 helicopter, with additional side view (top) of Lynx HAS. Mk 2 *(Pilot Press)*

First of six Navy Lynx Mk 86 for the Royal Norwegian Air Force

mounted on sponsons near rear of main fuselage, are fixed at 27° toe-out for deck landing, and can be manually turned into line and locked fore and aft for movement of aircraft into and out of ship's hangar. Twin-wheel nose unit can be castored hydraulically through 90° by the pilot. Designed for high shock-absorption to facilitate take-off from, and landing on, small decks under severe sea and weather conditions. Sprag brakes (wheel locks) fitted to each wheel prevent rotation on landing or inadvertent deck roll. These locks are disengaged hydraulically and will automatically re-engage in the event of hydraulic failure. Friction brakes may be fitted for shore use. Flotation gear, and hydraulically-actuated harpoon deck-lock securing system, optional.

Power Plant: Two Rolls-Royce Gem 2 turboshaft engines in Lynx AH. 1, HAS. Mk 2 and early export variants. Each has a max continuous rating of 559 kW (750 shp), a take-off and inter-contingency rating of 619 kW (830 shp), and a max contingency rating (2½ min) of 671 kW (900 shp). Uprating modifications have been incorporated progressively into the Gem engine to meet requirements for the increase in Lynx AUW to 4,763 kg (10,500 lb) and for operation in high ambient temperatures. The most recent engine version is the Gem 41-1, which has a max continuous rating of 663·5 kW (890 shp), a T-O rating of 746 kW (1,000 shp), an inter-contingency rating of 790·5 kW (1,060 shp), and a max contingency rating of 835 kW (1,120 shp). Engines mounted side by side on top of the fuselage upper decking, aft of the main rotor shaft and gearbox, and separated from fuselage, transmission area and each other by firewalls. Engine air intakes de-iced electrically. Fuel in five crashproof bag-type tanks, all within the fuselage structure, comprising two main tanks each of 204 kg (450 lb) capacity, two side-by-side collector tanks each of 93 kg (204·5 lb) capacity, and a 148 kg (326 lb) capacity underfloor tank at the forward end of the cabin. Total fuel capacity 733 kg (1,616 lb). Cross-feed system allows fuel to be supplied from both collector tanks to one engine or from one tank to both engines. If required, ferry range can be increased by installing in rear of cabin two metal auxiliary tanks with a combined capacity of 654 kg (1,442 lb). Single-point pressure refuelling (3·79 bars; 55 lb/sq in max) and defuelling; two points for gravity refuelling. A removable 114 litres (25 Imp gallons)/min pressure refuelling/defuelling pack can be fitted in the cabin and, with port engine running, can be used to refuel aircraft from dump stocks on ground or containers suspended from hoist. It is also possible to raise fuel about 5 m (15 ft) while the aircraft is hovering. Fuel jettison capability for main and forward tanks. Provision for self-sealing of both collector tanks (except in Royal Navy version) to provide protection against small-arms fire. Engine oil tank capacity 6·8 litres (1·5 Imp gallons). Main rotor gearbox oil capacity 18 litres (4 Imp gallons).

Accommodation: Pilot and co-pilot or observer on side-by-side seats which can accommodate back-type dinghies and are adjustable fore and aft and for height. Inertia-reel shoulder harness for pilot and co-pilot. Dual controls optional. Additional crew members (eg, gunner, hoist operator) according to role. Individual forward-hinged cockpit door and large rearward-sliding cabin door on each side; all four doors jettisonable. Main cabin doors manufactured by Hawker Pacific in Australia. Cockpit accessible from cabin area. Maximum high-density layout (general-purpose version) for one pilot and 10 armed troops or paratroops, on lightweight bench seats in soundproof cabin. Alternative VIP layouts for four to seven passengers, with additional cabin soundproofing. Seats can be removed quickly to permit the carriage of up to 907 kg (2,000 lb) of freight internally. Tiedown rings are provided at approx 51 cm (20 in) intervals on main cabin floor, which is stressed for loads of up to 976 kg/m² (200 lb/sq ft). Alternatively, loads of up to 1,360 kg (3,000 lb) can be carried externally on freight hook mounted below the cabin floor and fitted, in naval version, with electrically-operated emergency release system. In the casualty evacuation role, with a crew of two, the Lynx can accommodate three standard stretchers and a medical attendant; electrically-heated casualty bags can be provided. Both versions have secondary capability for search and rescue (up to nine survivors) and other roles (see introductory copy and 'Equipment' paragraphs).

Systems: Two independent hydraulic systems in all versions, pressure 141 bars (2,050 lb/sq in). Pumps powered by accessory drive from main rotor gearbox, enabling full power to be drawn from both main systems in event of engine failure. If either No. 1 or No. 2 main system fails, the other maintains adequate flying control. No. 1 system, additionally, actuates tail rotor yaw control and rotor brake. Tail rotor operation reverts to mechanical control if No. 1 system fails. A third hydraulic system, at the same pressure, is provided in the naval version when Alcatel sonar equipment, MAD or a hydraulic winch system is installed. When this third hydraulic system is installed, the deck-lock harpoon is also operated by this system. When the Bendix AN/AQS-13 sonar is installed, a 207 bar (3,000 lb/sq in) 'utility' hydraulic system is used. No pneumatic system. 28V DC electrical power supplied by two 6kW engine-driven starter/generators and an alternator. Engines can also be started from external 28V DC power source. 24V 23Ah (optionally 40Ah) nickel-cadmium battery fitted for essential services and emergency engine starting. 200V three-phase AC power available at 400Hz from two 15kVA transmission-driven alternators. AC and DC external ground power sockets on starboard side of fuselage. Graviner Triple FD engine fire detection system; two separate fire suppression systems fitted. Optional cabin heating and ventilation system, using mixing unit combining engine bleed air with outside air. Optional supplementary cockpit heating system. Electrical anti-icing and demisting of windscreen, and electrically-operated windscreen wipers, standard; windscreen washing system optional.

Avionics and Flight Equipment: Main equipment bays are in nose (under upward-hinged door) and at rear of cabin. All versions equipped as standard with navigation, cabin and cockpit lights; adjustable landing light under nose; anti-collision beacon; first aid kit(s); and hand-type fire extinguishers for cabin. Avionics common to all roles (general-purpose and naval versions) include Marconi duplex three-axis automatic stabilisation equipment; Sperry GM9 Gyrosyn compass system; Decca tactical air navigation system (TANS); Decca 71 Doppler, E2C standby compass; and S.G. Brown intercom system. Optional role equipment for both versions includes Marconi automatic flight control system (AFCS); Plessey PTR 377 UHF/VHF with homing; Collins ARC-159 UHF with homing; Plessey PTR 1751 UHF; Ultra D 403M standby UHF; Collins ARC-182 VHF/UHF; AM/FM with homing; Collins VHF-20B VHF/AM; Marconi AD 120 VHF/FM; Chelton 7 homer; Collins VOR/ILS; DME; Collins ARN-118 Tacan; I-band transponder (naval version only); Plessey PTR 446, APX-72, Siemens STR 700/375 or Italtel APX-77 IFF; Marconi AD 370 and AD 380 radio compass; and vortex-type sand filter for engine air intakes. Additional units are fitted in naval version, when sonar is fitted, to provide automatic transition to the hover and automatic Doppler hold in the hover. Other optional equipment (both versions) includes signal pistol and cartridges, Aldis lamp and stowage.

Armament and Operational Equipment: For armed escort, anti-tank or air-to-surface strike missions, general-purpose version can be equipped with one 20 mm Oerlikon or similar cannon mounted in the cabin with 1,500 rds; or two 20 mm cannon mounted externally so as to permit the carriage also of anti-tank missiles or a pintle-mounted 7·62 mm GEC Minigun inside the cabin; or a Minigun beneath cabin, in Emerson Minitat installation, with 3,000 rds. External pylon can be fitted on each side of cabin for a variety of stores, including two Minigun or other self-contained gun pods; two pods each carrying eighteen 68 mm SNEB, twelve 80 mm SURA, or nineteen 2·75 in rockets, the 2·75 in rockets containing illuminating flares if required; or up to six Aérospatiale AS.11, or eight Aérospatiale/MBB Hot or Hughes TOW, or similar air-to-surface missiles. An additional six or eight missiles can be carried in cabin, for rearming in forward areas, and a stabilised sight is fitted for target detection and missile direction. British Army Lynx aircraft equipped with TOW missiles have roof-mounted Hughes sight manufactured under licence by Dynamics Group of British Aerospace. The Lynx can also transport mobile anti-tank teams of three gunners with missiles and launchers. For search and rescue role, with a crew of three, both versions can be fitted with a waterproof floor, eight 4 in flares in utility version (or six 4·5 in flares in naval version), and a 272 kg (600 lb) capacity electrically-operated 'clip-on' hoist in starboard side of cabin. Alternative option of hydraulically-operated hoist in naval version when third hydraulic system is installed. Hoist, which can lift a load through 76 m (250 ft) at 30·5 m (100 ft)/min, can be swung back into cabin when not in use, permitting sliding door to be closed. General-purpose version can also be equipped for several other duties, including firefighting and crash rescue, reconnaissance, military command post, liaison, customs and border control, and pilot and operational training. Optional equipment, according to role, can include lightweight sighting system with alternative target magnification, vertical and/or oblique cameras, up to six 4·5 in flares for night operation, low light level TV, infra-red linescan, searchlight, and specialised communications equipment. Naval version can carry out a number of these roles, but has specialised equipment for its primary duties. For ASW role, this includes two Mk 44 or Mk 46 homing torpedoes, one each on an external pylon on each side of fuselage, and six marine markers; or two Mk 11 depth charges. Detection of submarines is by means of either Alcatel DUAV 4 or Bendix AN/AQS-18 dipping sonars or Texas Instruments AN/ASQ-81 magnetic anomaly detector. The dipping sonars are operated by a hydraulically-powered winch and cable hover mode facilities within the AFCS. Ferranti Sea Spray lightweight search and tracking radar, for detecting small surface targets in low visibility/high sea conditions. Armament includes four BAe Sea Skua semi-active homing missiles for attacking light surface craft; alternatively, four AS.12 or similar wire-guided missiles can be employed in conjunction with AF 530 or APX-334 lightweight stabilised optical sighting system.

Dimensions, external (A: general-purpose version; N: naval version):

Diameter of main rotor (A, N)	12·802 m (42 ft 0 in)
Diameter of tail rotor (A, N)	2·21 m (7 ft 3 in)
Main rotor blade chord (A, N, constant, each)	0·395 m (1 ft 3½ in)
Tail rotor blade chord (A, N, constant, each)	0·18 m (7·1 in)
Length overall, both rotors turning (A, N)	15·163 m (49 ft 9 in)
Length overall:	
A, main rotor blades folded	13·165 m (43 ft 2·3 in)
N, main rotor blades and tail folded	10·618 m (34 ft 10 in)
Length of fuselage, nose to tail rotor centre:	
A	12·06 m (39 ft 6·8 in)
N	11·92 m (39 ft 1·3 in)
Width overall, main rotor blades folded:	
A	2·94 m (9 ft 7¾ in)
N	3·75 m (12 ft 3¾ in)
Height overall, both rotors turning:	
A	3·66 m (12 ft 0 in)
N	3·60 m (11 ft 9¾ in)
Height overall, both rotors stopped:	
A	3·504 m (11 ft 6 in)
N	3·48 m (11 ft 5 in)
Height to top of rotor hub: A	2·964 m (9 ft 8·7 in)
Height overall, main rotor blades and tail folded:	
N	3·20 m (10 ft 6 in)
Tail rotor ground clearance: A	1·41 m (4 ft 7½ in)
N	1·38 m (4 ft 6·3 in)
Tailplane half-span (from fuselage c/l):	
A, N	1·776 m (5 ft 9·9 in)
Skid track: A	2·032 m (6 ft 8 in)
Wheel track: N	2·778 m (9 ft 1·4 in)
Wheelbase: N	2·94 m (9 ft 7¾ in)
Cabin door openings (A, N, each):	
Mean width	1·372 m (4 ft 6 in)
Height	1·194 m (3 ft 11 in)

DIMENSIONS, INTERNAL:
Cabin, from back of pilots' seats:
Min length 2·057 m (6 ft 9 in)
Max width 1·778 m (5 ft 10 in)
Width at rear 1·409 m (4 ft 7½ in)
Max internal floor width 1·715 m (5 ft 7½ in)
Max height 1·422 m (4 ft 8 in)
Floor area 3·72 m² (40·04 sq ft)
Volume 5·21 m³ (184 cu ft)

WEIGHTS (A: general-purpose version, N: naval version):
Manufacturer's bare weight: A 2,578 kg (5,683 lb)
N 2,740 kg (6,040 lb)
Manufacturer's basic weight: A 2,658 kg (5,860 lb)
N 3,030 kg (6,680 lb)
Operating weight empty, equipped:
A, troop transport (pilot and 10 troops) 2,787 kg (6,144 lb)
A, anti-tank strike (incl weapon pylons, firing equipment and sight) 3,072 kg (6,772 lb)
A, search and rescue (crew of three) 2,963 kg (6,532 lb)
N, anti-submarine strike 3,343 kg (7,370 lb)
N, reconnaissance (crew of two) 3,277 kg (7,224 lb)
N, anti-submarine classification and strike 3,472 kg (7,654 lb)
N, air to surface vessel search and strike (crew of two and four Sea Skuas) 3,414 kg (7,526 lb)
N, search and rescue (crew of three) 3,416 kg (7,531 lb)
N, dunking sonar search and strike 3,650 kg (8,047 lb)
Max T-O weight: A 4,535 kg (10,000 lb)
N 4,763 kg (10,500 lb)

PERFORMANCE (at normal max T-O weight at S/L, ISA, except where indicated. A: general-purpose version; N: naval version):
Max continuous cruising speed:
A 140 knots (259 km/h; 161 mph)
N 125 knots (232 km/h; 144 mph)
A (ISA + 20°C) 130 knots (241 km/h; 150 mph)
N (ISA + 20°C) 114 knots (211 km/h; 131 mph)
Max continuous cruising speed (1 h), one engine out:
A 134 knots (248 km/h; 154 mph)
N 122 knots (225 km/h; 140 mph)
A (ISA + 20°C) 114 knots (211 km/h; 131 mph)
N (ISA + 20°C) 99 knots (184 km/h; 114 mph)
Speed for max endurance:
A, N (ISA and ISA + 20°C) 70 knots (130 km/h; 81 mph)
Min flying speed (max contingency rating, one engine out): A 19 knots (35·5 km/h; 22 mph)
N 23 knots (43 km/h; 26·5 mph)
A (ISA + 20°C) 32 knots (60 km/h; 37 mph)
N (ISA + 20°C) 39 knots (73 km/h; 45 mph)
Max forward rate of climb: A 756 m (2,480 ft)/min
N 661 m (2,170 ft)/min
A (ISA + 20°C) 536 m (1,760 ft)/min
N (ISA + 20°C) 469 m (1,540 ft)/min
Max forward rate of climb (1 h power), one engine out:
A 277 m (910 ft)/min
N 223 m (730 ft)/min
A (ISA + 20°C) 72 m (235 ft)/min
N (ISA + 20°C) 64 m (210 ft)/min
Max vertical rate of climb:
A 472 m (1,550 ft)/min
N 351 m (1,150 ft)/min
A (ISA + 20°C) 390 m (1,280 ft)/min
N (ISA + 20°C) 244 m (800 ft)/min
Hovering ceiling OGE: A 3,230 m (10,600 ft)
N 2,575 m (8,450 ft)
Typical range, with reserves:
A, troop transport 292 nm (540 km; 336 miles)
Radius of action, out and back at max sustained speed, allowances for T-O and landing, 30 min loiter in search area, 3 min hover for each survivor, and 10% fuel reserves at end of mission:
N, search and rescue (crew of 3 and 2 survivors) 115 nm (212 km; 132 miles)
N, search and rescue (crew of 3 and 7 survivors) 96 nm (178 km; 111 miles)
Time on station at 50 nm (93 km; 58 miles) radius, out and back at max sustained speed, with 2 torpedoes, smoke floats and marine markers, allowances for T-O and landing and 10% fuel reserves at end of mission:
N, anti-submarine classification and strike, loiter speed on station 2 h 0 min
N, anti-submarine strike, loiter on station 2 h 29 min
N, dunking sonar search and strike, 50% loiter speed and 50% hover on station 1 h 5 min
Time on station at 50 nm (93 km; 58 miles) radius, out and back at max sustained speed, with crew of 2 and 4 Sea Skuas, allowances and reserves as above:
N, air to surface vessel strike, en-route radar search and loiter speed on station 1 h 36 min
Max range: A 340 nm (630 km; 392 miles)
N 320 nm (593 km; 368 miles)
A (ISA + 20°C) 339 nm (628 km; 390 miles)
N (ISA + 20°C) 320 nm (593 km; 368 miles)
Max endurance: A 2 h 57 min
N (ISA + 20°C) 2 h 50 min
Max ferry range with auxiliary cabin tanks:
A 724 nm (1,342 km; 834 miles)
N 565 nm (1,046 km; 650 miles)

WESTLAND WG. 30 LYNX and WESTLAND 30

Westland Helicopters first undertook studies leading to this enlarged, twin-engined development of the Lynx helicopter as a private venture in early 1976. Detail design of the bigger fuselage began about a year later, and the construction of two prototypes was authorised in January 1978. The first of these, registered G-BGHF, made its first flight on 10 April 1979, two weeks ahead of schedule. By February 1981 it had accumulated a total of 432 flight hours. The second prototype made its public debut at the Paris Air Show in June 1979.

Main differences from the Lynx are the completely new and more spacious fuselage; retractable landing gear, increased-diameter main and (on the production version) tail rotor; a dynamic system developed from that of the Lynx and retaining more than 85% of the proven system in that aircraft; increased fuel capacity; new automatic flight control system; and simplified electrical system. Payload/range capability is increased, and the manoeuvrability of the Lynx is retained.

The original designation of this new helicopter was WG 30, but in early 1981 this was changed to WG.30 Lynx for the military version, and Westland 30 for the civil version. Initially, the WG. 30 Lynx is planned to meet military needs for a tactical transport, battlefield support and aeromedical helicopter. The Westland 30 for the civil market can be equipped for such roles as passenger and/or cargo transport, executive and VIP use, offshore rig support, and Arctic operation. Military approvals and full Category A civil certification were expected to be obtained during 1981. Two Westland 30s were ordered by British Airways Helicopters in February 1981: one was scheduled for delivery in late 1981, the second in early 1982. At the 1981 Paris Air Show, it was announced that Airspur had ordered 18 (three with Gem 41 engines, the remainder with Gem 60s) for a network of scheduled services centred on Los Angeles International Airport, California. At that time, orders were also anticipated from Helicopter Hire (two) and British Caledonian (one). A production line of 45 Westland 30/WG. 30 Lynx has been laid down by the company.

TYPE: Twin-turboshaft general-purpose military and civil helicopter.

ROTOR SYSTEM: Four-blade semi-rigid main rotor and four-blade tail rotor. Main rotor blades, which can be folded for stowage, are of constant chord and cambered section; each has a stainless steel spar and a bonded GRP skin. Forged titanium hingeless main rotor head. The aircraft has a tail rotor of slightly larger diameter than that of the Lynx, with composite blades, which rotates in the opposite direction and is quieter.

ROTOR DRIVE: Engines drive directly into standard Lynx conformal main gearbox, thence by driveshafts to intermediate and tail rotor gearboxes.

FUSELAGE: Main cabin, which is of basically rectangular cross-section, is a conventional semi-monocoque structure of light alloy frames and stringers, with stringer spacing constant throughout the airframe. Roof panels, fuel tank surrounds and bulkheads are of aluminium honeycomb, floor panels of Ciba-Geigy Fibrelam GRP.

LANDING GEAR: Hydraulically retractable tricycle type, with oleo-pneumatic shock-absorber in each unit. Fairey Hydraulics main units each have a single Goodyear wheel and tyre of the size fitted to the Westland Sea King; they retract into fairings on the fuselage sides at the rear of the cabin, the wheels remaining partly exposed when retracted. Fairey nose unit, which is castoring, is fully retractable rearwards, and is fitted with twin Goodyear wheels and tyres.

POWER PLANT: Two Rolls-Royce Gem 41-1 turboshaft engines mounted side by side above cabin, each rated at 835 kW (1,120 shp) max contingency, 790 kW (1,060 shp) intermediate contingency, 746 kW (1,000 shp) for T-O, and 671 kW (900 shp) max continuous. Gem 60 engines available optionally. Engine intakes as for Lynx. Fuel in two FPT tanks, each of 499 kg (1,100 lb) capacity, one under front seats and one under rear seats in cabin. Intertechnique booster pumps, with provision for crossfeed to either engine.

ACCOMMODATION: Crew of two on flight deck, with provision for one-pilot operation. Large flight deck windows provide excellent field of view for crew. Windscreen

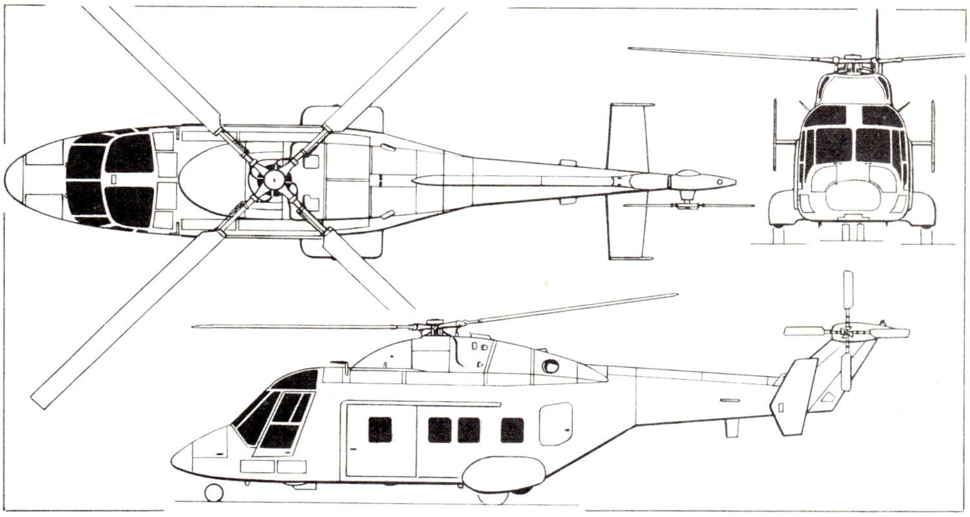

Westland 30 (two Rolls-Royce Gem 41-1 turboshaft engines) *(Pilot Press)*

Prototype of the Westland 30 twin-turboshaft general-purpose helicopter *(Brian M. Service)*

washers and wipers on both front transparencies. Main cabin can accommodate, in military version, 14 troops each weighing 127 kg (280 lb) including full equipment; 17 troops with less equipment; or a maximum of 22 troops in high-density configuration; or, in cargo configuration, ammunition, anti-tank missile launch teams, fuel, and supplies for battlefield support. Aeromedical version can accommodate six stretchers plus 8/10 sitting casualties/medical attendants. Civil passenger version can provide three/four-abreast seating for up to 17 persons in airline standards of comfort. This version, too, has a high-density layout, in which up to 22 passengers can be accommodated in two forward-facing rows of five and two inward-facing rows of six. Various other civil layouts include executive/VIP transport (six/eight seats, toilet and galley unit), offshore oil support, all-cargo, and mixed passenger/cargo. Large rearward-sliding door on each side of cabin. Four large square cabin windows each side, including one in each door. Step each side to facilitate access to flight deck. Baggage compartment in port side of rear fuselage, aft of cabin, with external access door. Stowage for safety equipment, liferafts, lifejackets and aircraft flotation gear for overwater operation.

Systems: Two independent hydraulic systems (each 141 bars; 2,050 lb/sq in), as in Lynx; No. 1 system actuates landing gear. Electrical compartment in starboard side of rear fuselage, aft of cabin. DC system as in Lynx (28V from two 6kW engine-driven starter/generators, alternator and battery); AC power provided by two 500V static inverters, but alternators are available at customer's option. Air-conditioning system optional.

Avionics and Equipment: Nose compartment for avionics and radio. Standard aircraft is VFR equipped, but IFR package (to include VOR/ILS, DME and ADF) is under development. Louis Newmark duplex automatic flight control system for all three axes, with computer-based control and duplex heading hold. SFENA spring feel units (one each for pitch and roll). Other flying controls essentially similar to Lynx; rods connecting hand and foot controls to operating jacks are of the type fitted to Wessex and Sea King. Duplicated three-axis automatic stabilisation equipment. Communications and security systems to customer's requirements.

Dimensions, external:
- Diameter of main rotor 13·31 m (43 ft 8 in)
- Diameter of tail rotor 2·44 m (8 ft 0 in)
- Length overall, rotors turning 15·90 m (52 ft 2 in)
- Length overall, main rotor blades folded 14·46 m (47 ft 5¼ in)
- Width overall, main rotor blades folded 3·25 m (10 ft 8 in)
- Height overall, rotors turning 4·39 m (14 ft 5 in)
- Height overall (minimum) 4·04 m (13 ft 3 in)
- Cabin doors (each): Height 1·37 m (4 ft 6 in)
  - Width 1·22 m (4 ft 0 in)
  - Height to sill 0·58 m (1 ft 10¾ in)

Dimensions, internal:
- Cabin, excl flight deck: Length 4·42 m (14 ft 6 in)
  - Width 1·98 m (6 ft 6 in)
  - Height 1·68 m (5 ft 6 in)
  - Volume 13·03 m³ (460 cu ft)
- Baggage compartment volume (rear) 1·05 m³ (37 cu ft)

Areas:
- Main rotor disc 139·14 m² (1,497·7 sq ft)
- Tail rotor disc 4·68 m² (50·4 sq ft)

Weights and Loading:
- Manufacturer's bare weight, with basic minimum equipment 3,030 kg (6,680 lb)
- Max fuel load 998 kg (2,200 lb)
- Typical operating weights (tactical troop transport):
  - 1 crew, VFR equipped 3,192 kg (7,037 lb)
  - 2 crew, IFR equipped 3,362 kg (7,412 lb)
- Max weight for hovering OGE, ISA + 10°C 5,443 kg (12,000 lb)
- Max T-O weight 5,443 kg (12,000 lb)
- Max cabin floor loading 976 kg/m² (200 lb/sq ft)

Performance (estimated, at max T-O weight except where indicated):
- Max level and cruising speed at S/L:
  - ISA 130 knots (241 km/h; 150 mph)
  - ISA + 20°C 119 knots (220 km/h; 137 mph)
- Max level speed at S/L, ISA, one engine out 98 knots (182 km/h; 113 mph)
- Min level speed at S/L, one engine out 35 knots (65 km/h; 40·5 mph)
- Hovering ceiling IGE 1,920 m (6,300 ft)
- Hovering ceiling OGE 1,190 m (3,900 ft)
- Radius of action (offshore oil support, IFR), 45 min hold, 5% fuel reserves 145 nm (269 km; 167 miles)
- Range with 1,815 kg (4,000 lb) internal payload 123 nm (228 km; 142 miles)
- Max ferry range (S/L, ISA), with reserves nearly 350 nm (648 km; 403 miles)

WESTLAND/AÉROSPATIALE PUMA

Following the completion of 40 Puma HC.Mk 1s for the RAF in 1972, Westland is building Puma component sets for the French production line. The first of an additional batch of eight for the RAF (ZA934) was delivered on 23 May 1980. These aircraft have composite main rotor blades, new multi-purpose air intakes which prevent ice and sand ingestion, and a max T-O weight of 7,000 kg (15,432 lb).

The Puma is described fully under the Aérospatiale heading in the French section.

WESTLAND/AÉROSPATIALE GAZELLE

The Gazelle, described fully under the Aérospatiale heading in the French section, is in joint production in Britain and France under the same Anglo-French agreement as the Puma. The Gazelle has been ordered by the British Army (158 **AH. Mk 1** initially; 29 more in 1978), Royal Navy (30 **HT. Mk 2**) and Royal Air Force (20 **HT. Mk 3**, one **HCC. Mk 4**). A further 36, for all three British services, were ordered in 1980; deliveries of these are due to begin in April 1982.

The first Gazelle HT. Mk 2 (XW845) was flown on 6 July 1972. The first HT. Mk 3 was delivered to the RAF Central Flying School on 16 July 1973. The AH. Mk 1 entered service on 6 July 1974, with No. 660 Squadron of the Army Air Corps at Soest, Germany; and the HT. Mk 2 with No. 705 Squadron at RNAS Culdrose on 10 December 1974.

British Army Gazelles began to be equipped in 1977 with Decca Doppler 80 radar and automatic chart display.

In 1981 Ferranti received a contract to supply roof-mounted gyro-stabilised periscopic sights for British Army Gazelles, to improve their reconnaissance capability. Design of the sight includes provision for including later, if required, such facilities as laser target designation, weapon aiming, and night vision.

WHE

W. H. EKIN (ENGINEERING) CO LTD

158 King's Gate, Aberdeen AB2 6BR, Scotland
Telephone: Aberdeen (0224) 33463
Registered Office: c/o 4 Malone Hill Park, Belfast BT9 6RL, Northern Ireland
Directors:
Dr William H. Ekin, PhD, BSc(Hons), CEng, MIMechE, DMS
Mrs M. J. H. Ekin

This company was formed in March 1969 to undertake the production of six McCandless Mk IV Gyroplanes (see 1972-73 *Jane's*), and the first of these made its first flight in February 1972. An extensive redesign was embarked upon by WHE in Autumn 1971 and a new prototype (G-AXXN) flew for the first time on 1 February 1973. The modified aircraft is known as the WHE Airbuggy.

WHE AIRBUGGY

The first production Airbuggy (G-AXYX) was delivered to an English customer in October 1975, and four more had been delivered by June 1980. The fourth Airbuggy, delivered in November 1979, was fitted with an 1,800 cc engine instead of the standard 1,600 cc unit. A sixth example was under construction in early 1981, and scheduled to fly late in the year, or in early 1982.

Type: Single-seat light autogyro.

Rotor System and Drive: Two-blade semi-rigid teetering rotor with an offset gimbal head, through the centre of which runs the rotor spin-up drive. Blades are secured to hub by bolts, and are of Rotordyne bonded metal construction. Rotor spin-up effected via V-belt drive to 9·667 : 1 worm reduction gearbox and universal and sliding joints. No rotor brake or blade folding.

Fuselage: Space-frame of T35 and T45 steel tube, assembled by sifbronze welding.

Tail Unit: Fin, rudder and tailplane formed from plywood sandwich. Ground-adjustable trim tab on rudder. Endplate auxiliary fins on tailplane of prototype.

Landing Gear: Non-retractable tricycle type. All three wheels have rubber bungee for shock-absorption. Nosewheel steerable. Main wheels with tyres sized from 12 × 2·5 to 13·5 × 5 and nosewheel from 12 × 2·5 to 12 × 3·8 according to surface from which aircraft operates. Internal expanding drum brake on nosewheel only.

Power Plant: One 56 kW (75 hp) 1,600-1,800 cc modified Volkswagen engine, driving a Hoffmann two-blade pusher propeller via a 1·5 : 1 reduction drive. Drive transmitted by 10 V-belts. Fuel capacity 29·5 litres (6·5 Imp gallons), in gravity tank mounted above engine, with refuelling point on top of tank. Oil capacity 2·5 litres (4·5 Imp pints).

Accommodation: Single seat in open cockpit behind large windscreen. Door on starboard side.

System: A 12V 25Ah ground rechargeable battery is installed to power an electric engine starter motor and to supply instruments and radio. The starters have been removed from the first two aircraft, as the engines have been found to start satisfactorily without them, using only the hand lever.

Avionics: Prototype had Parkair Nipper 24-channel com transceiver.

Dimensions, external:
- Diameter of rotor 6·63 m (21 ft 9 in)
- Rotor blade chord 0·188 m (7·4 in)
- Propeller diameter 1·45 m (4 ft 9 in)
- Length overall, excl rotor 3·51 m (11 ft 6 in)
- Height overall 2·21 m (7 ft 3 in)
- Wheel track 1·63 m-1·68 m (5 ft 4 in-5 ft 6 in)
- Wheelbase 1·32 m (4 ft 4 in)

Areas:
- Rotor blades (each) 0·52 m² (5·6 sq ft)
- Rotor disc 37·63 m² (405 sq ft)
- Fin 0·31 m² (3·3 sq ft)
- Rudder 0·49 m² (5·3 sq ft)

Weights and Loadings:
- Weight empty 161 kg (355 lb)
- Max T-O and landing weight 295 kg (650 lb)
- Max disc loading 8·5 kg/m² (1·75 lb/sq ft)
- Max power loading 5·27 kg/kW (8·7 lb/hp)

Performance (at max T-O weight):
- Never-exceed speed 69 knots (128 km/h; 80 mph)
- Max level speed at S/L 69 knots (128 km/h; 80 mph)
- Max cruising speed 61 knots (113 km/h; 70 mph)
- Econ cruising speed 52 knots (97 km/h; 60 mph)
- Max rate of climb at S/L 305 m (1,000 ft)/min
- T-O run 46-92 m (150-300 ft)
- Landing run, still air 9 m (30 ft)
- Landing run in 13 knot (24 km/h; 15 mph) wind Nil
- Range with max fuel, no allowances 121 nm (225 km; 140 miles)
- Range with max payload, no allowances 86 nm (161 km; 100 miles)

WHE Airbuggy single-seat autogyro (modified Volkswagen engine)

UNITED STATES OF AMERICA

AHRENS

AHRENS AIRCRAFT INC

WORKS: Hangar 403 (PO Box 432), Ramey, Puerto Rico 00604
Telephone: (809) 891 2033
Telex: 385 9451 AHRENS PR
PRESIDENT: Peter W. Ahrens
VICE-PRESIDENTS:
Kim K. Ahrens (Chief Engineer)
Edd H. Ahrens (Production)

AHRENS AIRCRAFT CORPORATION

HEAD OFFICE: 2800 Teal Club Road, Oxnard, California 93030
Telephone: (805) 985 2000
Telex: 659 240 Ahrens Air ONX

Ahrens Aircraft Corporation provides subassembly and shipping services for the main production and assembly plant of Ahrens Aircraft Inc, located at the former Ramey AFB, Puerto Rico. Current labour force at Ramey is approximately 180, but with full production this is expected to increase to 450-500 personnel.

AHRENS AR 404

Ahrens Aircraft Corporation initiated the design of the AR 404 in January 1975; the aim was to evolve a simple and robust multi-purpose transport aircraft that would be easy to operate and maintain. A constant-section square fuselage was adopted to provide maximum volume and simplify the loading of containers; the wing was mounted on the fuselage upper surface, eliminating the need for any wing carry-through structure within the fuselage. A modular concept was chosen for all systems to simplify maintenance, and it is claimed that an engine can be removed for replacement within 20 minutes.

Construction of a prototype began in August 1975, and this aircraft (N404AR) made its first flight on 1 December 1976. The first production aircraft (N1028G), built wholly in Puerto Rico, made its initial flight on 26 October 1979; it was used throughout 1980 for a continuation of ground handling tests, and for modification and improvement of the aircraft. Several internal refinements have been introduced, but the major changes involve a 0·46 m (1 ft 6 in) extension of the nose, to provide a 0·37 m³ (13 cu ft) forward baggage compartment to improve cargo loading flexibility, and a longer cabin.

The first and second aircraft are being used for the certification programme, and type certification was expected in January 1982. The third and fourth airframe will be used for structural testing. A production commitment was anticipated before the end of 1981, with initial deliveries following immediately after certification. Production estimates are for 12-16 aircraft in 1982, and 24-30 in 1983. The company has received letters of intent to purchase more than 100 aircraft. No deposits have been accepted, but line positions have been allocated. More than half of these aircraft are for the North American continent.

A twin-engined version of the AR 404, which would have a generally similar airframe and Garrett TPE331-11 or Pratt & Whitney Aircraft of Canada PT6A-45 turboprop engines, was under consideration in early 1981. No production decision was anticipated before the end of the year, and work on this aircraft would not start until the AR 404 has been certificated.

The following description applies to the AR 404 production version:

TYPE: Passenger/cargo transport.

WINGS: Cantilever high-wing monoplane. Wing section NACA 64_3-418. Dihedral 0°. Incidence 0°. Three-spar fail-safe light alloy structure, with light alloy skins. Electrically-operated two-section single-slotted trailing-edge flaps of light alloy construction on each wing. Manually-actuated plain ailerons of light alloy construction, with electrically-operated trim tab in each. Modified Hoerner-type wingtips.

FUSELAGE: Semi-monocoque square constant-section fail-safe structure of light alloy.

TAIL UNIT: Cantilever structure of light alloy. Manually-actuated control surfaces. Trim tabs in elevator and rudder.

LANDING GEAR: Hydraulically-retractable tricycle type, all units retracting forward to simplify free-fall extension in emergency. Main units retract into sponson on each side of fuselage. Oleo-pneumatic shock-absorber and twin wheels on each unit. Hydraulically-steerable nose unit has wheels and tyres size 6·00-6. Main wheels and tyres size 7·00-8. Hydraulically-operated disc brakes.

POWER PLANT: Four 314 kW (420 shp) Allison 250-B17C turboprop engines, each driving a Hartzell three-blade fully-feathering constant-speed metal propeller. Four wing fuel tanks, capacity 1,892 litres (500 US gallons). Single-point refuelling standard. Two optional leading-edge auxiliary tanks, capacity 946 litres (250 US gallons). Propeller blades de-iced electrically.

ACCOMMODATION: Crew of two side by side on flight deck, with dual controls. Up to 30 passengers can be accommodated in commuter version, with a 'two and one' seating arrangement having an aisle width of 0·38 m (1 ft 3 in). Most operators are expected to select a 27-passenger layout, with one flight attendant and full baggage load. Baggage can be carried in a rear-loaded container with capacity of 4·5 m³ (160 cu ft), and in a nose baggage compartment of 0·37 m³ (13 cu ft) capacity. Five-track seat/cargo restraint system in cabin floor may be used for seat attachment, direct cargo tiedown, or for the fitting of rollers to facilitate the handling of palletised cargo. Four standard D-3 containers can be accommodated. Twelve cabin windows on each side. Passenger door aft of wing on each side. Crew door on port side; communicating door between flight deck and cabin. Split cargo door forms undersurface of aft fuselage, lower half serving also as a loading ramp, which can be left open to permit the carriage of outsize cargo or to allow the airdropping of paratroops and supplies. Accommodation air-conditioned and heated.

SYSTEMS: Hydraulic system for landing gear retraction, brakes and nosewheel steering. Electrical system powered by four starter/generators, each rated at 150kVA; one 24V 36Ah lead-acid battery. A small turbine APU is being considered as optional equipment. A standby hydraulic pump is housed in one sponson for emergency hydraulic power. Sundstrand air cycle air-conditioning unit.

AVIONICS AND EQUIPMENT: Collins FD-112V flight director and HSI-84 horizontal situation indicator standard. A wide range of optional avionics by Collins is available to customer requirements. Standard equipment includes angle of attack indicator, clock, encoding altimeter, instant vertical speed indicators, outside air temperature gauge, dual turn and slip indicators, inertia-reel shoulder harnesses for pilot and co-pilot, alternate static source, dual heated pitot tubes, audible stall warning system, interior corrosion proofing; map, instrument panel post, overhead and floodlights; cabin, anti-collision, dual landing, navigation and taxi lights; engine fire-extinguishing system, jack pads and choice of exterior colour scheme.

DIMENSIONS, EXTERNAL:	
Wing span	20·12 m (66 ft 0 in)
Wing chord, constant	1·98 m (6 ft 6 in)
Wing aspect ratio	10·15
Length overall	16·69 m (54 ft 9 in)
Fuselage width	1·98 m (6 ft 6 in)
Height overall	5·64 m (18 ft 6 in)
Tailplane span	7·32 m (24 ft 0 in)
Wheel track	4·27 m (14 ft 0 in)
Propeller diameter	2·29 m (7 ft 6 in)
Propeller ground clearance	1·45 m (4 ft 9 in)
Cabin doors (each): Height	1·52 m (5 ft 0 in)
Width	0·76 m (2 ft 6 in)
Rear cargo doors (upper and lower):	
Width	1·63 m (5 ft 4 in)
DIMENSIONS, INTERNAL:	
Cabin: Length	8·53 m (28 ft 0 in)
Max width	1·85 m (6 ft 1 in)
Max height	1·83 m (6 ft 0 in)
Floor area	15·8 m² (170 sq ft)
Volume	28·55 m³ (1,008 cu ft)
AREAS:	
Wings, gross	39·20 m² (422 sq ft)
Ailerons (total, incl tabs)	2·79 m² (30 sq ft)
Trailing-edge flaps (total)	5·95 m² (64 sq ft)
Vertical tail surfaces (incl tab)	4·83 m² (52 sq ft)
Horizontal tail surfaces (incl tab)	11·15 m² (120 sq ft)
WEIGHTS AND LOADINGS:	
Weight empty, equipped	4,536 kg (10,000 lb)
Max T-O weight	7,938 kg (17,500 lb)
Max wing loading	202·1 kg/m² (41·4 lb/sq ft)
Max power loading	6·32 kg/kW (10·4 lb/shp)
PERFORMANCE (estimated, at max T-O weight):	
Max level speed at 1,525 m (5,000 ft)	190 knots (352 km/h; 219 mph)
Max cruising speed at 1,525 m (5,000 ft)	170 knots (315 km/h; 195 mph)
Stalling speed, flaps down	75 knots (139 km/h; 86·5 mph)
Max rate of climb at S/L	366 m (1,200 ft)/min
Rate of climb at S/L, one engine out	183 m (600 ft)/min
Service ceiling	5,500 m (18,000 ft)
T-O run	488 m (1,600 ft)
Landing run	396 m (1,300 ft)
Balanced field length, T-O and landing	915 m (3,000 ft)
Range with standard fuel, no reserves	850 nm (1,574 km; 978 miles)
Range with auxiliary fuel, no reserves	1,200 nm (2,222 km; 1,380 miles)

First production example of the Ahrens AR 404 passenger/cargo transport

AIC

AMES INDUSTRIAL CORPORATION

55 Orville Drive, Bohemia, NY 11716
Telephone: (516) 567 3780
President: M. Berlon
Vice-President and General Manager: H. A. Iversen

This company was awarded a NASA contract to develop and build a small manned aircraft embodying the pivoting-wing concept. This aircraft, which is designated **AD-1,** flew for the first time on 21 December 1979. Details can be found under the NASA entry in this section.

In early 1981 the company was involved in a new programme, building for an undivulged contractor a new low-cost proof-of-concept aircraft of advanced design. The only available information is that its unmoulded composite structure will incorporate graphite material, and that it is to be powered by two turbojet engines. (See Fairchild NGT in Addenda).

AIR TRACTOR

AIR TRACTOR INC

PO Box 485, Municipal Airport, Olney, Texas 76374
Telephone: (817) 564 5641
President: Leland Snow

The Air Tractor series of agricultural aircraft are new designs embodying more than 25 years' experience by their designer, Mr Leland Snow, who designed, developed, certificated and put into production the earlier Snow S-2 series of agricultural aircraft. The latter, which later became the Rockwell S-2R Thrush, have been described in previous editions of *Jane's*.

Five versions of the Air Tractor series were available in 1981: the Models AT-301 and AT-301A with Pratt & Whitney radial engine, the Models AT-302 and AT-302A, each with an Avco Lycoming turboprop engine, and the AT-400 with a Pratt & Whitney Aircraft of Canada PT6A-15AG turboprop engine.

Air Tractor Model AT-302 Air Tractor (Avco Lycoming LTP 101-600A1A turboprop engine)

AIR TRACTOR MODEL AT-301/301A AIR TRACTOR

Design of the Model **AT-301** Air Tractor was initiated in January 1971, and construction of the first prototype/pre-production aircraft started in August 1972. This aircraft flew for the first time in September 1973, and received FAA certification under FAR Pt 23 in November of that year. At that time the aircraft was also flight tested to meet FAR Pt 8 requirements. The Model **AT-301A**, introduced in 1980, is identical to the AT-301 except for the installation of a larger hopper, with a 0·97 m (3 ft 2 in) wide gatebox for high application rates of dry chemicals.

By 1 January 1981, a total of 421 AT-301/301As had been ordered, of which 326 had been completed, and production was continuing at a rate of eight aircraft per month.

Type: Single-seat agricultural aircraft.

Wings: Cantilever low-wing monoplane. Wing section NACA 4415. Dihedral 3° 30′. Incidence 2°. No sweepback. Conventional two-spar structure of 2024-T3 light alloy. Ailerons of light alloy construction, interconnected with trailing-edge flaps to droop 10° at maximum flap deflection of 30°. Electrically-operated Fowler-type trailing-edge flaps of light alloy construction. No trim tabs. Wing ribs and skins zinc chromated before assembly. Wing roots and skin overlaps sealed against chemical entry.

Fuselage: Welded structure of 4130N steel tube, oven stress relieved and oiled internally. Quickly detachable skins of 2024-T3 light alloy, with Camloc fasteners. Rear fuselage lightly pressurised to prevent chemical ingress.

Tail Unit: Light alloy structure, with cantilever fin and strut-braced tailplane. Fabric-covered rudder and elevators. Trim tab in each elevator.

Landing Gear: Non-retractable tailwheel type. Cantilever spring steel main gear; flat spring suspension for castoring and lockable tailwheel. Cleveland main wheels with tyres size 8·50-10 (8-ply), pressure 2·83 bars (41 lb/sq in). Tailwheel tyre size 12·50-4, pressure 2·42 bars (35 lb/sq in). Cleveland type 30-89 hydraulic disc brakes.

Power Plant: One 447 kW (600 hp) Pratt & Whitney R-1340 aircooled radial engine, driving a Hamilton Standard 12D40/6101A-12 two-blade constant-speed metal propeller. Fuel in two integral wing tanks with combined capacity of 288 litres (76 US gallons). Refuelling points on upper surface of wings at root. Oil capacity 30·3 litres (8 US gallons).

Accommodation: Single seat in enclosed cabin which is sealed to prevent chemical ingress. Downward-hinged window/door on each side. Baggage compartment in bottom of fuselage, aft of cabin, with door on port side. Cabin ventilation by 0·08 m (3 in) diameter airscoop.

Systems: Agricultural dispersal system comprises a 1,211 litre (320 US gallon) glassfibre hopper, mounted in the forward fuselage for the AT-301, 1,325 litre (350 US gallon) hopper for the AT-301A; Transland gatebox; Agrinautics valve and strainer; 5 cm (2 in) Root Model 67 pump; 5 cm (2 in) stainless steel plumbing; and up to 72 nozzles in spraybars mounted below and just aft of wing trailing-edges. 24V electrical system, supplied by 60A engine-driven alternator.

Avionics and Equipment: Optional avionics include Narco Com 120 radio and Nav 121/Com 120 nav/com. Optional equipment includes 600W retractable landing light in port wingtip; night-flying package of strobe/navigation, instrument, post, flexible cord utility and dome lights; night working lights; directional and attitude gyro instrument package; glassfibre engine cowling ring; new Hartzell propeller blades; Cleveland 29-11 wheels and high-flotation tyres; external power socket; and ferry fuel system. Alternative agricultural equipment includes Transland high volume and extra high volume spreader systems, and six- or eight-unit Micronair installations.

Dimensions, external:
- Wing span: 13·75 m (45 ft 1¼ in)
- Wing chord, constant: 1·83 m (6 ft 0 in)
- Wing aspect ratio: 7·5
- Length overall: 8·23 m (27 ft 0 in)
- Height overall: 2·59 m (8 ft 6 in)
- Propeller diameter: 2·77 m (9 ft 1 in)

Areas:
- Wings, gross: 25·08 m² (270 sq ft)
- Ailerons (total): 3·55 m² (38·2 sq ft)
- Trailing-edge flaps (total): 3·34 m² (36·0 sq ft)
- Fin: 0·90 m² (9·7 sq ft)
- Rudder: 1·30 m² (14·0 sq ft)
- Tailplane: 2·42 m² (26·0 sq ft)
- Elevators (incl tabs): 2·36 m² (25·4 sq ft)

Weights and Loadings (A: AT-301; B: AT-301A):
- Weight empty, spray equipped:
  - A: 1,723 kg (3,800 lb)
  - B: 1,746 kg (3,850 lb)
- Certificated gross weight (FAR 23):
  - A, B: 2,268 kg (5,000 lb)
- Typical operating weight (CAM 8):
  - A: 3,175 kg (7,000 lb)
  - B: 3,311 kg (7,300 lb)
- Max wing loading: A: 126·5 kg/m² (25·9 lb/sq ft)
  - B: 131·8 kg/m² (27·0 lb/sq ft)
- Max power loading: A: 7·10 kg/kW (11·67 lb/hp)
  - B: 7·40 kg/kW (12·17 lb/hp)

Performance (at max T-O weight unless stated otherwise):
- Max level speed at S/L:
  - A: 146 knots (270 km/h; 168 mph)
  - B: 142 knots (264 km/h; 164 mph)
- Max level speed at S/L with dispersal equipment:
  - A: 139 knots (257 km/h; 160 mph)
  - B: 135·5 knots (251 km/h; 156 mph)
- Cruising speed at 1,825 m (6,000 ft), spraypump removed for ferrying:
  - A: 135 knots (249 km/h; 155 mph)
  - B: 132 knots (245 km/h; 152 mph)
- Typical working speed: A, B: 104-121·5 knots (193-225 km/h; 120-140 mph)
- Stalling speed, flaps up, at 2,268 kg (5,000 lb) AUW:
  - A, B: 63·5 knots (117·5 km/h; 73 mph)
- Stalling speed, flaps down, at 2,268 kg (5,000 lb) AUW: A, B: 53 knots (98 km/h; 61 mph)
- Stalling speed as usually landed:
  - A, B: 49 knots (90 km/h; 56 mph)
- Max rate of climb at S/L, dispersal equipment installed, AUW of 2,268 kg (5,000 lb):
  - A: 488 m (1,600 ft)/min
  - B: 457 m (1,500 ft)/min
- T-O run at AUW of 2,268 kg (5,000 lb):
  - A, B: 244 m (800 ft)
- T-O run at AUW of 3,175 kg (7,000 lb):
  - A, B: 396 m (1,300 ft)
- Landing run at normal landing weight:
  - A: 91 m (300 ft)
- Range with max fuel, no allowances:
  - A: 304 nm (563 km; 350 miles)

Air Tractor Model AT-301 Air Tractor (Pratt & Whitney R-1340 engine)

AIR TRACTOR MODEL AT-302/302A AIR TRACTOR

Design of the basic **AT-302** turboprop-powered version of the Air Tractor was initiated in September 1976, and construction of the prototype started two months later. It flew for the first time in June 1977. The first flight of a production aircraft was made in November 1977, and certification under FAR Pt 23 was awarded during the following month. Apart from installation of the Avco Lycoming turboprop engine, and the structural modifications necessary to accept this power plant, the Model AT-302 is basically similar to the piston-engined AT-301. A total of 15 had been ordered by 1 January 1981, and all had been delivered.

The Model **AT-302A**, introduced in 1979, is identical to the AT-302 except for the installation of a 1,514 litre (400 US gallon) hopper, with a 0·97 m (3 ft 2 in) wide gatebox for high application rates of dry chemicals.

Wings, Fuselage, Tail Unit, Landing Gear: As for Model AT-301.

Power Plant: One 447 kW (600 shp) Avco Lycoming LTP 101-600A1A turboprop engine, driving a Hartzell three-blade constant-speed non-reversing metal propeller. Fuel system as for AT-301, except total capacity 477 litres (126 US gallons). Oil capacity 7·5 litres (2 US gallons). Air intake has two large dry paper type automobile air filters to avoid ingestion of chemicals.

Accommodation, Systems: As for Model AT-301, except electrical system supplied by 24V 150A engine-driven starter/generator.

Equipment: Optional equipment includes Transland high volume (AT-302) or extra high volume (AT-302A) dispersal system.

Dimensions, external: As for Model AT-301, except:
- Length overall: 8·99 m (29 ft 6 in)
- Propeller diameter: 2·74 m (9 ft 0 in)

Dimensions, internal, and Areas: As for Model AT-301

Weights and Loadings (A: AT-302; B: AT-302A):
- Weight empty:
  - A, spray equipped: 1,474 kg (3,250 lb)
  - B, duster equipped: 1,520 kg (3,350 lb)
- Max T-O weight: A: 2,994 kg (6,600 lb)
  - B: 3,266 kg (7,200 lb)
- Max wing loading: A: 119·4 kg/m² (24·44 lb/sq ft)
  - B: 130·2 kg/m² (26·67 lb/sq ft)
- Max power loading: A: 6·7 kg/kW (11·0 lb/shp)
  - B: 7·3 kg/kW (12·0 lb/shp)

PERFORMANCE (AT-302, at max T-O weight unless stated otherwise):
Never-exceed speed 155 knots (286 km/h; 178 mph)
Max level speed at S/L 148 knots (274 km/h; 170 mph)
Max cruising speed at 2,440 m (8,000 ft) 143 knots (266 km/h; 165 mph)
Econ cruising speed at 2,440 m (8,000 ft) 130 knots (241 km/h; 150 mph)
Stalling speed, flaps down, at normal landing weight 44·3 knots (82 km/h; 51 mph)
Max rate of climb at S/L (hopper empty) 610 m (2,000 ft)/min
Rate of climb at S/L (hopper full) 366 m (1,200 ft)/min
T-O run 238 m (780 ft)
Landing run at normal landing weight 91 m (300 ft)
Range with max fuel, no allowances 347 nm (644 km; 400 miles)

Air Tractor Model AT-400 Turbo Air Tractor (P&WC PT6A-15AG turboprop engine)

AIR TRACTOR MODEL AT-400 TURBO AIR TRACTOR

This version of the Air Tractor incorporates the larger hopper and wider gatebox of the AT-302A; is powered by a 507 kW (680 shp) Pratt & Whitney Aircraft of Canada PT6A-15AG turboprop engine, driving a Hartzell three-blade constant-speed reversible-pitch propeller; has size 29-11 wheels and high-flotation tyres as standard; and has a 250A starter/generator and two 24V 21Ah batteries. Fuel capacity is as for the AT-302/302A. Optional equipment includes a Transland extra high volume dispersal system.

Three Turbo Air Tractors were ferried from Olney, Texas, via the Arctic Circle to delivery points specified by the customer, Flug Service GmbH of Kapfenberg, Austria. One was delivered to the Austrian Civil Aviation Authority at Graz, Austria, the other two to Al-Minya, Egypt. Flight time for each of the two latter aircraft was 53·5 hours for the 7,295 nm (13,520 km; 8,400 mile) journey from Texas.

WEIGHTS AND LOADINGS:
Weight empty, spray equipped 1,610 kg (3,550 lb)
Certificated gross weight (FAR 23) 2,721 kg (6,000 lb)
Typical operating weight (CAM 8) 3,538 kg (7,800 lb)
Max wing loading 141·1 kg/m² (28·9 lb/sq ft)
Max power loading 6·98 kg/kW (11·47 lb/shp)

PERFORMANCE (at max T-O weight unless stated otherwise):
Max level speed at S/L, 'clean' 174 knots (322 km/h; 200 mph)
Max level speed at S/L with dispersal equipment 160 knots (298 km/h; 185 mph)
Cruising speed, 55% power at 2,440 m (8,000 ft), spraypump removed for ferrying 143 knots (265 km/h; 165 mph)
Typical working speed 113-126 knots (209-233 km/h; 130-145 mph)
Stalling speed, flaps up, at 2,721 kg (6,000 lb) AUW 66 knots (122 km/h; 76 mph)
Stalling speed, flaps down, at 2,721 kg (6,000 lb) AUW 57·5 knots (106 km/h; 66 mph)
Stalling speed as usually landed 48 knots (88·5 km/h; 55 mph)
Max rate of climb at S/L, dispersal equipment installed, AUW of 2,721 kg (6,000 lb) 457 m (1,500 ft)/min
T-O run at AUW of 2,721 kg (6,000 lb) 305 m (1,000 ft)
Landing run as usually landed 152 m (500 ft)

ARCTIC
ARCTIC AIRCRAFT COMPANY

PO Box 6-141, Anchorage, Alaska 99502
Telephone: (907) 243 1580
SECRETARY: C. J. Diehl

Arctic Aircraft is constructing and marketing, as the S1B2 Arctic Tern, an updated and improved version of the Interstate S1A, first flown more than 30 years ago. Built to CAR.04a (aerobatic) standard, it was certificated for operation on optional Edo floats on 20 January 1981.

Arctic Aircraft Interstate S1B2 Arctic Tern with optional float landing gear

ARCTIC AIRCRAFT INTERSTATE S1B2 ARCTIC TERN

TYPE: Two-seat sporting and general utility aircraft.

WINGS: High-wing monoplane, with V bracing struts each side and auxiliary struts. Wing section NACA 23012. Dihedral 2°. Incidence 4°. Composite structure with Sitka spruce spars, light alloy ribs and Dacron covering. Hoerner-type wingtips of glassfibre. Semi-Fowler-type single-slotted trailing-edge flaps. Plain inset ailerons.

FUSELAGE: Welded structure of 4130 chrome-molybdenum steel tube, Dacron covered. Two-piece engine cowling of glassfibre.

TAIL UNIT: Wire-braced structure of welded 4130 steel tube with Dacron covering. Trim tab in elevator; ground-adjustable tab on rudder.

LANDING GEAR: Non-retractable tailwheel type. Main wheels carried in two side Vs and half-axles hinged to fuselage. Shock-absorption by hydraulic/coil spring oleo unit. Cleveland main wheels with US Uniroyal tyres size 8·50-6. Maule tailwheel, diameter 203 mm (8 in). Scott tailwheel optional. Scott toe-operated brakes. Parking brake optional. Peekay 1800 or Edo 2000 floats or 2500 skis optional.

POWER PLANT: One 112 kW (150 hp) Avco Lycoming O-320 flat-four engine, driving a McCauley two-blade fixed-pitch metal propeller. One fuel tank in each wing, total capacity 151 litres (40 US gallons). Oil capacity 7·5 litres (2 US gallons). Belly-mounted auxiliary fuel tank optional.

ACCOMMODATION: Two seats in tandem; rear seat removable to provide additional space for cargo. Cabin door on starboard side, beneath wing. Cabin step. Baggage space in rear fuselage, with external door on starboard side, capacity 45 kg (100 lb). Tinted windows and cabin skylight optional. Safety belts and fittings standard. Dual controls standard. Cabin soundproofed, heated and ventilated. Windscreen de-icing by hot air. Cabin floor carpeted.

SYSTEM: Electrical system includes 55A engine-driven alternator, 12V DC storage battery, engine starter and navigation lights.

AVIONICS AND EQUIPMENT: A range of radios is available. Standard features include epoxy priming, ground manoeuvring handles and engine quick oil drain. Optional items include Alcor exhaust gas analyser, folding front seat, shoulder harness, dome light, instrument lights, landing light, three-colour external paint scheme, belly-mounted cargo pack, lumber rack and salt water corrosion proofing.

DIMENSIONS, EXTERNAL:
Wing span 11·18 m (36 ft 8 in)
Wing chord (constant) 1·57 m (5 ft 2 in)
Length overall: landplane 7·01 m (23 ft 0 in)
seaplane 7·32 m (24 ft 0 in)
Height overall 2·13 m (7 ft 0 in)
Tailplane span 3·35 m (11 ft 0 in)

DIMENSIONS, INTERNAL:
Cabin: Max width 0·66 m (2 ft 2 in)
Volume 1·38 m³ (48·7 cu ft)
Baggage volume 0·84 m³ (29·63 cu ft)

AREAS:
Wings, gross 17·30 m² (186·2 sq ft)
Flaps (total) 1·77 m² (19·02 sq ft)
Fin 0·74 m² (7·98 sq ft)
Rudder, incl tab 0·75 m² (8·10 sq ft)
Horizontal tail surfaces (total, incl tab) 2·67 m² (28·75 sq ft)

WEIGHTS AND LOADINGS (A: landplane; B: seaplane with Edo 2000 floats):
Weight empty: A 487 kg (1,073 lb)
B 521·5 kg (1,150 lb)
Max T-O weight: A 862 kg (1,900 lb)
B 965 kg (2,127 lb)
Max wing loading: A 49·8 kg/m² (10·2 lb/sq ft)
B 55·7 kg/m² (11·4 lb/sq ft)
Max power loading: A 7·7 kg/kW (12·67 lb/hp)
B 8·62 kg/kW (14·18 lb/hp)

PERFORMANCE:
Never-exceed speed:
A 152 knots (282 km/h; 175 mph)
Max cruising speed at S/L, 75% power:
A 102 knots (188 km/h; 117 mph)
B 91 knots (169 km/h; 105 mph)
Cruising speed, 65% power at optimum altitude:
A 96 knots (178 km/h; 111 mph)
Stalling speed, flaps down:
A, B 30 knots (55 km/h; 34 mph)
Max rate of climb at S/L, at max T-O weight:
A 389 m (1,275 ft)/min
B 305 m (1,000 ft)/min
Service ceiling: A, B 5,790 m (19,000 ft)
Absolute ceiling: A 6,400 m (21,000 ft)
T-O run at max T-O weight: A 99 m (325 ft)
T-O to 15 m (50 ft) at max T-O weight:
A 152 m (500 ft)
Landing from 15 m (50 ft): A 137 m (450 ft)
Range with max fuel, 45 min reserves:
A, 75% power 479 nm (888 km; 552 miles)
A, 65% power 566 nm (1,049 km; 652 miles)
Range with max fuel, no reserves:
B 499 nm (925 km; 575 miles)

ASTEC
ADVANCED SYSTEMS TECHNOLOGY INC

HEAD OFFICE AND WORKS: Snohomish County Airport, Everett, Washington 98204
Telephone: (206) 355 8700
Telex: Astec Evt 15 2954

Advanced Systems Technology (ASTEC) was founded in 1971 to provide consulting services in aerodynamics and aircraft structural design. It has provided technical support for such diversified types as the FMA IA 58A Pucará, Gates Learjet 50 series, LearStar (now Canadair CL-600 Challenger), Raisbeck/Rockwell Mark Five

The 19th production ASTEC/Cessna Eagle, a conversion of the Citation 500 with an advanced wing aerofoil section

System, and the Trident TR-1 Trigull. In 1976 the company initiated studies for a major aircraft modification, resulting in the ASTEC/Cessna Eagle.

Since July 1980, portions of ASTEC's engineering and manufacturing facility at Everett, Washington, have been leased to Robertson Aircraft Corporation (which see). Robertson has moved its total manufacturing capability to Everett, and in addition to its well-established line of STOL modifications to a variety of piston-engined general aviation aircraft has, under contract to ASTEC, taken on the manufacturing responsibilities of the ASTEC Eagle and Eagle SP.

ASTEC is continuing to maintain an active research, development and flight test department, which in early 1981 was involved in the certification programme of an Eagle with a Pratt & Whitney Aircraft of Canada JT15D-4 turbofan power plant. To be known as the **Super Eagle**, it was scheduled for certification in September 1981. ASTEC is working also on a project to retrofit Gates Learjet 25s with Pratt & Whitney Aircraft of Canada JT15D-5 engines, to offer increased range and reduced noise levels. Certification of this modification, identified currently as the **TF25**, is scheduled for 1983.

ASTEC/CESSNA EAGLE and EAGLE SP

ASTEC selected for its first major aircraft modification programme the Cessna Citation 500. It was chosen with the initial object of increasing cruising speed, but a market survey showed the major requirement to be extended range. Using computerised design and analysis techniques, ASTEC developed a wing modification to reduce drag and increase fuel capacity. To test the flight characteristics of this wing, a Citation 500 was flown with the new wing contours fabricated by the addition of polyurethane foam and glassfibre; a metal production prototype wing was then constructed and flight tested. FAA approval and issue of an STC for modification of Cessna Citation 500 and Citation I aircraft to the same standard was awarded on 4 August 1978.

Changes to the basic Citation 500 wing include use of an advanced technology aerofoil section of increased thickness and length over the inboard portion of the wing; wingtip extensions of 0·51 m (1 ft 8 in); and the addition of cove seals to the trailing-edge flaps. The thickened wing and the wingtip extensions contain an additional 392 kg (865 lb) of fuel, accounting for some 77% of the approximate 600 nm (1,112 km; 691 mile) increase in range over the basic Citation 500. The remainder of the increase is due to improved aerodynamic efficiency of the ASTEC wing.

An STC was received in May 1979 for similar modification of the Cessna Citation 501, this being known as the ASTEC/Cessna **Eagle SP** after conversion. It is certificated under FAR Pt 23 and approved for single-pilot operation.

In early 1981 production conversions were being carried out at a rate of one per month. The description of the Citation I under Cessna's entry in this edition applies also to the Eagle, except as follows:

WINGS: Thickness/chord ratio of inboard portions increased from 14% to 19%, and supercritical technology incorporated to improve wing/fuselage airflow. Span increased, by comparison with Citation 500, by addition of wingtip extensions.

POWER PLANT: As for Citation 500, except increased fuel capacity provided in the thicker wing centre-section and wingtip extensions.

DIMENSIONS, EXTERNAL:
Wing span 14·35 m (47 ft 0½ in)
Wing chord at root 2·95 m (9 ft 8⅛ in)
Wing chord at tip 0·93 m (3 ft 0½ in)
Wing aspect ratio 7·92
Length overall 13·26 m (43 ft 6 in)
Height overall 4·36 m (14 ft 3¾ in)
Wheel track 3·84 m (12 ft 7⅛ in)
Wheelbase 4·78 m (15 ft 8¼ in)

WEIGHTS:
Weight empty 2,971 kg (6,550 lb)
Max fuel weight 2,045 kg (4,510 lb)
Max T-O weight 5,670 kg (12,500 lb)
Max zero-fuel weight 4,309 kg (9,500 lb)
Max landing weight 5,148 kg (11,350 lb)

PERFORMANCE (at max T-O weight except as indicated, A: JT15D-1; B: JT15D-1A power plant):
Max cruising speed, AUW of 4,309 kg (9,500 lb), at 10,670 m (35,000 ft):
A 341 knots (631 km/h; 392 mph)
B 357 knots (661 km/h; 411 mph)
Cruising speed, AUW of 4,309 kg (9,500 lb), at 12,500 m (41,000 ft): A 317 knots (587 km/h; 365 mph)
B 336 knots (623 km/h; 387 mph)
Stalling speed at max landing weight:
A, B 78 knots (145 km/h; 90 mph)
Max certificated altitude: A 10,670 m (35,000 ft)
B 12,500 m (41,000 ft)
Balanced T-O field length: A, B 907 m (2,975 ft)
FAA landing field length at max landing weight:
A, B 722 m (2,370 ft)
Max range, with 227 kg (500 lb) fuel reserves:
A 1,800 nm (3,336 km; 2,073 miles)

AYRES

AYRES CORPORATION

PO Box 3090, Albany, Georgia 31706
Telephone: (912) 883 1440
Telex: 547629

AYRES THRUSH

Ayres Corporation purchased the manufacturing and world marketing rights of the Rockwell Thrush Commander-600 and -800 from Rockwell International's General Aviation Division in late November 1977, and is now offering the following versions of this large agricultural aircraft:

Thrush-600. Basic version with Pratt & Whitney R-1340 Wasp or Pezetel PZL-3S aircooled radial engine.

Turbo-Thrush S-2R. As Thrush-600, but with a Pratt & Whitney Aircraft of Canada PT6A-34 , -15 or -11 turboprop engine. Described separately.

Ayres took over the facility in Albany covering 23,226 m^2 (250,000 sq ft) at which Rockwell formerly built the Thrush, and retained the supervisory management and production personnel, amounting to 150 persons, who worked previously for Rockwell's General Aviation Division.

Ayres Thrush-600 agricultural aircraft (PZL-3S engine), showing optional two-seat accommodation

The Thrush-600 has a 1·50 m^3 (53 cu ft) hopper able to contain up to 1,514 litres (400 US gallons) of liquid or 1,487 kg (3,280 lb) of dry chemicals. Corrosion-proofing is of activated Copon, and it is certificated to both CAR 3 Normal category and CAM 8 Restricted category requirements.

The following details refer to the basic single-seat Thrush-600:

TYPE: Single-seat agricultural aircraft.

WINGS: Cantilever low-wing monoplane. Dihedral 3° 30′. Two-spar structure of light alloy throughout, except for main spar caps of heat-treated SAE 4000 Series steel. Leading-edge formed by heavy main spar and the nose-skin. Light alloy plain ailerons. Electrically-operated flaps. Wing roots sealed against chemical entry.

FUSELAGE: Welded chrome-molybdenum steel tube structure covered with quickly-removable light alloy panels. Underfuselage skin of stainless steel.

TAIL UNIT: Wire-braced welded chrome-molybdenum steel tube structure, fabric-covered. Streamline-section heavy-duty stainless steel wire bracing and heavy-duty stainless steel attachment fittings. Light alloy controllable trim tab in each elevator. Deflector cable from cockpit to fin-tip.

LANDING GEAR: Non-retractable tailwheel type. Main units have rubber-in-compression shock-absorption and 29 × 11·00-10 wheels with 10-ply tyres. Hydraulically-operated disc brakes. Parking brakes. Wire cutters on main gear. Steerable, locking tailwheel, size 12·5 × 4·5 in.

POWER PLANT: One 448 kW (600 hp) Pratt & Whitney R-1340 Wasp nine-cylinder aircooled radial engine, driving a Hamilton Standard type 12D40/EAC AG-100-2 two-blade constant-speed metal propeller, is standard. Available optionally is a 448 kW (600 hp) Pezetel PZL-3S seven-cylinder aircooled radial engine, driving a Dowty-Rotol type CR289-3 three-blade constant-speed metal propeller. One 200·5 litre (53 US gallon) integral tank in each wing, giving total fuel capacity of 401 litres (106 US gallons), of which 378·5 litres (100 US gallons) are usable.

ACCOMMODATION: Single adjustable seat in 'safety pod' sealed cockpit enclosure is standard, with steel tube overturn structure. (Second, tandem seat, with or without dual controls, is optional, to provide accommodation for additional crew member, passenger, or flying instructor.) Two overhead windows for improved view in turns. Downward-hinged door on each side. Tempered safety-glass windscreen. Dual inertia-reel safety harness standard. Baggage compartment. Windscreen wiper and washer.

SYSTEM: Electrical system powered by a 24V 50A engine-driven alternator; 70A alternator optional. Lightweight 24V 35Ah battery.

AVIONICS AND EQUIPMENT: Hopper forward of cockpit with capacity of 1·50 m^3 (53 cu ft) or 1,514 litres (400 US gallons). Hopper has a 0·33 m^2 (3·56 sq ft) lid, openable by two handles. Standard equipment includes Universal spray system with external 50 mm (2 in) stainless steel plumbing, 50 mm Root Model 67 pump with wooden fan, Transland gate, 50 mm valve, quick-disconnect pump mount and strainer. Streamlined spraybooms with outlets for 68 nozzles, 36 nozzles installed. Micro-adjust valve control (spray) and calibrator (dry). A 63 mm (2·5 in) side-loading system is installed on the port side. Navigation lights, instrument lights and two rotating beacons. Optional equipment includes a rear cockpit to accommodate aft-facing crew member, or forward-facing seat for passenger, or flying instructor if optional dual controls installed; space can be used alternatively for cargo. Other optional items are a Transland high-volume spreader, agitator installation, extra-high-density spray configuration with 70 nozzles installed; Agrinautics electrically-operated three-way valve, emergency shut-off valve, pump in lieu of Root pump and strainer in lieu of Transland strainer; Agevenco 6520 pump in lieu of Root pump. Six- or

Ayres Turbo-Thrush S-2R (Pratt & Whitney Aircraft of Canada PT6A-34 turboprop engine) *(R. Kunert)*

eight-unit AU3000 Micronair installation in lieu of standard booms and nozzles; Transland S-2 Boom-master with Q-D flange in lieu of standard gate and Root pump; night working lights including wingtip turn lights, cockpit fire extinguisher, and water bomber configuration. Optional avionics include Bendix T-12C ADF or T-12D digital ADF; King KX 170B, KX 170BE, KX 175B or KX 175BE nav/com; KI 201C converter indicator; Narco Com-11A or Com-11B com transceiver; and Nav-11 nav receiver.

DIMENSIONS, EXTERNAL:
Wing span 13·54 m (44 ft 5 in)
Length overall (tail up) 8·34 m (27 ft 4½ in)
Height overall 2·79 m (9 ft 2 in)
Tailplane span 4·86 m (15 ft 11½ in)
Wheel track 2·72 m (8 ft 11 in)
Propeller diameter 2·74 m (9 ft 0 in)

AREA:
Wings, gross 30·34 m² (326·6 sq ft)

WEIGHTS AND LOADINGS:
Weight empty, equipped 1,678 kg (3,700 lb)
Max T-O weight (CAR 3) 2,721 kg (6,000 lb)
Max T-O weight (CAM 8) 3,130 kg (6,900 lb)
Max wing loading 103·0 kg/m² (21·1 lb/sq ft)
Max power loading 6·99 kg/kW (11·5 lb/hp)

PERFORMANCE (with spray equipment installed and at CAR 3 max T-O weight, unless indicated otherwise):
Max level speed 122 knots (225 km/h; 140 mph)
Max cruising speed, 70% power 108 knots (200 km/h; 124 mph)
Working speed, 70% power 91-100 knots (169-185 km/h; 105-115 mph)
Stalling speed, flaps up 61 knots (113 km/h; 70 mph)
Stalling speed, flaps down 57·5 knots (107 km/h; 66 mph)
Stalling speed at normal landing weight, flaps up 50 knots (92 km/h; 57 mph)
Stalling speed at normal landing weight, flaps down 48 knots (89 km/h; 55 mph)
Max rate of climb at S/L 274 m (900 ft)/min
Service ceiling 4,575 m (15,000 ft)
T-O run 236 m (775 ft)
Landing run 152 m (500 ft)
Ferry range with max fuel at 70% power 350 nm (648 km; 403 miles)

AYRES TURBO-THRUSH S-2R

This turbine-engined version of the Thrush has as standard a 559 kW (750 shp) Pratt & Whitney Aircraft of Canada PT6A-34 turboprop, driving a three-blade constant-speed and reversible-pitch metal propeller. The 507 kW (680 shp) PT6A-15 or 373 kW (500 shp) PT6A-11 engines are available optionally. To compensate for the small size and light weight of the turboprop, it is mounted well forward of the firewall, in a slender cowling. Current installations have cowlings which comprise individual panels to improve access. Innovations include an oil cooler with an airscoop on top of the cowling, new exhaust outlets for the PT6A, and a special air induction system utilising large perforated aluminium panels at the aft end of the cowling side panels. Air entering via these panels is ducted to large centrifugal air filters, able to remove 98 per cent of all foreign matter in the airflow before it enters the engine.

A 1,514 litre (400 US gallon) hopper is standard for the Turbo-Thrush, but a 1,893 litre (500 US gallon) hopper is available optionally, and is suitable for installation with either the standard or optional power plants. Standard and optional equipment is generally as detailed for the basic Thrush, except that the two-seat conversion is not available.

Advantages claimed for this conversion include greatly improved take-off and climb performance; improved short landing capability; a 454 kg (1,000 lb) increase in payload due to reduced power plant weight; ability to operate on aviation turbine fuel, avgas or diesel fuel; a TBO of more than 3,000 h; quieter operation; and the ability to feather the propeller during fuelling and loading operations without shutting down the engine, because of the free-turbine configuration.

First flight of the prototype was made on 9 September 1975, and more than 125 Turbo-Thrushes are in operation.

DIMENSIONS, EXTERNAL:
Wing span 13·54 m (44 ft 5 in)
Length overall 10·06 m (33 ft 0 in)
Height overall 2·79 m (9 ft 2 in)
Tailplane span 4·86 m (15 ft 11½ in)
Wheel track 2·72 m (8 ft 11 in)

AREA:
Wings, gross 30·34 m² (326·6 sq ft)

WEIGHTS AND LOADINGS (A: standard hopper; B: optional 1,893 litre: 500 US gallon hopper):
Weight empty: A 1,633 kg (3,600 lb)
B 1,769 kg (3,900 lb)
Max T-O weight (CAR 3):
A, B 2,721 kg (6,000 lb)
Typical operating weight (CAM 8):
A 3,719 kg (8,200 lb)
B 3,856 kg (8,500 lb)
Max wing loading 127·1 kg/m² (26·0 lb/sq ft)
Max power loading 7·6 kg/kW (14·17 lb/shp)

PERFORMANCE (A and B, at max T-O weight unless specified otherwise):
Max level speed, with spray equipment 138 knots (256 km/h; 159 mph)
Cruising speed, 50% power 130 knots (241 km/h; 150 mph)
Working speed, 30-50% power 82·5-130 knots (153-241 km/h; 95-150 mph)
Stalling speed, flaps up 61 knots (113 km/h; 70 mph)
Stalling speed, flaps down 57 knots (106 km/h; 66 mph)
Stalling speed at normal landing weight, flaps up 51 knots (95 km/h; 59 mph)
Stalling speed at normal landing weight, flaps down 49·5 knots (92 km/h; 57 mph)
Max rate of climb at S/L 530 m (1,740 ft)/min
Service ceiling 7,620 m (25,000 ft)
T-O run 183 m (600 ft)
Landing run 152 m (500 ft)
Landing run with propeller reversal 91 m (300 ft)
Ferry range at 40% power 391 nm (724 km; 450 miles)

BEECHCRAFT

BEECH AIRCRAFT CORPORATION (Subsidiary of Raytheon Company)

HEAD OFFICE AND MAIN WORKS: Wichita, Kansas 67201
Telephone: (316) 681 7111
BRANCH DIVISIONS: Liberal, Kansas; Salina, Kansas; Selma, Alabama; and Boulder, Colorado
CHAIRMAN OF THE BOARD: Mrs O. A. (Walter H.) Beech
VICE-CHAIRMAN OF THE BOARD: Frank E. Hedrick
PRESIDENT: Edward C. Burns
SENIOR VICE-PRESIDENTS:
Stewart M. Ayton (Marketing)
Charles W. Dieker (Treasurer)
Glenn Ehling (Operations)
Michael G. Neuburger (International Division)
C. A. Rembleske (Engineering)
VICE-PRESIDENTS:
Harold W. Deets (Materiel and Production)
Max P. Eaton (Facilities)
Gary M. Hanssen (Industrial Relations)
E. C. Nikkel (Aerospace Programmes)
Austin Rising
William G. Robinson (Corporate Communications)
George D. Rodgers (Domestic Commercial Marketing)
William G. Rutherford (Government Relations)
W. D. Wise (Commuter Division)
CORPORATE SECRETARY: Mrs Ila A. Alumbaugh
ASSISTANT TREASURER AND CONTROLLER: L. R. Damon
ASSISTANT TREASURER AND CHIEF ACCOUNTANT:
D. C. Cullinane
ASSISTANT TREASURER: Wey D. Kenny
DIRECTOR, ADVERTISING AND SALES PROMOTION:
R. James Yarnell
DIRECTOR, PUBLIC RELATIONS: Stephen M. Caine

Beech Aircraft Corporation was founded jointly in 1932 by Mrs Olive Ann Beech and the late Walter H. Beech, pioneer designer and builder of light aeroplanes in the USA. On 8 February 1980, following approval by the stockholders of each company, Raytheon Company and Beech Aircraft Corporation completed the signing of closing merger documents under which Beech became a wholly-owned subsidiary of Raytheon. Beech continues to be operated as a separate entity under its former management, at its original locations; but the separate Board of Directors for Beech includes Mr D. Brainerd Holmes and Mr Thomas L. Phillips, respectively the President and the Chairman and Chief Executive Officer of Raytheon, as representatives of the parent company.

Beech Aircraft Corporation is currently engaged in the production of civil and military aircraft, missile targets, aircraft and missile components, and cryogenic equipment for spacecraft.

Deliveries by Beech in 1980 were made up of 368 King Airs, 33 Dukes, 298 Barons, 378 Bonanzas, 78 Duchesses, 40 Sierras, 55 Sundowners and 144 Skippers. By 1 March 1981 Beech had delivered 3,719 pressurised aircraft since introducing the King Air 90 in 1964. Total production of Beechcraft aeroplanes exceeded 44,000 by the beginning of 1981. Delivery of the 3,000th King Air, a C90, was made on 17 April 1981.

Deliveries of the Beechcraft UC-12B twin-turboprop transport to the US Navy continued in 1980. A $25 million contract received in 1980 extends deliveries of UC-12Bs for the US Navy until April 1982; procurement for the Navy now totals 66. An $8·5 million contract received in early 1981 covered the supply of six C-12Ds to the US Army; procurement for this service now totals 27, of which six had been delivered by 1 March 1981. In addition, the Army has already received 61 C-12As and 14 C-12Cs, and the US Air Force 30 C-12As.

During 1980 deliveries were resumed of the US Navy T-34C turboprop trainer for student primary training at NAS Milton, Florida, under a $9 million contract which extended deliveries to May 1981. A total of 184 T-34Cs has been ordered for the Navy.

Production of airframes for the Bell Helicopter Textron JetRanger, continuous at Beech since 1967, was extended until December 1981 by receipt of a $14·5 million contract in January 1980. Contracts for the commercial JetRanger helicopter; the US Army OH-58A observation helicopter, built from 1967-74; and spares, now exceed $168 million. Beech builds the complete airframe, including fuselage, cowling, tailboom and skid landing gear, at Wichita and the Salina, Kansas, division, and delivered the 3,000th commercial JetRanger airframe in March 1981.

In February 1980 Beech announced that it had formed a team with Grumman Aerospace Corporation to compete in the US Navy's VTX/TS jet trainer competition.

In early 1977 Beech established a wholly-owned subsidiary, Beech Aerospace Services Inc (BASI), to expand worldwide support of Beechcraft aerospace products. BASI headquarters for administration, spare parts, publications central supply and training is based at Jackson, Mississippi.

Beech Aircraft occupies 315,391 m² (3,394,841 sq ft) of plant area at its five major facilities in Wichita, Liberal and Salina, Kansas; Boulder, Colorado; and Selma, Alabama. Aircraft built in Wichita are flown to Selma Division for specialised work. Selma is also the production site for the Commuter C99, the company's initial product for re-entry into the commuter aircraft market.

The Salina division supplies all wings used in Wichita production and is responsible for manufacture and final assembly of the Beechcraft Duke, the pressurised Baron 58P, and the T-34C trainer.

Assembly, flight testing and delivery of the Duchess 76, Sierra and Sundowner are carried out at the Liberal Division, which also manufactures control surfaces for all Beech aircraft.

Work at Boulder involves space vehicle or missile applications. The company's contracts on the Space Shuttle programme alone totalled $41 million by the beginning of 1981 (details in 1980-81 *Jane's*).

Boulder also produces aircraft assemblies for other Beech divisions and AQM-37A and MQM-107 (VSTT) missile target systems for the US Army and Navy (see RPVs and Targets section). In 1979, it announced the development of a system using liquified methane gas (LMG) as an efficient, safe, and environmentally-clean alternative fuel for motor vehicles. In August 1980, Beech announced the establishment of an Alternative Energy Division, to manage the company's programmes for operating motor vehicles on LMG fuel, and other applications of LMG technology. During the year the company demonstrated two compact cars equipped with a Beech-developed LMG fuel system and expected in 1981 to flight-test a similar system in a Sundowner aircraft.

Wholly-owned subsidiaries of the parent company include Beech Acceptance Corporation Inc, which is engaged in business aircraft retail finance and leasing; Beechcraft AG, which has its headquarters in Zurich, Switzerland, and supports in Europe the sales, liaison and other activities of the parent company; Travel Air Insurance Company Ltd, a Bermuda-based company organised during 1972 to provide aircraft liability insurance; Beech Holdings Inc, which provides marketing support to the parent company; Beech International Sales Corporation, Wichita, through which all Beech export sales are made;

Beech Aerospace Services Inc, which provides world-wide support of Beechcraft military aircraft, missile targets and related products; Fuel and Line Service Inc, Wichita; and the following product sales outlets: Hedrick Beechcraft, Colorado Springs, Colorado; Houston-Beechcraft Inc, Houston, Texas; Denver-Beechcraft Inc, Denver, Colorado and Aircraftco Division, Broomfield, Colorado; United Beechcraft Inc, Wichita, Kansas; Beechcraft West Hayward, Van Nuys, Bakersfield and Fresno, California; Mission Beechcraft, Santa Ana, California; Indiana Beechcraft Inc, Indianapolis, Indiana; Beechcraft East Inc, Farmingdale, New York, and Teterboro, New Jersey; Baton Rouge Aircraft, Baton Rouge, Louisiana; and Thompson Beechcraft, Salt Lake City, Utah.

BEECHCRAFT MODEL T-34C

US Navy designation: T-34C

In March 1953 the US Air Force selected the Beechcraft Model 45 as its new primary trainer and, under the designation T-34A Mentor, 450 were eventually acquired. Power plant consisted of a 168 kW (225 hp) Continental O-470-13 flat-six engine. Just over a year later the US Navy reached a similar decision, and 423 T-34B Mentors were built for that service.

In 1973 Beech received a US Navy R & D contract to modify two T-34Bs to see whether the type could be upgraded for a continuing training role. This involved the installation of a turboprop engine and the latest avionics equipment, the primary object being to let student pilots have experience of operating turbine-powered aircraft from the beginning of their flight training. The power plant selected was the PT6A-25, which has a torque limiter in this application to restrict engine output to 298 kW (400 shp), ensure long engine life, and provide constant performance over a wide range of temperature and altitude.

Design of the modifications to update the aircraft began in March 1973, and conversion of two T-34Bs (140784 and 140861) started in May 1973. Designated YT-34C, the first of these aircraft (described in previous editions of *Jane's*) flew for the first time on 21 September 1973.

Beech received US Navy contracts totalling $89·5 million for 184 new-production T-34Cs, and the provision of engineering services and support. Deliveries to Naval Air Training Command at Whiting Field, Milton, Florida, began in November 1977, and student training in the T-34C started in January 1978. An export civil version, known as the **Turbine Mentor 34C**, is in service at the Algerian national pilot training school, which received six in 1979.

A **T-34C-1** armament systems trainer version is also available and, in addition to its basic role, is capable of carrying out forward air control (FAC) and tactical strike training missions. Contracts were received for the supply of 12 T-34C-1s to the Air Force of Morocco, 20 for the Air Force of Ecuador, 3 for the Ecuadorean Navy, 16 for the Air Force of Indonesia, 6 for the Peruvian Navy, 3 for the Uruguayan Navy and 15 for the Navy of Argentina. Delivery of all of these has been completed.

Type: Two-seat turbine-powered primary training and light strike training aircraft.

Wings: Cantilever low-wing monoplane. Wing section NACA 23016·5 (modified) at root, NACA 23012 at tip. Dihedral 7°. Incidence 4° at root, 1° at tip. No sweepback. Conventional box beam structure of light alloy. Ailerons of light alloy construction. Single-slotted trailing-edge flaps of light alloy. Manually operated trim tab in port aileron. Servo tabs in both ailerons.

Fuselage: Semi-monocoque light alloy structure.

Tail Unit: Cantilever structure of light alloy. Fixed-incidence tailplane. Manually-operated trim tabs in elevators and rudder. Twin ventral fins under rear fuselage.

Landing Gear: Electrically-retractable tricycle type. Main units retract inward, nosewheel aft. Beech oleo-pneumatic shock-absorbers. Single wheel on each unit. Main wheels size 7·00-8. Nosewheel and tyre size 5·00-5. Goodyear multiple-disc hydraulic brakes.

Power Plant: One 533 kW (715 shp) Pratt & Whitney Aircraft of Canada PT6A-25 turboprop engine, torque limited to 298 kW (400 shp), driving a Hartzell three-blade constant-speed fully-feathering metal propeller with spinner. Version of same engine derated to 410 kW (550 shp) is available optionally. Two bladder-type fuel cells in each wing, in inboard leading-edge and aft of main spar outboard of landing gear; total usable capacity 492 litres (130 US gallons). Oil capacity 15 litres (4 US gallons).

Accommodation: Instructor and pupil in tandem beneath rearward-sliding cockpit canopy. Cockpit ventilated, heated by engine bleed air and air-conditioned. Dual controls standard. All armament controls in forward cockpit of T-34C-1.

Systems: Hydraulic system for brakes only. Pneumatic system for emergency opening of cockpit canopy. Diluter demand gaseous oxygen system, pressure 103·5 bars (1,500 lb/sq in). Electrical power supplied by 250A starter/generator. Freon-type air-conditioner for cockpit cooling.

Avionics and Equipment: Standard avionics can include UHF or VHF com, VOR or Tacan nav, DME, transponder, angle of attack indicator, ADF, marker beacon receiver, compass and intercom system. Area-Nav, Loran, HF and specialised tactical systems available to customer's requirements. US Navy T-34C has ARC-159V UHF com, VIR-30A VOR/Omni, dual 255Y-1 ICS/audio, TCN-40 Tacan and PN-101 remote compass, all by Collins; two TDR-950 transponders and a CIR-11-2 emergency locator transmitter. Blind-flying instrumentation standard. Electrically-heated pitot.

Beechcraft Model T-34C-1 turboprop armed trainer for the Ecuadorean Air Force

Armament (T-34C-1): CA-513 fixed-reticle reflector gunsight. Four underwing hardpoints are provided for the carriage of stores. The inboard stations are rated at 272 kg (600 lb) each, the outboard stations at 136 kg (300 lb) each, with a maximum load of 272 kg (600 lb) each side and 544 kg (1,200 lb) total. Weapons which can be carried on MA-4 racks include AF/B37K-1 bomb containers with practice bombs or flares, LAU-32 or LAU-59 rocket pods, Mk 81 bombs, SUU-11 Minigun pods, BLU-10/B incendiary bombs, AGM-22A wire-guided anti-tank missiles and TA8X towed target equipment.

Dimensions, external:
- Wing span: 10·16 m (33 ft 3⅞ in)
- Wing chord at root: 2·55 m (8 ft 4½ in)
- Wing chord at tip: 1·05 m (3 ft 5¼ in)
- Wing aspect ratio: 6·22
- Length overall: 8·75 m (28 ft 8½ in)
- Height overall: 2·92 m (9 ft 7 in)
- Tailplane span: 3·71 m (12 ft 2⅛ in)
- Wheel track: 2·95 m (9 ft 8 in)
- Wheelbase: 2·41 m (7 ft 11 in)
- Propeller diameter: 2·29 m (7 ft 6 in)
- Propeller ground clearance: 0·29 m (11¼ in)

Dimensions, internal:
- Cabin: Length: 2·74 m (9 ft 0 in)
  - Max width: 0·86 m (2 ft 10 in)
  - Max height: 1·22 m (4 ft 0 in)

Areas:
- Wings, gross: 16·69 m² (179·6 sq ft)
- Ailerons (total): 1·06 m² (11·4 sq ft)
- Trailing-edge flaps (total): 1·98 m² (21·3 sq ft)
- Fin: 1·20 m² (12·9 sq ft)
- Rudder, incl tab: 0·64 m² (6·9 sq ft)
- Tailplane: 3·46 m² (37·2 sq ft)
- Elevators, incl tabs: 1·26 m² (13·6 sq ft)

Weights and Loading:
- Weight empty: T-34C: 1,342 kg (2,960 lb)
  - T-34C-1: 1,356 kg (2,990 lb)
- Max T-O and landing weight:
  - T-34C: 1,950 kg (4,300 lb)
  - T-34C-1, strike role: 2,494 kg (5,500 lb)
- Max ramp weight: T-34C: 1,962 kg (4,325 lb)
- Max wing loading: T-34C: 108·3 kg/m² (22·2 lb/sq ft)

Performance (T-34C, at T-O weight of 1,910 kg; 4,210 lb, except where indicated):
- Never-exceed speed: 280 knots (518 km/h; 322 mph)
- Max cruising speed at 5,180 m (17,000 ft): 214 knots (396 km/h; 246 mph)
- Stalling speed, flaps down, power off, at typical landing weight of 1,588 kg (3,501 lb): 53 knots (98 km/h; 61 mph)
- Max rate of climb at S/L: 451 m (1,480 ft)/min
- Service ceiling: over 9,145 m (30,000 ft)
- T-O run: 352 m (1,155 ft)
- T-O to 15 m (50 ft): 586 m (1,920 ft)
- Landing from 15 m (50 ft): 547 m (1,795 ft)
- Landing run: 226 m (740 ft)
- Range with max fuel:
  - at 181 knots (335 km/h; 208 mph) at 305 m (1,000 ft): 427 nm (790 km; 491 miles)
  - at 202 knots (374 km/h; 232 mph) at 3,050 m (10,000 ft): 523 nm (968 km; 601 miles)
  - at 180 knots (333 km/h; 207 mph) at 6,100 m (20,000 ft): 708 nm (1,311 km; 814 miles)
- *g* limits: +6; −3

Performance (T-34C-1 with 410 kW; 550 shp engine, estimated. A with two stores at AUW of 2,222 kg; 4,900 lb. B with four stores at AUW of 2,494 kg; 5,500 lb, except where indicated):
- Max level speed at 5,500 m (18,000 ft):
  - A: 209 knots (387 km/h; 241 mph)
  - B: 206 knots (382 km/h; 237 mph)
- Stalling speed, flaps down, idle power:
  - A: 65 knots (120 km/h; 75 mph) CAS
  - B: 69 knots (128 km/h; 80 mph) CAS
- Max rate of climb at S/L: A: 540 m (1,771 ft)/min
  - B: 436 m (1,431 ft)/min
- Typical combat radius:
  - FAC mission at AUW of 2,429 kg (5,355 lb), with four stores and optional max fuel, incl 2·6 h loiter over target and 20 min +5% reserves: 100 nm (185 km; 115 miles)

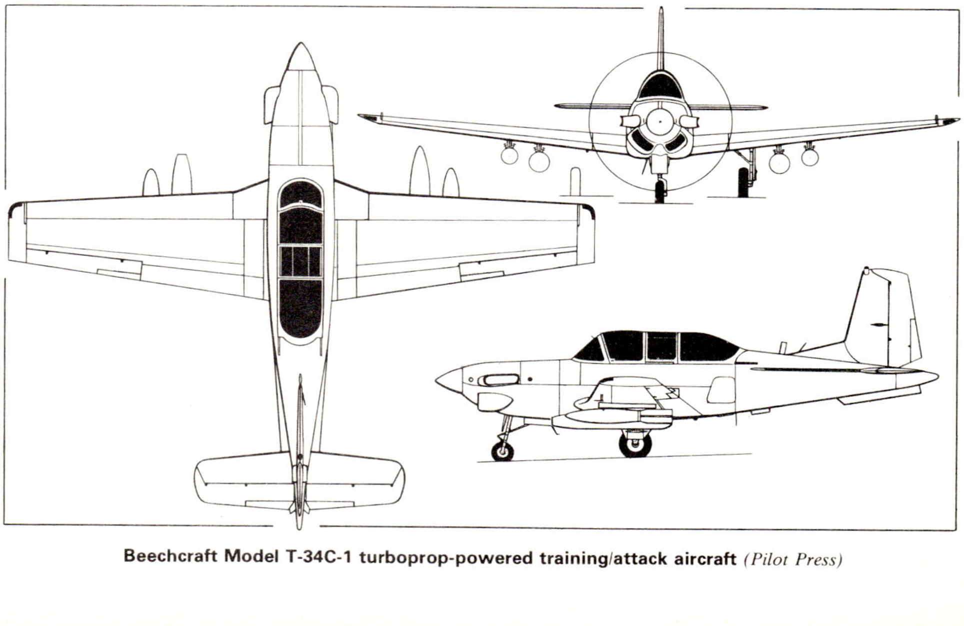

Beechcraft Model T-34C-1 turboprop-powered training/attack aircraft *(Pilot Press)*

Strike mission at AUW of 2,473 kg (5,452 lb), with four stores and optional max fuel, incl 20 min +5% reserves 300 nm (555 km; 345 miles)

BEECHCRAFT MODEL 77 SKIPPER

The prototype of this aircraft, designated PD 285, flew for the first time on 6 February 1975, and was described briefly in the 1975-76 *Jane's*. Beech announced on 17 April 1979 that the production Model 77 Skipper had received FAA certification in the Utility category, and deliveries began in May 1979. The first aircraft were allocated to Beech Aero Centers for flying training duties.

A total of 211 Skippers had been delivered by 1 January 1981, and the following description applies to the then-current production model. In July 1981 Beech announced a temporary suspension of Skipper production, which had been at the rate of ten per month, pending an improvement in market conditions.

TYPE: Two-seat training aircraft, approved for six-turn spins.

WINGS: Cantilever low-wing monoplane, of NASA GAW-1 section. Dihedral 6° from roots. Tubular spar formed of wrapped light alloy sheet and glassfibre adhesive to ensure required bonding stiffness at all wing stations; auxiliary rear spar, light alloy ribs and bonded light alloy skins. Plain ailerons with corrugated skins. Ground-adjustable tab on each aileron. Electrically-operated plain trailing-edge flaps with corrugated skins. Flaps and ailerons actuated via torque tubes.

FUSELAGE: Semi-monocoque structure of light alloy. Lower cabin section is of bonded construction.

TAIL UNIT: Cantilever T tail of light alloy. Fixed-incidence tailplane, leading-edge of fin, large portion of tailplane, elevator trim tabs and rudder are of bonded construction. Trim tab in each elevator; rudder trim standard.

LANDING GEAR: Non-retractable tricycle type with single wheel on each unit. Cantilever tubular spring main legs. Steerable and self-centering nosewheel with oleo shock-absorber. Main-wheel tyres size 15 × 6·00-6; nosewheel tyre size 5·00-5. Toe-operated hydraulic brakes. Parking brake.

POWER PLANT: One 85·5 kW (115 hp) Avco Lycoming O-235-L2C flat-four engine, driving a Sensenich fixed-pitch two-blade metal propeller with spinner. Fuel in integral wing tanks with combined usable capacity of 110 litres (29 US gallons). Refuelling point on upper surface of each wing. Oil capacity 5·7 litres (1·5 US gallons).

ACCOMMODATION: Two adjustable seats side by side in enclosed cabin, with inertia-reel harness. Door on each side. Tinted cabin windows. Carpeted floor. Dual controls standard. Ventilation, heating and windscreen defrosting standard. Baggage space.

SYSTEM: 14V electrical system, with 60A alternator and 25Ah battery.

DIMENSIONS, EXTERNAL:

Wing span	9·14 m (30 ft 0 in)
Wing area, gross	12·06 m² (129·8 sq ft)
Length overall	7·35 m (24 ft 1½ in)
Height overall	2·11 m (6 ft 11 in)
Tailplane span	3·00 m (9 ft 10 in)
Wheel track	2·53 m (8 ft 3½ in)
Wheelbase	1·57 m (5 ft 2 in)
Propeller diameter	1·83 m (6 ft 0 in)
Door: Max width	0·91 m (3 ft 0 in)
Height	0·95 m (3 ft 1½ in)

DIMENSIONS, INTERNAL:

Cabin: Length, panel to lower rear bulkhead	1·93 m (6 ft 4 in)
Max width	1·08 m (3 ft 6¾ in)
Max height	1·23 m (4 ft 0¾ in)
Baggage space	0·57 m³ (20·1 cu ft)

WEIGHTS:

Weight empty	501 kg (1,103 lb)
Max baggage	54 kg (120 lb)
Max T-O and landing weight	760 kg (1,675 lb)
Max ramp weight	762 kg (1,680 lb)

PERFORMANCE (at T-O weight of 726 kg (1,600 lb):

Max level speed at S/L	106 knots (196 km/h; 122 mph)
Max cruising speed, 80% power at 1,370 m (4,500 ft)	105 knots (195 km/h; 121 mph)
Econ cruising speed, 53% power at 1,370 m (4,500 ft)	88 knots (163 km/h; 101 mph)
Cruising speed, 61% power at 2,590 m (8,500 ft)	96 knots (177 km/h; 110 mph)
Econ cruising speed, 51% power at 2,590 m (8,500 ft)	85 knots (158 km/h; 98 mph)
Stalling speed, flaps up	48·2 knots (89·3 km/h; 55·5 mph) CAS
Stalling speed, flaps down	47 knots (87 km/h; 54 mph) CAS
Max rate of climb at S/L	219 m (720 ft)/min
Service ceiling	3,930 m (12,900 ft)
Absolute ceiling	4,570 m (15,000 ft)
T-O run	238 m (780 ft)
T-O to 15 m (50 ft)	390 m (1,280 ft)
Landing from 15 m (50 ft)	400 m (1,313 ft)
Landing run	204 m (670 ft)
Range, max cruising speed at 1,370 m (4,500 ft), with reserves	327 nm (606 km; 376 miles)
Range, econ cruising speed at 1,370 m (4,500 ft), with reserves	410 nm (760 km; 472 miles)
Range, cruising speed (61% power) at 2,590 m (8,500 ft), with reserves	388 nm (719 km; 447 miles)
Range, econ cruising speed at 2,590 m (8,500 ft), with reserves	413 nm (765 km; 475 miles)

Beechcraft Model 77 Skipper lightweight training aircraft (Avco Lycoming O-235 engine)

BEECHCRAFT SIERRA 200 and SUNDOWNER 180

In December 1971, Beech introduced a new light aircraft marketing programme centred around three models, which were given individual exterior paint schemes and renamed from their previous Musketeer designations (see 1971-72 *Jane's*).

In 1974 these designations were changed again to indicate the engine horsepower rating, so becoming Beechcraft Sierra 200 (formerly Model A24R Musketeer Super R), Sundowner 180 (Model C23, formerly Musketeer Custom), and Sport 150 (Model B19, formerly Musketeer Sport). The fourth aircraft in the former Musketeer line, the Super, was discontinued at the end of 1971 after a total of 368 had been built; and production of the Sport 150 (Model B19) ended in September 1978 after 905 had been built. Details of the Sport 150 can be found in the 1978-79 *Jane's*.

The Sierra 200 was recertificated in 1974 and redesignated Model B24R, due to the installation of a new engine, improved cowling and redesign of control features. Designated Model C24R in 1977, improvements included a propeller of increased diameter to enhance performance, and wheel well fairings to reduce drag when the wheels are retracted.

Details of the two current models are as follows:

Sundowner 180. Basic four-seat version with 134 kW (180 hp) Avco Lycoming O-360-A4K engine, driving a Sensenich Type 76EM8S5-0-60 two-blade fixed-pitch metal propeller with spinner, and non-retractable landing gear. Aerobatic version is approved for rolls, Immelmann turns, loops, spins, chandelles and other manoeuvres, carrying two persons. Three windows standard on each side of cabin.

Sierra 200. Generally similar to the Sundowner but with accommodation for four to six persons, a 149 kW (200 hp) Avco Lycoming IO-360-A1B6 engine, driving a Hartzell Type HC-M2YR-1BF/F7666A two-blade constant-speed metal propeller with spinner, and retractable tricycle landing gear. Electrically-actuated hydraulic system based on a self-contained unit in the rear fuselage, comprising electrically-driven hydraulic pump, fluid reservoir and valves. An emergency valve, sited adjacent to the pilot's feet, allows selection of the landing gear to free-fall within three seconds. Main wheels retract outward into wings; nosewheel turns through 90° as it retracts rearwards. Four windows standard on each side of cabin.

Factory-installed optional equipment packages are as follows:

Weekender. Includes sun visors; tinted windscreen and windows; dual controls and pedal-operated brakes for co-pilot; lighting group comprising rotating beacon, navigation, cabin dome, overhead instrument and map lights; cabin boarding steps; and acrylic enamel paint; adding 12·2 kg (27 lb) to basic empty weight.

Holiday. As above, plus instrument group comprising 3 in horizon and directional gyros with vacuum system, electric clock, outside air temperature gauge, rate of climb indicator and turn co-ordinator; coat hook and garment hanger in baggage compartment; wing-mounted landing light; and two 12V 25Ah batteries; adding 26·5 kg (58·4 lb) to basic empty weight.

Professional. As above, plus true airspeed indicator; two headrests; instrument post lights; wing-mounted taxi light; and heated pitot tube; adding 30·2 kg (66·6 lb) to basic empty weight.

Eleven factory-installed avionics packages are available optionally, and include com transceiver, nav receiver-converter, ADF, transponder and audio panel from Collins, King or Edo-Aire Mitchell, plus cabin speaker, microphone with jack and related antennae. An extensive selection of additional avionics is also available optionally.

Production is centred in Beech's Liberal, Kansas, plant. A total of 5,099 Musketeers, Sundowners, Sports and Sierras had been delivered by 1 January 1981. They included 20 aircraft supplied to the Mexican government for military training, 25 for the Canadian Armed Forces, 21 for the Indonesian Department of Transportation, Communication and Tourism for its primary training programme, and 19 for the University of Illinois Institute of Aviation. Three Sierra 200s were delivered in 1979 to the national pilot training school of Algeria.

The following details apply to both current models:

TYPE: Four- or four/six-seat cabin monoplane.

WINGS: Cantilever low-wing monoplane. Wing section NACA 63_2A415. Dihedral 6° 30′. Incidence 3° at root, 1° at tip. Single extruded main spar at 50% chord. Aluminium skin and stringers are bonded to honeycomb Trussgrid ribs on forward 50% of wing; rear 50% of wing is riveted. Slotted all-metal riveted ailerons and mechanically-controlled (optionally electrically-actuated) flaps have corrugated skin. No trim tabs. Plastics wingtips.

Beechcraft Sierra 200 four/six-seat light aircraft

Beechcraft Sundowner 180 four-seat light aircraft

FUSELAGE: Cabin section has basic keel formed by floor and lower skin, with rolled skin side panels, stringers, a minimum number of bulkheads and structural top. Conventional semi-monocoque rear fuselage.

TAIL UNIT: Cantilever all-metal structure, with swept vertical surfaces. One-piece all-moving horizontal surface with full-span anti-servo tab. Optional electric tailplane trim. Rudder and aileron controls interconnected for easy cross-country flying.

LANDING GEAR (Sundowner): Non-retractable tricycle type. Beech rubber-disc shock-absorbers. Nosewheel with tube-type tyre size 15 × 6·00-6, pressure 2·76 bars (40 lb/sq in). Main wheels have tube-type tyres size 17·5 × 6·00-6, pressure 1·52 bars (22 lb/sq in). Cleveland disc-type hydraulic brakes with toe-operated control. Steerable nosewheel. Parking brake.

POWER PLANT: One flat-four engine (details given under model listings). Two fuel tanks in inboard wing leading-edges, with usable capacity of 216 litres (57 US gallons). Refuelling points above tanks. Oil capacity 7·5 litres (2 US gallons).

ACCOMMODATION: Pilot and three or five passengers (Sierra); pilot and three passengers (Sundowner); in pairs, in enclosed cabin with door on each side. Compartment for 122 kg (270 lb) baggage, with external door on port side. In-flight-adjustable seats, pilot's storm window, windscreen defroster, instrument panel glareshield, air vents, map stowage, wall-to-wall carpeting. Optional aerobatic kit for Sundowner includes *g* meter and quick-release door.

SYSTEMS: Electrical system includes a 28V 60A alternator; 24V 15·5Ah battery standard, two 12V 25Ah batteries optional. Hydraulic system for brakes only, except on Sierra which has electro-hydraulic actuation system for landing gear. Vacuum system for instruments optional.

AVIONICS AND EQUIPMENT: Optional avionics as listed earlier. Standard equipment includes sensitive altimeter, fore and aft adjustable front seats with reclining backs, shoulder harness and lap belts, control locks, map stowage, pilot's storm window, emergency locator transmitter, stall warning device, and towbar. Optional equipment includes exhaust gas temperature gauge, Hobbs hour meter, outside air temperature gauge, true airspeed indicator, headrests, tinted windscreen and windows, alternate static source, heated pitot, rear cabin 'family' seat, internal corrosion proofing, instrument post lights, two-light strobe system, wing-mounted taxi light, external power socket and acrylic enamel exterior finish. An aerobatic kit is available for the Sundowner, and a 'Magic Hand' landing gear safety system for the Sierra.

DIMENSIONS, EXTERNAL:

Wing span	9·98 m (32 ft 9 in)
Wing chord, constant	1·34 m (4 ft 4¾ in)
Wing aspect ratio	7·4
Length overall	7·85 m (25 ft 9 in)
Height overall: Sundowner	2·51 m (8 ft 3 in)
Sierra	2·46 m (8 ft 1 in)
Tailplane span	3·30 m (10 ft 10 in)
Wheel track: Sundowner	3·61 m (11 ft 10 in)
Sierra	3·86 m (12 ft 8 in)
Wheelbase: Sundowner	1·93 m (6 ft 4 in)
Sierra	1·83 m (6 ft 0¼ in)
Propeller diameter	1·93 m (6 ft 4 in)
Propeller ground clearance:	
Sundowner	0·36 m (1 ft 2 in)
Sierra	0·33 m (1 ft 1 in)
Cabin doors: Height	0·91 m (3 ft 0 in)
Width	0·91 m (3 ft 0 in)
Baggage compartment door:	
Sundowner: Height	0·47 m (1 ft 6½ in)
Width	0·60 m (1 ft 11¾ in)
Sierra: Height	0·84 m (2 ft 9 in)
Width	0·56 m (1 ft 10 in)

DIMENSIONS, INTERNAL:

Cabin, aft of instrument panel:	
Length	2·41 m (7 ft 11 in)
Max width	1·18 m (3 ft 8 in)
Max height	1·22 m (4 ft 0 in)
Floor area	2·4 m² (25·84 sq ft)
Volume	2·92 m³ (103·2 cu ft)
Baggage compartment	0·55 m³ (19·5 cu ft)

AREAS:

Wings, gross	13·57 m² (146 sq ft)
Ailerons (total)	1·25 m² (13·47 sq ft)
Flaps (total)	1·74 m² (18·76 sq ft)
Fin	0·93 m² (9·98 sq ft)
Rudder	0·43 m² (4·62 sq ft)
Tailplane, incl anti-servo tab	2·52 m² (27·08 sq ft)

WEIGHTS AND LOADINGS:

Weight empty (incl oil and unusable fuel):	
Sundowner	683 kg (1,505 lb)
Sierra	771·5 kg (1,701 lb)
T-O weight, Utility category:	
Sundowner	920 kg (2,030 lb)
Max T-O and landing weight:	
Sundowner	1,111 kg (2,450 lb)
Sierra	1,247 kg (2,750 lb)
Max wing loading:	
Sundowner	81·9 kg/m² (16·78 lb/sq ft)
Sierra	91·9 kg/m² (18·84 lb/sq ft)
Max power loading:	
Sundowner	8·29 kg/kW (13·61 lb/hp)
Sierra	8·37 kg/kW (13·75 lb/hp)

PERFORMANCE (at max T-O weight):

Max level speed at S/L:	
Sundowner	123 knots (228 km/h; 141 mph)
Sierra	142 knots (262 km/h; 163 mph)
Cruising speed:	
Sundowner, 84% power at 1,370 m (4,500 ft)	123 knots (228 km/h; 141 mph)
Sundowner, 59% power at 1,370 m (4,500 ft)	98 knots (182 km/h; 113 mph)
Sierra, 75% power at 3,050 m (10,000 ft)	137 knots (254 km/h; 158 mph)
Sierra, 55% power at 3,050 m (10,000 ft)	115 knots (213 km/h; 132 mph)
Stalling speed, flaps down, power off:	
Sundowner	51 knots (94·5 km/h; 59 mph) IAS
Sierra	60 knots (111 km/h; 69 mph) IAS
Max rate of climb at S/L:	
Sundowner	241 m (792 ft)/min
Sierra	283 m (927 ft)/min
Service ceiling: Sundowner	3,840 m (12,600 ft)
Sierra	4,690 m (15,385 ft)
Absolute ceiling: Sundowner	4,390 m (14,400 ft)
Sierra	5,315 m (17,430 ft)
Min ground turning radius	7·29 m (23 ft 11 in)
T-O run: Sundowner	344 m (1,130 ft)
Sierra	324 m (1,063 ft)
T-O to 15 m (50 ft): Sundowner	596 m (1,955 ft)
Sierra	476 m (1,561 ft)
Landing from 15 m (50 ft):	
Sundowner	452 m (1,484 ft)
Sierra	446 m (1,462 ft)
Landing run: Sundowner	214 m (703 ft)
Sierra	249 m (816 ft)
Range with max fuel, allowances for warm-up, T-O, climb and 45 min reserves:	
Sundowner, 2,300 rpm at 1,370 m (4,500 ft)	597 nm (1,106 km; 687 miles)
Sierra, 2,400 rpm at 3,050 m (10,000 ft)	686 nm (1,271 km; 790 miles)

BEECHCRAFT BONANZA MODEL V35B

Beechcraft's V-tail V35B Bonanza four/five-seat light aircraft

The prototype Bonanza flew for the first time on 22 December 1945 and the type went into production in 1947. In February 1977 the 10,000th V-tail Bonanza Model 35 was completed, and this aircraft entered its 35th year of production in 1981. The current version is designated Model V35B.

The Bonanza Model A36 and A36TC utility aircraft, and Model F33 series, with conventional tail unit, are described separately. Total production of Bonanzas (all series) had exceeded 14,800 by mid-1981.

Bonanzas are equipped with a dual-duct fresh air system to increase cabin airflow. Safety features include single diagonal strap shoulder harness with inertia reel for all occupants as standard equipment. Five optional factory-installed IFR avionics packages include dual communication, navigation, marker beacon receiver, glideslope, DME, and transponder. Four packages meet FAA Technical Standard Order (TSO). Beech was in 1972 the first general aviation manufacturer to acquire approval for factory installation of area navigation equipment on production aircraft with IFR equipment.

In 1981 Beech was offering five 'Super Utility' packages of optional equipment, which includes a large cargo door on the Models V35B and F33A and club seating arrangement on the Models A36 and A36TC, gyro horizon and directional gyro with pressure system, heated pitot, alternate static source, super soundproofing and a choice of Edo-Aire Mitchell Century I, III or 41, Bendix FCS-870 or King KFC 200 autopilot.

Other optional extras available on the Bonanza include the Beech-designed 'Magic Hand' landing gear safety system. Designed to eliminate the possibility of wheels-up landing or inadvertent retraction of the landing gear on the ground, it lowers the gear automatically on approach when the engine manifold pressure falls below approximately 508 mm (20 in) and airspeed has been reduced to 104 knots (193 km/h; 120 mph). On take-off, it keeps the gear down until the aircraft is airborne and has accelerated to 78 knots (145 km/h; 90 mph) IAS. The system can be switched off by the pilot at will.

An optional 12,000 BTU refrigeration-type air-conditioning system was introduced during 1975, comprising an evaporator located beneath the pilot's seat, condenser mounted on the lower fuselage and an engine-mounted compressor. Air outlets are located on the centre console, and a two-speed blower is provided for air distribution (except on A36TC).

TYPE: Four/five-seat light cabin monoplane.

WINGS: Cantilever low-wing monoplane. Wing section Beech modified NACA 23016·5 at root, modified NACA 23012 at tip. Dihedral 6°. Incidence 4° at root, 1° at tip. Sweepback 0° at quarter-chord. Each wing is a two-spar semi-monocoque box-beam of conventional aluminium alloy construction. Symmetrical-section ailerons and single-slotted three-position flaps of aluminium alloy construction. Ground-adjustable trim tab in each aileron.

FUSELAGE: Conventional aluminium alloy semi-monocoque structure. Hat-section longerons and channel-type keels extend forward from cabin section, making the support structure for the engine and nosewheel an integral part of the fuselage.

TAIL UNIT: Cantilever V tail, with tailplane and elevators set at 33° dihedral angle. Semi-monocoque construction. Fixed surfaces have aluminium alloy structure and skin. Control surfaces, aft of the light alloy spar, are primarily of magnesium alloy, with large controllable trim tab in each. Tail surfaces are interchangeable port and starboard, except for tabs and actuator horns. Electrically-operated elevator trimming optional.

LANDING GEAR: Electrically-retractable tricycle type, with steerable nosewheel. Main wheels retract inward into wings, nosewheel aft. Beech oleo-pneumatic shock-absorbers in all units. Cleveland main wheels, size 6·00-6, and tyres, size 7·00-6, pressure 2·28-2·76 bars (33-40 lb/sq in). Cleveland nosewheel and tyre, size 5·00-5, pressure 2·76 bars (40 lb/sq in). Cleveland ring-disc hydraulic brakes. Parking brake. 'Magic Hand' landing gear system optional.

POWER PLANT: One 212·5 kW (285 hp) Continental IO-520-BB flat-six engine, driving a McCauley two-blade constant-speed metal propeller with spinner. Three-blade propeller optional. Manually-adjustable engine cowl flaps. Two standard fuel tanks in wing leading-edges, with total usable capacity of 280 litres (74 US gallons). Refuelling points above tanks. Oil capacity 11·5 litres (3 US gallons).

ACCOMMODATION: Enclosed cabin seating four or five persons on individual seats. Centre windows open for ventilation on ground and have release handles to permit their use as emergency exits. Pilot's storm window, port side. Cabin structure reinforced for protection in turnover. Space for up to 122·5 kg (270 lb) of baggage aft of seats. Passenger door and baggage access door both on starboard side. Cabin heated and ventilated.

SYSTEMS: Electrical system supplied by 28V 60A alternator, 24V 15·5Ah battery; a 100A alternator is available as an option, as is a standby generator. Hydraulic system for brakes only. Pneumatic system for instrument gyros and refrigeration-type air-conditioning system optional. Oxygen system optional.

AVIONICS AND EQUIPMENT: Standard avionics comprise King KX 170B 720-channel com transceiver, 200-channel nav receiver with KI 208 VOR/LOC converter-indicator, microphone, headset, cabin speaker, and nav/com/GS antenna. A wide range of optional avionics is available, by Bendix, Collins, Edo-Aire Mitchell, King and Narco. Standard equipment includes electric clock, outside air temperature gauge, rate of climb indicator, sensitive altimeter, turn co-ordinator, four fore and aft adjustable and reclining seats, armrests, headrests, shoulder harness and lap-belts, pilot's storm window, sun visors, ultraviolet-proof windscreen and windows, emergency locator transmitter, stall warning device, carpeted floor, coat hooks, glove compartment, in-flight storage pockets, utility shelf, cabin dome light, reading lights, instrument panel floodlights, electroluminescent sub-panel lighting, landing light, taxi light, full-flow oil filter, urethane exterior paint, and towbar. Optional equipment includes control wheel clock, 3 in horizon and directional gyros, exhaust gas temperature gauge, dual controls, co-pilot's wheel brakes, alternate static source, heated pitot, fifth seat, large cargo door, super soundproofing, control wheel map lights, entrance door courtesy light, instrument post lights, internally-lighted instruments, fresh air vent blower, rotating beacons, three-light strobe system, external power socket, and static wicks.

Dimensions, external:
- Wing span 10·21 m (33 ft 6 in)
- Wing chord at root 2·13 m (7 ft 0 in)
- Wing chord at tip 1·07 m (3 ft 6 in)
- Wing aspect ratio 6·2
- Length overall 8·05 m (26 ft 5 in)
- Height overall 2·31 m (7 ft 7 in)
- Tailplane span 3·10 m (10 ft 2 in)
- Wheel track 2·92 m (9 ft 7 in)
- Wheelbase 2·13 m (7 ft 0 in)
- Propeller diameter: two-blade 2·13 m (7 ft 0 in)
  - three-blade 2·03 m (6 ft 8 in)
- Passenger door: Height 0·91 m (3 ft 0 in)
  - Width 0·94 m (3 ft 1 in)
- Baggage compartment door:
  - Height 0·47 m (1 ft 6½ in)
  - Width 0·57 m (1 ft 10½ in)

Dimensions, internal:
- Cabin, aft of firewall: Length 3·07 m (10 ft 1 in)
  - Max width 1·07 m (3 ft 6 in)
  - Max height 1·27 m (4 ft 2 in)
  - Volume 3·31 m^3 (117 cu ft)
- Baggage space 0·99 m^3 (35 cu ft)

Areas:
- Wings, gross 16·80 m^2 (181 sq ft)
- Ailerons (total) 1·06 m^2 (11·4 sq ft)
- Trailing-edge flaps (total) 1·98 m^2 (21·3 sq ft)
- Fixed tail surfaces 2·20 m^2 (23·8 sq ft)
- Movable tail surfaces, incl tabs 1·34 m^2 (14·4 sq ft)

Weights and Loadings:
- Weight empty, standard 955 kg (2,106 lb)
- Max T-O and landing weight 1,542 kg (3,400 lb)
- Max ramp weight 1,547 kg (3,412 lb)
- Max wing loading 91·8 kg/m^2 (18·80 lb/sq ft)
- Max power loading 7·26 kg/kW (11·93 lb/hp)

Performance (at max T-O weight, except cruising speeds at mid-cruise weight):
- Max level speed at S/L 182 knots (338 km/h; 209 mph)
- Cruising speed:
  - 75% power at 1,830 m (6,000 ft) 172 knots (319 km/h; 198 mph)
  - 66% power at 3,050 m (10,000 ft) 168 knots (311 km/h; 193 mph)
  - 55% power at 3,660 m (12,000 ft) 157 knots (291 km/h; 181 mph)
  - 45% power at 2,440 m (8,000 ft) 136 knots (253 km/h; 157 mph)
- Stalling speed, power off, flaps up 64 knots (118 km/h; 74 mph) IAS
- Stalling speed, power off, 30° flap 51 knots (94 km/h; 59 mph) IAS
- Max rate of climb at S/L 356 m (1,167 ft)/min
- Service ceiling 5,445 m (17,860 ft)
- T-O run 305 m (1,002 ft)
- T-O to 15 m (50 ft) 539 m (1,769 ft)
- Landing from 15 m (50 ft) 404 m (1,324 ft)
- Landing run 233 m (763 ft)
- Range with 280 litres (74 US gallons) usable fuel, allowances for engine start, taxi, T-O, climb and 45 min reserves at 45% power:
  - 75% power at 1,830 m (6,000 ft) 716 nm (1,326 km; 824 miles)
  - 66% power at 3,050 m (10,000 ft) 777 nm (1,440 km; 894 miles)
  - 55% power at 3,660 m (12,000 ft) 838 nm (1,553 km; 964 miles)
  - 45% power at 2,440 m (8,000 ft) 889 nm (1,648 km; 1,023 miles)

BEECHCRAFT BONANZA MODEL F33A/C

The **F33A** version of the Bonanza is a four/five-seat single-engined executive aircraft, similar in general configuration to the Bonanza Model V35B, but distinguished by a conventional tail unit with sweptback vertical surfaces. The prototype flew for the first time on 14 September 1959, and the production models were known as Debonairs until 1967.

The **F33C** is generally similar to the A, but approved for both aerobatic and utility operation.

The Model G33 was discontinued in early 1973 after 49 had been produced.

A total of 2,362 Model 33s had been built by 1 January 1981. Twenty-one were bought for pilot training by Lufthansa in West Germany; and Pacific Southwest Airlines acquired 12 F33As for airline crew training. Deliveries of F33As and aerobatic F33Cs to foreign air forces were as follows: Imperial Air Force of Iran, 16 F33Cs; Mexican Navy, 5 F33Cs; Netherlands Government Flying School, 16 F33Cs; and Spanish Air Ministry and Air Force, 74 F33s.

Optional extras include the 'Magic Hand' automatic landing gear control system, air-conditioning system and other items described under the Model V35B Bonanza entry, except that the large cargo door, air-conditioning and fifth seat are not available for the Bonanza F33C.

Beechcraft Model F33A Bonanza four/five-seat cabin monoplane

Type: Four/five-seat cabin monoplane.

Wings: As for V35B Bonanza.

Fuselage: As for Bonanza V35B.

Tail Unit: Conventional cantilever all-metal stressed-skin structure, primarily of aluminium alloy but with corrugated magnesium skin on elevators. Large trim tab in each elevator. Fixed tab in rudder.

Landing Gear: As for Bonanza V35B series. Main wheels size 6·00-6, with tyres size 7·00-6, pressure 2·28-2·76 bars (33-40 lb/sq in); nosewheel size 5·00-5, tyre pressure 2·76 bars (40 lb/sq in).

Power Plant: As for Bonanza V35B.

Accommodation: Enclosed cabin with four individual seats in pairs as standard, plus optional forward-facing fifth seat (F33A only). Baggage compartment and hat shelf aft of seats. Passenger door and baggage compartment door on starboard side. Heater standard. Large cargo door, on starboard side of fuselage, optional on F33A.

Systems: As for V35B.

Avionics and Equipment: As for V35B.

Dimensions, internal: As for V35B
- Length overall 8·13 m (26 ft 8 in)
- Height overall 2·51 m (8 ft 3 in)
- Tailplane span 3·71 m (12 ft 2 in)
- Propeller diameter 2·13 m (7 ft 0 in)

Dimensions, internal: As for V35B.

Areas: As for V35B except:
- Fin 0·85 m^2 (9·1 sq ft)
- Rudder, incl tab 0·43 m^2 (4·6 sq ft)
- Tailplane 1·75 m^2 (18·82 sq ft)
- Elevators, incl tabs 1·24 m^2 (13·36 sq ft)

Weights and Loadings:
- Weight empty 964 kg (2,125 lb)
- Max T-O and landing weight 1,542 kg (3,400 lb)
- Max ramp weight 1,548 kg (3,412 lb)
- Max wing loading 91·8 kg/m^2 (18·8 lb/sq ft)
- Max power loading 7·26 kg/kW (11·93 lb/hp)

Performance: As for V35B

BEECHCRAFT BONANZA MODEL A36

This version of the Bonanza, introduced in mid-1968, is a full six-seat utility aircraft developed from the Bonanza Model V35B. It is generally similar to the V35B, but has a conventional tail unit with sweptback vertical surfaces, similar to that of the Bonanza F33 series. In addition, the A36 has large double doors on the starboard side of the fuselage aft of the wing root, to facilitate loading and unloading of bulky cargo when used in a utility role. The cabin volume is increased by 0·54 m^3 (18·9 cu ft) compared with the V35B, due to a fuselage extension of 0·25 m (10 in), and an increase of 0·28 m^3 (10 cu ft) in the baggage compartment volume.

Like all Bonanzas, the Model A36 is licensed in the FAA Utility category at full gross weight, with no limitation of performance.

The current version of the Bonanza Model A36 introduced options which include a club-seating interior layout with rear-facing third and fourth seats, executive writing desk, reading lights and fresh air outlets for fifth and sixth seats. Optional extras include the 'Magic Hand' automatic landing gear control system, refrigeration-type air-conditioning system and all other items mentioned under the Model V35B Bonanza entry, except for the large cargo door.

More than 2,100 Model 36 Bonanzas had been delivered by mid-1981.

Type: Four/six-seat utility light cabin monoplane.

Wings: As for Model V35B.

Fuselage: As for Model V35B but lengthened by 0·25 m (10 in).

Tail Unit: Conventional cantilever all-metal stressed-skin structure, primarily of aluminium alloy but with corrugated magnesium skin on elevators. Large trim tab in each elevator. Fixed tab in rudder.

Landing Gear: Electrically-retractable tricycle type, similar to that of Baron. Main units retract inward into wings, nosewheel rearward. Beech oleo-pneumatic shock-absorbers. Steerable nosewheel. Cleveland main wheels, size 6·00-6, with tyres size 7·00-6, pressure 2·28-2·76 bars (33-40 lb/sq in). Cleveland nosewheel and tyre size 5·00-5, pressure 2·76 bars (40 lb/sq in). Cleveland ring-disc hydraulic brakes. Parking brake.

Power Plant: As for Model V35B.

Accommodation: Enclosed cabin seating four to six persons on individual seats. Two rear removable seats and two folding seats permit rapid conversion to utility configuration. Optional club-seating layout with rear-facing third and fourth seats, executive writing desk, headrests for third and fourth seats, reading lights and fresh air outlets for fifth and sixth seats. Double doors of bonded aluminium honeycomb construction on starboard side facilitate loading of cargo. As an air ambulance, one stretcher can be accommodated with ample room for a medical attendant and/or other passengers. Extra windows provide improved view for passengers. Stowage for 181 kg (400 lb) of baggage.

Systems: Electrical system supplied by 28V 60A alternator, 24V 15·5Ah battery; a 100A alternator is available as an option, as is a standby generator. Hydraulic system for brakes only. Pneumatic system for instrument gyros, and refrigeration-type air-conditioning system, optional.

Avionics and Equipment: Standard avionics include King KX 170B 720-channel nav/com, with KI 208 VOR/LOC Omni converter/indicator and Beechcraft

Beechcraft four/six-seat A36 Bonanza (Continental IO-520-BB flat-six engine)

Beechcraft Duchess 76 light aircraft (two Avco Lycoming O-360-A1G6D engines)

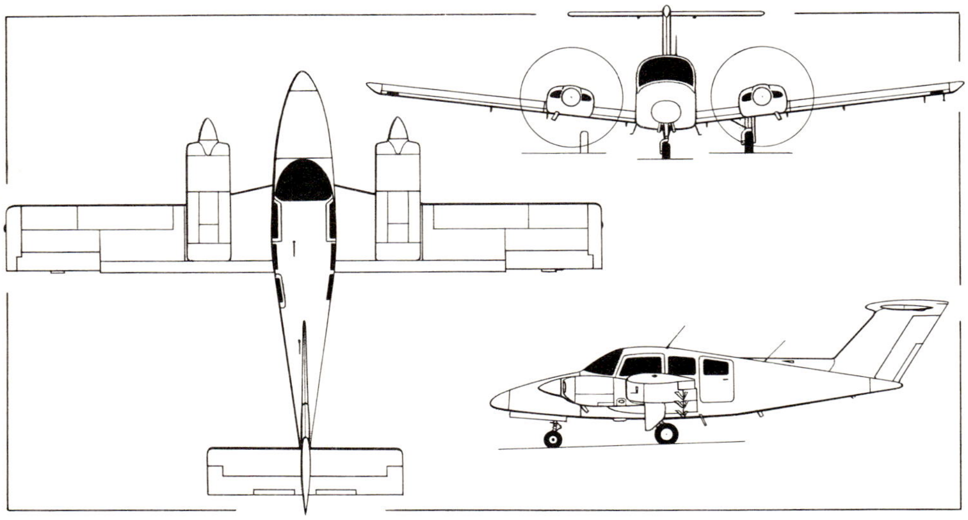

Beechcraft Duchess 76 four-seat light twin *(Pilot Press)*

antenna, but a wide range of optional avionics is available. Optional items of equipment are as detailed for the V35B Bonanza, except as noted.

Dimensions, external: As for V35B except:
- Length overall — 8·38 m (27 ft 6 in)
- Height overall — 2·57 m (8 ft 5 in)
- Wheelbase — 2·39 m (7 ft 10¼ in)
- Rear passenger/cargo door:
  - Height — 1·02 m (3 ft 4 in)
  - Width — 1·14 m (3 ft 9 in)

Dimensions, internal:
- Cabin, aft of firewall: Length, incl extended baggage compartment — 3·84 m (12 ft 7 in)
  - Max width — 1·07 m (3 ft 6 in)
  - Max height — 1·27 m (4 ft 2 in)
  - Volume — 3·85 m³ (135·9 cu ft)

Area:
- Wings, gross — 16·8 m² (181 sq ft)

Weights and Loadings:
- Weight empty, standard — 996 kg (2,195 lb)
- Max T-O weight — 1,633 kg (3,600 lb)
- Max ramp weight — 1,638 kg (3,612 lb)
- Max wing loading — 97·2 kg/m² (19·9 lb/sq ft)
- Max power loading — 7·68 kg/kW (12·6 lb/hp)

Performance (at max T-O weight, except cruising speeds at mid-cruise weight):
- Max level speed — 179 knots (332 km/h; 206 mph)
- Cruising speed:
  - 75% power at 1,830 m (6,000 ft) — 168 knots (311 km/h; 193 mph)
  - 66% power at 3,050 m (10,000 ft) — 163 knots (303 km/h; 188 mph)
  - 55% power at 3,660 m (12,000 ft) — 150 knots (278 km/h; 173 mph)
  - 45% power at 2,440 m (8,000 ft) — 127 knots (235 km/h; 146 mph)
- Stalling speed, flaps up, power off — 62 knots (115 km/h; 71 mph) IAS
- Stalling speed, 30° flap, power off — 52 knots (96·5 km/h; 60 mph) IAS
- Max rate of climb at S/L — 314 m (1,030 ft)/min
- Service ceiling — 5,060 m (16,600 ft)
- T-O run — 347 m (1,140 ft)
- T-O to 15 m (50 ft) — 622 m (2,040 ft)
- Landing from 15 m (50 ft) — 442 m (1,450 ft)
- Landing run — 256 m (840 ft)
- Range with 280 litres (74 US gallons) usable fuel, allowances for engine start, taxi, T-O, climb and 45 min reserves at 45% power:
  - 75% power at 1,830 m (6,000 ft) — 697 nm (1,292 km; 802 miles)
  - 66% power at 3,050 m (10,000 ft) — 748 nm (1,386 km; 861 miles)
  - 55% power at 3,660 m (12,000 ft) — 790 nm (1,464 km; 909 miles)
  - 45% power at 2,440 m (8,000 ft) — 824 nm (1,527 km; 948 miles)

BEECHCRAFT TURBO BONANZA MODEL A36TC

Beech introduced for 1979 a turbocharged version of the A36 Bonanza, following FAA certification on 7 December 1978. This addition to the Beechcraft range is generally similar to the A36, except as detailed:

Wings, Fuselage, Tail Unit and Landing Gear: As for Model A36.

Power Plant: One 223·7 kW (300 hp) Continental TSIO-520-UB turbocharged flat-six engine, driving a three-blade constant-speed metal propeller with spinner. Two fuel tanks in wing leading-edges, with total usable capacity of 280 litres (74 US gallons). Refuelling points above tanks. Oil capacity 11·5 litres (3 US gallons). Propeller anti-icing optional.

Accommodation and Systems: As for Model A36, except that air-conditioning is not available as an option.

Avionics and Equipment: As detailed for Model V35B, except that rear cabin utility shelf and exhaust gas temperature gauge are not available. A turbine inlet temperature gauge is standard.

Dimensions, external: As for Model A36, except:
- Propeller diameter — 2·03 m (6 ft 8 in)
- Propeller ground clearance — 0·24 m (9½ in)

Dimensions, internal, and Areas: As for Model A36

Weights and Loadings:
- Weight empty, standard — 1,033 kg (2,278 lb)
- Max T-O and landing weight — 1,655 kg (3,650 lb)
- Max ramp weight — 1,663 kg (3,666 lb)
- Max wing loading — 98·6 kg/m² (20·17 lb/sq ft)
- Max power loading — 7·40 kg/kW (12·17 lb/hp)

Performance (at max T-O weight):
- Never-exceed speed — 214 knots (396 km/h; 246 mph) IAS
- Max cruising speed (79% power) at 7,620 m (25,000 ft) — 199 knots (369 km/h; 229 mph)
- Max operating altitude — above 7,620 m (25,000 ft)

BEECHCRAFT DUCHESS 76

The Duchess 76, a four-seat twin-engined light aircraft, flew for the first time on 24 May 1977, and received FAA certification on 24 January 1978. A testbed version of this aircraft, designated PD 289, had been undergoing a comprehensive flight test programme since September 1974. Following certification in the Normal category for day and night VFR and IFR, the first production deliveries were made in May 1978. A total of 356 Duchesses had been delivered by 1 January 1981.

The Duchess 76 was planned for use by Beech Aero Centers, and is designed for the personal light twin, light charter and multi-engine flight trainer markets. Emphasis has been placed on good low-speed flight and single-engine handling characteristics, and counter-rotating propellers are fitted.

Factory-installed optional equipment packages are available as follows:

Weekender. Comprising sun visors; tinted windscreen and windows; landing, navigation, strobe, cabin dome, map, overhead instrument and instrument post lights; cabin boarding steps; propeller unfeathering accumulators; and acrylic enamel paint; adding 17·6 kg (38·8 lb) to basic empty weight.

Holiday. As above, plus coat hook and garment hanger in baggage compartment; instrument group comprising electric clock, 3 in horizon and directional gyros and pressure system, outside air temperature gauge, rate of climb indicator, turn co-ordinator, and two 12V 25Ah batteries; adding 32·75 kg (72·2 lb) to basic empty weight.

Professional. As above, plus true airspeed indicator; two seat headrests; heated pitot tube; and wing-mounted taxi lights; adding 35·9 kg (79·1 lb) to basic empty weight.

Eleven factory-installed avionics packages are also available optionally, and include com transceiver; nav receiver-converter, with VOR/LOC indicator; ADF; transponder; and audio panel; from Collins, Edo-Aire Mitchell and King. An extensive selection of additional avionics, including autopilot, DME, glideslope and marker beacon receiver systems, are available optionally from these manufacturers, and also from Bendix.

Type: Four-seat cabin monoplane.

Wings: Cantilever low-wing monoplane. Wing section NACA 63_2A415 with modified root section. Dihedral 6° 30′. Incidence 3° at root, 0° 38′ 28″ at tip. Wings and modified Frise-type ailerons of light alloy bonded honeycomb construction. Electrically-operated single-slotted trailing-edge flaps of light alloy.

Fuselage: Semi-monocoque structure of light alloy.

Tail Unit: Conventional cantilever T-tail structure of light alloy with swept vertical surfaces. Fixed-incidence tailplane. Trim tab in rudder and each elevator.

Landing Gear: Hydraulically-retractable tricycle type, with single wheel on each unit. Self-centering steerable nosewheel retracts forward, main units inward. Oleo-pneumatic shock-absorbers. Main wheels with tube type tyres size 17·5 × 6·00-6. Nosewheel with tube type tyre size 5·00-5. Hydraulic brakes. Parking brake.

Power Plant: Two 134·2 kW (180 hp) Avco Lycoming O-360-A1G6D counter-rotating flat-four engines, each driving a Hartzell Type HC-M2YR-2C(L)EUF/F(J)C 7666A two-blade constant-speed fully-feathering metal propeller with spinner. One fuel tank in each wing, with a combined usable capacity of 378·5 litres (100 US gallons). Refuelling point on upper surface of each wing. Oil capacity 15 litres (4 US gallons).

Accommodation: Two individual front seats, adjustable fore and aft, with reclining seatbacks. Rear bench seat for two passengers. Shoulder harness and lap belt for each seat. Door on each side of cabin. Baggage compartment (capacity 90 kg; 200 lb) with external access door on port side. Pilot's storm window. Accommodation heated and ventilated. Windscreen defroster.

Systems: 28V electrical system supplied by two 60A engine-driven alternators, 24V 15·5Ah battery. Electrically-driven hydraulic pump for landing gear retraction. Separate hydraulic system for brakes. 45,000 BTU heater.

Avionics and Equipment: Optional avionics as noted earlier. Standard equipment includes instrument panel glareshield, sensitive altimeter, control lock, map and storage pockets, emergency locator transmitter, stall warning device, armrests, carpeted floor, super soundproofing, utility shelf, tiedown rings, and towbar. Optional equipment includes Hobbs hour meter, ventilation blower, third and fourth split seat in lieu of bench seat, internal corrosion proofing, and external power socket.

Dimensions, external:
- Wing span — 11·58 m (38 ft 0 in)

Wing chord at root 1·71 m (5 ft 7½ in)
Wing chord at tip 1·36 m (4 ft 5¾ in)
Wing aspect ratio 7·973
Length overall 8·86 m (29 ft 0½ in)
Height overall 2·90 m (9 ft 6 in)
Tailplane span 3·81 m (12 ft 6 in)
Wheel track 3·20 m (10 ft 6 in)
Wheelbase 2·13 m (7 ft 0 in)
Propeller diameter 1·93 m (6 ft 4 in)
Propeller ground clearance 0·25 m (10 in)
Cabin doors (port, stbd): Height 0·91 m (3 ft 0 in)
Width 0·91 m (3 ft 0 in)
Baggage door (port): Height 0·84 m (2 ft 9 in)
Width 0·56 m (1 ft 10 in)

DIMENSIONS, INTERNAL:
Cabin: Length 2·41 m (7 ft 11 in)
Max width 1·12 m (3 ft 8 in)
Max height 1·22 m (4 ft 0 in)
Baggage compartment 0·55 m^3 (19·5 cu ft)

AREAS:
Wings, gross 16·81 m^2 (181 sq ft)
Ailerons (total) 0·99 m^2 (10·7 sq ft)
Trailing-edge flaps (total) 2·29 m^2 (24·7 sq ft)
Fin 1·70 m^2 (18·27 sq ft)
Rudder (incl tab) 0·68 m^2 (7·3 sq ft)
Tailplane 3·66 m^2 (39·4 sq ft)
Elevators (incl tabs) 1·28 m^2 (13·77 sq ft)

WEIGHTS AND LOADINGS:
Weight empty 1,116 kg (2,460 lb)
Max T-O and landing weight 1,769 kg (3,900 lb)
Max ramp weight 1,776 kg (3,916 lb)
Max zero-fuel weight 1,587 kg (3,500 lb)
Max wing loading 105·2 kg/m^2 (21·5 lb/sq ft)
Max power loading 6·59 kg/kW (10·8 lb/hp)

PERFORMANCE (at max T-O weight):
Never-exceed speed 194 knots (359 km/h; 223 mph)
Max level speed 171 knots (317 km/h; 197 mph)
Max cruising speed at 1,830 m (6,000 ft) 166 knots (308 km/h; 191 mph)
Recommended cruising speed at 3,050 m (10,000 ft) 158 knots (293 km/h; 182 mph)
Econ cruising speed at 3,660 m (12,000 ft) 151 knots (280 km/h; 174 mph)
Stalling speed, power off, flaps up 70 knots (130 km/h; 81 mph) IAS
Stalling speed, power off, flaps down 60 knots (111 km/h; 69 mph) IAS
Max rate of climb at S/L 380 m (1,248 ft)/min
Rate of climb at S/L, one engine out 72 m (235 ft)/min
Service ceiling 5,990 m (19,650 ft)
Service ceiling, one engine out 1,880 m (6,170 ft)
T-O run 310 m (1,017 ft)
T-O to 15 m (50 ft) 646 m (2,119 ft)
Landing from 15 m (50 ft) 573 m (1,881 ft)
Landing run 305 m (1,000 ft)
Range with max fuel, incl allowances for start, taxi, T-O, climb, and 45 min reserves at econ cruise power:
max cruising speed at 1,830 m (6,000 ft) 623 nm (1,155 km; 717 miles)
recommended cruising speed at 3,050 m (10,000 ft) 711 nm (1,317 km; 818 miles)
econ cruising speed at 3,660 m (12,000 ft) 780 nm (1,445 km; 898 miles)

BEECHCRAFT BARON MODEL 95-B55

US Army designation: T-42A Cochise

The original Baron Model 95-55 was a four/five-seat cabin monoplane developed from the earlier Travel Air but with more power, better all-weather capability and airframe refinements that included a swept tail-fin. It first flew in prototype form on 29 February 1960 and was licensed in the FAA Normal category in November 1960. The Baron Model 95-B55 was similarly licensed in September 1963.

The current Barons are optional four-, five- or six-seaters, with interior features as described for the Bonanza.

In February 1965 the US Army selected the Model 95-B55 as winner of its competition for a twin-engined fixed-wing instrument trainer. Beech identified the military trainer as the Model 95-B55B, and this received FAA Type Approval in the Normal and Utility categories in August 1964. The US Army ordered 65, which were delivered under the designation T-42A. During 1971 Beech delivered five more T-42As to the US Army, for service with the army of Turkey, under the Military Assistance Program. Export deliveries of the standard Model 95-B55 have included 19 for the Spanish Air Ministry and six for the Civil Air Bureau of Japan. These aircraft are used as instrument trainers.

A total of 2,442 civil and military Barons of the Model 95-55 series had been delivered by 1 January 1981.

TYPE: Four/six-seat cabin monoplane.

WINGS: Cantilever low-wing monoplane. Wing section NACA 23016.5 at root, NACA 23010·5 at tip. Dihedral 6°. Incidence 4° at root, 0° at tip. No sweepback. Each wing is a two-spar semi-monocoque box beam of conventional aluminium alloy construction. Symmetrical-section ailerons of light alloy construction, with beaded skins. Electrically-operated single-slotted light alloy trailing-edge flaps, with beaded skins. Manually-operated trim tab in port aileron. Pneumatic rubber de-icing boots optional.

FUSELAGE: Semi-monocoque aluminium alloy structure. Hat-section longerons and channel-type keels extend forward from the cabin section, making the support structure for the forward nose section and nosewheel gear an integral part of the fuselage.

TAIL UNIT: Cantilever all-metal structure. Elevators have smooth magnesium alloy skins. Manually-operated trim tab in each elevator and in rudder. Pneumatic rubber de-icing boots optional.

LANDING GEAR: Electrically-retractable tricycle type. Main units retract inward into wings, nosewheel aft. Beech oleo-pneumatic shock-absorbers in all units. Steerable nosewheel with shimmy damper. Cleveland wheels, with main-wheel tyres size 6·50-8, pressure 3·45-3·72 bars (50-54 lb/sq in). Nosewheel tyre size 5·00-5, pressure 3·31-3·59 bars (48-52 lb/sq in). Cleveland ring-disc hydraulic brakes. Parking brake.

POWER PLANT: Two 194 kW (260 hp) Continental IO-470-L flat-six engines, each driving a Hartzell two-blade constant-speed fully-feathering propeller with spinner. Optional Hartzell three-blade propellers. Manually-operated cowl flaps. Standard fuel system comprises two interconnected tanks in each wing leading-edge, with total usable capacity of 378 litres (100 US gallons). Optional interconnected fuel tanks may be added in each wing to provide a total usable capacity of 515 litres (136 US gallons). Single refuelling point in each wing for the standard or optional fuel systems. Optional fuel system includes a mechanical sight gauge in each wing leading-edge to give partial fuelling information. Oil capacity 23 litres (6 US gallons). Propeller de-icing, unfeathering accumulator and propeller synchrophaser optional.

ACCOMMODATION: Standard model has four individual seats in pairs in enclosed cabin, with door on starboard side. Single diagonal strap shoulder harness with inertia reel standard on all seats. Optional wider door for cargo. Folding airline-style fifth and sixth seats optional, complete with shoulder harness and inertia reel. Baggage compartments aft of cabin and in nose, both with external doors on starboard side and with capacity of 181 kg (400 lb) and 136 kg (300 lb) respectively. An extended rear compartment providing for an additional 54 kg (120 lb) of baggage is optional. Pilot's storm window. Openable windows adjacent to the third and fourth seats are used for ground ventilation and as emergency exits. Cabin heated and ventilated. Windscreen defrosting standard. Alcohol de-icing for port side of windscreen optional.

SYSTEMS: Cabin heated by Janitrol 50,000 BTU heater, which serves also for windscreen defrosting. Oxygen system of 1·41 m^3 (49·8 cu ft) or 1·87 m^3 (66 cu ft) capacity optional. Electrical system includes two 24V 50A engine-driven alternators and one 24V 15·5Ah battery. Two 12V 25Ah batteries optional. Hydraulic system for brakes only. Pneumatic pressure system for air-driven instruments, and optional wing and tail unit de-icing system. Oxygen system optional.

AVIONICS AND EQUIPMENT: Standard avionics comprise King KX 170B 720-channel com transceiver and 200-channel nav receiver with KI 208 VOR/LOC converter indicator, microphone, headset, cabin speaker and com and nav/GS antennae. RCA WeatherScout II weather radar, and a wide range of optional avionics by Bendix, Collins, King, Edo-Aire Mitchell and Narco, are available to customer's requirements. Standard equipment includes blind-flying instrumentation, clock, outside air temperature gauge, sensitive altimeter, turn co-ordinator, pilot's storm window, sun visors, ultraviolet-proof windscreen and windows, armrests, adjustable and retractable starboard side rudder pedals, emergency locator transmitter, heated pitot, carpeted floor, glove compartment, hatshelf, headrests for passenger seats, cabin dome light, instrument panel floodlights, map light, trim tab position indicator lights, passenger reading lights, navigation lights, position lights, dual landing lights, soundproofing, heated fuel vents, towbar, and engine winterisation kit. Optional equipment includes control wheel clock or chronograph, engine and flight hour recorders, instantaneous vertical speed indicator, exhaust gas temperature gauge, dual tachometers with synchroscope, dual controls, alternate static source, cabin fire extinguisher, super soundproofing, entrance door courtesy light, instrument post lights, internally illuminated instruments, rotating beacon, strobe lights, taxi light, wing ice lights, external power socket, and static wicks.

DIMENSIONS, EXTERNAL:
Wing span 11·53 m (37 ft 10 in)
Wing chord at root 2·13 m (7 ft 0 in)
Wing chord at tip 0·90 m (2 ft 11·6 in)
Wing aspect ratio 7·16
Length overall 8·53 m (28 ft 0 in)
Height overall 2·92 m (9 ft 7 in)
Tailplane span 4·19 m (13 ft 9 in)
Wheel track 2·92 m (9 ft 7 in)
Wheelbase 2·13 m (7 ft 0 in)
Propeller diameter: two-blade 1·98 m (6 ft 6 in)
three-blade 1·93 m (6 ft 4 in)
Passenger door: Height 0·91 m (3 ft 0 in)
Width 0·94 m (3 ft 1 in)
Height to step 0·41 m (1 ft 4 in)
Baggage door (fwd): Height 0·56 m (1 ft 10 in)
Width 0·64 m (2 ft 1 in)
Baggage door (rear):
Standard: Height 0·57 m (1 ft 10½ in)
Width 0·47 m (1 ft 6½ in)
Height to sill 0·71 m (2 ft 4 in)
Optional: Height 0·57 m (1 ft 10½ in)
Width 0·97 m (3 ft 2 in)

DIMENSIONS, INTERNAL:
Cabin: Length 3·07 m (10 ft 1 in)
Max width 1·07 m (3 ft 6 in)
Max height 1·27 m (4 ft 2 in)
Baggage compartment (fwd) 0·34 m^3 (12 cu ft)
Baggage compartment (rear) 0·99 m^3 (35 cu ft)

AREAS:
Wings, gross 18·5 m^2 (199·2 sq ft)
Ailerons (total) 1·06 m^2 (11·40 sq ft)
Trailing-edge flaps (total) 2·39 m^2 (25·70 sq ft)
Fin 1·02 m^2 (11·00 sq ft)
Rudder, incl tab 1·08 m^2 (11·60 sq ft)
Tailplane 4·46 m^2 (48·06 sq ft)
Elevators, incl tabs 1·51 m^2 (16·20 sq ft)

WEIGHTS AND LOADINGS:
Weight empty 1,468 kg (3,236 lb)
Max T-O and landing weight 2,313 kg (5,100 lb)
Max ramp weight 2,322 kg (5,121 lb)
Max wing loading 120·5 kg/m^2 (25·6 lb/sq ft)
Max power loading 5·96 kg/kW (9·8 lb/hp)

PERFORMANCE (at max T-O weight, except cruising speeds at average cruise weight):
Max level speed at S/L 201 knots (372 km/h; 231 mph)
Max cruising speed, 77% power at 1,830 m (6,000 ft) 188 knots (348 km/h; 216 mph)
Cruising speed, 66% power at 3,050 m (10,000 ft) 184 knots (341 km/h; 212 mph)
Econ cruising speed, 56% power at 3,660 m (12,000 ft) 173 knots (320 km/h; 199 mph)
Stalling speed, flaps up, power off 79 knots (146 km/h; 91 mph) IAS
Stalling speed, flaps down, power off 73 knots (135 km/h; 84 mph) IAS
Max rate of climb at S/L 516 m (1,693 ft)/min
Rate of climb at S/L, one engine out 121 m (397 ft)/min
Service ceiling 5,880 m (19,300 ft)
Service ceiling, one engine out 1,950 m (6,400 ft)
Min ground turning radius 9·00 m (29 ft 6 in)
Runway LCN 2
T-O run 427 m (1,400 ft)
T-O to 15 m (50 ft) 657 m (2,154 ft)
Landing from 15 m (50 ft) 655 m (2,148 ft)
Landing run 447 m (1,467 ft)

Beechcraft Baron Model E55 (foreground) and B55 four/six-seat cabin monoplanes

Range with 515 litres (136 US gallons) usable fuel, with allowances for engine start, taxi, T-O, climb and 45 min reserves at econ cruise power:
max cruising speed at 1,830 m (6,000 ft) 798 nm (1,479 km; 918 miles)
cruising speed at 3,050 m (10,000 ft) 907 nm (1,680 km; 1,044 miles)
econ cruising speed at 3,660 m (12,000 ft) 991 nm (1,836 km; 1,141 miles)

BEECHCRAFT BARON MODEL E55

The Baron E55 had its origin in the Baron 95-C55 when that model was added to the Baron series of twin-engined aircraft in August 1965. Compared to the B55, the 95-C55 had Continental IO-520-C engines, a pneumatic vacuum system for instrument gyros and the optional wing and tail unit de-icing system, two 24V 50A engine-driven alternators, increased tailplane span, swept vertical surfaces and an extended nose baggage compartment. It was followed by the D55 in October 1967, this model introducing a pneumatic pressure system in place of the pneumatic vacuum system. The subsequent Model E55, which has an improved interior and systems accessory refinements, was licensed in the FAA Normal category on 12 November 1969.

Beech had delivered a total of 1,178 of this Baron series by 1 January 1981.

Type: Four/six-seat cabin monoplane.

Wings: As for Model 95-B55.

Fuselage: As for Model 95-B55, except nose extended by 0·305 m (1 ft 0 in).

Tail Unit: As for Model 95-B55, except tailplane span increased.

Landing Gear: As for Model 95-B55, except main-wheel tyre pressure 3·59-3·86 bars (52-56 lb/sq in); nosewheel tyre pressure 3·79-4·14 bars (55-60 lb/sq in).

Power Plant: Two 212·5 kW (285 hp) Continental IO-520-CB flat-six engines, each driving a Hartzell two-blade constant-speed fully-feathering metal propeller with spinner. Hartzell three-blade propellers optional. Fuel system as for Model 95-B55, except optional total usable capacity 628 litres (166 US gallons). Oil capacity 23 litres (6 US gallons). Full-flow oil filters standard; propeller de-icing optional.

Accommodation: As for Model 95-B55, except that extended rear compartment, providing for an additional 54 kg (120 lb) of baggage, is standard.

Systems: As for Model 95-B55, except standard electrical system supplied by two 24V 60A alternators with alternator failure lights and one 24V 15·5Ah battery. Two 100A alternators and two 12V 25Ah batteries optional. Cabin air-conditioning system optional.

Avionics and Equipment: Standard and optional avionics as for Model 95-B55, except that a King KR 87 ADF with KI 227 indicator is standard. Standard and optional equipment as for Model 95-B55, except cabin door courtesy light and extended rear baggage compartment standard. Electric elevator trim optional.

Dimensions, external: As for Model 95-B55, except:
Length overall 8·84 m (29 ft 0 in)
Height overall 2·79 m (9 ft 2 in)
Tailplane span 4·85 m (15 ft 11 in)
Wheelbase 2·46 m (8 ft 1 in)

Dimensions, internal: As for Model 95-B55, except:
Cabin: Length (incl extended rear baggage compartment) 3·58 m (11 ft 9 in)
Baggage compartment (fwd) 0·51 m^3 (18 cu ft)
Baggage compartment (rear) 0·99 m^3 (35 cu ft)
Extension to rear baggage compartment 0·28 m^3 (10 cu ft)

Areas: As for Model 95-B55, except:
Tailplane 4·95 m^2 (53·30 sq ft)
Elevators, incl tabs 1·84 m^2 (19·80 sq ft)

Weights and Loadings:
Weight empty 1,483 kg (3,269 lb)
Max T-O and landing weight 2,405 kg (5,300 lb)
Max ramp weight 2,415 kg (5,324 lb)
Max wing loading 130·0 kg/m^2 (26·6 lb/sq ft)
Max power loading 5·66 kg/kW (9·3 lb/hp)

Performance (at max T-O weight, except cruising speeds at average cruise weight):
Max level speed at S/L 208 knots (386 km/h; 239 mph)
Max cruising speed, 77% power at 1,830 m (6,000 ft) 200 knots (370 km/h; 230 mph)
Cruising speed, 66% power at 3,050 m (10,000 ft) 195 knots (362 km/h; 224 mph)
Econ cruising speed, 56% power at 3,660 m (12,000 ft) 184 knots (341 km/h; 212 mph)
Stalling speed, flaps up, power off 83 knots (154 km/h; 96 mph) IAS
Stalling speed, flaps down, power off 73 knots (135 km/h; 84 mph) IAS
Max rate of climb at S/L 513 m (1,682 ft)/min
Rate of climb at S/L, one engine out 118 m (388 ft)/min
Service ceiling 5,820 m (19,100 ft)
Service ceiling, one engine out 2,010 m (6,600 ft)
T-O run 401 m (1,315 ft)
T-O to 15 m (50 ft) 625 m (2,050 ft)
Landing from 15 m (50 ft) 671 m (2,202 ft)

Beechcraft Baron Model 58 four/six-seat cabin monoplane

Landing run 377 m (1,237 ft)
Range with 628 litres (166 US gallons) usable fuel, with allowances for engine start, taxi, T-O, climb and 45 min reserves at econ cruise power:
max cruising speed at 1,830 m (6,000 ft) 933 nm (1,728 km; 1,074 miles)
cruising speed at 3,050 m (10,000 ft) 1,032 nm (1,912 km; 1,188 miles)
econ cruising speed at 3,660 m (12,000 ft) 1,135 nm (2,103 km; 1,306 miles)

BEECHCRAFT BARON MODEL 58

In late 1969 Beech introduced a new version of the Baron, designated Model 58. Developed from the Baron D55, it differed by having the forward cabin section extended by 0·254 m (10 in), allowing the windscreen, passenger door, instrument panel and front seats to be moved forward and so provide a more spacious cabin. This change was made without affecting the wing main spar location, but the wheelbase was extended by moving the nosewheel forward, to improve ground handling. New features included double passenger/cargo doors on the starboard side of the cabin, extended propeller hubs, redesigned engine nacelles to improve cooling, and a fourth window on each side of the cabin. The Model 58 Baron was licensed by the FAA in the Normal category on 19 November 1969.

Beech had delivered 1,213 of this Baron series (including Baron 58Ps and 58TCs) by 1 January 1981.

Type: Four/six-seat cabin monoplane.

Wings: As for Model 95-B55.

Fuselage: As for Model E55, except forward cabin section extended by 0·254 m (10 in).

Tail Unit: As for Model E55.

Landing Gear: As for Model E55, except wheelbase extended by 0·254 m (10 in).

Power Plant: As for Model E55, except that the standard and optional Hartzell propellers have extended hubs, and the engine nacelles are lengthened to accommodate these. The standard fuel system has a usable capacity of 514 litres (136 US gallons), with optional usable capacity of 628 litres (166 US gallons). Optional 'wet wingtip' installation also available, increasing usable capacity to 734 litres (194 US gallons).

Accommodation: As for Model E55, except that folding fifth and sixth seats, or club seating comprising folding fifth and sixth seats and aft-facing third and fourth seats, are optional. Executive writing desk available as option with club seating. Double passenger/cargo doors on starboard side of cabin provide access to space for 181 kg (400 lb) of baggage or cargo behind the third and fourth seats.

Systems: As for Model E55.

Avionics and Equipment: Standard and optional avionics as for Model E55. Equipment as for Model E55, except double passenger/cargo doors with door-ajar warning light standard. Optional equipment includes club seating arrangement as described above, executive writing desk, and Hartzell three-blade propellers.

Dimensions, external, and Areas:
As for Model E55, except:
Length overall 9·09 m (29 ft 10 in)
Height overall 2·90 m (9 ft 6 in)
Wheelbase 2·72 m (8 ft 11 in)
Rear passenger/cargo doors:
Max height 1·02 m (3 ft 4 in)
Width 1·14 m (3 ft 9 in)

Dimensions, internal: As for Model E55, except:
Cabin, incl rear baggage area:
Length 3·84 m (12 ft 7 in)
Floor area 3·72 m^2 (40 sq ft)
Volume 3·85 m^3 (135·9 cu ft)

Weights and Loadings:
Weight empty 1,524 kg (3,361 lb)
Max T-O and landing weight 2,449 kg (5,400 lb)
Max ramp weight 2,460 kg (5,424 lb)
Max wing loading 132·3 kg/m^2 (27·1 lb/sq ft)
Max power loading 5·76 kg/kW (9·5 lb/hp)

Performance (at max T-O weight, except cruising speeds at average cruise weight):
Max level speed at S/L 208 knots (386 km/h; 239 mph)
Max cruising speed, 77% power at 1,830 m (6,000 ft) 200 knots (370 km/h; 230 mph)
Cruising speed, 66% power at 3,050 m (10,000 ft) 195 knots (362 km/h; 224 mph)
Econ cruising speed, 56% power at 3,660 m (12,000 ft) 184 knots (341 km/h; 212 mph)
Stalling speed, flaps up, power off 84 knots (156 km/h; 97 mph) IAS
Stalling speed, flaps down, power off 74 knots (137 km/h; 85 mph) IAS
Max rate of climb at S/L 506 m (1,660 ft)/min
Rate of climb at S/L, one engine out 119 m (390 ft)/min
Service ceiling 5,670 m (18,600 ft)
Service ceiling, one engine out 2,135 m (7,000 ft)
T-O run 407 m (1,336 ft)
T-O to 15 m (50 ft) 640 m (2,101 ft)
Landing from 15 m (50 ft) 761 m (2,498 ft)
Landing run 439 m (1,439 ft)
Range with 734 litres (194 US gallons) usable fuel, with allowances for engine start, taxi, T-O, climb and 45 min reserves at econ cruise power:
max cruising speed at 1,830 m (6,000 ft) 1,108 nm (2,054 km; 1,275 miles)
cruising speed (66% power) at 3,050 m (10,000 ft) 1,224 nm (2,268 km; 1,409 miles)
econ cruising speed at 3,660 m (12,000 ft) 1,339 nm (2,482 km; 1,541 miles)

BEECHCRAFT BARON MODEL 58P

Design of this pressurised version of the Model 58 Baron started in June 1972; the first flight of the prototype was made in August 1973. Certification under FAR Part 23 was received in May 1974; the first production aircraft flew later the same year. Examples of the Model 58P produced prior to 1979 are powered by two 231 kW (310 hp) Continental TSIO-520-L (or -LB) engines. Current production aircraft have more powerful TSIO-520-WB engines, and introduced propeller synchrophasers as standard equipment in 1981.

Deliveries of production aircraft began in late 1975, and a total of 328 Baron 58Ps had been delivered by 1 January 1981. Twelve have been delivered to the US Forest Service for use as lead aircraft in smoke-jumping operations, as well as for reconnaissance, administration and cargo missions.

Type: Four/six-seat cabin monoplane.

Wings: As for Model 95-B55.

Fuselage: As for Model 58, except structural reinforcement to cater for pressurisation.

Tail Unit: As for Model 58.

Landing Gear: Main gear as for Duke, except main-wheel tyre pressure 5·24 to 5·66 bars (76 to 82 lb/sq in). Nosewheel unit as for Model 58. Goodrich single-disc hydraulic brakes. Parking brake.

Power Plant: Two 242 kW (325 hp) Continental TSIO-520-WB turbocharged flat-six engines, each driving a Hartzell three-blade constant-speed fully-feathering metal propeller with spinner. Propeller synchrophasers standard, unfeathering accumulators optional. Electrically-operated engine cowl flaps. Integral fuel tanks in wings, with standard capacity of 651 litres (172 US gallons) of which 628 litres (166 US gallons) are usable. Optional maximum capacity of 742 litres (196 US gallons) of which 719 litres (190 US gallons) are usable. Refuelling points in outboard leading-edge of wings and, for optional maximum fuel, in wingtips. Oil capacity 22·7 litres (6 US gallons). Electrical anti-icing for propellers optional.

ACCOMMODATION: Standard accommodation has four individual seats in pairs, facing forward, with shoulder harness and inertia reel. Fifth and sixth seats optional, as is club layout. Doors on starboard side, adjacent to co-pilot, and at trailing-edge of wing on port side. Baggage space in aft cabin and in fuselage nose, with door on starboard side of nose. Openable storm window for pilot on port side. Cabin heated and pressurised. Air-conditioning optional. Windscreen defrosting by hot air. Windscreen electrical anti-icing optional.

SYSTEMS: Garrett pressurisation system with max differential of 0·26 bars (3·7 lb/sq in) in aircraft manufactured prior to 1979, providing a 3,050 m (10,000 ft) cabin environment to an altitude of 6,460 m (21,200 ft). Aircraft manufactured subsequently have Garrett pressurisation system with max differential of 0·27 bars (3·9 lb/sq in), giving a 3,050 m (10,000 ft) cabin environment to a height of 6,705 m (22,000 ft). Beechcraft 14,000 BTU air-conditioning optional. Janitrol 35,000 BTU heater. Engine-driven compressors supply air for flight instruments, pressurisation control and optional pneumatic de-icing boots. Electrical system powered by two 24V 60A alternators, with two 12V 25Ah storage batteries. Two 24V 100A alternators optional. Hydraulic system for brakes only. Oxygen system of 0·42 m^3 (15 cu ft) optional.

AVIONICS AND EQUIPMENT: Standard avionics package comprises King KX 170B nav/com (720-channel com transceiver and 200-channel nav receiver) with KI 208 VOR/LOC converter-indicator, KR 87 ADF with KI 227 indicator, King combined loop/sense antenna, microphone, headset, cabin speaker and B38 nav and B6-1 com antennae. Optional avionics by Bendix, Collins, King, Narco, Edo-Aire Mitchell, Sperry and RCA. Standard equipment as for Model E55, plus emergency locator transmitter, heated stall warning transmitter, nose baggage compartment light, door-ajar warning light, step light, dual rotating beacons, and exterior urethane paint. Optional equipment includes engine and flight hour recorders, control wheel chronometer, dual controls, electrically-operated elevator trim, cabin fire extinguisher, executive writing desk, instrument post lights, internally illuminated instruments, strobe lights, wing ice lights, and static wicks.

DIMENSIONS, EXTERNAL:

Wing span	11·53 m (37 ft 10 in)
Wing chord at root	2·13 m (7 ft 0 in)
Wing chord at tip	0·90 m (2 ft 11½ in)
Length overall	9·12 m (29 ft 11 in)
Height overall	2·79 m (9 ft 2 in)
Tailplane span	4·85 m (15 ft 11 in)
Wheel track	2·92 m (9 ft 7 in)
Wheelbase	2·72 m (8 ft 11 in)
Propeller diameter	1·98 m (6 ft 6 in)
Propeller ground clearance	0·28 m (10¾ in)
Passenger door (stbd, fwd):	
Height	0·91 m (3 ft 0 in)
Width	0·94 m (3 ft 1 in)
Height to sill	0·51 m (1 ft 8 in)
Passenger door (port, aft):	
Height	0·89 m (2 ft 11 in)
Width	0·58 m (1 ft 11 in)
Height to sill	0·79 m (2 ft 7 in)
Baggage door (nose, stbd):	
Height	0·38 m (1 ft 3 in)
Width	0·64 m (2 ft 1 in)

DIMENSIONS, INTERNAL: As for Model 58

AREAS: As for Model 58

WEIGHTS AND LOADINGS:

Weight empty, equipped	1,822 kg (4,018 lb)
Max T-O and landing weight	2,812 kg (6,200 lb)
Max ramp weight	2,830 kg (6,240 lb)
Max zero-fuel weight	2,585 kg (5,700 lb)
Max wing loading	161·1 kg/m^2 (33 lb/sq ft)
Max power loading	5·81 kg/kW (9·54 lb/hp)

PERFORMANCE (at max T-O weight, except cruising speeds at average cruise weight):

Max level speed	261 knots (483 km/h; 300 mph)
Max cruising speed at approx 77% power:	
at 4,570 m (15,000 ft)	222 knots (412 km/h; 256 mph)
at 6,100 m (20,000 ft)	232 knots (430 km/h; 267 mph)
at 7,620 m (25,000 ft)	241 knots (447 km/h; 277 mph)
Cruising speed at approx 75% power:	
at 4,570 m (15,000 ft)	220 knots (407 km/h; 253 mph)
at 6,100 m (20,000 ft)	229 knots (425 km/h; 264 mph)
at 7,620 m (25,000 ft)	237 knots (439 km/h; 273 mph)
Cruising speed at approx 65% power:	
at 4,570 m (15,000 ft)	210 knots (389 km/h; 242 mph)
at 6,100 m (20,000 ft)	215 knots (399 km/h; 248 mph)
at 7,620 m (25,000 ft)	222 knots (412 km/h; 256 mph)
Econ cruising speed at approx 53% power:	
at 4,570 m (15,000 ft)	186 knots (344 km/h; 214 mph)
at 6,100 m (20,000 ft)	194 knots (359 km/h; 223 mph)
at 7,620 m (25,000 ft)	202 knots (375 km/h; 233 mph)
Stalling speed, flaps up, power off	84 knots (156 km/h; 97 mph)
Stalling speed, flaps down, power off	78 knots (145 km/h; 90 mph)
Max rate of climb at S/L	450 m (1,475 ft)/min
Rate of climb at S/L, one engine out	82 m (270 ft)/min
Service ceiling	above 7,620 m (25,000 ft)
Service ceiling, one engine out	4,110 m (13,490 ft)
T-O run	474 m (1,555 ft)
T-O to 15 m (50 ft)	806 m (2,643 ft)
Landing from 15 m (50 ft)	740 m (2,427 ft)
Landing run	420 m (1,378 ft)

Range with 719 litres (190 US gallons) usable fuel, and allowances for engine start, taxi, T-O, climb and 45 min reserves at econ cruising speed:

at approx 77% power:	
at 4,570 m (15,000 ft)	916 nm (1,697 km; 1,054 miles)
at 6,100 m (20,000 ft)	958 nm (1,775 km; 1,102 miles)
at 7,620 m (25,000 ft)	1,008 nm (1,868 km; 1,160 miles)
at approx 75% power:	
at 4,570 m (15,000 ft)	930 nm (1,723 km; 1,071 miles)
at 6,100 m (20,000 ft)	971 nm (1,799 km; 1,118 miles)
at 7,620 m (25,000 ft)	1,019 nm (1,888 km; 1,173 miles)
at approx 65% power:	
at 4,570 m (15,000 ft)	1,027 nm (1,904 km; 1,183 miles)
at 6,100 m (20,000 ft)	1,062 nm (1,968 km; 1,223 miles)
at 7,620 m (25,000 ft)	1,093 nm (2,026 km; 1,259 miles)
at approx 53% power:	
at 4,570 m (15,000 ft)	1,198 nm (2,220 km; 1,379 miles)
at 6,100 m (20,000 ft)	1,217 nm (2,255 km; 1,400 miles)
at 7,620 m (25,000 ft)	1,229 nm (2,277 km; 1,414 miles)

BEECHCRAFT BARON MODEL 58TC

This turbocharged version of the Baron Model 58 is generally similar to the Model 58P, sharing the same power plant, but is unpressurised and has detail differences in the airframe and equipment. The design originated in July 1974, and construction of a prototype to production standard began in February 1975. The first flight of this aircraft was made on 31 October 1975 and FAA certification in the Normal category was granted on 23 January 1976. Deliveries began in June 1976, and a total of 131 had been delivered by 1 January 1981.

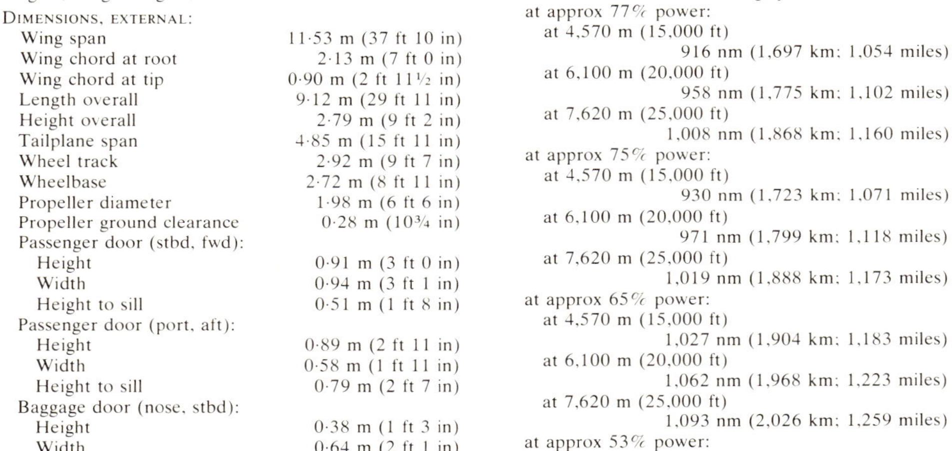

Beechcraft Baron Model 58TC, turbocharged version of the Baron 58

TYPE: Four/six-seat cabin monoplane.

WINGS, TAIL UNIT, LANDING GEAR, POWER PLANT: Generally similar to Model 58P.

FUSELAGE: Generally similar to Model 58.

ACCOMMODATION: Generally similar to Model 58, except windscreen electrical anti-icing optional. Baggage/cargo space aft of 3rd and 4th seats for 272 kg (600 lb). By removing these seats 408 kg (900 lb) baggage/cargo can be stowed aft of pilot and co-pilot seats with restraining net.

SYSTEMS: Generally similar to Model 58P, except Beech freon air-conditioning system optional. 50,000 BTU heater standard. Hydraulic system for brakes and propeller unfeathering only. Electrical power from two 24V 60A alternators, with 24V 15·5Ah storage battery. Two 100A alternators and two 12V 25Ah storage batteries optional. Oxygen system to supply crew and passengers.

AVIONICS AND EQUIPMENT: As for Baron 58P.

DIMENSIONS, EXTERNAL: As for Model 58P except:

Utility double door (stbd, aft):	
Height	1·02 m (3 ft 4 in)
Width	1·14 m (3 ft 9 in)
Baggage door (nose, stbd):	
Height	0·56 m (1 ft 10 in)
Width	0·64 m (2 ft 1 in)
Emergency exit window (port and stbd):	
Height	0·53 m (1 ft 9 in)
Width	0·61 m (2 ft 0 in)

DIMENSIONS, INTERNAL: As for Model 58

AREAS:

Wings, gross	17·47 m^2 (188·1 sq ft)
Ailerons (total, incl tabs)	1·08 m^2 (11·58 sq ft)
Trailing-edge flaps (total)	1·98 m^2 (21·3 sq ft)
Fin	1·46 m^2 (15·67 sq ft)
Rudder, incl tab	1·08 m^2 (11·6 sq ft)
Tailplane	5·11 m^2 (55·05 sq ft)
Elevators, incl tabs	1·84 m^2 (19·8 sq ft)

WEIGHTS AND LOADINGS:

Weight empty, equipped	1,720 kg (3,793 lb)
Max T-O and landing weight	2,812 kg (6,200 lb)
Max ramp weight	2,830 kg (6,240 lb)
Max zero-fuel weight	2,585 kg (5,700 lb)
Max wing loading	160·96 kg/m^2 (32·96 lb/sq ft)
Max power loading	5·81 kg/kW (9·54 lb/hp)

PERFORMANCE (at max T-O weight, except cruising speeds at average cruise weight):

As for Baron Model 58P except:

Max rate of climb:	
at 4,575 m (15,000 ft)	366 m (1,201 ft)/min
Rate of climb, one engine out:	
at 1,525 m (5,000 ft)	67 m (220 ft)/min

BEECHCRAFT DUKE B60

Design work on the original version of this 4/6-seat pressurised and turbocharged light twin-engined transport started in early 1965. Construction of the prototype began in January 1966, and the first flight was made on 29 December 1966. FAA Type Approval was granted on 1 February 1968.

The current version of the Duke B60 has a Garrett Lexan pressurisation system, with a mini controller that allows selection of cabin altitude prior to take-off or landing. This system can also change the aircraft cabin altitude at any desired rate from 15-610 m (50-2,000 ft)/min.

A total of 553 Dukes had been produced by 1 January 1981.

TYPE: Four/six-seat cabin monoplane.

WINGS: Cantilever low-wing monoplane. Wing section NACA 23016·5 at root, NACA 23010·5 at tip. Thickness/chord ratio 13·7% at root, 10·5% at tip. Dihedral 6°. Incidence 4° at root, 0° at tip. Each wing is a two-spar semi-monocoque box beam of conventional aluminium alloy construction. Overhang-balance ailerons constructed of aluminium alloy. Conventional hinged trim tab in port aileron. Electrically-operated single-slotted aluminium alloy flaps. Pneumatic rubber de-icing boots optional.

FUSELAGE: Semi-monocoque aluminium alloy structure. Heavy-gauge chemically-milled aluminium alloy skins.

TAIL UNIT: Cantilever all-metal structure. Aluminium spars and end ribs; magnesium alloy skins reinforced with metal bonded honeycomb stiffeners running chordwise. Dorsal fin. Swept vertical and horizontal surfaces. Tailplane dihedral 10°. Trim tabs in rudder and port elevator. Pneumatic rubber de-icing boots optional.

LANDING GEAR: Electrically-retractable tricycle type. Main units retract inward, nosewheel aft; all three units have fairing doors. Beechcraft oleo-pneumatic shock-absorbers. Goodrich main wheels and tyres size 19·50 × 6·75-8 10-ply rating, pressure 5·52 bars (80 lb/sq in). Goodyear steerable nosewheel with shimmy damper, tyre size 15 × 6·00-6, pressure 3·45 bars (50 lb/sq in). Goodrich single-disc hydraulic brakes. Parking brake.

Power Plant: Two 283 kW (380 hp) Avco Lycoming TIO-541-E1C4 turbocharged flat-six engines, each driving a Hartzell three-blade constant-speed fully-feathering metal propeller with spinner. Propeller synchrophaser, unfeathering accumulators, and electrical de-icing, optional. Electrically-operated engine cowl flaps. Two interconnected fuel cells in each wing containing 269 litres (71 US gallons); total usable capacity 538 litres (142 US gallons). Optionally, four interconnected fuel cells in each wing containing 382 litres (101 US gallons); total usable capacity 764 litres (202 US gallons); or five interconnected fuel cells in each wing containing 439 litres (116 US gallons), with total usable capacity of 878 litres (232 US gallons). Refuelling points in each leading-edge, near wingtip. Oil capacity 24·5 litres (6·5 US gallons).

Accommodation: Standard model has four individual seats in pairs, each complete with headrest and inertia-reel shoulder harness, in enclosed cabin with centre aisle. Door, hinged at forward edge, on port side at rear of cabin. Baggage hold in the nose, capacity 0·91 m^3 (32 cu ft), with external access door on port side of nose. Additional 0·80 m^3 (28·25 cu ft) of baggage space at rear of cabin. Optional extras include fifth and sixth seats, rearward-facing third and fourth seats, curtain separating passenger and pilot seating, writing desks, refreshment cabinets, toilet, windscreen electrical anti-icing and cabin fire extinguishers.

Systems: Cabin pressurisation system, differential 0·32 bars (4·6 lb/sq in), supplied by engine turbocharger bleed air, maintains cabin altitude equivalent to 3,050 m (10,000 ft) at 7,560 m (24,800 ft). Combustion heater of 45,000 BTU standard. Ventilation system has provisions for optional engine-driven vapour-cycle air-conditioning system of 14,000 BTU. Automatic altitude controller for cabin pressurisation system standard. Oxygen system optional, with 0·31 m^3 (11 cu ft), 0·62 m^3 (22 cu ft) or 1·39 m^3 (49 cu ft) bottle. Hydraulic system for brakes only. Pneumatic system for pressure-operated instruments and de-icing boots only. 24V 125A generators standard; two 12V 25Ah batteries.

Avionics and Equipment: Standard avionics comprise Collins VHF-251 VHF transceiver with PWC-150 power adaptor and B3 com antenna, Collins VIR-351 Omni receiver with IND-351A VOR/ILS indicator and B38 antenna, Collins AMR-350 audio panel, Collins marker beacon receiver incorporated in AMR-350, with single set marker lights and B16 antenna, Collins ADF-650A ADF with IND-650A indicator and ANT-650A antenna, Collins GLS-350 glideslope receiver with A-326A antenna, Collins TDR-950 transponder with B18 antenna, Beech metal radio panel, radio accessories, static wicks, microphone key button in pilot's control wheel, white lighting, dual microphones and headsets, single cockpit speaker, and avionics master switch. Optional avionics include an extensive range of Bendix, Collins, King, Edo-Aire Mitchell, Sperry and RCA equipment. Standard equipment includes blind-flying instrumentation, LCD digital clock/chronometer, outside air temperature gauge, turn co-ordinator, pilot's storm window, sun visors, alternate static source, heated stall warning device, heated pitot, armrests, headrests, in-flight storage pockets, carpeted floor, tinted cabin side windows, baggage straps, super soundproofing, cabin dome light, entrance light, instrument floodlights, instrument post lights, map lights, nose baggage compartment light, reading lights, dual landing lights, navigation lights, rotating beacons, taxi light, retracting step, heated fuel vents, full-flow oil filters, external power socket, towbar, and exterior urethane paint. Optional equipment includes de luxe instrument panel with duplicated blind-flying instrumentation for co-pilot, instantaneous vertical speed indicator, tachometer with synchroscope, flight and engine hour recorders, pilot's control wheel eight-day clock, pilot's relief tube, co-pilot's toe brakes, co-pilot's control wheel map light, internally lighted instruments, wing ice light, and strobe lights.

Dimensions, external:

Wing span	11·97 m (39 ft 3¼ in)
Wing chord at fuselage c/l	2·80 m (9 ft 2⅛ in)
Wing chord at tip	0·90 m (2 ft 11⅝ in)
Wing aspect ratio	7·243
Length overall	10·31 m (33 ft 10 in)
Height overall	3·76 m (12 ft 4 in)
Tailplane span	5·18 m (17 ft 0 in)
Wheel track	3·43 m (11 ft 3 in)
Wheelbase	2·82 m (9 ft 2⅜ in)
Propeller diameter	1·88 m (6 ft 2 in)
Passenger door: Height	1·21 m (3 ft 11½ in)
Width	0·67 m (2 ft 2½ in)
Height to sill	0·81 m (2 ft 8 in)
Baggage compartment door:	
Height	0·60 m (1 ft 11½ in)
Width	0·95 m (3 ft 1½ in)
Height to sill	0·95 m (3 ft 1½ in)

Dimensions, internal:

Cabin: Length	3·61 m (11 ft 10 in)
Max width	1·27 m (4 ft 2 in)
Max height	1·32 m (4 ft 4 in)
Floor area	3·36 m^2 (36·2 sq ft)
Volume	4·80 m^3 (169·6 cu ft)

Beechcraft Duke B60 four/six-seat pressurised transport

Areas:

Wings, gross	19·78 m^2 (212·9 sq ft)
Ailerons (total)	1·06 m^2 (11·4 sq ft)
Trailing-edge flaps (total)	2·76 m^2 (29·7 sq ft)
Fin	1·52 m^2 (16·38 sq ft)
Rudder, incl tab	1·15 m^2 (12·4 sq ft)
Tailplane	4·24 m^2 (45·6 sq ft)
Elevators, incl tab	1·52 m^2 (16·4 sq ft)

Weights and Loadings:

Weight empty, equipped	2,006 kg (4,423 lb)
Max T-O and landing weight	3,073 kg (6,775 lb)
Max ramp weight	3,093 kg (6,819 lb)
Max wing loading	155·3 kg/m^2 (31·8 lb/sq ft)
Max power loading	5·43 kg/kW (8·9 lb/hp)

Performance (at max T-O weight, except cruising speeds at average cruise weight):

Max level speed at 7,010 m (23,000 ft)
246 knots (455 km/h; 283 mph)

Max cruising speed:
approx 78% power at 7,620 m (25,000 ft)
239 knots (443 km/h; 275 mph)
approx 78% power at 6,100 m (20,000 ft)
231 knots (428 km/h; 266 mph)
approx 78% power at 4,570 m (15,000 ft)
220 knots (407 km/h; 253 mph)

Cruising speed:
approx 74% power at 7,620 m (25,000 ft)
233 knots (431 km/h; 268 mph)
approx 74% power at 6,100 m (20,000 ft)
225 knots (417 km/h; 259 mph)
approx 74% power at 4,570 m (15,000 ft)
214 knots (396 km/h; 246 mph)
approx 68% power at 7,620 m (25,000 ft)
225 knots (417 km/h; 259 mph)
approx 68% power at 6,100 m (20,000 ft)
215 knots (399 km/h; 247 mph)
approx 68% power at 4,570 m (15,000 ft)
205 knots (380 km/h; 236 mph)
approx 63% power at 7,620 m (25,000 ft)
217 knots (402 km/h; 250 mph)
approx 63% power at 6,100 m (20,000 ft)
209 knots (388 km/h; 241 mph)
approx 63% power at 4,570 m (15,000 ft)
199 knots (369 km/h; 229 mph)

Stalling speed, wheels and flaps up, power off
81 knots (150·5 km/h; 93 mph)

Stalling speed, wheels and flaps down, power off
73 knots (135 km/h; 84 mph)

Max rate of climb at S/L	488 m (1,601 ft)/min
Rate of climb at S/L, one engine out	94 m (307 ft)/min
Service ceiling	9,145 m (30,000 ft)
Service ceiling, one engine out	4,600 m (15,100 ft)
Runway LCN	4
T-O run	632 m (2,075 ft)
T-O to 15 m (50 ft)	800 m (2,626 ft)
Landing from 15 m (50 ft)	934 m (3,065 ft)
Landing run	402 m (1,318 ft)

Range with max optional fuel and allowances for engine start, taxi, take-off, climb to altitude and 45 min fuel reserves at 45% power, ISA:
approx 78% power at 7,620 m (25,000 ft)
1,045 nm (1,936 km; 1,203 miles)
approx 78% power at 6,100 m (20,000 ft)
967 nm (1,793 km; 1,113 miles)
approx 78% power at 4,570 m (15,000 ft)
916 nm (1,698 km; 1,054 miles)
approx 74% power at 7,620 m (25,000 ft)
1,072 nm (1,986 km; 1,234 miles)
approx 74% power at 6,100 m (20,000 ft)
1,010 nm (1,872 km; 1,163 miles)
approx 74% power at 4,570 m (15,000 ft)
964 nm (1,786 km; 1,110 miles)
approx 68% power at 7,620 m (25,000 ft)
1,112 nm (2,060 km; 1,280 miles)
approx 68% power at 6,100 m (20,000 ft)
1,070 nm (1,983 km; 1,232 miles)
approx 68% power at 4,570 m (15,000 ft)
1,028 nm (1,905 km; 1,183 miles)
approx 63% power at 7,620 m (25,000 ft)
1,168 nm (2,165 km; 1,344 miles)
approx 63% power at 6,100 m (20,000 ft)
1,122 nm (2,079 km; 1,291 miles)
approx 63% power at 4,570 m (15,000 ft)
1,083 nm (2,007 km; 1,247 miles)

BEECHCRAFT KING AIR MODEL C90

USAF designation: VC-6B

Introduced in September 1970, the King Air C90 is a pressurised 6/10-seat twin-turboprop business aircraft which superseded the original Models 90, A90 and B90 King Air. It is powered by Pratt & Whitney Aircraft of Canada PT6A-21 turboprop engines, which provide improved performance over a wide range of altitudes and temperatures. Increases in take-off and climb power offer improvements in high altitude and hot weather operation and, since these engines also run cooler, increases in useful life and lower overhaul costs result.

The C90 King Air utilises the more advanced cabin pressurisation and heating system of the King Air 100. This comprises a dual engine bleed air system for cabin pressurisation, with a max differential of 0·32 bars (4·6 lb/sq in). It also has as standard supplementary electric heating, air-conditioning, super soundproofing, a full anti-icing system, four cabin seats in club arrangement, forward cabin partition with curtain, aft starboard cabin partition with curtain, polarised cabin windows, exterior urethane paint scheme and a comprehensive avionics package which includes dual nav/com, transponder, DME, ADF, marker beacon receiver, glideslope and dual blind-flying instrumentation.

Beechcraft King Air C90 six/ten-seat business aircraft

A total of 952 commercial and 165 military King Air 90/A90/B90/C90s had been delivered by 17 April 1981. One was provided for the US Air Force's 1254th Special Air Missions Squadron at Andrews AFB, Maryland, for VIP transport duties under the designation **VC-6B.** Beech has delivered 10 King Air C90s to the Spanish Air Force and Civil Aviation School for instrument training and liaison. Delivery of five to the FAA, to replace Queen Airs which had been in service since 1963, began in early 1980.

TYPE: Six/ten-seat twin-turboprop business aircraft.

WINGS: Cantilever low-wing monoplane. Wing section NACA 23014·1 (modified) at root, NACA 23016·22 (modified) at outer end of centre-section, NACA 23012 at tip. Dihedral 7°. Incidence 4° 48' at root, 0° at tip. No sweepback at quarter-chord. Two-spar aluminium alloy structure. All-metal ailerons of magnesium, with adjustable trim tab on port aileron. Single-slotted aluminium alloy flaps. Automatic pneumatic de-icing boots on leading-edges are standard.

FUSELAGE: Aluminium alloy semi-monocoque structure.

TAIL UNIT: Cantilever all-metal structure with sweptback vertical surfaces. Fixed-incidence tailplane, with 7° dihedral. Trim tabs in rudder and each elevator. Automatic pneumatic de-icing boots on leading-edges of fin and tailplane are standard.

LANDING GEAR: Electrically-retractable tricycle type. Nosewheel retracts rearward, main wheels forward into engine nacelles. Main wheels protrude slightly beneath nacelles when retracted, for safety in a wheels-up emergency landing. Steerable nosewheel with shimmy damper. Beech oleo-pneumatic shock-absorbers. Goodrich main wheels with tyres size 8·50-10, pressure 3·79 bars (55 lb/sq in). Goodrich nosewheel with tyre size 6·50-10, pressure 3·59 bars (52 lb/sq in). Goodrich heat-sink and aircooled multi-disc hydraulic brakes. Parking brakes.

POWER PLANT: Two 410 kW (550 ehp) Pratt & Whitney Aircraft of Canada PT6A-21 turboprop engines, each driving a Hartzell three-blade constant-speed fully-feathering propeller with spinner. Propeller electro-thermal anti-icing, auto ignition system, environmental fuel drain collection system, and magnetic chip detector standard. Automatic propeller feathering, and propeller synchrophaser, optional. Fuel in two tanks in engine nacelles, each with capacity of 231 litres (61 US gallons), and bladder type auxiliary tanks in outer wings, each with capacity of 496 litres (131 US gallons). Total fuel capacity 1,454 litres (384 US gallons). Refuelling points in top of each engine nacelle and in wing leading-edge outboard of each nacelle. Oil capacity 13·2 litres (3·5 US gallons). Engine anti-icing system standard. Engine fire detection and extinguishing system optional.

ACCOMMODATION: Two seats side by side in cockpit with dual controls standard. Normally, four reclining seats are provided in the main cabin, in pairs facing each other fore and aft. Standard furnishings include cabin forward partition, with fore and aft partition curtain and coat rack, hinged nose baggage compartment door, seat belts and inertia-reel shoulder harness for all seats. Optional arrangements seat up to eight persons, some with two- or three-place couch, lateral tracking chairs, and refreshment cabinets. Baggage racks at rear of cabin on starboard side, with optional toilet on port side. Door on port side aft of wing, with built-in airstairs. Emergency exit on starboard side of cabin. Entire accommodation pressurised and air-conditioned. Electrically-heated windscreen, windscreen defroster and windscreen wipers standard.

SYSTEMS: Pressurisation by dual engine bleed air system with pressure differential of 0·32 bars (4·6 lb/sq in). Cabin heated by 45,000 BTU dual engine bleed air system and auxiliary electrical heating system. Electrical system includes two 28V 250A starter/generators, 24V 45Ah aircooled nickel-cadmium battery with failure detector. Complete de-icing and anti-icing equipment. Oxygen system, 0·62 m³ (22 cu ft), 1·39 m³ (49 cu ft) or 1·81 m³ (64 cu ft) capacity, optional. Vacuum system for flight instruments.

AVIONICS AND EQUIPMENT: Standard avionics package comprises dual Collins VHF-251 VHF transceivers with PWL-150 power adaptors and B3 antennae; dual Collins VIR-351 Omni nav receivers, one with Collins 331A-3G indicator and B17 antenna, the other with IND-351A on B17 antenna; Collins AMR-350 audio system; Collins ADF-650A ADF, with IND-650A indicator and ANT-650 antenna; Collins marker beacon receiver integral with AMR-350, plus marker lights and B16 antenna; dual Collins GLS-350 glide-slope receivers with B35 antenna; Collins DME-451, with IND-451 indicator, Nav1/Nav 2 switching, DME hold and B18 antenna; Collins PN-101 compass system (pilot); Standard Electric gyro horizon (pilot); CF gyro horizon and directional gyro (co-pilot); dual Flite-Tronics PC-125 125VA inverters with failure light; avionics transient protection; dual flight instrumentation; sectional instrument panel; white lighting; radio accessories, static wicks and Beech metal radio panel; microphone key button in pilot and co-pilot control wheels; dual microphones, headsets and cockpit speakers; and avionics master switch. Optional avionics include a wide range of equipment by Bendix, Collins, King, Edo-Aire Mitchell, RCA and Sperry. Standard equipment includes dual blind-flying instrumentation with sensitive altimeters, standby magnetic compass, outside air temperature gauge, LCD digital clock/chronometer, vacuum gauge, de-icing pressure gauge, cabin rate of climb indicator, cabin altitude and pressure differential indicators, pilot and co-pilot four-way adjustable seats with shoulder harness, map pockets, control locks, storm windows, sun visors, automatic fuel heater system, emergency locator transmitter, heated pitots, heated stall warning transmitter, stall warning device, cabin windows with adjustable polarised shades, carpeted floor, internal corrosion proofing, 'No smoking—Fasten seat belt' sign, fresh air outlets, dual map lights, primary and secondary instrument light systems, indirect cabin lighting, two overhead cabin spotlights, entrance door light, adjustable reading lights, aft compartment lights, dual landing lights, taxi light, position lights, dual rotating beacons, wing ice lights, heated fuel vents, external power socket, static wicks, and external urethane paint. Optional equipment includes flight hour recorder, instantaneous vertical speed indicator, cockpit and cabin fire extinguishers, a range of cabin seats, cabinets, storage drawers and toilets, entrance door step lights, tail floodlights, strobe lights, and wingtip recognition lights.

DIMENSIONS, EXTERNAL:

Wing span	15·32 m (50 ft 3 in)
Wing chord at root	2·15 m (7 ft 0½ in)
Wing chord at tip	1·07 m (3 ft 6 in)
Wing aspect ratio	8·57
Length overall	10·82 m (35 ft 6 in)
Height overall	4·34 m (14 ft 3 in)
Tailplane span	5·26 m (17 ft 3 in)
Wheel track	3·89 m (12 ft 9 in)
Wheelbase	3·75 m (12 ft 3½ in)
Propeller diameter	2·36 m (7 ft 9 in)
Passenger door: Height	1·30 m (4 ft 3½ in)
Width	0·69 m (2 ft 3 in)
Height to sill	1·17 m (3 ft 10 in)

DIMENSIONS, INTERNAL:

Total pressurised length	5·43 m (17 ft 10 in)
Cabin: Length	3·86 m (12 ft 8 in)
Max width	1·37 m (4 ft 6 in)
Max height	1·45 m (4 ft 9 in)
Floor area	6·50 m² (70 sq ft)
Volume	8·89 m³ (314 cu ft)
Baggage compartment, aft	1·51 m³ (53·5 cu ft)

AREAS:

Wings, gross	27·31 m² (293·94 sq ft)
Ailerons (total)	1·29 m² (13·90 sq ft)
Trailing-edge flaps (total)	2·72 m² (29·30 sq ft)
Fin	2·20 m² (23·67 sq ft)
Rudder, incl tab	1·30 m² (14·00 sq ft)
Tailplane	4·39 m² (47·25 sq ft)
Elevators, incl tabs	1·66 m² (17·87 sq ft)

WEIGHTS AND LOADINGS:

Weight empty	2,615 kg (5,765 lb)
Max T-O weight	4,377 kg (9,650 lb)
Max ramp weight	4,402 kg (9,705 lb)
Max landing weight	4,159 kg (9,168 lb)
Max wing loading	160·1 kg/m² (32·8 lb/sq ft)
Max power loading	5·34 kg/kW (8·8 lb/ehp)

PERFORMANCE (at max T-O weight, except where indicated):

Max cruising speed at 3,660 m (12,000 ft)
222 knots (412 km/h; 256 mph)

Max cruising speed at 4,880 m (16,000 ft) at AUW of 3,794 kg (8,365 lb)
219 knots (406 km/h; 252 mph)

Max cruising speed at 6,400 m (21,000 ft) at AUW of 3,794 kg (8,365 lb)
216 knots (401 km/h; 249 mph)

Stalling speed, wheels and flaps up, power off
89 knots (164 km/h; 102 mph) IAS

Stalling speed, wheels and flaps down, power off
76 knots (140 km/h; 87 mph) IAS

Max rate of climb at S/L	596 m (1,955 ft)/min
Rate of climb at S/L, one engine out	164 m (539 ft)/min
Service ceiling	8,565 m (28,100 ft)
Service ceiling, one engine out	4,587 m (15,050 ft)
Min ground turning radius	11·58 m (38 ft 0 in)
Runway LCN	4
T-O run	497 m (1,629 ft)
T-O to 15 m (50 ft)	689 m (2,261 ft)
Accelerate/stop distance	1,066 m (3,498 ft)
Landing from 15 m (50 ft) at max landing weight:	
without propeller reversal	613 m (2,010 ft)
with propeller reversal	510 m (1,672 ft)
Landing run at max landing weight:	
without propeller reversal	328 m (1,075 ft)
with propeller reversal	225 m (737 ft)

Range with max fuel at max cruising speed, incl allowance for starting, taxi, take-off, climb, descent and 45 min reserves at max range power, ISA, at:

6,400 m (21,000 ft)
1,202 nm (2,227 km; 1,384 miles)

4,875 m (16,000 ft)
1,059 nm (1,962 km; 1,219 miles)

3,660 m (12,000 ft)
957 nm (1,773 km; 1,102 miles)

Max range at econ cruising power, allowances as above, at:

6,400 m (21,000 ft)
1,281 nm (2,374 km; 1,474 miles)

4,875 m (16,000 ft)
1,159 nm (2,147 km; 1,334 miles)

3,660 m (12,000 ft)
1,068 nm (1,979 km; 1,229 miles)

BEECHCRAFT KING AIR E90

On 1 May 1972 Beech announced an addition to the King Air range of business aircraft. Designated King Air E90, this combines the airframe of the King Air C90 with 507 kW (680 ehp) Pratt & Whitney Aircraft of Canada PT6A-28 turboprop engines, each flat rated to 410 kW (550 ehp).

A total of 346 commercial King Air E90s had been delivered by 17 April 1981.

The description of the King Air C90 applies also to the King Air E90, except as follows:

LANDING GEAR: As King Air C90, except main-wheel tyre pressure 3·93 bars (57 lb/sq in).

POWER PLANT: Two 507 kW (680 ehp) Pratt & Whitney Aircraft of Canada PT6A-28 turboprop engines, flat rated to 410 kW (550 ehp), each driving a Hartzell three-blade fully-feathering and reversible-pitch constant-speed metal propeller with spinner. Standard fuel capacity 1,794 litres (474 US gallons).

AVIONICS AND EQUIPMENT: Standard avionics include dual Collins VHF-20A VHF transceivers with Gables controls and B3 antennae; Collins VIR-30 MGM Omni No. 1 receiver, with 331A-3G indicator, Gables control and B17 antenna; Collins VIR-30MG Omni No. 2 receiver with 331H-3G indicator and Gables control; Collins 356C-4 isolation amplifier and 356F-3 speaker amplifier with single set of audio switches; Collins ADF-650A ADF with IND-650A indicator, Gables control, voice range filter and ANT-650 antenna; Collins marker beacon receiver integral with VIR-30 No. 1, with marker lights and B16 antenna; Collins glideslope receiver integral with VIR-30 No. 1, with B35 antenna; dual flight instrumentation; Collins DME-40 with 339F-12 indicator, Nav 1/Nav 2 switching, DME hold, and 237Z-1 antenna; Collins TDR-950 transponder with 237Z-1 antenna; Collins PN-101 compass system (pilot); Standard Electric gyro horizon (pilot); CF gyro horizon and directional gyro (co-pilot); dual Flite-Tronics PC-125 125VA inverters with failure light; avionics transient protection; Beech edge-lighted radio panel, radio accessories and static wicks; white lighting; microphone key button in pilot and co-pilot control wheels; dual microphones, headsets and cockpit speakers; and avionics master switch. Optional avionics include a wide range of equipment by Bendix, Collins, King, Edo-Aire Mitchell, RCA and Sperry. Standard and optional equipment generally as listed for King Air C90.

WEIGHTS AND LOADINGS:

Weight empty	2,720 kg (5,996 lb)
Max T-O weight	4,581 kg (10,100 lb)
Max ramp weight	4,608 kg (10,160 lb)
Max landing weight	4,400 kg (9,700 lb)
Max wing loading	168·0 kg/m² (34·4 lb/sq ft)
Max power loading	5·59 kg/kW (9·18 lb/ehp)

PERFORMANCE (at max T-O weight, except where indicated):

Max cruising speed at 3,660 m (12,000 ft)
249 knots (462 km/h; 287 mph)

Cruising speed at max recommended cruise power:

at 4,875 m (16,000 ft)
248 knots (460 km/h; 286 mph)

at 6,400 m (21,000 ft)
245 knots (454 km/h; 282 mph)

Cruising speed for max range
197 knots (365 km/h; 227 mph)

Stalling speed, power off, wheels and flaps up
86 knots (159 km/h; 99 mph) IAS

Stalling speed, power off, wheels and flaps down
77 knots (143 km/h; 89 mph) IAS

Max rate of climb at S/L	570 m (1,870 ft)/min
Rate of climb at S/L, one engine out	143 m (470 ft)/min
Service ceiling	8,420 m (27,620 ft)
Service ceiling at 3,629 kg (8,000 lb) AUW	9,420 m (30,910 ft)
Service ceiling, one engine out	4,385 m (14,390 ft)
Service ceiling, one engine out, at 3,629 kg (8,000 lb) AUW	6,220 m (20,400 ft)
Min ground turning radius	11·58 m (38 ft 0 in)
Runway LCN	4·5
T-O run	473 m (1,553 ft)
T-O to 15 m (50 ft)	617 m (2,024 ft)
Landing distance, 5° approach angle, full flap, at max landing weight:	
landing from 15 m (50 ft)	643 m (2,110 ft)
landing run	314 m (1,030 ft)

Accelerate/stop distance, incl 2 s failure recognition time 1,139 m (3,736 ft)
Cruising range at max recommended cruise power:
at 4,875 m (16,000 ft) 1,125 nm (2,084 km; 1,295 miles)
at 6,400 m (21,000 ft) 1,309 nm (2,425 km; 1,507 miles)
Cruising range at max range power:
at 4,875 m (16,000 ft) 1,480 nm (2,742 km; 1,704 miles)
at 6,400 m (21,000 ft) 1,625 nm (3,011 km; 1,871 miles)
Range at max T-O weight, max recommended power at 6,400 m (21,000 ft), 45 min reserves, five occupants, 37 kg (82 lb) baggage and 1,440 kg (3,176 lb) fuel before engine start 1,309 nm (2,425 km; 1,507 miles)

BEECHCRAFT T-44A (KING AIR 90)

In 1976 Beech Aircraft won an industry-wide competition for a twin-turboprop advanced pilot training aircraft to meet the US Navy's VTAM(X) requirement. The aircraft selected was a version of the King Air 90, which has the US Navy designation T-44A. Production totalled 61 aircraft, and completion of delivery was announced on 19 June 1980. Brief details of this version of the King Air can be found in the 1980-81 *Jane's*.

BEECHCRAFT SUPER KING AIR MODEL F90

Deliveries of the Super King Air F90 began in mid-1979, shortly after FAA certification of this sixth member of the King Air range of corporate transport aircraft. Basically, it combines the pressurised fuselage of the King Air 90 with reduced-span wings similar to those of the King Air 100, and a T tail assembly similar to that of the Super King Air 200.

New Beechcraft assemblies and technology are utilised throughout its construction, and the PT6A-135 turboprop engines drive slow-turning four-blade propellers to reduce airport and in-flight noise. Cabin pressurisation is increased to 0·34 bars (5·0 lb/sq in) to give a sea level environment at 3,375 m (11,065 ft), a 2,440 m (8,000 ft) environment at 7,050 m (23,120 ft), and a 3,050 m (10,000 ft) environment at an altitude of 8,075 m (26,500 ft).

A total of 120 commercial King Air F90s had been delivered by 17 April 1981.

TYPE: Seven/ten-seat twin-turboprop business aircraft.

WINGS: Similar to King Air 100. De-icing system standard.

FUSELAGE: Similar to King Air 90.

TAIL UNIT: Similar to Super King Air 200. Tailplane de-icing standard.

LANDING GEAR: Retractable tricycle type, with twin-wheel main units and single steerable nosewheel. Electrical retraction, nosewheel rearward, main units forward into engine nacelles. Beech oleo-pneumatic shock-absorbers. Main-wheel tyres size 18 × 5·5. Nosewheel tyre size 22 × 6·75-10. Single-disc hydraulic brakes. Optional high-flotation gear for use on unimproved airstrips.

POWER PLANT: Two 559 kW (750 shp) Pratt & Whitney Aircraft of Canada PT6A-135 turboprop engines, each driving a Hartzell FT 101 73 four-blade constant-speed metal propeller with spinner. Propellers available optionally with reversible pitch. Usable fuel capacity 1,779 litres (470 US gallons). Automatic fuel transfer system, engine anti-icing, propeller de-icing, and ice-free fuel venting system, are standard.

ACCOMMODATION: Two seats side by side on flight deck, with dual controls. Seats for five to eight persons in main cabin, in deep-cushioned chairs. Passengers screened from flight deck and toilet by partitions at front and rear of cabin. Space for 172 kg (380 lb) of baggage. Windscreen anti-icing standard.

SYSTEMS: Pressurisation system, differential 0·34 bars (5·0 lb/sq in); 16,000 BTU air-conditioning system. Electrical system includes two 28V 250A starter/generators and 45Ah aircooled nickel-cadmium battery. Oxygen system, 0·62 m³ (22 cu ft) capacity, with eight automatically-deployed passenger masks and one first aid mask.

AVIONICS: Standard avionics package, by Collins, includes two VHF-20A transceivers, VIR-30MGM and VIR-30MG manual omnis, DB system Model 415 audio amplifier, ADF-650A ADF, marker beacon indicator, dual glideslope, DME-40, TDR-950 transponder, and PN-101 compass system. Full blind-flying instrumentation for pilot and co-pilot. Collins APS-80 autopilot, and large range of optional avionics available.

DIMENSIONS, EXTERNAL:
Wing span 13·99 m (45 ft 10¾ in)
Wing area, gross 25·98 m² (279·7 sq ft)
Length overall 12·13 m (39 ft 9½ in)
Height overall 4·60 m (15 ft 1¼ in)
Tailplane span 5·61 m (18 ft 4¾ in)
Propeller diameter 2·34 m (7 ft 8 in)
Passenger door: Height 1·31 m (4 ft 3½ in)
Width 0·69 m (2 ft 3 in)

DIMENSIONS, INTERNAL:
Cabin, excl flight deck: Length 3·86 m (12 ft 8 in)
Width 1·37 m (4 ft 6 in)
Height 1·45 m (4 ft 9 in)
Avionics compartment volume 0·45 m³ (16 cu ft)

WEIGHTS AND LOADINGS:
Weight empty 2,971 kg (6,549 lb)
Max T-O and landing weight 4,966 kg (10,950 lb)
Max ramp weight 5,003 kg (11,030 lb)
Max wing loading 190·8 kg/m² (39·1 lb/sq ft)
Max power loading 4·4 kg/kW (7·3 lb/shp)

PERFORMANCE (A: at max T-O weight; B: at 4,309 kg/9,500 lb AUW; C: at 4,082 kg/9,000 lb AUW):
Max cruising speed (B):
at 3,660 m (12,000 ft) 267 knots (495 km/h; 307 mph)
at 5,490 m (18,000 ft) 260 knots (483 km/h; 300 mph)
at 7,925 m (26,000 ft) 251 knots (465 km/h; 289 mph)
Take-off speed (A) 107 knots (198 km/h; 123 mph)
Approach speed (A) 108 knots (200 km/h; 124 mph)
Stalling speed (A), power off:
flaps up 94 knots (174 km/h; 108 mph)
32·5% flap 84 knots (156 km/h; 97 mph)
100% flap 77 knots (143 km/h; 89 mph)
Max rate of climb at S/L: A 725 m (2,380 ft)/min
C 947 m (3,108 ft)/min
Rate of climb at S/L, one engine out:
A 183 m (600 ft)/min
C 289 m (947 ft)/min
Service ceiling: A 9,084 m (29,800 ft)
C above 9,450 m (31,000 ft)
Service ceiling, one engine out: A 4,395 m (14,420 ft)
C 5,920 m (19,420 ft)
T-O run (A), flaps up 637 m (2,090 ft)
T-O to 15 m (50 ft) (A), flaps up 871 m (2,856 ft)
Landing from 15 m (50 ft) (A):
without propeller reversal 907 m (2,977 ft)
with propeller reversal 694 m (2,275 ft)
Landing run (A):
without propeller reversal 578 m (1,895 ft)
with propeller reversal 364 m (1,194 ft)
Cruising range at max cruise power, with reserves:
at 3,660 m (12,000 ft) 960 nm (1,779 km; 1,105 miles)
at 5,490 m (18,000 ft) 1,168 nm (2,165 km; 1,345 miles)
at 7,925 m (26,000 ft) 1,439 nm (2,667 km; 1,657 miles)
Cruising range at max range power, with reserves:
at 3,660 m (12,000 ft) 1,179 nm (2,185 km; 1,357 miles)
at 5,490 m (18,000 ft) 1,368 nm (2,535 km; 1,575 miles)
at 7,925 m (26,000 ft) 1,576 nm (2,920 km; 1,814 miles)

Beechcraft Super King Air F90 T-tailed seven/ten-seat business aircraft

BEECHCRAFT COMMUTER C99

Beech announced on 7 May 1979 the company's intention to re-enter the commuter airliner market, initially with two aircraft designated Commuter C99 and Commuter 1900. Since that time the company's efforts have been concentrated on the Commuter C99, of which a prototype, converted from a B99 Airliner airframe and powered by P&WC PT6A-34 turboprop engines, flew for the first time on 20 June 1980. Construction of production aircraft began in September 1980, and deliveries started on 30 July 1981, following FAA certification earlier in the same week.

By comparison with the B99 Airliner, this new aircraft has more powerful turboprop engines, the incorporation as standard of several items that formerly were optional, and some systems changes.

TYPE: Twin-turboprop commuter/cargo transport.

WINGS: Cantilever low-wing monoplane. Wing section NACA 23018 at root, NACA 23016·5 at centre-section joint with outer panel, NACA 23012 at tip. Dihedral 7°. Incidence 4° 48′ at root, 0° at tip. Two-spar all-metal structure of aluminium alloy. Single-slotted trailing-edge flaps of aluminium alloy. All-metal ailerons of magnesium alloy. Trim tab in port aileron. Optional pneumatic de-icing boots for wing leading-edges.

FUSELAGE: All-metal semi-monocoque structure.

TAIL UNIT: Cantilever all-metal structure with sweptback vertical surfaces and a ventral stabilising fin. Variable-incidence tailplane. Trim tab in rudder. Pneumatic de-icing boots for fin and tailplane leading-edges optional.

LANDING GEAR: Hydraulically retractable tricycle type with single steerable nosewheel and twin wheels on each main unit. Nose unit retracts aft, main units forward into engine nacelles. Beech oleo-pneumatic shock-absorbers. Goodrich wheels and tyres. Main-wheel tyres size 18 × 5·5, pressure 6·34-6·62 bars (92-96 lb/sq in). Nosewheel tyre size 6·50-10, pressure 3·45-3·79 bars (50-55 lb/sq in). Goodrich heat-sink and aircooled single-disc hydraulic brakes. Parking brake. Shimmy damper on nosewheel.

POWER PLANT: Two Pratt & Whitney Aircraft of Canada PT6A-36 turboprop engines, each flat rated to 533 kW (715 shp) and driving a Hartzell three-blade constant-speed fully-feathering reversible-pitch propeller. Five rubber fuel cells in each wing; total fuel capacity 1,393 litres (368 US gallons). Refuelling points on each engine nacelle and at each wingtip. Oil capacity 26·5 litres (7 US gallons).

ACCOMMODATION: Crew of two side by side on flight deck, with dual controls and full blind-flying instrumentation as standard. Half-curtain or bulkhead between flight deck and cabin. Standard high density seating arrangement accommodates 15 passengers two-abreast with centre aisle; single seat opposite cabin door. Baggage compartment aft of rear seats, with external door. Nose baggage compartment with two external doors. An underfuselage baggage/cargo pod with a volume of 1·68 m³ (59·4 cu ft) is available as an option. Airstair door at rear of cabin on port side, with aft-hinged cargo door adjacent and to the rear, to provide a large unobstructed opening for cargo loading. Emergency exit on each side at forward end of cabin. Emergency exit for flight deck optional.

SYSTEMS: Engine bleed air heating system of 45,000 BTU output. Freon air-conditioning system of 32,000 BTU optional. Hydraulic system, pressure 114 bars (1,650 lb/sq in), with duplicated lines and alternative reservoir for operation of landing gear and brakes. Electrical system includes two 28V 200A generators, 40Ah nickel-cadmium battery with failure detector, and dual solid state inverters.

AVIONICS AND EQUIPMENT: Dual nav/com, dual glideslope receivers and transponders, ADF, DME, marker beacon receiver, Bendix radar, encoding altimeter, and ELT.

DIMENSIONS, EXTERNAL:
Wing span 13·98 m (45 ft 10½ in)
Wing chord at root 2·15 m (7 ft 0½ in)
Wing chord at tip 1·07 m (3 ft 6 in)
Wing aspect ratio 7·51
Length overall 13·58 m (44 ft 6¾ in)
Height overall 4·37 m (14 ft 4¼ in)
Tailplane span 6·82 m (22 ft 4½ in)
Wheel track 3·96 m (13 ft 0 in)
Wheelbase 5·48 m (17 ft 11¾ in)
Propeller diameter 2·37 m (7 ft 9½ in)
Propeller ground clearance 0·34 m (1 ft 1½ in)
Cabin door (port, aft): Height 1·31 m (4 ft 3½ in)
Width 0·69 m (2 ft 3 in)
Height to sill 1·19 m (3 ft 11 in)
Cargo door (port, rearmost):
Height 1·31 m (4 ft 3½ in)
Width 0·53 m (1 ft 9 in)
Height to sill 1·19 m (3 ft 11 in)
Emergency exits (port and stbd, fwd):
Height 0·71 m (2 ft 4 in)
Width 0·53 m (1 ft 9 in)

DIMENSIONS, INTERNAL:
Cabin, incl flight deck and aft baggage compartment:
Length 7·73 m (25 ft 4½ in)

Prototype of the 15-passenger Beechcraft Commuter C99 light transport

Max width 1·40 m (4 ft 7 in)
Max height 1·45 m (4 ft 9 in)
Volume 13·71 m^3 (484·3 cu ft)
Baggage compartment volume:
aft of cabin 0·48 m^3 (17·0 cu ft)
nose 1·24 m^3 (43·9 cu ft)

Areas:
Wings, gross 25·98 m^2 (279·7 sq ft)
Vertical tail surfaces 4·17 m^2 (44·9 sq ft)
Horizontal tail surfaces 9·29 m^2 (100·0 sq ft)

Weights and Loadings:
Basic operating weight 2,946 kg (6,494 lb)
Max payload 1,474 kg (3,250 lb)
Max fuel weight 1,118 kg (2,466 lb)
Max T-O and landing weight 5,125 kg (11,300 lb)
Max ramp weight 5,162 kg (11,380 lb)
Max zero-fuel weight 4,420 kg (9,744 lb)
Max wing loading 197·25 kg/m^2 (40·4 lb/sq ft)
Max power loading 4·81 kg/kW (7·9 lb/shp)

Performance (estimated, at max T-O weight except where stated otherwise):
Never-exceed speed 282 knots (522 km/h; 324 mph)
Max level speed at 2,440 m (8,000 ft) 268 knots (496 km/h; 308 mph)
Cruising speed at AUW of 4,536 kg (10,000 lb), at:
2,440 m (8,000 ft) 249 knots (461 km/h; 287 mph)
3,660 m (12,000 ft) 248 knots (460 km/h; 286 mph)
4,875 m (16,000 ft) 245 knots (454 km/h; 282 mph)
Max rate of climb at S/L 677 m (2,221 ft)/min
Rate of climb at S/L, one engine out 164 m (539 ft)/min
Service ceiling 8,560 m (28,080 ft)
Service ceiling, one engine out 4,375 m (14,360 ft)
Range with max fuel, IFR reserves 910 nm (1,686 km; 1,048 miles)

BEECHCRAFT COMMUTER 1900

Development of this 19-passenger pressurised commuter transport is proceeding on schedule. First flight is planned for early 1982, with deliveries to begin during 1983. Power plant comprises two Pratt & Whitney Aircraft of Canada PT6A-65 turboprop engines, each flat rated to 746 kW (1,000 shp) and driving a 2·79 m (9 ft 2 in) three-blade Hartzell propeller with composite blades. A 1·32 m (4 ft 4 in) square cargo door will be optional.

Weights:
Max T-O weight 6,915 kg (15,245 lb)
Max zero-fuel weight 5,670 kg (12,500 lb)

Performance (estimated):
Max cruising speed more than 260 knots (483 km/h; 300 mph)
Range with max fuel, IFR reserves 555 nm (1,028 km; 639 miles)

BEECHCRAFT KING AIR A100

US Army designation: U-21F

Beech Aircraft announced on 26 May 1969 the addition of a new version of the King Air to its fleet of corporate transport aircraft. Designated King Air 100, this is a pressurised transport with increased internal capacity and more powerful engines, enabling it to carry a useful load of more than two short tons. By comparison with the King Air 90 series, it has a fuselage 1·27 m (4 ft 2 in) longer, reduced wing span, larger rudder and elevator and twin-wheel main landing gear. It is available in a variety of interior configurations, seating six to eight in executive versions, or up to 13 in high-density arrangement, plus crew of two.

The King Air 100 has been approved for Category 2 landing minima by the FAA. Initial deliveries were made in August 1969, following FAA certification. A total of 241 commercial and five military King Air A100s had been delivered by 17 April 1981, at which time the A100 remained in limited production only.

First deliveries of the advanced Model A100, comprising five U-21Fs for the Department of the Army, began in October 1971. Supplied under a \$2·5 million contract, they represented the first pressurised aircraft in the Army's inventory. Two A100s were supplied to the Spanish Air Force.

Two aircraft equipped with a Beech-developed UNACE package (Universal Aircraft Com/Nav Evaluation) have been delivered to Belgium and Indonesia. UNACE-configured aircraft, which provide an economical means of inspecting and calibrating aviation navigation aids, are operating also in Algeria, Canada, Malaysia, Mexico and the USA. Beech is able to modify King Airs for aerial photography, and deliveries of camera-equipped aircraft have been made to Canada, Chile, France, Jamaica, Saudi Arabia and Thailand, as well as to various US organisations.

Type: Twin-turboprop light passenger, freight or executive transport.

Wings: Cantilever low-wing monoplane. Wing section NACA 23018 at root, NACA 23016·5 at centre-section joint with outer panel, NACA 23012 at tip. Dihedral 7°. Incidence 4° 48′ at root, 0° at tip. Two-spar all-metal light alloy structure. All-metal ailerons of magnesium. Trim tab in port aileron. Single-slotted light alloy trailing-edge flaps. Pneumatic de-icing boots standard.

Fuselage: All-metal light alloy semi-monocoque structure.

Tail Unit: Cantilever all-metal structure with swept vertical surfaces and a ventral stabilising fin. Trim tab in rudder. Electrically-operated adjustment of tailplane incidence. Pneumatic de-icing boots standard.

Landing Gear: Retractable tricycle type with single steerable nosewheel and twin wheels on each main unit. Electrical retraction, nosewheel rearward, main units forward into engine nacelles. Hydraulic retraction system optional. Beech oleo-pneumatic shock-absorbers. Goodrich main wheels and tubeless tyres size 18 × 5·5, pressure 7·10 bars (103 lb/sq in). Nosewheel with tubeless tyre size 6·50 × 10, pressure 3·93 bars (57 lb/sq in). Goodrich heat sink and aircooled single-disc hydraulic brakes.

Power Plant: Two 507 kW (680 ehp) Pratt & Whitney Aircraft of Canada PT6A-28 turboprop engines, each driving a Hartzell four-blade fully-feathering and reversible-pitch constant-speed metal propeller with spinner. Automatic feathering, and propeller synchrophaser, optional. Rubber fuel cells in wings, with total capacity of 1,779 litres (470 US gallons). Automatic fuel heating systems; inertial engine inlet de-icing system; engine inlet lips de-iced by electro-thermally heated boots; auto ignition system; environmental fuel drain collector system; magnetic chip detector. Goodrich electrical propeller anti-icing system.

Accommodation: Crew of two side by side on flight deck, with full dual controls and instruments. Easily removable partition with sliding door between flight deck and cabin. Six fully-adjustable individual cabin chairs standard, with removable headrests, with a variety of alternative layouts, for up to 13 passengers in commuter role. Seat belts and inertia-reel shoulder harness for all seats. Polarised cabin windows. Dual storm windows. Fully-carpeted floor. External access door to forward radio compartment. Aft fuselage maintenance access door. Plug-type emergency exit at forward end of cabin on starboard side. Passenger door at rear of cabin on port side, with integral airstair. Easily removable aft cabin partition with sliding doors. Lavatory installation and stowage for up to 186 kg (410 lb) baggage in aft fuselage. Other standard cabin equipment includes

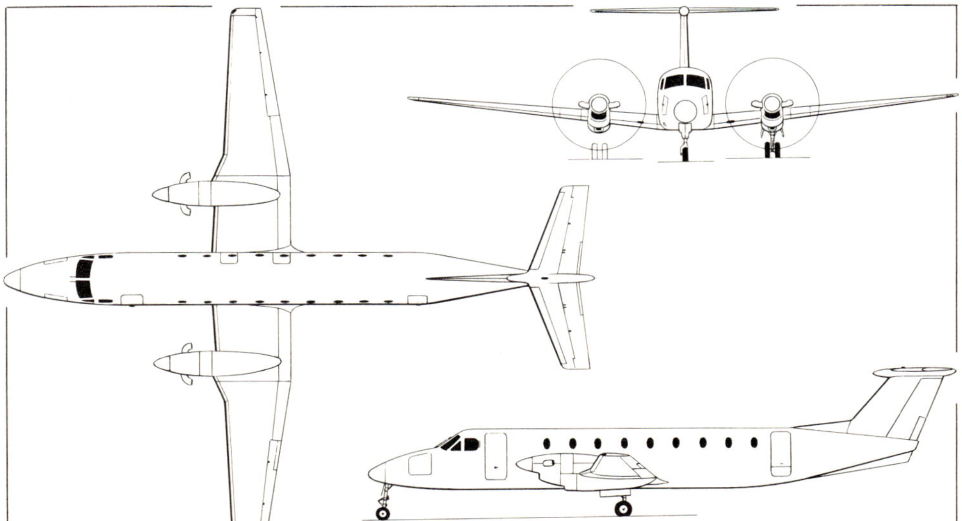

Beechcraft Commuter 1900 twin-turboprop commuter airliner *(Pilot Press)*

reading lights and fresh air outlets for all passengers, cabin coat rack and dual 'No smoking—Fasten seat belt' signs. Electrothermally heated windscreen, hot air windscreen defroster and windscreen wipers standard. Optional equipment includes cabin fire extinguisher, additional cabin window, flush toilet and a variety of interior cabinets.

SYSTEMS: Cabin pressurisation by engine bleed air with a differential of 0·32 bars (4·6 lb/sq in). Cabin heated by 27,000 BTU electrical heating system. Oxygen system for flight deck and 0·62 m³ (22 cu ft) oxygen system for cabin standard. Cabin oxygen system of 1·39 m³ (49 cu ft), or 1·81 m³ (64 cu ft) optional. Dual vacuum system for instruments. Hydraulic system for brakes only. Pneumatic system for wing and tail unit de-icing only. Electrical system includes two 28V 300A starter/generators, and two aircooled nickel-cadmium 28V 45Ah batteries with failure detector. Engine fire detection system.

AVIONICS AND EQUIPMENT: Standard avionics comprise dual King KTR 905 VHF transceivers, with Gables controls and B3 antennae; King KNR 630 Omni No. 1, with Collins 331A-3G indicator, Gables control and B17 antenna; King KNR 630 Omni No. 2, with Collins 331H-3G indicator and Gables control; dual Omni range filters; dual Collins 356F-3 audio amplifiers, each with 356C-4 isolation amplifiers and audio switches; King KDF 805 ADF less indicator, with KFS 580B control, voice range filter, and Beech flush sense antenna; King KMR 675 marker beacon receiver, with dual marker lights and B16 antenna; dual glideslope receivers integral with No. 1 and No. 2 Omni, with B35 antenna; Bendix RDR-130 weather radar, with phased array antenna and digital scope; Sperry C-14-43 compass system with servo amplifier (pilot); King KNI 581 RMI with VOR-1/ADF on single needle, VOR-2/ADF on double needle; King KXP 755 transponder, with Gables control and B18 antenna; King KDM 706 DME with KDI 572 indicator, Nav-1/Nav-2 switching, DME hold and B18 antenna; dual Flite-Tronics PC-250 250VA inverters with failure light; avionics transient protection; sectional instrument panel; Standard Electric gyro horizon (pilot); CF gyro horizon and directional gyro (co-pilot); Beech edge-lighted radio panel, radio accessories, microphone button in pilot and co-pilot control wheels, static wicks and white lighting; dual microphones, headsets and cockpit speakers; and avionics master switch. A wide range of optional avionics equipment is available, by Bendix, Collins, King, RCA, Sperry and SunAir. Heated main landing gear brake de-icing and tail floodlight offered as optional equipment. Other standard and optional equipment generally as listed for King Air C90.

DIMENSIONS, EXTERNAL:
- Wing span 14·00 m (45 ft 11 in)
- Wing chord at root 2·15 m (7 ft 0½ in)
- Wing chord at tip 1·07 m (3 ft 6 in)
- Wing aspect ratio 7·51
- Length overall 12·17 m (39 ft 11 in)
- Height overall 4·70 m (15 ft 5 in)
- Tailplane span 6·83 m (22 ft 5 in)
- Wheel track 3·96 m (13 ft 0 in)
- Wheelbase 4·55 m (14 ft 11 in)
- Propeller diameter 2·29 m (7 ft 6 in)
- Propeller ground clearance 0·34 m (1 ft 1½ in)

DIMENSIONS, INTERNAL:
- Cabin: Length (excl flight deck) 5·08 m (16 ft 8 in)
  - Max width 1·37 m (4 ft 6 in)
  - Max height 1·45 m (4 ft 9 in)
- Volume, avionics compartment in nose 0·45 m³ (16 cu ft)
- Volume, aft baggage compartment 1·51 m³ (53·5 cu ft)

WEIGHTS AND LOADINGS:
- Weight empty 3,083 kg (6,797 lb)
- Max T-O weight 5,216 kg (11,500 lb)
- Max ramp weight 5,247 kg (11,568 lb)
- Max zero-fuel weight 4,354 kg (9,600 lb)
- Max landing weight 5,084 kg (11,210 lb)
- Max wing loading 201 kg/m² (41·2 lb/sq ft)
- Max power loading 5·14 kg/kW (8·46 lb/ehp)

PERFORMANCE (at max T-O weight, unless otherwise quoted):
- Max cruising speed at 4,762 kg (10,500 lb) AUW:
  - at 6,400 m (21,000 ft) 235 knots (436 km/h; 270 mph)
  - at 4,875 m (16,000 ft) 243 knots (450 km/h; 280 mph)
  - at 3,050 m (10,000 ft) 248 knots (459 km/h; 285 mph)
- Stalling speed, power off, wheels and flaps up 89 knots (164 km/h; 102 mph)
- Stalling speed, power off, wheels and flaps down 75 knots (139 km/h; 86 mph)
- Max rate of climb at S/L 598 m (1,963 ft)/min
- Rate of climb at S/L, one engine out 138 m (452 ft)/min
- Service ceiling 7,575 m (24,850 ft)
- Service ceiling, one engine out 2,835 m (9,300 ft)
- Min ground turning radius 12·2 m (40 ft 0 in)
- Runway LCN 4·5
- T-O run: flaps up 628 m (2,060 ft)
  - 30% flap 565 m (1,855 ft)
- T-O to 15 m (50 ft): flaps up 989 m (3,245 ft)
  - 30% flap 817 m (2,681 ft)
- Landing from 15 m (50 ft) at max landing weight, without propeller reversal 897 m (2,944 ft)
- Landing run at max landing weight, without propeller reversal 545 m (1,787 ft)
- *Accelerate/stop distance, flaps up 1,303 m (4,275 ft)
- *Accelerate/stop distance, 30% flap 1,182 m (3,877 ft)
- Range at high cruise power, with 1,779 litres (470 US gallons) fuel, incl allowances for starting, taxi, take-off, climb, descent and 45 min reserves:
  - at 6,400 m (21,000 ft) 1,212 nm (2,247 km; 1,395 miles)
  - at 4,875 m (16,000 ft) 1,064 nm (1,971 km; 1,225 miles)
  - at 3,050 m (10,000 ft) 900 nm (1,667 km; 1,036 miles)
- Range at long range cruise power, fuel and allowances as above:
  - at 6,400 m (21,000 ft) 1,340 nm (2,483 km; 1,542 miles)
  - at 4,875 m (16,000 ft) 1,272 nm (2,358 km; 1,464 miles)
  - at 3,050 m (10,000 ft) 1,152 nm (2,136 km; 1,326 miles)

**Includes allowance for failure recognition*

BEECHCRAFT KING AIR B100

On 20 March 1975 Beech recorded the first flight of a new version of the King Air. Designated King Air B100, it is generally similar to its predecessors, except for the installation of two 533 kW (715 shp) Garrett TPE331-6-252B turboprop engines, giving improved performance.

A total of 110 commercial King Air B100s had been delivered by 17 April 1981.

The description of the King Air A100 applies also to the B100, except as follows:

TYPE: Twin-turboprop light passenger, freight or executive transport.

WINGS, FUSELAGE, TAIL UNIT, LANDING GEAR: As for Model A100.

POWER PLANT: Two 533 kW (715 shp) Garrett TPE331-6-252B turboprop engines, each driving a four-blade constant-speed fully-feathering metal propeller with spinner. Fuel system and anti-icing system as for Model A100.

ACCOMMODATION, SYSTEMS, AVIONICS AND EQUIPMENT: As for Model A100, except for slight variations in ancillary equipment associated directly with the power plant.

DIMENSIONS, EXTERNAL: As for Model A100, except:
- Propeller ground clearance 0·39 m (1 ft 3½ in)

DIMENSIONS, INTERNAL, AND AREAS: As for Model A100

WEIGHTS AND LOADINGS:
- Weight empty, equipped 3,212 kg (7,082 lb)
- Max T-O weight 5,352 kg (11,800 lb)
- Max ramp weight 5,386 kg (11,875 lb)
- Max zero-fuel weight 4,354 kg (9,600 lb)
- Max landing weight 5,085 kg (11,210 lb)
- Max wing loading 206·0 kg/m² (42·2 lb/sq ft)
- Max power loading 5·02 kg/kW (8·25 lb/shp)

PERFORMANCE (at max T-O weight):
- Max level speed and max cruising speed:
  - at 3,660 m (12,000 ft) 267 knots (495 km/h; 307 mph)
  - at 6,400 m (21,000 ft) 262 knots (486 km/h; 302 mph)
- Stalling speed, flaps up 93 knots (172·5 km/h; 107 mph) IAS
- Stalling speed, flaps down 83 knots (154·5 km/h; 96 mph) IAS
- Max rate of climb at S/L 652 m (2,139 ft)/min
- Rate of climb at S/L, one engine out 152 m (501 ft)/min
- Service ceiling 8,575 m (28,140 ft)
- Service ceiling, one engine out 3,695 m (12,120 ft)
- T-O run, flaps up 579 m (1,898 ft)
- T-O run, 30% flap 535 m (1,755 ft)
- T-O to 15 m (50 ft), flaps up 899 m (2,951 ft)
- T-O to 15 m (50 ft), 30% flap 821 m (2,694 ft)
- Landing from 15 m (50 ft) without propeller reversal 1,040 m (3,413 ft)
- Landing from 15 m (50 ft) with propeller reversal 817 m (2,679 ft)
- Landing run without propeller reversal 503 m (1,651 ft)
- Landing run with propeller reversal 393 m (1,290 ft)
- Range with 1,779 litres (470 US gallons) usable fuel, with allowances for start, taxi, T-O, climb, descent and 45 min reserves at max range power, ISA:
  - at max cruising power at:
    - 3,660 m (12,000 ft) 1,015 nm (1,881 km; 1,168 miles)
    - 4,875 m (16,000 ft) 1,119 nm (2,074 km; 1,288 miles)
    - 6,400 m (21,000 ft) 1,264 nm (2,343 km; 1,455 miles)
  - at econ cruising power at:
    - 3,660 m (12,000 ft) 1,108 nm (2,053 km; 1,275 miles)
    - 4,875 m (16,000 ft) 1,205 nm (2,232 km; 1,387 miles)
    - 6,400 m (21,000 ft) 1,325 nm (2,456 km; 1,525 miles)

BEECHCRAFT SUPER KING AIR 200 and B200

US military designations: C-12 and RU-21J

Design of the **Super King Air 200** began in October 1970, construction of the first prototype and first pre-production aircraft starting simultaneously a year later. The first prototype, c/n BB1, flew for the first time on 27 October 1972, followed by the second aircraft, BB2, on 15 December 1972. While the flight tests and testing of a static fuselage were under way, construction of the first production aircraft began in June 1973. FAA certification under FAR Part 23 was awarded on 14 December 1973, the aircraft satisfying also the icing requirements of FAR Part 25.

By comparison with the King Air 100, the Super King Air 200 has increased wing span, basically the same fuselage, a T tail, more powerful engines, additional fuel capacity, increased cabin pressurisation and a higher gross weight. The cargo door fitted to some military versions became available as an option on civil Super King Airs in 1979; first deliveries were for air ambulance use in Libya and commuter operations in Australia.

In August 1974, Beech received a military contract to build and support 34 modified versions of the Super King Air, designated C-12A. Orders for military Super King Airs had increased to 201 by 17 April 1981, of which 156 had been delivered. These are in five versions, as follows:

C-12A. Initial version for US Army (61) and US Air Force (30), with two 559 kW (750 shp) Pratt & Whitney Aircraft of Canada PT6A-38 turboprop engines, each driving a Hartzell three-blade constant-speed fully-feathering reversible-pitch propeller. Weights, loadings and performance given in 1980-81 *Jane's*. Total of 91 ordered by 17 April 1981.

UC-12B. US Navy/Marine Corps version, with 634 kW (850 shp) PT6A-41 turboprop engines, cargo door and high-flotation landing gear. Total of 66 ordered by 17 April 1981, which will extend deliveries into 1982.

C-12C. As C-12A, for US Army (14), but with PT6A-41 engines.

C-12D. As C-12C, for US Army, but with cargo door. Total of 27 ordered by 17 April 1981.

RU-21J. During 1974 the US Army added three Super King Airs of this version to its fleet of special mission aircraft. Under an R and D contract, Beech had modified these antenna-laden aircraft for the US Army's Cefly Lancer programme. They are approved for take-off at a special AUW of 6,804 kg (15,000 lb).

Worldwide deployment of the C-12s began in July

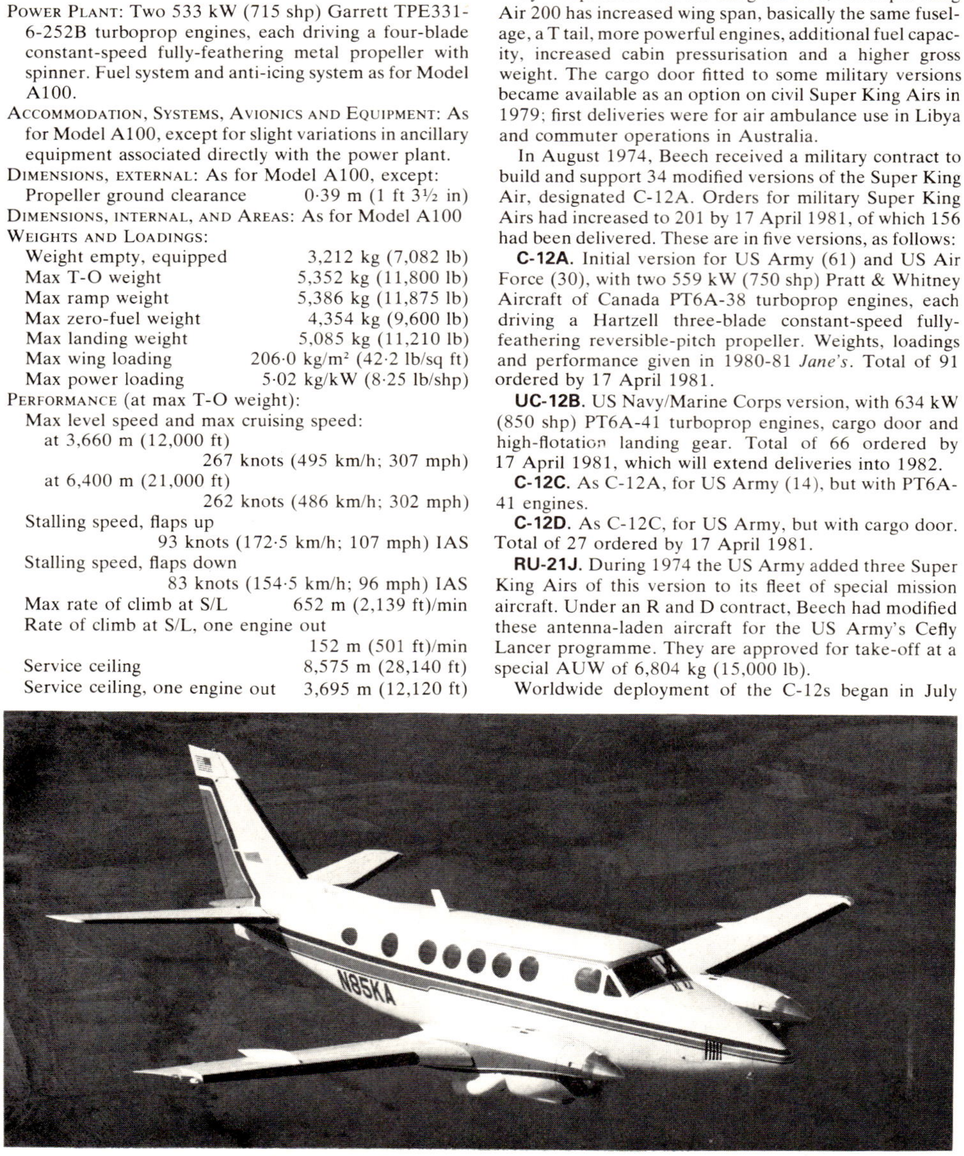

Beechcraft King Air B100 twin-turboprop light passenger, freight or executive transport

1975. They are described as "standard off-the-shelf Super King Air types, modified slightly to meet military flight requirements and to orient the control systems for two-pilot operation which is standard military practice". Accommodation is provided for eight passengers, plus two pilots, with easy conversion to cargo missions. The large baggage area has provisions for storing survival gear.

In February 1977 Beech delivered to the French Institut Géographique National two specially-modified Super King Airs. These have twin Wild RC-10 Superaviogon camera installations and Doppler navigation systems, and were the first Super King Airs to be equipped with optional wingtip fuel tanks, which increase the total usable fuel capacity from 2,059 litres (544 US gallons) to 2,457 litres (649 US gallons) to provide a max endurance of 10·3 h. Designated **Model 200T**, they are fitted with high-flotation main landing gear, and are being operated under a special French airworthiness certificate which allows max T-O and landing weights of 6,350 kg (14,000 lb) and 6,123 kg (13,500 lb) respectively. The aircraft can be operated with or without the wingtip tanks, for high-altitude photographic and weather observation missions.

Beech announced on 25 April 1977 the company's intention to introduce a specially-equipped maritime patrol version of the Super King Air, the **Maritime Patrol 200T**, which is described separately.

During 1978 Beech announced supply to the Egyptian government of a Super King Air which is being used to continue water, uranium and other natural resources exploration in the Sinai and Egyptian deserts which was originated by US ERTS-1 and Landsat satellites. This aircraft is equipped with remote sensing equipment, specialised avionics, and sophisticated cameras. In June 1978, Beech delivered to the government of Taiwan a Super King Air equipped to check ground-based navigation systems. A second special-mission aircraft was delivered to Taiwan's Ministry of the Interior in May 1979. In late 1978 Beech won an industry-wide competition for the lease of four Super King Airs to the border patrol fleet of the US Customs Service, and these were delivered in Spring 1979.

By 17 April 1981 Beech had delivered 844 Super King Airs to commercial and private operators and 156 military C-12s and RU-21Js to the US Air Force, US Navy and US Army.

The **Super King Air B200**, introduced in April 1981, is generally similar to the Super King Air 200, except for the installation of Pratt & Whitney Aircraft of Canada PT6A-42 turboprop engines, which provide better cruise and altitude performance than the PT6A-41s in the original Super King Air 200. In addition, max zero-fuel weight is increased by 272 kg (600 lb) and cabin pressure differential is increased from 0·41 bars (6·0 lb/sq in) to 0·44 bars (6·5 lb/sq in). Four versions are available:

Super King Air B200. Basic version, as detailed.

Super King Air B200C. As Super King Air B200, but with a 1·32 × 1·32 m (4 ft 4 in × 4 ft 4 in) cargo door as standard.

Super King Air B200T. Generally similar to Maritime Patrol 200T, with standard provision to carry removable wingtip tanks to increase maximum fuel capacity by 401 litres (106 US gallons), to a total of 2,460 litres (650 US gallons).

Super King Air B200CT. Version with both cargo door and wingtip tank provisions as standard.

Design of the Super King Air B200 began in March 1980, the prototype being a modified Super King Air 200, c/n BB343. Manufacture of production aircraft began in May 1980, and FAA certification of all four versions was granted on 13 February 1981.

US Army RU-21J, a Beech-modified version of the Super King Air 200, equipped for the Army's Cefly Lancer programme

A full description of the original Super King Air 200 can be found in the 1980-81 *Jane's*. The following description applies to the B200, which is to replace it in production:

Type: Twin-turboprop passenger or executive light transport.

Wings: Cantilever low-wing monoplane. Wing section NACA 23018·5 (modified) at root, NACA 23011·3 at tip. Dihedral 6°. Incidence 3°48′ at root, −1° 7′ at tip. No sweepback at quarter-chord. Two-spar light alloy structure. Conventional ailerons of light alloy construction, with trim tab in port aileron. Single-slotted trailing-edge flaps of light alloy construction. Pneumatic de-icing boots standard.

Fuselage: Light alloy semi-monocoque structure of safe-life design.

Tail Unit: Conventional cantilever T-tail structure of light alloy with swept vertical and horizontal surfaces. Fixed-incidence tailplane. Trim tab in each elevator. Anti-servo tab in rudder. Pneumatic de-icing boots standard, on leading-edge of tailplane only.

Landing Gear: Electrically-retractable tricycle type, with twin wheels on each main unit. Single wheel on steerable nose unit, with shimmy damper. Main units retract forward, nosewheel aft. Beech oleo-pneumatic shock-absorbers. Goodrich main wheels and tyres size 18 × 5·5, pressure 7·25 bars (105 lb/sq in). Oversize and/or 10-ply main-wheel tyres optional. Goodrich nosewheel size 6·50 × 10, with tyre size 22 × 6·75-10, pressure 3·93 bars (57 lb/sq in). Goodrich hydraulic multiple-disc brakes. Parking brake.

Power Plant: Two 634 kW (850 shp) Pratt & Whitney Aircraft of Canada PT6A-42 turboprop engines, each driving a Hartzell three-blade constant-speed fully-feathering reversible-pitch metal propeller with spinner. Bladder type fuel cells in each wing, with main system capacity of 1,461 litres (386 US gallons) and auxiliary system capacity of 598 litres (158 US gallons). Total fuel capacity 2,059 litres (544 US gallons). Two refuelling points in upper surface of each wing. Wingtip tanks optional, providing an additional 401 litres (106 US gallons) and raising maximum capacity to 2,460 litres (650 US gallons). Oil capacity 29·5 litres (7·8 US gallons). Anti-icing of engine air intakes by hot air from engine exhaust is standard. Electrothermal anti-icing for propellers standard; automatic feathering and synchrophaser optional.

Accommodation: Pilot only, or crew of two side by side, on flight deck, with full dual controls and instruments as standard. Six cabin seats standard, each equipped with seat belts and inertia-reel shoulder harness; alternative layouts for a maximum of 13 passengers in cabin and 14th beside pilot. Partition with sliding door between cabin and flight deck, and partition at rear of cabin. Door at rear of cabin on port side, with integral airstair. Large cargo door optional. Inward-opening emergency exit on starboard side over wing. Lavatory and stowage for up to 186 kg (410 lb) baggage in aft fuselage. Maintenance access door in rear fuselage; radio compartment access doors in nose. Standard equipment includes reading lights and fresh air outlets for all passengers, triple cabin windows with polarised glare control, fully-carpeted floor, 'No smoking—Fasten seat belt' sign, cabin coat rack, fluorescent cabin lighting, aisle and door courtesy lights. Electrically-heated windscreens, hot air windscreen defroster, dual storm windows, sun visors, map pockets and windscreen wipers. Cabin is air-conditioned and pressurised, and can be provided with optional radiant heat panels.

Systems: Cabin pressurisation by engine bleed air, with a maximum differential of 0·44 bars (6·5 lb/sq in). Cabin air-conditioner of 34,000 BTU capacity. Auxiliary cabin heating by radiant panels optional. Oxygen system for flight deck, and 0·62 m^3 (22 cu ft) oxygen system for cabin, with automatic drop-down face masks; standard system of 1·39 m^3 (49 cu ft); 1·81 m^3 (64 cu ft) or 2·15 m^3 (76 cu ft) optional. Dual vacuum system for instruments. Hydraulic system for brakes only. Pneumatic system for wing and tailplane de-icing. Electrical system has two 250A 28V starter/generators and a 24V 45Ah aircooled nickel-cadmium battery with failure detector. AC power provided by dual 250VA inverters. Engine fire detection system standard; engine fire extinguishing system optional.

Avionics and Equipment: Standard avionics include dual Collins VHF-20A VHF transceivers, with Gables controls and B3 antennae; Collins VIR-30AGM automatic Omni No. 1, with 331A-3G indicator, Gables control

Beechcraft Super King Air B200 eight/fifteen-seat pressurised transport (two Pratt & Whitney Aircraft of Canada PT6A-42 turboprop engines)

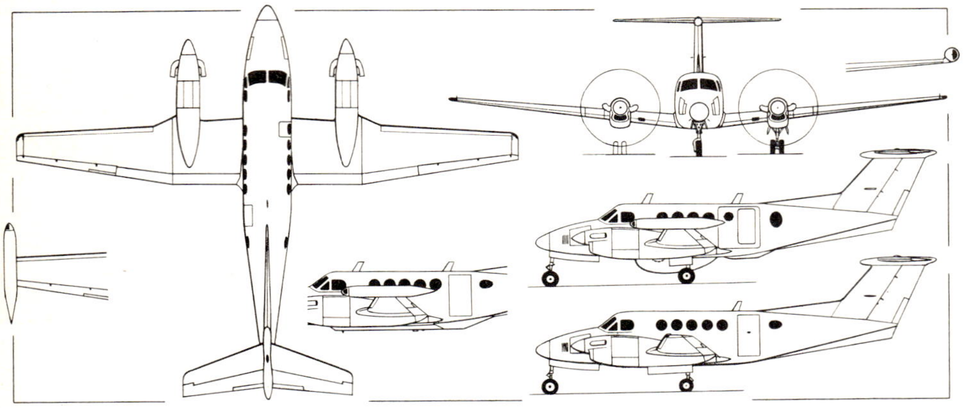

Beechcraft Super King Air 200 twin-turboprop transport, with additional side view of Maritime Patrol 200T (centre right); scrap views of wingtip tanks (left) and centre-fuselage of photo survey aircraft for IGN *(Pilot Press)*

and B17 antenna; Collins VIR-30AG automatic Omni No. 2, with 331H-3G indicator and Gables control; dual Omni range filters; Collins dual 356-F3 audio amplifiers, each with 356C-4 isolation amplifiers and audio switches; Collins ADF-60A ADF less indicator, with Gables control, voice range filter and ANT-60 antenna; Collins marker beacon receiver, integral with VIR-30 No. 1, with dual marker lights and B16 antenna; dual Collins glideslopes, integral with VIR-30 No. 1 and No. 2, with B35 glideslope antenna; Bendix RDR-130 weather radar, with phased array antenna and digital scope; Sperry C-14-43 compass system, with servo amplifier (pilot); Collins 332C-10 RMI, with Nav 1/ADF on single needle, Nav 2/ADF on double needle; Collins TDR-90 transponder, with Gables control and 237Z-1 antenna; Collins DME-40 with 339F-12 indicator, Nav 1/Nav 2 switching, DME hold and 237Z-1 antenna; dual Flite-Tronics PC-250 250VA inverters with failure light; sectional instrument panel; dual flight instrumentation; Standard Electric gyro horizon (pilot); CF gyro horizon and directional gyro (co-pilot); Beech edge-lighted radio panel, radio accessories, microphone key button in pilot's and co-pilot's control wheels, static wicks, and white lighting; dual microphones, headsets and cockpit speakers; and avionics master switch. A wide range of optional avionics by Bendix, Collins, King, RCA, Sperry and SunAir is available to customer's requirements. Standard equipment is generally as listed for King Air C90, plus dual max allowable airspeed indicators, control wheel mounted chronographs, toilet, fluorescent cabin lighting instead of indirect lighting, aisle courtesy light, transistor controlled blue/white cockpit lighting, passenger door light, rudder boost system, and yaw damper system. Optional equipment includes a flight hour recorder, instantaneous vertical speed indicator, cockpit and cabin fire extinguishers, a range of cabin chairs, cabinets and table, flushing toilet, aft cabin air-conditioning installation, passenger door step lights, wingtip recognition lights, strobe lights, and fin illumination lights.

DIMENSIONS, EXTERNAL:
- Wing span — 16·61 m (54 ft 6 in)
- Wing chord at root — 2·18 m (7 ft 1¾ in)
- Wing chord at tip — 0·90 m (2 ft 11⅝ in)
- Wing aspect ratio — 9·8
- Length overall — 13·34 m (43 ft 9 in)
- Height overall — 4·57 m (15 ft 0 in)
- Tailplane span — 5·61 m (18 ft 5 in)
- Wheel track — 5·23 m (17 ft 2 in)
- Wheelbase — 4·56 m (14 ft 11½ in)
- Propeller diameter — 2·50 m (8 ft 2½ in)
- Propeller ground clearance — 0·37 m (1 ft 2½ in)
- Distance between propeller centres — 5·23 m (17 ft 2 in)
- Passenger door: Height — 1·31 m (4 ft 3½ in)
  - Width — 0·68 m (2 ft 2¾ in)
  - Height to sill — 1·17 m (3 ft 10 in)
- Cargo door (optional):
  - Height — 1·32 m (4 ft 4 in)
  - Width — 1·32 m (4 ft 4 in)
- Nose avionics service doors (port and stbd):
  - Max height — 0·57 m (1 ft 10½ in)
  - Width — 0·63 m (2 ft 1 in)
  - Height to sill — 1·37 m (4 ft 6 in)
- Emergency exit (stbd):
  - Height — 0·66 m (2 ft 2 in)
  - Width — 0·50 m (1 ft 7¾ in)

DIMENSIONS, INTERNAL:
- Cabin (from forward to aft pressure bulkhead):
  - Length — 6·71 m (22 ft 0 in)
  - Max width — 1·37 m (4 ft 6 in)
  - Max height — 1·45 m (4 ft 9 in)
  - Floor area — 7·80 m² (84 sq ft)
  - Volume — 11·10 m³ (392 cu ft)
- Baggage hold, rear of cabin:
  - Volume — 1·51 m³ (53·5 cu ft)

AREAS:
- Wings, gross — 28·15 m² (303 sq ft)
- Ailerons (total) — 1·67 m² (18·0 sq ft)
- Trailing-edge flaps (total) — 4·17 m² (44·9 sq ft)
- Fin — 3·46 m² (37·2 sq ft)
- Rudder, incl tab — 1·40 m² (15·1 sq ft)
- Tailplane — 4·52 m² (48·7 sq ft)
- Elevators, incl tabs — 1·79 m² (19·3 sq ft)

WEIGHTS AND LOADINGS:
- Weight empty — 3,419 kg (7,538 lb)
- Max fuel load — 1,653 kg (3,645 lb)
- Max T-O and landing weight — 5,670 kg (12,500 lb)
- Max ramp weight — 5,710 kg (12,590 lb)
- Max zero-fuel weight — 4,990 kg (11,000 lb)
- Max wing loading — 201·6 kg/m² (41·3 lb/sq ft)
- Max power loading — 4·47 kg/kW (7·35 lb/shp)

PERFORMANCE (at max T-O weight ISA, except where indicated):
- Never-exceed speed — 260 knots (482 km/h; 299 mph) IAS
- Max operating Mach No. — 0·52
- Max level speed at 7,620 m (25,000 ft), average cruise weight — 294 knots (545 km/h; 339 mph)
- Max cruising speed at 7,620 m (25,000 ft), average cruise weight — 289 knots (536 km/h; 333 mph)
- Econ cruising speed at 7,620 m (25,000 ft), average cruise weight, normal cruise power — 282 knots (523 km/h; 325 mph)
- Stalling speed, flaps up — 99 knots (183 km/h; 114 mph) IAS
- Stalling speed, flaps down — 75·5 knots (140 km/h; 87 mph) IAS
- Max rate of climb at S/L — 747 m (2,450 ft)/min
- Rate of climb at S/L, one engine out — 226 m (740 ft)/min
- Service ceiling — over 10,670 m (35,000 ft)
- Service ceiling, one engine out — 6,625 m (21,735 ft)
- T-O run — 592 m (1,942 ft)
- T-O to 15 m (50 ft), 40% flap — 786 m (2,579 ft)
- Landing from 15 m (50 ft) — 867 m (2,845 ft)
- Landing run — 536 m (1,760 ft)
- Range with max fuel, allowances for start, taxi, climb, descent, and 45 min reserves at max range power, ISA:
  - max cruising power at:
    - 5,485 m (18,000 ft) — 1,192 nm (2,209 km; 1,372 miles)
    - 7,620 m (25,000 ft) — 1,461 nm (2,707 km; 1,682 miles)
    - 9,450 m (31,000 ft) — 1,766 nm (3,273 km; 2,033 miles)
    - 10,670 m (35,000 ft) — 1,972 nm (3,654 km; 2,271 miles)
  - econ cruising power at:
    - 5,485 m (18,000 ft) — 1,517 nm (2,811 km; 1,747 miles)
    - 7,620 m (25,000 ft) — 1,802 nm (3,339 km; 2,075 miles)
    - 9,450 m (31,000 ft) — 1,974 nm (3,658 km; 2,273 miles)
    - 10,670 m (35,000 ft) — 2,027 nm (3,756 km; 2,334 miles)

BEECHCRAFT MARITIME PATROL 200T

After investigating the potential market for a maritime patrol version of its Super King Air 200 twin-turboprop light transport, Beech announced on 9 April 1979 that it had begun to flight test such an aircraft for FAA certification as the Maritime Patrol 200T.

In production form, the 200T can be equipped for missions such as monitoring exclusive economic zones, detecting pollution, inspecting offshore installations, and conducting search and rescue flights. Special missions for which it could also be used include aerial photography, environmental and ecological research, airways and ground-based navigation equipment checks, and ambulance duties.

Modifications to the standard Super King Air to adapt it to Maritime Patrol 200T configuration include fitting new outboard wing assemblies, with mountings for a 200·5 litre (53 US gallon) removable fuel tank at each wingtip; strengthened landing gear to cater for higher take-off and landing weights; two bubble observation windows in the aft cabin for visual search and photography; a hatch for dropping survival equipment; and a search radar with full 360° scan in a radome beneath the fuselage. Advanced navigation equipment is available, especially for maritime patrol use; standard avionics include VLF/Omega which provides ground stabilisation and is coupled with the autopilot. This permits a search pattern to be programmed before take-off or en route.

Thirteen Maritime Patrol 200Ts bought by Japan's Maritime Safety Agency had been delivered by 22 September 1981, together with one for the Uruguayan Navy.

The description of the Super King Air B200 applies also to the 200T, except as follows:

TYPE: Twin-turboprop maritime patrol or multi-mission aircraft.

WINGS: As for Super King Air B200, except for new outboard wing panels redesigned to permit mounting of removable wingtip tanks.

LANDING GEAR: Strengthened to cater for higher operating weights.

POWER PLANT: As for Super King Air B200, plus provision for removable wingtip tanks to increase maximum fuel capacity by 401 litres (106 US gallons), to a total of 2,460 litres (650 US gallons).

AVIONICS AND EQUIPMENT: Standard items as detailed in introductory description. Optional avionics include INS, VHF/FM com, HF and VHF com, FLIR, LLLTV, multispectral scanner, tactical navigation computer, and two alternative search radar systems, both with 360° scan and weather avoidance capability.

DIMENSIONS, EXTERNAL: As for Super King Air B200, except:
- Wing span over tip-tanks — 17·25 m (56 ft 7 in)
- Wing aspect ratio — 10·5

DIMENSIONS, INTERNAL: As for Super King Air B200, except:
- Cabin: Length (excl flight deck) — 5·08 m (16 ft 8 in)

WEIGHTS (A: Normal category; B: Restricted category):
- Weight empty: A, B — 3,744 kg (8,255 lb)
- Max T-O weight: A — 5,670 kg (12,500 lb)
  - B — 6,350 kg (14,000 lb)
- Max landing weight: A — 5,670 kg (12,500 lb)
  - B — 6,123 kg (13,500 lb)

PERFORMANCE (at max T-O weight):
- Max cruising speed, AUW of 4,990 kg (11,000 lb) at 4,265 m (14,000 ft) — 265 knots (491 km/h; 305 mph)
- Typical patrol speed — 140 knots (259 km/h; 161 mph)
- Range with max fuel, patrolling at 227 knots (420 km/h; 261 mph) at 825 m (2,700 ft), 45 min reserves — 1,790 nm (3,317 km; 2,061 miles)
- Typical endurance at 140 knots (259 km/h; 161 mph), at 610 m (2,000 ft), 45 min reserves — 6 h 36 min

Rear view of Beechcraft Maritime Patrol 200T operated by the Uruguayan Navy, showing clearly the under-fuselage radome

BELL

BELL HELICOPTER TEXTRON (Division of Textron Inc)

HEAD OFFICE: PO Box 482, Fort Worth, Texas 76101
Telephone: (817) 280 2011
PRESIDENT: James F. Atkins
EXECUTIVE VICE-PRESIDENT: L. M. Horner
SENIOR VICE-PRESIDENTS:
Dwayne K. Jose (Marketing)
Robert R. Lynn (Research & Engineering)
Richard K. May (Operations)
Charles R. Rudning (Programmes)
Hans Weichsel Jr (Gen Manager, Customer Support and Service Division)
VICE-PRESIDENTS:
John A. Buyers (Model 222 Programme)
Charles E. Davis (Product Engineering)
Jan M. Drees (Technology)
George Galerstein (Law)
Webb F. Joiner (Finance)
Clifford J. Kalista (International Marketing)
Gainor J. Lindsey (Asst Gen Manager, Customer Support and Service Division)
C. E. McGuire (Commercial Marketing)
C. M. McKeen Jr (Procurement)
Stanley Martin (Development Engineering)
Phil C. Norwine (Market Development)
Edmund Szol (Employee Relations)
George G. Troutman (Washington Operations)
DIRECTOR, PUBLIC AFFAIRS: Carl L. Harris
GENERAL MANAGER, AMARILLO FACILITY: Jack A. Tarpley

Available details of the range of military and commercial helicopters in current production, or under development, by Bell Helicopter Textron are published in this entry. Several models are also built under licence by Agusta in Italy and Fuji in Japan (which see).

During 1972 Bell achieved a major breakthrough in the elimination of vibration in helicopters with what is known as the nodalisation concept, details of which can be found in the 1980-81 and earlier editions of *Jane's*. Flight tests of a Model 206 JetRanger with its fuselage suspended from a nodalised beam were so convincing that Bell decided to utilise this 'Noda-Matic' technique on new production helicopters, including the Models 206L LongRanger, 214, 214ST, 222 and 412.

Approximately 9,000 people were employed by Bell in 1981. The 25,000th Bell helicopter, a Model 222, was delivered on 18 January 1981.

BELL MODEL 205

US military designations: UH-1D/H, EH-1H and HH-1H Iroquois

Canadian military designation: CH-118 Iroquois

Although basically similar to the earlier Model 204 (see 1971-72 *Jane's*), the Model 205 introduced a longer fuselage, increased cabin space to accommodate a much larger number of passengers, and other changes. Details of the full range of military variants have been recorded in previous editions of *Jane's*. The final production version is the **UH-1H**. Deliveries of this version to the US Army began in September 1967 and 1,242 had been built by 1976, including nine for the RNZAF. Production of the UH-1H continued subsequently to satisfy export orders and a new contract for 27 for the US Army. In addition, under a licensing agreement, 118 UH-1Hs were produced for the Nationalist Chinese Army, with much of the manufacturing and assembly carried out by AIDC/CAF (which see) at Taichung, Taiwan.

Up to FY 1980, a total of $47 million was provided to modify UH-1Hs to **EH-1H** electronic countermeasures configuration, with the Quick Fix I airborne communications interception, emitter location and jamming system, including an AN/APR-39V2 radar warning receiver, XM130 chaff/flare dispenser and AN/ALQ-144 infra-red jammer. The FY 1981 budget added $5·1 million to convert initial Quick Fix IA systems in the EH-1H to Phase IB configuration, plus survivability equipment to protect the aircraft against known and postulated threats, including hot metal/plume suppression. By April 1981, three of the EH-1Hs had been delivered, with seven more to follow.

Four UH-1H aircraft have been modified for field evaluation of the US Army Stand-Off Target Acquisition System (I²SOTAS). General Dynamics' Electronics Division is prime contractor for this system, which will give battlefield commanders real-time radar moving target information on the battle situation, relayed from the helicopters. Added equipment includes a large plank-like antenna under the fuselage and Sperry coupled four-axis autopilot and three-cue flight director for precise helicopter stabilisation at very low speeds. In addition, the landing gear skids have been made retractable to allow for operation of the rotating antenna.

The US Army plans to retain at least 2,700 improved UH-1Hs in service beyond the year 2000 to perform such operations as resupply, troop transport, command and control, electronic warfare, medical evacuation and minefield emplacement. To make such a plan realistic, the current product improvement programme for the Army's UH-1H fleet introduces improved or new avionics and equipment that include an AN/ALQ-144 infra-red jammer, AN/APN-209 radar altimeter, AN/APR-39 radar warning receiver, AN/ARC-164 UHF/AM radio, AN/ARN-124 DME, XM130 chaff/flare dispenser, NOE communications (FM/HF), communications security, infra-red suppressor (hot metal and plume), altimeter lighting (5V), crashworthy auxiliary fuel system, closed-circuit refuelling, fuel tank vent, improved main input driveshaft, and main rotor mast plug. In addition to the above, it is planned to introduce, as a minimum, new composite main rotor blades; improved stabiliser bar, tail rotor hub and servo cylinders; a split engine deck, and improved oil filtration; a night-vision compatible cockpit; built-in Vibrex connections; an improved AN/ASN-43 gyro magnetic compass; and Doppler navigation.

The following details refer specifically to the current military UH-1H. The commercial Model 205A-1, developed from it, is described separately.

TYPE: Single-rotor general-purpose helicopter.

ROTOR SYSTEM: Two-blade all-metal semi-rigid main rotor with interchangeable blades, built up of extruded aluminium spars and laminates. Stabilising bar above and at right angles to main rotor blades. Underslung feathering axis hub. Two-blade all-metal tail rotor of honeycomb construction. Blades do not fold.

ROTOR DRIVE: Shaft-drive to both main and tail rotors. Transmission rating 820 kW (1,100 shp). Main rotor rpm 294-324.

FUSELAGE: Conventional all-metal semi-monocoque structure.

TAIL SURFACE: Small synchronised elevator on rear fuselage is connected to the cyclic control to increase allowable CG travel.

LANDING GEAR: Tubular skid type. Lock-on ground handling wheels and inflated nylon float-bags available.

POWER PLANT: One 1,044 kW (1,400 shp) Avco Lycoming T53-L-13 turboshaft engine, mounted aft of the transmission on top of the fuselage and enclosed in cowlings. Five interconnected rubber fuel cells, total capacity 844 litres (223 US gallons), of which 799 litres (211 US gallons) are usable. Overload fuel capacity of 1,935 litres (511 US gallons) usable, obtained by installation of kit comprising two 568 litre (150 US gallon) internal auxiliary fuel tanks interconnected with the basic fuel system.

ACCOMMODATION: Cabin space of 6·23 m³ (220 cu ft) provides sufficient room for pilot and 11-14 troops, or six litters and a medical attendant, or 1,759 kg (3,880 lb) of freight. Crew doors open forward and are jettisonable. Two doors on each side of cargo compartment; front door is hinged to open forward and is removable, rear door slides aft. Forced air ventilation system.

AVIONICS AND EQUIPMENT: FM, UHF, VHF radio sets, IFF transponder, Gyromatic compass system, direction finder set, VOR receiver and intercom standard. Optional nav/com systems. Standard equipment includes bleed air heater and defroster, comprehensive range of engine and flight instruments, power plant fire detection system, 30V 300A DC starter/generator, navigation, landing and anti-collision lights, controllable searchlight, hydraulically-boosted controls. Optional equipment includes external cargo hook, auxiliary fuel tanks, rescue hoist, 150,000 BTU muff heater.

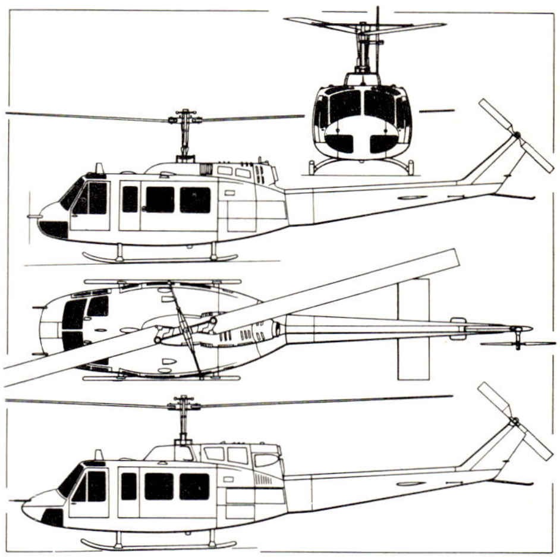

Bell UH-1H Iroquois, with additional side view (bottom) of UH-1N, described on page 299 *(pilot Press)*

Bell UH-1H in the US Army's I²SOTAS configuration

DIMENSIONS, EXTERNAL:
Diameter of main rotor 14·63 m (48 ft 0 in)
Diameter of tail rotor 2·59 m (8 ft 6 in)
Main rotor blade chord 0·53 m (1 ft 9 in)
Tail rotor blade chord 0·213 m (8·4 in)
Length overall (main rotor fore and aft) 17·62 m (57 ft 9⅝ in)
Length of fuselage 12·77 m (41 ft 10¾ in)
Height overall, tail rotor turning (excl fin-tip antenna) 4·41 m (14 ft 5½ in)
Height to top of main rotor hub 3·60 m (11 ft 9¾ in)
Stabiliser span 2·84 m (9 ft 4 in)
Width over skids 2·91 m (9 ft 6½ in)

DIMENSIONS, INTERNAL:
Cabin: Max width 2·34 m (7 ft 8 in)
Max height 1·25 m (4 ft 1¼ in)
Volume (excl flight deck) approx 6·23 m³ (220 cu ft)

AREAS:
Main rotor disc 168·06 m² (1,809 sq ft)
Tail rotor disc 5·27 m² (56·7 sq ft)

WEIGHTS AND LOADINGS:
Weight empty, equipped 2,363 kg (5,210 lb)
Basic operating weight (troop carrier mission) 2,520 kg (5,557 lb)
Mission weight 4,100 kg (9,039 lb)
Max T-O and landing weight 4,309 kg (9,500 lb)
Max zero-fuel weight 3,660 kg (8,070 lb)
Max disc loading 25·6 kg/m² (5·25 lb/sq ft)
Max power loading 4·13 kg/kW (8·63 lb/shp)

PERFORMANCE (at max T-O weight):
Never-exceed speed 110 knots (204 km/h; 127 mph)
Max level and cruising speed 110 knots (204 km/h; 127 mph)
Econ cruising speed at 1,735 m (5,700 ft) 110 knots (204 km/h; 127 mph)
Max rate of climb at S/L 488 m (1,600 ft)/min
Service ceiling 3,840 m (12,600 ft)
Hovering ceiling IGE 4,145 m (13,600 ft)
Hovering ceiling OGE 1,220 m (4,000 ft)
Range with max fuel, no allowances, no reserves, at S/L 276 nm (511 km; 318 miles)

BELL MODEL 205A-1

The Model 205A-1 is a fifteen-seat commercial utility helicopter developed from the UH-1H, with a 1,044 kW (1,400 shp) Avco Lycoming T5313B turboshaft engine, derated to 932 kW (1,250 shp) for take-off. It is designed for rapid conversion for alternative air freight, flying crane, ambulance, rescue and executive roles. Total cargo capacity is 7·02 m³ (248 cu ft) including baggage space in tailboom, with 2·34 m (7 ft 8 in) by 1·24 m (4 ft 1 in) door openings on each side of the cabin to facilitate loading of bulky freight. External load capacity in flying crane role is 2,268 kg (5,000 lb). The ambulance version can accommodate six litter patients and one or two medical attendants.

Normal fuel capacity is 814 litres (215 US gallons); optional capacity is 1,495 litres (395 US gallons).

The Model 205A-1 is also produced under licence in Italy by Agusta (which see) as the AB 205A-1. By mid-1980, more than 9,000 Model 205s had been built for commercial and military use by Bell and its licensees. One (c/n 30038, PK-HBK), operated by PT Bristow Masayu

Helicopters of Indonesia, had logged 21,036 flying hours by 1 May 1980.

The description of the Bell UH-1H applies also to the Model 205A-1, except for the following details:

TYPE: Fifteen-seat commercial utility helicopter.

AVIONICS AND EQUIPMENT: Standard avionics comprise 360-channel VHF transceiver and intercom system. Extensive range of optional nav/com systems available. Standard equipment includes vertical gyro system, 5 in gyro attitude indicator, gyro compass, master caution panel, force trim hydraulic boost controls, twin windscreen wipers, soundproof headliner, cabin fire extinguisher, map case, passenger retractable boarding steps, engine fire extinguishing system and bleed air heating. Optional items include dual controls, customised interiors, high-output cabin heater, litter installations, auxiliary fuel tanks, rescue hoist, rotor brake, float landing gear, external cargo suspension, and protective covers.

DIMENSIONS, EXTERNAL:
Length of fuselage 12·65 m (41 ft 6 in)
Height overall 4·39 m (14 ft 4¾ in)

WEIGHTS:
Weight empty, equipped 2,414 kg (5,323 lb)
Normal T-O weight 4,309 kg (9,500 lb)
Max T-O weight, external load 4,763 kg (10,500 lb)

PERFORMANCE (at normal T-O weight):
Max level speed from S/L to 915 m (3,000 ft) 110 knots (204 km/h; 127 mph)
Max cruising speed at S/L 110 knots (204 km/h; 127 mph)
Max cruising speed at 2,440 m (8,000 ft) 96 knots (179 km/h; 111 mph)
Max rate of climb at S/L 512 m (1,680 ft)/min
Max vertical rate of climb at S/L 259 m (850 ft)/min
Service ceiling 4,480 m (14,700 ft)
Hovering ceiling IGE 3,170 m (10,400 ft)
Hovering ceiling OGE 1,830 m (6,000 ft)
Range at S/L, at max cruising speed 270 nm (500 km; 311 miles)
Range at 2,440 m (8,000 ft) at max cruising speed, no reserves 298 nm (553 km; 344 miles)

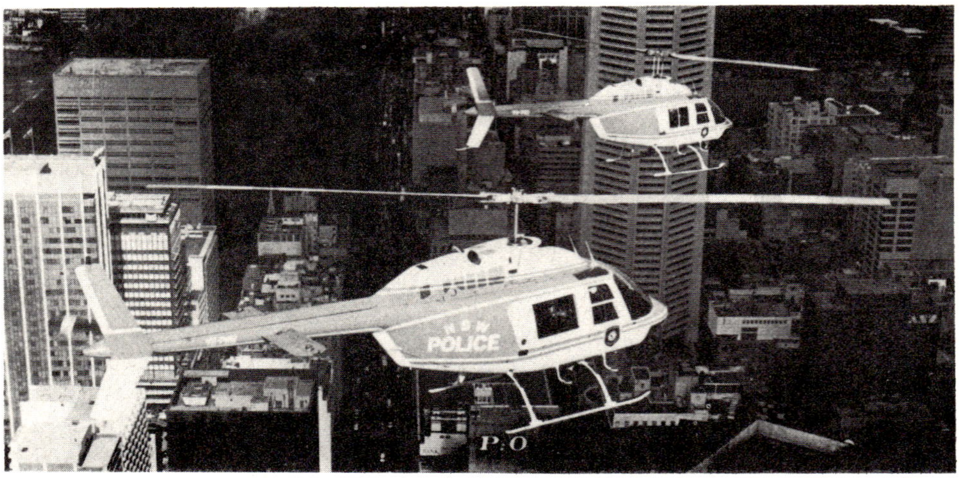

Bell Model 206B JetRanger IIIs of New South Wales police patrolling over Sydney

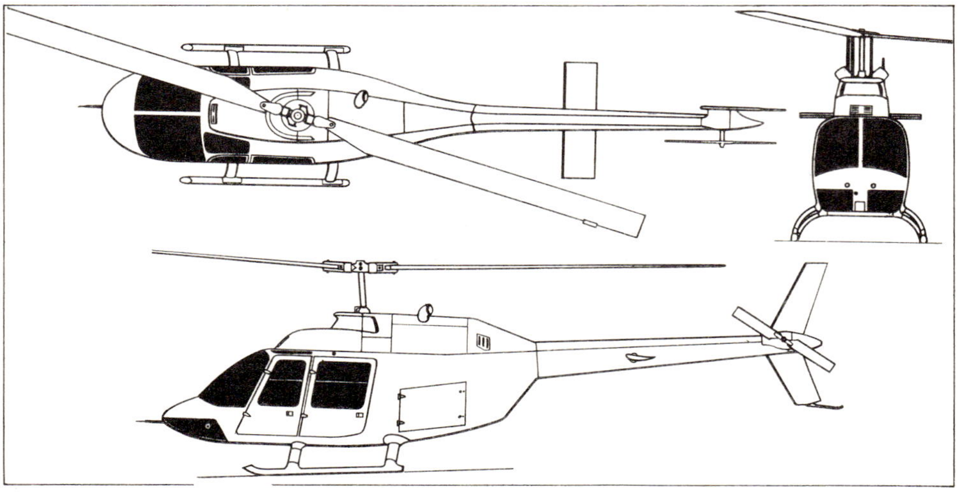

Bell 206B JetRanger III light utility helicopter *(Pilot Press)*

BELL MODEL 206B JETRANGER III

In the Summer of 1977, Bell began delivery of the Model 206B JetRanger III, which subsequently replaced in production the lower-powered JetRanger II, of which 1,619 were delivered.

Power plant of the JetRanger III is the Allison 250-C20B turboshaft, which Bell was able to install with minimal modification of the original airframe to meet requests for higher performance under hot-day/high-altitude conditions. This enables Bell to offer modification kits to convert JetRanger IIs to JetRanger III standard.

Under a succession of major contracts, Beech Aircraft has produced airframes for all the commercial and military versions of the JetRanger, the first airframe being delivered to Bell on 1 March 1968. The work involves manufacture of the fuselage, skid gear, tailboom, spars, stabiliser and two rear fairing assemblies.

By January 1980, Bell and its licensees had manufactured well over 6,000 helicopters of the Model 206 series, more than 2,300 of them for commercial customers.

TYPE: Turbine-powered general-purpose light helicopter.

ROTOR SYSTEM: Two-blade semi-rigid see-saw type main rotor, employing pre-coning and underslinging to ensure smooth operation. Blades are of standard Bell 'droop-snoot' section. They have a D-shape aluminium spar, bonded aluminium alloy skin, honeycomb core and a trailing-edge extension. Each blade is connected to the hub by means of a grip, pitch-change bearings and a tension-torsion strap assembly. Two tail rotor blades have bonded aluminium skin but no core. Main rotor blades do not fold, but modification to permit manual folding is possible. Rotor brake available as optional kit.

ROTOR DRIVE: Rotors driven through tubular steel alloy shafts with spliced couplings. Initial drive from engine through 90° spiral bevel gear to single-stage planetary main gearbox. Shaft to tail rotor single-stage bevel gearbox. Freewheeling unit ensures that main rotor continues to drive tail rotor when engine is disengaged. Main rotor/engine rpm ratio 1 : 15; main rotor rpm 374-394. Tail rotor/engine rpm ratio 1 : 2·3.

FUSELAGE: Forward cabin section is made up of two aluminium alloy beams and 25 mm (1 in) thick aluminium honeycomb sandwich. Rotor, transmission and engine are supported by upper longitudinal beams. Upper and lower structures are interconnected by three fuselage bulkheads and a centrepost to form an integrated structure. Intermediate section is of aluminium alloy semi-monocoque construction. Aluminium monocoque tailboom.

TAIL UNIT: Fixed stabiliser of aluminium monocoque construction, with inverted aerofoil section. Fixed vertical tail-fin in sweptback upper and ventral sections, made of aluminium honeycomb with aluminium alloy skin.

LANDING GEAR: Aluminium alloy tubular skids bolted to extruded cross-tubes. Tubular steel skid on ventral fin to protect tail rotor in tail-down landing. Special high skid gear (0·25 m; 10 in greater ground clearance) available for use in areas with high brush. Inflated bag-type pontoons or stowed floats, capable of in-flight inflation, available as optional kits.

POWER PLANT: One 313 kW (420 shp) Allison 250-C20B turboshaft engine, flat rated to 236 kW (317 shp). Fuel tank below and behind rear passenger seat, capacity 288 litres (76 US gallons). Refuelling point on starboard side of fuselage, aft of cabin. Oil capacity 5·2 litres (5·5 US quarts).

ACCOMMODATION: Two seats side by side in front and rear bench seat for three persons. Dual controls optional. Two forward-hinged doors on each side, made of formed aluminium alloy with transparent panels. Baggage compartment aft of rear seats, capacity 113 kg (250 lb), with external door on port side.

SYSTEMS: Hydraulic system, pressure 41·5 bars (600 lb/sq in), for cyclic, collective and directional controls. Electrical supply from 150A starter/generator. One 24V 13Ah nickel-cadmium battery.

AVIONICS AND EQUIPMENT: Full range of avionics available in form of optional kits, including VHF communications and omni navigation kit, Collins APS-841H autopilot, glideslope kit, ADF, DME, marker beacon receiver, transponder and intercom and speaker system. Standard equipment includes cabin fire extinguisher, first aid kit, door locks, night lighting, and dynamic flapping restraints. Optional items include clock, engine hour meter, turn and slip indicator, stability and control augmentation system, custom seating, internal litter kit, cabin heater, environmental control system, camera access door, high-intensity night lights, engine fire detection system, and external cargo sling of 545 kg (1,200 lb) capacity.

DIMENSIONS, EXTERNAL:
Diameter of main rotor 10·16 m (33 ft 4 in)
Diameter of tail rotor 1·57 m (5 ft 2 in)
Main rotor blade chord 0·33 m (1 ft 1 in)
Distance between rotor centres 5·96 m (19 ft 6½ in)
Length overall, rotors turning 11·82 m (38 ft 9½ in)
Length of fuselage, incl tailskid 9·50 m (31 ft 2 in)
Height over tail fin 2·54 m (8 ft 4 in)
Height overall 2·91 m (9 ft 6½ in)
Stabiliser span 1·97 m (6 ft 5¾ in)
Width over skids 1·92 m (6 ft 3½ in)

DIMENSIONS, INTERNAL:
Cabin: Length 2·13 m (7 ft 0 in)
Max width 1·27 m (4 ft 2 in)
Max height 1·28 m (4 ft 3 in)
Baggage compartment volume 0·45 m³ (16 cu ft)

AREAS:
Main rotor blades (total) 3·35 m² (36·1 sq ft)
Tail rotor blades (total) 0·22 m² (2·37 sq ft)
Main rotor disc 81·1 m² (873 sq ft)
Tail rotor disc 1·95 m² (20·96 sq ft)
Stabiliser 0·90 m² (9·65 sq ft)

WEIGHTS:
Weight empty, standard configuration 730 kg (1,610 lb)
Max T-O weight 1,451 kg (3,200 lb)

PERFORMANCE (at max T-O weight, ISA):
Never-exceed speed at S/L 122 knots (225 km/h; 140 mph)
Max cruising speed at 1,525 m (5,000 ft) 116 knots (216 km/h; 134 mph)
Max cruising speed at S/L 115 knots (214 km/h; 133 mph)
Max rate of climb at S/L 384 m (1,260 ft)/min
Vertical rate of climb at S/L 91 m (300 ft)/min
Service ceiling 4,115 m (13,500 ft)
Hovering ceiling IGE 3,900 m (12,800 ft)
Hovering ceiling OGE 2,680 m (8,800 ft)
Range with max fuel and max payload at S/L, no reserves 296 nm (549 km; 341 miles)
Range with max fuel and max payload at 1,525 m (5,000 ft), no reserves 328 nm (608 km; 378 miles)

BELL KIOWA and MODEL 406 (AHIP)

US Army designation: OH-58
US Navy designation: TH-57A SeaRanger
Canadian military designation: CH-136

On 8 March 1968 the US Army named Bell as winner of its reopened light observation helicopter competition, and awarded the company the first increment of a total order for 2,200 **OH-58A** Kiowa aircraft, generally similar to the Model 206A and each powered by a 236·5 kW (317 shp) Allison T63-A-700 turboshaft engine. Major difference concerns the main rotor, that of the Kiowa having an increased diameter. There are also differences in the internal layout and avionics.

The first OH-58A was delivered to the US Army on 23 May 1969 and deployment in Viet-Nam began in the early Autumn of 1969. During the previous year, the US Navy took delivery of 40 **TH-57A SeaRanger** primary training helicopters, also based on the Model 206A; seven more were ordered in FY 1981.

Seventy-four similar COH-58As were delivered to the Canadian Armed Forces, and are now designated **CH-136**. Fourteen of them, used for basic helicopter pilot training at Portage la Prairie, Manitoba, were replaced by 14 new JetRanger IIIs from May 1981, and reissued to other units.

Under a co-production agreement with the Australian government 56 Model 206B-1 Kiowa military light observation helicopters (similar to the OH-58A) were delivered over an eight-year period. The initial 12 206B-1s were built by Bell. Commonwealth Aircraft Corporation was prime Australian licensee, with responsibility for final assembly of the remainder. Only the engines and avionics were supplied from US sources. Delivery of 12 **OH-58Bs** to the Austrian Air Force was completed in 1976.

Under a US Army development qualification contract placed on 30 June 1976, Bell converted an OH-58A to an improved standard, under the designation **OH-58C**. This involved installation of a flat glass canopy to reduce glint, an uprated Allison T63-A-720 turboshaft engine, and an IR reduction package. Two additional OH-58As were modified to OH-58C configuration, for pre-production flight testing by Bell and the US Army, and production modification of 275 OH-58As to OH-58C standard began in March 1978 at Bell Helicopter's Amarillo plant. The final configuration includes a new instrument panel, modifications to reduce vulnerability in combat, CONUS (continental US) navigation equipment, day optics, improved avionics and improved maintenance features. The additional power significantly improves high-altitude, hot-weather performance.

A prototype mast-mounted sight, manufactured by Rockwell, was installed and flight tested on an OH-58C in 1979-80. The sight consists of a TV camera with automatic tracking, and a laser designator/rangefinder.

In February 1981, Bell notified the US Army Aviation Research and Development Command that it would participate in the Army's competitive programme for a near-term scout helicopter, as outlined in the Army's helicopter improvement programme (AHIP) request for proposals. Its winning **Model 406** entry will modify 720 existing OH-58s from the Army's inventory to have a four-blade composite main rotor and larger, slower-turning tail rotor, plus a McDonnell Douglas Astronautics Company mast-mounted sight, with Northrop TV telescope, FLIR and laser designator. Sperry Flight Systems will be responsible for the helicopter visual control, display and multiplex system. Power plant will be an Allison 250-C30R turboshaft engine. (See Addenda).

The following details apply specifically to the OH-58C:

TYPE: Turbine-powered light observation helicopter.

ROTOR SYSTEM: Two-blade semi-rigid see-saw type main rotor, employing pre-coning and underslinging to ensure smooth operation. Blades of standard Bell 'droop-snoot' section, with D-shape aluminium spar, bonded light alloy skin, honeycomb core and trailing-edge extension. Each blade is connected to the hub by means of a grip, pitch-change bearings and a tension-torsion strap assembly. The two tail rotor blades have bonded aluminium skin but no core. Main rotor blades do not fold, but modification to permit manual folding is possible. Rotor brake available as optional kit.

ROTOR DRIVE: Rotors driven through tubular steel alloy shafts with spliced couplings. Initial drive from engine through 90° spiral bevel gear to single-stage planetary main gearbox. Transmission improvements include a four-pinion upper planetary, with new thrust bearing and 'fly dry' capability. Shaft to tail rotor single-stage bevel gearbox protected by cover. Freewheeling unit ensures that main rotor continues to drive tail rotor when engine is disengaged. Redundant tail rotor control for improved safety if primary system is disabled. Main rotor/engine rpm ratio 1 : 17·44; main rotor rpm 354. Tail rotor/engine rpm ratio 1 : 2·353.

FUSELAGE: Forward cabin section is made up of two aluminium alloy beams and 25 mm (1 in) thick aluminium honeycomb sandwich. Rotor, transmission and engine are supported by upper longitudinal beams. Upper and lower structures are interconnected by three fuselage bulkheads and a centrepost to form an integrated structure. Intermediate section is a light alloy semi-monocoque. Aluminium monocoque tailboom. A low-glare canopy design reduces the solar glint signature. The windscreens are slightly convex to assist rain removal and increase their strength.

TAIL UNIT: Fixed stabiliser of aluminium monocoque construction, with inverted aerofoil section. Fixed vertical fin in sweptback upper and ventral sections, constructed of aluminium honeycomb with light alloy skins.

LANDING GEAR: Light alloy tubular skids bolted to extruded cross-tubes. Tubular steel skid on ventral fin to protect tail rotor in tail-down landing. Special high skid gear available, with 0·25 m (10 in) greater ground clearance, for use in areas with high brush. Inflated bag-type pontoons, or stowed floats capable of in-flight inflation, available as optional kits.

POWER PLANT: One 313 kW (420 shp) Allison T63-A-720 turboshaft engine. 'Black Hole' exhaust stacks and hot metal shroud for infra-red suppression. Fuel tank below and behind aft passenger seat, total usable capacity 276 litres (73 US gallons). Refuelling point on starboard side of fuselage, aft of cabin. Oil capacity 5·6 litres (1·5 US gallons).

ACCOMMODATION: Forward crew compartment seats pilot and co-pilot/observer side by side. Entrance to this compartment is provided by single door on each side of fuselage. The cargo/passenger compartment, which has its own doors, one on each side, provides approximately 1·13 m³ (40 cu ft) of cargo space, or provision for two passengers by installation of two seat cushions, seat belts and shoulder harnesses. A redesigned instrument panel houses new avionics, and all flight instruments have been modified for night operations using night vision goggles. An improved defrost/defog air circulation system increases the aircraft's mission readiness.

Bell Model 206L-1 LongRanger II of Houston Helicopters, equipped with pop-out floats

SYSTEMS: Hydraulic system, pressure 41·5 bars (600 lb/sq in) for cyclic and collective controls. Electrical supply from 150A starter/generator. One 24V 13Ah nickel-cadmium battery.

AVIONICS: C-6533/ARC intercommunication subsystem, AN/ARC-114 VHF/FM, AN/ARC-115 VHF/AM, AN/ARC-116 UHF/AM, AN/ARN-89 ADF, AN/ASN-43 gyro magnetic compass, AN/APX-100 IFF transponder, TSEC/KY-28 communications security set, C-8157/ARC control indication, MT-3802/ARC mounting, TS-1843/APX transponder test set and mounting, KIT-1A/TSEC computer and mounting, duplicate AN/ARC-114, AN/APR-39 radar warning, ID-1351 C/A HBI, ID-1347 C/ARN CDI; and provisions for AN/ARN-123(V)1 CONUS nav, AN/APN-209 radar altimeter and YG-1054 proximity warning.

ARMAMENT: Standard equipment is the M-27 armament kit, utilising the 7·62 mm Minigun.

DIMENSIONS, EXTERNAL: As JetRanger III, except:
Diameter of main rotor 10·77 m (35 ft 4 in)
Length overall, rotors turning 12·49 m (40 ft 11¾ in)
Length of fuselage 9·93 m (32 ft 7 in)

AREAS: As JetRanger III, except:
Main rotor blades (total) 3·55 m² (38·26 sq ft)
Main rotor disc 90·93 m² (978·8 sq ft)

WEIGHTS AND LOADINGS (A: OH-58A; C: OH-58C):
Weight empty: A 664 kg (1,464 lb)
C 719 kg (1,585 lb)
Operating weight: A 1,049 kg (2,313 lb)
C 1,104 kg (2,434 lb)
Max T-O and landing weight: A 1,360 kg (3,000 lb)
C 1,451 kg (3,200 lb)
Max zero-fuel weight: A 1,145 kg (2,525 lb)
C 1,200 kg (2,646 lb)
Max disc loading: A 14·9 kg/m² (3·07 lb/sq ft)
C 15·9 kg/m² (3·27 lb/sq ft)

PERFORMANCE (OH-58A at observation mission gross weight of 1,255 kg; 2,768 lb, ISA, except where indicated):
Never-exceed speed at S/L 120 knots (222 km/h; 138 mph)
Cruising speed for max range 102 knots (188 km/h; 117 mph)
Loiter speed for max endurance 49 knots (90·5 km/h; 56 mph)
Max rate of climb at S/L 543 m (1,780 ft)/min
Service ceiling 5,760 m (18,900 ft)
Hovering ceiling IGE 4,145 m (13,600 ft)
Hovering ceiling OGE 2,680 m (8,800 ft)
Hovering ceiling OGE (armed scout mission at 1,360 kg; 3,000 lb) 1,830 m (6,000 ft)
Max range at S/L, 10% reserves 259 nm (481 km; 299 miles)
Max range at S/L, armed scout mission at 1,360 kg (3,000 lb), no reserves 264 nm (490 km; 305 miles)
Endurance at S/L, no reserves 3 h 30 min

BELL MODEL 206L-1 LONGRANGER II, III and IV

Announced on 25 September 1973, Bell's LongRanger was intended to satisfy a requirement for a turbine-powered general-purpose light helicopter in a size and performance range between the five-seat JetRanger II and 15-seat Model 205A-1. The current LongRanger II, with more powerful engine and detail improvement, was certificated on 17 May 1978 and replaced the original model on the production line. FAA certification for single-pilot IFR operation was granted in December 1978. The LongRanger II has a 373 kW (500 shp) Allison 250-C28B engine with a max continuous rating of 365 kW (489 shp), and includes in its design an integral particle separator. The transmission is rated at 324 kW (435 shp) for take-off, its continuous rating remaining unchanged at 276 kW (370 shp). Main rotor rpm is 394. Fuel capacity remains the same, at 371 litres (98 US gallons). Deliveries were expected to total more than 600 by the end of 1980.

The LongRanger II incorporates Bell's Noda-Matic cabin suspension system, which gives a substantial reduction in rotor-induced vibration and results in a standard of comfort comparable with that of turboprop-powered fixed-wing aircraft.

Cabin volume is 2·35 m³ (83 cu ft), representing a considerable increase over that of the JetRanger III, and utility is enhanced by innovations that allow maximum use of this space. For example, the port forward passenger seat has a folding back to allow loading of a container measuring 2·44 × 0·91 × 0·30 m (8 × 3 × 1 ft), making possible the carriage of such items as survey equipment, skis, and long components that cannot be accommodated in any other light helicopter. Double doors on the port side of the cabin provide an opening 1·52 m (5 ft 0 in) wide, for easy straight-in loading of litter patients or utility cargo; in an ambulance or rescue role two litter patients and two ambulatory patients/attendants may be carried. With a crew of two, the standard cabin layout accommodates five passengers in two canted aft-facing seats and three forward-facing seats. An optional executive cabin layout has four individual passenger seats.

Detail improvements introduced in the LongRanger II include a redesigned rear cabin to provide 5 cm (2 in) more headroom for passengers in aft cabin seats; new cowlings, firewall, engine mountings, and engine deck area structure; new freewheeling unit, input shaft, forward tail rotor drive shaft, and increased diameter tail rotor; new engine oil system, oil tank, cooler, and transition duct; deletion of the water/alcohol system required formerly; and increased capacity 17Ah battery. Optional kits include emergency flotation gear, a 907 kg (2,000 lb) cargo hook, and an engine bleed air environmental control system. FAA certification was gained in December 1978 for the LongRanger II equipped with a Collins AP-107H autopilot, this providing single-pilot IFR capability. The Collins APS-841H autopilot is also available, featuring automatic heading, altitude navigation, approach and basic stabilisation modes of operation. A special turbulence mode improves response to handle rough air.

Under development in 1981 were a **LongRanger III**, with an Allison 250-C30 power plant, and a **LongRanger IV** with the same power plant plus a four-blade composite-construction main rotor and slower-turning tail rotor.

The following details apply to the LongRanger II:

DIMENSIONS, EXTERNAL:
Diameter of main rotor 11·28 m (37 ft 0 in)
Diameter of tail rotor 1·65 m (5 ft 5 in)
Length overall, both rotors turning 13·02 m (42 ft 8½ in)
Height over tail-fin 2·75 m (9 ft 0¼ in)
Height to top of rotor hub 3·05 m (10 ft 0 in)
Fuselage: Max width 1·32 m (4 ft 4 in)
Stabiliser span 1·98 m (6 ft 6 in)

WEIGHTS:
Weight empty, standard configuration 978 kg (2,156 lb)
Max T-O weight 1,882 kg (4,950 lb)

PERFORMANCE (ISA at T-O weight of 1,837 kg; 4,050 lb):
Never-exceed speed at S/L 130 knots (241 km/h; 150 mph)
Max cruising speed at 1,525 m (5,000 ft) 116 knots (215 km/h; 134 mph)
Max rate of climb at S/L 463 m (1,520 ft)/min
Service ceiling at max cruise power 5,945 m (19,500 ft)
Hovering ceiling IGE 4,025 m (13,200 ft)

Hovering ceiling OGE 2,470 m (8,100 ft)
Range at S/L, no reserves 297 nm (550 km; 342 miles)
Range at 1,525 m (5,000 ft), no reserves
335 nm (621 km; 386 miles)

BELL MODEL 206L TEXASRANGER

In October 1980, Bell announced that it was developing for the export market, as the TexasRanger, a multi-mission military version of the commercial LongRanger. A demonstration prototype has since completed a successful TOW missile firing programme in the Mojave Desert, and was scheduled to make sales tours throughout Europe, the Middle and Far East during 1981. Delivery dates for production helicopters had not been projected at the time of the initial announcement; but Bell considered that a version armed with folding-fin rockets and guns could be made available by the end of 1981, followed by TOW-equipped versions one year later.

The basic airframe of the TexasRanger is similar to that of the LongRanger, with Noda-Matic suspension to provide a vibration-reducing interface between the fuselage and rotor system. Quick-change design features enable the aircraft to be used for a variety of mission roles, including troop transport, battlefield resupply, armed reconnaissance and surveillance, search and rescue, medical evacuation, and command control. For example, the TOW missile avionics have a special rear-seat pallet mounting, which enables the cabin to be converted for utility missions in 15 minutes. The TOW system is also claimed to be the first that does not require re-boresighting in the field after removal and reinstallation.

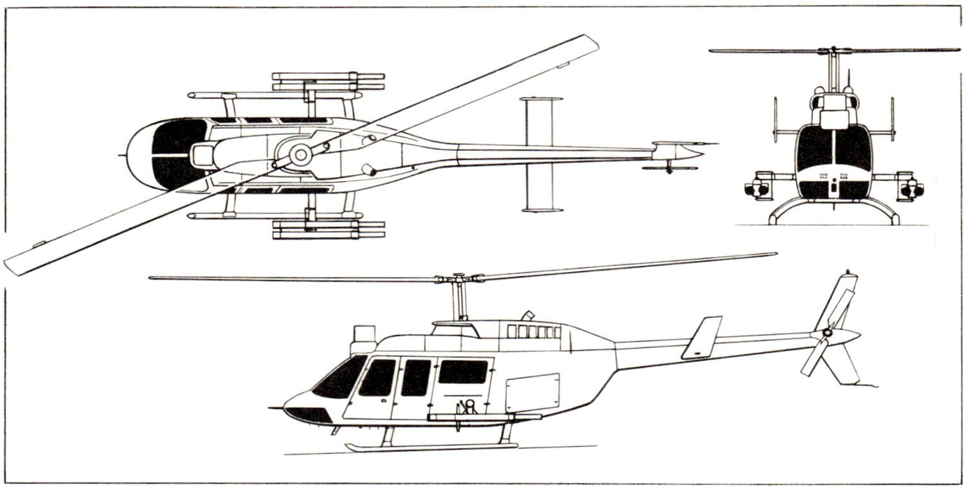

Bell Model 206L TexasRanger with TOW missile armament *(Pilot Press)*

Changes by comparison with the LongRanger are as follows:

TYPE: Light multi-mission military helicopter.

FUSELAGE: Light alloy tailboom is strengthened by comparison with that of LongRanger, by use of thicker skins and heavier longerons. Glassfibre blanket mounted externally on each side of rear fuselage to offset blast pressures of missile launch.

POWER PLANT: Demonstration aircraft has one 373 kW (500 shp) Allison 250-C28B turboshaft engine; production versions will have one 485 kW (650 shp) Allison 250-C30P turboshaft. Fuel carried by demonstrator comprises the standard 371 litre (98 US gallon) system of the LongRanger II; production aircraft will have a 416 litre (110 US gallon) crash-resistant system.

ACCOMMODATION: Pilot and weapons operator on side-by-side armoured seats for armed missions. Pilot and up to six passengers for personnel transport duties. Dual controls standard. Forward-hinged doors on each side are jettisonable.

SYSTEMS: Added stability augmentation and control system enhances capability as a weapons platform.

AVIONICS AND EQUIPMENT: Standard military nav/com. Pilot's steering indicator. TOW missile control avionics, missile firing and control system, and Hughes Aircraft Company TOW missile sight with low- and high-power optics. External portion of TOW sight is roof-mounted, so that target sighting can be carried out with most of the helicopter screened by terrain or trees. Sight has provisions for future installation of a FLIR receiver and a laser rangefinder/designator.

ARMAMENT: Can include four TOW missiles in two launchers, plus four TOW missiles carried internally for reloading during a brief landing; or two pods each containing seven 2·75 in folding-fin rockets; or two pods each containing two Fabrique Nationale 7·62 mm machine-guns with 500 rds/gun; or a total of four air-to-air missiles on launchers. All weapon pods are attached by three-point standard NATO shackles, and installations have been so designed that all weapon systems are jettisonable in flight.

DIMENSIONS, EXTERNAL:
Diameter of main rotor 11·28 m (37 ft 0 in)
Diameter of tail rotor 1·65 m (5 ft 5 in)
Length overall 12·92 m (42 ft 4¾ in)
Height overall 3·15 m (10 ft 4 in)
Width over armament 3·05 m (10 ft 0 in)

WEIGHTS:
Weight empty, standard configuration
999 kg (2,203 lb)
Max T-O weight, utility (internal load)
1,882 kg (4,150 lb)
Max T-O weight, armed mission (external jettisonable load) 1,927 kg (4,250 lb)

PERFORMANCE (at max T-O weight at S/L, ISA: A, armed mission; U, utility):
Max cruising speed:
A 102 knots (188 km/h; 117 mph)
U 112 knots (207 km/h; 129 mph)
Econ cruising speed: A 102 knots (188 km/h; 117 mph)
U 110 knots (204 km/h; 127 mph)
Rate of climb at S/L: A 404 m (1,325 ft)/min
U 445 m (1,460 ft)/min
Hovering ceiling, IGE: A 3,650 m (12,000 ft)
U 4,900 m (16,100 ft)
Range with max fuel: A 298 nm (552 km; 343 miles)
U 333 nm (616 km; 383 miles)

BELL MODEL 209 HUEYCOBRA and SEACOBRA

US Army designations: AH-1G, AH-1Q and AH-1R
US Navy/Marine Corps designations: AH-1J and AH-1T

Bell Helicopter Textron initiated the Model 209 in March 1965 as a company-funded development of the UH-1B/C Iroquois intended specifically for armed helicopter missions. The original design combined the basic transmission and rotor system and (in its standard form) the power plant of the UH-1C with a new, streamlined fuselage designed for maximum speed, armament load and crew efficiency. Relatively small, its low silhouette and narrow profile make it easy to conceal with small camouflage nets or to move under cover of trees. Tandem seating provides the best possible field of view for the crew of two.

Bell Model 206L TexasRanger, a multi-mission military version of the LongRanger

The Model 209 prototype made its first flight on 7 September 1965, and was sent to Edwards AFB in December 1965 for US Army evaluation. The Army's intention to order the aircraft was announced on 11 March 1966, the initial model being known as the AH-1G HueyCobra. Total orders to date exceed 1,800.

Versions announced so far are as follows:

AH-1G HueyCobra. Original version for US Army, powered by a single 1,044 kW (1,400 shp) Avco Lycoming T53-L-13 turboshaft engine, derated to 820 kW (1,100 shp) for T-O and max continuous rating. Development contract for two pre-production aircraft placed on 4 April 1966, followed on 13 April by an initial order for 110 aircraft plus long-lead-time spares. Subsequent contracts raised the total US Army order to 1,078 by mid-1971. Deliveries began in June 1967, and two months later the AH-1G was deployed operationally in Viet-Nam; it played a particularly important part in the Tet offensive in 1968, and in Laos in the Spring of 1971. The US Marine Corps acquired 38 AH-1Gs during 1969, for transition training and initial deployment pending deliveries of the AH-1J; these are included in the above total. The Spanish Navy received eight (designated **Z.14**), for anti-shipping strike duties, and six were supplied to Israel in 1974. A number of AH-1Gs have been converted to **TH-1G** dual-control trainers. Following the decision to equip the HueyCobra with TOW missiles, 93 AH-1Gs were converted to interim AH-1Q standard; all of these, and 197 other AH-1Gs, have been updated to Mod AH-1S standard. One AH-1G was converted to **JAH-1G** as a testbed aircraft, initially for the Hellfire laser-guided air-launched missile. More recent tests with this aircraft have included demonstrations of a General Electric M197 three-barrel 20 mm gun with an improved rate of fire of 1,500 rds/min. An increase to 3,000 rds/min is possible with this gun, and GE is also developing fast-firing 25 mm and 30 mm weapons for helicopter applications, with the emphasis on air-to-air anti-helicopter combat.

AH-1J SeaCobra. Initial twin-turboshaft version for US Marine Corps, powered by a 1,342 kW (1,800 shp) Pratt & Whitney Aircraft of Canada T400-CP-400 coupled free-turbine turboshaft engine, a military version of the PT6T-3 Turbo Twin Pac. Engine and transmission flat rated at 820 kW (1,100 shp) continuous output, with increase to 932 kW (1,250 shp) available for T-O or 5 min emergency power. Following an initial US Marine Corps order for 49, placed in May 1968, a pre-production example was displayed to representatives of the US armed forces at Enless, Texas, on 14 October 1969. Deliveries of these 49 began in mid-1970 and were completed in 1971; a further 20, ordered in early 1973, were delivered between April 1974 and February 1975. The last two of this batch were converted later as prototypes for the AH-1T; 58 USMC AH-1Js remained operational in May 1981. Under a $38.5 million contract announced on 22 December 1972, 202 TOW-capable AH-1Js were supplied to the Imperial Iranian Army Aviation from 1974, the US Army acting as purchasing agent.

AH-1Q HueyCobra. Interim anti-armour version for US Army, converted from AH-1G to fire Hughes TOW anti-tank missiles. First of eight 'pre-production' examples delivered in early 1973; first 'production' deliveries on 10 June 1975. Total of 93 converted; subsequently upgraded to Mod AH-1S standard.

AH-1R HueyCobra. As AH-1G, but with 1,342 kW (1,800 shp) T53-L-703 turboshaft engine. No TOW missile installation.

AH-1S HueyCobra. Advanced and modernised TOW-capable version for US Army; described separately.

AH-1T Improved SeaCobra. Improved version of twin-engined AH-1J for US Marine Corps. Last two of 69 AH-1Js modified as prototypes under a US Army Aviation Systems Command contract, with uprated components for significantly increased payload and performance. Incorporates features of AH-1J airframe, but embodies dynamic system of Bell Model 214, some technology developed for Bell Model 309 Kingcobra, and upgraded power plant (1,469 kW; 1,970 shp T400-WV-402) and transmission. Lengthened fuselage. Initial contract for 10 announced on 23 June 1975; total of 57 built, of which 23 are currently equipped to carry TOW missiles, with 28 more being modified to TOW configuration. First AH-1T (USN serial number 59228) flew on 20 May 1976, and was delivered to US Marine Corps on 15 October 1977.

Under US Navy contract, Hughes is adapting its Black Hole infra-red suppression system (developed for the YAH-64) for interchangeable use on the AH-1J and AH-1T.

During 1980, Bell flight tested successfully an AH-1T powered by two General Electric T700-GE-700 turboshaft engines with a combined output in excess of 2,386 kW (3,200 shp). This installation was made in an AH-1T loaned by the US Marine Corps, as part of an R & D programme to establish the specification of a similar helicopter with enhanced capability for future procurement. Improvements proposed for a qualification configuration, and suitable for retrofit to all existing AH-1Ts, include corrosion-free ballistically-tolerant composite main rotor blades; improved tail rotor hub; General Electric T700-GE-401 turboshafts with a combined output of 2,520 kW (3,380 shp); improved transmission with 1,678 kW (2,250 shp) 30-min fly-dry capability; a new combining gearbox; hydraulically-driven fan cooling system for main rotor and combining gearbox lubrication system; uprated electrical system; plus a number of detail improvements.

The following description applies primarily to the AH-1G and AH-1Q, except where indicated otherwise:

Type: Single-engined (AH-1G/Q/R/S) and twin-engined (AH-1J/T) close support and attack helicopters.

Rotor System and Drive (AH-1G/J/Q/R): Model 540 two-blade wide-chord 'door-hinge' main rotor, similar to that of UH-1C. Interchangeable blades, built up of extruded aluminium spars and laminates. Rotor brake fitted. Blades do not fold. Two-blade all-metal flex-beam tractor tail rotor on starboard side, of honeycomb construction; blade chord increased on AH-1J, which also has push/pull tail rotor controls. Shaft drive to both main and tail rotors. Main rotor rpm 294-324.

Rotor System and Drive (AH-1T): Similar to that of Bell Model 214, with strengthened main rotor hub incorporating Lord Kinematics Lastoflex elastomeric and Teflon-faced bearings. Main rotor blades have increased chord, and swept tips which reduce noise and improve high-speed performance. Tail rotor also similar to that of Model 214, with increased diameter and blade chord. Main rotor/engine rpm ratio: 1 : 21·288. Tail rotor/engine rpm ratio 1 : 4·52.

Wings: Small mid-mounted stub-wings, to carry armament and offload rotor in flight.

Bell AH-1S HueyCobra with flat plate canopy and missile launchers

Fuselage: Conventional all-metal semi-monocoque structure, with low silhouette and narrow profile. AH-1T has forward fuselage lengthened by insertion of a 0·305 m (1 ft 0 in) plug, to accommodate tankage for additional 181·5 kg (400 lb) of fuel, and tailboom lengthened by 0·79 m (2 ft 7 in).

Tail Unit: Sweptback vertical fin/tail rotor pylon, strengthened on twin-engined models to cater for increased power. Elevator, of inverted aerofoil section, mid-mounted on tailboom forward of fin.

Landing Gear: Non-retractable tubular skid type. Ground handling wheels optional.

Power Plant: Single or twin turboshaft engines, as detailed under model listings. Fuel capacity: G and J, 1,014 litres (268 US gallons); T, two fuselage tanks, total capacity 1,158 litres (306 US gallons). (Fuel loads, where known, are given under 'Weights' heading.) Refuelling point in port side of fuselage, aft of cockpits.

Accommodation: Crew of two in tandem, with co-pilot/gunner in front seat and pilot at rear. Crew are protected by seats and side panels of Norton Co 'Noroc' armour; other panels protect vital areas of aircraft.

Systems: Hydraulic system, with Abex pumps, for flight controls and other services. Battery-powered 28V DC electrical system. Environmental control and fire detection systems.

Avionics (AH-1G): Communications equipment includes AN/ARC-54/131 FM radio; AN/ARC-51 and AN/ARC-134 voice com; KY-28 secure voice system.

Avionics (AH-1T): AN/ARC-159(V)1 UHF command set, AN/ARC-114A FM tactical set, AN/AIC-18 intercom, AN/ARN-84(V) Tacan, AN/ARA-50 UHF DF, AN/ASN-75B gyrosyn compass, AN/ARN-83 DF, AN/APN-171(V) radar altimeter, AN/APX-72 IFF transponder and AN/APN-154(V) radar beacon. Provision for TSEC/KY-28 com security unit and KIT-1A/TSEC Mk XII computer.

Armament and Operational Equipment (AH-1G): Initial production AH-1Gs were fitted with GAU-2B/A 7·62 mm Minigun in Emerson Electric TAT-102A undernose turret (see 1978-79 *Jane's*). This was superseded by an M-28 turret, able to mount either two Miniguns (each with 4,000 rds), or two M-129 40 mm grenade launchers (each with 300 rds), or one Minigun and one M-129. The Miniguns in these turrets have two rates of fire, controlled by the gunner's trigger: 1,600 rds/min for searching or registry fire, or 4,000 rds/min for attack. The M-129 fires at a single rate of 400 rds/min. Four external stores attachments under stub-wings can accommodate seventy-six 2·75 in rockets in four M-159 launchers, 28 similar rockets in four M-157 launchers, or two M-18E1 Minigun pods. An initial batch of six AH-1Gs was delivered to the US Army in December 1969 equipped with a Bell/General Electric M-35 armament subsystem. This unit consists of an M-61 six-barrel 20 mm automatic cannon on the port inboard wing station, having a firing rate of 750 rds/min. Two ammunition boxes faired flush to the fuselage below the stub-wings each accommodate 500 rds, and total installed weight of the system is 531 kg (1,172 lb). A total of 350 M-35 kits was ordered subsequently by the US Army. All wing stores are symmetrically or totally jettisonable. In normal operation, the co-pilot/gunner controls and fires the turret armament, and the pilot (aided by an M-73 adjustable reflex rocket sight) normally fires the wing stores. The pilot can fire the turreted weapons only in the stowed (ie, dead ahead) position; the turret returns to the stowed position automatically when the gunner releases his grip on the slewing switch. The gunner also has the capability to fire the wing stores if required. Other operational equipment on the AH-1G includes an M-130 chaff dispenser.

Armament (AH-1J): Electrically operated General Electric undernose turret, housing an M-197 three-barrel 20 mm weapon (a lightweight version of the M-61 cannon). A 750-rd ammunition container is located in the fuselage directly aft of the turret; firing rate is 750 rds/min, but a 16-round burst limiter is incorporated in the firing switch. Barrel length of 1·52 m (5 ft) makes it imperative that the M-197 is centralised before wing stores are fired. Gun can be tracked 110° to each side, 18° upward, and 50° downward. Four attachments under stub-wings for various loads, including LAU-68A/A (seven-tube) or LAU-61A/A (19-tube) 2·75 in rocket launchers, or M-18E1 Minigun pods. Total possible armament load 245 kg (542 lb) internal, 998 kg (2,200 lb) external.

Armament (AH-1Q): M-28 turreted weapons as for AH-1G. Anti-armour configuration involves installation of eight Hughes TOW missile containers, disposed as two two-round pods on each of the outboard underwing stations. The inboard wing stations remain available for other stores, as listed for AH-1G. In the TOW configuration, a Sperry Univac helmet sight is used by both crew members to cue the turreted weapon or the TOW stabilised sight. In addition, the co-pilot/gunner may use the 2x or 13x magnification offered by the M-65 TOW system's telescopic sight unit for turret weapon engagements.

Armament (AH-1T): Chin turret as AH-1J. Underwing attachments for four LAU-61A, LAU-68A, LAU-68A/A, LAU-68B/A or LAU-69A 2·75 in rocket pods; or two CBU-55B fuel-air explosive weapons; four SU-44 flare dispensers; two M118 grenade dispensers; parachute flares; or two Minigun pods.

Dimensions, external:

Diameter of main rotor: G, J, Q, R	13·41 m (44 ft 0 in)
T	14·63 m (48 ft 0 in)
Main rotor blade chord: G, J, Q, R	0·69 m (2 ft 3 in)
T	0·84 m (2 ft 9 in)
Diameter of tail rotor: G, J, Q, R	2·59 m (8 ft 6 in)
T	2·96 m (9 ft 8½ in)
Tail rotor blade chord: G, Q, R	0·21 m (8·4 in)
J	0·29 m (11½ in)
T	0·305 m (1 ft 0 in)
Wing span (all)	3·23 m (10 ft 7 in)

Bell AH-1T improved version of the Model 209 SeaCobra for the US Marine Corps

US Marine Corps Bell AH-1T serving as testbed for General Electric T700-GE-700 power plant

Length overall, main rotor fore and aft:
G, Q, R 16·14 m (52 ft 11½ in)
J 16·26 m (53 ft 4 in)
T 17·68 m (58 ft 0 in)
Length of fuselage: G, J, Q, R 13·59 m (44 ft 7 in)
T 14·68 m (48 ft 2 in)
Width of fuselage: G, Q, R 0·965 m (3 ft 2 in)
J, T 0·98 m (3 ft 2½ in)
Height overall: G, Q, R 4·12 m (13 ft 6¼ in)
J 4·15 m (13 ft 8 in)
T 4·32 m (14 ft 2 in)
Elevator span (all) 2·11 m (6 ft 11 in)
Width over skids (all) 2·13 m (7 ft 0 in)
Width over TOW missile pods:
G, Q 3·26 m (10 ft 8¾ in)

AREAS:
Main rotor disc: G, J, Q, R 141·26 m^2 (1,520·53 sq ft)
T 168·11 m^2 (1,809·56 sq ft)
Tail rotor disc: G, J, Q, R 5·27 m^2 (56·75 sq ft)
T 6·88 m^2 (74·03 sq ft)

WEIGHTS:
Operating weight empty, incl amounts shown for crew, fluids, avionics and armour:
G (404 kg; 891 lb) 2,754 kg (6,073 lb)
J (398 kg; 877 lb) 3,294 kg (7,261 lb)
Weight empty: T 3,642 kg (8,030 lb)
Operating weight empty: T 3,904 kg (8,608 lb)
Mission fuel load:
G (871 litres; 230 US gallons) 680 kg (1,500 lb)
J 725 kg (1,600 lb)
T 944 kg (2,081 lb)
Max useful load (fuel and disposable ordnance):
J 1,144 kg (2,523 lb)
T 2,445 kg (5,392 lb)
Mission weight: G 4,266 kg (9,407 lb)
J 4,523 kg (9,972 lb)
Max T-O and landing weight:
G, Q, R 4,309 kg (9,500 lb)
J 4,535 kg (10,000 lb)
T 6,350 kg (14,000 lb)

PERFORMANCE (at max T-O weight, ISA):
Never-exceed speed:
G, Q, R 190 knots (352 km/h; 219 mph)
J 180 knots (333 km/h; 207 mph)
Max level speed at S/L:
G, Q, T 149 knots (277 km/h; 172 mph)
J 180 knots (333 km/h; 207 mph)
Max crosswind speed for hovering:
J 40 knots (74 km/h; 46 mph)
Vertical rate of climb at S/L:
T 92 m (301 ft)/min
Max rate of climb at S/L:
G, Q, normal rated power 375 m (1,230 ft)/min
J, normal rated power 332 m (1,090 ft)/min
T 544 m (1,785 ft)/min
Service ceiling:
G, Q, normal rated power 3,475 m (11,400 ft)
J, normal rated power 3,215 m (10,550 ft)
T, max cont power 2,255 m (7,400 ft)
Hovering ceiling IGE: G, Q 3,015 m (9,900 ft)
J 3,794 m (12,450 ft)
Hovering ceiling OGE: T 365 m (1,200 ft)
Combat radius at 138 knots (255 km/h; 158 mph) at S/L: T 108 nm (200 km; 124 miles)
Range with max fuel:
G, Q, both at S/L, 8% reserves 310 nm (574 km; 357 miles)
J, no reserves 311 nm (577 km; 359 miles)
T 227 nm (420 km; 261 miles)

BELL MODEL 209 HUEYCOBRA (MODERNISED VERSION) and MODEL 249

US Army designation: AH-1S

The AH-1S is an advanced version of the single-engined TOW-capable HueyCobra for the US Army, with upgraded power plant, gearbox, transmission and many other improvements. The supply of an undisclosed number to Israel has been authorised, and two have been delivered to Japan, for evaluation by the JGSDF for potential large-scale procurement.

The US Army's AH-1S fleet is being formed by procurement of new-production aircraft and by modification of AH-1Gs and other earlier models. Funds were provided in FY 1981 both to complete the new-production programme and to continue the conversions. A force of 1,000 AH-1Ss is currently planned to be achieved in FY 1983. At that time there will be a residue of approx 60 AH-1Gs; current plans do not envisage the conversion of these aircraft.

The first of a succession of US Army contracts was placed in 1975, and orders to the beginning of 1981 were as follows:

Mod AH-1S. This designation (the 'Mod' in this case indicating 'Modified') applies to 290 AH-1Gs retrofitted with the TOW missile system, and with a 1,342 kW (1,800 shp) Avco Lycoming T53-L-703 turboshaft engine, and the same rotor system dynamics as the Production AH-1S. No other changes were made to this Mod AH-1S version. The total includes the 93 AH-1Gs previously converted to AH-1Qs, which were further modified by Bell and Dornier to Mod AH-1S.

Bell Model 249, a version of the Modernised AH-1S with the advanced rotor system of the Model 412

Production AH-1S. Under Step 1 of a three-step new-production programme, 100 Production AH-1S HueyCobras were built and delivered to the US Army between March 1977 and September 1978. These aircraft have a new flat-plate canopy, improved nap-of-the-earth (NOE) instrument panel layout, continental United States (CONUS) navigation equipment, radar altimeter, improved communication radios, uprated engine and transmission, push/pull anti-torque controls, and (from the 67th aircraft onwards) new Kaman-developed composite rotor blades. First unit to receive this version, in August 1977, was the 82nd Airborne Division at Fort Bragg, North Carolina.

Up-gun AH-1S. The next 98 new-production aircraft (Step 2) have all the improvements detailed for the Production AH-1S, plus a new universal 20/30 mm gun turret, an improved wing stores management system for the 2·75 in rockets, automatic compensation for off-axis gun firing, and a 10kVA alternator to provide the necessary additional electric power. Deliveries of this version began in September 1978 and were completed in October 1979.

Modernised AH-1S. This version, not to be confused with the 'Mod AH-1S' referred to earlier, represents the fully-upgraded AH-1S, and became standard from the 199th new-production aircraft. To the improvements already mentioned for the two preceding stages are being added, as Step 3, a new fire control subsystem (comprising a laser rangefinder and tracker, ballistics computer, low-airspeed sensor, and pilot's head-up display), air data system, Doppler navigation system, IFF transponder, infra-red jammer, hot-metal and plume infra-red suppressor, closed-circuit refuelling, new secure voice communications, and new composite rotor blades developed by Kaman. Deliveries of the 196 Modernised AH-1S so far ordered are scheduled to be completed in February 1984.

Under a $13 million contract awarded by the US Army Missile Command, Hughes Aircraft Company began the manufacture in early 1980 of 157 Laser Augmented Airborne TOW (LAAT) stabilised sights for installation in Modernised AH-1S aircraft. The very small (13 × 13 × 4 cm; 5 × 5 × 1·5 in) laser transmitter has been developed to fit within the existing sight turret of the AH-1S.

Three engineering development models of the LAAT were used for flight testing by Bell at Yuma Proving Grounds, Arizona. They demonstrated that the LAAT can significantly improve first-burst accuracy of gun and rocket fire. In use, the gunner sights a target and fires the laser. Reflected from the target, the returning beam provides accurate and almost instantaneous range information, enabling the aircraft's fire control computer to integrate range, wind and ammunition ballistics data to direct weapon firing with great accuracy.

Hughes Aircraft has also developed, as part of the AH-1S TOW system, a FLIR-augmented Cobra TOW sight (FACTS). This telescopic sight, mounted under the nose of the helicopter, enables the gunner to 'see' through darkness, smoke and haze, and considerably improves round-the-clock combat capability.

The US Army hopes eventually to bring all of its AH-1S HueyCobras up to the full Modernised AH-1S standard over a period of about five years. Current plans envisage, first, the conversion of a further 425 AH-1Gs to Modernised AH-1S in 1979-82; the 290 'Mod AH-1S' aircraft would then be upgraded to Modernised AH-1S in 1982-83; finally, the 100 Step 1 aircraft (in 1983-84) and 98 Step 2 aircraft (in 1984) would be brought up to the full Step 3 standard. The US Army National Guard began to receive 27 Modernised AH-1S from April 1981, under a US Army contract announced in July 1980. These are in service with the Attack Helicopter Group of the 163rd Armored Cavalry Regiment, based in Utah.

Under the company designation **Model 249**, one Modernised AH-1S was tested in 1979-80 with the standard two-blade rotor replaced by a four-blade advanced rotor similar to that used on the Bell Model 412 transport helicopter, but with the diameter reduced by 0·61 m (2 ft 0 in). It has achieved speeds of up to 170 knots (315 km/h; 195 mph).

The major differences between the standard AH-1S and earlier single-engined HueyCobras may be summarised as follows:

TYPE: Anti-armour attack helicopter.

ROTOR SYSTEM AND DRIVE: Upgraded gearbox and transmission, the latter rated at 962 kW (1,290 shp). From 67th new-production AH-1S onward, new main rotor blades of composite construction are fitted, developed by Kaman Aerospace Corporation and equipped with tungsten carbide bearing sleeves. The outer 15% of these blades, which are tolerant of damage by weapons of up to 23 mm calibre, is tapered in both chord and thickness.

FUSELAGE: Tailboom strengthened to increase survivability against weapons of up to 23 mm calibre. Entire airframe has an anti-infra-red paint finish.

POWER PLANT: One 1,342 kW (1,800 shp) Avco Lycoming T53-L-703 turboshaft engine. Closed-circuit refuelling on Modernised AH-1S.

ACCOMMODATION: New flat-plate canopy has seven planes of viewing surfaces, designed to minimise glint and reduce possibility of visual detection during nap-of-the-earth (NOE) flying; it also provides increased headroom for pilot. Improved instrument layout and lighting, compatible with use of night vision goggles. Improved, independently-operating window/door ballistic jettison system to facilitate crew escape in emergency.

SYSTEMS: 10kVA AC alternator added to electrical system. Battery-driven Abex standby pump, for use in event of main hydraulic system failure, can be used for collective pitch control and for boresighting turret and TOW missile system. Improved environmental control and fire detection systems.

AVIONICS AND EQUIPMENT: Standard lightweight avionics equipment (SLAE) includes AN/ARC-114 FM, AN/ARC-164 UHF/AM voice com, and E-Systems (Memcor Division) AN/ARC-115 VHF/AM voice com (compatible with KY-58 single-channel secure voice system). Other avionics include AN/ARN-123 CONUS navigation system with VOR/ILS receivers, glideslope, marker beacon receiver and indicator lights (Doppler nav system in Modernised AH-1S); HSI; VSI; radar altimeter; push/pull anti-torque controls for tail rotor; co-pilot's standby magnetic compass.

ARMAMENT AND OPERATIONAL EQUIPMENT: M65 system with eight TOW missiles on outboard underwing stations, as in AH-1Q. Beginning with the 101st new-production AH-1S (the first 'Up-gun' example), the M28 (7·62/40 mm) turret in earlier HueyCobras is replaced by a new electrically-powered General Electric universal turret, designed to accommodate either a 20 mm or a 30 mm weapon and to improve stand-off capability. Initially, the 20 mm M197 three-barrel Vulcan (with 750 rds) is mounted in this turret. Rate of fire is 730 rds/min. Turret position is controlled by the pilot or co-pilot/gunner through helmet sights, or by the co-pilot using the M65 TOW missile system's telescopic sight unit. Field of fire is up to 110° to each side of aircraft, 20·5° upward and 50° downward. Also from the first 'Up-gun' AH-1S, the helicopter is equipped with a new Baldwin Electronics XM138 wing stores management subsystem, providing the means to select and fire, singly or in groups, any one of five types of external 2·75 in rocket store. These are mounted in launchers each containing from 7 to 19 tubes, and are additional to the TOW missile capability.

In addition to these installations the 199th new-built AH-1S (the first to full 'Modernised' standard) introduced a fire control subsystem which includes a Kaiser head-up display for the pilot, Teledyne Systems digital fire control computer for the turreted weapon and underwing rockets, omnidirectional airspeed system to improve cannon and rocket accuracy, Hughes laser rangefinder (accurate to 10,000 m; 32,800 ft), and AN/AAS-32 airborne laser tracker. Other operational equipment includes Hughes LAAT and FACTS sights (see introductory copy), a Marconi Avionics air data subsystem, AN/APX-100 solid-state IFF transponder, Sanders AN/ALQ-144 infra-red jammer, suppressor for infra-red signature from engine hot metal and exhaust plume, AN/APR-39 radar warning receiver, AN/ALQ-136 radar jammer (with M130 chaff system as backup), Perkin-Elmer laser warning receiver.

DIMENSIONS, EXTERNAL: As AH-1G except:
Main rotor blade chord (from 67th new-production AH-1S) 0·76 m (2 ft 6 in)
Tail rotor blade chord 0·29 m (11½ in)
WEIGHTS:
Operating weight empty 2,939 kg (6,479 lb)
Mission weight 4,524 kg (9,975 lb)
Max T-O and landing weight 4,535 kg (10,000 lb)
PERFORMANCE (at max T-O weight, ISA):
Never-exceed speed (TOW configuration) 170 knots (315 km/h; 195 mph)
Max level speed (TOW configuration) 123 knots (227 km/h; 141 mph)
Max rate of climb at S/L, normal rated power 494 m (1,620 ft)/min
Service ceiling, normal rated power 3,720 m (12,200 ft)
Hovering ceiling IGE 3,720 m (12,200 ft)
Range at S/L with max fuel, 8% reserves 274 nm (507 km; 315 miles)

Bell UH-1N of the US Marine Corps

Bell Model 212 twin-turbine helicopter of Aeroleo Taxi, Brazil

BELL MODEL 212 TWIN TWO-TWELVE

US military designation: UH-1N
Canadian military designation: CH-135

Bell announced on 1 May 1968 that the Canadian government had approved development of a twin-engined UH-1 helicopter to be powered by a Pratt & Whitney Aircraft of Canada PT6T-3 power plant. Subsequently, the Canadian government ordered 50 of these aircraft (designated CUH-1N) for the Canadian Armed Forces, with options on 20 more. Simultaneously, orders totalling 141 aircraft for the US services were announced, comprising 79 for the US Air Force, 40 for the US Navy and 22 for the US Marine Corps, all having the designation **UH-1N**. Subsequent orders covered the delivery of 159 more UH-1Ns to the US Navy and Marine Corps in 1973-78.

Initial deliveries for the US Air Force began in 1970, and the first CUH-1N for the Canadian Armed Forces was handed over officially at Uplands Airport, Ottawa, on 3 May 1971; the Canadian order was completed one year later. Deliveries to the US Navy and US Marine Corps began during 1971. Canadian aircraft are now designated **CH-135**. Six were delivered to the air force of Bangladesh in early 1977, and the Argentinian Air Force ordered eight in 1978.

A commercial version, known as the Twin Two-Twelve, is also in full-scale production. This received FAA type certification with PT6T-3 power plant in October 1970, and FAA Transport Type Category A certification on 30 June 1971. The Model 212 has the capability of carrying an external load of 2,268 kg (5,000 lb), and the military UH-1N a load of 1,814 kg (4,000 lb). The PT6T-3B engine was introduced in June 1980, offering improved single-engine performance and, consequently, additional safety margins.

Bell announced in January 1973 that two Twin Two-Twelves had been modified and flown in a programme to gain IFR certification from the UK's CAA and America's FAA. Conversion of the Model 212 from VFR to IFR configuration requires a new avionics package, new instrument panel and aircraft stabilisation controls. The Model 212 has also qualified for IFR certification by the Norwegian DCA and the Canadian DoT. In June 1977, it became the first helicopter FAA-certificated for single-pilot IFR operations with fixed floats.

An order for eight Model 212s, to support energy and natural resources development in China, was received in early 1979, and all were delivered by the end of the year. Operated by the Civil Air Authority of China (CAAC), they represented the first order received from the People's Republic of China by a US helicopter manufacturer.

TYPE: Twin-turbine utility helicopter.

ROTOR SYSTEM: Two-blade all-metal semi-rigid main rotor with interchangeable blades, built up of extruded aluminium spars and laminates. Stabilising bar above and at right angles to main rotor blades. Underslung feathering axis hub. Two-blade all-metal tail rotor. Main rotor blades do not fold. Rotor brake standard.

ROTOR DRIVE: Shaft drive to both main and tail rotors.

FUSELAGE: Conventional all-metal semi-monocoque structure.

TAIL SURFACE: Small fixed stabiliser on rear fuselage.

LANDING GEAR: Tubular skid type. Lock-on ground handling wheels, fixed floats and inflatable nylon float-bags optional.

POWER PLANT: Pratt & Whitney Aircraft of Canada PT6T-3B Turbo Twin Pac, comprising two PT6 turboshaft engines coupled to a combining gearbox with a single output shaft. Producing 1,342 kW (1,800 shp), the Twin Pac is flat rated to 962 kW (1,290 shp) for T-O and 842 kW (1,130 shp) for continuous operation. In the event of an engine failure, the remaining engine can deliver 764 kW (1,025 shp) for 2·5 min, 723 kW (970 shp) for 30 min, or 596 kW (800 shp) continuously. Five interconnected rubber fuel cells, total usable capacity 814 litres (215 US gallons). Auxiliary fuel tanks optional, to provide a max total capacity of 1,495 litres (395 US gallons). Single-point refuelling on starboard side of cabin. Oil capacity 11·5 litres (3 US gallons) for engines, 8·5 litres (2·25 US gallons) for transmission.

ACCOMMODATION: Pilot and up to 14 passengers. In cargo configuration there is a total internal volume of 7·02 m³ (248 cu ft), including baggage space in tailboom. Forward door on each side of fuselage, opening forward. Two doors on each side of cabin; forward door hinged to open forward, rear door sliding aft. Accommodation heated and ventilated. Garrett air-cycle environmental control unit available optionally.

SYSTEMS: Dual hydraulic systems. 28V DC electrical system supplied by two completely independent 30V 200A starter/generators. Secondary AC power supplied by two completely independent 250VA single-phase solid-state inverters. A third inverter can acquire automatically the load of a failed inverter.

AVIONICS AND EQUIPMENT: Optional IFR avionics include dual King KTR 900A com transceivers; dual King KNR 660A VOR/LOC/RMI receivers; King KDF 800 ADF; King KMD 700A DME; King KXP 750A transponder; King KGM 690 marker beacon/glideslope receiver; dual Sperry Tarsyn-444 three-axis gyro units; stability control augmentation system; and an automatic flight control system. Optional equipment includes a cargo sling, rescue hoist, emergency pop-out flotation gear and high skid gear.

DIMENSIONS, EXTERNAL:
Diameter of main rotor (with tracking tips) 14·69 m (48 ft 2¼ in)
Diameter of tail rotor 2·59 m (8 ft 6 in)
Main rotor blade chord 0·59 m (1 ft 11¼ in)
Tail rotor blade chord 0·292 m (11½ in)
Length overall (main rotor fore and aft) 17·46 m (57 ft 3¼ in)
Length of fuselage 12·92 m (42 ft 4¾ in)
Height to top of rotor hub 3·91 m (12 ft 10 in)
Height overall 4·53 m (14 ft 10¼ in)
Width over skids 2·64 m (8 ft 8 in)
Width overall (main rotor fore and aft) 2·86 m (9 ft 4½ in)
Stabiliser span 2·86 m (9 ft 4½ in)
Rear sliding doors (each):
Height 1·24 m (4 ft 1 in)
Width 1·88 m (6 ft 2 in)
Height to sill 0·76 m (2 ft 6 in)
Baggage compartment door:
Height 0·53 m (1 ft 9 in)
Width 1·71 m (2 ft 4 in)
Emergency exits (centre cabin windows, each):
Height 0·76 m (2 ft 6 in)
Width 0·97 m (3 ft 2 in)
DIMENSIONS, INTERNAL:
Cabin, excl flight deck:
Length 2·34 m (7 ft 8 in)
Max width 2·44 m (8 ft 0 in)
Max height 1·24 m (4 ft 1 in)
Volume 6·23 m³ (220 cu ft)
Baggage compartment volume 0·78 m³ (28 cu ft)
AREAS:
Main rotor disc 168·06 m² (1,809 sq ft)
Tail rotor disc 5·27 m² (56·74 sq ft)
WEIGHTS:
VFR empty weight plus usable oil 2,787 kg (6,143 lb)
Max T-O weight and mission weight 5,080 kg (11,200 lb)
PERFORMANCE (at max T-O weight):
Never-exceed speed at S/L 140 knots (259 km/h; 161 mph)
Max cruising speed at S/L 124 knots (230 km/h; 142 mph)
Max rate of climb at S/L 402 m (1,320 ft)/min
Service ceiling 4,330 m (14,200 ft)
Hovering ceiling IGE 3,350 m (11,000 ft)
Max range with standard fuel at S/L, no reserves 227 nm (420 km; 261 miles)

BELL MODEL 412

Bell announced on 8 September 1978 its intention to develop a variant of the twin-turbine Model 212 with a four-blade main rotor of advanced design. The new aircraft, designated Model 412, is the first production helicopter with a four-blade rotor to be manufactured by Bell, although the company has flown many helicopters with multi-blade rotors for research purposes.

Two new fully-certificated Model 212s were modified for use in the development and certification programme for the Model 412. The first of these began its flight trials in early August 1979, and the second in December 1979. FAA type approval, in accordance with FAR Pt 29, for VFR operation was received on 9 January 1981, and IFR certification on 13 February 1981. The first delivery to a customer was made to ERA Helicopters of Anchorage, Alaska, on 18 January 1981. Two were due for delivery to the Venezuelan Air Force by August 1981.

Production is being undertaken also by Bell's Italian licensee, Agusta. Deliveries of Italian-built AB 412s were due to begin in early 1982. Agusta is also developing a multi-role military version.

Introduction of the new rotor has not only improved performance and reduced noise, but has reduced vibration significantly without requiring a costly redesign of the fuselage structure to introduce nodal suspension.

The description of the Model 212 applies also to the Model 412, except as follows:

ROTOR SYSTEM: Four-blade flex-beam soft-in-plane advanced technology main rotor. Blades are of similar construction to those described for the Model 214ST, but have lightning-protection mesh moulded into the structure and there are provisions for inclusion of de-icing heater elements. New-design main rotor hub of steel and light alloy construction, with elastomeric bearings and dampers. Main rotor can be folded. Rotor brake standard. Two-blade tail rotor of all-metal construction. Main rotor rpm 314.

ROTOR DRIVE, FUSELAGE, TAIL SURFACE AND LANDING GEAR: As for Model 212, except for shorter main rotor

Bell Helicopter's Model 412 advanced technology four-blade variant of the twin-turbine Model 212

mast. Transmission rating 1,007 kW (1,350 shp) for T-O, 846 kW (1,134 shp) max continuous.

POWER PLANT: Pratt & Whitney Aircraft of Canada PT6T-3B-1 Turbo Twin Pac, comprising two PT6 turboshaft engines coupled to a combining gearbox with a single output shaft. Producing 1,342 kW (1,800 shp), the Twin Pac is flat rated to 975 kW (1,308 shp) for take-off and 843 kW (1,130 shp) for continuous operation. In the event of an engine failure the remaining engine can deliver up to 764 kW (1,025 shp) for 2½ min, or 723 kW (970 shp) for 30 min. Five interconnected rupture-resistant fuel cells, with automatic shut-off valves (breakaway fittings), have a combined capacity of 821 litres (217 US gallons). Optional auxiliary fuel tanks provide a maximum total capacity of 1,431 litres (378 US gallons). Single-point refuelling on starboard side of cabin.

ACCOMMODATION AND SYSTEMS: As for Model 212.

AVIONICS AND EQUIPMENT: Optional IFR avionics include dual King KTR 900A com transceivers, dual King KNR 660A VOR/LOC/RMI receivers, King KDF 800 ADF, King KMD 700A DME, King KXP 750A transponder, King KGM 690 marker beacon/glideslope receiver, and dual three-axis gyro units. A Sperry autopilot with stabilisation and attitude hold capability is standard; a flight director/navigation coupler is optional. Optional equipment includes a cargo sling, rescue hoist, emergency pop-out flotation gear and high skid gear.

DIMENSIONS, EXTERNAL:

Diameter of main rotor	14·02 m (46 ft 0 in)
Diameter of tail rotor	2·59 m (8 ft 6 in)
Main rotor blade chord: at root	0·40 m (1 ft 3·9 in)
at tip	0·22 m (8½ in)
Tail rotor blade chord	0·29 m (11½ in)
Length overall (rotors turning)	17·07 m (56 ft 0 in)
Length of fuselage (excl rotors)	12·92 m (42 ft 4¾ in)
Height to top of rotor hub	3·29 m (10 ft 9½ in)
Height overall, tail rotor turning	4·32 m (14 ft 2¼ in)
Stabiliser span	2·86 m (9 ft 4½ in)
Width over skids	2·59 m (8 ft 6 in)

AREAS:

Main rotor disc	154·40 m² (1,662 sq ft)
Tail rotor disc	5·27 m² (56·75 sq ft)

WEIGHTS:

Weight empty plus usable oil	2,823 kg (6,223 lb)
Max T-O weight	5,262 kg (11,600 lb)

PERFORMANCE (at max T-O weight except where indicated):

Never-exceed speed at S/L	140 knots (259 km/h; 161 mph)
Max cruising speed at S/L	124 knots (230 km/h; 143 mph)
Max rate of climb at S/L	442 m (1,450 ft)/min
Service ceiling	4,330 m (14,200 ft)
Hovering ceiling IGE	3,290 m (10,800 ft)
Hovering ceiling OGE, AUW of 4,762 kg (10,500 lb)	2,165 m (7,100 ft)
Max range with standard fuel at S/L, no reserves	227 nm (420 km; 261 miles)

BELL MODEL 214B BIGLIFTER

Bell announced its intention to develop the Model 214B on 4 January 1974, claiming that it would have a lift capability better than that of any existing commercial helicopter in the medium category. FAA certification was received on 27 January 1976.

Powered by a 2,185 kW (2,930 shp) Avco Lycoming T5508D turboshaft engine, the 214B has the same rotor drive and transmission system as the military Model 214A that was exported to Iran (see Model 214ST entry in 1980-81 and earlier editions of *Jane's*). The engine is flat rated to a maximum of 1,678 kW (2,250 shp) and the transmission is rated at 1,529 kW (2,050 shp) for take-off, with a maximum continuous power output of 1,379·5 kW (1,850 shp). The main rotor has a Wortmann blade section, swept tips, and an advanced rotor hub with elastomeric bearings on the flapping axis. The tail rotor hub requires no lubrication. Other features include an automatic flight control system, with stability augmentation and attitude retention; nodalised suspension; separate dual hydraulic systems; and a large engine deck that serves also as a maintenance platform. Differences from the military Model 214A include the addition of an engine fire extinguishing system, push-out escape windows in the cargo doors, and commercial avionics.

The standard **Model 214B** can cruise at 140 knots (259 km/h; 161 mph) with an internal load of 1,814 kg (4,000 lb). A passenger configuration provides seats for 14 persons, in addition to the crew of two. It is able to carry external loads in excess of 3,175 kg (7,000 lb) on its cargo hook, which is certificated for a maximum of 3,629 kg (8,000 lb). In an agricultural role this allows nearly four US tons of chemicals to be lifted. The firefighting version shown in the accompanying illustration carries one tank for 1,362 litres (360 US gallons) of liquid fire retardant in its cabin and another under its belly. Normally, the internal tank empties into the belly tank, but a gate between the two can be opened for a full 2,724 litre (720 US gallon) dump.

The **Model 214B-1** differs only in having its max T-O weight for internal load-carrying limited to 5,670 kg (12,500 lb) to meet different certification standards.

A new main rotor blade developed for the Model 214B became the first glassfibre blade of US manufacture to receive FAA certification on 24 July 1978. Assembled as a two-blade rotor for this helicopter, it has a diameter of 15.24 m (50 ft 0 in) and chord of 0·84 m (2 ft 9 in). Testing had exceeded 400 flying hours at the date of certification, and current FAA approval is for a retirement life of 7,200 h. Bell is confident that a retirement life of at least 10,000 h will be achieved when in-plant and service testing have been completed.

Production glassfibre blades were delivered initially as replacements for conventional blades in service on Model 214Bs. Since full-scale manufacture was established, they have become standard on all 214Bs coming off the assembly line.

The production version of the blade has a spar consisting entirely of machine-made elements. An orbital machine winds the spar caps, which are of spanwise-orientated S_2 glass fibres that carry bending loads and centrifugal force. The fibres of these spar caps wrap around the sleeve of the attachment bolt to the hub, forming integral attachment lugs to the hub. Torsional loads in the spar are carried by layers of filament-wound crossply material located inside and outside the spanwise spar caps.

The blade skins consist of layers of non-woven crossply E-glass. A layer of woven cloth is applied to the outside of the skin to minimise foreign object damage, and the skins are supported by a Nomex non-metallic honeycomb core. The leading-edge of the blade is protected by a full-length titanium abrasion strip. The paint finish incorporates a semi-conductive graphite layer to aid the dissipation of static electricity.

Tests have shown that blades built by this method do not dent as easily as metal ones. A fatigue crack usually will not grow from a small hole or puncture, and skin patches can be applied with less risk of subsequent cracking. Blades tested by the Lightning and Transient Research Institute of St Paul, Minnesota, were not damaged structurally by lightning strikes of 200,000 ampères, which is equivalent to the highest strikes recorded on aircraft. In ballistic tolerance tests, one blade virtually 'swallowed' a 23 mm high-explosive impact round rather than permitting it to exit.

DIMENSIONS, EXTERNAL:

Main rotor diameter	15·24 m (50 ft 0 in)
Main rotor blade chord	0·84 m (2 ft 9 in)
Tail rotor diameter	2·95 m (9 ft 8 in)
Tail rotor blade chord	0·305 m (1 ft 0 in)

WEIGHTS:

Max T-O weight:	
internal loading, 214B	6,260 kg (13,800 lb)
internal loading, 214B-1	5,670 kg (12,500 lb)
external loading, 214B and 214B-1	7,257 kg (16,000 lb)

BELL MODEL 214ST

The Model 214ST was developed originally for major production and service in Iran. It was expected to serve primarily as a military transport helicopter; but Bell is now developing and producing the 214ST as a commercial transport with multi-mission capability. Originally, the suffix ST indicated Stretched Twin: Bell is retaining these initials to represent Super Transport.

The prototype 214ST built by Bell flew for the first time in February 1977, and construction of three pre-production examples began in 1978. The decision to manufacture an initial series of 100 production 214STs was announced in November 1979. In October 1980, it was announced that Mitsui and Company would co-produce the 214ST in Japan. FAA certification for two-pilot IFR operation was expected in the fourth quarter of 1981. Deliveries by Bell are due to begin in early 1982, at a rate of three per month. It was reported in 1981 that an option for an undisclosed number, for SAR duties, had been taken by the Spanish Air Force.

TYPE: Twin-turboshaft commercial transport helicopter.

ROTOR SYSTEM: Two-blade advanced technology main rotor. Each blade has a unidirectionally laid glassfibre spar, with a ±45°-wound torque casing of glassfibre cloth. The trailing-edge is also of unidirectional glassfibre, and the space between spar and trailing-edge is filled by a Nomex honeycomb core. The entire blade is then bonded together by glassfibre wrapping, with the leading-edge protected by a titanium abrasion strip and

Bell Model 214B BigLifter equipped as a fire-bomber *(Howard Levy)*

the tip by a replaceable stainless steel cap. Two-blade tail rotor; interchangeable blades, each with a stainless steel leading-edge spar and covering, aluminium honeycomb core and glassfibre trailing-edge strip. Main rotor hub incorporates elastomeric bearings. Second-generation Noda-Matic nodal suspension system. Nodal beam requires no lubrication. Main rotor brake standard.

Rotor Drive: Main transmission, which has a one-hour run-dry capability, has a maximum rating of 1,678 kW (2,250 shp), maximum continuous rating of 1,380 kW (1,850 shp), and single-engine rating of 1,212 kW (1,625 shp). Combining, intermediate and tail rotor gearboxes, each with one-hour run-dry capability.

Fuselage: Conventional all-metal semi-monocoque structure, incorporating roll-over protection ring.

Tail Surface: Electronically-controlled elevator, which minimises trim changes with alterations of power and CG, and improves longitudinal stability.

Landing Gear: Energy-absorbing non-retractable tubular skid type. Future options will include tricycle-type wheeled landing gear.

Power Plant: Two 1,212 kW (1,625 shp) General Electric CT7-2A turboshaft engines, connected to a combining gearbox. In the event of an engine failure, the remaining engine is capable of developing 1,286 kW (1,725 shp) to provide continued flight capability. Standard fuel capacity 1,560 litres (412 US gallons), contained in seven interconnected cells, arranged to provide two independent fuel systems as required by FAR Pt 29. Single-point refuelling. Auxiliary fuel system optional, consisting of two tanks in rear of cabin.

Accommodation: Standard seating for pilot, co-pilot and 16 passengers, with alternative layout for 17 passengers. Dual controls under development. Crew seats adjustable. All seats have lap belt and shoulder harness. Jettisonable crew door each side. Large cabin door on port side for passengers or easy loading of cargo. Three emergency exits on each side. Baggage space aft of cabin, capacity 1·84 m³ (65 cu ft). Passenger seating removable to provide 8·95 m³ (316 cu ft) of cargo capacity. Cabin heated and ventilated.

Systems: Dual engine-driven hydraulic pumps for fully redundant hydraulic system. Redundant electrical system with dual engine-driven generators. Stability and control augmentation system (SCAS). Main rotor blade in-flight tracking system. Attitude/altitude retention system (AARS). Computer-controlled fly-by-wire automatic elevator trim system.

Avionics and Equipment: Avionics to provide full IFR capability, radar, and VLF navigation system, are under development. Equipment under development includes emergency flotation gear, external cargo suspension system, and internal rescue hoist.

Dimensions, external:
- Diameter of main rotor: 15·85 m (52 ft 0 in)
- Diameter of tail rotor: 2·95 m (9 ft 8 in)
- Main rotor blade chord: 0·84 m (2 ft 9 in)
- Tail rotor blade chord: 0·36 m (1 ft 2 in)
- Length overall, rotors turning: 18·95 m (62 ft 2¼ in)
- Length of fuselage: 15·24 m (50 ft 0 in)
- Height overall: 4·84 m (15 ft 10½ in)
- Skid track: 2·64 m (8 ft 8 in)

Dimensions, internal:
- Cabin: Length, instrument panel to centre rear bulkhead: 3·42 m (11 ft 2¾ in)
  - Length, max: 4·13 m (13 ft 6¾ in)
  - Max width: 2·41 m (7 ft 11 in)
  - Volume: 7·73 m³ (273 cu ft)

Areas:
- Main rotor disc: 197·32 m² (2,124 sq ft)
- Tail rotor disc: 6·82 m² (73·39 sq ft)

Weights:
- Max T-O weight: internal or external load: 7,802 kg (17,200 lb)

Performance (at max T-O weight, except where indicated):
- Normal cruising speed at S/L, at average cruise weight: 138 knots (256 km/h; 159 mph)
- Max cruising speed at 1,220 m (4,000 ft): 135 knots (250 km/h; 155 mph)
- Max rate of climb at S/L, ISA: 564 m (1,850 ft)/min
- Service ceiling, one engine out, ISA: 2,135 m (7,000 ft)
- Hovering ceiling IGE: 3,170 m (10,400 ft)
- Range at 1,220m (4,000 ft) with max standard fuel, VFR reserves: 353 nm (654 km; 406 miles)
- Range, ISA, VFR, standard fuel, no reserves: 421 nm (780 km; 485 miles)
- Ferry range with auxiliary fuel, pilot only, no payload, no reserves: over 550 nm (1,019 km; 633 miles)

BELL MODEL 222

In April 1974, Bell announced its intention of developing the Model 222, described as the first commercial light twin-engined helicopter to be built in the USA. Construction of five prototypes began on 1 September 1974 and the first of these flew for the first time on 13 August 1976. FAA certification for a Model 222 in pre-production configuration was received on 16 August 1979. The production model received approval for VFR operation on 20

Bell Model 214ST twin-turbine transport helicopter, scheduled for delivery in 1982

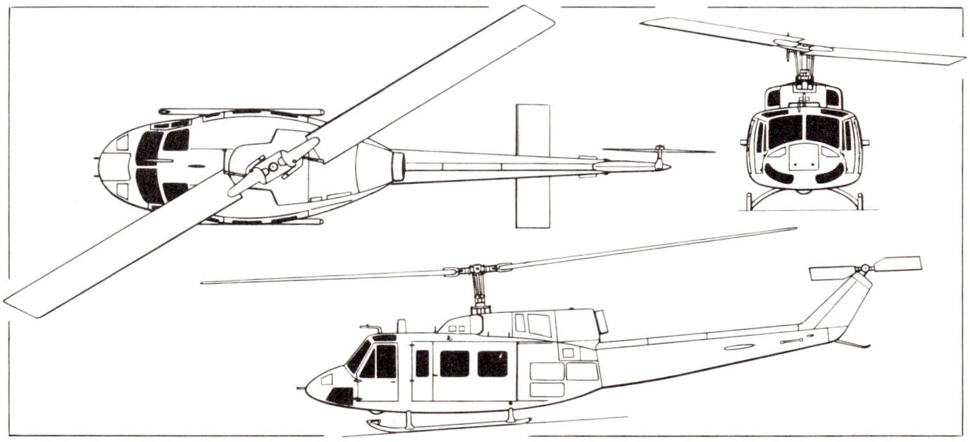

Bell Model 214B medium-size commercial heavy-lift helicopter *(Pilot Press)*

December, and the first delivery, to Petroleum Helicopters Inc, was made on 16 January 1980. Orders exceeded 150 at that date. FAA certification for single-pilot IFR operations in Category I weather conditions was granted on 15 May 1980. A Model 222 delivered to Omniflight Helicopters on 18 January 1981 was the 25,000th Bell helicopter built.

Production aircraft are offered in three configurations, as follows:

Basic 222. Standard model, as described.

222 Executive. Fully equipped for both single and dual pilot IFR flight. Sperry coupled automatic flight control system to provide stability augmentation and automatic hold for attitude, altitude, heading and airspeed, plus VOR/LOC course and glideslope hold during approach. Collins Proline avionics include dual VHF com, dual VOR nav with glideslope, ADF, marker beacon receiver, transponder, DME and area navigation. Luxury accommodation for five or six passengers, with automatic temperature control, fluorescent and reading lights, window curtains and ceiling speakers. Optional stereo system and refreshment cabinet.

222 Offshore. Equipped for dual pilot IFR operations over water, with emergency flotation system, auxiliary fuel tanks and Collins Microline avionics.

Type: Twin-turbine light commercial helicopter.

Rotor System: Two-blade main rotor. Blade section Wortmann 090. Thickness/chord ratio 8%. Each blade comprises a stainless steel spar with bonded glassfibre safety straps to retard crack propagation and offer secondary load path; replaceable stainless steel leading-edge; and afterbody of Nomex honeycomb covered with glassfibre skin. Each blade is attached to the hub by two chordwise bolts. Small trim tab on each blade. Completely dry titanium main rotor hub has conical elastomeric bearings. Two-blade tail rotor of stainless steel construction, with preconing, underslung feathering axis and skewed flapping axis. Rotor blades do not fold. A rotor brake is standard.

Rotor Drive: Rotors shaft-driven through gearbox with two spiral bevel reductions and one planetary reduction. Main rotor/engine rpm ratio 1 : 27·4; tail rotor/engine rpm ratio 1 : 5·08.

Sponsons: Short-span cantilever sponson set low on each side of fuselage, serving also as work platform. Section NACA 0035. Dihedral 3° 12′. Incidence 5°. Sweepback at quarter-chord 3° 30′. All-metal structure of light alloy sheet and honeycomb. No movable surfaces.

Fuselage: Semi-monocoque structure of light alloy, with limited use of light alloy honeycomb panels. Fail-safe structure in critical areas. One-piece nosecone tilts forward and down for access to avionics and equipment bay.

Tail Unit: Cantilever structure of light alloy. Fixed vertical fin in sweptback upper and lower sections. Tailplane, with slotted leading-edge and endplate fins, mounted midway along rear fuselage. Small skid below ventral fin for protection in tail-down landing.

Landing Gear: Hydraulically-retractable tricycle type. All units retract forward, main wheels into sponsons. Free-fall extension in emergency. Oleo-pneumatic shock-absorbers, with scissored yoke. Self-centering nosewheel, swivelling through 360°. Single wheel and tyre on each unit. Main-wheel tyres size 6·00-6, pressure 5·18 bars (75 lb/sq in). Nosewheel tyre size 5·00-5, pressure 4·14 bars (60 lb/sq in). Hydraulic disc brakes. New-type water-activated emergency 'pop-out' floats on Offshore version, which eliminate the need for tips on the sponsons and thus save 40 kg (88 lb) in installed weight.

Power Plant: Two Avco Lycoming LTS 101-650C-2 turboshaft engines, mounted in a streamline housing above the cabin and aft of the rotor pylon. Each engine is rated at 503 kW (675 shp) max for 2·5 min, 485 kW (650 shp) for 30 min and 446 kW (598 shp) max continuous; transmission rated at 652 kW (875 shp) from two engines or 519 kW (696 shp) from one engine. Bell focused pylon with nodalisation. Fuel contained in three crash-resistant internal bladders, in fuselage and sponsons, with total capacity of 710 litres (187·5 US gallons). Single-point refuelling on starboard side of fuselage. Oil capacity 3·2 litres (6·85 US quarts) per engine.

Accommodation: Pilot and seven passengers in standard 2-3-3 layout; alternatively pilot, co-pilot and six passengers. Two additional passengers can be accommodated in a high-density 2-2-3-3 arrangement. Energy-attenuating seats, all with shoulder harness. Crew door at forward end of cabin on each side; cabin door on each

Bell Model 222 light commercial helicopter in the insignia of Helikopter Service AS

side immediately forward of wing. Space for 1·05 m³ (37 cu ft) of baggage aft of cabin, with external door on starboard side. Ventilation standard; air-conditioning and heating optional.

SYSTEMS: Dual hydraulic systems, pressure 103·5 bars (1,500 lb/sq in). Electrical system supplied by dual 150A DC generators. Dual 250VA AC inverters, and 13Ah nickel-cadmium storage battery.

AVIONICS AND EQUIPMENT: VHF radio and Sperry IFR equipment standard. Other avionics, blind-flying instrumentation and equipment, including Bendix RDR-1400 weather radar and 1,588 kg (3,500 lb) capacity cargo hook kit, to customer's requirements.

DIMENSIONS, EXTERNAL:

Diameter of main rotor	12·12 m (39 ft 9 in)
Diameter of tail rotor	1·98 m (6 ft 6 in)
Main rotor blade chord	0·73 m (2 ft 4·6 in)
Tail rotor blade chord	0·254 m (10 in)
Sponson chord at root	1·55 m (5 ft 1 in)
Sponson chord at tip	1·49 m (4 ft 10¾ in)
Length of fuselage	10·98 m (36 ft 0¼ in)
Width overall	3·18 m (10 ft 5 in)
Height overall	3·51 m (11 ft 6 in)
Wheel track	2·77 m (9 ft 1 in)
Wheelbase	3·70 m (12 ft 1¾ in)
Passenger doors (each): Height	1·30 m (4 ft 3 in)
Width	0·99 m (3 ft 3 in)
Height to sill	0·46 m (1 ft 6 in)
Baggage door (stbd, aft): Height	0·62 m (2 ft 0½ in)
Width	0·89 m (2 ft 11 in)
Height to sill	1·14 m (3 ft 9 in)

DIMENSIONS, INTERNAL:

Cabin (passenger area): Length	2·01 m (6 ft 7 in)
Max width	1·41 m (4 ft 7½ in)
Max height	1·30 m (4 ft 3 in)
Volume, incl crew area	5·52 m³ (195 cu ft)
Baggage hold	1·05 m³ (37 cu ft)
Hat box (aft of cabin seats)	0·14 m³ (5 cu ft)

AREAS:

Main rotor blades (each)	4·40 m² (47·37 sq ft)
Tail rotor blades (each)	0·21 m² (2·29 sq ft)
Main rotor disc	115·29 m² (1,241 sq ft)
Tail rotor disc	2·15 m² (23·18 sq ft)
Vertical tail surfaces (total)	1·72 m² (18·53 sq ft)
Horizontal tail surfaces (total)	1·37 m² (14·8 sq ft)

WEIGHTS AND LOADING:

Weight empty, equipped	2,204 kg (4,860 lb)
Max T-O and landing weight	3,560 kg (7,850 lb)
Max disc loading	30·9 kg/m² (6·33 lb/sq ft)

PERFORMANCE (at max T-O weight):

Never-exceed speed	150 knots (278 km/h; 172 mph)
Max rate of climb at S/L, ISA	487 m (1,600 ft)/min
Service ceiling	6,100 m (20,000 ft)
Service ceiling, one engine out, ISA	1,585 m (5,200 ft)
Hovering ceiling IGE, ISA	1,280 m (4,200 ft)
Hovering ceiling OGE, ISA	1,400 m (4,600 ft)
Range with max fuel, 20 min reserves	282 nm (523 km; 325 miles)

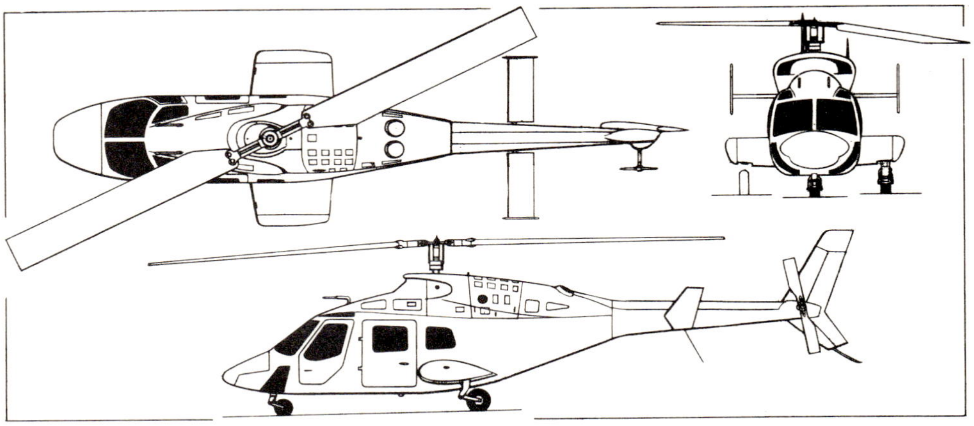

Bell Model 222 twin-turbine light commercial helicopter *(Pilot Press)*

BELL MODEL 301

US Army designation: XV-15

Bell Helicopter announced in May 1973 that it had been chosen by NASA and the US Army to build and test two twin-engined tilt-rotor research aircraft. Estimated cost of the six-year programme was $45 million.

The company has been working on tilt-rotor technology since the mid-1950s, proving the concept feasible with its XV-3 prototype, described in the 1962-63 *Jane's*. Since that time development of tilt-rotor systems has progressed steadily, leading to the Model 301 which Bell proposed to meet the NASA/Army requirement. The two research aircraft, on which design work was started in July 1973, have the Army designation XV-15. The fuselages and tail units were built under subcontract by Rockwell International's Tulsa Division.

The airframe structure is basically that of a conventional fixed-wing aircraft. However, the hover lift and cruise propulsive force is provided by low-disc-loading rotors located at each wingtip. These rotors, together with their wingtip-mounted turboshaft engines, rotate from a vertical position for hover and helicopter flight to horizontal for the conventional propeller-driven flight mode. Hover control is provided by rotor-generated forces and moments. In the conventional mode of flight, control is provided primarily by use of normal aerodynamic control surfaces of the kind fitted to fixed-wing aircraft. A cross-shafting system interconnecting the rotors precludes a complete loss of power to either rotor due to failure of one engine, permits power transfer for transient conditions, and achieves rotational speed synchronisation. A conversion system interconnect shaft caters for rotor axis tilt synchronisation.

The Avco Lycoming turboshaft engines are mounted in rotatable wingtip nacelles to minimise the operational loads on the cross-shaft system. The use of a free-turbine engine permits the reduction of rotor turning speed for conventional forward flight, thus improving rotor performance and reducing cruise noise. The rotors have elastomeric flapping restraints to increase helicopter mode control power and damping. The forward-swept fixed wings provide blade clearance which is adequate to cater for blade flexion resulting from gusts or manoeuvres while operating in an aeroplane flight mode. Wing/rotor/pylon stability is accomplished by selecting a stiff wing and pylon-to-wing attachment, and by minimising the distance of the rotor hub from the wing.

For hover flight the wing trailing-edge flaps and flap/ailerons (flaperons) are deflected downward to reduce the wing download, thereby increasing hovering efficiency. Hover roll control is provided by differential rotor collective pitch, pitch control by cyclic pitch, and yaw control by differential cyclic pitch. Dual controls for use in the helicopter mode are similar to those of a conventional rotating-wing aircraft. Thus, dual collective control sticks provide power and collective pitch for height control, and dual control columns provide longitudinal and lateral control: dual rudder pedals provide directional control.

In the aeroplane flight mode, the control columns and rudder pedals are employed conventionally, while the collective stick/power lever continues in use for power management. An H-tail configuration, with twin endplate fins and rudders, was selected to provide optimum aircraft directional stability around a zero yaw angle. Control authority for the power lever, blade pitch governor, cyclic, differential cyclic, differential collective, and flap/flaperon relationship, are phased with rotor mast angle by mechanical mixing linkages.

At intermediate rotor axis tilt angles (between 60° and 75°) the aircraft can perform STOL operations at weights above the maximum VTOL gross weight of 5,897 kg (13,000 lb).

The XV-15 is fitted with a stability and control augmentation system to improve the handling qualities and enhance pilot efficiency. Ejection seats are installed as a safety feature during flight trials.

Future commercial and military aircraft which might be derived from the XV-15 would have a two-, three- or four-engine power plant. One project, designated D303 by Bell, would have three General Electric T700 turboshafts, and would carry 20 troops in military service or 16 passengers as a civil transport.

Initially, the programme was funded and managed jointly by the NASA Ames Research Center and the US Army's Air Mobility Research and Development Laboratory, but US Navy funding was also provided in 1979 and 1980. The two XV-15s are being used in a research programme to prove the concept, explore the limits of the operational flight envelope and assess its application to military and civil transport needs.

The first aircraft (702) made its first free hovering flight on 3 May 1977. Flight tests of the second XV-15 in helicopter mode began on 23 April 1979. The first full in-flight conversion to aeroplane mode was made by this second XV-15 (N703NA) on 24 July 1979.

Bell Helicopter announced in July 1979 the receipt of a contract from NASA-Ames covering preliminary design study of an advanced rotor blade for the XV-15. The purpose of the study is to select a design concept for a composite rotor blade of 0·43 m (17 in) blade chord, that will offer improved performance and extended life expectancy compared with the existing metal blades of the XV-15.

The contractor flight test programme ended in August 1980; details of the earlier stages of this programme can be found in the 1980-81 and previous editions of *Jane's*. Test achievements by mid-1981 included:

Helicopter mode: 45° angle of bank; autorotative descents; speeds of up to 30 knots (55·5 km/h; 34·5 mph) sideways and 20 knots (37 km/h; 23 mph) rearward; insensitivity to wind direction; low rotor downwash; and excellent handling qualities.

Conversion mode: steady-state flight at all angles; more than 100 full conversions; a 70 knot (130 km/h; 81 mph) wide speed corridor; roll-on landings and take-offs; low pilot workload; and single-engined operation.

Aeroplane mode: a TAS range from 100-301 knots (185-557 km/h; 115-346 mph); 60° bank angle; 915 m (3,000 ft)/min climbs and descents; cruising altitude of 6,400 m (21,000 ft); low noise and vibration levels; and good ride quality.

In 1981 both aircraft were being used by NASA, in California, for the evaluation of tilt-rotor applications.

TYPE: Tilt-rotor research aircraft.

ROTOR SYSTEM: Two three-blade rotors, spring restrained, stiff in plane and gimballed, with an elastomeric hub spring to increase control power and damping. Stainless steel blades of high-twist design, suitable for both helicopter and high-speed aircraft flight modes. Blade section is Bell-modified NACA 6-series. Blades attached to titanium hub by tension-torsion straps and roller pitch bearings. Blades do not fold. No rotor brake.

ROTOR DRIVE: Each rotor is driven by individual engine via reduction gear, engine coupling, rotor planetary gear and shaft centrebox. Rotor/engine rpm ratio 1 : 35·11. Interconnected drive shafts and redundant tilting mechanisms permit single-engine operation and fail-operative tilt capability.

WINGS: Cantilever high-wing monoplane. Wing section Bell-modified NACA 64A223. Dihedral 2°. Incidence 3°. Forward sweep at quarter-chord 6° 30'. All-metal conventional structure, with light alloy ribs and honeycomb panels. Flap/aileron of light alloy construction on outer two-thirds of each wing trailing-edge, powered by HRT hydraulic actuators. Plain light alloy trailing-edge flap on inboard third of each wing, operated by Curtiss-Wright power hinges. No tabs.

FUSELAGE: Semi-monocoque fail-safe structure of light alloy.

TAIL UNIT: Cantilever structure of light alloy, with endplate fin and rudder mounted at each tailplane tip. Tailplane incidence ground-adjustable. Elevators and rudders powered by HRT hydraulic actuators. No tabs.

Bell XV-15 tilt-rotor research aircraft in state of partial conversion

Landing Gear: Hydraulically-retractable tricycle type, as used originally on Canadair CL-84, with twin wheels on each unit. Main units retract forward into fuselage-mounted landing gear pods, nose unit aft. Menasco oleo-pneumatic shock-absorbers. Nosewheel unit of self-centering type. Goodyear magnesium main wheels with Goodyear tyres size 6·50-8, pressure 3·8 bars (55 lb/sq in). Goodyear magnesium nosewheels with Goodyear tyres size 5·00-4, pressure 3·8 bars (55 lb/sq in). Goodyear hydraulically-operated magnesium/steel disc brakes.

Power Plant: Two 1,156 kW (1,550 shp) Avco Lycoming LTC1K-4K turboshaft engines, each with a two-minute contingency rating of 1,343 kW (1,800 shp), wingtip-mounted with tilt mechanism operated by SPECO interconnected double ballscrew actuators. Two fuel tanks in each wing, total capacity 867 litres (229 US gallons). Refuelling point on upper surface of each wing. Oil capacity 11·4 litres (3 US gallons).

Accommodation: Pilot and co-pilot on Rockwell-Columbus LW-3B ejection seats, side by side on flight deck, with access to cabin. Currently in austere test configuration for research equipment, cabin could accommodate nine personnel. Door on starboard side. Flight deck (only) is heated, ventilated and air-conditioned. Overhead and side windows jettisonable ballistically in emergency.

Systems: Garrett air-cycle environmental control unit. No pressurisation. Triplex hydraulic system, pressure 207 bars (3,000 lb/sq in): dual system for rotor and flight controls, with utility system as backup. Pneumatic system, pressure 207 bars (3,000 lb/sq in), for emergency actuation of landing gear. DC electrical system supplied by two 30V 300A generators. Two 28V 13Ah nickel-cadmium storage batteries. Oxygen system at pressure of 124 bars (1,800 lb/sq in). Engine inlet strut anti-icing.

Avionics and Equipment: Stability control augmentation system and FFS. King VHF, UHF, VOR, ILS, marker beacon receiver, and DME. Blind-flying instrumentation fitted.

Dimensions, external:	
Diameter of rotors (each)	7·62 m (25 ft 0 in)
Distance between rotor centres	9·80 m (32 ft 2 in)
Rotor blade chord	0·36 m (1 ft 2 in)
Wing span over engine nacelles	10·72 m (35 ft 2 in)
Wing aspect ratio	6·12
Wing chord, constant	1·60 m (5 ft 3 in)
Width overall, rotors turning	17·42 m (57 ft 2 in)
Length of fuselage	12·50 m (41 ft 0 in)
Length overall	12·83 m (42 ft 1 in)
Height over tail-fins	3·86 m (12 ft 8 in)
Height overall, nacelles vertical	4·67 m (15 ft 4 in)
Tail unit span (incl fins)	3·91 m (12 ft 10 in)
Wheel track, c/l of shock-absorbers	2·64 m (8 ft 8 in)
Wheelbase	4·80 m (15 ft 9 in)
Nacelle ground clearance, nacelles vertical	0·965 m (3 ft 2 in)
Fuselage ground clearance	0·305 m (1 ft 0 in)
Cabin door (stbd): Height	1·37 m (4 ft 6 in)
Width	0·81 m (2 ft 8 in)
Height to sill	0·56 m (1 ft 10 in)
Dimensions, internal:	
Cabin (excl flight deck):	
Length	4·53 m (14 ft 10½ in)
Max width	1·52 m (5 ft 0 in)
Max height	1·52 m (5 ft 0 in)
Floor area	5·40 m² (58·1 sq ft)
Volume	8·50 m³ (300 cu ft)
Areas:	
Rotor blades (each)	1·36 m² (14·6 sq ft)
Rotor discs (each)	45·61 m² (491 sq ft)
Wings, gross	15·70 m² (169 sq ft)
Flap/ailerons (total)	1·88 m² (20·2 sq ft)
Trailing-edge flaps (total)	1·02 m² (11·0 sq ft)
Fins (total)	3·99 m² (43·0 sq ft)
Rudders (total)	0·70 m² (7·5 sq ft)
Tailplane	3·46 m² (37·25 sq ft)
Elevators (total)	1·21 m² (13·0 sq ft)
Weights and Loading:	
Weight empty	4,341 kg (9,570 lb)
Max payload (STOL)	1,542 kg (3,400 lb)
Max fuel weight	676 kg (1,490 lb)
Design T-O weight (VTO)	5,897 kg (13,000 lb)
Max T-O weight (STO)	6,804 kg (15,000 lb)
Max disc loading	74·2 kg/m² (15·2 lb/sq ft)
Performance (estimated, at design T-O weight):	
Never-exceed speed	364 knots (674 km/h; 419 mph)
Max level speed at 5,180 m (17,000 ft)	332 knots (615 km/h; 382 mph)
Max cruising speed at 4,970 m (16,300 ft)	303 knots (561 km/h; 349 mph)
Econ cruising speed at 6,100 m (20,000 ft)	200 knots (371 km/h; 230 mph)
Max rate of climb at S/L	960 m (3,150 ft)/min
Service ceiling	8,840 m (29,000 ft)
Service ceiling, one engine out	4,570 m (15,000 ft)
Min ground turning radius	6·83 m (22 ft 5 in)
Hovering ceiling IGE	3,200 m (10,500 ft)
Hovering ceiling OGE	2,635 m (8,650 ft)
Range with max fuel	445 nm (824 km; 512 miles)

Bell XV-15 tilt-rotor research aircraft in conventional flight mode

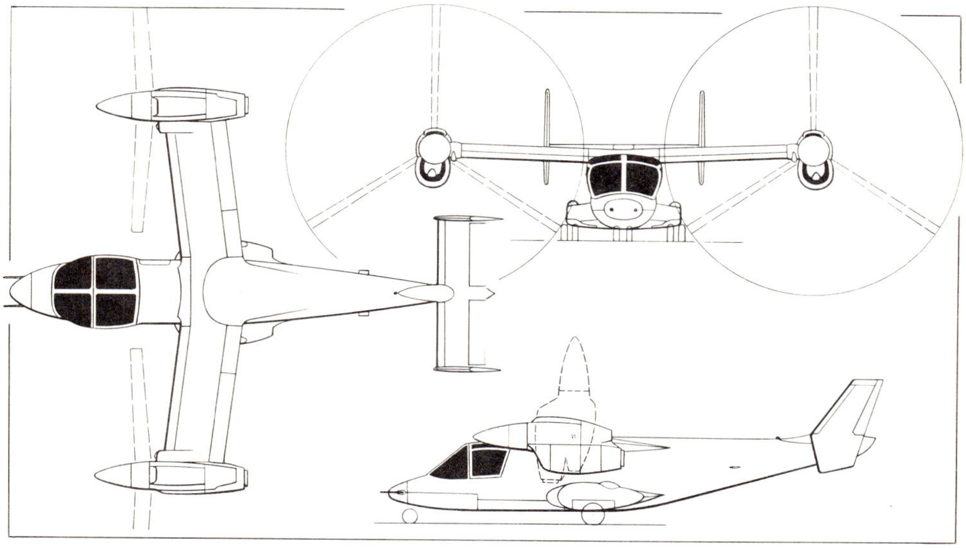

Bell XV-15 tilt-rotor research aircraft *(Pilot Press)*

BELLANCA

BELLANCA AIRCRAFT CORPORATION (subsidiary of Anderson, Greenwood and Co)

Head Office and Works: PO Box 69, Municipal Airport, Alexandria, Minnesota 56308

Known originally as International Aircraft Manufacturing Inc (Inter-air), Bellanca Sales Company (a subsidiary of Miller Flying Service) acquired the assets of Champion Aircraft Corporation on 30 September 1970. Following the merger, the name Bellanca Aircraft Corporation was adopted, and Bellanca subsequently marketed both its own products and those of Champion Aircraft.

For economic reasons, Bellanca halted production of its Viking four-seat light business aircraft, and suspended manufacture of the Citabria, Scout and Decathlon two-seat light aircraft, in the first months of 1980. Its remaining 60 production employees continued work on the Aries, of which details can be found in the 1980-81 *Jane's*, but no recent news of the company or its products has been received.

BELLANCA

BELLANCA AIRCRAFT ENGINEERING INC

Head Office and Works: PO Box 70, Scott Depot, nr Charleston, West Virginia 25560

Telephone: (304) 755 4354

President: August T. Bellanca
Managing Director: Henry E. Payne

The original Bellanca Aircraft Corporation of New Castle, Delaware, merged with companies not engaged in aircraft manufacture and lost its identity in 1959. The present company, formed by Mr August Bellanca and his father, the late G. M. Bellanca, bought all of the original Bellanca aircraft designs with the exception of the Model 14-19. On 1 December 1971 the company was reorganised and acquired corporate offices and production plant at Scott Depot, West Virginia. Development work has been concentrated on the Model 19-25 Skyrocket II, and details of this prototype aircraft can be found in the 1979-80 and earlier editions of *Jane's*.

BOEING

THE BOEING COMPANY

Head Office: PO Box 3707, Seattle, Washington 98124

Established: July 1916

Chairman of the Board and Chief Executive Officer:
T. A. Wilson

President: Malcolm T. Stamper

Senior Vice-Presidents:
H. W. Haynes (Executive Vice-President, Chief Financial Officer)
W. M. Maulden
C. E. Skeen

Vice-Presidents:
R. R. Albrecht (Counsel and Secretary)
R. E. Bateman (General Manager, Boeing Marine Systems)
D. P. Beighle (Contracts)
R. B. Light (Manager, Washington DC Office)
S. M. Little (Industrial and Public Relations)
J. E. Steiner (Corporate Product Development)
R. W. Tharrington (President, Boeing Computer Services)

B. M. Wheat (Senior Vice-President, Boeing Computer Services)
Other Vice-Presidents are listed under individual company headings
TREASURER: J. B. L. Pierce
PUBLIC RELATIONS AND ADVERTISING DIRECTOR: R. P. Bush

Operating components of The Boeing Company are as follows:

Boeing Commercial Airplane Company
See following entry

Boeing Aerospace Company
See pages 314-316 of this section

Boeing Military Airplane Company
See pages 316-317 of this section

Boeing Vertol Company
See pages 317-321 of this section

In May 1961 The Boeing Airplane Company changed its proprietary name to The Boeing Company as a recognition of its diversified interests. On 19 December 1972 it was announced that three of the company's operating organisations had been designated as companies, comprising Boeing Commercial Airplane Company, Renton, Washington; Boeing Aerospace Company, Kent, Washington; and Boeing Vertol Company, Philadelphia, Pennsylvania.

The Boeing Military Airplane Company at Wichita, Kansas, was formed in Autumn 1979 to replace the former Boeing Wichita Company and take over some aircraft programmes from Boeing Aerospace Company. Its responsibilities include modification programmes, 707, 727, 737 and 747 parts fabrication, research, programmes on military aircraft currently in use with the armed forces (B-52 and KC-135), and other support functions.

A factory with an area of 75,000 m² (807,300 sq ft) at St James-Assiniboia Airport, near Winnipeg, produces 747 components.

BOEING COMMERCIAL AIRPLANE COMPANY (BCAC)

PO Box 3707, Seattle, Washington 98124
PRESIDENT: E. H. Boullioun
EXECUTIVE VICE-PRESIDENT: Richard W. Welch
VICE-PRESIDENTS:
J. A. Blue (Product and Customer Support)
W. W. Buckley (Operations)
Ernest V. Fenn (General Manager, 757 Division)
Kenneth F. Holtby (New Programmes)
V. C. Moe (Contracts)
H. Carl Munson (Strategic Planning)
George D. Nible (Customer Services)
Lynn M. Olason (General Manager, 747 Division)
O. M. Roetman (International Sales)
William L. Shineman (General Manager, Fabrication Division)
Frank A. Shrontz (General Manager, 707/727/737 Division)
Joseph F. Sutter (Operations and Product Development)
John M. Swihart (Domestic and Canadian Sales)
R. W. Taylor (Special Assistant to President)
Dean D. Thornton (General Manager, 767 Division)
C. F. Wilde (Sales and Marketing)
H. W. Withington (Engineering)
Brien S. Wygle (Flight Operations)

The Boeing Commercial Airplane Company, with headquarters at the company's Renton, Washington, facility just south of Seattle, has five divisions. The 707/727/737 Division at Renton and the 747 Division at Everett continue to manufacture aircraft of those series, although production of commercial transport 707s has ended; the 767 Division was established to develop this new wide-body transport, which will be built at Everett; the 757 Division handles that programme at the Renton facility; and the Fabrication Division serves the other operating groups with its massive NC machine capability. A separate Engineering Organisation, reporting to company headquarters, is responsible for such functions as technology, quality control and flight operations.

Including military derivatives, 3 Model 707s, 131 Model 727s, 92 Model 737s and 73 Model 747s were delivered in 1980. Orders for 361 new air transports, valued at approximately $10·2 billion, were announced by Boeing customers during the year. Production in 1981 was scheduled at 2 Model 707s (for AWACS programme), 99 Model 727s, 109 Model 737s and 58 Model 747s.

BOEING MODEL 707

USAF designation: VC-137

The prototype for the Boeing Model 707, designated Model 367-80, was the first jet transport designed as such to be completed and flown in the United States. It made its first flight on 15 July 1954, and a developed version was ordered in large numbers as a flight refuelling tanker/-transport for the US Air Force under the designation KC-135.

On 13 July 1955 Boeing was given clearance by the US Air Force to build commercial developments of the prototype concurrently with the production of KC-135s. These transport aircraft, with the basic designations of Boeing 707 and 720, were manufactured in many versions. Details can be found in the 1980-81 and earlier editions of *Jane's*.

Manufacture was continuing in 1981 solely to provide airframes for the USAF/NATO military AWACS programme, details of which can be found in the Boeing Aerospace Company section of this entry.

BOEING MODEL 727

Boeing announced its intention to supplement the four-jet 707/720 series with the three-engined short/medium-range 727 on 5 December 1960. The 727 switched to a rear-engined configuration, but has an upper fuselage section identical with that of the 707/720 and many parts and systems are interchangeable between the three types.

The original short-fuselage 727-100, 727-100C, 727-100QC and 727-100 Business Jet versions of the Model 727 are no longer in production; details of these can be found in the 1973-74 *Jane's*.

The only version currently available is the advanced model of the lengthened 727-200 announced by Boeing on 12 May 1971, at an initial ramp weight of 86,635 kg (191,000 lb), and first delivered in June 1972. This has a much greater fuel capacity and range than earlier 727-200s; the interior features the 'Superjet-look', and a large 'Carry-all' compartment is available at no extra cost. Availability of successively more powerful engines has made possible weight increases up to a current max ramp weight of 95,254 kg (210,000 lb) for aircraft with JT8D-15/17/17R turbofans. The current 727-200 is certificated for Category IIIA 15 m (50 ft) decision height, 213 m (700 ft) RVR landing with optional autopilot systems. Improved overhead duct air-conditioning is now standard. A performance data computer system has been added as standard equipment to provide onboard information for optimisation of flight profile and fuel consumption.

A total of 1,815 Model 727s had been sold by 1 October 1981, of which 1,768 had been delivered. The 727 is the only commercial transport aircraft of which more than 1,500 have been delivered, the 1,500th example being handed over to United Air Lines on 2 July 1979. The number of individual customers for the 727 reached 100 when Wistair International of Texas ordered an executive 727-200 on 24 September 1980. More than 1,200 of the 727s sold by that time were of the 727-200 series. The 1,761st Model 727, a -200, handed over to Ansett Airlines of Australia on 30 July 1981, was the 4,000th Boeing jet transport to be delivered.

Production is expected to diminish as the new Models 757 and 767 enter production, from a total of 131 delivered in 1980 to 99 in 1981, 25 to 30 in 1982 and 12 in 1983.

TYPE: Three-turbofan airliner.

WINGS: Cantilever low-wing monoplane. Special Boeing aerofoil sections. Thickness/chord ratio from 9% to 13%. Dihedral 3°. Incidence 2°. Sweepback at quarter-chord 32°. Primary structure is a two-spar aluminium alloy box with conventional ribs. Upper and lower surfaces are of riveted skin-stringer construction. There are no chordwise splices in the primary structure from the fuselage to the wingtip. Advanced 727-200s at gross weight options have modified stringers and in-spar webs, as well as upper and lower surface wing skins of increased gauge. Structure is fail-safe. Hydraulically-powered aluminium ailerons, in inboard (high speed) and outboard (low speed) units, operate in conjunction with flight spoilers. Triple-slotted trailing-edge flaps constructed primarily of aluminium and aluminium honeycomb. Four aluminium leading-edge slats on outer two-thirds of wing. Three Krueger leading-edge flaps on inboard third of wing, made from magnesium or aluminium castings. Seven spoilers on each wing, consisting of five flight spoilers outboard and two ground spoilers inboard. Spoilers function also as airbrakes. Balance tab in each outboard aileron; control tab in each inboard aileron. Controls are hydraulically-powered dual systems with automatic reversion to manual control. Actuators manufactured primarily by Weston, National Water Lift and Bertea. Thermal anti-icing of wing leading-edges by engine bleed air.

FUSELAGE: Semi-monocoque fail-safe structure, with aluminium alloy skin reinforced by circumferential frames and longitudinal stringers.

TAIL UNIT: Cantilever structure, built primarily of aluminium alloys, with tailplane mounted near tip of fin. Dual-powered variable-incidence tailplane, with direct manual reversion. Hydraulically-powered dual elevator control system with control tab manual reversion. Hydraulically-powered rudders, utilising two main systems with backup third system for lower rudder. Anti-balance tabs; rudder trim by displacing system neutral.

LANDING GEAR: Hydraulically-retractable tricycle type, with twin wheels on all three units. Nosewheels retract forward, main gear inward into fuselage. Boeing oleo-pneumatic shock-absorbers. Goodrich nose-gear wheels, tyres and brakes are standard on all models. Goodrich and Bendix are both approved suppliers of main-gear wheels, tyres and brakes for all Model 727s. Nosewheels and tyres are size 32 × 11·5 Type VIII. Main-gear wheels size 49 × 17, with tyres size 50 × 21 Type VII, are standard.

POWER PLANT: Three Pratt & Whitney JT8D-9A turbofan engines with thrust reversers and full sound attenuation, each flat rated at 64·5 kN (14,500 lb st) to 29°C, are standard. Optionally, JT8D-15s rated at 68·9 kN (15,500 lb st), JT8D-17s rated at 71·2 kN (16,000 lb st), or JT8D-17Rs with automatic performance reserve (APR) and rated at 77·4 kN (17,400 lb st), can be fitted. The JT8D-17R engines are normally operated at an alternative rating of 72·9 kN (16,400 lb st); APR senses any significant loss in thrust by an engine during take-off

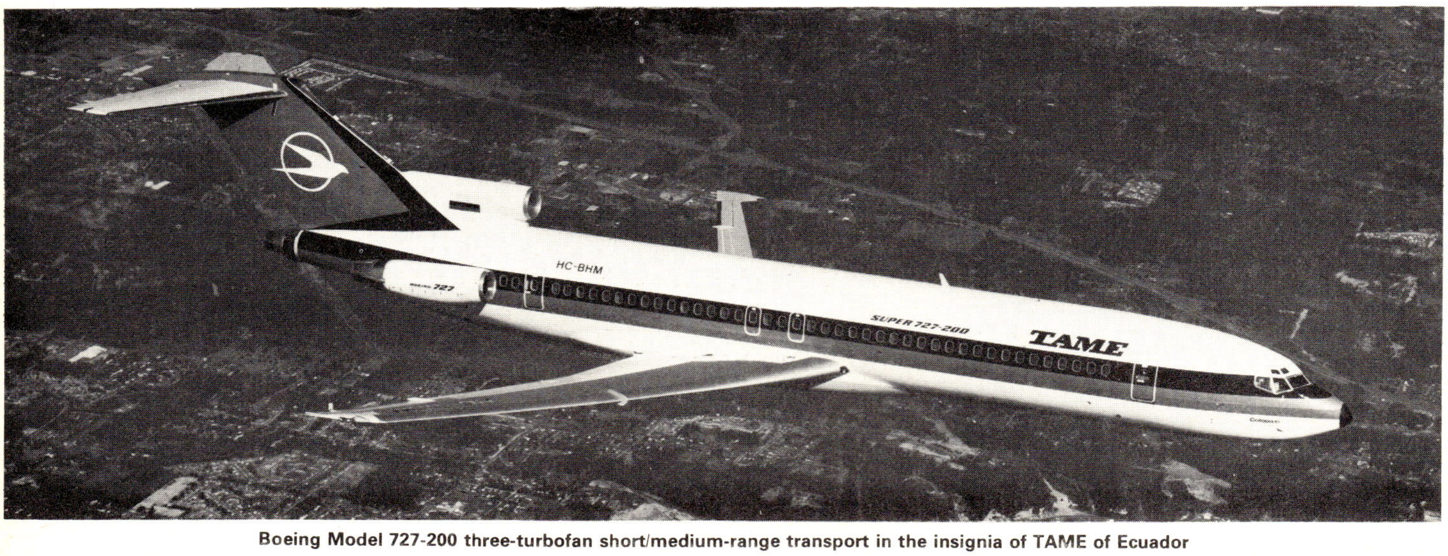

Boeing Model 727-200 three-turbofan short/medium-range transport in the insignia of TAME of Ecuador

and initial climb, automatically increasing thrust on the other engines to 77·4 kN (17,400 lb). This feature significantly improves performance from hot/high airports. Each engine has individual fuel system fed from integral tanks in wings, but all three tanks are interconnected. Optional fuselage fuel tanks can be installed, displacing forward and/or aft cargo compartment volume. Standard total fuel capacity 30,623 litres (8,090 US gallons). Modular design bladder cell tanks with dual fuel barrier can be installed to contain up to approximately 9,387 litres (2,480 US gallons). Single pressure fuelling point, rated at 2,271 litres (600 US gallons)/min, near leading-edge underside of starboard wing at mid-span. Total usable oil capacity 45·5 litres (12 US gallons).

Accommodation: Crew of three on flight deck. Basic accommodation for 145 passengers (14 first class, 131 tourist class, 4-6 abreast). Max capacity 189 passengers. Two galleys forward and two aft. One toilet forward and two aft. Other layouts to customer's specification. A 'Superjet-look' passenger interior design is standard. The wide-body effect is achieved (without any changes in cross-section dimensions) by lighting and architectural redesign. Retrofit kits for the 'Superjet-look' are offered, as are kits for larger 'Carry-all' compartments. Entry via hydraulically-operated integral aft stairway under centre engine and door at front on port side with optional Weber Aircraft electrically-operated airstairs. Two Type III emergency exits in mid-cabin on each side and aft service door on each side. The starboard forward service door is opposite the port forward passenger door. Two heated and pressurised baggage and freight compartments under floor, forward and aft of main landing gear bay. Each compartment has one outward-opening cargo door; a second cargo door is optional for the aft compartment.

Systems: Garrett air-conditioning and pressurisation system, using engine bleed air combined with air-cycle refrigeration. Pressure differential 0·59 bars (8·6 lb/sq in). Three independent 207 bar (3,000 lb/sq in) hydraulic systems, utilising Boeing Material Specification BMS 3-11 hydraulic fluid, provide power for flying controls, landing gear and aft airstairs. Electrical system includes three 40kVA 400Hz constant-frequency AC generators, three 50A transformer-rectifier units, one 22Ah battery. Garrett APU provides electrical power and compressed air for engine starting and air-conditioning on ground.

Avionics and Equipment: Standard avionics, available from alternative sources, include flight and service attendants' intercom, passenger address system, ARINC 531 Selcal, dual VHF com, ARINC 594 ground proximity warning system, ARINC 557 voice recorder, ARINC 542 flight recorder and remote encoder, marker beacon receiver, ARINC 552 radio altimeter, ARINC 572 ATC transponder, dual DME, ARINC 570 ADF, dual ARINC 547 VHF nav systems, dual radio direction/distance measuring indicators with dual directional gyros and compass systems, dual FD110 flight director systems, single-channel Mod Blk V autopilot, dual vertical gyros, dual yaw dampers, performance data computer system, dual ARINC 545 digital air data computer systems, push-button audio selector panels, and VHF nav/com frequency preselect. Optional avionics, available from several alternative sources, include a third VHF com, single or dual HF com, third VHF nav, Omega, single or dual ARINC 564 X-band weather radar, second ADF, Z-15 flight director systems, dual autopilot channels, roll monitor and flare coupler, third vertical gyro, and full range auto-throttle/speed control system.

Dimensions, external:
Wing span 32·92 m (108 ft 0 in)
Wing chord at root 7·70 m (25 ft 3 in)
Wing chord at tip 2·34 m (7 ft 8 in)
Wing aspect ratio 7·07
Length overall 46·69 m (153 ft 2 in)
Length of fuselage 41·51 m (136 ft 2 in)
Height overall 10·36 m (34 ft 0 in)
Tailplane span 10·90 m (35 ft 9 in)
Wheel track 5·72 m (18 ft 9 in)
Wheelbase 19·28 m (63 ft 3 in)
Passenger door (ventral): Length 1·93 m (6 ft 4 in)
Width 0·81 m (2 ft 8 in)
Passenger door (fwd): Height 1·83 m (6 ft 0 in)
Width 0·86 m (2 ft 10 in)
Height to sill 2·67 m (8 ft 9 in)
Service door (fwd): Height 1·65 m (5 ft 5 in)
Width 0·84 m (2 ft 9 in)
Service doors (aft, each): Height 1·52 m (5 ft 0 in)
Width 0·76 m (2 ft 6 in)
Baggage hold door (fwd): Height 1·07 m (3 ft 6 in)
Width 1·37 m (4 ft 6 in)
Baggage hold door (aft): Height 1·12 m (3 ft 8 in)
Width 1·37 m (4 ft 6 in)

Dimensions, internal:
Cabin (aft of flight deck to rear pressure bulkhead):
Length 28·24 m (92 ft 8 in)
Max width 3·55 m (11 ft 8 in)
Max height 2·11 m (6 ft 11 in)
Floor area 91·05 m² (980 sq ft)
Volume 188·4 m³ (6,652 cu ft)

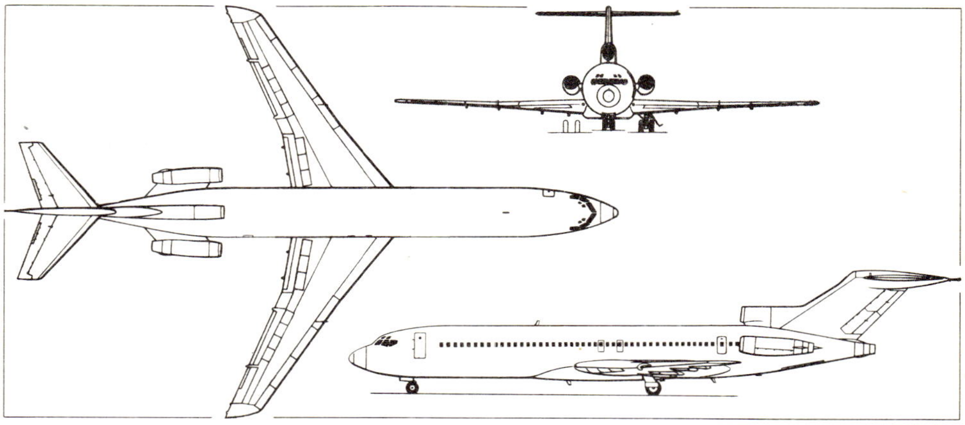

Boeing 727-200 three-turbofan short/medium-range transport *(Pilot Press)*

Baggage hold (fwd) 20·1 m³ (710 cu ft)
Baggage hold (aft): standard 23·1 m³ (815 cu ft)
with optional 2nd door 21·1 m³ (745 cu ft)

Areas:
Wings, gross 157·9 m² (1,700 sq ft)
Ailerons (total) 5·30 m² (57 sq ft)
Trailing-edge flaps, retracted (total) 26·10 m² (281 sq ft)
Trailing-edge flaps, extended (total) 36·04 m² (388 sq ft)
Flight spoilers (total) 7·41 m² (79·8 sq ft)
Fin 33·07 m² (356 sq ft)
Rudder, incl tabs 6·13 m² (66 sq ft)
Tailplane 34·93 m² (376 sq ft)
Elevators, incl tabs 8·83 m² (95 sq ft)

Weights and Loadings (A: brake release weight of 83,820 kg; 184,800 lb, B: brake release weight of 86,405 kg; 190,500 lb, C: brake release weight of 95,027 kg; 209,500 lb):
Operating weight empty (basic specification):
A 44,633 kg (98,400 lb)
B 45,178 kg (99,600 lb)
C 46,675 kg (102,900 lb)
Operating weight empty (typical airline):
A 45,360 kg (100,000 lb)
Max payload (structural, based on airline operating weight empty): A 18,144 kg (40,000 lb)
Max T-O weight: A 83,820 kg (184,800 lb)
B 86,405 kg (190,500 lb)
C 95,027 kg (209,500 lb)
Max ramp weight: A 84,275 kg (185,800 lb)
B 86,635 kg (191,000 lb)
C 95,254 kg (210,000 lb)
Max zero-fuel weight: A 62,595 kg (138,000 lb)
B 63,500 kg (140,000 lb)
C 65,315 kg (144,000 lb)
Max landing weight: A, B 70,080 kg (154,500 lb)
C 73,028 kg (161,000 lb)
Max wing loading: A 530·7 kg/m² (108·7 lb/sq ft)
B 546·8 kg/m² (112 lb/sq ft)
C 600·5 kg/m² (123 lb/sq ft)
Max power loading: A 433·2 kg/kN (4·2 lb/lb st)
B 446·5 kg/kN (4·4 lb/lb st)
C 491·1 kg/kN (4·8 lb/lb st)

Performance (A at brake release weight of 83,820 kg; 184,800 lb, B at brake release weight of 86,405 kg; 190,500 lb, and C at brake release weight of 95,027 kg; 209,500 lb, except where indicated):
Max operating speed Mach 0·90
Max level speed: A at 6,585 m (21,600 ft) 549 knots (1,017 km/h; 632 mph)
B, C at 6,250 m (20,500 ft) 539 knots (999 km/h; 621 mph)
Max cruising speed: A at 6,705 m (22,000 ft) 514 knots (953 km/h; 592 mph)
B at 7,530 m (24,700 ft) 520 knots (964 km/h; 599 mph)
Econ cruising speed at 9,145 m (30,000 ft) 495 knots (917 km/h; 570 mph)
Stalling speed at S/L, flaps down:
at 72,575 kg (160,000 lb) 104 knots (193 km/h; 120 mph)
Initial cruise altitude 10,060 m (33,000 ft)
Min ground turning radius 24·49 m (80 ft 4 in)
Runway LCN at max weight of 86,635 kg (191,000 lb), optimum tyre pressure and 0·51 m (20 in) flexible pavement:
50 × 21 tyres 70
CAR T-O distance to 10·7 m (35 ft):
A 2,880 m (9,450 ft)
B 2,554 m (8,380 ft)
C 2,804 m (9,200 ft)
CAR landing distance from 15 m (50 ft):
at 71,668 kg (158,000 lb) 1,430 m (4,690 ft)
Range at long-range cruising speed, with fuel load as specified and payload of 12,474 kg (27,500 lb), ATA domestic reserves:
A with 30,623 litres (8,090 US gallons) 2,000 nm (3,706 km; 2,303 miles)
B with 33,878 litres (8,950 US gallons) 2,180 nm (4,040 km; 2,510 miles)
C with 36,831 litres (9,730 US gallons) 2,370 nm (4,392 km; 2,729 miles)
Range with 18,144 kg (40,000 lb) payload, at long-range cruising speed, ATA domestic reserves:
A 1,530 nm (2,835 km; 1,762 miles)
B 1,615 nm (2,993 km; 1,860 miles)
C 2,160 nm (4,003 km; 2,487 miles)

Operational Noise Levels (with JT8D-15 engines, FAR Pt 36):
T-O at brake release weight of 86,405 kg (190,500 lb) 100 EPNdB
Approach at 70,080 kg (154,500 lb) landing weight and 30° flap 100·4 EPNdB
Sideline 102·2 EPNdB

BOEING MODEL 737

USAF designation: T-43A

The decision to build this short-range transport was announced by Boeing on 19 February 1965. Simultaneously, a first order for 21 aircraft was placed by Lufthansa.

The original Model 737 was designed to utilise many components and assemblies already in production for the Boeing 727. Design began on 11 May 1964, and the first Model 737 flew on 9 April 1967. Deliveries began before the end of 1967, following FAA certification on 15 December. Sales of the 737 totalled 951 by 6 August 1981, of which 779 had been delivered, including 19 Model 737-200s modified as T-43A navigation trainers for the US Air Force (see 1975-76 *Jane's*). The number of individual customers reached 100 in September 1980, when Air Pacific of Fiji ordered a 737-200.

Boeing announced on 30 April 1981 the receipt of an order from the Indonesian Air Force for three specially-equipped 737-200s for use primarily in a maritime surveillance role. Each will have 14 first class and 88 tourist class seats so that they can be used also for government transport purposes. The specialised equipment will consist of a Motorola high-resolution side-looking airborne multi-mission radar (SLAMMR), its console linked to two 5 m (16 ft) long blade antennae, mounted one on each side of the upper rear fuselage. This installation will make it possible to spot a small ship in heavy sea conditions at a range of 87 nm (160 km; 100 miles) on each side of the aircraft's flight path from a height of 9,150 m (30,000 ft). The aircraft are scheduled for delivery in May 1982, and June and September 1983. The first will be delivered without the multi-mission radar, but will be returned for retrofit of the SLAMMR installation when it has been certificated on the 737.

Details of the early production versions of the Model 737, and of subsequent design development, can be found in the 1974-75 *Jane's*. Versions currently available are as follows:

Advanced 737-200. Current standard model, with max ramp weight of 52,615 kg (116,000 lb) and max T-O weight of 52,390 kg (115,500 lb). JT8D-9A engines (each 64·5 kN; 14,500 lb st) standard; JT8D-15 (68·9 kN; 15,500 lb st); JT8D-17 (71·2 kN; 16,000 lb st); or JT8D-17R with Automatic Performance Reserve (75·6 kN; 17,000 lb st) engines optional; basic fuel capacity of 19,532 litres (5,160 US gallons). Accommodation for 115 passengers and baggage, with 86 cm (34 in) pitch seating, or up to 130 passengers in 74 cm (29 in) pitch seating with no reduction in cabin facilities. A gross weight option with a max ramp weight of 53,295 kg (117,500 lb) and max T-O weight of 53,070 kg (117,000 lb) is available.

Advanced 737-200C/QC. Standard convertible passenger/cargo model with strengthened fuselage and floor, and a large two-position upper-deck cargo door with effective opening of 2·15 m × 3·40 m (7 ft 0½ in × 11 ft 2 in). The quick-change (QC) feature allows more rapid conversion by using palletised passenger seating and other special interior furnishings. A gross weight option with a max ramp weight of 53,295 kg (117,500 lb) and max T-O weight of 53,070 kg (117,000 lb) is available.

Advanced 737-200 Executive Jet. Same as standard Advanced 737-200, except interiors are adapted to special business and executive luxury requirements. Executive interiors can be obtained from vendors. A limited kit selection is available from Boeing. Additional fuel capacity offered by installation of fuel cells in lower cargo compartments. With max fuel this model can carry 15 passengers up to 3,000 nm (5,560 km; 3,455 miles).

Advanced 737-200 High Gross Weight Structure. Higher gross weight models of the Advanced 737-200/200C, for longer-range use, are available in two versions. One has a maximum taxi weight of 56,700 kg (125,000 lb) and a maximum T-O weight of 56,472 kg (124,500 lb) with JT8D-15, -17, or -17R engines, and a fuel capacity of either 20,951 litres (5,535 US gallons) or 22,598 litres (5,970 US gallons). The additional capacity for increased range capability is provided by a 1,419 litre (375 US gallon) or a 3,066 litre (810 US gallon) fuel tank installed in the aft lower cargo compartment. The second version, with a maximum taxi weight of 58,332 kg (128,600 lb), maximum T-O weight of 58,105 kg (128,100 lb), design landing weight of 48,534 kg (107,000 lb), and maximum zero-fuel weight of 43,091 kg (95,000 lb), has approximately 650 nm (1,204 km; 748 miles) greater range capability than the standard Advanced 737-200 (130 passenger payload). Sectors of 2,300 nm (4,262 km; 2,648 miles) can be served with a 130 passenger payload and typical fuel reserves. Aircraft is identical to the current Advanced 737-200 except for the auxiliary fuel tank, new wheels, tyres and brakes, and strengthened landing gear and wing structure.

All higher gross weight versions meet FAR Pt 36 and ICAO Annex 16 in respect of noise characteristics.

An FAA-certificated kit is available which enables the Model 737 to operate from unpaved or gravel runways. The kit includes a vortex dissipator for each engine, consisting of a short hollow boom that protrudes from under each engine's forward edge. The boom is capped by a plug with downward-facing orifices. Pressurised engine bleed air forced through these orifices destroys any ground-level vortex and prevents small pieces of gravel being ingested by the engines. Other items include a gravel deflection 'ski' on the nosewheel, deflectors between the main landing gear wheels, protective shields over hydraulic tubing and speed brake cable on the main gear strut, glassfibre reinforcement of lower inboard flap surfaces, application of Teflon-base paint to fuselage and wing undersurfaces and provision of more robust DME, ATC and VHF antennae.

On 10 February 1980 British Airways began operations with the first of 28 Advanced 737s which the airline calls 'Super 737s'. These aircraft are equipped with advanced flight deck avionics, including a Sperry SP-177 digital automatic flight control system. Lufthansa has ordered 32 Advanced 737s with similar equipment; these will also have a Sperry RD-800J horizontal situation indicator and an engine noise reduction 'mixer'. Category IIIA certification of these AFCS versions was expected by late Spring 1981.

737-300. Developed from Advanced 737-200, for entry into service in 1984. Described separately.

The description which follows applies to the versions of the 737-200 in current production:

TYPE: Twin-turbofan short-range transport.

WINGS: Cantilever low-wing monoplane. Special Boeing wing sections. Average thickness/chord ratio 12·89%. Dihedral 6°. Incidence 1° at root. Sweepback at quarter-chord 25°. Aluminium alloy dual-path fail-safe two-spar structure. Ailerons of aluminium honeycomb construction. Boeing-developed triple-slotted trailing-edge flaps, all of aluminium with trailing-edges of aluminium honeycomb. Aluminium alloy Krueger flaps on leading-edge, inboard of nacelles. Three leading-edge slats of aluminium alloy with aluminium honeycomb trailing-edge on each wing from engine to wingtip. Two-section aluminium honeycomb flight spoilers on each outer wing serve both as airbrakes in the air and for lateral control, in association with ailerons. Two-section aluminium honeycomb ground spoilers on each wing, inboard of engine, are used only during landing. Ailerons are hydraulically powered by two hydraulic systems with manual reversion. Trailing-edge flaps are hydraulically powered, with electrical backup. Leading-edge slats and Krueger flaps are symmetrically powered by one hydraulic system normally, and by a second hydraulic system for alternate extension. Flight spoilers are symmetrically powered by the two main individual hydraulic systems. Engine bleed air for anti-icing supplied to engine nose cowls and all wing leading-edge slats.

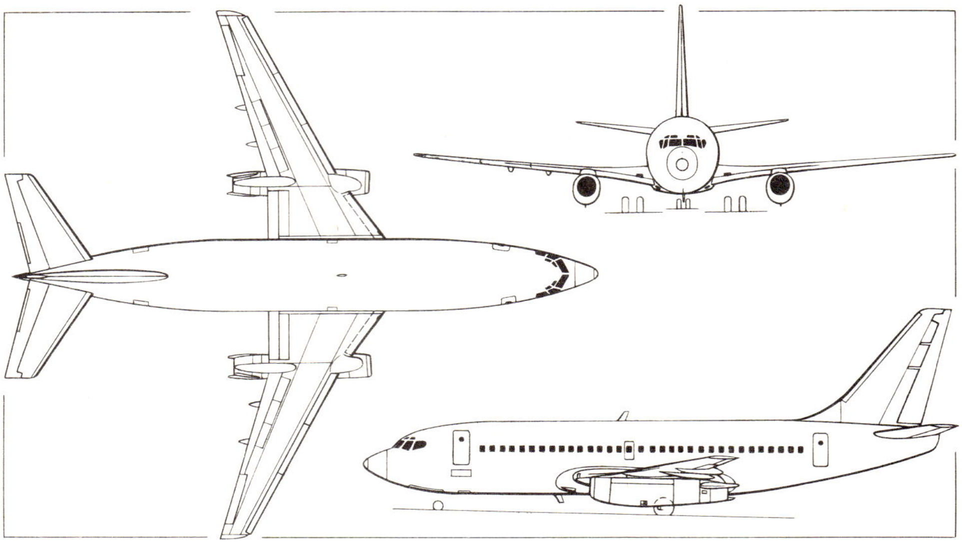

Boeing 737-200 twin-turbofan short-range transport *(Pilot Press)*

FUSELAGE: Aluminium alloy semi-monocoque fail-safe structure.

TAIL UNIT: Cantilever aluminium alloy multi-spar structure. Variable-incidence tailplane. Elevator has dual hydraulic power, with manual reversion. Rudder is powered by a dual actuator from two main hydraulic systems, with a standby hydraulic actuator and system. Tailplane trim has dual electric drive motors, with manual backup. Elevator control tabs for manual reversion are locked out during hydraulic actuation.

LANDING GEAR: Hydraulically-retractable tricycle type, with free-fall extension. Nosewheels retract forward, main units inward. No main gear doors: wheels form wheel-well seal. Twin wheels on each main and nose unit. Boeing oleo-pneumatic shock-absorbers. Main wheels and tyres size 40 × 14-16 (low-pressure 40 × 18-17 tyres, or C40 × 14-21/H40 × 14·5-19 tyres with heavy-duty wheel brakes, are available optionally). Nosewheels and tyres size 24 × 7·7 (low-pressure 24·5 × 8·5 tyres available optionally). Bendix or Goodrich multi-disc brakes. Hydro-Aire Mk III anti-skid units and automatic brakes standard.

POWER PLANT: Two Pratt & Whitney JT8D turbofan engines (details under individual model listings), in underwing pods. High-performance target-type thrust reversers, with full sound attenuation quiet nacelles. All models have standard fuel capacity of up to 19,532 litres (5,160 US gallons), with integral fuel cells in wing centre-section as well as two integral wing tanks. Long-range version has auxiliary fuel tank in aft lower cargo compartment, giving max fuel capacity of 22,598 litres (5,970 US gallons). Single-point pressure refuelling through leading-edge of starboard wing. Fuelling rate 1,135 litres (300 US gallons)/min. Auxiliary overwing fuelling points. Total oil capacity 41·5 litres (11 US gallons).

ACCOMMODATION: Crew of two side by side on flight deck. Details of passenger accommodation given under individual model descriptions. Passenger versions are equipped with forward airstair; an aft airstair is optional. Convertible passenger/cargo versions have the aft airstair as standard and forward airstair optional. One plug-type door at each corner of cabin, with passenger doors on port side and service doors on starboard side. Overwing escape hatches on each side. Basic passenger cabin has one lavatory and one galley at each end. Large-volume hand baggage overhead bins. Provision for a large variety of interior arrangements. Freight holds forward and aft of wing, under floor.

SYSTEMS: Air-conditioning and pressurisation system utilises engine bleed air. Max differential 0·52 bars (7·5 lb/sq in). Two independent hydraulic systems, using fire-resistant hydraulic fluid, for flying controls, flaps, slats, landing gear, nosewheel steering and brakes; pressure 207 bars (3,000 lb/sq in). No pneumatic system. Electrical supply provided by engine-driven generators.

Boeing Model 737-200 twin-turbofan short-range transport in the insignia of Guyana Airways

Garrett APU for air supply and electrical power in flight and on the ground, as well as engine starting.

Avionics and Equipment: Equipment to satisfy FAA Category II low weather minimum criteria is standard, as well as a Lear Siegler performance data computer system. Autopilot, specially designed for ILS localiser and glideslope control, with control wheel steering. Optional equipment will permit Category IIIA capability. Very Low Frequency (VLF-Omega) navigation systems, and a range of flight management systems with various levels of automation, including autothrottle, automatic flight control, and a performance navigation computer system, are available as options.

Dimensions, external:

Wing span	28·35 m (93 ft 0 in)
Wing chord at root	4·71 m (15 ft 5·6 in)
Wing chord at tip	1·60 m (5 ft 3 in)
Wing aspect ratio	8·83
Length overall	30·53 m (100 ft 2 in)
Length of fuselage	29·54 m (96 ft 11 in)
Height overall	11·28 m (37 ft 0 in)
Tailplane span	10·97 m (36 ft 0 in)
Wheel track	5·23 m (17 ft 2 in)
Wheelbase	11·38 m (37 ft 4 in)
Main passenger door (port, front):	
Height	1·83 m (6 ft 0 in)
Width	0·86 m (2 ft 10 in)
Height to sill	2·62 m (8 ft 7 in)
Passenger door (port, rear):	
Height	1·83 m (6 ft 0 in)
Width	0·76 m (2 ft 6 in)
Width with airstair	0·86 m (2 ft 10 in)
Height to sill	2·72 m (8 ft 11 in)
Galley service door (stbd, front):	
Height	1·65 m (5 ft 5 in)
Width	0·76 m (2 ft 6 in)
Height to sill	2·62im (8 ft 7 in)
Service door (stbd, rear):	
Height	1·65 m (5 ft 5 in)
Width	0·76 m (2 ft 6 in)
Height to sill	2·72 m (8 ft 11 in)
Freight hold door (stbd, fwd):	
Height	1·30 m (4 ft 3 in)
Width	1·22 m (4 ft 0 in)
Height to sill	1·30 m (4 ft 3 in)
Freight hold door (stbd, rear):	
Height	1·22 m (4 ft 0 in)
Width	1·22 m (4 ft 0 in)
Height to sill	1·45 m (4 ft 9 in)

Dimensions, internal:

Cabin, incl galley and toilet:	
Length	20·88 m (68 ft 6 in)
Max width	3·52 m (11 ft 6½ in)
Max height	2·18 m (7 ft 2 in)
Floor area	63·8 m² (687 sq ft)
Volume	131·28 m³ (4,636 cu ft)
Freight hold (fwd) volume	10·48 m³ (370 cu ft)
Freight hold (rear) volume	14·30 m³ (505 cu ft)

Area:

Wings, gross	91·04 m² (980 sq ft)

Weights and Loadings (standard aircraft at brake release weight of 52,390 kg; 115,500 lb except where indicated):

Operating weight empty:	
200	27,691 kg (61,050 lb)
200C all passenger	28,962 kg (63,850 lb)
200C all cargo	27,420 kg (60,450 lb)
200QC all passenger	30,467 kg (67,170 lb)
200QC all cargo	27,873 kg (61,450 lb)
Max payload:	
200	15,399 kg (33,950 lb)
200C all passenger	14,129 kg (31,150 lb)
200C all cargo	17,486 kg (38,550 lb)
200QC all passenger	12,623 kg (27,830 lb)
200QC all cargo	17,032 kg (37,550 lb)
200 Executive Jet	2,268 kg (5,000 lb)
Max T-O weight:	
All models, basic	52,390 kg (115,500 lb)
Optional:	53,070 kg (117,000 lb)
	or 56,472 kg (124,500 lb)
	or 58,105 kg (128,100 lb)
Max ramp weight:	
All models, basic	52,615 kg (116,000 lb)
Optional:	53,295 kg (117,500 lb)
	or 56,700 kg (125,000 lb)
	or 58,332 kg (128,600 lb)
Max zero-fuel weight:	
All models, basic	43,091 kg (95,000 lb)
Optional for 200C	up to 44,906 kg (99,000 lb)
Max landing weight:	
All models, basic	46,720 kg (103,000 lb)
Optional:	47,627 kg (105,000 lb)
	or 48,534 kg (107,000 lb)
Wing loading:	
All models, basic	575·5 kg/m² (117·9 lb/sq ft)
Max optional	638·2 kg/m² (130·7 lb/sq ft)
Power loading (JT8D-17):	
All models, basic	368 kg/kN (3·61 lb/lb st)
Max optional	408 kg/kN (4·00 lb/lb st)

Weights and Loadings (at brake release weight of 56,472 kg; 124,500 lb):

Operating weight empty	27,986 kg (61,700 lb)
Max payload	15,104 kg (33,300 lb)
Max T-O weight	56,472 kg (124,500 lb)
Max ramp weight	56,700 kg (125,000 lb)
Max zero-fuel weight	43,091 kg (95,000 lb)
Max landing weight	48,534 kg (107,000 lb)
Max wing loading	620·24 kg/m² (127·04 lb/sq ft)
Max power loading (JT8D-17)	397 kg/kN (3·9 lb/lb st)

Artist's impression of Boeing 737-200 equipped with Motorola side-looking airborne multi-mission radar (SLAMMR)

Performance (ISA, with JT8D-15 engines):

Max operating speed, all models: Mach 0·84

Max level speed, all models, at 7,165 m (23,500 ft): 509 knots (943 km/h; 586 mph)

Max cruising speed, 737-200 at an average cruise weight of 40,823 kg (90,000 lb) at 6,890 m (22,600 ft): 500 knots (927 km/h; 576 mph)

Econ cruising speed at 9,145 m (30,000 ft): Mach 0·73

Stalling speed, flaps down, at max landing weight: 99 knots (184 km/h; 114 mph)

Runway LCN (at max taxi weight of 52,615 kg; 116,000 lb, optimum tyre pressure and 20 in flexible pavement):

40 × 14-16 tyres	51
40 × 14-21 tyres	51
40 × 18-17 tyres	36

FAR T-O distance to 10·7 m (35 ft), 737-200 at 49,435 kg (109,000 lb) AUW and 28·9°C:

JT8D-9 engines	2,073 m (6,800 ft)
JT8D-17 engines	1,615 m (5,300 ft)

FAR landing distance from 15 m (50 ft), 737-200 at max landing weight: 1,341 m (4,400 ft)

Min ground turning radius: 17·58 m (57 ft 8 in)

Range, cruising at 10,060 m (33,000 ft), including 3,016 kg (6,650 lb) reserve fuel, 737-200 at 52,615 kg (116,000 lb) taxi weight with 115 passengers: 1,900 nm (3,521 km; 2,188 miles)

Range, all conditions as above, except 58,332 kg (128,600 lb) taxi weight: 2,400 nm (4,448 km; 2,764 miles)

Range, conditions as above, 737-200 with 130 passengers: 2,300 nm (4,262 km; 2,648 miles)

Operational Noise Levels (JT8D-9 engines and nacelle acoustic treatment, FAR Pt 36):

T-O at 52,390 kg (115,500 lb) brake release weight: 95·3 EPNdB

Sideline at 52,390 kg (115,500 lb) brake release weight: 100·6 EPNdB

Approach at 46,720 kg (103,000 lb) max landing weight: 101·9 EPNdB

BOEING MODEL 737-300

Work on this new short-range transport was started in early 1980. Its airframe is about 80 per cent common with that of the Advanced 737-200. Lengthening of the fuselage, to accommodate additional passengers and underfloor freight, and the installation of new-generation turbofan engines, offer much reduced fuel consumption per seat-mile and lower noise levels compared with the earlier model. Over a 500 nm (927 km; 576 mile) range, the 737-300 is expected to carry 20 more passengers than the Advanced 737-200, with a 16 per cent lower fuel consumption.

On 5 and 18 March 1981 respectively, USAir and Southwest Airlines each placed an order for ten 737-300s, with options on ten more; Southwest stated that it intended to negotiate an increase of its option to 30 aircraft. On this basis, Boeing announced on 26 March that the 737-300 had been committed to full development and production.

The manufacturing plan, announced at the time of production committal, schedules the completion of a Class II mockup, and initiation of tool design, in 1982; the beginning of major assembly in mid-1983; and rollout, first flight, certification and initial deliveries during 1984. Boeing does not regard the 737-300 as a replacement for the 737-200, but as a complement to the existing Boeing range of aircraft.

The description of the 737-200 applies also to the 737-300, except as follows:

Wings: Generally similar to the 737-200 except: modified aerofoil section for leading-edge slats outboard of engine nacelles; revised trailing-edge flap sections and flap track fairings aft of engines; wing structure strengthened; and each wingtip extended by 0·28 m (11 in), with wingtip flutter boom.

Fuselage: As for 737-200, but lengthened by a total of 2·64 m (8 ft 8 in), by insertion of a 1·12 m (3 ft 8 in) fuselage plug forward of the wing, and a 1·52 m (5 ft 0 in) plug aft of the wing carry-through structure. In addition to providing increased passenger capacity, this

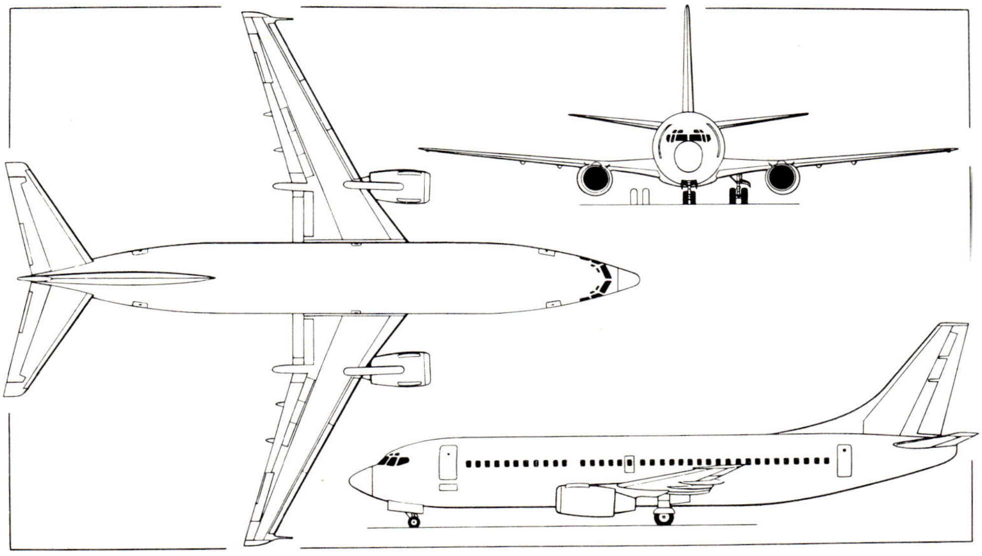
Boeing Model 737-300 (two CFM International CFM56-3 turbofan engines) *(Pilot Press)*

'stretch' gives a lower freight hold volume which is greater by 5·47 m³ (193 cu ft) than that of the standard 737-200.

Tail Unit: As for 737-200, except dorsal fin area and tailplane span increased.

Landing Gear: Generally as for 737-200, but nose unit repositioned and modified to ensure adequate ground clearance for larger engine nacelles. Twin nosewheels have tyres size 29 × 7·7. Main units have heavy-duty wheels, H40 × 14·5-19 heavy-duty tyres, and Bendix 4-rotor or Goodrich 5-rotor heavy-duty wheel brakes as standard. Main-wheel tyre pressure 11·7-12·2 bars (170-177 lb/sq in).

Power Plant: Two 89 kN (20,000 lb st) CFM International CFM56-3 turbofan engines, pylon-mounted one on each wing. Rolls-Royce/Japan Aero Engines RJ500 turbofans of similar thrust rating optional. Nacelles are forward of wings, and higher than those of 737-200; each is fitted with two aerodynamic fences. Standard fuel capacity up to 20,290 litres (5,360 US gallons), with integral fuel cells in wing centre-section and integral wing tanks. Single-point pressure refuelling through leading-edge of starboard wing.

Accommodation: Crew of two side by side on flight deck (unchanged from 737-200). Alternative cabin layouts seat from 121 to 149 passengers. Typical arrangements offer 8 first class seats four-abreast at 96·5 cm (38 in) pitch and 114 or 120 tourist class seats six-abreast at 86 cm (34 in) or 81 cm (32 in) respectively in mixed class; and 132, 140 or 148 all-tourist class at seat pitches of 86 cm (34 in), 81 cm (32 in) or 76 cm (30 in) respectively. One plug-type door at each corner of cabin, with passenger doors on port side and service doors on starboard side. Airstair for forward cabin door standard. Overwing emergency exit on each side. One galley and one lavatory forward, and one or two galleys and lavatories aft, depending on configuration. 'New look' large-volume interior or reduced-volume 'carry-all' interior optional: former has overhead baggage stowage capacity of 3·20 m³ (113 cu ft), latter 6·43 m³ (227 cu ft). Underfloor freight holds, forward and aft of wing, with access doors on starboard side.

Systems: Generally as for 737-200.

Avionics and Equipment: Equipped to FAA Category II low weather minimum criteria as standard. AFCS includes digital Category II autopilot, 12·7 cm (5 in) electro-mechanical flight displays, 10 cm (4 in) electrical air data displays, dual digital air data computer, and full-range digital autothrottle. Other items include a performance data computer, dual nav/com, VHF nav, colour digital radar, and digital autobrake. Optional equipment confers Category IIIA capability for the AFCs, a VLF/Omega nav system, and dual INS. A performance navigation computer system, with an associated dual electronic control display unit, is under study and, if satisfactory, will be available as an option.

Dimensions, external:

Wing span	28·91 m (94 ft 10 in)
Wing chord at root	4·71 m (15 ft 5·6 in)
Length overall	33·40 m (109 ft 7 in)
Height overall	11·13 m (36 ft 6 in)
Tailplane span	12·80 m (42 ft 0 in)
Wheel track	5·23 m (17 ft 2 in)
Wheelbase	12·45 m (40 ft 10 in)
Main passenger door (port, fwd):	
Height	1·83 m (6 ft 0 in)
Width	0·86 m (2 ft 10 in)
Passenger door (port, rear):	
Height	1·83 m (6 ft 0 in)
Width	0·76 m (2 ft 6 in)
Emergency exit (overwing, port and stbd):	
Height	0·97 m (3 ft 2 in)
Width	0·51 m (1 ft 8 in)
Service doors (stbd, fwd and rear):	
Height	1·65 m (5 ft 5 in)
Width	0·76 m (2 ft 6 in)
Freight hold door (stbd, fwd):	
Height	1·22 m (4 ft 0 in)
Width	1·30 m (4 ft 3 in)
Freight hold door (stbd, rear):	
Height	1·22 m (4 ft 0 in)
Width	1·22 m (4 ft 0 in)

Dimensions, internal:

Cabin, incl galley and toilet:	
Length	23·52 m (77 ft 2 in)
Max width	3·45 m (11 ft 4 in)
Max height	2·11 m (6 ft 11 in)
Freight hold (fwd) volume	12·03 m³ (425 cu ft)
Freight hold (rear) volume	18·21 m³ (643 cu ft)

Weights (estimated):

Operating weight empty	32,500 kg (71,650 lb)
Max T-O weight: standard	56,472 kg (124,500 lb)
optional	58,967 kg (130,000 lb)
Max ramp weight: standard	56,700 kg (125,000 lb)
optional	59,195 kg (130,500 lb)
Max zero-fuel weight	47,625 kg (105,000 lb)
Max landing weight	51,710 kg (114,000 lb)

Performance (estimated, A: at brake release weight of 56,472 kg; 124,500 lb. B: at optional BRW of 58,967 kg; 130,000 lb):

T-O field length, S/L, at 29° C:	
A	2,173 m (7,130 ft)
B	2,384 m (7,820 ft)
Wet landing field length, 40° flap, at typical landing weight	1,549 m (5,083 ft)
Still-air range, T-O at S/L:	
A	1,387 nm (2,570 km; 1,597 miles)
B	1,810 nm (3,354 km; 2,084 miles)

BOEING MODEL 747

USAF designation: E-4

First details of this wide-body commercial transport were announced on 13 April 1966, simultaneously with the news that Pan American had placed a $525 million contract for 25 Boeing 747s, including spares. Programme go-ahead date was officially 25 July 1966.

The first 747 (designated RA001) flew for the first time on 9 February 1969 and was retained by Boeing as an experimental flight test vehicle for airframe, avionics and systems technology development, evaluation, and certification. FAA certification of the basic 747 was granted on 30 December 1969. The first 747 to be delivered was received by Pan American on 12 December 1969, and this company inaugurated commercial service with the type on its New York/London route on 22 January 1970.

Orders for all versions of the 747 totalled 575 by 1 August 1981, of which 522 had been delivered. The 500th 747 was rolled out on 9 December 1980, for subsequent delivery to SAS. The total 747 fleet operating in Spring 1981 carried about 4·6 million passengers each month, and had flown more than 9 million hours in revenue service.

Versions of the Boeing 747 currently available are as follows:

747-100B. The original 747-100 was introduced into commercial service in January 1970. This aircraft, in versions with gross weights up to 334,751 kg (738,000 lb), more than doubled passenger and cargo payload capabilities by comparison with any previous commercial air transport, and 167 were sold. The current -100B incorporates strengthened wing, fuselage and landing gear structure. Initial order, with JT9D-7F engines, was placed by Iran Air in 1978, and this aircraft was delivered on 5 July 1979. Current versions, with max taxi weights of 323,411 kg (713,000 lb), 334,751 kg (738,000 lb) and 341,555 kg (753,000 lb), allow for the installation of a variety of optional engines in addition to the basic 208·8 kN (46,950 lb st) Pratt & Whitney JT9D-7A. These include the 216 kN (48,570 lb st) JT9D-7AW, 213·5 kN (48,000 lb st) JT9D-7F, 222·4 kN (50,000 lb st) JT9D-7FW, 222·4 kN (50,000 lb st) JT9D-7J, 233·5 kN (52,500 lb st) General Electric CF6-50E/E1/E2, 206·8 kN (46,500 lb st) CF6-45A2 and CF6-45B2, 222·8 kN (50,100 lb st) Rolls-Royce RB.211-524B2, and 229·5 kN (51,600 lb st) Rolls-Royce RB.211-524C2. Total of 8 ordered by 1 July 1981.

747SP. Lighter-weight, shorter-bodied derivative of the 747-100B. Described separately.

747SR. This short-range version of the 747-100B embodies structural changes required for high take-off and landing cycles. The 747SR is available at max taxi weights up to 341,555 kg (753,000 lb) with the same engines as available for the 747-100B. The initial purchase of four 747SRs by Japan Air Lines (later increased to seven) was announced on 30 October 1972. The JAL aircraft have max taxi weights of 273,515 kg (603,000 lb) and 237,225 kg (523,000 lb). The first 747SR flew on 4 September 1973 and was delivered on 26 September 1973. Total of 24 ordered by 1 July 1981.

747-200B. Passenger version, with same accommodation as 747-100B. First flown on 11 October 1970 and certificated on 23 December 1970; deliveries began on 15 January 1971. Basic version had max T-O weight of 351,535 kg (775,000 lb) and increased fuel capacity. Available now with 216 kN (48,570 lb st) JT9D-7AW engines and max T-O weight of 356,070 kg (785,000 lb); 222·4 kN (50,000 lb st) JT9D-7FW and max T-O weight of 365,140 kg (805,000 lb); 222·4 kN (50,000 lb st) JT9D-7J engines and max T-O weight of 362,875 kg (800,000 lb); 235·7 kN (53,000 lb st) JT9D-7Q engines and max T-O weight of 377,840 kg (833,000 lb); 240·2 kN (54,000 lb st) JT9D-7R4G2 engines and max T-O weight of 377,840 kg (833,000 lb); 233·5 kN (52,500 lb st) General Electric CF6-50E engines and max T-O weight of 371,945 kg (820,000 lb); 233·5 kN (52,500 lb st) CF6-50E1/E2 engines and max T-O weight of 377,840 kg (833,000 lb); 222·8 kN (50,100 lb st) Rolls-Royce RB.211-524B2 engines and max T-O weight of 371,945 kg (820,000 lb); 229·5 kN (51,600 lb st) RB.211-524C2 engines and max T-O weight of 377,840 kg (833,000 lb); and 236·25 kN (53,110 lb st) RB.211-524D4 engines and max T-O weight of 377,840 kg (833,000 lb). Total of 203 ordered by 1 July 1981.

During certification tests for the Pratt & Whitney JT9D-7Q engine, on 23 May 1979, a 747-200B took off from an airport near Glasgow, Montana, at a gross weight of 385,970 kg (850,920 lb). This is the heaviest T-O weight ever recorded by an aeroplane, though not certificated as an official record by Boeing. It was one of a series of take-offs at more than 385,553 kg (850,000 lb) made by this aircraft during heavy-weight noise certification tests.

747-200B Combi. A version of the basic 747-200B, incorporating a 3·12 × 3·40 m (123 × 134 in) cargo door in the port side of the fuselage, aft of the wing. This permits main deck layouts for passengers only, or for passengers and up to 12 main deck pallets/containers, with passenger and cargo areas separated by removable bulkhead. The first modification to Combi configuration was carried out on a Sabena 747-100, and redelivery was made in February 1974. The first 747-200B production Combi was delivered to Air Canada in March 1975. Total of 60 ordered by 1 July 1981.

747-200C Convertible. Version of 747-200 which can be converted from all-passenger to all-cargo, or five combinations of both. The first 747-200C flew on 23 March 1973, was certificated on 17 April, and was delivered to World Airways on 27 April 1973. Max T-O weight of 356,070 kg (785,000 lb) with Pratt & Whitney JT9D-7AW engines; 365,140 kg (805,000 lb) with JT9D-7FW engines; 362,875 kg (800,000 lb) with JT9D-7J engines; 371,945 kg (820,000 lb) with General Electric CF6-50E or Rolls-Royce RB.211-524B2 engines; and 377,840 kg (833,000 lb) with JT9D-7Q, JT9D-7R4G2, General Electric CF6-50E1/E2 or Rolls-Royce RB.211-524C2/D4 engines. Total of 13 ordered by 1 July 1981.

747-200F Freighter. Version of the 747-200 capable of delivering 90,720 kg (200,000 lb) of palletised cargo over a range of 4,100 nm (7,600 km; 4,720 miles). Certification of the first 747-200F, which is described separately, was awarded by the FAA on 7 March 1972. T-O weights as detailed for 747-200C. Total of 45 ordered by 1 July 1981.

747SUD. Version with 'stretched' upper deck. Described separately.

E-4. Advanced Airborne Command Post version of 747, developed for US Air Force. Described separately under Boeing Aerospace heading.

Boeing Model 747-200B four-turbofan commercial transport in the insignia of Philippine Airlines

The following details apply specifically to the basic Model 747 passenger airliner:

TYPE: Four-turbofan heavy commercial transport.

WINGS: Cantilever low-wing monoplane. Special Boeing wing sections. Thickness/chord ratio 13·44% inboard, 7·8% at mid-span, 8% outboard. Dihedral 7°. Incidence 2°. Sweepback 37° 30′ at quarter-chord. Aluminium alloy dual-path fail-safe structure. Low-speed outboard ailerons; high-speed inboard ailerons. Triple-slotted trailing-edge flaps. Six aluminium honeycomb spoilers on each wing, comprising four flight spoilers outboard and two ground spoilers inboard. Ten variable-camber leading-edge flaps outboard and three-section Krueger flaps inboard on each wing leading-edge. All controls fully powered.

FUSELAGE: Conventional semi-monocoque structure, consisting of aluminium alloy skin, longitudinal stiffeners and circumferential frames. Structure is of fail-safe design, utilising riveting, bolting and structural bonding.

TAIL UNIT: Cantilever aluminium alloy dual-path fail-safe structure. Variable-incidence tailplane. No trim tabs. All controls fully powered.

LANDING GEAR: Hydraulically-retractable tricycle type. Twin-wheel nose unit retracts forward. Main gear comprises four four-wheel bogies: two, mounted side by side under fuselage at wing trailing-edge, retract forward; two, mounted under wings, retract inward. Cleveland Pneumatic oleo-pneumatic shock-absorbers. All 18 wheels and tubeless tyres of Model 747-100B are size 46 × 16 Type VII. Tyre pressure: main wheels 14·49 bars (210 lb/sq in), nosewheels 13·11 bars (190 lb/sq in). Main wheels and tyres size 49 × 17 on 747-200B model, pressure 14·15 bars (205 lb/sq in). The high gross weight aircraft has 49 × 19 tyres at a pressure of 13·46 bars (195 lb/sq in). Disc brakes on all main wheels, with individually-controlled anti-skid units.

POWER PLANT: Four Pratt & Whitney, General Electric or Rolls-Royce turbofan engines, as detailed in model listings, in pods pylon-mounted on wing leading-edges. Fuel in seven integral tanks. Capacity of centre wing tank varies according to version: 747-100B: 49,966 litres (13,200 US gallons); 747-200: 64,973 litres (17,164 US gallons). Remaining tanks common to all versions: two inboard main tanks, each 47,492 litres (12,546 US gallons); two outboard main tanks, each 16,966 litres (4,482 US gallons); two inboard reserve tanks, each 2,021 litres (534 US gallons). Outboard mains are reduced by 1,234 litres (326 US gallons) when CF6-50E engines are installed. 747-200 also available with two outboard reserve tanks, each 2,983 litres (788 US gallons). Total capacity, including manifolds, 747-100B: 183,380 litres (48,445 US gallons); 747-200: 198,385 litres (52,409 US gallons). Refuelling point on each wing between inboard and outboard engines. Total usable oil capacity 19 litres (5 US gallons).

ACCOMMODATION: Normal operating crew of three, on flight deck above level of main deck. Observer station and provision for second observer station are provided. Basic accommodation for 452 passengers, made up of 32 first class and 420 economy class, which includes a 32-passenger upper deck. Alternative layouts accommodate 447 economy class passengers in nine-abreast seating or 516 ten-abreast, with 32 passengers on upper deck. All versions have two aisles. Five passenger doors on each side, of which two forward of wing on each side are normally used. Freight holds under floor, forward and aft of wing, with doors on starboard side. One door on forward hold, two on rear hold. Aircraft is designed for fully-mechanical loading of baggage and freight. An optional side cargo door is available for passenger, convertible and freighter versions of the Model 747. Installed aft of door 4 on the port side of the fuselage, it allows the carriage of main-deck cargo on passenger versions. Addition of this door to the freighter allows loads up to 3·05 m (10 ft) in height to be accommodated aft of the flight deck, and also makes possible simultaneous nose and side cargo handling.

SYSTEMS: Air-cycle air-conditioning system. Pressure differential 0·61 bars (8·9 lb/sq in). Four independent hydraulic systems, pressure 207 bars (3,000 lb/sq in), each with one engine-driven and one pneumatically-driven pump. The latter pumps supplement or substitute for engine-driven pumps. A small AC-powered electric pump is installed to charge the brake accumulator during towing of the aircraft. Electrical supply from four aircooled 60kVA generators mounted one on each engine. Two 60kVA generators (supplemental cooling allows 90kVA each) mounted on APU for ground operation and to supply primary electrical power when engine-mounted generators are not operating. Three-phase 400Hz constant-frequency AC generators, 115/200V output. 28V DC power obtained from transformer-rectifier units. 24V 36Ah nickel-cadmium battery for selected ground functions and as in-flight backup. Gas turbine APU for pneumatic and electrical supplies.

AVIONICS AND EQUIPMENT: Standard avionics include two ARINC 566 VHF communications systems, two ARINC 533A HF communications systems, one ARINC 531 Selcal, three ARINC 547 VOR/ILS navigation systems, two ARINC 570 ADF, marker beacon receiver, two ARINC 568 DME, two ARINC 572 ATC, two ARINC 552 low-range radio altimeters, two ARINC 564 weather radar units, three ARINC 561 inertial navigation systems, two heading reference systems, ARINC 412 interphone, passenger address system, passenger entertainment system, ARINC 573 flight recorder, ARINC 557 cockpit voice recorder, integrated electronic flight control system with autothrottle and rollout guidance to provide automatic stabilisation, path control and pilot assist functions for Category II and III landing conditions, two ARINC 565 central air data systems, stall warning system, central instrument warning system, ground proximity warning system, attitude and navigation instrumentation, and standby attitude indication.

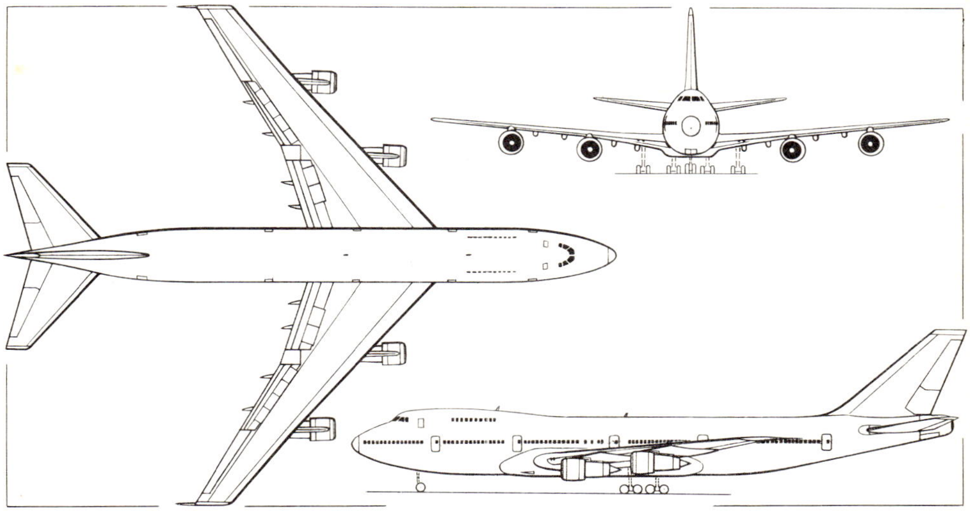

Boeing 747-200B four-turbofan heavy transport aircraft *(Pilot Press)*

DIMENSIONS, EXTERNAL:

Wing span	59·64 m (195 ft 8 in)
Wing chord at root	16·56 m (54 ft 4 in)
Wing chord at tip	4·06 m (13 ft 4 in)
Wing aspect ratio	6·96
Length overall	70·66 m (231 ft 10 in)
Length of fuselage	68·63 m (225 ft 2 in)
Height overall	19·33 m (63 ft 5 in)
Tailplane span	22·17 m (72 ft 9 in)
Wheel track	11·00 m (36 ft 1 in)
Wheelbase	25·60 m (84 ft 0 in)
Passenger doors (ten, each):	
Height	1·93 m (6 ft 4 in)
Width	1·07 m (3 ft 6 in)
Height to sill	approx 4·88 m (16 ft 0 in)
Baggage door (front hold):	
Height	1·68 m (5 ft 6 in)
Width	2·64 m (8 ft 8 in)
Height to sill	approx 2·64 m (8 ft 8 in)
Baggage door (forward door, aft hold):	
Height	1·68 m (5 ft 6 in)
Width	2·64 m (8 ft 8 in)
Height to sill	approx 2·69 m (8 ft 10 in)
Bulk loading door (rear door on aft hold):	
Height	1·19 m (3 ft 11 in)
Width	1·12 m (3 ft 8 in)
Height to sill	approx 2·90 m (9 ft 6 in)
Optional cargo door (port):	
Height	3·05 m (10 ft 0 in)
Width	3·40 m (11 ft 2 in)

DIMENSIONS, INTERNAL:

Cabin, incl toilets and galleys:	
Length	57·00 m (187 ft 0 in)
Max width	6·13 m (20 ft 1½ in)
Max height	2·54 m (8 ft 4 in)
Floor area, passenger deck	327·9 m^2 (3,529 sq ft)
Volume, passenger deck	789 m^3 (27,860 cu ft)
Baggage hold (fwd, containerised) volume	78·4 m^3 (2,768 cu ft)
Baggage hold (aft, containerised) volume	68·6 m^3 (2,422 cu ft)
Bulk volume	28·3 m^3 (1,000 cu ft)

AREAS:

Wings, reference area	511 m^2 (5,500 sq ft)
Ailerons (total)	20·6 m^2 (222 sq ft)
Trailing-edge flaps (total)	78·7 m^2 (847 sq ft)
Leading-edge flaps (total)	48·1 m^2 (518 sq ft)
Spoilers (total)	30·8 m^2 (331 sq ft)
Fin	77·1 m^2 (830 sq ft)
Rudder	22·9 m^2 (247 sq ft)
Tailplane	136·6 m^2 (1,470 sq ft)
Elevators	32·5 m^2 (350 sq ft)

WEIGHTS (letters are used to denote engine installations as follows: (A) JT9D-7A/7AW; (B) JT9D-7F/7FW/7FJ; (C) JT9D-7Q; (D) JT9D-7R4G2; (E) CF6-50E; (F)

Assembly-line photographs showing difference between stretched upper deck (top) and standard versions of Boeing 747

CF6-50E1/E2; (G) CF6-45A2 or -45B2; (H) RB.211-524B2; (J) RB.211-524C2; and (K) RB.211-524D4):

Operating weight empty (approx) for max available gross weights:

747SR (550 pass):
A, B 162,840 kg (359,000 lb)
E, F, G 164,655 kg (363,000 lb)
H, J 167,375 kg (369,000 lb)

747-100B (452 pass):
A, B 171,460 kg (378,000 lb)
E, F, G 173,270 kg (382,000 lb)
H, J 175,995 kg (388,000 lb)

747-200B (452 pass):
A, B 172,820 kg (381,000 lb)
C 173,725 kg (383,000 lb)
D 173,270 kg (382,000 lb)
E, F 175,085 kg (386,000 lb)
H, J 177,810 kg (392,000 lb)
K 177,355 kg (391,000 lb)

747-200B Combi (238 pass and 12 pallets):
A, B 172,365 kg (380,000 lb)
C 173,725 kg (383,000 lb)
D 173,270 kg (382,000 lb)
E, F 175,085 kg (386,000 lb)
H, J 177,810 kg (392,000 lb)
K 177,355 kg (391,000 lb)

747-200C (452 pass):
A, B 177,355 kg (391,000 lb)
C 178,715 kg (394,000 lb)
D 178,260 kg (393,000 lb)
E, F 180,075 kg (397,000 lb)
H, J 182,800 kg (403,000 lb)
K 182,345 kg (402,000 lb)

747-200C (28 pallets):
A, B 164,200 kg (362,000 lb)
C 165,560 kg (365,000 lb)
D 165,105 kg (364,000 lb)
E, F 166,920 kg (368,000 lb)
H, J 169,645 kg (374,000 lb)
K 169,190 kg (373,000 lb)

Max payload:

747SR (550 pass):
A, B 75,975 kg (167,500 lb)
E, F, G 74,160 kg (163,500 lb)
H, J 71,440 kg (157,500 lb)

747-100B (452 pass):
A, B 67,360 kg (148,500 lb)
E, F, G 65,545 kg (144,500 lb)
H, J 62,820 kg (138,500 lb)

747-200B (452 pass):
A, B 66,000 kg (145,500 lb)
C 65,090 kg (143,500 lb)
D 65,545 kg (144,500 lb)
E, F 63,730 kg (140,500 lb)
H, J 61,010 kg (134,500 lb)
K 61,460 kg (135,500 lb)

747-200B Combi (238 pass and 12 pallets):
A, B 74,840 kg (165,000 lb)
C 73,480 kg (162,000 lb)
D 73,935 kg (163,000 lb)
E, F 72,120 kg (159,000 lb)
H, J 69,400 kg (153,000 lb)
K 69,855 kg (154,000 lb)

747-200C (452 pass):
A, B 90,265 kg (199,000 lb)
C 88,905 kg (196,000 lb)
D 89,360 kg (197,000 lb)
E, F 87,545 kg (193,000 lb)
H, J 84,820 kg (187,000 lb)
K 85,275 kg (188,000 lb)

747-200C (28 pallets):
A, B 103,420 kg (228,000 lb)
C 102,060 kg (225,000 lb)
D 102,510 kg (226,000 lb)
E, F 100,695 kg (222,000 lb)
H, J 97,975 kg (216,000 lb)
K 98,430 kg (217,000 lb)

Max T-O weight:

747SR (initial aircraft):
A, B, E, F, G, H, J 235,870 kg (520,000 lb)
or 272,155 kg (600,000 lb)

747-100B, 747SR (optional):
A, B, E, F, G, H, J 322,050 kg (710,000 lb)
or 333,390 kg (735,000 lb)
or 340,195 kg (750,000 lb)

747-200B, -200B Combi, -200C:
A 351,535 kg (775,000 lb)
or 356,070 kg (785,000 lb)
B 362,875 kg (800,000 lb)
or (wet) 365,140 kg (805,000 lb)
E, H 362,875 kg (800,000 lb)
or 371,945 kg (820,000 lb)
C, D, F, J, K 362,875 kg (800,000 lb)
or 371,945 kg (820,000 lb)
or 377,840 kg (833,000 lb)

Max ramp weight:

747-100B, 747SR:
A, B, E, F, G, H, J 341,555 kg (753,000 lb)

747-200B, -200B Combi, -200C:
A 357,430 kg (788,000 lb)

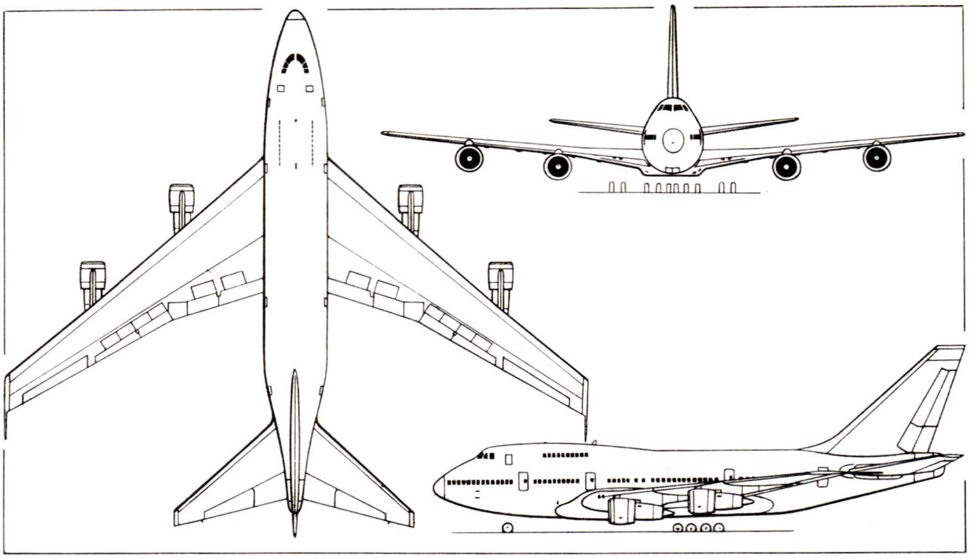

Boeing 747SP short-fuselage long-range version of the 747 *(Pilot Press)*

B 366,500 kg (808,000 lb)
E, H 373,305 kg (823,000 lb)
C, D, F, J, K 379,200 kg (836,000 lb)

Max zero-fuel weight:

747SR (initial aircraft):
A, B, E, F, G, H, J 215,455 kg (475,000 lb)
or 219,990 kg (485,000 lb)

747-100B, 747SR (optional):
A, B, E, F, G, H, J 238,815 kg (526,500 lb)

747-200B:
A, B, C, D, E, F, H, J, K 238,815 kg (526,500 lb)

747-200B Combi:
A, B, C, D, E, F, H, J, K 247,205 kg (545,000 lb)

747-200C:
A, B, C, D, E, F, H, J, K 267,620 kg (590,000 lb)

Max landing weight:

747SR (initial aircraft):
A, B, E, F, G, H, J 229,065 kg (505,000 lb)
or 238,135 kg (525,000 lb)

747-100B, 747SR (optional):
A, B, E, F, G, H, J 255,825 kg (564,000 lb)
or 265,350 kg (585,000 lb)

747-200B:
A, B, C, D, E, F, H, J, K 255,825 kg (564,000 lb)
or 265,350 kg (585,000 lb)
or 285,765 kg (630,000 lb)

747-200B Combi:
A, B, C, D, E, F, H, J, K 265,350 kg (585,000 lb)
or 285,765 kg (630,000 lb)

747-200C:
A, B, C, D, E, F, H, J, K 285,760 kg (630,000 lb)

Performance (at max T-O weight except where indicated):

Max level speed:

747-100B, JT9D-7A engines and AUW of 272,160 kg (600,000 lb) at 9,145 m (30,000 ft) 522 knots (967 km/h; 601 mph)

747-200B, CF6-50E engines and AUW of 317,515 kg (700,000 lb) at 9,145 m (30,000 ft) 523 knots (969 km/h; 602 mph)

Cruise ceiling, all versions 13,715 m (45,000 ft)

Min ground turning radius 22·86 m (75 ft 0 in)

Runway LCN (A: 334,750 kg; 738,000 lb, B; 341,560 kg; 753,000 lb, C: 366,500 kg; 808,000 lb, D: 379,200 kg; 836,000 lb max taxi weight on h=0·51 m; 20 in flexible pavement): A 81
B 83
C 86
D 88

Runway LCN (weights as above, on l=1·02 m; 40 in rigid pavement): A 87
B 89
C 93
D 95

FAR T-O distance to 10·7 m (35 ft) at S/L, ISA:

747-100B, JT9D-7A engines at AUW of 332,480 kg (733,000 lb) 2,880 m (9,450 ft)

747-200B, JT9D-7FW engines at AUW of 365,142 kg (805,000 lb) 3,338 m (10,950 ft)

747-200B, C, F, CF6-50E1/E2 engines at AUW of 377,840 kg (833,000 lb) 3,298 m (10,820 ft)

747-200B, C, JT9D-7Q engines at AUW of 377,840 kg (833,000 lb) 3,225 m (10,580 ft)

FAR landing field length, at max landing weights:

747-100B, -200B at 255,826 kg (564,000 lb) 1,881 m (6,170 ft)

747-100B, -200B at 265,352 kg (585,000 lb) 1,942 m (6,370 ft)

747-200B, C, F at 285,763 kg (630,000 lb) 2,109 m (6,920 ft)

Range (long-range cruise, FAR 121.645 reserves):

747-100B at T-O weight of 340,195 kg (750,000 lb), with 452 passengers and baggage 4,500 nm (8,339 km; 5,182 miles)

747-200B at T-O weight of 365,140 kg (805,000 lb), with 452 passengers and baggage 5,200 nm (9,636 km; 5,988 miles)

747-200B at T-O weight of 377,840 kg (833,000 lb), with 452 passengers and baggage 5,700 nm (10,562 km; 6,563 miles)

Ferry range (long-range cruise, FAR 121.645 reserves):

747-200B 6,800 nm (12,600 km; 7,830 miles)

Operational Noise Levels (As per FAR Pt 36, A: JT9D-7A engines at brake release weight (BRW) of 332,480 kg; 733,000 lb and landing weight of 255,826 kg; 564,000 lb, B: JT9D-7FW at BRW of 365,140 kg; 805,000 lb and landing weight of 255,826 kg; 564,000 lb, C: CF6-50E at BRW of 371,950 kg; 820,000 lb and landing weight of 285,763 kg; 630,000 lb):

T-O: A 106 EPNdB
B 109 EPNdB
C 107 EPNdB

Approach: A 106 EPNdB
B 107 EPNdB
C 106 EPNdB

Sideline: A 98 EPNdB
B 99 EPNdB
C 98 EPNdB

BOEING MODEL 747SP

The Boeing Company announced on 3 September 1973 that it intended to proceed 'incrementally' with development of a lower-weight longer-range version of the basic Model 747, for use on lower-density routes. A week later came the news that Pan American had placed an order for 10 747SP (Special Performance) aircraft, with an option on 15 more. Other orders, bringing the total to 42 by 1 July 1981, have been received from Braniff (4), China Air (3), Iran Air (4), South African Airways (6), Syrian Arab (2), Saudi Arabian government (1), Saudia (2), Korean Air Lines (2), Qantas (2), TWA (3) and CAAC (3).

Retaining a 90 per cent commonality of components with the standard Model 747, the major change is a reduction in overall length of 14·35 m (47 ft 1 in). Construction of the first production aircraft began in April 1974, with rollout on 19 May 1975, first flight on 4 July 1975, and FAA certification on 4 February 1976. First delivery was made on 5 March that year.

On 23-24 March 1976, taking off at a gross weight of 323,547 kg (713,300 lb) with 50 passengers, the first 747SP for South African Airways made a delivery flight from Paine Field, Washington, to Cape Town of 8,936 nm (16,560 km; 10,290 miles), a world record for nonstop distance flown by a commercial aircraft. The aircraft landed with fuel remaining for a further 2 h 27 min of flight.

A 747SP of Pan American, commanded by Capt Walter H. Mullikin, set a round-the-world speed record of 436·95 knots (809·24 km/h; 502·84 mph), by circumnavigating the globe in 1 day 22 h 0 min 50 s on 1-3 May 1976. The same pilot circumnavigated the world via the North and South Poles in a 747SP on 28-31 October 1977, covering 22,926 nm (42,459 km; 26,382·75 miles) in 2 days 6 h 7 min 12 s at an average speed of 423·49 knots (784·31 km/h; 487·35 mph). Start and finish were at San Francisco, with en route landings at London, Cape Town and Auckland, New Zealand.

The description of the basic Model 747 applies also to the 747SP, except for the following details:

Wings: As Model 747, except that trailing-edge flaps are of single-slotted variable pivot type, and wing structural materials are of reduced gauge. Large flap track fairings replaced by small link fairings. New wing/body fairings and leading-edge fillets.

Fuselage: As Model 747, except length reduced.

Tail Unit: Similar to 747, but tailplane span increased by 3·05 m (10 ft). Two-segment elevators. Height of fin increased by 1·52 m (5 ft 0 in). Double-hinged rudder.

Landing Gear: As Model 747, except structural weight reduced. Main-wheel tyres size 46 × 16, pressure 12·63 bars (183 lb/sq in). Nosewheel tyres size 49 × 17, pressure 13·8 bars (200 lb/sq in). Higher gross weight aircraft uses 747-100 wheels and brakes. Modified 747-100 steel brakes by Bendix.

Power Plant: Four Pratt & Whitney JT9D-7A turbofan engines, each 208·8 kN (46,950 lb st); or JT9D-7F of 213·5 kN (48,000 lb st); JT9D-7AW of 216 kN (48,570 lb st); JT9D-7FW of 222·4 kN (50,000 lb st); JT9D-7J of 222·4 kN (50,000 lb st); or four General Electric CF6-45A2/B2 turbofan engines, each of 206·8 kN (46,500 lb st); four Rolls-Royce RB.211-524B2 turbofan engines, each of 222·8 kN (50,100 lb st), or RB.211-524C2 engines, each of 229·5 kN (51,600 lb st). Fuel system and oil capacity as for Model 747-100B, except Model 747SP has an additional 5,966 litres (1,576 US gallons) reserve fuel, providing a total capacity of 190,625 litres (50,359 US gallons).

Accommodation: Normal operating crew of three on flight deck above level of main deck. Observer station and provision for second observer station are provided. Accommodation for 299 passengers on main deck, with 28 first class seats in forward area and ten-abreast seating throughout the major part of the main cabin. Seating for 32 passengers on upper deck, giving total capacity of 331 passengers. Max high-density accommodation for 440 passengers. Four doors on each side, two forward and two aft of the wing. Crew door on starboard side giving access to upper deck. Freight holds under floor, forward and aft of wing box, each with one door on starboard side.

Systems, Avionics and Equipment: As for Model 747.

Dimensions, external: As for Model 747 except:

Length overall	56·31 m (184 ft 9 in)
Height overall	19·94 m (65 ft 5 in)
Tailplane span	25·22 m (82 ft 9 in)
Wheelbase	20·52 m (67 ft 4 in)

Dimensions, internal:

Cabin, incl toilets and galleys:	
Length	42·27 m (138 ft 8 in)
Max width	6·13 m (20 ft 1½ in)
Max height	2·54 m (8 ft 4 in)
Floor area, passenger deck	253·2 m^2 (2,725 sq ft)
Volume, passenger deck	613·34 m^3 (21,660 cu ft)
Baggage hold volume (fwd)	48·99 m^3 (1,730 cu ft)
Baggage hold volume (aft, containerised)	48·99 m^3 (1,730 cu ft)
Bulk compartment volume (aft)	11·33 m^3 (400 cu ft)

Areas: As for Model 747 except:

Ailerons (total)	20·37 m^2 (219·3 sq ft)
Trailing-edge flaps (total)	78·78 m^2 (848 sq ft)
Fin	82·22 m^2 (885 sq ft)
Tailplane	142·51 m^2 (1,534 sq ft)

Weights: (letters are used to denote engine installations as follows: (A) JT9D-7A/7AW; (B) JT9D-7F/7FW/7J; (C) CF6-45A2 or -45B2; (D) RB.211-524B2; and (E) RB.211-524C2):

Operating weight empty (approx, with 331 passengers):	
A, B	147,870 kg (326,000 lb)
C	150,140 kg (331,000 lb)
D, E	152,860 kg (337,000 lb)
Max T-O weight (dry engines):	
A, B, C, D, E	285,765 kg (630,000 lb)
	or 299,370 kg (660,000 lb)
	or 303,905 kg (670,000 lb)
	or 312,980 kg (690,000 lb)
	or 315,700 kg (696,000 lb)
	or 317,515 kg (700,000 lb)
Max ramp weight:	
A, B, C, D, E	288,485 kg (636,000 lb)
	or 302,095 kg (666,000 lb)
	or 306,630 kg (676,000 lb)
	or 315,700 kg (696,000 lb)
	or 318,875 kg (703,000 lb)
Max zero-fuel weight:	
A, B, C, D, E	185,973 kg (410,000 lb)
	or 192,777 kg (425,000 lb)
Max landing weight:	
A, B, C, D, E	204,117 kg (450,000 lb)
	or 210,920 kg (465,000 lb)

Performance (at max T-O weight, except where indicated):

Never-exceed speed	Mach 0·92
Max level speed, AUW of 226,795 kg (500,000 lb) at 9,145 m (30,000 ft)	529 knots (980 km/h; 609 mph)
Service ceiling	13,745 m (45,100 ft)
Min ground turning radius over outer wingtip	22·25 m (73 ft 0 in)
Runway LCN (Y: 302,095 kg; 666,000 lb, Z: 317,515 kg; 700,000 lb max taxi weight on h = 0·51 m; 20 in flexible pavement):	
Y	70
Z	75
Runway LCN (weights as above, on l = 1·02 m; 40 in rigid pavement):	
Y	76
Z	80
FAR T-O distance to 10·7 m (35 ft) at S/L, ISA (X: 299,370 kg; 660,000 lb, Z: 317,515 kg; 700,000 lb max T-O weight):	
(A) X	2,393 m (7,850 ft)
Z	2,780 m (9,120 ft)
(B) X	2,164 m (7,100 ft)
Z	2,469 m (8,100 ft)
(C) X	2,362 m (7,750 ft)
Z	2,743 m (9,000 ft)
(D) X	2,118 m (6,950 ft)
Z	2,438 m (8,000 ft)
FAR landing field length:	
at max landing weight	1,594 m (5,230 ft)
at optional max landing weight	1,646 m (5,400 ft)
Range (long-range step cruise, FAR 121.645 reserves, with 331 passengers and baggage at X: 299,370 kg; 660,000 lb, Z: 317,515 kg; 700,000 lb):	
X	5,150 nm (9,544 km; 5,930 miles)
Z	5,850 nm (10,841 km; 6,736 miles)
Ferry range (long-range step cruise, FAR 121.645 reserves)	7,350 nm (13,620 km; 8,463 miles)

BOEING MODEL 747 WITH STRETCHED UPPER DECK

On 12 June 1980, Boeing announced a new option for the Model 747, which incorporates structural changes to the aircraft's upper deck area to increase passenger-carrying capacity. The bulged upper forward fuselage is extended aft by 7·11 m (23 ft 4 in) to increase upper-deck accommodation from 32 to 69 passengers in a standard economy class configuration. Seating is six abreast, with a single aisle, and panniers between the outer seats and cabin wall are provided for hand baggage. Seven additional seats can be accommodated on the main deck as a result of deleting the standard circular stairway. It is replaced by a new straight stairway at the rear of the upper deck area. Two new doors 1·83 m (6 ft 0 in) high and 1·07 m (3 ft 6 in) wide replace the existing 1·22 × 0·61 m (4 ft 0 in × 2 ft 0 in) upper-deck exits. Other structural changes include the provision of a new emergency exit and additional windows. The stretched upper deck option is available initially on existing aircraft of the 747-200 series; maximum take-off weights are unchanged, but operating weight empty is increased by about 3,630 kg (8,000 lb).

The stretched upper deck will be available to order on new 747-100B, 747SR, 747-200B and 747-200B Combi aircraft. It is not applicable to the 747SP, 747-200C or 747-200F.

Five 747s with stretched upper deck and 240·2 kN (54,000 lb st) JT9D-7R4G2 engines have been ordered by Swissair, and two by South African Airways. Deliveries are due to begin in March 1983.

BOEING MODEL 747-200F FREIGHTER

The Boeing Model 747-200F Freighter is a version of the standard Model 747-200, capable of delivering 90,720 kg (200,000 lb) of containerised or palletised main deck cargo over a range of 4,300 nm (7,969 km; 4,951 miles).

The first 747-200F flew for the first time on 30 November 1971. It was certificated on 7 March 1972 and delivered to Lufthansa two days later.

To ensure maximum utilisation, the 747-200F has a special loading system that enables two men to handle and stow the maximum load of up to 113,400 kg (250,000 lb) in 30 min. This system was fully described in the 1977-78 *Jane's*.

The 747-200F can carry up to 29 containers measuring 3·05 m × 2·44 m × 2·44 m (10 ft long, 8 ft high and 8 ft wide), plus 30 lower-lobe containers, each of 4·90 m^3 (173 cu ft) capacity, and 22·65 m^3 (800 cu ft) of bulk cargo. The main deck can accommodate ANSI/ISO containers of up to 12·2 m (40 ft) in length, and many combinations of pallets and igloos. The lower hold can accommodate combinations of IATA-A1 or -A2, and ATA LD-1 or -3 half-width containers, full-width or main-deck baggage containers, and many combinations of pallets and igloos.

The nose loading door, which is hinged just below the flight deck to allow it to swing forward and upward, gives clear access to the main deck to facilitate the handling of long or large loads. A side cargo door is available as an option, allowing simultaneous nose and side loading. The side cargo door will accept palletised loads up to 3·05 m (10 ft 0 in) in height.

The description of the Model 747-200B applies also to the Model 747-200F except as follows:

Type: Four-turbofan heavy commercial freighter.

Fuselage: As for Model 747-200B, except nose cargo loading door, which is hinged at the top and opens forward and upward.

Accommodation: Normal operating crew of three on flight deck. Nose cargo loading door, hinged at top. Lower lobe cargo doors, on starboard side, one forward and one aft of wing. Bulk compartment cargo door, on starboard side, aft of lower lobe cargo door. Two doors for crew on port side of aircraft. Aircraft is designed for fully-mechanical loading of freight.

Boeing Model 747SP four-turbofan special performance long-range transport in Korean Air Lines insignia

Boeing Model 747-200F freighter in service with the Korean Air Cargo division of Korean Air Lines

Dimensions, external: As for Model 747-200B except:
- Crew doors (two, each): Height — 1·93 m (6 ft 4 in)
  - Width — 1·07 m (3 ft 6 in)
  - Height to sill — approx 4·88 m (16 ft 0 in)
- Nose cargo loading door: Height — 2·49 m (8 ft 2 in)
  - Width at top (min) — 2·67 m (8 ft 9 in)
  - Max width — 3·81 m (12 ft 6 in)
  - Height to sill — approx 4·90 m (16 ft 1 in)

Dimensions, internal:
- Main cargo deck: Height — 2·54 m (8 ft 4 in)
  - Max width at floor level — 5·92 m (19 ft 5 in)
- Lower lobe: Width at floor level — 3·18 m (10 ft 5 in)
- Total cargo volume — 687·0 m^3 (24,260 cu ft)

Areas: As for Model 747-200B

Weights (letters are used to denote engine installations as follows: (A) JT9D-7A/7AW; (B) JT9D-7F/7FW/7J; (C) JT9D-7Q; (D) JT9D-7R4G2; (E) CF6-50E; (F) CF6-50E1/E2; (G) RB.211-524B2; (H) RB.211-524C2; and (J) RB.211-524D4):

Operating weight empty (approx) for max available gross weights:
- A, B — 155,130 kg (342,000 lb)
- C — 156,490 kg (345,000 lb)
- D — 156,035 kg (344,000 lb)
- E, F — 158,305 kg (349,000 lb)
- G, H — 161,025 kg (355,000 lb)
- J — 160,570 kg (354,000 lb)

Max payload (29 pallets):
- A, B — 112,490 kg (248,000 lb)
- C — 111,130 kg (245,000 lb)
- D — 111,585 kg (246,000 lb)
- E, F — 109,315 kg (241,000 lb)
- G, H — 106,595 kg (235,000 lb)
- J — 107,050 kg (236,000 lb)

Max T-O weight:
- A — 351,535 kg (775,000 lb) or 356,070 kg (785,000 lb)
- B — 362,875 kg (800,000 lb) or (wet) 365,140 kg (805,000 lb)
- E, G — 362,875 kg (800,000 lb) or 371,945 kg (820,000 lb)
- C, D, F, H, J — 362,875 kg (800,000 lb) or 371,945 kg (820,000 lb) or 377,840 kg (833,000 lb)

Max ramp weight:
- A — 357,430 kg (788,000 lb)
- B — 366,500 kg (808,000 lb)
- E, G — 373,305 kg (823,000 lb)
- C, D, F, H, J — 379,200 kg (836,000 lb)

Max zero-fuel weight:
- A, B, C, D, E, F, G, H, J — 267,620 kg (590,000 lb)

Max landing weight:
- A, B, C, D, E, F, G, H, J — 285,765 kg (630,000 lb)

Performance (at max T-O weight except where indicated):
- Max level speed at AUW of 272,160 kg (600,000 lb), at 9,145 m (30,000 ft) — 528 knots (978 km/h; 608 mph)
- Cruise ceiling — 13,715 m (45,000 ft)
- Min ground turning radius — 22·86 m (75 ft 0 in)
- FAR T-O distance to 10·7 m (35 ft) at S/L, ISA:
  - JT9D-7FW engines, at AUW of 365,140 kg (805,000 lb) — 3,338 m (10,950 ft)
  - CF6-50E/E1/E2 engines, at AUW of 377,840 kg (833,000 lb) — 3,298 m (10,820 ft)
  - JT9D-7Q engines, at AUW of 377,840 kg (833,000 lb) — 3,225 m (10,580 ft)
- FAR landing field length, at max landing weight — 2,103 m (6,900 ft)
- Range, long-range step cruise, FAR 121.645 reserves, with 90,720 kg (200,000 lb) payload:
  - T-O weight of 356,070 kg (785,000 lb):
    - A, B — 3,450 nm (6,390 km; 3,970 miles)
  - T-O weight of 377,840 kg (833,000 lb):
    - C, F — 4,150 nm (7,690 km; 4,780 miles)
    - D — 4,350 nm (8,061 km; 5,009 miles)
    - G, H — 4,050 nm (7,505 km; 4,663 miles)
    - J — 4,250 nm (7,876 km; 4,894 miles)
- Ferry range with max fuel, long-range cruise, FAR 121.645 reserves — 7,300 nm (13,525 km; 8,405 miles)

BOEING MODEL 757

In the early months of 1978, The Boeing Company announced a proposal to develop a new family of advanced technology commercial aircraft, to which it gave the Model designations 757, 767 and 777. The short/medium-range 757 was intended to differ considerably from the other two, being based on the Boeing 727 fuselage. Improved performance would come from two new high bypass engines and an advanced technology wing, with less sweepback than that of the Model 727.

On 31 August 1978 Eastern Air Lines and British Airways announced their intention to purchase 21 (since increased to 27) and 19 Model 757s respectively, the former taking an option on an additional 24, and the latter on 18. Following the signature in early 1979 of formal contracts by both of these airlines, Boeing announced on 23 March 1979 that the company had initiated full production of the Model 757; the first metal was cut for the new aircraft on 10 December 1979, and final assembly began in September 1981. First flight is scheduled for February 1982, with first deliveries in January 1983. Aircraft for both British Airways and Eastern Air Lines will be designated **Model 757-200** and will be powered by two 166·4 kN (37,400 lb st) Rolls-Royce RB.211-535C turbofan engines. This was the first time that Boeing had launched a new airliner with a non-American engine. On a typical 400 nm (740 km; 460 mile) stage, the Model 757-200 with RB.211-535 engines is expected to use up to 45 per cent less fuel per passenger than current medium-range aircraft.

By 1 August 1981, orders totalled 135, the additional customers being Transbrasil (9), Delta (60), American (15), Monarch (2) and Air Florida (3). Aircraft for Delta and American will have PW2037 engines.

The Model 757 retains the same fuselage cross-section as the 707/727/737 family. The original design provided for a T-tail configuration, but as a result of wind tunnel testing it was decided to mount the tailplane on the fuselage.

On 11 October 1979, more than $1,000 million worth of subcontracts for 757 components were awarded by Boeing, in each case for 200 ship-sets, with follow-on options for 200 more. Subcontracts for the wing centre-section and adjacent fuselage keel beam went to Avco Aerostructures; Fairchild Republic was selected to manufacture the overwing passenger cabin section (and, under a later subcontract, the wing leading-edge slats); Rockwell International's subcontract covered fuselage sections immediately forward and aft of the wing; Vought Corporation was made responsible for the tailplane, fin and extreme rear fuselage. Earlier subcontracts had gone to Short Brothers of Northern Ireland for inboard trailing-edge flaps; to Hawker de Havilland of Australia for wing in-spar ribs; to CASA of Spain for outboard trailing-edge flaps; and to Menasco of California for main and nose landing gear units. Other major subcontractors include Grumman (wing spoilers), Rohr Industries (engine pylons), and Heath Tecna (wing/fuselage fairings).

Type: Twin-turbofan short/medium-range transport aircraft.

Wings: Cantilever low-wing monoplane. Special Boeing wing sections. Dihedral 5°. Incidence 3° 12′. Sweepback at quarter-chord 25°. All aluminium alloy two-spar fail-safe wing-box structure. Centre-section continuous through fuselage. All-speed fully-powered ailerons outboard of double-slotted inboard and outboard trailing-edge flaps. High-lift leading-edge slats. Four flight spoilers and one ground spoiler on wing upper surface forward of trailing-edge flaps. Lateral control provided by ailerons, plus flight spoilers operated differentially. Spoilers operated collectively as speed brakes. Extensive use of honeycomb, and graphite composites and laminates in construction of ailerons, flaps and spoilers. Thermal anti-icing of wing leading-edges.

Fuselage: All-metal semi-monocoque fail-safe structure, its cross-section based on two circular arcs of different radii, the larger above, and faired into a smooth contoured oval.

Tail Unit: Tailplane consists of full-span torque boxes of conventional light alloy construction, and is attached at pivot points and to an actuator. The fin comprises a three-spar dual-cell light alloy torque box attached to the fuselage. Elevators and rudder have graphite/epoxy honeycomb skins, supported by honeycomb and laminated spar and rib assemblies.

Landing Gear: Retractable tricycle type, with main and nose units manufactured by Menasco. Each main unit comprises a four-wheel bogie, fitted with Goodrich wheels and brakes, and Dunlop tyres. Twin-wheel nose unit.

Power Plant: Two 166·4 kN (37,400 lb st) Rolls-Royce RB.211-535C or two 170 kN (38,200 lb st) Pratt & Whitney PW2037 turbofan engines, mounted in underwing pods. Fuel capacity 41,185 litres (10,880 US gallons) standard, 42,320 litres (11,180 US gallons) optional.

Accommodation: Crew of two on flight deck, with provision for an observer. Typical accommodation for 178 mixed class passengers in four/six-abreast seating at 96·5/86 cm (38/34 in) seat pitch, with central aisle. All-tourist configuration seats 196 at 86 cm (34 in). Max capacity 224 passengers at 76 cm (30 in) seat pitch.

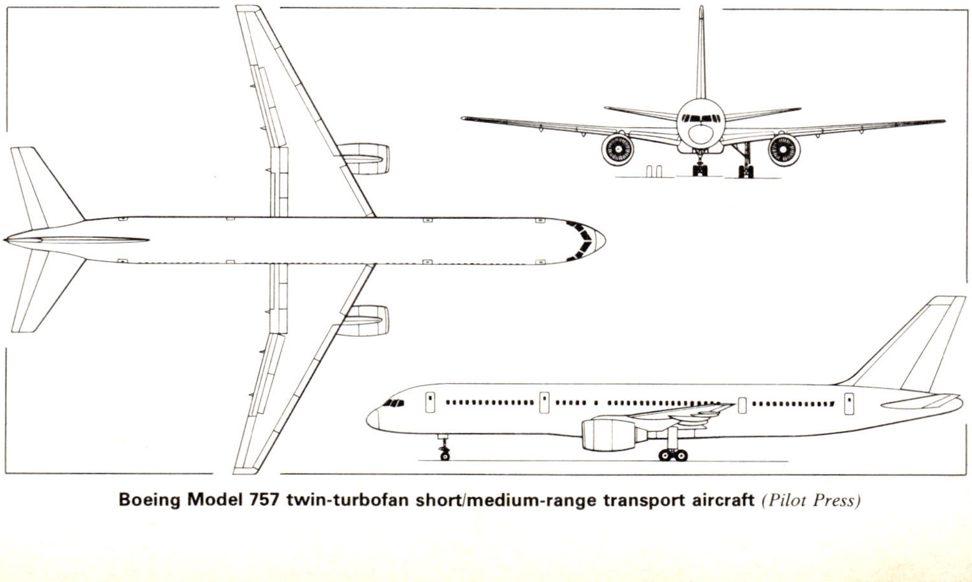

Boeing Model 757 twin-turbofan short/medium-range transport aircraft *(Pilot Press)*

Large overhead bins provide approximately 0·06 m³ (2 cu ft) of stowage per passenger. Two forward and one rear passenger door on port side. Four overwing emergency exits, two each side, are standard. With maximum capacity configuration, two additional emergency exits are added aft of wing, one each side.

Systems and Avionics: Garrett environmental control system; General Electric engine thrust management system; Sundstrand electrical power generating system and ram-air turbines; and Garrett GTCP331-200 APU. Honeywell inertial reference system (IRS); Sperry flight management computer system (FMCS) and digital air data computer (DADC); Collins FXS-700 flight control system, EFIS-700 electronic flight instruments system, RMI-743 radio distance magnetic indicator (RDMI) and optional radio magnetic indicator (RMI). Avionics also include a Honeywell inertial reference system (IRS). In this IRS, conventional mechanical gyroscopes are replaced by laser gyroscopes, and utilisation, in both the Models 757 and 767, represents their first commercial application. The IRS provides position, velocity and attitude information to flight deck displays, and for the flight management computer system (FMCS) and digital air data computer (DADC), to be supplied by Sperry Flight Systems. The FMCS provides automatic en-route and terminal navigation capability, and also computes and commands both lateral and vertical flight profiles for optimum fuel efficiency, maximised by electronic linkage of the FMCS with automatic flight control and thrust management systems.

Dimensions, external:

Wing span	37·95 m (124 ft 6 in)
Wing area, gross	181·25 m² (1,951 sq ft)
Length overall	47·32 m (155 ft 3 in)
Length of fuselage	46·89 m (153 ft 10 in)
Height overall	13·56 m (44 ft 6 in)
Tailplane span	15·21 m (49 ft 10¾ in)
Wheel track	7·32 m (24 ft 0 in)
Passenger doors (two, fwd, port):	
Height	1·83 m (6 ft 0 in)
Width	0·84 m (2 ft 9 in)
Passenger door (rear, port): Height	1·83 m (6 ft 0 in)
Width	0·76 m (2 ft 6 in)
Service door (fwd, stbd): Height	1·65 m (5 ft 5 in)
Width	0·76 m (2 ft 6 in)
Service door (stbd, opposite rear passenger door):	
Height	1·83 m (6 ft 0 in)
Width	0·84 m (2 ft 9 in)
Service door (rear, stbd): Height	1·83 m (6 ft 0 in)
Width	0·76 m (2 ft 6 in)
Emergency exits (four, overwing):	
Height	0·97 m (3 ft 2 in)
Width	0·51 m (1 ft 8 in)
Emergency exits, additional (two, aft of wings):	
Height	1·32 m (4 ft 4 in)
Width	0·61 m (2 ft 0 in)

Dimensions, internal:

Underfloor cargo volume (bulk loading):	
fwd	19·4 m³ (685 cu ft)
rear	30·58 m³ (1,080 cu ft)

Weights (with 178 passengers. A: RB.211-535C engines; B: PW2037s):

Operating weight empty: A	59,430 kg (131,020 lb)
B	59,021 kg (130,120 lb)
Max basic T-O weight: A, B	99,790 kg (220,000 lb)
Max T-O weight (medium range)	104,325 kg (230,000 lb)
Max T-O weight (optional)	108,860 kg (240,000 lb)
Max landing weight: A, B	89,810 kg (198,000 lb)
Max zero-fuel weight	83,460 kg (184,000 lb)

Performance (estimated, with 178 passengers, US mixed class operations; at max basic T-O weight and with engines as above):

Cruising speed: A, B	Mach 0·80
Approach speed:	
A, B	134 knots (248 km/h; 154 mph)
Initial cruising height: A	11,700 m (38,400 ft)
B	11,490 m (37,700 ft)
T-O field length (S/L, 29°C): A	2,095 m (6,870 ft)
B	2,054 m (6,740 ft)
Max range: A	2,150 nm (3,984 km; 2,476 miles)
B	2,490 nm (4,614 km; 2,867 miles)

BOEING MODEL 767

On 14 July 1978, Boeing announced its intention to launch full-scale development of the twin-turbofan Model 767, following receipt of an order from United Air Lines for 30, with initial deliveries scheduled for August 1982. This airline had participated actively in defining the design of the Model 767, as it did with the design of the Model 727 in 1959-60. It is estimated that when these aircraft enter service with United, powered by Pratt & Whitney JT9D-7R4D engines, they will be 35 per cent more fuel-efficient than the aircraft they will replace, and will comply also with the stringent noise regulations which are foreseen for 1984.

Construction of the first Model 767 began on 6 July 1979. It has a completely new airframe, with a fuselage 1·24 m (4 ft 1 in) wider than that of the Model 757, permitting a two-aisle seating layout. It was proposed initially in two forms, as the 767-100 with accommodation for approximately 180 passengers, and the 767-200MR with accommodation for up to 255 passengers. Since that time, a 211-passenger version known as the **767-200** has been finalised as the basic model. The prototype (N767BA) was rolled out on 4 August 1981, and made its first flight on 26 September 1981. One airframe has been completed for static testing. Another is being used for fatigue testing. It is intended to market an optional medium-range version with reduced fuel capacity, and a later option will cater for a version with a higher gross weight, although this is not expected to be available until late 1983. Delivery of the first 767, to United Air Lines, is scheduled for August 1982.

Alternative engines for the basic model are the Pratt & Whitney JT9D-7R4D, rated at 212·6 kN (47,800 lb st), and General Electric CF6-80A, rated at 213 kN (47,900 lb st). The same engines are projected for the later models. Boeing is also studying the Rolls-Royce RB.211 as an alternative option. The basic interior arrangements of all versions are identical.

By 1 August 1981, orders for the Model 767 totalled 173, with options on 138 more. Firm orders were as follows:

		Engines
Air Canada	12	JT9D-7R4D
Alaska International	2	JT9D-7R4D
All Nippon (Japan)	25	CF6-80A
American Airlines	30	CF6-80A
Ansett (Australia)	5	CF6-80A
Avianca (Colombia)	3	JT9D-7R4D
Braathens SAFE (Norway)	2	JT9D-7R4D
Britannia Airways (UK)	2	CF6-80A
China Airlines (Taiwan)	2	JT9D-7R4D
CP Air (Canada)	4	CF6-80A
Delta Air Lines (USA)	20	CF6-80A
El Al (Israel)	4	JT9D-7R4D
Pacific Western (Canada)	4	JT9D-7R4D
Transbrasil	3	CF6-80A
Trans World Airlines (USA)	10	JT9D-7R4D
United Air Lines (USA)	39	JT9D-7R4D
Western Airlines (USA)	6	JT9D-7R4D

Boeing has awarded manufacturing subcontracts to Grumman Aerospace Corporation (wing centre-section, an adjacent lower fuselage section, and fuselage bulkheads); to Vought Corporation (horizontal tail); and to Canadair Ltd (rear fuselage). In two other major work-sharing programmes announced in August and September 1978, Boeing gave details of co-production agreements whereby Aeritalia of Italy and member companies of the Civil Transport Development Corporation (CTDC) of Japan will participate as follows:

Aeritalia: Wing control surfaces, flaps and leading-edge slats; wingtips; elevators; fin, rudder; and nose radome.

Fuji: Wing fairings and main landing gear doors.

Kawasaki: Centre-fuselage body panels, exit hatches and wing in-spar ribs.

Mitsubishi: Rear-fuselage body panels, stringers, passenger and cargo doors, and dorsal fin.

The following details apply to the basic Model 767-200:

Type: Twin-turbofan medium-range commercial transport.

Wings: Cantilever low-wing monoplane. Special Boeing wing sections. Thickness/chord ratio 15·1% at root, 10·3% at tip. Dihedral 6°. Incidence 4° 15′. Sweepback 31° 30′ at quarter-chord. Fail-safe structure of aluminium alloy. Plain inboard and outboard ailerons with extensive use of graphite hybrid composites. Single-slotted linkage-supported aluminium outboard trailing-edge flaps; double-slotted inboard flaps. Conventional inboard and outboard spoilers of graphite composite construction are provided for roll control, to act as airbrakes, and for lift dumping. Track-mounted leading-edge slats of light alloy construction. Roll trim through spring feel system. All control surfaces are powered hydraulically. Anti-icing of outboard wing leading-edges.

Fuselage: Conventional semi-monocoque structure of aluminium alloy, consisting of skin, longitudinal stringers and circumferential frames. Structure is of fail-safe design, and is pressurised except for tailcone aft of passenger cabin, landing gear wheel wells and air cycle machine wells.

Tail Unit: Cantilever fail-safe structure of aluminium alloy and aluminium honeycomb. Variable-incidence tailplane. Elevators of single-hinge type with redundant parallel actuators. No trim tabs. All controls are powered hydraulically. Yaw trim through spring feel system. No tail unit anti-icing.

Landing Gear: Hydraulically-retractable tricycle type. Menasco twin-wheel nose unit retracts forward. Cleveland Pneumatic main gear, comprising two four-wheel bogies which retract inward. Oleo-pneumatic shock-absorbers. Bendix wheels and brakes. Main-wheel tyres size 45 × 17-20, pressure 12·6 bars (183 lb/sq in). Nosewheel tyres size 37 × 14-15, pressure 10·0 bars (145 lb/sq in). Steel disc brakes on all main wheels. Electronically-controlled anti-skid units.

Power Plant: Two high bypass turbofan engines in the 213·5 kN (48,000 lb st) class, in pods pylon-mounted on the wing leading-edges. Pratt & Whitney JT9D-7R4D and General Electric CF6-80A engines had been specified by customers up to August 1981, and Boeing has alternative Rolls-Royce RB.211 engines under study. Fuel in one integral tank in each wing, and in a centre tank, with total capacity of 58,900 litres (15,560 US gallons). Refuelling point in port outer wing. Anti-icing of engine air inlets.

Accommodation: Normal operating crew of two or three on flight deck. Basic accommodation for 211 passengers, made up of 18 first class passengers forward in six-abreast seating at 96·5 cm (38 in) pitch, and 193 tourist class in mainly seven-abreast seating at 86 cm

Rollout ceremony for the prototype Boeing 767 twin-turbofan transport aircraft on 4 August 1981

(34 in) pitch. Type A inward-opening plug doors are provided at both the front and rear of the cabin on each side of the fuselage, with a Type III emergency exit over the wing on each side. A total of five toilets is installed, two centrally in the main cabin, two aft in the main cabin, and one forward in the first class section. Galleys are situated at forward and aft ends of the cabin. With the two central toilets moved to the rear of the cabin, the number of tourist passengers can be increased to 198 in this mixed class, one meal service model. With an additional rear galley for one and one-half meal service, the rear centre row of three tourist class seats is removed in each configuration. Alternative single class layouts provide for 230 tourist passengers, seated mainly seven-abreast at 86 cm (34 in) pitch; 242 passengers seated seven-abreast at 81 cm (32 in) pitch; or 255 passengers mainly seven-abreast (two-three-two) at 76 cm (30 in) pitch, or eight-abreast (two-four-two) at 81 cm (32 in) pitch. Max seating capacity (requiring additional overwing emergency exit) for 289 passengers, mainly eight-abreast, at 76 cm (30 in) pitch. Underfloor cargo holds can accommodate, typically, up to 22 LD-2 or 11 LD-3 containers. Forward and rear cargo doors of equal size are standard, but a larger (1·75 by 3·40 m; 5 ft 9 in by 11 ft 2 in) forward cargo door is optional, to permit loading of Type 2 pallets. Bulk cargo door at rear on port side. Overhead stowage for carry-on baggage. Cabin is air-conditioned, cargo holds heated.

SYSTEMS: Garrett dual air-cycle air-conditioning system. Pressure differential 0·59 bars (8·6 lb/sq in). Electrical supply from two engine-driven 90kVA three-phase 400Hz constant-frequency AC generators, 115/200V output. 90 kVA generator mounted on APU for ground operation or for emergency use. Three hydraulic systems at 207 bars (3,000 lb/sq in), for flight control and utility functions, supplied from Garrett bleed air powered hydraulic pump or APU. Nitrogen chlorate oxygen generators in passenger cabin, plus gaseous oxygen for flight crew. Anti-icing for air data sensors and windscreen. APU, to provide ground and in-flight electrical power and pressurisation, in tailcone.

AVIONICS AND EQUIPMENT: Standard avionics include ARINC 700 Series equipment (Bendix in All Nippon aircraft), Collins caution annunciator, dual digital flight management systems, and triple digital flight control computers, including FCS-700 flight control system, EFIS-700 electronic flight instruments system and RMI-743 radio distance magnetic indicator. Honeywell IRS, and Sperry Flight Systems FMCS and DADC, as described in Boeing Model 757 entry.

DIMENSIONS, EXTERNAL:

Wing span	47·57 m (156 ft 1 in)
Wing chord at root	8·57 m (28 ft 1¼ in)
Wing chord at tip	2·29 m (7 ft 6 in)
Wing aspect ratio	7·9
Length overall	48·51 m (159 ft 2 in)
Length of fuselage	47·24 m (155 ft 0 in)
Fuselage: Max width	5·03 m (16 ft 6 in)
Height overall	15·85 m (52 ft 0 in)
Tailplane span	18·62 m (61 ft 1 in)
Wheel track	9·30 m (30 ft 6 in)
Wheelbase	19·69 m (64 ft 7 in)
Passenger doors (two, fwd and rear, port):	
Height	1·88 m (6 ft 2 in)
Width	1·07 m (3 ft 6 in)
Galley service door (two, fwd and rear, stbd):	
Height	1·83 m (6 ft 0 in)
Width	1·07 m (3 ft 6 in)
Emergency exits (two, each):	
Height	0·97 m (3 ft 2 in)
Width	0·51 m (1 ft 8 in)
Cargo doors (two, fwd and rear, stbd):	
Height	1·75 m (5 ft 9 in)
Width	1·78 m (5 ft 10 in)
Optional cargo door (fwd, port):	
Height	1·75 m (5 ft 9 in)
Width	3·40 m (11 ft 2 in)

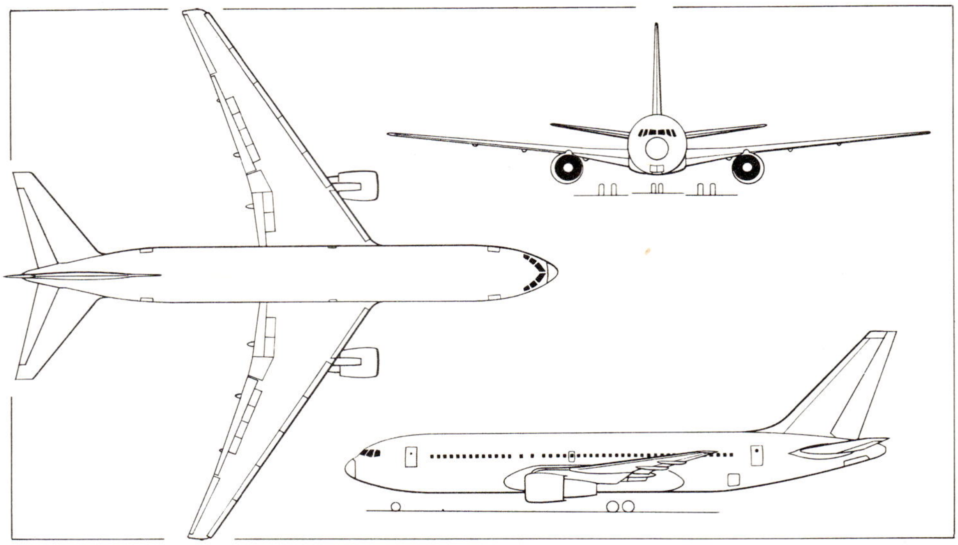

Boeing Model 767-200 wide-bodied medium-range commercial transport aircraft *(Pilot Press)*

DIMENSIONS, INTERNAL:

Cabin, excl flight deck:	
Length	33·93 m (111 ft 4 in)
Max width	4·72 m (15 ft 6 in)
Max height	2·87 m (9 ft 5 in)
Floor area	157·6 m^2 (1,696 sq ft)
Volume	428·2 m^3 (15,121 cu ft)
Volume, flight deck	13·5 m^3 (478 cu ft)
Baggage holds (containerised) volume	74·8 m^3 (2,640 cu ft)
Bulk cargo hold volume	12·2 m^3 (430 cu ft)

AREAS:

Wings, gross	283·3 m^2 (3,050 sq ft)
Ailerons (total)	11·87 m^2 (127·8 sq ft)
Trailing-edge flaps (total)	36·19 m^2 (389·5 sq ft)
Leading-edge slats (total)	28·37 m^2 (305·4 sq ft)
Spoilers (total)	15·83 m^2 (170·4 sq ft)
Fin	30·17 m^2 (324·7 sq ft)
Rudder	15·98 m^2 (172·0 sq ft)
Tailplane	77·69 m^2 (836·2 sq ft)
Elevators	17·47 m^2 (188·0 sq ft)

WEIGHTS (estimated, A: 767-200 basic/JT9D-7R4D engines; B: 767-200 basic/CF6-80A; C: medium-range version/JT9D-7R4D; D: medium-range version/CF6-80A; E: higher gross weight version/JT9D-7R4D; F: higher gross weight version/CF6-80A):

Manufacturer's weight empty:	
A, E	74,548 kg (164,352 lb)
B, F	74,208 kg (163,602 lb)
C	74,467 kg (164,172 lb)
D	74,127 kg (163,422 lb)
Operating weight empty:	
A, E	81,230 kg (179,082 lb)
B, F	80,890 kg (178,332 lb)
C	81,058 kg (178,702 lb)
D	80,717 kg (177,952 lb)
Max T-O weight: A, B	136,080 kg (300,000 lb)
C, D	127,910 kg (282,000 lb)
E, F	140,615 kg (310,000 lb)
Max ramp weight: A, B	136,985 kg (302,000 lb)
C, D	128,820 kg (284,000 lb)
E, F	141,520 kg (312,000 lb)
Max zero-fuel weight:	
A, B, E, F	112,490 kg (248,000 lb)
C, D	109,770 kg (242,000 lb)
Max landing weight:	
A, B, E, F	122,470 kg (270,000 lb)
C, D	116,575 kg (257,000 lb)

PERFORMANCE (estimated, with 211 passengers in US mixed-class operation, at max T-O weight except where indicated):

Normal cruising speed, all versions	Mach 0·80
Approach speed at max landing weight:	
A, B, E, F	134 knots (248 km/h; 154 mph)
C, D	131 knots (243 km/h; 151 mph)
Service ceiling: A	11,885 m (39,000 ft)
B	12,000 m (39,400 ft)
C	12,285 m (40,300 ft)
D	12,390 m (40,650 ft)
E	11,705 m (38,400 ft)
F	11,825 m (38,800 ft)
Service ceiling, one engine out (weight for 1,000 nm; 1,850 km; 1,150 miles range):	
A, E	6,505 m (21,350 ft)
B, D, F	6,460 m (21,200 ft)
C	6,525 m (21,400 ft)
T-O field length (S/L, 29°C): A	2,075 m (6,800 ft)
B	2,040 m (6,700 ft)
C	1,815 m (5,950 ft)
D	1,800 m (5,900 ft)
E	2,210 m (7,250 ft)
F	2,180 m (7,150 ft)
Design range: A	2,780 nm (5,152 km; 3,201 miles)
B	2,845 nm (5,272 km; 3,276 miles)
C	1,995 nm (3,697 km; 2,297 miles)
D	2,050 nm (3,799 km; 2,360 miles)
E	3,180 nm (5,893 km; 3,662 miles)
F	3,245 nm (6,013 km; 3,736 miles)

BOEING AEROSPACE COMPANY

Kent, Washington

PRESIDENT: H. K. Hebeler

VICE-PRESIDENTS:
R. L. Brock (Advanced Tactical Missiles)
C. P. Ekas (Advanced Strategic Missiles)
J. H. Goldie (Executive Vice-President, Department Management)
D. E. Graves (External Affairs)
R. W. Hager (Engineering)
D. B. Jacobs (Advanced Space Systems)
M. K. Miller (Space and Information Systems)
Howard N. Stuverude (Missile Systems)

The Boeing Aerospace Company has its headquarters at the company's space centre at Kent, Washington, some 12 miles south of Seattle. It consists of four major organisations: Space and Information Systems, Missile Systems, Business Management, and Department Management. Major programmes and activities include the E-3A Sentry Airborne Warning And Control System; Minuteman ICBM; MX ICBM ground support system; Inertial Upper Stage; E-4 Airborne Command Post; Roland: Air Launched Cruise Missile; Space Defense and Re-entry Systems Launch Branch; Technical Services; Automated Transportation Systems; and Electronic Support. Responsible for much of Boeing's military, space and diversification efforts, it has a labour force of approximately 17,000.

BOEING E-3 SENTRY

USAF designations: EC-137D and E-3A

The E-3A Sentry AWACS (Airborne Warning And Control System) is effectively a mobile, flexible, survivable and jamming-resistant high-capacity radar station, command, control and communications centre, installed within a Boeing 707 airframe. It offers the potential of long-range high- or low-level surveillance of all air vehicles, manned or unmanned, and provides detection, tracking and identification capability within its surveillance capacity during all weathers and above all kinds of terrain. The radar system of later production aircraft also incorporates a maritime surveillance mode. Each of these aircraft is able to support a variety of tactical and/or air defence missions with no change in configuration. Its data storage and processing capability can provide real-time assessment of enemy action, and of the status and position of friendly resources.

The USAF's first 23 Sentries, equipped to what is known as 'core E-3A' standard, each have 13 available communication links (seven UHF, three VHF/AM, one VHF/FM, and two HF/SSB), many of them in clear voice. In a new US/NATO standard configuration, introduced on the 24th E-3A delivered to the USAF during 1981, and fitted to all subsequent aircraft for the USAF and NATO, this communication system is replaced by the newly-developed joint tactical information distribution system (JTIDS). This high-capacity system, operating over a single secure communications channel to prevent enemy 'eavesdropping', is capable of providing links with from two to 98,000 participants, depending upon the proportions of an access time cycle allocated to each. Additional advantages of JTIDS are the high speed of communication, and the fact that it is far less vulnerable to enemy jamming than are more conventional communication systems. It is assumed that the first 23 E-3As will be retrofitted to this standard at some future date.

In US Air Force service, the E-3A has a dual use: as a command and control centre to support quick-reaction deployment and tactical operations by Tactical Air Command units; and as a survivable early-warning airborne command and control centre for identification, surveillance and tracking of airborne enemy forces, and for the command and control of NORAD (North American Air Defense) forces over the continental USA. The E-3A provides comprehensive surveillance out to a range of more than 200 nm (370 km; 230 miles) for low-flying targets, and still further for targets flying at higher altitudes.

Boeing's Aerospace Group, as it was then named, was awarded an initial contract as prime contractor and systems integrator for the AWAC system on 23 July 1970. In order to ensure that maximum effort and finance were devoted to the design and development of the most advanced radar and associated onboard systems, Boeing's design submission was based on the airframe of the Model

707-320B commercial jet transport. The only major change proposed for production E-3As was the installation of more powerful Pratt & Whitney TF33 turbofan engines, in lieu of the commercial turbofans then standard for the civil transport models. Two of these aircraft, with the prototype designation EC-137D, were modified initially for comparative trials with prototype downward-looking radars designed by Hughes Aircraft Company and Westinghouse Electric Corporation. After several months of airborne tests, the Westinghouse radar was selected on 5 October 1972.

On 26 January 1973 the US Air Force announced that, following satisfactory completion of the initial stage of the programme, approval had been given for full-scale AWACS development, and production received Congressional approval in the Spring of 1975, although by then the planned purchase had been cut to 34 instead of the 64 aircraft requested originally. The full-scale development test programme, which was completed at the end of 1976, involved a fleet of three aircraft completely equipped with mission avionics, and a fourth aircraft equipped for airworthiness testing.

In December 1976 Boeing awarded Westinghouse a contract to develop a maritime surveillance capability that could be incorporated into the E-3A radar system. An aircraft began flight testing this in June 1979 and all E-3As, beginning with the 22nd production aircraft, incorporate this maritime surveillance mode.

The first production E-3A Sentry was delivered on 24 March 1977 to Tactical Air Command's 552nd Airborne Warning and Control Wing, based at Tinker AFB, Oklahoma. Five had been delivered by the end of 1977, followed by ten more during 1978, five in 1979, three during 1980, and one during the first half of 1981. E-3As achieved initial operational status in April 1978, and have since completed deployments to Alaska, Iceland, Saudi Arabia, the Mediterranean area, and the Pacific. In December 1980 four E-3As were despatched to Ramstein AFB, West Germany, when tension in East Europe was at its height due to internal problems in Poland.

E-3As had begun to assume a role in US continental air defence on 1 January 1979, when NORAD personnel started to augment E-3A flight crews from TAC on all operational NORAD missions from Tinker AFB. The 552nd AWAC Wing reports directly to TAC Headquarters at Langley AFB, Virginia, and consists of several subordinate units. At Tinker, these include the 963rd and 964th AWAC Squadrons, the 966th AWAC Training Squadron, the 552nd Aircraft Generation Squadron (systems support), the 552nd Component Repair Squadron, and the 8th Tactical Deployment Control Squadron (flying EC-135/WC-135 aircraft). Overseas detachments of the 552nd AWACW include the 960th and 961st AWAC Support Squadrons. Based respectively at NAS Keflavik, Iceland, and Kadena AB, Okinawa, Japan, they provide command and control capability to CINCLANT (through the Commander, Iceland Defence Force) and CINCPAC. A third AWAC Support Squadron, the 962nd, is to be activated in 1982 at Elmendorf AFB, Alaska.

Funding for a total of 30 E-3As has been provided through FY 1982. Four more will be requested in FY 1983 to complete the currently planned force of 34 USAF AWACS. In addition, NATO has approved the acquisition of 18, under a cost-sharing agreement. Full basing details of these aircraft have yet to be announced, but their initial main operating base will be at Geilenkirchen in West Germany; some may later be based in Norway. For these NATO E-3As, much of the avionics is being produced in West Germany, with Dornier as systems integrator. NATO has funded a third HF radio, to cover the maritime environment; a new data analysis and programming group; and underwing hardpoints for self-defence system stores. The new data group, which handles the information gathered by the surveillance radar, IFF and other systems, is based on a new IBM CC-2 computer with a 665,360-word memory unit (compared with 114,688 words in the CC-1 computer of the earlier aircraft). Coupled with the increased data processing speed and capability is a greater tracking capacity, and the ability to initiate tracks automatically. Also funded by NATO, for its own E-3As only (although there are provisions for it in USAF aircraft), is a radio teletype to link the AWACS with the Organisation's maritime forces and commands. The first NATO production E-3A flew for the first time on 18 December 1980, from the manufacturing plant at Renton to Boeing Field, Seattle, where its rotodome and associated equipment were installed subsequently. It was delivered to Dornier's factory at Oberpfaffenhofen on 19 March 1981, and installation and checkout of mission avionics began on 1 April. Initial deliveries of operational E-3As to NATO are scheduled for 1982; attainment of limited initial operational capability is estimated in the second quarter of 1983.

Type: Airborne early-warning and command post aircraft.

Wings: Cantilever low-wing monoplane. Dihedral 7°. Incidence 2°. Sweepback at quarter-chord 35°. All-metal two-spar fail-safe structure. Centre-section continuous through fuselage. Normal outboard aileron, and small inboard aileron on each wing, built of aluminium honeycomb panels. Two Fowler flaps and one fillet flap of aluminium alloy on each wing. Full-span leading-edge flaps. Four hydraulically-operated aluminium alloy spoilers on each wing, forward of flaps. Primary flying controls are aerodynamically balanced and manually operated through spring tabs. Lateral control at low speeds by all four ailerons, supplemented by spoilers which are interconnected with the ailerons. Lateral control at high speeds by inboard ailerons and spoilers only. Operation of flaps adjusts linkage between inboard and outboard ailerons to permit outboard operation with flaps extended. Spoilers may also be used symmetrically as speed brakes. Thermal anti-icing of wing leading-edges.

Fuselage: All-metal semi-monocoque fail-safe structure with cross-section made up of two circular arcs of different radii, the larger above, faired into smooth-contoured oval. Structure strengthened by comparison with that of the commercial Model 707-320.

Tail Unit: Cantilever all-metal structure. Electrically and manually operated variable-incidence tailplane. Powered rudder. Anti-balance tab and trim tab in rudder. Trim and control tabs in each elevator.

Landing Gear: Hydraulically-retractable tricycle type. Main units are four-wheel bogies which retract inward into underside of thickened wing-root and fuselage. Twin-wheel nose unit retracts forward into fuselage. Landing gear doors close when legs fully extended. Boeing oleo-pneumatic shock-absorbers. Main wheels and tyres size 46 × 16. Nosewheels and tyres size 39 × 13. Multi-disc brakes by Goodyear. Hydro-Aire flywheel detector type anti-skid units.

Power Plant: Four Pratt & Whitney TF33-PW-100/100A turbofans, each rated at 93·4 kN (21,000 lb st), mounted in pods beneath the wings. Fuel contained in integral wing tanks. Provision for in-flight refuelling.

Accommodation: Basic operational crew of 17 includes a flight crew complement of four plus 13 AWACS specialists, though this latter number can vary for tactical and defence missions. Aft of flight deck, from front to rear of fuselage, are communications, data processing and other equipment bays; multi-purpose consoles; communications, navigation and identification equipment; and crew rest area.

Systems: A liquid cooling system provides protection for the radar transmitter. An air-cycle pack system, a draw-through system, and two closed-loop ram-cooled environmental control systems ensure a suitable environment for crew and avionics equipment. Electrical power generation has a 600kVA capability. The distribution centre for mission equipment power and remote avionics is located in the lower forward cargo compartment. The aft cargo compartment houses the radar transmitter and an APU. External sockets allow intake of power when the aircraft is on the ground. Two separate and independent hydraulic systems power flight-essential and mission-essential equipment, but either system has the capability of satisfying the requirements of both equipment groups in an emergency.

Avionics and Equipment: Prominent above the fuselage is the elliptical cross-section rotodome of 9·14 m (30 ft) diameter and 1·83 m (6 ft) max depth. It comprises four essential elements: a turntable, strut-mounted above the rear fuselage, that supports the rotary joint assembly to which are attached sliprings for electrical and waveguide continuity between rotodome and fuselage; a structural centre section of aluminium skin and stiffener construction which supports the AN/APY-1 surveillance radar and IFF/TADIL C antennae, radomes, auxiliary equipment for radar operation and environmental control of the rotodome interior; liquid cooling of the radar antennae; and two radomes constructed of multi-layer glassfibre sandwich material, one for the surveillance radar and one for the IFF/TADIL C array. For surveillance operations the rotodome is hydraulically driven at 6 rpm, but during non-operational flights it is rotated at only ¼ rpm, to keep the bearings lubricated. The Westinghouse radar operates in the S band and can function both as a pulse and/or a pulse-Doppler radar for detection of aircraft targets. A similar pulse radar mode with additional pulse compression and sea clutter adaptive processing is used to detect maritime/ship traffic. The radar is operable in six modes: PDNES (pulse-Doppler non-elevation scan), when range is paramount to elevation data; PDES (pulse-Doppler elevation scan), providing elevation data with some loss of range; BTH (beyond the horizon), giving long-range detection with no elevation data; Maritime, for detection of surface vessels in various sea states; Interleaved, combining available modes for all-altitude longer-range aircraft detection, or for both aircraft and ship detection; and Passive, which tracks enemy ECM sources without transmission-induced vulnerability. The radar antennae, spanning about 7·32 m (24 ft), and 1·52 m (5 ft) deep, scan mechanically in azimuth, and electronically from ground level up into the stratosphere. Heart of the data processing capability of the first 23 aircraft is an IBM 4 Pi CC-1 high-speed computer, the entire group consisting of arithmetic control units, input/output units, main storage units, peripheral control units, mass memory drums, magnetic tape transports, punched tape reader, line printer, and an operator's control panel. Processing speed is in the order of 740,000 operations/s; main memory size is 114,688 words (expandable to 180,224), and mass memory size 802,816 words (expandable to 1,204,224). An interface adapter unit developed by Boeing is the key integrating element interconnecting functional data between AWACS avionics subsystems, the data processing functional group, radar, communications, navigation/guidance, display, azimuth and identification, and also provides the central timing system. From the 24th production aircraft, the new and improved IBM CC-2 computer is installed, with a main storage capacity of 665,360 words. Data display and control are provided by Hazeltine Corporation multi-purpose consoles (MPC) and auxiliary display units (ADU); in present configuration

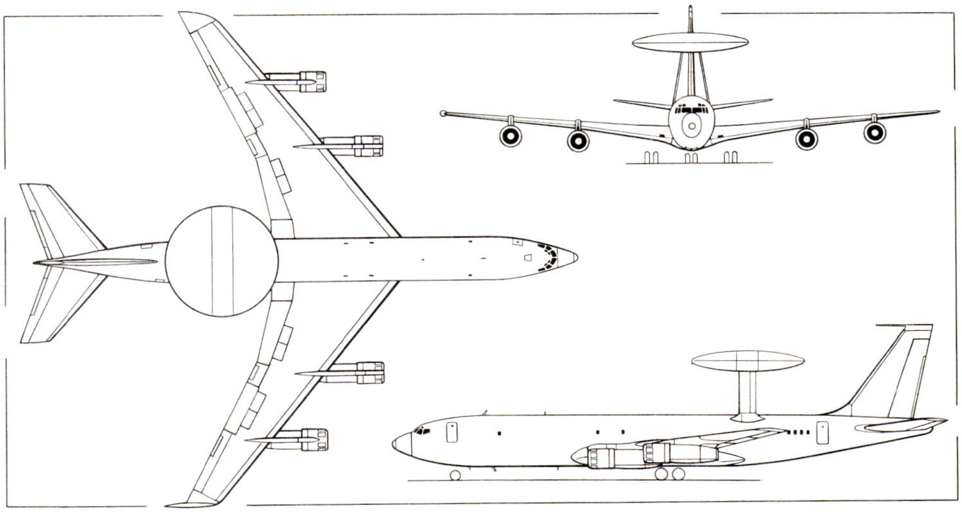

Three-view drawing *(Pilot Press)* **and photograph of the first of NATO's Boeing E-3A Sentry AWACS aircraft**

Boeing Canada (which see) is currently upgrading a number of Canadian Armed Forces CH-113/113As under a programme known as SARCUP.

In December 1980 the Naval Air Systems Command awarded Boeing Vertol an initial contract for a multi-year US Navy/Marine Corps helicopter improvement programme. Known as the Safety, Reliability and Maintainability (SR&M) programme, this will extend the effective service life of HH-46A, H-46D and CH-46E helicopters until the end of this century at significantly reduced operating costs. Improvements to these helicopters will involve an aircraft retrofit kit which Boeing Vertol will manufacture for installation by the Navy. During this eight-year programme, 368 modification kits are to be supplied. SR&M is a two-phase programme, the first involving design, development and prototype testing. The production phase is scheduled to begin during the first quarter of 1984, with delivery to the Navy of the first production kit in mid-1985. Production of kits will continue into 1988.

BOEING VERTOL MODELS 114 and 414

US Army designation: CH-47 Chinook
Canadian Armed Forces designation: CH-147
Royal Air Force designation: Chinook HC. Mk 1

Development of the CH-47 Chinook series of helicopters began in 1956, and the first of five YCH-47As (formerly YHC-1B) made its initial hovering flight on 21 September 1961. Since then, effectiveness of the CH-47 has been increased by successive improvement programmes. A total of 944 Chinooks had been ordered by 1 February 1981, of which 885 had been delivered; of these, 139 orders and 112 deliveries were from licence production by the Agusta Group in Italy (which see).

The CH-47 was designed to meet the US Army's requirement for an all-weather medium transport helicopter and, depending upon the series model, is capable of transporting specified payloads under severe combinations of altitude and temperature conditions. The primary mission radius criterion established by the US Army is 100 nm (185 km; 115 miles). The primary mission take-off gross weight is based on the capability of hovering out of ground effect at 1,830 m/35°C (6,000 ft/95°F). The CH-47C has demonstrated its ability to hover out of ground effect with a useful load of 11,453 kg (25,250 lb) at sea level under standard atmospheric conditions.

The following versions of the Chinook have been produced:

CH-47A. Initial production version, powered by two 1,640 kW (2,200 shp) Avco Lycoming T55-L-5 or 1,976 kW (2,650 shp) T55-L-7 turboshaft engines. Operation of the CH-47A by the Viet-Nam Air Force (VNAF) began in 1971. Four delivered to Royal Thai Air Force. Production completed. Transmissions uprated to CH-47C standard under 1978 Army contract.

CH-47B. Developed version with 2,125 kW (2,850 shp) T55-L-7C turboshaft engines, redesigned rotor blades with cambered leading-edge, blunted rear rotor pylon, and strakes along rear ramp and fuselage for improved flying qualities. First of two prototypes flew for the first time in early October 1966. Deliveries began on 10 May 1967. Production completed. Transmissions uprated to CH-47C standard under 1978 Army contract.

CH-47C. Increased performance from a combination of strengthened transmissions, two 2,796 kW (3,750 shp) T55-L-11A engines and increased integral fuel capacity. First flight of original CH-47C made 14 October 1967; production deliveries began in Spring 1968. First deployed in Viet-Nam in September 1968. Total of 210 US Army CH-47Cs currently undergoing retrofit with glassfibre rotor blades.

A crashworthy fuel system and an integral spar inspection system (ISIS) were made available during 1973.

Boeing Vertol updated CH-46E, for the US Marine Corps, with General Electric T58-GE-16 turboshaft engines

Boeing Vertol CH-47D utility transport helicopter of the US Army

CH-47C CHINOOK WEIGHTS AND PERFORMANCE

	Condition 1	Condition 2	Condition 3	Condition 4
Weight empty	9,736 kg (21,464 lb)	9,736 kg (21,464 lb)	9,812 kg (21,633 lb)	9,599 kg (21,162 lb)
Payload	5,284 kg (11,650 lb)	2,903 kg (6,400 lb)	9,843 kg (21,700 lb)	—
T-O weight	17,463 kg (38,500 lb)	14,968 kg (33,000 lb)	20,593 kg (45,400 lb)	20,865 kg (46,000 lb)
Max speed, S/L, ISA at normal rated power	155 knots (286 km/h; 178 mph)	164 knots (304 km/h; 189 mph)	127 knots (235 km/h; 146 mph)	—
Average cruising speed	139 knots (257 km/h; 160 mph)	137 knots (254 km/h; 158 mph)	114 knots (211 km/h; 131 mph)	133 knots (246 km/h; 153 mph)
Max rate of climb, S/L, ISA at normal rated power	649 m (2,130 ft)/min	878 m (2,880 ft)/min	421 m (1,380 ft)/min	402 m (1,320 ft)/min
Service ceiling, ISA, normal rated power	3,290 m (10,800 ft)	4,570 m (15,000 ft)	2,560 m (8,400 ft)	2,440 m (8,000 ft)
Hovering ceiling OGE, ISA, max power	2,805 m (9,200 ft)	4,145 m (13,600 ft)	Sea level	—
Mission radius	100 nm (185 km; 115 miles)	100 nm (185 km; 115 miles)	20 nm (37 km; 23 miles)	—
Ferry range	—	—	—	1,156 nm (2,142 km; 1,331 miles)

Condition 1 Criteria: Take-off gross weight equals gross weight to hover OGE at 1,830 m/35°C (6,000 ft/95°F). Radius of action 100 nm (185 km; 115 miles). Fuel reserve 10%. Payload carried internally.

Condition 2 Criteria: Take-off gross weight equals design gross weight. Radius of action 100 nm (185 km; 115 miles). Fuel reserve 10%. Payload carried internally.

Condition 3 Criteria: Take-off gross weight equals gross weight to hover OGE at S/L ISA. Radius of action 20 nm (37 km; 23 miles). Fuel reserve 10%. Payload carried externally. Except for the mission average cruising speed, all other performance is predicated on internal loading of cargo.

Condition 4 Criteria: Take-off gross weight represents alternative design gross weight. Max ferry range (integral and internal auxiliary fuel only), cruise at optimum altitude and standard temperature, no payload, 10% fuel reserves.

Incorporation of the crashworthy fuel system on US Army CH-47Cs was accomplished by retrofit kits, deliveries of which began in March 1973. Chinooks delivered to Australia have this system, which provides a total fuel capacity of 3,944 litres (1,042 US gallons).

CH-147. Designation of nine CH-47Cs delivered to Canada from September 1974. These aircraft have T55-L-11C engines, ISIS, crashworthy fuel system, forward door rescue hoist, ferry range tank kit, up to 44 troop seats, advanced flight control system, rear ramp with water dam, 12,700 kg (28,000 lb) cargo hook, T-O weight of 22,680 kg (50,000 lb), and weight for water operations of 16,330 kg (36,000 lb) normal or 20,865 kg (46,000 lb) emergency.

CH-47D. Three Chinooks (one each of the CH-47A, CH-47B, and CH-47C models) were modernised to CH-47D configuration under a $75 million contract received from the US Army in June 1976. The three aircraft were stripped down to their basic airframes, and were then rebuilt with improved components and systems to provide three CH-47D prototypes.

Conversion included the installation of Avco Lycoming T55-L-712 engines, which have an emergency power rating; the introduction of improved rotor transmissions with a rating of 5,593 kW (7,500 shp); the introduction of integral lubrication and cooling for the transmission systems, and of glassfibre rotor blades. Other improvements include a redesigned cockpit to reduce the pilot's workload; the introduction of redundant and improved electrical systems, modularised hydraulic systems, an advanced flight control system, and improved avionics equipment. In addition, a T62T-2B APU and a triple cargo hook suspension system were installed.

The first CH-47D, converted from a CH-47A airframe, was rolled out four months ahead of schedule on 6 March 1979. After flight testing had been completed by Boeing Vertol, this aircraft was accepted by the US Army, at Fort Rucker, Alabama, on 13 December 1979, and by mid-1980 the three prototypes had completed a 920-hour programme of flight testing, as well as an extensive series of ground tests, by Boeing Vertol and the US Army. Two of the CH-47D prototypes were allocated to an operational test programme, conducted by the 101st Airborne Division (Air Assault) at Fort Campbell, Kentucky, which took place in April and May 1980. The third aircraft was used for climatic tests. In October 1980 the first CH-47 modernisation programme contract was awarded to Boeing Vertol by the US Army. This contains funds for the remanufacture of nine CH-47As to CH-47D configuration; non-recurring production costs; and the procurement of long lead items needed for the Army's planned follow-on remanufacture of 19 Chinooks during FY1982. Work has started on remanufacturing the CH-47As as CH-47Ds, with delivery of the first example scheduled for the Spring of 1982.

Chinook HC.Mk 1. Version for Royal Air Force, which ordered 33. First example (ZA670) made its initial flight on 23 March 1980. The first example was handed over in August 1980, and this arrived in the UK in November 1980, where it was allocated to No. 240 OCU, RAF Odiham. Deliveries are to be completed by Spring 1982. Generally similar to Canadian CH-147, with Avco Lycoming T55-L-11E turboshaft engines, but with provision for glassfibre/carbonfibre rotor blades and three external cargo hooks (capacity 12,700 kg; 28,000 lb on centre hook, or 9,072 kg; 20,000 lb total on forward and rear hooks); accommodation for up to 44 troops or 24 standard NATO stretchers; engine and windscreen de-icing; provision for two self-ferry fuel tanks in cabin; and amphibious capability in sea states of up to 3. Intended for use on logistic support, tactical troop lift, casualty evacuation, air-mobility, and external load-carrying duties. Will equip one squadron based in the UK, and No. 18 Squadron at RAF Gutersloh in West Germany. Extensive range of British avionics and equipment, including Decca tactical navigation system, Doppler Mk 71 radar and Mk 19 area navigation; Marconi Avionics ARC340 VHF/FM com and homing system, AD120 VHF/AM com, AD380 ADF, AD2770 Tacan and AD27733 interface unit; Lucas 40kVA generators; and UHF/AM, VOR/ILS, HF/SSB, radar altimeter, UHF/VHF homing, and IFF transceiver.

Model 234. Commercial version, described separately.

Model 414. International military version, described in detail in this entry.

CH-47 helicopters are in service in many places, including Alaska, Australia, West Germany, Hawaii, Iran, Italy, Korea, Spain and Thailand, as well as at numerous US National Guard and US Army installations within the continental United States. Total US Army procurement of CH-47A/B/C models had reached 733 by 1 February 1981. Other customers for US-built CH-47Cs include Argentina (5), Australia (12), Canada (9), Spain (9), Thailand (4), and the UK (33). Deliveries of Agusta/EM-built CH-47Cs have been made to the air forces of Italy (26), Iran (60 delivered by start of US embargo), Libya (20), Morocco (6), Egypt (15), Tanzania (2) and Greece (10).

Details of the CH-47A and -47B can be found in the 1974-75 *Jane's*, and the CH-47C was described in the 1980-81 edition. The following data apply specifically to the current Model 414 international military Chinook:

TYPE: Twin-engined medium transport helicopter.

ROTOR SYSTEM: Two three-blade rotors, rotating in opposite directions and driven through interconnecting shafts which enable both rotors to be driven by either engine. Rotor blades, of glassfibre, have VR7 and VR8 aerofoil sections, cambered leading-edge, D-shaped glassfibre spar, and a fairing assembly consisting of Nomex honeycomb core with crossply glassfibre laminates for skin. All blades can be folded manually. Rotor heads are fully articulated with pitch, flapping and drag hinges. All bearings are submerged completely in oil. Provisions for optional rotor brake.

ROTOR DRIVE: Power is transmitted from each engine through individual overrunning clutches, into the combiner transmission, thereby providing a single power output to the interconnecting shafts. Rotor/engine rpm ratio 1 : 64.

FUSELAGE: Semi-monocoque all-metal fuselage structure comprising cockpit, cabin, aft fuselage, and pylon sections. The cabin is of constant section, with the lower half sealed during manufacture to form a watertight compartment that provides water landing capability. The aft fuselage contains the cargo ramp and door.

LANDING GEAR: Non-retractable quadricycle type, with twin wheels on each forward unit and single wheels on each rear unit. Oleo-pneumatic shock-absorbers in all units. Rear units fully castoring and steerable; power steering installed on starboard rear unit. All wheels are size 24 × 7·7-VII, with tyres size 8·50-10-III, pressure 4·62 bars (67 lb/sq in). Two single-disc hydraulic brakes. Provision for fitting detachable wheel-skis.

POWER PLANT: Two Avco Lycoming T55-L-712 turboshaft engines, mounted one on each side of the rear pylon, and each with a standard power rating of 2,796 kW (3,750 shp) and emergency rating of 3,356 kW (4,500 shp). Combined transmission continuous rating 5,054 kW (6,777 shp) at 225 rpm and 95 per cent torque; single-engine transmission emergency rating of 3,430 kW (4,600 shp) at 225 rpm and 129 per cent torque. Self-sealing pressure refuelled crashworthy fuel tanks in external pods on sides of fuselage. Total fuel capacity 3,929 litres (1,038 US gallons). Oil capacity 14 litres (3·7 US gallons).

ACCOMMODATION: Two pilots on flight deck, with dual controls. Jump seat for crew chief or combat commander. Jettisonable door on each side of flight deck. Depending on seating arrangement, 33 to 44 troops can be accommodated in main cabin, or 24 litters plus two attendants, or vehicles and freight. Rear loading ramp can be left completely or partially open, or can be removed to permit transport of extra-long cargo and in-flight parachute or free-drop delivery of cargo and equipment. Main cabin door, at front on starboard side, comprises upper hinged section which can be opened in flight, and lower section with integral steps. Lower section is jettisonable. Triple external cargo hook system, as on Model 234, with centre hook able to carry max load of 12,700 kg (28,000 lb). Provisions are installed for a power-down ramp and water dam which will permit ramp operation on water; for forward and aft cargo hooks, ferry fuel tanks, external rescue hoist, and windscreen washers.

SYSTEMS: Cabin heated by 200,000 BTU heater-blower. Hydraulic system provides pressures of 207 bars (3,000 lb/sq in) for flying controls, and 276 bars (4,000 lb/sq in) for engine starting. Electrical system includes two 40kVA air-cooled alternators driven by transmission drive system. Solar T62 APU runs accessory gear drive, thereby operating all hydraulic and electrical systems.

AVIONICS AND EQUIPMENT: Standard avionics include ARC-102 HF, ARC-186 UHF-AM/FM, ARC-164 UHF-AM com; C-6533 intercom; APX-100 IFF; APN-209 radar altimeter; ARN-89B ADF; ARN-123 VOR/glideslope/marker beacon receiver; and ASN-43 magnetic compass. Flight instruments are standard for IFR, and include an AQU-6A horizontal situation indicator. AFCS maintains helicopter stability, eliminating the need for constant small correction inputs by the pilot to maintain desired attitude. The AFCS is a redundant system using two identical control units and two sets of stabilisation actuators. Standard equipment includes a

Boeing Vertol Chinook HC. Mk 1 for the Royal Air Force *(Denis Calvert)*

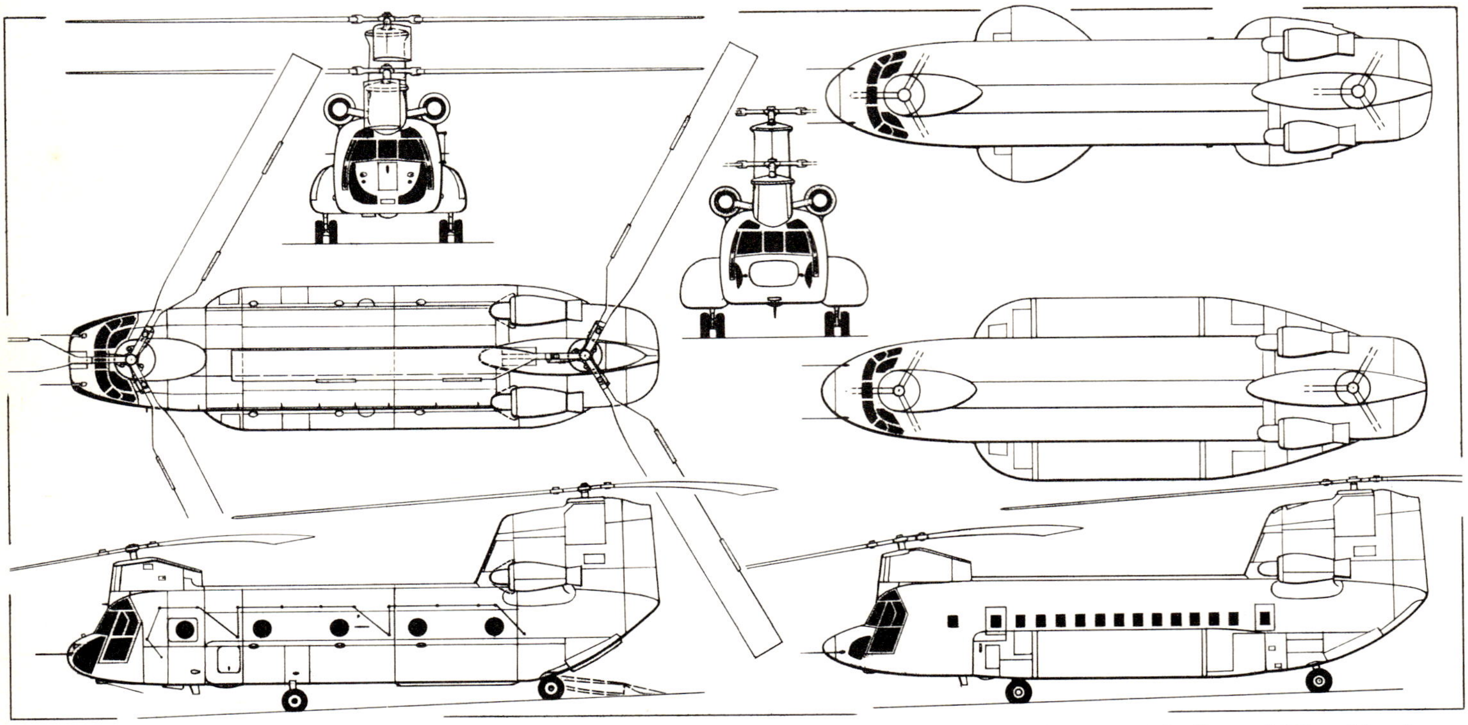

Three views on left depict the Boeing Vertol CH-47D military helicopter. Plan view at top right shows the Model 234 commercial utility model. Remaining three views depict the long-range commercial model *(Pilot Press)*

hydraulically powered winch for rescue and cargo handling, rearview mirror, integral work stands and step for maintenance, and provisions for static lines, and maintenance davits for removal of major components.

DIMENSIONS, EXTERNAL:

Diameter of rotors (each)	18·29 m (60 ft 0 in)
Main rotor blade chord	0·81 m (2 ft 8 in)
Distance between rotor centres	11·94 m (39 ft 2 in)
Length overall, rotors turning	30·18 m (99 ft 0 in)
Length of fuselage	15·54 m (51 ft 0 in)
Width, rotors folded	3·78 m (12 ft 5 in)
Height to top of rear rotor hub	5·68 m (18 ft 7·8 in)
Wheel track (c/l of shock-absorbers)	3·20 m (10 ft 6 in)
Wheelbase	6·86 m (22 ft 6 in)
Passenger door (fwd, stbd):	
Height	1·68 m (5 ft 6 in)
Width	0·91 m (3 ft 0 in)
Height to sill	1·09 m (3 ft 7 in)
Rear loading ramp entrance:	
Height	1·98 m (6 ft 6 in)
Width	2·31 m (7 ft 7 in)
Height to sill	0·79 m (2 ft 7 in)

DIMENSIONS, INTERNAL:

Cabin, excl flight deck: Length	9·20 m (30 ft 2 in)
Width (mean)	2·29 m (7 ft 6 in)
Width at floor	2·51 m (8 ft 3 in)
Height	1·98 m (6 ft 6 in)
Floor area	21·0 m² (226 sq ft)
Usable volume	41·7 m³ (1,474 cu ft)

AREAS:

Rotor blades (each)	7·43 m² (80 sq ft)
Rotor discs (total)	525·3 m² (5,655 sq ft)

WEIGHTS (CH-47C: see accompanying table. CH-47D as follows, A: guaranteed; B: estimated, based on whirl test results):

Internal payload over 100 nm (185 km; 115 miles) at 1,220 m (4,000 ft), hovering OGE at T-O:

A	5,896 kg (13,000 lb)
B	6,496 kg (14,322 lb)

External payload over 30 nm (55·5 km; 34·5 miles) at 1,220 m (4,000 ft), 61 m (200 ft)/min vertical climb at T-O, 35°C:

A	6,803 kg (15,000 lb)
B	7,155 kg (15,775 lb)

Gross weight, hovering OGE at S/L, ISA:

A	22,680 kg (50,000 lb)
B	24,267 kg (53,000 lb)

WEIGHTS (Model 414):

Weight empty	10,405 kg (22,939 lb)
Internal payload over 100 nm (185 km; 115 miles) at 1,525 m (5,000 ft), hovering OGE at 20,616 kg (45,450 lb), ISA + 20°C	8,255 kg (18,200 lb)
External payload over 30 nm (55·5 km; 34·5 miles), other conditions as above	9,435 kg (20,800 lb)
Design gross weight	14,968 kg (33,000 lb)
Alternative design gross weight	20,865 kg (46,000 lb)
Max T-O weight	22,679 kg (50,000 lb)

PERFORMANCE (CH-47C: see accompanying table. CH-47D: as follows):

Max level speed at S/L, ISA, at AUW of 14,968 kg (33,000 lb):

A	155 knots (287 km/h; 178 mph)
B	161 knots (298 km/h; 185 mph)

Single-engine service ceiling at 14,968 kg (33,000 lb) AUW, ISA:

A	3,050 m (10,000 ft)
B	4,270 m (14,000 ft)

PERFORMANCE (Model 414 at max continuous power, ISA, at gross weights of A: 22,679 kg: 50,000 lb; B: 20,865 kg: 46,000 lb; C: 14,968 kg: 33,000 lb):

Max level speed at S/L:	
A	145 knots (269 km/h; 167 mph)
B	150 knots (278 km/h; 173 mph)
C	160 knots (297 km/h; 184 mph)
Service ceiling: A	2,745 m (9,000 ft)
B	3,610 m (11,850 ft)
Hovering ceiling OGE: A	1,036 m (3,400 ft)
B	2,680 m (8,800 ft)

BOEING VERTOL MODEL 234 COMMERCIAL CHINOOK

Announced in the late Summer of 1978, this development of the military CH-47 Chinook is being produced for use as a commercial passenger transport, as a cargo carrier, and for specialised tasks such as servicing offshore oil and natural gas rigs, remote resources exploration and extraction, logging, and construction work.

The airframe of the Model 234 is based on that of the latest military Chinook, but has many new features. These include the use of wide-chord glassfibre rotor blades, instead of the usual metal blades; redesign of the fuselage-side fairings in two different forms; a lengthened nose to accommodate the weather radar antenna; and movement further forward of the front landing gear units.

Two basic versions of the commercial Model 234 are offered:

Long-range model. Identified by continuous fuselage-side fairings, approximately twice as large as those of the military Chinook and containing large fuel tanks. Equipped to airline standards as a passenger, passenger/freight 'combi' or all-cargo transport.

Utility model. Fuselage-side fuel tanks replaced by two drum-shape internal tanks, mounted longitudinally side by side at the front of the cabin. Fuselage-side fairings removed, leaving only an individual streamlined blister around each landing gear mounting. As well as reducing weight, this enhances the helicopter's lifting capability by reducing the airframe surface area on which the rotor downwash impinges.

Conversion from one configuration of the Model 234 to another is estimated to take eight hours, and requires four persons to handle the fuel tanks of the utility model and the ramp baggage bins of the passenger-carrying helicopter. Initial cost of the glassfibre blades is significantly greater than that of metal blades, but manufacturing time is reduced by 25% and in-service maintenance costs by 71%. Mean time between blade removal is estimated at 3,200 h. The three blades of any one rotor are interchangeable, but blades cannot be interchanged from one hub to another. Current engine TBO is 1,800 h.

Initial order for the Model 234 was placed by British Airways Helicopters, which ordered three in 1978 (since increased to six), primarily for North Sea oil rig support operations. Assembly of the first aircraft began in mid-1979. First flight was made on 19 August 1980, and FAA certification was received on 19 June 1981. Following CAA certification on 26 June, the first BAH Model 234 (which had been delivered in January) went into service on 1 July 1981. Three had been delivered to BAH by that date; Asahi Helicopters of Japan has an option for two Model 234s.

TYPE: Twin-turbine commercial transport helicopter.

ROTOR SYSTEM: Two three-blade rotors in tandem, turning in opposite directions and driven through interconnecting shafts which enable both rotors to be driven by either engine. Wide-chord glassfibre blades, with VR7 section over inboard 85% of span, and VR8 section on outer 15% of span; thickness/chord ratio 12% and 8% respectively. Overall blade twist 12°. Each blade comprises a laminated glassfibre skin over a glassfibre D spar, forming the front half of the section, and with the rear half filled with Nomex honeycomb. An aluminium screen inserted in the skin provides lightning protection, discharging strikes via the titanium leading-edge. Outboard 25% of leading-edge capped with replaceable nickel section. Blade balancing by tracking weights in tips. Two blades of each rotor can be folded manually. Hubs fully articulated, with pitch, flapping and drag hinges. All bearings submerged completely in oil. Blades embody electrical de-icing blankets, permitting addition of a de-icing kit if required. Rotor rpm 225.

ROTOR DRIVE: Power is transmitted from each engine through individual overrunning clutches, into the combiner transmission, thereby providing a single power output to the interconnecting shafts. Auxiliary transmission lubrication system enables flight to be completed after total loss of oil in primary system.

FUSELAGE: All-metal semi-monocoque structure of basically square section. Loading ramp forms undersurface of upswept rear fuselage. External fuel pods of long-range model made of advanced composites, including glassfibre, graphite/epoxy and Nomex nylon honeycomb. These fairing pods provide flotation capability adequate to meet British airworthiness requirements applicable to a sea state seven (9·15 m; 30 ft waves. Wave length-to-height ratio 15) without added flotation gear.

LANDING GEAR: Non-retractable quadricycle type, with twin wheels on each forward unit and single wheels on each rear unit. All tyres size 8·50-10, Type III (12 ply), pressure 8·55 bars (124 lb/sq in) on forward gear, 7·20 bars (104·4 lb/sq in) on aft gear. Oleo-pneumatic shock-absorbers in all units. Rear units fully castoring and steerable. Hydraulic disc brakes.

POWER PLANT: Two Avco Lycoming AL 5512 turboshaft engines, pod-mounted on each side of rear rotor pylon. Each engine has max T-O rating of 3,039 kW (4,075 shp), max continuous rating of 2,218·5 kW (2,975 shp), and 30 min contingency rating of 3,246 kW (4,353 shp). Transmission rated at 5,033 kW (6,750 shp) at 225 rotor rpm, and 3,430 kW (4,600 shp) for single-engine operation. Long-range model has two fuel tanks, one in each fuselage-side fairing, with total capacity of 7,911 litres (2,090 US gallons). Utility model has two drum-shape internal tanks, with total capacity of 3,785 litres (1,000 US gallons). Single-point pressure refuelling.

ACCOMMODATION: Two pilots side by side on flight deck, with dual controls. Passenger cabin of long-range model seats up to 44 persons four-abreast, with centre aisle. Each seat has overhead bin and underseat stowage for carry-on baggage; larger items are stowed over the rear ramp in the main baggage compartment. Galley, with

Boeing Vertol Model 234 Commercial Chinook in the insignia of British Airways Helicopters

cabin attendant's seat, and toilet, are standard, between flight deck and cabin. Typical 'combi' configuration accommodates 18 passengers forward and 7,250 kg (16,000 lb) of freight, loaded via rear ramp. All passenger facilities can be removed for freight-only service. Passenger door at front of cabin on starboard side. Crew door on each side of flight deck. Cabin floor supported by dynamically tuned fittings to reduce vibration. Hydraulically powered cargo ramp can be stopped at any intermediate position to match the level of the loading vehicle being used. Single central cargo hook is standard on utility model for carrying external loads of up to 12,700 kg (28,000 lb). Optional dual tandem hooks for precision operations and for load stability in high-speed flight; or three tandem hooks for delivering multiple loads.

Systems: Heating and ventilation systems maintain comfortable flight deck/cabin temperature in ambient temperatures down to −32°C. Duplicated flying control, hydraulic and electrical systems. Solar T62T-2B APU, rated at 71 kW (95 shp), drives auxiliary gearbox on rear transmission to start engines and provide power for two flying control system hydraulic pumps and two alternators. All critical systems heated to inhibit ice build-up.

Avionics and Equipment: Duplicated full blind-flying instrumentation, weather radar, and dual four-axis automatic flight control system with built-in test equipment, provide all-weather capability. Optional equipment includes passenger interior furnishings for the utility model, 'combi' interior, downward-shining cargo load light, rescue hoist of 272 kg (600 lb) capacity, glassfibre wheel-skis, an ice detector probe, and ditching equipment that includes two liferafts, each with an overload capacity of 36 persons. Standard items include integral work platforms, and a maintenance panel that allows 26 separate checks to be made from a single ground-level position.

Dimensions, external:
Rotor diameter (each) 18·29 m (60 ft 0 in)
Rotor blade chord (constant) 0·813 m (2 ft 8 in)
Length overall, rotors turning 30·18 m (99 ft 0 in)
Length of fuselage 15·87 m (52 ft 1 in)
Height overall 5·68 m (18 ft 7·8 in)
Width over fuselage-side fairings 4·78 m (15 ft 8 in)
Wheel track:
fwd gear 3·20 m (10 ft 6 in)
aft gear 3·40 m (11 ft 2 in)
Wheelbase 7·87 m (25 ft 9·9 in)

Dimensions, internal:
Passenger cabin: Length 9·19 m (30 ft 2 in)
Max width 2·51 m (8 ft 3 in)
Max height 1·98 m (6 ft 6 in)
Baggage compartment volume 4·42 m³ (156 cu ft)
Utility model, cargo hold volume 41·03 m³ (1,449 cu ft)

Weights (L, long-range model; U, utility model):
Manufacturer's weight empty:
L 11,090 kg (24,449 lb)
U 9,369 kg (20,655 lb)
Operating weight empty:
L 11,566 kg (25,500 lb)
U 9,576 kg (21,112 lb)
Fuel load: L 6,361 kg (14,024 lb)
U 1,826 kg (4,026 lb)
Max payload: L, internal 10,229 kg (22,551 lb)
U, internal 11,843 kg (26,109 lb)
L, U, external 12,700 kg (28,000 lb)
Max T-O weight:
L and U, internal load 21,318 kg (47,000 lb)
L, external load 22,226 kg (49,000 lb)
U, external load 23,133 kg (51,000 lb)

Performance (estimated: L, long-range model; U, utility model):
Never-exceed speed:
L, U 165 knots (305 km/h; 190 mph)
Max cruising speed at 610 m (2,000 ft):
L, U, internal load, at 21,318 kg (47,000 lb) AUW 145 knots (269 km/h; 167 mph)
Cruising speed for optimum range, at 610 m (2,000 ft):
L, U, internal load, at 18,150 kg (40,000 lb) AUW 137 knots (253 km/h; 157 mph)
L, U, internal load, at 21,318 kg (47,000 lb) AUW 135 knots (250 km/h; 155 mph)
Max rate of climb at S/L:
L, U, internal load 410 m (1,350 ft)/min
Service ceiling:
L, U, internal load 4,570 m (15,000 ft)
Service ceiling, one engine out:
L, U, internal load 457 m (1,500 ft)
Hovering ceiling IGE:
L 2,790 m (9,150 ft)
U 3,155 m (10,350 ft)
Hovering ceiling OGE:
L 1,495 m (4,900 ft)
U 2,195 m (7,200 ft)
Range with 45 min IFR reserves:
L with 44 passengers 545 nm (1,010 km; 627 miles)
L with max fuel 740 nm (1,371 km; 852 miles)
U with max internal load 180 nm (333 km; 207 miles)
U with max external load 135 nm (250 km; 155 miles)
Max endurance:
L 5 h 6 min
U, external load 2 h 24 min

BOLEN

RALPH BOLEN INC

1311 US 40 South West, London, Ohio 43140
Telephone: (614) 852 1990

BOLEN 'TAILDRAGGER' CONVERSIONS

Ralph Bolen Inc has designed a tailwheel landing gear conversion for application to Cessna 150/152, 172/Skyhawk and 175 aircraft. Flight testing has indicated an average 8·5-10·5 knots (16-19·5 km/h; 10-12 mph) increase in speed resulting from removal of the nosewheel unit; but this change in landing gear configuration is intended also to improve performance from short fields and rough surfaces, to permit a tighter turn radius on narrow strips, and to simplify operation on floats or skis when this is desirable.

The conversion is carried out by the addition of two bulkheads in the front fuselage and a new main landing gear attachment box, just forward of the existing box, to which new cantilever main-gear legs are installed in the case of Cessna Models 172 and 175. Cessna 150 series F,

Bolen's landing gear conversion, showing new tailwheel and new main gear on a Cessna Model 175

G, H, J and K, and Cessna 152s, retain their existing main-gear legs, but these are removed from their original mounting and attached to the new forward box. Earlier versions require new legs of the design fitted to Cessna 150s of the above series, so that the now-standard wheel with 15 × 6·00-6 tyre and wheel brakes can be installed. The remainder of the conversion covers the removal of the nose gear, and the installation of a new tailwheel unit that includes a Scott 3200 wheel. This unit is attached by stress plates and stringers, so that the tailwheel assembly becomes an integral part of the fuselage structure. No skin removal is required for the conversion.

The 'Taildragger' modification of the landing gear system has been so designed that on completion there is no shift in the CG position. The STC for most of the Cessna Model 150/152 series, and for the Model 172, covers operation on Fluidyne snow skis, and similar approval is being sought for the Model 175. No weight penalty results from conversion of Model 150/152 aircraft; the longer and stronger main gear units introduced on the Models 172/175, and accompanying structure, increase weight by 13·6 kg (30 lb).

The Bolen company can carry out conversions of customers' aircraft at its facility at Springfield Municipal Airport, Ohio, and also markets conversion kits under the name 'Taildragger' which include all components and fastenings, an instruction manual and blueprints. Company installations carry a full STC; conversions carried out by owners, or their mechanics, qualify after inspection and approval by the appropriate civil aviation authority.

BRANTLY-HYNES

BRANTLY-HYNES HELICOPTER INC

HEAD OFFICE AND WORKS: PO Box 697, Frederick, Oklahoma 73542

Telephone: (405) 335 2256

PRESIDENT: Michael K. Hynes

This company, formed on 1 January 1975, replaced Brantly Operators Inc which acquired all rights in Brantly helicopters in late 1970. Mr M. K. Hynes acquired also ownership of the Type Certificates for the Brantly B-2, B-2A, B-2B and Model 305.

Brantly-Hynes has put the two-seat B-2B back into production, and was scheduled to produce 18 of these aircraft by the Spring of 1982. It is planning to introduce an agricultural version of the Model 305 in early 1982.

BRANTLY-HYNES MODEL B-2B

TYPE: Two-seat light helicopter.

ROTOR SYSTEM: Three-blade main rotor. Articulated inboard flapping hinges offset 0·07 m (2·67 in) from hub, and coincident flap and lag hinges offset 1·31 m (4 ft 3¾ in) from hub. Symmetrical blade section with 29% thickness ratio on inboard portion; NACA 0012 section outboard of hinge. Inboard portion of each blade is rigid, built around a steel spar blade. Outboard portion is flexible, with an extruded aluminium leading-edge spar and polyurethane core; aluminium skin is bonded to core and riveted to spar. Blades are attached to hub by flapping links and do not fold. A rotor brake is standard equipment. Two-blade all-metal anti-torque tail rotor.

ROTOR DRIVE: Through centrifugal clutch and planetary reduction gears. Bevel gear take-off from main transmission with flexible coupling to tail rotor drive-shaft. Main rotor/engine rpm ratio 1 : 6·158. Tail rotor/engine rpm ratio 1 : 1.

FUSELAGE: Stressed-skin all-metal structure with conical tail section. Tail rotor on swept-up boom extension.

LANDING GEAR: Alternative skid, wheel or float gear. Skid type has small retractable wheels for ground handling, fixed tailskid and four shock-absorbers with rubber in compression. Tyres size 10 × 3½, pressure 4·14 bars (60 lb/sq in). Alternative non-retractable tricycle landing gear has oleo-pneumatic shock-absorbers in all units, with single wheels on main units and twin nosewheels. Inflatable pontoons, which attach to standard skids, are available to permit operation from water.

POWER PLANT: One 134 kW (180 hp) Avco Lycoming IVO-360-A1A flat-four engine, mounted vertically, with dual fan cooling system. Rubber bag-type fuel tank under engine, capacity 117 litres (31 US gallons). Refuelling point on port side of fuselage. Oil capacity 5·7 litres (1·5 US gallons).

Brantly-Hynes Model B-2B light helicopter (Avco Lycoming IVO-360-A1A engine)

Brantly-Hynes Model 305 light helicopter carrying 181 kg (400 lb) underslung load

ACCOMMODATION: Totally-enclosed circular-section cabin for two persons seated side by side. Forward-hinged door on each side. Dual controls, cabin heater and demisting fan standard. Compartment for 22·7 kg (50 lb) baggage in forward end of tail section.

AVIONICS AND EQUIPMENT: Provision for all standard nav/com radios. Blind-flying instrumentation available as an option, but the Model B-2B is not certificated for instrument flight. Twin landing lights in nose.

DIMENSIONS, EXTERNAL:	
Diameter of main rotor	7·24 m (23 ft 9 in)
Main rotor blade chord: inboard	0·225 m (8·85 in)
outboard	0·203 m (8·0 in)
Diameter of tail rotor	1·29 m (4 ft 3 in)
Length overall, rotor turning	8·53 m (28 ft 0 in)
Length of fuselage	6·62 m (21 ft 9 in)
Height overall	2·06 m (6 ft 9 in)
Skid track	1·73 m (5 ft 8¼ in)
Passenger doors (each): Height	0·79 m (2 ft 7 in)
Width	0·86 m (2 ft 9¾ in)
Baggage compartment door:	
Mean height	0·25 m (9¾ in)
Length	0·55 m (1 ft 9¾ in)
DIMENSIONS, INTERNAL:	
Max width of cabin	1·27 m (4 ft 2 in)
Baggage compartment	0·17 m³ (6 cu ft)
AREAS:	
Main rotor blades (each)	0·69 m² (7·42 sq ft)
Main rotor disc	41·06 m² (442 sq ft)
Tail rotor disc	1·21 m² (13 sq ft)
WEIGHTS AND LOADINGS:	
Weight empty with skids	463 kg (1,020 lb)
Weight empty with floats	481 kg (1,060 lb)
Max T-O weight	757 kg (1,670 lb)
Max disc loading	18·4 kg/m² (3·77 lb/sq ft)
Max power loading	5·65 kg/kW (9·27 lb/hp)
PERFORMANCE (at max T-O weight):	
Max level speed at S/L	87 knots (161 km/h; 100 mph)
Max cruising speed (75% power)	78 knots (145 km/h; 90 mph)
Max rate of climb at S/L	580 m (1,900 ft)/min
Service ceiling	3,290 m (10,800 ft)
Hovering ceiling IGE	2,040 m (6,700 ft)
Range with max fuel, with reserves	217 nm (400 km; 250 miles)

BRANTLY-HYNES MODEL 305

The Model 305 is a five-seat helicopter of similar configuration to the Model B-2B, but larger in every respect. The prototype of the original Model 305 flew for the first time in January 1964, and FAA Type Approval was received on 29 July 1965.

The following description applies to the certificated passenger-carrying version:

TYPE: Five-seat light helicopter.

ROTOR SYSTEM: Three-blade main rotor. Articulated inboard flapping hinges, offset 0·09 m (3·625 in) from hub, and coincident flap and lag hinges outboard. Inboard portion of each blade is rigid, built around a steel spar blade. All-metal outboard portion has a D-spar and is foam-filled. Two-blade all-metal tail rotor. Each blade has a forged aluminium leading-edge spar, ribs and riveted aluminium skin. Main rotor blades do not fold. Rotor brake is standard.

ROTOR DRIVE: Main rotor shaft-driven through centrifugal clutch and planetary reduction gears. Bevel gear take-off from main transmission, with flexible coupling, through tail rotor drive-shaft and intermediate gearbox to tail gearbox. Main rotor/engine rpm ratio 1 : 6·666. Tail rotor/engine rpm ratio 1 : 0·998.

FUSELAGE: Stressed-skin all-metal structure, with conical tail section. Tail rotor carried on swept-up boom extension.

TAIL UNIT: Small variable-incidence horizontal stabiliser of all-metal stressed-skin construction.

LANDING GEAR: Alternative skid, wheel or float gear. Skid landing gear has four oleo struts, two on each side, and small retractable ground handling wheels. The wheel gear has two main wheels and twin nosewheels, all with oleo-pneumatic shock-absorbers. Goodyear main wheels and tyres size 6·00-6, pressure 2·07 bars (30 lb/sq in). Goodyear nosewheels and tyres size 5·00-5, pressure 1·93 bars (28 lb/sq in). Goodyear single-disc hydraulic brakes on main wheels.

POWER PLANT: One 227·5 kW (305 hp) Avco Lycoming IVO-540-A1A flat-six engine, mounted vertically, with dual cooling fans. One rubber-cell fuel tank under engine, capacity 163 litres (43 US gallons). Refuelling point in port side of fuselage. Oil capacity 9·5 litres (2·5 US gallons).

Accommodation: Two individual seats side by side with dual controls. Rear bench seat for three persons. Door on each side. Rear compartment for 113 kg (250 lb) of baggage, with downward-hinged door on starboard side.

Avionics and Equipment: King or Narco radio, to customer's specification. Blind-flying instrumentation is available, but helicopter is not certificated for instrument flight.

Dimensions, external:
Diameter of main rotor 8·74 m (28 ft 8 in)
Main rotor blade chord (constant) 0·254 m (10 in)
Diameter of tail rotor 1·30 m (4 ft 3 in)
Length overall, rotor turning 10·03 m (32 ft 11 in)
Length of fuselage 7·44 m (24 ft 5 in)
Height overall 2·44 m (8 ft 0⅛ in)
Wheel track 2·10 m (6 ft 10¾ in)
Wheelbase 2·15 m (7 ft 0½ in)
Passenger doors (each): Height 0·82 m (2 ft 8⅛ in)
Width 1·02 m (3 ft 3⅞ in)
Baggage compartment door:
Mean height 0·30 m (1 ft 0¼ in)
Width 0·69 m (2 ft 3 in)

Dimensions, internal:
Cabin: Length 2·30 m (7 ft 6½ in)
Max width 1·39 m (4 ft 6¾ in)
Max height 1·22 m (4 ft 0½ in)
Baggage compartment volume 0·47 m³ (16·7 cu ft)

Areas:
Main rotor blades (each) 1·09 m² (11·79 sq ft)
Tail rotor blades (each) 0·05 m² (0·50 sq ft)
Main rotor disc 3·33 m² (35·8 sq ft)
Tail rotor disc 1·32 m² (14·18 sq ft)

Weights and Loadings:
Weight empty 817 kg (1,800 lb)
Max T-O and landing weight 1,315 kg (2,900 lb)
Max zero-fuel weight 1,224 kg (2,700 lb)
Max disc loading 22·7 kg/m² (4·65 lb/sq ft)
Max power loading 5·78 kg/kW (9·84 lb/hp)

Performance (at max T-O weight):
Max level speed at S/L 104 knots (193 km/h; 120 mph)
Max cruising speed at S/L 96 knots (177 km/h; 110 mph)
Max rate of climb at S/L 297 m (975 ft)/min
Service ceiling 3,660 m (12,000 ft)
Hovering ceiling IGE 1,245 m (4,080 ft)
Range with max fuel and max payload, with 15 min reserves 191 nm (354 km; 220 miles)

CAC
COMMUTER AIRCRAFT CORPORATION

Head Office: 2223 Avenida de la Playa, La Jolla, California 92037
Telephone: (714) 454 0467/8
Chairman: Dr Lynn L. Bollinger
President: Kornel Feher
Vice-President, Marketing: Kenneth C. Gordon

Commuter Aircraft Corporation was formed to develop and manufacture a new 50/60-seat commuter airliner under the designation CAC-100. The project had its origins in the GAC-100 programme of the former General Aircraft Corporation, of which details can be found in the 1970-71 *Jane's*. The CAC-100 is, however, a slightly larger aircraft, embodying an improved power plant and the latest technology. Its production is supported by substantial loan guarantees by the US Department of Commerce and the Ohio State administration. Production will be centred in a new factory that was under construction in mid-1981 at Youngstown Municipal Airport, Ohio.

COMMUTER AIRCRAFT CORPORATION CAC-100

Commuter Aircraft Corporation announced at the 1981 Paris Air Show that rollout of the CAC-100 prototype was planned for early 1983, and that it was hoped to begin deliveries of production aircraft later in that year, following certification to FAR Pt 25 standards.

Type: Four-turboprop commuter/cargo transport.

Wings: Cantilever low-wing monoplane. Wing section NACA 23020 (modified) at root, NACA 23015·5 at tip. Incidence 3°. Dihedral 7° at 40% chord. Sweepback at quarter-chord 1° 30′. All-metal two-spar fail-safe structure of 2024 ST light alloy. Full-span double-slotted trailing-edge flaps of light alloy construction. Full-span hydraulically-actuated segmented spoilers on wing upper surface, immediately forward of flaps. Full-span aerodynamically-actuated leading-edge slats. Trim tab in each outboard flap section. Pneumatic de-icing boots on wing leading-edges.

Fuselage: All-metal semi-monocoque fail-safe structure of circular cross-section with stringers and skins of 2024 ST light alloy. Flight deck, cabin and aft baggage compartment pressurised.

Tail Unit: Cantilever all-metal structure of 2024 ST light alloy with small dorsal fin and swept vertical surfaces. Trim tabs in rudder and elevators; geared servo tab in rudder. Pneumatic de-icing boots for fin and tailplane leading-edges.

Landing Gear: Hydraulically-retractable tricycle type, with oleo-pneumatic shock-absorbers and twin wheels on each unit. All units retract forward. Main-wheel tyres size 29 × 11-10, pressure 3·65 bars (53 lb/sq in); nosewheel tyres size 18·75 × 7-6, pressure 2·90 bars (42 lb/sq in). Multiple disc brakes. Modulated anti-skid units.

Power Plant: Four 875 kW (1,173 shp) Pratt & Whitney Aircraft of Canada PT6A-45A turboprop engines, each driving a four-blade constant-speed reversible-pitch metal propeller with spinner. Fuel in four integral wing tanks with combined capacity of 4,353 litres (1,150 US gallons). Overwing gravity refuelling points, and single-point fuel/de-fuel manifold in starboard inner engine nacelle. Electrical de-icing boots for propellers; air intakes de-iced by engine bleed air.

Accommodation: Crew of two on flight deck, with dual controls as standard. Main cabin accommodates 50 passengers as standard at 81 cm (32 in) seat pitch, with seat for stewardess, coat compartment, toilet and galley at rear of cabin. Max accommodation for 60 passengers at 76 cm (30 in) seat pitch, with seat for stewardess, and toilet at rear of cabin. Type 1 door with built-in airstairs at rear of cabin on port side. Type 1 door at forward end of cabin, port side, optional on 50-seat version. Service door at rear of cabin on starboard side. Pressurised baggage compartment aft of cabin, with door on port side of fuselage.

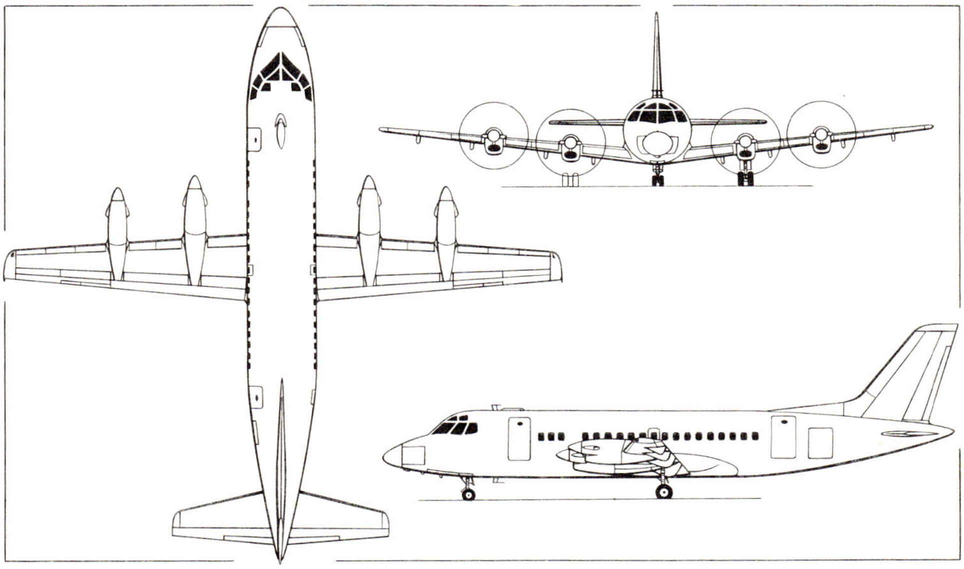

Commuter Aircraft Corporation CAC-100 four-turboprop commuter/cargo transport *(Pilot Press)*

Systems: Air-cycle pressurisation system using engine bleed air, max differential 0·45 bars (6·5 lb/sq in), to provide a sea level cabin altitude to 4,570 m (15,000 ft). Dual hydraulic systems at pressure of 207 bars (3,000 lb/sq in), powered by one electrically-driven and four engine-driven pumps. Pneumatic system, using engine bleed air, for flight instruments and de-icing boots. Electrical system includes four 300A 24-28V DC starter/generators, two 10kVA 120/208V 400Hz alternators driven by hydraulic motors, a standby solid state inverter, and two 24V 22Ah nickel-cadmium batteries. Oxygen system of 1·81 m³ (64 cu ft) capacity standard; 2·15 m³ (76 cu ft) system optional. APU optional.

Avionics and Equipment: Nav/com avionics to customer's requirements. Blind-flying instrumentation standard.

Dimensions, external:
Wing span 21·34 m (70 ft 0 in)
Wing chord at root 2·92 m (9 ft 7 in)
Wing chord at tip 1·09 m (3 ft 7 in)
Wing aspect ratio 10·6
Length overall 22·30 m (73 ft 2 in)
Fuselage basic diameter 2·82 m (9 ft 3 in)
Height overall 7·57 m (24 ft 10 in)
Tailplane span 8·38 m (27 ft 6 in)
Wheel track 6·71 m (22 ft 0 in)
Propeller diameter 2·74 m (9 ft 0 in)
Propeller ground clearance (inner) 0·46 m (1 ft 6 in)

Dimensions, internal:
Cabin, excl flight deck and aft baggage compartment:
Length 12·85 m (42 ft 2 in)
Max width 2·67 m (8 ft 9 in)
Max height 1·93 m (6 ft 4 in)
Volume 55·9 m³ (1,974 cu ft)
Aft baggage compartment:
Volume 6·80 m³ (240 cu ft)

Areas:
Wings, gross 42·83 m² (461 sq ft)
Vertical tail surfaces (total) 10·68 m² (115 sq ft)
Horizontal tail surfaces (total) 13·94 m² (150 sq ft)

Weights and Loadings (estimated):
*Operating weight empty 8,901 kg (19,623 lb)
Max payload 5,443 kg (12,000 lb)
Max fuel load 3,547 kg (7,820 lb)
Max T-O weight 15,422 kg (34,000 lb)
Max ramp weight 15,649 kg (34,500 lb)
*Max zero-fuel weight 13,437 kg (29,623 lb)
Max landing weight 14,969 kg (33,000 lb)
Max wing loading 361·3 kg/m² (74 lb/sq ft)
Max power loading 4·41 kg/kW (7·25 lb/shp)
* *50-passenger configuration*

Performance (estimated, at max T-O weight):
Max cruising speed at 4,570 m (15,000 ft) 300 knots (556 km/h; 345 mph)
T-O to 15 m (50 ft) FAR, ISA 1,066 m (3,500 ft)
Landing from 15 m (50 ft), at max landing weight, FAR, ISA 1,310 m (4,300 ft)
Range with max fuel, with reserves for 115 nm (213 km; 132 miles), plus 45 min 1,250 nm (2,316 km; 1,439 miles)
Range with max payload, with reserves for 115 nm (213 km; 132 miles), plus 45 min 625 nm (1,158 km; 720 miles)

CALSPAN
CALSPAN ADVANCED TECHNOLOGY CENTER

Head Office: PO Box 400, Buffalo, New York 14225
Telephone: (716) 632 7500
Telex: 91-270

Calspan Advanced Technology Center, a division of Arvin Industries' Applied Technology Group since 1978, carries out advanced research and development covering many technical disciplines. Its Flight Research Department has specialised in the design, construction and operation of variable-stability systems and research/simulation/training aircraft. These modified aircraft are used primarily to investigate flying qualities, control systems and displays, using in-flight simulation techniques. This is made possible by the installation of programmable computer systems to vary stability, control, guidance, and electronic displays.

The aircraft operated by Calspan in 1981 included:

X-22A. Since 1970, under contract to the US Naval Air Systems Command, Calspan has been operating the Bell/Navy X-22A V/STOL research aircraft, last described in the 1970-71 *Jane's*. In this continuing programme, research is focused on the evaluation of desirable V/STOL flight qualities and display formats. The X-22A is being used currently in a programme to investigate advanced control systems for helicopters which are required to land on the decks of ships in heavy seas.

NT-33A. From 1958, under contract to the US Air Force Flight Dynamics Laboratory, Calspan has been operating the Lockheed/US Air Force NT-33A for varied studies into flying qualities and training simulations. These programmes have investigated recently the effects of digital flight control systems on handling qualities, and the development of side stick controllers. The NT-33A has been used also to familiarise pilots with the characteristics of new aircraft, prior to their first flights in types such as the F-15, YF-16, YF-17 and F-18.

Bell/US Navy X-22A V/STOL research aircraft, in use for helicopter landing investigation

Lockheed/US Air Force NT-33A DEFT for investigation of display characteristics and handling qualities

Under a joint US Navy/Air Force programme, the NT-33A has been used since 1977 in a project known as Display Evaluation Flight Test (DEFT). The DEFT system, operational since September 1979, provides the necessary link between any new class of aircraft display/information hardware and software, and the pilot's needs. In this configuration the NT-33A is equipped with a programmable head-up display and associated sensors, in order to examine the interaction of display characteristics and handling qualities. Growth potential has been included for head-down vertical situation and energy manoeuvrability displays. The DEFT system can be used also to investigate terminal area problems, manoeuvring performance, energy management, ground attack, and air-to-air combat.

TIFS. Since 1969, under contract to the US Air Force Flight Dynamics Laboratory, Calspan has been operating this extensively modified Convair/US Air Force NC-131H. It is called the Total In-Flight Simulator (TIFS) as it has an independent control for each degree of freedom, with computer-controlled side force surfaces, direct-lift flaps, and throttle, in addition to similarly controlled ailerons, elevators and rudder. It is used primarily to investigate the flying qualities of larger aircraft, and is involved currently in programmes to study for NASA the Space Shuttle Orbiter's landing control system, and advanced supersonic cruise transport handling qualities. Other current tasks for the TIFS include study of aircraft motion effects on pilots and passengers, and gathering of data on human orientation perception in space. Installation of a multi-purpose digital computer in 1979 has expanded greatly the aircraft's simulation capabilities.

Learjet. For the past 20 years, Calspan has been operating two variable-stability Douglas B-26s for training and demonstration of aircraft stability and control at the US Air Force and Navy Test Pilot Schools. In 1979, Calspan was awarded a contract to modify a Learjet 24F to replace these B-26s. The aircraft, purchased by Calspan, was delivered to the company in December 1979. During 1980, installation of the Calspan-designed variable-stability system, including three-axis moment control, variable feel system, and digital computer configuration control system, was completed. First flight of the modified Learjet 24F was planned to take place during the first half of 1981, permitting operational use later in the year.

Convair NC-131H TIFS, studying currently the Space Shuttle Orbiter's landing control system

CESSNA

CESSNA AIRCRAFT COMPANY

Head Office and Works: Wichita, Kansas 67201
Telephone: (316) 685 9111

President, Chairman of the Board and Chief Executive Officer: Russell W. Meyer Jr

Senior Vice-Presidents:
R. P. Bauer (Finance)
William A. Boettger (Pawnee Division)
Charles B. Husick (Fluid Power and Aircraft Systems Divisions)
R. L. Lair (Commercial Aircraft Marketing)
Lloyd R. Leavitt Jr (Wallace Division)
J. Derek Vaughan (Commercial Jet Marketing Division)
William L. Worford (Personnel and Community Relations)

Vice-Presidents:
Pierre Clostermann (President, Reims Aviation)
John C. Dussault (McCauley Accessory Division)
Richard B. Foster (Aircraft Radio and Control Division)
L. C. Gartin (Commercial Aircraft Product Support)
Homer G. Nester (Controller)
Thaine L. Woolsey (Fluid Power Division)

Secretary: David R. Edwards

Cessna Aircraft Company was founded by the late Clyde V. Cessna, a pioneer in US aviation in 1911, and was incorporated on 7 September 1927. It has four plants in Wichita engaged on production of commercial and military aircraft, and the Fluid Power Division in Hutchinson, Kansas, which manufactures fluid power systems.

Subsidiary companies owned by Cessna are Aircraft Radio and Control Division at Boonton, New Jersey, the McCauley Accessories Division of Dayton, Ohio, Cessna Fluid Power Ltd of Glenrothes, Fife, Scotland, Cessna Finance Corporation and Cessna International Finance Corporation in Wichita. It has a 49 per cent interest in Reims Aviation of France.

By 1 January 1981 the company had produced a total of 166,344 aircraft. During 1980, its commercial sales totalled 6,393 aircraft, including units delivered by Reims Aviation of France (which see).

Military subcontract programmes include manufacture of crew door subassemblies for Bell helicopters.

Cessna Model 152 for 1981 (Avco Lycoming O-235-L2C engine)

CESSNA MODEL 152

During 1977 Cessna introduced a new two-seat cabin monoplane to replace the Model 150, which had been in production for almost 20 years. Designated Model 152, it differed primarily in having a more powerful engine using low-lead 100 octane fuel, an improved 'gull wing' propeller, and power plant installation and cowling changes to reduce engine noise and vibration.

Improvements introduced on the 1981 Model 152 include a spin-on oil filter for cleaner engine oil, improved battery contactors, main-wheel tyres of reduced frontal area and weight, and an avionics cooling system.

Four versions are available:

Model 152. Standard version, as described.

Model 152 II. As Model 152, but including as standard a factory installed avionics and equipment package which includes a Cessna Series 300 nav/com, dual controls, true airspeed indicator, navigation light detectors, heated pitot, courtesy lights, omni-flash beacon, alternate static source, and emergency locator transmitter.

Model 152 Trainer. Introduced in 1981 model range, with the most frequently ordered avionics and accessories installed as standard, at a package price. These include a cockpit intercom system with lightweight headset and 'hot' microphone, EGT indicator, all-purpose control wheel, oil quick-drain, flight hour recorder, refuelling steps and handles, tinted windows, dual controls, sun visors, Series 300 nav/com and transponder, basic avionics kit, attitude and directional gyros, outside air temperature indicator, vertical speed indicator, omni-flash beacon, cowl-mounted landing light, and variable-intensity light for radio stack.

Model 152 Aerobat. Described separately.

A total of 6,008 Model 152s had been built by 1 January 1981, including 413 F 152s built by Reims Aviation in France.

Type: Two-seat cabin monoplane.

Wings: Braced high-wing monoplane. Wing section NACA 2412 (tips symmetrical). Dihedral 1°. Incidence 1° at root, 0° at tip. All-metal structure of light alloy. Conical camber glassfibre wingtips optional. Modified Frise-type ailerons and electrically-actuated NACA single-slotted trailing-edge flaps of light alloy construction. No trim tabs.

Fuselage: Conventional semi-monocoque structure of light alloy.

Tail Unit: Cantilever structure of light alloy with swept vertical surfaces. Trim tab in starboard elevator. Ground-adjustable rudder tab.

Landing Gear: Non-retractable tricycle type. Land-O-Matic cantilever main legs, each comprising a one-piece machined conically-tapered spring steel tube. Steerable nosewheel on oleo-pneumatic shock-absorber strut. Main-wheel tyres size 15 × 6·00-6. Nosewheel size 5·00-5 with nylon tube-type tyre, pressure 2·07 bars (30 lb/sq in). Toe-operated single-disc hydraulic brakes. Rudder pedal extensions and wheel fairings optional.

Power Plant: One 82 kW (110 hp) Avco Lycoming O-235-L2C flat-four engine, driving a McCauley two-blade fixed-pitch metal propeller with spinner. Fuel tanks in wings, with total capacity of 98 litres (26 US gallons), of which 92·75 litres (24·5 US gallons) are usable. Optional long-range tanks have a total capacity of 147·5 litres (39 US gallons), of which 142 litres (37·5 US gallons) are usable. Refuelling points on upper surface of wing. Oil capacity 6·6 litres (1·75 US gallons).

Accommodation: Enclosed cabin seating two side by side. Vertically-adjustable seats for pilot and co-pilot; inertia-reel shoulder harness and dual controls optional on standard model. Baggage compartment behind seats, backs of which hinge forward. Baggage capacity 54 kg (120 lb). Optional 'family seat' can be fitted in baggage space, for two children not exceeding 54 kg (120 lb) in combined weight. Door, with opening window, on each side. Heating and ventilation standard. Dual windscreen defrosters standard. Cabin skylight windows optional.

Systems: Hydraulic system for brakes only. Electrical system includes a 28V 60A alternator and 28V battery.

Avionics and Equipment: Optional avionics include Cessna Series 300 nav/com (standard on 152 II and Trainer), Series 300 transceiver, Series 300 nav/com with remote VOR/LOC or VOR/ILS indicator, Series 300 ADF, marker beacon with three lights and aural signal, transponder with 4096 code capability (standard on Trainer), single unit avionics control panel, slimline microphone, intercom, padded headset with attached microphone and control wheel operating button, and a Telex 5X5 Mk IIA headset with distortion-free miniaturised microphone and adjustable headband. Standard equipment includes safety belts, map compartment, control locks, stall warning device, baggage retaining net, cabin dome lights, variable-intensity instrument panel red floodlights, navigation lights and spin-on oil filter. Optional equipment (standard on 152 II) includes true airspeed indicator, dual controls, alternate static source, emergency locator transmitter, heated pitot, courtesy lights, navigation light detectors, and omni-flash beacon. Other optional equipment includes artificial horizon, directional gyro, electric clock, economy mixture indicator, exhaust gas temperature indicator, outside air temperature gauge, flight hour recorder, rate of climb indicator, sensitive altimeter, turn and bank indicator, turn co-ordinator, cabin fire extinguisher, tinted windows, sun visors, control wheel map light, instrument post lights, cowl-mounted landing light, white strobe lights, anti-precipitation static kit, engine winterisation kit, external power socket, refuelling steps and handles, full-flow oil filter, and quick-drain oil valve.

Dimensions, external:

Wing span: standard	9·97 m (32 ft 8½ in)
with optional conical wingtips	10·11 m (33 ft 2 in)
Wing chord at root	1·63 m (5 ft 4 in)
Wing chord at tip	1·13 m (3 ft 8½ in)
Wing aspect ratio	6·7
Length overall	7·34 m (24 ft 1 in)
Height overall	2·59 m (8 ft 6 in)
Tailplane span	3·05 m (10 ft 0 in)
Wheel track	2·32 m (7 ft 7¼ in)
Wheelbase	1·47 m (4 ft 10 in)
Propeller diameter	1·75 m (5 ft 9 in)
Passenger doors (each): Width	0·86 m (2 ft 10 in)

Areas:

Wings, gross: standard	14·59 m² (157·0 sq ft)
with optional conical wingtips	14·82 m² (159·5 sq ft)

Weights and Loadings (A: standard 152; B: 152 II):

Weight empty: A	502 kg (1,107 lb)
B	518 kg (1,141 lb)
Max T-O and landing weight: A, B	757 kg (1,670 lb)
Max ramp weight: A, B	760 kg (1,675 lb)
Max wing loading	51·3 kg/m² (10·5 lb/sq ft)
Max power loading	9·24 kg/kW (15·2 lb/hp)

Performance (at max T-O weight, ISA):

*Max level speed at S/L
110 knots (204 km/h; 127 mph)

*Max cruising speed, 75% power at 2,440 m (8,000 ft)
107 knots (198 km/h; 123 mph)

Stalling speed, flaps up, power off
48 knots (89·5 km/h; 55·5 mph) CAS

Stalling speed, flaps down, power off
43 knots (80·5 km/h; 50 mph) CAS

Max rate of climb at S/L	218 m (715 ft)/min
Service ceiling	4,480 m (14,700 ft)
T-O run	221 m (725 ft)
T-O to 15 m (50 ft)	408 m (1,340 ft)
Landing from 15 m (50 ft)	366 m (1,200 ft)
Landing run	145 m (475 ft)

Range, recommended lean mixture, with allowance for start, taxi, T-O, climb and 45 min reserves at 75% power:

standard fuel, 75% power at 2,440 m (8,000 ft)
320 nm (592 km; 368 miles)

max fuel, 75% power at 2,440 m (8,000 ft)
545 nm (1,009 km; 627 miles)

Range, allowances as above, but with 45 min reserves at 45% power:

standard fuel, econ cruising power at 3,050 m (10,000 ft) 415 nm (769 km; 478 miles)

max fuel, econ cruising power at 3,050 m (10,000 ft)
690 nm (1,278 km; 794 miles)

**With wheel fairings which increase speeds by approximately 2 knots (3·7 km/h; 2·3 mph)*

CESSNA MODEL 152 AEROBAT

The Model 152 Aerobat combines the economy and versatility of the standard Model 152 with aerobatic capability. Structural changes allow the Aerobat to perform 'unusual attitude' manoeuvres and it is licensed in the Aerobatic category for load factors of +6*g* and —3*g* at full gross weight, permitting the performance of barrel and aileron rolls, snap rolls, Immelmann turns, Cuban eights, spins, vertical reversements, lazy eights and chandelles.

Equipment of the aircraft differs only slightly from that of the standard 152. Quick-release cabin doors, removable seat cushions and backs, quick-release lapstraps, and shoulder harnesses, are standard, as are two tinted skylights in the cabin roof which offer extra field of view. Distinct external styling provides immediate recognition of the Aerobat's role. The improvements for 1981 and the 152 II package detailed for the Model 152 apply also to the Aerobat.

Dimensions and Areas: As for Model 152

Weights and Loadings: As for Model 152 except:

Weight empty	512 kg (1,129 lb)

Performance (at max T-O weight, ISA): As for Model 152 except:

*Max level speed at S/L
109 knots (202 km/h; 126 mph)

*Max cruising speed, 75% power at 2,440 m (8,000 ft)
106 knots (196 km/h; 122 mph)

Range, recommended lean mixture, with allowance for engine start, taxi, T-O, climb and 45 min reserves at 75% power:

standard fuel, 75% power at 2,440 m (8,000 ft)
315 nm (583 km; 362 miles)

max fuel, 75% power at 2,440 m (8,000 ft)
540 nm (1,000 km; 621 miles)

Range, allowances as above, but with 45 min reserves at 45% power:

standard fuel, econ cruising power at 3,050 m (10,000 ft) 410 nm (760 km; 472 miles)

max fuel, econ cruising power at 3,050 m (10,000 ft)
680 nm (1,260 km; 783 miles)

**With wheel speed fairings which increase speeds by approximately 2 knots (3·7 km/h; 2·3 mph)*

CESSNA SKYHAWK

Two versions of the Skyhawk are currently available:

Skyhawk. This improved version of the Skyhawk, powered by a 119 kW (160 hp) engine able to operate on 100 octane low-lead fuel, was introduced in 1977. Improvements for 1981 include an increase of 40 kg (89 lb) in useful load; a new engine which offers improved performance; a modified elevator with rounded leading-edge which improves handling; addition of an avionics cooling fan which eliminates an external airscoop; a changed max flap setting for better balked landing performance; larger battery contactors; a new anti-precipitation static nav antenna which suppresses undesired signals; availability of an extra-long-range integral wing fuel tank system; and optional intercom.

Skyhawk II. As Skyhawk, but including as standard a 300 Series nav/com with 360-channel com and 160-channel nav, dual controls, true airspeed indicator, navigation light detectors, heated pitot, courtesy lights, omni-flash beacon, alternate static source and emergency locator transmitter. It can be equipped optionally with a Nav Pac which adds a second nav/com, a 300 Series ADF and a transponder.

The Skyhawk is certificated for operation as a floatplane, and can be fitted with skis. A version designated F 172 is produced in France by Reims Aviation.

A total of 33,588 commercial aircraft in the Model 172/Skyhawk series had been built by 1 January 1981, including 1,940 F 172s built in France. In addition, 237 were built in 1966-67 as T-41A Mescalero military basic trainers.

Type: Four-seat cabin monoplane.

Wings: Braced high-wing monoplane. NACA 2412 wing section. Dihedral 1° 44'. Incidence 1° 30' at root, −1° 30' at tip. All-metal structure, except for conical-camber glassfibre wingtips. Single bracing strut on each side. Modified Frise all-metal ailerons. Electrically-controlled NACA all-metal single-slotted flaps inboard of ailerons.

Fuselage: All-metal semi-monocoque structure.

Tail Unit: Cantilever all-metal structure. Sweepback on fin 35° at quarter-chord. Trim tab in starboard elevator. Ground-adjustable tab in rudder, in-flight adjustable trim tab optional.

Landing Gear: Non-retractable tricycle type. Cessna Land-O-Matic cantilever main legs, each comprising a one-piece machined conically-tapered spring steel tube. Nosewheel is carried on an oleo-pneumatic shock-strut and is steerable with rudder up to 10° and controllable up to 30° on either side. Cessna main wheels size 6·00-6 and nosewheel size 5·00-5 (optionally 6·00-6), with nylon cord tube-type tyres. Tyre pressure: main wheels 1·59 bars (23 lb/sq in), nosewheel 1·79 bars (26 lb/sq in). Hydraulic disc brakes. Optional wheel fairings. Alternative float and ski gear.

Power Plant: One 119 kW (160 hp) Avco Lycoming O-320-D2J flat-four engine, driving a two-blade fixed-pitch metal propeller. One fuel tank in each wing, total capacity 163 litres (43 US gallons). Usable fuel 151·4 litres (40 US gallons). Provision for long-range tanks, giving total capacity of 204 litres (54 US gallons), of which 189 litres (50 US gallons) are usable; or extra-long-range system, using integral tanks in wings to provide total capacity of 257 litres (68 US gallons), of which 234 litres (62 US gallons) are usable. Oil capacity 7·5 litres (2 US gallons). Oil cooler and full-flow oil filter standard.

Accommodation: Cabin seats four in two pairs, with optional fully-articulating front seats. Baggage space aft of rear seats, capacity 54 kg (120 lb). An optional fold-away seat can be fitted in baggage space, for one or two children not exceeding 54 kg (120 lb) total weight. Door on each side of cabin, giving access to all seats, simplifies loading if rear seats are removed and cabin used for freight. Pilot's window opens; co-pilot's opening side window and dual controls optional on Skyhawk; dual controls standard on Skyhawk II. Baggage door on port side. Combined heating and ventilation system; air-conditioning optional. Dual windscreen defrosters. Glassfibre soundproofing. Optional overhead skylights.

Systems: Electrical system of 28V includes electric engine starter and 24V battery. Air-conditioning system of

Cessna Skyhawk II four-seat lightplane for 1981 (Avco Lycoming O-320-D2J engine)

14,000 BTU capacity optional. Vacuum system for blind-flying instruments optional.

Avionics and Equipment: Optional avionics include Cessna Series 300 720-channel transceiver, 720-channel nav/com with remote VOR indicator, 720-channel nav/com with remote VOR/LOC indicator or VOR/ILS indicator, ADF, marker beacon receiver with three lights and aural signal, transponder with 4096 code capability, DME, 10-channel HF transceiver, Nav-O-Matic autopilot with heading control plus VOR, Series 400 glideslope receiver, intercom, boom microphone, with control-wheel switch, padded headset with boom microphone, a Telex 5X5 Mk IIA headset with distortion-free miniaturised microphone and adjustable headband, and speaker sidetone facility. Optional equipment (standard on Skyhawk II) includes true airspeed indicator, alternate static source, emergency locator transmitter, heated pitot, navigation light detectors, courtesy lights, and omni-flash beacon. Other optional equipment includes artificial horizon, directional gyro with movable heading index, carburettor air temperature gauge, electric clock, exhaust gas and outside air temperature gauges, flight hour recorder, rate of climb indicator, sensitive altimeter, turn and bank indicator, turn co-ordinator, reclining and vertically adjustable front seats, inertia reel shoulder harness, headrests, overhead skylights, rudder pedal extensions, starboard side storm window, sun visors, child's foldaway seat, fire extinguisher, portable stretcher, rear seats with individual reclining backs, safety belts for third and fourth seats, internal corrosion proofing, rear seat ventilation system, tinted windows, utility shelf, ventilation fan, control wheel map light, map and instrument panel lights, dual cowl-mounted landing lights, beacon, wingtip strobe lights, anti-precipitation static kit, engine priming system, floatplane kit, full-flow oil filter, glider tow hook, external power socket, quick-drain oil valve, tailplane abrasion boots, towbar, wing-strut and fuselage steps and handles for refuelling, and winterisation kit.

Dimensions, external (L: landplane; F: floatplane):
Wing span 10·92 m (35 ft 10 in)
Wing chord at root 1·63 m (5 ft 4 in)
Wing chord at tip 1·12 m (3 ft 8½ in)
Wing aspect ratio 7·52
Length overall: L 8·20 m (26 ft 11 in)
F 8·13 m (26 ft 8 in)
Height overall: L 2·68 m (8 ft 9½ in)
F 3·63 m (11 ft 11 in)
Tailplane span 3·45 m (11 ft 4 in)
Wheel track: L 2·53 m (8 ft 3½ in)
Wheelbase: L 1·63 m (5 ft 4 in)
Propeller diameter: L 1·91 m (6 ft 3 in)
F 2·03 m (6 ft 8 in)
Passenger doors (each): Height 1·01 m (3 ft 3¾ in)
Width 0·89 m (2 ft 11 in)

Areas:
Wings, gross 16·17 m² (174 sq ft)
Ailerons (total) 1·70 m² (18·3 sq ft)
Trailing-edge flaps (total) 1·97 m² (21·20 sq ft)
Fin 1·04 m² (11·24 sq ft)
Rudder, incl tab 0·69 m² (7·43 sq ft)
Tailplane 2·00 m² (21·56 sq ft)
Elevators, incl tab 1·35 m² (14·53 sq ft)

Weights and Loadings (Skyhawk landplane: L; floatplane: F):
Weight empty, equipped: L 642 kg (1,414 lb)
F 723 kg (1,593 lb)
Skyhawk II L 653 kg (1,440 lb)
Skyhawk II F 735 kg (1,619 lb)
Max T-O and landing weight: L 1,089 kg (2,400 lb)
F 1,007 kg (2,220 lb)
Max ramp weight: L 1,092 kg (2,407 lb)
F 1,010 kg (2,227 lb)
Max wing loading: L 67·3 kg/m² (13·8 lb/sq ft)
F 62·0 kg/m² (12·7 lb/sq ft)
Max power loading: L 9·15 kg/kW (15·0 lb/hp)
F 8·44 kg/kW (13·9 lb/hp)

Performance (L: Skyhawk and Skyhawk II landplane; F: floatplane, at max T-O weight, ISA):
Never-exceed speed:
L 151 knots (280 km/h; 174 mph)
Max level speed at S/L:
L 123 knots (228 km/h; 141 mph)
F 96 knots (178 km/h; 111 mph)
Max cruising speed (75% power):
L at 2,440 m (8,000 ft)
120 knots (222 km/h; 138 mph)
F at 1,220 m (4,000 ft)
95 knots (176 km/h; 109 mph)
Stalling speed, flaps up:
L 51 knots (95 km/h; 59 mph) CAS
F 48 knots (89 km/h; 55 mph) CAS
Stalling speed, flaps down:
L 46 knots (85 km/h; 53 mph)
F 44 knots (82 km/h; 51 mph) CAS
Max rate of climb at S/L: L 213 m (700 ft)/min
F 226 m (740 ft)/min
Service ceiling: L 3,960 m (13,000 ft)
F 4,570 m (15,000 ft)

Cessna Model R172 Hawk XP II, a version of the four-seat Skyhawk with more powerful engine

T-O run: L 272 m (890 ft)
F 427 m (1,400 ft)
T-O to 15 m (50 ft): L 556 m (1,825 ft)
F 658 m (2,160 ft)
Landing from 15 m (50 ft): L 390 m (1,280 ft)
F 410 m (1,345 ft)
Landing run: L 165 m (540 ft)
F 180 m (590 ft)
Range, at recommended lean mixture, with allowances for engine start, taxi, T-O, climb and 45 min reserves at 75% power:
max cruising speed:
L, standard fuel 440 nm (815 km; 506 miles)
F, standard fuel 360 nm (666 km; 414 miles)
L, 62 US gallons fuel
755 nm (1,398 km; 868 miles)
F, 50 US gallons fuel
475 nm (879 km; 546 miles)
Range, allowances as above, but with 45 min reserves at 45% power:
econ cruising speed at 3,050 m (10,000 ft):
L, standard fuel 520 nm (963 km; 598 miles)
F, standard fuel 435 nm (806 km; 501 miles)
L, 62 US gallons fuel
875 nm (1,620 km; 1,007 miles)
F, 50 US gallons fuel
565 nm (1,046 km; 650 miles)

CESSNA MODEL R172 HAWK XP

On 23 June 1976, Cessna introduced the Model R172 Hawk XP, and a similar model is being produced by Reims Aviation in France (which see), known as the Model FR 172/Reims Hawk XP. A total of 1,484 had been sold by 1 January 1981, this total including 75 Reims-built FR 172s. Eighteen Hawk XPs were supplied to the Chilean Army in 1978. Production ended in 1981.

Two versions were available:

Hawk XP. Standard version, powered by one 145·4 kW (195 hp) Continental IO-360-K engine. Built also as floatplane.

Hawk XP II. Version of the above equipped in a VFR and IFR configuration. It could be equipped optionally with Nav Pac which added a second nav/com system and other avionics.

The 1981 versions of the Hawk XP introduced as standard a modified elevator with rounded leading-edge which improved handling; a changed max flap setting for better balked landing performance; larger battery contactors; and a new anti-precipitation static nav antenna. A fan-operated avionics cooling system was available optionally.

The description of the Skyhawk applies also to the Hawk XP, except as follows:

Power Plant: One 145·4 kW (195 hp) Continental IO-360-KB flat-six fuel-injection engine, driving a McCauley type ZA34C203/90DCA-14 two-blade constant-speed metal propeller with spinner. Two fuel tanks in wings with a combined capacity of 197 litres (52 US gallons), of which 185 litres (49 US gallons) are usable. Optional integral fuel cells provide a maximum usable capacity of 250 litres (66 US gallons). To permit increased cabin loads on short-range flights, the long-range fuel cells can be filled to the bottom of a standpipe in the filler collar to provide 189·25 litres (50 US gallons) of usable fuel.

Accommodation: As for Skyhawk, except that Hawk XP has restyled sidewalls and overhead console to improve appearance. Overhead console change provides better access to overhead floodlighting. New control yokes and addition of control for constant-speed propeller. Baggage space has capacity of 91 kg (200 lb).

Dimensions, external (L: landplane; F: floatplane):
As for Skyhawk except:
Length overall: L 8·28 m (27 ft 2 in)
F 8·18 m (26 ft 10 in)
Height overall: F 3·78 m (12 ft 5 in)
Propeller diameter: L 1·93 m (6 ft 4 in)

Weights and Loadings (L: Hawk XP landplane; F: floatplane):
Weight empty: L 701 kg (1,546 lb)
F 820 kg (1,808 lb)
Hawk XP II: L 713 kg (1,572 lb)
F 832 kg (1,834 lb)
Max T-O and landing weight 1,157 kg (2,550 lb)
Max ramp weight 1,160 kg (2,558 lb)
Max wing loading 71·8 kg/m² (14·7 lb/sq ft)
Max power loading 7·96 kg/kW (13·08 lb/hp)

Performance (at max T-O weight, ISA):
Max level speed at S/L:
L 133 knots (246 km/h; 153 mph)
F 118 knots (219 km/h; 136 mph)
Max cruising speed (80% power at 1,830 m; 6,000 ft):
L 130 knots (241 km/h; 150 mph)
F 116 knots (215 km/h; 134 mph)
Stalling speed, flaps up, power off:
L 52 knots (97 km/h; 60 mph) CAS
F 50 knots (93 km/h; 58 mph) CAS
Stalling speed, flaps down, power off:
L 47 knots (87 km/h; 54 mph) CAS
F 44 knots (82 km/h; 51 mph) CAS
Max rate of climb at S/L:
L, F 265 m (870 ft)/min
Service ceiling: L 5,180 m (17,000 ft)
F 4,725 m (15,500 ft)
T-O run: L 244 m (800 ft)
F 346 m (1,135 ft)
T-O to 15 m (50 ft): L 415 m (1,360 ft)
F 564 m (1,850 ft)
Landing from 15 m (50 ft): L 410 m (1,345 ft)
F 424 m (1,390 ft)
Landing run: L 194 m (635 ft)
F 206 m (675 ft)
Range (recommended lean mixture, allowances for engine start, taxi, T-O, climb, and 45 min reserves at 45% power):
80% power at 1,830 m (6,000 ft) with 185 litres (49 US gallons) usable fuel:
L 440 nm (815 km; 506 miles)
F 395 nm (731 km; 454 miles)
80% power at 1,830 m (6,000 ft) with 250 litres (66 US gallons) usable fuel:
L 635 nm (1,176 km; 730 miles)
F 570 nm (1,055 km; 656 miles)
econ cruising power at 3,050 m (10,000 ft) with 185 litres (49 US gallons) usable fuel:
L 575 nm (1,065 km; 662 miles)
F 495 nm (916 km; 569 miles)
econ cruising power at 3,050 m (10,000 ft) with 250 litres (66 US gallons) usable fuel:
L 815 nm (1,510 km; 938 miles)
F 705 nm (1,305 km; 811 miles)

CESSNA CUTLASS RG

Cessna announced on 15 August 1979 the introduction of a new four-seat lightplane with retractable landing gear which it claimed to be the lowest-priced aircraft in this category. Known as the Cutlass RG, it combines the airframe of the Model 172 Skyhawk with the retractable landing gear developed for the Skylane RG. Power is provided by a 134 kW (180 hp) Avco Lycoming O-360-F1A6 flat-four engine, driving a two-blade constant-speed metal propeller. Two fuel tanks, one in each wing, have a combined usable capacity of 235 litres (62 US gallons).

The Cutlass RG became available in late 1979, together with a Cutlass RG II with avionics and equipment as detailed for the Skyhawk II, plus the optional Nav Pac. Equivalent versions designated F Cutlass RG and F Cutlass RG II were assembled under licence by Reims Aviation in France. A total of 671 had been built by 1 January 1981.

Improvements introduced as standard on the 1981 version of the Cutlass RG include an avionics cooling fan, a new oil pressure gauge, an improved anti-precipitation static navigation antenna, an alternator circuit breaker to protect the avionics, audio panel with marker beacon receiver incorporating a dimming switch and 30 s automatic-muting, separate nav and com audio switches to assist identification of localiser or VOR, a single avionics power switch, and fuel gauges that read in pounds and gallons. New options include an intercom system, and a Telex 5X5 Mk IIA headset and microphone.

Dimensions, external: As for Skyhawk except:
- Length overall 8·36 m (27 ft 5 in)
- Tailplane span 3·43 m (11 ft 3 in)
- Wheel track 2·59 m (8 ft 6 in)
- Propeller diameter 1·94 m (6 ft 4½ in)

Weights and Loadings (A: Cutlass RG; B: Cutlass RG II):
- Weight empty: A 706 kg (1,555 lb)
- B 721 kg (1,590 lb)
- Max T-O and landing weight: A, B 1,202 kg (2,650 lb)
- Max ramp weight: A, B 1,206 kg (2,658 lb)
- Max wing loading: A, B 74·4 kg/m² (15·2 lb/sq ft)
- Max power loading: A, B 8·97 kg/kW (14·7 lb/hp)

Performance (at max T-O weight, ISA):
- Max level speed at S/L 145 knots (269 km/h; 167 mph)
- Max cruising speed (75% power) at 2,740 m (9,000 ft) 140 knots (259 km/h; 161 mph)
- Stalling speed, flaps up, power off 54 knots (100 km/h; 62 mph) CAS
- Stalling speed, flaps down, power off 50 knots (93 km/h; 58 mph) CAS
- Max rate of climb at S/L 244 m (800 ft)/min
- Service ceiling 5,120 m (16,800 ft)
- T-O run 323 m (1,060 ft)
- T-O to 15 m (50 ft) 541 m (1,775 ft)
- Landing from 15 m (50 ft) 408 m (1,340 ft)
- Landing run 191 m (625 ft)
- Range with max fuel at max cruising speed, recommended lean mixture, with allowances for engine start, taxi, T-O, climb and 45 min reserves 720 nm (1,334 km; 829 miles)
- Range with max fuel at econ cruising speed, allowances as above 840 nm (1,557 km; 967 miles)

CESSNA MODEL 180 SKYWAGON

The Model 180 Skywagon has a typical Cessna braced high-wing monoplane layout, but with a tailwheel type of landing gear. Two commercial versions are available:

Model 180 Skywagon. Basic model, as described, which introduced as standard improvements for 1981 a new anti-precipitation static navigation antenna when factory-installed radios are ordered, and an avionics cooling fan. New options include an intercom system, a lightweight headset, and DME and R/Nav equipment which together provide an IFR-certificated area navigation system.

Model 180 Skywagon II. As Skywagon, plus factory-installed avionics package which includes Series 300 nav/com with remote VOR indicator, ADF and an emergency locator transmitter. Optional Nav Pac provides a second nav/com, with Series 400 glideslope and marker beacon receiver. Standard equipment is as Skywagon, plus long-range fuel tanks with a total usable capacity of 284 litres (75 US gallons), artificial horizon, clock, directional gyro, exhaust gas temperature gauge, outside air temperature gauge, turn co-ordinator, vertical speed indicator, dual controls, sun visor, heated pitot, map light, quick-drain oil plug, and external power socket.

A total of 6,161 Model 180s had been built by 1 January 1981. Production is ending after delivery of the 1981 quota.

Type: One/six-seat cabin monoplane.

Wings: Generally similar in construction to those of Skyhawk. Dihedral 1° 44′.

Fuselage: All-metal semi-monocoque structure. Identical to fuselage of Cessna 185, except for firewall and mounting brackets for dorsal fin.

Tail Unit: Unswept cantilever all-metal structure with adjustable-incidence tailplane. Normally no trim tabs; manually-operated rudder trim available optionally.

Landing Gear: Non-retractable tailwheel type. Cessna cantilever spring steel main legs. Tailwheel has tapered tubular spring. Main wheels and nylon tube-type tyres size 6·00-6 (optionally 8·00-6). Cessna tailwheel size 8·00 × 2·80. Tyre pressure, main wheels 2·07 bars (30 lb/sq in), tailwheel 3·79-4·48 bars (55-65 lb/sq in) according to load. Hydraulic disc brakes. Parking brake. Wheel and brake fairings optional. Alternative Edo Model 628-296 floats, snow skis or amphibious gear.

Power Plant: One 171·5 kW (230 hp) Continental O-470-U flat-six engine, driving a McCauley 2A34C203/90DA-8 two-blade constant-speed metal propeller. Two fuel tanks in wings, with total standard capacity of 333 litres (88 US gallons), of which 318 litres (84 US gallons) are usable. Oil capacity 11·5 litres (3 US gallons).

Accommodation: Standard seating is for a pilot only, with a choice of three optional arrangements. Maximum seating is for six persons in three pairs, without baggage space. With fewer seats there is space at rear of cabin for up to 181 kg (400 lb) of baggage. Door on each side of

Cessna Cutlass RG four-seat lightplane with retractable landing gear

cabin, plus optional cargo door and baggage compartment door on port side. Starboard door has quick-release hinge pins so that it can be removed when loading bulky cargo. Fifth and sixth passenger seats, attached to aft wall of cabin, can be folded when space is required for cargo. Hinged window each side. Instrument lighting controls are transistorised. Heating, ventilation, and windscreen defroster standard. Fully-articulating seats for pilot and co-pilot, child's foldaway seat for the rear cabin and safety belts for rear-seat passengers are available optionally. Dual controls optional (standard on Skywagon II).

Systems: Hydraulic system for brakes only. Electrical system powered by 28V 60A alternator. 24V 33Ah battery. Oxygen system, 1·36 m³ (48 cu ft) capacity, optional.

Avionics and Equipment: Optional avionics include Cessna Series 300 360-channel com transceiver, 360-channel nav/com with 160-channel nav and remote VOR indicator, 720-channel com and 200-channel nav with remote VOR/LOC indicator or VOR/ILS indicator, ADF with digital tuning, marker beacon receiver with three lights and aural signal, transponder with 4096 code capability, Collins R/Nav and DME-450C Nav-O-Matic single-axis autopilot with heading control and VOR intercept and track; or Series 400 720-channel com transceiver, 720-channel nav/com with remote VOR/LOC or VOR/ILS indicator, transponder with 4096 code capability, glideslope receiver and ADF with digital tuning. Standard avionics for Skywagon II include Cessna 300 Series nav/com with remote VOR indicator, ADF and an emergency locator transmitter. Standard equipment includes control locks, audible stall warning, baggage restraint net, cabin dome light, instrument panel red floodlights, and landing and taxi lights. Optional equipment (standard on Skywagon II) includes long-range fuel tanks, artificial horizon, clock, directional gyro, exhaust gas temperature gauge, outside air temperature gauge, turn co-ordinator, vertical speed indicator, dual controls, sun visor, heated pitot, map light, quick-drain oil valve and external power socket. Other optional equipment includes carburettor air temperature gauge, economy mixture indicator, flight hour recorder, true airspeed indicator, turn and bank indicator, co-pilot's seat installation, pilot and co-pilot headrests, rudder pedal extensions, stowable starboard side rudder pedals, inertia reel shoulder harness, twin beverage pack, tailwheel lock, alternate static source, emergency locator transmitter, navigation light detectors, cargo tiedown fittings, bubble and tinted windows, de luxe interior, internal corrosion proofing, cabin fire extinguisher, photographic provisions, stretcher installation, courtesy lights, boom microphone, control wheel with map light and microphone switch, auxiliary instrument lights, instrument panel post lights, omni-flash beacon, strobe light, amphibious kit, floatplane kit, ski kit, engine winterisation kit, agricultural sprayer system, non-congealing oil cooler, oil dilution system, overall paint scheme, and tailplane abrasion boots.

Dimensions, external:
- Wing span 10·92 m (35 ft 10 in)
- Wing chord at root 1·63 m (5 ft 4 in)
- Wing chord at tip 1·12 m (3 ft 8 in)
- Wing aspect ratio 7·52
- Length overall: Landplane 7·81 m (25 ft 7½ in)
  - Skiplane 8·47 m (27 ft 9½ in)
  - Floatplane 8·23 m (27 ft 0 in)
  - Amphibian 8·38 m (27 ft 6 in)
- Height overall:
  - Landplane, skiplane 2·36 m (7 ft 9 in)
  - Floatplane 3·71 m (12 ft 2 in)
  - Amphibian 3·86 m (12 ft 8 in)
- Tailplane span 3·35 m (11 ft 0 in)
- Wheel track: Landplane 2·26 m (7 ft 5 in)
- Propeller diameter:
  - Landplane, skiplane 2·08 m (6 ft 10 in)
  - Floatplane, amphibian 2·29 m (7 ft 6 in)
- Passenger doors (each): Height 1·01 m (3 ft 3¾ in)
  - Width 0·89 m (2 ft 11 in)

Areas:
- Wings, gross 16·16 m² (174 sq ft)
- Ailerons (total) 1·70 m² (18·3 sq ft)
- Trailing-edge flaps (total) 1·97 m² (21·23 sq ft)
- Fin 0·84 m² (9·01 sq ft)
- Dorsal fin 0·19 m² (2·04 sq ft)
- Rudder 0·68 m² (7·29 sq ft)
- Tailplane 1·94 m² (20·94 sq ft)
- Elevators 1·40 m² (15·13 sq ft)

Weights and Loadings:
- Weight empty, equipped:
  - Skywagon landplane 749 kg (1,650 lb)
  - Skywagon II landplane 772 kg (1,701 lb)
  - Floatplane 888 kg (1,958 lb)
  - Skiplane 813 kg (1,792 lb)
  - Amphibian 1,004 kg (2,213 lb)
- Max T-O weight:
  - Landplane, skiplane 1,270 kg (2,800 lb)
  - Floatplane, amphibian 1,338 kg (2,950 lb)
- Max ramp weight:
  - Landplane, skiplane 1,275 kg (2,810 lb)
  - Floatplane, amphibian 1,343 kg (2,960 lb)
- Max wing loading:
  - Landplane, skiplane 78·6 kg/m² (16·1 lb/sq ft)
  - Floatplane, amphibian 83·0 kg/m² (17·0 lb/sq ft)
- Max power loading:
  - Landplane, skiplane 7·41 kg/kW (12·2 lb/hp)
  - Floatplane, amphibian 7·80 kg/kW (12·8 lb/hp)

Performance (at max T-O weight, ISA):
- Never-exceed speed:
  - Landplane 167 knots (309 km/h; 192 mph)
- Max level speed at S/L:
  - *Landplane 148 knots (274 km/h; 170 mph)
  - Floatplane, amphibian, skiplane 129 knots (240 km/h; 149 mph)
- Max cruising speed (75% power) at 2,440 m (8,000 ft):
  - *Landplane 142 knots (264 km/h; 164 mph)

**These speeds are 1 knot (1·9 km/h; 1·2 mph) higher with optional speed fairings installed*

Cessna Model 180 Skywagon one/six-seat cabin monoplane (Continental O-470-U engine)

Floatplane, amphibian
123 knots (228 km/h; 142 mph)
Skiplane 124 knots (230 km/h; 143 mph)
Econ cruising speed at 3,050 m (10,000 ft):
Landplane 105 knots (195 km/h; 121 mph)
Floatplane, amphibian
99 knots (183 km/h; 114 mph)
Skiplane 88 knots (162 km/h; 101 mph)
Stalling speed, flaps up, power off:
all versions 53 knots (98·5 km/h; 61 mph) CAS
Stalling speed, flaps down, power off:
all versions 48 knots (88·5 km/h; 55 mph) CAS
Max rate of climb at S/L:
Landplane 335 m (1,100 ft)/min
Floatplane, amphibian 296 m (970 ft)/min
Skiplane 277 m (910 ft)/min
Service ceiling: Landplane 5,395 m (17,700 ft)
Floatplane, amphibian 4,663 m (15,300 ft)
Skiplane 4,480 m (14,700 ft)
T-O run: Landplane 190 m (625 ft)
Floatplane 354 m (1,160 ft)
Amphibian, on land 213 m (700 ft)
Amphibian, on water 354 m (1,160 ft)
T-O to 15 m (50 ft): Landplane 367 m (1,205 ft)
Floatplane 579 m (1,900 ft)
Amphibian, on land 401 m (1,315 ft)
Amphibian, on water 579 m (1,900 ft)
Landing from 15 m (50 ft):
Landplane 416 m (1,365 ft)
Floatplane 524 m (1,720 ft)
Amphibian, on land 442 m (1,450 ft)
Amphibian, on water 524 m (1,720 ft)
Landing run: Landplane 146 m (480 ft)
Floatplane 224 m (735 ft)
Amphibian, on land 226 m (740 ft)
Amphibian, on water 224 m (735 ft)
Range, at recommended lean mixture, with allowances for start, taxi, T-O, climb and 45 min reserves at 45% power:
max fuel, max cruising speed at 2,440 m (8,000 ft):
Landplane 785 nm (1,454 km; 903 miles)
Floatplane, amphibian
680 nm (1,260 km; 783 miles)
Skiplane 685 nm (1,270 km; 789 miles)
max fuel, econ cruising speed at 3,050 m (10,000 ft):
Landplane 1,010 nm (1,872 km; 1,163 miles)
Floatplane, amphibian, skiplane
815 nm (1,510 km; 938 miles)

CESSNA SKYLANE and TURBO SKYLANE

There were originally three variants of the Skylane, of which the Model 182 was the basic standard version. This was discontinued in 1976, but two new turbocharged versions were introduced for 1981; so, four versions are currently available.

Skylane. Standard version, powered by a 171·5 kW (230 hp) Continental O-470-U flat-six engine.

Skylane II. As Skylane, plus factory installed avionics package which includes Cessna Series 300 nav/com with 720-channel com and 200-channel nav with remote VOR/LOC, ADF, transponder and 200A Nav-O-Matic autopilot with VOR/LOC track and intercept. It is available optionally with a Nav Pac, which adds a second 300 nav/com plus VOR/ILS, Series 400 glideslope receiver with VOR/ILS indicator and Series 400 marker beacon receiver. Standard equipment includes exhaust gas temperature gauge, true airspeed indicator, dual controls, alternate static source, heated pitot, navigation light detectors, emergency locator transmitter, courtesy lights, omni-flash beacon, and external power socket.

Turbo Skylane. Generally similar to Skylane, but powered by a 175 kW (235 hp) Avco Lycoming O-540-L3C5D flat-six turbocharged engine.

Turbo Skylane II. As Turbo Skylane, but with standard avionics and equipment, and optional Nav Pac as detailed for Skylane II.

The 1981 versions of the Skylane have as standard improvements a new cabin door latch system, better ventilation, an avionics cooling fan, an anti-precipitation static nav antenna, and larger battery contactors. New options include Collins R/Nav and DME-450C which together provide an IFR-certificated area navigation system, new intercom system, and a lightweight microphone and headset. A three-blade propeller is available for the Turbo Skylane.

A version designated F 182 Skylane was produced for the European market by Reims Aviation in France (which see).

A total of 18,996 Model 182/Skylanes had been built by 1 January 1981, including 158 F 182s built by Reims Aviation.

Type: Four-seat cabin monoplane.

Wings: Braced high-wing monoplane. Wing section NACA 2412, modified. Incidence at root 0° 47′, at tip −2° 50′. Dihedral 1° 44′. Wing structure similar to Skyhawk, except metal-to-metal bonded leading-edge.

Fuselage: All-metal semi-monocoque structure.

Tail Unit: Cantilever all-metal structure with swept fin and rudder. Trim tab in starboard elevator. Electrically-operated elevator trim optional.

Cessna Turbo Skylane four-seat cabin monoplane (Avco Lycoming O-540-L3C5D engine)

Landing Gear: Non-retractable tricycle type. Land-O-Matic cantilever main legs, each comprising a one-piece machined conically-tapered spring steel tube. Steerable nosewheel with oleo-pneumatic shock-absorption. Cessna main wheels and tyres size 6·00-6, pressure 2·90 bars (42 lb/sq in). Cessna nosewheel and tyre size 5·00-5, pressure 3·38 bars (49 lb/sq in). Cessna hydraulic disc brakes. Parking brake. Optional wheel fairings.

Power Plant: As detailed in model listings. Two-blade constant-speed metal propeller standard; three-blade propeller optional for Turbo Skylane. Standard fuel capacity 348 litres (92 US gallons), of which 333 litres (88 US gallons) are usable. Refuelling point on upper surface of each wing. Oil capacity 11·5 litres (3 US gallons).

Accommodation: Generally similar to Skyhawk, with standard seating for four; four seat belts and two shoulder harnesses standard. Optional child's seat. Baggage space aft of rear seats and hatshelf with total capacity of 91 kg (200 lb), with external baggage door. Cargo tiedown net standard. Front seat inertia-reel shoulder harness, rear seat shoulder harness, leather seating, air vent for rear-seat passengers, and openable starboard window optional. Dual controls optional on Skylane. Heating, ventilation and windscreen defrosters standard.

Systems: Electrical system powered by 60A 28V engine-driven alternator. 24V battery. Hydraulic system for brakes only. Vacuum system optional. Oxygen system of 1·36 m³ (48 cu ft) capacity optional.

Avionics and Equipment: Standard avionics for Skylane II, optional on Skylane, include Cessna 200 Series 200A Nav-O-Matic autopilot, 300 Series 720-channel nav/com with remote VOR/LOC or VOR/ILS indicator, ADF with digital tuning, marker beacon receiver with three lights and aural signal, and transponder with 4096 code capability. Other optional avionics include Collins R/Nav and DME-450C, 10-channel HF transceiver, 300A Nav-O-Matic single-axis autopilot with heading control plus VOR intercept and track, 400 Series glideslope receiver, ADF with digital tuning, and transponder with 4096 code capability. Standard equipment includes control locks, armrests, tinted windscreen, audible stall warning, baggage restraint net, tinted windows, cabin dome light, adjustable cabin ventilators, variable-intensity instrument panel red floodlights, pedestal lights, navigation lights, and cabin steps. Optional equipment (standard for Skylane II) includes exhaust gas temperature gauge, true airspeed indicator, dual controls, alternate static source, heated pitot, emergency locator transmitter, navigation light detectors, courtesy lights, omni-flash beacon, and external power socket. Other optional equipment includes artificial horizon, carburettor air temperature gauge, directional gyro, electric clock, flight hour recorder, outside air temperature gauge, rate of climb indicator, sensitive altimeter, turn co-ordinator, inertia-reel shoulder harness for front seats, skylights, sun visors, heated stall warning transmitter, child's seat, headrests, shoulder harness for rear seats, leather seating, cabin fire extinguisher, internal corrosion proofing, rear window curtain, stretcher installation, utility shelf, writing table, stereo cassette entertainment centre, control wheel map light, instrument post lights, omni-flash beacon, wingtip strobe lights, engine priming system, engine winterisation kit, full-flow oil filter, non-congealing oil cooler, quick-drain oil valve, glider tow hook, overall paint scheme, tailcone lift handles, towbar, and tailplane abrasion boots.

Dimensions, external:
Wing span 10·92 m (35 ft 10 in)
Wing chord at root 1·63 m (5 ft 4 in)
Wing chord at tip 1·09 m (3 ft 7 in)
Length overall: Skylane 8·53 m (28 ft 0 in)
Turbo Skylane 8·66 m (28 ft 5 in)
Height overall 2·82 m (9 ft 3 in)
Tailplane span 3·55 m (11 ft 8 in)
Wheel track 2·74 m (9 ft 0 in)
Wheelbase 1·69 m (5 ft 6½ in)
Propeller diameter:
two-blade 2·08 m (6 ft 10 in)
three-blade 2·01 m (6 ft 7 in)
Passenger doors (each): Height 1·02 m (3 ft 4¼ in)
Width 0·90 m (2 ft 11¼ in)

Areas:
Wings, gross 16·16 m² (174 sq ft)
Ailerons (total) 1·70 m² (18·3 sq ft)
Trailing-edge flaps (total) 1·97 m² (21·20 sq ft)
Fin 1·08 m² (11·62 sq ft)
Rudder 0·65 m² (6·95 sq ft)
Tailplane 2·13 m² (22·96 sq ft)
Elevators 1·47 m² (15·85 sq ft)

Weights and Loadings (A: Skylane; B: Skylane II; C: Turbo Skylane; D: Turbo Skylane II):
Weight empty, equipped: A 780 kg (1,720 lb)
B 805 kg (1,775 lb)
C 782 kg (1,725 lb)
D 808 kg (1,781 lb)
Max T-O weight: all versions 1,406 kg (3,100 lb)
Max landing weight: all versions 1,338 kg (2,950 lb)
Max ramp weight: A, B 1,410 kg (3,110 lb)
C, D 1,411 kg (3,112 lb)
Max wing loading: all versions
86·9 kg/m² (17·8 lb/sq ft)
Max power loading: A, B 8·20 kg/kW (13·5 lb/hp)
C, D 8·02 kg/kW (13·2 lb/hp)

Performance (at max T-O weight, ISA):
Max level speed at S/L:
A, B 146 knots (271 km/h; 168 mph)
Max level speed at 6,100 m (20,000 ft):
C, D 168 knots (311 km/h; 193 mph)
Cruising speed, 75% power at 2,440 m (8,000 ft):
A, B 142 knots (263 km/h; 163 mph)
Cruising speed, 75% power at 3,050 m (10,000 ft):
C, D 145 knots (269 km/h; 167 mph)
Cruising speed, 75% power at 6,100 m (20,000 ft):
C, D 155 knots (293 km/h; 182 mph)
Stalling speed, flaps up, power off:
all versions 54 knots (100 km/h; 62·5 mph) CAS
Stalling speed, flaps down, power off:
all versions 49 knots (91 km/h; 56·5 mph) CAS
Max rate of climb at S/L:
A, B 264 m (865 ft)/min
C, D 294 m (965 ft)/min
Service ceiling: A, B 4,540 m (14,900 ft)
Max certificated operating altitude:
C, D 6,100 m (20,000 ft)
T-O run: A, B 245 m (805 ft)
C, D 241 m (790 ft)
T-O to 15 m (50 ft): A, B 462 m (1,515 ft)
C, D 450 m (1,475 ft)
Landing from 15 m (50 ft):
all versions 411 m (1,350 ft)
Landing run: all versions 180 m (590 ft)
Range with max fuel, recommended lean mixture, with allowances for start, taxi, T-O, climb and 45 min reserves at 45% power:
75% power at 2,440 m (8,000 ft):
A, B 820 nm (1,520 km; 944 miles)
75% power at 3,050 m (10,000 ft):
C, D 725 nm (1,343 km; 835 miles)
75% power at 6,100 m (20,000 ft):
C, D 745 nm (1,380 km; 858 miles)
econ cruising power at 3,050 m (10,000 ft):
A, B 1,025 nm (1,900 km; 1,180 miles)
C, D 920 nm (1,705 km; 1,059 miles)
econ cruising power at 6,100 m (20,000 ft):
C, D 885 nm (1,640 km; 1,019 miles)

CESSNA SKYLANE RG and TURBO SKYLANE RG

Introduced in late 1977, the retractable landing gear version of the Skylane is available in four models:

Skylane RG. Standard version, powered by Avco Lycoming O-540-J3C5D flat-six engine, derated to 175 kW (235 hp) and driving a two-blade constant-speed metal propeller with spinner.

Skylane RG II. As Skylane RG, plus additional avionics and equipment detailed for the Skylane II.

Turbo Skylane RG. Generally similar to Skylane RG, but powered by a 175 kW (235 hp) Avco Lycoming O-540-L3C5D flat-six engine with turbocharger, driving a two-blade constant-speed metal propeller.

Turbo Skylane RG II. As Turbo Skylane RG, plus additional avionics and equipment detailed for the Skylane II, except that the Series 300 transponder is replaced by the Series 400 model. Optional Nav Pac provides IFR capability with addition of a second Series 300 nav/com and indicator, and Series 400 glideslope and marker beacon receiver.

All versions of the Skylane RG for 1981 have the standard and optional improvements detailed for the 1981 Skylane models. Other options include the Cessna 400B autopilot which incorporates a new electrical trim system; a 95A alternator for aircraft with equipment that adds large electrical loads; and refuelling steps and handles. By 1 January 1981 a total of 1,694 Skylane RGs had been built, including 54 assembled by Reims Aviation in France as Reims 182 Skylane RGs.

The description of the Cessna Skylane applies also to the Skylane RG and Turbo Skylane RG, except as follows:

LANDING GEAR: Hydraulically-retractable tricycle type. Tubular spring steel main gear struts, retracting rearward into fuselage. Nosewheel, which retracts forward, is carried on a short-stroke oleo-pneumatic shock-absorber with hydraulic damper, and is steerable. Nosewheel enclosed by doors when retracted. Hydraulic brakes. Parking brake.

POWER PLANT: One Avco Lycoming flat-six engine, as detailed in model listings. Standard fuel capacity 348 litres (92 US gallons), of which 333 litres (88 US gallons) are usable. Refuelling point on upper surface of each wing. Oil capacity 8·5 litres (2·25 US gallons).

SYSTEMS: As described for Skylane, except self-contained electro-hydraulic system for operation of landing gear and brakes, plus optional air-conditioning system for all versions, and anti-icing system for Turbo Skylane RG versions. Optional 28V 95A alternator.

DIMENSIONS, EXTERNAL: As for Skylane except:
Length overall 8·72 m (28 ft 7½ in)
Height overall 2·72 m (8 ft 11 in)

WEIGHTS AND LOADINGS (A: Skylane RG; B: Skylane RG II; C: Turbo Skylane RG; D: Turbo Skylane RG II):
Weight empty: A 795 kg (1,752 lb)
B 819 kg (1,805 lb)
C 815 kg (1,797 lb)
D 839 kg (1,850 lb)
Max T-O and landing weight:
all versions 1,406 kg (3,100 lb)
Max ramp weight:
all versions 1,412 kg (3,112 lb)
Max wing loading:
all versions 86·9 kg/m² (17·8 lb/sq ft)
Max power loading:
all versions 8·03 kg/kW (13·2 lb/hp)

PERFORMANCE (at max T-O weight, ISA):
Max level speed at S/L:
A, B 160 knots (296 km/h; 184 mph)
Max level speed at 6,100 m (20,000 ft):
C, D 187 knots (346 km/h; 215 mph)
Max cruising speed, 75% power at 2,285 m (7,500 ft):
A, B 156 knots (290 km/h; 180 mph)
Max cruising speed, 75% power at 6,100 m (20,000 ft):
C, D 173 knots (320 km/h; 199 mph)
Max cruising speed, 75% power at 3,050 m (10,000 ft):
C, D 158 knots (293 km/h; 182 mph)
Stalling speed, flaps up, power off:
all versions 54 knots (100 km/h; 62 mph)
Stalling speed, flaps down, power off:
all versions 50 knots (92·5 km/h; 57·5 mph)
Max rate of climb at S/L:
A, B 347 m (1,140 ft)/min
C, D 317 m (1,040 ft)/min
*Service ceiling: A, B 4,360 m (14,300 ft)
Max certificated operating altitude:
C, D 6,100 m (20,000 ft)
T-O run: all versions 250 m (820 ft)
T-O to 15 m (50 ft): all versions 479 m (1,570 ft)
Landing from 15 m (50 ft):
all versions 402 m (1,320 ft)
Landing run: all versions 183 m (600 ft)
Range with max fuel, recommended lean mixture, with allowances for start, taxi, T-O, climb and 45 min reserves at 45% power:
75% power at 2,285 m (7,500 ft):
A, B 845 nm (1,566 km; 973 miles)
75% power at 6,100 m (20,000 ft):
C, D 825 nm (1,529 km; 950 miles)
75% power at 3,050 m (10,000 ft):
C, D 800 nm (1,482 km; 921 miles)
econ cruising power at 3,050 m (10,000 ft):
A, B 1,135 nm (2,103 km; 1,307 miles)
econ cruising power at 6,100 m (20,000 ft):
C, D 1,010 nm (1,871 km; 1,163 miles)
econ cruising power at 3,050 m (10,000 ft):
C, D 1,030 nm (1,908 km; 1,186 miles)

* *Service ceiling is 5,485 m (18,000 ft) if optional EGT gauge is used to set best power mixture*

Cessna Model 185 Skywagon one/six-seat cabin monoplane (Continental IO-520-D engine)

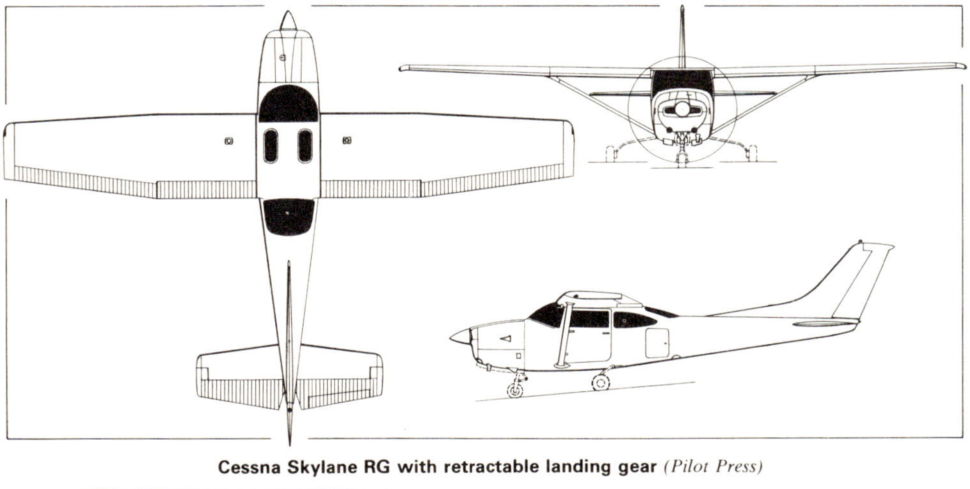

Cessna Skylane RG with retractable landing gear *(Pilot Press)*

Cessna Skylane RG II, a retractable landing gear version of the extensively built Model 182/Skylane

CESSNA MODEL 185 SKYWAGON

US military designation: U-17

The prototype of the Model 185 Skywagon flew for the first time in July 1960 and the first production model was completed in March 1961. It is generally similar to the Model 180 Skywagon, except for installation of a 224 kW (300 hp) Continental IO-520 engine.

Two versions are available:

Model 185 Skywagon. Standard version.

Model 185 Skywagon II. As Skywagon, plus factory-installed avionics package which includes Series 300 nav/com, ADF, and emergency locator transmitter. Optional Nav Pac provides a second nav/com with Series 400 glideslope and marker beacon receiver. Standard equipment as for basic Model 185, plus artificial horizon, clock, directional gyro, exhaust gas temperature gauge, outside air temperature gauge, sensitive altimeter, turn co-ordinator, vertical speed indicator, dual controls, sun visor, heated pitot and stall warning transmitter, map light, quick-drain oil valve, and external power socket.

The Model 185 Skywagon can be fitted with Edo 628-2960 floats, or Edo Model 597 amphibious floats, or Fli-Lite skis, and can be used in an agricultural role, for aerial application or agricultural pilot training, with optional 571 litre (151 US gallon) capacity Sorensen fan-driven spray system. It can carry under its fuselage a detachable glassfibre Cargo-Pack, more than 2·75 m long and 0·79 m wide (9 ft × 2 ft 7 in), with a volume of 0·61 m³ (21·5 cu ft) and capacity of 136 kg (300 lb). The Pack incorporates loading doors on the side and at the rear.

The 1981 version has the improvements and new options detailed for the Model 180 Skywagon.

Cessna has received important contracts to supply U-17A/B/C Skywagons to the US Air Force for delivery to overseas countries, under the US Military Assistance Program.

A total of 4,069 Model 185 Skywagons, including 497 U-17A/B/Cs, had been built by 1 January 1981. The 4,000th aircraft was delivered on 22 February 1980.

TYPE: One/six-seat cabin monoplane.

WINGS AND FUSELAGE: Similar to Model 180.

TAIL UNIT: Same as for Model 180, except for fin of increased area and manually-operated rudder trim as standard equipment.

LANDING GEAR: Similar to Model 180, except for tyre pressures: main wheels (6·00-6) 2·41 bars (35 lb/sq in), main wheels (8·00-6) 1·72 bars (25 lb/sq in), tailwheel 3·79-4·83 bars (55-70 lb/sq in) depending on load. Manual tailwheel lock standard. Wheel and brake fairings optional. Optional amphibious, float or ski gear.

POWER PLANT: One 224 kW (300 hp) Continental IO-520-D flat-six engine, driving a McCauley three-blade constant-speed metal propeller. Fuel in two tanks in wings, total capacity 333 litres (88 US gallons), of which 318 litres (84 US gallons) are usable. Oil capacity 11·5 litres (3 US gallons).

ACCOMMODATION, AVIONICS AND EQUIPMENT: Generally as for Model 180, except omni-flash beacon and manual tailwheel lock standard, plus availability of optional agricultural kit comprising 571 litre (151 US gallon) capacity Sorensen fan-driven spraygear with electric spray valve, deflector cable, windscreen and landing gear wire cutters, and external finish of Jet-flo polyurethane paint to provide extra corrosion protection.

DIMENSIONS: As for Model 180, except:
Propeller diameter 2·03 m (6 ft 8 in)

AREAS: As for Model 180, except:
Fin 1·29 m² (13·86 sq ft)

WEIGHTS AND LOADINGS:
Weight empty, equipped:
Skywagon landplane 769 kg (1,696 lb)
Floatplane 906 kg (1,998 lb)
Agplane option 858 kg (1,891 lb)
Amphibian 1,022 kg (2,253 lb)
Skiplane 834 kg (1,838 lb)
Skywagon II landplane 792 kg (1,747 lb)
Max T-O and landing weight:
Landplane, skiplane, Agplane option 1,519 kg (3,350 lb)
Floatplane 1,506 kg (3,320 lb)
Amphibian, land take-off 1,481 kg (3,265 lb)
Amphibian, water take-off 1,406 kg (3,100 lb)

Max ramp weight:
Landplane, skiplane, Agplane option 1,525 kg (3,362 lb)
Amphibian, land take-off 1,486 kg (3,277 lb)
Floatplane 1,511 kg (3,332 lb)
Max wing loading:
Landplane, skiplane, Agplane option 94·2 kg/m² (19·3 lb/sq ft)
Floatplane 93·3 kg/m² (19·1 lb/sq ft)
Amphibian 91·8 kg/m² (18·8 lb/sq ft)
Max power loading:
Landplane, skiplane, Agplane option 6·78 kg/kW (11·2 lb/hp)
Floatplane 6·72 kg/kW (11·1 lb/hp)
Amphibian 6·61 kg/kW (10·9 lb/hp)

Performance (at max T-O weight, ISA):
Never-exceed speed:
Landplane 182 knots (338 km/h; 210 mph)
Max level speed at S/L:
*Landplane 154 knots (285 km/h; 177 mph)
Floatplane 140 knots (259 km/h; 161 mph)
Agplane option 129 knots (239 km/h; 149 mph)
Amphibian 135 knots (249 km/h; 155 mph)
Skiplane 136 knots (252 km/h; 157 mph)
Max cruising speed (75% power) at 2,135 m (7,000 ft):
*Landplane 147 knots (272 km/h; 169 mph)
Floatplane 133 knots (246 km/h; 153 mph)
Agplane option 121 knots (224 km/h; 139 mph)
Amphibian 129 knots (240 km/h; 149 mph)
Skiplane 131 knots (243 km/h; 151 mph)
Stalling speed, flaps up, power off:
Landplane, skiplane, floatplane, Agplane option 56 knots (104 km/h; 64·5 mph) CAS
Amphibian 55 knots (102 km/h; 63 mph) CAS
Stalling speed, flaps down, power off:
Landplane, skiplane, Agplane option 49 knots (91 km/h; 56 mph) CAS
Amphibian 51 knots (94 km/h; 58 mph) CAS
Floatplane 52 knots (96 km/h; 60 mph) CAS
Max rate of climb at S/L:
Landplane 328 m (1,075 ft)/min
Floatplane 293 m (960 ft)/min
Agplane option 258 m (845 ft)/min
Amphibian 290 m (950 ft)/min
Skiplane 262 m (860 ft)/min
Service ceiling: Landplane 5,455 m (17,900 ft)
Floatplane 5,000 m (16,400 ft)
Agplane option 4,085 m (13,400 ft)
Amphibian 4,907 m (16,100 ft)
Skiplane 4,725 m (15,500 ft)
T-O run: Landplane 251 m (825 ft)
Floatplane 436 m (1,430 ft)
Agplane option 259 m (850 ft)
Amphibian, on land 238 m (780 ft)
Amphibian, on water 343 m (1,125 ft)
T-O to 15 m (50 ft): Landplane 436 m (1,430 ft)
Floatplane 648 m (2,125 ft)
Agplane option 463 m (1,520 ft)
Amphibian, on land 404 m (1,325 ft)
Amphibian, on water 521 m (1,710 ft)
Landing from 15 m (50 ft):
Landplane, Agplane option 427 m (1,400 ft)
Floatplane 477 m (1,565 ft)
Amphibian, on land 378 m (1,240 ft)
Amphibian, on water 450 m (1,480 ft)
Landing run:
Landplane, Agplane option 186 m (610 ft)
Floatplane 253 m (830 ft)
Amphibian, on land 174 m (570 ft)
Amphibian, on water 236 m (775 ft)
Range with max fuel (recommended lean mixture, with allowances for engine start, taxi, T-O, climb and 45 min reserves at 45% power):
75% power at 2,135 m (7,000 ft):
Landplane 645 nm (1,196 km; 743 miles)
Floatplane 585 nm (1,085 km; 674 miles)
Agplane option 530 nm (982 km; 610 miles)
Amphibian 570 nm (1,056 km; 656 miles)
Skiplane 575 nm (1,065 km; 662 miles)
econ cruising power at 3,050 m (10,000 ft):
Landplane 850 nm (1,575 km; 979 miles)
Floatplane 745 nm (1,381 km; 858 miles)
Amphibian 715 nm (1,324 km; 823 miles)
Skiplane 665 nm (1,233 km; 766 miles)

* *These speeds are 1 knot (1·9 km/h; 1·2 mph) higher with optional speed fairings installed*

Cessna Ag Wagon, the basic model of the company's series of agricultural aircraft

CESSNA AG WAGON and AG TRUCK

On 8 December 1971, Cessna introduced three new agricultural aircraft based on the earlier Ag Wagon low-wing monoplane. Of these, the Ag Pickup (53 built) was discontinued in 1976.

The current Ag Wagon and Ag Truck are of all-metal construction and have special corrosion proofing, heavy-duty spring steel Land-O-Matic landing gear and Cessna's Camber-Lift wing to provide better control during low-speed operations. Wing fences are used to smooth airflow over the wing. Special attention has been paid to safety features, and these include ensolite padding on the upper instrument panel, urethane padding on tubular structures in the cabin area and around doors, safe flush switch and control locations and quick-release door hinges. Other standard features include wide wing walks, large hopper loading doors, and fresh-air scoops that slightly pressurise the cockpit and tailcone to prevent the ingress of dust and fumes.

Optional equipment includes a special package to provide brilliant illumination for night operations. This comprises a 100A 24V alternator, taxi/landing lights, instrument panel lights, overhead floodlight, two 600W retractable spray lights, lighting angle control for spray lights, wingtip turning lights, hopper quantity light and a control stick grip incorporating light switches.

Improvements introduced for 1981 include dual-latching cabin doors for more positive sealing, a new high-volume dispersal system with a flow increase of up to 25 per cent, a redesigned emergency dump control, a new two-blade propeller, and new exterior and interior styling.

Differences between the two models are as follows:

Ag Wagon. Basic model, powered by a 224 kW (300 hp) Continental IO-520-D flat-six engine, driving a constant-speed propeller. Standard equipment includes a 757 litre (200 US gallon) hopper, a liquid and dry material dispersal control system, cockpit canopy with all-round view, tailplane abrasion boots, oversize 8·00-8 × 22 main-wheel tyres, 10 in tailwheel tyre, wire cutters, cable deflector, tailcone lift handles, hopper side-loading system on port side, navigation lights, pilot's four-way adjustable seat, control stick lock, quick oil drain, auxiliary fuel pump, steerable tailwheel and remote fuel strainer drain control. Options include an automatic flagman, hydraulic or fan-driven dispersal system, electric or fan-driven agitator, night lighting, and air-conditioning.

Ag Truck. As Ag Wagon, except for 1,060 litre (280 US gallon) hopper. Additional standard equipment includes a 22-nozzle engine-driven hydraulic spray system with manually-controlled spray valve and gatebox without agitator, wing fuel tanks, extended conical-cambered wingtips, automatic inertia reel for the safety belt system, sensitive altimeter, pilot's foul weather windows, strobe lights, instrument panel lights, landing and taxi lights, three-colour exterior styling, and oversize 10 in main and tailwheel tyres. Options include air-conditioning and a six-way articulating seat.

By 1 January 1981 deliveries totalled 1,581 Ag Wagons and 1,843 Ag Trucks. Production of the Ag Wagon is ending after delivery of the 1981 quota.

Type: Single-seat agricultural monoplane.

Wings: Braced low-wing monoplane, with single streamline-section bracing strut each side. Wing section NACA 2412, modified. Dihedral 9°. Incidence 1° 30′ at root, −1° 30′ at tip. All-metal structure with NACA all-metal single-slotted flaps inboard of Frise all-metal ailerons. Aileron leading-edge gaps sealed. Wing fences immediately outboard of bracing strut attachment points. Conical-cambered wingtips, extended on Ag Truck.

Fuselage: Rectangular-section welded steel tube structure with removable metal skin panels forward of cabin. All-metal semi-monocoque rear fuselage.

Tail Unit: Cantilever all-metal structure. Fixed-incidence tailplane. Trim tab in starboard elevator.

Landing Gear: Non-retractable tailwheel type. Land-O-Matic cantilever main legs of heavy-duty spring steel. Tapered tubular tailwheel spring shock-absorber. Main wheels and tyres size 8·00-8 × 22 on Ag Wagon, with oversize 10 in main-wheel tyres on Ag Truck. Ag Wagon has 10 in tailwheel tyre and Ag Truck an oversize 10 in tailwheel tyre. Hydraulic disc brakes and parking brake. Wheel fenders optional.

Power Plant: One 224 kW (300 hp) Continental IO-520-D flat-six engine, driving a two-blade constant-speed metal propeller. Three-blade propeller optional. Two fuel tanks in wings with combined capacity of 204 litres (54 US gallons), of which 197 litres (52 US gallons) are usable. Oil capacity 11·5 litres (3 US gallons).

Accommodation: Pilot only, on vertically and longitudinally adjustable seat, in enclosed cabin. Steel overturn structure. Combined window and door on each side, hinged at bottom. Heating and ventilation standard.

System: Electrical system has a 60A 12V alternator and 12V 24Ah battery as standard. A 60A 24V or a 100A 24V alternator is available optionally.

Equipment: Standard equipment is as detailed in model listings. Optional equipment includes fan-driven or engine-driven hydraulic spray systems with 22-nozzle spraybooms (standard on Ag Truck); 44- or 64-nozzle spraybooms; two spreader systems for either medium or high-volume applications; electric spray control valve; a Cessna-designed gatebox; wing leading-edge repair kit, a manual locking-type inertia-reel shoulder harness, and main gear and tailcone jacking points.

Dimensions, external (A: Ag Wagon; B: Ag Truck):
Wing span: A 12·41 m (40 ft 8½ in)
B 12·70 m (41 ft 8 in)
Wing chord at root 1·63 m (5 ft 4 in)
Wing chord at tip 1·12 m (3 ft 8 in)
Length overall 7·90 m (25 ft 11 in)
Height overall 2·49 m (8 ft 2 in)
Tailplane span 3·35 m (11 ft 0 in)
Wheel track 2·16 m (7 ft 1 in)
Propeller diameter: two-blade 2·18 m (7 ft 2 in)
three-blade 2·03 m (6 ft 8 in)

Areas:
Wings, gross: A 18·77 m² (202 sq ft)
B 19·05 m² (205 sq ft)

Weights and Loadings:
Weight empty, with liquid dispersal system gatebox and engine-driven hydraulic pump:
A 988 kg (2,179 lb)
B 1,012 kg (2,230 lb)
T-O weight, Normal category 1,496 kg (3,300 lb)
Max T-O weight, Restricted category:
A 1,814 kg (4,000 lb)
B 1,905 kg (4,200 lb)
Max landing weight 1,496 kg (3,300 lb)
Wing loading, Normal category:
A 79·6 kg/m² (16·3 lb/sq ft)
B 78·6 kg/m² (16·1 lb/sq ft)
Max wing loading, Restricted category:
A 96·6 kg/m² (19·8 lb/sq ft)
B 100·0 kg/m² (20·5 lb/sq ft)
Power loading, Normal category:
A, B 6·68 kg/kW (11·0 lb/hp)
Max power loading, Restricted category:
A 8·10 kg/kW (13·3 lb/hp)
B 8·50 kg/kW (14·0 lb/hp)

Performance (A: Ag Wagon, B: Ag Truck. At max T-O weight, ISA, with liquid dispersal equipment):
Max level speed at S/L:
A 107 knots (198 km/h; 123 mph)
B 106 knots (196 km/h; 122 mph)
Max cruising speed, 75% power at 1,980 m (6,500 ft):
A 102 knots (190 km/h; 118 mph)
B 101 knots (187 km/h; 116 mph)
Stalling speed, flaps up, power off:
A 58 knots (108 km/h; 67 mph) CAS
B 60 knots (111 km/h; 69 mph) CAS
Stalling speed, flaps down, power off:
A 55 knots (101 km/h; 63 mph) CAS
B 56 knots (104 km/h; 65 mph) CAS
Max rate of climb at S/L: A 160 m (525 ft)/min
B 142 m (465 ft)/min
Service ceiling: A 2,650 m (8,700 ft)
B 2,375 m (7,800 ft)
T-O run: A 381 m (1,250 ft)
B 430 m (1,410 ft)
T-O to 15 m (50 ft): A 599 m (1,965 ft)
B 686 m (2,250 ft)
Landing from 15 m (50 ft): A, B 386 m (1,265 ft)
Landing run: A, B 128 m (420 ft)
Range, recommended lean mixture, allowances for engine start, taxi, T-O, climb, and 30 min reserves at 45% power:
max fuel, 75% power at 1,980 m (6,500 ft):
A 256 nm (475 km; 295 miles)
B 252 nm (466 km; 290 miles)

CESSNA AG HUSKY

Cessna introduced for 1979 an addition to its line of agricultural aircraft. Named the Ag Husky, it is a turbocharged version of the Ag Truck, the 231 kW (310 hp) engine providing improved performance at varying operational altitudes. It introduces for 1981 the standard improvements as detailed for the Ag Wagon/Ag Truck.

By 1 January 1981 a total of 251 Ag Husky aircraft had been delivered.

The description of the Ag Truck applies also to the Ag Husky, except as follows:

Power Plant: One 231 kW (310 hp) Continental TSIO-520-T turbocharged flat-six engine, driving a three-blade constant-speed metal propeller.

Equipment: Standard equipment includes a 1,060 litre (280 US gallon) hopper, a Cessna hydraulic dispersal system with electric spray valve and 22-nozzle sprayboom, full-flow air filter, and a special lighting package.

Dimensions, external: As for Ag Truck, except:
Length overall 8·08 m (26 ft 6 in)
Propeller diameter 2·03 m (6 ft 8 in)

Weights and Loadings:
Weight empty, liquid dispersal system installed 1,053 kg (2,322 lb)
Max T-O weight, Restricted category 1,996 kg (4,400 lb)
Max landing weight 1,497 kg (3,300 lb)
Max wing loading 104·78 kg/m² (21·5 lb/sq ft)
Max power loading 8·64 kg/kW (14·2 lb/hp)

Performance (at max T-O weight, ISA):
Max level speed at S/L 113 knots (209 km/h; 130 mph)
Max cruising speed, 75% power at 1,980 m (6,500 ft) 106 knots (196 km/h; 122 mph)
Stalling speed, flaps up, power off 62 knots (114 km/h; 71 mph) CAS
Stalling speed, flaps down, power off 58 knots (108 km/h; 67 mph) CAS
Max rate of climb at S/L 155 m (510 ft)/min
Max certificated operating altitude 4,265 m (14,000 ft)
T-O run 393 m (1,290 ft)
T-O to 15 m (50 ft) 628 m (2,060 ft)
Landing from 15 m (50 ft) 386 m (1,265 ft)
Landing run 128 m (420 ft)
Range with max fuel, 75% power at 1,980 m (6,500 ft), recommended lean mixture, allowances for engine start, taxi, T-O, climb, and 30 min reserves at 45% power 217 nm (402 km; 250 miles)

CESSNA STATIONAIR 6 and TURBO STATIONAIR 6

Cessna first renamed the former U206 Skywagon and TU206 Turbo Skywagon as the Stationair and Turbo Stationair respectively. In 1978 a further name change to Stationair 6 and Turbo Stationair 6 highlighted the six-seat capacity of these cargo/utility aircraft. Among their important features are double cargo doors on the starboard side of the fuselage which permit the easy loading and unloading of a crate more than 1·22 m long, 0·91 m wide and 0·91 m deep (4 ft × 3 ft × 3 ft). Differences between the two versions are as follows:

Stationair 6. Standard cargo utility model with 224 kW (300 hp) Continental IO-520-F engine, as described in detail.

Turbo Stationair 6. Similar to the Stationair 6 but with 231 kW (310 hp) Continental TSIO-520-M turbocharged engine in modified cowling and with a manifold pressure relief valve to prevent overboost.

The 'II' package of optional avionics and equipment, as detailed for the Skylane, is available for both versions of the Stationair 6.

A utility version of the Stationair is also available, with a single seat for the pilot as standard, vinyl floor covering, two-colour paint scheme and no wheel fairings. Up to five passenger seats can be supplied optionally.

The 1981 models of the Stationair 6/Turbo Stationair 6 introduced a number of improvements as standard, including a new avionics cooling fan, better cabin ventilation, improved cabin heating heat-exchanger, an anti-precipitation static nav antenna, and larger battery contactors. New options include an intercom, a lightweight microphone headset, a 95A alternator for aircraft with equipment that adds large electrical loads, refuelling steps and handles, and Collins R/Nav and DME-450C which combined provide an IFR-certificated area navigation system.

A total of 6,635 Model 206 Skywagons and Stationairs had been built by 1 January 1981, including 643 de luxe Super Skylanes of similar basic design.

Cessna Stationair 6 II one/six-seat cargo/utility aircraft

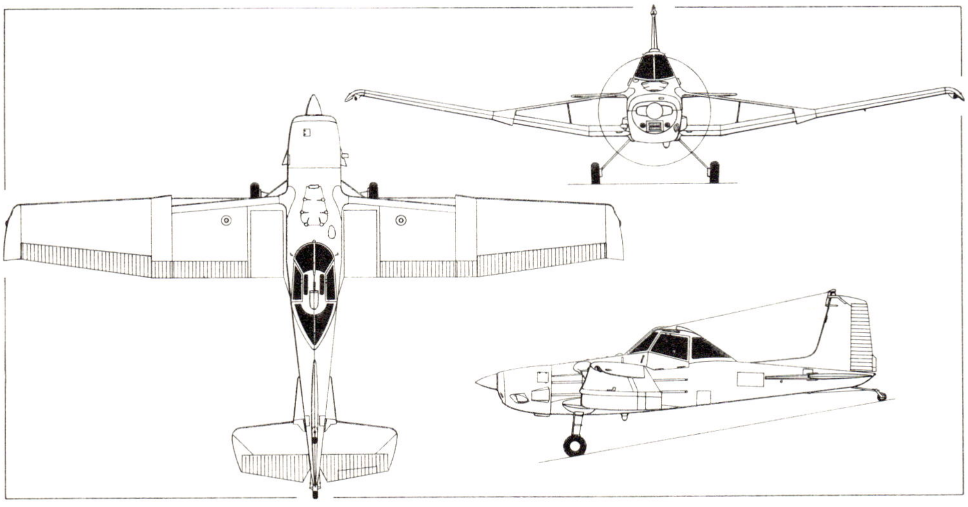

Photograph and three-view drawing *(Pilot Press)* of the Cessna Ag Husky, a turbocharged version of the extensively-built Ag Truck

Type: Single-engined cargo/utility aircraft.

Wings: Braced high-wing monoplane. Single streamlined-section bracing strut each side. Wing section NACA 2412, modified. Dihedral 1° 44'. Incidence 1° 30' at root, —1° 30' at tip. All-metal structure. Glassfibre conical camber tips. Modified Frise-type wide-chord ailerons. Electrically-operated wide-span NACA single-slotted flaps. No tabs.

Fuselage: Conventional all-metal semi-monocoque structure.

Tail Unit: Cantilever all-metal structure, with sweptback vertical surfaces. Large trim tab in starboard elevator. Electrical operation of trim tab optional.

Landing Gear: Non-retractable tricycle type. Cessna Land-O-Matic cantilever spring steel main legs. Steerable nosewheel with oleo-pneumatic shock-absorber. Cessna wheels, tubeless tyres and hydraulic disc brakes. Parking brake. Main wheels and tyres size 6·00-6, pressure 2·90 bars (42 lb/sq in). Nosewheel and tyre size 5·00-5, pressure 3·10 bars (45 lb/sq in). Main-wheel tyres size 8·00-6, nosewheel tyre size 6·00-6 and oversize wheel fairings optional. Floats and wheel-skis optional.

Power Plant: One Continental flat-six engine (details given under model listings), driving a McCauley three-blade constant-speed metal propeller type D2A32C90/82NC-2 (Stationair) or D3A-32C88/82NC-2 (Turbo Stationair). Two fuel cells in wings, with total standard capacity of 348 litres (92 US gallons), of which 333 litres (88 US gallons) are usable. Oil capacity 11·5 litres (3 US gallons).

Accommodation: Standard seating for pilot, co-pilot and up to four passengers, front seats with inertia safety belts. Club seating arrangement optional, with centre row of seats facing aft. Utility version has only pilot's seat as standard. Pilot's door on port side. Large double cargo doors on starboard side; forward door hinged to open forward, rear door hinged to open rearward. Aircraft can be flown with cargo doors removed for photography, airdropping of supplies or parachuting. Openable starboard window optional. Fully articulating seats for pilot and co-pilot and safety harness for four rear seats optional. Cabin heated and ventilated.

Systems: Electrical system powered by an engine-driven 28V 60A alternator. 24V 33Ah battery. 28V 95A alternator optional. Hydraulic system for brakes and optional wheel-skis. Oxygen system of 2·10 m³ (74 cu ft) capacity standard on Turbo Stationair 6; 1·36 m³ (48 cu ft) system optional for Stationair 6. Vacuum system optional.

Avionics and Equipment: Optional avionics as detailed for the Skylane, plus Series 400 Nav-O-Matic two-axis autopilot with heading control, VOR intercept and track and altitude control, Series 400 nav/com, a non-slaved HSI (with autopilot installation), audio/marker beacon receiver panel, Collins R/Nav and DME-450C, padded headset and electro-luminescent panel lighting, plus the optional avionics detailed for the Skylane II. Standard equipment as for the Skylane, plus electric clock, glareshield, outside air temperature gauge, sensitive altimeter, turn co-ordinator, sun visors, and overall paint scheme. Optional equipment, less the above items, is as detailed for the Skylane, plus ambulance kits, casket kit, writing desk, AM/FM stereo player, photographic provisions, glider towing provisions and skydiving kit. The child's seat and skylights are not available for the Stationair 6. The Turbo Stationair 6 has an overboost control valve, absolute pressure controller, pressurised fuel system, turbine access door, pilot's all-purpose control wheel, non-congealing oil cooler, full flow oil filter and alternate static source as standard.

Dimensions, external (L: landplane; F: floatplane):
Wing span 10·92 m (35 ft 10 in)
Wing chord at root 1·63 m (5 ft 4 in)
Wing chord at tip 1·09 m (3 ft 7 in)
Wing aspect ratio 7·63

Length overall:
Stationair 6: L 8·61 m (28 ft 3 in)
Stationair 6: F 9·04 m (29 ft 8 in)
Turbo Stationair 6 8·61 m (28 ft 3 in)
Height overall: L 2·83 m (9 ft 3½ in)
F 4·31 m (14 ft 1½ in)
Tailplane span 3·96 m (13 ft 0 in)
Wheel track 2·46 m (8 ft 1 in)
Propeller diameter: L 2·03 m (6 ft 8 in)
F 2·18 m (7 ft 2 in)
Pilot's door (port):
Height, mean 1·03 m (3 ft 4 in)
Cargo double door (stbd):
Height 0·98 m (3 ft 2½ in)
Width 1·13 m (3 ft 8½ in)
Height to sill 0·64 m (2 ft 1 in)

Dimensions, internal:
Cabin: Length 3·66 m (12 ft 0 in)
Max width 1·12 m (3 ft 8 in)
Max height 1·26 m (4 ft 1½ in)
Volume available for payload 2·87 m³ (101·2 cu ft)

Areas:
Wings, gross 16·17 m² (174·0 sq ft)
Ailerons (total) 1·60 m² (17·32 sq ft)
Trailing-edge flaps (total) 2·63 m² (28·35 sq ft)
Fin 1·08 m² (11·62 sq ft)
Rudder, incl tab 0·65 m² (6·95 sq ft)
Tailplane 2·31 m² (24·84 sq ft)
Elevators, incl tab 1·86 m² (20·08 sq ft)

Weights and Loadings (L: landplane; F: floatplane; A: Turbo Stationair 6 amphibian):
Weight empty:
Utility Stationair 6: L 854 kg (1,882 lb)
F 1,016 kg (2,241 lb)
Utility Stationair 6 II: L 881 kg (1,942 lb)
F 1,044 kg (2,301 lb)
Stationair 6: L 874 kg (1,928 lb)
F 1,029 kg (2,268 lb)
Stationair 6 II: L 902 kg (1,989 lb)
F 1,056 kg (2,329 lb)
Turbo Utility Stationair 6: L 889 kg (1,960 lb)
F 1,051 kg (2,317 lb)
A 1,183 kg (2,609 lb)
Turbo Utility Stationair 6 II: L 916 kg (2,020 lb)
F 1,078 kg (2,377 lb)
A 1,211 kg (2,669 lb)
Turbo Stationair 6: L 910 kg (2,006 lb)
F 1,063 kg (2,344 lb)
A 1,196 kg (2,636 lb)
Turbo Stationair 6 II: L 937 kg (2,066 lb)
F 1,090 kg (2,404 lb)
A 1,223 kg (2,696 lb)
Max T-O and landing weight:
Stationair 6: L 1,633 kg (3,600 lb)
F 1,587 kg (3,500 lb)
Turbo Stationair 6: L, F, A 1,633 kg (3,600 lb)
Max ramp weight:
Stationair 6: L 1,638 kg (3,612 lb)
F 1,593 kg (3,512 lb)
Turbo Stationair 6: L, F, A 1,640 kg (3,616 lb)
Max wing loading:
Stationair 6: L 101·1 kg/m² (20·7 lb/sq ft)
F 98·1 kg/m² (20·1 lb/sq ft)
Turbo Stationair 6:
L, F, A 101·1 kg/m² (20·7 lb/sq ft)
Max power loading:
Stationair 6: L 7·29 kg/kW (12·0 lb/hp)
F 7·08 kg/kW (11·7 lb/hp)
Turbo Stationair 6:
L, F, A 7·07 kg/kW (11·6 lb/hp)

Performance (at max T-O weight, ISA; L: landplane; F: floatplane; A: amphibian):
Max level speed:
Stationair 6 at S/L:
L 156 knots (290 km/h; 180 mph)
F 138 knots (256 km/h; 159 mph)
Turbo Stationair 6 at 5,180 m (17,000 ft):
L 174 knots (322 km/h; 200 mph)
F 155 knots (287 km/h; 178 mph)
A 150 knots (278 km/h; 173 mph)
Max cruising speed (75% power):
Stationair 6 at 1,980 m (6,500 ft):
L 147 knots (272 km/h; 169 mph)
F 132 knots (245 km/h; 152 mph)
Turbo Stationair 6, 80% power at 6,100 m (20,000 ft):
L 167 knots (309 km/h; 192 mph)
F 147 knots (272 km/h; 169 mph)
A 141 knots (261 km/h; 162 mph)
Turbo Stationair 6, 80% power at 3,050 m (10,000 ft): L 152 knots (282 km/h; 175 mph)
F 135 knots (249 km/h; 155 mph)
A 130 knots (241 km/h; 150 mph)
Stalling speed, flaps up, power off:
Stationair 6: L 62 knots (115 km/h; 71·5 mph) CAS
F 56 knots (104 km/h; 64·5 mph) CAS
Turbo Stationair 6:
L 62 knots (115 km/h; 71·5 mph) CAS
F, A 57 knots (106 km/h; 65·5 mph) CAS
Stalling speed, flaps down, power off:
Stationair 6: L 54 knots (101 km/h; 62·5 mph) CAS
F 51 knots (95 km/h; 59 mph) CAS
Turbo Stationair 6:
L 54 knots (101 km/h; 62·5 mph) CAS
F, A 52 knots (96·5 km/h; 60 mph) CAS
Max rate of climb at S/L:
Stationair 6: L 280 m (920 ft)/min
F 282 m (925 ft)/min
Turbo Stationair 6: L 308 m (1,010 ft)/min
F 255 m (835 ft)/min
A 247 m (810 ft)/min
Service ceiling:
Stationair 6: L 4,511 m (14,800 ft)
F 4,237 m (13,900 ft)
Turbo Stationair 6: L 8,230 m (27,000 ft)
F 7,800 m (25,600 ft)
A 7,650 m (25,100 ft)
T-O run:
Stationair 6: L 274 m (900 ft)
F 559 m (1,835 ft)
Turbo Stationair 6: L 255 m (835 ft)
F, A water 552 m (1,810 ft)
A land 288 m (945 ft)
T-O to 15 m (50 ft):
Stationair 6: L 543 m (1,780 ft)
F 860 m (2,820 ft)
Turbo Stationair 6: L 500 m (1,640 ft)
F, A water 850 m (2,790 ft)
A land 561 m (1,840 ft)
Landing from 15 m (50 ft):
Stationair 6: L 425 m (1,395 ft)
F 511 m (1,675 ft)
Turbo Stationair 6: L 425 m (1,395 ft)
F, A water 533 m (1,750 ft)
A land 430 m (1,410 ft)
Landing run:
Stationair 6: L 224 m (735 ft)
F 238 m (780 ft)
Turbo Stationair 6: L 224 m (735 ft)
F, A water 258 m (845 ft)
A land 229 m (750 ft)
Range, Stationair 6, recommended lean mixture, with allowances for start, taxi, T-O, climb, and 45 min reserves at 45% power:
max cruising speed at 1,980 m (6,500 ft) with max fuel: L 680 nm (1,260 km; 783 miles)
F 615 nm (1,139 km; 708 miles)
econ cruising speed at 3,050 m (10,000 ft) with max fuel: L 900 nm (1,667 km; 1,036 miles)
F 770 nm (1,427 km; 887 miles)
Range, Turbo Stationair 6, allowances and reserves as above:
max cruising speed at 6,100 m (20,000 ft) with max fuel: L 640 nm (1,186 km; 737 miles)
F 550 nm (1,019 km; 633 miles)
A 530 nm (981 km; 610 miles)
max cruising speed at 3,050 m (10,000 ft) with max fuel: L 610 nm (1,130 km; 702 miles)
F 535 nm (991 km; 616 miles)
A 515 nm (954 km; 593 miles)
econ cruising speed at 6,100 m (20,000 ft) with max fuel: L 775 nm (1,435 km; 892 miles)
F 660 nm (1,223 km; 760 miles)
A 630 nm (1,167 km; 725 miles)
econ cruising speed at 3,050 m (10,000 ft) with max fuel: L 805 nm (1,492 km; 927 miles)
F 690 nm (1,278 km; 794 miles)
A 660 nm (1,223 km; 760 miles)

CESSNA STATIONAIR 8 and TURBO STATIONAIR 8

For 1980, Cessna replaced the seven-seat Stationair 7 and Turbo Stationair 7 (see 1979-80 *Jane's*) with the eight-seat Stationair 8 and Turbo Stationair 8 respectively. Differences between the new models are as follows:

Stationair 8. Standard passenger/cargo utility model with 224 kW (300 hp) Continental IO-520-F engine.

Turbo Stationair 8. Generally similar to Stationair 8, but with 231 kW (310 hp) Continental TSIO-520-M turbocharged engine, driving a McCauley three-blade constant-speed metal propeller. Absolute pressure controller, pressurised fuel system, non-congealing oil cooler, full-flow oil filter, overboost control valve, alternate static source and oxygen system standard.

The 'II' package of optional avionics and equipment, as detailed for the Skylane, is available for both versions of the Stationair 8.

A utility version of the Stationair 8 is also available, having as standard a single seat for the pilot. Up to seven passenger seats can be supplied optionally.

The 1981 Stationair 8s introduced a number of improvements as standard, including a new avionics cooling fan, better cabin ventilation, improved cabin heating heat-exchanger, an anti-precipitation static nav antenna, and larger battery contactors. New options include Collins R/Nav and DME-450C which combine to provide an IFR-certificated area navigation system, intercom system, lightweight microphone headset, a 95A alternator for aircraft with equipment that adds large electrical loads, and refuelling steps and handles.

When production of the Stationair 7 and Turbo Stationair 7 ended, a total of 565 had been built, including early Model 207 Skywagons and Turbo Skywagons. By 1 January 1981 production of Stationair 8s and Turbo Stationair 8s totalled 105.

The following description applies to the Stationair 8, except where stated otherwise:

Type: Single-engined utility aircraft.

Wings: Braced high-wing monoplane. Single streamline-section bracing strut each side. Wing section NACA 2412 from root to just inboard of tip; wingtip is symmetrical. Dihedral 1° 44′. Incidence 1° 30′ at root, —1° 30′ at tip. All-metal structure. Glassfibre conical-camber

Cessna Stationair 8 II, a one/eight-seat version of this utility aircraft

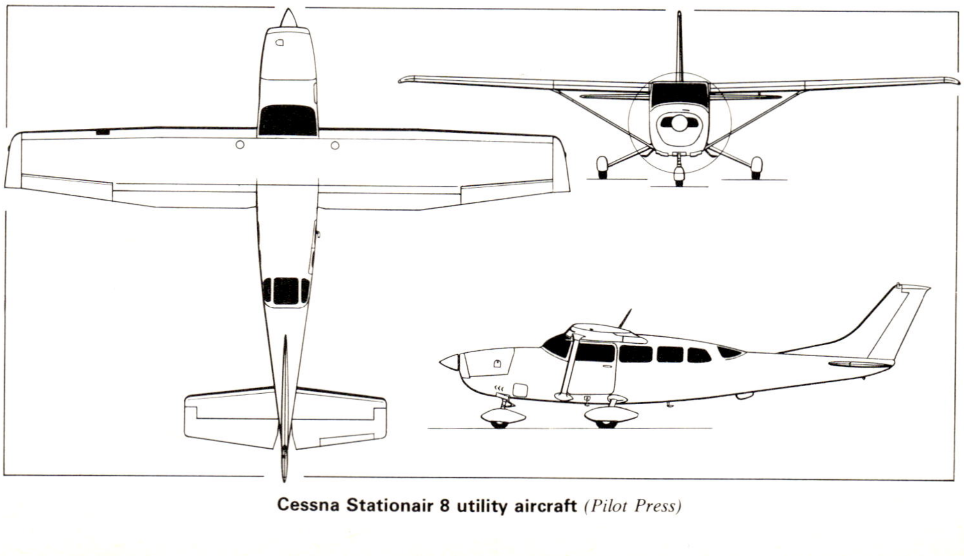

Cessna Stationair 8 utility aircraft *(Pilot Press)*

tips. Modified Frise-type all-metal wide-chord ailerons. Electrically-operated long-span NACA single-slotted all-metal flaps. No trim tabs.

Fuselage: Conventional all-metal semi-monocoque structure.

Tail Unit: Cantilever all-metal structure, with sweptback vertical surfaces. Tailplane fixed with —3° incidence. Large trim tab in starboard elevator. Electrical operation of trim tab optional. Rudder trimmed by adjustment of bungee.

Landing Gear: Non-retractable tricycle type. Improved Cessna Land-O-Matic cantilever main legs of one-piece tapered steel tube. Steerable nosewheel with Cessna oleo-pneumatic shock-absorber and hydraulic shimmy damper. Cessna wheels, tubeless tyres and hydraulic disc brakes. Main wheels and tyres size 6·00-6, pressure 3·79 bars (55 lb/sq in). Nosewheel and tyre size 5·00-5, pressure 3·38 bars (49 lb/sq in). Optional 8·00-6 main-wheel tyres, pressure 2·41 bars (35 lb/sq in), nosewheel tyre size 6·00-6, pressure 2·00 bars (29 lb/sq in). Wheel fairings standard; oversize wheel fairings optional.

Power Plant: One 224 kW (300 hp) Continental IO-520-F flat-six engine, driving a McCauley three-blade constant-speed metal propeller. A bladder-type fuel tank, capacity 115·5 litres (30·5 US gallons), is located in the inboard section of each wing. Total fuel capacity 231 litres (61 US gallons), of which 204 litres (54 US gallons) are usable. Optional tankage increases capacity to 151·5 litres (40 US gallons) in each wing, giving a total capacity of 303 litres (80 US gallons), of which 276 litres (73 US gallons) are usable. Refuelling point in upper surface of each wing. Oil capacity 11·5 litres (3 US gallons).

Accommodation: Standard seating is for a pilot and seven passengers, with individual seats arranged in four pairs. A six-seat club arrangement is optional. Seats removable individually to increase cargo space. Pilot's door on port side, co-pilot's door on starboard side at front. Large double cargo doors on starboard side at rear of cabin; forward door hinged to open forward, rear door hinged to open rearward. Aircraft can be flown with cargo doors removed for photography, airdropping of supplies or parachuting; optional equipment includes a spoiler for use when the aircraft is flown in this configuration. Openable window, port side; openable window starboard side optional. Separate baggage compartment, forward of cabin, capacity 54 kg (120 lb), accessible through top-hinged door on starboard side. External glassfibre cargo pack, capacity 136 kg (300 lb), carried beneath the fuselage, is available as an optional extra.

Systems: Hydraulic system for brakes. Electrical system powered by a 28V 60A engine-driven alternator. 24V 33Ah battery. 28V 95A alternator optional. Oxygen system of 2·15 m³ (76 cu ft) capacity standard on Turbo Stationair 8; 1·36 m³ (48 cu ft) system optional on Stationair 8.

Avionics and Equipment: Avionics detailed as standard and optional for the Stationair 6/Turbo Stationair 6 apply also to the Stationair 8/Turbo Stationair 8 respectively. Standard equipment is as described for the Model 180, plus electric clock, outside air temperature gauge, rate of climb indicator, sensitive altimeter, turn co-ordinator, sun visors, tinted windscreen and windows, emergency locator transmitter, baggage tiedown rings, electro-luminescent lights for switch and comfort control panels, instrument panel glareshield light, cabin dome lights, dual-beam landing lights, cabin steps and towbar. Additional optional items include articulating and vertically adjustable front seats, inertia-reel shoulder harness for two front seats, openable starboard window, heated pitot and stall warning transmitter, centre armrests, shoulder harness for six passenger seats, ambulance kit, skydiving kit, photographic provisions, instrument post lights, anti-precipitation static kit, flush com antenna, and glider tow hook.

Dimensions, external:
Wing span 10·92 m (35 ft 10 in)
Wing chord at root 1·63 m (5 ft 4 in)
Wing chord at tip 1·09 m (3 ft 7 in)
Wing aspect ratio 7·46
Length overall 9·80 m (32 ft 2 in)
Height overall 2·92 m (9 ft 7 in)
Tailplane span 3·96 m (13 ft 0 in)
Wheel track 3·09 m (10 ft 1¾ in)
Wheelbase 2·11 m (6 ft 11¼ in)
Propeller diameter 2·03 m (6 ft 8 in)
Forward cabin doors (each):
Height 1·05 m (3 ft 5½ in)
Width 0·89 m (2 ft 11½ in)
Height to sill 0·71 m (2 ft 4 in)
Cargo double doors (stbd):
Height 0·97 m (3 ft 2 in)
Width 1·13 m (3 ft 8½ in)
Height to sill 0·76 m (2 ft 6 in)
Baggage door (stbd): Height 0·61 m (2 ft 0 in)
Width 0·34 m (1 ft 1½ in)
Height to sill 1·02 m (3 ft 4 in)

Dimensions, internal:
Cabin: Length 4·27 m (14 ft 0 in)

Cessna Centurion II six-seat cabin monoplane (Continental IO-520-L engine)

Max width 1·13 m (3 ft 8½ in)
Max height 1·24 m (4 ft 1 in)
Floor area 4·38 m² (47·1 sq ft)
Volume 4·40 m³ (155·5 cu ft)
Forward baggage compartment:
Length 0·43 m (1 ft 5 in)
Max width 1·05 m (3 ft 5½ in)
Max height 0·69 m (2 ft 3 in)
Floor area 0·46 m² (4·9 sq ft)
Volume 0·27 m³ (9·5 cu ft)
Underfuselage cargo pack 0·34 m³ (12·0 cu ft)

Areas:
Wings, gross 16·17 m² (174·0 sq ft)
Ailerons (total) 1·60 m² (17·32 sq ft)
Trailing-edge flaps (total) 2·66 m² (26·60 sq ft)
Fin 0·84 m² (9·04 sq ft)
Rudder 0·65 m² (6·95 sq ft)
Tailplane 2·31 m² (24·84 sq ft)
Elevators, incl tab 1·86 m² (20·08 sq ft)

Weights and Loadings:
Weight empty: Utility Stationair 8 939 kg (2,070 lb)
Utility Stationair 8 II 969 kg (2,136 lb)
Stationair 8 957 kg (2,110 lb)
Stationair 8 II 987 kg (2,177 lb)
Turbo Utility Stationair 8 973 kg (2,146 lb)
Turbo Utility Stationair 8 II 1,004 kg (2,213 lb)
Turbo Stationair 8 992 kg (2,187 lb)
Turbo Stationair 8 II 1,022 kg (2,254 lb)
Max T-O and landing weight:
Stationair 8 and Turbo Stationair 8
1,723 kg (3,800 lb)
Max ramp weight: Stationair 8 1,729 kg (3,812 lb)
Turbo Stationair 8 1,731 kg (3,816 lb)
Max wing loading:
Stationair 8 and Turbo Stationair 8
106·44 kg/m² (21·8 lb/sq ft)
Max power loading:
Stationair 8 7·69 kg/kW (12·7 lb/hp)
Turbo Stationair 8 7·46 kg/kW (12·3 lb/hp)

Performance (at max T-O weight, ISA, and with optional wheel fairings, which increase speed by 3-4 knots; 5·5-7·5 km/h; 3·5-4·5 mph, with corresponding difference in range. A: Stationair 8; B: Turbo Stationair 8):
Never-exceed speed:
A, B 182 knots (338 km/h; 210 mph)
Max level speed:
A at S/L 150 knots (278 km/h; 173 mph)
B at 5,180 m (17,000 ft)
170 knots (315 km/h; 196 mph)
Max cruising speed:
A, 75% power at 1,980 m (6,500 ft)
143 knots (266 km/h; 165 mph)
B, 80% power at 6,100 m (20,000 ft)
161 knots (298 km/h; 185 mph)
B, 80% power at 3,050 m (10,000 ft)
148 knots (274 km/h; 170 mph)
Stalling speed, flaps up, power off:
A, B 65 knots (121 km/h; 75 mph) CAS
Stalling speed, flaps down, power off:
A, B 58 knots (108 km/h; 67 mph) CAS
Max rate of climb at S/L: A 247 m (810 ft)/min
B 270 m (885 ft)/min
Service ceiling: A 4,054 m (13,300 ft)
B 7,925 m (26,000 ft)
T-O run: A 335 m (1,100 ft)
B 314 m (1,030 ft)
T-O to 15 m (50 ft): A 600 m (1,970 ft)
B 567 m (1,860 ft)
Landing from 15 m (50 ft): A, B 457 m (1,500 ft)
Landing run: A, B 233 m (765 ft)
Range, A at recommended lean mixture, with allowances for engine start, taxi, T-O, climb and 45 min reserves at 45% power:
max cruising speed at 1,980 m (6,500 ft) with standard fuel 350 nm (649 km; 403 miles)
max cruising speed at 1,980 m (6,500 ft) with max fuel 525 nm (972 km; 604 miles)
econ cruising speed at 3,050 m (10,000 ft) with standard fuel 470 nm (871 km; 541 miles)
econ cruising speed at 3,050 m (10,000 ft) with max fuel 690 nm (1,279 km; 795 miles)
Range, B with allowances as above:
max cruising speed at 3,050 m (10,000 ft) or 6,100 m (20,000 ft) with standard fuel
300 nm (556 km; 345 miles)
max cruising speed at 6,100 m (20,000 ft) with max fuel 475 nm (880 km; 547 miles)
max cruising speed at 3,050 m (10,000 ft) with max fuel 460 nm (853 km; 530 miles)
econ cruising speed at 6,100 m (20,000 ft) with standard fuel 380 nm (703 km; 437 miles)
econ cruising speed at 3,050 m (10,000 ft) with standard fuel 415 nm (769 km; 478 miles)
econ cruising speed at 6,100 m (20,000 ft) with max fuel 580 nm (1,075 km; 668 miles)
econ cruising speed at 3,050 m (10,000 ft) with max fuel 610 nm (1,130 km; 702 miles)

CESSNA CENTURION and TURBO CENTURION

The original prototype Model 210 (now Centurion), which flew in January 1957, followed the general formula of the Cessna series of all-metal high-wing monoplanes, but was the first to have a retractable tricycle landing gear.

Later versions of the Model 210/Centurion have a fully-cantilever wing, eliminating the bracing struts used on earlier models. Their design was started on 24 October 1964 and construction of a prototype began on 29 November 1964. The first T210 (now Turbo Centurion) with the new wing flew on 18 June 1965.

On 3 December 1970 Cessna announced the introduction of two new versions known as Centurion II and Turbo Centurion II. These differ from the Centurion and Turbo Centurion by having as standard equipment a factory-installed IFR avionics package which offers a cost saving on avionics equipment, plus a gyro panel, dual controls, articulating front seats and all-purpose control wheel. A Pressurised Centurion was introduced in late 1977.

Standard improvements for the 1981 versions of the Centurion include a new avionics cooling fan, cabin overhead ventilators with movable shutters, an improved anti-precipitation static nav antenna, and larger battery contactors. New options include Collins R/Nav and DME-450C which, when combined, provide an IFR-certificated area navigation system, an intercom system, a lightweight microphone headset, an Alcor combustion analyser, and an exchange dual loop nav antenna.

The six current production versions of the Centurion are as follows:

Centurion. Standard model, with 224 kW (300 hp) Continental IO-520-L flat-six engine, driving a McCauley D3A32C88/82NC-2 three-blade constant-speed metal propeller.

Centurion II. Identical to Centurion but with a 720-channel Cessna Series 300 nav/com, ADF, transponder, artificial horizon, directional gyro, economy mixture indicator, true airspeed indicator, reclining and vertically adjustable co-pilot's seat, all-purpose control wheel, dual controls, alternate static source, emergency locator transmitter, heated pitot and stall warning transmitter, navigation light detectors, courtesy lights, instrument post lights, omni-flash beacon, and external power socket as standard. Optional avionics include Nav-O-Matic 200A autopilot, Bendix RDR 160 colour weather radar, a second Srs 400 glideslope receiver, this last-named item included in the optional Nav Pac which includes also a second Series 300 nav/com.

Turbo Centurion. Generally similar to Centurion, but powered by a 231 kW (310 hp) Continental TSIO-520-R turbocharged engine, driving a McCauley three-blade constant-speed metal propeller. Absolute pressure controller, full-flow oil filter, pressurised fuel system, oxygen system, non-congealing oil cooler and overboost control valve standard.

The Turbo Centurion holds an international altitude record for aircraft of this class with a height of 12,906·5 m (42,344 ft).

Turbo Centurion II. Identical to Turbo Centurion but with additional standard equipment as detailed for Centurion II, except that transponder is Series 400. The same optional avionics are available also.

Pressurised Centurion/Pressurised Centurion II. Introduced in late 1977, these pressurised versions of the Centurion are described separately.

The original versions received FAA Type Approval on 23 August 1966. A total of 7,631 Model 210/Centurions, plus an additional 611 Pressurised Centurions, had been delivered by 1 January 1981.

Type: Six-seat cabin monoplane.

Wings: Cantilever high-wing monoplane. Wing section

NACA 64_2A215 at root, NACA 64_1A412 (A=0·5) at tip. Dihedral 1° 30′. Incidence 1° 30′ at root, −1° 30′ at tip. All-metal structure, except for glassfibre conical-camber tips. All-metal Frise-type ailerons. Electrically-actuated all-metal Fowler-type flaps. Ground-adjustable tab in each aileron. Pneumatic de-icing system optional.

FUSELAGE: All-metal semi-monocoque structure.

TAIL UNIT: Cantilever all-metal structure with 36° sweep-back on fin. Fixed-incidence tailplane. Controllable trim tabs in rudder and starboard elevator. Electrical operation of elevator tab optional. Pneumatic de-icing system optional.

LANDING GEAR: Hydraulically-retractable tricycle type with single wheel on each unit. Nose unit retracts forward, main units aft and inward. Chrome vanadium tapered steel tube main legs. Steerable nosewheel with oleo-pneumatic shock-absorber. Cessna main wheels and tube-type tyres, size 6·00-6, pressure 2·90 bars (42 lb/sq in). Cessna nosewheel and tyre, size 5·00-5, pressure 3·10 bars (45 lb/sq in). Cessna hydraulic disc brakes. Parking brake.

POWER PLANT: One flat-six engine, as described under model listings. Electrical de-icing system for propeller optional. Integral fuel tanks in wings, with max total capacity of 340 litres (90 US gallons). Refuelling points above wing. Oil capacity 9·5 litres (2·5 US gallons) in Centurions, 10·5 litres (2·75 US gallons) in Turbo Centurions.

ACCOMMODATION: Six persons in pairs in enclosed cabin. Front two seats of fully-articulating type on Centurion II and Turbo Centurion II (pilot's seat only on other versions). Fifth and sixth seats have folding backs to accommodate articles up to 2·01 m (6 ft 7 in) long. Openable window on port side standard; optional for starboard side. Dual controls standard on Centurion II and Turbo Centurion II (optional on other models). Forward-hinged door on each side of cabin. Baggage space aft of rear seats, capacity 109 kg (240 lb), with outside door on port side. Combined heating and ventilation system. Windscreen defroster standard; electrical anti-icing optional.

SYSTEMS: Integral hydraulic-electric unit for landing gear operation. Hydraulic system for brakes. Electrical power supplied by 24V 60A engine-driven alternator. 24V battery; standby electric generator available optionally for Turbo Centurions. Oxygen system standard on Turbo Centurion, optional for Centurion. Vacuum system and air-conditioning system optional.

AVIONICS AND EQUIPMENT: Optional avionics as for Stationair, except that Series 300 or 400 integrated flight control system is available when the Series 200A, 300A, or 400 Nav-O-Matic autopilot is replaced by the Series 400A two-axis autopilot, which has automatic pitch trim and an optional ILS coupler. Automatic Radial Centering (ARC) is optional with Series 300 nav indicators. Series 400 avionics, including 400B integrated flight control system, and a second glideslope receiver, Series 800 encoding altimeter with altitude alert system, Bendix RDR-160 weather radar, and an exchange dual loop nav antenna, are also available as options. Standard equipment includes electric clock, outside air temperature gauge, rate of climb indicator, sensitive altimeter, turn co-ordinator, armrests, control locks, tinted windscreen and windows, sun visors, stall warning device, baggage restraint net, adjustable fresh air vents, dome lights, electro-luminescent lights for switch and comfort control panels, glareshield and map lights, variable-intensity instrument panel red flood-lights, landing lights, taxi light, navigation lights, quick fuel drains and sampler cup, overall paint scheme, cabin steps and towbar. Optional equipment for all versions includes an Alcor combustion analyser, cabin fire extinguisher, headrests, writing desk, stereo entertainment centre, leather seats, internal corrosion proofing, stretcher installation, ice detector light, wingtip-mounted strobe lights, fin light, engine priming system, glider tow hook, tailplane abrasion boots, and static dischargers. Equipment detailed under Centurion II/Turbo Centurion II model listings is available optionally for Centurion/Turbo Centurion. Optional for the Centurion and Centurion II only are a full-flow oil filter, non-congealing oil cooler, and engine winterisation kit.

DIMENSIONS, EXTERNAL:
- Wing span: 11·20 m (36 ft 9 in)
- Wing chord at root: 1·68 m (5 ft 6 in)
- Wing chord at tip: 1·22 m (4 ft 0 in)
- Wing aspect ratio: 7·66
- Length overall: 8·59 m (28 ft 2 in)
- Height overall: 2·95 m (9 ft 8 in)
- Tailplane span: 3·96 m (13 ft 0 in)
- Wheel track: 2·64 m (8 ft 8 in)
- Wheelbase: 1·75 m (5 ft 9 in)
- Propeller diameter: 2·03 m (6 ft 8 in)
- Passenger doors (each):
  - Height: 1·02 m (3 ft 4¼ in)
  - Width: 0·90 m (2 ft 11¼ in)
  - Height to sill: 0·91 m (3 ft 0 in)
- Baggage compartment door:
  - Height: 0·57 m (1 ft 10½ in)
  - Width: 0·74 m (2 ft 5 in)

Cessna Pressurised Centurion, with underwing weather radar

DIMENSIONS, INTERNAL:
- Cabin: Length: 3·50 m (11 ft 6 in)
  - Max width: 1·08 m (3 ft 6½ in)
  - Max height: 1·23 m (4 ft 0½ in)
  - Floor area: 2·69 m² (29·0 sq ft)
  - Volume: 3·96 m³ (139·9 cu ft)
- Baggage space: 0·46 m³ (16·25 cu ft)

AREAS:
- Wings, gross: 16·25 m² (175 sq ft)
- Ailerons (total): 1·75 m² (18·86 sq ft)
- Trailing-edge flaps (total): 2·74 m² (29·50 sq ft)
- Fin, incl dorsal fin: 0·95 m² (10·26 sq ft)
- Rudder, incl tab: 0·65 m² (6·95 sq ft)
- Tailplane: 1·73 m² (18·57 sq ft)
- Elevators, incl tab: 1·87 m² (20·08 sq ft)

WEIGHTS AND LOADINGS:
- Weight empty: Centurion: 970 kg (2,139 lb)
  - Centurion II: 998 kg (2,201 lb)
  - Turbo Centurion: 1,011 kg (2,228 lb)
  - Turbo Centurion II: 1,038 kg (2,289 lb)
- Max T-O and landing weight:
  - Centurion, Centurion II: 1,724 kg (3,800 lb)
- Max T-O weight:
  - Turbo Centurion, Turbo Centurion II: 1,814 kg (4,000 lb)
- Max landing weight:
  - Turbo Centurion, Turbo Centurion II: 1,724 kg (3,800 lb)
- Max ramp weight:
  - Centurion, Centurion II: 1,729 kg (3,812 lb)
  - Turbo Centurion, Turbo Centurion II: 1,822 kg (4,016 lb)
- Max wing loading:
  - Centurion, Centurion II: 106 kg/m² (21·7 lb/sq ft)
  - Turbo Centurion, Turbo Centurion II: 111·8 kg/m² (22·9 lb/sq ft)
- Max power loading:
  - Centurion, Centurion II: 7·69 kg/kW (12·7 lb/hp)
  - Turbo Centurion, Turbo Centurion II: 7·85 kg/kW (12·9 lb/hp)

PERFORMANCE (at max T-O weight, ISA):
- Max level speed:
  - Centurion, Centurion II at S/L: 175 knots (325 km/h; 202 mph)
  - Turbo Centurion, Turbo Centurion II at 5,180 m (17,000 ft): 204 knots (378 km/h; 235 mph)
- Max cruising speed:
  - Centurion, Centurion II, 75% power at 1,980 m (6,500 ft): 171 knots (317 km/h; 197 mph)
  - Turbo Centurion, Turbo Centurion II, 80% power:
    - at 6,100 m (20,000 ft): 196 knots (364 km/h; 226 mph)
    - at 3,050 m (10,000 ft): 180 knots (333 km/h; 207 mph)
- Stalling speed, flaps up, power off:
  - Centurion, Centurion II: 65 knots (121 km/h; 75 mph) CAS
  - Turbo Centurion, Turbo Centurion II: 67 knots (124 km/h; 77 mph) CAS
- Stalling speed, flaps down, power off:
  - Centurion, Centurion II: 56 knots (104 km/h; 64·5 mph) CAS
  - Turbo Centurion, Turbo Centurion II: 58 knots (108 km/h; 67 mph) CAS
- Max rate of climb at S/L:
  - Centurion, Centurion II: 290 m (950 ft)/min
  - Turbo Centurion, Turbo Centurion II: 283 m (930 ft)/min
- Service ceiling:
  - Centurion, Centurion II: 5,275 m (17,300 ft)
  - Turbo Centurion, Turbo Centurion II: 8,230 m (27,000 ft)
- T-O run:
  - Centurion, Centurion II: 381 m (1,250 ft)
  - Turbo Centurion, Turbo Centurion II: 396 m (1,300 ft)
- T-O to 15 m (50 ft):
  - Centurion, Centurion II: 619 m (2,030 ft)
  - Turbo Centurion, Turbo Centurion II: 658 m (2,160 ft)
- Landing from 15 m (50 ft):
  - all versions: 457 m (1,500 ft)
- Landing run: all versions: 233 m (765 ft)
- Range, Centurion and Centurion II with max fuel, at recommended lean mixture, with allowances for engine start, taxi, T-O, climb and 45 min reserves at 45% power:
  - max cruising speed at 1,980 m (6,500 ft): 805 nm (1,492 km; 927 miles)
  - econ cruising speed at 3,050 m (10,000 ft): 1,065 nm (1,973 km; 1,226 miles)
- Range, Turbo Centurion and Turbo Centurion II with max fuel, allowances as above:
  - max cruising speed at 3,050 m (10,000 ft): 725 nm (1,344 km; 835 miles)
  - max cruising speed at 6,100 m (20,000 ft): 755 nm (1,398 km; 869 miles)
  - econ cruising speed at 3,050 m (10,000 ft): 960 nm (1,778 km; 1,105 miles)
  - econ cruising speed at 6,100 m (20,000 ft): 940 nm (1,741 km; 1,082 miles)

CESSNA PRESSURISED CENTURION and PRESSURISED CENTURION II

On 10 November 1977, Cessna announced the introduction of a pressurised version of the Centurion. It is available in two forms:

Pressurised Centurion. Standard model, generally similar to Centurion, except for installation of 231 kW (310 hp) Continental TSIO-520-P engine with a special high-capacity turbocharger to support the pressurisation system, and pressure cabin. Easily identified by four smaller windows on each side of cabin.

Pressurised Centurion II. Identical to Pressurised Centurion, with additional standard avionics and equipment as described for Centurion II, except that autopilot is Nav-O-Matic 300A and transponder is Series 400.

These Centurions have a cabin pressure differential of 0·23 bars (3·35 lb/sq in), providing a cabin altitude of 3,695 m (12,127 ft) at 7,010 m (23,000 ft). Cabin heat is provided by a double heat exchange system using exhaust system heat. Unlike the average combustion heating unit, this system uses no fuel and is reported to offer outstanding performance even in extremely cold temperatures. A total of 611 Pressurised Centurions had been delivered by 1 January 1981.

The description of the Centurion applies also to the Pressurised Centurion, except as follows:

FUSELAGE: Conventional semi-monocoque structure of light alloy, with fail-safe structure in the pressurised section.

POWER PLANT: One 231 kW (310 hp) Continental TSIO-520-P flat-six engine; otherwise as for Centurion. Oil capacity 10·5 litres (2·75 US gallons).

ACCOMMODATION: As for Centurion, except forward-hinged door on port side of cabin. Large emergency exit on starboard side. Baggage space aft of cabin area, capacity 91 kg (200 lb). Four windows each side of cabin, two overhead windows above the rear seats. Cabin pressurised, heated and ventilated.

SYSTEMS: As for Centurion, except cabin pressurisation system by engine bleed air, max differential 0·23 bars (3·35 lb/sq in). Cabin heated by double heat exchange system using exhaust system heat.

DIMENSIONS, INTERNAL: As for Centurion except:
- Baggage space: 0·52 m³ (18·3 cu ft)

WEIGHTS AND LOADINGS (A: Pressurised Centurion; B: Pressurised Centurion II):
- Weight empty: A: 1,069 kg (2,357 lb)
  - B: 1,097 kg (2,419 lb)
- Max T-O weight: A, B: 1,814 kg (4,000 lb)
- Max landing weight: A, B: 1,723 kg (3,800 lb)

Max ramp weight: A, B 1,821 kg (4,016 lb)
Max wing loading: A, B 111·8 kg/m² (22·9 lb/sq ft)
Max power loading: A, B 7·85 kg/kW (12·9 lb/hp)

Performance (at max T-O weight, ISA, except where indicated):
Max level speed at 5,180 m (17,000 ft) 192 knots (356 km/h; 221 mph)
Max cruising speed, 75% power at 6,100 m (20,000 ft) 181 knots (335 km/h; 208 mph)
Max cruising speed, 75% power at 3,050 m (10,000 ft) 167 knots (309 km/h; 192 mph)
Stalling speed, flaps up, power off 67 knots (124 km/h; 77·5 mph) CAS
Stalling speed, flaps down, power off 58 knots (108 km/h; 67 mph) CAS
Max rate of climb at S/L 283 m (930 ft)/min
Rate of climb at 3,050 m (10,000 ft) 242 m (795 ft)/min
Rate of climb at 6,100 m (20,000 ft) 143 m (470 ft)/min
Max certificated operating altitude 7,010 m (23,000 ft)
T-O run 396 m (1,300 ft)
T-O to 15 m (50 ft) 658 m (2,160 ft)
Landing from 15 m (50 ft) 457 m (1,500 ft)
Landing run 233 m (765 ft)
Range with max fuel, recommended lean mixture, with fuel allowance for engine start, taxi, T-O, climb and 45 min reserves at 45% power:
75% power at 6,100 m (20,000 ft) 645 nm (1,195 km; 743 miles)
75% power at 3,050 m (10,000 ft) 615 nm (1,140 km; 708 miles)
econ cruising speed at 6,100 m (20,000 ft) 830 nm (1,538 km; 955 miles)
econ cruising speed at 3,050 m (10,000 ft) 860 nm (1,593 km; 990 miles)

CESSNA MODEL T303 CRUSADER

Cessna announced on 17 February 1978 the first flight, three days earlier, of a new lightweight twin-engined aircraft known as the Model 303. At that time it was a four-seat aircraft, with 119 kW (160 hp) engines, and made use of bonded structures and a supercritical wing section. It has been superseded by the new Model T303, with six seats, 186 kW (250 hp) turbocharged engines and conventional construction. To be certificated to the latest FAR Pt 23 regulations, this aircraft was undergoing certification flight testing, with deliveries of production aircraft scheduled to begin during 1981. The name Clipper was applied temporarily to the T303 but was changed to Crusader because of a trademark dispute.

Only preliminary details of the T303's specification and performance had been released by mid-1981. Cessna plans to produce 280 Crusaders during its 1982 model year.

Type: Twin-engined cabin monoplane.

Wings: Cantilever low-wing monoplane. Wing section of NACA 23000 series. Structure is primarily of light alloy. Conventional trailing-edge flaps and ailerons. Trim tab in starboard aileron.

Fuselage: Oval-section semi-monocoque structure of light alloy.

Tail Unit: Cantilever structure, primarily of light alloy, with horizontal surfaces mounted partway up fin. Fin and rudder sweptback. Long dorsal fin. Trim tab in rudder and starboard elevator.

Landing Gear: Retractable tricycle type, with single wheel on each unit. Main wheels retract inward, nosewheel forward. Main units of articulated (trailing-link) type. All wheels of same size, with tyres size 6·00-6.

Power Plant: Two 186 kW (250 hp) Continental TSIO-520-AE flat-six turbocharged engines, each driving a McCauley three-blade constant-speed metal propeller with spinner. Propellers are counter-rotating. Fuel in integral wing tanks with combined usable capacity of 549 litres (145 US gallons). Refuelling point on upper surface of each wing.

Accommodation: Standard seating for pilot and five passengers; or pilot, co-pilot and four passengers. Six individual forward-facing seats can be replaced by optional club arrangement. Outboard armrests are standard; inboard retractable armrests and adjustable headrests are normally optional, but are standard with club seating. Wide range of optional cabin furniture and equipment. Clamshell-type two-piece cabin door, with integral airstair, on port side at aft end of cabin. Emergency exit at front of cabin on starboard side. Baggage stowage in nose compartment (with door on port side), wing lockers, and at rear of cabin, with combined capacity of 267 kg (590 lb). Accommodation heated and ventilated.

Avionics and Equipment: Standard avionics include two ARC 485B nav/coms, 400B autopilot, slaved directional gyro, ADF, glideslope and marker beacon receivers and transponder. Optional equipment includes club seating with armrests and adjustable headrests, writing tables, refreshment units, storage drawer, ice chest, and stereo equipment.

Dimensions, external:
Wing span 11·90 m (39 ft 0½ in)

Cessna's new six-seat Model T303 Crusader

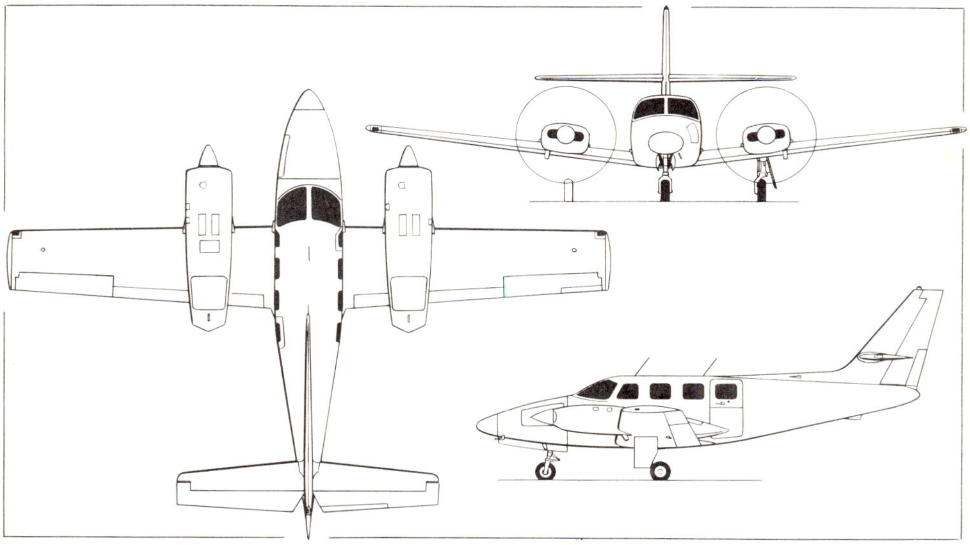

Cessna Model T303 Crusader (two Continental TSIO-520-AE engines) *(Pilot Press)*

Length overall 9·27 m (30 ft 5 in)
Height overall 4·06 m (13 ft 4 in)
Tailplane span 5·18 m (17 ft 0 in)
Wheel track 3·81 m (12 ft 6 in)
Wheelbase 2·29 m (7 ft 6 in)
Propeller diameter 1·98 m (6 ft 6 in)
Propeller ground clearance 0·25 m (10 in)

Area:
Wings, gross 17·6 m² (189·2 sq ft)

Weight (estimated):
Max useful load more than 860 kg (1,900 lb)

Performance (provisional, at max T-O weight, ISA, and in excess of figure quoted unless otherwise stated):
Max level speed at 6,100 m (20,000 ft) 215 knots (398 km/h; 247 mph)
Max cruising speed (72% power at 3,050 m; 10,000 ft) 180 knots (334 km/h; 207 mph)
Max rate of climb at S/L 427 m (1,400 ft)/min
Rate of climb at S/L, one engine out 61 m (200 ft)/min
Certificated operational ceiling 7,620 m (25,000 ft)
Service ceiling, one engine out 3,960 m (13,000 ft)
T-O to 15 m (50 ft) less than 610 m (2,000 ft)
Landing from 15 m (50 ft) less than 610 m (2,000 ft)
Range with max fuel 890 nm (1,648 km; 1,024 miles)

CESSNA MODEL 310 and 310 II

The Model 310 is a twin-engined five/six-seat cabin monoplane, the prototype of which flew on 3 January 1953. It went into production in 1954. The Turbo 310 was added in late 1968, and the first production model was delivered in December 1968. On 21 December 1973 Cessna announced two new versions of the Model 310, known as the 310 II and the Turbo 310 II, which have factory-installed IFR avionics plus other comfort and convenience features as standard. A total of 5,214 commercial examples of the Model 310 had been completed by 1 January 1981, plus 196 delivered to the US Air Force as U-3A/Bs.

The four final versions of the Model 310 were as follows:

310. Standard model, as described in detail, powered by two 212·5 kW (285 hp) Continental IO-520-MB flat-six engines, driving McCauley three-blade fully-feathering constant-speed metal propellers.

310 II. Identical to 310, but having as standard equipment dual 300 Series nav/com with 720-channel com, 200-channel nav, VOR/LOC and VOR/ILS indicators; 300 Series ADF with digital tuning; 400 Series glideslope receiver; marker beacon receiver; transponder; 400B Nav-O-Matic autopilot with approach coupler; associated antennae; avionics cooling kit and panel; economy mixture indicator, outside air temperature gauge, dual controls, six individual seats, emergency locator transmitter, large baggage door, starboard landing light, rotating beacon, taxi light, auxiliary fuel system of 238·5 litres (63 US gallons) capacity, nosewheel fender, static wicks, and external power socket.

Turbo T310. Similar to 310, but with two 212·5 kW (285 hp) Continental TSIO-520-BB turbocharged engines, with automatic propeller synchronisation, full-flow oil filters, absolute and pressure ratio controllers, overboost control valves and engine cowl flaps as standard.

Turbo T310 II. Identical to T310, but with the additional standard equipment as detailed for the 310 II.

The 1981 versions of the Model T310 introduced as standard new threadless blade propellers which are each 2·75 kg (6 lb) lighter in weight than those fitted previously. Production of all four models in the 310 series has now ended.

Type: Twin-engined five- or six-seat monoplane.

Wings: Cantilever low-wing monoplane. Wing section NACA 23018 at centreline, NACA 23009 at tip. Dihedral 5°. Incidence 2° 30′ at root, −0° 30′ at tip. All-metal structure. Electrically-operated split flaps. Trim tab in port aileron. Pneumatic de-icing system optional.

Fuselage: All-metal semi-monocoque structure.

Tail Unit: Cantilever all-metal structure, with 40° sweepback on fin at quarter-chord. Small ventral fin. Trim tabs in rudder and starboard elevator. Electrically-operated elevator trim optional. Pneumatic de-icing system optional.

Landing Gear: Retractable tricycle type. Electromechanical retraction. Cessna oleo shock-absorber struts. Nosewheel steerable to 15° and castoring from 15° to 55° each side. Main wheels size 6·50-10, tyre pressure 4·14 bars (60 lb/sq in). Nosewheel size 6·00-6, tyre pressure 2·76 bars (40 lb/sq in). Goodyear single-disc hydraulic brakes. Hydraulic parking brake.

Power Plant: Two flat-six engines, as described under individual model listings, driving three-blade constant-speed fully-feathering metal propellers. Automatic propeller unfeathering system and propeller de-icing optional; automatic propeller synchrophaser standard for T310 and T310 II, optional for 310 and 310 II. Standard fuel in two permanently attached canted wing-tip tanks, each holding 193 litres (51 US gallons), of which 189 litres (50 US gallons) are usable. Cross-feed fuel system. Optional fuel in two 77·5 litre (20·5 US gallon) rubber fuel cells installed between the wing spars outboard of each engine nacelle, two 43·5 litre (11·5 US gallon) rubber fuel cells further outboard in each wing, and two 77·5 litre (20·5 US gallon) wing locker fuel tanks, providing a maximum fuel capacity of 783 litres (207 US gallons), of which 768 litres (203 US gallons) are usable. Oil capacity 24·6 litres (6·5 US gallons).

Accommodation: Cabin normally seats five, two in front and three on cross-bench behind. Four alternative seating arrangements are available, with up to six individual seats in pairs, all of which can tilt and have fore and aft adjustment, individual air vents, reading lights and magazine pockets. Dual controls optional. Inertia seat-belts for two front seats (optional for rear seats). Pilot's storm window, port side. Cabin windows are double-glazed to reduce noise level. Large door on starboard side giving access to all seats. Cargo door, 1·02 m (3 ft 4 in) wide, for loading of bulky items, standard on 310 II and T310 II, optional on 310 and T310. Baggage compartment at rear of cabin, capacity 163 kg (360 lb), with internal and external access; locker for a further 54·5 kg (120 lb) of baggage in the rear of each engine nacelle; and baggage compartment in extended nose with capacity of 158 kg (350 lb). Total baggage capacity 430 kg (950 lb). Optional cabin accessories include writing desk, window curtains, vertical adjustment of pilot and co-pilot seats, all-leather seats, oxygen system and photographic survey provisions. Windscreen defrosting standard; windscreen alcohol de-icing system, and electrically-heated windscreen panel optional.

Systems: Electrical system powered by two 28V 50A engine-driven alternators and 24V 25Ah battery. 100A alternators optional. Oxygen system of 2·17 m³ (76·6 cu ft) or 1·37 m³ (48·3 cu ft) capacity optional; an automatic altitude compensating regulator is standard with this installation. Janitrol 45,000 BTU thermostatically-controlled blower-type heater for cabin heating and windscreen defrosting. Cabin air-conditioning system rated at 12,000 BTU optional. Vacuum system supplied by two engine-driven pumps with adequate capacity to cater for the pneumatic de-icing boots and flight instruments. Hydraulic system for brakes only.

Avionics and Equipment: Optional avionics include Series 300 nav/com transceiver with 720-channel com and 200-channel nav with remote VOR/LOC or VOR/ILS indicator, ADF with digital tuning, 10-channel HF and flight director system; Series 400 nav/com transceiver with 720-channel com and 200-channel nav with remote VOR/LOC or VOR/ILS indicator, 40-channel glideslope, ADF with digital tuning and BFO, transponder with 4096 code capability, encoding altimeter, Nav-O-Matic 400B two-axis autopilot and integrated flight control system with optional RMI or HSI; or Series 1000 com transceiver, nav receiver, ADF, and glideslope receiver, with Series 800 DME and R/Nav system. Additional avionics options include Bendix RDR-160 colour weather radar, KNC 610 area nav, AVQ-75 DME, KN 65 DME, radar altimeter, locator beacon, yaw damper, boom microphone and headset. Standard equipment includes blind-flying instrumentation, quartz crystal clock, sensitive altimeter, control locks, pilot and co-pilot safety belts, sun visors, alternator failure warning, heated pitot, heated stall warning transmitter, heater overheat warning, landing gear warning device, armrests, cabin radio speaker, baggage straps, adjustable air vents, emergency exit window, aft omni-vision window, tinted dual-pane windows, hatshelf, super soundproofing, map light, instrument post lights, variable-intensity emergency floodlight, reading lights, navigation light detectors, landing light, navigation lights, full-flow oil coolers, heated fuel vents, nosewheel fender, overall paint scheme, quick-drain fuel valves, retractable cabin step, and towbar. Optional equipment for Model 310/Turbo T310, but standard for 310 II/Turbo T310 II, as detailed in model listings. Optional equipment for all versions includes angle of attack indicator, co-pilot's blind-flying instrumentation, digital clock, instantaneous rate of climb indicator, flight hour recorder, synchronous tachometer, true airspeed indicator, turn co-ordinator, heated static sources, vertically adjustable seats with inertia-reel shoulder harness for pilot and co-pilot, pilot's relief tube, rudder pedal locks, cabin curtain, rear window curtains, cabin fire extinguisher, 'total flood' cabin fire extinguisher, eight-track stereo with cabin speakers and stereo headsets, internal corrosion proofing, electro-luminescent instrument panel lighting, baggage compartment courtesy lights, courtesy light timer, ice detection light, anti-collision light, second landing light, taxi light, three-light strobe system, carpet for nose baggage area, fuselage ice protection plates, photographic provisions, anti-icing kit, engine fire detection and extinguishing system, and radome nose. Additional optional items for the Model 310 and 310 II include automatic propeller synchrophaser and partial oxygen system plumbing.

Dimensions, external:
Wing span 11·25 m (36 ft 11 in)
Wing chord at root 1·71 m (5 ft 7½ in)
Wing chord at tip 1·16 m (3 ft 9½ in)
Wing aspect ratio 7·3
Length overall 9·74 m (31 ft 11½ in)
Height overall 3·25 m (10 ft 8 in)
Tailplane span 5·18 m (17 ft 0 in)
Wheel track 3·59 m (11 ft 9½ in)
Wheelbase 2·80 m (9 ft 2¼ in)
Propeller diameter: 310, 310 II 1·94 m (6 ft 4½ in)
T310, T310 II 1·98 m (6 ft 6 in)

Dimensions, internal:
Baggage compartment (cabin) 1·26 m³ (44·6 cu ft)
Baggage compartments (nacelles, total) 0·52 m³ (18·5 cu ft)
Baggage compartment (nose) 0·59 m³ (21 cu ft)

Areas:
Wings, gross 16·63 m² (179 sq ft)
Ailerons (total) 1·32 m² (14·17 sq ft)
Trailing-edge flaps (total) 2·14 m² (23·06 sq ft)
Fin, incl ventral fin 1·74 m² (18·70 sq ft)
Rudder, incl tab 1·09 m² (11·76 sq ft)
Tailplane 2·99 m² (32·15 sq ft)
Elevators, incl tab 2·05 m² (22·10 sq ft)

Weights and Loadings:
Weight empty: 310 1,523 kg (3,358 lb)
310 II 1,635 kg (3,604 lb)
T310 1,573 kg (3,467 lb)
T310 II 1,685 kg (3,714 lb)
Max ramp weight: all versions 2,511 kg (5,535 lb)
Max T-O weight: all versions 2,495 kg (5,500 lb)
Max landing weight: all versions 2,449 kg (5,400 lb)
Max zero-fuel weight:
310, 310 II 2,223 kg (4,900 lb)
T310, T310 II 2,275 kg (5,015 lb)
Max wing loading:
all versions 150 kg/m² (30·73 lb/sq ft)
Max power loading:
all versions 5·87 kg/kW (9·65 lb/hp)

Performance (at max T-O weight, ISA, except speeds at mid-cruise weight):
Never-exceed speed:
all versions 227 knots (420 km/h; 261 mph) CAS
Max level speed:
310 at S/L 207 knots (383 km/h; 238 mph)
T310 at 4,875 m (16,000 ft) 237 knots (439 km/h; 273 mph)
Max cruising speed:
310, 75% power at 2,285 m (7,500 ft) 195 knots (361 km/h; 225 mph)
T310, 73·6% power at 3,050 m (10,000 ft) 201 knots (372 km/h; 231 mph)
T310, 73·6% power at 6,100 m (20,000 ft) 223 knots (413 km/h; 257 mph)
Econ cruising speed with max fuel:
310 at 3,050 m (10,000 ft) 139 knots (258 km/h; 160 mph)
T310 at 6,100 m (20,000 ft) 171 knots (317 km/h; 197 mph)
Min control speed (Vmc):
all versions 81 knots (150 km/h; 93 mph)
Stalling speed, flaps and landing gear up, power off:
all versions 78 knots (145 km/h; 90 mph) CAS
Stalling speed, flaps and landing gear down, power off:
all versions 70 knots (130 km/h; 81 mph) CAS
Max rate of climb at S/L:
310 507 m (1,662 ft)/min
T310 518 m (1,700 ft)/min
Rate of climb at S/L, one engine out:
310 113 m (370 ft)/min
T310 119 m (390 ft)/min
Service ceiling: 310 6,020 m (19,750 ft)
T310 8,350 m (27,400 ft)
Service ceiling, one engine out:
310 2,255 m (7,400 ft)
T310 5,245 m (17,200 ft)
T-O run: 310 407 m (1,335 ft)
T310 398 m (1,306 ft)
T-O to 15 m (50 ft): 310 518 m (1,700 ft)
T310 507 m (1,662 ft)
Landing from 15 m (50 ft):
all versions, at 2,449 kg (5,400 lb) 546 m (1,790 ft)
Landing run:
all versions, at 2,449 kg (5,400 lb) 195 m (640 ft)
Range, recommended lean mixture, allowances for start, taxi, T-O, climb and 45 min reserves at selected cruise power:
310, 310 II, max cruising speed at 2,285 m (7,500 ft) with 272 kg (600 lb) usable fuel 440 nm (816 km; 507 miles)
310, 310 II, as above with 552 kg (1,218 lb) usable fuel 1,078 nm (1,997 km; 1,241 miles)
310, 310 II, econ cruising speed at 3,050 m (10,000 ft) with 272 kg (600 lb) usable fuel 631 nm (1,170 km; 727 miles)
310, 310 II, as immediately above with 552 kg (1,218 lb) usable fuel 1,534 nm (2,842 km; 1,766 miles)
T310, T310 II, max cruising speed at 6,100 m (20,000 ft) with 272 kg (600 lb) usable fuel 469 nm (869 km; 540 miles)
T310, T310 II, as immediately above with 552 kg (1,218 lb) usable fuel 1,197 nm (2,218 km; 1,378 miles)
T310, T310 II, econ cruising speed at 6,100 m (20,000 ft) with 272 kg (600 lb) usable fuel 586 nm (1,086 km; 675 miles)
T310, T310 II, as immediately above with 552 kg (1,218 lb) usable fuel 1,451 nm (2,689 km; 1,671 miles)

CESSNA MODEL 318E DRAGONFLY

Under the above Model designation, Cessna supplied a total of 577 A-37B COIN aircraft to the US and other air forces. During 1981, Nos. 45 and 47 Squadrons of the US Air Force Reserve were re-equipping from A-37Bs on to Fairchild A-10A Thunderbolt IIs. The Dragonflies are being converted for forward air control duty by the USAF, under the new designation **OA-37B**. They are to serve with Nos. 103 and 172 Squadrons of the Air National Guard, replacing O-2As.

CESSNA MODEL 335

Cessna introduced in 1979 this six-seat twin-engined aircraft, which was essentially a lighter-weight unpressurised version of the Model 340A. After 45 had been sold, production was ended during 1980. Details of the Model 335 can be found in the 1980-81 *Jane's*.

CESSNA MODEL 337 SKYMASTER and SKYMASTER II

USAF designation: O-2

This unorthodox 4/6-seat business aircraft was marketed for 18 years before production was terminated in 1980. A total of 2,054 Skymasters and 332 Pressurised Skymasters had been built by Cessna, plus 67 and 27 respectively by Reims Aviation in France. In addition, 513 O-2A military Skymasters were built, of which 31 were modified to O-2B standard. Details of these aircraft can be found in the 1980-81 and earlier editions of *Jane's*. Versions of the Model 337 for various military applications have been modified by Summit Aviation (which see).

CESSNA MODEL 340A

Cessna announced on 8 December 1971 the introduction of a pressurised twin-engined business aircraft designated Model 340. Developed from the Model 310, it had a wing and landing gear generally similar to those of the Model 414, a pressurised fuselage of fail-safe design, a tail unit similar to that of the Model 310, and 212·5 kW (285 hp) Continental TSIO-520-K engines. The Model 340 II followed, with factory-installed avionics as standard, and in 1978 a Model 340A III was introduced.

Versions of the Model 340 available in 1981 are as follows:

340A. Standard model, as described in detail.

340A II. Identical to Model 340A, but with dual Series 300 nav/coms with 720-channel com and 200-channel nav and VOR/LOC and VOR/ILS indicators, ADF, Series 400 DME, marker beacon receiver, glideslope, transponder and 400B two-axis Nav-O-Matic autopilot, basic avionics kit, avionics cooling kit, audio system, marker beacon receiver audio muting switch, and all necessary

Cessna Model 310 II twin-engined five/six-seat cabin monoplane

antennae for onboard avionics as standard. Other standard equipment includes economy mixture indicator, outside air temperature gauge, dual controls, electric elevator trim, emergency locator transmitter, cabin pressure control system, rear baggage compartment door, starboard landing light, taxi light, strobe lights, 238·5 litre (63 US gallon) auxiliary fuel system, nosewheel fender, static dischargers, and external power socket.

340A III. As Model 340A, plus an avionics package comprising dual Series 400 nav/coms, one with HSI and the other with Automatic Radial Centering (ARC), Series 400 dual glideslope, R/Nav, ADF, marker beacon receiver, DME, transponder, 400B IFCS, Series 800 encoding altimeter and altitude alerter, Series 1000 RMI, Bendix RDR-160 weather radar, yaw damper, basic avionics kit, avionics cooling, audio system, all essential antennae, hand fire extinguisher, and 100A alternators added and/or substituted for the equipment installed on the 340A II. Standard equipment is as detailed for the Model 340A II.

All 1981 models have as standard new threadless blade propellers which are each 2·75 kg (6 lb) lighter in weight than the type fitted previously. The airframe warranty of the Model 340 has been extended to one year.

A total of 1,155 Model 340s had been delivered by 1 January 1981.

Type: Six-seat pressurised business aircraft.

Wings: Cantilever low-wing monoplane, with 'Stabila-tip' fixed wingtip fuel tanks. Wing section NACA 23018 (modified) at aircraft centreline, NACA 23015 (modified) at centre-section/outer wing junction, NACA 23009 (modified) at tip. Dihedral 5° on outer panels. Incidence 2° 30′ at root, −0° 30′ at tip. All-metal two-spar structure. All-metal ailerons of single-spar construction; controllable trim tab in starboard aileron. Electrically-actuated all-metal trailing-edge split flaps, of single-spar construction with lower skin, comprising an inboard and outboard panel on each wing. Optional pneumatic de-icing system.

Fuselage: All-metal semi-monocoque structure. The pressurised cabin section, extending from station 100·00 aft to station 252·00, is of fail-safe construction. All openings are reinforced with doublers and frame members, and longitudinal continuity is provided by lightweight extruded T-section stringers.

Tail Unit: Cantilever all-metal structure with swept vertical surfaces. Fixed-incidence tailplane of conventional two-spar construction. Elevators of single-spar construction, with controllable trim tab in starboard elevator. Rudder, built up on a formed channel spar and transverse ribs, has a controllable trim tab. Optional pneumatic de-icing system.

Landing Gear: Retractable tricycle type, with single wheel on each unit. Electro-mechanical retraction, main units inward into wings and faired by doors when retracted, nose unit rearward into the fuselage nose and faired by two doors when retracted. Mechanically-operated emergency gear extension system. Cessna oleo-pneumatic shock-absorbers. Steerable nosewheel with shimmy damper and self-centering device. Mainwheel tyres size 6·50-10 (8-ply); nosewheel tyre size 6·00-6 (6-ply). Single-disc hydraulic brakes. Parking brake.

Power Plant: Two 231 kW (310 hp) Continental TSIO-520-NB flat-six turbocharged fuel-injection engines, each driving a McCauley three-blade constant-speed and fully-feathering metal propeller. Propeller anti-icing system optional. Fuel system, max usable capacity 768 litres (203 US gallons), as described for Model 310. Manifold pressure relief valves to prevent engines from overboosting are standard equipment. Oil capacity 24·6 litres (6·5 US gallons).

Accommodation: Standard seating for pilot and co-pilot on tilting and individually adjustable seats. Individual seats for four passengers, two forward-facing on the port side, one aft-facing and one forward-facing on starboard side. Door, on port side aft of wing, is two-piece type with built-in airstairs in bottom portion. Plug-type emergency escape hatch on starboard side of cabin, over wing. Foul-weather window for pilot. Baggage accommodated in nose compartment with external access doors, capacity 159 kg (350 lb), two wing lockers, capacity 54·5 kg (120 lb) each, and in rear cabin area, capacity 154 kg (340 lb). Total baggage capacity 422 kg (930 lb). Cabin pressurised, heated and ventilated. Air-conditioning optional. Windscreen defroster standard; windscreen alcohol de-icing system optional.

Systems: Electrical system powered by two 28V 50A engine-driven alternators and 24V 25Ah battery. 100A alternators optional, standard on 340A III. Vacuum system supplied by two engine-driven pumps. Hydraulic system for brakes only. Cabin pressurised by engine bleed air, max differential 0·29 bars (4·2 lb/sq in). Cabin heated by Stewart Warner 45,000 BTU gasoline heater. Lightweight air-conditioning system optional. Oxygen system, 0·31 m³ (11·0 cu ft) or 2·17 m³ (76·6 cu ft) capacity, optional.

Avionics and Equipment: Optional avionics for the Model 340A are as detailed for the Model 310. Standard equipment of Model 340A includes attitude and directional gyros, quartz crystal clock, cabin altitude and differential pressure gauge, rate of climb indicator, sensitive altimeter, control locks, pilot and co-pilot safety belts, sun visors, 'Door not locked' warning light, heater overheat warning light, stall warning device, heated pitot, heated stall warning transmitter, cabin speaker, super soundproofing, courtesy lights, instrument post lights, variable intensity floodlights, reading lights, retractable landing light in port wing, navigation lights with flasher unit, two rotating beacons, full-flow oil filters, quick-drain fuel valves, heated fuel vents, propeller synchrophaser, towbar, and all-over paint scheme. Optional items, except as detailed in model listings, include angle of attack indicator, blind-flying instrumentation for co-pilot, digital clock, economy mixture indicator, flight hour recorder, instantaneous rate of climb indicator, true airspeed indicator, inertia-reel shoulder harness for pilot and co-pilot, rudder pedal lock, emergency locator transmitter, heated dual static source, flight deck divider curtain, all-leather seats, headrests, tinted double-pane cabin windows, window curtains, refreshment centre, writing desk, stereo system, internal corrosion proofing, cabin fire extinguisher, 'total flood' cabin fire extinguisher system, baggage compartment courtesy light, courtesy light timer, ice detection light, taxi light, white strobe lights, engine fire detection and extinguishing system, propeller unfeathering system, fuselage ice impact panels, radome nose and dual pitot system.

Cessna Model 340A III six-seat pressurised business aircraft

Dimensions, external:
- Wing span — 11·62 m (38 ft 1·3 in)
- Wing chord at root — 1·714 m (5 ft 7·4 in)
- Wing chord, mean aerodynamic — 1·57 m (5 ft 1·68 in)
- Wing chord at tip — 1·156 m (3 ft 9·6 in)
- Wing aspect ratio — 7·2
- Length overall — 10·46 m (34 ft 4 in)
- Height overall — 3·84 m (12 ft 7 in)
- Tailplane span — 5·18 m (17 ft 0 in)
- Wheel track — 3·93 m (12 ft 10·7 in)
- Wheelbase — 3·11 m (10 ft 2·6 in)
- Propeller diameter — 1·94 m (6 ft 4½ in)
- Passenger door: Height — 1·25 m (4 ft 1¼ in)
  - Width — 0·53 m (1 ft 9 in)
- Emergency hatch: Height — 0·50 m (1 ft 7½ in)
  - Width — 0·66 m (2 ft 2 in)

Dimensions, internal:
- Cabin: Length, incl baggage compartment — 3·86 m (12 ft 8 in)
  - Max width — 1·18 m (3 ft 10½ in)
  - Max height — 1·24 m (4 ft 1 in)
  - Volume (total) — 4·6 m³ (162·4 cu ft)
- Baggage space: Cabin — 0·52 m³ (18·5 cu ft)
  - Nose — 0·44 m³ (15·5 cu ft)
  - Engine nacelles (each) — 0·13 m³ (4·625 cu ft)

Areas:
- Wings, gross — 17·09 m² (184 sq ft)
- Ailerons (total) — 1·06 m² (11·44 sq ft)
- Trailing-edge flaps (total) — 2·14 m² (23·06 sq ft)
- Fin — 1·51 m² (16·20 sq ft)
- Ventral fin — 0·54 m² (5·80 sq ft)
- Rudder, incl tab — 1·09 m² (11·76 sq ft)
- Tailplane — 2·99 m² (32·15 sq ft)
- Elevators, incl tab — 2·05 m² (22·10 sq ft)

Weights and Loadings:
- Weight empty: 340A — 1,778 kg (3,921 lb)
  - 340A II — 1,885 kg (4,155 lb)
  - 340A III — 1,926 kg (4,246 lb)
- Max T-O and landing weight — 2,717 kg (5,990 lb)
- Max ramp weight — 2,733 kg (6,025 lb)
- Max zero-fuel weight — 2,554 kg (5,630 lb)
- Max wing loading — 158·9 kg/m² (32·55 lb/sq ft)
- Max power loading — 5·88 kg/kW (9·66 lb/hp)

Performance (at max T-O weight, ISA, except speeds are those at mid-cruise weight):
- Never-exceed speed — 234 knots (433 km/h; 269 mph) CAS
- Max level speed at 6,100 m (20,000 ft) — 244 knots (452 km/h; 281 mph)
- Max cruising speed, 74·8% power:
  - at 7,470 m (24,500 ft) — 229 knots (425 km/h; 264 mph)
  - at 3,050 m (10,000 ft) — 198 knots (367 km/h; 228 mph)
- Econ cruising speed:
  - at 7,620 m (25,000 ft) with 272 kg (600 lb) usable fuel — 170 knots (315 km/h; 196 mph)
  - at 7,620 m (25,000 ft) with 552 kg (1,218 lb) usable fuel — 172 knots (319 km/h; 198 mph)
- Stalling speed, flaps up, power off — 79 knots (146 km/h; 91 mph) CAS
- Stalling speed, flaps down, power off — 71 knots (132 km/h; 82 mph) CAS
- Max rate of climb at S/L — 503 m (1,650 ft)/min
- Rate of climb at S/L, one engine out — 96 m (315 ft)/min
- Service ceiling — 9,085 m (29,800 ft)
- Service ceiling, one engine out — 4,815 m (15,800 ft)
- T-O run — 492 m (1,615 ft)
- T-O to 15 m (50 ft) — 663 m (2,175 ft)
- Landing from 15 m (50 ft) — 564 m (1,850 ft)
- Landing run — 235 m (770 ft)
- Range, recommended lean mixture, with allowances for start, taxi, T-O, climb, descent and 45 min reserves at selected cruise power:
  - 74·8% power at 7,470 m (24,500 ft) with 272 kg (600 lb) usable fuel — 418 nm (774 km; 481 miles)
  - 74·8% power at 7,470 m (24,500 ft) with 552 kg (1,218 lb) usable fuel — 1,106 nm (2,049 km; 1,273 miles)
  - 74·8% power at 3,050 m (10,000 ft) with 272 kg (600 lb) usable fuel — 391 nm (724 km; 450 miles)
  - 74·8% power at 3,050 m (10,000 ft) with 552 kg (1,218 lb) usable fuel — 986 nm (1,827 km; 1,135 miles)
  - econ cruising power at 7,620 m (25,000 ft) with 272 kg (600 lb) usable fuel — 530 nm (982 km; 610 miles)
  - econ cruising power at 7,620 m (25,000 ft) with 552 kg (1,218 lb) usable fuel — 1,377 nm (2,551 km; 1,585 miles)
  - econ cruising power at 3,050 m (10,000 ft) with 272 kg (600 lb) usable fuel — 579 nm (1,073 km; 667 miles)
  - econ cruising power at 3,050 m (10,000 ft) with 552 kg (1,218 lb) usable fuel — 1,405 nm (2,603 km; 1,618 miles)

CESSNA MODEL 402C

The original Model 402 was intended for the third-level airline market, with a convertible cabin and reinforced cabin floor of bonded crushed honeycomb construction, enabling it to be changed quickly from a ten-seat commuter to a light cargo transport. On 8 December 1971 Cessna renamed the original Model 402 as the Model 402 Utiliner and introduced a version designated Model 402 Businessliner. On 29 October 1975 Mk II versions of both aircraft were made available, each including a package of factory-installed equipment and avionics as standard; Model 402 III versions of both aircraft were introduced in 1978.

The 1981 Model 402C versions introduced as standard new threadless blade propellers which are each 2·75 kg (6 lb) lighter in weight than the type fitted previously. A 'Fasten seat belt, no smoking' sign, with instructions in international symbols, is available for Utiliners.

The five versions available for 1981 were as follows:

Model 402C Utiliner. Basic version, as described in detail.

Model 402C Businessliner. As basic version, except six/eight seats and optional side-hinged door, next to standard cabin door, to provide a total loading door width of 1·02 m (3 ft 4 in). Other options include folding business desks, stereo equipment, refreshment centre and cabin dividers.

Model 402C II Utiliner. As basic version, plus the

following factory-installed avionics and equipment as standard: Cessna Series 300 nav/com with 720-channel com, 200-channel nav and VOR/ILS and VOR/LOC indicators, Series 300 ADF, Series 400 marker beacon receiver, glideslope and transponder, Series 400B Nav-O-Matic autopilot, basic avionics kit, antennae, avionics cooling kit and audio system; economy mixture indicator, flight hour recorder, dual controls, electric elevator trim, marker beacon receiver audio muting switch, emergency locator transmitter, cabin fire extinguisher, starboard landing light, taxi light, nosewheel fender, external power socket, and static wicks.

Model 402C II Businessliner. As Model 402C Businessliner, plus standard factory-installed equipment and avionics detailed for Model 402C II Utililiner.

Model 402C III Businessliner. As Model 402C II Businessliner, except that the avionics package comprises Series 400 nav/com with HSI; second Series 400 nav/com with ARC; dual Series 400 glideslope; Series 400 ADF, DME, marker beacon receiver, R/Nav, transponder; Series 800 encoding altimeter and altitude alerter; Series 1000 RMI; Bendix RDR-160 weather radar; Series 400B IFCS; yaw damper; and basic avionics kit, audio system, avionics cooling and antennae. Standard equipment includes 100A instead of 50A engine-driven alternators.

The same prototype served for Models 401 and 402, and the FAA Type Certificate, awarded on 20 September 1966, also covered both types. A total of 1,346 Model 402s had been built by 1 January 1981. Twelve were delivered in 1975 to the Royal Malaysian Air Force, which uses ten of them for multi-engine training and the other two for photographic and liaison missions.

TYPE: Ten-seat (optional nine-seat) convertible passenger/freight transport (Utililiner) or six/eight-seat business aircraft (Businessliner).

WINGS: Cantilever low-wing monoplane. Wing section NACA 23018 (modified) at aircraft centreline, NACA 23015 (modified) at centre-section/outer wing junction, NACA 23009 (modified) at tip. All-metal two-spar structure of light alloy with stamped ribs and surface skins reinforced with spanwise stringers. Outer wing panels of bonded construction. All-metal ailerons and electrically-actuated split flaps. Trim tab in port aileron. Optional pneumatic de-icing system.

FUSELAGE: All-metal semi-monocoque structure.

TAIL UNIT: Cantilever all-metal structure, with 40° sweepback on fin at quarter-chord. Fixed-incidence tailplane. Trim tabs in rudder and starboard elevator. Electrical operation of trim tabs optional. Optional pneumatic de-icing system.

LANDING GEAR: Hydraulically-retractable tricycle type. Main units retract inward into wings, nosewheel unit rearward. No doors over main wheels when retracted. Emergency extension system. Oleo-pneumatic shock-absorbers. Steerable nosewheel. Cleveland heavy-duty wheels. Main-wheel tyres size 6·50-10, nosewheel tyre size 6·00-6. Cleveland heavy-duty hydraulic brakes. Parking brakes.

POWER PLANT: Two 242 kW (325 hp) Continental TSIO-520-VB flat-six turbocharged engines, each driving a three-blade constant-speed fully-feathering metal propeller. Propeller synchrophaser, automatic unfeathering and electrical de-icing optional. Integral wing fuel tanks with total capacity of 808 litres (213 US gallons), of which 780 litres (206 US gallons) are usable. Oil capacity 24·6 litres (6·5 US gallons).

ACCOMMODATION: Two seats side by side in pilot's compartment. Dual controls standard on Model 402C II/III versions, optional for Model 402C versions. The Utililiner cabin has four individual seats in pairs and two double seats. Passenger seats are 'Enviro-form' moulded honeycomb, glassfibre reinforced. Businessliner has four individual seats as standard, two additional seats optional, in the main cabin. Refreshment centre at aft end of cabin. Passenger reading lights standard on Businessliner, optional on Utililiner. Door with built-in airstair on port side of cabin at centre. Storm windows for pilot and co-pilot. Tinted cabin windows. An emergency escape hatch is provided on the starboard side of the cabin. Optional cargo door and crew door available. Baggage contained in area at rear of cabin, nose compartment, and wing locker at rear of each engine nacelle, with combined capacity of 680 kg (1,500 lb). Rear double seat on Utililiner can have fold-down back to facilitate access to rear baggage shelf. Cabin heated and ventilated. Windscreen defrosting standard. Electrical anti-icing of pilot's window or alcohol anti-icing of pilot's and co-pilot's windows optional.

SYSTEMS: Electrical system powered by two 24V 50A alternators. 24V 25Ah battery. Battery can be sited optionally in nose baggage area. 100A alternators optional, standard on 402C III version. Hydraulic system for brakes only. Vacuum system provided by two engine-driven pumps. Oxygen system of 1·25 m³ (44 cu ft) or 3·25 m³ (115 cu ft) capacity optional. Air-conditioning system optional. Heating and ventilation system with 45,000 BTU gasoline heater standard.

AVIONICS AND EQUIPMENT: Optional avionics as detailed for Model 310, plus radio telephone. Standard equipment includes attitude and directional gyros, quartz crystal clock, outside air temperature gauge, sensitive altimeter, pilot and co-pilot safety belts, control locks, sun visors, alternator failure lights, door 'Not locked' light, heater overheat warning light, emergency locator transmitter, stall warning device, armrests, cabin speaker, cabin air ventilators, super soundproofing, courtesy lights, map light, variable-intensity floodlight and instrument post lights, navigation light detectors, navigation lights, retractable landing light, towbar, and all-over paint scheme. Optional equipment, unless part of Model 402C II or III packages, includes co-pilot's blind-flying instrumentation, digital clock, economy mixture indicator, flight hour recorder, instantaneous rate of climb indicator, true airspeed indicator, pilot's and co-pilot's inertia-reel harness, rudder lock, dual heated static source, cabin hand fire extinguisher, 'total flood' cabin fire extinguisher, stereo system, Utililiner or Businessliner interiors (including flight deck divider curtains, window curtains, headrests, reading lights, 'Seat belt' and 'No smoking' signs and various arrangements of seats, tables, refreshment units and toilets), internal corrosion proofing, ice detection light, second retractable landing light, taxi light, three-light strobe system, vertical tail floodlight and photographic provisions.

DIMENSIONS, EXTERNAL:
- Wing span: 13·45 m (44 ft 1½ in)
- Wing chord at root: 1·77 m (5 ft 9¾ in)
- Wing chord at tip: 1·05 m (3 ft 5½ in)
- Length overall: 11·09 m (36 ft 4½ in)
- Height overall: 3·49 m (11 ft 5½ in)
- Tailplane span: 5·18 m (17 ft 0 in)
- Wheel track: 5·48 m (17 ft 11½ in)
- Wheelbase: 3·18 m (10 ft 5¼ in)
- Propeller diameter: 1·94 m (6 ft 4½ in)
- Passenger door (standard):
  - Height: 1·27 m (4 ft 2 in)
  - Width: 0·61 m (2 ft 0 in)
- Cargo door (optional): Height: 1·26 m (4 ft 1½ in)
  - Width: 1·05 m (3 ft 5¼ in)
- Nose baggage doors (each):
  - Height: 0·51 m (1 ft 8 in)
  - Width: 0·80 m (2 ft 7½ in)
- Nacelle baggage doors (each):
  - Length: 0·61 m (2 ft 0 in)
  - Width: 0·62 m (2 ft 0½ in)

DIMENSIONS, INTERNAL:
- Cabin: Length: 4·83 m (15 ft 10 in)
  - Max width: 1·42 m (4 ft 8 in)
  - Max height: 1·30 m (4 ft 3 in)
  - Volume: 6·30 m³ (222·4 cu ft)

AREAS:
- Wings, gross: 20·98 m² (225·8 sq ft)
- Fin: 2·46 m² (26·50 sq ft)
- Rudder, incl tab: 1·84 m² (19·85 sq ft)
- Tailplane: 4·02 m² (43·32 sq ft)
- Elevators, incl tab: 1·64 m² (17·63 sq ft)

WEIGHTS AND LOADINGS:
- Weight empty: Businessliner: 1,845 kg (4,069 lb)
  - Utililiner: 1,859 kg (4,098 lb)
  - Businessliner II: 1,912 kg (4,215 lb)
  - Utililiner II: 1,922 kg (4,238 lb)
  - Businessliner III: 1,960 kg (4,322 lb)
- Max T-O and landing weight: 3,107 kg (6,850 lb)
- Max ramp weight: 3,123 kg (6,885 lb)
- Max zero-fuel weight: 2,955 kg (6,515 lb)
- Max wing loading: 148·1 kg/m² (30·3 lb/sq ft)
- Max power loading: 6·42 kg/kW (10·5 lb/hp)

PERFORMANCE (at max T-O weight, ISA, except speeds are those at mid-cruise weight):
- Never-exceed speed: 231 knots (428 km/h; 266 mph) CAS
- Max level speed at 4,875 m (16,000 ft): 231 knots (428 km/h; 266 mph)
- Max cruising speed, 72% power:
  - at 6,100 m (20,000 ft) 213 knots (394 km/h; 245 mph)
  - at 3,050 m (10,000 ft) 194 knots (359 km/h; 223 mph)
- Econ cruising speed:
  - at 6,100 m (20,000 ft) with 272 kg (600 lb) usable fuel: 164 knots (304 km/h; 189 mph)
  - at 6,100 m (20,000 ft) with 560 kg (1,236 lb) usable fuel: 166 knots (307 km/h; 191 mph)
  - at 3,050 m (10,000 ft) with 272 kg (600 lb) usable fuel: 141 knots (261 km/h; 162 mph)
  - at 3,050 m (10,000 ft) with 560 kg (1,236 lb) usable fuel: 142 knots (263 km/h; 164 mph)
- Stalling speed, flaps up, power off: 78 knots (145 km/h; 90 mph) CAS
- Stalling speed, flaps down, power off: 68 knots (126 km/h; 78 mph) CAS
- Max rate of climb at S/L: 442 m (1,450 ft)/min
- Rate of climb at S/L, one engine out: 92 m (301 ft)/min
- Service ceiling: 8,200 m (26,900 ft)

Photograph and three-view drawing ***(Pilot Press)*** **of the Cessna Model 402C, current version of the Businessliner/Utililiner**

Service ceiling, one engine out 4,510 m (14,800 ft)
T-O run 537 m (1,763 ft)
T-O to 15 m (50 ft) 669 m (2,195 ft)
Landing from 15 m (50 ft) 757 m (2,485 ft)
Landing run 322 m (1,055 ft)
Range, recommended lean mixture, allowances for start, taxi, T-O, climb, descent and 45 min reserves at selected cruise power:
72% power at 6,100 m (20,000 ft) with 272 kg (600 lb) usable fuel 362 nm (671 km; 417 miles)
72% power at 6,100 m (20,000 ft) with 561 kg (1,236 lb) usable fuel 983 nm (1,822 km; 1,132 miles)
72% power at 3,050 m (10,000 ft) with 272 kg (600 lb) usable fuel 349 nm (647 km; 402 miles)
72% power at 3,050 m (10,000 ft) with 561 kg (1,236 lb) usable fuel 915 nm (1,695 km; 1,053 miles)
econ cruising power at 6,100 m (20,000 ft) with 272 kg (600 lb) usable fuel 459 nm (850 km; 528 miles)
econ cruising power at 6,100 m (20,000 ft) with 561 kg (1,236 lb) usable fuel 1,233 nm (2,285 km; 1,420 miles)
econ cruising power at 3,050 m (10,000 ft) with 272 kg (600 lb) usable fuel 499 nm (924 km; 574 miles)
econ cruising power at 3,050 m (10,000 ft) with 561 kg (1,236 lb) usable fuel 1,273 nm (2,359 km; 1,466 miles)

CESSNA MODEL 414A CHANCELLOR

Cessna introduced the pressurised twin-engined Model 414 on 10 December 1969 as a 'step-up' aircraft for owners of Cessna or other light unpressurised twins. It combined the basic fuselage and tail unit of the Model 421 with the wing of the Model 402 and had 231 kW (310 hp) turbocharged Continental engines.

It was replaced in 1978 by the similar but much improved Model 414A Chancellor. Major changes included a new bonded 'wet' wing of increased span, extended nose and baggage area, and introduction of an external access door to the tailcone. Three versions are available for 1981, as follows:

Chancellor. Standard version, as described in detail.

Chancellor II. As standard Chancellor, but with the following avionics and equipment as standard: dual Series 400 nav/com, with ARC, VOR/ILS and VOR/LOC; Series 400 ADF, DME, transponder, glideslope, marker beacon receiver; marker beacon receiver audio muting switch; 400B Nav-O-Matic autopilot; basic avionics kit, cooling kit, audio system, and all essential antennae; economy mixture indicator, flight hour recorder, co-pilot's blind-flying instrumentation, emergency locator transmitter, cabin pressure control system, cabin hand fire extinguisher, starboard landing light, taxi light, high-intensity strobe light, external power socket, nosewheel fender and static dischargers.

Chancellor III. As standard model, with standard equipment of Chancellor II plus cabin air-conditioning system and 100A alternators, and an all-weather avionics package which includes dual Series 1000 com, 1000 nav with HSI, 1000 nav with ARC, dual 1000 glideslope, 1000 ADF, RMI, Series 800 DME, R/Nav, transponder, encoding altimeter and altitude alerter; 800B IFCS; Series 400 marker beacon receiver; AA-100 radio altimeter; Bendix RDR-160 weather radar; Series 1000 audio panel, basic avionics kit, cooling kit, and all essential antennae.

The versions for 1981 introduce as standard improvements new threadless blade propellers which are each 2·75 kg (6 lb) lighter in weight than the type used previously; an engine oil heated fuel manifold valve for protection against icing, and frost-eliminating double-glazed panes on cockpit side windows. Polished propeller spinners are available as a new option.

A total of 513 of the original Model 414s were built before introduction of the more advanced Model 414A, and a combined total of 918 Model 414/414As had been delivered by 1 January 1981.

Type: Six/eight-seat pressurised light transport.

Wings: Cantilever low-wing monoplane. Wing section NACA 23018 (modified) at aircraft centreline, NACA 23015 (modified) at centre-section/outer wing junction, NACA 23009 (modified) at tip. Dihedral 5° on outer panels. Incidence 2° 30′ at root, —0° 30′ at tip. All-metal two-spar structure of light alloy with stamped ribs and surface skins reinforced with spanwise stringers. Outer wing panels of bonded construction. All-metal ailerons and electrically-actuated split flaps. Trim tab in port aileron. Optional pneumatic de-icing system.

Fuselage: Conventional all-metal semi-monocoque structure, with fail-safe construction in the pressurised section.

Tail Unit: Cantilever all-metal structure, with sweptback vertical surfaces. Fixed-incidence tailplane. Trim tabs in rudder and starboard elevator. Optional pneumatic de-icing system.

Landing Gear: Hydraulically-retractable tricycle type, main units retracting inward, nosewheel unit aft. Emergency extension by means of a 138 bar (2,000 lb/sq in) rechargeable nitrogen bottle. Oleo-pneumatic shock-absorbers. Steerable nosewheel. Main-wheel tyres size 6·50-10 (8-ply), nosewheel tyre size 6·00-6 (6-ply). Goodyear single-disc hydraulic brakes. Parking brakes.

Cessna Chancellor light transport with turbocharged engines

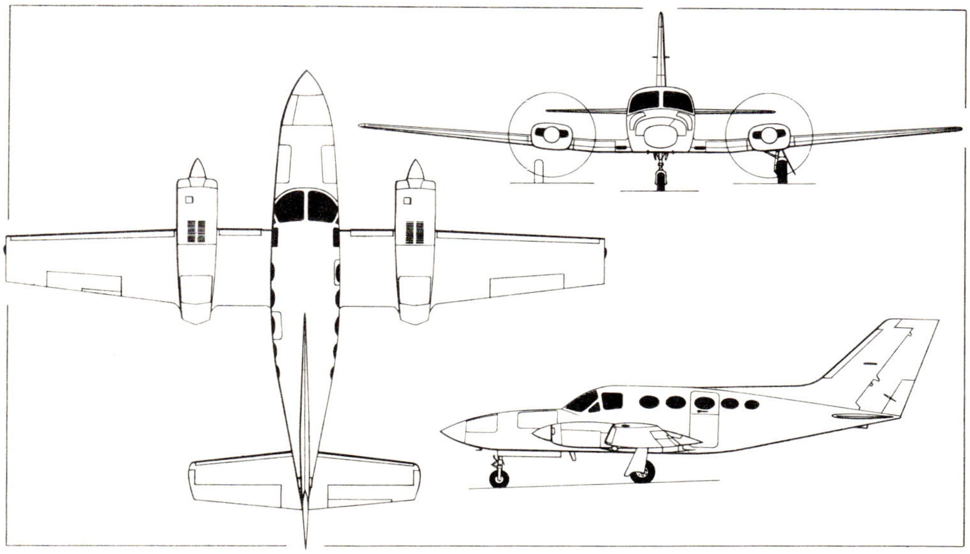

Cessna Chancellor six/eight-seat pressurised light transport *(Pilot Press)*

Power Plant: Two 231 kW (310 hp) Continental TSIO-520-NB flat-six turbocharged engines, each driving a McCauley 3AF32C93M/82NC-5·5 three-blade constant-speed fully-feathering metal propeller. Propeller synchrophasers standard; unfeathering pressure accumulator and electrical blade de-icing system optional. Fuel system with max usable capacity of 808 litres (213 US gallons). Oil capacity 24·6 litres (6·5 US gallons).

Accommodation: Two seats side by side in pilot's compartment. Optional curtain, or solid divider with curtain, to separate pilot's compartment from main cabin. Standard seating arrangement for four forward-facing passenger seats. Optional arrangements provide for front passenger seats to face aft and forward-facing seventh and eighth seats. Individual consoles each include reading light and ventilator. Optional items include executive writing desk, tables, hatshelf, stereo equipment, vertically-adjustable pilot's and co-pilot's seats, refreshment and Thermos units, fore and aft cabin dividers, electric shaver converter, all-leather seats, passenger instrument console (clock, true airspeed indicator and altimeter) and intercom. Door is two-piece type with built-in airstairs in bottom portion, on port side of cabin at rear. Plug-type emergency escape hatch on starboard side of cabin. Double-pane cabin windows. Foul-weather windows for pilot and co-pilot, on each side of fuselage. Windscreen defroster standard, electrical or alcohol windscreen de-icing optional. Baggage accommodated in nose compartment with external access doors, capacity 159 kg (350 lb), two wing lockers, capacity 54·5 kg (120 lb) each, and in rear cabin area, capacity 226 kg (500 lb). Total baggage capacity 494 kg (1,090 lb). External access door to tailcone on starboard side.

Systems: Cabin pressurisation system, max differential 0·34 bars (5·0 lb/sq in), maintains sea level cabin conditions to an altitude of 3,350 m (11,000 ft), and a 3,050 m (10,000 ft) cabin altitude to a height of 8,075 m (26,500 ft). Electrical system powered by two engine-driven 28V 50A alternators. 24V 25Ah battery. 28V 100A alternators optional, standard on Chancellor III version. Hydraulic system for operation of landing gear and brakes. Vacuum system for blind-flying instrumentation and optional wing and tail unit de-icing system. Oxygen system of 3·25 m³ (114·9 cu ft) capacity, or emergency oxygen system of 0·31 m³ (11·0 cu ft) capacity optional. Air-conditioning system optional, standard on Chancellor III.

Avionics and Equipment: The various versions of the Model 414A have avionics as detailed in the model listings, plus radio telephone with cockpit control. Optional avionics available for the basic Chancellor include those detailed for the Chancellor II/III, and alternative items from the Cessna Series 400, 800 and 1000 range are available for all versions. Standard equipment includes blind-flying instrumentation, quartz crystal clock, outside air temperature gauge, sensitive altimeter, control locks, dual controls, sun visors, alternator failure lights, aircraft systems monitoring device, heater overheat warning light, heated pitot and stall warning transmitter, stall warning device, door 'Not locked' warning light, armrests, cabin air ventilators, window curtains, super soundproofing, instrument post lights, aft cabin light, courtesy lights, reading lights, navigation light detectors, retractable landing light, navigation lights, rotating beacon, full-flow oil filters, non-congealing oil coolers, quick-drain fuel valves, heated fuel vents, towbar, and overall paint scheme. Equipment detailed as standard for the Chancellor II/III is available optionally for the basic Chancellor. Other optional items for all versions include angle of attack indicator, digital clock, digital fuel flow gauge with computer, instantaneous rate of climb indicator, true airspeed indicator, rudder pedal lock, electric elevator trim, flight deck/cabin divider or curtain, table, refreshment centre, tinted windows, 7th and 8th seats, 'Fasten seat belt' and 'Oxygen' signs, toilet with privacy curtain, internal corrosion proofing, ventilation fan, 'total flood' cabin fire extinguisher, 8-track stereo installation, timer for courtesy lights, ice detection lights, fuselage ice impact panels, dual pitot system, radome nose, and engine fire detection and extinguishing system.

Dimensions, external:
Wing span 13·45 m (44 ft 1½ in)
Wing chord at root 1·77 m (5 ft 9¾ in)
Wing chord at tip 1·05 m (3 ft 5½ in)
Length overall 11·09 m (36 ft 4½ in)
Height overall 3·49 m (11 ft 5½ in)
Tailplane span 5·18 m (17 ft 0 in)
Wheel track 5·47 m (17 ft 11¾ in)
Wheelbase 3·18 m (10 ft 5¼ in)
Propeller diameter 1·94 m (6 ft 4½ in)
Passenger door: Height 1·30 m (4 ft 3¼ in)
Width 0·63 m (2 ft 0¾ in)
Height to sill 1·21 m (3 ft 11½ in)

Dimensions, internal:
Cabin: Length 4·42 m (14 ft 6 in)
Max width 1·40 m (4 ft 7 in)
Max height 1·29 m (4 ft 3 in)
Volume 6·11 m³ (215·6 cu ft)

AREAS:
- Wings, gross — 20·98 m^2 (225·8 sq ft)
- Fin — 2·37 m^2 (25·53 sq ft)
- Rudder, incl tab — 1·46 m^2 (15·72 sq ft)
- Tailplane — 4·15 m^2 (44·62 sq ft)
- Elevators, incl tab — 1·49 m^2 (16·08 sq ft)

WEIGHTS AND LOADINGS:
- Weight empty: Chancellor — 1,976 kg (4,356 lb)
  - Chancellor II — 2,053 kg (4,526 lb)
  - Chancellor III — 2,160 kg (4,763 lb)
- Max T-O and landing weight — 3,062 kg (6,750 lb)
- Max ramp weight — 3,078 kg (6,785 lb)
- Max zero-fuel weight — 2,955 kg (6,515 lb)
- Max wing loading — 145·94 kg/m^2 (29·89 lb/sq ft)
- Max power loading — 6·63 kg/kW (10·89 lb/hp)

PERFORMANCE (at max T-O weight, ISA, except speeds are those at mid-cruise weight):
- Never-exceed speed — 232 knots (430 km/h; 267 mph) CAS
- Max level speed at 6,100 m (20,000 ft) — 235 knots (436 km/h; 271 mph)
- Cruising speed, 74·8% power at 7,470 m (24,500 ft) — 224 knots (415 km/h; 258 mph)
- Cruising speed, 74·8% power at 3,050 m (10,000 ft) — 193 knots (357 km/h; 222 mph)
- Econ cruising speed at 7,620 m (25,000 ft) with 408 kg (900 lb) usable fuel — 183 knots (339 km/h; 211 mph)
- Econ cruising speed at 3,050 m (10,000 ft) with 408 kg (900 lb) usable fuel — 143 knots (265 km/h; 165 mph)
- Stalling speed, flaps up, power off: all versions — 82 knots (152 km/h; 95 mph) CAS
- Stalling speed, flaps down, power off: all versions — 72 knots (133 km/h; 83 mph) CAS
- Max rate of climb at S/L — 463 m (1,520 ft)/min
- Rate of climb at S/L, one engine out — 88 m (290 ft)/min
- Service ceiling — 9,390 m (30,800 ft)
- Service ceiling, one engine out — 6,050 m (19,850 ft)
- T-O run — 666 m (2,185 ft)
- T-O to 15 m (50 ft) — 791 m (2,595 ft)
- Landing from 15 m (50 ft) at max landing weight — 729 m (2,393 ft)
- Landing run at max landing weight — 309 m (1,013 ft)
- Range, recommended lean mixture with allowances for start, taxi, T-O, climb, descent and 45 min reserves at selected cruise power:
  - 74·8% power at 7,470 m (24,500 ft) with 272 kg (600 lb) usable fuel — 404 nm (748 km; 465 miles)
  - 74·8% power at 7,470 m (24,500 ft) with 561 kg (1,236 lb) usable fuel — 1,099 nm (2,036 km; 1,265 miles)
  - 74·8% power at 3,050 m (10,000 ft) with 272 kg (600 lb) usable fuel — 382 nm (708 km; 440 miles)
  - 74·8% power at 3,050 m (10,000 ft) with 561 kg (1,236 lb) usable fuel — 984 nm (1,823 km; 1,133 miles)
  - econ cruising power at 7,620 m (25,000 ft) with 272 kg (600 lb) usable fuel — 482 nm (893 km; 555 miles)
  - econ cruising power at 7,620 m (25,000 ft) with 561 kg (1,236 lb) usable fuel — 1,293 nm (2,396 km; 1,489 miles)
  - econ cruising power at 3,050 m (10,000 ft) with 272 kg (600 lb) usable fuel — 532 nm (986 km; 613 miles)
  - econ cruising power at 3,050 m (10,000 ft) with 561 kg (1,236 lb) usable fuel — 1,327 nm (2,459 km; 1,528 miles)

CESSNA MODEL 421 GOLDEN EAGLE

On 28 October 1965, Cessna announced a pressurised twin-engined business aircraft designated Model 421, the prototype of which had flown for the first time on 14 October 1965. FAA type approval was received on 1 May 1967 and deliveries began in the same month.

Two developed versions of the Model 421 were produced subsequently as the 421B Golden Eagle and 421B Executive Commuter, remaining in production until replaced by the Model 421C Golden Eagle in 1976. Three versions were available for 1981, as follows:

Model 421C Golden Eagle. Standard version, as described in detail.

Model 421C Golden Eagle II. As basic model above, but with the following avionics and equipment as standard: dual Series 400 nav/com, one with VOR/ILS, the other with VOR/LOC, ADF, transponder, DME, glideslope, marker beacon receiver, 400B Nav-O-Matic autopilot, basic avionics and avionics cooling kits, associated antennae, audio system, marker beacon receiver audio muting switch, and slaved directional gyro; co-pilot's blind-flying instrumentation, propeller synchrophaser, economy mixture indicator, flight hour recorder, emergency locator beacon, cabin pressure control system, cabin hand fire extinguisher, starboard landing light, taxi light, high intensity strobe lights, external power socket, 100A alternators, static dischargers and nosewheel fender.

Model 421C Golden Eagle III. As basic model, with standard equipment of Golden Eagle II, plus cabin air-conditioning system and 100A alternators, and an all-weather avionics package which includes dual Series 1000 com and glideslope, Series 1000 nav with HSI, nav with ARC, ADF, RMI and audio panel; Series 800 DME, R/Nav, transponder, encoding altimeter and altitude alerter; 800B IFCS; Series 400 marker beacon receiver; AA-100 radio altimeter; RDR-160 weather radar; basic avionics and cooling kits, and all associated antennae.

The 1981 versions of the Golden Eagle introduce as standard improvements an engine oil heated fuel manifold valve for protection against icing, a pneumatic lower cabin door extender, an improved fuel injection system, and frost-eliminating double-glazed panes on cockpit side windows. Polished propeller spinners are available as a new option.

A total of 1,740 Model 421s had been delivered by 1 January 1981. Recent orders include three for the Royal New Zealand Air Force.

The description which follows applies to the Model 421C Golden Eagle:

TYPE: Six/eight-seat pressurised light transport.

WINGS, FUSELAGE, TAIL UNIT: As for Model 414A.

LANDING GEAR: Hydraulically-retractable tricycle type, main units retracting inward, nosewheel unit aft. Emergency extension by means of a 138 bar (2,000 lb/sq in) rechargeable nitrogen bottle. Oleo-pneumatic shock-absorbers. Main units of articulated (trailing-link) type. Steerable nosewheel. Main-wheel tyres size 6·50-10 (8-ply), nosewheel tyre 6·00-6 (6-ply). Goodyear single-disc hydraulic brakes. Parking brake.

POWER PLANT: Two 280 kW (375 hp) Continental GTSIO-520-N flat-six geared and turbocharged engines, each driving a McCauley three-blade fully-feathering constant-speed metal propeller. McCauley propeller synchrophaser optional. Standard total fuel capacity is 808 litres (213 US gallons), of which 780 litres (206 US gallons) are usable, contained in 'wet' wing. Optional wing locker tanks provide a maximum usable capacity of 991 litres (262 US gallons). Oil capacity 26·5 litres (7 US gallons).

ACCOMMODATION: Generally the same as for Model 414A; seats have tapered backs and headrests. The nose compartment can contain a total of 272 kg (600 lb) of baggage and avionics, and two wing lockers an additional 91 kg (200 lb) each, plus 226 kg (500 lb) in the rear cabin area, making a total capacity of 680 kg (1,500 lb). Dual-setting electrically-heated windscreen optional.

SYSTEMS, AVIONICS AND EQUIPMENT: Generally as for Model 414A. Hydraulic system for landing gear operation supplied by dual engine-driven pumps, pressure 103·5 bars (1,500 lb/sq in). Rechargeable nitrogen bottle for emergency extension of landing gear, pressure 138 bars (2,000 lb/sq in).

DIMENSIONS, EXTERNAL:
- Wing span — 12·53 m (41 ft 1½ in)
- Wing chord at root — 1·77 m (5 ft 9¾ in)
- Wing chord at tip — 1·14 m (3 ft 8½ in)
- Length overall — 11·09 m (36 ft 4½ in)
- Height overall — 3·49 m (11 ft 5½ in)
- Tailplane span — 5·18 m (17 ft 0 in)
- Wheel track — 5·28 m (17 ft 4 in)
- Wheelbase — 3·20 m (10 ft 6 in)
- Propeller diameter — 2·29 m (7 ft 6 in)

AREA:
- Wings, gross — 19·97 m^2 (215 sq ft)

WEIGHTS AND LOADINGS:
- Weight empty: Golden Eagle — 2,105 kg (4,640 lb)
  - Golden Eagle II — 2,194 kg (4,837 lb)
  - Golden Eagle III — 2,290 kg (5,048 lb)
- Max T-O weight — 3,379 kg (7,450 lb)
- Max ramp weight — 3,402 kg (7,500 lb)
- Max zero-fuel weight — 3,054 kg (6,733 lb)
- Max landing weight — 3,266 kg (7,200 lb)
- Max wing loading — 169·4 kg/m^2 (34·7 lb/sq ft)
- Max power loading — 6·03 kg/kW (9·9 lb/hp)

PERFORMANCE (at max T-O weight, ISA, except speeds are those at mid-cruise weight):
- Never-exceed speed — 238 knots (441 km/h; 274 mph) CAS
- Max level speed at 6,100 m (20,000 ft) — 258 knots (478 km/h; 297 mph)
- Max cruising speed, 73·5% power at 7,620 m (25,000 ft) — 241 knots (447 km/h; 278 mph)
- Max cruising speed, 73·5% power at 3,050 m (10,000 ft) — 208 knots (386 km/h; 240 mph)
- Econ cruising speed at 7,620 m (25,000 ft) with 637 kg (1,404 lb) usable fuel — 192 knots (356 km/h; 221 mph)

Photograph and three-view drawing *(Pilot Press)* **of the Cessna Model 421C Golden Eagle pressurised light transport**

Econ cruising speed at 3,050 m (10,000 ft) with 637 kg (1,404 lb) usable fuel
153 knots (283 km/h; 176 mph)
Stalling speed, flaps up, power off:
all versions 83 knots (154 km/h; 96 mph) CAS
Stalling speed, flaps down, power off:
all versions 74 knots (137 km/h; 85 mph) CAS
Max rate of climb at S/L 591 m (1,940 ft)/min
Rate of climb at S/L, one engine out
107 m (350 ft)/min
Service ceiling 9,205 m (30,200 ft)
Service ceiling, one engine out 4,540 m (14,900 ft)
T-O run 544 m (1,786 ft)
T-O to 15 m (50 ft) 708 m (2,323 ft)
Landing from 15 m (50 ft) 699 m (2,293 ft)
Landing run 219 m (720 ft)
Range, recommended lean mixture, with allowances for start, taxi, T-O, climb, descent and 45 min reserves at selected cruise power:
73·5% power at 7,620 m (25,000 ft) with 561 kg (1,236 lb) usable fuel
882 nm (1,633 km; 1,015 miles)
73·5% power at 7,620 m (25,000 ft) with 713 kg (1,572 lb) usable fuel
1,197 nm (2,218 km; 1,378 miles)
73·5% power at 3,050 m (10,000 ft) with 561 kg (1,236 lb) usable fuel
790 nm (1,464 km; 910 miles)
73·5% power at 3,050 m (10,000 ft) with 713 kg (1,572 lb) usable fuel
1,060 nm (1,963 km; 1,220 miles)
econ cruising power at 7,620 m (25,000 ft) with 561 kg (1,236 lb) usable fuel
1,088 nm (2,016 km; 1,253 miles)
econ cruising power at 7,620 m (25,000 ft) with 713 kg (1,572 lb) usable fuel
1,483 nm (2,748 km; 1,708 miles)
econ cruising power at 3,050 m (10,000 ft) with 561 kg (1,236 lb) usable fuel
1,107 nm (2,052 km; 1,275 miles)
econ cruising power at 3,050 m (10,000 ft) with 713 kg (1,572 lb) usable fuel
1,485 nm (2,752 km; 1,710 miles)

CESSNA MODEL 425 CORSAIR

Cessna introduced for 1980 a twin-turboprop business aircraft known as the Model 425 Corsair, which is based on the airframe of the Model 421 Golden Eagle. Design began on 1 November 1977, and construction of a prototype was initiated three months later, on 30 January 1978. This flew for the first time on 12 September 1978, and construction of a pre-production example was started during 1979. FAA certification was gained by mid-1980. Initial deliveries of production aircraft were made in November 1980; and seven Corsairs had been delivered by 1 January 1981.

Type: Six/eight-seat pressurised light transport.

Wings: Cantilever low-wing monoplane. Wing section NACA 23018-63 (modified) at root, NACA 23009-63 (modified) at tip. Dihedral 5° on outer panels. Incidence 2° 30′ at root, −0° 30′ at tip. All-metal two-spar structure of light alloy, with stamped ribs and surface skins reinforced with spanwise stringers. Outer wing panels of bonded construction. All-metal ailerons and electrically-operated trailing-edge split flaps. Trim tab in port aileron. Optional pneumatic de-icing of wing leading-edges.

Fuselage: Conventional all-metal semi-monocoque structure, with fail-safe construction in the pressurised section.

Tail Unit: Conventional all-metal cantilever structure, with sweptback vertical surfaces. Tailplane has dihedral of 12°. Trim tab in starboard elevator, with dual heavy-duty actuator. Trim tab in rudder. Optional pneumatic de-icing of fin and tailplane leading-edges.

Landing Gear: Hydraulically-retractable tricycle type, main units retracting inward, nosewheel aft. Oleo-pneumatic shock-absorbers, with main units of articulated (trailing-link) type. Steerable nosewheel. Single wheel on each unit. Main-wheel tyres size 6·50 × 10, 8-ply rating, pressure 4·83 bars (70 lb/sq in). Nosewheel tyre size 6·00 × 6, 6-ply rating, pressure 2·42 bars (35 lb/sq in).

Power Plant: Two Pratt & Whitney Aircraft of Canada PT6A-112 turboprop engines, flat rated to 335·5 kW (450 shp), each driving a Hartzell wide-chord three-blade constant-speed fully-feathering and reversible-pitch metal propeller. Propellers autofeather when engines are at rest. Fuel contained in integral tanks in outer wing panels, nacelle cells, and inboard collector tanks, with combined capacity of 1,411 litres (372·8 US gallons), of which 1,385 litres (366 US gallons) are usable. Refuelling point above each engine nacelle. Oil capacity 19·75 litres (5·25 US gallons). Engine inlet ducts have a separator mechanism to prevent ingestion of water. Propeller de-icing and synchrophaser standard. Engine fire detection system standard.

Accommodation: Two seats side by side in pilot's compartment, with dual controls. Optional curtain, or solid divider with curtain, to separate pilot's compartment from main cabin. Standard seating provides for four passengers, but optional arrangements have the front passenger seats facing aft and forward-facing seventh and eighth seats. Optional equipment includes storage drawers, refreshment centre, tables, toilet, radio telephone, stereo system, and aft cabin divider. Door is of two-piece type, with built-in airstairs in bottom portion, on port side of cabin at rear. Plug-type emergency escape hatch overwing on starboard side of cabin. Foul weather windows on each side of fuselage for pilot and co-pilot. Baggage accommodated in nose with external access doors, capacity 272 kg (600 lb), and in rear of cabin area, capacity 227 kg (500 lb). Total baggage capacity 499 kg (1,100 lb). Accommodation is pressurised, heated and air-conditioned. Electrically heated windscreen optional.

Systems: Freon air-conditioning system of 17,500 BTU capacity, plus engine bleed air and electric boost heating. Pressurisation system with max differential of 0·35 bars (5·0 lb/sq in) provides a 3,050 m (10,000 ft) cabin altitude to 8,075 m (26,500 ft), or 3,625 m (11,900 ft) cabin altitude to 9,145 m (30,000 ft). Electrical system includes a 28V 250A starter/generator on the starboard engine and a 40Ah nickel-cadmium battery. Hydraulic system for operation of landing gear and brakes. Vacuum system for blind-flying instrumentation and optional wing and tail unit de-icing. Oxygen system standard.

Avionics and Equipment: Standard avionics include Cessna Series 400 dual nav/coms, dual glideslope, ADF, DME, marker beacon receiver, transponder, and encoding altimeter; Series 1000 autopilot with 3 in HSI, and yaw damper, audio panel, basic avionics kit, avionics cooling kit, and all associated antennae. Optional avionics include Cessna Series 1000 equipment or Collins Pro Line. Standard equipment includes flight hour recorder; co-pilot's blind-flying instrumentation; cabin pressure control system; emergency locator transmitter; cabin fire extinguisher; starboard landing light; navigation, taxi and strobe lights; external power socket, nosewheel fender, and static wicks.

Dimensions, external:
Wing span 13·45 m (44 ft 1½ in)
Wing chord at root 1·77 m (5 ft 9¾ in)
Wing chord at tip 1·05 m (3 ft 5½ in)
Wing aspect ratio 8·65
Length overall 10·93 m (35 ft 10¼ in)
Height overall 3·84 m (12 ft 7¼ in)
Tailplane span 5·82 m (19 ft 1 in)
Wheel track 5·30 m (17 ft 4½ in)
Wheelbase 3·20 m (10 ft 6 in)
Propeller diameter 2·37 m (7 ft 9½ in)
Propeller ground clearance 0·23 m (9¼ in)
Passenger door: Height 1·30 m (4 ft 3¼ in)
Width 0·63 m (2 ft 8 in)
Height to sill 1·21 m (3 ft 11½ in)

Dimensions, internal:
Cabin: Length 4·42 m (14 ft 6 in)
Max width 1·40 m (4 ft 7 in)
Max height 1·29 m (4 ft 3 in)
Volume 6·11 m³ (215·6 cu ft)
Nose baggage compartment volume 0·93 m³ (33 cu ft)

Areas:
Wings, gross 20·90 m² (224·98 sq ft)
Fin 2·37 m² (25·53 sq ft)
Rudder (incl tab) 1·46 m² (15·72 sq ft)
Horizontal tail surfaces (total) 5·76 m² (61·99 sq ft)

Weights and Loadings:
Weight empty, equipped 2,209 kg (4,870 lb)
Max fuel weight 1,115 kg (2,459 lb)
Max T-O weight 3,720 kg (8,200 lb)
Max ramp weight 3,753 kg (8,275 lb)
Max zero-fuel weight 3,057 kg (6,740 lb)
Max landing weight 3,629 kg (8,000 lb)
Max wing loading 178·0 kg/m² (36·45 lb/sq ft)
Max power loading 5·54 kg/kW (9·11 lb/shp)

Performance (at max T-O weight, ISA, except where indicated):
Never-exceed speed
Mach 0·52 (229 knots; 424 km/h; 263 mph) CAS
Max cruising speed at 5,400 m (17,700 ft) at AUW of 3,175 kg (7,000 lb) 264 knots (489 km/h; 304 mph)
Stalling speed, flaps and landing gear up, engines idling 88 knots (163 km/h; 102 mph) CAS
Stalling speed, flaps and landing gear down, engines idling 79 knots (147 km/h; 91 mph) CAS
Max rate of climb at S/L 618 m (2,027 ft)/min
Rate of climb at S/L, one engine out
132 m (434 ft)/min
Service ceiling 10,575 m (34,700 ft)
Service ceiling, one engine out 5,640 m (18,500 ft)
T-O run 624 m (2,047 ft)
T-O to 15 m (50 ft) 714 m (2,341 ft)
Landing from 15 m (50 ft) 654 m (2,145 ft)
Landing run 290 m (952 ft)

Cessna Model 425 Corsair six/eight-seat twin-turboprop pressurised transport

Cessna Corsair (two Pratt & Whitney Aircraft of Canada PT6A-112 turboprop engines) *(Pilot Press)*

Range with max fuel at max cruising power, allowances for start, taxi, T-O, climb to cruise altitude, descent, and 45 min reserves at max cruise power:
at 6,100 m (20,000 ft)
1,020 nm (1,889 km; 1,173 miles)
at 9,145 m (30,000 ft)
1,406 nm (2,604 km; 1,618 miles)
Range with max fuel at max range power, allowances as above, and 45 min reserves at max range power:
at 6,100 m (20,000 ft)
1,370 nm (2,537 km; 1,576 miles)
at 9,145 m (30,000 ft)
1,646 nm (3,048 km; 1,854 miles)

CESSNA TITAN

On 16 July 1975 Cessna Aircraft Company announced that it was developing this twin-engined business/commuter/cargo aircraft, designated Model 404 Titan. The model number was deleted subsequently, and the aircraft is known currently as the Cessna Titan. It was designed to carry a nominal 1,588 kg (3,500 lb) useful load out of a 771 m (2,530 ft) airstrip. A prototype flew for the first time on 26 February 1975. Deliveries began in October 1976, and a total of 330 had been delivered by 1 January 1981.

The Titan offers an increase of more than 30 per cent in ton-miles per gallon by comparison with the Cessna 402. Its cabin, which is almost 5·79 m (19 ft 0 in) long, is designed for rapid conversion to satisfy cargo, commuter and executive transport roles.

The Cessna Titan is available in seven versions for 1981, as follows:

Titan Ambassador. Standard version, configured and equipped for passenger carrying, as described in detail.

Titan Ambassador II. As standard version, but with the following avionics and equipment: dual Series 300 nav/com with VOR/ILS and VOR/LOC; Series 300 ADF; Series 400 transponder, glideslope, marker beacon receiver; 400B Nav-O-Matic autopilot; basic avionics kit, marker beacon receiver audio muting switch, audio system, avionics cooling, and all associated antennae; economy mixture indicator, flight hour recorder, dual controls, emergency locator beacon, cabin fire extinguisher, starboard landing light, taxi light, external power socket, static dischargers, and nosewheel fender.

Titan Ambassador III. As Titan Ambassador, with standard equipment of Ambassador II plus heavy-duty brakes and 100A alternators, and an avionics package which includes Series 400 nav/com with HSI; 400 nav/com with ARC; dual 400 glideslope; 400 ADF, DME, marker beacon receiver, R/Nav, transponder; Series 800 encoding altimeter and altitude alert; Series 1000 RMI; 400B IFCS; Bendix RDR-160 weather radar; yaw damper, basic avionics kit, audio system, avionics cooling, and all associated antennae.

Titan Courier. Standard utility version for passenger/cargo role.

Titan Courier II. As standard Courier, but with the avionics and equipment detailed for the Ambassador II.

Titan Freighter. Cargo version with specially designed cabin walls and ceiling of impact resistant polycarbonate material and floor stressed to withstand loading of 976 kg/m^2 (200 lb/sq ft). Cargo handling facilities include the retention of floor tracks to provide tiedown attachment points, floor pallets to provide durable flooring flush with the tracks, five high-capacity cargo retaining nets. Cargo door 1·26 m (4 ft 1½ in) high and 1·24 m (4 ft 1 in) wide replaces normal cabin door on port side, and crew door is provided on port side over wing.

Titan Freighter II. As Titan Freighter, but with avionics and equipment detailed for Ambassador II.

The 1981 versions of the Titan have the TBO of the Continental GTSIO-520-M engines extended to 1,600 hours, and this applies retrospectively to aircraft delivered prior to introduction of the 1981 models. New for 1981 on Courier models is an optional 'Fasten seat belts, no smoking' sign with instructions in international symbols.

Type: Two/ten-seat passenger/executive/cargo transport.

Wings: Cantilever low-wing monoplane. Wing section NACA 23018 at root, NACA 23012 at tip. Dihedral 3° 30′ on wing centre-section, 4° 55′ on outer panels. Incidence 2° at root, −1° at construction tip. All-metal three-spar centre-section structure to meet FAR 23 fail-safe requirements: two-spar structure for outer wing panels. Hydraulically-operated Fowler-type trailing-edge flaps of light alloy construction. Plain ailerons of light alloy construction. Trim tab in port aileron. Pneumatic de-icing system optional.

Fuselage: All-metal semi-monocoque structure.

Tail Unit: Cantilever structure of light alloy with swept surfaces. Fixed-incidence tailplane, with dihedral of 12°. Trim tab in rudder and each elevator. Electrical operation of trim tabs, and pneumatic de-icing system, optional.

Landing Gear: Hydraulically-retractable tricycle type with single wheel on each unit. Main units retract inward into wing, nosewheel aft. Emergency extension by means of a 138 bar (2,000 lb/sq in) rechargeable nitrogen bottle. Cessna oleo-pneumatic shock-absorbers. All legs of articulated (trailing-link) type. Cleveland main wheels with tyres size 22 × 7·75-10, pressure 4·83 bars (70 lb/sq in). Cleveland nosewheel with tyre size

Cessna Titan business/commuter/cargo transport

6·00-6, pressure 3·45 bars (50 lb/sq in). Cleveland single-disc hydraulic brakes. Heavy duty brakes optional, standard on Ambassador III. Parking brake.

Power Plant: Two 280 kW (375 hp) Continental GTSIO-520-M flat-six engines, each driving a McCauley three-blade constant-speed fully-feathering metal propeller. Integral fuel tanks in wings, with combined usable capacity of 1,287 litres (340 US gallons). Oil capacity 26·5 litres (7 US gallons). Manifold pressure relief valves to prevent overboosting are standard. Propeller synchrophaser, automatic unfeathering system and electrical de-icing system optional.

Accommodation: Two seats side by side for pilot and co-pilot. Titan Courier has six individual seats and a two-passenger bench seat as standard; Titan Ambassador as Courier, except bench seat is optional. Passenger seats are 'Enviro-form' moulded honeycomb with glassfibre reinforcement. Passenger reading lights standard for Ambassador, optional for Courier. Door with built-in airstair on port side of fuselage, aft of wing. Emergency escape hatch on starboard side of cabin. Storm windows for pilot and co-pilot. Tinted cabin windows. Cargo door and crew access door optional for Courier, standard for Freighter. Baggage area at rear of cabin, capacity 227 kg (500 lb). Nose baggage compartment, with optional carpeting, can accommodate articles up to 1·96 m (6 ft 5 in) in length and has capacity of 159 kg (350 lb), accessible from each side. Avionics/baggage compartment in nose, separate from above baggage compartment and accessible through an 'over the top' 180° cam-lock door, has a capacity of 113 kg (250 lb); side access door optional. Wing lockers, at rear of each engine nacelle, each have capacity of 91 kg (200 lb). If no avionics carried in forward nose compartment, total baggage capacity is 680 kg (1,500 lb). Dual controls optional, standard on Titan II versions. Cabin heated and ventilated; air-conditioning optional. Windscreen defrosting standard. Electrical anti-icing of pilot's window, or alcohol anti-icing of pilot's and co-pilot's windows, optional.

Systems: Electrical system powered by two 28V 50A alternators; 100A alternators optional, standard on III versions. Battery, 24V 25Ah, can be sited optionally in nose baggage compartment. Hydraulic system for landing gear and flap operation, and for wheel brakes. Vacuum system provided by two engine-driven pumps. Oxygen system, capacity 1·25 m^3 (44 cu ft) or 3·25 m^3 (115 cu ft), optional. Heating and ventilation system standard, with 45,000 BTU gasoline heater. Air-conditioning system optional.

Avionics and Equipment: The various versions of the Titan have avionics as detailed in the model listings. Optional avionics available for the basic Titan versions include those detailed for the II and III versions, and alternative items from the Cessna Series 400, 800 and 1000 are available for all versions. Standard equipment includes blind-flying instrumentation, electric clock, outside air temperature gauge, sensitive altimeter, alternator failure lights, cabin door 'Not locked' warning light, heater overheat warning light, navigation light detectors, stall warning device, armrests, pilot and co-pilot safety belt system, control locks, sun visors, cabin speaker, adjustable air ventilators, shoulder restraint system, super soundproofing, courtesy lights, variable-intensity instrument post lights and floodlights, retractable landing light, navigation lights, three-light strobe system, towbar, and all-over polyurethane paint. Optional equipment for all versions includes GMT clock, high-efficiency exhaust gas temperature gauges, instantaneous rate of climb indicator, true airspeed indicator, digital fuel flow gauge with computer, dual heated static source, 'total flood' cabin fire extinguisher, stereo system, courtesy light timer, ice detection light, and anti-icing kit; items detailed as standard for II and III versions are also available as required.

Dimensions, external:
Wing span 14·12 m (46 ft 4 in)
Wing chord at root 1·78 m (5 ft 10 in)
Wing chord at tip 1·23 m (4 ft 0¼ in)
Wing aspect ratio 9·0
Length overall 12·04 m (39 ft 6¼ in)
Height overall 4·04 m (13 ft 3 in)
Tailplane span 5·82 m (19 ft 1 in)
Wheel track 4·28 m (14 ft 0½ in)
Wheelbase 3·82 m (12 ft 6¼ in)
Propeller diameter 2·29 m (7 ft 6 in)
Passenger door (port): Height 1·28 m (4 ft 2½ in)
Width 0·61 m (2 ft 0 in)
Height to sill 1·28 m (4 ft 2½ in)
Cargo door (optional, port; standard Freighter):
Height 1·28 m (4 ft 2½ in)
Width 1·24 m (4 ft 1 in)
Height to sill 1·28 m (4 ft 2½ in)
Nose baggage doors (each):
Height 0·41 m (1 ft 4 in)
Width 0·86 m (2 ft 9¾ in)
Nacelle baggage doors (each):
Length 0·62 m (2 ft 0¼ in)
Width 0·64 m (2 ft 1 in)

Dimensions, internal:
Cabin: Length 5·72 m (18 ft 9 in)
Max width 1·42 m (4 ft 8 in)
Max height 1·31 m (4 ft 3¼ in)
Volume 8·97 m^3 (316·6 cu ft)

Areas:
Wings, gross 22·48 m^2 (242 sq ft)
Fin 2·54 m^2 (27·33 sq ft)
Rudder, incl tab 1·50 m^2 (16·15 sq ft)
Tailplane 4·21 m^2 (45·31 sq ft)
Elevators, incl tab 1·68 m^2 (18·07 sq ft)

Weights and Loadings:
Weight empty: Ambassador 2,192 kg (4,834 lb)
Ambassador II 2,255 kg (4,972 lb)
Ambassador III 2,302 kg (5,075 lb)
Courier 2,205 kg (4,861 lb)
Courier II 2,267 kg (4,999 lb)
Freighter 2,133 kg (4,702 lb)
Freighter II 2,195 kg (4,840 lb)
Max T-O weight 3,810 kg (8,400 lb)
Max ramp weight 3,833 kg (8,450 lb)
Max zero-fuel and landing weight 3,674 kg (8,100 lb)
Max wing loading 169·5 kg/m^2 (34·71 lb/sq ft)
Max power loading 6·8 kg/kW (11·2 lb/hp)

Performance (at max T-O weight, ISA, unless otherwise indicated, except speeds are those at mid-cruise weight):
Never-exceed speed
238 knots (441 km/h; 274 mph) CAS
Max level speed at 4,875 m (16,000 ft)
232 knots (430 km/h; 267 mph)
Max cruising speed, 74·5% power at 6,100 m (20,000 ft) 217 knots (402 km/h; 250 mph)
Max cruising speed, 75·5% power at 3,050 m (10,000 ft) 199 knots (369 km/h; 229 mph)
Econ cruising speed at 6,100 m (20,000 ft) with 454 kg (1,000 lb) usable fuel
163 knots (302 km/h; 188 mph)
Econ cruising speed at 3,050 m (10,000 ft) with 454 kg (1,000 lb) usable fuel
140 knots (259 km/h; 161 mph)

Stalling speed, all versions:
flaps up, power off
83 knots (154 km/h; 96 mph) CAS
flaps down, power off, at max landing weight
70 knots (130 km/h; 81 mph) CAS
Max rate of climb at S/L 480 m (1,575 ft)/min
Rate of climb at S/L, one engine out
70 m (230 ft)/min
Service ceiling 7,925 m (26,000 ft)
Service ceiling, one engine out 3,080 m (10,100 ft)
T-O run 545 m (1,788 ft)
T-O to 15 m (50 ft) 721 m (2,367 ft)
Landing from 15 m (50 ft) at max landing weight
649 m (2,130 ft)
Landing run at max landing weight 335 m (1,100 ft)
Range, recommended lean mixture, allowances for start, taxi, T-O, climb, descent, and 45 min reserves at selected cruise power:
74·5% power at 6,100 m (20,000 ft) with 454 kg (1,000 lb) usable fuel
584 nm (1,081 km; 672 miles)
74·5% power at 6,100 m (20,000 ft) with 936 kg (2,064 lb) usable fuel
1,466 nm (2,717 km; 1,688 miles)
75·5% power at 3,050 m (10,000 ft) with 454 kg (1,000 lb) usable fuel
553 nm (1,025 km; 637 miles)
75·5% power at 3,050 m (10,000 ft) with 936 kg (2,064 lb) usable fuel
1,350 nm (2,501 km; 1,554 miles)
econ cruising power at 6,100 m (20,000 ft) with 454 kg (1,000 lb) usable fuel
726 nm (1,345 km; 836 miles)
econ cruising power at 6,100 m (20,000 ft) with 936 kg (2,064 lb) usable fuel
1,840 nm (3,410 km; 2,119 miles)
econ cruising power at 3,050 m (10,000 ft) with 454 kg (1,000 lb) usable fuel
756 nm (1,400 km; 870 miles)
econ cruising power at 3,050 m (10,000 ft) with 936 kg (2,064 lb) usable fuel
1,837 nm (3,404 km; 2,115 miles)

CESSNA CONQUEST

Cessna announced on 15 November 1974 that it was developing a twin-turboprop business aircraft designated Model 441, with initial deliveries scheduled for 1977. This was designed to slot into the market gap between existing twin piston-engined aircraft and turbofan-powered business aircraft.

Now marketed as the Cessna Conquest, this aircraft is powered by Garrett TPE331-8-401 turboprop engines, which have been developed specially to meet the high-altitude high-speed requirements set by Cessna for this aircraft. Its high performance stems in part from use of a new high aspect ratio bonded wing, and from the high-strength trailing-link-type hydraulically retractable tricycle landing gear.

Production aircraft beginning with c/n 116 introduced as standard an automatic engine torque and temperature limiting system, propeller synchrophaser, fuselage ice protection plates, pneumatic actuator for cabin upper door, new low-pressure fuel boost pumps, and fuel pressure switch. From and including c/n 173, the certificated altitude is increased to 10,670 m (35,000 ft); quick-donning pilot and co-pilot oxygen masks have been introduced, and additional options include Collins radios with electronic display, a dual inverter installation, Bendix RDR 160 colour radar display, and underseat storage drawers. At c/n 195 the original Hartzell propellers were replaced by McCauley equivalents that offer a weight saving of 5·2 kg (11·5 lb) each. At c/n 200, a Collins FIS-70 flight director and gyro system are added to the optional Collins avionics package.

Standard improvements introduced for 1981 include dual battery switches to allow independent battery checks; redesigned seat tracks on the port side, permitting seat movement to simplify passenger boarding; a pneumatic lower cabin door extender; failure indicator lights for the starter relay; and improved start control circuitry.

The prototype of the Conquest flew for the first time on 26 August 1975 and 192 aircraft had been delivered by 1 January 1981.

Type: Five/eleven-seat pressurised executive transport.

Wings: As Cessna Titan, except wing span and area increased by addition of wingtip extensions.

Fuselage: All-metal semi-monocoque structure of light alloy.

Tail Unit: Cantilever structure with sweptback vertical surfaces. Dihedral 12° on horizontal surfaces. Large tab in each elevator and rudder.

Landing Gear: As Cessna Titan.

Power Plant: Two Garrett TPE331-8-401S/402S turboprop engines, each flat rated to 474 kW (635·5 shp) to 4,875 m (16,000 ft). Hartzell Type HC-B3TN-5E/T10178B-11 constant-speed fully-feathering and reversible-pitch three-blade propellers up to and including Conquest c/n 194; McCauley propellers subsequently. Total fuel capacity 1,823 litres (481·5 US gallons), of which 1,798 litres (475 US gallons) are usable. Refuelling point on upper surface of each wing. Oil capacity 14·2 litres (3·75 US gallons).

Accommodation: Seats for four to ten persons, and pilot, in pressurised and air-conditioned cabin. Various optional seating arrangements. Door aft of wing on port side, with upward-hinged top portion and downward-hinged lower portion with integral airstairs. Emergency exit over wing on starboard side. Baggage door on each side of nose. Max baggage capacity 680 kg (1,500 lb). Optional items include aft cabin divider, refreshment centre, toilet, writing tables and stereo system.

Systems: Pressurisation system max differential 0·43 bars (6·3 lb/sq in). Freon air-conditioning systems of 8,000 BTU or 16,000 BTU capacity optional. Hydraulic system for operation of flaps, landing gear and brakes, pressure 103·5 bars (1,500 lb/sq in). Emergency extension of landing gear by means of 138 bar (2,000 lb/sq in) rechargeable gas bottle. Electrical power supplied by two 28V 200A starter/generators; two 24V 22Ah nickel-cadmium batteries housed in nose compartment. Electronic fuel control system.

Avionics: Standard avionics include dual Cessna 1038A com, dual 1048A nav, dual 1043A glideslope, 1046A ADF, 402A marker beacon receiver, 876A DME, 859A transponder, RMI, 400 encoding altimeter, Bendix RDR-160 weather radar, basic avionics kit. Collins avionics package and Bendix RDR-160 colour radar optional.

Dimensions, external:
Wing span, over tip-lights 15·04 m (49 ft 4 in)
Wing chord at root 1·78 m (5 ft 10 in)
Wing chord at tip 1·23 m (4 ft 0¼ in)
Wing aspect ratio 9·6
Length overall 11·89 m (39 ft 0¼ in)
Height overall 4·01 m (13 ft 1¾ in)
Tailplane span 5·81 m (19 ft 1 in)
Wheel track 4·28 m (14 ft 0½ in)
Wheelbase 3·81 m (12 ft 5⅞ in)
Propeller diameter 2·29 m (7 ft 6 in)

Dimensions, internal:
Cabin: Length 5·71 m (18 ft 9 in)
Max width 1·41 m (4 ft 7½ in)
Max height 1·29 m (4 ft 3 in)

Areas:
Wings, gross 23·56 m^2 (253·6 sq ft)
Ailerons (total) 1·37 m^2 (14·74 sq ft)
Trailing-edge flaps (total) 3·90 m^2 (42·00 sq ft)
Fin 2·54 m^2 (27·33 sq ft)
Rudder (incl tab) 1·50 m^2 (16·15 sq ft)
Tailplane 4·21 m^2 (45·31 sq ft)
Elevators (incl tabs) 1·68 m^2 (18·07 sq ft)

Weights and Loadings:
Weight empty, approx 2,588 kg (5,706 lb)
Max usable fuel 1,444 kg (3,183 lb)
Max ramp weight 4,502 kg (9,925 lb)
Max T-O weight 4,468 kg (9,850 lb)
Max landing weight 4,246 kg (9,360 lb)
Max zero-fuel weight 3,855 kg (8,500 lb)
Max wing loading 189·6 kg/m^2 (38·8 lb/sq ft)
Max power loading 4·79 kg/kW (7·88 lb/shp)

Performance (at max T-O weight, ISA, unless otherwise indicated, except speeds are those at mid-cruise weight of 3,788 kg; 8,350 lb):
Never-exceed speed
Mach 0·55 (243 knots; 450 km/h; 280 mph) CAS
Max level speed at 4,875 m (16,000 ft)
295 knots (547 km/h; 340 mph)
Max cruising speed at 7,315 m (24,000 ft)
293 knots (543 km/h; 337 mph)
Stalling speed, gear and flaps up, engines idling
90 knots (167 km/h;104 mph) CAS
Stalling speed, gear and flaps down, engines idling
76 knots (141 km/h; 87·5 mph) CAS
Stalling speed, gear and flaps down, power off, at max landing weight 75 knots (139 km/h; 86·5 mph) CAS
Max rate of climb at S/L 742 m (2,435 ft)/min
Rate of climb at S/L, one engine out
218 m (715 ft)/min
Service ceiling 11,275 m (37,000 ft)
Service ceiling, one engine out 6,515 m (21,380 ft)
Max certificated operating altitude
10,670 m (35,000 ft)
T-O run 544 m (1,785 ft)
T-O to 15 m (50 ft) 751 m (2,465 ft)
Landing from 15 m (50 ft) at max landing weight
572 m (1,875 ft)
Landing run 334 m (1,095 ft)
Range at max T-O weight with 1,444 kg (3,183 lb) fuel, at max cruising power, allowances for engine start, taxi, T-O, climb, descent and 45 min reserves at max cruise power:
at 5,180 m (17,000 ft)
1,199 nm (2,220 km; 1,379 miles)
at 7,620 m (25,000 ft)
1,571 nm (2,909 km; 1,807 miles)
at 10,060 m (33,000 ft)
2,063 nm (3,820 km; 2,374 miles)
at 10,670 m (35,000 ft)
2,193 nm (4,064 km; 2,525 miles)
Range at max T-O weight and at max range power, allowances as above:
at 5,180 m (17,000 ft)
1,471 nm (2,724 km; 1,692 miles)

Cessna Conquest (two Garrett TPE331-8-401S/402S turboprop engines)

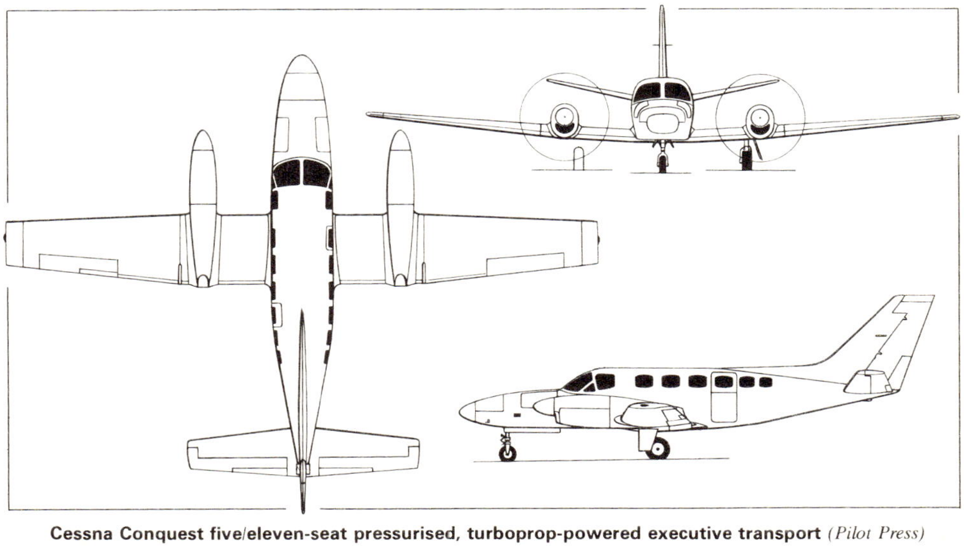

Cessna Conquest five/eleven-seat pressurised, turboprop-powered executive transport *(Pilot Press)*

Cessna Citation I seven/nine-seat twin-turbofan executive transport

at 7,620 m (25,000 ft)
1,816 nm (3,363 km; 2,090 miles)
at 10,060 m (33,000 ft)
2,212 nm (4,096 km; 2,545 miles)
at 10,670 m (35,000 ft)
2,291 nm (4,245 km; 2,638 miles)

CESSNA CITATION I

On 7 October 1968 Cessna announced that it was developing an eight-seat pressurised executive turbofan aircraft named Fanjet 500, which would be able to operate from most airfields used by light and medium twin-engined aircraft. After the first flight of the prototype, on 15 September 1969, the aircraft's name was changed to Citation. Subsequently, the gross weight was increased for the first time, and several other changes were made. These included a lengthened front fuselage, movement of the engine nacelles further aft, larger vertical tail, and resiting of, and introduction of dihedral on, the tailplane.

On 1 July 1971 Cessna announced that the first production Citation 0001 (N502CC) had made its first flight. Final FAA certification under FAR Part 25 was awarded on 9 September 1971, and has been followed by certification in many other countries.

The increase in take-off gross weight, to a maximum of 5,375 kg (11,850 lb), and the use of optional Rohr Industries thrust reversers, received FAA certification in February 1976. The improved Citation I became available in December 1976, with a wing of increased span and JT15D-1A turbofan engines. It superseded the earlier model on the production line from c/n 350 onward and was certificated on 15 December 1976. The first Citation I was delivered on 21 December 1976. Two versions are currently available:

Citation I. Basic version, as described in detail.

Citation I/SP Model 501. Basically the same as the Citation I, but certificated to FAR 23 for single-pilot operation. Type Certification granted on 7 January 1977, first aircraft delivered on 25 January 1977.

A total of 222 Citation Is, of both models, had been delivered by 1 January 1981.

Type: Seven/nine-seat twin-turbofan executive transport.

Wings: Cantilever low-wing monoplane without sweepback. Wing section at c/l NACA 23014 (modified), at wing station 247·95 NACA 23012. Incidence 2° 30′ at c/l, −0° 30′ at wing station 247·95. Dihedral 4°. All-metal fail-safe structure with two primary spars, an auxiliary spar, three fuselage attachments points, and conventional ribs and stringers. Manually-operated ailerons, with manual trim on port aileron. Electrically-operated single-slotted trailing-edge flaps. Hydraulically-operated aerodynamic speed brakes.

Fuselage: All-metal pressurised structure of circular section. Fail-safe design, providing multiple load paths.

Tail Unit: Cantilever all-metal structure. Horizontal surfaces have dihedral of 9°. Large dorsal fin and smaller ventral fin. Manually-operated control surfaces. Electrical elevator trim with manual override; manual rudder trim.

Landing Gear: Hydraulically-retractable tricycle type with single wheel on each unit. Main units retract inward into the wing, nose gear forward into fuselage nose. Free-fall and pneumatic emergency extension systems. Goodyear main wheels and 22 in diameter tyres, pressure 6·90 bars (100 lb/sq in). Steerable nosewheel with Goodyear wheel and tyre of 457 mm (18 in) diameter, pressure 8·27 bars (120 lb/sq in). Goodyear hydraulic brakes. Parking brake and pneumatic emergency brake system. Anti-skid system optional.

Power Plant: Two Pratt & Whitney Aircraft of Canada JT15D-1A turbofan engines, each rated at 9·77 kN (2,200 lb st) for take-off, mounted in pod on each side of rear fuselage. Rohr thrust reversers optional. Integral fuel tanks in wings, with combined usable capacity of 2,135 litres (564 US gallons).

Accommodation: Crew of two on separate flight deck. Fully-carpeted main cabin equipped with two individual forward-facing seats aft, one single forward-facing seat centre port, one single aft-facing seat centre starboard and a fifth aft-facing corner lounge chair at front of cabin on starboard side, all with headrests. Refreshment unit at front of cabin. Toilet compartment and main baggage area at rear of cabin. Second baggage area in nose. Total baggage capacity 454 kg (1,000 lb). Cabin is pressurised, heated and air-conditioned. Individual reading lights and air inlets for each passenger. Drop-out constant-flow oxygen system for emergency use. Plug-type door with integral airstair at front on port side and one emergency exit on starboard side. Doors on each side of nose baggage compartment. Tinted windows, each with curtains. Optional layouts for crew of two and six or seven passengers, with executive table, flush toilet replacing standard toilet, and choice of interior trims. Birdproof windscreen with de-fog system, anti-icing, and standby alcohol de-icing.

Systems: Pressurisation system supplied with engine bleed air, max pressure differential 0·59 bars (8·5 lb/sq in). Hydraulic system, pressure 103·5 bars (1,500 lb/sq in), with two pumps to operate landing gear and speed brakes. Electrical system supplied by two 28V 400A DC starter/generators, with two 300VA inverters and 24V 39Ah nickel-cadmium battery. Oxygen system of 0·62 m³ (22 cu ft) capacity includes two crew demand masks and five dropout constant-flow masks for passengers. High-capacity oxygen system optional.

Avionics and Equipment: Standard Category II avionics package on aircraft subsequent to c/n 275 comprises Sperry SPZ 500 flight control system with choice of single or double-cue command bars, including Sperry 500 autopilot, Sperry altimeter with altitude alerting and reporting functions, complete vertical navigation capability, air data computer, Sperry Model 600 (port)/Model 044 (starboard) horizontal situation indicator, Sperry ADI Model 300 or Model 600 command and control computer and autopilot servos, Bendix RDR 1200 (RDR 1100 subsequent to c/n 349, and Collins WXR-250C subsequent to c/n 550) continuous vision weather radar, two Collins VHF-20A com transceivers, two Collins VIR-30A nav receivers, Collins TDR-90 transponder, Collins DME-40, two Collins 332-C10 radio magnetic indicators, and ARC-846A ADF (Collins ADF-60 subsequent to c/n 525). Provision for advanced instrumentation and avionics to customer's specification. Standard equipment includes blind-flying instrumentation, generator load ammeters, standby magnetic compass, control locks, storm window, battery temperature and low fuel level warning lights, high Mach/airspeed warning, stall warning device, baggage tiedown kit, cabin fire extinguisher, individual life vests, 'No smoking, fasten seat belt' sign, internally lighted instruments, instrument standby lights, map light, entry light, tailcone compartment light, emergency exit lights, strobe lights, landing light, navigation lights, storm lights, taxi light, wing ice light, automatic engine start system, engine fire warning and extinguishing system, emergency battery pack, external power socket, inlet anti-icing, and surface de-icing system. Optional items include angle of attack indicator, radio telephone, cockpit voice recorder, flight data recorder, ground proximity warning system, emergency locator transmitter, navigation chart case, flush toilets, refreshment cabinets, storage drawers, tables, engine fan synchroniser, anti-collision beacon recognition lights, gravel runway kit, and drag-chute.

Dimensions, external:
Wing span 14·35 m (47 ft 1 in)
Wing aspect ratio 6·6
Length overall 13·26 m (43 ft 6 in)
Height overall 4·37 m (14 ft 4 in)
Tailplane span 5·74 m (18 ft 10 in)
Wheel track 3·84 m (12 ft 7 in)
Wheelbase 4·78 m (15 ft 8¼ in)
Cabin door (port): Height 1·29 m (4 ft 2¾ in)
Width 0·60 m (1 ft 11½ in)
Emergency exit (starboard):
Height 0·95 m (3 ft 1¼ in)
Width 0·56 m (1 ft 10 in)

Dimensions, internal:
Cabin:
Length, front to rear bulkhead 5·33 m (17 ft 6 in)
Max width 1·50 m (4 ft 11 in)
Max height 1·32 m (4 ft 4 in)
Baggage space: Cabin 1·10 m³ (39 cu ft)
Nose 0·45 m³ (16 cu ft)

Areas:
Wings, gross 25·9 m² (278·5 sq ft)
Horizontal tail surfaces 6·56 m² (70·6 sq ft)
Vertical tail surfaces 4·73 m² (50·9 sq ft)

Weights (from aircraft c/n 525 onward):
Weight empty (incl avionics) 2,974 kg (6,557 lb)
Max fuel weight 1,715 kg (3,780 lb)
Max T-O weight 5,375 kg (11,850 lb)
Max ramp weight 5,443 kg (12,000 lb)
Max landing weight 5,148 kg (11,350 lb)
Max zero-fuel weight 3,810 kg (8,400 lb)
Optional max zero-fuel weight 4,309 kg (9,500 lb)

Performance (at max T-O weight, ISA, except where indicated):
Max operating speed Mach 0·70
Cruising speed at average cruising weight
352 knots (652 km/h; 405 mph)
Stalling speed at max landing weight
82 knots (152 km/h; 94·5 mph) CAS
Max rate of climb at S/L 817 m (2,680 ft)/min
Rate of climb at S/L, one engine out
244 m (800 ft)/min
Max certificated altitude 12,495 m (41,000 ft)
Service ceiling, one engine out 6,400 m (21,000 ft)
T-O to 15 m (50 ft) 751 m (2,463 ft)
Landing run at max landing weight 692 m (2,270 ft)
Balanced field length:
at S/L, ISA 893 m (2,930 ft)
at 1,525 m (5,000 ft), 27°C 1,800 m (5,900 ft)
Range with 6 passengers, 45 min reserves
1,326 nm (2,454 km; 1,525 miles)

CESSNA CITATION II

First announced on 14 September 1976, this version of the Citation introduced several new features, including a fuselage lengthened by 1·14 m (3 ft 9 in), an increased-span high aspect ratio wing, increased fuel and baggage capacity, and the installation of Pratt & Whitney Aircraft of Canada JT15D-4 turbofan engines. The prototype (N550CC) flew for the first time on 31 January 1977, and certification was received in late March 1978; deliveries began immediately afterwards and a total of 300 had been delivered by June 1981.

In addition to the standard six/ten-passenger Model 550 **Citation II**, which is certificated for a crew of two, the Model 551 **Citation II/SP** is available for single-pilot operation, with up to 10 passengers, at a max T-O weight of 5,670 kg (12,500 lb).

In October 1977, the US Customs Service ordered a special-purpose Citation II/SP, fitted with government-furnished radar and sensing equipment. Sale of three aircraft to the People's Republic of China was confirmed in

The eight/twelve-seat Cessna Citation II executive transport *(Austin J. Brown)*

Prototype Cessna Citation III ten/fifteen-seat executive transport

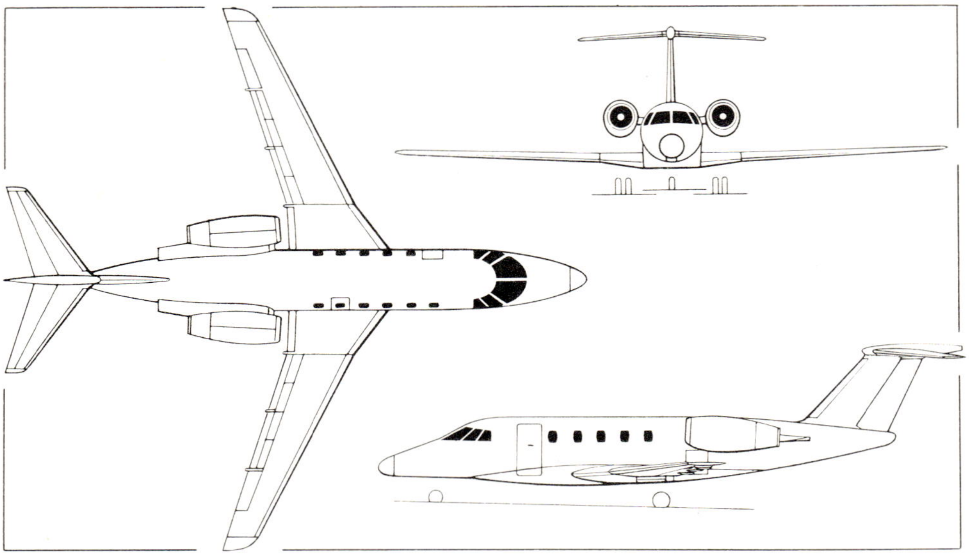

Cessna Citation III long-range executive transport *(Pilot Press)*

June 1981. The first will be equipped for flight inspection and calibration of air navigational aids, and the second for topographical mapping and survey photography. The third will be delivered in standard passenger-carrying configuration.

The description of the Citation I applies basically to the Citation II, except as follows:

Power Plant: Two Pratt & Whitney Aircraft of Canada JT15D-4 turbofan engines, each rated at 11·12 kN (2,500 lb st) for take-off, mounted in pod on each side of rear fuselage. Integral fuel tanks in wings, with usable capacity of 2,808 litres (742 US gallons).

Accommodation: As for Citation I, except seating for six to ten passengers in main cabin, with toilet in six/eight-seat versions, and increased baggage capacity of up to 522 kg (1,150 lb).

Dimensions, external:

Wing span	15·75 m (51 ft 8 in)
Wing aspect ratio	8·3
Length overall	14·38 m (47 ft 2 in)
Height overall	4·50 m (14 ft 9 in)
Wheel track	5·36 m (17 ft 7 in)
Wheelbase	5·54 m (18 ft 2 in)

Dimensions, internal:

Cabin: Length, front to rear bulkhead	6·38 m (20 ft 11 in)
Max height	1·45 m (4 ft 9 in)
Baggage capacity	1·84 m^3 (65 cu ft)

Weights (A: Citation II; B: Citation II/SP):

Weight empty (incl avionics): A, B	3,257 kg (7,181 lb)
Max fuel weight: A, B	2,272 kg (5,009 lb)
Max T-O weight: A	6,033 kg (13,300 lb)
B	5,670 kg (12,500 lb)
Max ramp weight: A	6,123 kg (13,500 lb)
B	5,760 kg (12,700 lb)
Max zero-fuel weight (standard): A, B	4,309 kg (9,500 lb)
Max zero-fuel weight (optional): A, B	4,989 kg (11,000 lb)
Max landing weight: A	5,760 kg (12,700 lb)
B	5,443 kg (12,000 lb)

Performance (at max T-O weight, ISA, except where indicated):

Max operating speed: A, B	Mach 0·705
Cruising speed at average cruising weight: A	385 knots (713 km/h; 443 mph)
B	387 knots (718 km/h; 446 mph)
Stalling speed at max landing weight: A	82 knots (152 km/h; 94·5 mph) CAS
B	80 knots (149 km/h; 92·5 mph) CAS
Max rate of climb at S/L: A	1,027 m (3,370 ft)/min
B	1,105 m (3,625 ft)/min
Rate of climb at S/L, one engine out: A	322 m (1,055 ft)/min
B	357 m (1,170 ft)/min
Max certificated altitude: A, B	13,105 m (43,000 ft)
Service ceiling, one engine out: A	7,680 m (25,200 ft)
B	8,290 m (27,200 ft)
T-O to 15 m (50 ft): A	727 m (2,385 ft)
B	639 m (2,095 ft)
Balanced field length (FAR Pt 25): A	912 m (2,990 ft)
B	808 m (2,650 ft)
Landing run at max landing weight: A	692 m (2,270 ft)
B	674 m (2,210 ft)
Range with six passengers and 45 min reserves: A	1,710 nm (3,169 km; 1,969 miles)
B	1,469 nm (2,721 km; 1,691 miles)

CESSNA CITATION III

While retaining some general similarity to earlier members of the Citation family, the Citation III is a very different aeroplane. The original design underwent major refinement during 1978. In particular, a T tail and an additional fuel tank in the rear fuselage were introduced.

First flight of the first prototype (N650CC) was made on 30 May 1979. The second prototype flew for the first time on 2 May 1980. Orders for more than 140 Citation IIIs had been received by July 1981; deliveries are scheduled to begin in the late Autumn of 1982.

Flight testing totalled 400 h in 372 flights by September 1980, including climb to the planned certificated ceiling of 15,545 m (51,000 ft) and max diving speeds up to Mach 0·88 (400 knots CAS). These speeds exceeded the designed operating limit of Mach 0·81 (320 knots CAS).

Type: Twin-turbofan 10/15-seat long-range executive transport.

Wings: Cantilever low-wing monoplane. NASA-developed supercritical section. Sweepback at quarter-chord 25°. Conventional two-spar structure of light alloy, utilising bonded and riveted construction. Anti-icing of wing leading-edges. Hydraulically-actuated trailing-edge flaps and spoilers.

Fuselage: Conventional semi-monocoque light alloy structure of circular cross-section. Fail-safe in pressurised area.

Tail Unit: Cantilever T-tail structure of light alloy, with swept horizontal and vertical surfaces. Anti-icing of tailplane leading-edges.

Landing Gear: Hydraulically-retractable tricycle type. Twin wheels on each main unit, single wheel on hydraulically-steerable nose unit. Oleo-pneumatic shock-absorbers. Hydraulic anti-skid braking system with pneumatic backup.

Power Plant: Two Garrett TFE731-3B-100S turbofan engines, each rated at 16·24 kN (3,650 lb st) for take-off, mounted in pod on each side of rear fuselage. Hydraulically-operated thrust reversers standard. Two independent fuel systems, with integral tanks in each wing; usable capacity 2,826 kg (6,230 lb). Additional fuel cell behind rear fuselage bulkhead. Single-point pressure refuelling. Fuel heaters optional. Engine intake anti-icing system.

Accommodation: Crew of two on separate flight deck, and up to thirteen passengers. Standard interior has eight individual seats, with toilet at rear of cabin. Storage in fuselage nose for crew baggage. Main baggage space in tailcone. Airstair door forward of wing on port side. Overwing emergency escape hatch on starboard side. Cabin is pressurised, heated and air-conditioned. Windscreen anti-icing.

Systems: Environmental control system, with separate control of flight deck and cabin conditions. Direct engine bleed pressurisation system, with nominal pressure differential of 0·66 bars (9·5 lb/sq in), provides 2,440 m (8,000 ft) cabin altitude to max certificated altitude. Dual parallel electrical buses. Hydraulic system of 207 bars (3,000 lb/sq in) for operation of flaps, spoilers, brakes, landing gear, nosewheel steering, and optional thrust reversers, with backup system to provide emergency power. Oxygen systems for crew and passengers.

Avionics: Standard avionics package provides full Category II capability, and includes a Sperry SPZ-650A automatic flight control system, digital central air data computer, vertical navigation (V/Nav) system with dual waypoints, AA-215 radio altimeter, dual C-14 slaved compass system and VG-14A vertical gyro; dual nav/com and RMI, transponder, DME and weather radar. A wide range of optional avionics is available to customer's requirements.

Dimensions, external:

Wing span	16·26 m (53 ft 4¼ in)
Wing aspect ratio	8·94
Length overall	16·90 m (55 ft 5½ in)
Height overall	5·27 m (17 ft 3½ in)
Tailplane span	5·60 m (18 ft 4½ in)
Wheel track	2·86 m (9 ft 4½ in)
Wheelbase	6·48 m (21 ft 3 in)

Dimensions, internal:

Cabin: Length, front to rear bulkhead	7·01 m (23 ft 0 in)
Length, aft of cockpit divider	5·66 m (18 ft 7 in)
Max width	1·73 m (5 ft 8 in)
Max height	1·78 m (5 ft 10 in)
Baggage capacity (aft)	2·10 m^3 (74 cu ft)
Crew baggage compartment (nose)	0·28 m^3 (10 cu ft)

Area:

Wings, gross	29·00 m^2 (312 sq ft)

Weights (estimated):

Weight empty	4,968 kg (10,951 lb)
Max fuel	3,245 kg (7,155 lb)

Max T-O weight 9,072 kg (20,000 lb)
Max landing weight 7,484 kg (16,500 lb)

PERFORMANCE (estimated, at T-O weight of 8,845 kg; 19,500 lb, ISA, unless stated otherwise):
- Max level speed at 10,060 m (33,000 ft) at average cruise weight of 6,395 kg (14,100 lb) 469 knots (869 km/h; 540 mph)
- Max cruising speed at average cruise weight of 6,214 kg (13,700 lb):
  - at 10,060 m (33,000 ft) 469 knots (869 km/h; 540 mph)
  - at 12,500 m (41,000 ft) Mach 0·8 (459 knots; 850 km/h; 528 mph)
  - at 13,715 m (45,000 ft) Mach 0·77 (442 knots; 818 km/h; 509 mph)
  - at 14,325 m (47,000 ft) 420 knots (779 km/h; 484 mph)
  - at 14,935 m (49,000 ft) 384 knots (711 km/h; 442 mph)
- Stalling speed, flaps and gear down, at max landing weight 90 knots (167 km/h; 104 mph) CAS
- Max rate of climb at S/L 1,365 m (4,475 ft)/min
- Rate of climb at S/L, one engine out 408 m (1,340 ft)/min
- Time to 12,500 m (41,000 ft) 21 min
- Certificated ceiling 15,545 m (51,000 ft)
- Ceiling, one engine out 8,780 m (28,800 ft)
- FAA T-O field length at S/L 1,375 m (4,510 ft)
- Landing distance at max landing weight 844 m (2,770 ft)
- Range, with two crew and six passengers, allowances for T-O, climb, descent and 45 min reserves 2,500 nm (4,630 km; 2,877 miles)

COLEMILL
COLEMILL ENTERPRISES INC

PO Box 60627, Cornelia Fort Air Park, Nashville, Tennessee 37206
Telephone: (615) 262 0456

Colemill Enterprises is a company which specialises in performance improvement conversions of light twin-engined aircraft. Its most recent programme led to certification of a re-engined version of the Piper Navajo, to which it has given the name Panther Navajo. Representing the company's first conversion of a 'cabin-class' aircraft, the Navajo was selected because, with more than 2,000 in service worldwide, there was a considerable market potential. Testing of the prototype showed that a significant improvement in performance had been achieved and since that time Panthers have set nine world records in the FAI's Classes C1d and C1e, six for declared distance flights, and three for time to height.

Colemill Panther Navajo with starboard propeller feathered *(Flight International)*

COLEMILL PANTHER NAVAJO

Colemill's conversion of the Navajo involves more than a new engine installation. In addition to the power plant change, there are redesigned nacelles, additional continuous-running fuel pumps, a digital fuel totaliser, heavy-duty brakes, and wingtip-mounted landing lights. Conversion normally takes 10-14 days. The basic description of the Navajo, which can be found under the Piper entry in this edition, applies also to the Panther Navajo, except as follows:

WINGS: As for Piper Navajo except for new wingtips with marked under camber, which increase wing span and area.

LANDING GEAR: As for Piper Navajo except for the introduction of Cleveland four-spot heavy-duty disc brakes.

POWER PLANT: Two 261 kW (350 hp) Avco Lycoming TIO-540-J2BD turbocharged engines, each driving a Hartzell four-blade constant-speed fully-feathering metal propeller with 'Q' tips. Pressurised magnetos, Woodward propeller governors, synchrophasers and unfeathering accumulators standard. Fuel system as for basic Navajo, except for the addition of continuous-running electrically-operated fuel pumps.

EQUIPMENT: Generally as for standard Navajo, but existing fuel-flow gauges are replaced by a Hoskins CFS 2000 fuel management computer giving digital readout of fuel remaining/fuel consumed. Supplemental wingtip landing lights can be operated independently of the standard nosewheel-mounted landing light, prior to lowering of landing gear.

DIMENSIONS, EXTERNAL: As for Piper Navajo except:
- Wing span 13·00 m (42 ft 8 in)

LOADING: As for Piper Navajo except:
- Max power loading 5·65 kg/kW (9·3 lb/hp)

PERFORMANCE (at max T-O weight):
- Max level speed 269 knots (498 km/h; 309 mph)
- Max cruising speed, 75% power at optimum altitude 248 knots (459 km/h; 285 mph)
- Cruising speed, 65% power at 7,300 m (24,000 ft) 235 knots (435 km/h; 270 mph)
- Cruising speed, 65% power at 3,650 m (12,000 ft) 206 knots (381 km/h; 237 mph)
- Max rate of climb at S/L 610 m (2,000 ft)/min
- Rate of climb at S/L, one engine out 122 m (400 ft)/min
- Short-field T-O run 229 m (750 ft)
- T-O to 15 m (50 ft) 458 m (1,500 ft)
- Landing from 15 m (50 ft) 427 m (1,400 ft)

COLEMILL PANTHER NAVAJO WITH WINGLETS

Colemill Enterprises has completed and flown a version of the Panther Navajo fitted with winglets. The prototype, which was a standard 1972 Piper Navajo, has been refurbished completely. In addition to improvements detailed for the Panther Navajo, and incorporation of winglets, it has its baggage floor area reduced by 0·28 m² (3 sq ft) to make possible a ten-seat interior with new lightweight seats.

The winglets improve stability in the lower-speed flight zone, down to stalling speed, and also provide cruising speed increases of 4-9 knots (8-16 km/h; 5-10 mph) at altitudes between 4,000 and 7,600 m (13,000 and 25,000 ft), at engine power settings between 45% and 65% of maximum power. This new version is, therefore, able to match the cruising speed of the standard Panther Navajo (without winglets) at power settings which offer greater fuel economy.

The flight test programme to gain a Supplemental Type Certificate was being flown during 1980.

COLEMILL EXECUTIVE 600

The Executive 600 is a Cessna 310, of Model G to U series, re-engined by Colemill to have two 224 kW (300 hp) Continental IO-520-E flat-six engines, each driving a McCauley three-blade propeller. Dimensions are unchanged; empty weight is increased by about 14 kg (30 lb); other data are as follows:

WEIGHT:
- Max T-O weight 2,358 kg (5,200 lb)

PERFORMANCE (at max T-O weight):
- Max cruising speed (75% power) 205 knots (379 km/h; 236 mph)
- Cruising speed (65% power) 195 knots (361 km/h; 224 mph)
- Stalling speed, wheels and flaps down 64 knots (119 km/h; 74 mph)
- Max rate of climb at S/L 762 m (2,500 ft)/min
- Service ceiling 5,940 m (19,500 ft)
- T-O to 15 m (50 ft) 518 m (1,700 ft)
- Landing from 15 m (50 ft) unchanged
- Range with max fuel, 45 min reserves 1,050 nm (1,944 km; 1,208 miles)

COLEMILL CENTURY 600

The Century 600 is a Cessna 320 re-engined by Colemill to have two 224 kW (300 hp) Continental IO-520-E flat-six engines, each driving a McCauley three-blade propeller. Dimensions are unchanged; empty weight is increased by about 14 kg (30 lb); other data are as follows:

WEIGHT:
- Max T-O weight 2,404 kg (5,300 lb)

PERFORMANCE (at max T-O weight):
- Max cruising speed (75% power) 202 knots (374 km/h; 232 mph)
- Cruising speed (65% power) 192 knots (355 km/h; 221 mph)
- Stalling speed, wheels and flaps down 64 knots (119 km/h; 74 mph)
- Max rate of climb at S/L 777 m (2,550 ft)/min
- Service ceiling 5,940 m (19,500 ft)
- T-O to 15 m (50 ft) 518 m (1,700 ft)
- Landing from 15 m (50 ft) unchanged
- Range with max fuel, 45 min reserves 1,060 nm (1,963 km; 1,220 miles)

COLEMILL PRESIDENT 600

The President 600 is a Beechcraft B55 Baron re-engined by Colemill to have two 224 kW (300 hp) Continental IO-520-E flat-six engines, each driving a three-blade propeller. Dimensions are unchanged; empty weight is increased by about 14 kg (30 lb); other data are as follows:

WEIGHT:
- Max T-O weight 2,313 kg (5,100 lb)

PERFORMANCE (at max T-O weight):
- Max cruising speed (75% power) 203 knots (376 km/h; 233 mph)
- Cruising speed (65% power) 193 knots (357 km/h; 222 mph)
- Stalling speed, wheels and flaps down 66 knots (123 km/h; 76 mph)
- Max rate of climb at S/L 823 m (2,700 ft)/min
- Service ceiling 5,940 m (19,500 ft)
- T-O to 15 m (50 ft) 497 m (1,631 ft)
- Landing from 15 m (50 ft) unchanged
- Range with max fuel, 45 min reserves 1,050 nm (1,944 km; 1,208 miles)

Colemill's Panther Navajo conversion with winglets *(Howard Levy)*

COMPOSITE AIRCRAFT CORPORATION
c/o DIETRICK SALES AND SERVICE

523 Ridgeview Drive, Florence, Kentucky 41042
Telephone: (606) 371 7247
PRESIDENT: Gerald P. Dietrick
CHIEF ENGINEER: George A. Alther

Mr Gerald P. Dietrick, the owner of Dietrick Sales and Service, has acquired the type certificate, the plant to manufacture non-woven unidirectional glassfibre fabric, moulds and engineering data, for construction of the Windecker Eagle 1, designed and developed from 1959 by Drs L. J. and F. M. Windecker. He has acquired also two of the five production Eagle 1s built by Windecker Industries Inc, and has an exclusive lease on a third aircraft. With these aircraft he is currently developing minor modifications before putting the Eagle back into production. It is planned to build 250 aircraft over a three-year period, with initial deliveries scheduled for 1982; Composite Aircraft Corporation has been registered for this purpose.

The five production Eagle 1s built by Windecker Industries had accumulated in excess of 8,000 flight hours by mid-January 1978. Details of the development of these aircraft can be found under the Windecker Industries entry in the 1974-75 *Jane's*.

In early 1979, aircraft N4198G was equipped with an autopilot, additional fuel tankage and navigation equipment, prior to being flown nonstop from Gander, Newfoundland, to Paris to participate in the 1979 Paris Air Show. This resulted in the Dietrick/Eagle combination setting two new world speed records in the FAI's Class C1d, with a flight which originated from Cincinnati, Ohio. The total distance covered was 3,622 nm (6,708 km; 4,168 miles), and the FAI accredited speeds were Cincinnati-Paris 164·21 knots (304·12 km/h; 188·97 mph), and New York-Paris 168·85 knots (312·72 km/h; 194·31 mph). Mr Dietrick flew N4198G to the Hanover Air Show in 1980, and designed and installed an 817·5 litre (216 US gallon) fuel tank in the cabin to make possible a non-stop flight from Cincinnati to Munich at a record speed of 161·68 knots (299·44 km/h; 186·06 mph). At the termination of that flight, N4198G had logged total flight hours exceeding 2,800. Its T-O weight at Cincinnati was 2,250 kg (4,960 lb).

DIETRICK/WINDECKER EAGLE

During 1980 further development of the Windecker design led to an improved single-spar wing with winglets, and three versions are planned:

Eagle I. Basic version, as described in detail.

Eagle II. Generally similar to the Eagle I, but powered by an Allison 250-B17C turboprop engine flat rated to 236·5 kW (317 shp).

Eagle III. Projected version with a lengthened, pressurised fuselage to accommodate six persons, and powered by a turboprop engine.

TYPE: Four-seat cabin monoplane.

WINGS: Cantilever low-wing monoplane. NACA 64_2415 wing section. Dihedral 4° 30′. Incidence 2° 30′. No sweepback. Fail-safe single-spar structure of GRP, and incorporating wingtip winglets. Frise-type ailerons of GRP with pilot-controlled bungee trimming. Plain trailing-edge flaps constructed of GRP.

FUSELAGE: Fail-safe monocoque structure of GRP.

TAIL UNIT: Conventional cantilever structure of GRP. Fixed-incidence tailplane. Trim tabs on starboard elevator and rudder. Electrically actuated elevator trim optional.

LANDING GEAR: Retractable tricycle type with single wheel on each unit. Electro-hydraulic retraction, main wheels inward, nosewheel rearward. Main wheels faired by landing gear doors when retracted. Nosewheel door functions as a cowl flap when the gear is retracted. Windecker oleo-pneumatic shock-absorbers. Cleveland 6·00-6 main wheels, 5·00-5 nosewheel; all tyre pressures 3·09 bars (45 lb/sq in). Cleveland hydraulic disc brakes. Parking brake.

POWER PLANT: One 212·5 kW (285 hp) Continental IO-520-C flat-six engine, driving a McCauley two-blade constant-speed propeller. Fuel capacity 318 litres (84 US gallons), in two 159 litre (42 US gallon) integral wing tanks. Refuelling point outboard on upper surface of each wing. Oil capacity 11·5 litres (3 US gallons).

ACCOMMODATION: Four persons in pairs in enclosed cabin. Forward-hinged door on each side of cabin. Compartment aft of rear seats for 54 kg (120 lb) baggage with external access door on port side; hatshelf on bulkhead at rear of cabin. Cabin ventilated and heated by ram air over exhaust. Windscreen defrosting standard.

SYSTEMS: Hydraulic system, pressure 103·5 bars (1,500 lb/sq in), for landing gear. Vacuum system for instruments. 12V 70A alternator. 12V 35Ah battery. Four-outlet oxygen system optional.

AVIONICS AND EQUIPMENT: Standard avionics comprise one Narco MK-16 VHF, one Narco VOA-40 VOR/LOC converter-indicator and an antennae package which includes VOR, communications, ADF sense, transponder, DME, glideslope and marker beacon receiver antennae, headset, microphone and all wiring harnesses. Optional avionics include the Brittain B5 autopilot, and a wide range of Narco units, available singly or as a complete package. Standard equipment includes full blind-flying instrumentation; 8-day clock; outside air temperature gauge; fully-reclining track-mounted seats with seat belt, headrest, fresh-air outlet, and map or magazine storage pocket for each seat; floor console between front seats for map and pencil storage; openable port side storm window; ultraviolet absorbent windscreen; shoulder harness for front seats; alternate static source; stall warning device; baggage net and attachments; soundproofing; tinted cabin windows; dome, radio, and map lights, and instrument panel floodlights; anti-collision beacons; landing, navigation and taxi lights; external power socket; handholds; jack pads; retractable steps; towbar; and choice of four interior and exterior trims, the latter with polyurethane paint finish. Optional equipment includes control wheel mounted chronograph; heated pitot; heated stall warning transmitter; cabin fire extinguisher; cargo tiedown kit; instrument post lights; reading lights for rear seats; white strobe lights; engine primer; engine winterisation kit; and glider tow provisions.

DIMENSIONS, EXTERNAL:	
Wing span	9·75 m (32 ft 0 in)
Wing chord, constant	1·68 m (5 ft 6 in)
Wing aspect ratio	5·82
Length overall	8·69 m (28 ft 6 in)
Height overall	2·90 m (9 ft 6 in)
Tailplane span	3·42 m (11 ft 3 in)
Wheel track	2·13 m (7 ft 0 in)
Wheelbase	1·98 m (6 ft 6 in)
Propeller diameter	2·13 m (7 ft 0 in)
Propeller ground clearance	0·28 m (11 in)
Cabin doors: Height	0·89 m (2 ft 11 in)
Width	0·97 m (3 ft 2 in)
Height to sill	0·40 m (1 ft 4 in)
Baggage door (port side):	
Height	0·60 m (1 ft 11½ in)
Width	0·39 m (1 ft 3½ in)
DIMENSIONS, INTERNAL:	
Cabin: Length	3·30 m (10 ft 10 in)
Max width	1·28 m (4 ft 2⅜ in)
Max height	1·07 m (3 ft 6 in)
AREAS:	
Wings, gross	15·51 m^2 (167 sq ft)
Ailerons (total)	1·42 m^2 (15·28 sq ft)
Trailing-edge flaps (total)	1·50 m^2 (16·1 sq ft)
Fin	0·93 m^2 (9·97 sq ft)
Rudder, incl tab	0·59 m^2 (6·31 sq ft)
Tailplane	1·70 m^2 (18·3 sq ft)
Elevators, incl tab	1·48 m^2 (15·97 sq ft)
WEIGHTS AND LOADINGS:	
Weight empty	975 kg (2,150 lb)
Max T-O and landing weight	1,542 kg (3,400 lb)
Max wing loading	94·7 kg/m^2 (19·3 lb/sq ft)
Max power loading	7·26 kg/kW (11·9 lb/hp)
PERFORMANCE (at max T-O weight):	
Max level speed at S/L	more than 182 knots (338 km/h; 210 mph)
Cruising speed, 75% power at 2,135 m (7,000 ft)	177 knots (328 km/h; 204 mph)
Cruising speed, 65% power at 3,660 m (12,000 ft)	175 knots (325 km/h; 202 mph)
Stalling speed, flaps and landing gear up, power off	61·5 knots (114 km/h; 71 mph)
Stalling speed, flaps and landing gear down, power off	57·5 knots (106 km/h; 66 mph)
Max rate of climb at S/L	372 m (1,220 ft)/min
Service ceiling	5,485 m (18,000 ft)
T-O run	261 m (855 ft)
T-O to 15 m (50 ft)	515 m (1,690 ft)
Range with max fuel at 3,050 m (10,000 ft), 45 min reserves	1,068 nm (1,979 km; 1,230 miles)

Dietrick/Windecker Eagle in original form; the developed versions will have an improved wing with winglets *(Brian M. Service)*

CONTINENTAL COPTERS
CONTINENTAL COPTERS INC

PRESIDENT: John L. Scott

Continental Copters developed and produced a series of specialised single-seat agricultural conversions of various versions of the Bell Model 47 helicopter, under the name El Tomcat. Design work on the original conversion began in 1959 and the prototype El Tomcat Mk II flew in April of that year, receiving an FAA Supplemental Type Certificate shortly afterwards. Improved Mk III, IIIA, IIIB and IIIC versions followed, and later conversions embraced the Mk V and VI versions.

By the beginning of 1980 Continental had delivered more than 80 El Tomcats to customers in the USA, Panama, Portugal, Puerto Rico, Switzerland and Turkey.

By the Spring of 1981 the company had left its former headquarters in Fort Worth, Texas, and had left no forwarding address. Details of its most recent conversions, the El Tomcat Mks V-A, VI-B and VI-C, can be found in the 1980-81 *Jane's*.

DEE HOWARD
THE DEE HOWARD COMPANY

International Airport, PO Box 17300, San Antonio, Texas 78217
Telephone: (512) 828 1341

DEE HOWARD XR LEARJET

The Dee Howard Company, in conjunction with the former Raisbeck Group, shared the development of a high-lift system for retrofit to Learjet Model 23/24/25 aircraft. Since that time, Dee Howard has developed an overall performance improvement system that is suitable for retrofit on all Model 24 Learjets which have a power plant comprising General Electric CJ610-6 or -8A turbojet engines.

The improvements incorporated in the XR Learjet include the provision of a new centre-section glove which reduces drag, improves spanwise lift, and accommodates an additional 245 kg (540 lb) of fuel; a new engine pylon/nacelle configuration that eliminates adverse flow pressure characteristics and channel flow Mach problems, and also improves engine bay cooling; the addition of a small span flow limiter and stall turbulator, at the junction of the inboard and outboard leading-edge on each wing, which produces stall buffet and improves overall stall performance; the introduction of a new leading-edge profile to optimise cruise drag and low-speed stall characteristics; improved ailerons, trailing-edge flaps, outer wing panels, and new tip-tank fin cuffs to improve cruise performance; introduction of a new engine exhaust nozzle that improves specific fuel consumption; and installation of a Teledyne angle of attack system. A new Sperry SPZ-500 LR autopilot is to be optional.

Conversion of a suitable Model 24 to XR Learjet configuration provides an extra 400 nm (741 km; 460 miles) of range at a constant cruising speed of Mach 0·78, plus an increase of 680 kg (1,500 lb) in maximum take-off weight, and increased payload.

The Dee Howard Company and Gates Learjet Corporation announced jointly, in September 1980, the intention to co-operate on several aircraft development projects. The first of these will involve the Learjet Model 25G, incorporating improvements similar to those of the XR Learjet. Following certification, these modifications will become available on new production aircraft after approximately one year, and will be retrofitted to existing aircraft by Dee Howard.

EAGLE

EAGLE AIRCRAFT COMPANY

PO Box 4127, Boise, Idaho 83704
Telephone: (208) 375 7500
President: Dwain Griggs
Vice-President: Wayne Mittleider
Chief Designer: Dean Wilson
Sales Manager: Chet Bowers

EAGLE AIRCRAFT EAGLE

Eagle Aircraft Company is marketing a new agricultural aircraft, with large-span narrow-chord wings based on advanced glider technology, to which it has given the name Eagle. Other innovations include spraybooms that form the trailing-edge of the lower wings, and spoilers for roll control mounted outboard on each lower wing.

The prototype flew originally with a Continental radial engine. This was replaced subsequently by an Avco Lycoming IO-540 flat-six engine, which is now available as an alternative power plant. The two versions are known as the **Eagle 220** and **Eagle 300** respectively, denoting the horsepower of their engine.

The Eagle 220 version, with modified Continental W-670 engine

Eagle Aircraft Eagle 300 agricultural aircraft (Avco Lycoming IO-540 engine)

The company announced on 7 June 1979 the signature of an agreement with Bellanca Aircraft Corporation to manufacture the Eagle on its behalf, and built a 2,954 m² (31,800 sq ft) factory adjacent to Bellanca's Alexandria, Minnesota, plant in which the aircraft is being produced. In January 1981 the manufacturing contracts with Bellanca were cancelled: Eagle Aircraft then engaged the 72 Bellanca employees who had been building these aircraft, and is continuing production of the Eagle in its factory at Alexandria.

First aircraft delivered, to Precissi Flying Service of Lodi, California, in June 1980, was the radial-engined N8800R. A total of 11 Eagles had been delivered by 1 March 1981.

Type: Single-seat agricultural aircraft.

Wings: Wire-braced single-bay biplane, with N-type interplane and centre-section struts. Narrow-chord wooden wings of advanced design, with spraybooms that form the trailing-edge of the lower wings. Spoilers, each with booster tab, in upper surface of outboard section of each lower wing. Conventional ailerons with booster tabs on upper wings. Dacron fabric covering, with butyrate dope and polyurethane finish.

Fuselage: Conventional welded steel tube structure; rear part covered by Dacron fabric, and the remainder by quickly-removable side panels.

Tail Unit: Conventional wire-braced structure, with Dacron fabric covering, butyrate dope and polyurethane finish. Booster tab in elevator and rudder. Trim tab in elevator. Wire deflector cable between tip of fin and centre-section of upper wing.

Landing Gear: Non-retractable tailwheel type. Two side Vs and half-axles. Firestone rubber-in-compression shock-absorbers. Cleveland main wheels and high-flotation tyres size 7·50-10. Cleveland hydraulically operated disc brakes. Parking brake. Wire cutter on leading-edge of each main gear V.

Power Plant: Alternative power plants include one 164 kW (220 hp) Gulf Coast conversion of the Continental W-670-6N radial engine, driving a Hartzell two-blade adjustable-pitch propeller; or a 224 kW (300 hp) Avco Lycoming IO-540-M1B5D flat-six engine, driving a Hartzell three-blade constant-speed propeller. Gravity-feed fuel tank in centre-section of upper wing, capacity 151·5 litres (40 US gallons). Wingtip fuel tanks for upper wing optional, with combined capacity of 75·5 litres (20 US gallons). Maximum fuel capacity 227 litres (60 US gallons). Refuelling point on top surface of upper wing.

Accommodation: Pilot only, beneath one-piece quick-release transparent canopy, which can be hinged to port or starboard by single lever, or can be removed entirely for open-cockpit flight. Pilot's seat has shoulder harness with inertia reel. Chemical hopper, capacity 946 litres (250 US gallons) in front fuselage, with lightweight bottom loading valve. Optional baggage compartment for pilot, with external access door.

Systems: 24V electrical system powered by 28V 70A alternator and heavy-duty 24V battery. 100A alternator optional. Hydraulic system for brakes only.

Equipment: Standard equipment includes sensitive altimeter, pilot's headrest, adjustable rudder pedals, stall warning device, windscreen cleaning access windows, stainless steel control cables, quick oil-drains, handholds on aft fuselage, tiedown rings, and complete wet dispersal system with 15·5 m (51 ft 0 in) boom. Internal and external polyurethane anti-corrosion protection. Optional items include air-conditioning and heating systems, dry dispersal system, and a night operations package which includes instrument panel lights, overhead floodlights; navigation, landing, taxi, strobe and hopper lights, and a heavy-duty battery.

Dimensions, external (A: Eagle 220; and B: Eagle 300):
- Wing span: both versions 16·76 m (55 ft 0 in)
- Wing aspect ratio 15
- Length overall: A 7·92 m (26 ft 0 in)
  - B 8·38 m (27 ft 6 in)
- Height overall: both versions 3·23 m (10 ft 7 in)

Area:
- Wings, gross 35·86 m² (386 sq ft)

Weights and Loadings:
- Weight empty: both versions 1,202 kg (2,650 lb)
- Max T-O weight: both versions 2,449 kg (5,400 lb)
- Max wing loading, Restricted category:
  - both versions 68·3 kg/m² (14·0 lb/sq ft)
- Max power loading, Restricted category:
  - A 14·93 kg/kW (24·5 lb/hp)
  - B 10·95 kg/kW (18·0 lb/hp)

Performance (at max T-O weight, Restricted category):
- Max working speed:
  - A 83 knots (153 km/h; 95 mph)
  - B 100 knots (185 km/h; 115 mph)
- Min working speed:
  - both versions 56·5 knots (105 km/h; 65 mph)
- Stalling speed, power off:
  - both versions 48 knots (88·5 km/h; 55 mph)
- Stalling speed, power on:
  - both versions 44 knots (82 km/h; 51 mph)

ECTOR

ECTOR AIRCRAFT COMPANY

414 East Hillmont Road, Odessa, Texas 79762
Telephone: (915) 362 1841
President: Timothy H. Parker

ECTOR L-19 MOUNTAINEER and SUPER MOUNTAINEER

The Ector Aircraft Company has in production a civil version of the Cessna L-19 Bird Dog (last described in the 1964-65 *Jane's*), to which it has given the name Mountaineer. This is available in two models:

Mountaineer. Standard model, with 159 kW (213 hp) Continental O-470-11 engine and fixed-pitch propeller. Continental O-470-11-13 (cruising speed increased to 108 knots; 201 km/h; 125 mph at 55% power) or O-470-11-13-15 engine and constant-speed propeller (104 knots; 193 km/h; 120 mph at 57% power) optional.

Super Mountaineer. More powerful version with a 186 kW (250 hp) Avco Lycoming O-540-A4B5 engine, driving a Hartzell Type HC-C2YK-1B two-blade constant-speed metal propeller, and with additional equipment as standard. In 1978 Ector obtained a Supplemental Type Certificate for the Super Mountaineer at a maximum T-O weight of 1,270 kg (2,800 lb), and delivered an aircraft for operation at this weight.

Generally similar to the original Cessna L-19, Ector's Mountaineers are rebuilt completely from new off-the-shelf or serviceable components. The entire airframe is corrosion-proofed with zinc chromate before assembly; mounting brackets for floats are built into the basic airframe; and all four side windows can be opened in flight. The rear seat is removable to permit the carriage of cargo.

More than 30 Mountaineers have been produced for service with various organisations as glider tugs and for patrol and general-purpose duties. Others are flown as sporting aircraft.

Type: Two-seat (optional three-seat) cabin monoplane.

Wings: Braced high-wing monoplane. Single streamline-section bracing strut each side. Wing section NACA 2412. Dihedral 2° 8′. Incidence 1° 30′ at root, −1° 30′ at tip. All-metal single-spar structure, with metal skin. Frise-type all-metal ailerons. Fowler all-metal trailing-edge flaps. No tabs.

Fuselage: Conventional all-metal semi-monocoque structure.

Tail Unit: Cantilever all-metal structure. Trim tab in elevator. Small auxiliary fins are attached to tailplane tips of floatplane version.

Landing Gear: Non-retractable tailwheel type. Cantilever spring steel main legs. Goodyear main wheels with tyres size 6·00-6. Scott steerable tailwheel. Single-disc hydraulic brakes. Floats, skis or tandem landing gear for rough terrain optional.

Power Plant: One Continental or Avco Lycoming flat-six engine as detailed in model listings, driving a two-blade fixed-pitch or constant-speed propeller. One fuel tank in each wing root, total capacity 151·4 litres (40 US gallons). Optionally, fuel cells in each wing with a total capacity of 246 litres (65 US gallons). Refuelling points on wing upper surface. Oil capacity 9·4 litres (2·5 US gallons).

Accommodation: Normally two seats in tandem in enclosed cabin with 360° field of view. Door on starboard side. All four cabin side windows can be opened fully. Six skylights in roof. Space for baggage behind rear seats. With rear seat removed, 0·85 m³ (30 cu ft) of space is available for freight. Optional conversion to provide three seats. Cabin heated and ventilated.

Systems: Hydraulic system for brakes only. Electrical system powered by 24V 50A engine-driven generator. Super Mountaineer has a 12V electrical system.

Avionics and Equipment: Radio equipment available to customer's requirements. Navigation and landing lights,

Ector Super Mountaineer, a civil version of the Cessna L-19 Bird Dog

and heated pitot, standard. Stall warning indicator, Hobbs hour meter and Whelen three-light strobe system standard on Super Mountaineer. External power socket and wing racks optional.

DIMENSIONS, EXTERNAL:

Wing span	10·97 m (36 ft 0 in)
Wing chord at root	1·63 m (5 ft 4 in)
Wing chord at tip	1·09 m (3 ft 7 in)
Wing aspect ratio	7·35
Length overall	7·86 m (25 ft 9½ in)
Height overall	2·29 m (7 ft 6 in)
Tailplane span	3·21 m (10 ft 6½ in)
Propeller diameter	2·29 m (7 ft 6 in)
Propeller ground clearance	0·23 m (9 in)
Door: Height	0·64 m (2 ft 1 in)
Width	0·81 m (2 ft 8 in)
Height to sill	1·12 m (3 ft 8 in)

WEIGHTS (A: Mountaineer with O-470-11 engine and fixed-pitch propeller; B: Super Mountaineer):

Weight empty, equipped: A	658 kg (1,450 lb)
B	673 kg (1,483 lb)
Max T-O weight: A	1,043 kg (2,300 lb)
B	1,270 kg (2,800 lb)

PERFORMANCE (at max T-O weight):

Max level speed at S/L:
- A: 87 knots (161 km/h; 100 mph)
- B: 112 knots (208 km/h; 129 mph) IAS

Max cruising speed: B at 2,135 m (7,000 ft) 139 knots (257 km/h; 160 mph)

Econ cruising speed: B at 2,895 m (9,500 ft) 109 knots (201 km/h; 125 mph) IAS

Stalling speed, flaps down:
- B: 45·5 knots (84 km/h; 52 mph)

Max rate of climb at S/L: A 366 m (1,200 ft)/min
- B: 549 m (1,800 ft)/min

Service ceiling: A 6,980 m (22,900 ft)

T-O run: A 122 m (400 ft)

Landing run: A 98 m (320 ft)

Range with max fuel, no reserves:
- A: 651 nm (1,207 km; 750 miles)

EMAIR

EMAIR (a division of Emroth Co)

Hangar 38, Industrial Airpark, Harlingen, Texas 78550

Telephone: (512) 425 6363

PRESIDENT: George A. Roth

The development history of the original Emair MA-1, of which 25 were built before production ended in January 1976, can be found in earlier editions of *Jane's*. Flight testing of the more powerful MA-1B Diablo 1200 began in August 1975, and this version received FAA Type Approval in May 1976. By early 1980, a total of 48 Diablo 1200s had been delivered.

In January 1981 the company stated that production had been suspended until the market improved. Details of the MA-1B Diablo 1200 can be found in the 1980-81 *Jane's*.

ENSTROM

THE ENSTROM HELICOPTER CORPORATION

HEAD OFFICE AND WORKS: PO Box 277, Twin County Airport, Menominee, Michigan 49858

Telephone: (906) 863 9971

Telex: 263451

CHAIRMAN OF THE BOARD: John M. Ferris

PRESIDENT: Herbert Moseley

SENIOR VICE-PRESIDENT: Paul L. Shultz

EXECUTIVE VICE-PRESIDENT: J. F. Hansen

VICE-PRESIDENTS:
D. E. Brandt (Engineering)
Robert Kroll (Quality Control)
Curt Larsen (Finance)
Alan C. Nitchman (Sales and Service)

In its original form, as the R. J. Enstrom Corporation, this company began in 1959 to develop a homebuilt light helicopter built by Rudolph J. Enstrom, which flew for the first time on 12 November 1960. There followed a design and development programme on a new helicopter, designated F-28, and this flew for the first time in May 1962. A limited number of F-28s were built, and were followed by the improved Model F-28A in 1968.

The company was acquired by the Purex Corporation in October 1968 and was operated for a time as part of the Pacific Airmotive Aerospace group. Under this ownership, a turbocharged F-28B version was developed, as well as a Model T-28 turbine-powered version.

The activities of this group ended in February 1970; but in January 1971 the Purex shares were acquired by F. Lee Bailey, and manufacture of the F-28A was resumed. In January 1980 these shares were purchased by Bravo Investments BV, of the Netherlands, which company is now the owner of Enstrom. A total of 715 Enstrom helicopters had been built by 1 February 1981.

ENSTROM MODELS F-28 and 280

Production of the basic Model F-28A and the Model 280 Shark ended during 1978; details of these can be found in the 1978-79 *Jane's*. They were replaced by turbocharged versions of both models, under the designations F-28C and 280C respectively. These received FAA certification on 8 December 1975.

Models available in 1981 were as follows:

F-28C-2. Basic version, with 153 kW (205 hp) Avco Lycoming HIO-360-E1BD engine and Rajay 301-E-10-2 turbocharger. Semi-circular guard-tube to protect tail rotor in tail-down attitude. Two-piece windscreen of original F-28C replaced by one-piece windscreen, and new pedestal type central instrument console for improved forward view.

Model 280C Shark. Business and sporting version of F-28C-2. Refined aerodynamic contours and revised tail stabilising surfaces. Two-piece windscreen.

F-28F Falcon. As F-28C-2, but with 168 kW (225 hp) Avco Lycoming HIO-360-F1AD engine and Rajay 325-E-10-2 turbocharger. Received FAA certification in January 1981.

Model 280F Shark. Refined version of the F-28F, incorporating features detailed for the Model 280C Shark. Received FAA certification in January 1981.

Enstrom F-28C-2, latest version of the basic F-28 helicopter

Enstrom Model 280C, with an Avco Lycoming HIO-360-E1BD turbocharged engine

In addition, the following version is under development:

Model 280L Hawk. Four-seat version of Model 280C with HIO-360-F1AD engine, first flown on 27 December 1978. Described separately.

An Enstrom wet or dry dispersal agricultural kit is available for fitment to the F-28C-2/280C. It comprises two side-mounted hoppers with large quick-fill openings, and spraybooms with a normal span of 9·04 m (29 ft 8 in), but extendable to 11·07 m (36 ft 4 in). A manually operated clutch provides positive control of the centrifugal pump, which has a liquid capacity of 227 litres (60 US gallons)/min. Dry discharge rate is variable from 0 to 272 kg (600 lb)/min. Weight of the entire quickly-removable dispersal system is 48 kg (105 lb). Hopper capacity is 340 litres (90 US gallons) of liquid or 0·5 m³ (17·4 cu ft) of dry chemicals.

The following description applies to the current production versions:

TYPE: Three-seat light helicopter.

ROTOR SYSTEM: Fully-articulated metal three-blade main rotor. Blades are of bonded light alloy construction, each attached to rotor hub by retention pin and drag link. Blade section NACA 0013·5. Two-blade teetering tail rotor, with blades of bonded light alloy construction. Tail rotor on port side. Blades do not fold. No rotor brake.

ROTOR DRIVE: Poly V-belt drive system. Right-angle drive reduction gearbox. Main rotor/engine rpm ratio 1 : 7·154; tail rotor/engine rpm ratio 1 : 1·156.

FUSELAGE: Glassfibre and light alloy cab structure, with welded steel tube centre-section. Semi-monocoque aluminium tailcone structure.

Tail Unit: All models have horizontal stabiliser forward of tail rotor. On F-28C-2, F-28F and 280L, this carries small endplate fins. The 280C and 280F have a dorsal fin, and sweptback ventral fin with tail rotor guard skid at the tip.

Landing Gear: Skids carried on Enstrom oleo-pneumatic shock-absorbers. Air Cruiser inflatable floats available optionally.

Power Plant: One Avco Lycoming flat-four engine with Rajay turbocharger, as detailed in model listings. Two fuel tanks, each of 75·7 litres (20 US gallons). Total fuel capacity 151·4 litres (40 US gallons). Oil capacity 9·5 litres (2·5 US gallons).

Accommodation: Pilot and two passengers, side by side on bench seat; centre place removable. Fully-transparent removable door on each side of cabin. Baggage space aft of engine compartment, capacity 49 kg (108 lb), with external access door. Cabin heated and ventilated.

Systems: Electrical power provided by 12V 70A engine-driven alternator.

Avionics and Equipment: Nav/com to customer's requirements. Cargo hook, floats, spraygear and litters optional.

Dimensions, external (A: F-28C-2/F; B: 280C/F):
- Diameter of main rotor: 9·75 m (32 ft 0 in)
- Diameter of tail rotor: 1·45 m (4 ft 9¼ in)
- Distance between rotor centres: 5·56 m (18 ft 3 in)
- Main rotor blade chord: 0·24 m (9½ in)
- Length overall: A 8·94 m (29 ft 4 in)
- B 8·43 m (27 ft 8 in)
- Height to top of rotor hub: 2·79 m (9 ft 2 in)
- Skid track: 2·24 m (7 ft 4 in)
- Cabin doors (each): Height 1·04 m (3 ft 5 in)
- Width 0·84 m (2 ft 9 in)
- Height to sill 0·64 m (2 ft 1 in)
- Baggage door: Height 0·55 m (1 ft 9½ in)
- Width 0·39 m (1 ft 3½ in)
- Height to sill 0·86 m (2 ft 10 in)

Dimensions, internal:
- Max width of cabin: 1·47 m (4 ft 10 in)
- Volume of baggage hold: 0·20 m³ (7 cu ft)

Areas:
- Main rotor disc: 74·69 m² (804 sq ft)
- Tail rotor disc: 1·66 m² (17·88 sq ft)

Weights and Loadings (A: F-28C-2 and 280C; B: F-28F and 280F Normal category; C: F-28F and 280F Utility category):
- Weight empty, equipped: A, B, C 680 kg (1,500 lb)
- Max T-O weight: A, B 1,066 kg (2,350 lb)
- C 1,179 kg (2,600 lb)
- Max disc loading: A, B 14·25 kg/m² (2·92 lb/sq ft)
- C 15·77 kg/m² (3·23 lb/sq ft)
- Max power loading: A, B 6·97 kg/kW (11·46 lb/hp)
- C 7·71 kg/kW (12·68 lb/hp)

Performance (at AUW of 1,066 kg; 2,350 lb, except where indicated):
- Max level speed, S/L to 915 m (3,000 ft):
  - F-28C-2, F-28F 97 knots (180 km/h; 112 mph) IAS
  - 280C/F 102 knots (188 km/h; 117 mph) IAS
- Max cruising speed:
  - F-28C-2 93 knots (172 km/h; 107 mph)
  - F-28F 97 knots (180 km/h; 112 mph)
  - 280C 95·5 knots (177 km/h; 110 mph)
  - 280F 102 knots (188 km/h; 117 mph)
- Econ cruising speed:
  - F-28C-2 74 knots (137 km/h; 85 mph)
  - F-28F 83 knots (154 km/h; 96 mph)
  - 280C 82·5 knots (153 km/h; 95 mph)
  - 280F 88 knots (163 km/h; 101 mph)
- Max rate of climb at S/L:
  - F-28C-2, 280C 350 m (1,150 ft)/min
  - F-28F, 280F 442 m (1,450 ft)/min
- Certificated operating ceiling:
  - F-28C-2, 280C 3,660 m (12,000 ft)
  - F-28F, 280F 3,050 m (10,000 ft)
- Hovering ceiling IGE:
  - F-28C-2, 280C 2,680 m (8,800 ft)
  - F-28F, 280F 3,780 m (12,400 ft)
- Hovering ceiling IGE at AUW of 1,179 kg (2,600 lb):
  - F-28F, 280F 1,555 m (5,100 ft)
- Hovering ceiling OGE:
  - F-28C-2, 280C 1,250 m (4,100 ft)
  - F-28F, 280F 2,285 m (7,500 ft)
- Max range, standard fuel, no reserves:
  - F-28C-2 234 nm (434 km; 270 miles)
  - 280C 231 nm (428 km; 266 miles)
  - F-28F 228 nm (423 km; 263 miles)
  - 280F 240 nm (445 km; 277 miles)

Prototype of the Enstrom Model 280L Hawk four-seat light helicopter *(Howard Levy)*

ENSTROM MODEL 280L HAWK

Design of the Model 280L Hawk, a four-seat version of the Model 280C, began in January 1978; construction of two prototypes started in August. The first flight was made on 27 December 1978. Further development of this aircraft has been deferred until late 1981.

The Model 280L is generally similar to the Model 280C, except as follows:

Type: Four-seat light helicopter.

Rotor System: As for Model 280C, except for increased-diameter main rotor blades, with sheet metal tabs at tip and mid-span.

Rotor Drive: As Model 280C, except that engine is installed with drive shaft forward.

Fuselage: Glassfibre and light alloy cab structure, with welded steel tube and light alloy centre-section. Semi-monocoque tailcone structure of light alloy. Fail-safe construction in major areas.

Tail Unit: Fixed horizontal surface of light alloy, with small endplate fins.

Landing Gear: As Model 280C.

Power Plant: One 168 kW (225 hp) Avco Lycoming HIO-360-F1AD flat-four engine with Rajay 325-E-10-2 turbocharger. Two fuel tanks, one each side of main transmission, with combined capacity of 170 litres (45 US gallons). Refuelling point on each side of fuselage upper surface. Oil capacity 7·5 litres (2 US gallons).

Accommodation: Individual, adjustable seats for pilot and co-pilot forward, two passengers aft, with baggage hold at rear of cabin. Double doors on each side of cabin. Accommodation is heated and ventilated as standard. Air-conditioning optional.

System: Electrical system powered by 12V 70A engine-driven alternator.

Avionics and Equipment: As for Model 280C.

Dimensions, external:
- Diameter of main rotor: 10·36 m (34 ft 0 in)
- Diameter of tail rotor: 1·42 m (4 ft 8 in)
- Distance between rotor centres: 5·87 m (19 ft 3 in)
- Main rotor blade chord: 0·24 m (9½ in)
- Length overall: 9·86 m (32 ft 4 in)
- Height overall: 2·79 m (9 ft 2 in)
- Cabin doors (each): Height 0·97 m (3 ft 2 in)
- Width 1·30 m (4 ft 2 in)
- Height to sill 0·71 m (2 ft 4 in)
- Baggage door: Height 0·48 m (1 ft 7 in)
- Width 0·58 m (1 ft 11 in)
- Height to sill 0·71 m (2 ft 4 in)

Dimensions, internal:
- Cabin: Length 1·78 m (5 ft 10 in)
- Max width 1·47 m (4 ft 10 in)
- Max height 1·22 m (4 ft 0 in)
- Floor area 2·60 m² (28 sq ft)
- Volume 3·06 m³ (108 cu ft)
- Volume of baggage hold 0·34 m³ (12 cu ft)

Areas:
- Main rotor disc: 84·35 m² (908 sq ft)
- Tail rotor disc: 1·58 m² (17·06 sq ft)

Weights and Loadings:
- Weight empty, basic: 708 kg (1,560 lb)
- Max T-O and landing weight: 1,179 kg (2,600 lb)
- Max disc loading: 13·98 kg/m² (2·86 lb/sq ft)
- Max power loading: 7·02 kg/kW (11·56 lb/hp)

Performance (at max T-O weight):
- Never-exceed speed: 104 knots (193 km/h; 120 mph)
- Max level and cruising speed at S/L: 95·5 knots (177 km/h; 110 mph)
- Max rate of climb at S/L: 349 m (1,144 ft)/min
- Service ceiling: 3,660 m (12,000 ft)
- Hovering ceiling IGE: 3,110 m (10,200 ft)
- Hovering ceiling OGE: 1,890 m (6,200 ft)
- Range with max fuel at 1,220 m (4,000 ft): 263 nm (488 km; 303 miles)

EXCALIBUR
EXCALIBUR AVIATION COMPANY

PO Box 32007, San Antonio, Texas 78216

Telephone: (512) 927 6201

President: William C. Hickey

Excalibur Aviation is engaged in the production of improved versions of the Beechcraft Queen Air. These modified aircraft, named Queenaire 800 and Queenaire 8800, are being completed at the rate of one per month. The conversion provides improved speed and range, together with reduced operating costs.

EXCALIBUR QUEENAIRE 800 and 8800

The Excalibur modification of Queen Air 65s and 80s includes installation of two 298 kW (400 hp) Avco Lycoming IO-720-A1B eight-cylinder engines, each driving a Hartzell three-blade constant-speed and fully-feathering metal propeller; new engine mountings; new exhaust system; new low-drag engine nacelles; new (or zero-time overhauled and certificated) accessories; and Excalibur fully-enclosed wheel-well doors. Modifications of the Beechcraft Queen Air 65, A65 and 80 of all serial numbers are designated Queenaire 800; similar modifications to the Queen Air A80 and B80 of all serial numbers have the designation Queenaire 8800.

Typical Excalibur Queenaire 800/8800 conversion (two Avco Lycoming IO-720-A1B engines)

Weights (A: Queenaire 800; B: Queenaire 8800):
- Weight empty, equipped (average):
  - A 2,449 kg (5,400 lb)
  - B 2,631 kg (5,800 lb)
- Max T-O weight: A 3,628 kg (8,000 lb)
- B 3,991 kg (8,800 lb)
- Max landing weight: A 3,447 kg (7,600 lb)
- B 3,792 kg (8,360 lb)

Performance (at max T-O weight):
- Cruising speed, 75% power:
  - A, B at 2,530 m (8,300 ft) 201 knots (372 km/h; 231 mph)

Cruising speed, 65% power:
A, B at 3,050 m (10,000 ft) 195 knots (362 km/h; 225 mph)
Cruising speed, 45% power at 3,050 m (10,000 ft):
A, B 172 knots (319 km/h; 198 mph)
Stalling speed, gear and flaps up:
A 80 knots (148 km/h; 92 mph)
B 86 knots (160 km/h; 99 mph)
Stalling speed, gear and flaps down:
A 68 knots (126 km/h; 78 mph)
B 70 knots (129 km/h; 80 mph)
Max rate of climb at S/L: A 468 m (1,535 ft)/min
B 454 m (1,490 ft)/min
Rate of climb at S/L, one engine out:
A 110 m (360 ft)/min
B 76 m (250 ft)/min
Service ceiling: A 6,005 m (19,700 ft)
B 5,700 m (18,700 ft)
Service ceiling, one engine out:
A 3,595 m (11,800 ft)
B 3,110 m (10,200 ft)
T-O to 15 m (50 ft): A 520 m (1,706 ft)
B 625 m (2,050 ft)
Landing from 15 m (50 ft): A 663 m (2,176 ft)
B 747 m (2,450 ft)
Range with max fuel at 3,050 m (10,000 ft), with 113·5 litres (30 US gallons) reserves:
A 1,322 nm (2,451 km; 1,523 miles)
B 1,547 nm (2,867 km; 1,782 miles)

FAIRCHILD INDUSTRIES

FAIRCHILD INDUSTRIES INC

Corporate Offices: 20301 Century Boulevard, Germantown, Maryland 20767
Telephone: (301) 428 6000
Chairman of the Board and Chief Executive Officer: Edward G. Uhl
President and Chief Operating Officer: John F. Dealy
Executive Vice-President: Charles Collis
Senior Corporate Vice-Presidents:
Emanuel Fthenakis (Chairman & Chief Executive Officer, Fairchild Space & Electronics Company)
W. Daniel Heidt (President, Fairchild Swearingen Corporation; General Manager, Fairchild Aircraft Service Division)
William H. Buckland (Treasurer)
Corporate Vice-Presidents:
George S. Attridge (Chief Financial Officer)
Robert J. Dixon (President, Fairchild Republic Company)
James J. Foody (Aerospace Development)
Richard R. Molleur (General Counsel)
J. Gilbert Nettleton Jr (Marketing)
Richard G. Orr (Chairman, Industrial Products Division)
Dr John W. Townsend Jr (President, Fairchild Space & Electronics Company)
Robert C. Woods
Secretary: John D. Jackson

Fairchild Republic Company
Divisional Office and Works: Farmingdale, Long Island, New York 11735
President: Robert J. Dixon
Hagerstown Facility: Hagerstown, Maryland 21740
Site Administrator: Paul Schmidt

Fairchild Aircraft Service Division
Divisional Office and Works: PO Drawer 1177, Crestview, Florida 32536
General Manager: W. Daniel Heidt

Fairchild Space and Electronics Company
Germantown, Maryland 20767
President: Dr John W. Townsend Jr

Fairchild Swearingen Corporation
San Antonio, Texas 78284
President: W. Daniel Heidt

Fairchild Stratos Division
Manhattan Beach, California 90266
General Manager: Donald B. Rassier

Fairchild Industrial Products Division
Winston-Salem, North Carolina 27105
Chairman: Richard G. Orr
General Manager: Robert L. Slater

Fairchild Burns Company
Winston-Salem, North Carolina 27105
General Manager: Anthony J. Spuria

American Satellite Company
Germantown, Maryland 20767
President: William S. Wheatley

VSI Corporation
Pasadena, California
President: Mason Phelps

Fairchild Industries is an aerospace, commercial/industrial products and communications company with business interests in civil and military aircraft, aerospace fasteners, aircraft subsystems, tooling for the plastics industry, industrial and electronic products, and domestic satellite communications systems.

The Fairchild Republic Company (see following entry) combines the corporation's military aircraft research, design, manufacture and personnel into one division. Facilities are located at Farmingdale, Long Island, New York; and Hagerstown, Maryland.

Based in Crestview, Florida, Fairchild Aircraft Service Division is involved in aircraft maintenance, repair and modification, as well as production of aircraft components. Current programmes include support equipment and fire-suppression foam for the A-10, components for Swearingen aircraft, and seats for military aircraft.

The Fairchild Space and Electronics Company directs the corporation's efforts in the design, development and manufacture of spacecraft, spacecraft subsystems, telecommunications ground equipment, and avionics systems; details of some of these programmes can be found in the Spaceflight section. The division also builds avionics systems, including armament control systems for the Fairchild A-10, Grumman F-14A Tomcat, and the Navy's Light Airborne Multi-Purpose System (LAMPS III) helicopter; and aircraft data annotation systems.

The Fairchild Stratos Division specialises in the development and manufacture of aerospace and commercial aircraft subsystems, and components. It is now manufacturing valves and the re-entry cooling system for the Space Shuttle, valves for space launch vehicles and missiles, bleed air turbines for Boeing 747 aircraft, avionics cooling systems for aircraft, aircraft pods, and ground support of missile systems.

Fairchild Swearingen Corporation (which see), a wholly owned Fairchild subsidiary, manufactures the Merlin and Metro series of executive, commuter and special-mission aircraft.

Fairchild Burns Company manufactures aircraft passenger seating used by more than 50 airlines worldwide, including the Airest 2000 line of lightweight seats.

FAIRCHILD REPUBLIC COMPANY

Divisional Office and Works: Farmingdale, Long Island, New York 11735
Telephone: (516) 531 0105
Telex: 96-7735

Founded on 17 February 1931, as the Seversky Aircraft Company, Republic operated as Republic Aviation Corporation from 1939 until September 1965, when it became a division of Fairchild Hiller Corporation, now Fairchild Industries Inc.

Its current work includes manufacture of the A-10 Thunderbolt II aircraft, structural components for the Boeing 747 and 757 airliners, tail fins and rudders for the Grumman F-14 Tomcat, and wings for Swearingen aircraft. It is developing a new transport aircraft jointly with Saab-Scania of Sweden (see International section).

FAIRCHILD REPUBLIC THUNDERBOLT II

USAF designation: A-10A

Fairchild Republic and Northrop each built two prototypes for evaluation under the US Air Force's A-X programme, initiated in 1967, for a close-support aircraft. The first Fairchild Republic prototype (71-1369), designated YA-10A, flew for the first time on 10 May 1972. On 18 January 1973 Fairchild was announced as the winner of the competitive evaluation of the prototypes, and received a contract for six A-10A DT and E aircraft, the first of which flew on 15 February 1975. Details of them can be found in the 1977-78 *Jane's*.

The first flight by a production A-10A Thunderbolt II (75-00258) was made on 21 October 1975. Purchase of a total of 747 aircraft for the USAF is planned (including the six DT and E aircraft). Of the latest increment of 60 A-10s ordered in FY 1981, 30 will be two-seat **combat-ready trainers**. Although resembling the prototype two-seat night/adverse weather (N/AW) attack aircraft built as a private venture by Fairchild (see 1980-81 *Jane's*), the trainer will be generally similar to the A-10A. The night/adverse weather capability of the single-seater is expected to be enhanced considerably by the addition of a Martin Marietta LANTIRN (Low-Altitude Navigation Targeting Infra-Red for Night) fire control pod in due course. Features of LANTIRN include a forward-looking infra-red sensor, automatic target recognition, automatic hand-off for Maverick IR missile delivery, automatic tracking and designation for laser-guided bombs, and terrain-following navigation for low-level day/night operations.

The first combat-ready A-10A wing was the 354th Tactical Fighter Wing, based at Myrtle Beach, South Carolina, to which deliveries began in March 1977. In May 1977 several A-10As were assigned to the 422nd Fighter Weapons Squadron of the 57th Tactical Training Wing at Nellis AFB, Nevada, for operational training and testing purposes. On 15 October 1977 the 356th Tactical Fighter Squadron of 354 TFW became the first combat-ready A-10 squadron.

In August 1977, six A-10As flew to Europe for the first live firing of the aircraft's 30 mm cannon and the anti-armour Maverick air-to-ground missile in the NATO theatre, expending 7·5 tons of conventional 500 lb bombs and 9,000 rounds of 30 mm ammunition during 117 close support sorties.

The first overseas deployment of the A-10 was made on 25 January 1979, with the arrival of 14 Thunderbolt IIs at RAF stations Bentwaters and Woodbridge in the UK, to equip the US Air Force's 81st Tactical Fighter Wing. Six squadrons will eventually be based at the two UK stations.

Deliveries of new production A-10As to one Tactical Fighter Wing (174th) and four Tactical Fighter Groups (103rd, 104th, 128th and 175th) of the Air National Guard are under way, the A-10 being the first first-line aircraft to be assigned to ANG units. Also currently receiving A-10As are the 434th TFW and 927th TFG of the Air Force Reserve.

With production at the peak rate of 12 per month in 1981, more than 500 A-10As had then been delivered to the US Air Force. Current USAF A-10A units are the 23rd, 81st, 354th and 355th Tactical Fighter Wings, and the 66th Fighter Weapons Squadron.

Type: Single-seat close-support aircraft.

Wings: Cantilever low-wing monoplane, with wide-chord, deep aerofoil section (NACA 6716 on centre-section and at start of outer panel, NACA 6713 at tip) to provide low wing loading. Incidence −1°. Dihedral 7° on outer panels. Aluminium alloy three-spar structure, consisting of one-piece constant-chord centre-section and tapered outer panels with integrally stiffened skins and drooped (cambered) wingtips. Outer panel leading-edges and cores of trailing-edges are of honeycomb sandwich. Four-point attachment of wings to fuselage, at front and rear spars. Two-segment,

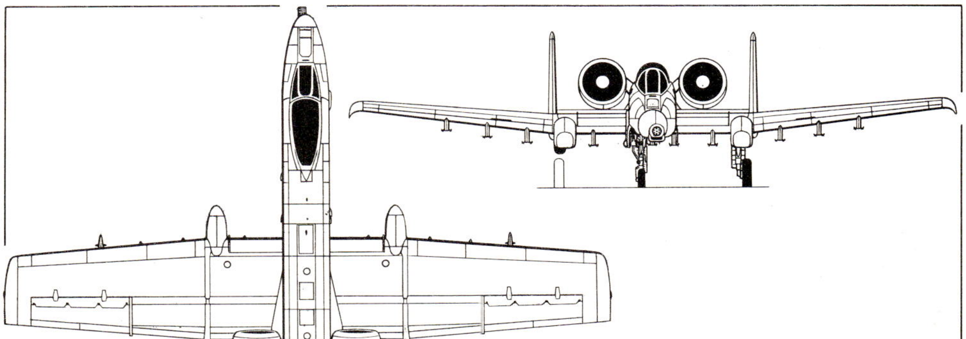

Fairchild Republic A-10A Thunderbolt II single-seat twin-engined close-support aircraft *(Pilot Press)*

Fairchild Republic A-10A Thunderbolt II single-seat close-support aircraft

three-position trailing-edge slotted flaps, interchangeable right with left. Wide-span ailerons, made up of upper and lower surfaces that separate to serve as airbrakes. Flaps, airbrakes and ailerons actuated hydraulically. Ailerons pilot-controlled by servo tab during manual reversion. Small leading-edge slat inboard of each main-wheel fairing. Redundant and armour-protected flight control system.

FUSELAGE: Semi-monocoque structure of aluminium alloy (chiefly 2024 and 7075), with four main longerons, multiple frames, and lap-jointed and riveted skins. Built in front, centre and aft portions. Single-curvature components aft of nose portion, interchangeable right with left. Centre portion incorporates wing box carry-through structure.

TAIL UNIT: Cantilever aluminium alloy structure, with twin fins and interchangeable rudders mounted at the tips of constant-chord tailplane. Interchangeable elevators, each with an electrically-operated trim tab. Rudders and elevators actuated hydraulically. Redundant and armour-protected flight control system.

LANDING GEAR: Menasco retractable tricycle type with single wheel on each unit. All units retract forward, and have provision for emergency gravity extension. Interchangeable main-wheel units retract into non-structural pod fairings attached to the lower surface of the wings. When fully retracted approximately half of each wheel protrudes from the fairing. Steerable nosewheel is offset to starboard to clear firing barrel of gun. Main wheels size 36 × 11, Type VII; nosewheel size 24 × 7·7-10, Type VII.

POWER PLANT: Two General Electric TF34-GE-100 high bypass ratio turbofan engines, each rated at 40·3 kN (9,065 lb st), enclosed in separate pods, each pylon-mounted to the upper rear fuselage at a point approximately midway between the wing trailing-edges and the tailplane leading-edges. Fuel is contained in two tear-resistant and self-sealing cells in the fuselage, and two smaller, adjacent integral cells in the wing centre-section. Maximum internal fuel capacity 4,853 kg (10,700 lb). All fuel cells are internally filled with reticulated foam, and all fuel systems pipework is contained within the cells except for the feeds to the engines, which have self-sealing covers. Three 2,271 litre (600 US gallon) jettisonable auxiliary tanks can be carried on underwing and fuselage centreline pylons. Provision for in-flight refuelling using universal aerial refuelling receptacle slipway installation (UARRSI).

ACCOMMODATION: Single-seat enclosed cockpit, well forward of wings, with large transparent bubble canopy to provide all-round view. Bulletproof windscreen. Canopy is hinged at rear and opens upward. Douglas ejection seat operable at speeds from 450 knots (834 km/h; 518 mph) down to zero speed at zero height. Entire cockpit area is protected by an armoured 'bathtub' structure of titanium, capable of withstanding projectiles of up to 23 mm calibre.

SYSTEMS: Redundant control system incorporates two 207 bar (3,000 lb/sq in) primary hydraulic flight control systems, each powered by an engine-driven pump, and a manual backup. Hydraulic systems actuate flaps, flying control surfaces, landing gear, brakes and nosewheel steering. Two independent hydraulic motors, either of which is sufficient to sustain half-rate firing, supply drive for 30 mm gun barrel rotation. Electrical system includes two 30/40kVA 115/200V AC engine-driven generators and a standby battery and inverter. Auxiliary power unit. Environmental control system, using engine bleed air for cockpit pressurisation and air-conditioning, pressurisation of pilot's *g* suit, windscreen anti-icing and rain clearance, fuel transfer, gun compartment purging, and other services.

AVIONICS AND EQUIPMENT: Head-up display giving airspeed, altitude and dive angle; weapons delivery package with dual reticle optical sight for use in conjunction with underfuselage Pave Penny laser designation pod; target penetration aids; associated equipment for Maverick and other missile systems; IFF/SIF (AIMS); UHF/AM; VHF/AM; VHF/FM; Tacan; UHF/ADF; ILS/FDC; X-band transponder; INS; heading and attitude reference system (HARS); radar homing and warning (RHAW); secure voice communications; active or passive electronic countermeasures (ECM); armament control panel; and gun camera. Space provisions for HF/SSB, and other 'growth' avionics and equipment.

ARMAMENT: General Electric GAU-8/A Avenger 30 mm seven-barrel cannon, mounted in nose with 2° depression and offset slightly to port so that as the barrels rotate the firing barrel is always on the aircraft centreline. Gun and handling system for the linkless ammunition are mechanically synchronised and driven by two motors fed from the aircraft's hydraulic system. The single-drum magazine has a capacity of 1,174 rounds, and has a dual firing rate of either 2,100 or 4,200 rds/min. Four stores pylons under each wing (one inboard and three outboard of each main-wheel fairing), and three under fuselage, for max external load of 7,250 kg (16,000 lb). External load with full internal fuel is 6,505 kg (14,341 lb). The centreline pylon and the two flanking fuselage pylons cannot be occupied simultaneously. The centreline pylon has a capacity of 2,268 kg (5,000 lb); the two fuselage outer pylons and two centre-section underwing pylons 1,587 kg (3,500 lb) each; the two innermost outer-wing pylons 1,134 kg (2,500 lb) each; and the four outermost wing pylons 453 kg (1,000 lb) each. These allow carriage of a wide range of stores, including twenty-eight 500 lb Mk 82 LDGP or Mk 82 retarded bombs; six 2,000 lb Mk 84 general-purpose bombs; eight BLU-1 or BLU-27/B incendiary bombs; four SUU-25 flare launchers; twenty Rockeye II cluster bombs, sixteen CBU-52/71 dispenser weapons; six AGM-65A Maverick missiles; Mk 82 and Mk 84 laser-guided bombs; Mk 84 electro-optically-guided bombs; two SUU-23 gun pods; ALE-40 chaff/flare system; ALQ-119 ECM pods, or other jammer pods; or up to three drop-tanks.

DIMENSIONS, EXTERNAL:

Wing span	17·53 m (57 ft 6 in)
Wing chord at root	3·04 m (9 ft 11½ in)
Wing chord (mean)	2·73 m (8 ft 11·32 in)
Wing chord at tip	1·99 m (6 ft 6·4 in)
Wing aspect ratio	6·54
Length overall	16·26 m (53 ft 4 in)
Height overall	4·47 m (14 ft 8 in)
Tailplane span	5·74 m (18 ft 10 in)
Wheel track	5·25 m (17 ft 2½ in)
Wheelbase	5·40 m (17 ft 8¾ in)

AREAS:

Wings, gross	47·01 m² (506·0 sq ft)
Ailerons (total, incl tabs)	4·42 m² (47·54 sq ft)
Trailing-edge flaps (total)	7·99 m² (85·99 sq ft)
Leading-edge slats (total)	0·98 m² (10·56 sq ft)
Airbrakes (total)	8·06 m² (86·78 sq ft)
Fins (total)	7·80 m² (83·96 sq ft)
Rudders (total)	2·18 m² (23·50 sq ft)
Tailplane	8·31 m² (89·40 sq ft)
Elevators (total, incl tabs)	2·69 m² (29·00 sq ft)

WEIGHTS AND LOADINGS:

Manufacturer's empty weight	9,771 kg (21,541 lb)
Operating weight empty	11,321 kg (24,959 lb)
*Basic design weight, equipped	14,438 kg (31,831 lb)
**Forward airstrip weight	14,865 kg (32,771 lb)
Internal fuel load	4,853 kg (10,700 lb)
Max external ordnance	7,250 kg (16,000 lb)
Max external ordnance with full internal fuel	6,505 kg (14,341 lb)
Max T-O weight	22,680 kg (50,000 lb)
Max wing loading	482·4 kg/m² (98·81 lb/sq ft)
Max power loading	281·39 kg/kN (2·76 lb/lb st)
Thrust/weight ratio	0·4

**incl six 500 lb bombs, 750 rds of ammunition, and 1,134 kg (2,500 lb) of fuel*

***with four Mk 82 bombs, 750 rounds of ammunition, and 2,041 kg (4,500 lb) of fuel*

PERFORMANCE: (at max T-O weight except where indicated):

Never-exceed speed 450 knots (834 km/h; 518 mph)
Max level speed at S/L, 'clean'
381 knots (706 km/h; 439 mph)
Combat speed at 1,525 m (5,000 ft), with six Mk 82 bombs 380 knots (704 km/h; 438 mph)
Cruising speed at S/L
300 knots (555 km/h; 345 mph)
Cruising speed at 1,525 m (5,000 ft)
336 knots (623 km/h; 387 mph)
Stabilised 45° dive speed below 2,440 m (8,000 ft), AUW of 15,932 kg (35,125 lb)
260 knots (481 km/h; 299 mph)
Max rate of climb at S/L at basic design weight
1,828 m (6,000 ft)/min
T-O distance:
at max T-O weight 1,220 m (4,000 ft)
at forward airstrip weight 442 m (1,450 ft)
Landing distance:
at max T-O weight 610 m (2,000 ft)
at forward airstrip weight 396 m (1,300 ft)
Operational radius, 20 min reserve:
close air support, 1·7 h loiter
250 nm (463 km; 288 miles)
deep strike 540 nm (1,000 km; 620 miles)
Ferry range, headwind of 50 knots (93 km/h; 58 mph)
2,131 nm (3,949 km; 2,454 miles)

FAIRCHILD REPUBLIC NIGHT/ADVERSE WEATHER A-10

This two-seat version of the A-10A Thunderbolt II is a night/adverse weather (N/AW) attack aircraft, with secondary use as a trainer retaining operational capability. The company-funded prototype was converted from one of the six single-seat DT & E A-10s built for the US Air Force in 1975, and made its first flight as a two-seater on 4 May 1979. The Fairchild test programme continued until the end of September 1979, followed by several months of testing by US Air Force pilots at Eglin AFB, Florida, where more than 100 flying hours were logged by the beginning of 1980. Details of the aircraft can be found in the 1980-81 *Jane's*.

The general appearance of the N/AW A-10 two-seater is shown in an accompanying illustration. There are no current plans to build more, although the A-10 two-seat combat-ready trainer (see main A-10 entry) will have a similar cockpit configuration. Instead, it is expected that single-seat A-10s will be given more effective N/AW capability by the addition of a LANTIRN fire control pod.

Fairchild Republic two-seat night/adverse weather version of the A-10

FAIRCHILD SWEARINGEN CORPORATION
(a subsidiary of Fairchild Industries)

PO Box 32486, San Antonio, Texas 78284
Telephone: (512) 824 9421
Telex: 767-315
President: W. Daniel Heidt
Executive Vice-President: Edward J. Swearingen
Vice-Presidents:
R. D. Busey (Administration)
D. E. Howard (Asst to the President)
R. E. McKelvey (Engineering)
J. A. Pike (New Product Development)
R. N. Robinson (Marketing)
R. C. Woods (Operations)
Directors:
P. D. Bartles (Manufacturing and Industrial Engineering)
R. E. Duncan (Contracts)
H. C. Henny (Quality Assurance)
J. N. Karamanian (Product Support)
D. T. Orseth (Materiel)
J. M. Rountree (Manufacturing)
Public Relations Manager: Marilyn A. Zimmer

As detailed in earlier editions of *Jane's*, the former Swearingen Aviation Corporation became a wholly-owned subsidiary of Fairchild Industries in 1979. Since January 1981 it has been known as Fairchild Swearingen Corporation.

Fairchild Swearingen has been engaged in manufacture of the Merlin series of twin-turboprop pressurised executive transport aircraft and the Metro airliner since 1966. A total of 512 Merlins and Metros of all models had been delivered throughout the world by January 1981. Production was then at a rate of over seven aircraft per month. This rate is to be increased to 12 a month, and the company's manufacturing facilities are being expanded by about 50 per cent to permit this. Three new models were introduced during 1980, with certification and first deliveries in the Spring of 1981. In addition, Fairchild Swearingen will be responsible for marketing in North America the new Saab-Fairchild 340 airliner, of which details can be found in the International section of this edition.

SWEARINGEN MODEL SA227-AC METRO III

The Metro III is a 19/20-passenger all-weather pressurised, air-conditioned airliner, certificated under the FAA's Special Federal Air Regulation 41B, which provides for compliance with ICAO Annex 8 specifications for operation at a gross weight in excess of 5,670 kg (12,500 lb). The Metro III is certificated for operation at a max T-O weight of 6,350 kg (14,000 lb), providing a useful load increase of 491 kg (1,082 lb) compared with the Metro II which it supersedes. The changes incorporated in the Metro III include a 3·05 m (10 ft 0 in) increase in wing span; more powerful engines, each driving a slow-turning four-blade Dowty Rotol propeller; new main landing gear doors, to improve take-off and landing performance and provide better access for maintenance; new streamline nacelle cowlings, with quick-action latches and hinged at the rear to improve engine access; improved handling characteristics; and new fire prevention and containment features, with all inflammable fluid pipework isolated physically from all electrical current-carrying components and wiring.

Nearly 200 Metro/Metro II/Metro IIIs were on order or in service with 26 airlines in the USA and 18 airlines in Africa, Australia, Europe, the Middle East and South America in mid-1981.

Type: Twin-turboprop 19/20-passenger commuter airliner.

Wings: Cantilever low-wing monoplane. Wing section NACA 65_2A215 at root, NACA 64_2A415 at tip. Dihedral 5°. Incidence 1° at root, −2° 30′ at tip. Sweepback at quarter-chord 0·9°. All-metal two-spar fail-safe structure of aluminium alloy, constructed in one piece. The main spar beams have laminated caps and these, in the centre-section, have titanium laminations. Hydraulically-operated double-slotted trailing-edge flaps. Manually-controlled trim tab in each aileron. Goodrich pneumatic de-icing boots on wing leading-edges, with automatic bleed air cycling system.

Fuselage: All-metal cylindrical semi-monocoque fail-safe structure of 2024 aluminium alloy, flush-riveted throughout. All but the nose section is pressurised. Glassfibre honeycomb nosecap can accommodate a 0·45 m (18 in) weather radar antenna.

Tail Unit: Cantilever all-metal structure with sweptback surfaces and dorsal fin. Small ventral fin. Electrically-adjustable variable-incidence tailplane. Manually-controlled rudder trim. Goodrich pneumatic de-icing boots on tailplane leading-edges, with automatic bleed air cycling system.

Landing Gear: Retractable tricycle type with twin wheels on each unit. Hydraulic retraction, with dual actuators on each unit. All wheels retract forward, main gear into engine nacelles, nosewheels into fuselage. Ozone Aircraft Systems oleo-pneumatic shock-absorber struts. Nosewheel steerable. Free-fall emergency extension system, with backup of hand-operated hydraulic pump. Goodrich main wheels with low-pressure tubeless tyres, size 18 × 5·50, type VII. Jay-Em nosewheels and Goodyear low-pressure tubeless tyres, size 16 × 4·40, type VII. Goodrich self-adjusting hydraulically-operated disc brakes and anti-skid system.

Power Plant: Two 745·5 kW (1,000 shp) dry/820 kW (1,100 shp) wet Garrett TPE331-11U-601G turboprop engines with continuous alcohol-water injection system, each driving a Dowty Rotol four-blade constant-speed fully-feathering reversible-pitch metal propeller with spinner. Automatic propeller synchrophasing, and full Beta control reversing, standard. In-flight windmill start capability. Integral fuel tank in each wing, each with a usable capacity of 1,226 litres (324 US gallons). Total usable fuel capacity 2,452 litres (648 US gallons). Refuelling point on each outer wing panel. Oil capacity 15·1 litres (4 US gallons). Engine inlet de-icing by bleed air. Electrical oil-cooler inlet anti-icing. Electrical propeller de-icing. Flush-mounted fuel vents. Single-point rapid defuelling provisions. Negative torque sensing, single red line/autostart, automatic engine temperature limiting, and engine fire extinguishing systems.

Accommodation: Crew of two on flight deck, each with four-way adjustable seat with folding armrests and shoulder harness, separated from passenger/cargo area by partial bulkhead on port side and armrest height curtain on starboard side. Dual controls standard. Bulkhead between cabin and flight deck optional. Standard accommodation for 19-20 passengers seated two abreast, on each side of centre aisle. 'No smoking' and 'Fasten seat belt' signs. Stowing fold-up seats for rapid conversion to cargo or mixed passenger/cargo configuration. Movable bulkhead between passenger and cargo sections. Snap-in carpeting. Self-stowing aisle filler. Tiedown fittings for cargo at 0·76 m (30 in) spacing. Integral-step passenger door on port side of fuselage, immediately aft of flight deck. Large cargo loading door on port side of fuselage at rear of cabin, hinged at top. Three window emergency exits, one on the port, two on the starboard side. Forward baggage/avionics compartment in nose, capacity 363 kg (800 lb). Pressurised rear cargo compartment, capacity 272 kg (600 lb). Cabin air-conditioned and pressurised. Electrical windscreen de-icing. Windscreen wipers.

Systems: Garrett automatic cabin pressure control system: max differential 0·48 bars (7·0 lb/sq in). Engine bleed air heating, dual air-cycle cooling system, with

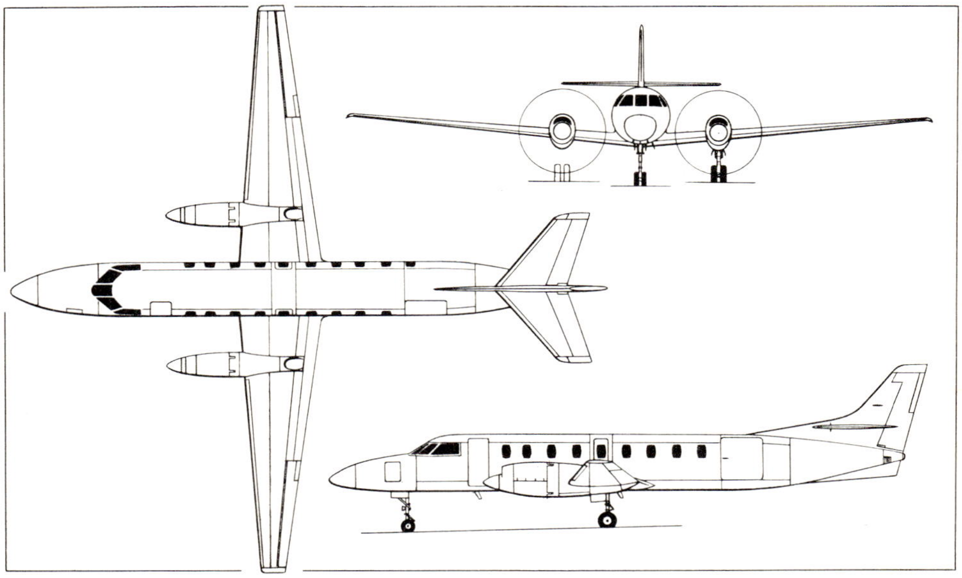

Swearingen Metro III commuter airliner (two Garrett TPE331 turboprop engines) *(Pilot Press)*

First production Fairchild Swearingen Metro III 19/20-passenger airliner, ordered by Crossair of Switzerland

Artist's impression of the Metro IIIA with PT6A-46 turboprop engines

automatic temperature control. Air blower system for on-ground ventilation. Independent hydraulic system for brakes. Dual engine-driven hydraulic pumps, using fire-resistant MIL-H-83282 hydraulic fluid, provide 138 bars (2,000 lb/sq in) to operate flaps, landing gear actuators and nosewheel steering. Electrical system supplied by two 200A 28V DC starter/generators. Fail-safe system with overload and over-voltage protection. Redundant circuits for essential systems. Two 350VA solid-state inverters supply 115V and 26V AC. Two 24V 25Ah nickel-cadmium batteries for main services. Engine fire detection system and fire extinguishing system standard. Wing overheat detection system. Oxygen system of 1·39 m³ (49 cu ft) capacity with flush outlets at each seat; system with capacity of 3·26 m³ (115 cu ft) optional. Stall avoidance system comprising angle indicator, visual and aural warning.

Avionics and Equipment: Two flight deck and four cabin speakers standard; provisions for installation of remotely-mounted or panel-mounted avionics, customer-furnished antennae, weather radar and autopilot. Standard equipment includes pilot and co-pilot foot warmers; edge-lighted consoles, pedestal and switch panels; integrally lighted instruments; annunciator panel with 48 indicators; internally-operated control locks, individual reading lights and air vents for each passenger; heated pitot; heated static sources; baggage compartment, cargo compartment, entrance, map and instrument panel, ice inspection, retractable landing, navigation, rotating beacon and taxi lights; automatic engine-start cycle; external power socket; and static wicks.

Dimensions, external:
Wing span 17·37 m (57 ft 0 in)
Wing mean aerodynamic chord 1·84 m (6 ft 0·33 in)
Wing aspect ratio 10·5
Length overall 18·09 m (59 ft 4¼ in)
Height overall 5·08 m (16 ft 8 in)
Tailplane span 4·86 m (15 ft 11½ in)
Wheel track 4·57 m (15 ft 0 in)
Wheelbase 5·83 m (19 ft 1½ in)
Propeller diameter 2·69 m (8 ft 10 in)
Passenger door (fwd): Height 1·35 m (4 ft 5 in)
Width 0·64 m (2 ft 1 in)
Cargo door (rear): Height 1·30 m (4 ft 3¼ in)
Width 1·35 m (4 ft 5 in)
Height to sill 1·30 m (4 ft 3¼ in)
Forward baggage doors (two, each):
Height 0·64 m (2 ft 1 in)
Width 0·46 m (1 ft 6 in)
Emergency exits (three, each):
Height 0·71 m (2 ft 4 in)
Width 0·51 m (1 ft 8 in)

Dimensions, internal:
Cabin, excl flight deck and rear cargo compartment:
Length 7·75 m (25 ft 5 in)
Max width 1·57 m (5 ft 2 in)
Max height (aisle) 1·45 m (4 ft 9 in)
Floor area 13·01 m² (140 sq ft)
Volume 13·88 m³ (490 cu ft)
Rear cargo compartment (pressurised):
Length 2·34 m (7 ft 8 in)
Max width 1·57 m (5 ft 2 in)
Max height 1·32 m (4 ft 4 in)
Volume 3·85 m³ (136 cu ft)
Nose cargo compartment (unpressurised):
Length 1·75 m (5 ft 9 in)
Volume 1·27 m³ (45 cu ft)

Areas:
Wings, gross 28·71 m² (309 sq ft)
Ailerons (total) 1·31 m² (14·12 sq ft)
Trailing-edge flaps (total) 3·78 m² (40·66 sq ft)
Fin, incl dorsal fin 3·40 m² (36·62 sq ft)
Rudder, incl tab 1·80 m² (19·38 sq ft)
Tailplane 5·08 m² (54·70 sq ft)
Elevators 1·98 m² (21·27 sq ft)

Weights and Loadings:
Operating weight empty 3,963 kg (8,737 lb)
Max fuel weight 1,969 kg (4,342 lb)
Max T-O and landing weight 6,350 kg (14,000 lb)
Max ramp weight 6,441 kg (14,200 lb)
Max zero-fuel weight 5,670 kg (12,500 lb)
Max wing loading 221·2 kg/m² (45·31 lb/sq ft)
Max power loading 3·87 kg/kW (6·36 lb/shp)

Performance (at max T-O weight, ISA, except where indicated):
Design diving speed
311 knots (576 km/h; 358 mph) CAS
Max operating speed
248 knots (459 km/h; 285 mph) CAS
Max operating Mach No. 0·52
Max cruising speed at mid-cruise weight of 5,670 kg (12,500 lb):
at 4,570 m (15,000 ft)
271 knots (501 km/h; 312 mph)
at 6,100 m (20,000 ft)
266 knots (492 km/h; 306 mph)
at 7,620 m (25,000 ft)
254 knots (470 km/h; 292 mph)
Stalling speed, flaps and wheels up
97 knots (180 km/h; 112 mph) IAS
Stalling speed, flaps and wheels down
87 knots (161 km/h; 100 mph) IAS
Max rate of climb at S/L 759 m (2,490 ft)/min
Rate of climb at S/L, one engine out
194 m (635 ft)/min
Service ceiling 8,380 m (27,500 ft)
Service ceiling, one engine out 3,960 m (13,000 ft)
T-O to 15 m (50 ft), dry power 953 m (3,125 ft)
Landing from 15 m (50 ft) 953 m (3,125 ft)
Range with max standard fuel, two crew, 19 passengers and baggage, 45 min reserves
620 nm (1,149 km; 714 miles)

SWEARINGEN METRO IIIA

Fairchild Swearingen announced on 20 August 1981 its intention to offer a new version of the Metro, with Pratt & Whitney Aircraft of Canada engines, to those operators who might wish to standardise on a common family of P&WC power plants throughout a multi-type fleet of aircraft. Designated Metro IIIA, the new model will have PT6A-46 turboprop engines. Deliveries are expected to begin in late 1983.

Performance data were still being studied by Fairchild Swearingen at the time of this announcement.

SWEARINGEN SA227-TT MERLIN IIIC

The original Merlin III received FAA certification on 27 July 1970, and was described in the 1974-75 *Jane's*. The later Merlin IIIA was described in the 1977-78 edition, and the Merlin IIIB in the 1980-81 edition.

A new Merlin IIIC was introduced for 1981, and was certificated under the FAA's Special Federal Air Regulation 41 for operation at a max T-O and landing weight of 6,001 kg (13,230 lb). Its design incorporates the stronger wing introduced on the Metro III, as well as the latter's streamlined nacelles and new main landing gear doors. It also has improved flight handling characteristics, integrally lighted instruments, an improved control panel quadrant, redesigned cabin with new passenger seats, and other detail refinements.

The description of the Metro III applies also to the Merlin IIIC, except as detailed below:

Type: Eight/eleven-seat twin-turboprop executive transport.

Wings: Generally as for Metro III, but without the increase in wing span, and with wing incidence at tip −1°.

Fuselage: All-metal cylindrical semi-monocoque fail-safe structure of 2024 aluminium alloy, flush-riveted throughout. Glassfibre honeycomb nosecap will accommodate a 0·38 m (15 in) weather radar antenna.

Tail Unit: As for Metro III.

Landing Gear: As for Metro III.

Power Plant: Two 671 kW (900 shp) Garrett TPE331-10U-503G turboprop engines, each driving a four-blade fully-feathering and reversible-pitch metal propeller with synchrophaser. Continuous alcohol/water injection system optional. Fuel system, engine inlet de-icing, and other power plant systems generally as for Metro III.

Accommodation: Crew of two on flight deck, each on four-way adjustable seat with shoulder harness; dual controls standard. Bulkhead with sliding door divides flight deck from cabin. Standard accommodation is for seven passengers with seats disposed on each side of a central aisle. Rapid relocation of seats and couches is made possible by continuous tracks recessed into floor, permitting layout to be varied according to mission. Passenger door at rear of cabin on port side, with integral airstair. Emergency exit on starboard side of cabin. Sliding door at rear end of cabin to separate it from the entrance vestibule. Baggage space in vestibule to accommodate 136 kg (300 lb); baggage and avionics in nose compartment, capacity 181·5 kg (400 lb). Accommodation pressurised, air-conditioned and ventilated. Electrically-heated windscreen. Two-speed windscreen wipers.

Systems: Generally as for Metro III, except electrical system has two 28V 300A starter/generators, and oxygen system has a capacity of 0·62 m³ (22 cu ft).

Avionics and Equipment: Optional avionics, and standard and optional equipment, are generally as for Metro III.

Dimensions, external:
Wing span 14·10 m (46 ft 3 in)
Wing mean aerodynamic chord 1·94 m (6 ft 4½ in)
Wing aspect ratio 7·71
Length overall 12·85 m (42 ft 2 in)
Height overall 5·13 m (16 ft 10 in)
Tailplane span 4·86 m (15 ft 11½ in)
Wheel track 4·57 m (15 ft 0 in)
Wheelbase 3·23 m (10 ft 7 in)
Propeller diameter 2·69 m (8 ft 10 in)
Passenger door (port, rear): Height 1·35 m (4 ft 5 in)
Width 0·64 m (2 ft 1 in)

Dimensions, internal:
Cabin, excl flight deck and rear compartment:
Length 3·23 m (10 ft 7 in)
Max width 1·57 m (5 ft 2 in)
Max height 1·45 m (4 ft 9 in)
Volume 9·09 m³ (321 cu ft)
Rear baggage compartment (pressurised):
Volume 2·12 m³ (75 cu ft)
Nose baggage/avionics compartment (unpressurised):
Volume 0·85 m³ (30 cu ft)

Areas: As for Metro III, except
Wings, gross 25·78 m² (277·50 sq ft)

Weights and Loadings:
Weight empty, equipped 3,696 kg (8,150 lb)

Swearingen Merlin IIIC eight/eleven-seat twin-turboprop executive transport

Fuel load 1,969 kg (4,342 lb)
Max T-O and landing weight 6,001 kg (13,230 lb)
Max ramp weight 6,046 kg (13,330 lb)
Max zero-fuel weight 5,670 kg (12,500 lb)
Max wing loading 232·8 kg/m² (47·68 lb/sq ft)
Max power loading 4·47 kg/kW (7·35 lb/shp)

Performance (at max T-O weight, ISA, except where indicated):
Max cruising speed, at mid-cruise weight of 4,990 kg (11,000 lb):
at 3,050 m (10,000 ft) 300 knots (556 km/h; 345 mph)
at 4,570 m (15,000 ft) 298 knots (552 km/h; 343 mph)
at 6,100 m (20,000 ft) 294 knots (545 km/h; 339 mph)
at 7,620 m (25,000 ft) 285 knots (528 km/h; 328 mph)
at 9,140 m (30,000 ft) 258 knots (478 km/h; 297 mph)
Stalling speed, flaps and gear up 106 knots (197 km/h; 122 mph)
Stalling speed, flaps and gear down 92 knots (171 km/h; 106 mph)
Max rate of climb at S/L 792 m (2,600 ft)/min
Rate of climb at S/L, one engine out 177 m (580 ft)/min
Service ceiling 8,230 m (27,000 ft)
Service ceiling, one engine out 4,265 m (14,000 ft)
T-O to, and landing from, 15 m (50 ft) 960 m (3,150 ft)
Range with max standard fuel, max cruise power at 7,925 m (26,000 ft), 45 min reserves:
six occupants 1,938 nm (3,591 km; 2,231 miles)
eight occupants 1,706 nm (3,161 km; 1,964 miles)
Ferry range, conditions and allowances as above, but with two flight crew only, no passengers 2,239 nm (4,149 km; 2,578 miles)

SWEARINGEN SA227-AT MERLIN IVC

Swearingen Merlin IVC, a corporate version of the Metro III commuter airliner

The Merlin IVC is a corporate version of the Metro III commuter airliner. It incorporates the same design and engineering improvements, differing primarily in its internal configuration, which provides more luxurious accommodation for 11 passengers as standard. Compared with the earlier Merlin IV/IVA, the IVC has a completely redesigned interior, with new reclining passenger seats, couches, interior furniture, decor and lighting. A new refreshment and entertainment centre includes a large buffet cabinet with beverage and food storage and preparation facilities, television and stereo equipment. At the rear of the cabin, separated by a bulkhead and hinged door, are a toilet and baggage compartment.

The Merlin IVC is certificated under the FAA's SFAR Pt 41B and ICAO Annex 8 specifications, and has a T-O and landing weight of 6,350 kg (14,000 lb). Its large cabin volume, and the availability of movable bulkheads and interchangeable cabin furnishings, makes it easily convertible to meet a company's airlift requirements in virtually any arrangement of passengers and/or cargo.

The description of the Metro III applies also to the Merlin IVC, except as follows:

Type: Thirteen/sixteen-seat corporate transport.

Weights and Loadings: As Metro III, except:
Weight empty, equipped 4,128 kg (9,100 lb)

Performance (at max T-O weight, ISA, except where indicated):
Max cruising speed, at mid-cruise weight of 5,670 kg (12,500 lb):
at 3,050-4,570 m (10,000-15,000 ft) 277 knots (513 km/h; 319 mph)
at 6,100 m (20,000 ft) 273 knots (505 km/h; 314 mph)
at 7,620 m (25,000 ft) 260 knots (482 km/h; 299 mph)
Stalling speed, flaps and gear up 97 knots (180 km/h; 112 mph)
Stalling speed, flaps and gear down 87 knots (161 km/h; 100 mph)
Max rate of climb at S/L 759 m (2,490 ft)/min
Rate of climb at S/L, one engine out 194 m (635 ft)/min
Service ceiling 8,380 m (27,500 ft)
Service ceiling, one engine out 3,960 m (13,000 ft)
T-O to 15 m (50 ft) 869 m (2,850 ft)
Landing from 15 m (50 ft) at AUW of 6,124 kg (13,500 lb) 884 m (2,900 ft)
Range with max standard fuel, max cruise power at 7,925 m (26,000 ft), 45 min reserves:
eight occupants 1,526 nm (2,828 km; 1,757 miles)
thirteen occupants 974 nm (1,805 km; 1,121 miles)
Ferry range, conditions and allowances as above, but with two flight crew only 2,117 nm (3,923 km; 2,438 miles)

FOXJET

FOXJET INTERNATIONAL INC (a subsidiary of Fox Industries Inc)

Head Office: 6701 West 110th Street, Minneapolis, Minnesota 55438
Telephone: (612) 941 8870 or (612) 944 2255

President: Tony Fox

In the Spring of 1977, details were released of a lightweight four-seat twin-turbofan transport named Foxjet F600 which was expected to operate at 20 per cent of the fuel costs of the most economical business jet then in service. Subsequently there was considerable development of the original design, with plans to fly a prototype during 1980. No recent information has been received from the company; details of the Foxjet F600 as proposed in early 1980 can be found in the 1980-81 *Jane's*.

FRONTIER

FRONTIER-AEROSPACE INC

Head Office: 2751 Temple Avenue, Long Beach, California 90806
Telephone: (213) 595 1261
Telex: 656485
President: John Halstead
Marketing Director: Doug Smith

Frontier-Aerospace assembles and markets the Fletcher agricultural and utility aircraft manufactured in New Zealand by New Zealand Aerospace Industries Limited (which see). Its current models, listed in this entry, include two military variants of which first details were released on 3 November 1980.

FRONTIER-AEROSPACE TASKMASTER

TaskMaster is the name under which Frontier-Aerospace markets the Aerospace Fletcher FU-24-954 piston-engined agricultural and utility aircraft. Full details can be found in the New Zealand section of this edition.

FRONTIER-AEROSPACE PEGASUS I

This is the military version of the TaskMaster, offered by Frontier-Aerospace for cargo and personnel transport, patrol, reconnaissance, medevac, defoliation, parachute and paramedic, aerial mapping, surveillance, tactical ground support and light attack duties. When operated at the same max T-O weight as the agricultural TaskMaster, the useful load of the Pegasus I is 1,270 kg (2,800 lb).

Frontier-Aerospace planned to conduct a demonstration tour of Latin America with a Pegasus I during 1981, and to display it at the Latin America National Security Expo in Panama in November 1981.

FRONTIER-AEROSPACE AGMASTER

AgMaster is the name under which Frontier-Aerospace markets the Aerospace Cresco 600 turboprop agricultural and utility aircraft. Full details can be found in the New Zealand section of this edition.

FRONTIER-AEROSPACE PEGASUS II

This is the military version of the AgMaster, offered by Frontier-Aerospace for the same variety of duties as those listed for the Pegasus I. It differs from the AgMaster in being available with a 462 kW (620 shp) Pratt & Whitney Aircraft of Canada PT6A-28 turboprop engine as an alternative to the 447 kW (600 shp) Avco Lycoming LTP 101. Useful load is quoted at 2,041 kg (4,500 lb), and length of the rear cabin as 3·65 m (12 ft) compared with 3·05 m (10 ft) in the Pegasus I.

GACC

GULFSTREAM AMERICAN CORPORATION OF CALIFORNIA (a subsidiary of Gulfstream American Corporation)

Head Office and Works: 7701 Woodley Avenue, Van Nuys, California 91406
Telephone: (213) 988 9900
Telex: 662461

Chairman, President and Chief Executive Officer: Allen E. Paulson
Senior Vice-President: Samuel J. Saso (Parts Sales)
Vice-Presidents:
Robert H. Cooper (Marketing)
Douglas A. Potter (Operations—Miami)
Secretary: John P. Innes II

This is the former American Jet Industries Inc, which was renamed following its acquisition of Grumman American Corporation.

At Van Nuys, California, GACC continues to develop the Hustler twin-turbine business aircraft, and to carry out cargo conversions of civil transport aircraft, and is the western support centre for Gulfstream II and III executive transports. In addition, the programme involving retrofit of the new Gulfstream III wing to Gulfstream II aircraft will be carried out by GACC at Van Nuys.

GULFSTREAM AMERICAN CV-880 AIRLIFTER

Current programmes at Gulfstream American's Van Nuys plant include the conversion of ex-airline Convair 880 passenger transports for operation as convertible all-cargo, or cargo/passenger aircraft. Work involved in this conversion includes the installation of a hydraulically-operated freight door in the port side of the fuselage, forward of the wing, measuring 3·66 × 2·07 m (12 ft 0 in × 6 ft 9½ in). New flooring is fabricated from extruded light alloy planking, able to accept a loading of 732·4 kg/m² (150 lb/sq ft), and six cargo tiedown tracks stressed to 9*g* are installed in the floor. Windows are plugged to save weight, a fireproof glassfibre liner installed throughout the main cabin/cargo hold, and a smoke detection system and 'door open' warning light are installed. Full cabin pressurisation integrity is retained.

The converted aircraft, known as the Airlifter, will carry its maximum payload of 22,725 kg (50,100 lb) over a range of 1,800 nm (3,336 km; 2,073 miles) at a block speed of nearly 400 knots (740 km/h; 460 mph). Maximum range with reduced payload is 3,000 nm (5,555 km; 3,450 miles).

The first two conversions were completed in 1979, and this programme was continuing in 1981.

Details of the standard Convair 880 can be found in the 1965-66 *Jane's*; the following details apply specifically to Gulfstream American's Airlifter conversion:

Dimension, internal:
Total cargo volume 180·0 m³ (6,358 cu ft)

Weights:
Operating weight empty, average 36,285 kg (80,000 lb)
Max payload 22,725 kg (50,100 lb)
Max T-O weight 83,685 kg (184,500 lb)
Max ramp weight 83,915 kg (185,000 lb)
Max zero-fuel weight 59,420 kg (131,000 lb)
Max landing weight 70,305 kg (155,000 lb)

GULFSTREAM AMERICAN HUSTLER 500

The former American Jet Industries announced in 1975 its plans to build a new general aviation aircraft. The design included a pressurised cabin, to permit cruising altitudes of up to 10,670 m (35,000 ft). STOL characteristics were to be provided by the use of a supercritical wing with full-span Fowler trailing-edge flaps, and spoilers, instead of ailerons, for lateral control. Most unusual feature was the power plant, comprising a nose-mounted

Gulfstream American Hustler 500 twin-engined business/utility aircraft *(Howard Levy)*

Pratt & Whitney Aircraft of Canada PT6A turboprop, with a small Williams turbofan standby engine mounted in the rear fuselage. Details of this aircraft, designated Hustler 400, can be found in the 1977-78 *Jane's*. The prototype, registered N400AJ, made its first flight on 11 January 1978 and had accumulated more than 250 hours' flying by mid-1979.

The desirability of obtaining certification of the Hustler as a twin-engined aircraft, coupled with a likely delay of two years before the Williams standby turbofan would be available, brought the decision to change to a Pratt & Whitney JT15D-1 turbofan in the rear fuselage. This, in turn, made necessary further airframe changes, which were described in detail in the 1980-81 *Jane's*. The turboprop engine in the nose was also replaced, with a Garrett TPE331-10-501, the single low-mounted exhaust of which seemed likely to minimise gas ingestion by the rear engine.

The redesigned aircraft (N501GA) was given the designation Hustler 500. Taxi trials were completed in December 1980, and it made a number of test flights from Gulfstream's Van Nuys facility and a base at Mojave. The programme was reported to be dormant in mid-1981. A full description of the Hustler 500 prototype can be found in the 1980-81 *Jane's*.

GULFSTREAM AMERICAN PEREGRINE

On 25 July 1979, Gulfstream American announced its intention to build a prototype of the Peregrine two-seat primary and basic military trainer. Based on the Hustler 500, it differs primarily by deletion of the forward engine and wingtip tanks. Seating is side by side in the prototype (N600GA), which flew for the first time on 22 May 1981 and is powered by a single 11·12 kN (2,500 lb st) Pratt & Whitney Aircraft of Canada JT15D-4 turbofan engine, mounted in the rear fuselage. Options available for production Peregrines will include a single 13·34 kN (3,000 lb st) JT15D-5, or a different power plant, consisting of two Williams WR44 turbofans (each rated at 6·67 kN; 1,500 lb st); and tandem seating, which will involve further changes to the forward fuselage, transparent cockpit canopy and windscreen. In most other respects, the Peregrine is similar to the Hustler 500.

Wings: Cantilever low/mid-wing monoplane. Conventional light alloy two-spar fail-safe structure with ribs, stringers and chemically-milled skins, flush riveted. Double-slotted hydraulically actuated light alloy Fowler trailing-edge flaps extending over two-thirds span. Conventional ailerons of light alloy construction. Small winglet above each tip.

Fuselage: Semi-monocoque light alloy structure.

Tail Unit: Cantilever light alloy structure with swept vertical surfaces. Trim tab in rudder and port elevator. Ventral fin.

Landing Gear: Hydraulically-retractable tricycle type. Main units retract inward, nose unit aft. Oleo-pneumatic shock-absorbers. Single wheel and tyre on each unit.

Dimensions, external:

Wing span	10·50 m (34 ft 5½ in)
Length overall	11·68 m (38 ft 4 in)
Height overall	4·09 m (13 ft 5 in)
Tailplane span	4·67 m (15 ft 4 in)

Weights (production version, estimated, A: JT15D-5; B: WR 44s):

Max T-O and landing weight: A	2,812 kg (6,200 lb)
B	2,903 kg (6,400 lb)
Mission weight: A, B	2,383 kg (5,254 lb)
Normal landing weight: A	1,997 kg (4,402 lb)
B	2,124 kg (4,683 lb)

Performance (production version, estimated, at max T-O weight except where indicated):

Max level speed at 6,100 m (20,000 ft):	
A	394 knots (730 km/h; 454 mph)
B	374 knots (693 km/h; 431 mph)
Stalling speed, power off:	
A	66 knots (122 km/h; 76 mph)
B	67 knots (124 km/h; 77 mph)
Max rate of climb at S/L: A	1,585 m (5,200 ft)/min
B	1,204 m (3,950 ft)/min
Service ceiling: A	14,630 m (48,000 ft)
B	12,620 m (41,400 ft)
T-O run: A	307 m (1,008 ft)
B	361 m (1,186 ft)
T-O to 15 m (50 ft): A	489 m (1,604 ft)
B	592 m (1,943 ft)
Landing from 15 m (50 ft) at max landing weight:	
A	799 m (2,621 ft)
B	880 m (2,888 ft)
Landing from 15 m (50 ft) at normal landing weight:	
A	639 m (2,095 ft)
B	814 m (2,670 ft)
Landing run at max landing weight:	
A	442 m (1,450 ft)
B	473 m (1,552 ft)
Landing run at normal landing weight:	
A	366 m (1,202 ft)
B	437 m (1,435 ft)
Mission range at 12,190 m (40,000 ft):	
A at 369 knots (683 km/h; 424 mph)	1,080 nm (2,001 km; 1,243 miles)
B at 319 knots (590 km/h; 367 mph)	1,315 nm (2,437 km; 1,514 miles)

GULFSTREAM AMERICAN PEREGRINE II

In early 1981, Gulfstream American announced its intention of developing a twin-turbofan executive transport, known as the Peregrine II, derived from the Peregrine trainer. Generally similar in overall configuration, it differs by having taller winglets at the wingtips, a pressurised fuselage to accommodate a pilot and five passengers, and two Williams WR44 turbofan engines mounted side by side in the rear fuselage, each developing 6·67 kN (1,500 lb st). Luxury interiors will include adjustable and side-tracking seats, foldaway tables, refreshment centre, toilet, and pressurised baggage space in the aft portion of the cabin, with a volume of 0·99 m³ (35 cu ft). The accommodation will be pressurised to a maximum differential of 0·59 bars (8·6 lb/sq in), and heated or cooled as required. Electrical, hydraulic and pressurisation systems will be fully redundant, the electrical system having solid-state controls. Anti-icing, de-icing, and full IFR instrumentation and avionics will be available.

A prototype is expected to fly by mid-1982.

Dimensions, external:

Wing span	11·00 m (36 ft 1¼ in)
Length overall	11·95 m (39 ft 2½ in)
Height overall	4·09 m (13 ft 5 in)

Prototype Gulfstream American Peregrine trainer, with side-by-side seating

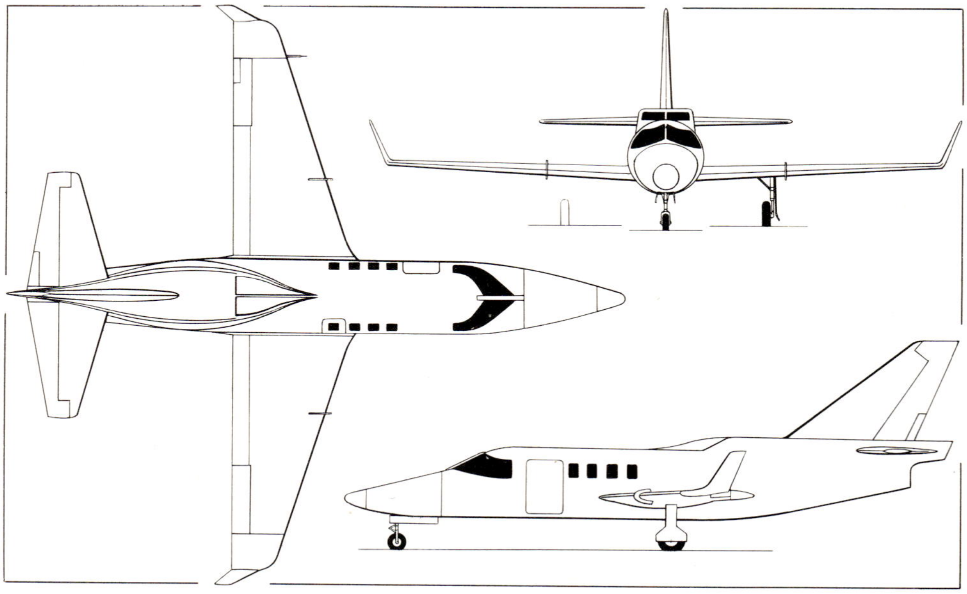

Gulfstream American Peregrine II twin-turbofan executive transport *(Pilot Press)*

Tailplane span 4·67 m (15 ft 4 in)
AREA:
Wing area, gross 19·88 m² (214 sq ft)
WEIGHTS AND LOADINGS (estimated):
Weight empty 1,937 kg (4,270 lb)
Max T-O weight 3,629 kg (8,000 lb)
Max landing weight 3,447 kg (7,600 lb)
Max wing loading 182·6 kg/m² (37·4 lb/sq ft)
Max power loading 272·0 kg/kN (2·67 lb/lb st)
PERFORMANCE (estimated, at max T-O weight, ISA, except where indicated):
Max cruising speed at 10,670 m (35,000 ft), at average cruise weight 403 knots (746 km/h; 464 mph)
Stalling speed at max landing weight 70 knots (129 km/h; 80 mph)
Max rate of climb at S/L 1,090 m (3,570 ft)/min
Max operational ceiling 13,410 m (44,000 ft)
T-O run 355 m (1,166 ft)
T-O to 15 m (50 ft) 532 m (1,745 ft)
Landing from 15 m (50 ft) 457 m (1,500 ft)
Range, max cruising speed at 10,670 m (35,000 ft), 45 min reserves 1,429 nm (2,649 km; 1,646 miles)
Range, max cruising speed at 12,200 m (40,000 ft), 45 min reserves 1,675 nm (3,104 km; 1,929 miles)

GATES LEARJET

GATES LEARJET CORPORATION

CORPORATE OFFICES, AIRCRAFT DIVISION: Mid-Continent Airport, PO Box 7707, Wichita, Kansas 67277
Telephone: (316) 946 2000
Telex: 417441
MARKETING HEADQUARTERS, TUCSON OPERATIONS: Tucson International Airport, PO Box 11186, Tucson, Arizona 85734
Telex: 666408
COMBS-GATES FIXED BASE OPERATIONS: Hangar 7, Stapleton International Airport, Denver, Colorado 80207
JET ELECTRONICS AND TECHNOLOGY INC: 5353 52nd Street, Grand Rapids, Michigan 49508
CHAIRMAN OF THE BOARD: Charles C. Gates
PRESIDENT: Harry B. Combs
GENERAL MANAGER, AIRCRAFT DIVISION: B. S. Stillwell
EXECUTIVE VICE-PRESIDENTS:
A. J. Brizzolara (J.E.T.)
R. E. Cloughley (Combs-Gates)
SENIOR VICE-PRESIDENTS:
James S. Medeiros (Product Support)
Salvatore J. Mira (Asst General Manager)
Ronald D. Neal (Programme Management and Technology)
R. C. Scott (Operations)
VICE-PRESIDENTS:
P. J. Baker (Flight Operations)
J. Benson (International Marketing)
M. Berger (Aircraft Sales) (Combs-Gates)
W. R. Edgar (Government Relations)
G. D. Gilmore (Government Programmes)
J. R. Greenwood (Corporate Affairs)
D. J. Grommesh (Engineering & Research)
R. E. Hamlin (Plant Manager, Wichita)
D. E. Howard (Plant Manager, Tucson)
B. Isaacman (Styling and Design Services)
E. C. Mandenberg (Materiel)
M. F. Mastin (Marketing Programmes)
Edward R. Miller (Employee Relations)
M. D. Mott (Domestic Marketing)
D. B. Norton (Technical and Service Facilities)
A. G. Ponte (Quality Assurance)
TREASURER: W. H. Webster
CONTROLLER: A. M. Hain
SECRETARY, GENERAL COUNSEL: R. C. Troll
DIRECTOR, PUBLIC RELATIONS: J. B. Meyer

Founded in 1960 by the late William P. Lear Sr, this company was known originally as the Swiss American Aviation Corporation, which was formed to manufacture a high-speed twin-jet executive aircraft known as the Learjet 23 (formerly SAAC-23). Most of the tooling for production of this aircraft was completed in Europe and then, in 1962, all company activities were relocated at Wichita, Kansas; shortly afterwards the company became known as Lear Jet Corporation. In 1967 all of Mr Lear's interests in the company (approximately 60 per cent) were acquired by The Gates Rubber Company of Denver, Colorado, and in January 1970 the company name was changed to Gates Learjet Corporation.

On 28 October 1975 Gates Learjet announced improved models of its Learjet 24, 25 and 35/36, then in production, under the designation Century III series. All Century III Learjets incorporate a cambered wing and other changes to reduce stall and approach speeds and balanced field length. This wing has a new contour, extending from the leading-edge aft to the second wing spar, and, coupled with other advances including an improved stall warning/prevention system, improves lateral stability and handling qualities in the slower flight regimes. Performance increases from Century III improvements include approach speed reductions of 16-18 knots (30-34 km/h; 18·5-21 mph), and increases in cruise range of as much as seven per cent. Century III improvements are available as a factory retrofit modification for earlier Learjet 24, 25, 35 and 36 models, and can be completed in approximately one week.

From 1 July 1979, all newly-delivered Learjets have embodied a 'Softflite' handling package to improve stall characteristics. Available for retrospective fit on earlier Century III series aircraft, this comprises a full-chord shallow fence on each wing, small devices on the inboard leading-edge, and two rows of boundary-layer energisers forward of each aileron, to energise the airflow and delay the onset of compressibility. Vortex generators are removed; the stick shaker/pusher is retained but unlikely to be required.

In June and September 1977 Gates Learjet announced a new family of advanced models, with wings of increased span fitted with supercritical winglets. Designated Learjet 28/29 Longhorn and Learjet 55/56 Longhorn, these have subsequently joined the earlier models in production.

Gates Learjet 25D executive transport, now the basic model of the Learjet range

In 1980, for the 16th consecutive year, Gates Learjet led in cumulative deliveries of business aircraft, with 120 Learjets of all models delivered (80 to customers in the USA, the balance for export). Of this total, 29 were Model 25s, 90 were Model 35/36s, and 1 was a Model 29. The company delivered its 1,000th Learjet on 28 March 1980.

On 23 September 1980, Gates Learjet and The Dee Howard Company (which see) announced jointly the intention to co-operate on several aircraft development projects. The first of these will involve the extended-range Learjet Model 25G, incorporating improvements similar to those of the Dee Howard XR Learjet. Following certification, these modifications will become available on new production aircraft, and will be available from The Dee Howard Company as a retrofit for existing aircraft.

Learjets are produced at the company's main facility at Wichita, Kansas. A completion centre at Tucson, Arizona, was opened in 1976, and a Customer Service Center was opened in 1978. During 1981 two new worldwide product support facilities were completed. The first, at Wichita, has an area of 14,270 m² (153,600 sq ft); that at Tucson is 5,017 m² (54,000 sq ft) in area. Other fixed-base operations, under the name of Combs-Gates, are located at Denver, Colorado; Fort Lauderdale, Florida; Indianapolis, Indiana; and Palm Springs, California. A new assembly plant, for production of the Learjet 55/56 Longhorn, was completed at Tucson in the Spring of 1979.

In January 1981 employees at the company's Aircraft Division facilities numbered approximately 5,300. Combined operations employment, including a wholly-owned subsidiary, Jet Electronics and Technology Inc, totals 6,348. Company facilities now cover more than 167,225 m² (1·8 million sq ft).

GATES LEARJET 24F

The prototype Lear Jet twin-jet executive transport flew for the first time on 7 October 1963 and deliveries of production Learjet 23 aircraft began on 13 October 1964. After a total of 104 of this version had been delivered, it was superseded by the Learjet 24, which was certificated under Federal Air Regulations Part 25 (formerly CAR 4B), as have been all subsequent models produced by the company. Production of the most recent model, the Learjet 24F, ended during 1980. Details of the Learjet 24 series can be found in the 1980-81 and earlier editions of *Jane's*.

GATES LEARJET 25D

First flown on 12 August 1966 as the Learjet 25, this version accommodates eight passengers and a crew of two. FAA certification in the air transport category (FAR Pt 25) was obtained on 10 October 1967 and the initial delivery was made in November 1967. British CAA certification was received on 26 June 1974.

The current basic model, embodying Century III improvements and the 'Softflite' handling package (see introductory notes), has the designation Learjet 25D. In addition, 8° flap settings for take-off are approved for this version, improving the high altitude/hot day take-off performance. Factory-installed thrust reversers for the General Electric CJ610-8A turbojet engines are optional.

For 1981, the TBO of these engines, in the Learjet 25D installation, was increased from 4,000 to 5,000 hours. It was planned also to incorporate during 1981 a smoother-operating freon rotary compressor.

TYPE: Twin-jet light executive transport.

WINGS: Cantilever low-wing monoplane. Wing section NACA 64A 109. Dihedral 2° 30′. Incidence 1°. Sweepback 13° at quarter-chord. All-metal eight-spar structure with milled alloy skins. Manually-operated, aerodynamically-balanced all-metal ailerons. Hydraulically-actuated all-metal single-slotted flaps. Hydraulically-actuated all-metal spoilers ahead of flaps. Electrically-operated trim tab in port aileron. Balance tab in each aileron. Anti-icing by engine bleed air ducted into leading-edges.

FUSELAGE: All-metal flush-riveted semi-monocoque fail-safe structure.

TAIL UNIT: Cantilever all-metal sweptback structure, with electrically-actuated variable-incidence T tailplane and small ventral fin. Conventional manually-operated control surfaces. Electrically-operated trim tab in rudder. Electrically-heated de-icing of tailplane leading-edges.

LANDING GEAR: Retractable tricycle type, with twin wheels on each main unit and single steerable nosewheel. Hydraulic actuation, with backup pneumatic extension. Oleo-pneumatic shock-absorbers. Main wheels fitted with Goodyear 18 × 5·50 10-ply tyres, pressure 7·93 bars (115 lb/sq in). Nosewheel fitted with Goodyear Dual Chine tyre size 18 × 4·40 10-ply rating, pressure 7·24 bars (105 lb/sq in). Goodyear multiple-disc hydraulic brakes. Pneumatic emergency braking system. Parking brakes. Fully-modulated anti-skid system.

POWER PLANT: Two General Electric CJ610-8A turbojet engines, each rated at 13·1 kN (2,950 lb st), pod-mounted on sides of fuselage aft of wings. Fuel in integral wing and wingtip tanks, and a bladder-type cell in the fuselage, with a total standard fuel capacity of 3,445 litres (910 US gallons). Oil capacity 3·75 litres (1 US gallon) per engine. Engine inlet anti-icing by bleed air.

ACCOMMODATION: Two seats side by side on flight deck, with dual controls. Up to eight passengers in cabin, with one on inward-facing bench seat on starboard side at front, then four on swivel seats which face fore and aft for take-off and landing, with centre aisle, and three on forward-facing couch. Toilet and stowage space under front inward-facing seat, which can be screened from remainder of cabin by curtain. Two refreshment cabinets, and two folding tables. Plexiglas divider between flight deck and cabin. Baggage compartment aft of cabin. With back of rear bench seat folded down, baggage compartment and rear of cabin can be used to carry cargo or stretchers. In full cargo version, the rearward-facing armchair seats are also removed. Two-piece door, with upward-hinged portion and downward-hinged portion with integral steps, on port side of cabin at front. Emergency exit on starboard side. Cargo door optional. Windscreen anti-icing by engine bleed air with liquid methyl alcohol backup.

SYSTEMS: Air-conditioning by freon R12 vapour cycle system, supplemented by ram-air heat exchanger during pressurised flight. Cabin pressurisation, by engine bleed air, has max differential of 0·65 bars (9·4 lb/sq in). Electrical system powered by dual 400A 30V DC starter/generators with AC power supplied from dual 1,000 VA solid-state inverters. Dual 24V 41Ah lead-acid batteries; nickel-cadmium batteries optional. Emergency battery pack optional. Dual engine-driven hydraulic pumps, each capable of maintaining full system pressure of 103·5 bars (1,500 lb/sq in). Auxiliary electrically-driven hydraulic pump. Pneumatic system at pressure of 124-207 bars (1,800-3,000 lb/sq in) for emergency extension of landing gear and operation of wheel brakes. Engine fire detection and extinguishing system. Oxygen system for emergency use has crew demand masks and passenger dropout masks. Alcohol anti-icing system for radome.

AVIONICS AND EQUIPMENT: Standard avionics, by Collins, include dual VHF-20A com, No. 1 with preselect control; dual VIR-30A VOR/ILS nav; dual marker lamps;

DME-40, with 339F-12A indicator and readout on pilot's HSI; TDR-90 transponder; ADF-60 ADF; FIS-84 flight director system (pilot); and Collins/J.E.T. PN-101/RAI-302 flight indicator (co-pilot). Bendix RDR-1200B weather radar; dual Allen 3137 RMIs; dual J.E.T. VG-206D vertical gyros; dual J.E.T. DN-104A directional gyros; IDC electric-encoding altimeter, with altitude alerter and static defect correction module; dual Avtech audio controls; and Gables controls for com, nav, DME and transponder. Standard equipment includes birdproof windscreen; dual clocks; control locks; depressurisation warning lights; engine fire warning lights; Mach warning; stall warning device; heated pitot tubes and static ports; ice detector; fire axe; cabin fire extinguisher; stereo system; baggage compartment, cabin dome, courtesy, instrument panel flood, reading, anti-collision, landing, navigation, taxi, and strobe lights; and lightning protection system.

Dimensions, external:
Span over tip-tanks 10·84 m (35 ft 7 in)
Wing chord at root 2·74 m (9 ft 0 in)
Wing chord at tip 1·40 m (4 ft 7 in)
Wing aspect ratio 5·01
Length overall 14·50 m (47 ft 7 in)
Length of fuselage 13·82 m (45 ft 4 in)
Height overall 3·73 m (12 ft 3 in)
Tailplane span 4·47 m (14 ft 8 in)
Wheel track (c/l shock-absorbers) 2·51 m (8 ft 3 in)
Wheelbase 5·84 m (19 ft 2 in)
Cabin door: Height 1·57 m (5 ft 2 in)
Standard width 0·61 m (2 ft 0 in)
Optional width 0·91 m (3 ft 0 in)
Emergency exit: Height 0·71 m (2 ft 4 in)
Width 0·48 m (1 ft 7 in)

Dimensions, internal:
Cabin, between pressure bulkheads:
Length 6·27 m (20 ft 7 in)
Max width 1·50 m (4 ft 11 in)
Max height 1·32 m (4 ft 4 in)
Volume, incl baggage compartment 8·47 m³ (299·0 cu ft)
Baggage compartment 1·13 m³ (40·0 cu ft)

Areas:
Wings, gross 21·53 m² (231·77 sq ft)
Ailerons (total) 1·08 m² (11·70 sq ft)
Trailing-edge flaps (total) 3·42 m² (36·85 sq ft)
Spoilers 0·66 m² (7·05 sq ft)
Fin 3·47 m² (37·37 sq ft)
Rudder, incl tab 0·67 m² (7·18 sq ft)
Tailplane 5·02 m² (54·00 sq ft)
Elevators 1·31 m² (14·13 sq ft)

Weights and Loadings:
Weight empty, equipped 3,606 kg (7,950 lb)
*Max payload, cargo configuration 1,587 kg (3,500 lb)
Max fuel weight 2,766 kg (6,098 lb)
Max T-O weight 6,804 kg (15,000 lb)
Max ramp weight 7,030 kg (15,500 lb)
Max landing weight 6,033 kg (13,300 lb)
Max wing loading 315·9 kg/m² (64·7 lb/sq ft)
Max power loading 259·7 kg/kN (2·54 lb/lb st)
Certification pending, February 1981

Performance (at max T-O weight, ISA, unless stated otherwise):
Never-exceed speed Mach 0·81
Max operating speed at 7,620 m (25,000 ft) 475 knots (880 km/h; 547 mph)
Max cruising speed, mid-cruise weight, at 14,325 m (47,000 ft) 451 knots (835 km/h; 519 mph)
Econ cruising speed, mid-cruise weight, at 14,325 m (47,000 ft) 428 knots (793 km/h; 493 mph)
Stalling speed, wheels and flaps up, engines idling, at max landing weight 113 knots (209 km/h; 130 mph) IAS

Gates Learjet 28/29 Longhorn, with increased-span wing and supercritical winglets

Stalling speed, wheels and flaps down, engines idling, at max landing weight 97 knots (180 km/h; 111·5 mph) IAS
Max rate of climb at S/L 2,080 m (6,830 ft)/min
Rate of climb at S/L, one engine out 582 m (1,910 ft)/min
Service ceiling 15,545 m (51,000 ft)
Service ceiling, one engine out 7,165 m (23,500 ft)
Min ground turning radius 11·43 m (37 ft 6 in)
T-O balanced field length, FAR Pt 25 1,200 m (3,937 ft)
Landing distance, FAR Pt 25, at max landing weight 859 m (2,817 ft)
Range with 4 passengers, max fuel and 45 min reserves 1,430 nm (2,650 km; 1,647 miles)

Operational Noise Levels (FAR Pt 36):
T-O 90·9 EPNdB
Approach 103·7 EPNdB
Sideline 95·2 EPNdB

GATES LEARJET 25G

Announced on 23 September 1980, this version of the Learjet 25 represents the first product of the co-operative agreement between Gates Learjet and The Dee Howard Company. It is similar in design to the Dee Howard XR Learjet (which see), and has a range more than 20 per cent better than that of other Model 25s. Features include an inboard-section glove on each wing, carrying additional fuel; a new nacelle pylon configuration that improves cruise performance; a wingtip tank fin cuff of new design; pressure-tuned wing leading-edges; and a new span flow limiter. Range capability is enhanced further by resulting drag reduction.

The Learjet 25G was expected to become available during 1981, following certification.

GATES LEARJET 28 and 29 LONGHORN

Displayed for the first time at the US National Business Aircraft Association's annual convention on 27-29 September 1977, the prototype of the Learjet 28/29 series was generally similar to the 10-seat Learjet 25D, except that it introduced a wing of much increased span, fitted with supercritical winglets. Combined with a cambered leading-edge, this is claimed to improve climb performance, reduce drag, offer improved high-altitude cruise efficiency, reduce runway requirements and reduce approach speed. The 'Softflite' handling package (see introductory notes) has been standard since 1 July 1979. No wingtip fuel tanks are fitted.

Like the Learjet 25D the 28/29 is powered by two General Electric CJ610-8A turbojet engines, each rated at 13·1 kN (2,950 lb st). Total usable fuel capacity of the Model 28 is 2,646 litres (699 US gallons), and of the Model 29 is 3,035 litres (802 US gallons). The Model 28 will accommodate a crew of two and eight passengers, the Model 29 a crew of two and six passengers. In other respects the 28/29 are similar to the 25D.

FAA certification of the Learjet 28/29 was received on 30 January 1979, and the first production deliveries were made shortly after.

On 19-20 February 1979, former astronaut Neil A. Armstrong set five records for jet aircraft in FAI categories C1f (6,000-8,000 kg) and C1e (3,000-6,000 kg). They were for altitude and sustained altitude of 15,584·6 m (51,130 ft) in each category, and climb to 15,000 m (49,213 ft) in 12 min 26·7 s.

Dimensions, external:
Wing span 13·35 m (43 ft 9·5 in)
Length overall 14·51 m (47 ft 7·6 in)
Height overall 3·73 m (12 ft 3·1 in)

Dimensions, internal:
Cabin, excl flight deck: Length:
28 4·37 m (14 ft 4 in)
29 3·86 m (12 ft 8 in)
Max width 1·49 m (4 ft 11 in)
Max height 1·32 m (4 ft 4 in)
Volume: 28 5·80 m³ (205 cu ft)
29 4·67 m³ (165 cu ft)
Baggage compartment: 28 0·85 m³ (30 cu ft)
29 0·76 m³ (27 cu ft)

Area:
Wings, gross 24·57 m² (264·5 sq ft)

Weights and Loadings:
Weight empty: 28 3,750 kg (8,268 lb)
29 3,730 kg (8,224 lb)
Max payload: 28 1,058 kg (2,332 lb)
29 1,078 kg (2,376 lb)
Max fuel weight: 28 2,124 kg (4,684 lb)
29 2,437 kg (5,373 lb)
Max T-O weight 6,804 kg (15,000 lb)
Max ramp weight 7,030 kg (15,500 lb)
Max landing weight 6,486 kg (14,300 lb)
Max wing loading 276·9 kg/m² (56·71 lb/sq ft)
Max power loading 246·52 kg/kN (2·42 lb/lb st)

Performance (at max T-O weight, except where indicated):
Never-exceed speed Mach 0·81
Max level speed at 7,620 m (25,000 ft) 477 knots (883 km/h; 549 mph)
Max cruising speed, mid-cruise weight, at 14,325 m (47,000 ft) 452 knots (836 km/h; 520 mph)
Econ cruising speed, mid-cruise weight, at 15,545 m (51,000 ft) 408 knots (756 km/h; 470 mph)
Stalling speed, flaps and wheels down, engines idling 89 knots (165 km/h; 102·5 mph) IAS
Service ceiling 15,545 m (51,000 ft)
Service ceiling, one engine out 8,840 m (29,000 ft)
T-O to 9 m (30 ft) 927 m (3,040 ft)
Landing distance from 15 m (50 ft) at max landing weight 833 m (2,734 ft)
Range with 4 passengers, max fuel and 45 min reserves:
28 1,137 nm (2,107 km; 1,309 miles)
29 1,376 nm (2,550 km; 1,584 miles)

Gates Learjet 28/29 Longhorn, with supercritical winglets *(Pilot Press)*

GATES LEARJET 35A and 36A

Although generally similar in basic configuration to the Learjet 25, the Learjet 35 and 36 are slightly larger in size and powered by turbofan engines. A prototype (known originally as the Learjet Model 26) made its first flight with Garrett TFE731-2 engines on 4 January 1973. The production 35 and 36, announced in May 1973, are almost identical, differing in fuel capacity and accommodation. FAA certification was awarded in July 1974, and customer deliveries began later that year. French and UK certification were gained during 1979.

Century III improvements, the 'Softflite' handling package (see introductory notes) and engine synchronisers are standard for both models; options include a higher max T-O weight of 8,300 kg (18,300 lb), and an Aeronca Inc thrust reverser package, available both as a retrofit kit and on new production aircraft. Improvements introduced as standard on the Learjet 30 series for 1981 include an increase of 136 kg (300 lb) in T-O weight of the 36A and of 454 kg (1,000 lb) in landing weight for both models; a new flap preselect system; and a more effective windscreen defogging system (available also as a retrofit kit). A smoother-operating freon rotary compressor was introduced during 1981; and a new option offers improved crew seats. Five Learjet 35As, including two specially equipped for high-altitude photography, have been delivered to the Argentinian Air Force; the Uruguayan Air Force has one 35A.

A special-mission version is available, and is described separately.

The description of the Learjet 25D applies also to the Learjet 35A and 36A, except in the following details:

Type: Twin-turbofan light executive transport.

Wings: As for Learjet 25D, except span increased.

Fuselage: As for Learjet 25D, except length increased.

Tail Unit and Landing Gear: As for Learjet 25D.

Power Plant: Two Garrett TFE731-2-2B turbofan engines, each rated at 15·6 kN (3,500 lb st), pod-mounted on sides of rear fuselage. Fuel in integral wing and wingtip tanks and a fuselage tank, with a combined usable capacity (Learjet 35A) of 3,524 litres (931 US gallons). Learjet 36A has a larger fuselage tank, giving a combined usable total of 4,201 litres (1,110 US gallons). Refuelling point on upper surface of each wingtip tank. Fuel jettison system. Engine nacelle leading-edges anti-iced by engine bleed air.

Accommodation: Crew of two on flight deck, with dual controls. Up to eight passengers in Learjet 35A; one on inward-facing bench seat on starboard side at front, then two pairs of swivel seats which face fore and aft for take-off and landing, with centre aisle, and three on forward-facing couch at rear of cabin. Alternative 'mid-cabin' arrangement, available optionally, places a refreshment area in the middle of the cabin, accessible from fore and aft club seating areas, each for four passengers. Learjet 36A can accommodate up to six passengers, one pair of swivel seats being removed. Toilet and stowage space under front inward-facing seat which can be screened from remainder of cabin. Refreshment cabinet opposite this seat, aft of passenger door. Baggage compartment with capacity of 226 kg (500 lb) aft of cabin. Two-piece clamshell door at forward end of cabin on port side, with integral steps built into lower half. Emergency exit on starboard side of cabin. Birdproof windscreens.

Systems: Environmental control system comprises cabin pressurisation, ventilation, heating and cooling. Heating and pressurisation are provided by engine bleed air, with a maximum pressure differential of 0·65 bars (9·4 lb/sq in), maintaining a cabin altitude of 1,980 m (6,500 ft) to an actual altitude of 13,715 m (45,000 ft). Freon R12 vapour cycle cooling system supplemented by a ram-air heat exchanger. Flight control system includes dual yaw dampers, dual stick pushers, dual stick shakers and Mach trim. Anti-icing system includes distribution of engine bleed air for wing, tailplane and engine nacelle leading-edges and windscreen; electrical heating of pitot heads, stall warning vanes and static ports; and alcohol spray on windscreen and nose radome. Hydraulic system supplied by two engine-driven pumps, each pump capable of maintaining alone the full system pressure of 103·5 bars (1,500 lb/sq in), for operation of landing gear, brakes, flaps and spoilers. Electrically-driven hydraulic pump for emergency operation of all hydraulic services. Pneumatic system of 124 to 207 bars (1,800 to 3,000 lb/sq in) pressure for emergency extension of landing gear and operation of brakes. Electrical system powered by two 30V 400A brushless generators, two 1kVA solid-state inverters to provide AC power, and two 24V 41Ah lead-acid batteries. Oxygen system for emergency use, with crew demand masks and drop-out masks for each passenger.

Avionics: Standard avionics, by Collins, include dual VHF-20A com, No. 1 with preselect control; dual VIR-30A VOR/ILS nav receivers; dual marker lamps; DME-40, with 339F-12A indicator, and readout on pilot's HSI; dual TDR-90 transponders; ADF-60 ADF, with preselect control; FIS-84 flight indicator, integrated with FC-200 FCS (pilot); Collins/J.E.T. PN-101/RAI-302 flight indicator (co-pilot); and ALT-50A radio altimeter; dual Allen 3137 RMIs; Bendix RDR-1200B weather radar; IDC electric encoding altimeter, with altitude alerter, and static defect correction module; IDC barometric altimeter (co-pilot); IDC electric rate of climb indicator (pilot); J.E.T. PS-823B and AI-804 emergency battery and attitude gyro; dual Avtech audio controls; and Gables controls for com, nav, DME and transponders.

Equipment: Standard equipment includes dual angle of attack indicators, dual battery temperature gauges, dual clocks, engine synchronisation meter, cabin differential pressure gauge, cabin rate of climb indicator, exhaust gas temperature and turbine temperature gauges, wing temperature indicator, alternate static source, dual battery overheat warning, depressurisation warning, engine fire warning lights, Mach warning system, dual stall warning systems, fire axe, cabin fire extinguisher, flotation jackets for crew and passengers, soundproofing; baggage compartment, courtesy, instrument panel, flood, map, and reading lights; dual anti-collision, landing, navigation, recognition, strobe, and taxi lights; dual engine fire extinguishing systems with 'systems armed' and fire warning lights, engine synchronisation system, control lock, external power socket, and lightning protection system.

Gates Learjet 35A Sea Patrol demonstration aircraft

Dimensions, external:

Wing span over tip-tanks	12·04 m (39 ft 6 in)
Wing chord at root	2·74 m (9 ft 0 in)
Wing chord at tip	1·55 m (5 ft 1 in)
Wing aspect ratio	5·74
Length overall	14·83 m (48 ft 8 in)
Height overall	3·73 m (12 ft 3 in)
Tailplane span	4·47 m (14 ft 8 in)
Wheel track	2·51 m (8 ft 3 in)
Passenger door:	
Standard: Height	1·57 m (5 ft 2 in)
Width	0·61 m (2 ft 0 in)
Optional: Height	1·57 m (5 ft 2 in)
Width	0·91 m (3 ft 0 in)
Emergency exit: Height	0·71 m (2 ft 4 in)
Width	0·48 m (1 ft 7 in)

Dimensions, internal (A: Learjet 35A; B: Learjet 36A):

Cabin: Length, incl flight deck:	
A	6·60 m (21 ft 8 in)
B	5·77 m (18 ft 11 in)
Max width	1·50 m (4 ft 11 in)
Max height	1·32 m (4 ft 4 in)
Volume, incl flight deck: A	7·99 m³ (282 cu ft)
B	6·49 m³ (229 cu ft)
Baggage compartment: A	1·13 m³ (40 cu ft)
B	0·76 m³ (27 cu ft)

Area:

Wings, gross	23·53 m² (253·3 sq ft)

Weights and Loadings (A: Learjet 35A; B: Learjet 36A):

Weight empty, equipped: A	4,342 kg (9,571 lb)
B	4,341 kg (9,570 lb)
*Max payload: A, B	1,587 kg (3,500 lb)
Max T-O weight: A	7,711 kg (17,000 lb)
B	8,300 kg (18,300 lb)
Max wing bending weight: A, B	6,123 kg (13,500 lb)
Max landing weight: A, B	6,940 kg (15,300 lb)
Max wing loading: A	327·6 kg/m² (67·1 lb/sq ft)
B	347·1 kg/m² (71·1 lb/sq ft)
Max power loading: A	247·1 kg/kN (2·43 lb/lb st)
B	261·7 kg/kN (2·57 lb/lb st)

* *Certification pending, February 1981*

Performance (at max T-O weight, except where indicated otherwise, A: Learjet 35A; B: Learjet 36A):

Never-exceed speed: A, B	Mach 0·83
Max level speed at 7,620 m (25,000 ft):	
A, B	471 knots (872 km/h; 542 mph)
Max cruising speed, mid-cruise weight, at 12,500 m (41,000 ft): A, B	460 knots (852 km/h; 529 mph)
Econ cruising speed, mid-cruise weight, at 13,700 m (45,000 ft): A, B	418 knots (774 km/h; 481 mph)
Stalling speed, wheels and flaps down, engines idling:	
A, B	96 knots (178 km/h; 111 mph) IAS
Max rate of climb at S/L: A	1,451 m (4,760 ft)/min
B	1,350 m (4,431 ft)/min
Rate of climb at S/L, one engine out:	
A	448 m (1,470 ft)/min
B	401 m (1,315 ft)/min
Service ceiling: A, B	13,715 m (45,000 ft)
Service ceiling, one engine out:	
A	7,710 m (25,300 ft)
B	7,165 m (23,500 ft)
T-O balanced field length, FAR Pt 25:	
A	1,287 m (4,224 ft)
B	1,515 m (4,972 ft)
Landing distance, FAR Pt 25, at max landing weight:	
A, B	937 m (3,075 ft)
Range with 4 passengers, max fuel and 45 min reserves:	
A	2,289 nm (4,242 km; 2,636 miles)
B	2,720 nm (5,040 km; 3,132 miles)

Operational Noise Levels (FAR Pt 36):

T-O: A	83·7 EPNdB
B	83·6 EPNdB
Approach: A	91·2 EPNdB
B	91·3 EPNdB
Sideline: A	86·9 EPNdB
B	87·4 EPNdB

GATES LEARJET SPECIAL-MISSIONS/ SEA PATROL VERSION

A prototype of this special-missions version of the Learjet, equipped for sea patrol duties, was exhibited at the 1979 Paris Air Show, before making a worldwide demonstration tour. This aircraft (N80SM) is a modified Learjet 35A, with a slightly larger cabin than the equally suitable 36A, enabling more observers to be carried during demonstrations.

Equipment available for the Sea Patrol version of the special-missions Learjet includes Litton AN/APS-504(V)2 sea surveillance radar, with 360° sweep from the underbelly radome and digital CFAR clutter suppression; low light level TV with video tape and scan conversion; forward-looking infra-red sensors; Daedalus DS-1210 multi-spectral infra-red and ultra-violet line scanner with tape data storage and hard copy printer; mini-computers for data processing, with tape input/output and graphic display capability; ASW sonobuoy drop and detection equipment; radar monitoring equipment (ESM); a hardpoint under each wing with an Alkan 165B ejector for survival equipment, flares or up to 220 kg (485 lb) of other stores; drop hatch for rescue gear; high-intensity searchlight; reconnaissance, mapping or LOROP cameras; HF, VHF and UHF homers; GNS-500A VLF Omega navigation system; side-looking airborne radar; Bendix RDR 1300B weather radar; and hand held cameras with position information printout.

Items fitted to the demonstrator make it suitable for location and detection of surface vessels, identification of targets, determination of target activities, and storage of permanent video tape and photographic records of target image, with position, time and date. Its performance ranges from a dash capability of nearly 485 knots (900 km/h; 560 mph) to manoeuvring speeds below 110 knots (200 km/h; 125 mph).

The company announced on 2 December 1980 the receipt of an order for three special-missions Learjet 35As to serve with the Finnish Air Force. Scheduled for delivery from mid-1982, they will be equipped for a variety of operations, including aerial mapping, air ambulance, air pollution control, oblique photography, rescue, and sea patrol. Primary role will be target towing for the Air Force's BAe Hawk combat trainers.

Performance (Learjet 36A):

Operating speed at 11,275-12,500 m (37,000-41,000 ft)	415 knots (769 km/h; 478 mph)
Operating speed at 4,575-7,620 m (15,000-25,000 ft)	319 knots (590 km/h; 367 mph)
Operating speed, S/L to 610 m (2,000 ft)	250 knots (463 km/h; 288 mph)
Rate of climb at S/L	1,380 m (4,525 ft)/min
Range at high altitude	2,249 nm (4,168 km; 2,590 miles)
Range at medium altitude	1,617 nm (2,996 km; 1,862 miles)
Range at low altitude	1,060 nm (1,964 km; 1,220 miles)

GATES LEARJET 55 and 56 LONGHORN

Gates Learjet announced at the Paris Air Show in June 1977 the company's decision to develop a new series of

Gates Learjet 35A ten-seat twin-turbofan light executive transport

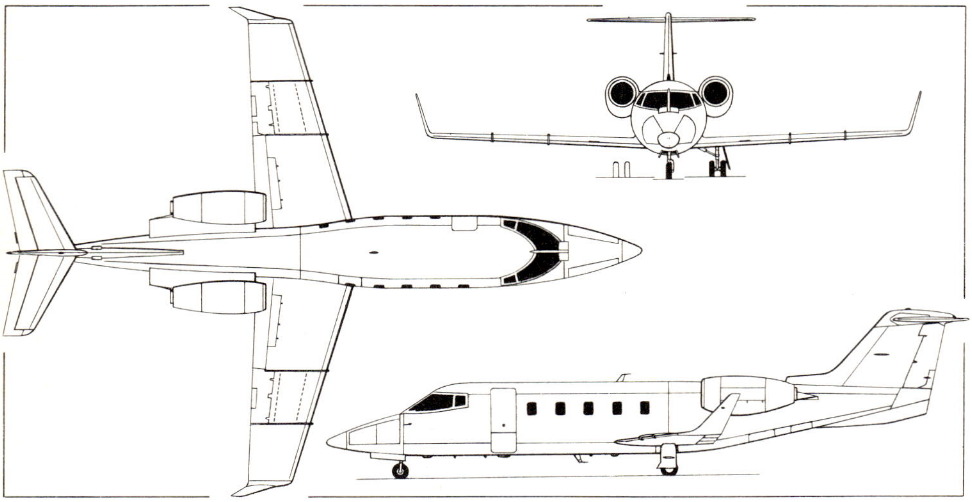

Gates Learjet 55 Longhorn eight/twelve-seat executive transport *(Pilot Press)*

The first prototype of the Gates Learjet 55 Longhorn in flight over Kansas

wide-body business aircraft known as the Learjet 50 series. Stemming from exhaustive engineering and marketing studies, these new aircraft supplement the earlier range of Learjets, providing a wide-body 'stand-up' cabin, and having accommodation for a maximum of nine passengers in the Learjet 55, and eight passengers in the Learjet 56.

The first of two Learjet 55 prototypes (N551GL) flew for the first time on 19 April 1979; the second prototype (N552GL) flew on 15 November 1979. The first was used to prove handling and performance characteristics, the second to verify aircraft systems. The first production example flew on 11 August 1980, and the second on 3 February 1981. At that time the first prototype had logged more than 400 flying hours, and the company had received more than 150 firm orders for these aircraft. Certification of the Model 55 under FAR Pt 25 was received on 18 March 1981. The initial customer delivery was made on 30 April 1981. Production is planned to reach a rate of seven 50-series Learjets per month.

Type: Twin-turbofan light executive transport.

Wings: Cantilever low-wing monoplane. Sweepback 13° at quarter-chord. All-metal multi-spar structure with cavity-milled wing skins. Wing upper surface skin tapers in thickness from wing root to wingtip to save weight. The design incorporates an advanced cambered leading-edge, the 'Softflite' handling package (see introductory notes to company entry) and supercritical winglets. Manually-operated ailerons. Electrically-operated trim tab in port aileron. Hydraulically-operated all-metal single-slotted trailing-edge flaps. Hydraulically-operated all-metal spoilers mounted on wing upper surface just forward of flaps. Anti-icing by engine bleed air ducted into leading-edges.

Fuselage: All-metal flush-riveted semi-monocoque fail-safe structure.

Tail Unit: Cantilever all-metal sweptback structure, with electrically-actuated variable-incidence T tailplane. Small ventral fin. Electrically-heated de-icing of tailplane leading-edge.

Landing Gear: Hydraulically-retractable tricycle type, with twin wheels on each main unit and single steerable nosewheel. High-pressure pneumatic system for emergency extension. Oleo-pneumatic shock-absorbers. Chined nosewheel tyre. High-energy hydraulic braking system, with pneumatic backup. Fully-modulated anti-skid units.

Power Plant: Two Garrett TFE731-3A-2B turbofan engines, each rated at 16·46 kN (3,700 lb st), pod-mounted on sides of fuselage aft of wing. Fuel in integral wing tanks and a bladder type fuselage tank, with a combined capacity of (Model 55) 3,774 litres (997 US gallons), (Model 56) 4,607 litres (1,217 US gallons). Refuelling point on upper surface of each outer wing. Engine nacelle leading-edges and fan hubs anti-iced by engine bleed air.

Accommodation: Crew of two on flight deck, with dual controls. Model 55 can seat from six to ten passengers in differing interior layouts. Model 56 seats six to eight passengers. Two folding tables, galley refreshment cabinet. Carpeted floor. Toilet. Baggage space at rear of cabin, and in fuselage nose and tailcone. Two-piece clamshell door at forward end of cabin on port side, with integral steps built into the lower section. Emergency exit on starboard side of cabin.

Systems: Environmental control system comprises cabin pressurisation, ventilation, heating and cooling. Heating and pressurisation are provided by engine bleed air, with a maximum pressure differential of 0·65 bars (9·4 lb/sq in), maintaining a cabin altitude of 2,440 m (8,000 ft) to an actual altitude of 15,545 m (51,000 ft). Freon vapour-cycle cooling system, supplemented by a ram-air system. Anti-icing system includes distribution of engine bleed air to wing leading-edges, engine nacelle leading-edges and fan hubs, and pilot and co-pilot windscreens; electrical anti-icing of tailplane leading-edge, pitot heads, stall warning vanes and static ports; and alcohol anti-icing of windscreens and radome. Hydraulic system supplied by two engine-driven variable-volume constant-pressure pumps, one on each engine, each capable of maintaining alone the full system pressure of 103·5 bars (1,500 lb/sq in) for operation of landing gear, brakes, flaps and spoilers. Electrically-driven hydraulic pump for emergency operation of all hydraulic services. Pneumatic system of 124 to 207 bars (1,800 to 3,000 lb/sq in) pressure for emergency extension of landing gear and operation of brakes. Electrical system powered by two 28V 400A engine-driven brushless generators, either of which is capable of maintaining adequate DC power to operate all electrical services; two 1kVA solid-state inverters to provide AC power; and two 24V 40Ah nickel-cadmium batteries. Oxygen system of 1·08 m³ (38 cu ft) capacity, with crew demand masks; dropout mask for each passenger, which is presented automatically if cabin altitude exceeds 4,265 m (14,000 ft).

Avionics: Standard avionics, by Collins, include dual VHF-20A com; dual VIR-30A nav; ADF-60 ADF; ALT-50A radio altimeter; DME-40 DME; dual marker beacon receivers; dual TDR-90 transponders; FIS-84 flight indicator integrated with a J.E.T. FC-550 FCS with altitude preselect (pilot); FIS-84 FDS (co-pilot); dual Allen 3137 RMIs; IDC electric-encoding altimeter, with altitude alerter (pilot); IDC barometric altimeter (co-pilot); dual J.E.T. VG-206D vertical gyros; dual J.E.T. DN-104A directional gyros; J.E.T. PS-823B and AI-804 emergency battery and attitude gyro; RCA Primus 300SL colour weather radar; Teledyne IVSI (pilot); and dual Avtech audio systems.

Equipment: Standard equipment includes sun visors, fire extinguisher and map lights on flight deck. Divider between flight deck and cabin. Cabin equipment includes two folding tables, galley refreshment cabinet, window shades, fire extinguisher and axe, soundproofing and insulation, carpeted floor; a lighting control panel and aisle, entrance step and indirect cabin lights.

Dimensions, external:	
Wing span	13·35 m (43 ft 9½ in)
Length overall	16·80 m (55 ft 1¼ in)
Height overall	4·48 m (14 ft 8¼ in)
Wheelbase	2·51 m (8 ft 3 in)
Dimensions, internal (A: Model 55; B: Model 56):	
Cabin: Length, excl flight deck: A	5·08 m (16 ft 8 in)
B	4·39 m (14 ft 5 in)
Max width	1·80 m (5 ft 11 in)
Max height	1·74 m (5 ft 8½ in)
Volume, incl flight deck: A	13·35 m³ (472 cu ft)
B	12·70 m³ (449 cu ft)
Baggage capacity, total: A	1·70 m³ (60 cu ft)
B	1·41 m³ (50 cu ft)
Area:	
Wings, gross	24·57 m² (264·5 sq ft)
Weights and Loadings (B estimated):	
Weight empty: A	5,502 kg (12,130 lb)
B	5,151 kg (11,356 lb)
Max payload: A	1,531 kg (3,375 lb)
B	1,653 kg (3,644 lb)
Max fuel weight: A	3,042 kg (6,707 lb)
B	3,699 kg (8,156 lb)
Max T-O weight: A	8,845 kg (19,500 lb)
B	9,298 kg (20,500 lb)
Max ramp weight: A	8,958 kg (19,750 lb)
B	9,412 kg (20,750 lb)
Max zero-fuel weight:	
A, B	6,804 kg (15,000 lb)
Max landing weight:	
A, B	7,711 kg (17,000 lb)
Max wing loading: A	359·3 kg/m² (73·6 lb/sq ft)
B	378 kg/m² (77·4 lb/sq ft)
Max power loading: A	272·3 kg/kN (2·67 lb/lb st)
B	286·3 kg/kN (2·81 lb/lb st)
Performance (B estimated, both at max T-O weight except where indicated):	
Never-exceed speed: A, B	Mach 0·81
Max level speed at 9,150 m (30,000 ft):	
A, B	470 knots (871 km/h; 541 mph)
Max cruising speed at 13,700 m (45,000 ft):	
A, B	439 knots (814 km/h; 506 mph)
Econ cruising speed at 14,935 m (49,000 ft):	
A, B	401 knots (743 km/h; 462 mph)
Stalling speed, flaps down, engines idling:	
A, B	98 knots (182 km/h; 113 mph)
Max rate of climb at S/L: A	1,417 m (4,650 ft)/min
B	1,335 m (4,380 ft)/min
Max certificated ceiling: A, B	15,545 m (51,000 ft)
T-O balanced field length, FAR Pt 25:	
A	1,509 m (4,950 ft)
B	1,670 m (5,480 ft)
Landing distance, FAR Pt 25, at max landing weight:	
A, B	948 m (3,109 ft)
Range with 4 passengers, max fuel and 45 min reserves:	
A	2,164 nm (4,010 km; 2,492 miles)
B	2,315 nm (4,290 km; 2,665 miles)

GENERAL DYNAMICS
GENERAL DYNAMICS CORPORATION

Head Office: Pierre Laclede Center, St Louis, Missouri 63105

Telephone: (314) 862 2440

Chairman and Chief Executive Officer: David S. Lewis

President: Oliver C. Boileau

Executive Vice-Presidents:
- James M. Beggs (Aerospace)
- Lester Crown
- Guy W. Fiske (Commercial)
- Gorden E. MacDonald (Finance)
- P. Takis Veliotis (Marine)

Vice-Presidents:
- Richard E. Adams (General Manager, Fort Worth Division)
- Leonard F. Buchanan (General Manager, Convair Division)
- James J. Cunnane (Controller)
- Otto J. Glasser (International, Washington)
- Ralph E. Hawes (General Manager, Pomona Division)
- Lyman C. Josephs (International)
- Edward J. LeFevre (Government Relations)
- Edward E. Lynn (General Counsel)
- Warren G. Sullivan (Industrial Relations)

Wayne Wells (Treasurer)
Robert H. Widmer (Science and Engineering)
Frederick S. Wood (Contracts and Pricing)
SECRETARY: John P. Maguire

Convair Division:
PO Box 80847, San Diego, California 92138
VICE-PRESIDENTS:
G. E. Blackshaw (Marketing)
Leonard F. Buchanan (General Manager)
D. E. Da Pra (Research and Engineering)
W. C. Dietz (Programme Director, Cruise Missiles)
B. R. Foushee (Programme Director, Aircraft)
Jay E. Hawley (Contracts and Estimating)
W. F. Rector (Programme Director, Space Programmes)
S. C. Wilkinson (Operations)

Fort Worth Division:
PO Box 748, Fort Worth, Texas 76101
VICE-PRESIDENTS:
Richard E. Adams (General Manager)
Charles A. Anderson (Research and Engineering)
Forrest E. Armstrong (111 Programmes)
W. D. Buntin (F-16 Engineering)
Edward E. Hatchett (Finance)
Rolf Krueger (Logistics and Support)
Robert W. McGuffee (Operations)
Herbert F. Rogers (Deputy General Manager, and Director, F-16 International Programmes)
Sterling V. Starr (Director, F-16/79 Export Fighter Programme)
D. J. Talley (Quality Assurance)
Theodore S. Webb Jr (Director, F-16 Programmes)
D. J. Wheaton (Marketing)

General Dynamics conducts its US aerospace activities at four divisions: Convair Division, with operations at San Diego, California; Fort Worth Division, with operations at Fort Worth, Texas; Pomona Division, at Pomona, California; and Electronics Division, with headquarters in San Diego. Convair Division is responsible for the design, development and production of aircraft structures and of systems for space exploration. Current programmes include development of the Tomahawk sea-launched cruise missile for the US Navy, and the ground-launched version for the US Air Force. Convair is also engaged in work on a medium-range air-to-surface missile (MRASM) for the US Air Force. Fort Worth Division is engaged in the design, development and production of military aircraft and avionics. Pomona Division is engaged in tactical missile and other ordnance programmes. Electronics Division is a leader in new technology to support the development and production of advanced electronics systems. Major programmes include sophisticated navigation positioning systems, tactical data and command control systems, automatic test equipment for high performance aircraft, and range measuring systems.

Fort Worth is currently responsible for production of the F-16 Fighting Falcon multi-role fighter; spares, support and modification/update for the F-111 fighter-bomber; and various ground-based radar systems. Convair Division is responsible for production of a major portion of the fuselage for the McDonnell Douglas DC-10 commercial transport aircraft and the KC-10A Extender tanker/cargo aircraft, the mid-fuselage section of the Space Shuttle Orbiter vehicle, engine mounting struts for the Boeing Model 767, the Atlas-Centaur launch vehicle used to boost unmanned spacecraft and satellites, and cruise missiles.

Convair Division also retains detailed tooling for high-usage spares for the Convair-Liner 240/340/440 series of piston-engined transports, and Convair 880 and 990 jet transports, and continues to manufacture components for these types.

GENERAL DYNAMICS F-111

Production of this two-seat variable-geometry tactical fighter was completed in 1976, but the various versions described in the 1978-79 and earlier editions of *Jane's* will remain important elements of US Air Force and Royal Australian Air Force strength for many more years, as will Strategic Air Command's FB-111A medium-range bomber variant. In addition, Grumman (which see) is developing and producing an ECM jamming version of the F-111A, under the designation EF-111A.

Important avionics update programmes are continuing to maintain the effectiveness of Tactical Air Command F-111s; details of some of these can be found in the 1980-81 *Jane's*.

The Australian government has announced its intention to purchase four USAF F-111As, to offset the cumulative loss of four of its fleet of F-111C strike aircraft. These will be refurbished and modified virtually to F-111C standard before delivery. It has also been announced that the Pave Tack infra-red target acquisition, designation and tracking system is to be acquired for service on the RAAF force of F-111Cs from 1984.

GENERAL DYNAMICS F-16 FIGHTING FALCON

USAF designations: F-16A/C and F-16B/D

The F-16 had its origin in the US Air Force's Lightweight Fighter (LWF) prototype programme, in 1972. The history of this programme and a description of the YF-16 prototypes can be found in the 1978-79 and 1977-78 editions of *Jane's* respectively.

General Dynamics F-111 armed with GBU-15 cruciform wing guided glide-bombs

The first of two YF-16 prototypes (72-01567) made its official first flight on 2 February 1974. A level speed of Mach 2 at 12,200 m (40,000 ft) was attained on 11 March 1974. The second YF-16 (72-01568) flew for the first time on 9 May 1974. During subsequent weapon trials, this aircraft extended the planned operational capability of the design by completing successfully the initial launch of Sparrow missiles on 6 October 1977 and of a Sky Flash missile in November 1978. Also in 1978, an F-16 prototype, with a Martin Marietta ATLIS II (Automatic Tracking and Laser Illumination System) pod, became the first single-seat fighter to hit ground targets with GBU-10 and GBU-16 laser-guided bombs without assistance from air/ground locators. Subsequent development milestones include the successful flight testing during 1980 of a pair of increased-capacity drop-tanks, each of 2,271 litres (600 US gallons; 500 Imp gallons). On 20 February 1981 an F-16 made the first launching of the new advanced medium range air-to-air missile (AMRAAM).

On 13 January 1975 the Secretary of the US Air Force announced that the F-16 had been selected for full-scale engineering development. The original YF-16 requirement for an air superiority day fighter was expanded, to give equal emphasis to the air-to-surface role, including provision of radar and all-weather navigation capabilities. The manufacture of eight pre-production aircraft, comprising six single-seat **F-16As** and two two-seat **F-16Bs**, began in July 1975. The first development F-16A made its first flight on 8 December 1976, and the first F-16B on 8 August 1977. The last of the eight development aircraft was the second two-seater, which made its first flight in June 1978. Meanwhile, the US Air Force had indicated its intention to procure a total of 1,388 F-16s, including 204 two-seaters. The name **Fighting Falcon** was adopted in 1980.

On 7 June 1975 a joint announcement by the four NATO countries of Belgium, Denmark, the Netherlands and Norway confirmed their selection of the F-16 to replace F-104s in current service. The initial order was for 348 aircraft (Belgium 116, Denmark 58, the Netherlands 102 and Norway 72), of which 58 will be two-seaters. Under co-production agreements, final assembly lines for these aircraft were established in Belgium and the Netherlands. About 30 European companies are producing F-16 components, avionics and equipment.

In August 1977, Israel announced plans to acquire a minimum of 75 F-16s (including eight two-seaters), and the first of these was delivered on 31 January 1980. In July 1980, Egypt formally agreed to acquire 40 F-16s, and delivery of these is scheduled to begin in early 1982. The sale of 36 F-16s to the Republic of Korea Air Force has also been approved; others are expected to be supplied to Pakistan. In January 1981, the Netherlands announced its intention to increase its original order for 102 F-16s by 22 aircraft (18 single-seat and four two-seat).

The first production F-16A (78-0001) flew for the first time on 7 August 1978, and was delivered to the US Air Force ten days later. The first F-16 to enter service was delivered to the US Air Force's 388th Tactical Fighter Wing at Hill AFB, Utah, on 6 January 1979. The first F-16 assigned to the 56th Tactical Fighter Wing, MacDill AFB, Florida, was delivered during October 1979. During 1980, the F-16 completed its first formal operational readiness inspection with the 388th TFW. The successful demonstration of reliability, sortie generation and bombing accuracy resulted in the F-16 achieving combat-ready status in October 1980, with the 4th Tactical Fighter Squadron of the 388th TFW. The first USAF F-16s to be stationed in the Pacific were ferried to Kunsan AFB, South Korea, in May 1981. Three squadrons of F-16s are expected to join the 50th TFW with the US Air Forces, Europe, in early 1982 at Hahn AFB, West Germany.

The first F-16 for Europe was delivered to the Belgian Air Force on 26 January 1979. The Royal Netherlands Air Force received its first two F-16s on 6 June 1979, and initial deliveries to the air forces of Denmark and Norway were made respectively on 28 and 25 January 1980. The 100th F-16 produced in Europe (by Fokker and SABCA/Sonaca) was delivered to the Royal Netherlands Air Force in mid-March. The 400th from combined US and European production became the 239th F-16 accepted by the USAF on 15 May 1981.

First operational NATO F-16 unit (from 1 January 1981) was No. 349 Squadron of the Belgian Air Force, at Beauvechain. The first Dutch squadron (No. 322) became operational on the F-16 at Leeuwarden on 1 May 1981. First Danish squadron is No. 727.

First combat use of the F-16 was by the Israeli Air Force, which used eight aircraft to destroy Iraq's Osirak nuclear reactor on 7 June 1981, with a top cover of six F-15s.

In February 1980, the US Air Force implemented a Multinational Staged Improvement Programme (MSIP) for the F-16, intended to assure the aircraft's capability to accept future systems now under development by the Air

General Dynamics F-16A Fighting Falcon multi-role fighters of the US Air Force's 388th Tactical Fighter Wing

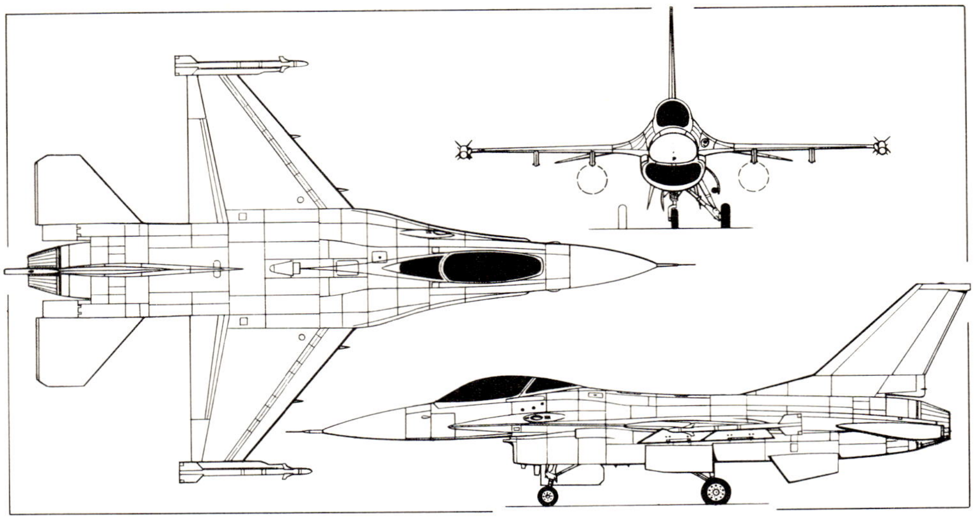

Three-view drawing of General Dynamics F-16A Fighting Falcon *(Pilot Press)*

Force. As a first stage, aircraft delivered after November 1981 will have built-in structural and wiring provisions, and system architecture, that will expand the single-seat F-16's multi-role flexibility to perform precision strike, night attack, and beyond-visual-range interception missions. Advanced cockpit displays and controls, and an improved fire control radar, will enable F-16s to launch AMRAAM air-to-air missiles at multiple targets in rapid succession. The enhancement programme will also qualify the F-16 for the US Air Force's LANTIRN nav/attack system and ASPJ jamming system, currently under development. Initial operational capability (IOC) is scheduled for December 1984. Designations will be **F-16C** (single-seat) and **F-16D** (two-seat).

The following description applies to the F-16A and F-16B, as indicated:

TYPE: Single-seat lightweight air combat fighter (F-16A) and two-seat fighter/trainer (F-16B).

WINGS: Cantilever mid-wing monoplane, of blended wing/body design and cropped-delta planform. The blended wing/body concept is achieved by flaring the wing/body intersection, thus not only providing lift from the body at high angles of attack but also giving less wetted area and increased internal fuel volume. In addition, thickening of the wing root gives a more rigid structure, with a weight saving of some 113 kg (250 lb). Basic wing is of NACA 64A-204 section, with 40° sweepback on leading-edges. Structure is mainly of aluminium alloy, with 11 spars, 5 ribs and single upper and lower skins, and is attached to fuselage by machined aluminium fittings. Variable wing camber is achieved by the use of leading-edge manoeuvring flaps that are programmed automatically as a function of Mach number and angle of attack. The increased wing camber maintains effective lift coefficients at high angles of attack. These flaps are one-piece bonded aluminium honeycomb sandwich structures, actuated by a drive system using rotary actuators. The trailing-edges carry large flaperons (flaps/ailerons), which are interchangeable left with right and are actuated by National Water Lift integrated servo-actuators. The maximum rate of flaperon movement is 80°/s.

FUSELAGE: Semi-monocoque all-metal structure of frames and longerons, built in three main modules: forward (to just aft of cockpit), centre and aft. Nose radome built by Brunswick Corporation. Highly-swept vortex control strakes along the fuselage forebody increase lift and improve directional stability at high angles of attack.

TAIL UNIT: Cantilever structure with sweptback surfaces. Fin is multi-spar multi-rib aluminium structure with graphite-epoxy skins, aluminium tip, and glassfibre dorsal fin and root fairing. Interchangeable all-moving tailplane halves, constructed of graphite-epoxy composite laminate skins mechanically attached to a corrugated aluminium substructure. Each tailplane half has an aluminium pivot shaft, and a removable full-depth bonded honeycomb leading-edge. Ventral fins are bonded aluminium honeycomb core with aluminium skins. Split speed-brake inboard of rear portion of each horizontal tail surface to each side of nozzle, each deflecting 60° from the closed position. National Water Lift servo-actuators for rudder and tailplane.

LANDING GEAR: Menasco hydraulically-retractable type, nose unit retracting aft and main units forward into fuselage. Nosewheel is located aft of intake, to reduce the risk of foreign objects being thrown into the engine during ground operation, and rotates 90° during retraction to lie horizontally under engine air intake duct. Oleo-pneumatic struts in all units. Goodyear main wheels and brakes; Goodrich main-wheel tyres, size 25·5 × 8-14. Steerable nosewheel with Goodrich tyre, size 18 × 5·5-8. All but two main unit components interchangeable. Brake-by-wire system on main gear, with Goodyear anti-skid units. Runway arrester hook under rear fuselage.

POWER PLANT: One Pratt & Whitney F100-PW-200 turbofan engine, rated at approx 111·2 kN (25,000 lb st) with afterburning, mounted within the rear fuselage. Fixed-geometry intake, with boundary layer splitter plate, beneath fuselage. A variable-geometry intake can be fitted later, without difficulty, if desirable to improve high-speed performance. The underfuselage intake position was chosen because here the airflow suffers least disturbance throughout the entire range of aircraft manoeuvres, and because it eliminates the problem of gun gas ingestion. Standard fuel contained in wing and five fuselage cells which function as two tanks; internal fuel weight is 3,162 kg (6,972 lb) in F-16A, and approx 17 per cent less in F-16B. In-flight refuelling receptacle in top of centre-fuselage, aft of cockpit. Auxiliary fuel can be carried in drop-tanks on underwing and under-fuselage hardpoints.

ACCOMMODATION: Pilot only in F-16A, in air-conditioned cockpit. McDonnell Douglas Aces II zero-zero ejection seat. Texstar transparent bubble canopy, made of polycarbonate, an advanced plastics material. The windscreen and forward canopy are an integral unit without a forward bow-frame, and are separated from the aft canopy by a simple support structure which serves also as the break-point where the forward section pivots upward and aft to give access to the cockpit. A redundant safety-lock feature prevents canopy loss. Windscreen/canopy design provides 360° all-round view, 195° fore and aft, 40° down over the side, and 15° down over the nose. Supersonic drag penalty is considered to be more than offset by the improved rearward view afforded to the pilot. To enable the pilot to sustain high-*g* forces, and for pilot comfort, the seat is inclined 30° aft and the heel-line is raised. In normal operation the canopy is pivoted upward and aft by electrical power; the pilot is also able to unlatch the canopy manually and open it with a backup handcrank. Emergency jettison is provided by explosive unlatching devices and two forward-mounted rockets. A limited-displacement, force-sensing control stick is provided on the right hand console, with a suitable armrest, to provide precise control inputs during combat manoeuvres. The F-16B has two cockpits arranged in tandem and equipped with all controls, displays, instruments, avionics and life-support systems required to perform both training and combat missions. The layout of the F-16B second station is essentially the same as that of the F-16A, and is fully systems-operational. A single-enclosure polycarbonate transparency, made in two pieces and spliced aft of the forward seat with a metal bow-frame and lateral support member, provides outstanding view from both cockpits.

SYSTEMS: Regenerative bootstrap air-cycle environmental control system by United Technologies' Hamilton Standard Division, using engine bleed air, for pressurisation and cooling. Two separate and independent hydraulic systems supply power for operation of the primary flight control surfaces and the utility functions. Electrical system powered by engine-driven Westinghouse 40kVA and Lear Siegler 5kVA generators and ground control units, with Sundstrand constant-speed drive. Four dedicated, sealed-cell batteries provide transient electrical power protection for the fly-by-wire flight control system. Application of the control configured vehicle (CCV) principle of relaxed static stability produces a significant reduction in trim drag, especially at high load factors and supersonic speeds. The aircraft centre of gravity is allowed to move aft, reducing both the tail drag and the change in drag on the wing due to changes in lift required to balance the down-load on the tail. Relaxed static stability imposes a requirement for a highly-reliable, full-time-operating, stability augmentation system, including reliable electronic, electrical and hydraulic provisions. The signal paths in this quad-redundant system are used to control the aircraft, replacing the usual mechanical linkages. Direct electrical control is employed from pilot controls to surface actuators. An onboard Sundstrand/Solar jet fuel starter is provided for engine self-start capability. Hamilton Standard turbine compressor, and Sundstrand accessory drive gearbox. Simmonds fuel measuring system. Garrett emergency power unit automatically drives a standby generator and pump to provide uninterrupted electrical and hydraulic power for control in the event of the engine or primary power system becoming inoperative.

General Dynamics F-16B two-seat fighter/trainer

AVIONICS AND EQUIPMENT: Westinghouse pulse-Doppler range and angle track radar, with planar array in nose. The radar has a lookdown range, in ground clutter, of 20-30 nm (37-56 km; 23-35 miles), and a lookup range of 25-40 nm (46-74 km; 29-46 miles). Forward avionics bay, immediately forward of cockpit, contains radar, air data equipment, inertial navigation system, flight control computer, and space and structural provisions for radar altimeter. Rear avionics bay contains ILS, Tacan and IFF, with space for future equipment. A Dalmo Victor ALR-69 radar warning system, with AEL antennae, is installed. Communications equipment includes Magnavox AN/ARC-164 UHF transceiver; provisions for a Magnavox KY-58 secure voice system; Collins AN/ARC-186 VHF AM/FM transceiver; government furnished AN/AIC-18/25 intercom; and Novatronics interference blanker. Sperry Flight Systems central air data computer. Singer-Kearfott modified SKN-2400 inertial navigation system; Collins AN/ARN-108 ILS; Collins AN/ARN-118 Tacan; Teledyne Electronics AN/APX-101 air-to-ground IFF transponder with a government furnished IFF control; government furnished National Security Agency KIT-1A/TSEC cryptographic equipment; Lear Siegler stick force sensors; Marconi Avionics electronic head-up display set; a government-furnished horizontal situation indicator; Teledyne Avionics angle of attack transmitter; Gull Airborne angle of attack indicator; Clifton Precision attitude director indicator; Delco fire control computer; Photo-Sonics gun camera; Kaiser radar electro-optical display. Landing/taxying light on each main landing gear strut. Essential structure and wiring provisions are built into the airframe to allow for easy incorporation of an advanced programmable signal processor radar, and future avionics systems under development for the F-16 by the US Air Force. These include advanced cockpit displays and communications/navigation controls, and advanced communications and data processing equipment.

ARMAMENT: General Electric M61A-1 20 mm multi-barrel cannon in the port-side wing/body fairing, equipped with a General Electric ammunition handling system and a 'snapshoot' gunsight (part of the head-up display system) and 500 rounds of ammunition. There is a mounting for an air-to-air missile at each wingtip, one underfuselage centreline hardpoint, and six underwing hardpoints for additional stores. For manoeuvring flight at 5·5*g* the underfuselage station is stressed for a load of up to 1,000 kg (2,200 lb), the two inboard underwing stations for 2,041 kg (4,500 lb) each, the two centre underwing stations for 1,587 kg (3,500 lb) each, the two outboard underwing stations for 318 kg (700 lb) each, and the two wingtip stations for 193 kg (425 lb) each. For manoeuvring flight at 9*g* the underfuselage station is stressed for a load of up to 544 kg (1,200 lb), the two inboard underwing stations for 1,134 kg (2,500 lb) each, the two centre underwing stations for 907 kg (2,000 lb) each, the two outboard underwing stations for 204 kg (450 lb) each, and the two wingtip stations for 193 kg (425 lb) each. The total possible external weapon load is 9,276 kg (20,450 lb), and for 9*g* manoeuvre capability 5,420 kg (11,950 lb). There are mounting provisions on each side of the inlet shoulder for the specific carriage of sensor pods (electro-optical, FLIR, etc); each of these stations is stressed for 408 kg (900 lb) at 5·5*g*, and 250 kg (550 lb) at 9*g*. Typical stores loads can include two wingtip-mounted AIM-9J/L Sidewinders, with up to four more on the outer underwing stations; Sargent-Fletcher 1,400 litre (370 US gallon; 308 Imp gallon) drop-tanks on the inboard underwing stations; a 1,136 litre (300 US gallon; 250 Imp gallon) drop-tank or a 2,200 lb bomb on the underfuselage station; a Martin Marietta Pave Penny laser tracker pod along the starboard side of the nacelle; and single or cluster bombs, air-to-surface missiles, or flare pods, on the four inner underwing stations. Stores can be launched from Aircraft Hydro-Forming MAU-12C/A bomb ejector racks, Hughes LAU-88 launchers, or Orgen triple or multiple ejector racks. Westinghouse AN/ALQ-119 and AN/ALQ-131 ECM (jammer) pods can be carried on the centreline and two underwing stations. Modified Tracor ALE-40 internal pyrotechnic/chaff dispensers have been specified. Current capabilities include air-to-air combat with gun and Sidewinder missiles; and air-to-ground attack with gun, rockets, conventional bombs, special weapons, laser-guided and electro-optical weapons. Specific structure and wiring provisions, and system architecture, are built-in to ensure the capability to accept future sensor and weapon systems, including electro-optical and FLIR pods, and advanced beyond-visual-range missiles. Weapons already launched successfully from F-16s, in addition to Sidewinders, include radar-guided Sparrow and Sky Flash air-to-air missiles, and TV-guided Maverick air-to-surface missiles.

DIMENSIONS, EXTERNAL:
- Wing span over missile launchers 9·45 m (31 ft 0 in)
- Wing span over missiles 10·01 m (32 ft 10 in)
- Wing aspect ratio 3·0
- Length overall 15·09 m (49 ft 5·9 in)
- Height overall 5·09 m (16 ft 8½ in)
- Tailplane span 5·64 m (18 ft 6 in)
- Wheel track 2·36 m (7 ft 9 in)
- Wheelbase 4·00 m (13 ft 1·44 in)

AREA:
- Wings, gross 27·87 m² (300·0 sq ft)

WEIGHTS AND LOADINGS:
- Weight empty: F-16A 7,070 kg (15,586 lb)
  - F-16B 7,374 kg (16,258 lb)
- Internal fuel load: F-16A 3,162 kg (6,972 lb)
  - F-16B 2,624 kg (5,785 lb)
- Max external load 9,276 kg (20,450 lb)
- Structural design gross weight (9·0*g*) with full internal fuel 10,205 kg (22,500 lb)
- Max symmetrical design load factor with full internal fuel at 10,205 kg (22,500 lb) gross weight:
  - F-16A/B 9·0
- Max T-O weight:
  - F-16A, air-to-air, no external tanks 10,800 kg (23,810 lb)
  - F-16B, air-to-air, no external tanks 10,568 kg (23,298 lb)
  - F-16A/B with external load 16,057 kg (35,400 lb)
- Wing loading:
  - at 9,979 kg (22,000 lb) AUW 356 kg/m² (73 lb/sq ft)
  - at 14,968 kg (33,000 lb) AUW 537 kg/m² (110 lb/sq ft)
- Thrust/weight ratio ('clean') 1·1 to 1

PERFORMANCE (F-16A):
- Max level speed at 12,200 m (40,000 ft) above Mach 2·0
- Service ceiling more than 15,240 m (50,000 ft)
- Radius of action more than 500 nm (925 km; 575 miles)
- Ferry range, with drop-tanks more than 2,100 nm (3,890 km; 2,415 miles)

GENERAL DYNAMICS F-16/79

This aircraft is essentially an F-16 tactical fighter powered by a General Electric J79-GE-119 afterburning turbojet engine. It was produced in response to US government authorisation of an FX programme, to develop aircraft for export to nations in which a first-line US fighter might not be required. The F-16/79 meets the FX requirements of being primarily an air defence fighter, while retaining the F-16's multi-role systems capability. It was initiated as a company-funded programme by General Dynamics, which leased back from the US Air Force the second F-16B development aircraft (75-0752) for conversion, testing and certification as the F-16/79 export fighter. The first flight was accomplished on 29 October 1980, and company certification flight testing was completed on 19 December 1980. By mid-1981 the prototype had accumulated 122 flying hours in 131 flights.

Replacement of the 111·2 kN (25,000 lb st) F100-PW-200 by the 80·1 kN (18,000 lb st) J79-GE-119 offers a considerable reduction in aircraft cost. The basic air combat effectiveness of the F-16 is retained, with an increase of some 817 kg (1,800 lb) in the aircraft's empty weight. Changes include internal modification of the engine air intake, to reduce airflow and increase pressure recovery, and lengthening of the rear fuselage fairing, increasing the length of the fuselage (but not overall length) by 0·45 m (1 ft 5½ in). Two versions are available: the single-seat F-16/79A and two-seat F-16/79B.

WEIGHTS:
- Weight empty:
  - A 7,730 kg (17,041 lb)
  - B 8,035 kg (17,714 lb)
- Internal fuel load:
  - A 3,162 kg (6,971 lb)
  - B 2,624 kg (5,785 lb)

Prototype General Dynamics F-16/79, armed with wingtip Sidewinder air-to-air missiles

General Dynamics F-16/101 testbed for the F101-DFE derivative fighter engine

Artist's impression of the AFTI-16 technology testbed aircraft produced by modifying an F-16A

Artist's impression of General Dynamics F-16E advanced technology development of the F-16 Fighting Falcon

Max external load: A, B	6,894 kg (15,200 lb)
Normal max T-O weight:	
A	11,633 kg (25,646 lb)
B	11,389 kg (25,108 lb)
Max T-O weight: A, B	16,057 kg (35,400 lb)

GENERAL DYNAMICS F-16/101

The first F-16 development aircraft was returned to Fort Worth by the US Air Force, to be modified by installation of a General Electric F101-DFE (derivative fighter engine). Developed from the F100-GE-100 engine that was designed by General Electric to power the Rockwell International B-1 strategic bomber, this advanced augmented turbofan has a rating of approximately 116 to 125 kN (26,000 to 28,000 lb st).

The F-16/101 programme called for flight testing of the engine over a five-month period, to determine its suitability as a power plant for advanced military aircraft. The contract required completion of approximately 100 flight hours, during which engine controls and response, emergency controls, air start capability, acceleration, afterburner operation, and performance in simulated combat, were to be evaluated.

The F-16/101 was first flown on 19 December 1980, and was scheduled to complete its flight test programme by the end of May 1981. General Dynamics was required to submit the final detailed report of its evaluation to the US Air Force's F-16 System Program Office at Wright-Patterson AFB, Ohio, during September 1981.

GENERAL DYNAMICS AFTI-16 PROGRAMME

In December 1978, the US Air Force selected the F-16 as a testbed with which to explore promising new fighter aircraft technologies, under the Advanced Fighter Technology Integration (AFTI-16) programme directed by the Air Force Systems Command's Flight Dynamics Laboratory at Wright-Patterson AFB, Ohio. General Dynamics was awarded a $34·3 million prime contract, under which it has modified an F-16A returned to the company by the US Air Force on 6 March 1980.

Changes visible in the accompanying artist's impression of the modified AFTI-16 testbed include the addition of two large manoeuvring canards for direct force control, under the engine air intake duct, and a dorsal fairing to house flight test equipment. Systems to be developed and evaluated when flight testing begins at Edwards AFB, California, include triplex digital fly-by-wire flight controls, advanced displays, system management computers, and integrated flight and weapons fire controls. Flight tests were scheduled to start during the Autumn of 1981. Information generated by the programme will be made available generally to the US aircraft industry for application to the design of future fighter aircraft.

GENERAL DYNAMICS F-16E

Under the above designation, General Dynamics has begun the company-funded development of an advanced version of the F-16 that will incorporate new aerodynamic and systems technologies. In March 1981, a design team began the preparation of engineering drawings, design analyses, and manufacturing planning that will lead to the construction of two flight demonstration aircraft, one of single-seat and the other of two-seat configuration. In this programme the company is receiving support from the US Air Force, which is supplying two single-seat full-scale development F-16 airframes for conversion to the new configuration, their Pratt & Whitney F100 turbofan engines, funding for one new two-seat cockpit, and the provision of flight test support facilities at Edwards AFB, California.

As can be seen in the accompanying illustration, the F-16E will have a new, highly-swept 'cranked-arrow', wing. This has been developed during some years of close collaboration between the company's Fort Worth Division and NASA. It will have an area of 60·01 m² (646 sq ft), more than double that of the standard F-16, and will incorporate graphite polyimide composite wing skins to provide the strength and rigidity essential for maximum wing performance. This wing is being combined with an F-16 fuselage that is lengthened by 1·42 m (4 ft 8 in), the additional volume being used to increase the internal fuel capacity by 82 per cent, and to provide an extra 1·13 m³ (40 cu ft) of storage space to cater for future avionics and sensor growth.

Preparatory work on this advanced version of the F-16 Fighting Falcon has been in progress at Fort Worth for a considerable time, originally under the designations F-16 Scamp and F-16XL, and the F-16E will capitalise on the extensive experience gained by the company since 1972 in the design and development of an air combat fighter that is currently the subject of a major international production programme. Its modular construction and electronic fly-by-wire control system will simplify the F-16E modification process. Wind tunnel and computer analyses have shown that the new configuration will extend the F-16's capabilities. By comparison, it will take off and land in only two-thirds of the distance, carry double the weapons load, and have a 45 per cent greater combat radius for both air-to-air and air-to-ground missions.

The first of the US Air Force's full-scale development aircraft was delivered to Fort Worth at the beginning of March 1981, and structural modification is in progress: the second aircraft was received during the Summer. First flight is scheduled for Summer 1982.

GREAT LAKES
GREAT LAKES AIRCRAFT COMPANY

Head Office and Works: Drawer A, Eastman, Georgia 31023
Telephone: (912) 374 5535
Sales Director: Richard Ivey

Following its acquisition by Mr R. Dean Franklin, Great Lakes Aircraft Company was relocated from Enid, Oklahoma, and Wichita, Kansas, to Eastman, Georgia. The Great Lakes Sport Trainer Model 2T-1A-2 is again in production, at the Eastman plant, from where deliveries began in March 1980. A complete parts department has been established to provide a service for Sport Trainer owners.

GREAT LAKES SPORT TRAINER MODEL 2T-1A-2

The Great Lakes Sport Trainer was produced with Cirrus and Menasco engines by the original Great Lakes Company, founded on 2 January 1929.

Certification of the Model 2T-1A-1 with 104 kW (140 hp) Avco Lycoming engine was obtained in May 1973, and delivery of production aircraft began in October 1973, but production of this version (described in the 1977-78 *Jane's*) has ended. Current production model is the 2T-1A-2 with 134 kW (180 hp) engine, constant-speed propeller, inverted fuel and oil systems, ailerons on both upper and lower wings, and with manifold pressure and fuel flow gauges as standard. Approved aerobatic manoeuvres are spins, chandelles, lazy eights, loops, barrel rolls, primary rolls, point rolls, slow rolls, snap rolls, Cuban 8s, hammerhead turns, Immelmann turns and split 'S'.

A total of 137 examples of the Great Lakes Sport Trainer were built at Enid and Wichita after production resumed in 1973. Production, at a projected rate of one aircraft every seven working days, is now centred at Eastman, Georgia.

Type: Two-seat sporting biplane.

Wings: Braced biplane, with N-type interplane struts, wire bracing and N-type centre-section support struts. Dual streamline-section landing and flying wires. Wing section M-12. No dihedral on upper wing, 2° on lower wings. Sweepback on upper wing 9° 13′. Composite structure, with Douglas fir spars, metal ribs and overall fabric covering. Ailerons on upper and lower wings. No flaps or tabs.

Fuselage: Welded chrome-molybdenum steel tube. Warren girder structure, with fabric covering.

Tail Unit: Wire-braced welded chrome-molybdenum steel tube structure, fabric-covered. Tailplane incidence manually adjustable from both cockpits. No controllable tabs.

Landing Gear: Non-retractable type, with steerable Scott tailwheel. Divided main legs with spring-oleo shock-absorbers standard. Main wheels size 6·00-6 with hydraulic disc brakes. Parking brake. Size 7·00-6 tyres optional. Leg fairings and wheel fairings standard.

Power Plant: One 134 kW (180 hp) Avco Lycoming AEIO-360-B1G6 flat-four engine, driving a Hartzell two-blade constant-speed metal propeller type HC-C2YK-4F/FC7666A-2 with spinner. Fully inverted fuel system with 98·5 litre (26 US gallon) centre-section tank in upper wing, and 5·3 litre (1·4 US gallon) inverted header tank. Refuelling point in upper wing surface. Fully inverted oil system.

Accommodation: Two seats in tandem in open cockpits. Dual controls, with adjustable rudder pedals, standard. Compass, airspeed indicator, altimeter and engine speed indicator standard in rear cockpit, optional for front cockpit. Cockpit heating standard. Baggage compartment aft of rear cockpit, capacity 18 kg (40 lb), optional.

Great Lakes Sport Trainer Model 2T-1A-2 (Avco Lycoming AEIO-360-B1G6 engine)

SYSTEMS: Engine-driven generator and sealed storage battery for electrical supply to navigation lights, two rotating beacons, and rear cockpit instrument lights, standard. Hydraulic system for brakes only.

AVIONICS AND EQUIPMENT: Emergency locator transmitter standard. Edo-Aire Mitchell 551 transceiver, intercom, and Edo-Aire Mitchell RT-553 with or without auto omni, optional. Standard equipment includes map case, seat belts and shoulder harness, cockpit lights, navigation lights, two omni-flash beacons, and a choice of four paint trims. Optional equipment includes turn and bank indicator, rate of climb indicator, cylinder head temperature gauge and accelerometer for rear cockpit, glider tow hitch, Hobbs meter, and cockpit covers.

DIMENSIONS, EXTERNAL:	
Wing span	8·13 m (26 ft 8 in)
Length overall	6·45 m (21 ft 2 in)
Height overall	2·34 m (7 ft 8 in)
Wheel track	1·78 m (5 ft 10 in)
AREAS:	
Wings, gross	17·43 m² (187·6 sq ft)
Ailerons (total)	2·70 m² (29·10 sq ft)
Fin	0·55 m² (5·87 sq ft)
Rudder	0·63 m² (6·81 sq ft)
Tailplane	1·43 m² (15·44 sq ft)
Elevators	0·99 m² (10·68 sq ft)
WEIGHTS AND LOADINGS:	
Weight empty	558 kg (1,230 lb)
Max T-O and landing weight	816 kg (1,800 lb)
Max wing loading	47·0 kg/m² (9·63 lb/sq ft)
Max power loading	6·09 kg/kW (10·0 lb/hp)
PERFORMANCE (at max T-O weight):	
Max level speed at S/L	114 knots (210 km/h; 131 mph)
Max level speed at 3,050 m (10,000 ft)	108 knots (201 km/h; 125 mph)
Max cruising speed	102 knots (190 km/h; 118 mph)
Approach speed	65 knots (121 km/h; 75 mph)
Stalling speed	50 knots (92 km/h; 57 mph)
Max rate of climb at S/L	350 m (1,150 ft)/min
Service ceiling	5,180 m (17,000 ft)
T-O run	145 m (475 ft)
T-O to 15 m (50 ft)	252 m (825 ft)
Landing from 15 m (50 ft)	252 m (825 ft)
Landing run	122 m (400 ft)
Range with max fuel, no reserves	260 nm (482 km; 300 miles)
Endurance with max fuel (75% power)	2 h 51 min
g limits	+5·38; −4

GRUMMAN

GRUMMAN CORPORATION

HEAD OFFICE: 1111 Stewart Avenue, Bethpage, New York 11714

Telephone: (516) 575 0574

CHAIRMAN OF THE BOARD AND CHIEF EXECUTIVE OFFICER: John C. Bierwirth

PRESIDENT AND CHIEF OPERATING OFFICER: Joseph G. Gavin Jr

VICE-CHAIRMAN OF THE BOARD: John F. Carr

VICE-PRESIDENTS:
- George Brown (Regional Offices)
- Nat P.Busi (Controller)
- John J. Carroll (Community Affairs)
- Howard Dunn (Audit)
- Robert G. Freese (Treasurer)
- Thomas L. Genovese (General Counsel)
- Weyman B. Jones (Public Affairs)
- Rolf J. Larson (Development)
- Ross S. Mickey (International Affairs)
- Gordon Ochenrider (Washington)

SECRETARY: Robert W. Bradshaw

GRUMMAN AEROSPACE CORPORATION

Entry follows

GRUMMAN ALLIED INDUSTRIES INC

HEAD OFFICE AND WORKS: 445 Broad Hollow Road, Melville, New York 11747

Telephone: (516) 454 8400

CHAIRMAN OF THE BOARD: Robert G. Freese

PRESIDENT: Robert W. Somerville

GRUMMAN DATA SYSTEMS CORPORATION

HEAD OFFICE AND WORKS: Bethpage, New York 11714

Telephone: (516) 575 0574

CHAIRMAN OF THE BOARD: John F. Carr

PRESIDENT: Robert A. Nafis

GRUMMAN ENERGY SYSTEMS INC

HEAD OFFICE AND WORKS: Bethpage, New York 11714

Telephone: (516) 575 0574

CHAIRMAN OF THE BOARD AND PRESIDENT: Ronald B. Peterson

GRUMMAN INTERNATIONAL INC

HEAD OFFICE: Bethpage, New York 11714

Telephone: (516) 575 1101

PRESIDENT: Michael Pelehach

The Grumman Aircraft Engineering Corporation was incorporated on 6 December 1929. Important changes in the corporate structure of the company were announced in 1969, resulting in the formation of Grumman Corporation, a small holding company, with Grumman Aerospace Corporation, Grumman Allied Industries Inc and Grumman Data Systems Corporation. A new organisation known as Grumman Ecosystems Corporation (now Grumman Energy Systems Inc) was brought into operation in January 1971, and on 2 January 1973 a merger took place with American Aviation Corporation of Cleveland, Ohio, to create Grumman American Aviation Corporation as a new subsidiary of Grumman Corporation. In 1978, Grumman Corporation's major holding in Grumman American Aviation was purchased by AJI, which renamed the latter company Gulfstream American Corporation (which see).

GRUMMAN AEROSPACE CORPORATION

HEAD OFFICE AND WORKS: South Oyster Bay Road, Bethpage, New York 11714

Telephone: (516) 575 0574

CHAIRMAN OF THE BOARD AND PRESIDENT: George M. Skurla

EXECUTIVE VICE-PRESIDENT: John O'Brien (Operations)

SENIOR VICE-PRESIDENTS:
- Dan F. Huebner (Marketing and Advanced Technology Development)
- Lawrence M. Mead Jr (Departmental Operations)
- Gordon H. Ochenrider (Washington Operations)
- Peter B. Oram (Aircraft Programmes)
- Carl A. Paladino (Treasurer)
- Philip S. Vassallo (Corporate Procurement Operations)

VICE-PRESIDENTS:
- Alexander D. Alexandrovich (Asst to Executive Vice-President Operations)
- John M. Buxton (President, Grumman Houston Corporation)
- Renso L. Caporali (Development)
- Edward Dalva (Stuart Operations)
- Fred W. Haise Jr (Space Programmes)
- Thomas J. Kane Jr (Business Development)
- Thomas J. Kelly (Engineering)
- Daniel E. Knowles (Personnel and Administration)
- Frank J. Messina (Manufacturing)
- Robert C. Miller (Presidential Asst for Corporate Development)
- Robert J. Myers (Resources)
- William E. Voorhest (Security and Corporate Services)
- William M. Zarkowsky (Milledgeville Operations)

SECRETARY AND GENERAL COUNSEL: Raphael Mur

Current products of this subsidiary of Grumman Corporation include versions of the A-6 Intruder, EA-6B Prowler, E-2C Hawkeye and F-14 Tomcat for the US Navy, and a tactical jamming version of the General Dynamics F-111, designated EF-111A, for the US Air Force. The Milledgeville plant manufactures wing skins and glassfibre wingtips for the Fairchild A-10A Thunderbolt II attack aircraft. A contract for Grumman Aerospace to build major components for the Sikorsky CH-53E heavy-lift helicopter was announced in the Spring of 1978. Grumman later received a contract from The Boeing Company to manufacture the wing centre-section of the Model 767 civil transport.

In 1979, the Corporation received a $3 million contract from the US Defense Advanced Research Projects Agency to proceed with design and wind tunnel testing of forward-swept wing (FSW) models. Advantages claimed for the FSW aircraft include reduced fuel consumption and greater manoeuvrability at supersonic speeds; greater flexibility in fuselage design; and reductions in aircraft cost, size and weight. Following satisfactory results from wind tunnel testing, DARPA has ordered a small manned FSW technology demonstrator.

Grumman/Resorts International commuter conversion of the Albatross military amphibian

GRUMMAN/RESORTS INTERNATIONAL G-111 ALBATROSS

Grumman's Stuart, Florida, facility completed the first modification of an Albatross amphibian to 28-passenger commuter configuration on behalf of Chalks International Airlines, a division of Resorts International. The first stage involved the inspection and replacement of parts to produce the equivalent of a zero-time airframe, followed by modernisation of the flight deck, and the provision of passenger cabin doors and a 28-seat interior. At the same time, the 1,100 kW (1,475 hp) Wright R-1820-982C9HE3 engines were removed and overhauled, and new fire detection and autofeathering systems installed, together with mufflers to bring noise levels down to FAR Pt 36 standards. During inspection, Grumman found that a rear spar capstrip of 7075-T6 light alloy had suffered deterioration, and it was replaced by a titanium capstrip. Subsequent conversions will have all four centre-section capstrips replaced by titanium to give the airframe an unlimited service life. Grumman modifications have included conversion of the port main passenger door to open outward, and the lower portion now mounts a drop-down ladder.

Twenty-eight non-reclinable seats are provided at 81 cm (32 in) pitch, with facilities for a flight attendant, and there is provision for a toilet at the rear of the cabin. New lightweight solid-state avionics include two Collins VHF-20A com, two VIR-30 nav, ADF-30 ADF, and RCA WeatherScout 2 radar. Equipment removed included the autopilot, JATO and drop fuel tank provisions.

The prototype conversion, of a UF-2 Albatross, flew for the first time on 13 February 1979, and FAA certification was received on 29 April 1980. Resorts International plans to have 11 HU-16s similarly converted. Its commuter carrier, Chalks International, has two of the first five conversions, and another has been sold to Pelita Air Service of Indonesia. Grumman has purchased 57 HU-16s for civil conversion, and believes that there is a potential market for conversions of approximately 200 Albatross aircraft which remain in worldwide use. A water bomber version, and a version with Garrett TPE331-15 turboprop engines and Dowty Rotol four-blade propellers, are under consideration.

Details of the HU-16 Albatross last appeared in the 1964-65 *Jane's*, and apply generally to this conversion, except as noted above and in the changed specification details which follow:

DIMENSIONS, INTERNAL:	
Cabin: Length	7·95 m (26 ft 1 in)
Max height	1·88 m (6 ft 2 in)
Max width	2·26 m (7 ft 5 in)
Floor area	13·47 m² (145 sq ft)
Baggage compartment	7·93 m³ (280 cu ft)
WEIGHTS AND LOADINGS:	
Operating weight empty	10,659 kg (23,500 lb)
Max fuel load	2,920 kg (6,438 lb)
Max T-O weight:	
land	13,970 kg (30,800 lb)
water	14,129 kg (31,150 lb)
Max landing weight:	
land	13,226 kg (29,160 lb)
water	14,129 kg (31,150 lb)
Max wing loading	146·9 kg/m² (30 lb/sq ft)
Max power loading	14·15 kg/kW (10·5 lb/hp)

Grumman E-2C Hawkeye airborne early-warning aircraft in service with the Israeli Air Force *(Denis Hughes)*

PERFORMANCE (at max T-O weight, ISA at S/L except where indicated):
- Never-exceed speed: 229 knots (424 km/h; 263 mph) IAS
- Normal operating speed: 206 knots (382 km/h; 237 mph) IAS
- Stalling speed, flaps and landing gear down, power off: 72 knots (134 km/h; 83 mph) IAS
- Max rate of climb, METO power: 381 m (1,250 ft)/min
- Max operating altitude: 2,440 m (8,000 ft)
- T-O to 15 m (50 ft): land 1,341 m (4,400 ft); water 1,349 m (4,425 ft)
- Landing from 15 m (50 ft): land 924 m (3,030 ft); water 966 m (3,170 ft)
- Accelerate/stop distance: land 1,825 m (5,990 ft); water 1,832 m (6,010 ft)
- Range at 1,525 m (5,000 ft), cruising speed of 162 knots (300 km/h; 186 mph) TAS, crew of 3 + 28 passengers, 45 min reserves:
  - land: 273 nm (506 km; 314 miles)
  - water: 405 nm (750 km; 466 miles)
- Max ferry range, height and speed as above, no reserves:
  - land, water: 1,480 nm (2,742 km; 1,704 miles)

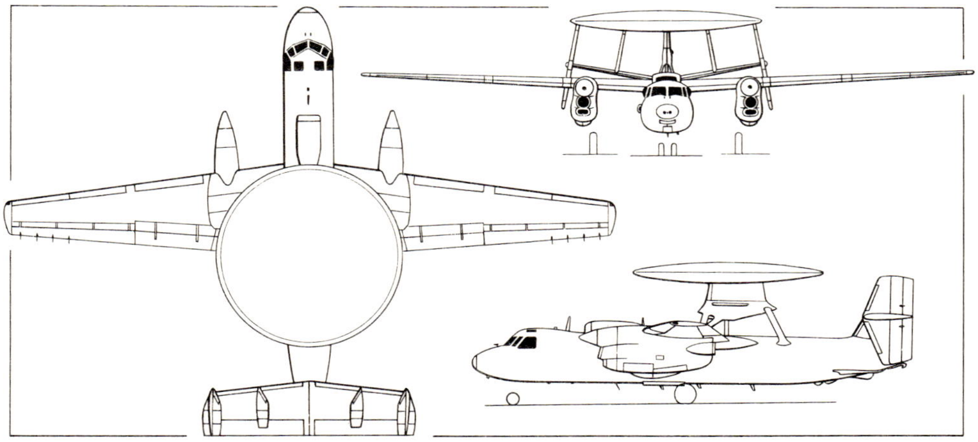

Grumman E-2C Hawkeye twin-turboprop airborne early-warning aircraft *(Pilot Press)*

GRUMMAN HAWKEYE

US Navy designation: E-2

The E-2 Hawkeye was evolved as a carrier-borne early-warning aircraft, but is suitable also for land-based operations. The prototype flew for the first time on 21 October 1960, since when the following versions have been built:

E-2A (formerly W2F-1). Initial production version. First flight on 19 April 1961, and first delivery to US Navy on 19 January 1964. Total of 62 built, including three prototypes.

E-2B. Prototype flew for the first time on 20 February 1969. Differs from E-2A by having a Litton Industries L-304 microelectronic general-purpose computer and reliability improvements. A retrofit programme, completed in December 1971, updated all operational E-2As to E-2B standard.

E-2C. First of two E-2C prototypes flew on 20 January 1971. Production began in mid-1971 and the first flight of a production aircraft was made on 23 September 1972. Orders from the US Navy now cover 95 aircraft, of which 58 had been delivered by the beginning of 1981. It is planned for production to continue at the rate of six per year until 1986, with the 95th aircraft delivered in 1987. Israel has four; Japan has ordered eight for delivery in 1982-85; and the US Navy has issued a letter of offer for four to the Egyptian Air Force. The French Air Force conducted an operational evaluation of the 58th production E-2C, from the CEAM, Mont-de-Marsan, in mid-1980.

The E-2C entered service, with airborne early-warning squadron VAW-123 at NAS Norfolk, Va, in November 1973, and went to sea on board the USS *Saratoga* in late 1974. Nine other squadrons have since received E-2C aircraft, and two **TE-2C** training aircraft are also in service.

The Hawkeye can maintain patrol on naval task force defence perimeters in all weathers, at an operating height of about 9,150 m (30,000 ft), and can detect and assess any threat from approaching enemy aircraft over ranges approaching 260 nm (480 km; 300 miles). The radar is capable of detecting airborne targets anywhere in a three million cubic mile surveillance envelope while simultaneously monitoring maritime traffic. Improvements compared with earlier radars provide increased reliability and easier maintenance. Long-range detection, automatic target track initiation, and high-speed processing, combine to enable each E-2C to track, automatically and simultaneously, more than 250 targets and to control more than 30 airborne intercepts. A Randtron Systems AN/APA-171 antenna system is housed in a 7·32 m (24 ft) diameter saucer-shaped rotodome mounted above the rear fuselage of the aircraft, which revolves in flight at 6 rpm, and can be lowered 0·64 m (1 ft 10¼ in) to facilitate aircraft stowage on board ship. The Yagi type radar arrays within the rotodome are interfaced to the onboard avionic systems, providing radar sum and difference signals plus IFF.

The AN/APS-125 search radar can detect targets as small as a cruise missile at ranges in excess of 100 nm (185 km; 115 miles). The AN/ALR-59 passive detection system (PDS) alerts operators to the presence of electronic emitters at distances up to twice the detection range of the radar system, thus expanding significantly the surveillance capability of the E-2C. Functions of these and other key elements of the E-2C's avionics systems were described more fully in the 1979-80 *Jane's*.

The following details apply to the E-2C Hawkeye:

TYPE: Airborne early-warning aircraft.

WINGS: Cantilever high-wing monoplane of all-metal construction. Centre-section is a structural box consisting of three beams, ribs and machined skins. Hinged leading-edge is non-structural and provides access to flying and engine controls. The outer panels fold rearward about skewed-axis hinge fittings mounted on the rear beams, to stow parallel with the rear fuselage on each side. Folding is done through a double-acting hydraulic cylinder. Trailing-edges of outer panels and part of centre-section consist of long-span ailerons and hydraulically-actuated Fowler flaps. When flaps are lowered, ailerons are drooped automatically. All control surfaces of E-2C are power-operated and incorporate devices to produce artificial feel forces. Automatic flight control system (AFCS) can be assigned sole control of the system hydraulic actuators, or AFCS signals can be superimposed on the pilot's mechanical inputs for stability augmentation. Pneumatically-inflated rubber de-icing boots on leading-edges.

FUSELAGE: Conventional all-metal semi-monocoque structure.

TAIL UNIT: Cantilever structure, with four fins and three double-hinged rudders. Tailplane dihedral 11°. Portions of tail unit made of glassfibre to reduce radar reflection. Power control and artificial feel systems as for ailerons. Pneumatically-inflated rubber de-icing boots on all leading-edges.

LANDING GEAR: Hydraulically-retractable tricycle type. Pneumatic emergency extension. Steerable nosewheel unit retracts rearward. Main wheels retract forward, and rotate to lie flat in bottom of nacelles. Twin wheels on nose unit only. Oleo-pneumatic shock-absorbers. Mainwheel tyres size 36 × 11 Type VII 24-ply, pressure 17·93 bars (260 lb/sq in) on ship, 14·48 bars (210 lb/sq in) ashore. Hydraulic brakes. Hydraulically-operated retractable tailskid. A-frame arrester hook under tail.

POWER PLANT: Two 3,661 kW (4,910 ehp) Allison T56-A-425 turboprop engines, driving Hamilton Standard type 54460-1 four-blade fully-feathering reversible-pitch constant-speed propellers. These have foam-filled blades which have a steel spar and glassfibre shell. Spinners and blades incorporate electrical anti-icers.

ACCOMMODATION: Normal crew of five on flight deck and in ATDS compartment in main cabin, consisting of pilot, co-pilot, combat information centre officer, air control officer and radar operator. Downward-hinged door, with built-in steps, on port side of centre-fuselage.

AVIONICS: AN/APA-171 rotodome (radar and IFF antennae), AN/APS-125 advanced radar processing system (ARPS) with overland/overwater detection capability, RT-988/A IFF interrogator with Hazeltine OL-76/AP IFF detector processor, Litton AN/ALR-59 passive detection system, Hazeltine AN/APA-172 control indicator group, Litton OL-77/ASQ computer programmer (L-304), ARC-158 UHF data link, ARQ-34 HF data link, ASM-440 in-flight performance monitor, ARC-51A UHF com, ARQ-34 HF com, AIC-14A intercom, Litton AN/ASN-92 (LN-15C) CAINS carrier aircraft inertial navigation system, Conrac Corporation CP-1085/AS air data computer, APN-153 (V) Doppler, ASN-50 heading and attitude reference system, ARN-52 (V) Tacan, ARA-50 UHF ADF, ASW-25B ACLS and APN-171 (V) radar altimeter.

DIMENSIONS, EXTERNAL:
- Wing span: 24·56 m (80 ft 7 in)
- Width, wings folded: 8·94 m (29 ft 4 in)
- Length overall: 17·54 m (57 ft 6¾ in)
- Height overall: 5·58 m (18 ft 3¾ in)
- Diameter of rotodome: 7·32 m (24 ft 0 in)
- Propeller diameter: 4·11 m (13 ft 6 in)

AREA:
- Wings, gross: 65·03 m² (700 sq ft)

WEIGHTS:
- Weight empty: 17,211 kg (37,945 lb)
- Max fuel (internal): 5,624 kg (12,400 lb)
- Max T-O weight: 23,503 kg (51,817 lb)

PERFORMANCE (at max T-O weight):
- Max level speed: 325 knots (602 km/h; 374 mph)
- Max cruising speed: 317 knots (587 km/h; 365 mph)

Cruising speed (ferry)	269 knots (499 km/h; 310 mph)
Approach speed	102 knots (189 km/h; 117 mph)
Stalling speed (landing configuration)	74 knots (138 km/h; 85·5 mph)
Service ceiling	9,390 m (30,800 ft)
T-O run	580 m (1,900 ft)
T-O to 15 m (50 ft)	793 m (2,600 ft)
Min landing run	381 m (1,250 ft)
Ferry range	1,394 nm (2,583 km; 1,605 miles)
Time on station, 175 nm (320 km; 200 miles) from base	3-4 h
Endurance with max fuel	6 h 6 min

GRUMMAN INTRUDER

US Navy designations: A-6, EA-6 and KA-6

The basic A-6A (originally A2F-1) Intruder was conceived as a carrier-borne low-level attack bomber equipped specifically to deliver nuclear or conventional weapons on targets completely obscured by weather or darkness. Of more than 600 A-6s built, in successive versions, approximately 250 were in US Navy service in early 1981. They currently equip 17 operational US Navy/Marine Corps squadrons, and three readiness training squadrons.

Competition for the original A-6 contract was conducted from May to December 1957, Grumman's contender being selected on 31 December 1957. Seven variants of the basic design have been built, of which the A-6A, A-6B and A-6C (described in 1978-79 *Jane's*) are no longer operational. The current operational versions are as follows:

EA-6A. First flown in 1963. Retains partial strike capability, but is equipped primarily to support strike aircraft and ground forces by suppressing enemy electronic activity and obtaining tactical electronic intelligence within a combat area. Total of 27 built for US Marine Corps, including six A-6As modified into EA-6As. Details in 1980-81 *Jane's*.

EA-6B Prowler. Advanced electronics development of the EA-6A, described separately.

KA-6D Intruder. Flight refuelling tanker, flown for first time on 23 May 1966. Could also act as a control aircraft for air/sea rescue operations or as a day bomber. Total of 62 A-6As modified to KA-6D configuration. Details in 1980-81 *Jane's*.

A-6E Intruder. An advanced conversion of the A-6A with multi-mode radar and an IBM computer similar to that first tested in the EA-6B. First flight of an A-6E was made on 27 February 1970. First squadron deployment was made in September 1972, and the A-6E was approved officially for service use in December 1972. It was planned to acquire a total of 318 A-6Es. Procurement is continuing beyond original termination date, and A-6E is now expected to remain in production, at rate of about 12 per year, until 1986.

An **A-6E/TRAM** (target recognition and attack multisensor) version of the A-6E flew for the first time in October 1974. TRAM adds an undernose precision-stabilised turret, with a sensor package containing both infra-red and laser equipment, to a full-system Intruder, updates the inertial navigation system with AN/ASN-92 CAINS, provides a new communications-navigation-identification (CNI) system and automatic carrier landing capability. The sensor package is integrated with the multi-mode radar, providing the capability of detecting, identifying and attacking a wide range of targets (as well as viewing the terrain) under adverse weather conditions, and with an improved degree of accuracy, using either conventional or laser-guided weapons. The bombardier/navigator operates the TRAM system by first acquiring the target on his radar screen. He then switches to the FLIR (forward-looking infra-red) system, using an optical zoom to enlarge the target's image. After identifying and selecting his targets, the bombardier uses a laser designator to mark the target with a laser spot, on which his own laser-guided weapons, or those from another aircraft, will home. Using TRAM's laser spot detector, the A-6E can also acquire a target that is being illuminated from another aircraft, or designated by a forward air controller on the ground. Delivery of fully provisioned TRAM aircraft began in February 1979, and the first carrier deployment of the A-6E/TRAM was completed successfully in May 1980.

Under plans current in early 1981, 50 A-6Es were scheduled to be equipped to carry McDonnell Douglas Harpoon anti-shipping missiles (six per aircraft), and to become operational in this form by the end of that year. It is planned eventually to give all A-6Es Harpoon capability.

Intruder training is carried out using Grumman TC-4C (modified Gulfstream I) aircraft, eight of which are in US Navy/US Marine Corps service. Grumman updated all eight to A-6E/TRAM standard during 1978-80.

The following description applies to the standard A-6E:

TYPE: Two-seat carrier-based attack bomber.

WINGS: Cantilever mid-wing monoplane, with 25° sweepback at quarter-chord. All-metal structure. Hydraulically-operated almost-full-span leading-edge and trailing-edge flaps, with inset spoilers (flaperons) of same span as flaps forward of trailing-edge flaps. Trailing-edge of each wingtip, outboard of flap, splits to form speed-brakes which project above and below wing when extended. Two short fences above each wing. Outer panels fold upward and inward.

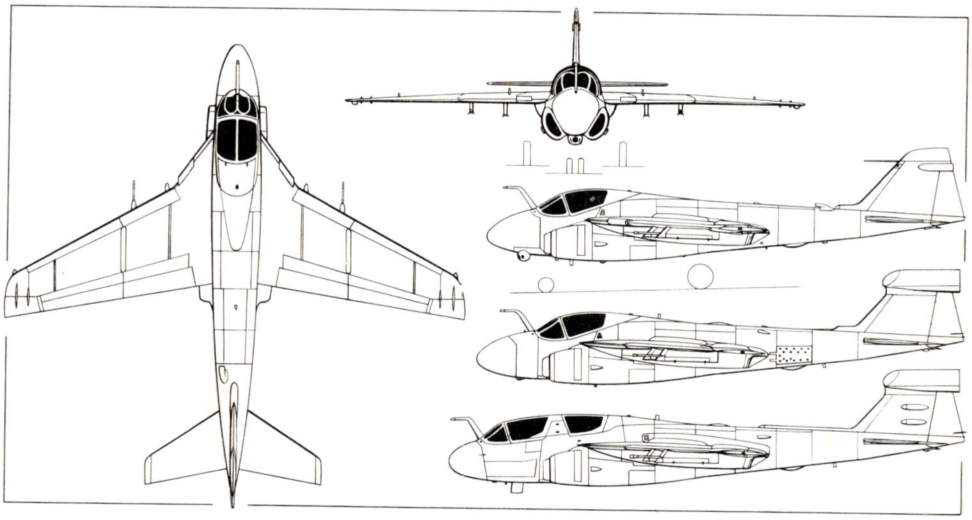

Grumman A-6E/TRAM, with additional side views of EA-6A (centre) and EA-6B (bottom) *(Pilot Press)*

Grumman A-6E/TRAM of the US Navy's VA-65 Tiger Squadron

FUSELAGE: Conventional all-metal semi-monocoque structure. Bottom is recessed between engines to carry semi-exposed store.

TAIL UNIT: Cantilever all-metal structure. All-moving tailplane, without separate elevators. Electronic antenna in rear part of fin, immediately above rudder.

LANDING GEAR: Hydraulically-retractable tricycle type. Twin-wheel nose unit retracts rearward. Single-wheel main units retract forward and inward into air intake fairings. A-frame arrester hook under rear fuselage.

POWER PLANT: Two 41·4 kN (9,300 lb st) Pratt & Whitney J52-P-8B turbojet engines. Max internal fuel 7,230 kg (15,940 lb), 9,028 litres (2,385 US gallons). Provision for up to four external fuel tanks under wings, each of 1,136 litres (300 US gallons) capacity. Removable flight refuelling probe projects upward immediately forward of windscreen.

ACCOMMODATION: Crew of two on Martin-Baker GRU7 ejection seats, which can be reclined to reduce fatigue during low-level operations. Bombardier/navigator slightly behind and below pilot to starboard. Hydraulically-operated rearward-sliding canopy.

SYSTEMS: Garrett environmental control system for cockpit and avionics bay. Dual hydraulic systems for operation of flight controls, leading-edge and trailing-edge flaps, wingtip speed-brakes, landing gear brakes and cockpit canopy. One electrically-driven hydraulic pump provides restricted flight capability by supplying the tailplane and rudder actuators only. Electrical system powered by two Garrett constant-speed drive starters that combine engine starting and electrical power generation, each delivering 30kVA. A Garrett ram-air turbine, mounted so that it can be projected into the airstream above the port wing-root, provides in-flight emergency electrical power for essential equipment.

AVIONICS AND EQUIPMENT: Development of the A-6E began with the substitution of a single simultaneous multi-mode navigation and attack radar, developed by the Norden division of UTC for the two earlier radar systems in the A-6A. Following the concepts of the EA-6B, the IBM Corporation and Fairchild Camera and Instrument Corporation have supplied a new attack and navigation computer system and an interfacing data converter. Conrac Corporation has designed an armament control unit and RCA has developed a video tape recorder for post-strike assessment of attacks.

The Norden division's AN/APQ-148 multi-mode radar provides simultaneous ground mapping; identification, tracking, and rangefinding of fixed or moving targets; and terrain-clearance or terrain-following manoeuvres. It can also detect, locate and track radar beacons used by forward air controllers when providing close support for ground forces. The APQ-148 has mechanical scanning in azimuth and utilises a specially-developed avionics system for simultaneous vertical scanning. During 1981-83, it is being updated by the addition of AMTI (airborne moving target indication) to enhance its ability to detect moving targets. There are two cockpit displays, one for the pilot and one for the bombardier/navigator, and terrain data is also presented on a vertical display indicator ahead of the pilot.

IBM's AN/ASQ-133 solid-state digital computer is coupled to the A-6E's radar, inertial and Doppler navigational equipment, communications and automatic flight control system. As mission data is measured in flight by onboard aerodynamic and electronic sensors, the computer compares the data with the programmed information, computes any differences, and provides corrective data that can be used to alter the parameters of the mission.

Fairchild Camera and Instrument Corporation's signal data converter for the A-6E accepts analogue input data from up to sixty sensors, and converts that information to a digital output that is fed into the computer of the navigation and attack system.

Conrac Corporation's armament control unit (ACU) for the A-6E provides in a single unit all the inputs and outputs necessary to select and release the Intruder's weapons. The master arming switch has a 'practice' position that allows the ACU to be cycled up to the point of firing command.

The multi-mode AN/AVA-1 display developed by Kaiser Aerospace and Electronics Corporation serves as a primary flight aid for navigation, approach, landing and weapons delivery. The basic vertical display indicator (VDI) is a 0·20 m (8 in) cathode-ray tube which shows a synthetic landscape, sky, and electronically-generated command flight path that move to simulate the motion of these features as they would be seen by the pilot through the windscreen of the aircraft. Symbols are superimposed to augment the basic attitude data, and for attack a second set of superimposed information provides a target symbol, steering symbol, and release and pull-up markers. A solid-state radar data scan converter can provide on the same display an apparent real-world perspective of terrain, ten shades of grey defining terrain elevation at ten different segmented contour intervals up to 8·7 nm (16 km; 10 miles) ahead of the aircraft. This makes it possible for the pilot to fly the Intruder in either a terrain-following or terrain-avoidance mode at low altitude. Flight path and attack symbols can be superimposed over the terrain elevation data on the VDI, enabling the pilot to make his attack while avoiding or following terrain in the target area. Kaiser has also developed a micromesh filter to prevent 'washout' of the data displayed on the VDI in sunlight conditions. Naval pilots use the VDI as a primary flight instrument, for precise steering in navigation, and for weapons cues, progress, and status information during an attack. For carrier landing the unit is used as a flight director and, linked to the APQ-148 radar, it presents steering information, allowing the pilot to select a descent angle for the final approach.

ARMAMENT: Five weapon attachment points, each with a 1,633 kg (3,600 lb) capacity (max external stores load 8,165 kg; 18,000 lb). Typical weapon loads are thirty 500 lb bombs in clusters of six, or three 2,000 lb general purpose bombs plus two 1,135 litre (300 US gallon) drop-tanks.

DIMENSIONS, EXTERNAL:
Wing span 16·15 m (53 ft 0 in)
Wing mean aerodynamic chord 3·32 m (10 ft 10¾ in)
Width, wings folded 7·72 m (25 ft 4 in)
Length overall 16·69 m (54 ft 9 in)
Height overall 4·93 m (16 ft 2 in)
Tailplane span 6·21 m (20 ft 4½ in)
Wheel track 3·32 m (10 ft 10½ in)

AREAS:
Wings, gross 49·1 m² (528·9 sq ft)
Flaperons (total) 3·81 m² (41·0 sq ft)
Trailing-edge flaps (total) 9·66 m² (104·0 sq ft)
Leading-edge slats (total) 4·63 m² (49·8 sq ft)
Fin 5·85 m² (62·93 sq ft)
Rudder 1·52 m² (16·32 sq ft)

WEIGHTS AND LOADING:
Weight empty 12,000 kg (26,456 lb)
Fuel load: internal 7,230 kg (15,939 lb)
external (four tanks) 3,638 kg (8,020 lb)
Max external load 8,165 kg (18,000 lb)
Max T-O weight:
catapult 26,580 kg (58,600 lb)
field 27,397 kg (60,400 lb)
Max zero-fuel weight 20,166 kg (44,460 lb)
Max landing weight:
carrier 16,329 kg (36,000 lb)
field 20,411 kg (45,000 lb)
Max wing loading 557·6 kg/m² (114·2 lb/sq ft)

PERFORMANCE (no stores, except where stated):
Never-exceed speed 700 knots (1,297 km/h; 806 mph)
Max level speed at S/L 560 knots (1,037 km/h; 644 mph)
Cruising speed at optimum altitude 412 knots (763 km/h; 474 mph)
Approach speed 110 knots (204 km/h; 127 mph)
Stalling speed, flaps up 142 knots (264 km/h; 164 mph)
Stalling speed, flaps down 98 knots (182 km/h; 113 mph)
Max rate of climb at S/L 2,307 m (7,570 ft)/min
Rate of climb at S/L, one engine out 652 m (2,140 ft)/min
Service ceiling 12,955 m (42,500 ft)
Service ceiling, one engine out 6,400 m (21,000 ft)
Min T-O run 1,185 m (3,890 ft)
T-O run to 15 m (50 ft) 1,380 m (4,530 ft)
Landing from 15 m (50 ft) 771 m (2,530 ft)
Min landing run 521 m (1,710 ft)
Combat range with max external fuel 2,823 nm (5,231 km; 3,250 miles)
Range with max external load 880 nm (1,631 km; 1,013 miles)

GRUMMAN EA-6B PROWLER

The EA-6B is an advanced electronics development of the EA-6A for which Grumman received a prototype design and development contract in the Autumn of 1966. Except for a 1·37 m (4 ft 6 in) longer nose section and large fin pod, the external configuration of this version is the same as that of the basic A-6.

The longer nose section provides accommodation for a total crew of four, the two additional crewmen being necessary to operate the more advanced ECM equipment. This comprises high-powered electronic jammers and modern computer-directed receivers, which provided the US Navy with its first aircraft designed and built specifically for tactical electronic warfare. The prototype EA-6B flew for the first time on 25 May 1968.

The FY 1969 defence budget allocated $139 million for the initial purchase of eight EA-6Bs, and the total programme is expected to cover the supply of 102 aircraft (including the four pre-production and one R and D aircraft), to equip Navy and Marine Corps squadrons. Deliveries of production aircraft began in January 1971

Grumman EA-6B Prowler tactical jamming system aircraft

and are expected to continue at a rate of six per year until 1986.

An ICAP (Increased Capability) version of the EA-6B, which increases substantially the jamming efficiency of the aircraft, has been developed. Current production aircraft are delivered with ICAP as standard, and the first 21 production EA-6Bs have been modified by Grumman to ICAP configuration. Modifications include an expanded onboard tactical jamming system with eight frequency bands, reduced response time, and a new multi-format display. In addition, an automatic carrier landing system (ACLS) to permit carrier recovery in zero-zero weather, a new defensive electronic countermeasures system (DECM) and new communications-navigation-identification (CNI) equipment are installed. Another advanced version, ICAP-2, is under evaluation; the prototype made its first flight on 24 June 1980. To follow this, an ACAP (Advanced Capability) programme is under development.

Ten US Navy squadrons (VAQ-129, 130, 131, 132, 133, 134, 135, 136, 137 and 138) were equipped with the Prowler by mid-1977. The first detachment of US Marine Corps Prowler squadron VMAQ-2 began training on the EA-6B in September 1977 at NAS Whidbey Island, Washington, and the detachment deployed in late 1978. Two additional detachments have since completed training, and at least one is deployed at all times.

Under a $4·8 million contract awarded in 1978, Norden Systems is to supply for the EA-6B an advanced navigation radar system designated AN/APS-130. The contract calls for one prototype and three pre-production systems before the start of full production. This is expected to run to approximately 100 systems.

The description of the standard A-6E Intruder applies also to the EA-6B, except as follows:

TYPE: Four-seat carrier- or land-based advanced ECM aircraft.

WINGS: As for A-6E, but reinforced to cater for increased gross weight, fatigue life and 5·5g load factor.

FUSELAGE: As for A-6E, but reinforcement of underfuselage structure in areas of arrester hook and landing gear attachments, and lengthened by 1·37 m (4 ft 6 in).

TAIL UNIT: As for A-6E, except for provision of a large fin-tip pod to house ECM equipment.

LANDING GEAR: As for A-6E, except for reinforcement of attachments, A-frame arrester hook, and upgrading of structure to cater for increased gross weight.

POWER PLANT: Two Pratt & Whitney J52-P-408 turbojet engines, each rated at 49·8 kN (11,200 lb st).

ACCOMMODATION: Crew of four under two separate upward-opening canopies. Martin-Baker GRUEA 7 ejection seats for flight crew. The two additional crewmen are ECM Officers to operate the ALQ-99 equipment from the rear cockpit. Either ECMO can independently detect, assign, adjust and monitor the jammers. The ECMO in the starboard front seat is responsible for communications, navigation, defensive ECM and chaff dispensing.

SYSTEMS: Generally as for A-6E.

AVIONICS: ALQ-99 advanced electronic countermeasures (ECM) to enable the EA-6B to fulfil a tactical jamming role. Five integrally powered pods, with a total of 10 jamming transmitters, can be carried. Each pod covers one of eight frequency bands. Sensitive surveillance receivers in the fin-tip pod for long-range detection of radars; emitter information is fed to a central digital computer that processes the signals for display and recording. Detection, identification, direction-finding and jammer-set-on sequence can be performed automatically or with manual assistance from crew.

DIMENSIONS, EXTERNAL:
As for A-6E, except:
Width, wings folded 7·87 m (25 ft 10 in)
Length overall 18·11 m (59 ft 5 in)
Height overall 4·95 m (16 ft 3 in)
Wheelbase 5·23 m (17 ft 2 in)

AREAS:
As for A-6E

WEIGHTS AND LOADING:
Weight empty 14,588 kg (32,162 lb)
Internal fuel load 6,995 kg (15,422 lb)
Max external fuel load 4,547 kg (10,025 lb)
T-O weight in stand-off jamming configuration 24,703 kg (54,461 lb)
T-O weight in ferry range configuration 27,492 kg (60,610 lb)
Max T-O weight, catapult or field 29,483 kg (65,000 lb)
Max zero-fuel weight 17,708 kg (39,039 lb)
Max landing weight, carrier or field 20,638 kg (45,500 lb)
Max wing loading 600·5 kg/m² (123 lb/sq ft)

PERFORMANCE (A: no stores; B: 5 ECM pods):
Never-exceed speed 710 knots (1,315 km/h; 817 mph)
Max level speed at S/L:
A 566 knots (1,048 km/h; 651 mph)
B 533 knots (987 km/h; 613 mph)
Cruising speed at optimum altitude:
A 418 knots (774 km/h; 481 mph)
B 420 knots (777 km/h; 483 mph)
Stalling speed, flaps up, max power:
A 124 knots (230 km/h; 143 mph)
Stalling speed, flaps down, max power:
A 84 knots (156 km/h; 97 mph)
Max rate of climb at S/L: A 3,932 m (12,900 ft)/min
B 3,057 m (10,030 ft)/min
Rate of climb at S/L, one engine out:
A 1,189 m (3,900 ft)/min
Service ceiling: A 13,565 m (44,500 ft)
B 12,500 m (41,000 ft)
Service ceiling, one engine out:
A 8,930 m (29,300 ft)
T-O to 15 m (50 ft): A 869 m (2,850 ft)
B 1,065 m (3,495 ft)
Landing from 15 m (50 ft): A 823 m (2,700 ft)
Landing run: A 579 m (1,900 ft)
B 655 m (2,150 ft)
Combat range with max external fuel: 2,085 nm (3,861 km; 2,399 miles)
Range with max external load, 5% reserves plus 20 min at S/L: B 955 nm (1,769 km; 1,099 miles)

GRUMMAN MOHAWK

In a continuing modernisation programme, under which more than 80 OV-1B/OV-1C Mohawks have already been upgraded to OV-1D standard, the US Army's Aviation Material Readiness Command has placed a further contract with Grumman, worth approx $16 million, to convert a further seven early-model Mohawks to OV-1D standard. The contract provides for four others to be converted to RV-1D electronic intelligence/emitter locating aircraft.

Grumman is also overhauling and modifying a total of about 20 OV-1s for the Philippines and another Far East country.

GRUMMAN TOMCAT

US Navy designation: F-14

Grumman announced on 15 January 1969 that it had been selected as winner of the design competition for a carrier-based fighter for the US Navy. Known as the VFX during the competitive phase of the programme, this aircraft was later designated F-14. First flight of the F-14A Tomcat prototype took place on 21 December 1970. It was lost in a non-fatal accident, and flight testing was resumed on 24 May 1971 with the second aircraft.

The F-14 is designed to fulfil three primary missions. The first of these, fighter sweep/escort, involves clearing contested airspace of enemy fighters and protecting the strike force, with support from early-warning aircraft, surface ships and communications networks to co-ordinate penetration and escape. Second mission is to defend carrier task forces via combat air patrol (CAP) and deck launched intercept (DLI) operations. Third role is secondary attack of tactical targets on the ground, supported by electronic countermeasures and fighter escort. It has also been reported that in tests with targets simulating anti-ship missiles, Phoenix-armed F-14s have proved effective against high, fast, relatively small radar cross-section threats such as the Soviet AS-4/6 air-to-surface missiles.

The configuration of the F-14 is unique, with variable-geometry wings, small foreplanes (glove vanes) which are extended automatically at supersonic speeds to control centre-of-pressure shift, manoeuvring slats and flaps to

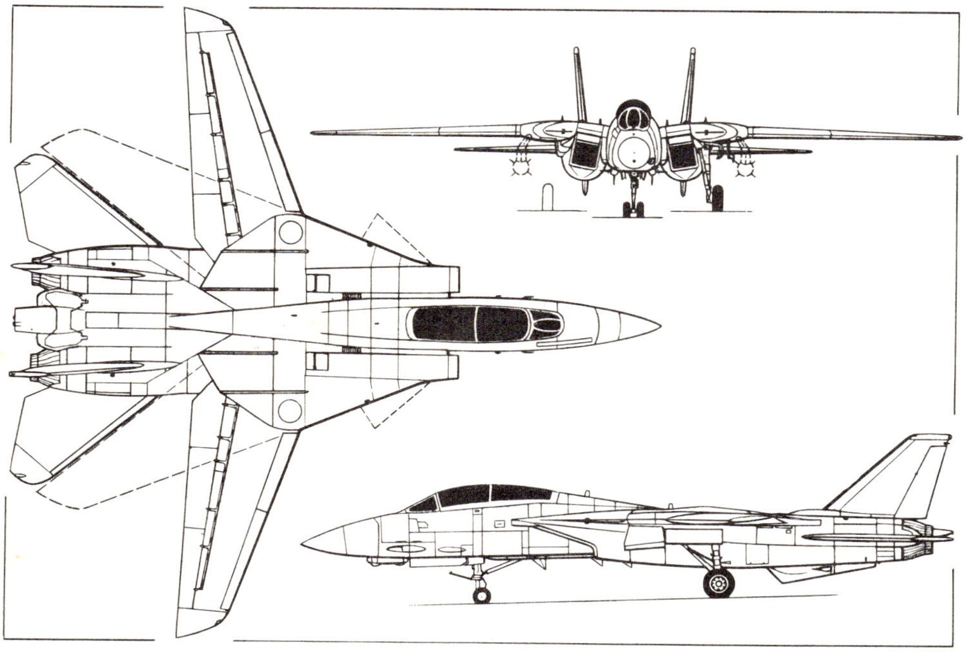

Grumman F-14A Tomcat carrier-based multi-mission fighter *(Pilot Press)*

create a lower effective wing loading, and twin outward-canted fins and rudders. Optimum sweep of the wing is controlled automatically by a Mach sweep programmer, which relates sweep to Mach number and altitude.

Under the initial contracts, Grumman was required to provide the US Navy with a mockup of the F-14A in May 1969, and to build 12 research and development aircraft. Subsequently, the US Navy ordered an initial series of 26 production F-14As, and is scheduled currently to acquire a total of 491 Tomcats, including the 12 development aircraft; the total may be increased to nearly 850 by the Reagan administration.

Carrier trials started in June 1972, and initial deployment with the fleet began in October 1972. Replacement Training Squadron (RTS) VF-124 at Miramar NAS, San Diego, California, was responsible for working up ground and air crews for the new aircraft, and the first two operational squadrons, VF-1 and VF-2, were serving on board the USS *Enterprise* in the Western Pacific in September 1974. Replacement Training Squadron VF-101 was established in July 1977 to train ground and flight crews for deployment from NAS Oceana, Virginia. A total of 386 F-14As, including the 12 R and D aircraft, had been delivered by 1 January 1981, and 30 more were scheduled for delivery during 1981.

In mid-1979 the US Navy awarded a $4 million contract to the Northrop Corporation to manufacture pre-production television camera sets (TCSs) for installation on F-14s. Developed by Northrop's Electro-Mechanical Division, the TCS is a closed-circuit TV system, offering both wide-angle (acquisition) and telescopic (identification) fields of view. Mounted beneath the nose of the F-14, the TCS automatically searches for, acquires and locks on to distant targets, displaying them on monitors for the pilot and flight officer. By allowing early identification of targets, the system permits crews of high-speed aircraft to make combat decisions earlier than was possible previously. A further $3 million initial production contract for TCSs was placed in early 1981.

The Naval Air Test Center at Patuxent River, Maryland, conducted test and evaluation of a 798 kg (1,760 lb) reconnaissance pod for the F-14 in 1979, and 49 F-14s were being fitted with this pod in 1980-81. Designated TARPS (Tactical Air Reconnaissance Pod System), it is mounted 0·38 m (1 ft 3 in) off the centreline of the F-14's underfuselage, in the tunnel between the two engine nacelles. It has four main compartments: the nose carries a CAI KS-87B frame camera for either forward oblique or vertical photographs; the second contains a Fairchild KA-99 low/medium-altitude horizon-to-horizon panoramic camera; the third has a Honeywell AN/AAD-5 infra-red reconnaissance set; and the fourth is an equipment bay containing the ground check maintenance panel and a sensor control data display set.

Without the TARPS pod installed, the F-14 retains its full weapon system capability. The F-14/TARPS aircraft will fulfil an important tactical reconnaissance role pending the evolution of new photo-reconnaissance aircraft.

The Iranian Air Force ordered 80 F-14As, all of which had been delivered by the end of 1978; first flight by one of these aircraft was made on 5 December 1975, and the first three aircraft were ferried to Khatami AFB, Isfahan, in January 1976. Iranian F-14As retain the Phoenix weapon system, but have slightly different ECM equipment from US Navy aircraft.

On 14 July 1981 Grumman initiated a short flight test programme (24 flights) with an F-14 (the seventh development aircraft, originally the prototype F-14B) refitted with two General Electric F101-DFEs (derivative fighter engines) as power plant. There are no plans at present to install this engine in production Tomcats, but Grumman has indicated that it is studying, for the mid-1980s, a possible 'Super Tomcat', which would make use of advanced technology engines. In the meantime, preparations have begun to introduce the **F-14C** version, with TF30-P-414A engines and improved avionics and radar, on to the Tomcat production line in late 1983.

TYPE: Two-seat carrier-based multi-role fighter.

WINGS: Variable-geometry mid-wing monoplane, with 20° of leading-edge sweep in the fully-forward position and 68° when fully swept. Oversweep position of 75° for carrier stowage. Wing position is programmed automatically for optimum performance throughout the flight regime, but manual override is provided. A short movable wing outer panel, needing only a comparatively light pivot structure, results from the wide fuselage and fixed centre-section 'glove', with pivot points 2·72 m (8 ft 11 in) from the centreline of the airframe. The inboard wing sections, adjacent to the fuselage, arc upward slightly to minimise cross-sectional area and wave-drag, and consist basically of a one-piece electron beam-welded titanium assembly, 6·70 m (22 ft) in span, made from Ti-6Al-4V titanium alloy. Small canard surfaces, known as glove vanes, swing out from the leading-edge of the fixed portion of the wing, to a maximum of 15° in relation to the leading-edge, as sweep of outer panels is increased. Spoilers on upper surfaces of wing. Stabilisation in pitch, provided by the canard surfaces, leaves the differential tailplane free to perform its primary control function. Trailing-edge control surfaces extend over almost entire span. Leading-edge slats.

FUSELAGE: The centre-fuselage section is a simple, fuel-carrying box structure; forward fuselage section comprises cockpit and nose. The aft section has a tapered aerofoil shape to minimise drag, with a fuel dump pipe projecting from the rear. Speed brakes located on the upper and lower surfaces, between the bases of the vertical tail fins.

TAIL UNIT: Twin vertical fins, mounted at the rear of each engine nacelle. Outward-canted ventral fin under each nacelle. The all-flying horizontal surfaces have skins of boron-epoxy composite material.

LANDING GEAR: Retractable tricycle type. Twin-wheel nose unit and single-wheel main units retract forward and upward. Existing beryllium brakes were replaced with Goodyear lightweight carbon brakes from Spring 1981. Arrester hook under rear fuselage, housed in small ventral fairing. Nose-tow catapult attachment on nose unit.

ENGINE INTAKES: Straight two-dimensional external compression inlets. A double-hinged ramp extends down from the top of each intake, and these are programmed to provide the correct airflow to the engines automatically under all flight conditions. Each intake is canted slightly away from the fuselage, from which it is separated by some 0·25 m (10 in) to allow sufficient clearance for the turbulent fuselage boundary layer to pass between fuselage and intake without causing turbulence within the intake. Engine inlet ducts and aft nacelle structures are manufactured by Rohr Corporation. The inlet duct, constructed largely of aluminium honeycomb, is about 4·27 m (14 ft) long, while the aft nacelle structure, of bonded aluminium honeycomb and conventional aluminium, is about 4·88 m (16 ft) in length.

POWER PLANT: Two Pratt & Whitney TF30-P-412A turbofan engines of 93 kN (20,900 lb st) with afterburning, mounted in ducts which open to provide 180° access for ease of maintenance. Garrett ATS200-50 air-turbine starter. An external auxiliary fuel tank can be carried beneath each intake trunk.

ACCOMMODATION: Pilot and naval flight officer seated in tandem on Martin-Baker GRU7A rocket-assisted zero-zero ejection seats, under a one-piece bubble canopy, hinged at the rear and offering all-round view.

AVIONICS: Hughes AN/AWG-9 weapons control system, with ability to detect airborne targets at ranges of more than 65-170 nm (120-315 km; 75-195 miles) according to their size, and ability to track 24 enemy targets and attack six of them simultaneously at varied altitudes and

Grumman F-14A Tomcat with a full complement of six AIM-54A Phoenix missiles

Northrop TCS under nose of an F-14

Grumman F-14 Tomcat fitted experimentally with General Electric F101-DFE augmented turbofan engines

distances. Kaiser Aerospace AN/AVG-12 vertical and head-up display system.

Armament: One General Electric M61A-1 Vulcan 20 mm gun mounted in the port side of forward fuselage. Four Sparrow air-to-air missiles mounted partially submerged in the underfuselage, or four Phoenix missiles carried on special pallets which attach to the bottom of the fuselage. Two wing pylons, one under each fixed wing section, can carry four Sidewinder missiles or two additional Sparrow or Phoenix missiles with two Sidewinders. Various combinations of missiles and bombs to a max external weapon load of 6,577 kg (14,500 lb). ECM equipment includes Goodyear AN/ALE-39 chaff and flare dispensers, with integral jammers.

Dimensions, external:
- Wing span: unswept — 19·54 m (64 ft 1½ in)
- swept — 11·65 m (38 ft 2½ in)
- overswept — 10·15 m (33 ft 3½ in)
- Wing aspect ratio — 7·28
- Length overall — 19·10 m (62 ft 8 in)
- Height overall — 4·88 m (16 ft 0 in)
- Tailplane span — 9·97 m (32 ft 8½ in)
- Distance between fin tips — 3·25 m (10 ft 8 in)
- Wheel track — 5·00 m (16 ft 5 in)
- Wheelbase — 7·02 m (23 ft 0½ in)

Areas:
- Wings, gross — 52·49 m^2 (565·0 sq ft)
- Horizontal tail surfaces (total) — 13·01 m^2 (140·0 sq ft)
- Vertical tail surfaces (total) — 10·96 m^2 (118·0 sq ft)

Weights:
- Weight empty — 18,036 kg (39,762 lb)
- Fuel (usable): internal — 7,348 kg (16,200 lb)
- external — 1,724 kg (3,800 lb)
- Normal T-O weight — 26,553 kg (58,539 lb)
- T-O weight with 4 Sparrow — 26,930 kg (59,372 lb)
- T-O weight with 6 Phoenix — 31,944 kg (70,426 lb)
- Max T-O weight — 33,724 kg (74,348 lb)
- Design landing weight — 23,510 kg (51,830 lb)

Performance:
- Max design speed — Mach 2·4
- Max cruising speed — 400-550 knots (741-1,019 km/h; 460-633 mph)
- Landing speed — 122 knots (226 km/h; 140 mph)
- Service ceiling — above 15,240 m (50,000 ft)
- Min T-O distance — 396 m (1,300 ft)
- Min landing distance — 762 m (2,500 ft)

GRUMMAN (GENERAL DYNAMICS) EF-111A

The programme to convert General Dynamics F-111As into EF-111A electronic warfare prototypes, and to evaluate their ability to provide ECM jamming coverage for air attack forces, was initiated in 1972-73. Operational deployment of the F-111A in Southeast Asia, from March 1968, had revealed shortcomings, despite special preparation under the Harvest Reaper programme to provide these aircraft with advanced ECM equipment that would facilitate penetration of enemy airspace. Subsequent enquiry revealed that many factors contributed to the limited success of the F-111A in Southeast Asia; lack of adequate and effective ECM jamming was responsible for many of its problems, as well as those of all other types of combat aircraft in that theatre of operations.

Because of the growing potential of Soviet-built air defence systems, which stretch across Eastern Europe, NATO anti-invasion forces must have the capability of suppressing literally thousands of radar 'eyes', able to locate precisely the route and speed of counter-attacking air strikes. In addition, updated SAM systems and new interceptors with sophisticated ECM equipment are being introduced regularly by the Soviet Union, providing its armed forces with a now-acknowledged lead in electronic warfare, both ground and airborne.

Senior US Air Force officials consider that utilisation of the EF-111 as a tactical jamming system, in combination with the E-3 AWACS, is vital to help offset this Soviet lead. Because of its vast masking power, the EF-111 is essential to provide cover for air-to-ground operations along the forward lines, and for support of penetrating allied strike forces. Should some future circumstances make it necessary to launch a counter-strike against Soviet penetration of NATO territory, EF-111s operating on the friendly side of the FEBA (forward edge of the battle area) could blind the other side's electronic 'eyes', making it possible for NATO strike forces to attack the armoured spearhead, as well as resupply areas, reserves and SAM installations 17-35 nm (32-64 km; 20-40 miles) behind the opposing lines, with something less than half the anticipated losses that could be expected without use of the EF-111s' jamming systems.

Three basic modes of deployment are foreseen for the EF-111: standoff, penetration, and close air support. In the standoff role, jamming aircraft would operate within their own airspace, at the FEBA. Out of range of the enemy's ground-based weapons, orbiting EF-111s would use their jamming systems to screen the routes of friendly strike aircraft. In the penetration role, the EF-111s would accompany strike aircraft to high-priority targets, their Mach 2 capability making them ideal escort aircraft for such a task. The close air support requirement calls for EF-111 escorts to neutralise anti-air radars while the strike force delivers its attack on enemy armour.

Design study contracts were awarded to General Dynamics and Grumman by the US Air Force in 1974, and in January 1975 it was announced that Grumman had been awarded an $85·9 million contract to convert two existing F-111As to EF-111A prototype configuration. Basic equipment of these prototypes comprises the AN/ALQ-99E tactical jamming system, comprising ten transmitters, five exciters, numerous receivers, computers, display systems, and one RF calibrator per aircraft. In addition, each has a modified AN/ALQ-137 self protection system, and a modified AN/ALR-62 terminal threat warning system. The ALQ-99E jammers are mounted in the weapons bay, with their antennae covered by a 4·9 m (16 ft) long canoe-shape radome. The fin-tip pod, similar in shape to that of the EA-6B Prowler, houses the receiver system and antennae. Total weight of the new equipment is about three tons.

The two-man crew of an EF-111 comprises a pilot and an electronic warfare officer (EWO). All tactical jamming functions are managed by the EWO who can, through computer management, handle a tactical electronic warfare workload which required previously several operators and more equipment. In addition, the automated system of the EF-111 has exceptional capability for locating, identifying, and assigning jammers to enemy emitters over a wide range of frequencies.

The first flight of an aerodynamic prototype was made on 10 March 1977, and the complete system was flown for the first time on 17 May 1977, on the second prototype. Subsequent Grumman flight testing of the jamming system involved 84 flights totalling 215 flight hours, completed by the system aircraft during a period of 3½ months. US Air Force flight testing involved 78 flights totalling 258 flight hours during a six-month test programme. The US Air Force tests verified various mission operational concepts, flight formations, and the jammer's electromagnetic compatibility with other strike aircraft. These latter tests dispelled an earlier concern that the friendly strike force, as well as enemy threats, might be jammed by the powerful signals emanating from the EF-111. Structural flight tests under all operating conditions demonstrated an 'infinite' life for all modified areas of the aircraft's structure, and flying qualities were considered virtually identical to those of the F-111 strike aircraft.

US Air Force plans envisage the conversion of 42 F-111As as ECM jamming aircraft. The production contract for the first six was signed in April 1979, and the first of these (66049) flew for the first time on 26 June 1981. They will be delivered to the 366th Tactical Fighter Wing at Mountain Home AFB, Idaho.

The description of the F-111A in the 1976-77 *Jane's* applies also to the EF-111A, except for the following additional or amended details:

Type: ECM tactical jamming aircraft.

Wings: As detailed for F-111A. Wing section NACA 64A210.68 modified at pivot point, NACA 64A209.80 with modified leading-edge at tip. Dihedral, at 16° sweep, 1°. Incidence, at 16° sweep, 1° at root, −3° at tip.

Power Plant: Two Pratt & Whitney TF30-P-3 turbofan engines, each rated at 82·3 kN (18,500 lb st) with afterburning. Fuel tanks in wings and fuselage, total capacity 18,919 litres (4,998 US gallons). Oil capacity 30·3 litres (8 US gallons).

Avionics: AN/ALQ-99E tactical jamming system; AN/ARC-164 UHF command, AN/AJQ-20A INS, AN/APQ-160 attack radar, AN/APN-167 radar altimeter, AN/APQ-110 terrain-following radar, AN/ARN-58 ILS, AN/ARC-112 HF transceiver, AN/AIC-25 intercom, AN/APX-64 IFF(AIMS), AN/ARN-118 Tacan, AN/ARA-50 UHF/DF, AN/ALQ-137 (modified) self-protection system (SPS), AN/ALR-62 (modified) terminal threat warning system (TTWS), AN/ALR-23 CMRS, and AN/ALE-28 CMDS.

Armament: None.

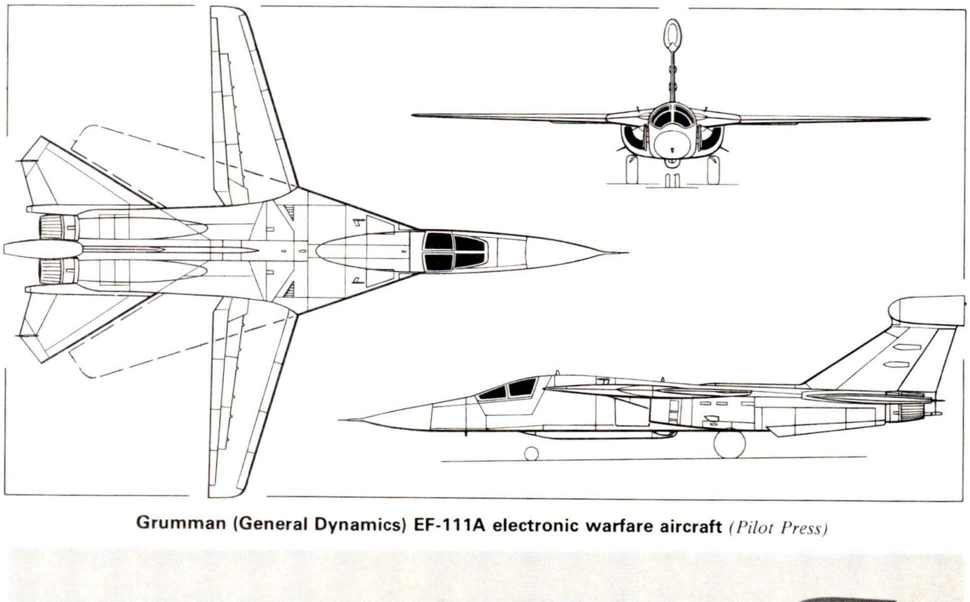

Grumman (General Dynamics) EF-111A electronic warfare aircraft *(Pilot Press)*

First Grumman-modified EF-111A production aircraft equipped with AN/ALQ-99E tactical jamming system

DIMENSIONS, EXTERNAL:

Wing span, spread	19·20 m (63 ft 0 in)
Wing span, fully swept	9·74 m (31 ft 11·4 in)
Wing mean aerodynamic chord	2·76 m (9 ft 0 in)
Wing aspect ratio (16° sweep)	7·56
Length overall	23·16 m (76 ft 0 in)
Height overall	6·10 m (20 ft 0 in)
Wheel track	3·19 m (10 ft 0·4 in)
Wheelbase	7·44 m (24 ft 4·8 in)

AREA:

Wings, gross (16° sweep)	48·77 m² (525 sq ft)

WEIGHTS:

Weight empty	25,072 kg (55,275 lb)
Max internal fuel	14,741 kg (32,500 lb)
Design T-O weight	33,000 kg (72,750 lb)
Combat T-O weight	31,751 kg (70,000 lb)
Max T-O weight	40,346 kg (88,948 lb)
Max landing weight	37,421 kg (82,500 lb)

PERFORMANCE (estimated for typical mission, A: basic standoff; B: penetration; C: close air support. At max T-O weight unless detailed otherwise):

Max combat speed at combat weight:	
A, B, C	1,196 knots (2,216 km/h; 1,377 mph)
Average speed, outbound:	
A, C	446 knots (826 km/h; 514 mph)
B	512 knots (949 km/h; 590 mph)
Average speed over combat area:	
A	321 knots (595 km/h; 370 mph)
B	507 knots (940 km/h; 584 mph)
C	462 knots (856 km/h; 532 mph)
Average speed, inbound:	
A, C	432 knots (800 km/h; 497 mph)
B	502 knots (930 km/h; 578 mph)
Stalling speed, power off:	
A, B, C	142·2 knots (263·5 km/h; 164 mph)
Rate of climb at S/L, intermediate power:	
A, B, C	1,006 m (3,300 ft)/min
Rate of climb at S/L, one engine out, with afterburning:	
A, B, C	1,021 m (3,350 ft)/min
Service ceiling with afterburning, at combat weight:	
A, B, C	13,715 m (45,000 ft)
T-O run: A, B, C	1,349 m (4,425 ft)
T-O to 15 m (50 ft): A, B, C	1,775 m (5,825 ft)
Landing from 15 m (50 ft) at weight of 26,968 kg (59,455 lb):	
A, B, C	945 m (3,100 ft)
Landing run at weight of 26,968 kg (59,455 lb):	
A, B, C	602 m (1,975 ft)
Combat radius, with reserves:	
A	200 nm (370 km; 230 miles)
B	807 nm (1,495 km; 929 miles)
C	623 nm (1,155 km; 717 miles)
Ferry range	2,022 nm (3,747 km; 2,328 miles)

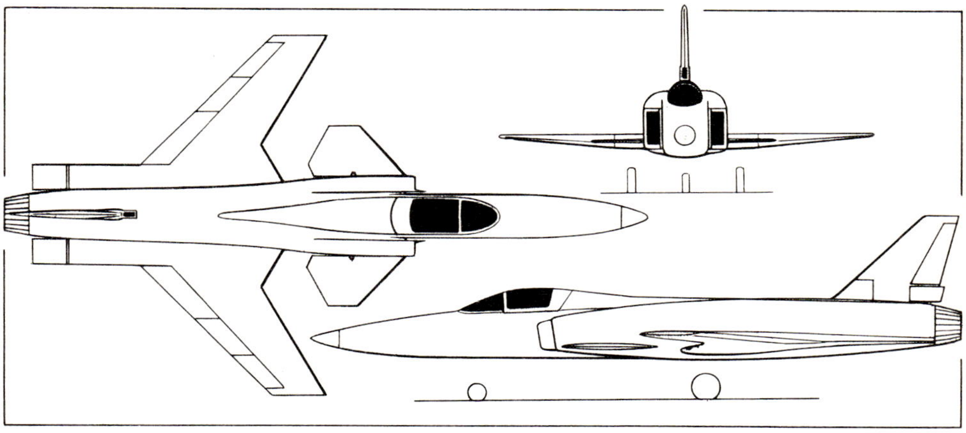

Grumman forward swept wing (FSW) demonstrator aircraft *(Pilot Press)*

GRUMMAN FORWARD SWEPT WING DEMONSTRATOR

Grumman has been exploring for some time the benefits offered by a forward swept wing design, including a series of wind tunnel test programmes funded by the Defense Advanced Research Projects Agency (DARPA). Monitored by the US Air Force, these programmes have verified the aerodynamic benefits of such a design, leading in January 1981 to the beginning of a programme to build a flight demonstration aircraft. The first flight of this vehicle is anticipated in late 1983.

Grumman's forward swept wing (FSW) design offers the promise of a new generation of tactical aircraft that will be smaller, lighter in weight, less costly, but more efficient than contemporary fighters. The concept is not new, as the aerodynamic advantages of forward wing sweep were recognised during the second World War. They include improved manoeuvrability, with virtually spin-proof characteristics; better low-speed handling; and reduced stalling speeds. In addition, such aircraft have the advantage of lower drag across their entire operational envelope, particularly at speeds approaching Mach 1, which will permit the use of a smaller engine. However, the achievement of a suitable structure to reap these benefits was found to be impracticable at the 1940s state of the art.

With an FSW aircraft of conventional construction, when aerodynamic stresses flex the wing in flight, this increases the angle of attack (and hence the lift) of the outer wing sections. This, in turn, increases the air loads and causes further deformation of the wings; higher speeds will raise these forces until they eventually exceed the strength of the wing structure. To compensate for this divergence problem, forward swept wings of metal construction had to be stiffened to the point where a weight penalty was incurred, negating any aerodynamic benefit. Grumman appreciated that the advent of advanced composite materials offered a solution. Exceptionally strong and light in weight, an FSW of graphite composite material can be tailored to eliminate twisting when the wing bends.

Grumman's design has a thin supercritical wing with a variable-camber trailing-edge that changes the shape of the wing to match flight conditions, and includes a close-coupled foreplane to reduce supersonic trim drag. Flight control will be by digital fly-by-wire. Sufficient flexibility will be built into the programme to allow for the flight testing of other advanced concepts, relating to cockpits, exhaust nozzles, weapons carriage, and techniques to reduce further the take-off and landing speed of FSW aircraft.

DIMENSIONS, EXTERNAL:

Wing span	8·23 m (27 ft 0 in)
Length overall	14·63 m (48 ft 0 in)
Height overall	4·27 m (14 ft 0 in)

WEIGHTS (estimated):

Weight empty	5,534 kg (12,200 lb)
Max fuel weight	1,814 kg (4,000 lb)
Max T-O weight	7,348 kg (16,200 lb)

GULFSTREAM AMERICAN
GULFSTREAM AMERICAN CORPORATION

HEAD OFFICE AND WORKS: PO Box 2206, Savannah, Georgia 31402
Telephone: (912) 964 3281
Telex: 804 705 Gulfjet Sav

CHAIRMAN, PRESIDENT AND CHIEF EXECUTIVE OFFICER:
Allen E. Paulson

SENIOR VICE-PRESIDENTS:
James L. Bradbury (Finance, and Chief Financial Officer)
Albert H. Glenn (Operations)
Charles G. Vogeley (Commercial Jet Aircraft Marketing)

VICE-PRESIDENTS:
Charles N. Coppi (Engineering)
Richard Kemper (Quality Control; New Products)
E. B. Pinkston (Government Relations)

TREASURER: J. Lamont Crosby

SECRETARY AND GENERAL COUNSEL: John P. Innes

On 1 September 1978 it was announced that American Jet Industries had purchased the 80% holding in Grumman American Aviation Corporation held by Grumman Corporation, and had made a cash offer to the holders of the remaining 20% stock. The company subsequently became known as Gulfstream American Corporation. Under the terms of the transfer of ownership, Grumman is continuing to develop the Gulfstream III executive jet transport under contract with Gulfstream American. It was to receive a royalty on all sales made after 31 December 1979.

On 9 October 1980, Gulfstream American announced that it had signed a letter of intent to purchase Rockwell International's General Aviation Division. The transaction was completed on 3 February 1981, and the facility is now known as Gulfstream American Corporation's Commander Division.

Gulfstream American is currently producing the Gulfstream I-C and Gulfstream III; and is continuing development of the Hustler/Peregrine concept originated by the former American Jet Industries, now known as GACC (Gulfstream American Corporation of California, which see). Its Commander Division is producing a family of three turboprop business aircraft known as the Gulfstream Commander Jetprop 840, 980 and 1000.

GULFSTREAM AMERICAN G159C GULFSTREAM I-C

During the first half of 1979, Gulfstream American conducted marketing and engineering studies to determine the feasibility of putting into production a 32/38-seat commuter version of the twin-turboprop G-159 Gulfstream I, which Grumman Corporation designed as a 19-seat executive aircraft and marketed from 1958 to 1969 (see 1968-69 *Jane's*). A total of 200 Gulfstream Is had been built when production was then terminated in favour of the twin-turbofan Gulfstream II. Gulfstream American acquired the tooling and fixtures for Gulfstream I production in September 1978.

Design changes introduced on the Gulfstream I-C include lengthening the fuselage by 3·25 m (10 ft 8 in) to accommodate a maximum of 37 passengers, with room for a toilet and carry-on baggage compartment at the front of the cabin, and a baggage compartment at the rear.

Gulfstream American lengthened the fuselage of an existing Gulfstream I during the Summer of 1979, for flight testing and demonstration. This aircraft (N5400C), which flew for the first time in its modified form on 25 October 1979, retains its original Rolls-Royce Dart turboprop engines and was certificated by the FAA under FAR Pt 121 on 30 October 1980.

One Gulfstream I-C was delivered to Air North on 11 November 1980, with a second scheduled in early 1981. One due for delivery to Air US in mid-1981 has Collins Pro-Line avionics and an APS-80T autopilot.

TYPE: Twin-turboprop commuter transport.

WINGS: Cantilever low-wing monoplane. Thickness/chord ratio 10%. Dihedral 6° 30′. Incidence 3° at root. Sweepback on leading-edge 4° 15′. Conventional structure of light alloy. Light alloy ailerons and trailing-edge flaps. Trim tab in port aileron. Pneumatic de-icing boots on leading-edges.

FUSELAGE: Conventional semi-monocoque structure of light alloy, with fail-safe construction in the pressurised section between fore and aft pressure bulkheads.

TAIL UNIT: Cantilever structure of light alloy. Trim tab in rudder and each elevator. Pneumatic de-icing boots on leading-edges of fin and tailplane.

LANDING GEAR: Hydraulically-retractable tricycle type, with twin wheels on each unit. Steerable nosewheels. All units retract forward. Oleo-pneumatic shock-absorbers. Single-disc brakes. Decelostat anti-skid unit in each brake.

POWER PLANT: Two 1,484 kW (1,990 ehp) Rolls-Royce Dart Mk 529-8X turboprop engines, each driving a Dowty Rotol four-blade constant-speed metal propeller. One fuel tank and one water/methanol tank in each wing. Total usable fuel capacity 5,867 litres (1,550 US gallons); total water/methanol capacity 166 litres (44 US gallons). Refuelling points on wing upper surface. Oil capacity 26·5 litres (7 US gallons).

ACCOMMODATION: Crew of three, comprising pilot, co-pilot and flight attendant. Cabin provides maximum seating for 37 passengers, three-abreast at 76 cm (30 in) pitch, with a toilet and compartment for carry-on baggage at the forward end of the cabin, plus an additional baggage compartment aft. Door with built-in airstair at forward end of cabin on port side. Three emergency exits, two starboard, one port. Accommodation air-conditioned, heated and pressurised. Windscreen wipers standard.

SYSTEMS: Pressurisation system max differential 0·45 bars (6·55 lb/sq in). Single hydraulic system, pressure 103·5 bars (1,500 lb/sq in), for operation of landing gear, brakes, nosewheel steering, trailing-edge flaps, door, airstairs and windscreen wipers. Pneumatic system, pressure 134·5 bars (1,950 lb/sq in), for emergency extension of landing gear. Electrical system includes two DC generators, three AC inverters and two storage batteries. Oxygen system for emergency use. Engine bleed air used for anti-icing and de-icing systems. APU optional.

AVIONICS AND EQUIPMENT: A wide range of optional items is available to customer's requirements.

DIMENSIONS, EXTERNAL:

Wing span	23·88 m (78 ft 4 in)
Wing chord at root	3·40 m (11 ft 2 in)
Wing chord at tip	1·35 m (4 ft 5 in)
Wing aspect ratio	10
Length overall	22·96 m (75 ft 4 in)
Diameter of fuselage	2·39 m (7 ft 10 in)
Height overall	7·01 m (23 ft 0 in)
Tailplane span	7·77 m (25 ft 6 in)
Wheel track	7·37 m (24 ft 2 in)
Propeller diameter	3·51 m (11 ft 6 in)
Propeller ground clearance	0·48 m (1 ft 7 in)

First production Gulfstream American G159C Gulfstream I-C commuter transport, delivered to Air North

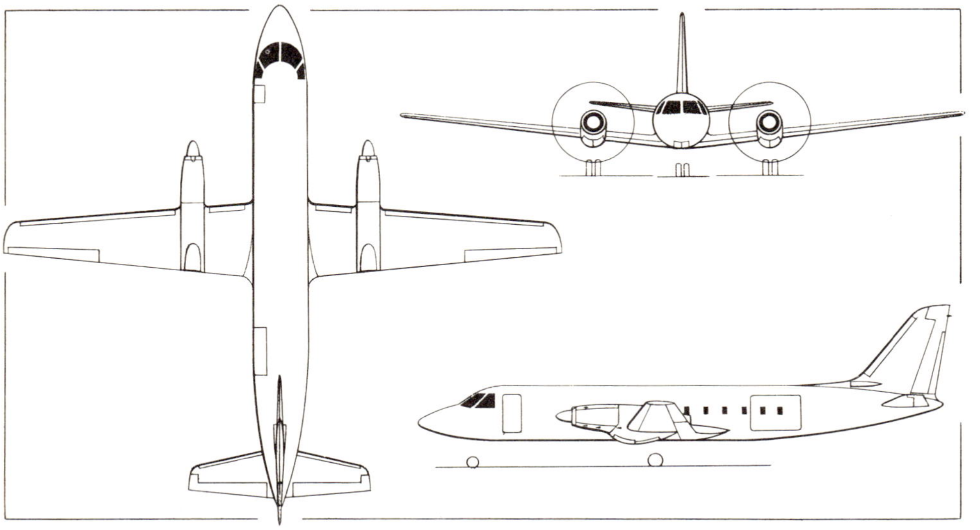

Gulfstream American G159C Gulfstream I-C commuter transport *(Pilot Press)*

Distance between propeller centres
7·37 m (24 ft 2 in)
Passenger door (fwd, port):
Height 1·57 m (5 ft 2 in)
Width 0·91 m (3 ft 0 in)
Baggage door (aft, stbd):
Height 0·99 m (3 ft 3 in)
Width 0·69 m (2 ft 3 in)

DIMENSIONS, INTERNAL:
Cabin (cockpit divider to aft baggage compartment, incl toilet and galley):
Length 13·21 m (43 ft 4 in)
Max width 2·24 m (7 ft 4 in)
Max height 1·85 m (6 ft 1 in)
Volume 42·76 m³ (1,510 cu ft)
Baggage hold, aft 4·08 m³ (144 cu ft)
Baggage hold, fwd 3·03 m³ (107 cu ft)

AREAS:
Wings, gross 56·7 m² (610·3 sq ft)
Ailerons (total) 3·44 m² (37·0 sq ft)
Trailing-edge flaps (total) 10·33 m² (111·2 sq ft)
Fin 8·21 m² (88·4 sq ft)
Rudder (incl tab) 2·67 m² (28·7 sq ft)
Tailplane 12·63 m² (136·0 sq ft)
Elevators (total, incl tabs) 3·25 m² (35·0 sq ft)

WEIGHTS AND LOADINGS:
Max payload (passengers) 3,356 kg (7,400 lb)
Max fuel weight 4,744 kg (10,460 lb)
Max T-O weight 16,329 kg (36,000 lb)
Max ramp weight 16,420 kg (36,200 lb)
Max zero-fuel weight 14,628 kg (32,250 lb)
Max landing weight 15,551 kg (34,285 lb)
Max wing loading 288·1 kg/m² (59 lb/sq ft)
Max power loading 5·50 kg/kW (9·04 lb/ehp)

PERFORMANCE (at max T-O weight):
Never-exceed speed 342 knots (634 km/h; 394 mph)
Max cruising speed 300 knots (555 km/h; 345 mph)
Stalling speed, flaps down, power off
87 knots (161 km/h; 100 mph)
Max rate of climb at S/L 580 m (1,900 ft)/min
Rate of climb at S/L, one engine out
165 m (540 ft)/min
Service ceiling 9,145 m (30,000 ft)
Service ceiling, one engine out 3,660 m (12,000 ft)
T-O to 15 m (50 ft) 1,480 m (4,850 ft)
Landing from 15 m (50 ft) 1,385 m (4,540 ft)
Range with max payload, IFR reserves
434 nm (804 km; 500 miles)
Range with max fuel, no reserves
2,171 nm (4,023 km; 2,500 miles)

GULFSTREAM AMERICAN GULFSTREAM II-B

The first production Gulfstream II (no prototype was built) was flown by Grumman on 2 October 1966. When production ended during 1979 a total of 256 had been manufactured, and details of this aircraft can be found in the 1979-80 and earlier editions of *Jane's*.

Following resumption of the Gulfstream III programme by Gulfstream American in early 1978, the company announced in the Autumn of that year that the Gulfstream III wing was to be offered as a retrofit for existing Gulfstream IIs. It was not until a year later, following certification of the Gulfstream III on 22 September 1980, that work was started to adapt the Gulfstream III wing for this retrofit application. It involved also structural modifications to permit increases in T-O weight, operating speeds and cruising heights.

Gulfstream II c/n 70, the property of Southland Corporation, Dallas, Texas, was used for the prototype conversion. The new wing has 27° 40′ of sweepback at quarter-chord, and incorporates winglets at the wingtips. Designated Gulfstream II-B in this new configuration, the prototype conversion (N711SC) was flown for the first time on 17 March 1981. This wing is expected to increase the range of the Gulfstream II by about 900 nm (1,668 km; 1,036 miles), and simultaneously to provide a significant improvement in fuel efficiency.

During the certification programme of the II-B, which was expected to be completed by the late Summer of 1981, Gulfstream American was endeavouring to gain certification at the same time for a modified Sperry SP-50 AFCS installation in this aircraft.

In April 1981 the company had concluded contracts for the conversion of 11 Gulfstream IIs to the II-B configuration, and nine more were under negotiation.

WEIGHTS:
Basic operating weight empty 17,735 kg (39,100 lb)
Max fuel weight 12,655 kg (27,900 lb)
Payload with max fuel 771 kg (1,700 lb)
Max ramp weight 31,160 kg (68,700 lb)
Max landing weight 26,535 kg (58,500 lb)
Max zero-fuel weight 19,050 kg (42,000 lb)

PERFORMANCE (estimated):
Cruising Mach number at max power at 9,145 m (30,000 ft), AUW of 24,947 kg (55,000 lb) 0·85
Typical cruising Mach number 0·775
FAA T-O distance 1,783 m (5,850 ft)
FAA landing distance 1,036 m (3,400 ft)
Range with NBAA IFR reserves
3,660 nm (6,782 km; 4,214 miles)

GULFSTREAM AMERICAN GULFSTREAM III

Grumman (now Gulfstream) American announced the resumption of the Gulfstream III programme in the Spring of 1978. It differs from the Gulfstream II primarily in having redesigned wings, with winglets at the tips, a lengthened fuselage, and increased fuel capacity combined with an increase of some 18 per cent in fuel economy and efficiency.

A prototype (N901G), converted from a production-line Gulfstream II (c/n 249), was rolled out on 21 September 1979, and made its first flight on 2 December 1979. The second prototype Gulfstream III (c/n 252) followed on 24 December 1979. FAA certification was received on 22 September 1980. Production deliveries by mid-1981 totalled 30 aircraft, and it was planned to deliver 13 more during 1981. Orders totalled more than 90 at that time, including one for the Ivory Coast Air Force. Other customers include the Royal Danish Air Force, which will use three (with options on five more) for fishing surveillance, search and rescue and personnel transport duties. This special-purpose maritime version is described separately.

The prototype Gulfstream III flew nonstop from Savannah, Georgia, to Hanover, West Germany, in April 1980, to participate in the Hanover Air Show. The 3,967·37 nm (7,347·58 km; 4,565·57 mile) flight, at an average speed of 449·00 knots (831·54 km/h; 516·70 mph), set a class C1k record for distance in a straight line and a speed record over this recognised course. On the return journey, the aircraft set further course records from Geneva, Switzerland, to Boston, New York and Washington, USA. It also set, on 3 May 1980, a class C1k record of 15,849·6 m (52,000 ft) for sustained altitude in horizontal flight.

On 15/16 February 1981, a production aircraft with 10

Prototype Gulfstream American Gulfstream II-B, incorporating the wing developed for the Gulfstream III

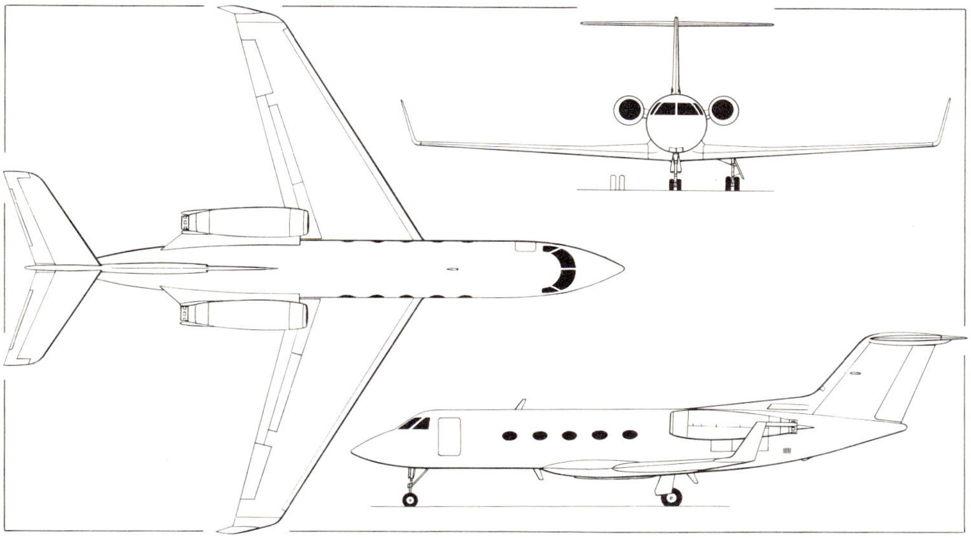

Gulfstream American Gulfstream III executive transport aircraft *(Pilot Press)*

persons on board completed the 4,152 nm (7,690 km; 4,778 mile) flight from Kona Airport, Hawaii, to Dulles International Airport, Washington, DC, in 8 h 50 min and this has been accepted by the FAI as a new world record in Class C1k, exceeding the earlier Savannah-Hanover distance.

The following details refer to the standard Gulfstream III transport:

Type: Twin-turbofan executive transport.

Wings: Cantilever low-wing monoplane of light alloy construction. Dihedral 3°. Incidence 3° 30′ at wing station 50, 1° 30′ at wing station 145, and −0° 30′ at wing station 414. Sweepback 27° 40′ at quarter-chord. By comparison with Gulfstream II, wing has extended-chord leading-edges (0·76 m; 2 ft 6 in at root, 0·13 m; 5 in at tip). NASA (Whitcomb) wingtip winglets. One-piece single-slotted Fowler-type trailing-edge flaps. Spoilers forward of flaps assist in lateral control and can be extended for use as airbrakes. All control surfaces actuated hydraulically. Trim tab in port aileron. Anti-icing by engine bleed air.

Fuselage: Conventional semi-monocoque structure of light alloy. Glassfibre nosecone hinged for access.

Tail Unit: Cantilever T-tail structure of light alloy, with swept horizontal and vertical surfaces. Trim tab in rudder and each elevator. Powered controls.

Landing Gear: Retractable tricycle type, with twin wheels on each unit. Inward-retracting main units; steerable nose unit retracts forward. Main-wheel tyres size 34 × 9·25-16, pressure 10·34 bars (150 lb/sq in). Nosewheel tyres size 21 × 7·25-10, pressure 6·55 bars (95 lb/sq in). Goodyear aircooled carbon brakes, with Goodyear fully-modulating anti-skid units.

Power Plant: Two Rolls-Royce Spey Mk 511-8 turbofan engines, each 50·7 kN (11,400 lb st), pod-mounted on sides of rear fuselage. Rohr target-type thrust reverser forms aft part of each nacelle when in stowed position. All fuel in integral tanks in wings, with total capacity of 16,656 litres (4,400 US gallons).

Accommodation: Crew of two or three. Standard seating for 19 passengers in pressurised and air-conditioned cabin. Large baggage compartment at rear of cabin, capacity 907 kg (2,000 lb). Integral airstair door at front of cabin on port side. Electrically-heated wraparound windscreen.

Systems: Cabin pressurisation system with max differential of 0·65 bars (9·45 lb/sq in). Two independent hydraulic systems, each 103·5 bars (1,500 lb/sq in). All flying controls hydraulically powered, with manual reversion. APU in tail compartment. Basic 28V DC electrical system, using two 300A generators and four 200A transformer-rectifiers. Two 20kVA alternators provide AC power for secondary and auxiliary systems. Third (APU-driven) 20kVA alternator for on-ground power. Three 2·5kVA inverters, powered by the transformer-rectifiers, provide 400Hz fixed-frequency power. Two 24V batteries.

Avionics and Equipment: Standard avionics include three Collins VHF com; two Collins VIR-31B VHF nav; two Collins ADF-60A; two Collins transponders; two Collins DME-40; cockpit voice recorders, Sperry SPZ-800 automatic flight guidance and control system; and Collins WXR-250A, or WXR-300; Bendix RDR-1200; or RCA Primus 40 WXD, or Primus 400 weather radar.

Dimensions, external:

Wing span	23·72 m (77 ft 10 in)
Wing area, gross	86·83 m² (934·6 sq ft)
Length overall	25·32 m (83 ft 1 in)
Fuselage length	22·66 m (74 ft 4 in)
Height overall	7·43 m (24 ft 4½ in)
Passenger door (fwd, port):	
Height	1·57 m (5 ft 2 in)
Width	0·91 m (3 ft 0 in)
Baggage door (aft): Height	0·72 m (2 ft 4½ in)
Width	0·91 m (2 ft 11¾ in)

Dimensions, internal:

Cabin: Length	12·60 m (41 ft 4 in)
Width	2·24 m (7 ft 4 in)
Height	1·85 m (6 ft 1 in)
Volume	42·53 m³ (1,502 cu ft)
Aft baggage compartment volume	4·44 m³ (157 cu ft)

Weights:

Manufacturer's bare weight	14,742 kg (32,500 lb)
Typical operating weight empty	17,372 kg (38,300 lb)
Max fuel load	12,825 kg (28,275 lb)
Typical payload	726 kg (1,600 lb)
Max T-O weight	30,935 kg (68,200 lb)
Max ramp weight	31,162 kg (68,700 lb)
Max zero-fuel weight	19,050 kg (42,000 lb)
Max landing weight	26,535 kg (58,500 lb)

Performance:

Max cruising speed	Mach 0·85
Long-range cruising speed	Mach 0·775
Approach speed	135 knots (250 km/h; 156 mph)
Stalling speed at max landing weight	105 knots (195 km/h; 121 mph)
Max operating altitude	13,720 m (45,000 ft)
FAA balanced T-O field length	1,738 m (5,700 ft)
FAA landing distance	1,040 m (3,400 ft)
NBAA range, IFR reserves	3,650 nm (6,760 km; 4,200 miles)

Operational Noise Levels (FAR Pt 36):

T-O	91 EPNdB
Approach	97 EPNdB
Sideline	104 EPNdB

GULFSTREAM AMERICAN GULFSTREAM III (MARITIME VERSION)

The special needs posed by the Royal Danish Air Force's fishery patrols, covering some 160,000 nm² (549,500 km²; 212,155 sq miles) around Greenland, and 85,000 nm² (291,912 km²; 112,708 sq miles) around the Faröe Islands, had to take into consideration that, in the event of bad weather prohibiting a landing at either of these places, an 800 nm (1,482 km; 921 mile) flight to an alternative landing field might be necessary. In addition to this primary role, the chosen aircraft was required to be suitable for airdrop, medevac, SAR, tactical air transport, and other special duties. The RDAF has ordered three Gulfstream IIIs (with options on five more) to replace eight Douglas C-47s currently in use, and to supplement the activities of three Lockheed C-130s. All three were scheduled for delivery by March 1982. It is anticipated that one will be operated from Sondrestrom AB, Greenland, the other two from Vaerloese, near Copenhagen.

In the primary role, for fishery patrol, the RDAF will operate the Gulfstream III with a crew of seven, comprising pilot, co-pilot, flight engineer, navigator, observer, photographer and radio operator. The radio operator's station is on the port side of the cabin, immediately aft of the door, and equipment will include dual HF and dual VHF/UHF com. This equipment is removable easily, and the seat can be turned 90° to port, so that the position may be used alternatively by an observer. A purpose-built observer's position is provided on the starboard side, directly opposite the radio operator's station. The standard Gulfstream III windows are retained, as adequate field of view is provided without recourse to special-purpose drag-inducing bubble windows. Aft of the radio operator's position, and also on the port side of the cabin, is the navigator's station. This is equipped with the master console for the Texas Instruments AN/APS-127 sea surveillance radar, dual control display units for the Litton 72R INS, a VHF navigation system, and basic flight instrumentation including rate of climb and airspeed indicators. Provision has been made for later installation of a VHF/Omega nav system if this is considered necessary.

The RDAF Gulfstream III has some structural differences to provide the essential multi-role capability. These include a 1·60 × 2·11 m (63 × 83 in) cargo door on the starboard side of the fuselage, forward of the wing; a cargo roller conveyor system in the aft cabin floor; an overhead cable system for the attachment of drop load parachute lanyards; and a flare launcher system in the aft fuselage, on the port side just aft of the wing trailing-edge. This last feature will permit the launch of a variety of pyrotechnic and/or signalling devices, including parachute flares as large as the LAU-2B. The existing 0·72 × 0·91 m (28·5 × 35·75 in) baggage door on the port side of the fuselage, aft of the wing, is to be used for the airdrop of emergency supplies and/or survival equipment, and a hydraulically-actuated air deflector will be mounted just forward of this door, opening automatically whenever the airdrop door is opened. It has not proved necessary to change the size of this baggage door, since the largest airdrop load for deployment by the RDAF will not exceed 0·60 × 0·60 × 0·90 m (23·5 × 23·5 × 35·4 in). The installation of AN/APS-127 radar antenna in the fuselage nose represents little more than an alternative to the weather radar, but substitution of the radar display for the weather radar display in the instrument panel has necessitated some

Gulfstream American Gulfstream III twin-turbofan executive transport aircraft

movement of the surrounding instruments and the inclusion of four sub-panels below the main pilot and co-pilot panels. Standby airspeed, altitude and attitude indicators have been installed on the glareshield panel, and other non-standard instrumentation includes duplicated Sperry AD650 Series attitude and RD650 horizontal situation indicators, SPZ-800 autopilot, and a Teledyne angle of attack indicator.

Performance figures will be very similar to those of the commercial Gulfstream III: max cargo weight will be 2,064 kg (4,551 lb) with reduced fuel load, or 848 kg (1,870 lb) with max fuel load. With a crew of three, max fuel, max ramp weight of 31,162 kg (68,700 lb), payload of 726 kg (1,600 lb) and NBAA VFR reserves, RDAF Gulfstream IIIs will have an estimated range of 4,205 nm (7,787 km; 4,839 miles) at a cruising speed of Mach 0·775.

GULFSTREAM AMERICAN SUPER AG-CAT

The prototype of the original Ag-Cat agricultural biplane flew for the first time on 27 May 1957. Series production was entrusted to Schweizer, under subcontract from Grumman. First deliveries were made in 1959, and more than 2,300 Ag-Cats (including 2,250 of the A and B models) had been built by 1 January 1980. Manufacturing then switched progressively from Elmira, NY, to Gulfstream American's plant at Savannah. During 1980, however, Gulfstream American Corporation sold the entire range of Ag-Cat agricultural aircraft to Schweizer Aircraft Corporation (which see), and production was resumed at Elmira during 1981.

GULFSTREAM AMERICAN CORPORATION, COMMANDER DIVISION

Executive Offices: 5001 North Rockwell Avenue, Bethany, Oklahoma 73008
Telephone: (405) 789 5000
Telex: 747193
President: Allen E. Paulson
Vice-President, Operations: Bill M. Humes
Division Executives:
Alvin F. Balaban (Manager, Communications)
Suresh Chandra (Director, Quality Assurance)
Jim Cobb (Director, Personnel)
Robert H. Cooper (Vice-President, Marketing)
Kenneth L. Hale (Director, Engineering)
Thomas R. Merry (Vice-President, Finance and Administration)
James M. O'Brien (Manager, Product Support)

Gulfstream American's Commander Division designs, manufactures, markets and supports twin-engined turboprop business aircraft. Its operations incorporate a worldwide network of individually contracted regional and district sales centres, and factory supervised product support facilities known as ServiCenters.

Since acquiring Rockwell International's General Aviation Division, Gulfstream American has ended production of the Shrike Commander 500S: details of this aircraft can be found in the 1980-81 *Jane's*.

GULFSTREAM COMMANDER JETPROP 840

The Commander Jetprop 840 is a pressurised business transport aircraft, powered by two Garrett turboprop engines. The prototype (c/n 11600) flew for the first time on 17 May 1979, and the first production aircraft (c/n 11601) on 23 October 1979. FAA certificated on 7 September 1979, the 840 has been in continuous production since that date, and 43 had been delivered by 1 January 1981. It was planned to produce 46 more during 1981.

The Commander Jetprop series introduced an aerodynamically refined wing incorporating winglets, and 'supercritical' propellers designed and developed by Dowty Rotol.

Type: Twin-engined light transport.

Wings: Cantilever high-wing monoplane. Wing section NACA 23012 (modified). Dihedral 4°. Incidence 3° at root, −0° 31′ at tip. Wing swept forward 3° 37′ at quarter-chord. All-metal two-spar flush-riveted structure of 2024 and 2014 light alloy. Shallow winglet above each wingtip. Frise statically-balanced all-metal ailerons. Hydraulically-operated all-metal slotted flaps. Electrically-operated trim tab in starboard aileron. Goodyear pneumatic de-icing boots standard, inboard and outboard of nacelles.

Fuselage: All-metal semi-monocoque structure of 2024 and 2014 light alloy with flush-riveted skin. Structure pressurised between forward bulkhead of flight deck, and aft bulkhead of cabin, under wing.

Tail Unit: Cantilever all-metal structure with 10° dihedral on fixed-incidence tailplane. Horn-balanced rudder; statically-balanced elevators. Trim tabs in each elevator and rudder. Electrically-actuated elevator trim when Collins or King autopilot installed. Goodyear pneumatic de-icing boots on fin and tailplane leading-edges. Electrical heaters on rudder horn slot, and leading-edge of rudder tab horn balance.

Landing Gear: Retractable tricycle type, with single wheel on each unit. All wheels retract rearward hydraulically, main wheels turning through 90° to stow horizontally in nacelles; wheels enclosed by doors when retracted. Oleo-pneumatic shock-absorbers manufactured by Ozone Aircraft Systems. Hydraulically-steerable nosewheel. Goodrich or Goodyear main-wheel tubeless tyres size 25·65 × 8·70-6·25, 10-ply rating, pressure 4·83 bars (70 lb/sq in). Goodyear nose-wheel tubed tyre size 14·90 × 6·14-4·96, 6-ply rating, pressure 4·69 bars (68 lb/sq in). Aircooled brakes of Goodyear multi-disc type, or Cleveland single-disc caliper type.

Power Plant: Two 626 kW (840 shp) Garrett TPE331-5-254K turboprop engines, each flat rated to a maximum of 535 kW (717·5 shp), and each driving a Dowty Rotol (C)R306/3-82-F/7-(C)VP2926 three-blade 'supercritical' constant-speed fully-feathering and reversible-pitch propeller. Propeller synchrophasers standard. Five bladder cells and one integral tank in each wing, inboard and outboard of nacelle respectively, with total usable capacity of 1,609 litres (425 US gallons). Optional system with usable capacity of 1,794 litres (474 US gallons). Refuelling points on upper surface of each wing. Oil capacity 21·2 litres (5·6 US gallons). Manually controlled oil temperature control doors on inboard side of each engine nacelle. Electrical propeller de-icing; engine intakes de-iced by engine bleed air.

Accommodation: Standard seating for pilot and seven passengers, on two adjustable seats on flight deck, a side-facing seat/toilet with privacy curtain aft of the co-pilot's seat, two aft-facing single seats and a three-place forward-facing bench seat in main cabin. Dual controls and co-pilot's instrumentation standard. A variety of optional seating layouts offer accommodation for a pilot and up to 10 passengers. Cabin pressurised, heated, ventilated and air-conditioned. Forward-hinged outward-opening cabin door on port side, with retractable cabin step. Plug-type pull-in emergency exit on starboard side of fuselage. Baggage compartment of 1·98 m³ (70 cu ft) with 272 kg (600 lb) capacity aft of rear pressure bulkhead with external door on port side of fuselage. Windscreen demisting and electrical windscreen anti-icing. Two-speed windscreen wipers.

Systems: Sundstrand bleed air air-conditioning, environmental control and pressurisation system, with normal differential of 0·36 bars (5·2 lb/sq in). Single hydraulic system with two engine-driven pumps, an emergency reservoir and an electrically-operated auxiliary pump; normal pressure between 62 and 74 bars (900 and 1,075 lb/sq in), with pressure relief valve set at 86 bars (1,250 lb/sq in), ±3·5 bars (50 lb/sq in); for operation of landing gear, brakes, nosewheel steering and trailing-edge flaps. Electrical system includes two 30V 300A starter/generators, and two 24V 48Ah lead-acid batteries. Emergency oxygen system of 0·6 m³ (22 cu ft) capacity, with automatic drop-down constant-flow outlets at each passenger position, and diluter-demand systems at crew stations. Pneumatic system using engine bleed air, for operation of wing and tail unit de-icing boots, cabin door seal, environmental control unit, and co-pilot's gyro horizon. Anti-icing and de-icing system includes bleed air heat for engine inlets, and electrical heaters for generator inlets, pitot heads, rudder horn slot, rudder tab horn, propeller boots, pitot heads, static ports, and windscreens.

Avionics and Equipment: Standard choice of Collins or King avionics package. Collins package includes two VHF-20A VHF com transceivers, No. 1 with dual heads; VIR-30 AGM nav receiver with GS and marker beacon indication; VIR-30AG nav with GS; FD-112V flight director with AP-106 autopilot; 331A-3F pictorial nav indicator (HSI) on co-pilot's panel; two Sperry C-14A slaved compasses; AAR-3137 RMI with VOR/ADF switching; DME-40 DME with IND-40A indicator; TDR-90 transponder; ALT-55B radio altimeter with DRI-55 indicator; ADF-60 ADF; Bendix RDR-1100 weather radar; and IDC 519-28704 series encoding altimeter. King package includes two KTR 905 VHF com transceivers, No. 1 with dual heads; two KNR 630 nav/GS receivers; KMR 675 marker beacon receiver; KFC 300 autopilot/flight director; AAR-3137 RMI with VOR/ADF switching; KDM 706 DME with KDI 572 indicator; KXP 755 transponder; ALT-55B radio altimeter with DRI-55 indicator; KDF 805 ADF; Bendix RDR-1100 weather radar; and Collins 331A-3F HSI for starboard instrument panel. Each package is accompanied by a basic installation kit which includes two Electret microphones, two Telex A610-1 headsets, two speakers, two 250VA inverters, static discharge wicks, and all associated equipment and wiring. A wide range of other optional items of avionics is available. Standard equipment includes aircraft hour meter, gyro pressure gauge, outside air temperature gauge, dual-intensity instrument panel floodlights, instrument lighting system with solid-state dimming, stall warning device, control lock, removable inner panes for crew side and eyebrow windows, stowable table, storage/refreshment console, individual reading lights, cabin lights, baggage compartment light, cabin fire extinguisher, tinted cabin windows, 'No smoking—fasten seat belt' sign, flashing beacon, landing and recognition lights, navigation lights, rotating beacon, taxi lights, Whelan three-light strobe system, external power socket, wing ice lights, fuselage ice impact shields and rudder gust lock. A wide range of optional cabin equipment, furnishings and seating is available to customer requirements.

Dimensions, external:

Wing span	15·89 m (52 ft 1½ in)
Wing chord at root	2·39 m (7 ft 10 in)
Wing chord at tip	0·69 m (2 ft 3 in)
Wing aspect ratio	9·772
Length overall	13·10 m (42 ft 11¾ in)
Fuselage height (constant section)	1·70 m (5 ft 7 in)
Fuselage width (constant section)	1·42 m (4 ft 8 in)
Height overall	4·55 m (14 ft 11½ in)
Tailplane span	6·03 m (19 ft 9¼ in)
Wheel track	4·70 m (15 ft 5 in)
Wheelbase	5·38 m (17 ft 7¾ in)
Propeller diameter	2·69 m (8 ft 10 in)
Propeller ground clearance	0·32 m (1 ft 0¾ in)
Distance between propeller centres	4·65 m (15 ft 3 in)
Cabin door: Height	1·19 m (3 ft 11 in)
Width	0·67 m (2 ft 2½ in)
Height to sill	0·47 m (1 ft 6½ in)
Baggage door: Height	0·79 m (2 ft 7¼ in)
Width	0·64 m (2 ft 1 in)
Height to sill	0·52 m (1 ft 8½ in)
Emergency exit: Height	0·48 m (1 ft 7 in)
Width	0·66 m (2 ft 2 in)

Dimensions, internal:

Cabin: Length, incl flight deck	4·34 m (14 ft 3 in)
Max width	1·26 m (4 ft 1½ in)
Max height	1·37 m (4 ft 5¾ in)
Floor area (incl flight deck)	5·22 m² (56·2 sq ft)
Volume (incl flight deck)	6·34 m³ (224 cu ft)
Baggage hold volume	1·98 m³ (70 cu ft)

Areas:

Wings, gross	25·95 m² (279·37 sq ft)
Ailerons (total)	1·90 m² (20·52 sq ft)
Trailing-edge flaps (total)	1·93 m² (20·80 sq ft)
Fin	2·50 m² (26·91 sq ft)
Rudder (incl tab)	1·59 m² (17·12 sq ft)
Tailplane	3·51 m² (37·83 sq ft)
Elevators (incl tab)	1·91 m² (20·54 sq ft)

Weights and Loadings:

Weight empty, equipped	3,007 kg (6,629 lb)
Max payload	832 kg (1,834 lb)
Max usable fuel weight:	
standard	1,292 kg (2,848 lb)
optional	1,441 kg (3,176 lb)
Max T-O weight	4,683 kg (10,325 lb)
Max ramp weight	4,706 kg (10,375 lb)
Max zero-fuel weight	3,839 kg (8,463 lb)
Max landing weight	4,389 kg (9,675 lb)
Max wing loading	180·5 kg/m² (36·96 lb/sq ft)
Max power loading	4·38 kg/kW (7·20 lb/shp)

Performance (at max T-O weight, except where indicated):

Never-exceed speed	Mach 0·52 (243 knots; 450 km/h; 280 mph CAS)
Max level and max cruising speed at 3,660 m (12,000 ft)	290 knots (537 km/h; 333 mph)
Econ cruising speed at 9,450 m (31,000 ft)	248 knots (460 km/h; 285 mph)
Stalling speed, flaps and landing gear up, engines idling	77 knots (143 km/h; 89 mph) CAS
Max rate of climb at S/L	860 m (2,824 ft)/min
Rate of climb at S/L, one engine out	306 m (1,003 ft)/min
Service ceiling	10,365 m (34,000 ft)
Service ceiling, one engine out	6,400 m (21,000 ft)
T-O run	392 m (1,285 ft)
T-O to 15 m (50 ft)	559 m (1,833 ft)
Landing from 15 m (50 ft), no propeller reversal	711 m (2,332 ft)
Landing run, no propeller reversal	472 m (1,550 ft)
Range with max payload, 45 min reserves	1,071 nm (1,985 km; 1,233 miles)
Range with standard fuel, 45 min reserves	1,780 nm (3,298 km; 2,050 miles)
Range with max optional fuel, 45 min reserves	2,040 nm (3,780 km; 2,349 miles)

GULFSTREAM COMMANDER JETPROP 980

The Commander Jetprop 980 differs from the 840 version primarily by having more powerful (730 kW; 980 shp) Garrett turboprop engines. The prototype (c/n 95000) flew for the first time on 14 June 1979, and the first production aircraft (c/n 95001) on 11 February 1980, following certification by the FAA on 1 November 1979. A total of 51 had been delivered by 1 January 1981; a further 55 were scheduled for delivery during 1981.

The description of the Commander Jetprop 840 applies also to this version, except as follows:

Power Plant: Two 730 kW (980 shp) Garrett TPE331-10-501K turboprop engines, each flat rated to 547 kW

Gulfstream American Corporation's Commander Jetprop 980 twin-engined light transport

(733 shp). Six bladder cells and one integral tank in each wing, with total usable capacity of 1,794 litres (474 US gallons).

WEIGHTS AND LOADING:

Weight empty, equipped	3,051 kg (6,727 lb)
Max payload	787 kg (1,736 lb)
Max power loading	4·28 kg/kW (7·04 lb/shp)

PERFORMANCE (at max T-O weight, except where indicated):

Never-exceed speed
Mach 0·52 (243 knots; 450 km/h; 280 mph CAS)
Max level and max cruising speed at 6,700 m (22,000 ft) 309 knots (573 km/h; 356 mph)
Econ cruising speed at 9,450 m (31,000 ft) 249 knots (461 km/h; 287 mph)
Stalling speed, flaps and landing gear up, engines idling 77 knots (143 km/h; 89 mph) CAS

Max rate of climb at S/L	846 m (2,777 ft)/min
Rate of climb at S/L, one engine out	299 m (982 ft)/min
Service ceiling	11,390 m (37,370 ft)
Service ceiling, one engine out	7,575 m (24,850 ft)
T-O run	396 m (1,299 ft)
T-O to 15 m (50 ft)	565 m (1,854 ft)
Landing from 15 m (50 ft), no propeller reversal	704 m (2,310 ft)
Landing run	472 m (1,549 ft)

Range with max payload, 45 min reserves
1,021 nm (1,892 km; 1,175 miles)
Range with max fuel, 45 min reserves
2,040 nm (3,780 km; 2,349 miles)

GULFSTREAM COMMANDER JETPROP 1000

This version of the Commander Jetprop range has overall dimensions which are unchanged from the 840 and 980, but has increased cabin volume as the result of a rearward shift of 0·92 m (3 ft 0 in) of the aft pressure bulkhead. Construction of the prototype (c/n 96000) began in October 1979, and this aircraft flew for the first time on 12 May 1980. FAA certification was gained on 30 April 1981, and deliveries began in July 1981. It was planned to produce 25 of these aircraft during 1981. An additional large-cabin model, the Jetprop 900, is planned for 1982.

The details of the Commander Jetprop 840/980 apply also to the Commander Jetprop 1000, except as follows:

LANDING GEAR: Goodrich or Goodyear main-wheel tubeless tyres size 24·15 × 7·65-5·50, 16-ply rating, pressure 6·21 bars (90 lb/sq in).

POWER PLANT: As for Commander Jetprop 980, except that the Garrett TPE331-10-501K engines are flat rated to a maximum of 611 kW (820 shp).

ACCOMMODATION: Generally as for Commander Jetprop 840/980, but with amended standard layout for a pilot and seven passengers, comprising two adjustable seats on flight deck, a single aft-facing seat behind co-pilot's position, two aft-facing single seats, two forward-facing single seats, and a combined seat/toilet with privacy curtain at the rear of the cabin. Maximum seating for 11 passengers. Extra cabin windows are provided on each side. Baggage area at rear of cabin to take 45·4 kg (100 lb), and baggage compartment aft of cabin rear pressure bulkhead, capacity 272 kg (600 lb).

SYSTEMS: Generally as for Commander Jetprop 840/980, except normal maximum cabin pressure differential of 0·46 bars (6·7 lb/sq in).

DIMENSIONS, EXTERNAL:

Baggage door: Height	0·79 m (2 ft 7 in)
Width	0·56 m (1 ft 10 in)
Height to sill	0·65 m (2 ft 1½ in)

DIMENSIONS, INTERNAL:

Cabin: Length (incl flight deck and toilet)	5·26 m (17 ft 3 in)
Max width	1·26 m (4 ft 1½ in)
Max height	1·45 m (4 ft 9¼ in)
Floor area (incl flight deck and toilet)	6·20 m^2 (66·7 sq ft)
Volume (incl flight deck and toilet)	8·44 m^3 (298 cu ft)
Baggage hold volume	1·30 m^3 (45·8 cu ft)

WEIGHTS AND LOADINGS:

Weight empty, equipped	3,183 kg (7,018 lb)
Max payload	899 kg (1,982 lb)
Max fuel weight	1,441 kg (3,176 lb)
Max T-O weight	5,080 kg (11,200 lb)
Max ramp weight	5,103 kg (11,250 lb)
Max zero-fuel weight	4,082 kg (9,000 lb)
Max landing weight	4,785 kg (10,550 lb)
Max wing loading	195·7 kg/m^2 (40·09 lb/sq ft)
Max power loading	4·16 kg/kW (6·83 lb/shp)

PERFORMANCE (at max T-O weight):

Never-exceed speed
Mach 0·6 (252 knots; 467 km/h; 290 mph CAS)
Max level and max cruising speed at 6,700 m (22,000 ft) 308 knots (571 km/h; 355 mph)
Econ cruising speed at 10,670 m (35,000 ft) 257 knots (476 km/h; 296 mph)
Stalling speed, flaps and landing gear up, engines idling 81 knots (150 km/h; 93·5 mph) CAS

Max rate of climb at S/L	855 m (2,802 ft)/min
Rate of climb at S/L, one engine out	283 m (929 ft)/min
Service ceiling	10,670 m (35,000 ft)
Service ceiling, one engine out	6,400 m (21,000 ft)
T-O run	429 m (1,407 ft)
T-O to 15 m (50 ft)	650 m (2,131 ft)
Landing from 15 m (50 ft)	814 m (2,670 ft)
Landing run	555 m (1,821 ft)

Range with max payload, 45 min reserves
1,311 nm (2,429 km; 1,510 miles)
Range with max fuel, 45 min reserves
2,080 nm (3,855 km; 2,395 miles)

Gulfstream American's increased-capacity Commander Jetprop 1000 light transport

HAMILTON
HAMILTON AVIATION

HEAD OFFICE: PO Box 11746, Tucson, Arizona 85734
Telephone: (602) 294 3481
PRESIDENT: Gordon B. Hamilton
VICE-PRESIDENT: Gordon D. Hamilton
CHIEF ENGINEER: Clayton C. Hamilton

Hamilton Aviation, which is engaged primarily in the procurement and overhaul of various types of military aircraft for foreign governments, is also building and marketing two Westwind turboprop conversions of the Beech Model 18 (last described in the 1969-70 *Jane's*).

HAMILTON WESTWIND III

The prototype Westwind III passenger/cargo aircraft flew for the first time in 1963, and FAA certification under CAR Part 3 was awarded in 1964. Passenger seats can be removed easily to make the whole cabin space available for cargo.

The Westwind III can have agricultural or military applications.

TYPE: Utility passenger/cargo commuter airliner.

WINGS, FUSELAGE, TAIL UNIT: As for Beech Model 18.

LANDING GEAR: Electrically-retractable tailwheel type. Oleo-pneumatic shock-absorbers in all units. Main wheels retract aft. Main-wheel tyre pressure 4·14 bars (60 lb/sq in). Tailwheel tyre pressure 5·52 bars (80 lb/sq in). Goodyear single-disc multi-puck hydraulic brakes. Main wheels fully enclosed by doors when retracted.

POWER PLANT: Two 432 kW (579 ehp) Pratt & Whitney Aircraft of Canada PT6A-20 turboprop engines standard. Optional power plants include two 533 kW (715 ehp) (derated to 470 kW; 630 ehp) PT6A-27s, two 533 kW (715 ehp) (derated to 470 kW; 630 ehp) PT6A-28s or two 455 kW (610 ehp) Avco Lycoming LTP 101s. Hartzell three-blade fully-feathering constant-speed metal propellers. Standard fuel capacity 1,544 litres (408 US gallons), contained in outer wing, inner wing and centre-section tanks. Optional total fuel in larger-capacity tanks 2,801 litres (740 US gallons). Refuelling points on wing upper surface. Engine air intakes have an inertial separator system, including foreign object and hail bypass and heating of the leading-edges by engine bleed air.

ACCOMMODATION: Pilot and co-pilot on flight deck, with cabin seating eight passengers. Door on port side, aft of wing, with built-in airstair, can be replaced by larger cargo door. Separate door to flight deck, on port side of fuselage, is optional. Emergency exit (push-out type) on starboard side of cabin. Passenger seating quickly removable for conversion to all-cargo role. Baggage or cargo space aft of cabin and in extended fuselage nose. Cabin is heated by bleed air, and can be cooled by a bleed air converter. Windscreen de-icing standard.

SYSTEMS: Cabin cooling by Garrett engine bleed air converter. Cabin heater manufactured by Hamilton Aviation. Pneumatic system, for flight instruments and wing and tail unit de-icing, supplied by engine bleed air. Electrical system powered by two 200A starter/generators and nickel-cadmium battery. Oxygen system optional.

AVIONICS AND EQUIPMENT: Radio com/nav and radar to customer's requirements. Blind-flying instrumentation standard.

ARMAMENT: Optional armament for military versions includes a cargo pod containing two General Electric Miniguns, and hardpoints on the wings for the carriage of bombs or rockets.

DIMENSIONS, EXTERNAL:

Wing span	14·02 m (46 ft 0 in)
Wing chord at root	4·19 m (13 ft 9 in)
Wing chord at tip	1·07 m (3 ft 6 in)
Wing aspect ratio	6·5
Length overall	10·85 m (35 ft 7¼ in)
Tailplane span	4·56 m (14 ft 11½ in)
Passenger door (port, aft): Height	1·22 m (4 ft 0 in)
Width	0·56 m (1 ft 10 in)
Height to sill	0·79 m (2 ft 7 in)

Cargo door (port, aft, optional):
Max height (forward edge) 1·52 m (4 ft 11¾ in)
Min height (rear edge) 1·19 m (3 ft 11 in)
Width 1·47 m (4 ft 9¾ in)
Height to sill 0·79 m (2 ft 7 in)
Emergency exit (stbd): Height 0·64 m (2 ft 1 in)
Width 0·48 m (1 ft 7 in)

Dimensions, internal:
Cabin (bare cargo configuration):
Length 4·57 m (15 ft 0 in)
Max width 1·32 m (4 ft 4 in)
Max height 1·55 m (5 ft 1 in)
Floor area 7·80 m² (84 sq ft)
Baggage/cargo hold (aft cabin) 0·85 m³ (30 cu ft)
Baggage/cargo hold (fuselage nose) 1·53 m³ (54 cu ft)

Areas:
Wings, gross 30·32 m² (326·4 sq ft)
Ailerons (total) 2·47 m² (26·6 sq ft)
Trailing-edge flaps (total) 3·49 m² (37·6 sq ft)
Fins (total) 3·03 m² (32·6 sq ft)
Rudders (total) 3·21 m² (34·56 sq ft)
Tailplane 6·08 m² (65·4 sq ft)
Elevator 2·53 m² (27·22 sq ft)

Weights and Loadings:
Weight empty 2,495 kg (5,500 lb)
Max payload 1,814 kg (4,000 lb)
Max T-O weight 5,094 kg (11,230 lb)
Max zero-fuel weight 4,854 kg (10,700 lb)
Max landing weight 4,763 kg (10,500 lb)
Max wing loading 167·9 kg/m² (34·4 lb/sq ft)
Max power loading 5·90 kg/kW (19·4 lb/ehp)

Performance (at max T-O weight. A: PT6A-20 engines; B: PT6A-27s):
Max level speed at 3,660 m (12,000 ft):
A 234 knots (435 km/h; 270 mph)
B 269 knots (499 km/h; 310 mph)
Max cruising speed at 3,660 m (12,000 ft):
A 217 knots (402 km/h; 250 mph)
B 252 knots (467 km/h; 290 mph)
Econ cruising speed at 3,050 m (10,000 ft):
A 204 knots (378 km/h; 235 mph)
B 234 knots (435 km/h; 270 mph)
Max rate of climb at S/L: A 549 m (1,800 ft)/min
B 823 m (2,700 ft)/min
Rate of climb at S/L, one engine out:
A 183 m (600 ft)/min
B 335 m (1,100 ft)/min
Service ceiling: A 7,315 m (24,000 ft)
B 8,535 m (28,000 ft)
Service ceiling, one engine out:
A 2,745 m (9,000 ft)
B 3,960 m (13,000 ft)
T-O run: A 549 m (1,800 ft)
B 366 m (1,200 ft)
T-O to 15 m (50 ft): A 1,005 m (3,300 ft)
B 731 m (2,400 ft)
Landing from 15 m (50 ft): A, B 549 m (1,800 ft)
Landing run: A, B 366 m (1,200 ft)
Range with max optional fuel:
A 3,240 nm (6,004 km; 3,731 miles)
Range with max payload:
A 810 nm (1,501 km; 933 miles)

Hamilton Westwind III conversion of the Beech Model 18 (two PT6A-27 derated turboprop engines)

HAMILTON WESTWIND II STD

The Westwind II STD is a 'stretched' version of the Beech 18, providing accommodation for a maximum of 17 passengers. Otherwise it is generally similar to the Westwind III except that a tricycle landing gear is available optionally. The Westwind II STD is intended primarily as a commuter airliner, but is convertible for freight carrying or military uses.

The description of the Westwind III applies also to the Westwind II STD, except as follows:

Landing Gear: Retractable tricycle type optional.

Power Plant: Two 626 kW (840 ehp) Pratt & Whitney Aircraft of Canada PT6A-34 turboprop engines, derated to 470 kW (630 ehp), are standard, each driving a Hartzell constant-speed fully-feathering reversible-pitch propeller. Two 579 kW (776 ehp) Garrett TPE331-6-251 turboprop engines, derated to 470 kW (630 ehp), available optionally.

Accommodation: Pilot and co-pilot or passenger on flight deck, with seating in main cabin for a maximum of 17 passengers. Two emergency exits on starboard side of fuselage.

Dimensions, external: As for Westwind III, except:
Length overall (standard) 13·72 m (45 ft 0 in)
Length overall (tricycle landing gear) 13·46 m (44 ft 2 in)

Dimensions, internal: As for Westwind III, except:
Cabin: Length 6·10 m (20 ft 0 in)

Weights and Loading (estimated):
Weight empty 2,712 kg (6,000 lb)
Max payload 2,041 kg (4,500 lb)
Max T-O weight 5,667 kg (12,495 lb)
Max zero-fuel weight 5,217 kg (11,500 lb)
Max landing weight 5,217 kg (11,500 lb)
Max wing loading 186·9 kg/m² (38·3 lb/sq ft)

Performance (at max T-O weight):
Max level speed at 4,265 m (14,000 ft) 278 knots (515 km/h; 320 mph)
Max cruising speed at 3,660 m (12,000 ft) 261 knots (483 km/h; 300 mph)
Econ cruising speed at 6,705 m (22,000 ft) 234 knots (435 km/h; 270 mph)
Stalling speed, flaps up 87 knots (161 km/h; 100 mph)
Stalling speed, flaps down 74 knots (137 km/h; 85 mph)
Max rate of climb at S/L 945 m (3,100 ft)/min
Rate of climb at S/L, one engine out 274 m (900 ft)/min
Service ceiling 9,755 m (32,000 ft)
Service ceiling, one engine out 5,485 m (18,000 ft)
T-O run 427 m (1,400 ft)
T-O to 15 m (50 ft) 1,036 m (3,400 ft)
Landing from 15 m (50 ft):
without propeller reversal 671 m (2,200 ft)
with propeller reversal 335 m (1,100 ft)
Range with max optional fuel 3,240 nm (6,004 km; 3,731 miles)
Range with max payload 810 nm (1,501 km; 933 miles)

Hamilton Westwind II STD of Connie Kalitta Services, with added ventral fin *(Austin J. Brown)*

HAWK

HAWK INDUSTRIES INC (Aircraft Division)

57430 Aviation Drive, Yucca Valley, California 92284

Telephone: (714) 365 9746

Telex: 656345

President: Ernest Hauk

Publicity Officer: W. C. Addison

Hawk Industries specialises in equipment for oil and water well-drilling, and fencing. Having experienced difficulties in transporting its products, the company's President designed a freight-carrying aircraft that might overcome both the slowness of road transport and the high cost, and loading/unloading difficulties, of conventional aircraft. He named his project the GafHawk 125, signifying general aviation freighter. It is intended to have STOL capability, and to be certificated initially at a gross weight of 5,670 kg (12,500 lb) under FAR Pt 23, although it would be capable of operating at a max T-O weight of 6,577 kg (14,500 lb).

HAWK GAFHAWK 125

Features considered of prime importance in design of the GafHawk included STOL capability for operation into and from small unprepared strips, a turboprop power plant for economic operations, a square-section fuselage for maximum utilisation of internal capacity, under-tail loading of bulk cargo at truckbed height, and a single engine for economy and for ease of certification and single-pilot operation. Construction of the prototype was still in progress in early 1981, with the first flight scheduled to take place during the year.

The GafHawk concept was tested initially in the form of a small-scale flying testbed known as the MiniHawk, which consisted of an extensively rebuilt Piper Tri-Pacer light aircraft. This made its initial flight in 1978, and was described and illustrated in the 1980-81 *Jane's*.

The following details apply to the prototype, known as the GafHawk 125:

Type: Single-engined turboprop freighter.

Wings: Strut-braced high-wing monoplane, with single bracing strut and short auxiliary strut on each side. NASA GAW-1 section, modified by use of a leading-edge cuff. No dihedral. Thickness/chord ratio 17%. Constant-chord structure of light alloy, made up of a leading-edge tubular spar, a box spar, a further tubular spar aft of the box, a total of 78 one-piece ribs, and light alloy skins. Electrically-actuated full-span trailing-edge flaps of similar construction. Half-span spoiler/aileron ('rolleron') hinged to top surface within slot between wing and flap in outer half of each wing, operating differentially through a range of 60° up and 10° down. Rolleron trimming by bungee.

Fuselage: Basic rectangular structure of welded square-section steel tubing, covered with non-structural corrugated Alclad light alloy skins. These are attached to the hermetically-sealed square tubing by clips, to facilitate the replacement of damaged sections.

Tail Unit: Cantilever structure of light alloy. Construction similar to that of wings, but with only fore and aft tubular spars in fin and tailplane. Horn-balanced rudder and one-piece elevator. Dorsal fin. Rudder and elevators trimmed by bungee.

Landing Gear: Non-retractable tricycle type, with single nosewheel, and twin wheels on main units. Shock-absorption by rubber in compression on all units. Goodrich wheels of the same size on each unit, with tubed tyres size 8·50-10. Hydraulic disc brakes. Parking brake.

Power Plant: Prototype has one 875 kW (1,173 shp) Pratt & Whitney Aircraft of Canada PT6A-45B turboprop engine, driving a Hartzell five-blade reversible-pitch low-speed propeller with spinner. Fuel tank, made of transparent material and with a capacity of 1,363 litres (360 US gallons), mounted above forward fuselage, directly over the wing, and providing gravity feed to engine. Refuelling point on upper surface of tank.

Accommodation: Pilot and co-pilot on flight deck. Dual controls and full blind-flying instrumentation for both pilots standard. Door to flight deck on each side of fuselage; communicating door between flight deck and cargo hold in forward bulkhead. Cabin door on port side, aft of wing. Electrically-actuated main cargo loading ramp/door, in undersurface of upswept aft fuselage, can be opened in flight. Heavy-duty corrugated light alloy floor in cargo hold, with cargo tiedowns along walls at each fuselage gusset frame. Main cabin volume augmented by usable space under flight deck, accommodating pipes and timber up to 6·1 m (20 ft) in length with aft loading door closed. Accommodation heated and ventilated.

Systems: Electrical system powered by 28V 250A Lear-Siegler starter/generator and 28V storage battery. Hydraulic system for brakes only. Vacuum system. De-icing system optional.

Avionics and Equipment: Standard avionics include King Goldcrown series dual com; dual nav with dual ILS, plus an HSI on the pilot's panel; ADF; DME; radar altimeter; transponder; switching panel; and VOR/localiser-coupled Century 11-B autopilot. Standard equipment includes dual blind-flying instrumentation, incl turn co-ordinator and rate of climb indicator; dual airspeed indicators; dual altimeters, one with encoding; eight-day clock; outside air temperature gauge; adjustable pilot/co-pilot seats with armrests; control locks; annunciator panel; cabin, compass, instrument post, landing, navigation, taxi and wingtip strobe lights; two rotating beacons; and heated pitot.

Full-size mockup of the Hawk GafHawk 125
(Henry Artof)

Dimensions, external:

Wing span	21·79 m (71 ft 6 in)
Wing chord (constant)	2·11 m (6 ft 11 in)
Wing area, gross	45·8 m^2 (493 sq ft)
Wing aspect ratio	10·4
Length overall	14·30 m (46 ft 11 in)
Height overall	5·49 m (18 ft 0 in)
Tailplane span	7·01 m (23 ft 0 in)
Wheel track (c/l outer tyres)	3·38 m (11 ft 1 in)
Wheelbase	4·39 m (14 ft 5 in)
Propeller diameter	2·74 m (9 ft 0 in)

Dimensions, internal:

Cabin: Length at floor level, excl flight deck	4·72 m (15 ft 6 in)
Max width	2·03 m (6 ft 8 in)
Max height	2·13 m (7 ft 0 in)
Volume	20·22 m^3 (714 cu ft)

Weights and Loadings (estimated):

Weight empty	2,835 kg (6,250 lb)
Max T-O weight (initial certification)	5,670 kg (12,500 lb)
Max wing loading	124·0 kg/m^2 (25·4 lb/sq ft)
Max power loading	6·48 kg/kW (10·66 lb/shp)

Performance (prototype, estimated, at max T-O weight):

Max cruising speed at 3,050 m (10,000 ft)	152 knots (282 km/h; 175 mph)
Econ cruising speed (55% power at 3,050 m; 10,000 ft)	126 knots (233 km/h; 145 mph)
Stalling speed, flaps down	44 knots (82 km/h; 51 mph)
Max rate of climb at S/L	290 m (950 ft)/min
Service ceiling	5,485 m (18,000 ft)
T-O run	189 m (620 ft)
T-O to 15 m (50 ft)	344 m (1,130 ft)
Landing from 15 m (50 ft)	296 m (970 ft)
Landing run	131 m (430 ft)
Range with max fuel	900 nm (1,668 km; 1,036 miles)

HIBBARD
HIBBARD AVIATION

Oakland International Airport, PO Box 2547, Oakland, California 94616

HIBBARD/VARGA 2150A-TG KACHINA

Under the above designation, Hibbard Aviation converted a Varga Model 2150A Kachina from tricycle to tailwheel landing gear. This involved removal of the standard nose unit, the installation of a castoring tailwheel, and the provision of speed fairings for the main units. The modification resulted in an 8·7 knot (16 km/h; 10 mph) increase in cruising speed, without any change in the useful load or gross weight, and the reduced drag has improved other performance figures.

Hibbard Aviation has gained an FAA Supplemental Type Certificate for this conversion, and is able to modify tricycle gear 2150As to this configuration. Varga Aircraft Corporation (which see) is marketing new aircraft with optional tricycle or tailwheel landing gear to customer's requirements.

Hibbard Aviation's tailwheel landing gear conversion for the Varga 2150A Kachina

HILLER
HILLER AVIATION INC

2075 West Scranton Avenue, Porterville, California 93257
Telephone: (209) 781 8000
Telex: 682454
President: Edwin L. Trupe
Domestic Sales Manager: Joe Campbell

Hiller Aviation, formed in January 1973, acquired from Fairchild Industries the design rights, production tooling and spares of the Hiller 12E piston-engined light helicopter. Initially, the company provided product support for UH-12 helicopters in service throughout the world, a total then estimated as being in excess of 2,200 aircraft. Service and repair facilities were added as a first move to expand the company's business. It was then decided to begin the manufacture of new aircraft from existing components, incorporating all modifications approved for the type since closure of the production line in the late 1960s.

It was decided also to develop a turbine-powered version of the UH-12, and in conjunction with Soloy Conversions of Chehalis, Washington, the company has developed such an aircraft. Its power plant is suitable not only for new production aircraft, but for retrospective installation in existing UH-12s.

More recently, in April 1980, the company announced that negotiations had been completed with Fairchild Industries to purchase all rights in the Fairchild FH-1100 five-passenger light turbine helicopter.

HILLER AVIATION FH-1100

The original FH-1100, the first production example of which was rolled out on 3 June 1966, was a refined development of the OH-5A helicopter which the former Hiller Aircraft Company designed for the US Army's LOH (Light Observation Helicopter) competition, and which was known initially as the Hiller Model 1100. Five HO-5 (later OH-5A) prototypes were ordered for evaluation in 1961, the first of them making its first flight on 26 January 1963. The OH-5A did not win the Army's LOH competition, but after the Hiller company became a subsidiary of Fairchild Stratos, in 1964, it was decided in early 1965 to develop this helicopter for the civil market as the Fairchild Hiller FH-1100. A total of 250 had been built when production ended in 1974.

Since acquiring the full rights of the FH-1100, Hiller Aviation has been working to improve the capability of this helicopter. Orders totalled 40 aircraft at 1 March 1981.

Type: Five-seat utility helicopter.

Rotor System: Two-blade semi-rigid main rotor of all-metal construction. Blade section NACA 63_2 015. Each blade attached to hub by single main retention bolt and drag link. The main rotor blades each have a rolled stainless steel leading-edge spar bonded to an aluminium trailing section with a honeycomb core. Two-blade tail rotor of stainless steel and honeycomb construction. Main rotor blades fold. Rotor brake optional. Electrically controlled trim system.

Rotor Drive: Mechanical drive through single-stage bevel and two-stage planetary main transmission, with intermediate and tail rotor gearboxes. Main rotor/engine rpm ratio 1:16·30. Tail rotor/engine rpm ratio 1:2·47.

Fuselage: Aluminium alloy semi-monocoque structure of pod and boom type.

Tail Unit: Vertical fin, and fixed horizontal surface, both of aluminium alloy and honeycomb construction. Tubular guard to protect rotor in tail-down landing.

Landing Gear: Skid type with torsion-tube suspension, with choice of standard or extended support struts. Extended struts necessary if optional inflatable float installation is required. Ground handling wheels standard.

Power Plant: One 313 kW (420 shp) Allison 250-C20B turboshaft engine, derated to 204 kW (274 shp). Single bladder fuel tank in bottom of centre-fuselage with usable capacity of 259 litres (68·5 US gallons). Refuelling point on starboard side of rear fuselage. Optional auxiliary fuel tank of 259 litres (68·5 US gallons) capacity can be installed in fuselage on centreline of main rotor mast. Oil capacity 2·6 litres (0·7 US gallons).

Accommodation: Pilot and co-pilot side by side with three passengers to rear, or pilot and four passengers. Four forward-hinged doors, two on each side of cabin. Stretcher kit optional. Baggage compartment to rear of cabin, capacity 0·30 m^3 (10·5 cu ft). Accommodation ventilated. Cabin heater and windscreen defroster optional.

Hiller Aviation FH-1100 five-seat turbine-powered utility helicopter

Systems: Hydraulic system for cyclic and collective pitch controls. Electrical system includes a 28V 60A DC starter/generator and nickel-cadmium battery.

Avionics: A range of nav/com systems is available to customer's requirements.

Equipment: Standard equipment includes clock, engine hour meter, outside air temperature gauge, transmission oil pressure gauge, edge-lighted instrument panel, seatbelts, provisions for shoulder harness, sliding side windows, tinted windows, hardpoint for optional external cargo hook, external power socket, and choice of exterior paint scheme and interior trim. Optional equipment includes stability augmentation system, dual controls, Mason cyclic control stick, inertia reel shoulder harness, cabin fire extinguisher, first aid kit, floor mats, night lighting system including two rotating beacons, strobe lights, rotor blade droop stops, engine auto relight, reverse scoop intake, particle separator, heated pitot, loudspeaker/siren, quick-release cargo hook, cargo racks, agricultural spray system, and hydraulic drive adapter for use with agricultural equipment.

Dimensions, external:

Diameter of main rotor	10·80 m (35 ft 5 in)
Diameter of tail rotor	1·83 m (6 ft 0 in)
Distance between rotor centres	6·29 m (20 ft 7½ in)
Main rotor blade chord	0·27 m (10·6 in)
Length overall, rotors turning	12·57 m (41 ft 3 in)
Length of fuselage	9·08 m (29 ft 9½ in)
Width, rotors folded	1·32 m (4 ft 4 in)
Height overall	2·83 m (9 ft 3½ in)
Skid track	2·20 m (7 ft 2¾ in)

Areas:

Main rotor blades (each)	2·80 m² (30·10 sq ft)
Tail rotor blades (each)	0·19 m² (2·04 sq ft)
Main rotor disc	91·0 m² (979 sq ft)
Tail rotor disc	2·63 m² (28·3 sq ft)

Weights:

Weight empty	680 kg (1,500 lb)
Max payload	467 kg (1,030 lb)
Max standard fuel weight	200·5 kg (442 lb)
Max T-O weight	1,247 kg (2,750 lb)

Performance (at max T-O weight):

Never-exceed speed at S/L	110 knots (204 km/h; 127 mph)
Econ cruising speed	106 knots (196 km/h; 122 mph)
Max rate of climb at S/L	740 m (2,430 ft)/min
Vertical rate of climb at S/L	612 m (2,010 ft)/min
Service ceiling	6,550 m (21,500 ft)
Hovering ceiling IGE	6,520 m (21,400 ft)
Hovering ceiling OGE	5,275 m (17,300 ft)
Range at S/L with max standard fuel	373 nm (692 km; 430 miles)
Max endurance with auxiliary fuel	8 h 48 min

HILLER AVIATION UH-12

Following the development of a turbine power plant for the UH-12, the following versions are available:

UH-12E. Basic model, with three-abreast accommodation for a pilot and two passengers.

UH-12E4. Four-seat version of UH-12E, with accommodation for the pilot on a single adjustable seat forward and three passengers side by side on a rear bench seat. Space for cargo in tailboom. An 'L-Cabin' version is available optionally, with wider cabin door for improved access.

UH-12ET. Turbine-powered version of the UH-12E, with the same three-seat accommodation. Power plant as for UH-12E4T.

UH-12E4T. Turbine-powered version with four-seat cabin configuration as in the UH-12E4. Power plant comprises one 313 kW (420 shp) Allison 250-C20B turboshaft engine, derated to 224 kW (301 shp). Certification achieved by Soloy Conversions in January 1976.

The description which follows applies generally to all versions, except as indicated:

Type: Three-seat or four-seat utility helicopter.

Rotor System: Two-blade main rotor mounted universally on driveshaft, with small servo rotor; the latter is connected directly to the pilot's cyclic control stick, through a universally-mounted transfer bearing and simple linkage. Movement of the control stick introduces positive or negative pitch changes to the servo rotor paddles. The resulting aerodynamic forces tilt the rotor head and produce cyclic pitch changes to the rotor blades. Main rotor blades are of bonded stainless steel (since April 1978), with aluminium honeycomb core, and wedge tips. Thickness/chord ratio 12%. Blades are interchangeable individually and are bolted to forks which are retained at the rotor head by tension-torsion bars. Blades do not fold. Rotor brake optional. Two-blade tail rotor of light alloy construction, mounted on port side of tailboom.

Rotor Drive: Mechanical drive through two-stage planetary main transmission. Bevel gear drive to auxiliaries. Tail rotor gearbox (and fan gearbox on piston-engined versions). Main rotor/engine rpm ratio 1 : 8·66 on piston-engined versions, 1:16·2 on turbine-powered versions. Tail rotor/engine rpm ratio 1 : 1·44 on piston-engined versions, 1:2·52 on turbine-powered versions.

Fuselage: Light alloy fully-stressed semi-monocoque platform structure supporting the non-stressed cabin enclosure, engine mounting and landing gear. Tailboom of beaded light alloy sheet with no internal stiffeners.

Tail Unit: Horizontal stabiliser on starboard side of tailboom of three-seat versions, with steel tube spar, light alloy ribs and skin. Incidence ground-adjustable. Inverted V stabilising surfaces forward of tail rotor on four-seat versions.

Landing Gear: Wide-track light alloy tube skids carried on spring steel cross members. Ground handling wheels standard. Optional 'zip-on' pontoons can be attached above the skids to permit water or land operations, but require alternative extended landing gear.

Power Plant: (UH-12E/E4), one 253·5 kW (340 hp) Avco Lycoming VO-540-C2A flat-six engine, installed vertically and derated to 227·5 kW (305 hp). (UH-12ET/E4T), one turboshaft engine as detailed in model listing, mounted diagonally aft of rotor pylon. Single bladder fuel cell, capacity 174 litres (46 US gallons), mounted in lower portion of rear fuselage, beneath engine. Two optional 72 litre (19 US gallon) auxiliary fuel tanks, mounted in fuselage on each side of engine. Oil capacity 12·5 litres (3·3 US gallons) on piston-engined versions.

Accommodation: Three persons side by side on bench seat; or pilot on single forward seat and three passengers side by side on bench seat. Seat belts for all occupants; provision for shoulder harness. Dual controls optional. Forward-hinged door on each side, with sliding window. Baggage compartment immediately aft of engine. Heater/defroster optional.

System: Electrical system includes a 72A alternator, nickel-cadmium battery, and battery temperature monitor.

Avionics and Equipment: A range of optional avionics is available. Standard equipment includes engine hour meter, edge-lighted instrument panel, outside air temperature gauge, eight-day clock, electrically-controlled trim system, tinted glazing, cargo hook hardpoint, external power socket, and polyurethane paint finish. Optional equipment includes Mason cyclic control grip, fire extinguisher, first aid kit, night lighting equipment including two rotating beacons, strobe lights, 454 kg (1,000 lb) capacity quick-release cargo hook, twin heavy-duty cargo racks, agricultural spray equipment, loudspeaker/siren, searchlight, tropical doors and (on E4) twin litters.

Dimensions, external:

Diameter of main rotor	10·80 m (35 ft 5 in)
Main rotor blade chord (constant)	0·33 m (1 ft 1 in)
Diameter of tail rotor	1·68 m (5 ft 6 in)
Distance between rotor centres	6·17 m (20 ft 3 in)
Length overall, rotors turning	12·41 m (40 ft 8½ in)
Length of fuselage: 3-seater	8·69 m (28 ft 6 in)
4-seater	9·08 m (29 ft 9½ in)
Height to top of rotor hub	3·08 m (10 ft 1¼ in)
Skid track	2·29 m (7 ft 6 in)
Cabin doors (standard, each):	
Height	1·13 m (3 ft 8½ in)
Max width	0·81 m (2 ft 8 in)
Height to sill	0·58 m (1 ft 11 in)
Cabin doors (L-cabin, each):	
As above except:	
Max width	0·99 m (3 ft 3 in)

Dimensions, internal:

Cabin: Length (3-seater)	1·52 m (5 ft 0 in)
Max width	1·50 m (4 ft 11 in)
Max height	1·35 m (4 ft 5 in)
Floor area (3-seater)	1·16 m² (12·5 sq ft)

Areas:

Tail rotor blades (each)	0·094 m² (1·01 sq ft)
Main rotor disc	91·97 m² (990 sq ft)
Tail rotor disc	2·57 m² (27·7 sq ft)

Weights and Loadings (A: UH-12E; B: UH-12E4; C: UH-12ET/E4T):

Weight empty: A	798 kg (1,759 lb)
B	833 kg (1,836 lb)
C	748·5 kg (1,650 lb)

Hiller UH-12ET, with an Allison 250-C20B turboshaft engine

Max T-O weight:
all versions 1,406 kg (3,100 lb)
Max disc loading:
all versions 15·28 kg/m² (3·13 lb/sq ft)
Max power loading:
A, B 6·18 kg/kW (10·16 lb/hp)
C 6·28 kg/kW (10·30 lb/shp)
PERFORMANCE (at max T-O weight):
Never-exceed and max level speed:
A, C 83 knots (154 km/h; 96 mph)
B 82·5 knots (153 km/h; 95 mph)
Cruising speed:
all versions 78 knots (145 km/h; 90 mph)
Max rate of climb at S/L: A 393 m (1,290 ft)/min
B 303 m (993 ft)/min
C 520 m (1,706 ft)/min
Vertical rate of climb at S/L:
A, B 225 m (740 ft)/min
C 446 m (1,464 ft)/min
Service ceiling: A, B 4,570 m (15,000 ft)
C 4,265 m (14,000 ft)
Hovering ceiling IGE: A, B 3,170 m (10,400 ft)
C 3,445 m (11,300 ft)
Hovering ceiling OGE: A, B 2,070 m (6,800 ft)
C 2,650 m (8,700 ft)
Range, no reserves:
A, B, with standard fuel
187 nm (346 km; 215 miles)
C, with auxiliary fuel
305 nm (565 km; 351 miles)
Endurance, no reserves:
A, B, with standard fuel 2 h 42 min
C with auxiliary fuel 3 h 32·5 min

HILLMAN
HILLMAN HELICOPTER ASSOCIATES

FLIGHT TEST CENTER: Stellar Air Park, Chandler, Arizona 85224
Telephone: (602) 838-8333
Telex: Hillman 669401 Air Cours
WORKS: Paso Robles Airport, PO Box 1014, Paso Robles, California 93446
PRESIDENT AND CHIEF EXECUTIVE OFFICER:
Douglas Hillman
VICE-PRESIDENT: Thomas H. Staten
CHIEF ENGINEER: Rudolph Enstrom
MARKETING DIRECTOR: Brian Kloos
EUROPEAN REPRESENTATION: Hillman Helicopters Europe, Leo Verschuren, PO Box 4, Beek en Donk, Netherlands, 5740 AA
Telephone: 4929-2486

In 1972 Mr Douglas Hillman began the design of the WankelBee, the first helicopter to be powered by a rotating-combustion engine, which flew successfully for the first time in July 1975. This aircraft was superseded by a two-seat helicopter named Hornet, which was powered by a 112 kW (150 hp) Avco Lycoming engine. The Hornet was first flown in February 1978, and won the Best Operational Helicopter award at the EAA Fly-in, Oshkosh, that year, followed by the Outstanding Design award at Oshkosh in 1979. In September 1979 Mr Rudolph Enstrom, designer of the Enstrom helicopter, joined Hillman's company as Chief Engineer, and this has led to development of the Hillman Model 360, first presented at the Paris Air Show in June 1981. Designed as a lightweight, sturdy, but mechanically simple helicopter requiring minimum maintenance, the prototype of this aircraft was undergoing pre-flight trials at Stellar Air Park near Phoenix, Arizona, in the Summer of 1981. Development is under the auspices of the FAA, with certification under FAR Pts 21 and 27 anticipated in late 1982.

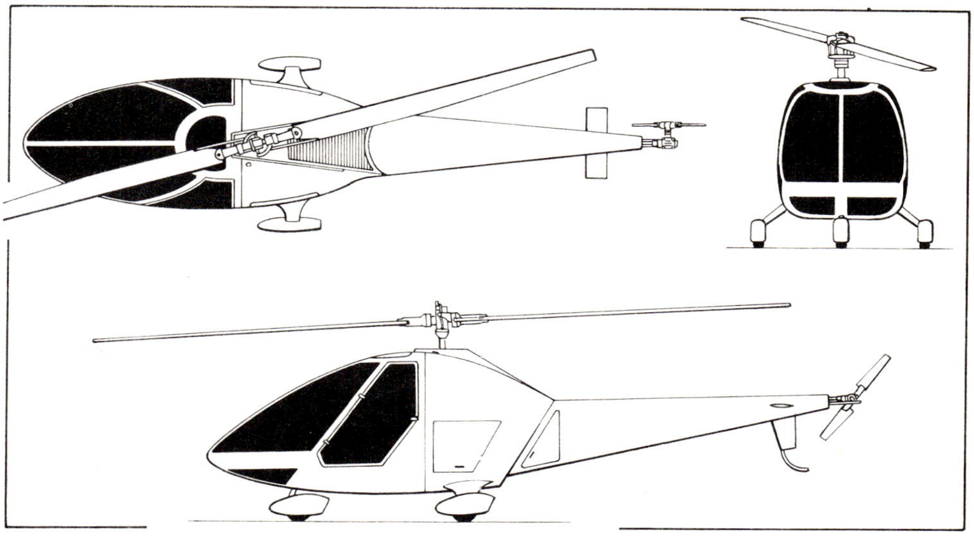

Hillman Model 360 three-seat light utility helicopter *(Michael A. Badrocke)*

HILLMAN MODEL 360

TYPE: Three-seat light utility helicopter.

ROTOR SYSTEM: Two-blade main rotor, with a semi-rigid underslung teetering hub to reduce rotor vibration and control force feedback. Tapered-chord main rotor blades are mounted 3° above horizontal to minimise blade flexing, and have 3° of twist. Construction is of glassfibre, with stainless steel leading-edge. Two-blade teetering tail rotor, with glassfibre blades which have leading-edges of stainless steel. No rotor brake at present.

ROTOR DRIVE: Eight-grooved common-back V-belt drive, with a sprag-type overrunning clutch in the driven pulley. Right-angle spiral bevel gear reduction in an aluminium gearcase. Tail rotor driven via spiral bevel gears. Main rotor/engine rpm ratio 1:5·5; tail rotor/engine rpm ratio 1:1.

FUSELAGE: Welded chrome-molybdenum steel tube centre-section and tailboom. Glassfibre cabin structure, and light alloy tailboom skin.

LANDING GEAR: Non-retractable tricycle type. Nose unit has oleo-pneumatic shock-absorption and a castoring and self-centering nosewheel. Main wheels carried on multiple spring steel leaves. Hydraulic brakes. Steel/aluminium tubular skid landing gear optional. Wheeled landing gear allows the Hillman 360 to take off with greater loads at higher density altitudes by making a short forward run.

POWER PLANT: One 153 kW (205 hp) Avco Lycoming HIO-360-C1A flat-four engine, mounted horizontally in the lower rear section of the fuselage pod. Exhaust is muffled, and sound-dampening foam is installed on firewall. Robertson crashworthy fuel system with single standard tank, capacity 265 litres (70 US gallons). Four optional tanks are being developed to provide a total max optional capacity of 591 litres (156 US gallons). Oil capacity 7·5 litres (2 US gallons).

ACCOMMODATION: Pilot and two passengers, side by side on contoured bench seat, with dual controls standard. Fully enclosed cabin with overhead eyebrow window. Removable door on each side of cabin. Baggage space at rear of cabin and around engine compartment. Cabin heating optional; ventilation standard.

SYSTEM: Electrical system includes a 12V 60A engine-driven alternator.

AVIONICS AND EQUIPMENT: Optional avionics include a King KY 197 com transceiver, KN 53 nav receiver, KT 76A transponder and K2 87 ADF, or similar installations by Edo or Narco. Standard equipment includes sensitive altimeter, low rpm warning lights, and anti-collision and navigation lights. Optional equipment includes amphibious floats, a lighting package, agricultural spray system, cargo racks, a cargo hook, and stretcher kits.

DIMENSIONS, EXTERNAL:
Main rotor diameter 8·15 m (26 ft 9 in)
Main rotor blade chord:
at root 0·289 m (11·375 in)
at tip 0·143 m (5·625 in)
Tail rotor diameter 1·22 m (4 ft 0 in)
Distance between rotor centres 4·88 m (16 ft 0 in)
Length overall, rotors turning 9·47 m (31 ft 1 in)
Length of fuselage 7·62 m (25 ft 0 in)
Height overall 2·54 m (8 ft 4 in)
Wheelbase 2·18 m (7 ft 2 in)
DIMENSIONS, INTERNAL:
Cabin: Max width 1·40 m (4 ft 7 in)
Baggage hold volume 0·17 m³ (6·0 cu ft)
AREAS:
Tail rotor blades (each) 0·05 m² (0·56 sq ft)
Main rotor disc 52·21 m² (562 sq ft)
Tail rotor disc 1·17 m² (12·57 sq ft)
WEIGHTS AND LOADINGS:
Weight empty 499 kg (1,100 lb)
Fuel weight, standard 185·5 kg (409 lb)
Max T-O and landing weight 998 kg (2,200 lb)
Max disc loading 19·05 kg/m² (3·91 lb/sq ft)
Max power loading 6·52 kg/kW (10·73 lb/hp)
PERFORMANCE (estimated at max T-O weight):
Never-exceed speed 113 knots (209 km/h; 130 mph)
Max cruising speed 100 knots (185 km/h; 115 mph)
Max rate of climb at S/L 457 m (1,500 ft)/min
Service ceiling 4,575 m (15,000 ft)
Hovering ceiling IGE 3,050 m (10,000 ft)
Hovering ceiling OGE 2,440 m (8,000 ft)
Range with max optional fuel and max payload
1,355 nm (2,511 km; 1,560 miles)

HUGHES
HUGHES HELICOPTERS, INC (Subsidiary of The Hughes Corporation)

HEAD OFFICE AND WORKS: Centinela and Teale Streets, Culver City, California 90230
Telephone: (213) 390 4451
Telex: 67-222
PRESIDENT AND CHIEF EXECUTIVE OFFICER:
J. G. Real
EXECUTIVE VICE-PRESIDENT:
C. D. Perry
VICE-PRESIDENTS:
W. J. Blackburn (Manufacturing Operations)
R. E. Brix (Ordnance)
William P. Brown (Engineering)
W. Ellis (Marketing)
M. F. Gerardis (Finance and Administration)
A. Haggerty (Senior V-P, Operations)
Norman B. Hirsh (Advanced Attack Helicopter Programme)
J. Kimmitt (Government Affairs)
William H. Murphy (Senior V-P and Chief Financial Officer)
C. E. Schaaf (Legal)
L. P. Sonsini (Product Assurance and Flight Operations)
F. C. Strible (Commercial Helicopters)
A. Taylor (General Counsel & Secretary)
PUBLIC RELATIONS MANAGER: Anthony J. Longo

Following reorganisation of Hughes Tool Company as the Summa Corporation, its former Aircraft Division became known as Hughes Helicopters. It has now separated from Summa Corporation, and is a wholly-owned subsidiary of The Hughes Corporation.

In 1981, nearly 4,500 Hughes helicopters were serving with civil and military operators in more than 100 countries worldwide, with production continuing of a series of advanced models. Research activities include work on composite rotor blades and tailbooms, metal insulation and IR suppression systems, and the Hughes Chain Gun and lockless ordnance systems for air or ground applications.

Licence manufacture of Hughes helicopters is undertaken by RACA in Argentina, Kawasaki in Japan, KAL in the Republic of Korea and BredaNardi in Italy.

HUGHES MODEL 300C

This is a developed version of the Model 300 (1976-77 *Jane's*), with improvements to allow a 45% increase in payload. The prototype made its first flight in August 1969, followed by the first production model in December 1969. FAA certification was received in May 1970. More than 550 Model 300Cs had been delivered by 1 January 1980. A two-month suspension of Model 300C final assembly, due to "softening of the market and a buildup of inventory", was announced on 17 July 1981.

The Model 300C is manufactured in Italy by BredaNardi (which see).

A specially equipped 300C is available for police patrol. Known as the **Sky Knight**, it has as standard equipment safety mesh seats with inertia-reel shoulder harness, ballistic glassfibre armour beneath each seat, a high-power public address/siren system, a high-intensity controllable searchlight system, an integrated communications system based on the King KY 195 VHF transceiver, a heavy-duty 28V 100A electrical system, cabin heater, night lights with strobe beacons, cabin utility light, external power socket, fire extinguisher, first aid kit and map case.

Hughes Model 300C supplied to Associated Industries of Lagos, Nigeria *(Howard Levy)*

Following the research that produced a modified version of the OH-6A known as 'The Quiet One', Hughes used similar techniques to develop and obtain full FAA certification of a quiet version of the Model 300, and this has the designation Model **300CQ**. In this configuration, emission of audible sound is 75 per cent less than with earlier models, and the necessary modifications can be retrofitted to existing 300Cs. Max T-O weight of the 300CQ for quiet operation is 873 kg (1,925 lb), with a useful load of 397 kg (875 lb), and there is little change in range and endurance by comparison with the standard Model 300C.

TYPE: Two- or three-seat light helicopter.

ROTOR SYSTEM: Fully-articulated metal three-blade main rotor. Blades are of bonded construction, with constant-section extruded aluminium spar, wraparound skin and a trailing-edge section. Blade section NACA 0015. Tracking tabs on blades at three-quarters radius. Electric cyclic trim. Two-blade teetering tail rotor, each blade comprising a steel tube spar with glassfibre skin. Limited blade folding. No rotor brake.

ROTOR DRIVE: Combination V-belt/pulley and reduction gear drive system. Main rotor and tail rotor gearbox have spiral bevel right-angle drive. Main rotor/engine rpm ratio 1 : 6·8. Tail rotor/engine rpm ratio 1 : 1·03.

FUSELAGE: Welded steel tube structure, with light alloy and Plexiglas cabin and one-piece light alloy tube tailboom.

TAIL UNIT: Horizontal and vertical fixed stabilising surfaces, made of light alloy ribs and skins.

LANDING GEAR: Skids carried on Hughes oleo-pneumatic shock-absorbers. Two cast magnesium ground handling wheels with 0·25 m (10 in) balloon tyres, pressure 4·14-5·17 bars (60-75 lb/sq in). Available optionally on floats made of polyurethane coated nylon fabric, 4·70 m (15 ft 5 in) long and with a total installed weight of 27·2 kg (60 lb).

POWER PLANT: One 142 kW (190 hp) Avco Lycoming HIO-360-D1A flat-four engine, mounted horizontally below seats. Aluminium fuel tank, capacity 103·5 litres (30 US gallons) mounted externally aft of cockpit. Crash-resistant fuel tank optional. Provision for aluminium auxiliary fuel tank, capacity 72 litres (19 US gallons), mounted opposite standard tank. Oil capacity 9·5 litres (2·5 US gallons).

ACCOMMODATION: Three persons seated side by side on sculptured and cushioned bench seat, in Plexiglas-enclosed cabin. Carpet and tinted canopy standard. Door on each side. Dual controls optional. Baggage capacity 45 kg (100 lb). Exhaust muff, or gasoline, heating and ventilation kits available.

SYSTEMS: Standard electrical system includes 28V 70A alternator, battery, starter and external power receptacle; 28V 100A alternator optional.

AVIONICS AND EQUIPMENT: Optional avionics include Collins VHF-251 or Bertea ML360 com transceiver and headsets. Standard equipment includes mapcase, first aid kit, fire extinguisher, engine hour meter, and main rotor blade tiedown kit. Optional equipment includes amphibious floats, litter kits, cargo racks with combined capacity of 91 kg (200 lb), external load sling of 408 kg (900 lb) capacity, agricultural spray or dry powder dispersal kits, 72 litre (19 US gallon) auxiliary fuel tank, night flying kit, dual controls, all-weather cover, heavy-duty skid plates, single or dual exhaust mufflers, door lock, dual oil coolers, tinted glass for cabin windows, gasoline or exhaust manifold cabin heating.

DIMENSIONS, EXTERNAL:
- Diameter of main rotor: 8·18 m (26 ft 10 in)
- Main rotor blade chord: 0·171 m (6¾ in)
- Diameter of tail rotor: 1·30 m (4 ft 3 in)
- Length overall, rotor blades fore and aft: 9·40 m (30 ft 10 in)
- Height over rotor hub: 2·67 m (8 ft 9 in)
- Width, rotor partially folded: 2·44 m (8 ft 0 in)
- Skid track: 1·98 m (6 ft 6 in)
- Passenger doors (each): Height: 1·09 m (3 ft 7 in)
  - Width: 0·97 m (3 ft 2 in)
  - Height to sill: 0·91 m (3 ft 0 in)

AREAS:
- Main rotor blades (each): 0·70 m^2 (7·55 sq ft)
- Tail rotor blades (each): 0·08 m^2 (0·86 sq ft)
- Main rotor disc: 52·5 m^2 (565·5 sq ft)
- Tail rotor disc: 1·32 m^2 (14·2 sq ft)
- Fin: 0·26 m^2 (2·8 sq ft)
- Horizontal stabiliser: 0·32 m^2 (3·44 sq ft)

WEIGHTS AND LOADINGS (A: 300C; B: 300CQ):
- Weight empty: 476 kg (1,050 lb)
- Max T-O weight: A: 930 kg (2,050 lb)
  - B: 921 kg (2,030 lb)
- Max disc loading: A: 17·67 kg/m^2 (3·62 lb/sq ft)
  - B: 17·48 kg/m^2 (3·58 lb/sq ft)

PERFORMANCE (at max T-O weight, ISA. A: 300C; B: 300CQ normal operation; C: 300CQ quiet operation at AUW of 873 kg; 1,925 lb):
- Never-exceed speed:
  - A, B, C: 91 knots (169 km/h; 105 mph)
- Max cruising speed at S/L:
  - A, B: 78 knots (145 km/h; 90 mph)
- Max level speed at S/L:
  - C: 61 knots (113 km/h; 70 mph)
- Max cruising speed at 1,220 m (4,000 ft):
  - A: 82 knots (153 km/h; 95 mph)
  - B: 78 knots (145 km/h; 90 mph)
- Max level speed at 1,220 m (4,000 ft):
  - C: 62·5 knots (116 km/h; 72 mph)
- Speed for max range, at 1,220 m (4,000 ft):
  - A, B: 67 knots (124 km/h; 77 mph)
- Max rate of climb at S/L: A: 229 m (750 ft)/min
  - C: 113 m (370 ft)/min
- Service ceiling: A: 3,110 m (10,200 ft)
  - B: 3,050 m (10,000 ft)
  - C: 2,440 m (8,000 ft)
- Hovering ceiling IGE: A: 1,830 m (6,000 ft)
  - B: 1,220 m (4,000 ft)
- Hovering ceiling OGE: A: 823 m (2,700 ft)
- Range at 1,220 m (4,000 ft), 2 min warm-up, max fuel, no reserves: 200 nm (370 km; 230 miles)

HUGHES MODEL 500 (CIVIL VERSIONS)

The Model 500, which entered full-scale production in November 1968, originated as a civil development of the OH-6A Cayuse military helicopter last described in the 1977-78 *Jane's*. From it have since been developed several new military export versions; these are described separately.

Civil versions of the Model 500 so far announced are as follows:

Model 500. Initial basic production version, with Allison 250-C18A turboshaft engine.

Model 500C. Similar to Model 500, but with Allison 250-C20 engine and improved 'hot and high' performance. Licence manufacture also undertaken by RACA (Argentina) and Kawasaki (Japan).

Model 500D. Announced in February 1975, the 500D is similar in size and general appearance to the Model 500C. It differs in having a 313 kW (420 shp) Allison 250-C20B engine; a five-blade main rotor; engine exhaust muffler; sound blanketing of the complete power plant assembly, including the engine air intake; and reshaping of the tips of the main rotor blades. It introduced also a small T tail which gives greater flight stability in both high and low speed regimes, as well as better handling characteristics in abnormal manoeuvres. Construction of the prototype and its first flight took place in August 1974, and the first flight of a production aircraft was made on 9 October 1975. Certificated by FAA on 8 December 1976. The 1,000th Model 500D was delivered in July 1981. Licence manufacture undertaken by BredaNardi (Italy), Kawasaki (Japan) and KAL (South Korea). Supplied also to air forces of Jordan (8) and Kenya (2) for training.

TYPE: Turbine-powered civil light helicopter.

ROTOR SYSTEM: Four-blade fully-articulated main rotor (five-blade on 500D), with blades attached to laminated strap retention system by means of quick-disconnect pins for folding. Each blade consists of an extruded aluminium spar hot-bonded to one-piece wraparound aluminium skin. Trim tab outboard on each blade. Main rotor blades can be folded. Two-blade tail rotor, each blade comprising a swaged steel tube spar and glassfibre skin covering (metal skin on 500D). Rotor brake optional.

ROTOR DRIVE: Three sets of bevel gears, three drive-shafts and one overrunning clutch. Strengthened transmission on 500D. Main rotor/engine rpm ratio 1 : 12·806 on 500/500C; 1 : 12·594 on 500D. Tail rotor/engine rpm ratio 1 : 1·987 on 500/500C; 1 : 1·956 on 500D.

FUSELAGE: Aluminium semi-monocoque structure of pod and boom type. Clamshell doors at rear of pod give access to engine and accessories.

TAIL UNIT: Fixed fin, horizontal stabiliser and ventral fin on Models 500/500C; Model 500D has T tail with horizontal stabiliser at tip of narrow-chord sweptback fin; small auxiliary fin at tip of tailplane on each side; narrow-chord sweptback ventral fin with integral tail-skid to protect tail rotor in tail-down attitude near ground.

LANDING GEAR: Tubular skids carried on Hughes single-acting (oleo-pneumatic on 500D) shock-absorbers. Utility floats, snow skis and emergency inflatable floats optional.

POWER PLANT: The 236·5 kW (317 shp) Allison 250-C18A turboshaft engine installed in the 500 is derated to 207 kW (278 shp) for T-O and has a max continuous rating of 181 kW (243 shp). Model 500C is powered by a 298 kW (400 shp) Allison 250-C20 turboshaft engine; this also is derated to 207 kW (278 shp) for T-O and has a max continuous rating of 181 kW (243 shp). Model 500D powered by a 313 kW (420 shp) Allison 250-C20B turboshaft engine. Two interconnected bladder fuel tanks with combined usable capacity of 240 litres (63·4 US gallons). Self-sealing fuel tank optional in 500D. Refuelling point on starboard side of fuselage. Auxiliary fuel system with 132 litre (35 US gallon) crashworthy internal fuel tanks, or two external glassfibre fuel cells with combined capacity of 167 litres (44 US gallons), optional. Oil capacity 5·7 litres (1·5 US gallons).

One of eight Hughes Model 500Ds for pilot training in the Royal Jordanian Air Force *(Howard Levy)*

Accommodation: Pilot and four passengers or equivalent freight in 500/500C. Optional accommodation for seven with litter kit in use or with four in passenger compartment. Model 500D has forward bench seat for pilot and two passengers, with two or four passengers, or two litter patients and two medical attendants, in aft portion of cabin. Baggage space, capacity 0·31 m³ (11 cu ft), under and behind rear seat in five-seat form. Clear space for 1·19 m³ (42 cu ft) of cargo or baggage with only three front seats in place. Two doors on each side.

System: Electrical system in 500D includes a 150A engine-driven generator and a nickel-cadmium battery.

Avionics and Equipment (500D): Optional avionics include dual King KY 195 com, KX 175 nav/com, KR 85 ADF, and KT 76 transponder; dual Collins VHF-251 com, VHF-251/351 nav/com, IND-350 nav indicator, ADF-650 ADF, and TDR-950 transponder; SunAir ASB 125; intercom system, headsets, microphones; public address system; stereo tape system; and flight management computer system. Standard equipment includes outside air temperature gauge, 8-day clock, engine hour meter, five sets inertia-reel shoulder harness, cargo tiedown fittings, fire extinguisher, first aid kit, passenger steps, external power socket, landing light, skid-tip position light, anti-collision strobe lights, cockpit utility light, aft cabin light, and instrument lights. Optional equipment includes dual controls, blind-flying instrumentation, electric hoist, cargo hook, external baggage pods, cargo racks, underfuselage cargo pod, nylon mesh seats, dual-strap shoulder harnesses, and heating/demisting system.

Equipment (500/500C): Standard equipment includes engine hour meter, navigation lights, clock, and ground handling wheels. Optional equipment includes shatterproof glass, heating system, radios and intercom, attitude and directional gyros, rate of climb indicator, inertia reels and shoulder harnesses for pilot and co-pilot, dual controls, cargo hook, hoist, auxiliary fuel system, fire extinguisher, heated pitot tube, extended landing gear, blade storage rack, litter kit, emergency inflatable floats, inflated utility floats, seating for four in passenger compartment, and first aid kit.

Dimensions, external:
- Diameter of main rotor:
  - 500/500C: 8·03 m (26 ft 4 in)
  - 500D: 8·08 m (26 ft 6 in)
- Main rotor blade chord: 0·171 m (6¾ in)
- Diameter of tail rotor: 1·30 m (4 ft 3 in)
- Distance between rotor centres:
  - 500/500C: 4·58 m (15 ft 0¼ in)
  - 500D: 4·62 m (15 ft 2 in)
- Length overall, rotors fore and aft:
  - 500/500C: 9·24 m (30 ft 3¾ in)
  - 500D: 9·30 m (30 ft 6 in)
- Length of fuselage: 7·01 m (23 ft 0 in)
- Height to top of rotor hub:
  - 500/500C: 2·48 m (8 ft 1½ in)
  - 500D: 2·59 m (8 ft 6 in)
- Skid track: 2·06 m (6 ft 9 in)
- Cabin doors (500/500C, fwd, each):
  - Height: 1·19 m (3 ft 11 in)
  - Width: 0·89 m (2 ft 11 in)
- Cabin doors (500D, each):
  - Height: 1·16 m (3 ft 9½ in)
  - Max width: 0·76 m (2 ft 6 in)
  - Height to sill: 0·76 m (2 ft 6 in)
- Cargo compartment doors (each):
  - Height: 1·04 m (3 ft 5 in)
  - Width: 0·88 m (2 ft 10½ in)
  - Height to sill: 0·57 m (1 ft 10½ in)

Dimensions, internal:
- Cabin: Length: 2·44 m (8 ft 0 in)
  - Max width: 500/500C: 1·37 m (4 ft 6 in)
    - 500D: 1·31 m (4 ft 3½ in)
  - Max height: 500/500C: 1·31 m (4 ft 3½ in)
    - 500D: 1·52 m (5 ft 0 in)

Areas:
- Main rotor blades (each):
  - 500/500C: 0·688 m² (7·41 sq ft)
  - 500D: 0·690 m² (7·43 sq ft)
- Tail rotor blades (each):
  - 500/500C: 0·079 m² (0·85 sq ft)
  - 500D: 0·09 m² (0·94 sq ft)
- Main rotor disc: 500/500C: 50·60 m² (544·63 sq ft)
  - 500D: 50·89 m² (547·81 sq ft)
- Tail rotor disc (all versions): 1·32 m² (14·19 sq ft)
- Fin: 500/500C: 0·52 m² (5·65 sq ft)
  - 500D: 0·56 m² (6·05 sq ft)
- Horizontal stabiliser: 500/500C: 0·72 m² (7·70 sq ft)
  - 500D: 0·61 m² (6·52 sq ft)

Weights and Loadings:
- Weight empty: 500: 493 kg (1,088 lb)
  - 500C: 501 kg (1,105 lb)
  - 500D: 598 kg (1,320 lb)
- Fuel load: 500D: 181 kg (400 lb)
- Max normal T-O weight:
  - 500/500C: 1,157 kg (2,550 lb)
- Max overload T-O weight:
  - 500/500C: 1,360 kg (3,000 lb)
- Max T-O and landing weight:
  - 500D: 1,360 kg (3,000 lb)
- Max disc loading: 500D: 26·76 kg/m² (5·48 lb/sq ft)
- Max power loading: 500D: 4·35 kg/kW (7·14 lb/shp)

Performance (500/500C at max T-O weight):
- Max level speed at 305 m (1,000 ft):
  - 500: 132 knots (244 km/h; 152 mph)
- Max cruising speed at S/L:
  - 500C: 125 knots (232 km/h; 144 mph)
- Max cruising speed at 1,220 m (4,000 ft):
  - 500C: 126 knots (233 km/h; 145 mph)
- Cruising speed for max range at S/L:
  - 500: 117 knots (217 km/h; 135 mph)
  - 500C: 124 knots (230 km/h; 143 mph)
- Max rate of climb at S/L:
  - 500, 500C: 518 m (1,700 ft)/min
- Service ceiling: 500: 4,390 m (14,400 ft)
  - 500C: 4,420 m (14,500 ft)
- Hovering ceiling IGE:
  - 500: 2,500 m (8,200 ft)
  - 500C: 3,960 m (13,000 ft)
- Hovering ceiling OGE:
  - 500: 1,615 m (5,300 ft)
  - 500C: 2,040 m (6,700 ft)
- Range at 1,220 m (4,000 ft):
  - 500: 327 nm (606 km; 377 miles)
- Range at 1,220 m (4,000 ft), 2 min warm-up with max fuel, no reserves:
  - 500C: 325 nm (603 km; 375 miles)

Performance (500D at max T-O weight, ISA):
- Never-exceed speed: 152 knots (282 km/h; 175 mph)
- Max cruising speed at S/L: 139 knots (258 km/h; 160 mph)
- Max cruising speed at 1,525 m (5,000 ft): 135 knots (249 km/h; 155 mph)
- Econ cruising speed at S/L: 130 knots (241 km/h; 150 mph)
- Econ cruising speed at 1,525 m (5,000 ft): 126 knots (233 km/h; 145 mph)
- Max rate of climb at S/L: 579 m (1,900 ft)/min
- Service ceiling: 4,570 m (15,000 ft)
- Hovering ceiling IGE: ISA: 2,590 m (8,500 ft)
  - ISA + 20°C: 1,830 m (6,000 ft)
- Hovering ceiling OGE: ISA: 2,285 m (7,500 ft)
  - ISA + 20°C: 1,370 m (4,500 ft)
- Range, 2 min warm-up, standard fuel, no reserves:
  - at S/L: 260 nm (482 km; 300 miles)
  - at 1,525 m (5,000 ft): 287 nm (531 km; 330 miles)

Hughes 500MD/TOW Defender of the Israeli Air Force during nap-of-the-earth exercises

HUGHES MODEL 500 (MILITARY VERSIONS)

Foreign military versions of the Hughes Model 500 so far announced are as follows:

Model 500M. Initial uprated version of OH-6A. Power plant as for civil Model 500, but fuel capacity of 227 litres (60 US gallons). First deliveries to Colombian Air Force in April 1968. Now in service also in Japan (Army and Navy), Argentina, Bolivia, Denmark (Navy), Spain (Navy), Mexico and the Philippines. The Model 500Ms delivered to the Spanish Navy for ASW duties have an AN/ASQ-81 magnetic anomaly detector installed on the starboard side of the fuselage, and can carry two Mk 44 torpedoes beneath the fuselage. Control boxes for the MAD equipment are mounted on the instrument panel and centre pedestal, and special instrumentation includes a 6 in attitude indicator and radar altimeter. Licence manufacture also undertaken by RACA (Argentina), and by Kawasaki in Japan as the Models 369HM/OH-6J (now ended) and 369D/OH-6D.

Model 500MD Defender. Multi-role military version. Structurally similar to civil Model 500D, from which it differs in having as standard or optional equipment self-sealing fuel cells, engine inlet particle separator, armour protection, Hughes 'Black Hole Ocarina' infra-red suppressor, and provisions for the carriage and deployment of a variety of weapons, including TOW missiles. Its diverse capabilities include training, command and control, scout, light attack, ASW, troop lift, and logistical support duties. It can carry up to seven people, including the pilot; or, in ambulance configuration, two stretcher patients with attendants in addition to a flight crew of two. Licence manufacture also undertaken by BredaNardi in Italy and (Scout and TOW versions) by KAL in Republic of Korea.

The versions available or proposed in 1981 were as follows:

500MD Scout Defender. Basic military version, able to carry a variety of alternative weapons, including fourteen 2·75 in rockets and either a 7·62 mm Minigun with 2,000 rounds of ammunition, a 40 mm grenade launcher, a 7·62 mm Chain Gun with 2,000 rounds of ammunition, or a 30 mm Chain Gun with 600 rounds of ammunition. Operators include Kenyan Army (15), Republic of Korea Air Force (over 100), and Royal Moroccan Air Force (24).

500MD Quiet Advanced Scout Defender. Basically similar to Scout, but with added quietening kit and Hughes

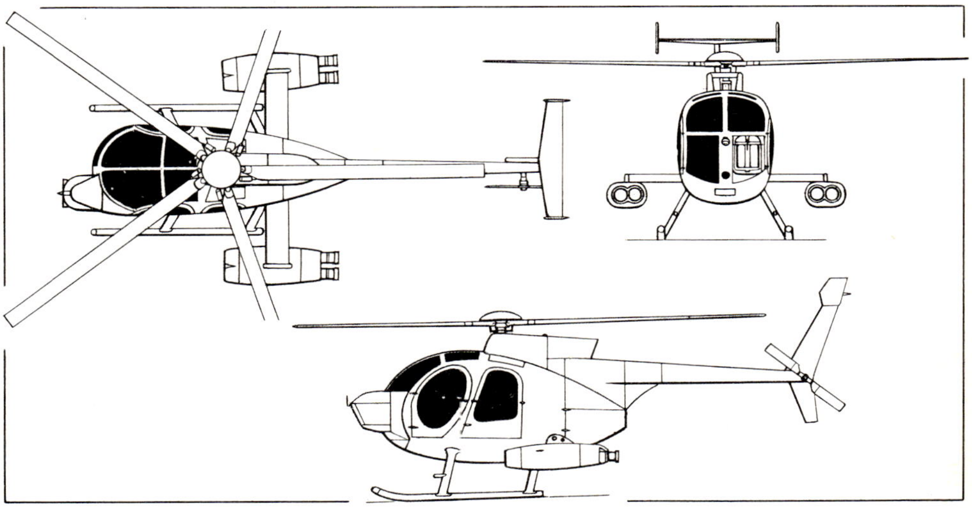

Hughes 500MD Defender with TOW missile launchers *(Pilot Press)*

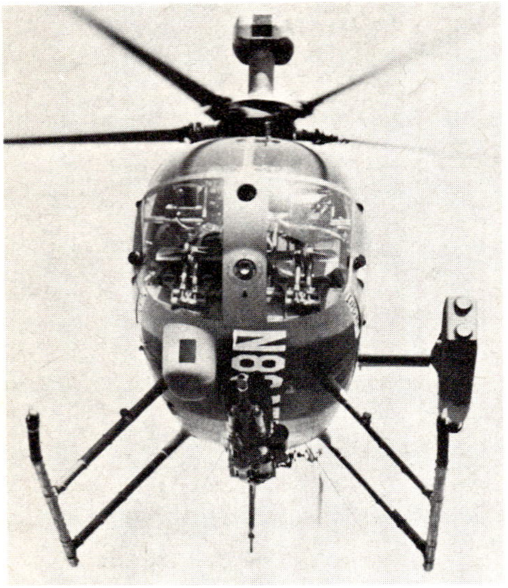

Hughes 500MD Defender II, with mast-mounted sight, underfuselage 30 mm Chain Gun, Stinger missile pod and other new features

Aircraft mast-mounted sight (MMS). Quietening kit features a slower-turning four-blade tail rotor, which imposes no reduction of performance. The MMS, described in the following Defender II paragraph, is mounted on a static mast 61 cm (2 ft) above the main rotor, and enables the aircraft to hover behind cover, using the small sight as a periscope to scan a large area out to a range of 3,000 m (9,840 ft). If employed to spot enemy armour, it is envisaged that Scouts would call in TOW Defenders to attack the targets.

500MD/TOW Defender. Anti-tank version armed with four TOW air-to-ground missiles. The TOW installation comprises four weapon pods, mounted two each side on a tubular mount carried through the lower aft fuselage, a stabilised telescopic sight mounted on the port side of the nose, sight control and armrest for the gunner, and a steering indicator for the pilot. Max T-O weight with four TOW missiles 1,360 kg (3,000 lb). In service with air forces of Israel (30), Kenya (15 ordered, of which 6 delivered by mid-1981), and Republic of Korea (25). Available also in **500MD/MMS-TOW** version, with Hughes Aircraft mast-mounted sight; empty weight of this version is reduced by 27 kg (60 lb).

500MD/ASW Defender. Version for anti-submarine warfare and surface search missions, with two crew, search radar on nose, AN/ASQ-81 towed MAD, smoke marker launchers, hauldown gear, emergency 'popout' floats and armament of two Mk 44 or Mk 46 homing torpedoes. Max T-O weight 1,610 kg (3,550 lb). Can remain on station for 1 h 48 min when operated at a typical ASW mission radius of 22-87 nm (40-160 km; 25-100 miles) from ship or shore base. Using its radar, 500MD/ASW could locate enemy destroyers and gunboats up to 150 nm (275 km; 172 miles) from its base-ship during a two-hour patrol. Twelve delivered to Taiwanese Navy.

500MD Defender II. Multi-mission version, introduced in Summer of 1980; available for delivery within two years. Five-blade main rotor standard; four-blade 'quiet' tail rotor optional: this turns at a rate 25 per cent slower than the standard two-blade rotor and is reported to be 47 per cent quieter in operation. Other options include Hughes Aircraft mast-mounted sight (MMS), two twin-round pods for four TOW anti-tank missiles, Hughes 30 mm Chain Gun (with firing rate reduced to 350 rds/min), Hughes 'Black Hole' infra-red suppression system, pod containing two Stinger or other air-to-air missiles, pilot's FLIR night vision system, AN/APR-39 (V-1) equipment to give warning that the helicopter is being tracked by hostile radar-directed weapon systems, self-sealing fuel tanks, auxiliary fuel tanks, and an advanced avionics/mission equipment package. The MMS uses a video link to TV displays for the crew, and includes a laser rangefinder. Use of the MMS enables the Defender II to hover virtually out of sight behind trees or natural terrain, while the crew surveys the battlefield over extended ranges.

Standard lightweight avionics equipment (SLAE) as developed for the OH-6A has been adapted for the 500MD with minimal changes. This equipment comprises AN/ARC-164 UHF/AM, AN/ARC-115 UHF/AM, AN/ARC-114 VHF/FM, ARN-89 ADF, APX-72 IFF transponder, AN/ASN-43 directional gyro, ID-1351 heading and bearing indicator, and C-6533/ARC intercom.

In addition to these export models, Hughes submitted to the US Army in 1981 a proposal, provisionally designated **OH-6D**, for consideration with the Bell Model 406 in the AHIP (Army Helicopter Improvement Program) competition for a 'near-term' scout helicopter. Basically, the OH-6D comprises an OH-6A airframe, upgraded with FAA-certificated components from the Model 500D, and fitted with graphite/Kevlar epoxy composite main rotor blades. First flight of the OH-6D took place in March 1981.

DIMENSIONS, EXTERNAL (500MD/TOW):

Diameter of main rotor	8·05 m (26 ft 4¾ in)
Diameter of tail rotor	1·40 m (4 ft 7¼ in)
Distance between rotor centres	4·63 m (15 ft 2½ in)
Length overall, rotors turning	9·39 m (30 ft 9½ in)
Length of fuselage	7·01 m (23 ft 0 in)
Height to top of rotor hub	2·65 m (8 ft 8½ in)
Height over tail (endplate fins)	2·71 m (8 ft 10¾ in)
Fuselage: Max width	1·40 m (4 ft 7¼ in)
Width over skids	1·95 m (6 ft 4¾ in)
Width over TOW pods	3·23 m (10 ft 7¼ in)
Tailplane span	1·68 m (5 ft 6 in)
Ventral fin ground clearance	0·67 m (2 ft 2½ in)

WEIGHTS (500M):

Weight empty	512 kg (1,130 lb)
Max normal T-O weight	1,157 kg (2,550 lb)
Max overload T-O weight	1,360 kg (3,000 lb)

PERFORMANCE (500M at max normal T-O weight):

Max level speed at 305 m (1,000 ft)	132 knots (244 km/h; 152 mph)
Cruising speed for max range at S/L	117 knots (217 km/h; 135 mph)
Max rate of climb at S/L	518 m (1,700 ft)/min
Service ceiling	4,390 m (14,400 ft)
Hovering ceiling IGE	2,500 m (8,200 ft)
Hovering ceiling OGE	1,615 m (5,300 ft)
Range at 1,220 m (4,000 ft)	318 nm (589 km; 366 miles)

HUGHES NOTAR HELICOPTER

On 20 May 1981, Hughes Helicopters announced that, following the award in September 1980 of a $1·4 million 24-month contract by the US Army Applied Technology Laboratory and the Defense Advanced Research Projects Agency, the company had designed and was to build a prototype no-tail-rotor (NOTAR) helicopter. This continues company-funded research and development efforts that began in 1975 and led, in 1977-78, to flight testing of an anti-torque tailboom which replaced a conventional tail rotor. Now, an Army OH-6 helicopter is to be modified to serve as the NOTAR prototype. The only change in this aircraft will be the installation of a completely new tailboom, manufactured by Aircraft Engineering Corporation of Paramount, California, under contract to Hughes.

Constructed of bonded aluminium honeycomb, this tailboom will have a variable-pitch fan mounted within the forward end to produce an airstream that will pressurise the hollow tailboom at just below 0·034 bars (0·5 lb/sq in). The fan will be driven by a shortened tail rotor driveshaft, and will absorb no more power from the engine than the OH-6's conventional tail rotor and its transmission system.

A slot, about 0·85 cm (0·33 in) wide, will run along the starboard lower side of the tailboom. Pressurised air venting through this slot will unite with the main rotor downwash, moving downward to follow the contour of the tailboom in the same way that an airstream travels over a conventional fixed wing: the resultant forces will counteract the torque of the main rotor.

To provide control in yaw, the aft end of the tailboom will incorporate a 'jet thruster', consisting of a fixed cone within a rotating cone. Air exits formed in both the port and starboard sides of the inner cone will allow pressurised air that has not vented through the slot to vent to atmosphere in controlled amounts, according to the position of a port in the outer cone, which will be rotated by conventional action of the pilot's rudder pedals. The standard OH-6 vertical fin, and stabiliser on the starboard side of the boom, will be mounted just forward of the jet thruster.

In hovering flight, the slotted tailboom will produce most of the necessary anti-torque forces; in forward or manoeuvring flight, when the main rotor downwash moves off the boom, the jet thruster and fixed tail surfaces will combine to produce the required forces for control. Hughes' engineers believe that a production version of NOTAR might also incorporate a controllable trim tab in the vertical fin, because there would not be any tail rotor flapping interference. A trim tab would lessen pilot workload, especially in crosswind conditions. A simple blipper-type button on the cyclic stick would activate the trim tab adjustments.

The first flight of the NOTAR prototype was scheduled for December 1981. Advantages claimed for the concept include elimination of such familiar tail rotor characteristics as safety problems, aerodynamic inefficiencies, noise, and maintenance costs. Most previous circulation control systems have required power in addition to that which is provided by the installed engine in a given aircraft.

HUGHES MODEL 600X

Hughes Helicopters announced at the Paris Air Show in June 1981 that it was "seriously considering" adding this six-seat development of the Model 500D to its product line, and would make a decision later that year.

The Model 600X retains the power plant and transmission of the Model 500D, but the airframe incorporates a 38 cm (15 in) widening of the standard cabin, providing 75 per cent more headroom. It is estimated that range and endurance would be 49 per cent greater than for the 500D.

HUGHES MODEL 77

US Army designation: AH-64

The Model 77 was designed by Hughes to meet the US Army's requirement for an Advanced Attack Helicopter (AAH) capable of undertaking a full day/night/adverse weather anti-armour mission, and of fighting, surviving and 'living with' troops in a front-line environment. Two flight test prototypes were built, for competitive evaluation against Bell's YAH-63, and these made their initial flights on 30 September and 22 November 1975. A ground test vehicle was also completed. The Hughes contract covered, in addition, development of the XM230 Chain Gun for installation in the Model 77 prototypes, which have the US Army designation YAH-64.

Selection of the YAH-64 was announced on 10 December 1976. This was followed by Phase 2 of the programme, which involved fitting the prototypes with advanced avionics, electro-optical equipment and weapon fire control systems, for further evaluation; continued development of the airframe; and the manufacture of three more aircraft.

These are identified as AVO.1 to AVO.3. In June 1979 the first was used by the US Army for continuing evaluation. AVO.3 was transferred to the Army's Yuma Proving Ground, Arizona, on 4 June, equipped with the Northrop TADS/PNVS. AVO.2, with the Martin Marietta TADS/PNVS, followed at the end of the same month. Company R & D with these two aircraft was continued until the end of the year, with Army fly-off testing of the competing TADS/PNVS systems beginning in January 1980, and scheduled for completion by August 1981.

Three additional development aircraft, designated AVO.4 to AVO.6, were flying by the Spring of 1980, as total systems aircraft.

Teledyne Ryan is responsible for the YAH-64 fuselage, wings, engine nacelles, avionics bays, canopy and tail unit. Martin Marietta and Northrop are developing competitive equipment to fulfil the TADS (target acquisition and designation sight) and PNVS (pilot's night vision sensor) tasks in the production AH-64. The former system includes direct-view optics, forward-looking infra-red, TV, laser designator/rangefinder, and a laser tracker. The PNVS consists of an advanced FLIR installation.

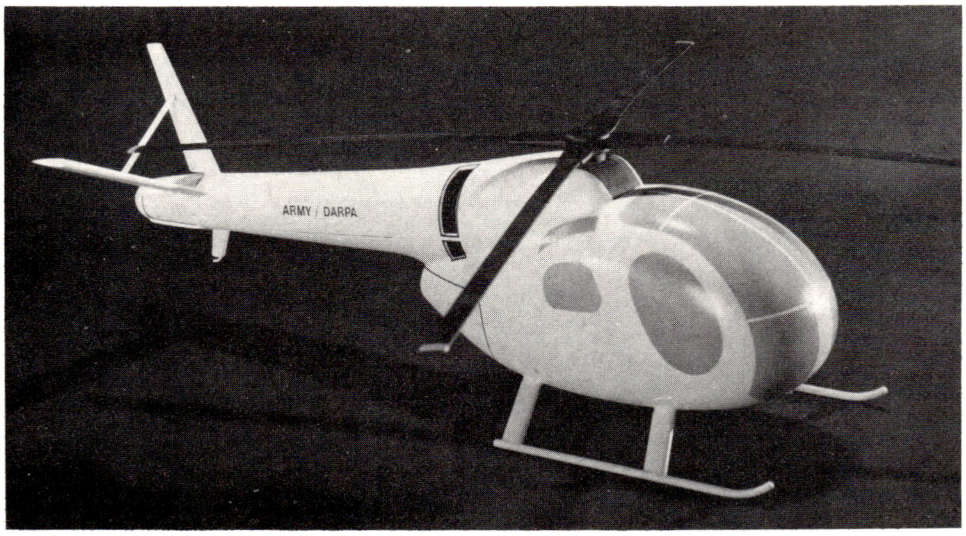

Model of Hughes prototype no-tail-rotor (NOTAR) helicopter

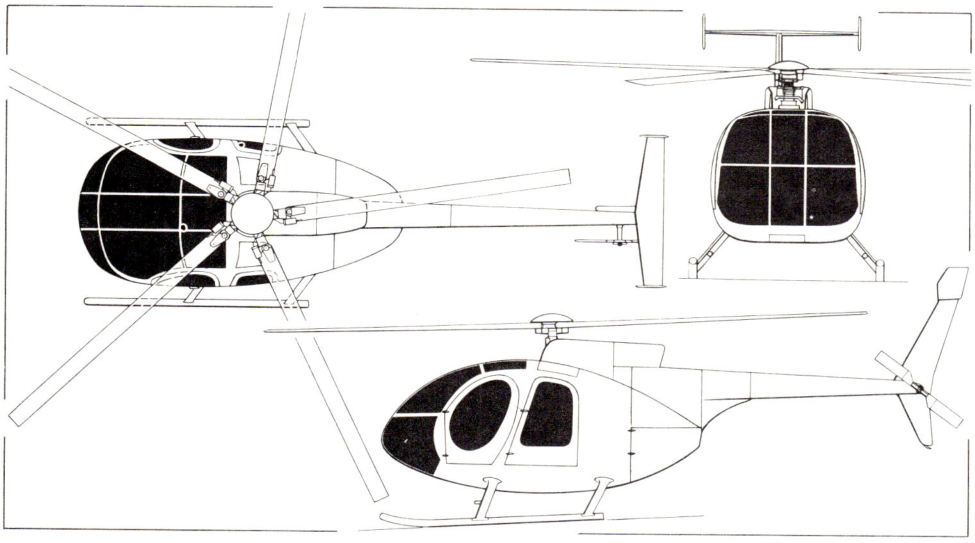

Hughes Helicopters' proposed six-seat Model 600X *(Pilot Press)*

In early 1978 the two Phase 2 helicopters began a new series of tests to evaluate planned design modifications known as Mod 1. These included swept tips on the main rotor blades; a Hughes-developed 'Black Hole' IR suppressor for each engine exhaust; a redesigned T tail; and a 76 mm (3 in) increase in tail rotor diameter. A Mod 2 programme, started later in that year, introduced final airframe improvements, as well as all mission equipment including armament, fire control and nav/com systems. These included cockpit windows of modified shape, with single-curvature side panels; and extending aft, to a point below the wing leading-edges, the fuselage side fairings over the forward avionics bay. The two prototype helicopters then completed a programme to confirm the airworthiness of the Mod 2 airframe changes and initial tests of the weapon system, including the Hellfire missile, 2·75 in rocket, 30 mm Chain Gun, and the fire control system. By mid-June 1979 five ballistic Hellfire missiles had been fired, without guidance system. The first fully-guided launch of a Hellfire was made successfully on 18 September 1979 at the Yuma, Arizona, proving grounds. Target for the laser-guided system was designated from the ground, and a direct hit was made. Northrop TADS/PNVS was installed and functional in the launch YAH-64, but was not used. Approximately one month later the first test firing was made with the Northrop TADS/PNVS providing illumination of the target. Other systems which were fully integrated and flight tested during 1979 included the Litton LR-80 HARS, and the Singer-Kearfott Doppler navigation system.

During 1980 AVO.4 was flown at a speed of almost 206 knots (382 km/h; 237 mph) during the flight envelope expansion testing, and had also demonstrated manoeuvring capability of more than 3*g* at speeds from 80 to 164 knots (148 to 304 km/h; 92 to 189 mph). All six YAH-64s have been converted to the Mod 1 configuration. Armament and fire control survey (AFCS) Part 3 was initiated by AVO.6, and joined later by AVO.2. This extends the test firings of 30 mm Chain Gun ammunition, 2·75 in air-to-ground rockets, and Hellfire missiles, as in AFCS Part 2, but covers all regimes and night operations, requiring use of the PNVS. During AFCS Part 2, AVO.2 established a record for the longest-range hit on a tank, scored by a Hellfire missile.

By early June 1981, YAH-64 prototypes had fired more than 50 Hellfire missiles, nearly 3,000 2·75 in rockets, and more than 25,000 rounds of 30 mm ammunition. Three YAH-64s were then about to begin a three-month Army exercise called OT-II (Operational Test II), in which all major weapons and systems, including the Martin Marietta TADS/PNVS and the Honeywell IHADSS (integrated helmet and display sight system), were to be field tested under operational conditions. The OT-II test area closely resembles the kind of terrain to be found in Western Europe, with low, tree-covered hills and gently rolling countryside.

The 56-month AAH development contract period was due to end in August 1981, with a production decision expected to be announced in December 1981. This means that initial delivery of production aircraft could begin in December 1983, with entry into service in 1984. Hughes announced in March 1981 the receipt of a $2·1 million contract for the procurement of long lead time items. The US Army's current stated requirement is for 536 AAHs.

The following description applies to the YAH-64 prototypes, except where indicated:

Type: Prototype armed helicopter.

Rotor System: Four-blade fully-articulated main rotor and four-blade tail rotor, with blades manufactured by Tool Research and Engineering Corpn (Advanced Structures Division). Main rotor blades are of high-camber aerofoil section and broad chord. Each blade has five stainless steel spars lined with structural glassfibre tubes, a laminated stainless steel skin and a composite aft section, bonded together. Blades are attached to hub by a laminated strap retention system similar to that of the OH-6A, and are fitted with elastomeric lead/lag dampers and offset flapping hinges. Four-blade tail rotor comprises two sets of two blades, mounted on port side of pylon/fin support structure at optimum quiet setting of approx 60°/120° to each other. Rotor mast similar to that of OH-6A, with driveshaft turning within a hollow, fixed outer shaft. Entire system capable of flight in negative *g* conditions.

Rotor Drive: Transmission to main rotor via Litton (Precision Gear Division) engine nose gearboxes, and to tail rotor via Aircraft Gear Corpn grease-lubricated intermediate and tail rotor gearboxes, with Bendix driveshafts and couplings. Garrett cooling fan for tail rotor gearbox. Redundant flight control system for both rotors. Selected dynamic components constructed of 70/49 aluminium and electro-slag remelt (ESR) steel; critical parts of transmission (eg, bearings) have ESR collars for protection against hits by 12·7 mm or 23 mm ammunition. Rotor/engine rpm ratios approx 1 : 66·7 for main rotor, approx 1 : 14·3 for tail rotor.

Wings: Cantilever mid-wing monoplane, of low aspect ratio, aft of cockpit. Trailing-edge flaps deploy automatically as function of control attitude and airspeed (max deflection 20°), and can be deflected 45° upward to offload wings in an emergency autorotative landing. Wings are removable, and attach to sides of cabin for transport and storage. Two hardpoints beneath each wing for the carriage of mixed ordnance.

Fuselage: Conventional semi-monocoque aluminium structure, built by Teledyne Ryan Aeronautical. Designed to survive hits by 12·7 mm and 23 mm ammunition.

Tail Unit: Hinged and foldable pylon structure with tail rotor mounted on port side. Low-mounted all-flying tailplane (replacing earlier fixed T tail).

Landing Gear: Menasco tailwheel type, with single wheel on each unit. Main legs fold rearward to reduce overall height for storage and transportation. Fully-castoring, self-centering tailwheel. Main-wheel tyres size 22 × 8; tailwheel tyre size 13 × 5. Hydraulic brakes.

Power Plant: Two 1,145 kW (1,536 shp) General Electric T700-GE-700 turboshaft engines, derated for normal operations to provide reserve power for combat emergencies. Engines mounted one on each side of fuselage, above wings. Two crashproof fuel cells, combined capacity 1,366 litres (361 US gallons).

Accommodation: Crew of two in tandem, co-pilot/gunner in front and pilot aft on 483 mm (19 in) elevated seat. Large, shaped transparent cockpit enclosure for optimum field of view. Lightweight boron armour shields in cockpit floor and sides. Cockpits separated by armour plating and a transparent plastics blast shield offering protection against 23 mm high-explosive and armour-piercing rounds.

Systems and Equipment: Large avionics bay adjacent to gunner's position, in lower fuselage. Parker-Bertea hydraulic control system, with hydraulic actuators ballistic tolerant to direct 12·7 mm hits. In the event of hydraulic control system failure, the system adjusts to secondary fly-by-wire control. Bendix electrical power system, with two fully-redundant engine-driven generators and standby DC battery. Sperry Flight Systems automatic stabilisation equipment. Litton LR-80 strapdown heading and attitude reference system (HARS). Singer-Kearfott lightweight Doppler navigation system. Garrett totally integrated pneumatic system includes a shaft-driven compressor, air turbine starters, pneumatic valves, temperature control unit and environmental control unit. Garrett also supplies the GTP 36-55(C) APU. Sperry Flight Systems all-raster symbol generator is under development to process TV

Hughes YAH-64 armed helicopter prototype, with 16 Hellfire missiles under its wings

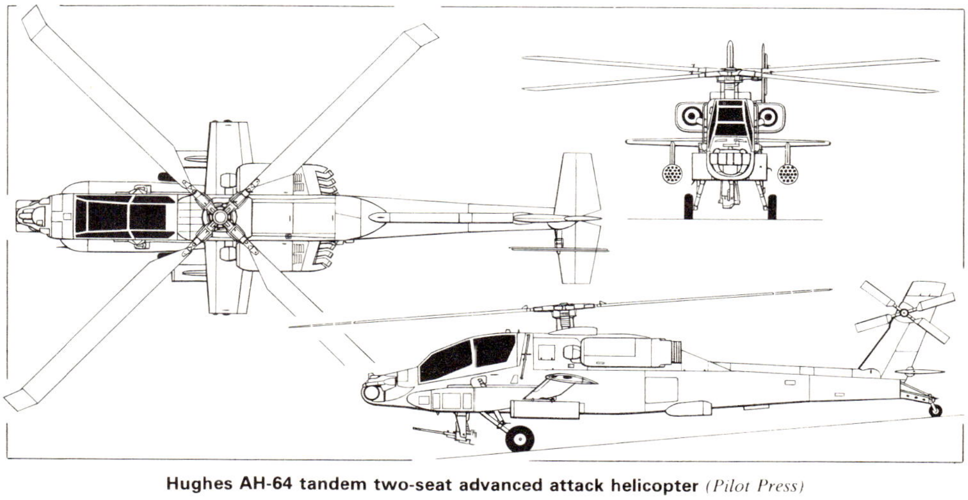

Hughes AH-64 tandem two-seat advanced attack helicopter *(Pilot Press)*

data from IR and other sensors, superimpose symbology and distribute the combination to CRT and helmet-mounted displays in the aircraft. BITE fault detection/location system. Hughes Black Hole IR suppression system, with no moving parts, to protect the aircraft from heat-seeking missiles. This eliminates an engine bay cooling fan, by operating from engine exhaust gas through ejector nozzles to lower the gas plume and metal temperatures.

ARMAMENT AND OPERATIONAL EQUIPMENT: Flexible armament consists of a Hughes-developed XM230E1 30 mm Chain Gun, mounted in an underfuselage turret between the main-wheel legs, and having a normal rate of fire of 800 rds/min of Honeywell TP, HE and HEDP ammunition. Ammunition max load is 1,200 rds, and is interoperable with NATO Aden/DEFA 30 mm rounds. Turret designed to collapse into fuselage between pilots in the event of a crash-landing. Four underwing hardpoints, on which can be carried up to sixteen Hellfire anti-tank missiles; or up to seventy-six 2·75 in folding-fin rockets in their launchers; or a combination of Hellfire missiles and rockets. CPG stabilised sight in forward fuselage, ahead of cockpit, incorporates day and night (FLIR: forward-looking infra-red) sighting equipment, laser ranger and target designator, and laser tracker equipment. Co-pilot/gunner has primary responsibility for firing guns and missiles, but pilot can override his controls to fire gun or launch missiles. Pilot's night vision sensor (PNVS) in extreme tip of nose. Integrated helmet and display sighting system (IHADSS) by Honeywell Avionics Division, used by both crew members, to enhance speed and flexibility of target acquisition. Forward bay includes avionics for stabilised sight, missiles and fire control computer provided by Teledyne Systems Inc.

DIMENSIONS, EXTERNAL (AH-64A):
- Diameter of main rotor: 14·63 m (48 ft 0 in)
- Diameter of tail rotor: 2·79 m (9 ft 2 in)
- Main rotor blade chord: 0·53 m (1 ft 9 in)
- Length of fuselage: 14·97 m (49 ft 1½ in)
- Length overall, rotors turning: 17·39 m (57 ft 0½ in)
- Wing span: 5·23 m (17 ft 2 in)
- Height to top of rotor hub: 4·22 m (13 ft 10 in)
- Height over tail fin: 3·54 m (11 ft 7½ in)
- Distance between c/l of pylons: 3·20 m (10 ft 6 in)
- Tailplane span: 3·40 m (11 ft 1¾ in)
- Wheel track: 2·03 m (6 ft 8 in)

AREAS:
- Main rotor disc: 168·11 m² (1,809·5 sq ft)
- Tail rotor disc: 6·13 m² (66·0 sq ft)

WEIGHTS:
- Weight empty: 4,657 kg (10,268 lb)
- Primary mission gross weight: 6,271 kg (13,825 lb)
- Structural design gross weight: 6,650 kg (14,660 lb)
- Max T-O weight: 8,006 kg (17,650 lb)

PERFORMANCE (at 6,316 kg; 13,925 lb AUW, ISA except where indicated):
- Never-exceed speed: 204 knots (378 km/h; 235 mph)
- Max level speed: 167 knots (309 km/h; 192 mph)
- Max cruising speed: 158 knots (293 km/h; 182 mph)
- Max vertical rate of climb at S/L: 878 m (2,880 ft)/min
- Max vertical rate of climb at 1,220 m (4,000 ft) at 35°C: 259 m (850 ft)/min
- Service ceiling: 6,250 m (20,500 ft)
- Service ceiling, one engine out: 3,505 m (11,500 ft)
- Hovering ceiling IGE: 4,633 m (15,200 ft)
- Hovering ceiling OGE: 3,780 m (12,400 ft)
- Max range, internal fuel: 330 nm (611 km; 380 miles)
- Ferry range, max internal and external fuel, still air: 974 nm (1,804 km; 1,121 miles)
- Endurance at 1,220 m (4,000 ft) at 35°C: 1 h 50 min
- Max endurance, internal fuel: 3 h 23 min

ICX—*See Lewis Aircraft Corporation*

ISHAM
ISHAM AIRCRAFT

PO Box 12172, Mid Continent Airport, Wichita, Kansas 67277
Telephone: (316) 755 0713
PRESIDENT: Brad E. Isham

ISHAM (PIPER) PA-28 CHEROKEE and (CESSNA) R172K SKYHAWK CONVERSIONS

Isham AirCraft is marketing, with FAA approval, modification kits for various models of the Piper Cherokee which are designed to improve the aircraft's performance and handling during approach and landing. They are also claimed to make slight improvements in take-off distance, rate of climb and cruising speeds.

Kits are available for the Piper PA-28-140, -150, -160 and -180, to provide light alloy wing and tailplane extensions, Hoerner-type wingtips, dorsal fins of ABS plastics, and a third window on each side. The third-window kit is available also for the Piper PA-28-235. A kit to provide the retractable-gear Piper PA-28R-180 and -200 with wing extensions, Hoerner wingtips and dorsal fin can be obtained from Isham AirCraft, as well as a kit to re-rate the Continental IO-360-K power plant of the Cessna R172K from 145 kW (195 hp) to 157 kW (210 hp) at take-off.

Piper PA-28R-180 Cherokee with Isham AirCraft wing and fin modifications

Performance and specification details of aircraft to which these kits have been applied are given in the accompanying table.

Dimensions/Performance	PA-28-140 PA-28-150	PA-28-180	PA-28R-180 PA-28R-200	R172K	Units
Wing span	9·81 (32·2)	9·81 (32·2)	9·81 (32·2)	N/C	m (ft)
Tailplane span	3·91 (12·83)	3·91 (12·83)	N/C	N/C	m (ft)
Wing area	15·7 (169)	15·7 (169)	15·7 (169)	N/C	m² (sq ft)
Max wing loading	62·1 (12·72)	69·3 (14·2)	72·2 (14·79)	N/A	kg/m² (lb/sq ft)
Max level speed	124 (230; 143)	135 (249; 155)	150 (278; 173)	N/A	knots (km/h; mph)
Max cruising speed (75% power)	119 (220; 137)	127 (235; 146)	143 (266; 165)	N/A	knots (km/h; mph)
Stalling speed	N/C	N/C	N/C	N/A	knots (km/h; mph)
Max rate of climb at S/L	247 (810)	274 (900)	302 (990)	351 (1,150)	m (ft)/min
T-O to 15 m (50 ft)	472 (1,547)	450 (1,477)	477 (1,565)	302 (990)	m (ft)
Landing from 15 m (50 ft)	N/C	N/C	N/C	N/A	m (ft)
Range, max fuel	687 (1,273; 791)	643 (1,191; 740)	757 (1,403; 872)	N/A	nm (km; miles)
Reduction in sink rate	27	26·4	27	N/A	%

Figures quoted are for S/L ISA at max T-O weight.
N/A: not available. N/C: no change.

JACOBS
JACOBS JETS

745 Emerson Street, Palo Alto, California 94301
Telephone: (415) 328 4672

JACOBS JETS PARIS JET IV

This company has developed a conversion of the Morane-Saulnier (Potez) MS 760 Paris military trainer, to produce a four-seat executive aircraft named Paris Jet IV. The conversion involves a refit of the cabin, the installation of new avionics, and the introduction of some aerodynamic refinements to improve performance. The original engine installation of two Turboméca Marboré VI turbojets, each of 4·71 kN (1,058 lb st), is retained, but it is reported that Jacobs has under consideration the installation of new Garrett engines.

In the Spring of 1981 the company had received orders for four conversions, and had about 20 aircraft available for early modification. About 250 of the military trainers are still in service, and it is believed that these may soon come on to the civil market. A description of the Morane-Saulnier MS 760 Paris can be found in the 1966-67 *Jane's*.

JETCRAFTERS
JETCRAFTERS INCORPORATED

PO Box 32622, San Antonio, Texas 78216
Telephone: (512) 657 2700
PRESIDENT: E. J. Swearingen
VICE-PRESIDENTS:
E. A. Banks
G. M. Loudermilk

Well-known as the designer of the Merlin and Metro twin-turboprop executive transports, 'Ed' J. Swearingen established Jetcrafters Inc to initiate a modification programme to enhance the performance and operating economy of the Beechcraft King Air 90 models.

The company's facility at San Antonio is also manufacturing engine nacelles for AiResearch Aviation Company's 731 JetStar conversions, and sheet metal assemblies for Learjet and Falcon Jet aircraft.

JETCRAFTERS TAURUS MODIFICATION

Any model of the Beechcraft King Air 90 series is suitable for incorporation of the Jetcrafters Taurus modification. This involves replacing the existing engines with Pratt & Whitney Aircraft of Canada PT6A-135 turboprop engines, each flat rated to 522 kW (700 shp); modification of the existing engine nacelles to reduce their height and cross-sectional area; provision of new cowls, engine air inlets, inertial water/ice separators, a new low-profile exhaust system to reduce drag and increase thrust, new oil cooler airflow control system, and a new integral nacelle fuel tank with full lightning strike protection; removal of the existing engine-driven cabin supercharger and provision of a new flow control system to provide dual bleed air pressurisation; and the supply of new engine torque indicators.

After embodiment of the Taurus modification, the King Air 90, A90, B90, C90 and E90 are redesignated respectively Taurus 90, A90, B90, C90 and E90.

Brief details of performance improvements follow:

PERFORMANCE (A: Taurus 90; B: A90; C: B90; D: C90; and E: E90):
- Max cruising speed at optimum altitude:
  - A, B, C, D, E — 278 knots (515 km/h; 320 mph)
- Stalling speed, flaps down:
  - A, B, C, D, E — 69 knots (127 km/h; 79 mph)
- Max rate of climb at S/L: A — 913 m (2,995 ft)/min
  - B — 884 m (2,900 ft)/min
  - C, D — 852 m (2,795 ft)/min
  - E — 814 m (2,670 ft)/min
- Service ceiling: A, B, C, D, E — 11,430 m (37,500 ft)
- T-O to 15 m (50 ft): A, B — 625 m (2,050 ft)
  - C — 533 m (1,750 ft)
  - D — 689 m (2,261 ft)
  - E — 617 m (2,024 ft)
- Landing from 15 m (50 ft): A, B — 716 m (2,350 ft)
  - C, D — 613 m (2,010 ft)
  - E — 643 m (2,110 ft)

Jetcrafters Taurus modification of a Beechcraft Model 90 King Air

KAMAN
KAMAN AEROSPACE CORPORATION
(a subsidiary of Kaman Corporation)

HEAD OFFICE: Old Windsor Road, Bloomfield, Connecticut 06002

Telephone: (203) 242 4461

PRESIDENT AND CHIEF EXECUTIVE: Fred L. Smith

VICE-PRESIDENTS:
- Wayne D. Hudson (Manufacturing)
- Eamon N. Kelly (Contracts)
- Walter R. Kozlow (Operations)
- Cornelia V. Lynn (Personnel)
- Donald W. Robinson (Marketing and Planning)
- Llewellyn C. Schuler (Engineering)

SECRETARY: J. S. Murtha

MANAGER, WASHINGTON OPERATIONS: Owen F. Polleys

The original Kaman Aircraft Corporation was founded in 1945 by Mr Charles H. Kaman, who continues as President and Chairman of the Board of Kaman Corporation. Its initial programme was to develop and test a novel servo-flap control system for helicopter rotors, and the Kaman K-125 of 1947 was the first in a series of 'synchropter' designs with intermeshing contra-rotating rotors and the servo-flap control system. The later H-2 Seasprite naval helicopter utilises the servo-flap control system on a single main rotor.

Current research and development programmes at Kaman Aerospace include evaluation of a Circulation Control Rotor (CCR) technology demonstrator aircraft for Naval Air Systems Command. In this concept a thin jet of air is ejected from a slot near the trailing-edge of each rotor blade, inducing a circulation which results in lift augmentation. Since conventional cyclic pitch is eliminated, the rotor hub and controls are simplified considerably. Full-scale whirl tests and wind tunnel tests were completed successfully in early 1978; flight tests on a Kaman NHH-2D helicopter began in September 1979 and the programme was continuing in early 1981.

Other R & D programmes are under sponsorship of the US Army Research and Technology Laboratories, Applied Technology Laboratory, Fort Eustis, Virginia. They include a controllable twist rotor (CTR); dynamic anti-resonant vibration isolator; composite rotor hub; repairable/expendable main rotor blade concept; new concepts in structural dynamic testing; maintainability of major helicopter components; elastic pitch beam tail rotor; and new design, fabrication and inspection techniques for helicopter structures. Kaman modified the CTR hardware to incorporate a multicyclic flap system at the request of NASA's Ames Research Center, Moffett Field, California, and testing in the 12 × 24 m (40 × 80 ft) wind tunnel has been completed. Flight testing of the CTR on NASA's Sikorsky-built Rotor Systems Research Aircraft (RSRA) is scheduled for 1983.

Kaman has been engaged in the development and testing of remotely piloted helicopters since 1952 (see RPVs and Targets section).

Kaman Aerospace is engaged in several major airframe programmes as a subcontractor, including the construction of flaps, slats, spoilers and cove doors for the US Navy's F-14 Tomcat fighter; engine access, tailpipe, and landing gear doors for the US Navy's A-6E and EA-6B aircraft; detail parts and tools on the US Air Force's A-10A Thunderbolt II close support aircraft; spare flaps and components for wing control surfaces of the C-5A heavy transport aircraft; components of the external fuel tank and solid rocket boosters of NASA's Space Shuttle Orbiter programme; acoustic structural components for jet engines; and structural components for various business and commercial aircraft, including the Boeing Model 767.

Kaman Aerospace is also a supplier of bonded components for C-130, B-52 and UH-1 aircraft for Army and Air Force government agencies. Kaman designed, and since mid-1977 has been producing, an all-composite rotor blade for Bell AH-1 HueyCobras in service with the US Army. The first AH-1 fitted with these blades made its first flight on 26 July 1976. Improved performance, life and operational features have been demonstrated.

Kaman Circulation Control Rotor (CCR) on NHH-2D flight test helicopter

Kaman SH-2F Mk I LAMPS ASW helicopter, showing sensors and stores

Kaman Sciences Corporation, with headquarters at Colorado Springs, Colorado, is engaged in nuclear research, weapons studies, advanced aerodynamics, computer programming and time sharing, neutron generators, advanced materials, measuring devices, systems analysis and solar energy. Kaman's general aviation subsidiaries comprise AirKaman Inc, a fixed-base operator at Bradley International Airport, Windsor Locks, Connecticut; Air Kaman of Omaha, Nebraska, and AirKaman of Jacksonville, Florida, all of which provide sales and service of light and twin-engined business aircraft, repairs, fuel sales, charter service, flight training and airline services.

KAMAN SEASPRITE

US Navy designations: HH-2 and SH-2

The prototype Seasprite flew for the first time on 2 July 1959, and many versions (described in previous editions of *Jane's*) were produced subsequently for the US Navy. Production of the SH-2F version was restarted in 1981.

From 1967, all of the original UH-2A/B Seasprites were converted progressively to UH-2C twin-engined configuration, with two 932 kW (1,250 shp) General Electric T58-GE-8B turboshaft engines in place of the former single T58. They have since undergone further modification, under the US Navy's Mk I LAMPS (Light Airborne Multi-Purpose System) programme, to provide helicopters for ASW (anti-submarine warfare) and ASST (anti-ship surveillance and targeting) operations.

The following versions remained available in early 1981:

HH-2D. Two aircraft, without LAMPS modifications, assigned to coast and geodetic survey work.

NHH-2D. Test aircraft assigned to the circulation control rotor (CCR) programme.

SH-2F. Deliveries of this developed Mk I LAMPS version began in May 1973 and the first unit became operational with squadron HSL-33, deployed to the Pacific, on 11 September 1973. Eighty-eight SH-2Fs were delivered, and earlier SH-2Ds are being uprated to SH-2F configuration, this programme being scheduled for completion in March 1982. An additional 18 new-production SH-2Fs were ordered on 21 April 1981; deliveries of these are due to begin in late 1983.

Features of the SH-2F include increased-strength landing gear; a shortened wheelbase by relocation of the tailwheel; and T58-GE-8F engines. In 1980 improvements were made to the LN-66HP radar, tactical navigation system, ESM, sonobuoys and data link.

In January-February 1973 Kaman flight-tested the prototype for qualification at a maximum gross weight of 6,033 kg (13,300 lb), which is 227 kg (500 lb) more than the current SH-2F. This could be utilised as increased payload, or in the form of additional fuel in larger auxiliary tanks to provide extended range and endurance in a new production version of the SH-2. US Navy tests have proved the SH-2 suitable for dipping sonar operations, air-to-surface missile firing, and equipment with various guns and rockets.

There are currently eight HSL LAMPS squadrons. Operational deployment began on 7 December 1971. By January 1980, 160 LAMPS SH-2D/F detachments had been deployed (not simultaneously) on long cruises, primarily in the Mediterranean and Pacific, on the following ship classes: FFG-7, DD-963, FFG-1, FF-1052, FF-1040, CGN-11, CG-25 and CG-26. The DD-963 and FFG-7 classes are designed to operate with two LAMPS helicopters per ship.

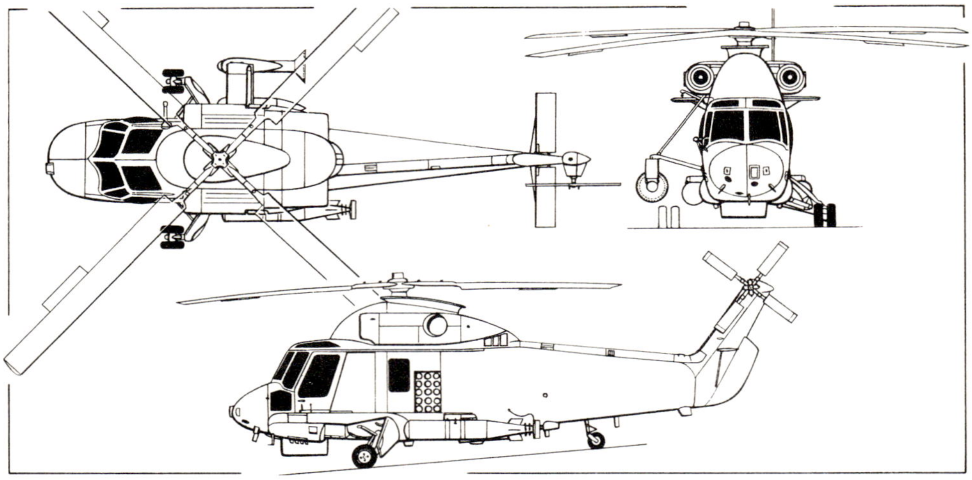

Kaman SH-2F Seasprite Mk I Light Airborne Multi-Purpose System (LAMPS) helicopter *(Pilot Press)*

The following details apply to the SH-2F version of the Seasprite:

Type: Naval anti-submarine warfare and anti-ship missile defence helicopter, with secondary capability for search and rescue, observation and utility missions.

Rotor System: Four-blade main and tail rotors. Kaman '101' main rotor utilises titanium hub and retention assemblies, reducing the number of control elements by two-thirds, and offering increased life for the entire rotor system. Blades of aluminium and glassfibre construction, with servo-flap controls. Blades folded manually. Main rotor rpm 287.

Fuselage and Tail Unit: All-metal semi-monocoque structure, with flotation hull housing main fuel tanks. Nose split on centreline, to fold rearward on each side to reduce stowage space required. Fixed horizontal stabiliser on tail rotor pylon.

Landing Gear: Tailwheel type, with forward-retracting twin main wheels and non-retractable tailwheel. Liquid spring shock-absorbers in main-gear legs; oleo-pneumatic shock-absorber in tailwheel unit, which is fully-castoring for taxying but locked fore and aft for T-O and landing. Main wheels have 8-ply tubeless tyres size 17·5 × 6·25-11, pressure 17·25 bars (250 lb/sq in); tailwheel 10-ply tubeless tyre size 5·00-5, pressure 11·04 bars (160 lb/sq in).

Power Plant: Two 1,007 kW (1,350 shp) General Electric T58-GE-8F turboshaft engines, mounted on each side of rotor pylon structure. Normal fuel capacity of 1,499 litres (396 US gallons), including external auxiliary tanks with a capacity of 454·6 litres (120 US gallons).

Accommodation: Crew of three, consisting of pilot, co-pilot and sensor operator. One passenger or litter patient with LAMPS equipment installed; four passengers or two litters with sonobuoy launcher removed. Provision for transportation of internal or external cargo.

Avionics and Equipment: Include Canadian Marconi LN-66HP surveillance radar; ASQ-81 magnetic anomaly detector; ALR-54 passive radiation detection receivers (being replaced by ALR-66); SSQ-41 passive and SSQ-47 active sonobuoys (being replaced by DIFAR and DICASS); smoke markers; one or two torpedoes; PT-429 plotting board system (being replaced by Teledyne Systems ASN-123 tactical navigation system with computer and CRT display); APN-182 Doppler radar; and AYK-2 analogue navigation computer. Cargo hook for external loads, capacity 1,814 kg (4,000 lb). Rescue hoist, capacity 272 kg (600 lb).

Dimensions, external:

Diameter of main rotor	13·41 m (44 ft 0 in)
Main rotor blade chord	0·55 m (21·6 in)
Diameter of tail rotor	2·49 m (8 ft 2 in)
Tail rotor blade chord	0·236 m (9·3 in)
Length overall (rotors turning)	16·03 m (52 ft 7 in)
Length overall, nose and blades folded	11·68 m (38 ft 4 in)
Height overall (rotors turning)	4·72 m (15 ft 6 in)
Height to top of rotor head	4·14 m (13 ft 7 in)
Stabiliser span	2·97 m (9 ft 9 in)
Wheel track (outer wheels)	3·30 m (10 ft 10 in)
Wheelbase	5·11 m (16 ft 9 in)

Weights:

Weight empty	3,193 kg (7,040 lb)
*Normal T-O weight	5,805 kg (12,800 lb)

**Although not yet certificated for a T-O gross weight of 6,033 kg (13,300 lb), all testing has been accomplished at that weight*

Performance (at normal T-O weight, except where indicated):

Max level speed at S/L	143 knots (265 km/h; 165 mph)
Normal cruising speed	130 knots (241 km/h; 150 mph)
Max rate of climb at S/L	744 m (2,440 ft)/min
Service ceiling	6,860 m (22,500 ft)
Hovering ceiling IGE	5,670 m (18,600 ft)
Hovering ceiling OGE	4,695 m (15,400 ft)
Normal range with max fuel	367 nm (679 km; 422 miles)

LAKE

LAKE AMPHIBIAN INC

Manufacturing and Sales: Laconia Airport, Laconia, New Hampshire 03246
Telephone: (603) 524 5868
Telex: 94-3554 Lake Air Lana
President: Armand Rivard
Vice-Presidents:
Bruce A. Rivard
Gordon Collins (Marketing)
Secretary: B. A. Sigsbee
Treasurer: Herbert P. Lindblad

This company has been reconstituted as Lake Amphibian Inc, and in 1981 was continuing production of the LA-4 amphibian.

Lake LA-4-200 Buccaneer four-seat light amphibian (Avco Lycoming IO-360-A1B flat-four engine)

LAKE LA-4-200 BUCCANEER

Design of the original C-1 Skimmer was started in August 1946. Construction of the prototype began in January 1947 and it flew for the first time in May 1948. Versions of the Lake LA-4 developed from the improved C-2 Skimmer IV have included the LA-4, LA-4A, LA-4P, LA-4S and LA-4T, as described in previous editions of *Jane's*.

The LA-4-200 current production version, described here, received FAA certification in 1970.

LA-4 c/n 1000 was completed on 1 January 1980.

Type: Single-engined four-seat amphibian.

Wings: Cantilever shoulder-wing monoplane with tapered wing panels attached directly to sides of hull. Wing section NACA 4415 at root, NACA 4409 at tip. Dihedral 5° 30′. Incidence 3° 15′. Structure consists of duralumin leading- and trailing-edge torsion boxes separated by a single duralumin main spar. All-metal ailerons and hydraulically-operated slotted flaps over 80% of span. Ground-adjustable trim tabs on ailerons. Wing balancer floats are light alloy monocoque structures.

Hull: Single-step all-metal structure, with double-sealed boat hull. Alodined and zinc chromated inside and out against corrosion, with polyurethane paint exterior finish.

Tail Unit: Cantilever all-metal structure. Outboard elevator section separate from inboard section and actuated hydraulically for trimming. Retractable water rudder in base of aerodynamic rudder.

Landing Gear: Hydraulically-retractable tricycle type. Consolidated oleo-pneumatic shock-absorbers in main gear, which retracts inward into wings. Long-stroke nosewheel oleo retracts forward. Gerdes main wheels with Goodyear tyres, size 6·00-6, pressure 2·41 bars (35 lb/sq in). Gerdes nosewheel with Goodyear tyre size 5·00-5, pressure 1·38 bars (20 lb/sq in). Gerdes disc brakes. Parking brake. Nosewheel is free to swivel 30° each side.

Power Plant: One 149 kW (200 hp) Avco Lycoming IO-360-A1B flat-four engine, mounted on pylon above hull and driving a Hartzell two-blade constant-speed metal pusher propeller. Rajay turbocharger, and Q-tip propeller, optional. US Rubber DL10 fuel tank in hull, capacity 151 litres (40 US gallons). Refuelling point above hull. Auxiliary fuel in stabilising floats, 28·4 litres (7·5 US gallons) each, optional. Total fuel capacity with optional tanks 208 litres (55 US gallons). Oil capacity 7·5 litres (2 US gallons).

Accommodation: Enclosed cabin seating pilot and three passengers. Front and rear seats removable. Front seats have inertia reel shoulder harness as standard. Dual controls standard; dual brakes for co-pilot optional. Entry through two forward-hinged windscreen sections. Baggage compartment, capacity 90·5 kg (200 lb), aft of cabin. Dual windscreen defroster system.

Systems: Vacuum system for flight instruments. Hydraulic system, pressure 86·2 bars (1,250 lb/sq in), for flaps, horizontal trim and landing gear actuation; handpump provided for emergency operation. Engine-driven 12V 60A alternator and 12V 30Ah battery. Stewart Warner 20,000 BTU heater optional.

Avionics and Equipment: Basic avionics installation includes com and nav antennae, cabin speaker, microphone and circuit breakers. An extensive range of avionics by Collins, King and Narco, and autopilots by Brittain and Edo-Aire Mitchell, are available to customers' requirements. Standard equipment includes full blind-flying instrumentation, electric clock, manifold pressure gauge, outside air temperature gauge, recording tachometer, stall warning device, control locks, instrument panel lighting, carpeted floor, four fresh air vents, tinted glass for all windows, map pocket on front seats, baggage tiedown straps, landing and taxi lights, navigation lights, strobe light, heated pitot, fuselage nose bumper, paddle, cleat, line, full-flow oil filter, quick fuel drains, and inboard and outboard tiedown rings. Optional equipment includes hour meter, true airspeed indicator, shoulder harness for rear seats, alternate static source, manual/automatic bilge pump, cabin fire extinguisher, and external metallic paint finish.

Dimensions, external:

Wing span	11·58 m (38 ft 0 in)
Wing chord, mean	1·35 m (4 ft 5·1 in)
Wing aspect ratio	8·67
Length overall	7·60 m (24 ft 11 in)
Height overall	2·84 m (9 ft 4 in)
Tailplane span	3·05 m (10 ft 0 in)
Wheel track	3·40 m (11 ft 2 in)
Wheelbase	2·69 m (8 ft 10 in)
Propeller diameter	1·88 m (6 ft 2 in)

Dimensions, internal:

Cabin: Length	1·57 m (5 ft 2 in)
Max width	1·05 m (3 ft 5½ in)
Max height	1·32 m (3 ft 11½ in)
Floor area	approx 1·53 m² (16·5 sq ft)
Volume	approx 1·70 m³ (60·0 cu ft)
Baggage hold	0·24 m³ (8·5 cu ft)

Areas:

Wings, gross	15·8 m² (170 sq ft)
Ailerons (total)	1·16 m² (12·5 sq ft)
Trailing-edge flaps (total)	2·28 m² (24·5 sq ft)
Fin	1·25 m² (13·5 sq ft)
Rudder	0·79 m² (8·5 sq ft)
Tailplane	1·45 m² (15·6 sq ft)
Elevators	0·78 m² (8·4 sq ft)

Weights and Loadings:

Weight empty, equipped	705 kg (1,555 lb)
Max T-O and landing weight	1,220 kg (2,690 lb)
Max wing loading	74·2 kg/m² (15·2 lb/sq ft)
Max power loading	8·19 kg/kW (13·45 lb/hp)

Performance (at max T-O weight):

Max level speed at S/L	134 knots (248 km/h; 154 mph) IAS
Max cruising speed, 75% power at 2,440 m (8,000 ft)	130 knots (241 km/h; 150 mph)
Stalling speed	39 knots (72·5 km/h; 45 mph)
Max rate of climb at S/L	366 m (1,200 ft)/min
Service ceiling	4,480 m (14,700 ft)
T-O run on land	183 m (600 ft)
T-O run on water	335 m (1,100 ft)
Landing run on land	145 m (475 ft)
Alighting run on water	183 m (600 ft)
Range with max fuel, at normal cruising speed, with reserves	564 nm (1,046 km; 650 miles)
Max range with max fuel, with reserves	716 nm (1,327 km; 825 miles)

LAS
LOCKHEED AIRCRAFT SERVICE COMPANY (Division of Lockheed Corporation)

Head Office and Works: PO Box 33, Ontario International Airport, Ontario, California 91761
Telephone: (714) 988 2411
Base: Luke Air Force Base, Arizona
Special Devices Division: Ontario, California
Lockheed Center for Marine Research: Carlsbad, California
JETPLAN Aviation Services: Los Gatos, California
President: Edward J. Shockley
Executive Vice-Presidents:
M. H. Greene
C. M. Schnepp (New Business Development)
Vice-Presidents:
B. H. Menke (Finance)
K. E. Neudoerffer (Operations)
I. S. Price (Marketing and New Business Development)
Director of Public Relations: John R. Dailey

Lockheed Aircraft Service Company is claimed to be the world's largest independent aircraft maintenance and modification company. It has designed and installed major modifications for such aircraft as the Boeing KC-135 and 707; Douglas DC-8; and Lockheed C-130, C-141, L-188 Electra, C-121, L-1649, L-1011 and P-3. In particular, LAS delivered 41 Electras in a cargo configuration; details of this conversion can be found in the 1970-71 *Jane's*. It has also designed and installed interiors for various transport aircraft.

LAS has diversified into many other fields, including aircraft maintenance training devices, aircraft maintenance recording systems and airborne integrated data systems. In addition, LAS and three other US companies were awarded a contract in November 1975 to construct a modern maintenance and manufacturing base for the Hellenic Aerospace Industry (HAI, which see) at Tanagra, some 60 km (37 miles) from Athens. This was dedicated officially on 18 December 1979, and provides airlines operating in the Mediterranean area with a base for contract maintenance and manufacturing facilities; it also offers depot level maintenance for the Greek Air Force and state services.

Lockheed EC-130E, modified by LAS for airborne battlefield command and control

In 1974 LAS began delivery of its Model 280 maintenance recorder, which records 50 h of flight data on an easily accessible cassette. During 1981 the company delivered its 1,000th Model 209 digital flight data recorder, which is suitable for use on all types of wide-body aircraft. Another LAS development is JETPLAN, a computerised flight planning and world-wide weather service for airlines and corporate jet operations, which has been available on a worldwide basis since 1971.

LOCKHEED C-130 and L-1011 CONVERSIONS

LAS specialises in complex aircraft modifications of all types, both of military aircraft such as the C-130 Hercules, and of commercial transports like the L-1011 TriStar. C-130 work included conversion of an HC-130H to DC-130H drone launch configuration for the US Air Force; and installation and flight testing of an in-flight refuelling system for the C-130, as described briefly in the 1978-79 *Jane's*.

In 1979 LAS delivered to the US Air Force the first C-130E modified to EC-130E configuration. Major exterior modifications added large blade antennae forward of the fin and beneath each wing; three trailing-wire antennae in bullet-shaped canisters, one beneath each wing and another under the tail unit; and two smaller horizontal blade antennae, on each side of the rear fuselage: all can be seen clearly in the accompanying illustration. The air-conditioning system has also been uprated to dissipate heat generated by the onboard electrical equipment, with heat exchanger pods located under each wing, above each main landing gear fairing. Lockheed EC-130Es have replaced EC-121s in US Air Force service.

In late 1979, LAS began a modification programme to convert to full passenger configuration the L-1011-500 TriStar testbed aircraft which was used in the type certification programme. This was completed and delivered in mid-1980. In a programme that will continue throughout 1981, LAS is modifying the first nine L-1011-500 TriStars to incorporate an active flight control system.

LEAR FAN
LEAR FAN CORPORATION

Head Office: PO Box 60000, Stead Airport, Reno, Nevada 89506
Telephone: (702) 972 0711
Telex: 35 4463
Chairman: Moya Olsen Lear
President: Samuel H. Auld
Executive Vice-President: Milton Weilenmann
Vice-Presidents:
J. S. Lewis (Sales)
W. W. Surley (Operations)
R. R. Tracy (Advanced Projects)

LEAR FAN MODEL 2100

The last aeroplane designed by Mr William P. Lear Sr, before his death on 14 May 1978, was a small twin-turbine business aircraft of advanced design, known originally as the Futura. Since that time it has undergone considerable modification and is known now as the Lear Fan Model 2100.

Of extremely clean appearance, the Model 2100 is built largely of graphite/epoxy composite materials, with other components made of boron, glassfibre, Kevlar and various resins. Design of this aircraft began in June 1977, and construction of the first prototype (N626BL) started in November 1978. This flew for the first time on 1 January 1981. One additional prototype and two static and fatigue test examples are being built at Reno. Certification to FAR Part 23 and BCAR Section K is scheduled for January 1983, with deliveries of production aircraft following immediately after certification. Orders for 203 Lear Fan 2100s had been received by June 1981.

Engineering, research and development of the Lear Fan 2100 is the responsibility of Lear Fan Corporation (USA), at Reno; the first 42 production aircraft are to be built in the USA, after which manufacture will be continued in Northern Ireland, by the newly-formed Lear Fan Ltd (see UK section).

Type: Twin-turbine business aircraft.

Wings: Cantilever low-wing monoplane. Thickness/chord ratio 13·5%. Dihedral 4°. Incidence 1° 30′. No sweepback. Three-spar bonded stressed-skin fail-safe structure of advanced graphite/epoxy composite materials. Each spar is made up of two channels, back to back, separated by a layer of honeycomb. Skins and spars each made in one piece, tip to tip. Hydraulically-actuated plain trailing-edge flaps, and manually-operated ailerons, of graphite/epoxy composites. Trim tab in port aileron. Pneumatic de-icing boots on leading-edges.

Fuselage: Semi-monocoque fail-safe pressurised structure of graphite/epoxy composites, comprising single frames and longerons bonded to the outer skin. Front, centre and rear fuselage sections each made of two shells, split on horizontal centreline, basically of four plies, increased to six to ten plies at cutouts for windows, etc.

Tail Unit: Cantilever Y-shaped structure of graphite/epoxy composites, comprising single-spar V tail and two-spar underfin, the latter stressed to withstand ground impact. Manually-operated elevators with trim tab in V tail, and rudder with trim tab on underfin. Pneumatic de-icing boots on V tail and fin leading-edges.

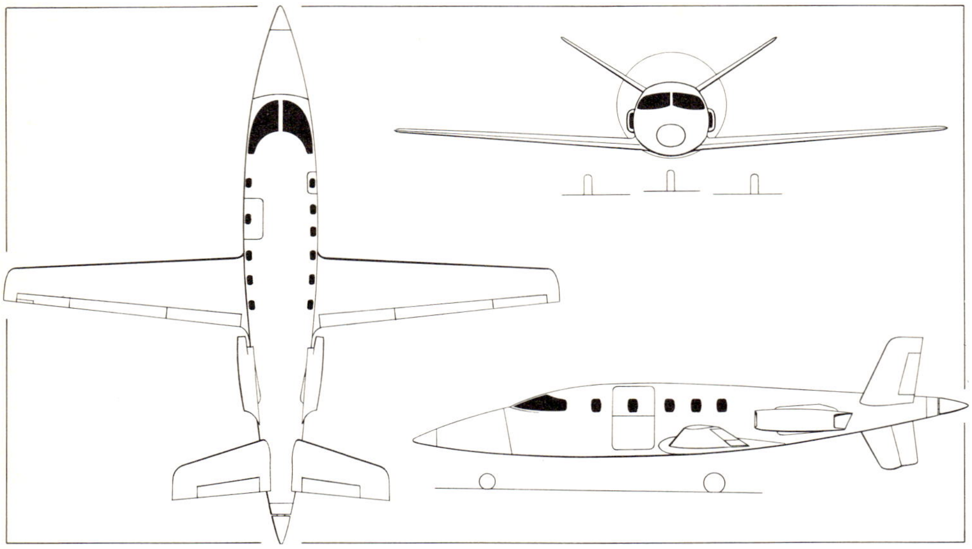

Lear Fan Model 2100 twin-turbine business aircraft *(Pilot Press)*

Lear Fan Model 2100 prototype photographed during its first flight

Landing Gear: Hydraulically-retractable tricycle type with single wheel on each unit; main units retract inward, nosewheel forward. Emergency extension by free fall, with pneumatic bottle backup. Nosewheel steering from rudder pedals. Oleo-pneumatic shock-absorbers. Main wheels have Goodrich tubeless tyres size 7·00-8, 8-ply rating. Nosewheel has Goodrich tyre size 6·00-6, 4-ply rating. Goodrich hydraulically actuated brakes, with optional anti-skid system.

Power Plant: Two 634 kW (850 shp) Pratt & Whitney Aircraft of Canada PT6B-35F turboshaft engines, each flat rated to 485 kW (650 shp), mounted in rear of fuselage and driving, via extension shafts and a combining/reduction gearbox (ratio 5·3 : 1) with separate clutches, a Hartzell four-blade constant-speed and reversible-pitch slow-turning pusher propeller constructed of Kevlar composite, with stainless steel leading-edges. Fuel in integral wing tanks with a usable capacity of 946 litres (250 US gallons). Refuelling points on wing upper surface.

Accommodation: Standard accommodation for a crew of two and six passengers, or crew of two with seven passengers, both layouts with refreshment cabinet and toilet. Alternative two crew/eight passenger high-density arrangement, which retains a toilet facility, or all-cargo version with crew of two. Seat tracks on each side of cabin simplify changes of interior layout, or removal of seats for use in a cargo role. Special optional ambulance version can accommodate two stretcher cases, each with attendant, and has biomedical facilities, therapeutic oxygen and toilet. Clamshell type door, with integral airstairs in lower half, on port side of cabin, forward of wing. Emergency exit on starboard side. Baggage space at rear of cabin, accessible in flight. Entire accommodation pressurised and air-conditioned. Windscreen defrosting and anti-icing.

Systems: Cabin pressurisation by engine bleed air, with max pressure differential of 0·59 bars (8·6 lb/sq in), can maintain a 2,440 m (8,000 ft) cabin altitude to max certificated altitude. Freon vapour-cycle cooling system. Electrical system powered by two 28V 200A starter/generators, with two 125VA 115V 400Hz solid-state inverters, and 24V nickel-cadmium battery. Hydraulic system of 103·5 bars (1,500 lb/sq in) pressure, provided by two engine-driven hydraulic pumps, either of which is capable of maintaining full system function for operation of trailing-edge flaps and landing gear. Oxygen system of 0·62 m³ (22 cu ft) capacity for emergency use by crew and passengers. Anti-icing system includes pneumatic de-icing of wing and tail unit leading-edges, and electrical or bleed air anti-icing of engine inlets, pitot tubes, propeller leading-edges, static ports, and windscreen.

Avionics and Equipment: Standard avionics by Collins include dual VIR-351 VHF nav receivers, dual VHF-251 VHF com transceivers, dual GLS-350 glideslope receivers,* TDR-950 ATC transponder,* ANS-352 area nav, ADF-650 ADF, AMR-350 marker beacon receiver/audio panel, WXR-250 weather radar, IND-351C VOR/ILS indicator and DME-451 DME; plus Sperry C14-D compass system, VG-14D attitude gyro, SPZ-650L AFCS with flight director,* radar altimeter and digital air data computer;* TBD angle of attack system and emergency locator transmitter. Items marked with an asterisk will be optional after first 50 aircraft. Other optional avionics include Collins HF-200 HF com and WXR-300 colour radar; Global Nav GNS-500A VLF/Omega; Sperry air data command display, co-pilot flight director system, and SAT/TAS indicator; TBD RMI/converter, and co-pilot slaved compass system and HSI. Standard equipment includes angle of attack indicator, encoding altimeter, Mach/airspeed indicators, vertical speed indicators, blind-flying instrumentation for co-pilot, cabin pressure indicator, crew seats with lapstraps and shoulder harness, chart/map holders, lap belts for all cabin seats, baggage net or straps, carpeted floor, 'Fasten seat belt—no smoking' sign, cabin fluorescent lighting, passenger reading lights, map and instrument panel lights, navigation lights, landing lights, strobe lights, engine fire detection and fire extinguishing systems.

Dimensions, external:	
Wing span	11·9 m (39 ft 4 in)
Wing aspect ratio	9·5
Length overall	12·07 m (39 ft 7 in)
Height overall	3·51 m (11 ft 6 in)
Wheel track	3·56 m (11 ft 8 in)
Wheelbase	4·90 m (16 ft 0¾ in)
Propeller diameter	2·29 m (7 ft 6 in)
Passenger door (port, fwd):	
Height	1·36 m (4 ft 5½ in)
Width	0·79 m (2 ft 7¼ in)
Height to sill	0·38 m (1 ft 3 in)
Emergency exit (stbd, fwd):	
Height	0·66 m (2 ft 2 in)
Width	0·48 m (1 ft 7 in)
Dimensions, internal:	
Cabin: Length, fwd to aft pressure bulkhead	5·54 m (18 ft 2 in)
Max width	1·47 m (4 ft 10 in)
Max height	1·42 m (4 ft 8 in)
Volume	6·63 m³ (234 cu ft)
Baggage compartment volume	1·19 m³ (42 cu ft)
Area:	
Wings, gross	15·13 m² (162·9 sq ft)
Weights and Loadings (estimated):	
Weight empty	1,746 kg (3,850 lb)
Max fuel weight	771 kg (1,700 lb)
Max T-O weight	3,266 kg (7,200 lb)
Max ramp weight	3,288 kg (7,250 lb)
Max zero-fuel weight	2,676 kg (5,900 lb)
Max landing weight	3,107 kg (6,850 lb)
Max wing loading	215·9 kg/m² (44·2 lb/sq ft)
Max power loading	3·65 kg/kW (6 lb/shp)
Performance (estimated, at max T-O weight unless indicated:	
Max level speed at 9,450 m (31,000 ft)	over 347 knots (643 km/h; 400 mph)
Max cruising speed at 9,145 m (30,000 ft)	347 knots (643 km/h; 400 mph)
Econ cruising speed at 12,495 m (41,000 ft)	304 knots (563 km/h; 350 mph)
Stalling speed, flaps down, power off	82 knots (151 km/h; 94 mph)
Max rate of climb at S/L	1,082 m (3,550 ft)/min
Rate of climb at S/L, one engine out	579 m (1,900 ft)/min
Service ceiling	12,500 m (41,000 ft)
Service ceiling, one engine out	8,840 m (29,000 ft)
T-O run	762 m (2,500 ft)
Balanced field length	823 m (2,700 ft)
Balanced field length at 2,722 kg (6,000 lb) AUW	579 m (1,900 ft)
Landing from 15 m (50 ft), at max landing weight	701 m (2,300 ft)
Range at max cruising speed with max payload and max fuel, 45 min reserves	1,415 nm (2,620 km; 1,630 miles)
Range at econ cruising speed with max payload and max fuel, 45 min reserves	2,000 nm (3,700 km; 2,300 miles)

LEWIS

LEWIS AIRCRAFT CORPORATION

This company, known formerly as ICX Aviation Inc, announced in December 1979 the conclusion of an agreement with Aviaexport of the USSR for exclusive world rights to manufacture and market a US re-engined derivative of the Yakovlev Yak-40 three-turbofan light transport aircraft. It negotiated for a production facility at Wheatfield, New York, and allocated the name X-Avia and designation LC-3 to the new aircraft.

LEWIS X-AVIA LC-3

As ICX Aviation, Lewis Aircraft Corporation began adapting the design of the Yak-40 for production in the USA in December 1975. It intended to develop three versions:

LC-3A. Basic version, to accommodate a crew of three and 30 passengers.

LC-3B. Version with high-density seating to accommodate a maximum of 40 passengers.

LC-3C. An all-cargo version, able to carry a load of up to 4,536 kg (10,000 lb).

No news of this programme has been received since early 1980. A full description of the proposed X-Avia LC-3 can be found in the 1980-81 *Jane's*.

LOCKHEED

LOCKHEED CORPORATION

Head Office: Burbank, California 91520
Telephone: (213) 847 6121

Chairman of the Board and Chief Executive Officer:
Roy A. Anderson

President and Chief Operating Officer:
Lawrence O. Kitchen

Directors:
Roy A. Anderson
Michael Berberian
Jack K. Horton
Lawrence O. Kitchen
Edward W. Carter
Joseph P. Downer
Gilbert R. Ellis
Houston I. Flournoy
Robert A. Fuhrman
John T. Gurash
Vincent N. Marafino
Joseph R. Rensch
Leslie N. Shaw
John E. Swearingen
Fred M. Vinson Jr

Corporate Senior Vice-Presidents:
Jack J. Catton (Business Development)
John E. Cavanagh (General Counsel)
R. F. Conley (Middle East)
Dr Edgar M. Cortright (President, Lockheed-California Co)
Robert A. Fuhrman (President, Lockheed Missiles and Space Co)
Vincent N. Marafino (Finance)
Robert B. Ormsby (President, Lockheed-Georgia Co)
Robert H. Wertheim (Science and Engineering)
William R. Wilson (Strategic Planning)

Corporate Vice-Presidents:
Harold L. Brownman (President, Lockheed Electronics Co)
F. A. Cleveland (Engineering)
Richard K. Cook (Washington Area)
H. David Crowther (Corporate Communications)
Charles de Bedts (International Marketing)
James Everington (Industrial Relations)
W. Paul Frech (General Manager, Operations, Lockheed-California Co)
Melvin H. Greene (Executive Vice-President, Lockheed Aircraft Service Co)
R. Richard Heppe (Advanced Programmes, Lockheed-California Co)
R. H. Hopps (General Manager, Commercial Programmes, Lockheed-California Co)
Fred S. Jacques (General Manager, Government Programmes, Lockheed-California Co)
H. Potter Kerfoot (General Manager, Advanced Systems Division, Lockheed Missiles and Space Co)
Alexander H. Lorch (Executive Vice-President, Lockheed-Georgia Co)
Robert R. McKirahan (Treasurer)
Robert H. Northcutt (Controller)
Frederick C. E. Oder (General Manager, Space Systems Division, Lockheed Missiles and Space Co)
Thomas J. O'Hara (Contracts and Pricing)
B. D. O'Laughlin (General Manager, Engineering, Lockheed-California Co)
James W. Plummer (Executive Vice-President, Lockheed Missiles and Space Co)
Ben R. Rich (General Manager, Advanced Development Projects, Lockheed-California Co)
James J. Ryan (Secretary and Assistant General Counsel)
Charles M. Schnepp (Executive Vice-President, Operations, Lockheed Aircraft Service Co)
Edward J. Shockley (President, Lockheed Aircraft Service Co)
David M. Simmons (President, Lockheed Air Terminal)
Lawrence A. Smith (President, Lockheed Shipbuilding and Construction Co)
Derald A. Stuart (General Manager, Missile Systems Division, Lockheed Missiles and Space Co)
Richard W. Taylor (Strategic Planning)
Joseph G. Twomey (Chief Counsel)
G. Graham Whipple (Executive Vice-President, Lockheed-California Co)

Corporate Officers:
Sam Pearce (Asst Chief Counsel and Asst Secretary)
A. Gene Otsea (Asst Treasurer)

DIVISIONS:

Lockheed-California Company:
Burbank, California 91520
President: Dr Edgar M. Cortright

Lockheed-Georgia Company:
Marietta, Georgia 30063
President: Robert B. Ormsby

Lockheed Aircraft Service Company:
Ontario, California 91761
President: Edward R. Shockley

SUBSIDIARIES:

Lockheed Air Terminal Inc: Burbank, California 91505
President: David M. Simmons
Lockheed Electronics Company Inc: Plainfield, New Jersey 07061
President: Harold L. Brownman
Lockheed Missiles and Space Company Inc: Sunnyvale, California 94086
President: Robert A. Fuhrman
Lockheed Shipbuilding and Construction Company: Seattle, Washington 98134
President: Lawrence A. Smith
Murdock Machine & Engineering Co of Texas: Irving, Texas 75060
President: Paul R. Holmes
Lockheed Engineering and Management Services Company Inc: Houston, Texas 77058
President: R. B. Young
Lockheed Finance Corporation: Burbank, California 91520
President: James T. West

Built by the brothers Allan and Malcolm Lockheed, the first Lockheed aircraft, a tractor seaplane, first flew in 1913. Three years later the brothers established a company at Santa Barbara, California, to manufacture a twin-engined flying-boat, two seaplanes for the Navy and a small sport biplane that was a forerunner of the true streamlined aeroplane. Lockheed Aircraft Co, formed in 1926, moved to Burbank, California, in 1928 and was reorganised as Lockheed Aircraft Corporation in 1932.

On 30 November 1943, the Vega Aircraft Corporation, which had been formed in 1937 as an affiliate and in 1941 became a wholly-owned subsidiary of the Lockheed Aircraft Corporation, was absorbed and the name Vega abandoned. In September 1977 the title Lockheed Aircraft Corporation was changed to Lockheed Corporation, to reflect the company's diversified activities.

Lockheed's aircraft and missile activities are now handled by three separate companies, which evolved from the former California, Georgia and Missiles and Space Divisions in the Summer of 1961. The current products of the Lockheed-California and Lockheed-Georgia companies are described hereafter under the individual company headings.

Lockheed has diversified into many fields of industry since 1959. Following the acquisition of Stavid Engineering Inc, it combined this company and its own Electronics and Avionics Division into Lockheed Electronics Company Inc. Effective in January 1980, certain activities of this company were transferred to a new subsidiary, Lockheed Engineering and Management Services Company Inc.

Lockheed Air Terminal Inc (LAT), a wholly-owned subsidiary, is operating and maintaining during a two-year transitional period the Hollywood-Burbank Airport, which now belongs to the cities of Burbank, Glendale and Pasadena, and provides fuelling and related services at 25 other locations in 11 states, the territory of Guam, and Panama. Lockheed Aircraft Service Co (LAS, which see) designs and manufactures products for the aerospace and marine industries, and carries out aviation maintenance, modification and management services in the USA and other nations. Much of its work outside the USA is done under subcontract for Lockheed Aircraft International AG, a separate subsidiary based in Geneva, Switzerland.

Since April 1959 Lockheed has also had an interest in shipbuilding and heavy construction, following its purchase of the Puget Sound Bridge and Dry Dock Company (now Lockheed Shipbuilding and Construction Co).

At the beginning of February 1981, Lockheed Corporation's total facilities covered more than 2,322,575 m^2 (25,000,000 sq ft), and it had approximately 75,000 employees in 25 US states and on worldwide company assignments.

LOCKHEED-CALIFORNIA COMPANY

Burbank, California 91520

Lockheed-California has responsibility for production of the P-3 Orion and CP-140 Aurora land-based ASW aircraft, L-1011 TriStar three-turbofan transport, and TR-1 high-altitude tactical surveillance and reconnaissance aircraft, which is a variant of the U-2R.

LOCKHEED F-104 STARFIGHTER

The Starfighter continues in production, under licence, by Aeritalia of Italy (which see).

LOCKHEED 'PROJECT STEALTH'

In the so-called 'Skunk Works' at Burbank, Lockheed-California is reported to have built, under a DARPA-funded contract from the USAF's Flight Dynamics Laboratory, a single-seat reconnaissance/strike aircraft, of which a primary feature is low radar, infra-red and optical signatures. The aircraft is said to be powered by two 53·4 kN (12,000 lb st) turbojet engines, and to have flown for the first time in 1977.

LOCKHEED U-2, ER-2 and TR-1

Development of the original U-2 began in the Spring of 1954 to meet a joint CIA/US Air Force requirement for a high-altitude strategic reconnaissance and special-purpose research aircraft. It took place in the Lockheed 'Skunk Works' at Burbank, California, where, after acceptance of the design in late 1954, two prototypes were hand-built in great secrecy by a small team of engineers. The aircraft's true purpose was cloaked under the US Air Force U-for-Utility designation U-2, and the first flight, by Lockheed test pilot Tony LeVier, took place on or about 1 August 1955 at Watertown Strip in the Nevada desert.

The configuration of the U-2 is basically that of a powered sailplane, which explains its unusual 'bicycle' landing gear, combined with underwing balancer units which provide stability during take-off and are then jettisoned. Range can, when necessary, be extended by shutting off the engine and gliding. Because of its configuration the U-2 requires unusually precise handling during take-off and landing—particularly the latter, since there is an extremely small margin between approach speed and stalling speed. After touchdown, the aircraft comes to rest on one of the down-turned wingtips, which serve as landing skids.

Initial quantities of 48 single-seat and five two-seat U-2s were ordered in FY 1956, but after about 30 of these had been completed the increasing weight of special equipment which the aircraft was required to carry had degraded performance to such an extent that a more powerful engine became necessary, and this was installed from 1959 onwards. At the same time a substantial increase in fuel capacity made possible a considerably greater range. In FY 1968 the U-2 was put back into production for a period, to replace some of the two dozen or more aircraft lost over hostile territory or in accidents. The production line was again reopened in 1980 to manufacture a new, tactical reconnaissance version, known as the TR-1. This has been selected as the preferred airborne relay vehicle for the PLSS (Precision Location Strike System), intended to locate and identify enemy radar emitters, and to direct strike aircraft against them.

In view of the continuing classified nature of much of its work, it is not possible to confirm officially many details concerning the U-2; but some information on its deployment, and on the versions built and subsequent conversions, can be found in the 1979-80 *Jane's*. Most recently built was the U-2R, and available details of this and the TR-1/ER-2 follow:

U-2R. Additional batch of 12 single-seat aircraft, ordered for strategic reconnaissance in FY 1968, by which time approx half of original U-2s had been lost through various causes. Serial numbers 68-10329 to 68-10340. Originally designated WU-2C. Bulged intakes, as on U-2C, but longer nose and fuselage, without dorsal spine fairing: increased wing span and internal fuel capacity: rear fuselage slightly bulged on top, just forward of fin; main-wheel unit further aft, tailwheel unit further forward, than on earlier models. Non-US bases have included Mildenhall, England.

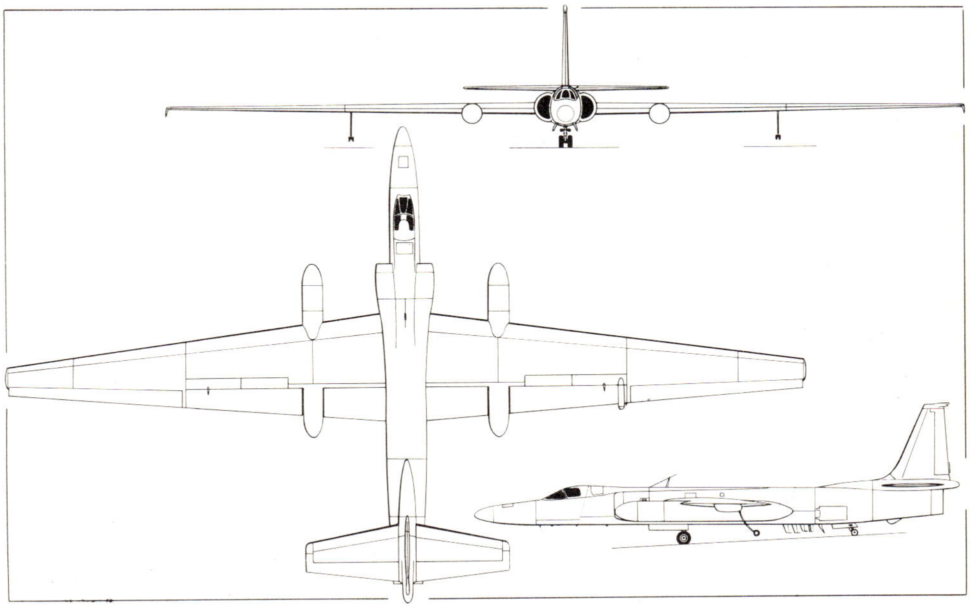

Lockheed U-2R high-altitude reconnaissance and research aircraft *(Pilot Press)*

TR-1. Single-seat tactical reconnaissance version, described by the Department of Defense as being "equipped with a variety of electronic sensors to provide continuously available, day or night, high-altitude all-weather stand-off surveillance of the battle area in direct support of the US and Allied ground and air forces during peace, crises, and war situations". Tooling for the U-2 had been kept in store at the USAF-owned Plant 42 at Palmdale, California, and the FY 1979 defence budget included $10·2 million to reopen the production line in FY 1980. The TR-1 has the same basic airframe as the U-2R, with a J75-P-13 engine, but with the significant addition of an 'advanced synthetic aperture' radar system (ASARS) in the form of a UPD-X side-looking airborne radar (SLAR) and modern electronic countermeasures (ECM). Seen as a replacement for the now-abandoned Compass Cope RPV (see 1977-78 *Jane's*), the TR-1 is intended primarily for use in Europe, where its SLAR will provide the capability to 'see' approximately 30 nm (55 km; 35 miles) into hostile territory without the need to overfly an actual or potential battle area. A further funding of $42·4 million was made under the FY 1980 budget, and a contract covering the construction of three aircraft was received on 16 November 1979. Two of these are standard **TR-1As,** which were scheduled for delivery to the US Air Force during 1981. The third aircraft (NASA 706) has been modified for use by NASA as an Earth resources research aircraft, with the designation **ER-2**, and was delivered to Ames Research Center in June 1981. Further details can be found under the NASA heading in this section. An additional 33 TR-1s, including two two-seat **TR-1Bs**, are planned for service with the US Air Force.

The following description applies primarily to the single-seat U-2R, but is applicable generally to the TR-1A:

Type: High-altitude reconnaissance and research aircraft.

Wings: Cantilever mid-wing monoplane, with wingtips turned down 90° for use as skids during landing. All-metal structure. Trailing-edge flaps (four segments on each wing, two inboard and two outboard of each underwing pod) occupy approx 70 per cent of each wing. Small tubular fuel vent fairing between each outermost flap segment and aileron, projecting slightly aft of the trailing-edge. Two small plate-type roll/lift spoilers on each wing, forward of outboard flap segments. Trim tab in each aileron.

Fuselage: All-metal semi-monocoque structure of circular cross-section, with thin-gauge skin. Fineness ratio approx 10:1. Forward-opening door-type airbrake on each side of fuselage aft of wings, used mainly as a landing aid. Large airscoop fairing on fuselage beneath rear of wing root: generally on starboard side, but sometimes on port side and sometimes on both. Since about 1974 some aircraft have had a modified tailpipe, to reduce the infra-red signature from the engine.

Tail Unit: Cantilever all-metal structure. Balanced rudder and elevators. Trim tab on rudder and in each elevator.

Landing Gear: Retractable bicycle type, with twin main wheels and twin small tailwheels in tandem, each unit retracting forward into fuselage. Balancer units under outer wings, each with twin small wheels, are jettisoned on take-off. Tailwheels and underwing wheels have solid tyres; castoring tailwheel unit aids manoeuvring on ground. Brakes on main wheels. Braking parachute in container under rudder.

Power Plant: One 75·6 kN (17,000 lb st) Pratt & Whitney J75-P-13B turbojet engine. All fuel in inboard and outboard main tanks filling each wing except for tip, each with overwing gravity refuelling point. Normal internal fuel capacity approx 4,448 litres (1,175 US gallons). Provision for 397·5 litre (105 US gallon) non-jettisonable auxiliary slipper tank on each wing leading-edge, between main tanks.

Lockheed TR-1 single-seat tactical reconnaissance aircraft for USAF, rolled out at Palmdale, California, on 15 July 1981

Accommodation: Pilot only in TR-1A/ER-2, on ejection seat. Rearward-sliding transparent canopy, protected internally against ultra-violet radiation. Accommodation is air-conditioned but not pressurised. Rearview periscope on most aircraft (positions vary). Food warmer, with spaceflight-type tubes of food. Liquid oxygen system for pilot.

Avionics and Equipment: Typical standard avionics include HF, UHF and VHF com, Tacan, ILS, autopilot, ADF, compass, and (for night flying) astro-compass. Equipment includes one vertical and two lateral cameras for training flights, or up to five 70 mm cameras (U-2) or side-looking airborne radar (TR-1) for operational missions. Panoramic camera(s), usually Model 73B or Perkin-Elmer Model 501, with ventral periscopic sight. Main avionics and equipment compartments of TR-1 are in detachable modular nose section, in a 'Q' bay aft of the cockpit, and in two large pods mounted underwing at approx one-third span. Each pod is approx 8·23 m (27 ft) long; has a volume of about 2·55 m^3 (90 cu ft), and weighs about 544 kg (1,200 lb) complete with sensors and/or equipment. There is a smaller 'E' bay between the 'Q' bay and main-wheel bay; additional small areas in the bottom of the rear fuselage and in the tailcone can also be used to house mission equipment.

Dimensions, external:

Wing span	31·39 m (103 ft 0 in)
Wing area, gross	approx 92·9 m^2 (1,000 sq ft)
Wing aspect ratio	approx 10·6
Length overall	19·20 m (63 ft 0 in)
Height overall	4·88 m (16 ft 0 in)

Weights:

Fuel and equipment payload:	
U-2R	approx 5,443 kg (12,000 lb)
Max T-O weight with slipper tanks:	
U-2R	13,154 kg (29,000 lb)
TR-1	18,143 kg (40,000 lb)

Lockheed ER-2, built for NASA to supplement two U-2s used by Ames Research Center for Earth resources survey

Performance (estimated for TR-1):

Max cruising speed: TR-1 at over 21,650 m (70,000 ft) more than 373 knots (692 km/h; 430 mph)
Operational ceiling: TR-1 27,430 m (90,000 ft)
Max range: TR-1 more than 2,605 nm (4,830 km; 3,000 miles)
Max endurance: U-2R, TR-1 12 h

LOCKHEED MODEL 185/285 ORION

US Navy designation: P-3
CAF designation: CP-140 Aurora

In 1958 Lockheed won a US Navy competition for an 'off-the-shelf' ASW aircraft with a developed version of its Electra four-turboprop commercial transport. An aerodynamic prototype flew for the first time on 19 August 1958. A second aircraft, designated YP-3A (formerly YP3V-1), with full avionics, flew on 25 November 1959.

Details of the P-3A initial production version (157 built; retired from operational use by active-duty USN patrol squadrons on 13 November 1978, but flown by Naval Reserve) and WP-3A can be found in the 1978-79 *Jane's*. Subsequent versions include the following:

P-3B. Follow-on production version with 3,661 kW (4,910 ehp) Allison T56-A-14 turboprop engines, which do not need water-alcohol injection. US Navy contracts covered 124 P-3Bs. In addition, five P-3Bs were delivered to the Royal New Zealand Air Force in 1966, ten to the Royal Australian Air Force during 1968 and five to Norway in the Spring of 1969. USN P-3Bs were modified retrospectively to carry Bullpup missiles. Others became **EP-3Bs.** The US Navy and Lear Siegler developed in 1976 a modification kit for retrofitting to P-3B aircraft. This includes a 32K Rom computer, ASN-84 inertial navigation system, Omega, ASA-66 displays and ASN-124 navigation controller, together with the necessary controls and equipment to integrate the new and existing systems. Kits have been made available for the above installation. New Zealand's five P-3Bs are currently undergoing modernisation by Boeing Aerospace, involving upgrading of their data handling and display systems, radar, infra-red detection system, and INS. It is planned to upgrade the ASW equipment in a subsequent programme.

P-3C. Advanced version with the A-NEW system of sensors and control equipment, built around a Univac digital computer that integrates all ASW information and permits retrieval, display and transmission of tactical data in order to eliminate routine log-keeping functions. This increases crew effectiveness by allowing them sufficient time to consider all tactical data and devise the best action to resolve problems. First flight of this version was made on 18 September 1968 and the P-3C entered service in 1969. A total of 192 of this version had been delivered to the US Navy by 1 January 1981. Production is continuing at the rate of about 12 per year, to complete delivery of currently planned total of 316 P-3Cs for the US Navy by 1990.

All of those delivered from January 1975 have been to **P-3C Update** standard, with new avionics and electronics software developed to enhance the effectiveness of the aircraft. Equipment includes a magnetic drum that gives a sevenfold increase in computer memory capacity, a new versatile computer language, Omega navigation system, improved acoustic processing sensitivity, a tactical display for two of the sensor stations, and an improved magnetic tape transport. A prototype with this equipment was handed over to the US Navy on 29 April 1974.

The US Navy and Lockheed began in 1976 a further P-3C avionics improvement programme known as **Update II**. This added an infra-red detection system (IRDS) and a sonobuoy reference system (SRS). The Harpoon missile and control system are included in Update II, which was incorporated into production aircraft from August 1977. The first Update II P-3C was delivered to the Naval Air Development Center in the same month. Ten Update II P-3Cs were delivered to the Royal Australian Air Force in 1978-79 for service with No. 10 Squadron. The Royal Netherlands Navy has ordered 13 Update II P-3Cs, for delivery from late 1981. Japan has ordered 45, of which four will be assembled and 38 licence-built in Japan by Kawasaki (which see), following delivery of three US-built aircraft.

Update III, of which development began in February 1978, involves mainly ASW avionics, including a new acoustic processor to analyse signals picked up from the sea, a new sonobuoy receiver which replaces DIFAR (directional acoustic frequency analysis and recording), an improved APU, and environmental controls to cater for increased heat from the avionics and to improve crew comfort further. A prototype Update III P-3C is scheduled for delivery to the US Navy for test and evaluation in April 1983, with full production to start in April 1984.

Two of the eight international records for turboprop aircraft set up in a P-3C by Cdr Donald H. Lilienthal, in early 1971, had not been beaten by mid-1981. They were a speed of 434·97 knots (806·10 km/h; 500·89 mph) over a 15/25 km course; and a time-to-height record, to 12,000 m in 19 min 42·24 s.

EP-3E. Ten P-3As and two EP-3Bs were converted to EP-3E configuration to replace Lockheed EC-121s in service with VQ-1 and VQ-2 squadrons. Identified by large canoe-shape radars on upper and lower surfaces of fuselage and ventral radome forward of wing.

P-3F. Six aircraft, similar to the US Navy's P-3Cs, for the Iranian Air Force. Used initially for long-range surface surveillance and subsequently also for ASW missions. Delivery completed in January 1975.

CP-140 Aurora. Version for Canadian Armed Forces. Described separately.

By mid-1980 Lockheed-California had delivered more than 500 P-3s of all versions. The following data refer to the P-3C, but are generally applicable to other versions, except for the details noted:

Type: Four-turboprop ASW aircraft.

Wings: Cantilever low-wing monoplane. Wing section NACA 0014 (modified) at root, NACA 0012 (modified) at tip. Dihedral 6°. Incidence 3° at root, 0° 30′ at tip. Fail-safe box beam structure of extruded integrally-stiffened aluminium alloy. Lockheed-Fowler trailing-edge flaps. Hydraulically-boosted aluminium ailerons. Anti-icing by engine bleed air ducted into leading-edges.

Fuselage: Conventional aluminium alloy semi-monocoque fail-safe structure.

Tail Unit: Cantilever aluminium alloy structure with dihedral tailplane and dorsal fin. Fixed-incidence tailplane. Hydraulically-boosted rudder and elevators. Leading-edges of fin and tailplane have electric anti-icing system.

Landing Gear: Hydraulically-retractable tricycle type, with twin wheels on each unit. All units retract forward, main wheels into inner engine nacelles. Oleo-pneumatic shock-absorbers. Main wheels have size 40-14 type VII 26-ply tubeless tyres. Nosewheels have size 28-7·7 type VII tubeless tyres. Hydraulic brakes. No anti-skid units.

Power Plant: Four 3,661 kW (4,910 ehp) Allison T56-A-14 turboprop engines, each driving a Hamilton Standard 54H60 four-blade constant-speed propeller. Fuel in one tank in fuselage and four wing integral tanks, with total usable capacity of 34,826 litres (9,200 US gallons). Four overwing gravity fuelling points and central pressure refuelling point. Oil capacity (min usable) 111 litres (29·4 US gallons) in four tanks. Electrically de-iced propeller spinners.

Accommodation: Normal ten-man crew. Flight deck has wide-vision windows, and circular windows for observers are provided fore and aft in the main cabin, each bulged to give 180° view. Main cabin is fitted out as a five-man tactical compartment containing advanced electronic, magnetic and sonic detection equipment, an all-electric galley and large crew rest area.

Systems: Air-conditioning and pressurisation system supplied by two engine-driven compressors. Pressure differential 0·37 bars (5·4 lb/sq in). Hydraulic system, pressure 207 bars (3,000 lb/sq in), for flaps, control surface boosters, landing gear actuation, brakes and bomb bay doors. Electrical system utilises three 60kVA generators for 120/208V 400Hz AC supply. 24V DC supply. Integral APU with 60kVA generator for ground air-conditioning, electrical supply and engine starting.

Avionics and Equipment: The ASQ-114 general-purpose digital computer is the heart of the P-3C system. Together with the AYA-8 data processing equipment and computer-controlled display systems, it permits rapid analysis and utilisation of electronic, magnetic and sonic data. Nav/com system comprises two LTN-72 inertial navigation systems; AN/APN-227 Doppler; ARN-81 Loran A and C; AN/ARN-118 Tacan; two VIR-31A VOR/LOC/GS/MB receivers; ARN-83 LF-ADF; ARA-50 UHF direction finder; AJN-15 flight director indicator for tactical directions; HSI for long-range flight directions; glideslope indicator; on-top position indicator; two ARC-161 HF transceivers; two ARC-143 UHF transceivers; ARC-101 VHF receiver/transmitter; AGC-6 teletype and high-speed printer; HF and UHF secure communication units; ACQ-5 data link communication set and AIC-22 interphone set; APX-72 IFF transponder and APX-76 SIF interrogator. Electronic computer-controlled display equipment includes ASA-70 tactical display; ASA-66 pilot's display; ASA-70 radar display and two auxiliary readout (computer-stored data) displays. ASW equipment includes two ARR-72 sonar receivers; two AQA-7 DIFAR (directional acoustic frequency analysis and recording) sonobuoy indicator sets; hyperbolic fix unit; acoustic source signal generator; time code generator and AQH-4(V) sonar tape recorder; ASQ-81 magnetic anomaly detector; ASA-64 submarine anomaly detector; ASA-65 magnetic compensator; ALQ-78 electronic countermeasures set; APS-115 radar set (360° coverage); ASA-69 radar scan converter; KA-74 forward computer-assisted camera; KB-18A automatic strike assessment camera with horizon-to-horizon coverage; RO-308 bathythermograph recorder. Additional equipment includes APN-194 radar altimeter; two APQ-107 radar altimeter warning systems; A/A24G-9 true airspeed computer and ASW-31 automatic flight control system. P-3Cs delivered from 1975 have the avionics/electronics package updated by addition of an extra 393K memory drum and fourth logic unit, Omega navigation, new magnetic tape transport, and an ASA-66 tactical display for the sonar operators. To accommodate the new systems a new operational software computer programme will be written in CMS-2 language. Marconi AQS-901 acoustic signal processing and display system in RAAF P-3Cs.

Armament: Bomb bay, 2·03 m wide, 0·88 m deep and 3·91 m long (80 in × 34·5 in × 154 in), forward of wing, can accommodate a 2,000 lb Mk 25/39/55/56 mine, three 1,000 lb Mk 36/52 mines, three Mk 57 depth bombs, eight Mk 54 depth bombs, eight Mk 43/44/46 torpedoes or a combination of two Mk 101 nuclear depth bombs and four Mk 43/44/46 torpedoes. Ten underwing pylons for stores: two under centre-section each side can carry torpedoes or 2,000 lb mines; three under outer wing each side can carry respectively (inboard to outboard) a torpedo or 2,000 lb mine (or searchlight on starboard wing), a torpedo or 1,000 lb mine or rockets singly or in pods; a torpedo or 500 lb mine or rockets singly or in pods. Torpedoes can be carried underwing only for ferrying; mines can be carried and released. Search stores, such as sonobuoys and sound signals, are launched from inside cabin area in the P-3A/B. In the P-3C sonobuoys are loaded and launched externally and internally. Max total weapon load includes six 2,000 lb mines under wings and a 3,290 kg (7,252 lb) internal load made up of two Mk 101 depth bombs, four Mk 44 torpedoes, pyrotechnic pistol and 12 signals, 87 sonobuoys, 100 Mk 50 underwater sound signals (P-3A/B), 18 Mk 3A marine markers (P-3A/B), 42 Mk 7 marine markers, two B.T. buoys, and two Mk 5 parachute flares. Sonobuoys are ejected from P-3C aircraft with explosive cartridge actuating devices (CAD), eliminating the need for a pneumatic system. Australian P-3Cs use BARRA sonobuoys.

Lockheed P-3C Orion in Update II configuration

Dimensions, external:

Wing span	30·37 m (99 ft 8 in)
Wing chord at root	5·77 m (18 ft 11 in)
Wing chord at tip	2·31 m (7 ft 7 in)
Wing aspect ratio	7·5
Length overall	35·61 m (116 ft 10 in)
Height overall	10·29 m (33 ft 8½ in)
Fuselage diameter	3·45 m (11 ft 4 in)
Tailplane span	13·06 m (42 ft 10 in)
Wheel track (c/l shock-absorbers)	9·50 m (31 ft 2 in)
Wheelbase	9·07 m (29 ft 9 in)
Propeller diameter	4·11 m (13 ft 6 in)
Cabin door: Height	1·83 m (6 ft 0 in)
Width	0·69 m (2 ft 3 in)

Dimensions, internal:

Cabin, excl flight deck and electrical load centre:	
Length	21·06 m (69 ft 1 in)
Max width	3·30 m (10 ft 10 in)
Max height	2·29 m (7 ft 6 in)
Floor area	61·13 m² (658 sq ft)
Volume	120·6 m³ (4,260 cu ft)

Areas:

Wings, gross	120·77 m² (1,300 sq ft)
Ailerons (total)	8·36 m² (90 sq ft)
Trailing-edge flaps (total)	19·32 m² (208 sq ft)

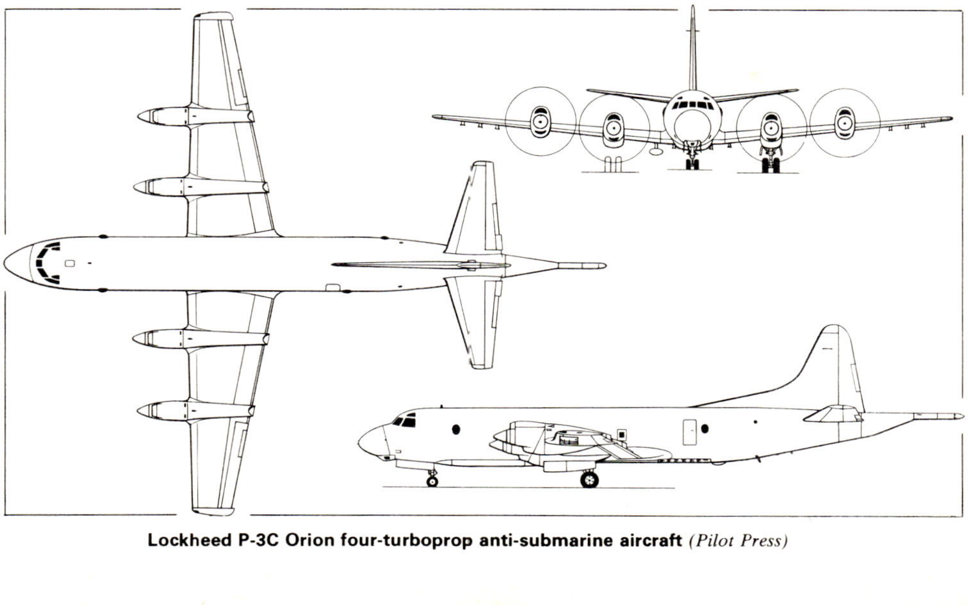
Lockheed P-3C Orion four-turboprop anti-submarine aircraft *(Pilot Press)*

Fin, incl dorsal fin 10·78 m² (116 sq ft)
Rudder, incl tab 5·57 m² (60 sq ft)
Tailplane 22·39 m² (241 sq ft)
Elevators, incl tabs 7·53 m² (81 sq ft)

WEIGHTS (P-3B/C):
Weight empty 27,890 kg (61,491 lb)
Max expendable load 9,071 kg (20,000 lb)
Max normal T-O weight 61,235 kg (135,000 lb)
Max permissible weight 64,410 kg (142,000 lb)
Design zero-fuel weight 35,017 kg (77,200 lb)
Max landing weight 47,119 kg (103,880 lb)

PERFORMANCE (P-3B/C, at max T-O weight, except where indicated otherwise):
Max level speed at 4,570 m (15,000 ft) at AUW of 47,625 kg (105,000 lb) 411 knots (761 km/h; 473 mph)
Econ cruising speed at 7,620 m (25,000 ft) at AUW of 49,895 kg (110,000 lb) 328 knots (608 km/h; 378 mph)
Patrol speed at 457 m (1,500 ft) at AUW of 49,895 kg (110,000 lb) 206 knots (381 km/h; 237 mph)
Stalling speed, flaps up 133 knots (248 km/h; 154 mph)
Stalling speed, flaps down 112 knots (208 km/h; 129 mph)
Max rate of climb at 457 m (1,500 ft) 594 m (1,950 ft)/min
Service ceiling 8,625 m (28,300 ft)
Service ceiling, one engine out 5,790 m (19,000 ft)
T-O run 1,290 m (4,240 ft)
T-O to 15 m (50 ft) 1,673 m (5,490 ft)
Landing from 15 m (50 ft) at design landing weight 845 m (2,770 ft)
Max mission radius (no time on station) at 61,235 kg (135,000 lb) 2,070 nm (3,835 km; 2,383 miles)
Mission radius (3 h on station at 457 m; 1,500 ft) 1,346 nm (2,494 km; 1,550 miles)

LOCKHEED AURORA

Canadian Armed Forces designation: CP-140

The purchase of 18 special variants of the Lockheed P-3 Orion maritime patrol aircraft for the Canadian Armed Forces, to replace CP-107 Argus maritime reconnaissance aircraft, was announced on 21 July 1976. Designated CP-140 Aurora, each of these aircraft combines the P-3 Orion airframe, power plant and basic aircraft systems with the avionics systems and data processing capability of the US Navy's carrier-based Lockheed S-3A Viking. It is able to perform missions involving a range of more than 4,000 nm (7,400 km; 4,600 miles), or flights of up to 17 hours' duration, and is being deployed initially for ASW duties; national sovereignty patrols; shipping, fisheries and Arctic surveillance; ice reconnaissance; and search and rescue. By the addition of a weapons bay sensors canister at a later date, the CP-140 will be able to undertake additional civilian tasks such as resources location, pollution control and aerial survey.

The first CP-140 (CAF serial number 140104) completed a successful first flight on 22 March 1979, and was handed over to the CAF on 29 May 1980. All 18 were due to be delivered by mid-1981, and are based at Greenwood, Nova Scotia, and Comox, British Columbia.

The cabin interior has been changed extensively to meet Canadian requirements: immediately aft of the flight deck are an observer's station on the port side and crew rest bunks on the starboard side. Moving aft, the tactical compartment comes next, with accommodation for the tactical navigator (tacnav), navigator/communicator (navcom), two acoustic sensor operators (ASO), and two non-acoustic sensor operators (NASO), all on the port side. Aft of the tactical compartment is the search stores and camera bay, with two more observer stations, one on each side. At the rear of the cabin are a galley, on the port side, a dinette area, and an airborne maintenance station on the starboard side. A toilet is located on the port side of the cabin, immediately aft of the forward observer's position.

On the flight deck, an IP-5043/A multi-purpose display (MPD) provides the pilots with a real-time presentation of the tactical situation and sensor information; directions from the tacnav and navcom are fed through the computer for display on both the MPD and the flight director indicators (FDIs). Cues and alerts, indicating required sequences of action, are displayed on the periphery of the MPD.

An AJN-501 flight director system supplies attitude, heading and fly-to-point references. For long-range navigation, data from the horizontal situation indicator are normally adequate. For precise, close-in tactical manoeuvring the FDI is used, and the automatic flight control system includes full-time attitude control and proportional control-wheel steering.

The three observer stations each have a fully-swivelling seat and are provided with intercom. Each of the observation windows gives full hemispherical view, and there are power and storage provisions for a hand-held camera. Each position is provided with isolation curtains, to screen observer and window from cabin lighting during night visual search. A fourth station can be made available on the starboard side, by removal of the crew rest bunks.

The tacnav has a console which includes an IP-5044/A MPD, a C5331/A keyset and trackball, and armament controls. With his keyset the tacnav can control, via the computer, the sonobuoy reference system (SRS), and can call up and display FLIR and other radar data on his MPD. The navcom also has an IP-5044/A and C5331/A, plus HF, VHF (FM) and UHF transceivers; inertial, VLF (Omega) and Doppler navigation sets; LF and UHF ADF, and VHF homer; a high-speed teleprinter and teletype keyboard; provisions for tactical satellite communications (tacsatcom); data link; control of reconnaissance photography; provisions for control of survey photography; and provisions for secure communications. The navcom's MPD and keyset serve as a backup for the tacnav in the event of equipment failure.

The two ASOs share a dual console and each has an IP-5044/A MPD and C5331/A keyset and trackball. They share also an IP-5045/A auxiliary readout unit (ARU), a time code generator, and an RD-5027/ASH 28-track tape recorder. Their MPDs can display acoustic data or the tactical plot, but the ARU is a dedicated acoustic display. The acoustic functions of receiving, processing, display and recording are controlled by the keysets through the computer.

The two NASOs also have a dual console, each with an IP-5044/A MPD and C5331/A keyset and trackball, the keysets being used to control radar, electronic support measures (ESM) and FLIR through the computer. Principal controls shared by these two operators, or available to only one of them, include ASQ-502 MAD, OA-5154/ASQ(FACS II) MAD compensator, video tape recorder, SIF and provisions for SLAR.

The heart of the entire control system is a Univac AN/AYK-502 navigation/tactical computer. Its two central processors function independently; both have coordinated access to a core memory of 65,536 words. There is growth capacity for an additional 32,000 words, and space has been allocated for a 127,000-word auxiliary memory in the acoustic system processor for the computer.

The search stores and camera bay has stowage for 'A' size sonobuoys, large and small marine markers, Signals Underwater Sound (SUS) and flares. Intercom controls and an ordnance status panel are provided for the ordnance crew member. The computer-controlled electrically-fired cartridge-actuated A-size launchers can all be operated with the aircraft pressurised. They comprise 36 underfloor launchers, loadable only on the ground, and three which can be loaded from the cabin with the aircraft pressurised or unpressurised. A C-size chute, just aft of the three cabin launch tubes, allows free-fall launch (with the aircraft unpressurised) of flares, small marine markers, SUS and mail, and airdrops to remote ships or stations.

A KA-501A day/night reconnaissance camera is installed beneath the floor in this area, and is accessible in flight through a floor hatch. The illuminator for night reconnaissance photography is located beneath the floor of the in-flight maintenance station. This position has a bench with 28V DC and 115V 400Hz AC power outlets, and there are provisions for a microfiche reader.

Aircraft operational support equipment for the CP-140 includes the ground-based data interpretation and analysis centre (DIAC), and a ground support computer complex (GSCC). The former provides operational support for the operating squadrons; the latter provides technical support for the operational software, and maintains software configuration records.

The aircraft's weapon bay, which has a maximum capacity of 2,177 kg (4,800 lb) on eight stations, can accommodate and drop a variety of ordnance. There are ten underwing hardpoints, with an individual capacity ranging from 277 kg (611 lb) to 1,111 kg (2,450 lb).

Canadair Ltd is manufacturing forward and aft radomes, rear fuselage sections, centre and outer wing sections, and main electrical load centres, both for the Aurora and for Lockheed-built P-3Cs.

TYPE: Four-turboprop long-range ASW and maritime patrol aircraft.

WINGS: As P-3C, with ailerons operated by dual hydraulic boosters, supplied from two independent hydraulic systems.

FUSELAGE: As for P-3C.

TAIL UNIT: As P-3C, with rudder and elevators each operated by dual hydraulic boosters, supplied from two independent hydraulic systems. Trim tabs in elevators and rudder.

LANDING GEAR: Hydraulically-retractable tricycle type with twin wheels on each unit. All units retract forward, main wheels into inner engine nacelles. Oleo-pneumatic shock-absorbers. All units can free-fall to the down and locked position in emergency. Hydraulically-powered steerable nose unit, controlled by handwheel on the pilot's side console. Hydraulically-operated dual segmented-disc brakes. Pneumatic emergency braking system.

POWER PLANT: Four 3,661 kW (4,910 ehp) Allison T56-A-14 turboprop engines, each driving a four-blade constant-speed fully-feathering and reversible metal propeller. Fuel in one fuselage and four wing integral tanks, with total usable capacity of 34,826 litres (9,200 US gallons). Single-point pressure refuelling, and four overwing gravity refuelling points, are provided. Fuel dump system. Propeller blade cuffs and spinners de-iced by electrical heating.

ACCOMMODATION: Normal ten-man crew, with seating for six additional passengers. Dual controls standard. Flight deck has wide-vision windows, and circular windows for up to four observers are provided in the main cabin, each bulged to give 180° visibility. Main cabin fitted out as detailed in introductory paragraphs. Door on port

CP-140 Aurora version of the Lockheed Orion for the Canadian Armed Forces

side, aft of wing. Overwing emergency exit on each side of cabin; others in side and ceiling of flight deck. Defogging and anti-icing of windscreens by electrical heating; windscreens have mechanical wipers, a washing system for the removal of salt deposits, and a rain-repellent spray system. Stowage for clothing, life jackets and parachute harness. Four floor tiedown areas, combined baggage/cargo capacity 442 kg (975 lb).

SYSTEMS: Air-conditioning and pressurisation system supplied by two engine-driven compressors, maintaining cabin temperatures between 15·6°C and 26·7°C (60°F and 80°F), and a cabin altitude of 2,440 m (8,000 ft) to a height of 9,145 m (30,000 ft). Two independent hydraulic systems, each at a pressure of 207 bars (3,000 lb/sq in) are powered by three interchangeable electrically-driven pumps, any two of which can maintain full hydraulic services. Pneumatic system at pressure of 207 bars (3,000 lb/sq in) for emergency braking. Electrical system of 120/208V 400Hz AC supplied by three 60/90kVA engine-driven generators, any one of which can maintain full normal load. DC power supplied by three 200A 24V transformer-rectifiers and one 31Ah storage battery. APU drives a 60/90kVA generator and provides power and bleed air for ground air-conditioning, weapons bay heating and engine starting; it can also provide emergency electrical power in flight. Oxygen system for crew of three on flight deck with 3·5 hour capacity. Individual portable chemical oxygen generators for emergency use by all crew members. Automatic flight control system (AFCS) with dual-channel fail-safe autopilot; includes tactical and airways nav modes and proportional control wheel steering.

AVIONICS AND EQUIPMENT: Univac AN/AYK-502 navigation/tactical computer; digital magnetic tape units; teleprinter; display generator units; APS-116 search radar; OR-5008/AA (modified) FLIR; video recorder for FLIR imagery; ARS-501 sonobuoy reference system; OL-5004/AXS acoustics data processor; OD-5006/A display group; ASN-505 dual inertial navigation systems; APN-510 Doppler; ARN-511 Omega; Tacan; revised airways/approach nav aids; dual VOR/ILS; communications sets comprising HF, UHF, VHF (AM), VHF guard receiver, VHF(FM); HF SIMOPS filters; RCVR homing; ASH-502 crash position indicator/flight data recorder; ASW-502 AFCS; ALR-502 ESM; RD-5027/ASH tape recorder; ASQ-502 MAD; OA-5154/ASQ (FACS II) MAD compensator; SLAR provisions; IFF; data link: airborne radiation thermometer (ART) provisions; and time coding generator. Equipment includes KA-501A day/night reconnaissance camera and night illuminator; provisions for civil sensors canister; galley with refrigerator and sink; blue-filtered white edge lighting for all console-mounted control panels and flight instruments; white cabin lighting; reading lights at all crew positions; white overhead lights; and aisle lights.

PERFORMANCE (with mission payload of 2,540 kg; 5,600 lb except where stated otherwise):
- Max transit speed at optimum altitude: 395 knots (732 km/h; 455 mph)
- Max level speed below cruise ceiling: 375 knots (695 km/h; 432 mph)
- FAR balanced field length: 2,408 m (7,900 ft)
- T-O to 15 m (50 ft): 1,829 m (6,000 ft)
- Landing from 15 m (50 ft) at 51,714 kg (114,000 lb) landing weight: 975 m (3,200 ft)
- Ferry range: 4,330 nm (8,024 km; 4,986 miles)
- Endurance on station at 1,000 nm (1,853 km; 1,151 miles) radius: 8 h 12 min

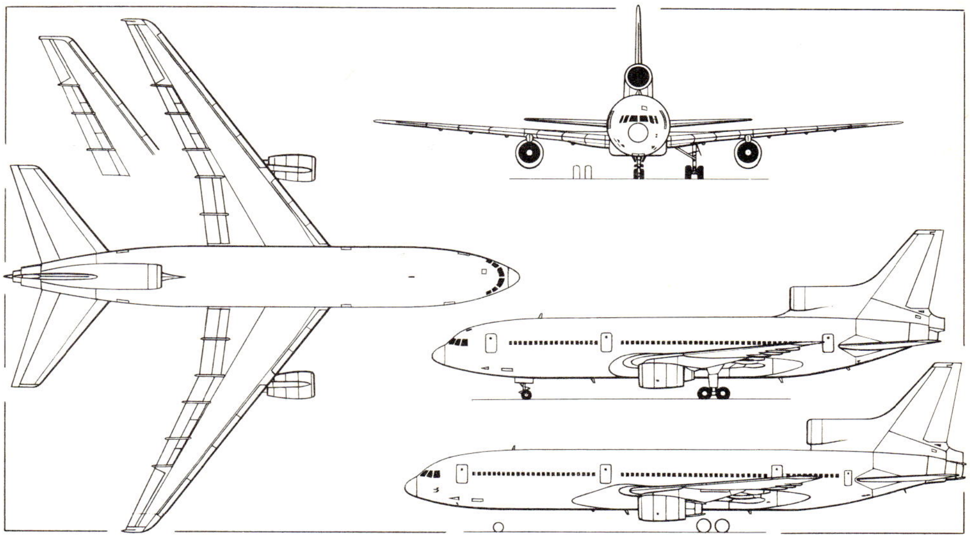

Lockheed L-1011-500 TriStar long-range wide-bodied transport, with added side view (bottom) and scrap view of wingtip of L-1011-100 *(Pilot Press)*

LOCKHEED VIKING

US Navy designation: S-3

Development of the **S-3A** Viking anti-submarine aircraft (1978-79 *Jane's*) was carried out by Lockheed in partnership with Vought Systems Division of LTV and Univac Federal Systems Division of Sperry Rand. Vought designed and built the wing, engine pods, tail unit and landing gear, and Univac was responsible for the digital computer. Lockheed built the fuselage, integrated the avionics, and was responsible for final assembly at Burbank, California, from where the first prototype flew on 21 January 1972.

Production of 187 S-3As for the US Navy ended in mid-1978. All tooling has been placed in storage at Burbank pending a US Navy decision on further orders, and in early 1980 demonstrator versions of the **US-3A** (COD) and **KS-3** (tanker) were being evaluated by the Navy. No production decision has been made, but it was expected that the US Navy would request a proposal for an advanced COD aircraft in 1981. Lockheed-California's submission for this competition is an S-3 derivative, with a wider fuselage that incorporates an aft loading door. Fleet introduction of this aircraft could begin in the late 1980s.

Lockheed announced on 18 August 1981 the receipt of a full-scale engineering development contract, from the US Naval Air Systems Command, for an improved avionics system for S-3A Vikings currently in service with the US Navy. With initial funding of $14·5 million, it follows a contract awarded by the Navy in 1980, under which Lockheed-California developed the specifications for an S-3A weapons system improvement programme (WSIP).

Improvements covered by the WSIP programme include increased acoustic processing capacity, expanded electronic support measure coverage, increased radar processing, a new sonobuoy receiver system, and provisions for the Harpoon missile. It is anticipated that a total of 160 S-3As could be retrofitted under the programme, after which they would be redesignated **S-3B**.

Top to bottom: Lockheed KS-3 (tanker), US-3A (COD) and S-3A Viking

LOCKHEED L-1011 (MODEL 385) TRISTAR

In January 1966, Lockheed-California began a study of future requirements in the short/medium-haul airliner market. The design which emerged, known as the L-1011 (Lockheed Model 385 TriStar), was influenced by the published requirements of American Airlines, which specified optimum payload/range performance over the Chicago-Los Angeles route, coupled with an ability to take off from comparatively short runways with full payload.

The original design centred around a twin-turbofan configuration. Discussions which followed with American domestic carriers led to the eventual selection of a three-engined configuration, and the Rolls-Royce RB.211 high bypass ratio turbofan was chosen as power plant.

In June 1968 the L-1011 TriStar moved to the production design stage. Construction of the first aircraft began in March 1969, and this was rolled out in September 1970. The first flight was made on 16 November 1970. On 22 December 1971 class II provisional type certification was received, permitting delivery of aircraft to customers for route proving and demonstration purposes.

The original version of the TriStar is now known as the L-1011-1. Three other versions were available in early 1981, the L-1011-100/-200/-500, which entered service respectively in 1975, 1977 and 1979. Details of current production versions, and available information on proposed new versions, follow:

L-1011-1. Basic TriStar, as described in detail. Initial delivery of the L-1011-1, to Eastern Air Lines for crew training, made on 6 April 1972, followed by a similar delivery to TWA. FAA certification was granted in the same month and the first passenger service with the TriStar was flown by Eastern on 15 April. Scheduled services began eleven days later.

L-1011-100. Extended-range version, with RB.211-22B engines (each 187 kN; 42,000 lb st). Outward configuration identical with that of L-1011-1. Max T-O weight of 211,375 kg (466,000 lb) with additional 8,165 kg (18,000 lb) of fuel in new centre-section tanks.

Lockheed L-1011-500 TriStar in the insignia of BWIA International

Ordered by Air Canada, Cathay Pacific, Gulf Air, Saudi Arabian Airlines and TWA.

L-1011-200. Extended-range version, with improved take-off and climb performance, offering particular benefits to operators serving 'hot or high' areas. Outward configuration identical with that of L-1011-1. Powered by RB.211-524 engines (each 213·5 kN; 48,000 lb st). Max T-O weight 211,375 kg (466,000 lb). First flight of TriStar testbed with RB.211-524 (one only) made on 10 April 1976. L-1011-200 certificated by FAA on 26 April 1977. Ordered by British Airways, Delta Air Lines, Gulf Air and Saudi Arabian Airlines.

L-1011-500. Long-range version, with a max T-O weight of 228,610 kg (504,000 lb) and max fuel capacity of 96,160 kg (212,000 lb) through added centre-section tankage. Fuselage is shortened by 4·11 m (13 ft 6 in). Three RB.211-524B or B4 engines (each 222·4 kN; 50,000 lb st). Galley located on main deck. Forward cargo hold accommodates 12 LD-3 containers or four pallets each measuring 2·24 m × 3·17 m (88 in × 125 in). Centre hold takes 7 LD-3 containers. In a mixed class configuration, with 24 first class passengers in six-abreast seating and 222 economy passengers in nine-abreast seating, the aircraft carries 246 passengers. Max accommodation for 330 passengers. Ordered by AeroPeru, Air Canada, Air Lanka, Alia, British Airways, BWIA, Delta Air Lines, LTU (West Germany), Pan Am and TAP/Air Portugal. New fuselage/centre engine fairing, installed first on the -500, may be adopted as standard for all TriStar models. Flight testing of this version began in October 1978, and -500s entered service with British Airways on 7 May 1979.

In November 1979, flight testing began of the first production L-1011-500 fitted with a 1·37 m (4 ft 6 in) extension to each wingtip. This was made possible by the introduction of automatically activated ailerons, and test results have shown a fuel efficiency improvement of 3 per cent. Extended wingtips and active controls of this kind were introduced on Pan Am -500s in 1980, and became standard in 1981. In early 1981 Lockheed initiated a modification programme to retrofit this wingtip extension to all earlier -500s.

On 17 June 1981, an L-1011-500 equipped with a Collins FCS-240 digital flight control system became the first aircraft to receive FAA certification for operation in Category IIIA conditions using a fully-digital autopilot system. Production L-1011-500 TriStars fitted with FCS-240 systems were delivered first to Pan Am and Air Canada in mid-1981.

Advanced TriStar. The original TriStar prototype (N1011) continues in use by Lockheed, under the name Advanced TriStar, to test and develop new ideas and systems that are under consideration for inclusion in future versions of this aircraft. Its equipment includes automatic brakes, automatic take-off thrust control, a flight management system fully coupled to the autopilot, extended wingtips, active aileron control, Autoland, direct lift control, all-moving tailplane, a digital autopilot, and an area navigation system with moving map display.

Automatic brakes operating in conjunction with anti-skid units ensure that braking is optimum at all times, in relation to load, speed and weather conditions. Automatic take-off thrust control allows the pilot to use reduced take-off power settings, and to operate from shorter field lengths than would be normal for such settings. With the throttles set to provide the requisite take-off power, in relation to field length, altitude and aircraft gross weight, failure of an engine during take-off would be offset by automatic advance of the remaining two engines to rated take-off thrust, or to a pre-set emergency power rating limit, thus minimising any time lag that might be experienced due to slow crew response.

The wingtip extensions now fitted on the -500 were first flown on the Advanced TriStar. It was appreciated that the added span, by itself, would generate an increased wing lift increment that would be unacceptable under certain manoeuvre or gust loads. This required the introduction of the now-standard active control system, to provide automatic aileron deflection to offset such loads, thus eliminating the need for wing structural redesign.

In mid-1980 Lockheed built for structure testing a TriStar fin made of advanced composite materials. Weighing 106 kg (235 lb) less than the standard metal fin, a similar fin may be flown eventually on the Advanced TriStar.

By June 1981, Lockheed had delivered 207 L-1011 TriStars to AeroPeru, Air Canada, All Nippon Airways, British Airways, BWIA, Cathay Pacific Airways, Court Line, Delta Air Lines, Eastern Air Lines, Gulf Air, Lufttransport Unternehmen, Pacific Southwest Airlines, Pan Am, Saudi Arabian Airlines and Trans World Airlines. Sales totalled 245, with options on 47 more, by the same date.

The description which follows applies to the L-1011-1 TriStar in its initial operational form, except where indicated. The basic structural details apply to all current production derivatives:

Type: Three-turbofan commercial transport.

Wings: Cantilever low-wing monoplane. Special Lockheed aerofoil sections. Dihedral at trailing-edge 7° 31′ on inner wings, 5° 30′ outboard. Sweepback at quarter-chord 35°. The wing consists of a centre-section, passing through the lower fuselage, and an outer wing panel on each side. It is of conventional fail-safe construction, with aluminium alloy surfaces, ribs and spars, and integral fuel tanks. Hydraulically-powered aluminium alloy ailerons of conventional two-spar box construction, with aluminium alloy honeycomb trailing-edge, in inboard and outboard sections on each wing, operate in conjunction with flight spoilers. The low-speed ailerons extend from approximately 80% of semi-span to within 0·25 m (10 in) of the wingtips, the high-speed ailerons extend from approximately WBL 387 to WBL 480 on each wing. Double-slotted Fowler trailing-edge flaps, constructed of aluminium alloy and aluminium alloy honeycomb. Each flap segment consists of a honeycomb trailing-edge, a front spar, ribs, skin panels, carriages, and tracks mounted on the forward segment to provide for extension and rotation of the aft segment. A sheet metal vane surface, actuated by a linkage system during flap rotation, forms the forward section of the extended flap. Four aluminium alloy leading-edge slats outboard of engine pylon on each wing. Each segment is mounted to two roller-supported tracks and extends in a circular motion down and forward for take-off and landing. Three leading-edge slats inboard of engine pylon on each wing, made of aluminium alloy honeycomb and sheet metal fairings. Six spoilers on the upper surface of each wing, two inboard and four outboard of the inboard aileron, constructed from bonded aluminium alloy tapered honeycomb. No trim tabs. Flight controls fully powered. Each control surface system is controlled by a multiple redundant servo actuator system that is powered by four independent and separate hydraulic sources. Thermal de-icing of outboard wing leading-edge slats by engine bleed air.

Fuselage: Semi-monocoque structure of aluminium alloy. Constant cross-sectional diameter of 5·97 m (19 ft 7 in) for most of the length. Bonding utilised in skin joints, for attaching skin-doublers at joints and around openings to improve fatigue life. Skins and stringers supported by frames spaced at 0·51 m (20 in) intervals, with fail-safe straps midway between frames. These frames, with the exception of main frames and door-edge members, are 0·076 m (3 in) deep at the sides of the cabin, increasing progressively to a depth of 0·15 m (6 in) at the top of the fuselage and below the floor. Fuselage length reduced on L-1011-500.

Tail Unit: Conventional cantilever structure, consisting of variable-incidence horizontal tailplane-elevator assembly and vertical fin and rudder. Primary loads of the fin are carried by a conventional box-beam structure, with ribs spaced at approx 0·51 m (20 in) centres. The rudder comprises forward and aft spars, glassfibre trailing-edges, hinge and actuator backup ribs, sheet metal formers, box surface panels and leading-edge fairings. Elevators are of similar construction. Truss members for the tailplane centre-section are built up from forged and extruded sections. Outboard of the centre-section, construction is similar to that of the fin box-beam, leading- and trailing-edges, except that the surface structure is integrally stiffened. The elevators are linked mechanically to the tailplane actuation gear, to modify its camber and improve its effectiveness. No trim tabs. Controls are fully powered, the hydraulic servo actuators receiving power from four independent hydraulic sources, under control of electronic flight control system. Control feel is provided, with the force gradient scheduled as a function of flight condition. No de-icing equipment.

Landing Gear: Menasco hydraulically-retractable tricycle type. Twin-wheel units in tandem on each main gear; twin wheels on nose gear, which is steerable 65° to each side. Nosewheels retract forward into fuselage. Main wheels retract inward into fuselage wheel-wells. Oleo-pneumatic shock-absorbers in all units. Goodrich forged aluminium alloy wheels of split construction. Main wheels have tubeless tyres size 50 × 20-20, Type VIII, pressure 10·34-11·38 bars (150-165 lb/sq in) for short- to medium-range operational weights, 12·41 bars (180 lb/sq in) for max-range weight. Nosewheels have tubeless tyres size 36 × 11-16, Type VII, pressure 12·76 bars (185 lb/sq in). Hydraulically-operated brakes, controlled by the rudder pedals. Anti-skid units, with individual wheel skid and modulated control, installed in the normal and alternative braking systems.

Power Plant (L-1011-1): Three Rolls-Royce RB.211-22B turbofan engines, each rated at 187 kN (42,000 lb st). Two engines mounted in pods on pylons under the wings, the third mounted in the rear fuselage at the base of the fin. Engine bleed air is used to anti-ice the engine inlet lips. Two integral fuel tanks in each wing: inboard tank capacity 30,581 litres (8,079 US gallons), outboard tank capacity 14,489 litres (3,828 US gallons). Total fuel capacity 90,140 litres (23,814 US gallons). Pressure refuelling points in wing leading-edges. Oil capacity approx 34 litres (9 US gallons) per engine. A detachable pylon can be fitted between the starboard engine nacelle and fuselage to permit carriage of a replacement engine for another TriStar. Alternative power plants for -100, -200 and -500 detailed under model listings. These four models each have provision for additional centre-section tankage, raising total fuel capacity to 100,317 litres (26,502 US gallons) in -100 and -200, and 119,774 litres (31,642 US gallons) in -500.

Accommodation: Basic flight crew of three, plus cabin attendants. First class and coach mixed accommodation for 256 passengers, with a maximum of 400 in all-economy configuration. Alternative intermediate seating capacities are provided by using eight seat-tracks which permit 6, 8, 9 or 10-abreast seating, with two full-length aisles. Underfloor galley. Seven lavatories are provided, two forward and five aft. Three Type A passenger doors of the upward-opening plug type on each side of the fuselage, one pair immediately aft of flight deck, one pair forward of wing, one pair aft of wing. Two Type I emergency exit doors, one each side of fuselage, at rear of cabin, replaced by two Type A doors for 10-abreast seating. Baggage and freight compartments beneath the floor able to accommodate 16 containers, totalling 71·58 m^3 (2,528 cu ft), and 19·8 m^3 (700 cu ft) bulk cargo (19 containers and 14·2 m^3; 500 cu ft in -500).

SYSTEMS: Air-conditioning and pressurisation system, using engine bleed air or APU air combined with air-cycle refrigeration. Pressurisation system maintains equivalent of 2,440 m (8,000 ft) conditions to 12,800 m (42,000 ft). Normal cabin pressure differential 0·582 bars (8·44 lb/sq in). Four independent 207 bar (3,000 lb/sq in) hydraulic systems provide power for primary flight control surfaces, normal brake power, landing gear retraction and nosewheel steering. Electrical system includes four 120/208V 400Hz alternators, one on each engine and one driven by the APU, which is sited in the aft fuselage. APU provides ground and in-flight power, to an altitude of 9,145 m (30,000 ft), producing both shaft and pneumatic power for utilisation by the electrical, environmental control and hydraulic systems. Integral electric heaters to anti-ice windscreens, pitot masts and total temperature probes.

AVIONICS AND EQUIPMENT: Standard equipment includes two ARINC 546 VHF communication transceivers, two ARINC 547 VHF navigation systems, two ARINC 568 interrogator units, an ARINC 564 weather radar system, two ARINC 572 air traffic control transponders, partial provision for a dual collision system, three vertical gyros, and full blind-flying instrumentation. Space is provided for installation of two ARINC 533A HF transceivers and a dual SATCOM system.

DIMENSIONS, EXTERNAL:
Wing span:
-1, -100, -200 47·34 m (155 ft 4 in)
-500 50·09 m (164 ft 4 in)
Wing chord at root 10·46 m (34 ft 4 in)
Wing chord at tip 3·12 m (10 ft 3 in)
Wing aspect ratio: -1, -100, -200 6·95
Length overall:
-1, -100, -200 54·17 m (177 ft 8½ in)
-500 50·05 m (164 ft 2½ in)
Height overall 16·87 m (55 ft 4 in)
Tailplane span 21·82 m (71 ft 7 in)
Wheel track 10·97 m (36 ft 0 in)
Wheelbase: -1, -100, -200 21·34 m (70 ft 0 in)
-500 19·71 m (64 ft 8 in)
Passenger doors (each): Height 1·93 m (6 ft 4 in)
Width 1·07 m (3 ft 6 in)
Height to sill 4·60 m (15 ft 1 in)
Emergency passenger doors (each):
Height 1·52 m (5 ft 0 in)
Width 0·61 m (2 ft 0 in)
Height to sill 4·60 m (15 ft 1 in)
Baggage and freight compartment doors (forward and centre): Height 1·73 m (5 ft 8 in)
Width 1·78 m (5 ft 10 in)
Height to sill 2·72 m (8 ft 11 in)
Baggage and freight compartment doors (aft):
Height 1·22 m (4 ft 0 in)
Width 1·12 m (3 ft 8 in)
Height to sill 2·92 m (9 ft 7 in)

DIMENSIONS, INTERNAL:
Cabin, excl flight deck and underfloor galley:
Length: -1, -100, -200 41·43 m (135 ft 11 in)
Max width 5·77 m (18 ft 11 in)
Max height 2·41 m (7 ft 11 in)
Floor area: -1, -100, -200 215·5 m^2 (2,320 sq ft)
-500 192·6 m^2 (2,073 sq ft)
Volume: -1, -100, -200 453 m^3 (16,000 cu ft)
Baggage/cargo holds, bulk capacity:
-1, -100, -200 110·4 m^3 (3,900 cu ft)
-500 118·9 m^3 (4,200 cu ft)

AREAS:
Wings, gross: -1, -100, -200 320·0 m^2 (3,456 sq ft)
Ailerons (total) 14·86 m^2 (160 sq ft)
Trailing-edge flaps (total) 49·80 m^2 (536 sq ft)
Leading-edge slats (total):
inboard slats 11·52 m^2 (124 sq ft)
outboard slats 21·93 m^2 (236 sq ft)
Spoilers (total) 19·88 m^2 (214 sq ft)
Fin 51·10 m^2 (550 sq ft)
Rudder 11·89 m^2 (128 sq ft)
Tailplane 119·10 m^2 (1,282 sq ft)

WEIGHTS:
Operating weight empty:
-1 109,633 kg (241,700 lb)
-100 111,674 kg (246,200 lb)
-200 112,670 kg (248,400 lb)
-500 111,311 kg (245,400 lb)
Max payload: -1 37,784 kg (83,300 lb)
-100 41,640 kg (91,800 lb)
-200 40,642 kg (89,600 lb)
-500 42,006 kg (92,608 lb)
Max T-O weight: -1 195,045 kg (430,000 lb)
-100, -200 211,375 kg (466,000 lb)
-500 228,610 kg (504,000 lb)
Max zero-fuel weight: -1 147,417 kg (325,000 lb)
-100, -200, -500 153,315 kg (338,000 lb)
Max landing weight: -1 162,385 kg (358,000 lb)
-100, -200, -500 166,920 kg (368,000 lb)

PERFORMANCE (A: L-1011-1 at max T-O weight of 195,045 kg: 430,000 lb: B and C: L-1011-100 and L-1011-200 respectively at max T-O weight of 211,375 kg: 466,000 lb: D: L-1011-500 at max T-O weight of 228,610 kg: 504,000 lb, except where indicated):
Never-exceed speed, all versions
Mach 0·95 (435 knots; 806 km/h; 501 mph) CAS
Max cruising speed, mid-cruise weight at 9,145 m (30,000 ft): A 520 knots (964 km/h; 599 mph)
B 515 knots (954 km/h; 593 mph)
C, D 525 knots (973 km/h; 605 mph)
Econ cruising speed, mid-cruise weight at 10,670 m (35,000 ft):
all versions 480 knots (890 km/h; 553 mph)
Stalling speed at max landing weight, flaps and gear up:
A 148 knots (274 km/h; 170 mph)
B, C 151 knots (280 km/h; 174 mph)
D 157 knots (291 km/h; 181 mph)
Stalling speed at max landing weight, flaps and gear down: A 108 knots (200 km/h; 124 mph)
B, C 110 knots (204 km/h; 127 mph)
D 114 knots (211 km/h; 131 mph)
Max rate of climb at S/L: A 838 m (2,750 ft)/min
B 731 m (2,400 ft)/min
C 945 m (3,100 ft)/min
D 908 m (2,980 ft)/min
Service ceiling: A, B, C 12,800 m (42,000 ft)
D 13,100 m (43,000 ft)
FAR T-O field length: A 2,440 m (8,000 ft)
B 3,260 m (10,700 ft)
C 2,485 m (8,150 ft)
D 2,620 m (8,600 ft)
FAR landing field length, at max landing weight:
A 1,734 m (5,690 ft)
B 1,775 m (5,820 ft)
C 1,960 m (6,430 ft)
D 2,065 m (6,770 ft)
Range with max passengers and baggage, international reserves: A 2,740 nm (5,077 km; 3,155 miles)
B 3,420 nm (6,338 km; 3,938 miles)
C 3,610 nm (6,690 km; 4,157 miles)
D 5,236 nm (9,697 km; 6,025 miles)
Range with max fuel, international reserves:
A 4,220 nm (7,815 km; 4,856 miles)
B 4,625 nm (8,565 km; 5,322 miles)
C 4,920 nm (9,111 km; 5,661 miles)
D 6,080 nm (11,260 km; 6,996 miles)

OPERATIONAL NOISE LEVELS (FAR Pt 36):
T-O: A 96 EPNdB
B, C, D 98 EPNdB
Approach: A, B 103 EPNdB
C 102 EPNdB
D 101 EPNdB
Sideline: A, B 95 EPNdB
C, D 98 EPNdB

LOCKHEED-GEORGIA COMPANY

86 South Cobb Drive, Marietta, Georgia 30063

Lockheed-Georgia's main building at Marietta is one of the world's largest aircraft production plants under a single roof. Aircraft in current production on its assembly lines are the C-130 Hercules turboprop transport and its commercial counterpart, the L-100, at a combined rate of three per month. Major modification programmes are under way on USAF C-141 StarLifter and C-5 Galaxy transports.

Lockheed-Georgia had approximately 11,000 employees at the beginning of 1981.

LOCKHEED MODEL 382 HERCULES

US Air Force designations: C-130, AC-130, DC-130, EC-130, HC-130, JC-130, MC-130, RC-130 and WC-130
US Navy designations: C-130, DC-130, EC-130 and LC-130
US Marine Corps designation: KC-130
US Coast Guard designation: HC-130
Canadian Armed Forces designation: CC-130
RAF designations: Hercules C Mk 1, W Mk 2 and C Mk 3

The C-130 was designed to a specification issued by the US Air Force Tactical Air Command in 1951. Lockheed was awarded its first production contract for the C-130A in September 1952, and 461 C-130As and C-130Bs were manufactured. Details of these basic versions and of many variants for special duties can be found in the 1967-68 and 1975-76 *Jane's*. Later military versions of the C-130 are as follows:

C-130E (Lockheed Model 382-44). Extended-range development of C-130B, with four 3,020 kW (4,050 ehp) T56-A-7 turboprop engines and two 5,145 litre (1,360 US gallon) underwing fuel tanks. Deliveries began in April 1962, and by February 1975 the planned production of a total of 503 C-130Es had been completed. Details of the basic C-130E can be found in the 1973-74 *Jane's*.

EC-130E. A version of the C-130E modified for USAF special mission applications of surveillance and communications. Equipped with ABCCC/USC-15 airborne battlefield command and control centre capsule (12·19 m; 40 ft long and weighing about 9,070 kg; 20,000 lb) which fits into cargo hold. Capsule accommodates 12-16 personnel and incorporates 20 different radios (four UF, eight UHF, four VHF and four FM), plus secure teletype and voice communications capability, and automatic radio relay. In service with 7th Airborne Command and Control Squadron, 552nd AWAC Wing, US Air Force. (See also the LAS entry in this section.)

MC-130E. Version of C-130E modified for USAF special tactical missions. ALQ-8 ECM pod under port wing. In service with 1st Special Operations Squadron, 3rd Tactical Fighter Wing.

EC-130G. The original configuration of the US Navy's TACAMO communications platform.

C-130H. Similar to earlier Hercules models except for more powerful engines: T56-A-15 turboprops rated at 3,661 kW (4,910 ehp) for take-off, but limited to 3,362 kW (4,508 ehp). Deliveries to USAF began in April 1975.

C-130H-MP. Maritime patrol, search and rescue version, based on C-130H. Max T-O weight 70,310 kg (155,000 lb). Max payload 18,630 kg (41,074 lb). Four 3,362 kW (4,508 ehp) T56-A-15 engines. Standard and optional equipment includes sea search radar, scanner seats with observation windows, computerised INS/Omega navigation system, crew rest and lavatory/galley slide-in module, flare launcher, loudspeaker system, rescue kit airdrop platform, side-looking airborne radar, passive microwave imager, low-light TV, infra-red scanner, camera with data annotation, and ramp equipment pallet which includes a station for an observer. Search time at an altitude of 1,525 m (5,000 ft) is 2 h 30 min at a radius of 1,800 nm (3,333 km; 2,070 miles); 16 h 50 min at radius of 200 nm (370 km; 230 miles). One aircraft on order for Indonesian Air Force, and three delivered to Royal Malaysian Air Force.

C-130H-30. 'Stretched' version, with structural changes similar to those of RAF Hercules C. Mk 3 (see C-130K paragraph). Seven ordered for Indonesian Air Force, of which the first two were delivered in December 1980.

HC-130H. Extended-range version for Aerospace Rescue and Recovery Service of the US Air Force for aerial recovery of personnel or equipment and other duties. The US Coast Guard subsequently ordered 12. Total of 55 delivered, of which the first one flew on 8

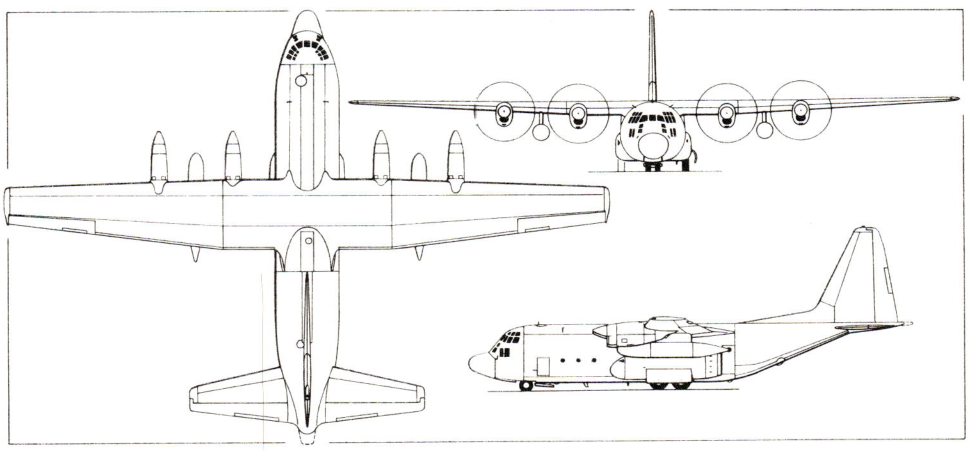

Lockheed C-130E Hercules four-turboprop medium/long-range combat transport *(Pilot Press)*

Fuel-saving strakes under rear fuselage of C-130E

December 1964. Details in 1979-80 *Jane's*. Four modified as **JHC-130H** with added equipment for aerial recovery of re-entering space capsules. One modified to **DC-130H** for drone control duties.

KC-130H. A tanker version of the C-130H, very similar to the KC-130R. Exported to Argentina (2), Brazil (2), Israel (2), Saudi Arabia (8) and Spain (8).

C-130K. This is basically a C-130H, modified to meet requirements of the Royal Air Force. Much of the avionics and instrumentation is of UK manufacture. Sixty-six delivered as **Hercules C. Mk 1**, of which the first flew on 19 October 1966. One modified by Marshall of Cambridge (Engineering) Ltd in the UK for use by the RAF Meteorological Research Flight, under the designation **Hercules W. Mk 2**. Thirty of the others are each being lengthened by 4·57 m (15 ft 0 in) (2·54 m; 8 ft 4 in plug forward of wing, 2·03 m; 6 ft 8 in aft of wing), equivalent to commercial L-100-30 standard, during 1978-82. This will increase payload capacity to seven cargo pallets instead of five, or 128 troops instead of 92, or 92 fully-equipped paratroops instead of 64, or 93 stretcher patients (and six attendants) instead of 70. The first aircraft was modified at Lockheed-Georgia in 1979; the remaining 29 are being lengthened by Marshall of Cambridge (which see). After modification, these aircraft are redesignated **Hercules C. Mk 3**.

HC-130N. Search and rescue version of the C-130H for recovery of aircrew and retrieval of space capsules after re-entry, using advanced direction-finding equipment. Fifteen delivered to US Air Force.

HC-130P. Version of the C-130H, modified to have capability of refuelling helicopters in flight and for mid-air retrieval of parachute-borne payloads. Details in 1979-80 *Jane's*.

EC-130Q. Similar to EC-130G, but with improved equipment and crew accommodation, for US Navy command communications (TACAMO) duties.

KC-130R. Tanker version of C-130H: 14 built for US Marine Corps. Major changes from earlier KC-130F include engines of 3,362 kW (4,508 ehp), increased T-O and landing weights, pylon-mounted fuel tanks to provide additional 10,296 litres (2,720 US gallons) of fuel, and removable 13,627 litre (3,600 US gallon) fuel tank in cargo compartment.

LC-130R. Basically a C-130H with wheel-ski gear for US Navy. Four converted, for service in the Antarctic. Two more acquired by National Science Foundation, for use in the Antarctic. Details in 1979-80 *Jane's*.

During 1980, a USAF C-130E was fitted with two light alloy and glassfibre strakes on the undersurface of the rear fuselage. Flight testing by Lockheed and the USAF showed that the resulting drag reduction offered fuel savings of more than 3·5 per cent at long-range cruising speeds, and an increase of more than 18 knots (33 km/h; 20 mph) in normal cruising speed at no cost in fuel consumption. The USAF was evaluating production strakes of Kevlar construction in Summer 1981, and was expected to begin a programme before the end of the year to retrofit most of its C-130s with strakes. If it does so, Lockheed is expected to introduce strakes on new production C-130s, and to offer retrofit kits for aircraft in service.

Commercial versions of the Hercules are described separately.

The C-130 is able to deliver single loads of up to 11,340 kg (25,000 lb) by the ground proximity extraction method. This involves making a flypast 1·2-1·5 m (4-5 ft) above the ground with the rear loading ramp open. The aircraft trails a hook which is attached by cable to the palletised cargo. The hook engages a steel cable on the ground and the cargo is extracted from the aircraft and brought to a stop on the ground in about 30 m (100 ft) by an energy absorption system manufactured by All American Engineering of Wilmington, Delaware. An alternative extraction technique involves deploying a 6·70 m (22 ft) ribbon parachute to drag the pallet from the cabin. Loads of up to 22,680 kg (50,000 lb) have been delivered by this method.

By January 1981 firm orders for all versions of the C-130 totalled 1,622 for 52 nations. This total comprised 1,042 C-130s for the US services, 495 for foreign military operators, and 85 commercial Hercules. Another 22 had been sold by June. The 1,600th Hercules, a C-130H-30 for the Indonesian Air Force, was delivered in December 1980.

The following details refer specifically to the C-130H, except where indicated otherwise:

Type: Medium/long-range combat transport.

Wings: Cantilever high-wing monoplane. Wing section NACA 64A318 at root, NACA 64A412 at tip. Dihedral 2° 30'. Incidence 3° at root, 0° at tip. Sweepback at quarter-chord 0°. All-metal two-spar stressed-skin structure, with integrally-stiffened tapered machined skin panels up to 14·63 m (48 ft 0 in) long. Conventional aluminium alloy ailerons have tandem-piston hydraulic boost, operated by either of two independent hydraulic systems. Lockheed-Fowler aluminium alloy trailing-edge flaps. Trim tab in each aileron. Leading-edges anti-iced by engine bleed air.

Fuselage: Semi-monocoque structure of aluminium and magnesium alloys.

Tail Unit: Cantilever all-metal stressed-skin structure. Fixed-incidence tailplane. Trim tab in each elevator and rudder. Elevator tabs use AC electrical power as primary source and DC as emergency source. Control surfaces have tandem-piston hydraulic boost. Hot-air anti-icing of tailplane leading-edge, by engine bleed air.

Landing Gear: Hydraulically-retractable tricycle type. Each main unit has two wheels in tandem, retracting into fairings built on to the sides of the fuselage. Nose unit has twin wheels and is steerable through 60° each side of centre. Oleo shock-absorbers. Main-wheel tyres size 56 × 20-20, pressure 6·62 bars (96 lb/sq in). Nose-wheel tyres size 39 × 13-16, pressure 4·14 bars (60 lb/sq in). Goodyear aircooled multiple-disc hydraulic brakes with anti-skid units. Retractable combination wheel-skis available.

Power Plant: Four 3,362 kW (4,508 ehp) Allison T56-A-15 turboprop engines, each driving a Hamilton Standard type 54H60 four-blade constant-speed fully-feathering reversible-pitch propeller. Fuel in six integral tanks in wings, with total capacity of 26,344 litres (6,960 US gallons) and two optional underwing pylon tanks, each with capacity of 5,146 litres (1,360 US gallons). Total fuel capacity 36,636 litres (9,680 US gallons). Single pressure refuelling point in starboard wheel well. Overwing gravity fuelling. Oil capacity 182 litres (48 US gallons).

Accommodation: Crew of four on flight deck, comprising pilot, co-pilot, navigator and systems manager. Provision for fifth man to supervise loading. Sleeping quarters for relief crew, and galley. Flight deck and main cabin pressurised and air-conditioned. Standard complements for C-130H are as follows: troops (max) 92, paratroops (max) 64, litters 74 and 2 attendants. Corresponding figures for C-130H-30 are 128 troops, 92 paratroops, and 97 litters. As a cargo carrier, loads can include heavy equipment such as a 12,080 kg (26,640 lb) type F.6 refuelling trailer or a 155 mm howitzer and its high-speed tractor, or up to five 463L pallets of freight (seven in C-130H-30). Hydraulically-operated main loading door and ramp at rear of cabin. Paratroop door on each side aft of landing gear fairing.

Systems: Air-conditioning and pressurisation system max pressure differential 0·52 bars (7·5 lb/sq in). Three independent hydraulic systems, utility and booster systems operating at a pressure of 207 bars (3,000 lb/sq in). Auxiliary system has handpump for emergencies. Electrical system supplied by four 40kVA AC alternators, plus one 40kVA auxiliary alternator driven by APU. Four transformer-rectifiers for DC power. Current production aircraft incorporate many systems and component design changes for increased reliability. There are differences between the installed components for US government and export versions.

Avionics: Dual 628T-1 HF com, dual 618M-3A VHF com, AN/ARC-164 UHF com, AN/AIC-13 PA system, AN/AIC-18 intercom, dual 621A-6A ATC transponders, DF-301E UHF nav, dual 51RV-4B VHF nav, CMA 771 Omega nav, LTN-72 INS, dual DF-206 ADF, 51Z-4 marker beacon receiver, dual 860E-5 DME, AL-101 radio altimeter, RDR-1F weather radar, dual C-12 compass systems, Mk II GPWS, AP-105V autopilot, and dual FD-109 flight directors.

Dimensions, external:

Wing span	40·41 m (132 ft 7 in)
Wing chord at root	4·88 m (16 ft 0 in)
Wing chord, mean	4·16 m (13 ft 8½ in)
Wing aspect ratio	10·09
Length overall:	
all except HC-130H and C-130H-30	29·79 m (97 ft 9 in)
C-130H-30	34·37 m (112 ft 9 in)
Height overall	11·66 m (38 ft 3 in)
Tailplane span	16·05 m (52 ft 8 in)
Wheel track	4·35 m (14 ft 3 in)
Wheelbase	9·77 m (32 ft 0¾ in)
Propeller diameter	4·11 m (13 ft 6 in)
Main cargo door (rear of cabin):	
Height	2·77 m (9 ft 1 in)
Width	3·05 m (10 ft 0 in)
Height to sill	1·03 m (3 ft 5 in)
Paratroop doors (each): Height	1·83 m (6 ft 0 in)
Width	0·91 m (3 ft 0 in)
Height to sill	1·03 m (3 ft 5 in)

Lockheed C-130H-30 'stretched' Hercules in service with the Indonesian Air Force

Lockheed L-100-30 Hercules commercial transport in service with Sfair of Bordeaux

DIMENSIONS, INTERNAL:
Cabin, excl flight deck:
Length without ramp:
C-130H 12·22 m (40 ft 1¼ in)
C-130H-30 16·79 m (55 ft 1¼ in)
Length with ramp: C-130H 15·73 m (51 ft 8½ in)
Max width 3·13 m (10 ft 3 in)
Max height 2·81 m (9 ft 2¾ in)
Floor area, excl ramp: C-130H 39·5 m² (425 sq ft)
Volume, incl ramp:
C-130H 127·4 m³ (4,500 cu ft)
C-130H-30 165·5 m³ (5,845 cu ft)

AREAS:
Wings, gross 162·12 m² (1,745 sq ft)
Ailerons (total) 10·22 m² (110 sq ft)
Trailing-edge flaps (total) 31·77 m² (342 sq ft)
Fin 20·90 m² (225 sq ft)
Rudder, incl tab 6·97 m² (75 sq ft)
Tailplane 35·40 m² (381 sq ft)
Elevators, incl tabs 14·40 m² (155 sq ft)

WEIGHTS AND LOADINGS:
Operating weight empty:
C-130H 34,356 kg (75,743 lb)
C-130H-30 36,067 kg (79,516 lb)
Max payload:
C-130H 19,685 kg (43,399 lb)
C-130H-30 17,974 kg (39,626 lb)
Max normal T-O and landing weight 70,310 kg (155,000 lb)
Max overload T-O weight 79,380 kg (175,000 lb)
Normal landing weight 58,965 kg (130,000 lb)
Max landing weight 79,380 kg (175,000 lb)
Max zero-fuel weight, 2·5*g* 54,040 kg (119,142 lb)
Max wing loading 434·5 kg/m² (89 lb/sq ft)
Max power loading 5·23 kg/kW (8·6 lb/ehp)

PERFORMANCE (C-130H at max T-O weight, unless indicated otherwise):
Max cruising speed 325 knots (602 km/h; 374 mph)
Econ cruising speed 300 knots (556 km/h; 345 mph)
Stalling speed 100 knots (185 km/h; 115 mph)
Max rate of climb at S/L 579 m (1,900 ft)/min
Service ceiling at 58,970 kg (130,000 lb) AUW 10,060 m (33,000 ft)
Service ceiling, one engine out, at 58,970 kg (130,000 lb) AUW 8,075 m (26,500 ft)
Min ground turning radius 19·2 m (63 ft)
Runway LCN at 70,310 kg (155,000 lb) AUW:
asphalt 37
concrete 42
T-O run 1,091 m (3,580 ft)
T-O to 15 m (50 ft) 1,573 m (5,160 ft)
Landing from 15 m (50 ft) at 45,360 kg (100,000 lb) AUW 731 m (2,400 ft)
Landing from 15 m (50 ft) at 58,967 kg (130,000 lb) AUW 838 m (2,750 ft)
Landing run at 58,967 kg (130,000 lb) AUW 518 m (1,700 ft)
Range with max payload, with 5% reserves and allowance for 30 min at S/L 2,160 nm (4,002 km; 2,487 miles)
Range with max fuel, incl external tanks, 7,293 kg (16,078 lb) payload and reserves of 5% initial fuel plus 30 min at S/L 4,100 nm (7,600 km; 4,721 miles)

LOCKHEED L-100 SERIES COMMERCIAL HERCULES

Details of initial versions of the commercial Hercules have appeared in previous editions of *Jane's*; current models are as follows:

L-100-20 (Model 382E). Certificated on 4 October 1968, this 'stretched' version of the Hercules has a 2·54 m (100 in) fuselage extension. A 1·52 m (60 in) fuselage plug is inserted aft of the forward crew door and a 1·02 m (40 in) plug aft of the paratroop doors. Allison 501-D22A engines. Operators include Southern Air Transport in the USA; Echo Bay Mines, and Pacific Western Airlines, in Canada; Safair Freighters in the Republic of South Africa; the Kuwait Air Force; the Peruvian Air Force; TAAG in Angola; the Philippine Aerospace Development Corporation; and the Republic of Gabon.

L-100-30 (Model 382G). Generally similar to the L-100-20, but with the fuselage extended a further 2·03 m (80 in). Rear cargo windows and provision for JATO eliminated; rear paratroop doors optional. Saturn Airways (now Transamerica Airlines) was the first operator of this model, in December 1970. Other operators include Air Botswana; Alaska International Air; Air Algérie; Bolivian Air Force; Air Force of Dubai; Indonesian Air Force; Northwest Territorial Airways; Pacific Western Airlines; Pelita Air Service (Indonesia); PEMEX (Mexico); Republic of Gabon; Safair Freighters; SCIBE (Zaïre); Sfair (France); Southern Air Transport; Uganda Airlines; United Trade International; and Wirtschaftsflug (West Germany).

A total of 85 commercial Hercules (all versions) had been delivered by January 1981. Details given for the C-130H apply also to the L-100-20 and L-100-30, except as follows:

TYPE: Medium/long-range transport.

LANDING GEAR: As for C-130H, except main-wheel tyre pressure 3·24-7·38 bars (47-107 lb/sq in) and nose-wheel tyre pressure 4·14 bars (60 lb/sq in).

POWER PLANT: Four 3,362 kW (4,508 ehp) Allison 501-D22A turboprop engines.

DIMENSIONS, EXTERNAL:
Length overall: L-100-20 32·33 m (106 ft 1 in)
L-100-30 34·37 m (112 ft 9 in)
Wheelbase: L-100-20 11·30 m (37 ft 1 in)
L-100-30 12·32 m (40 ft 5 in)
Crew door (integral steps):
Height 1·14 m (3 ft 9 in)
Width 0·76 m (2 ft 6 in)
Height to sill 1·04 m (3 ft 5 in)

DIMENSIONS, INTERNAL:
Cabin, excl flight deck:
Length: L-100-20 15·04 m (49 ft 4 in)
L-100-30, excl ramp 17·07 m (56 ft 0 in)
L-100-30, incl ramp 19·93 m (65 ft 4¾ in)
Max height 2·74 m (9 ft 0 in)
Floor area, excl ramp:
L-100-20 46·36 m² (499 sq ft)
L-100-30 52·30 m² (563 sq ft)
Floor area, ramp 9·57 m² (103 sq ft)
Volume, incl ramp:
L-100-20 150·28 m³ (5,307 cu ft)
L-100-30 171·5 m³ (6,057 cu ft)

WEIGHTS AND LOADINGS:
Operating weight empty:
L-100-20 32,934 kg (72,607 lb)
L-100-30 33,684 kg (74,262 lb)
Max payload: L-100-20 21,497 kg (47,393 lb)
L-100-30 23,014 kg (50,738 lb)
Max ramp weight 70,670 kg (155,800 lb)
Max T-O weight 70,308 kg (155,000 lb)
Max landing weight:
L-100-20 58,970 kg (130,000 lb)
L-100-30 61,235 kg (135,000 lb)

Newly-updated to C-141B standard, StarLifter *Golden Bear*, named for the State of California, had been the first all-jet freighter to enter USAF service in 1965 in its original form

A rewinged C-5A Galaxy takes off on its redelivery flight to a USAF base

Max zero-fuel weight:
L-100-20 54,430 kg (120,000 lb)
L-100-30 56,700 kg (125,000 lb)
Max wing loading 433·5 kg/m² (88·8 lb/sq ft)
Max power loading 5·23 kg/kW (8·6 lb/ehp)

PERFORMANCE (at max T-O weight except where indicated):
Max cruising speed at 6,100 m (20,000 ft) at 54,430 kg (120,000 lb) AUW 315 knots (583 km/h; 363 mph)
Landing speed:
L-100-20 126 knots (233 km/h; 145 mph)
L-100-30 128 knots (237 km/h; 147 mph)
Max rate of climb at S/L 579 m (1,900 ft)/min
Min ground turning radius: L-100-20 26·8 m (88 ft)
L-100-30 27·5 m (90 ft)
Runway LCN: asphalt 37
concrete 42
FAR T-O field length 1,830 m (6,000 ft)
FAR landing field length, at max landing weight:
L-100-20 1,450 m (4,760 ft)
L-100-30 1,478 m (4,850 ft)
Range with max payload, 45 min reserves:
L-100-20 2,100 nm (3,889 km; 2,417 miles)
L-100-30 1,742 nm (3,226 km; 2,005 miles)
Range with zero payload, 45 min reserves:
L-100-20 4,250 nm (7,871 km; 4,891 miles)
L-100-30 4,200 nm (7,778 km; 4,833 miles)

OPERATIONAL NOISE LEVELS (FAR Pt 36):
T-O 98·4 EPNdB
Approach 99·1 EPNdB
Sideline 93·9 EPNdB

LOCKHEED C-141B STARLIFTER

Operational experience with the Lockheed C-141A StarLifter by the USAF's Military Airlift Command showed that on many occasions the cargo compartment was physically packed to capacity without the aircraft's maximum payload capability being reached. Parametric studies carried out by Lockheed showed that lengthening the fuselage by 7·11 m (23 ft 4 in) would provide an optimum relationship between modification cost and payload improvement, while at the same time allowing the existing wings, landing gear and power plant to be retained.

Under USAF contract, Lockheed-Georgia began work on the conversion of a C-141 in 1976. Designated YC-141B, this prototype was rolled out on 8 January 1977 and made its first flight on 24 March. The fuselage had been lengthened by the insertion of a 4·06 m (13 ft 4 in) plug immediately forward of the wing, and by a similar 3·05 m (10 ft 0 in) plug immediately aft of the wing. These two plug sections are designed so that the lengthened aircraft retains the same operational features and mission versatility as the C-141A. As a result of this modification, the floor area of the cargo compartment is increased by 22·26 m² (239·6 sq ft), and its volume by 61·48 m³ (2,171 cu ft). Thirteen standard 463L cargo pallets can be accommodated in the C-141B StarLifter, instead of the ten carried by an unmodified C-141A. The four 93·4 kN (21,000 lb st) Pratt & Whitney TF33-P-7 turbofan engines are unchanged.

To satisfy another proven requirement, for flight refuelling capability, a Universal Aerial Refuelling Receptacle Slipway Installation (UARRSI) has been incorporated in the upper surface of the forward fuselage, just aft of the flight deck.

The first production C-141B was delivered to the USAF on 4 December 1979, ahead of schedule and below projected cost. Lockheed delivered 83 C-141B conversions during 1980, and is due to complete the modification of all of Military Airlift Command's 271 StarLifters by mid-1982.

DIMENSIONS, EXTERNAL:
Wing span 48·74 m (159 ft 11 in)
Length overall 51·29 m (168 ft 3½ in)
Height overall 11·96 m (39 ft 3 in)

DIMENSIONS, INTERNAL:
Cargo compartment: Length 28·44 m (93 ft 3½ in)
Max height 2·77 m (9 ft 1 in)
Max width 3·11 m (10 ft 2½ in)
Usable cargo volume 322·71 m³ (11,399 cu ft)

WEIGHTS:
Operating weight (MAC) 67,186 kg (148,120 lb)
Max payload (2·5*g*) 32,025 kg (70,605 lb)
Max payload (2·25*g*) 41,222 kg (90,880 lb)
Max T-O weight (2·5*g*) 146,555 kg (323,100 lb)
Max T-O weight (2·25*g*) 155,580 kg (343,000 lb)
Emergency war planning ramp weight 156,445 kg (344,900 lb)
Max zero-fuel weight (2·5*g*) 99,210 kg (218,725 lb)
Max zero-fuel weight (2·25*g*) 108,410 kg (239,000 lb)
Max landing weight 155,580 kg (343,000 lb)

PERFORMANCE (at max 2·5*g* T-O weight, except where indicated):
Max cruising speed 492 knots (910 km/h; 566 mph)
Long-range cruising speed 430 knots (796 km/h; 495 mph)
Max rate of climb at S/L 890 m (2,920 ft)/min
T-O to 15 m (50 ft) 1,768 m (5,800 ft)
Landing from 15 m (50 ft) at normal landing weight 1,128 m (3,700 ft)
Range with max payload 2,550 nm (4,725 km; 2,935 miles)
Ferry range 5,550 nm (10,280 km; 6,390 miles)

LOCKHEED C-5 GALAXY

USAF designation: C-5A

In early 1978 Lockheed received a USAF contract to manufacture two new sets of wings for the C-5A, of a design intended to reduce stress and increase service life to 30,000 h. Apart from the moving surfaces, these wings are of virtually new design, using 7175-T73511 aluminium alloy for greater strength and increased resistance to corrosion. One set was for ground testing, and one for flight trials, which were completed successfully on a prototype installation during 1980, the converted C-5A being redelivered to the USAF in early 1981. It is planned to fit similar new wings to the 77 Galaxies still in operational service with the USAF, between 1982 and 1987, and an incremental $157 million contract to begin production of the wings was received by Lockheed in January 1981.

Full structural and specification details of the C-5A can be found in the 1975-76 *Jane's*. An abbreviated entry appeared in the 1978-79 edition.

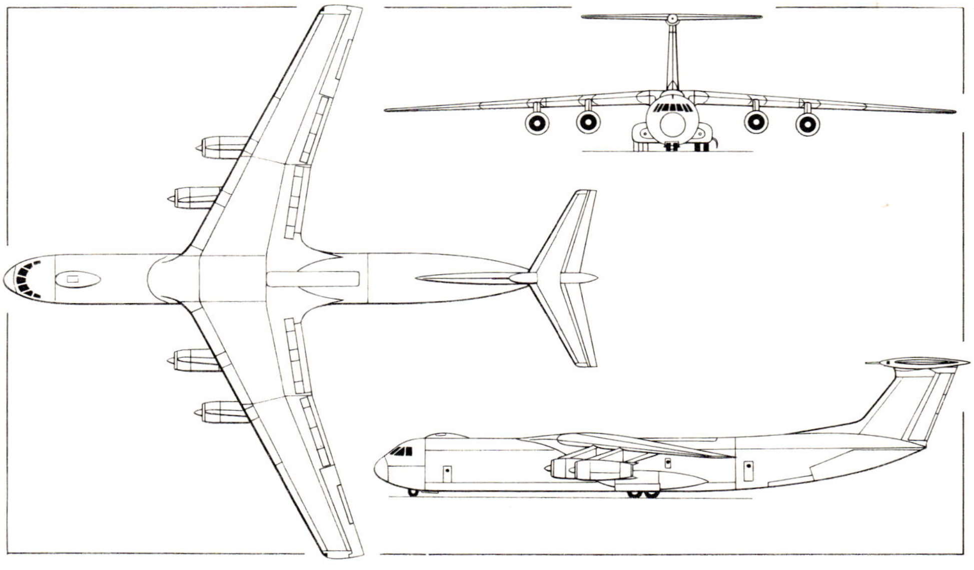

Lockheed C-141B lengthened version of the StarLifter logistics transport *(Pilot Press)*

MacCREADY

DR PAUL B. MacCREADY

AeroVironment Inc, 145 Vista Avenue, Pasadena, California 91107
Telephone: (213) 449 4392

Dr Paul MacCready headed the design team responsible for the Gossamer Condor which, on 23 August 1977, when propelled entirely by a man, became the first such aircraft to fly a figure-of-eight pattern around two pylons half a mile apart. This achievement won the £50,000 Kremer Prize that had been offered for the world's first man-powered aircraft to demonstrate such a capability. Since that time Dr MacCready has led the development of the man-powered Gossamer Albatross and Gossamer Penguin, the conversion of the Penguin for solar-powered flight, and the Solar Challenger.

MacCREADY GOSSAMER ALBATROSS

The Gossamer Albatross is similar in configuration to the Gossamer Condor, described in the 1978-79 *Jane's*, but has fewer external bracing wires and has structural members made of carbonfibre reinforced plastics (CFRP) tubing for optimum strength/weight ratio. The piloting position is upright rather than semi-reclined. Weight of the Albatross slightly exceeds that of the Condor, due mainly to the 2 litres (3·5 Imp pints) of drinking water provided for the pilot and the installation of a small Motorola Handi Talki communications radio with Bell Telephone earphone.

On 12 June 1979 the Gossamer Albatross completed successfully the first crossing of the English Channel by an aircraft propelled entirely by a man. Piloted by Mr Bryan Allen, who had gained the £50,000 prize in the Condor in 1977, it took off from Folkestone, England, at 05.51 h BST and landed on the beach at Cap Gris Nez, France, at 08.40 h BST. Straight-line distance covered was about 20 nm (37 km; 23 miles). Indicated airspeed varied from an initial 12-13 knots (22-24 km/h; 14-15 mph) to a later 9·5-10·5 knots (17·5-19·5 km/h; 11-12 mph). Average height above the sea was 0·5-2 m (2-6 ft), with a minimum of about 0·15 m (6 in) and maximum of 7·5 m (25 ft). This achievement gained a Kremer Prize of £100,000.

A second Gossamer Albatross, built as a backup for the cross-Channel flight, differed only in detail from the aircraft used on the successful attempt and described hereafter:

TYPE: Single-seat man-powered aircraft.

WINGS: Wire-braced high-wing monoplane, made in four panels, joined by light alloy tubes inside spar. Wing section Lissaman 7776. Thickness/chord ratio 11·5%. Moderate dihedral, constant from roots. Average sweepback at spar (28% chord) 7° 30′. Single spar of 50 mm (2 in) diameter CFRP tube; white expanded polystyrene ribs with CFRP capstrips and stiffeners; polystyrene leading-edge skin and Du Pont Mylar transparent heat-shrunk plastic film covering. Wingtips warped by starboard 'handlebar' control. No other control surfaces on wings. Kingpost of 25 mm (1 in) CFRP tubing provides overwing attachment for bracing wires.

FUSELAGE: Basic structure of CFRP tubing; formers of CFRP strip with Styrofoam core; leading-edge and bottom covering of polystyrene sheet, with inset window in line with pilot's head; Mylar film covering. Holes in leading-edge and vents in trailing-edge maintain flow of air through interior. Rhomboidal shape and aerofoil section of fuselage gondola enable it to act as vertical stabilising surface, requiring no additional fin area.

FOREPLANE: Wire-braced universally-pivoted structure of similar construction to wings, carried on a bowsprit of 38 mm (1·5 in) CFRP tubing. Fore and aft movement of pilot's port 'handlebar' grip controls foreplane pitch. Sideways movement operates servo-tabs on trailing-edge near tips, to provide roll and directional control.

LANDING GEAR: Two small fixed wheels at base of fore and aft gondola structural members, diameter 125 mm (5 in) and 100 mm (4 in) respectively.

MOTIVE POWER: Man-power on bicycle pedals, transmitted by chain drive to two-blade pusher propeller mounted on light alloy/CFRP shaft. Computer-designed propeller has Eppler 193 basic section; each blade is made in six pieces from Styrofoam, bonded on to a CFRP tube and covered with Kevlar cloth.

DIMENSIONS, EXTERNAL:

Wing span	28·60 m (93 ft 10 in)
Wing area	44·03 m² (474 sq ft)
Wing aspect ratio	18·6
Foreplane span	5·59 m (18 ft 4 in)
Foreplane area	5·95 m² (64 sq ft)
Propeller diameter	4·11 m (13 ft 6 in)

WEIGHTS AND LOADING:

Weight empty	31·75 kg (70 lb)
T-O weight with 63·5 kg (140 lb) pilot	97·5 kg (215 lb)
Wing loading	2·21 kg/m² (0·45 lb/sq ft)

Bryan Allen in Gossamer Albatross leaving the cliffs of Dover on 12 June 1979

MacCREADY SOLAR CHALLENGER

As a result of the successful flights of the Gossamer Condor and Gossamer Albatross, Dr MacCready and his team had gained an appreciation of the possibilities for improving the efficiency and energy conservation potential of lightweight vehicles, leading to the conclusion that it should be possible to build a practical solar-powered aircraft. To understand more fully the problems involved, battery-powered flights were made with two existing aircraft: Gossamer Albatross II (a backup that had been prepared for the cross-Channel flight) and the Gossamer Penguin, a three-quarter scale version of the Albatross, which had been almost completed as a second backup vehicle for the cross-Channel attempt. Dr MacCready's 13 year old son Marshall, weighing only 36·3 kg (80 lb), served as test pilot during approximately 50 battery-powered flights of the Penguin, his light weight being of considerable help in keeping the airspeed as low as possible.

This latter aircraft was used subsequently as a testbed for the initial exploration of manned solar-power flight, provided with a tilting solar cell panel that makes it possible to obtain maximum power from each cell with the Sun perpendicular, yet low on the horizon. This applies at the only times when flight was possible, in calm early morning conditions, for this early solar panel had only about 4,000 cells, and the Penguin itself is of delicate construction with only limited control, demanding calm conditions for successful flight. With this aircraft, on 18 May 1980, a short climbing flight was recorded on solar power alone, following an assisted take-off, the Penguin again being piloted by Marshall MacCready. A straight flight of about 1·7 nm (3 km; 2 miles) was made by this aircraft on 7 August 1980, piloted by Janice Brown, a 43 kg (95 lb) school teacher and glider pilot. In theory it should be possible for the aircraft to be flown over an out-and-return course, but the mid-turn would require it to exceed the 3 m (10 ft) height limit which, for safety reasons, is mandatory at the present time.

As a result of intensive studies, plus the initial success recorded by the Penguin, a decision was made to proceed with the construction and development of a purpose-designed solar-powered aircraft under the sponsorship of the Du Pont Company. Named appropriately Solar Challenger, it was required that this aircraft should be strong enough to cope with turbulence and permit safe flight at all heights; therefore, its configuration differs considerably from that of the Gossamer family which preceded it. As a start, it was necessary to eliminate drag-inducing bracing wires and the use of a kingpost which, at times, would shade the solar cells on the wing upper surface. This was important, because a slight shadow on one cell of a chain of cells can destroy the power output of the entire chain. A first battery-powered flight was recorded on 6 November 1980, with a number of battery-powered flights being made for test purposes before the first solar-powered flight was made on 20 November 1980. On 5 December a height of 1,220 m (4,000 ft) was gained during a flight of 1 h 32 min duration. On the following day Solar Challenger had been flown for a distance of 15·5 nm (29 km; 18 miles) between Tucson and Phoenix, Arizona, before a heavy rainstorm made it necessary for Janice Brown to land after being airborne for 2 hours. A cross-Channel flight of 163 nm (302 km; 188 miles) from Cormeilles-en-Vexin, near Paris, to Manston aerodrome, Kent, was made in 5 h 23 min on 7 July 1981, with Steve Ptacek as pilot. The design load factors of the Solar Challenger are +6*g* and −3*g*.

WINGS: Cantilever high-wing monoplane. Constant chord on centre-section, outer wing panels tapered. Aerofoil section AV 8017, in which 80% of the upper surface is flat, and the undersurface cambered. Thickness/chord ratio 13·5%. Dihedral on outer wing panels 3°. Incidence 9°. No sweepback. Wing spars consist of carbonfibre tubes, reinforced with Kevlar and Nomex. Rib doublers, internal bracing cables and control cables of Kevlar. Ribs and fairings of expanded polystyrene. Plain ailerons of similar construction. Wings and ailerons Mylar-covered. Solar cells on upper surface of wings.

FUSELAGE: Pod fuselage beneath wing. Structure of carbonfibre tubes, Nomex and Kevlar, with Mylar covering and a windscreen of Lucite.

TAIL UNIT: Tailboom, of carbonfibre, Kevlar and Nomex, carries on its upper surface, forward of the fin/rudder assembly, a fixed-incidence tailplane and an elevator. Fin and rudder assembly mounted at aft end of tailboom, at approximate midpoint of the fin. Construction of all tail surfaces similar to that of wings, except that no Nomex is incorporated. All surfaces Mylar-covered. Solar cells mounted on the tailplane upper surface.

LANDING GEAR: Non-retractable monowheel, plus nosewheel and tailwheel. Support structure of carbonfibre and Kevlar beneath the fuselage. Nylon nosewheel of 4 in diameter. Shock-absorption of monowheel by nylon webbing which restrains rear-swinging wheel mount. Monowheel is a Zytel plastic-rimmed bicycle wheel with spokes, and 18 in diameter tyre, pressure 1·03 bars (15 lb/sq in). Bicycle caliper wheel brake. Nylon tailwheel of 4 in diameter.

First solar-powered flight of Gossamer Penguin on 18 May 1980

MacCready Solar Challenger during first solar-powered flight on 20 November 1980 *(Martyn Cowley)*

POWER PLANT: One 2·05 kW (2·75 hp) Astro 25 electric motor with 23:1 geared reduction, driving an AeroVironment two-blade variable-pitch propeller which is constructed from blue Styrofoam and carbonfibre. This tractor propeller is mounted at the nose of a forward extension of the tailboom, and driven via a Kevlar motor drive belt and bicycle chain and sprocket. A total of 16,128 Spectrolab photovoltaic cells, on upper surface of wings and tailplane, provide maximum power of 2·25 kW (3 hp) at S/L. No batteries.

ACCOMMODATION: Pilot only, with seat harness and hang glider parachute of nylon. Rear swinging access door.

AVIONICS AND EQUIPMENT: Kraft transceiver. Instruments include altimeter; airspeed, rate of climb, and turn and bank indicators; ammeter; and Watt meter.

DIMENSIONS, EXTERNAL:
Wing span 14·33 m (47 ft 0 in)
Wing chord at root 1·78 m (5 ft 10 in)
Wing chord at tip 1·27 m (4 ft 2 in)
Wing aspect ratio 9·2
Length overall 9·27 m (30 ft 5 in)
Height overall 1·94 m (6 ft 4½ in)
Tailplane span 3·96 m (13 ft 0 in)
Propeller diameter 3·35 m (11 ft 0 in)
Propeller ground clearance 0·11 m (4½ in)

DIMENSIONS, INTERNAL (approx):
Cabin: Length 0·91 m (3 ft 0 in)
Max width 0·56 m (1 ft 10 in)
Max height 1·37 m (4 ft 6 in)
Floor area 0·46 m² (5 sq ft)
Volume 0·57 m³ (20 cu ft)

AREAS:
Wings, gross 23·7 m² (255 sq ft)
Ailerons (total) 1·58 m² (17 sq ft)
Fin 2·32 m² (25 sq ft)
Rudder 1·30 m² (14 sq ft)
Tailplane 9·48 m² (102 sq ft)
Elevator 1·21 m² (13 sq ft)

WEIGHTS:
Weight empty, incl solar cells 87·5 kg (193 lb)
Max T-O weight 160 kg (353 lb)

PERFORMANCE (at max T-O weight):
Max level speed at low altitudes 30·5 knots (56·5 km/h; 35 mph)
Max cruising speed at 9,150 m (30,000 ft) 61 knots (113 km/h; 70 mph)
Econ cruising speed 19 knots (36 km/h; 22 mph)
Stalling speed 15 knots (27·5 km/h; 17 mph)
Max rate of climb at S/L 61 m (200 ft)/min
Service ceiling 9,150-15,240 m (30,000-50,000 ft)
T-O run 152 m (500 ft)
T-O to 15 m (50 ft) 305 m (1,000 ft)
Landing from 15 m (50 ft) 91 m (300 ft)
Landing run 61 m (200 ft)
Range, dependent upon Sun condition, and limited by duration and intensity of Sun 217-260 nm (402-483 km; 250-300 miles)

MACHEN

MACHEN INC

Spokane, Washington

MACHEN SUPERSTAR

Under the name Superstar, Machen Inc has developed a conversion of the Piper (Ted Smith) Aerostar 601P to provide improved performance. The major element of this modification is the removal of the standard Avco Lycoming TIO-540 turbocharged engines. These are replaced by TIO-540 engines with Garrett TA-18 turbochargers, which are each flat rated to 261 kW (350 hp), and drive new Hartzell three-blade propellers with Q-tips. As an option to this conversion, a number of aerodynamic refinements are available as a single package; by reducing airframe drag they offer a considerable increase in cruising performance.

The description of the Aerostar 601P under the Piper entry in this edition of *Jane's* applies also to the Machen Superstar modification, except as noted. Changed full-conversion weight and performance figures include:

WEIGHTS: As for Piper Aerostar 601P except:
Weight empty, equipped plus 23 kg (50 lb)

PERFORMANCE (at max T-O weight, except where indicated):
Max cruising speed at 7,620 m (25,000 ft) 287 knots (531 km/h; 330 mph)
Cruising speed, average cruise weight at 7,620 m (25,000 ft):
75% power 273 knots (506 km/h; 314 mph)
65% power 258 knots (478 km/h; 297 mph)
Max rate of climb at S/L 714 m (2,342 ft)/min
Rate of climb at S/L, one engine out 152 m (500 ft)/min
Service ceiling 7,620 m (25,000 ft)
Service ceiling, one engine out 7,010 m (23,000 ft)
T-O to 15 m (50 ft) 541 m (1,775 ft)
Accelerate/stop distance 933 m (3,060 ft)
Range with max fuel at high-speed cruising power plus 250 nm (463 km; 287 miles)

MANCRO

MANCRO AIRCRAFT COMPANY

7716 East Alondra Boulevard, Paramount, California 90723
Telephone: (213) 774 1171 and 633 7114
Telex: 67 4865

MANCRO (FAIRCHILD) C-123T

Mancro Aircraft Company has completed the conversion of a Fairchild C-123B Provider to prototype C-123T configuration with a turboprop power plant. The company estimates that most of the C-123s in service have 10-15 years of useful life remaining, but their existing Pratt & Whitney R-2800 engines have been out of production for years, making these aircraft increasingly costly and difficult to maintain. The installation of current production turboprop engines not only simplifies maintenance and spares support, but potentially increases airframe life by at least ten years.

The conversion programme was initiated when the Thai Air Force sought to extend the life of its C-123B fleet by the installation of new engines. Mancro and the Thai government obtained US government support for the project by the temporary loan of engines and equipment. Since that time, budget restrictions have forced the Thai Air Force to withdraw from a conversion programme, but the Thai government is allowing Mancro to use the prototype as a demonstrator. It flew for the first time on 24 October 1980.

Structural modification of the prototype was limited to the wing section where the new 2,550 kW (3,420 shp) Allison T56-A-7B turboprops, with Hamilton Standard Type 56H60-91 propellers, are installed. It is planned, however, that production conversions would have 3,661 kW (4,910 shp) Allison T56-A-15 engines, flat rated to 3,132 kW (4,200 shp). Other modifications to be carried out simultaneously would include the provision of greater fuel capacity, the installation of an APU to provide self-start capability and auxiliary electrical power, and the introduction of powered control surfaces.

Mancro (Fairchild) C-123T prototype turboprop-engined conversion (two Allison T56-A-7B engines)
(Dave Menard)

Details of the Fairchild C-123K can be found in the 1970-71 *Jane's*. The data that follow apply to the C-123T prototype with Allison T56-A-7B engines:

WEIGHTS:
Weight empty 14,742 kg (32,500 lb)
Max T-O and landing weight 27,215 kg (60,000 lb)

PERFORMANCE (at AUW of 22,680 kg (50,000 lb), except where indicated):
Max cruising speed at 3,050 m (10,000 ft) 235 knots (436 km/h; 271 mph)
Max cruising speed at 1,525 m (5,000 ft) 225 knots (417 km/h; 259 mph)
Single-engine minimum control speed, landing gear and flaps up 104 knots (193 km/h; 120 mph)
Single-engine minimum control speed, landing gear and flaps down 98 knots (182 km/h; 113 mph)
Stalling speed, landing gear and flaps up 88 knots (163 km/h; 101·5 mph)
Stalling speed, landing gear and flaps down 73 knots (135 km/h; 84 mph)
Max rate of climb at S/L 780 m (2,560 ft)/min
Rate of climb at S/L, one engine out 277 m (910 ft)/min
Service ceiling 9,755 m (32,000 ft)
Service ceiling, one engine out 3,660 m (12,000 ft)
T-O run 351 m (1,150 ft)
T-O to 15 m (50 ft) 579 m (1,900 ft)
Landing from 15 m (50 ft) 549 m (1,800 ft)
Landing run 335 m (1,100 ft)
Range at AUW of 24,766 kg (54,600 lb), with 2,720 kg (6,000 lb) freight and 9,085 litres (2,400 US gallons) of fuel in integral wing tanks, econ cruising speed at 3,050 m (10,000 ft) 2,002 nm (3,710 km; 2,305 miles)

MARSH

MARSH AVIATION COMPANY

5060 East Falcon Drive, Mesa, Arizona 85205
Telephone: (602) 832 3770
Telex: 165 028
VICE-PRESIDENT: William G. Walker Jr

MARSH S2R-T TURBO THRUSH

Marsh Aviation Company has converted the piston-engined Rockwell Thrush Commander to turbine power by the installation of a Garrett TPE331-1-101 turboprop engine. Derated to 447 kW (600 shp) for this conversion, the full 580 kW (778 shp) output of the TPE331 is available in emergency. This engine drives a Hartzell constant-speed fully-feathering and reversible-pitch propeller. Single-cycle air-conditioning and cockpit heating provided by engine bleed air, and agricultural spraypump is also operated by bleed air. The empty weight of the Turbo Thrush is 227 kg (500 lb) less than that of the standard Thrush Commander, providing increased payload capability and improved speed and performance. For agricultural operators working in remote areas the TPE331 installation has the advantage that ordinary automotive diesel fuel can be used if jet fuel is not available.

Standard fuel capacity of the Turbo Thrush is 401 litres (106 US gallons). Standard hopper capacity is 1·50 m³ (53 cu ft) or 1,514 litres (400 US gallons). A larger hopper is available optionally, capacity 1·89 m³ (66·8 cu ft) or 1,892·5 litres (500 US gallons).

Following more than 600 h of flight by two prototypes, an FAA Supplemental Type Certificate was issued and the first production conversion was handed over in September 1976. Turbo Thrush deliveries totalled 67 by 1 March 1981, the majority for operators in Africa, Europe, Mexico, the Middle East and the USA.

DIMENSIONS, EXTERNAL: As for Thrush Commander except:
Length overall 9·27 m (30 ft 5 in)

WEIGHTS AND LOADINGS: As for Thrush Commander except:
Weight empty 1,633 kg (3,600 lb)
Typical operating weight (CAR Pt 8) 4,173 kg (9,200 lb)

PERFORMANCE (at 2,721 kg; 6,000 lb T-O weight, except where indicated):
Never-exceed speed 138 knots (256 km/h; 159 mph) IAS
Max level speed at 4,420 m (14,500 ft) 178 knots (330 km/h; 205 mph)
Econ cruising speed at 4,420 m (14,500 ft) 139 knots (257 km/h; 160 mph)
Cruising speed, 50% power 127 knots (235 km/h; 146 mph)
Working speed, 50% power 108·5 knots (201 km/h; 125 mph)
Stalling speed, flaps up 41·5 knots (77 km/h; 48 mph)
Stalling speed, flaps down 38 knots (71 km/h; 44 mph)

Stalling speed, flaps up at normal landing weight	39 knots (72·5 km/h; 45 mph)
Stalling speed, flaps down at normal landing weight	37 knots (69 km/h; 43 mph)
Max rate of climb at S/L	915 m (3,000 ft)/min
Service ceiling	7,620 m (25,000 ft)
T-O run	183 m (600 ft)
Landing run	91 m (300 ft)
Range with max payload	278 nm (515 km; 320 miles)
Ferry range, at 60% power	521 nm (966 km; 600 miles)

MARSH G-164 C-T TURBO CAT

Marsh Aviation has developed a turbine engine conversion for the Grumman/Gulfstream American/Schweizer G-164 Super Ag-Cat C, replacing the original Pratt & Whitney R-1340 piston engine with a 580 kW (778 shp) Garrett TPE331-1-101 turboprop. In this application, known as the Turbo Cat, the engine is derated to 447 kW (600 shp), and drives a Hartzell constant-speed fully-feathering and reversible-pitch propeller. The installation is approved for unlimited operation on automotive diesel fuel, as well as on all jet fuels, and for limited operation on 80/87/100 octane aviation gasoline.

Features of the Turbo Cat (which has an empty weight 263 kg; 580 lb below that of the standard G-164 Super Ag-Cat C) include automatic start sequencing, automatic fuel nozzle purging, engine inlet and fuel filter anti-icing, improved reliability and performance, and increased payload. Options available include air-conditioning, single-point refuelling, and hydraulic spray systems.

By 1 January 1980 the prototype had completed more than 100 h of flight testing, and certification was gained later in the year. Deliveries began shortly afterwards, and by 1 March 1981 six Turbo Cats had been completed. The description of the Schweizer Super Ag-Cat C in this edition applies also to the Marsh Turbo Cat conversion, except as follows:

DIMENSIONS, EXTERNAL:

Length overall	9·60 m (31 ft 6 in)

WEIGHTS:

Weight empty	1,424 kg (3,140 lb)
Typical operating weight (CAR Pt 8)	4,082 kg (9,000 lb)

PERFORMANCE (at max T-O weight, except where indicated):

Never-exceed speed	121 knots (223 km/h; 139 mph) IAS
Cruising speed at 4,420 m (14,500 ft)	139 knots (257 km/h; 160 mph)
Stalling speed	57 knots (106 km/h; 66 mph) IAS
Max rate of climb at S/L	914 m (3,000 ft)/min
T-O run	183 m (600 ft)
Landing run at max landing weight	91 m (300 ft)

Marsh Turbo Thrush, a turbine-engined conversion of the Rockwell International Thrush Commander

Marsh Turbo Cat, a turbine-engined conversion of the Grumman/Schweizer Super Ag-Cat C

Marsh Turbo Mentor, a turboprop conversion of the Beechcraft T-34A/B

MARSH/BEECHCRAFT T-34 TURBO MENTOR

In early 1980 Marsh Aviation was involved in the development of a turboprop conversion of the Beechcraft T-34A/B Mentor two-seat primary trainer (last described in the 1960-61 *Jane's*). This has involved replacement of the original 168 kW (225 hp) Continental O-470 piston engine that powered these aircraft, by a 580 kW (778 shp) Garrett TPE331-1 turboprop engine, flat rated to 335 kW (450 shp), and driving a Hartzell constant-speed fully-feathering and reversible-pitch propeller.

It is intended to market the Marsh Turbo Mentor as a low-cost, high-performance military trainer that will provide greatly improved performance and aerobatic capability by comparison with the piston-engined Beechcraft T-34A/B. The prototype conversion was flown for the first time in December 1979; since that time it has been upgraded to production status. In early 1981 negotiations were proceeding for the conversion of 20 aircraft overseas.

DIMENSIONS, EXTERNAL:

Wing span	10·00 m (32 ft 9½ in)
Length overall	8·05 m (26 ft 5 in)
Height overall	2·98 m (9 ft 9½ in)

WEIGHTS:

Weight empty	1,088 kg (2,400 lb)
Max T-O weight	1,497 kg (3,300 lb)

PERFORMANCE:

Never-exceed speed	243 knots (450 km/h; 280 mph) IAS
Cruising speed at 7,010 m (23,000 ft)	305 knots (565 km/h; 350 mph)
Normal operating speed	219 knots (406 km/h; 252 mph) IAS
Stalling speed, flaps and landing gear up	56 knots (104 km/h; 64·5 mph) IAS
Stalling speed, flaps and landing gear down	47 knots (88 km/h; 54·5 mph) IAS
Max rate of climb at S/L	762 m (2,500 ft)/min
Service ceiling	7,620 m (25,000 ft)
T-O and landing run	91 m (300 ft)
Range	912 nm (1,690 km; 1,050 miles)

MAULE

MAULE AIRCRAFT CORPORATION

HEAD OFFICE AND WORKS: Spence Air Base, Moultrie, Georgia 31768
Telephone: (912) 985 2045
PRESIDENT: B. D. Maule
VICE-PRESIDENT: Mrs B. D. (June) Maule (Treasurer)
SALES MANAGER: Dan Spader

This company was formed to manufacture the Maule M-4 four-seat light aircraft, production of which ended in 1975. It transferred to new facilities in Moultrie, Georgia, in September 1968, and is now concentrating on production of the uprated M-5 Lunar Rocket.

The company has also designed auxiliary fuel transfer tanks for installation in the outboard wing bays of M-4 and M-5 aircraft. Providing a total usable additional fuel capacity of 87 litres (23 US gallons), these tanks offer owners of the M-4 or M-5 a minimum-payload range of 650 nm (1,200 km; 750 miles). FAA approval of the modification was given on 31 October 1973.

MAULE M-5 LUNAR ROCKET

Developed from the M-4 Strata-Rocket, the M-5 Lunar Rocket series has a 30 per cent increase in flap area and enlarged tail surfaces to improve short-field performance and rate of climb. Two prototypes were built originally. First to fly, on 1 November 1971, was the M-5-220C prototype, powered by a 164 kW (220 hp) Franklin 6A-350-C1 engine. Manufacture of this version has since been discontinued as the Franklin engine is no longer in production in the USA.

Versions available currently are as follows:

M-5-180C. With 134 kW (180 hp) Avco Lycoming O-360-C1F flat-four engine, driving a Hartzell two-blade constant-speed metal propeller. Prototype first flown on 18 May 1978. One M-5-180C was completing its certification programme with an IO-360 fuel-injection engine in early 1981.

M-5-210C. Basic model with 156·5 kW (210 hp) Continental IO-360-D flat-six engine, driving a McCauley two-blade constant-speed metal propeller. Second prototype Lunar Rocket, flown on 16 October 1973, was of this version; FAA certification awarded on 28 December 1973.

M-5-210TC. Cargo version with 156·5 kW (210 hp) Avco Lycoming TO-360-C1A6D flat-four turbocharged engine, driving a Hartzell two-blade constant-speed metal propeller. Prototype first flown on 9 August 1978.

M-5-235C. Version with the more powerful 175 kW (235 hp) Avco Lycoming O-540-J1A5D flat-six engine, driving a larger-diameter Hartzell two-blade constant-speed metal propeller, certificated on 6 April 1976. It is available alternatively with the fuel-injection IO-540-W1A5D, with which it was certificated on 7 November 1980.

Maule Patroller. Civil patrol version of Lunar Rocket (any power plant), with any or all of the following modifications: Plexiglas-covered doors for improved view; port side rear observation window; 3·5 million candlepower belly-mounted manually-controlled searchlight; 100/200W siren and public address system; 28V electrical system; specialised radios to customer's requirements.

The M-5-210C/M-5-235C is a STOL aircraft, the 'C' in its designation implying that it has double aft doors on the starboard side to facilitate the loading of cargo. Production of all versions totalled approximately 1,100 by 1 January 1981.

TYPE: Four-seat light aircraft.

WINGS: Braced high-wing monoplane. Streamline-section V bracing strut each side. USA 35B (modified) wing section. Dihedral 1°. Incidence 0° 30′. All-metal two-spar structure with metal covering and glassfibre tips. All-metal ailerons and two-position flaps. Ailerons linked with rudder tab, so that aircraft can be controlled in flight by using only the control wheel in the cockpit. Cambered wingtips standard.

FUSELAGE: Welded 4130 steel tube structure. Covered with glassfibre, except for metal doors and aluminium skin around cabin.

TAIL UNIT: Braced steel tube structure with glassfibre covering. Trim tab in port elevator. Servo tab in rudder linked to aileron movement. Starboard rudder trim via spring to starboard rudder pedal.

LANDING GEAR: Non-retractable tailwheel type. Maule oleo-pneumatic shock-absorbers in main units. Maule steerable tailwheel. Cleveland main wheels with Goodyear or McCreary tyres size 17 × 6·00-6, pressure 1·79 bars (26 lb/sq in). Tailwheel tyre size 8 × 3·50-4, pressure 1·03-1·38 bars (15-20 lb/sq in). Cleveland hydraulic disc brakes. Parking brake. Oversize tyres, size 20 × 8·50-6 (pressure 1·24 bars; 18 lb/sq in), and fairings aft of main wheels optional. Provisions for fitting optional Edo Model 248B2440, Pee Kay Model 2300 or Aqua Model 2400 floats, or Federal skis Model C2200H or C3000H.

POWER PLANT: One flat-four or flat-six engine, driving a constant-speed propeller, as detailed in model listings. Two fuel tanks in wings with total usable capacity of 151 litres (40 US gallons). Optional auxiliary fuel tanks in outer wings, each with usable capacity of 43·5 litres (11·5 US gallons). Maximum usable fuel capacity 238 litres (63 US gallons). Refuelling points on wing upper surface. Oil capacity: M-5-210C 9·5 litres (2·5 US gallons); M-5-235C 11·5 litres (3 US gallons).

ACCOMMODATION: Pilot and three passengers on two front bucket seats and rear bench seat, or optional quickly-removed rear sling seat. One door on port side of fuselage, hinged at front edge and opening forward. Three doors on starboard side of fuselage, the forward and centre doors hinged at the front edge, the rear baggage door hinged at the rear edge. The centre and aft doors can be opened together to provide an opening 1·24 m (4 ft 1 in) wide to facilitate loading of bulky cargo. Accommodation heated and ventilated.

SYSTEMS: Hydraulic system for brakes only. Electrical system powered by 60A engine-driven alternator. 28V electrical system optional.

AVIONICS AND EQUIPMENT: A wide range of Collins Micro Line, King, Genave and Narco communication and navigation equipment is available to customer's requirements. Blind-flying instrumentation, autopilot, wing levelling system and automatic glideslope optional.

DIMENSIONS, EXTERNAL:

Wing span	9·40 m (30 ft 10 in)
Wing chord, constant	1·60 m (5 ft 3 in)
Wing aspect ratio	5·71
Length overall:	
M-5-180C, M-5-210C	6·93 m (22 ft 9 in)
M-5-210TC	7·26 m (23 ft 10 in)
M-5-235C	7·16 m (23 ft 6 in)
Height overall	1·89 m (6 ft 2½ in)
Tailplane span	3·28 m (10 ft 9 in)
Wheel track	1·83 m (6 ft 0 in)
Wheelbase	4·82 m (15 ft 10 in)
Propeller diameter: M-5-210C	1·88 m (6 ft 2 in)
M-5-235C	1·98 m (6 ft 6 in)
Cabin doors (fwd, each): Height	0·84 m (2 ft 9 in)
Width	0·76 m (2 ft 6 in)
Height to sill	0·94 m (3 ft 1 in)
Cabin door (centre, stbd):	
Height	0·75 m (2 ft 5½ in)
Width	0·69 m (2 ft 3 in)
Height to sill	0·76 m (2 ft 6 in)
Baggage door (aft, stbd): Height	0·58 m (1 ft 11 in)
Width	0·56 m (1 ft 10 in)
Height to sill	0·61 m (2 ft 0 in)

AREAS:

Wings, gross	14·67 m² (157·9 sq ft)
Ailerons (total)	1·19 m² (12·8 sq ft)
Trailing-edge flaps (total)	1·75 m² (18·8 sq ft)
Fin	1·22 m² (13·14 sq ft)
Rudder, incl tab	0·54 m² (5·83 sq ft)
Tailplane	1·32 m² (14·2 sq ft)
Elevators, incl tab	1·58 m² (17·0 sq ft)

WEIGHTS AND LOADINGS (A: M-5-180C; B: M-5-210C; C: M-5-210TC; D: M-5-235C with O-540-J1A5D engine):

Basic operating weight: B	601 kg (1,325 lb)
Weight empty: A	601 kg (1,325 lb)
B	612 kg (1,350 lb)
C, D	635 kg (1,400 lb)
Max T-O and landing weight:	
all versions	1,043 kg (2,300 lb)
Max wing loading: all versions	71·3 kg/m² (14·6 lb/sq ft)
Max power loading: A	7·78 kg/kW (12·78 lb/hp)
B, C	6·66 kg/kW (10·95 lb/hp)
D	5·96 kg/kW (9·79 lb/hp)

PERFORMANCE (at max T-O weight. A: M-5-180C; B: M-5-210C; C: M-5-210TC; D: M-5-235C with O-540-J1A5D engine):

Never-exceed speed:	
B, D	156 knots (290 km/h; 180 mph)
Max level speed at optimum altitude:	
A	117 knots (217 km/h; 135 mph)
B	137 knots (254 km/h; 158 mph)
C	165 knots (306 km/h; 190 mph)
D	149 knots (277 km/h; 172 mph)
Max cruising speed at optimum altitude:	
A	110 knots (204 km/h; 127 mph)
B	137 knots (254 km/h; 158 mph)
C	148 knots (274 km/h; 170 mph)
D	149 knots (277 km/h; 172 mph)
Econ cruising speed, 65% power at 2,440 m (8,000 ft):	
A	106 knots (196 km/h; 122 mph)
B	130 knots (241 km/h; 150 mph) CAS
C	135 knots (251 km/h; 156 mph)
D	139 knots (257 km/h; 160 mph) CAS
Stalling speed, flaps up:	
B	53 knots (98 km/h; 61 mph)
Stalling speed, flaps down:	
B	49 knots (90 km/h; 56 mph)
Max rate of climb at S/L: A	229 m (750 ft)/min
B	380 m (1,250 ft)/min
C, D	411 m (1,350 ft)/min
Service ceiling: A	4,570 m (15,000 ft)
B	5,485 m (18,000 ft)
C	7,620 m (25,000 ft)
D	6,100 m (20,000 ft)
T-O and landing run: B, D	122 m (400 ft)
T-O to and landing from 15 m (50 ft):	
B, D	183 m (600 ft)
Range with max standard fuel:	
A	426 nm (789 km; 490 miles)
B	521 nm (966 km; 600 miles)
C	391 nm (724 km; 450 miles)
D	478 nm (885 km; 550 miles)
Range with max fuel, 30 min reserves:	
A	651 nm (1,207 km; 750 miles)
B	760 nm (1,408 km; 875 miles)
C	625 nm (1,159 km; 720 miles)
D	695 nm (1,287 km; 800 miles)

Maule M-5-180C Lunar Rocket four-seat lightplane

MBB

MBB HELICOPTER CORPORATION (a subsidiary of Messerschmitt-Bölkow-Blohm GmbH)

HEAD OFFICE: Weston Way, PO Box 1507, West Chester, Pennsylvania 19380
Telephone: (215) 431 4150
Telex: 902041

PRESIDENT: C. W. Moore

DIRECTOR OF MARKETING: Samuel E. Rocray

Messerschmitt-Bölkow-Blohm of West Germany (which see) established this subsidiary company in the USA, in early 1979, to take over marketing and product support of the company's family of twin-turbine helicopters in North America, Central America and Mexico. These functions had been carried out since 1972 by Boeing Vertol, which maintained continuity of product support during the Summer of 1979 while the new company's 1,393 m² (15,000 sq ft) spares administration and maintenance training centre, which is near the West Chester head office, was being readied for operation. This facility has been fully operational since September 1979.

MBB Helicopter Corporation will market and support the entire range of MBB helicopters, including the 8/10-seat multi-purpose BK 117 developed jointly by MBB and Kawasaki of Japan (see International section), which was scheduled for certification in late 1981.

MCDONNELL DOUGLAS

MCDONNELL DOUGLAS CORPORATION

HEAD OFFICE AND WORKS: Box 516, St Louis, Missouri 63166
Telephone: (314) 232 0232
Telex: 44-857

CHAIRMAN OF THE BOARD OF DIRECTORS, AND CHIEF EXECUTIVE OFFICER: Sanford N. McDonnell

DIRECTORS:
John C. Brizendine
George H. Capps
Michael N. Chetkovich
William H. Danforth
Donald W. Douglas Jr
George S. Graff
Robert L. Johnson
Edwin S. Jones
Robert C. Little
Donald S. Macdonald
James S. McDonnell III
John F. McDonnell (President)
Sanford N. McDonnell (Chairman)
James T. McMillan
William R. Orthwein Jr
John T. Sant

PRESIDENT: John F. McDonnell

CORPORATE OFFICERS:
Harold D. Altis (Vice-President, Engineering and Research)
David C. Arnold (Vice-President)
Alvin L. Boyd (Vice-President)
Erwin F. Branahl (Vice-President)
John C. Brizendine (Vice-President)
Jerry G. Brown (Vice-President, Treasurer)
John E. Crosthwait (Vice-President, Far East)
Walter E. Diggs Jr (Secretary and Counsel)
Charles M. Forsyth (Vice-President)
George S. Graff (Vice-President)
Gordon M. Graham (Vice-President, Washington)
Robert L. Harmon (Vice-President)
Robert L. Johnson (Vice-President, Group Executive)
Warren E. Kraemer (Vice-President, Europe)
F. Mark Kuhlmann (Asst Secretary)
Robert C. Little (Vice-President, Operations and Marketing)
James H. MacDonald (Vice-President, Personnel)
James S. McDonnell III (Vice-President, Marketing)
James T. McMillan (Vice-President)
Donald Malvern (Vice-President)
Marvin D. Marks (Vice-President, Manager, C-X Transport Programme)
Gerald J. Meyer (Vice-President, External Relations)
William R. Orthwein Jr (Vice-President)
A. Joseph Quackenbush (Vice-President)
Albert J. Redway Jr (Vice-President, Eastern Region)
John T. Sant (Vice-President and General Counsel)
Stanley J. Sheinbein (Asst Treasurer)
Harry I. Sieferman (Tax Officer)
Albert H. Smith Jr (Vice-President, Contracts, and Controller)
John W. Walbran (Asst General Counsel)
John F. Yardley (Vice-President)

STAFF OFFICERS:
R. Joseph Alagna (Vice-President, Material)
Dr Donald P. Ames (Vice-President, McDonnell Douglas Research Laboratories)
Rowland G. Freeman III (Vice-President, Headquarters Executive)

Charles A. Gaskill (Vice-President, Properties and Facilities)
Arthur W. Hyland (Vice-President, Accounting)
Leo I. Mirowitz (Vice-President, Corporate Diversification)
Ernie H. Ridenhour (Vice-President, Information Resource Management)
William E. Schowengerdt (Vice-President, Auditing)
Howard C. Todt (Vice-President, Quality Assurance)
Michael Witunski (Vice-President)
CORPORATE DIRECTOR, EXTERNAL RELATIONS:
John L. Cooke

McDonnell Douglas Research Laboratories
Box 516, St Louis, Missouri 63166
STAFF VICE-PRESIDENT: Dr Donald P. Ames

McDonnell Douglas Corporation was formed on 28 April 1967, by the merger of the former Douglas Aircraft Company Inc and McDonnell Company. It encompasses both of the original companies and their subsidiaries.

Major operating components of McDonnell Douglas Corporation, are as follows:

Douglas Aircraft Company
See pages 407-413 of this section.
McDonnell Douglas Astronautics Company
See Air-launched Missiles and Spaceflight sections
McDonnell Aircraft Company
See following entry
McDonnell Douglas Automation Company
Box 516, St Louis, Missouri 63166
CHAIRMAN: William R. Orthwein Jr
PRESIDENT: A. Joseph Quackenbush
EXECUTIVE VICE-PRESIDENT, COMMERCIAL: Robert L. Harmon
McDonnell Douglas Electronics Company
St Charles, Missouri
PRESIDENT: David C. Arnold
McDonnell Douglas—Tulsa
Tulsa, Oklahoma
VICE-PRESIDENT AND GENERAL MANAGER:
Robert C. Lindstrom

Subsidiaries:

Subsidiaries of McDonnell Douglas Corporation include McDonnell Douglas Canada Ltd, Toronto, Ontario; McDonnell Douglas (Japan) Ltd, Tokyo; MDC Realty Company, Long Beach, California; McDonnell Douglas Finance Corporation, Long Beach, California; Microdata Corporation, Irvine, California; Vitek Systems Inc, Hazelwood, Missouri; CoaLiquid Inc, Louisville, Kentucky; and McDonnell Douglas Technical Services Company Inc, Huntsville, Alabama.

At 1 July 1981, McDonnell Douglas employed a total of 80,901 people, worldwide. Total office, engineering, laboratory and manufacturing floor area was 2,596,098 m^2 (27,945,077 sq ft).

MCDONNELL AIRCRAFT COMPANY (A Division of McDonnell Douglas Corporation)

HEADQUARTERS: Box 516, St Louis, Missouri 63166
Telephone: (314) 232 0232
PRESIDENT: George S. Graff
EXECUTIVE VICE-PRESIDENT: Donald Malvern
VICE-PRESIDENTS:
Askel R. Andersen (Avionics Engineering)
Robert C. Bartz (Material)
William J. Blatz (Engineering Technology)
Alvin L. Boyd (Fiscal Management)
Chester V. Braun (General Manager, F-15)
John J. Burns (Advanced Engineering)
Denver D. Clark (Marketing)
Paul T. Homsher (F-15 Saudi Arabian Operations)
Robert H. Koenig (Controller)
Edward B. Kuhlmann (Quality Assurance)
Richard A. Noyes (Aircraft Engineering)
Herbert Perlmutter (Manufacturing)
Madison L. Ramey (Engineering)
William S. Ross (Flight and Laboratory Development)
John N. Schuler (Contracts and Pricing)
Vernon E. Teig (Product Support)
Joseph C. Waldner (General Manager, F-18)
Darrell F. Waters (Personnel, East)
DIRECTOR, EXTERNAL RELATIONS: John J. McGrath

McDonnell Douglas F-15C Eagle of the German-based 36th TFW, US Air Force *(S. G. Richards)*

Development and production at St Louis continues to be concentrated on versions of the F-15 Eagle air superiority fighter, the AV-8B Advanced Harrier and F-18 Hornet naval strike fighter.

MCDONNELL DOUGLAS PHANTOM II

US Navy and USAF designations: F-4 and RF-4

The Phantom II was developed initially as a twin-engined two-seat long-range all-weather fleet air defence fighter for service with the US Navy. The change of primary mission, to missile fighter, was made in 1955. Production by McDonnell Douglas ended in 1979, and details of this extensively built fighter can be found in the 1980-81 and earlier editions of *Jane's*. The last Phantom to be completed was the 138th built by Mitsubishi in Japan, which was delivered on 20 May 1981. McDonnell Douglas continues to provide support for in-service aircraft, and is completing outstanding modification programmes.

MCDONNELL DOUGLAS F-15 EAGLE

The USAF requested development funding for a new air superiority fighter in 1965, and in due course design proposals were sought from three airframe manufacturers: Fairchild Hiller Corporation, McDonnell Douglas Corporation, and North American Rockwell Corporation. On 23 December 1969 it was announced that McDonnell Douglas had been selected as prime airframe contractor. The resulting contract called for the design and manufacture of 20 aircraft for development testing, these to comprise 18 single-seat **F-15As** and two TF-15A two-seat trainers. First flight of the F-15A was made on 27 July 1972, and the first flight of a two-seat TF-15A trainer (redesignated subsequently **F-15B**) on 7 July 1973.

A production go-ahead for the first 30 operational aircraft (FY 1973 funds) was announced on 1 March 1973. The FY 1974 Defense Procurement Bill authorised production of 62 aircraft, and subsequent Procurement Bills have authorised production of a further 622 aircraft through FY 1982. An F-15B (the 21st Eagle built) was the first Eagle delivered to the USAF, on 14 November 1974. Structural weight of the F-15B is approx 363 kg (800 lb) more than that of the single-seater.

Eagles delivered from mid-1979 are to **F-15C** and **F-15D** standard, which provides for 6,103 kg (13,455 lb) of internal fuel, and the ability to carry two low-drag fuel pallets known as FAST Packs (Fuel And Sensor Tactical Packs) developed specially for the F-15 by McDonnell Aircraft Company. Each FAST Pack contains approximately 3,228 litres (114 cu ft) of usable volume, which can accommodate 2,211 kg (4,875 lb) of JP-4 fuel. It attaches to the side of either the port or starboard engine air intake trunk (being made in handed pairs), is designed to the same load factors and airspeed limits as the basic aircraft, and can be removed in 15 minutes. FAST Packs can accommodate avionics such as reconnaissance sensors, radar detection and jamming equipment, a laser designator, low-light-level TV system, and reconnaissance cameras, in addition to fuel. All external stores stations remain available with the pallets in use. AIM-7F missiles and air-to-ground weapons can be attached to the corners of the FAST Packs.

Proposed Advanced Wild Weasel configuration for the McDonnell Douglas F-15D Eagle

The first F-15C (78-468) flew for the first time on 26 February 1979. Since 1980 the APG-63 radar of F-15C/D aircraft has been equipped with a Hughes Aircraft programmable signal processor, which enables changes to be incorporated in the radar earlier and more cheaply. An updated radar data processor increases memory capability from 24K to 96K. These added features enable the radar to operate in a high-resolution rate assessment mode which can identify clustered targets individually. F-15C and F-15D aircraft delivered prior to the availability of the programmable signal processor and expanded computer will be retrofitted to bring them up to standard.

Minor changes have been made to tyres, wheels and brakes to allow for an increased maximum T-O weight, which could be as high as 30,845 kg (68,000 lb) with full internal fuel, FAST Packs and external tanks. Landing gear and fuel system changes have added about 227 kg (500 lb) to the aircraft's dry weight.

By 1 February 1981 a total of 597 Eagles had been delivered, and were in operational service with the 57th TTW at Nellis AFB, Nevada, the 58th TTW at Luke AFB, Arizona, the 1st TFW at Langley AFB, Virginia, the 36th TFW at Bitburg, West Germany, the 49th TFW at Holloman AFB, New Mexico, the 33rd TFW at Eglin AFB, Florida, the 18th TFW at Kadena AB, Okinawa, and the 32nd Tactical Fighter Squadron based at Soesterburg in the Netherlands. The first US east coast air defence squadron to receive the Eagle, the 48th Fighter Interceptor Squadron at Langley AFB, Virginia, was due to complete conversion to the F-15 in early 1982.

It is planned to procure a total of 749 Eagles for the USAF by 1983, including the 20 R & D models. Twenty-five were delivered under an initial contract from the Israeli Air Force, with at least 15 more to follow; and 60 have been ordered by Saudi Arabia, of which deliveries are due to begin in January 1982. The JASDF plans to purchase 88 **F-15Js** and 12 **F-15DJs**, of which a total of 86 will be licence-built in Japan, with Mitsubishi as the prime contractor. The first of the 14 US-built aircraft was handed over on 15 July 1980, and the first two were flown to Japan in March 1981. The FY 1980 Japanese defence budget approved the purchase of the first 34 aircraft for the JASDF, which expects to form its first F-15J unit at Niutabaru AB in 1982.

Designed specifically as an air superiority fighter, the F-15 Eagle has proved equally suitable for air-to-ground missions without degradation of its primary role. It is able to carry a variety of air-to-air and air-to-ground weapons (see separate entry on Strike Eagle).

At the beginning of April 1981, development testing began of a US Air Force F-15 equipped with a Martin Marietta ATLIS II automatic tracking and laser illumination system pod, as part of a programme known as Integrated Flight Fire Control (IFFC)/Firefly III. The pod is mounted in the forward missile well position on the port side of the fuselage. Internal changes include installation of a modified central computer, a modified signal data processor for a pilot's head-up display, an added coupler interface unit to link flight and fire control systems, and new or modified flight/fire control software. During the 15-month, 150-flight test programme, the optical sensor/tracker pod is expected to enable air-to-air weapons to be fired accurately at simulated targets while the F-15 manoeuvres at high offset angles, for the first time in the case of a US Air Force fighter.

As a continuation of Advanced Wild Weasel studies which led to production of the F-4G version of the Phantom II, McDonnell Douglas has proposed to the US gov-

McDonnell Douglas F-15B Eagle two-seat trainer, fitted with FAST Packs

ernment a defence suppression version of the F-15D Eagle. In an accompanying illustration, an F-15 is shown with a HARM high-speed anti-radiation missile on its outboard underwing pylon, in addition to Snakeye bombs and Sidewinder missiles on the inner pylon and Sparrow missiles on the fuselage mountings. Its more sensitive ECM sensors and higher performance would make it capable of suppressing or destroying an increasing number of surface-to-air threats by comparison with the F-4G. Aerodynamic testing of a 'chin' pod, to house specialised avionics, was planned for late 1981.

Under a programme dubbed 'Streak Eagle', the F-15 demonstrated its climb capability by capturing eight time-to-height records, between 16 January and 1 February 1975. These records were listed in the 1979-80 *Jane's*; six of them remained unbeaten in mid-1981.

The following description applies to the F-15A:

TYPE: Single-seat twin-turbofan air superiority fighter, with secondary attack role.

WINGS: Cantilever shoulder-wing monoplane. Leading-edge swept back at 45°.

FUSELAGE: All-metal semi-monocoque structure.

TAIL UNIT: Cantilever structure with twin fins and rudders. All-moving horizontal tail surfaces outboard of fins, with extended chord on outer leading-edges. Rudder servo actuators by Ronson Hydraulic Units Corporation. Actuators for horizontal surfaces by National Water Lift Company. Boost and pitch compensator for control stick by Moog Inc, Controls Division.

LANDING GEAR: Hydraulically-retractable tricycle type, with single wheel on each unit. Nose and main landing gear by Cleveland Pneumatic Tool Company. Wheels and carbon brake assemblies by Goodyear Tire and Rubber Company. Main and nosewheel tyres by Goodrich Company. Wheel braking skid control system by Hydro-Aire Division of Crane Company. All units retract forward.

POWER PLANT: Two Pratt & Whitney F100-PW-100 turbofan engines, each rated at approx 106·4 kN (23,930 lb st) with afterburning for take-off. Internal fuel load 5,260 kg (11,600 lb). Fuel tanks by Goodyear Aviation Products Division. Fuel gauge system by Simmonds Precision Products Inc. FAST Pack conformal fuel pallets attached to side of engine air intakes, beneath wing, can be removed within 15 min. Each has usable volume of 3·23 m³ (114 cu ft) and can contain 2,268 kg (5,000 lb) of JP-4 fuel.

ENGINE INTAKES: Straight two-dimensional external compression inlets, on each side of the fuselage. Air inlet controllers by Hamilton Standard. Air inlet actuators by National Water Lift Company.

ACCOMMODATION: Pilot only, on ACES II ejection seat developed by Douglas. Stretched acrylic canopy and windscreen. Windscreen anti-icing valve by Dynasciences Corporation.

SYSTEMS: Electric power generating system by Lear Siegler Power Equipment Division; transformer-rectifiers by Electro Development Corporation; 40/50kVA generator constant-speed drive units by Sundstrand Corporation, Aviation Division. Three independent hydraulic systems (each 207 bars; 3,000 lb/sq in) powered by Abex engine-driven pumps; modular hydraulic packages by Hydraulic Research and Manufacturing Company. The oxygen system includes a liquid oxygen indicator by Simmonds Precision Products Inc. Air-conditioning system by Garrett. Automatic flight control system by General Electric, Aircraft Equipment Division. Auxiliary power unit for engine starting, and for the provision of electric or hydraulic power on the ground independently of the main engines, supplied by Garrett.

AVIONICS: Lightweight APG-63 pulse-Doppler radar developed by Hughes Aircraft Company provides long-range detection and tracking of small high-speed targets operating at all altitudes down to treetop level, and feeds accurate tracking information to the airborne central computer to ensure effective launch of the aircraft's missiles or the firing of its internal gun. For close-in dogfights, the radar automatically acquires the target on a head-up display. International Business Machines, Electronic Systems Center, is subcontractor for the central computer, and McDonnell Douglas Electronics Company for the head-up display. This latter unit projects all essential flight information in the form of symbols on to a combining glass positioned above the instrument panel at pilot's eye level. The display presents the pilot with all the information required to intercept and destroy an enemy aircraft without need for him to remove his eyes from the target. The display also provides navigation and other steering control information under all flight conditions. A transponder for the IFF system, developed by Teledyne Electronics Company, informs ground stations and other suitably equipped aircraft that the F-15 is a friendly aircraft. It also supplies data on the F-15's range, azimuth, altitude and identification to air traffic controllers. The F-15 carries an AN/APX-76 interrogator receiver-transmitter, built by Hazeltine Corporation, to inform the pilot if an aircraft seen visually or on radar is friendly. A reply evaluator for the IFF system, which operates with the AN/APX-76, was developed by Litton Systems Inc, Van Nuys. A vertical situation display set, that uses a cathode-ray tube to present radar, electro-optical identification and attitude director indicator formats to the pilot, has been developed by Sperry Rand Corporation, Sperry Flight Systems Division. This permits inputs received from the aircraft's sensors and the central computer to be visible to the pilot under any light conditions. This company has also developed an air data computer for the F-15, as well as an attitude and heading reference set to provide information on the aircraft's pitch, roll and magnetic heading that is fed to cockpit displays. This latter unit also serves as a backup to the inertial navigation set developed by Litton Guidance and Control Systems Division. This provides the basic navigation data and is the aircraft's primary attitude reference, enabling the F-15 to navigate anywhere in the world. In addition to giving the aircraft's position at all times, the inertial navigation system provides pitch, roll, heading, acceleration and speed information.

Other specialised equipment for flight control, navigation and communications includes a micro-miniaturised Tacan system by Collins Radio Company; a horizontal situation indicator to present aircraft navigation information on a symbolic pictorial display, by Collins Radio Company, which is also responsible for the ADF and ILS receivers. Magnavox provides the UHF transceiver and UHF auxiliary transceiver. The communications sets have cryptographic capability. Dorne and Margolin Aviation Products is responsible for the glideslope localiser antenna, and Teledyne Avionics Company for angle of attack sensors. An internal countermeasures set, designated AN/ALQ-135, which provides automatic jamming of enemy radar signals, is supplied by Northrop's Defense Systems Division; radar warning systems by Loral Electronic Systems; and an electronic warfare warning set by Magnavox.

EQUIPMENT: Tachometer, fuel and oil indicators by Bendix Corporation, Flight and Engine Instrument Division. Feel trim actuators by Plessey Airborne Corporation.

ARMAMENT: Provision for carriage and launch of a variety of air-to-air weapons over short and medium ranges, including four AIM-9L Sidewinders, four AIM-7F Sparrows, and a 20 mm M61A-1 six-barrel gun with 940 rounds of ammunition. A lead-computing gyro has been developed by the General Electric Co. To keep the pilot informed of the status of his weapons and provide for their management, an armament control set has been developed by Dynamic Controls Corporation. Five weapon stations allow for the carriage of up to 7,257 kg (16,000 lb) of bombs, rockets or additional ECM equipment.

DIMENSIONS, EXTERNAL:

Wing span	13·05 m (42 ft 9¾ in)
Length overall	19·43 m (63 ft 9 in)
Height overall	5·63 m (18 ft 5½ in)
Tailplane span	8·61 m (28 ft 3 in)
Wheel track	2·75 m (9 ft 0¼ in)
Wheelbase	5·42 m (17 ft 9½ in)

AREA:

Wings, gross	56·5 m² (608 sq ft)

WEIGHTS (A: F-15A; B: F-15C):

T-O weight (interceptor, full internal fuel and 4 Sparrows): A	18,824 kg (41,500 lb)
B	20,185 kg (44,500 lb)
T-O weight (incl three 2,271 litre; 600 US gallon drop-tanks): A	24,675 kg (54,400 lb)
B	26,035 kg (57,400 lb)
Max T-O weight: A	25,401 kg (56,000 lb)
B	30,845 kg (68,000 lb)

PERFORMANCE:

Max level speed	more than Mach 2·5 (800 knots; 1,482 km/h; 921 mph CAS)
Approach speed	125 knots (232 km/h; 144 mph) CAS
T-O run (interceptor)	274 m (900 ft)
Landing run (interceptor), without braking parachute	762 m (2,500 ft)
Absolute ceiling	30,500 m (100,000 ft)
Ferry range: without FAST Pack	more than 2,500 nm (4,631 km; 2,878 miles)
with FAST Pack	more than 3,000 nm (5,560 km; 3,450 miles)
g limits	+9·0/−3·0

MCDONNELL DOUGLAS F-15 STRIKE EAGLE

McDonnell Douglas has developed as a company-funded project an all-weather interdiction version of the F-15, identified as the F-15 Strike Eagle. Demonstrated at the 1980 Farnborough International Air Show, it is structurally generally similar to the F-15 Eagle, but has a synthetic aperture radar (SAR) configuration of the lightweight APG-63 pulse-Doppler radar which was developed by Hughes Aircraft Company for the standard F-15. SAR represents an advance in digital electronics which uses programmable signal processors with high-density storage elements in the radar. The SAR modification of the APG-63 provides a significant breakthrough in ground target resolution, making it possible for small targets to be 'seen' in any weather, or by night, from ranges of approximately 17 nm (32 km; 20 miles). It was anticipated that it would be possible for the radar to demonstrate a resolution of 3·05 m (10 ft) from a range of 8·5 nm (16 km; 10 miles).

A fully developed Strike Eagle will be capable of making night and all-weather attacks against ground targets from stand-off ranges. It was planned to demonstrate this capability by May 1981, at which time blind weapon-delivery techniques were to be simulated. It was hoped that within two months of that demonstration it would be possible to integrate SAR imagery with the FLIR (forward-looking infra-red) system of the Pave Tack targeting pod, leading to the first launching trials of live weapons, and the completion of all company-funded testing in late 1981. It was reported in early 1981 that the initial demonstrations were going well.

In the two-seat F-15 Strike Eagle, the rear cockpit is equipped for a specialist officer to handle both the SAR and the Pave Tack systems. He is provided with four displays, two of which are used for navigation to the target area. Of the others, one is dedicated to weapon selection, while the second monitors enemy tracking systems. Approach to the target area is made at low level, followed by a brief 'look-see' at a height of approximately 760 m

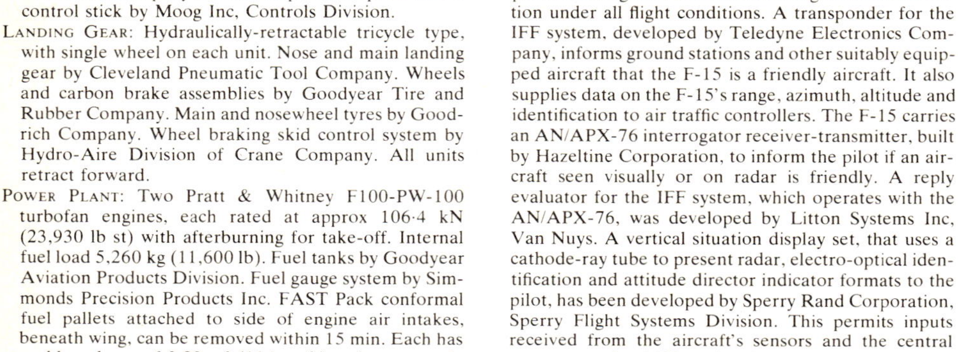

McDonnell Douglas F-15C Eagle single-seat air superiority fighter, with additional side view (top) of two-seat F-15B *(Pilot Press)*

(2,500 ft) to enable the radar officer to locate a specific target. After some 10 to 20 seconds the low-level track is resumed, with the onboard computer programmed with details of azimuth and range, and the FLIR system being used for pinpoint designation of the target.

The Strike Eagle retains the air superiority capability of the basic F-15, but is equally suitable for deployment in long-range all-weather interdiction roles. Three 30 mm gun pods can be carried on underfuselage and underwing racks, or up to 10,885 kg (24,000 lb) of other stores. These can include air-to-ground weapons such as HARM (High-speed Anti-Radiation Missile), Harpoon anti-ship missile, Mk 20 Rockeye bombs in an MER-200 dispenser which allows their delivery at supersonic speed, Mk 82 (500 lb) and Mk 84 (2,000 lb) bombs, Maverick television-guided missiles, and Pave Tack infra-red tracking and laser-spot designation pods. Strike Eagle can carry two low-drag conformal fuel tanks, known as FAST (Fuel And Sensor Tactical) Packs, to provide increased range. These can themselves be used to mount two AIM-7F Advanced Sparrow missiles, or 1,995 kg (4,400 lb) of air-to-ground weapons on each tank, or Pave Tack pods, or similar more advanced systems.

MCDONNELL DOUGLAS AV-8B

RAF designation: Harrier GR. Mk 5

In late 1973 and early 1974 the British and US governments received for approval various proposals for an advanced version of the BAe (HS) Harrier (see UK section), one of which was known by the US designation AV-16A. However, in March 1975, the British Secretary of State for Defence, Mr Roy Mason, stated that there was "not enough common ground on the Advanced Harrier for us to join in the programme with the US". Development studies for a US version were therefore continued primarily by McDonnell Douglas to meet requirements of the US Marine Corps.

Essentially, the objective of this programme was to develop an advanced V/STOL attack aircraft which, without too much departure from the existing Harrier airframe, would virtually double the aircraft's weapons load/combat radius. Initially McDonnell Douglas and the USMC modified two AV-8As as prototype **YAV-8Bs**. The first of these flew for the first time on 9 November 1978, and the second on 19 February 1979. Prototype demonstration was completed in Summer 1979, in 185 test flights totalling 173 flying hours, during which the two aircraft met or exceeded all performance requirements specified by the Marine Corps and Naval Air Systems Command.

Aim of the AV-8B is to achieve the improved performance capability required of the original AV-16A proposal by aerodynamic means, while retaining the same basic F402-RR-404 (Pegasus 11) engine, thus saving the cost of developing the Pegasus 15 engine that was originally considered necessary for the advanced version. However, a Pegasus 11F-35 engine, offering some 8·9 kN (2,000 lb st) more power than the Pegasus 11-21E currently specified, is being developed as a potential power plant for possible future developments of the AV-8B.

Aerodynamic changes include use of a supercritical wing; the addition of under-gun-pod strakes and a retractable fence panel forward of the pods, to increase lift for vertical take-off; larger wing trailing-edge flaps and drooped ailerons; and redesigned engine air intakes. The leading-edge root extensions (LERX) developed originally by British Aerospace for the UK-designed Big Wing Harrier will also be standard. The landing gear is strengthened to cater for the higher operating weights and greater external stores loads made possible by these changes.

Four full-scale development aircraft are being built, the first of which was scheduled to fly in October 1981, with Navy BIS (Bureau of Inspection and Survey) trials following in Spring 1983. The operational date planned for the AV-8B is mid-1985.

In addition to the two prototypes and four development aircraft, the USMC has a requirement for 336 AV-8Bs, and funding was provided in 1981 for a pilot production batch of 12 aircraft. Sixty AV-8Bs, which will be known as Harrier GR. Mk 5s, are to be acquired for the Royal Air Force, to enter service in 1986. On the combined USMC/RAF orders, approx 60 per cent of the airframe work will be undertaken by McDonnell Douglas; the other 40 per cent, including centre and rear fuselages, fins, rudders, tailplanes (for the RAF aircraft only), centreline pylons, reaction control system, and all major systems in these parts of the airframe, will be undertaken by British Aerospace, which will also be responsible for final assembly of the aircraft for the RAF. Under a similar industrial agreement Rolls-Royce will undertake 75 per cent of the engine work for the USMC, and Pratt & Whitney 25 per cent. In the event of exports to third countries, the McDonnell Douglas/British Aerospace work split will be 75/25 per cent.

The following description applies to the full-scale development and initial production AV-8B:

Type: Single-seat V/STOL combat aircraft.

Wings: Cantilever shoulder-wing monoplane, of broadly similar planform to Harrier/AV-8A but of supercritical section, approx 20% greater in span and 14% greater in area. Thickness/chord ratio 11·5% at root, 7·5% at tip. 10° less sweepback on leading-edges, and non-swept inboard trailing-edges. Composite construction, making extensive use of graphite epoxy in the main multi-spar torsion box, ribs, skins, outrigger fairings and wingtips. Trailing-edge single-slotted flaps, of substantially greater chord than those of AV-8A, with flap slot closure doors. Drooping ailerons, also of graphite epoxy construction.

McDonnell Douglas F-15 Strike Eagle, carrying four Sidewinder air-to-air missiles, a 30 mm gun pack, a Pave Tack infra-red sensor pod, two FAST Pack conformal fuel tanks, and eight Mk 82 500 lb bombs

McDonnell Douglas YAV-8B with leading-edge root extensions (LERX) and two underwing Sidewinder missiles

Fuselage: Generally similar to AV-8A, but longer. New forward fuselage, with raised cockpit, constructed from graphite epoxy composite material, and additional lift-augmenting surfaces. These comprise a fixed strake on each of the two underfuselage gun packs, and retractable forward fence just aft of the nosewheel unit. During VTOL modes the 'box' formed by the ventral strakes and the lowered nose flap augment lift by trapping the cushion of air bounced off the ground by the engine exhaust. This additional lift allows the AV-8B to take off vertically at a gross weight equal to its maximum hovering gross weight.

Tail Unit: Taller fin than that of AV-8A, resembling that of UK Sea Harrier. Graphite epoxy tailplane of new planform, with constant sweep on leading-edges and reduced sweep on trailing-edges.

Landing Gear: Main landing gear strengthened to cater for higher operating weights. Dowty Rotol/Cleveland outrigger wheels, moved inboard to approx mid-span beneath each wing between flaps and ailerons.

Power Plant: One Rolls-Royce Pegasus 11-21E (F402-RR-404) vectored-thrust turbofan engine rated at 95·64 kN (21,500 lb st). Zero-scarf front nozzles. Engine air intakes redesigned, with elliptical lip shape and double instead of single row of suction relief doors. Increased fuel tankage available in wings, raising total internal fuel capacity (fuselage and wing tanks) from approx 2,268 kg (5,000 lb) in the AV-8A to 3,402 kg (7,500 lb) in the AV-8B. Each of the four inner underwing stations capable of carrying a 1,135 litre (300 US gallon) auxiliary fuel tank. Probe for in-flight refuelling.

Accommodation: Pilot only, on ejection seat, in pressurised, heated and air-conditioned cockpit. In production configuration the cockpit is raised by comparison with the YAV-8B prototype, and has a redesigned windscreen and canopy to provide a better all-round view.

Avionics and Equipment: Improved attitude and heading reference system, AN/ARN-84 Tacan, AN/ARC-159 UHF, AN/APX-100 IFF, radar altimeter, radar warning receiver, visual landing aids, laser gyro inertial navigation system, dual combining glass head-up display, and CRT multi-purpose display. Marconi Avionics self-contained pitch and roll autostabilisation computer, with built-in rate gyroscopes and added electronic package to interface with forward reaction control nozzle. Garrett digital air data computer.

Armament and Operational Equipment: Two underfuselage gun/ammunition packs, mounting a General Electric GAU-12/U five-barrel 25 mm cannon or two 30 mm Aden guns. Single 454 kg (1,000 lb) stores point on fuselage centreline, between gun packs. Three stores stations under each wing, the inner one capable of carrying a 907 kg (2,000 lb) store, the centre one 454 kg (1,000 lb), and the outer one 286 kg (630 lb). The four inner wing stations are 'wet', permitting the carriage of auxiliary fuel tanks. Including fuel, stores, weapons and ammunition, and water injection for the engine, the maximum useful load for vertical take-off is approximately 3,175 kg (7,000 lb), and for short take-off nearly 7,710 kg (17,000 lb). Typical weapons include two or four Sidewinder or Maverick missiles; up to 16 general-purpose bombs, ten laser-guided bombs, 12 cluster bombs or ten fire bombs; up to ten rocket pods; or two underwing gun pods (in addition to the underfuselage gun packs). Main weapon delivery by Hughes Angle Rate Bombing System (ARBS), comprising a dual-mode (TV and laser) target seeker linked to a Smiths head-up display via an IBM digital computer. Passive ECM (chaff and flare) dispensers, and AN/ALQ-164 centreline defensive ECM pod.

Dimensions, external:

Wing span	9·25 m (30 ft 4 in)
Length overall	14·12 m (46 ft 4 in)
Height overall	3·56 m (11 ft 8 in)

Area:

Wings, gross	21·37 m² (230 sq ft)

Weights:

Basic operating weight empty	5,783 kg (12,750 lb)
Max external stores	4,173 kg (9,200 lb)
Max fuel: internal only	3,402 kg (7,500 lb)
internal and external	7,180 kg (15,829 lb)
Basic flight design gross weight for 7g operation	10,410 kg (22,950 lb)
Max T-O weight:	
366 m (1,200 ft) STO	13,494 kg (29,750 lb)
S/L VTO, 32°C	8,702 kg (19,185 lb)
Design max landing weight	8,799 kg (19,400 lb)
Max VL weight	8,096 kg (17,850 lb)

Performance (estimated):

Max Mach number in level flight 0·91

T-O run at max T-O weight less than 335 m (1,100 ft)

Operational radius with external loads shown:

short T-O (305 m; 1,000 ft), twelve Mk 82 Snakeye bombs, internal fuel, 1 h loiter
more than 150 nm (278 km; 172 miles)

short T-O (305 m; 1,000 ft), seven Mk 82 Snakeye bombs, external fuel tanks, no loiter
more than 650 nm (1,204 km; 748 miles)

Unrefuelled ferry range, with external tanks
2,500 nm (4,633 km; 2,879 miles)

BRITISH AEROSPACE/MCDONNELL DOUGLAS AV-8C HARRIER

The AV-8C is an updated conversion of the basic AV-8A Harrier that was produced by British Aerospace for the US Marine Corps. All 61 existing AV-8As are to be updated to the improved standard; the first conversions were made by McDonnell Douglas, the remainder will be carried out by the Marine Corps at Cherry Point, North Carolina. Details of the AV-8C can be found in the British Aerospace entry in the UK section.

MCDONNELL DOUGLAS F/A-18 HORNET

In the Spring of 1974 the US Department of Defense accepted a proposal from the US Navy to study a low-cost lightweight multi-mission fighter, then identified as the VFAX. In June 1974 the USN approached the US aircraft industry to submit critiques and comments on such an aircraft. Six companies responded, including McDonnell Aircraft Company; but in August of that year Congress terminated the VFAX concept, directing instead that the Navy should investigate versions of the General Dynamics YF-16 and Northrop YF-17 lightweight fighter prototypes then under evaluation for the USAF.

McDonnell Douglas made a study of the configuration of these two aircraft and concluded that Northrop's contender could be redesigned at minimum cost to meet the Navy's requirements. It then teamed with Northrop to propose a derivative of the YF-17 to meet the Navy's requirement, with McDonnell Douglas as the prime contractor. Identified as the Navy Air Combat Fighter (NACF), this received the name Hornet when selected for further development. Two single-seat versions were proposed originally, of which the F-18A was intended for fighter duties and the A-18 for attack missions. Except for a small amount of operational equipment and missile armament, the two proved so similar that the single designation F/A-18A now covers both configurations. The following production versions have been ordered:

F/A-18A. Single-seat escort fighter/interdictor to replace F-4, armed with fuselage-mounted Sparrows; also a single-seat attack aircraft to replace A-4 and A-7, with FLIR and a laser tracker, which are being developed as part of the Hornet programme, replacing the Sparrow missiles.

TF/A-18A. Tandem two-seat version of F/A-18A for training, with combat capability. Fuel capacity reduced by under 6 per cent.

CF-18. Version for Canadian Armed Forces, which plan to purchase 137, including two-seaters. Selection announced on 10 April 1980. Delivery to begin in October 1982, and end in 1988. CF-18s will replace CF-101s initially, and then CF-104s.

On 22 January 1976 it was announced that full-scale development had been initiated by the US Navy, with initial funding of $16 million. Total cost of the development programme was expected to be about $1·4 billion, including the production of 11 YF-18s for the flight test programme. A total of 1,377 Hornets, including the 11 development aircraft, is planned for construction by the end of the 1980s, for the US Navy and Marine Corps. More than 150 of those built will be two-seat trainers.

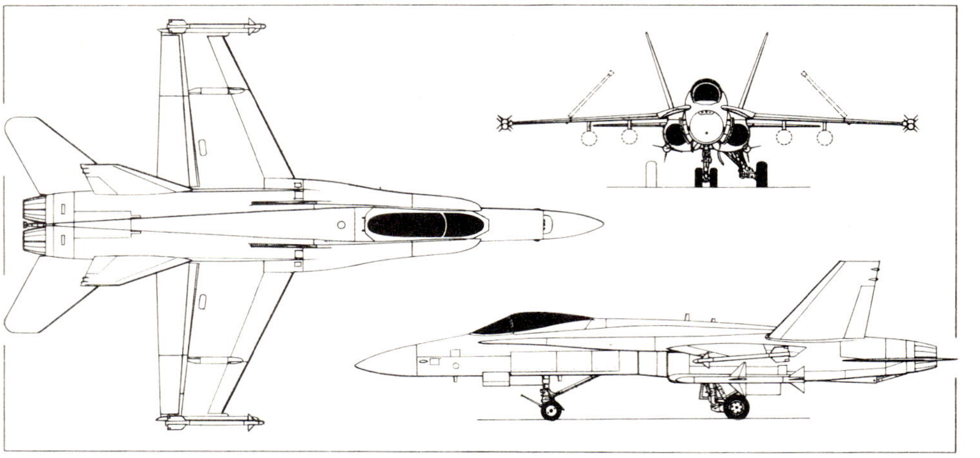

McDonnell Douglas F/A-18A Hornet (two General Electric F404-GE-400 turbofan engines) *(Pilot Press)*

The first Hornet (160775) was rolled out on 13 September 1978, and made its first flight on 18 November 1978; the second flew on 12 March 1979, and all 11 development aircraft were flying by March 1980, including two TF/A-18A two-seat combat-capable trainers. The first batch of nine production Hornets was authorised in FY 1979, followed by 25 in FY 1980 and 60 in FY 1981. In the fourth quarter of 1979, a Hornet became the first modern jet aircraft to complete initial sea trials within one year of its first flight, and the first production aircraft was delivered to the US Navy for operational evaluation in May 1980. Six more had been delivered by mid-June 1981. The first development squadron (VFA-125) was formed at NAS Lemoore, California, in November 1980. The F/A-18 is due to become operational with the US Navy in 1982.

The Hornet airframe differs from that of the YF-17 prototype by having increased wing area, a wider and longer fuselage to provide greater internal fuel capacity, an enlarged nose to accommodate the 0·71 m (28 in) radar dish to meet the Navy's search range requirement of over 30 nm (56 km; 35 miles), and strengthening of the airframe structure to cater for the increased loads caused by catapult launches and arrested landings. Approximately 2,000 kg (4,400 lb) of additional fuel is carried to meet Navy mission range requirements.

Ease of maintenance was given careful consideration in formulation of the F/A-18's design. An engine change can be effected within approx 20 min, and radar equipment is track-mounted so that it can be rolled out for maintenance. Electronics equipment is housed behind quick-release doors at chest height, and the windscreen is hinged to permit easy access behind the instrument panel. A built-in test panel mounted within the nosewheel well pinpoints system failures, and when the indicated access door is opened the assembly which has failed 'flags' confirmation that it needs repair or replacement. Groundcrew have access to 'go/no go' panel for rapid pre-flight check; this confirms levels of essential liquids, such as engine oil, hydraulic fluid, radar coolant, APU oil and oxygen. Safety features include self-sealing fuel tanks and fuel lines, fire suppressant foam within the fuel tanks, built-in fire extinguishers, filler foam in the fuselage for fire suppression, and a system which detects hydraulic fluid leaks and then isolates the relative section.

Conventional instrumentation has almost disappeared from the cockpit, replaced by three cathode ray tubes and an information control panel directly in front of the pilot. A head-up display is fitted and, so that the pilot will not be distracted by having to move his hands to different controls; every critical switch for air-to-air and air-to-surface engagements is either in the throttle in his left hand, or on the control stick in his right hand. During air-to-air combat the Hughes AN/APG-65 radar can track multiple targets, displaying up to eight target tracks while retaining up to ten in its memory. A raid assessment mode enables the pilot to discriminate between closely spaced targets. The radar information is displayed on a clutter-free scope in either lookup or lookdown attitude; it provides also range-while-search capability, long-range search and track, and several modes for close-in combat. Doppler beam sharpening (DBS) achieves greater resolution of radar signal returns during air-to-surface mapping.

McDonnell Douglas is prime contractor for the Hornet, with the centre of activities at St Louis, Missouri. Northrop builds the centre and aft fuselage, which is delivered totally assembled to McDonnell Douglas. Assembly is completed at St Louis and flight testing is carried out by VX-4 at NATC Patuxent River, Maryland.

The following information applies specifically to the single-seat F/A-18A:

TYPE: Single-seat carrier-based naval strike fighter.

WINGS: Cantilever mid-wing monoplane. Moderate-sweep multi-spar structure, primarily of light alloy and graphite/epoxy. Boundary layer control achieved by wing-root slots. Full-span leading-edge manoeuvring flaps have a maximum extension angle of 30°. Single-slotted trailing-edge flaps, actuated by Bertea hydraulic cylinders, deploy to a maximum of 45°. Ailerons, with Hydraulic Research actuators, can be drooped to 45°, providing the advantages of full-span flaps for low approach speeds. Leading- and trailing-edge flaps are computer programmed to deflect for optimum lift and drag in both manoeuvring and cruise conditions. Wing-root leading-edge extensions (LEX) permit flight at angles of attack exceeding 60°. Wings fold, by means of AiResearch mechanical drive, at the inboard end of each aileron.

FUSELAGE: Semi-monocoque basic structure, primarily of light alloy, with graphite/epoxy used for access doors/panels. Titanium firewall between engines. Air-brake in upper surface of fuselage between tail fins. Pressurised cockpit section of fail-safe construction.

TAIL UNIT: Cantilever structure with swept vertical and horizontal surfaces. Twin outward-canted fins and rudders, mounted forward of all-moving horizontal surfaces (stabilators), which are actuated collectively and differentially by National Water Lift servo-cylinder hydraulic units for pitch and roll control.

McDonnell Douglas TF/A-18A, a combat-capable two-seat training version of the Hornet

McDonnell Douglas F/A-18A Hornet with production-type wings, lacking the former extended-chord leading-edge on the outer panels *(Brian M. Service)*

Landing Gear: Retractable tricycle type, manufactured by Cleveland, with twin-wheel nose and single-wheel main units. Nose unit retracts forward, main wheels aft, turning 90° to stow horizontally inside the lower surface of the engine air ducts. Bendix wheels and brakes. Ozone nosewheel steering unit. Nose unit towbar for catapult launch. Arrester hook, for carrier landings, under rear fuselage.

Power Plant: Two General Electric F404-GE-400 low bypass turbofan engines, each producing approx 71·2 kN (16,000 lb thrust). Self-sealing fuel feed tanks and fuel lines; foam in wing tanks and fuselage voids. Internal fuel load approx 4,990 kg (11,000 lb); provision for up to three 1,192 litre (315 US gallon) external tanks, increasing total fuel capacity to more than 7,257 kg (16,000 lb). Simmonds fuel gauging system. Fixed-ramp air intakes.

Accommodation: Pilot only, on Martin-Baker US10S ejection seat in pressurised, heated and air-conditioned cockpit. Upward-opening two-part canopy, both sections hinged individually.

Systems: Two completely separate hydraulic systems. Quadruplex digital fly-by-wire flight control system, with direct electrical backup to all surfaces, and direct mechanical backup to stabilators. Garrett air-conditioning system. GEC electrical power system. Oxygen system. Fire detection and extinguishing systems.

Avionics and Equipment: Include an automatic carrier landing system (ACLS) for all-weather carrier operations; a Hughes AN/APG-65 multi-mode digital air-to-air and air-to-ground tracking radar, with air-to-air modes which include velocity search (VS), range while search (RWS), track while scan (TWS), which can track ten targets and display eight to the pilot, and raid assessment mode (RAM). Itek ALR-67 radar warning receiver; General Electric quadruple-redundant flight control system with two AYK-14 digital computers; Litton inertial navigation system; Kaiser multi-purpose cockpit display, including head-up display and three CRTs; Conrac communications system control; Normalair-Garrett digital data recorder for Bendix maintenance recording system; Smiths standby altimeter; and Kearflex standby airspeed indicator, standby vertical speed indicator, and cockpit pressure altimeter. Garrett APU for engine starting and ground pneumatic, electrical and hydraulic power.

Armament: Nine external weapon stations with a combined capacity of 7,710 kg (17,000 lb) of mixed ordnance at high *g*. These comprise two wingtip stations for AIM-9 Sidewinder air-to-air missiles; two outboard wing stations for an assortment of air-to-ground or air-to-air weapons, including AIM-7 Sparrows; two inboard wing stations for external fuel tanks or air-to-ground weapons; two nacelle fuselage stations for Sparrows or Martin Marietta AN/ASQ-173 laser spot tracker/strike camera (LST/SCAM) and Ford FLIR pods; and a centreline fuselage station for external fuel or weapons. An M61 20 mm six-barrel gun, with 570 rounds, is mounted in the nose and has a McDonnell Douglas director gunsight, with a conventional sight as backup.

Dimensions, external:

Wing span	11·43 m (37 ft 6 in)
Wing span over missiles	12·31 m (40 ft 4¾ in)
Width, wings folded	8·38 m (27 ft 6 in)
Length overall	17·07 m (56 ft 0 in)
Height overall	4·66 m (15 ft 3½ in)
Tailplane span	6·58 m (21 ft 7¼ in)
Wheel track	3·11 m (10 ft 2½ in)
Wheelbase	5·42 m (17 ft 9½ in)

Area:

Wings, gross	37·16 m² (400 sq ft)

Weights:

Fighter mission T-O weight	15,234 kg (33,585 lb)
Attack mission T-O weight	21,319 kg (47,000 lb)

Performance (estimated):

Max level speed	more than Mach 1·8
Max speed, intermediate power	more than Mach 1·0
Approach speed	130 knots (240 km/h; 150 mph)
Combat ceiling	approx 15,240 m (50,000 ft)
T-O run	less than 305 m (1,000 ft)
Combat radius, fighter mission	more than 400 nm (740 km; 460 miles)
Combat radius, attack mission	550 nm (1,019 km; 633 miles)
Ferry range, unrefuelled	more than 2,000 nm (3,706 km; 2,303 miles)

DOUGLAS AIRCRAFT COMPANY (a Division of McDonnell Douglas Corporation)

Headquarters: 3855 Lakewood Boulevard, Long Beach, California 90846
Telephone: (213) 593 5511

President: John C. Brizendine
Executive Vice-President:
Charles M. Forsyth
Senior Vice-Presidents:
Charles Conrad Jr (Marketing)
Edward Curtis (Fiscal Management)
William T. Gross (Operations)
Vice-Presidents:
Ray E. Bates (General Manager, New Commercial Programmes)
Jerald D. Burns (Business Management, New Commercial Programmes)
Larry S. Dickenson (Commercial Sales, Europe, Africa and Middle East)
James H. Douez (Manufacturing)
Eugene F. Dubil (Product Support)
N. Douglas Ingebretsen (Personnel, West)
Robert C. P. Jackson (Plans)
Robert H. Kinder (Government Marketing)
Ray J. Kleinberg (Controller)
Donald J. Krokus (Commercial Sales, Pacific and Asia)
Richard M. Randall (Commercial Contracts)
Roger D. Schaufele (Engineering)
John W. Stillwell (Quality Assurance)
Howell L. Walker (Commercial Sales, the Americas)
Wendell W. Way (Programme Management)
William R. Worrell (Material)
Chief Counsel: John H. Carroll
Director, External Relations: Donald N. Hanson

The Douglas Aircraft Company operates plants at Long Beach, Palmdale and Torrance, California.

MCDONNELL DOUGLAS DC-8 FREIGHTER CONVERSION

Under a programme launched at Tulsa works in February 1976, McDonnell Douglas is modifying DC-8 passenger transports into specialised freighters at a current rate of up to nine aircraft per year. First to be converted were two DC-8-43 series airliners, which were also re-engined with Pratt & Whitney JT3D turbofans, under contract from Frederick B. Ayer and Associates. Subsequent orders for a total of eight conversions have been received from Intercontinental Airways, International Air Leases Inc, International Air Service, Overseas National Airways and Transmeridian Air Cargo.

Modification includes removal of passenger installations and fitting a production freighter seven-track floor, and a 2·16 m × 3·56 m (85 × 140 in) main deck cargo door. Cabin windows are replaced by metal plugs, and a cargo loading system is installed. Conversion to turbofan power is optional for turbojet models.

MCDONNELL DOUGLAS DC-8 SERIES 71, 72 and 73

A brief paragraph in the 1979-80 *Jane's* gave preliminary details of the company's plans to re-engine DC-8 Srs 61, 62 and 63 aircraft with General Electric/SNECMA CFM56 or Pratt & Whitney JT8D-209 turbofan engines, under the respective designations of DC-8 Srs 71, 72 and 73. Since that time, engineering design for the DC-8/CFM56 conversion has continued. The first modification is that of a DC-8 Srs 61, and the resulting Srs 71 made its first flight on 15 August 1981. FAA certification was anticipated by the end of 1981. By that date a DC-8 Srs 63 was also to have been converted to the new power plant, and certification of this Srs 73 is programmed for April 1982. It is intended to carry out similar conversions to Srs 61F/61CF, 62F/62CF, and 63F/63CF Jet Traders. Up to February 1980, the company had received 56 firm orders and 44 options for conversions.

Overall management of the DC-8 re-engining programme is being handled by Cammacorp of Los Angeles. Other modifications include the optional incorporation of a Garrett APU and environmental control system.

Douglas states that Srs 71, 72 and 73 aircraft with CFM56-2-1C engines are expected to be the quietest large four-engined transports when they enter service, offering airport communities a true noise reduction in the order of 70 per cent. This means that these aircraft will not only be able to meet the requirements of FAR Pt 36-7 and -8, but will offer protection against future, more stringent regulations.

This considerable reduction in environmental noise pollution is gained without loss of aircraft performance. On the contrary, Douglas claims that the DC-8 Super Seventy series with CFM56 engines will offer improved performance, including reduced take-off run, increased range, and fuel savings over a 3,000 nm (5,560 km; 3,455 mile) route of as much as 7,711 kg (17,000 lb), 3,855 kg (8,500 lb), and 4,990 kg (11,000 lb) for the DC-8 Srs 71, 72 and 73 respectively.

Only limited specification estimates for these conversions were available by early 1981, as follows:

Weights and Loadings (estimated. A: Srs 71; B: Srs 72; C: Srs 73; D: Srs 71F Jet Trader; E: Srs 72F; and F: Srs 73F):

Operating weight empty: A	73,800 kg (162,700 lb)
B	69,220 kg (152,600 lb)
C	75,115 kg (165,600 lb)
Max payload: A, C	30,240 kg (66,665 lb)
B	19,315 kg (42,580 lb)
Fuel weight: A	71,093 kg (156,733 lb)
B	70,734 kg (155,942 lb)
C	73,773 kg (162,642 lb)
Max T-O weight: A, D	147,415 kg (325,000 lb)
B, E	151,955 kg (335,000 lb)
C	158,755 kg (350,000 lb)
F	161,025 kg (355,000 lb)
Max zero-fuel weight: A	101,605 kg (224,000 lb)
B	88,450 kg (195,000 lb)
C	104,325 kg (230,000 lb)
Max landing weight: A, B	108,860 kg (240,000 lb)
C	111,115 kg (245,000 lb)
D, E	117,025 kg (258,000 lb)
F	124,740 kg (275,000 lb)
Max wing loading:	
A, D	550·2 kg/m² (112·69 lb/sq ft)
B, E	558·8 kg/m² (114·45 lb/sq ft)
C	583·8 kg/m² (119·58 lb/sq ft)
F	592·1 kg/m² (121·28 lb/sq ft)

Performance (estimated, at max T-O weight with CFM56-2-1C engines):

Max level speed, all versions	521 knots (965 km/h; 600 mph)
Cruising speed, all versions	Mach 0·80

The first McDonnell Douglas DC-8 Srs 71, a version of the DC-8 Srs 61 re-engined with CFM56 turbofans

MCDONNELL DOUGLAS DC-9

USAF designations: C-9A and VC-9C
US Navy designation: C-9B

Design study data on the DC-9, then known as the Douglas Model 2086, were released in 1962. Preliminary design work began during that year. Fabrication was started on 26 July 1963 and assembly of the first airframe began on 6 March 1964. It flew for the first time on 25 February 1965 and five DC-9s were flying by the end of June 1965. These were of the basic version, known as the DC-9 Series 10, of which production has ended; details can be found in the 1978-79 *Jane's*. Other variants are as follows:

Series 20. For operation in hot climate/high-altitude conditions, combining long-span wings of Series 30 with short fuselage of original Series 10. Up to 90 passengers. Two 64·5 kN (14,500 lb st) JT8D-9 turbofans. The Series 20 flew for the first time on 18 September 1968, and was certificated on 11 December 1968. The first Series 20 was delivered to SAS on the same day and entered commercial service on 23 January 1969. Described in 1977-78 *Jane's*.

Series 30. Developed version, initially with 62·3 kN (14,000 lb st) JT8D-7s, increased wing span, longer fuselage accommodating up to 105 passengers (normal) or 119 (with reduced facilities), and new high-lift devices including full-span leading-edge slats and double-slotted flaps. First Srs 30 flew for first time on 1 August 1966. First delivery, to Eastern Air Lines, was made on 27 January 1967 and scheduled services began on 1 February 1967. Engine options available include JT8D-9 of 64·5 kN (14,500 lb st); JT8D-11 of 66·7 kN (15,000 lb st); JT8D-15 of 69 kN (15,500 lb st); and JT8D-17 of 71·2 kN (16,000 lb st). Newer engines have sound-treated nacelles that comply with FAA FAR Pt 36 noise regulations.

Series 40. As Series 30, but with 64·5 kN (14,500 lb st) JT8D-9, 69 kN (15,500 lb st) JT8D-15 or 71·2 kN (16,000 lb st) JT8D-17 turbofans, increased fuel capacity, longer fuselage accommodating up to 132 passengers, and greater AUW. First flight was made on 28 November 1967 and FAA certification was received on 27 February 1968. The first Series 40 was delivered to SAS on 29 February 1968 and entered commercial service with that airline on 12 March 1968.

Series 50. 'Stretched' short/medium-range development of the Series 30, announced on 5 July 1973. High-density seating for up to 139 passengers made possible by a 4·34 m (14 ft 3 in) fuselage extension. A 'new look' interior features enclosed overhead racks for carry-on baggage, sculptured wall panels, acoustically-treated ceiling panels and indirect lighting. Available with either Pratt & Whitney JT8D-15 or -17 turbofan engines, rated at 69 kN (15,500 lb st) and 71·2 kN (16,000 lb st) respectively, and embodying sound-absorption materials as developed for the engines and nacelles of the DC-10, the Series 50 meets FAR Pt 36 noise requirements. The engines are smokeless and have thrust reversers rotated 17° from the vertical to reduce the possibility of exhaust gas ingestion. The landing gear is fitted with an improved anti-skid braking system. First flight of a Series 50 was made on 17 December 1974. First deliveries were made to Swissair, with whom it entered service in August 1975.

Super 80. Increased capacity short/medium-range version of the DC-9, with accommodation for a maximum of 172 passengers. Described separately.

Current versions are offered in passenger, cargo (**DC-9F**), convertible (**DC-9CF**) or passenger-cargo (**DC-9RC**) configurations. The cargo and convertible models have a main cabin cargo door measuring 3·45 m (11 ft 4 in) wide and 2·06 m (6 ft 9 in) high. An executive transport version is also offered, with increased fuel, enabling up to 15 persons to be carried nonstop over 2,865 nm (5,300 km; 3,300 mile) transcontinental or transocean stages. First delivery of an all-cargo model, a DC-9 Srs 30F, was made to Alitalia on 13 May 1968. This model has 122·1 m³ (4,313 cu ft) of cargo space in main cabin, plus the underfloor hold, enabling it to carry eight full cargo pallets and two half-pallets with total weight of nearly 18,144 kg (40,000 lb).

There are also three military versions of the DC-9, as follows:

C-9A Nightingale. Aeromedical airlift transport, based on DC-9 Srs 30; 21 delivered to USAF between 1968-73. Details in 1977-78 *Jane's*.

C-9B Skytrain II. Fleet logistic support transport, of which 15 were ordered by the USN and two by Kuwait. Described in 1979-80 and earlier editions of *Jane's*.

VC-9C. Three DC-9-30 type aircraft with special configuration, delivered in 1975 to the USAF for service in the Special Air Missions Wing based at Andrews AFB, Maryland.

Orders for the DC-9 totalled 1,070 (with conditional orders and options for 31 others), including military versions, by 1 July 1981. A total of 984 DC-9s (including 21 C-9As, 17 C-9Bs and 3 VC-9Cs) had been delivered by that date. The 1,000th DC-9, a Super 81 for Swissair, was delivered on 3 September 1981.

In June 1966, the FAA certificated three Category 2 all-weather landing systems for the DC-9, comprising the Collins FD-108 flight director system, Sperry AD-200 flight director system, and coupled approach utilising the Sperry SP-50A autopilot.

The following structural details apply to the DC-9 Series 30:

TYPE: Twin-turbofan short/medium-range airliner.

WINGS: Cantilever low-wing monoplane. Mean thickness/chord ratio 11·0%. Dihedral 3°. Sweepback 24° 30′ at quarter-chord. All-metal construction, with three spars inboard, two spars outboard and spanwise stringers riveted to skin. Glassfibre trailing-edges on wings, ailerons and flaps. Manually-controlled aileron on each wing. Wing-mounted speed brakes. Full-span leading-edge slats (also on Srs 20/40/50). Hydraulically-actuated double-slotted trailing-edge flaps over 67% of semi-span. Single boundary-layer fence (vortillon) under each wing. Detachable wingtips. Thermal anti-icing of leading-edges.

FUSELAGE: Conventional all-metal semi-monocoque structure.

TAIL UNIT: Cantilever all-metal structure with electrically-actuated variable-incidence T tailplane. Manually-controlled elevators with servo tabs. Hydraulically-controlled rudder with manual override. Glassfibre trailing-edges on control surfaces.

LANDING GEAR: Retractable tricycle type of Menasco manufacture, with steerable nosewheel. Hydraulic retraction, nose unit forward, main units inward. Twin Goodyear wheels on each unit. Main-wheel tyres size 40 × 14. Nosewheel tyres size 26 × 6·60. Goodyear brakes. Hydro-Aire Hytrol anti-skid units.

POWER PLANT: Two Pratt & Whitney JT8D turbofan engines (details given under individual model listings), pod-mounted on each side of rear fuselage. Engines fitted with 40% target-type thrust reversers for ground operation only. Standard fuel capacity 13,937 litres (3,682 US gallons), also in Srs 10 and 20; 16,122 litres (4,259 US gallons) in Srs 40; and up to 19,074 litres (5,039 US gallons) in Srs 50. In Srs 30, 40 and 50 aircraft an additional 8,517 litres (2,250 US gallons) can be stored in auxiliary tanks.

ACCOMMODATION: Crew of two on flight deck, plus cabin attendants. Normal accommodation in main cabin is for 105 passengers, with seating for maximum of 119 with reduced facilities. Fully pressurised and air-conditioned. Toilets at rear of cabin. Provision for galley. Passenger door at front of cabin on port side, with electrically-operated built-in airstairs. Optional ventral stairway. Servicing and emergency exit door opposite on starboard side. Underfloor freight and baggage holds, with forward and rear doors on starboard side.

DIMENSIONS, EXTERNAL:

Wing span: Srs 30, 40, 50	28·47 m (93 ft 5 in)
Wing aspect ratio: Srs 30, 40, 50	8·71
Length overall: Srs 30	36·37 m (119 ft 3½ in)
Srs 40	38·28 m (125 ft 7¼ in)
Srs 50	40·72 m (133 ft 7¼ in)
Height overall: Srs 30	8·38 m (27 ft 6 in)
Srs 40, 50	8·53 m (28 ft 0 in)
Tailplane span:	
Srs 30, 40, 50	11·23 m (36 ft 10¼ in)
Wheel track: Srs 30, 40, 50	5·03 m (16 ft 6 in)
Wheelbase: Srs 30	16·22 m (53 ft 2½ in)
Srs 40	17·10 m (56 ft 1¼ in)
Srs 50	18·56 m (60 ft 11 in)
Passenger door (port, fwd):	
Height	1·83 m (6 ft 0 in)
Width	0·85 m (2 ft 9½ in)
Height to sill	2·13 m (7 ft 2 in)
Servicing door (stbd, fwd): Height	1·22 m (4 ft 0 in)
Width	0·69 m (2 ft 3 in)
Height to sill	2·18 m (7 ft 2 in)
Freight and baggage hold doors:	
Height	1·27 m (4 ft 2 in)
Width: fwd	1·35 m (4 ft 5 in)
rear	0·91 m (3 ft 0 in)
Height to sill	1·07 m (3 ft 6 in)

DIMENSIONS, INTERNAL:

Cabin: Max width	3·07 m (10 ft 1 in)
Floor width	2·87 m (9 ft 5 in)
Max height	2·06 m (6 ft 9 in)
Freight hold (underfloor):	
Srs 30	25·3 m³ (895 cu ft)
Srs 40	28·9 m³ (1,019 cu ft)
Srs 50	33·24 m³ (1,174 cu ft)

AREAS:

Wings, gross:	
Srs 30, 40, 50	92·97 m² (1,000·7 sq ft)
Ailerons (total): All versions	3·53 m² (38·0 sq ft)
Trailing-edge flaps (total):	
All versions	19·58 m² (210·8 sq ft)
Leading-edge slats (total):	
Srs 50	11·22 m² (120·8 sq ft)
Spoilers (total): All versions	3·22 m² (34·7 sq ft)
Fin: All versions	14·96 m² (161·0 sq ft)
Rudder: All versions	6·07 m² (65·3 sq ft)
Tailplane: All versions	25·60 m² (275·6 sq ft)
Elevators, incl tabs: All versions	9·83 m² (105·8 sq ft)

WEIGHTS:

Manufacturer's empty weight:	
Srs 30	25,940 kg (57,190 lb)
Srs 40	26,612 kg (58,670 lb)
Srs 50	28,068 kg (61,880 lb)
Max space-limited payload:	
Srs 30	14,118 kg (31,125 lb)
Srs 40	15,610 kg (34,415 lb)
Max weight-limited payload:	
Srs 30	12,743 kg (28,094 lb)
Srs 40	14,363 kg (31,665 lb)
Srs 50	15,617 kg (34,430 lb)
Max T-O weight:	
Srs 30, 40, 50	54,885 kg (121,000 lb)
Max ramp weight:	
Srs 30, 40, 50	55,338 kg (122,000 lb)
Max zero-fuel weight:	
Srs 30, 40, 50	44,678 kg (98,500 lb)
Max landing weight:	
Srs 30, 40, 50	49,895 kg (110,000 lb)

PERFORMANCE (at max T-O weight, except where indicated):

Never-exceed speed:	
Srs 50	537 knots (995 km/h; 618 mph)
Max level speed:	
all versions	500 knots (926 km/h; 575 mph)
Max cruising speed at 7,620 m (25,000 ft):	
Srs 30	490 knots (907 km/h; 564 mph)
Srs 40, 50	485 knots (898 km/h; 558 mph)
Average long-range cruising speed at 9,145-10,675 m (30,000-35,000 ft)	443 knots (821 km/h; 510 mph)
Max rate of climb at S/L:	
Srs 30	885 m (2,900 ft)/min
Srs 40	865 m (2,850 ft)/min
Srs 50	792 m (2,600 ft)/min
FAA T-O field length: Srs 30	1,685 m (5,530 ft)
Srs 40	2,088 m (6,850 ft)
FAA landing field length: Srs 30	1,425 m (4,680 ft)
Srs 40	1,440 m (4,720 ft)
Srs 50	1,485 m (4,880 ft)

Range at Mach 0·8, with reserves for 200 nm (370 km; 230 mile) flight to alternate and 60 min hold at 3,050 m (10,000 ft):

Srs 30 at 9,150 m (30,000 ft) with 64 passengers and baggage	1,160 nm (2,148 km; 1,335 miles)
Srs 40 at 7,620 m (25,000 ft) with 70 passengers and baggage	930 nm (1,723 km; 1,071 miles)

Range at long-range cruising speed at 9,150 m (30,000 ft), reserves for 200 nm (370 km; 230 mile) flight to alternate and 45 min continued cruise at 9,150 m (30,000 ft):

Srs 30 with 80 passengers and baggage	1,670 nm (3,095 km; 1,923 miles)
Srs 40 with 87 passengers and baggage	1,555 nm (2,880 km; 1,790 miles)
Srs 50 with 97 passengers and baggage	1,795 nm (3,326 km; 2,067 miles)
Ferry range, reserves as above:	
Srs 30	1,980 nm (3,669 km; 2,280 miles)
Srs 40	1,850 nm (3,428 km; 2,130 miles)
Srs 50	2,185 nm (4,049 km; 2,516 miles)

MCDONNELL DOUGLAS DC-9 SUPER 80

The McDonnell Douglas DC-9 Super 80 was developed specifically to meet the needs of operators on short/medium-range routes who require an aircraft of increased capacity, and the basic design has been modified to offer improved economy in operation, reduced fuel consumption, and far quieter engines. The following versions have been announced to date:

Super 81. Basic version, powered by two Pratt & Whitney JT8D-209 turbofan engines. Certificated at a max T-O weight of 63,500 kg (140,000 lb). Airlines from which firm orders have been received include Swissair (15), Austrian Airlines (10), Pacific Southwest Airlines (26) and Toa Domestic Airlines (8).

Super 82. Announced on 16 April 1979, with Pratt & Whitney JT8D-217 turbofan engines, each rated at 89 kN (20,000 lb st), and with an emergency thrust reserve of 3·78 kN (850 lb st). Certificated on 30 July 1981 at a max T-O weight of 66,680 kg (147,000 lb). Otherwise generally similar to Super 81, with same fuel capacity and max landing weight. Airlines from which firm orders have been received include Aeromexico (4), AirCal (5), Republic Airlines (14), and Inex Adria (3).

Super 80SF. Proposed short-field version, with the fuselage length and capacity of the DC-9 Srs 40, but with the new wings and more powerful engines of the Super 80 Series.

The new wings are increased in span by the insertion of wing root plugs and by a 0·61 m (2 ft 0 in) extension on each tip, giving a gross area 28% greater than that of the DC-9 Srs 50. The fuselage of the Super 81 is extended in length by insertion of a 3·86 m (12 ft 8 in) plug forward of the wing, and a 0·48 m (1 ft 7 in) segment aft of the wing. The cabin has 'wide look' decor, with large enclosed overhead baggage compartments, acoustic ceiling and soft fluorescent lighting.

Basic power plant of the Super 81 comprises two Pratt & Whitney JT8D-209 turbofan engines, each with an emergency thrust reserve which becomes available automatically in an engine-out situation. Refanned with a larger-diameter single-stage fan, this version of the JT8D engine has a bypass ratio of 1·68, by comparison with 1·00 of earlier versions, resulting in a lower specific fuel consumption and reduced noise emission. In addition, sound suppression materials applied to the inlet, fan duct and tailpipe duct of each nacelle, plus a cold/hot-stream exhaust mixer, reduce engine noise to levels below the requirements of FAR Pt 36 Stage 3, and also satisfy the more stringent requirements of ICAO Chapter 3 established for new aircraft designs. Standard fuel capacity is increased by 5,754 litres (1,520 US gallons) as a result of the larger wing.

The Super 82 with the more powerful Pratt & Whitney JT8D-217 turbofan engines, is regarded as being particularly suitable for operation from 'hot and high' airports. The higher thrust available makes possible an increased payload and range when operating from standard airports. At an airport such as Denver, Colorado, which is 1,525 m (5,000 ft) above sea level, the Super 82 is able to take off with 155 passengers and their baggage, and have a non-stop range of approximately 1,300 nm (2,409 km; 1,497 miles).

Systems improvements in the Super 80s include a new digital electronics integrated flight guidance and control system; a 'dial-a-flap' system to permit more accurate selection of flap angle for optimum take-off and landing performance; flow-through cooling of the aircraft's avionics compartment; a larger-capacity APU; a new recirculating system for ventilation air; and an advanced digital fuel quantity gauging system.

The first Super 81 made a first flight of 2 h 50 min on 18 October 1979. The second and third prototypes (N1002G and N1002W) flew on 6 December 1979 and 29 February 1980 respectively. FAA certification was granted on 25 August 1980. The initial example of the Super 82 made its first flight on 8 January 1981.

First order for the Super 80 series had been placed by Swissair, for 15 Super 81 aircraft, in October 1977. This airline was first to take delivery of a Super 81, on 12 September 1980, and started to operate the aircraft on its London-Zurich route from 5 October. First announced order (of three aircraft) for the Super 82 was placed by Aeromexico in September 1979. By 1 August 1981 firm orders totalled 110 aircraft, plus conditional firm orders and options for 26 more, to serve with 15 customers. Deliveries then totalled 35.

The following description applies to the Super 81, but except for the power plant applies generally also to the Super 82:

TYPE: Twin-turbofan short/medium-range airliner.

WINGS: Cantilever low-wing monoplane. Mean thickness/chord ratio 11·0%. Dihedral 3°. Incidence 1° 15′. Sweepback at quarter-chord 24° 30′. All-metal construction, with two spars and spanwise stringers riveted to skin. Glassfibre trailing-edges on wings, ailerons and flaps. Manually-controlled aileron on each wing. Wing-mounted speed brakes. Full-span leading-edge slats. Hydraulically-actuated double-slotted trailing-edge flaps over 67% of semi-span. Single boundary-layer fence (vortillon) under each wing. Three spoilers per wing; two outboard segments act as flight and ground spoilers, inboard segment is ground spoiler only. Detachable wingtips. Thermal anti-icing of leading-edges.

FUSELAGE: All-metal semi-monocoque fail-safe structure of heat-treated light alloy. Majority of cabin floor constructed of balsa or Nomex core composite. Engine pylons built by Calcor, fuselage panels by Aeritalia.

TAIL UNIT: Cantilever all-metal structure with electrically-actuated variable-incidence T tailplane. Manually-controlled elevators with servo tabs. Hydraulically-controlled rudder with manual override. Glassfibre trailing-edges on control surfaces.

LANDING GEAR: Retractable tricycle type of Cleveland Pneumatic manufacture, with steerable nosewheels. Hydraulic retraction, nose unit forward, main units inward. Twin Goodyear wheels and tyres on each unit. Main-wheel tyres size 44·5 × 16·5-20, pressure 11·38 bars (165 lb/sq in). Nosewheel tyres size 26 × 6·6-14, pressure 10·34 bars (150 lb/sq in). Goodyear disc brakes. Hydro-Aire Mk IIIA anti-skid units. Douglas ram-air brake cooling.

POWER PLANT (Super 81): Two Pratt & Whitney JT8D-209 turbofan engines, pod-mounted on each side of rear fuselage, and each rated at 82·3 kN (18,500 lb st), with emergency thrust reserve of 3·3 kN (750 lb st). Standard fuel capacity 21,876 litres (5,779 US gallons). Pressure refuelling point in starboard wing leading-edge. Overwing gravity refuelling points.

ACCOMMODATION: Crew of two on flight deck, plus cabin attendants. Seating arrangements are optional to meet specific airline requirements. Those selected by early 1981 range from 12 first class and 123 coach class, totalling 135 passengers, to 169 single class. Maximum optional seating capacity is for 172 passengers. Fully pressurised and air-conditioned. One toilet forward on port side, two at rear of cabin. Provisions for galley at both forward and aft ends of cabin. Passenger door at front of cabin on port side and a rear ventral stairway standard, both with built-in electrically-operated airstairs. Servicing and emergency exit doors at starboard forward end and port aft end of cabin. Three cargo doors for underfloor holds on starboard side. Overwing emergency exits, two each side.

SYSTEMS: Garrett dual air cycle air-conditioning and pressurisation system utilising engine bleed air, max differential 0·54 bars (7·77 lb/sq in). Two separate 207 bar (3,000 lb/sq in) hydraulic systems for operation of spoilers, flaps, slats, rudder, landing gear, nosewheel steering, brakes, rotated thrust reversers and ventral stairway. Pneumatic system, for air-conditioning/pressurisation, engine starting and ice protection, utilises 8th or 13th stage engine bleed air and/or APU. Electrical system includes three 40kVA 120/208V three-phase 400Hz alternators, two engine-driven, one driven by APU. Oxygen system of diluter-demand type for crew on flight deck; continuous-flow type with automatic mask presentation for cabin passengers. Anti-icing of wing and engine inlets, and de-icing of tailplane, by engine bleed air. Electrical de-icing of windscreen. APU provides pneumatic and electrical power on ground, and electrical power in flight.

AVIONICS AND EQUIPMENT: All-digital avionics, including dual Sperry integrated flight systems; Sperry Cat IIIA autoland; autopilot and stability augmentation system; speed command with digital autothrottles; thrust rating indicator system; dual air data systems; automatic reserve thrust; ADF system; and weather radar display. Sundstrand head-up display optional.

DIMENSIONS, EXTERNAL:
Wing span 32·87 m (107 ft 10 in)
Wing chord at root 7·05 m (23 ft 1½ in)
Wing chord at tip 1·10 m (3 ft 7½ in)
Wing aspect ratio 9·62
Length overall 45·06 m (147 ft 10 in)
Length of fuselage 41·30 m (135 ft 6 in)
Height overall 9·04 m (29 ft 8 in)
Tailplane span 12·24 m (40 ft 2 in)
Wheel track 5·08 m (16 ft 8 in)
Wheelbase 22·07 m (72 ft 5 in)
Passenger door (port, fwd):
Height 1·83 m (6 ft 0 in)
Width 0·86 m (2 ft 10 in)
Height to sill 2·24 m (7 ft 4 in)
Servicing door (stbd, fwd): Height 1·22 m (4 ft 0 in)
Width 0·69 m (2 ft 3 in)
Height to sill 2·24 m (7 ft 4 in)
Servicing door (port, rear): Height 1·52 m (5 ft 0 in)
Width 0·69 m (2 ft 3 in)
Height to sill 2·67 m (8 ft 9 in)
Freight and baggage hold doors:
Height 1·27 m (4 ft 2 in)
Width 1·35 m (4 ft 5 in)
Height to sill: fwd 1·17 m (3 ft 10 in)
centre 1·30 m (4 ft 3 in)
aft 1·52 m (5 ft 0 in)
Emergency exits (overwing, port and stbd):
Height 0·91 m (3 ft 0 in)
Width 0·51 m (1 ft 8 in)

DIMENSIONS, INTERNAL:
Cabin, excl flight deck, incl toilets:
Length 30·78 m (101 ft 0 in)
Max width 3·07 m (10 ft 1 in)
Max height 2·06 m (6 ft 9 in)
Floor area 89·65 m² (965 sq ft)
Volume 191·9 m³ (6,778 cu ft)
Freight holds (underfloor):
fwd 13·14 m³ (464 cu ft)
centre 9·80 m³ (346 cu ft)
aft 12·54 m³ (443 cu ft)

AREAS:
Wings, gross 118 m² (1,270 sq ft)
Ailerons (total) 3·53 m² (38·0 sq ft)
Fin, excl dorsal fin 9·51 m² (102·4 sq ft)
Rudder 6·07 m² (65·3 sq ft)
Tailplane 29·17 m² (314·0 sq ft)

WEIGHTS AND LOADINGS (A: Super 81; B: Super 82):
Weight empty: A 36,177 kg (79,757 lb)
B 36,534 kg (80,543 lb)
Fuel load: A, B 17,748 kg (39,128 lb)
Max T-O weight: A 63,500 kg (140,000 lb)
B 66,680 kg (147,000 lb)
Max zero-fuel weight: A, B 53,524 kg (118,000 lb)
Max landing weight: A, B 58,060 kg (128,000 lb)
Max wing loading: A 534·6 kg/m² (109·5 lb/sq ft)
B 561·0 kg/m² (114·9 lb/sq ft)
Max power loading: A 385·8 kg/kN (3·78 lb/lb st)
B 374·6 kg/kN (3·68 lb/lb st)

PERFORMANCE (at max T-O weight):
Max level speed:
A, B 500 knots (926 km/h; 575 mph)
Max cruising speed: A, B Mach 0·80
Normal cruising speed: A, B Mach 0·76
Landing speed:
A, B 124 knots (230 km/h; 143 mph)
FAA T-O field length: A 2,192 m (7,190 ft)
B 2,176 m (7,140 ft)
FAA landing field length: A, B 1,404 m (4,605 ft)
Range with max fuel:
A 2,657 nm (4,925 km; 3,060 miles)
B 2,596 nm (4,812 km; 2,990 miles)

OPERATIONAL NOISE LEVELS (FAR Pt 36):
T-O 90·4 EPNdB
Sideline 94·6 EPNdB
Approach 93·3 EPNdB

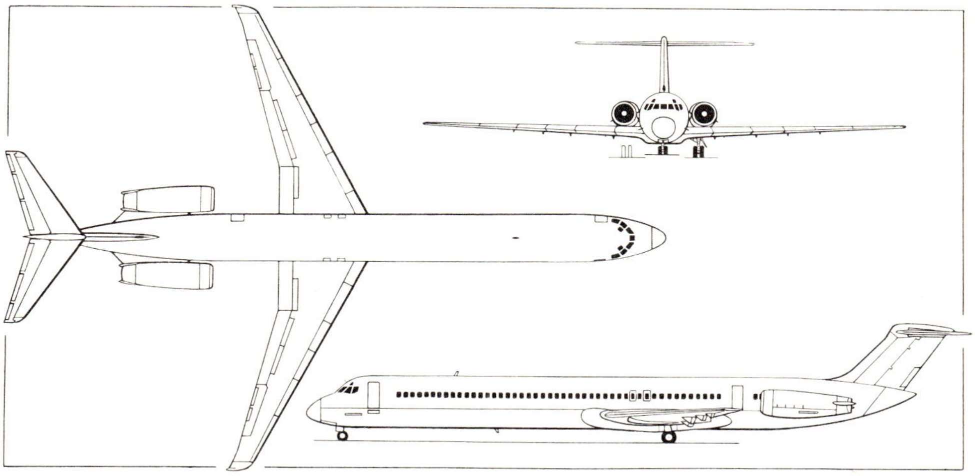

McDonnell Douglas DC-9 Super 80 'stretched' version of this twin-turbofan transport *(Pilot Press)*

McDonnell Douglas DC-9 Super 81 in the insignia of Japan's Toa Domestic Airlines

First flight take-off of a McDonnell Douglas DC-10 Series 15, powered by General Electric CF6-50C2F turbofan engines

MCDONNELL DOUGLAS DC-10

The DC-10 was developed by McDonnell Douglas as an all-purpose commercial transport able to operate economically over ranges from 260 nm to 5,300 nm (480 to 9,815 km; 300 to 6,100 miles), according to Series, and able to carry 270 mixed class passengers, or a maximum of 380 passengers in an all-economy configuration.

There are five basic production versions, as follows:

Series 10. Initial version, powered by three General Electric CF6-6D or -6D1 turbofan engines, each rated at 178 kN (40,000 lb st) or 182·4 kN (41,000 lb st) respectively. Intended for service on domestic routes of 260-3,125 nm (480-5,795 km; 300-3,600 miles). First ordered, by American Airlines, on 19 February 1968. First flight 29 August 1970. Type certificated by FAA on 29 July 1971. First scheduled passenger flight 5 August 1971, by American Airlines.

Series 15. Basically as Series 10, but with General Electric CF6-50C2F engines, each rated at 207 kN (46,500 lb st). Two each ordered by Aeromexico and Mexicana in Summer 1979, for delivery in 1981. Three more ordered subsequently by Mexicana. First flight 8 January 1981. Certificated on 12 June 1981, with deliveries beginning shortly afterwards.

Series 30. Extended-range version for intercontinental operations, powered by three General Electric CF6-50A or -50C turbofan engines, each rated at 218 kN (49,000 lb st) or 227 kN (51,000 lb st) respectively. Later versions have CF6-50C1 or C2 engines rated at 233·5 kN (52,500 lb st). Increased fuel capacity. Wing span increased by 3·05 m (10 ft 0 in). Landing gear supplemented by additional dual-wheel unit, mounted on the fuselage centreline between the four-wheel bogie main units. First flight 21 June 1972. FAA certification granted 21 November 1972, simultaneously with first deliveries of production aircraft to KLM and Swissair.

Series 30ER. Developed version of Series 30, with range extended still further by installing a 5,792 litre (1,530 US gallon) auxiliary fuel tank in rear of cargo compartment, giving total capacity of 144,224 litres (38,100 US gallons) and increasing range by up to 695 nm (1,287 km; 800 miles). More powerful (240·2 kN; 54,000 lb st) CF6-50C2B engines; max T-O weight 263,085 kg (580,000 lb); max range 5,730 nm (10,620 km; 6,600 miles). First order (for two) placed by Swissair in July 1980, plus kits to convert two of its existing DC-10s to Srs 30ER configuration. Since ordered by Finnair (one).

Series 40. Extended-range version for intercontinental operations, powered by three Pratt & Whitney turbofan engines. Twenty-two early models, built for Northwest Orient Airlines, had JT9D-20 turbofans, each rated at 220 kN (49,400 lb st) with water injection. First flight of Series 40 (known originally as Series 20) was on 28 February 1972; FAA certification received 20 October 1972. Later versions, built for Japan Air Lines, equipped with JT9D-59A turbofans, each rated at 236 kN (53,000 lb st). First flight of a Series 40 with this latter power plant was on 25 July 1975.

There are also convertible cargo versions designated Model **10CF** and **30CF**, of which the latter is described separately. On 19 December 1977 the USAF announced selection of the DC-10 Srs 30CF as its Advanced Tanker/Cargo Aircraft (ATCA), and this version also is described separately (see entry for **KC-10A Extender**).

Under consideration for possible future production are domestic and intercontinental versions with fuselages lengthened by 6·4 to 12·2 m (21 to 40 ft) compared with current series. The largest of the new models would accommodate a total of 395 passengers in mixed first class/coach seating, with improved seat/mile costs and reduced fuel consumption per seat. Winglets have been fitted to one DC-10, which began flight testing in this form on 31 August 1981, to evaluate the potential improvement in fuel efficiency.

DC-10 subassemblies and components are brought together at Long Beach for final assembly. Certain major subassemblies are produced at other divisions of McDonnell Douglas; Convair Division of General Dynamics Corporation at San Diego, California, is subcontractor for the fuselage, being responsible for five sections totalling 39·01 m (128 ft).

By 6 July 1981, McDonnell Douglas had received firm orders for 366 commercial DC-10s, plus conditional orders and options on 18 more, making a total of 384, of which 353 had been delivered. At that date, 47 airlines, plus the USAF, had ordered DC-10s: the worldwide fleet had then carried over 320 million passengers and had amassed more than 2,170 million nm (4,000 million km; 2,500 million miles) of flying.

Type: Three-turbofan commercial transport.

Wings: Cantilever low-wing monoplane of all-metal fail-safe construction. Several different wing sections of Douglas design are used between wing root and tip. Thickness/chord ratio varies from slightly more than 12·2% at root to less than 8·4% at tip. Dihedral 5° 14·4′ inboard, 3° 1·8′ outboard. Incidence ranges from positive at wing root to negative at tip. Sweepback at quarter-chord 35°. All-metal inboard and outboard ailerons, the former used conventionally, the latter only when the leading-edge slats are extended. Double-slotted all-metal trailing-edge flaps mounted on external hinges, with an inboard and outboard flap panel on each wing. Five all-metal spoiler panels on each wing, at the rear edge of the fixed wing structure, forward of the flaps. All spoilers operate in unison as lateral control, speed brake and ground spoilers. Full-span three-position all-metal leading-edge slats. Ailerons are powered by hydraulic actuators manufactured by Bertea Corporation, spoilers by hydraulic actuators manufactured by Parker-Hannifin Corporation. Each aileron is powered by either of two hydraulic systems; each spoiler is powered by a single system. All leading-edge slat segments outboard of the engines are anti-iced with engine bleed air.

Fuselage: Aluminium alloy semi-monocoque fail-safe structure of circular cross-section. Except for auxiliary areas the entire fuselage is pressurised.

Tail Unit: Cantilever all-metal structure. Variable-incidence tailplane, actuated by Vickers hydraulic motors. Longitudinal and directional controls are fully powered and comprise inboard and outboard elevators, each segment powered by a Bertea tandem actuator; upper and lower rudder each powered by a Bertea actuator. Rudder standby power supplied by two transfer motor pumps manufactured by Abex Corporation.

Landing Gear (Srs 10): Hydraulically-retractable tricycle type, with gravity free-fall for emergency extension. Nosewheel unit retracts forward, main units inward into fuselage. Twin-wheel steerable nose unit. Main gear comprises two four-wheel bogies. Oleo-pneumatic shock-absorbers on all units. Goodyear nosewheels and tyres size 37 × 14-14, pressure 11·03 bars (160 lb/sq in). Goodyear main wheels and tyres size 50 × 20-20, pressure 11·72 bars (170 lb/sq in). Goodyear disc brakes and anti-skid system, with individual wheel control.

Landing Gear (Srs 30 and 40): These versions have an additional dual-wheel main unit mounted on the fuselage centreline between the four-wheel bogie units, and this retracts forward. Goodyear nosewheels and tyres size 40 × 15·5-16, pressure 12·41 bars (180 lb/sq in). Four-wheel bogie main units and centreline unit have Goodyear wheels and tyres size 52 × 20·5-23. The former have a pressure of 11·38 bars (165 lb/sq in), the latter 9·65 bars (140 lb/sq in). Otherwise as Srs 10.

Power Plant: Three turbofan engines (details under Series descriptions), two of which are mounted on underwing pylons, the third above the rear fuselage at the base of the fin. All engines are fitted with thrust reversers for ground operation. Engine air inlets have load-carrying acoustically-treated panels for noise attenuation, and each engine fan case and fan exhaust is similarly treated. Three integral wing fuel tanks with a total capacity of approximately 82,142 litres (21,700 US gallons). Four standard pressure refuelling adapters, two in each wing outboard of the engine pylons. Series 30 and 40 aircraft have four integral wing fuel tanks and an auxiliary tank in the wing centre-section with a connected structural compartment fitted with a bladder cell, giving increased total capacity of approximately 138,165 litres (36,500 US gallons). Oil capacity, Series 10 and 30: 34·1 litres (9 usable US gallons); Series 40: 56·8 litres (15 usable US gallons).

Accommodation: Crew of three (pilot, first officer, flight engineer), with seating for two observers, plus cabin attendants. Standard seating for 255 or 270 in mixed class versions, with a maximum of 380 passengers in an economy class arrangement. Two aisles run the length of the cabin, which is separated into sections by cloakroom dividers. In the first class section, with three pairs of reclining seats abreast, the aisles are 0·78 m (2 ft 7 in) wide. In the coach class section, four pairs of seats, with a table between the centre pairs, also have two aisles, these being 0·51 m (1 ft 8 in) wide. One pair of seats is exchanged for a three-seat unit in the nine-abreast high-density layout. Up to nine lavatories located throughout the passenger cabin. Cloakrooms of standard and elevating type distributed throughout the cabin. Cabin windows, 0·28 × 0·41 m (11 in × 16 in), are spaced at 0·51 m (20 in) centres. Overhead stowage modules, fully enclosed and providing stowage for passengers' personal effects, are located on the sidewalls and extend the full length of the cabin. Optional centreline overhead baggage racks available. Eight passenger doors, four on each side, open by sliding inward and upward into the above-ceiling area. Containerised or bulk cargo compartments located immediately forward and aft of the wing, with outward-opening doors on the starboard side. A bulk cargo compartment is located in the lower aft section of the fuselage, with its door on the port side. Entire accommodation is fully air-conditioned, with five separate control zones for the standard below-floor galley configuration. Series 30 and 40 aircraft have an optional main cabin galley to replace the lower galley, and in this configuration there are four separate control zones for the air-conditioning. The lower-deck galley is provided with five to eight high-temperature ovens, and with refrigerators, storage space for linen, china and other accessories. Serving carts are taken to cabin level by two electric elevators, to a buffet service centre, from where stewardesses serve passengers. To permit quick turnround at terminals, without interference to passenger movement in the main cabin, the kitchen is provisioned through the cargo doors at ground level.

Systems: Three parallel continuously-operating and completely separate hydraulic systems supply the fully-powered flight controls and wheel brakes. Normally, one of the systems supplies power for landing gear actuation. Two reversible motor pumps, each sized to deliver power from one of the other two systems for standby operation of landing gear, can also power any other hydraulically operated unit. Each hydraulic system is powered by two identical engine-driven pumps, capable of delivering a total of 265 litres (70 US gallons)/min at 207 bars (3,000 lb/sq in) at take-off. All three hydraulic systems are applied to each primary control axis in a manner which ensures maximum control effectiveness in the event of single or dual hydraulic system failures. A Garrett TSCP-700-4 APU provides ground electrical and pneumatic power, including main engine starting, and auxiliary electrical power in flight.

AVIONICS AND EQUIPMENT: A dual fail-operative landing system is installed to meet Category IIIA weather minima. Digital air data computer meeting ARINC 576 requirements on Srs 10. Triple inertial navigation system meeting ARINC 561 requirements on Srs 30 and 40, with optional dual area navigation system capability.

DIMENSIONS, EXTERNAL:
Wing span: Series 10, 15 — 47·34 m (155 ft 4 in)
Series 30, 40 — 50·41 m (165 ft 4·4 in)
Wing chord at root — 10·71 m (35 ft 1·8 in)
Wing chord at tip: Series 10 — 3·21 m (10 ft 6½ in)
Series 30, 40 — 2·73 m (8 ft 11½ in)
Wing aspect ratio: Series 10 — 6·8
Series 30, 40 — 7·5
Length overall: Series 10 — 55·30 m (181 ft 5 in)
Series 15, 30, 40 — 55·50 m (182 ft 1 in)
Length of fuselage — 51·97 m (170 ft 6 in)
Height overall — 17·70 m (58 ft 1 in)
Tailplane span — 21·69 m (71 ft 2 in)
Wheel track — 10·67 m (35 ft 0 in)
Wheelbase: Series 10, 40 — 22·07 m (72 ft 5 in)
Series 30 — 22·05 m (72 ft 4 in)

DIMENSIONS, INTERNAL:
Cabin: Length, from aft bulkhead of flight deck to aft cabin bulkhead — approx 41·45 m (136 ft 0 in)
Max width — 5·72 m (18 ft 9 in)
Height (basic) — 2·41 m (7 ft 11 in)
Series 10, 30, 40 in lower-galley configuration:
Forward baggage and/or freight hold (forward of wing):
Containerised volume — 27·2 m³ (960 cu ft)
Bulk volume — 37·9 m³ (1,339 cu ft)
Centre baggage and/or freight hold (aft of wing):
Containerised volume — 36·2 m³ (1,280 cu ft)
Bulk volume — 43·9 m³ (1,552 cu ft)
Aft hold:
Bulk volume — 22·8 m³ (805 cu ft)
Series 30, 40 in upper-galley configuration:
Forward baggage and/or freight hold (forward of wing):
Containerised volume — 72·5 m³ (2,560 cu ft)
Bulk volume — 86·2 m³ (3,045 cu ft)
Centre baggage and/or freight hold (aft of wing):
Containerised volume — 45·3 m³ (1,600 cu ft)
Bulk volume — 54·8 m³ (1,935 cu ft)
Aft hold:
Bulk volume — 14·4 m³ (510 cu ft)

AREAS:
Wings, gross: Series 10 — 358·7 m² (3,861 sq ft)
Series 30, 40 — 367·7 m² (3,958 sq ft)
Ailerons, inboard (total) — 7·68 m² (82·7 sq ft)
Ailerons, outboard (total) — 9·76 m² (105·1 sq ft)
Trailing-edge flaps (total) — 62·1 m² (668·2 sq ft)
Leading-edge slats (total):
Series 10 — 42·05 m² (452·6 sq ft)
Series 30, 40 — 43·84 m² (471·9 sq ft)
Spoilers (total) — 12·73 m² (137·0 sq ft)
Fin — 45·92 m² (494·29 sq ft)
Rudders (total) — 10·29 m² (110·71 sq ft)
Tailplane — 96·6 m² (1,040·2 sq ft)
Elevators (total) — 27·7 m² (298·1 sq ft)

WEIGHTS AND LOADINGS:
Basic weight empty:
Series 10 — 111,086 kg (244,903 lb)
Series 15 — 111,832 kg (246,547 lb)
Series 30 — 121,198 kg (267,197 lb)
Series 40 — 122,951 kg (271,062 lb)
Max payload:
Series 10 — 44,678 kg (98,500 lb)
Series 30 — 48,330 kg (106,550 lb)
Series 40 — 46,243 kg (101,950 lb)
Max T-O weight:
Series 10, 15 — 206,384 kg (455,000 lb)
Series 30 — 259,450 kg (572,000 lb)
Series 40 (-20 engines) — 251,744 kg (555,000 lb)
Series 40 (-59A engines) — 259,450 kg (572,000 lb)
Max ramp weight:
Series 10 — 207,745 kg (458,000 lb)
Series 30 — 260,815 kg (575,000 lb)
Series 40 (-20 engines) — 253,105 kg (558,000 lb)
Series 40 (-59A engines) — 260,815 kg (575,000 lb)
Max zero-fuel weight:
Series 10 — 151,953 kg (335,000 lb)
Series 30, 40 — 166,922 kg (368,000 lb)
Max landing weight:
Series 10, 15 — 164,880 kg (363,500 lb)
Series 30, 40 — 182,798 kg (403,000 lb)
Max wing loading:
Series 10, 15 — 575·4 kg/m² (117·8 lb/sq ft)
Series 30 — 705·6 kg/m² (144·5 lb/sq ft)
Series 40 (-20 engines) — 684·6 kg/m² (140·2 lb/sq ft)
Series 40 (-59A engines) — 705·6 kg/m² (144·5 lb/sq ft)

PERFORMANCE (at max T-O weight unless specified):
Never-exceed speed — Mach 0·95
Max level speed at 7,620 m (25,000 ft) — Mach 0·88 (530 knots; 982 km/h; 610 mph)
Max cruising speed at 9,145 m (30,000 ft):
Series 10 (-6D engines) — 499 knots (925 km/h; 575 mph)
Series 10 (-6D1 engines) — 501 knots (928 km/h; 577 mph)
Series 30 — 490 knots (908 km/h; 564 mph)
Series 40 (-20 engines) — 489 knots (906 km/h; 563 mph)
Series 40 (-59A engines) — 498 knots (922 km/h; 573 mph)
Normal cruising speed, all versions — Mach 0·82
T-O speed (V_2):
Series 10 (-6D engines) — 181 knots (335 km/h; 208 mph)
Series 10 (-6D1 engines) — 175 knots (325 km/h; 202 mph)
Series 30 (-50C engines) — 189 knots (351 km/h; 218 mph)
Series 40 (-20 engines) — 187 knots (346 km/h; 215 mph)
Series 40 (-59A engines) — 178 knots (330 km/h; 205 mph)
Landing speed (with full load of passengers and baggage):
Series 10 — 128·5 knots (238 km/h; 148 mph)
Series 15 — 129 knots (240 km/h; 149 mph)
Series 30, 40 — 138 knots (256 km/h; 159 mph)
Max rate of climb at S/L:
Series 10 (-6D engines) — 817 m (2,680 ft)/min
Series 10 (-6D1 engines) — 838 m (2,750 ft)/min
Series 30 — 884 m (2,900 ft)/min
Series 40 (-20 engines) — 829 m (2,720 ft)/min
Series 40 (-59A engines) — 762 m (2,500 ft)/min
Service ceiling:
Series 10 (-6D engines) at 192,775 kg (425,000 lb) AUW — 10,605 m (34,800 ft)
Series 10 (-6D1 engines) at 192,775 kg (425,000 lb) AUW — 10,730 m (35,200 ft)
Series 30 at 249,475 kg (550,000 lb) AUW — 10,180 m (33,400 ft)
Series 40 (-20 engines) at 242,670 kg (535,000 lb) AUW — 9,660 m (31,700 ft)
Series 40 (-59A engines) — 9,965 m (32,700 ft)
En-route climb altitude, one engine out:
Series 10 at 195,045 kg (430,000 lb) AUW — 4,145 m (13,600 ft)
Series 30 at 251,744 kg (555,000 lb) AUW — 4,360 m (14,300 ft)
Series 40 (-20 engines) at 247,205 kg (545,000 lb) AUW — 3,565 m (11,700 ft)
Series 40 (-59A engines) at 254,010 kg (560,000 lb) AUW — 5,135 m (16,850 ft)
FAA T-O field length:
Series 10 (-6D1 engines) — 3,200 m (10,500 ft)
Series 15 (-50C2F engines) — 2,285 m (7,500 ft)
Series 30 (-50C2 engines) — 3,170 m (10,400 ft)
Series 40 (-59A engines) — 3,135 m (10,280 ft)
FAA landing field length:
Series 10 — 1,585 m (5,200 ft)
Series 15 — 1,600 m (5,250 ft)
Series 30, 40 — 1,630 m (5,350 ft)
Range with max fuel, no payload:
Series 10 — 5,514 nm (10,220 km; 6,350 miles)
Series 15 — 5,466 nm (10,130 km; 6,295 miles)
Series 30 — 6,504 nm (12,055 km; 7,490 miles)
Series 40 — 6,305 nm (11,685 km; 7,260 miles)
Range with max payload at max zero-fuel weight:
Series 10 — 2,350 nm (4,355 km; 2,706 miles)
Series 30 — 4,000 nm (7,413 km; 4,606 miles)
Series 40 (-20 engines) — 3,500 nm (6,485 km; 4,030 miles)
Series 40 (-59A engines) — 4,050 nm (7,505 km; 4,663 miles)

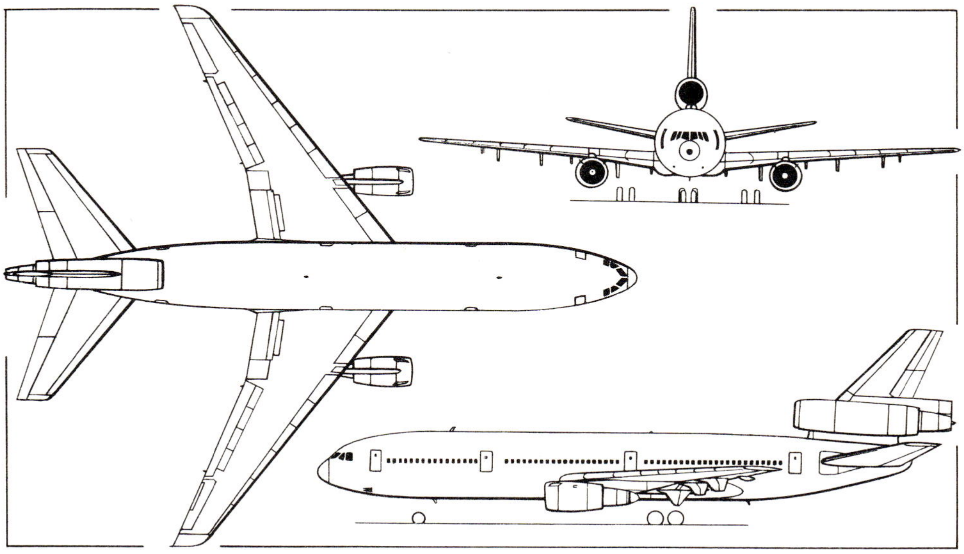

McDonnell Douglas DC-10 Series 30 high-capacity three-engined transport *(Pilot Press)*

First take-off of DC-10 fitted with drag-reducing winglets, to evaluate the effect on fuel consumption

MCDONNELL DOUGLAS DC-10 SERIES 30CF

The Series 30CF is a convertible freighter version of the McDonnell Douglas DC-10 transport. Generally similar to the basic DC-10 Series 30 and 40, it is designed for easy conversion to either passenger or cargo configuration. Its payload can consist of 380 passengers and baggage or 64,860 kg (143,000 lb) of cargo over full intercontinental range; or up to 70,626 kg (155,700 lb) of cargo on domestic transcontinental routes.

The first Series 30CF was powered by three General Electric CF6-50A turbofans. It flew for the first time on 28 February 1973 and initial deliveries were made to Trans International Airlines and Overseas National Airways on 19 April 1973 and 21 April 1973 respectively. In May and June 1977, Overseas National Airways took delivery of two of the later DC-10 Srs 30CFs, powered by General Electric CF6-50C1 engines rated at 233·5 kN (52,500 lb st).

In the passenger configuration, interior layout is generally similar to that of the DC-10, but the Series 30CF was designed to permit overnight conversion to an all-cargo configuration. This entails removal of seats, overhead baggage racks, forward food service centre, cloakrooms and carpeting from the main cabin, and installation of freight loading tracks and rollers, a cargo tiedown system and restraint nets. Coffee service fixtures and lavatories in the aft cabin may also be removed but are retained normally for regular cargo flights.

The cargo loading system for the Series 30CF is based on that in use in the DC-8 Super Sixty Series freighters. A two-channel network of roller conveyors, adjustable guide rails and pallet restraint fittings is installed in the seat tracks in the cabin floor by use of simple stud and locking pin devices. A 2·59 m high × 3·56 m wide (8 ft 6 in × 11 ft 8 in) cargo door in the side of the fuselage swings upward and allows easy loading of bulky freight.

A total of 30 standard 2·24 m × 2·74 m (7 ft 4 in × 9 ft) cargo pallets, or 22 larger pallets measuring 2·24 m × 3·18 m (7 ft 4 in × 10 ft 5 in) or 2·44 m × 3·05 m (8 ft × 10 ft), can be accommodated in the main cabin. The Series 30CF with upstairs galleys also has 132·2 m^3 (4,670 cu ft) of cargo space in the two lower cargo compartments for bulk freight, or for 26 half-size containers, or for five full-size pallets and 16 half-size containers.

A DC-10 Series 30CF delivered to Sabena Belgian World Airlines in 1973 is certificated for carrying combination loads of freight and passengers in the main cabin. Other DC-10 series aircraft are also offered in convertible versions. Continental Air Lines and Federal Express are both operating the DC-10 Series 10CF. More than 34 CF models had been ordered by 1 March 1981.

While weight and performance figures are generally similar to those of the DC-10 Series 30, the differing figures which follow should be noted for the Series 30CF:

WEIGHTS:

Weight empty	108,385 kg (238,948 lb)
Max T-O weight	267,620 kg (590,000 lb)
Max landing weight	190,962 kg (421,000 lb)

PERFORMANCE:

Landing speed at max landing weight
148·5 knots (275 km/h; 171 mph)

FAA T-O field length at max T-O weight
3,170 m (10,400 ft)

FAA landing field length at max landing weight
1,868 m (6,130 ft)

Max range with max cargo
3,581 nm (6,637 km; 4,124 miles)

MCDONNELL DOUGLAS EXTENDER

US Air Force designation: KC-10A

The USAF announced on 19 December 1977 that, following evaluation of the Boeing 747 and McDonnell Douglas DC-10 to meet its requirement for an Advanced Tanker/Cargo Aircraft (ATCA), the DC-10 ATCA was selected to fulfil this role. Subsequently, the aircraft was designated KC-10A and named Extender.

A force of KC-10As will greatly enhance the ability of the USAF to deploy combat aircraft, men and supplies on a global scale. This point was emphasised in a USAF submission to Congress in which the spokesman commented that 40 Boeing KC-135 tankers and a number of cargo aircraft would be needed to fuel an F-4 fighter squadron and carry its personnel and equipment from the USA to the Middle East. Just 17 of the proposed KC-10As could fulfil the same task, more economically and efficiently. USAF Military Airlift Command's (MAC) need for support by such aircraft was highlighted during the 1973 Arab-Israeli war, when many countries denied landing rights to MAC aircraft. From these circumstances came the decision to develop an ATCA to support the strategic airlift fleet, under the operational control of the USAF Strategic Air Command.

The initial $28 million contract awarded to McDonnell Douglas covered the funding for initial production planning, engineering and tooling. A second $429,000 contract was for initial planning of a logistics support programme covering the entire KC-10A military tanker fleet.

On 20 November 1978, the USAF authorised McDonnell Douglas to begin production of the KC-10A. This FY 1979 contract called for an expenditure of $148 million by the USAF, for the acquisition of two KC-10As, and in payment of the balance of the non-recurring engineering costs. In addition, under a separate $15·6 million contract, McDonnell Douglas was authorised to purchase the initial spare parts and support equipment for the KC-10A system. Four additional KC-10As were ordered by the USAF in November 1979, under the FY 1980 budget, and six were ordered in February 1981 under the FY 1981 budget. Quantities of KC-10As to be procured over subsequent years will be determined by available funding, but it is anticipated that at least 36 may be procured eventually by the USAF. The first KC-10A (USAF serial number 79-0433) made its first flight on 12 July 1980. The first to enter service (USAF serial number 79-0434) was delivered on 17 March 1981 to Barksdale AFB, Louisiana, for operation by SAC, and the second on 30 July 1981. Four more were due for delivery by the end of 1981 and the next six during 1982.

The commercial DC-10 Series 30CF convertible freighter, the basic airframe chosen for conversion to the ATCA role, is currently certificated for operation at a max T-O weight of 267,620 kg (590,000 lb).

The modifications necessary to convert the DC-10-30CF to the ATCA configuration include the installation of fuel cells in the lower fuselage compartment; the provision of a boom operator station, an aerial refuelling boom, a refuelling receptacle, an improved cargo handling system, and some military avionics systems. Various seating layouts are available in the forward area to permit the transport of a fighter squadron's essential support personnel. Seven bladder fuel cells are installed in the lower fuselage compartments, three forward and four aft of the wing, mounted within framework that restrains and supports the cells. These contain a total of 53,297 kg (117,500 lb) of fuel, equivalent to approx 68,420 litres (18,075 US gallons), and are interconnected with the aircraft's basic fuel system, comprising 108,211 kg (238,565 lb). All can be used for extended range, or fuel from the lower deck cells and the aircraft's basic fuel system can be used for in-flight refuelling. The KC-10A is designed to deliver 90,718 kg (200,000 lb) of fuel to a receiver 1,910 nm (3,540 km; 2,200 miles) from its home base.

The aerial refuelling operator's station, with access from the upper main deck, is sited in the lower aft fuselage and can accommodate the boom operator, an instructor and an observer, although only the boom operator is needed for a refuelling mission. The station has a rear window and a periscope observation system to give a wide field of view, and is pressurised and air-conditioned. The advanced aerial refuelling boom, which is the production version of a boom developed and tested in prototype form by McDonnell Douglas, provides greater capability than the type installed in the KC-135; in particular, it has a greater transfer flow rate, being rated at 5,678 litres (1,500 US gallons)/min. The boom operator 'flies' it by means of a digital fly-by-wire control system supplied by Sperry Flight Systems. A hose/reel unit for probe and drogue refuelling is also installed, so that the KC-10A can service USN, USMC and NATO aircraft, as well as older types of fighter still serving with Reserve and ANG units.

The provision of a refuelling receptacle, above the flight deck of the KC-10A, allows greater flexibility on long-range cargo or refuelling operations, extending the range beyond the nominal 6,000 nm (11,112 km; 6,905 miles) with a 45,400 kg (100,000 lb) payload. The improved

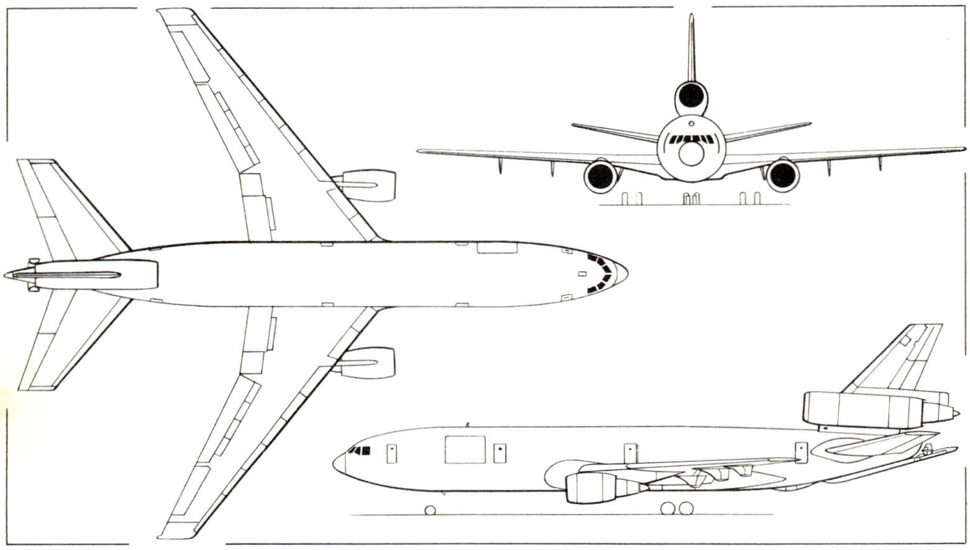

McDonnell Douglas KC-10A Extender advanced tanker/cargo aircraft for the US Air Force *(Pilot Press)*

McDonnell Douglas KC-10A Extender refuelling an F-15 Eagle

cargo handling system, by comparison with the basic DC-10-30CF, includes an increased floor area covered by omni-directional rollers, power rollers, and a portable winch to move cargo fore and aft.

Changes to the avionics are concerned chiefly with the deletion of equipment intended specifically for commercial operations, and the addition of UHF and secure com systems, Tacan, IFF, beacon transponder and a radar beacon mode.

The description of the DC-10 Series 30CF applies to the KC-10A, except as follows:

TYPE: Military flight refuelling/cargo aircraft.

WINGS, FUSELAGE, TAIL UNIT: As for DC-10-30CF, except for omission of most upper-deck cabin windows and lower-deck cargo doors.

LANDING GEAR: As for DC-10-30CF, except Goodyear nosewheels and tyres size 40 × 15·5-16, pressure 13·10 bars (190 lb/sq in). Four-wheel bogie main units and centreline unit have Goodyear wheels and tyres size 52 × 20·5-23. The former have a pressure of 13·79 bars (200 lb/sq in), the latter 10·69 bars (155 lb/sq in). Goodyear disc brakes and anti-skid system, with individual wheel control.

POWER PLANT: Three 233·53 kN (52,500 lb st) General Electric CF6-50C2 turbofan engines. Basic aircraft fuel system comprises three integral main wing fuel tanks, and an integral auxiliary tank in the wing centre-section with a connected structural compartment fitted with a bladder cell, giving a total capacity of approximately 132,331 litres (34,958 US gallons). Oil capacity 34·1 litres (9 US gallons).

ACCOMMODATION: Three crew on flight deck. Various seating arrangements for a limited number of essential support personnel at forward end of main cabin. Aerial refuelling station, with accommodation for boom operator, instructor and a student observer at aft end of lower fuselage compartment. Five passenger doors on main deck. A 2·59 m × 3·56 m (8 ft 6 in × 11 ft 8 in) cargo door on the port side of the fuselage permits loading of standard USAF 463L pallets, bulk cargo or wheeled vehicles. Maximum capacity for 25 pallets with access from both sides of the compartment, or 27 pallets with a single aisle on the starboard side.

SYSTEMS: As for DC-10-30CF.

AVIONICS AND EQUIPMENT: Include some additional military avionics, comprising navigation, communication, Tacan, IFF transponder, and modified commercial weather radar. Seven Goodyear Aerospace rubberised fabric fuel cells mounted in the lower fuselage compartments, with combined capacity of 53,446 kg (117,829 lb) fuel, equivalent to approx 68,610 litres (18,125 US gallons), which are interconnected into the aircraft's basic fuel system. Flight refuelling boom mounted under rear fuselage, plus hose/reel unit for probe and drogue refuelling. Director lights to guide receiver. Flight refuelling receptacle mounted on fuselage upper surface above flight deck.

DIMENSIONS, EXTERNAL: As for DC-10-30CF except:

Length overall 55·35 m (181 ft 7 in)
Height overall 17·70 m (58 ft 1 in)
Wheel track 10·57 m (34 ft 8 in)

AREAS: As for DC-10-30CF

WEIGHTS AND LOADING:

Operating weight empty:
tanker 108,874 kg (240,026 lb)
cargo 110,944 kg (244,591 lb)
Max cargo payload 76,842 kg (169,409 lb)
Design max T-O weight 267,620 kg (590,000 lb)
Max wing loading 727·8 kg/m² (149·06 lb/sq ft)

PERFORMANCE (estimated):

Critical field length 3,353 m (11,000 ft)
Max range with max cargo 3,797 nm (7,032 km; 4,370 miles)
Max ferry range, unrefuelled 9,993 nm (18,507 km; 11,500 miles)

MCDONNELL DOUGLAS SKYHAWK

US Navy designation: A-4

Construction of the XA-4A (originally XA4D-1) prototype Skyhawk began in September 1953, and the first flight of this aircraft took place on 22 June 1954.

Production of the A-4 ended on 27 February 1979, when the US Navy took delivery of the final aircraft off the assembly line, an A-4M for the US Marine Corps. Delivered to Marine Attack Squadron VMA-331, it was the 2,960th Skyhawk manufactured by McDonnell Douglas in 26 years of continuous production. The total was made up of 2,405 attack aircraft and 555 trainers, as follows:

XA4D-1	1	A4D-1 (A-4A)	146
YA4D-1	19	A4D-2 (A-4B)	542
A4D-2N (A-4C)	638	TA-4H	10
YA4D-5 (YA-4E)	2	TA-4J	291
A4D-5 (A-4E)	494	A-4K	10
A-4F	146	TA-4K	4
TA-4F	242	A-4KU	30
A-4G	8	TA-4KU	6
TA-4G	2	A-4M	162
A-4H	90	A-4N	117

In mid-1980 the Indonesian Air Force had begun to receive the first of 14 A-4Es and two TA-4Hs purchased from Israel. In mid-1981 the Royal Malaysian Air Force was due to receive the first 22 of a total of 88 surplus US Navy A-4Cs and A-4Ls. Under the designations A-4P and A-4Q, converted A-4Bs have been supplied respectively to the Argentine Air Force and Navy. Other A-4B conversions include 40 A-4S and seven two-seat TA-4S, this work being carried out by Lockheed Aircraft Service Company for the Republic of Singapore Air Force. Most TA-4Fs were converted subsequently to TA-4Js.

Details of Skyhawk variants can be found in earlier editions of *Jane's*.

MCDONNELL DOUGLAS DC-XX and MDF-100

To compete against the Boeing 757, McDonnell Douglas began studies for an advanced-technology medium-range (ATMR) transport aircraft. This was allocated the provisional designation DC-XX, and in 1980 was envisaged as a twin-aisle six-abreast 180-seat mixed-class airliner, as detailed in the 1980-81 *Jane's*.

In early 1981 the company was carrying out further studies, these involving a medium category aircraft of 150-160 seat capacity, with single and/or dual aisles, to be powered by two advanced turbofans in the 120-146·8 kN (27,000-33,000 lb st) range.

Since then, however, it has been announced that McDonnell Douglas is to team with Fokker BV of the Netherlands in developing a new joint project known as the MDF-100. Details of this programme are given in the International section.

MCDONNELL DOUGLAS C-17

The US Air Force announced on 29 August 1981 that it had selected McDonnell Douglas as prime contractor to develop its new C-17 long-range cargo aircraft. Further details can be found in the Addenda.

McKINNON—*See Canadian section*

MID-CONTINENT

MID-CONTINENT AIRCRAFT CORPORATION

Drawer L, Hayti, Missouri 63851
Telephone: (314) 359 0500
Telex: 447183
PRESIDENT: Richard Reade

MID-CONTINENT KING CAT

Mid-Continent Aircraft Corporation, an operator and distributor of Schweizer (Grumman) Ag-Cats, has obtained an FAA Supplementary Type Certificate for an Ag-Cat re-engined with an 895 kW (1,200 hp) Wright R-1820-202A radial engine, driving a three-blade metal propeller. The engineering work for this STC was carried out by Serv-Aero Engineering Inc (which see). After modification, the aircraft is known as a King Cat.

Based on the airframe of the Super Ag-Cat C, which has a hopper accommodating 1,893 litres (500 US gallons) of spray or 1,814 kg (4,000 lb) of dry chemicals, the King Cat offers improved high-altitude/high-temperature performance. It is available as a conversion of a new Super Ag-Cat C, or a kit is obtainable for installation by the customer or his maintenance organisation. Options which are available include Serv-O ailerons and increased fuel capacity of 431·5 litres (114 US gallons).

PERFORMANCE (at max T-O weight):

Working speed 87-113 knots (161-209 km/h; 100-130 mph)

Mid-Continent King Cat, a Wright-engined conversion of a Super Ag-Cat C *(Howard Levy)*

Ferry speed 117 knots (217 km/h; 135 mph)
T-O run, 32°C less than 365 m (1,200 ft)

MIT

MASSACHUSETTS INSTITUTE OF TECHNOLOGY

Department of Aeronautics and Astronautics, Chrysalis Human-Powered Flight Team, 77 Massachusetts Avenue, Cambridge, Massachusetts 02139
Telephone: (617) 253 4924
CHIEF DESIGNERS: Bob Parks and Harold Youngren
PROJECT MANAGER: John Langford
PROJECT ADVISER: E. Eugene Larrabee

MIT CHRYSALIS

A team of graduates and undergraduates of the MIT, with Professor E. Eugene Larrabee as Project Adviser, began in late 1978 the design and construction of a man-powered aircraft. Named Chrysalis, this aircraft achieved a successful first flight on 5 June 1979. When Chrysalis was dismantled in September 1979, and subsequently broken up, 44 pilots had made 345 flights in it during its brief life, including several successful turns.

TYPE: Single-seat man-powered aircraft.

WINGS: Braced biplane structure. Lissaman 7769 wing section. Dihedral 6° on outer panels of lower wing only. Each wing of two-spar construction. Spars of aluminium tube; ribs of Styrofoam and balsa; leading-edge of Styrofoam sheet; wingtip formed from laminated balsa strip; trailing-edge of wire; internal cross-bracing of Kevlar; and the entire structure covered by Mylar film. Interplane struts of aluminium tubing with Styrofoam fairing. No control surfaces, wings being warped for roll control.

FUSELAGE: Basic structure of aluminium tubing, lashed together with glassfibre and joints then saturated with epoxy; leading-edge and undersurface of pilot's enclosure formed of Styrofoam sheet, with inset transparent window to provide a forward view. Enclosure shaped by balsa ribs, and Mylar film covered. Aluminium tube tailbooms to carry tail unit.

TAIL UNIT: All-moving tailplane of similar construction to that of the wings. All-moving rudder of similar construction, except trailing-edge formed of carbonfibre and balsa.

LANDING GEAR: Fully-castoring and shock-absorbing monowheel mounted to fuselage basic structure, directly below pilot's seat.

MOTIVE POWER: Man-power on bicycle pedals, transmitted via nylon chain to two-blade tractor propeller mounted on aluminium tube shaft. Each blade constructed of foam material, hot-wired to shape, and supported by an aluminium root spar, carbonfibre surface spars, and a covering of Kevlar/epoxy.

ACCOMMODATION: Pilot only, on lightweight seat of aluminium tube with laced-on fabric cover. Seat position established by experimentation to allow the pilot to develop optimum power to the propeller via bicycle pedals and chain transmission. Lever for wing warping, and control stick for operation of tail control surfaces.

DIMENSIONS, EXTERNAL:

Wing span	21·95 m (72 ft 0 in)
Wing chord at root	1·83 m (6 ft 0 in)
Wing chord at tip	1·09 m (3 ft 7 in)
Length overall	11·89 m (39 ft 0 in)
Tailplane span	5·79 m (19 ft 0 in)
Propeller diameter	4·27 m (14 ft 0 in)
AREAS:	
Wings, gross	69·50 m² (748 sq ft)
Tailplane	5·02 m² (54 sq ft)
Rudder	4·09 m² (44 sq ft)
WEIGHTS AND LOADINGS:	
Weight empty	42 kg (93 lb)
Design gross weight	108 kg (238 lb)
*Gross weights at which flown	96·6-133 kg (213-293 lb)
Design wing loading	1·51 kg/m² (0·31 lb/sq ft)
Actual wing loading	1·37-1·90 kg/m² (0·28-0·39 lb/sq ft)

**Depending on pilot weight*

Chrysalis man-powered aircraft, designed and built at Massachusetts Institute of Technology *(Steve Finberg)*

MITSUBISHI

MITSUBISHI AIRCRAFT INTERNATIONAL INC (Subsidiary of Mitsubishi Heavy Industries Ltd)

HEAD OFFICE: One Lincoln Centre, 5400 LBJ Freeway, Suite 1500, Dallas, Texas 75240
Telephone: (214) 387 5600
Telex: 73 2575
PRESIDENT: Takaaki Yamada
VICE-PRESIDENT, INTERNATIONAL SALES: Bill Cole
VICE-PRESIDENT, MARKETING: George H. Scragg Jr
VICE-PRESIDENT, OPERATIONS: George M. Zieres Jr
MANAGER, PUBLIC RELATIONS: Patrick C. DeBlanc
WORKS: PO Box 3848, San Angelo, Texas 76901
Telephone: (915) 942 1511
Telex: 73 9438

This wholly-owned subsidiary of Mitsubishi Heavy Industries Ltd of Japan (which see) was established at San Angelo, Texas, in 1965 to undertake final assembly and flight testing of the MU-2 twin-turboprop STOL multi-purpose transport aircraft designed by the parent company. The worldwide headquarters of the company was moved to Dallas in 1977.

Mitsubishi Aircraft International is now responsible also for the new Diamond I business jet designed by the parent company.

MITSUBISHI MU-2

Details of early history and production of the MU-2 appear in the Japanese section of the 1980-81 and earlier editions of *Jane's*. By 1 January 1981, total orders for all versions had reached 676, including 49 for Japanese operators and 627 for customers outside Japan. The two current versions are:

Marquise. Basically similar to MU-2N (1978-79 *Jane's*), but with Garrett TPE331-10-501M turboprop engines, four-blade propellers and increased fuel capacity. Manufacturer's designation MU-2B-60. First flown 13 September 1977. Certificated 2 March 1978. Total of 74 delivered by 1 January 1981.

Solitaire. Basically similar to MU-2P (1978-79 *Jane's*), but with Garrett TPE331-10-501M turboprop engines, rated in this installation at 495·5 kW (665 shp) each. Propeller and fuel capacity details as for Marquise. Manufacturer's designation MU-2B-40. First flown 28 October 1977. Certificated 2 March 1978. Total of 36 delivered by 1 January 1981.

Wing, fuselage and tail unit components for the Marquise and Solitaire are manufactured in Japan by the parent company and shipped to the USA. At the San Angelo factory, they go on to a production line where American-built components (which include engines, propellers, tyres, brakes, avionics and interior furnishing) are installed. Examples of the MU-2 produced at San Angelo are in service in Africa, Australia, Canada, Europe, Mexico, the Middle East, South America and the USA.

TYPE: Twin-turboprop business aircraft.

WINGS: Cantilever high-wing monoplane. Wing section NACA 64A415 at root, NACA 63A212 (modified) at tip. No dihedral. Incidence 2°. Washout 3°. Sweepback 0° 21′ at quarter-chord. One-piece two-spar all-metal structure with chemically milled skins of 2024 and 7075 aluminium alloy. Spoilers for lateral control, between rear spar and flaps. Electrically-actuated full-span double-slotted Fowler-type flaps of aluminium alloy and plastics construction. Outboard flap section each side incorporates trim aileron. All primary controls manually-operated. Pneumatic de-icing rubber boots.

FUSELAGE: Circular-section aluminium alloy semi-monocoque structure.

TAIL UNIT: Cantilever structure of aluminium alloy, except for top of fin, which is of reinforced plastics. Small auxiliary fin beneath each side of rear fuselage on Marquise. Trim tab in rudder and each elevator. Pneumatic de-icing rubber boots.

LANDING GEAR (Marquise): Retractable tricycle type, with single wheel on each main unit and twin-wheel steerable nose unit. All wheels retract electrically, nosewheel forward, main wheels upward into fairings on fuselage sides. Manual backup system provided. Oleo-pneumatic shock-absorbers. Main-wheel tyres Type III, size 8·50-10 (10-ply). Nosewheel tyres Type III, size 5·00-5 (6-ply). Tyre pressure 2·76-4·34 bars (40-63 lb/sq in) on main units, 3·79 bars (55 lb/sq in) on nose unit. Goodrich double-disc hydraulic brakes. No anti-skid units or brake cooling.

LANDING GEAR (Solitaire): Retractable tricycle type, with single wheel on each main unit and twin-wheel steerable nose unit. All wheels retract rearward electrically. Manual backup system provided. Oleo-pneumatic shock-absorbers. Main-wheel tyres Type III 8·50-10 (10-ply). Nosewheel tyres Type III 5·00-5 (6-ply). Main-wheel tyre pressure 2·76-4·34 bars (40-63 lb/sq in), nosewheel tyre pressure 3·79 bars (55 lb/sq in). Goodrich double-disc hydraulic brakes. No anti-skid units or brake cooling.

POWER PLANT (Marquise): Two Garrett TPE331-10-501M turboprop engines, each flat rated to 533 kW (715 shp) and driving a Hartzell HC-B4TN-5DL/LT10282B-5·3R four-blade fully-feathering constant-speed reversible-pitch propeller. Fuel in five integral wing tanks, total usable capacity 844 litres (223 US gallons; 185·7 Imp gallons) and two fixed wingtip tanks with total usable capacity of 682 litres (180 US gallons; 150 Imp gallons). Max total usable fuel capacity of 1,526 litres (403 US gallons; 335·7 Imp gallons). Oil capacity 11·8 litres (3·1 US gallons; 2·6 Imp gallons).

POWER PLANT (Solitaire): Two Garrett TPE331-10-501M turboprop engines, each flat rated to 495·5 kW (665 shp). Propeller, fuel and oil details as for Marquise.

ACCOMMODATION (Marquise): Seats for pilot and co-pilot or passenger on flight deck. Seating in pressurised main cabin for seven to nine persons. Separate compartment at rear of cabin provides coat locker, toilet and baggage compartment. Door at rear of cabin on port side with built-in steps. Emergency exit under wing on starboard side.

ACCOMMODATION (Solitaire): Seats for pilot and co-pilot or passenger on flight deck. Standard seating for six or seven passengers in pressurised main cabin. Pressurised baggage compartment over main-wheel bays, capacity 100 kg (220 lb). Baggage compartments aft of main-wheel bays, capacity 70 kg (154 lb); port compartment pressurised, starboard compartment non-pressurised. Space for coats and small baggage at rear of cabin. Door under wing on port side. Emergency exit opposite main door.

Mitsubishi Solitaire business transport (two Garrett TPE331 turboprop engines)

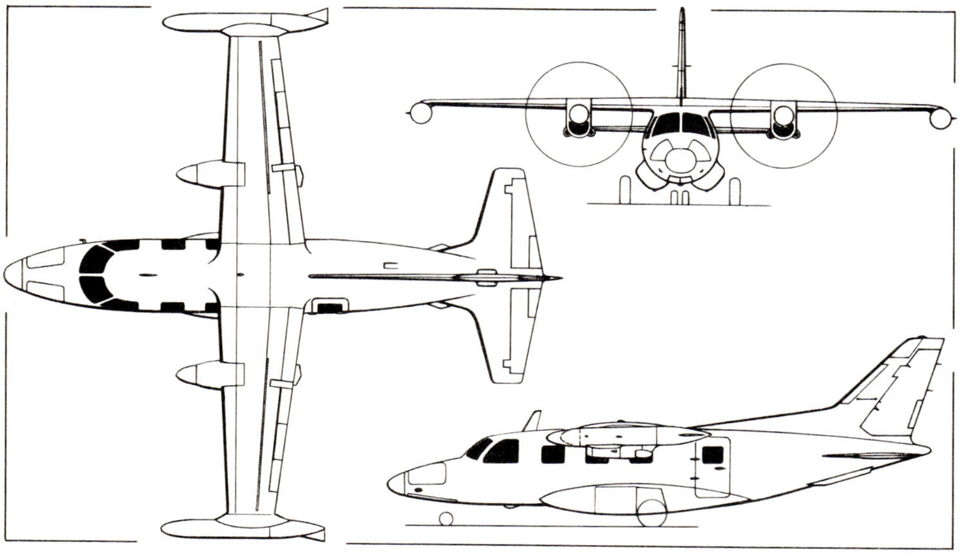

Mitsubishi Marquise twin-turboprop multi-purpose transport *(Pilot Press)*

Systems: Garrett pressurisation and Hamilton Standard air-conditioning systems, using engine bleed air. Differential 0·41 bars (6·0 lb/sq in). Hydraulic braking system. 28V DC primary electrical system, supplemented by 115V AC system (two static inverters) for instruments and avionics. DC power supplied by two 30V 200A starter/generators and two 24V 40Ah nickel-cadmium batteries. Scott automatic flow pressure-type oxygen system, with outlet for each seat. Rubber-boot pneumatic de-icing of wing and tail leading-edges; hot-air anti-icing of engine air intakes; heated windscreens; electrically heated propellers and oil cooler inlets.

Avionics and Equipment: Blind-flying instrumentation standard. Radio and radar to customers' requirements. Standard avionics include VOR/LOC, glideslope, ADF and marker beacon receivers; ATC transponder; DME; VHF or other communications systems; Sperry SPZ-500 integrated autopilot/flight director system; VG-14A vertical gyro; dual C-14 compass systems; and weather radar.

Dimensions, external:
Wing span over tip-tanks 11·94 m (39 ft 2 in)
Wing chord (mean) 1·54 m (5 ft 1 in)
Wing aspect ratio 7·71
Length overall: Marquise 12·02 m (39 ft 5 in)
Solitaire 10·13 m (33 ft 3 in)
Length of fuselage: Marquise 11·84 m (38 ft 10 in)
Solitaire 9·98 m (32 ft 9 in)
Fuselage: Max diameter 1·65 m (5 ft 5 in)
Height overall: Marquise 4·17 m (13 ft 8 in)
Solitaire 3·94 m (12 ft 11 in)
Tailplane span 4·80 m (15 ft 9 in)
Wheel track: Marquise 2·40 m (7 ft 11 in)
Solitaire 2·36 m (7 ft 9 in)
Wheelbase: Marquise 4·40 m (14 ft 5 in)
Solitaire 4·52 m (14 ft 10 in)
Propeller diameter 2·48 m (8 ft 2 in)
Distance between propeller centres 4·50 m (14 ft 9 in)
Propeller ground clearance:
Marquise 0·66 m (2 ft 2 in)
Solitaire 0·53 m (1 ft 9 in)
Cabin door: Height 1·17 m (3 ft 10 in)
Width 0·74 m (2 ft 5 in)
Height to sill: Marquise 0·38 m (1 ft 3 in)
Solitaire 0·51 m (1 ft 8 in)
Emergency exit: Height 0·71 m (2 ft 4 in)
Width 0·69 m (2 ft 3 in)

Dimensions, internal:
Cabin, incl flight deck and baggage compartment:
Length: Marquise 6·56 m (21 ft 6 in)
Solitaire 4·88 m (16 ft 0 in)
Max width 1·47 m (4 ft 10 in)
Max height 1·30 m (4 ft 3 in)
Volume: Marquise (usable) 9·83 m^3 (347 cu ft)
Solitaire (total) 6·94 m^3 (245 cu ft)
Baggage volume (total):
Marquise 1·95 m^3 (69·0 cu ft)
Solitaire 1·22 m^3 (43·0 cu ft)

Areas (both versions):
Wings, gross 16·55 m^2 (178 sq ft)
Flaps (total) 3·90 m^2 (42·00 sq ft)
Spoilers (total) 0·54 m^2 (5·82 sq ft)
Fin: Marquise 2·85 m^2 (30·70 sq ft)
Solitaire 2·77 m^2 (29·80 sq ft)
Rudder, incl tab 1·17 m^2 (12·60 sq ft)
Tailplane 4·01 m^2 (43·30 sq ft)
Elevators, incl tabs 1·40 m^2 (15·00 sq ft)

Weights and Loadings:
Weight empty, equipped:
Marquise 3,470 kg (7,650 lb)
Solitaire 3,180 kg (7,010 lb)

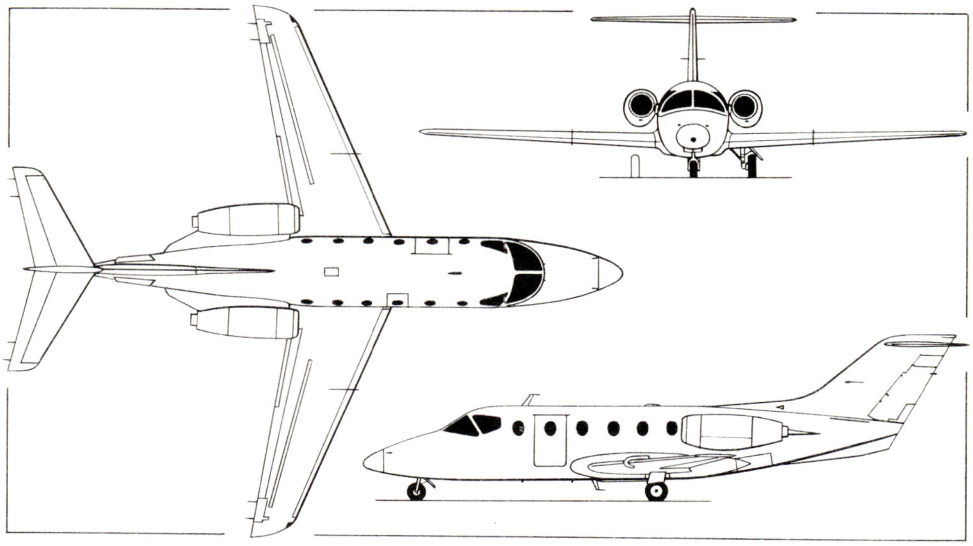

Mitsubishi Diamond I twin-turbofan business aircraft *(Pilot Press)*

Max payload: Marquise 1,043 kg (2,300 lb)
Solitaire 1,220 kg (2,690 lb)
Max ramp weight: Marquise 5,270 kg (11,625 lb)
Solitaire 4,770 kg (10,520 lb)
Max T-O weight: Marquise 5,250 kg (11,575 lb)
Solitaire 4,750 kg (10,470 lb)
Max landing weight: Marquise 5,000 kg (11,025 lb)
Solitaire 4,515 kg (9,955 lb)
Max zero-fuel weight: Marquise 4,513 kg (9,950 lb)
Solitaire 4,400 kg (9,700 lb)
Max wing loading:
Marquise 317·4 kg/m^2 (65·0 lb/sq ft)
Solitaire 287·9 kg/m^2 (59·0 lb/sq ft)
Max power loading:
Marquise 4·92 kg/kW (8·09 lb/shp)
Solitaire 4·79 kg/kW (7·87 lb/shp)

Performance (at max T-O weight, except where indicated):
Max operating speed (both, all weights)
Mach 0·57 (250 knots; 463 km/h; 288 mph) CAS
Max cruising speed:
Marquise at 4,880 m (16,000 ft)
308 knots (571 km/h; 355 mph)
Solitaire at 6,100 m (20,000 ft)
321 knots (595 km/h; 370 mph)
Econ cruising speed at 6,100 m (20,000 ft):
Marquise 295 knots (547 km/h; 340 mph)
Solitaire 313 knots (579 km/h; 360 mph)
Stalling speed, power off, flaps up:
Marquise, AUW of 4,626 kg (10,200 lb)
100 knots (185 km/h; 115 mph)
Solitaire, AUW of 4,173 kg (9,200 lb)
97 knots (180 km/h; 112 mph)
Stalling speed, power off, flaps down:
Marquise, AUW of 4,626 kg (10,200 lb)
76 knots (141 km/h; 87·5 mph)
Solitaire, AUW of 4,173 kg (9,200 lb)
73 knots (135 km/h; 84 mph)
Max rate of climb at S/L:
Marquise 670 m (2,200 ft)/min
Solitaire 716 m (2,350 ft)/min
Rate of climb at S/L, one engine out:
Marquise 125 m (410 ft)/min
Solitaire 145 m (475 ft)/min

Mitsubishi's seven/nine-passenger Marquise turboprop-powered business transport

Service ceiling: Marquise 9,070 m (29,750 ft)
Solitaire 10,210 m (33,500 ft)
Service ceiling, one engine out:
Marquise 4,510 m (14,800 ft)
Solitaire 5,150 m (16,900 ft)
Min ground turning radius:
Marquise (port) 25·30 m (83 ft 0 in)
Marquise (stbd) 22·20 m (72 ft 10 in)
Solitaire (port) 21·80 m (71 ft 6¼ in)
Solitaire (stbd) 25·90 m (84 ft 11¾ in)
T-O run: Marquise 564 m (1,850 ft)
Solitaire 472 m (1,550 ft)
T-O to 15 m (50 ft): Marquise 660 m (2,170 ft)
Solitaire 550 m (1,800 ft)
Landing from 15 m (50 ft):
Marquise 671 m (2,200 ft)
Solitaire 594 m (1,950 ft)
Landing run at max landing weight:
Marquise 402 m (1,320 ft)
Solitaire 357 m (1,170 ft)
Range with max fuel at 9,450 m (31,000 ft), 45 min reserves:
Marquise 1,395 nm (2,584 km; 1,606 miles)
Solitaire 1,600 nm (2,964 km; 1,842 miles)

MITSUBISHI DIAMOND I

Mitsubishi Heavy Industries Ltd had already completed a prototype of this twin-turbofan business aircraft (known in Japan as the MU-300) before releasing very brief details of it in the Summer of 1978. When this prototype (c/n 001) made its first flight on 29 August 1978, it was known only that it had two Pratt & Whitney Aircraft of Canada JT15D-4 turbofan engines, and that the major design objectives included low operating noise levels, a high standard of cabin comfort, fuel economy, and a high degree of reliability and maintainability. A second prototype (c/n 002) made its first flight on 13 December 1978. These two aircraft had logged more than 1,100 hours' flying by mid-1981.

Mitsubishi took the step of seeking US certification first, with domestic certification by the JCAB to follow later; the aircraft will be certificated in the Transport category of FAR Pt 25, for IFR operation in known icing conditions. The two prototypes, which by then had accumulated more than 350 hours' flying in Japan, were shipped in the Summer of 1979 to the USA, where they were reassembled by Mitsubishi Aircraft International (MAI) at San Angelo, Texas. First flight in the United States was made on 10 August 1979 by the second prototype, now registered N81DM. This was used as the systems and equipment test aircraft in the FAA certification programme, while the first prototype undertook general performance and handling trials. Certification was scheduled for September 1981.

The next four aircraft are Japanese-built production Diamond Is, dispatched to San Angelo for reassembly. The first production Diamond I made its initial flight on 21 May 1981. From the seventh aircraft onward, Mitsubishi (Japan) will deliver the aircraft in the form of major subassemblies to MAI, and the latter will be responsible for their assembly, installation of engines and avionics, furnishing, and flight test. Deliveries were expected to begin in January 1982.

By January 1981, MAI had reserved delivery positions for 114 Diamond Is.

The following description applies to the prototypes:

Type: Twin-turbofan business aircraft.

Wings: Cantilever low-wing monoplane. Mitsubishi computer-designed wing sections: thickness/chord ratio 13·2% at root, 11·3% at tip. Dihedral from roots. Incidence 3° at root. Washout 6° 30′. Sweepback 20° at quarter-chord. Wings are basically of aluminium alloy construction, and built in three portions: a centre-

Second prototype Mitsubishi Diamond I (two Pratt & Whitney Aircraft of Canada JT15D-4 turbofan engines)

section and two outer panels. Each wing has two primary box-beam spars, forming an integral fuel tank. Narrow-chord Fowler-type flaps over most of trailing-edges, actuated hydraulically and having double-slotted inboard and single-slotted outboard segments. Immediately forward of flaps are long-span narrow-chord spoilers, for roll control; these also serve as air-brakes, and can be used as lift dumpers to assist braking on touchdown. Outboard of each outer trailing-edge flap is a small, short-span aileron for roll trim. Leading-edges are anti-iced by hot air.

Fuselage: Pressurised, fail-safe fatigue-resistant semi-monocoque structure, of oval cross-section with flattened cabin floor. Construction is mainly of aluminium alloy, using multiple load paths, bonded doublers and small skin panels in the principal load-bearing members. Built in three main portions: forward (including flight deck), centre and rear.

Tail Unit: Cantilever T tail, with sweepback on all surfaces; construction generally similar to that of wings. Curved dorsal fin, plus small underfin. A small horizontal strake is fitted on each side of rear fuselage, abreast of main fin, to assist airflow control. Trim tab in base of rudder. Small yaw-damping control surface above rudder. Variable-incidence tailplane, with elevators.

Landing Gear: Sumitomo retractable tricycle type, with single wheel and oleo-pneumatic shock-absorber on each unit. Hydraulic actuation, controlled electrically. Emergency free-fall extension. Nosewheel, which is steerable by rudder pedals, retracts forward; main wheels retract inward into fuselage. Goodyear wheels, with Goodrich tyres, on all units. Goodyear brakes.

Power Plant: Two Pratt & Whitney Aircraft of Canada JT15D-4 turbofan engines, mounted in pod on each side of rear fuselage, each rated at 11·1 kN (2,500 lb st) for T-O and 10·4 kN (2,345 lb st) for max cruise. Aircraft is designed with capability to accept more powerful engines later, if required. Fuel in two 1,018 litre (224 Imp gallon; 269 US gallon) integral tanks in wings, plus a further 469 litres (103 Imp gallons; 124 US gallons) in aft tanks in fuselage. Total fuel capacity 2,505 litres (551 Imp gallons; 662 US gallons), of which 2,445 litres (538 Imp gallons; 646 US gallons) are usable. Fully-automatic fuel feed system. Refuelling points in upper surface of each wing and on starboard side of fuselage at rear.

Accommodation: Crew of two on flight deck. Standard layout seats seven passengers in pressurised cabin, with toilet and baggage compartment (capacity 295 kg; 650 lb) at rear. Crew/passenger door forward of wing on port side. Emergency exit forward of wing on starboard side.

Systems: Pressurisation system, with max differential of 0·62 bars (9·0 lb/sq in). Backup pressurisation system, using engine bleed air, for use in emergency. Hydraulic system, pressure 103·5 bars (1,500 lb/sq in), for actuation of flaps, landing gear and other services. All systems are, wherever possible, of modular conception: for example, entire hydraulic installation can be removed as a single unit. Stick shaker as backup stall warning device.

Avionics: Standard avionics include Sperry SPZ-900 integrated flight control system, with pilot's flight director, encoding altimeter, dual nav/com and audio systems, ADF, DME, ATC transponder, twin compasses and RMIs, and RCA colour weather radar.

Dimensions, external:
- Wing span: 13·23 m (43 ft 5 in)
- Wing aspect ratio: 7·54
- Length overall: 14·73 m (48 ft 4 in)
- Fuselage: Length: 13·16 m (43 ft 2 in)
  - Max width: 1·68 m (5 ft 6 in)
  - Max height: 1·85 m (6 ft 1 in)
- Height overall: 4·19 m (13 ft 9 in)
- Tailplane span: 5·00 m (16 ft 5 in)
- Wheel track: 2·84 m (9 ft 4 in)
- Wheelbase: 5·86 m (19 ft 3 in)
- Crew/passenger door: Height: 1·27 m (4 ft 2 in)
  - Width: 0·71 m (2 ft 4 in)

Dimensions, internal:
- Cabin: Max length, incl flight deck: 6·37 m (20 ft 11 in)
  - Length, excl flight deck: 4·77 m (15 ft 8 in)
  - Max width: 1·50 m (4 ft 11 in)
  - Max height: 1·47 m (4 ft 10 in)
  - Volume: incl flight deck: 11·33 m^3 (400 cu ft)
    - excl flight deck: 8·64 m^3 (305 cu ft)
- Baggage compartment volume: 1·4 m^3 (50 cu ft)

Areas:
- Wings, net: 22·43 m^2 (241·4 sq ft)
- Trailing-edge flaps (total): 2·62 m^2 (28·2 sq ft)

Weights and Loadings:
- Weight empty, equipped: 3,900 kg (8,600 lb)
- Basic operating weight empty: 4,082 kg (9,000 lb)
- Max fuel: 1,905 kg (4,200 lb)
- Max payload: 454 kg (1,000 lb)
- Max T-O weight: 6,395 kg (14,100 lb)
- Max landing weight: 5,987 kg (13,200 lb)
- Max zero-fuel weight: 4,900 kg (10,800 lb)
- Max wing loading: 285 kg/m^2 (58·4 lb/sq ft)
- Max power loading: 288 kg/kN (2·82 lb/lb st)

Performance (at max T-O weight except where indicated):
- Max operating Mach number (MMO) above 7,925 m (26,000 ft): Mach 0·78
- Max operating speed (VMO) between 4,875 m (16,000 ft) and 7,925 m (26,000 ft): 320 knots (593 km/h; 368 mph)
- Max level speed at 9,145 m (30,000 ft): 434 knots (805 km/h; 500 mph)
- Typical speed at 11,890 m (39,000 ft): 410 knots (760 km/h; 472 mph)
- Long-range speed at 11,890 m (39,000 ft): 375 knots (695 km/h; 432 mph)
- Stalling speed, flaps down, at AUW of 4,672 kg (10,300 lb): 77 knots (143 km/h; 89 mph) CAS
- Max rate of climb at S/L, ISA: 914 m (3,000 ft)/min
- Rate of climb at S/L, ISA, one engine out: 247 m (810 ft)/min
- Max operating altitude: 12,500 m (41,000 ft)
- FAA T-O field length, ISA, zero wind: 1,250 m (4,100 ft)
- FAA landing field length at max landing weight, ISA, zero wind: 823 m (2,700 ft)
- Range with four passengers at econ cruising speed, ISA, zero wind:
  - NBAA IFR reserves: 1,252 nm (2,320 km; 1,441 miles)
  - NBAA VFR reserves: 1,565 nm (2,897 km; 1,800 miles)

MOONEY
MOONEY AIRCRAFT CORPORATION
(Subsidiary of Republic Steel Corporation)

Head Office and Works: PO Box 72, Kerrville, Texas 78028

Telephone: (512) 896 6000

President: T. J. Smith

Vice-Presidents:
Don K. Cox (Marketing)
Vendal DiGiacinto (Finance)
Chris D. Hughes (Operations)
L. P. Lopresti (Engineering)

The original Mooney Aircraft Inc was formed in June 1948, in Wichita, Kansas, from where the single-seat Model M-18 Mooney Mite was produced until 1952. The company transferred to Kerrville, Texas, in 1953 and completed a merger with Alon Inc of McPherson, Kansas, in October 1967. Subsequently, in late 1969, Butler Aviation International acquired 100 per cent stock ownership of Mooney Aircraft, the company name being changed to Aerostar Aircraft Corporation on 1 July 1970. Production of Aerostar aircraft was suspended in early 1972. On 4 October 1973 came news that the Republic Steel Corporation of Cleveland, Ohio, had assumed control of the company, once again named Mooney Aircraft.

MOONEY 201 (M20J)

The Mooney 201, a faster development of the now discontinued Executive (last described in the 1977-78 *Jane's*), first flew in June 1976 and received FAA certification in September 1976. A total of 1,077 had been produced by 1 February 1981.

Type: Four-seat cabin monoplane.

Wings: Cantilever low-wing monoplane. Wing section NACA 63_2-215 at root, NACA 64_1-412 at tip. Dihedral 5° 30′. Incidence 2° 30′ at root, 1° at tip. Sweepforward 2° 29′. Light alloy structure with flush-riveted stretch-formed wraparound skins. Full-span main spar; rear spar terminates at mid-span of flaps. Sealed-gap differentially-operated light alloy ailerons. Electrically-operated single-slotted light alloy flaps over 70% of trailing-edge. No tabs.

Fuselage: Composite all-metal structure. Cabin section is of welded 4130 chrome-molybdenum steel tube with sheet light alloy covering. Rear section is of semi-monocoque construction, with sheet light alloy bulkheads and skin and extruded alloy stringers.

Tail Unit: Cantilever light alloy structure, with variable-incidence tailplane. All surfaces covered with wraparound metal skin.

Landing Gear: Electrically-retractable levered-suspension tricycle type. Nosewheel retracts rearward, main units inward into wings. Rubber disc shock-absorbers in main units. Delco hydraulic shock-absorber in nose unit. Cleveland main wheels, size 6·00-6, and steerable nosewheel, size 5·00-5. Tyre pressure: main wheels 2·07 bars (30 lb/sq in), nosewheel 3·38 bars (49 lb/sq in). Cleveland hydraulic single-disc brakes on main wheels. Parking brakes.

Power Plant: One 149 kW (200 hp) Avco Lycoming IO-360-A3B6D flat-four engine, driving a McCauley two-blade constant-speed metal propeller. Two integral fuel tanks in wings, with combined usable capacity of 242 litres (64 US gallons). Refuelling points in wing upper surface. Oil capacity 7·5 litres (2 US gallons).

Accommodation: Cabin accommodates four in two individual front seats and one side-by-side rear bench seat; front seats have reclining backs. Dual controls standard. Standard rudder pedals optionally removable to allow more legroom for passenger. Overhead ventilation system. Cabin heating and cooling system, with adjustable outlets and illuminated control. One-piece wraparound windscreen. Tinted Plexiglas windows. Starboard front seat removable for freight stowage. Single door on starboard side. Compartment for 54 kg (120 lb) baggage behind cabin, with access from cabin or through door on starboard side. Windscreen defrosting system standard.

Systems: Hydraulic system for brakes only. Electrical system includes 60A alternator, 12V 35Ah battery, voltage regulator and warning light, together with protective circuit breakers. Oxygen system optional.

Avionics and Equipment: An extensive range of optional avionics by Collins, Edo-Aire Mitchell and King, and RCA WeatherScout weather radar, are available to customer's requirements. Standard equipment includes sensitive altimeter, instrument panel glareshield, sun visors, shoulder harness, annunciator panel with low fuel warning, cabin lights, landing light, navigation lights, taxi light, full-flow oil filter, quick oil drain, zinc chromate anti-corrosion treatment, and streamlined propeller spinner. An optional operational package includes blind-flying instrumentation, electric clock, exhaust gas temperature gauge, outside air temperature gauge, turn co-ordinator, microphone jacks, radio master switch, emergency locator transmitter, alternate static source, heated pitot, three-light strobe system, com and nav antennae, and cabin speaker. Other optional equipment includes encoding altimeter, true airspeed indicator, hour meter, de luxe control wheel with map light and microphone button, dual brakes, headrests, all-leather interior furnishing, fuel management system, AM/FM cassette stereo system, cabin fire extinguisher, rotating beacon and external power socket.

DIMENSIONS, EXTERNAL:

Wing span	11·0 m (36 ft 1 in)
Wing chord, mean	1·50 m (4 ft 11¼ in)
Length overall	7·52 m (24 ft 8 in)
Height overall	2·54 m (8 ft 4 in)
Wheel track	2·76 m (9 ft 0¾ in)
Wheelbase	1·82 m (5 ft 11½ in)
Propeller diameter	1·88 m (6 ft 2 in)
Propeller ground clearance	0·24 m (9½ in)

DIMENSIONS, INTERNAL:

Cabin: Length	2·90 m (9 ft 6 in)
Max width	1·10 m (3 ft 7½ in)
Max height	1·13 m (3 ft 8½ in)
Baggage compartment	0·48 m^3 (17 cu ft)

WEIGHTS AND LOADINGS:

Weight empty	743 kg (1,640 lb)
Max T-O and landing weight	1,243 kg (2,740 lb)
Max wing loading	80·07 kg/m^2 (16·4 lb/sq ft)
Max power loading	8·34 kg/kW (13·7 lb/hp)

PERFORMANCE (at max T-O weight):

Never-exceed speed	196 knots (364 km/h; 226 mph)
Max level speed at S/L	175 knots (325 km/h; 202 mph)
Max cruising speed, 75% power at 2,470 m (8,100 ft)	169 knots (314 km/h; 195 mph)
Econ cruising speed, 55% power at 2,470 m (8,100 ft)	145 knots (269 km/h; 167 mph)
Stalling speed, flaps up	63 knots (117 km/h; 72·5 mph) IAS
Stalling speed, flaps down	55 knots (102 km/h; 63·5 mph) IAS
Max rate of climb at S/L	314 m (1,030 ft)/min
Service ceiling	5,730 m (18,800 ft)
T-O to 15 m (50 ft)	463 m (1,517 ft)
Landing from 15 m (50 ft)	491 m (1,610 ft)
Landing run	235 m (770 ft)
Range, 75% power, no reserves	974 nm (1,804 km; 1,121 miles)

TURBO MOONEY 231 (M20K)

Mooney began the design of this turbocharged version of the 201 in May 1976. Construction of the prototype began in the following month, and the first flight was recorded in October 1976. Certification in the Normal category was awarded by the FAA on 16 November 1978.

Generally similar to the 201, the Turbo Mooney 231 differs mainly in having a turbocharged engine. Production totalled 523 aircraft by 1 February 1981.

TYPE: Four-seat cabin monoplane.

WINGS: Cantilever low-wing monoplane. Wing section NACA 63_2-215 at root, NACA 64_1-412 at tip. Dihedral 5° 30′. Incidence 2° 30′. No sweepback. Conventional structure of light alloy. Plain ailerons and trailing-edge flaps of light alloy construction. No tabs.

FUSELAGE AND TAIL UNIT: As for 201.

LANDING GEAR: Electrically-retractable tricycle type. Nosewheel retracts rearward, main units inward into wings. All wheels faired by doors when retracted. Shock-absorption of nosewheel and main-wheel units by Lord rubber discs. Cleveland wheels, with main-wheel tyres size 6·00-6 6-ply, pressure 2·90 bars (42 lb/sq in). Nosewheel tyre size 5·00-5 6-ply, pressure 3·38 bars (49 lb/sq in). Cleveland hydraulic brakes on main wheels. Parking brake.

POWER PLANT: One 156·5 kW (210 hp) Continental TSIO-360-GB flat-six turbocharged engine, driving a McCauley constant-speed metal propeller. Goodrich propeller de-icing optional. Two integral fuel tanks in inner wings, with combined capacity of 297·5 litres (78·6 US gallons), of which 286 litres (75·6 US gallons) are usable. Refuelling points in upper surface of the inboard section of each wing. Oil capacity 7·5 litres (2 US gallons).

Mooney 201 (foreground) and turbocharged 231 four-seat cabin monoplanes

ACCOMMODATION: Cabin accommodates four on two individual front seats and a rear bench seat. Cabin door on starboard side. Baggage space aft of rear seat, accessible from cabin and via baggage door on starboard side. Accommodation heated and ventilated.

SYSTEMS: Hydraulic system for brakes only. Electrical system powered by a 64A engine-driven alternator. Oxygen system.

AVIONICS: Nav/coms by Collins and King, autopilot, flight director, R/Nav and RCA WeatherScout weather radar optional.

DIMENSIONS, EXTERNAL AND INTERNAL:

As Mooney 201 except:	
Wing aspect ratio	7·448
Length overall	7·75 m (25 ft 5 in)
Height overall	2·51 m (8 ft 3 in)
Tailplane span	3·58 m (11 ft 9 in)
Cabin door (stbd, over wing):	
Height	1·13 m (3 ft 8½ in)
Width	0·74 m (2 ft 5 in)
Baggage door (stbd, aft):	
Height	0·52 m (1 ft 8½ in)
Width	0·43 m (1 ft 5 in)
Height to sill	1·17 m (3 ft 10 in)

AREAS:

Wings, gross	16·24 m^2 (174·8 sq ft)
Ailerons (total)	1·06 m^2 (11·4 sq ft)
Trailing-edge flaps (total)	1·66 m^2 (17·9 sq ft)
Fin	0·72 m^2 (7·7 sq ft)
Tailplane	1·99 m^2 (21·42 sq ft)
Elevators	1·21 m^2 (13·0 sq ft)

WEIGHTS AND LOADINGS:

Weight empty	816 kg (1,800 lb)
Max T-O and landing weight	1,315 kg (2,900 lb)
Max zero-fuel weight	1,119 kg (2,468 lb)
Max wing loading	80·97 kg/m^2 (16·6 lb/sq ft)
Max power loading	8·4 kg/kW (13·8 lb/hp)

PERFORMANCE (at max T-O weight):

Max level speed	201 knots (373 km/h; 231 mph)
Max cruising speed, 75% power at 5,485 m (18,000 ft)	182 knots (338 km/h; 210 mph)
Econ cruising speed, 55% power at 7,315 m (24,000 ft)	162 knots (301 km/h; 187 mph)
Stalling speed, flaps down, idling power	57 knots (106 km/h; 66 mph) CAS
Max rate of climb at S/L	329 m (1,080 ft)/min
Certificated ceiling	7,315 m (24,000 ft)
T-O to 15 m (50 ft)	628 m (2,060 ft)
Landing from 15 m (50 ft)	695 m (2,280 ft)
Landing run	350 m (1,147 ft)
Range with max fuel, 55% power, no reserves	1,182 nm (2,190 km; 1,361 miles)

MOONEY M-30

Mooney Aircraft announced at the 1980 US Reading Air Show that the company had designed and planned to build a new six-seat pressurised aircraft designated M-30. Of cantilever low-wing configuration, it features a wing of advanced aerodynamic section, with two top-surface spoilers on each wing to supplement roll control, and Fowler-type trailing-edge flaps over 90% of the wing span. Entry to the cabin will be by means of a door with integral airstairs. Power plant will consist of one 261 kW (350 hp) Avco Lycoming turbocharged engine in the TIO-540 series; fuel capacity will be 380 litres (100 US gallons). Systems and equipment will include pressurisation (differential 0·32 bars; 4·7 lb/sq in), providing a 3,050 m (10,000 ft) cabin altitude to a height of 7,620 m (25,000 ft), anti-icing equipment, weather radar, and advanced avionics for communications and navigation. First flight of a prototype M-30 is planned for 1982, with certificated production aircraft to be available in 1985.

PERFORMANCE (estimated):

Max level speed	243-259 knots (450-480 km/h; 280-298 mph)
Cruising speed	210-235 knots (390-435 km/h; 242-270 mph)
Stalling speed	52-56·5 knots (96-104 km/h; 60-65 mph)
Range at 75% power	990 nm (1,840 km; 1,140 miles)
Range at 65% power	1,120 nm (2,080 km; 1,290 miles)

NASA

NATIONAL AERONAUTICS AND SPACE ADMINISTRATION

HEADQUARTERS: 400 Maryland Avenue SW, Washington, DC 20546

Telephone: (202) 755 2320

NASA has several research programmes of general aviation interest. Details of some of these are given in the following entry.

In March 1981, NASA announced that it was more than doubling the size of its '40 by 80' (implying a 12·2 by 24·4 m: 40 by 80 ft test section) wind tunnel at Ames Research Center. Prior to this reconstruction the tunnel was already regarded as the world's largest; the conversion will provide two tunnels: the existing '40 × 80', plus an '80 × 120' that will permit full-scale testing of aircraft with wing spans of up to 30·5 m (100 ft). Max airflow speed in the '40 × 80' has also been increased.

NASA/BOEING QSRA

Under a $21 million contract, Boeing carried out for NASA the conversion of a DHC C-8A Buffalo into a Quiet Short-haul Research Aircraft (QSRA). Details of the subsequent research programme, now successfully completed, and of the QSRA aircraft can be found in the 1980-81 and three previous editions of *Jane's*.

NASA AD-1 OBLIQUE-WING AIRCRAFT

On 20 February 1978, NASA announced the award of a $218,000 fixed-price contract to Ames Industrial Corporation of Bohemia, New York, for the construction of a small research aircraft to explore in piloted flight the viability of the pivoting-wing, or oblique-wing, concept that had emanated from NASA's Dryden Flight Research Center at Edwards, California.

Detail design of the aircraft, designated AD-1 (Ames/Dryden-1), was performed for NASA by Mr Burt Rutan, who was responsible for the well-known Vari-Viggen and VariEze designs for homebuilders. For operation at low speeds, and during take-off and landing, its wing is intended to be positioned conventionally. For flight at higher speeds, the wing is pivoted to form an oblique angle of up to 60° with the fore-and-aft centreline of the fuselage, reducing drag and promising increased speed and range for no increase in fuel consumption.

The AD-1 (N805NA) was delivered to Dryden Center in March 1979, and first flown on 21 December 1979. After eight flights with the wing in conventional configuration, during which the flight envelope was extended to 3,660 m (12,000 ft) altitude and maximum speed of 173 knots (322 km/h; 200 mph), the initial flight with the wing skewed was made in the first week of February 1980. The pivoting operation took place at an altitude of 3,810 m (12,500 ft), the wing being canted in stages of 5° to a maximum of 15°, with the starboard wing 15° forward and the port wing 15° aft of standard configuration. At each 5° stage of sweep, flutter control inputs were induced over a speed range of 110-150 knots (204-278 km/h; 127-173 mph) to determine flight characteristics. These procedures were repeated at a height of 2,285 m (7,500 ft) before the aircraft landed. Testing continued at this modest 15° angle through February and March 1980. NASA's programme called for the slew to be increased by increments of 15° up to the maximum of 60°, with at least two flights being made at each setting. No significant change in flight and control characteristics was anticipated until the wing setting exceeded 45°, the angle first adopted during the 14th flight of the AD-1 at the end of May 1980. The full 60° sweep angle was first flown successfully on 24 April 1981.

TYPE: Oblique-wing research aircraft.

WINGS: Pivoting cantilever high-wing monoplane, with wing constructed in one piece. Thickness/chord ratio 12%. No dihedral. Incidence 2° at root, 0° at tip. No sweepback. Composite structure, based on a foam core with glassfibre/epoxy laminate covering. Conventional manually-operated ailerons of similar construction. No trailing-edge flaps. Electrically-operated trim tab in starboard aileron.

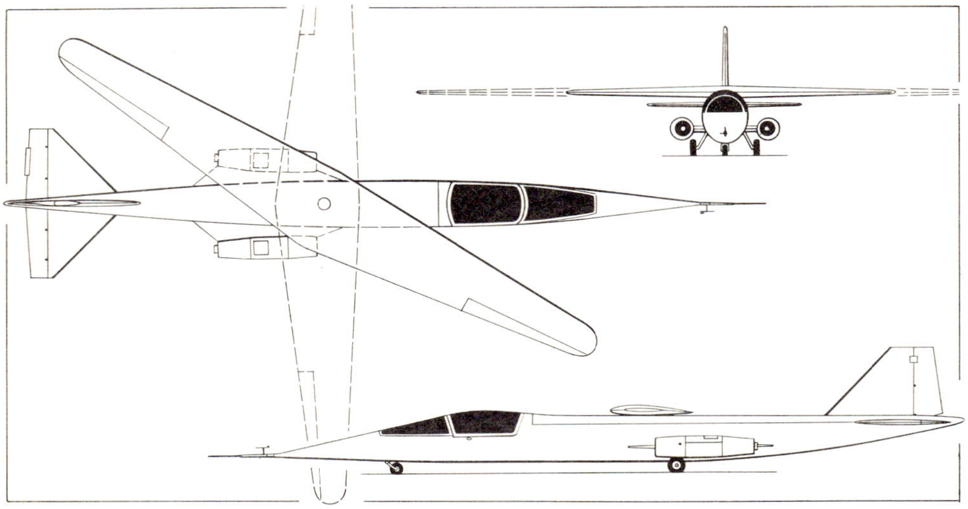

NASA/Ames AD-1 (two Ames/Microturbo TRS 18-046 turbojet engines) *(Pilot Press)*

During 1981, NASA Dryden test pilots have flown the AD-1 with the wing pivoted to its maximum of 60°

Fuselage: Composite semi-monocoque structure, utilising foam core with glassfibre/epoxy laminate covering for bulkheads and skin.

Tail Unit: Conventional tail unit, with cantilever fixed-incidence tailplane. Fixed surfaces and control surfaces constructed on a foam core with glassfibre/epoxy laminate covering. Electrically-operated trim tab in port elevator. All controls manually operated.

Landing Gear: Non-retractable tricycle type, with glassfibre/epoxy laminate cantilever struts manufactured by Jiran Glider Repairs. Single wheel on each unit. Cleveland 127 mm (5 in) main wheels and tyres, pressure 6·56 bars (95 lb/sq in). Nosewheel tyre pressure 3·11 bars (45 lb/sq in). Cleveland brakes.

Power Plant: Two 0·98 kN (220 lb st) Ames Industrial (Microturbo) TRS 18-046 turbojet engines, mounted on short mid-set stub wings on each side of fuselage, with their intakes just aft of the leading-edge of the pivoting wing. Two fuel tanks in fuselage; total capacity approx 303 litres (80 US gallons). Refuelling points on fuselage upper surface, one forward and one aft of the wing. Oil capacity 1·44 litres (0·38 US gallons).

Accommodation: Pilot only, beneath transparent cockpit canopy hinged on starboard side. Accommodation ventilated.

Systems: Electrical system powered by two 600W engine-driven DC generators; 28V SAFT nickel-cadmium battery. Scott Executive Mk II oxygen system.

Dimensions, external:	
Wing span	9·75 m (32 ft 0 in)
Wing chord at root	1·31 m (4 ft 3·4 in)
Wing chord at tip	0·47 m (1 ft 6·4 in)
Length overall, incl nose probe	11·68 m (38 ft 4 in)
Width, wing skewed	4·93 m (16 ft 2 in)
Height overall	1·98 m (6 ft 6 in)
Tailplane span	2·44 m (8 ft 0 in)
Wheel track	0·99 m (3 ft 3 in)
Wheelbase	4·27 m (14 ft 0 in)
Dimensions, internal:	
Cockpit: Length	1·52 m (5 ft 0 in)
Max width	0·57 m (1 ft 10·3 in)
Max height	0·84 m (2 ft 9 in)
Areas:	
Wings, gross	8·64 m² (93 sq ft)
Vertical tail surfaces (total)	1·29 m² (13·9 sq ft)
Horizontal tail surfaces (total)	2·46 m² (26·5 sq ft)
Weight:	
Max T-O weight	907 kg (2,000 lb)
Performance (demonstrated in early tests):	
Max level speed	173 knots (322 km/h; 200 mph)
Service ceiling	3,660 m (12,000 ft)

NASA TERMINAL CONFIGURED VEHICLE

The Terminal Configured Vehicle (TCV) programme was established to conduct research, and to develop and evaluate aircraft and flight management technology, that will benefit conventional take-off and landing operations in the terminal area.

The objectives and programme elements are:

1. To improve terminal area capacity and efficiency by:
   Systems and procedures for ATC evolution.
   Systems and procedures for runway capacity.
   Profiles and procedures for fuel conservation.
2. To improve approach and landing capability in adverse weather using:
   Human factor elements for effective flight management.
   Systems and information to minimise wind-shear hazard.
   Airborne sensors for weather penetration.
3. To reduce noise impact through:
   Development of profiles and configuration for noise reduction.

The research programme involves analyses, simulation and flight studies with a modified Boeing 737 aircraft, provided with highly flexible display and control equipment, and an aft flight deck for research purposes. A series of development and demonstration activities is being conducted to evolve practical systems and to solicit and encourage acceptance by flight crews and the airlines.

A Boeing 737-100 was obtained and a second (aft) flight deck and an array of computers and monitors were installed in the passenger cabin. The aircraft is designed to be flown from the forward flight deck with the conventional controls, and from the aft flight deck using a fly-by-wire, triple-redundant digital computer system. From the aft flight deck, the aircraft can be flown with advanced electronic displays and pilot selectable automatic navigation, guidance and control systems in simulated Cat III operations. Safety is assured through monitoring, and by takeover capability of the forward flight deck crew.

NASA AGRICULTURAL AIRCRAFT

NASA is undertaking studies, using an Ayres Thrush Commander 800, to improve the aerodynamics and application equipment and techniques of future agricultural aircraft.

NATURAL LAMINAR FLOW AIRCRAFT

In April 1979 it was reported that NASA had been gathering performance data on General Dynamics F-111 aircraft, prior to carrying out natural laminar flow testing of an F-111 at Dryden Flight Research Center in the late Summer of 1979. Study proved that it would be possible to fly the aircraft safely without flaps at the kind of take-off and landing speeds that would be encountered during the flight tests which were to be made.

It was known that significant improvements in the cruise efficiency of transport aircraft can be obtained if the flow over the wings is laminar. Active laminar flow control can prevent turbulent flow over most of the wing, but such systems add weight and cost, and complicate maintenance.

The F-111 which had been used earlier for the evaluation of a NASA supercritical wing, under a joint NASA/USAF Transonic Aircraft Technology (TACT) programme, was modified again for this new experiment to investigate natural laminar flow. This involved the installation of 'gloves' over the standard wing, these representing a supercritical aerofoil section, designed with favourable pressure gradients on both the upper and lower surfaces, to determine the extent of natural laminar flow that could be obtained with promising consistency over

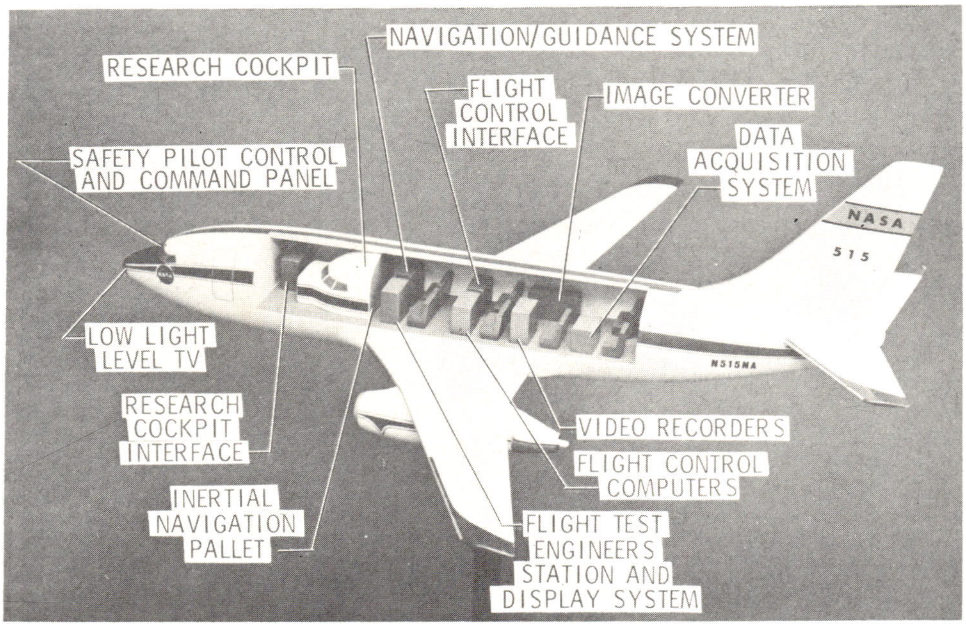

Cutaway model of NASA's TCV Boeing 737, showing internal layout

Boeing KC-135 fitted with NASA winglets, to evaluate their fuel-saving potential

both surfaces of this subsonic cruise aerofoil section. Following installation on the F-111, instrumentation was added to provide surface pressure data, as well as for determining transition location and boundary layer characteristics.

Nineteen flights were made with and without transition fixed at several locations, for wing leading-edge sweep angles which varied from 10° to 26°, at Mach numbers from 0·80 to 0·85, and at altitudes of 7,620 m (25,000 ft) to 9,145 m (30,000 ft). No details of the results had been announced at the time of going to press.

NASA/USAF KC-135 WINGLET TESTS

In February 1978 a US Air Force KC-135 cargo/tanker aircraft was delivered to NASA's Dryden Flight Research Center, where it was modified by the installation of NASA winglets and is being test flown to evaluate the fuel-saving potential of these supplementary aerofoil surfaces.

Under a US Air Force contract worth approximately $3 million, The Boeing Company constructed the winglets. Each has a chord of 0·61 m (2 ft 0 in) at the tip, 1·83 m (6 ft 0 in) at the base, and is 2·74 m (9 ft 0 in) in length. Constructed of light alloy, they weigh approximately 68 kg (150 lb) each. Boeing installed them on outer wing panels. These outer panels with winglets were then installed on the KC-135 by NASA personnel at Dryden, where the modified aircraft flew for the first time on 24 July 1979. It is possible to make adjustments to the incidence and cant angle of the winglets between flights, so that the optimum setting can be established.

Flight testing of the fully instrumented KC-135 was carried out prior to installation of the winglets to provide basic performance data for comparison. The flight test programme was completed in early 1981 after a total of 173 flight hours with the winglets installed. Preliminary results have shown that these provide a 5-7 per cent improvement in aircraft performance (see also KC-135A entry, under Boeing Military Airplane Company heading).

The winglet concept was developed in wind tunnels at NASA's Langley Research Center by Dr Richard T. Whitcomb, who was also responsible for the NASA supercritical wing.

GOSSAMER ALBATROSS

The short-term flight research programme with the MacCready (which see) Gossamer Albatross No. 2, of which details appeared in the 1980-81 *Jane's*, has been completed.

NASA/LOCKHEED U-2

NASA has been using an Ames Research Center U-2 Earth Resources Survey Aircraft to provide real-time infra-red images of smoke obscured landscapes. Equipped with a Daedalus multispectral scanner, and a Fairchild charge-coupled device linear array scanner, the U-2, flying at about 21,325 m (65,000 ft), transmits data directly to the Ames Center. There the digital information is processed into hard-copy images, and transferred immediately to US geological survey maps for transmission by telecopy machine direct to the fire control site. This entire process takes less than ten minutes, and can provide vital information on a fire's exact site and rate of propagation. Such information has not previously been obtainable when the site has been obscured by dense smoke, making difficult and dangerous the deployment of firefighters and their equipment. The US Forest Service, National Park Service, and California Department of Forestry, all regard this information as of great importance, and it was planned to expand the U-2's capability in 1981 to provide high-resolution imagery of near infra-red to reveal water sources close to the fire.

Natural laminar flow aerofoil sections superimposed at mid-span on the outer wings of NASA's F-111 test aircraft

NASA's F-15 test aircraft, with a section of Space Shuttle Orbiter tiles on its starboard wing

NASA/LOCKHEED ER-2

NASA took delivery, in June 1981, of the ER-2 high-altitude research aircraft which has been acquired for Earth resources studies, and for scientific measurements in the stratosphere. One of three single-seat Lockheed TR-1As contracted in November 1979, it has been specially modified for NASA to the ER-2 configuration. A description of the TR-1A and illustration of the ER-2 appear under the Lockheed entry in this section; the ER-2 (NASA 706) incorporates payload compartments in the nose, in a fuselage section behind the pilot which is identified as the Q-bay, and in two pods, carried one beneath each wing. The aircraft is designed to cruise at an altitude above 21,335 m (70,000 ft), at a speed of 415 knots (769 km/h; 478 mph), and with a range of more than 2,605 nm (4,828 km; 3,000 miles).

Based at Ames Research Center, the ER-2 is to be used for NASA missions under the direction of the Office of Space and Terrestrial Applications at NASA Headquarters. Payload areas within the ER-2 accommodate 272 kg (600 lb) in the nose, and 340 kg (750 lb) in the Q-bay: these, together with 340 kg (750 lb) in each of the underwing pods, provide a payload total of 1,292 kg (2,850 lb). Typical payloads include dual Wild-Heerbrugg RC-10 camera systems, a Daedalus infra-red scanner, linear array scanner, and radar system for a radar, photographic and infra-red mission; a similar installation plus LIDAR or MAPS system, and gas chromatograph or Itek Iris II high-resolution panoramic camera for a combined radar, photographic, infra-red and air sampling mission; and equipment that can include an aerosol particulate sampler, airborne gas chromatograph, cryogenic sampler, Decibi ozone monitor, frost point hygrometer, infra-red radiometer, infra-red spectrometer, INS stratospheric wind experiment, Knollenberg 2D probe, multi-filter sampler, SAS II stratospheric air sampler, and a water vapour hygrometer for stratospheric experiments. Much of the information gained by the ER-2 can be communicated on a real-time basis.

NASA's JetStar, modified for advanced propeller acoustics investigation

SHUTTLE TILE TESTS

In a twelve-month period to January 1981, NASA's Dryden Flight Research Center conducted a flight test programme to prove the integrity of the thermal protection tiles that had been specified for the Space Shuttle Orbiter vehicle. Approximately 60 flights were made by F-15 and F-104 test aircraft carrying on their wings tile sections from six different locations on the Orbiter. Speeds of up to Mach 1·4 and dynamic pressures up to 5,565 kg/m² (1,140 lb/sq ft) were achieved during the programme. After each flight the test section was inspected and measured precisely to identify any deformation or structural changes that may have occurred as a result of flight loads. The programme led to design changes to the thermal protection system installed on the Orbiter vehicles, notably to improve binding forces.

NASA ADVANCED PROPELLER ACOUSTICS TEST PROGRAMME

The Lockheed C-140 JetStar operated by NASA's Dryden Flight Test Center has been modified to measure and record propeller acoustical data which is unobtainable in wind tunnel tests. Modifications include the installation of a 91 cm (3 ft) high pylon on the upper surface of the fuselage, directly above the wing. This carries an Air Turbine Drive System (ATDS), to which is mounted the propeller under test. The ATDS is a three-stage air turbine motor, incorporating a flexible coupling between the propeller and turbine shaft, a brake system, and a slip-ring assembly. It provides a minimum of 145 kW (195 hp) at 7,636 rpm. Other equipment installed in the aircraft includes 28 flush-mounted microphones, and pylon and aircraft skin accelerometers for vibration measurements. The eight- and ten-blade propellers, or propfans, to be tested will be one-seventh scale models, each 61 cm (2 ft) in diameter. Their purpose is to make future aircraft significantly more fuel-efficient.

Initial flutter clearance flight testing of the first advanced turboprop propeller was under way in Spring 1981, to establish a safety and systems checkout before proceeding to gather acoustic test data. Much of this will be obtained in a Mach 0·8 and 6,100 to 9,145 m (20,000 to 30,000 ft) altitude regime, but terminal flight characteristics may also be examined. The current programme is expected to involve a total of twelve acoustic flights over a period of one to two years from June 1981.

USAF/NASA MISSION ADAPTIVE WING

Under a US Air Force contract awarded in 1979, Boeing Military Airplane Company is modifying the wings of NASA's General Dynamics F-111, which has been used most recently in a Transonic Aircraft Technology (TACT) programme. The wings were despatched to Boeing (which see) for modification in June 1981, and were expected to be returned to NASA's Dryden Flight Research Center in April 1982. When received they will have been converted to Mission Adaptive Wing (MAW) configuration, which is designed to alter wing camber to satisfy a variety of flight conditions and yet maintain a smooth, uninterrupted upper surface. Mechanisms built inside the MAW will act upon flexible leading- and trailing-edge surfaces to change wing camber during flight. The smooth, flexible surface of this unique wing will be made of glassfibre, and its contours will be altered by internal links driven by power hinges. The resulting variations in curvature will produce extremely efficient airflow over and under the wing. In addition to demonstrating the performance improvements of smooth variable-camber wings, the programme is intended to check and validate the deformable portion of the glassfibre structure, and to evaluate the agility and increased buffet-free envelope of the aircraft with the MAW.

NORTHROP
NORTHROP CORPORATION

Corporate Office: 1800 Century Park East, Los Angeles, California 90067
Telephone: (213) 553 6262
Chairman of the Board and Chief Executive Officer:
Thomas V. Jones
Vice-Chairman of the Board: Richard W. Millar
President and Chief Operating Officer:
Dr Thomas O. Paine
Senior Vice-Presidents:
William M. Elliott (General Counsel)
Welko E. Gasich (Advanced Projects)
Donald A. Hicks (Marketing and Technology)
James V. Holcombe (Asst to Chief Executive Officer for Government Relations)
Frank W. Lynch (Tactical and Electronic Systems Group)
William G. McGagh (Finance)
Division Vice-Presidents:
David N. Ferguson (Electronics)
George H. Hage (Electro-Mechanical)
Kent Kresa (Ventura)
F. J. Manzella (Aircraft Services)
Roy P. Jackson (Aircraft)
Wallace C. Solberg (Defense Systems)
Joseph Yamron (Precision Products)
Subsidiary Presidents:
Frank Caramelli (Wilcox Electric Inc)
D. A. McInnis (Northrop Services Inc)
James M. Ricketts (Page Communications Engineers Inc)
James E. Ware (Northrop Architectural Systems Inc)
Other Vice-Presidents:
J. R. Alison (Customer Relations)
L. F. Begin Jr (F/A-18L)
C. H. Bernstein (Asst to Chief Executive Officer for Analysis)
J. B. Campbell (Controller)
William J. Chalmers (Manager, Northrop Research and Technology Center)
W. B. Dennis (Forward Planning)
Stanley Ebner (Manager, Eastern Regional Office)
William A. Fifer (Senior Corporate Executive, Saudi Arabia)
C. Robert Gates (Northrop International, and F-5G Programme)
Manuel G. Gonzalez (F/A-18A and Asst General Manager, Aircraft Division)
D. G. Head (President, George A. Fuller Company)
R. P. Jackson (Programme Management)
J. C. Jones (Asst to Chief Executive Officer for Aeronautical Systems)
D. L. Lewis (Business Analysis)
T. A. McDougall (International Business Operations)
John J. Richardson (Industrial Relations)
Keith L. Robinett (Manager, Northrop Data Processing)
Donald D. Warner (Material, Facilities and Services)
Robert B. Watts Jr (Asst General Counsel)
G. Ronald Wenninger (Advanced Systems)
Corporate Treasurer: Richard B. Lohrer
Corporate Secretary: Sheila M. Gibbons
Asst to Chairman—Community Relations:
W. H. Habblett
Executive Asst to Chairman and Chief Executive Officer: Wesley B. Truitt
Director of Public Affairs—Europe: J. K. Corfield
Director of Information: A. W. Cantafio

This company was formed in 1939 by John K. Northrop and others to undertake the design and manufacture of military aircraft. During the second World War it built 1,131 aircraft of its own design and was engaged in extensive subcontract work. It also devoted considerable attention to the design and construction of aircraft of the 'flying wing' type.

Although continuing its activities in the design, development and production of aircraft, missiles and target drone systems, Northrop has broadened its scope of operation to include electronics, space technology, communications, construction, support services and commercial products. To reflect this changing character of its business, the company changed its name from Northrop Aircraft Inc to Northrop Corporation in 1959.

Divisions of Northrop now include Aircraft Division, specialising in the research, development, production and sale of military aircraft, commercial airframe assemblies and subassemblies; Aircraft Services Division, engaged primarily in worldwide total air base systems support, requirement analysis and planning, programme management, facilities development and maintenance, training, logistics, aircraft maintenance and overhaul, and base operation; Electronics Division, which handles the company's activities in electronics research, development, and manufacturing of navigation and guidance systems; Electro-Mechanical Division, which handles advanced missile programmes, electro-optical systems for target identification and target designation; Defense Systems Division, designer and manufacturer of electronics countermeasures systems, including the internal countermeasures set (ICS) for the US Air Force's F-15 Eagle; Precision Products Division, a leading supplier of precision gyroscopes and related guidance and control systems for communication and weather satellites, in key missile programmes, and on aircraft, submarines and ships; Ventura Division, designer and manufacturer of remotely piloted vehicles, target systems for anti-aircraft gunnery, surface-to-air and air-to-air missile crew training and weapons system evaluation, recoverable and expendable target aircraft, operation, maintenance and target range support services, torpedo modification kits, and mobile underwater targets; and George A. Fuller Company, a leading construction and construction-management company.

Northrop's communications activities are handled by two subsidiaries: Wilcox Electric Inc of Kansas City, Missouri, designers and producers of a wide range of ground-based navigation equipment, and Page Communications Engineers, Vienna, Virginia, which designs, engineers, installs and maintains total communications systems.

To further expand its research and development work, Northrop has divided its Research and Technology Center into three organisations: the Corporate Laboratories, Laser System Laboratories and Laser Technology Laboratories. Current programmes include research and development in such fields as information sciences, electronic devices and materials, nuclear radiation effects, high-energy laser development and laser systems applications.

The number of employees of Northrop Corporation totalled about 30,250 on 1 January 1981.

NORTHROP CORPORATION AIRCRAFT GROUP

3901 West Broadway, Hawthorne, California 90250
Telephone: (213) 970 2000
Senior Vice-President and Group Executive:
Ross F. Miller
Corporate Vice-President: R. P. Jackson (Manager, F-18 Programme)
Senior Vice-President: M. G. Gonzalez (Business Operations)
Vice-Presidents:
R. A. Graham (Marketing)
M. Kuska (General Manager Aircraft Programmes)
J. Mannion (Public Affairs)
G. S. Shackelford (Contracts and Pricing)
S. R. Smith (Iran Operations)
R. S. Taylor (Finance)
Robert W. Young (Administration)
Assts to V-P Finance:
W. G. Niemann
S. Zitter

Aircraft Division:
Corporate Vice-President and General Manager:
Ross F. Miller
Vice-Presidents:
C. W. Benson (Materiel)
R. D. Lovell (Manufacturing)
J. L. McCoy (Product Support)
B. T. Moser (Production Operations)
Thomas R. Rooney (Technical)

Northrop F-5E Tiger II light tactical fighter, with wingtip Sidewinder missiles

Aircraft Services Division:
Corporate Vice-President and General Manager:
F. J. Manzella
Vice-Presidents:
H. B. Gunther (Construction Operations)
L. J. Hunt Jr (Manager, Saudi Arabia Programme)
R. H. Madeira (Support Services)

Current production at Northrop's Aircraft Division is centred on the F-5E Tiger II fighter, F-5F two-seat fighter/trainer, F/A-18 Hornet multi-mission fighter, and major Boeing 747 subcontract work, which includes manufacture of the main fuselage section, the extra-large side-loading cargo door and the stretched upper deck. The company was also engaged in 1981 in the development of an advanced version of the F-5E, to which the designation F-5G Tigershark has been allocated.

Northrop's share of work on development and production of the US Navy's F/A-18 Hornet multi-mission fighter is proceeding on schedule. Northrop builds the centre and aft fuselage and the twin vertical tails of the Navy version of the Hornet. The company is responsible also for integration of all internal systems such as the environmental control system and auxiliary power system. McDonnell Douglas (which see) is prime contractor to the US Navy for the carrier-based Hornet, with Northrop responsible for about 40 per cent of production. The Navy has stated a requirement for 1,366 operational Hornets, to replace its existing A-7 and F-4 aircraft. Northrop's land-based version of the F-18 is known as the F-18L. As prime contractor for the land-based F-18, Northrop would carry out approximately 60 per cent of its production, McDonnell Douglas 40 per cent. Both carrier- and land-based versions use the same General Electric engines, and they are

80-90 per cent identical in high-usage spare parts. Design modifications to convert the carrier-designed F-18 airframe for land-based operation are intended to simplify it, and reduce the cost.

On 11 October 1979, Northrop announced the receipt of a $12·4 million contract from the US Air Force to design technology for highly loaded primary wing and fuselage structures, and to build and test graphite composite airframe components. Under the contract Northrop will build more than 800 test items, ranging from small components to 2·54 × 4·06 m (8 ft 4 in × 13 ft 4 in) structures. Each will be subjected to tests designed to prove material and structural integrity under a wide range of atmospheric conditions. The primary purpose of the five-year programme is to develop more cost-effective methods of testing airframe parts made from graphite composite.

In addition to its main factory at Hawthorne, the Aircraft Division has facilities at El Segundo, Carson, Torrance, Palmdale and Edwards Air Force Base, California. The 176,500 m^2 (1·9 million sq ft) facility at El Segundo, purchased in December 1978, now houses what are regarded as the most modern tactical fighter production lines in the USA, and has made it possible to double the machine tool capacity at Hawthorne.

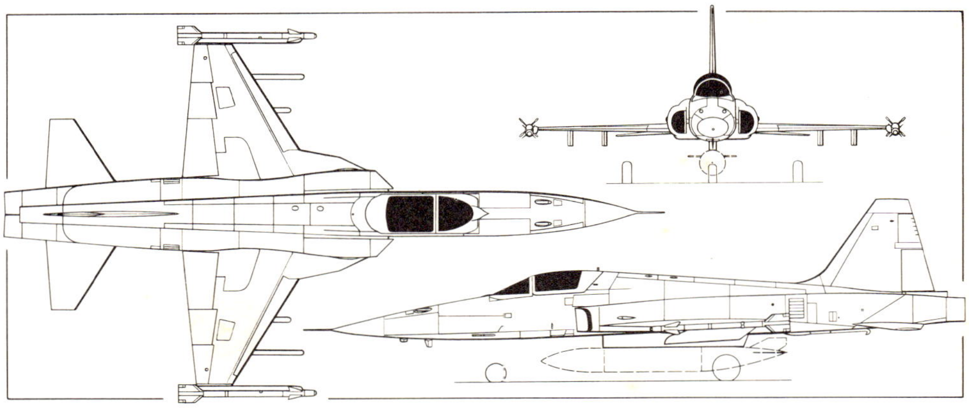

Northrop F-5E Tiger II single-seat twin-jet tactical fighter aircraft *(Pilot Press)*

Northrop F-5F tandem two-seat fighter-trainer, armed with wingtip-mounted Sidewinder missiles

NORTHROP TIGER II

USAF designations: F-5E and F-5F

The F-5E was selected in November 1970 by the US government as the winner of a competition to determine the single-seat International Fighter Aircraft (IFA) which was to succeed Northrop's F-5A. The two-seat F-5F was developed subsequently.

The F-5E design places particular emphasis on manoeuvrability rather than high speed, notably by the incorporation of manoeuvring flaps, based on the design of a similar system for the Netherlands Air Force's NF-5A/Bs. Full-span leading-edge flaps work in conjunction with conventional trailing-edge flaps, and are operated by a control on the pilot's throttle quadrant.

Wing loading on the F-5E is maintained at approximately the same value as on the F-5A, as the result of an increase in wing area to 17·30 m^2 (186 sq ft). This is due principally to the widened fuselage, which also increases wing span. The tapered wing leading-edge extension, between the inboard leading-edge and fuselage, was modified to enhance airflow over the wing at high angles of attack.

The F-5E incorporates other features developed for the Canadian, Dutch and Norwegian F-5s. These include two-position nosewheel gear, which increases wing angle of attack on the ground by 3° 22′ and which, in conjunction with the more powerful engines, has improved F-5E take-off performance some 30% by comparison with earlier F-5s. Arrester gear permits operation from short runways. It is qualified to carry two 1,040 litre (275 US gallon) underwing fuel tanks, in addition to the centreline 1,040 litre (275 US gallon) tank, and up to seven 500 lb Mk 82 bombs, following the addition of a multiple ejector rack (MER) on the centreline stores station.

The first F-5E was rolled out on 23 June 1972, and made its first flight on 11 August 1972. USAF Tactical Air Command, with assistance from Air Training Command, was assigned responsibility for training pilots and technicians of user countries. First deliveries of the F-5E, to the US Air Force's 425th Tactical Fighter Squadron, were made in the Spring of 1973. Twenty aircraft had been supplied for the US Air Force training programme by the end of September 1973, and deliveries to foreign countries began in late 1973. By 4 June 1981, a total of 973 F-5Es and 158 F-5Fs had been delivered, and production continues at an average rate of about five per month. Customers to date include the US Air Force (112 F-5Es), US Navy (10 F-5Es and 3 F-5Fs), Brazil, Chile, Iran, Jordan, Kenya, South Korea, Malaysia, Saudi Arabia, Singapore, Sudan, Switzerland, Taiwan, Thailand and the Yemen Arab Republic. It has been reported that 10 F-5Es and 2 F-5Fs were supplied to Mexico in late 1980.

In addition to their use as tactical fighters, F-5Es are operated by the US Air Force and US Navy in the 'aggressor' role, to simulate enemy aircraft at major air combat training schools in the USA, England and the Philippines.

Details of the current production versions of the Tiger II are as follows:

F-5E. Standard production version, to which the detailed description applies. In production also, under licence, by AIDC in Taiwan (which see), and being assembled by Swiss Federal Aircraft Factory. Can be fitted with an R-843A/ARN-58 localiser receiver and a reconnaissance nose containing four KS-121A 70 mm framing cameras and related equipment. Intended for low/medium-altitude photo-reconnaissance, the nose is similar to that of the RF-5A. New 'shark' profile nose developed for the F-5G (which see) is to be introduced on standard production F-5E and F-5F.

To extend the range of armament options, an F-5E completed a technology flying demonstration with a 30 mm underbelly gun pod developed by General Electric. More than 800 rounds of 30 mm GAU-8 type ammunition were fired on US Air Force test ranges at Edwards AFB, California.

The F-5Es for the Royal Saudi Air Force (RSAF) have a Litton LN-33 inertial navigation system, capable of accuracy exceeding 1·5 nm (2·7 km; 1·7 miles) CEP per flight hour, which provides attitude reference, range and bearing to ten pre-set destinations, as well as true ground track steering. The system is self-aligning in 10 min in the gyro compass mode, and can be aligned in 3 min to a stored heading. In-flight refuelling capability is also provided.

F-5Es of the Brazilian Air Force have a large dorsal fin to accommodate an ADF antenna.

F-5F. Tandem two-seat version of F-5E, with fuselage lengthened by 1·08 m (3 ft 6½ in). Fire control system retained, enabling aircraft to be used for both training and combat duties, but one M39 gun deleted. Development approved by US Air Force in early 1974. First flight was made on 25 September 1974. Two F-5Fs completed flight test and qualification in early 1976. Total of 118 ordered; deliveries began in the Summer of 1976.

On 31 March 1978, Northrop announced receipt of US government approval for a company-funded development and flight demonstration programme of an **RF-5E** specialised reconnaissance version of the F-5E. This has a modified forward fuselage with quick-change capabilities to accommodate a wide variety of reconnaissance equipment. Both day and night photo missions were demonstrated during the subsequent test programme. Modification of a production F-5E made possible the first flight of the RF-5E prototype in January 1979, and a brief description of this aircraft is given separately.

The following details refer to the F-5E, but are generally applicable to the F-5F also, except for details noted under model listings:

Type: Single-seat light tactical fighter.

Wings: Cantilever low-wing monoplane. Wing section NACA 65A004·8 (modified). No dihedral. No incidence. Sweepback at quarter-chord 24°. Multi-spar light alloy structure with heavy plate machined skins. Hydraulically-powered sealed-gap ailerons at approximately mid-span. Electrically-operated light alloy single-slotted trailing-edge flaps inboard of ailerons. Electrically-operated leading-edge manoeuvring flaps. No de-icing system.

Fuselage: Light alloy semi-monocoque basic structure, with steel, magnesium and titanium used in certain areas. Two hydraulically-actuated airbrakes of magnesium alloy construction, mounted on underside of fuselage forward of main-wheel wells. Avionics bay and cockpit pressurised; fail-safe structure in pressurised sections.

Tail Unit: Cantilever all-metal structure, with hydraulically-powered rudder and one-piece all-moving tailplane. Tailplane incidence varied by hydraulic actuators. No trim tabs. Dual hydraulic actuators of Northrop design for control of rudder and tailplane.

Landing Gear: Hydraulically-retractable tricycle type, main units retracting inward into fuselage, nosewheel forward. Oleo-pneumatic struts of Northrop design on all units. Two-position extending nose unit increases static angle of attack by 3° 22′ to reduce T-O distance, and is shortened automatically during the retraction cycle. Gravity-operated emergency extension. Main wheels and tyres size 24 × 8·00-13, pressure 14·48 bars (210 lb/sq in). Steerable nose unit with wheel and tyre size 18 × 6·50-8, pressure 8·27 bars (120 lb/sq in). All-metal multiple-disc brakes of Northrop design.

Power Plant: Two General Electric J85-GE-21A turbojet engines, each rated at 22·24 kN (5,000 lb st) with afterburning. Two independent fuel systems, one for each engine. Fuel for starboard engine supplied from two rubber-impregnated bladder-type nylon fabric cells, comprising a centre-fuselage cell of 803 litre (212 US gallon) capacity, and an aft-fuselage cell of 640 litre (169 US gallon) capacity. Port engine supplied from a forward fuselage cell of 1,120 litre (296 US gallon) capacity. Total fuel capacity 2,563 litres (677 US gallons). No fuel is carried in the wings. Fuel crossfeed system allows fuel from either or both cell systems to be fed to either or both engines. Auxiliary jettisonable fuel tanks of 568 or 1,041 litres (150 or 275 US gallons) can be carried on the fuselage centreline pylon and the inboard underwing pylons. Single refuelling point on lower fuselage for fuselage fuel cell and external tank installation. Provision for in-flight refuelling by means of a detachable probe. Oil capacity 4·5 litres (1·2 US gallons) per engine.

Engine Intakes: Intakes are supplemented with auxiliary air inlet doors for use during T-O and low-speed flight, to improve compressor face pressure recovery and to decrease distortion. Each door consists of a set of six pivot-mounted louvres in removable panels on each side of the fuselage. The doors are actuated by the pilot at T-O, and controlled automatically in flight by Mach sensor switches, and are maintained in the open position at airspeeds below Mach 0·35-0·4.

Accommodation: Pilot only in pressurised, heated and air-conditioned cockpit, on rocket-powered ejection seat (Martin-Baker RQ7A on aircraft for Iran and Brazil). Upward-opening canopy, hinged at rear.

Systems: Cockpit and avionics bay pressurised, heated and air-conditioned by engine bleed air, maximum pressure differential 0·34 bars (5 lb/sq in). Hydraulic power supplied by two independent systems at a pressure of 207 bars (3,000 lb/sq in). Flight control system provides power solely for operation of primary flight control surfaces. Utility system provides hydraulic backup power for the primary flight control surfaces and operating power for the landing gear, landing gear doors, airbrakes, wheel brakes, nosewheel steering, gun bay purge doors, gun gas deflectors and stability augmentation system. Electrical power supplied by two 13/15kVA 115/200V three-phase 320-480Hz non-paralleled engine-driven alternators. Each alternator has the capacity to accept full aircraft power load via an automatic transfer function. 250VA 115V 400Hz single-phase solid-state static inverter provides secondary AC source for engine starting. Two 33A 26-32V transformer-rectifiers and a 24V 11Ah nickel-cadmium battery provide DC power. Liquid oxygen system with capacity of 5 litres.

Avionics and Equipment (F-5E): AN/ARC-164 UHF command radio, 3,500-channel with 50kHz spacing. Emerson Electric AN/APQ-159 lightweight micro-miniature X-band pulse radar for air-to-air search and range tracking; target information, at a range of up to 20 nm (37 km; 23 miles), is displayed on a 0·13 m (5 in) DVST (direct vision storage tube) in cockpit. AN/ARA-50 UHF ADF; AN/AIC-18 intercom system; AN/APX-101 IFF/SIF system; AN/ARN-118 Tacan; attitude and heading reference system; angle of attack system; and central air data computer. Full blind-flying instrumentation. Optional avionics include Litton LN-33 inertial navigation system; AN/ARN-108 instrument landing system; CPU-129/A flight director computer; VHF; VOR/ILS with DME; LF ADF; CRT with scan converter for radar or electro-optical weapon (AGM-65 Maverick); and radar warning receiver. Optional equipment includes photo-reconnaissance nose.

Avionics and Equipment (F-5F): AN/ARC-164 UHF/AM command radio, AN/AIC-18 or -25 interphone set, AN/APX-101 IFF/SIF transponder, TS1843/APX transponder test set, AN/ARN-118 Tacan, AN/APQ-159 fire control radar, AN/ASG-29 lead computing optical sight, and AN/ARA-50 UHF ADF. Optional equipment includes photo-reconnaissance nose; in-flight refuelling system; pylon jettison conversion kits; anti-skid brakes; Northrop laser target designation set; and chaff/flare countermeasures package.

Armament (F-5E): Two AIM-9 Sidewinder missiles on wingtip launchers. Two M39A2 20 mm cannon mounted in fuselage nose, with 280 rounds per gun. Up to 3,175 kg (7,000 lb) of mixed ordnance can be carried on one underfuselage and four underwing stations, including M129 leaflet bombs; Mk 82 GP and Snakeye 500 lb bombs; Mk 36 destructors; Mk 84 2,000 lb bomb; BLU-1, -27 or -32 U or F napalm; LAU-68 (7) 2·75 in rockets; LAU-3 (19) 2·75 in rockets; CBU-24, -49, -52 or -58 cluster bomb units; SUU-20 bomb and rocket packs; SUU-25 flare dispensers; TDU-10 tow targets (Dart); and RMU-10 reel (Dart). Lead-computing optical gunsight uses inputs from airborne radar for air-to-air missiles and cannon, and provides a roll-stabilised manually-depressible reticle aiming reference for air-to-ground delivery. A 'snapshoot' capability is included for attack on violently manoeuvring and fleeting targets. The gunsight incorporates also a detachable 16 mm reticle camera with 15 m (50 ft) film magazine. Optional ordnance capability includes the AGM-65 Maverick; centreline multiple ejector rack; and laser guided bombs.

Armament (F-5F): Two AIM-9J Sidewinder missiles on wingtip launchers. One M39 20 mm cannon in port side of nose with 140 rounds. Underfuselage and underwing stores as detailed for the F-5E. Optional ordnance includes a laser designator in the rear cockpit.

Dimensions, external:
Wing span 8·13 m (26 ft 8 in)
Span over missiles 8·53 m (27 ft 11⅞ in)
Wing chord at root 3·57 m (11 ft 8⅝ in)
Wing chord at tip 0·68 m (2 ft 2⅞ in)
Wing aspect ratio 3·82
Length overall (incl nose-probe):
F-5E 14·68 m (48 ft 2 in)
F-5F 15·72 m (51 ft 7 in)
Height overall: F-5E 4·06 m (13 ft 4 in)
F-5F 4·01 m (13 ft 1¾ in)
Tailplane span 4·31 m (14 ft 1½ in)
Wheel track 3·80 m (12 ft 5½ in)
Wheelbase 5·17 m (16 ft 11½ in)

Areas:
Wings, gross 17·3 m² (186 sq ft)
Ailerons (total) 0·86 m² (9·24 sq ft)
Trailing-edge flaps (total) 1·95 m² (21·0 sq ft)
Leading-edge flaps (total) 1·14 m² (12·3 sq ft)
Fin 3·85 m² (41·42 sq ft)
Rudder 0·57 m² (6·10 sq ft)
Tailplane 5·48 m² (59·0 sq ft)

Weights and Loadings:
Weight empty: F-5E 4,392 kg (9,683 lb)
F-5F 4,793 kg (10,567 lb)
Max T-O weight: F-5E 11,193 kg (24,676 lb)
F-5F 11,442 kg (25,225 lb)
Max landing weight: F-5F 11,406 kg (25,147 lb)
Max zero-fuel weight: F-5E 7,953 kg (17,534 lb)
Max wing loading: F-5E 649·4 kg/m² (133 lb/sq ft)
Max power loading: F-5E 251·6 kg/kN (2·5 lb/lb st)

Performance (F-5E at combat weight of 6,055 kg; 13,350 lb, F-5F at combat weight of 6,375 kg; 14,055 lb, unless stated otherwise):
Never-exceed speed
710 knots (1,314 km/h; 817 mph) EAS
Max level speed:
F-5E at 10,975 m (36,000 ft) Mach 1·63
F-5F at 11,000 m (36,090 ft) Mach 1·55
Max cruising speed:
F-5E at 10,975 m (36,000 ft) Mach 0·98
Econ cruising speed Mach 0·80
Stalling speed, flaps down, power off:
F-5E 124 knots (230 km/h; 143 mph)
F-5F 136 knots (253 km/h; 157 mph)
Max rate of climb at S/L:
F-5E 10,516 m (34,500 ft)/min
F-5F 10,030 m (32,900 ft)/min
Service ceiling: F-5E 15,790 m (51,800 ft)
F-5F 15,484 m (50,800 ft)
Service ceiling, one engine out:
F-5E over 12,495 m (41,000 ft)
F-5F 12,285 m (40,300 ft)
T-O run:
F-5E at 7,053 kg (15,550 lb) 610 m (2,000 ft)
F-5F at 7,371 kg (16,250 lb) 701 m (2,300 ft)
T-O run at max T-O weight:
F-5E 1,737 m (5,700 ft)
F-5F 1,829 m (6,000 ft)
T-O to 15 m (50 ft):
F-5E at 7,053 kg (15,550 lb) 884 m (2,900 ft)
F-5F at 7,371 kg (16,250 lb) 975 m (3,200 ft)
Landing from 15 m (50 ft):
F-5E at 5,230 kg (11,530 lb), without brake-chute
1,417 m (4,650 ft)
F-5F at 5,554 kg (12,245 lb), without brake-chute
1,524 m (5,000 ft)
Landing run with brake-chute:
F-5E at 5,230 kg (11,530 lb) 762 m (2,500 ft)
F-5F at 5,554 kg (12,245 lb) 792 m (2,600 ft)
Range, F-5E:
with max fuel and reserves for 20 min max endurance at S/L: tanks retained
1,340 nm (2,483 km; 1,543 miles)
tanks dropped
1,545 nm (2,863 km; 1,779 miles)
Combat radius, F-5E:
with max fuel, two Sidewinder missiles, allowances as above and 5 min combat with max afterburning power at 4,570 m (15,000 ft)
570 nm (1,056 km; 656 miles)
with 2,358 kg (5,200 lb) ordnance load, two Sidewinder missiles, max fuel, allowances as above and 5 min combat at military power at S/L, lo-lo-lo mission 120 nm (222 km; 138 miles)
with max fuel, two Sidewinder missiles and two 530 lb bombs, allowances as above and 5 min combat at military power at S/L, hi-lo-hi mission
480 nm (890 km; 553 miles)
Range, F-5F:
ferry range with max fuel, allowances comprising 5 min at normal thrust, 1 min at max thrust, 20 min loiter at S/L, plus reserve of 5% of initial fuel: crew of two 1,270 nm (2,353 km; 1,462 miles)
Combat radius, F-5F:
with max internal fuel, and allowances comprising 2 min at normal thrust, 1 min at max thrust, 5 min max thrust for combat at 4,570 m (15,000 ft), 20 min loiter at S/L, plus reserve of 5% of initial fuel
520 nm (964 km; 599 miles)
with max fuel, two Sidewinder missiles and two 530 lb bombs, allowances as above, and 5 min combat at military power at S/L, hi-lo-hi mission
450 nm (834 km; 518 miles)

NORTHROP RF-5E

Northrop's RF-5E prototype made its international debut at the 1979 Paris Air Show, following a flight from Edwards AFB, California. First flown in January 1979, this prototype had completed more than 60 test and demonstration flights by the beginning of June 1979, during which missions had been flown at heights between 61 m (200 ft) and 14,325 m (47,000 ft), by day and night, at speeds of up to 600 knots (1,112 km/h; 691 mph). By mid-1981 it had been evaluated by 14 countries, and the US government had approved an order for two from Malaysia.

Basically similar to the F-5E Tiger II, the RF-5E differs by having a modified forward fuselage, and specialised equipment to enable it to fulfil a highly efficient reconnaissance role. The modified forward fuselage extends the overall length by 0·20 m (8 in), and provides 0·74 m³ (26 cu ft) of space to accommodate reconnaissance equipment. To allow maximum flexibility for differing reconnaissance roles, Northrop decided to group the various combinations of proposed cameras/sensors on portable pallets, any one of which could be loaded easily and quickly into this forward fuselage compartment. In addition to the selected pallet, a KS-87B oblique frame camera is mounted in a forward nose compartment and provided with lenses of 6 in and 12 in focal length.

Three pallets have been developed to date for the RF-5E, the first comprising a KA-95B medium-altitude panoramic camera, KA-56E low-altitude panoramic camera, and an RS-710 infra-red linescanner. Pallet 2 also has the KA-56E panoramic camera, with a KA-93B panoramic camera for horizon-to-horizon coverage at heights of 3,050-15,240 m (10,000-50,000 ft). Pallet 3 is configured for long-range oblique photo (LOROP) missions, and carries a single KS-147A LOROP camera of 66 in focal length. Other configurations are being studied, including one with a Zeiss mapping camera, two line-scanners and elint equipment.

The pilot has available advanced nav/com systems to complement the reconnaissance equipment, plus a video viewfinder system which enables him to view the terrain below the aircraft on a cathode ray tube (CRT) display in the cockpit. Using this system, the pilot can monitor and correct his line-of-flight during mapping runs, and can also update the INS when passing over recognisable terrain features. In addition, there is an integrated sensor control system (ISCS), which handles many operations automatically, thus reducing the pilot's workload. The ISCS takes inputs from the radar altimeter and INS to compute the angular velocity (V/H) for the terrain passing below the aircraft. V/H signals and the camera cycle rates are directed to individual cameras, with the appropriate scaling factor required for that particular sensor. The ISCS can cater for a wide variety of additional growth sensors.

Close-up of sensor windows beneath the forward fuselage of Northrop's RF-5E reconnaissance aircraft

The RF-5E retains the external stores stations of the F-5E, permitting the carriage of up to three external fuel tanks (each 1,041 litres; 275 US gallons) for maximum range and speed performance. It has essentially the same performance and armament capabilities as the F-5E tactical fighter, and on all missions is able to carry one M39 20 mm gun with 280 rounds, plus two AIM-9 missiles.

Performance:
Mission radius (A, with one external fuel tank and two AIM-9s; B, with three external fuel tanks and two AIM-9s):
low altitude throughout:
A 250 nm (463 km; 287 miles)
B 365 nm (676 km; 420 miles)
hi-lo-hi: A 410 nm (759 km; 471 miles)
B 560 nm (1,037 km; 644 miles)
hi-lo-lo-hi: A 365 nm (676 km; 420 miles)
B 520 nm (963 km; 598 miles)
high altitude throughout:
A 495 nm (916 km; 569 miles)
B 630 nm (1,166 km; 725 miles)

NORTHROP F-5G TIGERSHARK

Latest addition to Northrop's family of lightweight tactical aircraft, this export fighter has evolved from company studies which began more than four years ago. These showed that modification of the F-5E to have a single 71·17 kN (16,000 lb st) General Electric F404-GE-400 low bypass turbofan engine, with variable-geometry nozzle, in place of the two standard General Electric J85 turbojets, would result in a combat aircraft with much-improved performance at reasonable cost. Although the F404 engine is heavier than the two that it will replace, weight saving in other areas is expected to result in an aircraft empty weight only 17 per cent greater than that of the F-5E. Thus the additional 60 per cent of engine thrust will offer significant performance improvements.

Generally similar in appearance to, and with much the same avionics and weapon systems as, the F-5E, the basic **F-5G** has been shown by wind tunnel testing to retain the same handling characteristics, despite the much improved performance. Economies in purchase and operation will result from the interchangeability of many F-5E components, the ability to use existing F-5E support and training systems for the F-5G, and improved reliability and reduced maintenance associated with the General Electric F404 engine. For air forces requiring a more advanced version, the **F-5G-2** will be available in 1984, with new General Electric lookdown multi-mode radar, Honeywell laser gyro INS, head-up display and other refinements.

Following Presidential approval of the intermediate export fighter concept in January 1980, Northrop made the decision to proceed with the construction and development of several pre-production aircraft, and announced on 24 March 1980 the company's intention to engage an additional 200 employees to work on the project. By early February 1981 more than 600 employees were working on the development of this aircraft. The first of six pre-production aircraft is scheduled to make its first flight in September 1982, with customer deliveries to begin in July 1983.

The description of the F-5E applies also to the F-5G, except as follows:

Wings: As for F-5E, except that the tapered leading-edge extensions, between the inboard leading-edge and fuselage, are lengthened and modified as a result of engine inlet duct redesign. This changed leading-edge shape increases the maximum lift coefficient of the wings by about 12 per cent, for an increase in wing area of only 1·6 per cent. Strengthening of the structure by increasing the skin thickness of inboard sections will permit a manoeuvre load factor of 9*g*.

Fuselage: Basically as for F-5E, but nose flattened slightly and rounded in planform to enhance stability at high angles of attack. Area-ruling in the mid-fuselage section is reduced because of the higher thrust available. A 13 cm (5 in) fuselage plug aft of the cockpit permits rearrangement of the internal fuel cells and avionics, improving maintenance access to the latter. The rear fuselage is narrower as a result of the power plant change, but the longitudinal stability characteristics of earlier F-5s are retained by the addition of step fairings beneath the rear fuselage. A single variable-geometry exhaust nozzle replaces the twin nozzles of the former power plant. Weight-saving graphite composite skins are used in the rear fuselage.

Tail Unit: As for F-5E, except changed mounting of fin, and aerodynamic improvement of the drag-chute fairing. Forward end of this fairing incorporates an inlet to provide additional airflow to engine. As with the rear fuselage, weight-saving graphite composite skins are used on the tail surfaces.

Landing Gear: As for F-5E, except for heavier-duty Goodyear wheels and brakes to cater for the higher gross weight.

Power Plant: One 71·2 kN (16,000 lb st) General Electric F404-GE-400 afterburning turbofan engine installed in the rear fuselage. Fuel system, capacity 2,014 kg (4,440 lb), generally as for F-5E, but in Goodyear tanks rearranged to supply single engine.

Engine Intakes: Generally as for F-5E, but extended forward, enlarged slightly, and sited further from the fuselage to clear the thicker boundary layer airflow generated at higher airspeeds. Provisions for inlet enlargement to cater for growth versions of the F404 turbofan.

Accommodation: Generally similar to F-5E, with single zero-altitude ejection seat. Larger canopy, with 44 per cent more transparency area, plus improved headrest design and deletion of bulkhead behind pilot, considerably extend all-round field of view.

Systems: Hamilton Standard air-conditioning and demisting systems. Hydraulic system incorporates Ronson actuators. Electrical system utilises Westinghouse generators and Avtech transformer-rectifiers. Dual fly-by-wire longitudinal control system, with mechanical backup.

Avionics: To customer's requirements, ranging from latest available in F-5E (such as inertial navigation, angle-track radar and lead-computing gunsight) to new options. Latter include General Electric lockup/lookdown multi-mode radar, with advanced digital signal processing; Honeywell INS with laser gyro and solid-state digital mission computer; head-up display; radar warning receiver; and conformal countermeasures system.

Armament and Operational Equipment: Two 20 mm cannon in upper front fuselage. Centreline, four underwing and two wingtip stations for external stores, on which can be carried up to three General Electric 30 mm gun pods; six Sidewinder air-to-air or four Maverick air-to-surface missiles; seven Mk 82 bombs; laser-guided bombs; or chaff/flare dispensers.

Dimensions, external:

Wing span without missiles	8·13 m (26 ft 8 in)
Wing span over missiles	8·53 m (27 ft 11⅞ in)
Wing area, gross	17·28 m² (186 sq ft)
Length overall	14·19 m (46 ft 6⅔ in)
Height overall	4·22 m (13 ft 10¼ in)

Weights (estimated):

Weight empty	5,089 kg (11,220 lb)
Max internal fuel	2,014 kg (4,440 lb)
Max external weapons	3,175 kg (7,000 lb)
T-O weight 'clean'	7,820 kg (17,240 lb)
Max T-O weight	11,857 kg (26,140 lb)
Combat thrust/weight ratio	1·06

Performance (estimated):

Max level speed at height	Mach 2·1
Max rate of climb at S/L	15,330 m (50,300 ft)/min
Time to 12,190 m (40,000 ft)	2 min 18 s
Service ceiling	16,765 m (55,000 ft)
T-O run (S/L, ISA)	442 m (1,450 ft)

Combat radius with max internal fuel and two 1,041 litre (275 US gallon) external tanks, two Sidewinder missiles, seven Mk 82 bombs, 5 min combat at S/L military power, 20 min fuel reserves at S/L, hi-lo-hi mission 360 nm (667 km; 415 miles)

Combat radius with max internal fuel and three 1,041 litre (275 US gallon) external tanks, two Sidewinder missiles, 1 h 17 min on patrol, 20 min fuel reserves at S/L, combat air patrol mission 300 nm (556 km; 345 miles)

Ferry range with max internal and external fuel 1,490 nm (2,761 km; 1,715 miles)

g limit +9·0

Mockup of Northrop F-5G Tigershark with centreline 30 mm gun pod, wingtip Sidewinders, and underwing Maverick missiles and Mk 82 bombs

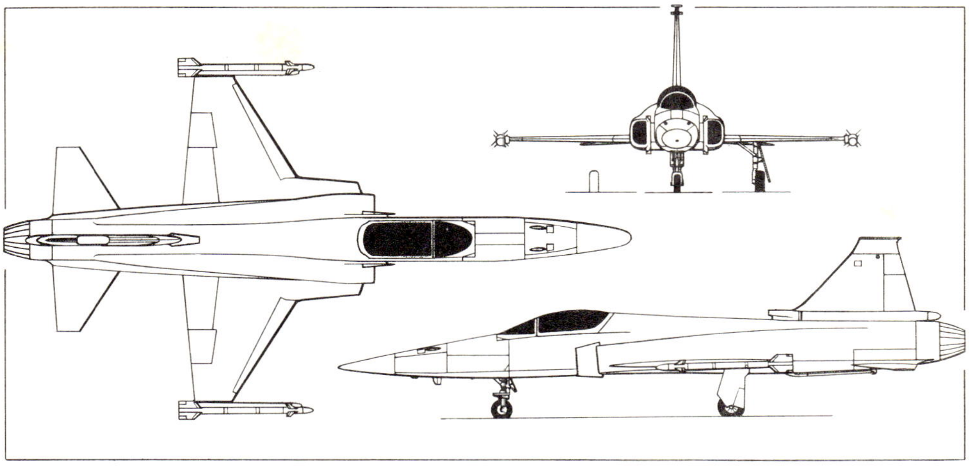

Northrop F-5G Tigershark intermediate export fighter (General Electric F404-GE-400 turbofan engine) *(Pilot Press)*

NORTHROP F/A-18L

The F/A-18L is a multi-role land-based version of the US Navy's F/A-18 Hornet carrier-suitable combat aircraft, for which McDonnell Douglas (which see) is prime contractor. Both aircraft are derivatives of the YF-17 twin-engined fighter prototypes, which represented Northrop's submission in the US Air Force's Lightweight Fighter (LWF) prototype programme. In the case of the F/A-18 Hornet production programme, Northrop's share of the work is approximately 40 per cent, that of McDonnell Douglas approximately 60 per cent. The percentages are reversed in the F/A-18L programme, for which Northrop would be prime contractor. The two F/A-18L prototypes had logged more than 600 flight hours by mid-1981, and had been flown by more than 50 pilots.

The land-based F/A-18L and carrier-suitable F/A-18 Hornet are about 80-90 per cent common in high-value/high-usage spare parts, with the same General Electric F404 engines. The F/A-18L is, however, considerably lighter in terms of empty weight as a result of the deletion of specifically-naval equipment and requirements, making it possible for the land-based version to carry up to 9,000 kg (19,840 lb) of stores on 11 weapon stations.

Major differences from the F/A-18 Hornet are a lighter and less complex landing gear, requiring 0·51 m³ (18 cu ft) less stowage space in the fuselage; a smaller and lighter internally-stowed arrester hook; deletion of the under-fuselage 'bird's nest' missile housings, to save weight and reduce drag; non-folding wings, with an extra stores station each side; replacement of the Hornet's ailerons and flaps by 'flaperons', which incorporate ride control for high-speed low-level dash and can be used also as lift dumpers during landing; and strengthened wingtips, able to carry Sparrow, Sky Flash, AMRAAM and ASRAAM missiles. Depending upon the options chosen, these refinements can virtually double the payload/range capability compared with the F/A-18 Hornet. If the F/A-18L's gun is removed, a further 454 kg (1,000 lb) of internal fuel can be carried, bringing total fuel capacity to approximately 5,443 kg (12,000 lb).

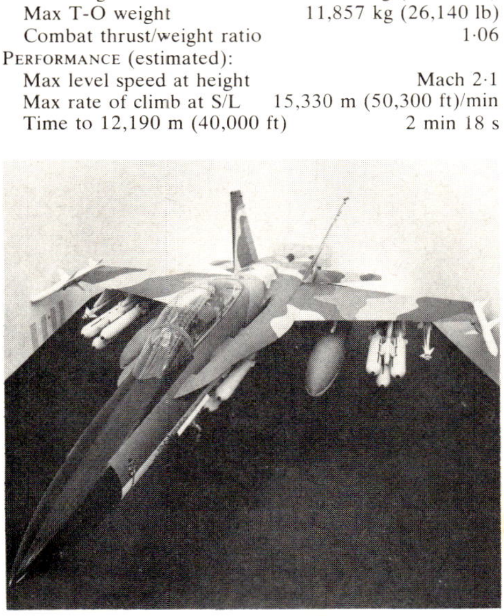

Mockup of two-seat version of F/A-18L at 1981 Paris Air Show

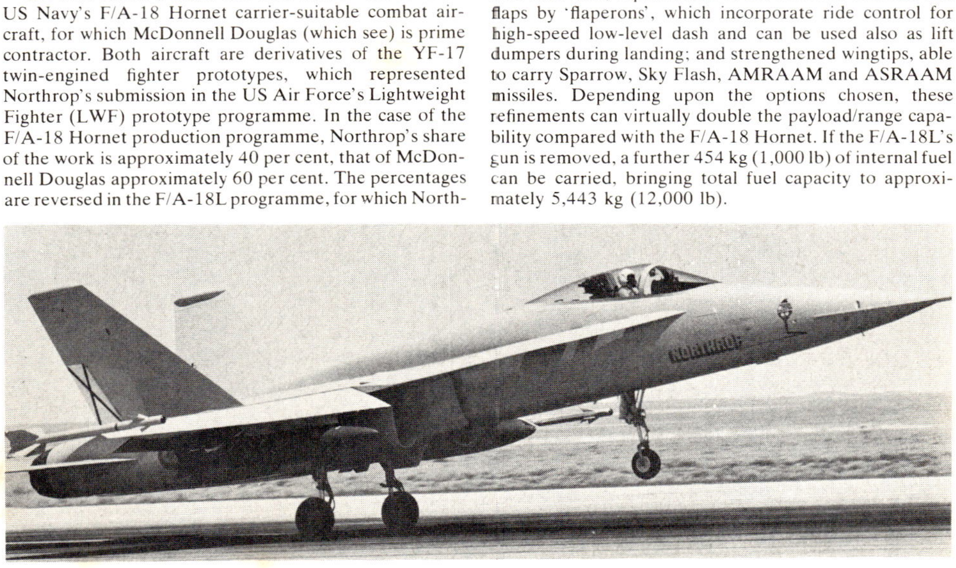

Prototype for the Northrop F/A-18L takes off at Edwards AFB on a Spanish Air Force evaluation flight

A two-seat version of the F/A-18L is also available, and would be capable of both transitional training and full tactical employment.

The description and dimensions given for the F/A-18 Hornet apply generally to the F/A-18L.

Performance (estimated, F-18L in basic air-to-air interception configuration):
Max level speed, with afterburning — in Mach 2 class
Max level speed, without afterburning, with two Sidewinders — above Mach 1
Approach speed — 115 knots (213 km/h; 132 mph)
Stalling speed at air-to-air landing weight — 92 knots (171 km/h; 106 mph)
Max rate of climb at S/L (50% fuel, two Sidewinders) — more than 18,290 m (60,000 ft)/min
T-O run (two Sidewinders):
with afterburning — 275 m (900 ft)
without afterburning — 396 m (1,300 ft)
T-O run with afterburning:
3,628 kg (8,000 lb) external load — 458 m (1,500 ft)
9,000 kg (19,840 lb) external load — 670 m (2,200 ft)
Landing run — 366 m (1,200 ft)
Combat radius — more than 700 nm (1,295 km; 805 miles)
Ferry range — more than 2,500 nm (4,630 km; 2,875 miles)
g limit (60% fuel) — +9

OMAC

OMAC INC

14325 Mt Lola Street, PO Box 60250, Reno, Nevada 89506
Telephone: (702) 972 3273 and 331 3954
President: Carl Parise
Vice-President: Larry Heuberger

This relatively new company was founded by Larry Heuberger and industrialist Carl Parise. Larry Heuberger was William Lear's project engineer in the early days of the Learjet 23 and 24, and chief engineer on the Lear Star 600. OMAC's aim is to develop a new low-cost high-performance and economical six/eight-seat turboprop-powered business aircraft of canard configuration. A full-size mockup was exhibited at the NBAA display at Atlanta, Georgia, in September 1979. All components and assemblies for production aircraft will be contracted out, leaving OMAC responsible for final assembly and flight testing.

OMAC I

Type: Lightweight business aircraft.

Wings: Cantilever high-wing monoplane, with 65° swept inboard leading-edge and 10° swept constant-chord outer wing panels. Aerofoil section NACA 64_1-212 (Mod B). Dihedral 1° 30′. Conventional two-spar structure of light alloy, with H-section extruded front spar. Leading-edge flaps on outer wing panels. Hydraulically-operated half-span trailing-edge plain flaps. Half-span cable-actuated ailerons. Trim tab in port aileron. Winglets fitted at wingtips extend above and below wing. Upper section canted outward 15°; lower section, which incorporates rudder with 25° of outward movement, canted outward 36°.

Fuselage: Conventional light alloy semi-monocoque pressurised structure.

Foreplane: Low-set constant-chord cantilever monoplane. Aerofoil section NACA 54_1-212 (Mod B). Conventional two-spar light alloy structure. Both spars of extruded H section. Full-span elevator each side. Trim tab in each elevator, electrically and manually actuated, and spring-interlocked with flaps for automatic trimming.

Landing Gear: Hydraulically-retractable tricycle type. Steerable nosewheel retracts forward. Oleo-pneumatic shock-absorber with shimmy damper on nosewheel. Main units have tubular spring steel legs. Nosewheel tyre size 6·00-6. Main wheels have heavy-duty tyres size 17·5-6·25 × 6. Heavy-duty brakes.

Power Plant: One 522 kW (700 shp) Avco Lycoming LTP 101-700A-A1 turboprop engine, driving a three-blade reversible-pitch metal pusher propeller, with 'Q' tips and Beta control. Fuel in integral wing tanks and rear-fuselage main feed tank, with combined capacity of 1,128 litres (298 US gallons). Refuelling points on wing upper surface.

Accommodation: Three alternative layouts for a pilot and one, six or seven passengers, with dual controls standard. Two forward-facing seats for pilot and co-pilot/passenger, with two or three individual reclining and swivelling seats, and three-seat bench at rear of cabin. Seats track-mounted to permit easy conversion for passenger/cargo use. Baggage compartment aft of rear seats, with space for toilet with privacy curtain. Split-type airstair door on port side, lower half incorporating folding step. Emergency exit on starboard side.

Systems: Air-conditioning and pressurisation system with max cabin pressure differential 0·38 bars (5·5 lb/sq in). Hydraulic system with electrically-powered pump for operation of flaps, landing gear and brakes. Electrical system. Oxygen system.

Avionics and Equipment: Standard avionics include dual Bendix VHF nav/com, 870 FCS, autopilot, ADF, DME, encoding altimeter, ELT, RDR160 colour radar, R/Nav, transponder, and VOR/LOC indicator. Standard equipment includes true airspeed indicator, full co-pilot instrumentation, instrument lights, stall warning device, pitot and stall warning heating systems, alternate static source, courtesy lights, navigation lights, strobe lights, dual landing and taxi lights, and external power socket.

Dimensions, external:
Wing span — 10·67 m (35 ft 0 in)
Foreplane span — 5·23 m (17 ft 2 in)
Length overall — 8·99 m (29 ft 6 in)
Height overall — 3·56 m (11 ft 8 in)
Wheel track — 2·84 m (9 ft 4 in)
Wheelbase — 4·69 m (15 ft 4¾ in)
Propeller diameter — 2·29 m (7 ft 6 in)

Dimensions, internal:
Cabin: Length — 4·45 m (14 ft 7 in)
Max height — 1·30 m (4 ft 3 in)
Max width — 1·42 m (4 ft 8 in)

Weights (estimated):
Weight empty — 1,175 kg (2,591 lb)
Max fuel weight — 906 kg (1,997 lb)
Max T-O weight — 2,585 kg (5,700 lb)

Performance (estimated, at max T-O weight):
Max level speed — 260 knots (482 km/h; 299 mph)
Cruising speed — 248 knots (460 km/h; 286 mph)
Econ cruising speed — 217 knots (402 km/h; 250 mph)
Stalling speed — 54 knots (100 km/h; 62·5 mph)
Max rate of climb at S/L — 610 m (2,000 ft)/min
Certificated ceiling — 7,620 m (25,000 ft)
T-O to 15 m (50 ft) — 482 m (1,583 ft)
Landing from 15 m (50 ft) — 488 m (1,600 ft)
Max range at econ cruising speed with allowances for 1 h hold — 2,950 nm (5,467 km; 3,397 miles)

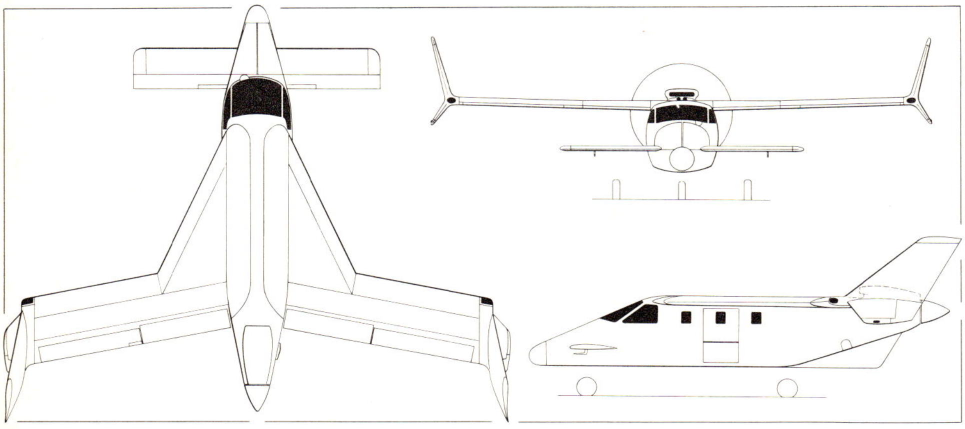
OMAC 1 lightweight turboprop-powered business aircraft *(Pilot Press)*

OMNIONICS

Meritt Island, Florida

In 1965, Omnionics began development of an all-glassfibre amphibian for commuter and business transport operation. This evolved as the Dolphinair, of which preliminary details can be found in the 1980-81 *Jane's*.

PIASECKI

PIASECKI AIRCRAFT CORPORATION

Head Office: Island Road, International Airport, Philadelphia, Pennsylvania 19153
Telephone: (215) 365 2222
Erection Works: PO Box 1776, Lakehurst, New Jersey 08733
Telephone: (201) 657 4222
Directors:
Virgil Kauffman
Donald N. Meyers
Arthur J. Kania
J. Micallef
Frank N. Piasecki
President: Frank N. Piasecki
Vice-President: Donald N. Meyers (Engineering)
Secretary: Arthur J. Kania
Industrial Engineering: K. R. Meenen

The Piasecki Aircraft Corporation was formed in 1955 by Mr Frank Piasecki, who was formerly Chairman of the Board and President of the Piasecki Helicopter Corporation (now the Boeing Vertol Company).

PIASECKI HELI-STAT

A Heli-Stat is a hybrid VTOL vehicle which links the envelope of a lighter-than-air craft with current-technology helicopters, the aerostat providing static lift to support approximately the full empty weight of the entire assembly. The helicopters furnish the lift to support the payload, as well as providing propulsion and control, with adequate control forces to enable the Heli-Stat to hover with precision, a characteristic with which conventional airships cannot comply.

Details of an early Heli-Stat project can be found in the 1977-78 *Jane's*. By 1979 Piasecki had started engineering work on a prototype, designated **Model 97-34J** and named **Logger**, and has since received a contract from the US Navy to design, build and fly this vehicle for use by the US Forest Service. In early 1981 the Heli-Stat was being erected at the company's works in Lakehurst, New Jersey. It is intended to demonstrate the advantages of such a vehicle for tasks such as the economic transportation of otherwise inaccessible timber over distances of up to 8·7 nm (16 km; 10 miles).

Fuselage: Frame structure of light alloy and composite materials to provide mountings for four Sikorsky SH-34J helicopters, on outriggers which will provide adequate clearance for their rotors and ensure that they are not blanketed by the aerostat envelope. In addition, this structure serves as the basic framework on which the cargo hold is built.

Artist's impression of a Heli-Stat hauling timber

TAIL UNIT: Comprises light alloy frame with fabric covering, and mounts a composite control surface which serves the dual roles of rudder and elevator.

LANDING GEAR: Four non-retractable units, each with twin wheels, are mounted one at each corner of the basic fuselage frame.

POWER PLANT: Basically four Sikorsky SH-34J helicopters without landing gear, each powered by a 1,137 kW (1,525 hp) Wright R-1820-84A radial aircooled engine, and each driving the conventional rotor system of the SH-34J, last described in the 1970-71 *Jane's*. Total combined fuel capacity 4,640 litres (1,226 US gallons).

ACCOMMODATION: Crew of four, comprising a master pilot in the port aft helicopter, and a flight engineer in each of the other helicopters. Controls of the four helicopters linked by a fly-by-wire system.

SYSTEMS: Duplicated hydraulic system in each helicopter, pressure 103·5 bars (1,500 lb/sq in), for operation of rotor controls. Electrical system includes four 28V 400A engine-driven DC generators, and eight 115V 400Hz AC inverters in the helicopters.

EQUIPMENT: Includes full blind-flying instrumentation.

DIMENSIONS, ENVELOPE:
- Length overall: 74·07 m (243 ft 0 in)
- Max diameter: 26·06 m (85 ft 6 in)

DIMENSIONS, EXTERNAL:
- Width overall, rotors turning: 60·05 m (197 ft 0 in)
- Width overall, rotor blades folded: 45·72 m (150 ft 0 in)
- Wheel track: 24·17 m (79 ft 3½ in)
- Wheelbase: 13·41 m (44 ft 0 in)

DIMENSION, INTERNAL:
- Freight hold will accept two containers, each 2·44 × 2·44 × 12·19 m (8 × 8 × 40 ft)
- Volume (total): 145·0 m^3 (5,120 cu ft)

AREA:
- Rotor discs (total): 915·3 m^2 (9,852 sq ft)

WEIGHTS (estimated):
- Weight empty: 24,895 kg (54,885 lb)
- Max payload: 23,586 kg (52,000 lb)
- Max fuel load: 4,646 kg (10,244 lb)
- Max hovering weight: 50,469 kg (111,265 lb)
- Max landing weight: 30,267 kg (66,729 lb)

PERFORMANCE (estimated, at max T-O weight):
- Max level speed at S/L: 71 knots (132 km/h; 82 mph)
- Max cruising speed at S/L: 61 knots (113 km/h; 70 mph)
- Max rate of climb at S/L: 290 m (950 ft)/min
- Ballonet ceiling: 3,810 m (12,500 ft)
- Range with max fuel, 10% reserves: 1,784 nm (3,307 km; 2,055 miles)
- Range with max payload, 10% reserves: 43 nm (80 km; 50 miles)

PIPER

PIPER AIRCRAFT CORPORATION (Subsidiary of Bangor Punta Corporation)

HEAD OFFICE AND WORKS: Lock Haven, Pennsylvania 17745

Telephone: (717) 748 6711

Telex: 841425

OTHER WORKS:
Vero Beach, Florida 32960
Lakeland, Florida 33801
Piper, Pennsylvania 16845
Renovo, Pennsylvania 17764

PRESIDENT AND CHIEF EXECUTIVE OFFICER: Max E. Bleck

EXECUTIVE VICE-PRESIDENT AND CHIEF OPERATING OFFICER: Robert D. Dickerson

SENIOR VICE-PRESIDENTS:
Thomas W. Gillespie Jr (Marketing and Sales)
Alfred J. Koontz Jr (Finance)

GROUP VICE-PRESIDENTS:
Vincent J. Montuoro (Vero Beach Division)
Harry M. Graham (Lakeland and Lock Haven Divisions)

VICE-PRESIDENTS:
Marion J. Dees Jr (Engineering)
Findley A. Estlick (Plant Facilities & Foreign Assembly Operations)
C. Raymond Johnson (Sales)
George W. Mabey (Personnel)
John S. McCollom (Special Projects)

CONTROLLER: Jack J. Cattoni

TREASURER: John R. Leeson

DIRECTOR, CORPORATE COMMUNICATIONS: Edwin C. Watson

In June 1981, Piper announced that its twin-engined Aerostar production was being transferred from Santa Maria, California, to Vero Beach, Florida, the move to be completed by October 1981. In addition to its other production of Piper single-engined and light twin-engined aircraft, Vero Beach is responsible for the experimental development of Piper aircraft and houses one of the company's Plastics Divisions. Lock Haven has R & D facilities for aircraft built at Lock Haven.

Piper operates three other plants. The first two are at Piper, Pennsylvania, where sheet metal parts are formed, and Renovo, Pennsylvania, which makes plastics components; the third is at Lakeland, Florida, where PA-31 aircraft are in production.

In 1971, 1974 and 1977 respectively, Piper announced agreements with Chincul SA (which see) for the manufacture of a broad range of Piper products in Argentina; with EMBRAER (which see) for the development, production and marketing of Piper aircraft in Brazil; and with PZL Mielec (which see) of Poland to assemble, manufacture and distribute the Piper Seneca III light twin-engined aircraft in that country.

Piper Super Cub two-seat light cabin monoplane (Avco Lycoming O-320 engine)

PIPER (PA-18-150) SUPER CUB

The original PA-18 with 67 kW (90 hp) Continental C90-12F engine received FAA Type Approval on 18 November 1949. The PA-18-150, PA-18A-150 agricultural aircraft and PA-18S and PA-18AS seaplanes were all approved on 1 October 1954.

The 1981 version of the Super Cub introduced as a standard improvement the zinc chromate treatment of all aluminium components.

Five factory-installed avionics packages are available optionally for the Super Cub, comprising nav/coms and associated equipment by Collins, Edo-Aire Mitchell, King and Narco. Optional items available from the same manufacturers, plus United, include altitude reporting/digitisers, encoding altimeters, and transponders.

The current international height record in Class C1b (aircraft with T-O weight of 500-1,000 kg) is held by Miss C. Bayley of the USA, who climbed to a height of 9,206 m (30,203 ft) in a Super Cub with 93·2 kW (125 hp) Lycoming engine, on 4 January 1951.

More than 40,000 examples of the PA-18 Cub and its predecessors have been delivered.

TYPE: Two-seat light cabin monoplane.

WINGS: Braced high-wing monoplane, with steel tube V bracing struts each side. Wing section USA 35B. Thickness/chord ratio 12%. Dihedral 1°. No incidence at mean aerodynamic chord. Total washout of 3° 18′. Aluminium spars and ribs, aluminium sheet leading-edge and aileron false spar, wingtip bow of ash, with Dacron covering overall and fire-resistant Duraclad plastic finish. Plain ailerons and trailing-edge flaps of light alloy construction.

FUSELAGE: Rectangular welded steel tube structure covered with Dacron. Fire-resistant Duraclad plastic finish.

TAIL UNIT: Wire-braced structure of welded steel tubes and channels, covered with Dacron. Fire-resistant Duraclad plastic finish. Tailplane incidence variable for trimming. Balanced rudder and elevators.

LANDING GEAR: Non-retractable tailwheel type. Two side Vs and half axles hinged to bottom of fuselage. Rubber cord shock-absorption. Main-wheel tyres size 6·00-6 four-ply, with 7·00-6 six-ply, or 8·00-6 four-ply, optional. Steerable leaf-spring tailwheel. Scott 20 cm (8 in) tailwheel optional. Dual expanding brakes. Parking brake.

POWER PLANT: One 112 kW (150 hp) Avco Lycoming O-320 flat-four engine, driving a Sensenich two-blade fixed-pitch metal propeller with spinner. Steel tube engine mounting is hinged at firewall, allowing it to be swung to port for access to rear of engine. One 68 litre (18 US gallon) metal fuel tank in each wing. Total fuel capacity 136 litres (36 US gallons), of which 135·5 litres (35·8 US gallons) are usable. Refuelling points on top of wing.

ACCOMMODATION: Enclosed cabin seating two in tandem with dual controls. Adjustable front seat. Rear seat quickly removable for cargo carrying. Inertia-reel shoulder harness and seat belts standard for front and rear seats. Heater and adjustable cool-air vent. Downward-hinged door on starboard side, and upward-hinged window above, can be opened in flight. Sliding windows on port side. Baggage compartments aft of rear seat, capacity 22 kg (50 lb).

SYSTEMS: Electrical system, comprising 14V 60A alternator, 12V 35Ah battery, optional. Vacuum system for gyro driven instruments optional.

AVIONICS AND EQUIPMENT: In addition to the optional factory-installed avionics packages mentioned, Edo-Aire Mitchell, King, Narco and United avionics items are available optionally. Standard equipment includes a recording tachometer, sensitive altimeter, provisions for emergency locator transmitter, stall warning device, seat belts and shoulder safety belts, engine quick oil drain, and tiedown rings. Optional equipment includes blind-flying instrumentation, clock, vertical speed indicator, outside air temperature gauge, emergency locator transmitter, control locks, cabin fire extinguisher, stainless steel control cables, dome light, instrument panel lights, landing light, navigation lights, strobe light, metallising and polyurethane finish for steel components, and glider tow hook.

DIMENSIONS, EXTERNAL:
- Wing span: 10·73 m (35 ft 2½ in)
- Wing chord (constant): 1·60 m (5 ft 3 in)
- Wing aspect ratio: 7
- Length overall: 6·88 m (22 ft 7 in)
- Height overall: 2·04 m (6 ft 8½ in)
- Tailplane span: 3·20 m (10 ft 6 in)
- Wheel track: 1·84 m (6 ft 0½ in)
- Propeller diameter: 1·88 m (6 ft 2 in)

DIMENSION, INTERNAL:
- Baggage compartment: 0·51 m^3 (18 cu ft)

AREAS:
- Wings, gross: 16·58 m^2 (178·5 sq ft)
- Ailerons (total): 1·75 m^2 (18·80 sq ft)
- Trailing-edge flaps (total): 1·07 m^2 (11·50 sq ft)
- Fin: 0·43 m^2 (4·66 sq ft)
- Rudder: 0·63 m^2 (6·76 sq ft)
- Tailplane: 1·40 m^2 (15·10 sq ft)
- Elevators: 1·09 m^2 (11·70 sq ft)

WEIGHTS AND LOADINGS (N: Normal category; R: Restricted, agricultural, category):
- Weight empty: N, R: 446 kg (983 lb)
- Max T-O and landing weight: N: 794 kg (1,750 lb)
  - R: 939 kg (2,070 lb)
- Max wing loading: N: 48·8 kg/m^2 (10·0 lb/sq ft)
  - R: 56·64 kg/m^2 (11·6 lb/sq ft)
- Max power loading: N: 7·09 kg/kW (11·6 lb/hp)
  - R: 8·38 kg/kW (13·8 lb/hp)

PERFORMANCE (at max T-O weight: N: Normal category; R: Restricted, agricultural, category):
- Never-exceed speed:
  - N, R: 132 knots (246 km/h; 153 mph)
- Max level speed: N: 113 knots (209 km/h; 130 mph)
- Max cruising speed, 75% power at 1,525 m (5,000 ft):
  - N: 100 knots (185 km/h; 115 mph)
  - R: 91 knots (169 km/h; 105 mph)
- Stalling speed, flaps down:
  - N: 37 knots (69 km/h; 43 mph)
  - R: 39 knots (73 km/h; 45 mph)
- Max rate of climb at S/L: N: 293 m (960 ft)/min
  - R: 232 m (760 ft)/min
- Service ceiling: N: 5,795 m (19,000 ft)
  - R: 5,180 m (17,000 ft)
- Absolute ceiling: N: 6,492 m (21,300 ft)
- T-O run: N: 61 m (200 ft)
  - R: 92 m (300 ft)
- T-O to 15 m (50 ft): N: 153 m (500 ft)
  - R: 290 m (950 ft)
- Landing from 15 m (50 ft): N: 270 m (885 ft)
  - R: 267 m (875 ft)
- Landing run: N: 107 m (350 ft)
  - R: 125 m (410 ft)
- Range with max fuel and max payload, 75% power at optimum altitude: N: 400 nm (741 km; 460 miles)
  - R: 312 nm (580 km; 360 miles)

Piper Turbo Aztec F six-seat twin-engined executive transport

PIPER (PA-23-250) AZTEC F

The 1981 version of the Aztec F introduced as standard improvements an additional 5 cm (2 in) of forward travel on the pilot's seat, and a heater thermal overheat reset button in the cockpit. New options include an alternative light grey instrument panel colour, and a wider range of avionics. The Aztec is available in basic form and in several optional configurations, as follows:

Custom Aztec. As basic model, described in detail, with addition of true airspeed indicator, alternate instrument static source and window curtains; adding 2·6 kg (5·8 lb) to basic empty weight.

Sportsman Aztec. As Custom model, with addition of external power socket; adding 4·5 kg (10 lb) to basic empty weight.

Professional Aztec. As Sportsman model, with addition of electrically heated windscreen panel, electrical propeller de-icing, pneumatic de-icing boots on wings and tail, and wing ice inspection light; adding 23·6 kg (52·1 lb) to basic empty weight.

Turbo Aztec. Turbocharged version, described separately.

Each of the above versions can be fitted with one of three factory-installed avionics packages, including nav/coms, ADF, DME, glideslope and marker beacon receivers, transponder, autopilot, and associated equipment by Bendix, Collins, Edo-Aire Mitchell, King, Narco, Piper and SunAir.

The Aztec received FAA Type Approval as a five-seat aircraft on 18 September 1959, and with six seats on 15 December 1961.

The prototype of a floatplane version of the Aztec was produced as a joint project by Melridge Aviation of Vancouver, Washington, and Jobmaster Company Inc of Seattle. Fitted with Edo 4930 floats, this aircraft can take off from calm water in 20 seconds at max T-O weight of 2,360 kg (5,200 lb). Useful load is 816 kg (1,800 lb), permitting a six-passenger load with 455 litres (120 US gallons) of fuel. To simplify docking and loading from either side, a door was designed for installation on the port side, by the pilot's seat, and is part of the kit offered by Melridge Aviation to permit conversion in the field.

TYPE: Six-seat twin-engined executive transport.

WINGS: Cantilever low-wing monoplane. Wing section USA 35-B (modified). Thickness/chord ratio 14%. Dihedral 5°. Incidence 0° at root, −1° 12′ at mean chord. All-metal stressed-skin structure, with heavy stepped-down main spar, front and rear auxiliary spars, ribs, stringers and detachable wingtips. Plain all-metal ailerons and hydraulically-actuated flaps. Optional Goodrich de-icing system.

FUSELAGE: Basic aluminium semi-monocoque structure with welded steel tube truss around cabin. Bob weight in nose to provide more balanced control forces in pitch.

TAIL UNIT: Cantilever all-metal structure with swept fin and all-moving horizontal surfaces which interconnect mechanically with the flap system to avoid need for pitch re-trim when flaps are lowered. Trim tab in rudder. Geared anti-servo tab in horizontal surfaces. Optional Goodrich de-icing system.

LANDING GEAR: Retractable tricycle type. Hydraulic retraction, with CO_2 emergency extension system. Nosewheel retracts rearward, main wheels forward. Wheel doors enclose landing gear fully when retracted. Electrol oleo shock-absorber struts. Cleveland main wheels, size 6·00-6, with size 7·00-6 8-ply tyres. Cleveland steerable nosewheel, size 6·00-6, with 6·00-6 4-ply tyre. Hydraulic disc brakes. Parking brake.

POWER PLANT: Two 186·5 kW (250 hp) Avco Lycoming IO-540-C4B5 flat-six engines, each driving a Hartzell HC-E2YK-2RB two-blade constant-speed fully-feathering metal propeller. Two rubber fuel cells in each wing with NACA type anti-icing non-siphoning vents. Total standard fuel capacity 544 litres (144 US gallons); 519 litres (137·2 US gallons) usable. Refuelling points above wings. Two internal wingtip fuel tanks optional, with total capacity of 151 litres (40 US gallons). Oil capacity 22·7 litres (6 US gallons). Propeller synchroniser and electrical de-icing system optional.

ACCOMMODATION: Six persons on two pairs of adjustable individual seats and rear bench seat. Shoulder harnesses for pilot's and co-pilot's seat; seat belts for all seats. Dual controls standard. Individual seat lights and controllable overhead ventilation. 35,000 BTU heater with four adjustable cool/warm air outlets and two windscreen defrosters. Heated windscreen optional. Double pane windows. Passenger step. Door at front of cabin on starboard side. Emergency exit at rear on port side. Centre and rear seats removable to provide space for stretcher, survey camera or up to 725 kg (1,600 lb) of freight. Rear cabin bulkhead removable for stretcher and cargo loading via rear baggage door. Baggage compartments at rear of cabin and in nose, with tiedown fittings, each with capacity of 68 kg (150 lb). Baggage doors on starboard side; rear one enlarged on current aircraft, for stretcher loading. Armrests, cabin dome light, individual reading lights, coat hooks, complete soundproofing and two sun visors. Seat headrests and leather upholstery optional.

SYSTEMS: Hydraulic system with dual engine-driven pumps, pressure 79 bars (1,150 lb/sq in), for landing gear and flaps. Manually operated secondary hydraulic pump is standard. Vacuum system with two engine-driven pumps. Two 70A 28V alternators. 24V 17Ah battery. 28V 25Ah battery optional. Oxygen system, with 3·23 m³ (114 cu ft) bottle and six outlets, optional.

AVIONICS AND EQUIPMENT: Optional avionics, in addition to items included in factory-installed packages, include automatic flight system/flight directors, encoding altimeters, radar and radio altimeters, radio telephone, R/Nav, and weather radar. Standard equipment includes full blind-flying instrumentation with 3 in pictorial rate of turn indicator; artificial horizon and directional gyro; clock; dual recording tachometers/hour meters; outside air temperature gauge; sensitive altimeter; dual vacuum gauges; sun visors; cabin and baggage door locks; nosewheel safety mirror; heated pitot head; door-ajar indicator lights; stall warning device; baggage compartment, cabin dome, instrument, map, reading, anti-collision, landing, navigation and taxi lights; quick oil drains; jack pads; tiedown rings; towbar; and a choice of exterior trims. Optional items as mentioned under model listings, plus co-pilot's blind-flying instrumentation and toe-brakes, exhaust gas temperature gauge, emergency locator transmitter, inertia reel shoulder safety belts for rear seats, cabin fire extinguisher, tinted windows, leather upholstery, alternative instrument static source, instrument panel coloured light, strobe tail light, propeller ice protection shields, wing ice inspection light, and zinc chromate treatment of aluminium parts.

DIMENSIONS, EXTERNAL:

Wing span	11·37 m (37 ft 3¾ in)
Wing chord (constant)	1·70 m (5 ft 7 in)
Wing aspect ratio	6·8
Length overall	9·52 m (31 ft 2⅝ in)
Height overall	3·07 m (10 ft 1 in)
Tailplane span	3·81 m (12 ft 6 in)
Wheel track	3·45 m (11 ft 4 in)
Wheelbase	2·29 m (7 ft 6 in)
Propeller diameter	1·96 m (6 ft 5 in)
Cabin door: Height	0·97 m (3 ft 2 in)
Width	0·84 m (2 ft 9 in)
Baggage compartment door (front):	
Height	0·49 m (1 ft 7½ in)
Width	0·77 m (2 ft 6½ in)
Baggage compartment door (rear):	
Height	0·76 m (2 ft 6 in)
Width	0·79 m (2 ft 7 in)

DIMENSIONS, INTERNAL:

Cabin: Length	2·59 m (8 ft 6 in)
Max width	1·14 m (3 ft 9 in)
Max height	1·28 m (4 ft 2½ in)
Baggage compartments: front	0·49 m³ (17·4 cu ft)
rear	0·65 m³ (23·0 cu ft)
Max cargo space, incl baggage compartments	3·45 m³ (122 cu ft)

AREAS:

Wings, gross	19·23 m² (207·0 sq ft)
Ailerons (total)	0·77 m² (8·38 sq ft)
Trailing-edge flaps (total)	1·54 m² (16·60 sq ft)
Fin	1·37 m² (14·80 sq ft)
Rudder	0·96 m² (10·30 sq ft)
Horizontal surfaces (total)	3·70 m² (39·80 sq ft)

WEIGHTS AND LOADINGS:

Weight empty	1,444 kg (3,183 lb)
Max T-O and ramp weight	2,360 kg (5,200 lb)
Max landing weight	2,240 kg (4,940 lb)
Max wing loading	122·7 kg/m² (25·12 lb/sq ft)
Max power loading	6·33 kg/kW (10·4 lb/hp)

PERFORMANCE (at max T-O weight, except where indicated):

Never-exceed speed	240 knots (446 km/h; 277 mph)
Max level speed	187 knots (346 km/h; 215 mph)
Normal cruising speed at 1,175 m (3,850 ft)	179 knots (332 km/h; 206 mph)
Intermediate cruising speed at 1,830 m (6,000 ft)	177 knots (328 km/h; 204 mph)
Econ cruising speed at 1,905 m (6,250 ft)	170 knots (315 km/h; 196 mph)
Long-range cruising speed at 3,290 m (10,800 ft)	162 knots (301 km/h; 187 mph)
Stalling speed, power off, flaps up	61 knots (113 km/h; 70 mph)
Stalling speed, power off, flaps down	54·5 knots (101 km/h; 63 mph)
Max rate of climb at S/L	426 m (1,400 ft)/min
Rate of climb at S/L, one engine out	72 m (235 ft)/min
Service ceiling	5,365 m (17,600 ft)
Service ceiling, one engine out	1,465 m (4,800 ft)
T-O run, normal	363 m (1,190 ft)
T-O run, short-field	309 m (1,015 ft)
T-O to 15 m (50 ft), normal	604 m (1,980 ft)
T-O to 15 m (50 ft), short-field	517 m (1,695 ft)
Landing from 15 m (50 ft) at max landing weight, normal	483 m (1,585 ft)
Landing from 15 m (50 ft) at max landing weight, short-field	400 m (1,310 ft)
Landing run at max landing weight, normal	290 m (951 ft)
Landing run at max landing weight, short-field	239 m (785 ft)

Range with standard max fuel, allowances for start, taxi, T-O, climb, and 45 min reserves at long-range cruise power:

Intermediate cruise power	790 nm (1,464 km; 910 miles)
Econ cruise power	830 nm (1,538 km; 956 miles)
Long-range cruise power at 2,135 m (7,000 ft)	985 nm (1,825 km; 1,134 miles)

Range with optional max fuel, allowances as above:

Intermediate cruise power	1,060 nm (1,963 km; 1,220 miles)
Econ cruise power	1,110 nm (2,055 km; 1,277 miles)
Long-range cruise power at 2,135 m (7,000 ft)	1,320 nm (2,444 km; 1,519 miles)

PIPER (PA-23T-250) TURBO AZTEC F

The Turbo Aztec F is identical in every way with the Aztec F, except that it has 186·5 kW (250 hp) Avco Lycoming TIO-540-C1A engines, with Garrett turbocharging system. These specially modified engines allow a turbo cruise setting at 2,400 rpm, providing a constant manifold pressure from sea level to 6,705 m (22,000 ft), and result in considerably improved performance.

Standard equipment includes a density controller to prevent inadvertent overboost of the engines at full throttle, and a differential pressure controller to provide constant manifold pressure during cruising flight. An exhaust gas temperature gauge is standard.

WEIGHTS AND LOADINGS:

As for Aztec F, except:

Weight empty (standard)	1,507 kg (3,323 lb)

PERFORMANCE (at max T-O weight):

As for Aztec F, except:

Max level speed at 5,640 m (18,500 ft)	220 knots (407 km/h; 253 mph)
Turbo cruising speed at 6,705 m (22,000 ft)	215 knots (399 km/h; 248 mph)
Intermediate cruising speed at 7,315 m (24,000 ft)	208 knots (384 km/h; 239 mph)
Econ cruising speed at 7,315 m (24,000 ft)	193 knots (357 km/h; 222 mph)
Long-range cruising speed at 7,315 m (24,000 ft)	174 knots (322 km/h; 200 mph)
Max rate of climb at S/L	448 m (1,470 ft)/min
Rate of climb at S/L, one engine out	68 m (225 ft)/min

Service ceiling over 7,315 m (24,000 ft)
Service ceiling, one engine out 4,055 m (13,300 ft)
Range with standard max fuel, allowances for start, taxi, T-O, climb, and 45 min reserves at long-range cruise power:
 Turbo cruising speed 695 nm (1,287 km; 800 miles)
 Intermediate cruising speed 740 nm (1,371 km; 852 miles)
 Econ cruising speed 780 nm (1,445 km; 898 miles)
 Long-range cruising speed 835 nm (1,548 km; 962 miles)
Range with optional max fuel, allowances as above:
 Turbo cruising speed 947 nm (1,753 km; 1,089 miles)
 Intermediate cruising speed 1,020 nm (1,889 km; 1,173 miles)
 Econ cruising speed 1,075 nm (1,990 km; 1,237 miles)
 Long-range cruising speed 1,145 nm (2,120 km; 1,317 miles)

PIPER (PA-25-235) PAWNEE D

The current version of the Pawnee D is available with five optional factory-installed avionics packages comprising nav/coms and associated equipment by Collins, Edo-Aire Mitchell, King and Narco.

Type: Single-seat agricultural monoplane.

Wings: Braced low-wing monoplane. Streamline V bracing struts on each side of fuselage, with additional short support struts. Wing section USA 35B (modified). Thickness/chord ratio 12%. Dihedral 7°. Incidence 1° 18' at mean aerodynamic chord. Wings are of fabric-covered aluminium construction, with fire-resistant Duraclad plastic finish. Trailing-edge flaps and ailerons of light alloy construction. No trim tabs.

Fuselage: Basically rectangular-section welded steel tube structure, with fabric covering and Duraclad plastic finish, except for removable metal underskin and removable metal top of rear fuselage. Glassfibre engine cowling.

Tail Unit: Wire-braced steel tube structure with fabric covering and Duraclad plastic finish. Fixed-incidence tailplane. Balanced rudder and elevators. No trim tabs. Cable from top of cockpit to top of rudder to deflect wires and cables.

Landing Gear: Non-retractable tailwheel type. Bungee rubber shock-absorbers. Main gear has two side Vs and half-axles hinged to centreline of underside of fuselage. Cleveland 40-84A main wheels, with 8·00-6 4-ply tyres. Cleveland type 30-41 toe-actuated hydraulic brakes. Parking brake. Wire-cutters on leading-edge of each side V. Scott 200 mm (8 in) steerable tailwheel.

Power Plant: One 175 kW (235 hp) (derated) Avco Lycoming O-540 flat-six engine, driving a McCauley Type 1A200/FA84 two-blade fixed-pitch metal propeller with spinner. Fuel tank in each wing, with combined capacity of 145·7 litres (38·5 US gallons), of which 136 litres (36 US gallons) are usable. Oil capacity 11·5 litres (3 US gallons).

Accommodation: Pilot on adjustable seat in specially-strengthened enclosed cockpit, with steel tube overturn structure. Heavy-duty safety belt and shoulder harness with inertia reel. Wire-cutter mounted on centre of windscreen. Combined window and door on each side, hinged at bottom. Window assemblies jettisonable for emergency exit; quick-release hinge pins in side windows. Cabin is heated and ventilated. Adjustable cool air vents. Utility compartment under seat.

Systems: Electrical system includes a 14V 60A alternator, 12V 35Ah battery and a battery charging diode. Hydraulic system for brakes only.

Avionics and Equipment: Optional avionics, additional to items included in factory-installed packages, add a transponder by each manufacturer. Standard equipment includes a non-corrosive hopper/tank, installed forward of cockpit and approximately on CG, volume of which is 0·59 m³ (21 cu ft) or 568 litres (150 US gallons), with capacity for 544 kg (1,200 lb) of dust; quick-change boom brackets, and quick dump valve; recording tachometer; sensitive altimeter; under-seat storage; low-fuel quantity warning light; provisions for emergency locator transmitter; full-flow oil filter; quick-drain oil sump; top-deck loading door; and tiedown rings. Optional high and low volume dispersal systems for wet or dry applications with associated equipment, side loading nozzle for liquid chemicals, electric turn and bank indicator, control lock, multi-directional inertia reel shoulder harness, emergency locator transmitter, hand held fire extinguisher, rotating beacon, landing light and navigation lights.

Dimensions, external:
 Wing span 11·02 m (36 ft 2 in)
 Wing chord (constant) 1·60 m (5 ft 3 in)
 Wing aspect ratio 7·15
 Length overall 7·53 m (24 ft 8½ in)
 Height overall 2·21 m (7 ft 3 in)
 Tailplane span 2·90 m (9 ft 6 in)
 Wheel track 2·13 m (7 ft 0 in)
 Wheelbase 5·52 m (18 ft 1¼ in)
 Propeller diameter 2·13 m (7 ft 0 in)

Areas:
 Wings, gross 17·0 m² (183 sq ft)
 Ailerons (total) 1·78 m² (19·2 sq ft)
 Trailing-edge flaps (total) 0·78 m² (8·4 sq ft)
 Fin 0·35 m² (3·8 sq ft)
 Rudder 0·64 m² (6·9 sq ft)
 Tailplane 1·21 m² (13·0 sq ft)
 Elevators 1·27 m² (13·7 sq ft)

Weights and Loadings:
 Weight empty:
  no dispersal equipment 725 kg (1,598 lb)
  duster 746 kg (1,645 lb)
  sprayer 749 kg (1,652 lb)
 Max T-O and landing weight 1,315 kg (2,900 lb)
 Max wing loading 77·15 kg/m² (15·8 lb/sq ft)
 Max power loading 7·51 kg/kW (12·3 lb/hp)

Performance (at max T-O weight, except where indicated):
 Never-exceed speed 135 knots (251 km/h; 156 mph)
 Max level speed at S/L:
  no dispersal equipment 108 knots (200 km/h; 124 mph)
  duster 96 knots (177 km/h; 110 mph)
  sprayer 102 knots (188 km/h; 117 mph)
 Max cruising speed (75% power):
  no dispersal equipment 99 knots (183 km/h; 114 mph)
  duster 87 knots (161 km/h; 100 mph)
  sprayer 91 knots (169 km/h; 105 mph)
 Stalling speed, flaps down 53 knots (98 km/h; 61 mph)
 Stalling speed at normal landing weight of 771 kg (1,700 lb) 40 knots (74 km/h; 46 mph)
 Max rate of climb at S/L:
  no dispersal equipment 213 m (700 ft)/min
  duster 152 m (500 ft)/min
  sprayer 192 m (630 ft)/min
 T-O run: no dispersal equipment 239 m (785 ft)
  duster 291 m (956 ft)
  sprayer 244 m (800 ft)
 T-O to 15 m (50 ft):
  no dispersal equipment 411 m (1,350 ft)
  duster 448 m (1,470 ft)
  sprayer 418 m (1,370 ft)
 Max landing run 259 m (850 ft)
 Range (75% power) with max fuel:
  no dispersal equipment 251 nm (467 km; 290 miles)
  duster 221 nm (410 km; 255 miles)
  sprayer 234 nm (434 km; 270 miles)

Piper Pawnee D agricultural aircraft (Avco Lycoming O-540 flat-six engine)

PIPER (PA-28-161) WARRIOR II

Piper Warrior II four-seat cabin monoplane

Design of the Warrior began in June 1972, an important feature being replacement of the earlier constant-chord wings of the Cherokee series by a longer-span wing with tapered outer panels. As a result of its introduction the Warrior, which at that time had essentially the same 112 kW (150 hp) engine as the discontinued Cherokee Cruiser, was certificated at a maximum T-O weight 79 kg (175 lb) greater. First flight of a prototype was made on 17 October 1972, and FAA certification of the original Model PA-28-151 was granted on 9 August 1973.

The current PA-28-161 Warrior II version, first flown on 27 August 1976, has a 119 kW (160 hp) engine which operates on 100 octane low-lead fuel.

Two groups of optional equipment are available for the current Warrior II as basic packages:

Custom. Includes blind-flying instrumentation with 3 in gyros; clock; outside air temperature gauge; rate of climb and true airspeed indicators; instrument panel white backlighting; overhead red spotlight; cabin dome, radio dimming, navigation, landing and taxi lights; rotating beacon; assist strap; aircraft step; engine primer system; engine-driven vacuum pump with vacuum gauge; quick oil drain; wheel fairings; and towbar; adding 19·7 kg (43·4 lb) to basic empty weight.

Executive. As Custom package less rotating beacon, plus pilot's vertically-adjustable seat, sun visors, alternate static source, heated pitot, emergency locator transmitter, strobe lights, external power socket, and 35Ah battery; adding 33·4 kg (73·7 lb) to basic empty weight.

In addition, ten optional avionics groups are available for factory installation in the Warrior II.

Type: Four-seat cabin monoplane.

Wings: Cantilever low-wing monoplane. Wing section NACA 65_2-415 on inboard panels; outboard leading-edge incorporates modification No. 5 of NACA TN 2228. Dihedral 7°. Incidence 2° at root, −1° at tip. Sweepback at quarter-chord 5°. Light alloy single-spar structure with glassfibre wingtips. Plain ailerons of light alloy construction. Four-position manually-actuated trailing-edge flaps constructed of light alloy with ribbed skins.

Fuselage: Light alloy semi-monocoque structure with glassfibre nose cowl and tailcone.

Tail Unit: Cantilever structure of light alloy, except for glassfibre tips on fin and tailplane. Fin and rudder have ribbed light alloy skins. One-piece all-moving tailplane, with combined anti-servo and trim tab. Rudder trimmable, but no trim tab in rudder.

Landing Gear: Non-retractable tricycle type. Steerable nosewheel. Piper oleo-pneumatic shock-absorbers; single wheel on each unit. Cleveland wheels with 4-ply tyres size 6·00-6 on main units, pressure 1·65 bars (24 lb/sq in). Cleveland nosewheel and 4-ply tyre size

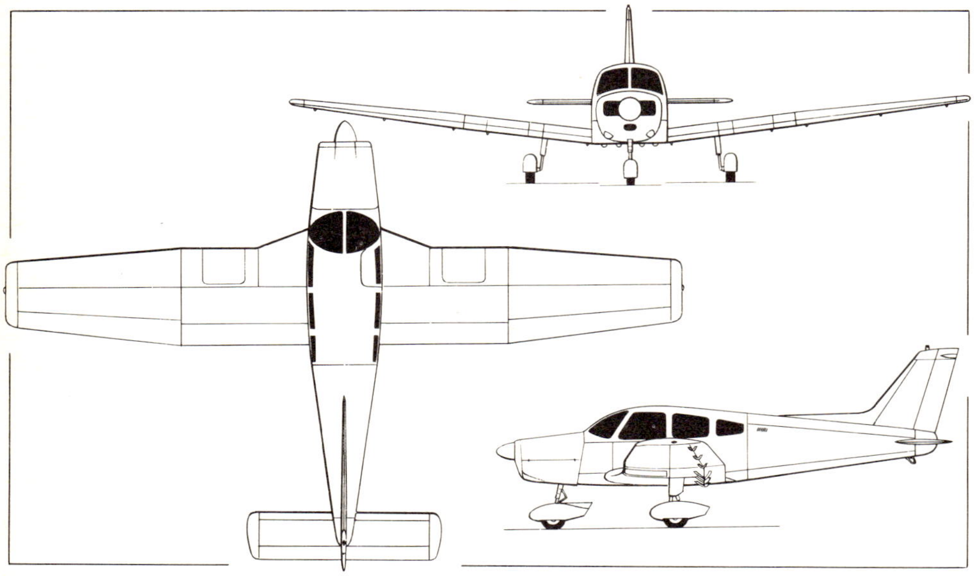

Piper Warrior II (Avco Lycoming O-320-D3G engine) *(Pilot Press)*

5·00-5, pressure 1·65 bars (24 lb/sq in). Cleveland disc brakes. Parking brake. Glassfibre wheel fairings optional.

POWER PLANT: One 119 kW (160 hp) Avco Lycoming O-320-D3G flat-four engine, driving a Sensenich two-blade fixed-pitch metal propeller type 74DM6-0-60 with spinner. Fuel in two wing tanks, with total capacity of 189 litres (50 US gallons), of which 181·5 litres (48 US gallons) are usable. Refuelling point on upper surface of each wing. Oil capacity 7·5 litres (2 US gallons).

ACCOMMODATION: Four persons in pairs in enclosed cabin. Individual adjustable front seats, bench type rear seat with seat belts. Shoulder harnesses with inertia reel on two front seats. Dual controls and brakes standard. Large door on starboard side. Baggage compartment and hatshelf at rear of cabin, with volume of 0·68 m^3 (24 cu ft) and capacity of 90 kg (200 lb). External access door on starboard side. Heating, ventilation and windscreen defrosting standard.

SYSTEMS: Hydraulic system for brakes only. Electrical system powered by 14V 60A engine-driven alternator. 12V 25Ah battery standard, 12V 35Ah battery optional. Vacuum system for optional blind-flying instrumentation is available, complete with vacuum gauge, regulator, filter, and annunciator light. Piper Aire air-conditioning system optional.

AVIONICS AND EQUIPMENT: Nav/coms, ADF, autopilot, glideslope and marker beacon receivers, and transponders, by Bendix, Collins, Edo-Aire Mitchell, King, Narco and Piper are available in ten optional groups, together with an extensive range of other optional avionics. Standard equipment includes recording tachometer, sensitive altimeter, armrests, map pockets, pilot's storm window, stall warning device, carpeted floor, soundproofing, provisions for emergency locator transmitter, provisions for air-conditioning, full-flow oil filter, tiedown rings and fuel tank quick drains. Optional equipment, in addition to that listed under Custom and Executive packages, includes encoding altimeter, digital clock, engine hour recorder, outside air temperature gauge, pilot and co-pilot headrests, cabin fire extinguisher, inertia reel shoulder safety belts for rear seats, floor and/or overhead ventilation with booster fan, stainless steel control cables, tinted windows, carburettor ice detection system, and zinc chromate paint for aluminium parts.

DIMENSIONS, EXTERNAL:

Wing span	10·67 m (35 ft 0 in)
Wing chord at root	1·60 m (5 ft 3 in)
Wing chord at tip	1·07 m (3 ft 6¼ in)
Wing aspect ratio	7·24
Length overall	7·25 m (23 ft 9½ in)
Height overall	2·22 m (7 ft 3½ in)
Tailplane span	3·96 m (12 ft 11¾ in)
Wheel track	3·05 m (10 ft 0 in)
Wheelbase	2·03 m (6 ft 8 in)
Propeller diameter	1·88 m (6 ft 2 in)
Propeller ground clearance	0·21 m (8¼ in)
Cabin door: Height	0·89 m (2 ft 11 in)
Width	0·91 m (3 ft 0 in)
Baggage door: Height	0·51 m (1 ft 8 in)
Max width	0·56 m (1 ft 10 in)
Height to sill	0·71 m (2 ft 4 in)

DIMENSIONS, INTERNAL:

Cabin: Length (instrument panel to rear bulkhead)	2·46 m (8 ft 1 in)
Max width	1·05 m (3 ft 5½ in)
Max height	1·24 m (4 ft 1 in)
Floor area	2·28 m^2 (24·5 sq ft)
Volume (incl baggage area)	3·00 m^3 (106·0 cu ft)

AREAS:

Wings, gross	15·8 m^2 (170 sq ft)
Ailerons (total)	1·23 m^2 (13·2 sq ft)
Trailing-edge flaps (total)	1·36 m^2 (14·6 sq ft)
Fin	0·69 m^2 (7·4 sq ft)
Rudder	0·38 m^2 (4·1 sq ft)
Tailplane, incl tab	2·46 m^2 (26·5 sq ft)

WEIGHTS AND LOADINGS:

Weight empty, standard	614 kg (1,353 lb)
Max T-O and landing weight	1,054 kg (2,325 lb)
Max wing loading	66·74 kg/m^2 (13·67 lb/sq ft)
Max power loading	8·86 kg/kW (14·53 lb/hp)

PERFORMANCE (at max T-O weight):

Never-exceed speed	153 knots (282 km/h; 176 mph)
*Max level speed at S/L	126 knots (233 km/h; 145 mph)
*Best power cruising speed, 75% power at 2,745 m (9,000 ft)	127 knots (235 km/h; 146 mph)
*Best power cruising speed, 65% power at 3,810 m (12,500 ft)	118 knots (219 km/h; 136 mph)
*Best power cruising speed, 55% power at 3,810 m (12,500 ft)	107 knots (198 km/h; 123 mph)
*Best econ cruising speed, 75% power at 2,745 m (9,000 ft)	123 knots (227 km/h; 141 mph)
*Best econ cruising speed, 65% power at 3,810 m (12,500 ft)	116 knots (215 km/h; 134 mph)
*Best econ cruising speed, 55% power at 3,810 m (12,500 ft)	103 knots (191 km/h; 119 mph)
Stalling speed, flaps up	56 knots (104 km/h; 65 mph) CAS
Stalling speed, flaps down	50 knots (93 km/h; 57·5 mph) CAS
Max rate of climb at S/L	216 m (710 ft)/min
Service ceiling	3,960 m (13,000 ft)
Absolute ceiling	4,600 m (15,100 ft)
T-O run	297 m (975 ft)
T-O to 15 m (50 ft)	454 m (1,490 ft)
Landing from 15 m (50 ft)	340 m (1,115 ft)
Landing run	181 m (595 ft)

*Range with max fuel with allowances for taxi, T-O, climb, and descent, and 45 min reserves at max range power:

at best power settings:	
75% power at 2,745 m (9,000 ft)	525 nm (972 km; 604 miles)
65% power at 3,810 m (12,500 ft)	553 nm (1,025 km; 637 miles)
55% power at 3,810 m (12,500 ft)	565 nm (1,047 km; 651 miles)
at best econ power settings:	
75% power at 2,745 m (9,000 ft)	590 nm (1,092 km; 679 miles)
65% power at 3,810 m (12,500 ft)	633 nm (1,173 km; 729 miles)
55% power at 3,810 m (12,500 ft)	640 nm (1,186 km; 737 miles)

**With optional wheel fairings*

PIPER (PA-28-181) ARCHER II

On 9 October 1972 Piper introduced the Cherokee Challenger as successor to the Cherokee 180. In 1974 this was superseded by the Cherokee Archer, with the same basic airframe and power plant, but with many additional equipment and avionics options. In 1976 this aircraft was redesignated PA-28-181 Cherokee Archer II, and in 1978 introduced the tapered wings of the Warrior II.

The Archer II can be fitted with generally similar optional Custom and Executive equipment packages, as well as the ten optional avionics packages, which are available for the Warrior II. The equipment packages vary as follows:

Custom. As for Warrior II, except for deletion of towbar and engine primer system.

Executive. As for Warrior II, plus exhaust gas temperature gauge, rotating beacon and strobe lights for wings only, and deletion of sun visors, towbar and engine primer system.

TYPE: Four-seat cabin monoplane.

WINGS: Cantilever low-wing monoplane. Wing section NACA 65_2-415 on inboard panels; outboard leading-edge has modification No. 5 of NACA TN 2228. Dihedral 7°. Incidence 2° at root, −1° at tip. Sweepback at quarter-chord 5°. Light alloy single-spar structure with glassfibre wingtips. Plain ailerons of light alloy construction. Trailing-edge flaps constructed of light alloy with ribbed skins.

FUSELAGE: Aluminium alloy semi-monocoque structure. Glassfibre engine cowling.

TAIL UNIT: Cantilever structure of aluminium alloy, except for glassfibre tips on fin and tailplane. Fin and rudder have corrugated metal skin. One-piece all-moving horizontal surface with combined anti-servo and trim tab. Trim tab in rudder.

LANDING GEAR: Non-retractable tricycle type. Steerable nosewheel. Piper oleo-pneumatic shock-absorbers. Cleveland wheels and Schenuit tyres, size 6·00-6, 4-ply rating, on all three wheels. Cleveland high capacity disc brakes. Parking brake. Wheel speed fairings optional.

POWER PLANT: One 134 kW (180 hp) Avco Lycoming O-360-A4M flat-four engine, driving a Sensenich two-blade fixed-pitch metal propeller with spinner. Fuel in two tanks in wing leading-edges, with total capacity of 189 litres (50 US gallons), of which 181·5 litres (48 US gallons) are usable. Oil capacity 7·5 litres (2 US gallons).

ACCOMMODATION: Four persons in pairs in enclosed cabin. Individual adjustable front seats, with dual controls; individual rear seats. Large door on starboard side. Baggage compartment aft of cabin, with volume of 0·68 m^3 (24 cu ft) and capacity of 90 kg (200 lb); door on

Piper Archer II, which has the tapered wings introduced on the Warrior

starboard side. Hatshelf. Rear seats removable to provide 1·25 m³ (44 cu ft) cargo space. Accommodation heated and ventilated. Windscreen defrosting.

Systems: Optional Piper Aire air-conditioning system. Electrical system includes 60A alternator and 12V 25Ah battery. 35Ah battery optional. Hydraulic system for brakes only. Vacuum system optional.

Avionics and Equipment: As for Warrior II, except that an engine priming system, sun visors and towbar are standard; an exhaust gas temperature gauge is optional for the Custom version.

Dimensions, external and internal:
As for Warrior II, except:
Tailplane span 3·92 m (12 ft 10½ in)
Wheelbase 2·00 m (6 ft 7 in)
Propeller diameter 1·93 m (6 ft 4 in)

Areas: As for Warrior II

Weights and Loadings:
Weight empty, equipped (standard) 641 kg (1,413 lb)
Max T-O and landing weight 1,156 kg (2,550 lb)
Max wing loading 73·2 kg/m² (15·0 lb/sq ft)
Max power loading 8·63 kg/kW (14·17 lb/hp)

Performance (at max T-O weight):
Never-exceed speed 148 knots (275 km/h; 171 mph) CAS
*Max level speed at S/L 128 knots (237 km/h; 147 mph)
*Best power cruising speed, 75% power at 2,440 m (8,000 ft) 129 knots (239 km/h; 148·5 mph)
*Best power cruising speed, 65% power at 3,660 m (12,000 ft) 125 knots (231 km/h; 144 mph)
*Best power cruising speed, 55% power at 3,810 m (12,500 ft) 111 knots (206 km/h; 128 mph)
*Best econ cruising speed, 75% power at 2,440 m (8,000 ft) 126 knots (233 km/h; 145 mph)
*Best econ cruising speed, 65% power at 3,660 m (12,000 ft) 122 knots (225 km/h; 140 mph)
*Best econ cruising speed, 55% power at 3,810 m (12,500 ft) 107 knots (198 km/h; 123 mph)
Stalling speed, flaps up 59 knots (109 km/h; 68 mph) CAS
Stalling speed, flaps down 53 knots (98 km/h; 61 mph) CAS
Max rate of climb at S/L 224 m (735 ft)/min
Service ceiling 4,160 m (13,650 ft)
Absolute ceiling 4,800 m (15,750 ft)
T-O run 265 m (870 ft)
T-O to 15 m (50 ft) 495 m (1,625 ft)
Landing from 15 m (50 ft) 424 m (1,390 ft)
Landing run 282 m (925 ft)
*Range with max fuel, allowances for taxi, T-O, climb and descent, and 45 min reserves at max range power:
at best power settings:
75% power at 2,440 m (8,000 ft) 529 nm (979 km; 608 miles)
65% power at 3,660 m (12,000 ft) 570 nm (1,055 km; 656 miles)
55% power at 3,810 m (12,500 ft) 586 nm (1,085 km; 674 miles)
at best econ power settings:
75% power at 2,440 m (8,000 ft) 600 nm (1,112 km; 691 miles)
65% power at 3,660 m (12,000 ft) 645 nm (1,196 km; 743 miles)
55% power at 3,810 m (12,500 ft) 670 nm (1,242 km; 772 miles)

With optional wheel fairings

PIPER (PA-28RT-201) ARROW IV

The Piper Arrow derives from the Cherokee Arrow II, which was generally similar to the Cherokee Archer II but had a retractable landing gear, more powerful engine, and the untapered wings of the 1975 PA-28-180 Archer. In 1977, Piper updated this model by fitting long-span tapered wings identical with those of the Archer II, but with increased fuel capacity, giving improved performance. The 1978 version of this aircraft was named Arrow III, the prototype of which flew for the first time on 16 September 1975, followed by the first production aircraft on 7 January 1977. Piper designation was PA-28R-201. The Turbo Arrow III differed by having a turbocharged engine, mounted in a streamline cowling, and the first production example of this version flew on 1 December 1976.

In 1979 Piper introduced an improved model with an all-moving T tailplane, and two versions are available currently:

Arrow IV. Basic model, as described; Piper designation **PA-28RT-201.**

Turbo Arrow IV. As Arrow IV, but power plant comprises one 149 kW (200 hp) Continental TSIO-360-FB flat-six turbocharged engine, mounted in a streamlined cowling, and driving a Hartzell two-blade constant-speed metal propeller with spinner. A three-blade propeller is optional (essential when built-in oxygen system is installed). Piper designation **PA-28RT-201T**.

Two groups of optional equipment are available for the Arrow IV as recommended packages:

Custom. Includes blind-flying instrumentation with 3 in gyros; electric clock; outside air temperature gauge; rate of climb and true airspeed indicators; vacuum system for gyro instruments; a basic lighting package comprising instrument panel white backlighting with overhead red lighting, cabin dome and radio dimming lights; landing, navigation, taxi and wing strobe lights; 35Ah battery; assist strap; and entrance step; adding 15·4 kg (34·0 lb) to aircraft basic weight.

Executive. As Custom package, plus alternate static source; heated pitot; landing gear sensor; emergency locator transmitter; pilot's vertically adjustable seat; and external power socket; adding 22·1 kg (48·7 lb) to aircraft basic weight.

In addition to the above equipment, the avionics packages which are available for the Warrior II are also available optionally for the Arrow IV.

The tricycle landing gear of both versions is retracted hydraulically, with an electrically-operated pump supplying the hydraulic pressure. In addition to the usual 'gear up' warning horn and red light, both Arrow IVs have an automatic extension system which drops the landing gear automatically if power is reduced and airspeed drops below 91 knots (169 km/h; 105 mph). The sensing system consists of a small probe mounted on the port side of the fuselage. Being located in the propeller slipstream, it can differentiate between a climb with power on and an approach to land with power reduced. A free-fall emergency extension system is also fitted. An 'anti-retraction' system guards against premature retraction of the landing gear below an airspeed of 74 knots (137 km/h; 85 mph) at take-off, or accidental retraction on the ground. There is also a manual override lever by which the pilot can hold the landing gear retracted as airspeed falls below 91 knots (169 km/h; 105 mph).

The description of the Archer II applies also to the Arrow IV, except for the following details:

Tail Unit: Cantilever T tail of light alloy construction. All-moving tailplane with trim tab. Rudder trim.

Landing Gear: Retractable tricycle type, with single wheel on each unit. Hydraulic retraction, main units inward into wings, nose unit rearward. Free-fall emergency extension system. All units fitted with oleo-pneumatic shock-absorbers. Main wheels and tyres size 6·00-6, 6-ply rating. Nosewheel and tyre size 5·00-5, 4-ply rating. High-capacity dual hydraulic disc brakes and parking brake.

Power Plant: Arrow IV has one 149 kW (200 hp) Avco Lycoming IO-360-C1C6 flat-four engine, driving a McCauley two-blade constant-speed metal propeller with spinner. Turbo Arrow IV as detailed in model listing. Both versions have fuel tanks in wing leading-edges with total capacity of 291 litres (77 US gallons), of which 273 litres (72 US gallons) are usable. Oil capacity 7·5 litres (2 US gallons).

Systems: Generally as for Archer II and Warrior II, plus electro-hydraulic system for landing gear actuation. An oxygen system is available optionally for the Turbo Arrow IV.

Avionics and Equipment: Optional avionics and standard equipment generally as detailed for Warrior II, plus cylinder head temperature gauge, exhaust gas temperature gauge, 65A electric generator, dual range electric auxiliary fuel pump, all standard on Turbo Arrow IV. Optional equipment as listed in Custom and Executive paragraphs and for Archer II and Warrior II, plus cold weather start kit for Turbo Arrow IV.

Dimensions, external:
Wing span 10·80 m (35 ft 5 in)
Wing chord at root 1·60 m (5 ft 3 in)
Wing chord at tip 1·07 m (3 ft 6¼ in)
Length overall: Arrow IV 8·23 m (27 ft 0 in)
Turbo Arrow IV 8·33 m (27 ft 3¾ in)
Height overall 2·52 m (8 ft 3¼ in)
Tailplane span 3·30 m (10 ft 10 in)
Wheel track 3·19 m (10 ft 5½ in)
Wheelbase 2·39 m (7 ft 10¼ in)
Propeller diameter 1·93 m (6 ft 4 in)
Cabin door (stbd): Width 0·91 m (3 ft 0 in)
Height 0·89 m (2 ft 11 in)
Baggage door (stbd): Width 0·56 m (1 ft 10 in)
Height 0·51 m (1 ft 8 in)

Dimensions, internal:
Cabin:
Length, panel to rear bulkhead 2·46 m (8 ft 1 in)
Max width 1·05 m (3 ft 5½ in)
Max height 1·24 m (4 ft 1 in)
Volume (incl baggage area) 3·00 m³ (106 cu ft)

Area:
Wings, gross 15·79 m² (170 sq ft)

Weights and Loadings (A: Arrow IV; B: Turbo Arrow IV):
Weight empty: A 742 kg (1,637 lb)
B 767 kg (1,692 lb)
Max T-O weight: A 1,247 kg (2,750 lb)
B 1,315 kg (2,900 lb)
Max wing loading: A 79·0 kg/m² (16·18 lb/sq ft)
B 83·29 kg/m² (17·06 lb/sq ft)
Max power loading: A 8·37 kg/kW (13·75 lb/hp)
B 8·83 kg/kW (14·5 lb/hp)

Performance (at max T-O weight):
Never-exceed speed 186 knots (344 km/h; 214 mph) CAS
Max level speed:
A at S/L 152 knots (282 km/h; 175 mph)
B at 4,265 m (14,000 ft) 178 knots (330 km/h; 205 mph)
Best power cruising speed, 75% power at optimum altitude: A 143 knots (265 km/h; 165 mph)
B 172 knots (319 km/h; 198 mph)
Best power cruising speed, 65% power at optimum altitude: A 138 knots (256 km/h; 159 mph)
B 167 knots (309 km/h; 192 mph)
Best power cruising speed, 55% power at optimum altitude: A 128 knots (237 km/h; 147 mph)
B 157 knots (291 km/h; 181 mph)
Best econ cruising speed, 75% power at optimum altitude: A 135 knots (250 km/h; 155 mph)
B 168 knots (311 km/h; 193 mph)
Best econ cruising speed, 65% power at optimum altitude: A 130 knots (241 km/h; 150 mph)
B 164 knots (304 km/h; 189 mph)
Best econ cruising speed, 55% power at optimum altitude: A 123 knots (228 km/h; 142 mph)
B 153 knots (284 km/h; 176 mph)
Stalling speed, flaps up:
A 60 knots (111 km/h; 69 mph) CAS
B 63 knots (117 km/h; 72 mph) CAS
Stalling speed, flaps down:
A 55 knots (102 km/h; 63·5 mph) CAS
B 58 knots (107·5 km/h; 67 mph) CAS
Max rate of climb at S/L: A 253 m (831 ft)/min
B 287 m (940 ft)/min
Service ceiling: A 4,940 m (16,200 ft)
*B 6,100 m (20,000 ft)
Absolute ceiling: A 5,610 m (18,400 ft)
*B 6,100 m (20,000 ft)
T-O run: A 312 m (1,025 ft)
B 338 m (1,110 ft)
T-O to 15 m (50 ft): A 488 m (1,600 ft)
B 494 m (1,620 ft)
Landing from 15 m (50 ft): A 465 m (1,525 ft)
B 474 m (1,555 ft)
Landing run: A 188 m (615 ft)
B 197 m (645 ft)
Range with max fuel, allowances for taxi, T-O, climb and descent, and 45 min reserves at max range power:
at best power settings:
75% power: A 730 nm (1,353 km; 840 miles)
B 695 nm (1,287 km; 800 miles)

Piper Turbo Arrow IV retractable-gear four-seat cabin monoplane with T tail

65% power: A 795 nm (1,473 km; 915 miles)
B 725 nm (1,343 km; 835 miles)
55% power: A 865 nm (1,603 km; 996 miles)
B 775 nm (1,435 km; 892 miles)
at best econ power settings:
75% power: A 785 nm (1,455 km; 904 miles)
B 790 nm (1,465 km; 910 miles)
65% power: A 850 nm (1,575 km; 979 miles)
B 830 nm (1,539 km; 956 miles)
55% power: A 935 nm (1,733 km; 1,076 miles)
B 900 nm (1,667 km; 1,036 miles)

**Max approved operating altitude*

PIPER (PA-28-236) DAKOTA

Piper introduced in 1978 an addition to the Warrior, Archer, Arrow line known as the PA-28-236 Dakota, which differs primarily by having a 175 kW (235 hp) Avco Lycoming engine to provide increased performance, and increased capacity fuel tanks to cater for this power plant. Like the related aircraft mentioned earlier, it is available with a variety of optional factory-installed avionics packages, and also two operational groups as follows:

Custom. As detailed for Warrior II, but with deletion of engine primer system, quick oil drain and towbar; adding 20·1 kg (44·4 lb) to basic empty weight.

Executive. As detailed for Warrior II, but with deletion of sun visors, engine primer system, quick oil drain and towbar; plus a tail-mounted rotating beacon instead of strobe light; adding 31·6 kg (69·7 lb) to basic empty weight.

Licence assembly of the Dakota is undertaken by the Chilean Air Force (Fuerza Aérea Chilena, which see); a total of 30 had been completed by mid-1981.

The description of the Archer II applies also to the Dakota, except as follows:

Power Plant: One 175 kW (235 hp) Avco Lycoming O-540-J3A5D flat-six engine, driving a Hartzell two-blade constant-speed metal propeller with spinner. Two integral fuel tanks in each wing, with a total capacity of 291 litres (77 US gallons), of which 273 litres (72 US gallons) are usable. Refuelling points on upper surface of each wing. Oil capacity 11·5 litres (3 US gallons).

Avionics and Equipment: As with the Archer II and Warrior II, ten optional factory-installed avionics packages are available, as well as a wide range of other avionics equipment to customer's requirements. Standard equipment as for Warrior II, plus quick oil drain valve. Optional equipment includes items provided in the Custom and Executive packages, plus engine hour recorder, exhaust gas temperature gauge, outside air temperature gauge, digital clock, co-pilot's vertically adjustable seat, cabin fire extinguisher, de luxe interior, shoulder safety belts with inertia reel for rear seats, super soundproofing, overhead air vents, Piper Aire air-conditioning, stainless steel control cables, tinted windows, ventilation fan, heavy-duty tyres and brakes, and zinc chromate treatment of aluminium parts.

Dimensions, external and internal: As for Archer II except:
Length overall 7·54 m (24 ft 8¾ in)
Height overall 2·18 m (7 ft 2 in)
Wheelbase 1·98 m (6 ft 6 in)

Weights and Loadings:
Weight empty 730 kg (1,608 lb)
Max T-O weight 1,361 kg (3,000 lb)
Max wing loading 85·93 kg/m² (17·6 lb/sq ft)
Max power loading 7·78 kg/kW (12·8 lb/hp)

Performance (at max T-O weight):
Max level speed at S/L 148 knots (274 km/h; 170 mph)
Best power cruising speed, 75% power at optimum altitude 144 knots (267 km/h; 166 mph)
Best power cruising speed, 65% power at optimum altitude 138 knots (256 km/h; 159 mph)
Best power cruising speed, 55% power at optimum altitude 130 knots (241 km/h; 150 mph)
Best econ cruising speed, 75% power at optimum altitude 139 knots (258 km/h; 160 mph)
Best econ cruising speed, 65% power at optimum altitude 134 knots (248 km/h; 154 mph)
Best econ cruising speed, 55% power at optimum altitude 126 knots (234 km/h; 145 mph)
Stalling speed, flaps up 63 knots (117 km/h; 72·5 mph) CAS
Stalling speed, flaps down 56 knots (104 km/h; 64·5 mph) CAS
Max rate of climb at S/L 338 m (1,110 ft)/min
Service ceiling 5,335 m (17,500 ft)
Absolute ceiling 5,945 m (19,500 ft)
T-O run 270 m (886 ft)
T-O to 15 m (50 ft) 371 m (1,216 ft)
Landing from 15 m (50 ft):
standard brakes 526 m (1,725 ft)
heavy-duty brakes 466 m (1,530 ft)
Landing run:
standard brakes 252 m (825 ft)
heavy-duty brakes 195 m (640 ft)
Range with max fuel, allowances for taxi, T-O, climb, cruise, descent, and 45 min reserves at max range power:
at best power settings at optimum altitude:
75% power 650 nm (1,205 km; 748 miles)
65% power 710 nm (1,315 km; 817 miles)
55% power 750 nm (1,390 km; 863 miles)
at best econ power settings at optimum altitude:
75% power 710 nm (1,315 km; 817 miles)
65% power 770 nm (1,427 km; 886 miles)
55% power 810 nm (1,501 km; 933 miles)

Piper Dakota, a four-seat aircraft with fixed landing gear

PIPER (PA-28-201T) TURBO DAKOTA

Piper introduced in 1979 a low-cost four-seat turbocharged aircraft with fixed landing gear, which was named Turbo Dakota and had the company designation PA-28-201T. Production of this aircraft ended during 1980; details can be found in the 1980-81 *Jane's*.

PIPER PA-28R-300XBT

Chilean Air Force name: Pillan

Piper PA-28R-300XBT Pillan basic/intermediate trainer for the Chilean Air Force

To meet the requirements of the Chilean Air Force for a basic/intermediate trainer with full aerobatic capability, Piper has developed a derivative of the Cherokee series to fulfil this role. In order to keep costs as low as possible, this incorporates an optimum number of components from various models within the series. Two examples had been completed by the company by mid-1981, when an initial batch of three more were to be supplied in kit form for assembly in Chile.

Type: Two-seat basic/intermediate military trainer.

Wings: Cantilever low-wing monoplane. Wing section NACA 65_2-415 on inboard panels, NACA 65_2-415 (modified) at tips. Dihedral 7°. Incidence 2° at root, −0° 30′ at tip. Single-spar structure of light alloy. Wingtips vacuum formed of Cycolac thermoplastics material. Slotted ailerons of light alloy construction, identical to those of the PA-32R-301 Saratoga. Single-slotted trailing-edge flaps of light alloy construction, identical to those of the PA-32-300 Lance.

Fuselage: Semi-monocoque structure of light alloy, based on that of PA-32R-301 Saratoga.

Tail Unit: Cantilever structure of light alloy with swept (38° 43′) vertical surfaces. One-piece all-moving horizontal surface with full-span anti-servo tab: this surface is a reduced-span version of the all-moving tailplane of the PA-28-236 Dakota.

Landing Gear: Hydraulically-retractable tricycle type, derived from PA-28R-200 Arrow and PA-32R-301 Saratoga landing gears, with single wheel on each unit. Main units retract inward, nosewheel rearward. All wheels enclosed by doors when retracted. Emergency free-fall extension system. Piper oleo-pneumatic shock-absorbers. Steerable nosewheel. High-capacity disc brakes. Parking brake.

Power Plant: One 224 kW (300 hp) Avco Lycoming AEIO-540 flat-six engine, driving a three-blade propeller with spinner. Fuel contained in integral wing tanks. Overwing gravity refuelling point on each wing. Fuel and oil systems suitable for inverted flight.

Accommodation: Two seats in tandem beneath sideways-opening (to starboard) one-piece transparent canopy. Dual controls standard. Accommodation is heated and ventilated.

Systems: Electrical system of 24V, powered by an engine-driven alternator. Hydraulic system powered by a self-contained unit incorporating an electric motor, gear pump, fluid reservoir, pressure regulator, flow control, and thermal expansion protection.

Dimensions, external:
Wing span 8·81 m (28 ft 11 in)
Wing chord at root 1·88 m (6 ft 2 in)
Wing chord at tip 1·26 m (4 ft 1½ in)
Wing aspect ratio 5·69
Length overall 7·97 m (26 ft 1¾ in)
Height overall 2·34 m (7 ft 8¼ in)
Tailplane span 3·05 m (10 ft 0 in)
Wheel track 3·02 m (9 ft 11 in)
Wheelbase 2·09 m (6 ft 10¼ in)

Dimensions, internal:
Cockpit: Length 3·24 m (10 ft 7½ in)
Max width 1·04 m (3 ft 5 in)
Max height 1·48 m (4 ft 10¼ in)

Areas:
Wings, gross 13·64 m² (146·8 sq ft)
Ailerons (total) 1·10 m² (11·84 sq ft)
Trailing-edge flaps (total) 1·36 m² (14·64 sq ft)
Fin 0·69 m² (7·43 sq ft)
Rudder 0·38 m² (4·09 sq ft)
Tailplane, incl tab 2·27 m² (24·43 sq ft)

Performance (at AUW of 1,315 kg; 2,900 lb):
Max level speed at S/L 173 knots (321 km/h; 199 mph)
Cruising speed:
75% power at 2,320 m (7,600 ft) 166 knots (308 km/h; 191 mph)
65% power at 3,450 m (11,300 ft) 161 knots (298 km/h; 185 mph)
55% power at 4,630 m (15,200 ft) 153 knots (284 km/h; 176 mph)
Stalling speed: flaps up 66 knots (122 km/h; 76 mph)
flaps down 60 knots (111 km/h; 69 mph)
Max rate of climb at S/L 378 m (1,240 ft)/min
Service ceiling 5,610 m (18,400 ft)
Absolute ceiling 6,100 m (20,000 ft)
T-O run 229 m (750 ft)
T-O to 15 m (50 ft) 408 m (1,340 ft)
Landing from 15 m (50 ft) 433 m (1,420 ft)
Landing run 207 m (680 ft)
Range with 45 min reserves:
75% power at 2,500 m (8,200 ft) 610 nm (1,130 km; 702 miles)
econ cruising power at 4,000 m (13,100 ft) 700 nm (1,297 km; 806 miles)

Range, no reserves:
75% power at 2,500 m (8,200 ft)
740 nm (1,371 km; 852 miles)
econ cruising power at 4,000 m (13,100 ft)
830 nm (1,538 km; 956 miles)

PIPER (PA-32-301) SARATOGA

On 17 December 1979, Piper announced that it had begun production of a new family of six/seven-seat single-engined aircraft known as Saratogas, to replace the PA-32 SIX 300 and T-tail Lance series (all described in 1979-80 *Jane's*). All Saratogas have a common airframe, with a conventional low-mounted tailplane and a semi-tapered wing of longer span than the wing of the aircraft they supersede.

Four versions of the Saratoga are available:

PA-32-301 Saratoga. Basic version, as described in detail.

PA-32-301T Turbo Saratoga. As Saratoga, but power plant comprises one 224 kW (300 hp) Avco Lycoming TIO-540-S1AD flat-six turbocharged engine, driving a Hartzell two-blade constant-speed metal propeller. Three-blade propeller optional. Additional standard equipment for the Turbo Saratoga includes cylinder head and exhaust gas temperature gauges, engine overboost warning light, and manually controlled cowl flap.

PA-32R-301 Saratoga SP. Retractable landing gear version of the Saratoga, described separately.

PA-32R-301T Turbo Saratoga SP. Retractable landing gear version of the Turbo Saratoga, described separately.

These aircraft are available with ten optional factory-installed avionics packages, and also two operational groups as follows:

Custom. Includes blind-flying instrumentation with 3 in attitude and directional gyros; electric clock; outside air temperature gauge; rate of climb and true airspeed indicators; basic lighting package which comprises avionics dimming, instrument panel white backlighting and overhead red lighting; cabin dome, map and reading lights; landing/taxi and navigation lights; rotating beacon; waterfall switch panel; assist straps; entrance step; wheel fairings; vacuum system with vacuum gauge, regulator, filter and warning lights; and 35Ah battery; adding 22·6 kg (49·8 lb) to basic empty weight.

Executive. As Custom package, plus alternate static source; emergency locator transmitter; heated pitot; pilot's vertically adjustable seat; courtesy lighting package comprising forward baggage compartment light, and forward and aft cabin door lights; wing strobe lights; and external power socket; adding 30·8 kg (68·0 lb) to basic empty weight.

TYPE: Six/seven-seat cabin monoplane.

WINGS: Cantilever low-wing monoplane. Light alloy single-spar structure with glassfibre wingtips. Plain ailerons of light alloy construction. Manually-operated four-position trailing-edge flaps of light alloy construction with ribbed skins.

FUSELAGE: Conventional semi-monocoque structure of light alloy. Glassfibre engine cowling.

TAIL UNIT: Cantilever structure of light alloy, except for glassfibre tips on fin and tailplane. Fin and rudder have ribbed metal skins. One-piece all-moving horizontal surface with combined anti-servo and trim tab. Trimmable rudder.

LANDING GEAR: Non-retractable tricycle type. Steerable nosewheel. Piper oleo-pneumatic shock-absorbers. Single wheel on each unit. Main wheels and tyres size 6·00-6, 8-ply rating. Nosewheel and tyre size 5·00-5, 6-ply rating. Nosewheel tyre size 6·00-6 optional. High-capacity disc brakes. Parking brake. Wheel fairings optional. Heavy-duty brakes and tyres optional.

POWER PLANT: One 224 kW (300 hp) Avco Lycoming IO-540-K1G5D flat-six engine, driving a Hartzell two-blade constant-speed metal propeller with spinner. Three-blade propeller optional. Turbo Saratoga as detailed in model listings. Two fuel tanks in each wing with combined capacity of 405 litres (107 US gallons), of which 386 litres (102 US gallons) are usable. Refuelling points on wing upper surface. Oil capacity 11·5 litres (3 US gallons).

ACCOMMODATION: Enclosed cabin, seating six people in pairs. Seventh seat, between two centre seats, optional. Dual controls and toe brakes standard. Two forward-hinged doors, one on starboard side forward, overwing; one on port side at aft end of cabin. Space for 45 kg (100 lb) baggage at rear of cabin, with external baggage/utility door on port side. Additional baggage space, capacity 45 kg (100 lb), between engine fireproof bulkhead and instrument panel, with external access door on starboard side. Pilot's storm window. Sun visors. Accommodation heated and ventilated. Windscreen defroster standard.

SYSTEMS: Piper Aire air-conditioning, vacuum and oxygen systems optional. Hydraulic system for brakes only. Electrical system includes a 14V 60A engine-driven alternator, and 12V 25Ah battery. 35Ah battery optional.

AVIONICS AND EQUIPMENT: A wide range of avionics is available to customer's requirements, including weather radar with internal wing-mounted antenna. Standard equipment includes recording tachometer, sensitive altimeter, dual cylinder head temperature and exhaust gas temperature gauges (EGT optional for Saratoga), fore and aft adjustable pilot and co-pilot seats with shoulder and safety belts and inertia reels, armrests, map pockets, glove compartment, stall warning device, provisions for emergency locator transmitter, alternator failure and low oil pressure warning lights, full-flow oil filter, fuel quick drains, oil quick drain, jack pads, stowable towbar, and tiedown rings. Optional equipment includes items detailed in Custom and Executive packages, plus digital clock; encoding altimeter; engine hour recorder; true airspeed indicator; de luxe interior groups including headrests, window curtains, club seating arrangement, refreshment console; shoulder safety belts with inertia reels for passenger seats; cabin fire extinguisher; tinted windows; stainless steel control cables; ventilation fan; super soundproofing; and zinc chromate treatment of aluminium parts.

Piper PA-32-301T Turbo Saratoga six/seven-seat cabin monoplane

Piper PA-32R-301 Saratoga SP, a version of the Saratoga with retractable landing gear

DIMENSIONS, EXTERNAL:
Wing span 11·02 m (36 ft 2 in)
Length overall: Saratoga 8·44 m (27 ft 8½ in)
Turbo Saratoga 8·59 m (28 ft 2 in)
Height overall 2·49 m (8 ft 2 in)
Tailplane span 3·94 m (12 ft 11 in)
Wheel track 3·23 m (10 ft 7 in)
Wheelbase 2·36 m (7 ft 9 in)
Cabin door (fwd, stbd): Height 0·89 m (2 ft 11 in)
Width 0·91 m (3 ft 0 in)
Cabin door (rear, port): Height 0·84 m (2 ft 9 in)
Width 0·74 m (2 ft 5 in)
Baggage door (fwd): Height 0·46 m (1 ft 6 in)
Width 0·61 m (2 ft 0 in)
Baggage/utility door (aft): Height 0·71 m (2 ft 4 in)
Width 0·51 m (1 ft 8 in)

DIMENSIONS, INTERNAL:
Cabin: Length (instrument panel to rear bulkhead)
3·18 m (10 ft 5 in)
Max width 1·24 m (4 ft 1 in)
Max height 1·24 m (4 ft 1 in)
Volume (incl aft baggage area) 5·53 m³ (195·3 cu ft)
Baggage compartment volume:
forward 0·20 m³ (7·0 cu ft)
aft 0·49 m³ (17·3 cu ft)

AREA:
Wings, gross 16·56 m² (178·3 sq ft)

WEIGHTS AND LOADINGS (A: Saratoga; B: Turbo Saratoga):
Weight empty: A 880 kg (1,940 lb)
B 908 kg (2,003 lb)
Max T-O weight: A, B 1,633 kg (3,600 lb)
Max ramp weight: A 1,640 kg (3,615 lb)
B 1,641 kg (3,617 lb)
Max wing loading: A, B 98·6 kg/m² (20·2 lb/sq ft)
Max power loading: A, B 7·30 kg/kW (12·0 lb/hp)

PERFORMANCE (at max T-O weight, except where indicated. A: Saratoga/two-blade propeller; B: Turbo Saratoga/two-blade propeller; C: Saratoga/three-blade propeller; D: Turbo Saratoga/three-blade propeller):
Max level speed at optimum altitude:
A 152 knots (282 km/h; 175 mph)
B 178 knots (330 km/h; 205 mph)
D 182 knots (337 km/h; 210 mph)
Best power cruising speed at optimum altitude:
at 75% power: A 150 knots (278 km/h; 173 mph)
B 165 knots (306 km/h; 190 mph)
at 65% power: A 146 knots (270 km/h; 168 mph)
B 154 knots (285 km/h; 177 mph)
at 55% power: A 136 knots (252 km/h; 156·5 mph)
B 139 knots (258 km/h; 160 mph)
Best econ cruising speed at optimum altitude:
at 75% power: A 148 knots (274 km/h; 170 mph)
B 160 knots (296 km/h; 184 mph)
at 65% power: A 144 knots (267 km/h; 166 mph)
B 148 knots (274 km/h; 170 mph)
at 55% power: A 133 knots (246 km/h; 153 mph)
B 132 knots (245 km/h; 152 mph)
Stalling speed, flaps up:
A 66 knots (122 km/h; 76 mph) CAS
B 67 knots (125 km/h; 77·5 mph) CAS
Stalling speed, flaps down:
A, B 60 knots (111 km/h; 69 mph) CAS
Max rate of climb at S/L: A 302 m (990 ft)/min
B 328 m (1,075 ft)/min
Service ceiling: A 4,300 m (14,100 ft)
*B 6,100 m (20,000 ft)
Absolute ceiling: A 4,845 m (15,900 ft)
*B 6,100 m (20,000 ft)

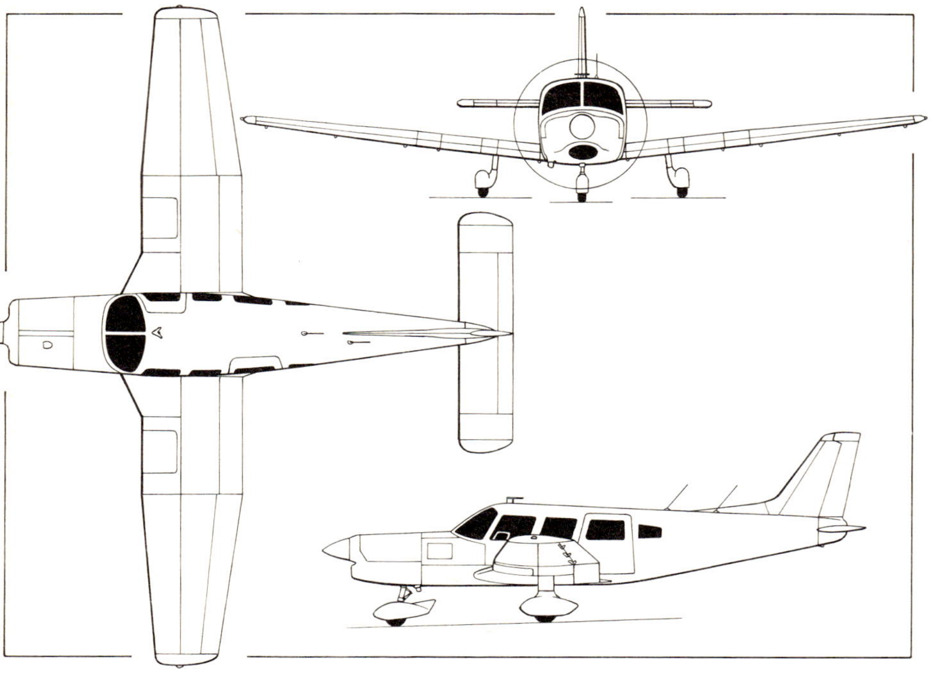

Piper PA-32-301 Saratoga (Avco Lycoming IO-540-K1G5D engine) *(Michael A. Badrocke)*

T-O run: A 361 m (1,183 ft)
B 338 m (1,110 ft)
C 309 m (1,013 ft)
D 293 m (960 ft)
T-O to 15 m (50 ft): A 536 m (1,759 ft)
B 485 m (1,590 ft)
C 479 m (1,573 ft)
D 433 m (1,420 ft)
Landing from 15 m (50 ft), standard brakes:
A 491 m (1,612 ft)
B 526 m (1,725 ft)
Landing from 15 m (50 ft), heavy-duty brakes:
A 466 m (1,530 ft)
B 500 m (1,640 ft)
Landing run, standard brakes: A, B 223 m (732 ft)
Landing run, heavy-duty brakes:
A, B 198 m (650 ft)
Range with max fuel, allowances for taxi, T-O, climb, descent, and 45 min reserves at max range power:
best power settings at optimum altitude:
75% power: A 745 nm (1,381 km; 858 miles)
B 684 nm (1,267 km; 788 miles)
65% power: A 805 nm (1,492 km; 927 miles)
B 730 nm (1,352 km; 841 miles)
55% power: A 849 nm (1,573 km; 978 miles)
B 772 nm (1,430 km; 889 miles)
best econ power settings at optimum altitude:
75% power: A 823 nm (1,525 km; 948 miles)
B 780 nm (1,445 km; 898 miles)
65% power: A 911 nm (1,688 km; 1,049 miles)
B 845 nm (1,566 km; 973 miles)
55% power: A 960 nm (1,778 km; 1,105 miles)
B 860 nm (1,593 km; 990 miles)

**Max certificated altitude*

PIPER (PA-32R-301) SARATOGA SP

This is a retractable landing gear model of the Saratoga, of which two versions were introduced for 1980:

PA-32R-301 Saratoga SP. Basic version, similar to the PA-32-301 Saratoga but with retractable landing gear and new options as described.

PA-32R-301T Turbo Saratoga SP. As Saratoga SP, but with turbocharged power plant and additional standard equipment as detailed for the Turbo Saratoga. Options available for this version include propeller de-icing and a built-in oxygen system.

Both models are available with ten optional factory-installed avionics packages, including a King flight director, and with the following Operational Groups:

Custom. As detailed for Saratoga/Turbo Saratoga, but with deletion of wheel speed fairings and rotating beacon, and addition of wing strobe lights; adding 10 kg (22·0 lb) to basic empty weight.

Executive. As detailed for Saratoga/Turbo Saratoga, except that a heated landing gear sensor is included, the entire package adding 22·4 kg (49·4 lb) to basic empty weight.

The description of the Saratoga/Turbo Saratoga applies also to the retractable landing gear versions, except as follows:

Landing Gear: Hydraulically-retractable tricycle type with single wheel on each unit. Main units retract inward, nosewheel aft. Integrated automatic system which extends the landing gear at 102 knots (189 km/h; 117 mph), unless overridden by pilot. Emergency free-fall extension system. Piper oleo-pneumatic shock-absorbers. Steerable nosewheel. Main wheels and tyres size 6·00-6, 8-ply rating. Nosewheel and tyre size 5·00-5, 6-ply rating. High-capacity disc brakes. Parking brake. Heavy-duty tyres and brakes optional.

Power Plant: As for Saratoga/Turbo Saratoga, except propeller de-icing optional for Turbo Saratoga SP.

Accommodation: As for Saratoga, but pilot's electrically heated windscreen plate optional for Turbo Saratoga SP.

Systems: As for Saratoga, but electrically-driven hydraulic pump for landing gear actuation; built-in oxygen system of 1·81 m³ (64 cu ft) capacity and 95A alternator available optionally for Turbo Saratoga SP.

Avionics and Equipment: Generally as listed for Saratoga/Turbo Saratoga, with amendments as noted in this entry, plus optional wing ice inspection light for Turbo Saratoga SP.

Dimensions, external: As for Saratoga, except:
Length overall: Saratoga SP 8·43 m (27 ft 8 in)
Turbo Saratoga SP 8·64 m (28 ft 4 in)
Height overall 2·59 m (8 ft 6 in)
Wheel track 3·39 m (11 ft 1½ in)
Wheelbase 2·43 m (7 ft 11½ in)

Weights and Loadings (A: Saratoga SP; B: Turbo Saratoga SP): As for Saratoga/Turbo Saratoga, except:
Weight empty: A 904 kg (1,994 lb)
B 937 kg (2,066 lb)

Performance (at max T-O weight, except where indicated. A: Saratoga SP/two-blade propeller; B: Turbo Saratoga SP/two-blade propeller; C: Saratoga SP/three-blade propeller; D: Turbo Saratoga SP/three-blade propeller):
Max level speed at optimum altitude:
A 164 knots (304 km/h; 189 mph)
B 191 knots (354 km/h; 220 mph)
D 195 knots (361 km/h; 225 mph)
Best power cruising speed at optimum altitude:
at 75% power: A 159 knots (295 km/h; 183 mph)
B 177 knots (328 km/h; 204 mph)
at 65% power: A 153 knots (283 km/h; 176 mph)
B 166 knots (307 km/h; 191 mph)
at 55% power: A 144 knots (267 km/h; 166 mph)
B 152 knots (282 km/h; 175 mph)
Best econ cruising speed at optimum altitude:
at 75% power: A 157 knots (291 km/h; 181 mph)
B 171 knots (317 km/h; 197 mph)
at 65% power: A 151 knots (280 km/h; 174 mph)
B 160 knots (296 km/h; 184 mph)
at 55% power: A 141 knots (261 km/h; 162 mph)
B 145 knots (269 km/h; 167 mph)
Stalling speeds, flaps up:
A 65 knots (121 km/h; 75 mph) CAS
B 63 knots (118 km/h; 72·5 mph) CAS
Stalling speed, flaps down:
A 59 knots (110 km/h; 68 mph) CAS
B 60 knots (111 km/h; 69 mph) CAS
Max rate of climb at S/L: A 308 m (1,010 ft)/min
B 341 m (1,120 ft)/min
Service ceiling: A 5,090 m (16,700 ft)
*B 6,100 m (20,000 ft)
Absolute ceiling: A 5,595 m (18,350 ft)
*B 6,100 m (20,000 ft)
T-O run, and T-O to 15 m (50 ft):
as for Saratoga and Turbo Saratoga
Landing from 15 m (50 ft), standard brakes:
A 491 m (1,612 ft)
B 526 m (1,725 ft)
Landing run, standard brakes: A, B 223 m (732 ft)
Range with max fuel, allowances for taxi, T-O, climb, descent, and 45 min reserves at max range power:
best power settings at optimum altitude:
75% power: A 784 nm (1,453 km; 903 miles)
B 730 nm (1,353 km; 840 miles)
65% power: A 828 nm (1,533 km; 953 miles)
B 790 nm (1,465 km; 910 miles)
55% power: A 869 nm (1,611 km; 1,001 miles)
B 843 nm (1,562 km; 971 miles)
best econ power settings at optimum altitude:
75% power: A 865 nm (1,603 km; 996 miles)
B 844 nm (1,564 km; 972 miles)
65% power: A 937 nm (1,736 km; 1,079 miles)
B 920 nm (1,704 km; 1,059 miles)
55% power: A 963 nm (1,785 km; 1,109 miles)
B 950 nm (1,760 km; 1,094 miles)

**Max certificated altitude*

PIPER (PA-31) NAVAJO

On 30 September 1964 Piper flew the first of what it described as a new series of larger executive aircraft for corporate and commuter airline service. Named Navajo, it was then available with normally-aspirated or turbocharged engines, the latter being known as the Turbo Navajo. Subsequently, three additional versions were introduced, the Navajo C/R, the Pressurised Navajo and Navajo Chieftain. The Pressurised Navajo has been discontinued; details can be found in the 1977-78 *Jane's*. The Navajo C/R and Chieftain are described separately.

Four optional factory-installed avionics packages are available for the Navajo, including nav/coms, ADF, AFCS, DME, glideslope and marker beacon receivers, transponder, and associated equipment by Bendix, Collins, King and Piper. There are, in addition, two optional equipment groups, as follows:

Co-pilot Flight Instrument Group. Includes blind-flying instrumentation with 3 in attitude and directional gyros, clock, rate of climb indicator, sensitive altimeter, true airspeed indicator, heated pitot, alternate static source, and individual rheostat controlled lighting. Group available with electrical or vacuum gyros; adding 5·6 kg (12·4 lb) or 5·9 kg (12·9 lb) respectively to basic empty weight.

De-icing Group. Pneumatic de-icing boot installation for wing and tail unit leading-edges; electrical propeller de-icing; ice inspection light; heated lift detector; fuselage ice protection shields; and electrical windscreen de-icing and windscreen wiper port side; adding 28·4 kg (62·7 lb) to basic empty weight.

Re-certification for flight into known icing conditions, to newer, more stringent standards, applies to aircraft equipped with the optional De-icing Group.

Type: Six/eight-seat corporate and commuter airline transport.

Wings: Cantilever low-wing monoplane. Wing section NACA 63_2415 at root, NACA 63_1212 at tip. 1° aerodynamic twist. 2° 30′ geometric twist. All-metal structure, with heavy stepped-down main spar, front and rear spars, lateral stringers, ribs and stressed skin. Wings spliced on centreline with heavy steel plates. Flush riveted forward of main spar. Wing-root leading-edge extended forward between nacelle and fuselage. Glassfibre wingtips. Balanced ailerons are interconnected with rudder. Trim tab in starboard aileron. Electrically-operated flaps. Pneumatic de-icing boots optional.

Fuselage: Conventional all-metal semi-monocoque structure.

Tail Unit: Cantilever all-metal structure, with sweptback vertical surfaces. Variable-incidence tailplane. Trim tabs in rudder and starboard elevator. Optional pneumatic de-icing boots.

Landing Gear: Hydraulically-actuated retractable tricycle type, with single wheel on each unit. Manual hydraulic emergency extension. Main wheels and tyres size 6·50-10, eight-ply rating, pressure 4·14 bars (60 lb/sq in). Steerable nosewheel and tyre size 6·00-6, six-ply rating, pressure 2·90 bars (42 lb/sq in). Toe-controlled hydraulic disc brakes. Heavy duty brakes and toe-operated brakes for co-pilot optional. Parking brake. Main-wheel doors close when gear is fully extended.

Power Plant: Two 231 kW (310 hp) Avco Lycoming TIO-540-A2C flat-six turbocharged engines. Hartzell three-blade constant-speed fully-feathering metal propellers with spinners. Propeller de-icing, and synchrophaser, optional. Four rubber fuel cells in wings; inboard cells each contain 212 litres (56 US gallons), outboard cells 151·5 litres (40 US gallons) each. Total fuel capacity 727 litres (192 US gallons), of which 709 litres (187·3 US gallons) are usable. Fuel cells equipped with NACA-type anti-icing non-siphoning fuel vents. Oil capacity 22·7 litres (6 US gallons). Two-piece glassfibre engine nacelles.

Accommodation: Six individual seats, with headrests and armrests, in pairs with centre aisle. Seventh and eighth seats optional. Dual controls standard. 'No smoking, fasten seat belt' sign. Thermostatically-controlled Janitrol 35,000 BTU combustion heater, windscreen defrosters and fresh air system standard. Double-glazed windows. Electrical de-icing and windscreen wiper for port side of windscreen optional. 'Dutch' door at rear of cabin on port side. Top half hinges upward; lower half hinges down and has built-in steps. Pilot's and co-pilot's storm windows. Sun visors. Super soundproofing. Divider curtain between cabin and flight deck. Window curtains. Emergency exit. Baggage compartments in nose, capacity 68 kg (150 lb), and in rear of cabin, capacity 91 kg (200 lb). Utility door and cockpit door optional.

Systems: Hydraulic system utilises two engine-driven pumps. 24V electrical system supplied by two engine-driven 28V 70A alternators and 24V 17Ah battery; 25Ah battery optional. External power socket standard. Oxygen system of 3·23 m³ (114 cu ft) capacity optional.

Avionics and Equipment: In addition to the optional factory-installed avionics packages, there is a wide range of alternative options by Aeronetics, Bendix, Collins, Foster, King, Piper, RCA and SunAir. These include encoding altimeter, HF transceiver, radar altimeter, radio altimeter, radio telephone, R/Nav, V/Nav, and weather radar. Standard equipment includes pilot's blind-flying instrumentation; clock; dual exhaust gas temperature gauges; flight hour recorder; dual manifold pressure gauges; outside air temperature gauge; rate of climb indicator; sensitive altimeter; true airspeed indicator; fully-adjustable pilot and co-pilot seats with shoulder safety belts and inertia reels; headrests; folding armrests; alternate static source; heated pitot; alternator and pneumatic failure, low fuel, door ajar, and heater overheat warning lights; stall warning device; corrosion proofing; cabin dome, courtesy, instrument panel, map, reading, landing, navigation, strobe and taxi lights; external power socket; tiedown rings; and stowable towbar. Optional equipment includes items covered by the operational groups as detailed, plus a wide range of cabin equipment including storage cabinets and drawers, tables, seventh and eighth seats, seventh seat with combined toilet and privacy curtain, shoulder safety belts and inertia reels for passenger seats, fore and aft cabin dividers, auxiliary cabin heater, cabin fire extinguisher, cabin chimes, tinted windows, Thermos unit and refreshment centre, emergency locator transmitter, ground and in-flight recognition lights, and fuselage ice protection shields.

Dimensions, external:
Wing span 12·40 m (40 ft 8 in)
Length overall 9·94 m (32 ft 7½ in)
Height overall 3·96 m (13 ft 0 in)
Tailplane span 5·52 m (18 ft 1½ in)
Wheel track 4·19 m (13 ft 9 in)
Wheelbase 2·64 m (8 ft 8 in)
Propeller diameter 2·03 m (6 ft 8 in)
Distance between propeller centres 3·86 m (12 ft 8 in)
Cabin door (port, rear): Height 1·14 m (3 ft 9 in)
Width 0·70 m (2 ft 3½ in)
Utility door (port, rear): Height 1·14 m (3 ft 9 in)
Width 0·43 m (1 ft 5 in)
Baggage door (port, fwd): Height 0·64 m (2 ft 1 in)
Width 0·71 m (2 ft 4 in)

Dimensions, internal:
Cabin: Length 3·33 m (10 ft 11 in)
Max width 1·27 m (4 ft 2 in)
Max height 1·31 m (4 ft 3½ in)
Baggage compartment: nose 0·40 m³ (14 cu ft)
rear 0·62 m³ (22 cu ft)

Area:
Wings, gross 21·3 m² (229 sq ft)

Weights and Loadings:
Weight empty 1,816 kg (4,003 lb)
Max T-O and landing weight 2,948 kg (6,500 lb)
Max ramp weight 2,965 kg (6,536 lb)
Max zero-fuel weight 2,812 kg (6,200 lb)
Max wing loading 138·7 kg/m² (28·4 lb/sq ft)
Max power loading 6·38 kg/kW (10·5 lb/hp)

Performance (at max T-O weight, except where indicated):
Max level speed at average cruise weight
227 knots (420 km/h; 261 mph)
Cruising speed at average cruise weight:
75% power at 6,705 m (22,000 ft)
215 knots (399 km/h; 248 mph)
75% power at 3,660 m (12,000 ft)
196 knots (363 km/h; 226 mph)
65% power at 7,315 m (24,000 ft)
204 knots (378 km/h; 235 mph)
65% power at 3,660 m (12,000 ft)
185 knots (343 km/h; 213 mph)
55% power at 7,315 m (24,000 ft)
186 knots (344 km/h; 214 mph)
55% power at 3,660 m (12,000 ft)
171 knots (317 km/h; 197 mph)
Stalling speed, flaps down
70 knots (130 km/h; 81 mph) IAS
Max rate of climb at S/L 372 m (1,220 ft)/min
Rate of climb at S/L, one engine out
75 m (245 ft)/min
Max certificated altitude 7,315 m (24,000 ft)

Piper Navajo C/R with counter-rotating engines and nacelle baggage compartments

Service ceiling, one engine out 4,635 m (15,200 ft)
Normal T-O run 524 m (1,720 ft)
Normal T-O to 15 m (50 ft) 698 m (2,290 ft)
*Normal landing from 15 m (50 ft) at max landing weight 554 m (1,818 ft)
*Short-field landing from 15 m (50 ft) 464 m (1,521 ft)
*Normal landing run 276 m (906 ft)
*Short-field landing run 274 m (900 ft)
Range with max fuel, allowances for start, taxi, T-O, climb, and 45 min reserves at long-range cruise power:
75% power at 6,100 m (20,000 ft)
1,005 nm (1,862 km; 1,157 miles)
75% power at 3,660 m (12,000 ft)
995 nm (1,844 km; 1,146 miles)
65% power at 6,100 m (20,000 ft)
1,055 nm (1,955 km; 1,215 miles)
65% power at 3,660 m (12,000 ft)
1,050 nm (1,945 km; 1,209 miles)
55% power at 6,100 m (20,000 ft)
1,065 nm (1,973 km; 1,226 miles)
55% power at 3,660 m (12,000 ft)
1,065 nm (1,973 km; 1,226 miles)

**With optional heavy-duty brakes*

PIPER (PA-31-325) NAVAJO C/R

Identical to the PA-31 Navajo except for having counter-rotating engines and nacelle baggage compartments as introduced on the Navajo Chieftain in 1972. The Navajo C/R is available in the same optional versions, and with the same optional avionics and operational package options, as the Navajo.

Power Plant: One 242·5 kW (325 hp) Avco Lycoming LTIO-540-F2BD and one 242·5 kW (325 hp) Avco Lycoming TIO-540-F2BD flat-six turbocharged counter-rotating engines. Standard usable fuel capacity 695 litres (183·5 US gallons). Optional fuel tank in each nacelle, reducing baggage capacity, to provide a total usable fuel capacity of 899 litres (237·5 US gallons).

Accommodation: As for Navajo, except for provision to stow 68 kg (150 lb) baggage in each engine nacelle. Baggage capacity reduced when optional fuel tanks installed in nacelles.

Dimensions, internal: As for Navajo, plus:
Baggage compartments:
Engine nacelles (each) 0·375 m³ (13·25 cu ft)

Weights and Loadings: As for Navajo, except:
Weight empty 1,859 kg (4,099 lb)
Max ramp weight 2,966 kg (6,540 lb)

Performance (at max T-O weight, except where indicated):
Max level speed at average cruise weight
228 knots (422 km/h; 262 mph)
Cruising speed at average cruise weight:
75% power at 6,100 m (20,000 ft)
220 knots (407 km/h; 253 mph)
75% power at 3,660 m (12,000 ft)
202 knots (375 km/h; 233 mph)

Photograph and three-view drawing *(Pilot Press)* **of the Piper Chieftain six/ten-seat executive/commuter cargo aircraft**

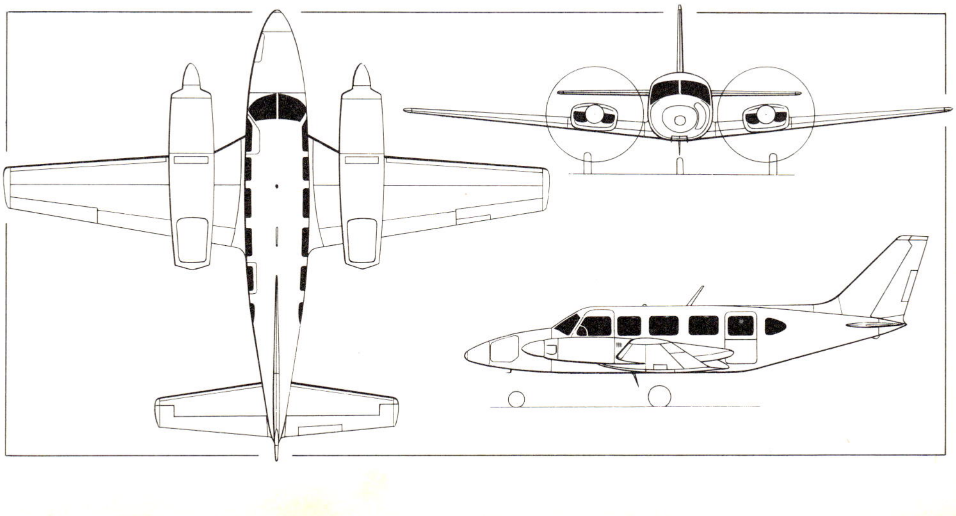

65% power at 6,100 m (20,000 ft)
208 knots (386 km/h; 240 mph)
65% power at 3,660 m (12,000 ft)
190 knots (352 km/h; 219 mph)
55% power at 4,875 m (16,000 ft)
180 knots (333 km/h; 207 mph)
55% power at 3,660 m (12,000 ft)
175 knots (325 km/h; 202 mph)
Stalling speed, flaps down
70 knots (130 km/h; 81 mph) IAS
Max rate of climb at S/L 366 m (1,200 ft)/min
Rate of climb at S/L, one engine out
78 m (255 ft)/min
Max certificated altitude 7,315 m (24,000 ft)
Service ceiling, one engine out 4,665 m (15,300 ft)
Normal T-O run 457 m (1,500 ft)
Normal T-O to 15 m (50 ft) 744 m (2,440 ft)
*Normal landing from 15 m (50 ft) 554 m (1,818 ft)
*Short-field landing from 15 m (50 ft) 464 m (1,521 ft)
*Normal landing run 276 m (906 ft)
*Short-field landing run 274 m (900 ft)
Range with max fuel, allowances for start, taxi, T-O, climb, and 45 min reserves at long-range cruise power:
75% power at 6,100 m (20,000 ft)
940 nm (1,741 km; 1,082 miles)
75% power at 3,660 m (12,000 ft)
910 nm (1,687 km; 1,048 miles)
65% power at 6,100 m (20,000 ft)
1,000 nm (1,854 km; 1,152 miles)
65% power at 3,660 m (12,000 ft)
970 nm (1,798 km; 1,117 miles)
55% power at 4,875 m (16,000 ft)
1,040 nm (1,928 km; 1,198 miles)
55% power at 3,660 m (12,000 ft)
1,025 nm (1,899 km; 1,180 miles)

**With optional heavy-duty brakes*

PIPER (PA-31-350) CHIEFTAIN

First announced on 11 September 1972, the PA-31-350 Chieftain is a lengthened version of the Navajo, with the fuselage extended by 0·61 m (2 ft 0 in) and with 261 kW (350 hp) counter-rotating turbocharged engines.

The main cabin floor is designed to carry heavy concentrated loads of up to 976 kg/m² (200 lb/sq ft) and, in addition to the 6·14 m³ (217 cu ft) of cargo space in the main cabin, 91 kg (200 lb) of cargo or baggage can be carried in the forward nose compartment, and 68 kg (150 lb) in the rear of each engine nacelle.

Two optional interior groups of equipment are available, depending upon the proposed use of the aircraft:

Standard Interior Group. Comprises six adjustable seats. Passenger seats in club arrangement with headrests, folding armrests, seat belts, oxygen mask stowage beneath each seat, and magazine storage pockets on each seat back. 'No smoking/Fasten seat belt' sign. Pull-curtain cockpit divider. Choice of nine interior colour schemes.

Commuter Interior Group. Comprises ten forward-facing seats. Eight adjustable and reclining passenger seats with oxygen mask stowage and magazine storage as above. 'No smoking/Fasten seat belt' sign. Pull-curtain cockpit divider. Choice of nine interior colour schemes. Adds 39·5 kg (87 lb) to basic empty weight.

Operational groups, comprising Co-pilot Instrument Group and De-icing Group, and avionics groups as for the Navajo, are available for the Chieftain, as well as a wide range of optional avionics.

The description of the Navajo applies also to the Chieftain, except as follows:

FUSELAGE: As for Navajo, except length increased by 0·61 m (2 ft 0 in).

POWER PLANT: Two 261 kW (350 hp) Avco Lycoming flat-six turbocharged engines, one TIO-540-J2BD and one LTIO-540-J2BD, each driving a Hartzell three-blade constant-speed fully-feathering metal propeller. Four rubber fuel cells in wings; inboard cells each contain 212 litres (56 US gallons), outboard cells each 151·5 litres (40 US gallons). Total standard fuel capacity 727 litres (192 US gallons), of which 689 litres (182 US gallons) are usable. Optional fuel, as for Navajo C/R, to provide total usable capacity of 893 litres (236 US gallons). Oil capacity 22·7 litres (6 US gallons).

ACCOMMODATION: Pilot and co-pilot on individually adjustable and reclining seats. Dual controls standard. Interior seating and equipment as detailed in optional interior groups. Cabin heated by thermostatically-controlled Janitrol 50,000 BTU combustion heater. Piper Aire 16,000 BTU air-conditioning system optional. Baggage/cargo compartments in nose and rear fuselage, capacity of each 91 kg (200 lb), and in the rear of each engine nacelle, each 68 kg (150 lb).

AVIONICS AND EQUIPMENT: A wide range of optional avionics is available, including a flight director system integrated with autopilot and weather radar. Standard and optional equipment is generally similar to that available for the Navajo. Also available optionally for all interior groups is a cargo kit which includes cargo barriers, tiedown rings and net, eight seat-track tiedown rings, four plug-in tiedown rings, and four tiedown straps.

DIMENSIONS, EXTERNAL:
Length overall 10·55 m (34 ft 7½ in)
Wheelbase 3·24 m (10 ft 7½ in)

DIMENSIONS, INTERNAL:
Cabin: Length 3·84 m (12 ft 7 in)
Baggage/cargo compartments:
Engine nacelles (each) 0·37 m³ (13·25 cu ft)
Engine nacelles (each) with optional fuel
0·17 m³ (6·0 cu ft)

WEIGHTS AND LOADINGS:
Weight empty (standard) 1,915 kg (4,221 lb)
Max T-O and landing weight 3,175 kg (7,000 lb)
Max ramp weight 3,196 kg (7,045 lb)
Max wing loading 149·4 kg/m² (30·6 lb/sq ft)
Max power loading 6·08 kg/kW (10·0 lb/hp)

PERFORMANCE (at max T-O weight, except where indicated):
Max level speed at average cruise weight
231 knots (428 km/h; 266 mph)
Cruising speed at average cruise weight:
75% power at 6,100 m (20,000 ft)
221 knots (409 km/h; 254 mph)
75% power at 3,660 m (12,000 ft)
205 knots (380 km/h; 236 mph)
65% power at 6,100 m (20,000 ft)
210 knots (389 km/h; 242 mph)
65% power at 3,660 m (12,000 ft)
191 knots (354 km/h; 220 mph)
55% power at 4,570 m (15,000 ft)
177 knots (328 km/h; 204 mph)
55% power at 3,660 m (12,000 ft)
173 knots (320 km/h; 199 mph)
Stalling speed, flaps up 80 knots (148 km/h; 92 mph)
Stalling speed, flaps down
74 knots (137 km/h; 85 mph)
Max rate of climb at S/L 341 m (1,120 ft)/min
Rate of climb at S/L, one engine out
70 m (230 ft)/min
Max certificated altitude 7,315 m (24,000 ft)
Service ceiling, one engine out 4,175 m (13,700 ft)
Normal T-O run 564 m (1,850 ft)
Normal T-O to 15 m (50 ft) 847 m (2,780 ft)
Normal landing from 15 m (50 ft) 573 m (1,880 ft)
Short-field landing from 15 m (50 ft) 491 m (1,610 ft)
Normal landing run 319 m (1,045 ft)
Short-field landing run 297 m (975 ft)
Accelerate/stop distance 640 m (2,100 ft)
Range, with allowances for start, taxi, T-O, climb, and 45 min reserves at long-range cruise power; A with max standard fuel, B with max optional fuel:
75% power at 6,100 m (20,000 ft):
A 885 nm (1,640 km; 1,019 miles)
B 1,210 nm (2,240 km; 1,392 miles)
75% power at 3,660 m (12,000 ft):
A 855 nm (1,585 km; 985 miles)
B 1,160 nm (2,148 km; 1,335 miles)
65% power at 6,100 m (20,000 ft):
A 925 nm (1,714 km; 1,065 miles)
B 1,260 nm (2,333 km; 1,450 miles)
65% power at 3,660 m (12,000 ft):
A 900 nm (1,667 km; 1,036 miles)
B 1,225 nm (2,268 km; 1,409 miles)
55% power at 4,570 m (15,000 ft):
A 950 nm (1,761 km; 1,094 miles)
B 1,290 nm (2,389 km; 1,484 miles)
55% power at 3,660 m (12,000 ft):
A 950 nm (1,761 km; 1,094 miles)
B 1,290 nm (2,389 km; 1,484 miles)

PIPER (PA-31T-1) CHEYENNE I

Introduced in 1978, the Cheyenne I is a low-cost version of the established Cheyenne PA-31T, which has been redesignated Cheyenne II, and which is described separately. It differs primarily by having less powerful engines. It is available with the optional Standard and Executive interiors, De-icing Group and Co-pilot Flight Group as detailed for the Cheyenne II. Also available optionally are three factory-installed avionics packages which include nav/coms, ADF, AFCS, DME, glideslope and marker beacon receivers, transponder, weather radar, and associated equipment by Bendix, Collins, King and Piper. Additional optional avionics items by Aeronetics, IDC, RCA, Smith and SunAir as well as the above manufacturers, include altitude alerter, encoding altimeter, radio telephone, radio altimeter, radar altimeter, and R/Nav.

FAA certification was received on 23 March 1978, and deliveries began at the end of April 1978. The 500th Cheyenne delivered, in August 1980, was a Cheyenne I.

The 1981 version of the Cheyenne I introduces a number of improvements as standard, including a new time delay circuit for the cabin entrance courtesy light system, a cabin temperature control, thermal overheat reset button located in the cockpit, improved sealing of engine and propeller controls to give smoother operation, a windscreen defroster on/off control, new wingtip strobe position light assembly, new split bus electrical system, inertia reel system for pilot's and co-pilot's shoulder restraints, and new interior and exterior trims. New options include an auto-ignition system, a light grey instrument panel, a four-point inertia shoulder belt system for pilot and co-pilot, plus new optional avionics.

The description of the Cheyenne II applies also to the Cheyenne I, except as follows:

LANDING GEAR: As Cheyenne II except: Cleveland Type 40-106 main wheels, with 6·50 × 10 10-ply tyres, pressure 5·52 bars (80 lb/sq in). Cleveland Type 40-120A nosewheel, with 18 × 4·40 6-ply tyre, pressure 5·52 bars (80 lb/sq in). Cleveland Type 30-106 brakes.

POWER PLANT: Two 372·6 kW (500 shp) Pratt & Whitney Aircraft of Canada PT6A-11 turboprop engines, each driving a Hartzell Type HC-BTN-3B three-blade constant-speed reversible-pitch and fully-feathering metal propeller. Each wing has three interconnected fuel cells, with total capacity of 1,166 litres (308 US gallons), of which 1,136 litres (300 US gallons) are usable. Optional wingtip tanks, each of 113·5 litres (30 US gallons) capacity, plus installational plumbing, provide a maximum total usable capacity of 1,446 litres (382 US gallons). Refuelling points on top of engine nacelles and on upper surface of each tip-tank. Oil capacity 24·6 litres (6·5 US gallons).

DIMENSIONS, EXTERNAL: As for Cheyenne II, except:
Wing span 12·40 m (40 ft 8 in)
Wing chord at root 2·61 m (8 ft 6¾ in)
Wing chord at tip 0·97 m (3 ft 2 in)
Wing aspect ratio 7·37
Passenger door (port, rear):
Height to sill 0·97 m (3 ft 2 in)
Baggage door (port, fwd):
Width 0·53 m (1 ft 9 in)

DIMENSIONS, INTERNAL:
Cabin (excl flight deck): Length 2·57 m (8 ft 5 in)
Max width 1·27 m (4 ft 2 in)
Max height 1·31 m (4 ft 3½ in)

AREAS: As for Cheyenne II, except:
Fin 1·37 m² (14·72 sq ft)
Rudder, incl tab 1·20 m² (12·88 sq ft)

WEIGHTS AND LOADINGS:
Weight empty 2,226 kg (4,907 lb)
Max T-O and landing weight 3,946 kg (8,700 lb)
Max ramp weight 3,969 kg (8,750 lb)
Max zero-fuel weight 3,266 kg (7,200 lb)
Max wing loading 185·5 kg/m² (38·0 lb/sq ft)
Max power loading 5·30 kg/kW (8·7 lb/shp)

Piper Cheyenne I pressurised six/eight-seat cabin monoplane, with optional wingtip tanks

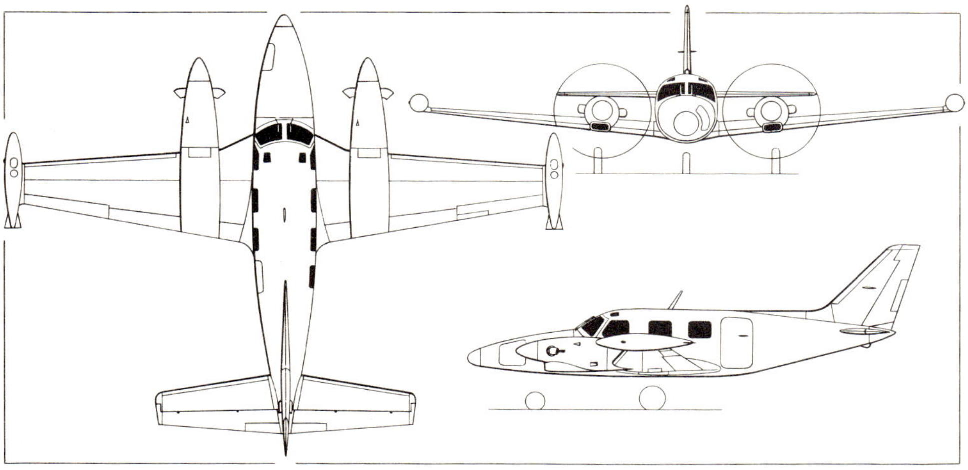

Original version of the Piper Cheyenne pressurised six/eight-seat cabin monoplane which is now designated Cheyenne II *(Pilot Press)*

Performance (at max T-O weight and without optional wingtip tanks installed, unless specified otherwise):
Cruising speed, max cruise power, average cruise weight of 3,400 kg (7,500 lb), at:
3,660 m (12,000 ft) 249 knots (461 km/h; 287 mph)
4,875 m (16,000 ft) 248 knots (460 km/h; 286 mph)
6,100 m (20,000 ft) 244 knots (452 km/h; 281 mph)
7,620 m (25,000 ft) 236 knots (437 km/h; 272 mph)
Stalling speed, wheels and flaps up, engines idling
84 knots (156 km/h; 97 mph) IAS
Stalling speed, wheels and flaps down, engines idling
72 knots (133 km/h; 83 mph) IAS
Rotation speed 89 knots (167 km/h; 104 mph)
Approach speed 102 knots (189 km/h; 117 mph)
Max rate of climb at S/L 533 m (1,750 ft)/min
Rate of climb at S/L, one engine out:
without tip-tanks 126 m (413 ft)/min
with tip-tanks 134 m (440 ft)/min
Service ceiling 8,595 m (28,200 ft)
Service ceiling, one engine out:
without tip-tanks 3,810 m (12,500 ft)
with tip-tanks 4,190 m (13,750 ft)
T-O run 436 m (1,429 ft)
T-O to 15 m (50 ft) 745 m (2,444 ft)
Landing from 15 m (50 ft) 690 m (2,263 ft)
Landing from 15 m (50 ft) with propeller reversal
507 m (1,663 ft)
Landing run 484 m (1,589 ft)
Landing run with propeller reversal 281 m (921 ft)
Accelerate/stop distance 1,016 m (3,334 ft)
Range with standard fuel, allowances for start, taxi, T-O, climb, descent, and 45 min reserves at max range cruising power:
max cruising power at:
3,660 m (12,000 ft)
715 nm (1,325 km; 823 miles)
4,875 m (16,000 ft)
780 nm (1,445 km; 897 miles)
6,100 m (20,000 ft)
850 nm (1,575 km; 978 miles)
7,620 m (25,000 ft)
940 nm (1,742 km; 1,081 miles)
max range power at:
3,660 m (12,000 ft)
775 nm (1,435 km; 892 miles)
4,875 m (16,000 ft)
845 nm (1,565 km; 972 miles)
6,100 m (20,000 ft)
910 nm (1,685 km; 1,047 miles)
7,620 m (25,000 ft)
980 nm (1,815 km; 1,128 miles)
Range with max optional fuel, allowances as above:
max cruising power at:
3,660 m (12,000 ft)
945 nm (1,751 km; 1,088 miles)
4,875 m (16,000 ft)
1,040 nm (1,927 km; 1,198 miles)
6,100 m (20,000 ft)
1,130 nm (2,094 km; 1,301 miles)
7,620 m (25,000 ft)
1,260 nm (2,335 km; 1,451 miles)
max range power at:
3,660 m (12,000 ft)
1,040 nm (1,927 km; 1,198 miles)
4,875 m (16,000 ft)
1,135 nm (2,102 km; 1,306 miles)
6,100 m (20,000 ft)
1,220 nm (2,260 km; 1,404 miles)
7,620 m (25,000 ft)
1,330 nm (2,463 km; 1,530 miles)

PIPER (PA-31T) CHEYENNE II

Design of the PA-31T began at the end of 1965. First flight of the prototype was made on 20 August 1969, and FAA certification was granted on 3 May 1972. The first production aircraft flew for the first time on 22 October 1973. Following the introduction of the low-cost Cheyenne I and 6/11-seat Cheyenne III, both of which are described separately, the original version was redesignated Cheyenne II.

It is available with Standard and Executive interior options and two operational group options as follows:

Standard. Six individual seats in pairs, with headrests and armrests. Pilot/co-pilot seats four-way adjustable with shoulder harness and inertia reels; third and fourth cabin seats aft-facing; and all cabin seats with seat belts; window curtains and wall to wall carpet. Rear cabin divider with clothes bar and baggage security net. Forward cabin divider curtain. 'No smoking/Fasten seat belt' sign. Oxygen outlets and masks at each seat position. Options available include a cabin instrumentation panel comprising digital readouts of altitude, outside air temperature, time, and true airspeed; pneumatic door extender; forward cabin combination unit; storage cabinets; folding tables; aft cabin combination unit, which includes side-facing seventh seat/toilet; seventh and eighth seats; tinted cabin windows; cabin fire extinguisher; stereo system; and all-leather seat covering.

Executive. Six individual seats, comprising two crew seats and four reclining chairs in the Standard arrangement. Other standard equipment as described, plus forward cabin combination unit which includes cabin dividers and curtain, electrically-heated Thermos unit, cup dispenser, storage for ice, beverages and manuals; two folding tables; pneumatic door extender and aft cabin combination unit which includes side-facing seventh seat/toilet, cabin divider with mirror, privacy curtain, refreshment centre; AC power outlet for electric razor; and baggage security net. Options are the same as for the Standard interior, unless included in the Executive package. Adds 55·3 kg (121·9 lb) to basic empty weight.

De-icing Group. Pneumatic de-icing boots for wing and tail unit leading-edges, and wing ice inspection light; adding 17·9 kg (39·4 lb) to basic empty weight.

Co-pilot Flight Group. Airspeed and rate of climb indicator, altimeter, electric turn rate indicator, attitude and directional gyros, clock, heated pitot, static system with alternate source, co-pilot's toe-brakes and windscreen wiper; adding 9·2 kg (20·3 lb) to basic empty weight.

Five optional factory-installed avionics packages are available for the Cheyenne II; other optional avionics generally as detailed for Cheyenne I.

In addition to the six/eight-seat passenger versions of the Cheyenne II, as described, Piper has developed a special multi-purpose model known as the **Maritime Surveillance Cheyenne II**. The prototype (N431PC), shown in an accompanying illustration, has a radar pod under the port wing. It can, however, be equipped for a wide range of special missions, including photo mapping, airways calibration, geological survey, rainmaking and hail suppression. A quick-change interior is standard, permitting conversion for corporate use in 30 min. Two aircraft of this type, each fitted with a Global GNS-500A navigation system, two 70 mm reconnaissance cameras (pod-mounted underwing), and a search radar were delivered to the Mauritanian Islamic Air Force in April 1981.

The 1981 versions of the Cheyenne II introduced the latest standard and optional improvements as detailed for the Cheyenne I.

Type: Six/eight-seat cabin monoplane.

Wings: Cantilever low-wing monoplane. Wing section NACA 63_2A415 at root, NACA 63_1A212 at tip. Dihedral 5°. Incidence 1° 30' at root, −1° at tip. Sweepback 0° at 30% chord. Three-spar structure of 2024ST light alloy. Balanced ailerons and single-slotted trailing-edge flaps of 2024ST light alloy. Trim tab in starboard aileron. Pneumatic de-icing boots on wing leading-edges optional.

Fuselage: Semi-monocoque structure of 2024ST light alloy, with fail-safe structure in the pressurised areas.

Tail Unit: Cantilever structure of 2024ST light alloy with sweptback vertical surfaces. Fixed-incidence tailplane. Trim tabs in elevators and rudder. Pneumatic de-icing of fin and tailplane leading-edges optional.

Landing Gear: Hydraulically-retractable tricycle type with single wheel on each unit, main units retracting inward and nosewheel aft. Nosewheel safety mirror. Piper oleo-pneumatic shock-absorbers. Main wheels and tyres size 6·50-10, 10-ply rating. Steerable nosewheel with Type VII tyre size 18 × 4·4, 6-ply rating. Goodyear disc-type hydraulic brakes. Parking brake.

Power Plant: Two 462 kW (620 ehp) Pratt & Whitney Aircraft of Canada PT6A-28 turboprop engines, each driving a Hartzell three-blade constant-speed reversible-pitch fully-feathering metal propeller type HC-BTN-3B. Propeller synchrophasers optional. Each wing has three interconnected fuel cells and a tip-tank, giving total fuel capacity of 1,476 litres (390 US gallons), of which 1,446 litres (382 US gallons) are usable. Refuelling points in engine nacelles and on upper surface of each tip-tank. NACA type anti-icing non-siphoning fuel tank vents with flame arresters. Oil capacity 24·6 litres (6·5 US gallons). Electrically heated air intake anti-icing boot, air intake ice deflection and air bypass doors. Electrical propeller de-icing.

Accommodation: Pilot and co-pilot on two individual adjustable seats. Dual controls standard. Pilot's storm window. Heated windscreen and windscreen wiper for pilot, optional for co-pilot. Cabin seating for four to six passengers on individual seats. Door with built-in airstair on port side, which has seven locking pins and inflatable pressurisation seal. Dual-pane windows. Emergency exit window on starboard side. Cabin heated and air-conditioned. Forward and aft cabin dividers. A wide range of options for cabin includes folding tables, beverage dispensers, pneumatic door extender, storage cabinets and tinted windows. Baggage compartments in nose, capacity 136 kg (300 lb), and rear of cabin, capacity 91 kg (200 lb). External access door to nose compartment.

Systems: Air-conditioning and pressurisation, with pressure differential of 0·38 bars (5·5 lb/sq in). Freon-type air-conditioner of 23,000 BTU capacity. Janitrol combustion heater of 35,000 BTU capacity with automatic windscreen defroster. Hydraulic system supplied by dual engine-driven pumps for landing gear retraction and brakes. Pneumatic system and vacuum system provided by engine bleed air. Electrical system supplied by two 28V 200A starter/generators and 24V 43Ah nickel-cadmium battery. External power socket standard. Oxygen system of 0·62 m^3 (22 cu ft) capacity. De-icing system comprises electric anti-icing boots for air intakes, heated pitot and electric propeller de-icing. Fire detection system with six sensors; engine fire extinguishing system optional.

Avionics: Six optional factory-installed avionics packages are available, these including nav/coms, ADF, AFCS, DME, glideslope and marker beacon receivers, R/Nav, transponder, and weather radar by Bendix, Collins, King and Piper. Options include alternatives to or duplication of the above equipment, plus altitude alerter, encoding altimeter, radar altimeter, radio altimeter,

Piper Cheyenne II (two P&WC PT6A-28 turboprop engines) equipped as maritime patrol demonstrator

radio telephone, global navigation system, and HF transceivers by manufacturers which include the above, plus Aeronetics, RCA, Smith, Sperry and SunAir.

EQUIPMENT: Installed standard equipment is extensive, and optional items include flight instrument group, toe brakes, heated windscreen and windscreen wiper for co-pilot, emergency locator transmitter, wing and tail pneumatic de-icing boots, engine fire extinguisher system, emergency electrical power supply system, fuselage ice protection plates, ice inspection lights and propeller synchrophasers.

DIMENSIONS, EXTERNAL:
Wing span over tip-tanks 13·01 m (42 ft 8¼ in)
Length overall 10·57 m (34 ft 8 in)
Height overall 3·89 m (12 ft 9 in)
Tailplane span 6·05 m (19 ft 10 in)
Wheel track 4·19 m (13 ft 9 in)
Wheelbase 2·64 m (8 ft 8 in)
Propeller diameter 2·36 m (7 ft 9 in)
Distance between propeller centres 3·85 m (12 ft 7½ in)
Propeller ground clearance 0·27 m (10½ in)
Passenger door (port, rear):
Height 1·17 m (3 ft 10 in)
Width 0·71 m (2 ft 4 in)
Height to sill 0·94 m (3 ft 1 in)
Baggage door (fwd): Height 0·53 m (1 ft 9 in)
Width 0·66 m (2 ft 2 in)
Height to sill 1·10 m (3 ft 7½ in)
Emergency exit (stbd, fwd):
Height 0·64 m (2 ft 1 in)
Width 0·48 m (1 ft 7 in)

DIMENSIONS, INTERNAL:
Cabin: Length, excl flight deck 2·57 m (8 ft 5 in)
Max width 1·27 m (4 ft 2 in)
Max height 1·31 m (4 ft 3½ in)
Floor area 4·37 m^2 (47 sq ft)
Volume 6·29 m^3 (222 cu ft)
Forward baggage compartment 0·57 m^3 (20 cu ft)
Rear baggage compartment 0·62 m^3 (22 cu ft)

AREAS:
Wings, gross 21·3 m^2 (229 sq ft)
Ailerons (total) 1·21 m^2 (13 sq ft)
Trailing-edge flaps (total) 3·12 m^2 (33·6 sq ft)
Fin 1·48 m^2 (15·9 sq ft)
Rudder, incl tab 0·98 m^2 (10·6 sq ft)
Tailplane 3·92 m^2 (42·2 sq ft)
Elevators, incl tab 2·63 m^2 (28·3 sq ft)

WEIGHTS AND LOADINGS:
Weight empty, standard, equipped 2,260 kg (4,983 lb)
Max T-O and landing weight 4,082 kg (9,000 lb)
Max ramp weight 4,105 kg (9,050 lb)
Max zero-fuel weight 3,265 kg (7,200 lb)
Max wing loading 191·9 kg/m^2 (39·3 lb/sq ft)
Max power loading 4·42 kg/kW (7·26 lb/ehp)

PERFORMANCE (at max T-O weight, unless specified otherwise):
Cruising speed, max cruise power, average cruise weight of 3,493 kg (7,700 lb) at:
3,350 m (11,000 ft) 283 knots (524 km/h; 325 mph)
4,875 m (16,000 ft) 278 knots (515 km/h; 320 mph)
6,400 m (21,000 ft) 270 knots (500 km/h; 311 mph)
8,840 m (29,000 ft) 253 knots (469 km/h; 291 mph)
Stalling speed, wheels and flaps up, engines idling 88 knots (162 km/h; 101 mph) IAS
Stalling speed, wheels and flaps down, engines idling 76 knots (141 km/h; 98 mph) IAS
Rotation speed 91 knots (167 km/h; 105 mph) IAS
Approach speed 98 knots (182 km/h; 113 mph)
Max rate of climb at S/L 826 m (2,710 ft)/min
Rate of climb at S/L, one engine out 201 m (660 ft)/min
Service ceiling 9,630 m (31,600 ft)
Service ceiling, one engine out 4,450 m (14,600 ft)
T-O run 430 m (1,410 ft)
T-O to 15 m (50 ft) 604 m (1,980 ft)
Landing from 15 m (50 ft) 756 m (2,480 ft)
Landing from 15 m (50 ft) with propeller reversal 567 m (1,860 ft)
Landing run 436 m (1,430 ft)
Landing run with propeller reversal 291 m (955 ft)
Accelerate/stop distance 1,006 m (3,300 ft)
Range with max fuel at max cruise power, with allowances for start, taxi, T-O, climb, descent, and 45 min reserves at long-range cruise power at:
3,660 m (12,000 ft) 905 nm (1,676 km; 1,041 miles)
4,875 m (16,000 ft) 1,020 nm (1,889 km; 1,173 miles)
6,400 m (21,000 ft) 1,155 nm (2,139 km; 1,329 miles)
8,840 m (29,000 ft) 1,380 nm (2,555 km; 1,588 miles)
Range at max range power, fuel and allowances as above, at:
3,660 m (12,000 ft) 1,090 nm (2,018 km; 1,254 miles)
4,875 m (16,000 ft) 1,195 nm (2,213 km; 1,375 miles)
6,400 m (21,000 ft) 1,330 nm (2,463 km; 1,530 miles)
8,840 m (29,000 ft) 1,510 nm (2,796 km; 1,737 miles)

PIPER (PA-31T2) CHEYENNE IIXL

On 17 October 1980, Piper announced a fourth member of the Cheyenne family of turboprop business aircraft. Known as the PA-31T2 Cheyenne IIXL, it is a new version of the Cheyenne II, with the fuselage lengthened by 0·61 m (2 ft 0 in) to increase cabin volume by 16 per cent and so provide more comfortable accommodation for passengers and crew. In addition, it has Pratt & Whitney Aircraft of Canada PT6A-135 turboprop engines, each rated at 559 kW (750 shp), but flat rated to only 462 kW (620 shp) in this installation, and driving large-diameter three-blade propellers at 1,900 rpm. This results in low noise levels, both internally and externally. By comparison with the Cheyenne II, the IIXL has a max T-O weight more than 180 kg (400 lb) greater. It uses the same type of environmental control unit as the Cheyenne III for cabin heating and air-conditioning, and has interior equipment options similar to those of the III.

The first six Cheyenne IIXLs were delivered to Piper sales centres on 2 June 1981.

The description of the Cheyenne II applies also to the Cheyenne IIXL, except as follows:

FUSELAGE: As for Cheyenne II, except lengthened by 0·61 m (2 ft 0 in).

POWER PLANT: Two 559 kW (750 shp) Pratt & Whitney Aircraft of Canada PT6A-135 turboprop engines, each flat rated to 462 kW (620 shp) and driving a Hartzell three-blade reversible-pitch constant-speed and fully-feathering metal propeller with spinner. Fuel, oil and de-icing system as for Cheyenne II.

DIMENSIONS, EXTERNAL: As for Cheyenne II except:
Length overall 11·18 m (36 ft 8 in)

DIMENSIONS, INTERNAL: As for Cheyenne II except:
Cabin: Length, excl flight deck 3·83 m (12 ft 7 in)
Forward baggage compartment 0·45 m^3 (16 cu ft)

WEIGHTS AND LOADINGS:
Weight empty, standard 2,325 kg (5,126 lb)
Max T-O weight 4,297 kg (9,474 lb)
Max ramp weight 4,327 kg (9,540 lb)
Max landing weight 4,082 kg (9,000 lb)
Max zero-fuel weight 3,447 kg (7,600 lb)
Max wing loading 202·0 kg/m^2 (41·37 lb/sq ft)
Max power loading 4·65 kg/kW (7·64 lb/shp)

PERFORMANCE (at max T-O weight, except where indicated):
Cruising speed, max cruise power, at average cruise weight of 3,720 kg (8,200 lb):
at 3,960 m (13,000 ft) 275 knots (510 km/h; 317 mph)
at 4,875 m (16,000 ft) 273 knots (506 km/h; 314 mph)
at 6,100 m (20,000 ft) 270 knots (500 km/h; 311 mph)
at 8,840 m (29,000 ft) 255 knots (473 km/h; 294 mph)
Stalling speed, flaps and landing gear up, engines idling 86 knots (160 km/h; 99 mph) IAS
Stalling speed, flaps and landing gear down, engines idling 77 knots (143 km/h; 89 mph) IAS
Rotation speed 101 knots (187 km/h; 116 mph) IAS
Approach speed 104 knots (193 km/h; 120 mph)
Max rate of climb at S/L 533 m (1,750 ft)/min
Rate of climb at S/L, one engine out 143 m (470 ft)/min
Service ceiling 9,875 m (32,400 ft)
Service ceiling, one engine out 4,540 m (14,900 ft)
T-O run 622 m (2,042 ft)
T-O to 15 m (50 ft) 896 m (2,940 ft)
Landing from 15 m (50 ft) 745 m (2,446 ft)
Landing from 15 m (50 ft), with propeller reversal 540 m (1,773 ft)
Landing run 479 m (1,571 ft)
Landing run, with propeller reversal 329 m (1,080 ft)
Accelerate/stop distance 1,276 m (4,186 ft)
Range with max fuel at max cruise power, allowances for start, taxi, T-O, climb, descent, and 45 min reserve at max cruise power, ISA:
at 3,660 m (12,000 ft) 785 nm (1,454 km; 904 miles)
at 4,875 m (16,000 ft) 887 nm (1,643 km; 1,021 miles)
at 6,100 m (20,000 ft) 1,001 nm (1,855 km; 1,152 miles)
at 8,840 m (29,000 ft) 1,233 nm (2,285 km; 1,420 miles)
Range with max fuel at max range power, allowances as above:
at 3,660 m (12,000 ft) 978 nm (1,812 km; 1,126 miles)
at 4,875 m (16,000 ft) 1,062 nm (1,968 km; 1,223 miles)
at 6,100 m (20,000 ft) 1,188 nm (2,201 km; 1,368 miles)
at 8,840 m (29,000 ft) 1,336 nm (2,476 km; 1,538 miles)

PIPER (PA-42) CHEYENNE III

Announced on 26 September 1977, the Cheyenne III differs from the I and II by having increased wing span, a lengthened fuselage, a T tail and more powerful PT6A-41 engines. The production prototype flew for the first time on 18 May 1979; FAA certification was gained in early 1980, and deliveries of production aircraft began on 30 June 1980.

The Cheyenne III is available with Standard and Executive interior options, and Co-pilot Flight Groups, as detailed for the Cheyenne II. Full de-icing equipment is standard. Three optional factory installed avionics packages are available for the Cheyenne III; other optional avionics are generally as detailed for the Cheyenne I.

TYPE: Six/eleven-seat corporate and commuter airline transport.

WINGS: Cantilever low-wing monoplane. Wing section NACA 63_2A415, modified, at root, NACA 63_1A212 at tip. Dihedral 5°. Incidence 1° 30′. No sweepback. Three-spar fail-safe structure of light alloy. Ailerons and trailing-edge flaps as for Cheyenne II. Goodrich pneumatic de-icing boots for wing leading-edges are standard.

FUSELAGE: Conventional semi-monocoque structure of light alloy, with fail-safe structure in the pressurised areas.

TAIL UNIT: Cantilever T tail of light alloy construction, with sweptback vertical surfaces. Fixed-incidence tailplane. Elevators and rudder of light alloy. Servo tab in rudder; anti-servo tab in elevator. Goodrich de-icing equipment standard for leading-edges of tailplane and fin.

LANDING GEAR: Hydraulically-retractable tricycle type with single wheel on each unit. Main units retract inward, nosewheel aft. Manually-operated hydraulic system for emergency extension. Piper oleo-pneumatic shock-absorbers. Cleveland main wheels with tyres size

Piper PA-31T2 Cheyenne IIXL, latest addition to the Cheyenne family of light transports

6·50-10 12-ply Type III, pressure 6·90 bars (100 lb/sq in). Cleveland steerable nosewheel with tyre size 17·5 × 6·25, 10-ply rating Type III, pressure 4·83 bars (70 lb/sq in). Cleveland hydraulically-operated disc brakes. Parking brake.

POWER PLANT: Two Pratt & Whitney Aircraft of Canada PT6A-41 turboprop engines, each flat rated to 537 kW (720 shp) and driving a Hartzell three-blade constant-speed feathering and reversible-pitch metal propeller with Q-tips. Automatic propeller feathering system and synchrophaser optional. Each wing has three interconnected fuel cells and a tip-tank, with a combined total capacity of 1,514 litres (400 US gallons), of which 1,476 litres (390 US gallons) are usable. Optional fuel tank, capacity 321·75 litres (85 US gallons), in each engine nacelle, to raise total usable capacity to 2,119·5 litres (560 US gallons). Refuelling points on upper surface of each tip-tank and engine nacelle. Oil capacity 24·6 litres (6·5 US gallons). Electrical intake anti-icing and propeller de-icing.

ACCOMMODATION: Pilot and co-pilot on four-way adjustable seats with armrests, headrests, shoulder safety belts with inertia reels, and stowage for oxygen mask beneath seats. To be certificated for single-pilot operation. Dual controls standard. Pilot's storm window. Cabin seats up to nine passengers, but standard interior includes six reclining and adjustable passenger seats with armrests, headrests, and magazine storage on seat back. Four optional executive interiors available, plus a wide range of options for cabin furnishing. Door with built-in airstair on port side, with seven locking pins and inflatable pressurisation seal. Emergency exit window on starboard side. Baggage compartments in nose and rear of cabin, each with capacity of 136 kg (300 lb). When aircraft is operated without engine nacelle fuel tank option, an additional 136 kg (300 lb) of baggage can be accommodated in each nacelle locker to give a maximum total baggage capacity of 544 kg (1,200 lb). Accommodation is pressurised, heated and air-conditioned. Pilot's windscreen heated; provisions for heating co-pilot windscreen. Pilot and co-pilot windscreen wipers standard.

SYSTEMS: Garrett pressurisation system with max differential of 0·43 bars (6·3 lb/sq in), maintaining a cabin altitude of 3,050 m (10,000 ft) to a height of 10,060 m (33,000 ft). Environmental control system, combining the functions of heater, air-conditioner and dehumidifier. Hydraulic system supplied by dual engine-driven pumps. Pneumatic system and vacuum system supplied by engine bleed air. Electrical system includes two 28V 250A engine-driven generators and 24V 43Ah storage battery. Oxygen system of 0·62 m³ (22 cu ft) capacity with ten outlets. De-icing system includes pneumatic wing and tailplane de-icing boots, electrical anti-icing of engine air intakes, heated pitots, electrical propeller de-icing, and windscreen heating.

AVIONICS AND EQUIPMENT: Generally as for Cheyenne II, including King 300 and Collins AP-106 autopilot/flight directors. Extensive standard installed equipment includes 'No smoking-fasten seat belt' sign; carpeted floor; tinted cabin windows; window curtains; curtain between flight deck and cabin; oxygen system with individual masks in storage compartments; indirect fluorescent lighting, individual reading lights, and courtesy lights. Optional equipment includes cabin chimes; stereo system; cabin instrument cluster giving digital readouts of altitude, outside air temperature, time, and true airspeed; cabin fire extinguisher; emergency locator transmitter; and engine fire extinguishing system.

DIMENSIONS, EXTERNAL: As for Cheyenne II, except:

Wing span over tip-tanks	14·53 m (47 ft 8⅛ in)
Wing chord at root	3·12 m (10 ft 3 in)
Wing chord at tip	0·97 m (3 ft 2 in)
Wing aspect ratio	7·82
Length overall	12·23 m (43 ft 4¾ in)
Height overall	4·50 m (14 ft 9 in)
Tailplane span	6·65 m (21 ft 10 in)
Wheel track	5·72 m (18 ft 9 in)
Wheelbase	3·23 m (10 ft 7¼ in)
Propeller diameter	2·41 m (7 ft 11 in)
Distance between propeller centres	5·38 m (17 ft 8 in)

DIMENSIONS, INTERNAL:

Cabin (incl flight deck and rear baggage area):	
Length	6·99 m (22 ft 11 in)
Max width	1·30 m (4 ft 3 in)
Max height	1·35 m (4 ft 5 in)
Volume	approx 9·91 m³ (350 cu ft)
Nose baggage compartment	0·46 m³ (16·25 cu ft)
Rear baggage compartment	0·88 m³ (31 cu ft)
Nacelle baggage locker (two, each)	0·41 m³ (14·5 cu ft)
Nacelle baggage locker with optional fuel (two, each)	0·19 m³ (6·7 cu ft)

AREAS:

Wings, gross	27·22 m² (293 sq ft)
Ailerons (total)	1·25 m² (13·5 sq ft)
Trailing-edge flaps (total)	3·98 m² (42·8 sq ft)
Fin	2·17 m² (23·36 sq ft)
Rudder, incl tab	1·88 m² (20·2 sq ft)
Tailplane	3·48 m² (37·5 sq ft)
Elevators, incl tab	2·26 m² (24·3 sq ft)

WEIGHTS AND LOADINGS:

Basic weight empty	2,898 kg (6,389 lb)
Max T-O weight	5,080 kg (11,200 lb)
Max ramp weight	5,119 kg (11,285 lb)
Max zero-fuel weight	4,241 kg (9,350 lb)
Max landing weight	4,685 kg (10,330 lb)
Max wing loading	186·6 kg/m² (38·22 lb/sq ft)
Max power loading	4·73 kg/kW (7·78 lb/shp)

PERFORMANCE (at max T-O weight, except where indicated):

Max level speed at average cruise weight of 4,218 kg (9,300 lb) — 296 knots (549 km/h; 341 mph)

Cruising speed at max cruise power, at average cruise weight of 4,218 kg (9,300 lb):
- at 6,100 m (21,000 ft) — 289 knots (536 km/h; 333 mph)
- at 7,620 m (25,000 ft) — 283 knots (524 km/h; 326 mph)
- at 9,145 m (30,000 ft) — 275 knots (509 km/h; 316 mph)
- at 10,060 m (33,000 ft) — 264 knots (489 km/h; 304 mph)

Stalling speed:
- flaps and gear up, power off — 95 knots (176 km/h; 109 mph) CAS
- flaps and gear down, power off — 83 knots (155 km/h; 96 mph) CAS

Rotation speed	98 knots (182 km/h; 113 mph) CAS
Approach speed	104 knots (193 km/h; 120 mph) CAS
Max rate of climb at S/L	682 m (2,236 ft)/min
Rate of climb at S/L, one engine out	162 m (530 ft)/min
Service ceiling	10,000 m (32,800 ft)
Service ceiling, one engine out	5,305 m (17,400 ft)
T-O run	768 m (2,520 ft)
T-O to 15 m (50 ft)	1,006 m (3,300 ft)
Landing from 15 m (50 ft)	930 m (3,050 ft)
Landing from 15 m (50 ft) with propeller reversal	791 m (2,595 ft)
Landing run	539 m (1,770 ft)
Landing run with propeller reversal	382 m (1,255 ft)
Accelerate/stop distance	1,646 m (5,400 ft)

Range, with max optional fuel, allowances for taxi, T-O, climb, descent, and 45 min reserves at max range power:

max cruising power at:
- 6,100 m (20,000 ft) — 1,360 nm (2,520 km; 1,566 miles)
- 7,620 m (25,000 ft) — 1,551 nm (2,874 km; 1,786 miles)
- 9,145 m (30,000 ft) — 1,670 nm (3,095 km; 1,923 miles)
- 10,060 m (33,000 ft) — 1,780 nm (3,298 km; 2,050 miles)

max range power at:
- 6,100 m (20,000 ft) — 1,755 nm (3,252 km; 2,021 miles)
- 7,620 m (25,000 ft) — 1,960 nm (3,632 km; 2,257 miles)
- 9,145 m (30,000 ft) — 2,120 nm (3,929 km; 2,441 miles)
- 10,060 m (33,000 ft) — 2,200 nm (4,077 km; 2,533 miles)

PIPER (PA-34-220T) SENECA III

On 23 September 1971, Piper announced a twin-engined light aircraft which had the company designation PA-34 and, following Piper tradition, the Indian name Seneca. Built at Piper's Vero Beach, Florida, factory, the aircraft was redesignated Seneca II from 1975. On 15 February 1981 Piper introduced an improved PA-34-220T Seneca III, with more powerful engines and improvements that include a redesigned one-piece windscreen, restyled window mouldings and curtains, a new all-metal instrument panel with improved layout, a newly designed glareshield, improved night-lighting and annunciator panel, improved braking system, and a number of detail refinements, plus a wider range of avionics.

The Seneca III has counter-rotating (C/R) engine and propeller installations. The retractable landing gear is operated by an electro-hydraulic system and includes an emergency extension system which allows the wheels to free-fall into the down and locked position. A dual-vane stall warning system provides warning by horn well in advance of the stall in either clean or gear/flaps-down configuration.

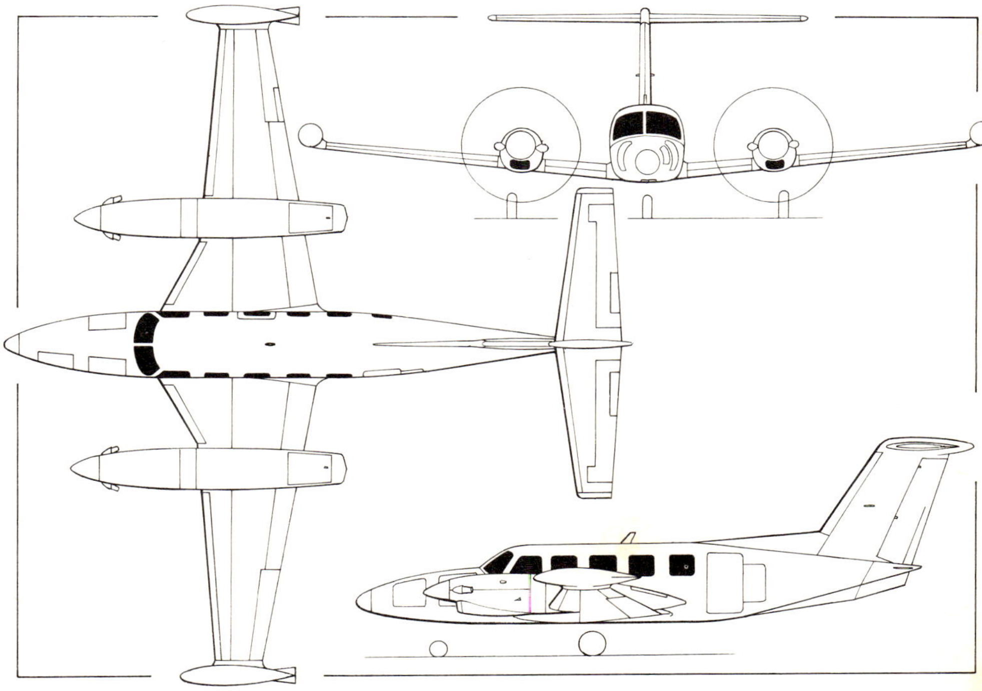

Photograph and three-view drawing *(Pilot Press)* **of Piper's T-tailed Cheyenne III**

Piper Seneca III, powered by two Continental TSIO-360-KB turbocharged counter-rotating engines

The following optional operational groups are available:

Custom. Comprising an instrument package which includes blind-flying instrumentation with 3 in attitude and directional gyros, electric clock, outside air temperature gauge, rate of climb indicator, true airspeed indicator, and dual engine-driven pressure pumps; a lighting package which includes instrument panel white lighting and overhead red lighting, avionics dimming, cabin dome, map, reading, waterfall switch panel, landing, navigation, taxi, and wing strobe lights; heated pitot and stall warning device; assist straps; and towbar; adding 16·6 kg (36·5 lb) to basic empty weight.

Executive. As Custom group, plus pilot's vertically adjustable seat, shoulder and safety belts with inertia reels for rear seats, tinted windows, emergency locator transmitter, baggage compartment and entrance door lights, external power socket, and quick oil drain valves; adding 23 kg (50·6 lb) to basic empty weight.

De-icing Group. Comprising pneumatic de-icing boots for wing and tail unit leading-edges, electrical propeller de-icing, luminous outside air temperature gauge, ice inspection light, and electrical windscreen de-icing plate; adding 24·5 kg (54·1 lb) to basic empty weight.

It was announced on 3 January 1977 that Piper had signed an agreement with Pezetel, the Polish foreign trade organisation, enabling PZL Mielec to assemble, manufacture and distribute the Seneca in Eastern Europe. These aircraft (several hundred are involved in the agreement) are powered by 164 kW (220 hp) PZL-Franklin engines and are named **M-20 Mewa** (Gull).

TYPE: Six/seven-seat twin-engined light aircraft.

WINGS: Cantilever low-wing monoplane. Dihedral 7°. Single-spar wings, Frise ailerons, and wide-span manually-operated slotted flaps, of light alloy construction. Glassfibre wingtips. Pneumatic de-icing boots for leading-edges optional.

FUSELAGE: Light alloy semi-monocoque structure.

TAIL UNIT: Cantilever structure of light alloy. One-piece all-moving horizontal surface with combined anti-balance and trim tab. Anti-servo tab in rudder. Pneumatic de-icing boots for fin and tailplane leading-edges optional.

LANDING GEAR: Hydraulically-retractable tricycle type. Main units retract inward, nose unit forward. Oleo-pneumatic shock-absorbers. Steerable nosewheel. Emergency free-fall extension system. Main wheels and tyres size 6·00-6, 8-ply rating; nosewheel and tyre size 6·00-6, 6-ply rating. Nosewheel safety mirror. High-capacity disc brakes. Parking brake. Heavy-duty tyres and brakes optional.

POWER PLANT: One 164 kW (220 hp) Continental TSIO-360-KB and one 164 kW (220 hp) Continental LTSIO-360-KB flat-six turbocharged counter-rotating engines, each driving a Hartzell two-blade constant-speed fully-feathering metal propeller with spinner. Three-blade propellers, propeller de-icing, and propeller synchrophasers optional. Fuel in two tanks in wings, with a total capacity of 371 litres (98 US gallons) of which 352 litres (93 US gallons) are usable. Optional 57 litre (15 US gallon) auxiliary tank in each wing to provide a max capacity of 485 litres (128 US gallons) of which 466 litres (123 US gallons) are usable. Oil capacity 7·5 litres (2 US gallons). Glassfibre engine cowlings.

ACCOMMODATION: Enclosed cabin, seating six people in pairs on individual seats with 0·25 m (10 in) centre aisle. Optional seventh seat between two centre seats. Dual controls standard. Pilot's storm window. Two forward-hinged doors, one on starboard side at front, the other on port side at rear. Large optional door adjacent to rear cabin door provides an extra-wide opening for loading bulky items. Passenger seats removable easily to provide different seating/baggage/cargo combinations. Space for 45 kg (100 lb) baggage at rear of cabin, and for 45 kg (100 lb) in nose compartment with external access door on port side. Cabin heated and ventilated. Windscreen defrosters standard. Electrically de-iced windscreen for pilot, and ice inspection light, optional.

SYSTEMS: Electro-hydraulic system for landing gear retraction. Electrical system powered by two 14V 65A alternators. 12V 35Ah battery. Oxygen system with six outlets, or built-in oxygen system of 1·78 m³ (63 cu ft), optional. Dual engine-driven vacuum pumps for flight instruments optional. Piper Aire air-conditioning system of 14,500 BTU capacity optional. Janitrol 45,000 BTU combustion heater standard.

AVIONICS AND EQUIPMENT: Seven optional factory-installed packages of avionics are available, these including nav/coms, ADF, autopilot, DME, glideslope and marker beacon receivers, and transponder, by Collins, King, Narco and Piper. Options include alternatives to or duplication of the above equipment, plus autopilot/flight director systems, encoding altimeter, HF transceiver, R/Nav, radio altimeter, and weather radar, by manufacturers which include the above, plus Bendix, Bonzer, Edo-Aire Mitchell, SunAir and United. Standard equipment includes two cylinder head and two exhaust gas temperature gauges, two manifold pressure gauges, two electric tachometers, and sensitive altimeter. Pilot's and co-pilot's seats are fore and aft adjustable and reclining, and have shoulder and safety belts with inertia reels, and armrests; pilot's storm window, sun visors, stall warning device, provisions for emergency locator transmitter, carpeted floor, soundproofing, jack pads, and tiedown rings. Items in operational groups are available optionally, plus co-pilot's blind-flying instrumentation, cabin fire extinguisher, headrests, window curtains, tables, refreshment console, shoulder safety belts with inertia reels for passenger seats, tinted windows, ventilation fan, stainless steel control cables, towbar, and zinc chromate finish for aluminium parts.

DIMENSIONS, EXTERNAL:

Wing span	11·85 m (38 ft 10¾ in)
Wing chord (constant)	1·60 m (5 ft 3 in)
Length overall	8·72 m (28 ft 7½ in)
Height overall	3·02 m (9 ft 10¾ in)
Tailplane span	4·14 m (13 ft 6¾ in)
Wheel track	3·37 m (11 ft 0¾ in)
Wheelbase	2·13 m (7 ft 0 in)
Propeller diameter	1·93 m (6 ft 4 in)
Distance between propeller centres	3·80 m (12 ft 5½ in)
Cabin door (port, rear): Height	0·84 m (2 ft 9 in)
Width	0·74 m (2 ft 5 in)
Baggage door (stbd, rear): Height	0·51 m (1 ft 8 in)
Width	0·74 m (2 ft 5 in)
Baggage door (stbd, aft): Height	0·51 m (1 ft 8 in)
Width	0·71 m (2 ft 4 in)
Baggage door (port, fwd): Height	0·46 m (1 ft 6 in)
Width	0·61 m (2 ft 0 in)

DIMENSIONS, INTERNAL:

Cabin (incl flight deck):	
Length	3·17 m (10 ft 5 in)
Max width and height	1·24 m (4 ft 1 in)
Volume	5·53 m³ (195·3 cu ft)
Forward baggage compartment	0·43 m³ (15·3 cu ft)
Rear baggage compartment	0·49 m³ (17·3 cu ft)

AREA:

Wings, gross	19·39 m² (208·7 sq ft)

WEIGHTS AND LOADINGS:

Weight empty, standard	1,304 kg (2,875 lb)
Max T-O weight	2,154 kg (4,750 lb)
Max ramp weight	2,165 kg (4,773 lb)
Max zero-fuel weight	2,027 kg (4,470 lb)
Max landing weight	2,047 kg (4,513 lb)
Max wing loading	111·1 kg/m² (22·76 lb/sq ft)
Max power loading	6·57 kg/kW (10·8 lb/hp)

PERFORMANCE (at max T-O weight, except where indicated):

Max level speed at optimum altitude, mid-cruise weight	196 knots (363 km/h; 226 mph)
Cruising speed at optimum altitude, mid-cruise weight:	
75% power	193 knots (357 km/h; 222 mph)
65% power	191 knots (354 km/h; 220 mph)
55% power	180 knots (333 km/h; 207 mph)
45% power	168 knots (311 km/h; 193 mph)
Cruising speed at 3,050 m (10,000 ft), mid-cruise weight:	
75% power	179 knots (332 km/h; 206 mph)
65% power	175 knots (324 km/h; 202 mph)
55% power	159 knots (295 km/h; 183 mph)
45% power	143 knots (265 km/h; 165 mph)
Stalling speed: flaps and landing gear up	66 knots (122 km/h; 76 mph) CAS
flaps and landing gear down	62 knots (115 km/h; 72 mph) CAS
Max rate of climb at S/L	427 m (1,400 ft)/min
Rate of climb at S/L, one engine out	73 m (240 ft)/min
Max certificated ceiling	7,620 m (25,000 ft)
Service ceiling, one engine out	3,750 m (12,300 ft)
T-O run	280 m (920 ft)
T-O to 15 m (50 ft)	369 m (1,210 ft)
Landing from 15 m (50 ft):	
standard brakes	658 m (2,160 ft)
heavy-duty brakes	603 m (1,978 ft)
Landing run: standard brakes	427 m (1,400 ft)
heavy-duty brakes	371 m (1,218 ft)
Accelerate/stop distance:	
standard brakes	732 m (2,400 ft)
heavy-duty brakes	636 m (2,088 ft)
Range with standard fuel, allowances for taxi, T-O, climb, descent, and 45 min reserves at max range power:	
at optimum altitude:	
75% power	462 nm (855 km; 531 miles)
65% power	550 nm (1,018 km; 633 miles)
55% power	630 nm (1,166 km; 725 miles)
45% power	670 nm (1,240 km; 771 miles)
at 3,050 m (10,000 ft):	
75% power	450 nm (833 km; 517 miles)
65% power	535 nm (990 km; 615 miles)
55% power	610 nm (1,129 km; 702 miles)
45% power	632 nm (1,170 km; 727 miles)
Range with max optional fuel, allowances as above:	
at optimum altitude:	
75% power	665 nm (1,232 km; 765 miles)
65% power	785 nm (1,454 km; 904 miles)
55% power	920 nm (1,705 km; 1,059 miles)
45% power	990 nm (1,834 km; 1,140 miles)
at 3,050 m (10,000 ft):	
75% power	640 nm (1,186 km; 737 miles)
65% power	760 nm (1,408 km; 875 miles)
55% power	860 nm (1,593 km; 990 miles)
45% power	900 nm (1,668 km; 1,036 miles)

PIPER (PA-36) BRAVE

On 9 October 1972 Piper Aircraft Corporation released details of a new agricultural aircraft named the Pawnee Brave, with a more powerful engine than the PA-25 Pawnee D and larger, with increased capacity for either liquid or dry chemicals.

The basic configuration seats the pilot well aft. The long nose is designed to collapse progressively in an emergency. The fuselage is a welded truss structure of chrome-molybdenum steel, which is graded in strength to provide high energy absorption and progressive collapse. A sturdy overturn pylon is an integral part of the fuselage structure. The wing is of conventional cantilever construction, with laminated spars to provide structural redundancy. The wing leading-edges each comprise two glassfibre sections, reinforced by a foam insert beam running spanwise. Normal impacts are absorbed by the leading-edge, more serious contacts by ribs designed to collapse with minimal impact transference to the basic wing structure.

The pilot is located in an isolated cockpit capsule which keeps him well clear of main structural members. The floor is 0·30 m (1 ft 0 in) above the lower longerons, and a cockpit width of 0·97 m (3 ft 2 in) allows for substantial deformation of the fuselage structure without hazard to the pilot. The seat is attached to the overturn pylon, and is articulated to allow the pilot's position to change with fuselage deformation. The cockpit capsule is sealed to prevent the ingress of toxic chemicals; and all protrusions, knobs and levers which might cause injury are eliminated. The instrument panel is equipped with a large energy-absorbing crash roll.

The cockpit capsule is ventilated via an airscoop in the top of the canopy, which filters the incoming air before discharge through two adjustable diffusers. A heating system is standard, and the inflow of ventilating and/or heated air has the effect of pressurising the cockpit, further discouraging any inflow of toxic fumes or chemicals.

The fuel tanks, located in the wing roots, are filled with reticulated polyurethane foam to serve both as a fire suppressant and as a constant baffle to reduce fuel surge. Fire-resistant fuel pipes are wire-reinforced at potential rupture points.

Two hopper sizes are available. The larger hopper has a maximum dry chemicals capacity of 862 kg (1,900 lb), and is compatible with applicators designed to spread chemicals at rates of up to 181 kg (400 lb) per acre.

Chincul Argentinian-built example of the Piper Brave 300 agricultural aircraft

Spray equipment for the Brave can dispense up to 863 litres (228 US gallons) per minute, which is the equivalent of 160 litres/hectare (17 US gallons/acre) at 117 knots (217 km/h; 135 mph) and with a 15·25 m (50 ft) swath width. The spray equipment consists of a quickly-removable pylon-mounted wind-driven spraypump, and sprayboom located just aft of the wing trailing-edges. This location reduces drag and allows the pilot to make visual checks of their operation.

All parts of the Brave's airframe are treated to prevent corrosion damage, with extensive use of polyurethane coating, selection of stainless steel for cables and other moving components in vulnerable areas, and internal oiling of lower truss sections. The design eliminates dust traps and inaccessible areas, and fuselage covering is spaced away from the frame to permit thorough hosing down. To facilitate washing, inspection and maintenance, the plastics side panels and entire belly covering are attached by quick-release fasteners.

Two versions of the Brave are available:

Brave 300. Basic version, equipped with one 224 kW (300 hp) Avco Lycoming IO-540-K1G5 flat-six engine, driving a Hartzell HC-C2YK-1BF two-blade constant-speed metal propeller. Three-blade propeller optional.

Brave 375. Introduced in 1978: basically as Brave 300, but equipped with 279·5 kW (375 hp) Avco Lycoming IO-720-D1CD flat-eight engine, driving a Hartzell three-blade constant-speed metal propeller, and with the 862 kg (1,900 lb) hopper as standard. FAA certification in Normal and Restricted categories was awarded on 4 October 1977.

The following description applies to both versions of the Brave:

TYPE: Single-seat agricultural aircraft.

WINGS: Cantilever low-wing monoplane. Wing section NACA 63_3-618. Dihedral 6°. Incidence 2° 30′ at root, 0° 30′ at tip. Conventional two-spar metal structure. Light alloy laminated spars with two-bolt main spar attachment to fuselage structure. Light alloy covering, except for detachable leading-edges of glassfibre, reinforced by foam inserts, and glassfibre wingtips. Conventional ailerons and trailing-edge flaps. Landing lights in wing leading-edges.

FUSELAGE: Welded chrome-molybdenum steel tube structure. Removable metal underskin and removable side panels of plastics material. Glassfibre engine cowling.

TAIL UNIT: Cantilever all-metal structure. Tailplane has glassfibre tips. Tab on rudder and in each elevator. Cable from top of cockpit structure to tip of fin to deflect cables.

LANDING GEAR: Non-retractable tailwheel type. Interchangeable cantilever spring steel main-gear struts, with wire-cutters on leading-edges. Cleveland main wheels type 40-101 with tyres size 8·50-10, 6-ply rating, pressure 1·93-2·21 bars (28-32 lb/sq in). Scott steerable tailwheel type 3450-21 with tyre of 0·25 m (10 in) diameter, pressure 2·41-3·10 bars (35-45 lb/sq in). Cleveland type 30-67B hydraulic brakes. Parking brake.

POWER PLANT: One engine as detailed in model listings. Three-blade constant-speed propeller optional for Brave 300. One fuel tank in each wing root, capacity 170·3 litres (45 US gallons). Total fuel capacity for both versions 337 litres (89 US gallons), of which 325·5 litres (86 US gallons) are usable. Refuelling point on upper surface of each wing. Fuel tanks filled with reticulated polyurethane safety foam (Safom). Oil capacity of Brave 300, 11·5 litres (3 US gallons); Brave 375, 16 litres (4·25 US gallons).

ACCOMMODATION: Pilot only, on adjustable seat in an isolated cockpit capsule, with steel tube overturn structure. Seat, equipped with double shoulder harness and inertia reel, is attached to overturn structure. Wire-cutter mounted in centre of windscreen. Combined window and door on each side, hinged at bottom. Cockpit capsule is ventilated and can be heated optionally. Entrance steps optional.

SYSTEMS: Electrical system of Brave 300 supplied by 28V 70A alternator, with 24V 17Ah battery. Brave 375 has 28V 70A alternator with two 12V 25Ah batteries in series. Hydraulic system for brakes only.

AVIONICS: The optional factory-installed avionics packages available for the Pawnee D can be obtained also for the Brave, as well as a range of transponders.

EQUIPMENT: Standard on Brave 300 is a non-corrosive hopper/tank of translucent GRP, installed forward of cockpit and approximately on CG, of 0·85 m^3 (30 cu ft) capacity, containing 852 litres (225 US gallons). Hopper/tank of 1·08 m^3 (38 cu ft) capacity, containing 1,041 litres (275 US gallons), optional for Brave 300, standard for Brave 375. The latter has a maximum capacity for dry chemicals of 862 kg (1,900 lb). Venturi-type dry material spreaders of either stainless steel or aluminium available, including a basic design capable of application rates of 5·6 to 224 kg/hectare (5 to 200 lb/acre). Spray system comprises an easily-removable wind-driven spraypump and 38 mm (1½ in) diameter spraybooms equipped with 60 nozzles. Other optional equipment includes 8-day clock; sensitive altimeter; turn coordinator; emergency locator transmitter; night working light package which includes two 600W cockpit adjustable/retractable working lights, two 450W ground-adjustable turning lights, hopper light, and related switching; landing and taxi lights; navigation, instrument panel, rotating beacon, and anti-collision lights; cockpit fire extinguisher; and heater.

DIMENSIONS, EXTERNAL:
Wing span 11·82 m (38 ft 9½ in)
Wing chord at root 2·03 m (6 ft 8 in)
Wing chord at tip 1·75 m (5 ft 9 in)
Wing aspect ratio 6·66
Length overall: Brave 300 8·17 m (26 ft 9½ in)
Brave 375 8·38 m (27 ft 6 in)
Height overall 2·29 m (7 ft 6 in)
Tailplane span 4·01 m (13 ft 1¾ in)
Wheel track 2·65 m (8 ft 8⅜ in)
Wheelbase: Brave 300 6·22 m (20 ft 4¾ in)
Brave 375 5·91 m (19 ft 4¾ in)
Propeller diameter: Brave 300 2·41 m (7 ft 11 in)
Brave 375 2·18 m (7 ft 2 in)
Propeller ground clearance:
Brave 300 0·25 m (10 in)
Brave 375 0·23 m (9 in)
Cockpit doors:
Both versions: Height 0·76 m (2 ft 6 in)
Width 0·71 m (2 ft 4 in)
Height to sill 1·57 m (5 ft 2 in)
Hopper loading door, Brave 375:
Length 1·27 m (4 ft 2 in)
Width 0·48 m (1 ft 7 in)

DIMENSIONS, INTERNAL:
Cockpit: Max width 0·97 m (3 ft 2 in)
Max height 1·32 m (4 ft 4 in)

AREAS:
Wings, gross 20·96 m^2 (225·65 sq ft)
Ailerons (total) 2·01 m^2 (21·6 sq ft)
Trailing-edge flaps (total) 2·32 m^2 (25·0 sq ft)
Fin 0·95 m^2 (10·2 sq ft)
Rudder 0·90 m^2 (9·7 sq ft)
Tailplane 2·11 m^2 (22·67 sq ft)
Elevators (incl tabs) 1·92 m^2 (20·66 sq ft)

WEIGHTS AND LOADINGS (A: Brave 300; B: Brave 375):
Weight empty: No dispersal equipment:
A 997 kg (2,198 lb)
B 1,118 kg (2,465 lb)
Sprayer: A 1,050 kg (2,314 lb)
B 1,154 kg (2,544 lb)
Duster: A 1,048 kg (2,310 lb)
B 1,152 kg (2,540 lb)
Max T-O weight, Normal category:
A, B 1,769 kg (3,900 lb)
Max T-O weight, Restricted category:
A 1,996 kg (4,400 lb)
B 2,177 kg (4,800 lb)
Max landing weight, Normal and Restricted categories:
A, B 1,769 kg (3,900 lb)
Max wing loading, Normal category:
A, B 84·4 kg/m^2 (17·3 lb/sq ft)
Max wing loading, Restricted category:
A 95·2 kg/m^2 (19·5 lb/sq ft)
B 103·9 kg/m^2 (21·3 lb/sq ft)
Max power loading, Normal category:
A 7·90 kg/kW (13·0 lb/hp)
B 6·33 kg/kW (10·4 lb/hp)
Max power loading, Restricted category:
A 8·91 kg/kW (14·7 lb/hp)
B 7·79 kg/kW (12·8 lb/hp)

PERFORMANCE (at Normal category max T-O weight, no dispersal equipment installed; A: Brave 300; B: Brave 375):
Max level speed at optimum altitude:
A 129 knots (238 km/h; 148 mph)
B 139 knots (257 km/h; 160 mph)
Cruising speed, best power mixture:
A, 75% power at 1,675 m (5,500 ft)
123 knots (228 km/h; 142 mph)
B, 75% power at 1,705 m (5,600 ft)
129 knots (240 km/h; 149 mph)
A, 65% power at 2,835 m (9,300 ft)
120 knots (222 km/h; 138 mph)
B, 65% power at 2,895 m (9,500 ft)
126 knots (233 km/h; 145 mph)
A, 55% power at 3,415 m (11,200 ft)
109 knots (203 km/h; 126 mph)
B, 55% power at 4,265 m (14,000 ft)
119 knots (220 km/h; 137 mph)
Cruising speed, best econ mixture:
A, 75% power at 1,675 m (5,500 ft)
123 knots (228 km/h; 142 mph)
B, 75% power at 1,705 m (5,600 ft)
128 knots (236 km/h; 147 mph)
A, 65% power at 2,835 m (9,300 ft)
118 knots (219 km/h; 136 mph)
B, 65% power at 2,895 m (9,500 ft)
124 knots (230 km/h; 143 mph)
A, 55% power at 3,415 m (11,200 ft)
108 knots (199 km/h; 124 mph)
B, 55% power at 4,265 m (14,000 ft)
117 knots (217 km/h; 135 mph)
Stalling speed, flaps up:
A, B 62·5 knots (116 km/h; 72 mph) CAS
Stalling speed, flaps down:
A 54 knots (100 km/h; 62 mph) CAS
B 57·5 knots (107 km/h; 66 mph) CAS
Max rate of climb at S/L: A 235 m (770 ft)/min
B 321 m (1,051 ft)/min
T-O run: A 293 m (960 ft)
B 218 m (715 ft)
T-O to 15 m (50 ft): A 465 m (1,525 ft)
B 368 m (1,208 ft)
Landing from 15 m (50 ft): A 503 m (1,650 ft)
B 564 m (1,850 ft)
Landing run: A 213 m (700 ft)
B 226 m (740 ft)
Range with max fuel, best econ mixture, with allowances for start, taxi, T-O climb, descent, and 45 min reserves:
A 75% power at 1,675 m (5,500 ft)
556 nm (1,030 km; 640 miles)
B 75% power at 1,705 m (5,600 ft)
465 nm (861 km; 535 miles)
A 65% power at 2,835 m (9,300 ft)
586 nm (1,086 km; 675 miles)
B 65% power at 2,895 m (9,500 ft)
495 nm (917 km; 570 miles)
A 55% power at 3,415 m (11,200 ft)
612 nm (1,135 km; 705 miles)
B 55% power at 4,265 m (14,000 ft)
525 nm (974 km; 605 miles)

PERFORMANCE: (at Restricted category max T-O weight, except where indicated. A: Brave 300 sprayer; B: Brave 300 duster; C: Brave 375 sprayer; D: Brave 375 duster):
Max speed at optimum altitude:
A 109 knots (203 km/h; 126 mph)
B 103 knots (191 km/h; 119 mph)
C 123 knots (229 km/h; 142 mph)
D 119 knots (220 km/h; 137 mph)
Cruising speed, best power mixture:
A 75% power at 1,675 m (5,500 ft)
102 knots (190 km/h; 118 mph)
A 65% power at 1,675 m (5,500 ft)
89 knots (166 km/h; 103 mph)
B 75% power at 1,675 m (5,500 ft)
94 knots (174 km/h; 108 mph)
C 75% power at 1,310 m (4,300 ft)
118 knots (219 km/h; 136 mph)
D 75% power at 1,310 m (4,300 ft)
113 knots (209 km/h; 130 mph)
Max rate of climb at S/L: A 114 m (375 ft)/min
B 97 m (320 ft)/min
C 168 m (550 ft)/min
D 116 m (380 ft)/min
T-O run: A 411 m (1,350 ft)
B 448 m (1,470 ft)
C 442 m (1,450 ft)
D 533 m (1,750 ft)
T-O to 15 m (50 ft): A 637 m (2,090 ft)
B 678 m (2,225 ft)
C 701 m (2,300 ft)
D 762 m (2,500 ft)
Landing from 15 m (50 ft), at Restricted category max landing weight: A 448 m (1,470 ft)
B 450 m (1,475 ft)
C, D 439 m (1,440 ft)

Landing run, at Restricted category max landing weight: A 232 m (760 ft)
B 216 m (710 ft)
C, D 140 m (460 ft)
Range with max fuel, cruising at best power mixture, allowances for start, taxi, T-O, climb, descent, and 45 min reserves at 45% power:
A (75% power) 399 nm (740 km; 460 miles)
A (65% power) 391 nm (724 km; 450 miles)
B (75% power) 360 nm (668 km; 415 miles)
C (75% power) 392 nm (727 km; 452 miles)
D (75% power) 372 nm (689 km; 428 miles)

PIPER (PA-38-112) TOMAHAWK II

In 1978 Piper introduced a completely new trainer named Tomahawk, which provided side-by-side seating in a roomy cabin with all-round view, and had a T tail and wide-track landing gear. On 14 February 1981 the company announced that this had been superseded by a new Tomahawk II, with improvements that include better soundproofing, improved door latches, larger throttle knob, new map pockets and seat adjustment assist grips, an improved nosewheel assembly, and more efficient windscreen defrosting. It is available with an optional factory-installed special training package or an optional factory-installed avionics package in lieu:

Special Training Package. Comprises advanced instrument package, true airspeed indicator, sun visors, tinted windscreen and side windows, dual toe-operated brakes, night lighting, entrance steps each side, towbar, and avionics including King KX 170B 720-channel transceiver with 200-channel nav, King KI 208 indicator with VOR/LOC converter, Telex 66T microphone, Telex headset, and Poly Planar cabin speaker; adding 21·3 kg (47·0 lb) to basic empty weight.

Group KS-1-38. Comprising King KX 170B 720-channel com transceiver and 200-channel nav receiver, King KI 208 VOR/LOC indicator with VOR/LOC converter, Telex 66T microphone, Telex headset, and Poly Planar cabin speaker; adding 5·3 kg (11·7 lb) to basic empty weight.

The STP and avionics package includes all associated antennae.

Recent Tomahawk deliveries have included eight to the Aero Club of Santiago, Chile, in April 1981.

FAA certification of the original Tomahawk trainer was gained on 20 December 1977, and well over 1,000 had been built before introduction of the improved Tomahawk II, to which the following description applies:

Type: Two-seat trainer/utility aircraft.

Wings: Cantilever low-wing monoplane. NASA GAW-1 aerofoil section. Thickness/chord ratio 17%. Dihedral 5°. Incidence 2° at root, 0° at tip. No sweepback. Conventional structure of light alloy with plastics wingtips. Plain ailerons of light alloy construction. Manually-operated plain trailing-edge flaps of light alloy construction. No trim tabs.

Fuselage: Conventional semi-monocoque structure, primarily of light alloy, but with some components of steel or plastics. High strength roll-over structure.

Tail Unit: Cantilever T tail of light alloy construction, with plastics fin and tailplane tips. Fixed-incidence tailplane. Spring-loaded elevator trim. Ground-adjustable tab on rudder.

Landing Gear: Non-retractable tricycle type. Main wheels carried on interchangeable cantilever steel leaf springs. Nose unit has Piper oleo-pneumatic shock-absorber. Cleveland main wheels Type 40-78B and nosewheel Type 40-77B, all fitted with 5·00-5 4-ply tyres and tubes, pressure 1·79 bars (26 lb/sq in). Wheels and tyres size 6·00-6 optional. Cleveland Type 30-9 hand-operated brakes. Dual toe-operated brakes optional. Parking brake.

Power Plant: One 83·5 kW (112 hp) Avco Lycoming O-235-L2C flat-four engine, driving a Sensenich two-blade fixed-pitch metal propeller with spinner. One integral fuel tank in each wing, total capacity 121 litres (32 US gallons), of which 113·5 litres (30 US gallons) are usable. Refuelling point in upper surface of each wing. Oil capacity 5·7 litres (1·5 US gallons).

Accommodation: Enclosed cabin seating two, side by side, with dual controls standard. High strength roll-over support structure in cabin area. Door on each side. Fore, aft, and vertically adjustable seats with safety belts and shoulder harnesses. Baggage space, capacity 45·4 kg (100 lb), behind seats. Accommodation heated and ventilated. Sun visors, tinted windscreen and side windows optional. Windscreen defroster standard.

System: Electrical system includes a 14V 60A alternator and 12V 25Ah battery. Hydraulic system for brakes only. Vacuum system optional.

Avionics and Equipment: Optional avionics, in addition to those detailed in the STP and KS-1-38 packages, include second nav/com, ADF, encoding altimeter, glideslope and marker beacon receiver, and transponder. Standard equipment includes instrument panel glareshield, outside air temperature gauge, recording tachometer, sensitive altimeter, fore/aft/vertically-adjustable pilot/co-pilot seats with safety belts and shoulder harness, dual controls, alternate static source, alternator failure warning light, stall warning device, baggage tiedown straps, large wraparound rear window with tinted glass, carpeted floor, soundproofing, engine priming system, full-flow oil filter, oil quick drain valve, tiedown rings, and winterisation kit. Optional equipment includes attitude and directional gyros, electric clock, flight hour recorder, outside air temperature gauge, true airspeed indicator, turn co-ordinator, vertical speed indicator, sun visors, tinted windscreen and windows, emergency locator transmitter, heated pitot, cabin fire extinguisher, corrosion proofing, stainless steel control cables, external power socket, and static wicks.

Dimensions, external:
Wing span 10·36 m (34 ft 0 in)
Wing chord, constant 1·12 m (3 ft 8 in)
Wing aspect ratio 9·27
Length overall 7·06 m (23 ft 2 in)
Height overall 2·77 m (9 ft 1 in)
Tailplane span 3·20 m (10 ft 6 in)
Wheel track 3·05 m (10 ft 0 in)
Wheelbase 1·45 m (4 ft 9 in)
Propeller diameter 1·83 m (6 ft 0 in)
Propeller ground clearance 0·18 m (7 in)
Doors (port and stbd, each):
Height 0·74 m (2 ft 5 in)
Width 0·58 m (1 ft 11 in)
Height to sill 1·02 m (3 ft 4 in)
Dimensions, internal:
Cabin, from instrument panel to rear bulkhead:
Length 1·74 m (5 ft 8½ in)
Max width 1·07 m (3 ft 6 in)
Max height 1·28 m (4 ft 2½ in)
Baggage space 0·57 m³ (20 cu ft)
Areas:
Wings, gross 11·59 m² (124·7 sq ft)
Ailerons (total) 1·02 m² (11·0 sq ft)
Trailing-edge flaps (total) 1·01 m² (10·9 sq ft)
Vertical tail surfaces (total) 1·42 m² (15·3 sq ft)
Horizontal tail surfaces (total) 2·03 m² (21·9 sq ft)
Weights and Loadings:
Weight empty 512 kg (1,128 lb)
Max T-O and landing weight 757 kg (1,670 lb)
Max wing loading 65·42 kg/m² (13·4 lb/sq ft)
Max power loading 9·07 kg/kW (14·9 lb/hp)
Performance (at max T-O weight):
Never-exceed speed 138 knots (256 km/h; 159 mph) IAS
Max level speed at S/L 109 knots (202 km/h; 126 mph)
Max cruising speed, 75% power at 2,165 m (7,100 ft) 108 knots (200 km/h; 124 mph)
Cruising speed, 65% power at 3,200 m (10,500 ft) 100 knots (185 km/h; 115 mph)
Stalling speed, flaps up 52 knots (96 km/h; 60 mph) IAS
Stalling speed, flaps down 49 knots (91 km/h; 56·5 mph) IAS
Max rate of climb at S/L 221 m (725 ft)/min
Service ceiling 3,960 m (13,000 ft)
Absolute ceiling 4,265 m (14,000 ft)
T-O run 245 m (805 ft)
T-O to 15 m (50 ft) 405 m (1,330 ft)
Landing from 15 m (50 ft) 471 m (1,544 ft)
Landing run 215 m (707 ft)
Range with max fuel, allowances for taxi, T-O, climb, descent, and 45 min reserves at 55% power:
75% power at 2,165 m (7,100 ft) 452 nm (838 km; 520 miles)
65% power at 3,200 m (10,500 ft) 468 nm (867 km; 539 miles)

Photograph and three-view drawing *(Pilot Press)* **of the Piper Tomahawk II two-seat trainer/utility aircraft**

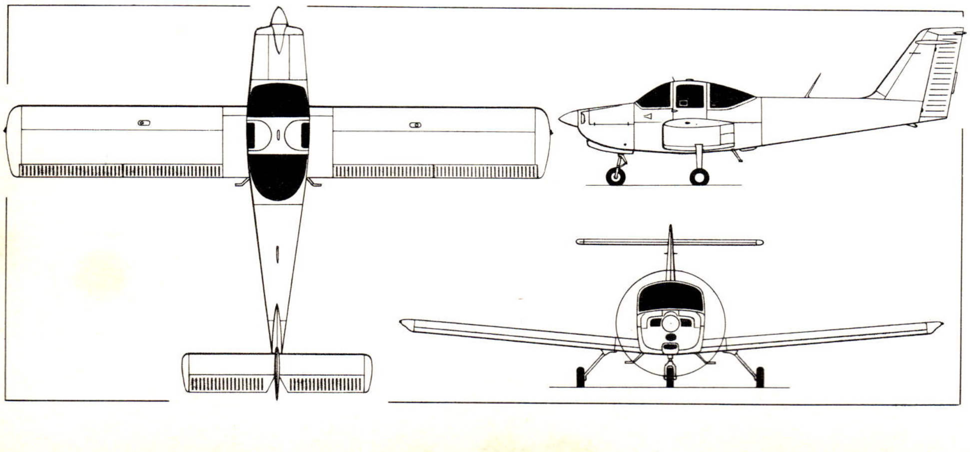

PIPER (PA-44-180) SEMINOLE

The Seminole lightweight twin-engined four-seat cabin monoplane was announced by Piper on 21 February 1978. Features of this economically-priced aircraft (first flown in prototype form in May 1976) are the T tail for improved controllability and stability, counter-rotating engines, a new fuel drain sump system with only two drain valves for the entire system, and a one-piece glassfibre nosecone to give easy access to the landing gear mechanism, combustion heater, battery and landing light.

Two optional equipment packages are available for the Seminole:

Custom. Comprising blind-flying instrumentation including 3 in attitude and directional gyros, electric clock, outside air temperature gauge, rate of climb indicator, pictorial turn rate indicator, and true airspeed indicator; avionics dimming, and instrument panel white and overhead red lighting; map, cabin dome, waterfall switch panel, landing/taxi, navigation, and wing strobe lights; assist strap; entrance step; vacuum system with dual engine-driven pumps; and quick oil drains; adding 15·8 kg (34·9 lb) to basic empty weight.

Executive. As Custom group, plus emergency locator transmitter, heated pitot and stall warning transmitter, pilot's vertically adjustable seat, and external power socket; adding 22·2 kg (48·9 lb) to basic empty weight.

Type: Twin-engined lightweight four-seat cabin monoplane.

Wings: Cantilever low-wing monoplane. Conventional structure of light alloy. Plain ailerons and manually-operated four-position trailing-edge flaps of light alloy construction.

Fuselage: Conventional semi-monocoque structure, primarily of light alloy.

Tail Unit: Cantilever T tail of light alloy construction. All-moving tailplane with full-span tab. Rudder anti-servo tab.

Landing Gear: Hydraulically retractable tricycle type. Free-fall emergency extension system. Piper oleo-pneumatic shock-absorbers. Main wheels and tyres size 6·00-6, 8-ply with tubes. Steerable nosewheel and tyre size 5·00-5, 6-ply with tube. Nosewheel safety mirror. Dual toe-operated high-capacity disc brakes. Parking brake. Heavy-duty brakes and tyres optional.

Power Plant: Two 134 kW (180 hp) Avco Lycoming flat-four counter-rotating engines (one O-360-E1A6D and one LO-360-E1A6D), each driving a Hartzell two-blade constant-speed fully-feathering metal propeller with spinner. Three-blade propellers and propeller synchrophasers optional. One bladder-type fuel tank in each engine nacelle, with total capacity of 416 litres (110 US gallons), of which 409 litres (108 US gallons) are usable. Refuelling point on upper surface of each nacelle. Oil capacity 11·5 litres (3 US gallons).

ACCOMMODATION: Cabin seats four in two pairs of individual seats. Pilot and co-pilot seats adjustable fore and aft and reclining, and fitted with seat belts and shoulder harness with inertia reels. Dual controls standard. Rear seats have safety belts and are reclinable. Cabin door on starboard side. Emergency exit window on port side. Pilot's storm window. Sun visors. Baggage compartment at rear of cabin, capacity 91 kg (200 lb). Accommodation heated and ventilated. Windscreen defrosters.

SYSTEMS: Electro-hydraulic system for landing gear actuation and brakes. Electrical system includes two engine-driven 14V 60A alternators and 12V 35Ah battery. Janitrol combustion heater of 45,000 BTU capacity. Vacuum system optional.

AVIONICS AND EQUIPMENT: A wide range of optional avionics is available to customer's requirements, including ten factory-installed avionics packages based on dual nav/com, ADF, DME, encoding altimeter, glideslope and marker beacon receivers, HSI, R/Nav, radar altimeter and transponder, by manufacturers which include Bendix, Collins, King, Narco and Piper. Standard equipment includes two cylinder head temperature and two manifold pressure gauges, two recording tachometers, sensitive altimeter, pilot and co-pilot's seats with fore/aft/recline adjustment and inertia reel shoulder and safety belts, map pockets, pilot's storm window, sun visors, armrests, alternate static source, alternator failure warning light, stall warning device, provisions for emergency locator transmitter, soundproofing, fuel tank quick drains, full flow oil filters, manually-controlled cowl flaps, engine priming system, jack pads, tiedown rings and stowable towbar. Optional equipment includes digital clock, engine hour recorders, exhaust gas temperature gauge, pilot and co-pilot's vertically-adjustable seats, cabin fire extinguisher, headrests, shoulder safety belts with inertia reels for rear seats, stainless steel control cables, tinted windows, ventilation fan, window curtains, carburettor ice detectors, and zinc chromate treatment of aluminium parts.

DIMENSIONS, EXTERNAL:
- Wing span — 11·77 m (38 ft 7¼ in)
- Length overall — 8·41 m (27 ft 7¼ in)
- Height overall — 2·59 m (8 ft 6 in)
- Wheel track — 3·20 m (10 ft 6 in)
- Wheelbase — 2·56 m (8 ft 4¾ in)
- Propeller diameter — 1·88 m (6 ft 2 in)
- Cabin door (stbd): Width — 0·91 m (3 ft 0 in)
  - Height — 0·89 m (2 ft 11 in)
- Baggage door: Width — 0·56 m (1 ft 10 in)
  - Height — 0·51 m (1 ft 8 in)

DIMENSIONS, INTERNAL:
- Cabin, instrument panel to rear bulkhead:
  - Length — 2·46 m (8 ft 1 in)
  - Max width — 1·05 m (3 ft 5½ in)
  - Max height — 1·245 m (4 ft 1 in)
  - Volume — 3·00 m³ (106 cu ft)

AREA:
- Wings, gross — 17·08 m² (183·8 sq ft)

WEIGHTS AND LOADINGS:
- Weight empty — 1,068 kg (2,354 lb)
- Max T-O and landing weight — 1,723 kg (3,800 lb)
- Max ramp weight — 1,731 kg (3,816 lb)
- Max wing loading — 100·9 kg/m² (20·67 lb/sq ft)
- Max power loading — 6·43 kg/kW (10·55 lb/hp)

PERFORMANCE (at max T-O weight, except where indicated):
- Max level speed at S/L — 168 knots (311 km/h; 193 mph)
- Cruising speed at optimum altitude, best power settings:
  - 75% power — 166 knots (308 km/h; 191 mph)
  - 65% power — 162 knots (300 km/h; 187 mph)
  - 55% power — 153 knots (284 km/h; 176 mph)
- Cruising speed at optimum altitude, best econ power settings:
  - 75% power — 161 knots (298 km/h; 185 mph)
  - 65% power — 157 knots (291 km/h; 181 mph)
  - 55% power — 149 knots (276 km/h; 171 mph)
- Stalling speed, flaps up — 63 knots (117 km/h; 72·5 mph) CAS
- Stalling speed, flaps down — 59 knots (109 km/h; 68 mph) CAS
- Max rate of climb at S/L — 408 m (1,340 ft)/min
- Rate of climb at S/L, one engine out — 66 m (217 ft)/min
- Service ceiling — 5,210 m (17,100 ft)
- Service ceiling, one engine out — 1,250 m (4,100 ft)
- Absolute ceiling — 5,670 m (18,600 ft)
- Absolute ceiling, one engine out — 1,600 m (5,250 ft)
- T-O run — 268 m (880 ft)
- T-O to 15 m (50 ft) — 427 m (1,400 ft)
- Landing from 15 m (50 ft) — 427 m (1,400 ft)
- *Landing from 15 m (50 ft) — 363 m (1,190 ft)
- Landing run — 180 m (590 ft)
- *Landing run — 117 m (383 ft)
- Accelerate/stop distance — 701 m (2,300 ft)
- *Accelerate/stop distance — 631 m (2,070 ft)
- Range at optimum altitude with max fuel at best power settings, allowances for taxi, T-O, climb, descent, and 45 min reserves at max range power:
  - 75% power — 690 nm (1,278 km; 794 miles)
  - 65% power — 725 nm (1,343 km; 834 miles)
  - 55% power — 760 nm (1,408 km; 875 miles)
- Range at optimum altitude with max fuel at econ power settings, allowances as above:
  - 75% power — 780 nm (1,445 km; 898 miles)
  - 65% power — 860 nm (1,593 km; 990 miles)
  - 55% power — 910 nm (1,685 km; 1,047 miles)

with optional heavy-duty brakes

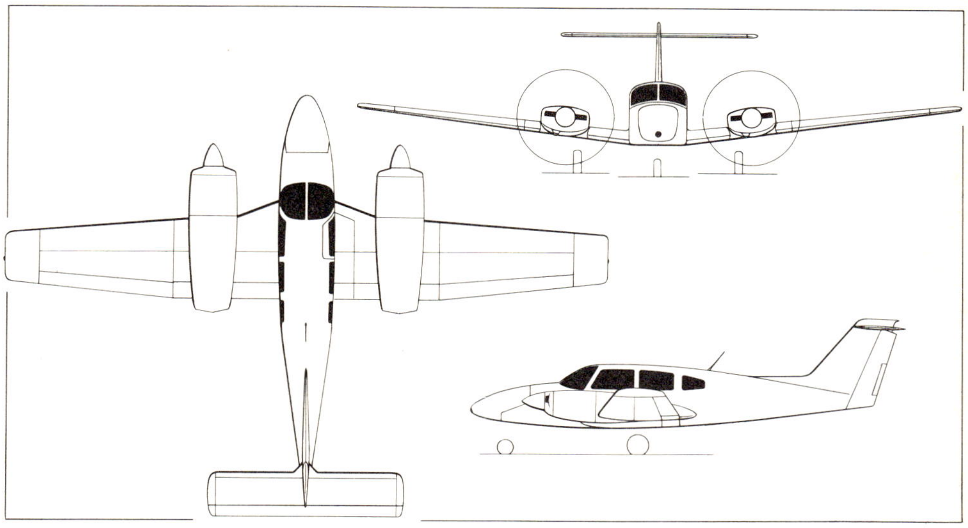

Piper Seminole lightweight twin-engined four-seat cabin monoplane *(Pilot Press)*

Piper Turbo Seminole with Avco Lycoming turbocharged engines

PIPER (PA-44-180T) TURBO SEMINOLE

Piper announced on 24 April 1980 the introduction of a turbocharged version of the twin-engined four-seat Seminole, from which it differs in its power plant and related equipment. It also includes provisions for a built-in optional oxygen system, to permit high-altitude operation.

The description of the PA-44-180 Seminole, including its optional packages, applies also to the Turbo Seminole, except as follows:

POWER PLANT: Two 134 kW (180 hp) Avco Lycoming flat-four counter-rotating (one TO-360-E1A6D and one LTO-360-E1A6D) turbocharged engines, each driving a Hartzell two-blade constant-speed fully-feathering metal propeller with spinner. Three-blade propellers and propeller synchrophasers optional. Propeller de-icing system optional. Fuel system and capacity as for Seminole.

ACCOMMODATION: Generally as for Seminole.

SYSTEMS: As for Seminole, except oxygen system optional.

AVIONICS: Generally as for Seminole.

EQUIPMENT: As for Seminole, except dual exhaust gas temperature gauges and engine hour recorders are standard, and wing ice inspection lights optional.

WEIGHTS AND LOADINGS:
- Weight empty — 1,102 kg (2,430 lb)
- Max T-O weight — 1,780 kg (3,925 lb)
- Max ramp weight — 1,788 kg (3,943 lb)
- Max zero-fuel and max landing weight — 1,723 kg (3,800 lb)
- Max wing loading — 104·2 kg/m² (21·35 lb/sq ft)
- Max power loading — 6·64 kg/kW (10·90 lb/hp)

PERFORMANCE (at max T-O weight):
- Max level speed at optimum altitude — 195 knots (361 km/h; 224 mph)
- Cruising speed at optimum altitude, best power settings:
  - 75% power — 183 knots (339 km/h; 211 mph)
  - 65% power — 172 knots (319 km/h; 198 mph)
  - 55% power — 158 knots (293 km/h; 182 mph)
- Cruising speed at optimum altitude, best econ power settings:
  - 75% power — 180 knots (334 km/h; 207 mph)
  - 65% power — 169 knots (313 km/h; 195 mph)
  - 55% power — 154 knots (285 km/h; 177 mph)
- Stalling speed, flaps and gear up — 64 knots (119 km/h; 74 mph) CAS
- Stalling speed, flaps and gear down — 61 knots (113 km/h; 70 mph) CAS
- Max rate of climb at S/L — 393 m (1,290 ft)/min
- Rate of climb at S/L, one engine out — 55 m (180 ft)/min
- Service ceiling, certificated — 6,100 m (20,000 ft)
- Service ceiling, one engine out — 3,800 m (12,500 ft)
- T-O run — 274 m (900 ft)
- T-O to 15 m (50 ft) — 457 m (1,500 ft)
- Landing from 15 m (50 ft) — 427 m (1,400 ft)
- *Landing from 15 m (50 ft) — 363 m (1,190 ft)
- Landing run — 180 m (590 ft)
- *Landing run — 117 m (383 ft)
- Accelerate/stop distance — 762 m (2,500 ft)
- *Accelerate/stop distance — 686 m (2,250 ft)
- Range at optimum altitude, with max fuel, at best power settings, allowances for taxi, T-O, climb, descent, and 45 min reserves at max range power:
  - 75% power — 710 nm (1,315 km; 817 miles)
  - 65% power — 725 nm (1,343 km; 835 miles)
  - 55% power — 745 nm (1,380 km; 858 miles)
- Range at optimum altitude, with max fuel, at econ power settings, allowances as above:
  - 75% power — 785 nm (1,454 km; 904 miles)
  - 65% power — 800 nm (1,482 km; 921 miles)
  - 55% power — 820 nm (1,519 km; 944 miles)

* *with optional heavy-duty brakes*

PIPER (PA-60) AEROSTAR 600A and 601P

Design work on this series was started by the Ted Smith Aircraft Company in November 1964 and the first Model 600/601 prototype made its first flight in October 1967. FAA Type Approval of the Model 600 was awarded in March 1968, and of the Model 601 in November 1968.

On 6 August 1975, an Aerostar 601A piloted by Jack F. Chrysler set a Class C1d 2,000 km closed-circuit speed record for piston-engined landplanes of 237·08 knots (439·07 km/h; 272·83 mph).

On 4-9 November 1977, an Aerostar 601P piloted by Philander Claxton III and Jack Cink recorded a new round-the-world speed record for piston-engined aircraft, completing the 19,974 nm (37,015 km; 23,000 mile) flight in 104 h 5 min 30 s, at a speed of 190·91 knots (353·57 km/h; 219·70 mph), so beating the previous record of 122 h set by two Australians in a Beechcraft Duke.

Current production versions of the Aerostar comprise:

Model 600A. Powered by two 216 kW (290 hp) Avco Lycoming IO-540-K1J5 flat-six engines. European models designated 600AE.

Model 601P. As Model 600A, but with increased wing span and two 216 kW (290 hp) Avco Lycoming IO-540-S1A5 flat-six engines, fitted with high flow-rate turbochargers to supply bleed air for cabin pressurisation. European models designated 601PE.

Model 602P. New model introduced in 1981. Described separately.

The 1981 versions of the Aerostar have a new exterior paint scheme as standard; new options include an alternative interior finish, shoulder safety belts with inertia reels for passenger seats, and an extended range of King avionics.

An optional group of factory-installed de-icing equipment is available for the Model 601P, which has received FAA approval for flight into icing conditions. It comprises de-icing boots for wing and tail unit leading-edges, electrically heated propeller boots, alcohol de-icing system or electrically-heated panel for windscreen, wing ice inspection light, and a manual alternative engine air control; adding 32·6 kg (71·9 lb) to basic empty weight.

Type: Twin-engined light transport aircraft.

Wings: Cantilever mid-wing monoplane. Wing section NACA 64_1A212. Dihedral 2°. Incidence 1°. No sweepback. All-metal structure using heavy gauge skins attached to three spars, several bulkheads and stringers. Entire wing assembly, excluding attachments for ailerons and flaps, contains fewer than 50 detail parts. Ailerons and hydraulically-actuated Fowler-type trailing-edge flaps each comprise a spar, ribs, nose skin and one-piece wraparound light alloy skin aft of spar. No trim tabs. De-icing for wing leading-edges optional.

Fuselage: All-metal fail-safe monocoque structure. Skin composed of large segments of light alloy sheet over stringers and frames. Entire fuselage contains fewer than 100 parts, including skin panels. All fuselage assemblies designed basically for pressurisation.

Tail Unit: Cantilever all-metal structure, with swept vertical and horizontal surfaces. Both fixed and control surfaces are interchangeable. Electrically-operated trim tab in rudder and each elevator. De-icing boots for fin and tailplane leading-edges optional.

Landing Gear: Hydraulically-retractable tricycle type. Main units retract inward, nosewheel forward. Hydraulically-powered nosewheel steering. Hydraulically-operated heavy-duty dual caliper brakes. Parking brake.

Power Plant: Two Avco Lycoming engines, as detailed in model listings, each driving a Hartzell three-blade constant-speed and fully-feathering metal propeller with spinner. Fuel in integral wing tanks and fuselage tank with total capacity of 657 litres (173·5 US gallons), of which 626 litres (165·5 US gallons) are usable. Oil capacity 22·7 litres (6 US gallons). Propeller synchrophasers and propeller de-icing optional. Engine fire detection system optional.

Accommodation: Cabin seats six people on track-mounted individual reclining seats, in pairs. Dual controls standard. Door on port side by pilot's seat; top half hinges upward, bottom half downward. Emergency escape window at rear of cabin. Tinted windscreen and dual-pane tinted cabin windows. Large utility shelf in rear of cabin. Baggage compartment, capacity 109 kg (240 lb), aft of cabin, with external access on port side. Individual air vents and reading lights for each seat. Cabin heated/ventilated in Model 600A, and also pressurised in 601P. Control locks. Windscreen defrosting; alcohol de-icing system, or electrically-heated windscreen plate, optional.

Systems: Heating/ventilation system includes a Janitrol 35,000 BTU heater. Pressurisation system for Model 601P supplied by engine bleed air; pressure differential 0·29 bars (4·25 lb/sq in). Hydraulic system for landing gear actuation, wheel brakes and flaps, powered by an engine-driven pump, pressure 69 bars (1,000 lb/sq in); auxiliary hydraulic system with electrically driven pump optional. Dual pneumatic systems for instrument gyros and optional de-icing boots. Electrical system powered by two 28V 70A engine-driven alternators with failure warning lights. Two 12V 24Ah batteries. External power socket. Model 601P has an oxygen system, with individual outlets at each seat, as standard; capacity 0·31 m³ (11 cu ft).

Avionics and Equipment: A wide range of optional avionics is available for all three versions of the Aerostar, including nav/coms, ADF, autopilots, DME, encoding altimeter, glideslope and marker beacon receivers, HF transceiver, R/Nav, radio altimeter, transponder and weather radar, by manufacturers which include Aeronetics, Aerosonics, Bendix, Bonzer, Collins, King, Piper and SunAir. Standard equipment includes blind-flying instrumentation with 3 in attitude and directional gyros, clock, flight hour meter, dual manifold pressure gauges (601P has also dual exhaust temperature gauges), outside air temperature gauge, sensitive altimeter, turn co-ordinator, and vertical speed indicator; control locks; inertia reel shoulder harness for pilot and co-pilot seats; storm windows; sun visors; alternator failure, heater failure, and low fuel warning lights; emergency locator transmitter; heated pitot (601P has also an alternate static source); cabin air ventilators, dual-pane tinted windows, carpeted floor, window curtains, super soundproofing; map, instrument panel red and white, reading, landing, navigation, taxi, and strobe lights; full-flow oil filters; jack pads and towbar. An extensive range of optional equipment is also available.

Dimensions, external:	
Wing span: 600A	10·41 m (34 ft 2 in)
601P	11·18 m (36 ft 8 in)
Wing chord at root	2·18 m (7 ft 2 in)
Wing chord at tip: 600A	0·87 m (2 ft 10⅜ in)
601P	0·76 m (2 ft 6 in)
Wing aspect ratio: 600A	6·83
601P	7·55
Length overall	10·61 m (34 ft 9¾ in)
Height overall	3·70 m (12 ft 1½ in)
Tailplane span	4·37 m (14 ft 4 in)
Wheel track	3·11 m (10 ft 2½ in)
Propeller diameter	1·98 m (6 ft 6 in)
Passenger door: Height	1·14 m (3 ft 9 in)
Width	0·71 m (2 ft 4 in)
Baggage compartment door:	
Height	0·61 m (2 ft 0 in)
Width	0·56 m (1 ft 10 in)
Dimensions, internal:	
Cabin: Length	3·81 m (12 ft 6 in)
Width	1·17 m (3 ft 10 in)
Height	1·22 m (4 ft 0 in)
Baggage space	0·85 m³ (30 cu ft)
Areas:	
Wings, gross: 600A	15·79 m² (170 sq ft)
601P	16·54 m² (178 sq ft)
Tailplane	4·20 m² (45·2 sq ft)
Weights and Loadings:	
Weight empty, equipped: 600A	1,695 kg (3,737 lb)
601P	1,840 kg (4,056 lb)
Max T-O and landing weight:	
600A	2,495 kg (5,500 lb)
601P	2,721 kg (6,000 lb)
Max ramp weight: 600A	2,506 kg (5,525 lb)
601P	2,733 kg (6,025 lb)
Max wing loading: 600A	158·2 kg/m² (32·4 lb/sq ft)
601P	164·5 kg/m² (33·7 lb/sq ft)
Max power loading: 600A	5·77 kg/kW (9·5 lb/hp)
601P	6·29 kg/kW (10·34 lb/hp)

Performance (at max T-O weight, A: 600A; B: 601P):

Cruising speed at average cruise weight:
- A 75% power at 2,285 m (7,500 ft): 220 knots (408 km/h; 253 mph)
- B 75% power at 7,620 m (25,000 ft): 257 knots (476 km/h; 296 mph)
- B 75% power at 4,570 m (15,000 ft): 233 knots (431 km/h; 268 mph)
- A 65% power at 3,050 m (10,000 ft): 213 knots (394 km/h; 245 mph)
- A 65% power at 2,285 m (7,500 ft): 208 knots (385 km/h; 239 mph)
- B 65% power at 7,620 m (25,000 ft): 238 knots (441 km/h; 274 mph)
- B 65% power at 4,570 m (15,000 ft): 218 knots (404 km/h; 251 mph)
- A 55% power at 3,050 m (10,000 ft): 197 knots (365 km/h; 227 mph)
- A 55% power at 2,285 m (7,500 ft): 193 knots (357 km/h; 222 mph)
- B 55% power at 7,620 m (25,000 ft): 218 knots (404 km/h; 251 mph)
- B 55% power at 6,100 m (20,000 ft): 209 knots (388 km/h; 241 mph)
- B 55% power at 4,570 m (15,000 ft): 201 knots (372 km/h; 231 mph)

Stalling speed, flaps down:	
A	74 knots (137 km/h; 85 mph) IAS
B	77 knots (143 km/h; 89 mph) IAS
Max rate of climb at S/L: A	549 m (1,800 ft)/min
B	445 m (1,460 ft)/min
Rate of climb at S/L, one engine out:	
A	110 m (360 ft)/min
B	73 m (240 ft)/min
Service ceiling: A	6,460 m (21,200 ft)
B	7,620 m (25,000 ft)
Service ceiling, one engine out:	
A	1,920 m (6,300 ft)
B	2,680 m (8,800 ft)
T-O run: A	472 m (1,550 ft)
B	579 m (1,900 ft)
T-O to 15 m (50 ft): A	594 m (1,950 ft)
B	759 m (2,490 ft)
Landing from 15 m (50 ft): A	561 m (1,840 ft)
B	619 m (2,030 ft)
Landing run: A	317 m (1,040 ft)
B	375 m (1,230 ft)

Range with max fuel, allowances for start, taxi, T-O, climb, and 45 min reserves at long-range cruise power:
- A 75% power at 1,830 m (6,000 ft): 983 nm (1,821 km; 1,132 miles)
- A 65% power at 2,745 m (9,000 ft): 1,079 nm (2,000 km; 1,242 miles)
- B 65% power at 4,570 m (15,000 ft): 1,024 nm (1,897 km; 1,179 miles)
- A 55% power at 3,050 m (10,000 ft): 1,201 nm (2,225 km; 1,383 miles)
- B 55% power at 4,570 m (15,000 ft): 1,173 nm (2,172 km; 1,350 miles)
- B 55% power at 6,100 m (20,000 ft): 1,198 nm (2,219 km; 1,379 miles)
- B 55% power at 7,620 m (25,000 ft): 1,101 nm (2,040 km; 1,268 miles)

Piper Aerostar 602P, the 1981 addition to this family of light transports

PIPER AEROSTAR 602P

On 14 February 1981, Piper announced an addition to the Aerostar family of twin piston-engined aircraft, designated Model 602P. The name Sequoya chosen for it was subsequently withdrawn. Generally similar to the Piper Aerostar 601P, the Model 602P is pressurised and has Avco Lycoming IO-540-AA1A5 low-compression engines with integral turbochargers, providing significantly improved performance, particularly at high altitudes. In addition, the 602P incorporates the improvements that were introduced on the other 1981 Aerostars, including an optional de-icing package.

The description of the Aerostar 601P applies also to the 602P, except as follows:

Wings, Fuselage, Tail Unit, Landing Gear: Generally similar to Aerostar 601P.

Power Plant: Two 216 kW (290 hp) Avco Lycoming IO-540-AA1A5 flat-six turbocharged engines, each driving a Hartzell three-blade constant-speed fully-feathering metal propeller with spinner. Fuel system and propeller options as detailed for other Aerostars.

Accommodation, Systems, Avionics and Equipment: Generally similar to Aerostar 601P.

Dimensions and Areas: As for Aerostar 601P

Weights and Loadings:	
Weight empty, equipped	1,848 kg (4,075 lb)
Max T-O and landing weight	2,721 kg (6,000 lb)
Max ramp weight	2,735 kg (6,029 lb)
Max wing loading	164·6 kg/m² (33·71 lb/sq ft)
Max power loading	6·30 kg/kW (10·34 lb/hp)

PERFORMANCE (at max T-O weight, except where indicated):
- Max level speed at average cruise weight of 2,495 kg (5,500 lb) 262 knots (486 km/h; 302 mph)
- Cruising speed, best power, at average cruise weight of 2,495 kg (5,500 lb):
  - 75% power at 7,620 m (25,000 ft) 245 knots (454 km/h; 282 mph)
  - 75% power at 7,010 m (23,000 ft) 241 knots (447 km/h; 278 mph)
  - 75% power at 4,570 m (15,000 ft) 226 knots (419 km/h; 260 mph)
  - 65% power at 7,620 m (25,000 ft) 226 knots (419 km/h; 260 mph)
  - 65% power at 7,010 m (23,000 ft) 222 knots (411 km/h; 256 mph)
  - 65% power at 4,570 m (15,000 ft) 210 knots (389 km/h; 242 mph)
  - 55% power at 7,620 m (25,000 ft) 204 knots (378 km/h; 235 mph)
  - 55% power at 7,010 m (23,000 ft) 202 knots (374 km/h; 233 mph)
  - 55% power at 4,570 m (15,000 ft) 192 knots (356 km/h; 221 mph)
- Stalling speed, flaps and landing gear down, power off 77 knots (143 km/h; 89 mph) IAS
- Max rate of climb at S/L 535 m (1,755 ft)/min
- Rate of climb at S/L, one engine out 92 m (302 ft)/min
- Service ceiling 8,535 m (28,000 ft)
- Service ceiling, one engine out 3,930 m (12,900 ft)
- T-O run 549 m (1,800 ft)
- T-O to 15 m (50 ft) 686 m (2,250 ft)
- Landing from 15 m (50 ft) 633 m (2,076 ft)
- Landing run 371 m (1,217 ft)
- Range with max fuel, allowances for start, taxi, T-O, climb, and 45 min reserves at long-range cruise power:
  - 65% power at 4,570 m (15,000 ft) 1,058 nm (1,960 km; 1,218 miles)
  - 65% power at 7,010 m (23,000 ft) 1,070 nm (1,983 km; 1,232 miles)
  - 65% power at 7,620 m (25,000 ft) 1,065 nm (1,973 km; 1,226 miles)
  - 55% power at 4,570 m (15,000 ft) 1,137 nm (2,107 km; 1,309 miles)
  - 55% power at 7,010 m (23,000 ft) 1,150 nm (2,131 km; 1,324 miles)
  - 55% power at 7,620 m (25,000 ft) 1,132 nm (2,098 km; 1,303 miles)

Piper Enforcer evaluated by USAF in the 1971 Pave Coin programme

PIPER ENFORCER

Piper completed two prototypes of this turboprop-powered light attack aircraft in the early 1970s, after acquiring the programme from Cavalier Aircraft Corporation (see 1971-72 *Jane's*). The Enforcer is a development of the wartime North American P-51D Mustang, with a 1,890 kW (2,535 shp) Avco Lycoming T55-L-9 engine and extensive armour protection. The first prototype made its initial flight on 29 April 1971, but was lost in a crash three months later. The second aircraft was evaluated in the US Air Force's Pave Coin programme in the same year, but was not selected for production.

The US Air Force will now re-evaluate the Enforcer, and in September 1981 Piper received a contract to produce and test two prototypes. For details see Addenda.

AIRLINE DIVISION

Formation of this new Division was announced by Piper on 4 June 1981. Under the direction of Mr C. Raymond Johnson, the company's Vice-President for Sales, it will support commuter airlines now operating Piper Chieftains, and those that will operate the new Piper T-1020 and T-1040 aircraft, specially designed for such operators.

The Airline Division provides its services under what is known as the T-1000 commuter transport system. It offers operators worldwide support, factory-direct prices, and the availability of a commuter support centre and parts warehouse at Tampa International Airport, Florida, that opened for 24-hours-per-day service on 1 September 1981. In addition to introducing on to the market the T-1020 and T-1040, the Division has announced a specialised training programme, developed by Flight Safety International at Lakeland, Florida, which will provide ground and airline-type simulator training for personnel of airlines operating these aircraft. Another feature of the T-1000 system is the availability of a computerised route segment and economic analysis programme. Financial aspects of using the T-1020 and T-1040 can be summarised for specific operations, with direct comparison of the operating profiles of those aircraft and other types, based on the individual carrier's load factors, costs and routes.

PIPER T-1020

This aircraft is based on the PA-31-350 Chieftain, of which approximately 500 are currently in service with commuter airlines. It retains the same basic airframe and power plant, comprising two 261 kW (350 hp) Avco Lycoming TIO-540 engines, with a special commuter TBO of 1,800 h. The interior is, however, 'hardened' to airline standards to accommodate two crew and nine passengers. The doors and landing gear are strengthened to withstand the high-frequency short-haul operations of a commuter line. Up to 317 kg (700 lb) of baggage can be carried, and stored in most cases entirely outside the cabin. Usable fuel capacity is 401 litres (106 US gallons) in a single tank in each wing. This is adequate for commuter operations, and permits a straightforward on/off/crossfeed cockpit fuel management system, with no necessity to switch tanks. Estimated full-payload range at 174 knots (322 km/h; 200 mph) at 1,525 m (5,000 ft), with reserves, is 310 nm (575 km; 355 miles).

The T-1020 can be equipped for flight into known icing conditions. Standard factory-installed avionics package includes Collins Micro Line or King Silver Crown com/nav equipment, with an extensive range of optional items. Optional equipment includes a crew door on the port side of the flight deck.

Deliveries were scheduled to begin in November 1981.

Prototype Piper T-1040, photographed before its first flight on 17 July 1981

Piper T-1040 commuter airliner (two P&WC PT6A-11 turboprop engines) *(Pilot Press)*

PIPER T-1040

The Piper T-1040, which has the same 11-seat capacity as the T-1020, combines the basic Chieftain fuselage with the nose, wings, tail unit and Pratt & Whitney Aircraft of Canada PT6A-11 engines of the PA-31T-1 Cheyenne I.

This represents an effective way of developing rapidly a reliable turboprop-powered commuter transport, eliminating most of the teething problems that can be anticipated with the production of a completely new design. With all seats installed, the T-1040 has more than 1·4 m³ (50 cu ft) of baggage space in nose and rear cabin baggage areas, and in engine nacelle lockers. Equipment for day or night IFR operations, and for flight into known icing conditions, is optional. Collins or King digital avionics packages are standard, including dual nav/com, ADF, audio panel with DME and marker beacon receiver, and a transponder. Other avionics and equipment, including a crew door on the port side of the flight deck, are available optionally.

The prototype T-1040 flew for the first time on 17 July 1981, and initial deliveries are scheduled for the Spring of 1982. It is estimated that this transport will have a maximum speed of 243 knots (450 km/h; 280 mph), a single-engine ceiling of 4,020 m (13,200 ft), a range of 600 nm (1,110 km; 690 miles) with full payload, and the ability to operate from airports with runways as short as 1,070 m (3,500 ft). It will be certificated for single-pilot operation.

WEIGHTS:
Max ramp weight 4,105 kg (9,050 lb)
Max zero-fuel weight 3,447 kg (7,600 lb)
PERFORMANCE (estimated, at max T-O weight):
Max cruising speed at 3,350 m (11,000 ft) 238 knots (440 km/h; 273 mph)
T-O to 15 m (50 ft) 808 m (2,650 ft)
Accelerate/stop distance 1,052 m (3,450 ft)
Range with max fuel at max cruising speed at 3,050 m (10,000 ft), 45 min reserves 670 nm (1,240 km; 770 miles)

PITTS

PITTS AEROBATICS

PO Box 547, Afton, Wyoming 83110
Telephone: (307) 886 3151
GENERAL MANAGER: E. H. Andersen Jr

One of the best-known US designers of high-performance sporting aircraft, Mr Curtis Pitts, announced on 1 January 1977 that he had sold the manufacturing, engineering and sales rights of the Pitts S-1 and S-2 aircraft to Doyle Child of Afton, Wyoming. A new company, Pitts Aerobatics, was formed to continue engineering and sales activities, being based at the same location as Aerotek Inc, which formerly produced the S-2 for Pitts Aviation Enterprises.

Detailed construction drawings for the single-seat S-1 version are available to amateur constructors (see Homebuilts section). For pilots who do not wish to build their own aircraft, Aerotek Inc can supply a ready-to-fly single-seater designated S-1S. The two-seat S-2 is available both as a factory-built aircraft and in kit form, and is produced by Aerotek for Pitts Aerobatics. The single-seat S-2S, of which the first production examples became available in late 1978, is also being marketed in kit form for amateur construction. Production S-2S aircraft are certificated currently in the Experimental-Exhibition category, but it is planned to gain full type certification.

Pitts Model S-2S single-seat biplane (Avco Lycoming AEIO-540-D4A5 engine)

PITTS S-1 SPECIAL

The original single-seat Pitts Special was designed in 1943-44. Construction of a prototype began in 1944 and it flew in September of that year. One of the most successful early models was *Little Stinker,* powered by a 67 kW (90 hp) Continental engine, and built by Mr Pitts in 1947 for Miss Betty Skelton, then an internationally-known aerobatic display pilot. The *Black Beauty* biplane, built by Pitts for Miss Caro Bailey, was of similar design, but powered by a 93 kW (125 hp) Avco Lycoming O-290-D engine.

Since then even more powerful engines have been installed in Pitts Specials built by both the designer and other people, and those versions of the single-seat Special for which drawings are available are designed to take an Avco Lycoming engine of up to 134 kW (180 hp), as noted in the Homebuilts section.

Current versions of the S-1 are as follows:

S-1D. Intended for homebuilders only, with plans available. Generally similar to the S-1S.

S-1E. Intended for homebuilders only; has wings of symmetrical aerofoil section. Plans available.

S-1S. Production aircraft, FAA type certificated and built at rate of 25 aircraft per year. It is available also in kit form, parts, materials and components being produced under an FAA Approved Production Certificate.

There is also an S-1C version, but Pitts can no longer supply plans for its construction.

Details of some of the major successes achieved by US pilots of Pitts Specials in national and international aerobatic competitions, since 1966, have been given in the 1972-73 and later *Jane's.*

The details which follow apply to the S-1S factory-built aircraft with 134 kW (180 hp) engine, but engines of 74·5-134 kW (100-180 hp) can be fitted to the S-1 Special.

TYPE: Single-seat sporting biplane.

WINGS: Braced biplane type, with single faired interplane strut each side and N-type cabane struts. Dual streamline flying and landing wires. Wing section M6. Thickness/chord ratio 12%. Dihedral 0° on upper wing, 3° on lower wings. Incidence 1° 30′ on upper wing, 0° on lower wings. Sweepback at quarter-chord 6° 40′ on upper wing only. Wooden structure, with fabric covering. Frise-type ailerons on both upper and lower wings, of similar construction to wings. No flaps or tabs.

FUSELAGE: Welded steel tube structure, fabric-covered.

TAIL UNIT: Wire-braced steel tube structure, fabric-covered. Fixed-incidence tailplane. Trim tab in each elevator.

LANDING GEAR: Non-retractable tailwheel type. Rubber-cord shock-absorption. Cleveland main wheels with 6-ply tyres, size 5·00-5, pressure 2·07 bars (30 lb/sq in). Cleveland hydraulic disc brakes. Steerable tailwheel. Glassfibre fairings on main wheels.

POWER PLANT: One 134 kW (180 hp) Avco Lycoming IO-360-B4A flat-four engine, driving a Sensenich type 76EM8-O-56/62 two-blade fixed-pitch metal propeller. Fuel tank aft of firewall, capacity 75 litres (20 US gallons). Refuelling point on upper surface of fuselage, forward of windscreen. Oil capacity 7·5 litres (2 US gallons). Inverted fuel and oil systems standard.

ACCOMMODATION: Single seat in open cockpit.

DIMENSIONS, EXTERNAL:
Wing span, upper 5·28 m (17 ft 4 in)
Wing chord (constant, both) 0·91 m (3 ft 0 in)
Wing aspect ratio 5·77
Length overall 4·71 m (15 ft 5½ in)
Height overall 1·92 m (6 ft 3½ in)
Tailplane span 1·98 m (6 ft 6 in)
Propeller diameter 1·93 m (6 ft 4 in)

AREA:
Wings, gross 9·15 m² (98·5 sq ft)

WEIGHTS AND LOADING:
Weight empty 326 kg (720 lb)
Max T-O weight 521 kg (1,150 lb)
Max power loading 3·89 kg/kW (6·38 lb/hp)

PERFORMANCE (at max T-O weight):
Never-exceed speed 176 knots (326 km/h; 203 mph)
Max level speed at S/L 153 knots (283 km/h; 176 mph)
Max cruising speed at S/L 122 knots (227 km/h; 141 mph)
Stalling speed 54 knots (100 km/h; 62 mph)
Max rate of climb at S/L 792 m (2,600 ft)/min
Service ceiling 6,795 m (22,300 ft)
T-O to 15 m (50 ft) 331 m (1,085 ft)
Range with max fuel, no reserves 273 nm (507 km; 315 miles)

Production version of the basic single-seat Pitts Special, the S-1S

PITTS S-2A SPECIAL

First flown in 1967, the S-2A is a two-seat version of the Pitts Special. It is similar to the single-seat S-1 in basic configuration and construction, but is slightly larger in overall dimensions, with no attempt at commonality of components. The increased size and power, coupled with aerodynamic changes, give the two-seater improved aerobatic and landing characteristics, and make it extremely stable in rough air conditions. Control responses are better than on the S-1. The ailerons are aerodynamically balanced for higher rate of roll at low speeds, and full vertical rolls can be made with ease. The different wing sections used on the S-2A provide inverted performance equal to conventional flight, and facilitate outside loops.

The S-2A is FAA type certificated in the Normal and Aerobatic categories. It is a production aeroplane, not intended for the homebuilder, and plans are not available.

Five S-2As, each fitted with a 149 kW (200 hp) Avco Lycoming engine, were supplied in early 1973 to the British aerobatic display team then financed by the Rothman Tobacco Company. During displays, the front cockpit of each aircraft was covered by a removable panel. Five similar aircraft were supplied to the Carling Black Label aerobatic team operating from Toronto, Canada.

S-2As have been exported to Australia, Brazil, Sweden and Venezuela. In mid-1976, the Peruvian Air Force announced an order for six Pitts Specials, for 'an unspecified training role'.

TYPE: Two-seat aerobatic biplane.

WINGS: Braced biplane type, with single faired interplane strut each side and N-type cabane. Wing section NACA 6400 series on upper wing, 00 series on bottom wings. Two-spar wooden (spruce) structure with fabric covering. Aerodynamically-balanced ailerons on both upper and lower wings. No flaps or tabs.

FUSELAGE: Welded steel tube structure with wooden stringers, covered with Dacron fabric except for metal top-decking.

TAIL UNIT: Wire-braced welded steel tube structure. Fixed surfaces metal-covered, control surfaces fabric-covered. Trim tab in each elevator.

LANDING GEAR: Non-retractable tailwheel type. Rubber-cord shock-absorption. Steerable tailwheel. Fairings on main wheels.

POWER PLANT: One 149 kW (200 hp) Avco Lycoming IO-360-A1A flat-four engine, driving a Hartzell type HC-C2YK-4/C7666A-2 two-blade constant-speed metal propeller with spinner. Fuel tank in fuselage, immediately aft of firewall, capacity 90·5 litres (24 US gallons). Refuelling point on fuselage upper surface forward of front windscreen. Oil capacity 7·5 litres (2 US gallons). Inverted fuel and oil systems standard.

ACCOMMODATION: Two seats in tandem cockpits with dual controls. Rear cockpit can be enclosed by a transparent canopy if required, and an extended one-piece canopy to cover both cockpits is also available. Space for 9 kg (20 lb) baggage aft of rear cockpit when flown in non-aerobatic category.

SYSTEM: Electrical system powered by 12V 40A alternator and non-spill 12V battery.

DIMENSIONS, EXTERNAL:
- Wing span, upper: 6·10 m (20 ft 0 in)
- Wing span, lower: 5·79 m (19 ft 0 in)
- Wing chord (constant, both): 1·02 m (3 ft 4 in)
- Length overall: 5·41 m (17 ft 9 in)
- Height overall: 1·94 m (6 ft 4½ in)

AREA:
- Wings, gross: 11·6 m² (125 sq ft)

WEIGHTS AND LOADINGS (A: Aerobatic; B: Normal category):
- Weight empty: A, B: 453 kg (1,000 lb)
- Max T-O weight: A: 680 kg (1,500 lb)
  - B: 714 kg (1,575 lb)
- Max wing loading: A: 58·6 kg/m² (12·0 lb/sq ft)
  - B: 61·5 kg/m² (12·6 lb/sq ft)
- Max power loading: A: 4·56 kg/kW (7·5 lb/hp)
  - B: 5·33 kg/kW (7·87 lb/hp)

PERFORMANCE (at max T-O weight. A: Aerobatic; B: Normal category):
- Never-exceed speed:
  - A, B: 176 knots (326 km/h; 203 mph)
- Max level speed at S/L:
  - A, B: 136 knots (253 km/h; 157 mph)
- Max cruising speed at S/L:
  - A, B: 132 knots (245 km/h; 152 mph)
- Stalling speed: A: 51 knots (94 km/h; 58 mph)
  - B: 51·5 knots (95 km/h; 59 mph)
- Max rate of climb at S/L: A: 579 m (1,900 ft)/min
  - B: 549 m (1,800 ft)/min
- Service ceiling: A: 6,125 m (20,100 ft)
  - B: 4,875 m (16,000 ft)
- T-O to 15 m (50 ft): A: 351 m (1,150 ft)
- Range with max fuel:
  - A: 297 nm (552 km; 343 miles)

PITTS MODEL S-2S

Pitts Aerobatics began production in late 1978 of this single-seat version of the S-2A. It has a modified forward fuselage to permit the installation of a 194 kW (260 hp) Avco Lycoming AEIO-540-D4A5 flat-six engine, driving a McCauley Type 1A/200 two-blade fixed-pitch metal propeller with spinner. The first flight of the prototype was made on 9 December 1977. Maximum level speed is 157 knots (291 km/h; 181 mph), max rate of climb at S/L 915 m (3,000 ft)/min. The Model S-2S is available either as a factory-built aircraft, or in kit form for amateur construction.

By early 1981 a total of 14 of these aircraft had been completed and certificated in the Experimental/Exhibition category. At that time the full certification programme was nearing completion, and the company anticipated that this would be attained by 1 May 1981.

RAISBECK
RAISBECK ENGINEERING

7536 Seward Park Avenue South, Seattle, Washington 98118

PRESIDENT: James D. Raisbeck

The Raisbeck Group, as included in the 1980-81 *Jane's*, is no longer in existence, and its former President, James D. Raisbeck, has formed Raisbeck Engineering to continue with the development of performance improvement systems for application to specific aircraft. In early 1981 the company was working on such a system that will eventually be suitable for modification of the entire Beechcraft King Air family. Known as the **Raisbeck Mark VI** Aerodynamic Performance Improvement System, this is being developed as a joint venture with Midcoast Aviation Services Inc of St Louis, Missouri. They, together with two other fixed base operators, will carry out the installation of these systems on in-service King Airs when certification has been gained.

It is intended to develop first a conversion for the King Air F90, and Raisbeck Engineering has come to an agreement with JetCrafters incorporated (which see) to supply for this programme the engine nacelles developed for the JetCrafters Taurus modification. These will be used by Raisbeck Engineering in conjunction with 15 other aerodynamic drag reduction improvements, including the use of winglets, to form the Mark VI System. Full certification is planned for the King Air F90 application in July 1982, and it is the intention to follow on with systems for the King Air E90 and Super King Air 200.

The Mark VI system was expected to fly on a prototype aircraft by the end of 1981. Improvements over the standard King Air F90 of more than 40 knots (74 km/h; 46 mph) in cruising speed, and a 30 per cent improvement in climb rate, were anticipated.

RILEY
RILEY AIRCRAFT CORPORATION

2016 Palomar Airport Road, Carlsbad, California 92008
Telephone: (714) 438 0660
Telex: 69 7936 Riley Jet
PRESIDENT: Jack M. Riley
VICE-PRESIDENT: Patrick M. Riley

Mr J. Riley was responsible for the Riley 55 Twin-Navion conversion scheme in 1952 and the Riley Rocket conversion of the Cessna 310 in 1962, as well as for Dove, Heron and Cessna 310/320 conversions known respectively as the Riley Turbo-Exec 400, Turbo Skyliner and Turbostream. His company is marketing currently a Cessna 310/320 conversion known as the Riley Super 310; two Cessna 340 conversions which are designated Riley Super 340 and Riley Rocket 340; a conversion of the Cessna 414 designated Riley Rocket Power 414, and two Cessna 421 conversions designated Riley Turbine Rocket 421 and Riley Turbine Eagle 421.

FAA-approved export conversion kits are available for the six aircraft in the current production programme, and assistance is available if required to gain approval by the civil aviation authority in the country to which the kit has been exported.

In early 1981 the company was involved in the certification programme of a Cessna 404 conversion, under the name Riley Turbine 404, and the company designation R404P. It was anticipated that certification would be gained in early 1982.

The growing demand for Riley conversions, with the company quoting a backlog of orders valued at $10 million, has required increased facilities, leading to the provision of a new hangar of 2,230 m² (24,000 sq ft) and a new ramp area of 4,180 m² (45,000 sq ft).

RILEY SUPER 310

The Super 310 (formal designation R310 Super) represents a similar conversion to that of the Super 340, the existing power plants in the Cessna 310/320 being replaced by 231 kW (310 hp) Continental TSIO-520-Ns. The conversion is FAA-approved for the Cessna 310I to 310R, 320B to 320F and T310P to T310R. Installation of the complete kit requires 12 working days.

A description of the current Cessna Model 310 appears under the Cessna entry in this edition. It applies also to the Riley Super 310, except as follows:

POWER PLANT: As described for Riley Super 340, except installed engines are TSIO-520-Ns instead of TSIO-520-Js.

WEIGHTS AND LOADING: As for the appropriate Cessna 310/320 model except as follows:
- Weight empty: increased by 10 kg (22 lb)
- Max payload: reduced by 10 kg (22 lb)
- Max power loading: 5·11 kg/kW (8·4 lb/hp)

PERFORMANCE (at max T-O weight):
- Max cruising speed, 75% power:
  - at 7,315 m (24,000 ft): 261 knots (483 km/h; 300 mph)
  - at 5,485 m (18,000 ft): 243 knots (451 km/h; 280 mph)
  - at 4,265 m (14,000 ft): 234 knots (435 km/h; 270 mph)
- Max rate of climb at S/L: 792 m (2,600 ft)/min
- Rate of climb at S/L, one engine out: 158 m (520 ft)/min
- Service ceiling: 10,670 m (35,000 ft)
- Service ceiling, one engine out: 7,620 m (25,000 ft)
- T-O to 15 m (50 ft): 503 m (1,650 ft)
- Landing from 15 m (50 ft): 546 m (1,790 ft)
- Range with max fuel, max cruising speed at 7,315 m (24,000 ft), 30 min reserves: 1,042 nm (1,931 km; 1,200 miles)
- Range with max fuel, econ cruising speed at optimum altitude, 30 min reserves: 1,737 nm (3,218 km; 2,000 miles)

RILEY SUPER 340

The Riley Super 340 (formal designation R340 Super) differs from the standard Cessna 340 primarily through replacement of the latter's TSIO-520-K engines by TSIO-520-Js. The most important change associated with this new engine installation is the addition of an intercooler which allows a higher power output and also improves specific fuel consumption, critical altitude, engine life and reliability.

The prototype Super 340 received FAA certification in June 1974. The standard Cessna 340A for 1976 was updated in a manner similar to that of the Riley Super 340. Consequently, only Cessna 340 aircraft constructed from 1972 until the last of the 1975 models were completed are suitable for this conversion, which is completed by Riley engineers in five working days.

A description of the current Cessna Model 340A appears under the Cessna entry in this edition. It applies also to the Riley Super 340, except as follows:

POWER PLANT: Two 231 kW (310 hp) Continental TSIO-520-J flat-six turbocharged and intercooled engines, each driving a McCauley three-blade propeller. Engine installation upgraded from Cessna 340 to Cessna 414 configuration. Fuel system as for Cessna 340. Optional aft nacelle fuel tanks, each of 75·7 litres (20 US gallons), provide max optional fuel capacity of 920 litres (243 US gallons) when used in conjunction with the Cessna optional 768 litre (203 US gallon) system.

WEIGHTS AND LOADING:
- Weight empty: 1,754 kg (3,868 lb)
- Max payload: 904 kg (1,993 lb)
- Max T-O weight: 2,710 kg (5,975 lb)
- Max power loading: 5·87 kg/kW (9·64 lb/hp)

PERFORMANCE (at max T-O weight):
- Max level speed at 7,315 m (24,000 ft): 252 knots (467 km/h; 290 mph)
- Max cruising speed, 75% power:
  - at 7,315 m (24,000 ft): 234 knots (435 km/h; 270 mph)
  - at 5,485 m (18,000 ft): 226 knots (418 km/h; 260 mph)
  - at 4,265 m (14,000 ft): 217 knots (402 km/h; 250 mph)
- Econ cruising speed at 7,315 m (24,000 ft): 208 knots (386 km/h; 240 mph)
- Max rate of climb at S/L: 549 m (1,800 ft)/min
- Cruise rate of climb at max cruising power: 305 m (1,000 ft)/min

Riley Rocket 340 conversion of the Cessna Model 340 (Avco Lycoming TIO-540-R engines)

Rate of climb at S/L, one engine out
107 m (350 ft)/min
Service ceiling 9,755 m (32,000 ft)
Service ceiling, one engine out 4,875 m (16,000 ft)
T-O run 457 m (1,500 ft)
T-O to 15 m (50 ft) 594 m (1,950 ft)
Landing from 15 m (50 ft) 561 m (1,840 ft)
Landing run 233 m (765 ft)
Range with max fuel, max cruising speed at 7,315 m (24,000 ft), 30 min reserves
1,042 nm (1,931 km; 1,200 miles)
Range with max fuel, econ cruising speed at 7,315 m (24,000 ft), 30 min reserves
1,737 nm (3,218 km; 2,000 miles)

RILEY ROCKET 340

This Riley conversion of the Cessna 340 is, like the Super 340 conversion, concerned primarily with the installation of a different power plant to improve performance. Thus, the Continental TSIO-520-J engines of the Super 340 are replaced by counter-rotating Avco Lycoming TIO-540-Rs of increased power. These engines feature the latest Garrett turbocharger system, with the wastegate and automatic manifold pressure controller integral with the engine turbocharger. First flight of a Rocket 340 (formal designation R340L) was made in December 1977.

A description of the current Cessna Model 340A appears under the Cessna entry in this edition. It applies as also to the Riley Rocket 340, except as follows:

Power Plant: One Avco Lycoming TIO-540-R2AD and one LTIO-540-R2AD flat-six engine, each rated at 254 kW (340 hp) for take-off and driving a Hartzell propeller with spinner. Fuel system as for Riley Super 340. Intake air intercoolers standard.

Avionics and Equipment: Standard equipment includes a propeller synchrophaser. A range of optional avionics is available to customer's requirements. Optional equipment includes a custom interior, seventh cabin seat, super soundproofing of nose baggage area adjacent to propeller arc, Cleveland heavy-duty wheels and brakes, J.B. Systems electrically-driven air-conditioner, automatic in-line fuel boost pumps, wing and tailplane de-icing boots, and polyurethane exterior paint finish.

Weight and Loadings:
Max T-O weight 2,853 kg (6,290 lb)
Max wing loading 158·7 kg/m² (32·5 lb/sq ft)
Max power loading 5·62 kg/kW (9·25 lb/hp)

Performance (at max T-O weight, ISA):
Max cruising speed at 7,315 m (24,000 ft)
261 knots (483 km/h; 300 mph)
Cruising speed, 65% power at 7,315 m (24,000 ft)
240 knots (444 km/h; 276 mph)
Max rate of climb at S/L 655 m (2,150 ft)/min
Rate of climb at S/L, one engine out
152 m (500 ft)/min

Service ceiling over 9,145 m (30,000 ft)
Service ceiling, one engine out 6,310 m (20,700 ft)
Minimum field length 561 m (1,840 ft)
Max range, 45 min reserves:
444 kg (978 lb) fuel
960 nm (1,778 km; 1,105 miles)
552 kg (1,218 lb) fuel
1,260 nm (2,335 km; 1,451 miles)

RILEY TURBINE 404

The Riley Turbine 404 (formal designation R404P) is a conversion of the Cessna Titan two/ten-seat passenger/executive/cargo transport. Improved performance with better safety margins results from the installation of Pratt & Whitney Aircraft of Canada PT6A-135 turboprops as used in the Riley Turbine Eagle 421 conversion, and the power plant and systems as described for this latter aircraft apply also to the R404P. It is planned to gain certification of this model also with PT6A-27 engines, each of 507 kW (680 shp), as options.

Following completion of the certification programme, which is scheduled for early 1982, it is anticipated that there will be considerable demand for this conversion. The Turbine 404 will operate at a higher gross weight than the Cessna Titan, with improved single-engine performance, range and serviceability. These factors, coupled with the versatility of the Titan, should prove an attractive proposition to commuter airlines.

Performance (provisional, at max T-O weight of 3,855 kg; 8,500 lb, ISA):
Max level speed at 8,230 m (27,000 ft)
290 knots (537 km/h; 334 mph)
Econ cruising speed at 3,050 m (10,000 ft)
235 knots (436 km/h; 271 mph)
Max rate of climb at S/L 1,000 m (3,280 ft)/min
Rate of climb at S/L, one engine out
213 m (700 ft)/min
Service ceiling 8,230 m (27,000 ft)
Service ceiling, one engine out 5,335 m (17,500 ft)
Range with max fuel at 3,050 m (10,000 ft), IFR reserves 1,050 nm (1,946 km; 1,209 miles)
Max range with max fuel at 8,230 m (27,000 ft), IFR reserves 1,250 nm (2,317 km; 1,439 miles)

RILEY ROCKET POWER 414

This Riley conversion of the Cessna 414, which has the formal designation R414L, involves installation of Avco Lycoming TSIO-720 flat-eight engines, with Riley/Garrett turbochargers that maintain a power output of 224 kW (300 hp) at 10,060 m (33,000 ft), to provide improved performance. It moves the empty aircraft CG forward, allowing more flexibility in loading, and introduces also Hartzell automatic propeller synchrophasing, an Aerosonic electronic fuel management system, and heavy-duty Cleveland brakes. The engines are mounted in new glassfibre cowlings which have louvres for engine cooling, making redundant the cowl flaps which were provided originally, and these are removed.

A description of the current Cessna Model 414 appears under the Cessna entry in this edition. It applies also to the Riley Rocket Power 414, except as follows:

Power Plant: Two 298 kW (400 hp) Avco Lycoming TSIO-720-B1B0 flat-eight engines, each driving a Hartzell four-blade constant-speed metal propeller with Q-tips. Riley intake air intercoolers standard. Fuel system as for Cessna 414, except that an Aerosonic fuel management system is provided. Hartzell automatic propeller synchrophasers standard.

Weights and Loadings:
Max T-O weight 3,250 kg (7,165 lb)
Max landing weight 3,096 kg (6,825 lb)
Max wing loading 154·9 kg/m² (31·73 lb/sq ft)
Max power loading 5·45 kg/kW (8·95 lb/hp)

Performance (at max T-O weight, ISA):
Max level speed, turbo cruise power:
at 7,315 m (24,000 ft)
261 knots (483 km/h; 300 mph)
at 3,660 m (12,000 ft)
217 knots (402 km/h; 250 mph)
Speed at intermediate cruise power:
at 7,315 m (24,000 ft)
243 knots (451 km/h; 280 mph)
at 3,660 m (12,000 ft)
200 knots (370 km/h; 230 mph)
Max rate of climb at S/L 686 m (2,250 ft)/min
Rate of climb at S/L, one engine out
122 m (400 ft)/min
Service ceiling 9,145 m (30,000 ft)
Service ceiling, one engine out 7,620 m (25,000 ft)
Max range, 45 min reserves:
490 kg (1,080 lb) fuel
771 nm (1,429 km; 888 miles)
653 kg (1,440 lb) fuel
1,136 nm (2,105 km; 1,308 miles)

RILEY TURBINE ROCKET 421

This Riley conversion of the pressurised Cessna 421 involves installation of Avco Lycoming LTP 101-700 turboprop engines, each rated normally at 522 ekW (700 ehp), but flat rated in this application to approximately 70 per cent of normal T-O power. Both the Cessna Model 421B and 421C are suitable for this conversion, under the formal designations of R421BL and R421CL respectively. First flight of a Turbine Rocket 421 was made on 10 January 1978, followed by the first production conversion on 4 June 1978. The R421BL was certificated in June 1981.

The description of the Cessna 421C Golden Eagle which appears under the Cessna entry in this edition applies also to the Riley Turbine Rocket 421, except that the landing gear is equipped with Cleveland 199-76 heavy-duty brakes; each engine drives a Hartzell propeller with Q-tips and Beta propeller reversal; and optional wing fuel tanks are available to add 363 litres (96 US gallons) to the max total capacity.

Weight and Loadings:
Max T-O weight 3,470 kg (7,650 lb)
Max wing loading 173·7 kg/m² (35·58 lb/sq ft)
Max power loading 4·75 kg/kW (7·81 lb/ehp)

Performance (at T-O weight of 3,453 kg; 7,612 lb, ISA):
Max cruising speed at 8,230 m (27,000 ft)
270 knots (500 km/h; 311 mph)
Max rate of climb at S/L 1,067 m (3,500 ft)/min
Rate of climb at S/L, one engine out
224 m (735 ft)/min
Service ceiling over 9,145 m (30,000 ft)
Service ceiling, one engine out 6,400 m (21,000 ft)
Min landing field length 686 m (2,250 ft)
Landing run with propeller reversal 152 m (500 ft)
Range with max fuel, IFR reserves
1,300 nm (2,407 km; 1,496 miles)

RILEY TURBINE EAGLE 421

Claimed by Riley Aircraft to be the world's fastest executive turboprop aircraft, this modification unites the Cessna 421C Golden Eagle airframe with Pratt & Whitney Aircraft of Canada PT6A turboprop engines, mounted in new low-drag nacelles. Each of these engines has more than double the power output of the engine it replaces, yet weighs some 40 per cent less. The prototype Turbine Eagle 421 flew for the first time in late November 1979, at Palomar Airport, and the certification programme was completed during 1980.

A description of the current Cessna 421C Golden Eagle appears under the Cessna entry in this edition; it applies also to the Riley Turbine Eagle 421 (R421CP), except as follows:

Power Plant: Two Pratt & Whitney Aircraft of Canada PT6A-135 turboprop engines, each flat rated to 559 kW (750 shp), and driving Hartzell three-blade constant-speed autofeathering reversible-pitch propellers with Q-tips and Hartzell spinners. Standard fuel as for Cessna 421C, but auxiliary tanks in rear of nacelles with combined capacity of 568 litres (150 US gallons). Advanced aerodynamic cowlings incorporate inertial separators. Electrical de-icing of propellers and air intakes.

Riley Rocket Power 414, a conversion of the Cessna 414 (two Avco Lycoming IO-720 engines)

Prototype of Riley's new Turbine Eagle 421, a conversion of the Cessna 421C Golden Eagle

SYSTEMS: Generally as for Cessna 421C, except bleed air unit for pressurisation system; electrical system with 200A starter/generators and heavy-duty battery; engine fire detection system; and pneumatic leading-edge de-icing boots for wings and tail unit.

WEIGHT:
Max T-O weight 3,492 kg (7,700 lb)

PERFORMANCE (at T-O weight of 3,453 kg; 7,612 lb, ISA):
Max cruising speed at 8,230 m (27,000 ft) 320 knots (593 km/h; 368 mph)
Max rate of climb at S/L 1,219 m (4,000 ft)/min
Rate of climb, one engine out 305 m (1,000 ft)/min
Service ceiling over 9,145 m (30,000 ft)
Service ceiling, one engine out 7,620 m (25,000 ft)
Range with max fuel, IFR reserves 1,500 nm (2,780 km; 1,727 miles)

ROBERTSON

ROBERTSON AIRCRAFT CORPORATION

Snohomish County Airport, North Complex C-72, Everett, Washington 98204
Telephone: (206) 355 8702
PRESIDENT: Leland R. Lynch
SENIOR VICE-PRESIDENT: Henry I. McKay
CHIEF ENGINEER: Sherman Hall
SALES MANAGER: Thomas Raschella
DIRECTOR OF PUBLIC RELATIONS: Merle E. Dowd

Robertson Aircraft Corporation was formed by the late Mr James L. Robertson, who had long been a pioneer in the development of STOL aircraft, having been responsible for the Skycraft Skyshark, Wren 460 and STOL modifications to the IMCO CallAir A-9 and B-1. It has designed, built and certificated a series of Hi-Lift advanced technology safety and performance systems for standard single- and twin-engined Beech, Cessna and Piper aircraft.

During 1980, Robertson transferred all its manufacturing, management, sales and engineering departments from its former facilities at Renton Municipal Airport to two buildings at the edge of Paine Field, near Everett. This has permitted expansion of its own manufacturing activities. It has also entered into an agreement with Astec Corporation (which see), to manufacture and install Astec's Eagle modification to Cessna Citation jet aircraft. Astec will continue to market and deliver the modified aircraft.

A new canard-like wing root airflow energiser was certificated on the Cessna 414A in early 1980. The Robertson Hi-Lift system for this aircraft includes also Fowler-type trailing-edge flaps of improved aerofoil section and with longer tracks. A similar programme has adapted the above system for installation on the Cessna 402C and 421C. The wing root airflow energiser consists of a flat plate 0·102 m (4 in) wide which is attached to the fuselage skin forward of each wing. With a length of 0·38 m (1 ft 3 in), each energiser is mounted directly ahead of the wing leading-edge and in line with the wing chord. The addition of this device reduces stalling speed by 6 knots (11 km/h; 7 mph) with flaps up, and by 7 knots (13 km/h; 8 mph) with the flaps extended.

Research contracts with Kansas University and NASA were completed in 1974, covering the design of an Advanced Technology Light Twin (ATLIT). A standard Piper Seneca fuselage was fitted with an experimental wing that had a new aerofoil section and 30% chord full-span Fowler flaps. There were no ailerons and roll control was achieved by the use of upper-surface spoilers. Construction of the aircraft was completed by Piper Aircraft Corporation and the first flight was recorded in late 1974.

Following completion of the ATLIT research programme, Robertson embarked on a project to develop spoilers and full-span flaps suitable for retrofit on the standard Piper Seneca I. FAA certification of the resulting system represented the first approval of spoilers for use on general aviation aircraft manufactured in the USA (see full description under this entry). This spoiler/full-span flap system has since been certificated for Beech Bonanza V35 and A36 series aircraft, and for the Piper PA-32-300 Cherokee SIX and Seneca II.

The most recent Robertson development has provided a high-differential aileron-droop Hi-Lift system for high-performance single-engined aircraft, such as the Cessna Model 210 Pressurised Centurion. In late 1980, three Cessna T210s equipped with this system were delivered for service with the Chilean Police, together with four T206s with Hi-Lift systems. They are used primarily for air surveillance to control smuggling and fishing, and for other activities that demand patrol speeds of 35-39 knots (64-72 km/h; 40-45 mph).

Beechcraft A36 Bonanza, newly certificated with Robertson Hi-Lift system including full-span flaps

R/STOL Cessna 340 tests Fowler flap extension at altitude

Robertson holds an FAA Supplemental Type Certificate covering the installation of RCA WeatherScout I radar inside the leading-edge of the starboard wing of a number of aircraft including the Cessna 182, 206F, T206F, TU206G, U206G and 210N series. The truncated 13 × 30 cm (5 × 12 in) antenna is housed inside a leading-edge cutaway, with the receiver/transmitter built integrally at the back of the antenna. The wing structure is reinforced around the cutaway. A reinforced epoxy/glassfibre radome encloses the opening. Total system weight is 10·5 kg (23·1 lb). WeatherScout I is basically weather radar, but provides limited ground-mapping capability.

ROBERTSON / CESSNA and ROBERTSON / PIPER SAFETY and STOL CONVERSIONS

Continuous product improvement has been made on Robertson's line of Hi-Lift systems for Cessna and Piper single-engined and smaller twin-engined aircraft. The Robertson modification, first applied to a Cessna 182, comprises full-span wing leading-edge and trailing-edge high-lift systems which greatly reduce the take-off and landing distances normally required by such aircraft.

The existing ailerons are used as an integral part of the full-span trailing-edge flap system. When the conventional inboard flaps are lowered for take-off or landing, the ailerons droop with them, virtually doubling the wing lift at low speeds. The ailerons retain their differential operation for roll control when drooped.

In addition, the wing is fitted with a full-span distributed-camber leading-edge to provide an optimum spanwise lift distribution for maximum cruise efficiency. The cambered leading-edge also reduces the aerofoil leading-edge pressure peak at high angles of attack, to impart maximum resistance to stall and to provide highly-responsive manoeuvrability at low airspeeds. Some types of Cessna single-engined aircraft built since 1972 include this Robertson leading-edge as a standard production feature.

The full-span flap system, in combination with Robertson's conical-cambered wingtips, dorsal and ventral fins, belly-mounted vortex generators, and flap/elevator automatic trim system, are combined in various models to offer increased performance.

To improve controllability at low speeds, stall fences are provided between flaps and ailerons, and to complete the Hi-Lift modification the aileron gap is sealed with a strip of aluminium sheet or rubberised canvas. These modifications permit safe STOL landings and take-offs by even novice pilots, and cruising speed and range are increased by 2-4%.

Maximum gross weight increases have accompanied the certification of Robertson modifications of twin-engined aircraft such as the Cessna Super Skymaster and Piper Twin Comanche. This is due primarily to their increased climb performance and slower take-off and landing speeds.

Development of a Hi-Lift system for the Cessna 400 series of twin-engined business aircraft entailed complete redesign of the wings from the rear spar aft to allow installation of 100% Fowler flaps and flap-actuated drooping ailerons. This system, in combination with Robertson's automatic pitch trim system and double-hinged rudder, led to FAA-certificated decreases in take-off and landing field lengths of approximately 40%.

In 1974, the FAA certificated the Robertson-equipped Piper Seneca I, with wing upper-surface spoilers for roll control, full-span slotted flaps, cambered wingtips and an anti-servo rudder tab. These modifications allow shorter take-off and landing distances; minimum control speed is reduced by 16%, the best rate of climb speed is lowered by 19%, single-engine service ceiling is raised by 213 m (700 ft) and roll response is increased greatly at all speeds and configurations. A similar modification of the Seneca II was certificated during 1976.

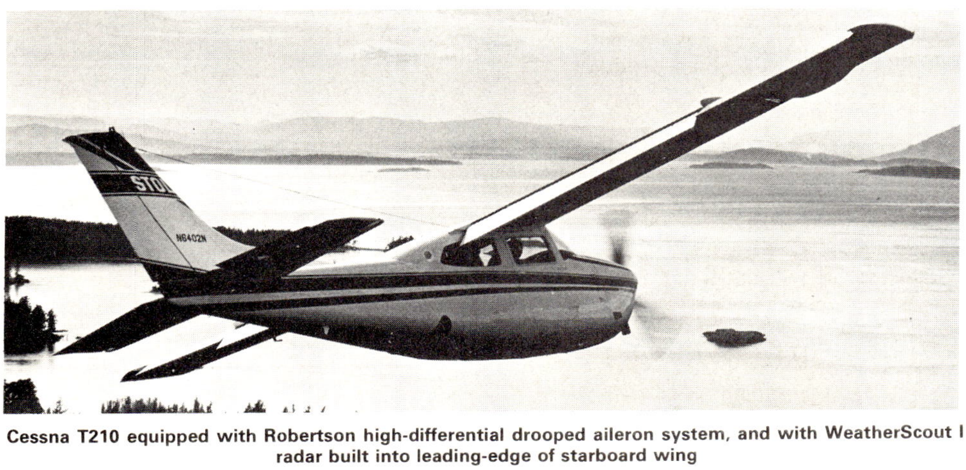

Cessna T210 equipped with Robertson high-differential drooped aileron system, and with WeatherScout I radar built into leading-edge of starboard wing

Robertson STOL conversion of the Cessna 414A; the Fowler-type trailing-edge flaps add 13 per cent to wing area. Also visible is the small airflow control device on the fuselage, forward of the wing leading-edge, which reduces stalling speed by 6 knots with flaps up

More recent activities have entailed fitting full-span trailing-edge flaps and spoilers for roll control on a Piper Cherokee SIX; and developing a high-differential aileron-droop Hi-Lift system for high-performance single-engined aircraft, such as the Cessna Model P210 Pressurised Centurion. In this application a new and improved control mechanism droops both ailerons symmetrically to 15° when the trailing-edge flaps are extended to 20° for take-off. For roll control the ailerons travel up and down in high-ratio differential: thus, with the control wheel fully to starboard, the starboard aileron moves up 42° to act as a spoiler (27° up from its normal faired position); at the same time the port aileron is deflected downward an additional 12° to a total of 27°. Ailerons are also drooped symmetrically by 15° when the trailing-edge flaps are extended to 30° for approach and landing.

Benefits from this advanced Hi-Lift system include a greatly-reduced adverse yaw tendency in turns, and a higher roll rate for more precise control when flaps are down and speed low. Control wheel forces are significantly less than with unmodified aircraft, due to a tailoring of aerodynamic forces on the up-moving aileron and a cross-over cable which minimises friction in the control runs.

Four stall strips, two on each wing, retain the desirable stall characteristics common to all Robertson Hi-Lift systems. For aircraft fitted with wing leading-edge de-icing boots, rubber strips are cemented to the rubber boots; when these are not fitted, aluminium stall strips are riveted to the wing leading-edge. A strip 0·13 m (5 in) long, close to the wing root, initiates an early-warning signal that the aircraft is approaching the stall. A longer strip, mounted further outboard and lower than the first, initiates the stall of the inboard and centre wing sections, while the outboard section and the wingtip are still unaffected by stall conditions. When the aircraft stalls it noses forward and down, under the complete control of ailerons which remain unstalled and effective for roll control.

The Robertson integrated high-lift and safety systems have been designed for easy field maintenance. They are designed to be applicable to almost the entire range of Cessna and Piper aircraft.

Full details of the basic Cessna and Piper airframes are given under the appropriate company headings in this and earlier editions of *Jane's*, and apply also to the Robertson versions, except for the added Hi-Lift systems as described. Weights and performance details of the entire range of Hi-Lift modifications are given in the accompanying tables. The conversions can be fitted as a retrospective modification to any of the models listed, irrespective of year.

The complete line of Robertson STOL systems is available throughout the world from 17 installation centres and more than 60 dealers.

R/STOL VERSIONS OF BEECH MODELS

	Weight empty equipped kg (lb)	Weight gross kg (lb)	Max level speed knots (km/h; mph)	Max cruising speed knots (km/h; mph)	Stalling speed, wheels and flaps down knots (km/h; mph)	Max rate of climb at S/L m (ft)/min	Service ceiling m (ft)	T-O run m (ft)	T-O to 15 m (50 ft) m (ft)	Landing from 15 m (50 ft) m (ft)	Landing run m (ft)	Max range** nm (km; miles)
V35B Bonanza	987 (2,176)	1,542 (3,400)	181 (335; 208)	176 (325; 202)	53 (98; 61)	343 (1,125)	5,443 (17,858)	B 239 (785)	B 420 (1,377)	B 279 (914)	B 146 (478)	875 (1,620; 1,007)
A36 Bonanza	1,009 (2,225)	1,633 (3,600)	177 (327; 204)	169 (313; 194)	54 (100; 62)	309 (1,015)	4,876 (16,000)	276 (904)	485 (1,590)	305 (1,000)	168 (550)	833 (1,543; 958)

B: Robertson Normal operation. **With optional long-range tanks fitted.

R/STOL VERSIONS OF CESSNA MODELS

	Weight empty equipped kg (lb)	Weight gross kg (lb)	Max level speed knots (km/h; mph)	Max cruising speed knots (km/h; mph)	Stalling speed, wheels and flaps down knots (km/h; mph)	Max rate of climb at S/L m (ft)/min	Single-engine rate of climb at S/L m (ft)/min	Service ceiling m (ft)	T-O run m (ft)	T-O to 15 m (50 ft) m (ft)	Landing from 15 m (50 ft) m (ft)	Landing run m (ft)	Max range** nm (km; miles)	Min control speed knots (km/h; mph)
Model 150 and Commuter	449 (990)	725 (1,600)	110 (204; 127)	105 (195; 121)	26 (48·3; 30)	213 (700)		3,930 (12,900)	A 129 (422) B 161 (527)	A 248 (815) B 273 (895)	A 193 (632) B 230 (755)	A 90 (295) B 106 (348)	790 (1,464; 910)	
Model 172 and Skyhawk†	572 (1,263)	1,043 (2,300)	126 (233; 145)	118 (219; 136)	28 (51·5; 32)	206 (675)		4,150 (13,600)	A 140 (460) B 175 (575)	A 274 (900) B 302 (990)	A 223 (730) B 267 (875)	A 92 (302) B 109 (356)	738 (1,367; 850)	
Model 172 and Skyhawk floatplane*	646 (1,425)	1,007 (2,220)	97 (180; 112)	94 (174; 108)	28 (51·5; 32)	191 (625)		3,765 (12,350)	A 256 (840) B 320 (1,050)	A 405 (1,330) B 451 (1,480)	A 267 (875) B 296 (970)	A 145 (475) B 171 (560)	477 (885; 550)	
Model 180 Skywagon†	707 (1,560)	1,270 (2,800)	152 (282; 175)	144 (267; 166)	32·2 (60; 37)	364 (1,195)		6,215 (20,400)	A 110 (360) B 137 (450)	A 216 (710) B 239 (785)	A 207 (680) B 254 (835)	A 88 (290) B 104 (342)	1,098 (2,035; 1,265)	
Model 180 Skywagon floatplane	850 (1,875)	1,338 (2,950)	142 (264; 164)	133 (246; 153)	32·2 (60; 37)	332 (1,090)		5,425 (17,800)	A 245 (805) B 307 (1,006)	A 369 (1,210) B 402 (1,320)	A 254 (832) B 308 (1,010)	A 145 (475) B 171 (560)	1,063 (1,971; 1,225)	

A, B, *, **, †, †† see notes at end of Cessna table.

CESSNA MODELS—*continued*

	Weight empty equipped kg (lb)	Weight gross kg (lb)	Max level speed knots (km/h; mph)	Max cruising speed knots (km/h; mph)	Stalling speed, wheels and flaps down knots (km/h; mph)	Max rate of climb at S/L m (ft)/min	Single-engine rate of climb at S/L m (ft)/min	Service ceiling m (ft)	T-O run m (ft)	T-O to 15 m (50 ft) m (ft)	Landing from 15 m (50 ft) m (ft)	Landing run m (ft)	Max range** nm (km; miles)	Min control speed knots (km/h; mph)
Model 182 and Skylane†	725 (1,599)	1,338 (2,950)	150 (278; 173)	143 (266; 165)	33 (62; 38)	288 (945)		5,610 (18,400)	A 131 (430) B 164 (537)	A 248 (815) B 270 (885)	A 237 (777) B 280 (920)	A 99 (325) B 117 (384)	1,050 (1,947; 1,210)	
Model 182 RG	800 (1,764)	1,412 (3,112)	187 (347; 215)	173 (321; 199)	42 (80; 48)	317 (1,040)		6,100 (20,000)	A 133 (435)	A 248 (815)	A 237 (777)	A 119 (389)	1,030 (1,909; 1,186)	
Model 185 Skywagon†	721 (1,590)	1,519 (3,350)	159 (295; 183)	149 (277; 172)	34 (63; 39)	320 (1,050)		5,440 (17,850)	A 114 (375) B 143 (469)	A 233 (763) B 265 (870)	A 230 (755) B 271 (890)	A 95 (310) B 111 (365)	972 (1,802; 1,120)	
Model 185 Skywagon floatplane	866 (1,910)	1,505 (3,320)	149 (277; 172)	140 (259; 161)	34 (63; 39)	311 (1,020)		5,210 (17,100)	A 182 (596) B 227 (745)	A 332 (1,090) B 364 (1,195)	A 265 (870) B 326 (1,070)	A 146 (480) B 173 (566)	903 (1,673; 1,040)	
Model 188 Ag Wagon 230	836 (1,844)	1,723 (3,800)	124 (230; 143)	116 (214; 133)	38·5 (71; 44)	245 (805)		4,330 (14,200)	A 207 (680) B 259 (850)	A 338 (1,110) B 431 (1,420)	A 186 (610) B 256 (840)	A 94 (308) B 111 (363)	303 (563; 350)	
Model A188 Ag Wagon 300†	843 (1,859)	1,814 (4,000)	135 (251; 156)	127 (235; 146)	40 (74; 46)	302 (990)		4,905 (16,100)	A 183 (600) B 229 (750)	A 293 (960) B 381 (1,250)	A 186 (610) B 256 (840)	A 94 (308) B 111 (363)	390 (724; 450)	
Model 206 Stationair 6†	785 (1,732)	1,633 (3,600)	155 (288; 179)	148 (274; 170)	36 (66; 41)	296 (970)		4,695 (15,400)	A 147 (482) B 184 (603)	A 302 (990) B 352 (1,155)	A 226 (740) B 270 (885)	A 92 (301) B 108 (355)	916 (1,697; 1,055)	
Model 206 Stationair 6 floatplane	943 (2,080)	1,587 (3,500)	144 (267; 166)	136 (253; 157)	36 (66; 41)	276 (905)		4,450 (14,600)	A 248 (815) B 311 (1,019)	A 454 (1,490) B 486 (1,595)	A 280 (917) B 343 (1,125)	A 148 (485) B 174 (572)	833 (1,544; 960)	
Model T206 Turbo Stationair 6	831 (1,832)	1,633 (3,600)	178 (330; 205)	163 (303; 188)	36 (66; 41)	322 (1,055)		8,260 (27,100)	A 148 (485) B 185 (606)	A 303 (995) B 358 (1,175)	A 226 (740) B 270 (885)	A 92 (301) B 108 (355)	963 (1,786; 1,110)	
Model T206 Turbo Stationair 6 floatplane	979 (2,160)	1,633 (3,600)	164 (304; 189)	148 (274; 170)	36 (66; 41)	311 (1,020)		7,650 (25,100)	A 241 (790) B 301 (987)	A 442 (1,450) B 475 (1,560)	A 283 (928) B 344 (1,130)	A 151 (495) B 178 (584)	911 (1,690; 1,050)	
Model 207 Skywagon and Stationair 7	862 (1,902)	1,723 (3,800)	150 (278; 173)	142 (264; 164)	38·5 (71; 44)	262 (860)		4,205 (13,800)	A 155 (510) B 194 (637)	A 332 (1,090) B 390 (1,280)	A 244 (800) B 298 (975)	A 97 (318) B 114 (375)	829 (1,536; 955)	
Model T207 Turbo Skywagon and Turbo Stationair 7	908 (2,002)	1,723 (3,800)	168 (312; 194)	156 (290; 180)	38·5 (71; 44)	277 (910)		7,650 (25,100)	A 155 (510) B 194 (637)	A 332 (1,090) B 390 (1,280)	A 244 (800) B 298 (975)	A 97 (318) B 114 (375)	812 (1,504; 935)	
Model 210 Centurion II	953 (2,102)	1,723 (3,800)	178 (330; 205)	167 (309; 192)	38·5 (71; 44)	274 (900)		4,905 (16,100)	A 155 (510) B 194 (637)	A 326 (1,070) B 376 (1,232)	A 239 (783) B 293 (960)	A 93 (305) B 110 (360)	1,137 (2,108; 1,310)	
Model T210 Turbo Centurion II	998 (2,202)	1,723 (3,800)	204 (378; 235)	192 (356; 221)	38·5 (71; 44)	293 (960)		9,020 (29,600)	A 160 (525) B 200 (656)	A 328 (1,075) B 401 (1,318)	A 239 (783) B 293 (960)	A 93 (305) B 110 (360)	1,133 (2,100; 1,305)	
Model P210 Pressurised Turbo Centurion II	1,008 (2,222)	1,723 (3,800)	204 (378; 235)	192 (356; 221)	49 (91; 57)	293 (960)		9,020 (29,600)	200 (656)	401 (1,318)	293 (960)	110 (360)	1,133 (2,100; 1,305)	
Model P210 Pressurised Centurion	1,064 (2,345)	1,822 (4,016)	206 (381; 237)	††	55 (102; 63)	283 (930)		7,010 (23,000)	366 (1,200)	607 (1,990)	271 (890)	424 (1,390)	925 (1,714; 1,065)	
Model 310R	1,701 (3,750)	2,495 (5,500)	237 (439; 273)	223 (414; 257)	64 (119; 74)	518 (1,700)	119 (390)	8,350 (27,400)	B 290 (950)	B 448 (1,470)	B 355 (1,165)	B 219 (720)	1,242 (2,301; 1,430)	69 (129; 80) CAS
Model 337 Super Skymaster	1,196 (2,638)	2,100 (4,630)	177 (328; 204)	168 (312; 194)	39 (72·5; 45)	369 (1,210)	99 (325)	6,125 (20,100)	A 130 (428) B 163 (535)	A 265 (870) B 322 (1,055)	A 273 (895) B 323 (1,060)	A 105 (343) B 123 (405)	†† ††	
Model T337 Turbo Super Skymaster	1,289 (2,843)	2,131 (4,700)	204 (378; 235)	200 (370; 230)	40 (74; 46)	353 (1,160)	93 (305)	9,265 (30,400)	A 136 (445) B 169 (556)	A 280 (920) B 332 (1,088)	A 273 (895) B 323 (1,060)	A 107 (352) B 126 (415)	†† ††	
Model T337 Pressurised Super Skymaster	1,315 (2,900)	2,132 (4,700)	217 (402; 250)	208 (385; 239)	40 (74; 46)	353 (1,160)	126 (415)	6,100 (20,000)	A 126 (413) B 157 (516)	A 280 (920) B 332 (1,088)	A 273 (895) B 323 (1,060)	A 107 (352) B 126 (415)	†† ††	

A, B, *, **, †, ††, see notes at end of Cessna table.

CESSNA MODELS—*continued*

	Weight empty equipped kg (lb)	Weight gross kg (lb)	Max level speed knots (km/h; mph)	Max cruising speed knots (km/h; mph)	Stalling speed, wheels and flaps down knots (km/h; mph)	Max rate of climb at S/L m (ft)/min	Single-engine rate of climb at S/L m (ft)/min	Service ceiling m (ft)	T-O run m (ft)	T-O to 15 m (50 ft) m (ft)	Landing from 15 m (50 ft) m (ft)	Landing run m (ft)	Max range** nm (km; miles)	Min control speed knots (km/h; mph)
Model 340	1,878 (4,140)	2,717 (5,990)	242 (447; 278)	228 (421; 262)	61 (114; 71)	503 (1,650)	96 (315)	9,085 (29,800)					1,372 (2,542; 1,580)	75 (140; 87) CAS
Model 401	1,673 (3,690)	2,858 (6,300)	226 (420; 261)	208 (386; 240)	65·5 (121; 75)	491 (1,610)	69 (225)	7,980 (26,180)	A 240 (786) B 300 (983)	A 378 (1,240) B 472 (1,550)	A 354 (1,160) B 442 (1,450)	A 155 (510) B 183 (600)	†† ††	72 (134; 83) 72 (134; 83)
Model 402C	1,849 (4,076)	3,107 (6,850)	††	††	65 (121; 75)	442 (1,450)	91 (300)	9,145 (30,000)	329 (1,080)	533 (1,750)	427 (1,400)	259 (850)	1,220 (2,261; 1,405)	74 (137; 85)
Model 411	1,764 (3,890)	2,948 (6,500)	233 (431; 268)	212 (393; 244)	61·5 (114; 71)	579 (1,900)	98 (320)	7,925 (26,000)	A 278 (912) B 347 (1,140)	A 372 (1,220) B 465 (1,525)	A 340 (1,115) B 425 (1,395)	A 183 (600) B 276 (905)	†† ††	76 (142; 88) 76 (142; 88)
Model 414A	1,975 (4,354)	3,062 (6,750)	††	††	65 (121; 75)	480 (1,575)	88 (290)	9,145 (30,000)	329 (1,080)	533 (1,750)	405 (1,330)	238 (780)	1,300 (2,409; 1,497)	73 (135; 84)
Model 421A	1,932 (4,260)	3,102 (6,840)	240 (444; 276)	224 (414; 258)	69 (128; 79)	512 (1,680)	88 (290)	8,230 (27,000)	A 307 (1,008) B 384 (1,260)	A 443 (1,452) B 553 (1,815)	A 419 (1,375) B 524 (1,720)	A 208 (683) B 245 (804)	1,488 (2,756; 1,713) ††	83 (153; 95) 83 (153; 95)
Model 421B	2,011 (4,435)	3,379 (7,450)	245 (454; 282)	230 (426; 265)	71 (132; 82)	564 (1,850)	93 (305)	9,450 (31,000)	A 313 (1,028) B 365 (1,196)	A 428 (1,403) B 535 (1,754)	A 427 (1,400) B 534 (1,752)	A 134 (440) B 158 (517)	1,490 (2,762; 1,716) ††	78 (145; 90) 78 (145; 90)
Model 421C	2,173 (4,790)	3,379 (7,450)	256 (475; 295)	240 (444; 276)	71 (132; 82)	591 (1,940)	107 (350)	9,200 (30,200)	B 396 (1,300)	B 558 (1,830)	B 518 (1,700)	B 317 (1,040)	1,487 (2,755; 1,712)	77 (143; 89) CAS

A: Robertson STOL operation. B: Robertson Normal operation. *Available also with engines of increased horsepower. **With optional long-range tanks fitted, if available. †Leading-edge already installed by Cessna on current models. ††No change from standard aircraft.

R/STOL VERSIONS OF PIPER MODELS

	Weight empty equipped kg (lb)	Weight gross kg (lb)	Max level speed knots (km/h; mph)	Max cruising speed knots (km/h; mph)	Stalling speed, wheels and flaps down knots (km/h; mph)	Max rate of climb at S/L m (ft)/min	Single-engine rate of climb at S/L m (ft)/min	Service ceiling m (ft)	T-O run m (ft)	T-O to 15 m (50 ft) m (ft)	Landing from 15 m (50 ft) m (ft)	Landing run m (ft)	Max range** nm (km; miles)	Min control speed knots (km/h; mph)
PA-28-140 Cherokee	558 (1,232)	975 (2,150)	126 (223; 145)	120 (222; 138)	29 (53·2; 33)	206 (675)		4,480 (14,700)	A 171 (560) B 189 (620)	A 354 (1,160) B 404 (1,325)	A 192 (630) B 221 (725)	A 94 (310) B 110 (360)	†† ††	
PA-28-160 Cherokee	576 (1,270)	997 (2,200)	128 (237; 147)	121 (224; 139)	31 (56·5; 35)	216 (710)		4,970 (16,300)	A 152 (500) B 177 (580)	A 341 (1,120) B 390 (1,280)	A 204 (670) B 226 (740)	A 104 (340) B 114 (375)	†† ††	
PA-28-180 Cherokee	638 (1,406)	1,111 (2,450)	129 (238; 148)	122 (227; 141)	36 (66; 41)	221 (725)		4,313 (14,510)	A 165 (540) B 186 (610)	A 351 (1,150) B 399 (1,310)	A 238 (780) B 262 (860)	A 131 (430) B 146 (480)	596 (1,104; 686) ††	
PA-28R-180 Cherokee Arrow II	611 (1,349)	1,134 (2,500)	149 (275; 171)	142 (262; 163)	35 (64·5; 40)	270 (885)		4,695 (15,400)	A 171 (560) B 195 (640)	A 347 (1,140) B 396 (1,300)	A 259 (850) B 299 (980)	A 145 (475) B 168 (550)	911 (1,690; 1,050)	
PA-28R-200 Cherokee Arrow II	693 (1,528)	1,202 (2,650)	152 (282; 175)	143 (266; 165)	37 (69; 43)	274 (900)		4,570 (15,000)	A 175 (575) B 198 (650)	A 344 (1,130) B 393 (1,290)	A 283 (930) B 319 (1,045)	A 168 (550) B 187 (615)	782 (1,448; 900)	
PA-28-235 Cherokee 235	712 (1,570)	1,361 (3,000)	140 (259; 161)	132 (245; 152)	39 (72; 45)	244 (800)		3,660 (12,000)	A 168 (550) B 191 (625)	A 265 (870) B 302 (990)	A 274 (900) B 312 (1,025)	A 149 (490) B 171 (560)	926 (1,716; 1,066)	
PA-32-260 Cherokee SIX	783 (1,726)	1,542 (3,400)	144 (267; 166)	137 (254; 158)	38 (71; 44)	259 (850)		4,420 (14,500)	A 180 (590) B 238 (780)	A 317 (1,040) B 341 (1,120)	A 247 (810) B 267 (875)	A 155 (510) B 171 (560)	964 (1,786; 1,110)	
PA-32-300 Cherokee SIX	825 (1,819)	1,542 (3,400)	151 (280; 174)	146 (270; 168)	38 (71; 44)	320 (1,050)		4,955 (16,250)	A 171 (560) B 226 (740)	A 299 (980) B 320 (1,050)	A 247 (810) B 267 (875)	A 158 (520) B 168 (550)	921 (1,706; 1,060)	
PA-24-180 Comanche	694 (1,530)	1,157 (2,550)	149 (277; 172)	143 (266; 165)	35 (64; 40)	293 (960)		5,850 (19,200)	A 183 (600) B 302 (990)	A 324 (1,065) B475 (1,560)	A 262 (860) B 326 (1,070)	A 128 (420) B 149 (490)	868 (1,609; 1,000)	

A: Robertson STOL operation. B: Robertson Normal operation.
**With optional long-range tanks fitted, if available. ††No change from standard aircraft.

PIPER MODELS—*continued*

	Weight empty equipped kg (lb)	Weight gross kg (lb)	Max level speed knots (km/h; mph)	Max cruising speed knots (km/h; mph)	Stalling speed, wheels and flaps down knots (km/h; mph)	Max rate of climb at S/L m (ft)/min	Single-engine rate of climb at S/L m (ft)/min	Service ceiling m (ft)	T-O run m (ft)	T-O to 15 m (50 ft) m (ft)	Landing from 15 m (50 ft) m (ft)	Landing run m (ft)	Max range** nm (km; miles)	Min control speed knots (km/h; mph)
PA-24-250 Comanche	776 (1,710)	1,315 (2,900)	168 (311; 193)	161 (298; 185)	35·5 (66; 41)	427 (1,400)		6,310 (20,700)	A 187 (615) B 290 (950)	A 296 (970) B 389 (1,275)	A 256 (840) B 347 (1,140)	A 140 (460) B 213 (700)	1,537 (2,848; 1,770)	
PA-24-260 Comanche	812 (1,792)	1,451 (3,200)	174 (322; 200)	164 (304; 189)	36 (66; 41)	410 (1,345)		6,355 (20,850)	A 200 (655) B 302 (990)	A 338 (1,110) B 401 (1,315)	A 268 (880) B 360 (1,180)	A 152 (500) B 226 (740)	1,137 (2,108; 1,310)	
PA-24-260 Turbo Comanche	821 (1,810)	1,451 (3,200)	213 (394; 245)	201 (372; 231)	36 (66; 41)	410 (1,345)		7,620 (25,000)	A 200 (655) B 302 (990)	A 338 (1,110) B 401 (1,315)	A 268 (880) B 360 (1,180)	A 152 (500) B 226 (740)	1,306 (2,422; 1,505)	
PA-24-400 Comanche	966 (2,130)	1,633 (3,600)	196 (364; 226)	189 (351; 218)	39 (72; 45)	506 (1,660)		6,155 (20,200)	A 126 (415) B 168 (550)	A 233 (765) B 271 (890)	A 303 (995) B 379 (1,245)	A 184 (605) B 216 (710)	1,568 (2,905; 1,805)	
PA-30 Twin Comanche*	1,022 (2,253)	1,724 (3,800)	158 (293; 182)	153 (283; 176)	45 (84; 52)	427 (1,400)	79 (260)	6,100 (20,000)	B 206 (675)	B 341 (1,120)	B 355 (1,165)	B 186 (610)	1,481 (2,744; 1,705)	69 (129; 80)
PA-30 Turbo Twin Comanche*	1,088 (2,399)	1,724 (3,800)	213 (394; 245)	197 (365; 227)	45 (84; 52)	427 (1,400)	69 (225)	7,620 (25,000)	B 206 (675)	B 341 (1,120)	B 355 (1,165)	B 186 (610)	1,528 (2,832; 1,760)	69 (129; 80)
PA-39 Twin Comanche C/R*	1,022 (2,253)	1,724 (3,800)	181 (336; 209)	174 (322; 200)	45 (84; 52)	445 (1,460)	79 (260)	6,100 (20,000)	B 206 (675)	B 320 (1,050)	B 355 (1,165)	B 186 (610)	1,468 (2,720; 1,690)	65 (121; 75)
PA-39 Turbo Twin Comanche C/R*	1,088 (2,399)	1,724 (3,800)	212 (393; 244)	197 (365; 227)	45 (84; 52)	427 (1,400)	69 (225)	7,620 (25,000)	B 158 (520)	B 323 (1,060)	B 355 (1,165)	B 189 (620)	1,515 (2,808; 1,745)	65 (121; 75)
PA-23-235 Aztec	1,241 (2,735)	2,177 (4,800)	182 (338; 210)	175 (325; 202)	40 (74; 46)	465 (1,525)	62 (205)	5,515 (18,100)	B 210 (690)	B 331 (1,085)	B 381 (1,250)	B 195 (640)	1,090 (2,020; 1,255)	54 (98; 62)
PA-E23-250 Aztec	1,339 (2,953)	2,266 (4,995)	197 (365; 227)	191 (354; 220)	41 (76; 47)	509 (1,670)	99 (325)	6,615 (21,700)	B 190 (625)	B 315 (1,035)	B 395 (1,295)	B 203 (665)	1,112 (2,060; 1,280)	56 (103; 64)
PA-23-250 Aztec	1,326 (2,925)	2,359 (5,200)	188 (348; 216)	179 (332; 206)	45·5 (84; 52)	491 (1,610)	85 (280)	6,035 (19,800)	B 195 (640)	B 323 (1,060)	B 395 (1,295)	B 203 (665)	916 (1,697; 1,055)	56 (105; 65)
PA-23-250 Turbo Aztec	1,397 (3,080)	2,359 (5,200)	222 (412; 256)	182 (388; 210)	43 (79; 49)	372 (1,220)	64 (210)	9,145 (30,000)	B 195 (640)	B 323 (1,060)	B 395 (1,295)	B 203 (665)	1,050 (1,947; 1,210)	56 (105; 65)
PA-34 Seneca I	1,160 (2,557)	1,905 (4,200)	170 (315; 196)	162 (300; 187)	58 (106; 66)	414 (1,360)	57 (190)	5,730 (18,800)	B 195 (640)	B 320 (1,050)	B 381 (1,250)	B 196 (645)	743 (1,378; 856)	57 (106; 65)
PA-34 Seneca II	1,280 (2,823)	2,073 (4,570)	195 (361; 225)	190 (352; 219)	66 (122; 76)	421 (1,380)	67 (220)	7,620 (25,000)	B 198 (650)	B 332 (1,090)	B 573 (1,880)	B 213 (700)	882 (1,633; 1,015)†	57 (106; 65)

A: Robertson STOL operation. B: Robertson Normal operation. *Available also with engines of increased horsepower.
**With optional long-range tanks fitted, if available. †1,192 (2,207; 1,371) with optional long-range tanks.

ROBINSON

ROBINSON HELICOPTER COMPANY INC

Head Office and Works: 24747 Crenshaw Boulevard, Torrance, California 90505
Telephone: (213) 539 0508
Chairman of the Board: C. K. LeFiell
President: Franklin D. Robinson
Vice-President: C. K. LeFiell
Marketing Co-ordinator: Karen L. Walling

Robinson Helicopter Company was formed to design and manufacture a lightweight helicopter which could be competitive in price with current two/four-seat fixed-wing light aircraft. The design of this aircraft, the Robinson R22, began in June 1973 with emphasis on efficiency, low noise emission and minimum maintenance. The first prototype flew for the first time on 28 August 1975, and the second was completed in early 1977. Both were employed in the programme which led to FAA certification on 16 March 1979. United Kingdom CAA certification was gained in June 1981.

Orders for a total of just over 800 R22s had been received by 1 January 1981. Deliveries began in October 1979, and had exceeded 180 by mid-1981; the production rate was then approximately five aircraft per week. Production was expected to increase to two R22s per working day by the end of 1981.

ROBINSON MODEL R22

Type: Two-seat lightweight helicopter.

Rotor System: Two-blade semi-articulated main rotor, with a tri-hinged underslung rotor hub to reduce blade flexing, rotor vibration and control force feedback. Main rotor blade section NACA 63-015 (modified). Rigid-in-plane, and free to flap, these blades are of bonded all-metal construction, with a stainless steel spar and leading-edge, light alloy skin and light alloy honeycomb core. One fixed trim tab on main rotor blades. The two-blade tail rotor, mounted on the port side, is of light alloy bonded construction.

Rotor Drive: V-belt drive with sprag-type overrunning clutch. The main and tail gearboxes each utilise spiral bevel gears. Maintenance-free flexible couplings of proprietary manufacture are used in both the main and tail rotor drive systems. Main rotor/engine rpm ratio 1:5. Tail rotor/engine rpm ratio 1·28:1.

Robinson Model R22 two-seat lightweight helicopter at the Westland Heliport in London

FUSELAGE: Welded steel tube and light alloy primary structure for cabin, rotor pylon and engine mounting, with full monocoque tailcone. Cabin skins of light alloy and glassfibre.

TAIL UNIT: Cruciform light alloy structure with fixed horizontal stabiliser and vertical fin. Small spring skid beneath lower half of fin to give protection in a tail-down landing.

LANDING GEAR: Welded steel tube and light alloy skid landing gear, with energy-absorbing crosstubes.

POWER PLANT: One 112 kW (150 hp) Avco Lycoming O-320-A2C flat-four engine (derated to 92·5 kW; 124 hp for T-O), mounted in the lower aft section of the main fuselage, and partially exposed to improve cooling and simplify maintenance. Light alloy fuel tank in upper rear section of the fuselage on port side, usable capacity 75·5 litres (20 US gallons). Oil capacity 7·5 litres (2 US gallons).

ACCOMMODATION: Two seats side by side in enclosed cabin. Cyclic control stick mounted between seats, with dual grips on yoke so that aircraft can be flown from either seat. Conventional dual collective and throttle controls mounted at the port side of each seat. Cyclic control pivots to either side to simplify entry and exit. Curved two-panel windscreen. Door, with window, on each side. Baggage space beneath each seat. Cabin heated and ventilated.

SYSTEM: Electrical system powered by 12V DC generator.

AVIONICS AND EQUIPMENT: Optional avionics include a King KY 197 com transceiver, KN 53 nav receiver, KT 76A transponder and KR 87 ADF. Standard equipment includes sensitive altimeter, low rpm warning horn, temperature and chip warning lights for main gearbox and chip warning light for tail gearbox, soundproofing, anti-collision light and ground handling wheels. Optional equipment includes cabin heater and lighting package.

DIMENSIONS, EXTERNAL:

Diameter of main rotor	7·67 m (25 ft 2 in)
Diameter of tail rotor	1·07 m (3 ft 6 in)
Main rotor blade chord	0·18 m (7·2 in)
Distance between rotor centres	4·39 m (14 ft 5 in)
Length overall (rotors turning)	8·76 m (28 ft 9 in)
Length of fuselage	6·30 m (20 ft 8 in)
Fuselage: Max width	1·12 m (3 ft 8 in)
Height overall	2·67 m (8 ft 9 in)
Skid track	1·93 m (6 ft 4 in)

DIMENSION, INTERNAL:

Cabin: Max width	1·12 m (3 ft 8 in)

AREAS:

Main rotor blades (each)	0·70 m² (7·55 sq ft)
Tail rotor blades (each)	0·037 m² (0·40 sq ft)
Main rotor disc	46·21 m² (497·4 sq ft)
Tail rotor disc	0·89 m² (9·63 sq ft)
Fin	0·21 m² (2·28 sq ft)
Stabiliser	0·14 m² (1·53 sq ft)

WEIGHTS AND LOADINGS:

Weight empty	361 kg (796 lb)
Fuel load	54·5 kg (120 lb)
Max T-O and landing weight	590 kg (1,300 lb)
Max zero-fuel weight	535 kg (1,180 lb)
Max disc loading	12·77 kg/m² (2·61 lb/sq ft)
Max power loading	6·38 kg/kW (10·48 lb/hp)

PERFORMANCE (at max T-O weight):

Never-exceed speed	102 knots (190 km/h; 118 mph)
Max level speed	97 knots (180 km/h; 112 mph)
Cruising speed, 75% power at 2,440 m (8,000 ft)	94 knots (174 km/h; 108 mph)
Econ cruising speed	82 knots (153 km/h; 95 mph)
Max rate of climb at S/L	366 m (1,200 ft)/min
Rate of climb at 1,525 m (5,000 ft)	323 m (1,060 ft)/min
Service ceiling	4,265 m (14,000 ft)
Hovering ceiling IGE	1,980 m (6,500 ft)
Hovering ceiling OGE	1,370 m (4,500 ft)
Range with max fuel and max payload, no reserves	208 nm (386 km; 240 miles)
Endurance at 65% power, no reserves	2 h 30 min

ROCKWELL INTERNATIONAL

ROCKWELL INTERNATIONAL CORPORATION

CORPORATE HEADQUARTERS: 600 Grant Street, Pittsburgh, Pennsylvania 15219

Telephone: (412) 565 2000

CORPORATE OFFICES: 2230 East Imperial Highway, El Segundo, California 90245

CHAIRMAN OF THE BOARD AND CHIEF EXECUTIVE OFFICER: Robert Anderson

PRESIDENT AND CHIEF OPERATING OFFICER: Donald R. Beall

CORPORATE VICE-PRESIDENTS:
Robert A. De Palma (Finance and Chief Financial Officer)
J. A. Earley (Corporate Development)
Charles Fazio (President, General Industries Operations)
Bastian J. Hello (President, Aircraft Operations)
George W. Jeffs (President, North American Aerospace Operations)
A. B. Kight (International)
Edward A. Loeser (Operations)
William L. Neely (Treasurer)
Carl J. Oles (Personnel)
Samuel Petok (Communications)
John J. Roscia (General Counsel)
C. E. Ryker (Controller)
W. F. Swanson Jr (Secretary)
Martin D. Walker (Senior V-P and President, Automotive Operations)
Donald J. Yockey (President, Electronics Operations)

North American Aviation Inc, incorporated in Delaware in 1928 and a manufacturer of aircraft of various kinds from 1934, and Rockwell-Standard Corporation of Pittsburgh, Pennsylvania, a manufacturer of automotive components and builder of the Aero Commander line of civilian aircraft, merged on 22 September 1967 to form North American Rockwell Corporation.

During 1971 the Corporation was reorganised into four principal parts: the North American Aerospace Group (formerly the Aerospace and Systems Office); the Industrial Products Group (formerly the Commercial Products Group); the Automotive Group; and the Electronics Group. There is, in addition, a further component known as the Utility and Consumer Products Group. The constitution of the Corporation was changed on 16 February 1973, when North American Rockwell and Rockwell Manufacturing Company merged to become Rockwell International Corporation.

In September 1978, as the culmination of a succession of organisational changes, the company's aerospace activities were consolidated into one organisation, North American Aerospace Operations, which currently comprises Aircraft Operations, Space Transportation and Systems Group, and Rocketdyne Division.

NORTH AMERICAN AEROSPACE OPERATIONS

EXECUTIVE OFFICES: 2230 East Imperial Highway, El Segundo, California 90245
Telephone: (213) 647 5000
PRESIDENT: George W. Jeffs

Rocketdyne Division
6633 Canoga Avenue, Canoga Park, California 91304
Telephone: (213) 334 4000
PRESIDENT: Norman J. Ryker

AIRCRAFT OPERATIONS

2230 East Imperial Highway, El Segundo, California 90245
Telephone: (213) 647 5000
PRESIDENT: Bastian J. Hello
EXECUTIVE VICE-PRESIDENT, AND PROGRAMME MANAGER, B-1: Sam F. Iacobellis
VICE-PRESIDENT, COMMERCIAL AIRCRAFT: James J. Edwards

North American Aircraft Division
Los Angeles Plant:
815 Lapham Street, El Segundo, California 90009
Telephone: (213) 647 1000
Columbus Plant:
4300 East Fifth Avenue, Columbus, Ohio 43216
Telephone: (614) 239 3344
VICE-PRESIDENT: Bastian J. Hello (Acting)

Sabreliner Division
6161 Aviation Drive, St Louis, Missouri 63134
Telephone: (314) 731 2260
PRESIDENT: Alan B. Kehlet

Tulsa Division
2000 North Memorial Drive, Tulsa, Oklahoma 74151
Telephone: (918) 835 3111
PRESIDENT: E. H. Ricketts

SPACE TRANSPORTATION AND SYSTEMS GROUP

EXECUTIVE OFFICES: 12214 Lakewood Boulevard, Downey, California 90241
Telephone: (213) 922 2111
PRESIDENT: George W. Jeffs

Shuttle Orbiter and Integration Division
12214 Lakewood Boulevard, Downey, California 90241
Telephone: (213) 922 2111
PRESIDENT: Charles H. Feltz

Space Operations and Satellite Systems Division
12214 Lakewood Boulevard, Downey, California 90241
Telephone: (213) 922 2111
VICE-PRESIDENT: William E. Dean

AIRCRAFT OPERATIONS

On 3 February 1981 it was announced that Gulfstream American Corporation (which see) had acquired Rockwell International's General Aviation Division. Current aircraft products of Aircraft Operations are as follows:

ROCKWELL INTERNATIONAL BUCKEYE

USN designation: T-2

After a design competition among several leading US manufacturers, what was then North American's Columbus Division was awarded a contract in 1956 to develop and build a jet training aircraft for the US Navy. The first T2J-1 flew on 31 January 1958. Five versions of the aircraft were produced subsequently, and have been described in the 1979-80 and earlier editions of *Jane's*.

Manufacture of the T-2 Buckeye ended after 609 production aircraft had been built, but tooling was retained in anticipation of additional foreign sales. A modified version, known as the T-2X, is a candidate in the US Navy's VTX/TS competition.

ROCKWELL INTERNATIONAL BRONCO

US military designation: OV-10

The development history of the Bronco and a detailed structural description can be found in the 1978-79 and earlier editions of *Jane's*.

The following production versions were built:

OV-10A. Initial production version, first flown on 6 August 1967. Used by US Marine Corps for light armed reconnaissance, helicopter escort and forward air control duties, and by US Air Force in the forward air control role, and for limited quick-response ground support pending the arrival of tactical fighters. Total of 271 built; six transferred to Royal Moroccan Air Force.

OV-10B. Generally similar to the OV-10A; six supplied to Federal German government for target towing.

OV-10B(Z). Structurally similar to OV-10B. Provision for General Electric J85-GE-4 turbojet of 13·12 kN (2,950 lb st) to be mounted above wing, to increase performance for target towing. Twelve delivered to Federal German government.

OV-10C. Version of OV-10A for Royal Thai Air Force; 32 delivered. A new contract valued at $58 million for the supply of an additional eight OV-10Cs for the Royal Thai Air Force was announced in late 1980.

OV-10D. In 1974, Rockwell received a US Navy contract to establish and test a production OV-10D configuration. This led to delivery of 17 US Marine Corps OV-10As to Rockwell's Columbus Aircraft Division, beginning in the Spring of 1978, for conversion to the Night Observation Surveillance (NOS) role. In addition to the NOS systems and the retention of basic OV-10A fuselage stores and external fuel capability, the OV-10D NOS has uprated (775·5 kW; 1,040 shp) engines, wing pylons capable of carrying rocket pods, flare pods, free-fall stores and 378 litre (100 US gallon) external fuel tanks when extended radius/loiter time is required. A Texas Instruments AN/AAS-37 FLIR sensor and laser target designator is installed in a rotating ball turret in the nose. The sensor turret can be linked to a turret-mounted General Electric M97 20 mm cannon, mounted beneath the fuselage, in lieu of normal operation with standard OV-10A armament sponsons and centreline station.

OV-10E. Version of OV-10A for Venezuelan Air Force. Sixteen delivered.

OV-10F. Version of OV-10A for Indonesian Air Force. Sixteen delivered.

An unconfirmed report in late 1980 suggested that the Philippine Air Force was seeking approval to purchase 18 Broncos under a contract valued at $113 million.

The following abbreviated description applies to the standard OV-10A, except where stated:

TYPE: Two-seat multi-purpose counter-insurgency aircraft.

POWER PLANT: Two 533 kW (715 ehp) Garrett T76-G-416/417 turboprop engines, each driving a Hamilton Standard three-blade constant-speed reversible-pitch and fully-feathering metal propeller. (OV-10D has 775·5 kW; 1,040 shp T76-G-420/421 engines, each driving a similar Hamilton Standard propeller but with glassfibre blades.) Five self-sealing bladder-type fuel tanks in wings, with combined capacity of 954 litres (252 US gallons). Gravity refuelling point above each tank on wing upper surface. Provision for carrying one 568 litre (150 US gallon) drop-tank on underfuselage pylon; OV-10D also has provisions for carrying one 378 litre (100 US gallon) drop-tank on each wing pylon. Oil capacity 11·5 litres (3 US gallons).

ACCOMMODATION: Crew of two in tandem, on LW-3B zero-zero ejection seats, under canopy with two large upward-opening transparent door panels on each side. Dual controls optional. Cargo compartment aft of rear seat, with rear-loading door at end of fuselage pod. Rear seat removable to provide increased space for up to 1,452 kg (3,200 lb) of freight, or for carriage of five paratroops, or two stretcher patients and attendant.

SYSTEMS: Heating and ventilation system combines engine bleed air and cold ram air to provide temperature controlled conditions. Engine bleed air is used also for windscreen defrosting and to supply crew's anti-*g* suits. Hydraulic system of intermittent-duty type powered by an electrically-driven hydraulic pump at a pressure of 103·5 bars (1,500 lb/sq in), for actuation of trailing-edge flaps, landing gear and nosewheel steering. Wheel brakes, which have two independent manually-driven brake units, are fed directly from the hydraulic system reservoir. Electrical system powered by two 30V 300A starter/generators and two 24V 22Ah nickel-cadmium batteries. (OV-10D has two 30V 400A starter/generators and two 24V 30Ah aircooled batteries.) AC power derived from two 750VA inverters which supply 115V at 400Hz three-phase; single-phase AC of 115V or 26V at 400Hz can be tapped from the bus system. (An additional 3,000VA inverter is installed in the OV-10D.) External power sockets for engine starting and utility services; the latter can be used to provide 28V DC to other aircraft for engine starting or servicing. Demand-regulated oxygen system supplied from two 0·008 m³ (0·3 cu ft) oxygen cylinders at a pressure of 124 bars (1,800 lb/sq in). Independent fire warning system for each engine, comprising control unit, sensing elements and warning lights. US Air Force aircraft (only) have an electrically-fired fire extinguishing system installed in each engine nacelle. No pneumatic system.

AVIONICS: US Air Force OV-10As are equipped with AN/AIC-18 intercom; AN/ARC-51BX UHF, Wilcox 807A VHF, dual FM-622A VHF, and HF-103 HF com radios; nav system includes AN/ASN-75 compass, AN/ARN-52(V) Tacan, AN/ARA-50 UHF-ADF, AN/ARN-83 LF-ADF, 51R-6 VOR, and 51V-4A ILS glideslope; identification system includes AN/APX-64(V) IFF/SIF, and SST-181-X radar beacon. US Marine Corps aircraft are equipped with AN/AIC-18 intercom; AN/ARC-51AX UHF, AN/ARC-54 VHF, and AN/ARC-120 HF com radios; nav system includes AN/ASN-75 compass, AN/APN-171 radar altimeter, AN/ARN-52(V) Tacan, and AN/ARA-50 UHF-ADF; AN/APX-64(V) IFF/SIF for identification. (OV-10D NOS aircraft have an AN/AAS-37 FLIR sensor system package, comprising FLIR, a laser target designator, and an automatic video tracker.)

ARMAMENT: Four weapon attachment points, each with capacity of 272 kg (600 lb), under short sponsons extending from bottom of fuselage on each side, under wings. Fifth attachment point, capacity 544 kg (1,200 lb) under centre fuselage. Two 7·62 mm M60C machine-guns, each with 500 rounds of ammunition, carried in each sponson. USMC OV-10A has provision also for carrying one AIM-9D Sidewinder missile under each wing. Stores which can be carried on the under-fuselage and sponson stations include Mk 81, 82 and 83 GP bombs, Mk 81 and 82 GP (Snakeye) bombs; Mk 77 Mod 2 and Mod 4 fire bombs; LAU-3/A, LAU-10/A, LAU-32/A, LAU-59/A, LAU-60/A, LAU-61/A, LAU-68/A and LAU-69/A rocket packages; SUU-11A/A (7·62 mm Minigun), Mk 4 Mod 0 (20 mm), and GPU-2/A (20 mm) gun pods; SUU-40/A and SUU-44/A with Mk 24 and Mk 45 flares; Mk 12 Mod 0 (Podeye) smoke tank; Mk 86A/A37B-3 MBR with Mk 76 and Mk 106 practice bombs; CBU-55/B cluster bomb. Max weapon load on fuselage stations 1,633 kg (3,600 lb). (OV-10D aircraft have wing pylons with stores capacity of 272 kg; 600 lb each, which can carry CBU-55/B cluster bombs; LAU-10/A, LAU-68/A and LAU-69/A rocket packages; SUU-40/A and SUU-44/A with Mk 24 and Mk 45 flares. In lieu of sponsons and fuselage centreline load, the OV-10D can accommodate a 20 mm gun turret kit installed on centreline station hardpoint.)

DIMENSIONS, EXTERNAL:

Wing span	12·19 m (40 ft 0 in)
Length overall: OV-10A	12·67 m (41 ft 7 in)
OV-10D	13·41 m (44 ft 0 in)
Height overall	4·62 m (15 ft 2 in)
Tailplane span	4·45 m (14 ft 7 in)
Wheel track	4·52 m (14 ft 10 in)
Wheelbase	3·56 m (11 ft 8 in)
Propeller diameter	2·59 m (8 ft 6 in)
Rear loading door: Height	0·99 m (3 ft 3 in)
Width	0·76 m (2 ft 6 in)

AREA:

Wings, gross	27·03 m² (291 sq ft)

WEIGHTS AND LOADING:

Weight empty	3,127 kg (6,893 lb)
Normal T-O weight	4,494 kg (9,908 lb)
Overload T-O weight	6,552 kg (14,444 lb)
Max wing loading	242·4 kg/m² (49·6 lb/sq ft)

PERFORMANCE (A: OV-10A/C/E/F; B: OV-10D; at weights stated):

Max level speed at S/L, without weapons:	
A	244 knots (452 km/h; 281 mph)
B	250 knots (463 km/h; 288 mph)
Max rate of climb at S/L at normal T-O weight:	
A	790 m (2,600 ft)/min
B	920 m (3,020 ft)/min
Rate of climb at S/L, one engine out, without weapons:	
A	58 m (190 ft)/min
B	168 m (550 ft)/min
Service ceiling at normal T-O weight:	
A	7,315 m (24,000 ft)
B	9,150 m (30,000 ft)
Service ceiling, one engine out, without weapons:	
A	3,505 m (11,500 ft)
B	3,810 m (12,500 ft)
T-O run at normal T-O weight	226 m (740 ft)
T-O to 15 m (50 ft):	
at normal T-O weight	341 m (1,120 ft)
at max T-O weight	853 m (2,800 ft)
Landing from 15 m (50 ft) at normal T-O weight	372 m (1,220 ft)
Landing run:	
at normal T-O weight	226 m (740 ft)
at max T-O weight	381 m (1,250 ft)
Combat radius with max weapon load, no loiter	198 nm (367 km; 228 miles)
Ferry range with auxiliary fuel	1,200 nm (2,224 km; 1,382 miles)

Rockwell International OV-10D, with production-configuration Night Observation Surveillance (NOS) equipment

ROCKWELL INTERNATIONAL B-1

The B-1 was the outcome of a succession of defence studies, begun in 1962 and leading to the AMSA (Advanced Manned Strategic Aircraft) requirement of 1965, for a low-altitude penetration bomber to replace the Boeing B-52s of USAF Strategic Air Command by 1980. It was to be the third and most flexible component of the US Triad defence system, comprising also land-based and submarine-launched ballistic missiles.

Research, development, test and evaluation contracts were awarded on 5 June 1970 to North American Rockwell's Los Angeles Division for the airframe and to General Electric for the F101 turbofan engine. The original contracts were for five flying prototypes, two structural test airframes and 40 engines; in January 1971, these quantities were reduced to three flight test aircraft, one ground test aircraft and 27 engines. Procurement of a fourth flight test aircraft, as a pre-production prototype, was approved under the FY 1976 budget. The US Air Force hoped to order 244 B-1s, including prototypes, to replace B-52s now in service.

The first B-1 prototype made its initial flight at Palmdale, California, on 23 December 1974. This occasion was also the first flight of the YF101 engine. The third B-1, used as a testbed for the avionics systems, made its first flight on 1 April 1976, and was followed by the first flight of the second B-1 on 14 June 1976. The fourth B-1 (76-0174), which flew for the first time on 14 February 1979, represented an operational configuration, with both offensive and defensive avionics systems installed.

The first prototype, after completing its test programme in 1978, was placed in storage. The second prototype completed its testing during 1979 and has been stored in flyable condition. A total of 687 h 48 min flying, in 139 flights, was achieved by these two prototypes.

The third prototype was modified by the addition of an advanced ECM system, and with a Doppler beam sharpening modification to the forward-looking attack radar. Continued testing of the third and fourth B-1s was concentrated on defensive system performance and advanced ECM development. Testing was carried out against simulated enemy threats, defence systems, and against US surrogate threats. By 30 April 1981, when the authorised test programme ended, the third prototype had made 138 flights, totalling 829 h 24 min in the air; the fourth B-1 had accumulated 378 h flying in 70 flights.

In 1978, B-1 derivative designs were included in US Department of Defense studies to evaluate various types of aircraft as cruise missile carriers. In November 1979, as a result of these studies, Rockwell was requested by the US Air Force to submit a proposal for the initial planning and design effort associated with flight demonstration of a prototype B-1 derivative aircraft. Identified as a strategic

Fourth Rockwell International B-1 strategic bomber, with wings in fully-forward position for take-off

ALCM launcher (SAL), it would have been produced by modification of the third B-1 prototype.

It was reported in May 1981 that authorisation for a new bomber aircraft based on the B-1, or a version of the B-1, had been recommended by the House Armed Services Committee, and that this might be developed to meet an IOC by 1 July 1987. In September 1981, President Reagan announced that the USAF would receive 100 **B-1Bs**, with max T-O weight of 216,360 kg (477,000 lb), wing sweep limited to about 59° 30′, ability to carry 30 ALCMs, uprated avionics, and some stealth qualities to reduce the likelihood of radar detection and successful attack by air defence weapons.

A full description of the B-1 prototypes can be found in the 1977-78 *Jane's*.

Dimensions, external:
Wing span: fully spread 41·67 m (136 ft 8½ in)
fully swept 23·84 m (78 ft 2½ in)
Wing area, gross approx 181·2 m^2 (1,950 sq ft)
Length overall: incl nose probe 45·78 m (150 ft 2½ in)
excl nose probe 44·70 m (146 ft 8 in)
Height overall 10·24 m (33 ft 7¼ in)
Wheel track (c/l of shock-absorbers) 4·42 m (14 ft 6 in)
Wheelbase 17·53 m (57 ft 6 in)

Weights:
Design max T-O weight 179,170 kg (395,000 lb)
Max landing weight approx 158,755 kg (350,000 lb)

Performance (estimated, with VG inlets):
*Max level speed at 15,240 m (50,000 ft) approx Mach 2·2 (1,260 knots; 2,335 km/h; 1,451 mph)
Max level speed at 152 m (500 ft) approx 650 knots (1,205 km/h; 750 mph)
Cruising speed at 15,240 m (50,000 ft) Mach 0·85 (562 knots; 1,042 km/h; 648 mph)
Max range without refuelling Intercontinental

**Highest speed reached in test programme, by second prototype on 5 October 1978, was Mach 2·22*

ROCKWELL INTERNATIONAL XFV-12A

The US Navy initiated the XFV-12A V/STOL Fighter/Attack Technology Prototype programme to develop the capability of V/STOL operation from comparatively small carrier decks that would have neither catapult nor arrester gear.

The single-seat prototype aircraft, roughly the size of a McDonnell Douglas A-4 Skyhawk, employs an augmentor-wing concept with forward canard and aft semi-delta wings, and is powered by a single, special version of the Pratt & Whitney F401-PW-400 advanced-technology turbofan engine.

Funding limitations continue to constrain severely the progress of this programme, a full description of which can be found in the 1979-80 *Jane's*.

SABRELINER DIVISION

Executive Offices: 6161 Aviation Drive, St Louis, Missouri 63134
Telephone: (314) 731 2260
Telex: 44-7227
President: Alan B. Kehlet
Executives:
E. J. Brandreth Jr (Vice-President, Marketing and Customer Service)
C. Castillo (Manager, Public Relations)
D. Denison (Manager, Pre-Owned Aircraft Sales)
H. L. Humble (Manager, Western Regional Sales)
A. J. Marcus (Manager, Marketing Services)
J. Medeiros (Manager, Sales Administration)
P. Picciano (Manager, Customer Support)
R. H. Reid (Vice-President, Finance, and Controller)
T. Reilly (Manager, Eastern Regional Sales)
F. Smith (Director, US Government Sales)
P. Wickham (Chief Engineer)
Carl G. Ziegler (Manager, International Sales)

On 4 March 1977 Rockwell International announced the intention to consolidate Sabreliner activities in St Louis, Missouri. Structures and components continued to be manufactured in Los Angeles and were shipped to Perryville Municipal Airport, Missouri, for final assembly and flight test.

Delivery of the 500th Sabreliner, a Model 75A, was made in July 1977. Production of this version, and of the Sabreliner 60, was completed in Spring 1979, after which manufacture concentrated on the Sabreliner 65.

ROCKWELL SABRELINER 40R

Plans for a remanufacturing programme for the Sabreliner 40 (see 1979-80 *Jane's*) were announced in mid-1981. This provides for extending the estimated airframe life to 30,000 h (or 30,000 missions); Pratt & Whitney JT12A-8 turbojet engines, with thrust reversers and no time since overhaul; an 11·33 m^3 (400 cu ft) cabin, with new interior similar to that of the Sabreliner 65; advanced avionics; and new exterior paint.

ROCKWELL SABRELINER 65

Construction of a prototype of Rockwell's Sabreliner 65, which has the company Model number NA-265-65, began in January 1977, and this aircraft made its first flight on 29 June 1977. Construction of a pre-production aircraft began in December 1977, and this entered the certification programme in February 1979. Certification was gained on 28 November 1979, and deliveries of production aircraft began during the following month.

On a demonstration flight from San Francisco to Paris, via Boston, a Sabreliner 65 piloted by Bill Hallock and Jud Brandreth of Rockwell's Sabreliner Division set seven world records for speed over recognised courses in FAI Class C1g. These include San Francisco to Boston, a distance of 2,342 nm (4,337 km; 2,695 miles), at 464·90 knots (860·99 km/h; 535·00 mph); and from Boston to Paris, 3,052 nm (5,652 km; 3,512 miles), at 449·47 knots (832·42 km/h; 517·24 mph).

Type: Twin-turbofan business transport.

Wings: Cantilever low-wing monoplane. Supercritical wing with dihedral. Incidence 1·68° at root, —4·64° at tip. Sweepback at quarter-chord 32·79°. Integrally stiffened milled skin light alloy structure. Conventional ailerons with single spar and ribbed skins of light alloy. Electrically-operated Fowler-type trailing-edge flaps of light alloy construction. Hydraulically-operated spoilers of light alloy construction, two on upper surface of each wing, forward of flaps, are operable in flight and on the ground. Electrically-operated trim tab in port aileron. Anti-icing of wing leading-edges by engine bleed air.

Fuselage: Conventional semi-monocoque safe-life structure of light alloy.

Tail Unit: Cantilever structure of light alloy. Electrically-operated variable-incidence tailplane. Electrically-operated trim tab in rudder.

Landing Gear: Hydraulically-retractable tricycle type, with single wheel on each main unit and twin nosewheels. Main units retract outward, nose unit forward; all wheels enclosed by fairings when retracted. Oleo-pneumatic shock-absorbers in all units. Main wheels have tyres size 26 × 6·75, 16-ply rating, pressure 17·86 bars (259 lb/sq in). Nosewheel tyres size 18 × 4·4 Type VII, 10-ply rating, pressure 5·17 bars (75 lb/sq in). Nose unit steerable. Hydraulically-operated disc brakes. Parking brake. Electrically-controlled hydraulically-actuated fully-modulating anti-skid units.

Power Plant: Two Garrett TFE731-3-1D turbofan engines, each 16·46 kN (3,700 lb st), pod-mounted on sides of rear fuselage. Hydraulically-actuated target-type thrust reversers. Integral fuel tanks in wings, with capacity of 4,063 litres (1,073·4 US gallons), one aft fuselage bladder tank, capacity 606 litres (160·2 US gallons), and one forward fuselage bladder tank, beneath cabin floor, with capacity of 213 litres (56·3 US gallons), providing total fuel capacity of 4,882 litres (1,289·9 US gallons). Single-point refuelling standard. Gravity refuelling points on wing upper surface, and on starboard side of fuselage, above engine.

Accommodation: Crew of two and eight passengers in pressurised and air-conditioned cabin with a variety of seating layouts, all with galley, toilet and baggage compartment. Dual controls standard. Bottom-hinged downward-opening door with built-in airstairs forward of wing on port side. Overwing emergency exit on each side of cabin, FAA Type IV, removable from inside or outside. Baggage compartment at forward end of cabin on starboard side; baggage space in aft toilet compartment; total baggage capacity 247 kg (545 lb).

Systems: Air-conditioning and pressurisation by engine bleed air, max differential 0·61 bars (8·8 lb/sq in) to provide cabin altitude of 2,440 m (8,000 ft) to a height of 13,715 m (45,000 ft); separate ducting and temperature controls for cabin and flight deck. Hydraulic system supplied by a single electrically-driven pump, pressure 207 bars (3,000 lb/sq in). Auxiliary hydraulic accumulator for emergency use provides through separate lines a limited number of cycles for anti-skid units, wheel brakes, nosewheel steering and spoilers. Electrical system includes two 400A engine-driven generators. Oxygen system comprising one 2·10 m^3 (74 cu ft) cylinder in aft cabin, with quick-donning crew masks and automatic dropout masks for passengers. Anti-icing of wing leading-edges and engine air intakes by engine bleed air; electrical de-icing of pitots, stall-warning vanes and windscreens. APU optional for air-conditioning and electrical/hydraulic power on the ground and for electrical/hydraulic power when airborne.

Avionics: Include dual Collins VHF com, dual VHF nav with VOR/ILS, dual FD 109 flight directors, APS-80 autopilot, WXR-250 weather radar, DME-40, ADF-60A, TRD-90 transponder, and dual MC-103 compasses. Full blind-flying instrumentation is standard.

Dimensions, external:
Wing span 15·37 m (50 ft 5⅛ in)
Length overall 14·30 m (46 ft 11 in)
Height overall 4·88 m (16 ft 0 in)
Tailplane span 5·91 m (19 ft 4⅝ in)
Wheel track 2·20 m (7 ft 2½ in)
Wheelbase 4·85 m (15 ft 10¾ in)
Cabin door (port, fwd): Height 1·19 m (3 ft 11 in)
Width 0·71 m (2 ft 4 in)
Emergency exits (port, stbd, each):
Height 0·66 m (2 ft 2 in)
Width 0·51 m (1 ft 8 in)

Dimensions, internal:
Cabin (excl flight deck): Length 5·79 m (19 ft 0 in)
Max width 1·60 m (5 ft 3 in)
Max height 1·60 m (5 ft 3 in)
Volume 13·59 m^3 (480 cu ft)
Baggage compartments (fore and aft):
Volume 1·22 m^3 (43 cu ft)

Areas:
Wings, gross 35·30 m^2 (380 sq ft)
Ailerons (total) 1·53 m^2 (16·42 sq ft)
Trailing-edge flaps (total) 3·87 m^2 (41·67 sq ft)
Spoilers (total) 1·11 m^2 (11·92 sq ft)
Fin 3·86 m^2 (41·58 sq ft)
Rudder, incl tab 0·83 m^2 (8·95 sq ft)
Tailplane 8·37 m^2 (90·08 sq ft)
Elevators 1·80 m^2 (19·43 sq ft)

Weights and Loadings:
Weight empty, basic (incl 2 crew) 6,420 kg (14,154 lb)
Max T-O weight 10,886 kg (24,000 lb)
Max zero-fuel weight 7,371 kg (16,250 lb)
Max landing weight 9,868 kg (21,755 lb)
Max wing loading 308·4 kg/m^2 (63·2 lb/sq ft)
Max power loading 330·7 kg/kN (3·24 lb/lb st)

Performance (at max T-O weight, unless specified otherwise):
Max operating speed Mach 0·83 IAS
High-speed cruise Mach 0·81
Recommended cruising speed Mach 0·77
Long-range cruising speed Mach 0·73
Stalling speed, flaps up 115 knots (212 km/h; 132 mph) IAS
Stalling speed, full flap, operating weight empty, plus 4 passengers and reserve fuel 82 knots (152 km/h; 95 mph) IAS
Max rate of climb at S/L 1,051 m (3,450 ft)/min
Rate of climb at S/L, one engine out 272 m (893 ft)/min
Service ceiling 13,715 m (45,000 ft)
Service ceiling, one engine out 6,100 m (20,000 ft)
FAA T-O field length, ISA 1,628 m (5,340 ft)
FAA (FAR Pt 91) landing field length at max landing weight 957 m (3,140 ft)
Range with 907 kg (2,000 lb) payload, NBAA VFR reserves 2,413 nm (4,469 km; 2,777 miles)
Range with max fuel and 395 kg (870 lb) payload:
NBAA VFR reserves 2,800 nm (5,185 km; 3,222 miles)
NBAA IFR reserves 2,400 nm (4,447 km; 2,763 miles)

Rockwell Sabreliner 65, with supercritical wing and other aerodynamic improvements

SCHAFER

SCHAFER AIRCRAFT MODIFICATIONS INC

HEAD OFFICE AND WORKS: PO Box 547, Clifton Airport, Clifton, Texas 76634
PRESIDENT: Earl Schafer

SCHAFER COMANCHERO

Under this designation, Schafer markets a conversion of the Piper Pressurised Navajo with 462 kW (620 shp) Pratt & Whitney Aircraft of Canada PT6A-135 turboprops replacing the standard 317 kW (425 hp) Avco Lycoming TIGO-541-E1A piston engines. The company received an FAA STC for the Comanchero in January 1981. By August 1981, three production conversions had been completed, with two more under way.

PERFORMANCE (provisional):
- Max cruising speed at 8,840 m (29,000 ft) 278 knots (515 km/h; 320 mph)
- Max rate of climb at S/L 1,372 m (4,500 ft)/min
- Rate of climb at S/L, one engine out 305 m (1,000 ft)/min
- Range with max fuel, 45 min reserves 1,355 nm (2,510 km; 1,560 miles)

SCHAFER COMANCHERO 500

Under the above designation, Schafer Aircraft is carrying out a conversion of the Piper Chieftain which, basically, replaces the standard piston engines with turboprops of greater power.

Design of the conversion originated in mid-1980, and a prototype Comanchero 500 was scheduled to begin flight testing in late August 1981. Subject to the programme progressing as planned, certification under SFAR 23 was expected by the end of 1981. Schafer Aircraft believes that the most likely market for the conversion is among commuter operators, and plans to carry out conversions of customer-supplied aircraft, the work taking approximately two months.

Conversion of the Chieftain involves replacement of its 261 kW (350 hp) Avco Lycoming TIO-540-J2BD piston engines by two 533 ekW (715 ehp) Pratt & Whitney Aircraft of Canada PT6A-27 turboprop engines, each flat rated to 431 ekW (578 ehp). Optionally, customers will be able to have lower-cost 410 kW (550 shp) PT6A-20s installed if they do not require the hot-day/high-altitude performance offered by the flat rated PT6A-27s. In addition to installation of the turboprop engines, the conversion adds a 132 litre (35 US gallon) supplementary fuel tank in each engine nacelle, and includes as standard an inspection of the airframe and replacement, if necessary, of such items as control surface and system bearings and bushings, landing gear bushings, and hydraulic components. Options include special interiors, avionics to customers' requirements, a detachable underfuselage cargo pod with capacity of 0·44 m³ (15·5 cu ft), and 341 litre (90 US gallon) supplementary nacelle tanks in lieu of the standard installation.

Schafer performance estimates for the Comanchero 500 are as follows:

PERFORMANCE (estimated):
- Max cruising speed at 3,660 m (12,000 ft) 240 knots (445 km/h; 276 mph)
- Max rate of climb at S/L 732 m (2,400 ft)/min
- Rate of climb at S/L, one engine out 259 m (850 ft)/min
- Service ceiling, one engine out 4,725 m (15,500 ft)
- Range with pilot and ten passengers, 13·6 kg (30 lb) baggage allowance per passenger, and 45 min reserves, with standard fuel 434 nm (805 km; 500 miles)
- Range with pilot and 181 kg (400 lb) payload 1,300 nm (2,410 km; 1,500 miles)

SCHAFER COMANCHERO 750

It is reported that, under this designation, Schafer Aircraft Modifications gained an STC during May 1981 for a re-engined Piper Cheyenne II. This modification involves replacement of the standard 462 ekW (620 ehp) Pratt & Whitney Aircraft of Canada PT6A-28 turboprops by PT6A-135 turboprops from the same manufacturer, which are flat rated to 559 kW (750 shp). No other details of this conversion have been received.

SCHAPEL

SCHAPEL AIRCRAFT COMPANY

PO Box 60039, Reno, Nevada 89506
Telephone: (702) 972 8937
PRESIDENT: Rodney E. Schapel

SCHAPEL S-525 SUPER SWAT

Believing that present-day agricultural aircraft designs are many years behind their potential, having regard to current technological advances in aerodynamics and structural materials, Schapel Aircraft Company began studies in 1977 which it was hoped would lead to the design of an advanced agricultural aircraft. By the Spring of 1979, wind tunnel testing of a model had been completed successfully. Construction of a prototype began in July 1979, and this aircraft was scheduled to make its first flight in October 1980. Since that time, however, the company has decided to give priority to two other projects, and further work on the S-525 has been deferred until late 1983. Details of this aircraft can be found in the 1980-81 *Jane's*.

Artist's impression of Schapel Model S-325 Mini-Swat advanced agricultural aircraft.

Artist's impression of the Schapel S-1080 Thunderbolt four-seat executive aircraft

SCHAPEL MODEL S-325 MINI-SWAT

The design of this agricultural aircraft was initiated in July 1980. Despite the similarity of designation to the earlier Super Swat, it is a very different aircraft, being, in configuration, a conventional mid-wing monoplane with tricycle landing gear. The unusual features of its design include the hopper, which is a major structural component of the fuselage to which the engine mounting, wings, main landing gear units, and cockpit structure are attached. In addition, the wing is of advanced aerofoil section and incorporates an innovative use of automatic camber-changing trailing-edge flaps. These are in three sections on each wing, the outer section serving as both aileron and flap. The flaps are positioned manually by the pilot to a predetermined deflection to produce the lowest drag relative to speed and load: in flight, the automatic camber-changing mechanism increases flap angle as the aircraft enters a turn, adjusting lift and drag characteristics to establish the optimum L/D ratio. Extensive use is made of advanced composite structural materials, and the aircraft can have, optionally, a newly designed spray system that utilises a hydraulic motor to give great flexibility in application techniques.

The first flight of a prototype is scheduled for August 1982.

TYPE: Single-seat agricultural aircraft.

WINGS: Cantilever mid-wing monoplane. Aerofoil section modified GAW-1. Dihedral 2°. Incidence 3° 6′. No sweepback. Three-spar structure composed largely of carbonfibre and glassfibre in an epoxy matrix. Spar caps, spar webs and ribs are of carbonfibre, skins of glassfibre reinforced by a honeycomb sandwich or hat-section stringers according to local loading. Full-span trailing-edge flaps in three sections on each wing, outer section on each side used differentially as aileron or collectively with two inboard sections as flap. Flaps constructed of glassfibre and epoxy with honeycomb sandwich. Wing leading-edge is of moulded glassfibre with polyurethane foam energy-absorbing reinforcement.

FUSELAGE: Comprises chrome-molybdenum steel tube engine support structure forward, attached to hopper which is a structural unit of the fuselage. Heavy-duty chrome-molybdenum steel tube pilot's protection structure bolted to aft end of hopper. Rear fuselage, attached to pilot's compartment structure, is a glassfibre/epoxy semi-monocoque.

TAIL UNIT: Cantilever structure of conventional configuration. Fin and tailplane are two-spar structures with glassfibre/epoxy skins stiffened by honeycomb sandwich. Elevators and rudder of similar materials. Servo and trim tabs on elevators.

LANDING GEAR: Non-retractable tricycle type with single wheel on each unit. Nose unit, attached to engine mounting structure, is of trailing link design and is steerable; it incorporates a hydraulic shock-absorber with integral coil spring. Main units, attached to base of hopper, have rubber-in-compression shock-absorption.

POWER PLANT: One 298 kW (400 hp) Avco Lycoming IO-720-B1B flat-eight engine, driving a Hartzell three-blade constant-speed propeller. One fuel tank in each wing, giving combined capacity of 378 litres (100 US gallons). Refuelling point in upper surface of each wing. Oil capacity 18 litres (4·75 US gallons).

ACCOMMODATION: Pilot only, in enclosed cabin with rearward-sliding transparent canopy. Heavy-duty roll-over structure of 4130 chrome-molybdenum steel tube, lined by insulated panels which are sealed to prevent chemical ingress. Removable glassfibre external skin panels. Accommodation can be air-conditioned and has a filtered ventilation system.

SYSTEMS: Electrical system of 24V. Air-conditioning system optional.

EQUIPMENT: Chemical hopper has volume of 1·33 m³ (46·8 cu ft) and capacity of 1,230 litres (325 US gallons). A variety of liquid and dry dispersal systems available optionally. Aerofoil-shape spray 'wings' are attached to hardpoints on wing undersurface. Specially designed optional spray system incorporates an engine-driven hydraulic pump, fluid reservoir, heat exchanger, and a hydraulic motor to drive the spraypump which has a maximum rating of 1,136 litres (300 US gallons)/min at a pressure of 0·83 bars (12 lb/sq in).

DIMENSIONS, EXTERNAL:
- Wing span 15·24 m (50 ft 0 in)
- Wing chord at root 2·12 m (6 ft 11½ in)
- Wing chord at tip 1·07 m (3 ft 6 in)
- Wing aspect ratio 9·55
- Length overall 9·21 m (30 ft 2½ in)
- Height overall 3·26 m (10 ft 8½ in)
- Tailplane span 4·08 m (13 ft 4¾ in)
- Wheel track 2·44 m (8 ft 0 in)
- Wheelbase 2·03 m (6 ft 8 in)
- Propeller diameter 2·29 m (7 ft 6 in)
- Propeller ground clearance 0·43 m (1 ft 5 in)

AREA:
- Wings, gross 25·55 m² (275·0 sq ft)

WEIGHTS AND LOADINGS (estimated):
- Weight empty, equipped, sprayer 1,153 kg (2,542 lb)
- Max fuel weight 272 kg (600 lb)
- Max T-O and landing weight 2,761 kg (6,088 lb)
- Max zero-fuel weight 2,489 kg (5,488 lb)
- Max wing loading 107·9 kg/m² (22·1 lb/sq ft)
- Max power loading 9·27 kg/kW (15·22 lb/hp)

PERFORMANCE (estimated, at max T-O weight):
- Max level speed at S/L 155 knots (287 km/h; 178 mph)
- Max cruising speed at 2,285 m (7,500 ft) 157 knots (291 km/h; 181 mph)
- Stalling speed, flaps up, power off 67 knots (124 km/h; 77·5 mph)
- Stalling speed, flaps down, power off 53 knots (98·5 km/h; 61 mph)
- Max rate of climb at S/L 184 m (605 ft)/min
- Service ceiling 4,570 m (15,000 ft)
- T-O to 15 m (50 ft) 396 m (1,298 ft)

Landing from 15 m (50 ft) 465 m (1,526 ft)
Range with max payload at 2,285 m (7,500 ft)
676 nm (1,253 km; 778 miles)

SCHAPEL MODEL S-1080 THUNDERBOLT

Schapel Aircraft's most recent project, of which the design was initiated in October 1980, is for a high-performance four-seat single-engined aircraft which has been allocated the designation S-1080 and the name Thunderbolt. Although varying in configuration from the Mini-Swat, it shares the same constructional materials and techniques, and has the full-span camber-changing trailing-edge flap system developed for that aircraft. No information regarding prototype construction and projected first flight date has been given by the company.

Type: Four-seat executive transport.

Wings: Cantilever mid-wing monoplane. Schapel laminar flow aerofoil section. Dihedral 2° 18′. Incidence 2° 48′. No sweepback. Three-spar structure of carbonfibre/glassfibre/epoxy, as described for Mini-Swat.

Fuselage: Fail-safe semi-monocoque structure of carbonfibre/glassfibre/epoxy.

Tail Unit: Cruciform cantilever structure with glassfibre/epoxy skins stiffened by honeycomb sandwich. Dorsal and ventral fins and rudders, with trim tab in dorsal rudder. Elevators have servo and trim tabs.

Landing Gear: Hydraulically retractable tricycle type, main units retracting inward and nose unit aft.

Power Plant: One 373 kW (500 shp) Pratt & Whitney Aircraft of Canada PT6A-112 turboprop engine, mounted in the rear fuselage and driving, via an extended shaft, a Hartzell three-blade constant-speed pusher propeller with spinner. Fuel in one 197 litre (52 US gallon) tank in each wing, and one 83 litre (22 US gallon) fuselage tank; total capacity 477 litres (126 US gallons). Refuelling point in upper surface of each wing.

Accommodation: Cabin seats four in two pairs. Baggage space aft of rear seats, and in nose. Clamshell-type two-piece cabin door on port side, forward of wing. The accommodation is pressurised and air-conditioned.

Systems: Air-conditioning and pressurisation system with max differential of 0·51 bars (7·4 lb/sq in). Hydraulic system for landing gear actuation and wheel brakes. Electrical system of 24V. Emergency oxygen system.

Avionics and Equipment: Nav/com system and full blind-flying instrumentation standard.

Dimensions, external:
Wing span 9·87 m (32 ft 4¾ in)
Wing chord at root 1·52 m (4 ft 11¾ in)
Wing chord at tip 0·71 m (2 ft 4 in)
Wing aspect ratio 8·75
Length overall 8·94 m (29 ft 4 in)
Height overall 2·98 m (9 ft 9½ in)
Tailplane span 3·94 m (12 ft 11 in)
Wheel track 2·50 m (8 ft 2½ in)
Wheelbase 3·86 m (12 ft 8 in)
Propeller diameter 2·13 m (7 ft 0 in)
Propeller ground clearance 0·74 m (2 ft 5 in)
Cabin door: Max width 0·83 m (2 ft 8½ in)

Dimensions, internal:
Cabin, incl flight deck and aft baggage space:
Length 2·74 m (9 ft 0 in)
Max width 1·22 m (4 ft 0 in)
Max height 1·30 m (4 ft 3 in)
Floor area 2·19 m² (23·61 sq ft)
Volume approx 2·24 m³ (79 cu ft)
Baggage compartment (aft) 0·26 m³ (9·25 cu ft)
Baggage compartment (nose) 0·44 m³ (15·63 cu ft)

Area:
Wings, gross 11·15 m² (120·0 sq ft)

Weights and Loadings (estimated):
Basic operating weight 1,020 kg (2,250 lb)
Max fuel weight 390 kg (860 lb)
Max T-O and landing weight 1,792 kg (3,950 lb)
Max zero-fuel weight 1,402 kg (3,090 lb)
Max wing loading 158·2 kg/m² (32·4 lb/sq ft)
Max power loading 4·80 kg/kW (7·9 lb/hp)

Performance (estimated, at max T-O weight):
Max level speed at 7,620 m (25,000 ft)
305 knots (565 km/h; 351 mph)
Max cruising speed at 7,620 m (25,000 ft)
300 knots (556 km/h; 345 mph)
Econ cruising speed at 10,670 m (35,000 ft)
296 knots (549 km/h; 341 mph)
Stalling speed, flaps up, power off
83 knots (154 km/h; 96 mph)
Stalling speed, flaps down, power off
60 knots (111 km/h; 69 mph)
Max rate of climb at S/L 606 m (1,989 ft)/min
Service ceiling 12,190 m (40,000 ft)
T-O to 15 m (50 ft) 431 m (1,413 ft)
Landing from 15 m (50 ft) 574 m (1,882 ft)
Range with max fuel at econ cruise power
1,456 nm (2,698 km; 1,676 miles)

SCHWEIZER

SCHWEIZER AIRCRAFT CORPORATION

Head Office and Works: PO Box 147, Elmira, New York 14902
Telephone: (607) 739 3821
Telex: 932459
President: William Schweizer
Vice-Presidents:
W. Stuart Schweizer (Manufacturing)
Paul Hardy Schweizer (Treasurer)
Lesle Schweizer (Engineering)
Donald Quigley (Production Engineering)

Schweizer Aircraft Corporation was established in 1939 by brothers Ernest and Paul Schweizer to design and construct sailplanes, becoming the leading US manufacturer in this field. From mid-1957 Schweizer built Grumman Ag-Cat agricultural aircraft under subcontract until the end of 1979, when manufacture was transferred progressively to Gulfstream American Corporation, which had acquired the former Grumman American Aviation Corporation, including all rights in the Ag-Cat. On 1 January 1981 Schweizer acquired from Gulfstream American the production rights, spare parts inventory and raw materials for the Ag-Cat line. Production tools are to be bought on a royalty basis over a seven-year period.

SCHWEIZER SUPER AG-CAT

The prototype of the original Ag-Cat agricultural biplane flew for the first time on 27 May 1957. First deliveries were made in 1959, and approximately 2,455 Ag-Cats (including 2,250 of the A and B models) had been built by 11 December 1980. Schweizer planned to have the basic Super Ag-Cat B/450 back in production in time to resume deliveries in September 1981.

When full production is re-established at Elmira, three versions of the current Super Ag-Cat will be available:

Super Ag-Cat B/450. Basic version, powered by a 335·5 kW (450 hp) Pratt & Whitney R-985 nine-cylinder radial aircooled engine, driving a Hamilton Standard two-blade constant-speed metal propeller with type 2D30 hub and AG-100-2 blades.

Super Ag-Cat C/600. As Super Ag-Cat B/450, but with 447·5 kW (600 hp) Pratt & Whitney R-1340 nine-cylinder radial aircooled engine, driving a Hamilton Standard two-blade constant-speed metal propeller with Type 12D40 hub and AG-100 blades. Larger-capacity hopper.

Turbo Ag-Cat D. As Super Ag-Cat C/600 generally, but with a Pratt & Whitney Aircraft of Canada PT6A-15AG turboprop engine. Described separately.

Turbo Ag-Cat D turboprop-powered agricultural aircraft *(J. M. G. Gradidge)*

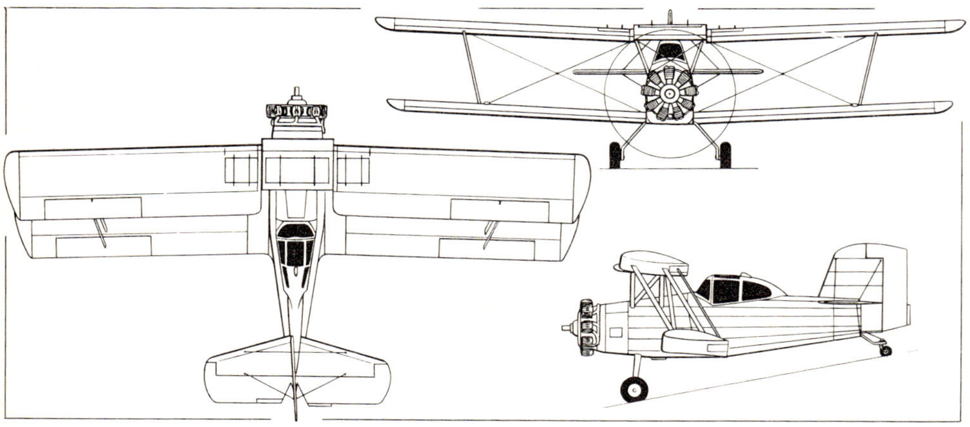

Three-view drawing of the Schweizer Super Ag-Cat B *(Pilot Press)*

Type: Single-seat agricultural biplane.

Wings: Single-bay staggered biplane. NACA 4412 (modified) wing section. Dihedral 3°. Incidence 6°. Aluminium alloy (6061-T6) two-spar structure with 6061-T6 skins on entire top surface, around leading-edge and back to front spar on undersurface. Remainder of undersurface fabric-covered. Each D leading-edge is made of five separate sections to facilitate replacement if damaged. Glassfibre wingtips. N-type interplane struts. Ailerons of light alloy construction, with fabric covering, on all four wings. Ground-adjustable tab in both ailerons of lower wing. No flaps.

Fuselage: Welded 4130 chrome-molybdenum steel tube structure, covered with duralumin sheet. Removable side panels.

Tail Unit: Welded 4130 chrome-molybdenum steel tube structure, covered with fabric and wire-braced. Cable deflector wire from tip of fin to top of cockpit canopy. Controllable trim tab in port elevator. Ground-adjustable tabs on rudder and starboard elevator.

Landing Gear: Non-retractable tailwheel type. Cantilever spring steel legs. Cleveland wheels with tyres size 8·50-10 6-ply, pressure 2·42 bars (35 lb/sq in), on the Super Ag-Cat B; size 29 × 11-10 10-ply rated at the same pressure on the Super Ag-Cat C. Steerable tailwheel with tyre size 12·4-4·5, pressure 3·45 bars (50 lb/sq in). Cleveland heavy duty aircooled disc brakes, Parking brake.

Power Plant: One Pratt & Whitney or Continental/Page nine-cylinder aircooled radial engine with Hamilton Standard constant-speed propeller, as detailed in model listings. Fuel tank in upper centre-section with standard usable capacity of 174 litres (46 US gallons). Optional tanks, installed in wings on one or both sides of centre-section, are available for total usable capacities of 241 or 302 litres (64 or 80 US gallons) on the Super Ag-Cat B; 302 litres (80 US gallons) standard on Super Ag-Cat C. Single-point refuelling on upper surface of upper wing centre-section. Oil capacity 32·2 litres (8·5 US gallons).

Accommodation: Single seat beneath enclosed cockpit canopy. Reinforced fairing aft of cockpit for turnover protection. Canopy side panels open outward and down, canopy top upward and to starboard, to provide access. Baggage compartment. Cockpit pressurised against dust ingress and ventilated by ram air. Air-conditioning optional. Safety-padded instrument panel. Air-conditioning by J.B. Systems optional.

Systems: Hydraulic system for brakes only. Optional electrical system with 24V alternator, navigation lights and/or strobe lights, external power socket, and electric engine starter.

Equipment: Radio installation optional. Standard equipment includes control column lock, instrument glareshield, seat belt and shoulder harness, tinted windscreen, stall warning light, refuelling steps and assist handles, tiedown rings, and urethane paint external yellow finish.

Agricultural Equipment: Forward of cockpit, over CG, is a 1·13 m³ (40 cu ft) glassfibre hopper, capacity (Super

Ag-Cat B) 1,136 litres (300 US gallons) of agricultural chemicals (dry or liquid) with distributor beneath fuselage. Low-volume, ULV or high-volume spray system, with leading- or trailing-edge booms. Super Ag-Cat C has 1,893 litre (500 US gallon) capacity hopper. Spray or dust distribution systems available. Emergency dump system for hopper load; can be used also for water-bomber operations.

DIMENSIONS, EXTERNAL (B: B/450; C: C/600):
Wing span 12·88 m (42 ft 3 in)
Wing chord (constant) 1·47 m (4 ft 10 in)
Wing aspect ratio: upper wing 8·74
biplane, effective mean 5·46
Length overall: B 7·90 m (25 ft 11 in)
C 9·14 m (30 ft 0 in)
Height overall: B 3·35 m (11 ft 0 in)
C 3·48 m (11 ft 5 in)
Tailplane span 3·96 m (13 ft 0 in)
Wheel track: B 2·44 m (8 ft 0 in)
C 2·54 m (8 ft 4 in)
Wheelbase: B 5·59 m (18 ft 4 in)
C 6·96 m (22 ft 10 in)
Propeller diameter (max) 2·74 m (9 ft 0 in)
Propeller ground clearance: B 0·27 m (10·8 in)
C 0·47 m (1 ft 6·7 in)

AREAS:
Wings, gross 36·42 m² (392 sq ft)
Ailerons (total) 2·93 m² (31·5 sq ft)
Fin: B 1·67 m² (17·97 sq ft)
C 0·84 m² (9·0 sq ft)
Rudder 1·12 m² (12·0 sq ft)
Tailplane 2·12 m² (22·8 sq ft)
Elevators 2·06 m² (22·2 sq ft)

WEIGHTS AND LOADINGS:
Weight empty, equipped, spray version:
B 1,404 kg (3,095 lb)
C 1,746 kg (3,850 lb)
Weight empty, equipped, duster version:
B 1,315 kg (2,900 lb)
C 1,692 kg (3,730 lb)
Max T-O weight (CAM.8):
B 2,755 kg (6,075 lb)
C (by formula) 3,855 kg (8,500 lb)
C (recommended) 3,629 kg (8,000 lb)
Max wing loading: B 75·68 kg/m² (15·50 lb/sq ft)
C 99·65 kg/m² (20·41 lb/sq ft)
Max power loading: B 8·21 kg/kW (13·5 lb/hp)
C 8·11 kg/kW (13·33 lb/hp)

PERFORMANCE (at CAM.8 max T-O weight):
Never-exceed speed:
B, C 128 knots (237 km/h; 147 mph)
Max cruising speed, 75% power at 1,525 m (5,000 ft):
B 102 knots (188 km/h; 117 mph)
Econ cruising speed, 50% power at 1,525 m (5,000 ft):
B 85 knots (158 km/h; 98 mph)
Stalling speed, power off:
B 52 knots (96 km/h; 60 mph) CAS
C 60 knots (111 km/h; 69 mph) CAS
Max rate of climb at S/L: B 323 m (1,060 ft)/min
T-O run: B 178 m (585 ft)
C 329 m (1,080 ft)
T-O to 15 m (50 ft): B 332 m (1,090 ft)
C 634 m (2,080 ft)
Landing from 15 m (50 ft): B 351 m (1,150 ft)
C 363 m (1,190 ft)
Landing run: B 172 m (565 ft)
C 180 m (590 ft)
Range with max standard fuel at 50% power:
B 216 nm (401 km; 249 miles)

MODEL G-164D TURBO AG-CAT D

This turboprop-powered version of the Ag-Cat is generally similar to the Super Ag-Cat C/600, but has a Pratt & Whitney Aircraft of Canada PT6A turboprop engine. It is available in three versions, as follows:

Turbo Ag-Cat D/T. With 507 kW (680 shp) PT6A-15 engine and 1,893 litre (500 US gallon) hopper/tank.

Turbo Ag-Cat D/ST. As D/T, but with 559 kW (750 shp) PT6A-34 engine.

Turbo Ag-Cat D/SST. As D/T, but with 634 kW (850 shp) PT6A-41 engine.

Schweizer began construction of a prototype in February 1978, and this flew for the first time on 19 July 1978. Construction of the first pre-production aircraft began in September 1978 and FAA certification under CAM 8 was received in early 1979, with production deliveries beginning shortly after. A total of 22 Turbo Ag-Cats had been built by 1 March 1981. Gulfstream American retains an unspecified number of this version which are for sale, and Schweizer intends to evaluate the potential market for the Turbo Ag-Cat before a decision is made to resume production.

The description of the Super Ag-Cat applies also to the Turbo Ag-Cat D except as follows:

WINGS AND FUSELAGE: As Super Ag-Cat.

TAIL UNIT: As for Super Ag-Cat, except ground-adjustable trim tab on rudder only.

LANDING GEAR: Non-retractable tailwheel type. Cantilever spring steel main-gear legs. Main-wheel tyres size 29 × 11-10, pressure 2·42 bars (35 lb/sq in). Steerable tailwheel with tyre size 12·5-4·5, pressure 3·80 bars (55 lb/sq in). Cleveland heavy-duty aircooled disc brakes. Parking brake.

POWER PLANT: One 507 kW (680 shp) Pratt & Whitney Aircraft of Canada PT6A-15AG turboprop engine, driving a Hartzell HC-B3TN-3D/T10282A+4 constant-speed fully-feathering reversible-pitch three-blade metal propeller with Beta control and spinner. Optional power plants as detailed in model listings. Fuel contained in upper centre-section and two wing tanks, with combined capacity of 302 litres (80 US gallons). Single-point refuelling on top surface of upper wing centre-section. Oil capacity 8·7 lires (2·3 US gallons).

ACCOMMODATION: As for Super Ag-Cat, except no optional air-conditioning system.

SYSTEMS: Hydraulic system for brakes only. Electrical system standard, with 24V 250A starter/generator and two 12V 85Ah batteries.

EQUIPMENT: Standard equipment includes sensitive altimeter, adjustable seat with seat belt and shoulder harness, control column lock, adjustable rudder pedals, tinted windscreen, stall warning light, engine oil chip detector with warning light, internal corrosion proofing, stainless steel control cables, external yellow urethane paint scheme. Optional equipment includes stall warning horn, navigation lights, strobe/navigation light system, and external power socket.

AGRICULTURAL EQUIPMENT: This equipment, which includes a variety of options, is generally similar to that described for Super Ag-Cat.

DIMENSIONS, EXTERNAL: As for Super Ag-Cat C, except:
Length overall 10·46 m (34 ft 4 in)
Propeller diameter 2·71 m (8 ft 10½ in)

DIMENSION, INTERNAL:
Max width 0·81 m (2 ft 8 in)
Max height 1·37 m (4 ft 6 in)

AREAS: As for Super Ag-Cat C

WEIGHTS AND LOADINGS (A: liquid dispersal; B: dust dispersal):
Weight empty, equipped: A 1,565 kg (3,450 lb)
B 1,520 kg (3,350 lb)
Max payload (CAM.8): A 2,291 kg (5,050 lb)
B 2,336 kg (5,150 lb)
Design T-O weight 2,857 kg (6,300 lb)
Max T-O weight (CAM.8): A, B 3,856 kg (8,500 lb)
Max wing loading 105·9 kg/m² (21·68 lb/sq ft)
Max power loading 9·85 kg/kW (16·19 lb/shp)

PERFORMANCE (D/T version with PT6A-15AG engine):
Max level speed 135 knots (249 km/h; 155 mph)
Typical working speed
82·5-117 knots (153-217 km/h; 95-135 mph)
Stalling speed at AUW of 2,857 kg (6,300 lb)
60 knots (110 km/h; 69 mph)
Stalling speed at usual landing weight
49 knots (90 km/h; 56 mph)
T-O run 223 m (730 ft)
T-O to 15 m (50 ft) 411 m (1,350 ft)
Landing from 15 m (50 ft) fully loaded
509 m (1,670 ft)
Landing from 15 m (50 ft) at usual landing weight
411 m (1,350 ft)
Landing run, fully loaded 259 m (850 ft)
Landing run at usual landing weight 150 m (490 ft)

SEAPLANES
SEAPLANES INC

PO Box 61586, Pearson Air Park, 1111 SE 5th Street, Vancouver, Washington 98666
Telephone: (206) 694 6287
PRESIDENT: L. W. Soukup

Seaplanes Inc is carrying out STOL and/or performance improvements to a variety of lightplanes manufactured by Cessna, Maule, Piper, Stinson and Waco. Performance increases result largely from replacing the original power plant by a Franklin engine of increased power, but the company has also designed a STOL kit for installation on Cessna 170B, 172 and 175 aircraft. This includes a stall fence, leading-edge cuff and aileron gap seals, which improve take-off and landing performance by approx 7 per cent.

Power plant replacements involve removal of the existing engine from Cessna 170B, 172 and 175, and Stinson 108 aircraft, and the installation of a 164 kW (220 hp) Franklin flat-six engine and constant-speed propeller. The 186 kW (250 hp) turbocharged Franklin, also with constant-speed propeller, can be installed instead in the Cessna 172, 175 and Stinson 108, and this power plant is used also for installation in the Maule M-5 Lunar Rocket

	Max level speed knots (km/h; mph)	Max cruising speed at 2,135 m (7,000 ft) knots (km/h; mph)	Max rate of climb at S/L m (ft)/min	Service ceiling m (ft)	T-O run m (ft)	Range nm (km; miles)
Cessna 170B	A 148 (274; 170)	A 137 (254; 158)	A 457 (1,500)	A 5,485 (18,000)	A 91 (300)	A 434 (804; 500)
Cessna 172	A 145 (269; 167) B 148 (274; 170)* C 126 (233; 145)	A 135 (251; 156) B 148+ (274; 170) C 117 (217; 135)**	A 427 (1,400) B 549 (1,800) C 335 (1,100)	A 5,485 (18,000) B 9,145 (30,000) C 4,875 (16,000)	A 91 (300) B 84 (275) C 320 (1,050)	A 521 (966; 600) B 521 (966; 600) C 412 (764; 475)
Cessna 175	A 145 (269; 167) B 148 (274; 170)*	A 135 (251; 156) B 148+ (274; 170)	A 427 (1,400) B 549 (1,800)	A 5,485 (18,000) B 9,145 (30,000)	A 91 (300) B 84 (275)	A 521 (966; 600) B 521 (966; 600)
Stinson 108	A 139 (258; 160) B 139 (258; 160)*	A 131 (243; 151) B 139 (258; 160)	A 488 (1,600) B 579 (1,900)	A 5,485 (18,000) B 9,145 (30,000)	A 91 (300) B 84 (275)	A 521 (966; 600) B 521 (966; 600)
Maule M-5	B 152 (282; 175)	B 142 (263; 164)	B 549 (1,800)	B 9,145 (30,000)	B 84 (275)	B 564 (1,045; 650)

A: 164 kW (220 hp) Franklin 6A-350-C2 normally aspirated engine
B: 186 kW (250 hp) Franklin 6AS-350-A1 turbocharged engine
C: Floatplane with 164 kW (220 hp) Franklin 6A-350-C2 engine
* max certificated speed
** at 1,370 m (4,500 ft)

in lieu of a 157 kW (210 hp) Continental or 164 kW (220 hp) Franklin engine. The 171·5 kW (230 hp) Continental O-470 is also available for the Stinson 108. Conversion is being carried out of the Piper Cherokee and Waco Vela to use a 164 kW (220 hp) and 186 kW (250 hp) Franklin engine respectively. During 1980 the company gained a Supplemental Type Certificate for a floatplane conversion of a Cessna 172 which had already been the subject of STOL/performance improvements and is now equipped with Edo Model 248-2440 floats. Performance figures for this aircraft, and other Seaplanes' conversions certificated by early 1981, are given in the table.

Seaplanes Inc modification of Cessna 172 with Franklin 6AS-350-A1 turbocharged engine

SERV-AERO

SERV-AERO ENGINEERING INC

Municipal Airport, Salinas, California 93901
Telephone: (408) 422 7866/7

PRESIDENT: Floyd N. Perry

Serv-Aero Engineering is involved in programmes to re-engine standard types of agricultural aircraft which are powered by Pratt & Whitney radial engines. The R-1340 radial, in particular, has been out of production since the mid-'fifties, and spare parts for this engine have become difficult to obtain.

THRUSH COMMANDER/R-1820

This conversion of the Thrush Commander involves removal of the Pratt & Whitney R-1340 or Wright R-1300 and its replacement by an 895 kW (1,200 hp) Wright R-1820-71 radial piston engine, driving a slow-turning large-diameter three-blade Hamilton Standard propeller. A Supplemental Type Certificate covering this airframe/engine combination was received on 8 October 1977, followed subsequently by Parts Manufacturing Approval covering the production of conversion kits.

It is reported that this engine installation provides a 396 m (1,300 ft) T-O run with full hopper (1,476-1,514 litres; 390-400 US gallons) at S/L in zero-wind ISA conditions. Working speed is 117-126 knots (217-233 km/h; 135-145 mph), resulting in a lower fuel consumption per acre worked. The slower-turning, quieter propeller permits operators to work closer to inhabited areas without creating noise problems.

Early operators of Thrush Commanders with the R-1820. In 1980, it was engaged in flight testing to achieve approval for a similar propeller on the quieter and smoother 16:9 series geared engines.
tion compared with that of the formerly-installed R-1340 engine.

Serv-Aero has FAA approval for a 3·05 m (10 ft) diameter propeller on the 3:2 geared version of the R-1820. In 1980, it was engaged in flight testing to achieve approval for a similar propeller on the quieter and smoother 16:9 series geared engines.

Thrush Commander with R-1820 engine and Serv-Aero's fuselage fairing and spring tailwheel modification

SERV-AERO/AYRES THRUSH MODIFICATION

This FAA-approved modification to the Ayres Thrush involves addition of a streamlined top fuselage fairing and a spring-type tailwheel leg. The fairing adds considerably to the airspeed and smooths airflow over the tail unit. By smoothing the aircraft's taxying, the new tailwheel leg reduces maintenance time, by reducing shock loads imposed on the fuselage structure, and improves directional stability on the ground. This modification is approved also for the Turbo-Thrush.

SIKORSKY

SIKORSKY AIRCRAFT, DIVISION OF UNITED TECHNOLOGIES CORPORATION

HEAD OFFICE AND WORKS: Stratford, Connecticut 06602
Telephone: (203) 386 4000

OTHER WORKS: South Avenue, Bridgeport, Connecticut; Sikorsky Memorial Airport, Stratford, Connecticut; and Development Flight Test Center, West Palm Beach, Florida

PRESIDENT AND CHIEF EXECUTIVE OFFICER:
Robert F. Daniell

EXECUTIVE VICE-PRESIDENT: William F. Paul

SENIOR VICE-PRESIDENT:
L. L. Allison (Finance)

VICE-PRESIDENTS:
John P. Balaguer (Bridgeport/Florida Manufacturing)
Eugene Buckley (Stratford Manufacturing)
J. Colin Green (Planning and Services)
Philip Locke (Contracts and Counsel)
Porter D. Lyke (Personnel and Industrial Relations)
Wayne B. Parker (Government Relations)
Allan K. Poole (Product Support—Government Programmes)
David O. Smith (Marketing and Customer Service)
Harvey I. White (Material)
Robert Zincone (Engineering)

DIRECTOR OF PUBLIC RELATIONS: Robert G. H. Carroll III

Founded on 5 March 1923 by the late Igor I. Sikorsky as the Sikorsky Aero Engineering Corporation, this company has been a division of United Technologies since 1929. Its main plant at Stratford, which has 120,775 m² (1,300,000 sq ft) of working space, produces the US Army's twin-turbine UH-60A Black Hawk helicopter. The Bridgeport facility, which has 55,740 m² (600,000 sq ft) of working space, houses the production line for the S-76 twin-turbine commercial helicopter, and is used for detail fabrication, overhaul and repair. The Development Flight Test Center is in Florida.

Sikorsky is involved currently in a number of important development programmes, including the three-engined CH-53E heavy transport helicopter for the US Navy and Marine Corps, the SH-60B Seahawk, also for the Navy, and the YEH-60B SOTAS helicopter for the US Army. Its ABC (advancing blade concept) helicopter, the S-69, has successfully completed flight tests using auxiliary propulsion for flight at speeds up to 263 knots (487 km/h; 303 mph).

Sikorsky licensees include Westland of Great Britain, Agusta of Italy, Aérospatiale of France, MBB in West Germany, Mitsubishi of Japan, and Pratt & Whitney Aircraft of Canada Ltd.

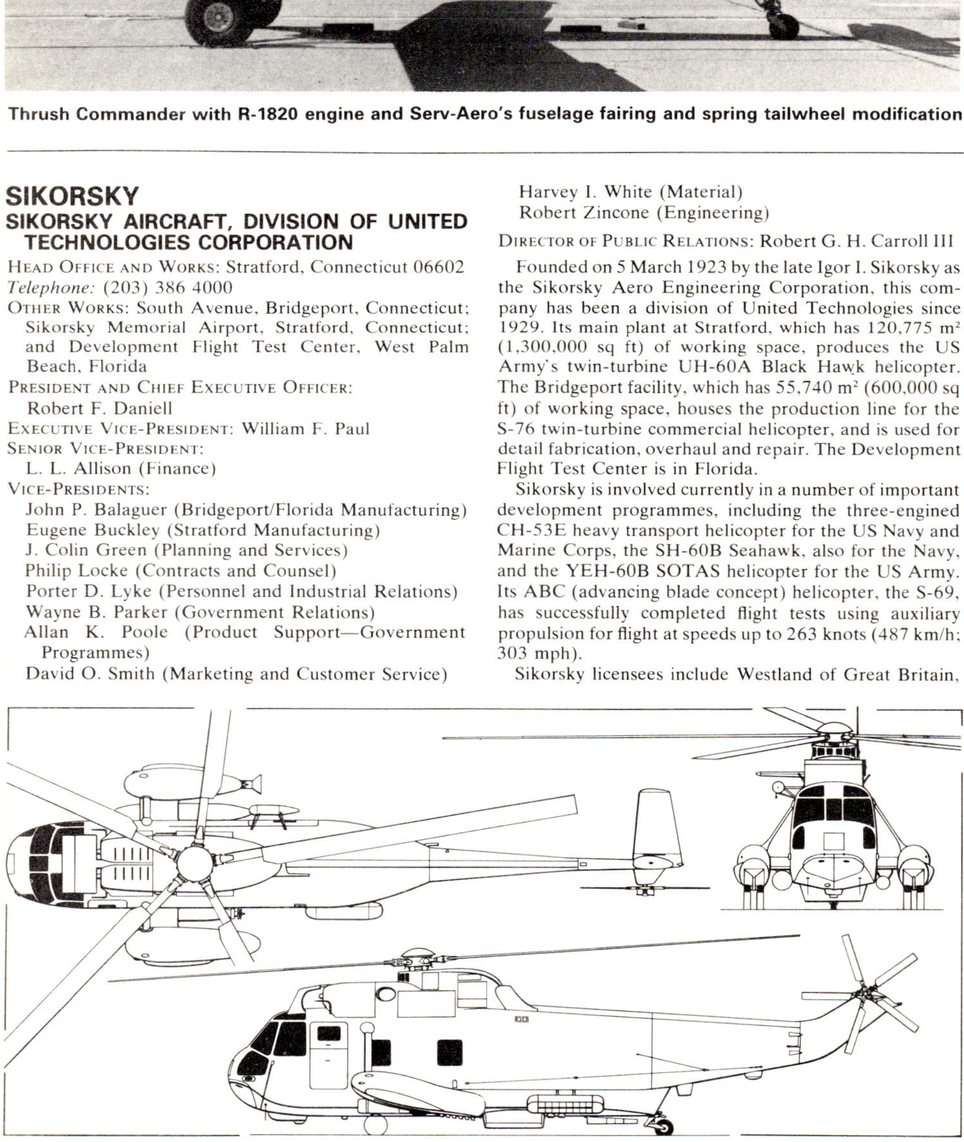

Sikorsky SH-3H twin-engined multi-purpose amphibious helicopter *(Pilot Press)*

SIKORSKY/US ARMY ACAP PROGRAMME

On 25 February 1981, it was announced that the US Army's Applied Technology Laboratory had selected Sikorsky as one of two contractors to negotiate a contract for the Army's Advanced Composite Airframe Program (ACAP). The objective is to develop, build and evaluate a helicopter fuselage made entirely of composite materials, in order to achieve a weight saving of 22 per cent, and cost saving of 17 per cent, compared with conventional metal airframes, while meeting established military requirements in terms of crashworthiness, ballistics tolerance, reliability, maintainability and reduced radar signature.

The ACAP contract involves a 42-month work programme, including fabrication of a vehicle with the composite airframe mated to the power plant and rotor system of the Sikorsky S-76. Subcontractors will include LTV Corporation, responsible for the lower tub section, and Hercules Corporation, which will fabricate the tailcone, tail pylon and stabiliser. All other fabrication and final assembly will be done by Sikorsky.

The ACAP aircraft proposed by Sikorsky will have a gross weight of 3,842 kg (8,470 lb) and accommodate six combat-equipped troops in addition to the pilot and co-pilot. Powered by twin Allison 250-C30 turboshaft engines, it will have an estimated max level speed of 140 knots (259 km/h; 161 mph) and a mission endurance of 2 h 18 min. The landing gear will be of non-retractable tricycle type. (See Addenda for further information.)

SIKORSKY S-61A and S-61B

US military designations: SH-3 Sea King, HH-3A, VH-3

CAF designation: CH-124

The first version of the S-61 ordered into production was the SH-3A (originally HSS-2) Sea King amphibious anti-submarine helicopter, of which the prototype flew for the first time on 11 March 1959. Deliveries to the US Navy began in September 1961. On 11 March 1979, Sikorsky drew attention to the 20th anniversary of the prototype's first flight, since when more than 900 military S-61s had accumulated over 3 million flight hours, and 130 commercial S-61s had logged a total of more than 815,000 hours.

The S-61 series includes the following military and commercial variants of which details can be found in the

Sikorsky SH-3H multi-purpose helicopter for ASW and fleet missile defence

1980-81 and earlier editions of *Jane's:* SH-3A Sea King, CH-124, HH-3A, VH-3D, S-61A, S-61A-4 Nuri (44 delivered), S-61D-4, S-61L, and S-61N; plus the SH-3D Sea King, SH-3G, SH-3H, and S-61R, of which details follow:

SH-3D Sea King. Standard anti-submarine helicopter of the US Navy, with T58-GE-10 engines and more fuel than SH-3A. First SH-3D, delivered in June 1966, was one of 22 for the Spanish Navy. Four were delivered to the Brazilian Navy and 72 to the US Navy. Versions with Rolls-Royce Gnome turboshaft engines and British anti-submarine equipment are manufactured by Westland Helicopters Ltd (which see). SH-3Ds have also been manufactured under licence by Agusta in Italy.

SH-3G. US Navy conversion of 105 SH-3As into utility helicopters, by removing anti-submarine warfare equipment. Six equipped with Minigun pods for search and rescue missions in combat conditions.

SH-3H. Multi-purpose version of SH-3A and SH-3G with T58-GE-10 engines. US Navy contracts, awarded from 1971, called for conversion to increase fleet helicopter capability against submarines and low-flying enemy missiles. New ASW equipment includes lightweight sonar, active and passive sonobuoys, and magnetic anomaly detection equipment. Electronic surveillance measurement (ESM) equipment enables the SH-3H to make an important contribution to the missile defence of the fleet. Also built under licence by Agusta in Italy.

S-61R. Development of S-61B for transport duties with US Air Force, under the designations **CH-3C** and **E**. Described separately.

By early 1981, more than 750 examples of the S-61 (all models) had been built by Sikorsky, and more than 350 by the company's foreign licensees.

The following details apply to the SH-3D Sea King, but are generally applicable to other versions except for accommodation and equipment:

Type: Twin-engined amphibious all-weather anti-submarine helicopter.

Rotor System: Five-blade main and tail rotors. All-metal fully-articulated oil-lubricated main rotor. Flanged cuffs on blades bolted to matching flanges on all-steel rotor head. Main rotor blades are interchangeable and are provided with an automatic powered folding system. Rotor brake standard. All-metal tail rotor.

Rotor Drive: Both engines drive through freewheel units and rotor brake to main gearbox. Steel drive-shafts. Tail rotor shaft-driven through intermediate and tail gearboxes. Accessories driven by power take-off on tail rotor shaft. Additional freewheel units between accessories and port engine, and between accessories and tail rotor shaft. Main rotor/engine rpm ratio 1 : 93·43. Tail rotor/engine rpm ratio 1 : 16·7.

Fuselage: Boat hull of all-metal semi-monocoque construction. Single step. Tail section folds to reduce stowage requirements.

Tail Surface: Fixed stabiliser on starboard side of tail section.

Landing Gear: Amphibious. Land gear consists of two twin-wheel main units, which are retracted rearward hydraulically into stabilising floats, and non-retractable tailwheel. Oleo-pneumatic shock-absorbers. Goodyear main wheels and tubeless tyres size 6·50-10 type III, pressure 4·83 bars (70 lb)sq in). Goodyear tailwheel and tyre size 6·00-6. Goodyear hydraulic disc brakes. Boat hull and pop-out flotation bags in stabilising floats permit emergency operation from water.

Power Plant: Two 1,044 kW (1,400 shp) General Electric T58-GE-10 turboshaft engines. Three bladder-type fuel tanks in hull: forward tank 1,314 litres (347 US gallons), centre tank 530 litres (140 US gallons), rear tank 1,336 litres (353 US gallons). Total fuel capacity 3,180 litres (840 US gallons). Refuelling point on port side of fuselage. Oil capacity 26·5 litres (7 US gallons).

Accommodation: Pilot and co-pilot on flight deck, two sonar operators in main cabin. Dual controls. Crew door at rear of flight deck on port side. Large loading door at rear of cabin on starboard side.

Systems: Primary and auxiliary hydraulic systems, pressure 103·5 bars (1,500 lb/sq in), for flying controls. Utility hydraulic system, pressure 207 bars (3,000 lb/sq in), for landing gear, winches and blade folding. Pneumatic system, pressure 207 bars (3,000 lb/sq in), for blow-down emergency landing gear extension. Electrical system includes one 300A DC generator, two 20kVA 115A AC generators and 24V 22Ah battery. APU optional.

Avionics and Equipment: Bendix AQS-13 sonar with 180° search beam width. Hamilton Standard autostabilisation equipment. Automatic transition into hover. Sonar coupler holds altitude automatically in conjunction with Teledyne APN-130 Doppler radar (Litton AN/APS-503 in CH-124) and radar altimeter. Provision for 272 kg (600 lb) capacity rescue hoist and 3,630 kg (8,000 lb) capacity automatic touchdown-release low-response cargo sling for external loads.

Armament: Provision for 381 kg (840 lb) of weapons, including homing torpedoes.

Dimensions, external:

Diameter of main rotor	18·90 m (62 ft 0 in)
Main rotor blade chord	0·46 m (1 ft 6¼ in)
Diameter of tail rotor	3·23 m (10 ft 7 in)
Distance between rotor centres	11·10 m (36 ft 5 in)
Length overall	22·15 m (72 ft 8 in)
Length of fuselage	16·69 m (54 ft 9 in)
Length, tail pylon folded	14·40 m (47 ft 3 in)
Width, rotors folded	4·98 m (16 ft 4 in)
Height to top of rotor hub	4·72 m (15 ft 6 in)
Height overall	5·13 m (16 ft 10 in)
Wheel track	3·96 m (13 ft 0 in)
Wheelbase	7·18 m (23 ft 6½ in)
Crew door (fwd, port): Height	1·68 m (5 ft 6 in)
Width	0·91 m (3 ft 0 in)
Height to sill	1·14 m (3 ft 9 in)
Main cabin door (stbd): Height	1·52 m (5 ft 0 in)
Width	1·73 m (5 ft 8 in)
Height to sill	1·14 m (3 ft 9 in)

Dimensions, internal (S-61A):

Cabin: Length	7·60 m (24 ft 11 in)
Max width	1·98 m (6 ft 6 in)
Max height	1·92 m (6 ft 3½ in)
Floor area	15·1 m² (162 sq ft)
Volume	28·9 m³ (1,020 cu ft)

Areas:

Main rotor blades (each)	4·14 m² (44·54 sq ft)
Tail rotor blades (each)	0·22 m² (2·38 sq ft)
Main rotor disc	280·5 m² (3,019 sq ft)
Tail rotor disc	8·20 m² (88·30 sq ft)
Stabiliser	1·86 m² (20·00 sq ft)

Weights:

Weight empty: S-61A	4,428 kg (9,763 lb)
S-61B	5,382 kg (11,865 lb)
Normal T-O weight: S-61A	9,300 kg (20,500 lb)
SH-3A (ASW)	8,185 kg (18,044 lb)
SH-3D (ASW)	8,449 kg (18,626 lb)
Max T-O weight: S-61A	9,750 kg (21,500 lb)
S-61B	9,300 kg (20,500 lb)
SH-3H	9,525 kg (21,000 lb)

Performance (at 9,300 kg; 20,500 lb AUW):

Max level speed	144 knots (267 km/h; 166 mph)
Cruising speed for max range	118 knots (219 km/h; 136 mph)
Max rate of climb at S/L	670 m (2,200 ft)/min
Service ceiling	4,480 m (14,700 ft)
Hovering ceiling IGE	3,200 m (10,500 ft)
Hovering ceiling OGE	2,500 m (8,200 ft)
Range with max fuel, 10% reserves	542 nm (1,005 km; 625 miles)

SIKORSKY S-61R

US military designations: CH-3 and HH-3 Jolly Green Giant

Although based on the SH-3A, this amphibious transport helicopter introduced many important design changes. They include provision of a hydraulically-operated rear ramp for straight-in loading of wheeled vehicles, a 907 kg (2,000 lb) capacity winch for internal cargo handling, retractable tricycle-type landing gear, pressurised rotor blades for quick and easy inspection, gas-turbine auxiliary power supply for independent field operations, self-lubricating main and tail rotors, and built-in equipment for the removal and replacement of all major components in remote areas.

The first S-61R flew on 17 June 1963, followed by the first CH-3C a few weeks later. FAA Type Approval was received on 30 December 1963, and the first delivery of an operational CH-3C was made on the same day, for drone recovery duties at Tyndall AFB, Florida. Subsequent deliveries made to USAF Aerospace Defense Command, Air Training Command, Tactical Air Command, Strategic Air Command and Aerospace Rescue and Recovery Service.

Production by Sikorsky has ended, but S-61R variants continue to be available from Agusta in Italy (which see).

There were four Sikorsky-built versions, as follows:

CH-3C. Two 969·5 kW (1,300 shp) T58-GE-1 turboshaft engines. After a total of 41 had been built for the US Air Force, production was switched to the CH-3E. All aircraft delivered as CH-3Cs were modified to CH-3E standard.

CH-3E. Designation applicable from February 1966, following introduction of uprated engines (1,118 kW; 1,500 shp T58-GE-5). A total of 42 were built as new aircraft to this standard.

HH-3E. For US Air Force Aerospace Rescue and Recovery Service. Additional equipment comprises armour, self-sealing fuel tanks, retractable flight refuelling probe, defensive armament and rescue hoist. Two 1,118 kW (1,500 shp) T58-GE-5 turboshafts. A total of 50 HH-3Es were converted from CH-3Es, and are known as **Jolly Green Giants**.

HH-3F. Similar to HH-3E, for US Coast Guard, which gave them the name **Pelican**. Advanced electronic equipment for search and rescue duties. No armour plate, armament or self-sealing tanks. First order announced in August 1965. Deliveries began in 1968 and 40 were built.

The following details apply to the CH-3E:

Type: Twin-engined amphibious transport helicopter.

Rotor System: Five-blade fully-articulated main rotor of all-metal construction. Flanged cuffs on blades bolted to matching flanges on rotor head. Control by rotating and stationary swashplates. Blades do not fold. Rotor brake standard. Conventional tail rotor with five aluminium blades.

Rotor Drive: Twin turbines drive through freewheeling units and rotor brake to main gearbox. Steel drive-shafts. Tail rotor shaft-driven through intermediate gearbox and tail gearbox. Main rotor/engine rpm ratio 1 : 93·43. Tail rotor/engine rpm ratio 1 : 16·7.

FUSELAGE: All-metal semi-monocoque structure of pod and boom type. Cabin of basic square section.

TAIL SURFACE: Horizontal stabiliser on starboard side of tail rotor pylon.

LANDING GEAR: Hydraulically-retractable tricycle type, with twin wheels on each unit. Main wheels retract forward into sponsons, each of which provides 2,176 kg (4,797 lb) of buoyancy and, with boat hull, permits amphibious operation. Oleo-pneumatic shock-absorbers. All wheels and tyres tubeless Type III rib, size 22·1 × 6·50-10, manufactured by Goodyear. Tyre pressure 6·55 bars (95 lb/sq in). Goodyear hydraulic disc brakes.

POWER PLANT: Two 1,118 kW (1,500 shp) General Electric T58-GE-5 turboshaft engines, mounted side by side above cabin, immediately forward of main transmission. Fuel in two bladder-type tanks beneath cabin floor; forward tank capacity 1,204 litres (318 US gallons), rear tank capacity 1,226 litres (324 US gallons). Total fuel capacity 2,430 litres (642 US gallons). Refuelling point on port side of fuselage. Total oil capacity 26·5 litres (7 US gallons).

ACCOMMODATION: Crew of two side by side on flight deck, with dual controls. Provision for flight engineer or attendant. Normal accommodation for 25 fully-equipped troops. Alternative arrangements for 30 troops, 15 stretchers or 2,270 kg (5,000 lb) of cargo. Jettisonable sliding door on starboard side at front of cabin. Internal door between cabin and flight deck. Hydraulically-operated rear loading ramp for vehicles, in two hinged sections, giving opening with minimum width of 1·73 m (5 ft 8 in) and headroom of up to 2·21 m (7 ft 3 in).

SYSTEMS: Primary and auxiliary hydraulic systems, pressure 103·5 bars (1,500 lb/sq in), for flying control servos. Utility hydraulic system, pressure 207 bars (3,000 lb/sq in), for landing gear, rear ramp and winches. Pneumatic system, pressure 207 bars (3,000 lb/sq in), for emergency blow-down landing gear extension. Electrical system includes 24V 22Ah battery, two 20kVA 115V AC generators and one 300A DC generator. APU standard.

DIMENSIONS, EXTERNAL:

Diameter of main rotor	18·90 m (62 ft 0 in)
Main rotor blade chord	0·46 m (1 ft 6¼ in)
Diameter of tail rotor	3·15 m (10 ft 4 in)
Distance between rotor centres	11·22 m (36 ft 10 in)
Length overall	22·25 m (73 ft 0 in)
Length of fuselage	17·45 m (57 ft 3 in)
Width over landing gear	4·82 m (15 ft 10 in)
Height to top of rotor hub	4·90 m (16 ft 1 in)
Height overall	5·51 m (18 ft 1 in)
Wheel track	4·06 m (13 ft 4 in)
Wheelbase	5·21 m (17 ft 1 in)
Cabin door (fwd, stbd): Height	1·65 m (5 ft 4¾ in)
Width	1·22 m (4 ft 0 in)
Height to sill	1·27 m (4 ft 2 in)
Rear ramp: Length	4·29 m (14 ft 1 in)
Width	1·85 m (6 ft 1 in)

DIMENSIONS, INTERNAL:

Cabin (excl flight deck):	
Length	7·89 m (25 ft 10½ in)
Max width	1·98 m (6 ft 6 in)
Max height	1·91 m (6 ft 3 in)
Floor area	approx 15·61 m² (168 sq ft)
Volume	approx 29·73 m³ (1,050 cu ft)

AREAS:

Main rotor blades (each)	3·71 m² (39·9 sq ft)
Tail rotor blades (each)	0·22 m² (2·35 sq ft)
Main rotor disc	280·5 m² (3,019 sq ft)
Tail rotor disc	7·80 m² (83·9 sq ft)
Stabiliser	2·51 m² (27·0 sq ft)

WEIGHTS:

Weight empty	6,010 kg (13,255 lb)
Normal T-O weight	9,635 kg (21,247 lb)
Max T-O weight	10,000 kg (22,050 lb)

PERFORMANCE (at normal T-O weight):

Max level speed at S/L	141 knots (261 km/h; 162 mph)
Cruising speed for max range	125 knots (232 km/h; 144 mph)
Max rate of climb at S/L	400 m (1,310 ft)/min
Service ceiling	3,385 m (11,100 ft)
Hovering ceiling IGE	1,250 m (4,100 ft)
Min ground turning radius	11·29 m (37 ft 0½ in)
Runway LCN at max T-O weight	approx 4·75
Range with max fuel, 10% reserves	404 nm (748 km; 465 miles)

SIKORSKY S-65A

US Navy designation: CH-53A Sea Stallion
US Air Force designations: HH-53B/C/H
US Marine Corps designations: CH-53A/D

Although no longer in production, this large twin-turboshaft helicopter continues in first-line service in the several versions described in the 1978-79 *Jane's*.

Eight HH-53Cs have been modified to HH-53H standard for night and adverse weather search and rescue operations under the US Air Force's Pave Low 3 programme, following evaluation of a prototype conversion which flew for the first time in June 1975.

The **Pave Low 3 HH-53H** has a stabilised Texas Instruments FLIR (forward-looking infra-red) installation mounted below the refuelling boom; a Litton inertial navigation system; a Canadian Marconi Doppler navigation system; an IBM computer; Systems Research symbol generator; and Texas Instruments terrain following/avoidance radar in an offset (to port) 'thimble' fairing on the nose.

The US Air Force accepted the first production Pave Low 3 HH-53H at Pensacola on 13 March 1979. All eight conversions were completed during 1980.

SIKORSKY SUPER STALLION

US Navy designation: CH-53E

The Sikorsky S-65 was chosen in 1973 for development with three engines to provide the US Navy and Marine Corps with a heavy-duty multi-purpose helicopter. Other changes to increase performance included installation of a new seven-blade main rotor of increased diameter, with blades of titanium/glassfibre construction, and an uprated transmission of 9,798 kW (13,140 shp) capacity to cater for future development.

Development was initiated by the award of a $1·7 million US Navy cost-plus-fixed-fee contract; in May 1973 Sikorsky announced that construction of two prototypes, designated YCH-53E, was to begin, with the first flight scheduled for April 1974. Bettering this by a month, the first YCH-53E made a successful half-hour flight on 1 March 1974. It was lost subsequently in an accident on the ground, but the programme was resumed on 24 January 1975 with the second YCH-53E. This aircraft flew subsequently at an AUW of 33,793 kg (74,500 lb), the highest gross weight achieved by any helicopter outside the USSR. It was used for preliminary evaluation and testing under Phase I of the development programme. Phase II covered the construction of a static test vehicle and two pre-production prototypes, the first of which flew on 8 December 1975. In February 1978 Sikorsky was awarded a contract to begin full-scale production, with initial approval for six aircraft. The first of these flew for the first time on 13 December 1980, and was accepted by the Navy four days later. A further 14 were ordered in FY 1979, 15 in FY 1980, and 14 in FY 1981. Of these 49 aircraft, 16 are for the US Navy and 33 for the Marine Corps. First unit to receive the CH-53E was HMH-464 (Marine Air Group 26) at New River, North Carolina, to which deliveries

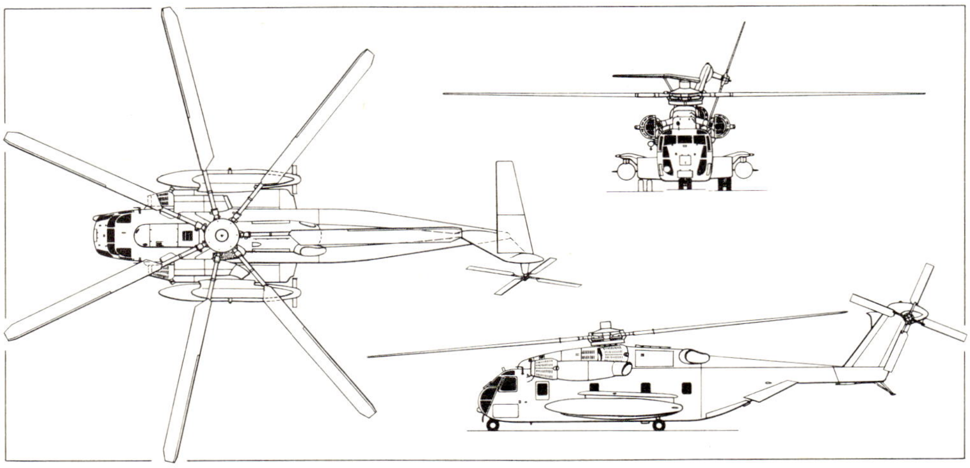

Sikorsky CH-53E Super Stallion heavy-duty helicopter *(Pilot Press)*

First production Sikorsky CH-53E Super Stallion heavy-lift helicopter (three General Electric T64-GE-416 turboshaft engines)

began on 16 June 1981. An additional 11 CH-53Es were requested in the FY 1982 budget. The eventual production total may exceed 100, including a proposed **MH-53E** minesweeping version, the equipment for which is to be fitted experimentally to the first production CH-53E for evaluation.

The US Navy plans to use the CH-53E for vertical onboard delivery operations, to support mobile construction battalions, and for the removal of battle-damaged aircraft from carrier decks. In amphibious operations, it would be able to airlift 93 per cent of a US Marine division's combat items, and would be able to retrieve 98 per cent of the Marine Corps' tactical aircraft without disassembly. Features include extended-range fuel tanks, in-flight refuelling capability, an onboard all-weather navigation system, and an advanced dual digital automatic flight control system.

The CH-53E is the largest helicopter capable of full operation from the Navy's existing and planned ships, requiring only 10 per cent more deck space than the twin-turbine H-53. It offers double the lift of the latter aircraft with an increase of only 50 per cent in engine power.

TYPE: Triple-turbine heavy-duty multi-purpose helicopter.

ROTOR SYSTEM AND TRANSMISSION: Seven-blade main rotor with blades of titanium/glassfibre construction. Titanium and steel main rotor head. Main rotor blades fold. Four-blade tail rotor mounted on pylon canted 20° to port. Rotor transmission is rated at 9,798 kW (13,140 shp) for 10 min, 8,628 kW (11,570 shp) for 30 min. Tail rotor pylon folds on starboard side of fuselage.

FUSELAGE: Conventional semi-monocoque structure of light alloy, steel and titanium.

TAIL SURFACE: Initial fixed tailplane on undersurface of fuselage, superseded successively by single high-mounted stabiliser on starboard side and lightweight gull-wing type.

LANDING GEAR: Retractable tricycle type, with twin wheels on each unit. Main units retract into rear of sponsons on each side of fuselage.

POWER PLANT: Three General Electric T64-GE-416 turboshaft engines, each with a max rating of 3,266 kW (4,380 shp) for 10 min, intermediate rating of 3,091 kW (4,145 shp) for 30 min and max continuous power rating of 2,756 kW (3,696 shp).

ACCOMMODATION: Crew of three. Main cabin will accommodate up to 55 troops in a high-density seating arrangement.

SYSTEM: Hamilton Standard automatic flight control system, using two digital onboard computers.

DIMENSIONS, EXTERNAL:

Main rotor diameter	24·08 m (79 ft 0 in)
Tail rotor diameter	6·10 m (20 ft 0 in)
Length overall, rotors turning	30·20 m (99 ft 1 in)
Length, rotor and tail pylon folded	18·44 m (60 ft 6 in)
Length of fuselage	22·35 m (73 ft 4 in)
Width of fuselage	2·69 m (8 ft 10 in)
Width, rotor and tail pylon folded	8·66 m (28 ft 5 in)
Height overall, tail rotor turning	8·66 m (28 ft 5 in)
Height, rotor and tail pylon folded	5·66 m (18 ft 7 in)
Wheel track (c/l of shock-struts)	3·96 m (13 ft 0 in)
Wheelbase	8·31 m (27 ft 3 in)

DIMENSIONS, INTERNAL:

Cabin: Length	9·14 m (30 ft 0 in)
Max width	2·29 m (7 ft 6 in)
Max height	2·01 m (6 ft 6 in)

WEIGHTS:

Weight empty	15,071 kg (33,226 lb)
Internal payload (100 nm; 185 km; 115 miles radius)	13,607 kg (30,000 lb)
External payload (50 nm; 92·5 km; 57·5 miles radius)	14,515 kg (32,000 lb)
Max T-O weight: internal payload	31,638 kg (69,750 lb)
external payload	33,339 kg (73,500 lb)

PERFORMANCE (ISA, at T-O weight of 25,400 kg; 56,000 lb):

Max level speed at S/L	170 knots (315 km/h; 196 mph)
Cruising speed at S/L	150 knots (278 km/h; 173 mph)
Max rate of climb at S/L	838 m (2,750 ft)/min
Service ceiling, at max continuous power	5,640 m (18,500 ft)
Hovering ceiling IGE, at max power	3,520 m (11,550 ft)
Hovering ceiling OGE, at max power	2,895 m (9,500 ft)
Range, at optimum cruise condition for best range	1,120 nm (2,075 km; 1,290 miles)

SIKORSKY S-69

US Army designation: XH-59A

On 7 February 1972 Sikorsky announced that the company was designing and building a research aircraft, designated S-69, to flight test the Advancing Blade Concept (ABC) rotor system, under a US Army contract. Subsequently, the value of the contract was increased to cover detail design changes and the construction of two demonstrator aircraft under the Army designation XH-59A. The first aircraft (21941) made its first flight on 26 July 1973.

In early 1981 the latest flight test programme, funded jointly by the US Army and Navy, was continuing. Scheduled for completion in July 1981, this was initiated to expand the altitude and CG flight envelopes, plus further investigation of general performance. Available details of the S-69 aircraft and programme can be found in the 1979-80 and 1980-81 *Jane's*.

Sikorsky UH-60A Black Hawk combat assault helicopter in service with the US Army's 101st Airborne Division

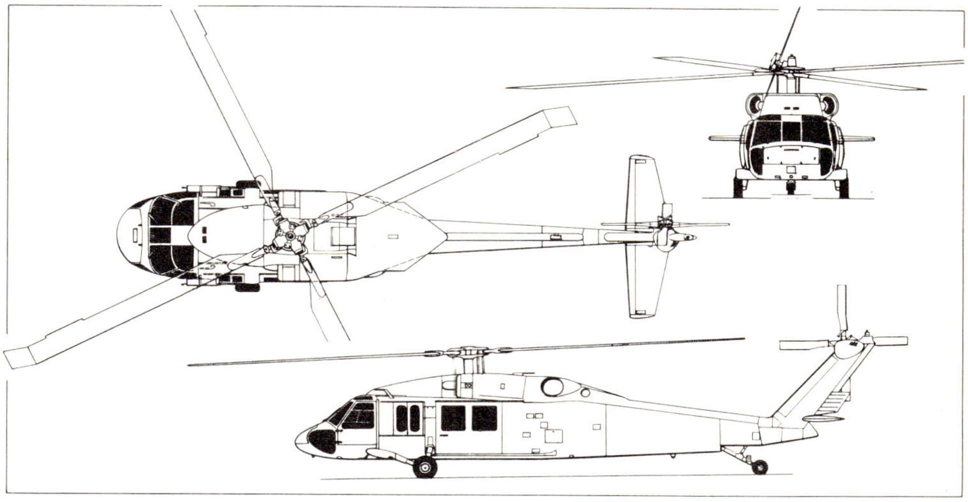

Sikorsky UH-60A Black Hawk combat assault helicopter *(Pilot Press)*

SIKORSKY S-70

US Army designations: UH-60A and EH-60A/B Black Hawk

At the end of August 1972, the US Army selected Sikorsky and Boeing Vertol as competitors to build three prototypes each, plus one ground test vehicle, of their submissions for the Utility Tactical Transport Aircraft System (UTTAS) requirement. Sikorsky's $61 million contract called for flight trials to begin in November 1974, but the first YUH-60A flew on 17 October 1974, six weeks ahead of schedule. The second prototype flew on 21 January 1975, followed by the third on 28 February 1975. Fly-off evaluation against Boeing Vertol's YUH-61A prototypes began in early 1976 and occupied a period of seven months. On 23 December 1976 Sikorsky's design was declared the winner, and it was subsequently named Black Hawk.

Designed to carry 11 fully-equipped troops plus a crew of three, the UH-60A has a large cabin which enables it to be used without modification for medical evacuation, reconnaissance, command and control purposes or troop resupply. For external lift missions its cargo hook has a capacity of up to 3,630 kg (8,000 lb). Design is compact, so that the helicopter itself can be airlifted over long ranges. One can be accommodated in a C-130, two in a C-141 and six in a C-5A.

The UH-60A is intended to serve as the US Army's primary combat assault helicopter, and the Army plans to procure a total of 1,107 by the mid-eighties. The basic production contract awarded to Sikorsky, plus options exercised by mid-1980, covered the construction of 255 aircraft during the first four years of production, which began in the Autumn of 1977. The FY 1982 defence budget increased the total so far ordered to 337, of which 171 had been delivered by 1 August 1981.

The first flight of a production aircraft was made in October 1978, and Black Hawks were delivered for pilot training to the US Army Aviation Center, Fort Rucker, Alabama, in April 1979. The first delivery of production aircraft to an operational unit was made on 19 June 1979, to Company D, 158th Assault Helicopter Battalion of the 101st Airborne Division at Fort Campbell, Kentucky. This Division had received its full complement of 90 UH-60As by early 1981. They were used initially by the 101st AD in an extensive force development test and experimentation (FDTE) programme under field conditions. Later, planned deployment of Black Hawks began to jungle areas of Panama, to desert regions in the southwest USA, and to the Arctic, for testing under a wide variety of temperatures and conditions. Second and third units equipped with Black Hawks are the 82nd Airborne Division at Fort Bragg, North Carolina, and the 9th Infantry Division, Fort Lewis, Washington. They will be followed by US Army, Europe; US Army, Korea; and US Forces Command aviation units.

In tests carried out during early 1979, the Black Hawk demonstrated its ability to sustain heavy landing impacts without damage. In a series of such tests, the helicopter sustained drop rates of 3·5 m (11·5 ft)/s at a forward speed of 63 knots (117 km/h; 73 mph) and gross weight of 7,632 kg (16,825 lb). Another aircraft, flown under artificial icing conditions, confirmed that safe flight can be made in moderate icing conditions by use of a specially developed de-icing kit.

Under an 11·4 million dollar contract awarded by the US Army in February 1981, Sikorsky is developing an external stores support system (ESSS) for the Black Hawk in a 24-month programme. This consists of a combination of fixed provisions built into the airframe, and removable external fittings for the carriage of auxiliary fuel tanks and a variety of weapons. With auxiliary fuel, the Black Hawk would have a significantly increased troop carrying range, and would be able to deploy over 1,000 nm (1,853 km; 1,151 miles) without refuelling. With the ESSS installed, and with the further development of a fire control system, external ordnance can be fired while the cabin remains usable to carry troops or cargo. Live weapon firings of an ESSS-equipped Black Hawk, at Redstone Arsenal, Huntsville, Alabama, will be a part of the programme. Among weapons to be evaluated will be a dispenser for 80

Sikorsky YEH-60A communications jamming version of the Black Hawk

anti-tank land mines, four of which can be carried on the ESSS stores points, and Hellfire missiles. Army plans anticipate the modification of all Black Hawks to accept the ESSS kit, if the programme is successful. Meanwhile, delivery of kits that will enable Black Hawks to perform medevac missions was scheduled to begin before the end of 1981.

Two special variants of the Black Hawk have been announced, as follows:

EH-60A. ECM variant, designed to intercept and jam enemy communications. During October 1980, Sikorsky received a $3·2 million contract from the US Army to initiate a prototype programme for the integration of Quick Fix II equipment into the Black Hawk. This involves the installation of 816 kg (1,800 lb) of electronics equipment that is designed to inhibit an enemy's battlefield communications. This EH-60A Quick Fix II represents but one component of the Army's special electronic mission aircraft (SEMA) programme. Subject to satisfactory results from the YEH-60A development programme, extending to February 1982, the Army plans to procure 36 of these aircraft.

EH-60B. Five YEH-60B prototypes are being produced under a $36·6 million US Army contract, announced in September 1979, for SOTAS (stand-off target acquisition system) missions. Each will be equipped with a rotating underbelly antenna, moving target indicating radar, and a data terminal in the main cabin, to detect the movement of enemy forces on the battlefield and relay the information to a ground station. This SOTAS data can be used also to cue RPVs assigned to the target acquisition role. Electronics packages are supplied by Motorola. First prototype, modified from a production UH-60A, made its first flight on 6 February 1981, with a fixed mockup antenna. The main landing gear of the EH-60B is designed to retract rearward, to enable the antenna to rotate in flight on operational aircraft. The US Army plans eventual production of 75-100 SOTAS helicopters.

The following description applies to the UH-60A:

TYPE: Twin-turbine combat assault squad transport.

ROTOR SYSTEM: Four-blade main rotor. Sikorsky SC-1095 blade section, with thickness/chord ratio of 9·5%. Middle section has leading-edge droop and trailing-edge tab to overcome vortex impingement from preceding blade in cruising flight. Blade twist 18°. Blade tips swept back 20°. Each blade consists of an oval titanium spar, Nomex honeycomb core, graphite trailing-edge and root, covered with glassfibre, and with plastics leading-edge counterweight and titanium leading-edge sheath. Blades are tolerant to small arms damage, and are pressurised and equipped with gauges providing fail-safe confirmation of blade structural integrity. Electrically-heated de-icing mat in leading-edge of each blade. C/R Industries elastomeric rotor hub bearings require no lubrication, reducing hub maintenance by 60%. Bifilar self-tuning vibration absorbers on rotor head. Manual blade folding. Canting of tail rotor (20° to port) increases vertical lift and allows greater CG travel. 'Cross beam' four-blade tail rotor of composite materials, eliminating all rotor head bearings.

ROTOR DRIVE: Conventional transmission system with both turbines driving through freewheeling units to main gearbox. This is of modular construction to simplify maintenance. Transmission can operate for 30 min following total oil loss. Intermediate and tail rotor gearboxes oil lubricated.

FUSELAGE: Conventional semi-monocoque light alloy crashworthy structure.

TAIL UNIT: Pylon structure with port-canted tail rotor mounted on starboard side. Large variable-incidence tailplane has a control system which senses airspeed, collective-lever position, pitch-attitude rate and lateral acceleration. Tailplane is set at about +34° incidence in the hover, and −6° for autorotation. Tailplane moved by dual electric actuators, with manual backup.

LANDING GEAR: Non-retractable tailwheel type with single wheel on each unit. Energy-absorbing main gear with a tailwheel which gives protection for the tail rotor in taxying over rough terrain or during a high-flare landing. Axle assembly and main gear shock-absorbers by General Mechatronics.

POWER PLANT: Two 1,151 kW (1,560 shp) General Electric T700-GE-700 advanced technology turboshaft engines; combined transmission rating 2,109 kW (2,828 shp). Two crashworthy, bulletproof fuel tanks, with combined capacity of 1,340 litres (354 US gallons), aft of cabin.

ACCOMMODATION: Pilot and co-pilot on armour protected seats. Main cabin area open to cockpit to provide good communication with flight crew and forward view for squad commander. Accommodation for 11 troops and crew of three. Eight troop seats can be removed and replaced by four litters for medevac missions, or to make room for internal cargo. Cabin heated and ventilated. External cargo hook, having a 3,630 kg (8,000 lb) lift capability. Large aft-sliding door on each side of fuselage for rapid entry and exit. Electrical windscreen de-icing.

SYSTEMS: Solar T-62T-40-1 APU; Garrett engine start system; Bendix 30/40kVA and 20/30kVA electric power generators; 17Ah nickel-cadmium battery. Engine fire extinguishing system.

AVIONICS: Include VHF/AM, VHF/FM and UHF/AM com, Singer Doppler, LF/ADF, AN/APR-39 radar warning receiver, SIF/IFF, and TSEC/KY-28 secure speech.

ARMAMENT AND OPERATIONAL EQUIPMENT: Provision for Hellfire missiles, one or two M60 side-firing machine-guns in forward area of cabin, infra-red jamming flares and XM130 chaff dispenser.

DIMENSIONS, EXTERNAL:	
Main rotor diameter	16·36 m (53 ft 8 in)
Main rotor blade chord	0·53 m (1 ft 8¾ in)
Tail rotor diameter	3·35 m (11 ft 0 in)
Length overall (rotors turning)	19·76 m (64 ft 10 in)
Length, rotors and tail pylon folded	12·60 m (41 ft 4 in)
Fuselage length	15·26 m (50 ft 0¾ in)
Fuselage width	2·36 m (7 ft 9 in)
Fuselage depth	1·75 m (5 ft 9 in)
Height overall, tail rotor turning	5·13 m (16 ft 10 in)
Height to top of rotor hub	3·76 m (12 ft 4 in)
Height in air-transportable configuration	2·67 m (8 ft 9 in)
Tailplane span	4·38 m (14 ft 4½ in)
Wheel track	2·705 m (8 ft 10½ in)
Wheelbase	8·84 m (29 ft 0 in)
Tail rotor ground clearance	1·98 m (6 ft 6 in)
Cabin doors (each): Height	1·37 m (4 ft 6 in)
Width	1·75 m (5 ft 9 in)
DIMENSION, INTERNAL:	
Cabin: Volume	10·90 m³ (385 cu ft)
AREAS:	
Main rotor blades (each)	4·34 m² (46·70 sq ft)
Tail rotor blades (each)	0·41 m² (4·45 sq ft)
Main rotor disc	210·05 m² (2,261 sq ft)
Tail rotor disc	8·83 m² (95·0 sq ft)
WEIGHTS:	
Weight empty	4,819 kg (10,624 lb)
Mission T-O weight	7,375 kg (16,260 lb)
Max alternative T-O weight	9,185 kg (20,250 lb)
PERFORMANCE (at mission T-O weight, except where indicated):	
Never-exceed speed	195 knots (361 km/h; 224 mph)
Max level speed at S/L	160 knots (296 km/h; 184 mph)
Max level speed at max T-O weight	158 knots (293 km/h; 182 mph)
Max cruising speed at 1,220 m (4,000 ft)	145 knots (269 km/h; 167 mph)
Single-engine cruising speed	105 knots (195 km/h; 121 mph)
Vertical rate of climb at S/L	over 137 m (450 ft)/min
Service ceiling	5,790 m (19,000 ft)
Hovering ceiling IGE (35°C)	2,895 m (9,500 ft)
Hovering ceiling OGE: ISA	3,170 m (10,400 ft)
35°C	1,705 m (5,600 ft)
Range at max T-O weight, 30 min reserves	324 nm (600 km; 373 miles)
Endurance	2 h 18 min

SIKORSKY S-70L

US Navy designation: SH-60B Seahawk

The S-70L, since designated SH-60B Seahawk, was Sikorsky's submission for the US Navy's LAMPS (Light Airborne Multi-Purpose System) Mk III competition, and was selected as the winner in September 1977.

Detail design of the Seahawk was initiated by a US Navy award to Sikorsky of a $2·7 million sustaining engineering contract. At the same time, General Electric was given a $547,000 contract for further development of the T700 advanced turboshaft engine to provide increased power and improved corrosion resistance, while a $17·9 million contract went to IBM Federal Systems to continue development of the avionics essential for the SH-60B to fulfil the LAMPS Mk III role. On 28 February 1978, it was announced that the US Department of Defense had authorised full-scale development of the SH-60B, and had awarded Sikorsky a $109·3 million contract for the development, manufacture and flight testing of five prototypes, plus a further airframe for ground testing.

Earlier, Sikorsky had updated the original UH-60A Black Hawk mockup to SH-60B configuration, and this

First of five Sikorsky YEH-60B SOTAS prototypes, with mockup underbelly radar antenna

was formally reviewed just prior to announcement of the contract award. In July and August 1978, this mockup was used for shipboard compatibility trials on board the frigate USS *Oliver Hazard Perry* (FFG-7) and the Spruance class destroyer USS *Arthur W. Radford* (DD-968).

In February 1979, the main transmission of the SH-60B completed qualification testing, during which it was run at up to 2,685 kW (3,600 shp). This is 447 kW (600 shp) in excess of the Navy's mission performance specification. On 29 March 1979 it was announced that final assembly of the first Seahawk prototype (US Navy serial number 161169) had begun, and the first flight was made on 12 December 1979. The remaining four prototypes were flown on 11 February, 17 March, 26 April and 14 July 1980.

During 1980 these prototypes were used as follows: No. 1 for preliminary evaluation at the Naval Air Engineering Center, Lakehurst, New Jersey; No. 2 for automatic flight control system checks by Sikorsky; Nos. 3 and 4 flown by the Naval Air Test Center, Patuxent River, Maryland, respectively for electromagnetic interference and compatibility tests, and for pilot training; No. 5 used by IBM Federal Systems Division at Owego, New York, for functional avionics testing. LAMPS operational evaluation by US Navy Squadron VX-1, including shipboard trials, began in May 1981. In the early flight test programme, the Seahawk has flown at 180 knots (334 km/h; 207 mph) in a dive, has demonstrated 153 knots (284 km/h; 176 mph) in level flight, and been flown at weights between 6,804 and 9,979 kg (15,000 and 22,000 lb).

The US Navy has indicated a requirement for 204 of these helicopters for deployment on board Spruance class ASW destroyers, Aegis-equipped guided missile destroyers, and guided missile frigates in the class of the FFG-7 USS *Oliver Hazard Perry*. Orders totalled 18 by mid-1981; deliveries are due to begin in 1983. In addition to the LAMPS Mk III primary missions of anti-submarine warfare (ASW) and anti-ship surveillance and targeting (ASST), the Seahawk is required also to perform secondary missions which include search and rescue (SAR), medical evacuation (Medevac) and vertical replenishment (Vertrep). It is intended for operational deployment in 1984.

Generally similar to the UH-60A Black Hawk, the Seahawk differs in the modifications necessary for shipboard compatibility and in the provision of avionics and equipment suitable for the naval mission. The former include a rotor brake, automatic blade folding, folding tailplane and tail rotor pylon, a simplified short-wheelbase landing gear, sliding cabin door, and the introduction of recovery assist, secure and traversing (RAST) equipment for rapid haul-down of the helicopter on to a small deck in rough sea conditions, hovering in-flight refuelling capability, and buoyancy features. Modifications necessary for the mission requirement include the provision of a sensor operator's station, rescue hoist, chin-mounted pods for ESM equipment, pylons for two Mk 46 torpedoes or auxiliary fuel tanks, a pylon on the starboard side to carry MAD equipment, a sonobuoy launcher on the port side, an increased capacity fuel system, and deletion of armour for the pilot's and co-pilot's seats.

TYPE: Twin-turbine ASW/ASST helicopter.

ROTOR SYSTEM: As for UH-60A, except that main rotor blades can be folded by electrical power, and a rotor brake is provided.

ROTOR DRIVE: As for UH-60A.

FUSELAGE: As for UH-60A, except for inclusion of flotation bags and sealing of tailboom to provide buoyancy.

TAIL UNIT: As for UH-60A, except that the pylon structure can be folded to port pneumatically, eliminating the necessity to fold the tail rotor, and the tailplane folds upward.

LANDING GEAR: Non-retractable tailwheel type, with single wheels on main units and twin wheels on tail unit. Wheelbase shortened by 46·6%. Landing gear structure is less complex since the SH-60B's vertical impact requirement is 71·5% below that of the UH-60A. Main-wheel tyres size 26 × 10·0-11; tailwheel tyres size 17·5 × 6·00-6. Multiple disc brakes.

POWER PLANT: Two 1,260 kW (1,690 shp) General Electric T700-GE-401 advanced technology turboshaft engines. Crash-resistant twin-cell fuel system in rear fuselage with total capacity of 2,241 litres (592 US gallons). Lower one-third of fuel cells is self-sealing. Single-point refuelling connection on port side. Hovering in-flight refuelling capability. Two auxiliary fuel tanks can be carried on fuselage pylons.

ACCOMMODATION: Pilot and co-pilot/ATO in cockpit, sensor operator in specially-equipped station. Sliding door with jettisonable window on starboard side. Accommodation is heated, ventilated and air-conditioned.

SYSTEMS: Generally as for UH-60A.

AVIONICS AND EQUIPMENT: Avionics include Collins AN/ARC-159 UHF and AN/ARC-174 HF com, Hazeltine AN/APX-76A active IFF, IBM AN/UYS-1 acoustic processor, Raytheon AN/ALQ-142 ESM, Teledyne Ryan AN/APN-217 Doppler, Texas Instruments AN/ASQ-81 MAD and AN/APS-124 search radar. Equipment includes a 25 tube pneumatic launcher for sonobuoys, and a rescue hoist for SAR operations.

First prototype Sikorsky SH-60B Seahawk, developed to meet the US Navy's LAMPS Mk III requirement

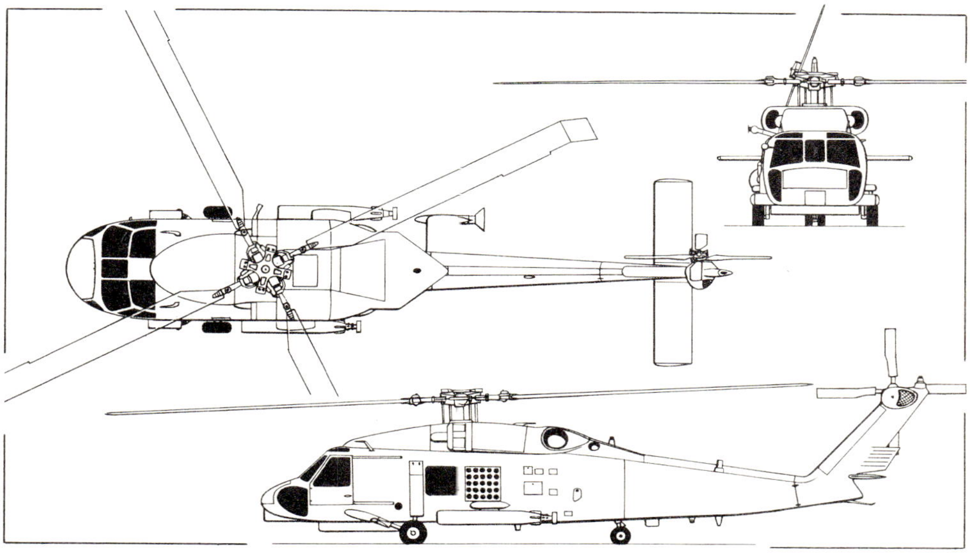

Sikorsky SH-60B Seahawk twin-turbine ASW/ASST helicopter *(Pilot Press)*

ARMAMENT: Includes two Mk 46 torpedoes.

DIMENSIONS, EXTERNAL: As UH-60A except:

Length overall (rotors and tail pylon folded)	12·51 m (41 ft 0½ in)
Width (rotors folded)	4·37 m (14 ft 4 in)
Height to top of rotor hub	3·63 m (11 ft 11 in)
Height overall, tail rotor turning	5·23 m (17 ft 2 in)
Height overall (pylon folded)	4·01 m (13 ft 2 in)
Tailplane span	4·37 m (14 ft 4 in)
Wheel track	2·79 m (9 ft 2 in)
Wheelbase	4·83 m (15 ft 10 in)

AREAS: As UH-60A

WEIGHTS (estimated. A, ASW mission; B, ASST mission):

Weight empty, equipped: A, B	6,191 kg (13,648 lb)
Desired mission T-O weight: A	8,983 kg (19,804 lb)
B	8,148 kg (17,963 lb)
Max T-O weight: A	9,908 kg (21,844 lb)
B	9,926 kg (21,884 lb)

PERFORMANCE (estimated, at mission T-O weight):

Max cruising speed at 1,525 m (5,000 ft), tropical conditions: A	135 knots (249 km/h; 155 mph)
Vertical rate of climb at S/L: A	363 m (1,192 ft)/min
Rate of climb at S/L, one engine out: A, B	235 m (770 ft)/min

SIKORSKY S-76

Sikorsky Aircraft announced on 19 January 1975 the company's decision to build a new 12-passenger twin-turbine commercial helicopter, as the first stage of a programme intended to give the company a bigger share of the civil aircraft market. The go-ahead for prototype construction followed a period of market research during

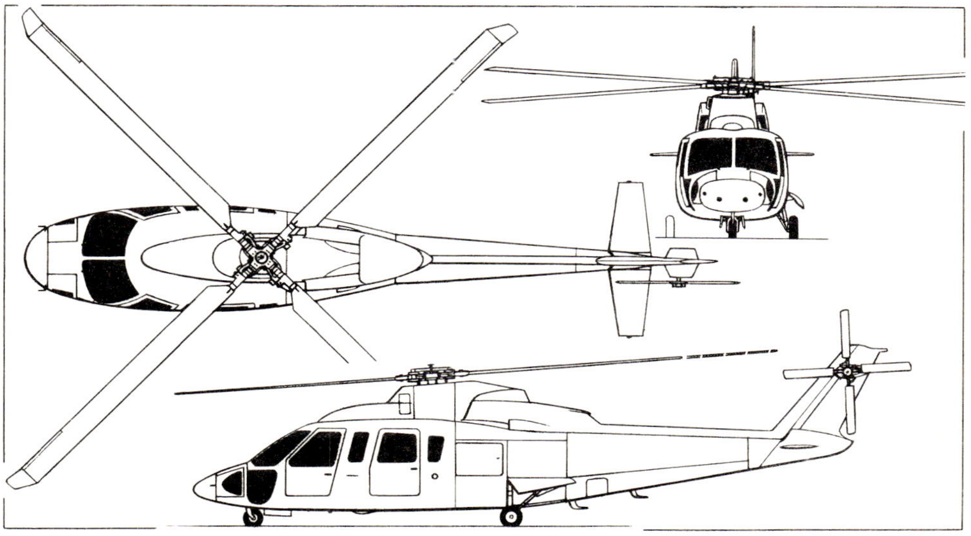

Sikorsky S-76 eight/twelve-passenger commercial transport helicopter *(Pilot Press)*

Sikorsky S-76 general-purpose commercial helicopter in corporate service

which firm contracts were signed with numerous commercial operators, both in the USA and abroad.

Construction of four prototypes began in May 1976, and the first flight was made, by the No. 2 aircraft (N762SA), on 13 March 1977. Second prototype to fly, in late April 1977, was the No. 3 aircraft, and a further prototype was flown about a month later.

Designated Sikorsky S-76, the design conforms with FAR Part 29 Category A IFR. By designing and building to this standard, Sikorsky has produced a rugged and reliable civil helicopter, and one which can be taken 'off the shelf' to satisfy a wide variety of air transport missions, with offshore oil support, corporate executive transport, and general utility operations regarded as primary markets.

The S-76 benefits from the design, research and development work carried out on the dynamic system of Sikorsky's UH-60A Black Hawk. The main rotor, for example, is a scaled-down version of that developed for the UH-60A. The power plant consists of two Allison 250-C30 turboshaft engines, a growth version of the current production Model 250-C20 which powers a number of important helicopters.

By 4 June 1981, a total of 444 S-76s had been ordered by 112 operators, half of them for offshore duties. Customers include the Icelandic Coast Guard (one) and Royal Brunei Malay Air Regiment (seven). Initial deliveries of fully certificated IFR production aircraft were made to Air Logistics of Lafayette, Louisiana, on 27 February 1979, and a total of 37 entered service during that year. Deliveries totalled 163, to 23 countries, by mid-July 1981. The 100th aircraft, one of 38 on order for Bristow Helicopters Ltd, was delivered in November 1980.

TYPE: Twin-turbine general-purpose all-weather helicopter.

ROTOR SYSTEM: Four-blade main rotor. Each blade consists of a titanium spar, titanium leading-edge cover, nickel leading-edge abrasion strip, and glassfibre/nylon honeycomb trailing-edge. Blades have swept tips (30° on leading-edges, 10° on trailing-edges). Elastomeric rotor hub bearings which need no lubrication. Bifilar vibration absorbers on rotor head. Cross-beam four-blade tail rotor of composite materials. Rotor brake optional.

ROTOR DRIVE: Conventional transmission system, with both turbines driving through freewheeling units to main gearbox. Intermediate and tail rotor gearboxes are oil lubricated.

FUSELAGE: Composite structure, comprising glassfibre nose, light alloy honeycomb cabin, semi-monocoque light alloy tailcone and Kevlar fairings.

TAIL UNIT: Pylon structure with tail rotor on port side. All-moving tailplane, which serves also to protect passengers or ground crew from contact with tail rotor.

LANDING GEAR: Hydraulically-retractable tricycle type, with single wheel on each unit. Nosewheel retracts aft, main units inward into rear fuselage; all three units are enclosed by wheel doors when retracted. Main-wheel tyres size 14·5 × 5·5-6, nosewheel tyre size 13 × 5·00-4. Hydraulic brakes; hydraulic main-wheel parking brake.

POWER PLANT: Two 485 kW (650 shp) Allison 250-C30 turboshaft engines, mounted above the cabin aft of the main rotor shaft. Standard fuel system has a capacity of 1,060 litres (280 US gallons). Extended range fuel tanks and closed circuit refuelling optional.

ACCOMMODATION: Pilot and co-pilot plus a maximum of 12 passengers. In this configuration passengers are seated on three four-abreast rows of seats, floor-mounted at a pitch of 79 cm (31 in). A number of executive layouts are available, including a four-passenger 'office-in-the-sky' configuration. Executive versions have luxurious interior trim, full carpeting, special soundproofing, radio-telephone, and co-ordinated furniture. Dual controls optional. Two large doors on each side of fuselage, hinged at their forward edges; sliding doors are available optionally. Baggage hold aft of cabin, with external access door on each side of the fuselage. Cabin heated and ventilated. Windscreen demisting and dual windscreen wipers. Windscreen heating optional. Optional external cargo hook with capacity of 1,814 kg (4,000 lb).

SYSTEMS: Hydraulic system at pressure of 207 bars (3,000 lb/sq in) supplied by two pumps driven from main gearbox. Electrical system comprises one Lucas 7·5kVA gearbox-driven generator, two 200A starter/generators, one 115V 600VA 400Hz static inverter and 24V 17Ah nickel-cadmium battery. Engine fire detection and extinguishing system.

AVIONICS AND EQUIPMENT: Standard equipment includes provision for dual controls; cabin fire extinguishers; cockpit, cabin, instrument, navigation and anti-collision lights; landing light; external power socket; first aid kit; and utility soundproofing. VHF com transceiver and intercom system standard. Optional equipment includes air-conditioning, cargo hook, rescue hoist, emergency flotation gear, engine air particle separators, full IFR instrumentation, litter installation and stability augmentation system. Wide range of optional avionics available, according to configuration, including VHF nav receivers, transponder, compass system, weather radar, flight director system, radar altimeter, ADF, DME, VLF nav system and ELT and sonic transmitters.

DIMENSIONS, EXTERNAL:
Diameter of main rotor 13·41 m (44 ft 0 in)
Main rotor blade chord 0·39 m (1 ft 3½ in)
Diameter of tail rotor 2·44 m (8 ft 0 in)
Tail rotor blade chord 0·16 m (6½ in)
Length overall, rotors turning 16·00 m (52 ft 6 in)
Height overall, tail rotor turning 4·41 m (14 ft 5¾ in)
Tail rotor ground clearance 1·97 m (6 ft 5¾ in)
Tailplane span 3·05 m (10 ft 0 in)
Length of fuselage 13·22 m (43 ft 4½ in)
Max width of fuselage 2·13 m (7 ft 0 in)
Depth of fuselage 1·83 m (6 ft 0 in)
Tail rotor ground clearance 1·97 m (6 ft 5¾ in)
Wheel track 2·44 m (8 ft 0 in)
Wheelbase 5·00 m (16 ft 5 in)

DIMENSIONS, INTERNAL:
Cabin: Length 2·46 m (8 ft 1 in)
Max width 1·93 m (6 ft 4 in)
Max height 1·35 m (4 ft 5 in)
Floor area 4·18 m² (45 sq ft)
Volume 5·78 m³ (204 cu ft)
Baggage compartment volume 1·08 m³ (38 cu ft)

AREAS:
Main rotor disc 116·77 m² (1,257 sq ft)
Tail rotor disc 4·67 m² (50·27 sq ft)
Tailplane 2·00 m² (21·5 sq ft)

WEIGHTS AND LOADING:
Weight empty, standard equipment 2,540 kg (5,600 lb)
Max T-O weight 4,536 kg (10,000 lb)
Max disc loading 38·9 kg/m² (7·96 lb/sq ft)

PERFORMANCE (A: at gross weight of 4,536 kg; 10,000 lb. B: at gross weight of 3,810 kg; 8,400 lb):
Max cruising speed:
A 145 knots (269 km/h; 167 mph)
B 155 knots (286 km/h; 178 mph)
Cruising speed for max range:
A 125 knots (232 km/h; 144 mph)
Max rate of climb at S/L 411 m (1,350 ft)/min
Service ceiling: A 4,570 m (15,000 ft)
Service ceiling, one engine out:
A 1,890 m (6,200 ft)
B 3,445 m (11,300 ft)
Hovering ceiling IGE: A 1,890 m (6,200 ft)
B 3,415 m (11,200 ft)
Range with 12 passengers, standard fuel, 30 min reserves 404 nm (748 km; 465 miles)
Range with 8 passengers, auxiliary fuel and offshore equipment 600 nm (1,112 km; 691 miles)

SMITH

MIKE SMITH AIRCRAFT INC

PO Box 430, Johnson, Kansas 67855
Telephone: (316) 492 6254

SMITH LIGHTNING MODEL 400

As a first step towards the development of a turboprop-powered four/six-seat light aircraft, which has since materialised as the Lightning, Mr Mike Smith decided in 1978 to design and build a two-seat lightplane with a turboprop power plant, to evaluate the speed and economics of an aircraft in this class. Named Super Interceptor, this aircraft was completed and flown but, at an early stage of flight testing, the failure of a fuel control unit caused an off-airport forced landing which resulted in substantial damage. The performance demonstrated by that time was sufficiently impressive for Mr Smith to proceed with design and construction of an advanced technology aircraft which has been given the name Lightning Model 400. Design began in July 1979, and construction of the prototype started in January 1980, with certification under FAA Part 23 anticipated by the Spring of 1982. A second prototype is to be built.

TYPE: Six-seat cabin monoplane.

WINGS: Cantilever mid-wing monoplane. Supercritical GST wing section of 12% thickness/chord ratio. Dihedral 1° 48′. Incidence 3° 12′. Sweepback at quarter-chord 7°. Conventional structure of light alloy. Plain trailing-edge flaps. Plain ailerons, hinged at lower surface and designed to droop with flaps. Servo tab on each aileron. Anti-icing of wing leading-edges by engine bleed air.

FUSELAGE: Semi-monocoque fail-safe structure of light alloy. Cabin area and part of baggage compartment pressurised.

TAIL UNIT: Cantilever structure of light alloy, comprising fixed-incidence tailplane and endplate fins and rudders. Servo tabs on elevators. Anti-icing of tailplane leading-edges by engine bleed air.

LANDING GEAR: Hydraulically-retractable tricycle type, with single wheel on each unit. Nosewheel retracts forward, main units aft. Cleveland brakes and brake cooling system.

POWER PLANT: One 633 kW (850 shp) Pratt & Whitney Aircraft of Canada PT6A-42 turboprop engine mounted in the aft fuselage, and driving a Hartzell three-blade constant-speed metal pusher propeller with spinner. One 227 litre (60 US gallon) fuel tank in each wing and one 113·5 litre (30 US gallon) fuselage tank, providing a total capacity of 567·5 litres (150 US gallons).

ACCOMMODATION: Pilot and five passengers in enclosed cabin, which is pressurised, heated and air-conditioned. Door of two-piece clamshell type on port side, forward of wing, with airstairs built into lower portion. Baggage compartments at rear of cabin and in nose.

SYSTEMS: Details not finalised. Air-conditioning and pressurisation system, with max differential of 0·59 bars (8·5 lb/sq in). Hydraulic system for landing gear operation. Electrical, oxygen and anti-icing systems.

EQUIPMENT: Blind-flying instrumentation is standard.

DIMENSIONS, EXTERNAL:
Wing span 9·04 m (29 ft 8 in)
Wing chord at root 1·42 m (4 ft 8 in)
Wing chord at tip 0·76 m (2 ft 5¾ in)

Wing aspect ratio	8·00
Length overall	9·51 m (31 ft 2½ in)
Height overall	2·98 m (9 ft 9½ in)
Tailplane span	3·48 m (11 ft 5 in)
Wheel track	2·29 m (7 ft 6 in)
Wheelbase	3·57 m (11 ft 8½ in)
Propeller diameter	2·29 m (7 ft 6 in)
Propeller ground clearance	0·68 m (2 ft 3 in)
Passenger door: Height	1·22 m (4 ft 0 in)
Width	0·61 m (2 ft 0 in)
Height to sill	0·92 m (3 ft 0 in)
DIMENSIONS, INTERNAL:	
Cabin: Max width	1·33 m (4 ft 4½ in)
Max height	1·26 m (4 ft 1½ in)
Volume	3·96 m³ (140 cu ft)
Baggage compartment: rear	0·41 m³ (14·4 cu ft)
nose	0·42 m³ (14·8 cu ft)
AREAS:	
Wings, gross	10·22 m² (110 sq ft)
Vertical tail surfaces (total)	0·94 m² (10·15 sq ft)
Horizontal tail surfaces (total)	2·94 m² (31·7 sq ft)
WEIGHTS AND LOADINGS (estimated):	
Weight empty	1,086 kg (2,394 lb)
Max fuel	459 kg (1,011 lb)
Max T-O and landing weight	1,905 kg (4,200 lb)
Max zero-fuel weight	1,446 kg (3,189 lb)
Max wing loading	186·4 kg/m² (38·18 lb/sq ft)
Max power loading	3·01 kg/kW (4·94 lb/shp)
PERFORMANCE (estimated, at max T-O weight):	
Max level speed at 7,620 m (25,000 ft)	400 knots (741 km/h; 460 mph)
Max cruising speed at 12,200 m (40,000 ft)	380 knots (703 km/h; 437 mph)
Econ cruising speed at 12,200 m (40,000 ft)	280 knots (518 km/h; 322 mph)
Stalling speed, flaps and gear up	91·8 knots (170 km/h; 105 mph)
Stalling speed, flaps and gear down	78 knots (145 km/h; 90 mph)
Max rate of climb at S/L	1,615 m (5,300 ft)/min
Service ceiling	12,200 m (40,000 ft)
T-O to 15 m (50 ft)	443 m (1,451 ft)
Landing from 15 m (50 ft)	695 m (2,281 ft)
Range with max fuel	2,700 nm (5,000 km; 3,107 miles)
Range with max payload	2,100 nm (3,890 km; 2,416 miles)

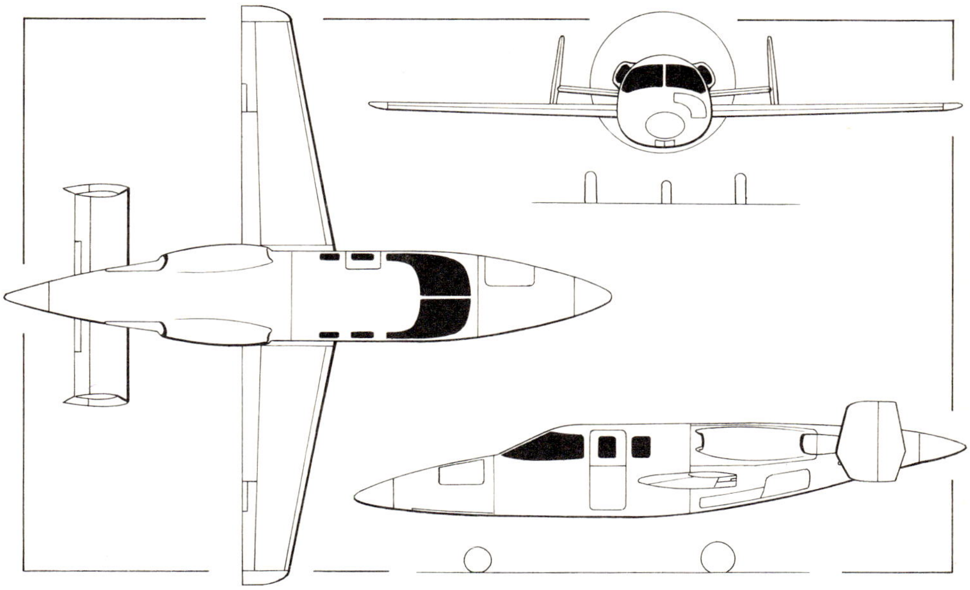

Smith Lightning Model 400 (one P&WC PT6A-42 turboprop engine) *(Michael A. Badrocke)*

SOLOY
SOLOY CONVERSION LTD

PO Box 60, Chehalis, Washington 98532
Telephone: (206) 748 0067
Telex: 510 786 0233
PRESIDENT: Joseph I. Soloy
SALES DIRECTOR: T. Koester

Soloy Conversion, which became known for its turbine engine conversions of Bell 47G and Hiller UH-12E helicopters, was reported in late 1980 to be working on a specialised turboprop conversion scheme that would be applicable to a number of different aircraft. To achieve this, Soloy is developing a Quick Engine Change unit built around Allison 250-C20B and -C30 turboshaft engines, and involving the introduction of a separate shaft-driven gearbox to provide the propeller drive. This gearbox is to be rated at more than 522 kW (700 shp) to allow for growth versions of the engine.

The company believes that such a QEC unit, requiring only minor modification for application to differing aircraft types, will speed both conversion and certification. It is planned to develop QEC units initially for the Cessna Ag Wagon and Piper PA-36 Brave, with other members of the Cessna single-engine line (Models 185, 206, 207 and 210) to follow. The application of Soloy QEC turboprop conversions to twin-engine installations is to be investigated at a later date.

SPITFIRE
SPITFIRE HELICOPTER COMPANY LTD

PO Box 61, Media, Pennsylvania 19063
Telephone: (215) 565 2986
Telex: 831750
PRESIDENT: John J. Fetsko

Spitfire Helicopter Company began in January 1975 the design of a new lightweight helicopter known as the Spitfire Mark I. Developed from the basic design of the Enstrom F-28A, this has a turbine power plant instead of the Avco Lycoming piston engine of the original aircraft. The drive from the turbine passes via a reduction gear, instead of through multiple V belts, showing a weight saving of 91 kg (200 lb), as well as freeing space which can be utilised for additional cargo or auxiliary fuel. Construction of the prototype began in January 1976, and of pre-production aircraft in February 1977.

The company has designed and built the prototype of an improved four-seat version designated Spitfire Mark II. It has also projected a completely new four-seat helicopter with a 224 kW (300 hp) Avco Lycoming piston engine and stub-wings, designated Spitfire Mark III; a Mark IIIA which differs by having a turboshaft engine; and an advanced version of the same design with stub-wings of increased span and area, to be powered by twin Allison turboshafts. In this last-mentioned version, which has the designation Spitfire Mark IV, reversible ducted fans, mounted at the tip of each stub-wing, will replace the tail rotor for torque control.

SPITFIRE/PZL SWIDNIK TAURUS

On 12 August 1978, Spitfire concluded an agreement with the Polish aircraft marketing organisation Pezetel, under which Spitfire has rights to sell a modified version of the PZL Swidnik Kania twin-turbine helicopter in Western markets. Spitfire's version, which is intended to have Allison 250-C28 instead of −C20B engines, will be known as the Taurus (see Polish section).

SPITFIRE MARK I

TYPE: Three-seat turbine-powered light helicopter.

ROTOR SYSTEM: Fully-articulated metal three-blade main rotor. Blades of bonded light alloy construction, each attached to rotor hub by retention pin and drag link. Blade section NACA 00135. Two-blade teetering tail rotor, with blades of bonded light alloy construction. Blades do not fold. Rotor brake optional.

ROTOR DRIVE: Shaft drive to both main and tail rotors through reduction gear. Main rotor/engine rpm ratio 1 : 18·2. Tail rotor/engine rpm ratio 1 : 2·54.

FUSELAGE: Glassfibre and light alloy cab structure, with welded steel tube centre-section. Semi-monocoque light alloy tailcone structure.

TAIL UNIT: Small horizontal surface with endplate fins.

LANDING GEAR: Skids, carried on oleo-pneumatic shock-absorbers. Ground handling wheels available.

POWER PLANT: One 313 kW (420 shp) Allison 250-C20B turboshaft engine, derated to 179 kW (240 shp) for take-off. Total standard fuel capacity 265 litres (70 US gallons).

ACCOMMODATION: Pilot and two passengers, side by side on bench seat. Removable door, with window, on each side of cabin. Baggage compartment of 0·57 m³ (20 cu ft) capacity forward of engine compartment, with external access door on each side of fuselage. Helicopter can be operated with these doors removed to permit loading of outsize cargo. Cabin heated and ventilated.

SYSTEM: Electrical system is supplied by a 28V DC engine-driven generator and 24V nickel-cadmium battery.

AVIONICS AND EQUIPMENT: Various nav/com systems available to customer's requirements. IFR provisions. Cargo hook and litters optional. It is intended to make available survey and agricultural equipment.

DIMENSIONS, EXTERNAL:	
Diameter of main rotor	9·75 m (32 ft 0 in)
Diameter of tail rotor	1·42 m (4 ft 8 in)
Distance between rotor centres	5·56 m (18 ft 3 in)
Main rotor blade chord	0·241 m (9½ in)
Length of fuselage	8·96 m (29 ft 4¾ in)
Height overall	2·74 m (9 ft 0 in)
Tailplane span	1·38 m (4 ft 6½ in)
Skid track	2·24 m (7 ft 4 in)
Cabin doors (each): Height	1·09 m (3 ft 7 in)
Width	1·02 m (3 ft 4 in)
Height to sill	0·61 m (2 ft 0 in)
AREAS:	
Main rotor blades (each)	2·22 m² (23·94 sq ft)
Main rotor disc	74·69 m² (804 sq ft)
Tail rotor disc	1·59 m² (17·1 sq ft)
WEIGHTS AND LOADINGS:	
Weight empty	567 kg (1,250 lb)
Max T-O weight	1,066 kg (2,350 lb)
Max disc loading	14·27 kg/m² (2·92 lb/sq ft)
Max power loading	5·95 kg/kW (9·79 lb/shp)
PERFORMANCE (at max T-O weight):	
Never-exceed speed	112 knots (208 km/h; 129 mph)
Max level speed	112 knots (208 km/h; 129 mph)
Cruising speed	95·5 knots (177 km/h; 110 mph)
Max rate of climb at S/L	472 m (1,550 ft)/min
Service ceiling	4,570 m (15,000 ft)
Hovering ceiling IGE	4,085 m (13,400 ft)

Spitfire Mark I, a three-seat turbine-powered development of the Enstrom F-28A *(Howard Levy)*

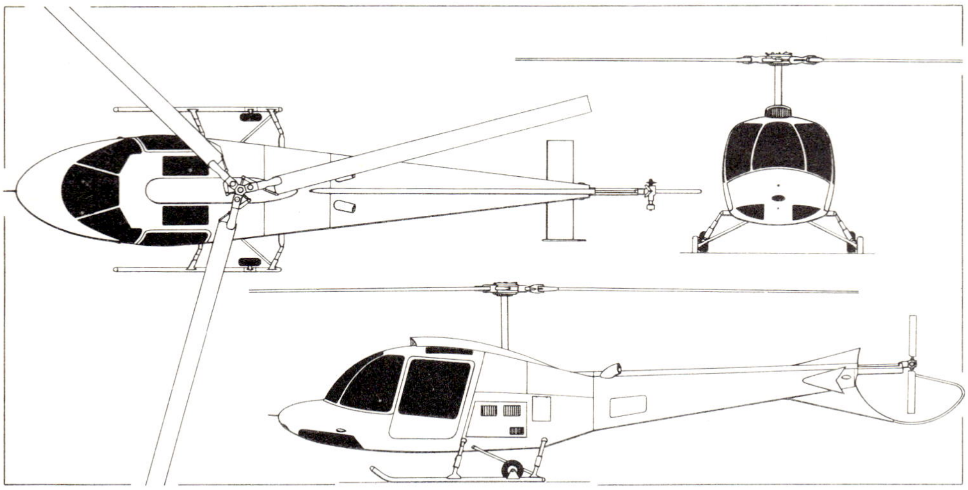

Spitfire Mark II (Allison 250-C20B turboshaft engine) *(Pilot Press)*

Model of the Spitfire Mark IV four/five-seat commercial helicopter

Hovering ceiling OGE	2,440 m (8,000 ft)
Range with max fuel	267 nm (495 km; 308 miles)
Endurance with max fuel	4 h

SPITFIRE MARK II

Information available indicates that the Mark II version of the Spitfire helicopter is generally similar to the production Mark I, except that more power is available at take-off from the derated Allison turboshaft engine, and that fuselage dimensions have been increased to provide room for an additional seat.

The description of the Spitfire Mark I applies also to the Mark II, except as follows:

TYPE: Four-seat turbine-powered light helicopter.

ACCOMMODATION: Pilot at front and three passengers side by side on rear bench seat. Cabin door on starboard side of fuselage. Baggage space in fuselage, aft of cabin, with external access door on each side; baggage volume 0·57 m³ (20 cu ft). Cabin heated and ventilated.

DIMENSIONS, EXTERNAL: As for Spitfire Mk I, except:

Length of fuselage	9·30 m (30 ft 6 in)
Height overall	2·86 m (9 ft 4½ in)

WEIGHTS AND LOADING (estimated):

Weight empty	601 kg (1,325 lb)
Max T-O weight	1,134 kg (2,500 lb)
Max disc loading	13·96 kg/m² (3·11 lb/sq ft)

PERFORMANCE (estimated, at max T-O weight):

Max level speed	117 knots (217 km/h; 135 mph)
Cruising speed	104 knots (193 km/h; 120 mph)
Max rate of climb at S/L	472 m (1,550 ft)/min
Service ceiling	4,570 m (15,000 ft)
Hovering ceiling IGE	4,085 m (13,400 ft)
Hovering ceiling OGE	2,440 m (8,000 ft)
Range with max fuel	267 nm (495 km; 308 miles)
Max endurance	4 h

SPITFIRE MARK IV

An accompanying illustration shows the four/five-seat Spitfire Mark IV, of which first details were given by Spitfire Helicopter Company at the Helicopter Association of America's 1979 convention in Las Vegas. The aircraft would be powered by two 313 kW (420 shp) Allison 250-C20B turboshaft engines, driving both the four-blade fully-articulated glassfibre rotor and a pair of reversible ducted fans, mounted at the tips of stub-wings and dispensing with the need for a tail rotor for torque control. Max fuel capacity is 455 litres (120 US gallons).

DIMENSIONS, EXTERNAL:

Diameter of main rotor	10·15 m (33 ft 3½ in)
Length overall	10·91 m (35 ft 9½ in)
Length of fuselage	9·15 m (30 ft 0 in)
Height overall	2·44 m (8 ft 0 in)

WEIGHTS:

Weight empty	1,044 kg (2,300 lb)
Max T-O weight	1,995 kg (4,400 lb)

PERFORMANCE (estimated, at max T-O weight):

Never-exceed speed	187 knots (346 km/h; 215 mph)
Max cruising speed	175 knots (323 km/h; 201 mph)
Max rate of climb at S/L	670 m (2,200 ft)/min
Service ceiling	5,180 m (17,000 ft)
Range with max fuel	418 nm (775 km; 482 miles)
Endurance with max fuel	4 h

SUMMIT

SUMMIT AVIATION INC

Summit Airpark, Middletown, Delaware 19709
Telephone: (302) 834 5400
Telex: 83-5499
MILITARY DIRECTOR: Patrick J. Foley

SUMMIT SENTRY 02-337

Summit Aviation was producing for sale to military customers a version of the Cessna Model T337 which is suitable for a wide range of missions. These include forward air control, helicopter escort, light air-to-ground attack, convoy protection, maritime patrol, six-seat personnel carrier, light cargo transport, aerial photography, psychological warfare and airborne discharge. Special configurations were available for VIP transport, medevac and high-altitude missions. In all configurations day or night capability could be provided.

Summit's modifications, to fit this aircraft for the specific role required by the customer, began with the purchase of a new T337 from Cessna. However, with production of this aircraft having been terminated by Cessna, and with no information received from Summit Aviation, the current status of this conversion programme is not known. All available details can be found in the 1980-81 *Jane's*.

TAYLORCRAFT

TAYLORCRAFT AVIATION CORPORATION

PO Box 243, 14600 Commerce Avenue NE, Alliance, Ohio 44601
Telephone: (216) 823 6675
PRESIDENT: Dorothy A. Feris

Taylorcraft Aviation Corporation, which was re-formed on 1 April 1968 primarily to provide product support, had in production a two-seat trainer/sporting aircraft designated Model F-19 Sportsman 100. Design originated in August 1967, based on the well-known Taylorcraft Model B of second World War origin; construction of pre-production and production aircraft started in 1973.

In early 1980, the Model F-19 (see 1979-80 *Jane's*) had been taken out of production, and the company became involved in the certification of a higher-powered version designated F-21.

Taylorcraft Model F-21 two-seat trainer/sporting aircraft

TAYLORCRAFT MODEL F-21

TYPE: Two-seat trainer/sporting aircraft.

WINGS: Braced high-wing monoplane with V bracing struts each side. Wing section NACA 23012. Dihedral 1°. Composite structure with spruce spars, stamped metal ribs and Dacron covering. Plain wide-span ailerons of similar construction. No flaps. No trim tabs.

FUSELAGE: Welded structure of 4130 steel tube with Dacron covering.

TAIL UNIT: Wire-braced welded steel tube structure with Dacron covering. Trim tab in port elevator.

LANDING GEAR: Non-retractable tailwheel type. Two side Vs and half-axles. Main wheels fitted with 6·00-6 4-ply tyres; swivelling tailwheel has pneumatic tyre of 20 cm (8 in) diameter. Cleveland hydraulic brakes. Parking brake. Wheel fairings, and Aero Skis Model M1500 or M2000, optional.

POWER PLANT: One 88 kW (118 hp) Avco Lycoming O-235-L2C flat-four engine, driving a Sensenich 72-0-50 two-blade fixed-pitch metal propeller with spinner. One fuel tank in each wing, with combined capacity of 45·4 litres (12 US gallons), and one fuel tank in fuselage, immediately aft of firewall, with capacity of 45·4 litres (12 US gallons). Total fuel capacity 90·8 litres (24 US gallons). Oil capacity 5·7 litres (1·5 US gallons).

ACCOMMODATION: Two seats, with shoulder harnesses, side by side in enclosed cabin. Dual controls standard. Carpeted floor. Metal door with sliding window each side. Baggage compartment aft of seats, capacity 37 kg (82 lb), with extension tube, length 4 ft (1·22 m), for fishing equipment. Cargo tiedown straps standard. Accommodation heated and ventilated.

SYSTEM: Electrical system powered by 12V 60A engine-driven alternator, with 12V storage battery. Wiring provisions for landing and strobe lights. Engine-driven vacuum pump when optional blind-flying instrumentation installed.

AVIONICS AND EQUIPMENT: Optional avionics include Genave Alpha 200B com transceiver, King KX 145 com transceiver, KR 86 ADF and KT 76A transponder, EBC-102A emergency locator transmitter, microphone, headsets, speakers and com antenna. Standard equipment includes recording tachometer, sensitive altimeter, glove compartment, wiring provisions for anti-collision and landing lights, navigation lights, and

tiedown rings. Optional equipment includes full blind-flying instrumentation with vacuum system; rate of climb indicator; electric clock; cockpit, Whelen strobe, and landing lights; and choice of Grade A or Dacron covering.

DIMENSIONS, EXTERNAL:

Wing span	10·97 m (36 ft 0 in)
Wing chord (constant)	1·60 m (5 ft 3 in)
Length overall	6·78 m (22 ft 2¾ in)
Height overall	1·98 m (6 ft 6 in)
Tailplane span	3·05 m (10 ft 0 in)
Wheel track	1·83 m (6 ft 0 in)
Propeller diameter	1·83 m (6 ft 0 in)

AREAS:

Wings, gross	17·07 m² (183·71 sq ft)
Ailerons (total)	1·86 m² (20·0 sq ft)
Fin	0·34 m² (3·7 sq ft)
Rudder	0·59 m² (6·3 sq ft)
Tailplane	1·21 m² (13·0 sq ft)
Elevators, incl tab	0·99 m² (10·66 sq ft)

WEIGHTS AND LOADINGS:

Weight empty	449 kg (990 lb)
Max T-O weight	680 kg (1,500 lb)
Max wing loading	39·9 kg/m² (8·17 lb/sq ft)
Max power loading	7·73 kg/kW (12·71 lb/hp)

PERFORMANCE (at max T-O weight):

Max level speed at S/L	108 knots (201 km/h; 125 mph)
Max cruising speed, 75% power at 2,440 m (8,000 ft)	102-106 knots (190-196 km/h; 118-122 mph)
Stalling speed	37·5 knots (69·5 km/h; 43 mph)
Max rate of climb at S/L	267 m (875 ft)/min
Service ceiling	5,485 m (18,000 ft)
T-O run	84 m (275 ft)
T-O to, and landing from, 15 m (50 ft)	107 m (350 ft)
Range with max fuel	347 nm (644 km; 400 miles)

TEXAS HELICOPTER

TEXAS HELICOPTER CORPORATION

1336 South Irving Heights Drive, Irving, Texas 75060
Telephone: (214) 445 0926
Telex: 73-2272

PRESIDENT: Gifton McCreary
CHIEF ENGINEER: Wallace Widtman

TEXAS HELICOPTER WASP

Texas Helicopter developed a series of high-performance agricultural/utility helicopters based on the Bell 47G. Details of these can be found in the 1980-81 *Jane's*.

VARGA

VARGA AIRCRAFT CORPORATION

12250 East Queen Creek Road, Chandler, Arizona 85224
Telephone: (602) 963 4914
PRESIDENT: George Varga Jr
GENERAL MANAGER: George Varga III

Varga Aircraft Corporation purchased from Shinn Engineering Inc in 1965 the full manufacturing rights, tooling and spare parts inventory for the Shinn Model 2150A, and put it into production as the Model 2150A Kachina. The design had originated in the plywood Morrisey Nifty of 1957, built by a former Douglas Aircraft chief test pilot. Ten examples of an improved all-metal version, designated Morrisey Model 2150, were produced by Morrisey Aviation Inc, and this version was described in the 1959-60 *Jane's*. Production rights were then acquired by Shinn Engineering Inc, which put into production the further improved Shinn Model 2150A, with changes to the engine installation, landing gear and interior arrangement. This version was certificated by the FAA on 31 July 1961 and was last described in the 1964-65 *Jane's*. The Varga Kachina is basically the same as the Shinn 2150A.

VARGA MODEL 2150A KACHINA

The current production Kachina is as described. A version with tailwheel landing gear has been developed and certificated by Hibbard Aviation (which see). That company will modify existing 2150As to this new 2150A-TG configuration, while new aircraft with optional tailwheel landing gear can be obtained from Varga.

In early 1981 Varga Aircraft Corporation announced that Canadian certification of the Model 2150A had been gained under Type Certificate 4A19 and the Bilateral Airworthiness Agreement.

TYPE: Two-seat lightweight sporting/training aircraft.

WINGS: Cantilever low-wing monoplane. Wing section NACA 43015. Dihedral 7°. Incidence 3°. No sweepback. Conventional two-spar stressed-skin structure of light alloy. Plain ailerons and two-position trailing-edge flaps of light alloy construction. Ground-adjustable tab in starboard aileron. Glassfibre wingtips.

FUSELAGE: Forward fuselage of welded 4130 steel tube with light alloy covering. Semi-monocoque light alloy structure aft of cockpit. Nose cowl and tailcone of glassfibre.

TAIL UNIT: Cantilever structure of light alloy. Fixed-incidence tailplane. Fin and tailplane have glassfibre tips. Ground-adjustable tab on rudder.

LANDING GEAR: Non-retractable tricycle type. Oleo-pneumatic shock-absorbers in all units. Main units attached directly to rear spar of wing. Steerable nosewheel. Main-wheel tyre pressure 1·66 bars (24 lb/sq in). Nosewheel tyre pressure 1·52 bars (22 lb/sq in). Cleveland hydraulic disc brakes. Parking brake.

Varga Model 2150A Kachina two-seat sporting/training aircraft

POWER PLANT: One 112 kW (150 hp) Avco Lycoming O-320-A2C flat-four engine, driving a Sensenich two-blade fixed-pitch metal propeller type M74DM. Fuel tank in each wing, total capacity 132·5 litres (35 US gallons). Refuelling points in upper surface of wings. Oil capacity 7·5 litres (2 US gallons).

ACCOMMODATION: Two seats in tandem under sideways-opening cockpit canopy, hinged on starboard side. Tinted windscreen and windows. Rear window of canopy completes 360° unrestricted view. Dual controls standard. Baggage space aft of rear seat, capacity 22·7 kg (50 lb). Utility shelf. Accommodation heated and ventilated.

SYSTEMS: Electrical system powered by 60A engine-driven alternator and 12V 25Ah battery. Hydraulic system for brakes only.

AVIONICS AND EQUIPMENT: Optional avionics equipment by Edo-Aire, King and Narco. Standard equipment includes sensitive altimeter, recording tachometer, stall warning device, carpet, soundproofing, navigation lights, landing light in nose cowl, electric starter, quick gascolator drain, and tiedown rings.

DIMENSIONS, EXTERNAL:

Wing span	9·14 m (30 ft 0 in)
Wing chord at root	1·52 m (5 ft 0 in)
Wing chord at tip	1·14 m (3 ft 9 in)
Length overall	6·45 m (21 ft 2 in)
Height overall	2·13 m (7 ft 0 in)
Tailplane span	2·79 m (9 ft 2 in)
Propeller diameter	1·88 m (6 ft 2 in)

AREAS:

Wings, gross	13·38 m² (144 sq ft)
Ailerons (total, incl tab)	2·60 m² (28 sq ft)
Trailing-edge flaps (total)	2·42 m² (26 sq ft)
Fin	0·33 m² (3·6 sq ft)
Rudder (incl tab)	0·53 m² (5·7 sq ft)
Tailplane	1·34 m² (14·4 sq ft)
Elevators	0·62 m² (6·7 sq ft)

WEIGHTS AND LOADINGS:

Weight empty	510 kg (1,125 lb)
Max T-O weight, Utility category	712 kg (1,570 lb)
Max T-O weight, Normal category	824 kg (1,817 lb)
Max wing loading	61·6 kg/m² (12·6 lb/sq ft)
Max power loading	7·36 kg/kW (12·1 lb/hp)

PERFORMANCE (at max T-O weight):

Never-exceed speed	147·5 knots (273 km/h; 170 mph)
Max level speed at S/L	128 knots (238 km/h; 148 mph)
Cruising speed, 75% power at optimum altitude	117 knots (217 km/h; 135 mph)
Cruising speed, 65% power at optimum altitude	110 knots (204 km/h; 127 mph)
Stalling speed, flaps up	49·5 knots (92 km/h; 57 mph)
Stalling speed, flaps down	45 knots (84 km/h; 52 mph)
Max rate of climb at S/L	442 m (1,450 ft)/min
Service ceiling	6,705 m (22,000 ft)
T-O to 15 m (50 ft)	134 m (440 ft)
Landing from 15 m (50 ft)	137 m (450 ft)
Range with max fuel	455 nm (845 km; 525 miles)

VARGA MODEL 2180 KACHINA

The Model 2180 Kachina, which gained FAA certification in early 1981, is generally similar to the Model 2150A Kachina except for the installation of a more powerful engine. In place of the standard Avco Lycoming O-320, a 134 kW (180 hp) Avco Lycoming O-360 is installed on dynafocal mounts, and provided with a larger capacity oil cooler and new cowlings. The 20 per cent increase in power is reported to have made a considerable improvement in performance, but no figures had been released by the company at the time of closing for press.

VOLPAR

VOLPAR INC

HEAD OFFICE AND WORKS: 7929 Hayvenhurst Avenue, Van Nuys, California 91406
Telephone: (213) 787 4393 and 873 5599
Telex: 65-1419
PRESIDENT: William E. Lindsey
EXECUTIVE VICE-PRESIDENT: Frank V. Nixon Jr
VICE-PRESIDENTS:
Robert A. Costley (Marketing)
Glade Johnson (Design)
Kevin Kuniyoshi (Maintenance)
Albert B. Seed (Operations)

Volpar Inc was formed in 1960 by Volitan Aviation Inc, a modification and repair company, and Paragon Tool, Die and Engineering, an established manufacturer of turbine engine components and other precision machined parts. Its original purpose was to design, develop and manufacture tricycle landing gear kits for the Beech Model 18, which Volitan was to install. In 1968, however, Volitan was merged into Volpar, which took over its aviation activities. The Volpar landing gear kit was certificated in 1960, becoming original equipment on new Beech aircraft.

As a follow-up to this modification, Volpar produces kits to convert the Model 18 to turboprop power, using Garrett TPE331 engines. The basic converted aircraft is known as the Turbo 18. After wide acceptance of this, Volpar introduced a 'stretched' version known as the Turboliner, and then the Turboliner II which is approved under SFAR 23 for commuter airline operation.

Using the nacelles that were developed for the Turbo 18 and Turboliner, Volpar next produced an engine installation which it markets under the name of Packaged Power. This is available with any of the Garrett TPE331 series of turboprop engines, with either over-engine or under-engine air intake as required by the particular installation. Packaged Power units have been fitted to such aircraft as the Beechcraft Model 18, de Havilland Dove, Grumman Goose and de Havilland Beaver. It is understood that they are being considered for installation in several aircraft currently in the development stage.

In February 1976 Volpar, in conjunction with Century Aircraft Corporation of Amarillo, Texas, became engaged in the installation of Garrett TPE331-3U-303 turboprop engines in Handley Page H.P.137 Jetstream aircraft. Other programmes have included modification of Cessna Model 402B and 421C aircraft to accept Wild camera systems and navigational sights, under contract to foreign governments.

Volpar was reorganised in 1979, under new ownership, to continue all previous activities and to expand its capabilities into the design and installation of commercial/executive aircraft interiors. To this end, a group of experienced professional designers has been formed, enabling the company to offer specialist services in the

Volpar Turbo 18 conversion of the Beechcraft Model 18, operated by the Environmental Protection Agency of the US government

design, fabrication and installation of such interiors.

VOLPAR (BEECHCRAFT) MODEL 18

The basic Volpar modification converts the Beechcraft Model 18 to a tricycle landing gear configuration, offering substantially lower approach speeds, greatly improved braking and easier ground handling. Cruising speed is improved, as all three wheels retract completely in flight. Futhermore, the aircraft can be kept in hangars with a lower roof clearance, since the overall height is reduced to 2·79 m (9 ft 2 in).

The Volpar kit, which passed all FAA static tests for a maximum landing weight of 4,433 kg (9,772 lb), utilises basic components of the existing main landing gear. The new nose gear is connected to the existing retraction system, where the tailwheel connection was removed. The complete modification can be made without removing the wings or stripping any of the wing skin. All cockpit controls and emergency procedures are unchanged, including the instrument panel wheel position indicator. Existing airstair doors can be retained with only minor modification.

Basically, the modification moves the main landing gear 1·22 m (4 ft 0 in) aft of the original position, attaching it to a welded tube truss that increases the torsional strength of the centre wing structure by 60 per cent in landing configuration. The nose assembly is completely new and includes a streamlined nose fairing which adds 0·67 m (2 ft 2½ in) to the fuselage length. Space inside the fairing can be used for additional equipment, including a weather radar dish of up to 0·305 m (12 in) diameter.

All three wheels are of aluminium and can be fitted with either Goodrich or Goodyear tubed or tubeless tyres, size 8·50-10, ten-ply rating. Main-wheel tyre pressure 4·48 bars (65 lb/sq in), nosewheel tyre pressure 3·10 bars (45 lb/sq in). Shock-absorption is provided by hydraulic oleo struts of Volpar manufacture. Goodrich multiple disc brakes. All three wheels retract forward in less than eight seconds. On the ground the cabin floor is only 1·07 m (3 ft 6 in) off the ground at the door. Wheelbase is 2·62 m (8 ft 7 in). The aircraft will turn on a 1·22 m (4 ft) radius of the inside wheel and a centering device is incorporated on the shimmy damper for take-off and landing.

The current Mk IV Volpar conversion incorporates Goodrich nine-piston full-circle brakes with twice the braking energy and three times the service life of the two-piston type fitted formerly. The new brakes fit on the original gear and are obtainable from either Volpar or Goodrich.

More than 400 sets of Volpar tri-gear have been delivered.

VOLPAR (BEECHCRAFT) TURBO 18

The Volpar Turbo 18 is a Beechcraft Model 18 fitted with the Volpar Mk IV tricycle landing gear and re-engined with two 526 kW (705 ehp) Garrett TPE331-1-101B turboprop engines, flat rated to 451 kW (605 ehp). The wing planform is changed, by extending forward the entire leading-edge inboard of each engine nacelle and carrying the new leading-edge line past the nacelle, so increasing the chord and sweepback to a point outboard of the nacelle. The rectangular wingtip panels of the standard Super 18 are replaced by smaller tips which decrease the wing span and maintain the normal leading-edge sweep to the tip.

Installation of TPE331 engines and Hartzell HC-B3TN-5 three-blade reversible-pitch propellers reduces the empty weight, permitting an increase in fuel or payload. Internal fuel capacity is increased by 379 litres (100 US gallons) by installing new integral tanks in the leading-edge immediately outboard of each engine nacelle. These become the main tanks, each delivering fuel directly to the adjacent engine. They increase the maximum fuel capacity to 2,385 litres (630 US gallons), with a normal capacity of 1,159 litres (306 US gallons).

Air-conditioning and heating installations are available, using engine bleed air. A large cargo door, 1·57 m (5 ft 2 in) wide, with a max height of 1·09 m (3 ft 7 in), can be provided, incorporating the existing airstair door.

The detailed description of the Turboliner (which follows), applies also to the Turbo 18, except that the Turbo 18 does not have the 'stretched' fuselage.

FAA Supplemental Type Approval of the Turbo 18 was received on 17 February 1966. Two aircraft were delivered to the US Environmental Protection Agency in 1966, and production of conversion kits was initiated. Customers include Air Asia of Taiwan, which converted 15 C-45H aircraft for the US government. These aircraft were acquired in 1977 by Ciba Pilatus Aerial Spraying Co Ltd of Switzerland.

In April 1977, Volpar received FAA Supplemental Type Approval to increase the max T-O weight of the Turbo 18 from 4,666 kg (10,286 lb) to 5,216 kg (11,500 lb). The required modification includes strengthening of the centre wing structure and replacement of some landing gear components, involving an increase of only 13·6 kg (30 lb) in the aircraft's structural weight.

DIMENSIONS, EXTERNAL:

Wing span	14·02 m (46 ft 0 in)
Length overall	11·40 m (37 ft 5 in)
Height overall	2·92 m (9 ft 7 in)
Wheelbase	2·62 m (8 ft 7 in)

DIMENSIONS, INTERNAL:

Cabin, excl flight deck:	
Length	3·87 m (12 ft 8½ in)
Max width	1·32 m (4 ft 4 in)
Max height	1·68 m (5 ft 6 in)
Volume	7·36 m^3 (260 cu ft)

WEIGHTS AND LOADINGS (A: basic Turbo 18; B Turbo 18 with Volpar increased gross weight conversion):

Weight empty, basic: A	2,495 kg (5,500 lb)
B	2,508 kg (5,530 lb)
Max payload: A	2,171 kg (4,786 lb)
B	2,708 kg (5,970 lb)
Max T-O weight: A	4,666 kg (10,286 lb)
B	5,216 kg (11,500 lb)
Max zero-fuel weight: A	4,082 kg (9,000 lb)
Max landing weight: A	4,433 kg (9,772 lb)
Max wing loading: A	134·3 kg/m^2 (27·51 lb/sq ft)
B	150·2 kg/m^2 (30·76 lb/sq ft)
Max power loading: A	5·17 kg/kW (8·50 lb/ehp)
B	5·78 kg/kW (9·50 lb/ehp)

PERFORMANCE (A, at max T-O weight):

Max cruising speed at 3,050 m (10,000 ft)	243 knots (451 km/h; 280 mph)
Econ cruising speed at 3,050 m (10,000 ft)	222 knots (412 km/h; 256 mph)
Stalling speed, wheels and flaps up, power off	80 knots (148 km/h; 92 mph)
Stalling speed, wheels and flaps down, power off	77 knots (142 km/h; 88 mph)
Max rate of climb at S/L	521 m (1,710 ft)/min
Service ceiling	7,925 m (26,000 ft)
Service ceiling, one engine out	4,265 m (14,000 ft)
T-O run	507 m (1,665 ft)
T-O to 15 m (50 ft)	725 m (2,380 ft)
Landing from 15 m (50 ft)	642 m (2,107 ft)
Landing run with reverse thrust	265 m (870 ft)
Range with max fuel at 222 knots (412 km/h; 256 mph), 45 min reserves	1,884 nm (3,492 km; 2,170 miles)
Range with max payload, 45 min reserves	400 nm (741 km; 461 miles)

VOLPAR (BEECHCRAFT) TURBOLINER

This is a 15-passenger version of the Volpar (Beechcraft) Turbo 18, with lengthened fuselage, intended for the third-level airline market. Design was started in August 1966 and construction of the prototype began in December 1966. This aircraft flew for the first time on 12 April 1967 and FAA certification was granted on 29 March 1968, the Turboliner being approved for operation at a gross weight of 5,216 kg (11,500 lb).

By the beginning of 1980 about 27 Turboliners had been delivered and were in service with small airlines throughout the world. In March 1970 a Turboliner (N353V), on a delivery flight from Los Angeles to Singapore, set six official international speed records. It carried on board during the flight all necessary spares for one year's normal operation, together with a 1,515 litre (400 US gallon) ferry tank in the fuselage, and was in operation with a commuter airline two days after arrival in Singapore.

TYPE: Twin-turboprop light transport aircraft.

WINGS: Cantilever low-wing monoplane. Wing section NACA 63-015 at station 28·0, NACA 23014 at station 144·5, NACA 23012 at station 260·4. Dihedral 6°. Incidence 5° 20′ at root, 1° at tip. Sweepback 16° 21′ on inner wings, 8° 23′ on outer panels. Steel truss centre-section spar; remainder of structure aluminium semi-monocoque. Plain differential ailerons and plain trailing-edge flaps of conventional aluminium construction. Trim tab in port aileron. Optional Goodrich pneumatic de-icing boots on leading-edges.

FUSELAGE: Conventional aluminium semi-monocoque structure.

TAIL UNIT: Cantilever aluminium semi-monocoque structure with twin endplate fins and rudders. Fixed-incidence tailplane. Trim tabs in rudder and elevators. Optional Goodrich pneumatic de-icing boots on leading-edges.

LANDING GEAR: Volpar electrically-retractable tricycle type. All units retract forward, main wheels into engine nacelles. Volpar hydraulic shock-absorbers. All three wheels size 8·50-10 with Goodrich or Goodyear tubeless or tubed tyres. Main-wheel tyre pressure 5·52 bars (80 lb/sq in); nosewheel tyre pressure 3·10 bars (45 lb/sq in). Goodrich multiple-disc brakes.

POWER PLANT: Two 526 kW (705 ehp) Garrett TPE331-1-101B turboprop engines, each driving a Hartzell HC-B3TN-5 three-blade reversible-pitch propeller with T10176H blades. Four to eight fuel tanks in wings, including new integral main tanks in wing leading-edges outboard of nacelles. Normal fuel capacity 1,159 litres (306 US gallons); max capacity 2,385 litres (630 US gallons). Refuelling points in upper surface of wings. Total oil capacity 11·5 litres (3 US gallons).

ACCOMMODATION: Crew of two and up to 15 passengers. Downward-hinged airstair door on port side at rear of cabin. Optional double-door for freight loading. Seats removable to enable aircraft to be used for freight-carrying. Heating and air-conditioning optional. Baggage space aft of cabin and in each wing.

SYSTEMS: Hydraulic system for brakes only. Electrical supply from two 200A starter/generators and two 24V batteries, for landing gear and flap operation, propeller anti-icing, landing lights, radio and lighting.

AVIONICS AND EQUIPMENT: Blind-flying instrumentation, radio and radar to customer's specification.

DIMENSIONS, EXTERNAL:

Wing span	14·02 m (46 ft 0 in)
Wing chord at root	4·15 m (13 ft 7·36 in)
Wing chord at tip	1·14 m (3 ft 8·94 in)
Wing aspect ratio	5·67
Length overall	13·47 m (44 ft 2½ in)
Height overall	2·92 m (9 ft 7 in)
Tailplane span	4·57 m (15 ft 0 in)
Wheel track	3·94 m (12 ft 11 in)
Wheelbase	3·84 m (12 ft 7 in)
Propeller diameter	2·46 to 2·57 m (8 ft 0⅜ in to 8 ft 5⅜ in)
Passenger door: Height	1·22 m (4 ft 0 in)
Width	0·69 m (2 ft 3 in)
Height to sill	1·07 m (3 ft 6 in)

DIMENSIONS, INTERNAL:

Cabin, excl flight deck: Length	5·94 m (19 ft 6 in)
Max width	1·32 m (4 ft 4 in)
Max height	1·68 m (5 ft 6 in)
Floor area	7·43 m^2 (80 sq ft)
Volume	11·16 m^3 (394 cu ft)
Freight hold (aft of cabin) volume	0·65 m^3 (23 cu ft)
Freight holds (wings) volume (total)	0·91 m^3 (32 cu ft)

AREAS:

Wings, gross	34·75 m^2 (374 sq ft)
Ailerons (total)	2·47 m^2 (26·6 sq ft)
Trailing-edge flaps (total)	2·62 m^2 (28·2 sq ft)
Fins (total)	1·51 m^2 (16·3 sq ft)
Rudders (total)	16·05 m^2 (17·28 sq ft)
Tailplane	35·49 m^2 (38·2 sq ft)
Elevators, incl tab	25·28 m^2 (27·22 sq ft)

WEIGHTS AND LOADINGS:

Weight empty: Cargo version	2,676 kg (5,900 lb)
Airliner	2,993 kg (6,600 lb)
Max T-O weight	5,216 kg (11,500 lb)
Max zero-fuel weight	4,762 kg (10,500 lb)
Max landing weight	4,989 kg (11,000 lb)
Max wing loading	150·1 kg/m^2 (30·75 lb/sq ft)
Max power loading	4·96 kg/kW (8·15 lb/ehp)

PERFORMANCE (at max T-O weight):

Max level and cruising speed at 3,050 m (10,000 ft)	243 knots (451 km/h; 280 mph)
Econ cruising speed at 3,050 m (10,000 ft)	222 knots (412 km/h; 256 mph)
Stalling speed, wheels and flaps up, power off	84 knots (154·5 km/h; 96 mph)
Stalling speed, wheels and flaps down, power off	80 knots (148·5 km/h; 92 mph)
Max rate of climb at S/L	463 m (1,520 ft)/min
Service ceiling	7,315 m (24,000 ft)
Service ceiling, one engine out	3,960 m (13,000 ft)
T-O run	570 m (1,870 ft)
T-O to 15 m (50 ft)	989 m (3,245 ft)
Landing from 15 m (50 ft)	762 m (2,500 ft)
Landing run	317 m (1,040 ft)
Range with max fuel, 45 min reserves	1,802 nm (3,340 km; 2,076 miles)
Range with max payload, 45 min reserves	300 nm (556 km; 346 miles)

VOLPAR (BEECHCRAFT) TURBOLINER II

The Turboliner II is basically a Turboliner modified to meet the requirements of SFAR 23. The prototype was completed in February 1970 and received certification in July 1970. Dimensions and performance are the same as those given for the Turboliner.

Recent conversions incorporate a number of improvements, including battery temperature indicators, a fail-safe Hydro-Aire Hytrol anti-skid braking system, installation of a 38,000 BTU Janitrol heater in the nose for ground heating of cockpit and engine nacelles, and modification to the standard Volpar side-opening cargo door. This now incorporates an inward-opening door 0·66 m (2 ft 2 in) in width and with a minimum height of 1·16 m (3 ft 9½ in), which may be opened in flight to permit the airdrop of firefighting personnel or cargo. During 1974 Volpar, in conjunction with Sierracin Manufacturing Company, developed electrically heated windscreens for the Turboliner II. Complete installation kits are available also for other Volpar conversions.

During 1977 Volpar developed and delivered a specially modified long-range Turboliner II to Société Interthon of Port Peche, France. This aircraft is identified as the 'Asterix' series VLR Turboliner, and is designed for long-range maritime reconnaissance. It can carry a complement of observers, offshore nav/com systems, proprietary marine search equipment and ocean survival gear. The standard 'Asterix' aircraft has 2,385 litres (630 US gallons) of fuel in wing tanks, plus 757 litres (200 US gallons) in two removable cabin fuel tanks; total capacity 3,142 litres (830 US gallons), providing an endurance of almost 12 h and range of 2,200 nm (4,076 km; 2,533 miles) with reserves. Special bubble-type windows are provided for both pilots; other equipment includes an inward/outward-opening cargo door for airdrops to surface vessels, lavatory, navigation tables, refreshment console, and provisions for camera and special search equipment. Certification by the FAA and French DGAC was awarded in December 1977.

VOUGHT

VOUGHT CORPORATION (a subsidiary of THE LTV CORPORATION)

HEAD OFFICE: PO Box 225907, Dallas, Texas 75265
Telephone: (214) 266 2011
PRESIDENT AND CHIEF EXECUTIVE OFFICER:
Robert L. Kirk
SENIOR VICE-PRESIDENTS:
E. F. Cvetko (Operations)
R. N. Parker (Research and Engineering)
B. M. Smith
J. J. Welch Jr
VICE-PRESIDENTS:
K. R. Chapman (Planning)
Michael Collins (Field Operations)
Philip R. Cowen (Finance)
Felix Fenter (Research & Advanced Technology)
H. M. Fish (International)
J. R. Grace (Materiel)
D. R. Hagler (General Counsel)
P. W. Hare (Miniature Systems)
C. H. McKinley (Missile Development Engineer)
Peter G. Paraskos (Domestic Operations, Programme Development)
R. J. Patton (Aircraft Development Engineer)
F. W. Randall (Aircraft Specialty Products)
J. J. Ryan
W. L. Shepard (Battlefield Interdiction Missile Systems)
C. E. Snyder (Human Resources)
R. W. Stoner (Operations & Development & Programme Support)
W. E. Stoney (European & Middle East Operations)
J. P. Woolnough (R & D and International Operations—MLRS)
Robert A. Zummo (Subcontract Programmes)
DIRECTOR, CORPORATE COMMUNICATIONS: Beal Box

The former Chance Vought Aircraft Inc, founded in 1917 and a leading producer of aircraft for the US Navy throughout its history, became the Chance Vought Corporation on 31 December 1960. On 31 August 1961, Chance Vought Corporation merged with Ling-Temco Electronics Inc, to form a combined company known as Ling-Temco-Vought Inc (now The LTV Corporation).

After a succession of reorganisations, what is now known as Vought Corporation took over responsibility for all LTV's aircraft and space activities, aerospace support and training equipment from 1 January 1976, including current construction of Boeing 747 tail assemblies, McDonnell Douglas DC-10 tailplanes and elevators, and Lockheed C-130 control surfaces.

On 23 February 1979 the company announced receipt of a multi-million dollar contract from The Boeing Company covering the construction of up to 300 complete tailplane assemblies for the Boeing Model 767. Deliveries began during November 1980, with planned production until June 1985 of six to nine units per month. On 11 October 1979 the company contracted also to build the complete tail unit for the Boeing 757, including the aft fuselage section.

Vought has in production for the US Air National Guard the A-7K, a two-seat trainer/fighter version of the US Air Force's A-7D Corsair II; the Scout launch vehicle for NASA; components for manned and unmanned space vehicles; advanced missile, guidance, control and environmental systems; Airtrans automatic transit systems; and advanced thermal protection systems.

Under an agreement signed in March 1980, Vought and MBB of West Germany have proposed jointly a modified version of the MBB Fantrainer aircraft to meet USAF requirements for a next-generation trainer (NGT).

In December 1974, Vought's Low Volume RamJet (LVRJ) propulsion system test vehicle made a successful first flight, covering a distance of more than 30 nm (56 km; 35 miles) and attaining a speed in excess of 1,259 knots (2,334 km/h; 1,450 mph). In four subsequent flights, extending into 1976, the vehicle demonstrated speeds well above 1,476 knots (2,736 km/h; 1,700 mph) and a range of more than 87 nm (161 km; 100 miles). Air-launched from a US Navy A-7 Corsair II, the flight test vehicle is 4·57 m (15 ft) long and 0·38 m (1 ft 3 in) in diameter. It can be scaled up or down for air-to-air, surface-to-air or surface-to-surface applications. Development is continuing as part of the US Navy's Supersonic Tactical Missile programme, and on 21 April 1979 an integral rocket/ramjet vehicle, featuring lower-cost propulsion system components and elements of a guidance system, was test flown with equal success.

Vought was selected in April 1980 as prime contractor for the Multiple Launch Rocket System for the US Army and NATO allies.

Vought A-7E Corsair II carrying FLIR pod and Mk 61 tactical nuclear weapon, on board USS *Kitty Hawk*

VOUGHT CORSAIR II

US military designation: A-7

On 11 February 1964 the US Navy named the former LTV Aerospace Corporation winner of a design competition for a single-seat carrier-based light attack aircraft. The requirement was for a subsonic aircraft able to carry a greater load of non-nuclear weapons than the A-4E Skyhawk. To keep costs to a minimum and speed delivery it had been stipulated by the Navy that the new aircraft should be based on an existing design; the LTV design study was based, therefore, on the F-8 Crusader. An initial contract to develop and build three aircraft, under the designation A-7A, was awarded on 19 March 1964; first flight was made on 27 September 1965.

Since that time several versions of the A-7 have been evolved as Corsair IIs, for the US Navy, the US Air Force and the Hellenic Air Force. Details of the A-7A, A-7B, A-7C and A-7D can be found in the 1979-80 and earlier editions of *Jane's*. The most recent versions are:

TA-7C. Sixty A-7Bs and A-7Cs have been converted into tandem two-seat trainers, with operational capability, under this designation. The first of them (154477) flew for the first time on 17 December 1976. Flight refuelling capability, gun and weapon pylons are retained, and TA-7Cs have the modern navigation/weapons delivery system used in the A-7E. Powered by non-afterburning 59·6 kN (13,400 lb st) Pratt & Whitney TF30-P-408 engine. Entered service with VA-122 and VA-174 in 1978.

A-7E. Developed version for the US Navy, equipped as a light attack/close air support/interdiction aircraft. All except first 67 aircraft (since redesignated A-7C) powered by Allison TF41-A-2 (Spey) non-afterburning turbofan engine, which provides 66·8 kN (15,000 lb st). First flight 25 November 1968; deliveries began 14 July 1969. The A-7E entered combat service in Southeast Asia with Attack Squadrons 146 and 147 in May 1970, and equipped 26 Navy squadrons in 1980. Production was at a rate of about two a month until the last of 596 A-7Es was delivered in March 1981.

In early 1977 production began of an **A-7E FLIR** version, called formerly TRAM (target recognition and attack multi-sensor system). This has a 327 kg (720 lb) pod under the starboard wing to house equipment which includes a Texas Instruments FLIR gimballed sensor, and a Marconi raster-HUD cockpit display, to provide improved night capability. Deliveries of new-production FLIR-equipped A-7Es to the Navy began on 15 September 1978. By the beginning of 1981, orders covered the production of 85 FLIR pods and 144 FLIR aircraft provisions. Plans call for eventual delivery of 91 pods and 231 FLIR installations.

A-7H. Land-based version of A-7E, retaining the folding wings. First A-7H flew for first time on 6 May 1975. Total of 60 delivered to the Hellenic Air Force. In service with Nos. 338, 340 and 345 Squadrons.

TA-7H. Two-seat version for the Hellenic Air Force, with an Allison TF41-A-400 engine. Configuration similar to TA-7C, but no in-flight refuelling capability. First TA-7H flew for first time on 4 March 1980. Five delivered between July and September 1980.

A-7K. Two-seat version of the US Air Force's A-7D, with fuselage lengthened by 0·86 m (2 ft 10 in). Total of 30 so far ordered, of a planned procurement of 42 for service with the US Air National Guard. Two will be assigned to each of the ANG's 13 A-7D units, in 11 States, and 16 to the 162nd Tactical Fighter Training Group in Tucson, Arizona. Basically two-seat trainers, these aircraft retain combat capability. They were preceded by a prototype, converted from an A-7D and delivered to the ANG in December 1980. The first production A-7K entered service in April 1981.

A-7P. Designation of 20 refurbished A-7As to be supplied to Portuguese Air Force from 1981. Refitted with TF30-P-408 engines and A-7E-standard avionics. Deliveries began on 18 August 1981, following a first flight during the preceding month.

Orders for all versions totalled 1,539 by 1 January 1980, of which more than 1,500 had been delivered by mid-1980.

The following description, which applies in particular to the A-7E, is generally applicable to other versions of the A-7 except as detailed under the individual model listings:

TYPE: Subsonic single-seat tactical fighter.

WINGS: Cantilever high-wing monoplane. Wing section NACA 65A007. Anhedral 5°. Incidence −1°. Wing sweepback at quarter-chord 35°. Outer wing sections fold upward for carrier parking and, in the A-7H, to allow best utilisation of revetments at combat airfields. All-metal multi-spar structure with integrally-stiffened aluminium alloy upper and lower skins. Plain sealed inset aluminium ailerons, outboard of wing fold, are actuated by fully-triplicated hydraulic system. Leading-edge flaps. Large single-slotted trailing-edge flaps. Spoiler above each wing forward of flaps.

FUSELAGE: All-metal semi-monocoque structure. Large door-type ventral speed-brake under centre-fuselage.

TAIL UNIT: Large vertical fin and rudder, swept back 44·28° at quarter-chord. One-piece all-moving tailplane, swept back 45° at quarter-chord and set at dihedral angle of 5° 25′. Tailplane is operated by triplicated hydraulic systems, and the rudder powered by two systems.

LANDING GEAR: Hydraulically-retractable tricycle type, with single wheel on each main unit and twin-wheel nose unit. Main wheels retract forward into fuselage, nosewheels aft. Main wheels and tyres size 28 × 9-12; nosewheels and tyres size 22 × 5·50. Nose gear launch bar for carrier catapulting. Sting-type arrester hook under rear fuselage for carrier landings, emergency landings or aborted take-offs. Anti-skid brake system.

POWER PLANT: One Allison TF41-A-2 (Rolls-Royce Spey) non-afterburning turbofan engine, rated at 66·7 kN (15,000 lb st). The A-7E has a pneumatic starter requiring ground air supply; A-7H, TA-7H and A-7K engines have self-start capability through the medium of battery-powered electric motor that actuates a small gas turbine engine (jet fuel starter) which, in turn, starts the main engine through the gearbox. The engine has self-contained ignition for start/airstart, automatic relight and selective ignition. Integral fuel tanks in wings and additional fuselage tanks. Maximum internal fuel 5,678 litres (1,500 US gallons). Maximum external fuel 4,542 litres (1,200 US gallons). The A-7E and A-7H have the fuselage sump tank filled with polyurethane fire-suppressing foam. Some fuselage tanks and fuel lines self-sealing. Flight refuelling capability of A-7E provided by a probe and drogue system; A-7K has boom receptacle above fuselage on port side in line with wing leading-edge. The A-7H and TA-7H do not have an air refuelling capability. Boron carbide (HFC) engine armour.

Accommodation: Pilot on McDonnell Douglas Escapac rocket-powered ejection system, complete with US Navy life support system on the A-7E/H. Escape system provides a fully-inflated parachute three seconds after sequence initiation; positive seat/man separation and stabilisation of the ejected seat and pilot. Boron carbide (HFC) cockpit armour.

Systems: Triple-redundant hydraulic system for flight controls; double-redundant system for flaps, brakes and landing gear retraction. Liquid oxygen system. An air-conditioning unit using engine bleed air provides pressurisation and cooling for the cockpit and cooling for certain avionics components. Automatic flight control system provides control-stick steering, altitude hold, heading hold, heading pre-select and attitude hold, which is coupled for automatic carrier landings. Ram-air turbine provides emergency hydraulic pressure and electrical power down to airspeeds below those used in normal landing approaches.

Avionics and Equipment: The navigation/weapon delivery system is the heart of the A-7E/H light attack aircraft. It performs continuously the computations needed for greatly increased delivery accuracy, and for manoeuvering freedom during navigation to a target and the attack, weapon release, pull up, and safe return phases of the mission. The system not only provides the pilot with a number of options during navigation and weapon delivery, but also relieves him of much of his workload. The AN/ASN-91(V) navigation/weapon delivery computer is the primary element of the system, in constant 'conversation' with basic electronic sensors, and computes and displays continuously present position, using computed position and stored data to calculate navigation and weapon delivery solutions, and monitors the reliability of data inputs and outputs. An AN/ASN-90(V) inertial measurement set is the basic three-axis reference system for navigation and weapon delivery. AN/APN-190(V) Doppler measures groundspeed and drift angle. AN/APQ-126(V) forward-looking radar provides the pilot with ten modes of operation: air-to-ground ranging; terrain following; terrain avoidance; ground mapping, shaped beam; ground mapping, pencil beam; beacon; cross-scan terrain avoidance; cross-scan ground mapping, pencil; TV; and Shrike integrated display system. An AN/AVQ-7(V) HUD receives and displays computed attack, navigation and landing data from the tactical computer; aircraft performance data from flight sensors: and discrete signals from various aircraft systems. CP-953A/AJQ air data computer is a solid-state servo-mechanical analogue computer which measures and computes continuously required altitude and airspeed information. The armament station control unit integrates and controls the weapon release system; it supplies electrical signals to arm and release or jettison external stores; controls and fires the Vulcan cannon; furnishes store-type information to the tactical computer; supplies weapon status information to the pilot; determines weapon release according to priority of stations; and determines compatibility of selected release mode with the stores on selected stations. Standard aeronautical charts reproduced on 35 mm film in full colour are stored in an AN/ASN-99 projected map display set which, as a subsystem of the tactical computer, provides a continuous display of the aircraft's geographical position. Other avionics include AN/ASN-54 approach power compensator; AN/ASW-30 AFCS; ARA-63 ACLS; dual AN/ARC-159 UHF com; AN/ARN-84 Tacan; AN/APX-72 IFF transponder; AN/APN-154 radar beacon; AN/ASW-25 data link; AN/ARA-50 ADF; and AN/AIC-25 audio system. ECM equipment includes ALR-45/50 internal homing and warning systems; ALQ-126 active ECM; chaff/flare dispensers; and external pod-mounted systems compatible with the aircraft's internal sytems.

Armament: A wide range of stores, to a total weight of more than 6,805 kg (15,000 lb), can be carried on six underwing pylons and two fuselage weapon stations, the latter suitable for Sidewinder air-to-air missiles. Two outboard pylons on each wing can each accommodate a load of 1,587 kg (3,500 lb). Inboard pylon on each wing can carry 1,134 kg (2,500 lb). Two fuselage weapon stations, one on each side, can each carry load of 227 kg (500 lb). Weapons carried include air-to-air and air-to-ground (anti-tank and anti-radar missiles); electro-optical (TV) and laser guided weapons; general-purpose bombs; bomblet dispensers; rockets; gun pods; and auxiliary fuel tanks. In addition, an M61A-1 Vulcan 20 mm cannon is mounted in the port side of the fuselage. This has 1,000-round ammunition storage and selected firing rates of 4,000 or 6,000 rds/min. Strike camera in lower rear fuselage for damage assessment.

Prototype of the Vought A-7K, a two-seat trainer/fighter for the US Air National Guard

Dimensions, external:

Wing span	11·80 m (38 ft 9 in)
Width, wings folded	7·24 m (23 ft 9 in)
Wing chord at root	4·72 m (15 ft 6 in)
Wing chord at tip	1·18 m (3 ft 10¼ in)
Wing aspect ratio	4
Length overall	14·06 m (46 ft 1½ in)
Height overall	4·90 m (16 ft 0¾ in)
Tailplane span	5·52 m (18 ft 1½ in)
Wheel track	2·90 m (9 ft 6 in)

Areas:

Wings, gross	34·83 m² (375 sq ft)
Ailerons (total)	1·85 m² (19·94 sq ft)
Trailing-edge flaps (total)	4·04 m² (43·48 sq ft)
Leading-edge flaps (total)	3·46 m² (37·24 sq ft)
Spoilers (total)	0·43 m² (4·60 sq ft)
Deflector	0·32 m² (3·44 sq ft)
Fin	10·70 m² (115·20 sq ft)
Rudder	1·40 m² (15·04 sq ft)
Tailplane	5·22 m² (56·19 sq ft)
Speed-brake	2·32 m² (25·00 sq ft)

Weights:

Weight empty	8,668 kg (19,111 lb)
Max T-O weight	19,050 kg (42,000 lb)

Performance:

Max level speed at S/L
600 knots (1,112 km/h; 691 mph)
Max level speed at 1,525 m (5,000 ft):
with 12 Mk 82 bombs
562 knots (1,040 km/h; 646 mph)
after dropping bombs
595 knots (1,102 km/h; 685 mph)
Sustained manoeuvring performance at 1,525 m (5,000 ft), at AUW of 13,047 kg (28,765 lb) with 6 pylons and 2 Sidewinder missiles
1,770 m (5,800 ft) turning radius at 4g and 500 knots (925 km/h; 575 mph)
T-O run at max T-O weight 1,830 m (6,000 ft)
Ferry range:
max internal fuel
1,981 nm (3,671 km; 2,281 miles)
max internal and external fuel
2,485 nm (4,604 km; 2,861 miles)

WEATHERLY

WEATHERLY AVIATION COMPANY INC

2304 San Felipe Road, Hollister, California 95023
Telephone: (408) 637 5534
President: John C. Weatherly

Weatherly Aviation developed in the early 1960s a conversion of the Fairchild M-62 for service in an agricultural role, and a total of 19 of the resulting Weatherly WM-62Cs was produced by the Autumn of 1965. Most of these were powered by Continental W670 radial engines, but a few had Pratt & Whitney R-985s. On the basis of this conversion programme, a somewhat larger aircraft, designated Model 201, was developed subsequently and remained in production in several successive versions until late 1979, when construction of the Model 201C ended after more than 100 examples of all versions had been built. A description of the Model 201C can be found in the 1979-80 *Jane's*.

Weatherly developed for it small sweptback vanes which can be attached to, or removed easily from, each wingtip. These have two primary functions: to increase the effective swath width; and to reduce the amount of spray materials lost from the swath area. With the vanes installed the swath width is increased by 1·80 to 2·75 m (6 to 9 ft), but because of reduced outboard airflow the vanes are not used when dispersing dust or other solid material. Normal flight characteristics are not impaired with the vanes installed. Take-off run is reduced by a few feet, and cruising speed is increased approximately 1·7 knots (3·2 km/h; 2 mph), due to a reduction in induced drag.

The turboprop Weatherly Model 620TP, fitted with wingtip vanes

WEATHERLY MODEL 620

For 1980 Weatherly introduced two new versions of a slightly larger, sturdier, and faster aircraft than the Model 201C, which has the designation Weatherly Model 620. Both are available with the optional small sweptback wingtip vanes which were developed for the Model 201. These can be folded back beneath the wing for hangar storage.

Two versions of the Model 620 are available:

Model 620. Basic version, with a 336 kW (450 hp) Pratt & Whitney R-985 radial engine, and with a 1,268 litre (335 US gallon) hopper for liquid or dry chemicals.

Model 620TP. Version with a 373 kW (500 shp) Pratt & Whitney Aircraft of Canada PT6A-11AG turboprop engine, and a 1,287 litre (340 US gallon) hopper.

Few details are available except that these two versions are basically the same from the engine firewall aft; it is believed that in most constructional details they are similar to the Model 201C.

Dimensions, external:

Wing span 12·50 m (41 ft 0 in)
Wing span, with optional wingtip vanes
14·33 m (47 ft 0 in)

Weights (A: Model 620; B: Model 620TP):

Weight empty, with spray system:

A	1,270 kg (2,800 lb)
B	1,134 kg (2,500 lb)
Design T-O weight: A, B	1,814 kg (4,000 lb)

Max T-O weight (Restricted category):

A	2,495 kg (5,500 lb)
B	2,449 kg (5,400 lb)

Performance (at design T-O weight):

Stalling speed: A, B 50·4 knots (93·3 km/h; 58 mph)

Endurance: A	2 h 30 min
B	2 h 0 min

WING

WING AIRCRAFT COMPANY

Head Office: 2925 Columbia Street, Torrance, California 90503
Telephone: (213) 533 0174
President: George S. Wing
Chief Engineer: Glenn Errington

Wing Aircraft Company was incorporated on 27 June 1966, when it became completely separated from its parent company, Hi-Shear Corporation. In 1971 the company's twin-engined Derringer was exhibited at the Paris Air Show, just prior to the start of planned production, but for several reasons this did not proceed, and no further progress was made until 1978. Mr George Wing then resigned from Chairmanship of Hi-Shear Corporation to concentrate on production of the Derringer. Production deliveries began in 1980.

First production example of the two-seat twin-engined Wing D-1 Derringer *(J. M. G. Gradidge)*

WING D-1 DERRINGER

Named after the well-known American compact pocket pistol, the twin-engined high-performance Derringer utilises manufacturing techniques that were new to the lightplane industry at the time of its inception. Butt-jointed, flush-riveted, chemically-milled and stretch-formed skins are used throughout the airframe. This simplifies the achievement of a flush surface finish and provides integral stiffness, since the skins are left thicker at the points where additional strength is needed. Assembly is simplified as, for example, each wing has a single stretch-formed chemically-milled skin that acts also as an integral fuel tank.

Design began originally in June 1960; the prototype flew for the first time on 1 May 1962, powered by two 86 kW (115 hp) engines. Used to prove the design concept, it was retired after logging more than 300 flying hours. The second aircraft, with 112 kW (150 hp) engines, was redesigned to production standards and flew for the first time on 19 November 1964, but was lost subsequently during flight testing. A third aircraft was used for static structural testing, and the fourth, with 119 kW (160 hp) engines, flew on 25 August 1965. FAA type certification under CAR Part 3 was gained on 20 December 1966.

Type: Two-seat twin-engined light aircraft.

Wings: Cantilever low-wing monoplane. Wing section NACA 65_2-415. Dihedral 6° from roots. Incidence 1°. All-metal two-spar structure of light alloy, with chemically milled skin. Plain ailerons of light alloy with piano-type hinges. Electrically-actuated slotted trailing-edge flaps of similar construction. Bungee-type trim control.

Fuselage: All-metal semi-monocoque light alloy structure, except for glassfibre nose.

Tail Unit: Cantilever structure of light alloy with swept vertical surfaces. Tailplane has variable incidence. Rudder and elevator trim tabs.

Landing Gear: Electrically-actuated retractable tricycle type with single wheel on each unit. Nosewheel retracts forward, main units upward into nacelles. Oleo-pneumatic shock-absorbers. Steerable nosewheel has tubeless tyre size 5·00-5, pressure 2·41 bars (35 lb/sq in); main wheels have low-profile tyres size 6·00-6, pressure 2·76 bars (40 lb/sq in). Cleveland single-disc hydraulic brakes. Parking brake.

Power Plant: Two 119 kW (160 hp) Avco Lycoming IO-320-B1C or -C1A flat-four engines, each driving a Hartzell HC-C2YL/8450-18 two-blade constant-speed fully-feathering metal propeller with spinner. Engine nacelles of glassfibre construction. Integral fuel tank in the leading-edge of each outer wing, with combined capacity of 333 litres (88 US gallons), of which 329 litres (87 US gallons) are usable. Refuelling point at each wingtip. Oil capacity 15 litres (4 US gallons).

Accommodation: Two adjustable seats, with seat belts and inertia reel shoulder harness, side by side under large upward-hinged canopy with opaque top. Tinted windscreen and windows. Dual controls standard. Baggage compartment with capacity of 113 kg (250 lb). Accommodation is heated and ventilated.

Systems: Electrical system includes two 60A alternators and a 12V 35Ah storage battery. 15,000 BTU heater/defroster. Vacuum system for flight instruments only. Hydraulic system for brakes only. Oxygen system optional.

Avionics and Equipment: Standard avionics include two 360-channel nav/coms; two VORs, one with glideslope indicator; transponder; ADF; three-light marker beacon receiver; related antennae, audio amplifier, cabin speaker, microphone and headset. Optional avionics include autopilot with three-axis electric trim, and DME. Standard equipment includes full IFR instrumentation; exhaust gas and outside air temperature gauges; turn and slip indicator; flap position indicator; rate of climb indicator; emergency locator transmitter; cabin soundproofing; map pockets; coat hooks; pilot's storm window; heated pitot; stall warning device; instrument lights; anti-collision, landing and navigation lights; corrosion proofing; and towbar. Optional equipment includes propeller unfeathering accumulators, Hartzell synchrophaser, inflatable canopy seal, and external power receptacle.

Dimensions, external:
Wing span 8·89 m (29 ft 2 in)
Wing chord, constant 1·27 m (4 ft 2 in)
Wing aspect ratio 7
Length overall 7·01 m (23 ft 0 in)
Height overall 2·44 m (8 ft 0 in)
Tailplane span 3·30 m (10 ft 10 in)
Wheel track 3·30 m (10 ft 10 in)
Wheelbase 1·64 m (5 ft 4½ in)
Propeller diameter 1·68 m (5 ft 6 in)
Propeller ground clearance 0·20 m (8 in)

Dimensions, internal:
Cabin: Length 2·54 m (8 ft 4 in)
Max width 1·12 m (3 ft 8 in)
Max height 1·22 m (4 ft 0 in)
Floor area 1·11 m² (12 sq ft)
Volume 1·59 m³ (56 cu ft)
Baggage hold volume 0·62 m³ (22 cu ft)

Areas:
Wings, gross 11·24 m² (121 sq ft)
Ailerons (total) 0·74 m² (8·00 sq ft)
Flaps (total) 1·11 m² (12·00 sq ft)
Fin 1·08 m² (11·65 sq ft)
Rudder, incl tab 0·48 m² (5·18 sq ft)
Tailplane 1·72 m² (18·56 sq ft)
Elevators, incl tabs 1·07 m² (11·51 sq ft)

Weights and Loadings:
Weight empty, standard 952 kg (2,100 lb)
Max T-O weight 1,383 kg (3,050 lb)
Max landing weight 1,315 kg (2,900 lb)
Max wing loading 123 kg/m² (25·2 lb/sq ft)
Max power loading 5·81 kg/kW (9·5 lb/hp)

Performance (at max T-O weight):
Max level speed at S/L 201 knots (373 km/h; 232 mph)
Max cruising speed, 75% power at 3,050 m (10,000 ft) 190 knots (352 km/h; 219 mph)
Econ cruising speed, 65% power at 3,050 m (10,000 ft) 182 knots (338 km/h; 210 mph)
Approach speed 82·5 knots (153 km/h; 95 mph)
Stalling speed, flaps and gear up 69·5 knots (129 km/h; 80 mph)
Stalling speed, flaps and gear down 63 knots (116 km/h; 72 mph)
Max rate of climb at S/L 518 m (1,700 ft)/min
Rate of climb at S/L, one engine out 128 m (420 ft)/min
Service ceiling 5,975 m (19,600 ft)
Service ceiling, one engine out 2,440 m (8,000 ft)
T-O run 280 m (920 ft)
T-O to 15 m (50 ft) 457 m (1,500 ft)
Landing from 15 m (50 ft) 640 m (2,100 ft)
Landing run 378 m (1,240 ft)
Range with max fuel, 65% power at 3,050 m (10,000 ft), no reserves 1,007 nm (1,866 km; 1,160 miles)
Endurance at econ cruising speed, 65% power 5 h 30 min

YUGOSLAVIA

SOKO

SOKO VAZDUHOPLOVNA INDUSTRIJA, RO VAZDUHOPLOVSTVO

79000 Mostar
Telephone: (088) 22-121, 25-120, 33-831, 35-244, 35-541, 37-943
Telex: 46-180 YU SOKOMO
Director, Aircraft Division: Dr Krsto Draca
Director, Aircraft Factory: Zijo Kreso, BSc
Director, Marketing: Novica Djurica, BSc
Director, Research and Development:
Dipl Ing Milenko Pjescić

Founded in 1951, this company manufactures aircraft of its own design and is participating, with Romania, in developing and producing the Orao strike aircraft described under the SOKO/CNIAR heading in the International section. Details of its SL-40 Liska motor glider can be found in the Sailplanes section.

Soko is also building under licence the Aérospatiale/Westland Gazelle helicopter, on behalf of the Yugoslav government.

SOKO G2-A GALEB (SEAGULL)

Design of the Galeb was started in 1957. Construction of two prototypes began in 1959 and the first of these flew for the first time in May 1961. Production began in 1963 and has continued since that time, to fulfil Yugoslav and export orders. First overseas operator was the Zambian Air Force, in early 1971.

There are two production versions of the Galeb:

G2-A. Standard version for Yugoslav Air Force. Progressive design improvements included optional cockpit air-conditioning system. Production completed.

G-2A-E. Export version, with updated equipment. First flown in late 1974. Series production began in 1975, reportedly for the Libyan Arab Air Force.

Production of the Galeb will continue into 1983 to meet current commitments.

Type: Two-seat armed jet basic trainer, designed for load factors of $+8g$ and $-4g$.

Wings: Cantilever low-wing monoplane. Wing section NACA 64A213·5 at root, NACA 64A212·0 at tip. Dihedral 1° 30′. No incidence. Sweepback at quarter-chord 4° 19′. Conventional light alloy two-spar stressed-skin structure, consisting of a centre-section, integral with the fuselage, and two outer panels which can be removed easily. Manually-operated light alloy ailerons. Trim tab on port aileron. Hydraulically-actuated Fowler flaps. No de-icing system.

Fuselage: Light alloy semi-monocoque structure in two portions, joined together by four bolts at frame aft of wing trailing-edge. Rear portion removable for engine servicing. Two hydraulically-actuated door-type airbrakes under centre-fuselage.

Tail Unit: Cantilever light alloy stressed-skin structure. Fixed-incidence tailplane. Rudder and elevators statically and dynamically balanced and manually operated. Manually-operated trim tab in each elevator. VHF radio aerial forms tip of fin.

Landing Gear: Hydraulically-retractable tricycle type, with single wheel on each unit. Nosewheel retracts forward, main units inward into wings. Oleo-pneumatic shock-absorbers manufactured by Prva Petoletka of Trstenik. Dunlop main wheels and tyres size 23 × 7·25-10, pressure 4·41 bars (64 lb/sq in). Dunlop nosewheel and tyre size 6·50-5·5 TC, pressure 3·43 bars (49·8 lb/sq in). Prva Petoletka hydraulic differential disc brakes, toe-operated from both cockpits.

Power Plant: One Rolls-Royce Viper 11 Mk 22-6 turbojet engine, rated at 11·12 kN (2,500 lb st). Two flexible fuel tanks aft of cockpits, with total capacity of 780 kg (1,720 lb). Two jettisonable wingtip tanks, each with capacity of 170 kg (375 lb). Refuelling point on upper part of fuselage aft of cockpits. Fuel system designed to permit up to 15 s of inverted flight. Oil capacity 6·25 litres (1·4 Imp gallons).

Accommodation: Crew of two in tandem on BAe (Folland) Type 1-B fully-automatic lightweight ejection seats. Separate sideways-hinged (to starboard) jettisonable canopy over each cockpit. Cockpit air-conditioning to special order only.

Soko G2-A Galeb two-seat basic training aircraft (Rolls-Royce Viper 11 turbojet engine)

Systems: Hydraulic system, pressure 58·5-69 bars (850-1,000 lb/sq in), for landing gear, airbrakes and flaps. Separate system for wheel brakes. Pneumatic system for armament cocking. Electrical system includes 6kW 24V generator, 24V battery, and inverter to provide 115V 400Hz AC supply for instruments. G2-A has low-pressure oxygen system, capacity 1,450 litres. G-2A-E has high-pressure oxygen system.

Avionics and Equipment (G2-A): Blind-flying instrumentation. Marconi radio compass (licence-built by Rudi Cajavec), intercom and STR-9Z1 VHF radio transceiver standard. Standard electrical equipment includes navigation lights, 250W landing light in nose, and 50W taxying light on nose landing gear. Camera, with focal length of 178 mm (7 in) and 125-exposure magazine, can be fitted in fuselage, under rear cockpit floor. Flares can be carried on the underwing bomb racks for night photography. Target towing hook under centre-fuselage.

Avionics and Equipment (G-2A-E): Full IFR instrumentation. Electronique Aérospatiale (EAS) Type TVU-740 VHF/UHF com radio transceiver. Marconi AD 370B radio compass, EAS RNA-720 VOR/LOC and ILS, Iskra 75R4 marker beacon receiver, and intercom. Vinten Type 360/140A camera with 3 in automatic exposure-control lens. Otherwise as G2-A.

Armament: All production aircraft have two 0·50 in machine-guns in nose (with 80 rds/gun); and underwing pylons for two 50 kg or 100 kg bombs and four 57 mm rockets or two 127 mm rockets; or clusters of small bombs and expendable bomblet containers of up to 150 kg (330 lb) weight (300 kg; 660 lb total).

Dimensions, external:

Wing span	10·47 m (34 ft 4½ in)
Wing span over tip-tanks	11·62 m (38 ft 1½ in)
Wing chord at root	2·36 m (7 ft 9 in)
Wing chord at tip	1·40 m (4 ft 7 in)
Wing aspect ratio	5·55
Length overall	10·34 m (33 ft 11 in)
Height overall	3·28 m (10 ft 9 in)
Tailplane span	4·27 m (14 ft 0 in)
Wheel track	3·89 m (12 ft 9 in)
Wheelbase	3·59 m (11 ft 9½ in)

Areas:

Wings, gross	19·43 m^2 (209·14 sq ft)
Ailerons (total)	2·36 m^2 (25·40 sq ft)
Trailing-edge flaps (total)	2·02 m^2 (21·75 sq ft)
Airbrake	0·34 m^2 (3·66 sq ft)
Fin	1·34 m^2 (14·42 sq ft)
Rudder, incl tab	0·56 m^2 (6·03 sq ft)
Tailplane	3·66 m^2 (39·40 sq ft)
Elevators, incl tabs	0·83 m^2 (8·93 sq ft)

Weights:

Weight empty, equipped	2,620 kg (5,775 lb)
Max T-O weight:	
Fully-aerobatic trainer ('clean')	3,374 kg (7,438 lb)
Basic trainer (no tip-tanks)	3,488 kg (7,690 lb)
Navigational trainer (with tip-tanks)	3,828 kg (8,439 lb)
Weapons trainer	3,988 kg (8,792 lb)
Strike version	4,300 kg (9,480 lb)

Performance (at normal T-O weight):

Max level speed at S/L	408 knots (756 km/h; 470 mph)
Max level speed at 6,200 m (20,350 ft)	438 knots (812 km/h; 505 mph)
Max cruising speed at 6,000 m (19,685 ft)	394 knots (730 km/h; 453 mph)
Stalling speed: flaps and airbrakes down	85 knots (158 km/h; 98 mph)
flaps and airbrakes up	97 knots (180 km/h; 112 mph)
Max rate of climb at S/L	1,370 m (4,500 ft)/min
Time to 3,000 m (9,840 ft)	2 min 24 s
Time to 6,000 m (19,685 ft)	5 min 30 s
Time to 9,000 m (29,520 ft)	10 min 12 s
Service ceiling	12,000 m (39,375 ft)
T-O run on grass	490 m (1,610 ft)
T-O to 15 m (50 ft)	640 m (2,100 ft)
Landing from 15 m (50 ft)	710 m (2,330 ft)
Landing run on grass	400 m (1,310 ft)
Max range at 9,000 m (29,520 ft), with tip-tanks full	669 nm (1,240 km; 770 miles)
Max endurance at 7,000 m (23,000 ft)	2h 30 min

UTVA

UTVA—SOUR METALNA INDUSTRIJA, RO FABRIKA AVIONA

Head Office and Works: Utve Zlatokrile 9, 26000 Pancevo

Telephone: 013 44 755 and 013 42 366

Telex: 131-16

General Manager: Dipl Ing Vladimir Nikolić

Manager of Aircraft Production:
Dipl Ing Miroslav Corbić

Manager of Aircraft Development:
Dipl Ing Dragoslav Dimić

UTVA-75

The UTVA-75 is a side-by-side two-seat training, glider towing and utility lightplane, which was projected, designed and built in partnership by UTVA-Pancevo, Prva Petoletka-Trstenik, Vazduhoplovnotehnicki Institut and Institut Masinskog Fakulteta of Belgrade. Design was started in 1974. Construction of two prototypes was undertaken in 1975; the first of these flew for the first time on 20 May 1976 and the second on 18 December 1976. Series production began immediately and is continuing.

Type: Two-seat light aircraft.

Wings: Cantilever low-wing monoplane, with short-span centre-section and two constant-chord outer panels. Wing section NACA 65_2415. Dihedral 0° on centre-section, 6° on outer panels. Conventional all-metal structure. Ailerons and flaps along entire trailing-edge of outer panels, except for tips. Flettner trim tab on each aileron.

Fuselage: Conventional all-metal semi-monocoque structure.

Tail Unit: Cantilever all-metal structure, with sweptback vertical surfaces. Fluted skin on fin and rudder. Elevator horn-balanced.

Landing Gear: Non-retractable tricycle type, with single wheel on each unit, and small tail bumper. Prva Petoletka-Trstenik oleo-pneumatic shock-absorbers. Dunlop tyres, size 6·00-6 on main wheels, 5·00-5 on nosewheel. Prva Petoletka-Trstenik hydraulic brakes.

Power Plant: One 134 kW (180 hp) Avco Lycoming IO-360-B1F flat-four engine, driving a Hartzell HC-C2YK-1BF/F7666A two-blade metal variable-pitch propeller. Two integral fuel tanks in wings, total capacity 160 litres (35 Imp gallons). Provision for carrying two 100 litre (22 Imp gallon) drop-tanks under wings, raising max total capacity to 360 litres (79 Imp gallons). Oil capacity 10 litres (2·2 Imp gallons).

Accommodation: Two seats side by side in enclosed cabin, with large upward-opening canopy door over each seat, hinged on centreline. Dual stick-type controls standard. Cabin heated and ventilated.

Systems: Dual hydraulic systems for brakes. 12V electrical system, with 35Ah battery, navigation lights, rotating beacon and landing lights as standard equipment.

Avionics and Equipment: King KY 195B radio optional. Standard equipment includes radio compass.

Armament and Military Equipment: Standard fittings for light weapon loads underwing.

Dimensions, external:

Wing span	9·73 m (31 ft 11 in)
Wing chord (constant)	1·55 m (5 ft 1 in)
Length overall	7·11 m (23 ft 4 in)
Height overall	3·15 m (10 ft 4 in)
Tailplane span	3·80 m (12 ft 5½ in)
Wheel track	2·58 m (8 ft 5½ in)
Wheelbase	1·99 m (6 ft 6¼ in)
Propeller diameter	1·93 m (6 ft 4 in)
Propeller ground clearance	0·295 m (11¾ in)

Areas:

Wings, gross	14·63 m^2 (157·5 sq ft)
Ailerons (total)	1·38 m^2 (14·85 sq ft)
Flaps (total)	1·61 m^2 (17·33 sq ft)
Horizontal tail surfaces (total)	3·34 m^2 (35·95 sq ft)

Weights:

Weight empty, equipped	650 kg (1,433 lb)
Max crew/military load	200 kg (441 lb)
Max fuel: standard	110 kg (242 lb)
with drop-tanks	256 kg (564 lb)
Max T-O weight	960 kg (2,116 lb)

Performance (at max T-O weight):

Max level speed	118 knots (220 km/h; 136 mph)
Econ cruising speed	89 knots (165 km/h; 102 mph)
Stalling speed, engine idling:	
flaps up	51·5 knots (95 km/h; 59 mph)
25° flap	43 knots (80 km/h; 50 mph)
Max rate of climb at S/L	270 m (885 ft)/min
Service ceiling	4,500 m (14,760 ft)
T-O run	125 m (410 ft)
T-O to 15 m (50 ft)	250 m (820 ft)
Landing from 15 m (50 ft)	350 m (1,150 ft)
Range with max standard fuel	432 nm (800 km; 497 miles)
Range with drop-tanks, no reserves	1,080 nm (2,000 km; 1,242 miles)

Assembly line of UTVA-75 training and utility aircraft (Avco Lycoming IO-360-B1F engine)

HOMEBUILT AIRCRAFT
(including racing aircraft)

AUSTRALIA

CORBY
JOHN C. CORBY

34 Coronet Court, North Rocks, Sydney 2151

Mr Corby, a consultant aeronautical engineer, has designed and is marketing plans for a single-seat wooden light aircraft known as the Starlet. By early 1981 at least nine Starlets had been completed and a further 40 were known to be under construction in Australia, Tasmania and New Zealand, including a metal version.

CORBY CJ-1 STARLET

The first Starlet (VH-ULV) was built by members of the Latrobe Valley division of the Australian Ultra Light Aircraft Association. Details of this aircraft were given in the 1974-75 *Jane's*.

Another Starlet (VH-WDJ), flown by Mr Peter Furlong and powered by a 36·5 kW (49 hp) Ardem XI engine, won an Australian aerobatic championship in 1973.

The following description applies to the standard Starlet, as built to current plans:

Type: Single-seat light aircraft.

Wings: Cantilever low-wing monoplane of wooden construction. Wing section NACA 43012A. Dihedral 6°. Incidence 2° 30′ at root, −1° at tip. Laminated main spar of solid spruce, subspars of spruce, built-up girder-type ribs and D-shaped nose section. Plywood covering from leading-edge to main spar, remainder fabric covered. Provision for dismantling into two equal halves. Ailerons, of spruce with birch plywood covering, deflect 15° up and down.

Fuselage: Plywood-covered spruce structure.

Tail Unit: Cantilever type, of similar construction to wings. Fixed-incidence tailplane. Plywood-covered fixed surfaces; fabric-covered rudder and elevators. Elevators deflect 30° up, 20° down; rudder deflects 25° to left and right.

Landing Gear: Non-retractable two-wheel type standard. Separate spring steel leaf-type shock-absorbing main legs, attached directly to fuselage via a solid spruce/ash beam which also serves as the wing leading-edge attachment member. Wheels, tyres and brakes of customer's choice, subject to main wheels of 89 mm (3½ in) min diameter. (Typically, 4·00-4 tyres and Olympic go-kart hubs.) Sturmey Archer cycle drum/shoe brakes may be used. Leaf-spring tailskid, or tailwheel at customer's option. Wheel fairings optional.

Power Plant: Any suitable engine of up to 56 kW (75 hp) and 72 kg (160 lb) weight, driving a two-blade propeller, including a 37·25 kW (50 hp) Volkswagen 1600 cc converted motorcar engine. Fuel tank, capacity 36-43 litres (8-9·5 Imp gallons), aft of engine firewall. Oil capacity 2·25 kg (5 lb).

Accommodation: Single seat. Sliding canopy optional. Baggage locker behind seat.

Dimensions, external:

Wing span	5·64 m (18 ft 6 in)
Wing chord at root	1·32 m (4 ft 4 in)
Wing area, gross	6·36 m² (68·50 sq ft)
Length overall	4·50 m (14 ft 9 in)
Fuselage: Max width	0·55 m (1 ft 9¾ in)
Height overall	1·47 m (4 ft 10 in)
Tailplane span	1·98 m (6 ft 6 in)
Wheel track	1·37 m (4 ft 6 in)
Propeller diameter	1·37 m (4 ft 6 in)

Weights:

Weight empty	183-212 kg (405-467 lb)
Max T-O weight (semi-aerobatic)	295 kg (650 lb)
Max T-O weight	322 kg (710 lb)

Performance (A: prototype with 36·5 kW; 49 hp engine at 295 kg; 650 lb AUW, B: with 37·25 kW; 50 hp VW engine at max T-O weight):

Never-exceed speed:	
A	138 knots (255 km/h; 159 mph) IAS
Max level speed: A	117 knots (217 km/h; 135 mph)
B	113 knots (209 km/h; 130 mph)
Max cruising speed:	
A	107 knots (198 km/h; 123 mph)
B	101 knots (186 km/h; 116 mph)
Stalling speed, power off:	
A	42 knots (79 km/h; 49 mph)
B	41 knots (76 km/h; 47 mph)
A, B	30 knots (57 km/h; 35 mph) IAS
Typical rate of climb at S/L:	
A	213-259 m (700-850 ft)/min
B	190 m (625 ft)/min
Service ceiling: A	4,420 m (14,500 ft)
T-O to, and landing from, 15 m (50 ft):	
A	305-335 m (1,000-1,100 ft)
Range with max fuel: B	231 nm (428 km; 266 miles)
g limits	±4·5

AUSTRIA

WESTERMAYER
OSKAR WESTERMAYER

A-2161 Poysbrunn, Hauptstrasse 11
Telephone: 02554 405

Herr Westermayer, who manufactures rotor blades for small aircraft, has built four single-seat rotorcraft. Details of the WE 01, WE 02 and WE 03 can be found in the 1980-81 *Jane's*. The WE 04 represents the fully-developed Westermayer rotorcraft, and information on this aircraft follows.

WESTERMAYER WE 04

Similar to the WE 03 in general configuration, the WE 04 (OE-AXR) was flown for the first time in early 1980. It embodies a number of design features new to aircraft in this category, including all-glassfibre rotor blades and a patented helicopter-type rotor head that provides cyclic and collective pitch to the blades, offering improved stability and safety. The blades are pre-spun to 500 rpm prior to take-off and pitched to +7° for flight.

The blades, built by Herr Westermayer, are made from Gevetex Glasrovings and cloth, with a Rohacell 31 foam core, bonded with CIBA epoxy resin in heated aluminium moulds. Blade section is NACA 8H-12.

The fuselage is made of metal, with GfK covering on the cabin and a side-hinged (to starboard) one-piece canopy/windscreen. A non-retractable tricycle landing gear is fitted, with steerable nosewheel and a small bumper wheel at the rear of the tailboom. The horizontal tail surfaces are fixed, but there is a large rudder with a ground-adjustable tab on its trailing-edge. All tail surfaces are wooden-covered.

Power plant is a 52 kW (70 hp) 2,600 cc Volkswagen engine, driving a two-blade wooden fixed-pitch pusher propeller. Normal autorotative rotor rpm in cruising flight is 380-400.

The WE 04 will be certificated to FAR Pt 27, and plans will be made available to amateur constructors.

Dimensions, external:

Rotor diameter	7·40 m (24 ft 3½ in)
Rotor blade chord, constant	18 cm (7·09 in)
Length of fuselage	3·92 m (12 ft 10½ in)
Height overall	2·20 m (7 ft 2½ in)
Propeller diameter	1·38 m (4 ft 6½ in)

Weights:

Weight empty	230 kg (507 lb)
Max T-O weight	340 kg (750 lb)

Performance:

Max level speed	81 knots (150 km/h; 93 mph)
Cruising speed	65 knots (120 km/h; 75 mph)
Max rate of climb at S/L	300 m (985 ft)/min
T-O run	10 m (33 ft)
Range with max fuel	135 nm (250 km; 155 miles)

BRAZIL

MOURA
MAURICIO IMPELIZIERI P. MOURA

Belo Horizonte, Minas Gerais State

MOURA ESQUALO (SHARK)

M Moura has designed and built, with some technical assistance from the CEA Universidade Federal de Minas Gerais, a two-seat sporting aircraft which he has named Esqualo. Its structure is stressed to +6*g* and −3*g*. The first flight of this aircraft was expected by early 1981. All available details follow:

Type: Two-seat sporting aircraft.

Wings: Cantilever low-wing monoplane of all-metal construction. Plain ailerons and trailing-edge flaps of similar construction. No trim tabs.

Fuselage: Conventional semi-monocoque structure of light alloy.

Tail Unit: Cantilever structure of light alloy. Horn-balanced control surfaces. Dorsal fin. No trim tab in rudder or elevators.

Landing Gear: Retractable tricycle type.

Power Plant: One 134 kW (180 hp) Avco Lycoming flat-four engine of the O-360 series, driving a two-blade variable-pitch propeller with spinner. Fuel capacity 140 litres (30·8 Imp gallons).

Accommodation: Two seats side by side beneath rearward-sliding transparent cockpit canopy.

Dimensions, external:

Wing span	7·98 m (26 ft 2¼ in)
Wing area, gross	8·86 m² (93·43 sq ft)
Length overall	6·68 m (21 ft 11 in)
Height overall	2·22 m (7 ft 3½ in)

Weights (estimated):

Weight empty	450 kg (992 lb)
Max T-O weight	804 kg (1,773 lb)

Performance (estimated at max T-O weight):

Max level speed	243 knots (450 km/h; 280 mph)
Cruising speed	183 knots (340 km/h; 211 mph)
Stalling speed, flaps and gear up	68 knots (125 km/h; 78 mph)
Stalling speed, flaps and gear down	49 knots (90 km/h; 56 mph)
Range with max fuel	755 nm (1,400 km; 870 miles)
Endurance at cruising speed	4 h 24 min

WEBER
WILLIBALD WEBER

c/o JATO-Aviação Sorocaba Ltda, Sorocaba Airport, São Paulo State

Austrian-born Mr Willibald 'Willi' Weber has been living and working in Brazil since 1949. During the 1950s he designed and built the twin-engined Casmuniz 5-2 (built in 1952; destroyed in an accident in 1968) and the W-141 (built in 1957). In 1962 he founded the CONAL aircraft maintenance company with Bertram Luiz Leopolz, and together they built the W-151 six-seater, which remains Mr Weber's personal aircraft.

Corby Starlet single-seat homebuilt aircraft

Westermayer WE 04 light rotorcraft

In the early 1970s Mr Weber left CONAL and set up a new company known as JATO-Aviação Sorocaba Ltda, for aircraft maintenance and repair work. More recently he designed a two-seat light aircraft known as the Perereca (Little Frog), of which a prototype has been built in the JATO workshop and was expected to fly for the first time in 1981. It is intended to be capable of limited aerobatics.

All available details follow.

WEBER PERERECA

Type: Two-seat light monoplane.

Wings: Cantilever low-wing monoplane, tapered towards tips. Airbrakes fitted in upper surface of constant-chord inner panels. Single aluminium spar, wooden ribs, wood and plastics covered, except for fabric-covered control surfaces. Wings can be removed for transportation.

Fuselage: Low-profile nacelle of glassfibre construction over a steel frame.

Tail Unit: Twin fins and rudders, supported on twin booms from the wing trailing-edges; high-mounted tailplane. Similar construction to wings.

Landing Gear: Non-retractable tricycle type, with nosewheel semi-recessed in streamline fairing.

Power Plant: One Volkswagen 1,800 cc modified motorcar engine, mounted to rear of cabin and covered by an aluminium cowling. Two-blade wooden pusher propeller.

Accommodation: Two seats side by side. Door each side of cabin. Dual controls.

Performance (estimated):

Max level speed 86 knots (160 km/h; 99 mph)

CANADA

FLETCHER

DANIEL & RICHARD FLETCHER

Apt 3112, 30 Exeter Road, Ajax, Ontario L1S 2J6
Telephone: (416) 683 0261

FLETCHER BUSHBIRD

Design of the Fletcher BushBird was started by Mr D. W. and Mr D. R. Fletcher in 1968. This was a spare-time project, intended to produce a practical sport/bushplane, and this is reflected by a first flight date of 14 August 1979. The aircraft is flown under a Department of Transport permit that, under Canadian amateur built regulations, limits maximum take-off weight to 900 kg (1,985 lb) and two-seat configuration. The BushBird was, however, designed as a four-seater with a gross weight of 1,089 kg (2,400 lb). The design includes provision for operation on floats, and it was the intention in early 1981 to complete the float installation for testing.

Type: Four-seat cabin monoplane (licensed for two-seat operation).

Wings: Braced high-wing monoplane, with single bracing strut each side. Aerofoil section modified Clark Y. Dihedral 0° 30′. Incidence 1°. No sweepback. Conventional structure of light alloy, with main spar at 30% chord and auxiliary rear spar. Glassfibre tips. Plain balanced ailerons of light alloy. Electrically-actuated slotted trailing-edge flaps of semi-Fowler type.

Fuselage: Forward fuselage and cabin area is a welded steel tube structure with light alloy skins. Tail cone is a monocoque structure of light alloy.

Tail Unit: Cantilever structure of light alloy, with both horizontal and vertical surfaces incorporating symmetrical aerofoil sections. Fixed-incidence tailplane. Trim tab in starboard elevator. Rudder has electrically controlled bungee trimming.

Landing Gear: Non-retractable tailwheel type. Steel leaf spring main legs, carrying Cleveland main wheels with tyres size 7·00 × 6, pressure 1·93 bars (28 lb/sq in). Scott steerable tailwheel carried on friction damped leaf springs with tyre size 2·50 × 4, pressure 2·07 bars (30 lb/sq in). Cleveland hydraulic disc brakes. Provisions for installation of floats.

Power Plant: One 112 kW (150 hp) Avco Lycoming O-320-E2D flat-four engine, driving a McCauley Type DTM 75 53 two-blade fixed-pitch propeller. Two integral wing fuel tanks, adjacent to wingtips, with combined capacity of 163·5 litres (36 Imp gallons). Refuelling point on upper surface of each wing, adjacent to wingtip. Oil capacity 7·5 litres (1·67 Imp gallons).

Accommodation: Pilot and co-pilot or passenger on individual front seats. Bench seat for two at rear of cabin, but aircraft licensed only for two-seat operation. Dual controls standard. Door at each side of cabin, hinged at upper edge. Baggage space aft of rear seats. Accommodation heated and ventilated. Windscreen and cabin windows of polycarbonate.

Systems: Electrical system includes 12V 60A engine-driven alternator, and 12V 35Ah battery. Hydraulic system for brakes only.

Avionics: Collins nav/com and transponder standard.

Dimensions, external:

Wing span	10·29 m (33 ft 9 in)
Wing chord (constant)	1·52 m (5 ft 0 in)
Wing aspect ratio	6·75
Wing area, gross	15·61 m² (168·0 sq ft)
Length overall	7·24 m (23 ft 9 in)
Height overall	2·29 m (7 ft 6 in)
Tailplane span	3·15 m (10 ft 4 in)
Wheel track	2·29 m (7 ft 6 in)
Wheelbase	5·18 m (17 ft 0 in)
Propeller diameter	1·91 m (6 ft 3 in)
Propeller ground clearance	0·23 m (9 in)

Weights:

Weight empty, equipped	637 kg (1,405 lb)
Max T-O and landing weight (licensed)	900 kg (1,985 lb)
Max T-O weight (design)	1,089 kg (2,400 lb)

Performance (at max licensed T-O weight):

Never-exceed speed	139 knots (257 km/h; 160 mph)
Max level speed at 915 m (3,000 ft)	118 knots (219 km/h; 136 mph)
Max cruising speed at 915 m (3,000 ft)	109 knots (203 km/h; 126 mph)
Econ cruising speed at 1,220 m (4,000 ft)	104 knots (193 km/h; 120 mph)
Stalling speed, flaps down, engine idling	32 knots (60 km/h; 37 mph)
Max rate of climb at S/L	320 m (1,050 ft)/min
Service ceiling (estimated)	4,265 m (14,000 ft)
T-O and landing run	107 m (350 ft)
Range with max fuel, 45 min reserves	499 nm (925 km; 575 miles)

JEAN ST-GERMAIN

CENTRE DE RECHERCHES JEAN ST-GERMAIN INC

924 St-Pierre, Drummondville, PQ J2C 3Y2
Telephone: (819) 477 1221
Designer: Jean Saint-Germain
General Manager: Gilbert Lapointe

JEAN ST-GERMAIN RAZ-MUT

The Raz-Mut was designed by Jean Saint-Germain as an easy-to-build and easy-to-fly light aircraft, suitable for home construction, for sport flying by amateurs, and for a variety of utility applications including agricultural spraying and survey work. Design of the aircraft was started in May 1976 and the original prototype, built at Mr Saint-Germain's research centre, flew for the first time on 24 October 1976, powered by a 33 kW (44 hp) Kohler K-440-2AS-M aircooled two-stroke engine. It proved to be underpowered, and the third Raz-Mut, demonstrated at the 1977 EAA Fly-in at Oshkosh and shown in the accompanying illustration, has a 1,700 cc Volkswagen engine. Also at the Fly-in was a then-unflown version with the alternative McCulloch MC-431 engine. Both embodied a number of aerodynamic refinements compared with the first Raz-Mut, including a drooped wing leading-edge, upper-surface fences and wide fairings over the wing bracing struts.

Kit production began in December 1976; by Spring 1981 many sets of plans and/or kits had been sold, and several amateur-built Raz-Muts were flying in Canada and elsewhere. The first example in Europe was built by M Jean-Pierre Marie, at Mantes in France.

The basic kit contains all raw materials and components needed to build the Raz-Mut, including the engine and propeller but excluding instruments, battery and seat cushion. A more expensive kit offers the materials and components in pre-finished state, so that only a simple tool kit is required to complete the aircraft. As an alternative, building and servicing instructions, and a list of materials, are available to those who are willing to purchase their own materials and perform all the machining and other work.

Type: Single-seat light aircraft.

Wings: Strut-braced high-wing monoplane, with single wide aerofoil-section bracing strut each side. Modified NACA 4400 wing section. Thickness/chord ratio 13%. Constant chord. Some dihedral. Plain aileron and three-position flap form entire trailing-edge of each wing. Entire wing and control surfaces of aluminium, including covering. Shallow fence above each wing between aileron and flap. Endplates standard.

Fuselage: Open structure of square-section aluminium tube, bolted together.

Tail Unit: All-moving horizontal surface with ground-adjustable tab at centre of trailing-edge. Small fin and large-area horn-balanced rudder with ground adjustable tab. All control surfaces cable-actuated. Entire structure of aluminium with fabric covering.

Landing Gear: Non-retractable tricycle type. Brake on nosewheel. Small tail bumper wheel in rear end of keel member. Skis and floats optional.

Power Plant: One 52 kW (70 hp) 1,700 cc Volkswagen flat-four four-stroke or 53·5 kW (72 hp) McCulloch MC-431 flat-four two-stroke engine, driving a two-blade fixed-pitch wooden pusher propeller through multiple belts. Fuel tank, capacity 23 litres (6 US gallons), aft of seat. Provision for second, similar, tank. Manual or electrical engine starting.

Accommodation: Pilot only, on open seat. Glassfibre cabin enclosure optional.

Dimensions, external:

Wing span	7·01 m (23 ft 0 in)
Wing area, gross	8·55 m² (92 sq ft)
Wing chord, constant	1·22 m (4 ft 0 in)
Length overall	5·58 m (18 ft 3½ in)
Height overall	1·77 m (5 ft 9½ in)

Moura Esqualo two-seat sporting aircraft, nearing completion

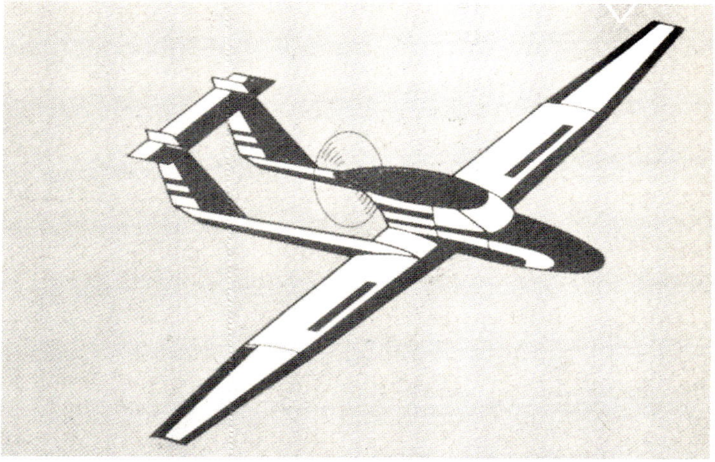

Artist's impression of Mr Willibald Weber's Perereca (Little Frog)

Raz-Mut with a glassfibre cabin enclosure and 48·5 kW (65 hp) Avco Lycoming O-145-B2 engine, built by Mr Joe Marczi of Welland, Ontario *(Neil A. Macdougall)*

Fletcher BushBird, designed as a four-seat monoplane

Jean St-Germain Raz-Mut (1,700 cc Volkswagen engine) *(Howard Levy)*

Macfam SA 102·5 Cavalier owned by Mr M. Johnson

Propeller diameter	1·09 m (3 ft 7 in)
WEIGHTS (VW engine):	
Weight empty	204 kg (450 lb)
Max T-O weight	363 kg (800 lb)
PERFORMANCE (VW engine):	
Max level speed	74 knots (136 km/h; 85 mph)
Max cruising speed	61 knots (112 km/h; 70 mph)
Landing speed	39 knots (73 km/h; 45 mph)
Max rate of climb at S/L	198 m (650 ft)/min
Service ceiling	2,745 m (9,000 ft)
T-O run	61 m (200 ft)
Landing run	54 m (175 ft)
Range with standard fuel	130 nm (240 km; 150 miles)

JONES

DANIEL G. JONES

Grosse Isle, Manitoba

JONES D-1

Designed and constructed by Daniel Jones, the D-1 (CF-UOZ) is a small single-seat biplane, powered by a 67 kW (90 hp) Continental engine. Large fairings cover the main wheels of the tailwheel-type landing gear. Wing span is reported to be 5·74 m (18 ft 10 in) and range 260 nm (480 km; 300 miles) at a cruising speed of 91 knots (169 km/h; 105 mph). No further information is available.

MACFAM

MACFAM WORLD TRADERS

4623 Fortune Road SE, Calgary, Alberta T2A 2A7
Telephone: (403) 272 3658

Macfam (formerly K & S Aircraft) is continuing to sell to amateur constructors plans for its popular SA 102·5 Cavalier, SA 103 and SA 104 Cavalier, and the SA 105 Super Cavalier with retractable landing gear. However, priority has shifted away from the SA 102·5, due to the ready availability of larger engines and the roomier cockpits and increased baggage capacity of the later models.

Each of the SA 103/104/105 models is available in **BCD** (bubble canopy with doors) and **BSC** (bubble sliding canopy) forms. BCD/BSC aircraft retain the original windscreen, but have the turtledeck aft of the seats lowered, giving improved rearward view and necessitating revision of the tail fin curve.

MACFAM SA 102·5 CAVALIER

TYPE: Two-seat or '2+2' light aircraft.

WINGS: Cantilever low-wing monoplane. Wing section NACA 23015 at root, NACA 23012 at tip. Dihedral 6°. Incidence 3° 30′ at root, washed out to 1° at tip. No sweepback at quarter-chord. Single wooden box-spar, plywood leading-edge, and auxiliary rear spar to carry aileron and flap loads. Diagonal I-section drag spar between front and rear spar in each wing. Entire centre-section is plywood-covered and contoured to serve as cabin seat. Outer panels covered with Dacron synthetic fabric, finished with a polyurethane compound. Single-slotted Frise-type ailerons of spruce and plywood. Cable-operated split flaps of spruce and birch ply. Optional plans available for building wing in three pieces and for electrical actuation of flaps.

FUSELAGE: Truss-type structure, with four main longerons, of spruce and birch plywood construction. Cockpit canopy and doors of moulded glassfibre, rear part of top-decking fabric-covered.

TAIL UNIT: Cantilever structure, with sweptback vertical surfaces. All-wood construction, with Dacron covering. Fixed-incidence tailplane. Elevators operated by pushrods. Trim tab in starboard elevator.

LANDING GEAR: Non-retractable tricycle type. All units have spring steel legs of Macfam design. Wheel size 5·00-5, tyre pressure 2·75 bars (40 lb/sq in), on all units. For rough-field operation, 6·00-6 main wheels with low-profile tyres are used optionally. Expanding-shoe brakes, operated hydraulically by dual toe-controls. Glassfibre wheel fairings. Alternative 680 kg (1,500 lb) capacity floats or ski landing gear optional.

POWER PLANT: Wide choice of four-cylinder engines available, including 63·5, 67 or 74·5 kW (85, 90 or 100 hp) Continental, 93-97 kW (125-130 hp) Franklin Sport 4A, 80·5 or 86 kW (108 or 115 hp) Avco Lycoming O-235, or 93 or 101 kW (125 or 135 hp) Avco Lycoming O-290. Choice of wood or metal, fixed-pitch or variable-pitch propellers, with diameters from 1·68 m (5 ft 6 in) to 1·83 m (6 ft 0 in). Whichever engine is used, an extension shaft 7·6 to 12·7 cm (3 to 5 in) long is fitted between the propeller and the crankshaft, permitting the use of more streamlined cowling panels and a reduction in the compression of airflow between the propeller and engine. All fuel in permanent wingtip tanks, with choice of two sizes: 64 or 75 litres (17 or 20 US gallons) each. Oil capacity 5·7-6·8 litres (1·5-1·8 US gallons), according to engine fitted.

ACCOMMODATION: Side-by-side seating for pilot and one adult passenger, with optional rear jump-seat for two small children in what would normally be the baggage area. Without this rear seat, up to 56 kg (125 lb) of baggage can be carried, depending upon engine and equipment installations. Forward-opening door on each side of heated and ventilated cabin.

AVIONICS AND EQUIPMENT: Standard nav/com equipment, including 1 or 2 VHF sets. ADF and transponder optional.

DIMENSIONS, EXTERNAL:	
Wing span over tip-tanks	8·33 m (27 ft 4 in)
Width, outer wing panels folded	3·63 m (11 ft 11 in)
Wing area, gross	10·96 m² (118·0 sq ft)
Wing aspect ratio	6·25
Length overall	6·71 m (22 ft 0 in)
Height overall	2·23 m (7 ft 4 in)
Wheelbase	1·37 m (4 ft 6 in)
Propeller diameter	see under 'Power Plant'
WEIGHTS AND LOADINGS:	
Basic operating weight empty	408 kg (900 lb)

Max T-O weight 680 kg (1,500 lb)
Max wing loading 2·60 kg/m² (12·7 lb/sq ft)
Max power loading (63·5 kW; 85 hp engine) 10·73 kg/kW (17·6 lb/hp)

PERFORMANCE (at max T-O weight, 93 kW; 125 hp Avco Lycoming engine):
Never-exceed speed 199 knots (370 km/h; 230 mph)
Max level speed at 2,135 m (7,000 ft) 160 knots (297 km/h; 185 mph)
Max cruising speed 143 knots (265 km/h; 165 mph)
Econ cruising speed 134 knots (249 km/h; 155 mph)
Stalling speed, flaps up 43·5 knots (81 km/h; 50 mph) IAS
Stalling speed, flaps down 35 knots (65 km/h; 40 mph) IAS
Max rate of climb at S/L over 518 m (1,700 ft)/min
Service ceiling 4,875 m (16,000 ft)
T-O run 107 m (350 ft)
T-O to 15 m (50 ft) 305 m (1,000 ft)
Landing from 15 m (50 ft) 396 m (1,300 ft)
Landing run 183 m (600 ft)
Max range, no reserves 720 nm (1,335 km; 830 miles)

MACFAM SA 103 CAVALIER

This aircraft combines a simplified version of the SA 105 airframe with a non-retractable tailwheel type landing gear. It provides a roomier cabin than the SA 102·5, for improved passenger comfort and increased baggage space, and allows use of the plentiful engines in the 93-134 kW (125-180 hp) category that are currently available, including the Javelin listed under the SA 105 entry.

Available in standard, BCD or BSC form (see introductory notes).

MACFAM SA 104 CAVALIER

The SA 104 differs from the SA 103 in having a non-retractable tricycle landing gear similar to that of the SA 102·5.

Available in standard, BCD or BSC form (see introductory notes).

MACFAM SA 105 SUPER CAVALIER

Although similar in general configuration to the SA 102·5, the SA 105 is an entirely new design, with only the ailerons and wingtip tanks interchangeable between the two types. Overall dimensions are slightly greater, offering improved comfort and more baggage space; and a retractable tricycle landing gear is standard.

The size of the elevator trim tab was increased in 1980; the plans have also been improved and simplified. First example to fly was the aircraft shown in the accompanying illustration (N17HJ), built by Mr Hank Q. Johnson of Memphis, Tennessee. This is powered by a 149 kW (200 hp) Avco Lycoming IO-360 flat-four engine in a Gulfstream American Traveler cowling, without a propeller extension, which compromises its appearance and performance by comparison with the basic Macfam design. Max cruising speed is 191 knots (354 km/h; 220 mph) at 2,135 m (7,000 ft) with a constant-speed propeller; econ cruising speed is 156-165 knots (289-305 km/h; 180-190 mph); and max rate of climb at S/L 760 m (2,500 ft)/min.

The following description applies to the standard SA 105 built to Macfam plans:

TYPE: Two-seat all-wood light aircraft.

WINGS: Cantilever low-wing monoplane. Wing section NACA 23015 at root, NACA 23010 at tip. Dihedral at chord line 6° from roots. Washout 2° 30′ between ribs 0 and 5. All-wood structure of Sitka spruce spars, fir/mahogany marine ply ribs, and leading-edge skin of Finnish birch plywood. Fabric covering aft of main spar, except for plywood walkways at roots, and tips. Ailerons of similar construction to wings. Split type mechanically-actuated (electrical or hydraulic actuation optional) flap system has been developed, effectively changing the wing aerofoil section from 230 series to 23 (sailplane) series. These flaps operate at 15°, 30° and 45°, and can be set at negative angles of up to 12° for cruising flight to increase cruising speed. Glassfibre canted wingtip tanks, which have similar effect to winglets, and increase wing area, aspect ratio and lift.

FUSELAGE: Truss structure, with four longerons, of Sitka spruce and Finnish birch plywood. Cockpit roof, side window/door frames and engine cowling of glassfibre. Rear fuselage and rear part of top-decking fabric covered.

TAIL UNIT: Cantilever wood structure, with sweptback vertical surfaces. Fin and tailplane plywood covered; control surfaces fabric covered. Trim tab in starboard elevator.

LANDING GEAR: Retractable tricycle type, with mechanical (optionally electric or hydraulic) actuation. Main units retract inward into wing roots, nosewheel rearward. Spring steel legs. All three wheels size 5·00-5. Brakes on main wheels, toe-operated from port seat.

POWER PLANT: One Avco Lycoming, Franklin or similar flat-four engine, with propeller extension shaft. Recommended are 112-119 kW (150-160 hp) Avco Lycoming O-320/IO-320, driving a variable-pitch or constant-speed propeller. Min recommended power plant is 93 kW (125 hp) Franklin 4A-235 or Avco Lycoming O-290/O-290-D2/O-290-G driving a fixed-pitch propeller. Max recommended is 149 kW (200 hp) Avco Lycoming O-360/IO-360. Other suitable engines include the Javelin Aircraft 2,300 cc turbocharged Ford conversion with ground-adjustable wooden propeller. Normal capacity of wingtip tanks 127 litres (28 Imp gallons). Optional sump tanks (with engines over 112 kW; 150 hp) raise total capacity to 152·75 litres (33·6 Imp gallons).

ACCOMMODATION: Two seats side by side in enclosed cabin. Dual controls. Forward-hinged window/door on each side in standard form. Available also in BCD or BSC form (see introductory notes). Baggage space aft of seats, capacity up to 172 kg (379 lb).

DIMENSIONS, EXTERNAL:
Wing span over tip-tanks 8·31 m (27 ft 3⅜ in)
Wing area, gross, flaps 0° 10·87 m² (117 sq ft)
Length overall 7·25 m (23 ft 9½ in)
Wheel track 2·58 m (8 ft 5½ in)
Wheelbase 1·33 m (4 ft 4¼ in)
Propeller diameter 1·83 m (6 ft 0 in)

WEIGHTS:
Weight empty 431-454 kg (950-1,000 lb)
Max T-O weight (dependent on engine) up to 885 kg (1,950 lb)

PERFORMANCE (estimated, at AUW of 816 kg; 1,800 lb. A = 93 kW; 125 hp. B = 112 kW; 150 hp. C = 134 kW; 180 hp):
Never-exceed speed 200 knots (370 km/h; 230 mph)
Cruising speed:
A 143-148 knots (265-273 km/h; 165-170 mph)
B 152-156 knots (281-289 km/h; 175-180 mph)
C 165-174 knots (305-322 km/h; 190-200 mph)
Stalling speed, flaps 0° 52-61 knots (96·5-105 km/h; 60-65 mph) IAS
Stalling speed, flaps down 15° 48-52 knots (89-96·5 km/h; 55-60 mph) IAS
Max rate of climb at S/L: A 183-213 m (600-700 ft)/min
B 274-305 m (900-1,000 ft)/min
C 365-488 m (1,200-1,600 ft)/min
T-O run: A 274 m (900 ft)
B 244 m (800 ft)
C 183 m (600 ft)

REPLICA PLANS
REPLICA PLANS

PO Box 94248, Richmond, BC V6Y 2A6

The SE-5A replica was designed to be an easy-to-build and inexpensive 85 per cent scale representation of the famous first World War fighter, although exact reproduction was waived in favour of making the aircraft simple to construct, using modern and more readily available materials.

Design of the aircraft began in 1969, in which year construction of the first prototype also started. The SE-5A prototypes were designed for Continental engines ranging from 48·5 to 74·5 kW (65-100 hp), but larger engines can be installed to the individual homebuilder's preference. The first prototype flew for the first time in 1970 and certification has been granted by the FAA in the Experimental (homebuilt) category.

Plans are available to amateur builders, and more than 300 sets had been sold by early 1981.

REPLICA PLANS SE-5A REPLICA

TYPE: Single-seat sporting biplane.

WINGS: Braced biplane wings of Clark CYH section. Dihedral 3°. Incidence 3°. Ailerons on lower wings only. Centre-section of upper wing houses a small tank which can be used as an auxiliary fuel tank or smoke tank, or can be left out at building stage. The centre-section is carried on four spruce cabane struts and is braced with stainless steel cables and turnbuckles. Spruce interplane struts, with 4130 steel end fittings. Stainless steel flying, landing and incidence wires. Wing ribs of mahogany plywood, with cap strips; spruce spars. From the front spar forward, the leading-edge is covered with glassfibre or aluminium. Wings are fabric-covered.

FUSELAGE: Ply-skinned box structure, with fabric-covered turtledeck and aluminium-covered forward top decking. Dummy Vickers machine-gun in housing on port side of fuselage, and gunsights on decking forward of windscreen.

TAIL UNIT: Tail surfaces built on spruce spars, with structure similar to that of wings except for drag bracing. Pushrod-operated elevators. Cable-operated rudder, with a horn for tailwheel steering.

LANDING GEAR: Non-retractable tailwheel type. Bungee cord shock-absorption. Converted motorcycle wheels on main units, size 3·25-16. Size 6·00-2 tailwheel. Mechanical brakes.

POWER PLANT: Various engines can be installed. Performance figures quoted relate to aircraft with 63·5 kW (85 hp) Continental C85 flat-four engine, driving a two-blade fixed-pitch wooden propeller. Fuel capacity 72 litres (19 US gallons). Oil capacity 3·8 litres (1 US gallon).

ACCOMMODATION: Single seat in open cockpit.

DIMENSIONS, EXTERNAL:
Wing span 6·96 m (22 ft 10 in)
Wing chord, constant 1·27 m (4 ft 2 in)
Wing area, gross 13·01 m² (140 sq ft)
Height overall 2·18 m (7 ft 2 in)
Wheel track 1·52 m (5 ft 0 in)
Propeller diameter 1·83 m (6 ft 0 in)

WEIGHTS:
Weight empty 358 kg (790 lb)
Max T-O weight 499 kg (1,100 lb)

PERFORMANCE:
Max level speed at S/L 78 knots (145 km/h; 90 mph)
Max cruising speed 74 knots (137 km/h; 85 mph)
Stalling speed 30·5 knots (57 km/h; 35 mph)
Max rate of climb at S/L 152 m (500 ft)/min

WESTERN
WESTERN AIRCRAFT SUPPLIES

623 Markerville Road NE, Calgary, Alberta T2E 5X1
Telephone: (403) 276 3087
DIRECTOR: Jean J. Peters

Western Aircraft Supplies markets materials for amateur aircraft constructors, and has sold plans for construction of the RL-3 Monsoon, a two-seat light aircraft originally designed in India. At least 12 sets of plans for the Monsoon and five kits of materials were sold. Details of this aircraft can be found in the 1977-78 *Jane's*.

Western has developed another aircraft, of basic wooden construction, known as the PGK-1 Hirondelle. Plans and wood kits are available to amateur constructors, as well as preformed engine cowlings, windscreens and fuel tanks.

PGK-1 HIRONDELLE

Design of the PGK-1 Hirondelle began in 1969 and construction of a prototype (C-GWYL) started in July 1970. First flight was achieved on 27 June 1976. It is said to be suited for cross-country and recreational flying, and simple enough for first-time constructors to build. All fabric covering is glued in place, rather than stitched.

TYPE: Two-seat light aircraft.

WINGS: Cantilever low-wing monoplane of constant chord. Wing section NACA 23012. Dihedral 4° from roots. Incidence 4° at root, 2° at tip. No sweepback. Wooden box spar and plank-type auxiliary rear spar. Plywood ribs. Dacron fabric covered, with dope and polyurethane finish. Ailerons of wood and fabric construction. No flaps or tabs.

FUSELAGE: Conventional structure of spruce bulkheads and spruce longerons, plywood covered to rear of cabin, with outer covering of Dacron. Cabin constructed of wood formers with polystyrene foam infill and glassfibre top. Rounded corners of turtledeck formed from foam and glassfibre. Rear fuselage Dacron-covered, with dope and polyurethane finish.

TAIL UNIT: Cantilever structure of capped spars and plywood ribs, all plywood-covered and with outer covering of Dacron. Tab in starboard elevator.

LANDING GEAR: Non-retractable tailwheel type. Tapered spring steel main struts; 5·00-5 tyres on main wheels. Industrial Dynamics brakes. Cessna 150 main-wheel fairings.

POWER PLANT: One 86 kW (115 hp) Avco Lycoming O-235-C1B flat-four engine, driving a Warnke three-blade ground-adjustable propeller. Two glassfibre fuel tanks in leading-edges of wings, each 54·5 litres (12 Imp gallons) capacity. Oil capacity 5·7 litres (1·25 Imp gallons).

ACCOMMODATION: Two seats side by side in enclosed cabin. Dual controls and heater standard.

SYSTEM: Electrical system includes 12V 20A generator for lights, radio and fuel pump.

AVIONICS: Genave Alpha 200B radio and VOR.

DIMENSIONS, EXTERNAL:
Wing span 7·92 m (26 ft 0 in)
Wing chord, constant 1·40 m (4 ft 7 in)
Wing area, gross 10·96 m² (118 sq ft)
Wing aspect ratio 5·7
Length overall 6·27 m (20 ft 7 in)
Height overall 2·29 m (7 ft 6 in)
Wheel track 2·03 m (6 ft 8 in)
Wheelbase 4·39 m (14 ft 5 in)
Propeller diameter 1·68 m (5 ft 6 in)

WEIGHTS:
Weight empty, equipped 428 kg (944 lb)
Max T-O weight 669 kg (1,475 lb)

Macfam SA 105 Super Cavalier built by Mr Hank Q. Johnson

Prototype PGK-1 Hirondelle two-seat light aircraft

Zénith-CH 200 *(Gordon S. Williams)*

Replica Plans SE-5A, an 85 per cent scale representation of a first World War fighter

Zénith-CH 250 with tailwheel type landing gear

PERFORMANCE (at max T-O weight):
Never-exceed speed 160 knots (297 km/h; 185 mph)
Max level speed 123 knots (228 km/h; 142 mph)
Max cruising speed 117 knots (217 km/h; 135 mph)
Stalling speed 52·5 knots (96·5 km/h; 60 mph)
Max rate of climb at S/L over 305 m (1,000 ft)/min
Service ceiling 3,840 m (12,600 ft)
T-O run approx 228 m (750 ft)
Endurance, with 45 min reserves 3 h 36 min

ZENAIR

ZENAIR LTD

236 Richmond Street, Richmond Hill, Ontario L4C 3Y8
Telephone: (416) 884 9044
WORKS: 25 King Road, Nobleton, Ontario
Telephone: (416) 859 4556
PRESIDENT AND DESIGNER: Christophe Heintz

M Heintz, a professional aeronautical engineer, participated in the design of several of the aircraft produced by Avions Pierre Robin. While in France, he also designed and built the prototype of a two-seat light aircraft named the Zénith, intended for amateur construction. More than 100 Zéniths are now flying, and others are under construction, including single-seat and three-seat models, as described in this entry. In addition, Mr Heintz' Zenair company is developing the four-seat Zénith-CH 400 for factory production (see main Aircraft section).

In 1974 the Zénith-CH 200 was granted the National Association of Sport Aircraft Designers (NASAD, USA) 'seal of quality' No. 108. In 1979 the Mono Z-CH 100 and Tri-Z CH 300 were granted NASAD 'seals of quality' Nos. 115 and 116 respectively; the Mono Z-CH 100, Zénith-CH 200 and CH 250, and Tri-Z CH 300 were granted NASAD kit 'seals of quality' Nos. 301, 302 and 303 respectively.

Zenair Ltd has thirteen full-time employees. Subsidiary manufacturing and distributing companies are Zenair Atlanta Inc (Georgia, USA) and Zenair Seattle Inc (Washington, USA).

ZÉNITH-CH 200

Work on the Zénith-CH 200 began in October 1968. The prototype, registered F-WPZY (later C-FEYC), flew for the first time on 22 March 1970 and was granted French CNRA (homebuilt experimental aircraft) certification. In October 1970 the original wing of NACA 64A315 (modified) section was replaced by one offering improved low-speed characteristics.

Sets of plans and a constructional manual for the Zénith-CH 200 are available to amateur builders, as follows:

French manual with metric measurements from D. Triques, 23 Ave Edouard Belin, Fontaine d'Ouche, F21 Dijon, France.

German manual and metric measurements from K. Arens, Rollstrasse 26, D 3392 Klausthal Zellerfeld-1, German Federal Republic.

English manual and drawings to US standards, with Imperial and metric measurements, from Zenair, which offers materials, parts and complete kits for all current Heintz designs. Zenair also designs and manufactures wooden propellers for engines of up to 134 kW (180 hp).

The Zénith-CH 200 is approved by the Australian DCA and the New Zealand airworthiness authorities. Several kits have been exported to these countries, as well as to 20 other countries, including Belgium, Botswana, Colombia, Iceland, Ireland, New Guinea, Norway, South Africa, Spain, Sweden, Switzerland and the UK.

During the 1976 eight-day EAA Fly-in at Oshkosh, Chris Heintz, assisted by volunteer workers, built a Zénith-CH 200 from standard pre-fabricated kit components (as available to the homebuilder) and flew it on the last day of the meeting.

The following description applies to the standard Zénith-CH 200, of which Zenair was producing five complete kits each month in early 1981:

TYPE: Two-seat all-metal light aircraft, with ultimate stress factor of 9*g*.

WINGS: Cantilever low-wing monoplane. Constant-chord wings, of NACA 64A515 (modified) section, are optionally detachable. Dihedral 6° from roots. Single-spar aluminium alloy structure, with blind riveted aluminium alloy skin. Hoerner wingtips. Aluminium alloy piano-hinged ailerons and electrically-actuated plain flaps on trailing-edge.

FUSELAGE: Conventional aluminium alloy stressed-skin structure, of basically rectangular section with rounded top-decking.

TAIL UNIT: Rectangular one-piece all-moving tailplane, with combined trim and anti-servo tabs. Plans show rudder only, with slight sweepback. Conventional fin and rudder can be fitted if desired. Tailplane and rudder are both single-spar structures with ribs and skin of aluminium alloy.

LANDING GEAR: Non-retractable tricycle type (tailwheel gear optional), with rubber-block shock-absorbers. Steerable nosewheel. All three Cleveland wheels and tyres size 6·00-6. Hydraulically-actuated disc brakes on main units. Streamlined glassfibre fairings over all three wheels and legs. Float kits are available. At least one aircraft has flown with skis for winter operation.

POWER PLANT: Design suitable for engines from 63·5 kW (85 hp) to 119 kW (160 hp). Fuel tank in fuselage, aft of seats, capacity 90 litres (20 Imp gallons). Optional fuel tanks in wing leading-edges, total capacity 72·5 litres (16 Imp gallons). Refuelling point aft of canopy on port side.

ACCOMMODATION: Side-by-side seating for pilot and one passenger under sideways-opening (to starboard) Plexiglas canopy. Forward-sliding canopy optional. Dual controls, with single control column located centrally between seats. Space for 35 kg (77 lb) of baggage aft of seats. Cabin heated and ventilated.

SYSTEMS: 12V battery and generator provide power for engine starting, fuel pump and flap actuation. VHF radio.

DIMENSIONS, EXTERNAL:
Wing span 7·00 m (22 ft 11¾ in)
Wing chord, constant 1·40 m (4 ft 7 in)
Wing area, gross 9·80 m² (105·9 sq ft)
Wing aspect ratio 5
Length overall 6·30 m (20 ft 8 in)
Height overall 1·85 m (6 ft 0¾ in)
Tailplane span 2·30 m (7 ft 6½ in)
Wheel track 2·25 m (7 ft 4½ in)
Wheelbase 1·42 m (4 ft 8 in)
Propeller diameter 1·83 m (6 ft 0 in)

DIMENSION, INTERNAL:
Cabin: Max width 1·01 m (3 ft 3¾ in)

WEIGHTS (with 74·5 kW; 100 hp engine):
Weight empty, equipped 400 kg (881 lb)
Normal T-O and landing weight 650 kg (1,433 lb)
Max T-O weight 680 kg (1,499 lb)

PERFORMANCE (at max T-O weight. A: 74·5 kW; 100 hp engine, B: 112 kW; 150 hp engine):
Max level speed at S/L:
A 126 knots (233 km/h; 145 mph)
B 143 knots (266 km/h; 165 mph)
Cruising speed (75% power) at S/L:
A 110 knots (205 km/h; 127 mph)
B 132 knots (245 km/h; 152 mph)
Cruising speed (75% power) at 2,750 m (9,000 ft):
A 116 knots (215 km/h; 134 mph)
B 135 knots (251 km/h; 156 mph)
Stalling speed, flaps down:
A, B 46 knots (85 km/h; 53 mph)
Max rate of climb at S/L:
A 240 m (787 ft)/min
B 426 m (1,400 ft)/min
Service ceiling:
A 4,600 m (15,100 ft)
B over 4,875 m (16,000 ft)
Range with max fuel, no reserves (75% power):
A 432 nm (800 km; 497 miles)
B 451 nm (836 km; 520 miles)

ZÉNITH-CH 250

The Zénith-CH 250 is an improved version of the Zénith-CH 200 with two 65 litre (14 Imp gallon) fuel tanks in the wings and a 0·71 m³ (25 cu ft) baggage area. A forward-sliding canopy and rear windows similar to those of the Tri-Z CH 300 are fitted. Recommended engines are in the 93-119 kW (125-160 hp) range. A tailwheel type or float landing gear can be fitted as alternatives to the standard tricycle type. Performance is the same as that of the Zénith-CH 200.

By early 1981, more than 300 sets of plans and 80 kits to construct the Zénith-CH 250 had been sold.

MONO Z-CH 100

The single-seat Mono Z-CH 100 is of generally similar all-metal construction to the two-seat Zénith, but is slightly smaller overall and is designed to be powered by engines in the 37·25 to 74·5 kW (50 to 100 hp) range. A forward-sliding canopy and float kit are not available optionally on this model. The prototype (C-GNYM), which has a Volkswagen engine, made its first flight on 8 May 1975. Like the two-seat Zénith, it is stressed to $\pm 9g$ ultimate at normal max T-O weight.

Construction drawings and manual, materials, parts and complete kits to build the Mono Zénith are available from Zenair, and 90 sets had been sold by early 1981.

TYPE: Single-seat light aircraft.

WINGS: Cantilever low-wing monoplane. Wing section GA(PC) 1. Thickness/chord ratio 15%. Dihedral 6°. Incidence 7° 30′. Single-spar aluminium alloy structure, with aluminium alloy skin, blind riveted. Aluminium alloy piano-hinged ailerons. Wings easily removable.

FUSELAGE: Conventional aluminium alloy stressed-skin structure, of basically rectangular section, with rounded top-decking.

TAIL UNIT: Rectangular one-piece all-moving tailplane, with automatic and controllable trim tab. Single-spar structures, with ribs and skins of aluminium.

LANDING GEAR: Non-retractable tricycle type, with rubber-block shock-absorbers. Tailwheel gear optional. All three Cleveland wheels and tyres size 5·00-5. Hydraulically-actuated disc brakes.

POWER PLANT: One 41 kW (55 hp) 1,700 cc converted Volkswagen motor-car engine in prototype, driving a Zenair wooden propeller. Continental engine optional. Fuel tank in fuselage, capacity 55 litres (12 Imp gallons).

ACCOMMODATION: Single seat under Plexiglas canopy. Baggage compartment aft of seat, capacity 11·3 kg (25 lb).

DIMENSIONS, EXTERNAL:

Wing span	6·71 m (22 ft 0 in)
Wing chord, constant	1·27 m (4 ft 2 in)
Wing area, gross	8·50 m^2 (91·5 sq ft)
Wing aspect ratio	5·27
Length overall	5·94 m (19 ft 6 in)
Tailplane span	2·26 m (7 ft 5 in)
Wheel track	2·13 m (7 ft 0 in)
Wheelbase	1·27 m (4 ft 2 in)
Propeller diameter	1·47 m (4 ft 10 in)

WEIGHTS:

Weight empty	263 kg (580 lb)
Max T-O weight:	
VW engine	413 kg (910 lb)
74·5 kW; 100 hp engine	444 kg (980 lb)

PERFORMANCE (A: 1,700 cc Volkswagen, B: 48·5 kW; 65 hp, C: 74·5 kW; 100 hp engine):

Max level speed:	
A	103 knots (190 km/h; 118 mph)
B	109 knots (200 km/h; 125 mph)
C	130 knots (240 km/h; 150 mph)
Cruising speed (75% power):	
A	91 knots (170 km/h; 105 mph)
B	97 knots (180 km/h; 112 mph)
C	118 knots (218 km/h; 135 mph)
Stalling speed:	
A, B, C	41 knots (76 km/h; 47 mph)
Max rate of climb at S/L:	
A	200 m (610 ft)/min
B	220 m (720 ft)/min
C	490 m (1,500 ft)/min
Service ceiling: A	3,050 m (10,000 ft)
T-O run: A	183 m (600 ft)
T-O to 15 m (50 ft): A	335 m (1,100 ft)
Landing run: A	152 m (500 ft)
Range with 55 litres (12 Imp gallons) fuel:	
A, B	350 nm (645 km; 400 miles)
C	312 nm (580 km; 360 miles)
Endurance with 55 litres (12 Imp gallons) fuel:	
A, B	4 h 0 min
C	2 h 30 min

TRI-Z CH 300

The three-seat Tri-Z CH 300 is a 'stretched' version of the two-seat CH 200, with a longer fuselage and enlarged cabin, to provide room for a rear bench seat able to carry a third adult, two children or 95 kg (210 lb) of baggage. It has a greater wing span and larger tailplane; a fin and rudder assembly is standard. The ailerons are aerodynamically balanced, and electrically-actuated slotted flaps are fitted. Recommended power is in the 93-134 kW (125-180 hp) range. Fuel is carried in two 65 litre (14 Imp gallon) tanks, one in each wing leading-edge. Extra fuel can be carried in similar tanks installed in the outer wing sections. A forward-sliding canopy is standard. Limiting load factors are $\pm 5{\cdot}7g$ ultimate.

The prototype (C-GQTR), built by Harold Allsop of Toronto and powered by a 112 kW (150 hp) Avco Lycoming O-320 engine, made its first flight on 9 July 1977.

Sets of plans and a constructional manual for the Tri-Z CH 300 are available to amateur builders (330 sold by early 1981), as are materials, prefabricated component parts and complete kits (including float kits).

A Tri-Z CH 300 (C-GOVK) was built by Robin 'Red' Morris, Gerry Boudreau and Doug Holtby in a little over six months, including modifications to the fuel system to provide 773 litres (170 Imp gallons) of fuel in two fuselage and six wing tanks. Max T-O weight of this aircraft is 1,134 kg (2,500 lb), with a 134 kW (180 hp) Avco Lycoming O-360-2F engine, and full IFR equipment. It was flown non-stop by 'Red' Morris from Vancouver International Airport to Halifax International Airport between 1 July and 2 July 1978, covering approximately 2,397 nm (4,440 km; 2,759 miles) in a flying time of 22 h 44 min. This set FAI Class C1c records between Vancouver and North Bay (103·831 knots; 192·295 km/h; 119·487 mph); Vancouver and Winnipeg (109·977 knots; 203·678 km/h; 126·560 mph); and Vancouver and Halifax (105·344 knots; 195·097 km/h; 121·228 mph).

DIMENSIONS, EXTERNAL:

Wing span	8·10 m (26 ft 6¾ in)
Wing chord, constant	1·48 m (4 ft 10½ in)
Wing area, gross	12·00 m^2 (129·2 sq ft)
Wing aspect ratio	5·48
Length overall	6·85 m (22 ft 5¾ in)
Tailplane span	2·60 m (8 ft 6¼ in)
Wheel track	2·25 m (7 ft 4½ in)
Wheelbase	1·45 m (4 ft 9 in)

WEIGHTS :

Weight empty	498 kg (1,100 lb)
Max T-O weight	840 kg (1,850 lb)

PERFORMANCE (at max T-O weight, with 112 kW; 150 hp engine):

Max level speed	137 knots (254 km/h; 158 mph)
Max cruising speed (75% power)	126 knots (233 km/h; 145 mph)
Stalling speed, flaps down	46 knots (85 km/h; 53 mph)
Max rate of climb at S/L	305 m (1,000 ft)/min
Range at max cruising speed	434 nm (804 km; 500 miles)
Endurance at max cruising speed	3 h 30 min

CH 50 MINI Z

Construction began in February 1978 of the prototype (C-GTZI) of this small single-seat monoplane, and the first flight was made in 1979.

TYPE: Single-seat light aircraft.

WINGS: Cantilever low-wing monoplane. Wing section NACA 4415. Dihedral 6°. Incidence 6°. Single-spar aluminium alloy structure. Elastic-hinged plain ailerons. No flaps. Wings removable.

FUSELAGE: Conventional aluminium alloy stressed-skin structure, of basically rectangular section with rounded top-decking.

TAIL UNIT: One-piece all-moving tailplane, with combined trim and anti-servo tabs. Vertical surfaces comprise rudder only. Tailplane and rudder are single-spar structures with ribs and skin of aluminium alloy.

LANDING GEAR: Non-retractable tailwheel type. Aluminium spring shock-absorption. Main wheels and tyres size 4·10 ×3·50-4. Tyre pressure 1·24 bars (18 lb/sq in).

POWER PLANT: One 18 kW (24 hp) Cuyuna two-cylinder two-stroke engine, driving a Zenair fixed-pitch wooden propeller. (A new power plant is under development.) Single fuel tank in fuselage, capacity 23 litres (5 Imp gallons). Refuelling point on top of fuselage.

ACCOMMODATION: Single seat in open cockpit. Space for 10 kg (22 lb) of baggage, aft of cockpit.

DIMENSIONS, EXTERNAL:

Wing span	5·70 m (18 ft 8¼ in)
Wing chord, constant	1·10 m (3 ft 7¼ in)
Wing area, gross	6·20 m^2 (66·74 sq ft)
Wing aspect ratio	5·2
Height overall	approx 1·60 m (5 ft 3 in)
Tailplane span	2·00 m (6 ft 7 in)
Wheel track	3·50 m (11 ft 5¾ in)
Wheelbase	1·35 m (4 ft 5 in)
Propeller diameter	1·16 m (3 ft 10 in)

WEIGHTS:

Weight empty	137 kg (303 lb)
Max T-O weight	252 kg (555 lb)

PERFORMANCE:

Max level speed at S/L	76·5 knots (142 km/h; 88 mph)
Max cruising speed at S/L	71 knots (132 km/h; 82 mph)
Stalling speed	40 knots (74 km/h; 46 mph)
Max rate of climb at S/L	128 m (420 ft)/min
T-O and landing run	183 m (600 ft)
T-O to, and landing from, 15 m (50 ft)	305 m (1,000 ft)
Range with max fuel	191 nm (354 km; 220 miles)

ACRO-ZÉNITH CH150

The prototype of this latest addition to the Zenair range made its first flight on 19 May 1980. Kits, complete except for engine and tyres, are available to amateur constructors. All of the more critical parts, such as wing spars and ribs, are pre-manufactured.

Developed from the Mono Z-CH 100, the Acro-Zénith CH150 has a tailwheel landing gear and is intended for aerobatic training and competition flying, with an engine of 74·5-134 kW (100-180 hp) and fuel and oil systems fully equipped for inverted flight. The airframe has been strengthened to withstand $\pm 12g$, and an NACA 0015 symmetrical wing section is employed. The cockpit length has been increased to allow room for a parachute, and the fuselage turtledeck has been lowered to accept a bubble canopy for improved rearward vision during aerobatics. The normal fuel capacity of 55 litres (12 Imp gallons) can be increased to 136 litres (30 Imp gallons) by use of a ferry tank. Other features include a battery-operated radio, quick-release canopy attachment, and five-point pilot's harness.

DIMENSIONS, EXTERNAL:

Wing span	6·15 m (20 ft 2 in)
Wing area, gross	7·83 m^2 (84·3 sq ft)
Wing aspect ratio	4·85
Length overall	6·17 m (20 ft 3 in)
Height overall	1·52 m (5 ft 0 in)
Tailplane span	2·26 m (7 ft 5 in)
Wheel track	1·47 m (4 ft 10 in)
Wheelbase	4·27 m (14 ft 0 in)
Propeller diameter	1·83 m (6 ft 0 in)

WEIGHTS (Avco Lycoming O-320 engine):

Weight empty	331 kg (730 lb)
Typical aerobatic T-O weight	440 kg (970 lb)
Max T-O weight	522 kg (1,150 lb)

PERFORMANCE (Avco Lycoming O-320 with fixed-pitch propeller):

Max level speed at S/L	150 knots (278 km/h; 173 mph)
Max cruising speed (75% power at S/L)	139 knots (257 km/h; 160 mph)
Stalling speed	48 knots (88·5 km/h; 55 mph)
Max rate of climb at S/L	670 m (2,200 ft)/min
Service ceiling	over 4,875 m (16,000 ft)
Range at 55% power, with ferry fuel	660 nm (1,223 km; 760 miles)

CZECHOSLOVAKIA

SIMŮNEK/KAMARÝT

JAN SIMŮNEK MSc/JAROSLAV KAMARÝT MSc

Rudý Letov Narodni Podnik, Praha 9, Letňany, Beranových 65, Post Code 199 02

SIMŮNEK/KAMARÝT SK-1 TREMPÍK (LITTLE TRAMP)

In 1968, Mr Jan Simůnek began the design of a light aircraft which eventually became the Trempík. In the following year, Mr Jaroslav Kamarýt joined the project to assist with the design and stress calculations. Construction began in Mr Simůnek's apartment in 1969, and continued there until 1975 when the aircraft outgrew the available space. The unfinished airframe was then transferred into the possession of the Rudý Letov company, with the understanding that completion would remain the responsibility of Mr Simůnek and Mr Kamarýt, with assistance from the SSM youth organisation and the Rudý Letov management. First flight of the Trempík (OK-006, now OK-JXA) was achieved on 19 October 1979, followed by the aircraft's public debut 12 days later. By the end of 1979, the first stages of certification under FAR 23 regulations had been completed. The Trempík is stressed to $+3{\cdot}8g$, $-1{\cdot}9g$.

TYPE: Tandem two-seat light monoplane.

WINGS: Strut-braced high-wing monoplane. Wing section NACA 23012 on constant-chord centre-section; NACA 4412 at wingtip. Washout 2°. Dihedral 1°. Incidence 4° 30′. Two-spar wooden structure, with leading-edge of birch plywood stiffened internally with styrofoam. Remainder of wing fabric covered. Wingtips of GRP; streamline-section bracing struts of steel tubing. Slotted and partially mass-balanced ailerons of wooden construction, fabric covered.

FUSELAGE: Composite structure, comprising welded steel tube forward section and wooden semi-monocoque rear section, made up of spruce longerons, plywood, spruce and pine bulkheads, and beech plywood covering. Side walls reinforced internally with styrofoam slabs.

TAIL UNIT: Conventional structure comprising all-wooden fin integral with fuselage, rudder of duralumin tubing

Mono Z-CH 100 (41 kW; 55 hp Volkswagen engine)

Tri-Z CH 300 (112 kW; 150 hp Avco Lycoming O-320)

Zenair Acro-Zénith CH 150 single-seat aerobatic and competition aircraft (112 kW; 150 hp Avco Lycoming O-320) *(Howard Levy)*

Simunek/Kamaryt SK-1 Trempik (Little Tramp) *(Letectvi + Kosmonautika)*

CH 50 Mini Z single-seat light aircraft

PIK-21 Super Sytky single-seat racing aircraft

with fabric covering, and all-wooden tailplane and elevator. Horizontal surfaces reinforced with styrofoam slabs. Trim tab on elevator.

LANDING GEAR: Non-retractable tailwheel type. Barum Aero main-wheel tyres size 420 × 152-52; Barum Aero tailwheel tyre size 260 × 85. Type B-2101 A3 mechanical drum brakes. Streamline fairings can be fitted to landing gear legs and wheels.

POWER PLANT: One 56 kW (75 hp) Praga D-1 flat-four engine, driving a two-blade fixed-pitch wooden propeller from a Praga E-117 Air Baby. Two fuel tanks, with total capacity of 60 litres (13·2 Imp gallons), fore and aft of cabin. Refuelling points in forward top-decking and on starboard side of fuselage aft of cabin door.

ACCOMMODATION: Two seats in tandem in fully-enclosed cabin. Forward-hinged jettisonable door on starboard side.

AVIONICS: Tesla LS-4/1 sailplane-type radio, battery-powered.

DIMENSIONS, EXTERNAL:

Wing span	9·294 m (30 ft 6 in)
Wing chord: at root	1·50 m (4 ft 11 in)
at tip	1·00 m (3 ft 3¼ in)
Wing area, gross	12·51 m² (134·7 sq ft)
Wing aspect ratio	6·7
Length overall	5·93 m (19 ft 5½ in)
Height overall	2·27 m (7 ft 5½ in)
Tailplane span	2·05 m (6 ft 8¾ in)
Wheel track	1·52 m (4 ft 11¾ in)
Wheelbase	4·12 m (13 ft 6¼ in)
Propeller diameter	1·60 m (5 ft 3 in)

WEIGHTS:

Weight empty	356 kg (785 lb)
Max payload	175 kg (386 lb)
Max T-O weight	575 kg (1,267 lb)

PERFORMANCE:

Never-exceed speed at S/L	116 knots (215 km/h; 133 mph) IAS
Max level speed at S/L	97 knots (180 km/h; 112 mph) IAS
Cruising speed at S/L	86 knots (160 km/h; 99 mph) IAS
Stalling speed	43 knots (80 km/h; 50 mph) IAS
Max rate of climb at S/L	210 m (690 ft)/min
Service ceiling	4,500 m (14,775 ft)
T-O run	120 m (394 ft)
Landing run	150 m (492 ft)
Max range (estimated)	324 nm (600 km; 373 miles)

FINLAND

PIK

POLYTEKNIKKOJEN ILMAILUKERHO

(The Flying Club of the Helsinki University of Technology)

Lepolantie 69D, 00660 Helsinki

PROJECT ENGINEER: Kai Mellén

Mr K. Mellén, an engineer with Finnair, has built, under the auspices of the Polyteknikkojen Ilmailukerho, a Formula V racing aircraft known as the Super Sytky. Design started at the end of 1973 and took 1,000 man-hours; construction of the prototype began on 15 March 1975.

The Super Sytky was rolled out on 20 January 1981, having taken 3,000 working hours to build, including engine conversion. It flew for the first time on 16 May 1981.

Latest known design by the students of the Flying Club is the PIK-23 Towmaster, a side-by-side two-seat sporting, training and glider-towing aircraft. This is to be manufactured in series by Valmet Oy, under whose heading it is described in the main Aircraft section of this edition.

PIK-21 SUPER SYTKY

TYPE: Single-seat racing monoplane.

WINGS: Cantilever constant-chord wooden wings, with no anhedral or dihedral. Wing section NACA 64₁212. Single box spar and truss ribs, plywood covered. Full-span narrow-chord ailerons of plywood-covered wood construction. No tabs.

FUSELAGE: Conventional semi-monocoque wooden structure, plywood covered.

TAIL UNIT: Cantilever wooden structure, plywood covered. Constant-chord tailplane and elevators. Fin integral with rear fuselage. Ground-adjustable trim tabs in rudder and elevator.

LANDING GEAR: Non-retractable tailwheel type. Steel leaf spring main legs, carrying Azuza wheels, with tyres size 5·00-5. Tapered rod tailspring with 12·7 cm (5 in) tailwheel. Reinforced plastics main-wheel fairings. Azuza mechanical brakes.

POWER PLANT: One 1,600 cc modified Volkswagen motor car engine, producing 37·25 kW (50 hp) at 3,400 rpm and driving a Ray Hegy two-blade fixed-pitch wooden propeller, with large spinner. One fuel tank in fuselage, capacity 40 litres (8·8 Imp gallons). Refuelling point in front of windscreen. Oil capacity 2·5 litres (0·55 Imp gallons).

ACCOMMODATION: Enclosed cabin seating pilot only. One-piece jettisonable windscreen/canopy, hinged on starboard side.

AVIONICS: Radio fitted.

DIMENSIONS, EXTERNAL:

Wing span	5·30 m (17 ft 4¾ in)
Wing chord, constant	1·37 m (4 ft 6 in)
Wing area, gross	7·20 m² (77·5 sq ft)
Wing aspect ratio	3·87
Length overall	5·16 m (16 ft 11¼ in)
Tailplane span	1·70 m (5 ft 7 in)
Wheel track	1·32 m (4 ft 4 in)
Wheelbase	3·93 m (12 ft 10¾ in)
Propeller diameter	1·32 m (4 ft 4 in)

WEIGHTS:

Weight empty, equipped	198·6 kg (438 lb)
Max T-O weight	320 kg (705 lb)

PERFORMANCE: No data available

TERVAMÄKI

JUKKA TERVAMÄKI

Aidasmäentie 16-20E, 00650 Helsinki 65

Mr Tervamäki, Technical Manager of a medical instrument company, first became interested in autogyros in 1956. In 1959 he worked briefly for the Bensen Aircraft Corporation in the USA. He obtained a Diploma in Aeronautical Engineering at the Helsinki Institute of Technology in 1963, and later served in the helicopter section of the Finnish Air Force. He was for two years project manager and chief designer of the PIK-19 Muhinu glider-towing aircraft. In 1974 he modified a Schleicher ASK 14 powered sailplane to make it capable of taxying on large airports; and designed the engine installation for the JT-6 prototype powered sailplane, which is in production as the PIK-20E (see Sailplanes section).

Autogyros designed by Mr Tervamäki were completed in 1958 (JT-1), 1965 (JT-2), 1968 (Tervamäki-Eerola ATE-3) and 1972 (JT-5).

TERVAMÄKI JT-5

The JT-5 is a development of the ATE-3, the major visible differences being the use of a triple tail assembly, to improve static and dynamic stability; a fully-enclosed cockpit; improved, low-drag fuselage contours; and extensive use of plastics materials in the basic structure and main components. Other features include an upward-directed exhaust, to reduce engine noise, and a simplified carburettor installation and heating system of Tervamäki design.

The prototype JT-5 (OH-XYS) was flown for the first time on 7 January 1973. It was later sold, together with production rights, tools and moulds, to Sr Vittorio Magni of Italy. In 1979 plans, component parts, materials and complete kits of the Magni-Tervamäki MT-5 were made available to amateur constructors by VPM (see Italy). By early 1981 Mr Tervamäki had sold about 30 sets of plans for the JT-5, and several aircraft were under construction.

The autogyro illustrated was completed in 1975 by Arvo and Hannu Taupila. It is a hybrid of the ATE-3 and JT-5, combining the steel tube structure, landing gear and rotor head of the former with the cockpit, tail surfaces and rotor blades of the JT-5:

The following details apply to the standard JT-5:

Type: Single-seat light autogyro.

Rotor System: Two-blade semi-rigid rotor of glassfibre-reinforced epoxy resin, with polyurethane plastics foam core. Blades, of constant chord and NACA 8-H-12 section, are each attached to hub by two bolts. A lead bar in each blade leading-edge forms the chordwise balance weight. Rotor mast of streamlined SAE 4130 steel tubing. Rotor head is of a compact offset-gimbal type with centrifugal teeter stops and rotor brake installed. There are two spiral springs for trim adjustment, which is effected via the control stick twist-grip handle. Normal rotor rpm is 400, maximum 600. Designed for the JT-5, but not yet fitted, is a modified Cierva-type inclined drag hinge which would allow the blades to move to zero pitch when pre-rotation torque is applied, permitting an increase of 100 rpm in pre-spin speed and, consequently, a shorter take-off.

Rotor Drive: Rotor spin-up by Vee-belt, clutch, 90° gearbox, sliding universal shaft and inertia-operated Bendix drive. Overall reduction ratio 8. Rotor spin-up of 300 rpm can be achieved.

Fuselage: Basic structure of welded 4130 steel tubing with a glassfibre/HFB honeycomb sandwich cockpit. All internal cockpit structures of glassfibre-reinforced epoxy resin. One-piece aluminium engine cowling.

Tail Unit: Central main fin and rudder, of glassfibre-reinforced epoxy resin, with rigid PVC foam ribs and Courtauld carbon fibre stiffeners. Horizontal tail and auxiliary endplate fins of glassfibre sandwich construction with honeycomb core. Tail assembly attached to fuselage by a single streamlined steel tube. Small tailwheel beneath base of fin.

Landing Gear: Non-retractable tricycle type. Main gear legs consist of 4 × 4 cm (1·6 × 1·6 in) glassfibre-reinforced epoxy resin springs, encased in streamline fairings of the same material. Cables inside these fairings to main gear drum brakes. Main-wheel tyres size 300 × 100. Compression rubber shock-absorption in nose gear. Nosewheel tyre size 260 × 80. Nosewheel steerable by rudder pedals.

Power Plant: One 56 kW (75 hp) 1·7 litre Volkswagen engine, converted for autogyro use by Limbach Motorenbau. No oil cooler, generator or electric starter. Two-blade pusher propeller, of glassfibre-reinforced epoxy. Glassfibre fuel tank, integrally built into fuselage aft of pilot's seat, capacity 50 litres (11 Imp gallons).

Accommodation: Single seat under sideways-opening Plexiglas canopy. Instrument panel cover and pilot's seat back (the latter also forming a firewall to the engine compartment) open together with the canopy.

Equipment: Standard equipment and controls are as listed in the 1974-75 *Jane's*. A 6-channel radio is installed.

Dimensions, external:	
Rotor diameter	7·00 m (22 ft 11½ in)
Rotor blade chord (constant, each)	0·18 m (7·1 in)
Length of fuselage	3·50 m (11 ft 5¾ in)
Height overall	2·00 m (6 ft 6¾ in)
Wheel track	1·70 m (5 ft 7 in)
Propeller diameter	1·20 m (3 ft 11¼ in)
Areas:	
Rotor blades (each)	0·63 m² (6·78 sq ft)
Rotor disc	38·50 m² (414·4 sq ft)
Weights:	
Weight empty, equipped	167 kg (368 lb)
Max T-O weight	290 kg (639 lb)
Performance (at max T-O weight):	
Never-exceed speed	97 knots (180 km/h; 111·5 mph)
Max level speed at S/L	92 knots (170 km/h; 106 mph)
Max cruising speed at S/L	81 knots (150 km/h; 93 mph)
Econ cruising speed	70 knots (130 km/h; 81 mph)
Min level speed	19 knots (35 km/h; 22 mph)
Max rate of climb at S/L	180 m (590 ft)/min
Service ceiling	4,000 m (13,125 ft)
T-O run	70 m (230 ft)
T-O to 15 m (50 ft)	120 m (394 ft)
Landing from 15 m (50 ft)	50 m (165 ft)
Landing run	5 m (16 ft)
Range with max fuel, no reserves	189 nm (350 km; 217 miles)

FRANCE

COLOMBAN

MICHEL COLOMBAN

37*bis* rue Lakanal, 92500 Rueil-Malmaison

Telephone: 751 88 76

Formerly with the Morane and Potez companies, and now an aerodynamicist with Aérospatiale, M Colomban designed and built a very small and unique twin-engined lightplane named the Cricri. Its construction required some 1,200 hours of work and cost only 5,000 francs (1971-72 prices), including the engines.

Plans and components for building the latest MC 12 version of the Cricri are available to amateur constructors from the Siravia company of Pons. Mr Colomban is thus free to concentrate on development of a new two-seat design.

COLOMBAN MC 12 CRICRI

The MC 10 prototype of the Cricri was powered by two Rowena 6507J single-cylinder two-stroke engines of 137cc, each giving 6·7 kW (9 hp) and weighing 6·5 kg (14·3 lb). It was claimed to be the smallest twin-engined aeroplane then flying, and the only one able to lift a useful load equivalent to 170 per cent of its own empty weight. Special constructional features permitted assembly or disassembly in only five minutes. Its light weight and small size made it particularly easy to transport on a trailer towed by car and to store in a garage or shed.

The Cricri was flown for the first time on 19 July 1973 by Robert Buisson, a 68-year-old pilot who had already logged 12,000 flying hours. A number of design refinements were made later that year, after which testing was resumed. Within fifteen days the Cricri had logged a total of 13 trouble-free flying hours, including rolls, reversements, 'split S' manoeuvres and inverted flight, made possible by its Tillotson diaphragm carburettor. Flight tests at up to 135 knots (250 km/h; 155 mph) confirmed that no special piloting skills were needed to fly the aircraft.

In particular, the Cricri handled like a single-engined design. This resulted from the fact that the two small engines were mounted close together, and from the carefully-conceived shape of the cockpit canopy which deflected the propeller slipstream over the tail surfaces in such a way that the failure of one engine would produce no dangerous handling problems. If one engine was throttled back fiercely, with hands and feet off the controls, the Cricri was said to do no more than begin a gentle turn.

Construction of a number of improved and re-engined Cricris has been undertaken by friends of M Colomban, and the first of these to be completed (No. 4, built by M Gérard Constant of Dreux) was exhibited in the static park at the 1977 Paris Salon. This aircraft, powered by two 125 cc McCulloch MC-101 single-cylinder two-stroke engines (each 8·9 kW; 12 hp), flew for the first time in July 1978 and performed entirely satisfactorily. Take-off was achieved in 80 m (263 ft), in a time of 7 s, in zero wind, followed by a stabilised climb at a rate of 360 m (1,180 ft)/min. On the other hand, the MC-101 engines were clearly too sophisticated for this application. So, although they proved generally satisfactory, they have been superseded by Valmet SM 160J engines on both M Constant's Cricri and on the version shown on the plans available to amateur constructors from the Siravia company. With Valmet engines the Cricri is designated MC 12.

A second amateur-built Cricri was flown in 1980, by M. J. Laurent of Rennes. Six MC 12s were expected to be flying during 1981.

Type: Twin-engined single-seat light aircraft, stressed to $+10g$ and $-5g$.

Wings: Cantilever low-wing monoplane of constant chord. Laminar-flow aerofoil derived from a Wortmann section. Thickness/chord ratio 21·7%. Dihedral 4° from roots. Incidence 1° at root, −30′ at tip. No sweep. Single-spar light alloy box structure. Spar comprises a web riveted to 2024 light alloy angle-section booms. Inboard end of spar in each wing is of 'forked-tongue' form, like that of many sailplanes, to permit rapid assembly and disassembly of wings (2 minutes). Closely-spaced Klégécel ribs are bonded fore and aft of the spar. Skin consists of a single sheet of 2017 light alloy, bonded to structure under pressure after its leading-edge has been formed. No rear spar. Wing box is closed at each end by a riveted metal rib. Entire trailing-edge is occupied by two-section external flaps of the kind fitted to many wartime Junkers aircraft, operating collectively as high-lift devices and differentially as ailerons. Flaps are spar-less, consisting of a metal monocoque structure, with four metal ribs per section (at each tip and each pivot point), filled with Klégécel over the entire span and over 20% of the chord. Flaps are each actuated via a ball-joint at the root. No controls pass through the wing box, which contains only a 2017 tube as provision for any future installation of fuel tanks on wingtips.

Fuselage: Simple metal box structure of 2017 sheet made in two parts to reduce space required for manufacture. Structure is stiffened by Klégécel stringers, bonded in place. Frames of 2017 light alloy riveted in position in line with the attachments for the wings, landing gear, tail unit and engine mountings.

Tail Unit: Cantilever T type, with sweptback vertical surfaces and all-moving constant-chord horizontal surface. Construction similar to that of wings. No tabs. Tailplane actuated by control rods, rudder by cables. Tailplane provided with artificial loading by bungee cord.

Landing Gear: Non-retractable tricycle type. Nosewheel fitted with bungee shock-absorption and linked to rudder bar for steering. Each main wheel carried on cantilever leg of glassfibre/epoxy laminations. Main-wheel tyres size 210-70, pressure 1·80 bars (28 lb/sq in). Nosewheel tyre size 200-50, pressure 0·80 bars (11·6 lb/sq in). Small diameter motorcycle drum brakes. Provision for fairing on all three wheels.

Power Plant: Two modified 160 cc Valmet SM 160J single-cylinder two-stroke engines, each giving 8·9 kW (12 hp) at 6,500 rpm. Tillotson diaphragm carburettors for all-attitude flight. Each engine drives an MCH 700-18 two-blade propeller, made of glassfibre. Plastics fuel tank in fuselage, capacity 23 litres (5 Imp gallons). Provision for tank on each wingtip, total capacity 24 litres (5·25 Imp gallons).

Accommodation: Single seat under large transparent canopy, hinged to open sideways, to starboard. Ventilation through port in side of fuselage. No heating.

Dimensions, external:	
Wing span, with or without tip-tanks	4·90 m (16 ft 0¾ in)
Wing chord, incl flap (constant)	0·63 m (2 ft 0¾ in)
Wing chord, excl flap (constant)	0·48 m (1 ft 6¾ in)
Wing area, gross	3·10 m² (33·4 sq ft)
Wing aspect ratio	7·75
Length overall	3·91 m (12 ft 10 in)
Height overall	1·20 m (3 ft 11¼ in)
Tailplane span	1·55 m (5 ft 1 in)
Wheel track	1·10 m (3 ft 7¼ in)
Wheelbase	1·15 m (3 ft 9¼ in)
Propeller diameter	0·69 m (2 ft 3 in)
Distance between propeller centres	0·95 m (3 ft 1½ in)
Dimensions, internal:	
Cabin: Length	1·30 m (4 ft 3¼ in)
Max width	0·55 m (1 ft 9½ in)
Max height	0·82 m (2 ft 8¼ in)
Weights:	
Weight empty	70 kg (154 lb)
Max T-O and landing weight	170 kg (375 lb)
Performance (at AUW of 160 kg; 352 lb):	
Max speed measured in flight	140 knots (260 km/h; 161 mph)
Max level speed	113 knots (210 km/h; 130 mph)
Max cruising speed (80% power)	103 knots (190 km/h; 118 mph)
Stalling speed, flaps down	39 knots (72 km/h; 45 mph)
Stalling speed, flaps up	50 knots (93 km/h; 58 mph)
Best glide ratio, engines off and T-O configuration, at 60 knots (110 km/h; 69 mph)	11

The hybrid ATE-3/JT-5 built by Arvo and Hannu Taupila of Kihniö, Finland

MC 12 Cricri (two Valmet SM 160J engines) built by Gérard Constant of Dreux
(Aviasport; M. Battarel)

Coupé-Aviation JC-01 two-seat lightplane

Croses EAC-3 Pouplume light aircraft (Moto engine)

Max rate of climb at S/L	336 m (1,100 ft)/min
Rate of climb at S/L, one engine out	80 m (262 ft)/min
Service ceiling	4,600 m (15,090 ft)
T-O run	110 m (360 ft)
T-O to 15 m (50 ft)	310 m (1,020 ft)
Landing from 15 m (50 ft)	350 m (1,150 ft)
Landing run	150 m (495 ft)
Range with max fuel	430 nm (800 km; 496 miles)

COUPÉ-AVIATION

COUPÉ-AVIATION

La Trute, Azay-sur-Cher, 37270 Montlouis
Telephone: (47) 55 31 84

M Jacques Coupé has designed and built a two-seat lightplane designated JC-01 (F-PXKV), which was flown for the first time on 16 March 1976. Design emphasis was placed on ensuring that homebuilders would find it easy to construct, and plans are available from Coupé-Aviation. In early 1980 the company had almost completed construction of a derivative of the JC-01. Designated JC-2, it is powered by a 67 kW (90 hp) engine and has tricycle landing gear. Its first flight was scheduled for 1980, but no confirmation of this has been received. Further JC-01s are being built, some powered by modified Volkswagen motor car engines of 1,600 or 1,700 cc.

COUPÉ-AVIATION JC-01

Type: Two-seat lightweight sporting aircraft.

Wings: Cantilever low-wing monoplane of constant chord. Conventional single-spar structure of wood, with lightweight lattice ribs and fabric covering. Leading-edge fixed slot. Slotted ailerons of similar construction to wings.

Fuselage: Lattice structure of wood, with wood and fabric covering. The fin is constructed integrally with the fuselage.

Tail Unit: Cantilever structure of wood; fin, rudder and elevators fabric-covered. Fixed-incidence tailplane has plywood skin. Manually-operated trim tab in elevator.

Landing Gear: Non-retractable tailwheel type. Shock-absorption of main units by rubber in compression, and of tailwheel by coil spring. Main-wheel tyres size 420 × 150. Cable-operated caliper brakes.

Power Plant: One 48·5 kW (65 hp) Continental A65-8F flat-four engine, driving a two-blade fixed-pitch wooden propeller with spinner. Fuel tank forward of cockpit, capacity 60 litres (13 Imp gallons).

Accommodation: Two seats side by side, beneath rearward-sliding transparent canopy. Dual controls. Accommodation is heated and ventilated.

Dimensions, external:

Wing span	8·35 m (27 ft 4¾ in)
Wing area, gross	11·69 m² (125·83 sq ft)
Wing aspect ratio	5·95
Length overall	6·40 m (21 ft 0 in)

Weights:

Weight empty	330 kg (728 lb)
Max T-O weight	580 kg (1,279 lb)

Performance (at max T-O weight):

Max level speed	108 knots (200 km/h; 124 mph)
Max cruising speed	81 knots (150 km/h; 93 mph)
Econ cruising speed	76 knots (140 km/h; 87 mph)
T-O speed	27 knots (50 km/h; 31 mph)
Approach speed	37-43 knots (70-80 km/h; 43-50 mph)
Stalling speed	24·5 knots (45 km/h; 28 mph)
T-O run	90 m (295 ft)

CROSES

EMILIEN CROSES

Route de Davayé (Aérodrome), 71000 Charnay les Macon
Telephone: (85) 38 07 31

Since M Emilien Croses began work as a designer/constructor in 1947, he has been responsible for nine different prototypes, all of which have been certificated. Addition of the initial 'B' in the designation of some of these reflects assistance given to M Croses by M R. Bujon, a specialist metal worker. Other assistance is given by MM J. Mottez with stressing and aerodynamics; Alain Croses with aerodynamic studies; and Y. Croses, an engineer specialising in applications of high-strength plastics and glassfibre.

CROSES EAC-3 POUPLUME

As in the familiar Mignet designs, the Pouplume single-seat tandem-wing biplane has a fixed rear wing and a pivoted forward wing which dispenses with the need for ailerons and elevators. A conventional rudder is fitted, with a large tailwheel built into its lower edge.

Construction is conventional, with spruce wing structure and a square-section spruce fuselage covered with okoumé ply. The main landing gear consists of Vespa scooter wheels carried on a wooden cross-member.

The power unit in the prototype (EAC-3-01) is a 7·8 kW (10·5 hp) Moto 232 cc two-stroke motorcycle engine, with chain reduction drive to the propeller shaft. The reduction ratio is 3·5 : 1, giving a propeller speed of 1,300 rpm. Fuel capacity is 10 litres (2·2 Imp gallons).

The EAC-3-01 Pouplume took 600 h to build and flew for the first time in June 1961. This machine was followed, in 1967, by a second prototype (EAC-3-02), with a 20 cm (8 in) longer fuselage. M Croses is offering sets of plans to other constructors, and at least 12 Pouplumes had flown by early 1981. The one shown in the accompanying illustration was built in France by an amateur constructor.

A version known as the Pouplume Sport differs in having a 1500 cc Volkswagen engine and reduced span of 6·40 m (21 ft 0 in). About 55 Pouplume Sports were under construction in early 1980.

The following data apply to the standard Moto-powered Pouplume:

Dimensions, external:

Span of forward wing	7·8 m (25 ft 7 in)
Span of rear wing	7·0 m (23 ft 0 in)
Wing area, gross	16·0 m² (172 sq ft)
Length overall	4·7 m (15 ft 3 in)
Height overall	1·8 m (5 ft 11 in)

Weights:

Weight empty	110-140 kg (243-310 lb)
Max T-O weight	220-260 kg (485-573 lb)

Performance (A: 7·8 kW; 10·5 hp engine. B: 13·4 kW; 18 hp engine):

Max level speed:	
A	38 knots (70 km/h; 43·5 mph)
B	65 knots (120 km/h; 75 mph)
Econ cruising speed:	
A	27 knots (50 km/h; 31 mph)
B	38 knots (70 km/h; 43·5 mph)
T-O speed:	
A	13·5 knots (25 km/h; 15·5 mph)
Landing speed:	
A	9·7 knots (18 km/h; 11 mph)
T-O run:	
A	60 m (200 ft)
B	40 m (131 ft)
Landing run:	
A	24 m (80 ft)
Fuel consumption:	
A	4·5 litres (1 Imp gallon)/h

CROSES EC-6 CRIQUET (LOCUST)

This design by Emilien Croses is a development of his earlier EC-1-02 prototype and is a side-by-side two-seater based on the familiar Mignet tandem-wing formula. Construction of the prototype was started in March 1964 and the EC-6-01 flew for the first time on 6 July 1965.

Plans of the wooden version of the Criquet, of which details follow, are available to amateur constructors. At least seven examples were flying by early 1981, with about 60 more under construction.

An all-plastics version, known as the LC-10, was described briefly in the 1977-78 *Jane's*. Plans of this are not available.

TYPE: Two-seat tandem-wing light aircraft.

WINGS: Forward wing built in one piece and pivoted on two streamlined supports, giving variable incidence between −2° and +12°. Fixed rear (lower) wing. Wing section NACA 23012 (modified). Both wings have two-spar wooden structure, with plywood leading-edge, overall fabric covering and some components of glassfibre. Ailerons optional.

FUSELAGE: Spruce structure, covered with plywood. Glassfibre engine cowling.

TAIL UNIT: Plywood-covered spruce fin and rudder. No tailplane or elevators.

LANDING GEAR: Non-retractable tailwheel type. Main wheels, size 420-150, carried on single cantilever arch structure made from ash wood on a forme and covered with glassfibre. Tailwheel, size 420-150, semi-enclosed in bottom of rudder.

POWER PLANT: One 67 kW (90 hp) Continental flat-four engine, driving a modified SIPA two-blade propeller. Fuel capacity originally 60 litres (13 Imp gallons); planned to be increased to 90 litres (20 Imp gallons).

ACCOMMODATION: Two seats side by side in enclosed cabin. Door on starboard side. Constructors can utilise either the special Mignet type of control system or a conventional system with ailerons and rudder bar.

DIMENSIONS, EXTERNAL:

Span of forward wing	7·80 m (25 ft 7 in)
Span of rear wing	7·00 m (22 ft 11½ in)
Wing chord (constant, each)	1·20 m (3 ft 11¼ in)
Wing area, gross	16·0 m² (172 sq ft)
Length overall	4·65 m (15 ft 3 in)

WEIGHTS:

Weight empty	290 kg (639 lb)
Max T-O weight	550 kg (1,213 lb)

PERFORMANCE (officially certificated, at max T-O weight):

Max level speed at S/L	115 knots (213 km/h; 132 mph)
Max cruising speed	92 knots (170 km/h; 106 mph)
Econ cruising speed	86 knots (160 km/h; 99 mph)
Min flying speed	22 knots (40 km/h; 25 mph)
Will not stall	
T-O time (max)	6 s
Climb to 2,000 m (6,560 ft)	6 min 14 s

CROSES EC-8 TOURISME

This three-seat touring aircraft is generally similar to the standard wooden Criquet but has an 'all-terrain' landing gear comprising two tandem pairs of main wheels. At least two EC-8s are flying.

CROSES B-EC-9 PARAS-CARGO

Unique in being a cargo transport for construction and operation by amateurs, the B-EC-9 employs the same Mignet tandem-wing configuration and simple wood construction as earlier Croses designs. Like them, it can be built with conventional three-axis controls or the special two-control system devised by Henri Mignet, with large 'tab' control surfaces on the trailing-edge of the rear wing.

It is intended to certificate the prototype, following tests of the aircraft's suitability as a transport for parachutists. It will be offered to amateur constructors and clubs in the form of both plans and kits of components.

TYPE: Light cargo transport and utility aircraft for amateur construction.

WINGS: Tandem-wing configuration of Mignet type, with forward wing supported, parasol-fashion, at three pivot points. Fixed cantilever aft wing mounted at top of rear fuselage. Both wings have same constant chord, with round and slightly upswept tips. No dihedral. No sweep. Wood structure, with fabric covering aft of main spar. Large inset tab in trailing-edge of each rear wing, but no ailerons, on prototype, which has Mignet two-control system.

FUSELAGE: Large-volume plywood box structure, with unobstructed interior.

TAIL UNIT: Fin, integral with fuselage, and rudder only, of plywood-covered wood construction. No tabs.

LANDING GEAR: Croses 'tous terrains' type. Two pairs of main wheels in tandem, carried on one-piece cantilever arch structure of wood covered with glassfibre. Two forward wheels larger than rear wheels. Large tailwheel semi-enclosed in base of rudder.

POWER PLANT: One 134 kW (180 hp) Avco Lycoming flat-four engine, driving a two-blade fixed-pitch propeller without spinner.

ACCOMMODATION: Crew of two side by side on flight deck, connected by a passage to the square-section main cabin. Accommodation for freight, four parachutists, two stretcher patients and seats for two attendants, or agricultural equipment. Very low floor for easy loading. Large door on port side of cabin opens inward and upward to permit airdropping of parachutists or cargo. Separate forward-hinged crew door on port side of flight deck. Use of a door at the rear of the cabin would permit the carriage of items such as metal tubes, boards or helicopter rotor blades up to 5·50 m (18 ft 0 in) long.

DIMENSIONS, EXTERNAL:

Wing span:	
front wing	9·60 m (31 ft 6 in)
rear wing	8·80 m (28 ft 10½ in)
Wing chord, constant	1·60 m (5 ft 3 in)
Wing gap (projected)	0·94 m (3 ft 1 in)
Length overall	7·50 m (24 ft 7¼ in)

DIMENSIONS, INTERNAL:

Cabin: Length	2·60 m (8 ft 6¼ in)
Max width	1·35 m (4 ft 5 in)
Max height	1·45 m (4 ft 9 in)
Usable volume	5 m³ (176 cu ft)

WEIGHTS:

Weight empty	650 kg (1,433 lb)
Max payload	450 kg (992 lb)

PERFORMANCE:

Normal cruising speed	94 knots (175 km/h; 108 mph)
Econ cruising speed	86 knots (160 km/h; 99 mph)
Speed for airdropping	32 knots (60 km/h; 37 mph)
Rate of climb at S/L	240 m (785 ft)/min
T-O run	150 m (490 ft)

DELEMONTEZ-CAUCHY
ALAIN CAUCHY

Les Bleuets Apt 24 rue A. France, 60230 Chambly

Among new amateur-built aircraft which took part in the RSA meeting at Brienne-le-Château in July 1979 was a small two-seater designated DC-1, with a configuration reminiscent of that of the well-known Jodel series of designs. It had been built over a period of 18 months by M Alain Cauchy, with design assistance from M Jean Delemontez, President-Director General of Avions Jodel SA, and had flown for the first time on 6 July 1979.

DELEMONTEZ-CAUCHY DC-1

TYPE: Low-cost two-seat light aircraft.

WINGS: Cantilever low-wing monoplane. Single-spar wing, with wide-span centre-section of constant chord and thickness, and two tapered outer panels set at a coarse dihedral (12°). All-wood structure. Plywood-covered leading-edge forward of spar; fabric covering aft of spar. Fabric-covered wooden ailerons.

FUSELAGE: Plywood-covered wood structure of basically rectangular section, with rounded top-decking.

TAIL UNIT: Cantilever wood structure. No fin. Tailplane plywood-covered; elevators and rudder fabric-covered.

LANDING GEAR: Non-retractable tailwheel type. Cantilever main legs with rubber-in-compression shock-absorption. Steerable tailwheel.

POWER PLANT: One 1,600 cc converted Volkswagen motorcar engine, driving a two-blade fixed-pitch wooden propeller made by Evra to Delemontez design.

ACCOMMODATION: Two seats side by side in enclosed cabin. Centrally-mounted control column and dual rudder pedals.

DIMENSIONS, EXTERNAL:

Wing span	7·40 m (24 ft 3½ in)
Wing area, gross	9·45 m² (101·7 sq ft)
Length overall	7·40 m (24 ft 3½ in)

DIMENSION, INTERNAL:

Cabin: Max width	1·00 m (3 ft 3¼ in)

WEIGHT:

Weight empty	216 kg (476 lb)

PERFORMANCE (measured on first flight, pilot only):

Max level speed	113 knots (210 km/h; 130 mph)
Max cruising speed (75% power)	86 knots (160 km/h; 99 mph)
Approach speed	43 knots (80 km/h; 50 mph)
Max rate of climb at S/L	240 m (785 ft)/min
T-O run	120 m (394 ft)
Landing run	100 m (328 ft)

DURUBLE
ROLAND DURUBLE

40 rue du Paradis, Les Essarts, 76530 Grand-Couronne
Telephone: 92 20 63

M Roland Duruble, with MM Guy Chanut and Legrand, of Rouen, designed and built a two-seat all-metal light aircraft named the RD-02 Edelweiss, which flew for the first time on 7 July 1962. Full details of this aircraft can be found in the 1972-73 *Jane's*.

Plans of an enlarged and improved version, known as the RD-03 Edelweiss, are available to other constructors.

DURUBLE RD-03 EDELWEISS

The RD-03 Edelweiss is designed to AIR 2052 (CAR 3) standards, in three versions, as follows:

RD-03A. With 74·5 kW (100 hp) Continental O-200 flat-four engine and fuel capacity of 100 litres (22 Imp gallons) in two wing tanks. Can be fitted with 67 kW (90 hp) engine or 100·5 kW (135 hp) Avco Lycoming O-320 engine. Side-by-side seats for pilot and one passenger (total weight 172 kg; 380 lb) in cabin.

RD-03B. With 100·5 kW (135 hp) Avco Lycoming O-320 or Franklin Sport 4B engine and same fuel capacity as RD-03A. Seating as RD-03A for Utility category operation, or in '2 + 2' arrangement for pilot and three passengers (154 kg; 340 lb on front seats, 110 kg; 240 lb on rear seats) in Normal category.

RD-03C. With 112 kW (150 hp) Avco Lycoming engine and additional wing tanks, increasing total fuel capacity to 150 litres (33 Imp gallons). In Utility two-seat form (as RD-03A) or with seating for a pilot and three passengers (total weight 308 kg; 680 lb) in Normal category.

Plans of the RD-03 have been available since the Autumn of 1970. By early 1981 about 48 sets of plans had been sold. Nine Edelweiss are known to be under construction in France, two in Belgium, eleven in Canada and the USA, and one in New Zealand. The first of these, constructed by M Duruble and M Gastan, was expected to fly for the first time in 1981.

TYPE: Two/four-seat light aircraft.

WINGS: Cantilever low-wing monoplane. Wing section NACA 23000 series. Thickness/chord ratio 18% at root, 12% at tip. Dihedral 6° 5' from roots. Incidence 3° at root, 0° at tip. No sweepback. All-metal two-spar duralumin structure, with metal slotted trailing-edge flaps and slotted ailerons. No trim tabs.

FUSELAGE: Conventional semi-monocoque duralumin structure.

TAIL UNIT: Cantilever all-metal structure, with sweptback vertical surfaces. Fixed-incidence tailplane. Trim tab in each elevator, one actuated by flap linkage and the other manually.

LANDING GEAR: Retractable tricycle type. Hydraulic retraction, nosewheel rearward, main units inward into wings. Duruble hydro-air shock-absorbers on all three units. Main-wheel tyres size 355 × 150, nosewheel tyre size 330 × 130. Pressure (all tyres) 1·24 bars (18 lb/sq in). Hydraulic disc brakes.

POWER PLANT: One 67, 74·5, 100·5 or 112 kW (90, 100, 135 or 150 hp) flat-four engine (details under individual model listings). Refuelling point above wing.

ACCOMMODATION: Side-by-side seats for two, three or four persons (details under individual model listings) in fully-enclosed cabin.

SYSTEM: Hydraulic system, pressure 69 bars (1,000 lb/sq in), for flap and landing gear actuation.

AVIONICS AND EQUIPMENT: Radio optional. Blind-flying instrumentation not fitted.

DIMENSIONS, EXTERNAL:

Wing span	8·82 m (28 ft 11¼ in)
Wing chord at root	1·70 m (5 ft 7 in)
Wing chord at tip	0·86 m (2 ft 10 in)
Wing area, gross	11·07 m² (119·2 sq ft)
Wing aspect ratio	7·05
Length overall (RD-03A)	6·875 m (22 ft 6¾ in)
Height overall	2·60 m (8 ft 6¼ in)
Tailplane span	3·05 m (10 ft 0 in)

DIMENSIONS, INTERNAL:

Cabin: Max length	2·44 m (8 ft 0 in)
Max width	1·10 m (3 ft 7¼ in)

WEIGHTS (estimated. A: RD-03A; B: RD-03B; C: RD-03C):

Weight empty, equipped: A	426 kg (940 lb)
B	436 kg (962 lb)
C	443 kg (976 lb)
Max T-O and landing weight:	
A (Utility)	689 kg (1,519 lb)
B (Utility)	705 kg (1,554 lb)
B (Normal)	817 kg (1,801 lb)
C (Utility)	731 kg (1,611 lb)
C (Normal)	887 kg (1,955 lb)

PERFORMANCE (estimated, at max T-O weight):

Never-exceed speed:	
A (Utility), B (Utility), C (Utility)	170 knots (316 km/h; 196 mph)
B (Normal), C (Normal)	182 knots (339 km/h; 210 mph)
Max level speed at S/L:	
A (Utility)	139 knots (257 km/h; 160 mph)
B (Utility)	143 knots (265 km/h; 165 mph)
B (Normal)	141 knots (262 km/h; 163 mph)
C (Utility)	149 knots (277 km/h; 172 mph)
C (Normal)	146 knots (270 km/h; 168 mph)
Max cruising speed at S/L:	
A (Utility)	126·5 knots (234 km/h; 145·5 mph)
B (Utility)	129 knots (240 km/h; 149 mph)
B (Normal)	128 knots (238 km/h; 148 mph)
C (Utility)	136 knots (252 km/h; 157 mph)
C (Normal)	133 knots (246 km/h; 153 mph)

Croses EC-6 Criquet *(Geoffrey P. Jones)*

Croses B-EC-9 Paras-Cargo light utility/cargo transport *(Geoffrey P. Jones)*

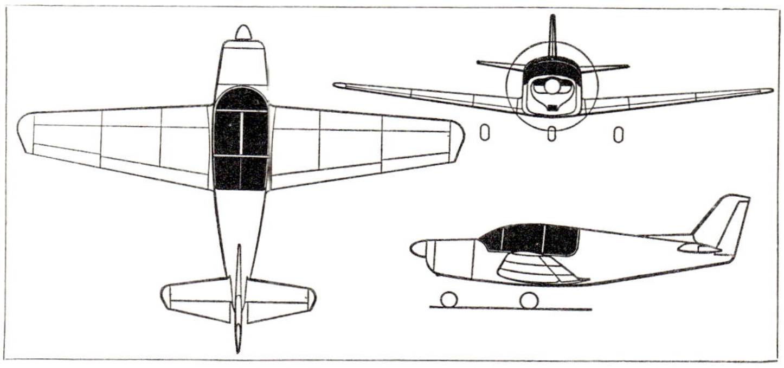

Duruble RD-03C Edelweiss two/four-seat light aircraft

Gatard AG 02 Poussin No. 19, built by M A. Hézard of Dijon

Delemontez-Cauchy DC-1 *(S. A. MacConnacher)*

Econ cruising speed at S/L:	
A (Utility), B (Utility)	121 knots (224 km/h; 139 mph)
B (Normal)	120 knots (222 km/h; 138 mph)
C (Utility)	128 knots (237 km/h; 147·5 mph)
C (Normal)	125 knots (233 km/h; 144·5 mph)
Stalling speed, flaps up:	
A (Utility), B (Utility)	50 knots (92 km/h; 57·5 mph)
B (Normal), C (Utility)	56 knots (104 km/h; 65 mph)
C (Normal)	59 knots (109 km/h; 68 mph)
Stalling speed, flaps down:	
A (Utility), B (Utility)	43 knots (79 km/h; 49·5 mph)
B (Normal)	48 knots (88 km/h; 55 mph)
C (Utility)	45 knots (83 km/h; 51·5 mph)
C (Normal)	48 knots (89 km/h; 55·5 mph)
Max rate of climb at S/L: A	198 m (650 ft)/min
B	213 m (700 ft)/min
C	300 m (985 ft)/min
Service ceiling: A, B	4,570 m (15,000 ft)
C	5,030 m (16,500 ft)
T-O to 15 m (50 ft): A, B	450 m (1,475 ft)
C	510 m (1,675 ft)
Landing from 15 m (50 ft): A	500 m (1,640 ft)
B, C	515 m (1,690 ft)
Range with max fuel, 30 min reserves:	
A, C	502 nm (930 km; 578 miles)
B	512 nm (950 km; 590 miles)

GATARD

AVIONS A. GATARD

Villa la Devallée, 17130 Montendres/Royan
Telephone: (46) 49 42 76

M Albert Gatard has developed a control system for aeroplanes which involves the use of a variable-incidence lifting tailplane of large area, and has built a series of aircraft, including the Alouette, Poussin and Pigeon, incorporating his ideas. Plans of the Poussin are available to amateur constructors. The Alouette and Pigeon have been described in previous editions of *Jane's*.

Instead of altering the wing angle of attack to increase lift on these aircraft, the pilot lowers full-span slotted aileron/flaps and adjusts the tailplane to maintain pitching equilibrium. In consequence, the aircraft climb with the fuselage datum at no more than 4° to the horizontal, which preserves a good forward view and low body drag.

GATARD STATOPLAN AG 02 POUSSIN (CHICK)

M Gatard built two prototypes of the Poussin, and the detailed description applies to the second of these, which introduced a number of design improvements, in its original form. Flight tests revealed excellent aerobatic qualities and the power plant is modified to permit up to 20 s of inverted flying.

This prototype was extensively flight-tested at the Centre d'Essais en Vol at Istres, and the performance figures quoted are those which were obtained during the tests. As a result of recommendations by the CEV, a 27 kW (36 hp) Rectimo (modified Volkswagen VW 1200) engine is suggested as the most suitable power plant for use by amateur constructors of the Poussin. The second prototype has been re-engined with a 1,200 cc Volkswagen, by M Mathevet of Mollard-Chateauneuf (Loirs), on behalf of M Gatard. Installation of this engine was expected to improve the CG position and make possible a max speed of approx 92 knots (170 km/h; 106 mph), a max cruising speed of approx 83 knots (155 km/h; 96 mph) and a rate of climb at S/L of 210 m (690 ft)/min.

Several Poussins are being built by amateur constructors, and three of these had been completed and flown by March 1979, one (F-PYBS) with a non-standard tricycle landing gear. Another of them is shown in an accompanying illustration.

The plans were refined during 1978-79, and the version of the Poussin now offered for amateur construction is designated **AG 02Sp**. The wingtips are now more rounded, with inset ailerons. The wing structure has been modified to permit an increased max T-O weight of 305 kg (672 lb); and the landing gear has been refined to give an increase of about 5 per cent in cruising speed.

The following details apply to the standard Poussins now flying:

Type: Single-seat light monoplane.

Wings: Cantilever low-wing monoplane. NACA 23012 wing section. Dihedral 4°. Incidence 3° 30′ at root, 2° at tip. Plywood-covered single-spar all-wood structure. Full-span slotted aileron/flaps, each in two sections which are moved together but at different angles (inboard sections up to 35°, outboard up to 20°) to give the effect of increased aerodynamic twist of the complete wing/aileron/flap assemblies. Aileron/flaps are linked with the variable-incidence tailplane.

Fuselage: Plywood-covered wood structure. Perforated airbrake, under fuselage, operates automatically when the main aileron/flaps are lowered at large angles, as during landing.

Tail Unit: Braced all-wood structure, with variable-incidence all-moving tailplane of NACA 2309 section. Endplates fitted to tailplane to increase vertical fin area and effective tailplane span. No elevators. Rudder trim tab actuated by lateral movement of control column, permitting full control by means of the control column alone in normal flight.

Landing Gear: Non-retractable tailwheel type. Cantilever levered-suspension main units with rubber-band shock-absorption. Modified Dunlop brakes. Steerable tailwheel.

Power Plant: One 18 kW (24 hp) modified Volkswagen flat-four engine, driving a Gatard two-blade fixed-pitch wooden propeller. Provision for fitting any alternative engine of up to 30 kW (40 hp) weighing between 50 and 60 kg (110-132 lb). Fuel tank aft of firewall, capacity 30 litres (6·6 Imp gallons). Oil capacity 2 litres (0·45 Imp gallons).

Accommodation: Single seat under large rearward-sliding transparent canopy. Baggage space aft of seat. Two map pockets.

Dimensions, external:	
Wing span	6·40 m (21 ft 0 in)
Wing chord, constant	1·00 m (3 ft 3¼ in)
Wing area, gross	6·15 m² (66·2 sq ft)
Length overall	4·53 m (14 ft 10½ in)
Height overall	1·50 m (4 ft 11 in)
Wheel track	1·50 m (4 ft 11 in)
Wheelbase	3·20 m (10 ft 6 in)
Weights:	
Weight empty	170 kg (375 lb)
Max T-O weight	280 kg (617 lb)
Performance (at max T-O weight):	
Never-exceed speed	116 knots (216 km/h; 134 mph)
Max cruising speed	77 knots (144 km/h; 89 mph)
Max speed for aerobatics	69 knots (130 km/h; 80 mph)
Stalling speed	35 knots (65 km/h; 40·5 mph)
Max rate of climb at S/L	132 m (435 ft)/min
T-O run	190 m (625 ft)
T-O to 15 m (50 ft)	435 m (1,425 ft)
Landing from 15 m (50 ft)	320 m (1,050 ft)
Landing run	200 m (655 ft)

GRINVALDS
JEAN GRINVALDS

5 rue Maillet, St Germain, 10120 St Andre les Vergers

Aircraft exhibited at the 1979 Paris Air Show under the auspices of the RSA (Réseau du Sport de l'Air) included the four-seat all-plastics Orion, designed and built by M Grinvalds with the assistance of M Calvel, whose Frelon motor glider is described in this edition. The Orion project had first attracted attention at the 1977 RSA meeting at Brienne-le-Château, where a radio-controlled scale model was demonstrated. The full-scale prototype was not quite complete, lacking its tail-mounted propeller, when shown at Le Bourget and Brienne in the Summer of 1979.

GRINVALDS ORION

The entire airframe of the Orion is made of glassfibre, reinforced locally with carbonfibre and Kevlar.

Type: Four-seat all-plastics light aircraft.

Wings: Cantilever low-wing monoplane, built in one piece tip-to-tip, and installed in housing moulded into under-surface of fuselage. Wing section NACA 43015 at root, NACA 43012 at tip. Dihedral 4° 18′ from root. Incidence 2° 18′ at root, 0° 18′ at tip. Aileron and slotted flap along entire trailing-edge of each wing. No tabs.

Fuselage: All-plastics structure of ovoid section, with integral tail fin and wing-root stubs and fillets.

Tail Unit: Cantilever T-tail, with fixed tailplane, horn-balanced elevator and inset rudder. Fixed ventral fin. Tab in elevator.

Landing Gear: Retractable tricycle type, with single wheel on each unit. Electrical retraction, nosewheel forward, main units inward under fuselage. Manual emergency actuation. Main gear fully enclosed when retracted, by doors attached to legs. All three legs of Wittman cantilever leaf-spring type. Hydraulic disc brakes on main wheels. Tailskid under ventral fin to protect propeller in taildown attitude.

Power Plant: Prototype has one Avco Lycoming flat-four engine, of 112-134 kW (150-180 hp), mounted inside rear fuselage and driving a tail-mounted four-blade propeller, with spinner, by means of Aerocar-type shafting. Access to engine via large door on each side, over wing trailing-edge. Cooling air enters via two underbelly scoops; hot air and exhaust gases ejected through ducts on each side of rear fuselage. After initial flight trials, and certification, the prototype is planned to flight test alternative power plants, including Wankel and 112 kW (150 hp) SACMA engines.

Accommodation: Four persons, in pairs, in enclosed cabin, or three persons, depending on engine power. Upward-opening window/door on each side, forward of wing. Baggage space at rear of cabin. Dual controls.

Dimensions, external:

Wing span	9·00 m (29 ft 6½ in)
Wing chord at root	1·50 m (4 ft 11 in)
Wing chord at tip	1·00 m (3 ft 3¼ in)
Wing area, gross	11·25 m² (121·1 sq ft)
Wing aspect ratio	7·2
Length overall	6·70 m (21 ft 11¾ in)
Tailplane span	3·30 m (10 ft 10 in)
Wheel track	2·90 m (9 ft 6 in)
Propeller diameter	1·50 m (4 ft 11 in)

Dimensions, internal:

Cabin: Length	2·30 m (7 ft 6½ in)
Max height	1·10 m (3 ft 7¼ in)
Max width	1·10 m (3 ft 7¼ in)

Weights (A, 112 kW; 150 hp: B, 134 kW; 180 hp):

Weight empty, equipped: A	535 kg (1,180 lb)
B	550 kg (1,213 lb)
Max T-O weight: A	975 kg (2,150 lb)
B	990 kg (2,183 lb)

Performance (estimated, at max T-O weight):

Max level speed, three occupants:	
A	162 knots (300 km/h; 186 mph)
Max level speed, three/four occupants:	
B	173 knots (320 km/h; 199 mph)
Max cruising speed: A	146 knots (270 km/h; 168 mph)
B	156 knots (290 km/h; 180 mph)

JODEL
AVIONS JODEL SA

Head Office: 36 Route de Seurre, 21200 Beaune

Design Office: 21-Darois

President-Director General: J. Delemontez

The Société des Avions Jodel was formed in March 1946, by MM Jean Delemontez and Edouard Joly, with the former acting as business and technical manager and the latter as test pilot.

Its first activities were concerned with the repair of gliders and light aircraft of the Service d'Aviation Légère et Sportive, on behalf of the State. Simultaneously, the company designed and built the D.9 Bébé Jodel single-seat light monoplane, which made its first flight in January 1948. This aeroplane, which is certificated with various power plants, proved ideal for amateur construction and can be built in as little as 500 man-hours.

As the result of official tests with the D.9, the French authorities placed an order for the development and construction of two prototypes of a two-seat model, the D.11 fitted with a 33·6 kW (45 hp) Salmson, and the D.111 with a 56 kW (75 hp) Minié engine. Subsequent developments of the D.11 are the D.112 and D.119, which have a 48·5 kW (65 hp) and 67 kW (90 hp) Continental engine respectively. These designs also have been built in large numbers, both commercially and by amateurs.

Avions Jodel now devotes its activities mainly to designing advanced developments of its established types and to acting as a consultant to those building and developing its designs.

JODEL D.9 and D.92 BÉBÉ

The type designation of the Bébé varies according to the type of engine fitted. The original version, with 18·6 kW (25 hp) Poinsard engine, was designated D.9; the D.92 has a modified Volkswagen engine.

The following details refer to all standard versions of the Bébé:

Type: Single-seat light monoplane.

Wings: Cantilever low-wing monoplane. Single-spar one-piece wing with wide-span centre-section of constant chord and thickness and two tapering outer portions set at a coarse dihedral angle (14°). Spar and ribs of spruce and plywood, with fabric covering. Ailerons similar in construction.

Fuselage: Rectangular spruce and plywood structure.

Tail Unit: Cantilever structure of spruce and plywood, with plywood covering on tailplane and fabric-covered rudder and elevators. No fin.

Landing Gear: Non-retractable cantilever main legs with rubber-in-compression springing. Leaf-spring tailskid or tailwheel. Cable brakes.

Power Plant: One 18·6 kW (25 hp) Poinsard (D.9) or modified Volkswagen (D.92) flat-four engine, but other engines of 18·5 to 48·5 kW (25 to 65 hp) may be fitted, including the 27 kW (36 hp) Aeronca JAP and Continental A40. Fuel tank in fuselage, capacity 25 litres (5·5 Imp gallons).

Accommodation: Single seat in open cockpit. Enclosed cockpit optional.

Dimensions, external:

Wing span	7·00 m (22 ft 11 in)
Wing chord (centre-section, constant)	1·40 m (4 ft 7 in)
Wing area, gross	9·0 m² (96·8 sq ft)
Wing aspect ratio	5·45
Length overall	5·45 m (17 ft 10½ in)

Weights:

Weight empty	190 kg (420 lb)
Max T-O weight	320 kg (705 lb)

Performance (30 kW; 40 hp engine, at max T-O weight):

Max level speed at S/L	87 knots (160 km/h; 100 mph)
Cruising speed	74 knots (137 km/h; 85 mph)
Stalling speed	35 knots (65 km/h; 40 mph)
Max rate of climb at S/L	180 m (590 ft)/min
T-O run	110 m (360 ft)
Landing run	100 m (330 ft)
Range with max fuel	217 nm (400 km; 250 miles)

JODEL D.11 and D.119

The original D.11, with 33·5 kW (45 hp) Salmson engine, was the basic model in the series of Jodel two-seaters for amateur and commercial production.

The version for amateur construction with 67 kW (90 hp) Continental engine is designated D.119.

A typical D.11 was built over an eight-year period by Wayne Nelson, an aeronautical engineer of Bountiful, Utah, at a cost of $2,000. The wing is of wood, covered with Dacron, the fuselage and tail unit of wood covered with glassfibre. Changes from the standard design include the fitting of a fixed tail-fin forward of the rudder, and of cantilever spring main landing gear legs. This D.11 spans 8·23 m (27 ft 0 in), has an empty weight of 340 kg (750 lb) and loaded weight of 562 kg (1,240 lb), and is powered by a 48·5 kW (65 hp) Continental A65-8 flat-four engine. Performance is as follows:

Performance:

Max level speed at S/L	93 knots (173 km/h; 108 mph)
Cruising speed	86 knots (161 km/h; 100 mph)
Landing speed	35 knots (64·5 km/h; 40 mph)
Max rate of climb at S/L	152 m (500 ft)/min
Service ceiling	4,875 m (16,000 ft)
T-O run	152 m (500 ft)
Landing run	244 m (800 ft)
Range with max fuel	260 nm (482 km; 300 miles)

JODEL D.112 CLUB

The D.112 is a two-seat dual-control version of the D.9. Except for increased overall dimensions, a wider fuselage and enclosed side-by-side cockpit, the D.112 conforms in layout and structure to the D.9, but is fitted normally with a 48·5 kW (65 hp) Continental flat-four engine. Fuel capacity is 60 litres (13 Imp gallons).

Dimensions, external:

Wing span	8·20 m (26 ft 10 in)
Wing area, gross	12·72 m² (137 sq ft)
Length overall	6·36 m (20 ft 10 in)
Dihedral on outer wings	19°

Weights:

Weight empty	270 kg (600 lb)
Max T-O weight	520 kg (1,145 lb)

Performance (at max T-O weight):

Max level speed at S/L	102 knots (190 km/h; 118 mph)
Max cruising speed	92 knots (170 km/h; 105·5 mph)
Econ cruising speed	81 knots (150 km/h; 93 mph)
Stalling speed	38 knots (70 km/h; 43 mph)
Max rate of climb at S/L	193 m (632 ft)/min
T-O run	137 m (450 ft)
Landing run	120 m (395 ft)
Range with max fuel	323 nm (600 km; 373 miles)

JURCA
MARCEL JURCA

2, rue des Champs Philippe, 92250 La Garenne-Colombe
Telephone: 242 9633 and 551 6306

M Marcel Jurca, an ex-military pilot and hydraulics engineer, has designed a series of high-performance light aircraft of which plans are available to amateur constructors.

A prototype of his first design, the M.J.1, was built but did not fly. To gain experience, M Jurca next built a two-seat Jodel light aircraft, with the help of members of the Aéro Club of Courbevoie, and this flew for the first time in 1954.

The same team then built a prototype of M Jurca's second design, the M.J.2 Tempête single-seat light aircraft, incorporating many Jodel components. It proved so successful that sets of plans were offered to amateur constructors and many Tempêtes are now flying or under assembly throughout the world.

M Jurca developed from the Tempête the two-seat M.J.5 Sirocco and the M.J.51 Sperocco, and has produced a further series of designs by scaling down the basic airframes of second World War fighters to two-thirds or three-quarters of the original size.

For the North American market, Jurca plans are available from Mr Ken Heit, 581 Helen Street, Mt Morris, Michigan 48458, USA. Representative for Australia and New Zealand is Mr Steve Rankin, RD 9, Whangarei, New Zealand.

JURCA M.J.2 and M.J.20 TEMPÊTE

The prototype Tempête was flown for the first time, by its designer, on 27 June 1956. It obtained its certificate of airworthiness very quickly, and a total of at least 28 Tempêtes are now flying, with 20 more under construction, in France, Denmark, Luxembourg, Portugal, the UK, the United States and Canada, all amateur built.

The type of engine fitted to a particular aircraft is indicated by a suffix letter in its designation. Suffix letters are A for the 48·5 kW (65 hp) Continental A65, B for the 56 kW (75 hp) Continental A75, C for the 63·5 kW (85 hp) Continental C85, D for the 67 kW (90 hp) Continental C90-14F, E for the 74·5 kW (100 hp) Continental O-200-A, F for the 78·5 kW (105 hp) Potez 4 E-20, G for the 86 kW (115 hp) Potez 4 E-30, and H for the 93 kW (125 hp) Avco Lycoming.

The standard version is the M.J.2A with A65 engine. The M.J.2D, with 67 kW (90 hp) C90-14F, cruises at 105 knots (195 km/h; 121 mph) and climbs to 1,000 m (3,280 ft) in 3 minutes. It can also perform aerobatics without loss of height. The Tempête built in Portugal is an M.J.2D with 67 kW (90 hp) Continental.

A version known as the **M.J.20** was under construction in Denmark with a 134 kW (180 hp) engine and strengthened airframe. It has been abandoned and its role as a prototype has been taken by a 112 kW (150 hp) Tempête being built in the Rouen area of France by M Yves Chopart.

The Tempête is basically a single-seat aircraft, but the 112 and 134 kW (150 and 180 hp) versions have provision for carrying behind the pilot on cross-country flights a second person weighing not more than 55 kg (121 lb). This is not permitted by the DGAC in France; but a two-seater has flown in the USA, built by Mr Don Kerkhof of Mankato, Minnesota.

The 112 and 134 kW versions are intended to have an aerobatic capability adequate to compete with the American Pitts Specials in international competitions.

The following details apply generally to all basic single-seat M.J.2 models:

Type: Single-seat light monoplane.

Wings: Cantilever low-wing monoplane. NACA 23012

Grinvalds Orion all-plastics light aircraft

Jodel F.11, a Canadian example of the D.11 built by Mr Anatel Yaremchuk of Vernon, BC *(Phil Hanson)*

Jurca M.J.2D Tempête built by M Sire of Pons *(Michel Bernard)*

Jodel Bébé single-seat homebuilt aircraft *(Roland Eichenberger)*

Prototype M.J.3H Dart, built by Mr Denis Jacobs of Dayton, Ohio

Jurca M.J.5EA2 Sirocco, with retractable landing gear *(Geoffrey P. Jones)*

wing section. Incidence varies according to engine power. The 48·5 kW (65 hp) version has an incidence of 4° at root, 2° at tip. No dihedral. All-wood one-piece single-spar structure with fabric covering. Fabric-covered wooden ailerons.

Fuselage: All-wood structure of basic rectangular section, plywood-covered.

Tail Unit: Cantilever all-wood structure. Tailplane and fin plywood-covered, elevators and rudder fabric-covered. Trim tab on starboard elevator.

Landing Gear: Non-retractable tailwheel type. Jodel D.112 cantilever legs with rubber-in-compression springing. Jodel D.112 wheels and Dunlop 420 × 150 tyres. Jodel D.112 tailskid or tailwheel.

Power Plant: One 48·5 kW (65 hp) Continental A65 flat-four engine, driving a Ratier two-blade wooden propeller with ground-adjustable pitch. Provision for fitting 56, 63·5, 67, or 74·5 kW (75, 85, 90 or 100 hp) Continental, 78·5 or 86 kW (105 or 115 hp) Potez or 93 kW (125 hp) Avco Lycoming engine. Jodel engine mounting and cowling. Jodel fuel tank, capacity 60 litres (13·2 Imp gallons), aft of firewall in fuselage.

Accommodation: Single seat under long rearward-sliding transparent canopy.

Dimensions, external:

Wing span	6·00 m (19 ft 8 in)
Wing chord (basic)	1·40 m (4 ft 7 in)
Wing area, gross	7·98 m² (85·90 sq ft)
Wing aspect ratio	4·5
Length overall	5·855 m (19 ft 2½ in)
Height overall	2·40 m (7 ft 10 in)
Tailplane span	2·50 m (8 ft 2 in)
Wheel track	2·30 m (7 ft 6½ in)

Weights:

Weight empty	90 kg (639 lb)
Max T-O weight	430 kg (950 lb)

Performance (48·5 kW; 65 hp engine):

Max level speed	104 knots (193 km/h; 120 mph)
Cruising speed	89 knots (165 km/h; 102 mph)
Landing speed	43 knots (80 km/h; 50 mph)
Max rate of climb at S/L	170 m (555 ft)/min
Service ceiling	3,500 m (11,500 ft)
T-O run	250 m (820 ft)
Endurance	3 h 20 min

JURCA M.J.3H DART

The M.J.3H Dart is a single-seat all-wood monoplane, with the fuselage of the M.J.2 Tempête and the wings of the M.J.5 Sirocco. The prototype, built by Denis Jacobs of Dayton, Ohio, made its first flight in 1977.

Type: Single-seat sporting aircraft.

Wings: As for M.J.5 Sirocco. Wing section NACA 23012. No dihedral. Incidence 3°. Washout 2° at half-span. Wood construction, including three-position split flaps and ailerons.

Fuselage: As for M.J.2 Tempête. All-wooden semi-monocoque structure.

Tail Unit: Conventional cantilever all-wooden structure, comprising fin and rudder, tailplane and elevators with trim tab on starboard side.

Landing Gear: Retractable tailwheel type. Main wheels retract inward.

Power Plant: One 86 kW (115 hp) Avco Lycoming O-235-C1 flat-four engine, driving a two-blade propeller. Fuel tanks in front and rear fuselage, total capacity 75·7 litres (16·6 Imp gallons).

Accommodation: Single seat under sliding transparent canopy.

Dimensions, external:

Wing span	6·35 m (20 ft 10 in)
Wing chord at tip	1·40 m (4 ft 7 in)
Wing area, gross	9·66 m² (104 sq ft)
Wing aspect ratio	4·25
Length overall	6·248 m (20 ft 6 in)
Tailplane span	2·54 m (8 ft 4 in)
Wheel track	2·44 m (8 ft 0 in)

Weights:

Weight empty	476 kg (1,050 lb)
Max T-O weight	612 kg (1,350 lb)

Performance (48·5 kW; 65 hp engine):

Never-exceed speed	173 knots (322 km/h; 200 mph)
Max level speed at 1,525 m (5,000 ft)	134 knots (248 km/h; 154 mph)
Max cruising speed (75% power) at 1,525 m (5,000 ft)	123 knots (229 km/h; 142 mph)
Econ cruising speed	117 knots (217 km/h; 135 mph)
Stalling speed, landing gear and flaps up	52 knots (97 km/h; 60 mph)
Stalling speed, landing gear and flaps down	48 knots (88·5 km/h; 55 mph)
Time to 1,000 m (3,280 ft)	3 min 15 s
Service ceiling (estimated)	3,660 m (12,000 ft)
T-O run, 15·5°C	229 m (750 ft)
Landing run	305 m (1,000 ft)
Endurance	2 h 45 min
g limits	+9, −4·5 (ultimate)

JURCA M.J.5 SIROCCO

The M.J.5 Sirocco is a tandem two-seat monoplane, developed from the M.J.2 Tempête as a potential club training and touring aircraft. It is fully aerobatic when flown as a two-seater.

The longer-span wings have an extended leading-edge inboard of the fence on each side and a completely new tip shape. A sweptback fin and rudder are standard.

The prototype M.J.5 flew for the first time on 3 August 1962, powered by a 78·5 kW (105 hp) Potez 4 E-20 engine. It was fitted originally with a non-retractable landing gear, but retractable landing gear and a 119·5 kW (160 hp) Avco Lycoming O-320 engine were fitted in 1966, followed by a 134 kW (180 hp) Avco Lycoming engine later. Its fuel capacity is 116 litres (25·5 Imp gallons).

By mid-February 1967, five more Siroccos were flying, one of them factory-built at Nancy. This aircraft, powered by a 74·5 kW (100 hp) Continental engine, concluded tests at Istres in January 1969. The French government then concluded an agreement with Constructions Aéronautiques Lorraines, François et Cie of Nancy, which built an airframe for static tests, in March 1971. These were required in view of the fact that the Sirocco is regarded as a basic trainer suitable for amateur construction; and it was awarded subsequently a certificate of airworthiness in the Utility category. Supplementary tests were conducted at the CEV with another Sirocco, powered by a 100·5 kW (135 hp) Avco Lycoming O-320 engine.

A full C of A, covering Aerobatic requirements and unlimited spinning, is applicable only when a power plant of 112 kW (150 hp) minimum rating is installed.

The version of the Sirocco for amateur construction is generally similar to the factory-built version, with optional retractable landing gear.

At least 40 Siroccos are reported to be flying or under construction by amateurs in France, Canada, West Germany, Switzerland, England and the USA, with various engines.

The type of engine fitted to a particular aircraft is indicated by a suffix letter in its designation. Suffix letters are A for the 67 kW (90 hp) Continental C90-8 or -14F, B for the 74·5 kW (100 hp) Continental O-200-A, C for the 78·5 kW (105 hp) Potez 4 E-20, D for the 86 kW (115 hp) Potez 4 E-30, E for the 78·5 kW (105 hp) Hirth, F for the 93 kW (125 hp) Avco Lycoming, G for the 100·5 kW (135 hp) Regnier, H for the 119·5 kW (160 hp) Avco Lycoming, K for the 134 kW (180 hp) Avco Lycoming and L for the 164 kW (220 hp) Franklin. Addition of the numeral 1 indicates a non-retractable landing gear and the numeral 2 indicates a retractable landing gear. Thus, the designation of the original prototype in its current form is M.J.5K2. The example built at Nancy for certification has a 74·5 kW (100 hp) Continental engine and so is designated M.J.5B1.

Two examples of the M.J.5K, flying in Vichy and Montelimar, France, are each fitted with a 134 kW (180 hp) Avco Lycoming engine and Christen inverted fuel system. Work on two M.J.5L Siroccos, with 164 kW (220 hp) Franklin engine, was started in 1975, in the USA and France, for use in international aerobatic championships.

A Sirocco with 86 kW (115 hp) Avco Lycoming O-235-C2B engine and 1·85 m (6 ft 0¾ in) diameter propeller was constructed by Luftsportgruppe Liebherr-Aero-Technik (LAT) in West Germany. This has a modified rudder of reduced height and greater chord, and a jettisonable, sideways-hinged cockpit canopy, and is intended for certification for aerobatic flying. The details which follow apply to this aircraft, but are generally typical of all versions.

A developed version, known as the M.J.51 Sperocco, is described separately.

Dimensions, external:
Wing span 7·00 m (23 ft 0 in)
Wing area, gross 10·00 m² (107·64 sq ft)
Wing aspect ratio 4·9
Length overall 6·15 m (20 ft 2 in)
Height overall, tail up:
standard model 2·80 m (9 ft 2¼ in)
LAT version 2·60 m (8 ft 6¼ in)
Tailplane span 3·24 m (10 ft 7½ in)
Wheel track 2·80 m (9 ft 2¼ in)

Weights:
Weight empty 430 kg (947 lb)
Max T-O weight 680 kg (1,499 lb)

Performance (at max T-O weight):
Max level speed 127 knots (235 km/h; 146 mph)
Cruising speed 116 knots (215 km/h; 134 mph)
Stalling speed 44 knots (80 km/h; 50 mph)
Climb to 1,000 m (3,280 ft) 4 min
Service ceiling 5,000 m (16,400 ft)
T-O run 250 m (820 ft)
Landing run 200 m (655 ft)
Endurance 4 h 20 min

JURCA M.J.5 SIROCCO (SPORT WING)

A special version of the Sirocco, with 86 kW (115 hp) engine and increased wing span, has been evolved for the New Zealand and Australian market. Known as a 'Sport' wing, the wing of this aircraft embodies one additional rib and inter-rib bay each side. The modification is available in the English-language set of Sirocco plans.

JURCA M.J.51 SPEROCCO

Using knowledge gained from flight experience with the M.J.5 and the Canadian prototype M.J.7, M Jurca developed, with the assistance of M. J. Lecarme, a design incorporating features of each aircraft. It is known as the M.J.51 Sperocco, the name being a contraction of 'Special Sirocco', and is intended for high-performance aerobatic and competition flying. Like other Jurca designs, the M.J.51 is suitable for amateur construction.

The wings, of Habib 64000 748 laminar-flow profile, are essentially those of the M.J.7 Gnatsum. They are without dihedral, and the angle of incidence varies according to the rating of the engine that is installed, as in the Tempête. The fuselage is of completely new design, with a basically triangular cross-section, but is of similar construction to the M.J.5. The tail unit consists of M.J.7 horizontal surfaces with a shorter and wider-chord fin and rudder. Landing gear is of the M.J.5 type and is fully retractable.

Any horizontally-opposed engine of 112-179 kW (150-240 hp) may be installed. Fuel is contained in two wing tanks, each of 55 litres (12 Imp gallons) capacity, and one fuselage tank of 45 or 100 litres (10 or 22 Imp gallons) capacity.

The M.J.51 seats two persons in tandem under a one-piece sliding canopy, the rear seat being 10 cm (3·9 in) higher than the front seat.

The first M.J.51, powered by a 134 kW (180 hp) Avco Lycoming AIO-360 engine, is under construction by M Serge Brillant at Melun.

Dimensions, external:
Wing span 7·623 m (25 ft 0 in)
Wing area, gross 11·00 m² (118 sq ft)
Length overall 7·24 m (23 ft 9 in)

Weight:
Max T-O weight 730 kg (1,653 lb)

Performance (estimated, with 112 kW; 150 hp Lycoming engine):
Max level speed 149 knots (275 km/h; 171 mph)
Max cruising speed (75% power) 135 knots (250 km/h; 155 mph)
Stalling speed 49 knots (90 km/h; 56 mph)
Time to 1,000 m (3,280 ft) 1 min 30 s

JURCA M.J.7 and M.J.77 GNATSUM

The Gnatsum is a scale replica, for amateur construction, of the North American P-51 Mustang single-seat fighter of the second World War. Its name 'Gnatsum' is 'Mustang' reversed.

Initially, M Jurca designed the wings, fuselage, tail surfaces and manually-retractable landing gear. The engine installation was deliberately not designed, to permit constructors to utilise any of the suitable Avco Lycoming, Continental, Ranger or other power plants that are available.

During construction of the M.J.7 prototype in Canada (see below), a number of modifications and improvements were made to the basic design. These were embodied in the drawings, which are available from M Jurca in two forms, as follows:

M.J.7. To two-thirds scale. Prototype (CF-XZI, now N51HR) built in the works of Falconar Aircraft Ltd on the Industrial Airport, Edmonton, Alberta, Canada, and first flown on 31 July 1969. Granted DoT type approval by early 1970. Described under the 'SAL' entry in the Canadian section of the 1973-74 *Jane's*. The second M.J.7 built in Canada was the one constructed by Captain W. T. Foster, to which the data given at the end of this entry apply. Two other Gnatsums were flown in 1977, under the SAL designation, built by Mr Bill Slater and Mr Ross Grady of the USA. The first example built in the UK, by Mr W. E. Wilks, was illustrated in the 1978-79 *Jane's*. Plastics fuselage shells for the M.J.7 are available from Boeve Brothers of Holland, Michigan, USA.

M.J.77. To three-quarters scale. Examples under construction by Mr Bob Aughton in Michigan, M Glorieux of Pau, M Piazola of Egletons and M Semenadisse of Hyères, near Toulon, France.

Unlike previous small-scale replicas of this aircraft, the Gnatsum is scaled down precisely. Use of an in-line engine, such as the 134 kW (180 hp) or 149 kW (200 hp) Ranger, permits the fuselage cowling lines to follow closely those of the original. Alternative installation of a 149 kW (200 hp) Avco Lycoming flat-four aircooled engine requires fairing blisters over the cylinders.

M Jurca's plans provide for alternative plywood-covered semi-monocoque fuselage construction or a wooden box structure covered with two plastics shells.

Dimensions, external:
Wing span 7·52 m (24 ft 8¼ in)
Wing area, gross 10·41 m² (112 sq ft)
Length overall 6·55 m (21 ft 6 in)

Weights (Capt Foster's M.J.7 with 149 kW; 200 hp Ranger engine):
Weight empty 673 kg (1,485 lb)
Max T-O weight 850 kg (1,875 lb)

Performance (Capt Foster's M.J.7 with 149 kW; 200 hp Ranger engine):
Max level speed 190 knots (351 km/h; 218 mph)
Cruising speed at 2,133 m (7,000 ft) 175 knots (325 km/h; 202 mph)
Take-off speed (flaps up) 70-75 knots (131-139 km/h; 81-86·5 mph)
Landing speed (flaps up) 65-70 knots (121-130 km/h; 75-81 mph)
Landing speed (flaps down) 63 knots (117 km/h; 72·5 mph)
Stalling speed (flaps down) 57 knots (106 km/h; 66 mph)
Max rate of climb at S/L 305 m (1,000 ft)/min
T-O run 457 m (1,500 ft)
Landing run 610-762 m (2,000-2,500 ft)

JURCA M.J.7S SOLO

Intended as a single-seat advanced trainer, the M.J.7S Solo is basically similiar to the M.J.7 Gnatsum but does not retain the underbelly scoop which the latter inherited from the original P-51 Mustang design. A prototype is under construction by M Duhamel of Strasbourg, with a 134 kW (180 hp) Avco Lycoming AIO-360 flat-four engine.

The wing section of the M.J.7S is quoted as Habib 64-000 748-MJ7-104.

Dimensions, external:
Wing span 7·523 m (24 ft 8½ in)
Wing area, gross 10·8 m² (116·2 sq ft)
Length overall, tail up 6·664 m (21 ft 10 in)
Tailplane span 3·00 m (9 ft 10 in)

JURCA M.J.8 and M.J.80 1-NINE-O

The M.J.8 is a single-seat sporting aircraft which has been designed by M Jurca by scaling down to three-quarters of the original dimensions the airframe of the Focke-Wulf Fw 190 fighter. Its general appearance is shown in the accompanying illustration.

The prototype, built by Mr Ronald Kitchen of Carson City, Nevada, flew for the first time on 30 March 1975. A second example was built by Mr J. Kiska of Norwalk, Connecticut; and Mr Evan Wolfe of California is constructing an enlarged M.J.8, in an attempt to produce a full-size 'Fw 190'. This has been designated **M.J.80.**

The M.J.8 prototype has a 216 kW (290 hp) Avco Lycoming IO-540 engine, but the design is suitable for the alternative use of any horizontally-opposed or radial engine in the 74·5-149 kW (100-200 hp) range. The landing gear is retractable.

Dimensions, external (M.J.8):
Wing span 7·87 m (25 ft 10 in)
Wing chord at root 1·70 m (5 ft 7 in)
Wing chord at tip 0·90 m (2 ft 11½ in)
Wing area, gross 10·2 m² (109·8 sq ft)
Length overall 6·63 m (21 ft 9 in)
Tailplane span 2·84 m (9 ft 4 in)

Weights (M.J.8, 119·5 kW; 160 hp engine):
Weight empty 400 kg (880 lb)
Max T-O weight 626 kg (1,380 lb)

Performance (M.J.8, estimated, with 119·5 kW; 160 hp engine):
Max level speed at S/L 139 knots (257 km/h; 160 mph)
Max cruising speed 124 knots (230 km/h; 143 mph)
Stalling speed 49 knots (90 km/h; 56 mph)
Max rate of climb at S/L 503 m (1,650 ft)/min

JURCA M.J.9 ONE-OH-NINE

A prototype three-quarter scale Messerschmitt Bf 109 fighter of the second World War is being constructed by Mr Werner Hohn of Carson City, Nevada. This has been designated M.J.9. A second example is being built by Mr Heyser of Tustin, California.

JURCA M.J.10 SPIT

The M.J.10 is a single-seat, three-quarter scale representation of the Supermarine Spitfire which can also be modified as a two-seater. It is suitable for any horizontally-opposed or in-line engine of 89·5-164 kW (120-220 hp), although some slight variations from the Spitfire's contours are necessary in the former case. Construction is entirely of wood, except for the glassfibre engine cowling and fabric covering on the control surfaces. The single-spar wing is similar in construction to that of the Sirocco. The manually-operated retractable landing gear is fitted with helicoidal spring shock-absorbers.

The basic plans adopted the Spitfire Mk IX as the standard M.J.10 version, but alternative detail plans are available for representing both Merlin- and Griffon-engined models, including the Mks VC and XIV, and for clipped, standard or extended-span wings.

A prototype is under construction by Mr Pendlebury of the Chesterfield Air Touring Group at West Bridgford, Nottingham, England. Another is being built by Mr Ed Storo of New York, USA.

Dimensions, external:
Wing span:
standard 8·40 m (27 ft 6¾ in)
clipped 7·46 m (24 ft 5½ in)
Wing area, gross 12·60 m² (135·6 sq ft)
Length overall 7·125 m (23 ft 4½ in)

Weights (119·5 kW; 160 hp engine):
Weight empty 658 kg (1,450 lb)
Max T-O weight 907 kg (2,000 lb)

Performance (estimated, with 119·5 kW; 160 hp engine):
Max level speed at S/L 139 knots (257 km/h; 160 mph)
Cruising speed 124 knots (230 km/h; 143 mph)
Stalling speed 49 knots (90 km/h; 56 mph)
Max rate of climb at S/L 503 m (1,650 ft)/min
T-O run 200 m (660 ft)

JURCA M.J.12 PEE-40

The M.J.12 is a three-quarter scale representation of the Curtiss P-40 single-seat fighter of the second World War. It spans 8·524 m (27 ft 11½ in) and has an overall length (tail up) of 7·62 m (25 ft 0 in)

Two M.J.12s are under construction in the USA, one by Mr Ron Kitchen of Carson City, Nevada, who built the prototype M.J.8.

JURCA M.J.14 FOURTOUNA

Designed in 1971, the M.J.14 Fourtouna (Tempête in Romanian) will be a small single-seat racing aircraft of unorthodox configuration, with a semi-reclining seat for the pilot. Construction of a prototype was undertaken by M Beliard in Coutances, France. Its general configuration is shown in the accompanying three-view drawing.

The standard tapered wings can be replaced by constant-chord wings of the same span if the aircraft is intended for 'rapid liaison'. Tailplane incidence is adjustable on the ground. The following data apply to the aircraft as illustrated, with a 67 kW (90 hp) Continental C90-8F flat-four engine. Fuel capacity is 70 litres (15·5 Imp gallons).

Dimensions, external:
Wing span 6·00 m (19 ft 8½ in)
Wing area, gross 6·1 m² (65·66 sq ft)
Length overall 5·68 m (18 ft 7½ in)
Tailplane span 2·97 m (9 ft 9 in)

Weights:
Weight empty 250 kg (550 lb)
Max T-O weight 420 kg (925 lb)

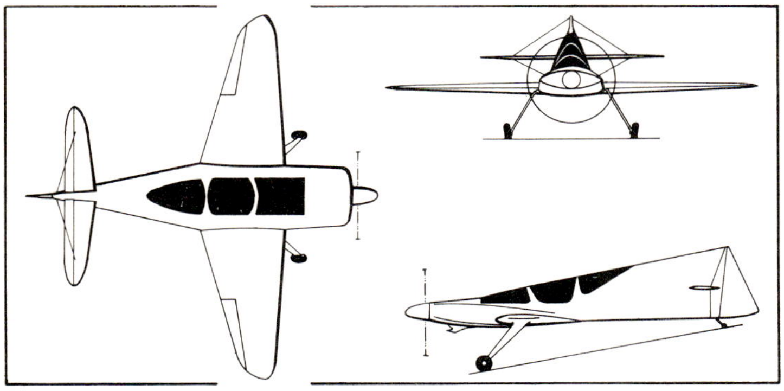

Jurca M.J.14 Fourtouna single-seat racing aircraft *(Roy J. Grainge)*

Jurca M.J.7H Gnatsum built in Canada by Capt W. T. Foster, CAF

Jurca M.J.8 1-Nine-O built by Mr Joseph Kiska of Connecticut

Landray GL03 tandem-wing light aircraft

Landray GL01 two-seat tandem-wing light aircraft (right) with the smaller GL02

LANDRAY

GILBERT LANDRAY

28 rue Remonteur, 91560-Crosne
Telephone: (6) 948 1019

M Landray has designed and constructed three tandem-wing light aircraft. The GL01 is based, presumably, on the general layout of the Croses EC-6 Criquet. However, there are significant differences, as can be noted from the following details. The GL02, as shown in the accompanying illustration with the GL01, is of much more basic design. The latest aircraft from M Landray is the GL03, of which brief details can be found below. It is believed to have flown first in August 1980.

LANDRAY GL01

TYPE: Two-seat tandem-wing light aircraft.

WINGS: Forward wing built in one piece and pivoted on two streamlined supports, giving variable incidence. Two forward wing control rods, and four pivot points. Fixed rear (lower) wing. Wooden structure, fabric covered. No ailerons. Two light alloy flaps.

FUSELAGE: All wooden structure.

TAIL UNIT: All wooden fin and rudder. No tailplane or elevators.

LANDING GEAR: Non-retractable tailwheel type. Main wheels carried on single cantilever arch structure of glassfibre and enclosed in streamline fairings. Mainwheel tyres size 420 × 150, pressure 1·8 bars (26 lb/sq in). Tailwheel, with tyre size 350 × 150, semi-enclosed in bottom of rudder. Landray brakes.

POWER PLANT: One 67 kW (90 hp) Continental C90-8F flat-four engine, driving a two-blade wooden propeller. Fuel capacity 80 litres (17·6 Imp gallons) in two tanks. Oil capacity 4 litres (0·9 Imp gallons).

ACCOMMODATION: Two seats side by side in enclosed cabin.

SYSTEM: Battery and alternator.

AVIONICS: VHF com; VOR nav.

DIMENSIONS, EXTERNAL:

Wing span: front wing	7·40 m (24 ft 3½ in)
rear wing	6·70 m (21 ft 11¾ in)
Wing chord, constant	1·30 m (4 ft 3¼ in)
Wing area, gross	15·50 m² (167 sq ft)
Length overall	5·00 m (16 ft 4¾ in)
Height overall	2·10 m (6 ft 10¾ in)

WEIGHTS:

Weight empty	361 kg (796 lb)
Baggage	13·6 kg (30 lb)
Max T-O weight	625 kg (1,378 lb)

PERFORMANCE:

Max level speed	119 knots (220 km/h; 137 mph)
Max cruising speed	97 knots (180 km/h; 112 mph)
Landing speed	44 knots (80 km/h; 50 mph)
T-O run	91 m (300 ft)
Landing run	61 m (200 ft)
Range with max fuel	378 nm (700 km; 435 miles)

LANDRAY GL02 AMI POU

The GL02 Ami Pou is a diminutive single-seat aircraft of 'Pou-du-Ciel' tandem-wing configuration, powered by a 20 kW (27 hp) converted Citroen Ami 8 motorcar engine. This retains the standard centrifugal clutch as a torque damper, and drives a two-blade fixed-pitch wooden propeller through a six-belt reduction.

The front wing is pivot-mounted. The landing gear comprises two main wheels, size 330 × 130, carried on cantilever light alloy spring legs, and a size 140 × 40 tailwheel.

DIMENSIONS, EXTERNAL:

Wing span: front wing	6·40 m (21 ft 0 in)
rear wing	5·40 m (17 ft 8½ in)
Wing chord, constant	1·10 m (3 ft 7¼ in)
Wing area, gross	12·0 m² (129·2 sq ft)
Propeller diameter	1·50 m (4 ft 11 in)

WEIGHTS:

Weight empty	180 kg (397 lb)
Max T-O weight	295 kg (650 lb)

PERFORMANCE:

Max level speed	70 knots (130 km/h; 80 mph)
Min flying speed	22 knots (40 km/h; 25 mph)

LANDRAY GL03

This new aircraft uses the same type of tandem wing arrangement as the GL01, but differs in other ways. It is a single-seater, with a rear-mounted 30 kW (40 hp) Citroen GS612 four-cylinder engine driving a two-blade pusher propeller, twin fins and rudders mounted on the rear wing, and a non-retractable tricycle landing gear. Like the GL01, it is of wooden construction, with fabric covering on the wings. These are of constant chord, except for a slight reduction near the tips.

DIMENSIONS, EXTERNAL:

Wing span:	
front wing	6·45 m (21 ft 2 in)
rear wing	6·35 m (20 ft 10 in)
Wing chord, basic	1·10 m (3 ft 7¼ in)
Wing chord at tips	1·00 m (3 ft 3¼ in)
Wing area, gross	13·50 m² (145 sq ft)
Length overall	4·05 m (13 ft 3½ in)
Wheel track	1·60 m (5 ft 3 in)
Propeller diameter	1·36 m (4 ft 5½ in)

WEIGHT:

Weight empty	220 kg (485 lb)

LEDERLIN

FRANÇOIS LEDERLIN

2 rue Charles Peguy, 38000 Grenoble

M Lederlin, an architect, designed and built a two-seat light aeroplane based on the familiar Mignet 'Pou-du-Ciel' formula. Although derived from the Mignet HM-380 and designated 380-L, it retains little of the original except for the wing section. First flight was made on 14 September 1965, a restricted C of A being granted in the following month.

Plans of the 380-L, annotated in English and with both English and metric measurements, are available to amateur constructors, and several examples are under construction or already flying.

LEDERLIN 380-L

TYPE: Two-seat amateur-built light aircraft.

WINGS: Tandem-wing biplane. Wing section 3·40-13. Dihedral 3° 30′ on outer sections only (both wings). Incidence variable from 0° to 12° (forward wing). Incidence of rear wing 6°. No sweepback. Each wing is made in two parts, bolted together at the centreline. Construction is conventional, with wooden box-spar and trellis ribs, plywood leading-edge and overall fabric covering. The variable-incidence front wing is pivoted on the cabane structure by ball-joints and on the bracing struts (one each side) by cardan-joints. No ailerons or flaps. Long-span tab on trailing-edge of rear wing, controllable in flight.

FUSELAGE: Welded steel tube structure, covered with light alloy to front of cabin and with fabric on rear fuselage, over light spruce formers.

TAIL UNIT: Fin and rudder only. Spruce and ply structure, covered with fabric. Ground-adjustable tab in rudder.

LANDING GEAR: Non-retractable tailwheel type. Cantilever main legs consist of conical spring steel rods, inclined rearward. Fournier main wheels and tyres, size 380 × 150, with mechanical brakes. Large tailwheel, carried on telescopic leg with spring shock-absorber, can be steered by the rudder controls through a linkage engaged by the pilot.

POWER PLANT: One 67 kW (90 hp) Continental C90-14F flat-four engine, driving a McCauley two-blade fixed-pitch metal propeller. Single fuel tank, capacity 85 litres (18·75 Imp gallons). Oil capacity 4·5 litres (1 Imp gallon).

ACCOMMODATION: Two seats side by side in enclosed cabin. Forward-hinged door on each side. Controls comprise a rudder bar for directional control and a stick, suspended from the roof of the cabin and free laterally, to control the incidence of the forward wing. A further lever, suspended from the roof, controls the tab on the rear wing. Baggage space aft of seats.

DIMENSIONS, EXTERNAL:

Wing span: forward	7·92 m (26 ft 0 in)
rear	6·00 m (19 ft 8¾ in)
Wing chord (constant, each)	1·30 m (4 ft 3¼ in)

Wing area, gross: forward	9·92 m² (106·8 sq ft)
rear	7·43 m² (80·0 sq ft)
Length overall	4·77 m (15 ft 7¾ in)
Height overall	2·08 m (6 ft 10 in)
Wheel track	2·05 m (6 ft 8¾ in)
Wheelbase	3·10 m (10 ft 2 in)
Propeller diameter	1·83 m (6 ft 0 in)
Doors (each): Height	0·90 m (2 ft 11½ in)
Width	0·75 m (2 ft 5½ in)
Height to sill	0·50 m (1 ft 7½ in)

Dimensions, internal:

Cabin: Max width	1·07 m (3 ft 6 in)
Max height	1·03 m (3 ft 4 in)
Baggage space	0·20 m³ (7 cu ft)

Weights:

Weight empty	360 kg (794 lb)
Max T-O weight	600 kg (1,323 lb)

Performance (at max T-O weight):

Never-exceed speed	126 knots (233 km/h; 145 mph)
Max level speed at 305 m (1,000 ft)	109 knots (201 km/h; 125 mph)
Max cruising speed	97 knots (180 km/h; 112 mph)
Econ cruising speed at 610 m (2,000 ft)	87 knots (161 km/h; 100 mph)
Stalling speed, power off	26 knots (49 km/h; 30 mph)
Max rate of climb at S/L	275 m (900 ft)/min
Service ceiling	over 3,660 m (12,000 ft)
T-O run	122 m (400 ft)
Landing run	153 m (500 ft)
Range at econ cruising speed	477 nm (885 km; 550 miles)

LEFEBVRE

ROBERT LEFEBVRE

CES A. Camus, rue Adeline, 76100 Rouen

M Lefebvre has built and flown a small single-seat racing aircraft named the Busard, assisted by pupils of the A. Camus technical school at Rouen. Basis of the design was the MP.204 prototype racer with 56 kW (75 hp) Minié engine, designed by Max Plan and first flown on 5 June 1952. By comparison with the MP.204, the Busard has been lightened, and simplified for construction by amateurs.

LEFEBVRE MP.205 BUSARD

The description below applies to the prototype Busard (F-PTXT) built and flown by M Lefebvre. This aircraft was powered originally with a 48·5 kW (65 hp) Continental engine and was illustrated in this form in the 1974-75 *Jane's*. After 20 flying hours it was re-engined with a 67 kW (90 hp) Continental and underwent several refinements, including the fitting of main-wheel fairings.

At least fifteen sets of plans have been sold to amateur constructors. One Busard is being built with a 48·5 kW (65 hp) Walter engine, two with versions of the Volkswagen, and the remainder with Continentals.

Type: Single-seat amateur-built racing aircraft.

Wings: Cantilever low-wing monoplane. Wing section NACA 23012. Constant incidence of 1°. Slight dihedral. Conventional single-spar wood structure, covered entirely with plywood. Fabric-covered wooden slotted ailerons, operated by control rods. Fabric-covered wooden slotted three-position trailing-edge flaps.

Fuselage: Conventional wooden structure, covered with plywood. Domed plywood decking. Plastics engine cowling, built in top and bottom sections.

Tail Unit: Cantilever wood structure. Fixed surfaces plywood-covered; control surfaces fabric-covered. Neither rudder nor elevators are aerodynamically balanced. Flettner tab in starboard elevator. Rudder is cable-operated.

Landing Gear: Non-retractable tailwheel type. Cessna-type aluminium leaf-spring cantilever main legs, with plastics wheel fairings. Steerable tailwheel. Brakes on main wheels.

Power Plant: One 67 kW (90 hp) Continental flat-four engine, driving a two-blade fixed-pitch propeller. Light alloy fuel tank aft of firewall in fuselage, capacity 40 litres (8·8 Imp gallons).

Accommodation: Single seat in enclosed cabin, with max width of 0·58 m (1 ft 10¾ in). Sideways-opening canopy, hinged on starboard side. Baggage space aft of seat.

Dimensions, external:

Wing span	6·00 m (19 ft 8¼ in)
Wing chord at root	1·50 m (4 ft 11 in)
Wing chord at tip	0·75 m (2 ft 5½ in)
Wing area, gross	6·00 m² (64·6 sq ft)
Wing aspect ratio	6·00
Length overall: Continental	5·35 m (17 ft 6¾ in)
Volkswagen	5·20 m (17 ft 0¾ in)
Height overall	1·50 m (4 ft 11 in)

Weights (A, 48·5 kW; 65 hp Continental; B, 67 kW; 90 hp Continental):

Weight empty: A	233 kg (514 lb)
B	239 kg (527 lb)
Max T-O weight: A	339 kg (747 lb)
B	345 kg (760 lb)

Performance (A, 48·5 kW; 65 hp Continental; B, 67 kW; 90 hp Continental; C, 1,600 cc Volkswagen):

Max level speed at S/L:	
A	127 knots (235 km/h; 146 mph)
B	156 knots (290 km/h; 180 mph)
C	113 knots (210 km/h; 130 mph)
Landing speed: A	43 knots (80 km/h; 50 mph)
Range with max fuel:	
A	242 nm (450 km; 279 miles)

LUCAS

EMILE LUCAS

Corbonod, 01420 Seyssel

Emile Lucas has built a two/three-seat light aircraft designated Lucas L5. Design began in 1969, and construction of the prototype was undertaken in the designer's living room. First flight was made on 13 August 1976. Twenty-five more L5s are under construction, some with non-retractable and others with retractable landing gear.

LUCAS L5

Type: Two/three-seat light monoplane.

Wings: Cantilever low-wing monoplane. Dihedral 3°. Incidence 3°. No sweepback. All-metal construction of AU2G and AU4G light alloy, including metal ailerons and two-position flaps. Chord sharply reduced at each wingtip by curved trailing-edge. Leading-edge of each wing extended forward at root to house main landing gear leg when retractable landing gear is fitted. No tabs.

Fuselage: Conventional all-metal structure of AU2G and AU4G light alloy.

Tail Unit: Cantilever structure of AU4G light alloy. Constant-chord tailplane, with one-piece elevator, horn-balanced at tips. Sweptback fin and horn-balanced rudder. No tabs.

Landing Gear: Non-retractable tricycle or mechanically-retractable tailwheel type, both having been fitted to prototype during development. Rubber shock-absorption. Wheel size 420 × 150 on nose and main units. Hydraulic brakes.

Power Plant: Prototype had one 85·75 kW (115 hp) Avco Lycoming O-235 flat-four engine, driving an EVRA two-blade fixed-pitch propeller with spinner. L5s built to plans can have 112 or 134 kW (150 or 180 hp) engines. One fuel tank standard in fuselage. Total fuel capacity 75 or 115 litres (16·5 or 25·3 Imp gallons).

Accommodation: Two seats side by side under transparent canopy. Provision for third seat when using engine of more than 85·75 kW (115 hp). Baggage hold, capacity 30 kg (66 lb).

Avionics: Jolliet ER400 VHF radio.

Dimensions, external:

Wing span	9·20 m (30 ft 2¼ in)
Wing area, gross	11·90 m² (128·1 sq ft)
Wing aspect ratio	7·2
Length overall	6·30 m (20 ft 8 in)
Height overall	2·10 m (6 ft 10¾ in)
Propeller diameter	1·70 m (5 ft 7 in)

Weights:

Weight empty, equipped	505 kg (1,113 lb)
Max T-O weight	746 kg (1,644 lb)

Performance (prototype):

Never-exceed speed	161 knots (300 km/h; 186 mph)
Max level speed:	
retractable landing gear	146 knots (270 km/h; 168 mph)
fixed gear	127 knots (235 km/h; 146 mph)
Max cruising speed:	
retractable landing gear	138 knots (255 km/h; 158 mph)
fixed gear	108 knots (200 km/h; 124 mph)
Econ cruising speed:	
retractable landing gear	116 knots (215 km/h; 134 mph)
fixed gear	97 knots (180 km/h; 112 mph)
Max rate of climb at S/L	300 m (985 ft)/min
T-O to 15 m (50 ft)	280 m (920 ft)
Landing from 15 m (50 ft)	380 m (1,245 ft)
Range with max fuel	539 nm (1,000 km; 621 miles)

NICOLLIER

HENRI NICOLLIER

69 rue des Cras, 25000 Besançon

M Nicollier, who began flying sailplanes and powered aircraft when he was 16 years old, has designed three light aircraft. Plans of two of them have been made available to other constructors.

NICOLLIER HN 433 MENESTREL

The prototype of this single-seat light aircraft (F-WKXO) flew for the first time on 25 November 1962 and continues in use. Plans are available, and Menestrels are currently under construction by amateurs in several countries. The following details refer to the prototype:

Type: Single-seat all-wood light aircraft.

Wings: Cantilever low-wing monoplane. Constant chord on inner half of each wing; semi-elliptical outer panels. All-wood single-spar structure, with plywood covering on leading-edge and inboard 25 per cent of each wing. Fabric-covered overall. Plain wooden ailerons, fabric-covered. No flaps or tabs.

Fuselage: Conventional wooden truss structure of basic rectangular section, with curved top-decking. Plywood-covered, except for sides of rear fuselage and top-decking. Fabric covering overall.

Tail Unit: Cantilever wood structure. Fin and tailplane plywood-covered; rudder and elevators fabric-covered. No tabs.

Landing Gear: Cantilever main units, with Vespa motor scooter wheels, and tailskid.

Power Plant: One 22 kW (30 hp) converted Volkswagen 1,300 cc motorcar engine, with aluminium pistons and dual ignition. Two-blade fixed-pitch propeller.

Accommodation: Single-seat in enclosed cockpit.

Dimensions, external:

Wing span	7·00 m (22 ft 11½ in)
Wing area	8·15 m² (87·7 sq ft)
Wing aspect ratio	6·4
Length overall	5·25 m (17 ft 2¾ in)
Height overall	1·48 m (4 ft 10¼ in)

Weights:

Weight empty	175 kg (386 lb)
Max T-O weight	285 kg (628 lb)

Performance (at max T-O weight):

Max level speed at S/L	92 knots (170 km/h; 105 mph)
Max cruising speed	81 knots (150 km/h; 93 mph)
Econ cruising speed	78 knots (145 km/h; 90 mph)
Stalling speed	35 knots (65 km/h; 41 mph)
Max rate of climb at S/L	180 m (590 ft)/min
Max range, with 20 min reserves	229 nm (425 km; 264 miles)

NICOLLIER HN 500 BENGALI

This side by side two-seat light aircraft (F-PXHN) was nearing completion in early 1981. M Nicollier does not intend to offer plans to other constructors.

Type: Two-seat all-wood light aircraft.

Wings: Cantilever low-wing monoplane of tapered planform, with increased sweepback towards roots. Wing section NACA 23000 series. Dihedral constant from roots. Washout 2° 10′. Single-spar all-wood structure made in one piece, with diagonal drag member at each wing root. Leading-edge and part of each root plywood-covered, with Dacron covering overall. Slotted ailerons of wooden construction with Dacron covering. Two-position (12° and 45°) trailing-edge flap between aileron and wing-root on each wing. No tabs.

Fuselage: All-wood semi-monocoque structure of trapezoidal section, with rounded top-decking. Four primary bulkheads, longerons and formers, with plywood stressed skin covering. Glassfibre engine cowling.

Tail Unit: Plywood-covered all-wood structure. Sweptback fin integral with fuselage. Semi-all-moving horizontal surfaces, with automatic tab in trailing-edge on each side.

Landing Gear: Non-retractable tricycle type. Cantilever spring steel main legs, each with wheel size 380 × 150. Self-centring but non-steerable cantilever nosewheel unit with helicoidal spring shock-absorption. Nosewheel size 330 × 130. Wheel fairings to be fitted. Hydraulic disc brakes. Parking brake.

Power Plant: One 74·5 kW (100 hp) Rolls-Royce Continental O-200-A flat-four engine, driving an EVRA two-blade fixed-pitch wooden propeller type D11.28.8C. Main fuel tank aft of cabin, under baggage space, capacity 67 litres (14·75 Imp gallons). Integral tank in leading-edge of each wing, with combined capacity of 54 litres (12 Imp gallons).

Accommodation: Two seats side by side in enclosed cabin, with dual controls. Tinted canopy over seats is hinged on centreline, allowing each half to open upward. Cabin

Lederlin 380-L two-seat light aircraft (Continental C90-14F engine) *(Peter J. Bish)*

Lefebvre MP.205 Busard (67 kW; 90 hp Continental engine) *(J. M. G. Gradidge)*

Lucas L5 prototype with non-retractable tricycle landing gear

Lucas L5 prototype with retractable tailwheel landing gear

Nicollier HN 433 Menestrel *(Photo Paille)*

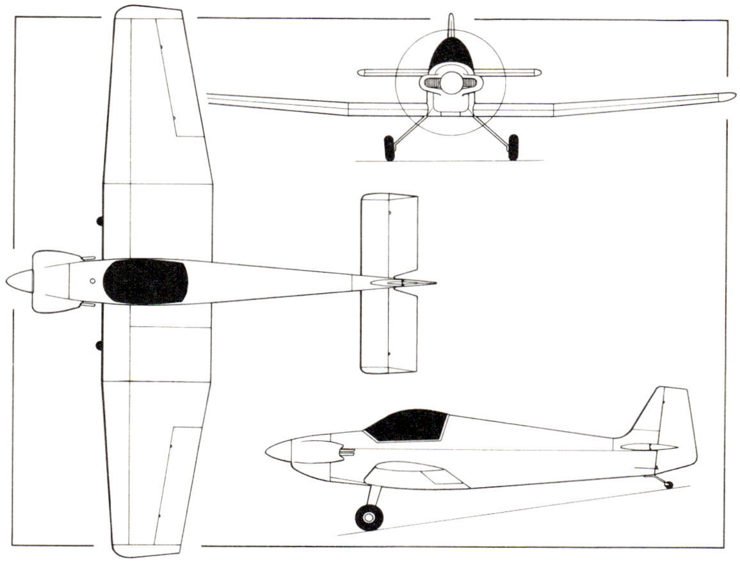

Nicollier HN 600 Week-end *(Michael A. Badrocke)*

Modified Piel Emeraude, with retractable landing gear *(J. M. G. Gradidge)*

heated and ventilated. Locker for 30 kg (66 lb) of baggage aft of seats.

Avionics and Equipment: VFR equipment standard, with provision for com/nav radio.

Dimensions, external:

Wing span	8·39 m (27 ft 6¼ in)
Wing area	11·25 m² (121·1 sq ft)
Wing aspect ratio	6·25
Length overall	6·20 m (20 ft 4 in)
Wheel track	2·52 m (8 ft 3¼ in)
Wheelbase	1·51 m (4 ft 11½ in)
Propeller diameter	1·76 m (5 ft 9¼ in)

Weights:

Weight empty	395 kg (871 lb)
Max T-O weight	650 kg (1,433 lb)

Performance (estimated, at max T-O weight):

Never-exceed speed	183 knots (340 km/h; 211 mph)
Max level speed at S/L	129 knots (240 km/h; 149 mph)
Normal cruising speed (with wheel fairings)	113 knots (210 km/h; 130 mph)
Approach speed	60 knots (110 km/h; 69 mph)
Stalling speed, flaps down	46 knots (85 km/h; 53 mph)
Max rate of climb at S/L	240 m (785 ft)/min
Range with max fuel	approx 540 nm (1,000 km; 620 miles)
g limits, Utility category	+4·4; −2·2

NICOLLIER HN 600 WEEK-END

The HN 600 Week-end is a single-seat light aircraft, designed for economical construction and operation by amateurs. The structure is simple, of wood and fabric, with some components of Klégècel foam. Plans were expected to be ready in time for display at the RSA meeting at Brienne in Summer 1981. A number of major components, such as the wing spar, ribs, cowlings and canopy, will be available in completed form from the Siravia company, of Pons.

The general appearance of the Week-end is shown in an accompanying three-view drawing. Recommended power plant is a converted Volkswagen 1,500 cc motorcar engine of 22-37 kW (30-50 hp).

Dimensions, external:

Wing span	7·00 m (22 ft 11½ in)
Wing area	8·20 m² (88·26 sq ft)
Length overall	5·25 m (17 ft 2¾ in)

Weights:

Weight empty	185 kg (408 lb)
Max T-O weight	310 kg (683 lb)

Performance (estimated, at max T-O weight):

Max level speed	108 knots (200 km/h; 124 mph)
Max cruising speed	92 knots (170 km/h; 106 mph)
Stalling speed	33 knots (60 km/h; 38 mph)
Max rate of climb at S/L	210 m (688 ft)/min
Range with max fuel at 86 knots (160 km/h; 99 mph)	345 nm (640 km; 397 miles)

PIEL

AVIONS CLAUDE PIEL

104 côte de Beulle, 78580 Maule
Telephone: 478 82 49

M Claude Piel has designed several light aircraft, including the Emeraude, Diamant and Beryl, of which sets of plans are available to amateur constructors. In addition, M Piel granted licence rights for their manufacture by several commercial concerns (see 1977-78 *Jane's*).

The authorised distributor for plans of all Piel designs available to amateur constructors is:

E. Littner, CP 272, Saint Laurent, Montreal, Quebec H4L 4V6, Canada.

PIEL EMERAUDE and SUPER EMERAUDE

There have been several factory-built versions of the Emeraude and Super Emeraude, but the aircraft are no longer being produced in this form. The designs continue to be available for amateur construction, and the following amateur-built versions have flown:

C.P.301. With 67 kW (90 hp) Continental engine.

C.P.304. With 63·4 kW (85 hp) Continental C85-12F engine and wing flaps.

C.P.305. With 86 kW (115 hp) Avco Lycoming engine.

C.P.308. With 56 kW (75 hp) Continental engine.

C.P.320. With Super Emeraude wings and 74·5 kW (100 hp) Continental engine. **C.P.320A** has sweptback fin.

C.P.321. As C.P.320, with 78·5 kW (105 hp) Potez engine.

C.P.323A. With 112 kW (150 hp) Avco Lycoming engine and sweptback fin. **C.P.323AB** has tricycle landing gear.

In addition, Mr Jason Petroelje of Holland, Michigan, brought to the 1979 EAA Fly-in an Emeraude (N78JP) with retractable tailwheel-type landing gear. This is shown in an accompanying illustration.

The Emeraude is one of the types approved by the Popular Flying Association for amateur construction in the United Kingdom.

The following details refer to the basic C.P.301

Emeraude and C.P.320 Super Emeraude, but are generally applicable to all versions:

Type: Two-seat light monoplane.

Wings: Cantilever low-wing monoplane. NACA 23012 wing section. Dihedral 5° 40′. Incidence 4° 10′. Inner half of each wing is rectangular in plan, outer half semi-elliptical. All-wood single-spar structure with fabric covering overall. Slotted ailerons and flaps.

Fuselage: Conventional wood structure, covered with fabric.

Tail Unit: Cantilever wood structure. Fin integral with fuselage. Single-piece all-wood tailplane. Elevators and rudder fabric-covered. Trim tab in starboard elevator.

Landing Gear: Non-retractable tailwheel type. Cantilever main legs have rubber-in-compression springing. Hydraulic brakes.

Power Plant (C.P.301): One 67 kW (90 hp) Continental C90-12F flat-four engine. Two-blade fixed-pitch wooden propeller. Fuel tank in fuselage, behind fireproof bulkhead, capacity 80 litres (17·6 Imp gallons). Provision for auxiliary tank, capacity 40 litres (8·8 Imp gallons).

Power Plant (C.P.320): One 74·5 kW (100 hp) Continental O-200 flat-four engine, driving a two-blade fixed-pitch wooden propeller. Fuel as for C.P.301.

Accommodation: Enclosed cockpit seating two side by side with dual controls. Sides of canopy hinge forward for access and exit. Heating and ventilation.

Dimensions, external:
- Wing span: 8·04 m (26 ft 4½ in)
- Wing chord at root: 1·50 m (4 ft 11 in)
- Wing chord at tip: 0·55 m (1 ft 9½ in)
- Wing area, gross: 10·85 m^2 (116·7 sq ft)
- Wing aspect ratio: 5·95
- Length overall:
  - C.P.301: 6·30 m (20 ft 8 in)
  - C.P.320: 6·45 m (21 ft 2 in)
- Height overall:
  - C.P.301: 1·85 m (6 ft 0¾ in)
  - C.P.320: 1·90 m (6 ft 2¾ in)
- Wheel track: 2·05 m (6 ft 8¾ in)
- Propeller diameter:
  - C.P.301: 1·80 m (5 ft 11 in)
  - C.P.320: 1·78 m (5 ft 10 in)

Weights:
- Weight empty:
  - C.P.301: 380 kg (838 lb)
  - C.P.320: 410 kg (903 lb)
- Max T-O weight:
  - C.P.301: 650 kg (1,433 lb)
  - C.P.320: 700 kg (1,543 lb)

Performance (at max T-O weight):
- Never-exceed speed:
  - C.P.301: 118·5 knots (220 km/h; 136·5 mph)
  - C.P.320: 149 knots (277 km/h; 172 mph)
- Max level speed:
  - C.P.301: 110 knots (205 km/h; 127 mph)
  - C.P.320: 124 knots (230 km/h; 143 mph)
- Max cruising speed (75% power) at 1,200 m (3,940 ft):
  - C.P.301: 108 knots (200 km/h; 124 mph)
  - C.P.320: 119 knots (220 km/h; 137 mph)
- Econ cruising speed (65% power) at 1,200 m (3,940 ft):
  - C.P.301: 101 knots (187 km/h; 116 mph)
  - C.P.320: 110 knots (205 km/h; 127 mph)
- Approach speed, flaps down:
  - C.P.301, C.P.320: 65 knots (120 km/h; 75 mph)
- Stalling speed, flaps up:
  - C.P.301: 51 knots (92 km/h; 58 mph)
  - C.P.320: 53 knots (97 km/h; 61 mph)
- Stalling speed, flaps down:
  - C.P.301: 46 knots (85 km/h; 53 mph)
  - C.P.320: 49 knots (90 km/h; 56 mph)
- Max rate of climb at S/L:
  - C.P.301: 168 m (551 ft)/min
  - C.P.320: 240 m (787 ft)/min
- Service ceiling:
  - C.P.301: 4,000 m (13,125 ft)
  - C.P.320: 4,300 m (14,100 ft)
- T-O run:
  - C.P.301: 250 m (820 ft)
  - C.P.320: 230 m (755 ft)
- T-O to 15 m (50 ft):
  - C.P.301: 440 m (1,443 ft)
  - C.P.320: 400 m (1,312 ft)
- Landing from 15 m (50 ft):
  - C.P.301: 475 m (1,558 ft)
  - C.P.320: 490 m (1,608 ft)
- Landing run:
  - C.P.301: 250 m (820 ft)
  - C.P.320: 260 m (853 ft)
- Range at econ cruising speed:
  - C.P.301, C.P.320: 538 nm (1,000 km; 620 miles)

PIEL C.P.1320

This aircraft combines the general characteristics of the Super Emeraude with the Diamant's three-seat cabin and fuel tanks in the wings. It can be fitted with engines of 112 kW (150 hp) to 149 kW (200 hp) and the prototype, built by Mr Lascoutounas of Blanquefort, near Bordeaux, has a 119·5 kW (160 hp) Avco Lycoming. Design started in January 1977 and construction of the prototype began in September of the same year. Although plans for amateur construction of the C.P.1320 had not been completed by January 1979, six sets had already been sold by that time and work on several aircraft had been started.

Normal load factors will be $+5g$ and $-2{\cdot}5g$. For aerobatics in two-seat form at a T-O weight of 720 kg (1,585 lb), the permissible load factors will be $+6g$ and $-3g$.

Type: Three-seat light monoplane.

Wings: Cantilever low-wing monoplane. NACA 23012 wing section. Dihedral 3°. Incidence 2°. Conventional all-wood structure. Slotted ailerons and slotted flaps of wood construction.

Fuselage: Conventional wooden semi-monocoque structure.

Tail Unit: Cantilever wooden structure. Sweptback fin and long dorsal fin integral with fuselage. All-wood constant-chord tailplane and elevator. Tab in elevator trailing-edge.

Landing Gear: Tailwheel type. Main units retract inward. Electrical retraction. Non-retractable tailwheel. Drum brakes on main wheels.

Power Plant: Prototype has one 119·5 kW (160 hp) Avco Lycoming engine, driving a two-blade EVRA fixed-pitch wooden propeller. Single fuel tank in fuselage, capacity 70 litres (15·4 Imp gallons), and two fuel tanks in wings with total capacity of 70 litres (15·4 Imp gallons). Max fuel capacity 140 litres (30·8 Imp gallons).

Accommodation: Three seats, under rearward-sliding transparent canopy.

Dimensions, external:
- Wing span: 7·90 m (25 ft 11 in)
- Wing chord, at root: 1·64 m (5 ft 4½ in)
- Wing chord, at tip: 1·14 m (3 ft 9 in)
- Wing area, gross: 11·10 m^2 (119·48 sq ft)
- Wing aspect ratio: 5·62
- Length overall: 6·60 m (21 ft 8 in)
- Height overall: 1·80 m (5 ft 11 in)
- Tailplane span: 2·50 m (8 ft 2½ in)
- Wheel track: 2·20 m (7 ft 2½ in)
- Propeller diameter: 1·80 m (5 ft 11 in)

Weights:
- Weight empty: 470 kg (1,036 lb)
- Max T-O weight: 800 kg (1,764 lb)

Performance:
- Never-exceed speed: 183 knots (340 km/h; 211 mph)
- Max level speed: 162 knots (300 km/h; 186 mph)
- Max cruising speed (75% power): 145 knots (270 km/h; 167 mph)
- Econ cruising speed (65% power): 132 knots (245 km/h; 152 mph)
- Approach speed, flaps down: 70 knots (130 km/h; 81 mph)
- Stalling speed, flaps up: 54 knots (100 km/h; 62 mph)
- Stalling speed, flaps down: 51·5 knots (95 km/h; 59 mph)
- Max rate of climb at S/L: 600 m (1,968 ft)/min
- Service ceiling: 5,000 m (16,400 ft)
- T-O run: 200 m (657 ft)
- T-O to 15 m (50 ft): 420 m (1,378 ft)
- Landing from 15 m (50 ft): 600 m (1,968 ft)
- Landing run: 300 m (984 ft)
- Range with max fuel (65% power): 593 nm (1,100 km; 683 miles)

PIEL SUPER DIAMANT

The Super Diamant is essentially a three/four-seat version of the Emeraude. It is fully certificated for commercial production and is available also in plan form for construction by amateurs. Current versions are as follows:

C.P.604. Prototype (F-PMEC) flown in Summer of 1964, with a 108 kW (145 hp) Continental engine. Current version has swept vertical tail surfaces. The C.P.604, built by retired aeronautical engineer Ken Mainzer of Norfolk, Virginia, USA, has a tricycle landing gear using Cessna 172 struts, wheels and tyres. Other changes from standard include increased cockpit area, and fuel tanks in the wings (total capacity 151·4 litres; 40 US gallons). First flown in July 1979, it is powered by a 119·5 kW (160 hp) Avco Lycoming IO-320-B1A engine. This aircraft is shown in an accompanying illustration.

C.P.605. Much-modified four-seat ('2+2') version, with 112 kW (150 hp) Avco Lycoming O-320-E2A engine. Fully certificated for commercial production, as well as for amateur construction. Details in 1973-74 *Jane's*.

C.P.605B. Version of C.P.605 with retractable tricycle landing gear.

Type: Three/four-seat light monoplane.

Wings: Cantilever low-wing monoplane. Wing section NACA 23012. Dihedral 5° 40′. Incidence 4° 10′. All-wood single-spar structure, made in one piece, with fabric covering. Slotted ailerons and slotted flaps of wood construction, with fabric covering.

Fuselage: Wood structure, covered with fabric.

Tail Unit: Cantilever wood structure, with sweptback fin and rudder. Fixed surfaces plywood-covered. Control surfaces fabric-covered. Ground-adjustable tab on each elevator.

Landing Gear (C.P.604): Non-retractable tailwheel type. Main wheels size 420 × 150. Hydraulic brakes. Wheel fairings. Steerable tailwheel, size 155 × 50.

Landing Gear (C.P.605B): Retractable tricycle type. Main wheels retract inward. All three wheels and tyres size 400 × 100.

Power Plant: One flat-four engine, driving an EVRA two-blade fixed-pitch wooden propeller. Fuel tank in fuselage, capacity 85 litres (18·7 Imp gallons). Provision for additional tankage to give total capacity of 160 litres (35 Imp gallons). Oil capacity 4 litres (0·9 Imp gallons).

Accommodation: Four seats ('2+2') in enclosed cabin under large rearward-sliding transparent canopy.

Dimensions, external (C.P.605B):
- Wing span: 9·20 m (30 ft 2¼ in)
- Wing chord at root: 1·50 m (4 ft 11 in)
- Wing area, gross: 13·30 m^2 (143·2 sq ft)
- Wing aspect ratio: 6·4
- Length overall: 7·00 m (22 ft 11¾ in)
- Height overall: 2·00 m (6 ft 6¾ in)
- Wheel track: 3·00 m (9 ft 10 in)
- Propeller diameter: 1·80 m (5 ft 11 in)

Weights (C.P.605B):
- Weight empty: 520 kg (1,146 lb)
- Max T-O weight: 850 kg (1,873 lb)

Performance (C.P.605B, at max T-O weight):
- Never-exceed speed: 151 knots (280 km/h; 174 mph)
- Max level speed: 141 knots (260 km/h; 162 mph)
- Max cruising speed (75% power) at 1,200 m (3,940 ft): 132 knots (245 km/h; 152 mph)
- Econ cruising speed (65% power) at 1,200 m (3,940 ft): 124 knots (230 km/h; 143 mph)
- Approach speed, flaps down: 68 knots (125 km/h; 78 mph)
- Stalling speed, flaps up: 49 knots (90 km/h; 56 mph)
- Stalling speed, flaps down: 45 knots (82 km/h; 51 mph)
- Max rate of climb at S/L: 330 m (1,082 ft)/min
- Service ceiling: 5,000 m (16,400 ft)
- T-O run: 160 m (525 ft)
- T-O to 15 m (50 ft): 380 m (1,247 ft)
- Landing from 15 m (50 ft): 600 m (1,969 ft)
- Landing run: 270 m (886 ft)
- Range at econ cruising speed: 620 nm (1,150 km; 714 miles)

PIEL C.P.70, C.P.750 and C.P.751 BERYL

The prototype **C.P.70 Beryl** was displayed publicly for the first time in August 1965. It combines the wing of the C.P.30 Emeraude, virtually unchanged, with a fuselage containing two seats in tandem, and with non-retractable tricycle landing gear. It is powered by a 48·5 kW (65 hp) Continental engine.

Intended for aerobatic flying, the **C.P.750 Beryl** is also similar in general appearance to the Emeraude but has a longer, steel tube fuselage seating two persons in tandem, slightly reduced span, non-retractable tailwheel-type landing gear, 112 kW (150 hp) Avco Lycoming engine and other changes.

The C.P.750 has so far been built principally by amateur constructors in Canada, but may also be built in France through the facilities offered by M Choisel at Abbeville. By early 1980, at least fifteen sets of plans had been sold.

A further variant is the **C.P.751**, illustrated in the 1980-81 *Jane's*. This aircraft (F-PYHM) was built by M Daniel Poulet and made its first flight at Les Mureaux on 2 March 1980. It introduced several new design features and has a 134 kW (180 hp) Avco Lycoming O-360-A engine.

Type: Two-seat aerobatic monoplane.

Wings: Cantilever low-wing monoplane. Wing section NACA 23012. Dihedral 5° 40′. Incidence 4° 10′. All-wood single-spar structure, made in one piece, with fabric covering. Slotted ailerons and slotted flaps of wood construction with fabric covering.

Fuselage: Fabric-covered structure of wood (C.P.70) or welded steel tube (C.P.750).

Tail Unit: Cantilever wood structure. Fixed surfaces plywood-covered, control surfaces fabric-covered. Ground-adjustable tab on each elevator.

Landing Gear (C.P.70): Non-retractable tricycle type.

Landing Gear (C.P.750): Non-retractable tailwheel type. Main wheels size 420 × 150, pressure 1·65 bars (24 lb/sq in). Hydraulic brakes. Wheel fairings. Steerable tailwheel.

Power Plant (C.P.70): One 48·5 kW (65 hp) Continental C65-8F flat-four engine, driving a two-blade wooden propeller. Fuel tank in fuselage, capacity 70 litres (15·4 Imp gallons).

Power Plant (C.P.750): One 112 kW (150 hp) Avco Lycoming O-320-E2A flat-four engine, driving an EVRA two-blade fixed-pitch wooden propeller. Fuel tank in fuselage, capacity 70 litres (15·4 Imp gallons), with provision for two auxiliary tanks in wings to give total capacity of 140 litres (30·75 Imp gallons). Oil capacity 5 litres (1·0 Imp gallon).

Accommodation: Two seats in tandem under rearward-sliding transparent canopy. Rear seat of C.P.70 is wide enough to accommodate one adult and a child, or two children.

Dimensions, external:
- Wing span: C.P.70: 8·25 m (27 ft 0¾ in)
  - C.P.750: 8·04 m (26 ft 4½ in)
- Wing chord at root: 1·50 m (4 ft 11 in)

Piel Super Diamant (112 kW; 150 hp Avco Lycoming engine) built by Mr Eric R. Glew of Agincourt, Ontario, Canada *(Neil A. Macdougall)*

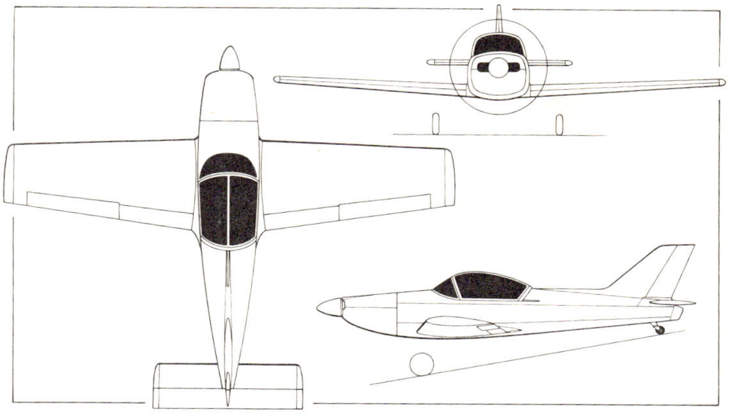

Piel C.P.1320 three-seat light aircraft *(Michael A. Badrocke)*

Piel C.P.750 Beryl built in the USA *(J. M. G. Gradidge)*

Piel C.P.80 No. 01 single-seat racing aircraft built by M Claude Piel

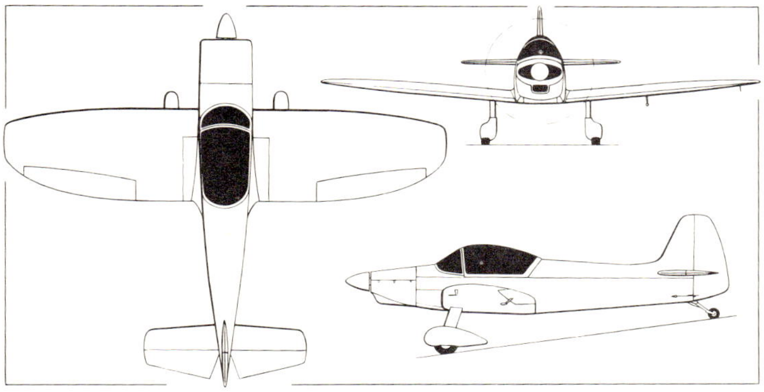

Piel C.P.90 Pinocchio single-seat light sporting aircraft *(Pilot Press)*

Piel C.P.604 Super Diamant with tricycle landing gear, built by Mr Ken Mainzer *(Howard Levy)*

Wing area, gross: C.P.70	10·85 m² (116·8 sq ft)
C.P.750	11·00 m² (118 sq ft)
Wing aspect ratio: C.P.70	5·95
C.P.750	5·85
Length overall: C.P.70	6·45 m (21 ft 2 in)
C.P.750	6·90 m (22 ft 7¾ in)
Height overall: C.P.70	1·60 m (5 ft 3 in)
C.P.750	2·10 m (6 ft 10¾ in)
Wheel track: C.P.70	2·00 m (6 ft 6¾ in)
C.P.750	2·40 m (7 ft 10½ in)
Propeller diameter	1·80 m (5 ft 11 in)
WEIGHTS:	
Weight empty: C.P.70	320 kg (705 lb)
C.P.750	480 kg (1,058 lb)
Max T-O weight: C.P.70	540 kg (1,190 lb)
C.P.750	760 kg (1,675 lb)
PERFORMANCE (at max T-O weight):	
Never-exceed speed:	
C.P.70	118·5 knots (220 km/h; 136·5 mph)
C.P.750	183 knots (340 km/h; 211 mph)
Max level speed:	
C.P.70	95 knots (175 km/h; 109 mph)
C.P.750	151 knots (280 km/h; 174 mph)
Max cruising speed (75% power) at 1,200 m (3,940 ft):	
C.P.70	84 knots (156 km/h; 97 mph)
C.P.750	143 knots (265 km/h; 165 mph)
Econ cruising speed (65% power) at 1,200 m (3,940 ft):	
C.P.70	78 knots (145 km/h; 90 mph)
C.P.750	135 knots (250 km/h; 155 mph)
Approach speed, flaps down:	
C.P.70	54 knots (100 km/h; 62·5 mph)
C.P.750	70 knots (130 km/h; 81 mph)
Stalling speed, flaps up:	
C.P.70	41 knots (75 km/h; 47 mph)
C.P.750	54 knots (100 km/h; 62·5 mph)
Stalling speed, flaps down:	
C.P.70	39 knots (70 km/h; 44 mph)
C.P.750	52 knots (95 km/h; 59 mph)
Max rate of climb at S/L:	
C.P.70	120 m (394 ft)/min
C.P.750	390 m (1,280 ft)/min
Service ceiling: C.P.70	3,000 m (9,850 ft)
C.P.750	5,200 m (17,060 ft)
T-O run: C.P.70	280 m (919 ft)
C.P.750	190 m (623 ft)
T-O to 15 m (50 ft): C.P.70	420 m (1,378 ft)
C.P.750	350 m (1,148 ft)
Landing from 15 m (50 ft): C.P.70	280 m (919 ft)
C.P.750	520 m (1,706 ft)
Landing run: C.P.70	140 m (459 ft)
C.P.750	280 m (919 ft)
Range at econ cruising speed:	
C.P.70	323 nm (600 km; 372 miles)
C.P.750	593 nm (1,100 km; 683 miles)

PIEL C.P.80

The C.P.80 was designed as a single-seat racing aircraft for amateur construction. The basic version is made of wood, as described; but M Calvel of l'Hospitalet du Larzac adapted the design to enable his C.P.80 Zef to be constructed of laminated plastics. This was the first C.P.80 to fly, followed in July 1974 by the C.P.80 Racer No. 01 (F-PVQF) built by M Claude Piel.

About 61 wooden C.P.80s are under construction by amateurs. The general appearance of the aircraft is shown in the accompanying illustration.

M Piel's prototype has confirmed the accuracy of the estimated weight and performance figures quoted. It has attained a maximum level speed of 162 knots (300 km/h; 186 mph) without main wheel fairings, suggesting an eventual maximum level speed of 172 knots (320 km/h; 199 mph) when these are fitted and development is completed.

TYPE: Single-seat amateur-built racing aircraft.

WINGS: Cantilever low-wing monoplane. Wing section NACA 23012. Dihedral 3°. Incidence 2° (constant). No sweep at quarter-chord. Conventional single-spar wood structure, plywood-covered and with polyester plastics tips. Ailerons mass-balanced and cable-actuated. No flaps or tabs.

FUSELAGE: Conventional plywood-covered wood structure of basic rectangular section, with four longerons, nine frames and domed rear decking. Polyester plastics engine cowling. Steel tube engine mounting attached to fireproof bulkhead.

TAIL UNIT: Cantilever plywood-covered all-wood structure, with vertical surfaces swept back at 50° on leading-edge. All-moving constant-chord horizontal surfaces, with centrally-positioned anti-balance and trim tab, and with mass-balance arm projecting forward inside fuselage. Horn-balanced rudder. Control surfaces cable-operated.

LANDING GEAR: Non-retractable tailwheel type. Main wheels carried on cantilever spring legs of treated AU4SG alloy. Steerable tailwheel carried on steel spring. Hydraulic brakes on main wheels.

POWER PLANT: One 67 kW (90 hp) Continental C90-8F flat-four engine, driving through a short extension shaft a two-blade fixed-pitch wooden propeller. Provision for other engines, including 48·5 kW (65 hp) Continental. Fuel tank of AG-3 alloy, capacity 40 litres (8·8 Imp gallons) aft of firewall, with refuelling point in top-decking.

ACCOMMODATION: Pilot only, in enclosed cockpit, under sideways-hinged transparent canopy.

DIMENSIONS, EXTERNAL:	
Wing span	6·00 m (19 ft 8¼ in)
Wing chord at aircraft centreline	1·35 m (4 ft 5¼ in)
Wing chord at tip	0·90 m (2 ft 11½ in)
Wing area, gross	6·20 m² (66·7 sq ft)
Wing aspect ratio	5·8
Length overall	5·30 m (17 ft 4¾ in)
Height overall	1·70 m (5 ft 7 in)
Tailplane span	1·58 m (5 ft 2¼ in)
Wheel track	1·60 m (5 ft 3 in)
Wheelbase	3·50 m (11 ft 5¾ in)
Propeller diameter	1·52 m (5 ft 0 in)
WEIGHTS (67 kW; 90 hp engine):	
Weight empty	260 kg (573 lb)
Max T-O weight	380 kg (837 lb)
PERFORMANCE (estimated, with 67 kW; 90 hp engine, at max T-O weight):	
Never-exceed speed	205 knots (380 km/h; 236 mph)
Max level speed	167 knots (310 km/h; 193 mph)
Max cruising speed (75% power) at 1,200 m (3,940 ft)	151 knots (280 km/h; 174 mph)
Econ cruising speed (65% power) at 1,200 m (3,940 ft)	129·5 knots (240 km/h; 149 mph)
Approach speed	70 knots (130 km/h; 81 mph)
Stalling speed	51·5 knots (95 km/h; 59 mph)
Max rate of climb at S/L	720 m (2,360 ft)/min
Service ceiling	6,000 m (19,685 ft)
T-O run	200 m (656 ft)
T-O to 15 m (50 ft)	400 m (1,312 ft)
Landing from 15 m (50 ft)	360 m (1,181 ft)

Landing run 200 m (656 ft)
Range at econ cruising speed 243 nm (450 km; 280 miles)
g limits +8; −6

PIEL C.P.90 PINOCCHIO

The C.P.90 Pinocchio is essentially a smaller, single-seat development of the basic Emeraude, intended for aerobatic and general sporting flying. At least six sets of plans had been sold by early 1981.

WINGS: Cantilever low-wing monoplane, of similar general planform and construction to Emeraude. Dihedral 5° 40′. Incidence 3°. Ailerons only, no flaps.
FUSELAGE: Fabric-covered wooden structure of basically rectangular cross-section with domed decking.
TAIL UNIT: Cantilever fabric-covered wooden structure, similar to that of Emeraude.
LANDING GEAR: Non-retractable tailwheel type. Streamline leg and wheel fairings on main units.
POWER PLANT: One 74·5 kW (100 hp) Continental O-200 flat-four engine, driving a two-blade wooden propeller. Fuel capacity 60 litres (13·2 Imp gallons).
ACCOMMODATION: Single seat under fully-transparent canopy.
DIMENSIONS, EXTERNAL:
Wing span 7·20 m (23 ft 7½ in)
Wing area, gross 9·65 m² (103·9 sq ft)
Wing aspect ratio 5·4
Length overall 6·00 m (19 ft 8¼ in)
Height overall 1·80 m (5 ft 11 in)
Wheel track 1·60 m (5 ft 3 in)
Propeller diameter 1·80 m (5 ft 11 in)
WEIGHTS:
Weight empty 335 kg (738 lb)
Max T-O weight 460 kg (1,014 lb)
PERFORMANCE (estimated, at max T-O weight):
Never-exceed speed 171 knots (320 km/h; 198 mph)
Max level speed 141 knots (260 km/h; 162 mph)
Max cruising speed (75% power) at 1,200 m (3,940 ft) 132 knots (245 km/h; 152 mph)
Econ cruising speed (65% power) at 1,200 m (3,940 ft) 124 knots (230 km/h; 143 mph)
Approach speed 59 knots (110 km/h; 68 mph)
Stalling speed 41 knots (75 km/h; 47 mph)
Max rate of climb at S/L 480 m (1,575 ft)/min
Service ceiling 6,000 m (19,685 ft)
T-O run 180 m (590 ft)
T-O to 15 m (50 ft) 400 m (1,312 ft)
Landing from 15 m (50 ft) 300 m (984 ft)
Landing run 160 m (525 ft)
Range at econ cruising speed 296 nm (550 km; 341 miles)

PIEL C.P.500

As can be seen in the accompanying illustration, the C.P.500 is a 'push and pull' twin-engined aircraft of staggered tandem-wing configuration. Although this gives it some similarity to the Mignet formula, the wings are fixed, and the pilot's controls conventional.

The strut-braced forward wing has four-section slotted trailing-edge flaps over 75% of the span and 25% of the chord, actuated electrically through 35°. The rear wing carries two elevons, actuated by control rods from 40° up to 35° down; these function differentially for roll control and collectively for pitch control. Endplate fins and rudders on the rear wing provide yaw control. The relative position of the wings is expected to permit steep 'parachute' descents of the kind possible with Mignet designs.

Wing section is NACA 23015. The front wing has a dihedral of 1° 30′ and incidence of 2° 30′ constant; the rear wing has a constant incidence of 4° 30′ but no dihedral.

Construction of the prototype C.P.500 has been started by the Aéro Club Jean Bertin. Although it was intended originally to be built of wood , this aircraft is now basically of metal construction, as specified for any future series production of the type. The engine cowlings, wingtips and fairings are of laminated plastics. Two 112/119·5 kW (150/160 hp) Avco Lycoming O-320 flat-four engines are specified, with the rear engine driving its propeller through an extension shaft, 15 cm (5·9 in) long. Fuel tanks in the tips of the forward wing have a combined capacity of 300 litres (66 Imp gallons).

A non-retractable tricycle landing gear is standard, with Wittman-type cantilever steel spring main legs and a steerable nosewheel. Each main wheel is fitted with a hydraulic brake.

Basic accommodation is for two persons side by side in front, with optional dual controls, and three passengers on a rear bench seat. Aft of the rear seat is space for a sixth person or a considerable quantity of baggage.

DIMENSIONS, EXTERNAL:
Wing span: front 8·80 m (28 ft 10½ in)
rear 6·43 m (21 ft 1¼ in)
Wing chord (constant): front 1·50 m (4 ft 11 in)
rear 1·10 m (3 ft 7¼ in)
Wing area, gross: front 13·20 m² (142·1 sq ft)
rear 7·10 m² (76·42 sq ft)
Wing aspect ratio: front 5·85
rear 5·88
Wing stagger 0·46 m (1 ft 6 in)
Length overall 6·10 m (20 ft 0 in)
Height overall 2·25 m (7 ft 4½ in)
Fuselage depth (max) 1·56 m (5 ft 1½ in)
Fuselage width (max) 1·40 m (4 ft 7 in)
Wheel track 2·56 m (8 ft 4¾ in)
Wheelbase 2·60 m (8 ft 6¼ in)
WEIGHTS:
Weight empty 866 kg (1,909 lb)
Max T-O weight 1,500 kg (3,307 lb)
PERFORMANCE (estimated):
Max level speed 162 knots (300 km/h; 186 mph)
Max level speed, one engine out 129 knots (240 km/h; 149 mph)
Max cruising speed (75% power) 143 knots (265 km/h; 165 mph)
Max cruising speed, one engine out (75% power) 108 knots (200 km/h; 124 mph)
Stalling speed, flaps down 49 knots (90 km/h; 56 mph)
Max rate of climb at S/L 540 m (1,770 ft)/min
Max rate of climb at S/L, one engine out 180 m (590 ft)/min
Service ceiling 6,800 m (22,300 ft)
Service ceiling, one engine out 3,000 m (9,850 ft)
Range with max fuel 647 nm (1,200 km; 745 miles)

POTTIER

JEAN POTTIER

4 rue de Poissy, 78130 les Mureaux
Telephone: 099 13-85

In addition to the light aircraft and sailplanes that he designed jointly with M Robert Jacquet, during his period as technical director at Société CARMAM, M Pottier is responsible for the purely amateur projects of which details follow.

POTTIER P.50 BOUVREUIL (BULLFINCH)

Designed by M Jean Pottier, the Bouvreuil is a single-seat racing monoplane, intended for construction by amateurs. Construction is entirely of wood, except for the plastics engine cowling and main-wheel fairings.

The Bouvreuil can be fitted with a variety of engines in the 48·5-86 kW (65-115 hp) category. It has also been designed from the start to have either a non-retractable (P.50) or retractable (P.50R) landing gear. Design load factors are ± 10.

The first Bouvreuil to be completed was number 04, built in Switzerland by M Sugnaux of Billens, which flew for the first time on 27 July 1979 and was displayed at the 1979 RSA meeting at Brienne in France. This aircraft (HB-YBF) has a 74·5 kW (100 hp) Continental O-200 engine, slotted flaps and electrically-retractable tailwheel-type landing gear. It differs from plans in having extended and cambered wingtips, and two underwing attachments for camera pods or external fuel tanks. Its load factors for aerobatics are calculated as +12/−7*g*. Equipment includes VHF radio and VOR. Number 04 has a wing span of 6·48 m (21 ft 3 in), wing area of 7·5 m² (80·7 sq ft), length of 5·25 m (17 ft 2¾ in), height of 1·70 m (5 ft 7 in), empty weight of 360 kg (794 lb) and max T-O weight of 450 kg (992 lb). Its max rate of climb at S/L is 360 m (1,180 ft)/min.

The French-built 02 has a 48·5 kW (65 hp) Continental engine and non-retractable landing gear.

TYPE: Single-seat racing monoplane.
WINGS: Cantilever low-wing monoplane. Wing section NACA 23015 at root, NACA 23012 at tip. Dihedral from roots. All-wood structure, with full-span ailerons and flaps. No tabs.
FUSELAGE: Conventional wood semi-monocoque structure, with plastics engine cowling.
TAIL UNIT: Cantilever all-wood structure, with swept vertical surfaces. Trim tab in each elevator.
LANDING GEAR: Alternative retractable or non-retractable tailwheel type. Wheel fairings standard on non-retractable main wheels. Steerable tailwheel. Independent main-wheel brakes.
POWER PLANT: Standard power plant is a 67 kW (90 hp) Continental C90 flat-four engine, driving a two-blade fixed-pitch propeller with spinner. Other engines of 48·5 to 86 kW (65-115 hp) are optional. Fuel capacity 60 litres (13 Imp gallons) for racing, 100 litres (22 Imp gallons) for touring. Provision for carrying one removable auxiliary fuel tank under each wing.
ACCOMMODATION: Single seat in enclosed cabin, under large rearward-sliding transparent canopy.
DIMENSIONS, EXTERNAL:
Wing span 6·20 m (20 ft 4 in)
Wing area, gross 7·50 m² (80·7 sq ft)
Wing aspect ratio 5·10
Length overall 5·65 m (18 ft 6½ in)
WEIGHTS (67 kW; 90 hp engine):
Weight empty 270 kg (595 lb)
Max T-O weight 400 kg (882 lb)
PERFORMANCE (estimated, with 67 kW; 90 hp engine and non-retractable landing gear):
Max level speed 167 knots (310 km/h; 192 mph)
Max cruising speed (75% power) 151 knots (280 km/h; 174 mph)
Min speed 43 knots (80 km/h; 50 mph)

POTTIER P.70S

This small sporting aircraft is derived from the P.70B, designed by M Pottier and built by M Alain Besneux (see 1976-77 *Jane's*). Design of the P.70S was started in January 1974; by January 1981 six were flying, with a further 25 under construction. First to fly, in 1977, was No. 23 (F-PYEF), powered by a 30 kW (40 hp) Volkswagen 1,500 cc engine and embodying some design changes, including a tailwheel-type landing gear (illustrated in 1978-79 *Jane's*).

The following details apply to the standard P.70S:

TYPE: Single-seat sporting aircraft.
WINGS: Cantilever mid-wing monoplane. Wing section NACA 4415. No dihedral. Incidence 2°. No sweep. Constant-chord all-metal structure of 2024 alloy, with I-beam main spar and channel-section rear spar. Entire trailing-edge of each wing formed by aileron hinged to upper surface and plain flap hinged to bottom surface. No tabs.
FUSELAGE: All-metal structure of 2024 alloy, built up on five frames.
TAIL UNIT: Cantilever all-metal structure, with sweptback vertical surfaces. Minimal fixed fin. No tabs.
LANDING GEAR: Non-retractable tricycle type. Cantilever main legs.
POWER PLANT: One 30/44·7 kW (40/60 hp) Volkswagen converted motor car engine, driving a two-blade fixed-pitch propeller. Single fuel tank in fuselage, aft of firewall, capacity 40 litres (8·75 Imp gallons).
ACCOMMODATION: Pilot only, in enclosed cockpit.
DIMENSIONS, EXTERNAL:
Wing span 5·85 m (19 ft 2¼ in)
Wing chord, constant 1·25 m (4 ft 1¼ in)
Wing area, gross 7·2 m² (77·5 sq ft)
Wing aspect ratio 4·8
Length overall 5·15 m (16 ft 10¾ in)
Height overall 1·60 m (5 ft 3 in)
Tailplane span 2·10 m (6 ft 10¾ in)
Wheel track 1·20 m (3 ft 11¼ in)
Propeller diameter 1·30 m (4 ft 3¼ in)
WEIGHTS:
Weight empty, equipped 215 kg (474 lb)
Max T-O and landing weight 325 kg (716 lb)
PERFORMANCE (A: standard P.70S with 30 kW; 40 hp engine, B: 44·7 kW; 60 hp engine, at max T-O weight, except where indicated):
Never-exceed speed at S/L 129 knots (240 km/h; 149 mph)
Max level speed at S/L:
A 97 knots (180 km/h; 112 mph)
B 116 knots (215 km/h; 133·5 mph)
No. 23 110 knots (205 km/h; 127 mph)
Max cruising speed at S/L:
A 89 knots (165 km/h; 103 mph)
B 108 knots (200 km/h; 124 mph)
No. 23 100 knots (185 km/h; 115 mph)
Econ cruising speed at S/L:
A 65 knots (120 km/h; 75 mph)
Stalling speed, flaps down 38 knots (70 km/h; 44 mph)
Max rate of climb at S/L:
A 150 m (490 ft)/min
B 330 m (1,080 ft)/min
Service ceiling: A 4,500 m (14,775 ft)
T-O run: A 350 m (1,150 ft)
B 200 m (657 ft)
Range 215 nm (400 km; 248 miles)
g limit +9 (ultimate)

POTTIER P.80S

The Pottier P.80S is a small single-seat sporting aircraft derived from the P.70S. Overall dimensions, structures, weights and performance are unchanged, but the P.80S is a low-wing monoplane and has a restyled cockpit hood. The prototype is powered by a 30 kW (40 hp) Rectimo 4 AR 1200 engine, itself a derivative of the Volkswagen motorcar engine, but engines of 30-44·7 kW (40-60 hp) can be fitted. Fuel capacity is 50 litres (11 Imp gallons).

Design of the P.80S began in January 1977 and construction of the prototype (F-PYEB) started in June 1977 at the Paris Air Show. Rollout was achieved at the Show on 12 June. By early 1981, three had flown, with a further 28 under construction by amateur builders.

POTTIER P.100TS and P.110TS

The P.100TS and P.110TS are small high-wing monoplanes of generally similar configuration, accommodating two and three persons respectively. One example of the P.100TS (F-WYJC) flew for the first time on 16 October 1980 and another is under construction, but plans are not available to amateur builders at the present time. Both aircraft have an NACA 4415 wing section. The prototype P.100TS has a 74·5 kW (100 hp) Continental engine; amateur-built examples can be fitted with Avco Lycoming

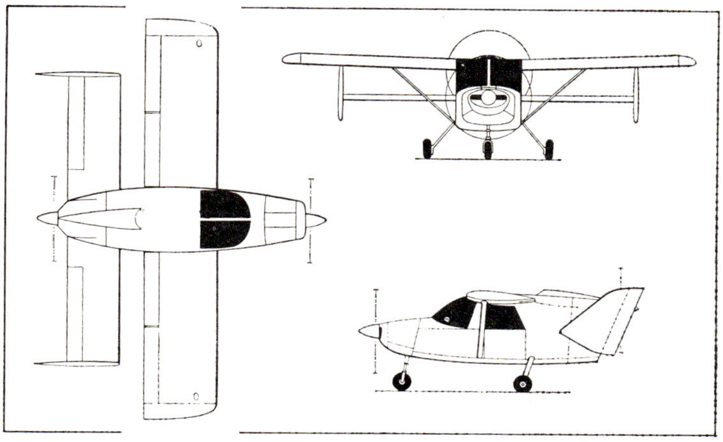

Piel C.P.500 tandem-wing twin-engined light aircraft *(Roy J. Grainge)*

Pottier P.70S with standard tricycle landing gear *(M. J. Hooks)*

Pottier P.50R Bouvreuil built in Switzerland by M Sugnaux *(Roland Eichenberger)*

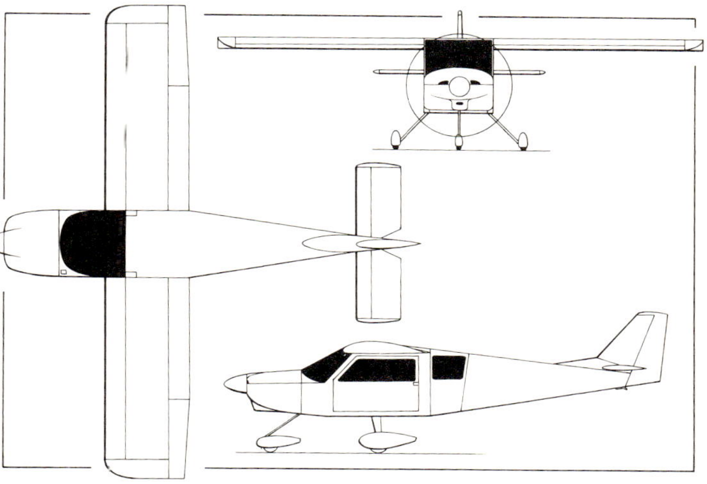

Pottier P.110TS three-seat light aircraft *(Michael A. Badrocke)*

The prototype Pottier P.80S, constructed at the 1977 Paris Air Show *(Geoffrey P. Jones)*

Prototype Pottier P.100TS (74·5 kW; 100 hp Continental engine)

or Continental engines of 48·5-93 kW (65-125 hp), the P.110TS with Avco Lycoming or Continental engines of 67-112 kW (90-150 hp). Fuel capacity of each aircraft is 90 litres (19·8 Imp gallons).

Type: Two-seat (P.100TS) or three-seat (P.110TS) all-metal light monoplane.

Dimensions, external (A: P.100TS, B: P.110TS):
Wing span: A 6·85 m (22 ft 5¾ in)
B 7·70 m (25 ft 3 in)
Wing chord, constant 1·35 m (4 ft 5¼ in)
Wing area, gross: A 9·25 m² (99·56 sq ft)
B 10·00 m² (107·64 sq ft)
Wing aspect ratio: A 5·2
B 5·9
Length overall: A 6·50 m (21 ft 4 in)
B 7·00 m (23 ft 0 in)
Height overall: A 2·20 m (7 ft 2½ in)
B 2·25 m (7 ft 4½ in)
Wheel track: A, B 1·70 m (5 ft 7 in)

Weights (A: P.100TS with 74·5 kW; 100 hp engine, B: P.110TS):
Weight empty: A 435 kg (959 lb)
Max baggage: A 33 kg (72·7 lb)
B 46 kg (101·4 lb)
Max T-O weight: A 680 kg (1,500 lb)
B 700 kg (1,543 lb)

Performance (estimated, A: P.100TS with 93 kW; 125 hp engine, B: P.110TS with 112 kW; 150 hp engine):
Max level speed:
A 135 knots (250 km/h; 155 mph)
B 143 knots (265 km/h; 165 mph)
Max cruising speed:
A 127 knots (235 km/h; 146 mph)
B 135 knots (250 km/h; 155 mph)
Stalling speed:
A, B 43·5 knots (80 km/h; 50 mph)
Max rate of climb at S/L: A 372 m (1,220 ft)/min
B 390 m (1,280 ft)/min
T-O run: A, B 230 m (755 ft)
Range with max fuel:
A, B 350 nm (650 km; 403 miles)
Endurance, with 45 min reserves:
A 3 h 30 min
B 3 h
g limit: A, B +5·7 (ultimate)

POTTIER P.105TS

The Pottier P.105TS is, as its designation suggests, a hybrid of the P.100TS and P.110TS. Power is provided by an Avco Lycoming or Continental engine of 48·5-104·5 kW (65-140 hp). Two P.105TS aircraft were under construction in 1981, but plans are not available to amateur constructors.

Type: Two-seat light monoplane.

Dimensions, external:
Wing span 7·70 m (25 ft 3 in)
Wing chord, constant 1·35 m (4 ft 5¼ in)
Wing area, gross 10·00 m² (107·64 sq ft)
Wing aspect ratio 5·9
Length overall 6·50 m (21 ft 4 in)
Height overall 2·20 m (7 ft 2½ in)
Wheel track 1·70 m (5 ft 7 in)

Weights:
Weight empty 350 kg (772 lb)
Max T-O weight 600 kg (1,323 lb)

Performance (A: 48·5 kW; 65 hp engine, B: 104·5 kW; 140 hp engine):
Max level speed:
A 94·5 knots (175 km/h; 109 mph)
B 135 knots (250 km/h; 155 mph)
Max cruising speed:
A 86·5 knots (160 km/h; 99·5 mph)
B 129 knots (240 km/h; 149 mph)
Stalling speed 41 knots (75 km/h; 47 mph)
Rate of climb at S/L: A 144 m (472 ft)/min
B 396 m (1,300 ft)/min
T-O run: A 320 m (1,050 ft)
B 180 m (590 ft)
Range with max fuel:
A, B 351 nm (650 km; 404 miles)
g limit +5·7 (ultimate)

POTTIER P.170S

The P.170S is a tandem two-seat version of the P.70S. It is of all-metal (AU4G) construction and has a retractable tricycle landing gear. Power is provided by a 37·25-52 kW (50-70 hp) Volkswagen modified motorcar engine. Plans are available to amateur constructors, and by early 1981 one P.170S was flying, with a further nine under construction.

Type: Two-seat sporting aircraft.

Dimensions, external:
Wing span 5·95 m (19 ft 6¼ in)
Wing chord, constant 1·25 m (4 ft 1¼ in)
Wing area, gross 7·40 m² (79·65 sq ft)
Wing aspect ratio 4·8
Length overall 5·70 m (18 ft 8½ in)
Height overall 1·65 m (5 ft 5 in)
Wheel track 1·20 m (3 ft 11¼ in)

Weights:
Weight empty 230 kg (507 lb)
Max T-O weight 445 kg (981 lb)

Performance (A: 37·25 kW; 50 hp engine, B: 52 kW; 70 hp engine)
Max level speed:
A 94·5 knots (175 km/h; 109 mph)
B 108 knots (200 km/h; 124 mph)
Max cruising speed:
A 84 knots (155 km/h; 96·5 mph)
B 102·5 knots (190 km/h; 118 mph)
Stalling speed 41 knots (75 km/h; 47 mph)
Max rate of climb at S/L: A 150 m (490 ft)/min
B 300 m (980 ft)/min
T-O run: A 380 m (1,250 ft)
B 220 m (722 ft)
Landing from 15 m (50 ft) 350 m (1,150 ft)
Range with standard 50 litres (11 Imp gallons) of fuel, 45 min reserves 215 nm (400 km; 248 miles)
g limits +6·6 (ultimate)

POTTIER P.180S

The P.180S is a side-by-side two-seat version of the P.80S. It is of all-metal (AU4G) construction and has a non-retractable tricycle landing gear with fairings over the wheels. Power is provided by a 41-67 kW (55-90 hp) Volkswagen modified motorcar engine. Plans are available to amateur constructors and, by early 1981, twenty-five P.180S were being built, one or two of which were expected to fly later in that year.

Type: Two-seat sporting aircraft.

Dimensions, external:

Wing span	6·50 m (21 ft 4 in)
Wing chord, constant	1·25 m (4 ft 1¼ in)
Wing area, gross	7·80 m² (83·96 sq ft)
Wing aspect ratio	5·3
Length overall	5·35 m (17 ft 6½ in)
Height overall	1·70 m (5 ft 7 in)
Wheel track	1·65 m (5 ft 5 in)

Weights:

Weight empty	240 kg (529 lb)
Max T-O weight	470 kg (1,036 lb)

Performance (estimated, A: 41 kW; 55 hp engine, B: 67 kW; 90 hp engine):

Max level speed:	
A	97 knots (180 km/h; 112 mph)
B	124 knots (230 km/h; 143 mph)
Max cruising speed:	
A	89 knots (165 km/h; 102·5 mph)
B	119 knots (220 km/h; 137 mph)
Stalling speed	41 knots (75 km/h; 47 mph)
Max rate of climb at S/L: A	168 m (550 ft)/min
B	330 m (1,080 ft)/min
T-O run: A	330 m (1,080 ft)
B	200 m (656 ft)
Landing from 15 m (50 ft)	350 m (1,150 ft)
Range	270 nm (500 km; 310 miles)
g limit	+5·7 (ultimate)

STARCK

ANDRÉ STARCK

M André Starck, who died on 16 December 1979, first left the ground in a Chanute-type hang glider. He next built a Mignet Pou-du-Ciel, followed by the first of his own designs, the AS-10. This was a tandem two-seat biplane, and led to the AS-20, first flown on 23 October 1942, with the sharply staggered narrow-gap biplane wing arrangement first conceived by an aerodynamicist named Nenadovitch.

Five fairly conventional monoplane designs were next, designated AS-70 Jac, AS-71, AS-57, AS-80 Holiday and AS-90 New-Look. In the AS-27 Starcky (see 1978-79 *Jane's*) and AS-37, described briefly below, M Starck reverted to the narrow-gap staggered biplane formula.

STARCK AS-37

As can be seen in the accompanying illustration, this design by the late M André Starck embodies the narrow-gap, sharply staggered biplane wing configuration that characterised several of his earlier products. The power plant is also unusual, as the engine is mounted behind the side by side two-seat cabin and drives, through timing belts, two pusher propellers mounted in the gap between the wings, on each side. This is claimed to enhance, by means of the propeller slipstream, the slot effect produced by the wing arrangement.

The prototype (F-WYBQ), built by M Rudy Nickel of Romans, flew for the first time on 13 January 1977. It was powered at that time by a 48·5 kW (65 hp) Citroën GS 1220 aircooled four-cylinder engine, with a 1:2 reduction in the belt drive, so that each propeller turned 2,200 rpm in cruising flight. Wingtip 'curtains', inclined at 45° to join the wingtips, stiffened the overall structure, making dihedral unnecessary on the main wings, and were claimed to improve stall characteristics and lateral control; the ailerons were attached to their trailing-edges. Construction of the aircraft was all-wooden, with spruce structure and acajou ply covering. Disc brakes were fitted to the main wheels of the non-retractable tricycle landing gear. Dual controls were fitted.

In that original form, as the **AS-37A,** M Nickel's aircraft logged 100 flying hours and 300 flights by October 1977. A second AS-37A (F-WXDU) was built and flown by M Léon Knoepfli of Sélestat, also with a GS 1220 engine. After a short period of flight testing, this was modified to **AS-37B** standard. The wingtip 'curtains' were removed and replaced by wide cantilever interplane struts, some distance inboard from extended wingtips. Conventional ailerons were installed on the upper wings, and flaps on the lower wings. Other changes included replacement of the original main landing gear legs, comprising side Vs and half-axles, with new cantilever legs of laminated glassfibre, with streamline fairings over all three wheels.

Flight trials of the AS-37B soon confirmed its stability in yaw and roll, as well as its manoeuvrability. However, the increase in empty weight, to 451 kg (995 lb), suggested that a more powerful engine might be desirable. Both the AS-37A of M Nickel and AS-37B of M Knoepfli have therefore been re-engined with a 74·5 kW (100 hp) Porsche 2 flat-four engine, with dual electronic ignition on the AS-37A. Other AS-37s, under construction by M Boucher of Saint-Marcellin and M Bourdreux of Cadarache, will have a 67 kW (90 hp) Renault 843-01 engine and a SACMA engine respectively. M Nickel's aircraft is also being tested with three-blade ground-adjustable wooden propellers.

At least 23 sets of plans for the AS-37 had been sold to amateur constructors by early 1981.

The following data apply to the AS-37A as originally flown:

Dimensions, external:

Wing span	6·30 m (20 ft 8 in)
Wing area, gross	13·60 m² (146·4 sq ft)
Wing aspect ratio	3·0
Length overall	6·00 m (19 ft 8 in)
Height overall	1·60 m (5 ft 3 in)

Weights:

Weight empty	400 kg (882 lb)
Max T-O weight	620 kg (1,366 lb)

Performance (estimated):

Max level speed at S/L	100 knots (185 km/h; 115 mph)
Cruising speed at S/L	91 knots (170 km/h; 105 mph)
Landing speed	38 knots (70 km/h; 43·5 mph)
Rate of climb at S/L	210 m (690 ft)/min
Service ceiling	4,500 m (14,750 ft)
T-O run	200 m (655 ft)
Landing run	140 m (460 ft)
Range with 90 litres (19·75 Imp gallons) fuel	810 nm (1,500 km; 930 miles)

STARCK SUPER NEW LOOK

This simple, low-cost, single-seat light aircraft had its origin in André Starck's AS-90 New Look, the first example of which was flown on 11 June 1950. At the request of French amateur constructors, M Starck updated the design, with a strut-braced wing of new section and powered by any one of a variety of modified motorcar engines. A prototype is under construction by M Parent of Cannes.

Type: Single-seat light monoplane.

Wings: Braced mid-wing monoplane. Single bracing strut each side, from main landing gear struts. Wing section GAW 1. Thickness/chord ratio 17%. Dihedral 3° from roots. Incidence 4°. Sweepback on leading-edge 6°. Constant-chord wood structure, with single plank-like main spar and trellis ribs, fabric-covered. Elliptical aileron articulated under each wingtip trailing-edge. No flaps.

Fuselage: Welded tube girder structure, fabric-covered.

Tail Unit: Cantilever structure of welded tube, fabric-covered. No tabs.

Landing Gear: Non-retractable two-wheel type, with tailskid. Each main wheel carried on a faired, welded tube V and a strut from the wing root.

Power Plant: One 24·5 kW (33 hp) Citroën Ami 6 converted motorcar engine (alternative engines from Volkswagen and DAF suitable). Up to 20 kg (44 lb) of fuel.

Accommodation: Single seat in enclosed cockpit, under sideways hinged fully transparent canopy.

Dimensions, external:

Wing span	8·00 m (26 ft 3 in)
Wing chord, constant	1·20 m (3 ft 11¼ in)
Wing area, excl ailerons	8·46 m² (91·06 sq ft)
Wing aspect ratio	7
Length overall	5·50 m (18 ft 0½ in)
Height overall	1·40 m (4 ft 7¼ in)

Weights:

Weight empty	154 kg (340 lb)
Max T-O weight	262 kg (578 lb)

Performance (estimated, with Citroën Ami 6 engine):

Max level speed at S/L	81 knots (150 km/h; 93 mph)
Cruising speed	73 knots (135 km/h; 84 mph)
Landing speed	27 knots (50 km/h; 31 mph)
Service ceiling	3,500 m (11,500 ft)

GERMANY
(FEDERAL REPUBLIC)

PFLUMM

MANFRED PFLUMM

Vahingerstrasse 60, D-7000 Stuttgart 80

Herr Manfred Pflumm, a member of the Oskar-Ursinus-Vereinigung (OUV), the West German Chapter of the EAA, has designed and constructed a simple single-seat unpowered rotor-kite. This needs to be towed to remain airborne, but can continue in limited autorotational flight after release of the towing cable.

As can be seen in the accompanying illustration, it is largely of tubular construction, with the fuselage structure welded together and the landing gear and rotor pylon attached by bolts. The two-blade rotor is of the simple teetering type; the strut-braced tail unit includes fin, fixed tailplane, and rudder. The landing gear comprises a steerable nosewheel, main units with spring-in-tension shock-absorption, plus a small bumper wheel beneath the tail. The towing coupling is incorporated in the triangular frame at the nose.

RICHTER

KLAUS J. RICHTER

Oeschwende 17, D-7900 Ulm

Telephone: 0731 60567

Mr Klaus J. Richter and Mr H. Dempfle have designed and built the prototype of a light single-seat aircraft known as the DeRic DE 13 P Delta-Ente. Design was started in 1978 and construction of the prototype began in 1980. First flight was scheduled for about September 1981. A military version of this aircraft has also been envisaged.

DeRIC DE 13 P DELTA-ENTE

Type: Single-seat light aircraft.

Wings: Basically delta wing of glassfibre, styrofoam and plywood construction. Three-quarter-span ailerons. Near vertical above-wing 'winglet' surface at each wingtip. Small swept ventral fin.

Foreplane: Cantilever structure under cockpit at nose. Similar construction to wings. Wide-chord elevators in trailing-edge.

Fuselage: Uniquely-shaped fuselage which sweeps upward and forward from centre-section of delta wing. Construction of glassfibre.

Landing Gear: Electrically-retractable tricycle type. Oleo-pneumatic shock-absorbers. Main wheel diameter 30·5 cm (12 in); nosewheel diameter 25 cm (10 in). DeRic brake.

Power Plant: One 41 kW (55 hp) DeRic engine, driving a two-blade variable-pitch DeRic pusher propeller via an extension shaft. One fuel tank in fuselage, capacity 55 litres (12 Imp gallons).

Accommodation: Pilot only, seated as on a motorcycle, his body stretching the length of the fuselage and his lower legs and feet inside the thick leading-edge of the wing.

Dimensions, external:

Wing span	5·04 m (16 ft 6½ in)
Wing area, gross	10·2 m² (109·8 sq ft)
Length overall	3·80 m (12 ft 5½ in)
Height overall	1·40 m (4 ft 7 in)

Weights:

Weight empty	150 kg (331 lb)
Max T-O weight	330 kg (728 lb)

Performance:

Max cruising speed	130 knots (240 km/h; 149 mph)
Econ cruising speed	119 knots (220 km/h; 137 mph)
Stalling speed	31·5 knots (58 km/h; 36 mph)
Max rate of climb at S/L	168 m (550 ft)/min
Service ceiling	4,500 m (14,775 ft)
T-O run	150 m (490 ft)
Landing run	120 m (395 ft)
Range	647 nm (1,200 km; 745 miles)

SCHOENENBERG

HEINRICH SCHOENENBERG

Cranachweg 2, D-5880 Lüdenscheid

Telephone: 02354 6282

SCHOENENBERG HS 9 BIPLANE

Herr Heinrich Schoenenberg, a member of the Oskar-Ursinus-Vereinigung (OUV), has built an aircraft similar to the EAA Biplane which he has designated HS 9. It differs primarily in being of smaller dimensions and lighter weight; but the main landing gear units with rubber cord shock-absorption, as detailed in the EAA plans, have been replaced by cantilever metal struts. Design began in

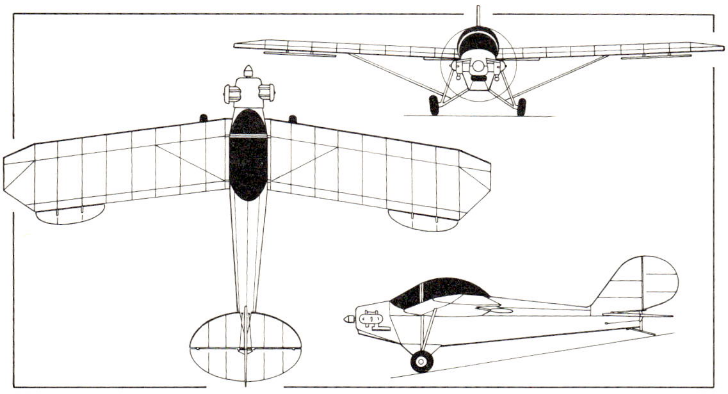

Starck Super New Look light monoplane *(Michael A. Badrocke)*

Starck AS-37A built by M Rudy Nickel, as originally flown *(Geoffrey P. Jones)*

Unpowered rotor-kite built by Manfred Pflumm *(Wolfgang Wagner)*

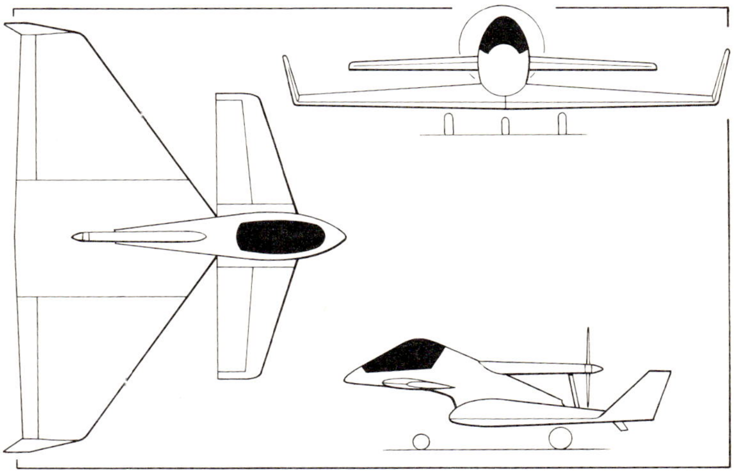

DeRic DE 13 P Delta-Ente single-seat light aircraft *(Michael A. Badrocke)*

Heinrich Schoenenberg's HS 9 biplane, derived from the EAA Biplane

Wallerkowski Hornisse single-seat monoplane with new streamline landing gear and canopy

1978, with construction starting in the following year. This aircraft (D-EHSE) flew for the first time during March 1980. Since that time Herr Schoenenberg has started production of aircraft of the same design, the first of them flown in early 1981, at which time five were on order.

Type: Single-seat sporting biplane.

Wings: Braced biplane, with N-shape streamline-section cabane and interplane struts. Standard aerofoil section NACA M6; available optionally with Wortmann FX 71/L150-20 symmetrical aerofoil section. Incidence 1° 30′. All-wood two-spar structure with fabric covering. Ailerons on all four wings. No flaps.

Fuselage: Welded steel tube structure, fabric-covered.

Tail Unit: Wire-braced structure of welded steel tube, with sheet metal ribs, fabric-covered. Trim tab in starboard elevator.

Landing Gear: Non-retractable tailwheel type. Cantilever aluminium alloy main legs, carrying wheels and tyres size 5·00 × 5. Cleveland brakes.

Power Plant: One 119 kW (160 hp) Avco Lycoming IO-320 flat-four engine (optionally an aerobatic AEIO-320-D1B of similar power) driving a Hoffmann Type HO 23-170 155 three-blade fixed-pitch wooden propeller with spinner. Aerobatic fuel system standard, with fuselage tank aft of firewall, capacity 91 litres (20 Imp gallons). Refuelling point on fuselage upper surface.

Accommodation: Single seat in open cockpit.

Dimensions, external:

Wing span	5·00 m (16 ft 5 in)
Wing area, gross	8·00 m² (86·1 sq ft)
Length overall	4·50 m (14 ft 9 in)
Height overall	1·40 m (4 ft 7 in)
Propeller diameter	1·55 m (5 ft 1 in)

Weights:

Weight empty	300 kg (661 lb)
Max T-O weight	480 kg (1,058 lb)

Performance (at max T-O weight):

Max cruising speed	173 knots (320 km/h; 199 mph)
Econ cruising speed	124 knots (230 km/h; 143 mph)
Stalling speed	43 knots (80 km/h; 50 mph)
Max rate of climb at S/L	660 m (2,165 ft)/min
Range with max fuel	378 nm (700 km; 435 miles)

WALLERKOWSKI

HEINZ WALLERKOWSKI

Rablstrasse 16/8, 8 Munich 80
Telephone: (089) 486678

Heinz Wallerkowski is an airline captain with Hapag Lloyd. He began the design of the Hornisse (D-EBXG) in January 1974; construction of the prototype was initiated one year later, and the first flight took place on 30 June 1978. By the end of October 1979, 80 flying hours had been logged, and the flight test programme to FAR Pt 23 had been completed. Captain Wallerkowski is a member of the Oskar-Ursinus-Vereinigung, the German Chapter of the EAA.

WALLERKOWSKI HORNISSE (HORNET)

Type: Single-seat monoplane.

Wings: Cantilever low-wing monoplane. NACA 23012 wing section. Dihedral 3° 30′. Incidence 1°. No sweepback. All-metal (2024-T3) construction, flush-riveted. All-metal Frise-type ailerons and all-metal flaps. No tabs.

Fuselage: All-metal semi-monocoque structure, flush-riveted.

Tail Unit: Conventional cantilever assembly of all-metal construction. Two-spar tailplane, with only two ribs each side and heavy skin. Flettner-type trim tabs in elevators.

Landing Gear: Non-retractable tailwheel type. Cantilever steel-tube main legs, with streamline fairings. Main wheels size 5·00-5. Wheel fairings. Steerable tailwheel. Gerdes hydraulic brakes.

Power Plant: One 112 kW (150 hp) Avco Lycoming O-320-E1C flat-four engine, driving a Hoffmann three-blade ground-adjustable propeller. Two fuel tanks in wings, total capacity 90 litres (20 Imp gallons).

Accommodation: Single seat under starboard-hinged transparent canopy. Baggage space behind seat.

System: 12V battery for engine starting.

Dimensions, external:

Wing span	6·14 m (20 ft 1¾ in)
Wing chord, constant	1·38 m (4 ft 6¼ in)
Wing area, gross	8·4 m² (90·4 sq ft)
Wing aspect ratio	4·45
Length overall	5·80 m (19 ft 1 in)
Height overall	1·60 m (5 ft 3 in)
Tailplane span	2·22 m (7 ft 3½ in)
Wheel track	1·80 m (5 ft 11 in)
Wheelbase	4·70 m (15 ft 5 in)

Propeller diameter 1·55 m (5 ft 1 in)
DIMENSIONS, INTERNAL:
Cockpit: Length 1·20 m (3 ft 11¼ in)
Max width 0·64 m (2 ft 1¼ in)
Max height 1·04 m (3 ft 5 in)
WEIGHTS:
Weight empty 334 kg (736 lb)
Max payload 95 kg (210 lb)
Max T-O weight: Normal 510 kg (1,124 lb)
Aerobatic 460 kg (1,014 lb)
PERFORMANCE (at max T-O weight):
Never-exceed speed 189 knots (350 km/h; 217 mph)
Max level speed at 2,440 m (8,000 ft) 188 knots (348 km/h; 216 mph)
Max cruising speed at 2,440 m (8,000 ft) 173 knots (320 km/h; 199 mph)
Econ cruising speed at 2,440 m (8,000 ft) 151 knots (280 km/h; 174 mph)
Stalling speed, flaps up, power off 48 knots (88 km/h; 55 mph)
Stalling speed, flaps down, power off 45 knots (82 km/h; 51 mph)
Max rate of climb at S/L 561 m (1,840 ft)/min
Service ceiling over 6,100 m (20,000 ft)
T-O run 150 m (490 ft)
T-O to 15 m (50 ft) 300 m (985 ft)
Landing from 15 m (50 ft) 300 m (985 ft)
Landing run 200 m (655 ft)
Range with max fuel, 45 min reserves 432 nm (800 km; 497 miles)

ITALY

VPM

VPM SnC

Via per Besnate 10, 21040 Jerago (Va)
Telephone: 0331 217328
WORKS MANAGER: Vittorio Magni

Sr Magni has built four small rotorcraft, of which the first and second were, respectively, single- and two-seat gyro-gliders of the Bensen type. The third aircraft was powered by a Volkswagen engine; the fourth had a 44·7 kW (60 hp) Franklin, and was displayed at the 1972 Turin Air Show. It was illustrated and briefly described in the 1973-74 *Jane's*.

Sr Magni has purchased from Mr Jukka Tervamäki of Finland the prototype, and all tooling, moulds and production rights, of the latter's JT-5 single-seat autogyro. In 1979 VPM SnC was established to produce plans, component parts and complete kits for amateur builders of the updated MT5 version of this autogyro. It is also developing a two-seat derivative, as the MT7.

VPM MT5

The description of the Tervamäki JT-5 that appears in the Finnish section of this edition is generally applicable also to the VPM MT5, the main difference being the use of a 1,700 cc Limbach engine to power the MT5. However, the MT5 has slightly greater dimensions, resulting in changes in all-up weight and performance, as follows:

DIMENSIONS, EXTERNAL:
Rotor diameter 7·19 m (23 ft 7 in)
Length of fuselage 3·61 m (11 ft 10 in)
Height overall 1·96 m (6 ft 5 in)
WEIGHTS:
Weight empty 167 kg (368 lb)
Max T-O weight 300 kg (661 lb)
PERFORMANCE:
Max level speed 81 knots (150 km/h; 93 mph)
Max cruising speed 70 knots (130 km/h; 81 mph)
Min level speed 22 knots (40 km/h; 25 mph)
Max rate of climb at S/L 183 m (600 ft)/min
Service ceiling 4,000 m (13,125 ft)
T-O run 80 m (262 ft)

VPM MT7

The MT7, the design of which began in early 1980, is basically an enlarged two-seat version of the MT5. Construction of a prototype was expected to begin at the end of the same year; but no plans or kits are yet available. Design of the airframe, control and rotor systems was entrusted to Mr Jukka Tervamäki of Finland, with VPM concentrating on the engine installation, instrumentation, landing gear and other details. Power plant will be a 112 kW (150 hp) Avco Lycoming O-320 flat-four engine. Design emphasis has been placed on aerodynamic refinement to give the aircraft the best possible cross-country performance at low power settings.

As can be seen from the accompanying drawing, the MT7 is generally similar in configuration to the MT5, the most obvious external differences being the selection of a twin-fin-and-rudder tail unit, with a forward-swept horizontal surface, and a more streamline landing gear incorporating wheel fairings. Provisional data are as follows:

DIMENSIONS, EXTERNAL:
Rotor diameter 8·70 m (28 ft 6½ in)
Length of fuselage 4·40 m (14 ft 5¼ in)
Width overall, excl rotor 2·10 m (6 ft 10¾ in)
Height overall 2·40 m (7 ft 10½ in)
WEIGHTS:
Weight empty 290 kg (639 lb)
Max T-O weight 480 kg (1,058 lb)
PERFORMANCE (estimated):
Max level speed 95 knots (175 km/h; 109 mph)
Max cruising speed 81 knots (150 km/h; 93 mph)
Max rate of climb at S/L 180 m (590 ft)/min
Range 161 nm (300 km; 186 miles)

JAPAN

ABE

KEÜCHI ABE

17-11-4 chome, Masuura, Kushiro-Shi, Hokkaido

Mr Keüchi Abe has designed a lightweight single-seat monoplane known as the Mizzet (Midget) No. III. Design work began in September 1977, and construction of the prototype was started in November of the same year. By early 1980 the aircraft was almost complete, but work has stopped temporarily to allow completion of Mr Abe's Motor Hang IV, details of which can be found in the section on Microlight aircraft.

ABE MIZZET No. III

TYPE: Single-seat lightweight monoplane.
WINGS: Strut-braced low-wing monoplane. Göttingen 387 wing section. Thickness/chord ratio 15%. Dihedral 3°. Incidence 0°. Constant-chord all-wood structure, with two box-spars, fabric covered. Ailerons, but no flaps.
FUSELAGE: Main structure of 6063T rectangular metal beams and tubes. Plywood and urethane used for sides of cockpit and nose, and to give form to rear fuselage. Fabric covering overall.
TAIL UNIT: Cantilever wooden structure, fabric covered. Constant-chord tailplane, with elevators. Small dorsal fin and large horn-balanced rudder.
LANDING GEAR: Non-retractable tricycle type, relying upon low-pressure tyres for shock-absorption. Mainwheel tyres size 3·40-10, nosewheel tyre size 2·50-4. Brakes fitted.
POWER PLANT: One 30 kW (40 hp) 1,200 cc modified Volkswagen motorcar engine, driving a two-blade fixed-pitch wooden propeller. Two glassfibre fuel tanks in wing leading-edge, with total capacity of 10 litres (2·2 Imp gallons). Oil capacity 2·5 litres (0·55 Imp gallons).
ACCOMMODATION: Single seat in open cockpit.
AVIONICS: Battery-powered VHF com radio.
DIMENSIONS, EXTERNAL:
Wing span 7·30 m (23 ft 11½ in)
Wing chord, constant 1·20 m (3 ft 11¼ in)
Wing area, gross 8·75 m² (94·2 sq ft)
Wing aspect ratio 6·3
Length overall 5·26 m (17 ft 3 in)
Height overall 2·05 m (6 ft 8¾ in)
Tailplane span 2·40 m (7 ft 10½ in)
Propeller diameter 1·35 m (4 ft 5 in)
WEIGHTS:
Weight empty 188 kg (414 lb)
Max fuel weight 14·4 kg (31·7 lb)
Max T-O weight 288 kg (635 lb)
PERFORMANCE (estimated):
Never-exceed speed 102 knots (189 km/h; 117 mph)
Max level speed 81 knots (150 km/h; 93 mph)
Max cruising speed 65 knots (120 km/h; 74·5 mph)
Stalling speed 38 knots (70 km/h; 43·5 mph)
Max rate of climb at S/L 366 m (1,200 ft)/min
Service ceiling 3,000 m (9,840 ft)
T-O to 15 m (50 ft) 116 m (381 ft)
Landing from 15 m (50 ft) 205 m (673 ft)
Range with max fuel 113 nm (210 km; 130 miles)

JAPAN AMATEUR BUILT AIRCRAFT LEAGUE

℅ JAA (Japan Aeronautical Association), 1-18-2 Shimbashi, Minato-ku, Tokyo
Telephone: (03) 502 1203

This organisation came into being in April 1978, superseding the former Japan Experimental Aircraft Association. Various fixed- and rotating-wing aircraft have been designed and/or built and flown by its members, and several of these have been described and illustrated in recent editions of *Jane's*. Others are examples of US homebuilt types such as the Baby Great Lakes, Flaglor Scooter, Hovey Whing Ding II, Stolp Starduster and Bensen Gyro-Copter as described in the US section.

MUKAI

ISAO MUKAI

KO 4, No. 1, Kisho, Utsumi-Machi, Shozu-Gun, Kagawa Prefecture

MUKAI OLIVE No. 2

The Olive No. 2 is Mr Mukai's second aircraft project. Design began in 1972 and the prototype was completed in March of the following year. Taxying trials started in July 1976, followed shortly afterwards by the first flight. At least one other Olive No. 2 has been constructed.

TYPE: Single-seat light seaplane.
WINGS: Braced high-wing monoplane. Constant chord. Clark Y wing section. No dihedral on centre-section; slight dihedral on outer wings. Two-spar all-wooden structure, fabric covered, with dope and enamel finish. V main bracing struts. Large-span wooden ailerons. No flaps.
FUSELAGE: Welded metal tube truss structure, fabric covered.
TAIL UNIT: Braced all-wood structure, fabric covered. Horizontal surfaces of constant chord. Ground-adjustable tab on starboard elevator. All-moving vertical tail surface.
LANDING GEAR: Twin floats of all-wooden construction, plywood covered.
POWER PLANT: One 48·5 kW (65 hp) 1,834 cc modified Volkswagen motorcar engine, driving a two-blade propeller. One fuel tank, aft of pilot's seat, capacity 50 litres (11 Imp gallons).
ACCOMMODATION: Single seat in enclosed cabin. Upward-hinged door on port side of fuselage. Space for baggage aft of pilot's seat.
DIMENSIONS, EXTERNAL:
Wing span 12·00 m (39 ft 4½ in)
Wing chord, constant 1·40 m (4 ft 7 in)
Wing area, gross 16·80 m² (180·8 sq ft)
Wing aspect ratio 8·6
Length overall 7·30 m (23 ft 11½ in)
Height overall 2·98 m (9 ft 9¼ in)
Tailplane span 3·20 m (10 ft 6 in)
Propeller diameter 1·35 m (4 ft 5 in)
DIMENSIONS, INTERNAL:
Cabin: Length 1·30 m (4 ft 3 in)
Max width 0·80 m (2 ft 7½ in)
Max height 1·00 m (3 ft 3½ in)
WEIGHTS:
Weight empty, incl oil and fuel 295 kg (650 lb)
Max T-O weight 365 kg (805 lb)
PERFORMANCE:
Max level speed at 1,000 m (3,280 ft) 59·5 knots (110 km/h; 68·5 mph)
Max cruising speed 49 knots (90 km/h; 56 mph)
Econ cruising speed 43·5 knots (80 km/h; 50 mph)
Stalling speed 24·5 knots (45 km/h; 28 mph)
Max rate of climb at S/L 150 m (490 ft)/min
T-O run 150 m (490 ft)
T-O to 15 m (50 ft) 300 m (980 ft)
Landing from 15 m (50 ft) 180 m (590 ft)
Range with max fuel 215 nm (400 km; 248 miles)

MUKAI OLIVE SMG III

Latest of Mr Mukai's aircraft is the Olive SMG III flying-boat motor glider. Design began on 3 May 1978,

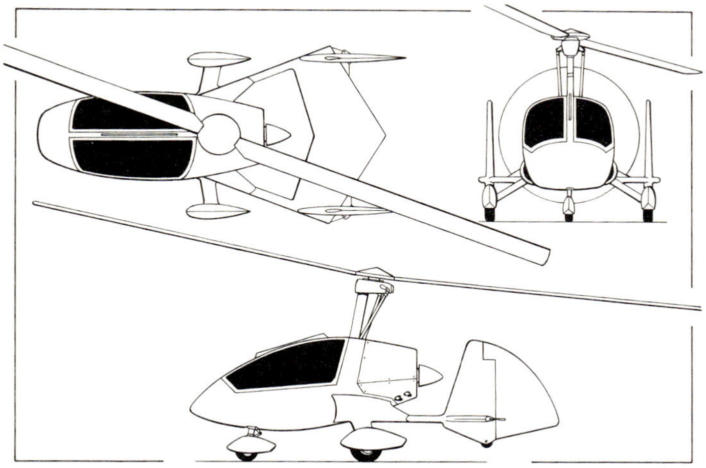

VPM MT7 two-seat autogyro *(Michael A. Badrocke)*

VPM MT5 single-seat autogyro

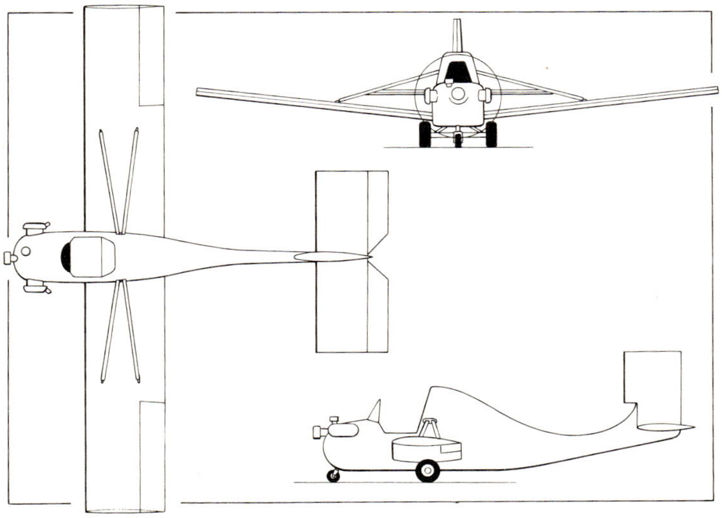

Abe Mizzet No. III *(Michael A. Badrocke)*

Abe Mizzet No. III under construction

Mukai Olive No. 2 seaplane

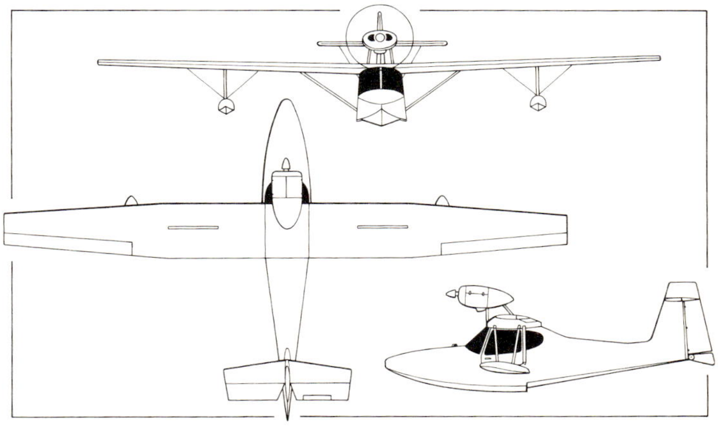

Mukai Olive SMG III *(Michael A. Badrocke)*

and construction of the prototype started on 10 October 1978. The first flight was originally scheduled for October 1980.

TYPE: Two-seat flying-boat motor glider.

WINGS: Strut-braced high-wing monoplane. Göttingen 549 wing section. Slight dihedral. Wooden box-spar and ribs, fabric covered. Large-span wooden ailerons. Airbrakes standard.

FUSELAGE: Conventional flying-boat hull. Welded stainless steel tube truss structure, glassfibre-covered on lower surface and Dacron fabric-covered on upper surface.

TAIL UNIT: Braced all-wood structure, fabric covered. High-mounted tailplane. Tab in starboard elevator. Fin integral with fuselage. Horn-balanced rudder, with water rudder beneath.

LANDING GEAR: Hull, with stabilising floats under wings.

POWER PLANT: One 45 kW (60 hp) modified Volkswagen motorcar engine, mounted on struts above wing and driving a two-blade propeller. Two fuel tanks, total capacity 30 litres (6·6 Imp gallons). Oil capacity 2·5 litres (0·55 Imp gallons).

ACCOMMODATION: Two seats in fully enclosed cabin.

DIMENSIONS, EXTERNAL:	
Wing span	14·00 m (46 ft 0 in)
Wing chord at root	1·40 m (4 ft 7 in)
Wing chord at tip	0·80 m (2 ft 7½ in)
Wing area, gross	16·80 m² (180·83 sq ft)
Wing aspect ratio	11·7
Length overall	7·80 m (25 ft 7 in)
Height overall	2·85 m (9 ft 4¼ in)
Tailplane span	3·20 m (10 ft 6 in)
WEIGHTS:	
Weight empty	430 kg (948 lb)
Max T-O weight	575 kg (1,268 lb)
PERFORMANCE (estimated):	
Best glide ratio	20
Max level speed	81 knots (150 km/h; 93 mph)
Max cruising speed	54 knots (100 km/h; 62 mph)
Stalling speed	30 knots (55 km/h; 34 mph)
Max rate of climb at S/L	150 m (492 ft)/min
Range with max fuel	215 nm (400 km; 248 miles)

NIHON UNIVERSITY

COLLEGE OF SCIENCE AND TECHNOLOGY (DEPARTMENT OF AEROSPACE ENGINEERING), NIHON UNIVERSITY

1-8 Kanda-Surugadai, Chiyoda-ku, Tokyo 101
Telephone: Tokyo (03) 293 3251
CHIEF PROFESSOR: Dr Hidemasa Kimura
CHAIRMAN: Dr Renzo Yokoi

Under the leadership of Dr Kimura, students of Nihon University have designed and built several aircraft, including the Okamura N-52 and N-58 Cygnet lightplanes and a STOL lightplane designated N-62, in collaboration with the Itoh company, which built the prototype. The N-62 was put into production by Itoh, as the Eaglet, and was described in the 1968-69 *Jane's*.

This design team has also built a series of successful man-powered aircraft named Linnet, Egret, Stork and Ibis A, all of which have been described fully in previous editions of *Jane's*. A motor glider, the N-70 Cygnus, was described in the 1974-75 edition; a new motor glider, known as N-75 Cygnus II, is currently under construction.

NEW ZEALAND

ARDC of NZ
AERONAUTICAL RESEARCH and DEVELOPMENT COMPANY OF NEW ZEALAND LTD

PO Box 11542, Wellington

DIRECTOR: James Lobet

TECHNICAL DIRECTOR: George Jacquemin

The original Ganagobie single-seat light aircraft was designed and built by the brothers William and James Lobet at Lille, France, and made its first flight in 1953, powered by an old Clerget engine. After modification and redesignation as Ganagobie 02 it flew for a further 30 h before being grounded by engine failure in 1954, and later became Ganagobie 2 when fitted with a two-stroke target drone engine.

Construction of a second aircraft, Ganagobie 03, was started in Alberta, Canada, by Mr La Rue Smith and completed by Mr Pierre Descamps. Built of birch plywood, it was somewhat heavier than the first aircraft and was powered by a 26 kW (35 hp) Poinsard engine. Engines fitted to later Ganagobie 03s were a 53·7 kW (72 hp) McCulloch and a 30 kW (40 hp) Continental; this version is also suitable for converted Volkswagen engines of 1,500 cc and above. The 'ultralight' version, known as the Ganagobie 04, is suitable for 35·8 kW (48 hp) Nelson and other small two-stroke engines. Very light okoumé mahogany, and other weight-saving features, may be used in its construction. Latest version is the Ganagobie 05, of which the prototype is being completed in New Zealand.

Mr Lobet decided to establish the present company in New Zealand for financial reasons, as it had proved difficult to produce kits at a competitive price in Australia, which had been the home of the Ganagobie programme for a number of years.

ARDC of NZ may develop a variety of aeronautical and engineering products, including energy systems using wind and solar power, and expects to market plans and kits for a number of aircraft in addition to the Ganagobies.

GANAGOBIE 05

The Ganagobie 05 is a small, high-wing single-seat aircraft, designed primarily for amateur construction. The basic homebuilt model is all-wooden; but improved production versions are under consideration, either in kit form with a fabric-covered steel tube fuselage and wooden wings and tail, or in factory-built form with all-metal fabric-covered wing and tail control surfaces. The 1979-80 edition of *Jane's* contains a description of the all-wood homebuilt version, of which a prototype (the property of ARDC of NZ) had been expected to fly in New Zealand by about 1981.

POLAND

JARZAB
KAZIMIERZ JARZAB

38-400 Krosno, ul. Staszica 11/22

JARZAB KLOS (EAR)

Mr Kazimierz Jarzab, with assistance from Mr S. Kustron and Mr S. Wojton, has designed and built an ultralight single-seat aircraft which has been named Klos. Design and construction began in 1974, and the aircraft was flown by Mr Jarzab for the first time in July 1977. All available details follow:

TYPE: Single-seat sporting aircraft.

WINGS: Braced high-wing monoplane, with single bracing strut each side. Wing section Clark Y. Constant-chord single-spar structure of wood, with plywood skins, and light alloy wingtips. Ailerons of wooden construction. No flaps or trim tabs.

FUSELAGE: Structure of pod and boom type. Forward fuselage has basic frame of welded steel tube, covered with light alloy. Tailboom consists of a light alloy tube.

TAIL UNIT: All-moving V tail, which has a light alloy tubular leading-edge spar, and is covered with fabric. Fixed triangular endplate fin at each tip of V tail. No trim tabs.

LANDING GEAR: Non-retractable tailwheel type with single wheel on each unit. Cantilever main legs of light alloy. Main-wheel tyres size 225 × 110 mm. Tailwheel diameter 90 mm (3½ in).

POWER PLANT: One 12 kW (16 hp) modified MZ-250 motorcycle engine, driving via V-belt reduction gear a two-blade metal pusher propeller with spinner. Fuel capacity 20 litres (4·4 Imp gallons).

ACCOMMODATION: Pilot only, in open cockpit.

DIMENSIONS, EXTERNAL:

Wing span	7·00 m (22 ft 11½ in)
Wing chord, constant	0·90 m (2 ft 11½ in)
Wing area, gross	6·00 m² (64·59 sq ft)
Length overall	4·20 m (13 ft 9¼ in)
Height overall	1·10 m (3 ft 7¼ in)
Tailplane span	1·76 m (5 ft 9¼ in)
Wheel track	0·90 m (2 ft 11½ in)

WEIGHTS:

Weight empty	95 kg (209 lb)
Normal T-O weight	190 kg (419 lb)

PERFORMANCE (at normal T-O weight):

Cruising speed	59 knots (110 km/h; 68 mph)
Max rate of climb at S/L	120 m (394 ft)/min
T-O run from grass	200 m (656 ft)
Range with max fuel	540 nm (1,000 km; 621 miles)
Endurance with reserves	5 h

SOUTH AFRICA

VAN ASWEGEN
C. H. J. VAN ASWEGEN

Box 88, Plettenberg Bay 6600

The 1979-80 *Jane's* contained details of a single-seat light monoplane named the Bergwind that was designed and built by Mr van Aswegen. This has been followed by a new four-seat lightplane named the Vansin.

VAN ASWEGEN VANSIN

Following completion of the Bergwind, Mr van Aswegen began in July 1977 the design of a lightplane named the Vansin. His aim was to evolve a four-seat aircraft of reasonable performance without the use of a high-power engine. As a result, he has had to rely upon producing a very clean aircraft of minimum frontal area and flush-riveted throughout. Construction also began in July 1977 and the Vansin was flown for the first time in May 1979. Later that year it was awarded a prize as the Best Homebuilt of 1979 by the EAA of South Africa.

TYPE: Four-seat cabin monoplane.

WINGS: Cantilever low-wing monoplane. Wing section NACA 2412 extensively modified. Thickness/chord ratio approximately 12%. Dihedral 7°. Incidence 1°. No sweepback. All-metal two-spar structure. Spars fabricated from 2024-T3 sheet and 7075-T6 extruded light alloy. Flush-riveted skins. Piano-hinged plain ailerons, and electrically-actuated Fowler-type trailing-edge flaps, of light alloy construction. Wingtip fuel tanks.

FUSELAGE: Conventional semi-monocoque structure of light alloy, with flush-riveted skins.

TAIL UNIT: Cantilever structure of light alloy, with flush-riveted skins. Fixed-incidence tailplane. Mechanically-actuated trim tab in elevator.

LANDING GEAR: Non-retractable tailwheel type. Cantilever main landing gear struts of spring steel. Cleveland main wheels, with tyres size 6·00-6. Maule 20 cm (8 in) diameter pneumatic tailwheel. Hydraulic disc brakes. Main-wheel fairings of GRP.

POWER PLANT: One 119 kW (160 hp) Avco Lycoming IO-320-C1A flat-four engine with turbocharger, driving a Hartzell Type HC-EZYL-2A two-blade constant-speed metal propeller with spinner. Two main fuel tanks in wings, each 70 litres (15·4 Imp gallons), and two wingtip tanks, each 50 litres (11 Imp gallons); total fuel capacity 240 litres (52·8 Imp gallons). Refuelling points on upper surface of each wing and each tip-tank. Oil capacity 8 litres (1·75 Imp gallons).

ACCOMMODATION: Pilot and three passengers in enclosed cabin. Upward-opening door on starboard side, hinged on fuselage centreline. Accommodation heated and ventilated.

SYSTEMS: Hydraulic system for brakes only. Electrical system includes engine-driven alternator and 25Ah storage battery, to operate engine starter and trailing-edge flaps.

AVIONICS AND EQUIPMENT: King KY 90 720-channel VHF com, and King ADF. Full IFR instrumentation.

DIMENSIONS, EXTERNAL:

Wing span over tip-tanks	9·22 m (30 ft 3 in)
Wing chord at root	1·80 m (5 ft 10¾ in)
Wing chord at tip	1·16 m (3 ft 9¾ in)
Wing area, gross	14·0 m² (150·7 sq ft)
Wing aspect ratio	6·1
Length overall	7·30 m (23 ft 11½ in)
Height overall	1·90 m (6 ft 2¾ in)
Tailplane span	2·80 m (9 ft 2¼ in)
Wheel track	2·10 m (6 ft 10¾ in)
Propeller diameter	1·80 m (5 ft 10¾ in)

WEIGHTS:

Weight empty, equipped	670 kg (1,477 lb)
Max fuel weight	170 kg (375 lb)
Max T-O weight	1,032 kg (2,275 lb)

PERFORMANCE (at max T-O weight):

Never-exceed speed	187 knots (346 km/h; 215 mph)
Max level speed at S/L	165 knots (306 km/h; 190 mph)
Max cruising speed at S/L	139 knots (257 km/h; 160 mph)
Econ cruising speed at 1,525 m (5,000 ft)	130 knots (241 km/h; 150 mph)
Stalling speed, flaps down, engine idling	42 knots (77 km/h; 48 mph)
Max rate of climb at S/L	244 m (800 ft)/min
Service ceiling	tested to 4,265 m (14,000 ft)
T-O run	220 m (722 ft)
Landing run	300 m (984 ft)
Range with max fuel, 1 h reserves	825 nm (1,529 km; 950 miles)
Range with max payload, 30 min reserves	521 nm (966 km; 600 miles)

SWEDEN

ANDREASSON
BJÖRN ANDREASSON

Collins Vag 22B, Halliviksnas

Mr Andreasson has designed eleven different types of light aircraft. Of these, the BA-7 was built in series by AB Malmö Flygindustri as the MFI-9B Trainer/Militrainer and by MBB in West Germany as the BO 208 C Junior (see 1970-71 *Jane's*). Plans of the MFI-9HB are available.

An earlier design, the BA-4 biplane, was modernised by Mr Andreasson for members of the Swedish branch of the Experimental Aircraft Association, and a prototype was built by apprentices of the MFI apprentice school as part of their training programme. To distinguish it from the original BA-4, it is designated BA-4B.

Mr Andreasson's latest design is the BA-11.

ANDREASSON BA-4B

The prototype BA-4B, built by MFI apprentices, was of all-metal construction. The design provides for alternative all-wooden wings.

World manufacturing rights in the BA-4B are held by Mr P. J. C. Phillips of Down House, Cocking, Midhurst, Sussex, and the aircraft has been built in small numbers in the UK by Crosby Aviation Ltd. Plans for homebuilders continue to be available from Mr Andreasson.

About three amateur-built BA-4Bs are flying in Sweden.

TYPE: Single-seat fully-aerobatic light biplane.

WINGS: Braced biplane type, with a single streamline-section interplane strut each side. A streamline-section bracing strut runs from the bottom fuselage longeron on each side to the top of the interplane strut, and an N-type cabane structure supports the centre-section. Incidence, upper wing 3°, lower wings 4°. Stagger 20°. Dihedral, upper wing 2°, lower wings 4°. Alternative all-metal structure or all-wood structure, with solid spars, covered with heavy plywood skin. Pop-riveted ailerons, of simplified sheet metal construction, on lower wings only. No flaps. Provision for fitting detachable plastics wingtips.

FUSELAGE: Sheet metal structure, with external stringers, making extensive use of pop-riveting. Turtledeck either sheet metal or reinforced plastics.

TAIL UNIT: Cantilever structure of pop-riveted sheet metal construction.

Mr Jarzab's Klos monoplane

van Aswegen Vansin four-seat cabin monoplane *(Steve Crutchley)*

Andreasson BA-11 tandem two-seat biplane *(Hans Strömberg)*

Andreasson BA-4B single-seat fully-aerobatic homebuilt biplane

Ekström Humlan 2 built by Mr Tommy Nilsson of Björklinge, Sweden

LANDING GEAR: Non-retractable tailwheel type. Cantilever spring steel main legs. Main wheels size 5·00-4 or 5·00-5. Hydraulic brakes. Steerable tailwheel carried on leaf spring.

POWER PLANT: Prototype has 74·5 kW (100 hp) Rolls-Royce Continental O-200-A flat-four engine. Provision for other engines, including Volkswagen conversions. Standard fuel tank, capacity 50 litres (11 Imp gallons), forward of cockpit. Provision for carrying external 'bullet' tank of 50 litres (11 Imp gallons) capacity under fuselage.

ACCOMMODATION: Single seat in open cockpit.

AVIONICS AND EQUIPMENT: Provision for battery, VHF radio and IFR instrumentation.

DIMENSIONS, EXTERNAL:

Wing span:	
upper	5·34 m (17 ft 7 in)
lower	5·14 m (16 ft 11 in)
Wing chord (upper and lower, constant)	0·80 m (2 ft 7½ in)
Wing area, gross	8·3 m² (90 sq ft)
Wing aspect ratio (upper and lower)	6
Length overall	4·60 m (15 ft 0 in)
Tailplane span	2·00 m (6 ft 6¾ in)

WEIGHT:

Max T-O weight	375 kg (827 lb)

PERFORMANCE (prototype, at max T-O weight):

Max level speed	122 knots (225 km/h; 140 mph)
Max cruising speed	104 knots (193 km/h; 120 mph)
Min flying speed	35 knots (64 km/h; 40 mph)
Max rate of climb at S/L	610 m (2,000 ft)/min
T-O and landing run	less than 100 m (330 ft)
Range with standard fuel	152 nm (280 km; 175 miles)

ANDREASSON BA-11

The BA-11 is an all-metal biplane, intended for single-seat aerobatic, two-seat training or competition flying. It is designed generally to FAR Pt 23 Appendix A category A (Aerobatic) requirements, but has enhanced limiting load factors of +9 to −6*g* as a single-seater, and in excess of +6 to −3*g* as a two-seater.

TYPE: Two-seat training and competition biplane.

WINGS: Biplane type, braced with dual sets of streamline tie-rods. Ailerons, of simplified pop-riveted sheet metal construction, on both upper and lower wings. Positive stagger.

FUSELAGE: Metal structure, with one-piece moulded glassfibre turtledeck.

TAIL UNIT: All-metal structure. Control surfaces of similar construction to ailerons.

LANDING GEAR: Non-retractable tailwheel type. Main legs consist of two steel leaf springs attached to bottom of fuselage. Size 5·50-5 main wheels, with hydraulic disc brakes. Tailwheel also uses leaf spring and is steerable.

POWER PLANT: Designed for one 149 kW (200 hp) Avco Lycoming fuel-injection engine, driving a 1·88 m (6 ft 2 in) diameter Hartzell constant-speed propeller. Main fuel tank in upper front fuselage, capacity approx 60 litres (13 Imp gallons). Auxiliary fuel tank, capacity approx 50 litres (11 Imp gallons), in upper wing centre-section.

ACCOMMODATION: Two seats in tandem, each designed to accommodate a back-type parachute. Prototype has enclosed cockpits, but open cockpits are optional. Basic instrumentation only in forward cockpit. Instrument panel of rear cockpit is large enough to accommodate a limited IFR panel in addition to the normal engine instruments. Electrical equipment, including starter, alternator and battery, can be fitted.

DIMENSIONS, WEIGHTS AND PERFORMANCE:
No details received for publication

EKSTRÖM
STAFFAN W. EKSTRÖM

Tivedsvägen 1, S-181 64 Lidingö
Telephone: (08) 766 34 48

EKSTRÖM HUMLAN 2

Mr Ekström began the design of this single-seat autogyro in June 1971. Construction began in April 1972, and it flew for the first time in June 1973. At least four more Humlans had been completed and flown in Sweden by early 1981, with others under construction.

The prototype (SE-HXE) was fitted initially with the standard tricycle landing gear, as described. It now has twin floats, each weighing 16 kg (35 lb), as illustrated in the 1978-79 *Jane's*.

TYPE: Single-seat homebuilt autogyro.

ROTOR SYSTEM: Single two-blade semi-rigid rotor, attached to hub by a single bolt. Ztan Zee rotor blades. Rotor brake added 1975.

ROTOR DRIVE: Flexible shaft for rotor spin-up only, via gearbox and two V-belts.

FUSELAGE: Cruciform chassis of 6061 T6 square-section aluminium tube, on which is mounted a pod-type nacelle.

TAIL UNIT: Conventional single fin and rudder, and fixed tailplane with dihedral, built of 0·4 mm and 0·8 mm aluminium sheet.

LANDING GEAR: Non-retractable tricycle type, with additional small wheel beneath tail. Rubber shock-absorption on tailwheel only. Go-kart wheels on main and nose units, tyre pressure 0·88 bars (12·8 lb/sq in). Nosewheel is steerable, self-centering, and is fitted with cycle-type brake.

POWER PLANT: One 67 kW (90 hp) McCulloch AF 100-X3 four-cylinder engine, driving a two-blade fixed-pitch pusher propeller. Fuel tank, capacity 32 litres (7 Imp gallons), behind pilot's seat. Fuel is a petrol/oil mixture, with 5% oil.

ACCOMMODATION: Single seat in open cockpit. One-piece curved windscreen. Shoulder harness fitted.

DIMENSIONS, EXTERNAL (wheel landing gear):

Rotor diameter	6·80 m (22 ft 3¾ in)
Length overall	3·42 m (11 ft 2¾ in)
Height overall	1·98 m (6 ft 6 in)
Width over wheels	1·65 m (5 ft 5 in)
Propeller diameter	1·20 m (3 ft 11¼ in)

WEIGHTS (A: wheel landing gear, B: floats):

Weight empty: A	145 kg (320 lb)
B	182 kg (401 lb)
Normal max T-O weight: A	260 kg (573 lb)
B	285 kg (628 lb)
Max T-O weight: A	295 kg (650 lb)

PERFORMANCE (A: wheel landing gear, B: floats, at 260 kg; 573 lb AUW):

Never-exceed speed:	
A, B	97 knots (180 km/h; 111·5 mph)
Max cruising speed:	
A	81 knots (150 km/h; 93 mph)
B	65 knots (120 km/h; 74 mph)
Econ cruising speed:	
A	64·5 knots (120 km/h; 74·5 mph)
B	54-59 knots (100-110 km/h; 62-68 mph)
Max rate of climb at S/L: A	300 m (984 ft)/min
B	180 m (590 ft)/min
T-O run: A	60 m (197 ft)
B	200 m (656 ft)
T-O to 15 m (50 ft): A	100 m (328 ft)
B	400 m (1,312 ft)
Landing from 15 m (50 ft), zero wind:	
A	30 m (98 ft)
Landing run, zero wind:	
A	5 m (16 ft)
Max range, no reserves:	
A	97 nm (180 km; 112 miles)
B	75 nm (140 km; 87 miles)

SWITZERLAND

BERGER

BERGER-HELICOPTER

CH 6573 Magadino TI

Telephone: (092) 64 21 71

DIRECTOR: Hans Berger

Mr Berger has built and flown prototypes of two light helicopters, of which available details follow:

BERGER BX-50A

TYPE: Single-seat light helicopter.

ROTOR SYSTEM: Two-blade main rotor and two-blade tail rotor.

FUSELAGE: All-metal frame of tubular alloy construction. Lower cabin fairing and forward-sliding transparent half-canopy.

TAIL UNIT: Small ventral fin.

LANDING GEAR: Skid type.

POWER PLANT: One 67 kW (90 hp) Continental C90 flat-four engine. Fuel in one or two tanks mounted aft of pilot.

ACCOMMODATION: Semi-enclosed cockpit for pilot only.

DIMENSIONS, EXTERNAL:
- Diameter of main rotor: 8·00 m (26 ft 3 in)
- Diameter of tail rotor: 1·10 m (3 ft 7¼ in)

WEIGHTS:
- Weight empty: 290 kg (639 lb)
- Max T-O weight: 410 kg (904 lb)

PERFORMANCE:
- Never-exceed speed: 75 knots (140 km/h; 87 mph)
- Max cruising speed: 65 knots (120 km/h; 75 mph)

BERGER BX-110

The BX-110 (HB-YAK) is a two-seat homebuilt light helicopter, powered by a Wankel rotating-piston engine. It flew for the first time on 3 June 1974, and has been awarded a permit for experimental flying by the Swiss Board of Aviation.

TYPE: Two-seat light helicopter.

ROTOR SYSTEM: Three-blade semi-rigid main rotor and two-blade tail rotor. Main rotor blades are of NACA 0012 section, and are foldable. Max pitch of main rotor 12°. Main and tail rotors of alloy construction. No rotor brake.

ROTOR DRIVE: By toothed belt, from specially designed gearbox. Main rotor/engine rpm ratio 0·09 : 1; tail rotor/engine rpm ratio 0·6 : 1.

FUSELAGE: All-metal frame, of square and circular tubular alloy construction. Large 'goldfish bowl' cabin, with framed transparencies.

TAIL UNIT: Half-tailplane on starboard side of tailboom, forward of tail rotor, with approx 45° dihedral.

LANDING GEAR: Original skid type replaced later by non-retractable tricycle wheeled gear, with main-wheel brakes.

POWER PLANT: One 134 kW (180 hp) Wankel rotating-piston engine. Fuel in two saddle tanks aft of cabin, total capacity 108 litres (23·75 Imp gallons). Refuelling point on top of each tank. Oil capacities: engine 4 litres (0·9 Imp gallons); gearbox 4 litres (0·9 Imp gallons); tail rotor 0·2 litres (0·04 Imp gallons).

ACCOMMODATION: Side-by-side seats for pilot and one passenger in 'goldfish bowl' cabin.

ELECTRICAL SYSTEM: 12V battery.

DIMENSIONS, EXTERNAL:
- Diameter of main rotor: 7·40 m (24 ft 3¼ in)
- Diameter of tail rotor: 1·20 m (3 ft 11¼ in)
- Distance between rotor centres: 4·35 m (14 ft 3¼ in)
- Length overall: 6·40 m (21 ft 0 in)
- Height to top of rotor hub: 2·52 m (8 ft 3¼ in)
- Wheel track: 1·85 m (6 ft 0¾ in)

DIMENSIONS, INTERNAL:
- Cabin: Length: 1·30 m (4 ft 3¼ in)
  - Max width: 1·25 m (4 ft 1¼ in)
  - Max height: 1·28 m (4 ft 2½ in)

AREAS:
- Main rotor disc: 43·00 m² (462·85 sq ft)
- Tail rotor disc: 1·13 m² (12·16 sq ft)

WEIGHTS:
- Weight empty: 460 kg (1,014 lb)
- Max T-O weight: 720 kg (1,587 lb)

PERFORMANCE (at max T-O weight):
- Never-exceed speed: 91 knots (170 km/h; 105 mph)
- Max cruising speed: 86 knots (160 km/h; 99 mph)
- Max rate of climb at S/L: 240 m (787 ft)/min

BEZZOLA

GION BEZZOLA

Croix-de-Pierre 742, 1470 Estavayer-le-Lac

Telephone: (037) 63 14 68

Mr Gion Bezzola, a member of the Amateur-bauvereinigung de Schweiz RSA, built and flew a single-seat lightweight monoplane designated GB-1 Luftibus (HB-YAY). This was described and illustrated in the 1978-79 *Jane's*. His latest aircraft is the **GB-2 Retro** (HB-YBH), which has rear-mounted wings, small fore-planes, and an engine at the rear, driving a pusher propeller. Further information is not available, but the accompanying illustration shows other details of the configuration.

BRÜGGER

MAX BRÜGGER

CH 1724 Zénauva

Telephone: (037) 33 29 20

Brief details of the Brügger Colibri 1 single-seat light aircraft, which flew for the first time on 30 October 1965, were given in the 1967-68 and 1971-72 *Jane's*.

More recent designs are the Colibri 2 and MB-3. Details and an illustration of the MB-3 can be found in the 1980-81 *Jane's*.

BRÜGGER MB-2 COLIBRI 2

Mr Brügger began design of the Colibri 2 in January 1966. Construction was started a year later, and the first of two prototypes flew for the first time on 1 May 1970. Plans are available to amateur constructors, and about 165 Colibri 2s were under construction or flying in Europe by early 1981.

TYPE: Single-seat light aircraft.

WINGS: Cantilever low-wing monoplane. Wing section NACA 23012. Dihedral from roots. Two-spar constant-chord wings. Wings and ailerons built of spruce with fabric covering. No flaps or tabs.

FUSELAGE: Plywood-covered wooden structure.

TAIL UNIT: Cantilever all-wood structure. Rudder only: no fin. All-moving horizontal surfaces, with Flettner-type elevators.

LANDING GEAR: Non-retractable tailwheel type, with coil spring shock-absorption on main units. Main wheels size 400 × 100, with streamline fairings. Tailwheel mounted on leaf spring. Mechanically-operated disc brakes.

POWER PLANT: One 30 kW (40 hp) 1,600 cc Volkswagen engine (Brügger modification), driving a Brügger two-blade fixed-pitch wooden propeller with plastics-coated blades. Fuel in single fuselage tank, capacity 33 litres (7·25 Imp gallons). Oil capacity 2·5 litres (0·55 Imp gallons).

ACCOMMODATION: Single seat under one-piece moulded transparent canopy, with quarter-lights to rear.

DIMENSIONS, EXTERNAL:
- Wing span: 6·00 m (19 ft 8¼ in)
- Wing chord, constant: 1·40 m (4 ft 7 in)
- Wing area, gross: 8·20 m² (88·25 sq ft)
- Length overall: 4·80 m (15 ft 9 in)
- Height overall: 1·60 m (5 ft 3 in)
- Tailplane span: 2·00 m (6 ft 6¾ in)
- Wheel track: 1·80 m (5 ft 11 in)
- Propeller diameter: 1·38 m (4 ft 6⅓ in)

WEIGHTS:
- Weight empty: 215 kg (474 lb)
- Max T-O and landing weight: 330 kg (727 lb)

PERFORMANCE (at max T-O weight):
- Max speed at 1,000 m (3,280 ft): 97 knots (180 km/h; 111 mph)
- Econ cruising speed (70% power) at 1,000 m (3,280 ft): 86 knots (160 km/h; 99 mph)
- Stalling speed: 32·5 knots (60 km/h; 37·5 mph)
- Max rate of climb at S/L: 180 m (590 ft)/min
- Service ceiling: 4,500 m (14,760 ft)
- T-O and landing run: 200 m (656 ft)
- Range with max fuel: 270 nm (500 km; 310 miles)

UNION OF SOVIET SOCIALIST REPUBLICS

KUYBYSHEV POLYTECHNIC

STUDENT DESIGN OFFICE

The Student Design Office of the Kuybyshev Polytechnic has been in existence for more than nine years, during which time several light aeroplanes, rotating-wing aircraft and gliders have been produced to give students practice in working as a team. Two of the newest aircraft from the SDO are described below.

STREKOZA (GRASSHOPPER)

This simple 'bathtub' single-seater is powered by an 18·4 kW (25 hp) Vichr-25 aircooled engine. Control surfaces comprise conventional rudder and elevators, and almost full-span auxiliary-aerofoil ailerons. The wings are the same as those fitted to the Szmiel, but are joined at the centre-section, and are supported on laminated struts. The tail unit is also of Szmiel type. Square-section tubes were used to construct the open girder fuselage, which originally had an exposed cockpit for the pilot. More recently a glassfibre nacelle has been fitted.

DIMENSIONS, EXTERNAL:
- Wing span: 7·20 m (23 ft 7½ in)
- Wing chord (constant): 1·1 m (3 ft 7¼ in)
- Wing area, gross: 9·8 m² (105·5 sq ft)
- Length overall: 6·07 m (19 ft 10¾ in)
- Height overall: 2·0 m (6 ft 6¾ in)
- Propeller diameter: 1·1 m (3 ft 7¼ in)

WEIGHTS:
- Weight empty: 150 kg (330 lb)
- Max T-O weight: 230 kg (507 lb)

PERFORMANCE:
- Max level speed: 54 knots (100 km/h; 62 mph)
- T-O speed: 27 knots (50 km/h; 31 mph)
- Landing speed: 25 knots (45 km/h; 28 mph)
- T-O run: 200 m (656 ft)
- Landing run: 180 m (590 ft)

SZMIEL (BUMBLE BEE)

The Szmiel is a single-seat strut-braced high-wing monoplane of conventional layout, except for its almost full-span auxiliary-aerofoil ailerons. The R-P-14 section single-spar wooden wings are fabric-covered, except for their leading-edge, which is plywood-covered. The fuselage is of girder construction, fabric covered. The M-73 motorcycle engine is rated at 28 kW (38 hp), and drives a two-blade fixed-pitch propeller. Two fuel tanks are carried in the wings.

The Szmiel was flight tested in 1977, and is said to be suitable for touring, liaison and forest fire patrol.

DIMENSIONS, EXTERNAL:
- Wing span: 7·60 m (24 ft 11¼ in)
- Wing chord (constant): 1·10 m (3 ft 7¼ in)
- Wing area, gross: 10·24 m² (110·2 sq ft)
- Length overall: 5·42 m (17 ft 9½ in)
- Height overall: 1·88 m (6 ft 2 in)
- Propeller diameter: 1·5 m (4 ft 11 in)

WEIGHTS:
- Weight empty: 220 kg (485 lb)
- Max T-O weight: 310 kg (683 lb)

PERFORMANCE:
- Max level speed: 65 knots (120 km/h; 74 mph)
- T-O speed: 30 knots (55 km/h; 34 mph)
- Landing speed: 27 knots (50 km/h; 31 mph)
- T-O run: 400 m (1,312 ft)
- Landing run: 300 m (985 ft)

UNITED KINGDOM

CLUTTON-TABENOR

ERIC CLUTTON

92 Newlands Street, Shelton, Stoke-on-Trent, Staffordshire ST4 2RF

CLUTTON-TABENOR FRED SERIES 2 and SERIES 3

FRED (Flying Runabout Experimental Design) was designed as a powered aircraft that could be flown by any reasonably experienced glider pilot without further training. Other aims were that it should be able to operate from small, rough fields and be roadable.

First flight of the prototype (G-ASZY) was made on 3

Berger BX-110 two-seat helicopter

Berger BX-50A single-seat helicopter in flight, with canopy open

Bezzola GB-2 Retro *(Roland Eichenberger)*

Brügger MB-2 Colibri 2 single-seat light aircraft

The Strekoza (left) and Szmiel homebuilt aircraft, made by students of the Kuybyshev Polytechnic

First 'plans-built' FRED Series 2, constructed by Mr Richard Yates

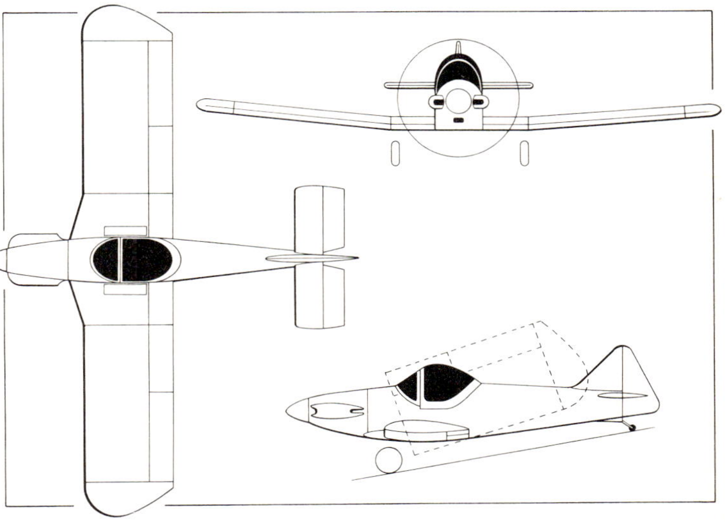

E.C.2 Easy Too, designed by Mr Eric Clutton *(Michael A. Badrocke)*

November 1963, with a 20 kW (27 hp) 500 cc Triumph 5T motorcycle engine; this was replaced later by a Scott A2S engine, and later still (1966) by a converted American-built Lawrance radial engine from an APU, at which stage the prototype was redesignated **FRED Series 2**.

Another change made after the first flights was replacement of the original bungee-in-tension landing gear shock-absorbers by steel springs.

During 1968 the Lawrance engine was replaced by a 49 kW (66 hp) Volkswagen 1,500 cc engine, modified by the provision of a toothed belt to drive the propeller at half engine speed. Next stage involved another engine change, to a Franklin 4AC-150.

The prototype has now been modified further, to **FRED Series 3** standard, with some dihedral on the wings, a shorter tailplane and increased fuel tankage. FRED Series 3 was complete in January 1980, when only adverse weather conditions were delaying the first flight in its new form.

FRED is described by its builders as being virtually unstallable with power on or off. Wings can be folded by one person unaided, and the tail unit is quickly detachable for easy transportation. Folded and tail-less width of FRED is 1·22 m (4 ft).

Plans for FRED have been available to amateur constructors since February 1970. Approximately 170 sets had been sold by early 1981, when at least four aircraft were flying, including one in New Zealand. Most are powered by direct-drive 1,600 cc Volkswagen engines.

The FRED Series 2 illustrated was built by Mr Richard Yates in nine months. It has a modified rudder shape, and this restyling is shown on current plans. First flight was made in November 1976. Powered by a 1,600 cc Volkswagen engine, Mr Yates' FRED has a cruising speed of 61·5 knots (114 km/h; 71mph) at 2,900 rpm and can take off to 61 m (200 ft) in 274 m (900 ft). Maximum nose-up attitude, power off, gives 35 knots (64 km/h; 40 mph) with no stall.

The following details apply basically to FRED Series 2 G-ASZY with Franklin engine:

TYPE: Single-seat light aircraft.

WINGS: Wire-braced parasol monoplane. Wing section Göttingen 535. Thickness/chord ratio 17·2%. No dihedral or incidence. 1° washout at tips. Spruce and plywood structure, with torsion-box leading-edge, auxiliary rear spar and drag spar, fabric-covered. Non-differential ailerons. No flaps or trim tabs.

FUSELAGE: Spruce longerons. Plywood covered to rear of cockpit, except for aluminium top-decking. Fabric covering on rear fuselage, except for plywood top-decking, front portion of which is removable for access to baggage locker.

TAIL UNIT: Cantilever structure of spruce and plywood. No fixed fin. Tailplane incidence adjustable on ground. Pushrod-operated elevators. No tabs.

LANDING GEAR: Non-retractable main wheels and tailskid. Main units sprung with motorcycle rear suspension springs. Industrial truck wheels. Tyre pressure 1·79 bars (26 lb/sq in). No brakes.

POWER PLANT: One 37·25 kW (50 hp) Franklin 4AC-150 engine in prototype, giving performance roughly equivalent to a geared 1,600 cc Volkswagen modified motorcar engine, driving a two-blade fixed-pitch propeller. Single fuel tank in centre-section, capacity 34 litres (7·5 Imp gallons). Provision for second centre-section tank. Oil capacity 3·5 litres (0·75 Imp gallons).

ACCOMMODATION: Single seat in open cockpit.

DIMENSIONS, EXTERNAL:

Wing span	6·86 m (22 ft 6 in)
Wing chord, constant	1·52 m (5 ft 0 in)
Wing area, gross	10·22 m² (110 sq ft)
Wing aspect ratio	4·4
Length overall	5·18 m (17 ft 0 in)
Height overall	1·83 m (6 ft 0 in)
Tailplane span	2·74 m (9 ft 0 in)
Wheel track	1·22 m (4 ft 0 in)
Wheelbase	3·20 m (10 ft 6 in)
Propeller diameter	1·83 m (6 ft 0 in)

WEIGHTS:

Weight empty	242 kg (533 lb)
Max T-O weight	350 kg (773 lb)

PERFORMANCE (at max T-O weight):

Max cruising speed	70 knots (130 km/h; 81 mph)
Econ cruising speed	55 knots (101 km/h; 63 mph)
Approach speed	40-45 knots (74-84 km/h; 46-52 mph)
Stalling speed	approx 35 knots (63 km/h; 40 mph)
Range with max fuel	173 nm (320 km; 200 miles)

CLUTTON-TABENOR E.C.2 EASY TOO

The Easy Too design was started in 1969 to utilise a geared Volkswagen power plant developed by Mr Clutton and Mr Tabenor. It is a single-seat all-wooden folding-wing aircraft, plywood-covered with a polyester resin finish, and is stressed for aerobatics. A full centre-section flap is fitted.

Completion of the prototype was delayed by other work, and in 1976-77 the Easy Too underwent some redesign, to enable it to qualify as an entrant for a PFA competition, on the lines of the Light Aeroplane Competitions organised at Lympne in the 1920s. By early 1978, only the wing outer panels were awaiting assembly; but completion was delayed by repair work on FRED after a heavy landing, by continued propeller research, and by the design of a new microlight aircraft of which no details are available.

The general appearance of the redesigned Easy Too can be seen in the accompanying three-view drawing. The prototype has a direct-drive 1,600 cc Volkswagen engine, but the aircraft is equally suited to 1,500 cc or larger-bore Volkswagen engines. Fuel capacity is 41 litres (9 Imp gallons).

The decision to use a direct-drive 1,600 cc VW engine meant that the engine mounting had to be lengthened by comparison with the original design (see 1976-77 *Jane's*), altering the aircraft's appearance considerably. The new engine drives a four-blade propeller. A further design change has been the adoption of a mechanically-retractable landing gear in place of the original fixed and trousered type.

The outer wing panels can be folded back by one person, by withdrawing pins and replacing them by an irreversible screwjack arrangement which locks them in position. The ailerons and flaps are coupled automatically. The folding hinge and support are entirely separate from the flying fittings, and the aeroplane can be towed on the road behind a motor car. The wingtips, of glassfibre, are of similar type and size to those fitted to the Druine Turbulent and the Taylor Monoplane. A one-piece sliding cockpit canopy is fitted. It is intended to make plans of the aircraft available after the conclusion of flight testing.

DIMENSIONS, EXTERNAL:

Wing span	7·11 m (23 ft 4 in)
Length overall	5·08 m (16 ft 8 in)
Height overall	1·37 m (4 ft 6 in)

WEIGHTS (estimated):

Weight empty	254 kg (560 lb)
Max T-O weight	363 kg (800 lb)

PERFORMANCE (estimated):

Normal cruising speed	100 knots (185 km/h; 115 mph)
Range with max fuel	347 nm (644 km; 400 miles)

COATES

J. R. COATES

The Spinney, Breachwood Green, Hitchin, Hertfordshire SG4 8PL

Mr Coates designed and built a two-seat light aircraft known as the S.A.II Swalesong, which was described fully in the 1975-76 *Jane's*. Drawings to allow amateur construction are not available, but a simplified version, the S.A.III, has been developed and will be suitable for homebuilding.

COATES S.A.III SWALESONG

This aircraft (G-BAAH) is a development of the S.A.II, and the best features of that aircraft have been maintained. Changes include a fuselage of more rounded form, a simplified wing of constant chord, and a vertical tail unit of reduced height. All control surfaces are mass and aerodynamically balanced.

The prototype S.A.III has been built in all-wood form. Following flight development, some parts may be changed to metal or structural foam, the latter being preferred as it enables weight reductions to be achieved and allows engines like the Volkswagen to be considered as adequate power sources.

TYPE: Two-seat light aircraft.

WINGS: Cantilever low-wing monoplane. Wing section NACA 63415. Dihedral 4°. Incidence 1° 30′. All-wood (spruce) structure with plywood covering, built in three pieces. All-wood slotted ailerons. Slotted all-wood flaps.

FUSELAGE: Semi-monocoque spruce structure with plywood covering.

TAIL UNIT: Cantilever structure, with sweptback vertical surfaces. Tailplane incidence adjustable on ground. One-piece fabric-covered wooden elevator, with tab.

LANDING GEAR: Non-retractable tricycle type. Cantilever light alloy main legs. Steerable nosewheel, with motor car shock-absorber and glassfibre fairing. Size 5·00-5 tyres on main wheels; 130-1300 tyre on nosewheel. Disc brakes.

POWER PLANT: One 63·4-80·5 kW (85-108 hp) Continental or Avco Lycoming flat-four engine, driving a fixed-pitch wooden propeller. Fuel capacity 72·7 litres (16 Imp gallons).

ACCOMMODATION: Two seats side by side in enclosed cockpit, with sliding one-piece hood. Baggage space aft of seats. Cockpit heated and ventilated.

DIMENSIONS, EXTERNAL:

Wing span	7·62 m (25 ft 0 in)
Wing chord, constant	1·37 m (4 ft 6 in)
Wing area, gross	10·41 m² (112 sq ft)
Wing aspect ratio	5·6
Length overall	5·64 m (18 ft 6 in)
Width, wings folded	1·73 m (5 ft 8 in)
Height overall	2·08 m (6 ft 10 in)
Tailplane span	2·44 m (8 ft 0 in)
Wheel track	1·68 m (5 ft 6 in)
Wheelbase	1·30 m (4 ft 3 in)
Propeller diameter	1·63 m (5 ft 4 in)

DIMENSION, INTERNAL:

Cockpit: Max width	1·07 m (3 ft 6 in)

WEIGHTS:

Weight empty	317 kg (700 lb)
Max T-O and landing weight	544 kg (1,200 lb)

PERFORMANCE (estimated):

Never-exceed speed at 305 m (1,000 ft)	165 knots (305 km/h; 190 mph)
Max level speed at 305 m (1,000 ft)	130 knots (241 km/h; 150 mph)
Max cruising speed at 305 m (1,000 ft)	113 knots (209 km/h; 130 mph)
Econ cruising speed at 305 m (1,000 ft)	95·5 knots (177 km/h; 110 mph)
Stalling speed, flaps down	44 knots (81 km/h; 50 mph)
Max rate of climb at S/L	260 m (850 ft)/min

ISAACS

JOHN O. ISAACS

23 Linden Grove, Chandler's Ford, Hampshire SO5 1LE
Telephone: 042 15 60885

Mr Isaacs designed and built a single-seat light aircraft, the airframe of which is basically a 7/10th scale wooden version of that of the Hawker Fury fighter of the 1930s. Constructional drawings are available to amateur builders.

He has also designed and built an all-wood scaled-down version of the Supermarine Spitfire single-seat fighter of the second World War.

ISAACS FURY II

Design of the Isaacs Fury was started in January 1961 and construction of the aircraft began in April 1961. It flew for the first time on 30 August 1963, powered by a 48·5 kW (65 hp) Walter Mikron engine (see 1965-66 *Jane's*).

In 1966-67 Mr Isaacs modified the Fury prototype to Mk II standard, by restressing the airframe and installing a 93 kW (125 hp) Avco Lycoming engine, and flew the aircraft in this form in the Summer of 1967. It was acquired subsequently by Mr W. Raper of Wrotham, Kent, who made further refinements, including the addition of blister fairings over the engine cylinders. It has since changed hands again, and is now based at Land's End airfield.

Fury plans have been supplied to amateur constructors in Australia, Canada, Denmark, Finland, West Germany, Jersey, the Netherlands, New Zealand, Norway, the UK, the USA and Yugoslavia. In addition to the prototype, two Furies are flying in the UK, three in New Zealand, one in Canada and two in the USA.

TYPE: Single-seat light biplane, stressed to 9*g* for aerobatics.

WINGS: Staggered biplane, with N type interplane strut each side and two N strut assemblies supporting centre-section of top wing above fuselage. Conventional wire bracing. Wing section RAF 28. Thickness/chord ratio 9·75%. Dihedral 1° on top wing, 3° 30′ on bottom wings. Incidence 3° 20′ on top wing, 3° 50′ on bottom wings. Spruce 'plank' spars and Warren girder ribs, with fabric covering. Fabric-covered spruce ailerons on top wing only. No flaps.

FUSELAGE: Spruce structure, covered with birch plywood.

TAIL UNIT: Strut-braced spruce structure of 'plank' spars and girder ribs, fabric-covered. Ground-adjustable tab in port elevator.

LANDING GEAR: Non-retractable type, with tailskid. Cross-axle tied to Vs with rubber-cord shock-absorption. Main wheels consist of WM.2 355 mm (14 in) rims spoked to home-made hubs. Dunlop tyre, size 3·25-14, pressure approx 2·28 bars (33 lb/sq in). Brakes optional.

POWER PLANT (prototype): One 93 kW (125 hp) Avco Lycoming O-290 flat-four engine. Two-blade fixed-pitch propeller. Fuel tank in fuselage, aft of fireproof bulkhead, capacity 45·5 litres (10 Imp gallons) or 54·5 litres (12 Imp gallons).

ACCOMMODATION: Single seat in open cockpit. Small door above top longeron on port side opens downward. Space for light baggage aft of seat. Radio optional.

DIMENSIONS, EXTERNAL:

Wing span:	
upper	6·40 m (21 ft 0 in)
lower	5·54 m (18 ft 2 in)
Wing chord (both, constant)	1·07 m (3 ft 6 in)
Wing area (total)	11·50 m² (123·8 sq ft)
Wing aspect ratio (upper)	6
Length overall	5·87 m (19 ft 3 in)
Height over tail (flying attitude)	2·16 m (7 ft 1 in)
Tailplane span	2·13 m (7 ft 0 in)
Wheel track	1·27 m (4 ft 2 in)

WEIGHTS (93 kW; 125 hp Avco Lycoming engine):

Weight empty	322 kg (710 lb)
Max permissible T-O weight	450 kg (1,000 lb)

PERFORMANCE (with uncowled 93 kW; 125 hp engine):

Max level speed	100 knots (185 km/h; 115 mph)
Stalling speed	33 knots (61 km/h; 38 mph)
Max rate of climb at S/L	488 m (1,600 ft)/min

ISAACS SPITFIRE

Construction of this prototype 6/10-scale Spitfire (G-BBJI) began in the Summer of 1969, and it flew for the first time on 5 May 1975. The airframe is stressed to meet the aerobatic requirements of +9*g* and −4·5*g* (factored) as laid down in BCAR. Plans for the Isaacs Spitfire are available to homebuilders, and about five sets had been supplied by early 1981.

TYPE: Single-seat sporting aircraft.

WINGS: Cantilever low-wing monoplane of semi-elliptical planform. Wing section NACA 2200 series. Thickness/chord ratio 13·2% at root, 6% at tip. Dihedral 6°. Incidence 2° at root, −30′ at tip. Two-spar wing built in one piece, mainly of spruce, with birch plywood covering, except for ailerons which are fabric-covered.

FUSELAGE: Spruce structure, covered with birch plywood.

TAIL UNIT: Cantilever structure of plywood-covered spruce.

LANDING GEAR: Non-retractable tailwheel type on prototype. Cantilever main legs. Dunlop 5·00-5 tyres, wheels and hydraulic disc brakes.

POWER PLANT: One 74·5 kW (100 hp) Continental O-200 flat-four engine, or alternative engine in same category. Two-blade ground-adjustable Ratier metal propeller. Fuel tank in fuselage, aft of fireproof bulkhead, capacity 45·5 litres (10 Imp gallons).

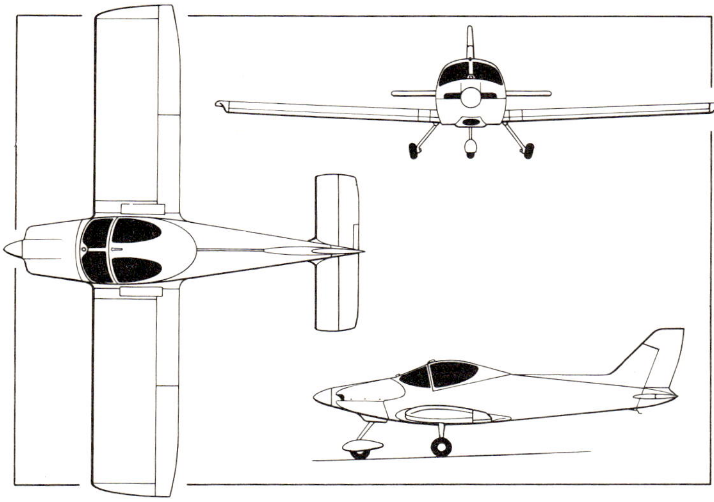

Coates S.A.III Swalesong *(Michael A. Badrocke)*

Prototype Isaacs Spitfire single-seat light sporting aircraft

Nipper Mk IIIb single-seat light aircraft

Isaacs Fury built in New Zealand

ACCOMMODATION: Single seat under blister-type transparent canopy. Space for light baggage aft of seat.

DIMENSIONS, EXTERNAL:

Wing span	6·75 m (22 ft 1½ in)
Wing chord at root	1·52 m (5 ft 0 in)
Wing area, gross	8·08 m² (87 sq ft)
Length overall	5·88 m (19 ft 3 in)
Height overall	1·73 m (5 ft 8 in)
Tailplane span	1·92 m (6 ft 3½ in)
Wheel track	1·80 m (5 ft 11 in)

WEIGHTS:

Weight empty	366 kg (805 lb)
Max T-O weight	499 kg (1,100 lb)

PERFORMANCE (at max T-O weight):

Max level speed	130 knots (240 km/h; 150 mph)
Cruising speed	116 knots (215 km/h; 134 mph)
Stalling speed, 'clean'	45-47 knots (84-87 km/h; 52-54 mph)
Stalling speed, with optional fuselage airbrake extended	41 knots (76 km/h; 47 mph)
Max rate of climb at S/L	336 m (1,100 ft)/min

NASH

NASH AIRCRAFT LTD

c/o Kinetrol Ltd, Trading Estate, Farnham, Surrey GU9 9NU

Telephone: 0252 723688

DIRECTORS:

Alan R. B. Nash, BSc, CEng, MICE, MRAeS
Richard C. Nash, BA (Cantab)
Roy G. Procter, CEng, MRAeS

In early 1978 a controlling interest in this company was obtained by Mr Alan Nash. It remains the intention of Nash Aircraft to exploit the commercial potential of both the single-seat Kittiwake (see 1978-79 *Jane's*) and the two-seat Petrel. However, Mr Nash's main interest is in using the aircraft as the basis for a manufacturing programme rather than in the amateur plans/kits market, although sympathetic to the latter.

A full description of the Nash Petrel can be found in the main Aircraft section of this edition.

NIPPER

NIPPER KITS AND COMPONENTS LTD

1 Ridgeway Drive, Bromley, Kent BR1 5DG

Telephone: 01-857 7821

DIRECTORS:

D. P. L. Antill (Chairman)
A. F. Ayles
A. S. Pearcey

Complete worldwide rights for the Nipper aircraft were purchased from Belgium in 1966, and it was marketed by Nipper Aircraft Ltd as a ready-to-fly lightplane and in the form of several stages of kits for amateur construction.

Nipper Aircraft Ltd went into receivership in May 1971. Prior to this Mr D. P. L. Antill, its Managing Director, acquired all rights in the Nipper, and on 20 October 1971 formed a new company, Nipper Kits and Components Ltd, to supply spares for existing aircraft and to encourage and support amateur construction of the Nipper. Plans and an advisory service for amateur constructors continue to be available, and a further new Nipper was completed in January 1981. Others are under construction.

NIPPER Mk IIIb

The Mk IIIb Nipper is powered by a 1,600 cc Rollason Ardem engine. Compared with the earlier Mk III and IIIA, it has a wider instrument panel to accept additional instruments to the owner's choice, and a new canopy of increased length and width to provide extra room inside the cockpit. It can be fitted with wingtip fuel tanks which almost double the standard fuel capacity. With these tanks fitted, but empty, the aircraft remains aerobatic. Flutter tests were completed satisfactorily at speeds up to 156 knots (290 km/h; 180 mph).

TYPE: Single-seat light monoplane.

WINGS: Cantilever mid-wing monoplane. Modified NACA 43012A wing section. Dihedral 5° 30'. Incidence 2°. All-wood one-piece single-spar structure, with plywood-covered leading-edge and overall fabric covering. Wooden ailerons with fabric covering. No flaps. Portion of port wing-root trailing-edge is made of light alloy and hinged, with built-in footrest, so that it can be folded down to assist access to cockpit. Wing is quickly removable, to permit aircraft to be towed behind a motorcar.

FUSELAGE: Welded steel tube structure. Underfuselage fairing of glassfibre. Rear fuselage fabric-covered.

TAIL UNIT: Braced tailplane and elevators of wood construction. No fin. Rudder of steel tube construction with fabric covering.

LANDING GEAR: Non-retractable tricycle type. Nieman transverse rubber-ring shock-absorbers. Steerable nosewheel. Continental tyres, size 4·00-4, pressure 1·79 bars (26 lb/sq in). Disc brakes.

POWER PLANT: One 41 kW (55 hp) Rollason Ardem XI flat-four engine, driving a two-blade fixed-pitch wooden propeller with glassfibre spinner. The Mk IIIb has new downward-facing stainless steel exhaust pipes for increased life and reduced noise level in cockpit. Fuel tank between engine and cockpit, capacity 34 litres (7·5 Imp gallons). Provision for two 16·5 litre (3·6 Imp gallon) wingtip fuel tanks. Oil capacity 3·5 litres (0·77 Imp gallons).

ACCOMMODATION: Single seat under blown Perspex canopy which hinges sideways to starboard. Small baggage space aft of seat.

AVIONICS: Standard radio is now the battery-operated 720-channel Dittel ATR-720, installed in port wing root.

DIMENSIONS, EXTERNAL:

Wing span (without tip-tanks)	6·00 m (19 ft 8 in)
Wing span (with tip-tanks)	6·25 m (20 ft 6 in)
Wing chord at c/l	1·40 m (4 ft 7¼ in)
Wing chord at tip	1·10 m (3 ft 7¼ in)
Wing area, gross	7·50 m² (80·70 sq ft)
Wing aspect ratio	4·8
Length overall	4·56 m (15 ft 0 in)
Height overall	1·91 m (6 ft 3 in)
Tailplane span	2·14 m (7 ft 0 in)
Wheel track	1·40 m (4 ft 7 in)
Wheelbase	1·13 m (3 ft 8 in)

WEIGHTS:

Weight empty	210 kg (465 lb)
Max T-O weight:	
Aerobatic	310 kg (685 lb)
Normal	340 kg (750 lb)

PERFORMANCE (at max T-O weight):

Never-exceed speed	126 knots (235 km/h; 146 mph)
Max level speed at S/L:	
without tip-tanks	93 knots (173 km/h; 107 mph)
with tip-tanks	83 knots (155 km/h; 96 mph)
Max cruising speed (75% power) at S/L:	
without tip-tanks	81 knots (150 km/h; 93 mph)
Econ cruising speed at S/L	78 knots (145 km/h; 90 mph)
Stalling speed, power off	33 knots (61 km/h; 38 mph)
Max rate of climb at S/L	198 m (650 ft)/min
Service ceiling	3,660 m (12,000 ft)

T-O run 85 m (280 ft)
T-O to 15 m (50 ft) 338 m (1,110 ft)
Landing from 15 m (50 ft) 457 m (1,500 ft)
Landing run 110 m (360 ft)
Range with max internal fuel, 30 min reserves
173 nm (320 km; 200 miles)
Range with tip-tanks 390 nm (720 km; 450 miles)

PFA
THE POPULAR FLYING ASSOCIATION

Terminal Building, Shoreham Airport, Shoreham-by-Sea, Sussex BN4 5FF
Telephone: 079 17 61616

Following the collapse of Phoenix Aircraft Ltd, the PFA assumed responsibility for marketing plans of the Luton L.A.4a Minor. Other types of which plans are obtainable from the Association include the Isaacs Fury, Evans VP-1 and Pazmany PL-4, described under their designers' names in this section of *Jane's*, and the Currie Wot and Druine Turbulent, described below with the L.A.4a Minor.

CURRIE WOT

This aircraft was designed by Mr J. R. Currie in 1937. Two examples were built at Lympne in that year, but were destroyed in a wartime bombing raid. Mr V. H. Bellamy took over the design after the war, at the Hampshire Aeroplane Club, and the first Wot built by members of this club (G-APNT) flew for the first time on 11 September 1958. The second example built at the club (G-APWT) was powered by a 45 kW (60 hp) Walter Mikron four-cylinder in-line engine.

Further Wots have since been completed, including G-ARZW with a 48·5 kW (65 hp) Walter Mikron III engine, built by Dr J. H. B. Urmston.

Dr Urmston purchased all rights in the design from Mr Bellamy, and the following details refer to G-ARZW, built to standard plans, as obtainable from the PFA. Data on the versions with Aeronca-JAP and Mikron engines can be found in the 1961-62 *Jane's*.

Type: Single-seat fully-aerobatic light biplane.

Wings: Braced biplane type, with two parallel interplane struts each side and N-type centre-section support struts. Wing section Clark Y. Dihedral (both wings) 3°. No incidence. Conventional spruce and plywood structure, with fabric covering. Fabric-covered ailerons on lower wings only. No flaps.

Fuselage: All-wood structure. Plywood-box construction, with overall fabric covering.

Tail Unit: Cantilever structure of spruce and plywood, with fabric covering. Fixed-incidence tailplane. Adjustable tab on rudder. Trim tab in port elevator.

Landing Gear: Non-retractable two-wheel type. Rubber-cord shock-absorption. Main wheels fitted with Dunlop tyres, size 400 × 8, pressure 1·24 bars (18 lb/sq in). No brakes.

Power Plant: One 48·5 kW (65 hp) Walter Mikron III four-cylinder inverted in-line engine, driving a two-blade fixed-pitch wooden propeller. Fuel tank aft of firewall, capacity 54·5 litres (12 Imp gallons). Oil capacity 7 litres (1·5 Imp gallons).

Accommodation: Single seat in open cockpit.

Dimensions, external:
Wing span (both) 6·73 m (22 ft 1 in)
Wing chord (both), constant 1·07 m (3 ft 6 in)
Wing area, gross 13·0 m² (140 sq ft)
Wing aspect ratio 6·3
Length overall 5·58 m (18 ft 3½in)
Height overall 2·06 m (6 ft 9 in)
Wheel track 1·38 m (4 ft 6½ in)

Weights:
Weight empty 250 kg (550 lb)
Max T-O weight 408 kg (900 lb)

Performance (at max T-O weight):
Never-exceed speed 112 knots (209 km/h; 130 mph)
Max level speed at 610 m (2,000 ft)
83 knots (153 km/h; 95 mph)
Max cruising speed at 610 m (2,000 ft)
78 knots (145 km/h; 90 mph)
Econ cruising speed at 610 m (2,000 ft))
69 knots (129 km/h; 80 mph)
Stalling speed 35 knots (65 km/h; 40 mph)
Max rate of climb at S/L 183 m (600 ft)/min
Range with max fuel 208 nm (385 km; 240 miles)

DRUINE D31 TURBULENT

The following data apply to the D31 Turbulent as factory-built in the UK for many years by Rollason Aircraft and Engines Ltd. Rollason's version was generally similar to the standard Druine design. Main differences were that it had wheels of slightly greater size and a tailskid instead of a tailwheel, although a tailwheel was available optionally.

The fitting of optional wheel fairings and a sliding canopy increases speed by about 7 knots (13 km/h; 8 mph).

Type: Single-seat light monoplane.

Wings: Cantilever low-wing monoplane. Wing section NACA 23012. Dihedral 4°. Incidence 3° 40′. All-wood two-spar structure, covered with fabric. Built-in leading-edge slot on outer 45% of half-span. Wooden slotted ailerons with fabric covering. No flaps or tabs.

Fuselage: Conventional rectangular four-longeron wood structure with domed decking. Plywood-covered.

Tail Unit: Cantilever wooden structure. Fixed surfaces plywood-covered, movable surfaces fabric-covered. No tabs.

Landing Gear: Non-retractable tailwheel type (optional tailskid). Compression-spring shock-absorbers. Dunlop or Goodyear main wheels and tyres, size 14 × 3, pressure 1·95 bars (28 lb/sq in). Vespa mechanical brakes. Wheel fairings, taxying and parking brakes optional. Skis or floats may be fitted as alternative to wheels.

Power Plant: One 33·5 kW (45 hp) Rollason Ardem 4CO2 Mk IV or 41 kW (55 hp) Ardem Mk V flat-four engine, driving a two-blade fixed-pitch wooden propeller. Fuel tank in fuselage forward of cockpit, capacity 39 litres (8·5 Imp gallons). Oil capacity 2·25 litres (0·5 Imp gallons).

Accommodation: Pilot only, in open cockpit (sliding canopy optional). Baggage locker aft of seat, capacity 11·5 kg (25 lb).

Avionics: Provision for lightweight radio.

Dimensions, external:
Wing span 6·58 m (21 ft 7 in)
Wing chord, constant 1·90 m (3 ft 11 in)
Wing area, gross 7·20 m² (77·5 sq ft)
Wing aspect ratio 5·4
Length overall 5·33 m (17 ft 6 in)
Height overall 1·52 m (5 ft 0 in)
Tailplane span 1·98 m (6 ft 6 in)
Wheel track 1·73 m (5 ft 8 in)
Wheel/tailskid base 3·81 m (12 ft 6 in)

Weights:
Weight empty 179 kg (395 lb)
Max T-O weight 281 kg (620 lb)

Performance (with 33·5 kW; 45 hp engine, at max T-O weight):
Never-exceed speed 108 knots (202 km/h; 125 mph)
Max level speed 95 knots (176 km/h; 109 mph)
Max cruising speed 87 knots (161 km/h; 100 mph)
Econ cruising speed 76 knots (141 km/h; 87 mph)
Stalling speed 39 knots (71 km/h; 44 mph)
Max rate of climb at S/L 137 m (450 ft)/min
Service ceiling 2,740 m (9,000 ft)
T-O run from grass 95 m (310 ft)
T-O to 15 m (50 ft) from grass 125 m (410 ft)
Landing from 15 m (50 ft) on grass 98 m (320 ft)
Landing run on grass 52 m (170 ft)
Range with max fuel, normal allowances
217 nm (400 km; 250 miles)

LUTON L.A.4a MINOR

The first Luton Minor flew in 1936 and proved entirely suitable for construction and operation by amateur builders and pilots. Examples were built pre-war in England and other parts of the world.

In 1960, the design was modernised and restressed completely to the latest British Airworthiness Requirements, allowing for a power increase to 41 kW (55 hp) and a maximum flying weight of 340 kg (750 lb).

Minors are under construction in many parts of the world, and a considerable number of amateur-built examples have been completed and flown successfully since mid-1962. At least one of them, built in Australia by R. A. Pearman and H. Nash, obtained a full Certificate of Airworthiness.

The following description applies to the aircraft in the form previously marketed by Phoenix:

Type: Single-seat light monoplane.

Wings: Strut-braced parasol monoplane. Wing section RAF 48. No dihedral. Wooden two-spar structure in two halves, attached to the fuselage by tubular centre-section pylons and braced by parallel lift struts of streamline-section steel tubing. Wings removable for ground transport and storage. Leading-edge and tips plywood-covered, remainder fabric-covered. Plain ailerons of wood construction, fabric-covered, hinged directly from rear spar. No flaps.

Fuselage: Rectangular all-wood structure. Sides and bottom plywood-covered. Curved decking aft of cockpit fabric-covered.

Tail Unit: Cantilever all-wood structure, fabric-covered. Fixed fin. Aerodynamically-balanced rudder.

Landing Gear: Non-retractable tailwheel type with divided main legs of tubular steel construction. Rubber disc shock-absorbers. Brakes and wheel fairings optional. Fully-castoring tailwheel.

Power Plant: One aircooled engine in the 27·5-41 kW (37-55 hp) range, driving a two-blade fixed-pitch wooden propeller. Fuel tank forward of cockpit, capacity 29·5 litres (6·5 Imp gallons). Provision for additional tanks in wings.

Accommodation: Single seat in open cockpit. Coupé top optional. Baggage space aft of seat.

Dimensions, external:
Wing span 7·62 m (25 ft 0 in)
Wing chord, constant 1·60 m (5 ft 3 in)
Wing area, gross 11·6 m² (125 sq ft)
Wing aspect ratio 5
Length overall 6·32 m (20 ft 9 in)
Height overall 2·29 m (7 ft 6 in)

Weights:
Weight empty 177 kg (390 lb)
Max T-O weight 340 kg (750 lb)

Performance (27·5 kW; 37 hp Aeronca-JAP J.99 engine, at normal T-O weight):
Max level speed at 450 m (1,500 ft)
60 knots (111 km/h; 69 mph)
Normal cruising speed 55 knots (102 km/h; 63 mph)
Stalling speed 25 knots (45 km/h; 28 mph)
Max rate of climb at S/L 76 m (250 ft)/min
T-O run 92 m (300 ft)
Landing run 36·5 m (120 ft)
Range with standard fuel
155 nm (290 km; 180 miles)
Range with auxiliary tanks
340 nm (645 km; 400 miles)

PLUMB
BARRY G. PLUMB

11 Copperbeech Way, Leighton Buzzard, Bedfordshire LU7 8BD
Telephone: 0525 376985

PLUMB BGP1 BIPLANE

In the Summer of 1981 Mr Barry Plumb was nearing the end of construction of a single-seat biplane of original design, with design load factors of +6·5*g* and −3·5*g*. It is to be powered by a Volkswagen motorcar engine, the modification of which has been carried out by Mr D. Burden. A first flight was anticipated during 1982. All available details follow:

Type: Single-seat sporting biplane.

Wings: Single-bay biplane, with I-type interplane struts, centre-section support struts, and a streamline-section bracing strut each side. Wing section N-71, of approx 12% thickness/chord ratio. Dihedral upper wing 0°, lower wing 4°. Incidence 2°. No sweepback. Unequal-span staggered constant-chord wings of conventional two-spar construction. Spruce spars and ribs of marine plywood, with fabric covering. Plain fabric-covered wooden ailerons on upper wings only.

Fuselage: All-wood rectangular-section structure, with spruce longerons, frames and bracing, plywood-covered. Turtleback aft of cockpit faired into fin.

Tail Unit: Wire-braced all-wood structure with fixed-incidence tailplane. Rudder and elevators fabric-covered. Trim tab in elevator.

Landing Gear: Non-retractable tailwheel type, with single wheel on each unit. Main units comprise steel tube legs with shock-absorption by rubber shock cord. Go-kart main wheels with tyres size 11 × 5·00-5, pressure 1·38 bars (20 lb/sq in). Tailwheel of 10 cm (4 in) diameter with solid tyre. Go-kart type brakes. Glassfibre wheel fairings.

Power Plant: One 1,834 cc modified Volkswagen motorcar engine, developing 30 kW (40 hp), and driving a two-blade fixed-pitch wooden propeller with spinner. Fuel tank in forward fuselage, immediately aft of firewall, capacity 36·4 litres (8 Imp gallons), and auxiliary tank in upper wing centre-section, capacity 9 litres (2 Imp gallons); total capacity 45·4 litres (10 Imp gallons). Refuelling point on upper surface of fuselage, forward of windscreen, and on top surface of upper wing.

Accommodation: Single seat in open cockpit. Baggage space in rear turtledeck.

System: Electrical system has a ground rechargeable battery to power radio.

Avionics: Narco Com III 360-channel VHF transceiver.

Dimensions, external:
Wing span, upper 4·88 m (16 ft 0 in)
Wing span, lower 4·27 m (14 ft 0 in)
Wing chord, constant (both) 1·07 m (3 ft 6 in)
Wing area, gross 8·83 m² (95 sq ft)
Wing aspect ratio 4·57
Length overall 4·57 m (15 ft 0 in)
Height overall 1·75 m (5 ft 9 in)
Tailplane span 1·98 m (6 ft 6 in)
Wheel track 1·22 m (4 ft 0 in)
Wheelbase 2·95 m (9 ft 8 in)
Propeller diameter 1·37 m (4 ft 6 in)

Druine D31 Turbulent, described under PFA entry

Currie Wot (Walter Mikron III engine), described under PFA entry

Luton L.A.4a Minor, described under PFA entry

Amateur-built Practavia Sprite *(J. M. G. Gradidge)*

Taylor J.T.1 single-seat fully-aerobatic aircraft built by Mr Ian Parker

Taylor J.T.2 Titch built by Mr Jim Miller of Missouri

WEIGHTS (estimated):

Weight empty	194 kg (428 lb)
Max T-O weight	329 kg (725 lb)

PERFORMANCE (estimated at max T-O weight):

Max level speed at S/L	90 knots (167 km/h; 104 mph)
Max cruising speed	70 knots (130 km/h; 81 mph)
Stalling speed, power off	43 knots (79 km/h; 49 mph)
Range with max fuel, 20% reserves	200 nm (370 km; 230 miles)

PRACTAVIA
PRACTAVIA LTD

Wycombe Air Park, Booker, near Marlow, Buckinghamshire
Telephone: 0494 35342
MANAGING DIRECTOR: C. B. Healey

This company was formed to market plans and kits of a two-seat all-metal aerobatic aircraft known as the Sprite, the design of which was initiated by *Pilot* magazine. It also manufactures a light aircraft engine.

PRACTAVIA SPRITE

Initial design work on the Sprite was started by the staff of *Pilot* magazine in early 1968, after consultation with many experienced light aircraft constructors, and a design and development panel was set up to foster the project. Detailed design began in November 1968 and construction of the first Sprite was started in 1969. Mr Brian Healey is project executive, and Mr Lloyd Jenkinson and Mr Peter Sharman, lecturers at Loughborough University, are the designers.

First flight of a Sprite, built by Mr Peter Burril of Snape, Yorkshire, took place on 16 June 1976. Plans and kits are available for amateur construction, and at least 150 Sprites were reported to be at various stages of completion by early 1981, when about eight were flying.

The Sprite is now available also from Practavia as a ready-to-fly production aircraft, as described in the main UK Aircraft section of this edition. The details given there are applicable also to the version shown on plans available to amateur constructors, with the following exceptions:

POWER PLANT: One 86 kW (115 hp) Lycoming O-235 or 104/112 kW (140/150 hp) Lycoming O-320 flat-four engine, driving a two-blade fixed-pitch propeller. Fuel contained in one fuselage tank, aft of firewall, capacity 54·5 litres (12 Imp gallons), and standard wingtip fuel tanks, capacity 54·5 litres (12 Imp gallons) each. Maximum total capacity 163·5 litres (36 Imp gallons). Oil capacity 4·5 litres (1 Imp gallon)

SYSTEMS: Hydraulic system for brakes only. 12V electrical system.

AVIONICS AND EQUIPMENT: Radio, blind-flying instrumentation and special equipment to individual builder's requirements.

DIMENSIONS, EXTERNAL:

Wing span over tip-tanks	8·23 m (27 ft 0 in)
Wing chord, constant	1·22 m (4 ft 0 in)
Wing area, gross	8·92 m² (96 sq ft)
Wing aspect ratio	6
Length overall	6·10 m (20 ft 0 in)
Width, outer wing panels removed	2·44 m (8 ft 0 in)
Height overall	2·51 m (8 ft 3 in)
Tailplane span	2·44 m (8 ft 0 in)
Wheel track	2·29 m (7 ft 6 in)
Wheelbase	1·40 m (4 ft 7 in)

DIMENSION, INTERNAL:

Cabin: Max width	1·17 m (3 ft 10 in)

WEIGHTS (typical):

Weight empty	454 kg (1,000 lb)
Max T-O weight	658 kg (1,450 lb)

TAYLOR
Mrs JOHN F. TAYLOR

25, Chesterfield Crescent, Leigh-on-Sea, Essex SS9 5PD
Telephone: 0702 521063

The late Mr John F. Taylor, AMIED, an amateur constructor, designed and built the prototype of a single-seat light sporting monoplane, designated J.T.1. His object was to produce the airframe for not more than £100 (at 1959 prices). Construction took about 14 months, and it flew for the first time on 4 July 1959.

A second design by Mr Taylor, the J.T.2 Titch, was awarded second prize in the Midget Racer Design Competition organised by Mr Norman Jones of the Rollason company in 1964. A prototype was built and flown successfully, but crashed on 16 May 1967, killing its designer.

Mrs J. F. Taylor is continuing to market plans of both of these aircraft to amateur constructors.

TAYLOR J.T.1 MONOPLANE

Plans of the Taylor monoplane have been sold to amateur constructors in the United Kingdom and about 20 other countries in all parts of the world. At least 80 J.T.1s are known to be flying in the UK, Canada, USA, Australia and New Zealand.

Aircraft currently flying or under construction are fitted with a variety of engines, including the 30 kW (40 hp) Aeronca E 113, 48·5 kW (65 hp) Continental A65, 48·5 kW (65 hp) Avco Lycoming, 53·7 kW (72 hp) two-stroke McCulloch and the modified Volkswagen series. Aircraft with the 48·5 kW (65 hp) engines have a 10 cm (4 in) longer nose and 25 cm (10 in) longer rear fuselage to maintain the correct CG position.

TYPE: Single-seat fully-aerobatic light monoplane.

WINGS: Cantilever low-wing monoplane. Wing section RAF 48. Constant chord. Dihedral on outer panels 4°. Incidence 3°. Wooden two-spar structure, comprising centre-section and outer panels. Plywood and fabric covering. Differential ailerons. Split trailing-edge flaps. No tabs.

FUSELAGE: Conventional plywood-covered wood structure of four main longerons and curved formers. Centre-section integral with fuselage.

TAIL UNIT: Cantilever fin and fixed-incidence tailplane are plywood-covered wood structures. Elevators and rudder are fabric-covered wood structures.

LANDING GEAR: Non-retractable two-wheel type. Cantilever main legs, with coil spring shock-absorption. Size 5 × 4 (305 mm; 12 in) main wheels, tyre pressure 1·65 bars (24 lb/sq in). Leaf spring tailskid with steerable skid-pad. One example, built by Mr R. Ladd in the USA (see 1972-73 *Jane's*), has manually-actuated inward-retracting main gear.

POWER PLANT (prototype): One 28·3 kW (38 hp) JAP two-cylinder engine, driving a modified Flottorp two-blade wooden fixed-pitch propeller. Fuel tank aft of firewall, capacity 27 litres (6 Imp gallons).

ACCOMMODATION: Single seat under transparent Perspex canopy. Aerobatic harness. Small locker aft of seat.

DIMENSIONS, EXTERNAL:
Wing span 6·40 m (21 ft 0 in)
Wing area, gross 7·06 m² (76 sq ft)
Wing aspect ratio 6
Length overall 4·57 m (15 ft 0 in)
Height over tail 1·47 m (4 ft 10 in)
Tailplane span 1·98 m (6 ft 6 in)
Wheel track 1·52 m (5 ft 0 in)

WEIGHTS:
Weight empty 186 kg (410 lb)
Max T-O weight 276 kg (610 lb)

PERFORMANCE:
Never-exceed speed 113 knots (209 km/h; 130 mph)
Max level speed at S/L 91 knots (169 km/h; 105 mph)
Econ cruising speed 78 knots (145 km/h; 90 mph)
Never-exceed speed with flaps down 56 knots (105 km/h; 65 mph)
Stalling speed, flaps up 40 knots (75 km/h; 46 mph)
Stalling speed, flaps down 33 knots (62 km/h; 38 mph)
Max rate of climb at S/L 305 m (1,000 ft)/min
Range 200 nm (370 km; 230 miles)
g limits +9; −9

TAYLOR J.T.2 TITCH

Construction of the prototype Titch was started in February 1965 and it flew for the first time on 22 January 1967.

Twenty-six Titches are known to have been completed from plans, and flown in the USA, New Zealand, France, Zimbabwe, and England; plans for Monoplanes and Titches have also been supplied to amateur constructors in Alaska, Argentina, Brazil, Denmark, Finland, Iceland, Ireland, Japan, Italy, Kenya, Mexico, Spain and Sweden.

The Titch illustrated was built by Jim Miller of Missouri, USA. Maximum speed is 165 knots (306 km/h; 190 mph) at 2,700 rpm, and cruising speed 130 knots (241 km/h; 150 mph) at 2,600 rpm with a metal propeller. When fitted with a wooden propeller, the rate of climb is stated to be 700 m (2,300 ft)/min. Empty weight is 255 kg (562 lb).

The following description refers to the prototype:

TYPE: Single-seat light monoplane.

WINGS: Cantilever low-wing monoplane. Taylor-modified NACA 23012 wing section. Dihedral 5° on top surface. Incidence 3°. Spruce structure with main box-spar and 'plank' auxiliary spar. Plywood and fabric covering. Plain manually-operated ply-covered flaps over half-span and fabric-covered differential ailerons.

FUSELAGE: All-wood structure, with four main longerons, four secondary longerons and double-curvature ply covering. Aluminium cockpit side panels.

TAIL UNIT: All-wood structure, with fixed-incidence tailplane. Fixed surfaces plywood-covered, control surfaces fabric-covered.

LANDING GEAR: Non-retractable tailwheel type. Steerable tailwheel. Chrome-vanadium compression coil-spring shock-absorbers. Wheels of constructor's manufacture with 4-ply tyres size 5·00-4 and drum brakes.

POWER PLANT: One 63·5 kW (85 hp) Continental C85 flat-four engine, driving a Hegy wooden two-blade scimitar propeller. Glassfibre fuel tank between firewall and instrument panel, capacity 45·5 litres (10 Imp gallons).

ACCOMMODATION: Single seat, with aerobatic harness, under bubble canopy hinged along starboard side.

DIMENSIONS, EXTERNAL:
Wing span 5·72 m (18 ft 9 in)
Wing chord at root 1·37 m (4 ft 6 in)
Wing chord at tip 0·91 m (3 ft 0 in)
Wing area, gross 6·32 m² (68 sq ft)
Wing aspect ratio 5·14
Length overall 4·91 m (16 ft 1½ in)
Height overall 1·42 m (4 ft 8 in)
Tailplane span 1·98 m (6 ft 6 in)
Wheel track 1·52 m (5 ft 0 in)
Propeller diameter 1·52 m (5 ft 0 in)

WEIGHTS (A: 63·5 kW; 85 hp Continental engine, B: Volkswagen engine):
Weight empty: A 227 kg (500 lb)
B 185 kg (410 lb)
Max T-O weight: A 338 kg (745 lb)
B 290 kg (640 lb)

PERFORMANCE (at max T-O weight except where indicated):
Never-exceed speed 195 knots (362 km/h; 225 mph)
Max level speed 174 knots (322 km/h; 200 mph)
Normal cruising speed 135 knots (250 km/h; 155 mph)
Econ cruising speed 95·5 knots (177 km/h; 110 mph)
Best approach speed 65 knots (121 km/h; 75 mph)
Stalling speed, flaps up:
A 50·5 knots (93·5 km/h; 58 mph)
B 46 knots (86 km/h; 53 mph)
Stalling speed, flaps down:
A 43·5 knots (80·5 km/h; 50 mph)
B 40 knots (74 km/h; 46 mph)
T-O speed 54 knots (100 km/h; 62 mph)
Touchdown speed 48 knots (89 km/h; 55 mph)
Max rate of climb at S/L 335 m (1,100 ft)/min

UNITED STATES OF AMERICA

ACE

ACE AIRCRAFT MANUFACTURING CO

106 Arthur Road, Asheville, North Carolina 28806
Telephone: (704) 252 4325
OWNER: Thurman G. Baird
SALES AND PUBLIC RELATIONS MANAGER: V. P. Baird

This company is a successor to the original Corben Aircraft Company, which was established in 1923 and began manufacturing the Baby Ace single-seat light monoplane in kit form in 1931. All rights in the current updated design were purchased by the present owner in 1965. Plans, kits and parts of the Baby Ace and Junior Ace are available to amateur builders from Ace Aircraft.

BABY ACE MODEL D

The prototype of the redesigned Baby Ace Model D flew for the first time on 15 November 1956. Large numbers have since been built by amateurs, some of whom have introduced authorised refinements to the basic design. At least one Baby Ace is flying with a float landing gear, and was illustrated in the 1975-76 *Jane's*. The floats were constructed of glassfibre, using a wood pattern and glassfibre mould, and have spruce struts wrapped in glassfibre, with cast aluminium end fittings. Twin water rudders are fitted for surface handling, and performance is claimed to be comparable with that of a Piper Cub seaplane with 48·5 kW (65 hp) engine.

TYPE: Single-seat light monoplane.

WINGS: Braced parasol monoplane. Wing section Clark Y (modified). Dihedral 1°. Incidence 1°. Fabric-covered two-spar wood structure. Fabric-covered wood ailerons. No flaps.

FUSELAGE: Welded steel tube structure, fabric-covered.

TAIL UNIT: Wire-braced steel tube structure, fabric-covered.

LANDING GEAR: Non-retractable tailwheel type. Combination special tubing and spring shock-absorption. Goodrich 8·00-4 main wheels. Scott hydraulic brakes. Wheel spats optional. Steerable tailwheel. Edo 1140 floats on seaplane version.

POWER PLANT: One Continental A65, A85, C65 or C85 flat-four engine of 48·5 or 63·5 kW (65 or 85 hp), driving a two-blade wooden fixed-pitch propeller. Fuel in tank aft of firewall with capacity of 63·6 litres (16·8 US gallons). Oil capacity 3·8 litres (1 US gallon).

ACCOMMODATION: Single seat in open cockpit. Wide door on starboard side. Space for 4·5 kg (10 lb) baggage.

DIMENSIONS, EXTERNAL:
Wing span 8·05 m (26 ft 5 in)
Wing chord, constant 1·37 m (4 ft 6 in)
Wing area, gross 10·43 m² (112·3 sq ft)
Wing aspect ratio 5·95
Length overall 5·40 m (17 ft 8¾ in)
Height overall 2·02 m (6 ft 7¾ in)
Tailplane span 2·13 m (7 ft 0 in)
Wheel track 1·83 m (6 ft 0 in)
Wheelbase 3·96 m (13 ft 0 in)

WEIGHTS:
Weight empty, equipped 261 kg (575 lb)
Max T-O weight:
48·5 kW (65 hp) 431 kg (950 lb)
63·5 kW (85 hp) landplane or seaplane 522 kg (1,150 lb)

PERFORMANCE (48·5 kW; 65 hp engine, at max T-O weight):
Max level speed at S/L 96 knots (177 km/h; 110 mph)
Max cruising speed 87-91 knots (161-169 km/h; 100-105 mph)
Stalling speed 30 knots (54·7 km/h; 34 mph)
Max rate of climb at S/L 365 m (1,200 ft)/min
Service ceiling 4,875 m (16,000 ft)
T-O run 60 m (200 ft)
Landing run 76 m (250 ft)
Range with max fuel 304 nm (560 km; 350 miles)

JUNIOR ACE MODEL E

The Junior Ace Model E differs from the Baby Ace Model D in being a side-by-side two-seater. It is powered usually by a 63·5 kW (85 hp) Continental C85 flat-four engine, and the details refer to an aircraft with this power plant that was built by Mr Louis C. Seno of Melrose Park, Illinois. First flown on 2 August 1966, it has a cockpit 7·5 cm (3 in) wider and 10 cm (4 in) deeper than that of the standard Model E, a full electrical system and increased fuel capacity of 85 litres (22·5 US gallons).

DIMENSIONS, EXTERNAL:
Wing span 7·92 m (26 ft 0 in)
Wing chord, constant 1·37 m (4 ft 6 in)
Wing aspect ratio 5·95
Length overall 5·50 m (18 ft 0 in)
Height overall 2·00 m (6 ft 7 in)

WEIGHTS:
Weight empty 367 kg (809 lb)
Max T-O weight 606 kg (1,335 lb)

PERFORMANCE:
Max level speed at S/L 113 knots (209 km/h; 130 mph)
Cruising speed 91 knots (169 km/h; 105 mph)
Landing speed 57 knots (105 km/h; 65 mph)
Service ceiling 3,050 m (10,000 ft)
T-O run 122 m (400 ft)
Landing run 183 m (600 ft)
Range with max fuel 304 nm (560 km; 350 miles)

AEROCAR

AEROCAR INC

PO Box 1171, Longview, Washington 98632
Telephone: (206) 423 8260
PRESIDENT AND GENERAL MANAGER: Moulton B. Taylor

In February 1948 Aerocar Inc began developing a flying automobile designed by Mr M. B. Taylor. The prototype Aerocar, with an Avco Lycoming O-290 engine, was completed in October 1949. It was followed by a pre-production Aerocar Model I, with an Avco Lycoming O-320 engine, and this was used for tests which led to FAA airworthiness certification of the Aerocar on 13 December 1956. Four additional Model I Aerocars were completed subsequently for demonstration tours of the United States and for sale to customers, and were followed by the prototype of a refined Model III Aerocar. Details of the development and history of these roadable aircraft can be found in the 1979-80 and earlier editions of *Jane's*.

Aerocar Inc, which is basically a three-person operation, decided that legislation and economics inhibited further development of the Aerocar, and has since devoted most of its activity to other projects, notably the single-seat Mini-Imp, which embodies features of the Aerocar and is suitable for amateur construction.

In 1966 Aerocar built the prototype of a light flying-boat for a private customer. It was followed by two further aircraft, named Coot, on similar lines but provided with tricycle landing gear for amphibious operation. Many sets of plans for this aircraft, now known as the Sooper-Coot, have been sold to amateur constructors.

Aerocar is also devoting much effort to the development of a low-cost and low-power aircraft engine, based on a Kawasaki motorcycle engine, which would be suitable for powering many types of light aircraft of the future. It originally flight tested for over 100 hours a 900 cc Kawasaki engine in a 1937 Taylorcraft. This was then replaced by a 1,000 cc Kawasaki engine, which had completed more than 200 flight hours by early 1980. During 1979 a turbocharger, Maloof three-blade constant-speed metal propeller, separate oil pump, and a Flexidyne dry fluid coupling in the Hy-Vo chain drive system have been incorporated in this power plant. Aerocar was investigating in 1980 the suitability of a solid state ignition system, with electronic spark advance, to facilitate engine starting and provide smoother running.

AEROCAR SOOPER-COOT MODEL A

The prototype of this aircraft flew for the first time in February 1971. It logged approximately 100 hours powered by an 89·5 kW (120 hp) Franklin 225 engine, driving a Sensenich fixed-pitch metal propeller. This was subsequently replaced by a 134 kW (180 hp) Franklin 335 engine, driving a Hartzell constant-speed metal propeller, and the aircraft has been flown extensively since 1972 with this more powerful engine.

The 'float-wing' configuration of the Sooper-Coot permits rough-water operation and, since the close proximity of the wings to the water forms a 'pressure wedge', unusually low take-off and landing speeds are possible without recourse to flaps or other lift-enhancing devices.

The structure is basically of wood, but the tailboom and tail unit can be of steel tube and fabric, wood monocoque or all-metal construction. The rearward-folding wings, of NACA 4415 section, can be folded by one person. The fabric-covered ailerons are of metal construction and statically balanced. There are no tip floats. Construction of the hull, which has only seven bulkheads, is straightforward, without the complication of wheel-well doors. Tailplane and elevators also fold and the elevators have trim tabs. All control surfaces are statically balanced. The tricycle-type landing gear is manually retractable into the wings, but an alternative powered retraction system is shown on the plans.

Baby Ace Model D (Continental A65 engine) built by Capt Lloyd Windh of Air Canada *(Neil A. Macdougall)*

Side-by-side two-seat Junior Ace Model E *(J. M. G. Gradidge)*

Aerocar Mini-Imp Model C built by Mr Pat Hart of Vancouver, Washington *(J. M. G. Gradidge)*

Aerocar Sooper-Coot Model A built by Mr Ed White of Seattle

Recommended power plant for the Sooper-Coot is a Franklin flat-six engine of either 134 or 164 kW (180 or 220 hp). However, many builders of the Sooper-Coot are fitting a 157 kW (210 hp) Continental IO-360. To minimise expenditure, others are fitting Avco Lycoming O-320 and O-320-B engines of 119 kW (160 hp). These are reported to be satisfactory if only a limited weight of optional equipment is installed in the aircraft. However, a constant-speed propeller has proved essential to maintain reasonable performance with the O-320/320-B. Maximum fuel capacity is 189 litres (50 US gallons).

Certain component parts (including the glassfibre engine cowls, glassfibre hull shell, foredeck, instrument panel, tail fairings, engine cooling-fan blades and spring steel main landing gear legs), and plans, are available to amateur constructors. The company also maintains a list of recommended suppliers of welded assemblies and machined components. More than 400 Sooper-Coots were under construction in 1980, and by 1 January 1981 more than 75 Coot-As and one twin-boom Coot-B were known to be flying in the USA and other countries.

DIMENSIONS, EXTERNAL:
- Wing span — 10·97 m (36 ft 0 in)
- Wing chord, constant — 1·52 m (5 ft 0 in)
- Wing area, gross — 16·72 m² (180 sq ft)
- Length overall — 6·10 to 6·71 m (20 ft 0 in to 22 ft 0 in)
- Height overall — 2·44 m (8 ft 0 in)
- Width folded — 2·44 m (8 ft 0 in)
- Tailplane span — 3·04 m (10 ft 0 in)

WEIGHTS:
- Weight empty — 499 kg (1,100 lb)
- Max T-O weight — 884 kg (1,950 lb)

PERFORMANCE (134 kW; 180 hp engine, at max T-O weight):
- Never-exceed speed — 120 knots (223 km/h; 139 mph)
- Max cruising speed — 113 knots (209 km/h; 130 mph)
- Econ cruising speed at 50% power — 95·5 knots (177 km/h; 110 mph)
- Max rate of climb at S/L — 381 m (1,250 ft)/min
- T-O run (land) — 61 m (200 ft)
- T-O (water) — 6-8 s

AEROCAR IMP

Design of the Imp (an acronym derived from Independently Made Plane) began in January 1972. Many design features of the Aerocar were embodied, including folding wings and a pusher propeller aft of the tail unit. All-metal construction, using pop-rivets, bolt-together assemblies and limited welding, was specified to reduce building time.

Construction of a two-seat prototype Imp began in August 1972, but completion of the test programme was delayed by the need for further investigation into many aspects of the design. Subsequent non-availability of the Franklin engine which was intended to power the aircraft led to the decision to give priority instead to the single-seat Mini-Imp (which see).

More recently, it was decided to resume the Imp programme by converting the original prototype to two-plus-two configuration, with the pilot and one passenger seated side by side at the front of the cabin and two folding seats at the rear for children. The necessary new skins for the modification had been built by early 1977. However, the aircraft was transferred to another builder in St Louis, Missouri, in 1978, and he is continuing the conversion. The modified prototype will not incorporate all changes shown on the current Imp information sheet supplied by Aerocar, such as use of the GA(PC)-1 aerofoil section, Kawasaki engine and a lengthened fuselage. Therefore, the details below refer to subsequent Imps and not the prototype.

The future of the Imp programme will be determined largely by the outcome of Aerocar's work to evolve a low-cost and low-power aero-engine from the Kawasaki motorcycle engine, but it was reported at the beginning of 1981 that development of this engine was progressing very well. Two 1,000 cc 74·5 kW (100 hp) four-cylinder engines of this type might be geared to drive a single propeller, with centreline thrust and 'either/or' engine drive capability, via a chain drive.

TYPE: Four-seat light aircraft.

WINGS: Cantilever high-wing monoplane. Wing section GA(PC)-1. No dihedral. Incidence 4°. No sweepback. All-metal structure of constant chord. Wings fold aft, alongside fuselage, for towing or storage. All-metal ailerons, each with trim tab. No flaps.

FUSELAGE: All-metal structure with glassfibre shell.

TAIL UNIT: Inverted V type, of all-metal construction, with variable-incidence tailplane surfaces.

LANDING GEAR: Electrically-retractable tricycle type, main units retracting outward. Special highway-type wheels and 6-ply tyres suitable for road towing. Tyre pressure 2·76 bars (40 lb/sq in). Rosenhan wheel brakes. Nosewheel retracts independently of main wheels for road towing.

POWER PLANT: Prototype may be powered by the existing 93 kW (125 hp) Franklin engine or a new 149 kW (200 hp) Franklin engine. Plans show installation of the Kawasaki twin-engine arrangement. One fuel tank of 53 litres (14 US gallons) capacity in each wing root. Total fuel capacity 106 litres (28 US gallons). Refuelling points in upper surface of wings, inboard of wing fold. Oil capacity 7·5 litres (2 US gallons).

ACCOMMODATION: Two seats, side by side, at front of enclosed cabin, with two folding seats at rear. Large baggage space if rear seats folded against bulkhead. Airstair doors.

SYSTEM: Electrical system powered by 12V 60A engine-driven alternator.

AVIONICS: Narco Escort 110 radio and Type 50A transponder.

DIMENSIONS, EXTERNAL:
- Wing span — 8·84 m (29 ft 0 in)
- Wing chord, constant — 1·22 m (4 ft 0 in)
- Wing area, gross — 10·41 m² (112 sq ft)
- Wing aspect ratio — 7
- Length overall — 6·71 m (22 ft 0 in)
- Width, wings folded — 2·13 m (7 ft 0 in)
- Height overall (propeller blades horizontal) — 1·68 m (5 ft 6 in)
- Tailplane span — 2·44 m (8 ft 0 in)
- Wheel track — 1·78 m (5 ft 10 in)
- Wheelbase — 2·67 m (8 ft 9 in)
- Propeller diameter — 1·83 m (6 ft 0 in)

DIMENSION, INTERNAL:
- Cockpit:
  - Max width — 1·12 m (3 ft 8 in)

WEIGHTS (estimated):
- Weight empty — 499 kg (1,100 lb)
- Max T-O and landing weight — 703 kg (1,550 lb)

PERFORMANCE (estimated):
- Max cruising speed — more than 130 knots (241 km/h; 150 mph)
- Stalling speed — 44 knots (81 km/h; 50 mph)
- Max rate of climb at S/L — 244 m (800 ft)/min

AEROCAR MINI-IMP

The Mini-Imp is a single-seat version of the Aerocar Imp, to which it is generally similar. The basic structure, which is stressed to $\pm 9g$, is all metal and is assembled with bolts and pop-rivets, thereby eliminating the need for welding skill. It is claimed that only a bench drill, sander and bandsaw are needed for home construction of the Mini-Imp, in addition to the usual hand tools and a pop-riveter. Average time taken for construction is 700 working hours, at a cost of $8,000, excluding engine, instruments and avionics.

By early 1980 the Mini-Imp prototype had completed about 250 flying hours, powered by a 51 kW (68 hp) Limbach 1,900 cc flat-four engine driving a wooden propeller. During the Winter of 1975/76 Aerocar built a folding wing for this aircraft, which folds in a similar way to that of the Imp. This permits the builder to construct the aircraft in small areas, and is a standard feature on the plans for the aircraft.

All drawings for the Mini-Imp have been available since 1978, together with parts kits. More than 200 amateur-built examples were being constructed during 1980 and at least five Mini-Imps were flying by 1 January 1981. Recommended power plant is the turbocharged Revmaster 2100D, driving a Maloof metal constant-speed propeller; Aerocar has built a prototype fitted with this propeller, and this is 30 cm (12 in) longer overall than the original prototype. Individual builders are fitting a variety of engines, including the 74·5 kW (100 hp) Continental O-200, 86 kW (115 hp) Avco Lycoming O-235, and two-cylinder 44·5 kW (60 hp) Franklin. Another Aerocar-built Mini-Imp will be fitted with the turbocharged Kawasaki engine and Maloof propeller, after testing in the Taylorcraft has been completed.

An accompanying illustration shows the Mini-Imp Model C constructed by Mr Pat Hart of Vancouver, Washington. This is powered by a 74·5 kW (100 hp) Continental O-200 flat-four engine driving a two-blade fixed-pitch propeller. It differs from standard plans by having a 0·30 m (1 ft 0 in) extension aft of the pilot's seat to increase baggage space, the addition of a small dorsal fin on the aft fuselage, and hydraulic retraction of the landing gear, using an electrically-powered hydraulic pump. Maximum level speed of Mr Hart's Mini-Imp is close on 174 knots (322 km/h; 200 mph) at 2,135 m (7,000 ft); cruising speed 143 knots (266 km/h; 165 mph) IAS at 1,220 m (4,000 ft); and the maximum rate of climb is 457 m (1,500 ft)/min.

Early in 1981, *Jane's* learned that Aerocar had received an enquiry regarding the possibility of installing two high rate of fire machine-guns in the Mini-Imp to provide a military capability. Design studies have shown that such weapons could be installed beneath the cockpit, with room in the baggage compartment for a considerable quantity of ammunition. The company has suggested that the nose of a military Mini-Imp could be skinned with Kevlar, as used for bulletproof clothing, to provide some protection against small-arms fire from the ground.

The description of the Imp applies also to the Mini-Imp, except as follows:

TYPE: Single-seat light sporting aircraft.

WINGS: As for Imp. Wing pivots 90° to a fore and aft position for towing by motor car.

FUSELAGE, TAIL UNIT AND LANDING GEAR: As for Imp.

POWER PLANT: Recommended engine is the turbocharged Revmaster 2100D, driving a Maloof constant-speed metal propeller via the standard extended driveshaft incorporating a Flexidyne dry fluid coupling. Alternative engines in the 44·5-74·5 kW (60-100 hp) range may be installed. Fuel tank in wing centre-section, capacity 45·4 litres (12 US gallons), plus tip-tanks, each of 34 litres (9 US gallons) capacity. Refuelling point on wing upper surface.

ACCOMMODATION: Single semi-reclining moulded glassfibre bucket seat beneath sideways-opening (to port) transparent canopy. Seat folds forward to provide access to space for baggage.

DIMENSIONS, EXTERNAL:

Wing span	7·77 m (25 ft 6 in)
Wing chord, constant	0·91 m (3 ft 0 in)
Length overall	4·88 m (16 ft 0 in)
Height over fuselage	1·24 m (4 ft 1 in)
Height over propeller	1·89 m (6 ft 2½ in)
Tailplane span	1·98 m (6 ft 6 in)
Propeller diameter	1·45 m (4 ft 9 in)

WEIGHTS:

Weight empty	236 kg (520 lb)
Max T-O weight	362 kg (800 lb)

PERFORMANCE (Limbach engine, except where indicated):

Never-exceed speed	260 knots (482 km/h; 300 mph)
Max cruising speed at S/L	130 knots (241 km/h; 150 mph)
Max cruising speed (Revmaster engine and Maloof propeller, estimated)	over 173 knots (322 km/h; 200 mph)
Stalling speed	43·5 knots (80·5 km/h; 50 mph)
Max rate of climb at S/L	366 m (1,200 ft)/min
T-O run	183 m (600 ft)
Range	over 434 nm (804 km; 500 miles)

AEROCAR MICRO-IMP

The Micro-Imp is of similar configuration to the Mini-Imp, but is constructed primarily of glassfibre-reinforced paper known as TPG (Taylor Paper Glass). Static testing has shown this to be far stronger than originally estimated, resulting in the redesign of many components to reduce weight. As an example, a main wing spar of TPG, weighing only 6 kg (13 lb), has been statically tested satisfactorily at a load of 589·5 kg (1,300 lb). Design had begun in August 1978, and construction of the prototype was started in December 1978. It was anticipated that the Micro-Imp with Citroen 2CV type engine would make its first flight during the Summer of 1981.

TYPE: Single-seat light sporting aircraft.

WINGS: Cantilever high-wing monoplane. Wing section NASA GA(PC)-1. Thickness/chord ratio 15%. No anhedral or dihedral. Incidence 3°. Glassfibre-reinforced paper structure, covered with rip-stop Dacron. Full-span constant-chord ailerons of similar construction to wings. Ailerons droop for flap function, and can be set in slight 'up' position for efficient cruise.

FUSELAGE: Glassfibre-reinforced paper structure. Non-structural and removable nosecone of moulded glassfibre.

TAIL UNIT: Inverted V type. Glassfibre-reinforced paper structure, covered with rip-stop Dacron. Variable-incidence tailplane surfaces.

LANDING GEAR: Manually-retractable tricycle type. Main units retract vertically. Cantilever spring steel tube main legs. Three plastics wheels, diameter 25·4 cm (10 in). Tyres size 3·00-4, pressure 3·45 bars (50 lb/sq in). Aerocar mechanical brakes.

POWER PLANT: One 18·6 kW (25 hp) modified and derated Citroen 2CV motor car engine, driving a tail-mounted Hendrickson two-blade wooden pusher propeller or a Maloof two-blade two-position propeller. One fuel tank aft of seat, capacity 26·5 litres (7 US gallons). Refuelling point on top of fuselage. NASA flush air intake ducts, plus cooling fan.

ACCOMMODATION: Single semi-reclining seat under sideways-hinged (to port) transparent canopy. Baggage space above engine.

AVIONICS: Radair 10A 10-channel radio, operated by nickel-cadmium battery.

DIMENSIONS, EXTERNAL:

Wing span	8·23 m (27 ft 0 in)
Wing chord, constant	0·91 m (3 ft 0 in)
Wing aspect ratio	9
Length overall	4·57 m (15 ft 0 in)
Max width of fuselage	0·69 m (2 ft 3 in)
Height overall	1·22 m (4 ft 0 in)
Tailplane span	2·13 m (7 ft 0 in)
Wheel track	1·22 m (4 ft 0 in)
Wheelbase	1·37 m (4 ft 6 in)
Propeller diameter	1·07 m (3 ft 6 in)

WEIGHTS:

Weight empty	118 kg (260 lb)
Max T-O weight	238 kg (525 lb)

PERFORMANCE (estimated):

Never-exceed speed	120 knots (223 km/h; 139 mph)
Max level speed	104 knots (193 km/h; 120 mph)
Max cruising speed	91 knots (169 km/h; 105 mph)
Econ cruising speed	87 knots (161 km/h; 100 mph)
Stalling speed	42 knots (78 km/h; 48 mph)
Max rate of climb at S/L	152 m (500 ft)/min
Service ceiling	3,050 m (10,000 ft)
T-O run	152 m (500 ft)
T-O to 15 m (50 ft)	290 m (950 ft)
Landing from 15 m (50 ft)	274 m (900 ft)
Landing run	122 m (400 ft)

AEROSPORT
AEROSPORT INC

Box 278, Holly Springs, NC 27540
Telephone: (919) 552 6375
PRESIDENT: E. B. Trent

Aerosport Inc was formed in 1970 by the late Mr H. L. Woods, former chief engineer of Bensen Aircraft Corporation, who designed and built a number of aircraft and air cushion vehicles, including flex-wing and rotating-wing aircraft, and a variety of gliders.

Since that time, the company has marketed plans and kits of a succession of light aircraft designed for construction by amateurs. Details of those which are still available follow. Construction plans for the Woody Pusher (see 1978-79 *Jane's*) are no longer available from Aerosport.

Aerosport also produces kits for a special agricultural cropspraying version of its single-seat Scamp, known as Scamp B. These kits are delivered to a company based in Colombia, South America, for commercial assembly and sale as completed aircraft. Details can be found in the main Aircraft section of this edition.

AEROSPORT RAIL Mk II

Plans and kits for this single-seat light aircraft are no longer being marketed. Full details can be found in the 1977-78 *Jane's*.

A new version was expected to be announced in 1980, but no details are currently available.

AEROSPORT QUAIL

Design of the Quail began in January 1970, and construction of the prototype was started in July 1971. It flew for the first time in December 1971, with an all-moving tailplane; but the designers incorporated a fixed-incidence tailplane with elevators before plans and construction kits were made available to amateur constructors. More than 375 sets of plans had been sold by 1981. Approximately 26 Quails are under construction and about 10 are flying.

TYPE: Single-seat lightweight cabin monoplane.

WINGS: Cantilever high-wing monoplane. Wing section NACA 23015. No sweepback. All-metal two-spar structure of 2024-T3 and 6061-T6 light alloy. Plain ailerons of light alloy construction with piano hinge at upper surface. Trailing-edge flaps. Endplates optional.

FUSELAGE: Semi-monocoque all-metal structure of 2024-T3 light alloy.

TAIL UNIT: Cantilever all-metal structure of light alloy, with swept vertical surfaces. Fixed-incidence tailplane. Trim tab on rudder and elevator of prototype; optional for later aircraft.

LANDING GEAR: Non-retractable tricycle type. Cantilever spring main gear struts of light alloy. Nosewheel shock-absorption by rubber in compression. Main wheels and tyres size 5·30 × 4·50-6; nosewheel and tyre size 2·80 × 2·50-4. Tyre pressure 1·38 bars (20 lb/sq in). Barrel-type wheel brakes. Nosewheel steerable. Optional wheel fairings.

POWER PLANT: One 1,600 cc modified Volkswagen motor-car engine, driving a 54-36 two-blade fixed-pitch wooden propeller. Provision for installation of Volkswagen engines from 1,500 cc to 1,800 cc capacity. Two fuel tanks in wings, and 7·5 litre (2 US gallon) meter tank; total capacity 38 litres (10 US gallons). Refuelling point in top of each wing.

ACCOMMODATION: Single seat in enclosed cabin, which is heated. Door on starboard side is hinged at top and opens upwards. Stowage for 9 kg (20 lb) baggage.

DIMENSIONS, EXTERNAL:

Wing span	7·32 m (24 ft 0 in)
Wing chord, constant	1·07 m (3 ft 6 in)
Wing area, gross	7·8 m² (84 sq ft)
Wing aspect ratio	6·87
Length overall	4·85 m (15 ft 11 in)
Height overall	1·69 m (5 ft 6½ in)
Tailplane span	1·83 m (6 ft 0 in)
Wheel track	1·52 m (5 ft 0 in)
Wheelbase	1·22 m (4 ft 0 in)
Propeller diameter	1·37 m (4 ft 6 in)

WEIGHTS:

Weight empty	242 kg (534 lb)
Normal T-O weight	345 kg (762 lb)
Max T-O weight	359 kg (792 lb)

PERFORMANCE (at max T-O weight):

Max level speed at S/L	113 knots (209 km/h; 130 mph)
Max cruising speed	100 knots (185 km/h; 115 mph)
Econ cruising speed	95·5 knots (177 km/h; 110 mph)
Max manoeuvring speed	78 knots (145 km/h; 90 mph)
Stalling speed, flaps down	42 knots (78 km/h; 48 mph)
Max rate of climb at S/L	259 m (850 ft)/min
Service ceiling (estimated)	3,660 m (12,000 ft)
T-O run	91 m (300 ft)
Landing run	122 m (400 ft)
Range with max fuel, no reserves	200 nm (370 km; 230 miles)

AEROSPORT SCAMP A

The prototype of the single-seat all-metal Scamp A flew for the first time on 21 August 1973. It was intended primarily for operation from grass strips, and tricycle landing gear was chosen as being more rational for a generation of amateur pilots who had received their initial flight training on aircraft equipped with landing gear of this configuration. Stressed to $+6g$ and $-3g$, the Scamp A can be used for limited aerobatics; and emphasis has been placed on simple construction techniques to make it an easy project for the homebuilder.

Plans and kits of parts, except for the engine, are available to amateur constructors. More than 640 sets of plans had been sold by 1981. At least 12 Scamps are flying and approximately 30 more are under construction from Aerosport kits.

An agricultural cropspraying version, known as Scamp B, is assembled in South America from Aerosport kits and marketed commercially (see under Colombia in main Aircraft section).

TYPE: Single-seat light aircraft.

WINGS: Braced biplane structure, with V-type interplane strut each side. Flying and landing wires of streamline section. Single 5 cm by 12·5 cm (2 in by 5 in) extruded section of 6061-T6 light alloy forms a pylon to support the centre-section of the upper wing. Dihedral 3° on lower wings only. All-metal two-spar structures of light alloy. Plain ailerons of light alloy construction, with piano hinge at upper surface, on upper wing only. No flaps. No trim tabs.

FUSELAGE: Semi-monocoque all-metal structure of light alloy.

TAIL UNIT: Braced T tail of light alloy construction. Single bracing strut on each side. Fixed-incidence tailplane. Ground-adjustable trim tab on rudder.

LANDING GEAR: Non-retractable tricycle type. Cantilever spring main-gear struts of light alloy. Wheel fairing on each unit optional.

POWER PLANT: Prototype has one 44·5 kW (60 hp) 1,834 cc Volkswagen modified motorcar engine. Design suitable for Volkswagen engines of 1,700 cc to 2,100 cc, driving a 56-38 two-blade fixed-pitch wooden propeller. Fuel tank in fuselage nose, aft of firewall, capacity 30·5 litres (8 US gallons). Refuelling point on fuselage upper surface, forward of windscreen.

ACCOMMODATION: Single seat in open cockpit.

DIMENSIONS, EXTERNAL:

Wing span	5·33 m (17 ft 6 in)
Wing chord, constant	0·91 m (3 ft 0 in)
Wing area, gross	9·52 m² (102·5 sq ft)
Length overall	4·27 m (14 ft 0 in)
Height overall	1·69 m (5 ft 6½ in)
Wheel track	1·52 m (5 ft 0 in)
Wheelbase	1·17 m (3 ft 10 in)

WEIGHTS (prototype):

Weight empty	236 kg (520 lb)
Max T-O weight	348 kg (768 lb)

PERFORMANCE (prototype):

Never-exceed speed	108 knots (201 km/h; 125 mph)
Max level speed	82 knots (153 km/h; 95 mph)
Cruising speed	74 knots (137 km/h; 85 mph)
Max manoeuvring speed	72 knots (134 km/h; 83 mph)
Stalling speed	39 knots (73 km/h; 45 mph)
Service ceiling (estimated)	3,660 m (12,000 ft)
T-O run	91 m (300 ft)
Landing run	122 m (400 ft)
Range at cruising speed	130 nm (241 km; 150 miles)

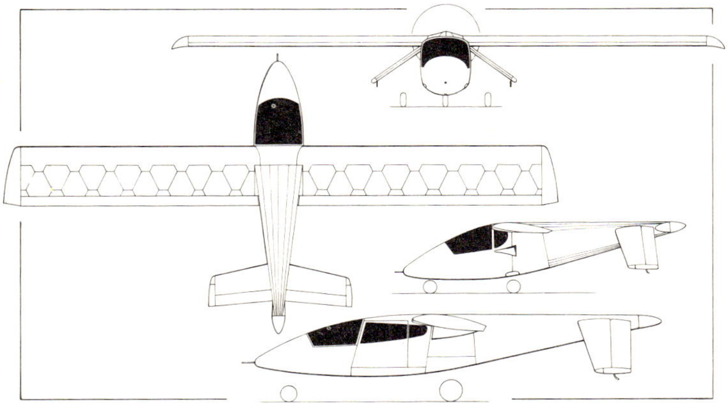

Aerocar Micro-Imp, with additional side elevation (bottom) of four-seat Imp *(Michael A. Badrocke)*

Aerosport Quail single-seat lightweight cabin monoplane

Aerosport Scamp A single-seat biplane *(Peter M. Bowers)*

Aero-Tech Boeing F4B/P-12 four-fifths scale replica

Bede BD-8 as completed by Mr Mike Huffman of Airplane Alley

AERO-TECH
AERO-TECH

Route 1, PO Box 27A, Hastings, Florida 32045
Telephone: (904) 692 1639
PRESIDENT: Alvin J. Jarvis

AERO-TECH BOEING F4B/P-12 REPLICA

Aero-Tech has designed and built a four-fifths scale replica of the Boeing F4B/P-12 biplane fighter, of which plans are available to amateur constructors. Design was initiated in June 1973, with construction beginning in the following month, and the prototype flew for the first time during August 1978. A total of 15 sets of plans had been sold by 1 February 1981.

TYPE: Single-seat sporting biplane.

WINGS: Braced single-bay biplane with N-type interplane and cabane struts. Aerofoil section NACA 64_1-212. Dihedral 0° upper wing, 2° lower wings. Incidence 0° all wings. No sweepback. Conventional two-spar structure of wood, with spruce spars, mahogany plywood ribs, birch plywood leading-edges, and Dacron fabric covering. Plain ailerons of plywood-covered wooden construction. No flaps. No trim tabs.

FUSELAGE: Composite structure with forward fuselage of welded steel tube and aft fuselage a wooden Warren truss. Fuselage forward of lower wings and upper fuselage decking covered by light alloy, remainder of fuselage part plywood and more extensively fabric-covered. Light alloy top-decking.

TAIL UNIT: Braced wooden structure with spruce spars, plywood ribs, plywood skins and polyurethane foam tips. Variable-incidence tailplane with electrical actuation. No trim tabs.

LANDING GEAR: Non-retractable tailwheel type. Main wheels on through axle, with shock absorption by spring in compression. Main wheels from Sunbeam Alpine motor car, with Goodyear tyres of 0·61 m (24 in) diameter, pressure 2·07 bars (30 lb/sq in). Castoring tailwheel with 20 cm (8 in) diameter solid tyre. Sunbeam Girling brakes.

POWER PLANT (prototype): One 182·7 kW (245 hp) Jacobs R-755-9 radial engine, driving a Hamilton Standard two-blade constant-speed metal propeller. Installation suitable for engines of 164-224 kW (220-300 hp). Fuselage fuel tank, immediately aft of firewall, capacity 151·5 litres (40 US gallons). Refuelling point on upper surface of forward fuselage. Oil capacity 23 litres (6 US gallons).

ACCOMMODATION: Single seat in open cockpit.

SYSTEM: Electrical system includes a 24V engine-driven generator.

DIMENSIONS, EXTERNAL:

Wing span, upper	7·32 m (24 ft 0 in)
Wing span, lower	6·40 m (21 ft 0 in)
Wing chord (constant), upper	1·27 m (4 ft 2 in)
Wing chord (constant), lower	1·02 m (3 ft 4 in)
Wing area, gross	14·99 m² (161·4 sq ft)
Wing aspect ratio (monoplane equivalent)	4·6
Length overall	5·41 m (17 ft 9 in)
Height overall	2·34 m (7 ft 8 in)
Tailplane span	2·95 m (9 ft 8 in)
Wheel track	1·83 m (6 ft 0 in)
Wheelbase	3·91 m (12 ft 10 in)
Propeller diameter	2·39 m (7 ft 10 in)

WEIGHTS:

Operating weight empty	694 kg (1,530 lb)
Max T-O weight	952·5 kg (2,100 lb)

PERFORMANCE (at max T-O weight):

Never-exceed speed	161 knots (298 km/h; 185 mph)
Max level speed at 1,067 m (3,500 ft)	143 knots (265 km/h; 165 mph) IAS
Max cruising speed at 1,067 m (3,500 ft)	117 knots (217 km/h; 135 mph) IAS
Stalling speed, power off	54 knots (100 km/h; 62·5 mph)
Max rate of climb at S/L	700 m (2,300 ft)/min
Range with max fuel, 45 min reserves	335 nm (621 km; 386 miles)

AIRPLANE ALLEY

Route 2, PO Box 1490, Owasso, Oklahoma 74055
Telephone: (918) 272 2775

BEDE-HUFFMAN BD-8

Mr Mike Huffman, who operates an aircraft building and restoration business under the above name, acquired the unfinished Bede BD-8 prototype and the only other incomplete BD-8, which was being built originally by Delmar Hostetler. For speed of construction, components and assemblies from both of these airframes have been combined to produce, after considerable additional construction by Mr Huffman, N88DH which was flown for the first time on 14 May 1980. It is planned to introduce refinements to the design, and to complete a second example from the somewhat less advanced assemblies that were left in abeyance during construction of N88DH.

TYPE: Single-seat aerobatic monoplane.

WINGS: Cantilever low-wing monoplane. Aerofoil section NACA 0012. Dihedral 2° 30′. Incidence 3°. No sweepback. Light alloy tubular spar with glassfibre 'panel ribs' and light alloy skins. Statically balanced plain ailerons of light alloy construction.

FUSELAGE: Light alloy box structure of angle section and gussets bolted together. Skins of light alloy.

TAIL UNIT: Cantilever structure incorporating light alloy spars, ribs and skins. All-moving tailplane, with anti-servo and trim tab.

LANDING GEAR: Non-retractable tailwheel type. Main units each consist of a glassfibre leaf spring, carrying a Cleveland wheel with a Goodyear 5·00-5 tyre, pressure 2·07 bars (30 lb/sq in). Tailwheel of 20 cm (8 in) diameter. Wheel fairings to be installed at a later date. Cleveland disc brakes.

POWER PLANT: One 149 kW (200 hp) Avco Lycoming IO-360-A1A flat-four engine, driving a Hartzell Type HC-C2YK-4F two-blade constant-speed propeller with spinner. Fuel tank in each wing, and in fuselage aft of firewall, with combined capacity of 170 litres (45 US gallons). Refuelling point on each wing and on upper surface of forward fuselage. Oil capacity 7·6 litres (2 US gallons).

ACCOMMODATION: Single seat beneath transparent canopy which opens upward and aft. Accommodation is heated.

SYSTEM: Electrical system includes 12V 60A engine-driven alternator and 12V battery.

DIMENSIONS, EXTERNAL:

Wing span	5·94 m (19 ft 6 in)
Wing chord, constant	1·52 m (5 ft 0 in)
Wing area, gross	9·01 m² (97·0 sq ft)
Wing aspect ratio	3·9
Length overall	5·31 m (17 ft 5 in)
Height overall	2·06 m (6 ft 9 in)
Tailplane span	2·44 m (8 ft 0 in)
Wheel track	2·13 m (7 ft 0 in)
Wheelbase	4·27 m (14 ft 0 in)
Propeller diameter	1·88 m (6 ft 2 in)

WEIGHTS:

Weight empty	442 kg (975 lb)
Max T-O weight	685 kg (1,510 lb)

PERFORMANCE (at max T-O weight):

Max level speed at S/L	187 knots (346 km/h; 215 mph)
Max cruising speed at 1,830 m (6,000 ft)	165 knots (306 km/h; 190 mph)
Stalling speed, power off	56·5 knots (105 km/h; 65 mph)
Max rate of climb at S/L	over 610 m (2,000 ft)/min
T-O run	152 m (500 ft)
T-O to 15 m (50 ft)	183 m (600 ft)
Landing from 15 m (50 ft)	457 m (1,500 ft)

ANDERSON

ANDERSON AIRCRAFT CORPORATION

c/o E. K. Morice Jr, PO Box 2048, Delray Beach, Florida 33444
Telephone: (305) 276 5207

Mr Earl Anderson, a Boeing 747 captain flying for Pan American World Airways, designed and built an original light amphibian which he named EA-1 Kingfisher. The project occupied a period of nine years from start of design to completion, and the first flight was made on 24 April 1969.

After a time, Mr Anderson replaced the original 74·5 kW (100 hp) Continental O-200 engine by an 86 kW (115 hp) Avco Lycoming O-235-C1, driving a Sensenich M76AM-4-44 propeller. With this power plant the Kingfisher has an empty weight of 495 kg (1,092 lb), a max T-O weight of 725 kg (1,600 lb) and improved performance. By January 1975 the prototype had accumulated a total of more than 600 flying hours.

Plans available to amateur constructors cover the increase in weight. The Kingfisher was designed originally to accept alternative power plants up to a maximum of 104 kW (140 hp), but on the basis of experience with the 86 kW (115 hp) Avco Lycoming engine, Mr Anderson is discouraging homebuilders from installing more powerful engines than this.

Mr Anderson formed Anderson Aircraft Corporation to market plans of the Kingfisher. By early 1981 well over 200 sets of plans had been sold, and more than 100 Kingfishers were under construction in the USA, Canada, Mexico, Sweden, West Germany and Panama. At least 10 homebuilt Kingfishers are known to be flying.

The following details apply to Mr Anderson's Kingfisher in its original configuration:

ANDERSON EA-1 KINGFISHER

TYPE: Two-seat light amphibian.

WINGS: Braced high-wing monoplane with streamline-section V bracing struts each side (standard J3, PA-11 or PA-12 Piper Cub wing). Stabilising floats mounted beneath wings, adjacent to wingtips, are constructed of ⅜ in square mahogany stringers, covered with $^1/_{16}$ in mahogany plywood coated with glassfibre. Each float weighs 4·1 kg (9 lb).

FUSELAGE: Conventional flying-boat hull of wooden construction with spruce frames and longerons, covered with $^1/_{16}$ in and ¼ in mahogany plywood coated with glassfibre.

TAIL UNIT: Conventional strut-braced tail unit of steel tubing, fabric covered.

LANDING GEAR: Retractable tailwheel type. Each main unit is retracted forward, manually and individually, with spring-loaded assist mechanism. Bungee shock-absorption.

POWER PLANT: One 74·5 kW (100 hp) Continental O-200 flat-four engine, driving a fixed-pitch two-blade tractor propeller. Alternatively an Avco Lycoming O-235-C1 of 86 kW (115 hp) or other engine of up to 112 kW (150 hp) can be fitted. Single fuel tank in hull, immediately forward of windscreen, capacity 76 litres (20 US gallons). Refuelling point on nose. Oil capacity 5·7 litres (1·5 US gallons).

ACCOMMODATION: Two seats, side by side, in enclosed cabin. Piper Tri-Pacer windscreen. Door on port side.

SYSTEM: 30A alternator for radios, lights and starter.

DIMENSIONS, EXTERNAL:

Wing span	11·00 m (36 ft 1 in)
Length overall	7·16 m (23 ft 6 in)
Height overall	2·44 m (8 ft 0 in)
Wheel track	1·52 m (5 ft 0 in)
Propeller diameter	1·83 m (6 ft 0 in)

WEIGHTS:

Weight empty	468 kg (1,032 lb)
Max T-O weight	680 kg (1,500 lb)

PERFORMANCE (at max T-O weight):

Never-exceed speed	104 knots (193 km/h; 120 mph)
Max level speed	104 knots (193 km/h; 120 mph)
Cruising speed at 305 m (1,000 ft)	74 knots (136 km/h; 85 mph)
Stalling speed	39 knots (72·5 km/h; 45 mph)
Max rate of climb at S/L	152-183 m (500-600 ft)/min
Service ceiling	3,050 m (10,000 ft)
T-O run	122-152 m (400-500 ft)
T-O to 15 m (50 ft)	305 m (1,000 ft)
Range with max fuel	173 nm (322 km; 200 miles)

BARNEY OLDFIELD

BARNEY OLDFIELD AIRCRAFT COMPANY

PO Box 5974, Cleveland, Ohio 44101
Telephone: (216) 449 6300
PRESIDENT: Harvey R. Swack
TECHNICAL DIRECTOR: Richard M. Lane

Barney Oldfield Aircraft Co markets plans and material kits for the 'Baby' Lakes, a scaled-down version of the Great Lakes Sport Trainer, the prototype of which was designed and built by Mr Andrew Oldfield, who died during 1970. More than 650 sets of drawings had been sold by early 1981, and many 'Baby' Lakes are under construction. About 40 were flying by 1981.

A variant of the 'Baby' Lakes is the 'Super Baby', with a more powerful engine and other modifications. Plans are available and at least 25 sets have been sold. Another new variant of the design is the 'Buddy Baby'.

Under development are inverted oil and fuel systems for the 'Baby' Lakes, so that it can be used as a low-cost aerobatic aircraft. It is stressed for $\pm 9g$ at the recommended 385 kg (850 lb) gross weight.

OLDFIELD 'BABY' LAKES

TYPE: Single-seat sporting biplane.

WINGS: Braced biplane, with N-type interplane struts, double landing and flying wires and N-type centre-section support struts. Wing section modified M6, tapering to USA 27 46 cm (18 in) from tips. Incidence 2° 30′ on top wing, 1° 30′ on bottom wing. Wood structure of spruce spars and Warren truss ribs, with overall fabric covering. Ailerons on lower wings only. No flaps.

FUSELAGE: Welded steel tube structure, fabric-covered.

TAIL UNIT: Wire-braced welded steel tube structure, fabric-covered.

LANDING GEAR: Non-retractable tailwheel type. Oleo main legs with size 5·00-4 wheels. Steerable tailwheel.

POWER PLANT: One 59·5 kW (80 hp) Continental A80 flat-four engine, driving a two-blade fixed-pitch propeller. Provision for alternative engines of between 37·25 and 74·5 kW (50 and 100 hp), and several aircraft now under construction will have 1,500 and 1,600 cc Volkswagen engines. Fuel tank in front fuselage, capacity 45 litres (12 US gallons).

ACCOMMODATION: Single seat, normally in open cockpit. Cockpit canopy optional.

DIMENSIONS, EXTERNAL:

Wing span: top	5·08 m (16 ft 8 in)
Wing chord (both wings, constant)	0·91 m (3 ft 0 in)
Wing area, gross	7·99 m² (86 sq ft)
Length overall	4·19 m (13 ft 9 in)
Height overall	1·37 m (4 ft 6 in)

WEIGHTS (A80 engine):

Weight empty	215 kg (475 lb)
Max T-O weight	385 kg (850 lb)

PERFORMANCE (A80 engine, at max T-O weight):

Max level speed at S/L	117 knots (217 km/h; 135 mph)
Cruising speed at S/L	102 knots (190 km/h; 118 mph)
Stalling speed	43·5 knots (81 km/h; 50 mph)
Max rate of climb at S/L	610 m (2,000 ft)/min
Service ceiling	5,200 m (17,000 ft)
T-O run	91 m (300 ft)
Landing run (no brakes)	122 m (400 ft)
Max range	217 nm (400 km; 250 miles)

OLDFIELD 'SUPER BABY' LAKES

Since 1976, a prototype 'Super Baby' Lakes (N362RB), based on the 'Baby' Lakes but with an 85·75 kW (115 hp) Avco Lycoming O-235 flat-four engine, has been under test by its constructors, Mr Ray Ball, Mr Alan Lane and Mr Richard Lane. Following completion of more than 100 flying hours by early 1977, plans were made available to other constructors, with provision for fitting Avco Lycoming engines of 80·5 to 93 kW (108-125 hp). At least two other plans-built 'Super Babies' were flying by early 1981.

The O-235 engine of the prototype drives a 69-54 Met-L-Prop two-blade metal propeller. Other modifications compared with the 'Baby' Lakes include use of a Burtch die spring landing gear, and routed wing spars which allow installation of wing fuel tanks without any appreciable increase in the empty weight by comparison with the wings of a standard 'Baby'. To prevent the possibility of fuel starvation, due to the high acceleration at take-off interupting forward fuel flow from the main tank, a 0·95 litre (1 US quart) header tank is installed forward of the carburettor. Covering is Dacron, finished with Stits Poly-Fiber and Imeron polyurethane colour.

Take-off distances are reduced by 20-25% compared with the 'Baby'. Other data measured with the prototype 'Super Baby' are as follows:

DIMENSIONS, EXTERNAL:

As for 'Baby' Lakes, except:

Length overall	4·34 m (14 ft 3 in)

WEIGHT:

Weight empty	218 kg (480 lb)
Max T-O weight	385 kg (850 lb)

PERFORMANCE:

Max level speed	135 knots (249 km/h; 155 mph) IAS
Cruising speed (75% power)	117 knots (217 km/h; 135 mph) IAS
Stalling speed	48 knots (88·5 km/h; 55 mph) IAS
Max rate of climb at S/L	915 m (3,000 ft)/min
T-O run	69 m (225 ft)
Landing run	130 m (425 ft)
Max range, with wing tanks	260 nm (483 km; 300 miles)

OLDFIELD 'BUDDY BABY' LAKES

The two-seat 'Buddy Baby' utilises standard 'Baby' wings and tail unit. To prevent a significant increase in wing loading, a 41 cm (16 in) upper wing centre-section has been introduced, together with short stub-wings built integrally with the fuselage to increase the lower span by a similar amount.

The fuselage is a completely new design, which has been widened and stretched only enough to permit seating positions for two average-sized (77 kg; 170 lb) people in a configuration much like that of a 'buddy seat' on a motorcycle. Dual throttle and rudder pedals are provided, with a single dual-position joystick. The aircraft is flown solo from the rear seat and dual from the front position.

BENSEN

BENSEN AIRCRAFT CORPORATION

Box 31047, Raleigh-Durham Airport, Raleigh, North Carolina 27622
Telephone: (919) 787-4224/0945
PRESIDENT: Dr Igor B. Bensen, PE

The Bensen Aircraft Corporation was formed by Dr Igor B. Bensen, formerly Chief of Research of the Kaman company, to develop a series of lightweight autogyros and rotary-wing gliders suitable for production in kit form for amateur construction as well as in ready-to-fly condition. More than 30,000 sets of plans to build these aircraft have been sold; the number flown to date is estimated at between 5,000 and 7,000.

Production is centred on the B-8M/V and Super Bug Gyro-Copter powered autogyros, and various land and waterborne versions of the B-8 rotor-kite. Research into new models and concepts continues, and Bensen has announced important new second-generation versions of the Gyro-Copter that are capable of 'jump' take-off and of hovering flight. The desirability of a Gyro-Copter capable of flying on leaded 'high-test' car fuel has led also to development of the Bensen B-80.

In 1980, the EAA named the Gyro-Copter as 'The World's Most Popular Homebuilt Aircraft'; and awarded one of ten Wright Brothers Invitational Awards to the individual builder of a Bensen B-80 at the annual Air Fair Airshow in Dayton, Ohio, dedicated to the memory of Orville and Wilbur Wright. In 1978 (last year for which figures are available), 772 Gyro-Copters were registered as 'active' by the FAA.

BENSEN MODEL B-8 GYRO-GLIDER

The Gyro-Glider is a simple unpowered rotor-kite which can be towed behind even a small motorcar and has achieved free gliding with the towline released. It is available as either a completed aircraft or kit of parts for amateur construction. Alternatively, would-be constructors can purchase a set of plans, with building and flying instructions. No pilot's licence is required to fly it in the United States and many hundreds of kits and plans have been sold. Capability of conversion into powered form, as the Gyro-Copter (which see), is an inherent feature of the design.

The original Model B-7 Gyro-Glider was described in the 1958-59 *Jane's*. It was followed by the Model B-8, which is offered as either a single-seater or two-seater, the latter version being suitable for use as a pilot trainer.

The Model B-8 consists basically of an inverted square-section tubular aluminium T-frame structure, of which the forward arm supports the lightweight seat, towing arm, rudder bar and landing gear nosewheel. The rear arm supports a large stabilising fin and rudder, made normally of plywood, but optionally of metal. The main landing gear wheels are carried on a tubular axle near the junction of the T-frame. The free-turning two-blade rotor is universally-mounted at the top of the T-frame and is normally operated directly by a hanging-stick control. A floor-type control column is available as optional equipment. Pedal controls for the rudder are standard.

The Gyro-Glider rotor is made normally of laminated plywood, with a steel spar. Factory-built all-metal rotor blades are available as optional items. By 1981, Bensen had produced more than 10,000 complete blades for the Gyro-Glider and Gyro-Copter, plus wood/steel-spar kits for homebuilt construction of another 4,000 blades.

The two-seat trainer version of the Gyro-Glider is fitted with castoring crosswind landing gear and has an extra-wide wheel track. It will maintain level flight down to 16·5 knots (30·5 km/h; 19 mph).

DIMENSIONS, EXTERNAL:

Diameter of rotor	6·10 m (20 ft 0 in)
Length of fuselage	3·45 m (11 ft 4 in)
Height overall	1·90 m (6 ft 3 in)

BENSEN MODEL B-8W HYDRO-GLIDER

The basic structure of this floatplane rotor-kite is similar to that of the B-8 Gyro-Glider and conversion from one to the other is simple. Main change is that the wheel

Anderson EA-1 Kingfisher amphibian *(Howard Levy)*

Oldfield 'Baby' Lakes with an enclosed cockpit and raised headrest for the pilot *(J. M. G. Gradidge)*

Bensen B-8MJ jump take-off Gyro-Copter

First Oldfield 'Super Baby' Lakes, built by Mr Ray Ball, Mr Alan Lane and Mr Richard Lane

Bensen B-18 Hover-Gyro, with co-axial rotors, being flight tested

Bensen Model B-8MW Hydro-Copter

landing gear is replaced by two floats. The original round-type floats have been superseded by flat-bottomed pontoons of polyurethane foam covered by glassfibre, which give better planing, with less spray.

The Hydro-Glider is towed by a motorboat.

BENSEN MODEL B-8M, B-8V and SUPER BUG GYRO-COPTERS and B-8MW HYDRO-COPTER

First flown on 6 December 1955, the Gyro-Copter is a powered autogyro conversion of the Gyro-Glider, designed for home construction from kits or plans. When fitted with floats it is known as a **Hydro-Copter**.

The current **B-8M** version of the Gyro-Copter has a more powerful engine than the original B-7M and can be equipped with an optional mechanical rotor drive. By engaging this drive, the rotor can be accelerated to flying speed while the aircraft is stationary. Then, by transferring the power to the pusher propeller, it is possible to take off in only 15 m (50 ft), with the rotor autorotating normally. Alternatively, a 1 hp Ohlsson & Rice Compact III two-stroke engine can be attached to the rotor for pre-rotation, automatically disengaging itself at take-off rpm.

Other non-standard items available optionally include a 67 kW (90 hp) engine instead of the normal 53·5 kW (72 hp) engine, a stronger aluminium bolted-together engine mounting, a larger-diameter rotor, an offset gimbal rotor head, a redundant mast consisting of a pair of closely matched alloy tubes instead of the normal square-tube mast, a floor-type control column instead of the normal overhead type of column, dual ignition, nosewheel arrester, Bensen-manufactured pontoons of polyurethane foam covered with glassfibre, and a detachable glassfibre open cabin nacelle for single-seat Gyro-Copters of 1976 or later type (only with redundant mast and aluminium engine mount of new design). All-metal rotor blades and tail surfaces are available as alternatives to the standard wooden components.

The prototype Model B-8M Gyro-Copter flew for the first time on 8 July 1957 and the first production model on 9 October 1957.

The B-8M is roadable, requiring no removal of, or changes in, its equipment for transition from air to ground travel. The rotor is merely stopped in a fore-and-aft position by a lock. Gyro-Copters have been driven on highways and have negotiated heavy city traffic with ease in a number of public demonstrations in the USA.

The **B-8V**, which flew for the first time in the Autumn of 1967, is basically a standard B-8M, but is powered by a 1,600 cc Volkswagen engine. In unmodified form the VW1600 yields just adequate flight performance at 272 kg (600 lb) gross weight. Since Bensen engineers considered that most Gyro-Copters would not have a gross weight as high as 600 lb, the VW engine justified inclusion as an alternative power plant to the standard McCulloch engine.

Kits and parts for the B-8V, excluding engine and mounting, are available, as are plans and an instruction manual for converting the B-8M to a B-8V, or for mounting a VW engine on a standard B-8 airframe.

In May 1971, Bensen announced introduction of the **Super Bug**, an advanced version of the standard Model B-8M. This features a twin-engine installation to spin up the rotor prior to take-off. Bensen claims this as an intermediate step towards full VTOL capability, as this more powerful pre-rotation enables the Super Bug to take off and clear a 15 m (50 ft) obstacle within 137 m (450 ft) of starting its T-O run in zero wind conditions at max T-O weight. Other standard equipment of the Super Bug includes rotor brake, parking brake on main wheels, single control of rudder and nosewheel steering, soft suspension of the auxiliary tailwheel and an increase of 45·5 kg (100 lb) in max T-O weight.

In July 1979 it was announced that Bensen was to begin marketing a 'Gold Seal' kit containing all the component parts, including engine, necessary to construct a standard Gyro-Copter. The price included a pre-flight checkout of the completed aircraft by an authorised Bensen dealer, and flying instruction for the builder (see separate item on Model B-80 Gyro-Copter).

The following description applies to the Models B-8M and B-8V:

TYPE: Single-seat light autogyro.

ROTOR SYSTEM: B-8M has single two-blade rotor of laminated plywood construction, with steel spar (optional all-metal rotor). Blade section Bensen G2. Teetering hub, with no lag hinges or collective pitch control. A similar rotor, of all-metal construction, is provided for the B-8V. A larger-diameter rotor is not available as an alternative for this latter model. No anti-torque rotor. Rotor speed 400 rpm.

ROTOR DRIVE (optional): An auxiliary 0·75 kW (1 hp) Ohlsson & Rice engine is available to spin up the rotor.

FUSELAGE: Square-section tubular 6061-T6 aluminium structure. Optional redundant twin rotor mast of alloy tubes.

TAIL SURFACES: Vertical fin and rudder of ¼ in plywood. Optional all-metal tail surfaces.

LANDING GEAR: Non-retractable tricycle type, with auxiliary tailwheel. No shock-absorbers. Steerable nosewheel. General Tire wheels, size 30·5-10·15 cm (12-4 in). Tyre pressure 0·69 bars (10 lb/sq in). Brake on nosewheel. Optional skis.

POWER PLANT: One 53·5 kW (72 hp) McCulloch Model 4318AX flat-four two-stroke engine (or, optionally, a 67 kW; 90 hp McCulloch 4318GX engine of similar weight and dimensions), driving a two-blade fixed-pitch Aero Prop Model BA 48-A2 wooden pusher propeller with leading-edges covered with stainless steel. Alternatively, one 47·5 kW (64 hp) Volkswagen 1,600 cc flat-four four-stroke engine, driving a Troyer Model 50-24-65 two-blade fixed-pitch wooden pusher propeller. Fuel tank under pilot's seat, capacity 22·75 litres (6·0 US gallons). Can be fitted with auxiliary tank for ferrying. Optional plastics tank shaped as pilot's seat, replacing standard tank and seat and increasing fuel capacity to 36·5 litres (9·7 US gallons).

ACCOMMODATION: Open seat. Overhead azimuth stick and rudder pedal controls. Optional floor-type control column. Safety belt. Optional glassfibre open cabin nacelle.

DIMENSIONS, EXTERNAL:

Diameter of rotor: standard	6·10 m (20 ft 0 in)
optional (on B-8M and B-8MW)	6·70 m (22 ft 0 in)
Rotor blade chord	0·18 m (7 in)
Length of fuselage	3·45 m (11 ft 4 in)
Height overall	1·90 m (6 ft 3 in)
Wheel track	1·52 m (5 ft 0 in)
Propeller diameter:	
53·5 kW (72 hp) McCulloch	1·22 m (4 ft 0 in)
47·5 kW (64 hp) Volkswagen	1·27 m (4 ft 2 in)

AREAS (standard rotor):

Rotor blades (each)	0·54 m² (5·83 sq ft)
Rotor disc	29·17 m² (314 sq ft)

WEIGHTS (standard rotor):

Weight empty: B-8M	112 kg (247 lb)
B-8V	158 kg (348 lb)
Max T-O weight: B-8M	227 kg (500 lb)
B-8V	272 kg (600 lb)

PERFORMANCE (at max T-O weight, with standard rotor):

Max level speed at S/L:	
B-8M	74 knots (137 km/h; 85 mph)
B-8V	52 knots (96·5 km/h; 60 mph)
Max cruising speed at S/L:	
B-8M	52 knots (96·5 km/h; 60 mph)
B-8V	43 knots (80·5 km/h; 50 mph)
Econ cruising speed:	
B-8M, B-8V	39 knots (72·5 km/h; 45 mph)
Min speed in level flight:	
B-8M	13 knots (24 km/h; 15 mph)
B-8V	17·5 knots (32 km/h; 20 mph)
T-O speed at S/L:	
B-8M	17·5 knots (32 km/h; 20 mph)
B-8V	22 knots (40 km/h; 25 mph)
Landing speed: B-8M	6 knots (11·5 km/h; 7 mph)
B-8V	9 knots (16 km/h; 10 mph)
Max rate of climb at S/L:	
B-8M	305 m (1,000 ft)/min
B-8V	198 m (650 ft)/min
Service ceiling: B-8M	3,800 m (12,500 ft)
B-8V	2,440 m (8,000 ft)
T-O run, unpowered rotor, zero wind:	
B-8M	92 m (300 ft)
B-8V	122 m (400 ft)
T-O run, powered rotor, zero wind:	
B-8M	15 m (50 ft)
Landing run in 9 knot (16 km/h; 10 mph) wind:	
B-8M, B-8V	0 ft
Landing run in zero wind: B-8M	6 m (20 ft)
B-8V	7·5 m (25 ft)
Normal range: B-8M	86 nm (160 km; 100 miles)
B-8V	130 nm (241 km; 150 miles)
Ferry range: B-8M	260 nm (482 km; 300 miles)
B-8V	345 nm (643 km; 400 miles)
Endurance: B-8M	1·5 h
B-8V	2·25 h

BENSEN MODEL B-8HD

This version of the Gyro-Copter is based on the Super Bug design, but uses hydraulic drive (hence the HD designation) to feed about 3 kW (4 hp) from the main engine to the rotor, instead of having a separate engine for pre-rotation. The rotor is thus engaged continuously during all aspects of flight. This is reported to give the aircraft a take-off run of less than 61 m (200 ft), and to improve climb and range, while allowing an increase in payload. A near-vertical landing is also possible.

The hydraulic drive is expected to be marketed as a retrofit kit for the current B-8M.

BENSEN MODEL B-8MJ

Bensen Aircraft Corporation announced in early 1977 that it was to market a Power Head to fit any standard B-8M Gyro-Copter, enabling the aircraft to take off without any ground roll. The Power Head is claimed to provide a 30° angle of climb. Hovering cannot be achieved.

Power Heads are offered only to builders or owners of Gyro-Copters with 50 or more hours of solo flight. The resulting aircraft are designated B-8MJ (for 'Jump' take-off).

BENSEN MODEL B-18 HOVER-GYRO

Demonstrated in public for the first time in 1976, the B-18 (known originally as the B-8MH) is described as a 'Hovering Gyro-Copter'.

Development dates from 1973. On 29 June that year, Dr Igor Bensen piloted a Gyro-Glider with two two-blade co-axial rotors during its first flight at Raleigh-Durham Airport. Nineteen further flights followed, during which the aircraft executed a full range of flight manoeuvres.

In the B-18 the lower two-blade rotor is driven by a vertically-mounted engine; the upper rotor autorotates. A small engine mounted immediately forward of the tail-fin drives a pusher propeller. The B-18 can fly forwards, sideways and backwards, as well as being able to hover. It is claimed to be safer than either an aeroplane or a helicopter, as it will not stall or spin and does not need any special skill to land with the engine stopped.

The following details apply to the prototype. In 1981, flight testing had been resumed after a period of inactivity, and a new twin-engine power plant was under development. Bensen planned to market the re-engined aircraft as a two-seater, in kit form, before the end of 1981.

POWER PLANT: One 52·2-82 kW (70-110 hp) modified watercooled outboard engine, driving the lower two-blade main rotor. Upper rotor autorotates. One 10·4 kW (14 hp) modified aircooled go-kart engine driving a two-blade pusher propeller mounted at rear. Fuel consumption at cruising speed, 19 litres (5 US gallons) per hour.

DIMENSIONS, EXTERNAL:

Rotor diameter (each)	8·53 m (28 ft 0 in)
Height overall	2·74 m (9 ft 0 in)
Width overall	2·13 m (7 ft 0 in)

WEIGHTS:

Max T-O weight	408 kg (900 lb)
Useful load	136 kg (300 lb)

PERFORMANCE:

Max cruising speed	56·5 knots (105 km/h; 65 mph)
Range at optimum cruising speed	104 nm (193 km; 120 miles)
Hovering endurance	45 min

BENSEN MODEL B-80 and B-80D GYRO-COPTER

In an attempt to reduce Gyro-Copter operating costs, and the problems associated with aviation fuel shortages at some airports, Bensen adapted the Gyro-Copter to fly on 100 octane low-lead fuel (100LL Avgas) instead of 110/130 octane aviation fuel. Further development allowed the Gyro-Copter to use Exxon leaded 'high-test' motorcar fuel. The test aircraft was designated Model B-80, and was flown with 53·5 kW (72 hp) and 67 kW (90 hp) McCulloch X engines, equipped with new carburettors, propellers and throttle controls to prevent knock and pre-ignition of the lower octane motorcar fuel.

Bensen first marketed the B-80 during 1980 in 'Gold Seal Kit' form, as a kit-only all-metal companion to the B-8M. The kit contained all hardware and instructions to complete the aircraft, including flight check, but no plans. Towards the end of that year, hybrid versions of the B-8M and B-80 were made available, to take advantage of interchangeability of certain components between the two models. By 1981, the dividing line between the B-8M and B-80 had become even less distinct, as customers may now choose which components they wish to build from plans and which from kits. Standard recommended power plant is the 53·5 kW (72 hp) McCulloch Model 4318AX, approved for use with automobile fuel, but other engines are being used successfully, as on the B-8M. A de luxe version of the aircraft is available as the **B-80D**.

In 1981, Bensen was converting a B-80 for evaluation as a radio-controlled pilotless drone, for patrolling areas considered too hazardous for piloted aircraft. Capability of manned operation is retained.

BOND

JOHN BOND

c/o Hollmann Aircraft, 11082 Bel Aire Court, Cupertino, California 95014

BOND SKY DANCER

Mr John Bond has designed and built the prototype of a gyroplane known as the Sky Dancer. Design was originated in 1971, construction beginning during 1972. Flown for the first time in 1973, its extremely low disc loading and high thrust-to-weight ratio has enabled this homebuilt aircraft to demonstrate outstanding performance. Plans of the Sky Dancer are available to amateur builders from Hollmann Aircraft at the above address.

By January 1981, twelve more Sky Dancers were under construction, including two in England.

TYPE: Single-seat lightweight gyroplane.

ROTOR SYSTEM: Two-blade teetering rotor of modified NACA 4412 blade section. Rotordyne blades of 6061-T6 light alloy retained to hub by steel straps. Blades can be folded. No rotor brake.

FUSELAGE: Welded structure of 4130 steel tube, integrating landing gear, rotor pylon, pilot's seat structure, engine mounting and tail support booms.

TAIL UNIT: Fixed horizontal surface and fin, with conventional rudder, all constructed of glassfibre with foam core.

LANDING GEAR: Non-retractable tricycle type, with additional small tail bumper wheel. Shock-absorption of main wheels by coil spring in compression; tailwheel carried on spring steel arm. Steerable nosewheel. Main-wheel tyres 33 cm (13 in) diameter, pressure 1·52 bars (22 lb/sq in). Nosewheel tyre 28 cm (11 in) diameter, pressure 1·38 bars (20 lb/sq in). Fairings over main wheels.

POWER PLANT: One 67 kW (90 hp) Revmaster modified Volkswagen 2,100 cc motorcar engine, driving a two-blade fixed-pitch pusher propeller manufactured by Troyer Boats. Fuel tank above engine, capacity 22·7 litres (6 US gallons). Refuelling point on tank. Oil capacity 2·8 litres (0·75 US gallons).

ACCOMMODATION: Pilot only, on exposed seat, immediately forward of rotor pylon.

DIMENSIONS, EXTERNAL:

Rotor diameter	7·01 m (23 ft 0 in)
Rotor blade chord, constant	0·17 m (6¾ in)
Rotor disc area	38·60 m² (415·5 sq ft)
Length of fuselage	3·30 m (10 ft 10 in)
Width, rotors folded	1·65 m (5 ft 5 in)
Height to top of rotor hub	2·02 m (6 ft 7½ in)
Wheel track	1·52 m (5 ft 0 in)
Propeller diameter	1·27 m (4 ft 2 in)

WEIGHTS:

Weight empty	127 kg (280 lb)
Max fuel weight	18 kg (40 lb)
Max T-O weight	245 kg (540 lb)

PERFORMANCE (at max T-O weight):

Never-exceed and max level speed	82 knots (153 km/h; 95 mph)
Max cruising speed at S/L	65 knots (121 km/h; 75 mph)
Econ cruising speed at S/L	48 knots (89 km/h; 55 mph)
Max rate of climb at S/L	366 m (1,200 ft)/min
Service ceiling	4,875 m (16,000 ft)
T-O run	46 m (150 ft)
Range with max fuel	104 nm (193 km; 120 miles)

BORCHERS

LOWELL J. BORCHERS

Mount Vernon, Ohio

BORCHERS DELTA STINGRAY

Mr Lowell J. Borchers has designed and built a two-thirds to three-quarter-scale single-seat derivative of the Dyke Aircraft JD-2 Delta. Mr John W. Dyke, designer of the JD-2 Delta, also worked on the project. Construction of the Delta Stingray took seven years, at a cost of $4,500. The first flight was achieved in mid-June 1980.

Unlike the JD-2 Delta, the Delta Stingray has a wooden

Bensen Model B-8HD Gyro-Copter

Bensen Model B-80, using 'high test' car fuel

Delta Stingray, designed and built by Mr Lowell J. Borchers *(Howard Levy)*

Bond Sky Dancer single-seat gyroplane

Replica Rose Parakeet built by Mr Richard R. Borg *(Howard Levy)*

structure. It uses plywood ribs and an I-beam spar, with foam filler forming the curvature of the wings and tailfin leading-edge. Mahogany and birch plywood skins are finished with epoxy, and painted to achieve high gloss. Power is provided by a 74·5 kW (100 hp) Continental O-200 flat-four engine. The tricycle landing gear is manually retractable, with hand-pumped hydraulic actuators for the main units. Small horizontal trimming surfaces near the top of the tailfin provide pitch control to maintain lift when elevons are up. The Delta Stingray is stressed to ±9g.

TYPE: Single-seat sporting aircraft.

DIMENSIONS, EXTERNAL:
- Wing span — 4·57 m (15 ft 0 in)
- Length overall — 4·52 m (14 ft 10 in)
- Height overall — 1·37 m (4 ft 6 in)

WEIGHTS:
- Weight empty — 336 kg (740 lb)
- Max T-O weight — 487 kg (1,075 lb)

PERFORMANCE:
- Max level speed — over 173 knots (322 km/h; 200 mph)
- Cruising speed — 148 knots (274 km/h; 170 mph)
- Landing speed — 70 knots (129 km/h; 80 mph)
- Max rate of climb at S/L — 427 m (1,400 ft)/min
- T-O and landing run — 457 m (1,500 ft)
- Range with 91 litres (24 US gallons) of fuel — 434 nm (804 km; 500 miles)

BORG

RICHARD R. BORG

San Jose, California

BORG ROSE PARAKEET REPLICA

Mr Richard R. Borg has built a replica of the Rose Parakeet. Construction took nearly eleven years and cost $8,000. It first flew on 11 July 1980. The original Parakeet was built by the Rose Aeroplane & Motor Company in the 1930s, as a light single-seat biplane for training or touring. Its 1935 cost was $975 (£245) with a 30 kW (40 hp) Continental A40 engine.

Like the original, the replica has fabric-covered wooden wings and fabric-covered steel-tube fuselage and tail unit. However, it differs by having a 63·5 kW (85 hp) Continental C85-12 engine in a full cowling, aluminium-covered wing leading-edges, a turtledeck fairing extended to the tailfin, full electrical system, two additional 15 litre (4 US gal) fuel tanks in the upper wing, and other changes. It won the EAA's 'Reserve Grand Champion Award' at Oshkosh in 1980.

TYPE: Single-seat sporting biplane.

DIMENSIONS, EXTERNAL:
- Wing span — 6·10 m (20 ft 0 in)
- Wing chord, constant — 0·97 m (3 ft 2 in)
- Length overall — 4·94 m (16 ft 2 in)
- Height overall — 1·88 m (6 ft 2 in)

WEIGHTS:
- Weight empty — 345 kg (760 lb)
- Max T-O weight — 526 kg (1,160 lb)

PERFORMANCE:
- Never-exceed speed — 156 knots (290 km/h; 180 mph)
- Cruising speed — 109 knots (201 km/h; 125 mph)
- Landing speed — 52 knots (96·5 km/h; 60 mph)
- Max rate of climb at S/L — 457 m (1,500 ft)/min
- T-O run — 122 m (400 ft)
- Landing run — 183 m (600 ft)
- Range — 347 nm (643 km; 400 miles)

BOWERS

PETER M. BOWERS

10458 16th Avenue South, Seattle, Washington 98168

Mr Peter Bowers, an aeronautical engineer with Boeing in Seattle, is a principal source of detailed information on vintage aircraft in the United States, and has provided much of the data for a number of replicas of first World War aircraft now under construction or flying.

Mr Bowers has built a number of aircraft himself. In particular, he designed and constructed prototypes of a single-seat light aircraft known as the Fly Baby, of which plans are available, and which is flying in very large numbers, in both monoplane and biplane forms.

BOWERS FLY BABY 1-A

The prototype Fly Baby monoplane was produced to compete in an Experimental Aircraft Association design contest, organised to encourage the development of a simple, low-cost, easy-to-fly aeroplane that could be built by inexperienced amateurs for recreational flying. It was built in 720 working hours, at a cost of $1,050, and flew for the first time on 27 July 1960. As only one other aircraft was completed by the specified closing date, the contest was postponed for two years.

When the EAA contest was finally held in the Summer of 1962, Fly Baby was placed first and won a prize of $2,500. Home construction plans of the aircraft are available and at least 3,925 sets had been sold by 1981. Construction of well over 700 Fly Babies is known to have been undertaken, of which at least 366 had been completed and flown by 1981, including some based on detailed drawings and instructions published in *Sport Aviation*, journal of the EAA.

The Fly Baby monoplane has been tested as a twin-float seaplane, in which configuration it has a max AUW of 454 kg (1,000 lb) and cruising speed of 84 knots (156 km/h; 97 mph). A version operated on skis was illustrated in the 1980-81 *Jane's*.

During 1968, Mr Bowers designed and built biplane wings for the Fly Baby, which are interchangeable with the monoplane wings. The monoplane version is known as the Fly Baby 1-A, and the biplane as the Fly Baby 1-B (described separately).

Mr Bowers amended his plans during 1973 to allow for construction of a two-seat version of the Fly Baby. The changes include a 0·97 m (3 ft 2 in) wide fuselage, a 1·52 m (5 ft 0 in) span wing centre-section to support a shock-absorbing landing gear similar to that of the Ryan ST/PT-22 of 1934-42, the use of heavier flying wires with swaged fork ends, a raised aft turtledeck to offset the drag of the larger cockpit and a recommendation to use an engine of 63·5 to 80 kW (85 to 107 hp). The outer wing panels are unchanged, giving the two-seat Fly Baby a span of 9·45 m (31 ft 0 in) and wing area of 12·4 m² (133·5 sq ft).

Mr Bowers feels that the use of side-by-side seating is a great performance handicap to the basic design, and approves the tandem arrangement worked out by Mr Victor F. Meznarsic (see accompanying illustration). At present, he is developing a 'stretched' tandem two-seater, to be fitted with longer two-bay biplane wings, in a deliberate attempt to resemble a first World War type such as the B.E.2c. This will be a low-performance 'fun flier' on engines up to 74·5 kW (100 hp), and will not be aerobatic.

Because of the shortage of aircraft engines in the 48·5-63·5 kW (65-85 hp) range, Mr Bowers has been asked repeatedly by would-be builders about the suitability of Volkswagen conversions for the Fly Baby. He has consistently recommended against this, but now that a Finnish constructor has had some success, first with a 1,600 cc and later with an 1,800 cc conversion, Mr Bowers approves the use of VW engines of over 1,800 cc.

The following description applies to the original single-seat Fly Baby 1-A:

Type: Single-seat light monoplane.

Wings: Wire-braced low-wing monoplane. Double ½ in 1 × 19 stainless steel bracing wires. Wing section NACA 4412. Wooden two-spar structure, covered with Dacron fabric and finished with two coats of nitrate dope and one coat of automotive enamel. Wings rotate about a special fitting to fold back alongside the fuselage for towing.

Fuselage: Conventional plywood-covered wood structure of rectangular section. Decking behind cockpit, including pilot's headrest, is removable and can be replaced with higher transparent section matched with a sliding transparent cockpit canopy for enclosed cockpit operation.

Tail Unit: Wire-braced wood structure, fabric-covered.

Landing Gear: Non-retractable tailwheel type. Main landing gear struts of laminated wood, braced by crossed steel wires. Steel tube straight-across axle faired with streamline-section steel tube. Ends of axles project beyond wheel hubs to serve as anchor points for wing bracing wires. Shock-absorption by low-pressure 8·00-4 tyres, carried on Piper Cub wheels, with hydraulic brakes.

Power Plant: One 63·5 kW (85 hp) Continental C75 flat-four engine, driving a two-blade fixed-pitch propeller. Fuel tank from Piper J-3 Cub, capacity 60·5 litres (16 US gallons).

Accommodation: Single seat in open or enclosed cockpit. Baggage in underfuselage 'tank' which can be removed and carried like a suitcase.

Dimensions, external:
- Wing span: 8·53 m (28 ft 0 in)
- Wing chord, constant: 1·37 m (4 ft 6 in)
- Length overall: 5·64 m (18 ft 6 in)
- Height, wings folded: 1·98 m (6 ft 6 in)

Weights:
- Weight empty: 274 kg (605 lb)
- Max T-O weight: 419 kg (924 lb)

Performance (at max T-O weight):
- Max level speed at S/L: over 104 knots (193 km/h; 120 mph)
- Cruising speed: 91-96 knots (169-177 km/h; 105-110 mph)
- Landing speed: 39 knots (72·5 km/h; 45 mph)
- Max rate of climb at S/L: 335 m (1,100 ft)/min
- T-O and landing run: 76 m (250 ft)
- Range with max fuel: 277 nm (515 km; 320 miles)

BOWERS FLY BABY 1-B

During 1968 Mr Bowers designed and built a set of interchangeable biplane wings for the original prototype Fly Baby and with these fitted it flew for the first time on 27 March 1969. About four more biplanes, designated Fly Baby 1-B, have been completed and flown.

The biplane wings have the same aerofoil section and incidence as those of the monoplane version, but the rib webs are made of $^1/_{16}$ in instead of ⅛ in plywood and the wingtip bows are formed from ½ in aluminium tube instead of laminated wood strips. This lightweight construction limits weight increase to only 21 kg (46 lb) for an increase of 2·79 m² (30 sq ft) in wing area. Span is reduced by 1·83 m (6 ft) and chord by 0·30 m (1 ft). Ailerons are fitted to the lower wings only.

To facilitate entry to the cockpit the upper wing has been located well forward, and in order to bring the centre of lift in line with the original CG, both planes have been given 11° of sweepback. Changeover from monoplane to biplane configuration can be accomplished by two people in approximately one hour.

The biplane is intended to use the same engines as the monoplane, for which Mr Bowers does not recommend anything heavier than the Continental O-200. Since some biplane builders have desired to use the 93 kW (125 hp) Avco Lycoming O-290, a modification has been authorised for the biplane, whereby the wing sweep is decreased by 5° for CG reasons.

Type: Single-seat light biplane.

Wings: Forward-stagger single-bay biplane with N-type interplane and centre-section struts. Landing and flying bracing wires. Sweepback 11°. Wooden structure with Dacron covering. Rib webs constructed of $^1/_{16}$ in plywood, wingtip bows formed of ½ in aluminium tube. Ailerons on lower wings only.

Power Plant: One 63·5 kW (85 hp) Continental C85 flat-four engine.

Dimensions, external:
- Wing span: 6·71 m (22 ft 0 in)
- Wing chord, both wings, constant: 1·07 m (3 ft 6 in)
- Wing area, gross: 13·94 m² (150 sq ft)
- Height overall: 2·08 m (6 ft 10 in)

Weights:
- Weight empty: 295 kg (651 lb)
- Max T-O weight: 440 kg (972 lb)

Performance (at max T-O weight):
- Cruising speed: 75·5 knots (140 km/h; 87 mph)
- Max rate of climb at S/L: 267 m (875 ft)/min

BOYD

GARY BOYD

2250 Judith Lane, Santa Ana, California 92706
Telephone: (213) 836 6580

BOYD G.B.1

Mr Gary Boyd has designed and built a two-seat light monoplane known as the G.B.1. Design and construction of the prototype took about two years, and a first flight was achieved in December 1979. By the beginning of 1981 more than 70 hours of flying had been logged.

The G.B.1 uses the RAF 48 wing section. Although the prototype is fitted with a 44·7 kW (60 hp) HAPI Volkswagen 1,835 cc modified motorcar engine driving a three-blade propeller, and a retractable tailwheel type landing gear, any modified Volkswagen motorcar engine in the 1,600-2,180 cc range can be fitted to the G.B.1, as well as a retractable tricycle landing gear or non-retractable landing gear. Virtually the entire aircraft is constructed from pre-moulded glassfibre components, bonded by polyester resin. This allows the G.B.1 to be constructed using only the most basic tools. The aircraft's outer wing panels are removable for transportation.

Sets of plans and glassfibre components are available to amateur builders. The following data apply to the prototype:

Type: Two-seat light monoplane.

Dimensions, external:
- Wing span: 6·40 m (21 ft 0 in)
- Length: 4·47 m (14 ft 8 in)

Weights:
- Weight empty: 249 kg (550 lb)
- Max T-O weight: 499 kg (1,100 lb)

Performance:
- Max level speed at 1,220 m (4,000 ft): 147 knots (273 km/h; 170 mph)
- Cruising speed: 130 knots (241 km/h; 150 mph)
- Stalling speed: 39 knots (73 km/h; 45 mph)
- T-O run: 137 m (450 ft)
- Endurance: 5 h

BROKAW

BROKAW AVIATION INC

2625 Johnson Point, Leesburg, Florida 32748
Telephone: (904) 787 2329
President: Bergon F. Brokaw, MD, FACFP

BROKAW BULLET

Dr B. F. Brokaw, a former US Navy pilot, and Dr Ernest R. Jones, who has a PhD in aeronautical engineering, combined their talents to design and build a low-wing monoplane which was claimed to be the world's fastest two-seat homebuilt. Dr Brokaw was concerned primarily with overall design and construction, Dr Jones with stress analysis and structural design.

Basic intention was to develop a high-speed all-weather two-seat homebuilt suitable for cross-country flying. Aerobatic potential was of secondary consideration, but the Bullet, which was known originally as the BJ-520, is stressed to ±6*g* for aerobatics and 9*g* ultimate, with a claimed 20 to 30 per cent safety margin.

Design began in August 1966 and construction of the prototype started six months later. First flight was made on 18 November 1972, and during the Summer of 1973 work was carried out to clean up the airframe to take better advantage of the design potential. This included the provision of wheel-well doors, and reduction of the drag of the engine cooling system, achieved by the use of baffles, ducts and direct ram-air cooling.

Further major redesign and modification were undertaken in 1976, and the prototype resumed flying in 1977 as the Bullet. By early 1980 it had logged more than 750 flying hours and had been flown in all weather conditions, including light ice. Subsequent refinements, nearing completion in early 1981, include changes to the wing span and section, and power plant, as described below. (For performance and other data relating to the original version, powered by a 212·5 kW; 285 hp Continental TSIO-520-B turbocharged engine, see 1975-76 *Jane's*.

Dr Brokaw formed Brokaw Aviation Inc to market plans, a construction manual and pilot handbook to amateur constructors. Although full kits are not available, it is intended also to supply certain components such as the canopy, cowling and, possibly, bulkheads and ribs, to simplify the task of the homebuilder. Thirteen Bullets were under construction in early 1981, one with a tailwheel landing gear and a 410 kW (550 hp) Ranger twelve-cylinder V engine.

Type: Two-seat sporting aircraft.

Wings: Cantilever low-wing monoplane. Laminar-flow wing section: NACA 64A412 at root, NACA 64A410 at tip. Dihedral 4° 30′. Incidence 1° at root, 0° at tip. Conventional structure of 2024-T3 light alloy. Ailerons and trailing-edge flaps of light alloy.

Fuselage: Semi-monocoque structure of light alloy.

Tail Unit: Cantilever light alloy structure. Fixed-incidence tailplane. Electrically-operated trim tabs in elevators and rudder.

Landing Gear: Hydraulically-retractable tricycle type. Main and nose units and associated hydraulic system are from a Navion aircraft. Main units retract inward and nosewheel rearward. Single wheel on each main unit, with 15 × 6·00-6 low-profile tyre, pressure 4·48 bars (65 lb/sq in). Nosewheel tyre size 14 × 5·00-4, pressure 3·10 bars (45 lb/sq in). Goodyear hydraulic brakes on main wheels.

Power Plant: One 283 kW (380 hp) Avco Lycoming TIO-541-E flat-six engine, driving a Sensenich three-blade constant-speed metal propeller with spinner. Alternative power plant is 213 kW (285 hp) Teledyne Continental TSIO-520-B flat-six engine, driving a McCauley three-blade constant-speed metal propeller with spinner, or other engine of 223-373 kW (300-500 hp). Four integral wing fuel tanks, with capacity of 363 litres (96 US gallons). Refuelling points at wingtips. Oil capacity 11·3 litres (3 US gallons).

Accommodation: Two seats in tandem beneath transparent individual canopies. Port half of each canopy hinged at centreline to open upwards. Baggage space, which accommodates 31·8 kg (70 lb), aft of rear seat. Cabin heated and ventilated.

Systems: Hydraulic system at 103·5 bars (1,500 lb/sq in) for landing gear retraction and brakes. 28V 70A alternator for electrical system. Oxygen system for pilot and passenger, with capacity of 5 hours.

Avionics and Equipment: Full IFR instrumentation, including DGO 10, DME 70, dual Narco 360-channel transceivers and dual VORs, ADF, transponder, marker beacon receiver and ILS. Landing gear warning lights, navigation lights and external power socket.

Dimensions, external:
- Wing span: 7·44 m (24 ft 4¾ in)
- Wing chord at root: 1·52 m (5 ft 0 in)
- Wing chord at tip: 0·76 m (2 ft 6 in)
- Wing area, gross: 8·79 m² (94·59 sq ft)
- Wing aspect ratio: 6
- Length overall: 6·25 m (20 ft 6 in)
- Height overall: 2·69 m (8 ft 10 in)
- Tailplane span: 2·69 m (8 ft 10 in)
- Wheel track: 2·54 m (8 ft 4 in)
- Wheelbase: 1·80 m (5 ft 11 in)
- Propeller diameter: 1·98 m (6 ft 6 in)

Dimensions, internal:
- Length: 2·54 m (8 ft 4 in)
- Max width: 0·91 m (3 ft 0 in)
- Max height: 1·37 m (4 ft 6 in)

Weights (TIO-541-E engine):
- Weight empty: 922 kg (2,033 lb)
- Max T-O weight: 1,425 kg (3,142 lb)

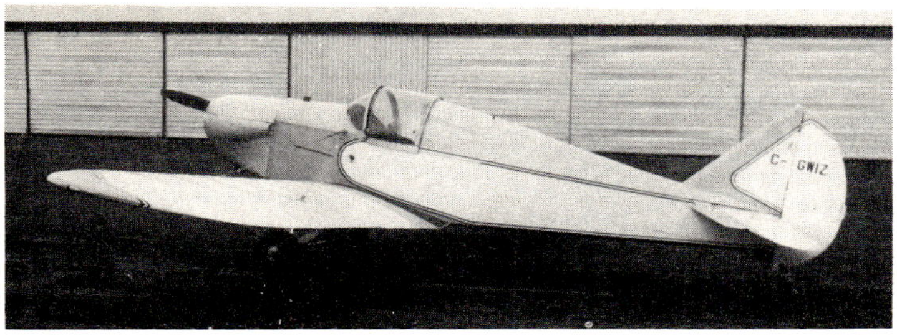

Bowers Fly Baby 1-A with a modified turtledeck (Continental A75 engine) *(Peter M. Bowers)*

Bowers Fly Baby 1-B with O-290-G engine and reduced sweep, built by Mr David Cronk *(Peter M. Bowers)*

Bowers Fly Baby 1-A (Continental A65) converted into a tandem two-seater by Mr Victor F. Meznarsic *(Howard Levy)*

Brokaw Bullet (with Continental TSIO-520-B engine)

Bushby M-II Mustang II (Avco Lycoming O-320 engine, driving Hartzell constant-speed propeller), built by Mr Charley H. Smith of Bolingbrook, Illinois

Bushby/Long Midget Mustang with wingtip fuel tanks, built in Managua, Nicaragua *(Neil A. Macdougall)*

PERFORMANCE (TIO-541-E engine):
Max level and normal cruising speed (75% power) at 7,315 m (24,000 ft)
303 knots (562 km/h; 349 mph)
Econ cruising speed (55% power) at 7,315 m (24,000 ft) 255 knots (473 km/h; 294 mph)
Stalling speed, clean 79 knots (147 km/h; 91 mph)
Stalling speed, wheels and flaps down
76 knots (142 km/h; 88 mph)
Max rate of climb at S/L 762 m (2,500 ft)/min
Service ceiling over 8,535 m (28,000 ft)
T-O run 396 m (1,300 ft)
T-O to 15 m (50 ft) 610 m (2,000 ft)
Landing from 15 m (50 ft) 853 m (2,800 ft)
Landing run 488 m (1,600 ft)
Range at econ cruising speed, 45 min reserves
1,600 nm (2,964 km; 1,842 miles)

BUSHBY

BUSHBY AIRCRAFT INC

Route 1, PO Box 13B, Minooka, Illinois 60447
Telephone: (815) 467 2346

Mr Robert W. Bushby, a research engineer with Sinclair Oil Co, began by building a Midget Mustang single-seat sporting monoplane, using drawings, jigs and certain components produced by the aircraft's designer, the late David Long. He has since produced the aircraft in kit form and also offers sets of plans of the Midget Mustang and a two-seat derivative known as the Mustang II to amateur constructors.

BUSHBY/LONG MM-1 MIDGET MUSTANG

The prototype of the Midget Mustang was completed in 1948 by David Long, then chief engineer of the Piper company. He flew it in the National Air Races that year, and in 1949 was placed fourth in the Continental Trophy Race at Miami.

Two basic versions were developed by Robert Bushby, as follows:

MM-1-85. Powered by 63·5 kW (85 hp) Continental C85-8FJ or -12 engine. Flew for the first time on 9 September 1959.

MM-1-125. Powered by 101 kW (135 hp) Avco Lycoming O-290-D2 engine. Otherwise similar to MM-1-85. Flew for first time in July 1963. New propeller introduced during 1973 has improved max speed and cruising speed.

At least 150 Midget Mustangs had been completed by the Spring of 1981, with 1,000 more under construction throughout the world. Several of those now flying have a 112 kW (150 hp) Avco Lycoming O-320 engine, providing a cruising speed of 234 knots (435 km/h; 270 mph) at 2,440 m (8,000 ft). Some have been fitted with retractable main landing gear.

The following details apply to the two basic versions:

TYPE: Single-seat fully-aerobatic sporting monoplane.

WINGS: Cantilever low-wing monoplane. Wing section NACA 64A212 at root, NACA 64A210 at tip. Dihedral 5°. Incidence 1° 30′. Two-spar flush-riveted stressed-skin aluminium structure. Aluminium statically-balanced ailerons and plain trailing-edge flaps.

FUSELAGE: Aluminium flush-riveted stressed-skin monocoque structure.

TAIL UNIT: Cantilever all-metal structure. Controllable trim tab in port elevator.

LANDING GEAR: Non-retractable tailwheel type. Cantilever spring steel main legs. Steerable tailwheel. Goodyear wheels and tyres, size 5·00-5, pressure 1·24 bars (18 lb/sq in). Goodyear hydraulic disc brakes.

POWER PLANT (MM-1-85): One 63·5 kW (85 hp) Continental C85-8FJ or -12 flat-four engine, driving a McCauley two-blade metal fixed-pitch propeller. Fuel tank aft of firewall, capacity 57 litres (15 US gallons). Optional integral wing fuel tanks, each with capacity of 57 litres (15 US gallons). Optional wingtip tanks, each with capacity of 13 litres (3·5 US gallons). Oil capacity 3·75 litres (1 US gallon).

POWER PLANT (MM-1-125): One 101 kW (135 hp) Avco Lycoming O-290-D2 flat-four engine, driving a Sensenich two-blade metal fixed-pitch propeller. Fuel tank aft of firewall, capacity 57 litres (15 US gallons). No provision for wingtip tanks. Oil capacity 5·75 litres (1·5 US gallons).

ACCOMMODATION: Single seat in enclosed cabin. Canopy hinged on starboard side. Space for 5·5 kg (12 lb) of baggage aft of seat. Room for back parachute.

AVIONICS AND EQUIPMENT: Radio optional. No provision for blind-flying instrumentation. Electrical system available on MM-1-85 only.

DIMENSIONS, EXTERNAL:
Wing span 5·64 m (18 ft 6 in)
Wing span over tip-tanks (MM-1-85)
5·99 m (19 ft 8 in)
Wing chord at root 1·53 m (5 ft 0 in)
Wing chord at tip 0·76 m (2 ft 6 in)
Wing area, gross 6·32 m² (68 sq ft)
Wing aspect ratio 4
Length overall 5·00 m (16 ft 5 in)
Height overall 1·37 m (4 ft 6 in)
Tailplane span 1·98 m (6 ft 6 in)
Wheel track 1·55 m (5 ft 1 in)

DIMENSIONS, INTERNAL:
Cabin:
Max width 0·56 m (1 ft 10 in)
Baggage space 0·057 m³ (2 cu ft)

WEIGHTS:
Weight empty:
MM-1-85 261 kg (575 lb)
MM-1-125 268 kg (590 lb)
Max T-O and landing weight:
MM-1-85 397 kg (875 lb)
MM-1-125 408 kg (900 lb)

PERFORMANCE (at max T-O weight):
Never-exceed speed 243 knots (450 km/h; 280 mph)
Max level speed at S/L:
MM-1-85 165 knots (306 km/h; 190 mph)
MM-1-125 195 knots (362 km/h; 225 mph)
Max cruising speed at 2,440 m (8,000 ft):
MM-1-85 171 knots (317 km/h; 197 mph)
MM-1-125 187 knots (346 km/h; 215 mph)
Econ cruising speed:
MM-1-85 129 knots (238 km/h; 148 mph)
MM-1-125 143 knots (265 km/h; 165 mph)
Stalling speed, flaps down:
MM-1-85 50 knots (92 km/h; 57 mph)
MM-1-125 53 knots (97 km/h; 60 mph)
Max rate of climb at S/L:
MM-1-85 533 m (1,750 ft)/min
MM-1-125 670 m (2,200 ft)/min
Service ceiling:
MM-1-85 over 4,875 m (16,000 ft)
MM-1-125 5,790 m (19,000 ft)
T-O run:
MM-1-85 137 m (450 ft)
MM-1-125 122 m (400 ft)
T-O to 15 m (50 ft):
MM-1-85 274 m (900 ft)
MM-1-125 213 m (700 ft)
Landing from 15 m (50 ft) 365 m (1,200 ft)
Landing run 152 m (500 ft)
Range with max fuel:
MM-1-85 347 nm (640 km; 400 miles)
MM-1-125 325 nm (603 km; 375 miles)
Range with max fuel and tip-tanks:
MM-1-85 651 nm (1,200 km; 750 miles)

BUSHBY M-II MUSTANG II

Design of this side-by-side two-seat derivative of the Midget Mustang was started in 1963. Construction of a prototype began in 1965 and it flew for the first time on 9 July 1966. During 1968 Mr Bushby designed an alternative non-retractable tricycle landing gear for the Mustang II, and amateur constructors have the option of either configuration. Another option is wing folding, to permit storage in a home garage. About 1,050 Mustang IIs were being built by amateurs in the Spring of 1981, at which time 100 were flying.

The description applies to the de luxe model, and the empty weight quoted includes IFR instrumentation and nav/com equipment. The M-II can also be operated as an aerobatic aircraft in what Bushby Aircraft calls the 'Sport' configuration. This is identical to the de luxe model except that the electrical system, radio and additional IFR instrumentation are deleted. The 'Sport' model has an empty weight of 340 kg (750 lb) and T-O weight of 567 kg (1,250 lb).

A Hartzell controllable-pitch propeller has replaced the original fixed-pitch propeller as standard, and the performance figures quoted for the 119 kW (160 hp) engine are obtained with this propeller. The figures quoted for the 93 kW (125 hp) engine are those obtained with the fixed-pitch propeller.

TYPE: Two-seat light sporting aircraft.

WINGS: Cantilever low-wing monoplane. Outer wings similar to those of Midget Mustang, attached to new constant-chord centre-section of short span. Wing section NACA 64A212 at root, NACA 64A210 at tip. Dihedral 5° on outer wings only. Incidence 1° 30'. Two-spar flush-riveted stressed-skin aluminium structure. Wing folding optional. Aluminium statically-balanced ailerons and plain trailing-edge flaps. No trim tabs.

FUSELAGE: Aluminium flush-riveted stressed-skin monocoque structure.

TAIL UNIT: Cantilever all-metal structure. Fixed-incidence tailplane. Controllable trim tab in starboard elevator.

LANDING GEAR: Standard version has non-retractable tailwheel type. Cantilever spring steel main legs. Goodyear 5·00-5 main wheels and tyres, pressure 1·38 bars (20 lb/sq in). Goodyear hydraulic disc brakes. Steerable tailwheel. Alternatively, non-retractable tricycle type. Cantilever spring steel main legs. Cleveland or Goodyear main wheels and tyres size 5·00-5. Non-steerable nosewheel, mounted on oleo-pneumatic shock-strut and free to swivel up to 16° either side. Goodyear nosewheel and tyre size 5·00-5. Goodyear or Cleveland hydraulic disc brakes. Wheel fairings optional on either type of landing gear.

POWER PLANT: Normally one 119 kW (160 hp) Avco Lycoming O-320 flat-four engine, driving a Hartzell two-blade controllable-pitch metal propeller. Provision for other engines including a 93 kW (125 hp) Avco Lycoming O-290, driving a two-blade fixed-pitch metal propeller. Fuel tank aft of firewall, capacity 94·6 litres (25 US gallons). Optional integral wing fuel tanks, each with a capacity of 45 litres (12 US gallons). Refuelling point on starboard side of fuselage aft of firewall. Provision for wingtip tanks. Oil capacity 7·5 litres (2 US gallons).

ACCOMMODATION: Two seats side by side, under large rearward-sliding transparent canopy. Dual controls. Baggage space aft of seats, capacity 34 kg (75 lb).

SYSTEMS: 12V electrical system, supplied by Delco-Remy 15A generator and Exide 33A battery.

AVIONICS AND EQUIPMENT: Provision for full IFR instrumentation and dual nav/com system.

DIMENSIONS, EXTERNAL:
Wing span 7·37 m (24 ft 2 in)
Width folded 2·44 m (8 ft 0 in)
Wing chord at root 1·47 m (4 ft 10 in)
Wing chord at tip 0·79 m (2 ft 7 in)
Wing area, gross 9·02 m² (97·12 sq ft)
Wing aspect ratio 5·5
Length overall 5·94 m (19 ft 6 in)
Height overall 1·60 m (5 ft 3 in)
Height folded 1·73 m (5 ft 8 in)
Tailplane span 2·29 m (7 ft 6 in)
Wheel track 2·08 m (6 ft 10 in)
Propeller diameter:
93 kW (125 hp) 1·73 m (5 ft 8 in)
119 kW (160 hp) 1·82 m (6 ft 0 in)

DIMENSIONS, INTERNAL:
Cabin:
Max width 1·02 m (3 ft 4 in)
Baggage space 0·16 m³ (5·5 cu ft)

WEIGHTS:
Weight empty, equipped (N: nosewheel, T: tailwheel landing gear):
N 93 kW (125 hp) engine 413 kg (911 lb)
T 93 kW (125 hp) engine 408 kg (900 lb)
N 119 kW (160 hp) engine 425 kg (938 lb)
T 119 kW (160 hp) engine 420 kg (927 lb)
*Max T-O and landing weight 680 kg (1,500 lb)

Except for countries that restrict max wing loading to 73·2 kg/m² (15 lb/sq ft), where T-O weight of 658 kg (1,450 lb) applies

PERFORMANCE (with tailwheel, at max T-O weight):
Never-exceed speed:
93 kW (125 hp) 173 knots (322 km/h; 200 mph)
119 kW (160 hp) 211 knots (391 km/h; 243 mph)
Max level speed at S/L:
93 kW (125 hp) 156 knots (290 km/h; 180 mph)
119 kW (160 hp) 200 knots (370 km/h; 230 mph)
Max cruising speed at 2,285 m (7,500 ft):
93 kW (125 hp) 152 knots (282 km/h; 175 mph)
119 kW (160 hp) 181 knots (335 km/h; 208 mph)
Stalling speed, flaps down:
93 kW (125 hp) 47 knots (87 km/h; 54 mph)
119 kW (160 hp) 51 knots (94 km/h; 58 mph)
Stalling speed, flaps up:
93 kW (125 hp) 51 knots (94 km/h; 58 mph)
119 kW (160 hp) 53 knots (96 km/h; 60 mph)
Max rate of climb at S/L:
93 kW (125 hp) 305 m (1,000 ft)/min
119 kW (160 hp) 670 m (2,200 ft)/min
Service ceiling:
93 kW (125 hp) 4,875 m (16,000 ft)
119 kW (160 hp) 6,400 m (21,000 ft)
T-O run:
93 kW (125 hp) 198 m (650 ft)
119 kW (160 hp) 137 m (450 ft)
T-O to 15 m (50 ft):
93 kW (125 hp) 320 m (1,050 ft)
119 kW (160 hp) 198 m (650 ft)
Landing from 15 m (50 ft):
93 kW (125 hp) 290 m (950 ft)
119 kW (160 hp) 259 m (850 ft)
Landing run:
93 kW (125 hp) 215 m (700 ft)
119 kW (160 hp) 168 m (550 ft)
Range with standard fuel (75% power):
93 kW (125 hp) 416 nm (770 km; 480 miles)
119 kW (160 hp) 373 nm (692 km; 430 miles)
Range with optional wingtip tanks:
119 kW (160 hp) 542 nm (1,005 km; 625 miles)

BUTTERWORTH
G. N. BUTTERWORTH

Richmond Airport, Heaton Orchard Road, West Kingston, Rhode Island 02892
Telephone: (401) 789 0384

BUTTERWORTH WESTLAND WHIRLWIND MARK II

Mr G. N. Butterworth has designed and built a ⅔-scale representation of a second World War Westland Whirlwind single-seat fighter. Design began on 20 October 1976, and construction of the prototype started simultaneously. Based on the wings and horizontal tail surfaces of a Grumman American AA-1A Trainer, construction took 900-950 working hours over a nine-month period. First flight was made in July 1977.

Plans are available to amateur constructors.

TYPE: Single-seat ⅔-scale replica fighter.

WINGS: Cantilever low-wing monoplane, based on those of a Grumman American AA-1A Trainer. Dihedral 2°. Incidence 2°. Aluminium structure, except for outer wing panels which are made of 29 mm (1·125 in) Sitka spruce spars and solid polyurethane foam filling, covered with Dynel and polyester resins. Outer wings removable in 15 minutes to permit road trailering. No flaps or tabs.

FUSELAGE: Conventional all-aluminium semi-monocoque structure, with ·02 in skins and ·03 in bulkheads.

TAIL UNIT: Cantilever aluminium structure. High-mounted horizontal surfaces based on those of a Grumman American AA-1A Trainer. Trim tab on starboard elevator.

LANDING GEAR: Hydraulically-retractable tailwheel type. Main wheels retract forward into engine nacelles. Cleveland wheels. Main-wheel tyres size 6·00-6. Tailwheel diameter 10 cm (4in). Oleo shock-absorbers. Cleveland disc brakes.

POWER PLANT: Two 48·5 kW (65 hp) 1,600 cc Volkswagen modified motor car engines, each driving a Hegy fixed-pitch two-blade propeller. Two fuel tanks, total capacity 76 litres (20 US gallons). Refuelling points in top of glassfibre engine cowlings. Oil capacity (total) 5·5 litres (1·5 US gallons).

ACCOMMODATION: Single seat under rearward-sliding transparent canopy. No heating or ventilation.

DIMENSIONS, EXTERNAL:
Wing span 8·53 m (28 ft 0 in)
Wing chord, constant 1·22 m (4 ft 0 in)
Length overall 5·94 m (19 ft 6 in)
Height overall 2·16 m (7 ft 1 in)
Tailplane span 2·49 m (8 ft 2 in)
Wheel track 2·44 m (8 ft 0 in)
Propeller diameter 1·35 m (4 ft 5 in)
Distance between propeller centres 2·13 m (7 ft 0 in)

WEIGHTS:
Weight empty 472 kg (1,042 lb)
Max T-O weight 635 kg (1,400 lb)

PERFORMANCE:
Max level and cruising speed 126 knots (233 km/h; 145 mph)
Econ cruising speed 87 knots (161 km/h; 100 mph)
Stalling speed 55·5 knots (103 km/h; 64 mph)
Max rate of climb at S/L 229 m (750 ft)/min
Service ceiling 2,440 m (8,000 ft)
T-O run 366 m (1,200 ft)
T-O distance to 15 m (50 ft) 549 m (1,800 ft)
Landing distance from 15 m (50 ft) 549 m (1,800 ft)
Landing run 427 m (1,400 ft)
Range with max fuel 608 nm (1,126 km; 700 miles)

CARMICHAEL
BOB CARMICHAEL

Roanoke, Texas

CARMICHAEL LASER 200

Mr Bob Carmichael, a member of the US Aerobatic Team, has built a single-seat sporting monoplane named the Laser 200 (N26SS), derived from the Stephens Akro (which see). It is basically similar to Leo Loudenslager's Laser 200, and to the Super Star built by Henry Haigh, US National Aerobatic Champion in 1979. It incorporates the original wing of the Super Star, which became surplus when Mr Haigh decided to rebuild this aircraft; is stressed to design load factors of $\pm 10g$ and to $\pm 13{\cdot}5g$ ultimate; and is powered by a 149 kW (200 hp) Avco Lycoming IO-360-A1A flat-four engine, driving a two-blade constant-speed metal propeller with spinner. Mr Carmichael first used this aircraft in competition during the 1979 US National Championships held at Sherman Texas.

Brief specification and performance details of the Laser 200 are as follows:

DIMENSIONS, EXTERNAL:
Wing span 7·47 m (24 ft 6 in)
Wing area, gross 8·70 m² (93·6 sq ft)
Length overall 6·10 m (20 ft 0 in)
Propeller diameter 1·93 m (6 ft 4 in)

WEIGHTS:
Weight empty 386 kg (850 lb)
Max T-O weight 499 kg (1,100 lb)

PERFORMANCE (at max T-O weight):
Max level speed 174 knots (322 km/h; 200 mph)
Cruising speed 148 knots (274 km/h; 170 mph)
Landing speed 56 knots (105 km/h; 65 mph)
Max rate of climb at S/L 915 m (3,000 ft)/min

CASSUTT

PLANS AND KITS FROM: National Aeronautics and Manufacturing Co Inc, 905 South 8th Street, Watertown, Wisconsin 53094

While employed as an airline pilot, Capt Tom Cassutt designed and built in 1954 a small single-seat racing monoplane known as the Cassutt Special I (No. 111), in which he won the 1958 National Air Racing Championships. In 1959, he completed a smaller aircraft on the same lines, known as the Cassutt Special II (No. 11). Twenty years later, Cassutt Specials took first, second and third places at the 1979 Cleveland National Air Races, and won the EAA 500 miles efficiency race at Oshkosh in all categories in both 1979 and 1980.

Plans and kits of parts of the original design, and plans of a sporting version of No. 111 with a larger cockpit, are available to amateur constructors through Mr James Adams of the National Aeronautics and Manufacturing Co Inc; about 2,000 sets of plans have been sold. More than 125 Cassutt Specials were flying, and many others were under construction by early 1981, in Australia, France, West Germany, New Zealand, South Africa, Sweden, the UK and elsewhere.

CASSUTT SPECIAL I

The following description applies to the Cassutt Special I as offered in Racing and Sport versions by the National Aeronautics and Manufacturing Co Inc. The Sport model, for sport flying and aerobatics, is essentially the same as the Racing model, except for having a greater wing span

Mr G. N. Butterworth's ⅔-scale replica of a Westland Whirlwind

Carmichael Laser 200 aerobatic aircraft *(Don Berliner)*

Cassutt Special owned by Mr Eldon McDaniel of Merrimac, Wisconsin

Christen Industries Eagle I flown by Mr Gene Soucy of the three-aircraft Eagles Aerobatic Flight

Christen Industries Eagle II two-seat aerobatic biplane built by Mr Dick Blair of Vincetown, NJ *(Howard Levy)*

and area, slightly raised seat for the pilot, and alternative open or enclosed cockpit.

A description of the original Cassutt Special I, *Jersey Skeeter*, can be found in the 1980-81 *Jane's*.

TYPE: Single-seat racing monoplane.

WINGS: Cantilever mid-wing monoplane. Wing section Cassutt 1107. No incidence or dihedral. All-wood two-spar structure with spruce ribs, solid spars and plywood skin, fabric-covered. Ailerons are of welded steel tube construction, fabric-covered. No flaps as standard, but a few constructors have added flaps. No tabs.

FUSELAGE: Steel tube structure, fabric-covered.

TAIL UNIT: Cantilever steel tube structure, with fabric covering. No tabs.

LANDING GEAR: Non-retractable tailwheel type. Wittman cantilever spring steel main legs. Main-wheel tyres size 5·00-5. Wheel fairings standard.

POWER PLANT: One Continental flat-four engine in the 63·5-74·5 kW (85-100 hp) range, driving a two-blade propeller. Fuel capacity 55 litres (14·5 US gallons). Oil capacity 3·8 litres (1 US gallon).

ACCOMMODATION: Single seat.

DIMENSIONS, EXTERNAL (A: Racing, B: Sport):
- Wing span: A — 4·57 m (15 ft 0 in)
- B — 5·18 m (17 ft 0 in)
- Wing chord (constant): A, B — 1·37 m (4 ft 6 in)
- Wing area, gross: A — 6·27 m² (67·5 sq ft)
- B — 7·11 m² (76·5 sq ft)
- Length overall: A, B — 4·88 m (16 ft 0 in)
- Height overall: A, B — 1·22 m (4 ft 0 in)
- Tailplane span — 1·19 m (3 ft 11 in)
- Wheel track — 1·37 m (4 ft 6 in)

WEIGHTS:
- Weight empty: A, B — 227 kg (500 lb)
- Max T-O weight: A, B — 363 kg (800 lb)

PERFORMANCE (63·5 kW; 85 hp engine):
- Max level speed: A, B — more than 174 knots (322 km/h; 200 mph)
- Max cruising speed: A, B — more than 156 knots (290 km/h; 180 mph)
- Landing speed: A — 61 knots (113 km/h; 70 mph)
- B — 52 knots (97 km/h; 60 mph)
- Stalling speed: A — 61 knots (113 km/h; 70 mph)
- Max rate of climb at S/L: A — 457 m (1,500 ft)/min
- B — 914 m (3,000 ft)/min
- Range: A, B — 425 nm (788 km; 490 miles)
- *g* limit: A, B — −6·5*g*

CASSUTT SPECIAL II

TYPE: Single-seat racing monoplane.

WINGS: Cantilever mid-wing monoplane. Wing section Cassutt 13106. No incidence or dihedral. All-wood structure. No flaps.

FUSELAGE: Steel tube structure, fabric-covered.

TAIL UNIT: Cantilever steel tube structure, fabric-covered. The prototype (No. 11) has small centre fin and auxiliary fins on the tailplane tips.

LANDING GEAR: Non-retractable tailwheel type. Wittman cantilever spring steel main legs. Main-wheel tyre size 5·00-5. Wheel fairings standard.

POWER PLANT AND ACCOMMODATION: As for Cassutt Special I.

DIMENSIONS, EXTERNAL:
- Wing span — 4·16 m (13 ft 8 in)
- Wing chord, constant — 1·47 m (4 ft 10 in)
- Wing area, gross — 6·13 m² (66 sq ft)
- Wing aspect ratio — 2·83
- Length overall — 4·88 m (16 ft 0 in)
- Height overall — 1·16 m (3 ft 10 in)
- Tailplane span — 1·14 m (3 ft 9 in)
- Wheel track — 0·97 m (3 ft 2 in)

WEIGHTS:
- Weight empty — 196 kg (433 lb)
- Max T-O weight — 363 kg (800 lb)

PERFORMANCE (at max T-O weight):
- Max level speed at S/L — 204 knots (378 km/h; 235 mph)
- Max cruising speed — 174 knots (322 km/h; 200 mph)
- Stalling speed — 54 knots (100 km/h; 62 mph)
- Max rate of climb at S/L — 915 m (3,000 ft)/min
- Endurance with max fuel — 3 h

CHRISTEN

CHRISTEN INDUSTRIES INC

1048 Santa Ana Valley Road, Hollister, California 95023
Telephone: (408) 637 7405
PRESIDENT: Frank L. Christensen

Two aircraft make up the current Christen Eagle series of aerobatic biplanes. The **Eagle I** is a special-purpose single-seater intended solely for unlimited class aerobatic competition and is structured to +9*g*, −6*g* operational. It is powered by a 194 kW (260 hp) Avco Lycoming AEIO-540-E4B5 flat-six engine, driving a Sensenich fixed-pitch propeller. The **Eagle II** is a less-austere two-seat sporting biplane that can be used for unlimited class aerobatics as well as for advanced aerobatic training and comfortable cross-country flying. It has the same structural limits as the Eagle I. All Eagles have a roll rate of 187° a second.

First version to become available was the Eagle II, which can be built by amateur constructors, using a series of 26 parts-kits supplied by Christen Industries. Each kit makes up a separate portion of the aircraft and is supported by a very detailed constructional manual. This allows the cost of building the aircraft to be spread over a period of time. Only common hand tools (no power tools) are required in construction, and no previous aircraft experience is thought necessary. Christen believes that a typical homebuilder could complete the aircraft in 1,400 to 1,600 man-hours.

CHRISTEN EAGLE I

Following completion of its flight test programme, the Eagle I was expected to become available in kit form in the Summer of 1981. It is generally similar to the Eagle II, with unchanged overall dimensions, but is a single-seater powered by a 194 kW (260 hp) Avco Lycoming engine. The firewall is moved aft for the larger engine, and a fuel tank replaces the forward seat. Aileron slave struts have been eliminated.

CHRISTEN EAGLE II

Design began in June 1974 and construction of a prototype started in August 1975. First flight was achieved in February 1977, when construction of other pre-production Eagles began. Manufacture of kits was initiated by Christen Industries in October 1977 and by January 1981 a total of 400 had been ordered.

TYPE: Two-seat unlimited class aerobatic biplane.

WINGS: Braced biplane with steel tube I-type main interplane struts. Symmetrical sections. Thickness/chord ratio 15%. Dihedral 0° 30′ on lower wings, 0° on upper wing. Incidence 0°. Sweepback at quarter-chord 7° 30′ on upper wing, 0° on lower wings. Wooden spars and ribs; metal leading- and trailing-edges, polyester fabric covered. Conventional ailerons on upper and lower wings, of similar construction to wings. No flaps or tabs.

FUSELAGE: Conventional structure of welded 4130N steel tubing. Covered with removable light alloy panels from firewall to back of rear seat. Rear fuselage fabric covered.

TAIL UNIT: Conventional wire-braced structure, comprising a tailplane and elevators fitted with boost tabs, and fin and rudder, all of welded steel tube, polyester fabric covered.

LANDING GEAR: Non-retractable tailwheel type. Aluminium spring main legs. Main tyres size 5·00-5, pressure 1·52 bars (22 lb/sq in). Cleveland hydraulic disc brakes. Streamline wheel fairings.

POWER PLANT: One 149 kW (200 hp) Avco Lycoming AEIO-360-A1D flat-four engine, driving a Hartzell HC-C2YK-4/C7666A-2 constant-speed propeller. One fuel tank in fuselage, capacity 98·4 litres (26 US gallons). Refuelling point in upper fuselage. Oil capacity 7·5 litres (2 US gallons). Fuel system allows unlimited inverted flight.

ACCOMMODATION: Two seats in tandem beneath one-piece side-hinged bubble canopy. Heated. Baggage hold in turtledeck, capacity 13·6 kg (30 lb).

SYSTEMS: 12V DC battery for starting and radio.

AVIONICS: Edo-Aire RT-563 or RT-553 radio.

DIMENSIONS, EXTERNAL:

Wing span	6·07 m (19 ft 11 in)
Wing chord, constant	1·02 m (3 ft 4 in)
Wing area, gross	11·61 m² (125 sq ft)
Length overall	5·64 m (18 ft 6 in)
Height overall	1·98 m (6 ft 6 in)
Tailplane span	2·13 m (7 ft 0 in)
Wheel track	1·83 m (6 ft 0 in)
Wheelbase	3·96 m (13 ft 0 in)
Propeller diameter	1·93 m (6 ft 4 in)

DIMENSIONS, INTERNAL:

Cabin: Length	2·13 m (7 ft 0 in)
Max width	0·71 m (2 ft 4 in)
Max height	0·99 m (3 ft 3 in)

WEIGHTS:

Weight empty	465 kg (1,025 lb)
Max T-O and landing weight	716 kg (1,578 lb)

PERFORMANCE:

Never-exceed speed	184 knots (341 km/h; 212 mph)
Max level speed at S/L	160 knots (296 km/h; 184 mph)
Max cruising speed at 1,825 m (6,000 ft)	143 knots (265 km/h; 165 mph)
Econ cruising speed at 1,825 m (6,000 ft)	137 knots (254 km/h; 158 mph)
Stalling speed	50·5 knots (94 km/h; 58 mph)
Max rate of climb at S/L	640 m (2,100 ft)/min
Service ceiling	5,180 m (17,000 ft)
T-O run	244 m (800 ft)
T-O to 15 m (50 ft)	381 m (1,250 ft)
Landing from 15 m (50 ft)	480 m (1,575 ft)
Range with max fuel and max payload	330 nm (611 km; 380 miles)

CONDOR

CONDOR AERO INC

PO Box 762, Vero Beach, Florida 32960

PRESIDENT: Landis G. Ketner

CONDOR SHOESTRING

The original Shoestring was designed and built by the Mercury Air Group, consisting of Mr Vincent Ast, Mr Carl Ast and Mr Rodney Kreimendahl, in 1949. As a racing monoplane (No. 16) it first competed at Miami in 1950 where, piloted by Bob Downey, it achieved third place at a speed of 157·474 knots (291·844 km/h; 181·334 mph). At San Jose, California, in the same year it finished first in the hands of Vincent Ast, at 152 knots (282 km/h; 175 mph). By 1974, when a new wing was fitted to Shoestring, the aircraft had achieved 14 first places in air races, three seconds and four thirds.

A full description of the original Shoestring appeared in the 1959-60 *Jane's*.

In 1965 Shoestring was purchased by John Anderson and taken to Miami, where a set of drawings of the aircraft was made. The first new racer to be built from these drawings was *Yellow Jacket*, constructed by Jim Strode, which flew for the first time on 15 July 1970. Although this aircraft was very similar to the original Shoestring, the fuselage structure was made 5 cm (2 in) wider and 10 cm (4 in) longer to provide a more comfortable cockpit. The power plant installation, canopy and fairings were also of Jim Strode's design.

Since then other Shoestrings have been built in modified form, and plans are available to amateur constructors.

The Shoestring illustrated was built by Ralph Moore of Titusville, Florida. It was the sixth plans-built aircraft to fly, on 16 December 1978. The weight and performance data which follow apply to this aircraft.

TYPE: Single-seat sporting monoplane.

WINGS: Cantilever shoulder-mounted monoplane of wood construction, built as one unit and fabric covered. Front spar built of three pieces of spruce bonded together. Rear spar can be cut from one piece of spruce or built up from several pieces. Plywood ribs with capstrips bonded to top and bottom. Skins bonded to spars and ribs. Ailerons built as part of wing and cut out after the top skin has been bonded in place.

FUSELAGE: Steel tube structure, with plywood formers and fabric covering from instrument panel aft. Thin-gauge aluminium skin forward of panel and on each side of cockpit above wing. Rollover structure of heavy gauge steel.

TAIL UNIT: Conventional cantilever spruce structure, with fabric-covered plywood skin.

LANDING GEAR: Non-retractable tailwheel type. Main wheels carried on one-piece formed aluminium member bolted directly to bottom of fuselage. Sheet aluminium fairings over legs; aluminium wheel fairings. Solid rubber tailwheel. Brakes fitted.

POWER PLANT: Originally one 63·5 kW (85 hp) Continental flat-four engine. Alternative 74·5 kW (100 hp) engine to meet current PRPA regulations. Recommended engines are Continental C85, C90 and O-200 flat-fours. Fuel tank aft of firewall, capacity 38 litres (10 US gallons). Provision for auxiliary tank behind seat.

ACCOMMODATION: Single seat under blown Plexiglas canopy.

DIMENSIONS, EXTERNAL:

Wing span	5·79 m (19 ft 0 in)
Wing area, gross	6·13 m² (66 sq ft)
Length overall	5·38 m (17 ft 7¾ in)
Height overall	1·42 m (4 ft 8 in)

WEIGHTS (aircraft illustrated):

Weight empty, equipped (with oil)	358 kg (788 lb)
Max T-O weight	476 kg (1,050 lb)

PERFORMANCE (aircraft illustrated):

Max level speed at S/L	200 knots (370 km/h; 230 mph)
Max cruising speed at 305 m (1,000 ft)	174 knots (322 km/h; 200 mph)
Econ cruising speed at 610 m (2,000 ft)	148 knots (274 km/h; 170 mph)
Stalling speed, power off	69·5 knots (129 km/h; 80 mph)
Service ceiling	5,485 m (18,000 ft)
T-O run	245 m (800 ft)
T-O to 15 m (50 ft)	305 m (1,000 ft)
Landing from 15 m (50 ft)	610 m (2,000 ft)
Landing run, without braking	488 m (1,600 ft)
Range with max fuel	295 nm (547 km; 340 miles)

CVJETKOVIC

ANTON CVJETKOVIC

5324 West 121 Street, Hawthorne, California 90250

Telephone: (213) 644 7931

When living in Yugoslavia, Mr Anton Cvjetkovic designed a single-seat light aeroplane designated CA-51 and powered by a modified Volkswagen engine. A prototype was built by members of Zagreb Aeroclub in 1951, and was followed by five more aircraft of the same type.

After moving to the USA, Mr Cvjetkovic began work, in May 1960, on the design of an improved light aircraft which he designated CA-61. Construction of a prototype was started in February 1961 and it flew for the first time in August 1962. Plans of both single-seat and two-seat all-wood versions are available to amateur constructors, together with plans of a two-seat all-wood aircraft designated CA-65, and an all-metal version, designated CA-65A.

By early 1981 more than 400 sets of plans of these aircraft had been sold, and completed aircraft were flying in Australia, Canada, South Africa and the USA.

CVJETKOVIC CA-61/-61R MINI ACE

The CA-61 can be built as a single-seat or side-by-side two-seat light aircraft, with any Continental engine of between 48·5 and 63·5 kW (65 and 85 hp). Alternatively, the single-seater can be fitted with a modified Volkswagen engine. Construction takes less than 1,000 h.

The design was modified during 1973 to allow for installation of retractable landing gear; when constructed in this form the aircraft is designated **CA-61R**.

The following details refer specifically to the single-seat CA-61 prototype:

TYPE: Single-seat light aircraft.

WINGS: Cantilever low-wing monoplane. Wing section NACA 4415. Dihedral 3°. No incidence. Structure consists of two spruce spars, each built in one piece, built-up spruce girder-type ribs and plywood-covered leading-edge torsion box, with fabric covering overall. Fabric-covered spruce ailerons. No flaps.

FUSELAGE: Conventional wooden structure of basic square section, plywood-covered.

TAIL UNIT: Cantilever wooden structure, covered with plywood. Fixed-incidence tailplane. Trim tab in elevator.

LANDING GEAR: Non-retractable tailwheel type. Cantilever main legs, with helical spring shock-absorption. Goodyear main wheels and tyres, size 5·00-5 Type III, and Model L5 brakes. Steerable tailwheel.

POWER PLANT: One 48·5 kW (65 hp) Continental A65 flat-four engine, driving a Flottorp 63-55 two-blade fixed-pitch propeller. Fuel in two steel tanks in fuselage, with capacities of 45 litres (12 US gallons) and 19 litres (5 US gallons) respectively. Total fuel capacity 64 litres (17 US gallons). Oil capacity 4·5 litres (1·25 US gallons).

ACCOMMODATION: Single seat in enclosed cockpit.

AVIONICS AND EQUIPMENT: Prototype fitted with Nova Star radio and Omni.

DIMENSIONS, EXTERNAL:

Wing span	8·38 m (27 ft 6 in)
Wing chord, constant	1·40 m (4 ft 7 in)
Wing area, gross	11·75 m² (126·5 sq ft)
Wing aspect ratio	6·0
Length overall	5·77 m (18 ft 11 in)
Height overall (in flying position)	2·08 m (6 ft 10 in)
Wheel track:	
Single-seat	2·49 m (8 ft 2 in)
Two-seat	2·62 m (8 ft 7 in)

WEIGHTS:

Weight empty:	
Single-seat	275 kg (606 lb)
Two-seat	363 kg (800 lb)
Max T-O weight:	
Single-seat	430 kg (950 lb)
Two-seat	590 kg (1,300 lb)

PERFORMANCE:

Max level speed at S/L	104 knots (193 km/h; 120 mph)
Normal cruising speed	87 knots (161 km/h; 100 mph)
Min flying speed:	
Single-seat	37 knots (67·5 km/h; 42 mph)
Two-seat	44 knots (80·5 km/h; 50 mph)
Range with max fuel:	
Single-seat	369 nm (685 km; 425 miles)
Two-seat	321 nm (595 km; 370 miles)

CVJETKOVIC CA-65

Design of this side-by-side two-seat light aircraft was started in September 1963. Construction of the prototype began in March 1964 and it flew for the first time in July 1965. Plans are available.

The CA-65 closely resembles the CA-61 in general appearance, but has a more powerful engine and retractable landing gear. A folding-wing version was introduced during 1967.

TYPE: Two-seat light aircraft.

WINGS: Cantilever low-wing monoplane. Modified NACA 4415 wing section. Dihedral 0° on centre-section, 3° on outer wings. Structure consists of two spruce spars, each built in one piece, and built-up spruce girder-type ribs, completely plywood-covered. Fabric-covered spruce ailerons. On the folding-wing version, the outer wings fold upward from their junction with the centre-section.

FUSELAGE: Conventional wooden structure of basically square section, plywood-covered. Manually-operated landing flap under fuselage.

TAIL UNIT: Cantilever wooden structure. Fixed surfaces covered with plywood. Elevator and rudder fabric-covered. Fixed-incidence tailplane.

LANDING GEAR: Mechanically-retractable tailwheel type. Main wheels retract inward. Goodyear main wheels and tyres, size 5·00-5 Type III. Goodyear type L5 brakes. Steerable tailwheel.

POWER PLANT: One 93 kW (125 hp) Avco Lycoming O-290-G flat-four engine, driving a Sensenich 66-68 two-blade fixed-pitch propeller. Two aluminium fuel tanks in fuselage, each with capacity of 53 litres (14 US gallons). Total fuel capacity 106 litres (28 US gallons).

ACCOMMODATION: Two seats side by side in enclosed cockpit, with dual controls; although hydraulic brakes can be operated only by the pilot. Forward-opening canopy.

RADIO: Bayside BE1-990 radio fitted in prototype.

DIMENSIONS, EXTERNAL:

Wing span	7·62 m (25 ft 0 in)
Wing area, gross	10·03 m² (108 sq ft)
Width, wings folded	2·74 m (9 ft 0 in)
Length overall	5·79 m (19 ft 0 in)
Height overall (in flying position)	2·24 m (7 ft 4 in)
Height, wings folded	3·05 m (10 ft 0 in)
Wheel track	2·11 m (6 ft 11 in)
Propeller diameter	1·73 m (5 ft 8 in)

WEIGHTS:

Weight empty	408 kg (900 lb)
Max T-O weight	680 kg (1,500 lb)

PERFORMANCE (at max T-O weight):

Max level speed	139 knots (257 km/h; 160 mph)
Normal cruising speed	117 knots (217 km/h; 135 mph)
Stalling speed	48 knots (89 km/h; 55 mph)
Max rate of climb at S/L	305 m (1,000 ft)/min
Service ceiling	4,575 m (15,000 ft)
T-O run	137 m (450 ft)
Landing run	183 m (600 ft)
Range with max fuel	434 nm (804 km; 500 miles)

Condor Shoestring built by Mr Ralph Moore of Titusville, Florida

First folding-wing version of the Cvjetkovic CA-65, built by Mr Graham Scarr of Duffy, Australia

The single-seat Cvjetkovic CA-61R (Continental A65 engine)

First Cvjetkovic CA-65A, built by students of the Kelowna Secondary School, Kelowna, British Columbia, Canada

CVJETKOVIC CA-65A

This aircraft is an all-metal version of the wooden CA-65, with swept vertical tail surfaces. It is designed for +9*g* and −6*g* ultimate loading.

The general description of the CA-65 applies also to the CA-65A, except in the following details:

Wings: The wing structure consists of a single main spar and an auxiliary wing spar, with aluminium sheet ribs and skin, riveted throughout. The main wing spar cap is made of extruded and bent-up sheet aluminium angles, tapered towards the tip to produce a wing of uniform bending strength. Ribs are formed from 0·025 in aluminium sheet. Wing skin is of 2024T-3 aluminium alloy sheet.

Fuselage: All-metal structure with four aluminium angle longerons and built-up frames. Fuselage skin is of 0·025-0·032 in 2024T-3 aluminium alloy sheet. To simplify formation of the curvature on the upper fuselage, the skins are broken up into small sections of flat panels.

Tail Unit: Cantilever all-metal structure, with swept vertical surfaces and dorsal fin. Construction similar to that of the wings.

Power Plant: The structure is designed to accommodate an Avco Lycoming engine of 80·5-112 kW (108-150 hp).

Dimensions, external: As for Model CA-65 except:

Wing span	7·75 m (25 ft 5 in)
Wing area, gross	10·16 m² (109·4 sq ft)
Length overall	5·99 m (19 ft 8 in)
Height overall	2·29 m (7 ft 6 in)

Performance (112 kW; 150 hp engine):

Max level speed	151 knots (280 km/h; 174 mph)
Normal cruising speed	130 knots (241 km/h; 150 mph)
Stalling speed	48 knots (89 km/h; 55 mph)
Max rate of climb at S/L	466 m (1,530 ft)/min
Service ceiling	4,570 m (15,000 ft)
T-O run	99 m (325 ft)
Landing run	183 m (600 ft)
Range with max fuel	460 nm (853 km; 530 miles)

D'APUZZO
NICHOLAS E. D'APUZZO

1029 Blue Rock Lane, Blue Bell, Pennsylvania 19422
Telephone: (215) 646 4792

Mr D'Apuzzo, who was formerly employed by the Naval Air Development Center, Warminster, Pennsylvania, as a project manager on specialised projects, retired from the Navy Department during 1973. He retains an association with the Navy on a consultant basis.

He has designed several sporting aircraft for amateur construction, among the best known of which are the Denight Special midget racer, described in the 1962-63 *Jane's,* and the PJ-260 single-seat aerobatic biplane described under the Parsons-Jocelyn heading in the 1974-75 *Jane's*. His other designs include the Senior Aero Sport, which is a two-seat version of the PJ-260, the smaller single-seat D-200 Freshman, and the D-201 Sportwing.

By early 1981 a total of 31 PJ-260s and Senior Aero Sports were known to have been completed by amateur constructors in the USA, with at least 33 more under construction.

D'APUZZO D-260/D-295 SENIOR AERO SPORT

There are six versions of the Senior Aero Sport, as follows:

D-260(1). With Avco Lycoming O-435 series engine.

D-260(2). With Continental O-470/E-185 series engine. Prototype first flown on 17 July 1965, with a 168 kW (225 hp) E-185 (modified) engine, driving an Aeromatic F-200H-O-85 propeller. Fuel in one 60·5 litre (16 US gallon) tank in fuselage and one 79·5 litre (21 US gallon) streamlined external tank under fuselage.

D-260(3). With Avco Lycoming GO-435 series engine. Construction of prototype started by Mr G. A. Shallbetter of Minneapolis in April 1961 and first flight made on 17 July 1965, with a 194 kW (260 hp) GO-435-C2 engine, driving a Hartzell controllable-pitch propeller of 2·29 m (7 ft 6 in) diameter. Four fuel tanks in fuselage, with total capacity of 136 litres (36 US gallons). Second D-260(3), completed by Mr Alfred Fessenden of Lafayette, New York, is fitted with a Hartzell constant-speed propeller.

D-260(4). With Ranger 6-440-C six-cylinder inverted aircooled engine.

D-260(5). With 224 kW (300 hp) Avco Lycoming R-680-E3 engine. Prototype built by Mr Henry Neys of Lake Stevens, Washington.

D-295. With 209 kW (280 hp) Avco Lycoming GO-480-G1D6 engine. First example built by Mr Ed Mahler; D-295-2 (illustrated in 1979-80 *Jane's*) completed in 1976 by Mr Geoffrey Geisz of Flint, Michigan.

Type: Two-seat sporting biplane.

Wings: Conventional braced biplane. Wing section NACA M-12 (modified). Dihedral 30′ on lower wings only. Incidence 2°. Sweepback 9° 15′ on upper wings. Two wood spars, metal ribs, fabric covering. Metal Frise ailerons on all four wings, with fabric covering. No flaps.

Fuselage: Steel tube structure, with aluminium alloy access panels forward of cockpit and fabric covering aft.

Tail Unit: Wire-braced fabric-covered steel tube structure. Trim tab in starboard elevator.

Landing Gear: Non-retractable tailwheel type. Cantilever spring steel main units. Goodyear 6·00-6 main wheels and tyres, pressure 1·38 bars (20 lb/sq in). Goodyear disc brakes.

Power Plant: One flat-six engine; details given under individual model listings. Fuel in 125 litre (33 US gallon) main tank and 41·5 litre (11 US gallon) aerobatic tank, both in fuselage. Total fuel capacity 166·5 litres (44 US gallons). Oil capacity 11 litres (3 US gallons).

Accommodation: Two seats in tandem in open cockpits. Baggage space behind headrest.

System: 12V electrical system, with optional starter and navigation lights.

Avionics: Prototypes fitted with two-way radio and Omni.

Dimensions, external:

Wing span	8·23 m (27 ft 0 in)
Wing chord, constant (both)	1·17 m (3 ft 10 in)
Wing area, gross	17·2 m² (185 sq ft)
Wing aspect ratio	4·22
Length overall	6·40 m (21 ft 0 in)
Height overall	2·32 m (7 ft 7½ in)
Tailplane span	3·10 m (10 ft 2 in)
Wheel track	2·57 m (8 ft 5 in)
Wheelbase	4·80 m (15 ft 9 in)

Weights:

Normal T-O weight:	
D-260 (2)	930 kg (2,050 lb)
D-260 (3)	975 kg (2,150 lb)
Max T-O and landing weight:	
All versions	975 kg (2,150 lb)

Performance (D-260 (3) at max T-O weight. D-260 (2) comparable):

Never-exceed speed	165 knots (305 km/h; 190 mph)
Max level speed at 2,135 m (7,000 ft)	135 knots (250 km/h; 155 mph)
Max cruising speed at 2,135 m (7,000 ft)	122 knots (225 km/h; 140 mph)
Econ cruising speed at 2,135 m (7,000 ft)	113 knots (209 km/h; 130 mph)
Stalling speed	48 knots (89 km/h; 55 mph)
Max rate of climb at S/L	610 m (2,000 ft)/min
Service ceiling	6,100 m (20,000 ft)
T-O run	122 m (400 ft)
T-O to 15 m (50 ft)	213 m (700 ft)
Landing from 15 m (50 ft)	275 m (900 ft)
Landing run	183 m (600 ft)
Range with max fuel and max payload	434 nm (805 km; 500 miles)

D'APUZZO D-200 FRESHMAN

The D-200 Freshman (formerly Junior Aero Sport) is a smaller single-seat version of the PJ-260. Its design was

Partially completed airframe of D'Apuzzo D-200 Freshman

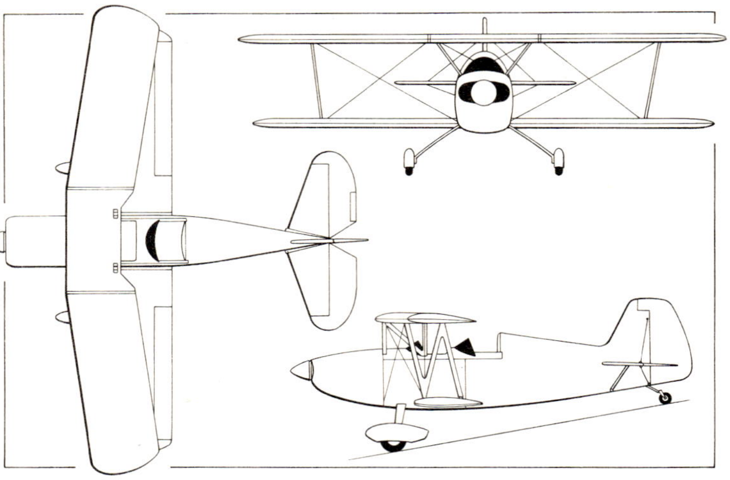

D'Apuzzo D-201 Sportwing, developed from the PJ-260/D-260 series
(Michael A. Badrocke)

D'Apuzzo D-260(1) Senior Aero Sport built by Mr Merrill Vallender of Downer's Grove, Illinois

D'Apuzzo D-201 Sportwing in January 1981

started in September 1963 and construction of two prototypes began in September 1964. The first of these was almost completed in early 1973 when the building in which it was being constructed was destroyed by fire. The Freshman was not damaged severely, but due to the delay in obtaining new workshop premises completion of this prototype was not resumed until early 1975. The aircraft has been inspected by the FAA and approved for final assembly. This is being deferred until initial testing of the Sportwing has been completed.

TYPE: Single-seat sporting biplane.

WINGS: Conventional braced biplane. Wing section NACA M-12 (mod). Dihedral 0° on top wing, 0° 30′ on bottom wing. Incidence 2° on both wings. No sweepback. Spruce spars, wooden ribs and light alloy nose skin, with fabric covering. Fabric-covered aluminium alloy ailerons. No flaps.

FUSELAGE: Welded steel tube structure, with aluminium alloy panels forward of cockpit and fabric covering aft.

TAIL UNIT: Wire-braced welded steel tube structure with fabric covering. Fixed-incidence tailplane. Trim tab in each elevator.

LANDING GEAR: Non-retractable tailwheel type. Cantilever spring steel main legs. Goodrich-Hayes main wheels and tyres, size 5·00-4, pressure 1·38 bars (20 lb/sq in). Goodrich-Hayes brakes.

POWER PLANT: One 134 kW (180 hp) Avco Lycoming O-360 flat-four engine, driving a Hartzell two-blade constant-speed metal propeller. Fuel tank in fuselage, capacity 75 litres (20 US gallons). Oil capacity 7·5 litres (2 US gallons).

ACCOMMODATION: Single seat in open cockpit.

DIMENSIONS, EXTERNAL:
Wing span (both) 6·60 m (21 ft 8 in)
Wing chord, constant 1·02 m (3 ft 4 in)
Wing area, gross 13·0 m² (140 sq ft)
Wing aspect ratio 4·2
Length overall 5·56 m (18 ft 3 in)
Height overall 1·93 m (6 ft 4 in)
Tailplane span 2·18 m (7 ft 2 in)
Wheel track 1·52 m (5 ft 0 in)
Wheelbase 4·11 m (13 ft 6 in)

WEIGHTS:
Weight empty 381 kg (840 lb)
Max T-O and landing weight 578 kg (1,275 lb)

PERFORMANCE (estimated, at max T-O weight):
Never-exceed speed 191 knots (354 km/h; 220 mph)
Max level speed at 2,135 m (7,000 ft) 139 knots (257 km/h; 160 mph)
Max cruising speed at 2,135 m (7,000 ft) 122 knots (225 km/h; 140 mph)
Stalling speed 48 knots (89 km/h; 55 mph)
Max rate of climb at S/L 762 m (2,500 ft)/min
Service ceiling 6,100 m (20,000 ft)
T-O run 122 m (400 ft)
T-O to 15 m (50 ft) 198 m (650 ft)
Landing from 15 m (50 ft) 260 m (850 ft)
Landing run 168 m (550 ft)
Range with max fuel 260 nm (480 km; 300 miles)

D'APUZZO D-201 SPORTWING

The D-201 is a completely redesigned development of the PJ-260/D-260 Senior Aero Sport series. Special attention has been given to reducing the cost and complexity of building the aircraft, while retaining the safety aspects of the previous models. The aircraft features long-span ailerons on the lower wings only, to reduce weight and to simplify construction; the front cockpit has been enlarged by 7·6 cm (3 in), without affecting the rear cockpit; and the wing panel structures and tailplane assembly have been redesigned. Many preformed parts are available to the homebuilder.

Construction of the prototype D-201 began in January 1977; by early 1981 it was at the final paint and trim stage, having been inspected by the FAA and covered. Rollout was expected to be in late Spring 1981.

TYPE: Two-seat sporting biplane.

WINGS: Conventional braced biplane. Wing section NACA M-12 (modified). Dihedral 0° on upper wings; 0° 30′ on lower wings. Incidence 2° on all wings. Sweepback 9° 15′ on upper wings only, outboard of centre-section. Long-span ailerons, on lower wings only. N-type interplane struts and cabane struts.

FUSELAGE: Conventional steel tube structure, with aluminium alloy panels forward of cockpit and fabric covering aft.

TAIL UNIT: Conventional wire-braced structure. Trim tab in starboard elevator. All control surfaces horn-balanced.

LANDING GEAR: Non-retractable tailwheel type. Cantilever spring steel main units. Wheel fairings.

POWER PLANT: One 119 kW (160 hp) Avco Lycoming IO-320-B1A flat-four engine driving a Hartzell constant-speed propeller in prototype. Design suitable for any Avco Lycoming engine from the 112 kW (150 hp) O-320 to the 149 kW (200 hp) O-360.

ACCOMMODATION: Two seats in tandem. Optional canopy over rear cockpit.

SYSTEM: Starter and generator standard.

DIMENSIONS, EXTERNAL:
Wing span, upper 8·23 m (27 ft 0 in)
Wing span, lower 7·86 m (25 ft 9½ in)
Wing chord, constant 1·17 m (3 ft 10 in)
Wing area, gross 17·14 m² (184·5 sq ft)
Length overall, tail up 6·59 m (21 ft 7½ in)
Height overall, tail down 2·34 m (7 ft 8 in)
Elevator span 3·15 m (10 ft 4 in)
Wheel track 2·57 m (8 ft 5 in)

WEIGHTS (estimated, with O-320 engine):
Weight empty 537·5 kg (1,185 lb)
Design max T-O weight 839 kg (1,850 lb)

PERFORMANCE:
No details yet available

DAVIS

LEEON D. DAVIS

PO Box 207, 405 North St Paul, Stanton, Texas 79782

Telephone: (915) 756 2100

Details of the first light aircraft designed by Mr Davis, the DA-1A five-seat high-wing monoplane, can be found in the 1960-61 *Jane's*. He completed subsequently the prototype of a two-seat low-wing monoplane designated DA-2A, of which plans are available to other builders, together with plans of the single-seat DA-5A of similar configuration.

DAVIS DA-2A

This side-by-side two-seat light aircraft was flown for the first time on 21 May 1966, after 18 months of spare-time work and an expenditure of $1,600. At the Experimental Aircraft Association's annual Fly-in a few weeks later, it gained the awards for both the most outstanding design and the most popular aircraft. Plans are available to other amateur constructors.

By early 1976 the prototype had flown a total of 860 hours, and the twenty DA-2As that were then flying had accumulated approximately 4,000 flying hours. It is believed that about 100 other DA-2As are under construction.

The DA-2A is of simple all-metal construction and has an all-moving V tail (included angle 100°) like Mr Davis's earlier DA-1A. The wings are of constant chord, without flaps. Dihedral is 5°. The non-retractable tricycle landing gear has cantilever spring steel main legs and a steerable nosewheel. Power plant is a 48·5 kW (65 hp) Continental A65-8 flat-four engine; but the DA-2A is stressed for engines of up to 74·5 kW (100 hp). Total fuel capacity is 75 litres (20 US gallons) and oil capacity 3·75 litres (1 US gallon).

There is baggage space aft of the side-by-side seats or,

Davis DA-5A single-seat lightweight sporting aircraft

Davis DA-2A side-by-side two-seat light aircraft *(J. M. G. Gradidge)*

alternatively, a child's seat may be located in this position.

DIMENSIONS, EXTERNAL:

Wing span	5·86 m (19 ft 2¾ in)
Wing chord, constant	1·31 m (4 ft 3½ in)
Wing area, gross	7·66 m² (82·5 sq ft)
Wing aspect ratio	4·48
Length overall	5·44 m (17 ft 10¼ in)
Height overall	1·65 m (5 ft 5 in)

DIMENSIONS, INTERNAL:

Cabin: Length	1·49 m (4 ft 6¾ in)
Max width	1·04 m (3 ft 5 in)
Max height	1·14 m (3 ft 8¾ in)

WEIGHTS:

Weight empty	277 kg (610 lb)
Max T-O weight	510 kg (1,125 lb)

PERFORMANCE (at max T-O weight):

Max level speed at S/L	104 knots (193 km/h; 120 mph)
Cruising speed	100 knots (185 km/h; 115 mph)
Landing speed	54 knots (100 km/h; 62 mph)
Range with max fuel	390 nm (725 km; 450 miles)

DAVIS DA-5A

Design of this aircraft began in October 1972, and Mr Davis and his son began construction of the prototype on 4 May 1974. It had been intended to power the aircraft with a two-cylinder Franklin Sport 2 engine, but non-availability of this power plant enforced a design change to utilise a 48·5 kW (65 hp) Continental A65. Despite the extra work involved, the first flight of the DA-5A was made on 22 July 1974.

The distance record set by a modified DA-5A on 6-7 August 1976 (see 1979-80 *Jane's*) has since been beaten.

Plans of the DA-5A are available to amateur constructors.

TYPE: Single-seat sporting aircraft.

WINGS: Cantilever low-wing monoplane. Wing section Clark Y. Thickness/chord ratio 12%. Dihedral 5°. Incidence 0°. Light alloy structure with single spar, ribs of 2024T-3 alloy, and stressed skins. Plain ailerons of light alloy construction. No flaps. No trim tabs.

FUSELAGE: Light alloy stressed-skin structure with four frames and stainless steel firewall. Steel tube overturn structure aft of pilot's seat.

TAIL UNIT: All-moving V tail (included angle 100°) with steel tube spar and light alloy ribs and skins. Anti-servo tabs in trailing-edges.

LANDING GEAR: Non-retractable tricycle type. Mainwheel legs of light alloy streamline section mounted rigidly to wing spar. Shock-absorption by tyres and rubber grommets. Main wheels have tyres size 14 × 5-4, pressure 1·03 bars (15 lb/sq in). Steerable nosewheel with tyre size 10 × 3-4, pressure 1·03 bars (15 lb/sq in). Rosenhan drum brakes.

POWER PLANT: One 48·5 kW (65 hp) Continental A65 flat-four engine, driving a Hegy type 60-70 two-blade fixed-pitch wooden propeller with spinner. Fuel tank in fuselage, immediately aft of firewall, capacity 64·3 litres (17 US gallons). Refuelling point on fuselage upper surface forward of windscreen. Oil capacity 3·75 litres (1 US gallon).

ACCOMMODATION: Single seat beneath canopy hinged on port side.

DIMENSIONS, EXTERNAL:

Wing span	4·76 m (15 ft 7¼ in)
Wing chord, constant	1·12 m (3 ft 8 in)
Wing area, gross	5·31 m² (57·20 sq ft)
Wing aspect ratio	4·26
Length overall	4·80 m (15 ft 9 in)
Height overall	1·35 m (4 ft 5¼ in)
Span over V tail	1·60 m (5 ft 3 in)
Wheel track	1·55 m (5 ft 1 in)
Wheelbase	1·12 m (3 ft 8 in)
Propeller diameter	1·52 m (5 ft 0 in)
Propeller ground clearance	0·203 m (8 in)

DIMENSION, INTERNAL:

Cockpit: Max width	0·53 m (1 ft 9 in)

WEIGHTS:

Weight empty	208 kg (460 lb)
Max T-O weight	351 kg (775 lb)

PERFORMANCE (at max T-O weight):

Never-exceed speed	147 knots (273 km/h; 170 mph)
Max level speed at S/L	139 knots (257 km/h; 160 mph)
Max cruising speed at S/L	122 knots (225 km/h; 140 mph)
Econ cruising speed at S/L	104 knots (193 km/h; 120 mph)
Stalling speed	52·5 knots (97 km/h; 60 mph)
Max rate of climb at S/L	244 m (800 ft)/min
Service ceiling	4,420 m (14,500 ft)
T-O run	183 m (600 ft)
T-O to 15 m (50 ft)	259 m (850 ft)
Landing from 15 m (50 ft)	335 m (1,100 ft)
Landing run	183 m (600 ft)
Range with max fuel	390 nm (724 km; 450 miles)

DURAND

DURAND ASSOCIATES INC

84th and McKinley Road, Omaha, Nebraska 68122
Telephone: (402) 571 7060

DURAND Mk V

Mr William H. Durand, a professional engineer, has designed and built a two-seat all-metal biplane known as the Durand Mk V (N100DV). Design began in 1967 and construction in 1970, and the first flight was achieved on 28 June 1978. Flight testing has shown that the Durand Mk V cannot be stalled.

Plans of the Durand Mk V are available to amateur builders, and 30 were known to be under construction in early 1981.

TYPE: Two-seat sporting biplane.

WINGS: Negative stagger biplane with wings of constant chord. Wing section NACA 23012. Dihedral 0° upper wing, 2° 30′ lower wings. Incidence 3° both wings. No sweepback. Strut-braced light alloy stressed-skin structure with spars of box form. Full-span flaps on upper and lower wings. Spoilers in place of ailerons, forward of flaps on lower wings only.

FUSELAGE: Semi-monocoque structure of 2024-T3 light alloy. Cabin area of fuselage skinned externally and internally, and insulated for temperature and sound.

TAIL UNIT: Cantilever structure of light alloy. All-moving tailplane has built-up box spar; fin and rudder have U-section spars. Full-span combined trim tab and anti-servo tab in tailplane.

LANDING GEAR: Non-retractable tricycle type. Cantilever main legs of reinforced glassfibre. All three wheels of Cleveland manufacture, with tyres size 5·00-5, pressure 2·07 bars (30 lb/sq in). It is intended to design fairings for the main wheels. Cleveland hydraulic brakes.

POWER PLANT: One 112 kW (150 hp) Avco Lycoming O-320-E2A flat-four engine, driving a Sensenich two-blade fixed-pitch metal propeller with spinner. Fuel tank in forward fuselage, aft of firewall, capacity 93 litres (24·5 US gallons). Two auxiliary wing tanks optional, each 22·5 litres (6 US gallons), to provide max fuel capacity of 138 litres (36·5 US gallons). Refuelling point for main tank on upper surface of forward fuselage. Oil capacity 7·5 litres (2 US gallons).

ACCOMMODATION: Two seats side by side beneath forward-sliding canopy. Full dual controls. Baggage space aft of seats, max capacity 58 kg (128 lb). Windscreen defroster. Cabin heated and ventilated.

SYSTEMS: Hydraulic system for brakes only. Electrical system includes 60A engine-driven alternator and 12V storage battery.

AVIONICS: Edo-Aire Mitchell nav/com can be installed optionally. Prototype has full IFR instrumentation, but is intended basically for VFR flying.

DIMENSIONS, EXTERNAL:

Wing span, both	7·47 m (24 ft 6 in)
Wing chord, constant (both)	0·914 m (3 ft 0 in)
Wing area, gross	13·38 m² (144 sq ft)
Wing aspect ratio, effective biplane	5·1
Length overall	6·17 m (20 ft 3 in)
Height overall	2·03 m (6 ft 8 in)
Tailplane span	2·71 m (8 ft 10½ in)
Wheel track	2·13 m (7 ft 0 in)
Wheelbase	1·83 m (6 ft 0 in)
Propeller diameter	1·83 m (6 ft 0 in)

DIMENSIONS, INTERNAL:

Cabin: Length	2·03 m (6 ft 8 in)
Max width	1·09 m (3 ft 7 in)
Max height	1·22 m (4 ft 0 in)

WEIGHTS:

Weight empty	549 kg (1,210 lb)
Max T-O weight	834 kg (1,840 lb)

PERFORMANCE (at max T-O weight):

Never-exceed speed	147 knots (272 km/h; 169 mph)
Max cruising speed at 2,285 m (7,500 ft)	117 knots (217 km/h; 135 mph)
Econ cruising speed at 2,285 m (7,500 ft)	109 knots (201 km/h; 125 mph)
Landing speed	48 knots (89 km/h; 55 mph)
Max rate of climb at S/L	366 m (1,200 ft)/min
Absolute ceiling, estimated	4,570 m (15,000 ft)
T-O run	168 m (550 ft)
T-O to 15 m (50 ft)	503 m (1,650 ft)
Landing from 15 m (50 ft)	274 m (900 ft)
Landing run	137 m (450 ft)
Range with standard fuel, no reserves	347 nm (644 km; 400 miles)
Range with max optional fuel, no reserves	451 nm (837 km; 520 miles)

DYKE

DYKE AIRCRAFT

2840 Old Yellow Springs Road, Fairborn, Ohio 45324
Telephone: (513) 878 9832

Mr John W. Dyke was the designer of a small delta-wing aircraft, designated JD-1 Delta, which he built with Mrs Jennie Dyke. This was described in the 1964-65 *Jane's*. Subsequently, Mr Dyke developed the design, and the first flight of the JD-2 Delta was made in July 1966.

DYKE AIRCRAFT JD-2 DELTA

Plans of the JD-2 are available to amateur constructors, and Mr Dyke has formed Dyke Aircraft to market them. A total of about 325 JD-2s were thought to be under construction in early 1981, in Australia, Brazil, Canada, France, West Germany, Japan, New Zealand, South Africa, the United Kingdom and the United States. About 25 had been completed and flown by then.

As a help to the homebuilder, hardware and tubing kits are available.

A feature of the Dyke Delta is that the outer wing panels can be folded flat above the cockpit for easy towing and to allow storage in a single-car garage.

The Dyke Delta illustrated (N78BS) was built by Mr Bernie Schaknowski of Clay, NY. First flown on 1 June 1980, it differs from the standard Dyke Delta by having a lengthened nose, giving an overall length of 22 ft (6·70 m); glassfibre covering on the underfuselage, control surfaces and fin; mass-balanced elevons; additional instruments; and other detail refinements. Power plant is a 134 kW (180 hp) Avco Lycoming O-360 flat-four engine. A standard Dyke Delta was illustrated in the 1980-81 *Jane's*.

TYPE: Delta-winged sporting aircraft.

WINGS: Delta wing of modified NACA 63012 and 66015 wing section. No dihedral. No incidence. Sweepback of wing centre-section 61°. Sweepback of outer wing

Durand Mk V all-metal two-seat biplane *(Howard Levy)*

Modified Dyke Delta built by Mr Bernie Schaknowski (Avco Lycoming O-360 engine) *(Howard Levy)*

panels 31°. Welded 4130 steel tube structure, with stainless steel capstrips to which the laminated glassfibre skins are secured by Dupont explosive rivets. Aluminium skin optional; attached by aluminium pop rivets. Trailing-edge elevons extending from centreline to approximately two-thirds span. Elevons have a basic structure of metal but are Polyfiber fabric covered. Trim tab at inboard edge of each elevon. Wing outer panels fold upward and lay above the fuselage so that the JD-2 can be towed on the landing gear. Folding and unfolding can be carried out by one person.

FUSELAGE: Welded steel tube structure with stainless steel capstrips and glassfibre skins, except for fuselage undersurface which has 4130 steel tube capstrips and polyfiber fabric covering.

TAIL UNIT: Welded steel tube structure with swept fin and rudder, both polyfiber fabric covered. Optional all-moving T tailplane, with steel tube main spar and aluminium skin, for improved low-speed trim when a heavier engine, such as the 149 kW (200 hp) Avco Lycoming, is installed.

LANDING GEAR: Manually retractable tricycle type, all units retracting aft. Shock-absorption by torsion bars of 6150 heat-treated steel. Main wheels have tyres size 14 × 5-6, pressure 3·10 bars (45 lb/sq in). Steerable nosewheel, tyre size 12 × 5-4, pressure 2·07 bars (30 lb/sq in). Firestone hydraulic brakes.

POWER PLANT: Prototype has one 134 kW (180 hp) Avco Lycoming O-360 flat-four engine, driving a McCauley two-blade fixed-pitch propeller with spinner. Constant-speed propeller and 149 kW (200 hp) engine installation optional. Glassfibre fuel tank in fuselage, immediately aft of cabin, capacity 178 litres (47 US gallons), with pilot and two passengers. Fuel capacity reduced to 87 litres (23 US gallons) as a four-seat aircraft. Refuelling point on port side of aft fuselage upper surface. Oil capacity 7·5 litres (2 US gallons).

ACCOMMODATION: Standard accommodation for pilot, on single forward seat, and three passengers on aft bench seat. Access by upward-opening canopy, hinged on starboard side. Accommodation is ventilated. Space for limited baggage in starboard wing centre-section.

SYSTEMS: Hydraulic system for brakes only. Vacuum system for instruments. Electrical supply from 12V DC generator; solid-state inverter provides 110V AC at 400Hz.

AVIONICS AND EQUIPMENT: Genave Series 200A 200-channel nav and com transceiver. Blind-flying instrumentation standard.

DIMENSIONS, EXTERNAL:
Wing span 6·87 m (22 ft 2½ in)
Wing chord at root (centre-section) 4·27 m (14 ft 0 in)
Wing chord at root (outer panels) 2·90 m (9 ft 6 in)
Wing chord at tip 0·51 m (1 ft 8 in)
Wing area, gross 17·00 m² (183 sq ft)
Wing aspect ratio 2·7
Length overall 5·79 m (19 ft 0 in)
Width, wings folded 2·24 m (7 ft 4 in)
Height overall 1·68 m (5 ft 6 in)
Height overall, wings folded 1·83 m (6 ft 0 in)
Tailplane span (T tail) 0·91 m (3 ft 0 in)
Wheel track 1·97 m (6 ft 5½ in)
Wheelbase 2·13 m (7 ft 0 in)
Propeller diameter 1·88 m (6 ft 2 in)

DIMENSIONS, INTERNAL:
Cabin: Length 1·88 m (6 ft 2 in)
Max width 1·22 m (4 ft 0 in)
Max height 1·02 m (3 ft 4 in)
Floor area 1·67 m² (18 sq ft)
Baggage space 0·17 m³ (6 cu ft)

WEIGHTS (with 134 kW; 180 hp or 149 kW; 200 hp engine):
Weight empty, basic equipped 435 kg (960 lb)
Weight empty, IFR equipped 483 kg (1,065 lb)
Max T-O weight 862 kg (1,900 lb)
Max landing weight 816 kg (1,800 lb)

PERFORMANCE (at max T-O weight with 134 kW; 180 hp engine, except where indicated):
Never-exceed speed 191 knots (354 km/h; 220 mph)
Max level speed at 2,285 m (7,500 ft) 165 knots (306 km/h; 190 mph)
Max cruising speed at 2,285 m (7,500 ft), fixed-pitch propeller 156 knots (290 km/h; 180 mph)
Max cruising speed, constant-speed propeller 165 knots (306 km/h; 190 mph)
Max cruising speed, with 149 kW (200 hp) engine and constant-speed propeller 174 knots (322 km/h; 200 mph)
Econ cruising speed at 2,285 m (7,500 ft) 135 knots (249 km/h; 155 mph)
Max rate of climb at S/L 305 m (1,000 ft)/min
Service ceiling 4,420 m (14,500 ft)
Service ceiling, with 149 kW (200 hp) engine 5,485 m (18,000 ft)
T-O run 213 m (700 ft)
T-O to 15 m (50 ft) 549 m (1,800 ft)
Landing from 15 m (50 ft) 549 m (1,800 ft)
Landing run 244-305 m (800-1,000 ft)
Range with max fuel and pilot only 625 nm (1,158 km; 720 miles)
Range with max payload 390 nm (724 km; 450 miles)

EAA
EXPERIMENTAL AIRCRAFT ASSOCIATION INC

PO Box 229, Hales Corners, Wisconsin 53130
Telephone: (414) 425 4860
PRESIDENT: Paul H. Poberezny
VICE-PRESIDENT: Ray Scholler
SECRETARY: S. H. Schmid
TREASURER: Arthur Kilps
EXECUTIVE DIRECTOR, INFORMATION SERVICES: Ben Owen

As a service to its members, the EAA decided in 1955 to develop a modern single-seat sporting biplane suitable for home construction by amateurs. The design drawings were prepared by Mr J. D. Stewart and Mr T. Seely of the Allison Division, General Motors Corporation, with the assistance of Mr Paul H. Poberezny, President of the EAA, and this aircraft became known as the EAA Biplane. It was followed in 1972 by the Acro-Sport, an aerobatic aircraft designed by Mr Poberezny. These two designs were supplemented in 1973 by a more advanced aerobatic aircraft, the Super Acro-Sport.

Latest designs to emanate from the EAA President are the lightweight Pober P-9 Pixie, which flew for the first time in July 1974, and the Acro-Sport II.

EAA BIPLANE

The prototype of the EAA Biplane was built between 1957 and May 1960 as a classroom project by students of St Rita's High School, Chicago, under the supervision of Mr Robert Blacker. It flew for the first time on 10 June 1960, powered by a 48·5 kW (65 hp) Continental A65 engine.

After this prototype had been taken over by the EAA, a number of modifications were made, including installation of a 63·5 kW (85 hp) engine. Mr Poberezny subsequently introduced further changes, including lighter wing spars and fittings, which reduced the empty weight of the aircraft, together with increases in the area of the tail surfaces and many other improvements. This modified version was known as the EAA Biplane Model P. Subsequent changes led successively to Models P1 and P2, with successively thinner spars and, in the P2, movement of the cockpit to a slightly more forward position.

More than 7,400 sets of plans of the EAA Biplane had been sold by the EAA Museum Foundation by early 1981, and many examples are flying and under construction.

The following details apply to the standard EAA Biplane Model P2:

TYPE: Single-seat sporting biplane.

WINGS: Braced biplane, with N-shape streamline-section cabane struts and interplane struts. Dihedral 0° on upper wing, 2° on lower wings. Incidence 2° on upper wing, 2° on lower wings. All-wood two-spar structure, with aluminium leading-edge and overall fabric covering. Spars 25·5 mm (1·0 in) thick. Ailerons of similar construction to wings, on lower wings only. No flaps.

FUSELAGE: Welded steel tube structure, fabric-covered.

TAIL UNIT: Wire-braced welded steel tube structure, fabric-covered.

LANDING GEAR: Non-retractable tailwheel type, modified from standard Piper J-3 components. Rubber cord shock-absorption. Brakes on main wheels. Wheel fairings optional. Piper J-3 steerable tailwheel.

POWER PLANT: Prototype Model P has a 63·5 kW (85 hp) Continental C85-8 flat-four engine, driving a two-blade fixed-pitch metal propeller. Provision for engines of up to 112 kW (150 hp); but most Model Ps have a 93 kW (125 hp) Avco Lycoming. Piper J-3 fuel tank aft of firewall in fuselage, capacity 68 litres (18 US gallons).

ACCOMMODATION: Single seat in open cockpit.

DIMENSIONS, EXTERNAL:
Wing span 6·10 m (20 ft 0 in)
Wing chord (both), constant 0·91 m (3 ft 0 in)
Wing area, gross 10·03 m² (108 sq ft)
Length overall 5·18 m (17 ft 0 in)
Height overall 1·83 m (6 ft 0 in)
Wheel track 1·71 m (5 ft 7½ in)
Propeller diameter: 85 hp 1·78 m (5 ft 10 in)
150 hp 1·88 m (6 ft 2 in)

WEIGHTS (63·5 kW; 85 hp engine):
Weight empty 322 kg (710 lb)
Max T-O weight 522 kg (1,150 lb)

PERFORMANCE (A, with 63·5 kW; 85 hp Continental engine; B, with 112 kW; 150 hp Avco Lycoming engine, at max T-O weight):
Max level speed at S/L:
A 109 knots (201 km/h; 125 mph)
Econ cruising speed:
A 96 knots (177 km/h; 110 mph)
Stalling speed: A 48 knots (89 km/h; 55 mph)
B 52 knots (97 km/h; 60 mph) CAS
Max rate of climb at S/L: A 305 m (1,000 ft)/min
B 370 m (1,210 ft)/min
Service ceiling: A 3,500 m (11,500 ft)
T-O run: A 91 m (300 ft)
Landing run: A 245 m (800 ft)
Range with max fuel:
A 304 nm (560 km; 350 miles)
g limits normal ±6, ultimate ±9

EAA ACRO-SPORT I

The Acro-Sport I was designed by Mr Paul Poberezny, President of the EAA, specifically for construction by school students, as a pupils' project. The EAA considers that such a project enables those who participate to discover their capabilities and potential in the manual crafts necessary to build an aircraft, possibly helping the individual to choose a career more wisely.

First flight of the prototype Acro-Sport (N1AC) was made on 11 January 1972, only 352 days after its design was started, although it represented a completely new design, unrelated to the EAA Biplane.

Plans and construction manuals are available to homebuilders from Acro-Sport Inc (see Super Acro-Sport for address), and about 850 sets had been sold by January 1981.

The following details apply to the prototype:

TYPE: Single-seat aerobatic biplane.

WINGS: Braced single-bay biplane, with single streamline-section interplane strut each side. N-type centre-section struts. Double streamline-section flying and landing wires. Wing section M-6. Dihedral: upper 0°, lower 2°. Incidence (both) 1° 30′. Conventional

EAA Super Acro-Sport built by Herr Manfred Krause (149 kW; 200 hp Avco Lycoming engine)

EAA Acro-Sport I aerobatic biplane built in the UK by Mr J. H. Kimber *(Geoffrey P. Jones)*

EAA Pober P-9 Pixie *(Howard Levy)*

EAA Biplane single-seat sporting aircraft

EAA Acro-Sport II two-seat aerobatic biplane *(Lee Fray)*

two-spar structure, with spruce spars and ribs, single wire drag and anti-drag truss, fabric-covered. Glassfibre wingtips. Ailerons, on all four wings, of wood construction with fabric covering. No flaps. Cutout in trailing-edge of upper wing.

FUSELAGE: Composite structure of welded steel tube, with wooden stringers, fabric-covered. Glassfibre nose cowl and light alloy engine cowlings.

TAIL UNIT: Wire-braced welded steel tube structure with fabric covering. Controllable trim tab on port side of elevator, servo tab on starboard side.

LANDING GEAR: Non-retractable tailwheel type, modified from Piper J-3 components. Two side Vs and half axles. Rubber bungee shock-absorption. Main-wheel tyres size 5·00-5, pressure 1·86-1·93 bars (27-28 lb/sq in). Cleveland or Goodyear hydraulic brakes. Glassfibre wheel fairings. Steerable tailwheel. Parking brake.

POWER PLANT: Prototype has a 134 kW (180 hp) Avco Lycoming engine, driving a Sensenich two-blade fixed-pitch metal propeller with spinner. Basic power plant is a 74·5 kW (100 hp) Continental O-200 flat-four engine. Single fuel tank immediately aft of firewall, capacity 104 litres (20 US gallons). Refuelling point on upper surface of fuselage. Small smoke oil tank, capacity 19 litres (5 US gallons), forward of instrument panel, could also be used for fuel.

ACCOMMODATION: Single seat in open cockpit, which is large enough to accommodate a pilot 1·95 m (6 ft 5 in) tall and weighing 115 kg (250 lb). Baggage space behind headrest, capacity 16 kg (35 lb).

DIMENSIONS, EXTERNAL:

Wing span, upper	5·97 m (19 ft 7 in)
Wing span, lower	5·82 m (19 ft 1 in)
Wing chord (both), constant	0·91 m (3 ft 0 in)
Wing area, gross	10·73 m² (115·5 sq ft)
Wing aspect ratio, upper	6·6
Length overall	5·33 m (17 ft 6 in)
Height overall	1·83 m (6 ft 0 in)
Tailplane span	2·16 m (7 ft 1 in)
Wheel track	1·78 m (5 ft 10 in)
Propeller diameter	1·93 m (6 ft 4 in)

WEIGHTS:

Weight empty, equipped	335 kg (739 lb)
Max T-O and landing weight	534 kg (1,178 lb)

PERFORMANCE (at max T-O weight):

Never-exceed speed	156 knots (289 km/h; 180 mph)
Max level speed	132 knots (245 km/h; 152 mph)
Max cruising speed	113 knots (209 km/h; 130 mph)
Econ cruising speed	91 knots (169 km/h; 105 mph)
Stalling speed	43·5 knots (80·5 km/h; 50 mph)
Max rate of climb at S/L	1,067 m (3,500 ft)/min
T-O run	46 m (150 ft)
T-O to 15 m (50 ft)	107 m (350 ft)
Landing from 15 m (50 ft)	267 m (875 ft)
Landing run	244 m (800 ft)
Range with max fuel	304 nm (563 km; 350 miles)

EAA SUPER ACRO-SPORT

Design of a developed version of the Acro-Sport was started in January 1971. Construction began in the following year and the first flight of this prototype was made on 28 March 1973. Known as the Super Acro-Sport, it is intended for unlimited International Class aerobatic competition at a world championship level. Generally similar in external appearance to the Acro-Sport, it has a more powerful engine, nearly symmetrical aerofoil section, a better rate of climb, and improved outside (negative *g*) aerobatic capability.

The differences by comparison with the standard Acro-Sport are covered in a supplement to the basic plans which are available from Acro-Sport Inc, PO Box 462, Hales Corners, Wisconsin 53130. The description of the Acro-Sport applies also to the Super Acro-Sport, except as follows:

TYPE: Single-seat advanced aerobatic biplane.

WINGS: As Acro-Sport, except wing section NACA 23012.

TAIL UNIT: As Acro-Sport, except controllable trim tab on starboard side of elevator, servo tab on port side.

LANDING GEAR: As Acro-Sport, except Cleveland hydraulic brakes. Steerable tailwheel has a solid tyre.

POWER PLANT: Prototype has a 149 kW (200 hp) Avco Lycoming IO-360-A2A flat-four engine, driving a Sensenich two-blade fixed-pitch metal propeller type 76EM8-0-60 with spinner. Fuel system and capacity as for Acro-Sport. Oil capacity 7·5 litres (2 US gallons).

DIMENSIONS, EXTERNAL: As for Acro-Sport, except

Length overall	5·30 m (17 ft 4½ in)

WEIGHTS:

Weight empty	401 kg (884 lb)
Max T-O weight	612 kg (1,350 lb)

PERFORMANCE (at max T-O weight):

Never-exceed speed	156 knots (289 km/h; 180 mph)
Max level speed at S/L	135 knots (251 km/h; 156 mph)
Max cruising speed	117 knots (217 km/h; 135 mph)
Stalling speed	43·5 knots (80·5 km/h; 50 mph)
Max rate of climb at S/L	1,128 m (3,700 ft)/min
Service ceiling	4,570 m (15,000 ft)
T-O run	38 m (125 ft)
T-O to 15 m (50 ft)	91 m (300 ft)
Landing from 15 m (50 ft)	274 m (900 ft)
Landing run	244 m (800 ft)
Range with max fuel	260 nm (482 km; 300 miles)

EAA ACRO-SPORT II

The Acro-Sport II is a two-seat aerobatic biplane derived from the Acro-Sport I, by Mr Paul Poberezny, for the pilot with only a small number of flying hours to his credit. Design began in 1976 and construction started in 1977. The first flight was achieved on 9 July 1978. Plans are available to the homebuilder from Acro-Sport Inc (see Super Acro-Sport for address), and about 500 sets had been sold by early 1981.

TYPE: Two-seat aerobatic biplane. Stress limits are +6·5*g*/−4·5*g*.

WINGS, FUSELAGE AND TAIL UNIT: As for Acro-Sport I.

LANDING GEAR: As for Acro-Sport I, except for main-wheel tyres size 6·00-6, pressure 1·86-1·93 bars (27-28 lb/sq in).

POWER PLANT: Prototype has one 134 kW (180 hp) Avco Lycoming O-360-A4B engine, driving a Sensenich two-blade fixed-pitch metal propeller with spinner. Can be powered by engines of 93-149 kW (125-200 hp). Fuel capacity 98·4 litres (26 US gallons).

ACCOMMODATION: Two seats in tandem in separate cockpits. Baggage space, capacity 13·6 kg (30 lb).

DIMENSIONS, EXTERNAL:

Wing span, upper	6·60 m (21 ft 8 in)
Wing span, lower	6·32 m (20 ft 9 in)
Wing chord, constant	1·09 m (3 ft 7 in)
Wing area, gross	14·12 m² (152 sq ft)
Length overall	5·75 m (18 ft 10¼ in)
Height overall	2·03 m (6 ft 7¾ in)
Wheel track	1·85 m (6 ft 0¾ in)

WEIGHTS:

Weight empty	397 kg (875 lb)
Max T-O weight	689 kg (1,520 lb)

PERFORMANCE (prototype):

Max cruising speed	107 knots (198 km/h; 123 mph)
Stalling speed	46 knots (86 km/h; 53 mph)
T-O run	91 m (300 ft)

EAA POBER P-9 PIXIE

Design and construction of this lightweight sporting aircraft began simultaneously in January 1974, and the first flight was made in July 1974.

It was conceived under 'Project Econoplane' to be economical in operation; and the prototype was originally powered by a converted Volkswagen motorcar engine. The Pixie made its public debut at the 1974 EAA Fly-in at Oshkosh, Wisconsin. Soon afterwards, the prototype was re-engined with a Limbach-VW engine, and a special pressure cowl was fitted to improve cooling and increase the cruising speed.

Plans are available to amateur constructors from Acro-Sport Inc (see Super Acro-Sport for address), and more than 400 sets had been sold by early 1981.

Type: Single-seat lightweight sporting aircraft.

Wings: Braced parasol monoplane. Two streamline-section struts and wire bracing each side. Wing section Clark Y. Dihedral 2°. Incidence 2°. Conventional wooden structure utilising spruce spars, mahogany plywood and fabric covering. Full-span Frise-type ailerons. No flaps. No trim tabs.

Fuselage: Welded structure of 4130 chrome molybdenum steel tubing with wood formers and fabric covering.

Tail Unit: Wire-braced structure of welded 4130 chrome molybdenum steel tubing, fabric covered. Fixed-incidence tailplane. No trim tabs.

Landing Gear: Non-retractable tailwheel type. Side Vs with half-axles carry main wheels. Shock-absorption by rubber bungee. Main-wheel tyres size 5·00-5, pressure 2·07 bars (30 lb/sq in). Tailwheel tyre of solid type. Cleveland brakes. Glassfibre fairings over main wheels.

Power Plant: One 44·5 kW (60 hp) Limbach SL 1700 EA flat-four engine, driving a Sensenich two-blade fixed-pitch propeller. Equally suited to Monnett Volkswagen engine conversions. Fuel tank in wing centre-section, capacity 46·6 litres (12·3 US gallons). Refuelling point in wing upper surface. Oil capacity 2·84 litres (0·75 US gallons).

Accommodation: Single seat in open cockpit. Door on starboard side. Baggage capacity 9 kg (20 lb).

Dimensions, external:

Wing span	9·09 m (29 ft 10 in)
Wing chord, constant	1·37 m (4 ft 6 in)
Wing area, gross	12·47 m² (134·25 sq ft)
Length overall	5·26 m (17 ft 3 in)
Height overall	1·88 m (6 ft 2 in)
Wheel track	1·60 m (5 ft 3 in)
Propeller diameter	1·35 m (4 ft 5 in)

Weights:

Weight empty	246 kg (543 lb)
Max T-O weight	408 kg (900 lb)

Performance (at max T-O weight):

Never-exceed speed	113 knots (209 km/h; 130 mph)
Max level speed at S/L	89 knots (166 km/h; 103 mph)
Max cruising speed	72 knots (134 km/h; 83 mph)
Stalling speed	26 knots (49 km/h; 30 mph)
Max rate of climb at S/L	213 m (700 ft)/min
Service ceiling	3,810 m (12,500 ft)
T-O and landing run	91 m (300 ft)
T-O to 15 m (50 ft)	305 m (1,000 ft)
Landing from 15 m (50 ft)	152 m (500 ft)
Range with max fuel	251 nm (466 km; 290 miles)

EAGLE
EAGLE HELICOPTER CORPORATION

Scottsdale, Arizona 85252

Eagle Helicopter Corporation developed and built prototypes of two lightweight helicopters. The Eagle II is a 'cold-jet' tip-driven helicopter; the Eagle III is powered by two 0·178 kN (40 lb st) Gluhareff G8 pressure jet engines, mounted at the rotor tips and burning liquid propane. A 151 litre (40 US gallon) fuel tank was expected to give the Eagle III an endurance of about two hours.

Details and an illustration of the Eagle II can be found in the 1980-81 *Jane's*. No later information has been received.

EICH
JAMES P. EICH

1820 W Grand Avenue, Alhambra, California 91801
Telephone: (213) 289 1983

After gaining considerable experience from flying a Barnett J-3M rotorcraft, Mr Eich designed and built a two-seat gyroplane of which details follow. Design started in 1975 and construction of the prototype began the following year. First flight was achieved in 1977.

EICH JE-2 GYROPLANE

During the past year, the JE-2 has been fitted with a system to pre-rotate the rotor and shorten the take-off run.

Type: Two-seat light autogyro.

Rotor System: Two-blade aluminium teetering rotor. Six struts forming the rotor pylon pick up the top of the fuselage structure at the corners of the front (passenger's) cockpit.

Rotor Drive: Pre-rotation system only, comprising a Poulan chain-saw motor which drives the rotor shaft through a V-belt slip-belt 'clutch'.

Fuselage: Square-section welded girder structure of chrome-molybdenum steel tubing, covered with Dacron and finished with Stits Poly-dope.

Tail Unit: Fixed horizontal stabiliser, triangular ventral fin and rudder. Fin is of Dacron, integral with fuselage covering. Stabiliser and rudder are of sheet aluminium, with simple ribs and spars of aluminium tubing.

Landing Gear: Non-retractable tricycle type. Spring-in-compression shock-absorption. Main-wheel tyres size 5·00-5, pressure 0·7 bars (10 lb/sq in). Steerable nose-wheel with tyre size 4·00-5, pressure 1·4 bars (20 lb/sq in). Go-kart brakes.

Power Plant: One 48·5 kW (65 hp) Continental A65 engine, driving a two-blade Eich fixed-pitch tractor propeller. One fuel tank aft of firewall, capacity 60·5 litres (16 US gallons).

Accommodation: Two seats in tandem in open cockpits. Dual controls.

Dimensions, external:

Diameter of main rotor	7·92 m (26 ft 0 in)
Rotor blade chord	0·187 m (7⅜ in)
Length of fuselage	4·27 m (14 ft 0 in)
Height overall	2·59 m (8 ft 6 in)
Wheel track	1·83 m (6 ft 0 in)
Wheelbase	1·65 m (5 ft 5 in)
Propeller diameter	1·78 m (5 ft 10 in)

Weights:

Weight empty	204 kg (450 lb)
Max T-O weight	385·5 kg (850 lb)

Performance:

Max level speed at 915 m (3,000 ft)	82 knots (153 km/h; 95 mph)
Max cruising speed at 915 m (3,000 ft)	69·5 knots (129 km/h; 80 mph)
Econ cruising speed at 915 m (3,000 ft)	52 knots (96·5 km/h; 60 mph)
Max rate of climb at 915 m (3,000 ft)	91·5 m (300 ft)/min
T-O run with pre-rotation	153 m (500 ft)
Landing run	15 m (50 ft)
Range with max fuel	156 nm (290 km; 180 miles)

EVANS
EVANS AIRCRAFT

PO Box 744, La Jolla, California 92037

Mr W. S. Evans, while employed as a design engineer with the Convair Division of General Dynamics Corporation, set out to design for the novice homebuilder an all-wood aircraft that would be easy to build and safe to fly. He was prepared to sacrifice both appearance and performance to achieve this aim. Two years of spare-time design and a year of construction produced a strut-braced low-wing monoplane with an all-moving tail unit, and powered initially by a 30 kW (40 hp) Volkswagen engine. Mr Evans named it the Volksplane but it was subsequently redesignated VP-1.

Mr Evans next developed a two-seat version of the VP-1, known as the VP-2. This is powered by a higher-rated Volkswagen engine, but in other respects has only minor constructional variations from the VP-1.

Plans of both models are available to amateur constructors, and approximately 6,000 sets had been sold by early 1981. VP-1 and VP-2 aircraft are currently flying in Australia, Belgium, Canada, France, West Germany, Italy, Ireland, Japan, New Zealand, the UK and the USA. VP-1s built in Italy and the Republic of Ireland were the first homebuilt aircraft licensed in those countries

EVANS VP-1

Type: Single-seat light aircraft.

Wings: Strut-braced low-wing monoplane. Two streamline-section bracing struts on each side. Wing section NACA 4412. Square tips. Dihedral 5°. Conventional wood structure with two rectangular spar beams, internal wooden compression struts and diagonal wire bracing, dispensing with the need for a complicated box spar. Fabric covering. Ailerons of wooden construction, fabric covered. No trim tabs. No flaps.

Fuselage: Rectangular-section all-wood stressed-skin structure, consisting essentially of three bulkheads, four longerons and plywood skin. Stressed-skin design eliminates the need for any diagonal bracing. Glassfibre fairing aft of pilot's seat.

Tail Unit: No fixed fin. The rudder is constructed of plywood ribs clamped to a 5 cm (2 in) aluminium tube which is mounted vertically through the rear fuselage and pivots in two nylon bushes. Leading- and trailing-edges are of wood and the whole unit is fabric-covered. The fabric-covered all-moving tailplane is a wooden cantilever structure, comprising ply ribs blocked and glued to a simple constant-section box spar. Both rudder and tailplane have anti-servo tabs.

Landing Gear: Non-retractable main wheels and tailskid. Main wheels carried on a bent section of heavy-gauge 24ST-3 aluminium bar, wire-braced by diagonal cables. Shock-absorption by low-pressure tyres. Main wheels and tyres size 6·00-6. Tyre pressure 0·83 bars (12 lb/sq in). Hydraulic brakes operated by single hand lever.

Power Plant: One 30 kW, 39·5 kW or 44·5 kW (40 hp, 53 hp or 60 hp) modified Volkswagen motorcar engine, driving a Hegy two-blade propeller, with pitch of 0·61 m (24 in) for 30 kW (40 hp) engine, 0·76 m (30 in) for 39·5 kW (53 hp) and 0·91 m (36 in) for 44·5 kW (60 hp). Glassfibre fuel tank aft of firewall and integral with the forward fuselage cowling, capacity 30 litres (8 US gallons). Filling point on top of fuselage, forward of windscreen.

Accommodation: Single seat in open cockpit. No baggage stowage.

Dimensions, external:

Wing span	7·32 m (24 ft 0 in)
Wing chord, constant	1·27 m (4 ft 2 in)
Wing area, gross	9·29 m² (100 sq ft)
Length overall	5·49 m (18 ft 0 in)
Height overall	1·56 m (5 ft 1½ in)
Tailplane span	2·13 m (7 ft 0 in)
Wheel track	1·50 m (4 ft 11 in)
Propeller diameter	1·37 m (4 ft 6 in)

Weights:

Weight empty	200 kg (440 lb)
Max T-O weight	340 kg (750 lb)

Performance (with 30 kW; 40 hp engine, at T-O weight of 295 kg; 650 lb):

Never-exceed speed	104 knots (193 km/h; 120 mph)
Cruising speed	65 knots (121 km/h; 75 mph)
Stalling speed	35 knots (65 km/h; 40 mph)
Max rate of climb at S/L	122 m (400 ft)/min
T-O run (average breeze)	137 m (450 ft)
Landing run (average breeze)	61 m (200 ft)

EVANS VP-2

Type: Two-seat light aircraft.

Wings: Generally similar to VP-1, except for NACA 4415 wing section and increased wing span and chord.

Fuselage: Similar to VP-1, but width increased by 0·305 m (1 ft 0 in).

Tail Unit: Similar to VP-1. No fin. Increased rudder area; all-moving tailplane of increased span and chord.

Landing Gear: Similar to VP-1. Wheel track increased by 0·23 m (9 in).

Power Plant: One 44·5 kW (60 hp) 1,834 cc or 48·5 kW (65 hp) 2,100 cc modified Volkswagen motorcar engine, driving a two-blade propeller. Glassfibre fuel tank aft of firewall, capacity 53 litres (14 US gallons).

Accommodation: Two seats side by side in open cockpit.

Dimensions, external:

Wing span	8·23 m (27 ft 0 in)
Wing chord, constant	1·47 m (4 ft 10 in)
Wing area, gross	12·08 m² (130 sq ft)
Length overall	5·87 m (19 ft 3 in)
Tailplane span	2·44 m (8 ft 0 in)
Wheel track	1·73 m (5 ft 8 in)
Propeller diameter	1·52 m (5 ft 0 in)

Weights:

Weight empty	290 kg (640 lb)
Max T-O weight	471 kg (1,040 lb)

Performance (44·5 kW; 60 hp engine, at max T-O weight):

Never-exceed speed	104 knots (193 km/h; 120 mph)
Max level speed	87 knots (161 km/h; 100 mph)
Max cruising speed	65 knots (121 km/h; 75 mph)
Stalling speed	35 knots (64·5 km/h; 40 mph)
Max rate of climb at S/L (pilot only)	213 m (700 ft)/min
Max rate of climb at S/L (pilot and passenger)	122 m (400 ft)/min

FLIGHT DYNAMICS *(see Sailplanes section)*

Eich JE-2 Gyroplane (Continental A65 engine) with pre-rotation gear fitted

Evans VP-1 single-seat light aircraft *(Roland Eichenberger)*

Evans VP-2 two-seat monoplane, built in England by pupils of Truro School

Frederick-Ames EOS/SFA single-seat sporting aircraft *(Howard Levy)*

FLSZ's Der Kricket DK-1 single-seat sporting biplane

FLSZ
FLIGHT LEVEL SIX-ZERO

PO Box 9980, Colorado Springs, Colorado 80932

FLSZ DER KRICKET DK-1

Flight Level Six-Zero is a company formed by John W. Dooley, Donald D. Miller, Cyril B. Smith and Roy E. Wheeler, to design, build, and market plans of an original biplane designated Der Kricket DK-1. Design emphasis has been to develop a 'good old-fashioned biplane', that would be easy to build, fly and maintain, and economical in operation. Design and construction of the prototype began in 1971, and this aircraft flew for the first time on 19 September 1978. A second prototype flew five days later.

TYPE: Single-seat sporting biplane.

WINGS: Braced cantilever biplane, with I-type streamline-section interplane struts, and N-shape cabane struts, but no flying wires. Aerofoil section NACA 4412. Wings moderately staggered. No dihedral, incidence or sweepback. Each wing is a two-spar structure of light alloy, including pop-riveted light alloy skins. Tips of Styrofoam covered with glassfibre. Plain ailerons on lower wings only. No flaps or trim tabs.

FUSELAGE: Basic structure of bolted and riveted light alloy angle, covered by light alloy skins. Can be built in flat-sided form or with more rounded contours, achieved by adding a single piece of angle each side.

TAIL UNIT: Cantilever light alloy structure comprising all-moving fin and all-moving tailplane. Tips of Styrofoam covered with glassfibre. Both surfaces have an anti-servo tab.

LANDING GEAR: Non-retractable tailwheel type. Main units consist of wire-braced light alloy struts. Azusa main wheels with tyres size 5·00-5. Azusa 8 in diameter tailwheel. Azusa mechanical brakes. Speed fairings on main units.

POWER PLANT: One modified Volkswagen motorcar engine of 37·3-48·5 kW (50-65 hp), driving a Cassidy two-blade fixed-pitch wooden propeller with spinner. Fuel tank in fuselage, aft of firewall, capacity 30 3 litres (8 US gallons). Refuelling point on upper surface of fuselage, just forward of upper wing leading-edge. Oil capacity 3·8 litres (1 US gallon).

ACCOMMODATION: Single seat in open cockpit.

DIMENSIONS, EXTERNAL:

Wing span (both)	4·88 m (16 ft 0 in)
Wing chord, constant (both)	0·91 m (3 ft 0 in)
Wing aspect ratio	5·33
Wing area, gross	8·92 m² (96·0 sq ft)
Length overall	4·88 m (16 ft 0 in)
Height overall	1·68 m (5 ft 6 in)
Tailplane span	1·98 m (6 ft 6 in)
Wheel track	1·52 m (5 ft 0 in)
Wheelbase	3·05 m (10 ft 0 in)
Propeller diameter	1·37 m (4 ft 6 in)

WEIGHTS:

Weight empty	249 kg (550 lb)
Max T-O weight	354 kg (780 lb)

PERFORMANCE (at max T-O weight):

Never-exceed speed	120 knots (222 km/h; 138 mph)
Max level speed	100 knots (185 km/h; 115 mph)
Max cruising speed	95·5 knots (177 km/h; 110 mph)
Econ cruising speed	82·5 knots (153 km/h; 95 mph)
Stalling speed, power off	45·5 knots (84 km/h; 52 mph)
Max rate of climb at S/L	253 m (830 ft)/min
Service ceiling	3,660 m (12,000 ft)
T-O run	213 m (700 ft)
T-O to 15 m (50 ft)	335 m (1,100 ft)
Landing from 15 m (50 ft)	290 m (950 ft)
Landing run	198 m (650 ft)
Range with max fuel	173 nm (322 km; 200 miles)

FREDERICK-AMES
FREDERICK-AMES RESEARCH CORPORATION

Anaheim, California
PRESIDENT: Frederick Smith

FREDERICK-AMES EOS/SFA

Mr Fred Smith, president of Frederick-Ames Research Corporation, designed a small single-seat sporting aircraft known as the EOS/SFA, which first appeared in an incomplete form at the 1973 EAA Fly-in at Oshkosh. After the meeting the EOS/SFA was completed; but on its maiden flight the Hirth engine seized and the aircraft crash landed, causing extensive damage. Although some restoration work was carried out, the aircraft was subsequently abandoned through lack of finance. In early 1978 the project was revived, with a Volkswagen modified motorcar engine in place of the original engine. First flight of the reworked aircraft took place in mid-1978, piloted by Mr Robert Bishop. By late 1980 the EOS/SFA had completed more than 100 hours of flight testing, including 20 hours with a Revmaster engine installed.

Kits to build the EOS/SFA are available to amateur constructors. All aluminium sheet is bent and flat wrapped, with no complicated forming. Structures are pop riveted and bonded, requiring no welding.

TYPE: Single-seat light sporting aircraft.

WINGS: Cantilever tapered low-wing monoplane. Wing section NACA 65212. Slight dihedral on outer panels only. Light alloy all-metal structure, except for glassfibre wing fillets, pop riveted and epoxy resin bonded. No flaps or tabs.

FUSELAGE: Conventional light alloy semi-monocoque structure, pop riveted and epoxy resin bonded. Glassfibre engine cowling.

TAIL UNIT: Cantilever light alloy structure, comprising triangular fin, integral with rear fuselage, rudder, and all-moving tailplane with projecting balance arms at roots. Tailplane actuated via servo tabs.

LANDING GEAR: Manually-retractable tricycle type. Main units retract inwards into wing roots; nosewheel rearward. Fully-castoring and steerable nosewheel. Rubber-in-compression shock-absorption. Disc brakes.

POWER PLANT: One 56 kW (75 hp) Revmaster or 55 kW (74 hp) 1,834 cc Volkswagen modified motorcar engine. Fuel capacity 56·75 litres (15 US gallons).

ACCOMMODATION: Single semi-reclined glassfibre seat under large sideways-hinged (to starboard) transparent canopy.

SYSTEM: Electrical system includes 45A alternator and starter.

DIMENSIONS, EXTERNAL:

Wing span	7·92 m (26 ft 0 in)
Wing area, gross	6·97 m² (75 sq ft)
Wing aspect ratio	9
Length overall	5·05 m (16 ft 7 in)
Height overall	1·98 m (6 ft 6 in)
Propeller diameter	1·27 m (4 ft 2 in)

WEIGHTS:

Weight empty	304 kg (670 lb)
Max T-O weight	454 kg (1,000 lb)

PERFORMANCE (Revmaster engine, constant-speed propeller):

Max level speed	174 knots (322 km/h; 200 mph)
Cruising speed (75% power)	165 knots (306 km/h; 190 mph)
Stalling speed at 1,370 m (4,500 ft)	50 knots (92 km/h; 57 mph) IAS
Max rate of climb at S/L	427 m (1,400 ft)/min
T-O and landing run	244 m (800 ft)

G/B AIRCRAFT

GLENN BEETS

460 Colorado Road, GVSR, Kingman, Arizona 86401
Telephone: (602) 565 3350

Mr Glenn Beets designed a lightweight sporting aircraft which he named the G/B Special. Construction of the prototype occupied two years and the first flight was made on 25 July 1973. Plans of the G/B Special, kits of components and materials are available to amateur constructors.

BEETS G/B SPECIAL

TYPE: Single-seat sporting aircraft.

WINGS: Braced parasol monoplane. Wing section Curtis 72. Sweepback 10°. V bracing struts each side with auxiliary struts. N-type cabane struts. Conventional wood structure with spruce spars and truss ribs, Dacron covered. Plain ailerons of similar construction. No trim tabs. Cutout in wing trailing-edge.

FUSELAGE: Welded structure of 4130 steel tube with Dacron covering.

TAIL UNIT: Cantilever structure of wood, with foam filling and covering of 1/16 in mahogany plywood.

LANDING GEAR: Non-retractable tailwheel type, with main wheels carried on braced tubular steel struts. Rubber-block shock-absorption; optional spring type or rigid gear. Glassfibre wheel fairings on main wheels.

POWER PLANT: One 52 kW (70 hp) Volkswagen 1,700 cc modified motorcar engine, driving a Hendrickson two-blade fixed-pitch laminated birch propeller through a 2½ : 1 reduction drive. Design will accept engines of 37·25-74·5 kW (50-100 hp). Fuel capacity 56·75 litres (15 US gallons).

ACCOMMODATION: Single seat in open cockpit.

DIMENSIONS, EXTERNAL:
Wing span 7·62 m (25 ft 0 in)
Wing chord 1·27 m (4 ft 2 in)
Wing area, gross 9·29 m² (100 sq ft)
Length overall 4·98 m (16 ft 4 in)
Height overall 1·83 m (6 ft 0 in)
Propeller diameter 1·98 m (6 ft 6 in)

WEIGHTS:
Weight empty 274 kg (603 lb)
Max T-O weight 420 kg (925 lb)

PERFORMANCE (at max T-O weight with cruise performance propeller):
Max level speed at S/L 135 knots (251 km/h; 156 mph)
Cruising speed 104 knots (193 km/h; 120 mph)
Landing speed 30·5 knots (56 km/h; 35 mph)
Max rate of climb at S/L 610 m (2,000 ft)/min
T-O run 61 m (200 ft)
Landing run 92 m (300 ft)
Range 520 nm (965 km; 600 miles)

GEIDE

RICHARD F. GEIDE

2323 S Crestway, Wichita, Kansas 67218

GEIDE SPORT

Mr Richard F. Geide has designed and built a tandem two-seat parasol-wing monoplane known as the Sport. Similar in configuration to the Pietenpol Air Camper, but constructed quite differently, the Sport is powered by a 63·4 kW (85 hp) Continental C85-12F engine driving a two-blade metal propeller. It uses a few Aeronca Champ components and the same 4412 wing section. Mr Geide claims that an important feature of the aircraft is its low-cost fabric covering, which uses medium weight Dacron fabric and cheaper non-aviation finishing materials. No further details were available at the time of writing.

GRADY

MELVIN GRADY

Jacksonville, Illinois

GRADY GRUMMAN/RAND MONOPLANE

Mr Melvin Grady has built a side-by-side two-seat light monoplane, using Rand KR-2 plans but incorporating components of Grumman American light aircraft. The aircraft (N2312N) took more than three years to construct and cost $5,600.

In order to save time in construction, Mr Grady incorporated into his aircraft a Grumman American tail unit, tailwheel-type landing gear, flaps (modified to suit his aircraft) and a few other components. He has also fitted a modified Thorp T-18 cockpit canopy. Power is provided by a 67 kW (90 hp) Continental C90 flat-four engine, driving a McCauley two-blade propeller.

TYPE: Two-seat monoplane.

DIMENSIONS, EXTERNAL:
Wing span 6·30 m (20 ft 8 in)
Length overall 5·49 m (18 ft 0 in)
Height overall, tail up 2·08 m (6 ft 10 in)

WEIGHTS:
Weight empty 426 kg (940 lb)
Max T-O weight 635 kg (1,400 lb)

PERFORMANCE:
Max level speed approx 104 knots (193 km/h; 120 mph)
Cruising speed 91 knots (169 km/h; 105 mph) IAS
Landing speed 52-56·5 knots (96·5-105 km/h; 60-65 mph)
T-O run 137 m (450 ft)

GREENAPPLES

GREENAPPLES AIRCRAFT

3100 6th Street, Boulder, Colorado 80302
Telephone: (303) 443 4371

GREENAPPLES AT-19

Mr L. Gale Abels, an architect of Boulder, Colorado, has designed and built a two-seat light monoplane known as the Greenapples AT-19. Construction took seven years, the aircraft beginning as a modified Thorp T-18. Constant redesign during construction led to only the landing gear and cockpit canopy of the finished aircraft being of Thorp type. First flight was made on 24 May 1979. The AT-19 received the Stan Dzik Memorial Award for 'Outstanding Design Contribution' at the 1979 EAA Fly-in. Used by Mr Abels for both business and pleasure, this aircraft had accumulated 220 flight hours by early 1981. At that time, no decision had been made as to whether plans would be made available to other homebuilders.

TYPE: Side-by-side two-seat light aircraft.

WINGS: Cantilever low-wing monoplane. Aerofoil section NACA 63,212 at root, NACA 65,415 at tip. Dihedral on outer wing panels 6°. Incidence 2°. No sweepback. Conventional flush-riveted structure of light alloy. Plain ailerons and plain trailing-edge flaps, all of light alloy construction. No trim tabs.

FUSELAGE: Semi-monocoque flush-riveted structure of light alloy.

TAIL UNIT: Cantilever V tail of flush-riveted light alloy construction. Fixed-incidence stabilisers. Electrically actuated trim tab in each movable surface.

LANDING GEAR: Non-retractable tailwheel type. Cantilever main legs, carrying Cleveland wheels. Maule steerable tailwheel. Main-wheel tyres size 5·00-5, pressure 2·07 bars (30 lb/sq in). Cleveland brakes. Glassfibre speed fairings for main wheels.

POWER PLANT: One 134 kW (180 hp) Avco Lycoming O-360-A1A flat-four engine, driving a Hartzell two-blade constant-speed propeller with spinner. Fuselage fuel tank, aft of firewall, capacity 125 litres (33 US gallons). Refuelling point on fuselage upper surface, forward of windscreen. Oil capacity 7·6 litres (2 US gallons).

ACCOMMODATION: Pilot and passenger side by side beneath rearward-sliding transparent canopy. Baggage space aft of seats. Accommodation heated and ventilated.

SYSTEM: Electrical system includes 12V battery. Vacuum system for blind flying instrumentation.

AVIONICS AND EQUIPMENT: Avionics include 720-channel nav/com, and transponder. Blind-flying instrumentation standard.

DIMENSIONS, EXTERNAL:
Wing span 7·42 m (24 ft 4 in)
Wing chord at root 1·27 m (4 ft 2 in)
Wing chord at tip 0·76 m (2 ft 6 in)
Wing aspect ratio 6·58
Wing area, gross 8·36 m² (90·0 sq ft)
Length overall 6·40 m (21 ft 0 in)
Height overall 1·57 m (5 ft 2 in)
Tailplane span 1·93 m (6 ft 4 in)
Wheel track 1·60 m (5 ft 3 in)
Propeller diameter 1·83 m (6 ft 0 in)

WEIGHTS:
Weight empty, equipped 472 kg (1,040 lb)
Max T-O and landing weight 796 kg (1,755 lb)

PERFORMANCE (at max T-O weight):
Never-exceed speed 204 knots (378 km/h; 235 mph)
Max level speed 189 knots (351 km/h; 218 mph)
Max cruising speed 187 knots (346 km/h; 215 mph)
Econ cruising speed at 3,660 m (12,000 ft) 172 knots (318 km/h; 198 mph)
Stalling speed, flaps down, power on 50·5 knots (94 km/h; 58 mph)
Max rate of climb at S/L 762 m (2,500 ft)/min
Range with max fuel, 30 min reserves 521 nm (965 km; 600 miles)

GREGA

JOHN W. GREGA

355 Grand Boulevard, Bedford, Ohio 44146

In addition to the standard Aircamper design, of which plans are available from Mr Bernard H. Pietenpol (see under 'Pietenpol'), a modernised version has been evolved by Mr John Grega. Plans of this are available to homebuilders, and by 1981 about 600 sets had been sold.

GN-1 AIRCAMPER

The prototype of this modernised version of the Aircamper two-seat light monoplane flew for the first time in November 1965. It used cut-down Piper J-3 Cub wings and bracing struts, and J-3 Cub landing gear. Other differences compared with the standard Aircamper (which see) are as follows:

POWER PLANT: One 48·5 kW (65 hp) Continental A65-8 flat-four engine in prototype, driving a two-blade 72/41 fixed-pitch metal propeller. Other engines of up to 63·5 kW (85 hp) can be installed. Fuel tanks aft of firewall, capacity 45 litres (12 US gallons), and in wing centre-section, capacity 22·5 litres (6 US gallons). Oil capacity 3·75 litres (1 US gallon).

DIMENSIONS, EXTERNAL:
Wing span 8·84 m (29 ft 0 in)
Wing chord, constant 1·52 m (5 ft 0 in)
Wing area, gross 13·94 m² (150 sq ft)
Wing aspect ratio 6
Length overall 5·51 m (18 ft 1 in)
Height overall 2·06 m (6 ft 9 in)
Tailplane span 2·29 m (7 ft 6 in)
Wheel track 1·60 m (5 ft 3 in)

WEIGHT:
Max T-O weight 499 kg (1,100 lb)

PERFORMANCE (at max T-O weight):
Max level speed at S/L 100 knots (185 km/h; 115 mph)
Max cruising speed at S/L 78 knots (145 km/h; 90 mph)
Stalling speed 31 knots (56 km/h; 35 mph)
Max rate of climb at S/L 152 m (500 ft)/min
T-O run 122 m (400 ft)
Landing run 76 m (250 ft)
Range with max fuel 347 nm (640 km; 400 miles)

HARMON

HARMON ENGINEERING COMPANY

PO Box 227, Rt 4, Sherman, Texas 75090
Telephone: (214) 893 2682
PRESIDENT: James B. Harmon

Harmon Engineering Company designed and built the prototype of a single-seat sporting aircraft named Der Donnerschlag (The Thunderclap), which flew for the first time in June 1974. Four months later, the aircraft suffered an engine failure while being flown by Mr James Harmon. He hit the ground at 52 knots (96 km/h; 60 mph) at a 45° impact angle, but was able to walk away from the accident. The aircraft was so little damaged that the decision was made to rebuild it as the prototype of a variant of Der Donnerschlag named Mister America. Design of the variant was started in November 1974, and first flight of Mister America took place on 31 October 1975.

Plans of Der Donnerschlag are no longer available, but full details of the aircraft can be found in the 1978-79 *Jane's*. Those for Mister America remain available and more than 30 sets had been sold by early 1981. Several aircraft are under construction and one, built by E. H. Holly Fletcher of Canby, Oregon, was flying by early 1981.

Mr Harmon is currently working on a number of new projects; no details of these were available at the time of going to press.

HARMON 1-2 MISTER AMERICA

TYPE: Single-seat lightweight sporting aircraft.

WINGS: Wire-braced shoulder-wing monoplane. Original wing section, with thickness/chord ratio of 14·5%. Dihedral 1°-2°. Incidence 0°. Wooden structure with

Beets G/B Special lightweight sporting aircraft (Volkswagen engine)

Grega GN-1 Aircamper two-seat light monoplane

Harmon Mister America sporting aircraft

Greenapples Aircraft Greenapples AT-19 *(Howard Levy)*

One of five Hatz CB-1s at the 1980 Oshkosh Fly-in *(J. M. G. Gradidge)*

fabric covering. Two simple beam spars; built-up ribs. Plain ailerons of similar construction. No flaps. Provision for ground-adjustable tabs in ailerons, if required. Dual flying and landing wires each side, landing wires attached to cabane structure at forward end of fuselage, which also forms windscreen support.

FUSELAGE: Welded structure of 4130 chrome-molybdenum steel tube with fabric covering. Wooden stringer turtledeck, with fabric covering. Aluminium and glassfibre engine cowling.

TAIL UNIT: Wire-braced welded structure of 4130 chrome-molybdenum steel tube with fabric covering. Tailplane incidence ground-adjustable. Elevators fitted. Provision for ground-adjustable trim tabs, if required.

LANDING GEAR: Non-retractable tailwheel type. Main units consist of two Vs, attached to fuselage structure, wire-braced and with a through axle. Shock-absorption by rubber bungee. Cleveland 5·00-5 main wheels and tyres, pressure 1·03 bars (15 lb/sq in). Aviation Products 10 cm (4 in) steerable tailwheel. Cleveland disc brakes with air cooling. Faired main legs and wheels.

POWER PLANT: One 44·7-48·5 kW (60-65 hp) 1,650 cc Volkswagen modified motorcar engine, driving a two-blade fixed-pitch 54-36 propeller. Fuel tank aft of engine firewall, capacity 34 litres (9 US gallons). Refuelling point forward of windscreen.

ACCOMMODATION: Single seat in open cockpit.

DIMENSIONS, EXTERNAL:

Wing span	5·99 m (19 ft 8 in)
Wing chord, constant	1·22 m (4 ft 0 in)
Wing area, gross	7·06 m² (76·0 sq ft)
Wing aspect ratio	4·916
Length overall	4·62 m (15 ft 2 in)
Height overall	1·52 m (5 ft 0 in)
Tailplane span	1·78 m (5 ft 10 in)
Wheel track	1·52 m (5 ft 0 in)
Wheelbase	3·51 m (11 ft 6 in)
Propeller diameter	1·37 m (4 ft 6 in)

DIMENSION, INTERNAL:

Cockpit: Max width	0·52 m (1 ft 8½ in)

WEIGHTS:

Weight empty	195 kg (430 lb)
Max T-O weight	295 kg (650 lb)

PERFORMANCE:

Never-exceed speed	143 knots (265 km/h; 165 mph)
Max level speed at S/L	109 knots (201 km/h; 125 mph)
Max cruising speed at S/L	95·5 knots (177 km/h; 110 mph)
Econ cruising speed at S/L	87 knots (161 km/h; 100 mph)
Stalling speed	42 knots (77·5 km/h; 48 mph)
Max rate of climb at S/L	244 m (800 ft)/min
Service ceiling	3,660 m (12,000 ft)
T-O run	61 m (200 ft)
T-O to 15 m (50 ft)	122 m (400 ft)
Landing from 15 m (50 ft)	152 m (500 ft)
Landing run	92 m (300 ft)
Range with max fuel	347 nm (643 km; 400 miles)

HATZ

HATZ AIRPLANE SHOP

Rt 2, Gleason, Wisconsin 54435

INFORMATION: Dudley Kelly, Rt 4, Versailles, Kentucky 40383

Mr John D. Hatz designed and built the prototype of a two-seat lightweight biplane, designated CB-1, of which plans are available to amateur constructors. Design and construction started in September 1959, and the first flight of the CB-1 was made on 19 April 1968. It was then powered by a 63·5 kW (85 hp) Continental C85-12 engine, but this was replaced subsequently by a 112 kW (150 hp) Avco Lycoming O-320.

The first plans-built CB-1 flew for the first time on 15 June 1975, powered by a Continental O-200 engine. Many more CB-1s are under construction, and at least 14 are flying.

HATZ CB-1 BIPLANE

The following amended details conform with the plans currently marketed by Hatz:

TYPE: Two-seat lightweight biplane.

WINGS: Braced single-bay biplane, with N-type interplane struts each side. N-type centre-section struts and streamline-section flying and landing wires. Wing section Clark Y. Dihedral 2°, on lower wings only. Incidence (both wings) 0°. Wooden two-spar structure with fabric covering. Cutout in trailing-edge of upper wing. Plain unbalanced ailerons of wood construction, with fabric covering, on both wings. No flaps. No trim tabs.

FUSELAGE: Welded steel tube structure, with fabric covering.

TAIL UNIT: Wire-braced welded steel tube structure, with fabric covering. Trim tab in elevator.

LANDING GEAR: Non-retractable tailwheel type. Two side Vs and half-axles hinged to fuselage structure. Steel spring shock-absorption. Main wheels size 6·50-6, with Goodyear tyres, pressure 1·03 bars (15 lb/sq in). Cleveland hydraulic wheel brakes. Glassfibre fairings on main wheels.

POWER PLANT: One 86 kW (115 hp) Avco Lycoming O-235 flat-four engine, driving a Sensenich two-blade fixed-pitch metal propeller. Single fuel tank in centre-section of upper wing, capacity 68 litres (18 US gallons). Refuelling point on upper surface of upper wing centre-section. Oil capacity 7·5 litres (2 US gallons).

ACCOMMODATION: Two seats in tandem in open cockpits.

SYSTEM: Electrical system, with 12V 12A DC engine-driven generator and 12V battery, for engine starting, navigation and instrument lights.

DIMENSIONS, EXTERNAL:

Wing span (both)	7·72 m (25 ft 4 in)

Wing chord, constant (both)	1·27 m (4 ft 2 in)
Wing area, gross	16·54 m² (178 sq ft)
Wing aspect ratio	6·06
Length overall	5·79 m (19 ft 0 in)
Height overall	2·39 m (7 ft 10 in)
Tailplane span	2·74 m (9 ft 0 in)
Wheel track	1·83 m (6 ft 0 in)
Propeller diameter	1·78 m (5 ft 10 in)
WEIGHTS:	
Weight empty	397 kg (875 lb)
Max T-O weight	726 kg (1,600 lb)
PERFORMANCE (at max T-O weight):	
Never-exceed speed	130 knots (241 km/h; 150 mph)
Max cruising speed	78 knots (145 km/h; 90 mph)
Stalling speed	35 knots (64·5 km/h; 40 mph)
Max rate of climb at S/L	244 m (800 ft)/min
T-O run	approx 122 m (400 ft)
Range with max fuel, 30 min reserves	174 nm (322 km; 200 miles)

HEADBERG
HEADBERG AVIATION INC

PO Box 557748, Miami, Florida 33155
Telephone: (305) 387 1704

The latest of a series of light aircraft designed and built in prototype form by Mr K. Flaglor is a light sporting monoplane named the Scooter. Design began in July 1965, and construction was started in November of the same year.

The prototype Scooter was powered originally by a 13·5 kW (18 hp) Cushman golf-kart engine, and it was with this power plant that the first flight was made in June 1967. Performance was marginal; as a result, Mr Flaglor replaced the Cushman with an 18·5-21 kW (25-28 hp) Volkswagen engine. Current power plant is a 1,500 cc Volkswagen engine developing 30 kW (40 hp). When flown to the 1967 EAA meet at Rockford, Illinois, the prototype Scooter won the 'Outstanding Ultra-light' and 'Outstanding Volkswagen-Powered Airplane' awards. Plans and kits of parts are available to amateur constructors from Headberg Aviation. More than 400 sets have been sold and many Scooters are flying.

FLAGLOR SCOOTER

The following description applies to the prototype Scooter in its current form:

TYPE: Light sporting monoplane.

WINGS: High-wing monoplane, braced by wires attached to fuselage and to kingpost mounted above centre-section. Wing section NACA 23012. Dihedral 2°. Incidence 3°. Two-spar structure with wood drag and anti-drag bracing. Aluminium leading-edge, spruce spars, plywood ribs and wingtips, and wooden trailing-edge, Ceconite covered. Conventional wooden ailerons. No flaps. No trim tabs.

FUSELAGE: Spruce wooden structure, plywood-covered in the forward cockpit area, Ceconite fabric-covered aft. Fuselage of triangular section aft of the wing. Wing centre-section and engine mounting constructed of 4130 steel tube.

TAIL UNIT: Spruce and plywood construction with strut bracing. No fixed fin. No trim tabs.

LANDING GEAR: Non-retractable tailwheel type. Cantilever spring steel main units. Steerable tailwheel. Main wheels of go-kart type, size 4·10 × 3·50-5. Tyre pressure 1·38 bars (20 lb/sq in). Vespa or Sears motor scooter brakes.

POWER PLANT: One 30 kW (40 hp) Volkswagen 1,500 cc flat-four engine, with Vertex magneto, driving a two-blade Troyer 54-28 propeller. Single fuel tank in fuselage nose, capacity 19 litres (5 US gallons). Refuelling point on top of fuselage forward of windscreen. Oil capacity 2·37 litres (2·5 US quarts).

ACCOMMODATION: Single seat in cockpit protected by deep windscreen.

DIMENSIONS, EXTERNAL:	
Wing span	8·48 m (27 ft 10 in)
Wing chord, constant	1·27 m (4 ft 2 in)
Wing area, gross	10·68 m² (115 sq ft)
Wing aspect ratio	6·7
Length overall	4·72 m (15 ft 6 in)
Height overall	2·13 m (7 ft 0 in)
Tailplane span	2·18 m (7 ft 2 in)
Wheel track	1·37 m (4 ft 6 in)
WEIGHTS:	
Weight empty	177 kg (390 lb)
Max T-O and landing weight	283 kg (625 lb)
PERFORMANCE:	
Max level speed	69 knots (129 km/h; 80 mph)
Max cruising speed	61 knots (112 km/h; 70 mph)
Econ cruising speed	56 knots (105 km/h; 65 mph)
Stalling speed	30 knots (55 km/h; 34 mph)
Max rate of climb at S/L	183 m (600 ft)/min
T-O and landing run	76 m (250 ft)
Range with max fuel, no reserves	152 nm (282 km; 175 miles)

HOLLMANN
HOLLMANN AIRCRAFT

11082 Bel Aire Court, Cupertino, California 95014
Telephone: (408) 255 2194

The HA-2M Sportster two-seat gyroplane has been developed by Mr Martin Hollmann, a senior design engineer in the aerospace industry. It is claimed to be the first aircraft of its type designed for the homebuilder who has access to a minimum of power tools. About 90 per cent of the structure is bolted and riveted together, with a minimum of machined parts.

Two average-size people can fly in the Sportster, which is suitable for pilot training and for flying on short cross-country journeys of up to 78 nm (145 km; 90 miles). It has been designed for towing behind a car, with the rotor stowed in a box attached to the car's roof. From the towed condition, the Sportster can be ready for its pre-flight walk-round inspection in ten minutes.

Plans, materials and many components are available to amateur constructors, and a network of distributors has been organised.

HOLLMANN HA-2M SPORTSTER

Design of this gyroplane began in June 1969, and construction of the first prototype was started in December 1972. The first flight of the prototype was made in October 1974, with FAA certification in the Experimental category. The first passenger was Dr Tom Butler, Vice-President Engineering and Research of AMF Incorporated, who flew in October 1975; test flying was completed by January 1976. The same year the prototype won a 'Best Original Design' award at the EAA Oshkosh meeting and an 'Outstanding New Design' award at the PRA Fly-in at Rockford. In 1977 it received the 'Best Autogyro' award at the EAA's Annual Antique Fly-in and air show at Watsonville, California. The prototype Sportster had logged over 230 flying hours by early 1981.

The Sportster utilises a mechanical pre-rotator which engages the engine, via a clutch, to pre-spin the rotor to 230 rpm prior to take-off. This allows the aircraft to take off after a very short run, and is expected to permit jump starts after brief further development. A rotor brake is utilised to slow down the rotor after landing and to stop the blades for taxying. A tension spring is adjusted from within the cockpit to trim the control stick in pitch.

By early 1981 more than 65 Sportsters were under construction in the USA, Canada, New Zealand and Sweden. Most of these will be powered by an Avco Lycoming O-320 engine, rated at 112 kW (150 hp); but the aircraft being built by Mr Bob Grayson, who lives at a height of 1,980 m (6,500 ft) in Montana, will have a 119 kW (160 hp) Avco Lycoming engine and larger rotor blades than standard. Twelve Sportsters were flying by January 1981.

TYPE: Two-seat gyroplane.

ROTOR SYSTEM: Two-blade rotor of NACA 8-H-12 section. Solidity ratio 0·035. Pre-cone angle 2°. Blade pitch +2½°. Metal blades, each made up of a 2024-T8511 leading-edge extrusion, aluminium formed ribs and Alclad skin, riveted and bonded together. Rotor rpm 380.

FUSELAGE: Square-tube 6061-T6 aluminium structure, bolted together. Aluminium skin. Glassfibre fairings. Two mast tubes. Large rear window for 360° view.

TAIL UNIT: Twin fins and balanced rudders carried on short tailbooms. Aluminium structure, with 2024-T3 Alclad skins pop-riveted in place. Glassfibre tips. Fixed horizontal surface between fins.

LANDING GEAR: Non-retractable tricycle type, with single wheel on each unit. Two small tailwheels. Main-wheel tyres size 18 × 6, pressure 1·8 bars (26 lb/sq in). Nose-wheel tyre of 0·25 m (10 in) diameter, pressure 1·8 bars (26 lb/sq in). Mechanical drum brake on nosewheel only.

POWER PLANT: One 97 kW (130 hp) Franklin Sport 4B flat-four engine, driving a Troyer 6635 pusher propeller, on prototype. One fuel tank of 45·4 litres (12 US gallons) capacity, with refuelling point inside cockpit. Oil capacity 5·7 litres (1·5 US gallons). Most other aircraft fitted with an Avco Lycoming O-320 engine, rated at 112 kW (150 hp), driving a Troyer 6648 propeller.

ACCOMMODATION: Two seats side by side in cabin with open sides.

SYSTEMS: Standard aircraft instruments and Narco Mk IV radio in prototype. 12V battery. Rotor pre-rotation and collective pitch system to spin rotor to 300 rpm.

DIMENSIONS, EXTERNAL:	
Rotor diameter	8·53 m (28 ft 0 in)
Rotor blade chord, constant	0·23 m (9 in)
Length overall	3·66 m (12 ft 0 in)
Height to top of rotor hub	2·34 m (7 ft 8 in)
Tailplane span	1·17 m (3 ft 10 in)
Wheel track	2·06 m (6 ft 9 in)
Wheelbase	1·42 m (4 ft 8 in)
Propeller diameter	1·68 m (5 ft 6 in)
DIMENSIONS, INTERNAL:	
Max width	0·914 m (3 ft 0 in)
Max height	1·45 m (4 ft 9 in)
WEIGHTS (A: prototype, B: 112 kW; 150 hp engine):	
Weight empty, equipped: A	281 kg (620 lb)
Max T-O and landing weight: A	476 kg (1,050 lb)
B	500 kg (1,100 lb)
PERFORMANCE (A: prototype, B: 112 kW; 150 hp engine):	
Never-exceed speed at S/L:	
A	78 knots (145 km/h; 90 mph)
Max cruising speed at S/L:	
A	65 knots (121 km/h; 75 mph)
Econ cruising speed at S/L:	
A	52 knots (97 km/h; 60 mph)
Stalling speed: A	24·5 knots (45 km/h; 28 mph)
Max rate of climb at S/L: A	213 m (700 ft)/min
B	274 m (900 ft)/min
Service ceiling	2,440 m (8,000 ft)
T-O run	107 m (350 ft)
Landing run	0-6 m (0-20 ft)
Range with max fuel	78 nm (145 km; 90 miles)
Range with max payload	61 nm (112 km; 70 miles)

HOVEY
ROBERT W. HOVEY

Aircraft Specialties Co, PO Box 1074, Saugus, California 91350
Telephone: (805) 252 4054

Mr Hovey designed and built an ultra-lightweight biplane of which plans are available to amateur constructors. His objective was to produce an aircraft which would require minimal construction time and have STOL performance, and which could be quickly disassembled for transportation. To achieve these ends the design has some unusual features, such as wing warping for roll control and use of an aluminium tube tailboom which has high-strength light alloy sheet and urethane foam stiffening. The original tail surfaces of craft paper over a styrofoam core have, however, been superseded by new horizontal and vertical tail surfaces of fabric-covered spruce construction.

Design began in October 1970, construction starting in the following month. First flight was made in February 1971, at which time the aircraft, known as Whing Ding II, received FAA certification in the Experimental category.

The original prototype was sold in Japan, where considerable interest was aroused among homebuilders. A second prototype was completed subsequently. Over 7,000 sets of plans had been sold and many Whing Dings were under construction or flying by early 1981. In January 1980 the original plans were revised to increase the wing span by 0·51 m (1 ft 8 in). This improves climb and load carrying performance, as a result of lower induced drag and reduced wing loading.

Mr Hovey's latest design to be available to amateur constructors is a monoplane derivative of the WD-II, known as the Beta Bird (originally Bushwacker), which was flown for the first time in April 1979. Few details are yet available on the new Hovey Delta Bird.

HOVEY WHING DING II (WD-II)

TYPE: Single-seat ultra-lightweight biplane.

WINGS: Braced single-bay biplane with parallel streamline-section interplane struts. Aircraft's fuselage, into which wing spars are socketed, gives location of inboard ends of wings. Landing and flying wires, the rear flying wires being used to control warping of upper wing. Wing section Hovey-10. Thickness/chord ratio 10%. Dihedral, both 1°. Incidence, both 4°. No sweepback. Wooden two-spar structure with ribs and wingtip bows of 9·5 mm (⅜ in) light alloy tube. Leading-edge faired in with rigid urethane foam. Wing structure fabric covered, tension of which retains the ribs in position. A plasticised fabric dope is used to ensure adequate flexibility for wing warping. No ailerons. No flaps. No trim tabs.

FUSELAGE: A closed box structure of 3·2 mm (⅛ in) mahogany plywood glued to 12·7 mm (½ in) square pine stringers, which is filled with urethane foam to stiffen and stabilise the plywood skin. This narrow fuselage provides attachment points for the seat, rudder bar and controls, and sockets for the wing spars. A reinforced extension at the top of the fuselage carries the engine. Aluminium tube tailboom is reinforced by

Flaglor Scooter light sporting monoplane

Hollmann HA-2M Sportster two-seat gyroplane

Hovey Beta Bird, a monoplane derivative of the Whing Ding II

Hovey Whing Ding II built by Mr Lin Bruty of Mt Emu Farm, Australia
(The Herald & Weekly Times, Melbourne)

Prototype Hovey Delta Bird *(Howard Levy)*

high-strength alloy sheet at the forward end, this being wrapped around the tube and bonded with epoxy resin. The entire tube is filled with free-foam urethane.

TAIL UNIT: Strut-braced structure. Horizontal and vertical surfaces constructed of 12·7 mm (½ in) square-section spruce, with fabric covering. All-moving tailplane, attached to tailboom by piano hinge. Rudder attached to fin by cloth hinges. No trim tabs.

LANDING GEAR: Non-retractable tailwheel type. Main wheels carried on spring-type strut of laminated fir covered with a layer of polyester glassfibre. Go-kart type main wheels with 28 cm (11 in) diameter tyres. Tyre pressure 1·38 bars (20 lb/sq in). Tailwheel has solid rubber tyre. Alternative steel tube landing gear available.

POWER PLANT: One 10·44 kW (14 hp) McCulloch MC-101A single-cylinder two-stroke aircooled go-kart engine, driving a two-blade hand-carved laminated birch or beechwood fixed-pitch pusher propeller. Original chain-drive reduction system now replaced by a double V belt drive. Fuel tank integral with engine, capacity 1·9 litres (0·5 US gallons).

ACCOMMODATION: Pilot only, on open seat.

EQUIPMENT: Basic instrumentation only, comprising airspeed and engine speed indicators and cylinder head temperature gauge.

DIMENSIONS, EXTERNAL:

Wing span (both)	5·69 m (18 ft 8 in)
Wing chord (both), constant	0·91 m (3 ft 0 in)
Wing area, gross	9·85 m² (106 sq ft)
Wing aspect ratio	5·9
Length overall	4·27 m (14 ft 0 in)
Height overall	1·68 m (5 ft 6 in)
Tailplane span	1·93 m (6 ft 4 in)
Wheel track	1·22 m (4 ft 0 in)
Wheelbase	2·97 m (9 ft 9 in)
Propeller diameter	1·22 m (4 ft 0 in)

WEIGHTS:

Weight empty, incl fuel	55·5 kg (123 lb)
Max T-O weight	140 kg (310 lb)

PERFORMANCE (at max T-O weight):

Never-exceed speed	52 knots (96·5 km/h; 60 mph)
Max level speed at S/L	43·5 knots (80·5 km/h; 50 mph)
Econ cruising speed at S/L	35 knots (64·5 km/h; 40 mph)
Stalling speed	23 knots (42 km/h; 26 mph)
Service ceiling	1,220 m (4,000 ft)
T-O run	76 m (250 ft)
T-O to 15 m (50 ft)	107 m (350 ft)
Landing from 15 m (50 ft)	76 m (250 ft)
Landing run	46 m (150 ft)
Range	17 nm (32 km; 20 miles)

HOVEY BETA BIRD

The Beta Bird is basically a monoplane version of the Whing Ding II. Design began in June 1977, and construction of the prototype started three months later. It was first flown in April 1979; the 40 h flight test programme to lift FAA test area restrictions was completed on 14 December 1979, and in early 1981 approximately 200 sets of plans and kits had been sold.

TYPE: Single-seat ultra-lightweight monoplane.

WINGS: Braced monoplane structure, with V strut on each side. Wing section Hovey BB-14. Thickness/chord ratio 14%. No dihedral. Incidence 3°. No sweepback. Constant-chord structure, with two spruce spars and ribs of single-piece aluminium tubes. Fabric covering. Full-span ailerons which can be drooped to a maximum of 16° to reduce take-off and landing runs. No trim tabs. Wings fold aft about a pin joint in the rear spar.

FUSELAGE: Generally the same as for WD-II.

TAIL UNIT: Wire-braced structure of aluminium tube, connected with pop-riveted gussets and fabric covered. Conventional tailplane with elevators. No trim tabs.

LANDING GEAR: Generally as for WD-II, except for Gerdes toe-operated brakes, and steerable tailwheel.

POWER PLANT: One 33·5 kW (45 hp) Volkswagen 1,385 cc modified motorcar engine, driving a Hovey two-blade fixed-pitch pusher propeller. Fuel tank within the vertical fuselage extension to which the engine is mounted, capacity 28·4 litres (7·5 US gallons). Oil capacity 3·8 litres (1 US gallon).

ACCOMMODATION: Pilot only, on exposed seat.

SYSTEM: Battery only, as source of electric power for basic instruments and radio.

AVIONICS: Six-channel VHF com radio.

DIMENSIONS, EXTERNAL:

Wing span	7·77 m (25 ft 6 in)
Width, wings folded	2·44 m (8 ft 0 in)
Wing chord, constant	1·07 m (3 ft 6 in)
Wing area, gross	8·08 m² (87 sq ft)
Wing aspect ratio	7·3
Length overall	5·33 m (17 ft 6 in)
Height overall	1·83 m (6 ft 0 in)
Tailplane span	2·44 m (8 ft 0 in)
Wheelbase	1·52 m (5 ft 0 in)
Propeller diameter	1·37 m (4 ft 6 in)

WEIGHTS:

Weight empty	184 kg (405 lb)
Max fuel weight	20·4 kg (45 lb)
Max T-O weight	295 kg (650 lb)

PERFORMANCE (at max T-O weight):

Never-exceed speed	82·5 knots (153 km/h; 95 mph) IAS
Max level speed at 915 m (3,000 ft)	74 knots (137 km/h; 85 mph)
Max cruising speed at 915 m (3,000 ft)	61 knots (113 km/h; 70 mph)
Econ cruising speed	52 knots (96·5 km/h; 60 mph)
Stalling speed, flaps down, power on	35 knots (64·5 km/h; 40 mph)
Max rate of climb at S/L	over 122 m (400 ft)/min
T-O and landing run	76 m (250 ft)
T-O to and landing from 15 m (50 ft)	152 m (500 ft)
Range with max fuel, no reserves	113 nm (209 km; 130 miles)

HOVEY DELTA BIRD

First flown in early 1981, the Hovey Delta Bird is a tractor biplane, powered by a 16·4 kW (22 hp) Onan four-stroke engine. Its structure is of aluminium tube, with pop-riveted gusset joints. Three-axis controls are used.

No other details are available, but the accompanying illustration shows the general appearance of the aircraft.

INTERNATIONAL HELICOPTERS

INTERNATIONAL HELICOPTERS INC

Dart Airport, PO Box 107, Mayville, New York 14757
Telephone: (716) 753 2113

PRESIDENT: Robert G. Dart
VICE-PRESIDENT: Winfield Babcock II

On 2 January 1979, this company announced that it had completed the purchase of Helicom Inc, which had produced and marketed kits for single-seat and two-seat homebuilt helicopters since the early 1960s. It transferred the entire operation from Long Beach, California, to the facilities of Aero Industries at Dart Airport (Detroit Sectional) in Mayville, NY, and began production of Commuter IIA kits immediately.

In 1980 International Helicopters and Aero Industries announced the intention to sell a ready-to-fly helicopter known as the Commuter IIB. This helicopter, which was expected to receive FAA certification in late 1981, is structurally identical to the IIA but has a roomier cabin. A 134 kW (180 hp) Avco Lycoming engine may be fitted in the prototype at a later stage. Orders for the IIB are being taken.

INTERNATIONAL HELICOPTERS COMMUTER IIA

A detailed description of the original Helicom Commuter Jr Model H-1A single-seat helicopter last appeared in the 1972-73 *Jane's*. The Commuter IIA is a more powerful two-seater, of generally similar configuration but embodying a number of refinements. The 69th kit of parts was delivered in late 1980. The Commuter IIB type of cabin (see introductory copy) is available in kit form for fitting to the Commuter IIA and older Commuter II.

TYPE: Two-seat light helicopter.

ROTOR SYSTEM: Two-blade main rotor and two-blade tail rotor. Main rotor blades each comprise a solid extruded 7075-T6 aluminium leading-edge spar, extending to 30% chord, solid aluminium alloy V section trailing-edge, cast ribs and Alclad skins, bonded together, with some rivets for added strength. Tail rotor has blades of stainless steel riveted to an aluminium alloy spar. All-metal drive system. Conventional main rotor gearbox has specially designed centrifugal clutch, and steel shaft drive to tail rotor gearbox. Main rotor/engine rpm ratio 1:5·5. Tail rotor/engine rpm ratio 1:1.

FUSELAGE: Basic truss structure of 4130 chrome-molybdenum steel tube. Glassfibre and Plexiglas cabin enclosure. Rear structure not covered.

TAIL UNIT: Small horizontal stabiliser and triangular ventral fin.

LANDING GEAR: Steel tube skids, with optional ground handling wheels.

POWER PLANT: One 112 kW (150 hp) Avco Lycoming O-320 flat-four engine, with dual ignition, mounted vertically aft of cabin. Two cylindrical fuel tanks, one each side of main rotor driveshaft, above engine, with combined capacity of 83 litres (22 US gallons). Cooling fan standard.

ACCOMMODATION: Pilot and passenger, side by side on bench seat. Cabin width 1·02 m (3 ft 4 in). Conventional helicopter controls.

DIMENSIONS, EXTERNAL:

Main rotor diameter	7·62 m (25 ft 0 in)
Main rotor blade chord	0·20 m (8 in)
Main rotor disc	45·6 m² (490·9 sq ft)
Tail rotor diameter	1·22 m (4 ft 0 in)
Length of fuselage	6·88 m (22 ft 6¾ in)
Length overall	8·84 m (29 ft 0 in)
Width of fuselage	1·02 m (3 ft 4 in)
Height overall	2·49 m (8 ft 2 in)

WEIGHTS:

Weight empty	318 kg (700 lb)
Max T-O weight	590 kg (1,300 lb)

PERFORMANCE (at max T-O weight):

Max level speed	87 knots (161 km/h; 100 mph)
Max cruising speed	78 knots (145 km/h; 90 mph)
Service ceiling	3,960 m (13,000 ft)
Hovering ceiling IGE	2,135 m (7,000 ft)
Range with max fuel	195 nm (362 km; 225 miles)

JAVELIN

JAVELIN AIRCRAFT COMPANY INC

Box 18486, Wichita, Kansas 67218
Telephone: (316) 733 1011
PRESIDENT AND CHIEF ENGINEER: David D. Blanton

Javelin Aircraft Company was founded on 1 March 1953 to manufacture a low-cost automatic pilot for small aircraft; this was followed by equipment manufacture, and aircraft development work. Following restoration of a Curtiss Robin between 1957 and 1961, which won the National Championship award for the best restored antique aircraft in 1961, Javelin sought an Arrow Sport biplane for similar treatment. Unable to find a suitable aircraft, the company began design and development of a new biplane on 1 January 1964.

The resulting aircraft (N71DB), named Wichawk, has structural geometry similar to a Stearman biplane, as well as some of its aerodynamic features, and is stressed for +12 and −6*g*. It flew for the first time on 24 May 1971 and has since logged more than 500 flying hours.

Javelin Aircraft does not produce any assembled Wichawks, but plans, wing ribs and fuel tanks are available to amateur constructors. Eventually, Javelin Aircraft intends to sell complete kits for amateur construction of the aircraft, and it has been reported that the type may go into production in a foreign country as a primary trainer. A total of at least 225 sets of plans have been sold, and more than 100 Wichawks are under construction or have flown in Canada, Australia and Argentina. The first plans-built Wichawk (N29JC), constructed by Mr Jim Crawford, a retired airline captain, made its first flight on 2 August 1975. A 168 kW (225 hp) Continental O-470-11 flat-six engine gives this aircraft maximum and cruising speeds very similar to those of the Avco Lycoming-engined prototype; max rate of climb is increased to 762 m (2,500 ft)/min.

Two other Wichawks are powered by 134 kW (180 hp) Avco Lycoming O-360 engines, one with a Continental E-225-8 (ex-Beechcraft Bonanza) engine, and one with a 172 kW (230 hp) Continental O-470-R engine taken from a Cessna 182. This last aircraft (illustrated) has a three-blade propeller. It is said to be very smooth to fly, and can take off in 2½ seconds, climb at 762 m (2,500 ft)/min, cruise on 65% power at 113 knots (209 km/h; 130 mph) IAS, and perform well aerobatically.

The eighth amateur-built Wichawk to fly (N75WT), in 1980, was the first three-seater, with a single cockpit aft of the normal side-by-side two-seat cockpit. With a Continental E-225-4 engine, this aircraft has an empty weight of 679 kg (1,496 lb).

The main current production work of Javelin Aircraft involves manufacture of 190 litre (50·2 US gallon) glassfibre slipper fuel tanks for the Rockwell Commander and Turbo-Commander 680/690 series. The company is also flight testing a turbocharged Ford Pinto motorcar engine in a 1956 Cessna 172; this develops 134 kW (180 hp), and Javelin will eventually sell plans for the conversion to amateur constructors.

JAVELIN WICHAWK

TYPE: Two/three-seat sporting biplane.

WINGS: Braced single-bay biplane, with N-shape streamline-section cabane and interplane struts. Streamline-section landing and flying wires. Wing section NACA 23015. 2° dihedral on lower wings only. Incidence 0°. No sweepback. Composite structure, with two wooden spars and 2024T3 light alloy ribs, fabric-covered. Simple sealed ailerons on lower wings only. No flaps. Geared trim tab.

FUSELAGE: Welded structure of 4130 chrome-molybdenum steel tube with light alloy tubular stringers, fabric-covered.

TAIL UNIT: Wire-braced welded structure of 4130 chrome-molybdenum steel tube with fabric covering. Fixed-incidence tailplane. Trim tab in starboard elevator.

LANDING GEAR: Non-retractable tailwheel type. Main wheels carried on side Vs hinged to lower fuselage longerons. Shock-absorption by automotive-type shock-struts, similar to those of Piper PA-20, and rubber shock cord. Main wheels and tyres size 6·00-6. Cleveland toe brakes. Steerable tailwheel.

POWER PLANT: Prototype has one 134 kW (180 hp) Avco Lycoming O-360 flat-four engine, driving a Sensenich two-blade fixed-pitch propeller type 76EM8-0-56. McCauley propellers optional. Provision for alternative horizontally-opposed or radial engines from 112 kW (150 hp) to 224 kW (300 hp), for which Javelin can provide installation drawings. Other engines currently being installed, other than those previously mentioned, include the 175 kW (235 hp) Avco Lycoming O-435, 224 kW (300 hp) Avco Lycoming O-540 and 224 kW (300 hp) Continental O-520. Fuel tank of 94·5 litre (25 US gallon) capacity in upper wing centre-section, and one of 56·8 litre (15 US gallon) capacity in fuselage aft of firewall. Refuelling points above tanks. Oil capacity of prototype 7·5 litres (2 US gallons).

ACCOMMODATION: Two seats, side by side, in open cockpit. Provision for tandem two-seat or three-seat configurations. Drawings available for rearward-sliding transparent cockpit canopy. Dual controls standard. Baggage compartment aft of seats, capacity 45·4 kg (100 lb). Baggage locker in turtleback, capacity 9 kg (20 lb).

SYSTEMS: Electrical system powered by 12V 50A engine-driven generator. Hydraulic system for brakes only.

DIMENSIONS, EXTERNAL:

Wing span (upper)	7·32 m (24 ft 0 in)
Wing chord (both), constant	1·27 m (4 ft 2 in)
Wing area, gross	17·2 m² (185 sq ft)
Wing aspect ratio	5·76
Length overall	5·87 m (19 ft 3 in)
Height overall	2·18 m (7 ft 2 in)
Tailplane span	2·44 m (8 ft 0 in)
Wheel track	1·87 m (6 ft 1½ in)
Propeller diameter	1·93 m (6 ft 4 in)

DIMENSIONS, INTERNAL:

Cockpit: Max width	0·93 m (3 ft 0½ in)
Baggage compartment	0·34 m³ (12 cu ft)

WEIGHTS (prototype with 134 kW; 180 hp engine):

Weight empty	599 kg (1,320 lb)
*Max T-O weight	907 kg (2,000 lb)

* *Max T-O weight is increased to 998 kg (2,200 lb) with high-powered engines*

PERFORMANCE (prototype with 134 kW; 180 hp engine):

Never-exceed speed	156 knots (289 km/h; 180 mph)
Max level speed at S/L	121·5 knots (225 km/h; 140 mph)
Max cruising speed	110 knots (204 km/h; 127 mph)
Landing speed	39 knots (72·5 km/h; 45 mph)
Stalling speed	50 knots (92 km/h; 57 mph) IAS
Max rate of climb at S/L	518 m (1,700 ft)/min
T-O run	46 m (150 ft)
g limits	+12, −6

JEFFAIR

JEFFAIR CORPORATION

PO Box 975, Renton, Washington 98055
Telephone: (206) 863 7992
PRESIDENT: Geoffrey L. Siers

This company was formed by Mr Geoffrey Siers, a former RAF fighter pilot and design engineer with BAC, who emigrated to the USA in 1964. In 1966 he began designing a high-performance all-wooden two-seat light aircraft, of which construction was started in June 1969. Now known as the Barracuda, this aircraft (N19GS) flew for the first time on 29 June 1975, and is certificated in the FAA's Experimental category. At the EAA Fly-in at Oshkosh in 1976, the Barracuda won the 'Most Outstanding New Design' award. Plans are available to amateur constructors, and about 350 Barracudas were being built by January 1981. An amateur-built example was illustrated in the 1979-80 *Jane's*.

JEFFAIR BARRACUDA

The prototype Barracuda flew originally with a 164 kW (220 hp) Avco Lycoming GO-435-2 engine. This was replaced with a more powerful O-540 engine in 1979, and the following details apply to the prototype in its current form.

TYPE: Two-seat all-wooden sporting monoplane.

WINGS: Cantilever low-wing monoplane, made in three pieces: two constant-chord outer panels and a centre-section which is integral with the fuselage. Wing section NACA $64_2$415. Upper surface of centre-section each side has slight anhedral, but bottom surface has neither anhedral nor dihedral. Dihedral of 5° on outer panels. No incidence. Basic structure of spruce, with mahogany and birch plywood covering. Box-section main spar with spruce booms and ply webs; truss-type ribs of spruce. Frise-type ailerons of wood and glassfibre. Electrically-actuated wooden plain flap on entire centre-section trailing-edge, extending under fuselage.

FUSELAGE: Conventional spruce structure, covered with birch plywood stressed skin.

TAIL UNIT: Cantilever spruce structure, plywood-covered. Glassfibre tips. Electrically-actuated trim tab in port elevator. Rudder has bungee trim.

LANDING GEAR: Retractable tricycle type. Electro-hydraulic retraction, main wheels inward, nosewheel rearward; all wheels fully enclosed by doors when retracted. Emergency extension by gravity. Legs made from 4130 steel tubing, with steel coil spring shock-absorption. Cleveland wheels, size 5·00-5, with 0·36 m (14 in) diameter tyres. Cleveland aircooled brakes.

POWER PLANT: One 194 kW (260 hp) Avco Lycoming O-540-E4B5 flat-six engine, driving a Hartzell three-blade constant-speed propeller. Two glassfibre fuel tanks in centre-section forward of spar, total capacity 166·5 litres (44 US gallons). Oil capacity 11·5 litres (3 US gallons).

ACCOMMODATION: Two armchair seats, with thigh support, side by side under upward-hinged individual 'gull-wing' canopy doors. Dual controls. Baggage space behind seats. Cabin heated and ventilated.

SYSTEMS: Electric pump supplies hydraulic system. Electrical supply via 12V DC system.

AVIONICS: Edo-Aire 360-channel nav/com.

DIMENSIONS, EXTERNAL:

Wing span	7·54 m (24 ft 9 in)
Wing chord at root	2·11 m (6 ft 11 in)
Wing chord, outer panels, constant	1·45 m (4 ft 9 in)
Wing area, gross	11·15 m² (120 sq ft)
Length overall	6·55 m (21 ft 6 in)
Wheel track	2·54 m (8 ft 4 in)
Propeller diameter	2·18 m (7 ft 2 in)

DIMENSION, INTERNAL:

Cabin: Max width	1·07 m (3 ft 6 in)

International Helicopters Commuter IIA two-seat light helicopter

Prototype Jeffair Barracuda two-seat sporting monoplane

Keleher Lark-1B single-seat sporting monoplane *(J. M. G. Gradidge)*

Amateur-built Javelin Wichawk (Continental O-470-R engine)

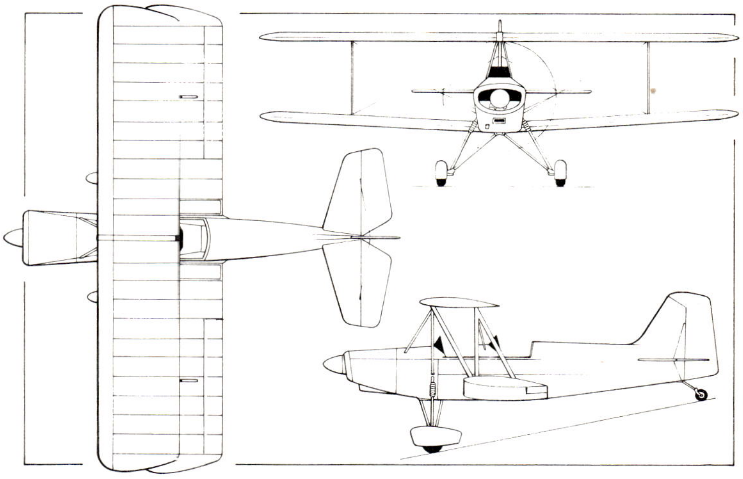

Kelly-D two-seat lightweight biplane *(Michael A. Badrocke)*

WEIGHTS:
Weight empty 678 kg (1,495 lb)
Baggage capacity 18 kg (40 lb)
Max T-O weight 1,043 kg (2,300 lb)
PERFORMANCE:
Never-exceed speed 260 knots (482 km/h; 300 mph)
Max level speed at 2,135 m (7,000 ft)
198 knots (367 km/h; 228 mph)
Max cruising speed at 2,135 m (7,000 ft)
182 knots (338 km/h; 210 mph)
Cruising speed (62% power) at 2,750 m (9,000 ft)
156 knots (290 km/h; 180 mph)
Stalling speed, flaps down
54 knots (100 km/h; 62 mph)
Max rate of climb at S/L 762 m (2,500 ft)/min
Range at max T-O weight, 65% power, with 30 min reserves 390 nm (724 km; 450 miles)

KELEHER

JAMES J. KELEHER

4321 Ogden Drive, Fremont, California 94538

In the early 1960s Mr J. Keleher designed and built a mid-wing sporting monoplane which he called the Lark. The design was revised in 1963, and the current model, for which plans are available to amateur constructors, is designated Lark-1B.

KELEHER LARK-1B

The following description applies to the Lark-1B with a 48·5 kW (65 hp) Continental engine. Examples are flying with 56 kW (75 hp) A75-8 and 74·5 kW (100 hp) O-200 Continental engines.

TYPE: Single-seat sporting monoplane.

WINGS: Braced mid-wing monoplane with streamline-section V bracing struts each side. Wing section NACA 2R₂12. No dihedral. Incidence 4° at root, 2° 30′ at tip. All-wood structure of Sitka spruce, with built-up I beam front spar and ribs, fabric-covered. Stressed to 6g plus. Fabric-covered wooden ailerons; no trim tabs or flaps.

FUSELAGE: Welded steel tube structure, fabric-covered, stressed to 6g plus.

TAIL UNIT: Wire-braced welded steel tube structure with sheet steel ribs and fabric covering. Adjustable-incidence tailplane. Swept fin. No trim tabs.

LANDING GEAR: Non-retractable tailwheel type. Divided main landing gear with shock-absorption by rubber cord in fuselage. Cleveland main wheels and tyres size 5·00-5, pressure 1·38-1·72 bars (20-25 lb/sq in). Cleveland disc brakes. Wheel fairings optional.

POWER PLANT: Provision for alternative flat-four engines of 48·5-74·5 kW (65-100 hp), driving a two-blade fixed-pitch metal propeller. One galvanised steel fuel tank in the fuselage, aft of the firewall, capacity 56 litres (15 US gallons). Refuelling point on top of cowl, forward of windscreen. Oil capacity 5·7 litres (1·5 US gallons).

ACCOMMODATION: Single seat in enclosed cockpit under sliding canopy. Lowered turtledeck and bubble canopy optional. Stowage for 9 kg (20 lb) baggage aft of seat.

DIMENSIONS, EXTERNAL:
Wing span 7·01 m (23 ft 0 in)
Wing chord, constant 1·22 m (4 ft 0 in)
Wing area, gross 7·48 m² (80·5 sq ft)
Wing aspect ratio 5·75
Length overall 5·18 m (17 ft 0 in)
Height overall 1·65 m (5 ft 5 in)
Tailplane span 2·03 m (6 ft 8 in)
Wheel track 1·57 m (5 ft 2 in)
Propeller diameter: A65-8 1·70 m (5 ft 7 in)
A75-8 1·66 m (5 ft 5½ in)

WEIGHTS (48·5 kW; 65 hp engine):
Weight empty 249 kg (550 lb)
Max T-O and landing weight 387 kg (855 lb)

PERFORMANCE (with 48·5 kW; 65 hp engine, at max T-O weight):
Never-exceed speed
160 knots (297·5 km/h; 185 mph)
Max level speed 115 knots (212 km/h; 132 mph)
Max cruising speed 103 knots (192 km/h; 119 mph)
Stalling speed 48 knots (89 km/h; 55 mph)
Max rate of climb at S/L 274 m (900 ft)/min
Service ceiling 5,950 m (19,500 ft)
T-O run 183 m (600 ft)
Range with max payload, with reserves
303 nm (563 km; 350 miles)

KELLY

DUDLEY R. KELLY

Route 4, Versailles, Kentucky 40383

Mr Dudley R. Kelly has designed and built, with Mr Jim Foster, the prototype of a biplane known as the Kelly-D, with the emphasis on improved efficiency and easier and cheaper construction than that of other designs in the same category. Design was initiated on 22 May 1977; construction began almost two years later, on 2 March 1979. The first flight was anticipated in the Spring of 1981.

KELLY-D

TYPE: Two-seat lightweight biplane.

WINGS: Braced single-bay biplane, with N-type interplane struts each side. Inverted V-type centre-section struts and streamline-section landing and flying wires. Wing section NACA 23012 (modified by Dr Robert Addoms). Dihedral 3°, on lower wings only. Incidence 3° on both wings. Constant chord, with no sweepback but considerable stagger. Wooden two-spar structure, with fabric covering. Aerodynamically-balanced ailerons of wooden construction, fabric-covered, on lower wing only.

FUSELAGE: Conventional welded steel tube structure, with fabric covering.

TAIL UNIT: Wire-braced structure of welded steel tube, fabric-covered. Fixed-incidence tailplane. Trim tab in port elevator.

LANDING GEAR: Non-retractable tailwheel type. Two side Vs and half-axles hinged to fuselage structure. Shock-absorption of main units by steel coil spring on each compression strut. Main-wheel tyres size 6·00-6, pressure 1·24 bars (18 lb/sq in). Tailwheel diameter 15 cm (6 in). Cleveland hydraulic disc brakes. Fairings over main wheels.

POWER PLANT: One 86 kW (115 hp) Avco Lycoming O-235 flat-four engine, driving a Sensenich two-blade fixed-pitch propeller. Flight testing will also be undertaken with a 48·5 kW (65 hp) Continental C65 engine.

Fuel tank in fuselage, aft of firewall, with capacity of 79·5 litres (21 US gallons). Refuelling point on upper surface of fuselage, forward of front cockpit. Oil capacity 5·7 litres (1·5 US gallons).

ACCOMMODATION: Two seats in tandem in open cockpits. Baggage space aft of pilot's seat in rear cockpit.

SYSTEMS: Hydraulic system for brakes only. Electrical system includes starter/generator and 12V storage battery.

AVIONICS AND EQUIPMENT: Genave Alpha 200B radio. Navigation lights.

DIMENSIONS, EXTERNAL:

Wing span (both)	7·42 m (24 ft 4 in)
Wing chord, constant (both)	1·27 m (4 ft 2 in)
Wing area, gross	19·54 m² (210·38 sq ft)
Wing aspect ratio	5·84
Length overall	5·87 m (19 ft 3 in)
Height overall	2·51 m (8 ft 3 in)
Tailplane span	2·82 m (9 ft 3 in)
Wheel track	1·83 m (6 ft 0 in)
Wheelbase	4·04 m (13 ft 3 in)
Propeller diameter	1·88 m (6 ft 2 in)

WEIGHTS (estimated, Avco Lycoming O-235 engine):

Weight empty, basic operating	352 kg (775 lb)
Max baggage	13·5 kg (30 lb)
Max fuel weight	54 kg (120 lb)
Max T-O and landing weight	635 kg (1,400 lb)

PERFORMANCE (estimated, Avco Lycoming O-235 engine, at max T-O weight):

Never-exceed speed	104 knots (193 km/h; 120 mph)
Max level speed at 1,525 m (5,000 ft)	91 knots (169 km/h; 105 mph)
Max cruising speed at 1,525 m (5,000 ft)	78 knots (145 km/h; 90 mph)
Econ cruising speed at 1,525 m (5,000 ft)	74 knots (137 km/h; 85 mph)
Stalling speed, power off	35 knots (64 km/h; 40 mph)
Max rate of climb at S/L	229 m (750 ft)/min
Service ceiling	3,050 m (10,000 ft)
T-O run	91 m (300 ft)
T-O to 15 m (50 ft)	274 m (900 ft)
Landing from 15 m (50 ft)	183 m (600 ft)
Landing run	76 m (250 ft)
Range with max fuel, 20 min reserves	234 nm (435 km; 270 miles)
Range with max payload, 20 min reserves	217 nm (402 km; 250 miles)

KIMBREL

MICHAEL KIMBREL

Oakville, Washington

KIMBREL DORMOY BATHTUB Mk 1

Mr Michael Kimbrel has built a single-seat light aircraft known as the Bathtub Mk 1, basing it on details published in a 1960 aviation magazine of the Dormoy Bathtub of 1924. Construction took seven years of part-time work and cost $700. The Bathtub Mk 1 flew for the first time in April 1978. Drawings are available to amateur builders and at least 140 sets have been sold.

The Bathtub Mk 1 has a braced parasol wing of wooden construction, Dacron fabric covered. The fuselage and tail have steel-tube structures, Dacron fabric covered, as shown in the accompanying illustration. Power is provided by a 30 kW (40 hp) Volkswagen 1,200 cc modified motor-car engine, driving a two-blade propeller.

DIMENSIONS, EXTERNAL:

Wing span	7·32 m (24 ft 0 in)
Wing chord, constant	1·07 m (3 ft 6 in)
Length overall	4·11 m (13 ft 6 in)
Height overall	2·13 m (7 ft 0 in)

WEIGHTS:

Weight empty	137 kg (302 lb)
Max T-O weight	245 kg (540 lb)

PERFORMANCE:

Max level speed	56·5 knots (105 km/h; 65 mph)
Cruising speed	48 knots (88·5 km/h; 55 mph)
Landing speed	30·5 knots (56·5 km/h; 35 mph)
Max rate of climb at S/L	122 m (400 ft)/min
T-O run	152 m (500 ft)
Landing run	91 m (300 ft)
Absolute ceiling	1,980 m (6,500 ft)
Range with max fuel	87 nm (161 km; 100 miles)

KINMAN

DUANE KINMAN

Rubidoux, California

KINMAN SUPER SIMPLE I

As its name suggests, the Super Simple I was designed by Mr Duane Kinman to be easy to construct. It is a small single-seater, the prototype of which was built in three weeks, at a cost of $2,680, from six sheets of 1·22 × 3·66 m (4 × 12 ft) aluminium. There are no compound curves, except in the glassfibre nose cowl, and no welding because of the extensive use of glue and pop rivets. Plans will be made available to amateur builders.

TYPE: Single-seat light monoplane.

WINGS: Cantilever low-wing monoplane. Wing section NACA 64212. Slight dihedral. No sweepback. Constant chord. Structure of 2024 T3 aluminium, bonded by Hysol 9410 glue. Full-span drooping ailerons. No tabs.

FUSELAGE: Conventional 2024 T3 aluminium structure, pop riveted. Tailcone is a 30 cm (12 in) propeller spinner.

TAIL UNIT: Conventional cantilever 2024 T3 aluminium structure, pop riveted. Constant-chord tailplane with full-span elevators. No tabs.

LANDING GEAR: Non-retractable tailwheel type. Cantilever main legs made from 1·9 cm (¾ in) thick 2024-T3 aluminium strip. Main wheels size 2·80-4. Hydraulic disc brakes.

POWER PLANT: One 42·5 kW (57 hp) 1,800 cc modified Volkswagen motorcar engine, driving a Bob Mende four-blade wooden propeller, with spinner. Fuel tank in nose, capacity 38 litres (10 US gallons).

ACCOMMODATION: Single semi-reclining seat under a bubble canopy.

DIMENSIONS, EXTERNAL:

Wing span	5·54 m (18 ft 2 in)
Wing chord, constant	1·27 m (4 ft 2 in)
Length overall	4·88 m (16 ft 0 in)
Propeller diameter	1·27 m (4 ft 2 in)

DIMENSIONS, INTERNAL:

Cockpit: Max width	0·51 m (1 ft 8 in)
Max height	0·76 m (2 ft 6 in)

WEIGHTS:

Weight empty	201 kg (443 lb)
Max T-O weight	306 kg (675 lb)

PERFORMANCE (estimated):

Max level speed over	139 knots (257 km/h; 160 mph)
Max cruising speed	130 knots (241 km/h; 150 mph)
Landing speed	39-42 knots (72·5-77 km/h; 45-48 mph)
T-O run	152 m (500 ft)
Range with max fuel	174 nm (322 km; 200 miles)

KOVACH

KIM KOVACH

31435 Alabama, Livonia, Minnesota 48150

KOVACH REPLICA IKE

Mr Kim Kovach has built a replica of Benny Howard's Ike racing monoplane of the 1930s. Powered by a 100 kW (134 hp) Menasco D4-87 in-line engine, it is 25% larger than the original Ike to make it suitable for fun flying.

As no plans of the original Ike were available, the replica has been built using details gleaned from contemporary photographs and articles. Construction of the replica Ike began in March 1975, and it was probably first flown in 1979. The wings are mostly of spruce construction and the fuselage of steel tubes with plywood formers and stringers, Dacron fabric covered. A few components, including the nose cowling and the headrest, are formed from glassfibre.

No further details were available at the time of writing.

TYPE: Single-seat replica monoplane.

DIMENSIONS, EXTERNAL:

Wing span	approx 7·62 m (25 ft 0 in)
Length	approx 6·40 m (21 ft 0 in)

KRAFT

KRAFT SYSTEMS INC

450 West California Avenue, PO Box 1268, Vista, California 92083

Telephone: (714) 724 7146

KRAFT K-1 SUPER FLI

Designed and built by Phil Kraft, with help from Paul White, the Super Fli was produced to model aircraft standards in terms of wing design, areas and moments, as Phil Kraft is a world champion model aircraft builder. After 1½ years of work, the Super Fli was first flown in December 1974 and was taken to the EAA Fly-in at Oshkosh in 1975. Aerobatic pilots at Oshkosh who flew the aircraft found it most satisfactory, and the prototype had logged more than 350 hours of almost continuous unlimited aerobatics by the Spring of 1977. Plans of the Super Fli are available to amateur constructors.

TYPE: Single-seat aerobatic monoplane.

WINGS: Cantilever low-wing monoplane. Spruce spars, plywood ribs and plywood covering. Horn-balanced ailerons.

FUSELAGE: Oval section fuselage of steel tube construction, aluminium covered.

TAIL UNIT: Wire-braced steel tube structure, fabric covered.

LANDING GEAR: Non-retractable tailwheel type. Faired main legs and faired wheels. Steerable tailwheel.

POWER PLANT: One 149 kW (200 hp) Avco Lycoming IO-360-A1D flat-four engine. Provision for fitting 134 kW (180 hp) Avco Lycoming.

ACCOMMODATION: Single seat under transparent canopy, hinged on starboard side.

DIMENSION, EXTERNAL:

Wing span	7·32 m (24 ft 0 in)

WEIGHTS:

Weight empty	445 kg (980 lb)
Max T-O weight	635 kg (1,400 lb)

PERFORMANCE:

Max level speed	174 knots (322 km/h; 200 mph)
Cruising speed	143 knots (265 km/h; 165 mph)
Landing speed	35 knots (65 km/h; 40 mph)
Max rate of climb at S/L	914 m (3,000 ft)/min
Service ceiling	3,660 m (12,000 ft)
Range with max fuel	260 nm (483 km; 300 miles)

LACO

LACO

PO Box 415, Desert Hot Springs, California 92240

Telephone: (714) 329 0955

LACO-125

Mr Joe Laven designed, built and has flown a two-seat light biplane known as the LACO-125. Design started in May 1971 and, following construction of two scale models to check the basic configuration, construction of a prototype began in April 1972. First flight was achieved on 29 May 1977 at the hands of Stolp Starduster test pilot Mr Eric Shilling. Recent modifications have included the fitting of a new nose cowl and louvred lower cowl discharge, an oil filter system, and a substantially modified windshield for the rear cockpit to eliminate slipstream buffeting and permit flying without the use of goggles. A second airframe has been completed, to the plans which are available to amateur constructors.

TYPE: Two-seat light biplane.

WINGS: Braced single-bay biplane. Wing section NACA 2412. Dihedral 2° on lower wings only. Incidence 2° on lower wings only. Conventional structure, with two spruce spars, spruce capstrips, mahogany ribs, aluminium leading- and trailing-edges and chrome-molybdenum steel fittings, all fabric covered. N-type interplane struts each side; two N-type struts support centre of upper wing above fuselage. Modified Frise-type ailerons on lower wings only, of similar construction to wings. No flaps or trim tabs.

FUSELAGE: Welded Warren truss structure of chrome-molybdenum steel tubing, with mahogany formers and spruce stringers, fabric covered.

TAIL UNIT: Conventional braced structure of welded chrome-molybdenum steel tube, fabric covered. Trim tab optional, but not fitted to prototype.

LANDING GEAR: Non-retractable tailwheel type, with steerable tailwheel. Polyurethane spring shock-absorption of LACO design. Cleveland main wheels, size 6·00-6. McKay Aero Research MF-7 main-wheel fairings. Cleveland hydraulic brakes.

POWER PLANT: One 93 kW (125 hp) Continental C-125-2 flat-four engine, driving a McCauley two-blade fixed-pitch propeller. Fuel tank in fuselage, capacity 91 litres (24 US gallons). Oil capacity 7·5 litres (2 US gallons).

ACCOMMODATION: Two seats in tandem in open cockpits.

SYSTEM: Engine-driven generator, lights and 30A starting system.

AVIONICS: Genave-100 channel transceiver.

Kimbrel Dormoy Bathtub Mk 1 single-seat light monoplane *(Howard Levy)*

Kinman Super Simple I single-seat monoplane *(Howard Levy)*

Kraft Super Flï aerobatic monoplane *(Howard Levy)*

LACO-125, designed and built by Mr Joe Laven

Langhurst 7/10th-scale replica of a Junkers Ju 87B-2 'Stuka' *(Howard Levy)*

Lasher Renegade I Formula V single-seat racer

DIMENSIONS, EXTERNAL:	
Wing span, upper	6·93 m (22 ft 8¾ in)
Wing span, lower	6·29 m (20 ft 7½ in)
Wing chord, upper (constant)	1·14 m (3 ft 9 in)
Wing chord, lower (constant)	1·07 m (3 ft 6 in)
Wing area, gross	13·94 m² (150 sq ft)
Length overall	5·94 m (19 ft 6 in)
Height overall	2·18 m (7 ft 2 in)
Tailplane span	2·64 m (8 ft 8 in)
Wheel track	1·88 m (6 ft 2 in)
Wheelbase	4·47 m (14 ft 8 in)
Propeller diameter	1·83 m (6 ft 0 in)
WEIGHTS:	
Weight empty	390 kg (860 lb)
Max T-O weight	635 kg (1,400 lb)
PERFORMANCE:	
Never-exceed speed	139 knots (257 km/h; 160 mph)
Max level speed	108 knots (200 km/h; 124 mph)
Max cruising speed	98 knots (182 km/h; 113 mph)
Stalling speed, power off	49 knots (90 km/h; 56 mph)
Stalling speed, power on	43·5 knots (80·5 km/h; 50 mph)
Max rate of climb at S/L	274 m (900 ft)/min
Range with max fuel	282 nm (523 km; 325 miles)

LANGHURST
LOUIS F. LANGHURST

Route 1, Box 315, Carriere, Mississippi 39426
Telephone: (601) 798 2880

LANGHURST REPLICA JUNKERS Ju 87B-2 'STUKA'

Mr Louis F. Langhurst, a retired design engineer, has designed and built a 7/10th-scale tandem two-seat replica of the German Junkers Ju 87B-2 'Stuka' dive-bomber of the second World War, finished in the personal markings of Major Hans-Ulrich Rudel of III Gruppe, Schlachtgeschwader 2 'Immelmann'. Research and design took two years and construction occupied 8,000 working hours over a six-year period. First flight was made on 19 July 1978. Construction is of steel tubing, with aluminium skinning. The control surfaces are identical in form with those of the original, with auxiliary aerofoil ailerons and flaps, but the dive-brakes remain fixed in flight. Power is provided by a 164 kW (220 hp) Avco Lycoming GO-435-B engine. Mr Langhurst was awarded the 'Replica Fighters of America Annual Award of Excellence' for the aircraft at the 1979 EAA Fly-in.

A designer-builder's guide has been prepared, and is available to amateur constructors.

DIMENSIONS, EXTERNAL:	
Wing span	9·91 m (32 ft 6 in)
Length overall	7·30 m (23 ft 11½ in)
WEIGHTS:	
Weight empty	794 kg (1,750 lb)
Max T-O weight	1,025 kg (2,260 lb)
PERFORMANCE:	
Max level speed	119 knots (220 km/h; 137 mph)
Cruising speed	102·5 knots (190 km/h; 118 mph)
Landing speed	54 knots (100 km/h; 62 mph)
Max rate of climb at S/L	over 305 m (1,000 ft)/min

LASHER
C. W. LASHER

320 Panama Circle, Winter Springs, Florida 32707

LASHER RENEGADE I

The Lasher-designed Renegade I Formula V racer (N73RL) was built by Mr M. Ricketts. First flight was made in 1974. It is owned currently by Mr Vernon Willingham of Sparta, Illinois, and has logged more than 500 flying hours. It took second and third places in Formula V at the 1978 and 1979 Cleveland National Air Races respectively, and is used for aerobatic displays. Structure is conventional, with all-wood wings, and fabric-covered tubular steel fuselage and tail unit. Power is provided by a 41 kW (55 hp) modified Volkswagen motorcar engine. Plans are available from the address shown.

DIMENSIONS, EXTERNAL:	
Wing span	5·18 m (17 ft 0 in)
Wing chord, constant	1·52 m (5 ft 0 in)
Length overall	4·57 m (15 ft 0 in)
Height overall	1·30 m (4 ft 3 in)
WEIGHTS:	
Weight empty	204·5 kg (451 lb)
Max T-O weight	304 kg (670 lb)
PERFORMANCE:	
Max level speed	over 148 knots (273 km/h; 170 mph)
Max cruising speed	126 knots (233 km/h; 145 mph)
Landing speed	43·5 knots (80·5 km/h; 50 mph)
Max rate of climb at S/L	259 m (850 ft)/min
Service ceiling	3,660 m (12,000 ft)
Range with max fuel	195 nm (362 km; 225 miles)

LUDWIG
LARRY LUDWIG

520 Bryden Avenue, Lewiston, Idaho 83501

LUDWIG PURE AIR MACHINE

Mr Larry Ludwig has designed and built a high-wing monoplane known as the Pure Air Machine. Construction of the prototype began in January 1978 and the first flight was recorded on 28 October the same year. Although flown initially as a single-seater, the Pure Air Machine was designed as a tandem two-seater and provision for a second seat has been made. Power is provided by a 56 kW (75 hp) Continental flat-four engine mounted on the open-frame fuselage just aft of the braced wing, driving a two-blade pusher propeller. No further details were available at the time of writing.

DIMENSIONS, EXTERNAL:
Wing span 9·12 m (29 ft 11 in)
Length overall 6·86 m (22 ft 6 in)

WEIGHT:
Weight empty 244 kg (538 lb)

MACE

MACE AIRCRAFT

30674 Pudding Creek Road, Fort Bragg, California 95437
PROPRIETOR: Harvey F. Mace

MACE M-102 SCORCHY

Veteran homebuilder Harvey Mace, whose earlier designs for the Macerschmitt, R-1, and R-2 Shark were described in the 1961-62, 1967-68, and 1971-72 *Jane's* respectively, has redesigned and completed a version of the Bede BD-5 as the Mace M-102 Scorchy. Modification of the original design to allow for the installation of a Revmaster engine with tractor propeller began during 1975. The M-102 flew for the first time on 25 June 1980, with unfaired engine cylinders and no propeller spinner. A streamlined cowling and spinner have since been installed, as illustrated.

TYPE: Single-seat sporting monoplane.

WINGS: Cantilever low-wing monoplane. Aerofoil section NACA 0009. Incidence 6°. No sweepback. Structure of light alloy. Plain ailerons and trailing-edge flaps of similar construction.

FUSELAGE: Semi-monocoque structure of light alloy. Aft fuselage terminates as a deep knife-edge section to enhance directional stability.

TAIL UNIT: Cantilever structure of light alloy, incorporating an all-moving tailplane. Anti-servo tab in tailplane.

LANDING GEAR: Manually-retractable type with fixed tailwheel. Main units retract inwards and are fitted with wheel doors. Shock-absorption of main units by glassfibre cantilever strut. Single wheel on each unit.

POWER PLANT: One 56 kW (75 hp) Revmaster 2100-D flat-four engine, driving a two-blade fixed-pitch wooden propeller of original design. Fuselage fuel tank immediately aft of firewall, capacity 49 litres (13 US gallons). Refuelling point on upper surface of fuselage, forward of windscreen.

ACCOMMODATION: Single seat beneath transparent canopy which opens upward and swings back. Accommodation is ventilated.

SYSTEM: Electrical system includes a 12V engine-driven alternator.

DIMENSIONS, EXTERNAL:
Wing span 6·55 m (21 ft 6 in)
Wing area, gross 4·46 m² (48 sq ft)
Length overall 4·11 m (13 ft 6 in)
Height overall 1·22 m (4 ft 0 in)
Tailplane span 1·98 m (6 ft 6 in)
Wheel track 1·19 m (3 ft 9 in)
Wheelbase 2·64 m (8 ft 8 in)
Propeller diameter 1·24 m (4 ft 1 in)

WEIGHTS:
Weight empty 207 kg (455 lb)
Max T-O weight 329 kg (725 lb)

PERFORMANCE (at max T-O weight):
Never-exceed speed 217 knots (402 km/h; 250 mph)
Max level speed at 1,525 m (5,000 ft) 152 knots (282 km/h; 175 mph)
Max cruising speed at 1,525 m (5,000 ft) 149 knots (275 km/h; 171 mph)
Econ cruising speed 122 knots (225 km/h; 140 mph)
Stalling speed, flaps down, power off 56·5 knots (105 km/h; 65 mph)
Max rate of climb at S/L 335 m (1,100 ft)/min
T-O run 229 m (750 ft)
T-O to and landing from 15 m (50 ft) 305 m (1,000 ft)
Landing run 213 m (700 ft)
Range with max fuel and reserves 300 nm (555 km; 345 miles)

MARCHETTI

FRANK MARCHETTI

Chicago, Illinois

MARCHETTI AVENGER

Mr Frank Marchetti has designed and built a two-seat autogyro known as the Avenger. Construction took three years and it made its first flight in September 1979. By Summer 1980 a total of 75 hours of flying had been completed. It received an EAA 'Rotorcraft Grand Champion' award at the 1980 Oshkosh Fly-in. Plans are available to amateur builders, plus kits including engine and instruments. Power is provided by a 112 kW (150 hp) Avco Lycoming O-320 engine, driving a Troyer propeller. The rotor blades came from a Hughes Model 269.

TYPE: Two-seat autogyro.

DIMENSIONS, EXTERNAL:
Rotor diameter 8·23 m (27 ft 0 in)
Length of fuselage 3·05 m (10 ft 0 in)
Height overall 2·13 m (7 ft 0 in)

WEIGHTS:
Weight empty 249 kg (550 lb)
Max T-O weight 544 kg (1,200 lb)

PERFORMANCE:
Max level speed 95·5 knots (177 km/h; 110 mph)
Econ cruising speed 52-56·5 knots (96·5-105 km/h; 60-65 mph)
Max rate of climb at S/L:
pilot only 487 m (1,600 ft)/min
two persons 366 m (1,200 ft)/min
T-O run 46 m (150 ft)
Endurance 1 h 30 min

McCARLEY

Mrs CHARLES E. McCARLEY

1119 27 Avenue, Hueytown, Alabama 35020

McCARLEY MINI-MAC

Designed by the late Mr Charles E. McCarley, the prototype Mini-Mac first flew in July 1970. It was totally destroyed, and Mr McCarley was fatally injured, in an accident on 28 October 1978. The aircraft was reportedly operating normally up to the time of the crash; Mr McCarley apparently was blinded by the sun and failed to see a powerline.

A modified version of the Mini-Mac (N75GH), built by George M. Harrison, was completed on 4 July 1974 and accumulated 122 flying hours in its first year. As well as having a shortened wing span, N75GH has 10° drooping ailerons.

The Mini-Mac is capable of limited aerobatics , and has a number of optional features, including a cockpit canopy, radio and an electrical system. Plans are available to amateur constructors and three more Mini-Macs were known to be under construction in 1980.

The following description generally applies to both aircraft mentioned above, except where indicated:

TYPE: Single-seat limited aerobatic monoplane.

WINGS: Cantilever low-wing monoplane. Clark Y wing section. All-metal construction. 10° drooping ailerons fitted to N75GH. Folding wings.

FUSELAGE: Conventional all-metal structure.

TAIL UNIT: Conventional cantilever all-metal structure. Tab in elevators.

LANDING GEAR: Non-retractable tricycle type. Wheels enclosed in streamline fairings. Small tail bumper skid.

POWER PLANT: One 1,834 cc (N152CM prototype) or 1,600 cc (N75GH) modified Volkswagen motorcar engine, driving a two-blade propeller. Conventional fuel and oil systems, not modified for inverted flight.

ACCOMMODATION: Single seat in open or enclosed cockpit, latter with rearward-sliding canopy.

DIMENSIONS, EXTERNAL (A: prototype; B: N75GH):
Wing span: A 6·25 m (20 ft 6 in)
B 5·79 m (19 ft 0 in)
Wing chord: A 1·02 m (3 ft 4 in)
B 1·04 m (3 ft 5 in)
Length overall: A 4·47 m (14 ft 8 in)
B 4·39 m (14 ft 5 in)
Height overall: A, B 1·73 m (5 ft 8 in)

WEIGHTS:
Weight empty: A 233 kg (514 lb)
B 208 kg (458 lb)
Max T-O weight: A 363 kg (800 lb)
B 318 kg (700 lb)

PERFORMANCE:
Max level speed:
A, B 139 knots (257 km/h; 160 mph)
Cruising speed: A 109 knots (201 km/h; 125 mph)
B 95 knots (177 km/h; 110 mph)
Landing speed: A 44 knots (81 km/h; 50 mph)
B 48 knots (89 km/h; 55 mph)
Max rate of climb at S/L: A 305 m (1,000 ft)/min
B 213 m (700 ft)/min
Service ceiling: A, B 2,440 m (8,000 ft)
T-O run: A 152 m (500 ft)
B 183 m (600 ft)
Landing run: A, B 152 m (500 ft)
Range with max fuel:
A 434 nm (804 km; 500 miles)
B 260 nm (483 km; 300 miles)

MEAD

MEAD ENGINEERING COMPANY

Box 354, Colwich, Kansas 67030
PROPRIETOR: George Mead

MEAD ADVENTURE

Mr George Mead, a design engineer with Piper Advanced Technology Inc, began in April 1977 the design of a small single-seat sporting monoplane. Construction began six months later, and the Adventure prototype (N36ME) flew for the first time in October 1978. In early 1981 Mr Mead was changing the landing gear from tricycle to tailwheel configuration for evaluation purposes. Similarly, the Continental A80 engine was being replaced by a converted Honda Civic motorcar engine. The first flight in this new form was scheduled for the Spring of 1981. Plans are available to amateur constructors.

TYPE: Single-seat sporting monoplane.

WINGS: Cantilever low-wing monoplane. Aerofoil section NACA 23018 at root, NACA 23012 at tip. Dihedral 4°. Incidence 0° 30′. Sweepback at quarter-chord 1° 18′. Composite structure of glassfibre/epoxy facings over a plastic foam core, and incorporating a built up box spar extending over 70% of span. Plain ailerons, inboard of wingtips, of glassfibre/foam sandwich. Wide-span single-slotted trailing-edge flaps of similar construction. Separate electrically-actuated roll trim surface set in trailing-edge of wing, at starboard wingtip.

FUSELAGE: Composite structure incorporating sandwich panels with glassfibre/epoxy facings and plastic foam core.

TAIL UNIT: Cantilever composite structure of full depth core glassfibre/epoxy/foam sandwich. Fixed-incidence tailplane. Bungee elevator trim. No trim tabs.

LANDING GEAR (original): Non-retractable tricycle type. Main legs comprise a glassfibre/epoxy spring formed as a one-piece inverted bow. Nosewheel carried on glassfibre/epoxy spring strut. Main-wheel tyres size 3·40 × 3·00-5, pressure 3·79 bars (55 lb/sq in). Rosenhan 13 cm (5 in) nosewheel, tyre pressure 4·14 bars (60 lb/sq in). Rosenhan hydraulic brakes. Speed fairings for main wheels optional.

POWER PLANT (original): One 59·5 kW (80 hp) Continental A80 flat-four engine, driving a two-blade fixed-pitch propeller with spinner. Two fuselage fuel tanks, one immediately aft of firewall, the other in turtledeck, with combined capacity of 87 litres (23 US gallons). Refuelling points on upper surface of forward and aft fuselage. Oil capacity 3·8 litres (1 US gallon).

ACCOMMODATION: Single seat in enclosed cockpit, beneath sideways-opening transparent canopy. Baggage space aft of seat. Accommodation is heated and ventilated.

SYSTEMS: Electrical system has small storage battery. Hydraulic system for brakes only.

DIMENSIONS, EXTERNAL:
Wing span 6·10 m (20 ft 0 in)
Wing chord at root 0·76 m (2 ft 6 in)
Wing chord at tip 0·46 m (1 ft 6 in)
Wing area, gross 3·72 m² (40·0 sq ft)
Wing aspect ratio 10·0
Length overall 3·96 m (13 ft 0 in)
Height overall 1·68 m (5 ft 6 in)
Tailplane span 2·56 m (8 ft 4¾ in)
Wheel track 1·52 m (5 ft 0 in)
Wheelbase 1·02 m (3 ft 4 in)
Propeller diameter 1·57 m (5 ft 2 in)

WEIGHTS:
Basic operating weight 209 kg (460 lb)
Max T-O weight 376·5 kg (830 lb)
Max landing weight 363 kg (800 lb)

PERFORMANCE (at max T-O weight):
Never-exceed speed 183 knots (338 km/h; 210 mph)
Max level speed at S/L 165 knots (306 km/h; 190 mph)
Max cruising speed at 2,440 m (8,000 ft) 156 knots (290 km/h; 180 mph)
Econ cruising speed at 3,500 m (11,500 ft) 130 knots (241 km/h; 150 mph)

Mace M-102 Scorchy, a modified version of the Bede BD-5

The Avenger two-seat autogyro built by Mr Frank Marchetti *(Howard Levy)*

K-Meyer Aero Model A, designed by Mr Les K. Meyer *(Howard Levy)*

Mead Adventure sporting monoplane as flown originally

Prototype WM-2 sport aircraft built by Mr William Miller

The prototype Mini-Mac (1,834 cc Volkswagen engine)

Stalling speed, flaps 45°, engine idling	49·5 knots (92 km/h; 57 mph)
Max rate of climb at S/L	488 m (1,600 ft)/min
Service ceiling	6,100 m (20,000 ft)
T-O run	183 m (600 ft)
T-O to 15 m (50 ft)	305 m (1,000 ft)
Landing from 15 m (50 ft)	457 m (1,500 ft)
Landing run	244 m (800 ft)
Range with max fuel at econ cruising speed, 45 min reserves	825 nm (1,529 km; 950 miles)

MEYER
LES K. MEYER

45105 244th Avenue SE, Enumclaw, Washington 98022

K-MEYER AERO MODEL A

Mr Les K. Meyer has designed a side-by-side two-seat high-wing monoplane known as the K-Meyer Aero Model A. By 1981 three Model As were flying, having accumulated a total of approximately 5,500 flight hours in 2½ years of development testing. Plans and some components are available to amateur constructors.

The Model A is currently powered by a 74·5 kW (100 hp) Continental O-200 flat-four engine, driving a two-blade fixed-pitch propeller, but is suited to other engines. The high-mounted wing has a wooden spar and metal ribs, and is covered with Ceconite. Ailerons and the tail unit are of aluminium construction. The fuselage has a 4130 steel tube structure, Ceconite covered. The engine cowling, wingtips and wheel fairings are all constructed from glassfibre.

Type: Two-seat light cabin monoplane.

Dimensions, external:

Wing span	7·62 m (25 ft 0 in)
Length overall	6·10 m (20 ft 0 in)

Weights:

Weight empty	389 kg (857 lb)
Max T-O weight	590 kg (1,300 lb)

Performance:

Max level speed	122 knots (225 km/h; 140 mph)
Max cruising speed	109 knots (201 km/h; 125 mph)
Stalling speed	42 knots (77·5 km/h; 48 mph)
Service ceiling	4,725 m (15,500 ft)
T-O and landing run	274 m (900 ft)
T-O to 15 m (50 ft)	335 m (1,100 ft)
Landing from 15 m (50 ft)	335 m (1,100 ft)
Range with 76 litres (20 US gallons) of fuel	347 nm (643 km; 400 miles)

MILLER
WILLIAM Y. MILLER

1439 West Second Street, Mesa, Arizona 85201

Mr William Y. Miller built the prototype of the WM-2 sporting aircraft, which had been designed by Mr W. Terry Miller for amateur construction.

MILLER WM-2

The WM-2 is a low-powered, high-performance aircraft, conceived originally for the exploration of wave soaring conditions, thermal soaring with the engine stopped, and high-altitude, economical powered sport flying. The prototype (N24832) was built by Mr William Miller between 1969 and 1972, and made its first flight in August 1972. Flight testing was undertaken during 1973-74, and by January 1978 the aircraft had completed 430 flying hours. Plans are available to amateur constructors, and approximately 12 WM-2s are under construction.

Type: Single-seat sport aircraft.

Wings: Cantilever low-wing monoplane. Modified NACA laminar-flow series wing sections. Thickness/chord ratio 15%. Dihedral 4°. Incidence 1°. Sweepback 0° 53′ at quarter-chord. Conventional structure of spruce spars, with birch plywood, glassfibre and fabric covering. Wooden ailerons. No flaps or tabs. Metal spoiler in each upper surface.

Fuselage: Conventional spruce structure, with birch plywood and glassfibre covering.

Tail Unit: Plywood- and fabric-covered spruce cantilever structure.

Landing Gear: Manually-retractable monowheel (wheel and tyre size 6·00-6) and steerable tailwheel. Hydraulic brake.

Power Plant: One 48·5 kW (65 hp) Continental flat-four engine, driving a two-blade fixed-pitch metal propeller. Fuel tank, capacity 37·8 litres (10 US gallons), aft of firewall. Fuel consumption at an econ cruising speed of 69·5 knots (129 km/h; 80 mph) is less than 7·5 litres (2 US gallons)/h.

Accommodation: Single seat under one-piece sideways-opening bubble canopy.

Dimensions, external:

Wing span	12·19 m (40 ft 0 in)
Wing chord at root	1·37 m (4 ft 6 in)
Wing chord at tip	0·76 m (2 ft 6 in)
Wing area, gross	13·38 m² (144·0 sq ft)
Wing aspect ratio	11·11

Length overall	6·10 m (20 ft 0 in)
Height over tail	1·60 m (5 ft 3 in)
Tailplane span	2·44 m (8 ft 0 in)
Propeller diameter	1·88 m (6 ft 2 in)

WEIGHTS:

Weight empty, equipped	351 kg (775 lb)
Max T-O weight	476 kg (1,050 lb)

PERFORMANCE (at max T-O weight):

Never-exceed speed	130 knots (241 km/h; 150 mph)
Max level speed at S/L	118 knots (219 km/h; 136 mph)
Normal cruising speed at 3,050 m (10,000 ft), 50% power	109 knots (203 km/h; 126 mph)
Stalling speed	39·5 knots (72·5 km/h; 45 mph)
Max rate of climb at S/L	271 m (890 ft)/min
Rate of climb at 4,575 m (15,000 ft)	152 m (500 ft)/min
Service ceiling (computed)	7,315 m (24,000 ft)
Range at normal cruising speed at 3,050 m (10,000 ft) with 30 min reserves	291 nm (540 km; 336 miles)
Best glide ratio at 54 knots (100 km/h; 62 mph), power off	15

MINI-HAWK

MINI-HAWK INTERNATIONAL INC

1930 Stewart Street, Santa Monica, California 90404
Telephone: (213) 828 4078
DIRECTOR OF MARKETING: E. Y. Treffinger

Mini-Hawk International was formed to market plans and kits for construction of an all-metal single-seat monoplane known as the Mini-Hawk TH.E.01 Tiger-Hawk. Three officers of the corporation, designer William B. Taylor, engineer Thomas E. Maloney and pilot E. Y. Treffinger, combined their efforts to design and construct the prototype. This aircraft flew for the first time during 1974 and the test programme was completed in 1976.

Mini-Hawk offers amateur constructors a complete set of plans and a construction manual or a complete kit package with or without an engine; by early 1981 at least five sets had been sold.

MINI-HAWK TH.E.01 TIGER-HAWK

TYPE: Single-seat lightweight aircraft.

WINGS: Cantilever low-wing monoplane of all-metal construction. Constant-chord wing. Dihedral 4° on outer panels. Full-span ailerons of all-metal construction, operated by push/pull rods. All-metal trailing-edge flaps. Wings removable in 10 min for towing or storage.

FUSELAGE: Built-up structure of light alloy.

TAIL UNIT: Cantilever all-metal structure. Fixed-incidence tailplane. Elevator of light alloy construction. No trim tabs.

LANDING GEAR: Non-retractable tricycle type with steerable nosewheel. Hurst/Airheart hydraulic disc brakes. Single wheel with speed fairing on each unit.

POWER PLANT: One 53·5 kW (72 hp) Revmaster Model 1831D modified Volkswagen motorcar engine, with dual ignition, driving an Eng. Duplicating 54/42 two-blade fixed-pitch propeller with spinner. Fuel tank in fuselage with capacity of 45·4 litres (12 US gallons). Suitable for Volkswagen engines of up to 67 kW (90 hp). Oil capacity 2·35 litres (5 US pints).

ACCOMMODATION: Single seat under transparent canopy.

AVIONICS: Genave radio optional.

DIMENSIONS, EXTERNAL:

Wing span	5·49 m (18 ft 0 in)
Wing chord, constant	0·99 m (3 ft 3 in)
Wing area, gross	5·30 m² (57 sq ft)
Length overall	4·04 m (13 ft 3 in)
Width overall, wings removed	1·83 m (6 ft 0 in)
Height overall	2·08 m (6 ft 10 in)
Tailplane span	1·83 m (6 ft 0 in)
Wheel track	1·78 m (5 ft 10 in)
Wheelbase	1·93 m (6 ft 4 in)
Propeller diameter	1·37 m (4 ft 6 in)

WEIGHTS:

Weight empty	238 kg (525 lb)
Max T-O weight	362 kg (800 lb)

PERFORMANCE (at max T-O weight):

Never-exceed speed	173 knots (321 km/h; 200 mph)
Max level speed	152 knots (282 km/h; 175 mph)
Max cruising speed	139 knots (257 km/h; 160 mph)
Stalling speed, flaps up	54 knots (100 km/h; 62 mph)
Stalling speed, flaps down	44 knots (81 km/h; 50 mph)
Max rate of climb at S/L	275-305 m (900-1,000 ft)/min
Service ceiling	3,050 m (10,000 ft)
Absolute ceiling	3,660 m (12,000 ft)
T-O and landing run	122 m (400 ft)
Range with max fuel	608 nm (1,126 km; 700 miles)

MONNETT

MONNETT EXPERIMENTAL AIRCRAFT INC

955 Grace Street, Elgin, Illinois 60120
Telephone: (312) 741 2223
PUBLIC RELATIONS: Betty Monnett

Mr John T. Monnett formed this company to market plans and certain components of an original-design Formula V racer. Known originally as the Monnett II Sonerai, this received the Best in Class Formula V Racer award at the EAA Fly-in at Oshkosh in 1971, as well as an award for its outstanding contribution to low-cost flying. Since that time Mr Monnett has designed a two-seat version of the Sonerai, with the result that the original single-seat model is now known simply as Sonerai, the two-seat model as Sonerai II. His latest aircraft is a high-performance single-seat monoplane named the Monex.

Details of the Monnett Monerai S homebuilt sailplane, and the Monerai P with auxiliary engine, can be found in the Sailplanes section.

Other products available from Monnett Experimental Aircraft include the company's Aero Vee light aero-engines. The two current models are the E-V and Super-Vee, supplied as either 1,600 cc or 1,700 cc units, with or without alternator. Both are stock Volkswagen direct-drive engines, producing 44·7-48·5 kW (60-65 hp) at 4,000 rpm. Customers can purchase either a conversion kit for any Volkswagen Type 3 case of 1969 or newer, or a ready-to-run engine.

MONNETT MONEX

Latest aircraft from Monnett is an all-metal single-seat high-performance monoplane named Monex. Design began in October 1979 and construction of a prototype started in the following month. This took approximately 600 working hours to complete and flew for the first time on 30 September 1980.

TYPE: Single-seat high-performance aircraft.

WINGS: Cantilever low-wing monoplane. Wing section NACA 64215. Dihedral 3°. Incidence 2°. No sweepback. Bonded light alloy structure, with Unispar extruded I-beam main spar, formed ribs and C-channel rear spar. Light alloy ailerons and flaps.

FUSELAGE: Light alloy tube structure, covered with light alloy sheet using stainless steel pop rivets. Flat sides and bottom; turtledeck has natural curves. Glassfibre engine cowling.

TAIL UNIT: Cantilever V structure of bonded light alloy. Combined rudders and elevators, known as 'ruddervators'. Bungee trim.

LANDING GEAR: Non-retractable tailwheel type. Cantilever spring main legs of aluminium alloy. Gerdes main wheels with tyres size 4·10 × 3·50-4. Tailwheel semi-enclosed in light alloy fairing. Gerdes hydraulic brakes. Moulded streamline fairings over main wheels.

POWER PLANT: One 60 kW (80 hp) Aero Vee 2,180 cc Volkswagen modified motorcar engine, driving a two-blade fixed-pitch Warnkee propeller. One fuel tank in fuselage, capacity 37·8 litres (10 US gallons). Refuelling point in fuselage forward of canopy. Oil capacity 2·8 litres (0·75 US gallon).

ACCOMMODATION: Single seat under long low-profile Plexiglas canopy, hinged at port side.

SYSTEM: 10A alternator.

AVIONICS: Alpha 100-channel com transceiver.

DIMENSIONS, EXTERNAL:

Wing span	5·08 m (16 ft 8 in)
Wing chord, constant	0·84 m (2 ft 9 in)
Wing area, gross	4·14 m² (44·6 sq ft)
Wing aspect ratio	5·9
Length overall	4·62 m (15 ft 2 in)
Height overall	1·09 m (3 ft 7 in)
Tailplane span	1·47 m (4 ft 10 in)
Wheelbase	1·52 m (5 ft 0 in)
Propeller diameter	1·32 m (4 ft 4 in)

WEIGHTS:

Weight empty	172 kg (380 lb)
Max T-O weight	295 kg (650 lb)

PERFORMANCE:

Never-exceed speed	217 knots (402 km/h; 250 mph)
Max level speed at S/L	182 knots (338 km/h; 210 mph)
Stalling speed	45·5 knots (84 km/h; 52 mph)
Max rate of climb at S/L (estimated)	over 610 m (2,000 ft)/min
Range (estimated)	over 348 nm (644 km; 400 miles)

MONNETT SONERAI

Mr Monnett began design of the Sonerai in September 1970, construction starting two months later. First flight was made in July 1971, with FAA certification in the Experimental category. Plans and certain components are available to amateur constructors, including glassfibre engine cowlings, clear or tinted Plexiglas cockpit canopy, main landing gear struts, formed aluminium ribs, tapered rod tail spring, fuel tanks, spar kits, instruments, injector carburettor and wheels and brakes. Estimated building time is 850 working hours. The aircraft is stressed to ±6*g*.

Approximately 550 sets of plans have been sold, and more than 400 Sonerais are under construction or flying.

TYPE: Single-seat Formula V racing aircraft.

WINGS: Cantilever mid-wing monoplane. Wing section NACA 64212. No dihedral, incidence or sweepback. Conventional light alloy structure. Full-span light alloy ailerons. No flaps or tabs. Wings fold on each side of the fuselage to allow the aircraft to be towed tail-first.

FUSELAGE: Welded chrome-molybdenum steel tube structure with fabric covering. Glassfibre engine cowling.

TAIL UNIT: Cantilever structure of welded chrome-molybdenum steel tube with fabric covering. Tailplane incidence ground-adjustable. No trim tabs.

LANDING GEAR: Non-retractable tailwheel type. Cantilever spring main gear of light alloy. Main wheels and tyres size 5·00-5. Caliper type wheel-brakes. Glassfibre fairings on main wheels.

POWER PLANT: One 44·7 kW (60 hp) Volkswagen 1,600 cc modified motorcar engine, driving a Hegy two-blade propeller with spinner. Fuel tank in fuselage, immediately aft of firewall, capacity 37·8 litres (10 US gallons). Refuelling point on fuselage upper surface forward of canopy. Oil capacity 2·82 litres (0·75 US gallons).

ACCOMMODATION: Single seat under jettisonable Plexiglas bubble canopy, hinged at the starboard side.

AVIONICS: Battery-powered 100-channel com transceiver.

DIMENSIONS, EXTERNAL:

Wing span	5·08 m (16 ft 8 in)
Wing chord, constant	1·37 m (4 ft 6 in)
Wing area, gross	6·97 m² (75 sq ft)
Length overall	5·08 m (16 ft 8 in)
Height overall	1·52 m (5 ft 0 in)
Tailplane span	1·98 m (6 ft 6 in)
Wheel track	1·22 m (4 ft 0 in)
Propeller diameter	1·27 m (4 ft 2 in)

WEIGHTS:

Weight empty	199 kg (440 lb)
Max T-O weight	340 kg (750 lb)

PERFORMANCE (at max T-O weight):

Max level speed at S/L	152 knots (281 km/h; 175 mph)
Max cruising speed	130 knots (241 km/h; 150 mph)
Econ cruising speed	109 knots (201 km/h; 125 mph)
Stalling speed	35 knots (64·5 km/h; 40 mph)
Max rate of climb at S/L	305 m (1,000 ft)/min
T-O and landing run	183 m (600 ft)
Range, with reserves	260 nm (482 km; 300 miles)

MONNETT SONERAI II and IIL

The success of the Sonerai encouraged Mr Monnett to begin the design and construction of the two-seat **Sonerai II** in December 1972. Generally similar to the Sonerai, it differs by being slightly larger and by having a more powerful Volkswagen engine. It is stressed to ±4·4*g* in the Utility category and to ±6*g* in a single-seat Aerobatic category. The prototype made its first flight in July 1973. Orders had been received for 843 sets of plans and at least 140 Sonerai IIs were flying by early 1981. Many components, complete kits for fuselage and wings, and materials, are available to amateur constructors. Estimated building time is 850 working hours.

A variant, first flown in June 1980, is the **Sonerai IIL**, with low-wing instead of mid-wing configuration and 3° of dihedral. The two designs are identical in all other respects. The IIL illustrated is powered by a 2,180 cc Aero Vee Volkswagen modified motorcar engine, which gives it a maximum cruising speed of 135 knots (249 km/h; 155 mph) and an econ cruising speed of 122 knots (225 km/h; 140 mph) at sea level.

The description of the Sonerai applies also to Sonerai II and IIL, except as follows:

TYPE: Two-seat high-performance sporting aircraft.

WINGS: Sonerai II as for Sonerai, except span increased. Sonerai IIL has low-wing configuration.

FUSELAGE: As for Sonerai, except length increased.

TAIL UNIT: As for Sonerai, except tailplane has fixed incidence and reduced span.

POWER PLANT: One 1,700 cc Volkswagen modified motorcar engine, developing 48·5-52·2 kW (65-70 hp), driving a two-blade ground-adjustable wooden propeller. Fuel capacity 37·8 litres (10 US gallons); oil as for Sonerai.

ACCOMMODATION: Two seats in tandem beneath transparent bubble canopy, hinged on starboard side.

AVIONICS: Prototype Sonerai II has alternator-powered 100-channel radio.

DIMENSIONS, EXTERNAL: As Sonerai, except:

Wing span	5·69 m (18 ft 8 in)
Wing area, gross	7·80 m² (84 sq ft)
Length overall	5·74 m (18 ft 10 in)
Tailplane span	1·83 m (6 ft 0 in)
Propeller diameter	1·32-1·37 m (4 ft 4 in to 4 ft 6 in)

DIMENSION, INTERNAL:

Cockpit: Max width	0·61 m (2 ft 0 in)

Mini-Hawk TH.E.01 Tiger-Hawk with current, revised canopy

Monnett Monex, designed for outstanding fuel economy

Mooney Mite single-seat light aircraft *(J. M. G. Gradidge)*

Monnett Sonerai (1,850 cc Volkswagen engine) built by Mr Zig Bergins of Montreal, Canada *(Neil A. Macdougall)*

The new low-wing Monnett Sonerai IIL two-seat sporting aircraft *(Howard Levy)*

Murphy VM-7 Competitor Unlimited class aerobatic monoplane

WEIGHTS:
- Weight empty 230 kg (506 lb)
- Max T-O weight 419 kg (925 lb)

PERFORMANCE (at max T-O weight):
- Max level speed at S/L approx 143 knots (266 km/h; 165 mph)
- Max cruising speed at S/L 122 knots (225 km/h; 140 mph)
- Econ cruising speed at S/L 113 knots (209 km/h; 130 mph)
- Stalling speed 39 knots (73 km/h; 45 mph)
- Max rate of climb at S/L 152 m (500 ft)/min
- T-O run, ISA 274 m (900 ft)
- Landing run 152 m (500 ft)
- Range, with reserves 217 nm (402 km; 250 miles)

MOONEY MITE
MOONEY MITE AIRCRAFT CORPORATION

PO Box 3999, Charlottesville, Virginia 22903

Mooney Mite Aircraft Corporation was formed to market to amateur constructors plans of the Mooney Mite, a version of the Mooney M-18 first designed and built as a production aircraft by Mooney Aircraft Inc.

MOONEY MITE

TYPE: Single-seat sporting aircraft.

WINGS: Cantilever low-wing monoplane. Wing section NACA 63215 at root, NACA 64415 at tip. Dihedral 5° 30′ from roots. Incidence 4°. Conventional single-spar structure of wood, with plywood D-type leading-edge torsion box, and fabric covering aft of the spar. Ailerons and trailing-edge flaps of welded steel tube construction with fabric covering.

FUSELAGE: Forward section to aft of cockpit of welded steel tube with light alloy skins. Aft fuselage of wood monocoque construction with fabric covering.

TAIL UNIT: Cantilever welded steel tube structure, fabric-covered. Tailplane incidence variable by 'Safe-Trim' system that interconnects tail trim with trailing-edge flaps to establish automatically the correct settings for take-off, climb, approach and landing.

LANDING GEAR: Manually retractable tricycle type. Shock-absorption of main units by rubber-in-compression. Nosewheel steerable. Main wheels of Cleveland, Firestone or Goodyear manufacture, with tyres size 5·00-5. Hydraulic brakes.

POWER PLANT: One 48·5 kW (65 hp) Avco Lycoming O-145-B2 flat-four aircooled engine, driving a Sensenich Type 66CB-54 two-blade fixed-pitch propeller with spinner. Fuel tank in fuselage, capacity 41·5 litres (11 US gallons). Provision for auxiliary tank, capacity 23 litres (6 US gallons). Oil capacity 3·75 litres (1 US gallon).

ACCOMMODATION: Single seat beneath aft-sliding transparent canopy. Space for 18 kg (40 lb) baggage aft of seat.

SYSTEM: Electric power supplied by wind-driven generator mounted on pylon on upper surface of aft fuselage.

DIMENSIONS, EXTERNAL:
- Wing span 8·19 m (26 ft 10½ in)
- Wing area 8·83 m² (95 sq ft)
- Length overall 5·37 m (17 ft 7¼ in)
- Height overall 1·89 m (6 ft 2½ in)
- Tailplane span 2·54 m (8 ft 4 in)
- Wheel track 1·55 m (5 ft 1 in)
- Wheelbase 1·22 m (4 ft 0 in)
- Propeller diameter 1·60 m (5 ft 3 in)

WEIGHTS:
- Weight empty 229 kg (505 lb)
- Max T-O weight 353 kg (780 lb)

PERFORMANCE (at max T-O weight):
- Max level speed at S/L 124 knots (230 km/h; 143 mph)
- Econ cruising speed, 50% power at S/L 99 knots (183 km/h; 114 mph)
- Min controllable speed, power on 33 knots (61·5 km/h; 38 mph)
- Landing speed, power off 39 knots (72 km/h; 45 mph)
- Stalling speed, power off 37·5 knots (69·5 km/h; 43 mph)
- Max rate of climb at S/L 332 m (1,090 ft)/min
- Service ceiling 5,915 m (19,400 ft)
- T-O run 213 m (698 ft)
- T-O to 15 m (50 ft) 305 m (1,000 ft)
- Landing from 15 m (50 ft) 366 m (1,200 ft)
- Landing run 73 m (240 ft)
- Range at econ cruising speed, standard fuel 382 nm (708 km; 440 miles)
- Range with auxiliary fuel 521 nm (965 km; 600 miles)

MURPHY
R. L. MURPHY

90 Pebble Woods Drive, Doylestown, Pennsylvania 18901

Mr Dick Murphy, a Captain with American Airlines, with 23 years of military and civil flying and engineering experience, designed and built a single-seat Unlimited class aerobatic aircraft known as the VM-7 Competitor (N5P). He had previously constructed a Van's RV-3, which was flown competitively during 1976; the VM-7 Competitor owes much to this aircraft and to ideas and suggestions from other designers and pilots. Design objectives for the VM-7 were that it should be easy to construct, should cost under $10,000, and should have good performance. Without the engine and propeller, it cost $6,000 and took 2,000 working hours to build. The prototype flew for the first time in March 1978, but was destroyed by fire during December 1979. Mr Murphy began construction of a second example immediately, with the intention of having it completed it time to compete in the 1980 World aerobatic contest. Plans may be made available to amateur builders when the aircraft has proved its capability in competitive aerobatics.

MURPHY VM-7 COMPETITOR

TYPE: Single-seat Unlimited class aerobatic monoplane.

WINGS: Cantilever mid-wing monoplane. Wing section NACA 23015. Constant chord. Light alloy structure, comprising a laminated main spar, passing through fuselage under pilot's knees, and twenty hydroformed ribs, covered with light alloy skin panels, bonded and pop-riveted in place. Symmetrical, balanced and sealed ailerons.

FUSELAGE: Light-alloy semi-monocoque structure, mostly bonded but with some pop rivets.

TAIL UNIT: Cantilever light alloy structure. Fin integral with rear fuselage. Trim tab in port elevator.

LANDING GEAR: Non-retractable tailwheel type. Wittman type cantilever spring steel main legs. Main-wheel tyres size 5·00-5. Cleveland brakes. Streamline fairings over main wheels. Steerable tailwheel.

POWER PLANT: One 149 kW (200 hp) Avco Lycoming O-360 flat-four engine, driving a Hartzell two-blade constant-speed propeller, with spinner. Engine fitted with IO-360 sump, manifold and fuel injector. Main and auxiliary fuel tanks in forward fuselage; total capacity 64·4 litres (17 US gallons). Second auxiliary tank optional. Oil capacity 7·6 litres (2 US gallons).

ACCOMMODATION: Single seat, inclined at about 15°, under forward-sliding canopy.

DIMENSIONS, EXTERNAL:
- Wing span: 5·94 m (19 ft 6 in)
- Wing chord (constant): 1·37 m (4 ft 6 in)
- Length overall: 6·02 m (19 ft 9 in)
- Propeller diameter: 1·93 m (6 ft 4 in)

WEIGHTS:
- Weight empty, with starter and battery: 361·5 kg (797 lb)
- Max T-O and landing weight: 567 kg (1,250 lb)

PERFORMANCE:
- Never-exceed speed: 186 knots (346 km/h; 215 mph)
- Max level speed at S/L: 174 knots (322 km/h; 200 mph)
- Max cruising speed (75% power): 152 knots (282 km/h; 175 mph)
- Econ cruising speed (50% power): 130 knots (241 km/h; 150 mph)
- Stalling speed, power off: 43·5 knots (80·5 km/h; 50 mph)
- Rate of climb at S/L: over 915 m (3,000 ft)/min
- Service ceiling, estimated: 5,485 m (18,000 ft)
- T-O run: 46 m (150 ft)
- T-O to 15 m (50 ft): 76 m (250 ft)
- Landing from 15 m (50 ft): 366 m (1,200 ft)
- Range with max fuel: 217 nm (402 km; 250 miles)
- *g* limits: normal: ±9
- ultimate: ±14

MUSSO
PAUL MUSSO
Mt Laurel, New Jersey

MUSSO REAL SPORTY

Mr Paul Musso designed and built a single-seat Formula I midget racing aircraft in 1978. Known as the Real Sporty, it is of metal, wood, glassfibre and fabric composite construction. Power plant is a 74·5 kW (100 hp) Continental O-200 flat-four engine.

Sponsored by the Sterling Model company of Philadelphia, Real Sporty is flown by Mr Charles 'Chuck' Andrews of Beckley, West Virginia, a retired USAF pilot and currently a pilot with Appalachian Tire Products. On 5 August 1980 it was flown at the Oshkosh, Lowers, Baker, Falck Memorial 500 Mile Race, during which it set FAI-approved world speed records for Class C1a over distances of 100 km and 500 km, with speeds of 192·743 knots (356·96 km/h; 221·804 mph) and 176·522 knots (326·92 km/h; 203·138 mph) respectively. Later, on 26 October 1980, the Real Sporty set a speed record of 205·010 knots (379·68 km/h; 235·922 mph) over a 3 km straight line course.

TYPE: Single-seat Formula I racer.

DIMENSIONS, EXTERNAL:
- Wing span: 5·79 m (19 ft 0 in)
- Length overall: 4·88 m (16 ft 0 in)

WEIGHT:
- Max T-O weight: 261 kg (575 lb)

PERFORMANCE:
- Max level speed: 213 knots (394 km/h; 245 mph)

OLSEN
GORDON L. OLSEN
4914 Alo Vera, Redding, California 96001

OLSEN NITE STAR

Dr Gordon L. Olsen has designed and built a two-seat cabin monoplane known as the Nite Star. Although owing something in design to the highly-successful Rand Robinson KR-2, the Nite Star is an original aircraft with several interesting features. The wings are of mostly wooden construction, with some aluminium components and glassfibre tips incorporating winglets. The fuselage uses a basic 4130 steel tube structure, covered with pre-moulded glassfibre shells to achieve a very streamline appearance. Full-span 'flaperons' are fitted to the detachable wings. The tail unit includes a high-mounted all-moving tailplane. Power is provided by a 67 kW (90 hp) turbocharged Revmaster 2100 engine, driving a Maloof two-blade constant-speed propeller.

Construction of the Nite Star began in late 1978 and completion took about four months. No further details were available at the time of writing.

OPTION AIR
OPTION AIR RENO
PO Box 20085, Reno, Nevada 89512
Telephone: (702) 825 0561

OPTION AIR RENO ACAPELLA

Mr Carl D. Barlow has formed Option Air Reno to market a kit that can be added to a Bede BD-5 fuselage, canopy, nosewheel, wings and tailfin to produce his twin-boom Acapella. The kit includes an engine mount, glassfibre cowling, new 2·44 m (8 ft) wing centre section, tailbooms, second tailfin, elevator, stabilator, main landing gear and all necessary plans for the conversion. The prototype Acapella has a 149 kW (200 hp) Avco Lycoming IO-360-C2B engine driving a Hartzell CS Q-Tip two-blade constant-speed pusher propeller, and is known as **Model 200S**. Other versions with a 74·5 kW (100 hp) Continental O-200 engine and increased wing span of 8·08 m (26 ft 6 in) are proposed as the **Model 100L**, with fuel capacity of 106 litres (28 US gallons), and **Model 100LR**, with fuel capacity of 182 litres (48 US gallons). The following data apply to the prototype Acapella 200S, with a fuel capacity of 128 litres (34 US gallons).

TYPE: Single-seat light monoplane.

DIMENSIONS, EXTERNAL:
- Wing span: 5·94 m (19 ft 6 in)
- Wing span, folded: 2·54 m (8 ft 4 in)
- Wing area, gross: 4·43 m² (47·7 sq ft)
- Wing aspect ratio: 8·0
- Length overall: 4·88 m (16 ft 0 in)

WEIGHTS: (A: Model 200S with 5·94 m; 19 ft 6 in wing span and IO-360 engine, B: Model 100L with 8·08 m; 26 ft 6 in wing span and O-200 engine):
- Weight empty: A: 422 kg (930 lb)
- B: 306 kg (675 lb)
- Max T-O weight: A: 612 kg (1,350 lb)
- B: 481 kg (1,060 lb)

PERFORMANCE (B, estimated):
- Max level speed:
  - A: over 247 knots (459 km/h; 285 mph)
  - B: 170 knots (315 km/h; 196 mph)
- Range (75% power):
  - A, B: 781 nm (1,448 km; 900 miles)

OSPREY
OSPREY AIRCRAFT
3741 El Ricon Way, Sacramento, California 95825
Telephone: (916) 483 3004

Osprey Aircraft was formed originally to market to amateur constructors plans of the Osprey I aircraft designed and built by Mr George Pereira. This was an unusual project for the homebuilder, being a flying-boat, intended for operation on and from enclosed waters rather than the open sea. The plans drawn up by Mr Pereira included drawings of a special trailer for carriage of the aircraft, which allowed the pilot to launch and recover the Osprey unassisted. Details of this aircraft can be found in the 1974-75 *Jane's*.

Mr Pereira subsequently completed the prototype of a two-seat amphibian version designated Osprey II.

PEREIRA GP3 OSPREY II

Design and construction of the Osprey II, a two-seat amphibian development of the Osprey I, began in January 1972.

Mr Pereira developed an unusual form of hull construction for this aircraft. When the all-wood fuselage structure had been completed and controls installed, the undersurface was given a deep coating of polyurethane foam. This was then sculptured to the requisite hull form before being covered with several protective layers of glassfibre cloth bonded with resin. The resulting structure is light, but extremely strong, with good shock resisting characteristics.

First flight of the Osprey II from water was made in April 1973, the amphibian becoming airborne in less than 244 m (800 ft), with no tendency to porpoise at any speed. In later tests from land it was found that with the landing gear retracted, and at a speed of about 104 knots (193 km/h; 120 mph), there was slight buffet aft of the cabin and the noise level was unacceptably high. Modifications carried out in early 1974 included lengthening of the cabin by 0·18 m (7 in), and installation of an Avco Lycoming O-320 engine in place of the original Franklin Sport, in a new cowling. Testing was resumed and completed satisfactorily during 1974, since when the shape of the tail fin has been changed. Sets of plans, as well as material and component kits, are available to amateur constructors. By early 1981, eighteen Ospreys were flying and several hundred more were under construction in 48 different countries.

TYPE: Two-seat lightweight amphibian.

WINGS: Cantilever mid-wing monoplane, of constant chord. Wing section NACA 23012. Dihedral 4° 30′. Incidence 5°. All-wood structure, with single box spar and auxiliary rear spar for aileron attachment. Forward of the main spar the wing is plywood-covered to form a rigid 'D' section. Aft of the spar the wing is fabric-covered. Conventional ailerons, 100% mass-balanced, will be fitted with a ground-adjustable tab if this proves desirable. No trailing-edge flaps. Wingtip stabilising floats of polyurethane foam covered with glassfibre.

HULL: All-wood structure of longerons and frames, covered with 2·5 mm ($^3/_{32}$ in) marine plywood. Hull undersurface contours formed from polyurethane foam, protected by several layers of glassfibre cloth bonded with resin.

TAIL UNIT: Cantilever all-wood structure, with swept vertical surfaces; tailplane mounted high on fin, which is integral with hull. Incidence of tailplane ground-adjustable. Controllable trim tab in starboard elevator. Water rudder, contained within the base of the aerodynamic rudder, is spring-loaded in the down position and retracted by cable.

LANDING GEAR: Retractable tricycle type, with single wheel on each unit. Main units retract inward into the wing roots, the wheel wells being covered by doors in the retracted position. Nosewheel retracts forward into the nosecone and is also enclosed by a door. Manual retraction system. Shock-absorption by coil springs. Cleveland main wheels and tyres size 5·00-5. Nosewheel, of industrial type with roller bearings, has a tyre of 10 in diameter. Cleveland hydraulic disc brakes.

POWER PLANT: One 112 kW (150 hp) Avco Lycoming O-320 flat-four engine, mounted on a steel tube pylon structure which is bolted to the wing truss. Hendrickson 66 × 52 three-blade fixed-pitch wooden pusher propeller. One glassfibre fuel tank standard, mounted beneath the main spar at the wing centre-section, usable capacity 98·4 litres (26 US gallons). Refuelling point on starboard side of hull, just aft of cabin. New wing tanks available to replace fuselage tank, allowing increased baggage area.

ACCOMMODATION: Two seats side by side beneath transparent canopy, which is hinged at rear and swings upward. Dual controls standard; but toe-operated wheel brakes on starboard side only. Baggage compartment aft of seats, capacity 41 kg (90 lb). With wing fuel tanks fitted, baggage area is large enough to store scuba diving tanks or camping gear.

SYSTEMS: Hydraulic system for brakes only. Electrical system powered by engine-driven generator.

DIMENSIONS, EXTERNAL:
- Wing span: 7·92 m (26 ft 0 in)
- Wing chord, constant: 1·52 m (5 ft 0 in)
- Wing area, gross: 12·08 m² (130 sq ft)
- Wing aspect ratio: 5·2
- Length overall: 6·25 m (20 ft 6 in)
- Height overall (wheels down): 1·83 m (6 ft 0 in)
- Tailplane span: 2·44 m (8 ft 0 in)
- Wheel track: 2·59 m (8 ft 6 in)
- Wheelbase: 2·13 m (7 ft 0 in)
- Propeller diameter: 1·68 m (5 ft 6 in)

WEIGHTS:
- Weight empty: 440 kg (970 lb)
- Max T-O weight: 707 kg (1,560 lb)

PERFORMANCE (at max T-O weight unless specified otherwise):
- Never-exceed speed: 130 knots (241 km/h; 150 mph)
- Max cruising speed at 75% power: 113 knots (209 km/h; 130 mph)
- Econ cruising speed at 55% power: 94 knots (175 km/h; 109 mph)
- Stalling speed: 53 knots (97 km/h; 60 mph)
- Max rate of climb at S/L, with pilot only: 365 m (1,200 ft)/min
- Rate of climb at S/L: 305 m (1,000 ft)/min
- T-O run, land: 122 m (400 ft)
- T-O run, water: 159 m (520 ft)
- Range with wing tanks: 313 nm (579 km; 360 miles)

Option Air Reno Acapella 200S *(Howard Levy)*

Mr Pereira's Osprey II prototype *(Scott Kemper)*

Vin-Del/Owl OR-71 *Lil Quickie* Formula One racing aircraft

Parker Teenie Two powered by a modified Volkswagen engine

OWL

GEORGE A. OWL Jr

17700 S. Western Avenue, Apartment 195, Gardena, California 90248

Telephone: (213) 323 3385

Mr George Owl, a member of the preliminary design staff of Rockwell International, is also the designer of two Formula One racing aircraft known as the Owl Racers OR-70 and OR-71. The former was designed to order for Bernadine and Jim Stevenson, both racing pilots, and was built by them (see 1977-78 *Jane's*). The OR-71 was produced in co-operation with Mr Vince DeLuca, proprietor of Vin-Del Aircraft, who built a prototype and is making plans available to amateur constructors.

A version of the OR-71 with a new wing has also been developed. This wing, of all-wood construction and utilising a special laminar flow aerofoil section, was expected to increase the maximum speed of the OR-71 by 5 knots (10 km/h, 6 mph). The designation of aircraft with the new wing is OR-71B.

Vin-Del Aircraft

29718 Knollview Drive, Miraleste, California 90732

Mr Vince DeLuca, proprietor of Vin-Del Aircraft, built the prototype of a Formula One racer of Mr Owl's design which he designated OR-71 *Lil Quickie*. Construction began on 2 December 1971 and the first flight was made on 6 June 1972. Plans are available to amateur constructors.

VIN-DEL/OWL OR-71 LIL QUICKIE

Type: Single-seat Formula One racing aircraft.

Wings: Cantilever mid-wing monoplane. Owl laminar flow aerofoil section, tapered in chord. Dihedral 0° 51′. Incidence 0°. Non-linear thickness distribution. Thickness/chord ratio is 13·7% at root, 10% on outer 60% of wing. All-wood structure. Laminated spruce one-piece main spar, plywood ribs with spruce caps, and plywood skins. Ailerons of spruce and plywood construction, mass-balanced at tip. No flaps. No trim tabs. Plans show optional high-lift leading-edge of larger radius and increased camber.

Fuselage: Welded steel tube structure, with light alloy fairings and Dacron covering. Glassfibre nose cowl incorporates an annular cooling inlet. Engine cowl has controllable air exit flap.

Tail Unit: Cantilever wooden structure, with spruce spars and mahogany plywood ribs and skins. Fixed-incidence tailplane. No trim tabs.

Landing Gear: Non-retractable tailwheel type. Prototype has cantilever light alloy main legs. Spring steel legs optional. Main wheels and tyres size 5·00-5, pressure 2·41 bars (35 lb/sq in). Cleveland hydraulic disc brakes. Glassfibre fairings on main wheels. Tailwheel is a ball-bearing castor.

Power Plant: One 74·5 kW (100 hp) Continental O-200 flat-four engine, driving an Anderson, McCauley or Sensenich two-blade fixed-pitch propeller. Fuel tank in fuselage, immediately aft of firewall, capacity 34 litres (9 US gallons). Auxiliary integral fuel tank in each wing leading-edge, with combined capacity of 22·7 litres (6 US gallons). Total fuel capacity 56·7 litres (15 US gallons). Refuelling point on upper surface of fuselage, forward of windscreen. Oil capacity 5·7 litres (1·5 US gallons).

Accommodation: Single seat beneath small transparent canopy, hinged on starboard side, opening upwards and to starboard.

Dimensions, external:
Wing span 6·10 m (20 ft 0 in)
Wing chord at root 1·27 m (4 ft 2 in)
Wing chord at tip 0·76 m (2 ft 6 in)
Wing area, gross 6·13 m² (66 sq ft)
Wing aspect ratio 6
Length overall 4·98 m (16 ft 4 in)
Height overall 1·40 m (4 ft 7 in)
Wheel track 1·30 m (4 ft 3 in)
Wheelbase 3·63 m (11 ft 11 in)
Propeller diameter (max) 1·52 m (5 ft 0 in)

Weights:
Weight empty 251 kg (553 lb)
Max T-O weight 386 kg (850 lb)

Performance (at max T-O weight):
Never-exceed speed
260 knots (482 km/h; 300 mph) IAS
Max level speed
more than 221 knots (410 km/h; 255 mph)
Stalling speed 60 knots (111 km/h; 69 mph)
g limits ±7·33 (±11 ultimate)

PARKER

C. Y. PARKER

PO Box 625, Coolidge, Arizona 85228
Telephone: (602) 723 5660

PARKER TEENIE TWO

Mr Cal Parker flew in 1969 an improved version of the lightweight all-metal homebuilt aircraft which he had designed and built earlier as Jeanie's Teenie. With completion of the new prototype, the original model became known as Teenie One. Plans for the Jeanie's Teenie were available for three years prior to Teenie Two plans being marketed.

Mr Parker's original aim had been to build an aircraft specifically to utilise the Volkswagen motorcar engine and, at the same time, to produce an all-metal design that would present few constructional problems even to homebuilders with virtually no metal-working experience. This was achieved, and no special tools or jigs are needed beyond a tool to close and form the cadmium-plated steel pop rivets that are used for practically all assembly. One gauge of aluminium sheet and one size of light alloy angle section are used for almost all of the structure, except for chromoly steel tube and sheet which are required for construction of the landing gear and control actuation tubes respectively. For simplicity and economy, push/pull tubes are used for all flying controls.

Teenie One conformed with these design objectives; Teenie Two is considerably refined to produce a much cleaner aeroplane. Its structure is stressed for full aerobatics, but the fuel and oil systems are not suitable for inverted flight.

Plans, complete kits of parts, and details of modifications for the Volkswagen engine are available to amateur constructors, and some Teenie Twos built from plans have completed well over 750 flying hours. During 1979 the Australian DoT gave approval for construction of the Teenie Two in that country.

Type: Single-seat light aircraft.

Wings: Cantilever low-wing monoplane. Wing section NACA 4415. All-metal two-spar structure, with detachable outer wing panels. Light alloy ribs and skin. Plain ailerons of metal construction. No flaps.

Fuselage: All-metal semi-monocoque structure with longerons of light alloy angle, three built-up bulkheads and light alloy skin.

Tail Unit: Cantilever all-metal structure with swept vertical surfaces. Small dorsal fin eliminates the need for a fourth bulkhead by carrying loads from fin leading-edge to centre bulkhead. Conventional rudder and elevators of metal construction.

Landing Gear: Non-retractable tricycle type. Shock-absorption provided by springs in compression and rubber hose. All three wheels same size, with tyres size 10·5 × 4·00-4, pressure 1·72 bars (25 lb/sq in). Mechanically-actuated wheel brakes.

Power Plant: One 31·5 kW (42 hp) 1,600 cc or 30 kW (40 hp) 1,500 cc Volkswagen modified motorcar engine (conversion parts sold by Parker), driving a two-blade fixed-pitch wooden propeller (a computer-designed propeller in the parts kit gives optimum performance for take-off, climb and cruise). Single fuselage fuel tank, immediately aft of firewall, capacity 34 litres (9 US gallons). Refuelling point on top of fuselage, forward of windscreen. Oil capacity 2·5 litres (0·66 US gallons).

Accommodation: Single seat in open cockpit. Drawings of optional canopy available; this increases max level speed to 122 knots (225 km/h; 140 mph).

System: Prototype now fitted with alternator, starter and battery to power lights and radio, giving an increase in empty weight of 18 kg (40 lb).

Dimensions, external:
Wing span 5·49 m (18 ft 0 in)
Wing chord, constant 1·02 m (3 ft 4 in)
Width, wings detached 1·83 m (6 ft 0 in)
Length overall 3·91 m (12 ft 10 in)

Weights:
Weight empty 140 kg (310 lb)
Max T-O weight 267 kg (590 lb)

Performance (at max T-O weight, 1,600 cc engine):
Max level speed 104 knots (193 km/h; 120 mph)
Max cruising speed (75% power)
95·5 knots (177 km/h; 110 mph)
Landing speed 43·5 knots (80·5 km/h; 50 mph)
Max rate of climb at S/L:
standard propeller 244 m (800 ft)/min
52 × 37 propeller 305 m (1,000 ft)/min
Service ceiling 4,575 m (15,000 ft)
Range 347 nm (643 km; 400 miles)

PAYNE

VERNON W. PAYNE

Route No. 2, PO Box 319M, Escondido, California 92025
Telephone: (714) 746 4465

Mr Vernon Payne is the designer of the Knight Twister, a light sporting biplane of which plans and kits are available for amateur construction. It exists in several versions. The original Payne Knight Twister was designed in 1928. The Knight Twister KT-85 single-seat sporting biplane flew in 1933. Details of this aircraft can be found in the 1980-81 *Jane's*. The developed and more powerful Sunday Knight Twister SKT-125 and the Knight Twister Junior KT-75 with a larger wing area (first flown in 1947) can also be found in the 1980-81 *Jane's*.

Recently Mr Payne changed the drawings to incorporate a new longer fuselage for all models, in which the pilot sits 0·23 m (9 in) further aft to give better view over the upper wing when rounding a pylon during air races. Details of current models, for which plans are available, follow. In late 1980 about six Knight Twister biplanes were under construction.

AERO KNIGHT TWISTER

The Aero Knight Twister is similar in most respects to the Knight Twister KT-85 (see 1980-81 *Jane's*), which it replaces.

Type: Single-seat light biplane.

Wings: Braced biplane type. Wing section NACA M-6. No dihedral. Incidence 1° 30′ at root; washed out at tip. All-wood two-spar structure, plywood-covered and with fabric covering overall. I-type interplane struts. Ailerons on lower wings only, of fabric-covered wood construction. No flaps.

Fuselage: Steel tube truss structure with wood stringers and fabric covering.

Tail Unit: Cantilever type. Vertical surfaces have fabric-covered steel tube structure. Horizontal surfaces have plywood-covered wood structure, with fabric covering overall.

Landing Gear: Non-retractable tailwheel type. Cantilever main units. Rubber cord or hydraulic shock-absorption. Wheels size 6·00-6 with Goodyear tyres, pressure 0·345-0·69 bars (5-10 lb/sq in). Goodyear disc brakes.

Power Plant: One 67 kW (90 hp) Continental C90 flat-four engine, driving a two-blade wood or metal fixed-pitch propeller. Alternatively any other Continental or Avco Lycoming flat-four engine of 63·5-108 kW (85-145 hp). Fuel tank aft of engine firewall, capacity 68 litres (18 US gallons). Oil capacity 3·7-5·7 litres (1-1·5 US gallons).

Accommodation: Single seat, normally in open cockpit. Baggage compartment capacity 9 kg (20 lb). Radio optional.

Dimensions, external:
- Wing span: upper 4·72 m (15 ft 6 in)
- lower 4·11 m (13 ft 6 in)
- Wing chord: upper at root 0·80 m (2 ft 7½ in)
- upper near tip 0·61 m (2 ft 0 in)
- lower at root 0·76 m (2 ft 6 in)
- lower near tip 0·61 m (2 ft 0 in)
- Wing area, gross 5·20 m² (56 sq ft)
- Length overall 4·91 m (16 ft 1½ in)
- Height overall, tail up 2·36 m (7 ft 8¾ in)
- Tailplane span 2·13 m (7 ft 0 in)
- Wheel track 1·52 m (5 ft 0 in)
- Wheelbase 5·23 m (17 ft 2 in)

Weight:
- Max T-O weight 408 kg (900 lb)

Performance:
- Max speed in dive 182 knots (338 km/h; 210 mph)
- Max level speed at S/L 149 knots (277 km/h; 172 mph)
- Max cruising speed 130 knots (241 km/h; 150 mph)
- Stalling speed 55 knots (102 km/h; 63 mph)
- g limit +9

HOLIDAY KNIGHT TWISTER SKT-1250

The Holiday Knight Twister has the same basic rib chords as the Aero Knight Twister, but the wing span is increased. This allows a more efficient performance and higher rate of climb at altitude, as well as shorter take-off runs.

Type: Single-seat light biplane.

Power Plant: One 93 kW (125 hp) Continental or Avco Lycoming flat-four engine.

Dimensions, external:
- Wing span: upper 5·94 m (19 ft 6 in)
- lower 5·33 m (17 ft 6 in)
- Wing chord: upper at root 0·88 m (2 ft 10½ in)
- upper near tip 0·61 m (2 ft 0 in)
- lower at root 0·84 m (2 ft 9 in)
- lower near tip 0·61 m (2 ft 0 in)
- Wing area, gross 7·53 m² (81 sq ft)
- Length overall, tail up 4·91 m (16 ft 1½ in)
- Height overall, tail up 2·05 m (6 ft 8¾ in)

Weight:
- Max T-O weight 441 kg (972 lb)

Performance (at max T-O weight):
- Max speed in dive 160 knots (296 km/h; 184 mph)
- Max level speed at S/L 129 knots (240 km/h; 149 mph)
- Max cruising speed 114 knots (211 km/h; 131 mph)
- Stalling speed 39 knots (72·5 km/h; 45 mph)
- g limit +7

IMPERIAL KNIGHT TWISTER M6

At the request of Don Fairbanks, owner of a flying training school at Lunken Airport, Cincinnati, Ohio, Mr Payne modified the design of the original Knight Twister to enable Mr Fairbanks to compete in US National Air Races in the Sport Biplane class. The resulting aircraft won the Silver Biplane Race at Reno, Nevada, in 1971. It also set a speed record for biplanes in its class, flying at 154·768 knots (286·814 km/h; 178·218 mph). Variations from the standard Knight Twister include a change in wing section, increased wing and tailplane span and increased fuel tankage. First flight was made on 19 June 1970.

Dubbed the Imperial Knight Twister by Mr Payne, this aircraft is generally similar to the Aero Knight Twister, except as follows:

Type: Single-seat light biplane.

Wings: Wing section NACA 21. Span and chord of both upper and lower wings increased. Flying and landing wires deleted. Incidence 0°.

Power Plant: One 101 kW (135 hp) Avco Lycoming O-290-D2 flat-four engine, driving a two-blade fixed-pitch propeller with spinner. Fuel contained in an upper tank of 85 litres (22·5 US gallons) capacity and a lower tank of 47 litres (12·5 US gallons) capacity, aft of firewall. Total fuel capacity 132 litres (35 US gallons).

Dimensions, external:
- Wing span: upper 5·33 m (17 ft 6 in)
- lower 4·72 m (15 ft 6 in)
- Wing chord: upper at root 0·91 m (3 ft 0 in)
- lower at root 0·88 m (2 ft 10½ in)
- Wing area, gross 7·06 m² (76 sq ft)
- Length overall, tail up 4·91 m (16 ft 1½ in)
- Height overall, tail up 2·01 m (6 ft 7 in)

Weights:
- Normal T-O weight 454 kg (1,000 lb)
- Racing weight 381 kg (840 lb)

Performance (at normal T-O weight):
- Max level speed 152 knots (282 km/h; 175 mph)
- Stalling speed 43·5 knots (80·5 km/h; 50 mph)
- Max rate of climb at S/L 488 m (1,600 ft)/min
- g limit +8

PAZMANY

PAZMANY AIRCRAFT CORPORATION

Box 80051, San Diego, California 92138
Telephone: (714) 276 0424

This company was formed by Mr Ladislao Pazmany, to develop and market a two-seat light aircraft known as the PL-1 Laminar, which he had designed. Some 5,000 design hours and 4,000 hours of construction went into the prototype PL-1, which was flown for the first time on 23 March 1962. Subsequent editions of *Jane's* recorded the history of the PL-1, of which about 375 sets of plans were sold, leading to construction of many amateur-built examples and quantity production in Taiwan for the Chinese Nationalist Air Force.

Pazmany Aircraft Corporation no longer markets plans of the PL-1, of which details can be found in the 1979-80 *Jane's*; instead, plans and instructions for building the improved PL-2 and the lightweight, low-cost single-seat PL-4A are available to amateur constructors and many aircraft of these types are being built. Mr Pazmany also contributed to the design of the Ryson Aviation Cloudster two-seat motor glider, described in the Sailplanes section of this edition.

PAZMANY PL-2

Shortly after flight trials of the PL-1 began, Mr Pazmany initiated a complete redesign of the aircraft. The resulting PL-2 is almost identical with the PL-1 in external configuration. Cockpit width is increased by 5 cm (2 in); wing dihedral is increased, and the internal structure is extensively changed, to simplify construction and reduce weight.

Static tests of every major assembly up to ultimate loads had been made by early 1967. The first PL-2 to be completed, by Mr H. Pio of Ramona, California, made its first flight on 4 April 1969, powered by an Avco Lycoming O-290-G engine. A total of 370 sets of plans had been sold by early 1981. Aircraft built and flown include several examples for evaluation and use by foreign military training centres (see 1977-78 *Jane's*), and a PL-2 was under construction by the Sri Lanka Air Force in 1979.

Type: Two-seat light aircraft.

Wings: Cantilever low-wing monoplane. Wing section NACA 63_2615. Dihedral 5°. Incidence −1° 20′. All-metal single-spar structure in one piece, with leading-edge torsion box. Plain piano-hinged ailerons and flaps of all-metal construction. No trim tabs.

Fuselage: Conventional all-metal semi-monocoque structure, with flat or single-curvature skins.

Tail Unit: Cantilever all-metal structure. One-piece horizontal surface, with anti-servo tab which serves also as a trim tab.

Landing Gear: Non-retractable tricycle type, with all three oleo-pneumatic shock-absorbers interchangeable. Goodyear wheels and tyres, size 5·00-5. Tyre pressure 2·14 bars (31 lb/sq in). Goodyear brakes. Steerable nosewheel.

Power Plant: One Avco Lycoming flat-four engine, driving a two-blade fixed-pitch metal propeller. Suitable power plants are the 80·5 kW (108 hp) O-235-C1, 93 kW (125 hp) O-290-G (ground power unit), 101 kW (135 hp) O-290-D2B or 112 kW (150 hp) O-320-A. Fuel in two glassfibre wingtip tanks, each of 47 litres (12·5 US gallons) capacity. Total fuel capacity 94 litres (25 US gallons). Oil capacity 4·5 litres (5 US quarts).

Accommodation: Two seats side by side under rearward-sliding transparent canopy. Dual controls. Space for 18 kg (40 lb) baggage aft of seats. Heater and airscoops for ventilation. VHF radio.

Dimensions, external:
- Wing span 8·53 m (28 ft 0 in)
- Wing chord, constant 1·27 m (4 ft 2 in)
- Wing area, gross 10·78 m² (116 sq ft)
- Wing aspect ratio 6·7
- Length overall 5·90 m (19 ft 3½ in)
- Height overall 2·44 m (8 ft 0 in)
- Tailplane span 2·44 m (8 ft 0 in)
- Wheel track 2·60 m (8 ft 5½ in)
- Wheelbase 1·30 m (4 ft 3 in)

Dimensions, internal:
- Cabin: Length 1·27 m (4 ft 2 in)
- Width 1·07 m (3 ft 6 in)
- Height 1·02 m (3 ft 4 in)

Weights (A with 80·5 kW; 108 hp engine, B with 93 kW; 125 hp, C with 101 kW; 135 hp, D with 112 kW; 150 hp):
- Weight empty: A 396 kg (875 lb)
- B, C 408 kg (900 lb)
- D 409 kg (902 lb)
- Max T-O weight: A 642 kg (1,416 lb)
- B, C 655 kg (1,445 lb)
- D 656 kg (1,447 lb)

Performance (at max T-O weight):
- Max level speed at S/L:
  - A 120 knots (222 km/h; 138 mph)
  - B 125 knots (232 km/h; 144 mph)
  - C 128 knots (238 km/h; 148 mph)
  - D 133 knots (246 km/h; 153 mph)
- Econ cruising speed:
  - A 103 knots (192 km/h; 119 mph)
  - B 111 knots (206 km/h; 128 mph)
  - C 113 knots (209 km/h; 130 mph)
  - D 118 knots (219 km/h; 136 mph)
- Stalling speed (flaps down):
  - A 45·5 knots (84 km/h; 52 mph)
  - B, C, D 47 knots (87 km/h; 54 mph)
- Max rate of climb at S/L: A 390 m (1,280 ft)/min
  - B 457 m (1,500 ft)/min
  - C 488 m (1,600 ft)/min
  - D 518 m (1,700 ft)/min
- Range at econ cruising speed:
  - A 427 nm (790 km; 492 miles)
  - B 422 nm (780 km; 486 miles)
  - C 428 nm (792 km; 493 miles)
  - D 330 nm (610 km; 381 miles)

PAZMANY PL-4A

This lightweight single-seat low-wing monoplane was designed for easy low-cost construction by amateur builders, and to provide a safe aircraft that would be economical in operation. The prototype flew for the first time on 12 July 1972 and had completed 335 hours of flight by January 1978. Sets of plans, kits of prefabricated components, glassfibre wingtips and fuel tank, and transparent cockpit canopy are available to amateur constructors.

By the beginning of 1981 more than 660 sets of plans had been sold, and the PL-4A had received approval in Australia for construction by amateurs.

Two PL-4As were built to evaluate different power plant installations, comprising a Volkswagen engine with 2¼:1 V-belt reduction; a Limbach SL 1700 E Volkswagen with direct drive to the propeller; and a Continental A65 aircooled engine.

Type: Single-seat lightweight sporting aircraft.

Wings: Cantilever low-wing monoplane. Wing section NACA 63_3418. Dihedral 5°. Incidence 3°. No sweepback. All-metal structure, with main spar, Z section rear beam, sheet metal ribs and skins. Wings fold alongside fuselage for towing or storage. Plain piano-hinged ailerons of all-metal construction. Cambered glassfibre wingtips. No flaps. No trim tabs.

Fuselage: All-metal structure, with bulkheads built up from bent sheet metal channels and standard extruded angles for longerons, and with sheet metal skins.

Tail Unit: All-metal cantilever T tail. All-moving tailplane with large anti-servo tab which serves also as a trim tab.

Landing Gear: Non-retractable tailwheel type. Spring steel cantilever main legs. Single go-kart type wheel on each main unit, with 4·10 × 3·50-6 four-ply tyre, pressure 4·48 bars (65 lb/sq in). Steerable and castoring tailwheel with solid tyre size 5 × 1·5-1·5. Go-kart type hydraulic disc brakes by Hurst-Airheart.

Power Plant: One 1,600 cc modified Volkswagen motor-car engine with Becar V-belt reduction of 2¼ : 1, developing approximately 37·5 kW (50 hp) and driving a

Knight Twister (Avco Lycoming O-235-C1 engine) built by Robert Ubel *(Howard Levy)*

Pazmany PL-2 built by Mr Hans Nielsen in Sweden

Pazmany PL-4A single-seat light sporting aircraft *(Peter M. Bowers)*

PDQ-2 Model E all-metal lightweight sporting aircraft

two-blade fixed-pitch wooden propeller of Pazmany design, manufactured by Ted Hendrickson. Glassfibre fuel tank immediately aft of firewall, usable capacity 45 litres (12 US gallons). Refuelling point on upper fuselage forward of windscreen. Oil capacity 2·8 litres (0·75 US gallons).

ACCOMMODATION: Single seat under transparent Plexiglas canopy, hinged on starboard side. Compartment aft of seat for 9 kg (20 lb) baggage. Cabin heated and ventilated.

SYSTEMS: Hydraulic system for brakes only. Electrical system powered by 12V 25Ah battery situated in baggage compartment.

DIMENSIONS, EXTERNAL:
- Wing span: 8·13 m (26 ft 8 in)
- Wing chord, constant: 1·02 m (3 ft 4 in)
- Wing area, gross: 8·27 m² (89·0 sq ft)
- Wing aspect ratio: 8·0
- Length overall: 5·04 m (16 ft 6½ in)
- Width, wings folded: 2·44 m (8 ft 0 in)
- Height overall: 1·73 m (5 ft 8 in)
- Tailplane span: 2·29 m (7 ft 6 in)
- Wheel track: 2·06 m (6 ft 9 in)
- Wheelbase: 3·56 m (11 ft 8 in)
- Propeller diameter: 1·73 m (5 ft 8 in)

WEIGHTS:
- Weight empty: 262 kg (578 lb)
- Max T-O and landing weight: 385 kg (850 lb)

PERFORMANCE (at max T-O weight):
- Never-exceed speed: 161 knots (299 km/h; 186 mph)
- Max level speed at S/L: 109 knots (201 km/h; 125 mph)
- Max cruising speed at S/L: 85 knots (158 km/h; 98 mph)
- Econ cruising speed at S/L: 78 knots (145 km/h; 90 mph)
- Stalling speed, power on: 40 knots (74 km/h; 46 mph)
- Stalling speed, power off: 42 knots (77·5 km/h; 48 mph)
- Max rate of climb at S/L: 198 m (650 ft)/min
- Service ceiling: 3,960 m (13,000 ft)
- Min ground turning radius: 3·05 m (10 ft 0 in)
- T-O run: 148 m (486 ft)
- Landing run: 133 m (436 ft)
- Range with max fuel, no allowances: 295 nm (545 km; 340 miles)

PDQ

PDQ AIRCRAFT PRODUCTS

28975 Alpine Lane, Elkhart, Indiana 46514
Telephone: (219) 264 2906

Mr Wayne Ison formed this company to market plans of the PDQ-2 lightweight sporting aircraft which he had designed and built. The PDQ-2 is intended to provide a cheap, robust, easily and quickly built aircraft which is easy for an average pilot to fly.

Design of the prototype began in September 1972, and construction was started on 5 January 1973. Excluding the 27 kW (36 hp) Rockwell (Venture) JLO-LB-600-2 engine that was originally fitted, the cost of construction was only $350 at 1973 prices, and the first flight was made on 30 May that year.

During 1975 it became clear that an alternative power plant was needed, owing to the increasing scarcity of the JLO engine. One amateur constructor fitted a BMW motorcycle engine; but Mr Ison decided to test the prototype PDQ-2 with a converted Volkswagen motorcar engine. The increased engine weight required a number of structural modifications, notably to the engine mounting, wing spars and landing gear; but flight testing proved satisfactory, and details of this engine installation have been sent to over 2,000 purchasers of sets of plans, in 35 countries. At least 50 to 60 PDQ-2s were flying by early 1981.

PDQ AIRCRAFT PRODUCTS PDQ-2

The version of this aircraft described below is the **PDQ-2 Model C** (unless otherwise stated), which has wings and tail constructed of wood and polyurethane foam. The **Model D** differs from the C only in having wings and tail of wood, with fabric covering; while the **Model E** (illustrated) has all-metal wings and tail, constructed from aluminium sheet. At least two aircraft have been fitted with wings of NASA GAW-1 section.

TYPE: Lightweight sporting aircraft.

WINGS: Wire-braced monoplane. Wing section NACA 63_2A615. Dihedral 5°. Incidence 3°. Composite structure with spruce spars, polyurethane foam ribs, sheet foam skins, covered with Sharkskin fabric. Full-span plain ailerons. No flaps or trim tabs. Wings quickly removable for towing and storage.

FUSELAGE: Basic structure consists of 0·05 m (2 in) square tubes of 6061-T6 light alloy. A lower forward tube carries the landing gear; attached to it is a vertical kingpost, to which the wing is attached and wire-braced. An aft horizontal tube attached to the kingpost carries the T tail.

TAIL UNIT: Strut-braced T tail, with swept vertical surfaces. Fixed-incidence tailplane. No trim tabs. Composite structure of polyurethane foam, spruce strips, plywood and Sharkskin fabric covering.

LANDING GEAR: Non-retractable tricycle type. Shock-absorption by spring steel leaf mounting struts. Go-kart type main wheels, diameter 0·28 m (11 in); steerable nosewheel, diameter 0·15 m (6 in). Go-kart type brakes.

POWER PLANT: Prototype flew originally with Rockwell (Venture) JLO-LB-600-2 engine, mounted on top of the kingpost and driving a two-blade fixed-pitch wooden pusher propeller. This engine remains suitable, but prototype now has a 1,385 cc converted Volkswagen motorcar engine, developing 30 kW (40 hp), and this is recommended for amateur-built PDQ-2s. Plastics fuel tank mounted alongside pilot's seat on port side, capacity 22·7 litres (6 US gallons). Refuelling point on tank upper surface.

ACCOMMODATION: Single open seat mounted immediately forward of kingpost.

DIMENSIONS, EXTERNAL:
- Wing span: 7·44 m (24 ft 5 in)
- Wing chord, constant: 1·07 m (3 ft 6 in)
- Wing area, gross: 7·90 m² (85 sq ft)
- Length overall: 4·42 m (14 ft 6 in)
- Height overall: 1·27 m (4 ft 2 in)
- Tailplane span: 1·83 m (6 ft 0 in)
- Wheel track: 1·27 m (4 ft 2 in)

WEIGHTS (Volkswagen engine):
- Weight empty: Model C: 164 kg (360 lb)
  - Model D: 145 kg (320 lb)
- Normal T-O weight: Model C: 257 kg (566 lb)
  - Model D: 239 kg (526 lb)
- Max T-O weight: Model C: 281 kg (620 lb)
  - Model D: 263 kg (580 lb)

PERFORMANCE (Model C and D, at max T-O weight, Volkswagen engine):
- Never-exceed speed: 78 knots (145 km/h; 90 mph)
- Max level speed at S/L: 69 knots (129 km/h; 80 mph)
- Max cruising speed at S/L: 61 knots (113 km/h; 70 mph)
- Stalling speed:
  - Model C: 37·5 knots (69·5 km/h; 43 mph)
  - Model D: 35 knots (65 km/h; 40 mph)
- Max rate of climb at S/L: 152 m (500 ft)/min
- Service ceiling: 3,050 m (10,000 ft)
- T-O run: 107 m (350 ft)
- Landing run: 122 m (400 ft)
- Range with max fuel: 173 nm (321 km; 200 miles)

PIETENPOL

DON PIETENPOL

215, 21st Street SE, Rochester, Minnesota 55901

The prototype of Mr Bernard H. Pietenpol's Aircamper two-seat parasol monoplane flew for the first time in 1929, powered by a 30 kW (40 hp) Ford Model A engine. Plans were published in the magazine *Modern Mechanics and Inventions* in the following year and large numbers of Aircampers, with a wide variety of power plants, were built by amateurs, either from the magazine plans or from kits of parts marketed by Mr Pietenpol.

The original Aircamper was of all-wood construction, with fabric covering, but some examples completed recently or currently being built have a steel tube fuselage and tail unit.

An illustration in the 1979-80 *Jane's* showed Mr Pietenpol's Aircamper, which he modified by extending the fuselage by 9 in (23 cm), fitting modified Piper J-3 Cub landing gear, strengthening the wings, and installing two 15 litre (4 US gallon) fuselage fuel tanks and a 38 litre (10 US gallon) fuel tank in the wing centre-section. This aircraft, which now has a converted Chevrolet Corvair motorcar engine, rated at 45 kW (60 hp), has completed over 600 flying hours.

The accompanying illustration shows another Aircamper, in the optional two-seat configuration. A further version, embodying many Piper Cub components, is described under the Grega entry in this section.

PIETENPOL B4 AIRCAMPER

Type: Single-seat light aircraft. (Basic design makes provision for second seat.)

Wings: Parasol monoplane, with two parallel bracing struts each side and centre-section cabane structure. Pietenpol special wing section. Thickness/chord ratio 16·6%. No dihedral. Incidence 2°. Fabric-covered two-spar wood structure. Tips may be rounded or square. Plain wooden ailerons with fabric covering. No flaps.

Fuselage: Wooden de Havilland truss structure of Sitka spruce, covered with birch plywood to back of rear cockpit and fabric on rear fuselage.

Tail Unit: Wire-braced wood and steel tube structure, covered with fabric.

Landing Gear: Divided main gear, with spring shock-absorption, modified from Piper J-3 Cub gear. Mainwheel tyres size 8·00-4, pressure 1·03 bars (15 lb/sq in). Piper hydraulic brakes. Steerable tailwheel.

Power Plant (Mr Pietenpol's aircraft): One 45 kW (60 hp) Chevrolet Corvair converted motorcar engine, driving a two-blade wooden propeller. A variety of other engines have been fitted in amateur-built Aircampers. One fuel tank in wing, capacity 38 litres (10 US gallons); two tanks in fuselage, total capacity 30 litres (8 US gallons). Oil capacity 4·75 litres (1·25 US gallons).

Accommodation: Single seat in open cockpit in Mr Pietenpol's Aircamper N7533U. Second seat is optional in basic design.

Dimensions, external:

Wing span	8·84 m (29 ft 0 in)
Wing chord, constant	1·52 m (5 ft 0 in)
Wing area, gross	13·47 m² (145 sq ft)
Wing aspect ratio	6
Propeller diameter	1·52 m (5 ft 0 in)

Weight:

Weight empty	282 kg (622 lb)

Performance:

Max level and never-exceed speed	95·5 knots (177 km/h; 110 mph)
Max cruising speed	74 knots (137 km/h; 85 mph)
Econ cruising speed	69 knots (129 km/h; 80 mph)
Stalling speed	33 knots (61·5 km/h; 38 mph)
Max rate of climb at S/L	152 m (500 ft)/min
T-O and landing run	76 m (250 ft)
Range, with 30 min reserves	330 nm (611 km; 380 miles)

PITTS

PITTS AEROBATICS

PO Box 547, Afton, Wyoming 83110
Telephone: (307) 886 3151
General Manager: E. H. Andersen Jr

In addition to marketing factory-built examples of the single-seat and two-seat Pitts Special biplanes, Pitts Aerobatics, which bought all rights to this range of aircraft from Mr Curtis Pitts, supplies plans of the S-1D and S-1E single-seaters to amateur constructors. It also began, in 1977, to supply kits for both the S-1E and the two-seat S-2E. No plans are available separately for the S-2 series.

PITTS S-1D/E SPECIAL

The versions of the single-seat S-1 for which plans are currently available are designated S-1D and S-1E. Details are generally similar to those given for the factory-built S-1S in the main US Aircraft section of this edition, but there are important differences in the two versions. The plans for the S-1E now include drawings for the four-aileron symmetrical wings which have not been available previously. The S-1D has four-aileron wings of M-6 aerofoil section, which are not symmetrical.

There is also an S-1C version, but Pitts Aerobatics does not supply plans for its construction. By comparison with earlier models, this has flat-bottomed wings, with ailerons on the lower wings only, and can be fitted with engines of between 74·5 and 134 kW (100 to 180 hp).

POLLIWAGEN

POLLIWAGEN INC

8782 Hewitt Place, Garden Grove, California 92644
Telephone: (714) 897 9852
President: Joseph P. Alvarez

POLLIWAGEN

The Polliwagen is a side-by-side two-seat sporting aircraft, stressed for aerobatics and designed for amateur construction. The first of two prototypes flew for the first time in July 1977. Plans are available, together with a wide range of component parts, including pre-moulded glassfibre cowlings, landing gear fairings, canopy and ribs, landing gear assemblies, brakes and instrument panel. By 1 January 1980, a total of 217 sets of plans had been sold; three Polliwagens had been built, and about 30 were known to be under construction.

Type: Two-seat light sporting aircraft.

Wings: Cantilever low-wing monoplane. Wortmann FX-67-K-150 wing section. Thickness/chord ratio 15%. Dihedral 4°. Incidence 1° 30′. No sweepback. Structure entirely of glassfibre epoxy composite, embodying sailplane technology. Ailerons and trailing-edge flaps of similar construction; both have an upward setting of 10° for higher cruising speed. Flaps have a maximum downward setting of 80°, so that they can be used for aerodynamic braking. Electrically-actuated trim tab in port aileron. Wingtip tanks increase effective span by reducing induced drag.

Fuselage: Glassfibre epoxy structure, consisting basically of one bottom and two side D spars, designed in such a way that the cockpit would break outward if subjected to impact loads.

Tail Unit: Cantilever T tail of similar construction to wings, with sweepback on all surfaces. Fin integral with fuselage. All-moving horizontal surfaces operated by electrically-actuated servo.

Landing Gear: Retractable tricycle type. Main units retract forward, nosewheel rearward. Polyurethane bumper shock-absorption. Rosenhan main wheels and nosewheel, all with tyres size 5·00-5. Main-wheel tyre pressure 2·62 bars (38 lb/sq in); nosewheel 1·38 bars (20 lb/sq in). Rosenhan hydraulic disc brakes. Main wheels remain semi-exposed to reduce damage in a wheels-up landing.

Power Plant: One 56 kW (75 hp) Revmaster 2100 D Turbo, which is a Revmaster-developed turbocharged Volkswagen engine, driving a Maloof two-blade constant-speed metal propeller. Total of more than 72 litres (19 US gallons) of fuel in two wingtip tanks (each 34 litres; 9 US gallons) and a gravity-feed tank in the fuselage nose. Refuelling point on upper surface of each tip-tank.

Accommodation: Two seats side by side under rearward-articulating transparent (clear or tinted) canopy. Accommodation is heated and ventilated. Baggage compartment.

Systems: All flying controls actuated through pushrods and bearings. Electrical system powered by 20A engine-driven alternator and storage battery. Pneumatic system for landing gear retraction is optional. Oxygen system of 0·62 m³ (22 cu ft) capacity.

Avionics and Equipment: Dual nav/coms. Provisions for three-axis electric trim adaptable to autopilot. Blind-flying instrumentation optional.

Dimensions, external:

Wing span over tip-tanks	7·92 m (26 ft 0 in)
Wing chord, constant	1·07 m (3 ft 6 in)
Wing area, gross	8·36 m² (90 sq ft)
Wing aspect ratio	7·43
Length overall	4·88 m (16 ft 0 in)
Height overall	1·70 m (5 ft 7 in)
Tailplane span	2·18 m (7 ft 2 in)
Wheel track	1·83 m (6 ft 0 in)
Wheelbase	1·09 m (3 ft 7 in)
Propeller diameter	1·45 m (4 ft 9 in)

Weights:

Weight empty	273 kg (600 lb)
Max fuel weight	54 kg (120 lb)
Max T-O weight	567 kg (1,250 lb)

Performance (at max T-O weight, except when stated otherwise):

Never-exceed speed	217 knots (402 km/h; 250 mph)
Max level speed	200 knots (370 km/h; 230 mph)
Max turbo cruising speed at 5,790 m (19,000 ft)	200 knots (370 km/h; 230 mph)
Econ cruising speed at 5,790 m (19,000 ft)	174 knots (322 km/h; 200 mph)
Cruising speed at 2,600 m (8,500 ft)	146 knots (270 km/h; 168 mph)
Stalling speed, flaps and landing gear down	39 knots (72·5 km/h; 45 mph)
Max rate of climb at S/L	198 m (650 ft)/min
T-O and landing run	122 m (400 ft)
Range with max payload and max fuel, 45 min reserves	955 nm (1,770 km; 1,100 miles)

POWELL

JOHN C. POWELL

4 Donald Drive, Middletown, Rhode Island 02840

John Powell, formerly a Commander in the US Navy, designed and built a two-seat parasol-wing monoplane, of which plans are available to amateur constructors. Known as the P-70 Acey Deucy, its design was started in 1966 and construction began during 1967. FAA certification in the Experimental homebuilt category was awarded on 19 June 1970 and the first flight was recorded on the following day. By the beginning of 1980 this prototype had accumulated 581 hours flying time. It was not flown during 1980, but was expected to be back in operation from the Spring of 1981.

More than 300 sets of plans have been sold, and the first aircraft built from plans was flying by the Autumn of 1973. At least seven Acey Deucys have been completed, with many more under construction.

One Acey Deucy, powered by a 74·5 kW (100 hp) Continental O-200 engine, has been fitted with twin floats and is said to perform well at a take-off weight of about 725 kg (1,600 lb).

POWELL P-70 ACEY DEUCY

Type: Two-seat light monoplane.

Wings: Braced parasol-wing monoplane with steel tube V bracing struts on each side, auxiliary bracing struts and N-type centre-section struts. Wing section NACA 4412. Dihedral 1° on outer panels. Incidence 2°. No sweepback. Composite structure of steel tube and wood, fabric covered. Frise-type ailerons of wooden construction, fabric covered. No flaps. No trim tabs.

Fuselage: Welded 4130 steel tube structure with wooden stringers, fabric covered.

Tail Unit: Wire-braced welded steel tube structure with U channel ribs. Tailplane incidence adjustable by screw-jack at leading-edge. No trim tabs.

Landing Gear: Non-retractable tailwheel type. Two side Vs and half axles hinged to fuselage structure. Shock-absorption by springs in compression. Goodyear main wheels and tyres size 8·00-4, pressure 0·83 bars (12 lb/sq in). Motor scooter type caliper brakes. At least one Acey Deucy is flying as a seaplane, with floats of 750 kg (1,650 lb) displacement.

Power Plant: Designer recommends use of engines from 48·5-67 kW (65 to 90 hp), although at least one Acey Deucy has a 74·5 kW (100 hp) Continental (see introductory copy). Prototype has one 48·5 kW (65 hp) Continental A65 flat-four engine, driving a McCauley two-blade fixed-pitch metal propeller. One fuel tank in fuselage, immediately aft of firewall, capacity 53 litres (14 US gallons). Refuelling point on top of fuselage, forward of front cockpit. Oil capacity 3·75 litres (1 US gallon).

Accommodation: Two persons in tandem, normally in open cockpits. Sliding canopy available. Small door by front cockpit on starboard side.

Dimensions, external:

Wing span	9·91 m (32 ft 6 in)
Wing chord, constant	1·52 m (5 ft 0 in)
Wing area, gross	14·4 m² (155 sq ft)
Wing aspect ratio	6·5
Length overall	6·32 m (20 ft 9 in)
Height overall	2·06 m (6 ft 9 in)
Tailplane span	2·59 m (8 ft 6 in)
Wheel track	1·83 m (6 ft 0 in)

Weights (prototype):

Weight empty	340 kg (750 lb)
Max T-O weight	578 kg (1,275 lb)

Performance (prototype):

Never-exceed speed	126 knots (233 km/h; 145 mph)
Max level speed at S/L	90 knots (167 km/h; 104 mph)
Max cruising speed at 610 m (2,000 ft)	80 knots (148 km/h; 92 mph)
Stalling speed	23·5-26 knots (44-48·5 km/h; 27-30 mph)
Max rate of climb at S/L	190 m (625 ft)/min
Service ceiling (approx)	3,050-3,355 m (10,000-11,000 ft)
T-O run	76 m (250 ft)
T-O to 15 m (50 ft)	approx 213 m (700 ft)
Landing run	approx 76 m (250 ft)
Range with 30 min reserves	217 nm (402 km; 250 miles)

QUICKIE

QUICKIE AIRCRAFT CORPORATION

PO Box 786, Mojave, California 93501 and Building 68, Mojave Airport, Mojave, California 93501
Telephone: (805) 824 4313

Quickie Aircraft Corporation is a company formed by Gene Sheehan and Tom Jewett to develop the Quickie aircraft and to sell complete kits, including engine, to amateur constructors who wish to build this type. Gene Sheehan has worked in the aerospace industry since 1964 and was involved in several previous homebuilt projects; Tom Jewett's career has been devoted to flight testing, and

Two-seat Pietenpol Aircamper built in Canada *(J. M. G. Gradidge)*

Powell P-70 Acey Deucy with new sliding canopy

Quickie Aircraft Corporation Quickie prototype *(Ian MacFarlane)*

Polliwagen two-seat light sporting aircraft *(Don Dwiggins)*

Quickie Aircraft Corporation's two-seat Quickie Q2

he was a Flight Test Engineer on board the Rockwell International B-1 supersonic bomber.

In mid-1981, Quickie Aircraft was well advanced with the development of a research aircraft which had been given the provisional name Big Bird, as well as an improved version of the Quickie under the name Super Quickie. A two-seat version of the Quickie has been developed under the designation Quickie Q2, and the delivery of complete kits for this version began in February 1981.

QUICKIE AIRCRAFT CORPORATION QUICKIE

The Quickie is a single-seat light aircraft of unusual configuration. The project began in 1975 when Mr Sheehan and Mr Jewett started looking for a low-power engine that could be installed in a highly-efficient single-seat sporting aircraft, the design of which had not then been initiated. Not until 1977 was the Onan engine found and tested. Having proved its reliability, Burt Rutan of VariViggen fame was contacted to assist in designing an aircraft that would make optimum use of the engine. Having established the basic configuration, Tom Jewett and Burt Rutan began detailed design work, while Gene Sheehan continued engine development.

On 13 August 1977 construction of the prototype Quickie (N77Q) was started, and it was completed in 400 working hours over a three-month period. First flight was achieved on 15 November 1977. Fifteen flights were made by the three designer/pilots during the first five days of the test programme.

The Quickie design was frozen on 14 April 1978, after 125 hours of flight testing. The full programme of air and ground testing included investigation of performance, stability and control, flutter testing to 156 knots (290 km/h; 180 mph) TAS, static load testing and landing gear drop testing to FAA Pt 23 certificated aircraft standards, fuel economy measurements, and taxying, take-off and landing in surface winds of over 50 knots. Extensive stall/departure/spin testing revealed that the Quickie could not be made to spin.

In August 1978, the Quickie was flown to the Experimental Aircraft Association's Fly-in at Oshkosh, Wisconsin, where it received the 'Outstanding New Design' award for 1978. In 1979, a company-owned Quickie was again flown to Oshkosh, covering the 3,650 nm (6,760 km; 4,200 mile) return journey at an average speed of 100 knots (185 km/h; 115 mph) and at a fuel consumption of 29·75 km/litre (70 miles/US gallon). Two homebuilt Quickies were also flown to the Oshkosh meeting.

Complete kits, including engines, have been available since June 1978. Each includes prefabricated cowling, canopy, all machined parts, all welded parts, and some of the tools, in order to reduce the building time to 400 man-hours for the inexperienced builder. More recently, Quickie has broken down the complete kit into three separate packages, making it possible to spread the capital cost over a longer period.

By mid-1981, more than 520 single-seat Quickie kits had been delivered, and 100 completed aircraft had flown.

Development work carried out during 1980 resulted in a turbocharged version of the Onan engine being installed in a prototype in early 1981. It is anticipated that this engine will develop between 18·6 and 22·4 kW (25 and 30 hp) from sea level to 3,660 m (12,000 ft), giving better performance than the optionally modified standard normally aspirated engine. It is intended that, subject to satisfactory testing, the turbocharged engine will be available as a retrofit to existing Quickies. A variable-pitch propeller was also scheduled for installation and testing with this new power plant at an early date.

TYPE: Single-seat light sporting aircraft of canard configuration.

WINGS: Tapered cantilever shoulder-wing monoplane. Dihedral from roots. Two unidirectional glassfibre spars and shaped low-density rigid foam core, covered with glassfibre. Constant-chord inboard ailerons. No flaps.

FOREPLANE: Tapered and slightly sweptback cantilever foreplane, mounted low on forward fuselage. Marked anhedral. Construction similar to that of wings. Mainwheel housings attached to tips. Full-span tapered elevator/flaps.

FUSELAGE: Semi-monocoque structure of banana shape, formed from 25 mm (1 in) thick foam, with glassfibre covering inside and out, and tapering towards rear. Foam/glassfibre sandwich bulkheads.

TAIL UNIT: Cantilever sweptback vertical fin and narrow-chord rudder of similar construction to wings.

LANDING GEAR: Non-retractable tailwheel type. Main wheels positioned at tips of foreplane in swept fairings. A kit to provide 50 per cent larger tyres for grass field operation is optional. Steerable tailwheel aft of fuselage.

POWER PLANT: One modified Onan horizontally-opposed two-cylinder four-stroke engine, developing originally 13·5 kW (18 hp) at 3,600 rpm and driving a two-blade fixed-pitch wooden propeller. Modifications to cylinder heads and exhaust system allow newly-supplied engines to develop 16·4 kW (22 hp) at 3,800 rpm: a retrofit kit is available to convert 13·5 kW (18 hp) engines to develop this increased output. Fuel capacity 30·3 litres (8 US gallons). Fuel consumption approx 44·2 km/litre (104 miles/US gallon).

ACCOMMODATION: Single seat under one-piece side-hinged canopy. Cockpit suitable for pilot up to 1·98 m (6 ft 6 in) tall and weighing up to 95 kg (210 lb). Max baggage capacity with lighter pilot 13·6 kg (30 lb). With engine developing 16·4 kW (22 hp), a 100 kg (220 lb) pilot, maximum fuel, and 5·4 kg (12 lb) of baggage can be carried.

DIMENSIONS, EXTERNAL:

Wing span	5·08 m (16 ft 8 in)
Wing area, gross	2·52 m² (27·08 sq ft)
Foreplane span	4·67 m (15 ft 4 in)
Foreplane area	2·47 m² (26·57 sq ft)
Length overall	5·28 m (17 ft 4 in)

DIMENSIONS, INTERNAL:

Cockpit: Length	1·63 m (5 ft 4 in)
Width	0·56 m (1 ft 10 in)
Height	0·86 m (2 ft 10 in)

WEIGHTS (A: 13·5 kW; 18 hp engine, B: 16·4 kW; 22 hp engine):

Weight empty: A, B	109 kg (240 lb)
Max T-O weight: A	218 kg (480 lb)
B	236 kg (520 lb)

PERFORMANCE:

Max level speed: A	109 knots (202 km/h; 126 mph)
B	122 knots (225 km/h; 140 mph)
Max cruising speed:	
A	105 knots (195 km/h; 121 mph)
B	115 knots (214 km/h; 133 mph)
Stalling speed, power off:	
A	46 knots (86 km/h; 53 mph)
Stalling speed, power on:	
A	43 knots (79 km/h; 49 mph)
Normal rate of climb at S/L: A	130 m (425 ft)/min
B	183 m (600 ft)/min
Max rate of climb at 1,525 m (5,000 ft):	
A	110 m (360 ft)/min
B	152 m (500 ft)/min
Service ceiling: A	3,750 m (12,300 ft)
B	4,665 m (15,300 ft)
T-O run: A	201 m (660 ft)
B	137 m (450 ft)
Landing run: A	255 m (835 ft)
B	183 m (600 ft)
Range at normal cruising speed:	
A	477 nm (885 km; 550 miles)
B	456 nm (845 km; 525 miles)
Range at econ cruising speed:	
A, B	712 nm (1,320 km; 820 miles)

QUICKIE AIRCRAFT CORPORATION QUICKIE Q2

Detail design of a two-seat version of the Quickie began in the Autumn of 1979, and the prototype was built by

Garry LeGare, the principal of Legair which is the Canadian distributor for Quickie Aircraft Corporation. This flew for the first time on 1 July 1980, before being transferred to Quickie's Mojave facility for development flying. Generally similar to the standard Quickie configuration, it differs primarily in having a wider fuselage, with a cockpit width of 1·10 m (3 ft 7½ in), to allow for side-by-side seating; a more powerful Revmaster engine; aerodynamic refinements; foreplane and vertical tail surfaces of increased area; and other improvements.

The Q2 is marketed in kit form, a complete kit containing all that is needed to build the aircraft, including the power plant and instruments. The fuselage is supplied as four prefabricated shells to simplify construction, and all components requiring machining or welding are also prefabricated. Delivery of kits began on 9 February 1981, at which time the company had a backlog of more than 60 Q2 orders. A composite construction starter kit is also available, to test the skills of a would-be amateur builder before he or she undertakes construction of a Q2.

The description of the Quickie applies also to the Quickie Q2, except as detailed:

TYPE: Two-seat light sporting aircraft of canard configuration.

WINGS AND FOREPLANE: Generally as for Quickie, except some modification of the aerofoil sections, and the incorporation of aerodynamic refinements. Foreplane increased in span and area.

FUSELAGE: Increased width and length, and built up from four prefabricated fuselage shells.

POWER PLANT: One 47·7 kW (64 hp) Revmaster 2100DQ flat-four engine, driving a two-blade fixed-pitch wooden propeller with spinner. Fuel capacity 76 litres (20 US gallons).

ACCOMMODATION: Pilot and passenger side by side in enclosed cockpit. Sideways opening transparent canopy, hinged on starboard side. Dual controls optional. Space for 18 kg (40 lb) baggage aft of seats.

DIMENSIONS, EXTERNAL:
Wing span 5·08 m (16 ft 8 in)
Foreplane span 5·08 m (16 ft 8 in)
Wing/foreplane area, gross 6·22 m² (67·0 sq ft)
Length overall 5·97 m (19 ft 7 in)

WEIGHTS:
Weight empty, equipped 243·5 kg (537 lb)
Max T-O weight 454 kg (1,000 lb)

PERFORMANCE (A: pilot only at 340 kg; 750 lb AUW. B: pilot and passenger at 454 kg; 1,000 lb max T-O weight):
Max level speed: B 156 knots (290 km/h; 180 mph)
Stalling speed: A 53 knots (98·5 km/h; 61 mph)
B 56 knots (103 km/h; 64 mph)
Max rate of climb at S/L: A 366 m (1,200 ft)/min
B 244 m (800 ft)/min
Service ceiling: A 5,790 m (19,000 ft)
B 4,570 m (15,000 ft)
T-O run: A 110-137 m (360-450 ft)
B 186-198 m (610-650 ft)
Landing run: A 219 m (720 ft)
B 241 m (790 ft)
Range with max fuel at max cruising speed, 45 min reserves: B 592 nm (1,097 km; 682 miles)
Range with max fuel at econ cruising speed, 45 min reserves: B 886 nm (1,641 km; 1,020 miles)

QUICKIE AIRCRAFT CORPORATION SUPER QUICKIE Q2

Generally similar in configuration to the Quickie, this new aircraft was redesigned to embody aerodynamic improvements, and to increase cockpit volume by 25%. Few details were available in mid-1981, but it is known that the fuselage is more rounded; the wing has a reduced area and embodies carbonfibre spars; and fuel capacity has been increased to 113 litres (30 US gallons). The Super Quickie is being used as a test aircraft, and is designed to achieve a fuel economy at maximum cruising speed of 32 km/litre (75 miles/US gallon) with a turbocharged Onan engine.

The following provisional specification details are estimated:

WEIGHTS:
Max useful load 141 kg (310 lb)
Baggage 18 kg (40 lb)

PERFORMANCE:
Max level speed 174 knots (322 km/h; 200 mph)
Stalling speed 49 knots (90 km/h; 56 mph)
Rate of climb at 3,050 m (10,000 ft) 335 m (1,100 ft)/min
Service ceiling 7,620 m (25,000 ft)
T-O and landing run 152 m (500 ft)
Range (normal) 1,042 nm (1,931 km; 1,200 miles)

QUICKIE AIRCRAFT CORPORATION 'BIG BIRD'

'Big Bird' (a temporary in-house name) is a single-seat single-engined T tail aircraft designed to fly long distances with outstanding efficiency. It is intended solely for special-purpose research and will attempt to set new world records for absolute distance. The only specification details available are as follows:

DIMENSION, EXTERNAL:
Wing span 15·00 m (49 ft 2½ in)

WEIGHT:
Max T-O weight approx 1,815 kg (4,000 lb)

PERFORMANCE (estimated):
Normal cruising speed 174 knots (322 km/h; 200 mph)
Service ceiling above 10,650 m (35,000 ft)
Range with max fuel more than 10,425 nm (19,310 km; 12,000 miles)
Endurance with max fuel at normal cruising speed more than 85 h
Fuel consumption:
normal 21·25-29·75 km/litre (50-70 miles/US gallon)
record attempts 32-42·5 km/litre (75-100 miles/US gallon)

RAND ROBINSON

RAND ROBINSON ENGINEERING INC

5842 K McFadden Avenue, Huntington Beach, California 92649

Telephone: (714) 898 3811

In 1974, the late Mr Kenneth Rand formed Rand Robinson Engineering Inc to market plans for the KR-1 single-seat lightweight sporting aircraft, and of a slightly larger two-seat version, designated KR-2. His last design was the KR-3 light amphibian, described in the 1978-79 *Jane's*.

RAND ROBINSON KR-1

Mr Kenneth Rand designed and built the prototype of a single-seat lightweight sporting aircraft known as the Rand KR-1. The design originated in 1969; construction of the prototype was started in 1970 and the first flight was made in February 1972. Plans are available to amateur constructors; about 6,000 sets had been sold by the beginning of 1981 and about 200 KR-1s were known to be flying.

The performance figures that are quoted relate generally to the re-engined prototype fitted with a 67 kW (90 hp) 2,074 cc Volkswagen engine. Performance figures with the original 27 kW (36 hp) 1,200 cc VW engine can be found in the 1975-76 *Jane's*.

TYPE: Single-seat lightweight sporting aircraft.

WINGS: Cantilever low-wing monoplane. Wing section RAF 48. Thickness/chord ratio 15%. Dihedral 5°. Incidence 5° at root, 2° at tip. No sweepback. Composite two-spar structure. Front spar of spruce; rear spar built of spruce and plywood. Most ribs formed from polyurethane foam, spaces between ribs being filled with polyurethane foam slab. Structure covered with Dynel reinforced epoxy. Outer wing panels removable for storage. Ailerons of polyurethane foam, with Dynel reinforced epoxy covering, over full span of outer panels. Trailing-edge flaps standard.

FUSELAGE: Composite structure, lower half of spruce longerons with plywood skin, upper surface of carved Styrofoam covered with Dynel epoxy. Firewall is a plywood, asbestos and stainless steel lamination.

TAIL UNIT: Cantilever structure with spruce spars, the remainder of the structure being carved polyurethane foam, Dynel epoxy covered. Fixed-incidence tailplane. Trim tabs in rudder and elevator.

LANDING GEAR: Tailwheel type. Main units retract aft manually into wing centre-section. Shock-absorption by flat spring crossbar to which main units are attached. Main-wheel tyres size 10½ × 4·00-5, pressure 2·07 bars (30 lb/sq in). Steerable tailwheel with solid tyre of 7·6 cm (3 in) diameter. Manual drum brakes.

POWER PLANT: One Volkswagen modified motorcar engine, driving a Maloof two-blade two-position metal propeller with spinner. Prototype has been re-engined with a Rajay-turbocharged 67 kW (90 hp) 2,074 cc VW. Fuel tankage with larger engine comprises one tank immediately aft of firewall, capacity 38 litres (10 US gallons), and one 76 litre (20 US gallon) tank in each wing, giving total capacity of 190 litres (50 US gallons). Refuelling point on fuselage upper surface, forward of windscreen. Oil capacity 2·8 litres (0·75 US gallons).

ACCOMMODATION: Pilot only, beneath rearward-sliding transparent cockpit canopy. Baggage space aft of seat.

SYSTEM: Electrical power supplied by 20A alternator and 12V 14Ah storage battery. Bosch electric starter.

AVIONICS AND EQUIPMENT: Edo-Aire 720-channel transceiver with 200-channel Auto Omni. Full blind-flying panel.

DIMENSIONS, EXTERNAL:
Wing span 5·18 m (17 ft 0 in)
Wing chord at root 1·22 m (4 ft 0 in)
Wing chord at tip 0·91 m (3 ft 0 in)
Wing area, gross 5·76 m² (62 sq ft)
Wing aspect ratio 4·5
Length overall 3·89 m (12 ft 9 in)
Width, wings removed 1·52 m (5 ft 0 in)
Height overall 1·07 m (3 ft 6 in)
Tailplane span 1·52 m (5 ft 0 in)
Wheel track 1·27 m (4 ft 2 in)
Propeller diameter 1·35 m (4 ft 5 in)

DIMENSIONS, INTERNAL:
Cockpit: Length 1·22 m (4 ft 0 in)
Max width 0·51 m (1 ft 8 in)
Max height 0·76 m (2 ft 6 in)
Baggage space 0·11 m³ (4 cu ft)

WEIGHTS (A: 27 kW; 36 hp. B: 67 kW; 90 hp):
Weight empty, equipped: A 154 kg (340 lb)
B 218 kg (480 lb)
Max T-O and landing weight: A 272 kg (600 lb)
B 408 kg (900 lb)

PERFORMANCE (A: 27 kW; 36 hp. B: 67 kW; 90 hp, at max T-O weight):
Never-exceed speed:
A 140 knots (259 km/h; 161 mph)
B 217 knots (402 km/h; 250 mph)
Max level speed at S/L:
A 130 knots (241 km/h; 150 mph)
B 191 knots (354 km/h; 220 mph)
Max cruising speed at 1,525 m (5,000 ft):
A 130 knots (241 km/h; 150 mph)
B 191 knots (354 km/h; 220 mph)
Econ cruising speed at 5,485 m (18,000 ft):
B 217 knots (402 km/h; 250 mph)
Stalling speed: A, B 39 knots (73 km/h; 45 mph)
Max rate of climb at S/L: A 182 m (600 ft)/min
B 457 m (1,500 ft)/min
Service ceiling: A 3,660 m (12,000 ft)
B 9,145 m (30,000 ft)
T-O run: A, B 122 m (400 ft)
T-O to 15 m (50 ft): A, B 244 m (800 ft)
Landing from 15 m (50 ft): A, B 305 m (1,000 ft)
Landing run: A, B 152 m (500 ft)
Range with max fuel:
B 2,600 nm (4,825 km; 3,000 miles)

RAND ROBINSON KR-2

The KR-2 is a slightly larger two-seat version of the KR-1, to which it is generally similar in construction. Design began in 1973 and the prototype flew for the first time in July 1974. Construction occupied approximately 800 man hours, at a cost of about $2,000. Plans and kits of parts are available to amateur constructors.

By early 1981, about 4,500 sets of plans and 2,800 kits had been sold; about 350 KR-2s were flying at that time.

TYPE: Two-seat lightweight sporting aircraft.

WINGS: As for KR-1, except span increased by 1·12 m (3 ft 8 in).

FUSELAGE: As for KR-1, except dimensions increased.

TAIL UNIT AND LANDING GEAR: As for KR-1, except wheel track increased.

POWER PLANT: Airframe designed to accept Volkswagen modified motorcar engines of 1,600 to 2,200 cc. Prototype has a Rajay turbocharged 2,074 cc Volkswagen engine, driving a Maloof two-position propeller with spinner. Fuel tank immediately aft of firewall, capacity 38 litres (10 US gallons). One fuel tank in each wing, capacity 53 litres (14 US gallons). Total fuel capacity 144 litres (38 US gallons). Refuelling point on fuselage upper surface, forward of windscreen.

ACCOMMODATION: Two persons, side by side, beneath transparent cockpit canopy.

AVIONICS AND EQUIPMENT: Provision for wide range of avionics and blind-flying instruments. Prototype has full IFR panel, dual nav/com transceivers, glideslope receiver, three-light marker beacon receiver, and radar transponder with altitude reporting.

DIMENSIONS, EXTERNAL:
Wing span 6·30 m (20 ft 8 in)
Wing chord at root 1·22 m (4 ft 0 in)
Wing chord at tip 0·91 m (3 ft 0 in)
Wing area, gross 7·43 m² (80 sq ft)
Wing aspect ratio 5·5
Length overall 4·42 m (14 ft 6 in)
Height overall 1·07 m (3 ft 6 in)
Tailplane span 1·52 m (5 ft 0 in)
Wheel track 1·52 m (5 ft 0 in)
Propeller diameter 1·32 m (4 ft 4 in)

DIMENSIONS, INTERNAL:
Cockpit: Length 1·22 m (4 ft 0 in)
Max width 0·91 m (3 ft 0 in)
Max height 0·76 m (2 ft 6 in)
Baggage space 0·11 m³ (4 cu ft)

WEIGHTS (prototype, A: without turbocharger, B: with turbocharger):
Weight empty, equipped: A 200 kg (440 lb)
B, with IFR and electrics 263 kg (580 lb)
Max T-O and landing weight: A 363 kg (800 lb)
B 499 kg (1,100 lb)

PERFORMANCE (prototype, at max T-O weight. A: without turbocharger, B: with turbocharger):
Never-exceed speed:
A, B 186 knots (346 km/h; 215 mph)

Rand Robinson KR-1 built by Mr K. Campbell *(J. M. G. Gradidge)*

A two-seat Rand Robinson KR-2 *(J. M. G. Gradidge)*

Breezy Model RLU-1 built by Mr R. J. Weller *(J. M. G. Gradidge)*

Rotor Sport Coupe RSH-1A two-seat helicopter

Max level speed at S/L:
A 156 knots (290 km/h; 180 mph)
B 161 knots (298 km/h; 185 mph)
Max cruising speed:
A at 1,525 m (5,000 ft)
156 knots (290 km/h; 180 mph)
B at 5,485 m (18,000 ft)
191 knots (354 km/h; 220 mph)
Econ cruising speed at 3,660 m (12,000 ft):
A 148 knots (274 km/h; 170 mph)
Stalling speed: A 39 knots (73 km/h; 45 mph)
Max rate of climb: A at S/L 244 m (800 ft)/min
B at 5,485 m (18,000 ft) 244 m (800 ft)/min
Service ceiling: A 4,875 m (16,000 ft)
B 7,925 m (26,000 ft)
T-O run: A, B 122 m (400 ft)
T-O to 15 m (50 ft): A, B 244 m (800 ft)
Landing from 15 m (50 ft): A, B 305 m (1,000 ft)
Landing run: A, B 152 m (500 ft)
Range with max fuel:
B 1,735 nm (3,215 km; 2,000 miles)

RLU

CHARLES ROLOFF, ROBERT LIPOSKY and CARL UNGER

c/o Charles B. Roloff, PO Box 358, Palos Park, Illinois 60464
Telephone: (312) 598 6210

Three professional pilots designed and built a unique light aircraft known as the Breezy Model RLU-1, the designation being made up of the initial letters of the surnames of the designers. Well over 700 sets of plans for the Breezy have since been sold, and many examples are flying, including some built in Australia, Canada, South Africa and Switzerland.

BREEZY MODEL RLU-1

Described as being of vintage configuration with all modern facilities, such as full radio, instruments and hydraulic brakes, the prototype Breezy was an open three-seat light aircraft powered by a 67 kW (90 hp) Continental engine. First flight was made on 7 August 1964.

First Breezy to be built from the published plans was that constructed by Airpark Aero of Santa Rosa, California, for Mr Jack Gardiner of Pandora, Ohio. It differed from the prototype only by having two bucket seats, one of these replacing the usual two-place bench seat behind the pilot.

The following description applies to the prototype Breezy:

Type: Three-seat parasol-wing monoplane.

Wings: Strut-braced parasol-wing monoplane. Prototype utilised standard Piper PA-12 wing, with V streamline-section bracing struts each side.

Fuselage: Triangular-section welded chrome-molybdenum steel tube structure, without any covering.

Tail Unit: Welded chrome-molybdenum steel tube braced structure; all surfaces fabric-covered.

Landing Gear: Non-retractable tricycle type. Main wheels and tyres size 6·00-6, 4-ply; nosewheel and tyre size 5·00-5. Cleveland hydraulic brakes.

Power Plant: One 67 kW (90 hp) Continental C90-8F-P flat-four engine, driving a Flottorp 72A50 two-blade pusher propeller. Single fuel tank, capacity 68 litres (18 US gallons), in wing centre-section. Oil capacity 4·5 litres (1·25 US gallons).

Accommodation: Pilot on single seat forward, two passengers on bench seat aft.

Dimensions, external:
Wing span 10·06 m (33 ft 0 in)
Wing area, gross 15·3 m² (165 sq ft)
Length overall 6·86 m (22 ft 6 in)
Height overall 2·59 m (8 ft 6 in)
Wheel track 1·83 m (6 ft 0 in)
Wheelbase 3·05 m (10 ft 0 in)

Weights:
Weight empty 317 kg (700 lb)
Max T-O weight 544 kg (1,200 lb)

Performance:
Never-exceed speed 91 knots (168·5 km/h; 105 mph)
Cruising speed, 70% power
65 knots (121 km/h; 75 mph)
Stalling speed 26 knots (49 km/h; 30 mph)
Service ceiling 4,572 m (15,000 ft)
T-O run (grass) 137 m (450 ft)
T-O to 15 m (50 ft) 335 m (1,100 ft)
Landing from 15 m (50 ft) 335 m (1,100 ft)
Landing run (grass) 91 m (300 ft)
Range with max fuel 217 nm (402 km; 250 miles)

ROTOR SPORT

ROTOR SPORT HELICOPTERS INC

PO Box 180, Milnesville, Pennsylvania 18239
Telephone: (717) 454 4530
President: Robert S. Kornishock

ROTOR SPORT COUPE RSH-1A

Under the above designation, Rotor Sport Helicopters has developed a two-seat light helicopter that is intended for amateur construction. It is claimed that it can be built as easily as any fixed-wing homebuilt, and it is powered by a low-cost secondhand V-6 motorcar engine. Design of the Coupe was initiated in 1969, but it was not until April 1976 that construction began, with a first flight recorded on 18 December 1978. Plans and kits of materials are available from Rotor Sport, which stated that in 1980 about 91 were under construction and orders for 440 outstanding.

Type: Two-seat light helicopter.

Rotor System: Two-blade semi-rigid main rotor. Blade section NACA 0012. Each blade is of wooden construction, incorporating a ground-adjustable trim tab, and is attached to the rotor hub by two retention straps. Two-blade wooden teetering tail rotor. Main and tail rotor hubs constructed of 2024-T3 light alloy. Rotor brake. Blades do not fold.

Rotor Drive: Simplified drive incorporating ring gear and pinion beneath main pylon housing, together with belts and drive shafts. Main rotor/engine rpm ratio 1:5 3. Tail rotor/engine rpm ratio 1:1.

Fuselage: Composite pod and boom type structure of steel tube, wood, and fabric.

Tail Unit: Small dorsal and ventral fin at end of tailboom.

Landing Gear: Rigid skid type.

Power Plant: Prototype powered by one 115·5 kW (155 hp) General Motors V-6 modified motorcar engine. Fuel tank in aft fuselage with standard capacity of 45·5 litres (12 US gallons); 91 litres (24 US gallons) optional. Refuelling point on upper port side of fuselage. Oil capacity 4·7 litres (1·25 US gallons).

Accommodation: Pilot and passenger side by side on individual seats. Dual controls standard. Forward-hinged door on each side. Baggage space aft of seats. Accommodation heated and ventilated.

System: Electrical system of 12V 60A DC.

Avionics and Equipment: Avionics comprise 360-channel nav/com, and ELT. Equipment includes a cargo sling.

Dimensions, external:
Diameter of main rotor 7·32 m (24 ft 0 in)
Diameter of tail rotor 1·22 m (4 ft 0 in)
Main rotor blade chord 0·20 m (8 in)
Length overall, rotors turning 8·84 m (29 ft 0 in)
Length of fuselage 6·10 m (20 ft 0 in)
Height to top of rotor hub 2·44 m (8 ft 0 in)

Weights:
Weight empty, equipped 508 kg (1,120 lb)
Max T-O weight 714 kg (1,575 lb)

Performance (at max T-O weight):
Max level speed at 3,050 m (10,000 ft)
95·5 knots (177 km/h; 110 mph)
Cruising speed at 762 m (2,500 ft)
69·5 knots (129 km/h; 80 mph)
Max rate of climb at S/L 366 m (1,200 ft)/min
Service ceiling 3,660 m (12,000 ft)
Range with max optional fuel
333 nm (618 km; 384 miles)

ROTORWAY

ROTORWAY AIRCRAFT INC

14805 S. Interstate 10, Tempe, Arizona 85284
Telephone: (602) 963 6652

Mr B. J. Schramm formed the Schramm Aircraft Company to market, in both ready-to-fly and prefabricated component form, a single-seat helicopter of his own design, named the Javelin. Details of this aircraft, which flew for the first time in August 1965, can be found in the 1967-68 *Jane's*.

Subsequently, a new company named RotorWay Aircraft Inc was formed to market to amateur constructors plans and kits of components to build Mr Schramm's Scorpion helicopter, described as a production version of the Javelin (see 1972-73 *Jane's*). After a period this aircraft was superseded by the two-seat Scorpion Too, now known as the Scorpion 133. An improved version of this aircraft, designated RotorWay Exec, was introduced in 1980.

RotorWay currently offers a package deal which includes Scorpion 133 construction kits, a RotorWay RW 133 engine, and a complete training programme for the constructor/pilot which covers flight training, operation and theory, and maintenance. Plans by themselves are no longer available.

During 1978, a turbocharged version of the RW 133 engine became available as an alternative to the basic, normally aspirated model, for owners who require improved performance at altitude.

ROTORWAY SCORPION 133

Type: Two-seat light helicopter.

Rotor System: Two-blade semi-rigid main rotor, incorporating Schramm Tractable Control rotor system. Blade section NACA 0015. All-metal D-spar thermal-bonded blades, which do not fold, are attached to aluminium teetering rotor hub by retention straps. Two-blade aluminium teetering tail rotor. Swashplate for cyclic pitch control. Cable through rotor shaft to blades for collective pitch control.

Rotor Drive: Drive from engine to vertical shaft via four power bands. Drive from vertical shaft to main rotor shaft via sprocketed triple-row chain. Tail rotor driven by V belt from first stage of reduction pulleys.

Fuselage: Basic 4130 steel tube structure of simplified form. Removable glassfibre body fairing.

Tail Unit: Braced steel tube tailboom only, to carry tail rotor.

Landing Gear: Tubular skid type.

Power Plant: One 99 kW (133 hp) RotorWay RW 133 watercooled flat-four engine, designed and produced for the Scorpion 133 by RotorWay and mounted aft of cabin area. Turbocharged version available, giving 119 kW (160 hp) at 70% boost. Standard fuel capacity 37·5 litres (10 US gallons) in tank mounted above drive chain, aft of main rotor shaft.

Accommodation: Two individual bucket seats, side by side, in enclosed cabin.

Dimensions, external:

Diameter of main rotor	7·62 m (25 ft 0 in)
Diameter of tail rotor	1·10 m (3 ft 7¼ in)
Length, nose to tail rotor axis	6·26 m (20 ft 6½ in)
Height to top of main rotor	2·22 m (7 ft 3½ in)
Width of cabin	1·22 m (4 ft 0 in)
Landing skid track	1·64 m (5 ft 4¾ in)

Weights (A without turbocharger, B with turbocharger):

Weight empty: A	365 kg (805 lb)
B	372 kg (820 lb)
Max T-O weight: A, B	560 kg (1,235 lb)

Performance (at max T-O weight. A without turbocharger, B with turbocharger):

Cruising speed: A, B	69·5 knots (129 km/h; 80 mph)
Max rate of climb at S/L: A	244 m (800 ft)/min
B	366 m (1,200 ft)/min
Service ceiling: A	3,050 m (10,000 ft)
B	4,875 m (16,000 ft)
Hovering ceiling IGE: A	1,675 m (5,500 ft)
B	2,745 m (9,000 ft)
Range with max fuel: A	104 nm (193 km; 120 miles)
B	100 nm (185 km; 115 miles)

ROTORWAY EXEC

Announced in 1980, the RotorWay Exec is a much-improved version of the Scorpion 133, and like that aircraft is available in kit form. The kit is broken down into six groups, so that the builder can, if he wishes, spread the cost of acquiring the complete kit over a period of time. The main differences between the Exec and the Scorpion 133 result from use of a more powerful engine, and an enclosed pod and boom type fuselage structure. By early 1981 it was reported that kits for about 50 of these helicopters had been delivered.

The description of the Scorpion 133 applies also to the RotorWay Exec, except as follows:

Rotor System: Generally similar to Scorpion 133, except for modified main rotor hub, increased-diameter tail rotor, and improved and simplified control system.

Tail Unit: A small horizontal stabiliser, and a one-piece sweptback dorsal/ventral fin, are mounted on the port side of the tailboom.

Fuselage: Basic 4130 steel tube structure for fuselage pod, with a monocoque tailboom of light alloy.

Power Plant: One 108 kW (145 hp) RotorWay RW 145 watercooled flat-four engine, mounted aft of cabin area. Standard fuel capacity 68 litres (18 US gallons).

Equipment: Optional equipment includes dual controls, passenger's seat belt and shoulder harness, ground handling wheels, engine-driven alternator, and seat upholstery to customer's requirements.

Dimensions, external:

Diameter of main rotor	7·62 m (25 ft 0 in)
Diameter of tail rotor	1·14 m (3 ft 9 in)
Length of fuselage	6·55 m (21 ft 6 in)
Height to top of main rotor	2·31 m (7 ft 7 in)
Landing skid track	1·55 m (5 ft 1 in)

Weights:

Weight empty	354 kg (780 lb)
Max T-O weight	573 kg (1,263 lb)

Performance (at max T-O weight):

Max level speed	108 knots (201 km/h; 125 mph)
Max rate of climb at S/L	366 m (1,200 ft)/min
Hovering ceiling IGE	2,135 m (7,000 ft)
Hovering ceiling OGE	1,220 m (4,000 ft)
Range with max fuel at optimum cruising power	174 nm (323 km; 201 miles)

RUTAN

RUTAN AIRCRAFT FACTORY

Building 13, Mojave Airport, Mojave, California 93501
Telephone: (805) 824 2645
President: Elbert L. Rutan

RUTAN VARIVIGGEN

Construction of the prototype VariViggen (N27VV) began during 1968, and the first flight was made in May 1972. By early 1976 it had accumulated a total flying time of nearly 600 hours, and flight testing had confirmed the spin-free characteristics demonstrated by an earlier free-flying scale model. The prototype had no conventional stall in its original form, and could climb, cruise, glide, turn and land with continuous full aft stick, giving a stable speed of 45 knots (83·5 km/h; 52 mph) throughout. Rate of climb at this speed proved to be 152 m (500 ft)/min.

The VariViggen's full-span ailerons provide a high rate of roll, and a 360° roll can be accomplished at a speed of only 80 knots (148 km/h; 92 mph) without loss of height. Manoeuvrability is such that the aircraft's turn radius is less than 61 m (200 ft) at speeds of 60-110 knots (111-204 km/h; 69-127 mph). Nosewheel rotation on take-off occurs at 50 knots (93 km/h; 58 mph), at which speed the aircraft will enter an immediate stable climb.

The full-span ailerons are described as 'reflexerons', since they serve both as ailerons and as an adjustable 'reflex' control for the main wing. Differential aileron motion is related mechanically to the stick, but the collective 'reflex' is electrically controlled. This is achieved by electrically controlled aileron droop, which causes the resulting nose-down trim to be countered by a nose-up deflection of the elevators on the forward canard surface. Thus both ailerons and elevators serve also as flaps. The ailerons are also adjusted at cruising speed to minimise trim drag and optimise the lift/drag ratio. Trim is achieved by an electrically-controlled bungee device on the mechanical elevator system.

In 1975 Mr Rutan began experimenting with a new SP (special performance) wing outer panel, constructed from urethane foam and unidirectional glassfibre, and with a Wortmann FX-60-126 section. With an increase of wing span to 7·23 m (23 ft 8½ in), and of area to 11·61 m² (125 sq ft), it was anticipated that this would provide a 25% increase in the max rate of climb and give a cruising speed 4·5-6 knots (8-11 km/h; 5-7 mph) greater than with the standard wing. Tests showed that the new wing increased rate of climb, but that the former 'no stall' characteristic had been sacrificed. With the SP wing, the VariViggen will 'roll off' if uncoordinated at full aft stick. It is, however, easier to build and stronger than the aluminium outer panel, and is covered on sets of plans currently available to amateur constructors.

A further refinement on the SP wing is the addition of NASA-developed 'winglets', designed by Dr Whitcomb, to enhance directional stability at no cost in performance. The SP wing panels attach to the standard inboard wing panels, without modification. Each SP outer wing houses a 28·4 litre (7·5 US gallon) fuel tank, giving an additional 56·8 litres (15 US gallons) of auxiliary fuel.

About 900 sets of VariViggen plans have been sold; it is believed that approximately 300 aircraft are under construction and several homebuilt examples were flying by early 1981. One aircraft assembled in France by M Chagnes is powered by two Microturbo TRS 18 turbojet engines, in which form it is known as the Microstar (see 1980-81 *Jane's*).

The following details apply to the VariViggen with the original type of outer wing panels:

Type: Two-seat (or 2+2) light aircraft.

Wings: Cantilever low-wing monoplane of cropped delta configuration. Rutan wing section. Thickness/chord ratio 7% at root, 9% at tip. Dihedral 3° on outer wing panels. Incidence 0°. Sweepback at quarter-chord 27°. Composite structure with spruce spars, plywood ribs and skin, Ceconite-covered, except for outboard aft wing panels which are of flush-riveted metal construction. Inward-canted fin and rudder each side at approximately one-third span. Full-span ailerons, extending between fins and wingtips, constructed as a shell of 0·016 in aluminium with foam filling. Cutout in inboard trailing-edges to accommodate pusher propeller.

Foreplane: Cantilever structure mounted high on the nose, forward of the windscreen. Aerofoil section NACA 4414 (modified). Slotted flap-type elevator in trailing edge.

Fuselage: Basically square-section fuselage of wooden construction, Ceconite-covered. Landing light in nosecone, which is hinged at top and opens upwards for access to equipment. Engine mounted in aft end of fuselage.

Landing Gear: Electrically-retractable tricycle type. Nosewheel retracts forward, main wheels inward into wings. Nosewheel mounted on oleo-pneumatic shock-strut. Shock-absorption of main units by rubber discs in compression. Goodyear main wheels, with tyres size 14 × 5·00-5, pressure 2·07 bars (30 lb/sq in). Scott nosewheel with 228 mm (9 in) diameter tyre, pressure 2·07 bars (30 lb/sq in). Goodyear caliper brakes.

Power Plant: One 112 kW (150 hp) Avco Lycoming O-320-A2A flat-four engine, mounted in rear fuselage and driving a Hegy two-blade fixed-pitch wooden propeller. One fuel tank in fuselage, capacity 87 litres (23 US gallons) and one external fuel tank, mounted on aircraft centreline under fuselage, capacity 45 litres (12 US gallons). Total capacity 132 litres (35 US gallons). Refuelling point on fuselage upper surface. Oil capacity 7·5 litres (2 US gallons).

Accommodation: Two seats in tandem in individual cockpits, beneath transparent canopies which are hinged on the starboard side. Space for two children, each weighing not more than 22·5 kg (50 lb), or 45 kg (100 lb) of baggage aft of rear seat.

Systems: Dual 12V electrical systems. Storage battery. Hydraulic system for brakes only.

Avionics and Equipment: ARC 360-channel VHF com transceiver, ARC 200-channel VHF nav receiver. Edo-Aire transponder. Angle of attack indicator.

Dimensions, external:

Wing span	5·79 m (19 ft 0 in)
Wing chord at root	2·26 m (7 ft 5 in)
Wing chord at tip	0·89 m (2 ft 11 in)
Wing area, gross	11·06 m² (119 sq ft)
Wing aspect ratio	3
Length overall	5·79 m (19 ft 0 in)
Foreplane span	2·44 m (8 ft 0 in)
Wheel track	2·20 m (7 ft 2½ in)
Wheelbase	2·44 m (8 ft 0 in)
Propeller diameter	1·78 m (5 ft 10 in)

Dimensions, internal:

Cabin: Length	2·54 m (8 ft 4 in)
Max width	0·64 m (2 ft 1 in)

Weights:

Weight empty, equipped	431 kg (950 lb)
Max T-O and landing weight	771 kg (1,700 lb)

Performance (A: standard wing; B: SP wing, at max T-O weight, except as indicated):

Never-exceed speed:	
A	156 knots (289 km/h; 180 mph)
Max level speed at S/L:	
A	142 knots (262 km/h; 163 mph)
Max cruising speed at 2,135 m (7,000 ft):	
A	130 knots (241 km/h; 150 mph)
B	137 knots (254 km/h; 158 mph)
Econ cruising speed at 2,135 m (7,000 ft):	
A	109 knots (201 km/h; 125 mph)
Max rate of climb at S/L: A	244 m (800 ft)/min
B	305 m (1,000 ft)/min
Service ceiling: A	4,265 m (14,000 ft)
T-O run: A	259 m (850 ft)
T-O to 15 m (50 ft): A	290 m (950 ft)
Landing from 15 m (50 ft) at max landing weight:	
A	183 m (600 ft)
Landing run at max landing weight: A	152 m (500 ft)
Range with max fuel, 30 min reserves:	
A	347 nm (643 km; 400 miles)

RUTAN VARIEZE

Mr Rutan designed and built two prototypes of a high-performance two-seat aircraft of canard configuration, named the VariEze, of which plans are available. The name stems from its simplicity of construction, the entire structure being a composite of high-strength, primarily unidirectional glassfibre with rigid foam as core material.

The configuration of the VariEze is based on that of the Rutan VariViggen, with a foreplane; but the cropped-delta wings of that aircraft are replaced by more conventional swept wings of high aspect ratio. At the tip of each wing is a vertical fin, known as a 'winglet'. Developed by

RotorWay Scorpion 133 with turbocharged engine

RotorWay's new Exec helicopter with enclosed fuselage structure *(Howard Levy)*

Rutan VariEze high-performance tail-first monoplane (background) flying with the Rutan VariViggen two-seat canard delta; the Vari-Eze is fitted with NASA 'winglets'

Rutan Long-EZ built by NASA engineer John C. Murphy Jr *(Howard Levy)*

Dr Richard Whitcomb at NASA's Langley Research Center, each 'winglet' consists of large above-wing and small below-wing surfaces. That beneath the wing extends aft from the leading-edge to 33% of the tip chord, is cambered inward and inclined outward at 30° from vertical. The above-wing surface, extending aft from 33% of the tip chord, is cambered outward and also inclined outward at 15° from vertical. This 'winglet' system has been shown to 'unwind' the wingtip vortex to a maximum, limiting induced drag and resulting in a fuel saving of 6%; the inclination of the upper and lower surfaces also offsets 40% of the parasitic drag of the vertical fins. These vertical surfaces include rudders, which are moved outward individually by single cables and centralised by return springs. The rudder control cables include in their run slotted bellcranks, so that extended movement of either rudder pedal causes the bellcrank to actuate its respective wheel-brake master cylinder. A conventional control column operates elevators mounted on the trailing-edges of the foreplane and ailerons on the trailing-edges of the rear wings.

The tricycle landing gear has another unusual feature: the main units are fixed but the nosegear retracts both on the ground and in the air. Retracting the nose unit on the ground, termed kneeled parking, not only facilitates access to the cockpits but raises the propeller to a convenient height for hand-swinging and, at the same time, dispenses with a need for wheel chocks.

Designed in late 1974, the first VariEze (N7EZ) was built over a ten-week period in the Spring of 1975 and made its first flight on 21 May, powered by a 47 kW (63 hp; 1,834 cc) Volkswagen engine. By early 1977 it had logged approximately 260 flying hours. Optimum economy cruise performance was a primary design aim, so that the prototype could be used to attack existing world distance records in the under 500 kg gross weight class C1a. On 4 August 1975 the aircraft set a closed-circuit distance record (since beaten) in this class, by covering 1,415·119 nm (2,620·80 km; 1,628·49 miles).

A second prototype (N4EZ), embodying some modifications and powered by a Continental O-200 engine, was built during the Winter of 1975-76. It represents the prototype of the version for which plans are available, and is described in detail below.

All raw materials and certain component parts of the VariEze are also available to homebuilders, including the Plexiglas canopy, moulded glassfibre nosewheel and main landing gear struts, and glassfibre cowling.

A large underfuselage drag device has been added to the VariEze to enable it to land in shorter distances. About 2,000 VariEzes are being constructed by homebuilders in many countries, and well over 200 were flying in early 1981.

Type: Two-seat sporting aircraft.

Wings: Cantilever mid-wing monoplane with sweptback surfaces. Single-spar structure of unidirectional glassfibre with rigid foam core. Inboard ailerons. Vertical above-wing 'winglet' surface at each wingtip includes rudder.

Foreplane: Cantilever structure of unidirectional glassfibre with rigid foam core. Trailing-edge elevators of similar construction.

Fuselage: Composite structure comprising large sheets of rigid urethane foam, with wood strips as corner fillers, and internal and external covering of unidirectional glassfibre. A standard-size light alloy extrusion is used for engine, landing gear, and control stick mounts, as well as for canopy latches and other parts.

Landing Gear: Tricycle landing gear, with fixed main units and mechanically-retractable nosewheel which is carried on a glassfibre strut moulded to conform to the outside contour of the fuselage, so eliminating need for a fairing door. Nose gear retracted by hand-lever. Main wheels carried on one-piece moulded glassfibre strut. Fairings on main wheels. Hydraulically-operated brakes.

Power Plant: One 74·5 kW (100 hp) Continental O-200-B flat-four engine, mounted in the rear fuselage and driving a two-blade fixed-pitch pusher propeller with spinner. Provision for Continental A75, C85 or C90 engines. Fuel tanks, of glassfibre/foam/glassfibre sandwich construction, form strakes on each side of the fuselage that fair the wing roots; total capacity 91 litres (24 US gallons). A small fuselage tank is also available. Refuelling points on each side of fuselage, on the upper surface of tanks.

Accommodation: Pilot and passenger on semi-reclining seats in individual cockpits. Side-stick controls. Dual controls limited to control stick at rear position. One-piece bubble canopy of moulded Plexiglas covers both cockpits and is hinged on starboard side. Roll-over structure. Space for 14 kg (30 lb) baggage in two specially designed suitcases which fit in rear seat area.

Systems: Hydraulic system for wheel brakes only. Can have simple electrical system with a single nav/com and gyro instrument.

Dimensions, external:

Wing span	6·77 m (22 ft 2½ in)
Wing chord at root	0·91 m (3 ft 0 in)
Wing chord at tip	0·41 m (1 ft 4 in)
Wing area, gross	4·98 m² (53·6 sq ft)
Foreplane span	3·81 m (12 ft 6 in)
Foreplane chord, constant	0·32 m (1 ft 0½ in)
Foreplane area	1·21 m² (13 sq ft)
Length overall	4·32 m (14 ft 2 in)

Weights (Continental O-200 engine):

Weight empty	254 kg (560 lb)
Max T-O weight	476 kg (1,050 lb)

Performance (at max T-O weight, Continental O-200 engine):

Max cruising speed	170 knots (313 km/h; 195 mph)
Econ cruising speed	143 knots (265 km/h; 165 mph)
Landing speed	60 knots (112 km/h; 70 mph)
Stalling speed	48 knots (90 km/h; 55·5 mph)
Max rate of climb at S/L	487 m (1,600 ft)/min
T-O and landing run	275 m (900 ft)
Range at 75% power	607 nm (1,126 km; 700 miles)
Range at econ cruising speed	738 nm (1,368 km; 850 miles)

RUTAN LONG-EZ

In late 1978, Mr Rutan began the design of an improved VariEze, to be powered by an Avco Lycoming O-235 engine that would have an electric starter and engine-driven alternator. The new model was intended specifically to have long range (hence the name Long-EZ), good forward field of view during landing, and lower approach and landing speeds than the VariEze, to make it more suitable for pilots of low proficiency. In its initial prototype form (N79RA), as first flown on 12 June 1979, it was generally similar to the VariEze; but in addition to the changed power plant, it had wings of increased span and sweep, a 'rhino' rudder on the nose, and no control surfaces on the winglets. Flight testing revealed several deficiencies in flight characteristics. As a result a new wing was designed, with an Eppler section similar to that used in the Rutan Defiant, reduced wing sweep, increased area, ailerons of increased span, and the eventual introduction of rudders in larger winglets, leading to deletion of the 'rhino' rudder.

When the Long-EZ was flown with the wing in this finalised form, the desired flight characteristics had been achieved. On 15 December 1979, piloted by Dick Rutan, it took off in an attempt to set a new closed-circuit distance record in FAI Class C1b. Landing 33 h 33 min and 41 s later, the Long-EZ had covered a distance of 4,171·3 nm (7,725·3 km; 4,800·3 miles) which ranked as a new world record, beating the former 20-year-old record by almost 1,602 nm (2,969 km; 1,845 miles).

On 5 June 1981, the same pilot took off from Anchorage, Alaska, for an attempt on the Class C1b straight-line distance record. This time he covered about 3,910 nm (7,240 km; 4,500 miles), landing at Turk Island, West Indies, after 30 h 8 min in the air. This Long-EZ had a 119 kW (160 hp) Avco Lycoming engine and fuel capacity of 556 litres (147 US gallons). The record awaited homologation in Summer 1981.

Generally similar in construction to the VariEze, the standard two-seat Long-EZ has an 85·75 kW (115 hp) Avco Lycoming O-235 flat-four engine, and a maximum fuel capacity of 197 litres (52 US gallons). Optional alternative power plant for the Long-EZ is the 74·5 kW (100 hp) Continental O-200 flat-four engine.

Plans for construction of this aircraft are available to homebuilders: by 1 January 1981 approximately 900 Long-EZs were under construction worldwide, and four examples had been completed and flown. The first of these (N21VE) is shown in an accompanying illustration. It was built in an eight-month period by Mr John C. Murphy Jr, a NASA engineer at Cape Canaveral. Powered by a 93 kW (125 hp) Avco Lycoming O-290-D2 flat-four engine and with an empty weight of 358 kg (789 lb) it is, in other respects, generally similar to Rutan plans.

Dimensions, external:
Wing span 7·96 m (26 ft 1¼ in)
Wing area, gross 7·62 m² (81·99 sq ft)
Foreplane span 3·59 m (11 ft 9½ in)
Foreplane area, gross 1·19 m² (12·8 sq ft)
Length overall 5·12 m (16 ft 9½ in)
Height overall 2·40 m (7 ft 10½ in)

Weights:
Weight empty, basic 322 kg (710 lb)
Weight empty, equipped 340 kg (750 lb)
Normal max T-O weight 601 kg (1,325 lb)

Performance (with O-235 engine at max T-O weight, A: solo; B: two-seat):
Max cruising speed, 75% power at 2,440 m (8,000 ft):
A 161 knots (298 km/h; 185 mph)
B 159 knots (295 km/h; 183 mph)
Cruising speed, 40% power at 3,660 m (12,000 ft):
A, B 125 knots (232 km/h; 144 mph)
Max rate of climb at S/L:
A 533 m (1,750 ft)/min
B 411 m (1,350 ft)/min
Service ceiling:
A 8,230 m (27,000 ft)
B 6,705 m (22,000 ft)
T-O run: A 168 m (550 ft)
B 253 m (830 ft)
Landing run: A 137 m (450 ft)
B 207 m (680 ft)
Range with max fuel at max cruising speed, 40 min reserves:
A 1,190 nm (2,205 km; 1,370 miles)
B 838 nm (1,553 km; 965 miles)
Range with max fuel at 40% power, 40 min reserves:
A 1,746 nm (3,235 km; 2,010 miles)
B 1,242 nm (2,301 km; 1,430 miles)

RUTAN DEFIANT

The Rutan Defiant (N78RA), which made its first flight in proof-of-concept prototype form on 30 June 1978, is a four-seat twin-engined light aircraft of unconventional design. It was intended that this aircraft should be built as a production model, but since late 1979 this programme has been in a state of suspension. Available details of the Defiant can be found in the 1980-81 *Jane's*.

SEQUOIA
SEQUOIA AIRCRAFT CORPORATION

900 West Franklin Street, Richmond, Virginia 23220
Telephone: (804) 353 1713
President: Alfred P. Scott

Sequoia Aircraft Corporation was formed by Mr Alfred P. Scott to develop the Sequoia cabin monoplane. Work on this aircraft was begun in July 1975, with all design and engineering by Mr David B. Thurston, a consultant whose other designs have included the Colonial Skimmer (now the Lake amphibian), the Teal and Marlin amphibians, and the TA-16 Trojan amphibian.

Two variations of the basic Sequoia design are available. The Model 300 Sequoia is a side-by-side two-seat aircraft, with two aft-facing rear seats optional and a sliding bubble canopy. The Model 302 Kodiak is a four-seat aircraft with an enclosed cabin.

These aircraft have a high degree of commonality of parts. Kits consisting of formed and machined parts have been produced and delivered to amateur constructors who are building the first examples of the aircraft. In January 1981, twenty aircraft were under construction; 40 sets of plans had been sold.

In addition to marketing plans and kits to amateur aircraft builders, Sequoia Aircraft Corporation has offered both aircraft as military trainers to countries that wish to assemble their own aircraft, using local labour. Aeronics (Pty) Ltd of the Republic of South Africa has signed an exclusive production agreement for manufacture of the aircraft for civilian use in South Africa, but start of manufacture is dependent on suitable financing being arranged.

In addition, Sequoia is marketing plans and kits for construction of the Italian Falco F.8L light aircraft.

MODEL 300 SEQUOIA

Type: Two-seat utility and aerobatic monoplane.

Wings: Cantilever low-wing monoplane. Wing section NACA 64_2A215 at root, NACA 64A210 at tip. Dihedral 3° from roots. Incidence 3° 30′ at root. Washout 2°. All-metal flush-riveted structure with single I-beam spar, 2024-T4 Alclad ribs and 2024-T3 Alclad skin. Glassfibre tips. Frise balanced ailerons of aluminium construction with rubber seals. Slotted aluminium flaps.

Fuselage: Welded 4130 steel tube structure. Entire fuselage, including wing fillets, covered with lightweight shell of glassfibre/PVC foam/glassfibre sandwich, attached to tubing with glassfibre and epoxy resin. Kevlar is optional alternative to glassfibre.

Tail Unit: Conventional cantilever unit of flush-riveted aluminium alloy construction. Elevators fitted, each with trim tab. Rudder trim is bungee system.

Landing Gear: Retractable tricycle type. Electro-hydraulic retraction. Oleo-pneumatic shock-absorbers.

Power Plant: One 224 kW (300 hp) Avco Lycoming TIO-540-S1AD flat-six engine, driving a Hartzell two-blade constant-speed propeller. Other Avco Lycoming engines of 175-224 kW (235-300 hp) may be used. Two integral fuel tanks in wings, with total capacity of 291·5 litres (77 US gallons). Provision for tip-tanks of approximately 75 litres (20 US gallons) capacity each; and two underwing attachment points for additional fuel tanks, radar or stores of up to 272 kg (600 lb) combined weight. Refuelling points in wingtips.

Accommodation: Two seats side by side, beneath sliding Plexiglas canopy. An aft-facing rear seat can be installed to provide a two-plus-two layout. Baggage compartment, capacity 45·4 kg (100 lb).

Avionics: A completely flush antenna system is being designed. The nav antennae will be in the glassfibre wingtips and the remainder will be under the fuselage shell.

Dimensions, external:
Wing span 9·14 m (30 ft 0 in)
Wing chord at root 1·68 m (5 ft 6 in)
Wing chord at tip 0·91 m (3 ft 0 in)
Wing area, gross 12·08 m² (130 sq ft)
Wing aspect ratio 6·92
Length overall 7·62 m (25 ft 0 in)
Height overall 2·90 m (9 ft 6 in)
Propeller diameter 2·03 m (6 ft 8 in)

Weights:
Weight empty 816 kg (1,800 lb)
Max T-O weight, utility 1,270 kg (2,800 lb)
Max T-O weight, aerobatic 1,088 kg (2,400 lb)

Performance (estimated):
Never-exceed speed 243 knots (450 km/h; 280 mph)
Max level speed at S/L
195 knots (362 km/h; 225 mph)
Max cruising speed at 2,440 m (8,000 ft)
185 knots (343 km/h; 213 mph)
Stalling speed 'clean' 75 knots (139 km/h; 86 mph)
Stalling speed, flaps and wheels down
60 knots (111 km/h; 69 mph)
Max rate of climb at S/L 664 m (2,180 ft)/min
Service ceiling 7,620 m (25,000 ft)
T-O run 457 m (1,500 ft)
T-O to 15 m (50 ft) 610 m (2,000 ft)
Landing run 548 m (1,800 ft)
Range at max cruising speed, with 45 min reserves
868 nm (1,609 km; 1,000 miles)

SEQUOIA MODEL 302 KODIAK

The Kodiak is a four-seat development of the Model 300 Sequoia, using the same wing assembly, tail unit, landing gear, power plant and mount, engine cowling, firewall, control system, instrument panel and windscreen. Wingtip fuel tanks can be fitted to increase range; gull-wing doors provide access to the enclosed cabin; and the aircraft is designed to be fully aerobatic.

Power plant will be in the 186-224 kW (250-300 hp) Avco Lycoming IO-540/TIO-540-S1AD range, driving a Hartzell two-blade constant-speed propeller with a diameter of 2·03 m (6 ft 8 in). Empty weight will be 839 kg (1,850 lb); max T-O weight 1,451 kg (3,200 lb); fuel capacity 341 litres (90 US gallons).

SEQUOIA FALCO F.8L

Sequoia Aircraft markets plans and kits to build the Falco F.8L high-performance low-wing monoplane, designed in Italy by Ing Stelio Frati and first flown on 15 June 1955. The aircraft was last marketed in Italy by Laverda SpA. In its current form, for amateur construction, it incorporates minor changes, and the performance data given below apply to the Sequoia Falco F.8L fitted with a 119 kW (160 hp) Avco Lycoming O-320-B3B flat-four engine, although Avco Lycoming engines of 100·5 kW (135 hp) and 112 kW (150 hp) can be installed.

The kits obtainable from Sequoia Aircraft comprise 21 separate packages which can be purchased collectively or, if the constructor wishes to spread the cost, be acquired as necessary. They do not include the engine, propeller, avionics, instruments, tyres and inner tubes. Approximately 174 Falcos were being built in January 1981.

Type: Two-seat light monoplane.

Wings: Cantilever low-wing monoplane. Laminar-flow NACA $64_2212{\cdot}5$ wing section at root, NACA 64_2210 at tip. Thickness/chord ratio 12·5% at root and 10% at tip. Dihedral 5°. Incidence at root 3°. Washout 3°. All-wood one-piece single-spar structure built integrally with forward portion of fuselage. Plywood-covered, with overall fabric covering. Frise-type ailerons and electrically-operated plain flaps can be of wood and fabric or all-metal construction. Optional leading-edge modification gives milder stall and slightly reduced cruising speed.

Fuselage: Plywood-covered semi-monocoque structure, with wooden frames and longerons, fabric covered overall. Normally built in two sections, comprising a centre-section integral with the wings, and a tailcone; but can be built in one piece.

Tail Unit: Cantilever type. All-wood tailplane and fin integral with rear portion of fuselage. Rudder and elevators can be of wood and fabric or all-metal construction. Controllable trim tab in starboard elevator.

Landing Gear: Retractable tricycle type. Manual or optional electrical retraction, with mechanical emergency actuation. Oleo-pneumatic shock-absorbers. Single-disc hydraulic brakes. Steerable nosewheel.

Power Plant: One Avco Lycoming O-290-D2B, O-320-A2A or O-320-B3B flat-four engine of 100·5 kW (135 hp), 112 kW (150 hp) or 119 kW (160 hp) respectively, driving a fixed-pitch or (except for O-290 engine) constant-speed two-blade propeller with spinner. Fuel in two metal tanks, capacity 118 litres (31·2 US gallons).

Accommodation: Enclosed cabin for two persons side by side. Rearward-sliding transparent canopy. Dual controls standard. Cabin heated and ventilated. Baggage locker behind seats. Provision for fitting seat for a child of not more than 40 kg (88 lb) weight in baggage space.

System: Electrical system of 12V for cabin lighting, landing lights, navigation lights, strobes, etc.

Dimensions, external:
Wing span 8·00 m (26 ft 3 in)
Wing area, gross 10·0 m² (107·5 sq ft)
Wing aspect ratio 6·4
Length overall 6·50 m (21 ft 4 in)
Height overall 2·29 m (7 ft 6 in)

Weights (O-320-B3B engine):
Weight empty 550 kg (1,212 lb)
Max aerobatic weight 748 kg (1,650 lb)
Max T-O weight 820 kg (1,808 lb)

Performance (O-320-B3B engine):
Never-exceed speed 208 knots (386 km/h; 240 mph)
Max level speed at S/L
184 knots (341 km/h; 212 mph)
Cruising speed at 1,830 m (6,000 ft), 75% power
165 knots (306 km/h; 190 mph)
Stalling speed, flaps and wheels down
54 knots (100 km/h; 62 mph)
Max rate of climb at S/L 347 m (1,140 ft)/min
Service ceiling 5,790 m (19,000 ft)
Range at econ cruising speed
755 nm (1,400 km; 870 miles)

SINDLINGER
SINDLINGER AIRCRAFT

37030 204th Avenue SE, Auburn, Washington 98002

Mr Fred G. Sindlinger began the design of a ⅝-scale replica of the second World War Hawker Hurricane IIC fighter in April 1969. Construction of the prototype was started three months later and the first flight was made in January 1972. By February 1977, this aircraft had accumulated a total of approximately 400 flying hours; it was subsequently sold.

Plans and certain component parts are available to amateur constructors; about 53 sets of drawings had been sold by early 1981, when 17 Sindlinger Hurricanes were under construction in Australia, France, West Germany and the United Kingdom. The first plans-built aircraft to

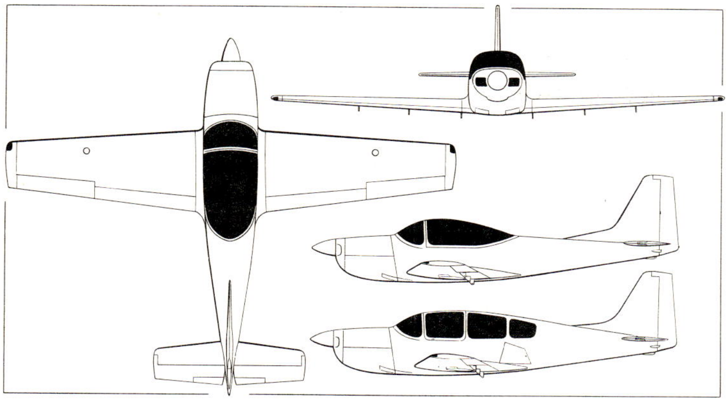

Model 300 Sequoia two-seat utility and aerobatic aircraft, with additional (lower) side view of Model 302 Kodiak *(Pilot Press)*

Italian-built Falco F.8L two-seat light monoplane *(James Gilbert)*

Sindlinger ⅝-scale replica of a Hawker Hurricane IIC

SF-2A Cygnet, designed and built by Captain A. M. Sisler of Bloomington, Minnesota

fly was constructed in South Africa and was flown in 1978; three more were nearing completion in 1980.

SINDLINGER HH-1 HAWKER HURRICANE

TYPE: Single-seat sporting aircraft.

WINGS: Cantilever low-wing monoplane. Wing section NACA 2418 in centre-section, with progressive change to NACA 2412 at tip. Dihedral 3° 30′ on outer panels only. Incidence 0° 36′. Sweepback at quarter-chord 3°. Two-spar structure of wood. Front spar is of I-beam construction in the centre-section and of built-up box section in the outer panels. Rear spar is an I-beam in the centre-section, and of U-channel form in the outer panels. Built-up truss ribs and 2·4 mm (3/32 in) plywood skin, fabric-covered overall. Frise-type ailerons of wood construction, fabric-covered and statically balanced. Split trailing-edge flaps of wood.

FUSELAGE: All-wood monocoque structure of 3·2 mm (⅛ in) plywood from firewall to aft of cockpit. Rear fuselage is a built-up box truss frame, with formers and stringers. Entire structure fabric-covered.

TAIL UNIT: All-wood cantilever structure. Two-spar tailplane and fin have plywood skins, covered with fabric overall. Rudder and elevators are of wood construction with fabric covering. Elevator trimmed by internal spring tension. Ground-adjustable trim tab on rudder.

LANDING GEAR: Manually-retractable tailwheel type. Main wheels retract inward. Shock-absorption of main and tail units by coil spring inside steel tubes. Tailwheel steerable. Goodyear main wheels and tyres size 5 00-5. Goodyear hydraulic brakes.

POWER PLANT: One 112 kW (150 hp) Avco Lycoming O-320 flat-four engine, driving a Hartzell two-blade constant-speed metal propeller with spinner. Up to three fuel tanks: one in fuselage aft of firewall with capacity of 54 litres (14 US gallons) and, optionally, one in each wing root, with capacity of 38 litres (10 US gallons) each. Max total fuel capacity 130 litres (34 US gallons). Refuelling point in fuselage upper surface, forward of windscreen. Oil capacity 7·5 litres (2 US gallons). Glassfibre engine cowlings.

ACCOMMODATION: Pilot only, beneath rearward-sliding transparent canopy. Cockpit heated and ventilated. Space for 18 kg (40 lb) baggage aft of pilot's seat.

SYSTEMS: Hydraulic system for brakes only. Electrical system powered by 12V DC engine-driven generator. 12V 35Ah battery.

AVIONICS AND EQUIPMENT: 90-channel VHF com transceiver, VOR Omni nav receiver. Partial IFR instrumentation. Wooden imitation cannon or machine-guns in wing leading-edges.

DIMENSIONS, EXTERNAL:
- Wing span: 7·65 m (25 ft 1 in)
- Wing chord on centre-section, constant: 1·52 m (5 ft 0 in)
- Wing chord at tip: 0·91 m (3 ft 0 in)
- Wing area, gross: 9·38 m² (101 sq ft)
- Wing aspect ratio: 6·2
- Length overall: 5·99 m (19 ft 8 in)
- Height overall, tail up: 1·78 m (5 ft 10 in)
- Tailplane span: 2·24 m (7 ft 4 in)
- Wheel track: 1·83 m (6 ft 0 in)
- Propeller diameter: 1·93 m (6 ft 4 in)

DIMENSION, INTERNAL:
- Cockpit: Max width: 0·66 m (2 ft 2 in)

WEIGHTS:
- Weight empty: 446 kg (984 lb)
- Max T-O weight: 624 kg (1,375 lb)

PERFORMANCE (at max T-O weight):
- Never-exceed speed: 208 knots (386 km/h; 240 mph)
- Max level speed at S/L: 174 knots (322 km/h; 200 mph)
- Max cruising speed, 65% power at 1,830 m (6,000 ft): 148 knots (273 km/h; 170 mph)
- Econ cruising speed, 60% power at 2,745 m (9,000 ft): 135 knots (249 km/h; 155 mph)
- Stalling speed, flaps up: 58 knots (108 km/h; 67 mph)
- Stalling speed, flaps down: 54 knots (100 km/h; 62 mph)
- Max rate of climb at S/L: 564 m (1,850 ft)/min
- Service ceiling: 6,400 m (21,000 ft)
- T-O run: 150 m (490 ft)
- Landing run: 168 m (550 ft)
- Range, 65% power, 30 min reserves: 500 nm (925 km; 575 miles)
- Max range, no reserves: 542 nm (1,005 km; 625 miles)
- g limits: +4·4 utility, +6·6 ultimate

SISLER
SISLER AIRCRAFT COMPANY

Box 20219, Bloomington, Minnesota 55420

Captain A. M. Sisler is an airline pilot who had logged more than 15,000 flying hours on 67 types of aircraft by the beginning of 1979. During his flying career, he has piloted seaplanes and sailplanes, instructed on single-engined and multi-engined aircraft, and served five years as a test pilot.

The first aircraft he designed and built was the Pipit two-seater, now owned by another pilot. It was followed by the SF-2 Whistler, which gained the 'Outstanding Design Contribution' award at the 1973 EAA Fly-in at Oshkosh. Capt Sisler has subsequently improved this design, making it easier to build, aerodynamically cleaner and 23 kg (50 lb) lighter. Plans of the improved version, now known as the SF-2A Cygnet, are available to other constructors.

SISLER SF-2A CYGNET

TYPE: Two-seat light aircraft.

WINGS: Strut-braced shoulder-wing monoplane of all-wood construction, swept forward 5° from roots. Wing section NACA 3413. Dihedral 3° 30′. Each wing consists of a main spar, light front and rear auxiliary spars, six plywood ribs, some stringers, geodetic upper and lower coverings of spruce strips, and an overall Dacron covering. Fabric-covered wooden ailerons, without tabs. No flaps.

FUSELAGE: Conventional welded chrome molybdenum steel tube structure, Dacron-covered.

TAIL UNIT: Conventional strut-braced unit, comprising tailplane and elevators, vertical fin and rudder. Structure similar to that of fuselage. Trim tab in port elevator. Ground-adjustable tab on rudder.

LANDING GEAR: Non-retractable tailwheel type. Two side Vs and half axles of welded steel tube. Bungee shock-absorption. Main-wheel tyres size 5·00-5 on prototype; size 6·00-6 tyres or skis optional. Steerable tailwheel.

POWER PLANT: One 46 kW (62 hp) 1,834 cc Volkswagen modified motorcar engine, driving a two-blade propeller. Fuel capacity 57 litres (15 US gallons).

ACCOMMODATION: Two seats side by side in enclosed cockpit. Windscreen made up of two door panels, hinged on centreline. Baggage capacity 23 kg (50 lb).

SYSTEM: Prototype has no electrical system, but lightweight alternator and battery are optional.

DIMENSIONS, EXTERNAL:
- Wing span: 9·14 m (30 ft 0 in)
- Wing chord, constant: 1·27 m (4 ft 2 in)
- Wing area, gross: 11·60 m² (124·8 sq ft)
- Wing aspect ratio: 7·1
- Length overall: 5·79 m (19 ft 0 in)
- Height overall: 1·78 m (5 ft 10 in)
- Tailplane span: 2·39 m (7 ft 10 in)
- Propeller diameter: 1·47 m (4 ft 10 in)

DIMENSION, INTERNAL:
- Cockpit width: 0·99 m (3 ft 3 in)

WEIGHTS:
- Weight empty: 265 kg (585 lb)
- Max T-O weight: 499 kg (1,100 lb)

PERFORMANCE (at max T-O weight):
- Max level speed: 94 knots (174 km/h; 108 mph)
- Cruising speed at 2,440 m (8,000 ft): 87 knots (161 km/h; 100 mph)
- Stalling speed: 42 knots (78 km/h; 48 mph)
- Max rate of climb at S/L: 177 m (580 ft)/min
- T-O and landing run: 213 m (700 ft)
- g limit: +4

SKYOTE
SKYOTE AEROMARINE LTD

PO Box 808, Clark, Colorado 80428

PRESIDENT: O. E. Bartoe Jr

This company was formed to sell sets of plans and component parts for the Skyote aerobatic biplane. Its President is the designer of the aircraft.

SKYOTE AEROMARINE SKYOTE

Construction of the prototype Skyote (N8XX), with a 67 kW (90 hp) Continental C90F engine, began in March

1975 and the first flight of this aircraft took place on 23 April 1976. FAA certification in the Aerobatic category was applied for on 17 February 1977. At that time, components for five other aircraft were being produced. The first of these was completed by Mr Duane Burnett of Boulder, and first flew in Summer 1978, powered by a 67 kW (90 hp) Continental C90-8 engine. Plans are available to amateur constructors, together with wing and other material kits. By early 1981 a total of 31 sets of plans had been sold.

TYPE: Single-seat aerobatic biplane.

WINGS: Braced biplane, with two parallel interplane struts each side and two pairs of centre-section support struts. Wing section NACA (1.8)412. No anhedral or dihedral. Incidence 2°. Sweepback on all wings at quarter-chord 7° 12′. Forward-staggered upper wing. All structure of 2024 aluminium, with hydroformed ribs, fabric covered. Plain ailerons on upper and lower wings, torque tube actuated.

FUSELAGE: Truss structure of welded 4130N chrome-molybdenum steel tubing, fabric covered.

TAIL UNIT: Wire-braced unit of 4130N chrome-molybdenum steel tube, fabric covered. Ground-adjustable tailplane incidence. No tabs.

LANDING GEAR: Non-retractable tailwheel type. Main wheels carried on side Vs and half-axles. Type 1280HD bungee cord shock-absorption. Cleveland main wheels and tyres size 18 × 6·00-6. Tyre pressure 1·24 bars (18 lb/sq in). Cleveland hydraulically-actuated disc brakes.

POWER PLANT: Aircraft used for FAA certification programme has one 88 kW (118 hp) Avco Lycoming O-235-K2A flat-four engine, driving a Sensenich two-blade fixed-pitch propeller. Other engines can be installed, including 67 kW (90 hp) Continental C85 or C90, or 74·5 kW (100 hp) O-200. Fuel tanks in wing centre-section and fuselage, each with capacity of 24·6 litres (6·5 US gallons). Total fuel capacity 49·2 litres (13 US gallons). Refuelling points in top of wing centre-section and top of fuselage. Oil capacity 7·5 litres (2 US gallons).

ACCOMMODATION: Single seat in open cockpit. Baggage compartment behind seat.

DIMENSIONS, EXTERNAL:	
Wing span	6·10 m (20 ft 0 in)
Wing chord, constant	0·97 m (3 ft 2 in)
Wing area, gross	11·43 m² (123 sq ft)
Wing aspect ratio	4·22
Length overall	4·95 m (16 ft 3 in)
Height overall	2·03 m (6 ft 8 in)
Tailplane span	2·03 m (6 ft 8 in)
Wheel track	1·22 m (4 ft 0 in)
Wheelbase	3·66 m (12 ft 0 in)
Propeller diameter	1·88 m (6 ft 2 in)
WEIGHTS (Continental C90-8F engine):	
Weight empty	269 kg (593 lb)
Max T-O weight	406 kg (895 lb)
PERFORMANCE (Continental C90-8F engine):	
Never-exceed speed	137 knots (253 km/h; 157 mph)
Cruising speed	97 knots (180 km/h; 112 mph)
Stalling speed	38 knots (71 km/h; 44 mph)
Max rate of climb at S/L	457 m (1,500 ft)/min
Service ceiling	5,030 m (16,500 ft)
Range with max fuel	180 nm (333 km; 207 miles)

SMITH

MRS FRANK W. (DOROTHY) SMITH

3502 Sunny Hills Drive, Norco, California 91760
Telephone: (714) 735 7152

The late Frank W. Smith built and flew in October 1956 the prototype of a single-seat fully-aerobatic sporting biplane which he designated the DSA-1 (Darn Small Aeroplane) Miniplane. This aircraft (N90P) is now in the EAA Museum. Plans continue to be marketed by Mrs Smith, and about 450 sets have been sold to constructors in several countries. Many Miniplanes have been completed and flown.

Mrs Smith's son designed a two-seat version of the DSA-1, which was provisionally designated Miniplane + 1; all available details of this aircraft appeared in the 1974-75 *Jane's*. It is expected to fly initially with a 93 kW (125 hp) Avco Lycoming engine.

SMITH DSA-1 MINIPLANE

The following details refer to the standard Miniplane, built according to Frank Smith's original plans:

TYPE: Single-seat sporting biplane.

WINGS: Braced biplane with N-type interplane struts each side and two N-type strut assemblies supporting centre of top wing above fuselage. NACA 4412 wing section. Dihedral 2° on lower wings only. Incidence 0° on top wing, 2° on lower wings. All-wood structure, fabric-covered. Fabric-covered wooden ailerons on lower wings only. No flaps.

FUSELAGE: Welded steel tube structure, fabric-covered.

TAIL UNIT: Wire-braced welded steel tube structure, fabric-covered. Adjustable-incidence tailplane.

LANDING GEAR: Non-retractable tailwheel type. Tripod streamlined-tube main legs. Compression-spring shock-absorbers optional (now fitted on prototype). Goodyear main wheels and tyres, size 7·00-4, pressure 1·38 bars (20 lb/sq in). Goodyear shoe-type brakes. Scott tailwheel.

POWER PLANT: Designed to take any engine in 48·5-93 kW (65-125 hp) category. Prototype has 80·5 kW (108 hp) Avco Lycoming O-235-C flat-four engine, driving a Sensenich two-blade fixed-pitch metal propeller. Most aircraft have a 48·5 kW (65 hp) Continental A65, 56 kW (75 hp) Continental A75 or 93 kW (125 hp) Avco Lycoming flat-four engine. Fuel in tank in fuselage, capacity 64·5 litres (17 US gallons). Oil capacity 5·7 litres (1·5 US gallons).

ACCOMMODATION: Single seat in open cockpit. Space for 27 kg (60 lb) baggage.

DIMENSIONS, EXTERNAL:	
Wing span (upper)	5·18 m (17 ft 0 in)
Wing span (lower)	4·80 m (15 ft 9 in)
Wing chord, constant (both)	0·91 m (3 ft 0 in)
Wing area, gross	9·29 m² (100 sq ft)
Length overall	4·65 m (15 ft 3 in)
Height overall	1·52 m (5 ft 0 in)
Wheel track	1·52 m (5 ft 0 in)
Propeller diameter	1·80 m (5 ft 11 in)
WEIGHTS (prototype):	
Weight empty, equipped	279 kg (616 lb)
Max T-O weight	454 kg (1,000 lb)
PERFORMANCE (prototype, at max T-O weight):	
Max level speed at S/L	117 knots (217 km/h; 135 mph)
Max cruising speed	102 knots (190 km/h; 118 mph)
Econ cruising speed	96 knots (177 km/h; 110 mph)
Stalling speed	48 knots (88·5 km/h; 55 mph)
Max rate of climb at S/L	380 m (1,250 ft)/min
Service ceiling	3,960 m (13,000 ft)
T-O run	107 m (350 ft)
Landing run	152 m (500 ft)
Endurance with max fuel	2 h 30 min

SMYTH

JERRY SMYTH

ADDRESS FOR PLANS AND KITS: George Blair, PO Box 815, Newbury Park, California 91320

In February 1958 Mr Smyth began the design of a sporting monoplane, setting out to develop an aircraft that would be reasonably easy to construct, easy to fly, stressed to 9g for limited aerobatics, of good appearance and offering economic operation. Construction of the prototype began in January 1967, and occupied two years before completion, at a cost of around $2,500. First flight of what Mr Smyth named the Model 'S' Sidewinder was made on 21 February 1969, and this aircraft received the 'Outstanding Design' award at the EAA Fly-in at Rockford, Illinois, in 1969. Plans and kits of parts are available to amateur constructors.

SMYTH MODEL 'S' SIDEWINDER

The following description applies to Mr Smyth's prototype:

TYPE: Two-seat sporting monoplane.

WINGS: Cantilever low-wing monoplane. Wing section NACA 64-612 at root, NACA 64-210 at tip. Dihedral 4°. Incidence 1° 30′. No sweepback. All-metal structure comprising a centre-section and two outer wing panels. Built-up main spar of ·04 in 2024-T3 aluminium U-sections, to which flat aluminium capstrips are riveted; secondary spar is of formed sections. Eleven equally-spaced ribs in each wing panel are made of ·025 in 6061-T4 aluminium. The wing skin, of ·025 in 2024-T3 aluminium, is in three sections: leading-edge, lower and upper skin, and is flush-riveted. Wings filled with epoxy. Simple sealed-gap ailerons of aluminium construction, attached to secondary spar by piano-type hinge. No trim tabs. No flaps.

FUSELAGE: Welded steel tube structure with aluminium formers and skin. Electrically-operated speed brake may be fitted on lower fuselage.

TAIL UNIT: Cantilever all-metal structure with swept vertical surfaces. All-moving tailplane with electrically-operated anti-servo tab.

LANDING GEAR: Non-retractable tricycle type. Wittman cantilever spring steel main gear. Main wheels and tyres size 5·00-5, pressure 1·72 bars (25 lb/sq in). Nose unit carries a 25 cm (10 in) diameter wheel and smooth tyre, free-castoring and non-steerable, pressure 1·72 bars (25 lb/sq in). Cleveland hydraulic brakes. Glassfibre fairings on all wheels. (Aircraft have been completed with retractable landing gear (see 1975-76 *Jane's*) and tailwheel type landing gear (see 1980-81 *Jane's*).)

POWER PLANT: Provision for installation of engines from 67-134 kW (90-180 hp). Prototype has a 93 kW (125 hp) Avco Lycoming O-290-G flat-four engine, driving a two-blade fixed-pitch aluminium propeller with spinner. Fuel tank in fuselage, forward of instrument panel, capacity 66·2 litres (17·5 US gallons). Refuelling point on top of fuselage, forward of windscreen. Provision for wingtip tanks. Oil capacity 7·5 litres (2 US gallons).

ACCOMMODATION: Pilot and passenger, seated side by side under rearward-sliding bubble canopy. Compartment for 40·8 kg (90 lb) of baggage aft of seats. Cabin heated and ventilated.

SYSTEMS: Hydraulic system for brakes and, optionally, for operation of aerodynamic speed brake. Engine-driven generator provides 35A 12V DC for instruments, lights, electrically-operated tailplane tab and optional electrically-driven hydraulic pump to operate aerodynamic speed brake.

AVIONICS: Simple 10-channel VHF com transceiver.

DIMENSIONS, EXTERNAL:	
Wing span	7·57 m (24 ft 10 in)
Wing chord at root	1·52 m (5 ft 0 in)
Wing chord at tip	0·91 m (3 ft 0 in)
Wing area, gross	8·92 m² (96 sq ft)
Wing aspect ratio	6·85
Length overall	5·89 m (19 ft 4 in)
Height overall	1·66 m (5 ft 5½ in)
Tailplane span	2·33 m (7 ft 7¾ in)
Wheel track	1·70 m (5 ft 7 in)
Wheelbase	1·30 m (4 ft 3 in)
Propeller diameter	1·70 m (5 ft 7 in)
DIMENSIONS, INTERNAL:	
Cabin: Max width	0·97 m (3 ft 2 in)
Baggage compartment	0·25 m³ (9 cu ft)
WEIGHTS:	
Weight empty	393 kg (867 lb)
Max T-O and landing weight	657 kg (1,450 lb)
PERFORMANCE (at max T-O weight):	
Never-exceed speed	173 knots (321 km/h; 200 mph)
Max level speed at 610 m (2,000 ft)	161 knots (298 km/h; 185 mph)
Max cruising speed, 75% power at 610 m (2,000 ft)	139 knots (257 km/h; 160 mph)
Stalling speed	48 knots (89 km/h; 55 mph)
Max rate of climb at S/L:	
at 0°C	366 m (1,200 ft)/min
at 24°C	274 m (900 ft)/min
Service ceiling	4,570 m (15,000 ft)
T-O run	244 m (800 ft)
T-O to and landing from 15 m (50 ft)	610 m (2,000 ft)
Landing run	457 m (1,500 ft)
Range with max fuel, no reserves	369 nm (684 km; 425 miles)

SORRELL

SORRELL AVIATION

16525 Tilley Road S, Tenino, Washington 98589
Telephone: (206) 264 2866

Sorrell Aviation designed and built a two-seat aerobatic biplane named the SNS-6 Hiperbipe, with the intention of providing a true HIgh PERformance BIPlanE that would be suitable for construction by amateurs. Flight testing confirmed that the aircraft had an outstanding aerobatic performance and, when demonstrated and displayed at the 1973 EAA Fly-in at Oshkosh, it received the 'Outstanding New Design' award. Plans for the fully-developed SNS-7 'production' version are not sold separately; but a basic kit package, containing construction drawings for the aircraft and certain completed components, is available. By early 1981 twelve Hiperbipes were flying, and more than 90 kits were under construction.

SORRELL SNS-7 HIPERBIPE

Design of the Hiperbipe began in 1964; construction of the SNS-6 first prototype started in June 1971. This aircraft made its initial flight in March 1973; a second prototype, modified to the SNS-7 standard offered to homebuilders, followed in March 1975.

The following details refer to the second prototype:

TYPE: Two-seat aerobatic biplane.

WINGS: Braced single-bay biplane, of modified NACA 0012 wing sections. Dihedral 1° 30′ on lower wings only. Incidence 1° 30′ on upper wing, 2° 30′ on lower wings. Sweepback 4° 30′ on lower wings only. Wide-chord welded 4130 steel I-type interplane struts and

Skyote Aeromarine Skyote aerobatic biplane built by Mr Duane Burnett

Smith DSA-1 Miniplane built by Mr Douglas Butzboch of Los Angeles, California

Smyth Sidewinder built by Mr Bill Uhl *(J. M. G. Gradidge)*

Sorrell Hiperbipe two-seat aerobatic biplane *(Peter M. Bowers)*

The prototype of the Spencer Amphibian Air Car *(Don Dwiggins)*

dual streamline-section landing and flying wires. Conventional structure of spruce spars, web ribs and stressed plywood skin, fabric covered overall. Centre-section of upper wing is skinned with transparent plastics to allow improved view for aerobatics. Cambered wingtips. Four full-span 'flaperons' of aluminium alloy sheet pop-riveted to aluminium torque tubes. No tabs.

FUSELAGE: Conventional structure of welded 4130 chrome-molybdenum steel tube, fabric-covered. Glassfibre engine cowling.

TAIL UNIT: Wire-braced structure of welded 4130 chrome-molybdenum steel tube, fabric-covered. Large dorsal fin. Tailplane and inset elevator extend aft of rudder trailing-edge. No tabs.

LANDING GEAR: Non-retractable tailwheel type. Wittman-type tapered spring steel rod main gear, with single wheel, glassfibre wheel fairing, and leg fairing on each unit. Cleveland 6·00-6 main wheels, with low-profile tyres. Maule 0·15 m (6 in) steerable tailwheel. Cleveland hydraulic spot disc brakes.

POWER PLANT: One 134 kW (180 hp) Avco Lycoming IO-360-B1E flat-four engine, driving a Hartzell HC-C2YK-4AF two-blade constant-speed metal propeller with spinner. Fuel tank in fuselage, capacity 147·5 litres (39 US gallons). Christen 801 inverted oil system, capacity 7·5 litres (2 US gallons).

ACCOMMODATION: Two seats side by side in enclosed cabin. Dual controls standard. Forward-hinged door on each side. Baggage capacity 36 kg (80 lb). Cabin heated and ventilated.

SYSTEM: Full electrical system, with lights.

AVIONICS AND EQUIPMENT: Narco Com 11A 360-channel radio.

DIMENSIONS, EXTERNAL:	
Wing span	6·96 m (22 ft 10 in)
Wing chord, constant	1·016 m (3 ft 3·9 in)
Wing area (projected)	13·9 m² (150 sq ft)
Length overall	6·35 m (20 ft 10 in)
Height overall	1·80 m (5 ft 10¾ in)
Tailplane span	2·87 m (9 ft 5 in)
Wheel track	2·16 m (7 ft 1 in)
Propeller diameter	1·93 m (6 ft 4 in)
DIMENSIONS, INTERNAL:	
Cabin: Length	1·40 m (4 ft 7 in)
Max width	1·07 m (3 ft 6 in)
WEIGHTS:	
Weight empty	561 kg (1,236 lb)
Max T-O weight: Aerobatic	766 kg (1,690 lb)
Normal	867 kg (1,911 lb)

PERFORMANCE (at max T-O weight):	
Never-exceed speed	195 knots (362 km/h; 225 mph)
Max level speed at S/L	149 knots (277 km/h; 172 mph)
Max cruising speed at S/L	139 knots (257 km/h; 160 mph)
Econ cruising speed	130 knots (241 km/h; 150 mph)
Stalling speed, flaperons down	43 knots (79 km/h; 49 mph)
Stalling speed, flaperons up	51 knots (94 km/h; 58 mph)
Max rate of climb at S/L	457 m (1,500 ft)/min
Service ceiling, estimated	6,100 m (20,000 ft)
T-O run	122 m (400 ft)
Landing run	181 m (595 ft)
Range	436 nm (807 km; 502 miles)

SPENCER

P. H. SPENCER

12780 Pierce Street, Pacoima, California 91331
Telephone: (213) 899 1010

Mr P. H. Spencer, who made his first solo flight in a powered aircraft on 15 May 1914, has been associated with the design of several single-engined amphibians, dating back to 1930, when Amphibians Inc of Garden City, Long Island, NY, put the Privateer amphibian into production. This was followed by the Spencer-Larsen, Spencer Air Car S-12, Republic Seabee RC 1, RC 2 and RC 3, and the more recent Canadian Trident TR-1.

All of these designs were variations of Mr Spencer's basic Air Car configuration, on which he was granted a patent on 3 January 1950. This was originally a two-seat amphibian powered by an 82 kW (110 hp) engine, and was developed into a four-seat version, known as the S-12-C. Mr Spencer then completed the design of the more advanced S-12-D, of which plans are available to homebuilders as well as certain glassfibre mouldings and metal assemblies. Since that time development has continued, the installation of a Teledyne Continental Tiara 6-285-B engine in the prototype (N111DA) resulting in a change of designation to S-12-E. The wing sweepback has also been increased, from 3° to 5°. By early 1978, this aircraft had accumulated a total of 800 hours' flying time, 450 of them with the Tiara engine.

By early 1981 at least 36 Air Cars were known to be under construction, with a variety of power plants ranging from 149 to 212·5 kW (200 to 285 hp), and more than 140 sets of plans had been sold. Air Car parts are currently being produced at a new facility at the address quoted.

The first S-12-E Air Car to be completed from Mr Spencer's plans made its first flight on 1 August 1974. Four more Air Cars flew for the first time in 1975, three in 1976, at least two in 1977, and a further two or three in 1978. Five are S-12-Es, with 212·5 kW (285 hp) Tiara engine, and five aircraft soon to fly are also powered by Tiara engines.

The Tiara 6-285-B engine has been removed from the prototype Air Car, and replaced by the 'C' model engine, of 224 kW (300 hp). Performance figures with this more powerful engine are not yet available. A second aircraft was fitted with a similar engine in 1980, together with a new constant-speed reversible-pitch propeller. Developed by Hartzell, this has no slip ring beta valve assembly, so offering a much refined installation.

During 1980-81 considerable effort has gone into the design and fabrication of a new two-seat version of the Air Car, known as the Air Car Jr, of which the prototype was scheduled to fly for the first time in July 1981.

SPENCER AMPHIBIAN AIR CAR MODEL S-12-E

TYPE: Four-seat amphibian.

WINGS: Braced high-wing monoplane with single streamline-section bracing strut on each side. Specially-designed STOL wing section. Thickness/chord ratio 15%. Dihedral 1°. Incidence 2°. Sweepback at quarter-chord 5°. Conventional two-spar structure of wood, steel and glassfibre. Frise-type ailerons of wooden construction. Electrically-operated trailing-edge flaps. Glassfibre stabilising float mounted on strut beneath each wing at approximately two-thirds span.

HULL: Conventional single-stepped hull with wood frames, longerons and skin. Welded steel tube structure to provide wing and engine mountings and attachment points for landing gear.

TAIL UNIT: Cantilever structure, comprising conventional fin and rudder, and all-moving tailplane set approximately midway up fin. Combined anti-servo and trim tab in tailplane. Retractable water rudder in base of aerodynamic rudder.

LANDING GEAR: Manually-retractable tricycle type. Main wheels retract aft to take up a near-vertical position on each side of the cabin. Nosewheel retracts forward through almost 180° and is partially housed in the nose of the hull, above the waterline, to form a nose fender. Cantilever spring steel main gear. Main wheels and tyres size 7·00-6. Nosewheel and tyre size 6·00-6. Cleveland hydraulic disc brakes.

POWER PLANT: One 212·5 kW (285 hp) Teledyne Continental Tiara 6-285-B flat-six engine, driving a Hartzell three-blade constant-speed reversible-pitch metal pusher propeller. Fuel tanks in fuselage and wing stabilising floats. Total fuel capacity 355 litres (94 US gallons). Oil capacity 8·5 litres (2·25 US gallons).

ACCOMMODATION: Four seats in pairs in enclosed cabin. Backs of front seats fold forward to improve access. Rear seats fold back against bulkhead to provide cargo or baggage space. Baggage space in rear fuselage, aft of rear cabin bulkhead. Door on each side of fuselage, hinged at forward edge. Bow access door on starboard side, hinged on centreline and opening upward. Dual controls standard. Accommodation heated and ventilated.

SYSTEMS: Hydraulic system for brakes only. Electrical system supplied by 24V 50A engine-driven alternator.

AVIONICS AND EQUIPMENT: Complete IFR instrumentation. Bendix 360 nav/com transceiver.

DIMENSIONS, EXTERNAL:

Wing span	11·38 m (37 ft 4 in)
Wing chord, constant	1·52 m (5 ft 0 in)
Wing area, gross	17·1 m² (184 sq ft)
Wing aspect ratio	7·4
Length overall	8·05 m (26 ft 5 in)
Height overall	2·90 m (9 ft 6 in)
Tailplane span	3·66 m (12 ft 0 in)
Wheel track	2·54 m (8 ft 4 in)
Wheelbase	3·10 m (10 ft 2 in)
Propeller diameter	2·13 m (7 ft 0 in)
Cabin doors (port and starboard, each):	
Height	0·97 m (3 ft 2 in)
Width	1·02 m (3 ft 4 in)
Height to sill	0·86 m (2 ft 10 in)
Cabin door (bow, starboard):	
Length	0·89 m (2 ft 11 in)
Width	0·51 m (1 ft 8 in)

DIMENSIONS, INTERNAL:

Cabin: Length	2·59 m (8 ft 6 in)
Max width	1·14 m (3 ft 9 in)

WEIGHTS:

Weight empty	993 kg (2,190 lb)
Max T-O weight	1,451 kg (3,200 lb)

PERFORMANCE (at max T-O weight):

Max level speed at S/L	128 knots (237 km/h; 147 mph)
Max cruising speed at 1,675 m (5,500 ft)	122 knots (225 km/h; 140 mph)
Econ cruising speed, 65% power at 2,315 m (7,600 ft)	117 knots (217 km/h; 135 mph)
Stalling speed, flaps up	46 knots (86 km/h; 53 mph)
Stalling speed, 35° flap	37·5 knots (70 km/h; 43 mph)
Max rate of climb at S/L	305 m (1,000 ft)/min
T-O time from calm water at S/L	16 s
Range, 65% power at 2,375 m (7,800 ft), 20 min reserves	695 nm (1,285 km; 800 miles)

SPENCER AMPHIBIAN AIR CAR Jr

TYPE: Two-seat amphibian.

WINGS: Braced high-wing monoplane, with V bracing struts each side. Conventional structure, with spruce spars and fabric covering. Trailing-edge flaps. Wings fold for towing and storage.

HULL: Conventional single-step hull of glassfibre, with birch and spruce bulkheads and reinforcements.

TAIL UNIT: Twin fins and rudders, and all-moving tailplane, carried on twin booms extending from the wing rear spar, and braced by struts from the aft point of the keel of the hull.

LANDING GEAR: Similar to that of Air Car S-12-E.

POWER PLANT: One 112 kW (150 hp) Avco Lycoming O-320 flat-four engine. Fuel capacity 151·4 litres (40 US gallons).

ACCOMMODATION: Two seats in enclosed cabin. Baggage capacity 18 kg (40 lb).

DIMENSIONS, EXTERNAL:

Wing span	10·06 m (33 ft 0 in)
Wing chord, constant	1·52 m (5 ft 0 in)
Wing area, gross	15·33 m² (165 sq ft)
Length overall, wings folded or spread	7·01 m (23 ft 0 in)
Height overall	2·21 m (7 ft 3 in)

WEIGHTS:

Weight empty	522 kg (1,150 lb)
Max T-O weight	816·5 kg (1,800 lb)

PERFORMANCE:

Max level speed at S/L	104 knots (193 km/h; 120 mph)
Cruising speed at 1,525 m (5,000 ft)	95·5 knots (177 km/h; 110 mph)
Max rate of climb at S/L	213 m (700 ft)/min
Range (65% power), with 30 min reserves	456 nm (845 km; 525 miles)

SPEZIO

WILLIAM EDWARDS

25 Madison Avenue, Northampton, Massachusetts 01060

Tony and Dorothy Spezio designed and built a two-seat light aircraft named the Tuholer, all rights of which were acquired by Mr Edwards in August 1973. He is continuing to market plans of the Tuholer to amateur constructors.

SPEZIO DAL-1 TUHOLER

Named Tuholer because of its two open cockpits, the prototype flew for the first time on 2 May 1961.

Folding wings enable the Tuholer to be kept in a normal home garage and it is towed behind a car on its own landing gear. It can be made ready for flight by two people in about 10 minutes or by one person in 20 minutes.

An accompanying photo shows the Spezio Special, which appears to be a Tuholer with the upper decking aft of the cockpits lowered and a single bubble canopy fitted over both cockpits. No further details were available at the time of writing.

The following description applies to the Tuholer built by Mr Edwards to the current plans:

TYPE: Two-seat sporting aircraft.

WINGS: Strut-braced low-wing monoplane, with streamline-section V bracing struts each side. Jury struts brace centre of these struts. Clark Y wing section. Dihedral 3°. Incidence 1°. Washout at wingtips 1°. Two-spar spruce structure, with plywood leading-edge and overall fabric covering. Conventional wooden ailerons. Drawings for Frise-type ailerons are available. No flaps. Wings fold back along sides of fuselage for stowage.

FUSELAGE: Steel tube structure with wood or light alloy stringers and fabric covering.

TAIL UNIT: Braced steel tube structure, fabric covered. Tailplane incidence adjustable by screwjack.

LANDING GEAR: Non-retractable tailwheel type. Coil spring shock-absorption. Main units fitted with Cleveland wheels and tyres, size 6·00-6. Cleveland brakes. Tyre pressure 2·76 bars (40 lb/sq in). Steerable tailwheel. Wheel fairings optional.

POWER PLANT: One 112 kW (150 hp) Avco Lycoming O-320 flat-four engine. Sensenich two-blade fixed-pitch metal propeller. Glassfibre fuel tank aft of firewall, capacity 90·5 litres (24 US gallons). Oil capacity 7·5 litres (2 US gallons).

ACCOMMODATION: Two persons in tandem in open cockpits. Small baggage compartment aft of rear seat.

EQUIPMENT: Nova-Tech TR-102 radio.

DIMENSIONS, EXTERNAL:

Wing span	7·55 m (24 ft 9 in)
Wing chord, constant	1·52 m (5 ft 0 in)
Wing area, gross	11·21 m² (120·7 sq ft)
Wing aspect ratio	5
Length overall	5·56 m (18 ft 3 in)
Height overall	1·57 m (5 ft 2 in)
Tailplane span	2·26 m (7 ft 5 in)
Wheel track	1·57 m (5 ft 2 in)

WEIGHTS:

Weight empty	408 kg (900 lb)
Max T-O weight	680 kg (1,500 lb)

PERFORMANCE (at max T-O weight):

Max level speed at S/L	130 knots (241 km/h; 150 mph)
Cruising speed	109-117 knots (201-217 km/h; 125-135 mph)
Stalling speed	48 knots (89 km/h; 55 mph)
Max rate of climb at S/L	732 m (2,400 ft)/min
T-O and landing run	61 m (200 ft)
Endurance with max fuel	3 h

SPORT AIR CRAFT

SPORT AIR CRAFT INTERNATIONAL

PO Box 1, Hillsboro, Oregon 97123

Telephone: (503) 985 7612

PROPRIETOR: Bill W. Johnson

Known originally as Chris Tena Aircraft Association, this concern is responsible for a lightweight all-metal single-seat sporting aircraft known as the Mini Coupe. Design originated in June 1968, and construction of the first prototype began in July 1970. This aircraft made its first flight in September 1971 and FAA certification in the Experimental category was awarded on 2 June 1972. Kits of components and materials for two alternative versions, less engine, are available to amateur constructors. Building of the Mini Coupe has now been approved also in Australia and Switzerland, where aircraft are being constructed.

SPORT AIR CRAFT MINI COUPE

Glassfibre wingtips have been added to the Mini Coupe since its early testing, to increase the wing area by 0·47 m² (5·1 sq ft). This improves the glide ratio, makes the aircraft more stable during banks, and has reduced stalling speed. At least 270 sets of plans of the Mini Coupe had been sold by early 1981, when about 60 aircraft were known to be flying.

There are two versions, as follows:

Mini Coupe. Basic model, similar to prototype, with tricycle landing gear and twin-fin tail unit. Most amateur-built Mini Coupes are of this type.

Mini Coupe TD. Generally similar to basic model, but with tailwheel landing gear and single fin and rudder. First TD was completed by Mr E. Argence of New Orleans, Louisiana, on 1 June 1977.

TYPE: Single-seat lightweight sporting aircraft.

WINGS: Cantilever low-wing monoplane. Wing section modified Clark Y. Thickness/chord ratio 7%. Incidence 0°. Conventional metal stressed-skin structure of constant chord, with glassfibre wingtips. Plain all-metal ailerons. No flaps. No trim tabs.

FUSELAGE: All-metal semi-monocoque structure.

TAIL UNIT: Cantilever all-metal structure with twin endplate fins and rudders of constant chord. Fixed-incidence tailplane. Manually controlled trim tab in centre of elevator. Alternative single fin and rudder configuration available on Mini Coupe TD.

LANDING GEAR: Non-retractable tricycle type. Shock-absorption on all units depends on use of oversize tyres. Main-wheel tyres size 14 × 6·00-6, pressure 0·55 bars (8 lb/sq in). Nosewheel tyre size 12 × 6·00-5, pressure 0·34 bars (5 lb/sq in). Azuza drum and band brakes. Alternative non-retractable tailwheel type, with faired main wheels, on Mini Coupe TD.

POWER PLANT: One 48·5 kW (65 hp) modified Volkswagen 1,600 cc motorcar engine, driving a Reese-Shores two-blade fixed-pitch wooden propeller. (Two aircraft fitted with 48·5 kW; 65 hp Continental.) One metal fuel tank in fuselage, immediately aft of firewall, capacity 49 litres (13 US gallons). Refuelling point on fuselage upper surface, forward of windscreen.

ACCOMMODATION: Single seat for pilot in open cockpit. Transparent cockpit canopy optional. Baggage compartment aft of headrest, volume 0·085 m³ (3 cu ft).

SYSTEM: Provision for electrical system, for engine starter and radio.

DIMENSIONS, EXTERNAL:

Wing span	7·32 m (24 ft 0 in)
Wing chord, constant	1·07 m (3 ft 6 in)
Wing area, gross	7·76 m² (83·5 sq ft)
Length overall	4·98 m (16 ft 4 in)
Height overall	1·80 m (5 ft 11 in)
Tailplane span	1·83 m (6 ft 0 in)
Wheel track	1·83 m (6 ft 0 in)
Wheelbase (tricycle landing gear)	0·97 m (3 ft 2 in)
Propeller diameter	1·37 m (4 ft 6 in)

WEIGHTS (A: tricycle landing gear; B: tailwheel landing gear):

Weight empty: A	225 kg (497 lb)
B	241 kg (532 lb)
Max T-O weight: A	385 kg (850 lb)
B	408 kg (900 lb)

PERFORMANCE (at max T-O weight with 1,600 cc engine; A and B as above):

Never-exceed speed:	
A, B	125 knots (231 km/h; 145 mph)
Max level speed at 610 m (2,000 ft):	
A	91 knots (169 km/h; 105 mph)
B	118 knots (219 km/h; 136 mph)
Max cruising speed at 610 m (2,000 ft):	
A	78 knots (145 km/h; 90 mph)
B	100 knots (185 km/h; 115 mph)
Stalling speed, power off:	
A	42 knots (78 km/h; 48 mph)
B	50·5 knots (94 km/h; 58 mph)
Stalling speed, power on:	
A, B	38 knots (70 km/h; 43 mph)
Max rate of climb at S/L: A	229 m (750 ft)/min
B	305 m (1,000 ft)/min
Service ceiling: A, B	3,810 m (12,500 ft)
T-O run: A, B	122 m (400 ft)
T-O to 15 m (50 ft): A, B	274 m (900 ft)
Landing from 15 m (50 ft): A, B	274 m (900 ft)
Landing run: A, B	152 m (500 ft)
Range with max fuel, 20 min reserves:	
A, B	260 nm (482 km; 300 miles)

SPRATT

SPRATT AND COMPANY INC

PO Box 351, Media, Pennsylvania 19063

Mr George G. Spratt, formerly a design engineer with The Boeing Company and Consolidated Vultee (now Convair Division of General Dynamics), has completed more than 30 years' work on developing a two-piece movable-wing control system, which he claims provides improved safety factors compared with the conventional aileron, elevator and rudder control system.

While he was with Consolidated Vultee, Mr Spratt designed a roadable aircraft which featured an earlier version of his wing control system, but this did not enter production. Since that time Mr Spratt has concentrated on perfecting his concept as a private venture.

To flight test the movable-wing control system, Mr Spratt built a lightweight experimental flying-boat (N910Z) constructed almost entirely of moulded plastics;

Model of the Spencer Amphibian Air Car Jr *(Bob Steele)*

Spezio Tuholer two-seat sporting aircraft

Spezio Special two-seat sporting aircraft *(J. M. G. Gradidge)*

Sport Air Craft Mini Coupe single-seat sporting aircraft

Spratt Model 107 two-seat movable-wing flying-boat

Statler Firefly homebuilt two-seat light monoplane *(Don Dwiggins)*

details of this can be found in the 1974-75 *Jane's*. Two further designs have followed, and plans of the Model 107 movable-wing flying-boat are available to amateur constructors. Details of the Model 105 landplane can be found in the 1979-80 *Jane's*.

SPRATT MODEL 107

Following construction and flight testing of his first experimental flying-boat, Mr Spratt designed and constructed a more advanced prototype known as the Model 107 (N2236) for public demonstration. Dimensions and weights are essentially the same as those of the test vehicle, but construction has been simplified to facilitate fabrication. The Mercury 800 modified outboard marine engine of the Model 107 is of slightly increased capacity, and produces greater horsepower with better fuel economy as a result of improved combustion chamber and inlet port design.

Mr Spratt claims that the Model 107 will neither stall nor spin, and displays 75% less reaction to turbulence than a conventional design.

Plans of this aircraft are available to amateur constructors, and by 1981 more than 80 sets had been sold. The first amateur-built Model 107 flew in the Autumn of 1975.

In Spring 1979 flight testing began of a new hull made from Airex PVC foam, with Kevlar skin, coated with epoxy resin. This is expected to prove more durable than the original sandwich structure of glassfibre and polyurethane foam. Twenty years' experience with the latter has shown that the foam deteriorates progressively, so that cells at the interface with the skin break down, destroying the integrity of the composite. Also, the weight of N2236 increased by 23·5 kg (52 lb) during the life of the original hull, due to water absorbed by the foam. Further deterioration resulted from abrasion of the glass fibre at junctures with metal fittings where vibration is severe.

The Model 107 prototype is now powered by a Mercury 850 modified outboard marine engine. The following data apply to the original version:

TYPE: Two-seat lightweight flying-boat.

WINGS: Pivoted controllable parasol wings, with inverted V bracing struts each side. Wing section NACA 23112. No dihedral. No sweepback. Reinforced plastics structure. No ailerons, flaps or trim tabs. Flying controls so arranged that the wings are allowed to move freely and collectively in incidence, while their incidence is controlled differentially by a steering wheel. The wings' angle of attack can be adjusted by a separate control.

HULL: Structure of polyurethane foam with reinforced plastics skin. Small water rudder interconnected with the control wheel.

TAIL UNIT: Butterfly-type tail unit, with no movable surfaces, constructed of reinforced plastics.

POWER PLANT: One 59·7 kW (80 hp) Mercury 800 modified outboard two-stroke marine engine, driving a two-blade plastics pusher propeller through an extended drive shaft, which locates the propeller between the butterfly tail surfaces. The pitch of the propeller, which is of Mr Spratt's design, is adjustable on the ground. Outboard engine type of fuel tank.

ACCOMMODATION: Two persons side by side, in open cockpit.

DIMENSIONS, EXTERNAL:	
Wing span	7·32 m (24 ft 0 in)
Wing chord, constant	1·22 m (4 ft 0 in)
Wing area, gross	8·92 m² (96 sq ft)
Wing aspect ratio	6
Length overall	5·18 m (17 ft 0 in)
Height overall	1·52 m (5 ft 0 in)
Propeller diameter	1·52 m (5 ft 0 in)
WEIGHTS:	
Weight empty	226 kg (500 lb)
Max T-O weight	453 kg (1,000 lb)

STATLER

WILLIAM H. STATLER

9300 Encino Avenue, Northridge, California 91325
Telephone: (213) 886 1854

While employed as an aeronautical engineer with Lockheed, from 1940 to 1972, Mr Statler built his first two homebuilt aircraft, Formula One racers known as *Ginny* and *Skeeter*. Following his retirement from Lockheed, he began work, in January 1973, on a two-seat homebuilt aircraft capable of limited aerobatic flying. Construction of what became the Firefly started in June 1973, with the help of Byrl Robinson and Claude Fiske. The Firefly was rolled out in May 1976, when taxying trials began. The first flight, piloted by Al Foss, took place on 8 October 1976.

Plans and a construction manual are available to amateur builders and at least 65 sets had been sold by early 1981. Meantime, work has begun on an Unlimited Class racing aircraft, of which no details are currently available.

STATLER FIREFLY

TYPE: Two-seat light monoplane.

WINGS: Cantilever low-wing monoplane. Wing section NACA 2412. Dihedral 5°. Incidence 2°. No sweepback. All-metal construction, with aluminium ribs and skins, flush-riveted. Glassfibre wingtips. All-metal ailerons and electrically-actuated plain trailing-edge flaps. Ground-adjustable tab on starboard aileron. Wings detachable for transportation.

FUSELAGE: All-metal semi-monocoque structure, with flush-riveted aluminium skin. Glassfibre engine cowl.

TAIL UNIT: Cantilever all-metal structure, with swept vertical surfaces. Fixed tailplane set at −3° incidence. Trim tab on elevator, controllable from either cockpit.

LANDING GEAR: Non-retractable tricycle type. Cantilever spring main units made from a single piece of aluminium. Ercoupe nosewheel, on oleo strut. Mainwheel tyres size 12 × 6·00-6, pressure 1·93 bars (28 lb/sq in). Nosewheel tyre size 10 × 5·00-5, pressure 1·93 bars (28 lb/sq in). Gerdes hydraulic brakes. Glassfibre wheel fairings.

POWER PLANT: One 63·5 kW (85 hp) Continental C85-12F flat-four engine, driving two-blade Wayne's Woodcraft fixed-pitch 62 × 54 propeller. Two interconnected fuel tanks in fuselage, total capacity 68 litres (18 US gallons). Refuelling point on forward tank. Oil capacity 5·7 litres (1·5 US gallons).

ACCOMMODATION: Two seats in tandem, under GB bubble canopy which is hinged on starboard side. Baggage compartment, capacity 22·6 kg (60 lb), in turtleback aft of rear cockpit.

SYSTEMS: Hydraulic system for brakes. Electrical system includes engine-driven generator and 12V 35Ah battery for engine starting, wing flap actuation and instruments.

AVIONICS: Bayside portable transceiver.

DIMENSIONS, EXTERNAL:

Wing span	5·84 m (19 ft 2 in)
Wing chord, constant	1·27 m (4 ft 2 in)
Wing area, gross	7·43 m² (80 sq ft)
Wing aspect ratio	4·6
Length overall	5·77 m (18 ft 11 in)
Height overall	2·15 m (7 ft 0½ in)
Tailplane span	2·82 m (9 ft 3 in)
Wheel track	1·37 m (4 ft 6 in)
Wheelbase	1·50 m (4 ft 11 in)
Propeller diameter	1·57 m (5 ft 2 in)

WEIGHTS:

Weight empty	379 kg (835 lb)
Max T-O weight	590 kg (1,300 lb)

PERFORMANCE:

Never-exceed speed	208 knots (386 km/h; 240 mph)
Max cruising speed at S/L	139 knots (257 km/h; 160 mph)
Econ cruising speed at 1,980 m (6,500 ft)	124 knots (230 km/h; 143 mph)
Stalling speed, flaps up	55 knots (102 km/h; 63 mph)
Max rate of climb at S/L	244 m (800 ft)/min
Service ceiling	4,875 m (16,000 ft)
T-O run	283 m (930 ft)
T-O to 15 m (50 ft)	354 m (1,160 ft)
Landing from 15 m (50 ft)	488 m (1,600 ft)
Landing run	152 m (500 ft)
Range with max fuel, no reserves	347 nm (643 km; 400 miles)

STEEN

STEEN AERO LAB INC

15623 De Gaulle Circle, Brighton, Colorado 80601
Telephone: (303) 659 7182

Mr Lamar Steen, an aerospace teacher in a Denver, Colorado, high school, designed a two-seat fully-aerobatic biplane named Skybolt which was built as a class project in the school. Simplicity of construction was a primary aim of the design, begun in June 1968, and it is stressed to +12 and −10*g*. Construction began on 19 August 1969, and the first flight was made in October 1970. The Skybolt received an EAA award for 'Best School Project'. Plans are available to amateur constructors, together with fuselage and wing kits; more than 2,000 sets of plans have been sold and many Skybolts are flying.

STEEN SKYBOLT

The following description applies to the prototype with 134 kW (180 hp) Avco Lycoming engine, built under Mr Steen's supervision:

TYPE: Two-seat aerobatic biplane.

WINGS: Braced biplane with single interplane strut each side. N-type centre-section struts. Streamline-section landing and flying wires. Wing sections: upper wing NACA 63_2A015, lower wing NACA 0012. Dihedral 0° on upper wing, 2° 30′ on lower wings. Incidence (both) 1° 30′. Sweepback 6° on upper wing only. Wooden two-spar structures with spruce spars, built-up ribs and fabric covering. Fabric-covered Frise-type ailerons on upper and lower wings. Cutout in centre-section trailing-edge of upper wing.

FUSELAGE: Welded structure of 4130 chrome-molybdenum steel tube, with fabric covering.

TAIL UNIT: Wire-braced welded structure of 4130 chrome-molybdenum steel tube, with fabric covering. Adjustable trim tab in port elevator.

LANDING GEAR: Non-retractable main wheels and tailwheel. Two side Vs and half axles hinged to fuselage structure. Shock-absorption by rubber bungee. Cleveland wheels with tyres size 6·00-6, pressure 1·72 bars (25 lb/sq in). Cleveland hydraulic disc brakes. Glassfibre fairings for main wheels.

POWER PLANT: One 134 kW (180 hp) Avco Lycoming HO-360-B1B flat-four engine, driving a McCauley two-blade fixed-pitch propeller with spinner. Provision for alternative engines of 93-194 kW (125-260 hp). Fuselage fuel tank, immediately aft of firewall, capacity 110 litres (29 US gallons). Optional tank of 37·8 litres (10 US gallons) capacity can be installed in centre-section of upper wing. Total optional fuel capacity 151·2 litres (40 US gallons). Refuelling points on fuselage upper surface, forward of windscreen, and on top surface of upper wing. Oil capacity 7·5 litres (2 US gallons).

ACCOMMODATION: Two seats in open cockpits. Provision for canopy over rear cockpit. Space for 13·6 kg (30 lb) baggage aft of rear seat.

SYSTEM: Hydraulic system for brakes only.

AVIONICS: Battery-powered Alpha 200 nav/com transceiver.

DIMENSIONS, EXTERNAL:

Wing span, upper	7·32 m (24 ft 0 in)
Wing span, lower	7·01 m (23 ft 0 in)
Wing chord, constant (both)	1·07 m (3 ft 6 in)
Wing area, gross	14·2 m² (152·7 sq ft)
Length overall	5·79 m (19 ft 0 in)
Height overall	2·13 m (7 ft 0 in)
Propeller diameter	1·88 m (6 ft 2 in)

WEIGHTS:

Weight empty	490 kg (1,080 lb)
Max T-O weight	748 kg (1,650 lb)

PERFORMANCE (at max T-O weight):

Max level speed	126 knots (233 km/h; 145 mph)
Cruising speed	113 knots (209 km/h; 130 mph)
Landing speed	43 knots (80·5 km/h; 50 mph)
Max rate of climb at S/L	762 m (2,500 ft)/min
Service ceiling	5,500 m (18,000 ft)
T-O run	122 m (400 ft)
Range with max fuel	390 nm (720 km; 450 miles)

STEPHENS

STEPHENS AIRCRAFT

ADDRESS FOR PLANS: Gerry Zimmerman, 8563 West Sixty-Eighth Place, Arvada, Colorado 80004

Mr C. L. Stephens designed a single-seat aerobatic monoplane specifically for homebuilders who wish to own an aircraft for competitive aerobatics. The prototype, designated Model A, was designed for Margaret Ritchie, US National Women's Aerobatic Champion in 1966, and the second aircraft, the Model B, for Dean S. Engelhardt of Garden Grove, California.

Stressed to +12*g* and −11*g*, it was the first US aircraft known to be designed around the Aresti Aerocriptografic System for competitive aerobatics. All control surfaces are fully static-balanced and the entire aircraft comes very close to being aerodynamically symmetrical. Design of the **Model A** started in July 1966 and construction of the prototype began a month later. First flight of this version was made on 27 July 1967, and of the **Model B**, with wings and ailerons of increased area and reduced fuel tankage, on 9 July 1969. Plans of the Stephens Akro are available to amateur constructors.

STEPHENS AKRO

TYPE: Single-seat sporting monoplane.

WINGS: Cantilever mid-wing monoplane. Wing section NACA 23012. No dihedral, incidence or sweepback. All-wood two-spar structure. One-piece wing, with solid spar passing through fuselage and positioned by means of removable top longeron sections. Rear spar in two pieces. No internal wires or compression struts. Wing covered with mahogany skin. Plain ailerons have a 4130 steel spar, and spruce ribs and trailing-edge, and are fabric-covered. Ground-adjustable trim tabs on ailerons, which are statically balanced. No flaps.

FUSELAGE: Welded 4130 steel tube structure, mostly of 0·75 in outside diameter tubing, with Ceconite covering.

TAIL UNIT: Wire-braced welded 4130 steel tube structure with swept surfaces, fabric-covered. Tailplane has variable incidence. Ground-adjustable trim tab on rudder; controllable trim tab in elevator. All control surfaces statically balanced.

LANDING GEAR: Non-retractable tailwheel type. Cantilever spring steel main gear. Goodyear main wheels and tyres size 5·00-5, pressure 1·93 bars (28 lb/sq in). Cleveland disc brakes. Maule steerable tailwheel. Glassfibre fairings on main wheels.

POWER PLANT: One 134 kW (180 hp) Avco Lycoming AIO-360-A1A flat-four engine, driving a Sensenich Type 7660 two-blade fixed-pitch metal propeller. Model A has fuel system for prolonged inverted flight, Model B has both fuel and oil system so modified. Model B can also have optional constant-speed propeller. Fuel tank in fuselage, forward of instrument panel. Model A has fuel capacity of 121 litres (32 US gallons), Model B has capacity of 102 litres (27 US gallons). Refuelling point on top of fuselage, forward of windscreen. Oil capacity 7·5 litres (2 US gallons).

ACCOMMODATION: Single seat for pilot under rearward-sliding bubble canopy. Large window in underfuselage, forward of control column. Model B has, in addition, a quarter window in each side of the fuselage, beneath the wings. Forced-air ventilation.

SYSTEM: Hydraulic system for brakes only.

AVIONICS: Battery-operated Bayside transceiver.

DIMENSIONS, EXTERNAL:

Wing span	7·47 m (24 ft 6 in)
Wing chord at root	1·60 m (5 ft 3 in)
Wing chord at tip: Model A	0·76 m (2 ft 6 in)
Model B	0·91 m (3 ft 0 in)
Wing area, gross: Model A	8·73 m² (94 sq ft)
Model B	9·29 m² (100 sq ft)
Length overall	5·82 m (19 ft 1 in)
Height overall	1·73 m (5 ft 8 in)
Tailplane span	2·44 m (8 ft 0 in)
Wheel track	1·37 m (4 ft 6 in)
Propeller diameter	1·93 m (6 ft 4 in)

WEIGHTS:

Weight empty: Model A	385 kg (850 lb)
Model B	431 kg (950 lb)
Max T-O weight: Model A	544 kg (1,200 lb)
Model B	589 kg (1,300 lb)

PERFORMANCE (at 544 kg; 1,200 lb T-O weight):

Never-exceed speed	191 knots (354 km/h; 220 mph)
Max level speed at 610 m (2,000 ft)	148 knots (274 km/h; 170 mph)
Max cruising speed at 610 m (2,000 ft)	139 knots (257 km/h; 160 mph)
Econ cruising speed at 610 m (2,000 ft)	109 knots (201 km/h; 125 mph)
Stalling speed	48 knots (89 km/h; 55 mph)
Max rate of climb at S/L	1,220 m (4,000 ft)/min
Service ceiling	6,705 m (22,000 ft)
T-O run	61 m (200 ft)
T-O to 15 m (50 ft)	122 m (400 ft)
Landing from 15 m (50 ft)	457 m (1,500 ft)
Landing run	183 m (600 ft)
Range with 121 litres (32 US gallons) of fuel	303 nm (563 km; 350 miles)

STEWART

STEWART AIRCRAFT CORPORATION

11420 State Route 165, Salem, Ohio 44460
Telephone: (216) 332 0865

Mr Donald Stewart formed this company to market plans of a simple single-seat light aircraft named the Headwind, of which he designed and built a prototype. During 1969 he designed a new wing for the Headwind and this is an integral part of the plans available to homebuilders. A two-seat version was under construction, but work on this has been suspended to allow full concentration on development of a new foot-launched microlight aircraft named the Puffin (see Microlights section).

Work on the original single-seat Headwind was followed by design and construction of the JD₂FF Foo Fighter, which is described separately.

STEWART JD₁ HW 1·7 HEADWIND

Built in only five months, the prototype Headwind flew for the first time on 28 March 1962, and has since undergone considerable refinement. Plans are available to amateur constructors, who can build the aircraft with either an open or enclosed cockpit. The wings can be removed or fitted in about 20 minutes by two people.

The power plant of the prototype Headwind has a belt-driven propeller reduction drive designed by Mr Stewart. Given the name Maximizer, this unit was put into production in the Spring of 1972 and is available to amateur constructors for use with Volkswagen power plants.

Several thousand sets of plans for the Headwind have been sold. About 250 aircraft are believed to be under construction, with approximately 30 already flying, including examples in Mexico and South Africa.

TYPE: Single-seat light aircraft.

WINGS: Strut-braced high-wing monoplane, with streamline-section V bracing strut each side. Wing section NACA 4412. Dihedral 2°. Incidence 2°. Two spruce spars, steel tube compression members, drag and anti-drag wires, plywood ribs, fabric covering. Frise-type ailerons of similar construction to wings. No flaps.

FUSELAGE: Welded steel tube structure, fabric-covered.

TAIL UNIT: Braced steel tube structure, fabric-covered. Ground-adjustable tailplane incidence. Fixed tabs on rudder and starboard elevator.

LANDING GEAR: Non-retractable tailwheel type. Shock-absorption by low-pressure tyres. Hayes main wheels with Goodyear tyres size 8·00-4. Tyre pressure 0·83 bars (12 lb/sq in). Alternatively, rubber-in-compression shock-struts, with Cleveland or Goodyear wheels and tyres size 6·00-6, tyre pressure 1·24 bars (18 lb/sq in). No brakes. Steerable tailwheel.

POWER PLANT: One 27 kW (36 hp) modified Volkswagen 1,192 cc motorcar engine, driving a special Stewart/Kirk two-blade fixed-pitch propeller via a Stewart Maximizer belt-driven reduction unit, ratio 1·6 : 1. Engines weighing up to 84 kg (185 lb) can be utilised. Fuel tank aft of firewall, capacity 19 litres (5 US gallons). Oil capacity 2·4 litres (5 US pints).

ACCOMMODATION: Single seat in open or enclosed cockpit, with door on starboard side. Provision for up to 4·5 kg (10 lb) baggage in net, directly aft of cockpit.

DIMENSIONS, EXTERNAL:

Wing span	8·61 m (28 ft 3 in)

White Der Jäger D.IX built by Mr B. W. Salyer *(J. M. G. Gradidge)*

Wittman W-8 Tailwind (93 kW; 125 hp Avco Lycoming engine) built in South Africa by Mr Nick Turvey *(R. Kunert)*

WITTMAN

S. J. WITTMAN

Box 2672, Oshkosh, Wisconsin 54901

Famous as a racing pilot since 1926, Steve Wittman has designed and built a large number of different racing and touring aeroplanes at Winnebago County Airport, of which he became manager in 1931.

Most popular current Wittman design is the W-8 Tailwind side-by-side two-seat light aeroplane. The prototype was built in 1952-53 and proved so successful that sets of plans and prefabricated components were made available to amateur builders. By early 1981 there were more than 325 Model W-8 Tailwinds flying, including a number built in foreign countries, and more than 100 were known to be under construction.

WITTMAN TAILWIND MODEL W-8

Some Tailwinds have been built with tricycle landing gear, retractable main wheels and other design changes. The following data refer to the standard W-8 Tailwind built to Mr Wittman's plans:

Type: Two-seat cabin monoplane.

Wings: Braced high-wing monoplane. Wing section is a combination of NACA 4309 (upper surface) and NACA 0006 (lower surface). Thickness/chord ratio 11·5%. No dihedral. Incidence 1°. Wooden structure with plywood and fabric covering. Single bracing strut each side. Ailerons and flaps of steel and stainless steel construction.

Fuselage: Steel tube structure, fabric-covered.

Tail Unit: Cantilever structure of steel and stainless steel. Ground-adjustable trim tabs in control surfaces.

Landing Gear: Non-retractable tailwheel type. Spring steel cantilever main legs. Goodyear 15 × 5 main wheels and tyres, pressure 2·21 bars (32 lb/sq in). Goodyear brakes.

Power Plant: Normally one 67 kW (90 hp) Continental C90-12F flat-four engine, driving a Sensenich or Flottorp two-blade fixed-pitch wood propeller. Alternative engines are the 63·5 kW (85 hp) Continental C85, 74·5 kW (100 hp) Continental O-200, 86 kW (115 hp) Avco Lycoming O-235 or 104·5 kW (140 hp) Avco Lycoming O-290. One fuel tank of 94·5 litres (25 US gallons) capacity in fuselage. Oil capacity 1·85-2·8 litres (1-1½ US gallons).

Accommodation: Two seats side by side in enclosed cabin, with door on each side. Space for 27 kg (60 lb) baggage.

Dimensions, external:

Wing span	6·86 m (22 ft 6 in)
Wing chord, constant	1·22 m (4 ft 0 in)
Wing area, gross	8·36 m² (90 sq ft)
Wing aspect ratio	5·5
Length overall	5·87 m (19 ft 3 in)
Height overall	1·73 m (5 ft 8 in)
Tailplane span	2·03 m (6 ft 8 in)
Wheel track	1·65 m (5 ft 5 in)
Propeller diameter	1·63 m (5 ft 4 in)

Weights (74·5 kW; 100 hp Continental engine):

Weight empty	318 kg (700 lb)
Max T-O weight	590 kg (1,300 lb)

Performance (74·5 kW; 100 hp Continental engine at max T-O weight):

Never-exceed speed	160 knots (297 km/h; 185 mph)
Max level speed at S/L	143 knots (265 km/h; 165 mph)
Max cruising speed	139 knots (257 km/h; 160 mph)
Econ cruising speed	113 knots (209 km/h; 130 mph)
Stalling speed, flaps down	48 knots (89 km/h; 55 mph)
Max rate of climb at S/L	275 m (900 ft)/min
Service ceiling	4,876 m (16,000 ft)
T-O run	245 m (800 ft)
T-O to 15 m (50 ft)	405 m (1,325 ft)
Landing from 15 m (50 ft)	350 m (1,150 ft)
Landing run	183 m (600 ft)

Range with max payload at 3,050 m (10,000 ft), no reserves:
at 139 knots (257 km/h; 160 mph) 521 nm (965 km; 600 miles)
at 122 knots (225 km/h; 140 mph) 607 nm (1,125 km; 700 miles)

SAILPLANES

(For specification data, see tables at end of Section)

AUSTRALIA

BETTERIDGE
DAVID BETTERIDGE

c/o Box 390, Morphett Vale, South Australia 5162

Mr Betteridge has designed an unusual motor glider, which was test-flown by Mr Colin Scott at Aldinga, South Australia, in early 1979, and was expected to be put into production by Free Flight Gliders towards the end of that year. No further news has been received of this aircraft, of which an illustration and brief description appeared in the 1980-81 *Jane's*.

BUCHANAN
JOHN BUCHANAN

c/o Murraycraft Fibreglass, David Muir Street, Mackay, Queensland 4740

BUCHANAN RICOCHET

Mr Buchanan is building the prototype of this sailplane, which he describes as "just a pod big enough to fit the pilot, with a wing attached". The wing, constructed of carbonfibre and glassfibre, is of swept configuration, with a 15 metre span (49 ft 2½ in), narrow chord and very high aspect ratio. It is mid-mounted on the small, streamline fuselage pod, at the rear of which is a small sweptback fin; there are no horizontal tail surfaces. A flush-fitting one-piece cockpit transparency encloses the pilot. Fuselage length is 3·70 m (12 ft 1¾ in).

Mr Buchanan, who is a graduate aeronautical engineer from the Cranfield Institute of Technology in England and a former RAAF pilot, is building the Ricochet under a $A10,000 grant from the Australian government.

SUNDERLAND
GARY SUNDERLAND

1 Nicholas Grove, Heatherton, Victoria 3202
Telephone: 551 2564

SUNDERLAND MOBA 2C

In 1972 the Australian *Gliding* magazine, with the Australian branch of the Royal Aeronautical Society, conducted a competition for a 13 metre sailplane design to be built in Australia. The MOBA 2B was one of the two finalists selected before the contest was abandoned. The MOBA 2A was the original 15 metre span version; the MOBA 2C represents an improved version of both earlier designs.

Construction of the MOBA 2C began in 1975, and it flew for the first time on 12 December 1979.

Certification flight testing was completed in 1980, and the award of a C of A was expected in 1981.

TYPE: Single-seat 15 metre Standard Class sailplane; *g* limits +4; −1·5.

AIRFRAME: Cantilever high-wing monoplane, with T tail. Box-spar wings, of Wortmann FX-67-K-150 section, built in three pieces. Dihedral 1° at tips. Plywood ribs, with urethane foam infill; spar of pop-riveted aluminium alloy sheet. Epoxy-resin glasscloth skin. Aluminium alloy plain flaps on trailing-edges, serving also as airbrakes; GRP-covered wooden ailerons. Nose and central fuselage are riveted aluminium alloy box structures, with sliding nose of balsa/GRP sandwich; centre fairing of foam with GRP skin; tailboom is of riveted aluminium alloy sheet. Tail unit of wood, with foam infill and GRP skin. Elevator has spring trim. Manually retractable monowheel and non-retractable tailwheel, both from a Standard Libelle. Entire nose section, including canopy, slides forward to provide access to cockpit. No water ballast provision.

SUNDERLAND MOBA 3 VIRAGO

As a result of experience gained with the MOBA 2C, work has begun on the design of a 15 metre span FAI racing class sailplane to suit Australian soaring conditions. Structure will be similar to that of the MOBA 2C, with a full-span flap providing variable chord and wing area.

TODHUNTER
R. W. TODHUNTER

5 Leemon Street, Condell Park, New South Wales 2200

Mr Reg Todhunter, designer and builder of the Skycycle man-powered aircraft (see Homebuilts section of the 1979-80 *Jane's*), is developing a very lightweight powered sailplane known as the Blue Wren. He is being assisted by Mr Milton Lalas, a senior aeronautical engineer, and a small group of friends with considerable aviation experience, led by Mr Bob Letson.

TODHUNTER T-5 BLUE WREN

Basic design and stressing of the Blue Wren had been completed by early 1980, and construction began in January of that year. It was expected that the aircraft would be completed by late 1981. In February of that year the fuselage pod skin moulds had been completed, the welded steel tube fuselage frame was approx half completed, and the tailboom and tail unit spars were under construction. The engine, in parallel development, was due to be run for the first time in March 1981. The Blue Wren has been designed to comply initially with the Australian Dept of Air Transport ANO Section 95-10.

TYPE: Single-seat lightweight powered sailplane; *g* limits +4·0/−1·5 (+6·0/−2·25 ultimate).

AIRFRAME: Cantilever tapered high-wing monoplane, with Wortmann FX-05-H-126 wing section. Dihedral 2° 30′. Incidence 2° 30′. Sweepforward 2° 6′ at quarter-chord. Wings are constructed of aluminium alloy, foam and glassfibre, with trailing-edge flap/spoilers and ailerons, and are detachable for transportation and storage. Turned-down wingtips. Pod-type fuselage of glassfibre over a steel tube frame. Tapered 2024-T3 aluminium alloy tailboom faired to bottom of pod. Engine, mounted above boom, is faired in behind streamlined cockpit canopy and drives a pusher propeller centred behind wing trailing-edge. Propeller can be stopped in horizontal position to reduce drag in soaring mode. Cantilever fixed-incidence T tailplane, with elevator, of plastics foam with plywood skin. Elevator has spring trim. Non-retractable unsprung monowheel (tyre size 10 × 3 in), with drum brake; tailskid. Single 40° reclining seat under one-piece single-curvature canopy.

POWER PLANT: One 11 kW (15 hp) TS 162 two-stroke flat-twin engine, driving (on prototype) a 0·91 m (3 ft) diameter fixed-pitch wooden pusher propeller via a reduction drive. Single fuselage fuel tank, capacity 7 litres (1·5 Imp gallons).

AUSTRIA

BRDITSCHKA
H. W. BRDITSCHKA OHG

A-4053 Haid, Dr Schärfstrasse 42, Postfach 12
Telephone: 07229/8355
Telex: 21909
DIRECTOR: Heinz W. Brditschka

BRDITSCHKA HB-21

The HB-21, which first flew on 22 March 1974, is essentially an enlarged version of the HB-3 (1977-78 *Jane's*), with increased-span wings, tandem seating for two persons, and a Volkswagen engine. Four prototypes were built. Certification was granted in February 1978, and 15 HB-21s had been built by early 1980.

TYPE: Two-seat motor glider; *g* limits +5·3; −2·7.

AIRFRAME: Fabric-covered wooden wings, with Wortmann sections (FX-61-184 at root, FX-60-126 at tip). Dihedral 2°. Incidence 3°. All-wood ailerons and upper-surface spoilers. Steel tube fuselage with glassfibre covering. Fixed-incidence all-wood tailplane. Non-retractable tricycle landing gear, with Tost mechanical main-wheel brakes.

POWER PLANT: One 35·75 kW (48 hp) VW-Westermayer 1600G flat-four piston engine, driving a Hoffmann HO 14-175 117 LD two-blade fixed-pitch pusher propeller. Aluminium fuel tank in wing, capacity 54 litres (11·9 Imp gallons). Fuel is 100LL avgas.

BRDITSCHKA HB-22

A much-redesigned development of the HB-21, the HB-22 has a more streamline fuselage, cantilever T tail and retractable landing gear. It is of GfK construction, and is powered by an 82 kW (110 hp) Renault four-cylinder in-line engine. For other details see table.

First flight was expected before the end of 1981.

BRAZIL

CTA
CENTRO TÉCNICO AEROESPACIAL

São José dos Campos, São Paulo State

IPD (Instituto de Pesquisas e Desenvolvimento)/PAR (Divisão de Aeronaves)

The IPD is the aeronautical research and development institute of the CTA (see Aircraft section). The PAR is its Aircraft Division, having responsibility for flight testing, evaluating, certificating and developing new aircraft projects. One of these is the Urubu two-seat all-metal training sailplane.

IPD/PAR PE-80367 URUBU

The PE-80367 programme is intended to develop a modern training sailplane which, after testing and certification, is to be handed over to the Brazilian aircraft industry for series production, as a replacement for the Czechoslovak Blanik sailplanes used at present. One static test prototype, and one (PP-ZUR) for flight test and certification, have been built.

TYPE: Tandem two-seat training sailplane; *g* limits: semi-aerobatic, +6·2/−3·4; fully aerobatic, +6·5/−3·5.

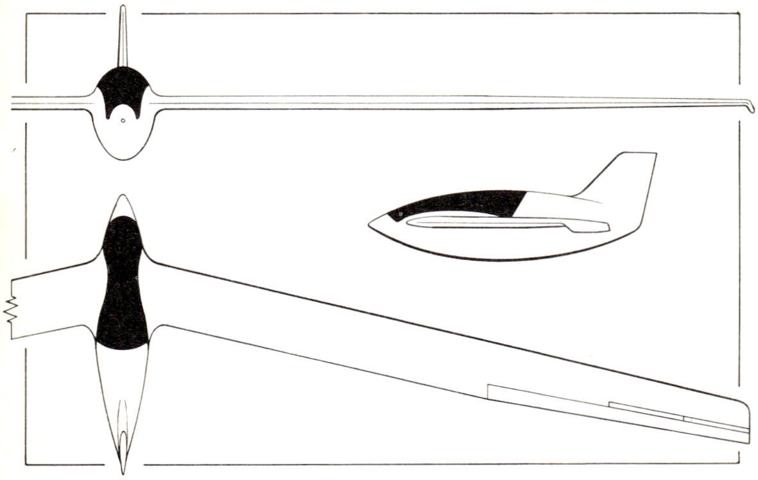

Buchanan Ricochet single-seat homebuilt sailplane *(Michael A. Badrocke)*

Sunderland MOBA 2C single-seat sailplane, showing nose section slid forward

AIRFRAME: Cantilever shoulder-wing monoplane. Wing section NACA 63_2A-615 at root, NACA 63_1A-412 at tip. Dihedral 3° on lower surface. Sweepforward 1° 51′ at quarter-chord. Incidence 3° at root. All-metal two-spar wings, with light alloy airbrakes in upper and lower surfaces, and all-metal ailerons. All-metal semi-monocoque oval-section fuselage. All-metal tail unit, with fixed-incidence tailplane; controllable trim tab in each elevator. Non-retractable rubber-sprung monowheel, with mechanical brake; non-retractable tailwheel.

EEUFMG (CEA)
ESCOLA DE ENGENHARIA DA UNIVERSIDADE FEDERAL DE MINAS GERAIS (Centro de Estudos Aeronáuticos)

Rua Espirito Santo 35, 30.000 Belo Horizonte, Minas Gerais State
Telephone: 222 4011
HEAD OF CEA: Prof Cláudio Pinto de Barros

CB-2/B MINUANO

The Aeronautical Research Centre of the Engineering School at Minas Gerais Federal University began the design of this single-seat sailplane in 1969. Its designation indicates that it is the second design by Professor Cláudio Barros. The CB-2 prototype (PP-ZPZ) flew for the first time on 20 December 1975 and was described in the 1977-78 *Jane's*.

A further four Minuanos were ordered, and the second aircraft was scheduled for completion by the end of 1979, though no information has been received from EEUFMG since that time. The second aircraft was intended to incorporate several modifications, and is designated CB-2/B; the description applies to this version. The Minuano is named after a strong, cold wind in southern Brazil.

TYPE: Single-seat high-performance sailplane; *g* limits (at safety factor of 1·72): normal, +5·3/−3·0; ultimate, +9·14/−5·19.

AIRFRAME: High-wing monoplane, with Wortmann FX-67-K-170/17 (root) and FX-60-126 (tip) wing sections. Single aluminium alloy main spar; wing skin of plywood/glassfibre honeycomb sandwich. Flaps and ailerons, similar except for wooden spars, are interconnected; flaps can be used also as airbrakes. Provision for 80 kg (176 lb) of water ballast. All-wood semi-monocoque fuselage. Plywood-covered tail surfaces, stiffened with foam plastics. All-moving tailplane, with trim tab in each half. Retractable unsprung monowheel (tyre pressure 2·45 bars; 35·5 lb/sq in), with internal shoe brake, and sprung tailskid.

IPE
INDÚSTRIA PARANAENSE DE ESTRUTURAS

J. Durski 357, 80.000 Curitiba, Paraná State
MANAGER: Eng J. C. Boscardin

IPE was in 1980 continuing to develop a two-seat training sailplane known as the IPE 02 Nhapecam. A licence has also been acquired for the construction of three examples of the Swiss Neukom Elfe 4, with wings and part of the fuselage of sandwich construction.

IPE 02 NHAPECAM

The general appearance of this aircraft can be seen from the accompanying illustration. The prototype (PP-ZQL) flew for the first time on 24 May 1979, and had accumulated more than 50 hours' flying by mid-February 1980.

Two more prototypes were then under construction, and an order for at least 30, for Brazilian flying clubs, was awaiting confirmation.

TYPE: Tandem two-seat training sailplane.

AIRFRAME: Shoulder-wing monoplane with Scheibe Spatz (modified) wing section. Ailerons and upper/lower surface spoilers. Fabric-covered elevators and rudder. Non-retractable monowheel, in streamline fairing, and tailwheel.

RIO CLARO
AÉRO CLUBE DE RIO CLARO

Rua Cinco 1152, 13.500 Rio Claro (Caixa Postal 147), São Paulo State

The Rio Claro aero club is building, under the direction of Eng Sylvio de Oliveira, an all-wood sporting glider known as the **Araponga.** General appearance can be seen in the accompanying three-view drawing.

SIDOU
ENG ANTONIO MENEZES SIDOU

Rua Luciana de Abreu 184, Moinhos de Vento, 90.000 Porto Alegre, Rio Grande do Sul State

SIDOU JOÃO GRANDE (STORK)

Eng Sidou is developing the João Grande at Passo Fundo, Rio Grande do Sul, currently with GRP construction instead of the original wooden-construction version. For general appearance, see accompanying three-view drawing.

TYPE: Tandem two-seat Open Class and advanced training sailplane (OSTIV aerobatic category A); *g* limits +9·75; −6·00 (ultimate).

AIRFRAME: Cantilever mid-wing monoplane, with cambered wingtips and T tail. Wortmann wing sections: FX-61-184 (modified) at root, FX-60-126 (modified) at tip. Dihedral 3° 36′. GRP structure, with wide-span ailerons and upper/lower surface Schempp-Hirth airbrakes. All wing and tail control surfaces statically and dynamically balanced. GRP semi-monocoque fuselage. Jettisonable cockpit canopy. Retractable rubber-sprung monowheel, with mechanical brake, and tail bumper.

CANADA

MARSDEN
PROF DAVID J. MARSDEN

c/o Dept of Mechanical Engineering, The University of Alberta, Edmonton, Alberta T6G 2G8
Telephone: (403) 432 3705

Prof Marsden of the University of Alberta acquired from Operation Sigma Ltd the **Sigma** variable-geometry sailplane last described in the 1974-75 *Jane's*; it now carries the Canadian registration C-GVJW-X. He replaced the former area-increasing flap system by full-span slotted flaps similar to those used on the Gemini, a two-seat sailplane of his own design. Modification work was carried out at the Cranfield Institute of Technology in the UK. After 16 hours of flight testing in 1979, handling and performance were considered to be very satisfactory, and a further 59 hours were flown in 1980. These included entry in the Canadian national competition (gaining 3rd place in the Open Class), and some performance flight testing. Results of the latter indicated a climbing and gliding performance equivalent to that of a Nimbus 2 with 100 kg (220 lb) of water ballast; this was confirmed by comparison flights during the Canadian nationals.

MARSDEN GEMINI

This successful variable-geometry sailplane flew for the first time in October 1973, and was fully described and illustrated in the 1980-81 and previous editions of *Jane's*.

The Gemini is still being flown regularly, but no further development is envisaged at present.

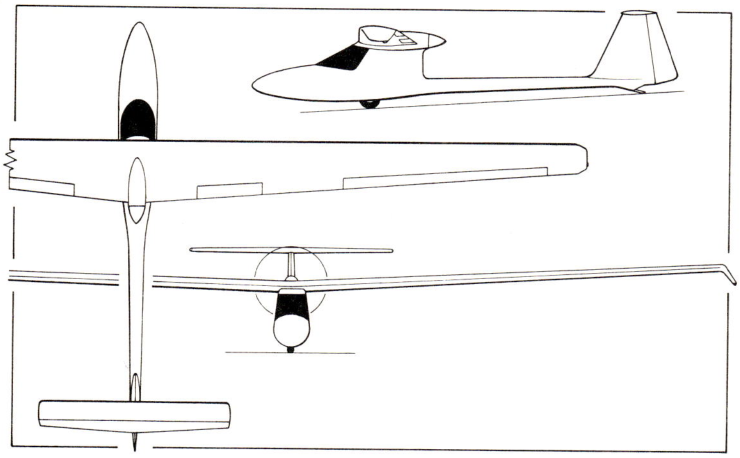

Todhunter T-5 Blue Wren powered sailplane *(Michael A. Badrocke)*

Brditschka HB-21 two-seat motor glider

CTA (IPD/PAR) PE-80367 Urubu *(Roberto Pereira de Andrade)*

First prototype of the IPE 02 two-seat training sailplane

Prototype EEUFMG CB-2 Minuano single-seat sailplane

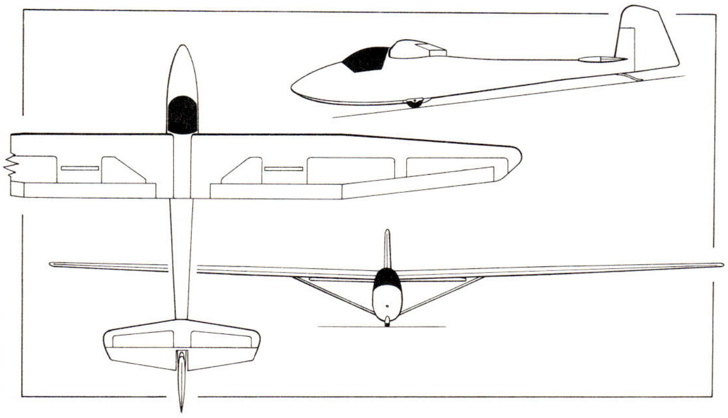

Rio Claro Araponga all-wood sporting glider *(Michael A. Badrocke)*

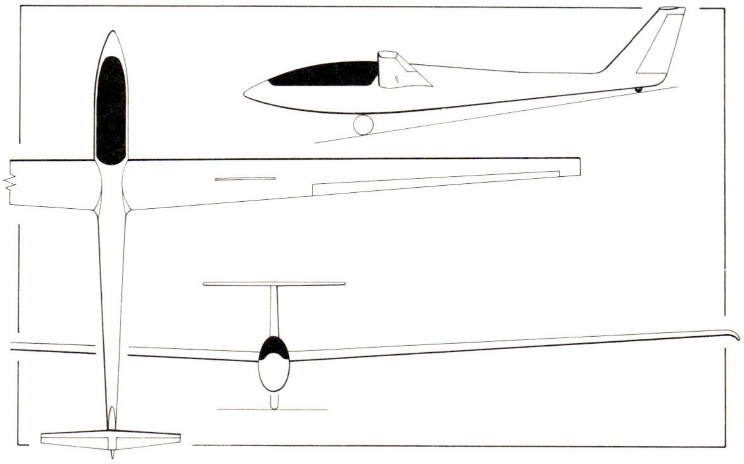

Sidou João Grande *(Michael A. Badrocke)*

Sigma variable-geometry sailplane (see Marsden entry) flying over Alberta

CHINA
(PEOPLE'S REPUBLIC)

STATE AIRCRAFT FACTORIES

WORKS: see Aircraft section

Sailplane development in China began in 1958, when a number of Polish gliding instructors were invited into the country to train air force cadets. Since that time the Chinese industry has manufactured more than 1,000 gliders of various types.

CHENDU X-7 JIAN FAN (SWORD POINT)

Built at the Chendu sailplane factory in Sichuan Province, the X-7 was flown for the first time in October 1966. A total of 130 had been built by the beginning of 1980.

TYPE: Tandem two-seat basic training glider; *g* limits +4·0; −2·0.

AIRFRAME: Braced high-wing monoplane, braced by single I strut on each side. Wing section Göttingen 535 (modified). Dihedral from roots. Single-spar constant-chord wings, comprising a glassfibre/honeycomb/epoxy sandwich torsion box, fabric-covered aft of spar. Glassfibre ailerons and upper-surface airbrakes. Semi-monocoque glassfibre pod-and-boom fuselage, moulded in two halves and joined at centreline. Fabric-covered glassfibre cruciform tail unit; fin built integrally with fuselage. Non-retractable monowheel, nose-skid and tailskid. Tandem open cockpits, with windscreen.

SHENYANG X-9

Used extensively in the Chinese People's Republic, the X-9 is a tandem two-seat training glider, said to have flying characteristics similar to those of the Schweizer SGS 2-33 (see US section). Of braced high-wing monoplane configuration, it is of wood and aluminium construction, and has a best glide ratio of approx 17. Design and first flight took place in 1977, and approx 150 had been built by the Autumn of 1980.

SHENYANG X-10 QIAN JIN (FORWARD)

Similar in general appearance to the Polish SZD-36 Cobra, this single-seat high-performance sailplane is of recent design and is of wood and glassfibre construction. It has a best glide ratio of approx 35.

SHENYANG X-11

The X-11, which was in the design stage in 1980, is a side-by-side two-seat motor glider, powered by a Revmaster engine.

CZECHOSLOVAKIA

OMNIPOL
OMNIPOL FOREIGN TRADE CORPORATION

Nekázanka 11, 11221 Prague 1
Telephone: 268261-8
Telex: 121299
SALES MANAGER: Ing Miloslav Branný

Omnipol handles all exports of products of the Czechoslovak aircraft industry.

LET
LET NÁRODNI PODNIK (Let National Corporation)

Uherské Hradiste-Kunovice

LET L-13 BLANÍK

The Blanik was designed for training in all categories from elementary to 'blind' flying and for high-performance flight. First flight was made in March 1956.

Production, which terminated at the end of 1978, totalled 2,616. Of these, over 2,000 had been exported to customers in more than 40 countries, including 1,289 to the USSR and 236 to the USA.

A description of the standard Blanik can be found in the 1978-79 *Jane's*. A number of Blaniks were converted into motor gliders, and all known details of these were given in the 1979-80 and 1980-81 *Jane's*.

A motorised version has also been developed by Mr Pentti Alanne of Finland (which see).

VSO
VYVOJOVÁ SKUPINA ORLICAN

c/o Orlican Národní Podnik, 565 37 Chocen
Telephone: Chocen 70 and 80
Telex: 0 196 210
CHIEF DESIGNER: Dipl Ing Jan Janovec

This group was formed by members of the former VSB (1973-74 *Jane's*) and some of the design staff of the Orlican National Works. Its first product is the VSO 10.

VSO 10

Design of the VSO 10 began in March 1972. Construction of three prototypes (one for structural test and two for flight test) began in 1975, and the first flight took place on 26 October 1976. Series production began in December 1978, with first deliveries being made to the Aeroclub of the Czechoslovak SSR.

By the beginning of 1981 a total of 30 had been delivered to the Czechoslovak Aeroclub, including a number of the VSO 10C Club Class version, with non-retractable monowheel.

TYPE: Single-seat high-performance sailplane; *g* limits +5·3/−3·5.

AIRFRAME: Cantilever shoulder-wing monoplane, with Wortmann wing sections: FX-61-163 at root, FX-60-126 at tip. Dihedral 3°. All-wood single-spar forward-swept wings with glassfibre sandwich skin. All-metal DFS airbrakes on upper surfaces. All-wood slotted ailerons. Glassfibre monocoque front and centre fuselage sections, latter reinforced by steel tube frame. Monocoque rear fuselage of aluminium alloy sheet. Metal T tail with fabric-covered elevators and rudder. Fixed-incidence tailplane. Retractable rubber-sprung monowheel (tyre pressure approx 2·45 bars; 35·5 lb/sq in), with drum brake. Semi-recessed unsprung tailwheel. Detachable cockpit canopy. Provision for 56 litres (12·3 Imp gallons) water ballast.

DENMARK

PROJEKT 8
PROJEKT 8 I/S

Fynsvej 56, DK-4000 Roskilde

DOLPHIN

The Projekt 8 I/S company, formed by Mr Helge Petersen and 10 other glider pilots, is building a two-seat motor glider known as the Dolphin. Construction is taking place at three separate sites near Copenhagen. The fuselage, power plant installation and tail unit have been completed, and successful initial taxying trials have been carried out using a spar-and-outrigger-wheels assembly in place of the wings. Construction of the wings was proceeding in 1980.

Dolphin is a testbed for the power system to be used in developing an improved motor glider known as Dolphin M2. The latter is described separately; the following description applies to the Dolphin prototype:

TYPE: Tandem two-seat motor glider; *g* limits +6; −4.

AIRFRAME: Cantilever mid-wing monoplane with T tail. Wortmann and NACA wing sections: FX-67-K-170 on centre-section; outer panels vary through FX-67-K-150 to NACA 64-212 at tip. Dihedral 4°. Aluminium wing centre-section; outer panels, including flaps and ailerons, of wood and glassfibre. Aluminium upper-surface airbrakes. Welded steel tube forward fuselage, covered by light glassfibre shell; wooden rear fuselage, reinforced by glassfibre. Wooden tail unit, with fixed-incidence tailplane. Central trim tab in elevator, inset tab at base of rudder. Semi-retractable rubber-sprung Tost main wheel; Tost nosewheel; steerable solid-tyre tailwheel; retractable wingtip balancer wheels. Cockpit canopy opens sideways.

POWER PLANT: One 40·25 kW (54 hp) VW 1600 engine, driving a two-blade tractor propeller, mounted on pylon which retracts rearward into top of fuselage when not in use.

DOLPHIN M2

Dolphin M2 is a developed version of the original Dolphin, from which it differs mainly in having a redesigned fuselage of improved aerodynamic shape, and sweptback fin and rudder. Design of the fuselage, of which the plug for the moulds was being made in early 1980, was by Mr Petersen's son Lars. Manufacture of the M2 will be undertaken by KFI (Kalundborg FlyIndustri) of Holbaekvej 109, DK-4400 Kalundborg.

TYPE: Side-by-side two-seat motor glider.

AIRFRAME: Cantilever mid-wing monoplane with sweptback fin and rudder. Construction generally similar to that of Dolphin prototype. Outer wing panels detachable, reducing span to 7·60 m (24 ft 11¼ in) for hangarage. Retractable twin monowheels; steerable tailwheel. Seats for two persons under two-piece upward-opening clamshell canopy; right-hand seat is staggered 25 cm (10 in) aft of left-hand seat. Towing hooks in nose and at CG for launching training.

POWER PLANT: One 45 kW (60 hp) VW engine, driving a two-blade tractor propeller, mounted on pylon which retracts rearward into top of fuselage when not in use. Propeller pylon movement is actuated by push-button electrical control from cockpit; engine clutch is operated automatically by pylon movement. When retracted, propeller is automatically aligned longitudinally in fuselage. Engine can be started before pylon is raised, to facilitate transition from gliding to powered flight with minimum loss of height.

FINLAND

ALANNE
PENTTI ALANNE

Orapihlajantie 9, 02620 Espoo 62

Mr Pentti Alanne has been concerned with a number of motor glider conversions since the second World War. In 1949, with Mr Vilho Swahn, he modified a Grunau Baby glider by installing a Poinsard engine in the nose and fitting a landing gear similar to that of a Piper Cub. The Motorlerche, described hereafter, is a Rhönlerche II glider fitted with similar landing gear and a modified Volkswagen engine. His latest conversion is of a Czechoslovak Blanik sailplane to receive a dorsally-mounted VW engine.

ALANNE MOTORLERCHE

The Motorlerche, which flew for the first time on 10 August 1973, is a conversion of the Rhönlerche II, a former training glider now readily obtainable at a low cost. The forward section of the fuselage is removed, and attachments welded on for a modified Volkswagen 1,500 cc engine. Landing gear consists of main-wheel and tail-wheel units similar to those of a Piper Cub, and the complete conversion can be accomplished in approx 200 h.

By 1 January 1981, eight Motorlerche were flying and four more were under construction.

TYPE: Single-seat motor glider. Aerobatics prohibited.

AIRFRAME: As for Rhönlerche II, except for modifications necessary to install power plant and non-retractable landing gear.

POWER PLANT: One 37·2 kW (50 hp) modified Volkswagen 1,500 cc engine (Ted Barker conversion), driving a two-blade propeller. Usable fuel capacity 26 litres (5·7 Imp gallons). Oil capacity 2·5 litres (0·5 Imp gallons).

ALANNE MOTORBLANIK

First flown on 19 May 1980, the Motorblanik is converted from a standard Czechoslovak Let L-13 Blanik tandem two-seat sailplane. It had accumulated 35 h flying by 1 January 1981.

AIRFRAME: As for L-13 Blanik (see 1978-79 *Jane's*), except for modifications necessary for power plant installation. Monowheel tyre size 350 × 135 mm, pressure 2·55 bars (37 lb/sq in).

POWER PLANT: One 37·2 kW (50 hp) modified Volkswagen 1,600 cc engine, driving a two-blade wooden propeller. Fuel capacity 28 litres (6·2 Imp gallons). Installation is pod-mounted on struts above fuselage, aft of cockpit, and is non-retractable.

Chinese X-5A tandem two-seat elementary training glider *(Charles M. Gyenes)*

Chendu X-7 Jian Fan two-seat glassfibre training glider *(Charles M. Gyenes)*

Shenyang X-10 Qian Jin single-seat sailplane *(Charles M. Gyenes)*

Shenyang X-9 two-seat training glider, with enclosed cockpits *(Charles M. Gyenes)*

VSO 10 single-seat high-performance sailplane

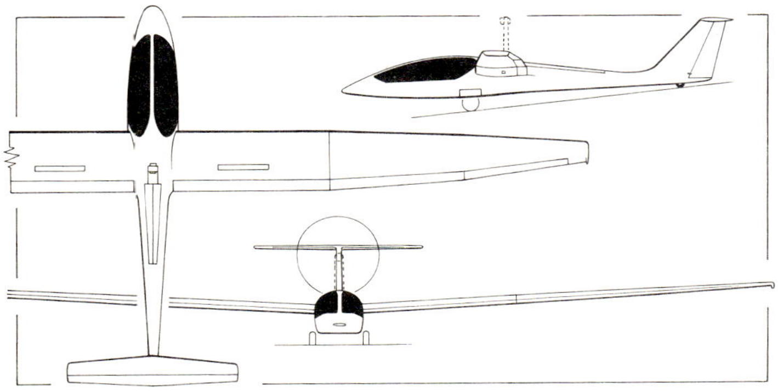

Projekt 8 Dolphin M2 side-by-side two-seat motor glider *(Michael A. Badrocke)*

Projekt 8 Dolphin tandem two-seat motor glider, partially complete and undergoing taxying trials

Alanne Motorlerche motor glider (modified VW engine)

Alanne Motorblanik, converted from an L-13 sailplane

EIRI

EINO RIIHELÄ KY (Eiriavion Dept)

Box 107, SF-15101, Lahti 10

Manufacture of the EIRI PIK-20E powered sailplane has been transferred to Siren SA of France, under which heading the description can now be found.

FRANCE

CALVEL

JACQUES CALVEL

Aérodrome Millau-Larzac, 12230 La Cavalerie

CALVEL FRELON (HORNET)

Illustrated in an accompanying photograph, the Frelon was built by M Jacques Calvel, a specialist in the use of laminated plastics materials.

The prototype, which made its public debut at the 1979 Paris Air Show, was expected to fly for the first time in March 1981. At that time, M Calvel had sold 30 sets of plans to amateur constructors, and intends in due course also to market the aircraft in kit form.

TYPE: Single-seat lightweight homebuilt powered glider; *g* limits +4·0/−2·0.

AIRFRAME: Braced high-wing monoplane, of wood and plastics construction. Wortmann FX-61-184 wing section. Dihedral 3° from roots. Incidence 1° 30′. Single wing bracing strut on each side. Wings have wood and plastics-tube basic structure, and shaped styrofoam core covered with plastics skin. Airbrake in upper surface of each wing, well inboard of aileron and close to trailing-edge. Main load-bearing structure of cabin is of plastics-coated plywood, faired at rear to a plastics tailboom. Cantilever tail unit comprises a sweptback fin with horn-balanced rudder, a small ventral fin, and a low-set, non-swept one-piece all-moving tailplane. Non-retractable tricycle landing gear, with Kevlar and carbonfibre epoxy laminated cantilever mainwheel legs, size 360 × 85 tyres, and cable-operated drum brakes. Steerable nosewheel, on metal leg. Streamline fairings on all three wheels. One-piece flush-fitting cockpit canopy.

POWER PLANT (prototype): One 26 kW (35 hp) Rockwell JLO two-cylinder two-stroke engine, driving a 0·80 m (2 ft 7½ in) diameter two-blade fixed-pitch wooden pusher propeller. Plastics fuel tank, capacity 15 litres (3·3 Imp gallons).

CARMAM

CARMAM SOCIÉTÉ ANONYME

Aérodrome d'Avermes, 03000 Moulins
Telephone: (70) 44 36 18
PRESIDENT: Xavier Laguette
DIRECTOR GENERAL: Jean Pottier
TECHNICAL DIRECTOR: Christian Rocheteau

Since July 1979, CARMAM has been a member of the Siren group, together with Issoire-Aviation (which see), Siren SA and Pelletier Exploitation. Its current products are the J.P.15-36A and AR, the Pottier Kit Club 15-34, the C.38 (formerly CARMAM 15-38), and a new sailplane known as the C.40.

CARMAM J.P.15-36A and AR AIGLON (EAGLET)

The Aiglon was designed as a private venture by M Robert Jacquet and M Jean Pottier. Design began in September 1971, and prototype construction started at the end of 1972. The prototype (F-WCAP) made its first flight on 14 June 1974.

Series production by CARMAM began in 1976, and the first production **J.P.15-36A** flew for the first time on 16 October 1976. With retractable monowheel, the aircraft is designated **J.P.15-36AR.**

Forty-one J.P.15-36A and J.P.15-36AR Aiglons had been sold by January 1981.

TYPE: Single-seat Standard Class sailplane; *g* limit +5·3.

AIRFRAME: Cantilever mid-wing monoplane, with Wortmann wing sections: FX-67-K-170 at root, FX-60-126 at tip. Dihedral 3°. Single glassfibre wing spar, with glassfibre/Rohacell/epoxy sandwich skin and steel-tipped wingtip 'salmons'. Plastics plain ailerons, which can be operated differentially or in unison. Schempp-Hirth upper/lower surface airbrakes. Provision for 55 kg (121 lb) water ballast. Semi-monocoque glassfibre fuselage, moulded in two halves and joined at centre-line. Single bulkhead forms cockpit backrest and shock-absorbing structure for monowheel. Sweptback fin, integral with fuselage, with fabric-covered wooden rudder. All-moving plastics tailplane. Non-retractable monowheel, with brake, and tail bumper, on J.P.15-36A. Retractable monowheel on J.P.15-36AR. Detachable cockpit canopy.

CARMAM (POTTIER) KIT CLUB 15-34

CARMAM produces factory-built examples of the Kit Club 15-34 homebuilt sailplane described under the Pottier heading in this section.

CARMAM C.38

Construction of this single-seat Standard Class sailplane, known originally as the 15-38, began in February 1977, and the prototype flew for the first time on 17 January 1979. It utilises the same wings as the J.P.15-36A, but with provision for 100 kg (220 lb) water ballast; the fuselage is of new design, but of similar construction to the J.P.15-36A; it has a T tail, with elevator spring trim, and a mechanically retractable unsprung monowheel.

Certification was under way in Spring 1981.

CARMAM C.40

Construction of the C.40 was due to begin in 1980; its general appearance can be seen in the accompanying three-view drawing.

TYPE: Single-seat 15/17 metre sailplane; *g* limit +6.

AIRFRAME: Cantilever mid-wing monoplane. Wings as for J.P.15-36, but of carbonfibre and epoxy construction, with detachable tips to extend span to 17 m. Fuselage is a thin-skin carbonfibre and epoxy structure, over a plywood frame in the centre-section. Sweptback fin and rudder, and cantilever fixed-incidence T tail, of similar construction to wings. Elevator spring trim. Retractable unsprung monowheel, tyre size 4·00-4, with Tost brake; tailskid. Provision for 180 kg (397 lb) water ballast.

CENTRAIR

SA CENTRAIR

Aérodrome Le Blanc, BP 44, 36300 Le Blanc
Telephone: (54) 37 06 91 and 37 07 96
Telex: 750272 F
PRESIDENT/DIRECTOR GENERAL: Marc Ranjon

CENTRAIR (SCHLEICHER) ASW 20

Centrair has since 1978 been producing under licence the Schleicher ASW 20 sailplane, a description of which can be found in the German section of this edition. Five versions were available in 1981, as follows:

ASW 20F. Licence version of the standard 15 m Contest Class ASW 20, for which French certification was received on 31 March 1978. Provision for 120 litres (26·5 Imp gallons) of water ballast. Standard version has fixed wingtips, but the ASW 20F can also be supplied with detachable wingtips to facilitate conversion to models with extended span and/or winglets. Production of the ASW 20F totalled approx 100 by May 1981, and was then continuing at the rate of about five per month.

ASW 20FL. Licence version of the Open Class ASW 20L, with detachable wingtips and add-on outer panels extending the span to 16·59 m (54 ft 5¼ in). No provision for water ballast.

ASW 20FP. Version of ASW 20F with carbonfibre NASA-type winglets fitted to the detachable wingtips. Developed by Centrair and test-flown August-December 1980. Available from late 1981.

ASW 20FLP. As ASW 20FP, but with the add-on outer wing panels of the FL.

CENTRAIR 101

Of all-French design, the Centrair 101 is a Standard Class sailplane, with a choice of retractable or non-retractable monowheel landing gear. A prototype was expected to fly in the Summer of 1981.

TYPE: Single-seat Standard Class sailplane.

AIRFRAME: Cantilever shoulder-wing monoplane, constructed of Conticell sandwich, glassfibre and epoxy resin. Onera COAP-01 (modified Wortmann FX-61-147) wing section. Dihedral 2° 18′. Double-plate upper surface airbrakes. Turned-down wingtips. Glassfibre/epoxy resin monocoque fuselage. Cantilever T tailplane, with elevator. Retractable or non-retractable monowheel (size 5·00-5), with drum brake; rubber tail bumper, with metal skid or recessed wheel. One-piece cockpit canopy. Provision for 120 litres (26·5 Imp gallons) water ballast in two wing leading-edge tanks.

ENSMA (SSRPPL)

ÉCOLE NATIONALE SUPÉRIEURE DE MÉCANIQUE ET D'AÉROTECHNIQUE (Société Scientifique de Recherche et de Promotion du Planeur Léger)

20 rue Guillaume VII, 86034 Poitiers Cédex
Telephone: (16) 49 41 48 05
PRESIDENT OF SSRPPL: Jean-Yves Piquereau

ENSMA FS-25 S CUERVO

Students and staff of the ENSMA have built this modified version of the Akaflieg Stuttgart FS-25 Cuervo (described and illustrated in the 1973-74 *Jane's*). Known as the FS-25 S, it has a retractable monowheel and modifications to the internal structure and wing/fuselage attachments. Construction began in 1976, and the aircraft (F-WRFM) made its first flight on 5 February 1981.

TYPE: Single-seat Standard Class sailplane; *g* limits +5·6/−2·3.

AIRFRAME: Cantilever shoulder-wing monoplane. Wortmann wing sections: FX-66-S-196 at root, varying through FX-61-184, -168 and -147 to FX-60-126 at tip. Three-spar wings, without ribs, with glassfibre/balsa sandwich skin. Dihedral 2° 42′. Incidence 0°. Sweep-forward 3° at quarter-chord. Glassfibre/balsa sandwich ailerons and DFS-type trailing-edge airbrakes. Pod-and-boom fuselage: forward portion of welded steel tube with glassfibre/balsa sandwich skin in cockpit area; duralumin rear portion. Cantilever T tail, with one-piece all-moving tailplane; construction similar to that of wings. Retractable unsprung monowheel (tyre size 300 × 100, pressure 2·5 bars; 36·25 lb/sq in), and tailskid. No brakes. Single seat under two-piece canopy.

FERRIER

HUBERT FERRIER

48 rue du Coq, 13001 Marseille
Telephone: 16 91 64 68 60

FERRIER CONDOR

M Ferrier, an engineer at the helicopter division of Aérospatiale, began in September 1975 the design of this 15 metre Class sailplane, the general appearance of which is shown in an accompanying three-view drawing. Construction of the prototype, which was expected to fly in June 1981, was undertaken by M Jacques Boulay.

TYPE: Single-seat 15 metre Class homebuilt sailplane.

AIRFRAME: Cantilever mid-wing monoplane. Wing centre-section is of Wortmann FX-67-K-170 section, modified on outer panels to introduce small washout. Dihedral 3° from roots. Incidence 1° at root. All-plastics construction, the wings having a carbonfibre main spar and a carbonfibre/Kevlar skin. Trailing-edge flaps, interconnected ailerons, and upper-surface airbrakes, all of carbonfibre. Moulded-shell fuselage, with forward and central sections of glassfibre; centre-fuselage has floor and seat of Nomex honeycomb. Carbonfibre rear fuselage (reinforced with glassfibre). Cantilever T tail, of carbonfibre/glassfibre construction. Mechanically retractable rubber-sprung monowheel (tyre size 330 × 130, pressure 2·0 bars; 29·0 lb/sq in), and tail bumper. One-piece forward-opening canopy. Provision for 100 litres (22 Imp gallons) water ballast in wings.

FOURNIER

FOURNIER AVIATION

Aérodrome de Nitray, 37270 Montlouis
Telephone: (47) 50 68 30
MANAGING DIRECTOR: René Caillet
CONSULTANT: René Fournier
SALES MANAGER: Patrice Caillet

This company was formed in the Spring of 1978 as a successor to Avions Fournier (see 1977-78 *Jane's*).

FOURNIER RF-9

The prototype RF-9 (F-WARF) flew for the first time on 20 January 1977. A second prototype made its first flight in April 1979, and French certification was granted on 20 December 1979. Ten production aircraft had been delivered by May 1981.

TYPE: Side-by-side two-seat training and sporting motor glider.

AIRFRAME: Cantilever low-wing monoplane, of all-wood construction. NACA 64_3-618 section wings, with dihedral from roots. Ailerons and upper-surface airbrakes. Outer portion of each wing can be folded inward for stowage (span folded 10·00 m; 32 ft 9½ in), with connect/disconnect aileron controls. Semi-monocoque fuselage, slightly-swept fin and rudder, and small dorsal fin. Retractable main wheels, with oleo-pneumatic suspension and hydraulic disc brakes; steerable tailwheel. One-piece cockpit canopy opens upward and rearward.

POWER PLANT: One 50 kW (68 hp) Limbach SL 1700 E flat-four engine, driving a Hoffmann two-blade three-position propeller. Fuel capacity 30 litres (6·6 Imp gallons).

FOURNIER RF-10

First flown on 6 March 1981, the side-by-side two-seat RF-10 is generally similar to the RF-9 but is of plastics construction and has a carbonfibre main spar. The wings are completely detachable in four sections for transporta-

Calvel Frelon ultralight powered glider *(J. M. G. Gradidge)*

CARMAM J.P.15-36A Aiglon Standard Class sailplane *(Peter F. Selinger)*

Centrair ASW 20FP, winglet version of the Schleicher ASW 20

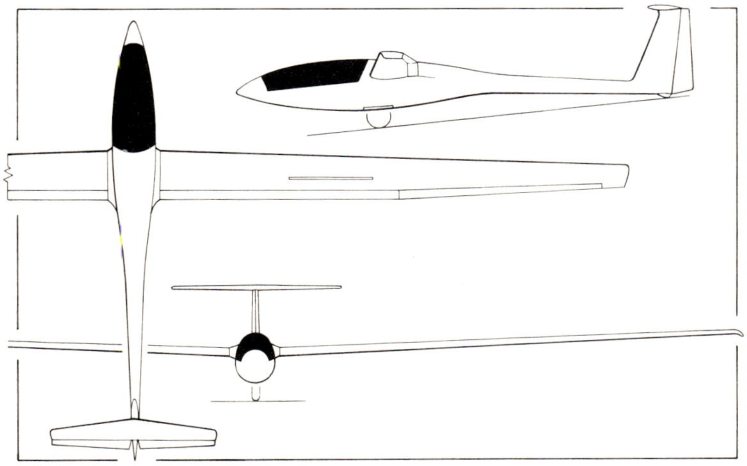

Ferrier Condor all-plastics homebuilt sailplane *(Michael A. Badrocke)*

Prototype Fournier RF-9 two-seat training motor glider

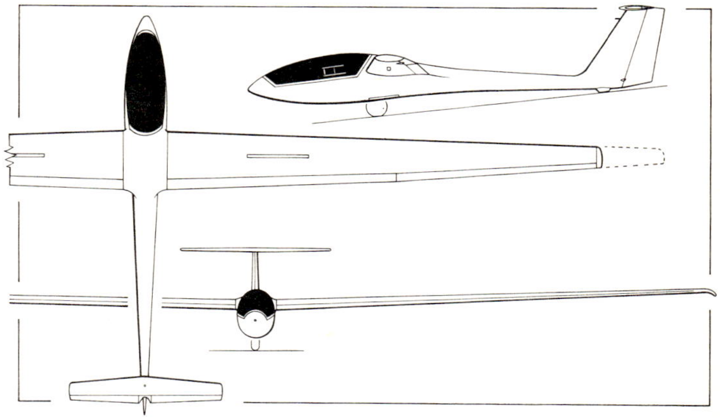

CARMAM C.40 single-seat carbonfibre sailplane *(Michael A. Badrocke)*

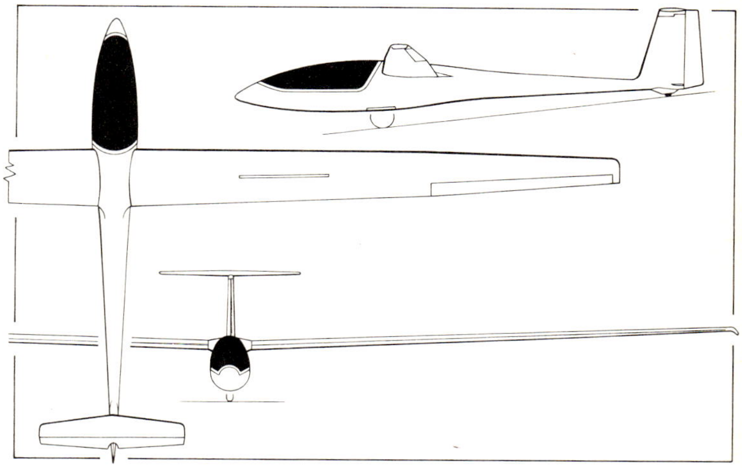

Centrair 101 Standard Class sailplane *(Michael A. Badrocke)*

ENSMA FS-25 S Cuervo single-seat Standard Class sailplane

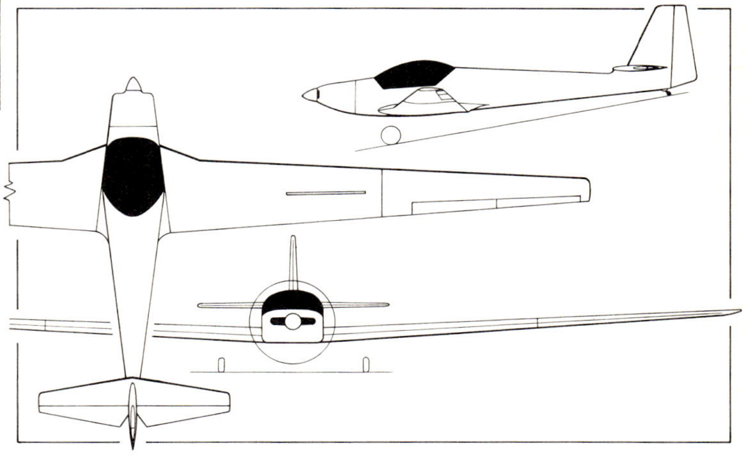

Fournier RF-10 all-plastics motor glider *(Michael A. Badrocke)*

tion and storage. In an alternative form, under development, the wings will be foldable to a span of 9·60 m (31 ft 6 in). A more powerful engine is fitted, fuel capacity is increased, all control surfaces are balanced, and the main landing gear has larger wheels (size 330 × 130). Production was scheduled to begin in the Autumn of 1981 at the rate of one aircraft per week.

Power Plant: One 59·5 kW (80 hp) Limbach L 2000 EO 1 flat-four engine, driving a Hoffmann two-blade three-position variable-pitch propeller. Fuel capacity 40 litres (8·8 Imp gallons) standard; provision for auxiliary tank, also of 40 litres capacity.

GROSS
FRANÇOIS and ROGER GROSS
Le Clairupt, 54120 Baccarat
Telephone: (83) 75 14 37

GROSS (FAUVEL) AV.451
This modified version of a Fauvel AV.45 self-launching sailplane (F-WRRM) was designed and built by M François Gross, with the help of his son, in less than a year. Principal changes include an increase in wing span; adoption of a new Wortmann aerofoil section; a longer fuselage, modified in the cockpit area to accept the canopy from an Astir sailplane; and the installation of a 28·5 kW (38 hp) Rotax engine aft of the cockpit. The wood-and-fabric construction of the standard AV.45 is retained, as are the regular control surfaces.

Redesign started in November 1976, construction began in September 1977, and the aircraft flew for the first time on 12 September 1979. It is the 13th AV.451 to be completed.

TYPE: Single-seat self-launching sailplane; *g* limit +10.

AIRFRAME: Cantilever mid-wing monoplane. Wing section Wortmann FX-66-H-159. Dihedral 6° 30′ on outer panels. Wooden wings (Finnish spruce) with fabric covering. Fabric-covered wooden differential ailerons, to counteract adverse yaw. Twin fins and rudders on wing trailing-edge at tips of centre-section. Plywood-covered fuselage, with bulkheads and four main longerons. Wooden elevons on inboard trailing-edge of wing; trim tab in port elevon. Upper/lower surface airbrakes. Monowheel and steerable nosewheel (linked to rudders), both non-retractable and semi-recessed. Tyre sizes 380 × 150 (main), 300 × 100 (nose); tyre pressures 1·80 bars (26 lb/sq in) and 0·80 bars (11·5 lb/sq in) respectively. Tost wheel brakes. Cockpit canopy from Grob Astir sailplane. Radio fitted.

POWER PLANT: One 28·5 kW (38 hp) Rotax 642E two-cylinder two-stroke engine, driving a Hoffmann 136/95/LD two-blade pusher propeller. Fuel in two wing tanks, total capacity 35 litres (7·7 Imp gallons).

ISSOIRE
ISSOIRE-AVIATION SA
Aérodrome d'Issoire-le-Broc (Puy-de-Dôme), BP No. 7, 63501 Issoire Cédex
Telephone: (73) 89 01 54
Telex: ISSAVIA 990 185 F
PRESIDENT-DIRECTOR GENERAL: Xavier Laguette
TECHNICAL DIRECTOR: Philippe Moniot
DESIGNER: Bruno Boulnois

This company was formed in late 1977 by the President of Siren SA, following the bankruptcy of Wassmer Aviation. It is responsible for the Siren (formerly CERVA CE 75) Silène and the Siren D 77 Iris sailplanes. It is now one of four companies forming the Siren group, the others being Siren SA and CARMAM (which see), and Pelletier Exploitation.

ISSOIRE (SIREN) D 77 IRIS
Design of the Iris began in 1973, and the prototype (F-WAQA) flew for the first time on 26 February 1977.

French certification was granted in early February 1979 and BGA certification in May 1979 (semi-aerobatic category).

By 1 January 1981 a total of 20 Iris had been ordered, of which five had been completed. Fuselages and tail units are manufactured by Siren; Issoire builds the wings, and undertakes final assembly and flight testing.

The D 77 is also available in kit form for amateur construction.

TYPE: Single-seat training sailplane; *g* limits +5·3/−2·65 (normal), +8·0/−4·0 (ultimate).

AIRFRAME: Cantilever mid-wing monoplane. Bertin E55-166 wing section. Dihedral 3°. Glassfibre/polyester/PMC sandwich construction, with glassfibre ailerons and Schempp-Hirth metal airbrakes on upper surfaces. Glassfibre/polyester monocoque fuselage, built in two halves. Cantilever tail unit, of similar construction to wings. Fixed-incidence tailplane. Spring tab in each elevator. Non-retractable unsprung monowheel, tyre size 330 × 130, pressure 1·8 bars (26 lb/sq in); tailskid. Siren hydraulic brake. Cockpit canopy opens sideways to starboard. Winch-launching and bungee-launching equipment available.

ISSOIRE (SIREN) E 78 SILÈNE
Design of the original E 75 Silène (by Siren SA) began on 1 January 1972. Construction by CERVA of a CE 75 prototype started on 1 February 1973, and this aircraft (F-CCFF) made its first flight at Argenton on 2 July 1974. French certification was granted on 3 February 1978, and BGA certification in March 1979 (semi-aerobatic category).

Fifteen E 78s had been delivered by 1 January 1981, of 22 then on order, including examples for customers in Australia, the Federal Republic of Germany, the UK and the USA. Production is on the same basis as for the Iris, with fuselages and tail units built by Siren; the wings are built by Issoire, which also carries out final assembly and flight testing.

Current versions of the Silène are designated as follows:

E 78. Cockpit improvements by comparison with E 75 include a larger canopy, lowered seats, increased width at shoulder level, and repositioned starboard rudder pedals. Retractable monowheel. Certificated 1 August 1978. In production.

E 78B. Basic version, with non-retractable monowheel. Certificated 1 August 1978.

The following description applies to the E 78 and E 78B:

TYPE: Side-by-side two-seat training sailplane; *g* limits +5·3/−2·65 (normal), +8·0/−4·0 (ultimate).

AIRFRAME: Cantilever mid-wing monoplane, built entirely of glassfibre/polyester/PMC sandwich. Bertin E55-166 wing section. Dihedral 2°. Sweepforward 2° at quarter-chord. Two-section ailerons, with spring tabs. Schempp-Hirth airbrakes above and below each wing. Semi-monocoque fuselage. Cantilever tail unit, with fixed-incidence tailplane. Spring tab in each elevator. Retractable or non-retractable rubber-sprung monowheel, tyre size 330 × 130, pressure 2·4 bars (35 lb/sq in). Siren hydraulic brake optional. Tailskid. Winch-launching and bungee-launching equipment available.

PONSOT
BERNARD PONSOT
5 rue Bicquilley, 54200 Toul

A French press report in September 1979 referred briefly to a Fauvel AV.45 built by M Ponsot, which had recently made its first flight powered by a Microturbo Eclair turbojet in place of the Nelson piston engine previously fitted. No further details had been received at the time of closing for press.

POTTIER
JEAN POTTIER
4 rue de Poissy, 78130 Les Mureaux
Telephone: 099 13 85

POTTIER KIT CLUB 15-34
This is essentially the same sailplane as the CARMAM J.P.15-36A Aiglon (which see), with some constructional simplification to make it suitable for amateur builders.

Prototype construction began in November 1975; first flight was made on 6 November 1976.

French certification was received on 15 February 1979. First kits for homebuilders became available in mid-March 1979, and by March 1981 ten examples had been flown in France, with five others under construction.

The Kit Club is also produced in factory-built form by CARMAM (which see).

The description of the Aiglon applies also to the Kit Club, except in the following respects:

TYPE: Single-seat homebuilt sailplane; *g* limits +8·3/−4·0.

AIRFRAME: As Aiglon, but of plywood-covered spruce construction except for glassfibre nosecone. Provision for 80 kg (176 lb) of water ballast. Rudder and rear part of tailplane fabric-covered. Non-retractable unsprung monowheel, with cable-operated drum brake for production aircraft.

SCAP
SOCIÉTÉ DE COMMERCIALISATION AÉRONAUTIQUE DU PLESSIS, SàRL
Aérodrome de Bailleau-Armenonville, 28320 Gallardon
Telephone: (37) 23 43 74 and 23 54 32

SCAP CIRRUS 78L
The Cirrus 78L is a version of the Standard Cirrus sailplane produced originally by Schempp-Hirth in Germany. It is manufactured in France by the holding company Dubigeon-Normandie, and distributed by SCAP. Thirty had been completed by 1 January 1979.

TYPE: Single-seat high-performance Standard Class sailplane; *g* limit +10.

AIRFRAME: Cantilever mid-wing monoplane with T tail. Wortmann wing section. Dihedral 3°. Sweepback 1°18′ at leading-edge. Glassfibre/foam sandwich wings, ailerons and tail surfaces. Schempp-Hirth glassfibre airbrakes on wing upper surface. Provision for 130 litres (28·6 Imp gallons) of water ballast. Glassfibre fuselage shell, stiffened with bonded-in foam rings. Fixed tailplane, with elevator. Retractable monowheel standard. Non-retractable faired monowheel optional. Tost wheel with drum brake and Continental tyre, pressure 3·45 bars (50 lb/sq in). New-design canopy, opening sideways to starboard.

SIREN
SIREN SA
Route des Chambons, BP 42, 36200 Argenton-sur-Creuse
Telephone: (54) 04 14 47
Telex: 750534 Chamco-Châteauroux Siren 200-1
DIRECTOR: Xavier Laguette

In addition to Siren SA, the Siren group has since July 1979 comprised CARMAM and Issoire-Aviation (which see), and Pelletier Exploitation.

The current activities of Siren SA include the manufacture of fuselages and tail units for the Issoire D 77 Iris and E 78 Silène sailplanes (which see). Under an agreement signed on 23 January 1981, Siren is now responsible also for the production and marketing of the PIK-20E powered sailplane built formerly by Eiriavion in Finland.

SIREN (EIRI) PIK-20E
Details of the PIK-20D 15 metre Contest Class sailplane and earlier versions of the PIK-20 can be found under the EIRI heading in the Finnish sections of the 1979-80 and previous editions of *Jane's*.

The prototype of the PIK-20E powered version flew for the first time on 2 October 1976 and was described in the 1977-78 *Jane's*. The production prototype, with a manually retractable Rotax 503 engine, made its first powered flight on 18 March 1978. Series production, with a Rotax 501 engine, longer nose and 25% more tailplane area than the PIK-20D, began in late 1978, and approx 50 were delivered by EIRI before production in Finland ended in 1980. A description can be found in the 1980-81 *Jane's*. The following description applies to the French-built version, which has a Rotax 505 engine and is designated PIK-20E2F:

TYPE: Single-seat self-launching 15 metre Class sailplane.

AIRFRAME: Cantilever shoulder-wing monoplane with T tail. Wortmann wing sections: FX-67-K-170 at root, FX-67-K-150 at tip. Dihedral 3°. Sweepback 1° 21·6′. Glassfibre/epoxy/PVC foam sandwich wings. Spars of carbonfibre-reinforced epoxy. Schempp-Hirth airbrakes standard. Plain flaps ('flaperons') function as both flaps and ailerons. Provision for 120 litres (26·4 Imp gallons) of water ballast. Glassfibre/epoxy monocoque fuselage, reinforced with ribs and carbonfibre. Optional Tost towing hook. T tail of similar construction to wings. Fixed-incidence tailplane, with one-piece elevator. Retractable sprung Tost monowheel with drum brake. Steerable rubber-sprung tailwheel. Non-retractable wingtip wheels. Forward-hinged one-piece cockpit canopy.

POWER PLANT: One 32 kW (43 hp) Rotax 505 two-cylinder two-stroke engine, driving a Hoffmann two-blade fixed-pitch wooden propeller, retracting manually into fuselage aft of cockpit when not in use. Electrical starter. Kevlar fuel tank, capacity 30 litres (6·6 Imp gallons).

Gross (Fauvel) AV.451 self-launching sailplane

Issoire (Siren) D 77 Iris single-seat training sailplane

Pottier Kit Club 15-34 prototype homebuilt sailplane *(Geoffrey P. Jones)*

Issoire (Siren) E 78 Silène two-seat training sailplane *(Peter F. Selinger)*

Akaflieg Braunschweig SB-12 *(Peter F. Selinger)*

Akaflieg Darmstadt D-39 single-seat motor glider *(Peter F. Selinger)*

Siren (Eiri) PIK-20E self-launching sailplane, with retractable power plant

GERMANY
(FEDERAL REPUBLIC)

AKAFLIEG BRAUNSCHWEIG
AKADEMISCHE FLIEGERGRUPPE BRAUNSCHWEIG eV

3300 Braunschweig, Flughafen Akafliegheim
Telephone: 0531 3952149
DIRECTOR: Martin Hansen

The students of Brunswick University have built a series of high-performance sailplanes (see 1977-78 and earlier editions of *Jane's*). The Open Class SB-10 was described in the 1974-75 *Jane's*, and the variable-geometry SB-11 Antares in the 1979-80 edition.

The latest design by Akaflieg Braunschweig of which details have been received is the SB-12. A new sailplane designated SB-13 was reported to be under construction in the Spring of 1981.

AKAFLIEG BRAUNSCHWEIG SB-12

The SB-12 is a Standard Class sailplane using a new wing aerofoil section designed for high cruising performance; its general appearance is shown in an accompanying illustration. Design began in the Summer of 1979, and the prototype construction started in the following November, utilising the wings and fuselage of a Glasflügel Hornet C as a basis. First flight was made on 9 April 1980. Production is not intended.

TYPE: Single-seat experimental Standard Class sailplane; *g* limits +5·3/−2·65.

AIRFRAME: Cantilever mid-wing monoplane. Specially-designed DFVLR HQ 014 wing section, with thickness/chord ratio of 18·43%. Dihedral 3°. Incidence 0° 30′. No sweep. Single I spar, with carbonfibre flanges and GRP shear web; outer skin of carbonfibre and plastics foam. GRP ailerons and wide-span trailing-edge airbrakes. No flaps or spoilers. GRP monocoque fuselage. Cantilever T tail, with fixed surfaces of GRP/plastics foam sandwich, rudder and elevators of GRP only. Retractable unsprung monowheel, size 5·00-5, with Tost internal drum brake; semi-recessed tailwheel. Canopy hinged at front to open upward. Provision for 150 kg (331 lb) water ballast.

AKAFLIEG DARMSTADT
AKADEMISCHE FLIEGERGRUPPE DARMSTADT eV

6100 Darmstadt, Technische Hochschule, Hochschulstrasse 1
Telephone: 06151 162790

The Fliegergruppe of Darmstadt University has been designing, building and flying sailplanes since 1921. Its postwar products have been described in previous editions of *Jane's*.

AKAFLIEG DARMSTADT D-39

The D-39 motor glider utilises the basic D-38 airframe, except for a low-mounted wing and a nose-mounted power plant.

The prototype flew for the first time on 28 June 1979. Series production is not intended.

TYPE: Single-seat motor glider; *g* limits +5·3/−3·6.

AIRFRAME: Cantilever low-wing monoplane with T tail. Wortmann wing sections: FX-61-184 at centreline, FX-60-126 at tip. Dihedral 4° from roots. Glassfibre/balsa sandwich wings, tail and monocoque fuselage, with glassfibre/Klégécel foam sandwich ailerons. All-moving one-piece swept tailplane, with Flettner tab. Retractable sprung monowheel, with Tost drum brake, and small tailwheel.

POWER PLANT: One 51 kW (68 hp) Limbach SL 1700 E engine, driving a two-blade propeller. Fuel capacity 35 litres (7·7 Imp gallons).

AKAFLIEG DARMSTADT D-40

A prototype of this new variable-geometry sailplane was completed in early 1981. The fuselage is that of a Rolladen-Schneider LS3. The wings, using the same aerofoil section as those of the Akaflieg München Mü 27 (which see), are fitted with variable-area Fowler-type trailing-edge flaps which, when extended, increase available wing area by more than 20 per cent. The wings are of Wortmann FX-67-VG-170 aerofoil section, and have 1° 28′ sweepforward at the leading-edge. Weights and performance figures had not been released at the time of closing for press.

DIMENSIONS, EXTERNAL:

Wing span	15·00 m (49 ft 2½ in)
Wing area, gross: flaps in	9·50 m² (102·25 sq ft)
flaps out	11·50 m² (123·8 sq ft)
Wing aspect ratio: flaps in	23·7
flaps out	19·6
Length overall	6·75 m (22 ft 1¾ in)

AKAFLIEG HANNOVER
AKADEMISCHE FLIEGERGRUPPE HANNOVER eV
Welfengarten 1a, 3000 Hannover

AKAFLIEG HANNOVER AFH 22
The AFH 22, under development in 1980, is a tandem two-seat sailplane with a mid-mounted sweptforward wing, T tail and retractable monowheel.

Specification:

Wing span	17·50 m (57 ft 5 in)
Length overall	8·50 m (27 ft 10¾ in)
Weight empty	390 kg (860 lb)
Max T-O weight	600 kg (1,322 lb)

AKAFLIEG KARLSRUHE
AKADEMISCHE FLIEGERGRUPPE KARLSRUHE eV
Technische Hochschule, 7500 Karlsruhe 1, Postfach 6380, Kaiserstrasse 12
Telephone: (0721) 608 2044
Information: Hartmut Walter

AKAFLIEG KARLSRUHE AK 2
Design of the AK 2 was begun in 1970, and construction of a prototype began two years later. This was still proceeding in early 1981.

The 1980-81 *Jane's* included a three-view drawing of the AK 2 as then envisaged, with a wing span of 20·00 m (65 ft 7½ in). This dimension has since been increased, and the latest available data are given in the table at the end of this section.

Type: Single-seat motor glider.

Airframe: Cantilever mid-wing monoplane. Three-piece wing, with carbonfibre centre-section. Trailing-edge flaps, ailerons, and upper-surface airbrakes. Fuselage of GfK/Conticell sandwich construction. Cantilever T tail, with all-moving tailplane of GfK sandwich. Retractable GfK-sprung Tost monowheel, diameter 380 mm, with Tost disc brake. Cockpit canopy, from a Glasflügel Hornet sailplane, opens upward. Provision for 72 litres (15·8 Imp gallons) water ballast.

Power Plant: One 48·5 kW (65 hp) Volvo three-cylinder two-stroke engine, driving a Hoffmann constant-speed propeller which retracts into fuselage when not in use. Fuel capacity 70 litres (15·4 Imp gallons).

AKAFLIEG MÜNCHEN
AKADEMISCHE FLIEGERGRUPPE MÜNCHEN eV
8000 München 2, Arcisstrasse 21, Postfach 202420
Telephone: (089) 28 61 11
Director: Klaus Raeder
Design Manager: Werner Pattermann

Recent designs by students at Munich University have included the Mü 27 and Mü 28.

AKAFLIEG MÜNCHEN Mü 27
The prototype Mü 27 (D-2827) flew for the first time on 24 February 1979 at Oberpfaffenhofen.

Type: Tandem two-seat high-performance sailplane.

Airframe: Cantilever shoulder-wing monoplane with T tail. Wortmann FX-67-VC-170/136 section, with electrically-actuated Fowler-type flaps which increase wing area by 36 per cent when fully extended. Aluminium alloy wing spar and metal webs, with glassfibre/foam sandwich wing and tail skins; all-glassfibre semi-monocoque fuselage. Double ailerons, linked to flaps when latter are extended. Carbonfibre airbrakes on wing upper surfaces. Retractable monowheel and fixed tailwheel. Front portion of canopy opens sideways to starboard; rear portion opens upward.

AKAFLIEG MÜNCHEN Mü 28
A prototype of the Mü 28 was expected to fly for the first time in 1980. Its general appearance is shown in the accompanying three-view drawing.

Type: Single-seat sailplane for full aerobatics and training.

Airframe: Cantilever mid-wing monoplane, with detachable tips to extend span from 12 m to 14 m. Wings are of Wortmann FX-71-L-150/20 symmetrical section, fitted with automatic trailing-edge flaps/ailerons ('flaperons'), and are of glassfibre/foam sandwich construction. Airbrakes in upper surfaces. All-glassfibre monocoque fuselage. Retractable monowheel; non-retractable tailwheel. Canopy opens forward and upward.

AKAFLIEG STUTTGART
AKADEMISCHE FLIEGERGRUPPE STUTTGART eV
Pfaffenwaldring 35, 7000 Stuttgart 80
Telephone: (0711) 784 2443
President: Andreas Haufler

Akaflieg Stuttgart's latest known design is the FS-31.

AKAFLIEG STUTTGART FS-31
This tandem two-seat sailplane has mid-mounted wings of Eppler E 603 section, a T tail and a retractable monowheel. Carbonfibre and Kevlar 49 are used in its construction, to save weight. The fuselage is made of these materials, although it retains the general shape of the fuselage of the University's earlier FS-28 and FS-29 sailplanes, to save time and expense.

General appearance of the FS-31 can be seen in the accompanying three-view drawing.

Specification:

Wing span	17·50 m (57 ft 5 in)
Length overall	8·70 m (28 ft 6½ in)
Weight empty	340 kg (750 lb)
Max T-O weight	560 kg (1,234 lb)

BLESSING
GERHARD BLESSING
205 Hamburg 80, Ochsenwerder Landstrasse 33
Telephone: (040) 737 2325

BLESSING STAFF REBELL
First flown on 3 June 1973, the original Rebell (D-KEBO) was a single/two-seat homebuilt glider, powered by a 40·5 kW (54 hp) Hirth engine driving a pusher propeller. In this form it has been described in previous editions of *Jane's*; it was subsequently flown with a modified VW engine.

Beginning in 1976, the aircraft was redesigned to accept a front-mounted engine and tractor propeller. In this form it is known as the Staff Rebell, and made its first flight in August 1980. The following description applies to the Staff Rebell in its current (1981) form:

Type: Single/two-seat homebuilt motor glider.

Airframe: Cantilever low-wing monoplane, with Wortmann wing sections. Dihedral on outer panels only. Incidence 3°. Single-spar wooden wings, outer panels of which can be folded inwards to reduce width to 7·50 m (24 ft 7¼ in) for storage. Steel tube fuselage, of basically rectangular cross-section, with Dacron covering. Wooden tail unit, with sweptback vertical surfaces. Ground-adjustable aluminium tabs on rudder and elevator. Non-retractable monowheel (tyre size 380 × 150 mm), fitted with handbrake connected to spoiler drive. Auxiliary outrigger wheel under each outer wing panel. Perspex cockpit canopy, with single-curvature surfaces.

Power Plant: One 44·5 kW (60 hp) Limbach SL 1700 EA engine, driving a Hoffmann two-blade variable-pitch propeller. Single fuel tank in fuselage, capacity 55 litres (12 Imp gallons).

FVA
FLUGWISSENSCHAFTLICHE VEREINIGUNG AACHEN (1920) EV
Templergraben 55, 5100 Aachen

Telephone: (0241) 806824

Telex: 832704 thac d (FVA)

Project Director: Jochen Ewald

FVA-20 F. B. SCHMETZ
Design of the FVA-20, to explore experimental methods of glassfibre construction, began in 1967 and work on the prototype began two years later. This aircraft (D-6020) flew for the first time on 28 November 1979; production is not intended.

Type: Single-seat Standard Class sailplane; *g* limit +6.

Airframe: Cantilever shoulder-wing monoplane. Wortmann wing sections: FX-61-168 at root, FX-60-126 at tip. Schempp-Hirth upper/lower-surface airbrakes. Wings are of GRP/Conticell sandwich construction; original GRP ailerons being replaced in 1981 by new, lighter-weight surfaces of CRP construction, to eliminate aileron flutter problems. GRP/balsa sandwich fuselage. Cantilever T tail, of similar construction to wings, has fixed-incidence tailplane; elevators fitted with Hähnle trim system. Retractable unsprung monowheel, with footbrake, and tail bumper. One-piece cockpit canopy, opening forward.

GLASER-DIRKS
GLASER-DIRKS FLUGZEUGBAU GmbH
7520 Bruchsal 4, Postfach 47, Im Schollengarten 19-20
Telephone: 07257 (1071)
Telex: 782241 gldg
Directors: Gerhard Glaser and Dipl-Ing Wilhelm Dirks

GLASER-DIRKS DG-100
Manufacture of the DG-100 series is now carried out under licence by Elan in Yugoslavia, under whose entry a description and illustration can be found.

The prototype of a new model, designated **DG-101**, was due to fly in March 1981, with deliveries to begin in May 1981. Improvements include a one-piece cockpit canopy, automatic self-correcting elevator, and sprung monowheel, and are applicable to both the DG-100 and DG-100G.

GLASER-DIRKS DG-200
The DG-200 (D-8200), which flew for the first time in prototype form on 22 April 1977, is an improved DG-100 fitted with wing flaps, and was developed for unlimited international competition.

Versions so far announced are as follows:

DG-200. High-performance 15 metre Class standard version. Total of 137 built by beginning of 1981.

DG-200A Acroracer. As DG-200, but with detachable wingtips to reduce span for aerobatics. No water ballast. First flown on 28 October 1978. One only built. Details in 1980-81 *Jane's*.

DG-200/17. As DG-200, but with add-on wingtips to extend span for Open Class competition. First flown 14 March 1979. Total of 35 built by beginning of 1981.

DG-200C. See 1980-81 *Jane's*. Not built.

DG-200/17C. Similar to DG-200/17, but with carbonfibre wing spar boom, wing skin, ailerons and flaps. Weight empty 228 kg (503 lb); max T-O weight and max wing loading of 450 kg (992 lb) and 42·5 kg/m² (8·71 lb/sq ft) with 17 m wings, 480 kg (1,058 lb) and 48·0 kg/m² (9·84 lb/sq ft) with 15 m span wings. Other data as given in table for DG-200/17. First flight 16 April 1980. Type certification granted in December 1980. Total of 10 built by beginning of 1981.

DG-202. Developed version, first flown on 30 April 1980. Improved cockpit and controls, and other improvements as for DG-101. Available as DG-202, DG-202/17 and DG-202/17C models.

The following description applies to the standard DG-200, unless indicated otherwise:

Type: Single-seat high-performance sailplane.

Airframe: Cantilever shoulder-wing monoplane. Wing section Wortmann FX-67-K-170-17 at root, FX-67-K-170 (15 m span) or FX-60-K-126 (17 m span) at tip. Dihedral 3° from roots. Incidence 0°. Glassfibre or carbonfibre roving main spar (see under model listings). Glassfibre/Conticell/foam sandwich wings, with glassfibre or carbonfibre flaps and ailerons. All-glassfibre semi-monocoque fuselage, fin and rudder. Glassfibre/foam sandwich T tailplane, with all-glassfibre elevator. Schempp-Hirth aluminium airbrakes on upper wing surfaces. Water ballast tank in each wing (except Acroracer), combined capacity 130 kg (286 lb). Manually-retractable monowheel, size 5·00-5, with Tost drum brake; size 200 × 50 tailwheel. One-piece cockpit canopy on DG-202, hinged at front to open upwards.

GLASER-DIRKS DG-400
The DG-400 is a self-launching development of the DG-202, from which it differs principally in having a slightly deeper rear fuselage to accommodate the power plant when retracted. Twenty-five had been ordered by early 1981; the prototype (D-KOLL) made its first flight in early May 1981. As with the unpowered versions, the DG-400 is available with either 15 m or 17 m span wings. Deliveries were expected to begin in late 1981.

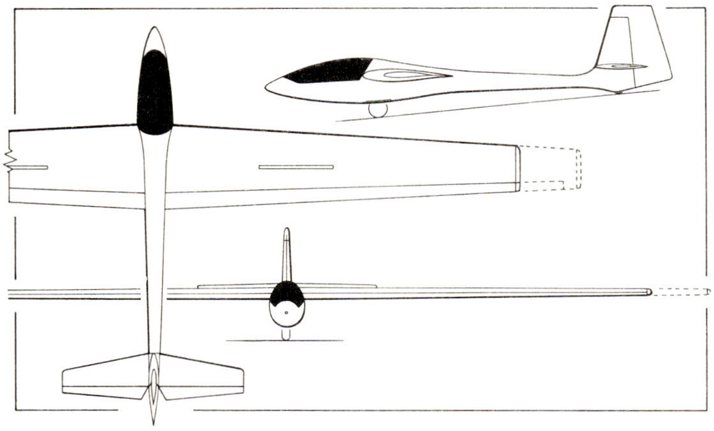

Akaflieg München Mü 28 single-seat sailplane *(Michael A. Badrocke)*

Akaflieg München Mü 27 prototype tandem two-seat sailplane

Blessing Staff Rebell homebuilt motor glider

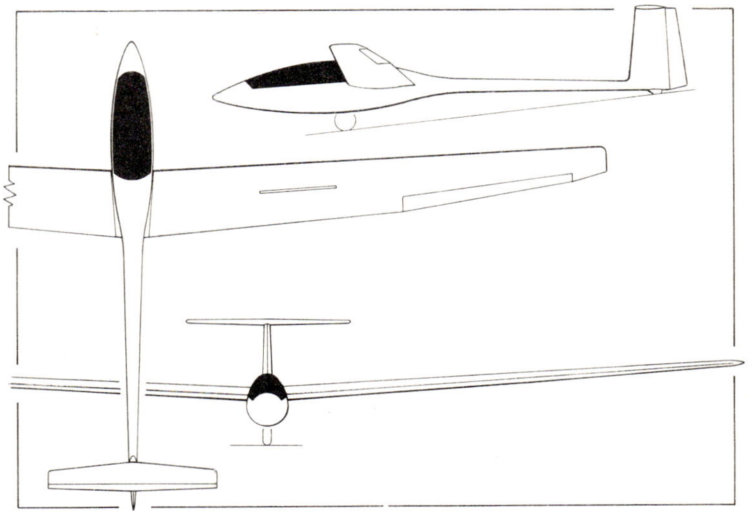

Akaflieg Stuttgart FS-31 tandem two-seat sailplane *(Michael A. Badrocke)*

FVA Aachen FVA-20 Standard Class sailplane *(Peter F. Selinger)*

17 metre span version of the Glaser-Dirks DG-202

Glaser-Dirks DG-101

Glaser-Dirks DG-200 standard 15 metre Class sailplane

Type: Single-seat powered sailplane; *g* limits +6/−4.

Airframe: Recontoured upper fuselage aft of cockpit. Wing incidence −1°. Water ballast provision 90 litres (19·8 Imp gallons). Carbonfibre elevators. Otherwise generally as described for DG-200/17C Version 202. Tyre pressures: main 3·0 bars (43·5 lb/sq in), tailwheel 2·0 bars (29 lb/sq in).

Power Plant: One 32 kW (43 hp) Rotax 501 two-stroke engine, pylon-mounted on fuselage aft of wing trailing-edge and driving a 1·40 m (4 ft 7 in) diameter Hoffmann two-blade fixed-pitch propeller. Installation retracts, electrically, rearward into fuselage when not in use. Single fuselage fuel tank standard, capacity 15 litres (3·3 Imp gallons). Optional 20 litre (4·4 Imp gallon) tank in each wing, in lieu of water ballast, raising total fuel capacity to 55 litres (12 Imp gallons).

Dimensions, external: As DG-200 (15 m and 17 m)

Weights and Loadings:

Weight empty, equipped:	
15 m, 17 m	290 kg (639 lb)
Max water ballast	90 kg (198 lb)
Max T-O weight: 15 m	480 kg (1,058 lb)
17 m	450 kg (992 lb)

Max wing loading: 15 m 48·0 kg/m² (9·84 lb/sq ft)
17 m 42·5 kg/m² (8·71 lb/sq ft)
Max power loading: 15 m 14·9 kg/kW (24·6 lb/hp)
17 m 14·0 kg/kW (23·1 lb/hp)
PERFORMANCE (power off): As DG-200 (15 m and 17 m) except:
Best glide ratio at 59·5 knots (110 km/h; 68·5 mph):
15 m 47
17 m 45
Min sinking speed at 43 knots (80 km/h; 50 mph)
15 m 0·60 m (1·97 ft)/s
17 m 0·45 m (1·45 ft)/s
Stalling speed: 15 m 35 knots (65 km/h; 40·5 mph)
17 m 34 knots (63 km/h; 39·5 mph)
PERFORMANCE (power on, both versions):
Never-exceed speed 146 knots (270 km/h; 168 mph)
Max cruising speed (60% power)
75·5 knots (140 km/h; 87 mph)
Econ cruising speed 70 knots (130 km/h; 81 mph)
Stalling speed 35 knots (65 km/h; 40·5 mph)
Max rate of climb at S/L 234 m (768 ft)/min
Range with max fuel 307 nm (570 km; 354 miles)

GLASFLÜGEL

GLASFLÜGEL GmbH (Deutsch-Brasilianische Flugzeug- und Fahrzeugbau GmbH)

7318 Lenningen 1 Schlattstall, Württ 1

Telephone: 07026 855
Telex: 7267785 gafud
DIRECTOR: Erich Weinzierl

The present company was formed on 4 April 1979. The products of the former Glasflügel company were described in the 1979-80 and earlier editions of *Jane's*. Sailplanes in production in 1981 were as follows:

GLASFLÜGEL HORNET

The Hornet is a derivative of the Club Libelle (1977-78 *Jane's*), from which it differs principally in having a retractable monowheel, provision for water ballast, and an enlarged flush-fitting canopy. The prototype (D-9432) flew for the first time on 21 December 1974. Approx 90 Hornets had been delivered by the beginning of 1980.

TYPE: Single-seat Standard Class sailplane; *g* limits +5·3/−3·25 (safety factor 1·5) .

AIRFRAME: Cantilever mid-wing monoplane with T tail. Wortmann FX-66-17AII-182 wing section. Entire structure of glassfibre monocoque and glassfibre/foam sandwich. Rotating airbrakes on wing trailing-edges; partially mass-balanced ailerons. Elevator spring trim. Retractable unsprung monowheel, with internally-expanding brake; tyre size either 5·00-5, pressure 3·5 bars (51 lb/sq in), or 4·00-4, pressure 5·0 bars (73 lb/sq in). Non-retractable tailwheel. Provision for 170 kg (375 lb) water ballast.

GLASFLÜGEL HORNET C

First flown on 6 April 1979, the Hornet C has a wing of carbonfibre construction, reducing the aircraft empty weight to 210 kg (463 lb). Five examples of this version had been built by the beginning of 1980.

GLASFLÜGEL MOSQUITO B

Glasflügel began building a prototype of this Mosquito version in September 1977. It differs from the original Mosquito (1979-80 *Jane's*) in having GRP ailerons, no fuselage/wing fairings, reduced wing weight and slimmer horizontal tail surfaces.

The Mosquito B flew for the first time on 24 March 1978; approx 90 had been delivered by January 1980.

TYPE: Single-seat 15 metre Class sailplane.

AIRFRAME: Cantilever mid-wing monoplane with T tail. Wortmann FX-67-K-150 wing section. Dihedral 3° on wing upper surface; sweepback 0° 9′ at quarter-chord. Construction generally similar to Hornet, but with GRP mass-balanced ailerons. Camber-changing trailing-edge flaps; spoiler forward of each flap combines with flap to act as trailing-edge airbrake. Flap, aileron and airbrake controls linked automatically. Fixed-incidence tailplane; elevator has spring trim. Retractable unsprung monowheel, with internally-expanding brake; non-retractable semi-recessed tailwheel. Provision for 120 kg (265 lb) water ballast.

GLASFLÜGEL 304

Design of the Glasflügel 304 single-seat 15 metre Class sailplane began in 1979 under the direction of Herr Martin Hansen, and the prototype (D-9304) flew for the first time on 10 May 1980. Deliveries of production aircraft began later that year.

In configuration and construction the 304 is very similar to its predecessor, the Glasflügel 303 Mosquito, the principal structural differences being in the wings, which have trailing-edge flaps of a new HQ 014-1642 aerofoil section; the nose, which is slimmer and more pointed; and the instrument panel, which is integral with the front-hinged, upward-opening canopy. The wings have provision for 115 kg (253 lb) of water ballast.

GLASFLÜGEL 402

A prototype (D-2611) was flown on 16 December 1980 of a version of the Glasflügel 304 fitted with removable wingtips which extend the span to 17 m (55 ft 9¼ in). Carbonfibre is used in construction of the wings, which have the same water ballast capacity as the 304. Fuselage of the 402 is slightly longer.

Production of the Glasflügel 402 was expected to begin in mid-1981.

GROB

GROB FLUGZEUGBAU GmbH & Co KG

Flugplatz Mindelheim-Mattsies, 8939 Mattsies
Telephone: 08268 411
Telex: 53 96 23
DIRECTOR: Ernst Grob

This company built the Schempp-Hirth Standard Cirrus under licence between 1972 and 1975. It is now manufacturing sailplanes and a motor glider of its own design.

GROB ASTIR II

Prototype construction of the Astir CS (Club Standard) single-seat Standard Class sailplane began in March 1974, and this aircraft (D-6102) made its first flight on 19 December 1974. German certification by the LBA was awarded in September 1975. The Astir CS77, first flown on 26 March 1977, introduced modified tailplane actuation.

The Astir is produced in two forms: the **Standard Astir II**, and the **Club Astir II**, the latter having a non-retractable monowheel and no water ballast provision.

TYPE: Single-seat Standard and Club Class sailplanes.

AIRFRAME: Cantilever mid-wing monoplane with T tail. Eppler E 603 wing section. Wing sweepback 1°. Carbonfibre main spar; wings and tail surfaces have glassfibre/epoxy resin sandwich skin. Ailerons are of glassfibre sandwich, with elastic gap seals. Schempp-Hirth upper-surface aluminium airbrakes. Glassfibre semi-monocoque fuselage, with towing/launching hook. Carbonfibre used to reinforce fuselage and fin, and for canopy frame. Retractable Tost monowheel standard (non-retractable on Club version), tyre pressure 2·5 bars (36·3 lb/sq in), with internally-expanding drum brake; rubber-sprung tailwheel. Canopy opens sideways to starboard. Provision for 90 kg (198 lb) water ballast in Standard Astir II.

GROB SPEED ASTIR IIB

The prototype Speed Astir II made its first flight in mid-April 1978. More than 50 were then on order; deliveries began in 1979. After the first 25 aircraft, the fuselage was lengthened from 6·60 m (21 ft 7¾ in) to 6·80 m (22 ft 3¾ in).

TYPE: Single-seat 15 metre Class sailplane.

AIRFRAME: Cantilever mid-wing monoplane with T tail. Generally similar to Standard Astir except for wings, which have reduced area, Eppler E 662 section, a carbonfibre spar and other load-bearing components, and 'elastic flaps' (ie, the gaps are elastically sealed with a thin glassfibre skin) which can be deflected upward or downward. Ailerons actuated by a similar system. Provision for 180 litres (39·5 Imp gallons) water ballast.

GROB G 102 SERIES III

Representing a further development of the Astir II series, the G 102 has Eppler E 603 aerofoil-section wings, lower wing loading, better take-off and landing characteristics, and a longer cockpit with a more upright seat. It is available in three versions:

G 102 Club III. Basic Club Class version, with non-retractable monowheel and tailwheel, no water ballast tanks, and a choice of drum or disc brakes. Wings and fuselage of GRP construction. First flown in late 1980.

G 102 Club IIIb. Monowheel moved further aft; nosewheel added. Otherwise as Club III.

G 102 Standard III. Standard Class version, having retractable monowheel with drum brake, non-retractable tailwheel, and water ballast tanks (capacity 100 kg; 220 lb).

GROB G 103 TWIN II

This tandem two-seat sailplane for training and club flying is the successor to the Twin Astir (1979-80 *Jane's*), from which it differs mainly in having lower-mounted wings, a more streamlined fuselage, modified landing gear, improved cockpit layout and reduced weight. General appearance is shown in an accompanying three-view drawing. First flight was made in late 1979.

TYPE: Tandem two-seat training and club sailplane.

AIRFRAME: Cantilever low/mid-wing monoplane, of similar construction to Astir. Wings have ailerons and trailing-edge 'elastic flaps', similar to those of Speed Astir. Cantilever T tail. Non-retractable nosewheel and monowheel, both semi-recessed; plus tailwheel. No water ballast provision.

GROB G 109

The prototype of the G 109 (D-KBGF) flew for the first time on 14 March 1980.

TYPE: Side-by-side two-seat motor glider.

AIRFRAME: Cantilever low-wing monoplane with T tail. Wings of E 572 (modified Eppler E 603) section. Construction mainly of glassfibre (GfK). Non-retractable tailwheel-type landing gear, with Scheibe hydraulic brakes. One-piece fully-transparent canopy.

POWER PLANT: One 59 kW (80 hp) Limbach L 2000 EJ engine, driving a Hoffmann two-blade variable-pitch wooden propeller. Fuel capacity 80 litres (17·5 Imp gallons).

HÄNLE

URSULA HÄNLE

Postfach 1112, 5438 Westerburg

HÄNLE H 101 SALTO

Last described in the 1978-79 *Jane's*, under the entry for the former Start + Flug company, the Salto continues to be available from Frau Ursula Hänle, and in 1980 won the Club Class competitions at both the central and regional gliding championships in West Germany.

Design of the Salto is based upon that of the Glasflügel Standard Libelle, from which it differs chiefly in having a V tail. It first flew in 1971, and is certificated by the LBA and FAA for both Utility and Aerobatic category flying. Deliveries had totalled 60 by the Spring of 1977, before the closure of the Start + Flug company.

TYPE: Single-seat Standard and Club Class sailplane.

AIRFRAME: Cantilever mid-wing monoplane. Wings have Conticell sandwich shell; HH-type glassfibre spar caps. Glassfibre ailerons and four flush-fitting airbrakes on trailing-edges. Glassfibre monocoque fuselage. Cantilever V tail (included angle 99°), with glassfibre fixed surfaces and glassfibre/honeycomb sandwich control surfaces. Non-retractable semi-recessed monowheel, size 300 × 100, with glassfibre shock-absorption and internally-expanding brake. Non-retractable semi-recessed tailwheel.

HOFFMANN

WOLF HOFFMANN FLUGZEUGBAU KG

8048 Haimhausen bei München, Hauptstrasse 13
Telephone: 08133/843
DIRECTOR: Dipl-Ing Wolf D. Hoffmann

HOFFMANN H-36 DIMONA

This two-seat motor glider flew for the first time on 9 October 1980. Series production is undertaken at Friesach in Austria. Deliveries began in Summer 1981.

TYPE: Side-by-side two-seat motor glider.

AIRFRAME: Cantilever low/mid-wing monoplane, constructed mainly of GfK. Wing section Wortmann FX-63-137. Upper-surface airbrakes. Wings are attached independently to fuselage by bolts, and can be folded back alongside fuselage for transportation and storage. Fuselage sidewalls and frames are strengthened with extensive GfK rovings. Double shell in cockpit area. Cantilever T tail. Non-retractable tailwheel-type landing gear, with 6 in main wheels and steerable fully-castoring tailwheel. Cantilever GfK mainwheel legs. Fairings on main legs and wheels. Cockpit canopy hinged at rear to open upwards. Baggage space aft of seats.

POWER PLANT: One 59·5 kW (80 hp) Limbach L 2000 EBI engine, driving a Hoffmann HO-V62/160 two-blade propeller. Fuel tank capacity 80 litres (17·5 Imp gallons).

Glasflügel Hornet C 15 metre competition sailplane *(Peter F. Selinger)*

Glasflügel Mosquito B single-seat 15 m Class sailplane *(Peter F. Selinger)*

Glasflügel 304 single-seat 15 metre Class sailplane *(Peter F. Selinger)*

Grob Club Astir II single-seat sailplane *(Peter F. Selinger)*

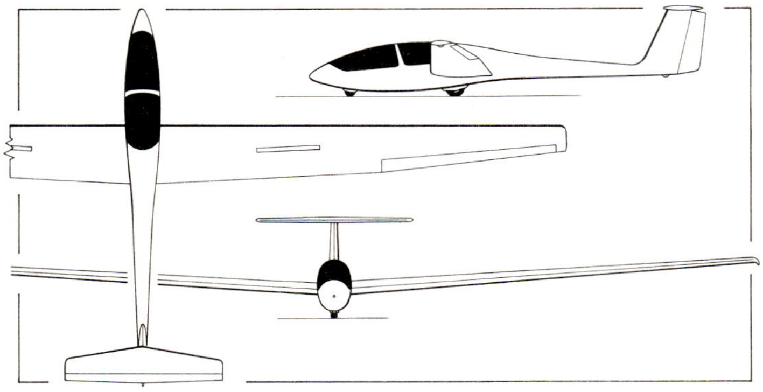

Grob G 103 Twin II two-seat training and club sailplane *(Michael A. Badrocke)*

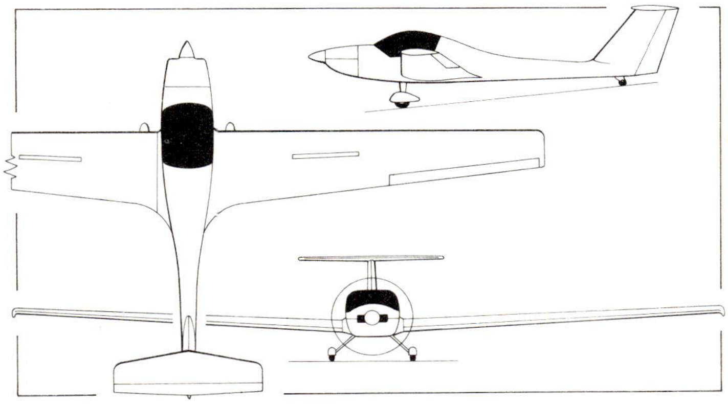

Grob G 109 two-seat motor glider *(Michael A. Badrocke)*

Rhönlerche IIM motorised glider built by Ing Wilhelm Knechtel *(Peter F. Selinger)*

Hänle H 101 Salto single-seat sailplane *(Peter F. Selinger)*

Hufnagel Zamberle single-seat homebuilt glider *(Peter F. Selinger)*

HUFNAGEL

BERNHARD HUFNAGEL

Bernhardswiesen 53, 8800 Ansbach

HUFNAGEL ZAMBERLE

This single-seat, open-cockpit primary glider, designed and built by Herr Hufnagel, made its first flight on 23 March 1979. It is of mid-wing configuration (see accompanying illustration), and fabric-covered plywood construction.

SPECIFICATION:

Wing span	8·00 m (26 ft 3 in)
Wing area, gross	12·00 m² (129·2 sq ft)
Length overall	5·00 m (16 ft 4¾ in)
Weight empty	48·6 kg (107 lb)
Max T-O weight	130 kg (286·5 lb)
Max speed (approx)	19·5 knots (36 km/h; 22 mph)
Best glide ratio	15-17

KNECHTEL

ING WILHELM-J. KNECHTEL

Raulinenstrasse 24, 6301 Rodheim

KNECHTEL RHÖNLERCHE IIM

Herr Knechtel has converted a Rhönlerche (Rhône Lark) glider into a motor glider by installing a 32 kW (43 hp) Volkswagen 1500 motorcar engine above the wings, driving two small two-blade propellers mounted on outriggers. This aircraft (D-KABQ) is shown in an accompanying illustration.

Herr Knechtel also has under construction a two-seat motor glider, powered by a Volkswagen minibus engine and utilising a fuselage of his own design in combination with the wings of a Grob Twin Astir sailplane.

Specification (Rhönlerche IIM):	
Wing span	13·00 m (42 ft 7¾ in)
Length overall	7·00 m (22 ft 11½ in)
Weight empty, equipped	405 kg (893 lb)
Max T-O weight	535 kg (1,179 lb)
Max level speed	86 knots (160 km/h; 99 mph)
Service ceiling	4,500 m (14,765 ft)
Range	135 nm (250 km; 155 miles)

LINDNER
FIBERGLAS-TECHNIK RUDOLF LINDNER GmbH & Co KG

Ortsstrasse 70, 7959 Walpertshofen

LINDNER PHOEBUS B 3

Features of the Phoebus B 3 are 'elastic' flaps and automatically-actuated plain flaps, developed jointly by Lindner and Prof Eppler and first flight tested in 1978. The aircraft is illustrated in an accompanying photograph.

MFB
MISTRAL-FLUGZEUGBAU (DIPL-ING STRAUBER GmbH & Co KG)

Flugplatz, 8728 Hassfurt
Telephone: (09521) 4730
Director: Dipl-Ing M. Strauber

MFB MODEL 2 MISTRAL-C

The original Mistral was described and illustrated under the Strauber heading in the 1976-77 *Jane's*. Design of a second model, known as the ISF Mistral-C, was started in October 1974, and this made its first flight on 21 October 1976. Twenty-five of this model had been completed by the beginning of 1980.

A new production facility was completed in 1980, which from January 1981 was continuing manufacture of the Mistral-C at the rate of five per month.

Type: Single-seat Club Class sailplane; *g* limits +5·8/−3·8.

Airframe: Cantilever shoulder-wing monoplane with T tail. Wortmann wing sections: FX-61-163 at root, FX-60-126 at tip. Dihedral 4° 18′. Sweepforward 1° at quarter-chord. Structure: main wings and tail of GRP/foam/Conticell CC60 sandwich, ailerons of GRP. Schempp-Hirth aluminium airbrakes on upper wing surfaces. GRP monocoque fuselage. Fixed-incidence tailplane, with spring trim. Non-retractable monowheel, tyre pressure 2·5 bars (36·25 lb/sq in), with Tost brake; tailskid.

ROLLADEN-SCHNEIDER
ROLLADEN-SCHNEIDER FLUGZEUGBAU GmbH

6073 Egelsbach, Mühlstrasse 10 (Postfach 1130)
Telephone: (06103) 4126
Officers: Walter Schneider and Dipl-Ing Wolf Lemke

All models of Rolladen-Schneider's LS1 sailplane are now out of production. A total of 224 examples of the models LS1-o/a/b/c/d/e/ef were built, as well as 240 examples of the LS1-f. Current (1981) production is concerned with two versions of the LS3, the LS4, and development of the new LS5.

ROLLADEN-SCHNEIDER LS3

The LS3 was developed from the LS1-f. Design and construction began in 1975, and the prototype flew for the first time on 4 February 1976. A total of 491 of all versions had been built by the beginning of 1981.

Three versions have been built, as follows:

LS3. Standard FAI 15 metre Class sailplane. Total of 155 built; no longer in production.

LS3-a. Similar to LS3, but with horizontal and vertical tail surfaces of greater area and different aerofoil sections, and reduced empty weight. First flown on 18 February 1978; deliveries began in Spring 1978. Total of 272 built by beginning of 1981.

LS3-17. Extended-span (17 metre) version of LS3; first flown on 10 February 1979. Total of 64 built by beginning of 1981.

The following description applies to the LS3-a:

Type: Single-seat FAI 15 metre Class sailplane.

Airframe: Cantilever mid-wing monoplane, with modified Wortmann wing section. Dihedral 4°. Ailerons and flaps over entire trailing-edge. Schempp-Hirth airbrakes on upper wing surfaces. Semi-monocoque fuselage. Cantilever T tailplane, with elevator. Entire structure of GfK sandwich. Retractable Tost rubber-sprung monowheel, tyre pressure 2·94 bars (43 lb/sq in), with drum brake; tailskid. Provision for 150 kg (330 lb) water ballast.

ROLLADEN-SCHNEIDER LS4

Design of this Standard Class sailplane began in the Winter of 1979/80, and the prototype (D-6680) flew for the first time on 28 March 1980. By early 1981 a total of 35 had been built, of 123 then on order.

The LS4 utilises a modified LS3-a fuselage combined with a new, thin-section wing, mid-mounted on the fuselage.

In the 1981 World Championships at Paderborn, West Germany, LS4s took eight of the first ten places in the Standard Class, including all of the first seven places.

Type: Single-seat Standard Class sailplane; *g* limits +5·3/−2·65 at 97 knots (180 km/h; 112 mph), +4·0/−1·5 at 146 knots (270 km/h; 168 mph).

Airframe: Cantilever mid-wing monoplane, of glassfibre/Conticell foam construction. Wings have thickness/chord ratio of 15% at root and 13% at tip (modified Wortmann sections), 3° 30′ dihedral and 0° incidence. Schempp-Hirth upper-surface airbrakes. No flaps. Provision for 140 litres (30·8 Imp gallons) water ballast in wings. Fuselage, tail unit and landing gear as described for LS3-a. Tost Kobold brake on main wheel. Canopy hinged at front to open upward.

ROLLADEN-SCHNEIDER LS5

The LS5 is a 22 m (72 ft 2¼ in) span Open Class sailplane of GRP construction, with a new carbonfibre wing of thin Wortmann section, fitted with four flaps/airbrakes, of which the inner segments can be deflected 60°. The aircraft will have a fuselage similar to that of the LS3-a, with a larger tail unit and modified landing gear. Performance is expected to be at least as good as that of the LS3-17.

SCHEIBE
SCHEIBE FLUGZEUGBAU GmbH

Head Office and Works: D-8060 Dachau, August-Pfaltz-Strasse 23, Postfach 1829, near Munich
Telephone: Dachau 4047, 6813 and 81886
Managers: Dipl-Ing Egon Scheibe and Ing Christian Gad

Scheibe Flugzeugbau GmbH was founded at the end of 1951 by Dipl-Ing Scheibe; its first type produced in quantity was the Mü-13E Bergfalke I.

Subsequently, Scheibe has built many new types of sailplane, and since 1957 has become the major producer of motor gliders in the Federal Republic of Germany. Currently in production are the SF-25C, C-S and K Falke, SF-25E Super-Falke, SF-28A Tandem-Falke and SF-36 motor gliders, in addition to the Bergfalke-IV and SF-34 sailplanes.

Scheibe has built more than 2,000 aircraft of various types, in addition to many kits for home construction by amateurs. Gliders of Scheibe design are being built under licence by gliding clubs as well as by foreign companies.

SCHEIBE SF-25C and C-S FALKE '80 (FALCON)

The SF-25C is an improved version of the SF-25B Falke, to which it is structurally similar. The primary difference is in the use of a more powerful engine, giving an enhanced performance.

By January 1981 a total of 312 SF-25C Falkes had been built by Scheibe; a further 50 were built under licence by Sportavia in Germany. Type certification was granted in September 1972.

With a Hoffmann feathering propeller, adjustable engine cowl flap and slightly modified fuselage, the aircraft is known as the **SF-25C-S** and has a best glide ratio of 25 at 43·5 knots (80 km/h; 50 mph). Twenty of this version had been built by January 1981.

Optional features include an additional exhaust outlet and a slower-turning propeller. With this installation the nominal noise level is reduced to less than 60 dB. In addition, Scheibe has developed an optional folding wing, which reduces the span to 9·50 m (31 ft 2 in) for easier hangar storage.

Current models, known as **C-Falke '80**, introduced a number of design improvements, as listed in the 1977-78 *Jane's*. With fully-folding wings, they are known as **K-Falke '80**.

Type: Side-by-side two-seat motor glider, particularly suitable for basic and advanced training.

Airframe: Cantilever low-wing monoplane. Forward-swept wooden wings, with airbrakes and aerodynamically balanced ailerons. Wing folding optional. Fabric-covered welded steel tube fuselage; forward section coated with laminated glassfibre. Wooden tail unit. Non-retractable rubber-sprung monowheel with brake and aerodynamic fairing; steerable tailwheel; sprung outrigger stabilising wheel under each wing. Alternative twin-wheel main gear available optionally, with streamline wheel fairings.

Power Plant: One 48·5 kW (65 hp) Limbach SL 1700 EA I modified Volkswagen engine, driving a two-blade propeller. Electric starter. Fuel in single fuselage tank, capacity 45 litres (9·9 Imp gallons) standard, 55 litres (12·1 Imp gallons) optional.

SCHEIBE SF-25E SUPER-FALKE

Developed from the SF-25C-S, the Super-Falke has an increased wing span and a rubber-sprung monowheel; a cabin heater is fitted as standard. Production aircraft have a tailwheel, and upper-surface Schempp-Hirth airbrakes.

A 48·5 kW (65 hp) Limbach SL 1700 EA I engine is fitted, with a 12V battery and alternator for electrical engine starting. Fuel capacity is 45 litres (9·9 Imp gallons). Optionally, the outer wing panels can be made foldable to facilitate transportation and storage.

The Super-Falke was flown for the first time in the Summer of 1974; 58 had been delivered by January 1981.

SCHEIBE SF-28A TANDEM-FALKE

The Tandem-Falke, as its name implies, is a further development of the Falke series of motor gliders in which the two seats are arranged in tandem. Design began in 1970, and the prototype (D-KAFJ) flew for the first time in May 1971, powered by a 33·5 kW (45 hp) Stamo MS 1500 engine. Details apply to the production version, of which 117 had been built by January 1981.

Type: Tandem two-seat motor glider.

Airframe: Cantilever low-wing monoplane. Wing section Gö 533. Single-spar wooden wings, with wooden ailerons. No flaps. Spoiler on upper surface of each wing. Fabric-covered steel tube fuselage. Wooden tail unit, with trim tab in elevator. Non-retractable monowheel, with internal brake, and steerable tailwheel. Nylon leg with outrigger stabilising wheel under each wing. Can be flown solo from front seat, with space for 90 kg (198 lb) of baggage on rear seat.

Power Plant: One 48·5 kW (65 hp) Limbach SL 1700 EAI engine, driving a Hoffmann two-blade feathering (optionally, fixed-pitch) propeller. 12V alternator standard, for engine starting. Fuel capacity 40 litres (8·8 Imp gallons).

SCHEIBE BERGFALKE-IV

The Bergfalke-IV is a developed version of the Bergfalke-III, with a new wing which provides improved performance. Construction of the prototype began in early 1969 and first flight was accomplished a few months later. A total of 70 had been built by the beginning of 1981.

Type: Tandem two-seat training and competition sailplane; *g* limit +8 (ultimate).

Airframe: Cantilever mid-wing monoplane, primarily of plywood-covered wooden construction except for welded steel tube fuselage, which is covered with moulded glassfibre shell (nose section) and fabric. Wortmann wing sections: SO 2 at root, SO 2/1 at tip. Dihedral 3°. Wooden ailerons and Schempp-Hirth airbrakes. Flettner trim tab on starboard elevator. Non-retractable monowheel and tailwheel.

SCHEIBE SF-34 DELPHIN

Design of the SF-34 sailplane began in 1978, and it flew for the first time on 28 October that year. Seven production-standard aircraft had been completed by the beginning of 1981.

Type: Tandem two-seat training and sporting sailplane.

Airframe: Cantilever mid-wing monoplane. Wortmann wing sections: FX-61-184 at root, FX-60-126 at tip. Glassfibre roving main spar. Wings and tail unit of GfK honeycomb sandwich construction; fuselage is a GfK shell. Schempp-Hirth airbrake in upper surface of each wing. Non-retractable semi-exposed monowheel and nosewheel, tyre sizes 5·00-4 and 265 × 85 respectively; tailskid. Tandem seats under frameless one-piece sideways-opening flush canopy. Towing hooks under nose and at CG.

Lindner Phoebus B 3 sailplane, with experimental flap system
(via Peter F. Selinger)

Prototype of the Rolladen-Schneider LS4 Standard Class sailplane

Scheibe Bergfalke-IV two-seat training and competition sailplane
(Peter F. Selinger)

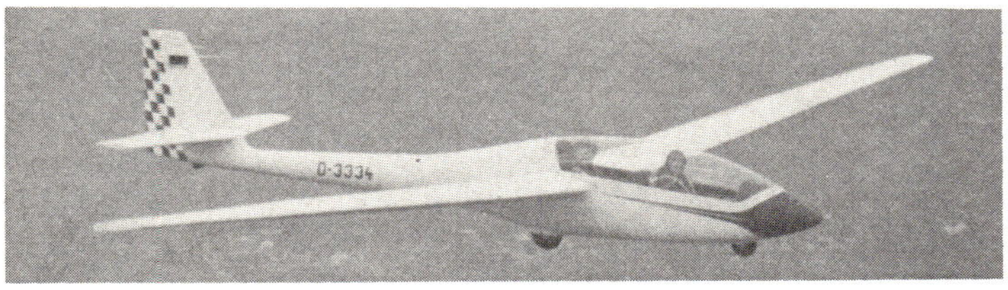

Scheibe SF-34 two-seat training and sporting sailplane

Scheibe SF-36 two-seat motor glider

MFB Model 2 Mistral-C single-seat Club Class sailplane

Rolladen-Schneider LS3-17 single-seat 17 metre sailplane

Rolladen-Schneider LS3-a 15 metre Class sailplane

Scheibe SF-25C Falke '80, version with twin-wheel main gear

Scheibe SF-25K Falke '80, with fully-folding wings

SCHEIBE SF-36

A prototype of this two-seat motor glider (D-KOOP) was flown for the first time in the Summer of 1980; production began in January 1981. It is of GfK construction, utilising the wings and tail unit of the SF-34 sailplane.

Type: Side-by-side two-seat motor glider.

Airframe: Cantilever low-wing monoplane. Wortmann wing sections: FX-61-184 at root, FX-60-126 at tip. Dihedral 1° 30′. Wings detachable for transportation and storage. Schempp-Hirth upper-surface airbrakes. Twin-wheel main landing gear similar to that of C-Falke '80, or single main wheel plus underwing outriggers. Forward-sliding cockpit canopy. Baggage space aft of seats.

Power Plant: One 59·5 kW (80 hp) Limbach L 2000 EI engine, driving a two-blade fixed-pitch propeller. Feathering propeller optional. Fuel capacity 95 litres (20·9 Imp gallons).

SCHEMPP-HIRTH

SCHEMPP-HIRTH FLUGZEUGBAU GmbH

7312 Kirchheim-Teck, Krebenstrasse 25, Postfach 1143
Telephone: (07021) 2441 and 6097
Telex: 7267817 hate
Directors:
Dipl-Ing Klaus Holighaus
Wilfried Müller

Schempp-Hirth specialises in the production of high-performance Open Class and 15 metre Class sailplanes. Dipl-Ing Klaus Holighaus is 100% shareholder.

Production by Schempp-Hirth of the original Cirrus ended in late 1971, after 120 had been built; manufacture of the Cirrus and Standard Cirrus is now undertaken by VTC in Yugoslavia (which see).

SCHEMPP-HIRTH NIMBUS 2C

Generally similar to the HS-3 Nimbus (1972-73 *Jane's*), from which it was developed, the Nimbus 2 differs principally by having reduced span and a wing built in four pieces to limit weight and dimensions for rigging, storage and trailer transport. The first flight was made in April 1971; in the 1976 World Championships at Räyskälä,

Finland, Nimbus 2s took no fewer than 14 of the first 25 places, and in the 1981 World Championships four of the first 12 places in the Open Class were taken by Nimbus 2/2Cs. The world distance record to a goal, for single-seat gliders, was set by a group of three Nimbus 2s flown by S. H. Georgeson, B. L. Drake and D. N. Speight, who covered 677·25 nm (1,254·26 km; 779·36 miles) in New Zealand on 14 January 1978.

Production of the original Nimbus 2, with all-moving tailplane, totalled 132. Aircraft from No. 133 onward (Nimbus 2B) introduced a fixed-incidence tailplane. The current model, the Nimbus 2C, is similar, but has a new brake/flap system, higher gross weight and (optionally) wings and horizontal tail surfaces of carbonfibre construction. A total of 100 Nimbus 2B/2C had been built by 1 January 1981.

The following description applies primarily to the Nimbus 2C:

Type: Single-seat Open Class sailplane; *g* limit +10·5.

Airframe: Cantilever mid-wing monoplane. Wortmann wing section. Dihedral 2°. Sweepback at leading-edge 1° on inner panels, 2° on outer panels. Glassfibre/foam sandwich (optionally carbonfibre) wings and tail. Combined flap/airbrake on each wing trailing-edge. Central tubular steel fuselage framework. Glassfibre shell, stiffened with bonded-in foam bulkheads. Cantilever T tail, with fixed-incidence tailplane. Retractable rubber-sprung monowheel, tyre pressure 3·45 bars (50 lb/sq in). Tost drum brake. Tailskid. Canopy opens sideways to starboard. Provision for up to 250 kg (550 lb) of water ballast.

SCHEMPP-HIRTH HS-7 MINI-NIMBUS C

As its name implies, the Mini-Nimbus is a scaled-down version of the Nimbus 2, the chief differences being a shorter wing, fitted with trailing-edge flaps/airbrakes, designed to qualify under the new FAI regulations for 15 metre Standard Class sailplanes. With optional fixed-incidence tailplane it is known as the Mini-Nimbus B; with the same new features as listed for the Nimbus 2C, it is known as the Mini-Nimbus C.

The prototype Mini-Nimbus (D-3119) flew for the first time on 18 September 1976. A total of 158 Mini-Nimbus had been delivered by 1 January 1981. The following details apply primarily to the Mini-Nimbus B/C:

Type: Single-seat 15 metre Class sailplane.

Airframe: Cantilever mid-wing monoplane. Wortmann wing section. Dihedral 3°. Glassfibre/foam sandwich (optionally carbonfibre) wings and tail; flaps/airbrakes and ailerons are glassfibre shells. Glassfibre fuselage shell, stiffened with bonded foam rings and a central tubular steel frame. Fixed-incidence T tailplane; fin integral with fuselage. Manually-retractable rubber-sprung monowheel, tyre pressure 3·45 bars (50 lb/sq in), with Tost drum brake. Provision for up to 190 kg (418 lb) of water ballast.

SCHEMPP-HIRTH NIMBUS 3

Except for the forward fuselage, which is of glassfibre, the Nimbus 3 is built extensively of carbonfibre, and has mid-mounted wings with Wortmann/Holighaus thin section (thickness/chord ratio 14%). Built in four segments, the wings are fitted with ailerons, trailing-edge flaps, airbrakes, and tanks for 310 kg (683 lb) of water ballast. The T tailplane has a separate elevator.

The prototype Nimbus 3 (D-2111) flew for the first time on 21 February 1981; by April 1981 the whole of the first year's output had been sold. The Nimbus 3 took the first three places in the Open Class of the World Championships at Paderborn, West Germany, in June 1981.

SCHEMPP-HIRTH JANUS

The Janus is of all-glassfibre construction. Original design work, begun by Dipl-Ing Holighaus in 1969, was continued from early 1972 onward, and the prototype made its first flight in the Spring of 1974.

A Janus flown by E. Mouat Biggs and S. Murray set two current speed records over triangular courses in South Africa in November 1977. On 17 November, it averaged 75·631 knots (140·068 km/h; 87·034 mph) over 500 km; on 21 November it averaged 79·476 knots (147·19 km/h; 91·46 mph) over 100 km. The record of 75·85 knots (140·48 km/h; 87·29 mph) over a 300 km course was set by Erwin Müller and Otto Schäffner of West Germany in a Janus on 30 November 1979, in South Africa.

Production began in January 1975 with the second, improved aircraft; a **Janus B** version became available in March 1978, with fixed-incidence tailplane; the **Janus C** has 20 m span carbonfibre wings and carbonfibre tailplane. A total of 125 Janus, Janus Bs and Janus Cs had been delivered by early 1981.

The **Janus CM** is a powered version of the C, first flown in 1978; it has a 37 kW (50 hp) engine, pylon-mounted aft of the cockpit and retracting into the top of the fuselage when not in use. Empty and max T-O weights are 465 kg (1,025 lb) and 680 kg (1,500 lb) respectively. Five Janus CMs had been completed by early 1981.

The following description applies to the Janus B:

Type: Tandem two-seat high-performance training sailplane; *g* limits +5·3/−2·65.

Airframe: Cantilever mid-wing monoplane. Wortmann wing sections: FX-67-K-170 at root, FX-67-K-15 at tip. Dihedral 4°. Sweepforward 2° on leading-edge. Glassfibre/foam sandwich wings, with glassfibre monocoque ailerons, trailing-edge flaps and Schempp-Hirth upper-surface airbrakes. Glassfibre monocoque fuselage with bonded-in foam bulkheads. Cantilever glassfibre/foam sandwich fixed-incidence T tailplane, with elevator. Non-retractable semi-recessed monowheel and nosewheel. Continental tyres: pressure 2·69 bars (39 lb/sq in) on main wheel; pressure 0·79 bars (11·5 lb/sq in) on nosewheel. Tost drum brake on main wheel. Bumper under rear fuselage. Tail drag 'chute. Canopy opens sideways to starboard. Provision for 200 kg (440 lb) of water ballast.

SCHEMPP-HIRTH VENTUS

First flown on 3 May 1980, the Ventus (Latin for wind) features a thin-section carbonfibre wing. It is available with a choice of two fuselages, the Ventus b having an overall length of 6·53 m (21 ft 5 in) and a slightly larger cockpit for pilots more than 1·75 m (5 ft 9 in) tall.

Ventus sailplanes took 3rd, 4th and 7th places in their class at the 1981 World Championships at Paderborn, West Germany.

Type: Single-seat 15 metre Class sailplane.

Airframe: Cantilever mid-wing monoplane, with newly developed Wortmann/Holighaus/Althaus thin-section wings of carbonfibre construction, including skins. Two-segment ailerons. Schempp-Hirth trailing-edge flaps/airbrakes, as on Nimbus 2C. Fuselage centre-frame of steel tube to carry main wing and landing gear loads. Cantilever fixed-incidence T tailplane, with elevators; rudder area increased on Ventus b. Retractable monowheel; tail bumper. One-piece cockpit canopy, opening sideways to starboard. Provision for 150 kg (331 lb) of water ballast.

SCHLEICHER

ALEXANDER SCHLEICHER SEGELFLUGZEUGBAU

Head Office and Works: D-6416 Poppenhausen/Wasserkuppe

Telephone: (06658) 225

This company is one of the oldest manufacturers of sailplanes in the world. Its founder, Alexander Schleicher, won a contest for training sailplanes in 1927 at the Wasserkuppe. In the same year he built at Poppenhausen a small factory for manufacturing gliders, two of his best-known pre-war products being the Rhönbussard and Rhönadler, designed by Hans Jacobs.

During the second World War, the factory was engaged in the repair of Baby IIb sailplanes. For a time afterwards it became a furniture factory, but it began producing sailplanes once more in 1951.

Descriptions of sailplanes in current production follow. Those designed by Ing Rudolf Kaiser are prefixed ASK; those with ASW designations are designed by Ing Gerhard Waibel. Schleicher also manufactures and markets spare parts, constructional materials and dust- and weather-proof covers for sailplanes.

SCHLEICHER ASK 13

This tandem-seat sailplane was developed from the K 7, which is in worldwide use by gliding clubs. The prototype first flew in July 1966.

Production by Schleicher, which totalled 617 of these aircraft, has ended, but the ASK 13 is still being manufactured, under licence, by another Federal German company.

Type: Tandem two-seat training and high-performance sailplane; *g* limit +4·0 (safety factor of 2).

Airframe: Cantilever mid-wing monoplane. Wing section developed from Göttingen 535 and 549. Sweepforward at quarter-chord 6°. Dihedral 5°. Single-spar wooden wings, with fabric covering. Wooden ailerons, with plywood covering; Schempp-Hirth metal airbrakes above and below each wing. Welded steel tube fuselage with spruce formers and fabric main covering. Nose made of glassfibre. Turtledeck aft of canopy is plywood shell. Cantilever wooden tail unit, plywood-covered except for fabric-covered rear portion of rudder and elevators. Flettner tab in starboard elevator. Non-retractable sprung monowheel, with Tost drum brake; skid in front of wheel; steel tailskid. Canopy opens sideways to starboard.

SCHLEICHER ASW 19

This sailplane, which flew for the first time on 23 November 1975, is essentially an improved version of the ASW 15 B (1977-78 *Jane's*), from which it differs in having a T tail unit. Production began in the Spring of 1976, and approx 350 had been built by the beginning of 1981. The ASW 19 took 1st place in the Standard Class at the 1978 World Championships at Châteauroux, France.

There are two current production versions:

ASW 19B. Standard Class version, with retractable monowheel, water ballast provision and, from October 1979, double-panelled airbrakes.

ASW 19 Club. Club Class version, with non-retractable unsprung monowheel, no water ballast provision and larger-area double-panelled airbrakes. Best glide ratio of 36; other data as for ASW 19B.

Type: Single-seat Standard Class and Club Class sailplane.

Airframe: Cantilever mid-wing monoplane. Wortmann wing sections: FX-61-163 at root, FX-60-126 at tip. Dihedral 2°. Glassfibre roving wing main spar. Wings and tail surfaces of glassfibre/foam sandwich, with Schempp-Hirth metal upper-surface airbrakes. Glassfibre/honeycomb sandwich fuselage, with reinforced keel. T tailplane with elevator. Retractable monowheel (non-retractable on Club version), size 5·00-5, with internal drum brake; rubber tailskid. Canopy hinged at front and opens upward. Provision for 100 kg (220 lb) water ballast (19B only). Towing hook at CG; optional second hook in nose.

SCHLEICHER ASW 20

The ASW 20 is intended to take advantage of the March 1975 CIVV regulations for unlimited 15 metre sailplanes, and is fitted with trailing-edge flaps as well as large upper-surface spoilers. It has an additional high-drag flap range incorporating a special mechanism to eliminate pitch and airspeed changes when changing flap position between 30° and 55°.

The prototype ASW 20 made its first flight on 29 January 1977, and aircraft of this type won 2nd, 3rd, 4th, 6th and 7th places in their class at the 1978 World Championships. Approx 350 had been built by the beginning of 1981. A version designated **ASW 20F** is one of five models being built under licence in France by Centrair (which see); this received French certification on 31 March 1978.

Seven of the first ten places in the 1981 World Championships, including 1st and 2nd, were won by ASW 20s.

Type: Single-seat 15 metre Class sailplane.

Airframe: As described for ASW 19B, but with Wortmann FX-62K-131 (modified) wing section at root, FX-60-126 at tip, no dihedral, addition of wing flaps, no option for nose towing hook, and max water ballast increased to 120 kg (265 lb).

SCHLEICHER ASW 20 L

The ASW 20 L (for Lang: long), first flown in May 1977, is a version of the ASW 20 with detachable outer wing panels, increasing the span to 16·59 m (54 ft 5¼ in) for Open Class competition. Production began in late 1978, and approx 100 had been built by the beginning of 1981. It is produced also by Centrair in France as the **ASW 20FL**.

Type: Single-seat Open Class sailplane; *g* limits +5·3/−2·65.

Airframe: Generally as for ASW 20 except for wings, which have detachable span-increasing outer panels and no provision for water ballast.

SCHLEICHER ASK 21

Designed by Ing Kaiser, the ASK 21 prototype flew for the first time on 6 February 1979; it received LBA certification on 18 April 1980. Approx 60 had been built by the beginning of 1981.

Type: Tandem two-seat competition and training sailplane.

Airframe: Cantilever mid-wing monoplane. Two-piece wings, primarily of GRP/foam sandwich construction with GRP roving spar. Wortmann wing sections: FX-S-02-196 from root to inboard of ailerons, FX-60-126 at tip. Dihedral 4°. Schempp-Hirth upper-surface metal airbrakes. Fixed T tailplane and separate elevator. Non-retractable semi-recessed nosewheel (size 4·00-4) and rubber-sprung monowheel (size 5·00-5), the latter with hydraulic disc brake; steel-shod rubber tailskid standard, pneumatic tailwheel optional. Front and rear canopies hinged at front and rear respectively, both opening upward.

SCHLEICHER ASW 22

Extensive use is made of CRP, GRP and synthetic fibre reinforced plastics in the construction of this new Open Class mid-wing, T tailed sailplane, which has a standard wing span of 22 m, extendable to 24 m by the addition of detachable tip extensions. A 21 m span version, with higher max T-O weight, is also to be made available for competition and record attempts in rough weather conditions. Combined flaps/airbrakes on the wing trailing-edge are similar to those fitted to the ASW 20. Main landing gear is a twin-wheel retractable unit, with shock-absorption and hydraulic disc brakes; the tailwheel is also retractable.

SCHLEICHER ASK 23

Announced in the Spring of 1981, the ASK 23 is essentially a successor to the K 8 as a single-seat training and club glider, but making use of GfK constructional materials. Its general appearance is shown in an accompanying three-view drawing.

Schempp-Hirth Janus two-seat all-glassfibre training sailplane

Schempp-Hirth Nimbus 2C high-performance Open Class sailplane

Schempp-Hirth HS-7 Mini-Nimbus C single-seat 15 metre sailplane *(Peter F. Selinger)*

Schempp-Hirth Ventus b 15 metre Class sailplane

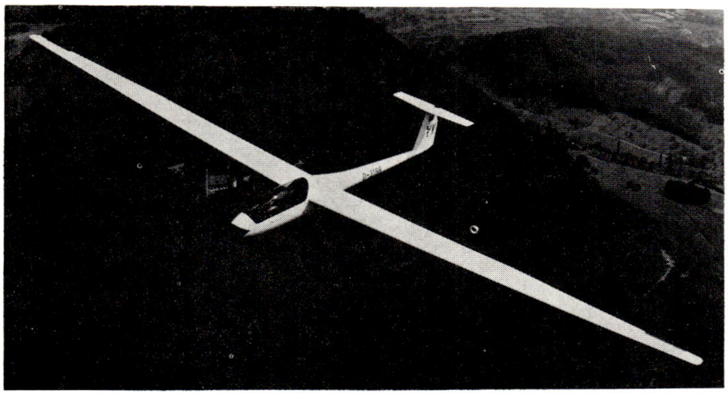

Schempp-Hirth Janus C, with 20 metre carbonfibre wings

Schleicher ASW 20 15 metre Class sailplane

Schleicher ASK 21 tandem two-seat sailplane

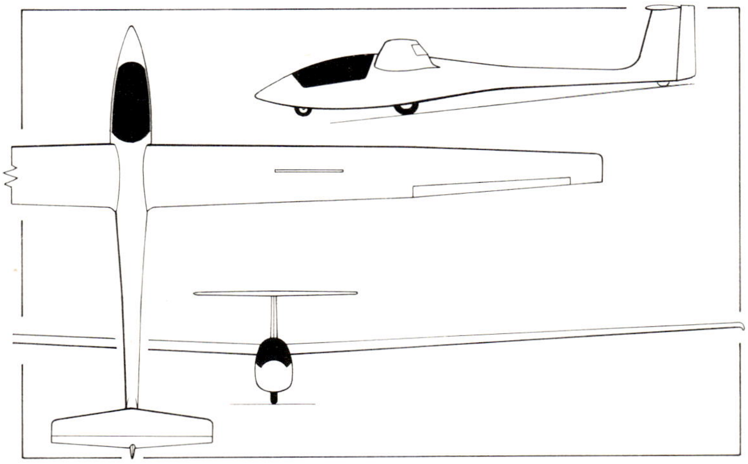

Schleicher ASK 23 training and club sailplane *(Michael A. Badrocke)*

Schleicher ASW 19 single-seat Standard Class sailplane

VALENTIN

VALENTIN GmbH GERÄTE- und MASCHINENBAU

8901 Königsbrunn, Germanenstrasse 2 (Postfach 1165)
Telephone: (08231) 4033/4
DESIGNER: Dipl-Ing Jörg B. Stieber

VALENTIN TAIFUN

The Taifun (typhoon) motor glider is proposed in four versions, as follows:

Taifun 15S. Basic model, with 15 m span wings and non-retractable tailwheel-type landing gear.

Taifun 15E. With 15 m span wings and retractable tricycle landing gear.

Taifun 17S. With 17 m span wings and non-retractable tailwheel-type landing gear.

Taifun 17E. With 17 m span wings and retractable tricycle landing gear.

Construction of the first prototype, to Taifun 17E configuration, began in July 1979, and this aircraft (D-KONO) made its first flight on 28 February 1981.

TYPE: Two-seat motor glider; *g* limits +5·3/−3·0.

AIRFRAME: Cantilever low-wing monoplane with T tail. Wortmann FX-67-K-170/17 wing section. Dihedral 3°. Incidence 0° 48′. Sweepforward 0° 38′ 24″ at quarter-chord. Wings and ailerons of glassfibre/foam sandwich construction. All-glassfibre trailing-edge flaps. Schempp-Hirth upper-surface airbrakes. Fuselage is a stressed-skin structure of glassfibre and foam sandwich. Fixed-incidence T tailplane, with horn-balanced elevator, of similar construction to wings. Flettner tab in port half of elevator. Non-retractable or mechanically retractable tricycle or tailwheel-type landing gear (see model listings), with Goodyear 5 in main-wheel tyres, pressure 2·5 bars (36 lb/sq in), and Cleveland disc brakes. Single nosewheel or tailwheel. Cockpit seats two persons side by side under rearward-sliding canopy.

POWER PLANT: One 59 kW (80 hp) Limbach L 2000 EO four-cylinder four-stroke engine, driving a Hoffmann HO-V62 two-blade propeller. Fuel in two wing tanks, total capacity 90 litres (19·8 Imp gallons).

WIEGAND
HEINZ WIEGAND

c/o Wiegand und Wisser, Industriestrasse 2, 5014 Kerpen

WIEGAND WM-1

The WM-1, which flew for the first time on 24 May 1980, is a Standard Class sailplane of conventional appearance, with a T tail unit and retractable monowheel. The wings have a Wortmann FX-61-184 section. Construction is of GfK sandwich. All known details are given in the table at the end of this section.

INDIA

CIVIL AVIATION DEPARTMENT
TECHNICAL CENTRE, CIVIL AVIATION DEPARTMENT

Civil Aviation Department, R. K. Puram, New Delhi 22

WORKS: Technical Centre, opposite Safdarjung Airport, New Delhi 110003

Telephone: 611504

DIRECTOR GENERAL: G. R. Kathpalia

DEPUTY DIRECTOR GENERAL: K. B. Ganesan

DIRECTOR (R & D): P. R. Chandrasekhar

The Technical Centre is the research and development establishment of the Indian Civil Aviation Department. It is engaged in the type certification of aircraft; development of design, airworthiness and operational standards; the study of operational problems; development testing and standardisation of aeronautical materials; laboratory investigation of accidents; fatigue research; and the evaluation of aircraft performance and economics.

Since 1950 the Technical Centre has undertaken the design and development of gliders utilising predominantly indigenous materials. The first of these gliders, of the open-cockpit primary type, was flown in November 1950. Since then the Technical Centre has built gliders of nine types for use at gliding clubs and centres in India, as listed in previous editions of *Jane's*. Of these, six have been original designs, the latest being the two-seat Ardhra training sailplane.

The Technical Centre does not undertake quantity production of gliders. Drawings of designs developed at the Centre are supplied to interested organisations with permission to manufacture them in series.

HS-II MRIGASHEER

This single-seat Standard Class sailplane was the first to be designed and developed at the Technical Centre by the team of designers and engineers led by Mr K. B. Ganesan. The original HS-I was described and illustrated in the 1973-74 *Jane's*.

The prototype of the further-developed HS-II flew for the first time in April 1973. A second prototype, fitted with trailing-edge slotted flaps instead of airbrakes, made its first flight in May 1977, and was described in the 1980-81 *Jane's*.

ATS-1 ARDHRA

This two-seat training sailplane was designed by the team responsible earlier for the Mrigasheer. The prototype (VT-GEJ) made its first flight on 5 March 1979 and, after extensive flight testing, was certificated by the Indian DGCA in November of that year.

The Ardhra has been found suitable for use by the National Cadet Corps, and is expected to enter large-scale production at the Kanpur Division of Hindustan Aeronautics Ltd. A second prototype, scheduled to fly in March 1981 and described below, differs from the first in having 2° geometric twist (washout) on the outer 50 per cent of each wing, and 3° dihedral.

TYPE: Tandem two-seat advanced training sailplane; *g* limits +5·3/−2·65.

AIRFRAME: Cantilever shoulder-wing monoplane. Wortmann FX-61-184 wing section from root to tip. Dihedral 3°. Incidence 3° at root. Sweepforward 3° at leading-edge; 2° geometric twist on outer half of each wing. Plywood-covered two-spar wooden wings, plywood covered on leading- and trailing-edges; plain wooden ailerons; wooden airbrakes in upper and lower wing surfaces. Tail unit of similar construction, but with rear portions of rudder and elevator fabric covered. Horn-balanced rudder; mass- and horn-balanced elevator. Plywood-covered tab in starboard elevator. Semi-monocoque wooden fuselage, with plywood covering. Nosecone of glassfibre. Non-retractable unsprung monowheel, tyre size 6·00-4. Rubber-sprung nose-skid, with replaceable steel shoe, and rubber-sprung tailskid.

ITALY

CAPRONI VIZZOLA
CAPRONI VIZZOLA COSTRUZIONI AERONAUTICHE SpA

Via Durini 24, 20122 Milan

Telephone: (02) 700826 and 781975

Telex: Caproni 332554 CAVIZ I

WORKS: Via Montecchio 2, CP 11 Gallarate, 21010 Vizzola Ticino (Varese)

Telephone: (0331) 230826

PRESIDENT: Dott Giovanni Caproni di Taliedo

The Caproni company, formed in 1910, is the oldest Italian aircraft manufacturer. Its works at Vizzola Ticino have approx 30,000 m² (322,917 sq ft) of covered space, and are adjacent to Malpensa Airport. They are equipped to manufacture complete structural subassemblies for helicopters and medium-sized fixed-wing aircraft. Caproni Vizzola also produces ground support equipment for General Electric T64/CT64, J79, J85 and CF6 and Turbo Union RB.199 turbojet and turbofan engines.

In 1969 Caproni Vizzola began producing a series of Calif sailplanes designed by Carlo Ferrarin and Livio Sonzio. Details of the earlier A-10 (one built), A-12 (two built), A-14 (one built), A-15 (one built), A-20 and A-21 have been given in the 1972-73 and subsequent *Jane's*.

Development of an improved two-seat version of the A-20, designated A-20S, has been suspended.

CAPRONI VIZZOLA CALIF A-21S

The original A-21 was a two-seat version of the A-14, from which it differed in having a wider (to accommodate two side-by-side seats) and slightly longer fuselage. This version (1975-76 *Jane's*) was superseded by the A-21S which has been certificated by the RAI, FAA, LBA and DGAC.

It was in an A-21S that, on 18 August 1974, Adele Orsi and Franca Bellingeri set the current Class D2 speed record for women over a 300 km closed circuit at 52·742 knots (97·741 km/h; 60·733 mph). Several other national and international records have been set by sailplanes of this type.

By January 1981 a total of 50 A-21S had been delivered.

TYPE: Side-by-side two-seat high-performance sailplane; *g* limit +5·0 (ultimate).

AIRFRAME: Cantilever mid-wing monoplane. Wing section Wortmann FX-67-K-170 at root, FX-60-126 at tip. Dihedral 0° at root, 1° 30′ on outer panels. Incidence 4°. No sweep. Single-spar wings, of aluminium alloy stressed-skin construction, with glassfibre tips. Extruded aluminium alloy ailerons are aerodynamically balanced. Extruded aluminium alloy plain trailing-edge flaps. Extruded aluminium alloy airbrakes on upper wing surface move in conjunction with flaps. Glassfibre/foam plastics semi-monocoque forward fuselage, with load-carrying light alloy frame; all-metal stressed-skin rear fuselage. T tail of aluminium alloy, with glassfibre tips on the fin, rudder and tailplane; spring-adjusted elevator trim. Mechanically retractable twin main wheels, with rubber-in-compression shock-absorbers, tyres size 3·50-5, pressure 3·54 bars (51·3 lb/sq in), and Tost drum brake; non-retractable steerable tailwheel, size 200 × 50; small wheel built into each wingtip. Canopy hinged to open upward and rearward. Space for baggage (0·05 m³; 1·76 cu ft) aft of seats.

CAPRONI VIZZOLA CALIF A-21SJ

There are three known motorised versions of the Calif A-21S. First of these to fly, in the USA in late January 1972, was an **A-21J** powered by a 0·90 kN (202 lb st) Microturbo TRS 18 turbojet. The engine, installed in the bottom of the fuselage, was provided with an NACA flush-type intake in each side of the fuselage, and a straight exhaust pipe. This aircraft had A-21S-type landing gear. A second A-21S was purchased by Lockheed-Georgia and modified to A-21J configuration for a joint research programme with Mississippi State University by the University's Raspet Flight Research Laboratory. Details of the early stages of this programme were given in the 1977-78 *Jane's*. The programme was continuing in 1980, the aircraft now being further modified and having an aluminium and glassfibre glove fitted over the inboard portion of the starboard wing, for laminar flow control tests. This glove replaced one of the two underwing 113·5 litre (25 Imp gallon; 30 US gallon) auxiliary fuel tanks normally carried by this aircraft. In another test, the aircraft evaluated a spanwise blowing system, mounted on each wing at approx mid-span and significantly increasing available lift. Other tests with this aircraft have included techniques for measuring airframe noise, command augmentation by means of an auxiliary side-stick controller to improve flying characteristics, identification of flight condition parameters, and the use of fibre optic pitch controls. To determine the effects of scale, the results of many of the tests were checked against a remotely controlled three-tenths scale model of the A-21J, embodying modifications corresponding to those being tested on the full-size aircraft.

One **Calif M**, powered by a Wankel rotating-piston engine, is known to be flying in the Federal Republic of Germany.

The jet-powered version produced by Caproni Vizzola, designated **A-21SJ**, has a more powerful engine, installed in the upper fuselage with a single NACA intake on top, aft of the cockpit, a reinforced main spar, and larger wheels and tyres. Two A-21SJs have been built in Italy, and production of this version is continuing.

The following description applies to the A-21SJ:

TYPE: Side-by-side two-seat jet-powered sailplane.

AIRFRAME: As A-21S, except for modifications necessary to incorporate power plant aft of cockpit. Main-wheel tyres size 5·00-5. Steerable tailwheel, tyre size 210 × 65.

POWER PLANT: One 1·08 kN (242·5 lb st) Microturbo TRS 18 turbojet engine (or 1·1 kN; 247 lb st KHD T317 turbojet), installed to rear of cockpit. Fuel load, in one tank in fuselage and two flexible wing tanks, 110 kg (242·5 lb). 24V/600W starter/generator and 24V 12Ah batteries for in-flight and ground starting.

GLASFASER
GLASFASER ITALIANA SrL

Via Ghiaie 3, 24030 Valbrembo (Bergamo)

Telephone: (035) 612617

PRESIDENT: Dott Ing Sergio Aldo Capoferri

TECHNICAL DIRECTOR: Giampaolo Ghidotti

This company's factory occupies an area of 1,600 m² (17,225 sq ft) and is insulated and heated to maintain a controlled temperature of 20°C, essential for work on glassfibre structures.

Glasfaser holds RAI licences to build glassfibre sailplanes and to repair, maintain and modify sailplanes and

Model of the Valentin Taifun two-seat motor glider

First prototype of the Civil Aviation Dept ATS-1 Ardhra advanced training sailplane

Caproni Vizzola Calif A-21S two-seat high-performance sailplane

Caproni Vizzola Calif A-21SJ jet-powered sailplane

First Japanese-built Nippi Pilatus B4-PC11AF Standard Class sailplane

Maggi MG3-15L Condor homebuilt sailplane

training aircraft.

Glasfaser built 25 single-seat 17 metre Kestrel sailplanes and 130 complete fuselage assemblies for the Kestrel, and assembled 10 Glasflügel Hornets. It has also completed 15 Glaser-Dirks DG-200s, and is an agency for Grob sailplanes (see Federal German section).

Other activities include construction of glassfibre trailers for road transportation of sailplanes, the repair and maintenance of sailplanes of glassfibre, wood and metal construction, installation and modification work on sailplanes of all types, and construction of glassfibre and carbonfibre windmill blades.

MAGGI
ANGELO MAGGI

Viale C. Troya 23, 20144 Milan

MAGGI MG3-15L CONDOR

This homebuilt sailplane, of which construction began in May 1975, made its first flight on 2 July 1980.

Type: Single-seat homebuilt sailplane.

Airframe: Cantilever shoulder-wing monoplane. Wing section NACA 65_3-618. Dihedral 4°. Incidence 4°. Single light alloy main spar; wing leading-edges of PVC resin, coated with glassfibre. NASA (Whitcomb) winglets at each wingtip. Fiat light alloy ailerons; airbrakes and plain trailing-edge flaps, also of light alloy. Fuselage has light alloy structure with glassfibre/polyurethane foam covering. Cantilever tailplane of light alloy, leading-edges of which are of balsa with glassfibre coating; separate elevators, with spring trim. Retractable unsprung monowheel, tyre size 5·00-5, pressure 3·2 bars (46·4 lb/sq in), with drum brake; non-retractable tailskid.

JAPAN

NIPPI
NIHON HIKOKI KABUSHIKI KAISHA (Japan Aircraft Manufacturing Co Ltd)

Head Office and Sugita Plant: 3175 Showa-machi, Kanazawa-ku, Yokohama 236

Telephone: Yokohama (045) 771 1251

Telex: 3822267 Nippi J

Chief, Sailplane Sales Section: Norio Kagata

Other Officers: see Aircraft section

NIPPI NP-100A ALBATROSS

Nippi began the design of this motor glider in late 1973, and the prototype (NP-100) made its first flight on 25 December 1975. Flight testing was carried out during 1976, as a result of which several modifications were made. Principal point of interest in the NP-100A is the power plant, which is of the ducted-fan type and is fully buried within the fuselage aft of the main landing gear.

A secondary flight test programme took place between March and July 1978, to determine the final specification for a production version. The planned production programme is still awaiting final selection of a type-certificated power plant to replace the Nippi M2 two-stroke flat-four engine which powers the prototype.

A description of the prototype in its current form can be found in the 1980-81 *Jane's*.

NIPPI PILATUS B4-PC11AF

Pilatus Flugzeugwerke of Switzerland designed and built this single-seat Standard Class sailplane for multi-purpose training and aerobatics. Swiss certification was granted on 12 June 1972, and by March 1978 more than 330 had been delivered by Pilatus to customers in 30 countries in all parts of the world. For these aircraft, Pilatus continues to be responsible for after-sales service.

Production and marketing rights in the aircraft were transferred to Nippi in June 1978, and it is now known as the Nippi Pilatus B4-PC11AF. The first Nippi-built example flew for the first time on 25 November 1979. Initial production, following JCAB certification on 13 May 1980, is at the rate of one per month.

Type: Single-seat Standard Class and Club Class sailplane; *g* limits +6·32/−4·32 (Normal); +7·0/−4·79 (Aerobatic).

AIRFRAME: Cantilever shoulder-wing monoplane with T tail. Wing section NACA 64_3-618. Dihedral 1°. Conventional light alloy wings with single U spar and hard PVC foam ribs between metal ribs; conventional ailerons of similar construction. Light alloy spoilers at 60% chord on wing upper surfaces. Semi-monocoque light alloy fuselage, with flush-riveted skin. Rear fuselage consists of two half-shells, riveted together. Cantilever light alloy tail unit, with PVC ribs. Fixed-incidence tailplane. Elevator spring trim. Retractable (optionally non-retractable) unsprung Tost monowheel, and fixed tailwheel, in tandem. Main wheel faired by doors when retracted. Mechanical independent drum brakes. Canopy jettisonable in flight.

TAINAN

TAINAN INDUSTRY CO LTD (Tainan Kogyo KK)

5139-3 1-chome, Komatsubara, Zama-shi, Kanagawa-ken 228
Telephone: 0462 54 2332/3/4

This company took over the manufacture of sailplanes from LADCO (see 1971-72 *Jane's*). Details of the former company's Mita III, which Tainan continued to build under licence, can be found in the 1979-80 *Jane's*.

Tainan's latest known design is the TN-1, described and illustrated in the 1980-81 edition.

No news of this company's activities, if any, has been received since early 1978.

POLAND

SZD

PRZEDSIEBIORSTWO DOSWIADCZALNO-PRODUKCYJNE SZYBOWNICTWA (Experimental and Production Concern for Gliders) PZL-BIELSKO

HEAD OFFICE AND WORKS: 43-300 Bielsko-Biala 1, ul. Cieszynska 325
Telephone: 250 21 to 250 26
Telex: 031-259 SZD PL
DIRECTOR: Ing Jerzy Cieśla
SALES REPRESENTATIVE: Pezetel, 00-991 Warszawa, 61 Aleja Stanow Zjednoczonych (PO Box 6)
Telephone: 10 80 01
Telex: 813314

The Instytut Szybownictwa (Gliding Institute), formed officially in January 1946 at Bielsko-Biala, has since undergone several changes of name, as detailed in the 1977-78 *Jane's*. The change to the present title took place in July 1975, but the well-known designation initials SZD are retained. This organisation is responsible for the design and development of all Polish gliders and sailplanes. Production plants are situated at Bielsko-Biala, Wroclaw and Jezów.

Between 1947 and 1980 the Polish aircraft industry produced 4,037 gliders of more than 100 different types, and SZD sailplanes have been exported all over the world in substantial numbers. The SZD-9 Bocian, last described in the 1978-79 *Jane's*, was in production for 25 years.

SZD-41A JANTAR STANDARD (AMBER)

The Jantar Standard was flown for the first time on 3 October 1973, and a total of 160 had been built by the beginning of 1980, for customers in Argentina, Australia, Austria, Belgium, Bulgaria, Canada, Denmark, Finland, Germany (Democratic Republic), Germany (Federal Republic), Hungary, Poland, Sweden, Switzerland, the UK, the USA, the USSR and Yugoslavia. Production has now ended. A description can be found in the 1980-81 *Jane's*.

SZD-42-2 JANTAR 2B

Designed by Dipl-Ing Adam Kurbiel, the SZD-42-2 Open Class sailplane is based on the SZD-42-1. The prototype flew for the first time on 13 March 1978, and 46 Jantar 2Bs had been built by the beginning of 1981.

TYPE: Single-seat Open Class sailplane; *g* limits +5·3/−2·65.

AIRFRAME: Generally similar to SZD-41A Jantar Standard (1980-81 *Jane's*), but with wings raised from mid-mounted to shoulder position, 2° dihedral, and incidence reduced by 1° 30′. Wing section NN-8. One-piece slotless ailerons; top-hinged 'elastic' flaps. Light alloy DFS-type airbrakes. Cantilever cruciform tail unit of glassfibre/epoxy resin. Monowheel, diameter 400 mm, tyre pressure 2·96 bars (43 lb/sq in), with two rubber shock-absorbers and disc brake. Tailwheel diameter 200 mm. Water ballast provision 170 litres (37·4 Imp gallons). Improvements made to elevator spring trim, monowheel retraction system and cockpit comfort. Hinged canopy and provision for CG tow hook.

SZD-48-1 JANTAR STANDARD 2

This Standard Class sailplane is a development of the SZD-41A Jantar Standard (which see), and was designed by Ing Wladyslaw Okarmus. The prototype flew for the first time on 10 December 1977, and 201 had been built by the beginning of 1981.

With simple modifications to the trim and speed measuring systems, the SZD-48 can be converted quickly to the **SZD-48-2** version, which has a never-exceed speed of 158 knots (285 km/h; 177 mph).

TYPE: Single-seat Standard Class sailplane.

AIRFRAME: Cantilever shoulder-wing monoplane, with NN-8 wing section. Dihedral 1° 30′. Single glassfibre roving main spar, with glassfibre/foam/glassfibre moulded skins (no ribs) and plain ailerons. Duralumin upper/lower surface airbrakes. Glassfibre fuselage, with steel tube central support structure, rear portion stiffened by half-frames and fin ribs. Cantilever T tail of similar construction to wings. Fin integral with fuselage. Mass-balanced elevator, with spring trim. Retractable monowheel, tyre size 350 × 235 mm, with disc brake; semi-recessed tailwheel, diameter 200 mm. Optional CG towing hook on main-wheel fork. Provision for 150 litres (33 Imp gallons) water ballast.

SZD-49 JANTAR K

The Jantar K is an FAI 15 metre Class version of the Jantar Standard, fitted with wing flaps. It was built only as a prototype (SP-2583), which flew for the first time on 10 October 1978.

SZD-50-3 PUCHACZ (EAGLE OWL)

The Puchacz, designed by Dipl-Ing Adam Meus, is a high-performance sailplane intended particularly for training and performance flying. It has been modified and developed from a prototype, designated SZD-50-1 Dromader, which first flew on 21 December 1976.

A second prototype made its initial flight on 20 December 1977, and the first production SZD-50-2 Puchacz (SP-3151) followed on 13 April 1979. A total of 12 had been built by the beginning of 1981. The current version is designated SZD-50-3.

TYPE: Tandem two-seat high-performance training sailplane; *g* limits +5·2/−2·65.

AIRFRAME: Cantilever mid-wing monoplane, mainly of glassfibre sandwich construction. Wortmann wing sections. Plain ailerons. Upper/lower surface airbrakes. Glassfibre fuselage, supported in central portion by two wooden frames, with integral fin. Glassfibre sandwich cruciform tail unit, with fabric-covered glassfibre rudder. Non-retractable semi-recessed nosewheel (size 255 × 110 mm) and sprung monowheel (size 350 × 135 mm, with disc brake), and tailskid. One-piece canopy opens sideways.

ROMANIA

ICA

INTREPRINDERE DE CONSTRUCTII AERONAUTICE (Aeronautical Construction Enterprise)

Brasov

As detailed in the Aircraft section, the current activities of the Romanian aircraft industry are divided between three main industrial centres, IAvB Bucharest, IAvB Bacau, and ICA at Brasov. In addition to its work on powered aircraft, the ICA is responsible for all sailplane development and production previously undertaken by URMV-3 (up to 1959) and IIL (Ghimbav) up to 1968. The principal Romanian designer of sailplanes was Dipl Ing Iosif Silimon, who died in February 1981. His designs are prefixed with the letters IS. Details of his earlier sailplanes appeared in the 1961-62, 1965-66 and 1972-73 *Jane's*.

IS-28B2

Despite its similar designation, this high-performance training sailplane represents a considerable advance over the IS-28 (1975-76 *Jane's*), from which it differs principally in having 17 m span all-metal wings, a longer and more slender fuselage, and reduced wing and tailplane dihedral. Design began in the Autumn of 1971, and the first IS-28B made its first flight on 26 April 1973.

There have been two production versions, as follows:

IS-28B1. Version without wing flaps. No longer in production. Described in 1976-77 *Jane's*.

IS-28B2. Current (since 1976) production version, with Schempp-Hirth type (instead of DFS type) airbrakes and trailing-edge split flaps. Description applies to this version, of which more than 300 have been sold, chiefly in Australia and the USA. The IS-28B2 has also been certificated in Canada, France, West Germany and the UK. Production in 1980 was continuing at a rate of 70-100 per year.

TYPE: Tandem two-seat high-performance training sailplane; *g* limits (single-seat) +6·4/−4·2; (two-seat) +5·3/−2·65.

AIRFRAME: Cantilever shoulder-wing monoplane. Wortmann wing sections: FX-61-163 at root, FX-60-126 at tip. Forward-swept all-metal wings, with L-section main spar booms and dural web, dural auxiliary spar, and dural ribs. Schempp-Hirth metal airbrake above and below each wing. Ailerons and split trailing-edge flaps fabric-covered. No tabs. All-metal semi-monocoque forward and centre fuselage. Rear fuselage is duralumin monocoque. Cantilever T tail, with moderate tailplane dihedral. Elevator trailing-edges and rudder fabric-covered. Trim tab in each elevator. Semi-retractable monowheel with oleo-pneumatic shock-absorber and disc brake; sprung tailskid. Canopy opens sideways (to starboard) and can be jettisoned in flight. Nose towing hook, with Tost cable release, is standard; CG towing hook optional.

IS-28M

Two motor glider versions of the IS-28B2 have been developed. The rear fuselage and main wings are virtually unchanged from the IS-28B2, but the powered versions are of low-wing configuration and have redesigned forward fuselages, cockpit canopies and main landing gear.

The two versions are designated as follows:

IS-28M1. Tandem two-seater, with main landing gear comprising a retractable central monowheel, with balloon tyre, and underwing outrigger wheels. In production. Deliveries scheduled to begin during 1981.

IS-28M2. Side-by-side two-seater, with main landing gear comprising two retractable wheels side by side under fuselage centre-section, with shock-absorbers. Steerable tailwheel, also with shock-absorber. Prototype (YR-1013) flew for the first time on 26 June 1976. The first 10 aircraft of this type were allocated to the UK, where they are marketed by Morisonics Ltd of The Parade, Frimley, Surrey. Certificated in Australia, Portugal, Sweden and the UK. Delivered to customers in Australia, Canada, Denmark, Hungary, Israel, the Philippines, Spain, Sweden, Switzerland and the USA. In production.

TYPE: Two-seat motor glider; *g* limits +4·0/−1·5 (M1), +5·3/−2·65 (M2).

AIRFRAME: Cantilever low-wing monoplane, of mainly metal construction. Wortmann wing sections: FX-61-163 at root, FX-60-126 at tip. Dihedral 2°. Sweepforward 2° 30′ at quarter-chord. Single-spar wings, with aluminium ribs and skin, fabric-covered metal ailerons and (optionally) all-metal split flaps on trailing-edges. Flaps can be set to a negative position. All-metal two-section Hütter airbrakes on upper surfaces. Conventional fuselage, in three parts: metal front portion, built up on two longerons and cross-frames and having glassfibre fairings and engine cowling panels; aluminium alloy monocoque centre portion; and rear portion of aluminium alloy frames and skin. Cantilever aluminium alloy T tail, with dihedral tailplane. Rudder and elevator trailing-edges fabric-covered. Trim tab in each elevator. Landing gear as described under model listings. Rearward-sliding canopy. Dual controls standard.

POWER PLANT: One 50·7 kW (68 hp) Limbach SL 1700 EI flat-four engine, driving a Hoffmann HO-V-62/R160T two-blade variable-pitch fully-feathering propeller. Single fuel tank aft of cockpit, capacity 40 litres (8·8 Imp gallons).

Model of the Valentin Taifun two-seat motor glider

First prototype of the Civil Aviation Dept ATS-1 Ardhra advanced training sailplane

Caproni Vizzola Calif A-21S two-seat high-performance sailplane

Caproni Vizzola Calif A-21SJ jet-powered sailplane

First Japanese-built Nippi Pilatus B4-PC11AF Standard Class sailplane

Maggi MG3-15L Condor homebuilt sailplane

training aircraft.

Glasfaser built 25 single-seat 17 metre Kestrel sailplanes and 130 complete fuselage assemblies for the Kestrel, and assembled 10 Glasflügel Hornets. It has also completed 15 Glaser-Dirks DG-200s, and is an agency for Grob sailplanes (see Federal German section).

Other activities include construction of glassfibre trailers for road transportation of sailplanes, the repair and maintenance of sailplanes of glassfibre, wood and metal construction, installation and modification work on sailplanes of all types, and construction of glassfibre and carbonfibre windmill blades.

MAGGI
ANGELO MAGGI

Viale C. Troya 23, 20144 Milan

MAGGI MG3-15L CONDOR

This homebuilt sailplane, of which construction began in May 1975, made its first flight on 2 July 1980.

Type: Single-seat homebuilt sailplane.

Airframe: Cantilever shoulder-wing monoplane. Wing section NACA 65_3-618. Dihedral 4°. Incidence 4°. Single light alloy main spar; wing leading-edges of PVC resin, coated with glassfibre. NASA (Whitcomb) winglets at each wingtip. Fiat light alloy ailerons; airbrakes and plain trailing-edge flaps, also of light alloy. Fuselage has light alloy structure with glassfibre/polyurethane foam covering. Cantilever tailplane of light alloy, leading-edges of which are of balsa with glassfibre coating; separate elevators, with spring trim. Retractable unsprung monowheel, tyre size 5·00-5, pressure 3·2 bars (46·4 lb/sq in), with drum brake; non-retractable tailskid.

JAPAN

NIPPI
NIHON HIKOKI KABUSHIKI KAISHA (Japan Aircraft Manufacturing Co Ltd)

Head Office and Sugita Plant: 3175 Showa-machi, Kanazawa-ku, Yokohama 236
Telephone: Yokohama (045) 771 1251
Telex: 3822267 Nippi J
Chief, Sailplane Sales Section: Norio Kagata
Other Officers: see Aircraft section

NIPPI NP-100A ALBATROSS

Nippi began the design of this motor glider in late 1973, and the prototype (NP-100) made its first flight on 25 December 1975. Flight testing was carried out during 1976, as a result of which several modifications were made. Principal point of interest in the NP-100A is the power plant, which is of the ducted-fan type and is fully buried within the fuselage aft of the main landing gear.

A secondary flight test programme took place between March and July 1978, to determine the final specification for a production version. The planned production programme is still awaiting final selection of a type-certificated power plant to replace the Nippi M2 two-stroke flat-four engine which powers the prototype.

A description of the prototype in its current form can be found in the 1980-81 *Jane's*.

NIPPI PILATUS B4-PC11AF

Pilatus Flugzeugwerke of Switzerland designed and built this single-seat Standard Class sailplane for multi-purpose training and aerobatics. Swiss certification was granted on 12 June 1972, and by March 1978 more than 330 had been delivered by Pilatus to customers in 30 countries in all parts of the world. For these aircraft, Pilatus continues to be responsible for after-sales service.

Production and marketing rights in the aircraft were transferred to Nippi in June 1978, and it is now known as the Nippi Pilatus B4-PC11AF. The first Nippi-built example flew for the first time on 25 November 1979. Initial production, following JCAB certification on 13 May 1980, is at the rate of one per month.

Type: Single-seat Standard Class and Club Class sailplane; *g* limits +6·32/−4·32 (Normal); +7·0/−4·79 (Aerobatic).

AIRFRAME: Cantilever shoulder-wing monoplane with T tail. Wing section NACA 64_3-618. Dihedral 1°. Conventional light alloy wings with single U spar and hard PVC foam ribs between metal ribs; conventional ailerons of similar construction. Light alloy spoilers at 60% chord on wing upper surfaces. Semi-monocoque light alloy fuselage, with flush-riveted skin. Rear fuselage consists of two half-shells, riveted together. Cantilever light alloy tail unit, with PVC ribs. Fixed-incidence tailplane. Elevator spring trim. Retractable (optionally non-retractable) unsprung Tost monowheel, and fixed tailwheel, in tandem. Main wheel faired by doors when retracted. Mechanical independent drum brakes. Canopy jettisonable in flight.

TAINAN

TAINAN INDUSTRY CO LTD (Tainan Kogyo KK)

5139-3 1-chome, Komatsubara, Zama-shi, Kanagawa-ken 228
Telephone: 0462 54 2332/3/4

This company took over the manufacture of sailplanes from LADCO (see 1971-72 *Jane's*). Details of the former company's Mita III, which Tainan continued to build under licence, can be found in the 1979-80 *Jane's*. Tainan's latest known design is the TN-1, described and illustrated in the 1980-81 edition.

No news of this company's activities, if any, has been received since early 1978.

POLAND

SZD

PRZEDSIEBIORSTWO DOSWIADCZALNO-PRODUKCYJNE SZYBOWNICTWA (Experimental and Production Concern for Gliders) PZL-BIELSKO

HEAD OFFICE AND WORKS: 43-300 Bielsko-Biala 1, ul. Cieszynska 325
Telephone: 250 21 to 250 26
Telex: 031-259 SZD PL
DIRECTOR: Ing Jerzy Cieśla
SALES REPRESENTATIVE: Pezetel, 00-991 Warszawa, 61 Aleja Stanow Zjednoczonych (PO Box 6)
Telephone: 10 80 01
Telex: 813314

The Instytut Szybownictwa (Gliding Institute), formed officially in January 1946 at Bielsko-Biala, has since undergone several changes of name, as detailed in the 1977-78 *Jane's*. The change to the present title took place in July 1975, but the well-known designation initials SZD are retained. This organisation is responsible for the design and development of all Polish gliders and sailplanes. Production plants are situated at Bielsko-Biala, Wroclaw and Jezów.

Between 1947 and 1980 the Polish aircraft industry produced 4,037 gliders of more than 100 different types, and SZD sailplanes have been exported all over the world in substantial numbers. The SZD-9 Bocian, last described in the 1978-79 *Jane's*, was in production for 25 years.

SZD-41A JANTAR STANDARD (AMBER)

The Jantar Standard was flown for the first time on 3 October 1973, and a total of 160 had been built by the beginning of 1980, for customers in Argentina, Australia, Austria, Belgium, Bulgaria, Canada, Denmark, Finland, Germany (Democratic Republic), Germany (Federal Republic), Hungary, Poland, Sweden, Switzerland, the UK, the USA, the USSR and Yugoslavia. Production has now ended. A description can be found in the 1980-81 *Jane's*.

SZD-42-2 JANTAR 2B

Designed by Dipl-Ing Adam Kurbiel, the SZD-42-2 Open Class sailplane is based on the SZD-42-1. The prototype flew for the first time on 13 March 1978, and 46 Jantar 2Bs had been built by the beginning of 1981.

TYPE: Single-seat Open Class sailplane; *g* limits +5·3/−2·65.

AIRFRAME: Generally similar to SZD-41A Jantar Standard (1980-81 *Jane's*), but with wings raised from mid-mounted to shoulder position, 2° dihedral, and incidence reduced by 1° 30'. Wing section NN-8. One-piece slotless ailerons; top-hinged 'elastic' flaps. Light alloy DFS-type airbrakes. Cantilever cruciform tail unit of glassfibre/epoxy resin. Monowheel, diameter 400 mm, tyre pressure 2·96 bars (43 lb/sq in), with two rubber shock-absorbers and disc brake. Tailwheel diameter 200 mm. Water ballast provision 170 litres (37·4 Imp gallons). Improvements made to elevator spring trim, monowheel retraction system and cockpit comfort. Hinged canopy and provision for CG tow hook.

SZD-48-1 JANTAR STANDARD 2

This Standard Class sailplane is a development of the SZD-41A Jantar Standard (which see), and was designed by Ing Wladyslaw Okarmus. The prototype flew for the first time on 10 December 1977, and 201 had been built by the beginning of 1981.

With simple modifications to the trim and speed measuring systems, the SZD-48 can be converted quickly to the **SZD-48-2** version, which has a never-exceed speed of 158 knots (285 km/h; 177 mph).

TYPE: Single-seat Standard Class sailplane.

AIRFRAME: Cantilever shoulder-wing monoplane, with NN-8 wing section. Dihedral 1° 30'. Single glassfibre roving main spar, with glassfibre/foam/glassfibre moulded skins (no ribs) and plain ailerons. Duralumin upper/lower surface airbrakes. Glassfibre fuselage, with steel tube central support structure, rear portion stiffened by half-frames and fin ribs. Cantilever T tail of similar construction to wings. Fin integral with fuselage. Mass-balanced elevator, with spring trim. Retractable monowheel, tyre size 350 × 235 mm, with disc brake; semi-recessed tailwheel, diameter 200 mm. Optional CG towing hook on main-wheel fork. Provision for 150 litres (33 Imp gallons) water ballast.

SZD-49 JANTAR K

The Jantar K is an FAI 15 metre Class version of the Jantar Standard, fitted with wing flaps. It was built only as a prototype (SP-2583), which flew for the first time on 10 October 1978.

SZD-50-3 PUCHACZ (EAGLE OWL)

The Puchacz, designed by Dipl-Ing Adam Meus, is a high-performance sailplane intended particularly for training and performance flying. It has been modified and developed from a prototype, designated SZD-50-1 Dromader, which first flew on 21 December 1976.

A second prototype made its initial flight on 20 December 1977, and the first production SZD-50-2 Puchacz (SP-3151) followed on 13 April 1979. A total of 12 had been built by the beginning of 1981. The current version is designated SZD-50-3.

TYPE: Tandem two-seat high-performance training sailplane; *g* limits +5·2/−2·65.

AIRFRAME: Cantilever mid-wing monoplane, mainly of glassfibre sandwich construction. Wortmann wing sections. Plain ailerons. Upper/lower surface airbrakes. Glassfibre fuselage, supported in central portion by two wooden frames, with integral fin. Glassfibre sandwich cruciform tail unit, with fabric-covered glassfibre rudder. Non-retractable semi-recessed nosewheel (size 255 × 110 mm) and sprung monowheel (size 350 × 135 mm, with disc brake), and tailskid. One-piece canopy opens sideways.

ROMANIA

ICA

INTREPRINDERE DE CONSTRUCTII AERONAUTICE (Aeronautical Construction Enterprise)

Brasov

As detailed in the Aircraft section, the current activities of the Romanian aircraft industry are divided between three main industrial centres, IAvB Bucharest, IAvB Bacau, and ICA at Brasov. In addition to its work on powered aircraft, the ICA is responsible for all sailplane development and production previously undertaken by URMV-3 (up to 1959) and IIL (Ghimbav) up to 1968. The principal Romanian designer of sailplanes was Dipl Ing Iosif Silimon, who died in February 1981. His designs are prefixed with the letters IS. Details of his earlier sailplanes appeared in the 1961-62, 1965-66 and 1972-73 *Jane's*.

IS-28B2

Despite its similar designation, this high-performance training sailplane represents a considerable advance over the IS-28 (1975-76 *Jane's*), from which it differs principally in having 17 m span all-metal wings, a longer and more slender fuselage, and reduced wing and tailplane dihedral. Design began in the Autumn of 1971, and the first IS-28B made its first flight on 26 April 1973.

There have been two production versions, as follows:

IS-28B1. Version without wing flaps. No longer in production. Described in 1976-77 *Jane's*.

IS-28B2. Current (since 1976) production version, with Schempp-Hirth type (instead of DFS type) airbrakes and trailing-edge split flaps. Description applies to this version, of which more than 300 have been sold, chiefly in Australia and the USA. The IS-28B2 has also been certificated in Canada, France, West Germany and the UK. Production in 1980 was continuing at a rate of 70-100 per year.

TYPE: Tandem two-seat high-performance training sailplane; *g* limits (single-seat) +6·4/−4·2; (two-seat) +5·3/−2·65.

AIRFRAME: Cantilever shoulder-wing monoplane. Wortmann wing sections: FX-61-163 at root, FX-60-126 at tip. Forward-swept all-metal wings, with L-section main spar booms and dural web, dural auxiliary spar, and dural ribs. Schempp-Hirth metal airbrake above and below each wing. Ailerons and split trailing-edge flaps fabric-covered. No tabs. All-metal semi-monocoque forward and centre fuselage. Rear fuselage is duralumin monocoque. Cantilever T tail, with moderate tailplane dihedral. Elevator trailing-edges and rudder fabric-covered. Trim tab in each elevator. Semi-retractable monowheel with oleo-pneumatic shock-absorber and disc brake; sprung tailskid. Canopy opens sideways (to starboard) and can be jettisoned in flight. Nose towing hook, with Tost cable release, is standard; CG towing hook optional.

IS-28M

Two motor glider versions of the IS-28B2 have been developed. The rear fuselage and main wings are virtually unchanged from the IS-28B2, but the powered versions are of low-wing configuration and have redesigned forward fuselages, cockpit canopies and main landing gear.

The two versions are designated as follows:

IS-28M1. Tandem two-seater, with main landing gear comprising a retractable central monowheel, with balloon tyre, and underwing outrigger wheels. In production. Deliveries scheduled to begin during 1981.

IS-28M2. Side-by-side two-seater, with main landing gear comprising two retractable wheels side by side under fuselage centre-section, with shock-absorbers. Steerable tailwheel, also with shock-absorber. Prototype (YR-1013) flew for the first time on 26 June 1976. The first 10 aircraft of this type were allocated to the UK, where they are marketed by Morisonics Ltd of The Parade, Frimley, Surrey. Certificated in Australia, Portugal, Sweden and the UK. Delivered to customers in Australia, Canada, Denmark, Hungary, Israel, the Philippines, Spain, Sweden, Switzerland and the USA. In production.

TYPE: Two-seat motor glider; *g* limits +4·0/−1·5 (M1), +5·3/−2·65 (M2).

AIRFRAME: Cantilever low-wing monoplane, of mainly metal construction. Wortmann wing sections: FX-61-163 at root, FX-60-126 at tip. Dihedral 2°. Sweep-forward 2° 30' at quarter-chord. Single-spar wings, with aluminium ribs and skin, fabric-covered metal ailerons and (optionally) all-metal split flaps on trailing-edges. Flaps can be set to a negative position. All-metal two-section Hütter airbrakes on upper surfaces. Conventional fuselage, in three parts: metal front portion, built up on two longerons and cross-frames and having glassfibre fairings and engine cowling panels; aluminium alloy monocoque centre portion; and rear portion of aluminium alloy frames and skin. Cantilever aluminium alloy T tail, with dihedral tailplane. Rudder and elevator trailing-edges fabric-covered. Trim tab in each elevator. Landing gear as described under model listings. Rearward-sliding canopy. Dual controls standard.

POWER PLANT: One 50·7 kW (68 hp) Limbach SL 1700 EI flat-four engine, driving a Hoffmann HO-V-62/R160T two-blade variable-pitch fully-feathering propeller. Single fuel tank aft of cockpit, capacity 40 litres (8·8 Imp gallons).

SZD-42-2 Jantar 2B single-seat Open Class sailplane

SZD-50-2 Puchacz two-seat training sailplane

ICA IS-32 two-seat Open Class sailplane

IS-29EM motor glider (Rectimo 4AR-1200 engine)

IS-28M2 side-by-side two-seat motor glider *(Brian M. Service)*

IS-29D2 single-seat Standard Class sailplane produced by ICA

SZD-48-1 Jantar Standard 2 single-seat sailplane

Close-up of cabin and main landing gear of IS-28M1 *(Brian M. Service)*

IS-30 two-seat sailplane *(J. M. G. Gradidge)*

IS-28B2 tandem two-seat high-performance training sailplane
(The Age, Melbourne)

IS-29 and IS-33

The IS-29 can be adapted to suit a variety of requirements or weather conditions. All versions have an identical fuselage and tail unit, and a choice of wings is available.

Descriptions can be found in the 1980-81 and earlier editions of *Jane's* of the IS-29B, D, D3, D4, E, E4, G, and IS-31(E3). None of these versions is now in production. Models current in 1981 were as follows:

IS-29D2. Current Standard Class version, with improved cockpit and controls, Hütter airbrakes, separate tailplane and elevator, and improved rigging system. More than 100 sold in Australia, the USA, and several other countries. More than 150 built by mid-1981. Production continuing in 1980-81 at rate of 20-25 a year.

The following are all variants of the IS-29D2 design:

IS-29D2 Club. Club Class version of D2, with flaps deleted and non-retractable monowheel. Certification pending in mid-1981.

IS-29DM. Powered version of D2; see separate description which follows.

IS-29E2. Open Class version, having 19 m span wings and interconnected flaps and ailerons, Schempp-Hirth airbrakes and integral tanks for 100 litres (22 Imp gallons) water ballast.

IS-29EM. Powered version of E2; see separate description which follows.

IS-33. Racing version, with 15 m span wings fitted with 'flaperons'. To be offered with 150 litres (33 Imp gallons) water ballast.

The following description applies to the IS-29D2 and E2, except where a specific model is indicated:

TYPE: Single-seat Standard Class (D2) and Open Class (E2) sailplanes.

AIRFRAME: Cantilever shoulder-wing monoplane, with T tail. Wortmann wing sections: FX-61-165 at root, FX-61-124 at tip on D2; FX-K-170 (root) and FX-K-150 (tip) on E2. All-metal wings, with main spar, false rear spar and riveted dural skin. Full-span trailing-edge flaps and ailerons, coupled to operate in unison but can be disconnected for separate operation during landing. Airbrakes (see under model listings for type) in upper and lower surface of each wing (upper surface only on D2). All-metal semi-monocoque fuselage identical on all versions except for local variations at wing attachment points. Detachable glassfibre nosecap. Cantilever all-metal T tail, with full-span elevator. Retractable sprung monowheel, with brake; non-retractable tailwheel. Canopy hinges sideways to starboard and can be jettisoned in flight. Provision for water ballast in E2.

IS-29DM and IS-29EM

These two motor gliders, of which prototypes have been flown, are respectively powered versions of the IS-29D2 and IS-29E2, fitted with a 29 kW (39 hp) Rectimo 4AR-1200 engine driving a Hoffmann HO-V42 propeller; fuel capacity is 27 litres (6 Imp gallons). A retractable rubber-sprung Tost monowheel, diameter 345 mm, is fitted; the Tost tailwheel has a diameter of 210 mm.

IS-30

The IS-30, which underwent flight testing in the Spring of 1978, is based on the IS-28B2. It has 18·00 m (59 ft 0¾ in) span wings, the redesigned tail of the IS-32, and is available (as the **IS-30M**) with an optional power plant and retractable propeller pylon aft of the cockpit. There is no water ballast provision; *g* limits are the same as for the IS-28B2.

IS-32

First shown publicly at the Paris Air Show in June 1977, the IS-32 tandem two-seat Open Class sailplane is developed from the IS-28B2. Of all-metal construction, it has 20 m wings of improved Wortmann FX-K-170 (root) and FX-K-150 (tip) sections, with interconnected flaps and ailerons and water ballast tanks, and redesigned tail surfaces; *g* limits are +4·0/−1·5. Certification has been awarded, and deliveries were intended to begin in 1981.

SPAIN

CASES
FRANCISCO CASES MASIÁ

Miguel Hernandez 44, Benetússer, Valencia
Telephone: (93) 3754380

Sr Cases designed the Libel-Lula lightweight glider in early 1977, and began building it later that year. In 1980 he was working, as time permitted, on the design of a Standard Class sailplane.

CASES LIBEL-LULA

This lightweight homebuilt glider, built by an experienced aeromodeller and glider pilot, was flown for the first time on 14 September 1979. A description and illustration appeared in the 1980-81 *Jane's*.

SWITZERLAND

EFF
ENTWICKLUNGSGEMEINSCHAFT FÜR FLUGZEUGBAU

c/o Dr H. K. Stiffler, Dufourstrasse 101, CH-8000 Zürich
Telephone: (01) 813 3244
PRESIDENT: Thomas Bircher

EFF PROMETHEUS

After four years of design work, construction of the Prometheus began in 1975, and it flew for the first time on 22 June 1978.

This original version, with 19·40 m (63 ft 7¾ in) wing span, is known as the Prometheus 19. A Prometheus 12 version, with 12·63 m (41 ft 5¼ in) span, flew for the first time on 17 September 1979. Another version, with tip-tanks and winglets, was planned. No production of the Prometheus is intended at present.

TYPE: Two-seat twin-jet motor glider; *g* limits ±4·0.

AIRFRAME: Cantilever mid-wing monoplane. Wing section Wortmann FX-67-170-17. Wings have aluminium main spar, with plywood skin and glassfibre/foam sandwich infill. Trailing-edge 'flaperons'. DFS upper-surface airbrakes. Forward fuselage built of wood, with glassfibre skin; rear fuselage is of glassfibre sandwich, as are the fixed-incidence T tailplane and elevator. Retractable tricycle landing gear, comprising twin main wheels with Goodyear tyres (diameter 127 mm; 5 in, pressure 4·3 bars; 62·4 lb/sq in) and Tost brake; and single nosewheel with General Tire and Rubber Co tyre (pressure 5 bars; 72·5 lb/sq in). Nosewheel is oleo-sprung, main wheels by rubber blocks. Seats for two persons side by side; VFR instrumentation standard, IFR optional.

POWER PLANT: Two 0·88 kN (198 lb st) Microturbo TRS 18 turbojet engines. Fuel in two 82 litre (18 Imp gallon) wing tanks and two 14 litre (3 Imp gallon) fuselage tanks in Prometheus 19, giving total capacity of 192 litres (42 Imp gallons). Prometheus 12 has two 35 litre (7·75 Imp gallon) tip-tanks, increasing total capacity to 262 litres (57·5 Imp gallons). Oil capacity 2 litres (0·4 Imp gallons).

FARNER
DIPL-ING ETH HANS ULRICH FARNER

c/o Aviafiber AG, Sagenrainstrasse 4, CH-8636 Wald
Telephone: (055) 952055

FARNER HF COLIBRI 1 SL

Herr Farner designed, built and flew this unusual experimental motor glider, a description and illustration of which can be found in the 1980-81 *Jane's*.

No news of the HF Colibri 1 SL has been received for the past two years.

NEUKOM
ALBERT NEUKOM SEGELFLUGZEUGBAU

Flugplatz Schmerlat, CH-8213 Neuenkirch

NEUKOM S-4A ELFE 15

The S-4A Elfe 15 flew for the first time in 1970, and 10 had been built by early 1973. Production continues, though at a slow rate. The Elfe 15 is also available in kit form for amateur construction.

TYPE: Single-seat Standard Class sailplane.

AIRFRAME: Cantilever high-wing monoplane. Wortmann wing sections: FX-61-163 at root, FX-60-126 at tip. Single wing spar of aluminium alloy, with plywood/foam sandwich skin. Plain ailerons. Schempp-Hirth airbrakes on upper surfaces. Forward fuselage of reinforced glassfibre; rear portion and tail unit of plywood/foam sandwich. Elevator spring trim. Retractable rubber-sprung monowheel, with brake. Detachable canopy. Universal aero-tow/winch coupling forward of monowheel.

NEUKOM ELFE 17

The Elfe 17 is an Open Class version of the S-4A, employing the same fuselage, but having a two-piece wing of 17 m span with provision for a tank in each leading-edge to contain a total of 60 kg (132 lb) water ballast. A braking parachute is carried on this version.

A total of 10 Elfe 17s had been built by the Spring of 1973. Production continues, though at a slow rate. The Elfe 17 is available also in kit form for amateur construction.

UNION OF SOVIET SOCIALIST REPUBLICS

LAK
LITOVSKAYA AVIATSIONNAYA KONSTRUKTSIYA (Lithuanian Aircraft Construction)

Prenaisk, Litovsk (Lithuania)
DIRECTOR: Vitautas Pakarskas
DEPUTY DIRECTOR: V. Myakshryunas
CHIEF DESIGNER: Boleslavas Karvyalis
CHIEF TEST PILOT: Vitautas Shliumba

This factory originated in the late 1960s with a 12-man 'initiative group', led by designers B. Oshkinis, B. Karvyalis and A. Paknis, and assisted by the Lithuanian Central Committee of DOSAAF. Its primary objective was the evaluation of a plastics-construction sailplane. A workshop, opened in the Summer of 1969, began by building the Oshkinis BRO-11 M elementary training glider. At that time it was known as the Experimentalnii Zavod Sportivnoi Aviatsii (Experimental Sports Aviation Factory). Its first (and the first Soviet) plastics sailplane, designed by Boleslavas Karvyalis, was the **BK-7 Lietuva**, which was first flown by Alexander Ionushas on 8 December 1972 and was described and illustrated in the 1975-76 *Jane's*. This was followed by the **BK-7A**, which had a 2·22 m (7 ft 3½ in) increase in wing span, and retractable airbrakes instead of the 60° landing flaps and brake parachute of the BK-7.

Two experimental versions were then built: the **BK-7V** and **BK-7S**. These had wings with a modified spar, interconnected flaps and ailerons, and increased water ballast; a longer cockpit canopy; modified main wheel, and a sprung tailskid instead of a tailwheel; a lighter-weight elevator; and no brake-chute. The BK-7V and BK-7S took part in an international competition between Socialist countries at Kishinev in April 1976. By that time a developed version of the BK-7A, known as the **LAK-9**, was already in production, and took part in the 1976 World Championships in Finland.

From 1972 further experience was gained by repairing Czechoslovak Blanik sailplanes. More than 300 Blaniks had been overhauled by mid-1977, and LAK is now the sole overhaul centre in the USSR for this type of sailplane. Repairs have also been undertaken of German MBB Phoebus and Schleicher ASW 15 plastics sailplanes. The factory has its own static test laboratory; design calculations and wind tunnel testing of LAK designs are carried out by the Central Aero-Hydrodynamic Institute (TsAGI) in Moscow. Branch factories now handle repairs and overhauls of Czechoslovak Blanik, and Polish Cobra and Foka gliders, and BRO-11 production. Future plans include repair and overhaul of Polish PZL-104 Wilga utility aircraft. A new, larger, sports aviation factory was due to be opened in 1981.

OSHKINIS BRO-11 M ZILE (BLUETIT)

The BRO-11, designed by Bronius Oshkinis, was first built in 1954, and has since been built in considerable numbers as an elementary training glider for use by the DOSAAF. Production of the slightly larger BRO-11 M by the LAK began in mid-1979. The 1977-78 *Jane's* contained brief details of a biplane seaplane version, the **BRO-16**, first flown in August 1973, and a shorter-span (experimental?) version of this, known as the **BRO-17U Utochka**.

The BRO-11 M was still in production by LAK in 1980.

TYPE: Single-seat basic training glider.

AIRFRAME: Strut-braced high-wing monoplane, of fabric- and plywood-covered wooden construction. Towing hook under nose. Semi-recessed monowheel landing gear.

EFF Prometheus twin-jet motor glider

Neukom S-4A Elfe 15 single-seat Standard Class sailplane *(Martin Fricke)*

Oshkinis (LAK) BRO-11 primary training glider (*reproduced from* Krilya Rodini)

LAK-12 Lietuva Open Class sailplane (*reproduced from* Krilya Rodini)

LAK-9 Lietuva single-seat Open Class sailplane (*reproduced from* Krilya Rodini)

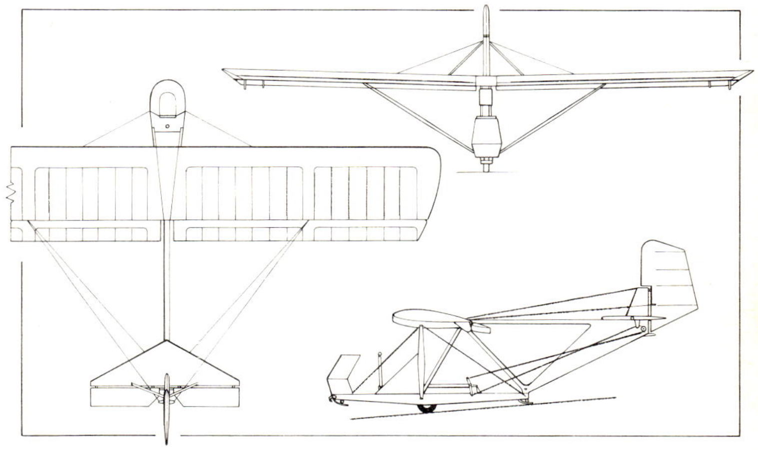
Oshkinis BRO-11 M basic training glider *(Michael A. Badrocke)*

LAK-5 NYAMUNAS

An experimental motor glider was produced by the LAK in 1975, by fitting an engine to a standard Czechoslovak L-13 Blanik sailplane. An illustration of this aircraft, which was designated LAK-6, appeared in the 1977-78 *Jane's*.

The designation LAK-5 was reserved for a motor glider of national design, and several prototypes of this aircraft were under construction in the Summer of 1980. The LAK-5, and its 56 kW (75 hp) engine, were designed by engineer Vintsas Lapenas. The aircraft, named after a river in Lithuania, is intended primarily for training pilots of sailplanes and powered aircraft for participation in national and international competitions, and for record attempts. Other possible applications include forest fire-spotting patrol, meteorological investigation, and geological reconnaissance.

LAK-9 LIETUVA (LITHUANIA)

As indicated in the introductory copy, the LAK-9 is a developed version of the Karvyalis BK-7A; its chief designer was Kiastutis Gechas. The first LAK-9, registered CCCP-6301, was flown by Oleg Pasechnik in the Open Class of the June 1976 World Championships at Räyskälä, Finland. It was reported that the three LAK-9s then built had not, at that time, fully completed flight testing.

A modified version, designated **LAK-9M**, was produced in 1978, but the improvement in performance was not considered to be up to contemporary requirements. The LAK-9/9M was therefore superseded, first by the LAK-10 and later by the LAK-12.

Type: Single-seat Open Class sailplane; *g* limits +6·0/−3·0.

Airframe: Cantilever shoulder-wing monoplane. Single-spar wings, of Wortmann FX-67-K-170 section, with interconnected flaps and ailerons on trailing-edges. Upper and lower surface airbrakes. Glassfibre/epoxy construction. All-moving tailplane. Retractable monowheel and tailwheel. Provision for 90 kg (198 lb) water ballast.

LAK-10 LIETUVA

Representing the fifth stage in the development of the Lietuva series, after the BK-7, BK-7A, LAK-9 and LAK-9M, the LAK-10 successfully completed its flight test programme in July 1978 and was then in production for one year.

Compared with the LAK-9, the wings of the LAK-10 have a 0·40 m (1 ft 3¾ in) increase in span, higher aspect ratio, and slightly less gross area. The Wortmann FX-67-K-170 section is retained only from the root to the junction with the outer panels, then tapering to an FX-67-K-150 section at the tip. The wing spars are of hot-bonded carbonfibre. Airbrakes are fitted to the upper surfaces only, and water ballast capacity is increased to 220 litres (48·5 Imp gallons). Flaps and ailerons are interconnected, as on the LAK-9/9M. Cockpit layout, and external airframe finish, are also improved; the cockpit canopy opens sideways to starboard.

On 31 May 1979 a LAK-10 flown by Oleg Pasechnik set a new Soviet out-and-return distance record of 335·6 nm (622 km; 386·5 miles), at an average speed of 60·4 knots (112 km/h; 69·6 mph). However, it was found that the use of an all-moving tailplane limited the sailplane's speed (to avoid the danger of flutter), and its maximum speed was only 97 knots (180 km/h; 112 mph). Production therefore ended in favour of the further-redesigned LAK-12.

LAK-11

Brief reference has been made in the Soviet aviation press to a Standard Class sailplane designated LAK-11, said to be designed by Yonas Bankauskas. It is not certain whether there is any connection between this design and that of the Nida sailplane described separately.

LAK-12 LIETUVA

This Open Class sailplane, now in production, was designed under the leadership of Kiastutis Gechas. It is intended as a training aircraft for high-class glider pilots participating in all-Soviet and international events, and for record attempts. Compared with the LAK-10 it has a more comfortable cockpit, with improved instrument layout, a better-positioned control column, and a one-piece canopy. Streamlining of the airframe is enhanced, the wings are lowered to a mid position, the fuselage and horizontal tail surfaces are of completely new design, and construction makes extensive use of composite materials, including carbonfibre for the wing spars and other components.

First flight of the LAK-12 took place on 21 December 1979. Flight testing was completed in 1980, and deliveries of production aircraft began later that year.

Type: Single-seat Open Class sailplane.

Airframe: Cantilever mid-wing monoplane. Wortmann wing sections: FX-67-K-170 from root to junction with outer panels, then tapering to FX-67-K-150 at tip. No wing fillets. Main wing spar is a carbonfibre T beam. Trailing-edge flaps and interconnected ailerons are attached to auxiliary rear spar. Airbrakes on upper surfaces only. Wings, flaps and ailerons have a skin of three-ply glasscloth, with foam plastics infill between each layer; carbonfibre tape is used to reinforce flap construction. Fuselage, and integral fin, are of glassfibre construction, reinforced with carbonfibre tape. Slightly-swept fin and rudder, and non-swept low-mounted tailplane and elevators (all of Wortmann FX-71-L-150/25 section), are of similar construction to wings. Rudder and elevators are mass-balanced. Retractable monowheel, with mechanical brake and oleo-pneumatic shock-absorber; non-retractable shock-absorbing tail bumper. One-piece glassfibre cockpit canopy. Provision for 250 litres (55 Imp gallons) water ballast in wing leading-edge tanks.

LAK NIDA

Under the supervision of engineer Yonas Bankauskas, work was proceeding in the Summer of 1980 on the construction of three prototypes of this 15 metre Contest Class sailplane. Construction is of carbonfibre and other composite materials. The aircraft is named after a tributary of the River Vistula.

UNITED KINGDOM

SLINGSBY

SLINGSBY ENGINEERING LTD (Aircraft Division)

Head Office and Works: Ings Lane, Kirkbymoorside, North Yorkshire YO6 6EZ
Telephone: 0751 31751
Telex: 57911
Directors:
R. J. Kendrick (Chairman)
J. S. Tucker, BSc (Managing)
R. Dobson
G. B. Robertson
Sales Manager, Aircraft Division: Roger Bull

Slingsby's present production includes the Vega 15 metre Class sailplane, the Sport Vega sport class sailplane, the T.61F Venture, which is a 'military' version of the T.61A built between 1971 and 1974, and the T.61G Falke, the latest civilian development of the type.

As described in the main Aircraft section, Slingsby also manufactures under licence, as the T.67A, the Fournier RF6B-120 light aircraft, of which French production has now ended.

SLINGSBY VEGA

The first sailplane entirely of Slingsby design to appear since the failure of the previous Slingsby company in 1969, the Vega was designed to take advantage of the change in the rules for Standard Class sailplanes permitting, after 1976, the installation of flaps in this class. Designed for optimum performance in its class, its wings have a unique flap and airbrake speed limiting system fitted to their trailing-edges, operated by a single control lever in the cockpit. The fuselage is gently 'waisted' to reduce the possibility of flow separation over the fuselage/wing junction.

The prototype Vega flew for the first time on 3 June 1977. A total of 43 had been delivered by the beginning of 1981.

Type: Single-seat 15 metre Class sailplane.

Airframe: Cantilever mid-wing monoplane with T tail. Wing section Wortmann FX-67-K-150 from root to tip. Wings of foam plastics sandwich, with single carbonfibre main spar. Combined flaps/airbrakes inboard, ailerons outboard, on trailing-edges; glassfibre ailerons can be operated independently or in conjunction with flaps. Turned-down wingtips. Conventional semi-monocoque plastics fuselage, with retractable tow hook. Cantilever T tail, with GRP/carbonfibre tailplane spar; elevator spring trim. Tost monowheel (size 5·00-5), with Tost drum brake, and tailwheel, both fully retractable. Canopy opens forward and upward, and is jettisonable. Provision for 159 kg (350 lb) water ballast.

SLINGSBY T.65C SPORT VEGA

A prototype of the Sport Vega flew for the first time on 18 December 1979. Eight had been delivered by the beginning of 1981.

Type: Single-seat sporting sailplane; *g* limit +5·3.

Airframe: Generally similar to Vega, but with glassfibre main spar, rotating trailing-edge airbrakes instead of flaps, and non-retractable monowheel and tailwheel. No provision for water ballast.

SLINGSBY T.61F VENTURE

RAF name: Venture T.Mk 2

Slingsby began producing the Scheibe SF-25B Falke motor glider under licence in 1970, and completed a total of 35.

The Venture is a special version, of which 40 have been ordered by the Ministry of Defence (Air) for the Air Training Corps. It features a special glassfibre spar, encased in plywood, which simultaneously reduces the empty weight and increases the max permissible T-O weight of the aircraft. Many other glassfibre components are used in the aircraft, including a new seat designed to improve comfort and to reduce the hazard of loose articles slipping under the seat into the control area. The power plant is a Rollason-modified Volkswagen 33·5 kW (45 hp) 1,600 cc motorcar engine, driving a two-blade fixed-pitch propeller.

First flight of a production Venture T.Mk 2 (XZ550) was made on 2 July 1977. Deliveries began in Autumn 1977, and 33 had been completed by the beginning of 1981.

SLINGSBY T.61G FALKE

The T.61G is a civil development of the T.61F Venture, differing mainly in having a 45 kW (60 hp) Limbach SL 1700 EA flat-four engine driving a Hoffmann two-blade fixed-pitch (optionally variable-pitch) propeller. Fuel capacity is 32 litres (7 Imp gallons).

SWALES

SWALES SAILPLANES

Long Street, Thirsk, North Yorkshire
Telephone: 0845 23096
Officers:
Bryan Swales, MBIM, Eng Dip, MIMI
G. L. Kemp, BEM (Chief Inspector)

SWALES SD3-15T

Construction of the prototype SD3 started in September 1974, and it flew for the first time in March 1975. This aircraft was designated SD3-13V. Production aircraft were designated SD3-15V (first flight July 1975) and SD3-15T (first flight December 1976), the suffix letter denoting the tail configuration. By mid-1979, six SD3-15s had been built. Construction is now to order, and is restricted to the T-tail version, of which one was nearing completion at that time. No later information has been supplied.

Type: Single-seat sailplane; *g* limits +3·5/−1·0.

Airframe: Cantilever mid-wing monoplane. Wortmann FX-61-168 wing section. Dihedral 3°. Wings built mainly of metal, with metal and polystyrene ribs and GRP wingtips. All-metal trailing-edge flaps/airbrakes. Metal plain ailerons, with foam ribs. Semi-monocoque fuselage with metal skin and GRP nosecone. Cantilever T tail, of metal construction with 50% foam ribs. Full-span elevator with spring trim. Non-retractable monowheel, tyre pressure 2·41 bars (35 lb/sq in), with internally-expanding brake. Canopy opens sideways to starboard.

WRIGHT

PETER W. WRIGHT

3 The Close, Burton Lazars, near Melton Mowbray, Leicestershire
Telephone: Melton Mowbray (0664) 4361

WRIGHT FALCON 15 m

Mr Wright began the design and construction of this single-seat 15 metre Class homebuilt sailplane in 1975. It flew for the first time on 16 September 1979, and in March 1980 was certificated for semi-aerobatic and cloud flying. Kits for amateur constructors may be made available at a later date.

Type: Single-seat homebuilt 15 metre Class sailplane; *g* limits +12/−8 (design), +6/−4 (operating).

Airframe: Cantilever mid-wing monoplane. Wings have a 16 per cent laminar-flow section, with low Reynolds number, and a constant thickness/chord ratio of 16 per cent from root to tip. Dihedral 3°. Incidence 3°. No sweepback. Wings are formed from a foam/glasscloth spar web, pultruded stringers and GRP spar caps, and a glasscloth skin. Wide-chord, 1·22 m (4 ft) span flap/airbrake on each trailing-edge. Mass-balanced ailerons, of glasscloth on foam ribs. Fuselage is a five-ply glasscloth shell, moulded in top and bottom halves, with foam bulkhead stiffeners and GRP/foam longitudinal top-hat stiffeners. Cantilever V tail surfaces (included angle 80°), fins being built integrally with fuselage. Mass-balanced 'ruddervators', with horn aerodynamic balance, have electrically-operated spring trim. Non-retractable unsprung monowheel, size 12 × 4½ in, semi-recessed in underfuselage fairing; 3 in diameter tailwheel, with leaf spring, in similar but smaller fairing under rear fuselage. Stainless steel band brake on main wheel. Semi-reclining seat under detachable framed canopy.

UNITED STATES OF AMERICA

AmEAGLE

AmEAGLE CORPORATION

841 Winslow Court, Muskegon, Michigan 49441
Telephone: (616) 780 4680
President: Larry Haig

This company was formed by Mr Larry Haig, a 1958 graduate engineer from Wayne State University, following seven years experience in gas turbine engine development with the research laboratories of General Motors; a further seven years in automotive and diesel engine design and application work for the Allison Division of General Motors and Teledyne Continental Motors; and two years experience with composite laminate structures with Brunswick Corporation.

Its first product is the American Eaglet homebuilt self-launching sailplane.

AmEAGLE AMERICAN EAGLET

Design of the American Eaglet began in September 1974. Construction of the prototype started in June 1975, and this aircraft (N101EA) made its first flight on 19 November of that year.

A total of 400 kits had been ordered by early 1980, of which 20 had then been completed.

Type: Single-seat homebuilt self-launching sailplane; *g* limits +4·4/−2·2 (safety factor 1·5).

Airframe: Shoulder-wing monoplane, with single aluminium tube bracing strut on each side. Wing section Wortmann FX-61-184. Dihedral 1° 30′ from roots. Sparless stressed-skin structure with urethane foam core, moulded glassfibre leading-edges and wingtips, and epoxy-bonded pre-cured glassfibre skin. 'Spoilerons' (for both roll and glidepath control) at 30 per cent chord on each upper surface. Pod-and-boom fuselage, the forward portion comprising two pre-formed glassfibre half-shells pop-riveted to tubular aluminium longerons. Tailboom is a thin-wall aluminium tube with a moulded glassfibre tailcone. Combined pitot tube/lifting handle in nose. Cantilever inverted-V tailplane and elevators, suspended from rear of tailboom. Tail surfaces have urethane foam core and epoxy/glassfibre skin. Manually-retractable monowheel, tyre pressure 2·41 bars (35 lb/sq in), with external friction-pad brake, forward of CG. Tailwheel under tip of each half-tailplane.

Power Plant: One 9 kW (12·2 hp at 8,000 rpm) McCulloch MC-101B single-cylinder two-stroke engine, installed aft of cockpit and driving a two-blade fixed-pitch pusher propeller with folding plastics blades. Recoil starting. Engine is for T-O and self-recovery only, and not for cross-country continuous operation. Fuel tank capacity 2 litres (0·5 US gallons). Normal fuel load of approx 1·81 kg (4 lb) (consumption at max power is 5·5 kg/h; 12 lb/h) is sufficient for one T-O and climb to 610 m (2,000 ft) and three airborne restarts and climbs from 150 m (500 ft) to 610 m (2,000 ft).

Performance (powered):

Max rate of climb at S/L	137 m (450 ft)/min
T-O run	305 m (1,000 ft)
Landing run	90 m (300 ft)

BRYAN

BRYAN AIRCRAFT INC

Head Office and Works: Williams County Airport, PO Box 488, Bryan, Ohio 43506
Telephone: (419) 636 1340
Director: R. E. Schreder

This company markets plans and kits for the RS-15 and HP-18 high-performance 15 metre Class sailplanes, and is developing the more recent HP-19.

BRYAN (SCHREDER) RS-15

The RS-15 was designed by Mr Schreder to meet OSTIV Standard Class specifications. It is designed for simple, rapid assembly by the homebuilder and is licensed in the amateur-built Experimental category. No jigs are

Slingsby T.61F Venture T.Mk 2 in service with UK Air Cadets

Slingsby Vega high-performance 15 metre sailplane

Known as the Airmate, this Bryan (Schreder) HP-18 was modified by its Canadian builders Don Brand and Peter Masak to have Wortmann FX-60-126 section winglets of glassfibre and balsa at the wingtips *(Neil A. Macdougall)*

Bryan (Schreder) RS-15 Standard Class sailplane built by Mr Gordon Anderson of Michigan *(Howard Levy)*

Swales SD3-15T single-seat sailplane

Wright Falcon 15 m single-seat homebuilt sailplane

AmEagle American Eaglet single-seat self-launching sailplane *(J. M. G. Gradidge)*

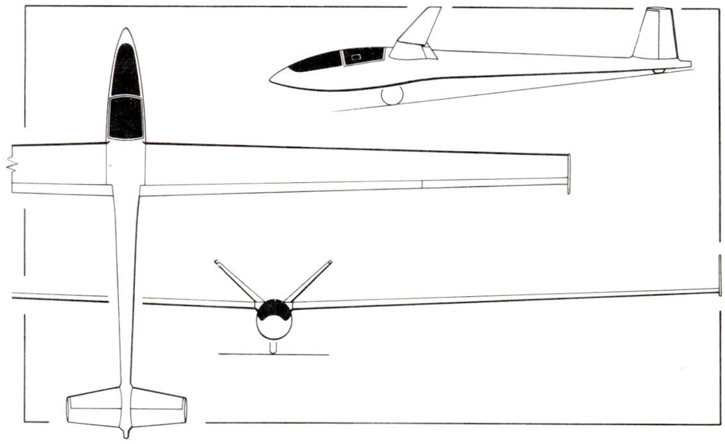

Bryan (Schreder) HP-19 high-performance sailplane *(Michael A. Badrocke)*

required, and most major components are prefabricated, to reduce assembly time to approx 500 man-hours for a builder with average mechanical aptitude.

TYPE: Single-seat 15 metre Class sailplane.

AIRFRAME: Cantilever shoulder-wing monoplane. Wing section Wortmann FX-67-K-150. Dihedral 2° 18′. All-metal wings except for polyurethane foam plastics ribs. Main wing spar caps of aluminium plate. Plain ailerons. Optional trailing-edge flaps/airbrakes, of aluminium sheet bonded to PVC foam ribs, which can be linked with ailerons. Monocoque fuselage: prefabricated glassfibre forward pod, complete with bulkheads, floorboards and finish; aluminium tube tailboom. All-metal V tail, which can be folded upward for towing or storage. Retractable monowheel, with hydraulic shock-absorber and brake, and non-retractable steerable tailwheel. Provision for 91 kg (200 lb) water ballast in wing box spars.

BRYAN (SCHREDER) HP-18

The HP-18 was flown for the first time in 1975. Approx 170 had been built or were under construction by early 1979. Assembly time is approx 700 man-hours.

TYPE: Single-seat 15 metre Class sailplane; *g* limits ±12·0.

AIRFRAME: Generally similar to the RS-15, the HP-18 has a slightly longer fuselage, with low profile circular instead of oval section, and other features designed to reduce drag and produce a superior competition aircraft. Incidence 2°. Improvements include better gap seals, new wingtips, faired tailwheel and better streamlining; ailerons and 90° flaps are coupled, to improve cross-country speed; use of full flap permits steeper approaches and slower landing speeds. NASA-type winglets can be added to increase glide ratio and improve aileron control. Fuselage comprises a prefabricated glassfibre pod and aluminium monocoque tailboom. The Tost monowheel has a tyre pressure of 2·07 bars (30 lb/sq in) and is fitted with a mechanically-expanding brake.

BRYAN (SCHREDER) HP-19

The HP-19 is an improved version of the HP-18. A carbonfibre main spar and other weight savings reduce wing weight by 32 kg (70 lb), while increasing water ballast capacity to 136 kg (300 lb). A new Schreder wing section with higher maximum lift coefficient, together with NASA-type winglets, decrease thermalling and landing speeds, decrease minimum sink rate, improve aileron control and increase glide ratios. Coupled flaps and ailerons improve performance at all speeds, while 90° flaps permit steeper approaches and slower landing speeds than were possible with conventional airbrakes. Simplified constructional materials and methods reduce assembly time to about 500 man-hours. A prototype was expected to fly during 1979, but no news has been received from the company since early that year.

DSK

DSK AIRCRAFT CORPORATION

13161 Sherman Way, North Hollywood, California 91605
VICE-PRESIDENT, MARKETING: Rodney E. Gage

In addition to the well-established Duster homebuilt sailplane, DSK is believed to be developing an improved version known as the DS-1. It is also an approved supplier of kits for the Woodstock One homebuilt sailplane designed by Mr Jim Maupin (which see).

DSK BJ-1b DUSTER

The Duster is available in the form of plans and/or component kits for amateur construction. Average construction time is approx 600-700 hours.

TYPE: Single-seat homebuilt sailplane; *g* limits +5·3/−3·1.

AIRFRAME: Cantilever shoulder-wing monoplane. NACA 4415 (modified) wing section, with 2° washout at tips. Single wooden main spar, with plywood ribs and plywood covering. Wing built in three portions: two outer panels, and a 2·13 m (7 ft) span centre-section integral with fuselage. Trailing-edge flaps/airbrakes; top-hinged ailerons. Straight-sided plywood fuselage shell, having an elongated hexagonal cross-section in the cockpit area, changing to triangular section aft of wings. Ventral plywood keel, reinforced by two bulkheads shaped to seat contour and supporting floor-mounted seat. Glassfibre nosecone and fairings. Cantilever tail unit, with tailplane mounted on top of fuselage. Construction similar to that of wings. All tail surfaces plywood skinned. Non-retractable semi-recessed monowheel beneath keel, and tailwheel. Aircraft can be flown with one-piece flush-fitting canopy, or with open cockpit and windscreen only.

POWER PLANT: Auxiliary power package, using a 19·4 kW (26 hp) engine, under development. Plans detail all hardware necessary to support an engine mount, which can be added at builder's option without further alteration of airframe.

EXPLORER

EXPLORER AIRCRAFT COMPANY INC (Sierra-Nevada Soaring Division)

PO Box 60036, 4895 Texas Avenue, Reno-Stead Airport, Reno, Nevada 89506
Telephone: (702) 972 7757

Explorer Aircraft markets plans of the biplane Aqua Glider. Designed by Colonel William L. Skliar, USAF (Retd), it is intended for tethered gliding by unlicensed pilots, towed behind any motorboat able to attain a speed of 30 knots (56 km/h; 35 mph). If the pilot has the necessary licence, he can cast off from the motorboat when airborne and make a free flight before landing back on the water.

EXPLORER PG-1 AQUA GLIDER

Design of the Aqua Glider began in September 1958, and the prototype made its first flight in July 1959. Plans are available to amateur constructors and approx 1,000 sets have been sold in more than 20 countries all over the world. About 12 Aqua Gliders are known to have flown, in the USA, the Bahamas, Brazil and Japan, and about 200 more are under construction.

The prototype, after making about 1,000 flights and being flown by about 60 pilots, was donated to the Experimental Aircraft Association Museum in 1970. In the previous year, as illustrated in the accompanying photograph, it had been converted to a landplane glider and ground-towed by a motorcar. Some homebuilders have converted their Aqua Gliders to similar configuration.

TYPE: Single-seat homebuilt waterborne glider; *g* limit +4·0.

AIRFRAME: Forward-stagger single-bay biplane with N interplane and parallel centre-section struts. Wing section (both) NACA 4412. Dihedral 2° 30′ on lower wing only. Conventional single-spar wooden structure with fabric covering. Spoiler-type light alloy ailerons on lower wing only, immediately aft of main spar. Balancer floats at lower wingtips. Unstepped watertight wooden hull of spruce with mahogany plywood bow, bottom skins and sides. Plywood is glassfibre-covered below waterline. Tow hook on nose. Wire-braced spruce tail unit, with plywood and fabric covering, carried on welded steel tube or wire-braced wooden boom. All-moving one-piece tailplane with bungee trim. Conventional rudder. Standard jumper skis, 1·83 m (6 ft) in length, attached to small wire-braced struts below hull. Open cockpit.

FLIGHT DYNAMICS

FLIGHT DYNAMICS INC

PO Box 5070, State University Station, Raleigh, North Carolina 27607
Telephone: (919) 781 6198 or 834 6806
PRESIDENT: Thomas H. Purcell Jr

FLIGHT DYNAMICS SEASPRITE

The Seasprite is the unpowered Stage I version of the aircraft which in a fully-developed powered form will be designated Seasprite II. The Seasprite is of simplified construction, and is available in plans form to amateur sailplane constructors.

Flight Dynamics has modified the original design, replacing the single tail unit by twin fins and rudders; the Seasprite has also been provided with non-reversing differential ailerons. It was planned to power this airframe with two West Bend engines, each of approximately 7·5 kW (10 hp).

The following description applies to the unpowered model:

TYPE: Single/two-seat homebuilt glider.

AIRFRAME: Fuselage, of aerofoil/hydrofoil cross-section, is built on a framework of aluminium tubing. Sides and top are covered with polyethylene foil, except for upper part of nose section, which has a transparent covering of polyester film and Plexiglas. Pilot sits in centre-section, between twin booms, each of which carries at the aft end a fin and rudder. The fins are linked by a tailplane, attached to the tops of the fins and carrying a full-span elevator. Underside of fuselage/hull has catamaran-type twin floats, built of plastics with a plywood covering. For only a minimal weight increase, structure can be covered instead with aluminium foil, which has a much longer life. Wings are of triangular planform, and are of sailwing type, with aluminium tube leading-edge, wire trailing-edge, and polyethylene covering. They are braced to fuselage sides, and can be folded rearward when not in use. Wingtips pivot about their leading-edges, to provide roll control.

HOLLMANN

HOLLMANN AIRCRAFT

11082 Bel Aire Court, Cupertino, California 95014
Telephone: (408) 255 2194
DIRECTOR: Martin Hollmann

HOLLMANN-KLEPEIS CONDOR

Designed jointly by Mr Eddie Klepeis and Mr Martin Hollmann, the Condor is a two-seat motor glider, of composite construction, powered by a Volkswagen motorcar engine fitted with a turbocharger. Design began in 1979. One prototype is under construction by each co-designer, in New York and California respectively, with first flights anticipated in 1982.

TYPE: Two-seat motor glider; *g* limits +3·8/−2·0.

AIRFRAME: Cantilever mid-wing monoplane with turned-down wingtips, a central fuselage nacelle containing the tandem two-seat cockpit and rear-mounted engine, and two short tailbooms which support twin, sweptback, inward-canted endplate fins and rudders bridged at the top by a fixed tailplane and elevator. Landing gear comprises a non-retractable semi-recessed nosewheel and main wheel. Wings are of Wortmann FX-67-K-170 section, with 0° dihedral and 0° 30′ incidence. Constant-chord centre-section; taper on outer leading-edges. Aluminium frame construction, with glassfibre skin and ailerons. No flaps, spoilers, airbrakes or tabs. Welded steel tube fuselage, with glassfibre outer shell. Tail unit of similar construction to wings. Landing gear is unsprung, fitted with tyres size 5·00-5 (main) and 4·00-4 (nose), pressure 1·93 bars (28 lb/sq in). Hydraulic caliper brake on main wheel. Cockpit seats two persons in tandem, under Plexiglas canopy identical to that of NASA/Ames AD-1. Standard equipment includes engine starter, alternator, lights, and nav/com transceiver.

POWER PLANT: One 41 kW (55 hp) Revmaster VW engine, driving a two-blade fixed-pitch wooden pusher propeller. Fuel tank capacity 75·5 litres (16·5 Imp gallons; 20 US gallons).

MAUPIN

JIM MAUPIN

PO Box 5127, San Pedro, California 90733

MAUPIN WOODSTOCK ONE

The Woodstock One was designed by Mr Jim Maupin and was flown for the first time in early 1978. It has been flight tested to include spins (four turns) and stalls; and has been dived at 110 per cent of its never-exceed speed. Performance and handling are claimed to be easier and better than those of a Schweizer SGS 1-26.

Plans, including an instructional booklet, are available to amateur constructors, and by early 1981 about 200 sets had been sold in the USA and some 20 other countries.

Approved kits for homebuilders are available from two sources:

DSK Aircraft Corporation, 13161 Sherman Way, North Hollywood, California 91605, USA; and

Aircraft Manufacturers, 16 Perth Street, Pietermaritzburg 3201, South Africa.

TYPE: Single-seat homebuilt sailplane; *g* limit +5.

AIRFRAME: Cantilever shoulder-wing monoplane, of Douglas fir and birch construction with some fabric covering. Culver wing sections (thickness/chord ratio 18% at root, 13% at tip). Airbrakes on upper wing surfaces. Non-retractable semi-recessed monowheel. Undernose towing hook. Plans depict both open and enclosed cockpits, and a powered version with a non-retractable pylon-mounted power plant aft of the cockpit.

MONNETT

MONNETT EXPERIMENTAL AIRCRAFT INC

955 Grace Street, Elgin, Illinois 60120
Telephone: (312) 741 2223

In addition to its Sonerai powered aircraft (see Homebuilts section), Monnett produces kits of parts for the Monerai single-seat homebuilt sailplane.

By January 1981 a total of 320 Monerai kits had been sold, and 40 aircraft had flown.

MONNETT MONERAI S

The Monerai, development of which began in 1976, was designed to fill a need for a low-cost homebuilt sailplane with better-than-average performance. It flew for the first time in March 1978. The kit includes many extruded or preformed components, to permit quick construction with a minimum of equipment or experience. Average construction time is 300 hours.

TYPE: Single-seat homebuilt sailplane; *g* limits ±6·0.

AIRFRAME: Cantilever mid-wing monoplane. Wing section Wortmann FX-61-192 (modified). Dihedral 4°. Aluminium main spar of extruded, modified I section; rear spar is an aluminium C channel. Constant chord allows all the preformed ribs to be identical, and single-piece aluminium skin to wrap around from trailing-edge to trailing-edge. Skin bonded to spars and ribs with Hysol epoxy. Glassfibre wingtips. Aileron and flap skins are preformed and ready for assembly. Fuselage is built up from chrome-molybdenum steel tube and is covered by a non-structural premoulded glassfibre shell. A 12·7 cm (5 in) diameter extruded aluminium tube is attached to the frame and extends 3·66 m (12 ft) to the rear where an upright, all-flying, all-aluminium V tail is attached. Non-retractable monowheel, with mechanical brake; tailskid. One-piece flush-fitting canopy. No provision for water ballast.

MONNETT MONERAI P

Exhibited at Oshkosh in 1978, the Monerai P is identical to the S version, but has a small auxiliary engine (15 kW; 20 hp 250 cc Zenoah single-cylinder two-stroke) to permit self-launch. Fuel capacity is 4·25 litres (1·13 US gallons). A combination turnover structure and power pod mount, aft of the cockpit, allows the sailplane to be converted to the P version in only five minutes. The engine and propeller do not fold away when not in use, but a streamline glassfibre fairing encloses the engine and pylon. The Monerai P has a rate of climb of 91-122 m (300-400 ft)/min. Approx 100 power pod kits had been delivered by the Summer of 1981.

DSK BJ-1b Duster single-seat homebuilt sailplane

Flight Dynamics Seasprite single/two-seat homebuilt glider

Maupin Woodstock One homebuilt sailplane

Monnett Monerai P, powered version of the S for self-launch

Prototype Explorer PG-1 Aqua Glider with water skis replaced by wheeled landing gear

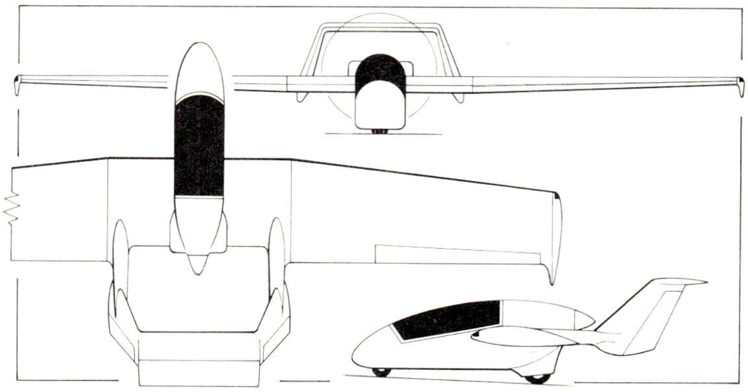

Hollmann-Klepeis Condor two-seat motor glider *(Michael A. Badrocke)*

Monnett Monerai S single-seat homebuilt sailplane

RAND ROBINSON

RAND ROBINSON ENGINEERING INC
5842 K McFadden Avenue, Huntington Beach, California 92649
Telephone: (714) 898 3811

RAND ROBINSON KR-1B

Under this designation, Rand Robinson was in late 1979 proposing to market plans for a motor glider version of the KR-1 (see Homebuilts section), with enlarged wing outer panels to increase the span to 8·23 m (27 ft 0 in).

Rand Robinson KR-1B motor glider *(Don Dwiggins)*

RYSON

RYSON AVIATION CORPORATION
548 San Fernando Street, San Diego, California 92106
Telephone: (714) 278 4791

PRESIDENT: T. Claude Ryan

EXECUTIVE VICE-PRESIDENT AND OPERATIONS MANAGER: Jerome D. Ryan

Ryson Aviation Corporation was founded by Mr T. Claude Ryan, until 1969 Chairman and Chief Executive Officer of the Ryan Aeronautical Company (now Teledyne Ryan Aeronautical). The aim of the Corporation is to develop aeronautical products and make them available for manufacture by other companies.

Since early 1975 the company has been working on the ST-100 (Soaring/Touring—100 hp) motor glider.

RYSON ST-100 CLOUDSTER

The ST-100, design of which was started on 18 March 1974, incorporates all-metal construction, fully-folding wings, power-operated flaps for air braking, and ailerons which operate with the flaps. It is designed to be aerobatic and to meet current FAR Pt 23 gust load requirements (15·25 m; 50 ft/s); is capable of performing as an aero-tow aircraft for unpowered sailplanes; and can be used as a conventional powered aircraft for cross-country flying.

The ST-100 prototype (N2RY) flew for the first time on 21 December 1976. By January 1981 it had accumulated about 800 h of flight testing and demonstration, including an economy record flight from California to Oshkosh, Wisconsin (1,456 nm; 2,697 km; 1,676 miles), on 28 of the available 32 US gallons of fuel (106 of 121 litres); and a 5,214 nm (9,656 km; 6,000 mile) flight around the perimeter of the United States.

FAA certification, expected originally in early 1980, was delayed because of difficulty in meeting the engine manufacturer's specified cylinder-head maximum temperatures. This problem, which developed as a result of the unusual conditions required by the capabilities of the aircraft (eg, full-power engine output in a climb), has now been solved, and final-phase flight testing by the FAA was resumed in January 1981. Type certification was expected to follow shortly afterwards, permitting completion of plans for series production of the ST-100 by a licensee.

Type: Tandem two-seat motor glider; *g* limits +6·75/−3·5.

Airframe: Cantilever low-wing monoplane. Wing section Wortmann FX-67-K-170/17 throughout, with max thickness at 40 per cent chord. Dihedral 4° from roots. All-metal safe-life structure, with some fail-safe features, built of extruded 2024-T4 and other alloys of aluminium, with sheet aluminium skin. Aluminium plain ailerons and plain trailing-edge flaps, with foam core. Flaps, powered by electrically-operated screwjack, can be deflected downward for use as airbrakes. Ailerons operate with flaps. No spoilers or tabs. Wings fold back alongside fuselage for stowage or transportation. Conventional all-metal semi-monocoque fuselage, with extruded aluminium longerons, and sheet metal frames, bulkheads and skins. Cantilever T tail, with sweptback fin and rudder and non-swept horizontal surfaces, latter comprising fixed-incidence tailplane and one-piece balanced elevator. Rudder and elevator have riveted sheet aluminium primary ribs, bonded intermediate foam ribs, and bonded sheet aluminium skin. Elevator tips removable for transportation or storage. Anti-servo and trim tab, of similar construction, in centre of elevator. Two-wheel main gear and steerable tailwheel, all non-retractable. Ryson oleo-pneumatic shock-absorbers. Cleveland main wheels, with Cleveland hydraulic disc brakes, and Goodyear tailwheel. Streamline glassfibre fairings on main gear legs, main wheels and tailwheel. Canopy opens sideways to starboard. Dual controls standard. Baggage space aft of rear seat.

Power Plant: One 74·5 kW (100 hp) Continental O-200 flat-four engine, driving a Hoffmann HO-V-62 variable-pitch fully-feathering propeller with two composite blades. Fuel in two integral tanks in leading-edges of wing centre-section, with combined capacity of 121 litres (26·6 Imp gallons; 32 US gallons). Oil capacity 5·7 litres (1·25 Imp gallons; 1·5 US gallons).

SCHWEIZER
SCHWEIZER AIRCRAFT CORPORATION

Head Office and Works: Box 147, Elmira, New York 14902
Telephone: (607) 739 3821
Telex: 932459
President: William Schweizer
Chief Engineer: Leslie E. Schweizer
Sales Manager: James B. Short

Schweizer Aircraft Corporation is a leading American designer and manufacturer of sailplanes. Its current products include the SGS 2-33A, SGS 1-35 series and SGS 1-36.

In December 1980 Schweizer and Gulfstream American Corporation approved an agreement under which Schweizer purchased the G-164 Ag-Cat agricultural aircraft, which will henceforth be manufactured and marketed by Schweizer.

Schweizer is also well known as an aircraft subcontract manufacturer, and is currently engaged in such work for Beech, Bell Helicopter, Boeing, Sikorsky, and several other companies.

About 250 people were employed by Schweizer in early 1981.

SCHWEIZER SGS 1-26

Production of the 1-26E ended in January 1980 with the completion of the 700th aircraft. During the 26-year production life of the 1-26 series, approx 200 of these aircraft were produced in kit form.

A description and illustration of the SGS 1-26E can be found in the 1980-81 *Jane's*.

SCHWEIZER SGS 2-33A

The SGS 2-33 was developed to meet the demand for a medium-priced two-seat sailplane for training and general family soaring. The prototype was first flown in the Autumn of 1966 and received FAA Type Approval in February 1967. Production began in January 1967, and 575 had been built by January 1981. The current 2-33A is also available in kit form.

Type: Tandem two-seat training sailplane.

Airframe: Strut-braced high-wing monoplane. Aluminium alloy structure with metal skin and all-metal ailerons. Airbrakes fitted. Welded chrome-molybdenum steel tube fuselage. Nose covered with glassfibre, remainder with Ceconite fabric. Steel tube tail unit, covered with Ceconite fabric. Braced tailplane. Non-retractable Cleveland monowheel immediately aft of nose-skid. Rubber-block shock-absorption for skid. Wingtip wheels.

SCHWEIZER SGS 1-35 SERIES

The SGS 1-35 is an all-metal single-seat high-performance 15 metre Class sailplane, the prototype of which was flown for the first time in April 1973. The FAA certification programme was completed in Spring 1974.

The 1-35 is claimed to have the widest wing loading range in the FAI 15 metre Class (28·22 to 43·75 kg/m²; 5·78 to 8·96 lb/sq ft), and carries 146·5 kg (323 lb) of water ballast. It can thus compete effectively in both light and strong soaring conditions. The 80° landing flaps are used instead of the more conventional airbrakes.

Two newer models are also available, as follows:

SGS 1-35A. Unrestricted 15 metre Class version. Monowheel forward of CG, large tailwheel and no nose-skid. Interconnected flaps and ailerons standard. Data as for standard 1-35 except for empty equipped weight of 222 kg (490 lb) and best glide ratio of 41.

SGS 1-35C (Club-35). Simplified, lower-cost version for club or syndicate ownership. Non-retractable Cleveland unsprung monowheel aft of CG, Cleveland hydraulic brake, nose-skid, and no water ballast provision. Flush rivets used on aft fuselage. Data as for standard 1-35 except for empty equipped weight of 192 kg (425 lb), max T-O weight of 310 kg (685 lb), max wing loading of 32·16 kg/m² (6·59 lb/sq ft), and best glide ratio of 36.

A total of 98 SGS 1-35s of all versions had been produced by January 1981. The following description applies to the standard SGS 1-35:

Type: Single-seat 15 metre Class sailplane; *g* limits +8·33/−5·33.

Airframe: Cantilever shoulder-wing monoplane. Wortmann wing sections: FX-67-K-170 at root, FX-67-K-150 at tip. Dihedral 3° 30′. Aluminium stressed-skin and stringer wing structure. Bottom-hinged trailing-edge flaps and top-hinged ailerons of aluminium torque cell construction. No airbrakes or spoilers. All-aluminium monocoque fuselage. Cantilever aluminium T tail, with fixed-incidence tailplane and fabric-covered elevator. No tabs. Mechanically-retractable unsprung monowheel, tyre pressure 2·41 bars (35 lb/sq in). Hydraulic brake. Nose-skid and tailwheel or tailskid. Detachable canopy. Provision for 146·5 kg (323 lb) water ballast.

SCHWEIZER SGS 1-36 SPRITE

The Sprite was designed to offer modern performance, handling and appearance to up to 99 per cent of the solo glider pilots in the USA and overseas as a personal, school or club sailplane. Although a new design, it utilises some components of the 1-34 and 1-35 in its construction. The Sprite is, however, aimed at a different market from these earlier types, and is intended to be a 'one-design' class, as was the 1-26. The Sprite is also intended to fulfil a worldwide need for a single-seat Club Class sailplane.

Two versions are available:

36903-1. With forward-positioned monowheel, sprung tailwheel, and no nose-skid.

36903-3. With monowheel further aft, unsprung tailwheel, and aluminium nose-skid. This version is recommended for school and club operation, where ruggedness and ease of ground handling are important considerations.

The prototype flew for the first time on 2 August 1979. FAA certification was received in September 1980, and by 1 March 1981 a total of 17 had been delivered.

Type: Single-seat multi-purpose sailplane.

Airframe: Cantilever mid-wing monoplane, of all-aluminium construction. Wortmann wing sections: FX-61-163 at root, FX-60-126 at tip. Incidence 1° at root, 0° 4′ at tip. Airbrakes in upper and lower wing surfaces. Cantilever T tail, with elevators. Non-retractable semi-recessed 5·00-5 monowheel, with hydraulic brake, and tailwheel; aluminium nose-skid on 36903-3 version.

VOLPAR
VOLPAR INC

7929 Hayvenhurst Avenue, Van Nuys, California 91406
Telephone: (213) 787 4393 and 994 5023
Officers: See Aircraft section

In February 1977 Volpar retained Mr Percival H. Spencer to design the Drag-N-Fly, a small but fully manoeuvrable water-based glider having simple controls that could be operated safely by an amateur or non-pilot.

The prototype was first flown, by Mr Spencer, on 20 April 1977, and in early 1979 was said to be ready for production. In January 1981 it was stated to be no longer in production.

A description and illustration of the Drag-N-Fly can be found in the 1980-81 *Jane's*.

YUGOSLAVIA

ELAN
ELAN TOVARNA SPORTNEGA ORODJA N.SOL.O

Begunje St 1, 64275 Begunje na Gorenjskem
Telephone: (064) 75 010, 75 218 and 75 560
Telex: 34518
Manager, Plastics Division: Tadej Lazar

Formed in 1944 by a group of 10 workers to make skis for Yugoslavia's partisan fighters, Elan is now well known as a manufacturer of sports equipment, glassfibre pleasure and fishing craft, and gliders. In early 1979 it assumed responsibility for licence production of the DG-100 sailplane built formerly by Glaser-Dirks of the Federal Republic of Germany (which see).

ELAN (GLASER-DIRKS) DG-100 ELAN

The DG-100 is a modified and lighter-weight development of the Akaflieg Darmstadt D-38 described in the 1973-74 *Jane's*. Its design, by Dipl-Ing Wilhelm Dirks, began in August 1973, and the prototype (D-7100) flew for the first time, in West Germany, on 10 May 1974. By the end of 1978 Glaser-Dirks had completed 105 DG-100s, including 16 examples of the DG-100G (first flight 11 June 1976) which differs in having a tailplane similar to that of the Glaser-Dirks DG-200.

From early 1979, the DG-100 entered licence production in Yugoslavia, and the two versions are now known as

Ryson ST-100 Cloudster (Continental O-200 engine)

Schweizer SGS 2-33A two-seat general-purpose sailplane

Schweizer SGS 1-35A single-seat 15 metre sailplane

Schweizer SGS 1-36 Sprite training and club sailplane (36903-1 version)

Schweizer SGS 1-35C Club version of the 1-35A

Elan-built Glaser-Dirks DG-100 Standard Class sailplanes

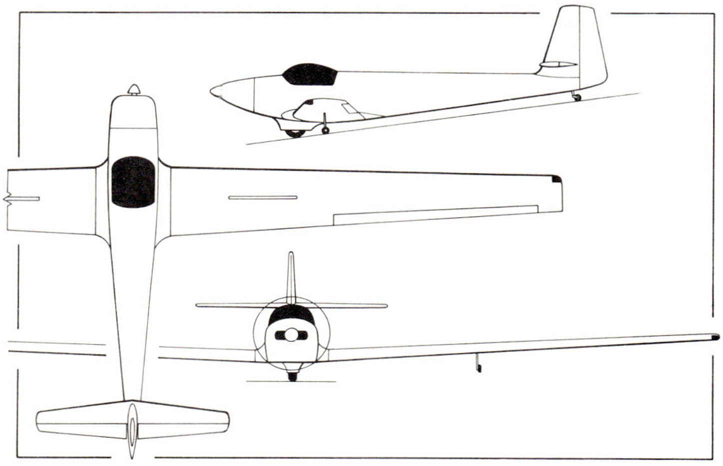

Jastreb SOLE-77 two-seat motor glider *(Michael A. Badrocke)*

the **DG-100 Elan** and **DG-100G Elan**. Ten of each version were built by Elan during 1979, and by the end of 1980 production totalled 21 DG-100s and 27 DG-100Gs. A **Club Elan** version of the DG-100, without water ballast, is also available; this may be fitted with either a conventional tailplane, or with an all-moving tailplane incorporating a trimmable anti-Flettner tab.

The following description applies to the DG-100 standard version:

TYPE: Single-seat Standard Class sailplane; *g* limit +6·1.

AIRFRAME: Cantilever shoulder-wing monoplane. Wortmann wing sections: FX-61-184 at centreline, FX-60-126 at tip. Dihedral 3° from roots. Glassfibre roving main spar. Glassfibre/Conticell/foam sandwich wings, ailerons and all-moving T tailplane; all-glassfibre semi-monocoque fuselage, fin and rudder. Schempp-Hirth duralumin airbrakes on wing upper surfaces. Water ballast tank in each wing (except on Club version), combined capacity 100 kg (220 lb). Full-span Flettner tab in tailplane. Manually-retractable unsprung monowheel, tyre pressure 2·0 bars (29 lb/sq in), with Tost drum brake, and tailwheel.

ELAN DG-101 (G)

The DG-101 (G) is an improved version of the DG-100 with a one-piece canopy, more comfortable cockpit, sprung monowheel, and automatically connected elevator. It flew for the first time in March 1981.

JASTREB
JASTREB AEROPLANE AND GLIDER FACTORY

Vrsac

Telephone: (013) 813157, 813051, 813639 and 815491

JASTREB SOLE-77

The Jastreb SOLE-77, illustrated in an accompanying three-view drawing, is a tandem two-seat motor glider for elementary and advanced training, and can be used for basic aerobatics. Other applications include atmospheric sampling, photogrammetry, and study of air currents. The monowheel is said to be retractable; the tailwheel, linked to the rudder pedals, is fully castoring. The cockpit canopy, which opens forward, is jettisonable. The engine, of unknown type, develops 54 kW (72 hp) and drives a two-blade propeller; there is provision for an auxiliary fuel tank. Without this tank, standard fuel load is 50 kg (110 lb) and powered flight endurance 5 h.

All other known details are given in the table.

MF
MASINSKI FAKULTET (Faculty of Mechanical Engineering, Aeronautical Institute ATI)

27 Marta 80, 11000 Belgrade
Telephone: (011) 329 362
SAILPLANE DESIGNER:
Prof Dr Tomislav Dragovic, Dipl Eng

The Faculty of Mechanical Engineering at the University of Belgrade is the leading research and educational institution for aircraft engineering in Yugoslavia. It has tackled many important light aviation projects in recent years, and collaborates closely with the national aerospace industry. Recent designs have included the UTVA-75, described in the Aircraft section.

MF VUK-T

The prototype of this single-seat all-plastics sailplane was designed and built at the MF in 1976, and made its first flight in 1977. It entered production in the Jastreb Aeroplane and Glider Factory at Vrsac (which see) in January 1979. At that time 110 had been ordered by the Yugoslav Aeronautical Union.

Chief features of the VUK-T are an entirely new technique of plastics stressed-skin construction and the use of a supercritical wing section. The sailplane is designed to OSTIV airworthiness requirements, and is cleared for cloud flying, aerobatics and spinning.

TYPE: Single-seat basic and advanced training sailplane; *g* limits +5·3/−2·65 (safety factor 1·5).

AIRFRAME: Cantilever mid-wing monoplane. NASA GAW-1 supercritical wing section from root to tip, with no twist. Thickness/chord ratio 17%. Dihedral 3°. Glassfibre roving main spar, integral with the upper and lower wing skins, which are of glassfibre/Conticell foam sandwich. Auxiliary rear spar carries the top-hinged plain ailerons. Schempp-Hirth airbrakes in upper surfaces at 50 per cent chord. Fuselage is of glassfibre sandwich construction forward, glassfibre monocoque aft. Cantilever T tailplane, with separate elevator, of similar construction to wings. Retractable monowheel, with brake, and non-retractable tailwheel. Semi-reclining seat under one-piece flush-fitting jettisonable canopy. No water ballast provision.

SOKO

SOKO VAZDUHOPLOVNA INDUSTRIJA, RADNA ORGANIZACIJA VAZDUHOPLOVSTVO

79000 Mostar
Telephone: (088) 22 121
Telex: 46 180 YU SOKOMO
DIRECTOR, RESEARCH AND DEVELOPMENT:
Dipl Ing Milenko Pjescić

SOKO SL-40 LISKA

Construction of the prototype Liska single-seat motor glider began at the end of 1975, design having started about a year earlier. It flew for the first time on 19 February 1981.

TYPE: Single-seat motor glider.

AIRFRAME: Cantilever low-wing monoplane. Wortmann wing sections: FX-61-184 at root, FX-61-163 at tip. Dihedral 4° from roots. Incidence 2°. Sweepforward 0° 22′ 48″ at quarter-chord. Single-spar light alloy wings, of stressed-skin construction, with light alloy slotted ailerons. Light alloy DFS-type spoiler on each wing, inboard of aileron. Light alloy semi-monocoque fuselage. Single-spar stressed-skin light alloy tail unit, with sweptback fin and rudder, and non-swept all-moving tailplane. Plain-hinged light alloy trim and balance tabs in tailplane. Manually-operated mechanically-retractable monowheel, retracting forward, is a Moravan-Otrokovice unit taken from a Let L-13 sailplane; it has SOKO rubber-in-compression shock-absorption, and a mechanically-actuated brake. Non-retractable steerable tailwheel, with solid rubber tyre. Non-retractable outrigger under each wing. Mainwheel tyre size 350 × 135 × 150 mm, pressure 2·45 bars (35·5 lb/sq in); tailwheel tyre size 100 × 28 × 50 mm. Cockpit canopy opens sideways to starboard. Compartment behind seat, accessible from cockpit, for 5 kg (11 lb) of baggage. Cockpit is heated and ventilated; 12V 34Ah battery and engine-driven alternator provide power for instruments (including standard blind-flying panel) and Dittel FSG-16 VHF com radio. Oxygen system, VOR/LOC and ADF are optional.

POWER PLANT: One 33·5 kW (45 hp) Pieper-Stark Stamo MS 1500/2 piston engine, driving a Hoffmann HO11-137B85L or HO11-150B70L two-blade fixed-pitch propeller. Single metal fuel tank aft of cockpit, capacity 40 litres (8·8 Imp gallons). Refuelling point immediately aft of cockpit on port side. Oil capacity 2·5 litres (0·55 Imp gallons).

VTC

VAZDUHOPLOVNO TEHNICKI CENTAR—VRSAC

Vrsac, 29 Novembra b.b. Guduricki put
Telephone: 80-111
DIRECTOR: Veselinovic Zivota
CHIEF DESIGNER: Dipl Ing Ivan Sostaric

The VTC at Vrsac is producing the Schempp-Hirth Cirrus and Standard Cirrus under licence. The first production Cirrus built at Vrsac was delivered in late 1971.

VTC (SCHEMPP-HIRTH) CIRRUS

This single-seat high-performance sailplane was designed in the Federal Republic of Germany by Dipl-Ing Klaus Holighaus. The prototype flew for the first time in January 1967 with a V tail unit. The second prototype had

DATA: SAILPLANES AND MOTOR GLIDERS (POWER OFF)

Manufacturer and Model	Dimensions: Wing span: m; ft in	Wing aspect ratio	Wing area (gross): m²; sq ft	Length overall: m; ft in	Height over tail: m; ft in	Weights and Loading: Empty: kg; lb	Max T-O: kg; lb	Max wing loading: kg/m²; lb/sq ft
AUSTRALIA								
Sunderland MOBA 2C	14·99;	24·7	9·08;	6·78;	1·36;	263;	361;	39·8;
	49 2		97·7	22 3	4 5½	580	796	8·15
Todhunter T-5 Blue Wren	10·97;	14·9	8·08;	5·49;	1·00;	106;	197;	24·42;
	36 0		87·0	18 0	3 3¼	235	435	5·00
AUSTRIA								
Brditschka HB-21	16·24;	13·9	18·98;	8·10;	2·60;	505;	710;	37·4;
	53 3½		204·3	26 7	8 6¼	1,113	1,565	7·66
Brditschka HB-22	15·00;	12·2	18·45;	8·50;	n.k.	520;	750;	40·6;
	49 2½		198·6	27 10½		1,146	1,653	8·33
BRAZIL								
CTA PE-80367 Urubu	17·80;	15·1	20·915;	8·96;	2·00;	370;	550;	26·3;
	58 4¾		225·1	29 4¾	6 6¾	816	1,212	5·39
EEUFMG (CEA) CB-2/B Minuano	15·00;	22·1	10·20;	7·00;	1·43;	230;	400;	39·22;
	49 2½		109·8	22 11¾	4 8¼	507	882	8·03
IPE 02 Nhapecam	17·20;	15·4	19·26;	8·40;	1·57;	370;	570;	29·6;
	56 5¼		207·3	27 6¾	5 1¾	816	1,256	6·06
Sidou João Grande	18·00;	25·0	12·96;	9·00;	2·52;	270;	450;	34·72;
	59 0¾		139·5	29 6¼	8 3¼	595	992	7·11
CHINA (PEOPLE'S REPUBLIC)								
Chendu X-7	13·07;	9·49	18·00;	7·06;	1·60;	220;	370;	20·55;
	42 10½		193·75	23 2	5 3	485	816	4·21
CZECHOSLOVAKIA								
VSO 10	15·00;	18·75	12·00;	7·00;	1·38;	250;	380;	31·67;
	49 2½		129·2	22 11¾	4 6¼	551	837	6·49
VSO 10C	15·00;	18·75	12·00;	7·00;	1·38;	250;	380;	31·67;
	49 2½		129·2	22 11¾	4 6¼	551	837	6·49
DENMARK								
Projekt 8 Dolphin	18·72;	16·8	20·80;	8·50;	1·30;	480;	750;	36·0;
	61 5		223·9	27 10¾	4 3¼	1,058	1,653	7·37
Projekt 8 Dolphin M2	18·60;	16·6	20·80;	8·40;	1·52;	520;	750;	36·0;
	61 0¼		223·9	27 6¾	4 11¾	1,146	1,653	7·37
FINLAND								
Alanne Motorblanik	16·20;	13·7	19·15;	8·40;	1·14;	384;	584;	30·5;
	53 1¾		206·1	27 6¾	3 9	846	1,287	6·25
Alanne Motorlerche	13·00;	10·3	16·34;	6·86;	1·58;	288;	400;	23·8;
	42 7¾		175·9	22 6	5 2¼	635	882	4·88
FRANCE								
Calvel Frelon	10·40;	13·5	7·80;	5·30;	1·60;	160;	240;	30·8;
	34 1½		83·96	17 4¾	5 3	352	529	15·12
CARMAM J.P. 15-36A Aiglon	15·00;	20·4	11·00;	6·18;	1·40;	229;	390;	35·0;
	49 2½		118·4	20 3¼	4 7	505	860	7·16
CARMAM C.38	15·00;	20·6	11·00;	6·87;	1·85;	230;	440;	40·0;
	49 2½		118·4	22 6½	6 0¾	507	970	8·19
CARMAM C.40	15·00;	20·6	11·00;	6·65;	1·40;	240;	370;	34·0;
	49 2½		118·4	21 9¾	4 7	529	816	6·97
or	17·00;	24·7	11·70;	6·65;	1·40;	250;	470;	40·0;
	55 9½		125·9	21 9¾	4 7	551	1,036	8·20

n.k. = not known; n.a. = not applicable

VUK-T all-plastics training sailplane, designed by the Belgrade University Faculty of Mechanical Engineering

SOKO SL-40 Liska single-seat motor glider

a conventional tail unit, as fitted to production models.

Schempp-Hirth built 120 Cirrus before ending production in late 1971. Since then the Cirrus has been produced under licence in Yugoslavia by VTC, which had completed about 60 by early 1979. According to Schempp-Hirth, production was continuing in 1981.

Type: Single-seat sailplane.

Airframe: Cantilever mid-wing monoplane. Wortmann FX-66 series section. Dihedral 3° at spar centreline. Wing shell is a glassfibre/foam sandwich structure, with an all-glassfibre box spar. Hinged ailerons of glassfibre/balsa sandwich. No flaps. Schempp-Hirth aluminium alloy airbrakes. Glassfibre fuselage shell, stiffened with foam rings, secured with resin. Cantilever tail unit of glassfibre/foam sandwich. Retractable rubber-sprung Tost monowheel, tyre pressure 3·38 bars (49 lb/sq in). Tost drum brake.

VTC (SCHEMPP-HIRTH) STANDARD CIRRUS 75-VTC

Designed by Dipl-Ing Klaus Holighaus, this Standard Class version of the Schempp-Hirth Cirrus entered production in the Federal Republic of Germany in the Summer of 1969, following the first flight of the prototype in March 1969.

By April 1977 a total of 700 Standard Cirrus had been built, including 200 under licence by Grob Flugzeugbau in West Germany. Production by Grob ended in July 1975, and by Schempp-Hirth in April 1977. Since then licence manufacture has been undertaken by SCAP in France, and from January 1979 by VTC in Yugoslavia. VTC had built 14 Standard Cirrus by the Spring of 1979. According to Schempp-Hirth, production was continuing in 1981.

Type: Single-seat Standard and Club Class sailplane; *g* limit +10.

Airframe: Cantilever mid-wing monoplane. Wortmann wing section. Dihedral 3°. Sweepback 1°18′ at leading-edge. Glassfibre/foam sandwich wings, ailerons and tail surfaces. Schempp-Hirth glassfibre airbrakes on upper surfaces. Glassfibre fuselage shell, stiffened with bonded-in foam rings. All-moving T tailplane. Retractable monowheel standard; non-retractable faired monowheel optional. Tost wheel with drum brake and Continental tyre, pressure 3·45 bars (50 lb/sq in). Canopy opens sideways to starboard. Provision for 60 kg (132 lb) of water ballast.

Performance (at max T-O weight except where indicated)									
Best glide ratio	at (speed): knots; km/h; mph	Min sinking speed: m (ft)/s	at (speed): knots; km/h; mph	Stalling speed: knots; km/h; mph	Max speed (smooth air): knots; km/h; mph	Max speed (rough air): knots; km/h; mph	Max aero-tow speed: knots; km/h; mph	Max winch-launching speed: knots; km/h; mph	Remarks
38	55; 102; 48	0·64; 2·10	48; 89; 42	*40; 74; 46	110; 204; 127 (IAS)	93; 172; 107 (IAS)	86; 159; 99 (IAS)	70; 130; 81 (IAS)	*Flaps up (35 knots; 65 km/h; 40·5 mph flaps down)
n.k.	50; 93; 58	n.k.	n.k.	35; 64; 40	n.k.	n.k.	n.a.	n.a.	Max power loading 17·6 kg/kW (29·0 lb/hp). No other performance details available until flight testing completed
24-26	56·5; 105; 65	1·20; 3·94	45; 84; 52	38; 70; 43·5	108; 200; 124	94; 174; 108	n.a.	n.a.	
24	n.k.	1·20; 3·94	n.k.	40·5; 75; 47	135; 250; 155	n.k.	n.a.	n.a.	
30·6	46; 85; 53	0·70; 2·30	39; 72; 45	35; 64; 40	139; 257; 160	87; 161; 100	81; 150; 93	n.k.	
39	51; 95; 59	0·60; 1·97	45; 83; 51·5	35; 65; 40·5	140; 260; 161	86; 160; 99	86; 160; 99	n.a.	
28	43; 80; 50	0·75; 2·46	37; 68; 42·5	30; 55; 35	108; 200; 124	81; 150; 93	65; 120; 75	n.a.	
43·9	51; 95; 59	0·57; 1·87	44; 81; 50·5	37·5; 69; 43	162; 300; 186	97; 180; 112	97; 180; 112	59·5; 110; 68·5	
12	30; 55; 34	1·40; 4·59	28·5; 53; 33	n.k.	81; 150; 93	n.k.	n.k.	n.k.	
36	49; 90; 56	0·64; 2·10	39; 73; 45·5	37; 68; 42·5	135; 250; 155	86; 160; 99	86; 160; 99	65; 120; 75	
34	51; 95; 59	0·72; 2·36	43; 79; 49	38; 70; 43·5	135; 250; 155	86; 160; 99	86; 160; 99	65; 120; 75	
32	54; 100; 62	0·70; 2·30	43·5; 80; 50	38; 70; 43·5	141·5; 263; 163	141·5; 263; 163	67·5; 125; 77·5	59; 110; 68	
33	54; 100; 62	0·70; 2·30	43·5; 80; 50	n.k.	n.k.	n.k.	n.k.	n.k.	
n.k.	43; 80; 50	n.k.	40·5; 75; 47	n.k.	76; 140; 87	70; 130; 81	59; 110; 68	n.a.	
17	42; 78; 48·5	1·00; 3·28	35; 65; 40·5	24·5; 45; 28	81; 150; 93	65; 120; 75	n.a.	n.a.	
n.k.	48·5; 90; 56	n.k.	27; 50; 31	24·5; 45; 28	65; 120; 75	48·5; 90; 56	n.a.	n.a.	Performance estimated
36·9	43·5; 80; 50	0·62; 2·03	39; 72; 45	33·5; 62; 39	119; 220; 136	89; 166; 103	89; 166; 103	n.k.	
38	54; 100; 62	0·60; 1·97	40; 75; 46·5	35; 64; 40	119; 220; 136	89; 166; 103	119; 220; 136	89; 166; 103	
39	49; 90; 56	n.k.	n.k.	31·5; 58; 36	146; 270; 168	97; 180; 112	97; 180; 112	97; 180; 112	Data for 15 m version without water ballast, 17 m version with water ballast
41	37; 68; 42	n.k.	n.k.	31·5; 58; 36	146; 270; 168	97; 180; 112	97; 180; 112	97; 180; 112	

	Dimensions					Weights and Loading		
Manufacturer and Model	Wing span: m; ft in	Wing aspect ratio	Wing area (gross): m²; sq ft	Length overall: m; ft in	Height over tail: m; ft in	Empty: kg; lb	Max T-O: kg; lb	Max wing loading: kg/m²; lb/sq ft
Centrair 101	15·00;	21·4	10·50;	6·82;	1·42;	238;	455;	43·3;
	49 2½		113·0	22 4½	4 8	525	1,003	8·88
Centrair (Schleicher) ASW 20FP	15·00;	21·4	10·50;	6·82;	1·45;	260;	454;	43·2;
	49 2½		113·0	22 4½	4 9	573	1,000	8·86
ENSMA FS-25 S Cuervo	15·00;	26·4	8·54;	6·48;	1·42;	183;	270;	31·6;
	49 2½		91·9	21 3	4 8	403	595	6·47
Ferrier Condor	15·00;	22·4	10·057;	6·95;	1·50;	245;	450;	44·7;
	49 2½		108·25	22 9½	4 11	540	992	9·17
Fournier RF-9	17·00;	16·1	18·00;	7·86;	1·93;	520;	750;	41·67;
	55 9½		193·75	25 9½	6 4	1,146	1,653	8·53
Fournier RF-10	17·47;	16·3	18·70;	7·89;	1·93;	530;	770;	41·2;
	57 3¾		201·3	25 10¾	6 4	1,168	1,697	8·44
Gross (Fauvel) AV.451	15·00;	13·2	17·017;	3·84;	1·84;	290;	420;	25·9;
	49 2½		183·17	12 7¼	6 0½	639	926	5·31
Issoire (Siren) D 77 Iris	13·50;	16·0	11·40;	6·37;	1·20;	220;	345;	30·3;
	44 3½		122·7	20 10¾	3 11¼	485	760	6·21
Issoire (Siren) E 78 Silène	18·00;	18·0	18·00;	7·95;	1·50;	365;	565;	31·3;
	59 0½		193·8	26 1	4 11	805	1,245	6·43
Pottier Kit Club 15-34	15·00;	20·6	11·00;	6·25;	1·40;	225;	420;	39·0;
	49 2½		118·4	20 6	4 7	496	926	7·99
SCAP Cirrus 78L	15·00;	22·5	10·00;	6·41;	1·32;	218;	450;	45·0;
	49 2½		107·6	21 0½	4 4¾	480	992	9·22
Siren (EIRI) PIK-20E2F	15·00;	22·5	10·00;	6·53;	1·47;	310;	470;	47·0;
	49 2½		107·6	21 5	4 10	683	1,036	9·63
GERMANY (FEDERAL REPUBLIC)								
Akaflieg Braunschweig SB-12	15·00;	22·5	10·02;	6·40;	1·40;	220;	450;	44·9;
	49 2½		107·8	21 0	4 7	485	992	9·20
Akaflieg Darmstadt D-39	15·00;	20·5	11·00;	7·15;	*1·02;	370;	470;	42·0;
	49 2½		118·4	23 5½	3 4¼	816	1,036	8·61
Akaflieg Karlsruhe AK 2	22·00;	29·8	16·23;	7·60;	1·67;	480;	670;	41·3;
	72 2¼		174·7	24 11¼	5 5¾	1,058	1,477	8·46
Akaflieg München Mü 27 (flaps in)	22·00;	27·5	17·60;	10·30;	2·10;	710;	900;	51·0;
	72 2¼		189·4	33 9½	6 10¾	1,565	1,984	10·45
Akaflieg München Mü 27 (flaps out)	22·00;	20·2	23·90;	10·30;	2·10;	710;	900;	37·6;
	72 2¼		257·3	33 9½	6 10¾	1,565	1,984	7·70
Akaflieg München Mü 28	12·00;	10·9	13·20;	6·78;	1·53;	250;	360;	27·27;
	39 4½		142·1	22 3	5 0¼	551	794	5·59
or	14·00;	13·2	14·56;	6·78;	1·53;	260;	370;	25·41;
	45 11¼		156·7	22 3	5 0¼	573	816	5·21
Blessing Staff Rebell	15·00;	13·2	17·00;	6·30;	2·05;	420;	600;	35·3;
	49 2½		183·0	20 8	6 8¾	926	1,322	7·23
FVA-20 F. B. Schmetz	15·00;	17·6	12·80;	7·00;	*0·90;	280;	380;	29·7;
	49 2½		137·8	22 11½	2 11½	617	838	6·08
Glaser-Dirks DG-200	15·00;	22·5	10·00;	7·00;	1·40;	238;	450;	45·0;
	49 2½		107·6	22 11¾	4 7	525	992	9·22
Glaser-Dirks DG-200/17	17·00;	27·3	10·57;	7·00;	1·40;	248;	450;	42·5;
	55 9¼		113·8	22 11¾	4 7	547	992	8·71
Glasflügel Hornet	15·00;	22·96	9·80;	6·40;	1·40;	227;	420;	42·9;
	49 2½		105·5	21 0	4 7	500	926	8·79
Glasflügel Hornet C	15·00;	22·96	9·80;	6·40;	1·40;	210;	450;	45·9;
	49 2½		105·5	21 0	4 7	463	992	9·40
Glasflügel Mosquito B	15·00;	22·8	9·86;	6·40;	1·40;	235;	450;	46·0;
	49 2½		106·1	21 0	4 7	518	992	9·42
Glasflügel 304	15·00;	22·78	9·88;	6·45;	1·36;	235;	450;	44·5;
	49 2½		106·35	21 2	4 5½	518	992	9·12
Glasflügel 402	17·00;	27·3	10·60;	6·75;	n.k.	250;	500;	47·2;
	55 9¼		114·1	22 1¾		551	1,102	9·66
Grob Standard Astir II	15·00;	18·2	12·40;	6·80;	1·26;	255;	450;	36·3;
	49 2½		133·5	22 3¾	4 1½	562	992	7·43
Grob Club Astir II	15·00;	18·2	12·40;	6·80;	1·26;	250;	380;	30·6;
	49 2½		133·5	22 3¾	4 1½	551	837	6·27
Grob Speed Astir IIB	15·00;	19·6	11·50;	6·80;	1·27;	260;	515;	44·8;
	49 2½		123·8	22 3¾	4 2	573	1,135	9·18
Grob G 102 Club III	15·00;	18·2	12·40;	6·70;	1·26;	245;	380;	30·6;
	49 2½		133·5	21 11¾	4 1½	540	838	6·27
Grob G 102 Club IIIb	15·00;	18·2	12·40;	6·70;	1·26;	248;	380;	30·6;
	49 2½		133·5	21 11¾	4 1½	546	838	6·27
Grob G 102 Standard III	15·00;	18·2	12·40;	6·70;	1·26;	255;	450;	36·3;
	49 2½		133·5	21 11¾	4 1½	562	992	7·44
Grob G 103 Twin II	17·50;	17·2	17·80;	8·18;	1·55;	370;	580;	32·6;
	57 5		191·6	26 10	5 1	815	1,278	6·68
Grob G 109	16·60;	13·5	20·4;	7·80;	1·80;	510;	780;	38·2;
	54 5½		219·6	25 7	5 10¾	1,124	1,719	7·83
Hänle H 101 Salto (Aerobatic)	13·30;	20·6	8·58;	5·70;	n.k.	182;	280;	32·6;
	43 7½		92·4	18 8½		401	617	6·68
Hänle H 101 Salto (Utility)	15·50;	26·4	9·10;	5·70;	n.k.	187;	310;	34·0;
	50 10¼		98·0	18 8½		412	683	6·97
Hoffmann H-36 Dimona	16·00;	16·8	15·20;	6·85;	n.k.	470;	740;	48·6;
	52 6		163·6	22 5¾		1,036	1,631	9·97
MFB Model 2 Mistral-C	15·00;	20·7	10·85;	6·73;	1·45;	230;	350;	32·3;
	49 2½		116·8	22 1	4 9	507	771	6·62
Rolladen-Schneider LS3-a	15·00;	21·4	10·50;	6·84;	1·20;	240;	472;	45·0;
	49 2½		113·0	22 5¼	3 11¼	529	1,040	9·22
Rolladen-Schneider LS4	15·00;	21·4	10·50;	6·84;	1·26;	240;	472;	44·9;
	49 2½		113·0	22 5¼	4 1½	529	1,040	9·21
Scheibe Bergfalke-IV	17·20;	17·4	17·00;	8·00;	1·50;	300;	500;	29·4;
	56 5¼		183·0	26 3	4 11	661	1,102	6·02
Scheibe SF-25C and C-S Falke '80	15·25;	12·8	18·20;	7·55;	1·85;	375;	610;	33·5;
	50 0¼		195·9	24 9¼	6 0¾	826	1,345	6·86

n.k. = not known; n.a. = not applicable

Performance (at max T-O weight except where indicated)									
Best glide ratio	at (speed): knots; km/h; mph	Min sinking speed: m (ft)/s	at (speed): knots; km/h; mph	Stalling speed: knots; km/h; mph	Max speed (smooth air): knots; km/h; mph	Max speed (rough air): knots; km/h; mph	Max aero-tow speed: knots; km/h; mph	Max winch-launching speed: knots; km/h; mph	Remarks
41	n.k.	0·68; 2·23	n.k.	35·5; 65; 40·5	135; 250; 155	97; 180; 112	n.k.	n.k.	Performance estimated
45	54; 100; 62	n.k.	n.k.	35; 64; 40	135; 250; 155	86; 160; 99	n.k.	n.k.	
38	49; 90; 56	0·60; 1·97	43; 80; 50	35·5; 65; 40·5	108; 200; 124	86; 160; 99	76; 140; 87	86; 160; 99	
40	56·5; 105; 65	0·57; 1·87	n.k.	39; 72; 45	156; 290; 180	n.k.	n.k.	n.k.	Performance estimated
28	49; 90; 56	0·80; 2·62	43·5; 80; 50	39·5; 73; 45·5	119; 220; 136	99; 183; 114	n.a.	n.a.	
30	46; 85; 53	0·80; 2·62	43·5; 80; 50	39; 72; 45	108; 200; 124	n.k.	n.a.	n.a.	
30·5	47·5; 88; 55	0·72; 2·36	39; 72; 45	38; 70; 43·5	121; 225; 140	86; 160; 99	n.a.	n.a.	
33	51; 95; 59	0·65; 2·13	43·5; 80; 50	32·5; 60; 37·5	121; 225; 140	91; 170; 105	91; 170; 105	68; 126; 78	
38	51; 95; 59	0·67; 2·20	43·5; 80; 50	35·5; 65; 40·5	119; 220; 136	91; 170; 105	91; 170; 105	63; 117; 73	
36	41·5; 77; 48	0·63; 2·07	39; 72; 45	34; 62; 39	135; 250; 155	135; 250; 155	135; 250; 155	65; 120; 75	
38·5	49; 90; 56	0·60; 1·97	40·5; 75; 47	34; 62; 39	135; 250; 155	135; 250; 155	81; 150; 93	65; 120; 75	
41	63; 117; 73	0·70; 2·30	47·5; 88; 55	40·5; 75; 47	154; 285; 177	119; 220; 136	105; 195; 121	67; 125; 78	
**42	53; 98; 61	**0·60; 1·97	42; 78; 48·5	n.k.	*135; 250; 155	*81; 150; 93	*81; 150; 93	*81; 150; 93	*Estimated **At 310 kg (683 lb) AUW
**36	56·5; 105; 65	**0·70; 2·30	45; 84; 52	43·5; 80; 50	151; 280; 174	151; 280; 174	97; 180; 112	59; 110; 68	*Height over fuselage **Without propeller
48	58; 107; 66·5	0·58; 1·90	50; 92; 57	42·5; 78; 48·5	135; 250; 155	135; 250; 155	97; 180; 112	75·5; 140; 87	All performance figures are with propeller retracted
47	62; 115; 71·5	0·65; 2·13	58; 108; 67	53; 98; 61	162; 300; 186	162; 300; 186	108; 200; 124	108; 200; 124	
36	50; 93; 58	0·66; 2·16	41·5; 77; 48	37; 68; 42	75·5; 140; 87	75·5; 140; 87	75·5; 140; 87	75·5; 140; 87	
28	51; 95; 59	0·84; 2·76	43·5; 80; 50	32·5; 60; 37·5	205; 380; 236	167; 310; 193	108; 200; 124	108; 200; 124	Performance estimated
32	51; 95; 59	0·72; 2·36	40·5; 75; 47	32; 59; 37	156; 290; 180	108; 200; 124	86; 160; 99	86; 160; 99	Performance estimated
24	54; 100; 62	1·10; 3·61	43; 80; 50	38; 70; 44	n.k.	n.k.	n.a.	n.a.	
37	49; 90; 56	0·60; 1·97	37; 68; 43	33·5; 62; 39	135; 250; 155	97; 180; 112	81; 150 93	70; 130; 81	*Height over fuselage
42·5	59·5; 110; 68·5	0·59; 1·94	43; 80; 50	*34; 62; 39	146; 270; 168	102; 190; 118	102; 190; 118	70; 130; 80·5	*At 300 kg (661 lb) AUW; other performance at 420 kg (926 lb)
44·6	59·5; 110; 68·5	0·53; 1·74	43; 80; 50	32·5; 60; 37·5	146; 270; 168	102; 190; 118	102; 190; 118	70; 130; 80·5	
38	41; 75; 47	*0·60; 1·97	41; 75; 47	39; 72; 45	135; 250; 155	135; 250; 155	81; 150; 93	65; 120; 74·5	*At 345 kg (760 lb) AUW
38·5	58; 107; 66·5	0·59; 1·94	40; 74; 46	35; 65; 40·5	135; 250; 155	135; 250; 155	81; 150; 93	81; 150; 93	
42	59·5; 110; 68·5	0·67; 2·20	51; 95; 59	42·5; 78; 48·5	135; 250; 155	108; 200; 124	81; 150; 93	81; 150; 93	
43	63; 116; 72	0·69; 2·26	50; 93; 58	40; 73; 46	135; 250; 155	n.k.	n.k.	n.k.	
45	62; 115; 71·5	*0·54; 1·77	43·5; 80; 50	*33·5; 62; 39	135; 250; 155	n.k.	n.k.	n.k.	*At wing loading of 31 kg/m² (6·35 lb/sq ft)
37·5	51; 95; 59	0·58; 1·90	41; 75; 47	32·5; 60; 37·5	135; 250; 155	135; 250; 155	91; 170; 105	65; 120; 74·5	
37	49; 90; 56	0·60; 1·97	40; 74; 46	32·5; 60; 37·5	135; 250; 155	135; 250; 155	91; 170; 105	65; 120; 74·5	
41·5	64·5; 120; 74·5	0·57; 1·87	41; 75; 47	33·5; 62; 39	146; 270; 168	108; 200; 124	91; 170; 105	65; 120; 74·5	
36	50; 92; 57	0·62; 2·03	41; 76; 47	32·5; 60; 37·5	135; 250; 155	n.k.	91; 170; 105	65; 120; 74·5	
35·5	49; 90; 56	0·65; 2·13	42; 78; 48·5	32·5; 60; 37·5	135; 250; 155	n.k.	91; 170; 105	65; 120; 74·5	
38	56; 105; 65	0·70; 2·30	46; 85; 53	32·5; 60; 37·5	135; 250; 155	n.k.	91; 170; 105	65; 120; 74·5	
37	56; 105; 65	0·64; 2·10	43; 80; 50	*33·5; 62; 39	135; 250; 155	91; 170; 105	91; 170; 105	65; 120; 74·5	*With pilot only
30	65; 120; 75	0·90; 2·95	n.k.	40·5; 75; 47	118; 220; 136	n.k.	n.k.	n.k.	
35	n.k.	*0·70; 2·30	39; 72; 45	38; 70; 43·5	151; 280; 174	151; 280; 174	n.k.	n.k.	*At 280 kg (617 lb) AUW; 0·60 m (1·97 ft)/s at 250 kg (551 lb) AUW
36	n.k.	*0·70; 2·30	39; 72; 45	33·5; 62; 39	135; 250; 155	135; 250; 155	n.k.	n.k.	*As above
27	56; 105; 65	0·90; 2·95	43; 80; 50	39; 72; 45	148; 275; 171	91; 170; 105	n.a.	n.a.	
37·5	51·5; 95; 59	0·66; 2·17	44·5; 82; 51	36·5; 67; 42	135; 250; 155	135; 250; 155	86; 160; 99·5	70; 130; 80·5	
41·8	54; 100; 62	0·60; 1·97	38; 70; 43·5	34; 62; 39	146; 270; 168	102·5; 190; 118	102·5; 190; 118	70; 130; 80·5	
40	54; 100; 62	0·60; 1·97	40·5; 75; 47	37; 68; 42·5	146; 270; 168	97; 180; 112	97; 180; 112	70; 130; 81	
34	51·5; 95; 59	0·75; 2·46	46; 85; 53	35·5; 65; 40·5	108; 200; 124	92; 170; 106	76; 140; 87	59·5; 110; 68·5	
23	41; 75; 47	1·00; 3·28	38; 70; 43·5	32·5; 60; 37·5	102·5; 190; 118	102·5; 190; 118	n.a.	n.a.	

Manufacturer and Model	Dimensions					Weights and Loading		
	Wing span: m; ft in	Wing aspect ratio	Wing area (gross): m²; sq ft	Length overall: m; ft in	Height over tail: m; ft in	Empty: kg; lb	Max T-O: kg; lb	Max wing loading: kg/m²; lb/sq ft
Scheibe SF-25K Falke '80	14·60; 47 10¾	12·5	17·10; 184·1	7·55; 24 9¼	1·85; 6 0¾	*410; 904	610; 1,345	35·7; 7·32
Scheibe SF-25E Super-Falke	18·00; 59 0¾	17·8	18·20; 195·9	7·50; 24 7¼	1·85; 6 0¾	440; 970	650; 1,433	35·0; 7·17
Scheibe SF-28A Tandem-Falke	16·30; 53 5¾	14·4	18·50; 199·1	8·10; 26 7	1·55; 5 1	410; 904	610; 1,345	33·0; 6·76
Scheibe SF-34 Delphin	15·80; 51 10	16·9	14·80; 159·3	7·50; 24 7¼	1·50; 4 11	306; 675	500; 1,102	33·8; 6·92
Scheibe SF-36	16·35; 53 7¾	17·1	15·60; 167·9	7·18; 23 6¾	n.k.	430; 948	630; 1,389	40·4; 8·27
Schempp-Hirth Nimbus 2C	20·30; 66 7¼	28·6	14·40; 155·0	7·33; 24 0½	1·45; 4 9	*355; 782	650; 1,433	45·0; 9·22
Schempp-Hirth Mini-Nimbus C	15·00; 49 2½	22·8	9·86; 106·1	6·41; 21 0½	1·32; 4 4	*235; 518	500; 1,102	51·0; 10·45
Schempp-Hirth Nimbus 3	22·90; 75 1½	32·3	16·20; 174·4	7·70; 25 3¼	n.k.	360; 793	750; 1,653	46·3; 9·49
Schempp-Hirth Janus B	18·20; 59 8½	19·95	16·60; 178·7	8·62; 28 3¼	1·45; 4 9	365; 805	620; 1,366	37·0; 7·58
Schempp-Hirth Janus C	20·00; 65 7½	23·0	17·40; 187·3	8·62; 28 3¼	1·45; 4 9	355; 783	700; 1,543	40·0; 8·20
Schempp-Hirth Ventus a	15·00; 49 2½	23·7	9·51; 102·4	6·35; 20 10	n.k.	220; 485	430; 948	45·2; 9·26
Schleicher ASK 13	16·00; 52 6	14·6	17·50; 188·4	8·18; 26 9½	1·60; 5 3	290; 640	480; 1,060	26·8; 5·49
Schleicher ASW 19 B	15·00; 49 2½	20·4	11·00; 118·4	6·82; 22 4½	1·45; 4 9	245; 540	454; 1,000	41·3; 8·46
Schleicher ASW 20	15·00; 49 2½	21·4	10·50; 113·0	6·82; 22 4½	1·45; 4 9	250; 551	454; 1,000	43·2; 8·86
Schleicher ASW 20 L	16·59; 54 5¼	24·9	11·05; 118·9	6·82; 22 4½	1·45; 4 9	*255; 562	380; 838	34·385; 7·05
Schleicher ASK 21	17·00; 55 9¼	16·1	17·95; 193·2	8·35; 27 4¾	1·55; 5 1	360; 794	600; 1,323	37·0; 7·58
Schleicher ASW 22	22·00; 72 2¼	32·5	14·88; 160·2	8·10; 26 7	n.k.	420; 926	750; 1,653	50·4; 10·33
	or 24·00; 78 9	37·2	15·48; 166·6	8·10; 26 7	n.k.	420; 926	600; 1,323	38·7; 7·94
Schleicher ASK 23	15·00; 49 2½	17·4	12·90; 138·8	7·10; 23 3½	n.k.	230; 507	380; 837	29·5; 6·04
Valentin Taifun 15	15·00; 49 2½	13·8	16·35; 176·0	7·785; 25 6½	2·40; 7 10½	480; 1,058	770; 1,697	47·1; 9·65
Valentin Taifun 17	17·00; 55 9¼	16·4	17·60; 189·4	7·785; 25 6½	2·40; 7 10½	490; 1,080	770; 1,697	43·75; 8·96
Wiegand WM-1	15·00; 49 2½	19·1	11·80; 127·0	6·91; 22 8	n.k.	230; 506	400; 882	33·9; 6·95
INDIA								
Civil Aviation Dept ATS-1 Ardhra (2nd prototype)	16·50; 54 1½	12·5	21·83; 235·0	8·61; 28 3	2·464; 8 1	328; 723	508; 1,120	23·28; 4·77
ITALY								
Caproni Vizzola Calif A-21S	20·38; 66 10¼	25·65	16·19; 174·3	7·737; 25 4½	1·84; 6 0½	436; 961	644; 1,419	39·8; 8·15
Caproni Vizzola Calif A-21SJ	20·38; 66 10¼	25·65	16·19; 174·3	7·737; 25 4½	1·84; 6 0½	528; 1,164	808; 1,781	49·9; 10·23
Maggi MG3-15L Condor	15·00; 49 2½	20·0	11·25; 121·1	6·80; 22 3¾	1·56; 5 1½	340; 750	510; 1,124	45·3; 9·28
JAPAN								
Nippi Pilatus B4-PC11AF	15·00; 49 2½	16·0	14·04; 151·1	6·57; 21 6¾	1·57; 5 1¾	230; 506	350; 770	25·0; 5·13
POLAND								
SZD-42-2 Jantar 2B	20·50; 67 3	29·2	14·25; 153·4	7·18; 23 6¾	1·76; 5 9¼	362; 798	*649; 1,430	45·6; 9·28
SZD-48-1 Jantar Standard 2	15·00; 49 2½	21·1	10·66; 114·7	6·71; 22 0¼	1·50; 4 11	265; 584	535; 1,179	50·2; 10·28
SZD-50-3 Puchacz	16·67; 54 8¼	15·3	18·16; 195·5	8·38; 27 6	1·92; 6 3½	380; 837	550; 1,212	30·3; 6·20
ROMANIA								
IS-28B2	17·00; 55 9¼	15·8	18·24; 196·3	8·17; 26 9¾	1·80; 5 11	360; 793	590; 1,300	32·34; 6·62
IS-28M1	18·00; 59 0¾	17·1	18·95; 204·0	7·96; 26 1½	1·90; 6 2¾	485; 1,069	730; 1,609	38·52; 7·89
IS-28M2	17·00; 55 9¼	15·8	18·24; 196·3	7·50; 24 7¼	2·15; 7 0¾	495; 1,091	745; 1,642	40·8; 8·36
IS-29D2	15·00; 49 2½	21·6	10·40; 111·9	7·30; 23 11½	1·68; 5 6¼	230; 507	360; 793	34·62; 7·09
IS-29DM	15·00; 49 2½	22·1	10·20; 109·8	7·00; 22 11½	1·75; 5 9	310; 683	440; 970	43·1; 8·83
IS-29E2	19·00; 62 4	26·5	13·60; 146·4	7·30; 23 11½	1·68; 5 6¼	320; 705	500; 1,102	36·76; 7·53
IS-29EM	20·00; 65 7½	27·2	14·68; 158·0	7·00; 22 11½	1·75; 5 9	400; 882	530; 1,168	36·1; 7·40
IS-30	18·00; 59 0¾	17·1	18·92; 203·6	8·36; 27 5¼	1·90; 6 2¾	365; 804	590; 1,300	31·71; 6·49
IS-32	20·00; 65 7½	27·2	14·68; 158·0	8·36; 27 5¼	1·90; 6 2¾	390; 860	590; 1,300	40·6; 8·31
IS-33	15·00; 49 2½	21·2	10·61; 114·2	6·35; 20 10	1·68; 5 6¼	240; 529	500; 1,102	47·1; 9·65
SWITZERLAND								
EFF Prometheus 12	12·63; 41 5¼	12·1	13·21; 142·2	8·30; 27 2¾	1·70; 5 7	*668/688; 1,472/1,516	n.k.	n.k.
EFF Prometheus 19	19·40; 63 7¾	21·3	17·64; 189·9	8·30; 27 2¾	1·70; 5 7	735; 1,620	999; 2,202	56·6; 11·59

n.k. = not known; n.a. = not applicable

Performance (at max T-O weight except where indicated)

Best glide ratio	at (speed): knots; km/h; mph	Min sinking speed: m (ft)/s	at (speed): knots; km/h; mph	Stalling speed: knots; km/h; mph	Max speed (smooth air): knots; km/h; mph	Max speed (rough air): knots; km/h; mph	Max aero-tow speed: knots; km/h; mph	Max winch-launching speed: knots; km/h; mph	Remarks
*22	41; 75; 47	*1·10; 3·61	97; 180; 112	35; 65; 40	102·5; 190; 118	102·5; 190; 118	n.a.	n.a.	*Approx
28/29	46; 85; 53	0·85; 2·79	41; 75; 47	37; 68; 42·5	102·5; 190; 118	102·5; 190; 118	n.a.	n.a.	
26/27	51·5; 95; 59	0·90; 2·95	38; 70; 43·5	32·5; 60; 37·5	102·5; 190; 118	102·5; 190; 118	n.a.	n.a.	
35	51·5; 95; 59	0·70; 2·30	41; 75; 47	35·5; 65; 40·5	135; 250; 155	135; 250; 155	88; 163; 101	67·5; 125; 78	
28	n.k.	0·90; 2·95	48·5; 90; 56	n.k.	n.k.	n.k.	n.a.	n.a.	
49	62; 115; 71	0·47; 1·54	43; 80; 50	46; 77; 48	146; 270; 168	146; 270; 168	97; 180; 112	81; 150; 93	*315 kg (694 lb) with carbonfibre wings and tail
42	65; 120; 75	0·70; 2·30	46; 85; 53	43·5; 80; 50	135; 250; 155	135; 250; 155	97; 180; 112	81; 150; 93	*215 kg (474 lb) with carbonfibre wings and tail
55	67; 125; 78	*0·52; 1·71	n.k.	42; 77; 48	146; 270; 168	n.k.	n.k.	n.k.	*At wing loading of 30 kg/m² (6·15 lb/sq ft)
*39·5	*59·5; 110; 68·5	*0·70; 2·30	*48·5; 90; 56	*38; 70; 43·5	118; 220; 136	118; 220; 136	91; 170; 105	65; 120; 74·5	*At wing loading of 36·5 kg/m² (7·48 lb/sq ft)
*43·5	*62; 115; 71·5	*0·60; 1·97	*48·5; 90; 56	*38; 70; 43·5	135; 250; 155	135; 250; 155	97; 180; 112	81; 150; 93	*At wing loading of 36·5 kg/m² (7·48 lb/sq ft)
44	65; 120; 75	0·66; 2·16	n.k.	38; 70; 43·5	135; 250; 155	135; 250; 155	n.k.	n.k.	Ventus b weights and performance similar
28	49; 90; 56	0·80; 2·62	38; 70; 43·5	33; 61; 38	108; 200; 124	75·5; 140; 87	75·5; 140; 87	54; 100; 62	Performance at 470 kg (1,036 lb) AUW
38·5	60·5; 112; 70	0·73; 2·40	48·5; 90; 56	36·5; 67; 42	137; 255; 158	137; 255; 158	92; 170; 106	67·5; 125; 78	
*42	52; 96; 60	0·59; 1·94	45·5; 84; 52·5	37·5; 69; 43	143; 265; 165	97; 180; 112	97; 180; 112	67·5; 125; 78	*At 350 kg (772 lb) AUW
46	52; 96; 60	0·55; 1·80	46·5; 86; 53·5	37; 68; 42·5	135; 250; 155	89; 165; 102·5	86; 160; 99·5	65; 120; 74·5	*Approx
34	46; 85; 53	0·65; 2·13	36; 67; 42	33·5; 62; 39	151; 280; 174	135; 250; 155	94; 175; 109	70; 130; 80·5	
n.k.	n.k.	n.k.	n.k.	n.k.	n.k.	n.k.	n.k.	n.k.	
55	56; 105; 65	0·45; 1·48	46; 85; 53	38; 70; 43·5	146; 270; 168	n.k.	n.k.	n.k.	
35	49; 90; 56	0·53; 1·74	38; 70; 43·5	35·5; 65; 40·5	119; 220; 137	97; 180; 112	n.k.	n.k.	
28	65; 120; 75	0·99; 3·25	47; 87; 54	n.k.	135; 250; 155	100; 185; 115	n.a.	n.a.	Performance data are for 15E
30	59·5; 110; 68·5	0·85; 2·79	43; 80; 50	39; 72; 45	135; 250; 155	100; 185; 115	n.a.	n.a.	Performance data are for 17E
*37·6	*47; 87; 54	*0·57; 1·87	*39; 72; 45	33·5; 62; 39	124; 230; 143	n.k.	n.k.	n.k.	*At wing loading of 27 kg/m² (5·53 lb/sq ft)
26	47; 87; 54	0·78; 2·56	39; 72; 45	33; 61; 38	108; 200; 124	68·5; 127; 79	59·5; 110; 68	59·5; 110; 68	Performance estimated
43	56·5; 105; 65	0·60; 1·97	43; 80; 50	34; 63; 39·5	136; 252; 156	91; 169; 105	75·5; 140; 87	70; 130; 81	
43	65; 120; 75	0·68; 2·23	65; 120; 75	34; 63; 39·5	136; 252; 156	n.k.	n.a.	n.a.	
42	62; 115; 71·5	0·66; 2·16	47; 87; 54	40·5; 74·5; 46·5	137; 254; 158	137; 254; 158	97; 180; 112	n.a.	
35	46; 85; 53	0·64; 2·10	39; 72; 45	33; 61; 38	129; 240; 149	129; 240; 149	88; 163; 101	70·5; 130; 81	
50·3	56; 103; 64	0·53; 1·74	52; 95; 59	43; 79; 49	135; 250; 155	108; 200; 124	75·5; 140; 87	62; 115; 71·5	Performance at 649 kg (1,430 lb) AUW; *458 kg (1,010 lb) without water ballast
38	70; 130; 81	0·77; 2·53	52·5; 97; 60	44·5; 82; 51	154; 285; 177	154; 285; 177	81; 150; 93	67·5; 125; 78	*366 kg (807 lb) without water ballast
30	45; 85; 53	0·70; 2·30	40·5; 75; 47	32·5; 60; 37·5	116; 215; 133	86; 160; 99	81; 150; 93	59; 110; 68·5	
34	49; 90; 56	0·67; 2·20	43·5; 80; 50	39; 72; 45	124; 230; 143	91; 169; 105	75·5; 140; 87	59·5; 110; 68·5	Performance at 540 kg (1,190 lb) AUW
32	54; 100; 62	0·81; 2·66	42; 78; 48·5	*35; 64; 40	132; 245; 152	89; 165; 102	n.a.	n.a.	*Flaps down
29	54; 100; 62	0·87; 2·85	40·5; 75; 47	*38; 70; 43·5	124; 230; 143	95; 177; 110	n.a.	n.a.	*Flaps down
37	43·5; 80; 50	0·55; 1·80	40·5; 75; 47	38; 70; 43·5	121; 225; 140	93; 172; 107	75·5; 140; 87	70; 130; 81	
30	46; 85; 53	0·65; 2·13	41·5; 77; 48	43·4; 80·3; 50	125; 231; 144	87; 160·5; 100	n.a.	n.a.	
48	49; 90; 56	0·50; 1·64	43·5; 80; 50	38·5; 71; 44·5	121; 225; 140	88·5; 164; 102	75·5; 140; 87	67·5; 125; 74·5	
42	46; 85; 53	0·55; 1·80	43·5; 80; 50	42; 77; 48	113; 210; 130	84; 155; 96	n.a.	n.a.	
36·5	51; 95; 59	0·62; 2·03	40·5; 75; 47	38; 70; 43·5	135; 250; 155	90; 166; 103	75·5; 140; 87	67·5; 125; 78	
46	46; 85; 53	0·53; 1·74	43·5; 80; 50	39; 72; 45	124; 230; 143	87; 161; 100	75·5; 140; 87	67·5; 125; 78	
41	54; 100; 62	0·64; 2·10	46·5; 86; 53·5	40·5; 75; 47	130; 240; 149	111; 205; 127	75·5; 140; 87	67·5; 125; 78	
n.k.	n.k.	n.k.	n.k.	n.k.	n.k.	n.k.	n.a.	n.a.	*Without/with tip-tanks
n.k.	n.k.	n.k.	n.k.	n.k.	n.k.	n.k.	n.a.	n.a.	

Manufacturer and Model	Dimensions: Wing span: m; ft in	Wing aspect ratio	Wing area (gross): m²; sq ft	Length overall: m; ft in	Height over tail: m; ft in	Weights and Loading: Empty: kg; lb	Max T-O: kg; lb	Max wing loading: kg/m²; lb/sq ft
Neukom S-4A Elfe 15	15·00; 49 2½	19·1	11·80; 127·0	7·10; 23 3½	1·50; 4 11	230; 507	350; 771	29·6; 6·06
Neukom Elfe 17	17·00; 55 9¼	21·8	13·20; 142·1	7·10; 23 3½	1·50; 4 11	255; 562	380; 837	28·8; 5·90
USSR								
Oshkinis BRO-11	7·28; 23 10½	5·05	10·50; 113·0	5·17; 16 11½	2·40; 7 10½	58; 128	118; 260	11·24; 2·30
Oshkinis BRO-11 M	7·80; 25 7	5·16	11·80; 127·0	5·47; 17 11½	2·50; 8 2½	65; 143	125; 275	10·59; 2·17
LAK-5 Nyamunas	n.k.	n.k.	n.k.	n.k.	n.k.	n.k.	750; 1,653	n.k.
LAK-9 Lietuva	20·02; 65 8¼	26·8	14·99; 161·35	7·20; 23 7½	1·53; 5 0¼	380; 837	*560; 1,234	37·36; 7·65
LAK-12 Lietuva	20·42; 67 0	28·5	14·63; 157·5	7·23; 23 8½	1·70; 5 7	340; 749	*650; 1,433	44·42; 9·10
LAK Nida	15·00; 49 2½	n.k.	n.k.	n.k.	n.k.	220; 485	480; 1,058	n.k.
UK								
Slingsby Vega	15·00; 49 2½	22·4	10·05; 108·2	6·72; 22 0½	1·50; 4 11	233; 515	508; 1,120	50·51; 10·35
Slingsby Sport Vega	15·00; 49 2½	22·4	10·05; 108·2	6·72; 22 0½	1·50; 4 11	233; 515	354; 780	35·2; 7·21
Slingsby Venture T.Mk 2	15·25; 50 0¼	13·8	18·20; 195·9	7·55; 24 9¼	1·85; 6 0¾	376·5; 830	612; 1,350	33·6; 6·88
Slingsby T.61G Falke	15·25; 50 0¼	13·8	18·20; 195·9	7·55; 24 9¼	1·85; 6 0¾	376·5; 830	612; 1,350	33·6; 6·88
Swales SD3-15T	15·00; 49 2½	23·7	9·48; 102·0	6·10; 20 0	1·30; 4 3	218; 480	331; 730	34·96; 7·16
Wright Falcon 15 m	15·00; 49 2½	25·7	8·76; 94·3	5·94; 19 6	1·02; 3 4	238; 525	363; 800	41·38; 8·48
USA								
AmEagle American Eaglet	10·97; 36 0	18·0	6·69; 72·0	4·88; 16 0	0·91; 3 0	72·5; 160	163; 360	24·41; 5·00
Bryan (Schreder) RS-15	15·00; 49 2½	21·4	10·50; 113·0	6·71; 22 0	1·17; 3 10	204; 450	431; 950	41·04; 8·41
Bryan (Schreder) HP-18	15·00; 49 2½	21·4	10·50; 113·0	7·06; 23 2	1·22; 4 0	213; 470	440; 970	41·87; 8·58
Bryan (Schreder) HP-19	15·00; 49 2½	21·4	10·50; 113·0	7·06; 23 2	1·22; 4 0	181; 400	453; 1,000	43·18; 8·85
DSK BJ-1b Duster	13·00; 42 7¾	17·4	9·72; 104·65	n.k.	n.k.	177; 390	281; 620	28·9; 5·92
Explorer PG-1 Aqua Glider	4·88; 16 0	5·0	8·73; 94·0	4·17; 13 8	1·52; 5 0	81; 180	181; 400	20·5; 4·20
Flight Dynamics Seasprite	10·36; 34 0	6·96	15·42; 166·0	6·10; 20 0	2·44; 8 0	75; 165	227; 500	14·70; 3·01
Hollmann-Klepeis Condor	9·75; 32 0	7·7	12·36; 133·0	4·88; 16 0	1·75; 5 9	227; 500	463; 1,020	37·43; 7·67
Maupin Woodstock One	11·89; 39 0	14·53	9·73; 104·7	n.k.	n.k.	106·5; 235	204; 450	20·9; 4·29
Monnett Monerai S	10·97; 36 0	16·6	7·25; 78·0	5·97; 19 7	1·32; 4 4	100; 220	204; 450	28·11; 5·76
Ryson ST-100 Cloudster	17·58; 57 8	15·6	19·79; 213·0	7·78; 25 6·4	1·78; 5 10	550; 1,212	794; 1,750	37·82; 7·75
Schweizer SGS 2-33A	15·54; 51 0	11·85	20·39; 219·48	7·85; 25 9	2·83; 9 3½	272; 600	472; 1,040	23·14; 4·74
Schweizer SGS 1-35	15·00; 49 2½	23·29	9·64; 103·8	5·84; 19 2	1·35; 4 5	199; 440	422; 930	43·75; 8·96
Schweizer SGS 1-36 Sprite	14·07; 46 2	15·15	13·07; 140·72	6·27; 20 7	1·45; 4 9	215; 475	322; 710	24·64; 5·05
YUGOSLAVIA								
Elan (Glaser-Dirks) DG-100/DG-100G Elan	15·00; 49 2½	20·5	11·00; 118·4	7·00; 22 11¾	1·40; 4 7	*230; 507	418; 921	38·0; 7·78
Elan (Glaser-Dirks) DG-100 Club Elan	15·00; 49 2½	20·5	11·00; 118·4	7·00; 22 11¾	1·40; 4 7	*230; 507	385; 848	35·0; 7·17
Jastreb SOLE-77	16·40; 53 9¾	n.k.	n.k.	7·67; 25 2	n.k.	500; 1,102	n.k.	n.k.
MF VUK-T	15·00; 49 2½	18·75	12·00; 129·2	6·50; 21 4	1·45; 4 9	245; 540	355; 783	29·7; 6·08
SOKO SL-40 Liska	15·00; 49 2½	14·78	15·225; 49 11½	6·40; 21 0	2·00; 6 6¾	358; 789	500; 1,102	32·8; 6·72
VTC (Schempp-Hirth) Cirrus	17·74; 58 2½	24·98	12·60; 135·6	7·20; 23 7¼	1·56; 5 1½	260; 573	400; 882	31·7; 6·49
VTC (Schempp-Hirth) Standard Cirrus 75	15·00; 49 2½	22·5	10·00; 107·6	6·41; 21 8½	1·32; 4 4¾	215; 474	330; 727	33·0; 6·76

n.k. = not known; n.a. = not applicable

Performance (at max T-O weight except where indicated)									
Best glide ratio	at (speed): knots; km/h; mph	Min sinking speed: m (ft)/s	at (speed): knots; km/h; mph	Stalling speed: knots; km/h; mph	Max speed (smooth air): knots; km/h; mph	Max speed (rough air): knots; km/h; mph	Max aero-tow speed: knots; km/h; mph	Max winch-launching speed: knots; km/h; mph	Remarks
37·5	49; 90; 56	0·60; 1·97	38; 70; 43·5	35·5; 65; 40·5	113; 210; 130	113; 210; 130	75·5; 140 87	54; 100; 62	
40·5	49; 90; 56	0·56; 1·84	40·5; 75; 46·5	35·5; 65; 40·5	113; 210; 130	113; 210; 130	75·5; 140 87	54; 100; 62	
11	n.k.	1·00; 3·28	n.k.	22; 40; 25	n.k.	n.k.	n.k.	n.k.	
12	n.k.	1·00; 3·28	n.k.	16·5; 30; 19	n.k.	n.k.	n.k.	n.k.	
42-43	n.k.	n.k.	n.k.	40·5; 75; 47	n.k.	n.k.	n.k.	n.k.	
48	54; 100; 62	**0·51; 1·67	40·5; 75; 46·5	37; 68; 42·5	113; 210; 130	113; 210; 130	75·5; 140 87	n.k.	*470 kg (1,036 lb) without water ballast; **at 470 kg (1,036 lb) AUW
48	61; 113; 70	**0·50; 1·64	46; 85; 53	35·5; 65; 40·5	135; 250; 155	n.k.	62; 115; 71·5	n.k.	*430 kg (948 lb) without water ballast; **at 430 kg (948 lb) AUW
42	n.k.	n.k.	n.k.	n.k.	146; 270; 168	n.k.	n.k.	n.k.	
42	62; 115; 71	0·67; 2·21	44; 82; 51	41; 76·5; 47·5	135; 250; 155	105; 195; 121	80; 148; 92	70; 129; 80	
39	50; 93; 58	0·70; 2·30	44; 82; 51	37; 69; 43	135; 250; 155	105; 195; 121	100; 185; 115	70; 129; 80	
20·5	50·5; 93·5; 58	1·00; 3·30	43·5; 80·5; 50	37; 69; 43	100; 185; 115	80; 148; 92	n.a.	n.a.	
21	47; 87; 54	1·00; 3·30	43·5; 80·5; 50	37; 69; 43	100; 185; 115	79; 146; 91	n.a.	n.a.	
36	48; 88·5; 55	0·73; 2·40	42; 78; 48·5	34; 63·5; 39·5	109; 201; 125	86; 159; 99	78; 145; 90	65; 121; 75	
39	55; 102; 63	0·61; 2·00	46; 85; 53	40; 74·5; 46·5	135; 250; 155	100; 185; 115	100; 185; 115	70; 130; 81	
27	45; 84; 52	0·76; 2·50	35; 64·5; 40	33; 61·5; 38	100; 185; 115	69·5; 129; 80	69·5; 129 80	n.a.	
38	59; 109; 68	*0·64; 2·10	43·5; 80; 50	**40; 74; 46	130; 241; 150	104; 193; 120	104; 193; 120	78; 145; 90	*At 284 kg (626 lb) AUW; **at 335 kg (740 lb) AUW
40	59; 109; 68	*0·52; 1·70	39; 73; 45	**35; 64·5; 40	130; 241; 150	104; 193; 120	104; 193; 120	78; 145; 90	*At 275 kg (606 lb) AUW; **at 326 kg (720 lb) AUW
42	n.k.	*0·46; 1·50	35; 64·5; 40	*30·5; 56·5; 35	130; 241; 150	104; 193; 120	104; 193; 120	78; 145; 90	*At 266 kg (586 lb) AUW
n.k.	n.k.	n.k.	n.k.	n.k.	n.k.	n.k.	n.k.	n.k.	
6·5	39; 72·5; 45	n.a.	n.a.	*30·5; 56·5; 35	56·5; 105; 65	56·5; 105; 65	56·5; 105 65	56·5; 105; 65	*At 149 kg (330 lb) AUW
n.k.	n.k.	n.k.	n.k.	n.k.	n.k.	n.k.	n.k.	n.k.	
n.k.	69; 129; 80	n.k.	n.k.	40; 74; 46	104; 193; 120	87; 161; 100	n.a.	n.a.	Performance estimated
24	n.k.	0·79; 2·60	n.k.	n.k.	n.k.	n.k.	n.k.	n.k.	
30	52; 97; 60	0·79; 2·60	48; 88·5; 55	33; 61·5; 38	104; 193; 120	78; 145; 90	78; 145; 90	–	
28	52; 97; 60	0·90; 2·95	47; 87; 54	36·5; 68; 42	139; 258; 160	139; 258; 160	n.a.	n.a.	
22-25	45; 84; 52	0·91; 3·00	33; 61; 38	30·5; 57; 35	85; 158; 98	85; 158; 98	85; 158; 98	60; 111; 69	
39	46; 85; 53	0·61; 2·00	43; 79; 49	28; 51·5; 32	121; 223; 139	121; 223; 139	121; 223; 139	n.a.	
31	46; 85; 53	0·68; 2·25	36·5; 68; 42	30·5; 57; 35	105; 195; 121	105; 195; 121	105; 195; 121	68; 126; 78	
39	56·5; 105; 65	0·59; 1·94	40; 74; 46	32·5; 60; 37·5	140; 260; 161	140; 260; 161	89; 165; 102·5	70; 130; 80·5	*With normal tailplane; 235 kg (518 lb) with all-moving tailplane
36	49; 90; 56	0·60; 1·97	40; 74; 46	32·5; 60; 37·5	140; 260; 161	140; 260; 161	89; 165; 102·5	70; 130; 80·5	*With normal tailplane; 235 kg (518 lb) with all-moving tailplane
29	n.k.	n.k.	n.k.	n.k.	n.k.	n.k.	n.k.	n.k.	
37·5	51; 95; 59	0·65; 2·13	42; 78; 48·5	32; 59; 37	129; 240; 149	81; 150; 93	81; 150; 93	59; 110; 68	
n.k.	n.k.	n.k.	n.k.	n.k.	n.k.	n.k.	n.k.	n.k.	
44	46; 85; 53	0·50; 1·64	39; 73; 45	33·5; 62; 39	119; 220; 137	119; 220; 137	76; 140; 87	59; 110; 68	
38·5	49; 90; 56	0·60; 1·97	39·5; 75; 47	33·5; 62; 39	119; 220; 137	119; 220; 137	81; 150; 93	65; 120; 75	

DATA: MOTOR GLIDERS (POWER ON)

	Loading	Performance at max T-O weight (powered)								
Manufacturer and Model	Max power loading: kg/kW: lb/hp:	Max cruising speed: knots; km/h; mph	Econ cruising speed: knots; km/h; mph	Stalling speed: knots; km/h; mph	Max rate of climb at S/L: m; ft/min	Service ceiling: m; ft	T-O run: m; ft	Landing run: m; ft	Range with max fuel: nm; km; miles	Remarks
AUSTRIA										
Brditschka HB-21	15·80; 26·00	86; 160; 99	70·5; 130; 81	39·5; 75; 47	198; 650	6,300; 20,675	100; 330	80; 262	431; 800; 497	
Brditschka HB-22	9·14; 15·03	108; 200; 124	n.k.	39·5; 75; 47	270; 886	n.k.	*300; 915	n.k.	432; 800; 497	*To 15 m (50 ft)
FINLAND										
Alanne Motorblanik	15·66; 25·74	75; 140; 87	65; 120; 75	35·5; 65; 40·5	120; 394	n.k.	200; 656	200; 656	162; 300; 186	
Alanne Motorlerche	12·07; 19·84	65; 120; 75	54; 100; 62	24·5; 45; 28	150; 492	n.k.	100; 328	150; 492	162; 300; 86	
FRANCE										
Calvel Frelon	9·19; 15·12	65; 120; 75	48·5; 90; 56	n.k.	210; 689	n.k.	n.k.	n.k.	n.k.	Performance estimated
Fournier RF-9	14·79; 24·31	86; 160; 112	n.k.	35·5; 65; 40·5	180; 590	6,000; 19,685	n.k.	n.k.	n.k.	
Fournier RF-10	12·91; 21·2	97; 180; 112	n.k.	39; 72; 45	210; 689	n.k.	n.k.	n.k.	540; 1,000; 621	
Gross (Fauvel) AV.451	14·82; 24·35	81; 150; 93	65; 120; 75	49; 90; 56	252; 827	n.k.	80; 262	200; 656	215; 400; 248	
Siren (EIRI) PIK-20E2F	14·62; 24·03	73; 135; 84	73; 135; 84	40·5; 75; 47	162; 531	5,200; 17,050	*500; 1,640	300; 85	156; 290; 180	*To 15 m (50 ft)
GERMANY (FEDERAL REPUBLIC)										
Akaflieg Darmstadt D-39	9·20; 15·12	*97; 180; 112	n.k.	43·5; 80; 50	240; 787	n.k.	150; 492	200; 656	324; 600; 373	†Height over fuselage *At 75% power
Akaflieg Karlsruhe AK 2	13·82; 22·71	n.k.	n.k.	n.k.	n.k.	n.k.	n.k.	n.k.	n.k.	
Blessing Staff Rebell	13·41; 22·03	81; 150; 93	73; 135; 84	38; 70; 44	180; 590	4,500; 14,760	120; 394	135; 443	402; 745; 463	
Grob G 109	13·07; 21·49	108; 200; 124	n.k.	40·5; 75; 47	*180; 590	n.k.	200; 656	n.k.	*540; 1,000; 621	*Approx
Hoffmann H-36 Dimona	12·40; 20·38	97; 180; 112	n.k.	39; 72; 45	210; 689	6,000; 19,685	180; 590	n.k.	540; 1,000; 621	
Scheibe SF-25C and C-S Falke '80	12·58; 20·68	86; 160; 99	70·5; 130; 81	35·5; 65; 40·5	138; 453	5,000; 16,400	180; 590	100; 328	404; 750; 466	
Scheibe SF-25K Falke'80	12·58; 20·68	86; 160; 99	70·5; 130; 81	35·5; 65; 40·5	138; 453	5,000; 16,400	*180; 590	100; 328	*378; 700; 435	*Approx
Scheibe SF-25E Super-Falke	13·41; 22·04	*97; 180; 112	81; 150; 93	37; 68; 42·5	150; 492	5,000; 16,400	150-200; 490-655	100; 328	323-600; 372	*Max level speed
Scheibe SF-28A Tandem-Falke	12·58; 20·68	*97; 180; 112	81; 150; 93	33·5; 62; 39	132; 435	5,000; 16,400	150-200; 490-655	150; 492	280; 520; 323	*Max level speed
Scheibe SF-36	10·56; 17·35	97; 180; 112	n.k.	39; 72; 45	180; 590	6,000; 19,685	200; 656	n.k.	Endurance 4-5 h	
Valentin Taifun 15	12·90; 21·22	121; 224; 139	114; 212; 132	40·5; 75; 47	162; 531	5,000; 16,400	200; 656	150; 492	675; 1,250; 777	Performance data are for Taifun 15E
Valentin Taifun 17	12·90; 21·22	118; 220; 137	112; 208; 129	39; 72; 45	168; 551	5,000; 16,400	200; 656	150; 492	675; 1,250; 777	Performance data are for Taifun 17E
ITALY										
Caproni Vizzola Calif A-21SJ	915 kg/kN; 8·97 lb/lb st	108; 200; 124	102·5; 190; 118	34; 63; 39·5	240; 787	11,000; 36,100	300; 985	200; 656	189; 350; 217	
ROMANIA										
IS-28M1	14·4; 23·6	95; 176; 109	86; 159; 99	38; 70; 44	192; 630	5,500; 18,050	180; 590	90; 295	242; 450; 280	
IS-28M2	14·7; 24·1	92; 170; 106	81; 150; 93	35·5; 65; 40·5	186; 610	5,000; 16,400	240; 788	90; 295	242; 450; 280	
IS-29DM	15·1; 24·9	92; 170; 106	75·5; 140; 87	43·4; 80·3; 50	180; 590	4,500; 14,765	*370; 1,215	200; 656	270; 500; 310	*To 15 m (50 ft)
IS-29EM	18·2; 29·9	86; 160; 99	70·5; 130; 81	42; 77; 48	150; 492	4,500; 14,765	*400; 1,315	200; 656	270; 500; 310	*To 15 m (50 ft)
SWITZERLAND										
EFF Prometheus 12	n.k.	n.k.	n.k.	n.k.	366+; 1,200+	n.k.	450; 1,476	n.k.	n.k.	
EFF Prometheus 19	567·6 kg/kN; 5·56 lb/lb st	*140; 260; 161	n.k.	n.k.	244+; 800+	n.k.	450; 1,476	n.k.	n.k.	*Max attained speed (88% power)
USSR										
LAK-5 Nyamunas	13·4; 22·0	*135; 250; 155	n.k.	40·5; 75; 47	n.k.	n.k.	n.k.	n.k.	n.k.	*Max level speed
UK										
Slingsby Venture T.Mk 2	18·3; 30·0	76; 140; 87	65; 121; 75	37·5; 69·5; 43	125; 410	4,267; 14,000	198; 650	100; 330	217; 402; 250	
Slingsby T.61G Falke	13·69; 22·5	90; 167; 104	75; 139; 86	36; 67; 41·5	140; 460	5,000; 16,400	*180; 590	100; 330	217; 402; 250	*To 15 m (50 ft)
USA										
Hollmann-Klepeis Condor	11·3; 18·5	78; 145; 90	69; 129; 80	40; 74; 46	183; 600	3,660; 12,000	305; 1,000	183; 600	495; 917; 570	Performance estimated
Ryson ST-100 Cloudster	10·0; 16·5	130; 241; 150	*69; 129; 80	37·5; 69·5; 43	256; 840	7,315; 24,000	290; 950	244; 800	595; 1,104; 686	*At 26% power
YUGOSLAVIA										
Jastreb SOLE-77	n.k.	92; 170; 106	n.k.	39; 72; 45	186; 610	4,500; 14,760	*350; 1,148	n.k.	378; 700; 435	*To 15 m (50 ft)
SOKO SL-40 Liska	14·9; 24·5	*94; 175; 109	n.k.	**39; 72; 45	204; 669	6,600; 21,650	254; 833	126; 413	n.k.	Performance estimated. *Max level speed at S/L; **engine idling

n.k. = not known

MICROLIGHT AIRCRAFT and HANG GLIDERS

CIVL
COMMISSION INTERNATIONALE DE VOL LIBRE

c/o Fédération Aéronautique Internationale, 6 rue Galilée, 75782 Paris, Cédex 16, France

PRESIDENT: Erwin Kjellerup (Sweden)

HONORARY PRESIDENTS:
Daniel F. Poynter (USA)
Mrs Ann Welch (UK)

VICE-PRESIDENTS:
Tom Hudson (Ireland)
Stein Arne Fossum (Norway)

SECRETARY: Tom Hudson

INTERNATIONAL REPRESENTATIVES:

Australia: W. T. J. Moyes, 173 Bronte Road, Waverley, Sydney, NSW 2024

Austria: Sepp Himberger, A-6345 Kössen

Belgium: Dimitri Caloussis, Aéro-Club Royal de Belgique, 1 rue Montoyer, 1040 Brussels

Bulgaria: Georgi Jankov; Janko Prodanov (addresses not known)

Canada: Willi Müller, Hang Gliding Association of Canada, PO Box 4063, PSC, Calgary, Alberta

Denmark: Palle J. Christensen (address not known)

France: René Coulon, 3 rue Ampère, 94200 Ivry-sur-Seine; alternate: Maurice Bourgueil (address not known)

Germany (Federal Republic): Rainer Kolm, Mitterharthansen 81b, 8441 Feldkirchen Ndb

Hungary: Dr M. Ordody, 11 Szilagyi E.f. 13-15, 1024 Budapest

Ireland: Tom Hudson, 60 Hillcourt Road, Glenageary, Co Dublin

Israel: Ziv Brosh, 18 Koumemiout Street, Ramat, Ahsharon

Italy: Heinz Kostner, 39033 Coruara; Alfio Caronti (address not known)

Japan: Asahi Miyahara, 2-27, Uehara Shibuya-ku, Tokyo 151

Korea (Republic): Kim Suk Whan (address not known)

Luxembourg: Jean Willems, 15 rue de la Sûre, Echternach, 62 Rabatt

Mexico: Miguel ; Alfonso Espinoza (addresses not known)

Netherlands: Floor C. G. Gremmen, Achterde, Uismarkt 41, 2801 NB Gouda; Maarten Brandt, 2e v.d. Heydenstraat 93 I, 1074 XT Amsterdam

New Zealand: Mark Nichols, 73 Bryndwr Road, Christchurch 5

Norway: Stein Arne Fossum, Brendöyveien 21, 6900 Florö; Werner Johannessen, Emil Stang s.v. 23B, 1346 Giettum

Poland: Dr Jerzy Wolf, Instytut Lotnictwa, Al. Krakowska 110/114, 02-256 Warsaw

South Africa: Eric Cornhill, PO Box 18.185, Dalbridge 4014

Spain: José Maria Garcia Planas (address not known)

Sweden: Erwin Kjellerup, Helgeredsvägen 30, S-430 50 Kallered; Hans Säfwenberg, Polhemsgatan 4, 112 36 Stockholm

Switzerland: Werner Müller, President, SHV Sportkommission, Weiherweg 14, 3053 Mibuchsee

USSR: E. N. Elizarov, FAS of the USSR, PO Box 395, Moscow D-362

UK: Mrs Ann Welch, 14 Upper Old Park Lane, Farnham, Surrey GU9 0AS; R. J. Spooner, Clifton House, Bath Road, Cowes, Isle of Wight; Barry Blore, 40 Castle Street, Steventon, Abingdon, Oxfordshire OX13 6SR; Roy D. Hill, 68 Besselsleigh Road, Abingdon, Oxfordshire OX13 6DX

USA: Harry W. Robb, 2909 Gulf to Bay 0203, Clearwater, Florida 33515; Richard Heckman, 5401 Lookout Drive, Huntsville, Alabama 35801

(Where personal addresses of representatives are not given, they may be contacted via the national Aero Club concerned)

One of the first actions of the CIVL, following its formation in mid-1975, was to define a hang glider as "a heavier-than-air, fixed-wing (not rotating) glider, capable of being carried, foot-launched and landed solely by the energy and use of the pilot's legs". It subsequently defined three Competition classes of hang glider. Other tasks of the Commission include safety and pilot ratings.

In recent years hang gliders have been joined by the fast-growing category of what are now almost universally known as microlight aircraft. By mid-1981 some 9,000 of these aircraft were estimated to be flying in 15 or more countries worldwide, including about 6,000 in the USA and 200 in the UK. The FAI formed a Committee in mid-1981, with provisional status for one year, to look after and encourage the development of microlight flying. The Fédération has also agreed an international definition of a microlight as a single- or two-seat powered aircraft having a dry (empty) weight (W) not exceeding 150 kg (330 lb), and a wing area in m² of not less than W÷10 and in no case less than 10 m² (107·6 sq ft). In other words, a microlight of 150 kg dry weight would be required to have a minimum wing area of 15 m² (161·5 sq ft).

Pending worldwide adoption of such an international definition, the regulation of microlight aircraft varies from country to country. For example, in the United States these aircraft have not, since 13 May 1977, required FAA certification, even if fitted with landing gear, provided that they retain a capability for foot-launch; but in 1981 this definition was reported to be losing credibility. At Oshkosh in August 1980 the EAA announced the formation of an Ultralight Division, and it is anticipated that a revised definition may emerge after consultation with the FAA. A similar situation exists in the UK, where discussions continue between the Civil Aviation Authority and the British Microlight Aircraft Association. Up to mid-1981 the CAA remained the controlling body, but had given temporary exemptions from the relevant Air Navigation Order until details of pilot licensing, registration and airworthiness requirements are worked out. In the meantime, British microlight pilots were obtaining the BMAA pilot certificate of competence issued by UK training schools. The CAA has recommended a wing loading of 10 kg/m² (2·05 lb/sq ft) for microlights; but that single-seaters not exceeding a dry weight of 70 kg (154 lb) be exempted from airworthiness requirements, provided that they are not used for public transport or aerial work. Those above 70 kg dry weight would still require a Permit to Fly. Microlight aircraft in Australia do not at present require certification or a pilot's licence, and may weigh up to 181 kg (400 lb) including the pilot (subject to a max wing loading of 19·5 kg/m²; 4·0 lb/sq ft), but all flying must be done below 91·5 m (300 ft) altitude, and must not approach within 5 km (3 miles) of a licensed airfield or 300 m (985 ft) of a metalled road.

The listing which follows includes data for all microlight aircraft and hang gliders of which details had been received at the time this edition closed for press; however, many manufacturers have failed to supply information for the present edition.

ARGENTINA

LADAS
LIGA ARGENTINA DE AERODESLIZADORES SUPERLIVIANOS

San Martin 359, 5539 Las Heras, Mendoza

PRESIDENT: Johan F. Byttebier

Recent activities concerned the Johan SH1, based on the Seagull Seahawk, of which LADAS had produced nine and sold seven by the end of 1979. Sr Byttebier has also test-flown a larger Johan SH2 version.

Details of the SH1 can be found in the table and in the 1980-81 Jane's. No 1981 details were received for either type.

AUSTRALIA

KIMBERLEY
GARY J. KIMBERLEY

255 Woniora Road, Blakehurst, New South Wales 2221
Telephone: (02) 546 4143

KIMBERLEY SKY-RIDER

Conforming to Australian Department of Transport Air Navigation Order 95.10, which covers powered hang gliders and 'minimum' aircraft, the Sky-Rider is a single-seat monoplane powered by a 9 kW (12 hp) modified McCulloch two-stroke engine. Construction is of wire-braced aluminium tubing, with Dacron covering on the wings and tail unit. Conventional control surfaces are fitted. The Sky-Rider can be dismantled for transportation on the roof of a car and for stowage. Plans are available to amateur constructors.

TYPE: Single-seat microlight aircraft.

DIMENSIONS:

Wing span	9·86 m (32 ft 4 in)
Wing area	13·38 m² (144·0 sq ft)
Length overall	5·79 m (19 ft 0 in)
Height overall	2·39 m (7 ft 10 in)

WEIGHTS AND LOADING:

Weight empty	88·5 kg (195 lb)
Max T-O weight	181·5 kg (400 lb)
Max wing loading	13·57 kg/m² (2·78 lb/sq ft)

PERFORMANCE:

Max level speed	43·5 knots (80·5 km/h; 50 mph)
Cruising speed	35 knots (64·5 km/h; 40 mph)
Stalling speed, flaps up	22 knots (40·5 km/h; 25 mph)
Stalling speed, flaps down	17·5 knots (32 km/h; 20 mph)
Stalling speed IGE	16 knots (29 km/h; 18 mph)
Max rate of climb at S/L	46 m (150 ft)/min
Max permitted flying height	91·5 m (300 ft)
T-O run	75 m (246 ft)
Landing run	50 m (164 ft)
Endurance	45 min

MOYES
MOYES DELTA GLIDERS PTY LTD

173 Bronte Road, Waverley, Sydney, NSW 2024
Telephone: 387 5114

Manufacturer of the Stinger and Mega ranges, of which the following models were available in 1981. Most recent production model is the Mega Mark III. Mega sales totalled 750 by early 1981; Stinger sales exceed 2,000. A Mega flown by Mr George Worthington set a 96·4 nm (178·6 km; 111 mile) distance record for Rogallo-type hang gliders on 25 July 1980.

Maxi Stinger Mark III. Class 2 competition model, for expert pilots only. Aluminium alloy airframe, with Dacron sail and vinyl-coated stainless steel cables. Kingpost standard. Features include floating keel pocket, keel pocket battens, folding crossbar, cambered sail, deflexers, and provisions for tow launch and passenger-carrying. Airframe stressed to $+7/-4g$.

Maxi Stinger Mark IV. Similar to Mk III, but with reduced sail area, shorter keel and higher aspect ratio.

Maxi Stinger SP. Similar to Maxi Mark III, but scaled down 5% for lighter-weight pilots; folding crossbar.

Redtail Stinger. In production since 1978. Basic Class 1 model, with single-batten roached tip and moderate speed range; easy to fly and suitable for learner pilots and experts.

Kimberley Sky-Rider microlight aircraft

Wheeler Scout microlight aircraft *(Maurice Allward)*

Mega Mark II. Competition model for advanced pilots only. Airframe stressed to +7/−4·5*g*. Features include feathered radial tip, pre-formed keel pocket battens, deep fin floating keel pocket, folding crossbar, and double-thickness covering on 40 per cent of sail area. Flown prone only.

Mega Mark III. Similar to Mega II but with double-thickness covering on 65 per cent of sail area and 11 ribs per side instead of eight. Can be flown prone or seated.

Mega Major. Larger Mega, for heavier pilots. Features include floating crossbar and increased double-surface covering. Eight ribs per side. Can be flown prone or seated.

For other details see table.

Moyes Maxi Stinger

WHEELER
RON WHEELER AIRCRAFT (SALES) PTY LTD

152 Bellevue Parade, Carlton, NSW 2218
Telephone: Sydney 546 2501
DIRECTOR: R. G. Wheeler

WHEELER SCOUT

Designed by Mr R. G. Wheeler, and first built and flown in May 1974, the Scout 'minimum aircraft' can be quickly dismantled and carried on the roof rack of a motorcar. It has been in production since August 1976, when the Australian Department of Transport issued Air Navigation Order 95.10 permitting its operation subject to certain exemptions and limitations. Worldwide sales totalled more than 400 by early 1981. The Scout is marketed in the UK by Skycraft (UK) Ltd of 79-81 Prestwick Road, Ayr KA8 8LH, Scotland.

AIRFRAME: Wire-braced aluminium alloy extrusions, with Dacron-covered (heavy-duty single-surface) wings and tail surfaces. Wings have a leading-edge spar, set at a dihedral which raises the tips 0·51 m (20 in) above the centreline, and are set at a mean incidence of 4°. No ailerons or wing warping. Tail surfaces are all-moving, comprising a rudder and elevators only, operated by the control column. No rudder bar. Non-retractable steel-leaf main landing gear forward of CG, below the A-frame which supports the pilot's glassfibre seat, with 12 in × 2·5 in wheels. Small alloy steerable tailwheel attached to base of rudder post.

POWER PLANT: One 10·4 kW (14 hp) 173 cc Pixie Major two-stroke piston engine, with twin polyflex V-belt reduction drive to a two-blade wooden tractor propeller. Fuel tank capacity 2·3 litres (0·5 Imp gallons).

DIMENSIONS:	
Wing span (nominal)	8·69 m (28 ft 6 in)
Wing area, gross	13·56 m² (146·0 sq ft)
Length overall	5·03 m (16 ft 6 in)
Height overall	1·57 m (5 ft 2 in)
WEIGHTS AND LOADING:	
Weight empty, equipped	55 kg (122 lb)
Max pilot weight	79 kg (175 lb)
Max T-O weight	135 kg (297 lb)
Max wing loading	9·91 kg/m² (2·03 lb/sq ft)
PERFORMANCE:	
Max level speed	42 knots (78 km/h; 48 mph)
Cruising speed	36 knots (67 km/h; 41·5 mph)
T-O and landing speed	24 knots (44·5 km/h; 28 mph)
Stalling speed	18 knots (33·5 km/h; 21 mph)
Rate of climb	61 m (200 ft)/min
T-O run	64 m (210 ft)
Landing run	28 m (90 ft)
Range	24 nm (44·5 km; 28 miles)
Endurance	40 min
Best glide ratio	7

WINTON
WINTON AIRCRAFT

D4/1 Campbell Parade, Manly Vale, Sydney, NSW 2093

This company markets two single-seat microlight aircraft, the Grasshopper and the Cricket, both designed by Mr Colin Winton.

WINTON GRASSHOPPER

AIRFRAME: Cantilever shoulder-wing monoplane, built mainly of aluminium alloy tube and glassfibre. Wing leading-edges, wingtips and ailerons are of glassfibre; remainder of wings, and all tail surfaces, are Dacron-covered. Wings detachable for transportation and stowage. Pod-shaped fuselage nacelle, to accommodate pilot and to support power plant, with slender boom to carry tail unit. Three-axis control by ailerons, elevators and rudder. Non-retractable tricycle-type sprung landing gear. Cockpit equipped with upholstered seat and windscreen.

POWER PLANT: One 17 kW (23 hp) 432 cc two-cylinder two-stroke engine, mounted on top of fuselage pod aft of cockpit and driving a two-blade pusher propeller.

DIMENSION:	
Wing span	approx 8·23 m (27 ft 0 in)
WEIGHT:	
Weight empty	115 kg (253 lb)
PERFORMANCE:	
Max level speed	65 knots (120 km/h; 74 mph)

WINTON CRICKET

AIRFRAME: Extremely simple minimum airframe comprising constant-chord 'plank' wings and a keel-type boom to support the power plant and a constant-chord cruciform fin/rudder/tailplane/elevator assembly. Lower (dependent) part of fin is strut-braced to 'fuselage' boom, as are the pilot's seat and tricycle sprung landing gear forward, below the wings. Construction is of aluminium alloy, with Dacron-covered glassfibre wings and tail surfaces built in same way as those of Grasshopper. Three-axis control by ailerons, elevators and rudder.

POWER PLANT: Two Victor aircooled piston engines, each driving a fixed-pitch propeller. Fuel capacity 13·6 litres (3 Imp gallons).

DIMENSIONS:	
Wing span	7·62 m (25 ft 0 in)
Length overall	4·42 m (14 ft 6 in)
WEIGHTS:	
Weight empty	93 kg (205 lb)
Max T-O weight	184 kg (405 lb)
PERFORMANCE:	
Max level speed	54 knots (100 km/h; 62 mph)
Cruising speed	43·5 knots (80·5 km/h; 50 mph)
T-O speed	26 knots (48 km/h; 30 mph)
Stalling speed	17·5 knots (32·5 km/h; 20 mph)
T-O run	46 m (150 ft)

BELGIUM

EMC2
EMC2 AVIONICS

EMC2 SAUTERELLE (GRASSHOPPER)

Broadly similar in overall configuration to the Weedhopper (see US section), the Sauterelle is constructed mainly of aluminium tube and Dacron, and has three-axis control. The aircraft can be fitted with any suitable engine from 9 to 26 kW (12-35 hp) which does not weigh more more than 25 kg (55 lb) or turn at more than 6,000 rpm. Typical power plants include the French JPX 340, German König (three-cylinder), Solo (single-cylinder) and Sachs 350. Both direct-drive and reduction-drive installations are available.

The Sauterelle has a wing span and area of 8·50 m (27 ft 10½ in) and 15·20 m² (163·6 sq ft) respectively. Empty weight is 85 kg (187 lb) and max pilot weight 110 kg (242·5 lb).

BULGARIA

ZMIEJ AERO CLUB

Sofia

ZMIEJ-4

All known details of this Bulgarian Rogallo, which has a Dacron sail, are given in the table at the end of this section.

Direct-drive version of the EMC2 Sauterelle
(J. M. G. Gradidge)

Birdman MJ-5 with engine

Talanchuk WT-9 Mars-Marta

Talanchuk WT-10 Mars-Beryl

Birdman MJ-6

CANADA

AFC
ADVENTURE FLIGHT CENTRES INC

PO Box 172, Station A, Winnipeg, Manitoba R3K 2A1
Telephone: (204) 632 4200

AFC CANADIAN SKYSEEKER

The Canadian Skyseeker microlight aircraft is a development of the Eipper-Formance Quicksilver (see US section), with design improvements by AFC. Tailwheel-type or tricycle landing gear can be fitted; skis or twin floats optional. Production totalled 200 by early 1981.

AIRFRAME: Aluminium alloy structure, with Dacron-covered wings, tailplane and rudder; anodised stainless steel control cables.

POWER PLANT: One 11·2 kW (15 hp) Yamaha engine, with reduction gear; two-blade pusher propeller. Fuel tank capacity 11·5 litres (3 US gallons). 15 kW (20 hp) engine available optionally.

DIMENSIONS:

Wing span	9·75 m (32 ft 0 in)
Wing area, gross	14·9 m² (160·0 sq ft)
Wing aspect ratio	6·4

WEIGHTS AND LOADING:

Weight empty, incl power package and landing gear	68 kg (150 lb)
Max T-O weight	179 kg (395 lb)
Max wing loading	12·1 kg/m² (2·47 lb/sq ft)

PERFORMANCE:

Best glide ratio	6
Min sinking speed, power off	67 m (220 ft)/min
Max level speed	approx 39 knots (72 km/h; 45 mph)
Cruising speed	22-26 knots (40-48 km/h; 25-30 mph)
T-O speed	15 knots (27·5 km/h; 17 mph)
Stalling speed	14 knots (26 km/h; 16 mph)
Endurance	2 h 30 min

BIRDMAN
BIRDMAN ENTERPRISES LTD

8027 Argyll Road, Edmonton, Alberta T6C 4A9

Producer in 1980 of the **Falcon I and II, MJ-5, MJ-6 and XC** hang gliders (see table for details). All airframes are of aluminium alloy, with vinyl-coated stainless steel cables, single-thickness Dacron sail, and a kingpost as standard. All are stressed to a *g* limit of +6·5. Both Falcon models are suitable for training; MJ-5, MJ-6 and XC are suitable for competition flying. Falcons have floating keel pocket, single deflexers and ribs, but no fin; and can be tow-launched. The two MJ models have a stabilising fin, floating keel pocket and ribs, but no deflexers; MJ-5 can be tow-launched. The XC has a fin, floating keel pocket, triple deflexers and ribs.

Birdman also markets power plant packs suitable for most high-performance hang gliders.

No 1981 details were received.

HIGH PERSPECTIVE
HIGH PERSPECTIVE INC

RR No. 2, Claremont, Ontario L0H 1E0

By early 1979, High Perspective had sold 50 examples of its **HP IV** Class 2 glider, details of which can be found in the 1980-81 *Jane's*. No more recent information has been received.

RFC
RECREATIONAL FLIGHT CENTRES INC

RR 2, Lucknow, Ontario N0G 2H0

RFC SKYRIDER

The Skyrider is a modified version of the US Eipper-Formance Quicksilver (which see). A choice of power plants includes a 7·5 kW (10 hp) Chrysler, 8·2 kW (11 hp) McCulloch go-kart engine, and 11·2 kW (15 hp) Komet go-kart engine. Fuel tank capacity is 6·5 litres (1·4 Imp gallons).

WEIGHTS:

Weight empty, incl power package and landing gear	54 kg (120 lb)
Max T-O weight	170 kg (375 lb)

PERFORMANCE:

Best glide ratio	6
Min sinking speed, power off	67 m (220 ft)/min
Max level speed	43 knots (80 km/h; 50 mph)
Cruising speed	26-30 knots (48-56 km/h; 30-35 mph)
T-O speed	15 knots (27·5 km/h; 17 mph)
Stalling speed	14 knots (26 km/h; 16 mph)

TALANCHUK
VLADIMIR TALANCHUK

c/o 13340 104th Street, Edmonton, Alberta T5E 4P1

Mr Talanchuk, a former member of the Aeroklub Wroclaw, was responsible for the Mars-Agat WT-8 competition glider, of which eight had been built in Poland by early 1979. His previous designs included the WT-6 Mars-S and WT-7 Mars-2S. Details of the WT-8 were published in the 1979-80 *Jane's*.

Mr Talanchuk's most recent designs are as follows:

WT-9 Mars-Marta. Competition hang glider, built in Poland in 1979. Duralumin airframe, with Dacron sail (double over 40 per cent of area) and anodised stainless steel cables. Kingpost standard. Other features include ribs of duralumin tube, pre-formed battens, floating keel pocket, adjustable camber (+2°), and single deflexers. Airframe stressed to +4/−3*g*; can be flown seated or prone. Other details in table.

WT-10 Mars-Beryl. Competition hang glider, built in 1980. Airframe as described for WT-9 except that it is stressed to +5/−3*g* and has plastics-tube ribs. For details see table.

ULTRAFLIGHT
ULTRAFLIGHT INC

6 George Street, Port Colborne, Ontario L3K 3S1
Telephone: (416) 835 1933

ULTRAFLIGHT LAZAIR

The Lazair twin-engined single-seat microlight aircraft was designed by Mr Dale Kramer in 1977-78, making its first flight in November 1978. It is available in kit form for amateur construction.

AIRFRAME: Braced high-wing monoplane, with inverted-V tail unit. Wings have an aluminium tube and foam plastics D-section leading-edge, with foam plastics ribs. Wings and tail surfaces are Mylar-covered. Three-axis control by ailerons and 'ruddervators'. Non-retractable tricycle landing gear. Twin-float gear available optionally.

POWER PLANT: Two 4·5 kW (6 hp) 100 cc Pioneer chainsaw engines, each driving a two-blade tractor propeller. Fuel tank capacity 4·5 litres (1·2 US gallons).

DIMENSIONS:

Wing span	11·07 m (36 ft 4 in)
Wing area, gross	13·19 m² (142·0 sq ft)
Wing aspect ratio	9·3
Length overall	4·27 m (14 ft 0 in)
Height overall	1·93 m (6 ft 4 in)
Propeller diameter	0·71 m (2 ft 4 in)

WEIGHTS AND LOADING:

Weight empty	approx 64·5 kg (142 lb)
Max pilot weight	100 kg (220 lb)
Max T-O weight	170 kg (375 lb)
Max wing loading	12·9 kg/m² (2·64 lb/sq ft)

PERFORMANCE:

Best glide ratio at 19 knots (35 km/h; 22 mph)	13
Min sinking speed	76 m (250 ft)/min
Never-exceed speed	47 knots (88 km/h; 55 mph)
Cruising speed	35 knots (64 km/h; 40 mph)
Manoeuvring speed	26 knots (48 km/h; 30 mph)
Stalling speed (AUW of 140 kg; 310 lb)	15 knots (27·5 km/h; 17 mph)
Max rate of climb at S/L:	
at 140 kg (310 lb) AUW	91 m (300 ft)/min
at max T-O weight	61 m (200 ft)/min
T-O run	61 m (200 ft)
Landing run	46 m (150 ft)
Range	35 nm (64 km; 40 miles)
g limits (140 kg; 310 lb AUW)	+4·0; −2·0

Ultraflight Lazair *(J. M. G. Gradidge)*

CHINA
(PEOPLE'S REPUBLIC)

BEIJING AERONAUTICS INSTITUTE

Beijing

FEIYAN (FLYING SWALLOW)

Led by Mr Wu Ji Zhong as its chief designer, the Beijing Aeronautics Institute has built a series of three Feiyan Rogallo-type hang gliders. The Feiyan II, illustrated in an accompanying photograph, has an aluminium tube airframe and nylon sail; control is by weight shift. At a T-O weight of 90 kg (198 lb), its T-O speed is 13·5 knots (25 km/h; 15·5 mph), and it can be used both for free flight and for towing behind a motorboat. All known details of the Feiyan II are given in the table; the Feiyan III has a nose angle of 110°.

MIFENG-1 (HONEYBEE-1)

The Mifeng-1, as the accompanying photographs show, is a single-seat microlight aircraft with a Rogallo-type wing strut-mounted well above the main airframe structure. It was designed and built at the Beijing Aeronautics Institute, and was intended to be powered by a 22·4 kW (30 hp) piston engine. However, its first flight, on 6 June 1979, was as a pilotless, radio-controlled aircraft, powered by a 11·2 kW (15 hp) engine. In this form it had a T-O weight of 100 kg (220 lb) and achieved a max level speed of 30 knots (55 km/h; 34 mph). The Mifeng-1 has a non-retractable tricycle-type landing gear, and is controlled in pitch and yaw by the rudder and elevator. The 45° swept wing has no separate controls.

The following data apply to the aircraft with 22·4 kW (30 hp) engine:

DIMENSIONS:

Wing span	8·20 m (26 ft 11 in)
Wing area, gross	24·00 m² (258·3 sq ft)
Length overall	5·45 m (17 ft 10½ in)
Height overall	3·30 m (10 ft 10 in)

WEIGHT:

Max T-O weight	170 kg (375 lb)

PERFORMANCE:

Max level speed	35 knots (65 km/h; 40 mph)
T-O run	20 m (66 ft)

SHENYANG AVIATION SCHOOL

Shenyang

SHENYANG AVIATION SCHOOL POWERED HANG GLIDER

An accompanying photograph shows a simple, single-seat powered Rogallo-type aircraft built by the Shenyang Aviation School. An 11·2 kW (15 hp) engine is mounted behind the pilot's seat. The craft is fitted with rudder, elevator, and a tricycle landing gear, and has an empty weight of 50-60 kg (110-132 lb).

CZECHOSLOVAKIA

SVAZARM
SVAZU PRO SPOLUPRÁCI S ARMÁDOU

Aprom

This local repair centre for aeroclub gliders and light aircraft began the production of sailwing-type hang gliders in 1980, with the objective of producing between 30-50 in that year and 100 in 1981. The gliders are of two types: the **Zk-1**, with a sail area of 24·0 m² (258·3 sq ft) and glide ratio of 8, and a training type known as the **Zk-2**.

FRANCE

AGRIPLANE
LA CULTURE DE L'AN 2000 SA

Groupe Roland Périnet et Cie, Route de Poitiers, 86110 Mirebeau
Telephone: (49) 50 44 34
Telex: 790334

This company displayed at the 1981 Paris Air Show a crop-spraying microlight aircraft known as the **Agriplane 2000**, powered by a Hirth two-cylinder vertically-opposed engine driving a four-blade propeller, and carrying a 90 litre (20 Imp gallon) chemical tank and spraybars.

No other details had been received at the time of closing for press.

AIR-CEFELEC
AIR-CEFELEC AÉRONAUTIQUE

Moisselles
Telephone: (Paris) 607 07 23 and (Moisselles) 991 32 78
DIRECTOR: Alain Bernet

AIR-CEFELEC ULM 811

The ULM 811 (ultra-léger motorisé) is a twin-engined microlight aircraft resembling, and to some extent based on, the US Eipper-Formance Quicksilver (which see). It has rudder and foreplane control surfaces, and the wing is braced with lightweight aluminium alloy struts instead of cables. Rudder pedals are linked to the nosewheel of the tricycle landing gear. The wing is based on that of the Quicksilver, but reinforced with aluminium alloy tube leading-edge; covering of all flying surfaces is of Dacron. Two small piston engines, each driving a two- or three-blade wooden tractor propeller, are mounted on the wing leading-edge; each develops about 9 kW (12 hp). First flight was made in the Spring of 1981.

A two-seat development, with dual controls, is under construction.

DIMENSION:

Wing area, gross	15·0 m² (161·5 sq ft)

WEIGHT:

Weight empty	70 kg (154·3 lb)

PERFORMANCE:

Max level speed	48·5-54 knots (90-100 km/h; 56-62 mph)
Cruising speed	27-32·5 knots (50-60 km/h; 31-37 mph)
Stalling speed	19 knots (35 km/h; 22 mph)
Max rate of climb at S/L	120 m (394 ft)/min
T-O run	less than 50 m (164 ft)
Landing run	15 m (49 ft)
Endurance (max)	1 h 30 min
Best glide ratio	7-8

DAHU PLANE
DAHU PLANE (ÉTABLISSEMENTS NOIN AÉRONAUTIQUE)

Le Logis-Neuf, RN85 Châteauvieux, 05130-Tallard

Models in production in 1979 were described in the 1980-81 *Jane's*. No more recent information has been received.

Beijing Aeronautics Institute Mifeng-1 microlight aircraft

Beijing Aeronautics Institute Feiyan III

Shenyang Aviation School powered hang glider (inset: Beijing Feiyan II)

Agriplane 2000 crop-spraying microlight aircraft *(J. M. G. Gradidge)*

Air Cefelec ULM 811 *(J. M. G. Gradidge)*

Danis single-seat microlight aircraft *(J. M. G. Gradidge)*

Danis Sabre *(J. M. G. Gradidge)*

DANIS
BERNARD DANIS

71 rue Roger François, 94700 Maisons Alfort
Telephone: 368 22 37
DIRECTOR: Bernard Danis

Details of earlier hang gliders by this pioneer French designer/manufacturer can be found in the 1979-80 and previous editions of *Jane's*. Models current in 1981 included the **Aiglon**, **Météor** and **Sabre**, in single- and tandem two-seat versions, and with optional tricycle landing gear and/or power plant. The Aiglon was available with 15 m² and 16 m² (161·5 and 172·2 sq ft) sail areas; Météor with 14, 16 and 17 m² (150·7, 172·2 and 183·0 sq ft); Sabre with 16 and 17 m² as a single-seater, 23 m² (247·6 sq ft) as a two-seater. Single engines of 9, 22·4 and 31·3 kW (12, 30 and 42 hp) were available, or an 18 kW (24 hp) twin-engine installation. No other details were known at the time of closing for press.

LA MOUETTE
LA MOUETTE

1 rue de la Petite Fin, 21121 Fontaine les Dijon

Manufacturer in 1980 of the **Atlas 14 and 16** (25 and 90 sold respectively by early 1979); **Lotus 15 and 17** (sales 5 and 5); and **Mouette 15, 17, 19 and 23B** (sales 25, 240, 780 and 90).

Details can be found in the table and in the 1980-81 *Jane's*. No more recent information has been received.

MOTO-DELTA
JEAN MARC GEISER

95 boulevard Saint-Michel, 75005 Paris
Telephone: 326 50 99

MOTO-DELTA G11

The Moto-Delta G11, as the accompanying photograph shows, is a single-seat powered flexwing aircraft. It is road-transportable on top of a family car.

The first prototype flew originally as the G10 with a completely open pilot's seat, as illustrated in the main Aircraft section of the 1976-77 *Jane's*. The single-seat G11 version, which entered production in April 1981, has a streamlined cockpit fairing of glassfibre, with a windscreen to protect the pilot.

AIRFRAME: Delta-shaped flexwing is made of Dacron, mounted on a tubular duralumin frame. An overhead control arm regulates the movement of the wing. Pedals actuate the rudder and nosewheel steering. The fuselage, landing gear legs and main beam of the wing are built entirely of laminated glassfibre, with some Klégécel and polyurethane foam plastics. The tail unit consists of a fin and rudder only.

POWER PLANT: One 13·5 kW (18 hp) flat-twin two-stroke

engine is mounted behind the pilot's seat and drives a pusher propeller.

Dimensions:

Wing span	8·00 m (26 ft 3 in)
Wing area, gross	14·00 m² (151 sq ft)
Length overall	3·40 m (11 ft 2 in)

Weights:

Weight empty	52 kg (115 lb)
Max T-O weight	140 kg (308 lb)

Performance:

Max level speed	49 knots (90 km/h; 56 mph)
Max cruising speed	38 knots (70 km/h; 43·5 mph)
Stalling speed	16 knots (30 km/h; 19 mph)
Max rate of climb at S/L	120 m (395 ft)/min
T-O run	25 m (82 ft)
Landing run	0-15 m (0-49 ft)
Endurance with max fuel	2 h

MOTO-DELTA G20

This two-seat version of the Moto-Delta was expected to fly during 1981. No details have yet been received.

SOFREC

SOFREC 'VÉLIPLANE' (Société Française de Représentation et d'Échanges Commerciaux)

16 rue Georges-Appay, 92150 Surèsnes

Models in production in 1979 were described in the 1980-81 *Jane's*. No more recent information has been received, although a powered flexwing with a Sachs single-cylinder engine has been reported.

ULM

ULTRA LÉGER MOTORISÉ SARL

Domaine du Montcel, 78350 Jouy-en-Josas
Telephone: (3) 956 46 46

This company is the French distributor for the Weedhopper and Sky Sports Vector 600, described in the US section.

GERMANY
(FEDERAL REPUBLIC)

EEL

ENTWICKLUNG UND ERPROBUNG VON LEICHTFLUGZEUGEN

Fichtenstrasse 7, 8077 Reichertshofen

ULF-1

The ULF-1 (Ultra Leicht Flugzeug) is a one-man rigid-wing foot-launched glider, designed by Dieter Reich and built by Heiner Neumann. First flight was made in November 1977, and certification was received in July 1980. Plans (but not kits or materials) are available to amateur constructors, and at least 15 are believed to be flying or under construction.

Airframe: Cantilever high-wing monoplane. Wings have Wortmann FX-63-137 section, and comprise a single wooden (spruce) spar, plywood nose section, plywood/balsa ribs and fabric covering. Thickness/chord ratio 18% at root, 15% at tip. Fuselage is a fabric-covered wooden frame, of triangular cross-section aft of cockpit. Cantilever tail unit, of similar construction to wings. Single sliding seat for pilot, who retracts his legs after foot-launch to operate rudder pedals. Landing is made on a nose-skid and glassfibre tube (fishing rod) tailskid. Three-axis aerodynamic control by ailerons, one-piece elevator, and rudder.

Dimensions:

Wing span	10·40 m (34 ft 1½ in)
Wing area, gross	13·40 m² (144·2 sq ft)
Wing aspect ratio	8
Length overall	5·55 m (18 ft 2½ in)
Height overall	2·55 m (8 ft 4½ in)
Tailplane span	2·90 m (9 ft 6¼ in)

Weights and Loading:

Weight of glider	46 kg (101·5 lb)
Pilot weight	60-90 kg (132-198 lb)
Max T-O weight	136 kg (300 lb)
Max wing loading	10·1 kg/m² (2·07 lb/sq ft)

Performance:

Best glide ratio at 30 knots (55 km/h; 34 mph)	15
Min sinking speed at 19-21·5 knots (35-40 km/h; 22-25 mph)	48 m (157 ft)/min
Max speed	38 knots (70 km/h; 43·5 mph)
Stalling speed	17·5 knots (32 km/h; 20 mph)

HIRTH

WOLF HIRTH GmbH

7312 Kirchheim/Teck-Nabern

Details of the W-17 Dracula series can be found in the 1980-81 *Jane's*. No details of this company's activities have been received since early 1979.

IKARUSFLUG

IKARUSFLUG BODENSEE

Forellenweg 1, 7770 Überlingen
Telephone: 07551 64700
Director: Ing Rudolf Schäfer

WINDSPIEL 2

The Windspiel 2 is a 'semi-rigid' hang glider (FAI Class 3), which is available also in powered form as a microlight aircraft. It made its first flight on 19 November 1980. Eighty had been sold by the Spring of 1981.

Airframe: Constant-chord non-swept wings, with ribs and marked dihedral; tailplane, underfin and rudder standard; ailerons optional. Main airframe of 2017TF duralumin, with Dacron-covered wing and tail surfaces and PVC-coated stainless steel cables. Kingpost standard. Airframe stressed to +6*g*. Can be flown seated or supine. Wheel or ski landing gear on powered version.

Power Plant: Can be fitted optionally with a 3·7 kW (5 hp) König three-cylinder two-stroke engine, two-blade tractor propeller and 20 litre (4·4 Imp gallon) fuel tank.

Dimensions: see table

Weights:

Weight of glider: without power plant	28 kg (62 lb)
with power plant (excl fuel)	51 kg (112·5 lb)

Performance (powered):

Cruising speed	19 knots (35 km/h; 22 mph)
Rate of climb	120 m (400 ft)/min

PFLUMM

MANFRED PFLUMM

Vahingerstrasse 60, 7000 Stuttgart 80

Herr Manfred Pflumm, a member of the Oskar-Ursinus-Vereinigung (OUV), has designed and built an unusual powered hang glider. Believing that many potential enthusiasts were deterred by the exposed position of the pilot, he attached a miniature aircraft fuselage beneath a Rogallo-type wing. This fuselage is of metal tube construction, fabric-covered, has a tail unit comprising fin and rudder only, and non-retractable tailwheel-type landing gear. A small piston engine, mounted conventionally in the nose of the fuselage, drives a two-blade fixed-pitch propeller, permitting powered take-off and flight.

No further details or dimensions are known.

RICHTER

KLAUS J. RICHTER INGENIEURBÜRO

Öschwende 17, 7900 Ulm-Lehr
Telephone: 0731 60567
Directors:
Ing K. J. Richter
H. Dempfle

In association with Herr Pflumm of Stuttgart, Ing Richter was responsible for the Faltente ultralight experimental powered glider, brief details of which were given in the Sailplanes section of the 1980-81 *Jane's*. More recent designs are the Dempfle-Richter DeRic Delta-Ente (see Homebuilt Aircraft section) and the AN 20B microlight aircraft, of which a description follows.

RICHTER AN 20B

The AN 20B, designed in 1979 and first flown in 1980, is an extremely lightweight motor glider based on the Neukom AN-20M (see Swiss section) and available in plans, kit or ready-to-fly form. Two prototypes and five production aircraft had been built by early 1981.

Airframe: Parasol monoplane configuration, supported by centre-section struts and single bracing strut on each side. Wortmann wing section FX-63-137 from root to tip. Dihedral 2°. Wings built of glassfibre, foam and plywood; glassfibre ailerons and trailing-edge flaps, and plywood upper-surface airbrakes. All-glassfibre fuselage. Cantilever T tail (fixed-incidence tailplane plus elevator), of glassfibre and foam. Non-retractable tricycle landing gear, with glassfibre sprung legs, 30·5 cm (12 in) diameter main wheels, 25·4 cm (10 in) diameter nosewheel, and wheel brakes. Single seat in open cockpit, with windscreen.

Power Plant: One 19 kW (25·5 hp) König three-cylinder piston engine, driving a 1·06 m (3 ft 5¾ in) diameter Dorn two-blade wooden pusher propeller. Fuel tank capacity 20 litres (4·4 Imp gallons).

Dimensions:

Wing span	12·60 m (41 ft 4 in)
Wing area, gross	10·08 m² (108·5 sq ft)
Wing aspect ratio	15·75
Length overall	5·40 m (17 ft 8½ in)
Height overall	1·20 m (3 ft 11¼ in)

Weights and Loading:

Weight empty	99 kg (218 lb)
Max T-O weight	200 kg (441 lb)
Max wing loading	18·45 kg/m² (3·78 lb/sq ft)

Performance (power off):

Best glide ratio at 44·5 knots (82 km/h; 51 mph)	20
Min sinking speed at 37 knots (68 km/h; 42·5 mph)	48 m (157 ft)/min
Max speed (smooth air)	81 knots (150 km/h; 93 mph)
Max speed (rough air)	70 knots (130 km/h; 81 mph)
Stalling speed	24·5 knots (45 km/h; 28 mph)

Performance (power on):

Max cruising speed	73 knots (135 km/h; 84 mph)
Econ cruising speed	59·5 knots (110 km/h; 68·5 mph)
Stalling speed	24·5 knots (45 km/h; 28 mph)
Max rate of climb at S/L	108 m (350 ft)/min
Service ceiling	3,500 m (11,475 ft)
T-O run	80 m (262 ft)
Landing run	120 m (395 ft)
Range with max fuel	189 nm (350 km; 217 miles)
g limits	+6·0; −3·0

Geiser Moto-Delta G11 powered flexwing-type aircraft

EEL (Reich & Neumann) ULF-1

Ikarusflug Windspiel 2 hang glider

Powered hang glider built by Herr Manfred Pflumm *(Wolfgang Wagner)*

Richter AN 20B ultra-lightweight motor glider

US-built example of the Wizard microlight aircraft *(Howard Levy)*

ROCHELT

GÜNTER ROCHELT

Josef-Schwarz-Weg 11, 8000 München 71

ROCHELT SOLAIR I

The Solair I, which made its initial flight on 17 December 1980, is the first German solar-powered aircraft to fly. It is modified from a Canard-2 FL (see Swiss section), and is fitted with extended-span wings. Along the upper surface of these are 2,499 solar cells, which are used to power an 88V 2·2 kW electric motor driving a two-blade large-diameter pusher propeller.

Dimensions:	
Wing span	16·00 m (52 ft 6 in)
Wing area, gross	22·32 m² (240·25 sq ft)
Wing aspect ratio	17
Foreplane span	3·90 m (12 ft 9½ in)
Length overall	5·40 m (17 ft 8½ in)
Height overall	2·00 m (6 ft 6¾ in)
Propeller diameter	2·65 m (8 ft 8¼ in)
Weights and Loading:	
Weight empty	120 kg (264·5 lb)
Max T-O weight	200 kg (441 lb)
Max wing loading	8·96 kg/m² (1·83 lb/sq ft)
Performance:	
Best glide ratio at 24 knots (45 km/h; 28 mph)	26
Min sinking speed at 21·5 knots (40 km/h; 25 mph)	0·42 m (1·38 ft)/s
Stalling speed	15·5 knots (28 km/h; 17·5 mph)
T-O run	60 m (200 ft)
Landing run	30 m (100 ft)
g limits	+4·0/−2·0

SCHÜRMANN

WERNER SCHÜRMANN

Hafenstrasse 7, 4500 Osnabrück

Telephone: 0541 65055 and 432334

In 1981 Herr Schürmann was advertising for sale a microlight aircraft known as the **Wizard,** of 'semi-rigid' monoplane form with tricycle landing gear and an 11·2 kW (15 hp) Yamaha single-cylinder two-stroke engine. It has a wing span of 9·90 m (32 ft 6 in), T-O run of 30 m (100 ft) and landing run of 15 m (50 ft). No other details were received for publication, but the aircraft is believed to be a US design. As an accompanying photograph shows, it bears a general resemblance to the Eipper-Formance Quicksilver.

TEMPLE

PROF DIPL-ING BERNHARD E. R. J. DE TEMPLE

Am Pfingstborn 23, 6501 Heidesheim/Rhein

TEMPLE-WING TE-F1, TE-F2 and TE-F3

The Temple-Wing (Temple-Flügel) TE-F1 rigid-wing hang glider uses pre-stressed sheets of foam plastics in the wings and tail unit to provide firm lifting surfaces which are virtually insensitive to dynamic loads. Because these surfaces are pre-stressed, they do not become deformed in flight. Other advantages include low constructional and repair costs. A full-size prototype of the TE-F1 was flown in Austria on 13 August 1978, and was described and illustrated in the 1980-81 *Jane's*.

Built in the Spring of 1979, the TE-F2 has a rigid wing, made of the same materials as that of the TE-F1, and a similar inverted-V tail unit. All pre-stressed sheets are secured firmly by thin strips of polypropylene. The glider can be dismantled and folded to half-span dimensions for transportation and stowage. The TE-F2 was tested by the DAeC-Testauto at Malmsheim in August 1979, and aerodynamic results were sufficiently encouraging to lead to a new design, the TE-F3, which is a powered hang glider with a multi-propeller propulsion system. It was hoped to fly the TE-F3 during 1980, but no news of this was received before closing for press.

The following details apply to the TE-F2:

DIMENSIONS:
- Wing span: 11·00 m (36 ft 1 in)
- Wing area, gross: 15·00 m² (161·5 sq ft)
- Length overall: 3·52 m (11 ft 6½ in)
- Height overall: 2·80 m (9 ft 2¼ in)

WEIGHTS:
- Weight empty: 35 kg (77 lb)
- Weight of pilot: 75 kg (165 lb)

PERFORMANCE:
- Best glide ratio: 12
- Min sinking speed: 48 m (157 ft)/min
- Max level speed: 27 knots (50 km/h; 31 mph)

JAPAN

ABE
KEIICHI ABE

17-11-4 Chome, Masuura, Kushiro City, Hokkaido

Mr Abe, whose Midget (Mizzett) III lightplane is described in the Homebuilts section, was in 1981 constructing a microlight aircraft known as the Motor Hang IV.

ABE MOTOR HANG IV

TYPE: Single-seat microlight aircraft.

AIRFRAME: Wire-braced constant-chord high-wing monoplane with T tail. Dihedral 2°. Main frame of 6061-T8 tube, with wooden wing ribs, leading-edge and auxiliary spar. Wing and tail surfaces fabric-covered; fuselage covering of glassfibre sheet. Wing section CTÖ 387, with thickness/chord ratio of 15%. Non-retractable tricycle landing gear, with bicycle wheels and 25·4 cm (10 in) diameter tyres. No brakes.

POWER PLANT: One 6·7 kW (9 hp) McCulloch engine initially, driving a two-blade tractor propeller. Intended power plant is a 9·7 kW (13 hp) Japanese-built 100 cc go-kart engine with 1:4 reduction gear.

DIMENSIONS:
- Wing span: 6·00 m (19 ft 8¼ in)
- Wing area, gross: 6·00 m² (64·6 sq ft)
- Wing aspect ratio: 6·0
- Length overall: 3·81 m (12 ft 6 in)
- Height overall: 2·52 m (8 ft 3¼ in)

WEIGHTS:
- Weight empty: 47 kg (104 lb)
- Max T-O weight: 127 kg (280 lb)

PERFORMANCE (estimated):
- Max level speed: 40·5 knots (75 km/h; 46·5 mph)
- Stalling speed: 28 knots (52 km/h; 32·5 mph)
- Rate of climb: 250 m (820 ft)/min
- Range with max fuel: 40·5 nm (75 km; 46·5 miles)

POLAND

CIESLAR/ŚWIGÓN
MARIAN CIESLAR and PIOTR ŚWIGÓN

Mr Cieslar, of Bielsko-Biala aero club, assisted by Mr Świgón of the Slasa aero club, has fitted a **Balans** hang glider with a 30 kg (66 lb) power plant installation and a tricycle landing gear, as shown in an accompanying illustration. The power plant is a WSK 175 motorcycle engine, driving a two-blade pusher propeller. First flight, by Mr Świgón, was made on 31 May 1981.

PW
POLITECHNIKI WARSZAWSKIEJ (Warsaw Polytechnic)

AKL

The students' aero club (Academicki Klub Lotniarski) of Warsaw Polytechnic has designed, built and flown a hang glider powered by a 3·3 kW (4·5 hp) 75 cc engine driving a two-blade ducted propulsion unit. General appearance of the aircraft is shown in an accompanying illustration; data include the following:
- Wing span: 9·60 m (31 ft 6 in)
- Wing area: 18·00 m² (193·75 sq ft)
- Nose angle: 110°
- Weight of glider: 20 kg (44 lb)
- Weight of power plant: 12 kg (26·5 lb)

ULS

General appearance of the ULS (Ultralekkiego Szybowca: ultralight glider) can be seen in an accompanying photograph. Construction is of glassfibre/epoxy (62%), foam infill (12%), polyurethane lacquer (7%), steel (5%), plywood (4%), duralumin (4%), synthetic fabric covering (4%) and other materials (2%). It was designed and built by a group of 40 students under the leadership of Prof Marek Dietrych, and was completed in the Summer of 1981.

DIMENSIONS:
- Wing span: 10·87 m (35 ft 8 in)
- Wing area, gross: 12·65 m² (136·2 sq ft)
- Wing aspect ratio: 9·34
- Length overall: 5·45 m (17 ft 10½ in)

WEIGHTS:
- Weight empty: 45 kg (99 lb)
- Max T-O weight: 125 kg (276 lb)

PERFORMANCE (estimated):
- Best glide ratio at 30 knots (55 km/h; 34 mph): 17
- Min sinking speed at 24 knots (45 km/h; 28 mph): 51 m (167 ft)/min
- Max level speed (smooth air): 67·5 knots (125 km/h; 78 mph)
- Stalling speed: 22 knots (40 km/h; 25 mph)
- *g* limits: +5·3; −2·65

WIERZBOWSKI
PAWEL WIERZBOWSKI

ul. Wroblewskiego 25/506, 51-627 Wroclaw

Details of Mr Wierzbowski's Vega 106 hang glider were published in the 1979-80 *Jane's*. More recent designs include the **Vega SST-A** (ten built by Spring 1980, plus 30 sets of plans sold); **Vega SST-B** (five built by the same date, plus 25 sets of plans sold); and the **Vega-Star** (four built by February 1981, plus 18 sets of plans sold).

All three types are suitable for competition flying; the SST-A is also suitable for private flying, and the Vega-Star for training. All can be flown prone; SST-A can be flown seated. All three are stressed to +5/−3*g*. Airframes are of duralumin, with a single-thickness nylon (A) or Dacron (A and B) sail (30 per cent double-thickness Dacron on Vega-Star) and vinyl-coated stainless steel cables. Kingpost, pre-formed battens and a floating keel pocket are standard on all models.

For other details see table.

SWITZERLAND

CANARD
CANARD AVIATION

Sagenrainstrasse 4, CH-8636 Wald
Telephone: (055) 952055
PRESIDENT: Dipl-Ing Hans U. Farner

CANARD AVIATION CANARD-2 FL

First flown on 7 September 1977, the Canard-2 FL was designed by Dipl-Ing Hans U. Farner, who also designed the Colibri 1 SL canard motor glider described in the Sailplanes section. It can be bungee-launched in a conventional manner, foot-launched, or rolled downhill on wheels for take-off. Production of the Canard-2 FL began in 1978. A two-seat version was under development in 1981.

TYPE: One-man foot-launched sailplane; *g* limits +5·3 (limit)/+7·9 (ultimate).

AIRFRAME: 'Parasol monoplane' configuration, with main wing at rear, supported by V-form pylons which act as both lifting surfaces and 'vertical' fins; fixed-incidence foreplane. Wings, pylons and canard surface have Wortmann FX-63-137 section. Wings have external aerofoil ailerons; trailing-edges of V-tail pylons movable to act as Venom-type airbrakes. Wing is a single-spar structure, with vacuum-formed shell of laminated resin (Du Pont Kevlar, CIBA-Geigy 2878) and glassfibre, and core of styrofoam plate; other structures have a similar shell, with cores of hard foam cells (ailerons) or solid hard foam (airbrakes). Leg and head doors for foot-launch, after which pilot swivels into prone position, on sliding board, and closes both doors. One-piece canopy/windscreen, open at rear. Retractable front skid for landing; monowheel, or jettisonable twin main- and tailwheels, optional for downhill rolling take-off.

DIMENSIONS:
- Wing span: 13·50 m (44 ft 3½ in)
- Wing area, gross: 11·00 m² (118·4 sq ft)
- Wing aspect ratio: 16·6
- Foreplane span: 3·74 m (12 ft 3¼ in)
- Foreplane area: 2·00 m² (21·5 sq ft)
- Foreplane aspect ratio: 7·0
- Length overall: 4·94 m (16 ft 2½ in)
- Height over tail: 1·80 m (5 ft 11 in)

WEIGHTS AND LOADING:
- Weight empty: 58 kg (128 lb)
- Max T-O weight: 170 kg (375 lb)
- Max wing loading: 12·5 kg/m² (2·56 lb/sq ft)

PERFORMANCE (at max T-O weight):
- Best glide ratio at 31 knots (57 km/h; 35·5 mph): 30
- Min sinking speed at 27 knots (50 km/h; 31 mph): 30 m (98 ft)/min
- Stalling speed: 21 knots (38 km/h; 24 mph)
- Max speed (rough and smooth air): 65 knots (120 km/h; 74·5 mph)

NEUKOM
ALBERT NEUKOM SEGELFLUGZEUGBAU

Flugplatz Schmerlat, CH-8213 Neuenkirch

NEUKOM AN-20 and AN-20M

The **AN-20** hang glider has an open-frame 'fuselage' beneath the strut-braced wing, fin and rudder extending above and below the slender steel-tube tailboom, and a T tailplane.

The **AN-20M** powered version has a tricycle landing gear, and is fitted with either a 9·7 kW (13 hp) Solo 220 cc motorcar engine or an 18·6 hp (25 hp) König three-cylinder engine. It is available optionally with a tailboom of GfK instead of steel tube.

DIMENSIONS (A: AN-20; B: AN-20M; C: AN-20M with GfK tailboom):
- Wing span: A, B: 12·60 m (41 ft 4 in)
  - C: 12·50 m (41 ft 0 in)
- Wing area, gross: A, B: 10·08 m² (108·5 sq ft)
  - C: 10·00 m² (107·6 sq ft)
- Wing aspect ratio: A, B: 15·75
  - C: 15·62
- Length overall: A: 5·50 m (18 ft 0½ in)
  - B: 6·50 m (21 ft 4 in)
  - C: 6·20 m (20 ft 4 in)

WEIGHTS:
- Weight empty: A: 70 kg (154 lb)
  - B: 96 kg (212 lb)
  - C: 110 kg (242 lb)
- Max T-O weight: A: 170 kg (375 lb)
  - B, C: 200 kg (441 lb)

PERFORMANCE:
- Best glide ratio:
  - A at 32·5 knots (60 km/h; 37 mph): 16
  - B at 35 knots (65 km/h; 40·5 mph): 14
  - C at 35 knots (65 km/h; 40·5 mph): 19
- Min sinking speed:
  - A at 26 knots (48 km/h; 30 mph): 60 m (197 ft)/min
  - B at 27 knots (50 km/h; 31 mph): 66 m (216 ft)/min
  - C at 27 knots (50 km/h; 31 mph): 54 m (177 ft)/min

Temple-Wing TE-F2, folded for road transportation

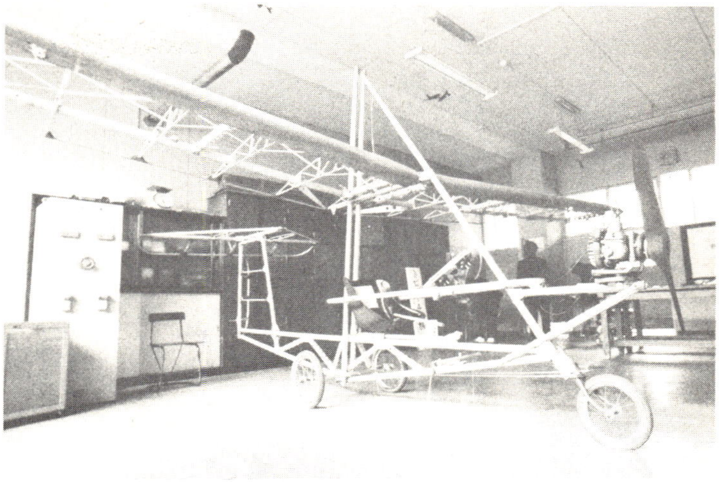

Abe Motor Hang IV under construction

Warsaw Polytechnic ULS sailplane *(Skrzydlata Polska)*

AKL powered hang glider built by students of Warsaw Polytechnic *(Skrzydlata Polska)*

Wierzbowski Vega SST

Cieslar/Świgó powered hang glider *(Skrzydlata Polska)*

Neukom AN-20M one-man microlight aircraft

Neukom AN-21 ultralight aircraft, in unpowered form *(Peter F. Selinger)*

Max level speed: B, C	81 knots (150 km/h; 93 mph)
Stalling speed: A	23 knots (42 km/h; 26·5 mph)
Rate of climb: B	90 m (295 ft)/min
C	150 m (492 ft)/min
T-O run (grass): B	110 m (361 m)
C	80 m (263 m)

NEUKOM AN-21R

Of canard configuration, the AN-21R was designed by Mr Albert Neukom, and the first completed example was constructed by Herr Klaus-Jurgen Richter at his factory at Ulm-Lehr in West Germany. It is powered by a 22·5 kW (30 hp) König three-cylinder engine, driving a two-blade pusher propeller, mounted aft of the pilot's seat.

The braced wings of the AN-21R have a Wortmann FX-63-137 section, 20° of sweepback, and are fitted with upper-surface spoilers, ailerons, and endplate fins and rudders at their tips. The non-swept, constant-chord foreplane is fitted with an elevator. The aircraft is of mixed construction (mainly metal and GfK), and has a non-retractable tricycle landing gear. The cockpit/fuselage can be an openwork frame, or covered.

DIMENSIONS:	
Wing span	12·60 m (41 ft 4 in)
Foreplane span	3·00 m (9 ft 10 in)
Wing area, gross	10·7 m² (115·2 sq ft)
Wing aspect ratio	14·8
Foreplane area	1·6 m² (17·2 sq ft)
Length overall	4·90 m (16 ft 1 in)
Fuselage length	3·40 m (11 ft 1¾ in)
Height overall	2·40 m (7 ft 10½ in)
WEIGHTS:	
Weight empty	100 kg (220 lb)
Max T-O weight	200 kg (441 lb)
PERFORMANCE:	
Max level speed	70 knots (130 km/h; 81 mph)
Stalling speed	23 knots (42 km/h; 26 mph)
Rate of climb	150 m (492 ft)/min
T-O run (grass)	75 m (246 ft)

NEUKOM AN-22

This semi-rigid hang glider was under development in 1981. It has sweptback wings, with an elevon at each tip, no tail surfaces, and a tailwheel-type landing gear.

Canard Aviation Canard-2 FL ultralight sailplane in foot-launch configuration

DIMENSIONS:	
Wing span	12·60 m (41 ft 4 in)
Wing area, gross	10·80 m² (116·25 sq ft)
Wing aspect ratio	14·7
WEIGHTS AND LOADING:	
Weight empty	50 kg (110 lb)
Max T-O weight	140 kg (308 lb)
Max wing loading	12·96 kg/m² (2·65 lb/sq ft)

UNION OF SOVIET SOCIALIST REPUBLICS

Details of hang gliders and microlights built in the USSR are somewhat elusive, but the following brief particulars have appeared in the Soviet and Polish press during the past year:

ANTONOV

The O.K. Antonov design bureau has designed a flex-wing hang glider known as the **Slavutitch** which is in series production. All known details are given in the table.

DMITRIYEV
VICTOR DMITRIYEV
DMITRIYEV X-12C

Mr Dmitriyev is a driver at the Kirghiz National University in Frunze, the capital of Kirghizia; he built his first aircraft in 1969. In designing and building the X-12C he was assisted by Nikolai Kitz (who built the engine), sports fliers Alexander Ugryumov and Yuri Tsybenko, and Aeroflot pilot Yuri Batsura.

DIMENSIONS:

Wing span	5·54 m (18 ft 2 in)
Length overall	3·35 m (11 ft 0 in)
Height overall	1·26 m (4 ft 1½ in)

WEIGHT:

Weight empty	51 kg (112·5 lb)

KAZUROV/POLUSHKIN/RUSAKOV

The **Shtyvnoplat** semi-rigid hang glider shown in an accompanying photograph is credited to J. Kazurov, I. Polushkin and V. Rusakov. Wing span 9·60 m (31 ft 6 in), wing area 14·0 m² (150·7 sq ft), length 5·20 m (17 ft 0¾ in), glide ratio 15-17.

LENINGRAD CENTRAL NII

This organisation has produced three known types of hang glider, known as the **Grach, Crechet** and **Sokol.** No details are known of the Grach-1, Grach-2 and Crechet, all said to be 1978 designs. The original Sokol, designed by V. Mikhailov, appeared in 1976, and has been followed by the Sokol-2 (1978) and Sokol-3 (see accompanying photograph). Known details of this latter series include the following:

Airframe of D16T alloy, with fabric-covered sail

Wing span: Sokol	8·62 m (28 ft 3¼ in)
Sokol-2	10·30 m (33 ft 9½ in)
Wing area: Sokol	17·00 m² (183·0 sq ft)
Sokol-2	18·00 m² (193·8 sq ft)
Wing aspect ratio: Sokol	4·4
Sokol-2	6·0
Pilot weight: Sokol-2 and -3	55-95 kg (121-209 lb)
T-O weight: Sokol	110 kg (243 lb)

PLETNYOV

The **Zhuravlic** microlight aircraft designed by Alexander Pletnyov is illustrated in an accompanying photograph. The following details are known:

Power plant	350 cc piston engine
Wing span	9·60 m (31 ft 6 in)
Weight empty	55 kg (121 lb)
Glide ratio	30
Max speed	32·5 knots (60 km/h; 37·5 mph)
Ceiling	2,000 m (6,560 ft)
T-O run	15-20 m (49-66 ft)
Landing run	10 m (33 ft)

SP
SINIAYA PTICA

Klub Siniaya Ptica, ul. Vavishova, Rostov on Don 344064

Accompanying photographs illustrate two Rogallo-wing microlight aircraft, both credited to Yevgeni Shevchenko.

The **Ptica-2** has a 'pilot pod' equipped with three-axis controls and a tricycle landing gear, and is powered by a 31·3 kW (42 hp) engine driving a 0·38 m (1 ft 3 in) diameter wooden propeller. Wing area is 22·5 m² (242·2 sq ft); T-O speed is 35 knots (65 km/h; 40·5 mph), and cruising speed 51·5-54 knots (95-100 km/h; 59-62 mph).

The **SPM** (Siniaya Ptica Modernisovanna) has a 34·3 kW (46 hp) engine, wing area of 20·5 m² (220·7 sq ft) and T-O weight of 182 kg (401 lb). Performance includes a T-O speed of 21·5 knots (40 km/h; 25 mph), T-O run of 30-40 m (100-130 ft), and landing run of 25-30 m (85-100 ft).

UNITED KINGDOM

BIRDMAN
BIRDMAN SPORTS LTD

Overtown House, Mildenhall, Marlborough, Wiltshire

Producer of the **Cherokee**, of which 250 had been sold by the beginning of 1980. No 1981 information received: details of Cherokee (one size) can be found in the table and in the 1980-81 *Jane's*.

CHARGUS
CHARGUS HANG GLIDERS

Wagland Engineering, Gawcott, Buckingham

This company, one of the leading UK hang glider concerns, has not yet responded to any *Jane's* request for information about its products.

In Spring 1980 it flew a prototype of the **Vortex** powered hang glider (see accompanying photograph), which has an 8·2/10·6 kW (11/19 hp) 244 cc Fuji Robin engine and three-blade pusher propeller. Stalling speed is approx 16 knots (29 km/h; 18 mph). No other details known at the time of closing for press.

COOK
DAVID COOK

Hillcrest, Aldeburgh Road, Aldringham, near Leiston, Suffolk

Mr David Cook acquired a Volmer VJ-23 Swingwing hang glider (which see) in the Spring of 1974, subsequently fitting it with a 6·7 kW (9 hp) McCulloch MC-101B piston engine and two 5 litre (1·1 Imp gallon) fuel tanks. In this form, Mr Cook piloted the aircraft across the Channel from Walmer Castle, Deal, to Blériot Plage, France, on 9 May 1978 in a flying time of 1 h 15 min. This VJ-23E thus became the lowest-powered aeroplane to make a cross-Channel flight up to that time. It was presented to the Shuttleworth Collection on 8 May 1981.

At that time Mr Cook was constructing a new microlight aircraft powered by a 125 cc engine.

ECLIPSE

7 Exbridge Road, Dulverton, Somerset

Markets the **Super Eagle** microlight powered hang glider shown in an accompanying photograph. No other details known.

EURO WING

Unit 20, Dixon Place, College Milton North, East Kilbride G74 5JF

Telephone: 03 552 46498

PARTNERS:

B. K. Harrison

P. M. Coppola

This company, formerly the distributor for US Catto and Electra Flyer hang gliders, no longer manufactures hang gliders. In 1981 it was producing under licence the Goldwing and CP 16 microlight aircraft (see US section).

FLEXI-FORM
FLEXI-FORM SKYSAILS

Unit 24, Nassau Mill, Cawdor Street, Patricroft, Manchester

Details of 1979 models can be found in the 1980-81 *Jane's*. No more recent information has been received.

HIWAY
HIWAY HANG GLIDERS LTD

Sirhowy Hill, Tredegar, Gwent NP2 4XP

Details of 1979 models can be found in the 1980-81 *Jane's*. No more recent information has been received, but press reports have referred to a powered version of the Super Scorpion known as **Skytrike** (see accompanying photograph), which has a landing gear/seat/power plant unit in which the engine drives a pusher propeller.

MARSHALL
FRANK MARSHALL

19 Station Road, Filton, Bristol BS12 7BZ

Mr Marshall, whose Fantail hang glider was described under the South African section in the 1978-79 *Jane's*, later designed and built the Heatwave and Madcap powered hang gliders. No recent news of these has been received; descriptions and illustrations can be found in the 1980-81 *Jane's*.

SKYHOOK
SKYHOOK SAILWINGS LTD

Vale Mill, Chamber Road, Hollinwood, Oldham, Lancashire OL8 3PX

Telephone: 061 624 8351

Telex: 667 849

DIRECTOR: L. Gabriels

Hang glider models current in 1981 were:

Cutlass. Improved version of Silhouette (which see), having a lower-twist wing and improved performance.

Dmitriyev X-12C microlight aircraft *(Tass)*

Pletnyov Zhuravlic microlight aircraft *(Tass)*

Soviet Shtyvnoplat *(Skrzydlata Polska)*

Leningrad (Mikhailov) Sokol-3 *(Krilya Rodiny)*

Siniaya Ptica (Shevchenko) SPM *(Skrzydlata Polska)*

SP (Shevchenko) Ptica-2 *(Skrzydlata Polska)*

Chargus Vortex *(Peter R. March)*

David Cook and the VJ-23E Swingwing powered glider on which he crossed the Channel in May 1978

Eclipse Super Eagle *(Peter R. March)*

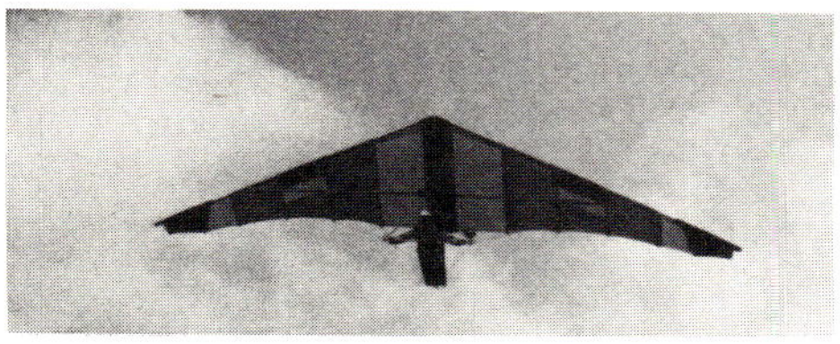

Skyhook Sailwings Silhouette

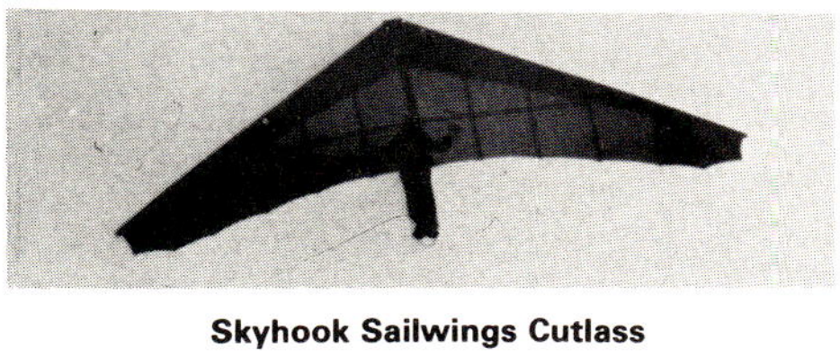

Skyhook Sailwings Cutlass

Hiway Skytrike *(Peter R. March)*

Double-thickness Dacron over 30 per cent of sail area. Anodised airframe, which can be broken down to 3·86 m (12 ft 8 in) or less. Sliding centre box for final sail tension. Otherwise generally as described for Silhouette.

Sabre. Developed version of Cutlass, with fully-enclosed floating crossboom and 55 per cent double surface sail. Airframe stressed to $+6/-3g$ (minimum); max speed over 43·5 knots (80·5 km/h; 50 mph).

Silhouette. Single-seater, for training, private and competition flying. Airframe of anodised aluminium alloy, with single-thickness Dacron sail and vinyl-coated stainless steel cables. Kingpost standard. Features include preformed aluminium battens, floating keel pocket and (optionally) provision for tow launch. Airframe is stressed to $+6/-5g$, and can be broken down to 4·1 m (13 ft 6 in) or less for transportation and stowage.

The company also markets a 41 kg (90 lb) Power Trike unit consisting of a 7·5 kW (10 hp) 210 cc aircooled two-stroke engine, two-blade pusher propeller, 8 litre (1·75 Imp gallon) fuel tank, frame-mounted tricycle landing gear with 30·5 cm (12 in) diameter go-kart wheels (front wheel steerable), and padded fabric seat with lap-straps. It can be attached to any Skyhook hang glider (and many others) by a single bolt and backup support strap.

UAC

THE ULTRALIGHT AVIATION CENTRE LTD

Wellesbourne Airfield, Wellesbourne, Warwick CV35 9EU

Telephone: 0789 841114

DIRECTORS: Paul Baker and David Garrison

The Ultralight Aviation Centre has since September 1979 assembled under licence the Pterodactyl Fledgling powered hang glider described under Pterodactyl Ltd in the US section.

UNITED STATES OF AMERICA

ACT

AIRCRAFT COMPOSITE TECHNOLOGY

4909 Looman, Wichita, Kansas 67220
Telephone: (316) 686 4193
DESIGNERS: J. G. Black and J. L. Long

ACT IBIS

As an accompanying photograph shows, the Ibis microlight aircraft is of canard configuration, similar to that of the Goldwing (which see), and has a tricycle-type landing gear. It was designed by Mr James G. Black, a power plant group design engineer with Gates Learjet, assisted by Mr Jerry Long. The Ibis is of glassfibre/foam/epoxy sandwich construction, reinforced with steel and aluminium, has a GAW-1 wing section, three-axis control (ailerons, elevator and rudders), and can be foot-launched. It first flew in May 1980 and is now available in kit form.

POWER PLANT: One McCulloch 91-101 piston engine, driving a two-blade pusher propeller.

DIMENSIONS:

Wing span (excl fins and rudders)	7·62 m (25 ft 0 in)
Foreplane span	2·44 m (8 ft 0 in)
Length overall	3·96 m (13 ft 0 in)
Height overall	1·68 m (5 ft 6 in)

WEIGHTS:

Weight empty	68 kg (150 lb)
Max T-O weight	159 kg (350 lb)

PERFORMANCE (estimated):

Cruising speed	over 43·5 knots (81 km/h; 50 mph)
T-O speed	17·5 knots (32 km/h; 20 mph)

AMERICAN AEROLIGHTS

AMERICAN AEROLIGHTS

700 Comanche NE, Albuquerque, New Mexico 87107
Telephone: (505) 344 6366
PRESIDENT: Larry Newman
VICE-PRESIDENT: Bryan L. Allen

Under its former name of Electra Flyer International, this company produced the hang gliders listed in the 1980-81 *Jane's*. It no longer manufactures hang gliders, but in 1981 was continuing to produce the highly successful Eagle microlight aircraft. By February 1981 a total of 600 had been built, of which 560 had been sold.

The Eagle was one of four types of microlight aircraft purchased in 1980 by the US Navy for evaluation at the Naval Weapons Center, China Lake, California.

AMERICAN AEROLIGHTS EAGLE

TYPE: One-man microlight aircraft, stressed to +8·5/−2·5*g*.

AIRFRAME: Sweptback flexible-wing monoplane (nose angle 148°, washout 14°), with non-swept foreplane and elevator; rudders at wingtips. Main frame is of 6061-T6 anodised aluminium alloy, with Dacron covering (double surface on foreplane and part of wing) and vinyl-coated stainless steel cables. Kingpost standard. Features include floating keel pocket and ribs. Tricycle coil-sprung landing gear, with steerable nosewheel, standard; twin-float gear available optionally. Control by weight shift and tip rudders.

POWER PLANT: Two 134 cc Chrysler Talon two-stroke engines (each approx 6·5 kW; 9 hp), driving a two-blade pusher propeller via a reduction gear. Alternative power plants include a single 17 kW (23 hp) 250 cc Fuji Robin FC-25P single-cylinder engine. Fuel tank capacity 9·5 litres (2·5 US gallons).

DIMENSIONS:

Wing span	10·97 m (36 ft 0 in)
Foreplane span	2·74 m (9 ft 0 in)
Wing area, gross	15·8 m² (170·0 sq ft)
Foreplane area	1·67 m² (18·0 sq ft)
Wing aspect ratio	7·6
Foreplane aspect ratio	4·5
Length overall	4·27 m (14 ft 0 in)
Propeller diameter	1·37 m (4 ft 6 in)

WEIGHTS AND LOADING:

Weight empty	69·5 kg (153 lb)
Pilot weight range	43-113 kg (95-250 lb)
Max T-O weight	193 kg (425 lb)
Max wing loading	10·64 kg/m² (2·18 lb/sq ft)

PERFORMANCE:

Never-exceed speed	43·5 knots (80 km/h; 50 mph)
Cruising speed	28 knots (52 km/h; 32 mph)
Manoeuvring speed	26 knots (48 km/h; 30 mph)
Stalling speed	16 knots (29 km/h; 18 mph)
Max rate of climb at S/L	137 m (450 ft)/min
Service ceiling (standard)	3,050 m (10,000 ft)
T-O and landing run	23 m (75 ft)
Typical endurance	1 h 30 min
Best glide ratio at 24·5 knots (45 km/h; 28 mph)	7
Min sinking speed at 23·5 knots (43·5 km/h; 27 mph), power off	91 m (300 ft)/min

BIRDMAN

BIRDMAN INC

217 Sandy Circle, South Daytona, Florida 32019
Telephone: (904) 761 1883
PRESIDENT: Leonard I. Roberts
ENGINEERING: J. G. Ladesic
PRODUCTION: Gerald Zimmick
SALES: J. Roberts

BIRDMAN TL-1A

Birdman Aircraft Inc has developed a microlight aircraft which, despite a wing span of 10·36 m (34 ft 0 in), has an empty weight of only 55·5 kg (122 lb) in its current form. The company's aim was to design a strong, lightweight and inexpensive aircraft that could be assembled easily by a novice builder. In achieving this aim, it introduced some unusual ideas and construction materials, as detailed in the Homebuilts section of the 1980-81 *Jane's*.

Design of what proved to be the world's lightest powered aircraft, known as the Birdman XTL-1, with an empty weight of 45 kg (100 lb) in prototype form, began in April 1969. The first flight was recorded on 25 January 1975, after considerable structural research and testing. The XTL-1 (N111ET) had full-span trailing-edge flaps. These are deleted on the TL-1A production version, which also has a V tail instead of the T tail fitted to the prototype. The company is marketing the aircraft in kit form, with all materials, components, engine and accessories provided, together with plans and a builder's manual.

First orders were taken in June 1975 and by early 1981 a total of 575 kits had been sold; 330 TL-1As were then flying.

TYPE: Single-seat microlight aircraft; *g* limits ±6.

AIRFRAME: Cantilever low-wing monoplane. Wings built in three sections, with detachable outer panels. Wing section USA 35B (modified). Thickness/chord ratio 14%. Dihedral on outer wing panels 5°. Incidence 3°. No sweepback. Composite structure, with single built-up spruce spar, wooden ribs, leading-edge D cell of reinforced expanded synthetic foam and aircraft Monokote covering. No ailerons. Sequentially-operated spoilers, immediately aft of main spar, for both yaw and roll control; the more the stick is moved, the more panels come into operation. Three-part fuselage: forward portion of riveted semi-monocoque 2024-T3 light alloy, centre section of spruce and plywood, and tapered plywood monocoque aft section. Braced V tail. Two-spar structure with laminated plywood ribs and aircraft Monokote covering. No rudders. Non-retractable main wheels, sited at point of balance. Castor at aft end of fuselage to support aircraft on ground. Shock-absorption by rubber bungee. Main-wheel tyres size 12 × 1·25, pressure 3·10 bars (45 lb/sq in). No brakes. Exposed seat of aluminium tube and cloth, with harness, adjustable to allow pilot to fly the aircraft in an upright or semi-reclining position.

POWER PLANT: One 8·6 kW (11·5 hp) Chrysler Power Bee two-stroke engine, pylon-mounted from centre-section structure aft of pilot, and driving a two-blade fixed-pitch laminated wood pusher propeller through a 2·65:1 reduction gear. Fuel contained in one (normal) or two (auxiliary) aluminium tanks with combined capacity of 15 litres (4 US gallons). Lubricating oil mixed in fuel at ratio of 1 : 16. Standard recoil starter.

DIMENSIONS:

Wing span	10·36 m (34 ft 0 in)
Wing area, gross	13·42 m² (144·5 sq ft)
Wing aspect ratio	8
Length overall	5·92 m (19 ft 5 in)
Height overall, tail up	2·18 m (7 ft 2 in)

WEIGHTS AND LOADING:

Weight empty	55·5 kg (122 lb)
Normal T-O and landing weight	131 kg (288 lb)
Max T-O and landing weight	159 kg (350 lb)
Max wing loading	11·82 kg/m² (2·42 lb/sq ft)

PERFORMANCE (at max T-O weight):

Never-exceed speed	78 knots (145 km/h; 90 mph)
Max level speed at S/L	60·5 knots (112 km/h; 70 mph)
Max cruising speed at 1,980 m (6,500 ft)	48 knots (88 km/h; 55 mph)
Econ cruising speed at 1,220 m (4,000 ft)	35 knots (64 km/h; 40 mph)
Stalling speed	13 knots (24 km/h; 15 mph)
Touchdown speed	10·5 knots (19·5 km/h; 12 mph)
Max rate of climb at S/L	152 m (500 ft)/min
Service ceiling	2,800 m (9,200 ft)
T-O run	23 m (75 ft)
T-O to 15 m (50 ft)	61 m (200 ft)
Landing from 15 m (50 ft)	30 m (100 ft)
Landing run	9 m (30 ft)
Min field length	91 m (300 ft)
Normal range	191 nm (354 km; 220 miles)
Endurance	4 h
Best glide ratio	15
Min sinking speed (power off)	44 m (144 ft)/min

CASCADE

CASCADE ULTRALITES

1750 12th Street NW, Issaquah, Washington 98027
Telephone: (206) 392 0388
DIRECTOR: Steve Grossruck

CASCADE KASPERWING

Design of the Kasperwing, originally using a Manta Fledge hang glider as basis, was begun by Mr Grossruck in order to flight test the theories of Mr Witold Kasper in regard to what the latter described as 'vortex lift'. Mr Kasper, formerly an aerodynamicist with the Boeing company, believes that it is possible to develop a wing configuration which, at angles of attack of 30° or more, can induce a spanwise vortex that would produce enough lift to bring about drastic reductions in aircraft stalling speed and rate of sink. Mr Kasper himself tested his theories on two tailless gliders in the 1960s, and subsequently on a powered glider, proving the concept and demonstrating near-vertical descents and zero-forward-speed landings.

The current Cascade Kasperwing, now bearing little resemblance visually or structurally to the Fledge on which it was based, is powered by a 15 kW (20 hp) Honda or 17 kW (23 hp) Yamaha 250 cc engine driving a two-blade pusher propeller, and has a fuel capacity of 9·5 litres (2·5 US gallons). It was an award winner at the 1980 Oshkosh Fly-in. An accompanying photograph illustrates the reflexed wingtips and endplate fins and rudders which are among the aircraft's design features.

DIMENSIONS:

Wing area, gross	16·72 m² (180 sq ft)
Propeller diameter	1·27 m (4 ft 2 in)

WEIGHTS AND LOADING:

Weight empty	59 kg (130 lb)
Max pilot weight	100 kg (220 lb)
Max T-O weight	168 kg (370 lb)
Max wing loading	10·00 kg/m² (2·05 lb/sq ft)

PERFORMANCE:

Never-exceed speed	52 knots (96 km/h; 60 mph)
Cruising speed	26-35 knots (48-65 km/h; 30-40 mph)
Stalling speed	13 knots (24·5 km/h; 15 mph)
Max rate of climb at S/L	198 m (650 ft)/min
Endurance	2 h
Best glide ratio at 20-22 knots (37-41 km/h; 23-25 mph)	10

CGS

CGS AVIATION INC

4252 Pearl Road, Cleveland, Ohio 44109

Known previously as Chuck's Glider Supplies (see earlier editions of *Jane's*), this company had two basic sailwing-type hang gliders in production in 1980: the Falcon V and Falcon 5+. It also marketed the Power Hawk, a reduction-drive power package suitable for various types of hang glider. The Power Hawk can be used in Mitchell and Birdman Inc homebuilts, and as a power plant for such types as the UFM Easy Riser and the Volmer VJ-23 and VJ-24.

Details of the Falcon V and Falcon 5+ can be found in the table and in the 1980-81 *Jane's*. No more recent information has been received.

Aircraft Composite Technology Ibis *(Howard Levy)*

American Aerolights Eagle

Birdman TL-1A homebuilt microlight aircraft

Cascade Kasperwing *(J. M. G. Gradidge)*

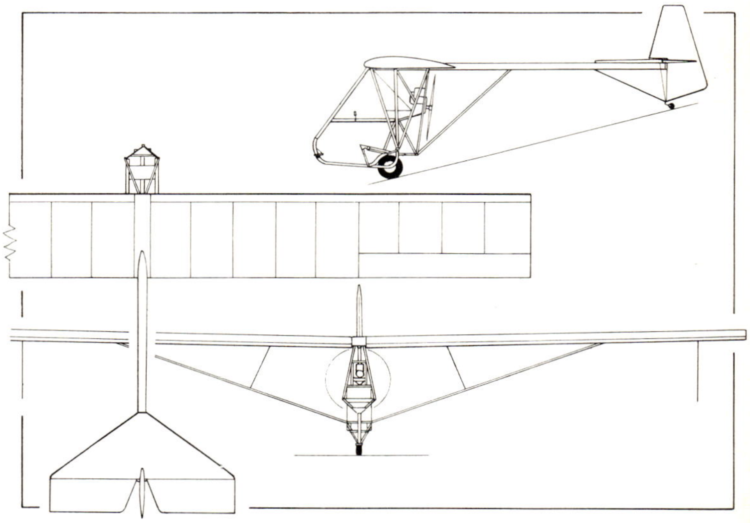

Delta Sailplane DS-26A Nomad
(Michael A. Badrocke)

DECHOW
CURT DECHOW
Bay City, Michigan

DECHOW SOJOURN I

The Sojourn I is a sweptwing rigid monoplane hang glider, with wingtip rudders. It has a 9·75 m (32 ft) wing span, weighs 41 kg (90 lb), and is built largely of 4 mm birch plywood with a corrugated cardboard core. No other details were known at the time of closing for press.

DELTA SAILPLANE
DELTA SAILPLANE CORPORATION
13161 Sherman Way, North Hollywood, California 91605

DELTA DS-26 NOMAD

Marketed as a fully prefabricated kit for homebuilders, the Nomad is a successor to the DSK Flatlander described in the 1979-80 *Jane's*. It is available in three versions: the standard DS-26A with 7·5 kW (10 hp) engine and monowheel landing gear; DS-26A with optional 13·4 kW (18 hp) engine; and DS-26B, in which the 18 hp engine and a tricycle landing gear are standard. The airframe is stressed to *g* limits of +4·67/−1·67 (normal) and +7/−2·5 (ultimate), and meets FAA sailplane Utility category requirements. Deliveries began in mid-March 1980.

Type: Single-seat microlight aircraft.

Airframe: Strut-braced high-wing monoplane. All-aluminium frame, assembled with bolts and pop-rivets. Tubular wing spars and fuselage boom. Auxiliary wing box-spar, of formed sheet metal, to support ailerons. All flying surfaces covered with polyester fabric and doped. Full aerodynamic control in all three axes by ailerons, elevators and rudder. Foldaway seat for pilot. Single main wheel and steerable tailwheel on DS-26A; tricycle gear plus tailwheel on DS-26B. Wings and tail detachable for transportation.

Power Plant: One 7·5 kW (10 hp) Unitrek C10M42 single-cylinder two-stroke engine standard in DS-26A; 13·4 kW (18 hp) Unitrek S18M42 optional for DS-26A, standard for DS-26B.

Dimensions:
Wing span 11·00 m (36 ft 1 in)
Wing area, gross 13·69 m² (147·34 sq ft)
Wing aspect ratio 8·84
Length overall: DS-26A 5·62 m (18 ft 5¼ in)
DS-26B 5·72 m (18 ft 9¼ in)
Height overall: DS-26A 1·73 m (5 ft 8 in)
DS-26B 2·82 m (9 ft 3 in)
Propeller diameter 1·07 m (3 ft 6 in)

Weights and Loadings:
Weight empty: DS-26A (10 hp) 69 kg (152 lb)
DS-26A (18 hp) 72 kg (158 lb)
DS-26B 75 kg (165 lb)
Max T-O weight:
10 hp 164 kg (362 lb)
18 hp 185 kg (408 lb)
Max wing loading:
10 hp 12·01 kg/m² (2·46 lb/sq ft)
18 hp 13·52 kg/m² (2·77 lb/sq ft)

Performance:
Never-exceed speed (all versions)
48 knots (88·5 km/h; 55 mph) IAS
Cruising speed (85% power):
DS-26A (10 hp) 30·5 knots (56 km/h; 35 mph)
DS-26A (18 hp) 42 knots (77 km/h; 48 mph)
DS-26B 39 knots (72 km/h; 45 mph)
Stalling speed:
10 hp 21 knots (39 km/h; 24 mph) IAS
18 hp 22 knots (40·5 km/h; 25 mph) IAS
Min landing speed (all versions)
19·5 knots (35·5 km/h; 22 mph) IAS
Max rate of climb at S/L:
DS-26A (10 hp) 56 m (185 ft)/min
DS-26A (18 hp) 104 m (340 ft)/min
DS-26B 98 m (320 ft)/min
Service ceiling (85% power):
10 hp 1,370 m (4,500 ft)
18 hp 1,980 m (6,500 ft)
Absolute ceiling: 10 hp 1,830 m (6,000 ft)
18 hp 3,660 m (12,000 ft)
T-O run: DS-26A (10 hp) 69 m (225 ft)
DS-26A (18 hp) 50 m (165 ft)
DS-26B 46 m (150 ft)
T-O to 15 m (50 ft): DS-26A (10 hp) 256 m (840 ft)
DS-26A (18 hp) 177 m (580 ft)
DS-26B 183 m (600 ft)
Range at cruising power:
DS-26A (10 hp) 92 nm (170 km; 106 miles)
DS-26A (18 hp) 87 nm (161 km; 100 miles)
DS-26B 81·5 nm (151 km; 94 miles)
Endurance at best L/D power:
DS-26A (10 hp) 2 h 56 min
DS-26A (18 hp) 3 h 40 min
DS-26B 3 h 20 min
Best glide ratio: DS-26A 14·5
DS-26B 14·3
Min sinking speed, power off:
DS-26A (10 hp) 51·2 m (168 ft)/min
DS-26A (18 hp) 55 m (180 ft)/min
DS-26B 56·7 m (186 ft)/min

DELTA DS-27A HONCHO and DS-28A SUPER HONCHO

Generally similar to the Nomad, except for a shorter fuselage and shorter-span wings, the DS-27A has as standard a tricycle landing gear (plus tailwheel) and the 13·4 kW (18 hp) engine of the DS-26B. The DS-28A Super Honcho, introduced in mid-1981, is fitted with a 20·9 kW (28 hp) 338 cc watercooled inline engine, with 12V electrical starting; the landing gear has high-flotation aircraft-type rib tyres and toe-actuated drum brakes.

Dimensions:
Wing span 9·80 m (32 ft 2 in)
Wing area, gross 12·20 m² (131·35 sq ft)
Wing aspect ratio 7·88
Length overall 5·31 m (17 ft 5¼ in)
Height overall: DS-27A 2·79 m (9 ft 2 in)
DS-28A 2·82 m (9 ft 3 in)
Propeller diameter: DS-27A 1·07 m (3 ft 6 in)
DS-28A 1·37 m (4 ft 6 in)

WEIGHTS AND LOADING:
Weight empty: DS-27A 72 kg (159 lb)
DS-28A 99 kg (218 lb)
Max T-O weight (both) 185 kg (407 lb)
Max wing loading (both) 15·14 kg/m² (3·10 lb/sq ft)

PERFORMANCE:
Never-exceed speed:
DS-27A 52 knots (96·5 km/h; 60 mph) IAS
DS-28A 74 knots (137 km/h; 85 mph) IAS
Cruising speed (85% power):
DS-27A 45 knots (84 km/h; 52 mph)
DS-28A 61 knots (113 km/h; 70 mph)
Stalling speed (both versions)
23 knots (42 km/h; 26 mph) IAS
Min landing speed (both versions)
20 knots (37 km/h; 23 mph) IAS
Max rate of climb at S/L:
DS-27A 104 m (340 ft)/min
DS-28A 625 m (2,050 ft)/min
Service ceiling:
DS-27A (85% power) 1,830 m (6,000 ft)
DS-28A (52% power) 3,660 m (12,000 ft)
Absolute ceiling: DS-27A 3,050 m (10,000 ft)
DS-28A 5,335 m (17,500 ft)
T-O run: DS-27A 46 m (150 ft)
DS-28A 15 m (50 ft)
T-O to 15 m (50 ft): DS-27A 183 m (600 ft)
DS-28A 66 m (215 ft)
Range at cruising power:
DS-27A 92 nm (170 km; 106 miles)
DS-28A 70·5 nm (130 km; 81 miles)
Endurance at best L/D power:
DS-27A 3 h 12 min
DS-28A 3 h 30 min
Best glide ratio (both versions) 13·6
Min sinking speed, power off (both versions)
62·2 m (204 ft)/min

DELTA WING

BILL BENNETT'S DELTA WING KITES AND GLIDERS INC

PO Box 483, Van Nuys, California 91408
Telephone: (213) 787 6600 and 785 2474
Telex: 65-1425

Current (1981) models were as follows:

Phoenix 6D. Single-place glider for training and private flying; 853 sold by 1 January 1980. Aluminium alloy (6061-T6) airframe, with single-thickness Dacron sail and vinyl-coated stainless steel cables. Kingpost standard. No deflexers. Features include floating keel pocket and provision for tow launch. Available in four sizes: details of one of these are given in the table; data for the other three were not received. Larger sizes can carry a passenger. All can be flown seated or prone.

Phoenix Lazor. Single-place glider for private and competition flying; 1,200 sold by 1 January 1980. Airframe construction similar to Phoenix 6D. Kingpost standard. Features include floating keel pocket, 'window' panels and provision for tow launch. Available in three sizes: details of two of these are given in the table; data for the other were not received. Largest size can carry a passenger. All can be flown seated or prone. New **Lazor II** series introduced in 1980.

Phoenix Viper. New model introduced in late 1980. Double-surface covering over 57 per cent of sail area; fully-floating, totally-enclosed crossbars; fully-cambered and reflexed metal battens; cantilevered leading-edges, without deflexers. All other known details given in table.

A Soarmaster tricycle landing gear/power plant package, with single or twin Chrysler engines, can be fitted to all current Phoenix models.

EIPPER

EIPPER-FORMANCE INC

1070 Linda Vista Drive, San Marcos, California 92069
Telephone: (714) 744 1514
PRESIDENT: Lyle Byrum
VICE-PRESIDENT: Larry Cook

Eipper's Flexi 2, Flexi 3 and Antares hang gliders were described in the 1980-81 *Jane's*. These are no longer in production.

Activities in 1981 were concentrated upon the highly successful Quicksilver powered hang glider/microlight aircraft, which is currently available in four models.

The Quicksilver was one of four types of microlight aircraft acquired by the US Navy in 1980 for evaluation at the Naval Weapons Center, China Lake, California.

EIPPER-FORMANCE QUICKSILVER

Designed in 1972 by Mr Bob Lovejoy, the Quicksilver has been in production by Eipper-Formance for several years, and some 3,500 had been sold by the beginning of 1981.

The four models available in 1981 were as follows:

Quicksilver. Basic model, to which the detailed description mainly applies. Powered by 11·2 kW (15 hp) 100 cc Yamaha KT100SD two-stroke engine with reduction gear.

Seaquick. Seaplane version, with twin-float landing gear. Choice of 11·2 kW (15 hp) 100 cc Yamaha or 22·4 kW (30 hp) 430 cc Cuyuna two-cylinder engine.

Doublequick. Standard power plant of 22·4 kW (30 hp) 430 cc Cuyuna two-cylinder engine; structural modifications to basic airframe; manually operated trim tab on horizontal stabiliser.

Quicksilver MX. Full multiple-axis aircraft-type flight control system; rigid seat for pilot; 22·4 kW (30 hp) Cuyuna engine and direct drive standard. Wheel/stick combination actuates horizontal stabiliser and air-balanced rudder. Independent differential spoiler system, applied by foot pedals, permits rapid roll rate and steep, slow-airspeed descent.

TYPE: Single-seat microlight aircraft; design *g* limits ±4·0.

AIRFRAME: High-wing monoplane, with rudder and one-piece fixed horizontal stabiliser supported on U-shaped rear twin booms. Wing dihedral 8°, incidence 2° 30′. Kingpost above centre-section. Framework of anodised aluminium alloy tubing. Stainless steel fittings; vinyl-coated stainless steel cables. Dacron-covered wings, tailplane and rudder. Control by weight shift and rudder.

POWER PLANT: One 11·2 kW (15 hp) or 22·4 kW (30 hp) piston engine, as detailed under model listings, driving a two-blade pusher propeller. Fuel capacity 13 litres (3·4 US gallons) standard on all models, of which 11·35 litres (3·0 US gallons) are usable. Provision for two optional 6·4 litre (1·7 US gallon) tanks, increasing total usable capacity to 22·7 litres (6·0 US gallons).

DIMENSIONS:
Wing span 9·75 m (32 ft 0 in)
Wing area, gross 14·86 m² (160·0 sq ft)
Wing aspect ratio 6·4
Length overall:
Quicksilver 5·33 m (17 ft 6 in)
Quicksilver MX 5·51 m (18 ft 1 in)
Height overall 2·95 m (9 ft 8 in)
Wheel track 1·65 m (5 ft 5 in)
Propeller diameter: max 1·32 m (4 ft 4 in)
min 0·91 m (3 ft 0 in)

WEIGHTS AND LOADINGS (A: standard Quicksilver; B: 15 hp Seaquick; C: 30 hp Seaquick; D: Doublequick; E: Quicksilver MX):
Weight empty: A 70 kg (155 lb)
B 86 kg (190 lb)
C 97·5 kg (215 lb)
D 84 kg (185 lb)
E 91 kg (200 lb)
Pilot weight range: A 54·5-91 kg (120-200 lb)
B 54·5-82 kg (120-180 lb)
C 66-91 kg (145-200 lb)
D 66-100 kg (145-220 lb)
E 45·5-91 kg (100-200 lb)
Max wing loading: A 10·25 kg/m² (2·1 lb/sq ft)
B, D 11·23 kg/m² (2·3 lb/sq ft)
C 12·21 kg/m² (2·5 lb/sq ft)
E 11·72 kg/m² (2·4 lb/sq ft)

PERFORMANCE:
Never-exceed speed (all versions)
48 knots (88·5 km/h; 55 mph)
Max level speed: A 36·5 knots (68 km/h; 42 mph)
B 32 knots (60 km/h; 37 mph)
C 39 knots (72 km/h; 45 mph)
D, E 43·5 knots (80 km/h; 50 mph)
Cruising speed (75% power): A
28 knots (51·5 km/h; 32 mph)
B 26 knots (48 km/h; 30 mph)
C 34 knots (63 km/h; 39 mph)
D, E 37·5 knots (69 km/h; 43 mph)
Stalling speed, power off:
A 15 knots (27·5 km/h; 17 mph)
B, D, E 16 knots (29 km/h; 18 mph)
C 16·5 knots (31 km/h; 19 mph)
Max rate of climb at S/L: A 76 m (250 ft)/min
B 61 m (200 ft)/min
C 152 m (500 ft)/min
D 229 m (750 ft)/min
E 183 m (600 ft)/min
Service ceiling: A, C 2,745 m (9,000 ft)
B 1,830 m (6,000 ft)
D, E 3,660 m (12,000 ft)
T-O run (calm wind): A, D 18 m (60 ft)
B 52 m (170 ft)
C 40 m (130 ft)
E 15 m (50 ft)
T-O to 15 m (50 ft): A 189 m (620 ft)
B 265 m (870 ft)
C 133 m (435 ft)
D 85 m (280 ft)
E 98 m (320 ft)
Landing from 15 m (50 ft): A, C 76 m (250 ft)
B 73 m (240 ft)
D 91 m (300 ft)
E 61 m (200 ft)
Landing run (calm wind): A, B, C 15 m (50 ft)
D 18 m (60 ft)
E 23 m (75 ft)
Range (75% power, standard fuel, no reserves):
A 55·5 nm (103 km; 64 miles)
B 42 nm (77 km; 48 miles)
C 44 nm (82 km; 51 miles)
D, E 56·5 nm (104 km; 65 miles)
Range (75% power, max fuel, no reserves):
A 111 nm (206 km; 128 miles)
B 83 nm (154 km; 96 miles)
C 88 nm (164 km; 102 miles)
D, E 113 nm (209 km; 130 miles)
Endurance (75% power, standard fuel, no reserves):
A 2 h 0 min
B 1 h 36 min
C 1 h 18 min
D, E 1 h 30 min
Endurance (75% power, max fuel, no reserves):
A 4 h 0 min
B 3 h 12 min
C 2 h 36 min
D, E 3 h 0 min
Best glide ratio (all versions) 7
Min sinking speed, power off: A, D 91 m (300 ft)/min
B 114 m (375 ft)/min
C 116 m (380 ft)/min
E 94 m (310 ft)/min

EIPPER-FORMANCE QUICKSILVER QM-1 and QM-2

The QM-1 and QM-2 are proposed military versions of the Quicksilver; details can be found in the Addenda.

FISHER

MICHAEL FISHER

Route 2, Box 282, South Webster, Ohio

FISHER FLYER

Completed in July 1980, the Flyer is basically a conventional single-seat light aircraft fuselage and tail unit combined with the biplane wings of a UFM Easy Riser hang glider. Airframe construction is of aluminium tube, with fabric covering, and a non-retractable tricycle landing gear is fitted. Power plant is a 22·4 kW (30 hp) Zenoah engine.

DIMENSIONS:
Wing span 7·92 m (26 ft 0 in)
Length overall 4·27 m (14 ft 0 in)
Height overall 1·83 m (6 ft 0 in)

WEIGHTS:
Weight empty 109 kg (240 lb)
Max T-O weight 190 kg (420 lb)

PERFORMANCE:
Cruising speed 39 knots (72 km/h; 45 mph)
Landing speed 26 knots (48·5 km/h; 30 mph)
T-O run 46-61 m (150-200 ft)
Landing run 46 m (150 ft)

FLIGHT DESIGNS

FLIGHT DESIGNS

PO Box 1503, Salinas, California 93902
Telephone: (408) 758 6896

Markets the **Lancer** and **SL** (Super Lancer) **180, 200** and **220** hang gliders. Airframe of the SL series is of anodised aluminium and stainless steel, with plastics-coated cables and a Dacron sail. Kingpost standard. Features include single deflexers, riveted spar caps and sliding crossbars. For other details see table.

Delta Wing Phoenix 6D

Delta Wing Phoenix Lazor

Eipper-Formance Seaquick

Eipper-Formance Quicksilver *(Howard Levy)*

Eipper-Formance Quicksilver MX

Gemini Hummingbird

Fisher Flyer *(Howard Levy)*

GEMINI

GEMINI INTERNATIONAL INC

75 Bank Street, No. 13, Sparks, Nevada 89431
Telephone: (702) 331 3638
PRESIDENT: Edward C. Sweeney Jr

GEMINI HUMMINGBIRD

The Hummingbird is a development of the Sky Sports (Klaus Hill) Humbug, the original prototype of which was acquired by Gemini. The standard production Hummingbird is illustrated in an accompanying photograph. Purchasers of this aircraft must be qualified pilots. A simpler, lighter and less expensive **Hummingbird Sport** is available for trainee pilots; there is also a 22·4 kW (30 hp) **Hummingbird GT** utility model with improved load-carrying and performance.

TYPE: Single-seat foot-launchable microlight aircraft.

AIRFRAME: Main frame of 6061-T6 aluminium tube. Wings have KH2 thick aerofoil section, two Manta-built spars, ribs, and double-surface Dacron covering. Kingpost standard. V tail unit (inverted-V optional), with double-surface Dacron covering. Three-axis control by ailerons and 'ruddervators'. Tricycle landing gear standard. Options include oversize wheels and twin floats.

POWER PLANT: Two 5·2 kW (7 hp) Gemini/A. B. Partner K-1200 piston engines, each with reduction drive to a two-blade propeller. Standard fuel tank capacity 5·7 litres (1·5 US gallons); larger tank optional.

DIMENSIONS:
Wing span 10·36 m (34 ft 0 in)
Wing area, gross 14·59 m² (157·0 sq ft)
Wing aspect ratio 7·36
Propeller diameter 1·07 m (3 ft 6 in)

WEIGHTS AND LOADING:
Weight empty 74 kg (163 lb)
Max pilot weight 113 kg (250 lb)
Max T-O weight 188 kg (415 lb)
Max wing loading 12·88 kg/m² (2·64 lb/sq ft)

PERFORMANCE:
Never-exceed speed 47·5 knots (88·5 km/h; 55 mph)
Max level speed 43·5 knots (80·5 km/h; 50 mph)
Cruising speed 30·5 knots (56·5 km/h; 35 mph)
Stalling speed:
power on 13 knots (24·5 km/h; 15 mph)
power off 17·5 knots (32·5 km/h; 20 mph)
Typical rate of climb at S/L 91 m (300 ft)/min
T-O run 61 m (200 ft)
Landing run 23 m (75 ft)
Best glide ratio at 23 knots (42 km/h; 26 mph) 11
Endurance 1 h

GLA

GLA INC

841 Winslow Court, Muskegon, Michigan 49441
Telephone: (616) 780 4680

GLA MINIBAT

GLA produces the Minibat, an extremely lightweight sailplane for amateur construction. A complete constructional manual is available, the cost of which is refunded upon purchase of a Minibat kit. By early 1981 GLA had sold 32 kits, and three Minibats had been flown. Assembly time is 40-60 man-hours. Optional extras include a 2·2 kW (3 hp) sustainer engine, wing extension panels, and de luxe instrumentation. The engine is restartable in the air, but the Minibat must be auto, bungee or winch launched to an altitude of at least 30·5 m (100 ft) before relying upon the engine to sustain flight.

TYPE: Single-seat extremely lightweight glider, with optional power plant; *g* limits ±6·0, with safety factors of 2·0 (standard wings) and 1·5 (extended wings).

AIRFRAME: Cantilever high-wing monoplane, with forward-swept wings and single fin and rudder. Three-axis control by ailerons, flaps and rudder; there are no horizontal tail surfaces. Wings have unidirectional glassfibre spar caps, moulded glassfibre/foam shear webs, and moulded upper and lower skins of woven epoxy glassfibre/foam sandwich. Fuselage, of elliptical cross-section, is of similar construction. Non-retractable semi-recessed unsprung monowheel (3·5 × 4 in), nose-skid and tailwheel. Semi-reclining seat for pilot, under fully-transparent canopy. Entire airframe comprises only 18 major components; it can be dismantled for transportation and stowage, and reassembled in 5 min.

POWER PLANT (optional): One 2·2 kW (3 hp) two-stroke chain-saw engine, driving a pusher propeller turning between fuselage sternpost and rudder. Engine is for boosting flying speed only, and does not develop enough power for self-launched take-off (aircraft is launched by bungee, tow or winch). Fuel capacity 2 litres (0·4 US gallons).

DIMENSIONS:
Wing span: standard 7·62 m (25 ft 0 in)
extended 10·00 m (32 ft 10 in)
Wing area, gross: standard 6·04 m² (65 sq ft)
extended 7·15 m² (77 sq ft)
Wing aspect ratio: standard 9·6
extended 14·0
Length overall 2·84 m (9 ft 4 in)
Height overall 1·52 m (5 ft 0 in)
Propeller diameter 0·51 m (1 ft 8 in)

WEIGHTS AND LOADINGS (A: standard wings; B: extended-span wings):
Weight empty: A 47·5 kg (105 lb)
B 57 kg (125 lb)
Max T-O weight: A 147 kg (325 lb)
B 159 kg (350 lb)
Max wing loading: A 24·4 kg/m² (5·0 lb/sq ft)
B 22·0 kg/m² (4·5 lb/sq ft)

PERFORMANCE:
Max level speed (A, B, smooth air) 109 knots (203 km/h; 126 mph)

GLA Minibat with standard wings and (right: *Don Dwiggins photo*) **with wingtip extension panels**

Manoeuvring speed (A, B) 76·5 knots (142 km/h; 88 mph)
Stalling speed (A, B) 34 knots (63 km/h; 39 mph)
Max rate of climb at S/L (A, B, still air) 30·5 m (100 ft)/min
Best glide ratio at 48 knots (88 km/h; 55 mph):
A 23
B 30
Min sinking speed at 37·5 knots (69 km/h; 43 mph), power off: A 55 m (180 ft)/min
B 42 m (138 ft)/min
Speed for sink rate of 120 m (390 ft)/min:
A 68 knots (126 km/h; 78 mph)
B 74 knots (137 km/h; 85 mph)

GOLDWING

GOLDWING LTD

PO Box 1123, Amador County Airport, Building No. 3, Jackson, California 95642
Telephone: (209) 223 0384
President: Brian Glenn
Designer: Craig Catto

Many hundreds of kits of Mr Catto's earlier CA 14 and CA 15 (1980-81 *Jane's*) were sold. Goldwing Ltd was formed in 1980, initially to market the Goldwing microlight aircraft, also designed by Mr Catto.

GOLDWING GOLDWING

The Goldwing was designed in 1978 and flew for the first time in January 1979. Five complete prototypes, six different wing sections, various planforms and wing loadings, and many detail modifications were tested before the aircraft became available in production kit form in 1980. Assembly time is 100 hours or less; the Goldwing can be dismantled for transportation, and can be reassembled for flight in 10 min. The Goldwing is also available as a factory-built aircraft.

Type: Single-seat microlight aircraft; *g* limits +6·0/−3·0.

Airframe: Canard monoplane configuration, with swept-back main wings and non-swept foreplane. Construction is primarily of wood, plywood, foam plastics and glassfibre. Wings have ailerons and interconnected outboard overwing spoilers, and a NASA-type winglet, with rudder, at each tip. Foreplane is fitted with an elevator. Wedge-shaped fuselage nacelle. Non-retractable tricycle landing gear, with shock-absorption. Ski and float gear optional. Open cockpit, with semi-reclining seat and windscreen.

Power Plant: Kits for US customers are supplied with 'best available' snowmobile engine, which may vary from 9·7 to 22·4 kW (13 to 30 hp). A 13·4 kW (18 hp) Fuji Robin 250 cc engine is fitted in European models. Two-blade wooden pusher propeller (direct or reduction drive). Standard fuel tank capacity 7·5 litres (2 US gallons). Long-range tank (22·7 litres; 5 Imp gallons; 6 US gallons) optional.

Dimensions:
Wing span 9·14 m (30 ft 0 in)
Wing and foreplane area (total) 11·9 m² (128·0 sq ft)
Length overall 3·66 m (12 ft 0 in)

Weights:
Weight empty 84 kg (185 lb)
Max T-O weight approx 197 kg (435 lb)

Performance (typical):
Max level speed 46 knots (137 km/h; 85 mph)
Cruising speed over 52 knots (97 km/h; 60 mph)
Stalling speed 21 knots (39 km/h; 24 mph)
Max rate of climb at S/L 183 m (600 ft)/min
T-O run 46 m (150 ft)
Landing run 23 m (75 ft)
Best glide ratio at 37 knots (69 km/h; 43 mph) 16
Min sinking speed at 33 knots (61 km/h; 38 mph) 76 m (250 ft)/min

GOLDWING CA 15

Details of the CA 15 microlight aircraft appeared in the 1980-81 *Jane's*. It is believed that production has now ended, though confirmation from the company was not received.

GOLDWING CP 16

This new Catto design, illustrated in an accompanying photograph, is in production, including licence manufacture in the UK by Euro Wing. No details of dimensions or weights were received for publication.

Power Plant: One 13·4 kW (18 hp) 250 cc engine.

Performance:
Cruising speed 24 knots (72 km/h; 45 mph)
Stalling speed 10 knots (29 km/h; 18 mph)
Max rate of climb at S/L 122 m (400 ft)/min
T-O run 46 m (150 ft)
Range 108 nm (322 km; 200 miles)

HALL

LARRY HALL

Mr Hall is the designer of a quadruplane hang glider, all known details of which follow:

Dimensions:
Wing span: upper 6·35 m (20 ft 10 in)
centre upper 6·05 m (19 ft 10¼ in)
centre lower 5·74 m (18 ft 10 in)
lower 5·44 m (17 ft 10 in)
Wing dihedral (each) 6°
Wing stagger 45°
Wing area (total) 17·20 m² (185·1 sq ft)
Rudder area (each) 0·22 m² (2·37 sq ft)
Length overall 3·00 m (9 ft 10 in)

Weight:
Weight empty 29·5 kg (65 lb)

HIGHSTER

HIGHSTER AIRCRAFT INC

Produced the **Highster** intermediate-performance hang glider in 13·94, 15·79 and 17·65 m² (150, 170 and 190 sq ft) sizes and the **Goshawk** microlight aircraft. No other details known.

HUMMER

THE HUMMER

Box 1655, Salt Lake City, Utah 84110
Telephone: (801) 451 5622
Principal Manufacturer: Maxair Sports Inc, RD2, Glen Rock, Pennsylvania 17327

THE HUMMER

The Hummer was designed by the late Mr Klaus Hill, using hang glider technology. Design began in August 1977; the first prototype flew in November 1977, and the first production example on 17 March 1980.

Earlier, non-powered, aircraft designed and built by Mr Hill included two sailplanes, a Rogallo-type and Voyager hang gliders, and the Fledge hang glider currently in production by Manta Products Inc (which see).

Hummer kits and plans are being handled by several distributors, with the Maxair Sports Inc division of Franklin Manufacturing Corporation as the main supplier and manufacturer of component parts. By early 1981 at least 65 Hummers (nine prototypes and 56 production aircraft) had been completed. Two models (A and B) are available, depending on pilot weight.

Type: Single-seat microlight aircraft.

Airframe: Wire-braced high-wing monoplane. Dihedral 45·7 cm (18 in) from horizontal at tips. Incidence 5°. Leading- and trailing-edge spars of 6061-T6 aluminium tube. Twelve ribs of aluminium tubing. Structure bolted together. Dacron fabric covering sewn into shape and slipped over wing structure, avoiding use of dope, glue or rib stitching. No ailerons or spoilers. Wings fold back and attach to tail for transportation. Assembly and disassembly takes one man 15-20 min. Fuselage main boom comprises a 12·7 cm (5 in) diameter aluminium tube, to which wing and engine supporting truss structures are attached. Strut-braced V tail, with tailplane and elevators/rudders set at dihedral angle of about 33°. Construction similar to wings. Control by cables. Non-retractable tailwheel landing gear, with shock-absorption on main units; steerable tailwheel. Formed sheet metal main legs with two steel cross-tubes. Go-kart wheels. No brakes. Float and ski gear available optionally. Single seat in open position; pilot enclosure optional. Prototype of two-seat version has been flown.

Power Plant: One 14·9 kW (20 hp) 250 cc Zenoah G-25A two-stroke engine standard, with direct drive to a two-blade laminated wood pusher propeller. Various engines of up to 48·5 kW (65 hp) have been adapted. Belt-driven reduction gear and larger propeller optional. Plastics fuel tank aft of seat, capacity 19 litres (5 US gallons).

Dimensions:
Wing span 10·06 m (33 ft 0 in)
Wing area, gross 11·89 m² (128 sq ft)
Length overall 5·79 m (19 ft 0 in)
Height overall 2·51 m (8 ft 3 in)
Propeller diameter: standard 0·84 m (2 ft 9 in)
optional 1·37 m (4 ft 6 in)

Weights and Loading:
Weight empty 77 kg (170 lb)
Pilot weight: Hummer A up to 79·5 kg (175 lb)
Hummer B 79·5-100 kg (175-220 lb)
Max T-O weight 177 kg (390 lb)
Max wing loading 14·6 kg/m² (3·0 lb/sq ft)

Performance:
Max level speed 39-43·5 knots (72·5-80 km/h; 45-50 mph)
Cruising speed 26 knots (48 km/h; 30 mph)
T-O and landing speed 22-24·5 knots (40·5-45 km/h; 25-28 mph)
Stalling speed 19·5-21 knots (35·5-39 km/h; 22-24 mph)

Goldwing Goldwing

Goldwing CP 16 microlight aircraft

Hummer microlight aircraft

Jaeger J-Bird *(J. M. G. Gradidge)*

Hall Quadruplane hang glider

Kolb Flyer twin-engined microlight aircraft

Rate of climb at S/L	at least 122 m (400 ft)/min
Service ceiling	3,050 m (10,000 ft)
T-O run	91-122 m (300-400 ft)
Landing distance	18-46 m (60-150 ft)
Range with max fuel	173 nm (322 km; 200 miles)
Best glide ratio	10
Min sinking speed, power off	68·5 m (225 ft)/min

JAEGER
JIM JAEGER

The Jaeger **J-Bird** (see accompanying photograph) is essentially a UFM Easy Riser to which the builder has added a high-mounted horizontal tail surface. For other details see Addenda.

KICENIUK
TARAS KICENIUK JR

976 La Vuelta, Santa Paula, California 93060

No news of Mr Kiceniuk's activities, if any, has been received since early 1979. Details of his Icarus II and Icarus V hang gliders can be found in the 1980-81 *Jane's*.

KITTY HAWK
KITTY HAWK KITES INC

PO Box 340, Nags Head, North Carolina 27959
Telephone: (919) 441 6247 and 441 7575

In late 1980 this company acquired all physical assets of the former Seagull Aircraft Inc of Carpinteria, California. Its intention was to transfer the manufacture of Seagull hang glider components to a new facility in North Carolina.

Details of the most recent Seagull designs can be found in the 1980-81 *Jane's*. At the beginning of 1980 Seagull had completed prototypes of two new models, the Sierra and Sailwing. Kitty Hawk is also a distributor for the Eipper-Formance Quicksilver (which see).

KOLB
KOLB COMPANY INC

RD 3, Box 38, Phoenixville, Pennsylvania 19460
Telephone: (215) 948 3264
PRESIDENT: Homer Kolb

KOLB FLYER

First flown in 1970, this twin-engined microlight aircraft was then shelved for nearly ten years before being restored for the 1980 EAA Fly-in at Oshkosh, Wisconsin. The original twin McCulloch MC-101 engines have since been replaced by a pair of chain-saw engines; a new Flyer has also been completed, with single struts (instead of V struts), different wheels, more comfortable seating, and slightly more compact dimensions. Plans and kits of this version were due to become available during 1981

TYPE: Single-seat microlight aircraft.

AIRFRAME: High-wing monoplane, of 4130 steel and 6061-T6 aluminium tube construction with Dacron-covered wing and tail surfaces. Three-axis control (ailerons, elevators and rudder). Kolb original wing section. Non-retractable main wheels and tailskid.

POWER PLANT: Two 4·5 kW (6 hp) Chrysler 820 engines, each with direct drive to a two-blade pusher propeller. Other engines available optionally. Fuel tank capacity 7·5 litres (2 US gallons).

DIMENSION:

Wing area, gross	14·86 m² (160·0 sq ft)

WEIGHT AND LOADING:

Weight empty	70·5 kg (155 lb)
Max wing loading	9·76 kg/m² (2·0 lb/sq ft)

PERFORMANCE:

Landing run	15 m (50 ft)

KUBASEK
JOHN W. KUBASEK

5460 Antoinette Street, Sarasota, Florida 33580

KUBASEK CLOUDBUSTER

Mr Kubasek's Cloudbuster microlight aircraft is built of aluminium tube, with foam plastics covering for the wing and tail surfaces. It has 10·97 m (36 ft 0 in) span wings, and is powered by a 250 cc Yamaha engine. No other details were known at the time of closing for press.

LAFAYETTE
LAFAYETTE AVIATION & AIRCRAFT

PO Box 10139, Orlando, Florida 32809
Telephone: (305) 859 5551

LAFAYETTE HI-NUSKI

Very similar in appearance to the Eipper Quicksilver (which see), the Hi-Nuski microlight aircraft has an airframe of 6061-T6 aluminium tube, Dacron wing and tail surface covering, and stainless steel cables. The tricycle landing gear has coil-spring shock-absorption. Control is by weight shift, ailerons and rudder. Power plant comprises a single or twin engine pack, using 7·5 kW (10 hp) Chrysler Power Bee engine(s) driving wooden fixed-pitch propeller(s); or a 22·4 kW (30 hp) Cuyuna engine. Fuel capacity is 6·4 litres (1·7 US gallons) or 11·7 litres (3·1 US gallons). Flight is not recommended in winds of more than 17·5 knots (32 km/h; 20 mph).

DIMENSIONS:
- Wing span — 10·21 m (33 ft 6 in)
- Wing area, gross — 15·2 m² (164·0 sq ft)
- Wing aspect ratio — 6·84
- Propeller diameter (Cuyuna engine) — 0·91 m (3 ft 0 in)

WEIGHTS (A: single 10 hp engine; B: twin 10 hp engines; C: 30 hp engine):
- Weight empty: A — 62 kg (137 lb)
- B — 70 kg (154 lb)
- C — 86 kg (190 lb)
- Max T-O weight: A — 150 kg (330 lb)
- B — 191 kg (420 lb)
- C — 207 kg (458 lb)

PERFORMANCE:
- Never-exceed speed: A — 39 knots (72 km/h; 45 mph)
- B — 43·5 knots (80 km/h; 50 mph)
- C — 47 knots (87 km/h; 54 mph)
- Cruising speed: A — 30·5 knots (56 km/h; 35 mph)
- B — 35 knots (64 km/h; 40 mph)
- C — 39 knots (72 km/h; 45 mph)
- Stalling speed: A, B — 15 knots (27·5 km/h; 17 mph)
- C — 17 knots (32 km/h; 20 mph)
- Max rate of climb at S/L: A — 61 m (200 ft)/min
- B — 122 m (400 ft)/min
- C — 244 m (800 ft)/min
- Max power ceiling: A, B — 2,440 m (8,000 ft)
- T-O run: A, B — 12-33·5 m (40-110 ft)
- Min landing distance: A, B — 15 m (50 ft)
- Range: A — 39 nm (72 km; 45 miles)
- B — 56 nm (105 km; 65 miles)
- Best glide ratio: A, B — 7
- Min sinking speed, power off:
- A — 67 m (220 ft)/min
- B — 76 m (250 ft)/min

LEAF
LEADING EDGE AIR FOILS INC

331 South 14th Street, Colorado Springs, Colorado 80904

In late 1980 this company was offering a series of kits for converting standard Rogallo-type hang gliders to improved configuration, with higher aspect ratio sails, raised keel pockets, bridles and tip limiters. Details of six LEAF models are given in the table.

MANTA
MANTA PRODUCTS INC

1647 East 14th Street, Oakland, California 94606
Telephone: (415) 536 1500

MANTA FLEDGE 2B

This rigid sweptwing tail-less monoplane hang glider can be flown from either a seated or prone position, and is suitable for training, private or competition flying. Sales totalled 1,600 by February 1981.

The Fledge 1 and 2A, described in the 1980-81 *Jane's*, are no longer in production. Manufacture was continuing in 1981 of the Fledge 2B, to which the following description applies. Manta expected to release details of a Trike version in April 1981, and of a new Fledge 3 model in the following month. Another powered version, known as the Fledgling, is produced by Pterodactyl Ltd (which see).

TYPE: One-man Fledge class rigid-wing hang glider; *g* limits ±7·5.

AIRFRAME: Sweptback flying-wing monoplane, with 18° sweep on leading-edges, 2° washout, and dihedral which raises wingtips 46 cm (18 in) above keel. Each half-wing has two aluminium alloy spars, and compression struts; wings are linked by vinyl-coated stainless steel sweep/flying wires. Structure includes 14 pre-formed aluminium ribs and tip ribs. Kingpost standard. Swept and canted drag rudders above 'booster' wingtips which facilitate turning, take-off and landing. All surfaces covered with double-thickness Dacron. Provision for power plant and landing gear.

DIMENSIONS:
- Wing span (included) — 10·06 m (33 ft 0 in)
- Wing leading-edge length — 5·03 m (16 ft 6 in)
- Wing chord at root (keel length) — 1·68 m (5 ft 6 in)
- Wing chord at tip — 1·37 m (4 ft 6 in)
- Wing area, gross — 15·6 m² (168·0 sq ft)
- Wing aspect ratio — 6·8
- Nose angle — 144°
- Billow angle — not applicable

WEIGHTS:
- Weight empty — 28·5 kg (63 lb)
- Pilot weight — 72·5-100 kg (160-220 lb)

PERFORMANCE:
- Best glide ratio — 12
- Min sinking speed — 61 m (200 ft)/min
- Max speed — 48 knots (88 km/h; 55 mph)
- Cruising speed — 16-22 knots (29-40 km/h; 18-25 mph)
- Stalling speed — 12·5 knots (23 km/h; 14 mph)

MARSKE
MARSKE AIRCRAFT CORPORATION

130 Crestwood Drive, Michigan City, Indiana 46360
Telephone: (219) 879 7039
PRESIDENT: James J. Marske

MARSKE MONARCH C

The Monarch was designed and built by Mr Jim Marske, and made its first flight on 4 July 1974. Plans and kits are available to amateur constructors. A total of 71 Monarchs were under construction by the beginning of 1979, of which four were then known to have been completed. No later information has been received.

Mr Marske has successfully test-flown the Monarch prototype with a 9 kW (12 hp) McCulloch engine installed behind the pilot's seat, driving a 0·635 m (2 ft 1 in) diameter pusher propeller. It is considered, however, that an engine of approx 15 kW (20 hp) should be fitted to homebuilt aircraft.

The earlier Monarch B was described in the 1977-78 *Jane's*; the following description applies to the Monarch C version which replaced it:

TYPE: Single-seat ultralight homebuilt glider; *g* limits +8·0/−4·0 (ultimate).

AIRFRAME: Braced high-wing monoplane, with single steel strut each side. Wing section NACA 43012R or 43112. Dihedral 2°. Sweepforward 3° at quarter-chord. Moulded glassfibre D leading-edge, with glassfibre front spar web and booms, wood and foam plastics ribs, wooden rear spar and trailing-edge, and fabric covering. Ailerons (outboard) and elevators (inboard) have single Sitka spruce spars with foam plastics ribs. Fixed tab on each elevator. All control surfaces Dacron-covered. Aluminium spoiler above each wing. Simple minimal beam-type fuselage of laminated glassfibre, moulded in two halves and joined at centreline. Forward section supports pilot's seat, with nose fairing over instrument panel; rear section forms integral fin leading-edge. CG tow hook on each side of nose fairing. Fin and rudder, above and below level of wings. Glassfibre leading-edge (fin), wooden trailing-edge (rudder), foam ribs and fabric covering. No horizontal tail surfaces. Reinforced underfuselage landing skid. Single landing wheel, below pilot's open seat. Conventional floor-mounted control column.

DIMENSIONS:
- Wing span — 12·80 m (42 ft 0 in)
- Wing area, gross — 17·19 m² (185·0 sq ft)
- Wing aspect ratio — 9·5
- Length overall — 3·51 m (11 ft 6 in)
- Height over tail — 2·39 m (7 ft 10 in)

WEIGHTS AND LOADING:
- Weight empty — 100 kg (220 lb)
- Max T-O weight — 204 kg (450 lb)
- Max wing loading — 11·72 kg/m² (2·40 lb/sq ft)

PERFORMANCE:
- Best glide ratio at 35 knots (64 km/h; 40 mph) — 19
- Min sinking speed at 26 knots (48·5 km/h; 30 mph) — 49 m (162 ft)/min
- Stalling speed — 21 knots (39 km/h; 24 mph)
- Max speed (smooth air) — 61 knots (113 km/h; 70 mph)
- Max speed (rough air), max aero-tow and max winch-launching speed — 43·5 knots (80 km/h; 50 mph)

MITCHELL
MITCHELL AIRCRAFT CORPORATION

1900 South Newcomb, Porterville, California 93257
Telephone: (209) 781 8100
PRESIDENT: James M. Meade
GENERAL MANAGER: Richard T. Avalon

Mitchell Aircraft Corporation, which produces the Mitchell Wing hang glider, is also marketing plans and kits for the Mitchell Wing Model B-10, Mitchell Wing U-2, and P-38 Lightning, all of these aircraft being in the microlight category. Both B-10 versions utilise the basic hang glider wing, whereas those of the U-2 and P-38 are of different design.

MITCHELL WING MODEL B-10

Design of this aircraft originated in 1975, construction of the prototype following very soon after, and it flew for the first time in 1976, powered by a 9 kW (12 hp) McCulloch MC-101 engine. Two powered versions are available:

Model B-10F. Foot-launch powered version, with a tubular frame structure that serves as fuselage, engine mounting, and open cockpit for the pilot. Two main wheels only, at aft end of fuselage frame.

Model B-10. Powered version with tricycle-type landing gear that allows conventional powered take-offs and landings.

Both powered versions, as well as the hang glider, have standard attachments to unite wing and fuselage, the wing being common to all three.

A B-10 was one of four types of microlight aircraft acquired in 1980 by the US Navy for evaluation at the Naval Weapons Center, China Lake, California.

In 1981 Mitchell introduced what it calls a 'podule' (a cross between a hanging pod and a cockpit module), designed specifically for the B-10. Apart from its obvious advantage of protecting the pilot in adverse flying conditions, use of the podule is claimed to provide a 15 per cent improvement in fuel economy, cruising and high speeds, and glide ratio. The podule kit, which weighs only 6·8 kg (15 lb), includes pre-moulded glassfibre fairing, windscreen, matching wheel fairings, and nosewheel brake assembly. Application to other tricycle-gear microlights is being considered.

By early 1981 approximately 1,000 kits and plan sets for B-10/10F versions had been sold.

TYPE: Single-seat microlight monoplane; *g* limits ±6·0.

AIRFRAME: Tapered parasol monoplane. Wing section NACA 23015. Dihedral 6° on outer wing panels. Sweepback at quarter-chord 12°. Structure is of wood and fabric, with D section leading-edge, its form retained by foam ribs, and attached to a wooden web spar. Trailing-edge ribs are of built-up truss type, those in the centre-section being of heavier material and carrying the rear spar for accessory attachment points. Dacron covering, with a small section of the wing undersurface, immediately above the cockpit, left uncovered to simplify access to seat. Mitchell patented 'stabilators' occupy the entire trailing-edge of each outer wing panel and function differentially as ailerons and collectively as elevators. Rudders, mounted above each wingtip, are operated conventionally for yaw control; they can also be actuated differentially as airbrakes for increased descent rates. Outer wing panels fold over centre-section for transportation. Braced fuselage framework of aluminium tube, serving also as support structure for wing, landing gear and power plant. B-10F has two main wheels only; B-10 has non-retractable tricycle configuration with steerable nosewheel. All wheels of nylon, with tubed tyres size 4·10 × 3·50-4. No brakes.

Marske Monarch B single-seat ultralight glider

Mitchell 'podule' fitted to a B-10

Manta Fledge 2B in powered form *(Peter R. March)*

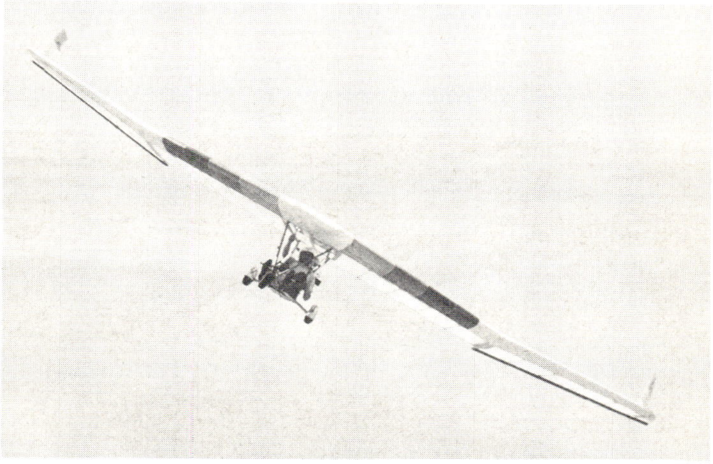

Mitchell Wing Model B-10 microlight aircraft

Lafayette Hi-Nuski microlight aircraft *(Howard Levy)*

Mitchell Wing U-2, with optional main-gear leg/wheel fairings

Mitchell P-38 Lightning microlight aircraft

Pilot seated within fuselage framework on reinforced canvas sling seat, with seat belt or full shoulder-harness, or in optional 'podule'.

POWER PLANT: One 15 kW (20 hp) Honda 250 cc two-stroke engine, driving a two-blade fixed-pitch pusher propeller via a 2·25:1 reduction gear. Fuel tank, which can be mounted within the wing or on the fuselage framework, has a capacity of 5·7 litres (1·5 US gallons).

DIMENSIONS:

Wing span	10·36 m (34 ft 0 in)
Wing area, gross	12·63 m² (136 sq ft)
Wing aspect ratio	·8
Width, wings folded	5·49 m (18 ft 0 in)
Height overall	1·22 m (4 ft 0 in)
Propeller diameter	1·22 m (4 ft 0 in)

WEIGHTS:

Weight empty	86 kg (190 lb)
Max T-O weight	195 kg (430 lb)

PERFORMANCE (B-10 at max T-O weight):

Max level and max cruising speed at S/L	48 knots (89 km/h; 55 mph)
Econ cruising speed at S/L	31 knots (58 km/h; 36 mph)
Stalling speed	17·5 knots (32·5 km/h; 20 mph)
Max rate of climb at S/L	122 m (400 ft)/min
Service ceiling	3,660 m (12,000 ft)
T-O run	61 m (200 ft)
T-O to 15 m (50 ft)	91 m (300 ft)
Landing from 15 m (50 ft)	61 m (200 ft)
Landing run	30 m (100 ft)
Range, with max fuel and payload	35 nm (64 km; 40 miles)
Best glide ratio	17

MITCHELL WING U-2

With the aim of developing a microlight powered aircraft superior in controllability, performance, and structure to powered hang gliders. Mr Don Mitchell designed the U-2, based on the Mitchell Wing. Initiation of the design, and construction and first flight of the prototype, were accomplished during 1979. By early 1981, orders for plans and/or kits totalled 400.

TYPE: Single-seat microlight monoplane; *g* limits ±10·0.

AIRFRAME: Wings of production kits are generally similar to those of the B-10, except for a modified Wortmann section of 19% thickness/chord ratio, and combined fin and rudder assemblies at the wingtips, the aft 65% of each being movable. Fuselage is a chrome-molybdenum steel tube structure within the wing centre-section. Two hardwood runners extend the full length of this structure, beneath the pilot's seat, for protection in a wheels-up landing. Manually-retractable tricycle landing gear, all units retracting aft. Optional non-retractable gear. Steerable nosewheel. Nylon wheels with tubed tyres size 4·10 × 3·50-4. Friction brake operating on surface of nosewheel tyre. Reinforced canvas sling seat, beneath one-piece bubble canopy secured by fore and aft locking pins, and easily removed for entry and exit. Mylar panels in cockpit floor to provide downward view. Cockpit ventilated.

POWER PLANT: As described for B-10. Plastics fuel tank mounted within wing has capacity of 5·7 litres (1·5 US gallons).

DIMENSIONS: As for B-10 except:

Length overall	2·84 m (9 ft 4 in)
Height overall	0·91 m (3 ft 0 in)

WEIGHTS:

Weight empty	113 kg (250 lb)
Max T-O weight	218 kg (480 lb)

PERFORMANCE (at max T-O weight):

Max level speed at S/L	82 knots (153 km/h; 95 mph)
Max cruising speed at S/L	65 knots (121 km/h; 75 mph)
Econ cruising speed at S/L	56 knots (105 km/h; 65 mph)
Stalling speed	28 knots (51·5 km/h; 32 mph)
Max rate of climb at S/L	152 m (500 ft)/min
Service ceiling	3,660 m (12,000 ft)
T-O run	61 m (200 ft)
T-O to 15 m (50 ft)	84 m (275 ft)
Landing from 15 m (50 ft)	76 m (250 ft)
Landing run	38 m (125 ft)
Range, with max fuel and max payload	39 nm (72 km; 45 miles)
Best glide ratio	27

MITCHELL P-38 LIGHTNING

Mr Mitchell's objective in designing the P-38 was to produce a conventionally-controlled microlight aircraft, with reasonable performance, which could be assembled by four people in one day. It flew for the first time in late 1980 and became available in plan/kit form in March 1981. Assembly time is quoted as 38 man-hours (from factory-supplied kit), or 110-150 man-hours if built from plans.

TYPE: Single-seat microlight aircraft; *g* limits ±6·0.

AIRFRAME: Braced monoplane, with twin tailbooms and twin fins and rudders. Main structure is of aluminium, with quick-fit ribs and Dacron-covered wing and tail surfaces. Full-span ailerons, set below and behind wings. One-piece elevator aft of tailbooms. Wings can be folded back alongside fuselage for transportation and stowage. Non-retractable tricycle landing gear, with glassfibre rod main-wheel legs, 10 in wheels, and nose-wheel drum braking. Open seat for pilot. Options include 20 in wheels for rough-field operation, main-wheel fairings, ski or float gear, cockpit enclosure, flight instrumentation, and night lights.

POWER PLANT: One 13·4-15 kW (18-20 hp) 250 cc Honda or Zenoah two-stroke engine, driving a two-blade wooden pusher propeller via a 2·25:1 reduction gear. Fuel tank capacity 5·7 litres (1·5 US gallons). Options include electric engine starting and 15 litre (4 US gallon) auxiliary fuel tank.

DIMENSIONS:

Wing span	7·92 m (26 ft 0 in)
Wing area, gross	10·22 m² (110 sq ft)
Length overall	5·18 m (17 ft 0 in)
Height overall	1·52 m (5 ft 0 in)

WEIGHTS:

Weight empty	91 kg (200 lb)
Max T-O weight	191 kg (420 lb)

PERFORMANCE (approx):

Max level speed	more than 52 knots (97 km/h; 60 mph)
Max cruising speed	48 knots (88 km/h; 55 mph)
Approach speed	30·5 knots (56 km/h; 35 mph)
T-O speed	28 knots (51 km/h; 32 mph)
Landing speed	less than 26 knots (48 km/h; 30 mph)
Stalling speed	24·5 knots (45 km/h; 28 mph)
Max rate of climb at S/L	76-91 m (250-300 ft)/min
T-O run	less than 61 m (200 ft)
Landing run	less than 30 m (100 ft)

MONARCH
MONARCH PRODUCTS INC

Route 2, Box 376-D, Pine, Colorado 80470
Telephone: (303) 838 0267

MONARCH II

This new flexible-wing hang glider received USHGMA certification in early 1981, after six months of development and testing. It has a Dacron sail (double-surfaced over one-third of its area) of unusual tightness, the camber being fixed by pre-formed 6061-T6 anodised aluminium and graphite ribs. Other characteristics include an extra-stiff leading-edge, shifting crossbar system, kingpost, double reflex bridles, no deflexers, and floating tips. All other known details are given in the table.

ODYSSEY
ODYSSEY INC

PO Box 299, Amherst, Massachusetts 01002

This company was licenced in the Spring of 1981 by Pacific Kites to build a new high-performance hang glider known as the **Vampyr**. No details of this aircraft were known at the time of closing for press.

PACIFIC GULL
PACIFIC GULL

150 H, Los Obreros Lane, San Clemente, California 92672

For details of 1979 Alpine series see the 1980-81 *Jane's*. No more recent information has been received.

PAUP
D. PAUP

802 NE Street, Carroll, Iowa 51401

PAUP PAPILLON

First flown in mid-1980, the Papillon is an unusual microlight aircraft (see accompanying photograph) with increased dihedral on the outer wings, and a V tail. The strut-braced wings are fitted with deflexers, but no ailerons; two-axis control is exercised in a similar manner to that of the Hummer (which see). The Papillon is powered by a 16·4 kW (22 hp) Polaris snowmobile engine, with direct drive to a two-blade pusher propeller, and is fitted with a tailwheel-type landing gear.

DIMENSION:

Wing span	10·82 m (35 ft 6 in)

WEIGHTS:

Weight empty	86 kg (189 lb)
Max T-O weight	175 kg (385 lb)

PERFORMANCE:

Cruising speed	26 knots (48 km/h; 30 mph)
Landing speed	22 knots (40 km/h; 25 mph)

PROGRESSIVE
PROGRESSIVE AIRCRAFT COMPANY

6474 East Sibley, Simi Valley, California 93063
PRESIDENT: R. Boone

This company specialises in high-performance hang gliders and associated products. Its 1981 model, known as the **Pro Air Series I**, has a metal-batten, zero billow sail, with an enclosed floating crossbar. No other details were known at the time of closing for press.

PTERODACTYL
PTERODACTYL LTD

PO Box 191, San Andreas Road, Watsonville, California 95076
Telephone: (408) 724 2233
PRESIDENT: Jack McCornack

Mr Jack McCornack designed and built the prototype of the Fledgling powered hang glider, which is based on the Manta Fledge 2B (which see). Mr McCornack and Mr Keith Nicely each flew one of these aircraft from Monterey to the EAA Fly-in at Oshkosh, Wisconsin, in 1979, averaging about 174 nm (322 km; 200 miles) per day. One of these aircraft has since been donated to the National Air and Space Museum in Washington. A Fledgling was one of four types of microlight aircraft acquired by the US Navy in 1980 for evaluation at the Naval Weapons Center.

PTERODACTYL FLEDGLING

The Fledgling is based upon the Manta Fledge 2B hang glider wing, modified by Mr Jack McCornack of Pterodactyl Ltd, and has been in production in the USA since 1978. In its current (1981) form, it is powered by a 336 or 430 cc two-cylinder two-stroke engine, driving a 0·91 m (3 ft) diameter pusher propeller. A joystick is standard, to actuate the overwing drag rudders. These produce yaw-induced roll; pitch control is by a combination of weight shift and drag rudders.

In 1980, Mr David Garrison and Mr John Leigh-Pemberton of the UK became the first pilots to fly microlight aircraft from Land's End to John O'Groats, completing the distance, in stages, in a total flying time of 17 h 5 min. In the United States, four Pterodactyls made a west-to-east coast-to-coast crossing in 1979, a total distance of approx 2,605 nm (4,830 km; 3,000 miles), during the course of which they overflew the 4,570 m (15,000 ft) Grand Teton Mountains.

On 24 May 1981 a Ptraveler called *No Turkey* was flown by Mr Jim Campbell at NAS Lakehurst, New Jersey, to an unofficial world altitude record for microlight aircraft of 6,465 m (21,210 ft). At that height the aircraft was still climbing, and the flight was terminated only because Mr Campbell—clad in thermal underwear, two snowsuits and a flight jacket—could no longer withstand the −46°C temperature.

In mid-1981 Mr Campbell and a partner were awaiting delivery of two special **Acro-Dactyl** aerobatic prototypes, to be flown in air display demonstrations.

Four versions of the Fledgling were available in 1981:

Fledgling OR (Oshkosh Replica). Basic version, similar to those at 1979 Fly-in. Powered by 336 cc single-cylinder snowmobile engine, built for Pterodactyl in West Germany. Rudders controlled by individual twist-grips.

Fledgling 430D. Similar to OR, but powered by a Pterodactyl-built Cuyuna 22·4 kW (30 hp) 429 cc two-cylinder snowmobile engine. Rudders controlled by single left or right joystick.

NFL (Not Foot Launchable). Basically a Fledgling 430D, fitted with a rear-mounted elevator which replaces pilot weight shift as the means of pitch control. Conventional joystick control system, using pushrod for elevator and cables for rudders.

Ptraveler. Basically a Fledgling 430D, but fitted with a canard elevator and retaining foot-launch capability. Control system as for NFL.

TYPE: Single-seat microlight aircraft.

AIRFRAME: Braced monoplane wing, with tubular leading-edge spar, tubular secondary spar and light-weight light alloy ribs. Sweepback 18°. Dihedral 6°. Nylon or Dacron covering on upper surface only. Wing wires braced to light alloy tube 'fuselage' and to king-post extending above wing. Controllable rudder above each wingtip. Elevator (NFL) and canard surface (Ptraveler) covered with Stitts fabric. Non-retractable tricycle landing gear.

POWER PLANT: One snowmobile engine (details under model listings), driving a two-blade fixed-pitch pusher propeller. Fuel in two plastics tanks, total capacity 19 litres (5 US gallons).

DIMENSIONS:

Wing span	10·06 m (33 ft 0 in)
Wing area, gross	15·0 m² (162·0 sq ft)
Wing aspect ratio	6·7
Length overall	2·44 m (8 ft 0 in)
Height overall	3·05 m (10 ft 0 in)

WEIGHTS:

Weight empty: OR	70·5 kg (155 lb)
430D	75 kg (165 lb)
NFL, Ptraveler	82 kg (180 lb)
Max T-O weight: OR	192 kg (425 lb)
430D, NFL, Ptraveler	204 kg (450 lb)

PERFORMANCE:

Never-exceed speed	48 knots (88 km/h; 55 mph)
Cruising speed range	30-39 knots (56-72 km/h; 35-45 mph)
Stalling speed	20 knots (37 km/h; 23 mph)
Max rate of climb at S/L: OR	91 m (300 ft)/min
430D, NFL, Ptraveler	122 m (400 ft)/min
Service ceiling	5,180 m (17,000 ft)
T-O run (430D, NFL, Ptraveler)	38 m (125 ft)
Landing run	6 m (20 ft)
Range with max fuel	174 nm (322 km; 200 miles)
Best glide ratio at 22-26 knots (41-48 km/h; 25-30 mph)	9

ROTEC
ROTEC ENGINEERING INC

PO Box 124, Duncanville, Texas 75116
Telephone: (214) 298 2505
PRESIDENT: William Adaska

Formerly a helicopter engineer with Bell and Aérospatiale, Mr Adaska has been developing the Rally series of microlight aircraft since 1977. The latest versions, available in kit form or ready to fly, are the Rally 2B and twin-float Rally Marine. Rally sales (all versions) totalled more than 2,000 by mid-1981.

ROTEC RALLY 2B and RALLY MARINE

TYPE: Single-seat microlight aircraft; *g* limits +4·0/−2·0.

AIRFRAME: Main frame of 6061-T6 aluminium tubing, with single- and (over 30 per cent of area) double-thickness Dacron-covered wing and tail surfaces. Vinyl-coated stainless steel cables. Kingpost standard.

Paup Papillon (16·4 kW; 22 hp snowmobile engine) *(J. M. G. Gradidge)*

Pterodactyl Fledgling

Pterodactyl NFL

Pterodactyl Ptraveler *No Turkey*, **in which Mr Jim Campbell set a microlight aircraft altitude record in May 1981** *(Howard Levy)*

Rotec Rally 2B *(Howard Levy)*

Rotec Rally Marine

Three-axis control by ailerons, elevators and rudder. Wings are sweptback 5° and fitted with endplates. Tricycle landing gear, with shock-absorption (twin-float gear on Rally Marine). Rigid seat and nose structure. Optional features include ski gear, second seat, windscreen, instrument panel, and agricultural spray kit.

Power Plant: One 12·7 kW (17 hp) Solo two-cylinder two-stroke engine in Rally 2B, driving a two-blade wooden pusher propeller via a reduction gear. Rally Marine has a 22·4 kW (30 hp) Chaparral engine, which is optional also for the Rally 2B. Fuel capacity 7·5 litres (2 US gallons) standard, 22·7 litres (6 US gallons) optional.

Dimensions:
- Wing span: 9·75 m (32 ft 0 in)
- Wing area, gross: 14·4 m² (155·0 sq ft)
- Wing aspect ratio: 6·4
- Length overall: 8·48 m (27 ft 10 in)
- Propeller diameter: 1·37 m (4 ft 6 in)

Weights and Loading:
- Weight empty: 61 kg (135 lb)
- Max pilot weight: 104 kg (230 lb)
- Max T-O weight: 166 kg (365 lb)
- Max wing loading: 11·5 kg/m² (2·35 lb/sq ft)

Performance (Rally 2B, 17 hp engine):
- Never-exceed speed: 39 knots (72 km/h; 45 mph)
- Cruising speed: 29 knots (54 km/h; 33 mph)
- T-O speed: 13-15 knots (24-27 km/h; 15-17 mph)
- Stalling speed:
  - power on: 12·5 knots (22·5 km/h; 14 mph)
  - power off: 14 knots (26 km/h; 16 mph)
- Max rate of climb at S/L: 106 m (350 ft)/min
- Best glide ratio at 12 knots (22 km/h; 14 mph): 7
- Min sinking speed, power off: 85·5 m (280 ft)/min

Performance (Rally Marine, 30 hp engine):
- Cruising speed: 39 knots (72 km/h; 45 mph)
- Max rate of climb at S/L: over 152 m (500 ft)/min
- T-O run: 27 m (90 ft)

SKY SPORTS

SKY SPORTS INC

PO Box 507 (Route 83), Ellington, Connecticut 06029
Telephone: (203) 872 7317
President: Ed Vickery

Details of Sky Sports' Bobcat IV and Osprey II hang gliders can be found in the 1980-81 *Jane's*. Models available for 1981 were the **Eaglet, Sirocco III** and **Peregrine**. All have aluminium tube airframes, stainless steel cables and double-surface Dacron sails. All other known details of these craft are given in the table at the end of this section.

Sky Sports acquired the rights to the Humbug microlight aircraft following the death in 1979 of its designer. Mr Klaus Hill. Production of this aircraft, assisted by finance from Vector Aircraft Corporation, began in 1980; it is now known as the Vector 600. A separate development of the Humbug has been made by Gemini International (which see) under the name Hummingbird.

SKY SPORTS VECTOR 600

Although the Vector 600 is very similar to Klaus Hill's Humbug, the landing gear has been improved to make it more suitable for operation from grass fields, the twin power plant has been changed from tractor to pusher configuration, and the wing design has been modified to incorporate 'spoilerons'. These are linked to an Ercoupe-like two-control system, allowing the 'spoilerons' and 'ruddervators' to be operated by a single control to produce smooth co-ordinated turns. The Vector 600 is foot launchable and, under FAA regulations current in 1981, can be flown without being licensed, and by an unlicensed pilot. It can be folded for transportation on the roof rack of a motorcar.

Type: Single-seat microlight aircraft; *g* limits +5·7/−2·8.

Airframe: Wire-braced high-wing monoplane, with 6° dihedral from roots. Wings similar to those of a flex-wing, with a leading-edge tubular spar, a similar spar at approx 70 per cent chord, and a wire-in-tension trailing-edge. The tubular spars fit into sockets of wingtip tubes, and the wing is tensioned by spring compression tubes between the tubular spars. Light alloy tubular ribs are contained within pockets sewn into the wing fabric. Double-surface Dacron covering over most of wing area. Wings braced by stainless steel landing wires running from triangular kingpost above fuselage structure, and by flying wires attached to fuselage. The fuselage itself has a rectangular base frame of 6061-T6 aluminium alloy tube, to carry landing gear and support pilot's sling seat. Light alloy tube pyramid structure above base frame supports forward end of wing mounting, kingpost and power plant. V tail of lightweight 6061-T6 tube with fabric covering, carried on twin light alloy tube tailboom and further tubes from each wing rear spar. Movable surfaces of V tail serve as both rudders and elevators ('ruddervators'). Non-retractable tricycle landing gear, with non-steerable nosewheel; larger-diameter main wheels with pneumatic tyres.

Small castoring wheel beneath tail unit. Float and ski landing gear available optionally. Adjustable-tension sling seat for pilot.

POWER PLANT: One 11·2 kW (15 hp) or two 6·7 kW (9 hp) single-cylinder two-stroke engines, with single-shaft reduction drive to a two-blade metal pusher propeller. Fuel tank capacity 15 litres (4 US gallons).

DIMENSIONS:

Wing span	10·51 m (34 ft 6 in)
Wing area, gross	12·40 m² (133·5 sq ft)
Wing aspect ratio	8·9
Length overall	5·49 m (18 ft 0 in)
Height overall	2·59 m (8 ft 6 in)
Propeller diameter	1·27 m (4 ft 2 in)

WEIGHTS AND LOADING:

Basic operating weight empty	81·5 kg (180 lb)
Max T-O weight	193 kg (425 lb)
Max wing loading	15·53 kg/m² (3·18 lb/sq ft)

PERFORMANCE:

Never-exceed speed	60·5 knots (112 km/h; 70 mph)
Max level speed	43·5 knots (80 km/h; 50 mph)
Econ cruising speed	24 knots (45 km/h; 28 mph)
Stalling speed, power off	16 knots (29 km/h; 18 mph)
Max rate of climb at S/L	122 m (400 ft)/min
Service ceiling	3,050 m (10,000 ft)
T-O run	35 m (115 ft)
Landing run	61 m (200 ft)
Range with max fuel	26 nm (48 km; 30 miles)
Best glide ratio	10

SPECTRA

SPECTRA AIRCRAFT CORPORATION

2151 Arnold Industrial Highway, Shop 5, Concord, California 94520

Telephone: (415) 798 9993

SPECTRA AOLUS

The Aolus, undergoing certification in early 1981, is a 'bowsprit' design semi-rigid hang glider with high aspect ratio wings, a large nose angle, and tail surfaces. It was being offered in three sizes, all known details of which are given in the table at the end of this section.

STEWART

STEWART AIRCRAFT CORPORATION

11420 State Route 165, Salem, Ohio 44460
Telephone: (216) 332 0865
DESIGNER: James Stewart

STEWART JD3 PUFFIN

Mr Stewart has devised an unusual power plant for microlight aircraft, consisting of two entirely separate Chrysler piston engines, mounted in tandem and driving, via concentric shafts, two separate counter-rotating two-blade pusher propellers. This unit, and an elevon control system, were being flight tested in early 1981 on a UFM Easy Riser hang glider. Subject to successful completion of these tests, Mr Stewart hoped to fly in the Spring of 1981 a new microlight design, the JD3 Puffin, incorporating these features.

TYPE: Single-seat microlight aircraft.

AIRFRAME: Mid-wing wire-braced sweptback tailless monoplane, with specially-developed aerofoil section and endplate winglets. Central pod nacelle for pilot, under flush-fitting canopy, with power plant mounted aft of pilot's seat. Ventral fin. Non-retractable tricycle landing gear, which can be removed if required (saving approx 6·4 kg; 14 lb weight) to permit foot-launch via 'bomb bay' doors in cockpit floor. Steerable nosewheel. Cockpit has adjustable seat and is heated by engine bleed air.

POWER PLANT: Two 6·7 kW (9 hp) Chrysler West Bend engines, mounted in tandem and driving, via concentric shafts, two separate two-blade counter-rotating pusher propellers. Fuel capacity 21 litres (5·5 US gallons).

DIMENSIONS:

Wing span	10·52 m (34 ft 6 in)
Wing area, gross	13·47 m² (145·0 sq ft)
Wing aspect ratio	8·2

WEIGHTS:

Weight empty	59 kg (130 lb)
Max T-O weight	181 kg (400 lb)

PERFORMANCE: No details known

STRATUS

STRATUS

The Old Schoolhouse, Montara, California 94037
Telephone: (415) 728 7655

STRATUS V

The Stratus V is a flexible-wing hang glider of 'bowsprit' configuration, with a fully-cambered double-surface sail and no deflexers. It was available in 1981 in two sizes, all known details of which are given in the table. A 17·19 m² (185 sq ft) version was due to enter production in the Summer of 1981.

STRIPLIN

STRIPLIN AIRCRAFT CORPORATION

PO Box 2001, Lancaster, California 93534
Telephone: (805) 942 0768
PRESIDENT: Kenneth Striplin

Striplin's original FLAC (Foot Launched Air Cycle), flown for the first time on 24 October 1978, was a foot-launchable microlight aircraft fitted with an auxiliary tricycle landing gear. A description and illustration appeared in the Homebuilts section of the 1980-81 *Jane's*.

Production of the FLAC has ended during the past year. Current models in 1981 were a developed version, the Super FLAC, and a new design known as the Lone Ranger.

STRIPLIN SUPER FLAC

The FLAC prototype was fitted with a non-retractable landing gear; the retractable nosewheel was introduced later, but one result of this was to reduce legroom for the pilot. Striplin therefore introduced the Super FLAC, with a roomier cockpit and a twin-engine power plant. A recovery parachute is available, as are snow skis for all three wheels.

Announced in the Spring of 1980, the Super FLAC retains the overall dimensions of the standard FLAC, but has an enlarged pilot pod which not only provides greater legroom but also has a compartment for up to 2·25 kg (5 lb) of baggage. The aircraft can also be flown by pilots heavier than those for whom the standard FLAC is suitable.

TYPE: Single-seat microlight aircraft.

AIRFRAME: Cantilever high-wing monoplane of constant chord. Wortmann FX-126 wing section. Thickness/chord ratio 12%. No dihedral. Incidence 3°. Sweepback at quarter-chord 8°. Structure of glassfibre and epoxy with foam core, fabric-covered aft of spar. Elevon control surfaces attached to trailing-edge near tips operate collectively as elevators, differentially as ailerons. (Outboard section of each elevon is used for upward deflection only, being picked up automatically by the inboard section when the stick is pulled back; otherwise, the outboard section does not move. The outboard section is also set at a specific reflexed position to provide an upward pitching moment; when the aircraft reaches a certain airspeed it will automatically pull out of a dive.) Large endplates on wingtips each carry a control surface at the rear, operated independently as rudders, collectively as airbrakes. Wings fold for transportation. Fuselage pod, consisting of a moulded shell formed from glassfibre over foam reinforcement, is suspended from wings, and has a large windscreen and open sides. 'Bomb-bay' type doors in bottom open for foot launching. Non-retractable tricycle type landing gear, with laminated glassfibre legs. Optional mainwheel fairings, skis, and floats. Single seat, with baggage space to rear. Elevons actuated by side-stick, rudders by foot pedals.

POWER PLANT: Two Soarmaster two-cylinder engines, geared to a common propeller shaft and driving a two-blade wooden pusher propeller. Two glassfibre tanks in wing, with a combined fuel capacity of 19 litres (5 US gallons).

DIMENSIONS:

Wing span	9·75 m (32 ft 0 in)
Wing span, folded	5·64 m (18 ft 6 in)
Wing area, gross	14·4 m² (155 sq ft)
Wing aspect ratio	7·1
Length of fuselage pod	2·13 m (7 ft 0 in)
Height overall	1·22 m (4 ft 0 in)
Propeller diameter	1·22 m (4 ft 0 in)

WEIGHTS AND LOADING:

Weight empty	91 kg (200 lb)
Max T-O weight	197 kg (435 lb)
Max wing loading	13·7 kg/m² (2·80 lb/sq ft)

PERFORMANCE:

Never-exceed speed	69 knots (129 km/h; 80 mph)
Max level speed	52 knots (96·5 km/h; 60 mph)
Cruising speed	48 knots (88·5 km/h; 55 mph)
Stalling speed	21 knots (39 km/h; 24 mph)
Best glide ratio at 33 knots (61 km/h; 38 mph), power off	20
Min sinking speed at 31 knots (58 km/h; 36 mph), power off	55 m (180 ft)/min

STRIPLIN LONE RANGER

First flown on 23 October 1980, the Lone Ranger is a single-seat, single-engined microlight aircraft with conventional three-axis controls, designed for amateur construction.

TYPE: Single-seat microlight aircraft; *g* limits ± 4·5 (without wing struts), +4·5/−3·9 (with struts).

AIRFRAME: Cantilever high-wing monoplane, with provision for wing bracing struts. Wings can be folded back alongside fuselage for transportation and stowage. Basic load-bearing wing box structure formed from an I-beam main spar and D-section leading-edge of wood and glassfibre, with glassfibre and foam ribs and Dacron covering. Leading-edge is of pre-formed foam with a glassfibre skin; trailing-edge is a wooden strip. Piano-hinged ailerons, also of glassfibre/foam construction. Pod-and-boom fuselage, the forward portion having a base-frame of unidirectional glassfibre, impregnated with epoxy resin, with a moulded outer shell of glassfibre. At top of pod is a U-shaped channel, in which is buried a lightweight metal tube to which the wings are bolted and which also supports the rear fuselage and tail unit. Tailboom is a triangular-section glassfibre and foam sandwich structure; tail surfaces are of similar construction to wings. Non-retractable tricycle landing gear, with laminated glassfibre legs and steerable nose unit. Float and ski gear available optionally. Cockpit similar to that of Super FLAC, including optional 'bomb bay' doors for foot launch.

POWER PLANT: Single 9-15 kW (12-20 hp) engine, mounted in overwing nacelle above cockpit, driving a two-blade tractor propeller via a reduction gear. Striplin prototypes have been flown with a variety of engines and reduction units. Electrical starting optional for larger engines. Fuel tank in each wing.

DIMENSIONS:

Wing span	9·75 m (32 ft 0 in)
Wing area, gross	13·4 m² (144·0 sq ft)
Wing aspect ratio	7·1
Width, wings folded	1·07 m (3 ft 6 in)
Length overall	4·78 m (15 ft 8 in)
Length, wings folded	5·74 m (18 ft 10 in)
Height overall	1·27 m (4 ft 2 in)

WEIGHTS AND LOADINGS (A: without; B: with optional wing struts):

Weight empty: A	80 kg (176 lb)
B	82·5 kg (182 lb)
Max T-O weight: A	175 kg (385 lb)
B	197 kg (435 lb)
Max wing loading: A	13·04 kg/m² (2·67 lb/sq ft)
B	14·74 kg/m² (3·02 lb/sq ft)

PERFORMANCE (typical):

Never-exceed speed	69 knots (129 km/h; 80 mph)
Max level speed	65 knots (121 km/h; 75 mph)
Cruising speed	48 knots (88 km/h; 55 mph)
Landing speed	20 knots (37 km/h; 23 mph)
Stalling speed	18·5 knots (34 km/h; 21 mph)
Max rate of climb at S/L:	
12 hp engine	91 m (300 ft)/min
20 hp engine	168 m (550 ft)/min
T-O run: 12 hp engine	61 m (200 ft)
20 hp engine	53 m (175 ft)
Best glide ratio at 33 knots (61 km/h; 38 mph)	17
Min sinking speed at 30 knots (55 km/h; 34 mph), power off	64 m (210 ft)/min

STRIPLIN SKY RANGER

Due to fly in April 1981, the Sky Ranger is a single-engined, side-by-side two-seat dual-control development of the Lone Ranger; other features include a T tail. Power plant is a 22·4 kW (30 hp) engine, driving two wing-mounted propellers; fuel capacity is 19 litres (5 US gallons).

DIMENSIONS:

Wing span	9·75 m (32 ft 0 in)
Length overall	4·27 m (14 ft 0 in)
Height overall	1·93 m (6 ft 4 in)

WEIGHTS:

Weight empty	118 kg (260 lb)
Max T-O weight	272 kg (600 lb)

PERFORMANCE (estimated):

Never-exceed speed	95·5 knots (177 km/h; 110 mph)
Cruising speed (75% power)	69 knots (129 km/h; 80 mph)
Stalling speed:	
flaps up	31·5 knots (58 km/h; 36 mph)
flaps down	22 knots (40·5 km/h; 25 mph)
Endurance	more than 2 h

Sky Sports Vector 600 microlight aircraft in current 'pusher' configuration *(J. M. G. Gradidge)*

Stewart JD3 Puffin, minus wings *(Howard Levy)*

Striplin Super FLAC

Striplin Lone Ranger

Teman Mono-Fly

Partially-completed Striplin Sky Ranger, with Fowler-type flaps and wing cutout showing position of propeller installation on each side *(Howard Levy)*

SUNBIRD

SUNBIRD ULTRALIGHT GLIDERS

12501 Gladstone Avenue, No. A4, Sylmar, California 91342
Telephone: (213) 361 8651

Sunbird's 1980 hang glider was the Nova, details of which can be found in the table and in the 1980-81 *Jane's*.

A new model in 1981 was the **Challenger**, reported to have a top speed of 48 knots (88 km/h; 55 mph). No details of this were received from Sunbird, but press reports describe it as having an anodised airframe, kingpost, preformed battens, floating crossbar, 132° nose angle, stalling speed of 14-15 knots (26-27 km/h; 16-17 mph) IAS, and min sinking speed of 17·5 knots (32 km/h; 20 mph) IAS. The sail is double-surfaced over 36 per cent of its area.

TEMAN

TEMAN AIRCRAFT CO

10092 Northampton Avenue, Westminster, California 92683
Telephone: (714) 531 2655
FOUNDER: Robert C. Teman

TEMAN MONO-FLY

Design of the Mono-Fly started in January 1976. prototype construction began two years later, and the first flight was made in July 1979. FAA certification (Experimental category) was received on 7 July 1980, and approx 60 sets of plans had been sold by February 1981. Kits for amateur construction were due to become available later.

TYPE: Single-seat microlight aircraft.

AIRFRAME: Strut-braced high-wing monoplane. Modified Clark Y wing section, with 17% thickness/chord ratio. Dihedral 3°. Incidence 6° at root, 3° at tip. Aluminium tube main spar, with glassfibre ribs and Dacron covering. Glassfibre and foam ailerons. Fuselage frame of 7·6 cm (3 in) aluminium tube. All-moving tailplane and rudder, of Dacron-covered aluminium tube. Non-retractable tricycle landing gear, with coil spring shock-absorption on nose unit, Teman bungee/cable system on main units. Single open seat for pilot.

POWER PLANT: One 13·4-15·7 kW (18-21 hp) modified Onan BGM-18 two-cylinder four-stroke engine, driving a two-blade wooden pusher propeller. Single polyurethane fuel tank, capacity 22·7 litres (6 US gallons). 15A generator and 12V battery for engine starting.

DIMENSIONS:	
Wing span	9·37 m (30 ft 9 in)
Wing area, gross	11·52 m² (124 sq ft)
Wing aspect ratio	7·5
Length overall	5·23 m (17 ft 2 in)
Height overall	2·01 m (6 ft 7 in)
Propeller diameter	1·22 m (4 ft 0 in)
WEIGHTS AND LOADING:	
Basic operating weight empty	136 kg (300 lb)
Max T-O weight	249 kg (550 lb)
Max wing loading	22·0 kg/m² (4·5 lb/sq ft)
PERFORMANCE:	
Never-exceed speed	43·5 knots (80 km/h; 50 mph)
Max level speed	39 knots (72 km/h; 45 mph)
Max cruising speed	35 knots (64 km/h; 40 mph)
Econ cruising speed	33 knots (61 km/h; 38 mph)
Stalling speed:	
power on	17·5 knots (32·5 km/h; 20 mph)
engine idling	19·5 knots (35·5 km/h; 22 mph)
Max rate of climb at S/L	91 m (300 ft)/min
Service ceiling	2,440 m (8,000 ft)
T-O run	61 m (200 ft)
T-O to 15 m (50 ft)	152 m (500 ft)
Landing from 15 m (50 ft)	122 m (400 ft)
Landing run	46 m (150 ft)
Range with max fuel, 15 min reserves	139 nm (257 km; 160 miles)

THEIS

JAMES THEIS

Mr Theis is the designer of the **Nighthawk**. No details of this aircraft were known at the time of closing for press.

UFM

ULTRALIGHT FLYING MACHINES OF WISCONSIN

PO Box 248, Kansasville, Wisconsin 53139

This company produces John Moody's power package and crosswind landing gear for the Easy Riser, a rigid-wing powered hang glider stressed to +11/−5*g*. More than 900 Easy Risers had been delivered by early 1981, many of them for construction by homebuilders.

UFM EASY RISER

AIRFRAME: Constant-chord sweptback (15°) flying-wing biplane, of LM 7610 aerofoil section and unequal span, built of aluminium alloy with double-thickness Dacron covering. No fuselage. Drag rudder aft of each wing, attached to outermost rear interplane strut. Crosswind tricycle landing gear optional.

POWER PLANT (optional): One Chrysler West Bend 820 single-cylinder two-stroke piston engine, driving a two-blade wooden pusher propeller via a 'Maximizer' reduction gear. Fuel capacity 4·25 litres (9 US pints) standard; additional tank of similar capacity is available optionally.

DIMENSIONS:

Wing span (trailing-edge)	9·14 m (30 ft 0 in)
Wing area, gross	15·98 m² (172 sq ft)
Aspect ratio (upper wing)	8·8
Length overall	2·44 m (8 ft 0 in)
Propeller diameter	1·22 m (4 ft 0 in)

WEIGHTS AND LOADING:

Weight of glider without engine or landing gear	22·7 kg (50 lb)
Weight of glider with engine:	
foot-launched	38·5 kg (85 lb)
with landing gear	47 kg (104 lb)
Pilot weight range	54·5-95 kg (120-210 lb)
Max wing loading	7·17 kg/m² (1·47 lb/sq ft)

PERFORMANCE (A: unpowered; B: with Maximizer unit):

Best glide ratio:	
A	over 10
B at 22-24 knots (40-45 km/h; 25-28 mph)	9·3
Min sinking speed:	
A	55 m (180 ft)/min
B, at 16·5-18 knots (31-34 km/h; 19-21 mph):	
foot-launched	70 m (230 ft)/min
with landing gear	76 m (250 ft)/min
Max speed:	
B	39-43 knots (72-80 km/h; 45-50 mph)
Cruising speed:	
B	22-30 knots (40-56 km/h; 25-35 mph)
Stalling speed: A	14 knots (26 km/h; 16 mph)
B	15-17 knots (27·5-31 km/h; 17-19 mph)
Typical rate of climb:	
B	76-106 m (250-350 ft)/min
Typical endurance: B	1 h

UFM EASY RISER ARROW

This version of the Easy Riser has an 11·2 kW (15 hp) Yamaha engine and fuel tank capacity of 9·5 litres (2·5 US gallons), giving it an endurance of up to 3 hours. Max T-O weight is increased to 156·5 kg (345 lb); speed performance is generally similar to earlier models, but max climb rate is increased to 152 m (500 ft)/min.

UFM SOLAR RISER

Mr Larry Mauro, designer of the Easy Riser biplane hang glider, has built a solar-powered aircraft, called the Solar Riser. Based on the hang glider, the Solar Riser has a power system incorporating solar panels with 300 cells, giving a total of 40V 10A power output. Associated equipment includes a battery and a 2·24 kW (3 hp) General Design electric motor of the type developed originally for the Apollo spacecraft programme. To convert the hang glider into a simple research aeroplane, a pilot nacelle and a tricycle landing gear are also fitted.

The Solar Riser is claimed to have been the first solar-powered aircraft to fly, and by mid-1979 had completed six flights, totalling 10 minutes flying time. The first flight, in late April 1979, covered a distance of about 805 m (2,640 ft), at an altitude of 12·2 m (40 ft). Normal flying time is from 1 to 3 minutes. The current research and development programme is aiming at a flight of one hour, but improvements to the aircraft and its power system could eventually lead to flights of about two hours.

TYPE: Single-seat solar-powered biplane.

AIRFRAME: Based on Easy Riser hang glider, with aluminium alloy wing spars, ribs, leading- and trailing-edges, stainless steel interplane struts, and Dacron skin. Upper wing top surface is covered with solar panels made up of 300 cells. Small and basic nacelle for pilot, with electric motor mounted at rear. Non-retractable tricycle landing gear.

POWER PLANT: One 2·24 kW (3 hp) General Design electric motor, driving a 1·04 m (3 ft 5 in) diameter two-blade Soarmaster pusher propeller.

WEIGHTS:

Weight empty	57 kg (125 lb)
Max T-O weight	131·5 kg (290 lb)

PERFORMANCE:

Max level speed approx	17·3 knots (32 km/h; 20 mph)
T-O run	23-61 m (75-200 ft)
Endurance (to date)	1 to 3 min

ULTIMATE HI
ULTIMATE HI

14328 Lolin Lane, Poway, California 92064

Producer in 1980 of the Starship series (three sizes), all known details of which are given in table. No more recent information has been received.

UP
UP INC (ULTRALITE PRODUCTS)

PO Box 582, 28011 Front Street, Rancho California, Temecula, California 92390
Telephone: (714) 676 5652
DIRECTOR: Peter Brock

Details of the UP Condor, Firefly and Spyder series can be found in the 1979-80 *Jane's*, and of the Mosquito in the 1980-81 edition.

UP's major model for 1980-81 was the **Comet**, which is available in three sizes; details are given in the table. The Mosquito has an airframe of 6061-T6 anodised aluminium alloy tube, with kingpost as standard, Dacron sail (double surface over 60 per cent of area), stainless steel cables, ribs, enclosed cross-spars, and a floating keel pocket. It can be flown from a prone or seated position. A total of 580 had been sold by February 1981.

The Comet 165 won most of the major US competition events in 1980, including the XC (cross-country) Classic and the American Cup, against strong national and international competition.

VOLMER
VOLMER AIRCRAFT

PO Box 5222, Glendale, California 91201
Telephone: (213) 247 8718
PRESIDENT: Volmer Jensen

VOLMER VJ-23 SWINGWING

This monoplane hang glider was designed in 1971 by Volmer Jensen and Irving Culver, making its first flight towards the end of that year. Subsequent modifications made the aircraft safer and more controllable. Volmer Aircraft markets plans (but not kits or materials) for the VJ-23; several hundreds have been sold.

Following the success of David Cook in the UK (which see) in installing an engine in his Swingwing, drawings for a **VJ-23E** powered version are now also available from Volmer Aircraft.

AIRFRAME: Cantilever high-wing monoplane, constructed of steel tube, spruce, mahogany and plywood, with Dacron or fabric covering. Control by means of ailerons, elevators and rudder. Airframe stressed to ±3·0*g*.

DIMENSIONS:

Wing span	9·93 m (32 ft 7 in)
Wing area, gross	16·63 m² (179·0 sq ft)
Length overall	5·31 m (17 ft 5 in)
Height overall	1·83 m (6 ft 0 in)

WEIGHTS:

Weight empty: VJ-23	45·5 kg (100 lb)
VJ-23E	63·5 kg (140 lb)
Pilot weight range	45-91 kg (100-200 lb)
Max T-O weight: VJ-23	136 kg (300 lb)

PERFORMANCE (VJ-23):

Best glide ratio at 17·5 knots (32 km/h; 20 mph)	9
Cruising speed	17 knots (32 km/h; 20 mph)
Stalling speed	13 knots (24 km/h; 15 mph)

VOLMER VJ-24 SUNFUN

The VJ-24 rigid-wing monoplane hang glider is essentially a simplified-construction version of the earlier VJ-23 Swingwing, from which it differs primarily in having wings and tail of fabric-covered aluminium tube instead of wooden spars and ribs; rectangular instead of tapered planform wings, of increased span, with V bracing struts; and a pair of ground handling wheels.

A **VJ-24E** is also available, with a 7·5 kW (10 hp) McCulloch go-kart engine, pusher propeller and fuel tank installed beneath the wing, aft of the pilot, to permit take-offs from level ground or uphill. The **VJ-24W** has a tractor engine installation, two-wheeled landing gear and pilot seat, which can easily be attached to the basic VJ-24 airframe.

The following details apply to the unpowered VJ-24, except where indicated:

DIMENSIONS:

Wing span	11·13 m (36 ft 6 in)
Wing area, gross	15·14 m² (163·0 sq ft)
Length overall: VJ-24	5·54 m (18 ft 2 in)
VJ-24W	5·64 m (18 ft 6 in)
Height overall: VJ-24	1·73 m (5 ft 8 in)
VJ-24W	1·75 m (5 ft 9 in)

WEIGHTS:

Weight empty: VJ-24	50 kg (110 lb)
VJ-24W	75 kg (165 lb)
Max T-O weight: VJ-24	140·5 kg (310 lb)
VJ-24W	156 kg (345 lb)

PERFORMANCE (VJ-24): As for VJ-23

PERFORMANCE (VJ-24W):

Cruising speed	24·5 knots (45 km/h; 28 mph)
Stalling speed	13 knots (24·5 km/h; 15 mph)

WASPAIR
WASPAIR CORPORATION

1881 Enterprise Boulevard, West Sacramento, California 95691
Telephone: (916) 372 5791
PRESIDENT: Robin Haynes

This is the former British company Waspair (see earlier editions of *Jane's*), which transferred its activities from the UK to the USA in 1979. Since then it has been developing a new microlight aircraft, the H.M.81 Tomcat.

WASPAIR H.M.81 TOMCAT

Design of the Tomcat began in 1979, and an H.M.80 first prototype was flown in May 1980. The second prototype, designated H.M.81, was completed in october of that year and incorporated a number of modifications. This version was available in 1981 in kit form for amateur construction.

TYPE: Single-seat microlight aircraft.

AIRFRAME: Configuration is that of a parasol monoplane with a boom-mounted all-moving foreplane. Main wing is a non-swept, constant-chord structure, supported above the aircraft keel by a pair of 'dihedral wings', set at approx 45° dihedral and having 13° of sweepback. In addition to performing the major lateral stability function of twin fins, by providing side area aft of the CG, they also tie the aircraft's yaw and roll reactions closely together, so providing a single co-ordinated yaw/roll response to a control input or gust. The all-moving canard surface, in addition to providing inherent longitudinal stability, can also be tilted forwards/backwards and side to side to provide pitch and yaw/roll control. The main wings are foldable for transportation and stowage. A tricycle landing gear is fitted. The pilot's open seat is located in the V formed by the dihedral wings.

POWER PLANT: The Tomcat will perform adequately with a 9 kW (12 hp) engine, but best prototype test results have been achieved with a 22·4 kW (30 hp) 430 cc Zenoah engine. This is mounted aft of the pilot's seat, driving a two-blade wooden pusher propeller. Fuel tank capacity 19 litres (5 US gallons).

DIMENSIONS:

Wing area, gross	16·26 m² (175 sq ft)
Propeller diameter	0·91 m (3 ft 0 in)

WEIGHTS AND LOADING:

Weight empty	77 kg (170 lb)
Max pilot weight	100 kg (220 lb)
Max T-O weight	186 kg (410 lb)
Max wing loading	11·43 kg/m² (2·34 lb/sq ft)

PERFORMANCE (30 hp engine):

Max level speed	48·5 knots (90 km/h; 56 mph)
Cruising speed	39 knots (72 km/h; 45 mph)
T-O speed	23·5 knots (43·5 km/h; 27 mph)
Stalling speed	20 knots (37·5 km/h; 23·5 mph)
Max rate of climb at S/L	183 m (600 ft)/min
Normal range	approx 130 nm (240 km; 150 miles)
Best glide ratio at 30 knots (56 km/h; 35 mph)	15

UFM powered Easy Riser *(Howard Levy)*

UFM Solar Riser solar-powered aeroplane *(Henry Artof)*

Volmer VJ-24E powered Sunfun

Volmer Jensen flying a VJ-23 Swingwing

UP Comet 165

David Cook (UK) on a Volmer VJ-24W

Waspair Tomcat with designer Mr Robin Haynes *(Howard Levy)*

Volmer VJ-23E powered Swingwing

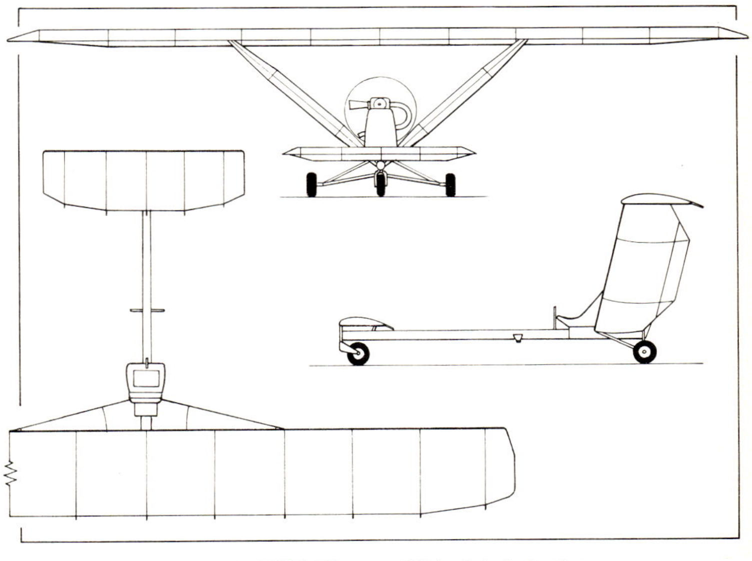

Waspair H.M.81 Tomcat *(Michael A. Badrocke)*

WEEDHOPPER

WEEDHOPPER OF UTAH INC

PO Box 2253, 1148 Century Drive, Ogden, Utah 84404
Telephone: (801) 621 3941
DESIGNER: John F. Chotia

WEEDHOPPER JC-24 WEEDHOPPER

This single-seat ultralight aircraft represents the 23rd design by Mr Chotia; the others comprised 18 hang gliders and four powered lightplanes. Design of the original wooden version of the Weedhopper (which is reminiscent in configuration of the 1909 Demoiselle of Alberto Santos-Dumont) began in 1975. Two prototypes were tested, after which the decision was taken to switch to metal structures. The first of two pre-production metal prototypes flew in February 1978. In November 1979 a prototype with an increased-area wing flew for the first time, and three versions of the JC-24 were available in 1981:

JC-24B Weedhopper. Basic version, as described in detail. Compared with original JC-24, has simplified structure and improved manoeuvrability. Full 360° turns have been completed in less than 8 s, with a turn radius of less than 21 m (70 ft). Stalls are preceded by greater warning buffet. Foot launch has been demonstrated, but is discouraged by Weedhopper. It is claimed that the JC-24B Weedhopper will recover from a stall, with power on, within an altitude loss of 3 m (10 ft). If the optional spoilers are not fitted to this version, bank and roll control are effected by using the rudder, or by optional wing warping.

JC-24BL Weedhopper Two. Soaring version, with wing of increased span and area that gives greater lift, and reduces both drag and stalling speed. Because this wing is more affected by gusts and turbulence, spoilers are mounted adjacent to the wingtips on the upper surface of the wings, to function as ailerons. In their neutral position they serve as leading-edge slots, inhibiting wingtip stalls. These supplementary aerofoils are available optionally for installation also on the standard wing of the JC-24B, and offer considerable aerodynamic improvement to the basic version.

JC-24C. New version, introduced in 1981. No details received for publication.

Kits of parts to build the Weedhopper first became available to homebuilders in July 1978. By 1 January 1980 orders for kits totalled 410, of which 330 had been supplied: about 100 Weedhoppers were then known to be flying.

TYPE: Single-seat microlight aircraft.

AIRFRAME: Strut-braced high-wing monoplane. Thickness/chord ratio 11%. Dihedral 12°. Incidence 3°. Washout 4° 30′. Sweepback 4° on JC-24B, 3° on JC-24BL. Aluminium tube structure, covered with Dacron fabric. Upper surface of JC-24B wing is fabric-covered, undersurface optional; but JC-24BL has both surfaces covered as standard. Aluminium tube V bracing struts each side, from landing gear axle to wing leading- and trailing-edges. Wingtip spoilers for roll control standard on JC-24BL, optional for JC-24B. Alternative wing warping system also available. Open triangulated fuselage of aluminium tubing, bolted together. Upper fuselage tube supports wing roots, tailplane and engine. Cantilever tailplane with elevators, ventral fin and rudder. Tailplane incidence and rudder trim ground-adjustable. Aluminium tube structure, Dacron fabric covered. Non-retractable tricycle type landing gear. Steerable nosewheel. All three tyres size 3·50-4. Brake on nosewheel optional. Ski landing gear optional. Single semi-reclined and adjustable open seat. Two-seat training version available optionally.

POWER PLANT: One Chotia 460 two-stroke engine, developing 18·6 kW (25 hp) at 4,500 rpm, with direct drive to a Weedhopper-built two-blade fixed-pitch wooden propeller. Single fuel tank of standard 3·8 litre (1 US gallon) capacity; optional 13·25 litre (3·5 US gallon) capacity on JC-24B, 9·5 litre (2·5 US gallon) capacity on JC-24BL. 6V battery for ignition system.

DIMENSIONS (A: JC-24B; B: JC-24BL):
- Wing span: A — 8·53 m (28 ft 0 in)
  - B — 10·46 m (34 ft 4 in)
- Wing area, gross: A — 15·61 m² (168 sq ft)
  - B — 19·32 m² (208 sq ft)
- Wing aspect ratio: A — 4·7
  - B — 5·7
- Length overall — 5·64 m (18 ft 6 in)
- Height overall — 1·98 m (6 ft 6 in)
- Propeller diameter — 1·12 m (3 ft 8 in)

WEIGHTS:
- Weight empty: A — 75 kg (165 lb)
  - B — 82 kg (180 lb)
- Max T-O weight: A — 175 kg (385 lb)
  - B — 204 kg (450 lb)

PERFORMANCE (at max T-O weight):
- Never-exceed speed and max level speed:
  - A, B — 43 knots (80 km/h; 50 mph)
- Max cruising speed at S/L:
  - A — 30 knots (56 km/h; 35 mph)
  - B — 26 knots (48 km/h; 30 mph)
- Econ cruising speed at S/L:
  - A — 26 knots (48 km/h; 30 mph)
  - B — 22 knots (40 km/h; 25 mph)
- Stalling speed, power on:
  - A — 19 knots (35·5 km/h; 22 mph)
  - B — 17 knots (31·5 km/h; 20 mph)
- Max rate of climb at S/L: A — 183 m (600 ft)/min
  - B — 198 m (650 ft)/min
- Service ceiling: A — over 3,050 m (10,000 ft)
- T-O run: A — 30·5 m (100 ft)
- T-O to 15 m (50 ft): A — 91 m (300 ft)
- Landing run: A — 18 m (60 ft)
- Range with optional fuel — 78 nm (145 km; 90 miles)

WEEDHOPPER CHOTIA GYPSY

This new (1980) design by Mr John Chotia had completed 20 hours' flying prior to its appearance at the 1980 Oshkosh Fly-in. Shown in an accompanying illustration, it has three-axis control and a 22·4 kW (30 hp) Chotia 460 direct-drive two-stroke engine. Standard fuel capacity is 7·5 litres (2 US gallons).

DIMENSIONS:
- Wing area, gross — 13·38 m² (144 sq ft)
- Propeller diameter — 1·12 m (4 ft 4 in)

WEIGHTS AND LOADING:
- Weight empty — 75 kg (165 lb)
- Max pilot weight — 106 kg (234 lb)
- Max T-O weight — 188 kg (415 lb)
- Max wing loading — 14·06 kg/m² (2·88 lb/sq ft)

PERFORMANCE:
- Never-exceed speed — 52 knots (96 km/h; 60 mph)
- Cruising speed — 35 knots (65 km/h; 40 mph)
- Stalling speed — 19 knots (35·5 km/h; 22 mph)
- Max rate of climb at S/L — 183 m (600 ft)/min
- Best glide ratio at 24 knots (44·5 km/h; 27·5 mph) — 16

WILLS WING

WILLS WING INC
1208-h, East Walnut, Santa Ana, California 92701
Telephone: (714) 547 1344

Formerly known as Sport Kites Inc, this company marketed in 1980 the Alpha, Omega and Omni hang gliders, of which details appeared in the 1980-81 *Jane's*.

No updated information was received from the company in 1981, when the major models in production appeared to be the **Harrier** and **Raven.** The latter is known

FLEXIBLE WING HANG GLIDER DATA

Manufacturer and Model	FAI class	Span: m/ ft-in	Leading-edge: m/ ft-in	Keel: m/ ft-in	Nose/ Billow angle: degrees	Wing area: m²/ sq ft	Wing aspect ratio
ARGENTINA							
LADAS *(1980: no 1981 details received)*							
Johan SH1	n.k.	9·20/30-2¼	5·60/18-4½	3·20/10-6	108/n.k.	16·00/172	n.k.
AUSTRALIA							
Moyes							
Maxi-Stinger Mark III	2	10·40/34-1½	6·10/20-0	4·00/13-1½	117/1·0	20·00/215	5·40
Maxi Stinger SP	2	10·00/32-9¾	5·79/19-0	4·00/13-1½	117/0·0	18·00/194	5·55
Redtail Stinger	1	9·00/29-6½	5·50/18-0½	4·00/13-1½	109/1·5	16·00/172	4·40
Mega Mark II	1	9·86/32-4	5·79/19-0	3·81/12-6	120/0·0	15·79/170	6·20
Mega Mark III	1	9·86/32-4	5·79/19-0	3·81/12-6	120/0·0	16·00/172	6·20
Mega Major	1	10·36/34-0	6·10/20-0	4·06/13-4	120/0·0	18·58/200	6·10
BULGARIA							
Zmiej-4	2	8·38/27-6	5·82/19-1	4·52/14-10	90/3·25	20·20/217·5	3·50
CANADA							
Birdman Enterprises *(1980: no 1981 details received)*							
Falcon I	1	8·23/27-0	5·49/18-0	3·96/13-0	96/1·5	15·33/165	4·40
Falcon II	1	8·84/29-0	5·79/19-0	4·27/14-0	96/1·5	16·72/180	4·60
MJ-5	2	9·60/31-6	5·49/18-0	3·05/10-0	120/0·5	16·72/180	5·50
MJ-6	3	10·06/33-0	5·61/18-5	3·35/11-0	115/0·0	17·00/183	5·70
XC	2	10·21/33-6	5·94/19-6	3·51/11-6	116/0·75	19·51/210	5·40
Talanchuk							
WT-9 Mars-Marta	1	12·60/41-4	6·00/19-8¼	4·57/15-0	142/1·0	17·80/191·5	8·90
WT-10 Mars-Beryl	1	11·20/36-9	5·62/18-5¼	4·57/15-0	142/0·0	17·00/183	7·30
CHINA							
Beijing Feiyan II	n.k.	9·00/29-6½	n.k.	4·50/14-9¼	100/n.k.	20·00/215·3	4·05
FRANCE							
La Mouette *(1980: no 1981 details received)*							
Atlas 14	2	9·30/30-6¼	5·36/17-7	3·50/11-5¾	120/0·5	13·80/148·5	6·25
Atlas 16	2	9·80/32-1¾	5·70/18-8½	3·50/11-5¾	120/0·5	15·50/166·8	6·20
Lotus 15	2	9·45/31-0	5·40/17-8½	3·50/11-5¾	120/2·0	15·50/166·8	5·70
Lotus 17	2	9·80/32-1¾	5·70/18-8½	3·50/11-5¾	120/2·0	17·00/183·0	5·60
Mouette 15	1	8·00/26-3	5·00/16-4¾	3·80/12-5½	102/2·0	15·00/161·5	4·30
Mouette 17	1	8·50/27-10½	5·21/17-1	4·00/13-1½	102/2·0	17·00/183·0	4·30
Mouette 19	1	9·00/29-6¼	5·40/17-8½	4·20/13-9¼	102/2·0	19·00/204·5	4·30
Mouette 23B	1	10·60/34-9¼	5·40/17-8½	4·50/14-9¼	102/2·0	23·00/247·6	4·30
GERMANY (Federal Republic)							
Ikarusflug							
Windspiel 2	3	10·00/32-9¾	5·00/16-4¾	4·00/13-1½	n.a.	15·00/161·5	6·70

Weedhopper Chotia Gypsy *(Howard Levy)*

Version of the Weedhopper JC-24B with stiffened wings *(Howard Levy)*

to have an airframe of aluminium alloy and composite materials, with a kingpost as standard; cables are of vinyl-coated stainless steel; double-surface sail is of Dacron. Other features include applied leading-edge pockets, adjustable camber, and washout control tips. It can be flown seated or prone, can be tow-launched, and (larger sizes only) can carry two persons. More than 1,000 Ravens have been sold.

Constructional details of Harrier are not known. All other known details for both models are given in the table.

WOLF
WILLIAM H. WOLF

18703 S.E. 44th Place, Issaquah, Wichita 98027

No news has been received of the Valkyrie rigid-wing hang glider since 1979. All known details appeared in the 1980-81 *Jane's*.

JC-24C version of the Weedhopper, new in 1981 *(Howard Levy)*

Weight: kg/lb	Glide ratio	Sink rate: m/ft per min	Pilot weight: kg (lb)	Stalling speed: knots (km/h; mph)	Remarks
21·0/46·3	n.k.	72/236	40-75 (88-165)	7 (13; 8·5)	
28·0/61·7	8·2	61/200	50-100 (110-220)	11 (20·5; 13)	Mk IV sail area 18 m² (194 sq ft); otherwise similar
26·0/57·3	8·5	61/200	45-86 (100-190)	10 (18; 11·5)	
25·0/55·1	5·5	87/285	55-75 (121-165)	10 (18; 11·5)	
31·8/70·0	7·5	61/200	66-100 (145-220)	16·5 (31; 19)	
31·8/70·0	10·0	61/200	63·5-118 (140-260)	17 (31·5; 19·5)	
34·0/75·0	8·0	56/185	63·5-127 (140-280)	15 (28·5; 17·5)	
n.k.	6·0/6·5	84/276	n.k.	n.k.	
20·4/45	6	87/285	45-72·5 (100-160)	10·5 (19·5; 12)	
20·4/45	6	87/285	61-91 (135-200)	10·5 (19·5; 12)	
22·2/49	8+	68·5/225	63·5-88·5 (140-195)	9 (16·5; 10)	
22·7/50	8·5	61/200	63·5-95 (140-210)	9 (16·5; 10)	
24·9/55	9+	61/200	82-104 (180-230)	9 (16·5; 10)	
27·0/59·5	10	60/197	55-85 (121-187)	10 (18; 11·5)	
26·0/57·3	11	60/197	55-85 (121-187)	11·5 (21; 13)	
17·0/19·0	10+	n.k.	approx 70 (154)	n.k.	
37·5-42					
24·5/54·0	9	54/177	65-75 (143-165)	10 (18; 11·5)	
25·0/55·1	9	54/177	65-100 (143-220)	10 (18; 11·5)	
23·0/50·7	7·5	66/217	40-75 (88-165)	11 (20; 12·5)	
24·0/52·9	7·5	66/217	65-95 (143-209)	11 (20; 12·5)	
15·0/33·1	6·5	78/256	40-55 (88-121)	10 (18; 11·5)	
17·0/37·5	6·5	78/256	50-68 (110-150)	10 (18; 11·5)	
19·0/41·9	6·5	78/256	65-100 (143-220)	10 (18; 11·5)	
30·0/66·1	6·5	78/256	100-180 (220-397)	11 (20; 12·5)	
28·0/61·7	11	54/175	60-95 (132-209)	13·5 (25; 16)	

Manufacturer and Model	FAI class	Span: m/ ft-in	Leading-edge: m/ ft-in	Keel: m/ ft-in	Nose/ Billow angle: degrees	Wing area: m²/ sq ft	Wing aspect ratio
POLAND							
Wierzbowski							
Vega SST-A	1	9·60/31-6	5·70/18-8½	3·50/11-5¾	116/0·8	18·60/200·2	4·9
Vega SST-B	1	10·10/33-1¾	6·00/19-8¼	3·13/10-3¼	116/0·5	18·90/203·4	5·4
Vega-Star	1	10·40/34-1½	5·90/19-4¼	2·70/8-10¼	126/0·0	15·80/170	6·8
USSR							
Antonov							
Slavutitch	n.k.	8·80/28-10½	5·60/18-4½	4·30/14-1¼	104/n.k.	17·50/188·4	4·45
UNITED KINGDOM							
Birdman *(1980: no 1981 details received)*							
Cherokee	1	10·06/33-0	5·49/18-0	3·35/11-0	120/0·0	18·58/200	n.k.
Skyhook Sailwings							
Cutlass A	n.k.	n.k.	n.k.	n.k.	n.k./0·0	14·68/158	n.k.
Cutlass B	n.k.	n.k.	n.k.	n.k.	n.k./0·0	16·26/175	n.k.
Cutlass C	n.k.	n.k.	n.k.	n.k.	n.k./0·0	18·39/198	n.k.
Silhouette A	2	n.k.	n.k.	n.k.	n.k.	16·72/180	5·72
Silhouette B	2	10·06/33-0	5·79/19-0	3·76/12-4	120/0·5	18·21/196	5·72
Silhouette C	2	n.k.	n.k.	n.k.	n.k.	20·25/218	5·72
UNITED STATES OF AMERICA							
CGS Aviation *(1980: no 1981 details received)*							
Falcon V	2	8·66/28-5	5·49/18-0	n.k.	105/1·65	14·31/154	5·25
Falcon V	2	9·63/31-7	6·10/20-0	n.k.	105/1·65	17·65/190	5·28
Falcon V	2	10·59/34-9	6·71/22-0	n.k.	105/1·65	21·18/228	5·30
Falcon 5+	2	8·66/28-5	5·49/18-0	n.k.	105/1·5	14·31/154	5·25
Falcon 5+	2	9·63/31-7	6·10/20-0	n.k.	105/1·5	17·65/190	5·28
Falcon 5+	2	10·59/34-9	6·71/22-0	n.k.	105/1·5	21·18/228	5·30
Delta Wing Kites and Gliders *(1980: no 1981 details received)*							
Phoenix 6D	n.k.	9·14/30-0	5·49/18-0	3·05/10-0	110/0·25	17·19/185	4·86
Phoenix Lazor 170	n.k.	9·45/31-0	5·26/17-3	2·44/8-0	130/0·0	15·79/170	5·70
Phoenix Lazor 190	n.k.	10·06/33-0	5·56/18-3	2·59/8-6	130/0·0	17·65/190	5·70
Phoenix Lazor II 155	n.k.	9·25/30-4	5·26/17-3	2·34/7-8	130/n.k.	14·49/156	5·90
Phoenix Lazor II 175	n.k.	9·80/32-2	5·56/18-3	2·44/8-0	130/n.k.	16·07/173	5·98
Phoenix Lazor II 195	n.k.	10·36/34-0	5·79/19-0	2·67/8-9	130/n.k.	18·21/196	5·90
Phoenix Viper	n.k.	10·06/33-0	5·49/18-0	2·51/8-3	n.k./n.k.	16·26/175	6·20
Flight Designs *(1980: no 1981 details received)*							
Lancer IVS	n.k.	9·91/32-6	n.k.	2·74/9-0	120/n.k.	n.k.	6·90
Lancer IVL	n.k.	10·36/34-0	n.k.	3·00/9-10	120/n.k.	n.k.	6·90
SL 180	n.k.	9·96/32-8	5·59/18-4	3·56/11-8	120/0·5	16·26/175	6·09
SL 200	n.k.	10·57/34-8	5·79/19-0	3·66/12-0	120/0·5	18·58/200	6·32
SL 220	n.k.	10·57/34-8	5·79/19-0	3·66/12-0	120/0·33	20·44/220	5·63
LEAF							
Model 140	n.k.	7·54/24-9	4·90/16-1	3·38/11-1	103/1·5	13·01/140	4·45
Model 155	n.k.	7·98/26-2	5·18/17-0	3·48/11-5	103/1·5	14·40/155	4·45
Model 170	n.k.	8·36/27-5	5·49/18-0	3·66/12-0	103/1·5	15·79/170	4·45
Model 185	n.k.	8·71/28-7	5·61/18-5	3·78/12-5	103/1·5	17·19/185	4·45
Model 200	n.k.	8·76/28-9	5·92/19-5	3·96/13-0	103/1·5	18·58/200	4·45
Model 215	n.k.	9·35/30-8	6·10/20-0	4·09/13-5	103/1·5	19·97/215	4·45
Monarch							
Monarch II	n.k.	9·91/32-6	n.k.	n.k.	121/0·35	16·72/180	5·87
Sky Sports							
Eaglet	1	8·46/27-9	5·79/19-0	3·96/13-0	105/1·5	13·10/141	4·00
Peregrine	2	10·36/34-0	5·79/19-0	2·90/9-6	130/1·0	15·79/170	6·00
Sirocco III	2	10·67/35-0	6·40/21-0	2·74/9-0	120/0·0	17·19/185	6·62
Spectra							
Aolus 150	n.k.	10·06/33-0	n.k.	n.k.	150/n.k.	13·94/150	7·26
Aolus 170	n.k.	10·36/34-0	n.k.	n.k.	150/n.k.	15·79/170	6·80
Aolus 190	n.k.	10·67/35-0	n.k.	n.k.	150/n.k.	17·65/190	6·45
Stratus							
Stratus V (small)	n.k.	9·91/32-6	n.k.	2·36/7-9	130/n.k.	13·75/148	7·26
Stratus V (large)	n.k.	10·67/35-0	n.k.	2·44/8-0	130/n.k.	15·24/164	7·46
Sunbird Ultralight Gliders *(1980: no 1981 details received)*							
Nova 150	2	8·92/29-3	4·88/16-0	3·66/12-0	130/1·0	13·94/150	5·70
Nova 170	2	9·45/31-0	5·18/17-0	3·81/12-6	130/1·0	15·79/170	5·70
Nova 190	2	10·00/32-9¾	5·49/18-0	3·96/13-0	130/1·0	17·65/190	5·70
Nova 210	2	10·52/34-6	5·79/19-0	4·27/14-0	130/1·0	19·51/210	5·70
Nova 230	2	10·97/36-0	6·10/20-0	4·42/14-6	130/1·0	21·37/230	5·70
Ultimate Hi *(1980: no 1981 details received)*							
Starship 170	n.k.	10·46/34-4	5·79/19-0	2·44/8-0	124/1·0	15·79/170	6·10
Starship 190	n.k.	10·67/35-0	5·94/19-6	2·59/8-6	124/1·0	17·65/190	6·00
Starship 220	n.k.	11·38/37-4	6·35/20-10	2·84/9-4	124/1·0	20·44/220	6·20
UP Inc							
Comet 135	1	8·86/29-1	5·22/17-1½	2·18/7-2	120/0·0	12·54/135	6·30
Comet 165	1	9·96/32-8	5·86/19-2¾	2·49/8-2	120/0·0	15·33/165	6·50
Comet 185	1	10·57/34-8	6·22/20-4¾	2·62/8-7	120/0·0	17·19/185	6·60
Wills Wing *(1980: no 1981 details received)*							
Harrier 147	n.k.	9·14/30-0	5·09/16-8½	n.k.	130/n.k.	13·66/147	6·10
Harrier 177	n.k.	10·16/33-4	5·64/18-6	n.k.	130/n.k.	16·44/177	6·30
Raven 149	n.k.	8·64/28-4	5·03/16-6	n.k.	n.k.	13·84/149	5·40
Raven 179	2	9·60/31-6	5·55/18-2½	3·75/12-3½	115/1·0	16·63/179	5·50
Raven 209	2	10·52/34-6	6·10/20-0	3·84/12-7¼	115/1·0	19·42/209	5·70
Raven 229	2	11·04/36-2½	6·40/21-0	4·11/13-6	115/1·0	21·27/229	5·70

n.k. = not known

Weight: kg/lb	Glide ratio	Sink rate: m/ft per min	Pilot weight: kg (lb)	Stalling speed: knots (km/h; mph)	Remarks
20·5/45·2	9	60/197	65-90 (143-198)	10 (18; 11·5)	
22·0/48·5	9+	60/197	68-97 (150-214)	10 (18; 11·5)	
24·0/53·0	n.k.	60/197	60-85 (132-187)	14 (26; 16·5)	
25·0/55·0	7	78/256	n.k.	13·5 (25; 16)	
24·9/55	9	61/200	55-95 (121-209)	13(24·5; 15)	Other sizes available
n.k.	10+	<61/200	61-82·5 (135-182)	16 (29; 18)	
n.k.	10+	<61/200	72·5-89 (160-196)	16 (29; 18)	
n.k.	10+	<61/200	86-98·5 (190-217)	16 (29; 18)	
n.k.	10	61/200	61-82·5 (135-182)	n.k.	
24·5/54	10	61/200	72·5-89 (160-196)	n.k.	
n.k.	10	61/200	86-98·5 (190-217)	n.k.	
17·7/39	8	61/200	45-63·5 (100-140)	16-19·5 (29-35·5; 18-22)	
18·6/41	8	61/200	61-84 (135-185)	16-19·5 (29-35·5; 18-22)	
19·5/43	8	61/200	77-113 (170-250)	16-19·5 (29-35·5; 18-22)	
19·1/42	n.k.	n.k.	45-63·5 (100-140)	n.k.	
20·0/44	n.k.	n.k.	61-84 (135-185)	n.k.	
20·9/46	n.k.	n.k.	77-113 (170-250)	n.k.	
25·4/56	n.k.	61/200	45-125 (100-275)	14 (26; 16)	Three other sizes available
n.k.	n.k.	n.k.	61-79 (135-175)	n.k.	
24·9/55	n.k.	n.k.	75-100 (165-220)	15 (27·5; 17)	
n.k.	n.k.	n.k.	54·5-72·5 (120-160)	n.k.	
n.k.	n.k.	n.k.	68-86 (150-190)	n.k.	
n.k.	n.k.	n.k.	81·5-100 (180-220)	n.k.	
28·6/63	n.k.	n.k.	68-91 (150-200)	n.k.	
20·9/46	n.k.	n.k.	n.k.	n.k.	
22·7/50	n.k.	n.k.	n.k.	n.k.	
n.k.	n.k.	n.k.	56·5-79·5 (125-175)	14 (26; 16)	
n.k.	n.k.	n.k.	77-95 (170-210)	13 (24·5; 15)	
n.k.	n.k.	n.k.	86-118 (190-260)	12·5 (23; 14)	
17·2/38	n.k.	n.k.	45·5-59 (100-130)	n.k.	
18·6/41	n.k.	n.k.	55-66 (121-145)	n.k.	
19·5/43	n.k.	n.k.	63-74·5 (139-164)	n.k.	
20·4/45	n.k.	n.k.	69-80 (152-176)	n.k.	
21·8/48	n.k.	n.k.	76-86·5 (167-191)	n.k.	
22·5/49·5	n.k.	n.k.	82·5-95 (182-210)	n.k.	
29·0/64	n.k.	n.k.	68-95 (150-210)	13·5 (24·5; 15)	
15·9/35	6	61/200	63·5 (86)	15 (27·5; 17)	
29·0/64	10+	55/180	77-100 (170-220)	12·5 (23; 14)	
24·5/54	10	55/180	72·5-91 (160-200)	13·5 (24·5; 15)	
24·5/54	n.k.	n.k.	54·5-72·5 (120-160)	n.k.	
24·5/54	n.k.	n.k.	63·5-81·5 (140-180)	n.k.	
26·3/58	n.k.	n.k.	72·5-91 (160-200)	n.k.	
23·5/52	n.k.	n.k.	54·5-77 (120-170)	n.k.	
25·4/56	n.k.	n.k.	63·5-95 (140-210)	n.k.	
20·4/45	8·5	58/190	45-63·5 (100-140)	15 (27·5;17)	
22·2/49	8·5	58/190	54-72·5 (120-160)	15 (27·5;17)	
23·6/52	8·5	58/190	63·5-86 (140-190)	15 (27·5;17)	
27·2/60	8·5	58/190	72·5-91 (160-200)	15 (27·5;17)	
28·1/62	8·5	58/190	79-100 (175-220)	15 (27·5;17)	
23·6/52	n.k.	n.k.	54-82 (120-180)	n.k.	
24·9/55	n.k.	n.k.	68-100 (150-220)	n.k.	
26·3/58	n.k.	n.k.	82-113 (180-250)	n.k.	
26/57	10+	55/180	45·5-84 (100-185)	13·5 (24·5;15) IAS	
29/64	10+	55/180	59-104·5 (130-230)	13·5 (24·5;15) IAS	
35·4/78	10+	55/180	68-113·5 (150-250)	13·5 (24·5;15) IAS	
24/53	n.k.	n.k.	47·5-93 (105-205)	n.k.	
28·6/63	n.k.	n.k.	68-113·5 (150-250)	n.k.	
20·4/45	n.k.	n.k.	41-63·5 (90-140)	n.k.	
22·2/49	10	+<61/200	52-77 (115-170)	16 (29;18)	
26·3/58	10	+<61/200	68-95 (150-210)	16 (29;18)	
28·1/62	10	+<61/200	77-104 (170-230)	16 (29;18)	

LIGHTER-THAN-AIR: AIRSHIPS

JAPAN

FUJI
FUJI MANUFACTURING CO LTD

16 Hotoku-cho, Kita-ku, Nagoya
Telephone: (052) 991 8171
President: Yasushi Koizumi

In the Autumn of 1974 Fuji Manufacturing Co began construction of a small remotely controlled pilotless research airship (construction number 2) which flew for the first time on 23 December 1974. (It is believed that No. 1 was the airship built by members of the Japan Experimental Aircraft Association and described under the JEAA heading in the 1975-76 *Jane's.)*

Airship No. 3 flew on 28 July 1975, this being a modified version of No. 2. No details are known of Nos 4, 5 and 6; but No. 7, which first flew on 21 November 1975, has an overall length of 6·0 m (19 ft 7½ in) and was supplied to Tohoku University Geographical Laboratory. Airship No. 12 began tests at the end of 1978, and was supplied to the Japan Geographical Survey Institute during February 1980.

FUJI MODEL 500 AERO-SHIP

Fuji Model 500 Aero-Ship, showing clearly the unique horizontal lifting surface

The word Aero-Ship, which forms a part of the designation of the Fuji Model 500, is an allusion to the unique aeroplane-type horizontal lifting surface which is attached to the lower structure of the gondola. The Aero-Ship's envelope is helium filled, and it is claimed that the wing not only provides additional lift but, in conjunction with the fairly large tail surfaces, helps to make the vessel highly manoeuvrable. Flown under radio control, the Aero-Ship can be utilised for such applications as aerial photography, meteorological and pollution measurement, seed and fertiliser distribution, and advertising.

The envelope of the Model 500 is of plastics, the gondola and cruciform tail surfaces of wood, with silk covering. Tailplane incidence is normally 3°, but is ground adjustable. Trim tabs in elevators. The wing is also of wood, with silk covering. Wing section is Göttingen Gö 535 and the wing is normally mounted without incidence. Two handling wheels, with tyres of 365 mm (14·4 in) diameter, are mounted beneath the wing, one at the aft end of the gondola and one beneath the lower fin. Two ENYA 60 IIIB aircooled engines, each developing 0·97 kW (1·3 hp), are mounted at the rear of the gondola, at each end of an outrigger. The forward part of the gondola is able to house a camera or other equipment. Radio control equipment and batteries are accommodated in the central area, and a fuel tank in the aft section.

Limited range of the radio control equipment used to date has prevented full evaluation of the Aero-Ship's performance, but a speed range of 5·4-43 knots (10-80 km/h; 6·2-50 mph) and ceiling of 1,000 m (3,280 ft) have been demonstrated.

Dimensions:	
Length overall	8·0 m (26 ft 3 in)
Wing span	3·26 m (10 ft 8½ in)
Wing chord, constant	0·68 m (2 ft 2¾ in)
Tailplane span	2·25 m (7 ft 4½ in)
Dimensions, envelope:	
Max diameter	1·90 m (6 ft 2¾ in)
Volume, gross	18·28 m³ (645 cu ft)
Areas:	
Wing, gross	1·01 m² (10·87 sq ft)
Rudders (total)	0·43 m² (4·63 sq ft)
Tailplane	0·86 m² (9·26 sq ft)
Elevators (incl tabs)	0·49 m² (5·27 sq ft)
Weights and Loading:	
Weight empty, equipped	22·7 kg (50 lb)
Weight, helium filled	4·0 kg (8·8 lb)
Max T-O weight	6·0 kg (13·2 lb)
Max wing loading	5·96 kg/m² (1·22 lb/sq ft)

UNITED KINGDOM

AIRSHIP INDUSTRIES
AIRSHIP INDUSTRIES LTD

Head Office: Meadows Court, Ramsey, Isle of Man
London Office: 2 York Street, London W1H 1FA
Telephone: (01) 486 3356
Telex: 299964 Bitco G

On 12 June 1980 Thermo-Skyships Ltd, an airship design company based on the Isle of Man, acquired the assets of Airship Developments Ltd (formerly Aerospace Developments, a CAA approved manufacturer) for the sum of £1·1 million: the resulting company is known as Airship Industries Ltd. It was the intention of this company to build initially two new Skyship 500s (formerly AD-500s), the first of which was to be flown during 1981 from new facilities, including an airship hangar, built on the Isle of Man. The original AD-500 airframe was to be used for static and structural testing.

In mid-1980, there was understood to be considerable interest in the Skyship 500 in Australia, and the Royal Navy and United States Navy were both willing to carry out flight test and evaluation of this vessel for such duties as maritime patrol, fishery protection, and anti-submarine warfare. As a result, Airship Industries planned to complete by 1981 the tooling to build to order a special-purpose AEW version known as the Skyship 5000. This would carry comprehensive avionics, possibly including the MSA (Mission System Avionics) which Marconi Avionics Ltd has developed for the British Aerospace Nimrod AEW.Mk 3.

Details of the Skyship 600 (formerly AD-600) appear in this entry, but the company will build examples of this vessel only to specific order.

On 11 July 1980, Redcoat Cargo Airlines, a Surrey-based freight operator, announced that it had concluded an agreement with Airship Industries covering the supply of four rigid cargo airships, with options for ten more. Allocated the provisional designation Model 4060 by Airship Industries, detail design was in progress in the Summer of 1980, but since that time the company's efforts have been concentrated on completion of the second Skyship 500, for extensive flight testing. It was nearing completion in early May 1981, and was expected to be flying in July.

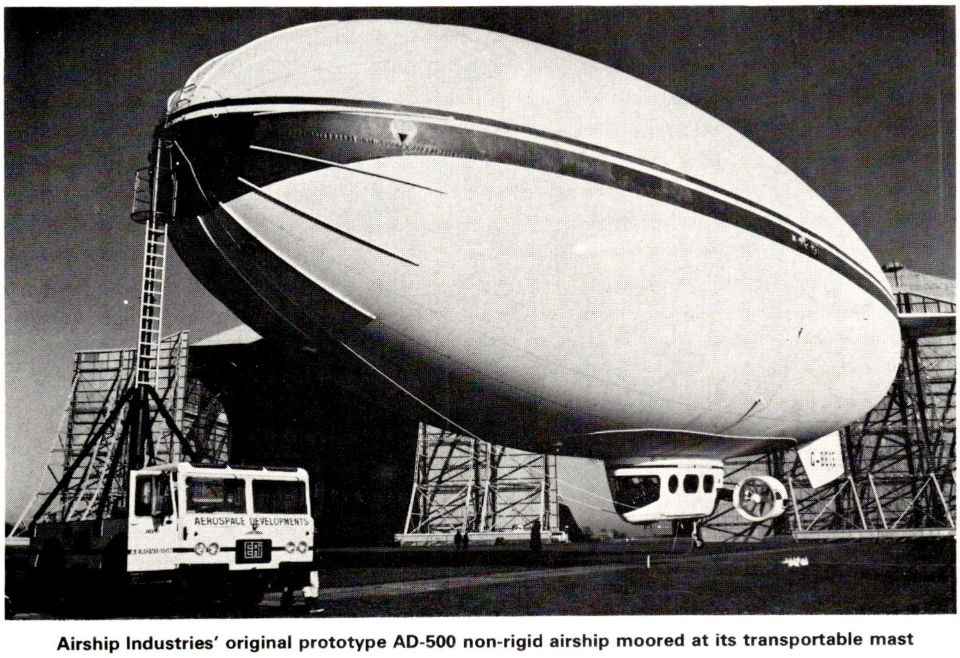

Airship Industries' original prototype AD-500 non-rigid airship moored at its transportable mast

AIRSHIP INDUSTRIES SKYSHIP 500

The envelope of the Skyship 500 (originally AD-500) is manufactured from a strong-in-weft single-ply polyester fabric, developed by Airship Industries, and coated with titanium dioxide polyurethane sealant to minimise loss of the helium lifting gas. The nose structure consists of a domed disc, moulded from Kevlar and carrying the fitting by which the airship is moored to its transportable mast. Two ballonets, which together comprise 25 per cent of the envelope volume, are installed fore and aft, so that differential inflation will provide some degree of fore and aft trim. Catenary curtains carry 12 main cables of Kevlar within the envelope for suspension of the gondola. The tail unit is of conventional cruciform layout, each surface being attached to the envelope at its root and braced by four wires on each side. All four surfaces are constructed from interlocking ribs and spars of Fibrelam and have GRP skins: their hinged rudder and elevator control surfaces are cable-operated and each has a trim tab.

The gondola is a one-piece moulding of Kevlar-reinforced plastics, with flooring and bulkheads of Fibrelam panels; those which form the engine compartment at the rear of the gondola are faced with stainless steel for fire protection. There is accommodation for a pilot and co-pilot, with dual controls, although the vessel is designed to be flown by a single pilot; there are seven individual seats for passengers. Door on port side, just aft of flight deck. Transparent plastics dome in ceiling of flight deck for visual internal inspection of envelope. Ballast in the form of lead shot is contained in a box situated below the crew seats, and water ballast totalling 513 kg (1,130 lb) is contained in tanks at the rear of the gondola. A single fully-castoring wheel with pneumatic tyre is mounted on an oleo-pneumatic strut beneath the gondola to ensure that the latter is not damaged by contact with the ground.

Power plant comprises two 149 kW (200 hp) Porsche six-cylinder aircooled motorcar engines. Each drives, via a modified Lynx helicopter tail rotor gearbox, a ducted propulsor designed by Airship Industries, consisting of a Hoffmann five-blade reversible-pitch metal propeller rotating within an annular duct constructed of GRP, reinforced with carbonfibre. Each propulsor can be rotated about its pylon attachment to the gondola through an arc of 210°: 90° upward and 120° downward. The 30° of movement past centre with the ducted propulsors facing downward is to maintain cable tension while the airship is being moored to its mast. The vectored thrust available throughout this range of 210° simplifies take-off and landing. A fuel tank, capacity 536 litres (118 Imp gallons), is mounted at the rear of the engine compartment. Max fuel capacity is 1,577 litres (347 Imp gallons). Engine modifications include provision of automatic mixture control, fuel injection and electronic ignition. The 28V electrical system is supplied by engine-driven alternators. Avionics installed in the prototype Skyship 500 include Bendix 2000 series dual nav/com, ADF, VOR/ILS and Omega, Bendix 1400 weather radar, and Decca Doppler 80 nav system with PBDI and automatic chart display.

Dimensions, envelope:

Length overall	50·00 m (164 ft 0½ in)
Max diameter	14·00 m (45 ft 11¼ in)
Volume, gross	5,130 m³ (181,160 cu ft)
Volume, ballonets	1,282·5 m³ (45,290 cu ft)

Dimensions, gondola:

Length overall	9·25 m (30 ft 4 in)
Height	2·26 m (7 ft 4⅞ in)
Max width	2·41 m (7 ft 10¾ in)

Weight:

Nominal payload to 610 m (2,000 ft) 2,540 kg (5,600 lb)

Performance (estimated):

Max level speed	62 knots (114 km/h; 71 mph)
Cruising speed	49 knots (90 km/h; 56 mph)
Pressure altitude	3,050 m (10,000 ft)
Range with max fuel, 15% reserves: at 15 knots (28 km/h; 17 mph)	6,375 nm (11,800 km; 7,335 miles)
at max speed	600 nm (1,110 km; 690 miles)

AIRSHIP INDUSTRIES SKYSHIP 600

Airship Industries has completed the design of what is basically a 'stretched' version of the Skyship 500. The increased volume of the envelope will provide approximately 1,016 kg (2,240 lb) more lift, and the lengthened gondola would have a usable floor area of 11·15 m² (120 sq ft). Developed to meet the needs of operators who require greater payload and/or endurance than are provided by the Skyship 500, it is considered to be suited to roles which require long-endurance patrols or extended time on station. With the maximum fuel that can be carried, patrols of up to three days' duration would be possible, and the large cabin would permit a high degree of crew comfort and efficiency.

Airship Industries believes that the size and capability of the Skyship 600 would make it particularly suitable for deployment in a surveillance role, but other obvious military applications include minesweeping, ASW and electronic warfare. It is understood that the Royal Navy and US Navy are interested in evaluating such a vessel for maritime applications.

Envelope: Of ellipsoidal form with parallel mid-body. Four load curtains and two ballonets. Envelope fabric comprises a Terylene strength layer, with a weather protection coating of titanium oxide-loaded polyurethane, and an internal helium-retention layer of Saran film.

Tail Fins: Of conventional cruciform layout, each aerofoil section has ribs and spars constructed from Ciba Geigy Fibrelam-bonded honeycomb structure panels. Control surfaces have setback hinges, balance tabs, and adjustable gearing to minimise control forces.

Gondola: A one-piece shell, moulded from Kevlar. Fireproof bulkheads in engine bay are honeycomb panels, faced with stainless steel skins. For patrol, or associated military duties, accommodation can be provided for a crew of up to seven, with bunks, galley, mess, and toilet facilities. In a passenger-carrying role, up to 20 passenger seats can be installed at a pitch of 91 cm (36 in). For use as a cargo carrier in an emergency, the gondola can be configured to provide a maximum disposable freight capacity of 28·32 m³ (1,000 cu ft).

Power Plant: Two 186·4 kW (250 hp) Porsche Turbo 930 turbocharged engines. Each drives, via a Westland bevel gearbox, a ducted propulsor designed by Airship Industries which houses a Hoffmann five-blade variable-pitch fan. Each propulsor can be rotated about its pylon attachment to the gondola through an arc of 180°, thus simplifying the tasks of take-off and landing. A fuel tank, mounted at the rear of the engine compartment, has a maximum capacity of 2,273 litres (500 Imp gallons).

Avionics and Equipment: Full equipment and instrumentation for IFR operation by day or night is standard. Optional avionics considered suitable for surveillance and patrol include Bendix RDR 1400 digital radar, Ferranti Sea Spray, or MEL Marec. Bendix radar and Omega VLF navigation are optional for civil versions of the Skyship 600, and are considered suitable also for a surveillance role.

Dimensions, envelope:

Length overall	56·00 m (183 ft 8¾ in)
Max diameter	14·00 m (45 ft 11¼ in)
Height overall	18·65 m (61 ft 2¼ in)
Tail fin span	17·00 m (55 ft 9¼ in)
Volume, gross	6,055 m³ (213,830 cu ft)
Volume, ballonets	1,332·1 m³ (47,040 cu ft)
Helium volume, at 85 per cent inflation	5,146 m³ (181,730 cu ft)
L/D ratio	4·0

Dimensions, gondola:

Length overall	12·10 m (39 ft 8¼ in)
Length between cabin bulkheads	7·00 m (22 ft 11½ in)
Height	1·96 m (6 ft 5¼ in)
Max width	2·41 m (7 ft 10¾ in)

Weights (estimated, A: patrol configuration; B: passenger configuration):

Weight empty, basic: A, B	3,331 kg (7,344 lb)
Fuel weight: A	1,633 kg (3,600 lb)
B	816 kg (1,799 lb)
Max T-O weight: A	6,220 kg (13,713 lb)
B	6,087 kg (13,420 lb)
Gross lift at 85 per cent inflation, ISA, plus vectored thrust	6,270 kg (13,823 lb)

Performance (estimated):

Max level speed	65 knots (120 km/h; 75 mph)
Max cruising speed	50 knots (93 km/h; 58 mph)
Cruising speed (43·3 kW; 58 hp rating per engine)	40 knots (74 km/h; 46 mph)
Cruising speed (26·1 kW; 39 hp rating per engine)	35 knots (64 km/h; 40 mph)
Max rate of climb at S/L	610 m (2,000 ft)/min
Rate of climb at S/L, at max cruising power	549 m (1,800 ft)/min
Cruising altitude	610 m (2,000 ft)
Pressure height (85 per cent inflation at S/L)	1,675 m (5,500 ft)
Max altitude pressure height, with ballonets full at S/L	2,440 m (8,000 ft)

AIRSHIP INDUSTRIES MODEL 4060

This airship, under detail design for Redcoat Cargo Airlines, is of rigid configuration and will have a cargo capacity of 1,070 m³ (37,755 cu ft), compared with the total cargo volume of 687 m³ (24,260 cu ft) for a Boeing 747-200F Freighter. Apart from offering considerably more economical operation than a fixed-wing aircraft, which it is believed will open up new export markets because of lower airfreight rates, it is anticipated that such a vessel will allow cargoes to be flown directly from factory to factory, since landing sites can be established at low cost. It is expected that construction of the prototype will begin in early 1982.

As envisaged prior to final design, the Model 4060 will have conventional lines for a rigid airship, except that the flight deck will be located in the nose of the vessel. It will have four cruciform tail fins, with control surfaces; the power plant will comprise four 835 kW (1,120 shp) Pratt & Whitney Aircraft of Canada PT6A-50 turboprop engines.

Dimensions:

Length overall	182·88 m (600 ft)
Diameter	36·58 m (120 ft)
Span over tail fins	51·21 m (168 ft)
Volume, gross	128,559 m³ (4,540,000 cu ft)
Volume, gas	119,498 m³ (4,220,000 cu ft)

Areas:

Hull surface area	16,201 m² (174,395 sq ft)
Total net fin area	826·8 m² (8,900 sq ft)

Weights:

Weight empty	51,030 kg (112,500 lb)
Aerostatic lift at pressure altitude	103,455 kg (228,080 lb)
Max vertical T-O weight	113,800 kg (250,890 lb)

Performance (estimated, at 915 m; 3,000 ft altitude):

Design diving speed	103·5 knots (192 km/h; 119 mph)
Max level speed at max continuous power	95·3 knots (177 km/h; 110 mph)
Cruising speed at cruising power	67·6 knots (126 km/h; 78 mph)
Ferry range, still air, at cruising speed of 50 knots (92·5 km/h; 57·5 mph)	10,420 nm (19,310 km; 12,000 miles)
Range with 58 ton payload	435 nm (805 km; 500 miles)
Range with 45 ton payload	1,735 nm (3,215 km; 2,000 miles)
Range with 40 ton payload	2,170 nm (4,020 km; 2,500 miles)

Artist's impression of Airship Industries Model 4060 rigid cargo airship

CAMERON
CAMERON BALLOONS LTD

1 Cotham Park, Bristol BS6 6BZ
Telephone: (0272) 41455

Directors:
D. A. Cameron, BSc (Aero Eng), MIE, MRAeS
Kim Cameron
R. I. M. Kerr, BSc, MRAeS
Tom Sage, ARPS, AIIP

Cameron Balloons Ltd, which designs and manufactures a wide range of hot-air balloons (which see), also designed and built the world's first hot-air airship, the D-96, of which production versions are available. A smaller airship, designated D-38, has now completed its development programme and was in production in 1981 to customer order.

CAMERON D-96 HOT-AIR AIRSHIP

First flight of the prototype (G-BAMK) was made at Wantage, Berkshire, on 7 January 1973. Since that time a considerable amount of work has been carried out to improve and develop the airship, and the current production model has two vertical and two horizontal stabilisers. An improved method of suspending the gondola has also been developed, eliminating distortion of the envelope which occurred in the early stages. During 1978, further development led to an increase in the length and volume of the envelope. Like the envelopes of Cameron hot-air balloons, this is made from a light but high-strength nylon fabric. A lightweight gondola carries the propane burner, gas supply, pilot and power plant. A maximum of two passengers can be accommodated.

First two examples of Cameron's new D-38 hot-air airship

Single-seat gondola of the D-38 airship

Power plant is a 33·5 kW (45 hp) 1,600 cc Volkswagen modified motorcar engine, using propane as fuel and driving a large-diameter semi-shrouded pusher propeller.

The first production airship was completed for a customer in the USA; several more have been constructed for customers in Australia, Belgium, Canada, France, the Netherlands, Sweden and the UK.

The following data apply to the current version, with the envelope lengthened by 3·66 m (12 ft):

DIMENSIONS, ENVELOPE:	
Length overall	34·14 m (112 ft 0 in)
Max diameter	13·72 m (45 ft 0 in)
Volume, gross	2,917 m³ (103,000 cu ft)
PERFORMANCE:	
Max speed	13 knots (24 km/h; 15 mph)
Turning radius at 8·7 knots (16 km/h; 10 mph)	30·5 m (100 ft)
Endurance	2 h

CAMERON D-38 HOT-AIR AIRSHIP

This hot-air airship, which benefits from experience gained during development of the larger D-96, was first flown in a kite balloon configuration, as illustrated in the 1980-81 *Jane's*. It was the company's original intention to market it in that form, under the designation C-38; but the decision has now been made to manufacture only the single-pilot D-38 airship, to complement the two-man D-96. The kite balloon (G-BGEP) served also as the prototype of the powered airship, which flew for the first time at Ashton Park, Bristol, on 25 September 1980.

Construction of the envelope is similar to that of the D-96, made from lightweight high-strength ripstop nylon, but with only one vertical and two horizontal stabilisers. A completely new gondola has been designed, with a basic structure of welded steel tube, a plywood floor and fabric enclosure. This accommodates the pilot and two cylinders of propane gas; and the steel frame provides mountings for the single standard Cameron burner and the power plant, which comprises a 250 cc Robin single-cylinder two-stroke engine. It drives, through a V-belt reduction gear, a two-blade pusher propeller; its petrol/oil fuel mixture is contained in a tank mounted alongside the engine.

The first production D-38 airship (G-CULT) was completed for Colt Cars, and further examples were under construction to customer order in early 1981.

DIMENSIONS, ENVELOPE:	
Length overall	22·86 m (75 ft 0 in)
Height, incl gondola	15·24 m (50 ft 0 in)
Max diameter	9·96 m (32 ft 8 in)
Volume, gross	1,076 m³ (38,000 cu ft)
DIMENSIONS, GONDOLA:	
Length	1·65 m (5 ft 5 in)
Propeller diameter	1·32 m (4 ft 4 in)
WEIGHTS:	
Envelope	89 kg (196 lb)
Gondola, empty	91 kg (200 lb)
Propane fuel	23·6 kg (52 lb)
PERFORMANCE:	
Max speed	12 knots (22 km/h; 14 mph)

THUNDERCOLT

THUNDERCOLT BALLOONS

SALES OFFICE: Thunder Balloons Ltd, 75 Leonard Street, London EC2A 4QS
Telephone: (01) 739 0775/0776
Telex: 261234 ref H5758D

WORKS: Colt Balloons Ltd, Maesbury Road, Oswestry, Shropshire SY10 8HA
Telephone: (0691) 2216
Telex: 35503
DIRECTORS: see Balloons subsection

THUNDERCOLT BALLOONS A.S.80

ThunderColt Balloons has been involved in the design and development of a hot-air airship during the past three years. A one-fifth-scale radio controlled model was first flown successfully, to confirm flight characteristics before construction of the full-size airship was initiated. The latter was completed in early 1981, and the first flight of the A.S.80, as it is designated, was made on 12 April 1981.

Comparatively few details of this airship have yet been released. It is powered by a Honda CX500 V twin-cylinder watercooled engine. The envelope length is 32·00 m (105 ft 0 in), its diameter 11·50 m (37 ft 8¾ in), and volume 2,209 m³ (78,000 cu ft). A crew of two can be carried in the gondola, which also contains four fuel cylinders for the two burners. Cruising speed is 15-20 knots (28-37 km/h; 17-23 mph), maximum speed 25 knots (46 km/h; 29 mph), and endurance 2-4 hours.

UNITED STATES OF AMERICA

BOLAND

BOLAND BALLOON

Pine Drive, RD 2, Burlington, Connecticut 06013
Telephone: (203) 673 1307

BOLAND A-1 ALBATROSS

Mr Brian Boland, an art and photographic teacher at Farmington High School, Connecticut, has, with his wife Kathy, designed and built several hot-air balloons, described in the Balloons section, as well as a hot-air airship named *Albatross*. Construction began in January 1975 and it was flown for the first time on 11 October 1975.

The envelope is made from 32 separate strips of polyurethane-coated rip-stop nylon. The gondola, which can accommodate four persons, has a basic structure of light alloy, around which wicker sidewalls have been woven. It has oak landing skids on its undersurface, and is mounted on a wheeled trailer for easy ground handling, this being left on the ground on lift-off. The gondola is suspended from the envelope by 28 flexible steel cables which run through its light alloy tubular framework and are attached to two catenary suspension curtains. The tubular framework of the gondola serves also as mounting for three 8 million BTU propane burners. The fuel supply comprises six low-pressure cylinders, with a combined capacity of 227 litres (60 US gallons) of liquid propane, and a 22·7 litre (6 US gallon) tank of fuel/oil mixture for the propulsion engine. This is a 29·8 kW (40 hp) Rockwell JLO two-stroke engine, driving a Banks Maxwell two-blade ducted pusher propeller.

On the basis of flight experience with *Albatross* Mr Boland designed a number of improvements, including shortening each of the original catenary cables by 2·29 m (7 ft 6 in); adding 7·32 m (24 ft) of catenary curtain aft of the original curtains; and improving the structural integrity of the gondola by heliarc welding of all metal fittings.

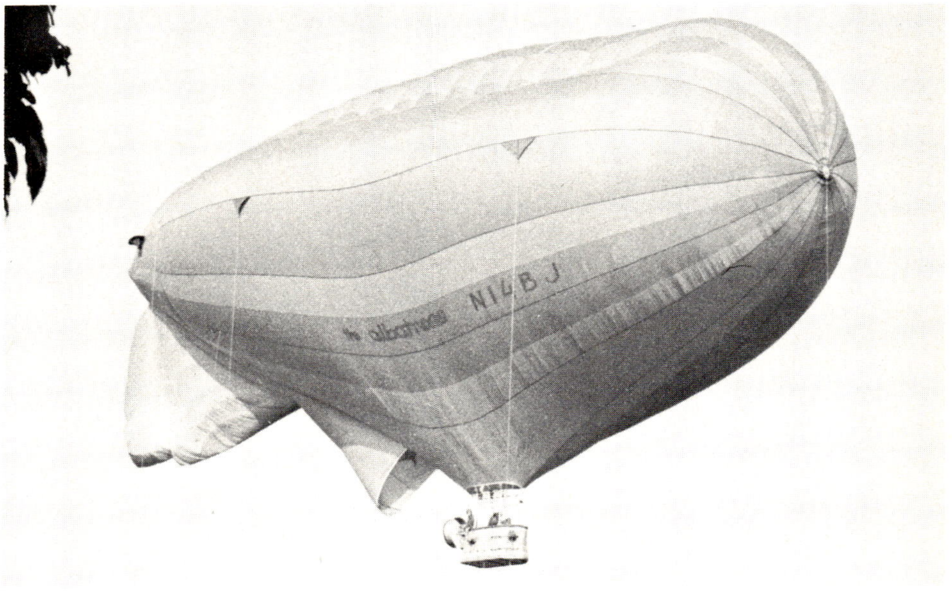

Boland A-1 *Albatross* in flight, giving clear view of fin and rudder pressurisation scoop

The rigid-frame fabric-covered fin and rudder, with which *Albatross* was flown originally, was replaced by a multi-chambered inflatable fin and rudder with three times the surface area. Initial flight testing showed that this could deflate during prolonged turns, and an enlarged-scoop fin pressurisation sleeve has been designed and installed. The *Albatross* was flown in this form for the first time at the end of August 1979. Testing has shown that it is possible to operate this vessel as a free balloon, with the power plant closed down.

A second hot-air airship, named *Roover*, was designed in 1978. This will have a volume of about 1,982-2,265 m³

(70,000-80,000 cu ft) and is described as being ultra-lightweight, responsive and compact. Construction was expected to start in 1980, but no recent information has been received from Mr Boland.

The following details apply to *Albatross:*

DIMENSIONS, OVERALL:	
Length	34·14 m (112 ft 0 in)
Height	19·81 m (65 ft 0 in)
Width	15·24 m (50 ft 0 in)
Volume	3,965 m³ (140,000 cu ft)
DIMENSIONS, GONDOLA:	
Length overall	3·35 m (11 ft 0 in)
Height	3·05 m (10 ft 0 in)
Width	1·22 m (4 ft 0 in)
WEIGHTS:	
Flight weight, fuelled	658 kg (1,450 lb)
Max design gross weight	1,089 kg (2,400 lb)
PERFORMANCE:	
Max speed	approx 8·7 knots (16 km/h; 10 mph)
Turning radius	approx 91 m (300 ft)

GOODYEAR

THE GOODYEAR TIRE & RUBBER COMPANY—AIRSHIP OPERATIONS

1144 East Market Street, Akron, Ohio 44316
Telephone: (216) 794 4045
CHAIRMAN OF THE BOARD AND CHIEF EXECUTIVE OFFICER:
Charles J. Pilliod Jr
VICE-PRESIDENT, PUBLIC RELATIONS:
William L. Newkirk

Goodyear has built over 300 airships, more than any other company in the world. Of these, over 250 were constructed under contract for the US Army and Navy, and included the USS *Akron* and USS *Macon,* the largest rigid airships constructed in the USA. The remainder have been commercial airships, of which the first was the *Pilgrim*, launched in 1925.

Goodyear operates currently four non-rigid airships for public relations and sales promotion activities: *America, Columbia, Europa* and *Enterprise*. *Europa* is based near Rome, Italy, the other three airships in the USA. Details of the newly constructed *Enterprise* follow. In size, equipment and performance it is generally the same as the three other airships in the fleet.

Goodyear has been studying the development of heavy-lift airships, and believes that existing technology can provide such a vehicle in the 1980s if adequate funding is available. The company suggests that a helium-filled airship, using helicopter-type rotors for lift and conventional propellers for forward flight, would prove more economical than helicopters in a flying-crane role, and would also have a very much greater lift capability. Studies indicate that by using such techniques a 135 m (443 ft) long, 73,625 m³ (2·6 million cu ft) airship would have a useful lifting capacity of 76·2 tonnes (75 tons).

The non-rigid airship *Enterprise*, constructed by Goodyear during 1979

ENTERPRISE

The *Enterprise* (N1A) was constructed during 1979 and erected at Houston, Texas, where it was flown for the first time in late October 1979. Some two weeks later it was flown to its base at Pompano Beach, Florida. Like its sister ships, it has a gross volume of 5,740 m³ (202,700 cu ft) and envelope surface area of 2,006 m² (21,600 sq ft). The envelope is made of two-ply Neoprene-coated Dacron and is helium-filled. On each side is a four-colour sign 32·00 m (105 ft 0 in) long and 7·47 m (24 ft 6 in) high, containing 3,780 lamps to flash static or animated messages. These can be read at a distance of 1·6 km (1 mile) when the airship is cruising at a height of 305 m (1,000 ft). A turbojet APU, mounted in a removable pod on the undersurface of the 'ship's gondola, drives a 500A 28V generator to supply electrical power for the signs and their control equipment. The turbojet is designed to operate without developing any appreciable amount of forward thrust for the airship.

The gondola, attached to the undersurface of the envelope, has accommodation for a pilot and six passengers, and has a single non-retractable landing wheel mounted beneath it.

Power plant consists of two 157 kW (210 hp) Continental IO-360-D flat-six engines, each driving a Hartzell two-blade metal reversible-pitch pusher propeller. Tankage is provided for 527 litres (138 US gallons) of fuel.

Model of heavy-lift airship studied by the Goodyear company

DIMENSIONS, OVERALL:	
Length	58·67 m (192 ft 6 in)
Width	15·24 m (50 ft 0 in)
Height	18·14 m (59 ft 6 in)
DIMENSIONS, ENVELOPE:	
Length	58·00 m (190 ft 3½ in)
Max diameter	14·00 m (45 ft 11½ in)
Fineness ratio	14·4
Volume, gross	5,740 m³ (202,700 cu ft)
Volume, ballonets	1,662·2 m³ (58,700 cu ft)
DIMENSIONS, GONDOLA:	
Length overall	6·93 m (22 ft 9 in)
Height	2·47 m (8 ft 1¼ in)
Height, incl landing gear	3·59 m (11 ft 9½ in)
Width at ceiling	2·13 m (7 ft 0 in)
Width at floor	1·31 m (4 ft 3½ in)
WEIGHTS:	
Weight empty	4,252 kg (9,375 lb)
Max design gross weight	5,824 kg (12,840 lb)
PERFORMANCE:	
Max speed	43·5 knots (80 km/h; 50 mph)
Normal cruising speed	30-35 knots (56-64 km/h; 35-40 mph)
Max rate of climb at S/L	732 m (2,400 ft)/min
Max rate of descent	427 m (1,400 ft)/min
Normal operational altitude	305-915 m (1,000-3,000 ft)
Service ceiling	2,285 m (7,500 ft)
Endurance at cruising speed	approx 10 h

HOV-AIR-SHIP

HOV-AIR-SHIP INC

1427 West 22nd Street, Miami Beach, Florida 33140
Telephone: (305) 531 8968/8969

HOV-AIR-SHIP HX-1

Designed by Mr Saul I. Slater, who formed Hov-Air-Ship Inc to hold the patents of this and other designs he has produced, the HX-1 is a small-scale remotely controlled airship 5·79 m (19 ft) long and 2·13 m (7 ft) in diameter, with cruciform tail surfaces. It has been test-flown tethered to a 110V electrical power source and piloted by a controller on the ground; a magnetic device was provided to moor the airship, without any need for groundcrew assistance.

A feature which at the present time is unique to this airship is an ability to hover over a fixed point, and ascend or descend above that point.

The HX-1 was acquired by the US Navy, and was delivered on 16 October 1978 for extensive tests at the Naval Air Development Center, Warminster, Pennsylvania. In late 1979, the Navy planned to exchange the electric

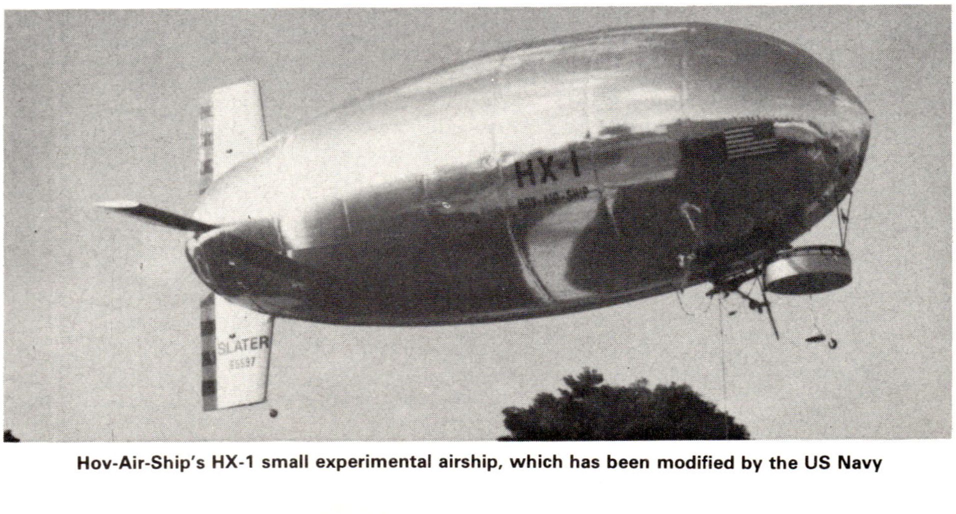

Hov-Air-Ship's HX-1 small experimental airship, which has been modified by the US Navy

motors which then powered the HX-1 for nitro-alcohol fuelled engines of the type used to power model aircraft. However, it was decided subsequently to retain electric power, and in 1980 the HX-1 was sent by the Navy to the California Polytechnic Institute for two additional electric motors of the same type to be installed. The resulting 'quad-rotor' airship is undergoing further tethered testing in a slightly 'heavy' form.

Mr Slater has proposed an enlarged HX-2 version to the US Navy and Coast Guard. This would be approximately 45·72 m (150 ft) long and 9·14 m (30 ft) in diameter, with an envelope volume of 2,294 m³ (81,000 cu ft), and would carry a crew of two. Power would be provided by two conventional aircraft engines, each of 168 kW (225 hp). The envelope would be of the super-pressure type, with the lifting gas maintained at approximately 3·5 in of water pressure. It is estimated that a maximum speed of 65 knots (121 km/h; 75 mph) could be attained by such an airship.

ILC

ILC DOVER (Division of ILC Industries Inc)

PO Box 266, Frederica, Delaware 19946
Telephone: (302) 335 3911

ILC Dover designs and manufactures a range of unmanned non-rigid aerostats of the kite balloon type. It has a variety of balloons available as standard, but the company offers specialist services to develop a balloon system to satisfy precisely a user's requirements. The standard range covers aerostats from as small as 1·70 m³ (60 cu ft) to 7,079 m³ (250,000 cu ft) radar platforms. Brief details of three examples follow:

ILC DOVER AG-60 BALLOON

ILC's AG-60 balloon is designed for any general use where a highly visible ground marker is required for such applications as aerial spraying and surveying. Its envelope is made from heat-sealed white urethane-coated nylon, and incorporates three inflatable fins forming an inverted 'Y' to ensure stable flight in wind velocities from 0 to 39 knots (72·5 km/h; 45 mph) on a 91 m (300 ft) nylon tether. The fins are designed so that no rigid members are required to increase stiffness or prevent flutter. It is equipped with an automatic self-closing fill port to provide easy inflation, and a pressure relief valve to prevent overpressurisation. Handled easily by one operator, it can be transported, flown and retrieved from a standard open pickup truck.

DIMENSIONS, ENVELOPE:	
Length overall	3·05 m (10 ft 0 in)
Max diameter	1·17 m (3 ft 10 in)
Volume, gross	1·70 m³ (60 cu ft)
WEIGHT:	
Unfilled	1·4 kg (3 lb)

ILC DOVER SSP-6000 TETHERED AEROSTAT

The SSP-6000 (Stable Sensor Platform) is a non-rigid aerostat based upon the design of the successful Family II D7 developed by the Advanced Research Projects Agency for the surveillance and communications relay requirements of the US Department of Defense. It consists of an elliptically-shaped envelope, containing a lenticular-shaped ballonet, and incorporating three fins forming an inverted 'Y'. Each fin has 13 spar webs to form an NACA 0018 aerofoil section. The SSP-6000 has a pressure control system which maintains the envelope's design shape, with redundant relief valves to relieve pressure resulting from superheat, atmospheric pressure drop, or increase in altitude. A battery-powered electric blower, with more than five days' sustained in-flight capability, maintains pressure during descent, atmospheric pressure increase, or temperature drop. The fins are air-inflated and coupled to the ballonet, so that the complete aerostat is maintained at the same pressure.

Flight loads are taken by 14 confluence lines and patches, seven port and seven starboard; four close-haul lines, two port and two starboard, are provided for ground handling. ILC has developed a tower and monorail system for mooring aerostats of this type, making them free to weathervane and to survive without damage winds in excess of 65 knots (120 km/h; 75 mph). To secure the aerostat to this mooring tower, eight load patches are distributed equally around the nose of the envelope.

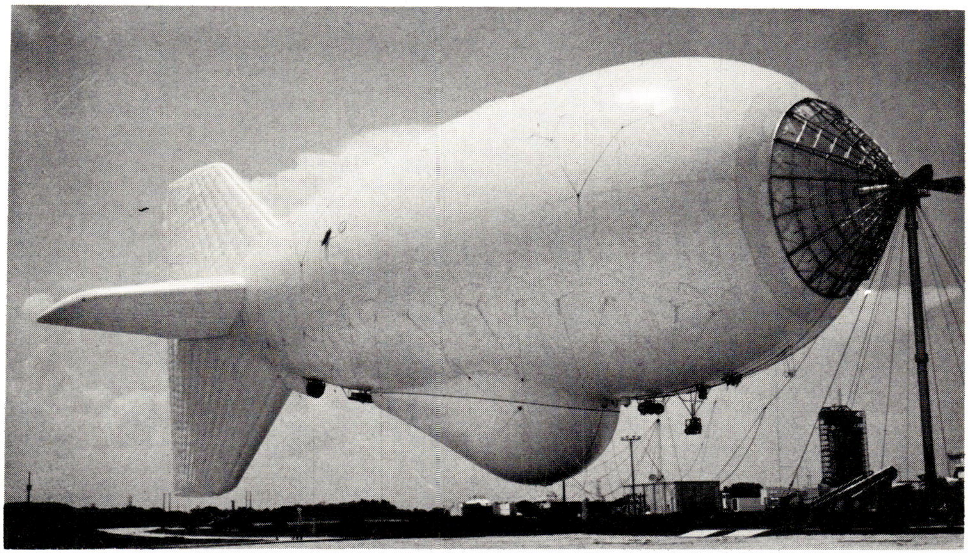

ILC Dover's 250,000 cu ft aerostat, carrying Seek Skyhook surveillance equipment, during tests at Cape Canaveral

DIMENSIONS, ENVELOPE:	
Length overall	15·35 m (50 ft 4¼ in)
Max diameter	4·98 m (16 ft 4 in)
Tail fin span, horizontal	5·06 m (16 ft 7¼ in)
Tail fin span, vertical	6·10 m (20 ft 0 in)
Max helium volume	170·0 m³ (6,000 cu ft)
Max ballonet volume	28·33 m³ (1,000 cu ft)
Nominal helium volume at S/L	147·25 m³ (5,200 cu ft)
Nominal ballonet volume at S/L	22·65 m³ (800 cu ft)
WEIGHTS:	
Gross lift, 95% helium purity	144 kg (317 lb)
Free lift	14·5 kg (32 lb)
Aerostat weight	72·6 kg (160 lb)
Net lift	56·7 kg (125 lb)
PERFORMANCE:	
Max design altitude	1,525 m (5,000 ft)
Max wind velocity, operational	45 knots (83·5 km/h; 52 mph)
Max wind velocity, moored	65 knots (120 km/h; 75 mph)

ILC DOVER 250,000 CU FT AEROSTAT

ILC has developed a large-volume aerostat able to lift a payload of 454 kg (1,000 lb) to an altitude of 3,660 m (12,000 ft). The shape of its envelope is a modification of the Class 'C' developed by the US Air Force as a radar surveillance platform. The envelope incorporates a large ballonet, and has four tail fins in cruciform configuration: these not only provide pitch and yaw stability, but generate additional lift as windspeeds increase. The avionics or other payload is carried beneath the envelope, and a pressurised cover is available to protect the payload and decrease aerodynamic drag. Pressure control is maintained by a system of valves and blowers which add or dump air from the ballonet, tail fins, and pressurised load cover. This pressure maintenance is totally automatic, controlled by pressure switches and logic circuits, but manual control from the ground via onboard telemetry can be used to override any pre-set limits. Onboard power supplies can be matched to the payload requirements but, typically, a 6 kW generator will meet radar payload and pressure system requirements. A balloon of this size, shown in an accompanying illustration during tests at Cape Canaveral, used Seek Skyhook equipment to detect an armed Soviet brigade in Cuba during 1979. In this particular operation, the balloon was operating from Cudjoe Key, Florida, at a height of 3,660 m (12,000 ft).

DIMENSIONS, ENVELOPE:	
Length overall	53·34 m (175 ft 0 in)
Max diameter	17·31 m (56 ft 9½ in)
Tail fin span	24·62 m (80 ft 9½ in)
Volume, gross	7,079 m³ (250,000 cu ft)
Volume, ballonet	3,030 m³ (107,000 cu ft)
PERFORMANCE:	
Design altitude, with 454 kg (1,000 lb) payload	3,660 m (12,000 ft)
Max wind velocity, operational	60 knots (111 km/h; 69 mph)
Max wind velocity, moored	90 knots (167 km/h; 103·5 mph)

TUCKER

TUCKER AIRSHIP COMPANY

13218 Lake Street, Los Angeles, California 90066
Telephone: (213) 798 3719; (213) 577 9289
PRESIDENT: Curtis E. Tucker

The Tucker Airship Company was founded in 1973 by Mr Curtis Tucker, with the object of building airships that would benefit from the latest technology and materials.

Production of three prototype airships was planned originally, but only the initial TX-1 was nearing completion in early 1981.

Tucker TX-1 airship *Silver Hawk* during inflation checks

TUCKER TX-1 SILVER HAWK

Design of the TX-1 originated in October 1972 and construction began in May 1973. During 1976 the company encountered a problem with its envelope sealing process, but this was subsequently resolved. Funding was reduced in 1977, but additional finance was made available during 1978 and 1979 so that the vessel could become operational and available for company use. In early 1980 the envelope had been coated with a 0·018 mm (0·0007 in) thickness of light alloy by means of a new metalising process, and testing of the envelope was being carried out. Installation of final wiring, and rigging checks, were being completed in early 1981, when the company was seeking sponsorship to carry out the flight test programme.

The TX-1 is very much an experimental vessel, intended to provide experience in design, construction and operation, and will be used to test new design ideas, materials and ground handling concepts. Should the future airship proposals develop, it is envisaged that the TX-1 type would serve as a low-cost primary trainer. It is intended to be available for research use by scientific, military and industrial organisations upon request.

A semi-rigid design was chosen to provide for multi-mission roles, permitting the carriage of external loads; all of the superstructure, including the tail control surfaces, is located beneath the envelope. Oversize ballast and trim tanks are provided to keep the airship level irrespective of load disposition, and to provide additional safety during pilot training. A unique feature of the design is its modular construction. The TX-1 breaks down into twelve units for transportation or storage, and can be assembled and inflated in an open field. A portable mooring mast, which can be erected in 30 minutes, has been designed and built, and this accepts the company's Mk 1 pneumatic mooring spindle which requires only one man on the ground to carry out mooring and castoff operations. When broken down for transit, the entire vessel can be accommodated in a standard-width trailer, 4·57 m (15 ft) in length, and there is sufficient space in the trailer for a power generating unit, air-to-ground communication system, and other essential supplies and equipment. During 1978 a portable double-wall air-inflated hangar was developed to accommodate the TX-1. Measuring 33·5 × 12·2 × 10·7 m (110 × 40 × 35 ft), it has an air-operated upward-opening clamshell door.

The forward-mounted gondola of the TX-1 airship provides accommodation for the pilot and one passenger, seated in tandem. Excellent view is provided by a wide and deep windscreen, plus large sliding side windows. Standard airship controls are provided and equipment includes VFR instrumentation and communication radio. A monowheel is mounted beneath the gondola. Tail control surfaces comprise a large fin and rudder, with a wide-span horizontal tail surface and elevator. Power plant consists of an 89 kW (120 hp) McCulloch O-150-O two-stroke six-cylinder engine, mounted beneath the keel structure at approximately its mid-point, and driving an AAF tractor propeller 1·12 m (3 ft 8 in) in diameter. Total fuel capacity is 227 litres (60 US gallons).

Three holds are provided within the keel structure, each with a capacity of 90·7 kg (200 lb), and are available for equipment or baggage. Hardpoints are provided for the carriage of external loads.

The envelope, which contains a single ballonet, is intended to be hydrogen filled for short test flights. For operational use the envelope would be inflated with helium. Special equipment includes pneumatic ballast controls, and there are provisions for full IFR instrumentation.

Details of proposals for the developed prototypes can be found in the 1979-80 *Jane's*.

Dimensions, overall:

Length	32·00 m (105 ft 0 in)
Height	7·92 m (26 ft 0 in)

Dimensions, envelope:

Length	27·74 m (91 ft 0 in)
Max diameter	6·10 m (20 ft 0 in)
Length of constant-diameter section	13·11 m (43 ft 0 in)
Volume	651 m³ (23,000 cu ft)

Dimensions, gondola:

Length overall	2·74 m (9 ft 0 in)
Height	1·37 m (4 ft 6 in)
Max width	0·91 m (3 ft 0 in)

Areas:

Fin	6·97 m² (75·0 sq ft)
Rudder	2·93 m² (31·5 sq ft)
Tailplane	7·43 m² (80·0 sq ft)
Elevator	5·39 m² (58·0 sq ft)

Weights (estimated):

Weight empty	368 kg (810 lb)
Max design gross weight	635 kg (1,400 lb)

Performance (estimated, with 89 kW; 120 hp engine):

Max speed	56 knots (105 km/h; 65 mph)
Service ceiling	1,370 m (4,500 ft)
Range with max fuel	434 nm (804 km; 500 miles)
Range with max payload	43 nm (80 km; 50 miles)

LIGHTER-THAN-AIR: BALLOONS

FRANCE

CHAIZE
BALLONS CHAIZE

48 rue Balay, 42000 Saint-Étienne
Telephone: (77) 33 43 76

M Maurice Chaize markets five hot-air balloons, one in the FAI AX-6 class, three in the AX-7 class and, the most recently introduced, a 4,000 m³ (141,260 cu ft) balloon in the AX-9 class. The envelopes are made from high-strength nylon, with the lower panels in the throat made from Nomex flame-resistant fabric. They are supplied in a ready-to-fly state, complete with basket, burners and gas cylinders, but without instruments. Several options are available to customer requirements.

The company holds the French Certificat de Navigabilité de Type for all five models, details of which can be found in the accompanying table.

The prototype balloon was flown for the first time in December 1972. Deliveries began in the Summer of 1977, and totalled 20 by late 1979. On 20 October 1979 two Chaize balloons, one of 2,000 m³ (70,630 cu ft) and one of 4,000 m³ (141,260 cu ft) capacity, crossed the Alps from France to Italy, flying directly over Mont Blanc.

GERMANY (FEDERAL REPUBLIC)

BALLONFABRIK
BALLONFABRIK SEE- UND LUFT-AUSRÜSTUNG GmbH und Co KG

Austrasse 35, 89 Augsburg 1, Postfach 101327
Telephone: (0821) 41 50 41

Ballonfabrik is currently manufacturing and marketing a range of seven gas balloons, all of which are within the FAI categories AX-3/-4/-5/-6. All available details of these balloons are given in the accompanying table.

UNITED KINGDOM

CAMERON
CAMERON BALLOONS LTD

1 Cotham Park, Bristol BS6 6BZ
Telephone: (0272) 41455

Directors: See Airships subsection

Cameron has been manufacturing hot-air balloons in Bristol since 1968, its products including the world's largest hot-air balloon, the *Gerard A. Heineken* of 14,158 m³ (500,000 cu ft). The company holds CAA, FAA, Certificat de Navigabilité de Type and the West German Musterzulassungsschein type certificates for its balloons.

Envelopes are made from rip-stop nylon which has been treated with a polyurethane sealant to reduce porosity. The lower section of the envelope is made from Nomex flame-resistant fabric. The envelopes are available with either a Velcro seal or parachute deflation system, and those with the Velcro seal have also a vertical flap-type vent. Cameron offers a wide range of balloons, details of which can be found in the accompanying table. Each is supplied complete with willow and cane basket, burner and gas cylinders and is ready to fly when delivered. Optional items include instrument pack, envelope thermometer, pressure scoop, Nomex skirt, rigid-suspension basket, trail rope and additional gas cylinders.

During 1976 the company initiated the design of specially shaped balloons for advertising purposes. Three new sporting balloons were introduced during 1979, comprising the N-65 and Viva-65 in the AX-7 class, and the O-105 in the AX-8 class. It also constructed the world's first hot-air airship, and is now manufacturing such aircraft to order (see Airships subsection).

In mid-1981, Cameron hot-air balloons held the world's absolute duration record of 24 h 7 min 58 s; and, subject to homologation, the absolute altitude record of 17,404 m (57,100 ft).

During 1978 an attempt was made by Don Cameron and Maj Christopher Davey to be the first balloonists to cross the North Atlantic, in the combination helium/hot-air balloon *Zanussi*. This flight had to be abandoned on 30 July, when they were little more than 87 nm (161 km; 100 miles) from the French coast. Further details can be found in the 1980-81 *Jane's*.

Experience gained during this attempted transatlantic flight contributed to the design of an advanced balloon that will be used for an attempt to achieve a round-the-world flight in the Autumn of 1981. Under the sponsorship of Imperial Chemical Industries, with backing from British aerospace companies that include Air Products Ltd, Racal Electronics Ltd, and British Airways, Don Cameron and Peter Bohanna have designed respectively the helium/hot-air balloon and gondola that will be used to carry a four-man crew on the planned 20-day flight, scheduled for lift-off from the South of France on 1 October 1981. The balloon, which will be about 61 m (200 ft) high when inflated, will have minimum and maximum volumes of approximately 21,235 m³ (750,000 cu ft) and 28,315 m³ (1,000,000 cu ft) respectively. Its circumnavigation will be carried out at an average height of about 12,200 m (40,000 ft), but it must be capable of considerable altitude changes to maintain a position within varying jet streams that will propel the balloon around the world. Because of the high altitudes that are involved, the gondola which contains the crew is, in effect, a self-contained life support system, able to sustain them for more than the scheduled duration of the flight. The accompanying sectional drawing of this gondola gives an appreciation of its comprehensive and sophisticated equipment.

THUNDERCOLT
THUNDERCOLT BALLOONS

Sales Office: Thunder Balloons Ltd, 75 Leonard Street, London EC2A 4QS
Telephone: (01) 739 0775/0776
Telex: 261234 ref H5758D

Works: Colt Balloons Ltd, Maesbury Road, Oswestry, Shropshire SY10 8HA
Telephone: (0691) 2216
Telex: 35503

Directors:
T. M. Donnelly
P. Lindstrand
K. F. Simonds
A. R. Wirth

ThunderColt Balloons was formed during 1980 by the amalgamation of Colt Balloons Ltd and Thunder Balloons Ltd. The resulting company is claimed to be the world's largest manufacturer of balloons outside of the USA. It manufactures a range of ready-to-fly hot-air balloons, details of which can be found in the accompanying table. In addition to those which are listed in the table, there are five balloons with a designation that has an A suffix, namely the 31A, 56A, 65A, 77A and 105A. These have the same specification as the balloons designated 31Z,

Maurice Chaize CS.4000 balloon in the FAI AX-9 class

A typical Ballonfabrik gas balloon *(Peter J. Bish)*

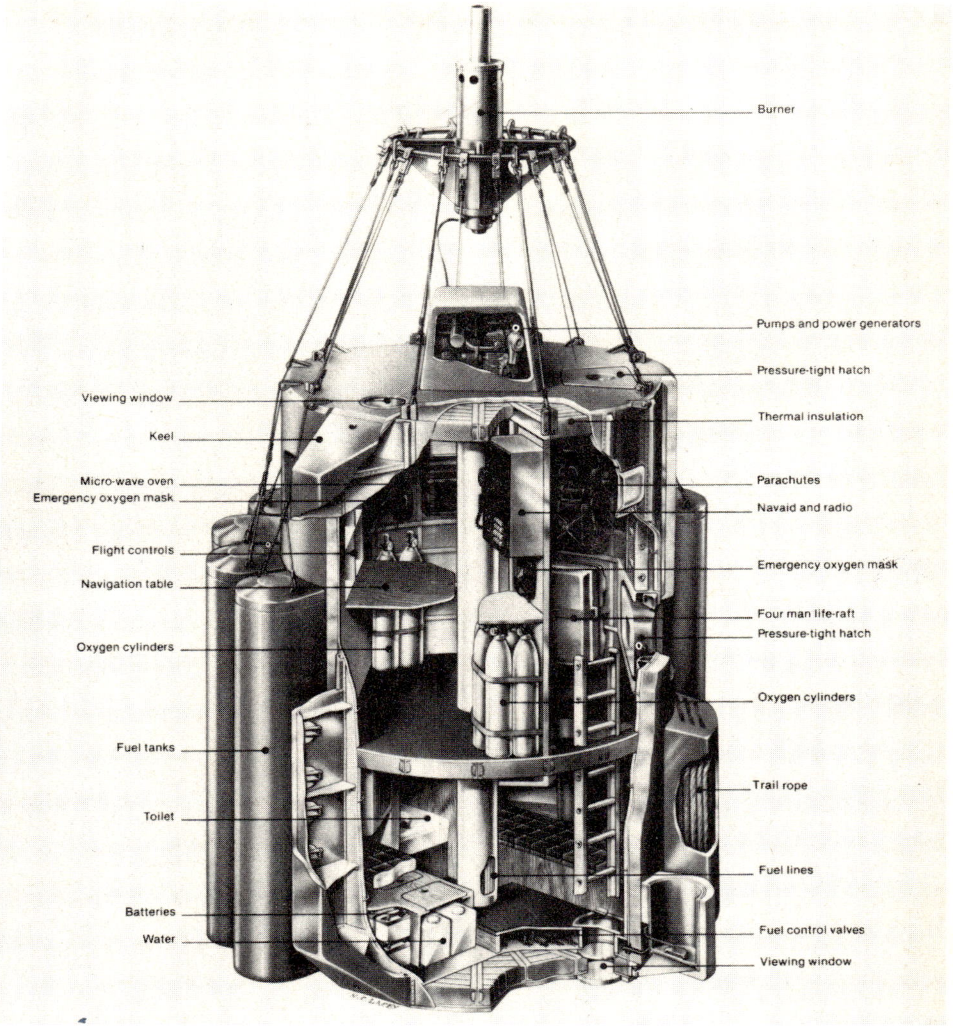

Sectional drawing of the gondola to be used under a Cameron balloon in an attempted round-the-world balloon flight *(Imperial Chemical Industries)*

56Z, 65Z, 77Z and 105 respectively, but differ in envelope construction. Those with an A suffix have vertical panels, and those with a Z suffix horizontal panels: the 105 also has horizontal panels.

Envelopes are made from coated polyamide or polyester fabrics, with lower panels of Nomex flame-resistant fabric. All balloons are supplied complete with basket, burner and fuel cylinders, and are certificated for use throughout the world. Series 1, 2 and Z balloons are available with an instrument dashboard which covers the fuel cylinders and provides more room within the basket. 'Bolt' type balloons, in a lower price range, were introduced during 1978. The British team competing in the World Championship events in 1979 flew Z-type Thunder balloons.

The company is developing also a hot-air airship, all available details of which are given in the Airships subsection.

Cameron Viva-77 balloon in the AX-7 class

British team's ThunderColt balloons at Uppsala, Sweden

UNITED STATES OF AMERICA

ADAMS

MIKE ADAMS' BALLOON LOFT INC

PO Box 12168, Atlanta, Georgia 30355
Telephone: (404) 452 8066

Mike Adams' Balloon Loft is manufacturing a range of hot-air balloons, of which details can be found in the accompanying table.

Envelopes are made from rip-stop nylon, which may be coated optionally with a polyurethane or aluminised sealant to reduce porosity. Nomex flame-resistant fabric is used to protect the envelope throat. Two manoeuvring vents are located in the crown of the balloon and there is a duffle bag type of deflation system, or an optional Paravent.

In 1978 Adams introduced the first two of a series of balloons named 'Little Devil' models. These feature a deflation system which the company calls MultiVent. A single control line creates 16 diamond-shape openings which snap shut immediately the control line is released, so providing an improved measure of control.

Adams balloons are supplied complete with a woven

One of the Mike Adams' Balloon Loft range of hot-air balloons

Avian Balloon 20-gore envelope in the AX-7 class

Boland *Dimples* AX-5 class balloon

rattan basket, burner, aluminium fuel cylinders, and instruments which include a sensitive altimeter, rate of climb indicator, envelope temperature gauge, bubble type compass and fuel gauges. Optional items include tank-mounted pressure gauge, Nomex skirt, Paravent or MultiVent deflation system, inflator fan and additional fuel cylinders.

AVIAN

AVIAN BALLOON COMPANY

South 3722 Ridgeview Drive, Spokane, Washington 99206
Telephone: (509) 928 6847

Avian Balloon is marketing currently three hot-air balloons. Identified as the Avian Sparrow, Falcon II and Skyhawk, these are in FAI classes AX-5, AX-7 and AX-8 respectively.

Avian balloon envelopes are made from rip-stop nylon, which can be treated with a sealant to reduce porosity, or can optionally be made from aluminised balloon fabric. Four design patterns and twelve fabric colours are available to customer's choice. A feature of the design is the incorporation of a mechanical rapid deflation system. The basket is of wicker and, in addition to the burner and gas cylinders, an instrument panel that includes an altimeter, rate of climb indicator and envelope temperature gauge is standard. Details of Avian's balloons are given in the accompanying table.

Boland *Peaches* AX-6 balloon, showing translucent semi-sphere

BALLOON WORKS

THE BALLOON WORKS

Rhyne Aerodrome—RFD2, Statesville, North Carolina 28677
Telephone: (704) 873 0503
DIRECTORS:
Tracy Barnes
Dodds Meddock

The Balloon Works is currently marketing four category sizes of hot-air balloons, details of which are given in the accompanying table.

The envelopes of this company's balloons are made of a polyester fabric, which tests have shown to be less affected by heat and sunlight exposure than other materials, and are coated with a urethane sealant to reduce porosity. Rapid deflation is made possible by use of a wide-diameter envelope valve, which is self-sealing immediately the valve line is released. Triangular wicker baskets are leather trimmed and are available in three sizes. One high-output burner is fitted.

Standard equipment includes the Barnes triple burner, one main tank and related fuel lines, apex handling line, thermometer, rate of climb indicator, sensitive altimeter, and a fuel pressure gauge.

Optional equipment includes Fire 2 supplementary quiet burner system, Nomex envelope skirt, electronic temperature and rate of climb indicators, extra fuel tanks and tethering kit. A wide variety of envelope artwork can be provided.

Group of balloons made by Raven Industries; largest is the S-66A

BOLAND

BOLAND BALLOON

Pine Drive, RD 2, Burlington, Connecticut 06013
Telephone: (203) 673 1307

Mr Brian Boland, designer and builder, with his wife Kathy, of the *Albatross* hot-air airship (described in the Airships section) has also designed and built a number of hot-air balloons. Early examples included the very small *Piccolo*, and the somewhat larger *Fred II* of which Mr Boland built six examples. A plan/construction manual and materials source book is available for *Fred II*, and details of both of these early balloons can be found in the 1978-79 *Jane's*.

During 1978 the Bolands built three new balloons, two of them in the lower-size FAI categories. These included the XXUS in the AX-1 class, *High Fred* in the AX-3 class, and *Spirit of Lake Garda* in the AX-8 class. Details of these are given in the accompanying table; additional information can be found in the 1979-80 *Jane's*.

During 1979 the Bolands originated two more designs, the first being a very lightweight balloon with the model name *Peaches*, which is categorised in the AX-6 class. Believed to be the lightest-weight balloon yet built in relation to volume, it has a duration of 2-3 h with a crew of two. A feature of its design is the use of translucent material above, and black opaque material below, the equator line to gain some benefit from solar heating at high altitude. The second new design, the AX-5 class *Dimples*, is intended for one-person operation, and has a duration of about 1 h 30 min with two gas cylinders. No recent information has been received from Mr Boland.

Available details of these balloons are given in the accompanying table.

GENERAL

GENERAL BALLOON CORPORATION

865 West 18th Street, Costa Mesa, California 92627
Telephone: (714) 642 3545

General Balloon Corporation, known formerly as Piccard Balloons, manufactures a range of hot-air balloons, of which eight different models are available, all classified within the FAI's AX-2 to AX-8 ranges. In addition, three gas balloons are available, and details of both categories can be found in the accompanying table.

RAVEN

RAVEN INDUSTRIES INC

Box 1007, Sioux Falls, South Dakota 57101
Telephone: (605) 336 2750

Raven Industries has been manufacturing hot-air balloons since the early 1960s, and also built the STAR hot-air airship of which details were given in the 1977-78 *Jane's*. It claims to be the world's largest manufacturer of hot-air balloons.

In 1979 Raven introduced a new balloon designated RX-7, which falls within the FAI class AX-7 and joins the RX-6 in Raven's Rally II economy balloon class. In addition, the company markets five sizes of custom design balloons. Details of these may be found in the accompanying table.

Raven envelopes are made from rip-stop nylon treated with ultra-violet inhibitors to prolong fabric life. They embody to the customer's choice either a parachute-style deflation system or a Velcro rapid deflation port. All models have woven rattan wicker baskets. Standard equipment includes the burner, gas cylinders, and an instrument case which includes a sensitive altimeter, rate of climb indicator and envelope temperature gauge. A range of options includes fire retardant Kynol skirt liner, Kynol throat liner, stainless steel or additional aluminium gas cylinders, electric variometer, special colours, designs and decorations for the envelope, banners, bunting and repair fabrics.

HOT-AIR AND GAS BALLOON DATA

Country	Company	Model	Volume m³/cu ft	Diameter m/ft	Height m/ft	Crew	Basket	Fuel Cylinders	Burners	Lift, S/L ISA kg/lb
FAI CLASS AX-1										
United States	Boland Balloon	XXUS	239/8,450	7·62/25	8·53/28	1	Harness seat	1	1	—
FAI CLASS AX-2										
United Kingdom	ThunderColt Balloons	Cloudhopper Junior	396·4/14,000	8·53/28	8·53/28	1	Harness seat	1	1	159/350
United States	Boland Balloon	Piccolo	345/12,200	8·53/28	9·75/32	1	Harness seat	1	1	136/300
United States	General Balloon	Piccard	396·4/14,000	9·75/32	11·28/37	1	—	1	1	127/280
FAI CLASS AX-3										
West Germany	Ballonfabrik	K 630/1-Ri	630/22,250	10·59/34·75	18·0/59	3-4	—	NA	NA	730/1,609
United Kingdom	Cameron Balloons	Viva-20	566/20,000	—	—	1	Harness	1	1	—
United Kingdom	ThunderColt Balloons	Sky Chariot (Mini)	481·4/17,000	—	—	1	Harness seat	1	1	159/350
United Kingdom	ThunderColt Balloons	Cloudhopper Super	481·4/17,000	9·45/31	9·45/31	1	Harness seat	1	1	193/425
United Kingdom	ThunderColt Balloons	Sky Chariot (Midi)	538/19,000	9·14/30	9·14/30	1	Harness seat	1+aux	1	190·5/420
United Kingdom	ThunderColt Balloons	Sky Chariot (Maxi)	623/22,000	10·06/33	10·06/33	1	Harness seat	2	1	211/465
United States	Boland Balloon	Fred II	566/20,000	10·4/34·25	12·2/40	1	Wicker	1	1	181/400
United States	Boland Balloon	High Fred	595/21,000	10·06/33	—	1	Harness seat	1	1	—
United States	General Balloon	Piccard A-3	594·7/21,000	9·14/30	13·72/45	3	—	NA	NA	454/1,000
United States	General Balloon	Piccup-3	595/21,000	10·67/35	12·2/40	1	Harness seat	1	1	200/440
FAI CLASS AX-4										
West Germany	Ballonfabrik	B 800/2-Ri	800/28,250	11·4/37·4	20·12/66	4	—	NA	NA	928/2,046
West Germany	Ballonfabrik	K 780/2-Ri	780/27,500	11·4/37·4	19·81/65	4	—	NA	NA	905/1,995
West Germany	Ballonfabrik	K 945/2-Ri	945/33,400	12·28/40·3	20·42/67	5-6	—	NA	NA	1,095/2,416
United Kingdom	Cameron Balloons	N-31	890/31,430	12·5/41	14·63/48	1	Willow/Cane	2	1	—
United Kingdom	Cameron Balloons	O-31	890/31,430	—	—	1	Willow/Cane	2	1	—
United Kingdom	Cameron Balloons	Viva-31	890/31,430	—	—	1	Willow/Cane	2	1	—
United Kingdom	ThunderColt Balloons	Air Chair	877·8/31,000	10·36/34	10·36/34	2	Harness seat	2	1	272/600
United Kingdom	ThunderColt Balloons	31Z	890/31,430	—	—	1	Rattan	2	1	—
United States	General Balloon	Piccard A-4	877·8/31,000	10·97/36	16·76/55	4	—	NA	NA	794/1,750
United States	General Balloon	Lightning 33	934/33,000	12·19/40	14·33/47	1-2	—	2	1	272/600
United States	Raven Industries	S-40	900/31,779	12·19/40	16·76/55	1	Aluminium chair	1	1	299/660
FAI CLASS AX-5										
West Germany	Ballonfabrik	K 1050/3-Ri	1,050/37,100	12·6/41·3	20·7/68	5-6	—	NA	NA	1,218/2,685
West Germany	Ballonfabrik	K 1260/3-Ri	1,260/44,500	13·4/44	22·25/73	6	—	NA	NA	1,460/3,219
United Kingdom	Cameron Balloons	O-42	1,190/42,024	—	—	2	Willow/Cane	2	1	—
United Kingdom	ThunderColt Balloons	42	1,190/42,024	13·4/44	14·1/46·4	2	Rattan	2	1	—
United Kingdom	ThunderColt Balloons	42 Bolt	1,190/42,024	—	—	2	Willow	2	1	—
United States	Adams' Balloon Loft	LD	964/34,029	12·95/42·5	15·54/51	1	Rattan	1/2	1	318/700
United States	Adams' Balloon Loft	LD-S	1,104/38,978	12·95/42·5	16·46/54	1-2	Rattan	1/2	1	318/700
United States	Avian	Sparrow	1,189/42,000	13·41/44	16·46/54	1	—	—	—	286/630
United States	Boland Balloon	Dimples	906/32,000	—	—	1	Wicker	2	1	272/600
United States	The Balloon Works	FireFly 5	1,182/41,742	13·9/45·6	14·8/48·4	2	Wicker	2	2	226/499
United States	General Balloon	Piccard A-5	1,189/42,000	11·89/39	17·68/58	6	—	NA	NA	1,134/2,500
United States	General Balloon	Ballymurray	1,189/42,000	14·02/46	14·63/48	1-2	—	2	1-2	363/800
United States	General Balloon	Lightning 45	1,274/45,000	13·72/45	15·24/50	1-2	—	2	1	363/800
FAI CLASS AX-6										
France	M. Chaize	CS. 1600	1,600/56,503	—	—	2	Rattan	2	—	—
West Germany	Ballonfabrik	K 1680/4-Ri	1,680/59,300	14·78/48·5	24·4/80	6	—	NA	NA	1,960/4,321
United Kingdom	Cameron Balloons	O-56	1,590/56,150	—	—	3	Willow/Cane	2	1	—
United Kingdom	Cameron Balloons	Viva-56	1,590/56,150	—	—	3	Willow/Cane	2	1	—
United Kingdom	Cameron Balloons	N-56	1,590/56,150	15·25/50	17·07/56	3	Willow/Cane	2	1	—
United Kingdom	ThunderColt Balloons	56	1,590/56,150	14·8/48·4	15·5/50·9	3	Rattan	2	1	—
United Kingdom	ThunderColt Balloons	56Z	1,590/56,150	14·51/47·6	15·7/51·5	3	Rattan	2	1	—
United Kingdom	ThunderColt Balloons	56 Bolt	1,590/56,150	—	—	3	Willow	2	1	—
United States	Adams' Balloon Loft	A50	1,557/54,985	15·24/50	17·68/58	2	Rattan	1-4	1	544/1,200
United States	Boland Balloon	Peaches	1,416/50,000	—	—	2	Wicker	2	1	454/1,000
United States	The Balloon Works	FireFly 6	1,576/55,656	15·3/50·1	16·1/52·8	3	Wicker	3	2	715/1,576
United States	The Balloon Works	FireFly 6B	1,586/56,000	15·24/50	16·46/54	1-3	—	—	—	715/1,576
United States	General Balloon	Le Mans	1,586/56,000	14·94/49	16·45/54	2-3	—	2	2	454/1,000
United States	Raven Industries	S-50A	1,597/56,400	15·24/50	17·68/58	2-4	Wicker	3	1	635/1,400
United States	Raven Industries	RX-6	1,597/56,400	15·24/50	17·68/58	2-3	Wicker	2	1	—

NA: Not Applicable

HOT-AIR AND GAS BALLOON DATA

Country	Company	Model	Volume m³/cu ft	Diameter m/ft	Height m/ft	Crew	Basket	Fuel Cylinders	Burners	Lift, S/L ISA kg/lb
FAI CLASS AX-7										
France	M. Chaize	CS. 1800	1,800/63,565	—	—	2-3	Rattan	2	—	—
France	M. Chaize	CS. 2000	2,000/70,629	—	—	3-4	Rattan	3	—	—
France	M. Chaize	CS. 2200	2,200/77,682	—	—	4	Rattan	3	—	—
United Kingdom	Cameron Balloons	O-65	1,840/64,980	—	—	3	Willow/Cane	3	1	—
United Kingdom	Cameron Balloons	O-77	2,190/77,339	—	—	4	Willow/Cane	4	1	—
United Kingdom	Cameron Balloons	N-65	1,840/65,000	16·15/53	18·0/59	3	Willow/Cane	3	1	—
United Kingdom	Cameron Balloons	Viva-65	1,840/65,000	—	—	3	Willow/Cane	2	1	—
United Kingdom	Cameron Balloons	Viva-77	2,190/77,339	—	—	4	Willow/Cane	2	1	—
United Kingdom	Cameron Balloons	N-77	2,190/77,339	17·07/56	18·9/62	4	Willow/Cane	4	1	—
United Kingdom	ThunderColt Balloons	65	1,840/64,980	15·5/50·8	16·3/53·5	3	Rattan	2	1	—
United Kingdom	ThunderColt Balloons	65Z	1,840/64,980	—	—	3	Rattan	2	1	—
United Kingdom	ThunderColt Balloons	65 Bolt	1,840/64,980	—	—	3	Willow	2	1	—
United Kingdom	ThunderColt Balloons	77	2,190/77,339	16·4/53·8	17·3/56·6	4	Rattan	2	1	—
United Kingdom	ThunderColt Balloons	77Z	2,190/77,339	15·97/52·38	16·66/54·66	4	Rattan	2	1	—
United Kingdom	ThunderColt Balloons	77 Bolt	2,190/77,339	—	—	4	Willow	2	1	—
United States	Adams' Balloon Loft	A50S	1,755/61,977	15·24/50	18·59/61	3	Rattan	1-4	1	544/1,200
United States	Adams' Balloon Loft	A55	2,123/74,973	16·76/55	19·51/64	3	Rattan	1-4	1	680/1,500
United States	Avian Balloon	Falcon II	1,700/60,000	15·39/50·5	18·6/61	1-3	Wicker	2	1	376/830
United States	The Balloon Works	FireFly 7	2,167/76,526	17·0/55·8	17·7/58·2	3	Wicker	3	2	517·5/1,141
United States	General Balloon	Stretch	1,982/70,000	16·15/53	17·68/58	3	—	4	2	549/1,210
United States	Raven Industries	S-55A	2,195/77,500	16·76/55	19·2/63	3-4	Wicker	3	1	657/1,450
United States	Raven Industries	RX-7	2,195/77,500	16·76/55	19·2/63	3	Wicker	3	1	657/1,450
FAI CLASS AX-8										
United Kingdom	Cameron Balloons	O-84	2,380/84,047	—	—	4	Willow/Cane	4	1	—
United Kingdom	Cameron Balloons	A-105	2,970/104,884	—	—	6	Willow/Cane	6	2	—
United Kingdom	Cameron Balloons	N-105	2,970/104,884	18·9/62	20·72/68	6	Willow/Cane	6	2	—
United Kingdom	Cameron Balloons	O-105	2,970/104,884	—	—	6	Willow/Cane	6	2	—
United Kingdom	ThunderColt Balloons	84	2,380/84,047	—	—	4	Rattan	2	1	—
United Kingdom	ThunderColt Balloons	90	2,550/90,050	17·5/57·5	17·5/57·5	5	Rattan	2	2	—
United Kingdom	ThunderColt Balloons	105	2,970/104,884	18·44/60·5	18·44/60·5	6	Rattan	2	2	—
United States	Adams' Balloon Loft	A55S	2,350/82,990	16·76/55	20·42/67	4	Rattan	1-4	1	680/1,500
United States	Avian Balloon	Skyhawk	2,265/80,000	16·76/55	20·1/66	3-4	Wicker	3	1	544/1,200
United States	The Balloon Works	FireFly 8	2,956/104,400	18·8/61·8	19·5/64	4-5	Wicker	6	2	651/1,436
United States	Boland Balloons	*Spirit of Lake Garda*	2,322/82,000	—	—	4	Wicker	3/6	2	—
United States	General Balloon	Newport	2,265/80,000	16·75/55	18·6/61	4	—	4	2	549/1,210
United States	Raven Industries	S-60A	2,973/105,000	18·29/60	21·03/69	3-4	Wicker	3	1	680/1,500
FAI CLASS AX-9										
France	M. Chaize	CS.4000	4,000/141,260	—	—	8	Rattan	—	2	—
United Kingdom	Cameron Balloons	A-140	3,960/139,846	—	—	8	Willow/Cane	6	2	—
United Kingdom	ThunderColt Balloons	140	3,965/140,022	20·36/66·8	20·36/66·8	8	Rattan	2	2	—
United Kingdom	ThunderColt Balloons	160A	4,530/160,000	—	—	9	Rattan	—	3	1,589/3,504
United Kingdom	ThunderColt Balloons	200A	5,663/200,000	—	—	12	Rattan	—	4	1,987/4,380
United States	Raven Industries	S-66A	4,000/141,260	—	—	4-6	Wicker	2	2	998/2,200

NA: Not Applicable

RPVs AND TARGETS

ARGENTINA

AMC

AREA DE MATERIAL CÓRDOBA, FUERZA AÉREA ARGENTINA (Fábrica Militar de Aviones)

Avenida Fuerza Aérea Argentina Km 5½, 5103 Guarnicion Aérea Córdoba
Telephone: 45011/15
Telex: 51965 AMCOR AR

HEAD OF AMC: Brigadier Guillermo Héctor Marotta
DIRECTOR OF IIAE: Comodoro Edgardo Carlos A. Stahl

FMA IA 59

Designed to Argentinian Air Force requirements by the IIAE (Instituto de Investigaciónes Aeronáuticas y Espaciales: Aerospace Research Institute), the IA 59 target drone was built by the Grupo Fabricación of the FMA as an experimental aerial target for anti-aircraft guns and surface-to-air missiles. Construction of a prototype vehicle, designated IA-X-59-1, began in 1975, and this made its first flight in December 1976. A second prototype flew in July 1977, and a third was completed for static tests.

Series production is not intended at present. Full details and an illustration of the IA 59 can be found in the 1980-81 *Jane's*.

AUSTRALIA

GAF

GOVERNMENT AIRCRAFT FACTORIES

Fishermen's Bend, Private Bag No. 4, Post Office, Port Melbourne, Victoria 3207
Telephone: 64 0661
Telex: AA 34397
OFFICERS: see Aircraft section

GAF's current products include the Jindivik weapons target drone.

GAF JINDIVIK Mk 3B

The Jindivik continues to be a standard weapons target in Australia and Great Britain.

Design began in March 1948. Construction of the prototype Jindivik Mk 1 started in December 1950, and it flew for the first time on 28 August 1952.

A total of 526 Jindiviks had been ordered by 1 January 1981, of which 480 had been delivered.

The 700 Series Mks 103BL and 203B versions are still in production; deliveries of these versions began in 1976. Customers include the UK (40 Mk 103BL) and Royal Australian Navy (12 Mk 203B). These models were designed to cater for low-level trials at speeds in excess of 500 knots (925 km/h; 575 mph). The autopilot is the improved Marconi L4 or L5, and system 'black boxes' incorporate printed-circuit techniques.

An improved Mk 4 Jindivik was in the early production stage in January 1981. The Mk 4 has a rationalised electrical/electronic system, increased manoeuvring capability (to +6*g*), and greater endurance. No other details were available at the time of going to press.

For low-altitude work, the standard-span Jindivik can be fitted either with Mk 5 wingtip camera pods, or with larger Mk 9 pods each containing two cameras, a Luneberg lens and a small amount of fuel. For high-altitude work (also with a choice of Mk 5 or Mk 9 wing pods), constant-chord wing extension panels can be added outboard of the pods. For extra high altitude flying (with Mk 5 pods only), these panels can be replaced by increased-span panels, tapered on the leading-edge. A ventral tail fin is also fitted in this configuration.

Up to 1 January 1981, Jindiviks had flown more than 6,500 sorties at the RAE, Llanbedr, North Wales, and in Australia; one particular Mk 3A drone (WRE 418) was eventually destroyed after successfully completing 285 sorties at Woomera.

TYPE: Pilotless target drone.

WINGS: Cantilever low/mid-wing monoplane. Wing section NACA 64A-106 with modified trailing-edge. Dihedral 2° 30′. Incidence 1°. Bonded multi-spar box structure of aluminium alloy. Interspar torsion box utilised as integral fuel tankage. Aluminium alloy monocoque flaps and ailerons. Ailerons fitted with inset geared tab and driven by GAF-designed twin-motor servo motor. Flaps operated pneumatically.

FUSELAGE: Aluminium alloy semi-monocoque structure, built in front, centre and rear sections. Front fuselage carries all control equipment, autopilot amd telemetry equipment. Pitot head and wave-guide boom mounted on permanent nose probe. Moulded glassfibre canopy, which lifts off for access to equipment, forms ram-type air intake. Rear end of front fuselage and front end of centre-fuselage form bay for all special trials equipment. Centre-fuselage also houses landing skid. Rear fuselage carries engine and jetpipe.

TAIL UNIT: Cantilever multi-spar light alloy tailplane. Elevators driven by GAF-designed twin-motor servo motor. Inset geared tabs. Fin of light alloy skin bonded to two spars, stabilised by metal honeycomb filling. No rudder. Ventral fin on extra high altitude version.

LANDING GEAR: Pneumatically-extended, manually retracted (on ground) central skid. Pneumatic jack acts as shock-absorber. Steel auxiliary skids at wingtips. See also paragraph on 'Take-off and Landing'.

POWER PLANT: One Rolls-Royce Viper Mk 201 turbojet engine, rated at 11·1 kN (2,500 lb st). Engine relight capability available in the event of flameout. Flexible rubber main fuselage fuel tank, capacity 291 litres (64 Imp gallons), and two integral wing tanks, total capacity 173 litres (38 Imp gallons). Oil capacity 4·5 litres (1 Imp gallon). Mk 4 has additional 54 litre (12 Imp gallon) fuselage tank.

LAUNCH AND RECOVERY: Take-off from aircraft trolley, steered by gyro and servo-controlled nosewheel which responds to signals from ground controller. Aircraft/trolley combination accelerates under normal jet power with flaps retracted and with aircraft set at negative incidence. When unstick speed (125 knots; 231 km/h; 144 mph) is reached, aircraft is rotated to take-off incidence and flaps are lowered rapidly. Rotation of aircraft initiates trolley release system and the aircraft climbs away. When Jindivik is in approach run, flaps and skid are selected down for landing. On touchdown, at approx 120 knots (222 km/h; 138 mph), a 'sting' extended below main skid rotates on impact and initiates rapid retraction of flaps. Fuel supply terminated by radio command.

GUIDANCE AND CONTROL: Radio control equipment comprises two receivers and GAF relay receiving set. Telemetry equipment consists of NIC transducers and Australian-designed transmitter and junction box.

SYSTEMS: No hydraulic system. Non-regenerative pneumatic system: air stored at 138 bars (2,000 lb/sq in) in power pack which supplies air to flaps and landing skid reduced to 39·6 bars (575 lb/sq in). Engine-driven 6 kW generator delivers 208V AC 3-phase electrical supply at 300-550Hz. In event of alternator failure, a 24V DC battery provides limited power for essential control functions. Automatic orbit and/or destruct systems provided, consistent with range safety requirements.

EQUIPMENT: Transponders and microwave reflectors for trials of active, semi-active or beam-riding missiles. Heat sources, including infra-red flare packs mounted on rear of fuselage, can be fitted to provide low-frequency IR output. Transponders in X, S and C bands can be fitted for target acquisition and to enable Jindivik to be tracked at greater range. Provision for Tonic Beam under each wing, on which can be carried recoverable Tonic Tow Bodies. These Tow Bodies can carry either active radar, in-flight-commanded IR flares, or forward-looking Luneberg lens. They can be towed at 15-150 m (50-500 ft) behind aircraft; recovery by electric winch mounted in centre of fuselage. Other types of special tow may also be carried. Cameras fitted with wide-angle lenses are carried in wingtip pods, with all-round viewing capability. Variants are Mk 5 pod with cameras only and Mk 9 with cameras, fuel and provision for fitment of microwave reflectors in leading-edge and trailing-edge radomes. By fitting rearward-looking prisms to the lower cameras they can be used to film and record missile performance when Tonic towed targets are used.

DIMENSIONS, EXTERNAL:
Wing span:
  short span, low altitude — 6·32 m (20 ft 8·99 in)
  extended, high altitude — 7·92 m (26 ft 6 in)
  extended, extra high altitude — 9·78 m (32 ft 1·4 in)
Wing area, gross: short span — 7·06 m² (76·0 sq ft)
  extended span, high altitude — 9·48 m² (102·0 sq ft)
  extended span, extra high altitude — 10·68 m² (115·0 sq ft)
Length overall: incl nose probe — 8·15 m (26 ft 8¾ in)
  excl nose probe — 7·11 m (23 ft 3¾ in)
Height overall, skid extended — 2·08 m (6 ft 9·85 in)

WEIGHTS:
Weight empty, equipped (min) — 1,315 kg (2,900 lb)
Max payload: short-span version — 249 kg (550 lb)
  extended-span versions — 181 kg (400 lb)
Max T-O weight:
  short span, Mk 5 wing pods — 1,451 kg (3,200 lb)
  short span, Mk 9 wing pods — 1,655 kg (3,650 lb)
  high altitude, Mk 5 wing pods — 1,474 kg (3,250 lb)
  high altitude, Mk 9 wing pods — 1,655 kg (3,650 lb)
  extra high altitude, Mk 5 wing pods — 1,496 kg (3,300 lb)

PERFORMANCE (A: short span, Mk 5 pods; B: short span, Mk 9 pods; C: high altitude, Mk 5 pods; D: high altitude, Mk 9 pods; E: extra high altitude, Mk 5 pods):
Max level speed at max operational ceiling:
  A, B, C, D — Mach 0·86 (490 knots; 908 km/h; 564 mph)
  E — Mach 0·82 (470 knots; 871 km/h; 541 mph)
Min operating height: A, B — 15 m (50 ft)
Max operational ceiling: A — 17,375 m (57,000 ft)
  B — 16,460 m (54,000 ft)
  C — 19,200 m (63,000 ft)
  D — 18,595 m (61,000 ft)
  E — 20,420 m (67,000 ft)
Time to max operational ceiling: A — 26 min
  B, C — 30 min
  D, E — 34 min
Typical max on-station endurance: A, C — 1 h 24 min
  B — 1 h 56 min
  D — 2 h 17 min
  E — 1 h 28 min
Max range: A — 560 nm (1,038 km; 645 miles)
  B — 800 nm (1,482 km; 921 miles)
  C — 740 nm (1,371 km; 852 miles)
  D — 1,100 nm (2,038 km; 1,266 miles)
  E — 900 nm (1,668 km; 1,036 miles)

CANADA

CANADAIR

CANADAIR LTD

PO Box 6087, Montreal, Quebec H3C 3G9
Telephone: (514) 744 1511
OFFICERS: see Aircraft section

In addition to its work on manned aircraft, Canadair is developing or producing various drone systems, including the CL-289 described in the International section.

CANADAIR CL-89

NATO designation: AN/USD-501

The Canadair CL-89 (AN/USD-501) airborne surveillance drone system evolved from a need of the western Allied armed forces for an intelligence-gathering device for battlefield commanders; details of its early history can be found in the 1977-78 and earlier editions of *Jane's*.

With a very high probability of survival against all known air defence systems, the CL-89 can acquire timely and accurate battlefield intelligence using its photographic and infra-red line-scanning equipment. The system, consisting of the air vehicles plus the related ground support and operational maintenance equipment, is totally integrated, mobile, and independent of such external services as electrical power supplies.

The CL-89 system has been produced for Canada and the Federal Republic of Germany, as described in previous editions of *Jane's*. It is shortly to enter service with the armed forces of France and Italy.

Type: Recoverable airborne surveillance drone system.

Airframe: Cylindrical metal body, with curved nosecone and tapering tailcone. Three detachable dorsal packs for forward and rear landing bags and flare container; two detachable ventral packs for sensor equipment and parachute recovery system. Four rectangular stub-wings at rear of body, in cruciform arrangement at 45° to horizontal and vertical centrelines. Upper pair fold out of way when landing airbags are inflated. Ailerons on port upper and starboard lower stub-wings. Two pairs of foreplanes aft of nosecone on horizontal and vertical centrelines, for pitch and yaw trim respectively.

Power Plant: One 0·56 kN (125 lb st) Williams International WR2-6 turbojet engine, with variable exhaust nozzle, installed in tailcone aft of wings. Air intake duct on each side of fuselage, forward of wings. Fuel and oil tanks in central body compartment, forward of air intakes. One 22·24 kN (5,000 lb st) average thrust BAJ Vickers Wagtail booster rocket motor, attached to body of drone by three V-shaped thrust arms and cable.

Launch and Recovery: Launched from truck-mounted zero-length ramp; booster separates automatically after 2·5 s of flight. For recovery, after final positioning by ground homing beacon, drone is slowed by drogue parachute until main parachute deploys. Drone is then inverted, and forward and rear airbags are inflated and deployed automatically to absorb landing impact.

Guidance and Control: Flight path, altitude and sensor on/off commands are controlled by preset programmer which receives information from onboard Air Distance Measuring Unit (ADMU) and combines this with preset programme to control flight path. Ground homing beacon positions drone in final stages of flight to ensure accuracy of landing.

Equipment: Engine-driven alternator for electrical power during flight. Two main sensor systems currently in use: Carl Zeiss KR68/24C camera system and British Aerospace Dynamics Type 201 infra-red linescan system. Aft of sensor pack is compartment for fuel and oil tanks. Compartment aft of tanks has ventral forward-hinged door providing access to engine start air connector, and dorsal pack containing 12 photoflares just forward of rear landing bag container. Final cylindrical compartment houses rear landing bag container and parachute recovery pack, between dorsal and ventral pairs of wings respectively. Other onboard equipment includes forward and rear landing airbags; Air Distance Measuring Unit (ADMU); programmer; static power converter; homing receiver; amplifier; flash detector; directional and vertical gyros; transponder antenna; and air bottle to inflate airbags.

Dimensions, external:
Length overall, excl nose probe:
with booster — 3·73 m (12 ft 3 in)
without booster — 2·60 m (8 ft 6½ in)
Body diameter — 0·33 m (1 ft 1 in)
Span of wings — 0·94 m (3 ft 1 in)
Span of foreplanes — 0·48 m (1 ft 7 in)

Weights:
Weight dry (excl fuel, oil and mission equipment) — 78·2 kg (172·4 lb)
Mission equipment — 17-20 kg (37·5-44 lb)
Max launching weight: with booster — 156 kg (343 lb)
without booster — 108 kg (238 lb)

Performance:
Max speed — 400 knots (741 km/h; 460 mph)
Max operating altitude — 3,050 m (10,000 ft)
Max range: standard — 65 nm (120 km; 74 miles)
with extended range fuel tank — 75 nm (140 km; 87 miles)

CANADAIR CL-227

The CL-227 is designed as a highly survivable real-time surveillance and target acquisition system for use at medium range. It has VTOL capability, can translate to horizontal flight and hover, and is launched and recovered using a small platform. First flight took place on 25 August 1978. Field flight demonstrations were scheduled to take place at Canadian Forces Base Suffield, Alberta, in October 1981.

It has a peanut-shaped body housing the power plant in the upper sphere, with the sensors and autopilot in the lower sphere. Two three-blade contra-rotating rigid propellers, mounted amidships, provide the lift and attitude control. It is powered by a Williams International WR-34 turboshaft engine. Sensors can include daylight or low light level TV, a laser designator, thermal imager, radiation detector or decoy equipment.

The CL-227 can be winched on to a landing device to provide an automatic landing capability. This platform can also act as the transportation platform and minimise ground handling of the air vehicle between flights.

Dimensions:
Height overall — 1·64 m (5 ft 4½ in)
Body diameter (max) — 0·64 m (2 ft 1 in)
Propeller diameter — 2·54 m (8 ft 4 in)

Weights:
Mission equipment — 36 kg (80 lb)
Fuel — 25·5 kg (56 lb)
Max T-O weight — 125 kg (275 lb)

Performance:
Max level speed — 70 knots (130 km/h; 81 mph)
Max operating altitude — 2,800 m (9,185 ft)
Typical operating radius — 27 nm (50 km; 31 miles)
Typical mission endurance — 2 h

FRANCE

AÉROSPATIALE
SOCIÉTÉ NATIONALE INDUSTRIELLE AÉROSPATIALE

Division des Engins Tactiques

2-18 rue Béranger, BP 84, 92320 Châtillon-sous-Bagneux
Telephone: 657 14 21
Telex: AIRSPA 250881 F
Officers: see Aircraft section

Aérospatiale (Division des Engins Tactiques) is currently producing the CT.20 target drone for the French armed forces. Approx 50 examples of the C.20, referred to in the 1978-79 *Jane's*, were produced. The C.22 is under development as a high-performance subsonic target.

AÉROSPATIALE CT.20

The CT.20 medium performance target is also used as a tug for a towed target. Series production began in 1958; by 1 January 1981 a total of 1,450 had been ordered, including more than 300 for export to NATO and non-NATO countries. Rate of production in 1980 was approx five to six per month.

It is standard equipment for training military units in the use of air-to-air and surface-to-air missiles. A version known as the **CT.20 TBA,** with very low altitude capability, is also operational.

Type: Turbojet-powered radio-controlled target.

Airframe: Body in three main sections. Forward section, of aluminium alloy, contains command guidance, autopilot, batteries and principal recovery parachute. Central section consists of structural steel tank divided into two parts, one for fuel and one containing chemicals for tracking smoke. Rear fuselage, of aluminium alloy, contains engine and carries aluminium alloy V-tail. Braking parachute in cone above jet nozzle. Mid-mounted swept wings, of aluminium alloy, have wingtip spoilers for lateral control; elevators controlled simultaneously by single jack. Ventral fin.

Power Plant: One Turboméca Marboré II (3·92 kN; 880 lb st) turbojet engine in CT.20 Version IV; Marboré VI (4·7 kN; 1,056 lb st) turbojet in CT.20 Version VII.

Launch and Recovery: Launched by booster rockets from nearly-zero-length vehicle-mounted ramp. Recovery by parachute and impact-absorbing airbag, deployed after engine is stopped by transmission of landing signal. Can be recovered from land or water.

Guidance and Control: Radio command guidance system, operated by controller on ground or in airborne director aircraft. Nine basic signals can be transmitted: turn right; turn left; nose up; nose down; increase power; decrease power; trace smoke; operate cameras; and land.

Equipment: Can be used to tow Dornier target which is pylon-mounted under starboard wing at launch; or can be used with trailed target system developed for the Centre d'Essais des Landes, primarily for training in use of air-to-air missiles with electro-magnetic (BEY target) or infra-red homing (EMIR target) systems. CT.20 TBA version has TRT AVH-6 radio altimeter for very low (about 30 m; 100 ft) altitude capability, and an improved remote guidance system.

Dimensions, external (A: standard version; B: extended version with trailed target):
Wing span — 3·16 m (10 ft 4½ in)
Wing area, gross — 3·20 m² (33·34 sq ft)
Length overall:
A — 5·45 m (17 ft 10½ in)
B — 5·60 m (18 ft 4½ in)
Body diameter (max) — 0·66 m (2 ft 2 in)

Weights (A: standard version; B: extended version with trailed target):
Weight empty: A — 490 kg (1,080 lb)
B — 610 kg (1,344 lb)
Fuel load: A — 164 kg (361 lb)
B — 192 kg (423 lb)
Max launching weight: A — 660 kg (1,455 lb)
B — 800 kg (1,763 lb)

Performance (A: standard version; B: extended version with trailed target):
Max speed at 10,000 m (32,800 ft):
Marboré II — 485 knots (900 km/h; 560 mph)
Marboré VI — 512 knots (950 km/h; 590 mph)
Max Mach number (A, B) — Mach 0·85
Service ceiling: Marboré II — 12,000 m (39,375 ft)
Marboré VI — 15,000 m (49,200 ft)
Max operating height: A — 14,000 m (45,925 ft)
B — 13,000 m (42,650 ft)
Time to max operating height: A, B — 15 min
Time to 10,000 m (32,800 ft) — 6 min
Endurance at max operating height: A — 50 min
B — 1 h 10 min
Min operating height: A, B — 30 m (100 ft)
Practical range of command guidance and tracking system — 135 nm (250 km; 155 miles)
Endurance at min operating height: A — 15 min
B — 21 min
Average endurance at 10,000 m (32,800 ft) — 45 min

AÉROSPATIALE C.22

The C.22, which first flew on 6 June 1980, was designed for use as a variable-speed target for anti-aircraft weapons and, especially, for training of fighter pilots and anti-aircraft system crews. Its dimensions, radar signature and high performance enable it to simulate combat aircraft flying at any altitude, and also sea-skimming missiles.

Type: High-performance variable-speed subsonic target.

Airframe: Moulded plastics wings, with symmetrical profile and no control surfaces. Fuselage of wound glassfibre, impregnated with epoxy resin, reinforced by metal inserts at attachment points. Nosecone and tailcone of moulded plastics. Cruciform light alloy tail-fins, with four control surfaces operated by electrical actuators.

Power Plant: One 3·4 kN (772 lb st) Microturbo TRI 60-2 turbojet engine, mounted in pod on top of fuselage; and two jettisonable solid-fuel booster rockets (each 28·3 kN; 6,360 lb st) attached beneath wings on sides of fuselage. Fuel tanks for TRI 60 in centre of fuselage, capacity 245 litres (54 Imp gallons).

Launch and Recovery: Launched by jettisonable rockets from ground or ship base. Parachute system for recovery from land or sea. Inflatable airbags beneath fuselage to absorb landing impact.

Guidance and Control: Radio command digital guidance system. Flight control, based upon a mini-computer, permits complex manoeuvres at more than 6*g*. Aerodynamic control by cruciform tail-fins indexed at 45° from vertical axis.

Special Equipment: Nose compartment available for equipment, including tow winch. Tail compartment includes smoke generator and recovery parachutes. Equipment/mission load in nosecone includes up to 50 kg (110 lb) of Luneberg lenses or infra-red augmentation devices; active or passive countermeasures; recording equipment for assessing the effectiveness of weapon systems during training; or equipment for towing successively two 30 kg (66 lb) secondary targets at speeds of more than 485 knots (900 km/h; 559 mph), with 800 m (2,625 ft) cable for each target.

Dimensions, external:
Wing span — 2·50 m (8 ft 2½ in)
Span over tail-fins — 1·20 m (3 ft 11¼ in)
Length overall — 5·25 m (17 ft 2¾ in)
Height overall — 1·15 m (3 ft 9¼ in)
Body diameter — 0·40 m (1 ft 3¾ in)

Weights:
Weight empty — 255 kg (562 lb)
Fuel load — 186 kg (410 lb)
Total mission load, incl towed targets — up to 135 kg (297·5 lb)
Max launching weight — 630 kg (1,389 lb)

Performance:
Max speed — Mach 0·95
Time to 12,000 m (39,375 ft) — less than 6 min
Service ceiling — 14,000 m (45,925 ft)
Minimum operating altitude — 15 m (50 ft)
Max endurance at 12,000 m (39,375 ft) — 2 h 30 min
g limit — more than +6

GAF Jindivik Mk 203B target drone, with Mk 1 in background

Canadair CL-227 surveillance/target acquisition RPV

Canadair CL-89 airborne surveillance/target acquisition drone system

Aérospatiale CT.20 target drone on launching ramp

Aérospatiale C.22 variable-speed subsonic target drone

EERM SAM-B meteorological drone

EERM

ÉTABLISSEMENT D'ÉTUDES ET DE RECHERCHES MÉTÉOROLOGIQUES (Direction de la Météorologie Nationale)

77 rue de Sèvres, 92106 Boulogne-Billancourt Cédex
Telephone: (1) 604 91 51
DIRECTOR: A. Villevieille

Several types of SAM (Sonde Aérologique Motorisée) have been used in considerable numbers by the EERM since 1976 as pilotless atmospheric sounding vehicles.

EERM SAM-B

As the accompanying photograph shows, SAM-B is an extremely simple vehicle, of disc-shaped planform with the trailing-edges hinged to act as control surfaces. It is powered by a 1·2 kW (1·6 hp) Webra 61 10 cc two-stroke piston engine, driving a two-blade fixed-pitch glassfibre propeller, and is used under line-of-sight radio control for low-altitude sounding flights up to 500 m (1,640 ft). No other details are known.

EERM SAM-C

SAM-C is an atmospheric sounding vehicle for work at higher altitudes than SAM-B, up to cloud level. Design began in January 1977, and the first of two prototypes made its initial flight on 5 May that year. This version, with the engine pod-mounted on a pylon above the centre-fuselage, is designated **SAM-C MP** (moteur en pylône). A later version, which began flight testing in July 1978, is known as the **SAM-C MAR** (moteur arrière) and has the engine mounted at the rear, driving a pusher propeller.

First high-altitude (3,200-4,000 m; 10,500-13,125 ft) tests with a SAM-C MP were made in May 1978 at Deux-Alpes, and in the following month one was flown over the 3,300 m (10,825 ft) Mount Etna in Sicily, to measure thermal conditions in the vicinity of the volcano and to take gas samples to determine their chemical composition.

Five production SAM-C MAR vehicles, to which version the following description applies, had been built by 1 February 1980; production was continuing at that time.

TYPE: Pilotless atmospheric sounding vehicle.

AIRFRAME: Cantilever shoulder-wing monoplane, with inverted-V tail surfaces. Wings, of Aérospatiale RA 16 RC3 section, have 10° dihedral and 3° incidence, and are of glassfibre/polystyrene foam construction without spars or ribs. Schempp-Hirth metal airbrakes, or plastics trailing-edge flaps, with upward travel only and actuated by electrical servo. Glassfibre frameless monocoque fuselage, in four sections: nose, front, centre and tail. Tail surfaces of similar construction to wings; elevators actuated by electrical servo. No landing gear.

POWER PLANT: One 1·2 kW (1·6 hp) Webra 61 10 cc two-stroke piston engine, mounted in rear fuselage and driving a two-blade fixed-pitch glassfibre pusher propeller. Fuel tank in centre fuselage section, capacity 0·5 to 2·0 litres (0·9 to 3·5 Imp pints). Refuelling point in fuselage tail section. NACA-type flush air intake above rear fuselage.

LAUNCH AND RECOVERY: Launched by catapult from 3·10 m (10 ft 2 in) ramp.

GUIDANCE AND CONTROL: Line-of-sight radio control. Aerodynamic control by airbrakes/flaps and elevators.

SYSTEM: 12V 0·5Ah nickel-cadmium battery for sounding systems; 4·8V 1·2Ah nickel-cadmium battery for radio controls.

EQUIPMENT: Temperature, humidity and atmospheric pressure sensors.

DIMENSIONS:

Wing span	3·15 m (10 ft 4 in)
Wing chord (constant)	0·35 m (1 ft 1¾ in)
Wing area, gross	1·10 m^2 (11·85 sq ft)
Wing aspect ratio	9
Length overall	2·10 m (6 ft 10¾ in)
Body diameter	0·16 m (6·3 in)
Height overall (on ground)	0·20 m (7·9 in)
Propeller diameter	0·30 m (11·8 in)

WEIGHTS AND LOADINGS:

Weight empty:	
MP with minimum 0·5 kg (1·1 lb) ballast in forward fuselage	6·5 kg (14·33 lb)
MAR with 2 kg (4·5 lb) ballast	8 kg (17·63 lb)
Max mission load	2·5 kg (5·5 lb)
Max fuel	1·3 kg (2·9 lb)
Max launching and landing weight	9·5 kg (20·9 lb)
Max wing loading	8·63 kg/m^2 (1·76 lb/sq ft)
Max power loading	7·9 kg/kW (13·1 lb/hp)

PERFORMANCE (at max launching weight except where indicated):

Never-exceed speed	68 knots (126 km/h; 78 mph)
Max level speed at S/L	50 knots (92 km/h; 57 mph)
Max cruising speed	30 knots (55 km/h; 34 mph)
Econ cruising speed at S/L	25 knots (46 km/h; 29 mph)
Stalling speed, power off	15 knots (28 km/h; 17·5 mph)
Max rate of climb at S/L at 8·5 kg (19 lb) normal operating weight	210 m (690 ft)/min
Service ceiling	5,000 m (16,400 ft)

EERM SAM-D

Design of SAM-D began at the same time as that of SAM-C, and the first of two prototypes made its initial flight in March 1978. Intended to obtain aerological measurements under difficult conditions (wind speeds of up to 29 knots; 54 km/h; 33 mph), it has a wide speed range of 17-64 knots (32-119 km/h; 20-74 mph), and entered production in 1979. Three production SAM-Ds had been completed by 1 February 1980.

TYPE: Pilotless atmospheric sounding vehicle.

AIRFRAME: Cantilever monoplane, with cropped-delta mid-mounted main wings at rear and shoulder-mounted foreplanes with elevons. Main wings have 15% thickness/chord ratio, 4° dihedral, 2° incidence and 20° sweepback at quarter-chord; they are of glassfibre/polystyrene foam construction, without spars or ribs, and have no movable control surfaces. Foreplanes are of similar construction, the elevons being actuated by electrical servos. Single sweptback fin; no rudder. Glassfibre frameless monocoque fuselage, of rectangular cross-section. No landing gear.

POWER PLANT: As for SAM-C. Single plastics fuel tank, capacity 0·5 to 1·0 litre (0·9 to 1·75 Imp pints), in fuselage centre-section beneath wing roots. Refuelling point in rear fuselage.

LAUNCH AND RECOVERY: No details received.

GUIDANCE AND CONTROL: No details received.

SYSTEM: Electrical power sources as for SAM-C.

EQUIPMENT: No details received.

DIMENSIONS, EXTERNAL:

Wing span	1·40 m (4 ft 7¼ in)
Wing chord at root	0·60 m (1 ft 11½ in)
Wing area, gross	0·63 m² (6·78 sq ft)
Wing aspect ratio	3·1
Length overall	1·55 m (5 ft 1 in)
Fuselage: Max width	0·14 m (5½ in)
Height overall	0·47 m (1 ft 6½ in)
Propeller diameter	0·30 m (11¾ in)

WEIGHTS:

Weight empty	4·4 kg (9·7 lb)
Max mission load	1 kg (2·2 lb)
Max fuel load	0·6 kg (1·3 lb)
Max launching and landing weight	6 kg (13·2 lb)

PERFORMANCE (at max launching weight):

Max level speed	60 knots (111 km/h; 69 mph)
Max cruising speed	50 knots (92 km/h; 57 mph)
Stalling speed, power off	15 knots (28 km/h; 17·5 mph)
Max rate of climb at S/L	360 m (1,180 ft)/min
Service ceiling	4,000 m (13,125 ft)

FRANCE-ENGINS
SOCIÉTÉ ANONYME FRANCE-ENGINS

c/o Microturbo SA, BP 2089, Chemin du Pont-de-Rupé, 31019 Toulouse Cédex
Telephone: (61) 70 07 77
Telex: 531442 miturbo
PRESIDENT-DIRECTOR GENERAL: André de Boysson

This company was formed, in early 1975, by Microturbo SA (see Aero-Engines section) and Société Soulé.

FRANCE-ENGINS MITSOUBAC

The Mitsoubac is a low-cost target drone for naval operation, designed to provide a target for ships' armament without the need for support by land-based installations.

Flight trials began in 1975 as a private venture, using a Bell 47 light helicopter and later an Alouette III as the launch aircraft. A low-cost launching system, using a Dornier Do 28 as the launch aircraft, has been developed, capable of operating 135 nm (250 km; 155 miles) offshore from its base. Details of the standard Mitsoubac can be found in the 1979-80 *Jane's*.

A long-range version is under development. This will have an engine thrust of 1·13 kN (253·5 lb st); payload of 30 kg (66 lb); launch weight of 200 kg (441 lb); max range of 324 nm (600 km; 373 miles) at 330 m (1,000 ft) or 450 nm (835 km; 519 miles) at 3,280 m (10,000 ft); and endurance of 35 min at 330 m (1,000 ft) at a max speed of 450 knots (835 km/h; 519 mph).

GERMANY
(FEDERAL REPUBLIC)

DORNIER
DORNIER GmbH

Postfach 2160, 8000 München 66
Telephone: München 8715480
Telex: 05-23543
OFFICERS: see Aircraft section

Current activities of Dornier GmbH include the development of drones, RPVs, reconnaissance systems, missiles and air target systems.

Dornier Aerial Target Systems (DATS) can be adapted to a wide variety of aircraft and target drones, and the original DATS was described briefly in the 1975-76 *Jane's*. Later systems include DATS 2, now in service with the German armed forces for surface-to-air and air-to-air gunnery training; DATS 3, for air-to-air gunnery in the high subsonic and transonic speed ranges; DATS 5, designed for operation with target drones but otherwise identical to DATS 2; and DATS 6, operable with either drones or piloted aircraft. Details of these were given in the 1980-81 *Jane's*.

Dornier developed the Do 34 Kiebitz mobile drone system, based on the Do 32 K Experimental Kiebitz; and the Aerodyne unmanned VTOL research vehicle, described in previous editions of *Jane's*. It is currently developing the LA-RPV, and is active in the fields of tactical RPVs, stand-off missiles, mini-RPVs and helicopter drones. Under subcontract to LMSC (see US section), it is supplying the net recovery system for the Aquila mini-RPV ordered by the US Army.

Jointly with Canadair and SAT, Dornier is developing the AN/USD-502, details of which can be found in the International section.

DORNIER Do 34 KIEBITZ (PEEWIT)

In August 1972 Dornier announced the receipt of a DM 7 million contract from the Federal Ministry of Defence for the first development phase of a prototype of an operational Kiebitz system, to be used for reconnaissance, fire control, communications and traffic monitoring duties. This followed the completion in mid-1972 of the design of a Do 34 operational model as a sensor platform with a payload of 140 kg (308 lb). Full-scale mockups of the Do 34 were followed by the construction of two prototype flight vehicles and two ground stations. Under a bilateral agreement between the West German and French governments, both flight vehicles are fitted with an advanced version of the French LCT Orphée II radar, to define, integrate and test a new battlefield reconnaissance system known as **Argus** (see International section). The first Argus prototype made its initial flight on 14 May 1979.

Flight testing of the first Do 34 Kiebitz (F 01) began on 1 February 1978, and on 3 March the maximum flight altitude of more than 300 m (985 ft) was reached, the Kiebitz staying aloft for 40 minutes. By February 1980 the prototypes had made more than 240 successful flights, totalling approx 90 flying hours.

The following details apply to the basic Do 34 flight vehicle:

AIRFRAME: Flight vehicle is roughly cone-shaped, to reduce radar signature. Approximately cylindrical payload compartment located beneath this, enabling sensors to be changed quickly and allowing space for large-volume radome. Two-blade rotor, with foldable blades attached by straps and driven by cold air expelled through blade-tip nozzles. Air to nozzles supplied by MTU radial compressor driven by engine.

POWER PLANT: One 313 kW (420 shp) Allison 250-C20B turboshaft engine, installed at an angle to give good air intake position.

LAUNCH AND RECOVERY: Not applicable. Kiebitz is cable-tethered, housed in vehicle which incorporates its landing platform, winch, control post, fuel tank and auxiliary equipment.

GUIDANCE AND CONTROL: After arrival on site, can be in position at operational height of 300 m (985 ft) in 8 min; limiting factors are wind speed of 14 m (46 ft)/s, added gusts of 8 m (26 ft)/s, available thrust reserves, and sensor requirements. Control system aligns drone according to airframe attitude and position in relation to ground. Electromagnetic sensor measures any drift from desired position.

EQUIPMENT: Bodenseewerk autopilot and Lear Siegler 3 kW starter/generator. Appropriate payload packages for reconnaissance, target acquisition, fire control, communications relay or traffic monitoring.

DIMENSIONS, EXTERNAL:

Diameter of rotor	8·00 m (26 ft 3 in)
Height overall	2·70 m (8 ft 10¼ in)
Body diameter (max) at bottom	2·00 m (6 ft 6¾ in)

WEIGHTS AND LOADING:

Weight without cable or mission load	350 kg (772 lb)
300 m (985 ft) cable	85 kg (187 lb)
Max mission load to 300 m (985 ft), ISA	140 kg (309 lb)
Max launching weight	550 kg (1,212 lb)
Disc loading	10·9 kg/m² (2·2 lb/sq ft)

DESIGN PERFORMANCE:

Reel-in/reel-out speed	3 m (9·8 ft)/s
Operating altitude	300 m (985 ft)
Mission endurance	more than 24 h

DORNIER MTC II

Details of this remotely piloted mini-helicopter were released on 8 May 1981, shortly after its first flight. Described as "a developed version of the smaller MTC I", it was designed as an equipment-carrying platform for Army and Navy use, and is capable of carrying a 60 kg (132 lb) load. The MTC II is fully stabilised, and is remotely operated from a ground-based console via a loose cable for initial testing; radio control is planned for subsequent flight test phases. Power plant is a 30 kW (40 hp) two-stroke engine, driving contra-rotating co-axial rotors.

SPECIFICATION:

Rotor diameter	3·20 m (10 ft 6 in)
Max T-O weight	190 kg (419 lb)
Endurance	approx 2 h

DORNIER LA-RPV

Formerly known as the UKF, the LA (Luft-Angriffs: air attack) RPV has been proposed by Dornier, under the Federal Defence Ministry's Component Air Testing Programme, as a means of examining the basic problems of target acquisition, armament and flight guidance for ground support RPVs.

The principal tasks of the LA-RPV are seen as air strike (especially battlefield interdiction) against highly defended fixed and transient ground targets; tactical reconnaissance; and electronic warfare including air defence suppression. Several different possible configurations have been studied, the one most favoured in 1979-80 (see accompanying photograph) being a ground-launched re-usable vehicle with retractable landing gear, an internal weapon load, an externally-mounted turbojet engine, and inertial navigation updated by satellite data or terrain comparison. This would take off and land conventionally, having provision for booster-assisted take-off and an arrester system for landing.

Particular attention is being paid to accuracy of navigation and weapons delivery, with or without terminal guidance or with remote-based ground control. Since most missions will be against preselected targets, there will be no onboard detection sensors for an enemy to jam. Most missions will be undertaken at high speed and low altitude, and will be pre-programmed out and back, with provision for the ground controller to override the mission programme during the take-off, formation, flight monitoring and recovery phases. Target attack would be by means of 'scatter' weapons ejected from an onboard munitions container.

An Aeritalia (Fiat) G91R aircraft is serving as a multipurpose RPV testbed, and will be used in co-operation with VFW in a remote target acquisition programme. This aircraft is manually piloted, and has been modified by Dornier to carry an AEG-Telefunken underwing television camera, an Eltro forward-looking infra-red (FLIR) sensor, a two-axis stabilisation system, and a TV transmitter, all housed in an underwing pod. The TV/infra-red experiments were completed by the beginning of 1979, and the G91 is being used by the two companies in other programmes.

DIMENSIONS:

Wing span	3·80 m (12 ft 5½ in)
Length overall	6·50 m (21 ft 4 in)
Height overall	1·90 m (6 ft 2¾ in)

DORNIER KDAR-LOCUST

Dornier has been studying for some years the development of mini-RPVs, and a vehicle of this type was displayed for the first time at the Paris Air Show in May/June 1977. A half-scale model has been flight-tested, as well as experimental full-scale prototypes of the Dornier/Texas Instruments KDAR (Kleindrohne Antiradar; anti-radiation mini-drone).

Typical missions for which the vehicle was intended include electro-optical reconnaissance, target acquisition and fire control, anti-radar operations, anti-tank or point target attack, and target presentation. For the reconnaissance and target acquisition and fire control missions, the integration of a sensor package comprising a stabilised TV camera, with autotrack facility coupled with a laser illuminator, was under investigation. Anti-radar missions and attacks on tanks or point targets would be 'kamikaze' missions in which the vehicle would home on to and attack the target after a search flight using, for example, passive or active radar and/or infra-red seeker heads. For target missions, it could be fitted with various augmentation devices, including smoke cartridges, flares or Luneberg lenses. It could also be used to test missile proximity fuses or homing heads.

The Dornier vehicle was one of three RPVs (others were by MBB and VFW, which see) designed for the joint

EERM SAM-D adverse-weather meteorological drone

EERM SAM-C MAR atmospheric sounding vehicle

Dornier KDAR-Locust mini-RPV

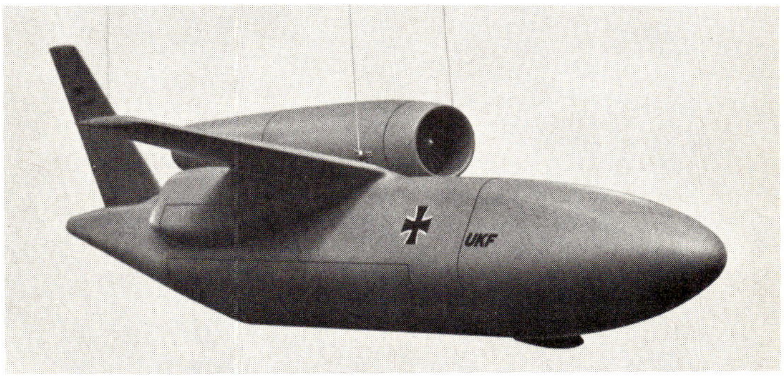

Model of possible Dornier LA-RPV strike RPV *(Brian M. Service)*

Dornier Spähplattform on truck-towed trailer

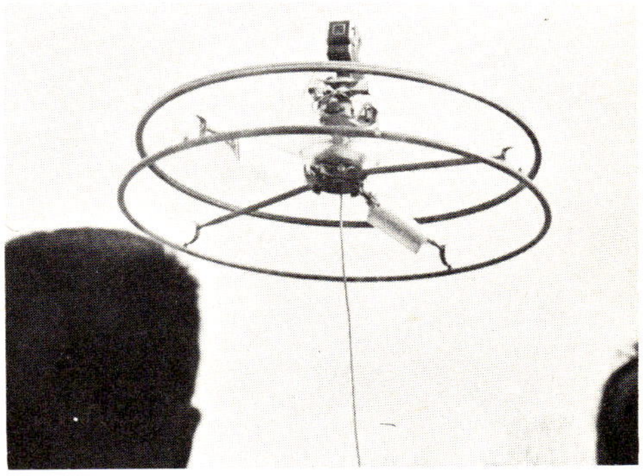

Dornier Spähplattform in flight

US-West German **Locust** harassment weapon system programme. German participation in this programme was suspended in 1981.

Airframe: Consists essentially of two glassfibre half-shells. Wings built integrally with fuselage, and fitted with elevons on trailing-edges. Tail surfaces comprise a main fin and an underfin, the latter also serving as a keel surface to protect the propeller during landing.

Power Plant: One 16·4 kW (22 hp) Limbach two-cylinder two-stroke piston engine, mounted in rear of fuselage and driving a four-blade pusher propeller. Fuel in two fuselage tanks; provision to install a third tank in centre of fuselage instead of main recovery parachute. Provision for booster engine for launch.

Launch and Recovery: During initial flight tests, drone was launched by catapult from hydraulically operated launching sled. When required for re-use, the RPV is fitted with an electro-mechanically deployed parachute recovery system (drogue parachute in fin-tip fairing, main parachute in centre-fuselage), activated by radio command, with automatic activation in the event of onboard systems failure. Inflatable airbags to absorb landing impact. For expendable missions, parachute can be replaced by an additional fuel tank or other equipment.

Guidance and Control: Automatic or remote radio command guidance system. Aerodynamic control by elevons.

Dimensions, external:
- Wing span 2·10 m (6 ft 10½ in)
- Length overall 2·00 m (6 ft 6¾ in)

Dimensions, internal:
- Mission load compartment:
  - Length 0·60 m (1 ft 11½ in)
  - Diameter 0·20 m (8 in)

Weights:
- Mission load 15 kg (33 lb)
- Max launching weight 70 kg (154 lb)

Performance:
- Max diving speed 194 knots (360 km/h; 223 mph)
- Max level speed at S/L, ISA 135 knots (250 km/h; 155 mph)
- Max operating height 3,000 m (9,850 ft)
- Max endurance 3 h

DORNIER SPÄHPLATTFORM

The Spähplattform (spotting platform) is an unpowered camera-carrying device being developed for battlefield surveillance. To improve mobility, the experimental system is being installed, under government contract, in a Unimog 1·4 ton truck with a two-wheeled trailer. Field tests, simulating military missions, were planned to take place during 1980.

Airframe: Small, domed mission load compartment, mounted on top of an upper gyroscopic ring encircling a four-blade rotor, and a lower stationary ring, in which four control vanes are mounted.

Power Source: 25 kW (33·5 hp) is available for initial rotor spin-up; 90 N is required to maintain the hover mode.

Launch and Recovery: Not applicable. The Spähplattform is vehicle-mounted, and tethered to its control vehicle by an umbilical cable, via which drive from the parent vehicle is transmitted to spin up the four-blade flywheel rotor to 4,000 rpm for take-off. The blade-tips of the rotor are attached to the upper gyroscopic ring, in which the energy from spin-up is stored; the four control vanes, mounted on cruciform tubes within the stationary lower ring, can be actuated to control the platform in azimuth. A conventional cyclic pitch control of the rotor determines the platform's position, and gyroscopic effect stabilises its attitude. The Spähplattform is started and reeled out from a transportation box with automatic covers, and hovers on the energy stored in the flywheel rotor.

Guidance and Control: Control and sensor signals are transmitted via the tethering cable from the ground control vehicle.

Special Equipment: Still camera of 12·5 mm to 75 mm focal length, or TV camera, in mission equipment compartment.

Dimension:
- Rotor diameter 1·20 m (3 ft 11¼ in)

Weights:
- Weight of twin-ring assembly 20 kg (44 lb)
- Weight of complete vehicle, without mission load 30 kg (66 lb)
- Payload 5 kg (11 lb)
- Max T-O weight 35 kg (77 lb)

Performance:
- Reel-in/reel-out speed 5 m (16·4 ft)/s
- Operating altitude 50 m (164 ft)
- Mission endurance 1 min

MBB

MESSERSCHMITT-BÖLKOW-BLOHM GmbH

Ottobrunn bei München, 8 München 80, Postfach 801220

OFFICERS: see Aircraft section

MBB's drone/RPV activities include studies for a modular-concept weapon-carrying RPV, for use against heavily defended, previously reconnoitred targets, or for electro-optical reconnaissance. All-weather navigation, and the ability to deliver weapons even under adverse conditions of visibility, would be achieved with the aid of a combined flight guidance and fire control system also under development by MBB. The RPV would be truck-launched, using a booster, and recovered either by arrester cable or in a horizontal, parachute-assisted glide landing.

MBB KDAR-LOCUST

This anti-radiation mini-RPV was developed under Federal German Ministry of Defence contract, to combat the radar equipment of enemy air defence installations. It was a competitor for US/German contracts intended to have Locust vehicles in service with field units by the end of FY 1983. Further details of the Locust, now a USAF-only programme, can be found under the US Department of Defense/US Air Force heading in this section.

The MBB vehicle is launched, by means of a solid (powder) propellant booster motor, directly from its storage and transport container. When inside the container the wings, tail control surfaces and propeller are folded; they flip out for use immediately upon launch.

Designed to search for, and attack, enemy air defence radars, this small RPV was evolved in collaboration with Teledyne Brown Engineering of Huntsville, Alabama, USA. Components for the terminal phase guidance system are manufactured by Motorola.

First launch was made in the Spring of 1980.

DIMENSIONS, EXTERNAL:

Wing span (deployed)	2·60 m (8 ft 6½ in)
Length	2·20 m (7 ft 2½ in)
Body diameter	0·183 m (7¼ in)

VFW

VEREINIGTE FLUGTECHNISCHE WERKE GmbH

Hünefeldstrasse 1-5, 28 Bremen 1, Postfach 1206

VFW RT 910 TUCAN

This mini-RPV, developed under contract to the Federal Ministry of Defence, was displayed in model form at the 1979 Paris Air Show and made its first flight in 1980. Intended for real-time target detection and acquisition, the Tucan has a broadly similar configuration to the prototype mini-RPV No. 2012 (see 1980-81 *Jane's*), but without the latter's canard control surfaces. The Tucan is of modular design, with a radio command guidance system, and telemetry for relaying visual and signal data. Power plant is a horizontally-opposed piston engine, driving a two-blade pusher propeller. Launch is by catapult, recovery by parachute. Mission equipment, accommodated in the long fuselage nose, can include cameras and transmitters; the CG of the vehicle can be adjusted, according to the weight of this equipment, by positioning the wings accordingly. Targets are detected by means of a TV or infra-red camera, the images from which are relayed via video or radio link to the ground control station.

WEIGHT:

Max launching weight	100 kg (220 lb)

PERFORMANCE:

Max level speed	135 knots (250 km/h; 155 mph)
Operational radius, incl 30 min loiter	38 nm (70 km; 43 miles)

VFW KDAR-LOCUST

VFW's contender for the US Air Force Locust anti-radiation mini-RPV programme, developed in association with General Dynamics, was launched for the first time on 1 April 1980 at Meppen. A progressive development of the RT 910 Tucan, it has cruciform wings, fitted with control surfaces, and is powered by a small piston engine driving a two-blade pusher propeller. The VFW vehicle is container-launched, and has a launching weight of 60 kg (132 lb).

INDONESIA

LAPAN

LEMBAGA PENERBANGAN DAN ANTARIKSA NASIONAL (National Aeronautics and Space Institute)

HEADQUARTERS: Jalan Pemuda Persil No. 1, PO Box 3048, Jakarta Timur

Telephone: (021) 465125, 482808 and 482653

Telex: 45675 LAPAN IA

OFFICERS: see Aircraft section

This Institute, established in 1963, designed and built an experimental mini-RPV known as the XTG-01, of which a description and illustration can be found in the 1980-81 *Jane's*.

INTERNATIONAL PROGRAMMES

AN/USD-502

PRIME CONTRACTOR:

Canadair Ltd, PO Box 6087, Montreal, Quebec H3C 3G9, Canada
Telephone: (514) 744 1511

PRINCIPAL SUBCONTRACTOR:

Dornier GmbH, Postfach 2160, 8000 München 66, Federal Republic of Germany
Telephone: München 8715480
Telex: 05-23543

AN/USD-502 / CL-289

The AN/USD-502 originated as a joint Canadair/Dornier programme, based on AN/USD-501 (Canadair CL-89, which see) technology, to meet a German military technical requirement for a longer-range drone for service from about 1983. It has the NATO designation AN/USD-502, and the Canadair designation **CL-289**.

The main contract was awarded in July 1976 to Canadair, as part of a shared-cost project between Canada and West Germany. Dornier is subcontracted for approximately half the work. France joined the programme on 29 March 1977, and assigned SAT as the industrial subcontractor. The SAT work represents approx 10 per cent of the total programme.

Resembling a scaled-up CL-89, the AN/USD-502 has greater mission equipment capacity and is equipped with both a Zeiss three-lens reconnaissance camera and an SAT infra-red linescan (IRLS) sensor with real-time data transmission, plus a ground recording system. The navigation/control system comprises a Canadian Marconi Doppler sensor and a Dornier System airborne digital computer. The mission programme is stored in the computer, which generates the command signals required for navigation, attitude control and sensor activation. A 1·07 kN (240 lb st) KHD T117 turbojet engine gives the drone a high-subsonic cruising speed; a 29 kN (6,500 lb st) BAJ Vickers solid-propellant booster rocket is used to accelerate the drone off the mobile zero-length launcher. Heading correction before landing, descent on the recovery parachute, and damping of the landing impact by inflatable landing bags, correspond to the well-proven concepts of the CL-89, which are characterised by their high landing accuracy and the avoidance of damage to the flight vehicle.

The first AN/USD-502/CL-289 flight was made on 3 March 1980. Contractor flight trials were completed successfully in March 1981. Customer evaluation and troop trials are scheduled to take place between December 1981 and August 1982; production is expected to begin in 1984.

DIMENSIONS, EXTERNAL:

Wing span	1·32 m (4 ft 4 in)
Length overall (excl ADMU):	
with booster	4·67 m (15 ft 4 in)
without booster	3·61 m (11 ft 10 in)
Body diameter (max)	0·38 m (1 ft 3 in)

ARGUS

Dornier GmbH, Postfach 2160, 8000 München 66, Federal Republic of Germany
Telephone: München 8715480
Telex: 05-23543

LCT (Laboratoire Central de Télécommunications), 18-20 rue Grange-Dame-Rose, 78140 Vélizy-Villacoublay, BP 40, France
Telephone: 946 96 15
Telex: 690892

ARGUS

Argus (Autonomes Radar Gefechtsfeld Uberwachungs System) is a battlefield surveillance system being developed under a German-French government agreement signed on 6 March 1974. Dornier has the main contract for co-ordinating, integrating and testing the system, and the programme is administered by a board of directors representing the two governments.

The Argus prototypes consist of the two prototypes of the Dornier Do 34 Kiebitz, each fitted with an advanced version of the LCT Orphée II radar as the primary sensor. The first Argus made its initial flight on 14 May 1979 and the second at the end of 1979. The first part of the troop demonstration was also completed in that year. The second part, using an improved Orphée II airborne radar and ground control station, was completed at Erprobungsstelle 91, Meppen, West Germany, at the end of 1980. During the three-month demonstration to representatives of the German and French procurement agencies, 34 flights (including one at night) were made, totalling 44 h in the air. The Argus system successfully demonstrated reconnaissance of individual targets and convoys over various distances, and reconnaissance of several vehicle convoys in an assumed tactical situation. At the end of this second stage the two Argus vehicles had accumulated 150 h flying time and made a total of 306 flights, of which 47 were at the design operating height.

The complete Argus system incorporates a mobile ground station and a tether cable. The ground station is installed in a container mounted on a cross-country 10 ton truck and provided with all equipment necessary for transportation and up to 24 h operation of the system. Operating height of the flight vehicle is 300 m (985 ft).

A description of the Kiebitz appears under the Dornier heading in the German section.

ISRAEL

IAI

ISRAEL AIRCRAFT INDUSTRIES

Ben Gurion International Airport, Lydda (Lod)
Telephone: (03) 973 111
Telex: Isravia 031102 and 031114

IAI SCOUT

Israel Aircraft Industries has designed and built a mini-RPV system known as Scout, primarily for real-time battlefield reconnaissance and surveillance. The complete system comprises the Scout aircraft, ground control station, launcher and retrieval net.

Military and civil applications include missile site reconnaissance, battlefield control, target identification, strike force control, artillery targeting, border patrol, coastal and waterway control, and damage assessment.

Scouts are in service with the Israeli Defence Forces,

MBB KDAR-Locust anti-radar RPV being launched from its container

Artist's impression of the VFW KDAR-Locust mini-RPV

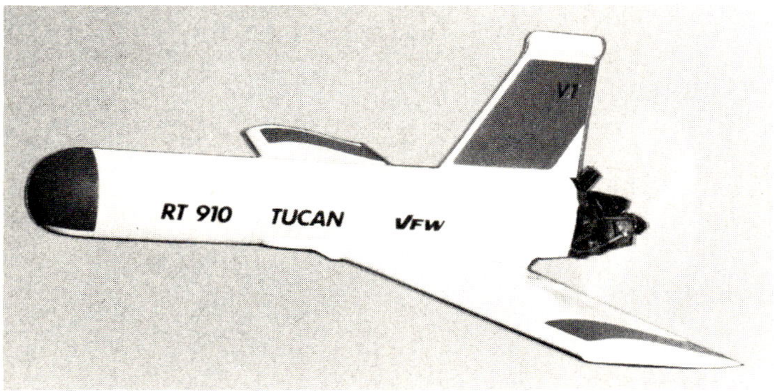

First prototype VFW RT 910 Tucan target acquisition mini-RPV

Launch of an AN/USD-502/CL-289 surveillance drone

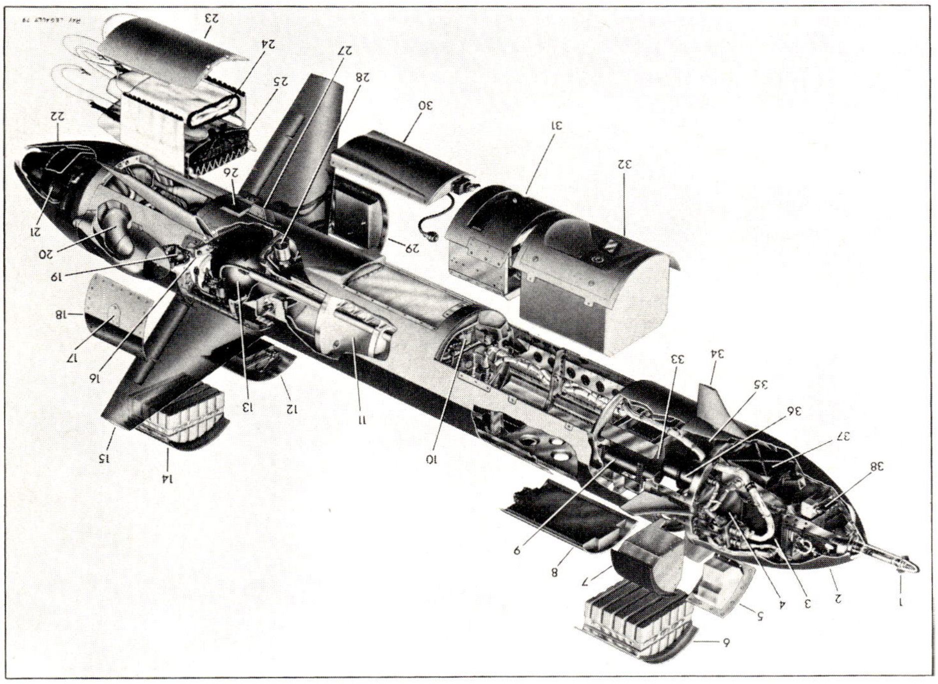

Exploded view of the Canadair/Dornier AN/USD-502 surveillance drone

Dornier Do 34 Kiebitz in Argus configuration

1 Air distance measuring unit (ADMU)
2 Nosecone
3 Heading reference unit
4 Airborne computer
5 Servo amplifier
6 Forward landing bag container
7 Power supply unit
8 Video data link
9 Barometric altitude reference unit
10 Lanyard valve
11 Fuel tank
12 Command receiver/decoder
13 Air reservoir
14 Aft landing bag container
15 Wing (one of four)
16 Transponder unit
17 Air charging door
18 Access panel
19 Pressure regulator
20 Engine air intake
21 T117 engine
22 Tailcone
23 Recovery parachute door
24 Drogue parachute
25 Main parachute
26 Transponder antenna
27 Data link antenna
28 Fuel expulsion unit
29 28V regulator
30 Doppler antenna
31 IRLS sensor
32 Optical sensor
33 Doppler electronic sensor unit
34 Yaw canard surface
35 Yaw actuator
36 Vertical gyro
37 Terminal guidance unit
38 Non-tactical telemetry unit for mission equipment

and are also operational with a number of export customers.

AIRFRAME: Cantilever high-wing monoplane. Fuselage is a rectangular-section nacelle; twin inward-canted fins and rudders, supported by twin tailbooms extending from wings outboard of fuselage. No landing gear on catapult-launched version, but a non-retractable tricycle gear can be fitted, if required, for operation from paved strips. Modular construction, with large access panels. Low detection signatures.

POWER PLANT: One 13·4 kW (18 hp) two-cylinder air-cooled piston engine, installed in rear of fuselage nacelle, driving a specially designed two-blade pusher propeller. Fuel is a 20:1 petrol/oil mixture.

LAUNCH AND RECOVERY: Truck-mounted pneumatic rail launcher, and trailer- or truck-mounted recovery net, are standard. Conventional runway T-O and landing available optionally, if tricycle landing gear is fitted.

GUIDANCE AND CONTROL: Real-time guidance by auto-tracker and two-way data link. Re-programmed dead reckoning capability. For recovery, aircraft is guided semi-automatically by an optical device into centre of retrieval net.

SPECIAL EQUIPMENT: TV camera, with telephoto lens, mounted in fuselage on stabilised platform servo-controlled for vibration damping. Large transparent hemispherical blister under centre of fuselage nacelle. Camera can rotate and scan through 360° in azimuth and 0-90° in pitch. Field of view adjustable, varying between 47·5° and 3·4°. The mini-RPV is also fitted with panoramic camera to scan an area within 60° on each side of flight path. Configuration permits development of other mission equipment packages, to customer's requirements.

DIMENSIONS, EXTERNAL:

Wing span	3·60 m (11 ft 9¾ in)
Length overall	3·68 m (12 ft 1 in)

WEIGHTS:

Mission equipment	22·7 kg (50 lb)
Fuel	14·5 kg (32 lb)
Launching weight	118 kg (260 lb)

PERFORMANCE (typical):

Max level speed	80 knots (148 km/h; 92 mph)
Speed for max range	55 knots (102 km/h; 63 mph)
Stalling speed	42 knots (78 km/h; 49 mph)
Rate of climb at S/L	152 m (500 ft)/min
Max operating altitude	3,050 m (10,000 ft)
Flight endurance	more than 4 h

TADIRAN

TADIRAN ISRAEL ELECTRONICS INDUSTRIES LTD

3 Hashalom Road, PO Box 648, Tel-Aviv 61000
Telephone: 267272
Telex: 03-3537
MANAGING DIRECTOR: E. Caspi
ADVERTISING MANAGER: N. Leshem
PLANNING AND DEVELOPMENT: I. Ish-Hurvitz

TADIRAN MASTIFF

This mini-RPV completed more than 100 test flights with great success, and is now in production for the Israeli armed forces and for export. The tractor-engined **Mk 1** configuration described and illustrated in the 1980-81 *Jane's* has been changed on **Mk 2** vehicles to the twin-boom, pusher-engined layout of which a description follows.

The Mastiff is designed for a wide range of intelligence and reconnaissance missions, including real-time reconnaissance, remotely controlled photography, decoy manoeuvres, electronic interference and jamming, and other uses within the limitations of payload size and weight.

AIRFRAME: High-wing braced monoplane, with high-mounted tailplane and single fin and rudder, supported by two slender tailbooms. Non-retractable tricycle landing gear. Nacelle-type fuselage, in which power plant and mission equipment are mounted. Modular construction, with wings and tailplane detachable to facilitate transportation. Low radar and infra-red signatures; low visibility and noise levels.

POWER PLANT: One 10·4 kW (14 hp) Kolbo Korp two-cylinder piston engine, driving a two-blade pusher propeller. Fuel capacity 22·7 litres (5 Imp gallons).

LAUNCH AND RECOVERY: Vehicle-mounted pneumatic launcher activated by a compressor and controlled automatically by an electronic system. Recovery by means of arrester wire stretched at near ground level between energy absorbers. On landing, RPV engages wire with tailhook.

GUIDANCE AND CONTROL: Remotely controlled from ground radio control station (GCS) or, independently, by a portable control station (PCS). Principal GCS functions are vehicle and mission equipment control; RPV tracking; video and telemetry signal reception; mini-computer processing of data received; and display for real-time or subsequent analysis. Use of PCS permits T-O and landing at sites distant from GCS; when terrain near GCS is unusable for RPV T-O and landing; and/or when extended mission range is required. All mission phases are under autopilot control. Aerodynamic control by ailerons, elevators and rudder.

EQUIPMENT: Up to 350W of electrical power can be supplied. This permits installation of a variety of mission equipment, such as: (1) TV camera on stabilised gimbals, operable in yaw (360°) and pitch (90° down, 10° up) movements, with controllable angle of view of camera lenses; (2) gimbal-mounted TV camera and miniature panoramic film camera for detail photography; (3) various electronic warfare and ECM packages; (4) customer-ordered items such as laser designator and miniature FLIR. The Mastiff can also deliver explosive charges, and simulate larger aircraft or ships by means of radar lenses.

DIMENSIONS, EXTERNAL:

Wing span	4·30 m (14 ft 1¼ in)
Wing area	2·00 m² (21·53 sq ft)
Length overall	2·60 m (8 ft 6¼ in)
Height overall	1·00 m (3 ft 3¼ in)

WEIGHTS:

Weight empty	52 kg (114·5 lb)
Max mission equipment load	15 kg (33 lb)
Max launching weight	75 kg (165 lb)

PERFORMANCE:

Max level speed	70 knots (130 km/h; 80·5 mph)
Typical cruising speed	40-60 knots (74-111 km/h; 46-69 mph)
Stalling speed	30 knots (55·5 km/h; 34·5 mph)
Max rate of climb at S/L	more than 152 m (500 ft)/min
Max operating altitude	3,050 m (10,000 ft)
Range:	
without PCS	38 nm (70 km; 43·5 miles)
with PCS	75·5 nm (140 km; 87 miles)
Endurance	more than 4 h

ITALY

METEOR

METEOR AIRCRAFT AND ELECTRONICS INDUSTRY SpA

146 Via Nomentana, 00162 Rome
Telephone: (06) 830991
Telex: 680136 Meteor
PRESIDENT: Comm Avv Furio Lauri
GENERAL MANAGER: Dott Antonio Castelli
PRODUCTION PLANT: 34074 Monfalcone, Stazione F.S. Ronchi Nord
Telephone: (0481) 778001
Telex: 460288 METMON
PLANT MANAGER: Ing Carlo Spanò
PRODUCTION AND TECHNICAL DIVISION MANAGER: P. I. Paolo Raiteri
ADMINISTRATIVE DIVISION MANAGER: Mrs Mirella Tiburzio

Meteor was established in Trieste as a joint stock company in 1947. Its present head office is in Rome, supported by facilities at Monfalcone (Trieste) and Cagliari (Sardinia). The former is a production factory, the latter being equipped for flight operations and for technical assistance to the users of the tri-service range at Salto di Quirra.

Meteor is producing for the Italian and foreign armed forces various propeller-driven and turbojet-powered radio-controlled drones covering a speed range from 323 knots (600 km/h; 372 mph) up to Mach 2·8, and altitudes from 10 m (35 ft) to 25,000 m (82,000 ft) above sea level. It also co-produces the Northrop-Ventura Meteor 1 and 2 (NVM-1 and NVM-2), Northrop-Ventura Meteor/USD-1, Aérospatiale Meteor 20 (AM/CT.20) target drones and the BM-1 (Beech Meteor 1) missile target system. Under licence from Canadair, Meteor produces 50 per cent of the AN/USD-501 (CL-89) reconnaissance systems ordered by Italy.

METEOR ANDROMEDA/MIRACH SYSTEM

The Andromeda system is designed to meet a basic operational requirement which synthesises and integrates the requirements of most potential operators. Such a multi-role system, equipped with various types of RPV for combat or training use by air, land and naval forces in natural and operational environments, could reduce the attrition of conventional piloted aircraft used in conjunction with them and/or replaced by them in missions operationally suited to the use of RPVs.

The Andromeda system, which is characterised by completely independent operation, maximum utilisation and considerable mobility, is composed of the following subsystems:

1st Subsystem (Alamak). Ground station for accomplishing and/or controlling the mission by the ground crew (Standard and RID versions). Produced for foreign markets.

2nd Subsystem. Equipment for launch, recovery and maintenance of the aerial vehicles. Produced for foreign markets.

3rd Subsystem. Equipment for preparation, recovery, maintenance and operation and/or evaluation of the payload and/or its results. Produced for foreign markets.

4th Subsystem. The aerial vehicle system, configured in a range of five types of vehicle, known as **Mirach**, compatible with the Andromeda system to guarantee maximum effectiveness.

The Andromeda system can be adapted, with only slight modification, to use other types of air vehicle which operators may already have at their disposal. It is being developed in collaboration with Aerosystem Electronic AG of Switzerland.

The following versions of the Mirach flight vehicle are available:

Mirach-20. RPV for target acquisition, location and designation; surveillance; 'kamikaze' missions; and enemy defence saturation. Powered by single 16·4 kW (22 hp) piston engine. Meteor radio command guidance system. Equipped with real-time TV, photo-reconnaissance cameras and laser designator. Under development for foreign markets.

Mirach-70. Target, or RPV for surveillance, target acquisition, electronic warfare, or decoy saturation. Powered by single 52 kW (70 hp) McCulloch piston engine. Meteor radio command guidance system. Various combat or training payloads. In production for foreign markets.

Mirach-100. Target, or RPV for surveillance; reconnaissance; target acquisition, location and designation; electronic warfare strike; and defence saturation. Powered by single 1·13 kN (253·5 lb st) Microturbo TRS 18-076 turbojet engine. Meteor radio command guidance, or programmed guidance with real-time updating. Various combat or training payloads. In production for Italy and foreign markets.

Mirach-300. Target, or RPV for surveillance, reconnaissance, target acquisition, location and designation, electronic warfare strike, and defence saturation. Powered by single 3·34 kN (750 lb st) Microturbo TRI 60 turbojet engine. Guidance as for Mirach-100. Various combat or training payloads. Under development for foreign markets.

Mirach-600. Target, or RPV for training and evaluation, area reconnaissance, electronic warfare strike, and defence suppression. Powered by two 3·34 kN (750 lb st) Microturbo TRI 60 turbojet engines. Guidance as for Mirach-100. Various combat or evaluation payloads. Under development for foreign markets.

DIMENSIONS, EXTERNAL:

Span overall (wings or fins):	
20	3·94 m (12 ft 11 in)
70	3·57 m (11 ft 8½ in)
100	2·80 m (9 ft 2¼ in)
300	2·83 m (9 ft 3½ in)
600	3·60 m (11 ft 9¾ in)
Length overall (excl booster):	
20	1·91 m (6 ft 3¼ in)
70	3·66 m (12 ft 0 in)
100	3·91 m (12 ft 10 in)
300	4·88 m (16 ft 0 in)
600	6·10 m (20 ft 0 in)

WEIGHTS:

Weight empty: *20	90 kg (198 lb)
70	200 kg (441 lb)
100	210 kg (463 lb)
*300	320 kg (705 lb)
*600	680 kg (1,499 lb)
Combat payload (internal): 70	20 kg (44 lb)
100	40 kg (88 lb)
300	80 kg (176 lb)
600	160 kg (352 lb)
Max T-O weight (incl booster): *20	70 kg (154 lb)
70	260 kg (573 lb)
100	310 kg (683 lb)
*300	450 kg (992 lb)
*600	1,000 kg (2,204 lb)

**estimated*

PERFORMANCE:

Max level speed:	
20	97 knots (180 km/h; 112 mph)
70	167 knots (310 km/h; 193 mph)
100	512 knots (950 km/h; 590 mph)
300, 600	Mach 0·9
Max endurance: 20	6 h 0 min
70	1 h 0 min
100	1 h 10 min

Israel Aircraft Industries Scout mini-RPV

Tadiran Mastiff mini-RPV (current configuration)

Meteor Mirach-300 target/RPV on launching ramp

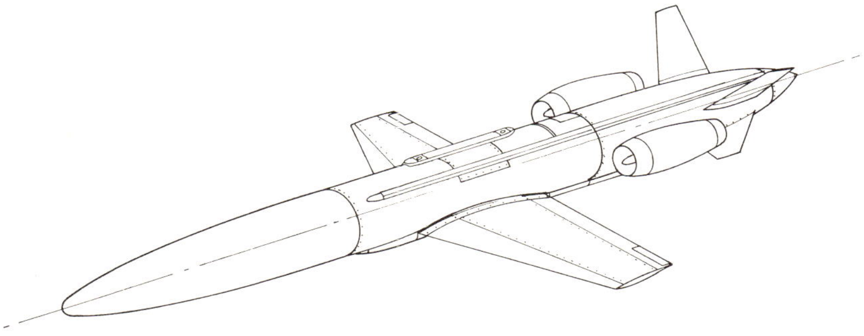

Drawing of Mirach-600

Mirach-70 RPV on its launch trailer

Mirach-100 on Italian Army Agusta A 109

JAPAN

FUJI

FUJI JUKOGYO KABUSHIKI KAISHA (Fuji Heavy Industries Ltd)

Subaru Building, 7-2, 1-chome, Nishi-Shinjuku, Shinjuku-ku, Tokyo

Telephone: Tokyo (03) 347 2505

Telex: 0-232-2268

OFFICERS: see Aircraft section

Under contract from the Japan Defence Agency, Fuji is building the Teledyne Ryan BQM-34A Firebee I subsonic target drone (see US section) for use in the training of Tartar missile and gunnery crews, and for the evaluation of air-to-air missile systems. The first shipboard launch of a Fuji-built drone was carried out successfully in 1970, and 200 flights had been made by the beginning of 1981.

Fuji-built BQM-34As include two fitted with RALACS (Radar Altimeter Low Altitude Control System: see Teledyne Ryan entry in US section), for training the crews of defensive weapons against attack by anti-shipping missiles.

Twenty-seven Fuji-built BQM-34As had been completed and delivered to the Japan Maritime Self-Defence Force, and two to the Technical Research and Development Institute of the Japan Defence Agency, by the beginning of 1981. A total of 36 was scheduled to be completed by the end of FY 1981 (March 1982).

The Japan Defence Agency has authorised the design, development and construction of the first all-Japanese RPV. Intended for use by all three Self-Defence forces, its missions will include reconnaissance, surveillance and target designation.

Fuji Heavy Industries has been selected as prime contractor, with Nippon Avionics providing the guidance system, Hitachi the onboard TV camera, and Nippon Electric the ground control equipment. Other Japanese participants are Nippon Aviation Electronics and Daicel; the power plant is of US origin.

NEC

NIPPON ELECTRIC CO LTD (Nippon Denki Kabushiki Kaisha)

33-1 Shiba Gochome, Minato-ku, Tokyo 108
Telephone: Tokyo (03) 454 1111
Telex: NECTOK J22686
PRESIDENT: Tadahiro Sekimoto
ASST GENERAL MANAGER, LEGAL AND ADMINISTRATION DIVISION: Kiyoshi Yamauchi

Under a technical aid agreement with Northrop Corporation, USA, NEC is responsible for production and repair of Northrop KD2R5 (Shelduck) target drones for the Japan Maritime Self-Defence Force and Ground Self-Defence Force. Delivery of these drones to the JMSDF and JGSDF began in 1961, and a total of 292 had been delivered by the beginning of 1981. Production was expected to continue at the rate of approx 15 per year.

SAUDI ARABIA

MCS

MID-CONTINENT SCIENTIFIC CO (A subsidiary of Qutub International Ltd)

PRESIDENT: Dr Musa Qutub, PhD

MCS PL-60

The PL-60 is a twin-engined mini-RPV capable of use for surveillance and other military duties, border patrol, forest fire surveillance, aerial photography, traffic patrol, air pollution and weather monitoring, and similar applications. It is claimed that it cannot be detected by radar.

Design began, in the USA, in 1971; a prototype PL-60 was flown in 1972, and production began in the same year. By March 1979 a total of 1,580 had been ordered and built in two versions, designated **PL-60A** and **PL-60B**, as described in the 1980-81 *Jane's*.

No news has been received from the company since that time.

SWEDEN

FOA

FÖRSVARETS FORSKNINGSANSTALT (National Defence Research Institute), Applied Electronics Dept (FOA 3)

Huvudavdelning 3, Box 1165, S-581 11, Linköping
Telephone: (013) 11 80 00
HEAD OF DEPARTMENT: Torsten Linell

FOA SKATAN

The Skatan system was designed for a feasibility study concerning short-range daylight-only reconnaissance and surveillance over zones immediately behind the forward edge of the battle area (FEBA). The complete system comprises miniature aircraft, a radio transmitter and telescopic antenna, two pairs of binoculars, and a case for cameras and film processing equipment. The first phase of the Skatan programme had been completed by the end of 1979; a second phase, under way in early 1981, includes tests with real-time optical sensors. No decision regarding series production had been taken up to that time.

TYPE: Experimental short-range battlefield reconnaissance mini-RPV.

AIRFRAME: Strut-braced shoulder-wing monoplane, with constant-chord wings and square-section fuselage. Non-retractable tailwheel-type landing gear. Construction of balsa wood proposed, which would give typical lifetime of 5-10 missions. Extremely small radar signature; not affected by ECM in programmed flight.

POWER PLANT: One 1·2 kW (1·6 hp) O.S. Max H80/RC 13·23 cc single-cylinder glow-plug engine, driving a two-blade propeller. Fuel, contained in a tank of 0·5 litres (0·11 Imp gallons) capacity, is a mixture of methanol (70%), castor oil (25%) and nitromethane (5%).

LAUNCH AND RECOVERY: Conventional T-O and landing, under radio control, is standard; alternative parachute recovery system available optionally.

GUIDANCE AND CONTROL: Radio command guidance. Aircraft can be guided by manual control throughout flight, controlled in line-of-sight by use of two pairs of binoculars (one of low magnification for very close range control, and a higher-magnification pair for use at longer range), mounted coaxially; manual guidance can also be effected by use of radar data and triangulation instead of binoculars. Of these, the distance-calculation method has proved to be the most effective. Alternatively, drone can be fitted with an automatic flight control system for pre-programmed flight. An aneroid altimeter is linked to an onboard control to maintain operating altitude within an accuracy of 20 m (66 ft). Course is controlled by means of a special device based on a magnetic terrain compass. Using this system, the onboard radio receiver can be switched off over the surveillance area, avoiding ECM while the aircraft photographs the area below. Aerodynamic control of the drone is by movement of conventional ailerons and tail control surfaces.

SPECIAL EQUIPMENT: Several onboard camera systems have been tested. The best results were obtained with a Hasselblad 500 EL (70 frames) and a Nikon F (36 frames), both cameras having an electric motor to advance the film. Various lenses, depending on mission configuration, have also been tested. Other potential payloads include radio relay equipment, jammers, smoke generators, laser target designator, TV camera, radar reflector, or (in 'kamikaze' role) small attack weapons.

DIMENSIONS, EXTERNAL:
Wing span 2·14 m (7 ft 0¼ in)
Length overall 1·61 m (5 ft 3½ in)

WEIGHTS:
Weight without camera 4·4 kg (9·7 lb)
T-O weight with 0·4 litres (0·09 Imp gallons) fuel:
with Canomatic M70 camera 6·0 kg (13·2 lb)
with Nikon F camera 7·4 kg (16·8 lb)

PERFORMANCE:
Max level speed 54 knots (100 km/h; 62 mph)
Optimum operating height 300-500 m (985-1,640 ft)
T-O run (on tarmac, zero wind) 7 m (23 ft)
Landing run 20 m (66 ft)
Line-of-sight control range 2·7 nm (5 km; 3·1 miles)
Typical mission times:
briefing and preparation 5-10 min
out and back flight 7-10 min
film removal and processing 5-10 min
total time 17-30 min

UNION OF SOVIET SOCIALIST REPUBLICS

UR-1

According to unconfirmed press reports appearing in early 1979, the designation UR-1 applies to a target drone observed during tests of late-model MiG-25 'Foxbat' interceptors equipped with AA-X-9 radar-guided air-to-air missiles. These tests were said to have been successful against drones flying at low level, in ground clutter, and at altitudes up to approx 21,335 m (70,000 ft). The drones, used to simulate targets of some 25·00 m² (269 sq ft) gross area, are air-launched from Tu-16 carrier aircraft. They are said to operate normally between 19,810 m (65,000 ft) and 30,500 m (100,000 ft), but to be capable of flying as high as 39,625 m (130,000 ft).

UNITED KINGDOM

AEL

AERO ELECTRONICS (AEL) LTD

Gatwick House, Horley, Surrey RH6 9SU
Telephone: 02934 5353
Telex: 87116
MANAGING DIRECTOR: M. H. Nicholas
MANAGER, AERIAL TARGET DIVISION: I. G. Matyear

This company manufactures small, low-cost radio- and optically-guided systems for target drone, reconnaissance or other RPV applications.

AEL 4111 SNIPE Mk II

The AEL Snipe aerial target system was developed as a dynamic target for use with air defence gun systems, of up to 40 mm calibre, and close-range missile systems. It combines simplicity of design with outstanding performance and ease of operation. Its high manoeuvrability enables it faithfully to reproduce any of the complex attack profiles which are normally executed by full-size, manned ground attack aircraft.

The Mk II Snipe is a development of the original aircraft (see 1978-79 *Jane's*) which is in service in seven countries (eight services). Negotiations were at an advanced stage for sale to a further five countries in Spring 1981.

AIRFRAME: High-wing monoplane configuration. Built of wood and plywood veneer covering a polystyrene foam core. Finish is to customer's specification, but is normally a high gloss red paint.

POWER PLANT: One 100 cc single-cylinder two-stroke air-cooled engine, with electronic ignition. Petrol/oil mixture (20:1), contained in a 2·3 litre (0·5 Imp gallon) tank.

LAUNCH AND RECOVERY: Launch by simple rubber bungee catapult mounted on two-wheel transporting trailer. Recovery by orthodox belly landing or by parachute. Fail-safe system automatically closes throttle and deploys parachute in event of loss of command signal, heavy radio interference or loss of internal power supplies.

GUIDANCE AND CONTROL: By specially-developed radio control equipment operating in VHF and UHF bands. Two transmitters normally supplied with each trailer, on frequencies specified by customer. Aerodynamic control by ailerons and elevators.

SPECIAL EQUIPMENT: For visual enhancement, wingtip pods can be fitted to carry four smoke flares which can be ignited sequentially. The 40 s burning time is adequate to allow full target engagement. For weapons employing infra-red homing techniques, heat from engine is normally sufficient to allow lock-on and full target engagement. For special applications, one infra-red flare with 25 s burning time can be fitted to each wingtip and fired sequentially. For radar enhancement, a special radar-reflective mesh is incorporated between foam core and skin of aircraft which is sufficient to provide a return on most radars operating between 2-10 GHz. For customers requiring a greater radar response, a corner reflector can be mounted in rear of fuselage. For special applications a robot camera can be supplied for aerial photography. Under development are a Doppler miss-distance indicating equipment (by an associate company); an airborne lightweight closed-circuit TV system; and an autopilot that will enable Snipe to be flown out of sight, at ranges up to 4·3 nm (8 km; 5 miles). For naval air defence training, target aircraft are specially waterproofed for recovery from sea. GSE (ground support equipment) comprises trailer, launcher, and all ground-based stores needed to fly aircraft. Trailer is matched to standard Land-Rover truck with Avonride suspension. Trailer can carry two target aircraft and has built-in workbench for ease of maintenance. For naval air defence training, the trailer body can be removed quickly and easily from the chassis, to permit deck mounting in limited spaces.

DIMENSIONS, EXTERNAL:
Wing span 2·44 m (8 ft 0 in)
Wing area 0·99 m² (10·64 sq ft)
Length overall 2·13 m (7 ft 0 in)

NEC-built Northrop Shelduck target drone

FOA Skatan experimental battlefield reconnaissance mini-RPV

Launch of a BAe Stabileye Mk 2

AEL Snipe miniature target aircraft

British Aerospace Stabileye Mk 3

AEL Streek trainer/air defence target aircraft

WEIGHT:
- Max launching weight 19 kg (42 lb)

PERFORMANCE:
- Max speed 113 knots (209 km/h; 130 mph)
- Average sortie 35 min

AEL 4600 STREEK

This radio-controlled trainer/all arms air defence target has been developed as a cost-effective and realistic target for small arms and machine-guns used in air defence roles. It is hand-launched by the operator, or the operator's assistant, and requires a minimum of ground support equipment. Recovery is by belly landing on any reasonably level terrain.

Radio control equipment, in either VHF or UHF wavebands, is available to customer's frequency requirements.

AIRFRAME: Fuselage constructed of GRP, wings of veneer-covered polystyrene foam. The fuselage is sturdy enough to withstand belly landings; the whole aircraft has been designed to facilitate component replacement and repair. All parts are interchangeable between aircraft.

GROUND SUPPORT EQUIPMENT (GSE): Comprises GSE box, including a fuel container (2·3 litres; 0·5 Imp gallon), electric fuel pump, electric engine starter, ammeter, 12V battery to power equipment, and selection of hand tools.

DIMENSIONS, EXTERNAL:
- Wing span 1·70 m (5 ft 7 in)
- Length overall 1·385 m (4 ft 6½ in)

WEIGHT:
- Max launching weight 2·7 kg (6 lb)

PERFORMANCE:
- Max level speed 78 knots (145 km/h; 90 mph)
- Average sortie 20 min

BAe

BRITISH AEROSPACE PUBLIC LIMITED COMPANY, DYNAMICS GROUP

STEVENAGE DIVISION: PO Box 19, Six Hills Way, Stevenage, Hertfordshire SG1 2DA
Telephone: Stevenage (0438) 2422
Telex: 825125/6
HEAD OF RPV RESEARCH (BRISTOL): R. Stephenson
PUBLIC RELATIONS MANAGER: Philip J. Birtles

BAe FLYBAC

This small RPV flew for the first time on 27 July 1976. Since then, many examples have been used in a programme of research into lightweight, low-cost command, control and sensor systems. The vehicles are instrumented to permit autonomous or remote pilot control, telemetry of the onboard systems, and parachute recovery for reasons of range safety.

POWER PLANT: One 1·34 kW (1·8 hp) Webra piston engine.

DIMENSIONS, EXTERNAL:
- Wing span 2·00 m (6 ft 6¾ in)
- Length overall 1·47 m (4 ft 9¾ in)
- Height overall 0·41 m (1 ft 4¼ in)

WEIGHTS:
- Max mission equipment load 3 kg (6·6 lb)
- Max launching weight 11 kg (24·25 lb)

PERFORMANCE:
- Cruising speed 39 knots (72 km/h; 45 mph)
- Max range 13 nm (24 km; 15 miles)
- Max endurance 20 min

BAe STABILEYE

The Stabileye family of mini-RPVs has been used for a number of years in various Ministry of Defence and BAe research programmes. The initial concept of Stabileye was to produce a simple, robust mini-RPV research vehicle, capable of flight testing a large variety of sensors and other mission equipment loads. The general configuration is that of a twin-tailboom monoplane, with the single engine mounted at the rear of the central nacelle, driving a pusher propeller. Stabileye is normally launched from a moving truck, but a static pneumatic powered launcher may be employed instead. Recovery is by skid landing or by parachute.

The following versions of Stabileye have been produced:

Mk 1. First prototype made its initial flight on 24 October 1974, and the type has been in almost continuous service since then. Twelve examples have been flown in the programmes, configured to accept various loads for investigation of the use of mini-RPVs in military roles. Powered by a 3 kW (4 hp) Rowena 60 cc piston engine. May be phased out during 1981.

Mk 2. Same basic aerodynamic configuration as Mk 1, but structurally uprated in order to carry test loads of up to 15 kg. (33 lb). Construction mainly of GRP and aluminium alloy, with a steel tube space-frame fuselage centre-section. The prototype, powered by a 5·6 kW (7·5 hp) Weslake T-116 engine, was equipped with wingtip extensions in order to make use of the truck method of launch; it made its first flight on 20 December 1979. Now superseded by Mk 3.

Mk 3. Designed to carry a max mission load of 25 kg (55 lb), making use of the five years' experience in the construction and operation of the Mks 1 and 2. Appearance differs in some respects from its predecessors. Fuselage is rectangular in section, and the shoulder-mounted wings are of constant chord instead of the previous tapered planform. A single central fin, of greater area, replaces the twin fins of the earlier models. First flight was made on 24 September 1980. A batch of Mk 3s was under construction in the Spring of 1981, to be powered by either 5·6 kW (7·5 hp) Weslake T.116 or 9·7 kW (13 hp) Weslake T.200 engines.

DIMENSIONS, EXTERNAL:

Wing span: Mk 1	3·10 m (10 ft 2 in)
Mk 2	3·70 m (12 ft 1¾ in)
Mk 3	3·65 m (11 ft 11¾ in)
Wing area: Mk 1	1·65 m² (17·76 sq ft)
Mk 2	1·68 m² (18·08 sq ft)
Mk 3	2·23 m² (24·00 sq ft)
Length overall: Mk 1, Mk 2	2·60 m (8 ft 6½ in)
Mk 3	2·87 m (9 ft 5 in)
Height overall: Mk 1, Mk 2	0·89 m (2 ft 11 in)
Mk 3	1·03 m (3 ft 4½ in)

WEIGHTS:

Max mission load: Mk 1	8 kg (17·6 lb)
Mk 2	15 kg (33 lb)
Mk 3	25 kg (55 lb)
Max launching weight: Mk 1	42 kg (92·5 lb)
Mk 2	55 kg (121 lb)
Mk 3	60 kg (132 lb)

PERFORMANCE:

Max level speed: Mk 1	78 knots (144 km/h; 90 mph)
Mk 2	93 knots (173 km/h; 107 mph)
Mk 3	86 knots (159 km/h; 99 mph)
Cruising speed: Mk 1	68 knots (126 km/h; 78 mph)
Mk 2	83 knots (155 km/h; 96 mph)
Mk 3	70 knots (130 km/h; 80·5 mph)
Stalling speed: Mk 1	39 knots (72 km/h; 45 mph)
Mk 2	43 knots (79 km/h; 49 mph)
Mk 3	41 knots (76 km/h; 47·5 mph)
Max endurance: Mk 1	30 min
Mk 2	1 h
Mk 3	2 h

FR

FLIGHT REFUELLING LTD

Wimborne, Dorset BH21 2BJ
Telephone: 0202 882121
Telex: 41247
CHAIRMAN: M. J. Cobham, CBE, MA, FRAeS, Barrister

For more than 25 years, Flight Refuelling Ltd has been engaged, inter alia, in the design and manufacture of control systems for pilotless aircraft. It was responsible for the conversion of more than 250 Gloster Meteor aircraft to pilotless or drone configuration. Some of these aircraft are still in service use for missile development trials and practice firings.

Since 1977 work has proceeded on fitting an initial batch of 25 Hawker Siddeley Sea Vixen D. Mk 3 aircraft with Flight Refuelling's Universal Drone Pack (UDP). This packaged system enables most types of aircraft to be readily converted to drone or pilotless configuration with only minor modifications to their conventional flying control system inputs, and without degrading the original manned performance. The pack is designed for installation on the existing seat rail in the aircraft's cockpit.

Flight Refuelling is continuing its work on RPVs and target drones, and is currently developing the ASAT target vehicle for the UK Ministry of Defence.

FR ASAT

Flight Refuelling Ltd designed this jet-propelled target drone to meet a Ministry of Defence requirement known as ASAT (Advanced Subsonic Aerial Target). The requirement calls for a level speed in excess of 400 knots (741 km/h; 461 mph) and a steady state manoeuvring capability of up to 6g. Design began in May 1979, and Flight Refuelling has a contract to build eight flight vehicles, of which the first was scheduled to fly during the second half of 1981. If approved, the drone is expected to be produced at the rate of about 50 per year.

The max fuel load of 60 kg (132 lb) permits sortie times in excess of 1 h. Approach and intercept legs of a typical mission total nearly 5·5 nm (10 km; 6·2 miles), over the whole of which the target vehicle is flown at more than 400 knots (741 km/h; 461 mph) at S/L. Ten such circuits can be repeated at 4·5 min intervals. During less demanding operations, or where more time is required between presentations, the rate of fuel usage will be lower and the vehicle can be loitered at 3,000 m (9,845 ft) altitude for 90 min, with full allowances for T-O, climb, descent and recovery.

TYPE: Subsonic target drone.

AIRFRAME: Cantilever low/mid-wing monoplane. Wings, tailplane and elevators folded from flat aluminium alloy sheet, without compound forming. Wings are each attached to fuselage by four bolts, are fitted with plain ailerons, and are interchangeable port/starboard. Tailplane held in place by GRP fin, via two studs. Aluminium alloy stressed-skin fuselage, of circular cross-section, with crushable nosecone to absorb nose-down ground impact. Nosecone attachment and body joints made by band-clamps. Nosecone is expendable, and is replaced after each flight. Cylindrical GRP canister in rear fuselage houses recovery parachutes. Engine pod attached under fuselage by two bolts.

POWER PLANT: One 1·08 kN (242·5 lb st) Microturbo TRS 18-075 turbojet engine, built under licence by Ames Industrial Ltd and mounted in pod under centre of fuselage. Single integral fuel tank in centre of fuselage, between wings; this has a capacity of 75 litres (16·5 Imp gallons) and is pressurised at 1·5 bars (22 lb/sq in). Refuelling point on top of fuselage. Aft of fuel tank is a smaller tank for oil, which can be injected into jetpipe, on command from ground, to produce smoke.

LAUNCH AND RECOVERY: Takes off under own power from three-wheeled trolley, running on a circular track 115 m (377 ft) in diameter and tethered to a pylon in the centre. Aircraft will reach T-O speed in three laps of this track, equivalent to a straight-line T-O run of more than 1,000 m (3,280 ft), and with this system can always be launched into wind. Day and night phased-release parachute recovery system, consisting of a cluster of three Irvin cruciform 'chutes, stowed in GRP canister which is rear-loaded into aft fuselage and covered by an ejector spring and drogue tailcone.

GUIDANCE AND CONTROL: Radio command PCM guidance system, with Skyleader receiver and Marconi Avionics autopilot. Programmable ground-based control option, using a microprocessor-based ground control station which utilises radar data as the basis for more sophisticated control and guidance functions. Aerodynamic control by ailerons and elevators, actuators for which are in rear fuselage, between fuel/oil tanks and parachute compartment.

SYSTEMS: Fuel and smoke/oil tanks pressurised by engine compressor bleed air. 400W direct-drive alternator/PCU.

EQUIPMENT: Large bay in forward fuselage, aft of nosecone, for avionics and optional target equipment such as miss-distance indicator and radio altimeter. Pylon provision for optical or infra-red flares, radar enhancement devices, subtargets, or other external stores.

DIMENSIONS, EXTERNAL:

Wing span	3·05 m (10 ft 0 in)
Wing area, gross	1·40 m² (15·1 sq ft)
Wing aspect ratio	6
Length overall	3·50 m (11 ft 6 in)
Body diameter (max)	0·387 m (1 ft 3¼ in)
Height overall	1·32 m (4 ft 4 in)
Tailplane span	1·32 m (4 ft 4 in)

WEIGHTS:

Basic weight empty	123 kg (271 lb)
Max fuel load	60 kg (132 lb)
Max launching weight	195 kg (430 lb)

PERFORMANCE (estimated):

Max level speed	450 knots (834 km/h; 518 mph)
Min loitering speed	150 knots (278 km/h; 173 mph)
Max rate of climb at S/L	2,600 m (8,530 ft)/min
Typical endurance	1 h 0 min
Max endurance	1 h 30 min
g limit	+6

INTERGARD

INTERGARD ELECTRONICS LTD

LONDON OFFICE: No. 5, 34 Frognal, Hampstead, London NW3
Telephone: 01 435 1622
Telex: 8951859 Basil G
WORKS: 3 Tayfen Road, Bury St Edmonds, Suffolk
Telephone: 0284 4382
Telex: 81503
MANAGING DIRECTOR: J. Van Herbert

This company has in production the Shrike small aerial target, or target tug, and the Sky-Eye real-time TV surveillance mini-RPV; it has also developed a third miniature target drone aircraft known as the Pipit.

INTERGARD IETS 7501 SHRIKE

This in-sight vehicle, for gunnery practice with small arms and anti-aircraft guns, has been ordered by two Middle Eastern countries; ten were ordered by the Malaysian Army in 1980. Shrike can be flown in a low-flying attack mode, to simulate an attack directly upon the operator's gun position; as a target towing system for live weapon practice; and as a target training system for ground-to-air missiles. The complete system includes the air vehicle, self-contained transporter and optional launcher, ground control and support equipment, operator spares pack, drogue assembly, and spare drogue.

TYPE: Miniature aerial gunnery target or target-towing vehicle.

AIRFRAME: Cantilever shoulder-wing monoplane, with dihedral wings, built of balsa wood, plywood, expanded plastics foam and glassfibre. Twin-skid metal landing gear.

POWER PLANT: One 32-56 cc spark ignition or glow ignition two-stroke piston engine, driving a small-diameter two-blade propeller. Petrol/oil or methanol/oil mixture fuel, depending upon engine chosen.

LAUNCH AND RECOVERY: Aircraft can be launched from a four-wheeled drop dolly or bungee launcher, or can take off conventionally with skid landing gear. Recovered by conventional landing on skids.

GUIDANCE AND CONTROL: Radio command guidance system, operating on one of 12 spot frequencies in 27MHz band with a bandwidth of approx 25kHz. Different frequency required for each system if more than one target is flown at the same time within radio control range of each other (approx 2·2 nm; 4 km; 2·5 miles). Aerodynamic control by rudder and elevators.

EQUIPMENT: Electronic fail-safe device automatically closes engine throttle if aircraft is accidentally flown beyond visual range or suffers a radio link malfunction.

DIMENSIONS, EXTERNAL:

Wing span	3·05 m (10 ft 0 in)
Wing chord (average)	0·425 m (1 ft 4¾ in)
Length overall	2·31 m (7 ft 7 in)

WEIGHT:

Launching weight	14 kg (31 lb)

PERFORMANCE:

Max speed (depending on engine fitted)	80-120 knots (148-222 km/h; 92-138 mph)
Cruising speed	60 knots (111 km/h; 69 mph)
Operating altitude, and range	limited only by operator's visual range
Flight duration	up to 50 min

INTERGARD GTS 7901 SKY-EYE

Sky-Eye ordered by RAE Farnborough and also in production for customers in two countries in the Middle East, was developed as a close-support information relay vehicle for use by forward and front-line troops. It is designed to provide, directly to a command headquarters or forward artillery control centre, information about the deployment and movement of hostile troops or armour. Alternatively, by removing the TV camera and transmitter, a chaff dispenser or radar jammer can be installed for use against enemy ground radar sites; or, for use in conjunction with close-support aircraft, a laser designator can be fitted for accurate target identification.

Other applications include TV surveillance of urban riot areas, border surveillance, and forest fire location and assessment. A new guidance system, to enable Sky Eye to travel up to 21·5 nm (40 km; 25 miles) and return on a pre-programmed flight pattern, was under development in 1981.

TYPE: Multi-purpose surveillance/ECM mini-RPV.

AIRFRAME: Cantilever high-wing monoplane, with pod-and-boom fuselage and sweptback fin and rudder; non-swept dihedral wings and non-swept tailplane. Construction of balsa wood, plywood, expanded plastics foam, glassfibre and aluminium alloy. Rudder and full-span elevator standard; ailerons optional.

POWER PLANT: As for Shrike, but installed at rear of fuselage pod to drive a pusher propeller.

LAUNCH AND RECOVERY: As for Shrike.

GUIDANCE AND CONTROL: Similar to that for Shrike. For flight programmes in areas of high air turbulence an electrostatic stabilisation system, using a slightly radioactive polonium element at each wingtip, can be

Mounted on its ground transporter, an FR Universal Drone Pack awaits installation in a Sea Vixen aircraft

Flight Refuelling ASAT (Advanced Subsonic Aerial Target)

Intergard Shrike aerial gunnery target

Intergard Sky-Eye RPV on four-wheeled launch dolly

installed for improved pitch and roll control. Ailerons are required if this system is fitted, since for it to be effective the RPV must initially be unstable in flight.

Equipment: TV camera and transmitter standard; these can be replaced, according to mission, by a multi-frame 35 mm still camera, 8 mm movie camera, chaff dispenser, radar jammer, laser designator or other sensor. Electronic fail-safe throttle closure/engine stop, as in Shrike.

Dimensions, external:
Wing span 2·44 m (8 ft 0 in)
Wing chord (constant) 0·38 m (1 ft 3 in)
Length overall 2·03 m (6 ft 8 in)

Weight:
Launching weight 12 kg (26·5 lb)

Performance:
Max speed (depending on engine fitted) 60-80 knots (111-148 km/h; 69-92 mph)
Cruising speed 30 knots (56 km/h; 35 mph)
Operating altitude limited only by operator's visual range
Range limited only by operator's visual range, but guidance system range is up to 5·4 nm (10 km; 6·2 miles) with high-power transmitter
Flight duration up to 45 min

INTERGARD IETS 7502 PIPIT

Pipit, designed to train radio control operators to use the larger Shrike, is of generally similar configuration, construction and operation to the latter vehicle. Being smaller, it has a less powerful engine (1·1 kW; 1·5 hp 10 cc glow plug ignition type), and has twin landing skids attached directly to the underside of the fuselage.

Dimensions, external:
Wing span 2·01 m (6 ft 7 in)
Length overall 1·22 m (4 ft 0 in)

Weight:
Launching weight 3 kg (6·6 lb)

Performance:
Max speed 60 knots (111 km/h; 69 mph)
Cruising speed 25 knots (46 km/h; 29 mph)
Operating altitude, and range limited only by operator's visual range
Flight duration up to 30 min

MARCONI

MARCONI AVIONICS LIMITED

Airport Works, Rochester, Kent ME1 2XX
Telephone: 0634 44400
Telex: 96333
Press Officer: Malcolm Moulton

Machan (pronounced Ma'shahn), a name derived from the old Hindi word for a treetop surveillance platform, represents one of several involvements by Marconi Avionics in unmanned aircraft. Others include the development of lightweight surveillance equipment, including TV systems, and associated control and data link equipment.

MARCONI AVIONICS MACHAN

The Machan programme was established to develop an unmanned flight research aircraft. It is funded by the Ministry of Defence (Procurement Executive) via the Radio Navigation Department (XR) at RAE Farnborough, with a substantial private-venture investment by Marconi Avionics; the Cranfield Institute of Technology is a major subcontractor to Marconi for the airframe, power plant and digital flight control system.

The main purposes of the programme are to investigate the surveillance role, including target recognition and designation; navigation; and environmental/operational aspects such as the relationship between power plant induced vibration, manoeuvre and performance; automation of launch and recovery; and the flight evaluation of RPV components and systems. The aim is to gain a fuller understanding of the operational role of RPVs, and to provide information that will lead to cost-effective procurement and deployment of vehicles with enhanced capability.

With the emphasis on research capability, Machan has been designed to simplify changes of onboard avionics and mission equipment. The air vehicle reflects not only the research aspect of this programme but also an inherent flexibility of design. For example, the 2·44 m (8 ft) span wing was designed without control surfaces, as the basic configuration was intended to provide a highly stable aircraft with low agility. An extended-span wing, with ailerons, has been installed only because of the restricted airspace in which early testing has to be conducted, but the ailerons can be disconnected during flight to permit evaluation of the intended wing configuration. The diamond-shaped fuselage cross-section presents a low radar signature, and provision exists for glassfibre fuselage and propeller construction to be used in the future. The emergency parachute system, used for test range recovery, is housed in a detachable pod; it is not likely to be required on an operational vehicle.

Marconi's Flight Automation Research Laboratory is overall programme manager, responsible for system integration and ground control equipment; other systems are provided by the company's Electro-Optical Surveillance Division (surveillance equipment and data links), and Gyro Division (strapdown body motion sensor package).

The first prototype Machan, to which the following description applies, made its first flight at RAE Bedford on 19 February 1981.

Type: Experimental research RPV.

Airframe: Cantilever shoulder-wing monoplane. Wings are of composite construction with a light alloy spar, urethane foam core and glassfibre skin. Ailerons, installed on first prototype, are for specific initial test conditions only, and can be disconnected in flight to allow evaluation of aircraft in its intended configuration. Diamond cross-section fuselage, of mainly metal construction on prototype, with aft-mounted propulsor duct; future version for more extensive production may be of glassfibre construction. Duct is a foam-filled double metal skin, carrying three identical all-flying control surfaces and a tailskid of GRP. The single vertical surface is a rudder; the two horizontal 'tailerons' combine the functions of ailerons and elevators, operating differentially for roll control and collectively for pitch control.

Power Plant: One 13·5 kW (18 hp) Weslake two-stroke flat-twin engine, mounted in rear of fuselage and driving a two-blade wooden pusher fan within a propulsor duct. Four-blade wooden and glassfibre fans are to be evaluated. Fuel, contained in a centre-fuselage tank, is a petrol/oil mixture.

LAUNCH AND RECOVERY: Launched currently by radio-controlled, steerable tricycle-type dolly; development of a pneumatic catapult launch system is planned. Normal recovery by conventional landing on GRP main skids and tailskid. Emergency landing during test phase by parachute contained in detachable dorsal canister.

SYSTEMS, AVIONICS AND EQUIPMENT: 28V DC electrical system, incorporating engine-driven alternator, with central rectifier and regulator, plus a sealed 28V lead-acid battery. Digital microprocessor-based flight control system with 68MHz command link. Stability augmentation and attitude reference from three-axis 'strapdown' sensor package. Command and telemetry facilities can be used in conjunction with the ground station computer to investigate outer loop control without modifying onboard equipment. Aircraft is designed to permit carriage of a variety of sensors in nose equipment bay. As first flown, the experimental surveillance equipment consists of a vibration-insulated TV camera which has remotely controlled zoom, focus and aperture, but is not stabilised or steerable. Later, a gimballed, stabilised and steerable camera will be installed, to permit a far more detailed investigation of the surveillance role. Navigational research is aimed ultimately at the development of advanced equipment that will offer easy pre-programming of a mission, using onboard navigational systems able to function effectively without any (or only minimal) communication with ground-based equipment.

DIMENSIONS, EXTERNAL (first prototype):
- Wing span: current — 3·66 m (12 ft 0 in)
- designed — 2·44 m (8 ft 0 in)
- Length overall — 2·13 m (7 ft 0 in)
- Height overall — 0·55 m (1 ft 9½ in)

WEIGHTS (first prototype):
- Basic weight empty — approx 57 kg (125 lb)
- Mission equipment — 15 kg (33 lb)
- Max T-O weight — 73 kg (161 lb)

PERFORMANCE (first prototype, at max T-O weight):
- Max level speed at S/L — 115 knots (213 km/h; 132 mph)
- Cruising speed — 64 knots (119 km/h; 74 mph)
- Stalling speed — 43 knots (80 km/h; 50 mph)
- Range — 130 nm (241 km; 150 miles)
- Endurance at cruising speed — 2 h

RCS
RCS GUIDED WEAPONS SYSTEMS (Division of Radio Control Specialists Ltd)

6 Wolsey Road, Ashford, Middlesex TW15 2RB
Telephone: 078 42 53661/2

DIRECTORS:
E. A. Falkner, CEng, MRAeS, FIED
P. R. Conway, BSc (Eng), CEng, MICE
K. E. Mackley, CEng, MIERE

This company manufactures a range of small, low-cost radio- and optically-guided systems for target drone, reconnaissance or other RPV applications. All types in full-scale production at the beginning of 1979 were described and illustrated in the 1980-81 *Jane's*; no information has been received from the company since that time.

SHORTS
SHORT BROTHERS LIMITED

PO Box 241, Airport Road, Belfast BT3 9DZ, Northern Ireland
Telephone: 0232 58444
Telex: 74688
OFFICERS: see Aircraft section

SHORTS SKEET

The Skeet, a development of Short Brothers' highly successful MATS-B drone (see 1979-80 *Jane's*) now in service with the British Army, provides a realistic and cost-effective aerial target for use with close-range weapon systems. Like MATS-B, it combines simplicity with high manoeuvrability and ease of operation, but offers the additional advantages of higher speed, greater range and better payload, and has a maximum endurance twice that of the parent drone. The Skeet completed acceptance trials in September 1979, and is now in production under MoD contract.

AIRFRAME: Modular construction cantilever high-wing monoplane. Fuselage of metal and GRP. Polyurethane foam-filled glassfibre wings, fin and tailplane.

POWER PLANT: One 13·4 kW (18 hp) Weslake 274-6 274 cc two-cylinder two-stroke aircooled engine. Fuel capacity 9·1 kg (20 lb).

LAUNCH AND RECOVERY: Launched pneumatically from self-contained two-wheel launcher having a capacity of six launches from one charging. Recovery by parachute or by conventional landing, using fuselage underfairing as a skid. Launches can be carried out in headwinds of up to 25 knots (46 km/h; 29 mph).

GUIDANCE AND CONTROL: Eight-channel (elevator, aileron, throttle, smoke generator firing, height hold, and three spare) radio command system (hand-held transmitter and 10W rf amplifier), in the 68MHz band. Automatic barometric height hold available. Visual control sight with binoculars to assist pilot in flying the aircraft visually to over 2·7 nm (5 km; 3·1 miles) range and to enable repeatable race tracks to be flown.

SPECIAL EQUIPMENT: X-band reflector in rear fuselage, and parachute in forward fuselage. Sixteen 40 s smoke generators fitted in landing skid. Two wing-mounted pods can carry, for example, Simflak for training, and the fuselage can accept either a miss-distance indicator or a transponder. An autopilot, complete with heading device, is under development for out-of-sight operation. An automatic parachute release can be provided. Without parachute, and with part fuel, Skeet can accept mission equipment loads of up to 18 kg (40 lb) of a non-target nature (eg, for surveillance and tactical aggression).

DIMENSIONS, EXTERNAL:
- Wing span — 3·35 m (11 ft 0 in)
- Wing area — 1·53 m² (16·5 sq ft)
- Length overall, incl MDI aerial — 2·72 m (8 ft 11 in)

WEIGHTS:
- Mission equipment (normal) — 8 kg (17·6 lb)
- Max launching weight — 63 kg (139 lb)

PERFORMANCE:
- Max level speed — 125 knots (232 km/h; 144 mph)
- Stalling speed — 50 knots (93 km/h; 58 mph)
- Max in-sight range — 2·7 nm (5 km; 3·1 miles)
- Max range under control — more than 5·4 nm (10 km; 6·2 miles)
- Endurance at full throttle — 1 h 15 min

WALLIS/VINTEN

Wallis Autogyros Ltd, Reymerston Hall, Norfolk NR9 4QY
Telephone: 0362 850418
W. Vinten Ltd, Western Way, Bury St Edmunds, Suffolk IP33 3TB
Telephone: 0284 2121
Telex: 81176

On 1 March 1981 a collaborative agreement was reached between Wallis Autogyros Ltd (see main Aircraft section) and W. Vinten Ltd for the purpose of exploiting Wallis-designed autogyros for military and non-military roles. Wg Cdr K. H. Wallis will act as consultant and technical adviser to Vinten; the latter will be responsible for development, manufacture and marketing of the autogyros on a worldwide basis, though it is stressed that the partnership does not involve any production of Wallis autogyros for the general aviation market.

The partnership aims to fulfil market needs for both manned and an unmanned reconnaissance autogyros, and Vinten plans to begin production of the manned version in Spring 1982, following initial deliveries of a new British-designed Weslake engine, a 56 kW (75 hp) flat-twin able to run on automobile fuel. Meanwhile, trials are to be conducted by Vinten with an unmanned autogyro; this will be capable of carrying 160 kg (353 lb) of mission equipment, compared with 75 kg (165 lb) in the manned version, and will be able to mount any combination of current production Vinten cameras. Eventually it will be equipped with a Vinten real-time reconnaissance pack, currently undergoing evaluation by RAE Farnborough.

PERFORMANCE (RPV version, anticipated):
- Cruising speed — approx 80 knots (148 km/h; 92 mph)
- Typical radius (battlefield surveillance) — 10-30 nm (18·5-55·5 km; 11·5-34·5 miles)
- Max range — 868 nm (1,610 km; 1,000 miles)

WESTLAND
WESTLAND HELICOPTERS LTD

Yeovil, Somerset BA20 2YB
Telephone: 0935 5222
Telex: 46277

OFFICERS: see Aircraft section

WESTLAND WIDEYE

Since the cancellation of the Supervisor surveillance system programme in December 1979, Westland Helicopters has continued development testing of Wideye remotely piloted helicopters on a private venture basis. An illustration and all known details of the Wideye can be found in the 1979-80 *Jane's*.

UNITED STATES OF AMERICA

APL
APPLIED PHYSICS LABORATORY, THE JOHNS HOPKINS UNIVERSITY

AERONAUTICS DIVISION: Johns Hopkins Road, Laurel, Maryland 20810
Telephone: (301) 953 7100
Telex: 89 548
DIRECTOR: Maynard L. Hill

APL was responsible for developing the RPD2 miniature target drone, a description of which can be found in the 1980-81 *Jane's*. These vehicles are now in storage, and are unlikely to be flown again as targets, although they may be used for further research and development testing.

A more recent programme is that for an unmanned meteorological research vehicle, all available details of which are given hereafter.

APL MAP/UV 8001

Design of the MAP/UV 8001 (Maneuverable Atmospheric Probe, Unmanned Vehicle) began in August 1978, and prototype construction started in June of the following year. By early 1981 four prototypes had been completed, the first of which made its initial flight in August 1979.

The RPV was designed to be low in cost, operable in hazardous regions, and capable of slow flight and tight manoeuvres, to provide a means of gathering data in more confined locations than would be possible with a manned aircraft. It is used to investigate meteorological and electrical characteristics of the lower atmosphere near a laser test facility at the White Sands Missile Range, in co-operation with the US Army Atmospheric Sciences Laboratory, the Army Research Office, and the University of Texas at El Paso (UTEP).

The method of vertical stabilisation employed (see following 'Guidance and Control' paragraph) has been used in several delta-planform mini-RPVs developed previously at APL, and test programmes have also been conducted using an instrumented full-size aircraft.

The main objective of the MAP/UV 8001 programme is to measure atmospheric processes and characteristics that might affect the performance of electro-optical devices in battlefield environments, such as obscuration by smoke, dust and aerosol particles; and diffraction caused by turbulence or other forms of temperature and density gradients. Another objective is to investigate meteorological factors that affect the electrical field (which does not remain vertical in the presence of some forms of adverse weather), and to define types of weather or other phenomena that may impair or prevent the operation of vehicles stabilised by the type of lightweight vertical reference system fitted to the MAP/UV vehicle.

Typical tests carried out to date have included attempts to define the rate of dispersal of dust clouds formed above exploding projectiles, at altitudes as low as 9·14 m (30 ft). In another series of tests, in July 1980, the RPV was flown close to the face of a cliff and at 2,440 m (8,000 ft) near the peak of a mountain, controlled during take-off and landing by a 'valley' pilot in the foothills and in mid-flight by a 'mountain' pilot at the peak of the mountain. The latter was able to operate the vehicle safely to within 91-182 m (300-600 ft) of the cliff face, by using visual inputs to control the flight path.

TYPE: Meteorological research mini-RPV; *g* limits ±15.

AIRFRAME: High-wing monoplane configuration. Wing section NACA 2417 (modified). Dihedral 4° from roots. Wing and tail surfaces of composite construction, consisting of a glassfibre/epoxy skin laid over polystyrene expanded foam. Fuselage consists of moulded

First prototype Marconi Avionics Machan experimental RPV on its launch dolly

Beechcraft Model 1019 AQM-37A target drone

Shorts Skeet target drone taking off from pneumatic launcher

Applied Physics Laboratory MAP/UV 8001 atmospheric research RPV

Beechcraft MQM-107B improved version of the MQM-107A Streaker

glassfibre/epoxy shells. Non-retractable tricycle landing gear. Underwing or wingtip sensor pods.

Power Plant: One 8·2 kW (11 hp) Herbrandson Dyad 180 two-cylinder two-stroke engine, driving a two-blade APL propeller. Single fuselage fuel tank, capacity 5·7 litres (1·25 Imp gallons; 1·5 US gallons).

Launch and Recovery: Conventional runway-type take-off and landing.

Guidance and Control: Radio command guidance and telemetry system. Two independently operable vertical stabilisation systems, both developed by APL. One of these, based on fluidic angular rate sensors, provides wing levelling and pitch stabilisation in all weather. The other is an advanced version of an electrostatic autopilot, previously invented at APL, which senses the attitude of the aircraft in relation to an electrical field in the atmosphere. In fair weather, and in certain types of bad weather, the electrical field is usually vertical, but in some adverse conditions (eg, thunderstorms) the device is not adequate to derive a vertical reference. One of the goals of the MAP/UV programme has been to investigate meteorological conditions that disturb the atmospheric electrical field, and to measure and record atmospheric characteristics that influence laser propagation in the lower atmosphere. Aerodynamic control of the RPV is by conventional ailerons, wing flaps, rudder and elevators.

Equipment: Sensors developed at APL include airspeed, engine tachometer, heading, altitude, three components of the electrical field, and yaw and pitch gust probes. The telemetry system and sensors for temperature, humidity, ozone concentration, and high frequency turbulence, were developed at UTEP. Devices for gathering aerosol samples and dust particles are carried in pods slung underneath the wing (or, optionally, located at the wingtips). A radar transponder assists in accurate tracking of the flight path.

Dimensions, external:

Wing span	3·05 m (10 ft 0 in)
Wing span over wingtip instrument pods	3·35 m (11 ft 0 in)
Wing chord: at root	0·53 m (1 ft 9 in)
at tip	0·41 m (1 ft 4 in)
Wing area, gross	1·44 m² (15·48 sq ft)
Wing aspect ratio	6·46
Length overall	2·41 m (7 ft 11 in)
Fuselage: Max diameter	0·23 m (9 in)
Height overall	0·81 m (2 ft 8 in)
Tailplane span	1·02 m (3 ft 4 in)
Propeller diameter	0·56 m (1 ft 10 in)
Sensor pod diameter	0·15 m (6 in)

Weights (as flown up to early 1981):

Weight empty, equipped	21·8 kg (48 lb)
Sensor load	11·3 kg (25 lb)
Fuel load	4·8 kg (10·5 lb)
T-O weight at 1,770 m (5,800 ft)	37·9 kg (83·5 lb)
T-O weight at S/L	43·5 kg (96 lb)

Performance (at T-O weight of 37·9 kg; 83·5 lb):

Max level speed at S/L	91 knots (169 km/h; 105 mph)
Econ cruising speed at S/L	54 knots (100 km/h; 62 mph)
Stalling speed, flaps down	27 knots (50 km/h; 31 mph)
Max rate of climb at S/L	579 m (1,900 ft)/min
Service ceiling (calculated)	7,620 m (25,000 ft)
T-O run	46 m (150 ft)

BEECHCRAFT
BEECH AIRCRAFT CORPORATION

Wichita, Kansas 67201
Telephone: (316) 689 7111
Officers: see Aircraft section

In addition to manufacturing piloted aircraft, Beech has been designing and producing pilotless target drones since 1955. By 1 January 1981 it had built more than 6,000 target drones, including more than 2,000 MQM-61As (now out of production) for the US Army, more than 3,600 AQM-37As, and more than 440 MQM-107As.

Beech has also carried out design studies to examine the suitability of its range of light aircraft for operation under remote control.

BEECHCRAFT MODEL 1019 SERIES
US military designation: AQM-37A

This normally non-recoverable, supersonic air-launched missile target system is designed to simulate invader aircraft and missile threats, and to provide defence weapon system evaluation and crew training. The target provides both active and passive radar augmentation for radar acquisition and tracking; a chemical flare is provided for infra-red homing. Two optional miss-distance systems are available.

The original Model 1019 won a 1959 US Navy/Air Force design competition, and has since entered service in a variety of versions with the US Navy (1962), US Air Force (1968) and US Army (1973), as well as the air forces of the UK (1966), Italy (1969), France (1974) and Israel (1975). Total deliveries of all versions had reached 3,620 by January 1981.

The Model 1019 is the basic AQM-37A target. Other versions are as follows:

Model 1019A. Twenty AQM-37As built to a 1975 US Navy contract. Capable of flying at heights of 10-300 m (33-985 ft) for weapon system evaluation. These 'Sea Skipper' vehicles were equipped with an autopilot which incorporated a homing device for longitudinal guidance and a radar altimeter for altitude control.

Model 1072 (Shorts SD.2 Stiletto). Version for UK, substantially re-engineered by Shorts to meet British requirements, including incorporation of telemetry and radar systems and control system changes. Vehicle is fitted by Beech with single-chamber rocket motor to give Mach

2 performance at 18,300 m (60,000 ft). Principal modifications made by Shorts were detailed in the 1978-79 *Jane's*. Current Stiletto system consists of basic target vehicle plus a number of optional mission kits which can either be installed by Shorts or be delivered separately for customer installation. Total of 150 ordered by late 1980.

Model 1088. Manufacturer's designation of 11 targets for Italian Air Force, used primarily for weapon systems evaluation. Pre-programmed dive manoeuvre to give supersonic presentation at medium altitudes. Radio controlled lateral manoeuvring capability to position target accurately.

Model 1094. Manufacturer's designation of 50 targets for French Air Force, used primarily for air-to-air weapon systems evaluation. French name **Vanneau** (Lapwing). Modified by Matra to be compatible with appropriate weapon system. Modifications include addition of scoring, radar augmentation, telemetry, and tracking beacon systems. Primary missions are performed at Mach 1·6 at 16,750 m (55,000 ft) and Mach 2 at 21,335 m (70,000 ft).

Model 1095. Manufacturer's designation of 40 targets ordered by British Ministry of Defence. Modified to MoD specifications for use in crew training exercises on Hebrides range.

Model 1098. Manufacturer's designation of 20 targets for Israeli Air Force.

Models 1100 and 1101. The US Army contracted with Beech in 1975 to provide 48 AQM-37As for weapon system evaluation. These were the first production vehicles to incorporate a two-stage parachute recovery system. Under the Army contract, Beech produced two versions of the AQM-37A. One (the Model 1101) is a supersonic, high-altitude target capable of operation at speeds up to 1,182 knots (2,188 km/h; 1,360 mph) and altitudes up to 21,335 m (70,000 ft). The other (the Model 1100) is a low-altitude version which can be flown to within 15 m (50 ft) of the terrain. Both models are equipped with a radio controlled guidance system permitting both longitudinal and vertical positioning.

Model 1102. Version of AQM-37A for US Navy, incorporating a solid-state autopilot, improved wing design, and other changes yielding improved performance. Initial contract for 100 placed in November 1976; total of 415 ordered by early 1980.

Model 1104. Special version of AQM-37A for US Navy. Ten modified in 1980 for altitude control up to 27,430 m (90,000 ft). Speed control for Mach 2·5 and 3·0 from 22,860 m (75,000 ft). Controlled dive angle upon command is available. Space in nose section for ECM, scoring and augmentation systems.

The following description applies to the standard Model 1019/AQM-37A:

Type: Air-launched expendable or recoverable target.

Airframe: Mid-wing monoplane, of canard configuration, with slim-delta main wings (of double-wedge section) at rear having 76° sweep on leading-edges, 0° dihedral, 0° incidence, and full-span ailerons. Movable foreplanes, of modified double-wedge section. Fixed endplate fin at tip of each main wing. Cylindrical centre-fuselage, with ogival nose section and tapering rear section over rocket chambers. Underbelly tunnel for rocket-engine cartridge-operated start valves, plumbing, infra-red flare and miss-distance scoring system antenna.

Power Plant: One Rocketdyne/AMF LR64 P-4 two-chamber liquid-propellant rocket engine (2·81 kN; 631 lb st). Three propellant tanks, for nitrogen pressurant, mixed amine fuel (MAF-4) and IRFNA oxidiser, form integral part of centre-fuselage.

Launch and Recovery: Air-launched from launcher which is adaptable to various fighter aircraft. Normally non-recoverable, but US Army Models 1100/1101 have parachute recovery system.

Guidance and Control: Programmed guidance system. Flight normally terminated by automatic destruct system, which also operates in event of a major failure. Command standby destruct system for added range safety.

Special Equipment: Approx 0·04 m^3 (1·4 cu ft) of space in nose section for optional scoring and augmentation systems (see introductory paragraph). Target is compatible with most non-co-operative scoring systems.

Dimensions, external:

Wing span	1·00 m (3 ft 3½ in)
Wing area (exposed)	0·87 m^2 (9·35 sq ft)
Length overall	3·82 m (12 ft 6½ in)
Height overall	0·51 m (1 ft 8 in)
Diameter of fuselage	0·33 m (1 ft 1 in)

Weight:

Max launching weight	256 kg (565 lb)

Performance (rated):

Operating speed	Mach 0·4 to Mach 3·0
Operating height	300-27,430 m (1,000-90,000 ft)
Range	more than 100 nm (185 km; 115 miles)
Endurance (powered)	8 min)

BEECHCRAFT MODEL 1089

US Army designation: MQM-107A Streaker
Swedish Air Force designation: RB06 Girun

After a performance and cost competition, Beech was awarded a multi-year production contract by the US Army in 1975 for the Model 1089 Variable-Speed Training Target, under the designation MQM-107A Streaker. Final assembly is performed at Beech's Boulder Division in Colorado. Beech designed, developed and now delivers ground support equipment, spares and ancillaries; and also provides operation of the system.

In addition to purchase for its own use, the US Army has supplied the MQM-107A system through Foreign Military Sales to Iran and Sweden for operational use. Direct sales have been made to Taiwan and Korea; contracts with other governments were pending in early 1981. The US Navy and Air Force have also completed evaluation programmes, involving firings of air-to-air and surface-to-air missiles. More than 430 Model 1089s had been ordered, and approx 400 delivered, by early 1981.

The principal function of the Streaker is to tow a variety of targets for missile training and evaluation, including gunnery banners, or sophisticated infra-red and radar-augmented tow targets with cable lengths of up to 2,440 m (8,000 ft).

The MQM-107A itself also serves as an aerial target for air defence systems such as Chaparral, Redeye, Hawk and improved Hawk, and is the primary subsonic missile training target for the US Army.

Programmes recently completed include a 0·38 m (1 ft 3 in) extension to the fuselage for greater internal payload capacity and easier access to augmentation and scoring components, improved control of high-*g* manoeuvres, and a digital flight control system.

Type: Re-usable variable-speed target.

Airframe: Low-wing monoplane configuration, with sweptback wings and tail surfaces. Engine suspended on pylon beneath centre of fuselage. Modular design throughout, with flat-section wings and tail surfaces of bonded honeycomb and foam-filled aluminium ailerons and elevators. Cylindrical body, with ogival nose and tail cones.

Power Plant: One 2·85 kN (640 lb st) Teledyne CAE 372-2 (J402-CA-700) turbojet engine in underfuselage pod, with 242 litres (64 US gallons) of fuel. M-60 or MARC-117 JATO bottle for surface launch boost.

Launch and Recovery: Lightweight surface launcher and checkout system, easily transportable in two suitcase-size containers. Drogue and main parachute command recovery system.

Guidance and Control: Radio command guidance of flight profile, with pre-programmed or commanded manoeuvres. In flight, guidance and control system automatically stabilises about roll, yaw and pitch attitudes, and provides altitude and speed hold modes.

Special Equipment: Principal function is to tow a variety of targets for missile training and evaluation. Two TRX-4 radar or two TA-8 infra-red augmentation targets can be carried on each mission and towed separately up to 2,440 m (8,000 ft) behind MQM-107A. Streaker serves as aerial target for such air defence systems as Chaparral, Redeye, Hawk, Roland and Stinger, and the Vulcan rapid-fire gun system. Internal payload capacity for avionics; wingtip payload capacity for decoy augmentation and additional avionics; capacity for up to 113 kg (250 lb) beneath wings.

Dimensions, external and internal:

Wing span	3·00 m (9 ft 10 in)
Length	5·23 m (17 ft 2 in)
Height (total)	1·47 m (4 ft 10 in)
Body diameter	0·38 m (1 ft 3 in)
Volume available for mission equipment and core avionics	0·092 m^3 (3·26 cu ft)

Areas:

Wings (total projected)	2·52 m^2 (27·16 sq ft)
Vertical tail surfaces	0·43 m^2 (4·63 sq ft)
Horizontal tail surfaces	0·79 m^2 (8·50 sq ft)

Weights:

Weight empty	218 kg (480 lb)
Usable fuel and visual smoke oil	191 kg (420 lb)
Launching weight (incl booster)	460 kg (1,014 lb)

Performance:

Operating speed range	247-500 knots (459-925 km/h; 285-575 mph)
Operating height range	S/L to 12,200 m (40,000 ft)
Endurance	more than 3 h

BEECHCRAFT MODEL 999F/MQM-107B

US military designation: MQM-107B

This improved version of the MQM-107A incorporates a new engine for higher performance, increased internal payload capacity, and improved manoeuvring capability with a new control system design. In 1981, Beech was delivering 10 to the US Army and 10 to the US Air Force; 21 similar units, with the standard-length MQM-107A fuselage, have been ordered by Taiwan, and are designated Model 999F.

Type and Airframe: As MQM-107A except for longer fuselage with increased payload/avionics volume. Impact-absorbing nosecone.

Power Plant: One 3·68 kN (827 lb st) Microturbo TRI 60-2 Model 074 turbojet engine; or one 3·25 kN (730 lb st) Teledyne CAE 372-11A (improved version of CAE 372-2). Fuel capacity and JATO details as for MQM-107A.

Launch and Recovery: Launch and control systems mobile and easily transportable. Drogue and main parachute recovery system. Water recovery kit available.

Guidance and Control: As described for MQM-107A. Autopilot provides for high-*g* manoeuvres. Terrain-following guidance system available optionally.

Special Equipment: Principal function is to provide a variety of threat simulations for missile and gunnery systems evaluation and combat crew training. A number of augmentation and scoring devices can be carried internally, or externally on pylons and/or wingtips, or can be towed up to 2,440 m (8,000 ft) behind the MQM-107B. Increased internal capacity compared with MQM-107A.

Dimensions, external and internal: As MQM-107A except:

Length	5·51 m (18 ft 1 in)
Volume available for mission equipment and core avionics	0·136 m^3 (4·79 cu ft)

Areas: As MQM-107A

Weights:

Weight empty	245 kg (540 lb)
Usable fuel and visual smoke oil	199 kg (438 lb)
Launching weight (incl booster)	494 kg (1,090 lb)

Performance: As MQM-107A except:

Operating speed range	275-535 knots (510-991 km/h; 317-616 mph)

BENSEN

BENSEN AIRCRAFT CORPORATION

Box 31047, Raleigh-Durham Airport, Raleigh, North Carolina 27612

Telephone: (919) 787 4224/0945

President: Dr Igor B. Bensen, PE

Bensen Aircraft Corporation (see Homebuilt Aircraft section) was in early 1981 converting a B-80 autogyro to a radio controlled pilotless aircraft. It is to be used by the military for patrolling hazardous areas where manned aircraft cannot be used. The converted B-80 remains suitable for manned flight.

BRUNSWICK

BRUNSWICK DEFENSE DIVISION

3333 Harbor Boulevard, Costa Mesa, California 92626

Telephone: (714) 546 8030

Marketing Director: A. H. Bickel

On behalf of the US Air Force Avionics Laboratory, Brunswick developed the small Maxi-Decoy glider, and the larger rocket-powered Propelled Decoy, to be carried by US strike aircraft. Equivalent to an 'iron' bomb, the decoys are carried with wings folded on the aircraft external stores points. Shortly after release the wings are opened by internal command. Flight profiles and manoeuvres are pre-programmed.

BRUNSWICK MAXI-DECOY

Unpowered decoy glider, having a square-section body and flip-out sweptback wings. Capable of high subsonic speed. Under development by US Air Force primarily as active or passive ECM carrier, launched from F-4 Phantom aircraft. Three Maxis can be accommodated in each of F-4's four underfuselage Sparrow missile recesses, and two on each 340 kg (750 lb) underwing hardpoint; several decoys can be deployed simultaneously. Electronic warfare packages are designed and manufactured by Automated Systems Division of RCA. Series of mission equipment test flights conducted at Eglin AFB during 1977. Typical loads include a C-band jammer; or a G-band jammer combined with radar signature augmentation. Further details are classified.

Dimensions:

Wing span	approx 0·76 m (2 ft 6 in)
Body width	0·13 m (5 in)
Length	1·14 m (3 ft 9 in)
Mission equipment volume	0·008 m^3 (500 cu in)

Weight:

Max launching weight	59 kg (130 lb)

Performance:

Max speed	Mach 0·8-0·9

BRUNSWICK PROPELLED DECOY

Larger than Maxi-Decoy, with circular-section body, flip-out sweptback wings and cruciform sweptback tail

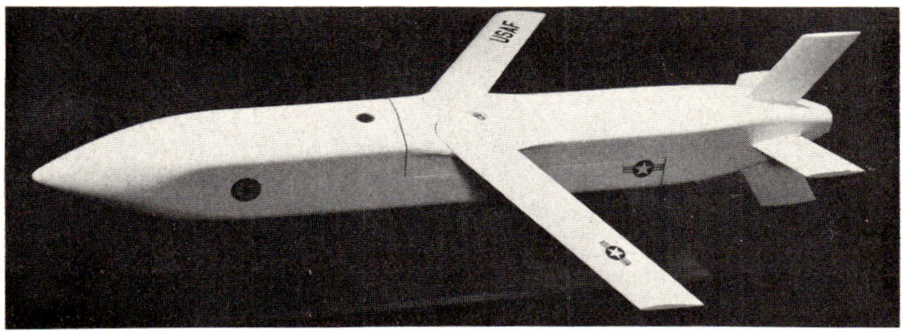

Brunswick Maxi-Decoy unpowered decoy, with flip-out wings

DSI Locomp high-subsonic multi-purpose target

DSI Gunsight low-cost target

DSI armed Sky Eye R-4D, with underwing rocket launch tubes

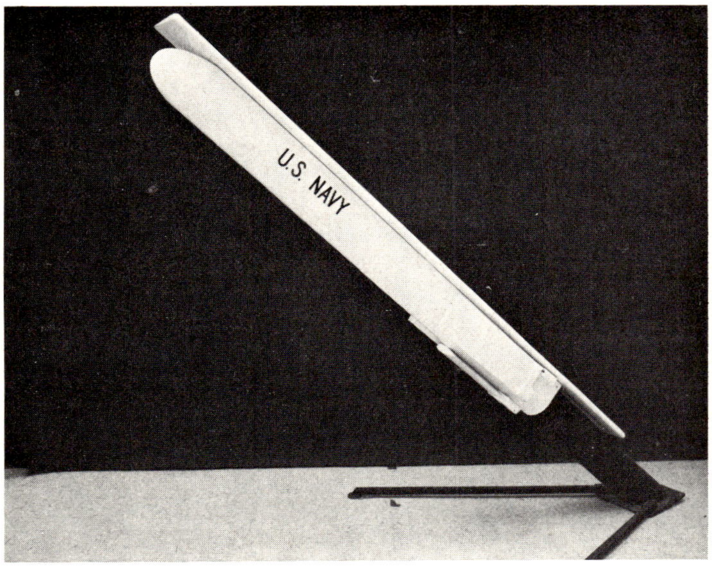

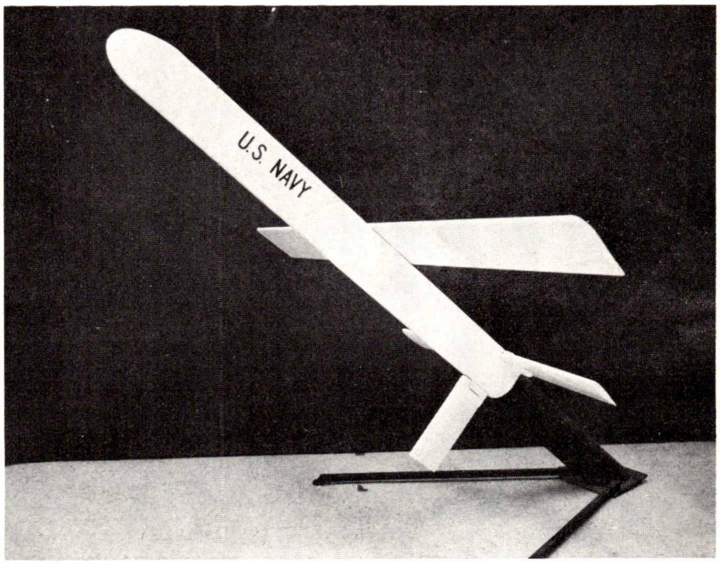

The active expendable decoy developed by DSI for the US Navy, showing the flying surfaces folded (left) and deployed (right)

surfaces; powered by Atlantic Research solid-propellant rocket motor. Flight testing began in 1974. Two-axis (roll and heading/altitude) autopilot allows decoy to manoeuvre at high cruising speeds as well as performing in glide modes; unlike Maxi, Propelled Decoy can cruise at high subsonic speeds. Six successful launchings from F-4 carrier aircraft made at Eglin AFB in 1974. One Propelled Decoy can be carried on each of Phantom's 340 kg (750 lb) underwing hardpoints. Further details are classified.

DIMENSIONS:

Wing span	1·40 m (4 ft 7 in)
Body diameter	0·25 m (10 in)
Length	2·24 m (7 ft 4 in)
Mission equipment volume	0·03 m³ (1,800 cu in)

WEIGHT:

Launching weight	136 kg (300 lb)

PERFORMANCE:

Burn time	5 min
Max speed	Mach 0·8-0·9

DSI
DEVELOPMENTAL SCIENCES INC

15757 East Valley Boulevard, PO Box 1264, City of-Industry, California 91749
Telephone: (213) 330 6865
PRESIDENT: Dr Gerald R. Seemann

DSI designed and built, under contract to NASA, a prototype oblique-wing RPRA which made its first flight in mid-1976 and was described and illustrated in the 1976-77 *Jane's*. The company also designed and built the Sky Eye and Scout mini-RPVs, and, for LMSC, the Aquila, based on the Sky Eye. In 1977 it undertook, for the US Army, a study of the feasibility of equipping mini-RPVs with 2·75 in FFAR rockets for attack duties. A further contract, for test and analysis of projectiles launched from small RPVs, was awarded in 1978 by the US Army Research and Technology Laboratories. Also in 1978, DSI completed, for the US Army, a recovery programme using LMSC Aquila drones fitted with a parachute and an inflatable airbag. Since early 1978 the company has been investigating, with promising results, the feasibility of a solar-powered aeroplane.

DSI is currently engaged in several classified RPV programmes, and in 1979 began a programme for the US Air Force concerned with an electrical propulsion system for micro- and mini-RPVs. This involves an 'exotic' primary battery as the energy source, a rare earth electric motor, and a composite-material propeller.

DSI has also initiated, for the Environmental Protection Agency, a concept study and preliminary design of an HMS-RCV (Hazardous Material Spill-Remotely Controlled Vehicle). No details have been received for publication, but the HMS vehicle is believed to be circular in shape, of hovercraft configuration, with twin ducted-fan propulsors.

DSI RPA-12 SKY EYE R-4D

The prototype Sky Eye first flew on 26 April 1973; details of this and early models can be found in previous editions of *Jane's*. More than 200 flights have been made by ten Sky Eye vehicles.

The latest known version is the Sky Eye R-4D, an upgraded version of the II-R (1980-81 *Jane's*) with improved wing; larger, flat-bottomed fuselage; no landing gear; and upgraded flight control system. More powerful engine, increased mission equipment capacity and higher gross weight. Endurance more than 8 h with 36 kg (80 lb) of mission equipment. The Sky Eye R-4D first flew in 1978, and in the following year DSI was engaged, with All American Engineering Inc, in an overseas programme involving this version.

AIRFRAME: Predominantly of Kevlar 49 skins with Nomex honeycomb core, giving a lightweight but robust vehicle capable of sustaining ±6*g* (normal) and 8*g* axial loading at max gross weight, and of surviving rough handling in operation. Sweptback high-wing monoplane, with servo-actuated trailing-edge elevons and turned-down wingtips. Wing section NACA 23-015 (modified). Incidence 3° at root. Aerodynamic twist (washout) 3°. Removal of wing and nosecone provides access to entire payload installation in nose and central bays. Fuselage consists of four basic components: nose fairing, central body, cowling and propeller shroud. Flat-bottomed fuselage has three quickly removable structural bottom panels, intended specifically for easy pallet-style installation and interchanging of a wide variety of loads. Rudder and/or dive-brakes optional.

POWER PLANT: One 11·9 kW (16 hp) or 16·4 kW (22 hp) DH Enterprises 220 or 280 two-stroke flat-twin engine, driving a two-blade wooden pusher propeller turning within a 0·69 m (2 ft 5·2 in) diameter annular duct. Fuel in two standard wing-root tanks, combined capacity 19 litres (4·2 Imp gallons; 5 US gallons). Optional tanks available to double this capacity for long-endurance missions.

Launch and Recovery: Two launch systems available: truck-mounted pneumatic launcher or simpler launch apparatus (also truck-mounted). With latter system, RPV is attached to launcher mounted on pickup truck; with RPV engine operating, truck is then accelerated to safe launch speed and RPV is released. Various recovery systems (eg skid landing gear) available, but usual method is by vertical barrier net in conjunction with blower-inflated air mattress. Emergency backup parachute system available optionally.

Guidance and Control: VHF radio command guidance system. Aerodynamic control by elevons.

Systems, Avionics and Equipment: Equipped initially with 600W electrical system consisting of a battery, engine-driven alternator and regulator. Alternators of 300 to 1,000W output (28V DC) available currently. Battery can provide 20 min of emergency power in event of alternator failure. Lear Siegler P/N R-4D-412-1 two-axis autopilot system (third axis optional when rudder is fitted), with variety of available control modes; inputs can be in form of vertical velocity and turn-rate commands, or heading and altitude commands. Tayburn Electronics data link system, with onboard TBD 400 encoder/decoder. Wide-band airborne transmitter, using sidebands for telemetry.

Dimensions, external (R-4D):
- Wing span: 3·78 m (12 ft 4¾ in)
- Wing area, gross: 2·991 m² (32·2 sq ft)
- Wing aspect ratio: 4·75
- Length overall: 2·12 m (6 ft 11½ in)
- Body: Max diameter: 0·355 m (1 ft 2 in)
- Propeller diameter: 0·64 m (2 ft 1 in)

Dimensions, internal:
- Equipment volume: more than 0·085 m³ (3 cu ft)

Weights:
- Weight empty, dry: 45·4 kg (100 lb)
- Fuel load (standard tanks): 13·6 kg (30 lb)
- Mission equipment: 36·3 kg (80 lb)
- Launching weight: 99·3 kg (219 lb)

Performance ('clean' at 99·3 kg; 219 lb gross weight, 22 hp engine):
- Max level speed:
  - at S/L: 130 knots (241 km/h; 150 mph)
  - at 1,525 m (5,000 ft): 129 knots (239 km/h; 148 mph)
  - at 3,050 m (10,000 ft): 128 knots (237 km/h; 147 mph)
  - at 4,575 m (15,000 ft): 127 knots (235 km/h; 146 mph)
- Truck launch speed: 48 knots (89 km/h; 55·5 mph)
- Stalling speed: 39 knots (72·5 km/h; 45 mph)
- Max rate of climb:
  - at S/L: 600 m (1,970 ft)/min
  - at 1,525 m (5,000 ft): 482 m (1,580 ft)/min
  - at 3,050 m (10,000 ft): 375 m (1,230 ft)/min
  - at 4,575 m (15,000 ft): 280 m (918 ft)/min
- Service ceiling (46 m; 150 ft/min climb): 6,100 m (20,000 ft)
- Endurance with 13·6 kg (30 lb) standard fuel: 6 h

DSI RPMB

Details of DSI's original 4·88 m (16 ft) long RPMB (Remotely Piloted Mini-Blimp) appeared in the 1977-78 *Jane's*; and of a later 7·92 m (26 ft) long version in the 1978-79 and 1979-80 editions.

For 1979, DSI designed a third, larger RPMB to what it described as a final configuration, powered by a 15 kW (20 hp) D & H two-stroke engine. A description of this version appeared in the 1980-81 edition.

DSI/US NAVY ACTIVE EXPENDABLE DECOY

Since 1976, DSI has been working with the Naval Research Laboratory and other US Navy agencies on a programme for an active expendable decoy vehicle. Cruising flight, and boost launch from a folded position, were demonstrated successfully in 1978 in the Mojave Desert and at the Naval Weapons Center, China Lake, California.

No other details of this programme are known, but two accompanying photographs show that the decoy is basically of long and narrow cylindrical shape, with three foldaway/flip-out tail control surfaces and a one-piece pivoting wing.

DSI GUNSIGHT

Gunsight is a low-cost target for tracking and actual weapon-firing use; it was in the feasibility demonstration phase in early 1980.

Power Plant: One 11·2 kW (15 hp) McCulloch flat-twin piston engine. Alternative power plant available, depending upon the speed desired.

Launch and Recovery: Rocket launch; parachute recovery.

Equipment: Two-axis autopilot. Visual (smoke), infra-red (flares) and R/F (three Luneberg lenses) augmentation available.

Dimensions, external:
- Wing span: 1·02 m (3 ft 4 in)
- Length overall: 2·03 m (6 ft 8 in)
- Body diameter: 0·22 m (8⅝ in)

Weight:
- Max launching weight: 34 kg (75 lb)

Performance (with 11·2 kW; 15 hp engine):
- Never-exceed speed at S/L: 300 knots (555 km/h; 345 mph)
- Max level speed at S/L: 240 knots (444 km/h; 276 mph)
- Service ceiling: 6,100 m (20,000 ft)
- Endurance at S/L: 40 min

DSI LOCOMP

Locomp, also in the feasibility demonstration phase in early 1980, is a multi-purpose system for operation at high subsonic speeds.

Power Plant: No details known.

Launch and Recovery: Air or ground launch; recoverable from land or water (floats with 50% fuel or less).

Dimensions:
- Wing span: 1·37 m (4 ft 6 in)
- Length overall: 2·31 m (7 ft 7 in)
- Body diameter: 0·33 m (1 ft 1 in)
- Mission equipment volume: 0·057 m³ (2 cu ft)

Weights:
- Weight empty (incl 52 kg; 115 lb equipment): 136 kg (300 lb)
- Max launching weight: 200 kg (440 lb)

Performance:
- Max level-flight Mach No. at S/L: 0·80
- Max level-flight Mach No. at 6,100 m (20,000 ft): 0·82
- Endurance at max Mach No. at S/L: 37 min
- Endurance at max Mach No. at 6,100 m (20,000 ft): 1 h 0 min

E-SYSTEMS

E-SYSTEMS INC

PO Box 6030, Dallas, Texas 75222

Melpar Division: 7700 Arlington Boulevard, Falls Church, Virginia 22046

Telephone: (703) 560 5000

E-Systems Inc's activities lie predominantly in aerospace systems development and manufacture, in addition to which it has carried out specialised conversion work on nearly 400 C-135 and KC-135 series aircraft for the US Air Force. Its Greenville Division designed and built the L450F, a single-engined aircraft capable of manned or unmanned operation and described in previous editions of *Jane's*. Melpar Division has designed a number of mini-RPVs, including those of which details follow.

E-SYSTEMS E-55

The E-55 is an improved version of the E-45 mini-RPV (1978-79 *Jane's*), with a more powerful engine and higher gross weight. All existing E-45 vehicles have been upgraded to the E-55 configuration.

The E-55 is designed to carry a variety of electronic packages for real-time reconnaissance, jamming, targeting, and homing missions. Several E-55s were employed successfully during a harassment weapons systems concept demonstration at Meppen in the Federal Republic of Germany in early 1977.

Three modified **E-55S** were undergoing evaluation in 1980 as part of the US Army's Superfly demonstration programme. These have side-force generator panels ('vertical wings') above and below each wing, to enable the aircraft to turn without banking during a terminal dive, and are described separately.

Airframe: High-wing twin-tailboom monoplane, with pod-shaped central nacelle. Dihedral on outer wing panels. Twin endplate fins and rudders; enclosed tailplane with full-span elevator.

Power Plant: One 2·76 kW (3·7 hp) single-cylinder two-stroke Roper engine, mounted in rear of fuselage nacelle and driving a two-blade pusher propeller. Fuel capacity 3·4 litres (0·9 US gallons).

Launch and Recovery: Launch by vehicle-mounted launcher, using jettisonable take-off dolly, or by catapult. Recovery by conventional landing, on skid beneath central nacelle, or by net.

Guidance and Control: Either by real-time tracking radar mounted on top of two-man mobile van, or by standard non-tracking radio control link. Radar system can control RPV effectively for approx 43·5 nm (80·5 km; 50 miles). This can be extended to 130 nm (241 km; 150 miles) by pre-programming flight and using airborne relay for video downlink. Aircraft can be controlled either manually or by combination of pre-programmed flight and manual landing.

Equipment: Melpar-designed five-axis autopilot weighing 0·9 kg (2 lb), which can be used for dead reckoning or Omega navigation. Melpar TV surveillance system provides means for remote manual control, target identification, and navigational assistance away from launch site.

Dimensions, external:
- Wing span: 2·41 m (7 ft 11 in)
- Wing area, gross: 0·68 m² (7·36 sq ft)
- Length overall: 2·36 m (7 ft 9 in)
- Height overall: 0·51 m (1 ft 8 in)
- Width over tailbooms: 0·71 m (2 ft 4 in)
- Fuselage diameter: 0·20 m (8 in)

Weights:
- Weight empty, equipped: 11·3 kg (25 lb)
- Mission equipment: 13·6 kg (30 lb)
- Fuel: 2·7 kg (6 lb)
- T-O weight: 25 kg (55 lb)

Performance:
- Cruising speed: 52 knots (97 km/h; 60 mph)
- Stalling speed: 39 knots (72 km/h; 45 mph)
- Rate of climb at S/L: 198 m (650 ft)/min
- Service ceiling: 3,050 m (10,000 ft)
- Endurance: 4 h

E-SYSTEMS E-55S

Airframe: Shoulder-wing twin-tailboom monoplane, with central fuselage nacelle for engine and avionics. No dihedral on outer wing panels. Twin fins and rudders. Side-force generator panel above and below each wing, outboard of tailboom/wing joint. Lower side-force panels are frangible.

Power Plant: As for E-55. Fuel capacity 3·8 litres (1 US gallon).

Launch and Recovery: Pneumatic catapult launch; recovery by radio-controlled conventional landing.

Guidance and Control: Either by autonomous, programmed autopilot; or by intermittently updated autopilot; or by continuous, real-time autopilot command. Manual control available for recovery. Addition of side-force generators provides increased guidance accuracy and permits a simplified electronic control system.

Dimensions, external: As E-55 except:
- Wing span: 2·44 m (8 ft 0 in)
- Wing area, gross: 0·69 m² (7·40 sq ft)
- Height overall: 0·61 m (2 ft 0 in)
- Side-force generator span: 0·61 m (2 ft 0 in)
- Side-force generator area (each): 0·31 m² (3·33 sq ft)

Weights and Performance: As E-55 except:
- Rate of climb at S/L: 137 m (450 ft)/min
- Service ceiling: 2,745 m (9,000 ft)

E-SYSTEMS E-90

The E-90 is a 41 kg (90 lb) version of the E-75 harassment vehicle (1978-79 *Jane's*). The use of a more powerful engine improves performance and allows an increased amount of mission equipment to be carried. It was first flown in October 1977, and development was continuing in 1980.

In addition to its potential mission as a passive radar homing system, future applications for the E-90 include loiter over enemy defences while carrying RV reconnaissance jamming, FLIR, or photographic payloads.

Airframe: High-wing single-boom monoplane, with side-force generator panel under each wing. Glassfibre wings, with high/low-density foam core; moulded polyethylene fuselage; internal phenolic resin coated cardboard tubes to stiffen tailboom and tail surfaces.

Power Plant: One 8·2 kW (11 hp) Herbrandson DH-160 flat-twin two-stroke piston engine, driving a two-blade pusher propeller. Standard fuel capacity of 5·7 litres (1·5 US gallons) can be increased by reducing mission load.

Equipment: Pre-programmed autonomous autopilot/navigation system using onboard microprocessor.

Dimensions, external:
- Wing span: 3·05 m (10 ft 0 in)
- Wing area, gross: 1·11 m² (12 sq ft)
- Length overall: 2·29 m (7 ft 6 in)
- Height overall: 0·51 m (1 ft 8 in)
- Fuselage: Max width: 0·25 m (10 in)

Weights:
- Weight empty, equipped: 22·7 kg (50 lb)
- Mission equipment (normal): 9 kg (20 lb)
- Fuel: 4·5 kg (10 lb)
- T-O weight: 41 kg (90 lb)

Performance (typical):
- Max level speed: 95 knots (177 km/h; 110 mph)
- Max cruising speed: 69 knots (128 km/h; 80 mph)
- Stalling speed: 44 knots (81 km/h; 50 mph)
- Rate of climb at S/L: 488 m (1,600 ft)/min
- Service ceiling: 3,660 m (12,000 ft)
- Range at 1,525 m (5,000 ft): 191 nm (354 km; 220 miles)
- Endurance at 1,525 m (5,000 ft): 4 h 0 min

E-SYSTEMS E-100X

Essentially, the E-100X is an enlarged development of the E-45 (1978-79 *Jane's*), having extended-span wings, greater engine power, and increased volume available for mission equipment. It was described in the 1980-81 *Jane's*. The programme ended in December 1979.

E-SYSTEMS E-130

The E-130 is essentially a scaled-up version of the E-90 (which see), utilising the proven very-low-cost constructional features and materials of the E-75/E-90 series but

E-Systems E-55 mini-RPV at a desert test site

E-Systems E-90 mini-RPV with underwing side-force panels

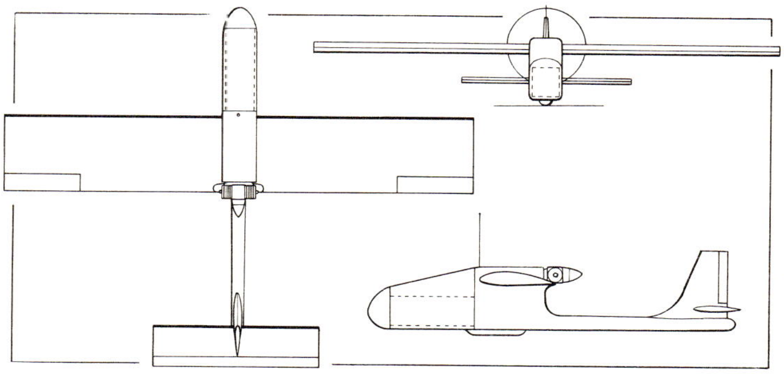

Provisional drawing of E-Systems E-200 testbed RPV *(Michael A. Badrocke)*

E-Systems E-150 TV-carrying RPV

E-Systems E-55S RPV, evaluated in the US Army Superfly programme

E-Systems E-130, with underwing side-force generator panels

able to carry a significantly greater load. A wind tunnel model (83·3% of full size) was tested at the University of Washington during 1979, and in early 1980 a batch of ten full-sized E-130s was under construction. A distinctive feature is the provision of side-force generator panels below each wing.

Airframe: Cantilever shoulder-wing monoplane with pod-and-boom fuselage. Foam plastics wings; moulded polyethylene fuselage, with pneumatic resin-coated cardboard tailboom stiffener. Side-force generator panel beneath each wing, at approx half-span.

Power Plant: One 13·4 kW (18 hp) Herbrandson DH 220 flat-twin two-stroke engine, driving a two-blade pusher propeller.

Equipment: Pre-programmed autonomous auto-pilot/navigation system, using an onboard microprocessor.

Dimensions, external:

Wing span	3·05 m (10 ft 0 in)
Wing area, gross	1·29 m² (13·89 sq ft)
Length overall	2·36 m (7 ft 9 in)
Height overall	0·61 m (2 ft 0 in)
Fuselage: Max width	0·305 m (1 ft 0 in)

Weights:

Weight empty, equipped	27 kg (60 lb)
Mission equipment	20 kg (45 lb)
Fuel	7 kg (15 lb)
T-O weight	59 kg (130 lb)

Performance:

Max level speed	116 knots (216 km/h; 134 mph)
Cruising speed	87 knots (161 km/h; 100 mph)
Stalling speed	52 knots (97 km/h; 60 mph)
Rate of climb at S/L	610 m (2,000 ft)/min
Service ceiling	3,660 m (12,000 ft)
Endurance	4 h

E-SYSTEMS E-150

The E-150, which flew for the first time on 4 August 1978, was designed to carry a 27·2 kg (60 lb) stabilised TV camera. It was constructed with as many standard Melpar parts as possible, but with new nose structure and fairings; extended wings, tailboom and horizontal tail; additional fin area; rudder control; and a new shock-absorbing landing gear.

Airframe: Derived from that of E-75 (1978-79 *Jane's*).

Power Plant: One 9·7 kW (13 hp) Herbrandson DH-174 horizontally-opposed two-stroke piston engine, driving a two-blade wooden pusher propeller. Fuel capacity 9·5 litres (2·5 US gallons).

Launch and Recovery: Launched from truck top, free in yaw and pitch. Lands on shock-absorbing, self-stabilising skid/castor gear.

Guidance and Control: Autopilot-guided, similar to E-45/55.

Dimensions, external:

Wing span	3·96 m (13 ft 0 in)
Wing area, gross	1·45 m² (15·6 sq ft)
Length overall	2·22 m (7 ft 3½ in)
Height overall	0·89 m (2 ft 11 in)
Fuselage width	0·25 m (10 in)

Weights:

Weight empty, equipped	37·6 kg (83 lb)
Mission equipment	27·2 kg (60 lb)
Fuel	6·8 kg (15 lb)
Max T-O weight	71·6 kg (158 lb)

Performance:

Max level speed	74 knots (137 km/h; 85 mph)
Cruising speed	56·5 knots (105 km/h; 65 mph)
Stalling speed	46 knots (86 km/h; 53 mph)
Endurance	2 h

E-SYSTEMS E-200

The E-200 is a scaled-up derivative of the E-75 (1978-79 *Jane's*) and E-55. Its 91 kg (200 lb) take-off gross weight will allow a 31·75 kg (70 lb) mission load to be flown at altitudes of up to 3,660 m (12,000 ft). The vehicle is configured for use as a flying testbed, permitting flight demonstration of airborne reconnaissance, jamming or navigation systems without incurring the costs of manned flight test programmes. Electrical power for the payload and the Melpar-designed digital autopilot/flight control system is provided by a 0·5kW engine-driven alternator.

Airframe: A formed and machined aluminium backbone joins the firewall, propulsion unit, tailboom and wing spars. Fuselage structure and mission equipment covered with cast plastics fairing. Firewall provides for structural mounting of equipment package. Wings and horizontal tail surfaces are shaped foam cores covered with glassfibre-reinforced skin. Wing spars and tailboom are structural tubes. Fin of cast plastics. Nosecone removable for ease of access to equipment; wings and horizontal tail surfaces removable for transportation. Lifting yoke at CG.

Power Plant: One 18 kW (24 hp) flat-twin two-stroke piston engine, driving a two-blade fixed-pitch pusher propeller. Fuel capacity 8·7 litres (2·3 US gallons).

Launch and Recovery: Vehicle-mounted launcher. Recovery by conventional landing on skid beneath central nacelle.

Guidance and Control: Radio command control for launch and recovery. In-flight guidance by three-axis Melpar-designed autopilot. Aerodynamic control by ailerons, elevator and rudder.

Equipment: Volume of 0·06 m³ (2·08 cu ft) available for user-defined mission equipment.

DIMENSIONS, EXTERNAL (provisional):

Wing span	4·27 m (14 ft 0 in)
Wing area, gross	3·07 m² (33·0 sq ft)
Length overall	3·35 m (11 ft 0 in)
Height overall	0·71 m (2 ft 4 in)
Fuselage: Max width	0·305 m (1 ft 0 in)

WEIGHTS (typical):

Weight empty, incl flight controls	52·1 kg (115 lb)
Mission equipment	31·75 kg (70 lb)
Fuel	6·8 kg (15 lb)
Max T-O weight	91 kg (200 lb)

PERFORMANCE (estimated):

Max level speed	96 knots (177 km/h; 110 mph)
Cruising speed	82 knots (153 km/h; 95 mph)
Stalling speed	39 knots (73 km/h; 45 mph)
Rate of climb at S/L	610 m (2,000 ft)/min
Service ceiling	3,660 m (12,000 ft)
Endurance	2 h

FSI
FLIGHT SYSTEMS INC

4000 Westerly Place, PO Box 2400, Newport Beach, California 92663
Telephone: (714) 833 9661
DIRECTOR, AERIAL TARGETS: S. C. Warrick

FSI QF-86E SABRE

FSI demonstrated to the US Army in 1975 two remotely controlled North American F-86 Sabre jet fighters, the converted aircraft actually being Canadair-built Sabre 5s, structurally similar to the US-built F-86E.

Forty-one QF-86Es had been ordered by 1 February 1981; deliveries began in mid-1977. They are being produced to test and evaluate US Army air defence systems. Manoeuvres of up to 8*g* can be performed, and the QF-86Es can also deploy stores, initiate jamming and provide other countermeasures, all under remote control.

A description of the basic F-86 airframe can be found in the 1959-60 *Jane's*; the following details apply to the FSI QF-86E/Sabre 5:

POWER PLANT: One Orenda 10 turbojet engine, rated at 28·15 kN (6,325 lb st). Fuel capacity 1,571 litres (415 US gallons) internal, plus provision for two 454 or 757 litre (120 or 200 US gallon) underwing drop-tanks.

LAUNCH AND RECOVERY: Conventional runway T-O and landing, on retractable tricycle landing gear. Aircraft programmed to come to a halt, or continue and make safe T-O and climb-out, if ground control link is lost during T-O run.

GUIDANCE AND CONTROL: Radio command guidance system. Primary mode of operation is NOLO (No Local Operator), over full range of pre-programmed flight manoeuvres including T-O and landing; but provision for onboard human pilot is retained (eg for manned practice presentations, or maintenance or ferry flights). Remote control exercised from one fixed and one mobile ground station, both manufactured by Vega Precision Laboratories; or by the drone formation control system manufactured by IBM Corporation.

SPECIAL EQUIPMENT: Onboard avionics and instrumentation comprise three basic installations: autopilot, with FSI avionics; Vega Precision Laboratories command/telemetry data system; and FSI interface coupler for processing uplink command and downlink telemetry data to and from drone aircraft. A fourth installation, the IBM drone formation control system, is optional. Radar altimeter optional, for simulated low-level attack presentations. Ancillary equipment, according to mission, may include scoring gear, infra-red flare dispenser, chaff dispensers or ECM pods. Television has been added as an optional T-O and landing aid.

DIMENSIONS, EXTERNAL:

Wing span	11·31 m (37 ft 1·2 in)
Length overall	11·43 m (37 ft 6 in)
Height overall	4·48 m (14 ft 8·4 in)

WEIGHTS:

Basic weight empty	4,921 kg (10,850 lb)
T-O weight 'clean'	6,123 kg (13,500 lb)
T-O weight with two 454 litre (120 US gallon) drop-tanks	6,894 kg (15,200 lb)

PERFORMANCE:

Max level speed above 11,000 m (36,000 ft)	Mach 0·92 (527 knots; 977 km/h; 607 mph)
Service ceiling	13,715 m (45,000 ft)
Max range at 9,145 m (30,000 ft) with two 454 litre (120 US gallon) drop-tanks	600 nm (1,112 km; 691 miles)
Max endurance, conditions as above	2 h 0 min
g limit	+7·0

ILC DOVER
ILC INDUSTRIES INC
Dover Division

PO Box 266, Harrington Road, Frederica, Delaware 19946
Telephone: (302) 335 3911

This company, which has for some years been a manufacturer of tethered balloons, and the spacesuits worn by America's Apollo, Skylab and Space Shuttle astronauts, has evaluated the application of inflatable surfaces to lightweight RPVs. Details of the Apteron, an inflatable-wing mini-RPV, can be found in the 1980-81 *Jane's*.

KAMAN
KAMAN AEROSPACE CORPORATION

Old Windsor Road, Bloomfield, Connecticut 06002
Telephone: (203) 242 4461
OFFICERS: see Aircraft section

KAMAN STAPL

Kaman began design of the STAPL (Ship Tethered Aerial Platform) under contract to the US Office of Naval Research in September 1972, and testing of two demonstration vehicles and their mobile launch and recovery platform and control system began in 1974; low-altitude testing was completed in 1979.

The test phase of the programme has been inactive, except for report preparation, since the Spring of 1980. In early 1981 the future of the programme was awaiting completion of Congress and Defense Department budget deliberations. It was not anticipated that programme tests would resume earlier than October 1981.

The demonstrators, which are of autogyro configuration, are equipped with automatic flight control systems and data recording equipment. A photograph of the first demonstrator appeared in the 1977-78 *Jane's*; modifications were later made to add a horizontal stabiliser with vertical fins at the tips. The self-contained AFCS, engineered by Kaman, provides three-axis stabilisation and automatic flight path control, and incorporates redundancy for mission reliability.

AIRFRAME: Welded steel tube fuselage and skid-type landing gear. Sheet metal vertical fin and aerodynamically balanced rudder. Electrically actuated variable-incidence plywood horizontal stabiliser, with plywood endplate fins. Two-blade Bensen G2 teetering and autorotating rotor, with metal blades bolted to hub. Blades do not fold. No rotor brake.

POWER PLANT: One 67 kW (90 hp) McCulloch 4318G piston engine, driving a Bensen BA-48-A8-70 two-blade fixed-pitch propeller. Single fuselage fuel tank, capacity 22·4 litres (6 US gallons).

LAUNCH AND RECOVERY: Tethered T-O and landing.

GUIDANCE AND CONTROL: Electrically-driven automatic flight control system (AFCS).

EQUIPMENT: Two 12V batteries to drive AFCS.

DIMENSIONS, EXTERNAL:

Rotor diameter	6·63 m (21 ft 9⅛ in)
Rotor blade chord	0·18 m (7 in)
Rotor disc area	34·56 m² (372 sq ft)
Length overall (excl rotor)	3·27 m (10 ft 8¾ in)
Width overall	1·75 m (5 ft 9 in)
Height to top of rotor hub	1·92 m (6 ft 3½ in)
Propeller diameter	1·22 m (4 ft 0 in)

WEIGHTS AND LOADING:

Weight empty	172 kg (380 lb)
Fuel load	17·7 kg (39 lb)
Max mission equipment load	34 kg (75 lb)
Max T-O weight	224 kg (494 lb)
Max disc loading	6·35 kg/m² (1·3 lb/sq ft)

PERFORMANCE: No details available

KAMAN HELICOPTER DRONE KITS

Kaman is providing, under US Army contracts, a number of helicopter drone kits for use in defence programmes. These kits provide complete radio remote control from the ground for take-off, performance of mission, return and landing. Kaman designs and manufactures the kits; installs and flight tests them, using an onboard safety pilot; and provides the ground controller to fly the target missions.

Ten drone kits, for installation in surplus Bell UH-1B target aircraft, were for use at Fort Bliss, Texas, in testing competing designs for the Division Air Defence (DIVAD) gun system for the US Army. Two kits, also for installation in UH-1Bs, will be used at White Sands missile range, New Mexico, in trials of the Patriot missile. Two others will be used at White Sands in a test programme for the Stinger man-portable surface-to-air missile.

LMSC
LOCKHEED MISSILES AND SPACE COMPANY INC

1111 Lockheed Way, Sunnyvale, California 94088
Telephone: (408) 742 4321
PUBLIC RELATIONS: Paul J. Binder

LMSC AQUILA
US Army designation: YMQM-105

The Aquila (Latin for 'eagle') mini-RPV was developed under US Army contract to demonstrate available technology for RPV use in surveillance, target acquisition, artillery adjustment, and laser designation for precision-guided munitions. Design and development utilised prior efforts by DSI (which see) with its Sky Eye RPV, and DSI supplied LMSC with technology demonstration airframe parts. Twenty-three XMQM-105 Aquilas were built under the initial system technology demonstrator programme, which was concluded in 1978. This vehicle was described and illustrated in the 1980-81 *Jane's*.

The US Army awarded LMSC contracts totalling $179·3 million in FY 1980-82 for a full-scale development version of the Aquila, designated YMQM-105, for target acquisition, laser designation and reconnaissance missions. The target acquisition system can be used for conventional artillery as well as for laser homing missiles such as Copperhead. It will also provide damage assessment data, and accuracy in locating a target is such that, at a range of 11 nm (20 km; 12·5 miles), it offers a circular area prediction of only 70 m (230 ft).

Under the US Army contract LMSC is to supply, over a 43-month period, a total of 22 YMQM-105s, together with four ground control stations, three launchers, three recovery units, three maintenance shelters, training simulators and training manuals. An operational section equipped with the system would consist of seven trucks, three trailers, 13 troops and an officer in charge, and would be air-transportable (incl five RPVs) in a Lockheed C-5A Galaxy transport aircraft. Set-up and launching can be done in 1 hour after arrival at a tactical site; stowage and make-ready for transport takes 30 min.

TYPE: Recoverable tactical mini-RPV.

AIRFRAME: Shoulder-wing monoplane of near-delta planform, built of Kevlar 49 honeycomb material, for low radar signature. Elevons on trailing-edges; turned-down wingtips. Airframe dismantles into four major components (centrebody, two wings, and propeller duct), and has quick-disconnect bladder fuel system.

POWER PLANT: One Herbrandson two-stroke piston engine (19·4 kW; 26 hp at 8,000 rpm), driving a two-blade fixed-pitch wooden pusher propeller within an annular duct. Fuel is a petrol/oil mixture.

LAUNCH AND RECOVERY: Launched from Fairchild Stratos launcher mounted on a 5 ton truck. Recovered by Dornier vertical ribbon net, raised on back of truck and capable of being lowered quickly after recovery in order to maintain low profile.

GUIDANCE AND CONTROL: Operates independently of ground control, inertial navigation being provided via an onboard computer.

EQUIPMENT: Target acquisition, laser designator, or real-time surveillance equipment, depending upon mission.

DIMENSIONS, EXTERNAL:

Wing span	3·89 m (12 ft 6 in)
Length overall	1·98 m (6 ft 6 in)
Propeller diameter	1·83 m (2 ft 0 in)

WEIGHT:

Max launching weight	100 kg (220 lb)

PERFORMANCE:

Range	more than 27 nm (50 km; 31 miles)

Flight Systems Inc QF-86E drone conversion of the Canadair Sabre 5

Kaman STAPL demonstrator on mobile launch and recovery platform

DAST ARW-1 being retrieved after flight

LMSC YMQM-105 Aquila mini-RPV, under full-scale development for the US Army *(Howard Levy)*

Northrop BQM-74C target being air-launched from a US Navy TA-4J Skyhawk

NASA

NATIONAL AERONAUTICS AND SPACE ADMINISTRATION

1520 H Street NW, Washington, DC 20546

HUGH L. DRYDEN FLIGHT RESEARCH CENTER

PO Box 273, Edwards, California 93523
Telephone: (805) 258 3311

NASA MINI-SNIFFER II

The first two prototype vehicles, designated Mini-Sniffer I, were described and illustrated in the 1975-76 *Jane's*. A third prototype, designated Mini-Sniffer II, made its first flight in the Summer of 1975; a description of this can be found in the 1980-81 edition.

The Mini-Sniffer has been considered as a potential RPV to collect rock samples and data on Mars.

LANGLEY RESEARCH CENTER

Hampton, Virginia 23665
Telephone: (804) 827 1110

NASA PROJECT DAST

Project DAST (Drones for Aerodynamic and Structural Testing) is conducted jointly by NASA's Langley and Dryden Flight Research Centers, utilising modified Firebee IIs as testbeds for high-risk evaluation of various research wing configurations. Primary object of the programme is to demonstrate the ability of advanced flight control systems to control wing flutter.

The first drone, designated ARW-1 (for aeroelastic research wing), began flight testing in 1979, but was lost as a result of wing structural problems during a flight on 12 June 1980. A second type of wing (ARW-2), also of supercritical type, is scheduled for testing later.

NORTHROP

NORTHROP CORPORATION—VENTURA DIVISION

1515 Rancho Conejo Boulevard, Newbury Park, California 91320
Telephone: (805) 498 3131
Telex: 659 220
General Manager: Kent Kresa
Vice-President, Aeronautical Systems: Vincent W. Howard
Public Relations Manager, Tactical and Electronic Systems Group: Park H. Irvine

Northrop's Ventura Division designs and manufactures pilotless target aircraft and related equipment. It also produces glassfibre wing fairings for the Boeing 747 transport aircraft, as well as various parts and subassemblies for Northrop F-5E/F and McDonnell Douglas F-18A combat aircraft.

Northrop Ventura (formerly Radioplane) undertook the design, development and construction of its first radio controlled target drone in the mid-thirties. Since then it has become a leader in the field of pilotless aircraft. More than 77,000 drones have been delivered to the US military services and 25 allied nations.

NORTHROP KD2R5 BASIC TRAINING TARGET (BTT)

US Army designation: MQM-33C

This target drone is currently in use by the armed forces of 18 countries as a training device for ground-to-air gunnery and is used as a training target for surface-to-air missiles such as Seacat, Tigercat, Redeye, Blowpipe, Sparrow, Chaparral and Hawk.

Design began in 1946 and the prototype flew for the first time in 1947. Since then more than 65,000 of this type, including early KD2R versions, have been built and production continues.

Northrop is delivering to the US Army a modified version of the BTT, known as the **MQM-33C**, incorporating a Northrop-developed G-band command control system. By early 1981, Northrop had received contracts totalling $5 million for 89 MQM-33Cs and 18 ground control stations.

The following description applies to the standard version:

TYPE: Remotely controlled aerial target.

AIRFRAME: High-wing monoplane, of aluminium alloy construction. No dihedral. Wing incidence 1° at root, −2° at tip. Ailerons and elevator servo-operated by D-9 actuators.

POWER PLANT: One 67 kW (90 hp) Northrop O-100-3 flat-four engine, driving a two-blade fixed-pitch wooden propeller. Steel integral fuel tank in mid-fuselage, capacity 44 litres (11·6 US gallons). Refuelling point in fuselage forward of wing.

LAUNCH AND RECOVERY: Surface launch from land or ship, by either rotary or zero-length launcher. Recovery by parachute released by radio command. Engine is stopped and parachute deploys automatically in event of serious damage by gunfire, or loss of radio control or electrical power.

GUIDANCE AND CONTROL: AN/ARW-79 radio command guidance system, with automatic altitude hold. Visual or radar tracking (radar or FM type tracking systems or equivalent).

EQUIPMENT: 28V battery for all electrical power. L-band tracking system, smoke generating kit, tow banner and many other accessories available to customer's requirements. For radar augmentation, two wingtip reflector pods are optional.

DIMENSIONS, EXTERNAL:

Wing span	3·50 m (11 ft 6 in)
Wing area, gross	1·74 m² (18·7 sq ft)
Length overall	3·85 m (12 ft 7½ in)
Height overall	0·76 m (2 ft 6 in)

WEIGHTS:

Weight empty	123 kg (271 lb)
Max launching weight	181 kg (400 lb)
Max zero-fuel weight	133 kg (292 lb)
Max landing weight	154 kg (340 lb)

PERFORMANCE:

Max level speed at S/L and max cruising speed	195 knots (360 km/h; 224 mph)
Stalling speed	58 knots (108 km/h; 67 mph)
Max rate of climb at S/L	1,341 m (4,400 ft)/min
Service ceiling	more than 8,230 m (27,000 ft)
Range at S/L with max fuel	183 nm (338 km; 210 miles)

NORTHROP CHUKAR II

US military designation: MQM-74C

The MQM-74C is an improved version of the MQM-74A (1974-75 *Jane's*), evolved via an MQM-74B developmental model to meet requirements for a 500 knot (926 km/h; 576 mph) target. Since early 1974 more than 1,400 MQM-74Cs have been delivered to the US Navy, making more than 3,800 MQM-74 Chukars manufactured in all. Production of the MQM-74C is continuing.

Modified versions of the MQM-74C have been tested and operated as reconnaissance, electronic warfare and strike RPVs under the US Navy's Persistent Anti-Radiation Missile (PARM) and US Air Force's TED programmes. More recently, Northrop has developed a tactical reconnaissance and target acquisition version, known as the **Chukar-R**, fitted with a modular nose section mounting a Perkin-Elmer 35 mm camera and a Teledyne Brown TV camera, and incorporating preprogrammed navigation. Under US Navy contract, Northrop is producing a new version of the MQM-74C known as the **BQM-74C** (described separately), for use by the US Navy as a cruise missile simulator and standard Navy aerial training target.

The basic Chukar II/MQM-74C target aircraft was designed to meet requirements for a small, lightweight target for anti-aircraft gunnery, surface-to-air missile training and weapon system evaluation. Chukar II is used at the NATO Missile Firing Installation (NAMFI) on the island of Crete in the Mediterranean. Meteor of Italy (which see) was selected by NATO's Hawk management office to provide Chukar II services at the Salto di Quirra range in Sardinia.

TYPE: Radio-controlled recoverable target.

AIRFRAME: Shoulder-wing monoplane, of aluminium construction. No-dihedral detachable wings, each with electrically-actuated aileron. Tapered circular-section body, with underslung air intake duct. Inverted-Y tail unit, with 30° anhedral on tailplane, fixed vertical fin and electrically-actuated elevators.

POWER PLANT: One Williams International Corporation Model WR24-7 (J400-WR-401) turbojet engine, rated at 0·80 kN (180 lb st). Fuel tank in centre of fuselage.

LAUNCH AND RECOVERY: Zero-length launching by means of two Mk 91 Mod 0 JATO rockets and a ZL-5 launcher. Normal recovery by automatic drone pull-up followed by main parachute deployment, initiated automatically in emergencies such as interruption of continuous radio signal or loss of parachute command channel. Alternative method consists of direct main parachute deployment and is initiated automatically on loss of electrical power. Main parachute housed in fuselage immediately aft of wing, with automatic disconnect on impact.

GUIDANCE AND CONTROL: Radio command guidance system. Out-of-sight control by automatic stabilisation and command, with radar tracking; in-sight control with visual acquisition aids. Proportional feedback stabilisation and control system for pitch and bank. Engine throttle position, altitude hold initiation and recovery system initiation controlled by audio tone signals. Components include receiver, decoder, autopilot, aileron and elevator servos, altitude hold and airspeed pressure transducers. Command control antenna in upper forward fuselage.

EQUIPMENT (target): Electrical power from engine-driven alternator through a rectifier-regulator. 28V nickel-cadmium battery secondary power source used during glide. Onboard acquisition and tracking aids include fore and aft Luneberg lenses for passive radar augmentation, four wingtip-mounted Mk 28 Mod 3 infra-red flares, pyrotechnic infra-red plume augmentors, active L-band augmentation, and a smoke system, designed to improve visual detection. Main payload compartment is in front fuselage between control equipment bay and fuel tank. Improved Manoeuvrability Package (IMP) successfully flight tested in 1976 has closed-loop control device installed in flight control system, enabling Chukar II to perform constant *g* manoeuvres at any of five selected levels up to and including 6*g*. A low-cost infra-red tow target, for use with the Chukar II system, was successfully flight tested in early 1978. One can be attached to each wingtip and towed approx 30 m (100 ft) behind the Chukar.

DIMENSIONS, EXTERNAL:

Wing span	1·76 m (5 ft 9¼ in)
Length overall	3·87 m (12 ft 8½ in)
Body diameter	0·38 m (1 ft 3 in)
Height overall	0·71 m (2 ft 4 in)

WEIGHTS:

Weight empty	128·4 kg (283 lb)
Max launching weight	223 kg (492 lb)

PERFORMANCE:

Max level speed at 6,100 m (20,000 ft)	515 knots (954 km/h; 593 mph)
Max level speed at S/L	475 knots (880 km/h; 547 mph)
Econ cruising speed at S/L	250 knots (463 km/h; 288 mph)
Max rate of climb at S/L with full fuel	1,780 m (5,840 ft)/min
Service ceiling	12,200 m (40,000 ft)
Range at max speed at S/L	205 nm (380 km; 236 miles)
Range at max speed at 6,100 m (20,000 ft)	330 nm (611 km; 380 miles)
Range at econ cruising speed at S/L	245 nm (454 km; 282 miles)

NORTHROP BQM-74C

US Navy designation: BQM-74C

The BQM-74C is a US Navy target version of the MQM-74C Chukar II (which see) with added air-launch capability. It can be used as a cruise missile simulator, in training pilots for air-to-air combat, and as a target for anti-aircraft gunnery and surface-to-air missiles. Design began in November 1977, and construction of 16 pre-production examples started in September 1978. Navy technical and operational evaluation tests of the basic BQM-74C configuration were completed successfully in 1980. Technical evaluation of the **MSR** (Mobile Sea Range) version was due for completion in the first half of 1981, and deliveries of production targets to the US Navy were scheduled to begin in the second half of that year.

TYPE: Radio controlled or programmable automatic recoverable target.

AIRFRAME: Shoulder-wing monoplane of aluminium construction. Northrop G-9224-080 wing section of 8% thickness/chord ratio. No-dihedral non-swept wings, each with electrically-actuated aileron. Semi-monocoque aluminium body, similar to that of MQM-74C, houses all equipment, power plant and fuel tank. Nose and tail skins removable for access to equipment and power plants. Underslung engine air intake. Inverted-Y aluminium tail unit, comprising fixed vertical fin, fixed tailplane halves (anhedral 30°) and two electrically actuated elevators.

POWER PLANT: One Williams International Corporation Model WR24-7A (J400-WR-402) turbojet engine, rated at 0·80 kN (180 lb st). Pressurised fuel tank in centre of fuselage.

LAUNCH AND RECOVERY: Can be launched from ground or shipborne launcher in same manner as MQM-74C. Can also be air-launched from Grumman A-6E Intruder or McDonnell Douglas TA-4J Skyhawk (one under each wing); is also compatible with underwing launchers of Lockheed DP-2 Neptune or DC-130A Hercules. Parachute recovery, on land or from water, as for MQM-74C.

GUIDANCE AND CONTROL: Out-of-sight control by automatic stabilisation and command; radar tracking in-sight control, with visual acquisition aids. Proportional feedback stabilisation and control system for pitch and bank. Engine throttle position, altitude hold initiation and recovery initiation are controlled by audio tone signals. Components include a Northrop digital avionics processor, vertical gyro, Motorola AN/DKW-3 integrated target control system, aileron and elevator servos, and altitude hold pressure transducer.

EQUIPMENT: Electrical power from engine-driven alternator through a rectifier-regulator. 28V nickel-cadmium battery secondary power source. Onboard acquisition and tracking aids include fore and aft Luneberg lenses, for passive radar augmentation, and a smoke system. Main payload compartment is in forward fuselage section. Equipment includes locator beacon, radar altimeter, seeker simulator (to duplicate cruise missile emissions), radar transponder for IFF, and scoring. Provisions include flight profile programmer with UHF command override; radar altimeter; active J-band, B-band, L-band and X-band radar augmentation; and Tacan receiver. System also includes payload kits for mobile sea range (MSR) operations.

DIMENSIONS, EXTERNAL:

Wing span	1·76 m (5 ft 9·4 in)
Wing area, gross	0·74 m² (8·0 sq ft)
Length overall	3·94 m (12 ft 11·4 in)
Height overall	0·72 m (2 ft 4·3 in)

WEIGHTS AND LOADING:

Basic operating weight empty	133 kg (294 lb)
Max payload	78·5 kg (173 lb)
Max launching weight	204 kg (451 lb)
Max wing loading	273 kg/m² (56 lb/sq ft)

PERFORMANCE (at max T-O weight):

Max level speed at 6,100 m (20,000 ft)	500 knots (927 km/h; 576 mph)
Econ cruising speed	300 knots (555 km/h; 345 mph)
Service ceiling	9,145 m (30,000 ft)
Max range	450 nm (833 km; 518 miles)

ROCKWELL INTERNATIONAL

ROCKWELL INTERNATIONAL CORPORATION, NORTH AMERICAN AIRCRAFT DIVISION

PO Box 92098, Los Angeles, California 90009

ROCKWELL INTERNATIONAL/NASA/US AIR FORCE HiMAT

HiMAT (Highly Manoeuvrable Aircraft Technology) is a programme evolved by NASA's Dryden Flight Research Center at Edwards AFB, California, and the US Air Force Flight Dynamics Laboratory (AFFDL) at Wright-Patterson AFB, Ohio. Its basic purposes are to speed up the progress of advanced design technology into the flight test phase; to assist designers in taking larger technological steps forward between generations of aircraft; and, more specifically, to provide a low-cost, low-risk means of testing the advanced manoeuvring capability of future aircraft. NASA began evolving control techniques for the HiMAT programme in 1975, first with ⅜-scale unpowered glassfibre models of the F-15 fighter and later with two modified BQM-34F Firebee IIs.

After receipt of programme proposals from Grumman Aerospace, McDonnell Aircraft Co and Rockwell International, NASA announced in October 1975 the award of an $11·8 million contract to Rockwell for the design and construction of two prototype HiMAT remotely piloted research vehicles (RPRVs). Major subcontractors are Teledyne Ryan Aeronautical (airborne flight control system) and Hydraulic Research Corpn (servo actuators).

To meet the requirements of the programme, HiMAT's design consists of a basic core vehicle which includes the engine and all essential subsystems. To the core vehicle are added, as modular units, the main wings, canard surfaces, tail surfaces, and engine intake and afterburner/exhaust structures. In this way the modular components can be replaced during the programme, at minimum cost, with others of alternative design. Among these is expected to be a so-called '2D' vectored-thrust exhaust nozzle; other features to be tested include advanced supercritical wings; variable-camber wings; deformable, self-trimming outer wings; CCV (control configured vehicle) techniques; a digital fly-by-wire system; and a variable-thrust engine control system.

The first HiMAT RPRV was delivered to NASA on 7 March 1978 and the second on 12 June 1978. The first free flight was made on 27 July 1979, initiating the 'envelope expansion' phase of the flight test programme.

During the first eight flights, HiMAT achieved speeds up to Mach 0·93 and a max load factor of 8*g*. Control system modifications were being made in the early Spring of 1981, prior to the next phase of testing. The second HiMAT was scheduled to become available by mid-1981, and flights up to Mach 1·4 will be attempted in the next phases of the programme.

Launch of a Northrop MQM-74C Chukar II at the Pacific Missile Test Range

Northrop Chukar-R, with interchangeable nosecones for tactical reconnaissance or target use

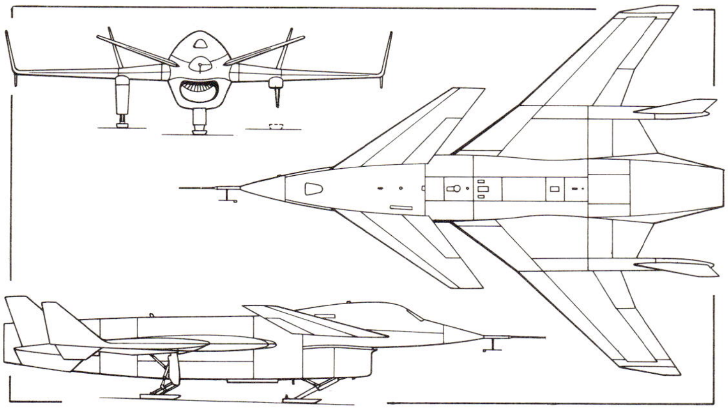

Rockwell International HiMAT research RPV *(Michael A. Badrocke)*

MQM-33C version of the Northrop Basic Training Target (BTT)

One of the two Rockwell HiMAT research RPV prototypes

The HiMAT is flown by a pilot in a ground-based control centre in which all of the aeroplane's flight parameters are displayed via telemetry from the RPRV. The ground-based cockpit is equipped with conventional aircraft controls. The pilot is aided in the landing manoeuvre by a televised view of the landing area taken with a TV camera mounted in the HiMAT.

The following description applies to the HiMAT prototypes:

Type: Remotely piloted research vehicle.

Airframe: Mid-wing monoplane, of roughly tandem-delta configuration, with sharply-swept main wings and foreplanes having graphite composite skins. Glassfibre outer leading-edges; remainder of wings of aluminium, titanium and other materials. Main wings have neither dihedral nor anhedral, and have ailerons, elevons and elevators on the trailing-edges, and NASA 'winglets' at tips. Canard surfaces have marked dihedral and are fitted with elevators. Semi-monocoque area-ruled fuselage, with graphite composite skin. Twin, swept, outward-canted pivoted fins on short booms extending from trailing-edges of main wings at approx mid-span. Graphite composite fin skins. Retractable tricycle landing gear, of skid type for landing on dry lake bed at Edwards AFB. All units retract rearward, main units to form continuous fairing with wing/tail booms, nose unit into underside of engine air intake trunk.

Power Plant: One General Electric J85-GE-21 turbojet engine (15·6 kN; 3,500 lb st dry and 22·25 kN; 5,000 lb st with afterburning), mounted centrally in fuselage. Fuel capacity (JP-5) 295 kg (650 lb).

Launch and Recovery: Air-launched from B-52 carrier aircraft. Recovered by conventional runway landing.

Guidance and Control: Primary control from ground console, by TV, telemetry, and radar link with onboard systems. If ground control is lost, backup inputs from the RPRV will be relayed to a TF-104G chase-plane. On occasions when the TF-104G is out of control range, the HiMAT has an onboard self-righting system that will bring the RPRV into constant-altitude, orbiting, subsonic flight until the former can resume control.

Equipment: Onboard equipment includes flight and control parameter sensors, 164 research instrumentation sensors (pressures, strain gauges, accelerometers), signal processors, data link to ground and chase-plane, backup flight control system computer, and TV camera.

Dimensions, external:
- Wing span: 4·755 m (15 ft 7¼ in)
- Length overall, incl probe: 6·86 m (22 ft 6 in)
- Height overall: 1·31 m (4 ft 3·6 in)

Weights:
- Weight empty: 1,200 kg (2,645 lb)
- Max air-launching weight: 1,528 kg (3,370 lb)
- Thrust/weight ratio: approx 1

Performance (estimated):
- Max level speed: Mach 1·6
- Touchdown speed: 180 knots (333 km/h; 207 mph)
- Average research flight duration: 30 min
- *g* limit: +12

RS
RS SYSTEMS
Beltsville, Maryland

RS RCMAT
US Army designation: FQM-117A

The US Army has ordered 14,000 examples of this miniature expendable aerial target, as a training vehicle for air defence gunners, tank and armoured vehicle gunners, air-to-air gunners and infra-red systems tracking; and for payload missions. At a distance of 200 m (650 ft) and a speed of 80 knots (148 km/h; 92 mph), the RCMAT simulates a full-size aircraft 1,200 m (3,935 ft) away, travelling at 480 knots (890 km/h; 550 mph).

Type: Radio controlled miniature aerial target.

Airframe: Simple styrofoam structure, of cropped-delta monoplane planform with V tail surfaces. No landing gear.

Power Plant: One 0·75 kW (1 hp) K & B 0·6 cu in two-stroke model aircraft engine.

Launch and Recovery: Hand-launched; non-recoverable (expendable).

Guidance and Control: Radio command guidance system.

Equipment: Aluminium tube along top of fuselage provides infra-red target signature.

Dimensions, external:
- Wing span: 1·60 m (5 ft 3 in)
- Length overall: 0·81 m (2 ft 8¹/₁₆ in)
- Height overall: 0·19 m (7⅝ in)

Weight:
- Mission equipment: 1·4 kg (3 lb)

Performance:
- Max level speed: 87 knots (161 km/h; 100 mph)
- Min flying speed: 25 knots (47 km/h; 29 mph)
- Service ceiling: 3,050 m (10,000 ft)
- Endurance at max speed: 10 min

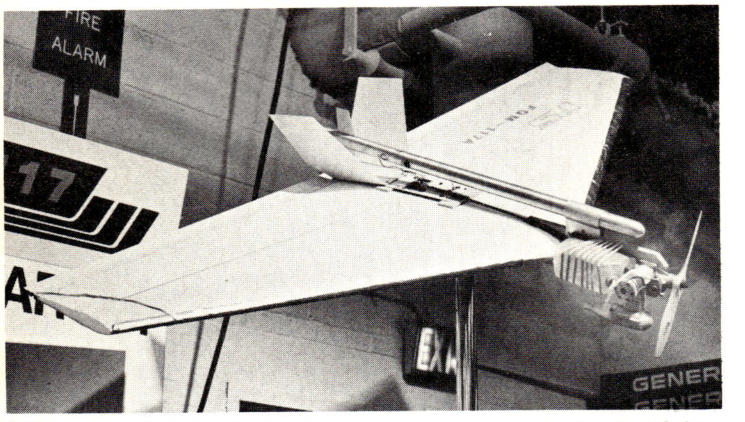

RS Systems RCMAT (FQM-117A) low-cost expendable target for the US Army *(Howard Levy)*

Sperry PQM-102B target drone with underwing pods

SPERRY

SPERRY FLIGHT SYSTEMS DIVISION, SPERRY CORPORATION

PO Box 21111, Phoenix, Arizona 85036
Telephone: (602) 869 2311
Communications Representative, Defense and Space Systems Division: Brent Hosage

SPERRY (CONVAIR/GENERAL DYNAMICS) PQM-102 DELTA DAGGER

Conversion of US Air Force F-102A fighters for remotely piloted operation, under the Pave Deuce programme, has resulted in three versions, as follows:

QF-102. Prototypes, with provision for manned operation. Five converted by Fairchild Aircraft Service Division. Used for captive missile evaluation, crew training and systems checkout.

PQM-102A. Initial unmanned version, of which 63 produced under subcontract to Sperry by Fairchild Aircraft Service Division. Used for development testing, including pre-programmed repeatable manoeuvres. Details in 1977-78 *Jane's*.

PQM-102B. Current unmanned version; 148 being converted by Sperry Flight Systems. Retains same operational characteristics as PQM-102A except for one low-altitude mode, which can be implemented for special missions by installing radar altimeter. Improvements made to scoring system; smoke, brake and explosive destruct systems; and lift compensation in altitude hold mode. Modification of basic aircraft; avionics and operational complexity; and throttle quadrant, are all simplified. Additional batteries are reduced to one; second transformer installation is deleted, a slightly modified form of the basic AC system in the F-102A being used; aircraft wiring is simplified; and engine start is done in cockpit, eliminating need for external control unit. Flight control avionics are located in the nose bay. Manoeuvres can be initiated and programmed from fixed ground control site, using new PQM-102B auxiliary panel.

The PQM-102s are the first-ever fighter aircraft converted for drone duties with no provision whatever for manned operation, and cannot be flown except under remote control. Missiles fired at them since operation began in October 1974 have included AIM-9J/L Sidewinder, AIM-7E/F Sparrow, Stinger and Patriot.

The following description applies to the PQM-102B:

Airframe: As F-102A (see 1961-62 *Jane's*).

Power Plant: One Pratt & Whitney J57-P-23A turbojet engine, rated at approx 45·37 kN (10,200 lb st) dry and 71·17 kN (16,000 lb st) with afterburning.

Launch and Recovery: Normal runway T-O and landing.

Guidance and Control: Dual Vega command guidance and telemetry systems. Fully redundant digital tracking and control system for command/telemetry link, in conjunction with AN/FPS-16 ground-based range radar. Simultaneous control of two targets has been demonstrated.

Special Equipment: Upper and lower fuselage avionics bays, missile bay, cockpit and two pylon-mounted underwing pods available for mission equipment. Digidops miss-distance scoring system standard (antennae aft of cockpit, under fuselage, and in fairing each side of tailpipe); installation and test of vector miss-distance indication (VMDI) is planned. Two types of scoring camera operational, covering forward and aft areas to provide missile approach angle, velocity and miss distance; these are mounted in front and at rear of cockpit. Flight control stabilisation system (FCSS), comprising flight reference computer, interface coupler and air data computer, provides eight longitudinal/vertical and four lateral/directional modes, and interface between aircraft systems and command/telemetry system. Automatic control modes are provided for take-off, loss of command carrier, take-off abort, and other safety modes. In case of primary FCSS failure, a backup system provides independent control in pitch, roll and yaw, and of thrust and other vital functions. This backup is operated by independent electrical and hydraulic power systems. Redundant AC and DC power systems, and redundant dual autopilots. Manoeuvre programmer can be pre-programmed for three or four manoeuvres, to run in any required sequence, and provides backup for FCSS. Command/telemetry system provides for four proportional (pitch, roll and two spare) and 48 discrete commands, 22 proportional and 40 discrete telemetry channels; 11 additional commands (nine discrete and two proportional) available for standard payload control. Manoeuvre destruct and explosive destruct systems incorporated. Visual augmentation (smoke) system, operable at any altitude or power setting. Radar and infra-red augmentation not required, due to size of aircraft.

Dimensions:
- Wing span: 11·62 m (38 ft 1½ in)
- Wing area, gross: 64·57 m² (695·0 sq ft)
- Length overall: 20·84 m (68 ft 4⅔ in)
- Height overall: 6·46 m (21 ft 2½ in)
- Payload volume (max, excl wing pods): 5·66 m³ (200 cu ft)

Weights:
- Basic mission weight: 9,076 kg (20,010 lb)
- Payload (max): 1,905 kg (4,200 lb)
- Mission operational T-O weight: 14,186 kg (31,276 lb)

Performance:
- Max speed at altitude: Mach 1·2 (688 knots; 1,274 km/h; 792 mph)
- Operating height range: 61 m (200 ft) to 17,070 m (56,000 ft)
- Range (dictated by effective control range of guidance radar): more than 174 nm (322 km; 200 miles)
- Normal endurance of mission: 40-55 min
- *g* limit: +8

SPERRY (NORTH AMERICAN) QF-86F SABRE

Under a $ 350,000 contract announced on 18 August 1980, Sperry Flight Systems is converting ten North American F-86F Sabre jet fighters into QF-86F target drones for the US Naval Weapons Center at China Lake, California. The conversions are due to be completed by October 1981. Follow-on orders to convert an additional 90 F-86Fs are in prospect.

The QF-86F drone programme is similar to other current Sperry programmes to convert F-100 and F-102A fighters (see accompanying entries) for the US Air Force.

Airframe: As F-86F Sabre (see 1961-62 *Jane's*).

Power Plant: One 26·56 kN (5,970 lb st) General Electric J47-GE-27 turbojet engine.

Dimensions, external:
- Wing span: 11·91 m (39 ft 1 in)
- Wing area, gross: 29·12 m² (313·4 sq ft)
- Length overall: 11·43 m (37 ft 6 in)
- Height overall: 4·49 m (14 ft 8¾ in)

Weights (F-86F):
- Weight empty: 5,046 kg (11,125 lb)
- T-O weight ('clean'): 6,893 kg (15,198 lb)

Performance (F-86F):
- Max level speed at S/L: 588 knots (1,091 km/h; 678 mph)
- Max rate of climb at S/L, AUW of 6,446 kg (14,212 lb): 2,987 m (9,800 ft)/min
- Service ceiling, AUW as above: 15,120 m (49,600 ft)

SPERRY (NORTH AMERICAN) QF-100 SUPER SABRE

Conversion of US Air Force /Air National Guard F-100 fighter-bombers for remotely piloted operation as QF-100 aerial targets is being undertaken by Sperry under a full-scale engineering development (FSED) contract from the US Air Force Armament Development and Test Center, Eglin AFB, Florida. This full-scale aerial target (FSAT) programme is for a multi-service interim target to provide air-to-air and ground-to-air missile evaluation and combat crew training. The QF-100 will fill the gap between the PQM-102 (which see) and the recommended target conversion of the McDonnel Douglas F-4 Phantom, which will not be available until well into the 1980s.

The initial full-scale engineering development programme involves the conversion of nine Super Sabres into four different configurations. Of these, configuration No. 2, converted from the single-seat F-100D, will be the standard US Air Force target version; No. 1 will incorporate additional cockpit controls to permit evaluation of system performance from within the cockpit; No. 3 is the same configuration as No. 2, except that it is converted from the two-seat F-100F; No. 4 is the target configuration for the US Army, and incorporates a drone formation control system (DFCS) for multiple-target missions. The FSED programme includes DT & E (development, test and evaluation) and IOT & E (initial operational test and evaluation). Conversions are carried out at Sperry's facility at Phoenix-Litchfield Airport (formerly the Naval Air Facility, Litchfield Park) in Goodyear, Arizona.

Deliveries began on 13 March 1981 to Tyndall AFB, Panama City, Florida, for US Air Force DT & E; that for the DFCS will be carried out at Holloman AFB, New Mexico. The development contract includes an option for up to 300 'production' QF-100D/F conversions, delivery of which would begin in March 1983. Initial operational capability is due to be achieved by June 1983.

The QF-100 uses current Sperry PQM-102 drone ground control and test equipment, as well as many PQM-102 airborne subsystems. While sharing a common conversion and operational scheme with the PQM-102 series, the QF-100 utilises an SDP-175 digital flight control computer instead of analogue computers for ease of testing and flexibility for future growth of operational modes.

Airframe: As F-100D/F (see 1961-62 *Jane's*).

Power Plant: One Pratt & Whitney J57-P-21A turbojet, rated at 75·40 kN (16,950 lb st) with afterburning.

Launch and Recovery: Normal runway T-O and landing.

Guidance and Control and Special Equipment: Generally similar to that of PQM-102B (which see), using existing drone tracking and control system for the command/telemetry link. Major difference from PQM-102 is the use of a digital flight control computer (FCC) system, which incorporates air data sensors, an SDP 175 processor, analogue/digital and digital/analogue converters, a power supply and the necessary interface electronics. Use of FCC permits automatic checkout of many primary autopilot functions; it also provides a flexible system for incorporating target system functions, or for adaptation to other target programmes. Digidops scoring system, as in PQM-102, is fitted; VMDI (vector miss-distance indicator) scoring system, currently under development, will provide a directional parameter to help further in evaluating missile performance. A DLQ-3B ECM pod and ALE-40 infra-red/chaff pod are incorporated to provide realistic evaluation of missile performance against anticipated countermeasures. Drone formation control system incorporated in US Army version to permit formation flight of two or more targets, to provide a realistic challenge for missiles equipped with a seeker head.

Dimensions, external:
- Wing span: 11·82 m (38 ft 9⅓ in)
- Wing area, gross: 35·79 m² (385·2 sq ft)
- Length overall, incl probe: 16·54 m (54 ft 3 in)
- Height overall: 4·95 m (16 ft 2⅔ in)

Weight:
- Mission operational T-O weight: 14,060 kg (31,000 lb)

Performance:
- Max speed at altitude: Mach 1.2 (688 knots; 1,274 km/h; 792 mph)
- Range, nominal (guidance radar range-limited): 120 nm (222 km; 138 miles)
- Normal mission endurance: 40-55 min
- *g* limit: +8

Sperry QF-100D target drone conversion of an F-100D Super Sabre

Sperry QF-86F conversion of an F-86F Sabre

Teledyne Ryan ZBQM-111A Firebrand naval supersonic target

Model of the Teledyne Ryan Model 305 Firebolt

BQM-34S Firebee I target drone of the US Navy

TELEDYNE RYAN

TELEDYNE RYAN AERONAUTICAL

2701 Harbor Drive, San Diego, California 92138
Telephone: (714) 291 7311

President: R. S. McCarter

Senior Vice-President, Administration: C. E. McGill

Vice-Presidents:

D. L. Arney (Administration and Public Relations)
E. C. Chapman Jr (Contracts)
F. E. Oldfield (Aerospace Engineering)
A. C. Richards (Subcontract Programmes)
R. R. Schwanhausser (RPV Programmes)
H. B. Starkey (Missiles and Simulator Programmes)
G. E. Timmons (Target Programmes)

The former Ryan Aeronautical Company was an indirect successor to Ryan Airlines Inc, which produced the aeroplane in which Charles A. Lindbergh made the first nonstop flight from New York to Paris in 1927. It was renamed Teledyne Ryan Aeronautical in December 1969.

Current activities include the design, production and field operation of high-performance aerial jet targets and RPV systems. Teledyne Ryan is also a major subcontractor to Hughes Helicopters on the AH-64 programme, for which it builds the complete airframe.

Major production items at Teledyne Ryan's plant for many years have been the Firebee jet-powered targets and special-purpose vehicles (pre-programmed and remotely piloted) for various types of reconnaissance mission. The supersonic Firebee II (286 built: see 1979-80 *Jane's*), although no longer in production, remains in service with the US Navy (BQM-34E and T) and Air Force (BQM-34F); a total of 167 remained in the US government inventory in February 1981.

TELEDYNE RYAN MODEL 124 FIREBEE I

US Air Force designation: BQM-34A
US Army designation: MQM-34D
US Navy designations: BQM-34A and BQM-34S

The Firebee I remotely piloted vehicle was developed as a joint US Air Force/Army/Navy project, in collaboration with the US Air Force Air Research and Development Command.

Glide flight tests of the original version began in March 1951, and the first powered flights were made that Summer at the US Air Force Holloman Air Development Center, Alamogordo, New Mexico. A total of 1,280 of these early Q-2A and KDA versions were built eventually for all three US services and for the Royal Canadian Air Force, and full details of these can be found in previous editions of *Jane's*.

Development of the current **BQM-34A** (originally Q-2C) improved Firebee began on 25 February 1958. Construction of the prototype started on 1 May and it flew for the first time on 19 December 1958. The first production model flew on 25 January 1960.

By February 1981, a total of 6,151 Firebee Is had been produced, including the early Q-2A and KDA versions. These included Firebees supplied under contract to NATO for use in a missile test and evaluation programme, and a contract from the Japan Defence Agency for Firebees to support the training of missile and gunnery crews. The latter are being built by Fuji (which see). The latest US contract, for 189 more Firebee Is, includes 127 powered by 10·9 kN (2,450 lb st) General Electric J85-GE-7 turbojets.

Current Firebee targets for the US Navy incorporate a Motorola integrated track and control system (ITCS) and have the designation **BQM-34S**; US Air Force BQM-34A targets have a Vega Drone Tracking and Control System (DTCS).

By 1 January 1981 Firebee I targets had provided more than 29,000 flights in support of weapon system and target research, development, test, evaluation, quality assurance, training and annual service practices conducted by the US Army, Navy and Air Force, and foreign governments. Since the current Firebee I and its predecessor models have been in operational use, target presentations have been made to virtually every surface-to-air and air-to-air weapon system in the US arsenal. To reduce the vulnerability of the target, and increase its cost-effectiveness, a 'non-kill' environment has been created and extensive use made of infra-red and/or radar augmented towed targets ('Towbees') or cloth banners, towed behind the Firebee on cables.

RPVs (remotely piloted vehicles) using airframes developed from that of the Firebee I were described in the 1980-81 and earlier editions of *Jane's*.

The following details refer to the standard BQM-34A target vehicle:

Type: Remotely piloted jet target vehicle.

Airframe: Cantilever mid-wing monoplane, of aluminium alloy semi-monocoque construction. Three-spar wings, incorporating leading-edge droop. No dihedral or incidence. Sweepback at quarter-chord 45°. Single-spar ailerons, with Lear servo-actuators. Wingtips detachable. Provision for wingtip extensions. Tapered, circular-section body, with chemical-etched components. Glassfibre tailcone and nose section. Keel under central portion, to absorb landing impact. All tail surfaces swept 45° at quarter-chord. Multi-spar fin, with glassfibre tip housing guidance and control antenna. Trim rudder operated electrically by Bendix actuator. Single-spar tailplane, with glassfibre tips housing radar echo enhancing antennae. Ventral fin under tailcone, aft of main tail unit. Magnesium elevators powered by Lear servo.

Power Plant: One 7·56 kN (1,700 lb st) Teledyne CAE J69-T-29 turbojet engine. Integral fuel tank in forward fuselage, capacity 378 litres (100 US gallons). Provision for one 94·5 litre (25 US gallon) auxiliary fuselage tank and one 378 litre (100 US gallon) drop-tank under each wing. Oil capacity 5·75 litres (1·5 US gallons).

Launch and Recovery: Either air-launching, from suitably-modified aircraft, or ground-launching, using 50·3 kN (11,300 lb st) (nominal) solid-propellant JATO bottle. US Navy has launched BQM-34As from ships under way at up to 15 knots (27·5 km/h; 17 mph). Two-stage parachute recovery system operates automatically in event of target hit, loss of radio wave carrier from remote control station, engine failure, or upon command by remote control operator. To prevent damage by dragging, recovery system incorporates disconnect which releases parachute from Firebee on contact with ground or water.

Guidance and Control: Remote control methods include choice of radar, radio, active seeker and automatic navigator, developed and designed by Teledyne Ryan. Normal method is through UHF radio link using AN/FRW-2 or SRW-4 ground transmitter and AN/DRW-29 airborne receiver. Target can be controlled either from manned aircraft or from surface station. Remote command includes activation of special scoring and augmentation equipment in target. Onboard beacon facilitates radar tracking; provision to install telemetry system to relay pertinent flight data to controller. Basic commands consist primarily of on/off functions, received by onboard radio receiver and relayed to appropriate subsystem. Motorola ITCS (integrated tracking and control system) in BQM-34S. BQM-34As currently being fitted with Vega DTCS (drone tracking and control system). Other types of remote command and tracking system can include microwave command and guidance system to control Firebee beyond line-of-sight from ground station through airborne relay station. Operational Firebees can be equipped with increased manoeuvrability flight control system for tactical air combat simulation which gives target capability to perform 4, 5 or 6*g* manoeuvres. Other systems include active and passive radar augmentation, and afterburning plume devices. Radar Altimeter Low Altitude Control System (RALACS), when added to Firebee I control system, permits precision low-altitude flights at 15 m (50 ft) over water and 30 m (100 ft) over land.

Systems: Electrical power only. Primary power furnished by 28V 200A DC engine-driven generator. Power for control systems furnished by 400Hz 115V 250W AC inverter; 28V 12·5Ah lead-acid battery provides power for electrical devices of recovery system and for control during pre-landing glide phase.

AVIONICS AND EQUIPMENT: AN/DRW-29 radio receiver with Dorsett TM-4-31A telemetry system, or DKW-2 guidance transponder, or DTCS Model 685-2 guidance system. A/A37G-3 or A/A37G-8 flight control system. Wide range of 'building block' operational equipment includes visual or radar-reflecting banner targets; radar or infra-red Towbee towed targets or tow target Doppler 'bird'; two underwing drop-tanks, 500 lb bombs or bomblet dispensers; AN/ALE-33 or other ECM containers; wingtip tow launchers, camera pods, scoring equipment, flares or other forms of infra-red augmentation, or reflector pods for radar augmentation. BQM-34A can be equipped with adjustable travelling wave tube amplifiers for use as radar echo enhancers in L, S, X and C frequency bands.

DIMENSIONS, EXTERNAL:

Wing span	3·93 m (12 ft 10·8 in)
Wing area, gross	3·34 m² (36·00 sq ft)
Length overall	6·98 m (22 ft 10·8 in)
Body diameter	0·94 m (3 ft 1·2 in)
Height overall	2·04 m (6 ft 8·4 in)

WEIGHTS:

Weight empty	680 kg (1,500 lb)
Basic gross weight	934 kg (2,060 lb)
Max launching weight	1,134 kg (2,500 lb)

PERFORMANCE:

Never-exceed speed Mach 0·96 (635 knots; 1,176 km/h; 731 mph at 15,240 m; 50,000 ft)

Max level speed at 1,980 m (6,500 ft) 600 knots (1,112 km/h; 690 mph)

Max cruising speed at 15,240 m (50,000 ft) at 816 kg (1,800 lb) AUW 547 knots (1,015 km/h; 630 mph)

Stalling speed, power on, at 816 kg (1,800 lb) AUW 101 knots (187 km/h; 116 mph)

Max rate of climb at S/L at 1,000 kg (2,200 lb) AUW 4,875 m (16,000 ft)/min

Operating height range 15 m-18,300 m (50 ft to more than 60,000 ft)

Endurance at 15,240 m (50,000 ft), incl 2 min 40 s glide after fuel expended 75 min 30 s

Max range 692 nm (1,282 km; 796 miles)

Flotation time with 25% fuel 24 h

TELEDYNE RYAN MODELS 147 and 255

US Air Force designation: AQM-34

The Model numbers 147 and 255, and the basic US Air Force designation AQM-34, encompass a large family of surveillance, reconnaissance and ECM RPVs evolved from the subsonic Firebee 1 target.

Details of all AQM-34s, and their uses from 1964-74, can be found in the 1975-76 and later *Jane's* up to and including the 1980-81 edition. They are no longer in USAF use, but are available for reactivation if required.

TELEDYNE RYAN FIREBRAND

US Navy designation: ZBQM-111A

Initial US Navy funding of approx $3 million was announced in May 1977 for this new supersonic target, designed to simulate the threat posed by Soviet anti-shipping missiles. The total value of the development and construction programme will be $41·7 million.

The initial contract calls for the completion of nine targets: six for research and development and three for special tests. The Firebrand will serve as a realistic 'enemy' in testing US Navy shipboard defence weapons under simulated combat conditions. Subject to the satisfactory completion of flight testing, which is due to begin in March 1983, production Firebrands are expected to be ordered in early 1984.

TYPE: Recoverable supersonic target drone.

AIRFRAME: Shoulder-wing monoplane, with small delta wings (approx 63° sweep on leading-edges) set well back on tapered, circular-section body. Built mainly of aluminium alloy, with wings of aluminium honeycomb and bonded stainless steel skins. Differentially-operating elevons and rudder. Ram-air intake in base of fin.

POWER PLANT: Two Marquardt ramjets, mounted externally one on each side of fuselage at rear. Fuel tank in centre of fuselage. Thiokol booster rockets.

LAUNCH AND RECOVERY: Air or surface launch. Fail-safe parachute recovery system.

GUIDANCE AND CONTROL: Pre-programmed radio command guidance system, using AN/AYK-14 flight control computer. Provision for manual override.

EQUIPMENT: Ram-air turbogenerator; batteries; Honeywell 7193 altimeter; Tacan; drag 'chute, recovery parachute and flotation bag; locator beacon; and destruct receiver. Mission equipment, in 0·17 m³ (6 cu ft) compartment in nose, includes ECM; active and passive radar augmentation; IR augmentation; expendable H-, I- and J-band emitter; and scoring system.

DIMENSIONS, EXTERNAL:

Wing span	2·74 m (9 ft 0 in)
Wing area, gross	3·72 m² (40·0 sq ft)
Length overall	10·36 m (34 ft 0 in)
Body diameter	0·71 m (2 ft 4 in)

WEIGHTS (surface launch):

Fuel	998 kg (2,200 lb)
Mission equipment	113 kg (250 lb)
Max gross weight (incl booster)	2,812 kg (6,200 lb)

TELEDYNE RYAN MODEL 305 FIREBOLT

US Air Force designation: AQM-81A

Teledyne Ryan is conducting, for the US Air Force Armament Division at Eglin AFB, Florida, full-scale development of the Firebolt High Altitude High Speed Target (HAHST). HAHST is a continuation and refinement of the High Altitude Supersonic Target (HAST) which was an advanced development programme conducted by Beech Aircraft Corporation. Teledyne Ryan was the winner in a competitive procurement bid for the full-scale engineering development of this target, and on 10 December 1979 was awarded a contract to produce and deliver nine test vehicles. The four-year programme includes a series of 30 test flights, which will begin in August 1982 and end in December 1983.

The flight performance envelope of the Firebolt covers a range from Mach 1·2 at 12,200 m (40,000 ft) to Mach 4·0 at 30,500 m (100,000 ft).

TYPE: Supersonic air-launched recoverable target drone.

AIRFRAME: Mid-wing monoplane, with slim-delta main wings swept 75° on leading-edges and of constant thickness except for leading-edges. Full-span aileron on each trailing-edge. Arrow-planform foreplanes for longitudinal control. Fixed endplate fin at tip of each main wing. Cylindrical body, with 3·5 calibre von Kármán nose section and conical boat-tail section. Ventral fin under rear fuselage.

POWER PLANT: CSD hybrid rocket engine. Propellant is polybutadiene and polymethyl-methacrylate, with IRFNA oxidiser. System is inherently safe, since propellants will not burn unless external ignition is applied. Engine is throttleable, with thrust variable from 0·53 to 5·34 kN (120-1,200 lb st). The 0·33 m (13 in) thrust chamber forms integral part of fuselage assembly. Oxidiser pressurisation and electrical power provided by ducted power unit. This unit, developed by Marquardt, is powered by ram-air turbine with air intake and exit on lower side of fuselage mid-section. Manoeuvring requirements dictate positive expulsion system for oxidiser.

LAUNCH AND RECOVERY: Air-launched from carrier aircraft at speeds between Mach 1·2 and 2·5. Recovery by 13·72 m (45 ft) diameter parachute from land, water or mid-air.

GUIDANCE AND CONTROL: Manoeuvres can be either pre-programmed or initiated via ground command radio link. Manoeuvres of between 5*g* at 10,670 m (35,000 ft) and 1·15*g* at 27,400 m (90,000 ft) are part of flight test programme; vehicle also capable of 'S' and 180° turns in horizontal plane and altitude changes in vertical plane.

SPECIAL EQUIPMENT: The flight test vehicles include provisions for a Scaler miss-distance scoring system and a point source radar augmentation system.

DIMENSIONS, EXTERNAL:

Wing span	1·02 m (3 ft 4 in)
Wing area (total exposed)	0·97 m² (10·44 sq ft)
Length	5·18 m (17 ft 0 in)
Height (stabiliser)	0·66 m (2 ft 2 in)
Body diameter	0·33 m (1 ft 1 in)

VOLUME:

Mission equipment	0·033 m³ (1,970 cu in)

WEIGHTS:

Launching weight	558 kg (1,231 lb)
Propellant	29·7 kg (65·5 lb)

PERFORMANCE:

Endurance at Mach 3	5 min

US DEPARTMENT OF DEFENSE

Major areas in which the Department of Defense has concentrated RPV research during recent years include the testing of a missile- or aircraft-launched mini-RPV; development and test of shipboard launch and recovery techniques; battery- and solar-powered RPVs; miniature infra-red and laser designator and rangefinder systems that will enable a mini-RPV to find targets and guide weapons at night; and data links resistant to hostile jamming. The use of mini-RPVs for hostile weapon and radar location is foreseen, using the air vehicle as a platform equipped with a lightweight high-performance radar, or a laser line-scanner, and a direct line-of-sight flash detector, or with an explosive charge to destroy a target. A small, low-cost system for locating a mortar projectile is being designed, and investigation of radar options for projectile tracking is in progress.

In addition to drone/RPV programmes for which hardware contracts have been awarded (described under the appropriate contractor's heading in this section), the following are among the more important programmes receiving attention by the three US services:

US AIR FORCE

Office of Public Affairs, Aeronautical Systems Division (AFSC), Wright-Patterson AFB, Ohio 45433

Telephone: (513) 253 7111

Locust. Mini-drone programme to develop a small, low-cost system to provide defence suppression of hostile air defences. Formerly known as LCEHV (Low Cost Expendable Harassment Vehicle) and later as HWS (Harassment Weapons System). Will be ground-launched, and pre-programmed to navigate to, and loiter in, the area of hostile air defences. Its presence will draw the defences' fire, provoke radar silence, or (if ignored) will home in to allow an explosive charge to damage a hostile radar.

The Locust vehicle will be able to re-loiter in the event of a hostile radar being turned off during a terminal attack. A concept validation flight test was conducted in early 1979, using a US Air Force Flight Dynamics Laboratory XBQM-106 vehicle (see next entry) with a DH 220 two-stroke engine and existing modified sensor and warhead. A Memorandum of Agreement was signed on 21 March 1980 for joint development of the Locust production vehicle. If full-scale development is completed within 30 months of the contract award, production systems should become operational in late FY 1984.

Sensor full-scale development contracts were awarded to Texas Instruments and General Dynamics in June 1978. Each contractor will deliver 20 sensors for use by the system integration contractors. The sensors, scheduled for delivery in early 1981, will provide the initial acquisition, target selection, and terminal homing on to emitting targets of hostile air defences, when incorporated into the system. Overall dimensions and performance characteristics include a diameter of 0·203 m (8 in) and length of 0·38 m (1 ft 3 in), including antenna; weight of 6·8 kg (15 lb); a probable performance reliability of 0·9; storage life of 10 years; an operational endurance of 2 h search and 3 h search; nominal power bus of 28V DC; and a go/no-go maintainability checkout with the Locust system.

Competing designs are described under the Dornier, MBB and VFW entries in the Federal German part of this section, but German participation was suspended in 1981 for budgetary reasons. Unilateral development by the USAF is continuing.

XBQM-106. This experimental mini-RPV was designed and built by the US Air Force Flight Dynamics Laboratory, and was first flown in 1975. It is designed for easily-obtainable flexibility; up to early 1981, 18 vehicles had been built and 10 variants had flown, incorporating various alternative wing, nose, tail and engine configurations, and with payloads ranging from 11·3 to 59 kg (25-130 lb).

The XBQM-106 in its 1980 configuration was described and illustrated in the 1980-81 *Jane's*.

US NAVY

Office of Public Affairs, Naval Air Systems Command, Department of the Navy, the Pentagon, Washington, DC 20361

Naval Air Development Center HI-SPOT. HI-SPOT (HIgh altitude Surveillance Platform for Over-the-horizon Targeting) is a programme to develop an unmanned platform for high altitude (18,500 to 24,000 m; 60,700 to 78,750 ft) ocean surveillance and communications relay. It will be designed to remain on station for periods of more than 30 days. Programme is sponsored by the Naval Air Systems Command, with technical management by the Naval Air Development Center. Model testing is scheduled to begin in 1982.

Naval Air Development Center QF-4B. QF-4B target version of McDonnell Douglas F-4B was developed by Naval Air Development Center for use at Pacific Missile Test Center. First aircraft was delivered in Spring 1972 to PMTC at Point Mugu.

Seven QF-4B conversions were produced by Naval Air Rework Facility, Cherry Point, North Carolina. Target is equipped for high-*g* manoeuvring and has provisions for photo and electronic scoring and ECM. Production continuing into 1980s. A follow-on version is under development to incorporate all-attitude manoeuvring.

Naval Weapons Center QF-86. Thirty-one North American F-86H Sabres were converted by Naval Weapons Center, China Lake, California, into QF-86H pilotless target aircraft, as described in 1977-78 *Jane's*. Three F-86F Sabres underwent prototype conversion at the NWC in 1979 into QF-86F pilotless target aircraft. Ten F-86Fs are being converted to QF-86F drones by Sperry (which see) under contract to the NWC; a further 90 may be converted later. The QF-86Fs and QF-86Hs are used as interim all-attitude manoeuvring targets.

AIR-LAUNCHED MISSILES

ARGENTINA

CITEFA
INSTITUTO DE INVESTIGACIONES CIENTIFICAS Y TECNICAS DE LAS FUERZAS ARMADAS

Zufriategui y Varela, 1603 Villa Martelli, Provincia de Buenos Aires

Weapons under development by the Argentinian Armed Forces Scientific and Technical Research Institute include a supersonic air-launched tactical missile, of which details follow:

MARTIN PESCADOR (KINGFISHER)

Since CITEFA completed development of this supersonic tactical missile in 1979, it has built a number of pre-production models prior to the start of series manufacture for the Argentinian armed forces. One was exhibited at the 1981 Paris Air Show, as a potential weapon to be carried by the Agusta A 109 helicopters ordered for the Argentinian Army. When launched from a hovering helicopter, the Martin Pescador has a range of 2·3 nm (4·3 km; 2·7 miles), with a target impact speed of Mach 1·1. It can also be launched from any fixed-wing aircraft capable of a speed of Mach 0·5 or more; performance in this form is given in the table at the end of this section.

Martin Pescador is guided by radio command from the launch aircraft, along a line of sight. Its layout is conventional, with canard control surfaces, a single-stage solid-propellant rocket motor, and a high-explosive warhead detonated by impact fuse.

DATA: See table

BRAZIL

PIRANHA MAA 1

The Brazilian press has published brief reports of an air-to-air missile named Piranha MAA 1 which is said to be under development in Brazil. It is expected to be ready for service by 1984.

DATA: See table

FRANCE

AÉROSPATIALE
SOCIÉTÉ NATIONALE INDUSTRIELLE AÉROSPATIALE

Division des Engins Tactiques

2 à 18 rue Béranger, BP 84, 92320 Châtillon sous Bagneux

Telephone: 657 11 80

Telex: AISPA 250881 F

When Nord-Aviation became part of Aérospatiale, responsibility for its guided missiles, target vehicles and test vehicles was allocated to Aérospatiale's Tactical Missiles Division. Subsequently, this Division developed and put into production a new generation of short-range battlefield weapons, in association with MBB of Federal Germany (see Euromissile entry in International section).

A total of 521,706 tactical missiles and 1,925 target and reconnaissance RPVs had been ordered from the Division and its overseas partners by January 1981. Employees totalled 6,200 at the beginning of 1981.

AÉROSPATIALE AS.11

The AS.11 is an air-to-surface version of the SS.11 line-of sight wire-guided battlefield missile. It is powered by a two-stage solid-propellant rocket motor. Directional control is achieved by varying the sustainer efflux through two side-nozzles.

In action, with a typical helicopter installation, the operator acquires the target by means of a stabilised and magnifying optical sight. As soon as the missile enters his field of vision after launch, he passes to it the signals needed to align it with the target, while keeping it above the terrain until impact. The signals are given by the operator by means of a control stick which makes it possible to send simultaneously up or down and port or starboard commands. The signals are transmitted to the missile over wires, wound on twin spools. Tracer flares are installed on the rear of the missile for visual reference.

The current AS.11 B.1 version is available with a variety of warheads, including an inert type for practice, the Type 140AC anti-tank warhead capable of perforating 60 cm (24 in) of armour plate, the Type 140AP02 explosive warhead (2·6 kg; 5·72 lb of explosive) which will penetrate an armoured steel plate 1 cm (0·4 in) thick at a range of 3,000 m (9,840 ft) and explode about 2·1 m (7 ft) behind the point of impact, and the Type 140AP59 high-fragmentation anti-personnel type with contact fuse.

The AS.11 B.1 and the basic surface-launched SS.11 B.1 have been supplied to all three French services and the armed forces of 27 other countries, including the USA and UK. Orders totalled 174,741 by January 1981, including surface-to-surface Harpons with automatic infra-red command guidance. Of these, 60 per cent were for export. Licence manufacture was undertaken in Federal Germany and India. Production by Aérospatiale is expected to continue on a reduced scale until about 1985.

DATA: See table

AÉROSPATIALE AS.12

This spin-stabilised missile was derived from the AS.11 and has a warhead weighing 28·4 kg (62·6 lb), which makes it suitable for use against fortifications as well as tanks, ships and other vehicles. The current OP.3C warhead can pierce 40 mm (1·5 in) of armour and explode on the other side.

The AS.12 has a two-stage solid-propellant rocket motor and a command to line-of-sight guidance system. It can be air-launched at speeds up to about 200 knots (370 km/h; 230 mph), in conjunction with an APX 260 or APX 334 gyro-stabilised sight, and can be used at night with target illuminating equipment.

The AS.12 arms seven types of fixed-wing aircraft and eight types of helicopter in service with 16 navies or air forces. They include Breguet Atlantic and P-2 Neptune aircraft of the Royal Netherlands Navy, the Breguet Alizé, and helicopters of the French Navy and Royal Navy.

A total of 9,223 AS.12s and surface-to-surface SS.12Ms had been ordered by January 1981, when manufacture was continuing. Of these, 79 per cent were for export.

DATA: See table

AÉROSPATIALE AS.15TT

The AS.15TT is being developed by Aérospatiale as a lightweight all-weather missile for attacking surface targets of all tonnages at sea. It is intended for operation from comparatively slow-flying aircraft, such as ship-based or shore-based helicopters and maritime patrol aircraft, or from surface ships or coastal defence batteries. A helicopter such as the SA 365 Dauphin 2 will be able to carry four, in pairs, on fuselage-side racks.

The AS.15TT uses an extremely accurate automatic guidance system, and carries a warhead identical to that of the AS.12 missile.

Associated surface-surveillance radar is the Thomson-CSF Agrion 15, which has a 360° scan; is designed for automatic target tracking and determination of the differential target/missile range and bearing; and is able to operate in an ECM environment. As well as forming part of the AS.15TT system, the Agrion 15 can be used to designate over-the-horizon targets for long-range anti-ship missiles in the Exocet class.

In operation, the target is detected; the radar is switched to automatic tracking and the missile launched. After a shallow descent, the AS.15TT follows a sea-skimming trajectory to the target, under entirely automatic guidance. The characteristics of the system are such that approach to the target and missile launch can be performed at very low altitude.

Initial firings, to check the aerodynamic characteristics of the AS.15TT, were made from a ground launcher at a test range in the Landes region of SW France. The decision to complete development of the weapon followed receipt of an order from Saudi Arabia for 20 SA 365F anti-ship helicopters and 200 AS.15TT missiles.

DATA: See table

AÉROSPATIALE AS.30 LASER

Developed as a successor to the original command-guided AS.30 (see 1976-77 *Jane's*), this tactical air-to-surface supersonic missile is intended to be compatible with all types of modern single-seat or multi-seat combat aircraft, for use against hardened targets on land or at sea. It has a cylindrical body, cruciform sweptback wings which are canted to impart spin stabilisation, cruciform 'flip-out' tail-fins indexed in line with the wings, and a two-stage (boost and sustainer) solid-propellant rocket motor. A two-phase guidance system is utilised, comprising a pre-guidance phase on gyro reference, followed by automatic (TV) target tracking combined with semi-active laser terminal homing, by means of the missile's Thomson-CSF Ariel homing head.

Standard equipment required in the launch aircraft consists of an armament control panel, pilot's sight and TV screen, a firing button on the control column and an emergency jettison switch. Equipment associated specifically with the AS.30 Laser missile comprises launchers with the appropriate firing circuits, and a Thomson-CSF Atlis target illuminating pod.

Tests of an Atlis II pod under a French Air Force Jaguar A were made successfully at the CEV, Cazaux, in 1977. Test firings of the AS.30 Laser missile were completed successfully in 1980.

DATA: See table

AÉROSPATIALE EXOCET AM39

Exocet is a missile that was devised originally to provide warships with all-weather attack capability against other surface vessels. The air-to-surface version is designated AM39. Propulsion is by a tandem two-stage solid-propellant motor, and the highly-destructive warhead is in the same order as that of a torpedo.

Compared with the surface-to-surface versions, the AM39 air-to-surface Exocet has a reduced launch weight and a new rocket motor that burns for 150 s, giving an increased range. Another modification to the propulsion system ensures a one-second delay in ignition after launch, allowing the missile to drop clear of the launch aircraft.

The missile's high subsonic flight profile consists of a pre-guidance phase, during which it travels towards the target, whose range and bearing have been determined by an airborne radar and fire control computer and set up in the missile pre-guidance circuits before launch; an inertially-guided midcourse phase; and a final guidance phase during which the missile flies directly towards the target under the control of its active homing head. Throughout its flight the missile is maintained at a very low altitude (reported to be 2 to 3 m; 6·5 to 10 ft) by an FM radio altimeter. Its homing head is reported to pick up the target over a range of up to 6·5 nm (12 km; 7·5 miles). Exocet is intended to operate efficiently in an ECM (electronic countermeasures) environment. Subcontractors to Aérospatiale include Electronique Marcel Dassault for the ADAC homing head, TRT for the AHV-7 radio altimeter, SNPE for the solid propellants, SERAT for the explosive charge, SFENA and SAGEM for accelerometers and gyroscopes, and Jaeger for control surface actuators.

Launch tests from a Super Frelon helicopter began in April 1973. The type of missile employed was an intermediate version of Exocet designated AM38, which consisted of a surface-to-surface MM38 fitted with the one-

CITEFA Martin Pescador air-to-surface missile *(Brian M. Service)*

Aérospatiale AS.12 air-to-surface missile on an Alouette III helicopter

second ignition delay that is embodied in the AM39. In each of the powered launches, the missile dropped about 10 m (33 ft) in a stable, horizontal attitude from its pylon before ignition, despite rotor downwash.

Firings of AM39 missiles from a Super Frelon began in December 1976, and were followed by the first launches from a Super Etendard in the first half of 1977.

Carrying two Exocet AM39s and 5,000 litres (1,100 Imp gallons) of fuel, a Super Frelon can carry out a 6 h patrol near its shore or ship base, or attack a specific target up to 350 nm (650 km; 400 miles) from base. Equipment added to the helicopter consists of two launch pylons, an operator's console, an Omera ORB-31D X-band search and tracking radar capable of acquiring targets the size of a fast patrol boat over a range of 28·5 nm (53 km; 33 miles) in sea state 4 or 5, an EMD Alto-2 Doppler radar altimeter, a SFIM CV-153 vertical gyro for a compass system, a compensated airspeed indicator and a Crouzet 70 computer.

Of the 140 production AM39s ordered by January 1980, more than 60 per cent were for export. The missile has been selected as armament for the French Navy's Atlantic NG and Super Etendard aircraft, Pakistani Sea King helicopters and Iraqi Super Frelon helicopters. It is being adapted for carriage by the Super Puma helicopter and several additional types of strike aircraft.

Range (estimated):
launched from a helicopter at 60 knots (110 km/h; 69 mph) at a height of 100 m (330 ft)
28 nm (52 km; 32·25 miles)
launched from an Atlantic aircraft at heights between 300 and 5,000 m (1,000 ft and 16,400 ft)
29-32·25 nm (54-60 km; 33·5-37·25 miles)
launched from a Super Etendard aircraft at heights between 100 m (330 ft) and 10,000 m (33,000 ft)
32-37 nm (60-70 km; 37-43 miles)
For other data see table

ASMP

As mentioned in the 1977-78 *Jane's*, the original missile that was being developed under the ASMP (Air-Sol Moyenne Portée) designation reverted to definition study status when the ACF (Avion de Combat Futur) programme was cancelled. However, on 6 April 1978 it was announced that a new design and development contract for an air-to-surface medium-range missile (ASMP) had been awarded to Aérospatiale's Tactical Missiles Division, which is collaborating with ONERA and SNPE. Intended as the vehicle for the French Air Force's latest tactical nuclear warhead, ASMP is stated to benefit from ramjet engine-research carried out by Aérospatiale and its predecessors, and from experience gained in developing the guidance systems for the Exocet and Pluton missiles. It will be carried by 15 Dassault Mirage IV bombers, Super Etendard naval fighters and Mirage 2000 tactical fighters, from about 1985.

The ASMP will be powered in supersonic cruising flight by a kerosene-burning ramjet, supplied with air by a pair of two-dimensional side intakes which will serve also as wings. Stability and control will be provided by two pairs of tail-fins and rudders, with guidance by a SAGEM inertial system. Range is expected to be in the order of 40-54 nm (75-100 km; 46-62 miles) depending on launch and cruising height. Yield of the CEA nuclear warhead is said to be between 100 and 150 kT, and ASMP is intended for stand-off use against heavily defended targets such as airfields, and command and communications centres.

It seems likely that the power plant will have a solid-propellant launch booster integrated in the ramjet combustion chamber, forming a two-stage rocket-ramjet.

MATRA
SA MATRA

Head Office: 4 rue de Presbourg, 75116 Paris
Management and Works, Postal Address: BP No. 1, 78140 Vélizy
Telephone: 946 96 00
Telex: ENMATRA 698077F
President Director-General: Jean-Luc Lagardère
Director of Public Relations: Philippe Chassagny
Press Department: 6 rond-point des Condamines, 78000 Versailles
Telephone: 953 00 05
Telex: DREMATRA 695 413 F
Chief of Press Department: Roland Sanguinetti

Since 1948, Matra has been engaged in extensive research and experimental work in the guided missile, propulsion and guidance fields. The wide range of weapon systems in current production includes a total of eight air-to-air, surface-to-air, air-to-surface and anti-ship missiles.

Other Matra weapon systems include the Durandal rocket-propelled penetration bomb for attacking runways; and the Beluga cluster weapon, developed in association with Thomson/Brandt, which dispenses 151 grenades or other munitions, lowered to the ground on individual brake-chutes. A kit able to adapt several standard types of bomb for laser guidance has been developed.

MATRA SUPER 530

The basic **Super 530** air-to-air weapon is deployed on Dassault Mirage F1 interceptors of the French Air Force; the Mirage 2000 will be armed with the improved **Super 530 D**, compatible with its Doppler radar.

The Super 530 is an all-weather and all-sector weapon, with the ability to attack targets flying at an altitude 9,000 m (29,500 ft) higher or lower than that of the launch aircraft. Range limits are described as being from "several hundreds of metres" to "several tens of km".

The missile has skin, wings and tail surfaces of sheet steel and steel honeycomb, with a ceramic radome over its EMD Super AD-26 semi-active pulse radar homing head. Its Angèle dual-thrust rocket motor, supplied by Thomson/Brandt, utilises Butalane propellant. The fragmenting warhead is supplied by Thomson/Brandt, with a proximity fuse by Thomson-CSF; and advanced ECM anti-jamming circuits are fitted. Control is by means of the cruciform tail-fins. Transverse acceleration limit is 20*g* on each axis up to 17,000-18,000 m (56,000-59,000 ft) and 6*g* from that height up to 25,000 m (82,000 ft).

Ramp-launch tests of the Super 530 began in January 1971, and the first controlled model was fired successfully on 27 February 1973. Five complete missiles had been fired successfully against airborne targets by early 1975; the first launch from a Mirage F1 was conducted successfully in January 1976. Manufacture for the Air Force was started at the end of 1977, reaching mass production levels by January 1979. The Super 530 has been operational on French home defence interceptors since December 1979. Orders, from several countries, covered the production of more than 1,000 operational and training missiles by early 1981.

Data: See table

MATRA R.550 MAGIC

This air-to-air weapon system was intended initially to meet a French Air Force requirement for a highly-manoeuvrable short/medium-range 'dogfight' missile. Development was started in 1967 as a private venture and has continued under official contract since 1969.

The R.550 Magic has a 'double-canard' configuration, with a set of movable foreplane control surfaces immediately behind and indexed in line with cruciform fixed surfaces. The missile has an SNPE Romeo single-stage solid-propellant motor, an AD-3601 infra-red homing head with a cooled cell, manufactured by SAT, and a 12·5 kg (27·5 lb) warhead with 6 kg (13·2 lb) of explosive, impact and proximity fuses. It can utilise the same aircraft launcher as the Sidewinder which it replaces.

The Magic has wide range limits, is stressed for 30*g* manoeuvres and can be fired from an aircraft in a 6*g* turn, singly or at one second interval between rounds. There is no minimum launch speed; maximum is more than 700 knots (1,300 km/h; 805 mph), against targets in a forward sector of 140°.

The first air-launch of an R.550 complete with guidance equipment, against a target, took place at the Landes Test Centre on 11 January 1972, from a Gloster Meteor test aircraft. Delivery of the first production missiles for operational testing and evaluation by the French Air Force began in 1974. Full series manufacture began in 1975 and orders for more than 6,000 training and operational rounds had been received from 15 countries by February 1981.

Adaptation to the Mirage III, Mirage 5, Mirage F1 and French Navy Crusader has been completed; adaptation to the Jaguar, Alpha Jet, Aermacchi M.B. 326K and Super Étendard is under way, and Magic will also be the basic weapon of the Mirage 2000. A report from the Middle East, in March 1981, claimed that Pakistani engineers had modified Iraqi MiG-21s to carry Magic, and that these MiGs had shot down an Iranian F-14 Tomcat and an F-4 Phantom.

Data: See table

GERMANY
(FEDERAL REPUBLIC)

BODENSEEWERK
BODENSEEWERK GERÄTETECHNIK GmbH

Head Office: Postfach 1120, 7770 Überlingen
Telephone: (07 551) 811
Telex: 07 33924
Liaison Office: Bundeskanzlerplatz HI 1002, 5300 Bonn
Telephone: (02 221) 214432
UK Office: 27-28 Crossway House, Bracknell, Berkshire RG12 1DA, England
Telephone: (0344) 26070 and 54782

Under the original NATO European production programme for the Sidewinder air-to-air missile, Bodenseewerk was prime contractor for the licence manufacture of 13,000 of the AIM-9B and FGW Mod 2 versions. The

Launch of an Aérospatiale AS.11 air-to-surface missile from an Alouette III helicopter

Aérospatiale AS.15TT missile installation on mockup of SA 365F Dauphin 2 anti-ship helicopter

Launch of an Aérospatiale AS.30 Laser air-to-surface homing missile from a Jaguar

AM39 Exocet air-to-surface missile on an Aéronavale Super Étendard

Matra Durandal runway penetration bombs and a wingtip-mounted Magic air-to-air missile on a Dassault Mirage F1

Matra R.550 Magic and Super 530 D air-to-air missiles under the wings of a Mirage 2000

One of Matra's laser-guided air-to-surface weapons

company is now prime contractor for European production of the advanced AIM-9L Sidewinder, described under the NWC heading in the US section. Participating countries, in addition to West Germany, are Italy, Norway and the UK.

In the late 1960s, Bodenseewerk began developing the Viper close-range air-to-air missile system, comprising the high-performance all-aspect Viper infra-red homing missile, a new launcher, and a launcher-mounted passive infra-red target acquisition unit. When Dornier entered the programme in 1972, a joint management team was set up. After cancellation of Viper in 1974, the joint US/FRG missile programme known as ALASCA (all aspect capability) was launched, combining the Viper IR seeker with the aft section of the AIM-9L/H Sidewinder. Under this programme, 12 ALASCA guidance and control sections were built for flight testing. The project was stopped later in favour of the AIM-9L, but flight tests of ALASCA seekers were performed during 1980.

Bodenseewerk has been selected as West German lead contractor to carry out with British Aerospace studies for a new generation of air-to-air missiles.

MBB

MESSERSCHMITT-BÖLKOW-BLOHM GmbH

HEAD OFFICE AND WORKS: Ottobrunn bei München, 8000 München 80, Postfach 801220
Telephone: (089) 60 00 25 90

In addition to its manufacture of piloted aircraft, MBB has engaged in the development and manufacture of guided weapons for many years. It has achieved considerable success with its short-range surface-to-surface and surface-to-air battlefield missiles, some of which result from partnership with Aérospatiale of France.

Weapons in current production include an air-launched anti-shipping missile known as the Kormoran.

MBB KORMORAN

This roll-stabilised all-weather anti-shipping missile can be carried by any aircraft which is able to maintain a speed between Mach 0·6 and 0·95 during the attack, and which is equipped with target detection radar and an autonomous navigation system such as an inertial platform. The aircraft's avionics must be supplemented by a position and homing indicator (PHI) and Kormoran firing equipment. Launch information is obtained from the aircraft's target detection radar, navigation system, and a navigation support device which adapts the signals representing velocity for the Kormoran airborne computer. In addition, if the missile is operated in an optical mode, instead of through target detection radar, the PHI has to be used in conjunction with computer co-ordinated data (CCD) and a vector addition unit (VAD).

Two built-in SNPE Prade boosters accelerate Kormoran to high subsonic cruising speed, which is then maintained by the SNPE Eole IV solid-propellant sustainer motor. The guidance system employs pre-guidance and homing phases, using a Thomson-CSF RE-576 radar terminal seeker. The new-type high-energy warhead, developed at MBB's Schrobenhausen works, weighs 160 kg (352 lb) and can penetrate 70-90 mm (2·75-3·5 in) of steel plate, making it effective against ships up to the size of a destroyer.

The launch aircraft can approach the target area at very low level, to escape detection by the ship's radar, by means of the PHI, with radar switched off. With the radar then turned on, the target is acquired and its position is fed into the PHI; the radar is again switched off while the aircraft takes up the optimum attack position. The radar is switched on, locked on to the target, and initial data are transmitted automatically to the missile. As soon as the missile has been launched under power, directly from the pylon, as little as 30 m (100 ft) above the water, the aircraft breaks away, outside the range of enemy anti-aircraft defences. The missile descends to its programmed flight level and locks on to the target, guided by inertial navigation aided by its radar altimeter. At a prescribed distance, the inertial system releases the active radar seeker. After this has locked on target, the inertial guidance is slaved and corrected in azimuth and range by the seeker head. At a short distance from the target the missile descends to its terminal flight level in order to hit the target just above the waterline.

The secondary optical firing mode, in which the pilot aims the missile by means of his optical weapon sight, is used against surprise targets, in the presence of extreme enemy ECM, or when the target is a small vessel, such as a fast patrol boat or minesweeper. Kormoran itself is immune to all known kinds of ECM.

In-plant testing of Kormoran was completed by early 1974, at which time official tests from F-104G Starfighter aircraft were under way. A preliminary production contract to equip West German naval air arm F-104Gs was placed in 1974; it was followed by a full contract in November 1976, for 350 missiles and 56 aircraft installations, for which MBB is prime contractor. Seven qualification test firings in Autumn 1977 resulted in six direct hits on target and a seventh which met specification requirements. Kormoran has been in series production since December 1977, and will arm West German Navy Tornados when these aircraft replace the F-104G Starfighters now in service. In October 1980 it was announced that the Italian Air Force will also use the Kormoran on its Tornados.

DATA: See table

INTERNATIONAL PROGRAMMES

EMDG
EUROMISSILE DYNAMICS GROUP

12 rue de la Redoute, 92260 Fontenay-aux-Roses, France
Telephone: 662 33 11
Telex: EUROM 204691 F
SUPERVISORY BOARD:
Michel Allier (Aérospatiale)
Admiral Sir Raymond Lygo (British Aerospace)
Gunther Kuhlo (MBB)

Formation of this guided weapons company was announced by Aérospatiale, British Aerospace and MBB on 2 January 1980, following the signing of a memorandum of understanding by the governments of France, the UK and the German Federal Republic. EMDG is subject to French law. Its supervisory board is made up of the Chief Executives of Aérospatiale's Division Engins Tactiques, MBB Unternehmensbereich Apparate and British Aerospace Dynamics Group.

Initially, EMDG will be involved in research, development, production and marketing of third-generation anti-tank guided weapon systems, both medium and long range. It is also the intention of the three participating companies that other weapon systems undertaken jointly by them, on behalf of their respective governments, shall be the responsibility of EMDG.

Aérospatiale, MBB and British Aerospace were already working together on an anti-ship missile system through a joint company known as Anti-Ship Euromissile (ASEM). It is intended to absorb ASEM into EMDG as soon as practicable.

EUROMISSILE
EUROMISSILE GROUPEMENT D'INTÉRÊT ÉCONOMIQUE

12 rue de la Redoute, 92260 Fontenay-aux-Roses, France
Telephone: 662 33 11
Telex: EUROM 204691 F
PRESIDENT: Marcel Morer
SALES DIRECTORS: Pierre Froget and Georg Erlenwein

Euromissile is a Groupement d'Intérêt Economique formed by Aérospatiale of France and MBB of Federal Germany. First products of the Euromissile team are three short-range battlefield weapons known as Milan, Hot and Roland, developed initially for the armed services of France and Federal Germany. Hot is in production for both air-launched and surface-launched operation.

HOT

The Hot (High-subsonic, optically-guided, tube-launched) is a tube-launched wire-guided anti-tank missile suitable for both surface-to-surface and air-to-surface use. It has fins which fold down against the body when it is in its launching tube, and open out in flight. The power plant is a two-stage solid-propellant rocket motor. A jet-vane is used to steer the missile. Guidance is by automatic command to line of sight, with infra-red tracking.

The HLVS (Hot, stabilised localiser-sight) system offers magnifications of 3·2 × 17° field and 10·8 × 5° field. Sight limits are ±118° in bearing and +28°/−20° in elevation. To engage a target, the aimer maintains a sighting cross on the target, switches on the firing installation and selects a missile on the control box.

Flight testing of an operational Hot installation on an Aérospatiale/Westland SA 341 Gazelle helicopter, in conjunction with the HLVS stabilised sight system, was performed in 1973. With the target hidden from the sight of the pilot and missile operator before take-off, the helicopter proved able to acquire the target and launch a missile within 4 s of lift-off. Hit probability proved to be 90 per cent over ranges varying from a minimum 400 m (1,310 ft) to a maximum of 4,000 m (13,125 ft) in hovering flight and 4,300 m (14,100 ft) in translational flight at speeds up to 100 knots (185 km/h; 115 mph). During missile flight, the helicopter was able to take evasive action at a turning speed of up to 6° per second.

Further helicopter trials, by the official services of France and Federal Germany, took place during 1974, from a Gazelle and a BO 105 respectively. Subsequently, France adopted Hot as armament for the Gazelle helicopter; Federal Germany selected it for the BO 105 M (PAH 1), and several other nations have ordered this weapon system for helicopter use. Series production began in 1976, and the first production units to equip helicopters were delivered in 1977. Hot was fired from Lynx and Dauphin helicopters during 1978.

A system known as Venus (Viseur Ecartometric Nuit Stabilisé), which includes a Hector infra-red sensor, is under development by Aérospatiale and the SFIM, TRT/SAT and Thomson-CSF companies, to permit firing at night. A prototype installation on a Dauphin helicopter was used to fire two Hot missiles at the Cazaux flight test centre, in France, on the nights of 20 and 21 January 1981. Using BNL binoculars, the pilot launched the missiles while hovering about 15 m (50 ft) above the ground. Despite extremely adverse weather (dense drizzle; approx 100% RH), both missiles hit the 2·3 m (7 ft 6 in) square targets, over successive ranges of 2,900 m (9,500 ft) and 2,350 m (7,700 ft).

DATA: See table

ISRAEL

RAFAEL
RAFAEL ARMAMENT DEVELOPMENT AUTHORITY

Ministry of Defence, POB 2082, 31021 Haifa
Telephone: 04 714168
Telex: 45173 VERED IL

SHAFRIR

Shafrir is a short-range infra-red homing dogfight missile developed for use against aircraft at heights up to 18,000 m (60,000 ft). Many rounds have been fired against enemy aircraft since the current Mk 2 version entered service in 1969, with considerable success. This was particularly evident during the Yom Kippur War of October 1973, when Shafrir is claimed to have destroyed more than 100 Arab aircraft in air combat, representing a success rate of 60% in terms of the number of missiles fired.

Relatively small in size and simple in conception, Shafrir has a solid-propellant rocket motor and is a solid-state weapon, fully transistorised and with guidance by proportional navigation, for optimum results against manoeuvring targets.

The missile and its launcher are mounted under the wing of the aircraft on a specially-designed adapter which is capable of carrying other types of weapon as an alternative to Shafrir. Attachment is mechanical and the missile requires no support from the aircraft except for the firing circuit. When a target is detected within firing range, an audio signal is heard and a light is switched on automatically on the pilot's control panel as an indication that the firing button should be pressed. After launch, the missile tracks the target entirely automatically, and the warhead is detonated either on impact or by the proximity fuse within optimum distance of the target.

Shafrir Mk 3, under development, is expected to offer greater manoeuvrability and all-aspect launch capability.

Sales of the Mk 2 version are reported to have been made to several overseas customers, including Chile and Taiwan.

DATA: See table

PYTHON 3

At the 1981 Paris Air Show, Rafael announced that it was developing the Python 3 to meet an Israeli Air Force requirement for a close-range air-to-air dogfight missile that would also be suitable for longer-range interception. The configuration of the new weapon is very like that of Shafrir, except that the infra-red homing head is slightly tapered and the tail surfaces have sweepback on both leading- and trailing-edges. Development was at an advanced stage in mid-1981, suggesting that Python 3 will be operational within two or three years. Its performance was claimed to be superior to that of the AIM-9L Sidewinder.

LUZ

First reported in the 1974-75 *Jane's*, Luz-1 is now confirmed as a TV-guided air-to-surface missile developed and produced by Rafael Armament Development Authority. No details have been released officially, but in mid-1977 the Israeli Air Force was reported to be modifying its F-4E Phantom and IAI Kfir-C2 fighters to carry Luz. The missile's main purpose is said to be the destruction of enemy surface-to-air weapons, with a conventional warhead. Impervious to jamming, Luz is believed to have a maximum range of 43 nm (80 km; 50 miles) and a launch weight of 200 kg (440 lb).

ITALY

SELENIA
SELENIA INDUSTRIE ELETTRONICHE ASSOCIATE SpA

HEAD OFFICE AND WORKS: Via Tiburtina km. 12.400, 00131 Rome
POSTAL ADDRESS: PO Box 7083, 00100 Rome
Telephone: 43601
Telex: 61106 Seleniat
PUBLIC RELATIONS: Gen Clemente Paolozzi

ASPIDE

This missile, which is in series production for use by nine navies in the Albatros naval surface-to-air defence system, is also under development as an all-weather all-aspect weapon for high-performance interceptors. It will improve their effectiveness in terms of maximum missile range, operation at very low altitudes, multiple target engagement and resistance to advanced ECM. In particu-

Launch of an MBB Kormoran missile from a Tornado

Launching a Hot missile from a Gazelle helicopter

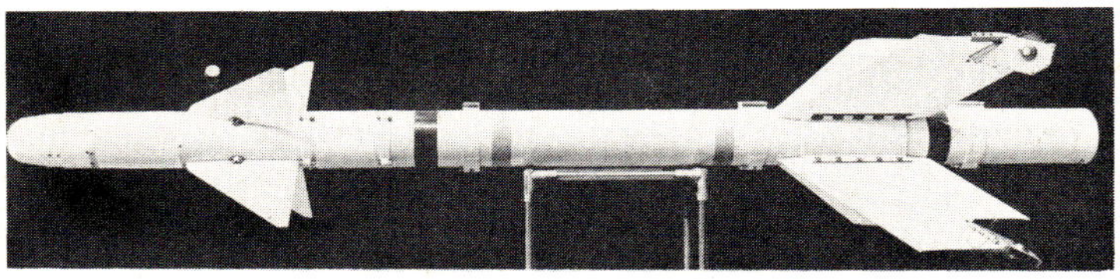

Python 3 air-to-air missile

Sistel Sea Killer Mk 2 missile installed on SH-3D helicopter as component of Marte air-to-surface weapon system

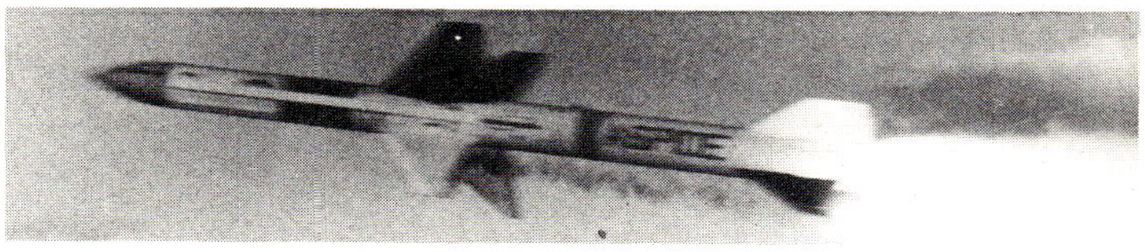

Ground-to-air firing of an Aspide missile in Sardinia

Shafrir under wing of a Kfir fighter *(Israir)*

lar, it is expected to enhance the dogfight and shootdown capabilities of the F-104S interceptors of the Italian Air Force.

Aspide has a single-stage solid-propellant rocket motor, supplied by Difesa e Spazio, which gives it a speed described as being "well in the hypersonic field". Its guidance system is of the semi-active radar type, developed by Selenia. The warhead is a fragmentation type produced by Difesa e Spazio.

The missile's final configuration is almost identical to that of the US Sparrow and the five missiles used in first-phase firing tests, concluded on 17 December 1975, were each powered by a licence-built Rocketdyne Mk 38 motor, as fitted in Sparrow. Second-phase testing involved five surface-to-air and five air-to-air firings, using missiles powered by the Difesa e Spazio motor developed especially for Aspide and fitted with an improved autopilot. Final air-to-air verification trials, phased in with Italian Air Force priorities, took place in 1979 and 1980.

Data: See table

SISTEL
SISTEL—SISTEMI ELETTRONICI SpA

Head Office: via Tiburtina 1210, 00131 Rome
Telephone: 43 69 41
Telex: 680112 Sistel I
Managing Director: Ing Aldo Barontini

MARTE

In 1967, the Italian Navy initiated a development programme, known as Project Marte, to enhance the capabilities of shore-based or shipborne helicopters by arming them with anti-shipping missiles. Sistel's Sea Killer family of missiles was chosen as most suitable for this application, because of the weapons' target discrimination capability, all-weather operability, automatic guidance system, inherent insensibility to ECM and sea-skimming flight profile. The system has been studied for installation on a wide range of helicopters, with take-off weights varying from 3,000 kg to more than 10,000 kg (6,600-22,050 lb), dependent on the number of missiles to be carried and the possible requirement for simultaneous anti-submarine and surface strike capability.

Equipment added to the Marte helicopter is lightweight and easily operated. The radar performs navigation, search and target tracking as well as guidance of the missile in azimuth. Thus an SH-3D helicopter, fully equipped for surface strike and anti-submarine duties, and with an autopilot for instrument flying, can be assigned a 4½ h maritime patrol at a search speed of 100 knots (185 km/h; 115 mph), carrying a crew of four. Total weight of the system is 1,320 kg (2,910 lb), including 158 kg (348 lb) for the radar, 600 kg (1,323 lb) for two Sea Killer Mk 2 missiles, 300 kg (661 lb) for electronic warfare equipment, and 22 kg (48 lb) for an optical sight.

In action, it can be assumed that the helicopter will locate a target in a few seconds of radar operation at the limit of radar range. To reduce the possibility of enemy recognition of the helicopter's radar interrogation, the airborne radar is then switched off and the aircraft descends in order to fly toward the target at the lowest practical height above the water. At an estimated distance just beyond missile range, the helicopter climbs again to missile launching height, re-acquires the target and launches a Sea Killer, which takes slightly more than one minute to reach the enemy ship over a range of 10·8 nm (20 km; 12·4 miles).

The Sistel radar altimeter in the missile can be pre-set before launch to control the cruising height at values down to 3 m (10 ft) or less, depending on factors such as sea state and target dimensions. The altitude can be changed by command signal during flight, if required. Control in azimuth can be achieved either by automatic radar mode in all-weather conditions or by a standby fair-weather system, using an optical sight and joystick controller.

Initial firing trials of the Marte system, in flight, had been made successfully from an AB 204 helicopter by the Spring of 1973. The system was put into production subsequently for installation on Agusta SH-3D helicopters of the Italian Navy, utilising two Sea Killer Mk 2/Marte missiles and chin-mounted APQ-706 X-band track-while-scan radar supplied by SMA. Installation on smaller helicopters, such as the Agusta-Bell 212ASW, is being studied.

SISTEL SEA KILLER Mk 2

This two-stage missile is a development of the Sea Killer Mk 1 (see 1977-78 *Jane's*), with a heavier warhead and extended range. It is operational in surface-to-surface form on board four Vosper Mk 5 frigates of the Iranian Navy. With a guidance modification, it is also the sea-skimming missile utilised in the Marte air-to-surface weapon system for helicopters.

Power Plant: SEP 299 solid-propellant booster, rated at 43·16 kN (9,702 lb st) for 1·6 s. SEP 300 solid-propellant sustainer, rated at 0·98 kN (220·5 lb st) for 73 s.

Guidance: Alternative all-weather beam-rider/command/radio altimeter guidance, or optical radio command/radio altimeter guidance.

Warhead: High-explosive semi-armour-piercing with impact/proximity fuse.

Data: See table

JAPAN

MITSUBISHI

MITSUBISHI DENKI KABUSHIKI KAISHA (Mitsubishi Electric Corporation)

HEAD OFFICE: 2-3, Marunouchi 2-chome, Chiyoda-ku, Tokyo 100
Telephone: 218 2111
PRESIDENT: Sadakazu Shindo

Mitsubishi Electric is prime contractor for licence production of Sparrow air-to-air missiles to arm F-4EJ fighters of the JASDF. This work began in FY 1972, and more than 800 Sparrows were delivered, for air-to-air use, between 1974 and 1979. Production is continuing, the current version being the advanced AIM-7F Sparrow (see Raytheon entry in US section).

MITSUBISHI JUKOGYO KABUSHIKI KAISHA (Mitsubishi Heavy Industries, Ltd)

HEAD OFFICE: 5-1 Marunouchi 2-chome, Chiyoda-ku, Tokyo 100

Mitsubishi Heavy Industries has Japanese government contracts to develop two important air-launched missiles, of which available details follow:

ASM-1

JASDF designation: Type 80

Development of the ASM-1 anti-shipping missile, to arm the Mitsubishi F-1 close-support fighter, was initiated in 1973 by Mitsubishi Heavy Industries, as prime contractor. In December 1980 the missile was accepted officially by the Japan Defence Agency, following the completion of evaluation tests. Production was initiated in FY 1980, and approximately 25-30 ASM-1s will be produced annually for the JASDF.

The missile is intended to travel to the target at a low altitude, using a Japan Aviation Electronics inertial system and TRT/Japan Radio ANV-7 radio altimeter for mid-course guidance, and a Mitsubishi Electronics active radar seeker for terminal guidance. Subcontractors include Nissan Motor Co for the rocket motors.

A surface-to-surface version of the ASM-1, designated SSM-2, is also being developed and produced by Mitsubishi, for the JGSDF.

DATA: See table

IR-AAM

As a result of its work on the AAM-1 and AAM-2 missiles, described briefly in earlier editions of *Jane's*, Mitsubishi has gained considerable experience in infra-red homing air-to-air missile design and potential. Under an agreement with the Technical Research and Development Institute of the Japan Defence Agency, it has begun the development and pre-series manufacture of a highly-manoeuvrable new dogfight missile known at present as the IR-AAM. No details are yet available.

NORWAY

KONGSBERG

A/S KONGSBERG VAAPENFABRIKK

HEAD OFFICE AND WORKS: Postboks 25, N-3601 Kongsberg
Telephone: (03) 73 82 50
Telex: 71491A vaapn n

Defence Products Division
MARKETING MANAGER, DEFENCE EQUIPMENT: Per Simonsen
PRODUCT INFORMATION: Jan E. Andersen

This government-owned company was prime contractor for European production of the Bullpup air-to-surface missile and participated in other NATO-co-ordinated programmes such as those for the Sidewinder air-to-air missile and NATO Sea Sparrow. It is manufacturer of the Norwegian-developed Penguin anti-ship missile system.

PENGUIN

This anti-ship missile system was developed by the Norwegian Defence Research Establishment and A/S Kongsberg Vaapenfabrikk. It is in quantity production, and can be installed on ships, helicopters and other platforms.

In its ship-to-ship version, Penguin is delivered in a container with integral launch-rail and can utilise most existing types of shipboard fire-control system. It embodies a solid-propellant rocket motor and a programmed inertial midcourse guidance system with infra-red terminal homing. Its warhead is similar to that of Bullpup, with a contact fuse.

The air-launched Penguin Mk 3 version was selected for carriage by F-16 fighters of the Norwegian Air Force, after evaluation in competition with the American Harpoon. Compared with the ship-launched version, it has an extended range, higher cruising speed and reduced wing span, and dispenses with the booster stage of the rocket motor. The missile's canard actuation system is being developed by Sperry Gyroscope of the UK; the launcher and power unit to be installed on the F-16s is being developed and produced by Saab-Scania of Sweden.

Penguin Mk 3 is expected to enter service in 1987.

DATA: See table

SOUTH AFRICA

ARMSCOR

ARMAMENTS CORPORATION OF SOUTH AFRICA

The Kentron branch of ARMSCOR is reported to have brought a new air-to-air missile, designated V3, beyond the development stage to the start of production.

V3

The V3 air-to-air missile was stated, in 1979, to have been totally developed and produced by Kentron, with support from the Council for Scientific and Industrial Research (CSIR), Atlas Aircraft Corporation and the South African Air Force. No details have been released, except that it is derived from the earlier Whiplash programme (see 1976-77 *Jane's*) and is claimed to be superior to the Matra 550 Magic. The V3 is said to be ready for operational use by the SAAF, to supplement and eventually replace the Sidewinders and Magics now in service.

SWEDEN

SBMC

SAAB BOFORS MISSILE CORPORATION

Stureplan 15, S-111 45 Stockholm
Telephone: 08 21 49 17
Telex: 12798 SBMC S
PRESIDENT: Arne Hult

This company was created jointly by Saab-Scania and Bofors, on 14 April 1978, to co-ordinate future missile activities within the two companies and to be responsible for marketing. The intention is that the work will continue to be located in the facilities of both parent companies, which remain responsible individually for weapons developed before the creation of SBMC. Details of three of these, the air-to-surface Saab RB04E and RB05A and Bofors RB53 Bantam wire-guided anti-tank missile, can be found in the 1977-78 *Jane's*.

In July 1979, SBMC was awarded a contract for the RBS 15 ship-to-ship weapon system for the Swedish Navy. Air-launched versions of both this missile and the surface-to-air RBS 70 are projected.

RBS 15

The RBS 15 anti-ship missile, shown in an accompanying illustration, is the first project undertaken by the Saab Bofors Missile Corporation. Saab-Scania is prime contractor, responsible for systems design and integration. Other Swedish companies are participating in the programme, which was launched after studies of the capability of existing foreign missiles such as Harpoon and Exocet to satisfy the requirements. They include Philips Elektronikindustrier for the frequency-agile radar seeker and fire control computer; Förenade Fabriksverken for the heavy warhead; and Bofors for the canister launcher.

As currently envisaged, the RBS 15 will require very little pre-launch preparation. All calculations will be performed automatically by a fire control computer, which will also control the status of the complete weapon system. The missile itself will be of the 'fire and forget' type, with a turbojet engine for long range at sea-skimming height, an ECM-resistant homing system, and a heavy and effective warhead.

The RBS 15 will be deployed initially as armament for Swedish torpedo boats of the 'Spica' class, with which it will become operational in 1985. Second stage of the programme will involve adapting the same weapon to meet Swedish Air Force requirements for its next type of attack aircraft.

DIMENSIONS:

Length overall	4·35 m (14 ft 3¼ in)
Body diameter	0·50 m (1 ft 7¾ in)
Wing span, folded	0·85 m (2 ft 9½ in)

WEIGHT:

Launching weight, without booster	560 kg (1,235 lb)

RBS 70

At the 1981 Paris Air Show, Saab Bofors exhibited a partial mockup of an attack helicopter carrying a four-round pack of RBS 70 missiles for anti-helicopter use. The basic RBS 70 is a portable or vehicle-mounted surface-to-air weapon employing optical beam-riding guidance on a laser beam. The missile weighs 24 kg (53 lb) in its container and has a max range of 2·7 nm (5 km; 3·1 miles).

UNION OF SOVIET SOCIALIST REPUBLICS

AIR-TO-AIR MISSILES

AA-1

NATO reporting name: Alkali

This Soviet air-to-air missile was produced as standard armament for the older generation of PVO-Strany interceptors, such as the Sukhoi Su-9 and all-weather versions of the MiG-19, and can be expected to disappear from service soon. Radars fitted to the interceptors are known to NATO as 'Spin Scan' and 'Scan Odd' respectively.

'Alkali' is a solid-propellant missile, with large delta cruciform wings at the rear and small cruciform foreplanes indexed in line with the wings. There appear to be control surfaces in the trailing-edges of the wings and 'Alkali' employs I/J-band semi-active radar homing. The warhead is carried immediately aft of the foreplanes.

DATA: See table

K-13A (AA-2)

NATO reporting name: Atoll

This missile, which has the Soviet designation K-13A, has long been standard equipment on home and export versions of the MiG-21, and is carried by export models of

Mitsubishi ASM-1 anti-shipping missile

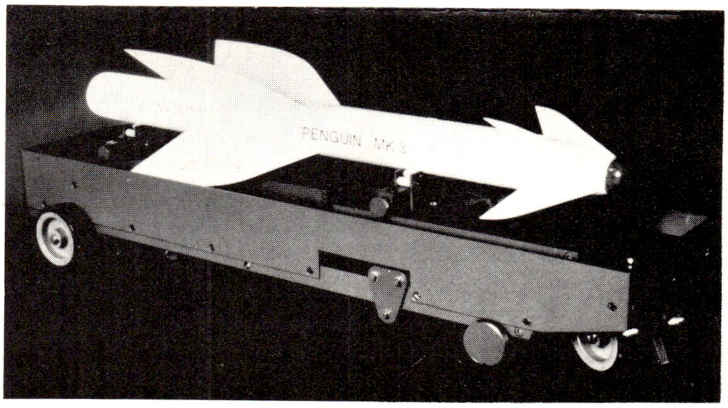

Penguin Mk 3 anti-ship missile on its transporter/loader

Artist's impression of the Swedish RBS 15 anti-ship missile

Mockup of RBS 70 anti-helicopter missile system *(Brian M. Service)*

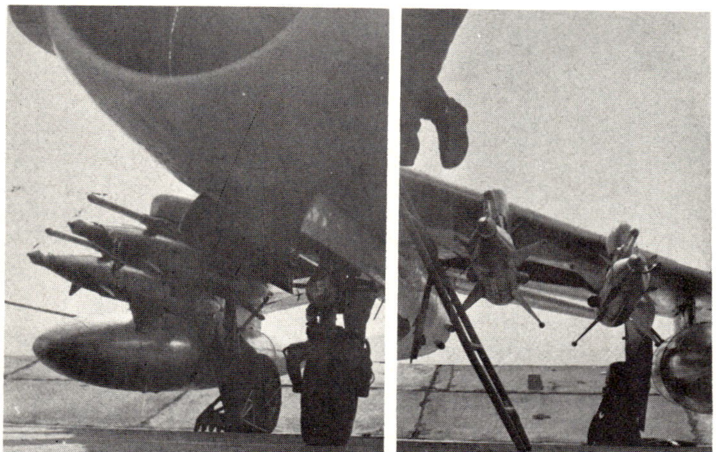

Two views of the missile known to NATO as 'Alkali' *(Tass)*

'Atoll' air-to-air missile under wing of MiG-21

'Anab' air-to-air missiles on a Yak-28P ('Firebar') fighter *(Flug Revue)*

the MiG-23 and Sukhoi Su-22. It is almost identical to the first-generation American Sidewinder (AIM-9B) in size and configuration and has similar infra-red guidance.

The body is cylindrical with cruciform control surfaces near the nose, indexed in line with the fixed cruciform tail-fins. There are no external cable or control conduits.

The triangular control surfaces have a compound sweep averaging about 60° on the leading-edge and 10° on the trailing-edge. They are linked in opposite pairs, with a maximum movement of 20-30°.

Leading-edge sweep on the tail-fins is about 40°, with straight trailing-edges. A small gyroscopically-controlled tab is inset in the trailing-edge of each fin, at the tip, presumably for anti-roll stabilisation but possibly with an added control function.

Nozzle diameter of the solid-propellant motor is 8 cm (3⅛ in). Weight and performance of 'Atoll' should be very similar to those of Sidewinder.

DATA: See table

AA-2-2

NATO reporting name: Advanced Atoll

The latest multi-role versions of the MiG-21 ('Fishbed-J, K, L and N') carry a mix of standard infra-red 'Atolls' and a version of this weapon with a radar homing head. The radar version is known as 'Advanced Atoll'.

AA-3

NATO reporting name: Anab

First seen as underwing armament on the Yakovlev Yak-28P fighter ('Firebar') in the 1961 Soviet Aviation Day display, 'Anab' is a standard air-to-air missile in the Soviet Air Force. It was carried by Yak-28, Sukhoi Su-11 and Sukhoi Su-15 interceptors taking part in the 1967 air display at Domodedovo. All three types are fitted with the radar known to NATO as 'Skip Spin'.

'Anab' has a cylindrical body, with small cruciform canard control surfaces indexed in line with very large cruciform tail-fins. Both infra-red and I/J-band semi-active radar homing versions are operational, and have been built in thousands.

DATA: See table

AA-5

NATO reporting name: Ash

'Ash' is the large air-to-air missile carried under the wings of Tupolev Tu-28P ('Fiddler') fighters. It has cruciform wings and tail surfaces indexed in line, and is operational in two versions, which have infra-red and I/J-band semi-active radar homing heads respectively.

The radar fitted to the Tu-28P is known to NATO as 'Big Nose'. It has been suggested that the MiG-25 ('Foxbat-A') also carries 'Ash' as one of its alternative weapons. Several thousand of these missiles have been built.

DATA: See table

Infra-red (left) and radar versions of 'Ash'

Launch of AS-2 ('Kipper') missile from a Tupolev Tu-16 ('Badger-C')

AA-6

NATO reporting name: Acrid

This air-to-air missile was identified during 1975 as one of the weapons carried by the 'Foxbat-A' interceptor version of the MiG-25, which has fire control radar known to NATO as 'Fox Fire'. Its configuration is similar to that of 'Anab' but it is considerably larger. Photographs suggest that the version of 'Acrid' with an infra-red homing head is normally carried on each inboard underwing pylon, with a radar-homing version on each outer pylon. The latter is about 6·10 m (20 ft) long, with an estimated range of at least 20 nm (37 km; 23 miles). The wingtip fairings on the fighter, different in shape from those of 'Foxbat-B', are thought to house continuous-wave target illuminating equipment for the radar-homing missiles.

'Fox Fire' radar is believed to be evolved from the Tu-28P's 'Big Nose', with a detection range of 43-54 nm (80-100 km; 50-62 miles) and ability to track targets at 27 nm (50 km; 31 mile) range.

AA-7

NATO reporting name: Apex

This long-range air-to-air missile is one of two types of missile known to be carried as standard armament by interceptor versions of the MiG-23. No details are yet available, except that 'Apex' has a solid-propellant rocket motor and exists in two versions, with infra-red and semi-active radar homing heads, the latter operating in conjunction with the MiG-23's 'High Lark' fire control radar. 'Apex' is also reported to be an alternative weapon for the MiG-25.

DATA: See table

AA-8

NATO reporting name: Aphid

Second type of air-to-air missile carried by the MiG-23, 'Aphid' is a close-range solid-propellant weapon with infra-red homing guidance.

The accompanying illustration suggests that it has a canard configuration, with small cruciform control surfaces immediately behind the hemispherical nose, indexed in line with the cruciform rear-mounted wings, which are similar in shape to those of 'Acrid'.

DATA: See table

AA-X-9

The missile known in the West as AA-X-9 is reported to have achieved successes against simulated cruise missiles, after 'lookdown/snapdown' launch from a modified MiG-25 interceptor. No details are available.

HELICOPTER MISSILES

AT-2

NATO reporting name: Swatter

This standard Soviet anti-tank weapon formed the original missile armament of the Mil Mi-24 ('Hind-A and D') helicopter gunship, and is carried by the 'Hip-E' version of the Mi-8. 'Swatter' is controlled by elevons on the trailing-edges of its rear-mounted cruciform wings, and embodies infra-red terminal homing, guiding the missile via two small, movable canard surfaces at the nose. The motor appears to exhaust through two vents diametrically opposed between the wings. Two more tubes, projecting rearward from opposite wings, probably house tracking flares.

DATA: See table

AT-3

NATO reporting name: Sagger

In conformity with the Soviet practice of not supplying advanced equipment on its export aircraft, the wire-guided 'Sagger' replaces 'Swatter' on the 'Hip-F' version of the Mi-8, as well as arming the Polish-built Mi-2. It has a short cylindrical body, with conical nose. Rear-mounted cruciform wings, which fold for storage, are swept near the body of the missile but unswept outboard. A hollow charge warhead is fitted.

DATA: See table

AT-6

NATO reporting name: Spiral

Unlike previous Soviet helicopter-launched anti-tank missiles, 'Spiral' does not appear to have a surface-launched application. Few details are yet available, except that it is tube-launched and homes on targets illuminated by a laser designator. It equips the 'Hind-E' version of the Mi-24, and is said to have a range of 3·75-5·3 nm (7-10 km; 4·3-6·2 miles).

AIR-TO-SURFACE MISSILES

AS-2

NATO reporting name: Kipper

First seen in the 1961 Soviet Aviation Day flypast, this aeroplane-configuration missile, with swept wings and underslung turbojet engine, was described by the commentator at Tushino Airport as an anti-shipping weapon. Radar is carried in the nose of the Tu-16 ('Badger-C') carrier aircraft, and guidance is believed to comprise initial beam-riding, subsequent pre-programmed flight under autopilot control, and infra-red terminal homing. A nuclear warhead can be fitted.

DATA: See table

AS-3

NATO reporting name: Kangaroo

Largest of the air-to-surface missiles first seen in the Soviet Aviation Day display at Tushino in 1961, and known still to equip more than half of the Tu-95s ('Bears') of the Soviet Long-Range Aviation bomber force, 'Kangaroo' is a winged missile with an airframe similar in size and shape to a sweptwing turbojet-powered fighter aircraft. The tail unit is conventional, with sweepback on all surfaces. The vertical surfaces, concealed inside the launch aircraft until the missile is dropped, are of rhomboid form. A nuclear or high-explosive warhead can be fitted.

What was believed originally to be a radome on the missile's nose was identified subsequently as either a duct through which air can be fed to start the missile's turbojet engine prior to launching or a fairing over the air intake. Radar is carried in the nose of the Tu-95 launch aircraft and guidance is assumed to be by initial beam-riding and subsequent pre-programmed flight under autopilot control.

DATA: See table

AS-4

NATO reporting name: Kitchen

Developed as a stand-off weapon for the Tu-95 and Tu-22 strategic bombers, and now carried also by the variable-geometry 'Backfire', the AS-4 was first seen on a single Tu-22 ('Blinder-B') in 1961. Most of the 22 Tu-22s which participated in the 1967 Aviation Day display at Domodedovo carried an AS-4, semi-submerged in the fuselage, and production by 1976 was stated by the UK Defence Minister to be around 1,000. The missile, which has been seen in more than one form, has an aeroplane configuration, with stubby delta wings and cruciform tail surfaces. Propulsion is believed to be by liquid-propellant rocket motor. Guidance is reported to be inertial, with infra-red terminal homing. The speed of 'Kitchen' is probably high, and a choice of nuclear or high-explosive warhead can be assumed.

DATA: See table

AS-5

NATO reporting name: Kelt

This air-to-surface missile, carried by the Tu-16 bomber, is externally similar to the earlier turbojet-powered 'Kennel' missile, which looked rather like a scaled-down unpiloted version of the MiG-15 fighter with a hemispherical radome above its air intake. On 'Kelt', both the ram-air intake and radome are replaced by a hemispherical nose fairing, probably housing a larger radar. 'Kelt' is liquid-propellant rocket-powered and appears to be longer than 'Kennel'. Guidance is said to be by autopilot on a pre-programmed flight path, with radar terminal homing which can be switched from active to passive as required. Well over 1,000 'Kelts' are thought to have been delivered, with high-explosive warheads.

According to Israeli reports, about 25 'Kelts' were launched against Israeli targets by Tu-16s from Egypt during the Arab-Israeli War of October 1973. Twenty of the missiles were claimed as destroyed by the air and ground defences; the others hit two radar sites and a supply centre in Sinai.

DATA: See table

AS-6

NATO reporting name: Kingfish

First sighting of this air-to-surface missile was by the pilot of a Japan Air Self-Defence Force F-86F in December 1977. When investigating a Tu-16 ('Badger') flying 43 nm (80 km; 50 miles) to the north of the Noto Peninsula, he was able to photograph the bomber, carrying a 'Kingfish' under its port wing. The missile has a cylindrical body with ogival nose, two short-span long-chord wings, and a cruciform tail unit with folding ventral fin. Propulsion is said to be by liquid-propellant rocket motor, with inertial midcourse guidance and active radar terminal homing, giving exceptional accuracy. Primary carrier was expected to be the variable-geometry 'Backfire'; there has been little evidence of this, but a Tu-16 was photographed near Japan in early 1980 with a 'Kingfish' under each wing. Others have since been photographed over the Baltic.

DATA: See table

AS-7

NATO reporting name: Kerry

This tactical air-to-surface missile is reported to be carried by the Sukhoi Su-24 ('Fencer') fighter-bomber and other current Soviet close-support aircraft. It is said to have a radio command guidance system, to weigh about 1,200 kg (2,650 lb) and to have a range of 5·4 nm (10 km; 6·2 miles).

AS-X-9

A reported anti-radiation missile, with a range of 43-48 nm (80-90 km; 50-56 miles), to arm the Sukhoi Su-24 ('Fencer').

NEW MISSILE

US administration officials are reported to have said, on 1 February 1979, that the Soviet Union had begun to test a missile in the category of the USAF's ALCM. They claimed that at least eight missiles had been test-launched from 'Backfire' bombers during preceding months, over ranges of about 650 nm (1,200 km; 750 miles). This was said to be the first suggestion that the Soviet Union was testing long-range cruise missiles from its bombers.

The same US officials were quoted as saying that the maximum range of the new Soviet missile could be as much as 1,300 nm (2,400 km; 1,500 miles), but that it did not appear to have the high accuracy of the US ALCM designs or their ability to approach targets at tree-top heights.

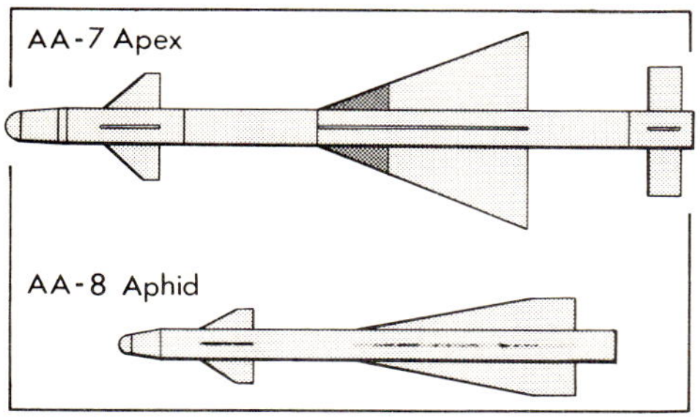

'Apex' and 'Aphid' missiles *(Michael A. Badrocke, with reference to data from Hasegawa Seisokusho Company)*

'Kangaroo' air-to-surface missile under the fuselage of a Tu-95 ('Bear-B') strategic bomber

'Swatter' anti-tank missiles on BRDM vehicle. Missiles of this type arm Soviet helicopters *(Tass)*

Soviet Tu-16 carrying a 'Kingfish' missile under each wing *(JASDF)*

One of several types of 'Kitchen' air-to-surface missile photographed under 'Backfire-B' bombers

The 'Badger-G' version of the Tupolev Tu-16, with two 'Kelt' missiles on underwing launchers

UNITED KINGDOM

BAe
BRITISH AEROSPACE PUBLIC LIMITED COMPANY

CORPORATE HEADQUARTERS: Brooklands Road, Weybridge, Surrey KT13 0SJ
Telephone: (0932) 45522
Telex: 27111

BRITISH AEROSPACE DYNAMICS GROUP

HEADQUARTERS: Six Hills Way, Stevenage, Herts SG1 2DA
Telephone: (0438) 2422
Telex: 825125
CHAIRMAN AND CHIEF EXECUTIVE: Admiral Sir Raymond Lygo
GROUP DEPUTY CHIEF EXECUTIVE: T. G. Kent
GROUP SALES AND MARKETING DIRECTOR: L. A. Sanson, OBE
SECRETARY: M. W. Plimley, LLB
HEAD OF PUBLIC RELATIONS: M. K. Hird

Stevenage Division (Army Weapons)
MANAGING DIRECTOR: R. J. Parkhouse, MSc, BSc, CEng

Hatfield Division (Air Weapons)
MANAGING DIRECTOR: L. G. Evans, MA, CEng, FRAeS

Bristol Division (Naval Weapons)
MANAGING DIRECTOR: H. Metcalfe, OBE, BSc, ARCS, CEng, FRAeS

Space and Communications Division (Stevenage Site B)
MANAGING DIRECTOR: P. L. V. Hickman

British Aerospace Dynamics Group was established on 1 January 1978 as a separate operating unit within British Aerospace, to continue the guided weapon and associated business of the predecessor companies. On 1 January 1980 it was reorganised to comprise the Group Board and Headquarters, and four business units (Divisions). Lostock works continues as a separate self-accounting centre and a main production unit of Dynamics Group. Total number of employees on 1 January 1980 was approximately 17,500.

Among weapons for which the Dynamics Group of British Aerospace is currently prime contractor are the air-launched missiles described in this section, the Rapier and Seawolf surface-to-air missiles, the Sea Dart ship-to-air and ship-to-surface missile, and the Swingfire series of wire-guided anti-tank missiles.

Dynamics Group produces components of the American TOW anti-tank weapon system, excluding the missile itself but including the roof sight, for installation on British Army Lynx helicopters; and items for the guidance and control units and other sections of AIM-9L Sidewinder air-to-air missiles built under a joint European manufacturing programme. It is the British 'daughter firm' for the Australian-developed Ikara long-range anti-submarine weapon used by the Royal Navy. Other defence work includes development and manufacture of infra-red linescan systems which equip the Jaguar aircraft and the CL-89 surveillance RPV (see RPVs section).

SEA SKUA

This lightweight all-weather sea-skimming anti-ship missile is designed for operation from the Royal Navy's Lynx helicopter, to meet the threat from missile-firing fast patrol craft and to provide a good stand-off capability for the helicopter, with consequent protection from counterattack. The complete helicopter-borne weapon system includes surveillance and tracking radar, fire control equipment, launching arrangements and up to four of the semi-active homing missiles. Main contractors are Dynamics Group of British Aerospace for the missile and fire control, Ferranti for the Sea Spray radar, and Westland for the Lynx helicopter.

In addition to its ability to engage fast patrol boat targets, Sea Skua has a significant effectiveness against larger warships. It offers high lethality, with the capability of firing salvoes. The radar provides continuous surveillance, and includes facilities for weapon control and target identification.

Sea Skua is a quick-reaction system, at immediate readiness throughout the helicopter sortie, with a high degree of system reliability and full capability in a complex ECM environment. Wherever possible, it embodies existing, proven techniques and components. The solid-propellant boost and sustainer motor provides a high subsonic speed. Sea-skimming capability is conferred by a licence-built TRT radio altimeter, with Marconi Space and Defence Systems semi-active radar for terminal homing and a high-efficiency warhead actuated by a direct-action fuse. Simple go/no-go testing is all that is required on board ship before the missile is mounted on the helicopter.

The weapon system is in full production and entered service on Lynx helicopters of the Royal Navy in 1981, as the Navy's principal all-weather light strike weapon for defence of the fleet. It can be fitted to other medium-sized helicopters and long-range maritime patrol aircraft.

DATA: See table

SEA EAGLE (P3T)

On 17 August 1977, Dr John Gilbert, UK Minister of State for Defence, announced that a sea-skimming 'fire and forget' anti-ship missile designated P3T was to be developed to replace Martel on RAF aircraft in the 1980s. Since given the name Sea Eagle, it is powered by a Microturbo TRI-60 turbojet engine instead of the rocket motor of the earlier missile, to give longer range. The configuration follows closely that of Martel, except for the addition of an underbelly air intake for the engine.

The Sea Eagle is an all-weather missile, with an active radar homing head developed by Marconi Space and Defence Systems. Prior to launch, the onboard microcomputer is supplied with target positional information from the carrier aircraft. The computer controls the flight path of the Sea Eagle until the target is acquired by the seeker during the final sea-skimming phase of the attack. The missile is designed to operate against targets protected by sophisticated ECM.

British Aerospace Dynamics Group announced on 31 July 1979 that it had completed the definition phase of the programme, as a result of which it had been awarded a contract to complete development and evaluation of Sea Eagle, and to supply an initial production quantity. The weapon will arm RAF Buccaneer and Tornado GR. Mk 1 aircraft and the Royal Navy's Sea Harriers. Air carriage and preliminary launch trials began in November 1980. The first flight demonstrating use of the turbojet and sea-skimming autopilot was made successfully from a Buccaneer on 24 April 1981.

DATA: See table

SKY FLASH

Swedish Air Force designation: RB71

On 17 April 1973 Hawker Siddeley Dynamics (now British Aerospace Dynamics Group) announced that it had received a prime contract from the Ministry of Defence for project definition and pre-development studies of a medium-range all-weather air-to-air missile based on the American AIM-7E Sparrow, which is itself in service with the RAF. The name Sky Flash was given subsequently to the new missile. Raytheon, the US Sparrow prime contractor, is participating in the programme through cross-licensing agreements with British Aerospace and Marconi Space and Defence Systems Ltd.

Project definition was completed in 1973, and it was announced on 26 February 1974 that Dynamics Group had received instructions from the Ministry of Defence to proceed with full development and initial production of Sky Flash. A contract for full production was awarded in 1975, initially to arm the RAF's Phantom and, later, the Tornado F. Mk 2.

The 'boost and coast' Sky Flash has the same general configuration and dimensions as the Raytheon AIM-7E, but is fitted with a semi-active radar homing head of inverse monopulse design, developed by Marconi Space and Defence Systems under a Ministry of Defence contract. This has been designed as a completely self-contained unit with modular construction, which provides considerable flexibility in the overall design to cater for a wide range of different applications. Microstrip circuitry and the latest types of solid-state technology and micro-circuits have offered reliability, reduced size and high performance. The complete homing head is only 283 mm (11 in) long and can be fitted in any appropriate missile body of 180 mm (7 in) internal diameter.

The advanced radar proximity fuse is designed by EMI Electronics Ltd, and is claimed to offer a high single-shot kill capability against targets at high, medium and low altitudes. Dynamics Group has updated the autopilot and power systems, and is responsible for building the missile structure, as well as for assembly and test.

The successful completion of Sky Flash firing trials was announced in December 1977, by which time 22 missiles had been launched from US Navy F-4J Phantom aircraft over the Pacific Missile Test Center, California, in a variety of attack situations against targets representative of a range of hostile aircraft types. These included targets flying at subsonic and supersonic speeds, singly and in formation, and in severe ECM environments. Several missiles scored direct hits and others were classified as partial successes, simple remedial modifications being made to subsequent missiles. During the trials the snapdown capability of Sky Flash was demonstrated when a missile launched from medium altitude achieved a target 'kill' against an unaugmented QF-86 drone flying only 150 m (500 ft) above the ground. Aircraft from which Sky Flash has since been launched successfully include the General Dynamics F-16 Fighting Falcon. The weapon has shown itself able to engage hostile aircraft at ranges in excess of 21 nm (40 km; 25 miles) at high altitudes or as low as 75 m (250 ft).

The first firing of a Sky Flash in the UK from an RAF aircraft was made on 15 August 1979 from a Phantom of No. 29 Squadron, which shot down a Meteor target aircraft flying at a lower level over the Aberporth test range.

The first export contract, valued at £60 million, was announced in December 1978 and is providing Sky Flash missiles to arm the Swedish Air Force's Saab JA 37 Viggen all-weather fighter. The Sky Flash integration programme for the Viggen, involving many air firings, with a 100 per cent success rate, was completed by the end of 1979, and delivery of production missiles began in October 1980. A follow-on contract, valued at £11 million, was placed by Sweden in June 1981.

Further work on an improved Sky Flash Mk 2 (see 1980-81 *Jane's*) was cancelled by the UK Ministry of Defence, as an economy measure, in January 1981.

DATA: See table

ASRAAM PROGRAMME

A new fully-guided short-range air-to-air missile developed by British Aerospace Dynamics Group has undergone a series of successful test firings, in which it has demonstrated outstanding manoeuvrability. Shown in an accompanying illustration, the missile is wingless, uses an infra-red seeker, is steered by swivel-nozzle thrust vector control, and has a motor bleed roll control system that vents exhaust gases through four tangential thrusters. Motor and actuation systems were developed by Imperial Metal Industries and Sperry Gyroscope.

Fired from a Hunter aircraft, the missile intercepted an airborne target over the Aberporth range on 18 August 1980. The trial was set up to provide severe interception conditions, using a missile that is the result of a substantial development programme. The technologies used and experience gained will make a valuable contribution to the European advanced short-range air-to-air missile (ASRAAM) programme in which British Aerospace is UK lead contractor.

In 1980, an ASRAAM pre-feasibility study was conducted jointly by British Aerospace Dynamics Group in the UK and Bodenseewerk Gerätetechnik, the lead contractor in West Germany. The missile is intended as a Sidewinder replacement, for eventual manufacture in both Europe and, under licence, in the USA. Parallel US development of an advanced medium-range air-to-air missile (AMRAAM) is intended to result in production of the weapon in the USA and, under licence, in Europe, so providing a modern NATO-standard air-to-air missile inventory.

UNITED STATES OF AMERICA

BOEING
BOEING AEROSPACE COMPANY

HEAD OFFICE AND WORKS: Seattle, Washington 98124

ALCM

USAF designation: AGM-86

The Air-Launched Cruise Missile (ALCM) is a small unmanned winged air vehicle capable of sustained subsonic flight following launch from an airborne carrier aircraft. It is propelled by a turbofan engine, incorporates a nuclear warhead for 'hard target' kill capability, and is programmed to strike a pre-determined surface target. Guidance is by a combination of inertial and terrain comparison (Tercom) techniques. In the terrain comparison mode the missile is kept on course by a computer which compares pre-programmed geographical features on its flight plan with the geography seen by its sensors during actual flight.

When launched in large numbers, each of the missiles would have to be countered, making defence against them difficult and costly. The missiles are intended both to serve as stand-off weapons and to dilute defences and improve the ability of manned aircraft to penetrate to major targets. Small radar signature and low-level flight capability would enhance their effectiveness.

In its initial form, designated **AGM-86A**, Boeing's ALCM design made maximum use of the air vehicle developed under the former AGM-86A Subsonic Cruise Armed Decoy (SCAD) programme, for which Boeing was prime contractor. It was similar in overall dimensions to

Sea Skua anti-ship missile on Lynx prior to successful first launch

Sea Eagle anti-ship missile on RAF Buccaneer

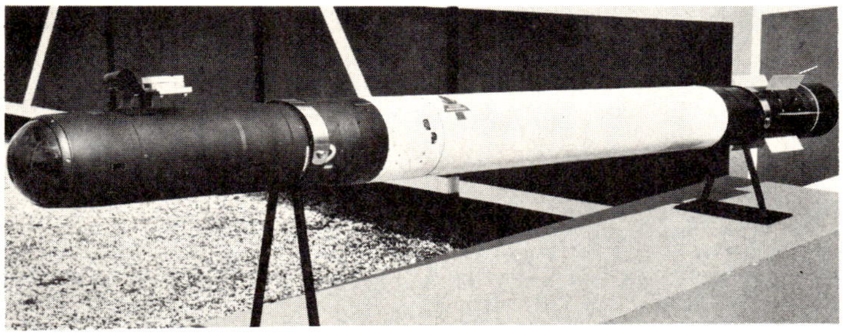

British Aerospace's wingless infra-red homing air-to-air development missile *(Brian M. Service)*

Sky Flash air-to-air missiles on inboard underwing hardpoints of JA 37 Viggen

Exhibition model of Boeing Wasp *(Brian M. Service)*

Boeing AGM-86B air-launched cruise missile in flight

SRAM (short-range attack missile; see 1977-78 *Jane's*) and was suitable for carriage on the rotary launcher developed for this latter weapon, with wings and tail folded and engine air intake retracted. A B-52G was expected to carry 12 AGM-86As externally and 8 internally, complementing the aircraft's SRAMs and free-fall weapons. When carried externally, the missile was to have an underbelly auxiliary fuel tank fitted to increase its range.

Six test flights (three performance and three navigation/guidance) were conducted with AGM-86As between March and November 1976. Development was then switched to the **AGM-86B**. This has a fuselage more than 30% longer than that of the AGM-86A, to house sufficient fuel for double the range of the basic 'A'. On 1 February 1978, Boeing Aerospace received the first increment of a contract calling for a competitive fly-off between the AGM-86B and an AGM-109 air-launched version of the General Dynamics Tomahawk SLCM. Department of Defense contracts funded a total of 24 Boeing/GD ALCMs in FY 1978, and a similar quantity in FY 1979. Ten examples of each missile were used in the competitive fly-off, which ended on 8 February 1980. Three of the AGM-86Bs crashed, and another flight suffered unscheduled termination; four of the AGM-109s crashed. Programme officials considered that, overall, the fly-off had achieved its targets, and Boeing was declared the winner on 25 March 1980. Reasons for the choice were given as better guidance system software, aerodynamic features that offered better terrain following in rough country, and promise of superior field maintenance characteristics. A follow-on series of up to 19 test and evaluation flights began on 12 June 1980 and was to extend into late 1981. Of the first nine missiles launched in this series, five completed their missions.

Current plans envisage procurement of 3,418 ALCMs between FY 1980 and FY 1987, with deliveries to be completed in FY 1989. Initial contract, placed on 2 May 1980, was for 225 missiles, at a cost of $365·4 million, plus $111 million for initial spares, RDT&E and military construction. The budget for FY 1981 provides a further increment of 480 missiles, and 440 are requested in FY 1982. Scheduled date for fitting an AGM-86B to the first operational B-52G, at Griffiss AFB, NY, was September 1981. Initial operational capability, comprising one squadron of 14 B-52Gs externally loaded with missiles (six under each wing) is expected to be achieved in December 1982. Subsequently, each bomber is intended to be modified to have a bomb-bay rotary launcher for eight more ALCMs, eight SRAMs, or a mixture of both.

Boeing manufactures AGM-86B airframes. The missile's F107-WR-100 turbofan engine (2·67 kN; 600 lb st) continues to be produced by Williams International Corporation, as prime contractor; but Teledyne CAE has been permitted to tender as second source contractor from FY 1981. McDonnell Douglas is responsible for supplying guidance systems, with inertial navigation platforms by Litton.

DATA: See table

WASP

Boeing announced in February 1980 that it had received a $43 million contract from the US Air Force Armament Division, covering its participation in the competitive validation phase of the Wasp anti-tank missile programme. A similar contract was awarded to Hughes Aircraft Company, under whose entry further details of the programme can be found. A further $11·2 million contract was awarded to each company in 1981 to compensate for an extension of the development time-scale.

The Boeing Wasp missile is approximately 1·37 m (4 ft 6 in) long and weighs 45 kg (100 lb). It is capable of being launched individually or in salvo from pod launchers on aircraft. Each missile carries its own seeker and guidance system for maximum accuracy in even adverse weather. The validation phase is planned to be completed within 42 months, after which Boeing or Hughes will be selected to continue full-scale engineering development of Wasp. Work on target seeker designs, under Boeing subcontract, is being carried out by Sperry Microwave Electronics and Raytheon Missile Systems Division.

FORD

FORD AEROSPACE & COMMUNICATIONS CORPORATION

HEAD OFFICE: 20th Floor, 300 Renaissance Center, PO Box 43342, Detroit, Michigan 48243

Aeronutronic Division

Ford Road, Newport Beach, California 92663
Telephone: (714) 759 5996

Ford Aerospace & Communications Corporation, and its Philco predecessor, have produced Sidewinder air-to-air missiles and Chaparral surface-to-air missiles for the US armed services for many years. Details of the Sidewinder can be found under the NWC entry in this section.

GENERAL DYNAMICS

GENERAL DYNAMICS CORPORATION

HEAD OFFICE: Pierre Laclede Center, St Louis, Missouri 63105

Convair Division

5001 Kearny Villa Road, San Diego, California 92123
VICE-PRESIDENT AND GENERAL MANAGER:
Dr Leonard F. Buchanan

Pomona Division

PO Box 2507, Pomona, California 91766
VICE-PRESIDENT AND GENERAL MANAGER:
Ralph E. Hawes

The Pomona Division of General Dynamics is responsible for development and production of a wide variety of tactical weapon systems for the US Navy, US Army, US Air Force and friendly foreign powers. Current and recent products include the Standard Missile-1 naval ship-launched weapon, the Standard ARM anti-radiation missile, the Stinger advanced man-portable air defence weapon, and Phalanx, the Navy's all-weather automatic gun fire-control system. In pilot production is the longer-range Standard Missile-2.

In addition, the US Naval Air Systems Command awarded a contract to Pomona for second-source production of the AIM-7F Sparrow air-to-air missile (see Raytheon entry).

Convair Division was selected in 1976 to develop the US Navy's Sea Launched Cruise Missile, under the designation BGM-109 Tomahawk. Variants include the USAF's Ground Launched Cruise Missile; and the new Tomahawk II medium-range air-to-surface missile (MRASM) is based on Tomahawk.

TOMAHAWK II (MRASM)

US military designation: AGM-109

In March 1980, the Department of Defense announced that a new medium-range air-to-surface missile (MRASM) for tactical operations was to be based on the General Dynamics Tomahawk cruise missile. Intended to provide tactical aircraft with a reasonable-cost subsonic stand-off weapon with which to attack heavily defended high-value targets, MRASM will be produced initially for US Air Force use in attacking airfields, using runway-cratering conventional submunitions.

Guidance for the Tomahawk II will be based on the Tercom inertial navigation/terrain contour matching system used in the land attack Tomahawk Sea Launched Cruise Missile (SLCM), with the addition of digital scene matching area correlation (DSMAC) for final target homing. Lower-cost hardware, such as ring laser gyro and tactical computers, may also be used.

For the USAF, MRASM will be 5·80 m (19 ft) long and will have a launch weight of approximately 1,315 kg (2,900 lb). It will be powered by a Teledyne CAE J402 turbojet engine, and will have a range of approximately 250 nm (460 km; 285 miles). A version that could meet the requirements of a stand-off weapon for the US Navy would be somewhat shorter (4·88 m; 16 ft), to allow for carrier operations.

HUGHES

HUGHES AIRCRAFT COMPANY

Corporate Office: Culver City, California 90230
Chairman of the Board and Chief Executive Officer:
Dr Allen E. Puckett
President: John H. Richardson

Missile Systems Group
Canoga Park, California 91304
Telephone: (213) 883 2400
Vice-President and Group Executive:
Dr Malcolm R. Currie

Electro-Optical & Data Systems Group
Culver City, California 90230
Vice-President and Group Executive: H. W. Boehmer

Hughes Aircraft began developing an air-to-air missile for the USAF in 1947, and this weapon has been in squadron use for many years as the Falcon. Currently, Hughes is producing an air-to-air missile named Phoenix for the US Navy, an air-to-surface missile named Maverick for US and foreign operators, and a wire-guided anti-tank missile named TOW for the US Army and other customers. It is also engaged in prototype validation of an advanced medium-range air-to-air missile (AMRAAM). The planar wing GBU-15 glide bomb programme, described in the 1980-81 *Jane's*, has been terminated.

PHOENIX

US Navy designation: AIM-54

Phoenix arms the Grumman F-14A Tomcat two-seat carrier-based fighter and is claimed to have capabilities exceeding those of any other air-to-air missile yet operational. It has a solid-propellant rocket motor, and is fitted with a large proximity-fused high-explosive warhead.

The launch aircraft's AWG-9 weapon control system is able to track an enemy target, at high or low altitude, in any kind of weather, and launch the Phoenix missile. The data from the radar is processed by a high-speed digital computer, the output of which is displayed to the missile control officer in the launch aircraft on two displays: a 254 mm (10 in) cathode-ray tube and a 127 mm (5 in) multimode storage tube. The long-range high-power pulse-Doppler radar has a 'lookdown' capability that enables it to pick out moving targets from the ground clutter that normally obscures them on a conventional radar. The AWG-9 has a track-while-scan radar mode that makes it possible to launch up to six missiles and keep them on course while searching for other possible targets, all in the presence of sophisticated enemy countermeasures.

All F-14s built for the US Navy are equipped to carry Phoenix. Production was initiated in late 1970, and continued missile purchases are envisaged until at least 1991. Details of early testing can be found in the 1979-80 *Jane's*. Phoenix entered first-line service in September 1974, when two squadrons of US Navy F-14s were deployed on board the nuclear-powered USS *Enterprise* for a six-month tour of duty in the Western Pacific. Eight US carriers were operating F-14 fighter squadrons in 1979.

There are two versions of Phoenix, as follows:

AIM-54A. Initial production version, to which all details above apply. Manufacture ended in last quarter of 1980; approx 2,500 built. Of 155 production rounds launched between May 1972 and Autumn 1980, 92 per cent were judged to have achieved target 'kills'.

AIM-54C. Improved version, with much enhanced ECCM capability and reliability. New target detecting device, digital electronics unit, solid-state radar transmitter and Northrop strapdown inertial reference system for improved accuracy and greater range. First of 15 engineering development models delivered in Summer 1980. Firing trials, started later that year, began with three passes well within the lethal distance from drone targets by unarmed missiles. Thirty pilot production missiles following in 1981-82, leading to series production of approximately 1,400 AIM-54Cs in 1982-90. AIM-54As in USN inventory may also be updated with AIM-54C electronics.

Data: See table

AMRAAM

On 5 February 1979, the USAF announced that it had selected Hughes Aircraft Company's Missile Systems Group and Raytheon Company's Missile Systems Division to begin prototype validation on an advanced medium-range air-to-air missile (AMRAAM). This weapon is intended to replace the AIM-7 Sparrow in USAF and US Navy service, as primary armament of the F-14, F-15, F-16 and F/A-18 fighters.

The AMRAAM programme was initiated by a 13-month USAF/USN/USMC study of the likely air-to-air threat for the next 30 years, and the kind of beyond-visual-range weapon that would best meet such a threat.

Basic parameters laid down by the AMRAAM Joint System Program Office included an upper weight limit of about 160 kg (350 lb), with 136 kg (300 lb) as the desirable goal; all-weather capability; snapdown capability to permit interception of low-flying targets over land or sea; a resistance to ECM at least as good as that offered by current monopulse techniques; all-aspect launch and tracking ability; a minimum range at least equal to that of the AIM-7F Sparrow and a maximum range as great as the state of the art permits. These requirements seemed likely to demand inertial midcourse guidance and self-contained terminal homing such as active radar; however, it seemed likely that a missile with these features would appproach its target so closely that a conventional proximity-fused warhead weighing only 14-22 kg (30-50 lb) would be adequate.

Few details of Hughes' AMRAAM missile have been released, except that it embodies a Northrop strapdown midcourse inertial reference system. Hughes claims that it will outperform the Sparrow, cost less to build and be one half as heavy as the earlier weapon. Its version of AMRAAM will offer maximum independence of the launching aircraft, and will enable the pilot to make launches against multiple targets with no changes in current cockpit procedures.

First launch of a powered test round, without seeker, warhead or control systems, was made sucessfully from an F-16 flying at Mach 0·85 at 6,100 m (20,000 ft) over White Sands Missile Range, New Mexico, on 20 February 1981. The AMRAAM validation phase is expected to last 33 months, after which the winning design will be taken through full-scale engineering development into production.

TOW

US Army designation: BGM-71

Hughes Aircraft was prime contractor for development of this high-performance surface-to-surface and air-to-surface anti-tank missile, the basic characteristics of which are indicated by its name, as TOW is an acronym for Tube-launched, Optically-tracked, Wire-guided.

The basic ground-fired TOW system consists of a glassfibre launch tube, a tripod, a traversing and sighting unit, an electronic package, and missiles encased in shipping containers. Total weight of the entire weapon system, including missile, is approximately 91 kg (200 lb).

The missile has low aspect ratio wings and tail control surfaces that remain folded while in the launcher and flick open as the missile leaves the launch tube. The wings flick forward during extension, the tail surfaces rearward.

TOW is inserted into the rear end of the tube in its container, which forms an extension of the tube. Electrical and mechanical connections to the missile are made automatically during this operation. There are two separate motors. The launch motor fires first to propel the missile from the launcher. The missile coasts for a short period after leaving the mouth of the tube, before the second (boost) motor fires.

The operator guides the missile by keeping the target centred in a telescopic sight. Movement of the sight generates electronic signals to correct the missile's course, the signals being passed through two wires. The warhead is a high-explosive shaped charge, developed under the Army Munitions Command, Picatinny Arsenal, New Jersey.

In this basic BGM-71A form, TOW is a heavy assault weapon for use against tanks, armoured vehicles and gun emplacements over ranges of more than 3,750 m (12,300 ft). To permit its use from helicopters, Hughes developed under US Army contract a gyro-stabilised sight that would eliminate the effects of aircraft vibration and manoeuvres. As part of the XM26 missile/launcher/sight subsystem, this was installed on a UH-1B helicopter, from which air-to-surface firing tests of TOW missiles were then made successfully at Redstone Arsenal, Alabama, with hits on moving tank targets over ranges of more than one mile.

Two UH-1Bs with early-model Hughes XM26 sighting systems were shipped to Viet-Nam on 24 April 1972. After a short training period, during which each pilot-gunner fired one TOW missile for the first time, the aircraft were committed at Kontum to meet an expected armour threat. By 27 June, in 77 combat launches, they had scored 62 hits on point targets and had destroyed 39 armoured vehicles, trucks and howitzers, without themselves being hit by hostile fire.

As a result of these successes, the US Army embarked on a programme to convert AH-1G HueyCobras to AH-1Q/S Cobra/TOW standard, with improved M65 missile system. In addition, TOW arms helicopters of various types selected by foreign services, including the Hughes 500 MD Defender, Italy's Agusta A 109, and the British Army Lynx.

TOW has been in production for the US Army since 29 November 1968. By the beginning of 1981, more than 275,000 missiles had been delivered from Hughes' factory at Tucson, Arizona, with production continuing at a rate of about 3,000 a month. Deployment to US troops in the USA and Europe began in November 1970, to replace the 106 mm recoilless rifle and the Entac and SS.11 missiles. Subsequently, TOW was ordered by more than 33 other countries. In over 10,500 firings, including combat, TOW has demonstrated an operational reliability better than 96 per cent, and 83 per cent of the missiles have hit their target.

During the late Summer of 1979, at Fort Hood, Texas, first tests were conducted of an AH-1S HueyCobra fitted with a modified TOW system that provides a FLIR (forward-looking infra-red) night vision capability. Known as the FLIR Augmented Cobra TOW Sight (FACTS), the system utilises a standard US Army thermal imaging night vision sensor. Following an initial firing using the sight's daylight viewing channel, two TOWs were fired in early evening and two in total darkness. All five missiles scored hits on stationary targets. US Army evaluation tests were to follow.

In September 1980 it was announced that the US Army had initiated a two-stage programme to improve the performance of TOW against advanced armour. First phase involves use of an improved 127 mm (5 in) diameter warhead, approximately the same size and weight as the current warhead, to increase the missile's armour penetrating capability. It will be capable of replacing the original type on all existing TOW missiles, without any changes to the launcher or guidance hardware, and will be applicable to all TOW system platforms, including helicopters and tracked vehicles. Development and testing of the warhead, which embodies an extensible probe to provide standoff detonation, were underway at the time of the announcement. DoD requested $105·2 million in FY 1981 to purchase 18,000 of the improved warheads to retrofit existing TOWs, and $76·6 million for 12,000 new missiles with this warhead. Deployment is expected to begin in the early 1980s.

The improved TOW will be followed later in the decade by a more extensively upgraded **TOW 2**. This will introduce a heavier 152 mm (6 in) warhead, occupying the full diameter of the missile body, and a microprocessor-based digital missile guidance set, which will provide greater flexibility in guidance programming and higher precision. The boost motor will be reloaded with an improved propellant, to provide a higher impulse that will offset the additional weight of the warhead and equipment changes. TOW 2 modifications will be applicable by retrofit to early missiles, but will require more sophisticated procedures than first-stage improved TOW upgrading. Any earlier TOW missile will, however, be able to be fired from a TOW 2 launcher. DoD funding of $20·6 million was requested in FY 1981 to continue development of TOW 2, which is focussed initially on infantry applications but is expected to be introduced into other launch platforms. Both warhead improvement programmes are being directed by the US Army's Research and Development Command, Picatinny Arsenal.

Data: See table

First engineering development model AIM-54C Phoenix air-to-air missile

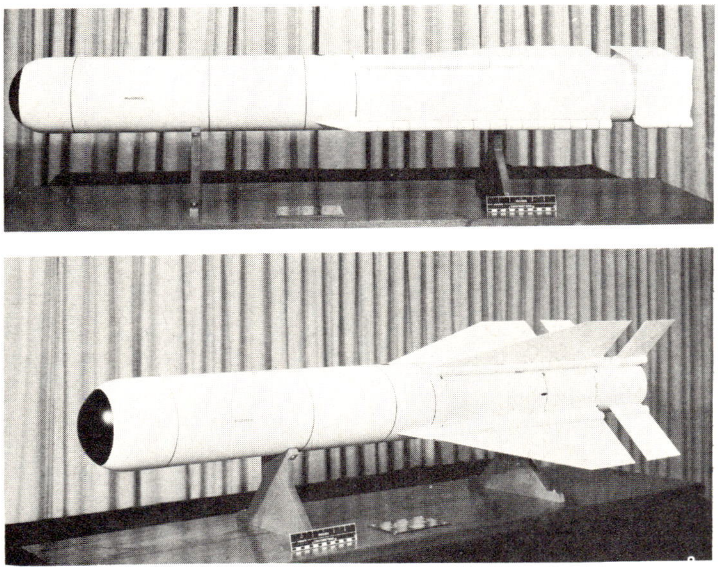

Hughes Aircraft Company's Wasp anti-armour missile with wings and fins folded (top) and spread

An extensible warhead probe characterises improved TOW

AGM-65E Maverick under wing of A-4M aircraft

MAVERICK

USAF designation: AGM-65

The Maverick is a precision-guided air-to-surface missile in the 500 lb class, which developed from a 1965 USAF concept. A three-year total package contract, covering development, testing, support equipment, launchers, and initial production, was put into effect in July 1968. The production option was exercised in July 1971 and first deliveries of the AGM-65A Maverick were made in August 1972. By the end of 1978, deliveries totalled more than 26,000, all TV-guided versions. Maverick is operational on the A-7D, A-10A, F-4D/E and F-5E/F and has been integrated with the F-16 and F-111. The A-4M Skyhawk II will be equipped with the new AGM-65E version of the missile, when development has been completed. Other carrier-based aircraft with which Maverick will be integrated are the F/A-18 Hornet and A-7E Corsair II; plans call for integration on the A-6E and AV-8B. It has been possible to adapt existing controls and displays in launch aircraft to operation of the Maverick weapon system. The minimum aircraft requirement is a raster-scan TV-type display in the cockpit and conventional control functions. A single-rail launcher has been developed for wing stations which cannot accommodate the standard three-round launcher.

More than 900 Mavericks have been launched during testing, training, operational evaluation, demonstrations, and actual combat, with over 85 per cent achieving direct hits on their targets. First combat employment was by the USAF in Southeast Asia. In October 1973, Mavericks were used operationally by the Israeli Air Force during the Yom Kippur War, with spectacular success. Others have been delivered to Greece, Iran, Saudi Arabia, Sweden and Turkey. Sweden has integrated Maverick with the AJ 37 version of the Viggen. Additionally, integration is planned on the Tornado and Alpha Jet.

There are five current versions of Maverick:

AGM-65A. Basic TV-guided model with 5° field of view camera in the seeker, centroid tracker and single-stage solid-propellant boost/sustain rocket motor. Warhead in central portion of airframe. When triggered by contact fuse in the nose, a shaped-charge jet fires through the axis of the guidance unit into the target. First employed by USAF in Viet-Nam. Production for USAF was terminated in favour of AGM-65B, but has been resumed for friendly foreign nations.

AGM-65B. 'Scene magnification' version of TV Maverick, with 2·5° field of view camera and associated minor changes to the guidance equations. Generally similar to AGM-65A, with same warhead. Manufacture of 6,000 for USAF completed, but production resumed for friendly foreign nations.

The basic 57 kg (125 lb) shaped-charge warhead is designed to defeat heavy armour and reinforced concrete, making it most suitable for use against tanks and pillbox-type targets. Missile reliability obviates the need for a missile check on aircraft, and operational use is in the 'fire and forget' mode. The pilot can slew the seeker for rapid lock-on to the target, using a TV visual display, launch the missile and immediately take evasive action while the missile guides itself to the target. Simplicity of operation was demonstrated by a successful firing from a Teledyne Ryan BGM-34 RPV.

AGM-65D. Has imaging infra-red (IIR) seeker, which tracks passively the natural thermal image of a tactical target, providing a pictorial display of the target to the pilot in darkness, haze and smoke. Lock-on, launch and 'fire and forget' homing flight as for AGM-65A/B. Full-scale engineering development for USAF started October 1978. Following captive flight trials on fixed-wing aircraft and helicopters, the first of 30 development and initial operational test launches was made from an F-4 at Eglin AFB, Florida, in late 1980. The missile tracked and destroyed a stationary tank at night. Production of 22,400 AGM-65Ds is anticipated, with an initial request for 490 in FY 1982.

AGM-65E. Fitted with a Rockwell laser seeker which homes on a target illuminated by a ground or airborne laser designator. Can be used by day or night, as the pilot of the launch aircraft does not need to acquire the target visually. Provision is made for rendering the warhead inert, to protect friendly troops in the event of a designator failure or other loss of lock-on during close-support missions. Contractor and USAF flight test launches were made against moving and stationary tanks, armoured personnel carriers and simulated bunkers, by day and night, from high and low altitudes, at various ranges, and at speeds from slow to supersonic. After 19 direct hits had been achieved in 22 firings, development was switched from the original AGM-65C model to the AGM-65E, combining the laser seeker with a new 135 kg (298 lb) blast/penetrator warhead. This is being developed to extend Maverick's capability against larger targets. Intended for US Navy and Marine Corps utilisation. The first three test launches all scored direct hits on target. The third, fired from a Marine Corps A-4M, impacted within 1 m (3 ft) of the aiming point over a range of more than 4·3 nm (8 km; 5 miles).

AGM-65F. This version combines the imaging infra-red seeker of the AGM-65D with the warhead and propulsion sections of the AGM-65E. Intended initially for the US Navy, it has a fine-tuned tracker, to enhance its effectiveness against ships at sea. Selectable delayed or contact fusing of the blast penetration warhead will enable the flight crew to attack ship or shore targets with optimum effect. Firings were to begin in mid-1981. Delivery of 7,000 production AGM-65F missiles is scheduled to begin in mid-1984, for deployment on A-7, A-6 and F/A-18 aircraft.

DATA: See table

WASP

The Statement outlining the FY 1981 Department of Defense Programme for Research, Development and Acquisition advised Congress that the DoD had "begun the development of a family of area munitions, dispensers, warheads and guidance systems in the Advanced Attack Weapons programme. The Wide Area Anti-armour Munitions (WAAM) programme will provide a system capable of multiple kills of armour targets per aircraft pass, even at night and in adverse weather. The four munitions concepts originally in development have been reduced to three: the anti-armour cluster munitions (ACM), the extended range anti-tank mine (ERAM), and the Wasp mini-missile. The Army and the Air Force will co-ordinate Wasp and Hellfire (see Rockwell entry in this section) developments to determine the opportunities to utilise common systems or subsystems to meet both Air Force and Army anti-armour requirements".

Announcing its involvement in Wasp development, in February 1980, Hughes Aircraft Company described it as "the first weapon with built-in ability to identify and aim itself at a target". Wasps will be fired from aircraft in clusters of ten or more, by day or night and in inclement weather, in a 'fire and forget' mode. Onboard computers will direct each mini-missile in the 'swarm' to a different target as long as there are more targets than missiles. Thus, it will not be necessary for the aircrew to see and designate a target for the missile before it is fired. Wasp will be programmed to fly to the area where enemy armour has been located. A millimetre wave radar terminal guidance seeker will then take over, identifying the armour and steering the Wasp to its individual target. Meanwhile, the launch aircraft can withdraw from the area.

Despite its potential, Wasp will be approximately one-fifth the weight of the Maverick air-to-surface missile currently in production by Hughes. Further details can be found in the entry for Boeing, the other competitor in the validation phase of the programme.

MARTIN MARIETTA

MARTIN MARIETTA CORPORATION

Corporate Headquarters: 6801 Rockledge Drive, Bethseda, Maryland 20034
Telephone: (301) 897-6000
Executives: See Spaceflight section

Martin Marietta Aerospace produces missiles, spacecraft, launch vehicles, communications systems and electronic systems. It also fabricates precision components for the US Department of Defense and NASA, and conducts research related to defence and space systems.

WALLEYE

US Navy designation: AGM-62A

When introduced into service, the original version of Walleye was described by the US Navy as "the most accurate and effective air-to-surface conventional weapon ever developed anywhere". It was designed and developed at the NOTS (now Naval Weapons Center), China Lake, California. A production contract awarded to Martin Marietta on 21 January 1966 covered manufacture of qualification weapons as well as an initial quantity for operational use, and carried options for continued production in succeeding years. These options were taken up, the work being performed at Orlando Division. Hughes Aircraft was for a period second source manufacturer.

Walleye is simple in configuration, comprising a torpedo-shaped body with cruciform wings of long-chord cropped-delta form. Control surfaces are hinged to the wing trailing-edges. A ram-air turbine provides the electrical and hydraulic power to operate the guidance/control system. The target seeker is housed behind a window in the nose, and a conventional high-explosive warhead is fitted.

Walleye has undergone considerable development, and the following versions may now be listed:

Walleye 1 (Guided Weapon Mk 1 Mod 0). Basic version with 385 kg (850 lb) warhead. Before launch, the pilot focussed the missile's gyro-stabilised TV camera on target, with the aid of a CRT monitor screen in the cockpit. Once the camera had been locked on target, the missile could be released. Mechanism inside Walleye then took over, using signals from the TV seeker to correct any deviation from the flight path to the target during the bomb's unpowered descent.

Walleye 2 (Guided Weapon Mk 5 Mod 4). Larger than Walleye 1, with warhead in 907 kg (2,000 lb) class for attacking semi-hard targets such as bridges, air base facilities, and ships. Development by Martin Marietta started in Spring 1971.

Extended-Range Data Link Walleye (Guided Weapon Mk 13 Mod 0). Larger wings, to extend glide range, and data link equipment. These improvements permit earlier release, and subsequent remote acquisition and lock-on of the Walleye's TV seeker by the pilot of the launch aircraft (or the pilot of a companion aircraft), after the initiation of evasive action and withdrawal from the target area. Initial batch of 170 Walleye 2s had been updated to ER/DL standard by October 1976, and work on a further batch of 430 then started. Conversion of Walleye 1s began in FY 1978. Current plans envisage conversion of a total of about 1,400 Walleye 1s and 2,400 Walleye 2s.

IR Walleye. Hughes Aircraft has developed an imaging infra-red seeker for Walleye, aimed at achieving guidance system commonality with the GBU-15 glide bomb and the AGM-65D version of the Maverick missile. This provides night and adverse weather capability, together with improved tracking precision. First IR Walleye was scheduled for delivery in Summer 1980.

Data: See table

MDAC

MCDONNELL DOUGLAS ASTRONAUTICS COMPANY (A Division of McDonnell Douglas Corporation)

5301 Bolsa Avenue, Huntington Beach, California 92647
Telephone: (714) 896 3311
President: John F. Yardley
Vice-President and General Manager: C. James Dorrenbacker
Vice-President, Engineering: Adrain P. O'Neal
Vice-President, Marketing: Paul L. Smith
McDonnell Douglas Astronautics Company-St Louis
Box 516, St Louis, Missouri 63166
Telephone: (314) 232 0232
Vice-President and General Manager: Erwin F. Branahl
Vice-President and Programme Manager, Harpoon: Clifford D. Marks
Vice-President, Advanced Missile Programmes: R. Wayne Lowe

McDonnell Douglas Astronautics Company was formed on 26 June 1968, by merging the former Douglas Missile and Space Systems Division and the McDonnell Astronautics Company into a single management structure.

In the missile field, McDonnell Douglas Astronautics Company has developed and is producing the Harpoon anti-shipping missile for the US Navy and other customers, and is developing cruise missile guidance systems for the USAF's ALCM and GLCM, US Navy's SLCM, and the Tomahawk II (MRASM).

HARPOON

US Navy designations: AGM-84A and RGM-84A

The US Navy selected McDonnell Douglas as prime contractor for development of the Harpoon all-weather anti-ship missile in June 1971. The work was allocated to McDonnell Douglas Astronautics Company-St Louis, with Texas Instruments Inc and Sperry Systems Management as major subcontractors.

The initial contract covered the development and demonstration of a number of engineering-model missiles over a two-year period. This phase of the programme was completed successfully on schedule, and was followed by a 30-month final development phase in which more than 30 missiles were launched from Harpoon-designated aircraft, ships and submarines. Production of 100 pilot missiles was authorised in 1974. A proportion of them was used for operational evaluation, which began in November 1975 and was completed in February 1977. The go-ahead for initial production had been given in July 1975, and authorisation for deployment on aircraft, surface ships and submarines was granted in July 1977.

The US Navy plans to equip all of its guided missile cruisers, guided missile destroyers, FF-1052 frigates, FFG-7 patrol frigates, DD-963 ASW destroyers and PHM-1 hydrofoil patrol craft; P-3A/B/C, A-6E and S-3A aircraft; and 594, 637, 688 and 700 class nuclear attack submarines with the Harpoon Weapon System (HWS). By the beginning of 1981, the missiles equipped 91 destroyers, frigates and cruisers, 28 nuclear attack submarines, and 33 P-3C aircraft of the US Navy, with deployment on the first 50 A-6E aircraft to begin later that year. In addition the HWS has been procured by 12 allied nations for launch from surface ships, submarines and aircraft.

By early 1981, a total of 2,587 of the missiles had been ordered. The 1,500th Harpoon produced by MDAC was delivered in late 1980, with production continuing at the rate of 40 per month. Manufacture will continue until at least 1986, and could be increased to 55 per month if justified by further orders.

The general configuration of the **AGM-84A** air-launched version of Harpoon is shown in the accompanying illustration. The Teledyne CAE J402-CA-400 turbojet power plant is housed in the rear of the body, with a ventral flush air intake.

Prior to launch, targeting data for Harpoon are provided by the command and launch subsystem, which interfaces with onboard systems. The Harpoon data processor, a general-purpose digital computer, receives targeting and attitude data from existing systems, and computes the necessary missile and launcher orders. After launch, guidance is provided by a midcourse guidance system consisting of a strapdown attitude reference assembly and digital computer. No inputs from the launch platform are required by the missile after launch. Cruise altitude is monitored by a radar altimeter, enabling the flight to the target to be made at low altitude, so offering both optimum target acquisition capability through reduction of clutter effects, and the ability to penetrate enemy defences. Offset launch capability is provided for all launch modes.

When the target comes within the search area of the active radar seeker, the high-resolution system detects and locks on to the target, even in rain and high sea states. Seeker lock-on is maintained until impact. Capability to perform high-*g* manoeuvres throughout flight permits successful operation against fast manoeuvring targets. A terminal 'pop-up' manoeuvre counters close-in defences and offers maximum warhead effectiveness. Counter-countermeasures devices are installed. The warhead is a penetration/high-explosive blast type.

The free-flight reliability rate for Harpoon stood at 87 per cent in 151 launches by February 1981, and 93 per cent in the last 73 launches by that date.

Advanced versions of Harpoon are under development, including some with an alternative guidance system, using an imaging infra-red seeker.

Data: See table

NWC

NAVAL WEAPONS CENTER

China Lake, California 93555
Telephone: (714) 939 3511

The Naval Weapons Center is located 250 km (155 miles) northeast of Los Angeles, in the Mojave Desert. Its current mission is to be the principal US Navy RDT&E centre for air warfare systems (except anti-submarine warfare systems) and missile weapon systems, and to operate the National Range/Facility for parachute T&E.

Weapons developed at the Center have included the Mighty Mouse 2·75 in folding-fin aircraft rocket, the 11·75 in Tiny Tim rocket, the Zuni 5 in folding-fin aircraft rocket, the Sidewinder air-to-air guided missile, the Snakeye 250/500 lb bomb with folding dive-brake retardation system to avoid fragmentation damage to the launch aircraft during low-level strikes, the Shrike and Standard ARM air-to-surface anti-radar missiles and the Walleye glide bomb. Production of many of these weapons was entrusted to commercial companies under whose entries they are, or were, described in *Jane's*.

SIDEWINDER

US military designation: AIM-9

The **AIM-9A** prototype version of the Sidewinder air-to-air missile was developed by the NWC and was first fired successfully on 11 September 1953. The **AIM-9B** first-generation production version was manufactured by Philco (now Ford Aerospace & Communications Corporation) and General Electric for the US Navy and the USAF, with a Naval Propellant Plant solid-propellant rocket motor and infra-red seeker, and was supplied to many foreign services. Production in the USA, from 1955, totalled more than 80,000 AIM-9Bs.

Although the US services no longer deploy the AIM-9B and its semi-active radar counterpart, the **AIM-9C**, other countries continue to utilise the 'B', of which licence manufacture was undertaken in Germany by Bodenseewerk, in association with subcontractors in the Netherlands, Denmark, Norway, Greece, Portugal and Turkey.

Also no longer deployed is the **AIM-9D**, which had higher speed and greater range than the 'B', from which it could be distinguished by its tapering nose, longer-chord nose fins and greater sweepback on the tail fins. This version had a Rocketdyne Mk 36 Mod 5 solid-propellant rocket motor, different guidance unit and continuous-rod warhead with proximity and impact fuses.

Of the developed versions of the infra-red homing Sidewinder produced for air-to-air use, to enhance and update the missile's capabilities, those in service with, or in production for, the US services in 1981 were as follows:

AIM-9E. Approximately 5,000 of this version were produced by Philco for the USAF, by modification of AIM-9Bs to have a new seeker section, Thiokol Mk 17 solid-propellant motor, refurbished electronics and wiring. Major increase in capability claimed. Production completed.

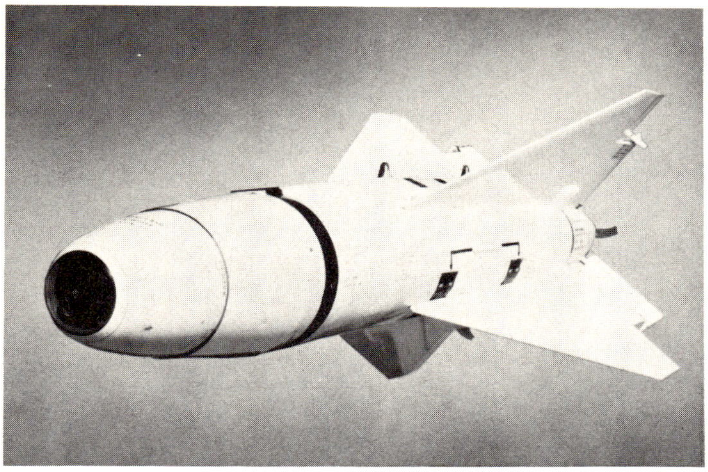

AGM-62A Walleye 1 TV-guided glide bomb

Extended-Range Walleye and data link pod on A-7 Corsair II

Grumman A-6E Intruder carrying two Harpoons on underwing pylons

AIM-9H Sidewinder on McDonnell Douglas F-4 Phantom II

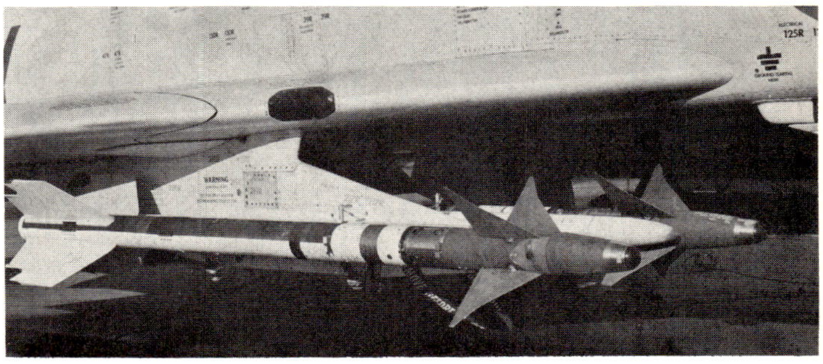

AIM-9L advanced version of Sidewinder on US Navy F-4

AIM-9G. Similar to AIM-9D but with off-boresight target acquisition and lock-on. Production by Raytheon for US Navy completed in FY 1970.

AIM-9H. Version for US Navy with improved close-range 'dogfight' capability. Basically similar to AIM-9G, but with solid-state guidance instead of vacuum tube electronics for improved reliability and maintainability, decreased minimum ranges and faster angle tracking rates. Production completed.

AIM-9J. Conversion of AIM-9Bs and 9Es, with new 'front end' to enhance 'dogfight' capability by improved manoeuvring ability. About 14,000 modified for USAF by Ford Aerospace to equip the F-15 Eagle and other types. Deliveries began in 1977 and have been completed.

AIM-9L. Third-generation version of Sidewinder with all-aspect detection and launch capability. Double-delta nose fins for improved inner boundary performance and better manoeuvrability. AM-FM conical scan for increased seeker sensitivity and improved tracking stability. Active optical fuse for increased lethality and low susceptibility to countermeasures. Annular blast fragmentation warhead. Series of 20 joint service technical evaluation firings completed on 1 March 1975, including successful firings against PQM-102 targets, and with particular emphasis on difficult shots on the beam or forward quarter of the target. Procurement between FY 1976 and FY 1980 totals 3,550 for US Navy and more than 5,000 for the USAF (to arm the F-15 and F-16). Eventual total for the two services is expected to be 11,707. In FY 1976, Raytheon was contracted to deliver 1,584 AIM-9Ls. Ford Aerospace received a contract in FY 1977. Total of 2,945 provided under FY 1978 budget, 3,494 in FY 1979 and 3,969 in FY 1980. The AIM-9L was chosen in August 1977 to arm Phantom, Tornado F. Mk 2 and Hawk aircraft of the Royal Air Force and Sea Harriers of the Royal Navy. It is also produced under licence by a consortium of manufacturers in NATO countries.

AIM-9N. Version of AIM-9P for Foreign Military Sales. Deliveries have been completed.

AIM-9P. Improved version of AIM-9J produced by Ford Aerospace for USAF. Increased lethality due to fuse improvements. Reduced-smoke rocket motor.

Versions of Sidewinder under development for proposed air-to-air use are as follows:

AIM-9J+ (J-3). Further improvement of AIM-9J, under development for USAF by Ford Aerospace. Increased target acquisition envelope, solid-state electronics and increased lethality due to seeker improvements. Proposed production by conversion of existing AIM-9Es and 9Js.

AIM-9M. Improved version of AIM-9L, under development for US Navy and USAF. Increased ECCM capability, improved background discrimination, reduced-smoke rocket motor. A pilot production contract for 50 units was awarded to Raytheon, for delivery in 1979-80. Series of 37 joint service firings was completed on 20 December 1980, including successful firings against PQM-102, QF-86 and QT-38 targets. Planned procurement of 480 in FY 1981, with substantially larger quantities in following years.

DATA: See table

RAYTHEON

RAYTHEON COMPANY

141 Spring Street, Lexington, Massachusetts 02173
Telephone: (617) 862 6600
Telex: 92-3455
CHAIRMAN OF THE BOARD AND CHIEF EXECUTIVE OFFICER:
Thomas L. Phillips
PRESIDENT: D. Brainerd Holmes

Raytheon Company is prime contractor for the US Army's Patriot and Improved Hawk surface-to-air missile systems, and the US Navy/USAF Sparrow air-to-air missile. It produces the Dragon anti-tank missile for the US armed services, and versions of the US Navy/USAF Sidewinder air-to-air missile (described under NWC heading), and is the major subcontractor on the US Navy's Aegis advanced surface missile system.

Raytheon Company's Equipment Division is prime contractor for the NATO Sea Sparrow system, under development for the US Navy, the Royal Norwegian Navy, the Royal Danish Navy, the Royal Netherlands Navy, the Royal Belgian Navy and the Italian Navy.

SPARROW

US military designations: AIM-7 and RIM-7H

The AIM-7 Sparrow, developed by Raytheon Company for the US Naval Air Systems Command, is a radar-homing air-to-air missile with all-weather all-altitude operational capability. It can also be used against shipping targets from aircraft or ships.

The Sparrow equips McDonnell Douglas F-4 Phantom II aircraft of the US Navy, USAF, Royal Air Force, Greek Air Force, Iranian Air Force, Israeli Air Force, Republic of Korea Air Force and Spanish Air Force. The Lockheed F-104S fighters licence-built in Italy are Sparrow-armed, and both the USAF's McDonnell Douglas F-15 Eagle and the US Navy's Grumman F-14 Tomcat employ this missile. Mitsubishi is manufacturing more than 800 in Japan for the Japan Air Self-Defence Force, under licence from Raytheon.

Without change, the Sparrow is used on US Navy ships as a surface-to-air and anti-shipping weapon in the Basic Point Defense Surface Missile System. It is also used in Canadian ships in the Close Range Missile System.

The advanced **AIM-7F** Sparrow, in production in 1981 by both Raytheon and General Dynamics Pomona Division, and by Mitsubishi of Japan, is powered by a Hercules Mk 58 Mod 0 solid-propellant motor and has a Raytheon semi-active Doppler radar homing system, with a smaller solid-state seeker than that of earlier versions. The heavier continuous-rod warhead is actuated by a proximity fuse or contact fuse, and is mounted forward of the wings, instead of aft as on earlier versions. Manoeuvrability has been improved, and Sparrow is now considered a good dogfight missile as well as a good medium-range weapon with all-aspect capability, including shootdown.

The AIM-7F is the fourth operational version of the Sparrow and is the only one that can be carried by the F-15. It has been fired successfully from a YF-16, but there is no current requirement for Sparrow armament on USAF F-16s. Production of the earlier AIM-7C, D and E models totalled 34,000 missiles, and the AIM-7F is being produced for both the US Navy and USAF, with General Dynamics as second-source contractor. Total procurement under the AIM-7F programme is expected to provide 9,804 missiles for the US Navy and 9,150 for the USAF. The missile was approved for deployment at the beginning of 1977. Production of 1,725 was funded under the FY 1978 budget, 1,910 in FY 1979, and 1,560 in FY 1980, with a further 1,680 proposed for FY 1981, and 1,965 for FY 1982, to meet USAF and USN requirements (numbers include AIM-7M missiles, below).

An advanced monopulse seeker has been developed for Sparrow, with the aims of reducing cost and overcoming present shortcomings in the ECM and lookdown/clutter areas. It has been fitted to a new version of the missile, designated **AIM-7M**, of which production began in FY 1980. In FY 1981 all Sparrow production was expected to switch to the AIM-7M.

A missile known as Sky Flash, evolved from Sparrow, is in production for the Royal Air Force and Swedish Air Force by Dynamics Group of British Aerospace (which see).

DATA: See table

AMRAAM

Raytheon's Missile Systems Division is one of the two suppliers selected by the USAF on 5 February 1979 to participate in the 33-month validation phase of the advanced medium-range air-to-air missile (AMRAAM) programme. Aim of the programme (described in more detail under the Hughes Aircraft entry) is to provide an advanced beyond-visual-range missile for operational use between 1985 and 2005.

Raytheon's AMRAAM utilises an inertial reference unit and microcomputer to project target co-ordinates obtained from the launch aircraft's radar, and an active radar seeker for terminal homing. The first rail launch from an F-15 in the Spring of 1981 demonstrated clean missile separation from the aircraft. Stable and controlled aerodynamic flight was also demonstrated by proper response to an onboard acceleration command system.

ROCKWELL INTERNATIONAL
ROCKWELL INTERNATIONAL CORPORATION

Missile Systems Division

4300 East Fifth Avenue, PO Box 1259, Columbus, Ohio 43216

Rockwell International's Columbus (Ohio) Missile Systems Division is playing a major role in several important US weapon programmes. Brief details are given of the GBU-15 air-to-surface weapons and the Hellfire anti-armour weapon system which the Division is developing and/or producing for the US Army and Air Force.

HELLFIRE

The US Army announced in October 1976 that Rockwell International had won the competition for engineering development of the Hellfire (HELicopter-Launched, FIRE and forget) modular missile system. The company was awarded a $66·7 million contract for full-scale engineering development, including design, manufacture, testing and evaluation of the missile, launcher, ground and logistics support equipment. Rockwell is performing most of the work at Columbus, but propellant loading and final assembly are done at Redstone Arsenal, Alabama, under subcontract with Thiokol Chemical Company.

The initial version of Hellfire uses a laser seeker as its guidance module, and will equip the US Army's AH-64 advanced attack helicopter and UH-60 Black Hawk utility helicopter. Use of laser homing in the initial application of Hellfire is based on successful testing by the Army Missile Research and Development Command in Huntsville, Alabama, using Rockwell International Hornet missile airframes and Rockwell laser seekers. Hellfire is designed to accept a variety of other terminal homing seeker modules, including TV, imaging infra-red (IIR), radio frequency (RF), and dual-mode RF/IIR.

In November 1974 two test missiles, fitted with laser seekers supplied by Rockwell International, scored direct hits on two tank targets in the first successful ripple fire demonstration of such weapons at Redstone Arsenal. A US Army crew, flying an AH-1G HueyCobra helicopter, 'popped up' from cover and fired the first test missile at a tank illuminated by a ground-based laser designator. Less than one second later, the second missile was launched against a tank situated about 20 m (65 ft) from the first target and illuminated by an airborne laser on board a second AH-1G.

The Hellfire system's demonstrated rapid, ripple and indirect fire performance and effectiveness, coupled with the capability of being launched from a variety of helicopters, ground vehicles and fixed-wing aircraft, provides a significant improvement over current systems. Its 178 mm (7 in) warhead has a high level of effectiveness against present and near-term future types of armour.

The first guided flight of a Hellfire from a YAH-64 helicopter took place on 18 September 1979, and a total of 21 guided firings had been completed by February 1980. Initial production of laser Hellfire was delayed for about one year to be consistent with the AH-64 production schedule. Full-scale development of an imaging infra-red seeker started in FY 1981, to provide the missile with true 'launch and leave' capability, and the US Army has been directed to pursue a parallel detector development approach, evaluating a focal plane array detector as the primary effort. $96·5 million was requested for Hellfire procurement under the initial FY 1982 budget, including a first batch of 502 laser-guided rounds. The seekers will be supplied by Martin Marietta. The US Army is expected to acquire an eventual total of 24,600 laser Hellfires and 18,500 of the IIR version. Others will equip US Marine Corps AH-1s.

DATA: See table

GBU-15

The GBU-15 cruciform wing weapon (CWW) is an air-launched glide bomb fitted with a guidance system which is claimed to give it pinpoint accuracy from altitudes below 60 m (200 ft) or over standoff ranges greater than 5 nm (9·25 km; 5·75 miles). Development began in 1974, with Rockwell as prime contractor, on the basis of experience gained in Viet-Nam with the earlier Pave Strike/GBU-8 HOBOS programme (see 1980-81 *Jane's*).

Like HOBOS, the GBU-15 is a modular weapon, intended for tactical use to suppress enemy defences and to destroy other high-value targets. It is normally built around a standard Mk 84 2,000 lb general-purpose bomb, minus fins, but is adaptable to the CBU-75 cluster munition as an alternative warhead. The target detecting device carried on the front of the warhead, via an adaptor, can utilise a TV or imaging infra-red (IIR) seeker, with a DME midcourse guidance system available for increased accuracy. The control module, with autopilot, and data link module attach to the rear of the warhead. The CWW aerofoil group of fixed foreplanes and large rear-mounted wings and control surfaces completes the weapon. The alternative planar wings described under the Hughes Aircraft heading in the 1980-81 *Jane's* are no longer intended for use on the GBU-15.

Two basic attack trajectories are available. For direct trajectories, the weapon is locked on target before launch and flies a near line-of-sight profile to impact. The indirect profile includes a midcourse glide phase which extends standoff capability. In this profile, the seeker can be locked on to the target after launch, or the operator can fly the weapon manually to impact, using guidance updates provided through the data link. Many launches have been made successfully from F-4, F-111 and B-52 aircraft.

The Missile Systems Division of Rockwell began full-scale production of the GBU-15(V)/B CWW, with TV guidance and Hughes Aircraft data link, in September 1980. The imaging infra-red seeker is under development.

DATA: See table

TEXAS INSTRUMENTS
TEXAS INSTRUMENTS INC

PO Box 6015, Dallas, Texas 75222
Telephone: (214) 238 2011
Telex: 7-3324

Texas Instruments has played a major part in many US missile programmes. The latest missile for which it is prime contractor is the AGM-88A HARM, of which brief details follow:

HARM

US Navy designation: AGM-88A

On 24 May 1974, Texas Instruments was named as prime contractor for HARM (High-speed Anti-Radiation Missile) by the US Naval Air Systems Command. The initial phase of the contract covered four months of basic design co-ordination and was followed by the Advanced Development Phase; during 1979 development continued in the company's Dallas facilities, where production of the Shrike anti-radiation missile took place for more than a decade.

Few details of HARM may yet be published. Its basic configuration is conventional, with a slim cylindrical body, ogival head, cruciform double-delta wings at mid-length for simplified roll control, and cruciform tail-fins indexed in line with the wings. The fixed antenna for proportional navigation is in the extreme nose, with the seeker. The smokeless solid-propellant dual-thrust rocket motor is provided by Thiokol.

The emphasis on high speed reflects experience gained in Viet-Nam, where Soviet surface-to-air missile radar systems sometimes detected the approach of US anti-radiation missiles such as the first-generation Shrike and ceased operation before the missile could lock on to them. In the case of the Navy's A-7E aircraft, HARM can cover a wide range of frequency spectra through the use of programmable digital processors in both the aircraft's AN/AWG-25 avionics equipment and the missile. The USAF will also utilise HARM on the F-4G 'Wild Weasel'. The missile has been launched successfully from F-4, A-6 and A-7 aircraft, and is readily adaptable to the F/A-18, B-52, F-15 and F-16.

For technical and budgetary reasons, the planned procurement of 80 pilot production HARMs for early Navy IOC (initial operational capability) was delayed. Operational evaluation by the US Air Force and Navy was scheduled to start during 1981. Production deliveries will begin in 1982, with full-scale procurement expected in FY 1982.

DATA: See table

PAVEWAY LASER GUIDED BOMBS

The Laser Guided Bomb (LGB) concept was introduced by Texas Instruments Incorporated in 1965. The US Air Force initiated an extensive development programme and the name Paveway was assigned to this. Texas Instruments builds the guidance units and rear wing assemblies, which are fitted to US Mk 80 series bomb bodies. With minor modifications, the Paveway units have been installed on other aircraft ordnance, including the British Mk 13/18 and 540 lb stores.

Paveway LGBs are semi-active laser-guided munitions which home on reflected laser energy from a target being illuminated by a laser designator. Day and night attacks can be accomplished against all types of tactical target, since no intrinsic target signature is required. LGBs have been used effectively against anti-aircraft gun and missile installations, tanks, tracks, bridges, power stations, railroads, buildings, ships and caves.

Conventional bomb handling equipment is used for loading and fusing. No electrical interface to the delivery aircraft is required. Paveway munitions are delivered in a similar manner to conventional unguided munitions.

Continued development and engineering upgrading has added Paveway II to the family of weapons proven in Southeast Asia. Folding rear wings on the Paveway II have increased weapon performance and aircraft load density, increasing the number of aircraft, worldwide, that are compatible with the Paveway system.

The same laser detector and guidance unit is used with all weapons. The rear wing and control canards vary in size according to the size of the weapon.

Both Paveway I and II are currently in production for customers that include the US Air Force and Navy, Royal Air Force, Royal Netherlands Air Force, Hellenic Air Force, Turkish Air Force, Canadian Armed Forces, Royal Saudi Air Force and Republic of Korea Air Force.

Sparrow (lower) and Sidewinder (upper, with protective nosecap) missiles on an F-14A Tomcat fighter of the US Navy *(Brian M. Service)*

Laser Hellfire modular air-to-surface missile

Cruciform Wing Weapon developed under GBU-15 programme

Paveway laser-guided weapons on a USMC AV-8A Harrier

AGM-88A HARM air-to-surface missile

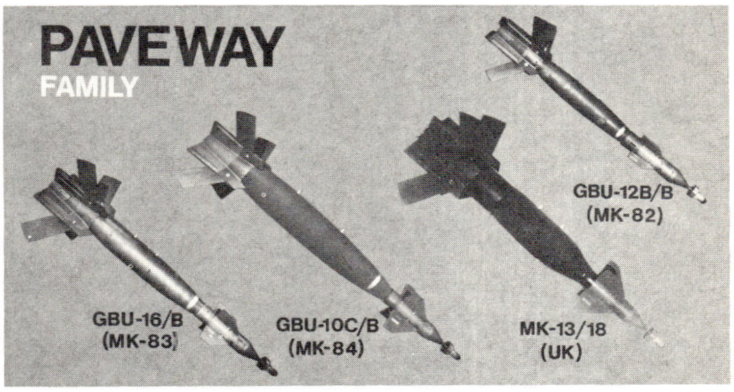

Paveway II laser-guided bombs

AIR-LAUNCHED MISSILES

Note: entries in italics are estimated

Country	Prime Contractor	Model	Length m (ft in)	Body diam cm (in)	Wing span m (ft in)	Launch weight kg (lb)	Cruising speed knots (km/h; mph)	Range	
Argentina	CITEFA	Martin Pescador	2·94 (9 7¾)	21·85 (8·6)	0·73 (2 4¾)	140 (308)	Mach 2·3	1·3-4·8 nm (2·5-9 km; 1·5-5·5 miles)	Warhead 40 kg (88 lb)
Brazil		Piranha MAA 1	2·72 (8 11)	15 (5·9)	NA	85·5 (188)	NA	5·4 nm (10 km; 6·2 miles)	Warhead 12 kg (26·5 lb)
France	Aérospatiale	AS.11	1·21 (3 11½)	16·4 (6½)	0·50 (1 7½)	29·9 (66)	214-370 (396-685; 246-425)	350-3,000 m (1,150-9,850 ft)	Time of flight (propelled) 18-20 s. Min turning radius 1,000 m (3,300 ft)
	Aérospatiale	AS.12	1·87 (6 2)	21 (8¼)	0·65 (2 1½)	76 (168)	at impact 505 (936; 581)	8,000 m (26,250 ft); launch speed 180 (333; 207)	Range in relation to surface. Time of flight 32 s. Warhead 30 kg (66 lb)
	Aérospatiale	AS.15TT	2·30 (7 6½)	18·8 (7·4)	0·56 (1 10)	100 (220)	540 (1,000; 620)	more than 8 nm (15 km; 9·3 miles)	Warhead 30 kg (66 lb) Time of flight (propelled) 45·2 s
	Aérospatiale	AS.30 Laser	3·65 (11 11¾)	34·2 (13½)	1·00 (3 3¼)	520 (1,146)	supersonic	1·6-5·4 nm (3-10 km; 1·8-6·2 miles)	Warhead 240 kg (529 lb) Time of flight 21 s
	Aérospatiale	Exocet AM39	4·69 (15 4½)	35 (13¾)	1·10 (3 7¼)	652 (1,438)	Mach 0·93	see text	Warhead 160 kg (352 lb)
	Matra	Super 530	3·54 (11 7¼)	26 (10¼)	0·64 (2 1¼)	250 (550)	Mach 4·5	*16-19 nm (30-35 km; 18·5-21·75 miles)*	Fin span 0·90 m (2 ft 11½ in). Operational ceiling above 21,350 m (70,000 ft). Warhead greater than 30 kg (66 lb)

NA: not available

Note: entries in italics are estimated

Country	Prime Contractor	Model	Length m (ft in)	Body diam cm (in)	Wing span m (ft in)	Launch weight kg (lb)	Cruising speed knots (km/h; mph)	Range	
	Matra	R.550 Magic	2·80 (9 2¼)	15·7 (6)	0·47 (1 6½)	88 (194)	Mach 2+	3·25 nm (6 km; 3·75 miles) at medium altitude; 5·4 nm (10 km; 6·2 miles) at high altitude	Fin span 0·65 m (2 ft 1½ in) Min range 500 m (1,640 ft)
Germany, Federal Rep	MBB	Kormoran	4·40 (14 5)	34 (13½)	1·00 (3 3¼)	600 (1,320)	*Mach 0·95*	*20 nm (37 km; 23 miles)*	
International	Euromissile	Hot	1·28 (4 2½)	14·4 (5¾)	*0·31 (1 0¼)	23·5 (51·8)	486 (900; 560)	400-4,000 m (1,310-13,125 ft)	Time of flight to 4,000 m 17 s. Warhead 6 kg (13·2 lb)
Israel	Rafael	Shafrir	2·47 (8 1¼)	16 (6¼)	0·52 (1 8½)	93 (205)	NA	2·7 nm (5 km; 3·1 miles)	Warhead 11 kg (24·25 lb)
Italy	Selenia	Aspide	3·70 (12 1½)	20·3 (8)	1·00 (3 3¼)	220 (485)	*Mach 4*	*27-54 nm (50-100 km; 31-62 miles)*	Warhead *33 kg (73 lb)*
	Sistel	Sea Killer Mk 2	4·70 (15 5)	20·6 (8⅛)	1·00 (3 3¼)	300 (660)	Mach 0·74	over 10·8 nm (20 km; 12·4 miles)	Warhead 70 kg (154 lb)
Japan	Mitsubishi	ASM-1	4·00 (13 1½)	35 (13¾)	1·20 (3 11¼)	610 (1,345)	Mach 0·9	43-48 nm (80-90 km; 50-56 miles)	Warhead 200 kg (440 lb)
Norway	Kongsberg	Penguin Mk 3	3·18 (10 5¼)	28 (11)	1·00 (3 3½)	347 (765)	Mach 0·8+	over 22 nm (40 km; 25 miles)	Warhead 120 kg (265 lb)
USSR		AA-1 'Alkali'	1·88 (6 2)	17·8 (7)	0·58 (1 10¾)	*90 (200)*		3·2-4·3 nm (6-8 km; 3·7-5 miles)	
		AA-2 'Atoll' (K-13A)	2·80 (9 2)	12·0 (4¾)	0·45 (1 5¾)	70 (154)	Mach 2·5	2·5-3·5 nm (5-6·5 km; 3-4 miles)	Fin span 0·53 m (1 ft 8¾ in)
		AA-3 'Anab'	*4·10 (13 5)*	*28·0 (11)*	*1·30 (4 3)*			over 8·75 nm (16 km; 10 miles)	Length for IR version; radar version is 4·0 m (13 ft 1 in) long
		AA-5 'Ash'	*5·50 (18 0)*					16 nm (30 km; 18·5 miles)	Length for IR version; radar version is 5·2 m (17 ft 0 in) long
		AA-7 'Apex'	*4·30 (14 1¼)*	*24·0 (9½)*	*1·05 (3 5½)*	*320 (705)*		*15 nm (27 km; 17 miles)*	
		AA-8 'Aphid'	*2·00 (6 6¾)*	*13·0 (5)*		*55 (121)*		*3-4 nm (5·5-8 km; 3·5-5 miles)*	*Warhead 6 kg (13·2 lb)*
		AT-2 'Swatter'	0·90 (2 11½)	15·0 (6)	0·65 (2 2)	25 (55)	290 (540; 335)	300-2,200 m (985-7,220 ft)	
		AT-3 'Sagger'	0·86 (2 10)	*12·0 (5)*	0·46 (1 6)	11·3 (24·9)		500-3,000 m (1,640-9,840 ft)	
		AS-2 'Kipper'	*9·50 (31 0)*	*90 (35·4)*	*4·88 (16 0)*	*4,200 (9,260)*	*Mach 1·2*	*115 nm (213 km; 132 miles)*	*Warhead 1,000 kg (2,200 lb)*
		AS-3 'Kangaroo'	14·90 *(48 11)*	*185 (72·8)*	9·15 (30 0)	*11,000 (24,250)*	Mach 1·8	350 nm (650 km; 400 miles)	*Nuclear or HE warhead 2,300 kg (5,070 lb)*
		AS-4 'Kitchen'	*11·30 (37 0)*	*90 (35·4)*	*3·00 (9 10)*	*6,000+ (13,225)*	*Mach 2+*	*160 nm (300 km; 185 miles)*	Range at low altitude. *Warhead 1,000 kg (2,200 lb)*
		AS-5 'Kelt'	*9·45 (31 0)*	*100 (39·5)*	*4·57 (15 0)*	*3,500 (7,715)*	*Mach 0·9 at low altitude; Mach 1·2 at 9,150 m (30,000 ft)*	over 85 nm *(160 km; 100 miles)*	Range at low altitude; doubled at height *HE warhead; 1,000 kg (2,200 lb)*
		AS-6 'Kingfish'	*10·50 (34 6)*		*2·50 (8 2½)*		*Mach 3*	*120 nm (220 km; 135 miles)*	Range at low altitude
UK	British Aerospace	Sea Eagle	4·10 (13 5½)	40 (15¾)	1·20 (3 11¼)	NA	NA	NA	
	British Aerospace	Sea Skua	2·50 (8 2½)	25 (9¾)	0·72 (2 4½)	*147 (325)*	NA	over 8 nm (15 km; 9·3 miles)	*Warhead 20 kg (44 lb)*
	British Aerospace	Sky Flash	3·66 (12 0)	20·3 (8)	1·02 (3 4¼)	193 (425)	Mach 4	27 nm (50 km; 31 miles)	
USA	Boeing Aerospace	AGM-86B ALCM	6·32 (20 9)	62 (24½)	3·66 (12 0)	1,282 (2,825)	approx 435 (805; 500)	approx 1,350 nm (2,500 km; 1,550 miles)	Height overall 1·19 m (3 ft 11 in)
	Hughes	AIM-54A Phoenix	3·95 (13 0)	38 (15)	0·91 (3 0)	447 (985)	Mach 5+	over 108 nm (200 km; 124 miles)	Data for AIM-54C similar
	Hughes	BGM-71A TOW	1·17 (3 10)	15 (6)	0·34 (1 1½)	19 (42)	High subsonic	500-3,750 m (1,640-12,300 ft)	Warhead 3·9 kg (8·6 lb) Max range increased by 25% in 1978
	Hughes	AGM-65A Maverick	2·49 (8 2)	30 (12)	0·72 (2 4½)	210 (462)	NA	*12 nm (22 km; 14 miles)*	Weight 288 kg (637 lb) with alternative 135 kg (298 lb) warhead. Data for AGM-65B/D/E/F similar
	Martin Marietta	AGM-62A Walleye 1	3·44 (11 3½)	38·1 (15)	1·14 (3 9)	499 (1,100)	Subsonic	—	Warhead 385 kg (850 lb)
	Martin Marietta	AGM-62 Walleye 2	4·04 (13 3)	45·7 (18)	1·30 (4 3)	1,061 (2,340)	Subsonic	—	Warhead 907 kg (2,000 lb)
	MDAC	AGM-84A Harpoon	3·84 (12 7)	34 (13½)	NA	526 (1,160)	High subsonic	over 50 nm (92 km; 57 miles)	
	NWC/ Raytheon	AIM-9G/H Sidewinder	2·87 (9 5)	13 (5)	*0·63 (2 0¾)	86 (190)	Mach 2·5	over 1·75 nm (3·35 km; 2 miles)	Warhead 11·4 kg (25 lb)
	Ford Aerospace	AIM-9J/J+/N/P Sidewinder	3·05 (10 0)	13 (5)	*0·56 (1 10)	77 (170)	Mach 2·5	over 1·75 nm (3·35 km; 2 miles)	Warhead 11·4 kg (25 lb)
	NWC/ Raytheon/ Ford Aerospace	AIM-9L Sidewinder	2·87 (9 5)	13 (5)	*0·63 (2 0¾)	86 (190)	Mach 2·5	over 3·75 nm (7 km; 4·35 miles)	Warhead 11·4 kg (25 lb)
	Raytheon	AIM-7F Sparrow	3·66 (12 0)	20 (8)	1·02 (3 4)	227 (500)	*Mach 3·5+*	*24 nm (44 km; 28 miles)*	
	Rockwell	GBU-15(V)/B	3·92 (12 10½)	45·7 (18)	1·50 (4 11)	1,187 (2,617)	Subsonic	—	Warhead 907 kg (2,000 lb)
	Rockwell	Hellfire	1·625 (5 4)	17·8 (7)	NA	43 (95)	NA	NA	
	Texas Instruments	AGM-88A HARM	4·18 (13 8½)	25·4 (10)	1·13 (3 8½)	362 (798)	Supersonic	over 8·5 nm (16 km; 10 miles)	Altitude limits S/L to 12,200 m (40,000 ft)

*Fin span NA: not available

SPACEFLIGHT

CANADA

SPAR

SPAR AEROSPACE LIMITED

Royal Bank Plaza, South Tower, PO Box 83, Toronto, Ontario M5J 2J2
Telephone: (416) 865 0480
Telex: 065-24240 Sparcal Tor

SPACE AND ELECTRONICS GROUP

1700 Ormont Drive, Weston, Ontario M9L 2W7
Telephone: (416) 745 9680
Telex: 065 27360

Chairman of the Board and Chief Executive Officer: L. D. Clarke

Senior Vice-President Corporate and Personnel Relations: E. P. Birch

Vice-President and Group Executive; Space & Electronics Group: J. D. MacNaughton

Director, Public Relations: Hugh A. MacLean

Spar is an independent Canadian public company, active in varied fields of advanced technology and serving commercial, scientific and military markets. Up to the beginning of 1981, it had participated in five scientific satellite programmes, and in 17 communications satellite programmes, involving a total of 60 spacecraft. Major hardware provided by Spar included transponder equipment, power supplies, telemetry, structure and digital electronics.

ANIK

Anik B, Canada's latest domestic communications satellite (see RCA entry in US section of 1979-80 *Jane's*), launched on 15 December 1978.

The **Anik C** series of three further satellites, to be launched during 1981 and 1982, will provide K-band services to the Trans-Canada Telephone System. Hughes Aircraft Company of California, UA, as prime contractor, has awarded to Spar major Anik C subcontracts, under which Spar's Mechanical and Electrical Group will supply antennae and transponders.

Spar will have prime contract management and will supply major hardware for the C-band **Anik-D** programme.

SHUTTLE REMOTE MANIPULATOR SYSTEM (RMS)

Under Canadian/US co-operative arrangements, a Canadian team led by Spar is developing the remote manipulator system for the Space Shuttle programme. The team has, in addition, established a computer-controlled facility capable of simulating under laboratory conditions the performance of remote manipulator systems for space, Arctic, underwater and nuclear applications.

The RMS consists of a 15 m (50 ft) long mechanical arm with joints similar to a human shoulder, elbow and wrist. It will be used to deploy satellites into space from the vehicle's cargo bay, and to retrieve orbiting satellites for onboard servicing or return to Earth. It will be capable of handling satellites weighing up to 29,480 kg (65,000 lb), with a length of up to 18·28 m (60 ft) and maximum diameter of 4·57 m (15 ft). Television cameras and lights on the RMS facilitate the handling of payloads.

NASA has placed a contract for three RMS systems, for delivery in 1982, 1983 and 1984. These, together with the first flight system developed by Spar for the National Research Council of Canada, will be installed on the NASA fleet of four Space Shuttle vehicles. The first RMS was delivered in 1981 and was to be tested on the second Shuttle flight.

Engineering model of RMS under test at Spar's Toronto facility

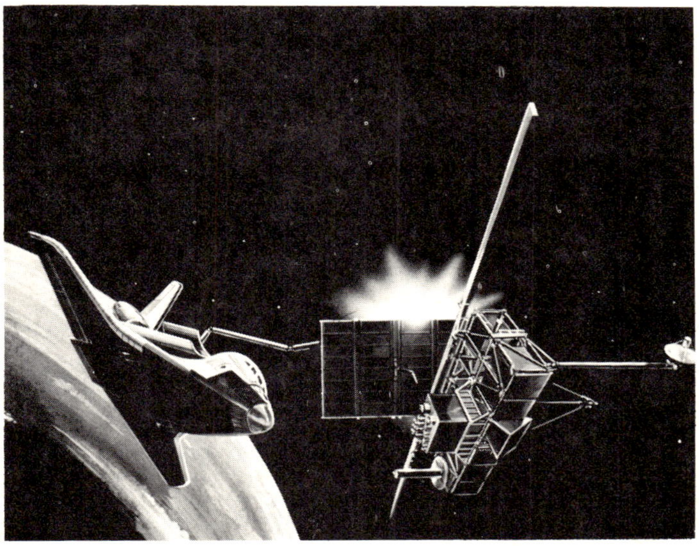

RMS deploying a satellite *(artwork by MBB)*

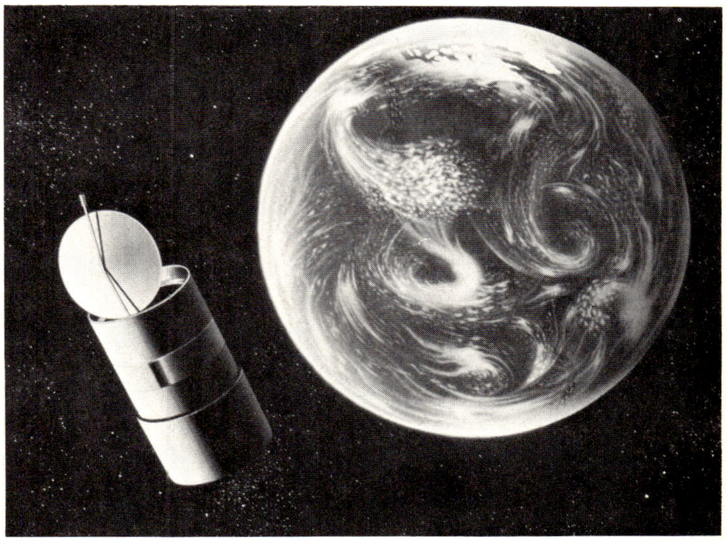

Artist's impression of Anik D in orbit

CHINA
(PEOPLE'S REPUBLIC)

China launched its first satellite (Norad designation *Chicom 1*) on 24 April 1970, and thus became the fifth country to orbit a payload using national resources, following the USSR, USA, France and Japan. The launch is believed to have taken place from the main Chinese rocket centre near Shuangzhengzu (Shuang Cheng Tsu), 800 km (500 miles) east of the nuclear test establishment of Lop Nur.

Brief details of the China 2, 3, 4 and 5 satellites were given in the 1976-77 *Jane's*. China 6 and 7 were briefly described in the 1978-79 *Jane's*. China 8, launched on 26 January 1978, and weighing an estimated 3,630 kg (8,000 lb), returned a re-entry module to Earth.

In 1979 a Japanese delegation was told that China was developing a new three-stage launch vehicle, named Long March III, capable of placing an 800 kg (1,764 lb) payload in geostationary orbit. It was reported that the first and second stages had been developed and tested, and that the third stage was being tested in 1979, with a possible first launch of the complete vehicle in 1981 or 1982. China is also reported to be developing a communications satellite and a weather satellite.

A Chinese spokesman has announced that China has long-range plans for launching manned spacecraft and space laboratories, and a documentary film has shown men in spacesuits undergoing stress tests and weightlessness.

In 1978 a Chinese delegation, led by Dr Ren Xinmin, President of the Chinese Academy of Space Technology, held discussions with US officials regarding possible US-Chinese co-operation in the peaceful utilisation of space technology. The discussions resulted in an informal agreement for purchase of a US satellite communications system.

The discussions also led to the signing of a Memorandum of Understanding between NASA and the Chinese Academy of Sciences in January 1980, covering China's participation in an experimental Landsat programme. Under the agreement, China agreed to establish a ground station in the Beijing (Peking) area to read out Earth resources data from Landsat D.

CZECHOSLOVAKIA

CESKOSLOVENSKA AKADEMIE VED, GEOFYZIKÁLNÍ ÚSTAV
(Geophysical Institute, Czechoslovak Academy of Sciences)

Bocni II, CP1401, 141 31 Praka 4 Sporilov
Telephone: 761941 and 762 541
Telex: 121546

Project Leader: Dr Pavel Triska
Experimenters:
Dr Frantisek Jiricek
Ing Jaroslav Vojta

MAGION

Magion, the first satellite to be designed and built in Czechoslovakia, was launched into orbit with the Soviet satellite Intercosmos 18 on 24 October 1978. The 15 kg (33 lb) satellite carried six experiments to study the Earth's magnetosphere and ionosphere. Magion was ejected into independant orbit on 14 November 1978, the purpose of the mission being to study ELF/VLF phenomena by obtaining simultaneous data from a pair of slowly diverging satellites. Telemetry data from both satellites were received by ground stations in the USSR and by the Observatory of the Academy of Sciences at Panska Ves. Owing to a power malfunction, except for a brief period after activation, the data acquisition rate has been substandard, using low-power telemetry only.

Magion is box-shaped, measuring 300 × 300 × 150 mm (11·8 × 11·8 × 5·9 in).

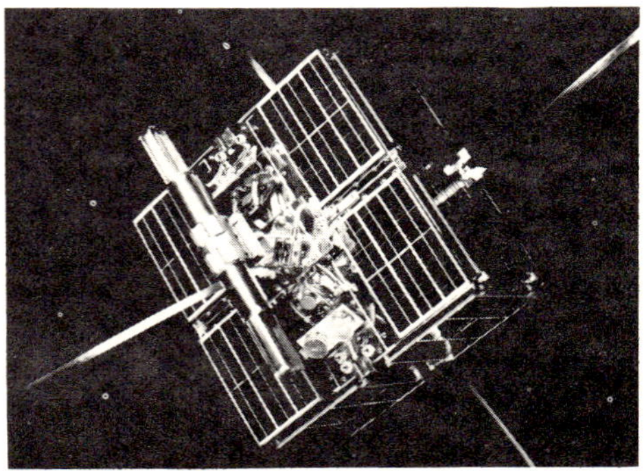

Magion satellite (courtesy *Letectvi + Kosmonautika***)**

FRANCE

AÉROSPATIALE
SOCIÉTÉ NATIONALE INDUSTRIELLE AÉROSPATIALE
Division Systèmes Balistiques et Spatiaux

Management: Route de Verneuil, BP 96, 78130 Les Mureaux
Telephone: 33 (1) 474 72 13
Telex: AISPA 696 759F
Director: P. M. Usunier

The Space and Ballistic Systems Division of Aérospatiale handles research, development, testing and engineering for entire missile and space vehicle systems, including a range of solid-propellant sounding rockets (see 1977-78 *Jane's*). Satellite programmes to which it is contributing include Intelsat V, Exosat, MARECS, Spot, Telecom 1, TDF-1, TV-Sat and Meteosat, for which it is prime contractor.

Aérospatiale is a member of the COSMOS international consortium (see International section) and has founded a joint subsidiary, named Eurosatellite, with AEG Telefunken, MBB, ETCA and Thomson-CSF, for the development and marketing of applications satellites. Studies are currently underway for direct-to-home TV broadcast satellites and communications satellites in both the 500 kg (1,102 lb) and 1,000 kg (2,204 lb) classes.

Second launch of Ariane, 23 May 1980

ARIANE

This heavy three-stage launch vehicle is being developed and produced in Europe as a co-operative project involving eleven countries. The programme is financed by the European Space Agency, with the French space agency CNES as prime contractor. Ariane's payload potential is 1,700 kg (3,748 lb) into a transfer orbit, or applications satellites of up to 970 kg (2,138 lb) weight in geostationary orbit.

Aérospatiale is responsible for the design of Ariane and for integration of the entire vehicle. It is producing major structural elements of the first stage and is responsible for the final delivery of all three stages and the fairings.

SEP is producing the engines and associated subsystems. Matra is building the vehicle equipment bay, and ETCA is responsible for the checkout and launch command facilities.

The first test flight was made successfully on 24 December 1979, with a payload consisting of a technological capsule (CAT) and a dummy payload weighing 1,600 kg (3,527 lb). The second attempted test flight, on 23 May 1980, was unsuccessful due to imperfect combustion in the Viking engines of the first stage. Improved injectors are being designed. The third test launch, using the modified injectors, with a payload consisting of Meteosat 2 and APPLE, was due to take place in the second half of 1981.

The revised schedule of follow-on flights is as follows:

L04	Oct 1981	MARECS A
L05	Dec 1981	Exosat
L06	Feb 1982	MARECS B/SIRIO and SYLDA
L07	Apr 1982	Intelsat V (6) or ECSI
L08	Jun 1982	ECSI or Intelsat V (6)
L09	Oct 1982	Intelsat V (7)
L10	Dec 1982	Intelsat V (8)

Projected developments include lengthening the third stage fuel tank (Ariane 2) and the addition of strap-on boosters (Ariane 3) to increase the transfer orbit payload respectively to 1,950 kg (4,299 lb) and 2,400 kg (5,291 lb).

Model of the Sonate communications satellite

SYLDA dual launch system

Details of the individual stages of the basic Ariane are as follows:

L 140 FIRST STAGE

This first stage is made up of two identical steel tanks for the 145,000 kg (319,670 lb) of UDMH and nitrogen tetroxide propellants, linked together by means of a cylindrical skirt. It is powered by four Viking engines (each 611·25 kN; 137,412 lb st), carried on a cylindrical thrust frame and protected by fairings with fins. Burn time is 145 s.

An interstage skirt joins the first and second stages.

L 33 SECOND STAGE

The second stage comprises two light alloy tanks for 34,000 kg (74,955 lb) of UDMH and nitrogen tetroxide propellants, separated by a common bulkhead. The single Viking engine (713 kN; 160,280 lb thrust in vacuum) is linked to the stage by a conical thrust frame. Burn time is 132 s.

A cylindrical interstage skirt connects the second and third stages.

H 8 THIRD STAGE

The third stage is made of light alloy and houses 8,200 kg (18,075 lb) of liquid hydrogen and liquid oxygen propellants for its single HM7 cryogenic engine (60 kN; 13,485 lb thrust). Burn time is 570 s.

A cylindrical vehicle equipment bay is situated above the third stage. The equipment platform is an annular plate, the inside flange of which supports the payload attachment fittings.

Dimensions:

Length: L 140 1st stage	18·40 m (60 ft 4½ in)
L 33 2nd stage	10·40 m (33 ft 1 in)
H 8 3rd stage	8·60 m (26 ft 2½ in)

Overall length	47·0 m (154 ft 2 in)
Body diameter: L 140	3·80 m (12 ft 5½ in)
L 33, H 8, equipment bay	2·60 m (8 ft 6½ in)
WEIGHTS:	
L 140 1st stage	165,000 kg (363,762 lb)
L 33 2nd stage	37,600 kg (82,895 lb)
H 8 3rd stage	9,400 kg (20,725 lb)
Total weight at lift-off	207,000 kg (456,355 lb)

SONATE

Sonate is a drum-shape telecommunications satellite, weighing 300 to 400 kg (660-880 lb). It is optimised for domestic or regional missions, involving point-to-point telecommunications, relay, data collection and transmission.

Utilising proven technologies and existing hardware, Sonate will be reliable and relatively inexpensive. Its configuration is based on the Meteosat spacecraft platform, and on Symphonie (4-6GHz) or OTS (12-14GHz) payloads, with a mechanically de-spun antenna. It will be spin-stabilised and will have more than 300W power at the end of its lifetime.

Sonate is configured for dual launch with Ariane/SYLDA, but users can choose between this method and the Space Shuttle with an upper stage (SSUS-D).

SYLDA

SYLDA (SYstem for Launching Dual payloads with Ariane) is designed for missions requiring the simultaneous launch of two large satellites. Its first use is scheduled for February 1982, on the sixth launch, with MARECS B and Sirio 2. SYLDA is an egg-shaped device of carbon-fibre/metal honeycomb material. Its height in the current version is 3·9 m (12 ft 9½ in) and its diameter 2·8 m (9 ft 2½ in), making SYLDA the largest carbonfibre structure yet made in Europe. It weighs only 180 kg (397 lb) fully equipped, but can place a total mass (two spacecraft) of 1,520 kg (3,351 lb) into transfer orbit with Ariane 1. A 'stretched' version of SYLDA will be ready for Ariane 2 and 3 by mid-1983, and will accommodate two satellites of 1,400 kg (3,086 lb) total mass, destined for transfer orbit.

TDF-1

This is a joint Franco-German programme for the development of direct-to-home TV broadcast satellites. Under a decision made by the heads of state in October 1979, two satellites are planned, TDF-1 for France and TV-Sat for West Germany, using the same spacecraft platform. Aérospatiale and Thomson-CSF in France, and MBB and AEG-Telefunken in West Germany, are developing jointly the two three-channel pre-operational satellites, and subsequent operational versions.

ARABSAT

On 25 May 1981 Aérospatiale received a $134·35 million contract from the Arab Satellite Communications Organisation to develop and manufacture three communications satellites (including one spare) for use by 22 member states of the Arab League. Compatible with both Ariane/SYLDA and Space Shuttle launch vehicles, they will be launched in late 1983.

The satellites will each have 25 C-band channels and one S-band (TV) channel, and a lifetime of seven years. Tip-to-tip length with solar panels deployed is 21 m (68 ft 10¾ in). The central body measures 2·2 × 1·5 × 1·6 m (7 ft 2½ in × 4 ft 11 in × 5 ft 3 in); mass in orbit will be 680 kg (1,500 lb).

Artist's impression of TDF-1 TV satellite

MATRA
SA MATRA

SPACE BRANCH: 37 avenue Louis Breguet, 78140 Vélizy
Telephone: 946 96 00
Telex: MATRA 698 077F
OFFICERS: See Missiles section

On 18 September 1979, Matra was appointed prime contractor for Telecom 1, the first French telecommunications satellite. It is also prime contractor for the French Earth observation satellite SPOT, which is being developed in association with Sweden and Belgium. Other current commitments include the command and data management system for Spacelab, as contractor for ESA, and the vehicle equipment bay for the Ariane European launch vehicle.

Following successful operation in orbit of the Orbital Test Satellite (OTS), for which Matra developed the attitude and orbit control system, the company is developing within the MESH international consortium the attitude and orbit control system of the European Communications Satellite (ECS) and its maritime version (MARECS), and will integrate ECS.

Matra is also working, under ESA contract, on the Faint Object Camera (FOC) for NASA's Space Telescope.

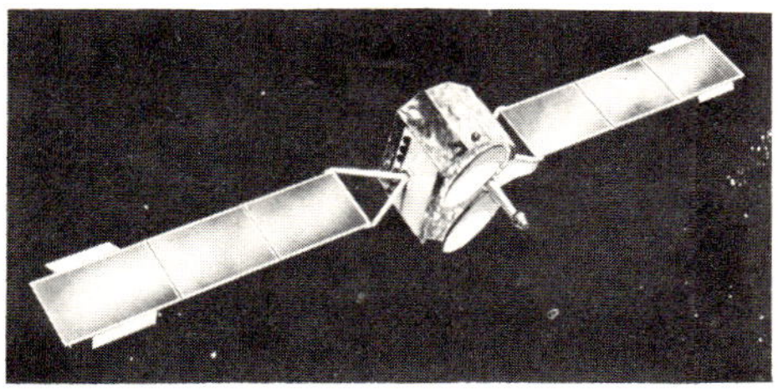

Artist's impression of Telecom 1

TELECOM 1

This telecommunications satellite utilises the platform of MARECS (see MESH entry in International section). The first Telecom 1 will be employed for inter-computer data transfer, video conferences, video transmissions, telecopying, and telephone and TV hook-up with overseas French territories. Its general appearance is shown in the accompanying illustration.

SPOT

Matra is prime contractor for the SPOT Earth observation satellite which, utilising advanced instrumentation, is expected to provide images of greater resolution than those obtained from the US Landsat series of spacecraft.

SPOT comprises two modules: a lower platform housing the stabilisation subsystem, power supplies, onboard management systems, telemetry and solar array; and an upper payload module housing the mission instrumentation. The solar panel array will provide 1,800W power. Overall length of the satellite is 7·35 m (24 ft 1½ in) and weight with maximum payload 1,750 kg (3,858 lb).

SPOT is scheduled to be launched by Ariane in 1984.

Mockup of SPOT Earth observation satellite

GERMANY
(FEDERAL REPUBLIC)

DORNIER
DORNIER SYSTEM GmbH

Postfach 6136048, 7990 Friedrichshafen
Telephone: Immenstaad (07545) 81
Telex: 073 4359
GENERAL MANAGERS:
Dipl-Ing Silvius Dornier
Dipl-Ing Dr Jr Karl-Wilhelm Schäfer
Dr Ing Bernhard Schmidt
Dr Ing Helmut Ulke
Dipl-Kfm Klaus-Peter Thomé
PUBLIC RELATIONS: Gerhard Patt

Dornier System, a member company of the Dornier group, is engaged in activities involving spaceflight, new technologies, electronics and research. It is a major European contractor for the development of equipment for satellites, and for associated equipment such as the communications terminals mounted on West German ships for use with the Marisat satellite system.

Within the STAR international consortium, Dornier System is leading the development programme for the ISPM (International Solar Polar Mission) scientific satellite. Two spacecraft, the second developed by NASA (which see), will be launched by Space Shuttle into planetary trajectories towards Jupiter, the gravitational field of which will deflect them out of the ecliptic plane in opposite directions for unique observations of our solar system.

Dornier System is also developing the environmental control and life support subsystem (ECLSS) and the instrument pointing system (IPS) for the Spacelab which is to be carried into orbit by the US Space Shuttle.

For Ariane, the European launch vehicle, Dornier has developed the second-stage tank structure.

ERNO
ERNO RAUMFAHRTTECHNIK GmbH

Hünefeldstrasse 1-5, Postfach 10 59 09, 2800 Bremen 1
Telephone: (0421) 5391
Telex: 024 5548
MANAGEMENT:
Dipl-Ing Hans Hoffmann
Fritz Sandermeier

ERNO is the prime contractor for Spacelab, which is to be carried into orbit by the Space Shuttle. The Company is also responsible for the second stage of the Ariane European launch vehicle and is involved in work on communications satellites such as the European Communications Satellite (ECS), MARECS, the French Telecom 1, and the Franco-German direct-broadcast TV-Sat project. The Company's main activities in satellite technology involve

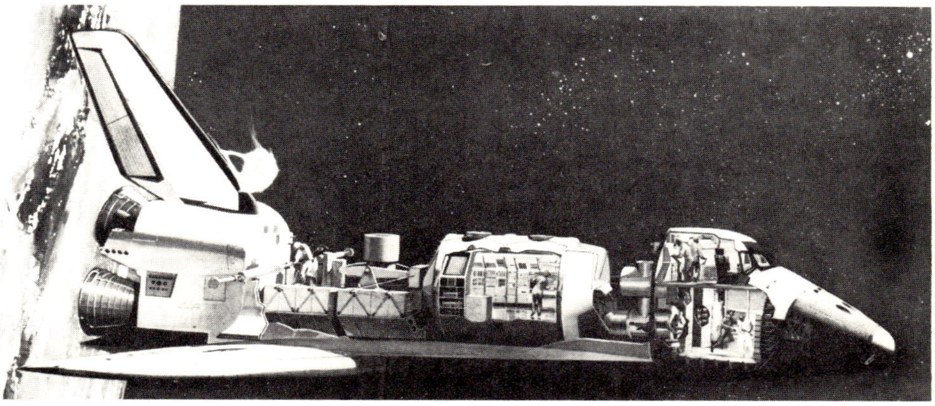

Artist's impression of Spacelab in orbit

subsystems and components for thermal and reaction control systems.

SPACELAB

Under development by a ten-nation consortium led by VFW/ERNO, Spacelab is the only manned and reusable payload for the American Space Shuttle. A typical payload will comprise a two-segment module and two pallets with an overall length of 13·80 m (45 ft 3 in) and a diameter of 4·06 m (13 ft 4 in).

The first launch of the Space Shuttle with a Spacelab is scheduled for mid-1983. Up to four payload specialists can work in two shifts for a period of seven days; extensions up to 30 days are envisaged. Spacelab will remain in the cargo bay of the Orbiter during the mission and will be brought back to Earth when the Orbiter lands. The payload specialists will be able to work under normal atmospheric conditions in the pressurised module.

MBB
MESSERSCHMITT-BÖLKOW-BLOHM GmbH

Space Division
8012 Ottobrunn bei München
Telephone: (089) 60 00 25 90

In addition to its manufacture of piloted aircraft, Messerschmitt-Bölkow-Blohm is engaged in the development of guided weapons and RPVs (see relevant sections of this edition), research rockets and satellites. Contributions to ten space programmes have been made since 1968, including the two Helios solar probes, the Symphonie communications satellites and the cryogenic engine for the third stage of the Ariane European launch vehicle. Currently the company is building the Exosat X-ray satellite, under contract to ESA, and is a major subcontractor for the Intelsat V communications satellite.

MBB is developing the propulsion module for the Galileo Jupiter probe, sponsored by NASA and the Federal German Ministry of Research and Technology, and subsystems for the SPAS (Shuttle Pallet Satellite) family of satellites for use with the Space Shuttle. The company is also engaged in the Franco-German programme for domestic TV-Sat/TDF-1 satellites, each designed to transmit three television programmes directly to every home TV receiver in France and Germany, from a geostationary orbit. These satellites, designed for a life of seven years, will be forerunners of future national operational TV satellites, each with a capacity of five TV channels.

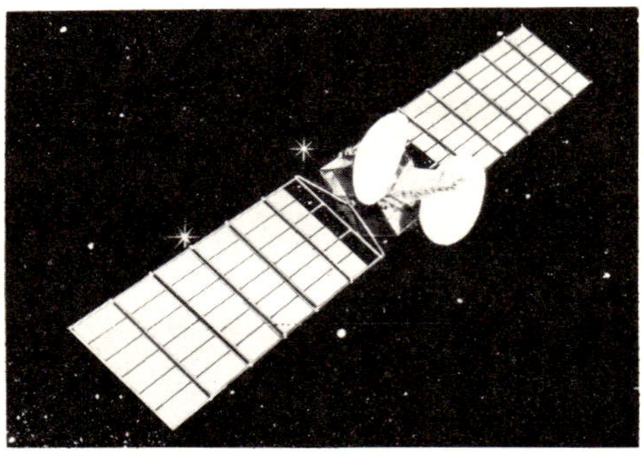

Artist's impression of TV-Sat/TDF-1

OTRAG
ORBITAL TRANSPORT- UND RAKETEN AG

Daimlerstrasse 19, 8046 Garching
Telephone: 089 320 1079
Telex: 528548 otragd
CHAIRMAN: Dipl-Ing Lutz T. Kayser
PRESIDENT: Dipl-Ing Frank K. Wukasch

OTRAG is developing as a private venture a low-cost satellite launch vehicle based on a building block concept, by which tank-plus-engine modules are clustered in increasing numbers to form larger vehicles.

The basic OTRAG building block is a two-tank/two-engine module, one tank containing white fuming nitric acid and the other kerosene. The two throttleable engines each develop 29·36 kN (6,600 lb st). Two such building blocks together form the smallest flight vehicle.

The tanks, which have a diameter of 27·0 cm (10·6 in), are made from cold-spun stainless steel and are built up from modular segments each 3·0 m (9 ft 10 in) long. Length of the tanks on the basic vehicle is 12·0 m (39 ft 4½ in), but this is increased for the larger vehicles. Each tank is filled with propellant to about two-thirds capacity. An adiabatic blow-down feed system utilising compressed air occupies the remaining space and forces the propellant into the engine during flight. Radial injection and ablative thrust chambers are used, lined with a coating of asbestos and phenolic plastics resin.

Engine throttling capability is provided by ball valves, which allow propellants to pass from the tanks to the engines, and are opened or closed by actuators to provide full, 50 per cent or zero thrust.

To achieve low cost, existing aerospace components and commercially available components are used wherever possible, examples of the latter being the fuel valves and their actuators.

Guidance is by an inertial strap-down platform and computer in the payload section. These send signals to each engine actuator unit, which in turn throttles the engines on one side or the other to control the attitude of the vehicle.

A basic flight vehicle, consisting of two modules, was successfully test-launched from North-Shaba in Zaïre on 17 May 1977, when an altitude of about 12,000 m (39,370 ft) was reached. A second launch, on 20 May 1978, utilised a cluster of four modules and reached a height of 30,000 m (98,400 ft). No. 3 flight test took place on 5 June 1978, with a cluster of four modules and tank length of 12·0 m (39 ft 4½ in).

Sub-orbital test flight No. 4 was made on 1 March 1981 from a new test range in the Central Sahara, about 800 km (497 miles) south of Tripoli, in Libya. During this test the propulsion system, as well as the newly-developed inertial strap-down guidance system, worked perfectly.

Further launches were planned for later in 1981.

From these basic flight vehicles OTRAG plans to evolve a series of efficient but relatively inexpensive launchers capable of undertaking a wide variety of duties. The first orbital insertion is planned for late 1982, with a 64-module vehicle; by 1984 it is hoped to launch a vehicle of more than 600 modules. On the larger vehicles the propellant tanks will be lengthened to a maximum of 40 m (131 ft), and the basic modules will be clustered in stages one inside the other. As one stage or 'layer' burns out, it will be jettisoned when the next outermost layer or stage ignites. The process will continue until only a slender core stage is left to insert the payload into orbit.

A 600-module vehicle could place a 10,000 kg (22,050 lb) payload in low Earth orbit, or 2,000 kg (4,410 lb) in geosynchronous orbit.

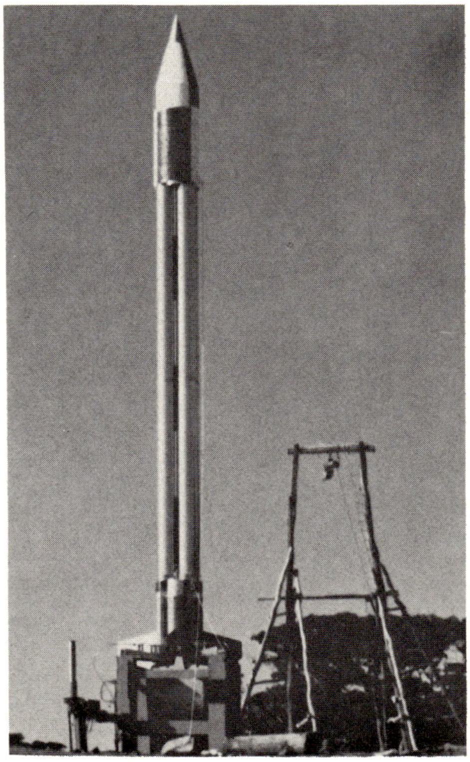

OTRAG 4-module launch vehicle

INDIA

ISRO
INDIAN SPACE RESEARCH ORGANISATION

F Block, Cauvery Bhavan, Kempe Gowda Road, Bangalore 560 009
Telephone: 27371/76
Telex: 0845-499 IN
CHAIRMAN: Prof S. Dhawan
DIRECTOR, PUBLICATIONS & PUBLIC RELATIONS UNIT: Dr S. P. Kosta

The government of India set up a Space Commission in June 1972, backed by the Department of Space (DOS), and entrusted DOS with responsibility for conducting India's space programme. ISRO functions under DOS as its research and development organisation and for executing the space research activities.

ISRO operates and maintains the UN-sponsored Thumba Equatorial Rocket Launching Station (TERLS) near Trivandrum, and encourages international collaboration in space research experiments using rockets.

The activities of ISRO, which has its Headquarters at Bangalore, are carried out at the following centres:

Vikram Sarabhai Space Centre (VSSC)
ISRO PO, Trivandrum 695 022
Telephone: 56-2444 and 56-2555
Telex: 0884-201 IN and 0884-202 IN
DIRECTOR: Dr V. R. Gowarikar

VSSC serves as the main research and development centre for space technology. It is engaged in the development of sounding rockets and satellite launch vehicles.

Space Applications Centre (SAC)
Jodhpur Tekra, Ahmedabad 380 053
Telephone: 442700 and 443296
Telex: 012-239 IN
DIRECTOR: Prof E. V. Chitnis

This Centre is concerned with space applications programmes of ISRO in the areas of satellite communications, TV broadcasting via satellite, and the survey of Earth resources and meteorological parameters from space/aerial platforms using remote sensing and satellite geodesy.

SHAR Centre
Sriharikota PO 524 124, Nellore District (AP)
Telephone: 2001 and 2345 (via Madras)
Telex: 041-394 and 041-7353 IN
DIRECTOR: N. Pant

This centre contains facilities for launching and flight testing large multi-stage sounding rockets. Facilities for launching satellites are under development. The centre also has facilities for the production of solid propellants and for the static testing of rocket motors. The main Indian ground station for tracking, commanding ISRO satellites and receiving telemetry data is located at this centre.

ISRO Satellite Centre (ISAC)
Peenya Industrial Estate, Bangalore 560 058
Telephone: 38261-69
Telex: 0845-325 IN
DIRECTOR: Prof U. R. Rao

This centre provides inputs for planning satellite programmes/missions of ISRO, and for the fabrication of satellites. It is the research centre for satellite technology.

The latest ISRO programmes are as follows:

SATELLITE LAUNCH VEHICLE (SLV-3)

This VSSC project has the objective of developing an indigenous satellite launch vehicle capable of putting a 40 kg (88 lb) Rohini satellite into a near-Earth orbit. The four-stage solid-propellant vehicle, designated SLV-3, has a length of 22·77 m (74 ft 8 in), with a maximum diameter of 1 m (39 in) and weighs 16,900 kg (37,258 lb) at lift-off. Thrust of the first stage is 422 kN (95,000 lb). Inertial guidance is employed.

The first experimental flight on 10 August 1979 was only partially successful; the second flight, on 18 July 1980, was successful.

APPLE

APPLE (Ariane Passenger Payload Experiment), weighing about 657 kg (1,448 lb) is India's first three-axis stabilised geostationary experimental communications satellite. Its communications payload is being built by the Space Applications Centre, Ahmedabad. Major subsystems, such as the Apogee Boost Motor, C-Band Antenna, Momentum Wheel and Solar Array Drive, are under development at VSSC. The satellite was due to be launched by the third Ariane vehicle of the European Space Agency in 1981. APPLE is a forerunner of future

Indian operational communications satellites and represents a significant advance in indigenous satellite technology.

BHASKARA

The second Indian satellite, known as Bhaskara (originally SEO) and designed primarily for Earth observation experiments, was launched from the Soviet Union on 7 June 1979. A malfunction of the TV camera system was overcome in May 1980, since when it and the radiometers have been providing useful ocean and land-surface data.

An improved Bhaskara was being prepared for launch in 1981 from the Soviet Union.

INSAT

To further the objective of the Indian Space Programme to use space technology for practical applications, the government approved establishment of the Indian National Satellite (INSAT) system. INSAT is a multipurpose geostationary satellite system for telecommunications and meteorology, and also has the capability for nationwide direct TV broadcast to community TV receivers in rural areas. INSAT involves the departments of Space, Communications and Meteorology, as well as the Ministry of Information and Broadcasting.

Each of the two first-generation INSAT (INSAT-1) satellites will combine capability for long-distance telephony, continuous meteorological Earth observations in the visible and infra-red bands, relay of meteorological data from unattended data platforms, disaster warning, direct TV broadcasting and programme distribution.

The Ford Aerospace & Communications Corporation, USA, has been awarded a contract for fabrication of the INSAT-1 spacecraft and associated equipment for the master control facility (MCF) that is being created near Hassan, Karnataka.

The first INSAT-1 is scheduled to be launched early in 1982.

ROHINI

The 35 kg (77 lb) Rohini Satellite (RS-1), designed to evaluate the performance of the fourth stage of the SLV-3 vehicle, was launched successfully on 18 July 1980. It was the first satellite to be made and launched in India, and in mid-1981 it was functioning well. A series of Rohini satellites, carrying scientific and applications payloads, is planned.

INTERNATIONAL PROGRAMMES

CIFAS

CONSORTIUM INDUSTRIEL FRANCO-ALLEMAND POUR LE SATELLITE SYMPHONIE (DEUTSCH-FRANZÖSICHES INDUSTRIE-KONSORTIUM FÜR DEN SATELLITEN SYMPHONIE)

This consortium was formed on 25 April 1968 to develop a telecommunications satellite named Symphonie. Following successful completion of the programme, described in the 1980-81 *Jane's*, the consortium has been disbanded.

COSMOS

ÉTUDES TECHNIQUES ET CONSTRUCTIONS AÉROSPATIALES SA (ETCA)

BP 97, 6000 Charleroi 1, Belgium

SOCIÉTÉ NATIONALE INDUSTRIELLE AÉROSPATIALE

BP 96, 78130 Les Mureaux, France

SOCIÉTÉ ANONYME DE TÉLÉCOMMUNICATIONS (SAT)

41 rue Cantagrel, 75624 Paris Cédex 13, France

MESSERSCHMITT-BÖLKOW-BLOHM GmbH (MBB)

8 München 80, Postfach 801 169, West Germany

CONSTRUCCIONES AERONAUTICAS SA (CASA)

Division Espacial, Getafe, Madrid, Spain

MARCONI SPACE & DEFENCE SYSTEMS LTD (MSDS)

The Grove, Stanmore, Middlesex HA7 4LY, England

SELENIA SpA

CP 7083, 00100 Rome, Italy

The COSMOS consortium was formed in November 1970. It is involved in several major space programmes, notably Exosat and Meteosat, of which details follow:

EXOSAT

In 1978, MBB of West Germany, acting as prime contractor for the COSMOS consortium, was awarded a contract by ESA for development of the European X-Ray Observatory Satellite (Exosat). This satellite will extend the understanding of both point source and distributed cosmic X-ray sources.

MSDS is responsible for the attitude control system, which is so precise (about 2 arc seconds peak-to-peak in fine pointing) that astronomers will be able to identify for the first time point sources with very faint visible stars. In addition, MSDS is responsible for the orbit control system which, by adjusting the satellite velocity, will cause selected X-ray objects to be occulted by the Moon, giving further definition of the location of point sources and the distribution of extended ones. Exosat was scheduled to be launched in December 1981 by a European Ariane.

Mechanical test model of Exosat

METEOSAT

Aérospatiale of France, acting as prime contractor for the COSMOS consortium, was awarded a $56·76 million contract by ESA for development of this geosynchronous meteorological satellite in December 1973. The work culminated in launch of Meteosat 1 by a Delta 2914 vehicle on 23 November 1977. Launch of the second Meteosat was scheduled for late 1981, by an Ariane vehicle.

Of the other COSMOS members, MBB is responsible for the spacecraft structure; Marconi Space & Defence Systems supplies the attitude and orbit control system and checkout equipment. MSDS is also responsible for the satellite's main communications package for all future missions. Ford Aerospace is a consultant for the programme.

Meteosat represents Europe's contribution to a worldwide Global Atmospheric Research Programme (GARP), intended to utilise also US and Japanese satellites to provide data for the World Weather Watch Organisation. Meteosat's main mission is to transmit images of the Earth and its cloud cover every half-hour, using both visible light and the infra-red spectrum. It also relays processed images from the main station in Darmstadt to user stations. Data are collected from land-based, marine or airborne platforms at various points in the coverage zone.

Meteosat meteorological satellite

MESH

AERITALIA SpA

Piazzale V. Tecchio 51, 80125 Naples, Italy

BRITISH AEROSPACE PUBLIC LIMITED COMPANY, DYNAMICS GROUP

Space and Communications Division, Gunnels Wood Road, Stevenage, Hertfordshire SG1 2AS, England

ERNO RAUMFAHRTTECHNIK GmbH

28 Bremen 1, Hünefeldstrasse 15, Postfach 1199, Federal Republic of Germany

FOKKER BV

PO Box 1065, 1000BB Amsterdam, Netherlands

INTA

Departamento de Equipo y Armamento, Paseo del Pintor Rosales 34, Torrejon de Ardoz, Madrid 8, Spain

SA MATRA

BP No. 1, 78140 Vélizy, France

SAAB-SCANIA AB

S-581 88 Linköping, Sweden

This international consortium was formed in October 1966, the word MESH being an acronym of the initial letters of the four original founder members.

In addition to the OTS and ECS satellite programmes, MESH is involved in the maritime communications satellite MARECS, which employs basic ECS technology.

ECS

Five European Communications Satellites (ECS) are being developed and built, with British Aerospace Dynamics Group as prime contractor. ECS, based on the technology of OTS, will offer a fully operational indigenous regional satellite communications system, and will be capable of carrying a significant proportion of future European telephone, telex and TV traffic. The first ECS will be placed in geostationary orbit in 1982, by the European Ariane launcher, and it is planned that this should be followed by others, which will contain a package dedicated to business systems communications, between then and 1990. Each satellite is designed for a minimum operational life of seven years. All launches will be made from the equatorial site at Kourou in French Guiana, using Ariane.

MARECS

This maritime communications programme, for which British Aerospace Dynamics Group is lead contractor, involves the initial procurement of two MARECS (the maritime version of ECS) satellites, together with components for a third. The two MARECS are to be leased by INMARSAT (International Maritime Satellite Organisation) for inclusion in a global maritime communications network which will provide merchant shipping with instant communications to shore terminals by links vastly superior to the present highly congested HF circuits.

The basic structure, control and power systems are those of ECS, to which will be fitted complete communications packages being developed by MSDS. The first satellite was scheduled to be launched into geostationary orbit by Ariane in October 1981. The system will be augmented by further satellites in different orbital locations, until INMARSAT has established a worldwide maritime satellite communications system.

To permit the use of simple shipborne and shore-based terminals, MARECS will be equipped with a high-efficiency 'shaped beam' antenna, coupled with a transistorised L-band power amplifier. Ships will use L-band frequencies to communicate to and from the satellite. As shore terminals will be able to use more sophisticated equipment, the shore-to-satellite frequency will be in the 6GHz band and satellite-to-shore communications will use the 4GHz band.

OTS

This Orbital Test Satellite was produced under a three-year contract, with British Aerospace Dynamics Group as prime contractor, responsible for both development and launch support. It is intended to demonstrate the operational capabilities of high-technology communications

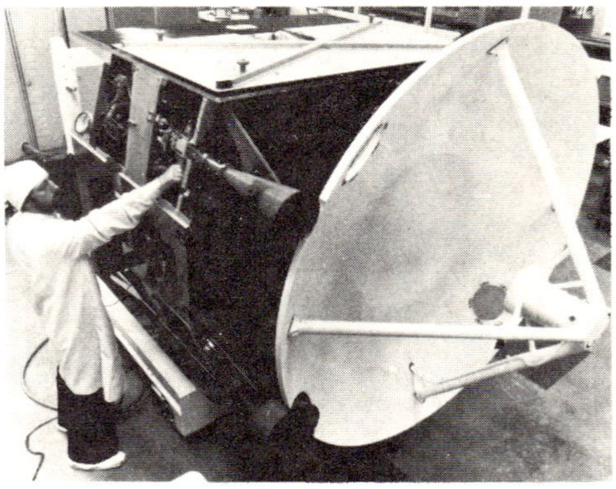

MARECS maritime communications satellite

systems, and to test experimental advanced communications concepts and propagation assumptions in space, as a preliminary to the ECS programme.

A failure in the Delta launch vehicle caused the first OTS to be lost in September 1977. The second OTS was launched successfully on 11 May 1978 and arrived on station at 10°E on 21 May.

OTS 2 has exceeded its design life, and the performance standards specified, and has demonstrated its ability to transmit TV signals to, and receive them from, small mobile ground stations.

Artist's impression of ECS satellite in orbit

STAR

BRITISH AEROSPACE PUBLIC LIMITED COMPANY, DYNAMICS GROUP

Space and Communications Division, Gunnels Wood Road, Stevenage, Hertfordshire SG1 2AS, England

CONTRAVES AG

Schaffhauserstrasse 580, CH-8052 Zurich 11, Switzerland

CGE-FIAR

Via G.B. Grassi 93, 20157 Milan, Italy

DORNIER SYSTEM GmbH

Postfach 648, 799 Friedrichshafen/Bodensee, Federal Republic of Germany

TELEFONAKTIEBOLAGET L. M. ERICSSON

126 25 Stockholm, Sweden

MONTEDEL (MONTECATINI EDISON ELETRONICA SpA)

Via E Bassini 15, 20133 Milan, Italy

SENER SA

Guzman el Bueno 133, Madrid 3, Spain

SOCIÉTÉ EUROPÉENNE DE PROPULSION

Tour Roussel-Nobel, 92080 Paris Défense Cédex 3, France

THOMSON-CSF

173 boulevard Hausmann, 75360 Paris Cédex 08, France

STAR (Satellites for Telecommunications, Applications and Research) companies are involved in several major space programmes, including development of the ISPM satellite, described under the NASA entry.

Details of STAR's earlier ISEE-2 satellite can be found in the 1979-80 *Jane's*.

GIOTTO

The European Space Agency (ESA) has selected British Aerospace Dynamics Group, Space and Communications Division, to develop the GIOTTO satellite that will investigate Halley's comet in 1986. The work will be led by the Division's Bristol factory. The object of the mission will be to intercept the comet with GIOTTO, carrying a variety of scientific instruments. These will provide data on the chemical composition of the 'coma' region surrounding the nucleus, and of the tail of the comet. A camera will take pictures of the comet's nucleus, and measurements may also be made of its magnetic field.

The design of Giotto will be based on that of the highly successful GEOS-1 and 2 satellites (see 1977-78 *Jane's*).

ITALY

AERITALIA

AERITALIA-SOCIETÀ AEROSPAZIALE ITALIANA p.A.

Space and Alternative Energies Group

HEAD OFFICE: Corso Marche 41, 10146 Turin
Telephone: (011) 33321
Telex: 221235 AERSPZ

GROUP DIRECTOR: Dott Stefano Abbà
DIRECTOR OF EXTERNAL AFFAIRS: Dott Ugo Sacerdote
DIRECTOR, SPACE SECTOR: Dott Ernesto Vallerani

Aeritalia is engaged in research and development of missile and space vehicle systems. It is a member of the ERNO Spacelab consortium and of the MESH satellite consortium. It is responsible for design and development of the Spacelab pressurised module, and the active and passive thermal control subsystem. Activities within the MESH consortium include design, development, construction and structural testing of the ECS communications satellite and of its MARECS maritime version; and manufacture of the structure of the Matra Telecom 1.

Aeritalia is responsible for the thermal control subsystem of the ESA L-Sat telecommunications satellite, and for the L-Sat payload module; it is also involved in AOCS, EMC and EGSE activities. For the Sirio-2 satellite, Aeritalia is producing the structure, thermal control, and VHF antenna.

IRIS

IRIS (Italian Research Interim Stage) is a propulsive upper stage (perigee stage) of the solid-spinning type (SSUS), designed to carry satellites in the 600 kg (1,323 lb) class from low orbit into a geosynchronous transfer orbit. Aeritalia is responsible for overall system activities and for several subsystems, collaborating with Difesa e Spazio (formerly SNIA) in programme management.

ITALSAT

Italsat is a pre-operational telecommunications satellite, designed to provide telephony and other services to national users. Aeritalia is responsible for structure, thermal control and AOCS subsystems.

TETHERED SATELLITE

This programme is for a satellite to be operated in low orbit from the Space Shuttle, through a tether about 120 km (74·5 miles) long. Aeritalia is prime contractor for a pre-development study, in preparation for later phases which may be carried out as a bilateral collaboration between Italy and NASA.

CNS

COMPAGNIA NAZIONALE SATELLITI per TELECOMUNICAZIONI SpA

Via Salaria Km 9·3, 00138 Rome
Telephone: (06) 840 2021
Telex: 610295 CNASPA I
DIRECTOR GENERAL: Ing Antonio Roolotà

This company succeeds the former Compagnia Nazionale Aerospaziale, which was formed in 1965 for the design and development of aerospace systems. Its current parent companies are Aeritalia, Bastogi Sistemi, Selenia and Difesa e Spazio.

The company was involved primarily in development of the Sirio-1 satellite, produced under the sponsorship of the Consiglio Nazionale Ricerche (CNR) of Rome and launched on 25 August 1977. Sirio-1 reached station at 15°W longitude on 8 September 1977, where it is still successfully involved in SHF propagation experiments. Brief details of this satellite were given in the 1977-78 *Jane's*.

CNS is currently integrating Sirio-2, under contract to the European Space Agency. It is also involved, as space segment prime contractor, in the Italsat programme. Italsat (see Aeritalia entry) is the pre-operational communications satellite envisaged by the Italian National Space Programme to carry out experimental point-to-point communications using 20/30GHz bands.

SIRIO-2

The Sirio-2 satellite is derived directly from Sirio-1, and will carry out two different missions:

MDD. (Meteorological Data Distribution). In support of the Meteosat 2 mission, Sirio-2 will distribute meteorological data, mainly in regions of Africa.

Lasso. Experiment for synchronisation of atomic clocks, using laser techniques.

Sirio-2 will be launched in about February 1982 by Ariane L5, equipped for the first time with the SYLDA double launch facility, from Kourou, the ESA launch site in Guiana.

CNS Sirio-2 prototype

JAPAN

MITSUBISHI

MITSUBISHI JUKOGYO KABUSHIKI KAISHA (Mitsubishi Heavy Industries Ltd)

AIRCRAFT AND SPACE VEHICLE DEPARTMENT:
5-1, Marunouchi, 2-chome, Chiyoda-ku, Tokyo 100

OFFICERS: See Aircraft section
MANAGER, SPACE SYSTEMS DEPARTMENT: M. Hamada

Mitsubishi Heavy Industries has co-operated with the National Space Development Agency of Japan in the programme for the N-I and N-II launch vehicles; with the National Aerospace Laboratory, which supports rocket research; and with the Institute of Space and Aeronautical Science, University of Tokyo, which is a major contributor to scientific satellite programmes. It is also a designer and manufacturer of space subsystems, and ground support equipment.

Mitsubishi had delivered six N-Is by early 1980 and is scheduled to produce two N-IIs before 1983. It is also developing the second stage of the H-1 launch vehicle for NASDA, and GS-1, Japan's first geosurvey satellite, in co-operation with Kawasaki and Canon Corporation. The GS-1 is to be launched by NASDA, using an H-I launch vehicle, in FY 1985.

MITSUBISHI ELECTRIC CORPORATION (MELCO)

2-3, Marunouchi, 2-chome, Chiyoda-ku, Tokyo 100

Mitsubishi Electric Corporation has designed and manufactured a succession of satellites for NASDA (which see), often in partnership with Ford Aerospace & Communications Corporation. Among the latest of these was Japan's second Experimental Communications Satellite, ECS-B. This was launched successfully on 22 February 1980, but was lost in orbit three days later during a boost motor burn towards geostationary orbit.

NASDA

NATIONAL SPACE DEVELOPMENT AGENCY OF JAPAN

2-4-1, Hamamatsu-cho, Minato-ku, Tokyo 105
Telephone: (03) 435 6111
Telex: J28424
PRESIDENT: Masao Yamanouchi
DIRECTOR, INTERNATIONAL AFFAIRS DEPARTMENT:
Hiroshi Uda

This Agency was established on 1 October 1969, with primary responsibility for Japanese development of satellites, launch vehicles, tracking systems, and associated facilities and equipment.

In addition to the programme of satellites to be launched by the N vehicle (which see), NASDA has another programme involving three geostationary satellites launched by US vehicles—Geostationary Meteorological Satellite (GMS-1), launched on 14 July 1977 and described in the 1977-78 *Jane's*; the experimental Communications Satellite (CS) Sakura, launched on 15 December 1977; and the experimental Broadcasting Satellite (BSE) Yuri, launched on 7 April 1978. CS and BSE were described in the 1978-79 *Jane's*.

The future NASDA programme calls for the launching of two more engineering satellites, a second GMS met-sat, a geosurvey satellite, second and third com-sats, the second and third TV-sats and a maritime observation satellite.

H-1 LAUNCH VEHICLE

This three-stage vehicle is being designed to launch heavy satellites in the mid-eighties. The H-1 will employ Japan's first liquid hydrogen/liquid oxygen rocket motor, developed by Mitsubishi, as its second stage, and will be able to launch satellites weighing 500 kg (1,102 lb) into geostationary orbit.

N LAUNCH VEHICLE

The three-stage N-I launch vehicle is capable of putting a 1,200 kg (2,645 lb) satellite into a 300 km (186 mile) circular orbit, or a 130 kg (286 lb) satellite into a geostationary orbit. The first stage consists of a Douglas DSV-3N-1 liquid-propellant (liquid oxygen/kerosene) rocket, with three Nissan (Thiokol TX-354-5) solid-propellant strap-on boosters (approx 1,471 kN; 330,600 lb st); the LE-3 second stage is powered by a storable liquid rocket (NTO/A-50), and the third stage by a Thiokol TE-364-14 solid rocket. The guidance system is of the strap-down radio command type. The N-II vehicle has an extended long-tank DSV-3P-1, nine strap-on boosters and inertial guidance, and is designed to put a 350 kg (772 lb) satellite into geostationary orbit. It was scheduled to be tested in 1981, and will have an Aerojet AJ10-118FJ second stage instead of the LE-3 of the N-I.

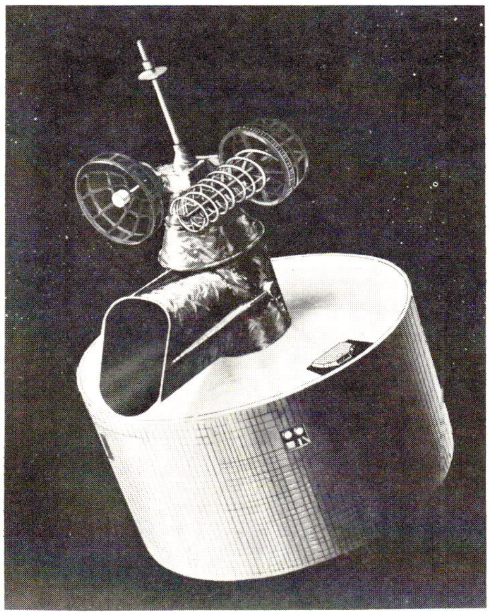

Artist's impression of GMS-2

With the N vehicle NASDA launched the Ionosphere Sounding Satellite-B (ISS-B) Ume on 16 February 1978, and the Experimental Communications Satellite (ECS-A) Ayame on 6 February 1979. ECS-B, the backup of Ayame, was launched successfully on 22 February 1980. ETS-III, which is being designed to establish the production and operational technology of three-axis stabilised satellites with a high-accuracy attitude control system, is to be launched in 1982. Other satellites planned for launch by N-11 vehicles are ETS-IV, designed to prove the operational technology required for later large-scale satellites; GMS-2, CS-2 and BS-2 (second-generation developments of Himawari, Sakura and Yuri); and AMES, which is to establish operational technology for aeronautical and maritime satellite systems. CS-2 is scheduled for launch in 1983 and AMES in 1986.

The main parameters of the N-I and N-II vehicles are:

DIMENSIONS (approx):	
Length overall: N-I	32·6 m (106 ft 11 in)
N-II	35·4 m (116 ft 2 in)
Diameter of first stage:	
N-I and N-II	2·4 m (8 ft 2½ in)
WEIGHT (approx): N-I	90,400 kg (199,298 lb)
N-II	134,700 kg (296,963 lb)

First launching of N rocket

NISSAN

NISSAN JIDOSHA KABUSHIKI KAISHA (Nissan Motor Co Ltd)

HEAD OFFICE: 17-1, 6-chome, Ginza, Chuo-ku, Tokyo
AERONAUTICAL AND SPACE DIVISION: 5-1, 3-chome, Momoi, Suginami-ku, Tokyo
Telephone: Tokyo (390) 1111
Telex: 232 2271
CHAIRMAN: Katsuji Kawamata
PRESIDENT: Takashi Ishihara
MANAGING DIRECTOR AND GENERAL MANAGER, AERONAUTICAL AND SPACE DIVISION: Kazuo Shibata

The Aeronautical and Space Division of Nissan has a major share of the market for solid rocket motors in Japan. It has produced many sounding rockets and space launch vehicles for the Institute of Space and Aeronautical Science, University of Tokyo (ISAS); National Aerospace Laboratory; Meteorological Agency; and Antarctic Research Centre. Nissan has also fabricated a number of strap-on boosters for the N satellite launch vehicles of the National Space Development Agency of Japan (NASDA), under licence from Thiokol Corporation, USA. The third-stage solid-propellant motors of the N vehicle are procured from Thiokol and serviced by Nissan.

Mu ROCKET

Details of Mu-4S, Mu-3C, Mu-3C-1 and Mu-3C-2 satellite launch rockets were given in the 1977-78 *Jane's*. The Mu-3H launch vehicle, described in the 1980-81 *Jane's*, was basically an Mu-3C with an improved M13 first stage. The latest development, Mu-3S, is basically an Mu-3H with thrust vectoring control (TVC) on both the first and second stages. The new rocket was used to launch the 196 kg (432 lb) ISAS scientific satellite Astro-A (Hi No Tori) on 21 February 1981. When developed, the Mu-3S will be able to place payloads of around 290 kg (639 lb) into a nominal 250 km (155 mile) orbit.

Nissan is also developing a new launch vehicle, similar in size to Mu-3S, but with improved second and third stages,

Upper-stage spherical motor and Hi No Tori satellite

and two large strap-on boosters for the first stage. This rocket will be able to place a 670 kg (1,477 lb) payload into a 250 km (155 mile) circular orbit, and is to be used to send Japan's first interplanetary spacecraft, Planet-A, to Halley's comet in FY 1985.

Brief details of the MS-T3 Tansei 3 and Exos-A satellites launched by Mu vehicles were given in the 1978-79 *Jane's*, and details of the Exos-B (Jikiken) and Corsa-B (Hakucho) satellites in the 1980-81 *Jane's*. The 178 kg (392 lb) engineering test satellite MS-T4 was launched on 17 February 1980 and placed in a 720 × 518 km (447 × 322 mile) orbit; when in orbit this satellite was named Tansei 4. The ISAS scientific satellites Astro-B and Exos-C will be launched by the Mu-3S in the 1980s.

The following details apply to the Mu-3S:

TYPE: Solid-propellant three-stage satellite launch vehicle.

POWER PLANT:

M13-TVC 1st stage	1,117 kN (251,325 lb st)
+ SB-310 (8) boosters	1,068 kN (240,300 lb st)
M22-TVC 2nd stage	367 kN (82,575 lb st)
M3A 3rd stage	66·8 kN (15,030 lb st)

DIMENSIONS:

Length overall:	
M13 1st stage	15·0 m (49 ft 2½ in)
M22 2nd stage	4·9 m (16 ft 1 in)
M3A 3rd stage	1·5 m (4 ft 11½ in)
Diameter:	
M13 1st stage	1·41 m (4 ft 7½ in)
M22 2nd stage	1·41 m (4 ft 7½ in)
M3A 3rd stage	1·13 m (3 ft 8½ in)

WEIGHTS:

At launch:	
M13 1st stage	32,440 kg (71,518 lb)
SB boosters (8)	3,872 kg (8,536 lb)
M22 2nd stage	8,686 kg (19,149 lb)
M3A 3rd stage	1,222 kg (2,694 lb)

Mu-3S-1 launch vehicle

SWEDEN

SAAB-SCANIA

SAAB-SCANIA AKTIEBOLAG

HEAD OFFICE: See Aircraft section

EXECUTIVES: See Aircraft section

Saab-Scania is engaged in the manufacture of telemetry, tracking and command equipment, and onboard computer subsystems for satellites. In September 1980 the company received a contract from the Swedish Space Corporation for the development of Sweden's first satellite, known as Viking.

VIKING

The Viking satellite will explore the Earth's magnetosphere at high latitudes at altitudes of up to 15,000 km (9,320 miles). The results are expected to provide data on the transfer of energy between the Sun and the magnetosphere, and on other processes behind auroral phenomena—an important field of research in Sweden.

Boeing Aerospace Company will participate as a subcontractor, with responsibility for the spacecraft platform, on which a boost motor and equipment will be mounted. Solar power will be obtained from arrays fixed to panels on the spacecraft body. Weight of the satellite will be 550 kg (1,212 lb).

Viking is scheduled to be launched in 1984 by the European space launcher Ariane, together with the French Earth resources satellite SPOT, in which Sweden also has a share.

Model of the Saab-Scania Viking satellite

UNION OF SOVIET SOCIALIST REPUBLICS

Details of Soviet satellites, space probes and spacecraft have appeared in *Jane's* since the 1959-60 edition, together with descriptions of various research and meteorological rockets.

In October 1976, the Soviet authorities announced the creation of the Morsvyazsputnik (Sea Communication Satellite) Organisation, which is responsible for the management of Soviet maritime communications satellites.

COSMOS SATELLITES

This series of satellites is continuing the Soviet programme of research into physical phenomena in space and the Earth's upper atmosphere; and into the technical problems involved in spaceflight and the development of spacecraft design and their systems; as well as conducting experiments of an applied nature of interest to science and the Soviet economy.

The wide terms of reference mean that a Cosmos satellite can vary from a small uninstrumented device to a large spacecraft capable of life support, and military applications. The majority of the scientific satellites are of a basic standard design in which various experiment payloads can be accommodated. They are cylindrical in shape, approximately 1·83 m (6 ft) long by 1·05 m (3 ft 6 in) in diameter and weigh about 360 kg (800 lb).

MILITARY COSMOS

As the satellite table at the end of this section indicates, Cosmos designations are given to Soviet reconnaissance satellites and other types of military spacecraft.

Cosmos reconnaissance satellites are launched from the bases at Plesetsk and Tyuratam, usually into orbits with inclinations of 52°, 62·8°, 65° or 72°. Most eject capsules after 8 or 13 days and these are recovered. Some have the capacity for in-flight frequency changing; others have a small manoeuvring capability for precise target coverage.

Cosmos scientific satellite *(Maurice Allward)*

Mockup of Cosmos 782 biological satellite, embodying components of Vostok manned spacecraft *(Tass)*

A number of test vehicles for the Soviet Union's Fractional Orbital Bombardment System were given Cosmos designations. Cosmos designations have also been given to spacecraft which have the capability of intercepting other satellites. Early 'interceptors' were Cosmos 249, 252, 374 and 375. Other Cosmos satellites used for intercept tests were described in the 1977-78 and 1979-80 *Jane's*.

Two successful tests were reported in the West in early 1981: the first on 2 February when the 'hunter-killer' Cosmos 1,243 intercepted the target Cosmos 1,241; the second on 14 March when Cosmos 1,258, in a 299 × 1,021 km (186 × 634 mile) orbit intercepted the same Cosmos 1,241 target in a 973 × 1,009 km (605 × 627 mile) orbit, the attack being made within two orbits. On the second occasion Cosmos 1,243 exploded. The Soviet system is said to utilise shrapnel-like pellets to destroy or disable its targets, and this test is reported to be the first involving an exploding anti-satellite weapon. The target Cosmos was not completely destroyed by the explosion, but its sensors and sensitive instruments would almost certainly have been disabled.

Three distinct attack procedures appear to be used: Eccentric Orbit mode, where the attack satellite is launched into an eccentric orbit in which it makes a flyby of the target; Circular Orbit mode, where the attack satellite is injected into a circular orbit similar to that of the target, the interception being made at a slower flyby speed; Pop-Up mode, where the attack satellite intercepts within one revolution, in contrast to the other two modes, which require several revolutions. The Pop-Up mode of attack is difficult to counter, as there is little advance indication of an intended target. It was reported in 1980 that at least one of the Soviet anti-satellite systems was then operational.

A number of Cosmos satellites have been launched into orbits suitable for ocean surveillance missions. The first of these was Cosmos 198, launched on 27 December 1967. Cosmos 651 and subsequent spacecraft in this series appear to be bigger, with a length of about 12 m (40 ft) compared with 6 m (20 ft) of the earlier satellites. All the satellites were launched into an initially low orbit of about 250 km (155 miles) but after a few weeks moved into an approximately circular orbit of 1,000 km (620 miles).

COSMOS LAUNCHERS

Two categories of launch vehicle appear to be used for Cosmos and Intercosmos satellites, and other Soviet spacecraft. One category is based on the structures and power plants of standard missiles, such as the SS-4 ('Sandal'), SS-5 ('Skean') and SS-9 ('Scarp'), with additional upper stages as required. The other combines the basic core vehicle developed originally for the Vostok manned spacecraft with a variety of upper stages. Examples are as follows:

SS-4 + Cosmos stage. First stage powered by 706·34 kN (158,800 lb st) RD-214 four-chamber liquid-propellant rocket engine, burning nitric acid and kerosene. Second stage powered by RD-119 single-chamber engine, burning liquid oxygen and dimethylhydrazine, and giving 107·87 kN (24,250 lb thrust) in vacuum. Typical early launch, on 26 June 1974, orbited Cosmos 662, a 408 kg (900 lb) ellipsoid, 1·83 m (6 ft) long with a diameter of 1·22 m (4 ft), intended for scientific research. (Referred to as B-1 Cosmos in drawing near end of Spaceflight section).

SS-5 + Restart stage. A typical application for the SS-5 is to orbit satellites like the early Cosmos 655 and 661. Shaped as cylinders, 1·83 m (6 ft) long and 0·91 m (3 ft) in diameter, with paddle-type solar panels, these were thought to have navigation and/or electronic intelligence missions. (Referred to as C-1 Cosmos in drawing near end of Spaceflight section).

Ekran geostationary TV relay satellite, displayed at 1977 Paris Air Show *(Flight International)*

SS-9 + FOBS stage. Frequent launches of this vehicle are expected to contribute to continued development of ocean surveillance missions.

Vostok core + Venus stage. This standard launch vehicle has many applications. It is used with an escape stage to orbit the 1,250 kg (2,750 lb) uprated Molniya 2 communications satellites. Typical early military payloads were Cosmos 639, a manoeuvrable reconnaissance satellite intended probably to study the breakup of Arctic pack ice; and Cosmos 658, a reconnaissance satellite in the form of a four-ton sphere-cylinder, 5·0 m (16 ft 6 in) long, which remained in orbit for 12 days.

EKRAN

Ekran (Screen) is a direct-broadcast geostationary satellite, relaying television signals directly to small community aerials throughout the entire Soviet Union. The spacecraft operates at 714MHz and, in addition to advanced relay equipment, embodies a three-axis Earth-directed orientation system, a power supply system using solar batteries which is independently orientated on the Sun, an orbital correction system, a radio telemetry system to transmit information on the functioning of the various systems, and a radio system for the accurate measurement of the orbital parameters and spacecraft control.

Ekran 1 (Statsionar T) was launched on 26 October 1976 into a geostationary orbit and then manoeuvred into its scheduled position at 90° E. Subsequent Ekran launchings have been recorded in the annual tables of satellites.

HORIZONT

Horizont (Horizon) is the name given to a series of communications satellites for telephone and television transmissions. The use of an inclined orbit gives better coverage for areas in the far north of the Soviet Union. The first Horizont satellite was launched on 19 December 1978 into an elliptical orbit with a significantly higher perigee than Molniya satellites.

INTERCOSMOS SATELLITES

Intercosmos is the designation given to a series of scientific satellites carrying experiments provided by countries with which the USSR has close political relationship. Intercosmos I, launched on 14 October 1969, involved the co-operation of seven countries: Bulgaria, Czechoslovakia, Germany (Democratic Republic), Hungary, Poland, Romania and the USSR. The satellite carried scientific instruments developed and made in Czechoslovakia and Germany as well as in the USSR.

Details of subsequent Intercosmos launches have been given in successive editions of *Jane's*.

METEOR

Meteor is the name given to the Soviet series of first-generation operational meteorological satellites, developed from Cosmos prototypes.

The Meteor system consists of three satellites in orbital planes at 90° and 180° to each other, so that they pass over a given area of the Earth on the northbound pass at intervals of about 6 h and 12 h, and again during the southbound pass.

The satellites, which provide information about the state of the atmosphere both on the 'daylight' and 'night' sides of the Earth, are stabilised so that the camera lenses and infra-red instruments always point towards the Earth.

Information received from Meteors is supplied to the Soviet hydro-meteorological service and to the World Meteorological Service. Cloud cover picture charts are transmitted to Washington, Geneva, Tokyo, Sydney and other foreign weather services.

Meteor 1 was launched on 26 March 1969, and launchings have continued at a current rate of three or four per year. An improved version, **Meteor 2**, is now in service. Launchings of both types are recorded in the annual satellite table.

MOLNIYA

Molniya 1 (Lightning), launched on 23 April 1965, was a communications satellite placed in a highly elliptical orbit designed to provide the longest possible communications sessions between Moscow and Vladivostok. It was the first Soviet communications satellite. On 24 November 1971, **Molniya 2A**, the first of a new uprated version, was launched from Plesetsk.

Launches of both versions have continued, and the current operational system consists of four pairs of each type of Molniya circling the Earth in orbital planes spaced at 90° intervals. The planes rotate at the rate of about 1° a day, in order to maintain stationary ground tracks over the Earth throughout the year, thus providing predictable communications coverage over the whole of the USSR.

Molniya 3 is the latest version, embodying improved communications equipment able to accommodate colour television and new communications frequencies. The first of the new spacecraft was launched on 21 November 1974, and was placed in a normal Molniya-type 12 h 17 min orbit, with its 650 km (400 mile) perigee over the southern hemisphere and its apogee of 40,690 km (25,283 miles) over the northern hemisphere.

Full-scale model of Molniya 2 communications satellite
(Brian M. Service)

Prognoz 2 solar research satellite *(Tass)*

The Molniya satellites, together with the Ekran and Raduga satellites (which see), operate in conjunction with 60 ground stations to provide the Orbita national communications system.

PROGNOZ

Launched on 14 April 1972, Prognoz 1 (Forecast) was the first of a series of satellites designed specifically to study processes of solar activity, and their influence on the interplanetary medium and the Earth's magnetosphere.

Prognoz is basically spherical, with four cruciform solar panels extended to provide power via an internal rechargeable chemical battery. The sphere, filled with an inert gas, contains telemetry equipment, temperature control components and the electrical system. Experiments on board are intended to measure: electromagnetic solar radiation generated simultaneously with solar flares; solar cosmic radiation; solar wind plasma; and radio waves.

Subsequent Prognoz launches have been recorded in the annual tables of satellites.

PROGRESS

Progress is a non-reusable unmanned spacecraft, designed to ferry supplies and equipment up to space stations of the Salyut type. A development of the basic two-man Soyuz spacecraft, it is lightened as much as possible, with all the normal Soyuz life support systems removed, to permit carriage of the maximum practicable payload. The propulsion module is almost unchanged; but the size of the instrument module is increased. The major change entails replacement of the standard Soyuz descent module with a tanker module carrying propellants.

Positioning, rendezvous, approach and docking can be effected automatically or controlled from Salyut. The freight compartment of Progress is hermetically sealed, so that, after the vehicle has docked, cosmonauts can work in comfort while transferring supplies to their space station.

After transfer of the supplies, the spacecraft is used as a receptacle for waste material such as spent oxygen regenerators and filters, empty food and water cartons, sewage and redundant flight documentation. It is then jettisoned, to burn up in the atmosphere on re-entry; recovery is not attempted.

Weight of the first Progress was announced as 7,020 kg (15,476 lb), of which 2,300 kg (5,071 lb) was cargo, 1,000 kg (2,204 lb) fuel and 1,300 kg (2,866 lb) food etc. Weight varies according to mission. Progress is 10·8 m (35 ft 5 in) long with a maximum diameter of 3·7 m (12 ft 2 in).

Progress 1. Launched on 20 January 1978, ferried supplies to Salyut 6, with which it docked on 22 January. While docked, the first-ever refuelling in space was conducted on 3 February, following checks by the crew of Soyuz 26. The spacecraft undocked on 6 February, and re-entered and burned up the next day.

Details of **Progress 2 to 7**, which supported the Soyuz 26-28 and Soyuz 29-34 missions to Salyut 6, were given in the 1979-80 *Jane's*. **Progress 8, 9 and 10,** which supported the Soyuz 35-37 and Soyuz T-2 missions to Salyut 6, were described in the Addenda to the 1980-81 *Jane's*.

RADUGA/STATSIONAR

Raduga (Rainbow) is the name of a Soviet system of geostationary-orbit communications satellites, the first operational version of which, **Statsionar 1**, was launched on 22 December 1975.

Statsionars are three-axis stabilised, and transmit at higher power levels than comparable US-launched Intelsat IV spacecraft. This enables them to communicate with smaller and therefore cheaper ground terminals. Currently, the spacecraft communicate with the Orbita network of ground stations set up for the Molniya programme.

Details of Statsionar launches can be found in the annual tables of satellites.

SALYUT

Salyut (Salute) spacecraft have served as orbital scientific stations for the crews of Soyuz manned spacecraft. Basically, the station is a stepped cylinder about 13 m (42 ft 8 in) long, from 2·13 m (7 ft) to 4·0 m (13 ft) in diameter, with a weight of 18½ tons.

Details of Salyut 1 and 2 can be found in the 1973-74 *Jane's*. Salyut 3 was described in the 1975-76 edition. Salyut 4 and 5 were described in the 1977-78 edition.

Salyut 6, launched on 29 September 1977, is a major development of Salyut 5, the main change being the embodiment of a second docking port so that two Soyuz-type spacecraft can be docked simultaneously. The additional port is located on the instrument section, opposite the original port on the transfer compartment. The two ports are identical, and the automatic rendezvous procedure is the same at both, but only the new port embodies refuelling facilities. Normal crew is two, increasing for short periods of time to four. Soviet sources indicate that the station can support six cosmonauts.

Internal modifications by comparison with early Salyuts include the installation of a folding shower cubicle, into which hot water is sprayed and recovered by an extractor

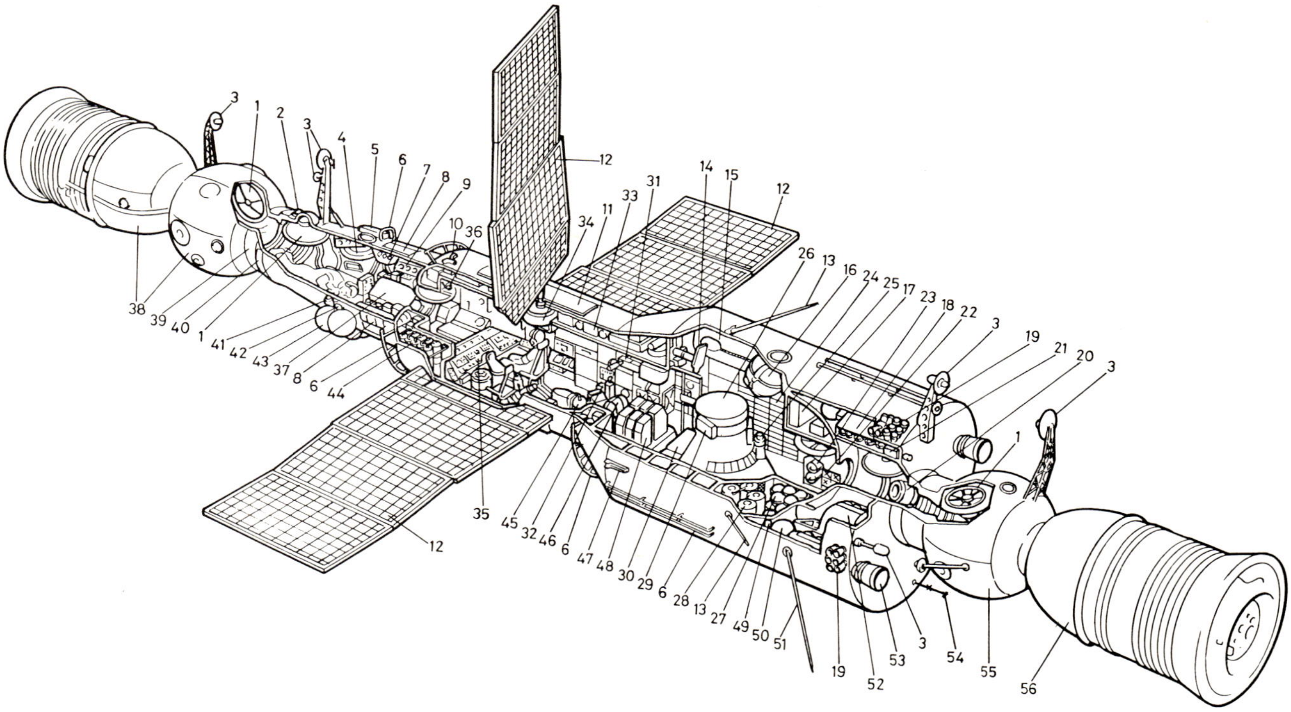

Salyut 6 Space Station *(M. A. Barnes)*

1 Forward and aft hatches
2 External TV camera mount
3 Rendezvous antenna
4 Exploration hatch?
5 External instrument package
6 EVA handrails
7 Airlock pneumatic controls
8 External thermal control panel
9 Airlock controls
10 Sun sensor
11 Protective screen
12 Rotating solar arrays
13 Telemetry antennae
14 Zero-gravity cosmonaut weighing scale
15 Sleeping berth
16 Airlock for debris ejection
17 Dust filter
18 Pneumatic and hydraulic systems for attitude control and main propulsion engines
19 Attitude control engines
20 Air ventilator
21 TV camera for aft docking operations
22 Sanitary facilities
23 Toilet
24 Food lockers
25 Sighting device (12 × magnification, for aligning the BST-1M telescope)
26 Container for scientific instruments (BST-1M telescope)
27 Fresh water storage
28 Garbage containers
29 Electronic control panels for instrument module
30 Running track for cosmonaut exercises
31 Veloergometer exerciser
32 Photographic apparatus
33 Electronics bay
34 Solar panel rotating mechanism
35 Commander's control panel
36 TV camera
37 EVA spacesuits (stowage)
38 Transport spaceship Soyuz
39 Active docking system of Soyuz
40 Passive docking system of the Salyut station
41 Sun sensor
42 Porthole
43 Compressed air cylinders
44 Oxygen cylinders for station's atmosphere
45 Fresh water tank
46 Vacuum cylinder
47 Attachment point for launch shroud
48 MFK-6M photographic apparatus
49 High-pressure airlines
50 High-pressure air storage cylinder
51 Communications antenna
52 Propellant tank
53 Main propulsion system engine
54 Visual docking target
55 Soyuz orbital module
56 Soyuz descent module

Salyut 6 in orbit with Soyuz spacecraft docked on aft port

pump. Sensors are fitted to register the impact of micrometeorites.

Major items of equipment installed in the working compartment include a BST-IM submillimetre telescope, an MKF-6M multi-spectral camera and a KT-140 high-resolution topographical camera system.

Three large solar arrays are installed, which can be rotated to any one of 16 positions to maintain solar contact whatever the attitude of the space station.

An initial programme of research involving Soyuz 26, 27 and 28, and Progress 1, was described in the 1978-79 *Jane's*. A second programme started on 16 June 1978 with the docking of Soyuz 29 (see 1979-80 *Jane's*).

A third programme started on 26 February 1979 with the docking of Soyuz 32 and the boarding of cosmonauts Lyakhov and Ryumin. During this mission, Soyuz 33 and 34, and Progress 5, 6 and 7 docked with the space station. The mission ended on 19 August 1979 with the return to Earth of Lyakhov and Ryumin in Soyuz 34.

The fourth programme started on 10 April 1980, with the docking of Soyuz 35 and the boarding of cosmonauts Leonid Popov and Valery Ryumin. During this mission, Soyuz 36, 37 and 38, Soyuz T-2, and Progress 8, 9 and 10 docked with the space station. The mission ended on 11 October 1980, when Popov and Ryumin returned safely to Earth in Soyuz 37, after a record endurance for manned spaceflight of 185 days in space.

The fifth programme started on 28 November 1980, with the docking of Soyuz T-3 and the boarding of cosmonauts Leonid Kizim, Oleg Makarov and Gennady Strekalov. The main task of this mission was to test and refurbish the space station, which had by this time been in orbit for more than three years. To assist this task, supplies were used from Progress 11, which had been put into orbit on 28 September. The mission ended on 10 December, when the three cosmonauts returned safely to Earth.

The sixth programme started on 14 March 1981, with the docking of Soyuz T-4 and the boarding of cosmonauts Vladimir Kovalenok and Victor Savinykh. This mission utilised supplies ferried up in Progress 12, which had docked on the station's aft docking port on 26 January. Progress 12 was separated on 19 March, to free the docking port for Soyuz 39 which docked on 23 March.

The activities of the crew of Soyuz T-4 during their first seven days on Salyut 6 were devoted almost entirely to repair and refurbishment operations, the most important of which involved the replacement of onboard components to free a stuck Salyut solar array. The inability of the panel to track the sun had seriously reduced the power available to the station.

Soyuz 40 (which see) also docked during this programme, which ended on 26 May when Kovalenok and Savinykh returned safely to Earth after 86 days in orbit, leaving Salyut 6 orbiting empty.

SOYUZ SPACECRAFT

Developed for the Soviet Earth-orbital space station programme, Soyuz spacecraft each comprise three basic sections or modules: a laboratory-cum-rest compartment (orbital module), a descent compartment (landing module) and a propulsion and instrument section (service module). The orbital module is mounted on the extreme nose of the craft, and communicates with the landing module via a hermetically-sealed hatch. The orbital and landing modules are pressurised to 1·01 bars (14·7 lb/sq in), have a combined internal volume of 9 m³ (318 cu ft) and can accommodate up to four cosmonauts.

The service module contains the main systems for orbital flight, together with a liquid-propellant propulsion system embodying two motors (one a standby) each with a thrust of 3·92 kN (880 lb). These allow midcourse manoeuvres, up to heights of 1,300 km (800 miles), and are used for the de-orbit manoeuvre. Another system provides attitude control. Attached to the service module is a solar-cell array having an area of about 14 m² (150 sq ft).

The landing module contains the parachutes and landing rockets. A backup parachute system is available in case of failure. The main parachute, preceded by a pilot 'chute, is deployed at 8,000 m (27,000 ft). Retro-rockets, operating at a height of about 1 m (3 ft) above the ground, ensure a landing velocity not exceeding 3 m/s (10 ft/s). The aerodynamic design of the landing module permits landing loads to be kept within 3-4*g*, although ballistic re-entries, involving loads of 8-10*g*, can be made if required. The overall length of the craft is about 9 m (30 ft), the diameter of the crew compartments about 2·1 m (7 ft) and the all-up weight about 6,000 kg (13,000 lb).

The Soyuz craft are equipped with an automatic control system for approach and docking manoeuvres, the technique and external aerials being similar to those employed on the Cosmos spacecraft 186 and 188, and 212 and 213.

The missions of Soyuz 1 to 37 were described in previous editions of *Jane's*, including the Addenda to the 1980-81 edition.

Soyuz 38. Launched on 18 September 1980, with Col Yuri Romanenko (commander) and Cuban cosmonaut Arnaldo Tamayo Mendez (engineer). Soyuz 38 docked with Salyut 6/Soyuz 37 on 19 September, to continue the fourth programme of research in company with Leonid Popov and Valery Ryumin. The spacecraft and its two cosmonauts returned safely to Earth on 26 September.

Soyuz 39. Launched on 22 March 1981, with cosmonaut Vladimir Dzhanibekov (commander) and Mongolian cosmonaut Jugderdemidiyn Gurragcha (engineer). Soyuz 39 docked with Salyut 6/Soyuz T-4 on 23 March. After a week of experiments and observations, the spacecraft and its two cosmonauts returned safely to Earth on 30 March.

Soyuz 40. Launched on 14 May 1981, with Leonid Popov (commander) and Romanian cosmonaut Dumitru Prunariu (engineer), Soyuz 40 docked with Salyut 6/Soyuz T-4 on 15 May to continue the sixth programme of

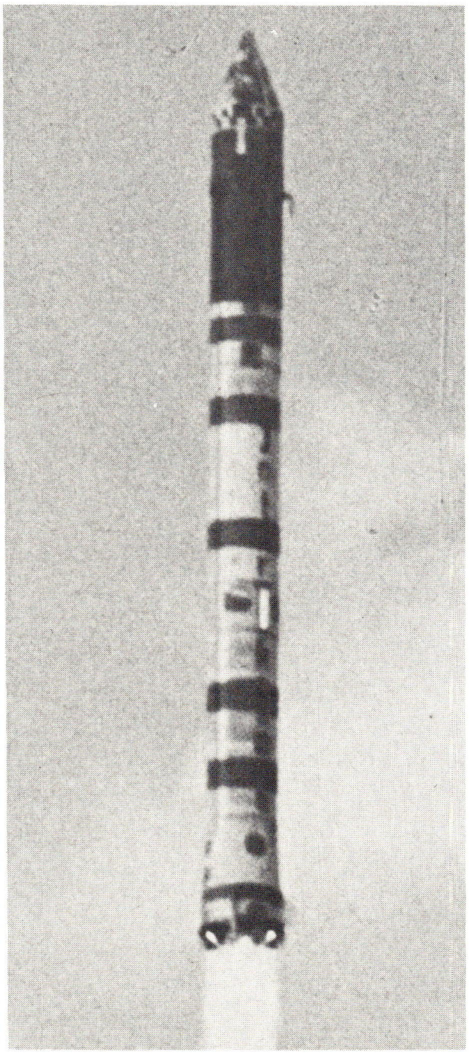

Launch vehicle for Intercosmos 16 satellite *(Tass,*

Launch of Soyuz 28 (Note: photograph could not be trimmed to show true vertical lift-off) *(Tass)*

research in company with cosmonauts Vladimir Kovalyonok and Victor Savinykh, who had been working in near-Earth orbit since 12 march. The spacecraft and its two cosmonauts returned safely to Earth on 22 May.

SOYUZ T SPACECRAFT

Soyuz T is a new version of the spacecraft used normally to orbit two-man crews. Dr Konstantin Feoktistov, the cosmonaut and spacecraft designer, described the spacecraft as being similar to earlier Soyuz craft in external appearance, but with many changes to internal layout and equipment. The onboard digital computer, which controls manoeuvring operations and the systems, also transmits processed data to Earth. For example, when docking, the computer displays on a visual display unit (VDU) at mission control how all the main systems are working. A new propulsion system is embodied, in which all engines (main propulsion, manoeuvre and attitude control) use a common source of propellants. The landing system has also been changed.

Soyuz T was first launched on 16 December 1979, unmanned, and docked with Salyut 6 (also unmanned) on 19 December on a test mission. Earlier tests of Soyuz T systems were carried out on Cosmos 1001, launched on 4 April 1978, and Cosmos 1074, launched on 31 January 1979.

Soyuz T-2, the first manned Soyuz T, was launched on 5 June 1980. This mission is described briefly in the Addenda to the 1980-81 *Jane's*.

Soyuz T-3. Launched on 27 November 1980, with Leonid Kizim (commander), Oleg Makarov (engineer) and Gennady Strekalov (researcher-engineer). Soyuz T-3 docked with Salyut 6/Progress 11 on 28 November. The new onboard computer was used for the link-up, which took place on schedule, and while in space the crew tested systems onboard their improved spacecraft. Soyuz T-3 returned to Earth and landed safely on 10 December.

Soyuz T-4. Launched on 12 March 1981, with Vladimir Kovalyonok (commander) and Victor Savinykh (engineer, and the 100th person to travel in space). Soyuz T-4 docked with Salyut 6/Progress 12 on 14 March 1981.

SOYUZ LAUNCHER

The vehicle used for launching Soyuz spacecraft appears to be a development of the booster used for launching the original Vostok spacecraft, with some 11·8 m (36 ft) of additional upper staging and structures. To cater for the increased weight and bending moment the interstage truss is strengthened. During launch, the Soyuz vehicle is surmounted by an escape tower.

It is not possible to identify the current engines in the launch vehicle, or give their individual ratings. However, official Soviet reports have stated that the vehicle has a total thrust of around 60 million horsepower, which is three times the power quoted for the original Vostok launcher. The basic configuration has not changed. Thus, the first stage consists of a central core, powered by an engine with four primary nozzles and four verniers. This is surrounded by four wrap-round boosters, each with four primary nozzles and two verniers, so that 32 rocket chambers are fired simultaneously during lift-off.

Data given in a brochure issued by Aérospatiale of France suggest that the Soyuz launcher can insert payloads of up to 7,500 kg (16,535 lb) into low circular orbit; 2,400 kg (5,290 lb) into geostationary transfer orbit; and satellites of up to 1,100 kg (2,425 lb) into geostationary orbit.

ZONDA LAUNCHER

A drawing purporting to show this largest known Soviet launch vehicle appeared in a brochure issued by Aérospatiale of France in 1977. The drawing, of which a copy appears on page 710, suggests that Zonda has the same configuration as the Soyuz vehicle. The first stage is shown as a large-diameter core, surrounded by six large boosters. The upper stages are attached by a truss structure.

Overall height of Zonda is shown as about 61·33 m (201 ft). It is stated to be able to insert a payload of 22,000 kg (48,500 lb) into low circular orbit; 5,000 kg (11,020 lb) into geostationary transfer orbit; and satellites weighing up to 1,600 kg (3,525 lb) into geostationary orbit.

UNITED KINGDOM

BAe

BRITISH AEROSPACE PUBLIC LIMITED COMPANY, DYNAMICS GROUP

Six Hills Way, Stevenage, Hertfordshire SG1 2DA
CHAIRMAN AND CHIEF EXECUTIVE:
Admiral Sir Raymond Lygo, KCB
HEAD OF PUBLIC RELATIONS: M. K. Hird

Space and Communications Division:
Gunnels Wood Road, Stevenage, Hertfordshire SG1 2AS
MANAGING DIRECTOR: P. L. V. Hickman
SALES DIRECTOR:
R. G. T. Munday, BSc, ARCS, CEng, FRAeS

Bristol Works:
PO Box No. 5, Filton, Bristol BS12 7QW

Following nationalisation of major units of the UK aerospace industry, British Aircraft Corporation's Guided Weapons Division and Hawker Siddeley Dynamics Ltd became the Dynamics Group of British Aerospace on 1 January 1978. All space projects are handled by the Space and Communications Division.

BAe Dynamics has been chosen as prime contractor for all the European Space Agency's (ESA) communications programmes, beginning with the OTS satellites. Currently, it is leading the MESH international consortium as prime contractor for ESA's European Communications Satellite (ECS) and MARECS, a maritime version of ECS. The ECS project aims to provide a fully operational satellite communications system for Europe in the 1980s, providing bulk telephone transmissions in 'time division multiple-access' mode, and TV distribution to complement existing terrestrial systems. MARECS (described in the MESH entry) will provide direct telephone and telex links between ships in distant oceans and shore stations in the UK and elsewhere.

BAe Dynamics has been appointed prime contractor by ESA for the L-Sat large multi-role communications satellite, capable of providing a direct TV broadcast service of up to 5 channels. L-Sat will be about twice the body size of ECS, but up to five times more powerful. Also for ESA, BAe Dynamics in 1979 completed a study of the payload for an Earth Resources Survey Satellite System, and is currently developing and manufacturing the 33 m² (356 sq ft) 4kW solar array and the photon detector assembly for NASA's Space Telescope being built by Lockheed Missiles and Space Company. In 1981 the Group signed a Memorandum of Understanding with Comsat General Corporation to initiate the leasing of military satellite communications services to user organisations.

BAe Dynamics is to develop the attitude and orbit control equipment for the interplanetary satellite being manufactured by the STAR consortium for the International Solar Polar Mission (ISPM). Also, as a member of this consortium, it has been nominated prime contractor for the Giotto scientific satellite which is to intercept Halley's comet in 1986. (For further details see the International Section.)

BAe has completed studies of the feasibility of Solar Power Satellites (SPS), and is monitoring and contributing to US and European work on the concept. Stationed in geosynchronous orbit, these satellites would receive energy via large arrays of photo-electric cells, convert it on board into microwave radio energy, and beam it towards receiving antennae on the Earth's surface. There, the microwave energy would be converted to usable electrical power and fed into a supply grid. Under contract to the UK Department of Industry, BAe Dynamics has led a study into the industrial implications to the UK of such satellites.

As a member of the nine-nation European Spacelab consortium led by ERNO of West Germany, British Aerospace is making the pallets (experiment-mounting platforms) for the Spacelab that is to be carried into orbit by the US Space Shuttle. By mid-1981, thirteen pallets had been completed and a further five were under construction. Study has been made to assess the feasibility of using the pallets as the basic structure of free-flying spacecraft and as elements of space stations.

Other space engineering work performed by BAe Dynamics includes the design, testing and supply of satellite power and electrical subsystems; attitude and orbit control systems; solar array drives; and satellite reaction and momentum control wheels; and studies of reusable spacecraft and 'Getaway Specials' for the Space Shuttle. These latter are small self-contained experiment packages orbited at reduced rates when space is available after the installation of the prime payload. BAe Dynamics also designed and built the hold-down launch release system for the Ariane European launcher.

Manufacture of the widely-used Skylark high-altitude sounding rocket continues and by mid-1981 more than 400 had been launched. For details see 1977-78 *Jane's*.

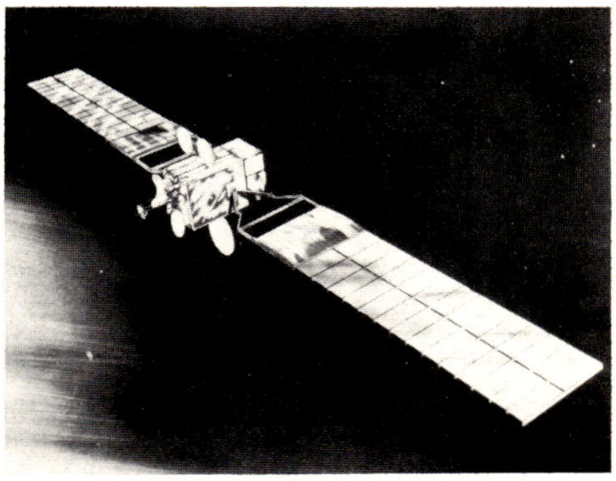

Artist's impression of the L-Sat communications satellite.

Artist's impression of the MARECS maritime communications satellite

L-SAT

BAe has been appointed by ESA as prime contractor for the L-Sat programme. L-Sat is a communications satellite, about twice the size of the European Communications Satellite (described in the International pages of this section) and several times more powerful, developing up to 7 kW of electrical power. In addition to transmitting TV broadcasts at high power, for direct reception on Earth by individual houses equipped with small rooftop-mounted antennae, L-Sat will also carry a significant volume of telephone and data traffic. Fokker is contributing to the L-Sat programme.

Configured for launch by either Ariane or the Space Shuttle, L-Sat is scheduled to be placed in geosynchronous Earth orbit in 1984, to provide communications coverage in Europe.

MSDS

MARCONI SPACE AND DEFENCE SYSTEMS LTD

The Grove, Stanmore, Middlesex HA7 4LY
Telephone: 01 954 2311
Telex: 22616
CHAIRMAN:
General Sir Harry Tuzo, GCB, OBE, MC, MA

Marconi Space and Defence Systems, a member of GEC-Marconi Electronics, assembled and tested the hydrazine propulsion subsystem used to keep the NATO III communications satellite on station in its synchronous orbit. This series of satellites also embodies local oscillators built by MSDS; these act as highly stable frequency sources with which incoming signals to the satellites are mixed to produce the intermediate frequency suitable for the onboard electronics.

As a member of COSMOS consortium, MSDS provided the attitude control system and other equipment for the first Meteosat which was launched on 23 November 1977. MSDS is providing similar equipment for the second Meteosat, scheduled to be launched by an Ariane vehicle in mid-1981, and for Exosat, a precision three-axis stabilised X-ray observatory. Under subcontract to Aérospatiale, it is responsible for onboard software for the Ariane European launch vehicle. In association with the MESH international consortium, it is totally responsible for the maritime payload modules which will be mounted on the ECS platform to form the MARECS spacecraft.

Details of the UK 6 (Ariel) scientific satellite, for which MSDS was prime contractor, can be found in the 1979-80 *Jane's*.

UOSAT
UNIVERSITY OF SURREY

Guildford, Surrey GU2 5XH
PROJECT MANAGER: Dr M. N. Sweeting

UOSAT

Research staff and academics at the University of Surrey, in conjunction with the UK branch of the Amateur Satellite Corporation (Amsat), are designing and building a satellite to provide radio enthusiasts with a means of studying radio propagation in the ionosphere. For this purpose the satellite, named Uosat, will carry high-frequency beacons, transmitting on 7, 14, 21 and 28MHz. A digitally synthesised speech experiment module, under the control of the onboard computer, will 'speak' telemetry, experiment data and spacecraft 'news'. Other equipment will include a three-axis magnetometer, supplied by a US enthusiast, and radiation counters. Four solar panels, made by Solarex (USA), will each provide around 27W. Stabilisation will be by gravity-gradient forces.

To assist the project, which has a budget of £85,000, BAe Dynamics carried out a structural analysis of Uosat. NASA agreed to launch the satellite, as a piggy-back payload, on the Delta vehicle scheduled to orbit the Solar Mesosphere Explorer in September 1981.

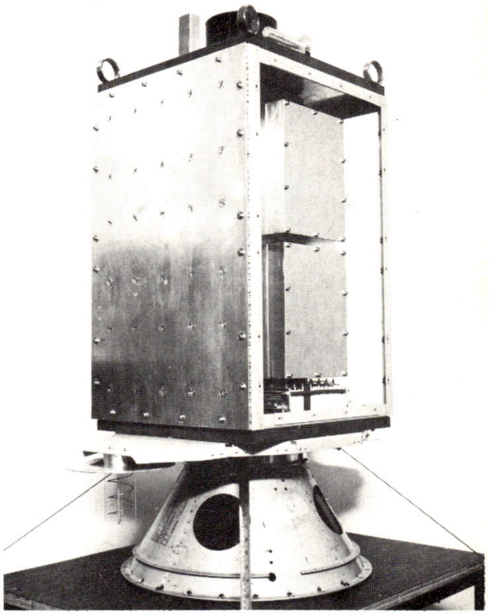

UOSAT 'amateur' satellite

UNITED STATES OF AMERICA

BOEING
THE BOEING COMPANY

Seattle, Washington 98124
Boeing Aerospace Company
PO Box 3999, Seattle, Washington 98124
OFFICERS: See Aircraft section

INERTIAL UPPER STAGE (IUS)

In August 1976 Boeing was selected by the US Air Force Space and Missile Systems Organization (SAMSO) to develop the Inertial Upper Stage (IUS) that will carry Space Shuttle payloads to orbits not attainable by the Shuttle Orbiter. The IUS has also been chosen as the upper stage for the Air Force's Titan 34-D rocket.

IUS vehicles are designed in two sizes, using common components.

The basic two-stage vehicle is 5·0 m (16 ft 4½ in) long and 2·9 m (9 ft 6 in) in diameter. It consists of an aft skirt, a 9,707 kg (21,400 lb) aft-stage solid rocket motor generating an average of 189·5 kN (42,600 lb) of thrust, an interstage, a 2,722 kg (6,000 lb) forward stage solid rocket motor generating an average of 77·5 kN (17,430 lb) of thrust, and an equipment support structure which contains electronics for guidance and navigation, reaction control subsystem and electrical power. Its weight is 14,515 kg (32,000 lb).

The NASA twin-stage IUS uses two 9,707 kg (21,400 lb) motors and, except for the larger forward-stage motor and minor modifications to the interstage and equipment support structure, is quite similar to the basic two-stage vehicle. Tooling is identical for both vehicles. The NASA two-stage version is 6·64 m (21 ft 9½ in) long and weighs 20,956 kg (46,200 lb).

Boeing's propulsion team member, Chemical Systems Division of United Technologies, will design and test all solid motors. Production deliveries were programmed to begin in 1981.

FAIRCHILD
FAIRCHILD SPACE & ELECTRONICS COMPANY

Germantown, Maryland 20767
Telephone: (301) 428 6000

Fairchild manufactures the Stage Vehicle System for use on the Atlas E/F launch vehicle. Used as an interstage between Atlas and the NavStar Global Positioning Satellite System, it contains two tandem-mounted TE-M-364-4 solid rocket motors for providing transfer orbit insertion of the satellites.

Fairchild was responsible for the integration and test, and the production of subsystems for the ISEE-3 satellite, described in the 1980-81 *Jane's*. Current contracts include manufacture of the Multimission Modular Spacecraft (MMS).

MMS

In 1978 Fairchild was selected by NASA as the integration and test contractor for the Multimission Modular Spacecraft (MMS), a standard bus for many future NASA and DoD missions. The MMS standardises into modules basic spacecraft functions such as power supply, attitude control, communications and data handling, and propulsion. The first launch using MMS was the Solar Maximum Mission, on 14 February 1980. The second mission was to be Landsat-D, scheduled for 1981. Shuttle-compatible, retrievable and refurbishable, the MMS is also being considered for such missions as GRO, UARS and NOSS.

Under separate contract, Fairchild builds the Communications and Data (C&DH) module for the MMS. This module provides full telemetry and command link between the host spacecraft and the ground facility; onboard control of all spacecraft functions; and retrieval of spacecraft housekeeping and scientific payload data. Besides its application in the MMS, the C&DH can be interjected into other systems, and Fairchild is under contract to build one version of the C&DH for Lockheed's Space Telescope Programme.

Solar Maximum Mission, first spacecraft to use the MMS

FORD
FORD AEROSPACE & COMMUNICATIONS CORPORATION

20th Floor, 300 Renaissance Center, PO Box 43342, Detroit, Michigan 48243
Telephone: (313) 568 7640
DIRECTOR, PUBLIC AFFAIRS: Don Flamm

In addition to satellites built for US agencies and NATO, Ford Aerospace manufactured the Sakura Experimental Communications Satellite, described under the NASDA entry in the Japanese section of the 1978-79 *Jane's*.

INTELSAT V

Ford Aerospace, under a contract from the 105-nation International Telecommunications Satellite Organisation (Intelsat) is building 12 Intelsat V communications satellites. The first Intelsat V was launched by an Atlas-Centaur vehicle on 6 December 1980, and was placed successfully in a geostationary orbit on 11 December. Nine Intelsat V satellites will be launched by the end of 1982. These will be followed by three improved Intelsat VA satellites, the first of which will be launched in 1984.

Intelsat V's capacity of 24,500 telephone half-circuits is twice that of Intelsat IVA. It is also the first satellite of the Intelsat series to be three-axis stabilised, using technology developed during the European Symphonie programme. The spacecraft contains 54,373 electronic components, 40,000 machined components and 17,576 solar cells. Launch weight of the 6.4 m (21 ft) high satellite is 1,950 kg (4,300 lb). Planned life span is seven years.

Currently, the Intelsat global system carries two-thirds of all international trans-oceanic communications traffic. Assisting Ford Aerospace in the Intelsat V programme are Aérospatiale and Thomson-CSF of France, Marconi Space and Defence Systems of England, MBB of West Germany, Mitsubishi Electric Corporation of Japan, and Selenia of Italy.

Artist's impression of Intelsat V in orbit

GENERAL DYNAMICS

GENERAL DYNAMICS CORPORATION

Pierre Laclede Center, St Louis, Missouri 63105
Convair Division
5001 Kearny Villa Road, PO Box 80847, San Diego, California 92138
Telephone: (714) 277 8900
OFFICERS: See Aircraft section

Convair Division devotes a major part of its activity to production and launch of Atlas and Centaur space launch vehicles, as well as production of DC-10 fuselages. Convair is also developing the Tomahawk cruise missile for the US Navy, and producing the Space Shuttle Orbiter mid-fuselage for NASA.

ATLAS E and F

Atlas E and F series rockets were formerly ICBMs deployed at Strategic Air Command bases as part of the US strategic missile force. At Vandenberg AFB the vehicles undergo extensive refurbishment, which includes the installation of re-certified components, new electronics and harnessing.

Twenty-three Atlas E/Fs remain to be launched. These vehicles are currently assigned to the Air Force Phase I and Phase II Global Positioning System programme, the Defense Meteorological Satellite programme, the Geostat programme and NOAA's Meteorological Satellite programme.

ATLAS SLV

The Atlas Standardised Launch Vehicle (SLV) had its inception in Atlas, the USA's first intercontinental ballistic missile (ICBM). Two versions are currently in service:

SLV-3A, for use with Agena.

SLV-3D, for use with the Centaur D-1A.

These vehicles differ from their immediate predecessor, the SLV-3, mainly in increased tank length and rocket engine thrust.

Atlas is a 'stage-and-a-half' vehicle, consisting of side booster and central sustainer sections. The sustainer section includes the propellant tanks and a single rocket engine. The booster engines receive fuel from the sustainer tanks and are jettisoned midway into flight. The main propellant tanks are currently being lengthened to provide more usable propellants for Intelsat VA missions.

The engine system is the Rocketdyne MA-5, using liquid oxygen and RP-1 propellants. Total thrust developed is 1,917·4 kN (431,040 lb), including 1,646 kN (370,000 lb) total from the two boosters, 266·9 kN (60,000 lb) from the sustainer, and 4·6 kN (1,040 lb) total axial thrust from the two vernier rockets. All engines are ignited at lift-off.

Most of the electronic command and control functions for the SLV-3D are generated by its Centaur D-1A upper-stage electronics system. The SLV-3A has its own systems, independent of the upper stage, including radio guidance.

Atlas has many space 'firsts' to its credit, including launch of the world's first communications satellite, in 1958, and the first US manned orbital flights (Mercury). All US planetary spacecraft have been launched by Atlas or Centaur.

DIMENSIONS:

Diameter	3·05 m (10 ft 0 in)
Length:	
SLV-3A	24·0 m (78 ft 11 in)
SLV-3A/Agena	36·0 m (118 ft 0 in)
SLV-3D	21·2 m (69 ft 6 in)
SLV-3D/Centaur	39·9 m (131 ft 0 in)

PERFORMANCE:
Atlas/Centaur: See Centaur entry
Atlas SLV-3A/Agena:
3,856 kg (8,500 lb) into 185 km (115 mile) orbit
1,238 kg (2,730 lb) into synchronous transfer orbit

CENTAUR

Centaur was the first US high-energy upper stage and the first to utilise liquid hydrogen as a propellant. The current version is combined with the Atlas SLV-3D.

In April 1973, the first Atlas/Centaur D-1A launched Pioneer 11 on a Jupiter fly-by mission. Three more Intelsat IVs, Mariner 10, seven Intelsat IVAs, one Intelsat V, four Comstars (domestic communication satellites leased to AT and T by Comsat), three High Energy Astronomy Observatories (HEAO) and four US Navy fleet satellite communications spacecraft have also been launched successfully by this vehicle. In early 1981, Atlas/Centaur D-1A had been assigned future missions extending into 1984. They include further launches of FLTSATCOMs and Intelsat V and VA satellites.

Centaur was also used as an upper stage for Titan IIIE, for Viking and Voyager missions, described in previous editions of *Jane's*.

NASA has selected Centaur to operate out of the Space Shuttle for the Galileo mission to Jupiter in 1985 and for the European spacecraft for the Solar Polar Mission in 1986. Use of the stage for the VOIR mission is also contemplated.

Centaur D-1A retains the same propulsion and structural features as its predecessor, Centaur D, with stainless steel pressurised tanks and two 73·4 kN (16,500 lb st) Pratt & Whitney RL10A liquid oxygen/liquid hydrogen rocket engines. Specific impulse with this propellant combination is 445 s, equalling that of the advanced main engines on the Space Shuttle Orbiter. Total propellants carried weigh 13,950 kg (30,750 lb). Attitude control is achieved by gimballing the two main engines or by clusters of small hydrazine rocket motors.

Several of the electronics components were redesigned for Centaur D-1A. The most significant addition is a 16,000 word capacity Teledyne digital computer. Navigation, guidance, vehicle stability, tank pressurisation, propellant management, telemetry formats and transmission, and event initiation are all controlled by the computer. Guidance, control and sequencing for the Atlas booster are provided by the Centaur D-1A electronics system.

Payloads are carried on adapters mounted on the forward end of the Centaur. A 3·05 m (10 ft) diameter fairing protects payloads for Centaur D-1A.

Atlas/Centaur AC-54 launch vehicle for Intelsat V

DIMENSIONS:

Centaur length	9·14 m (30 ft 0 in)
Centaur diameter	3·05 m (10 ft 0 in)
Atlas/Centaur length	39·9 m (131 ft 0 in)

PERFORMANCE:
Atlas/Centaur:
5,942 kg (13,100 lb) into 185 km (115 mile) circular orbit
2,222 kg (4,900 lb) into synchronous transfer orbit
1,179 kg (2,600 lb) to near planet (with kick motor)

GENERAL ELECTRIC

GENERAL ELECTRIC COMPANY SPACE DIVISION

Valley Forge Space Center, PO Box 8555, Philadelphia, Pennsylvania 19101

General Electric's Space Division is vehicle contractor for NASA's Landsat programme (formerly Earth Resources Technology Satellite, ERTS). Other satellites designed and manufactured by this Division have included Nimbus meteorological satellites, BSE (Broadcast Satellite Experimental) and DSCS III (Defense Satellite Communications Systems).

LANDSAT (ERTS)

Designed to study the resources of Earth from space, ERTS-1 (Earth Resources Technology Satellite) was launched by NASA on 23 July 1972. It was subsequently renamed Landsat-1 and was described in the 1977-78 *Jane's*. This spacecraft was retired by NASA on 16 January 1978, after operating for 5½ years and after revolutionising the technology of observing the Earth from space.

Landsat-2 (known formerly as ERTS-B) was launched on 22 January 1975, into an orbit similar to that of Landsat-1 but 180° out of phase from it to allow repetitive coverage of all portions of the Earth every nine days with two spacecraft, instead of once every 18 days with one spacecraft. Brief details of Landsat-2 were given in the 1977-78 *Jane's*. It was retired on 22 January 1980, because of wear-induced failure of its primary flight control system (FCS). Reactivation became possible on 22 May following restart of a yaw flywheel in the FCS. In Spring 1981, the spacecraft was providing prime multispectral scanner subsystem (MSS) coverage to all users in real time mode, but with no working video recorder.

A third spacecraft in the series, Landsat-3, was launched on 5 March 1978. This embodies refined sensors which are supplying data significantly improved over those obtained from Landsat -1 and -2. The MSS detects temperature differences in vegetation, bodies of water and urban areas by day and night. An improved return beam vidicon (RBV) sensor has increased the resolution of recorded images by 50 per cent, enabling areas as small as half an acre to be identified. A line start anomaly in the MSS results in loss of 30 per cent of the image; but Landsat-3 continues to provide useful data.

NIMBUS

Nimbus-7, the seventh satellite in an atmospheric research programme, was launched on 23 October 1978. Designed to gather data from the atmosphere and oceans, it is obtaining global information on air and water pollution for continuing research on climate and weather. In addition one of the sensors, called a Coastal Zone Colour Scanner (CZCS), is available to identify and track oil spills over eleven major ocean areas traversed by tankers.

Since Nimbus-1 was launched on 28 August 1964, and up to 1 January 1981, the spacecraft in the series returned more than 60,000 h of infra-red data, 45,700 h of ultraviolet data, 197,000 h of 'sounder' data on temperature and pressure, and more than 22,600 h of data on the Earth's heat balance. The CZCS had transmitted 20,000 two-minute scenes each year, in 670 h operation.

Landsat-3 Earth resources technology satellite

Details of earlier Nimbus spacecraft were given in the 1974-75 *Jane's*.

HUGHES

HUGHES AIRCRAFT COMPANY

Space and Communications Group
PO Box 92919, Los Angeles, California 90009
Telephone: (213) 648 0884

Hughes is prime contractor for a number of major space programmes, including the US Navy's LEASAT; Indonesia's Palapa B communications satellites; Satellite Business Systems' SBS satellite; the Geostationary Operational Environmental Satellites (GOES); and the Ku-band Integrated Radar and Communications Subsystem for NASA's Space Shuttle. The company is also engaged on development and manufacture of Earth stations for satellite telecommunications; classified military satellite systems; Visible/Infra-red Spin-Scan

Radiometers (VISSR), used as the primary sensor on GMS and, with atmospheric sounder (VAS), on GOES-D, E, and F; Multispectral Scanners (MSS) used on Landsat spacecraft and a Thermatic Mapper for Landsat-D; and on investigation and flight testing of experimental ion engines for advanced spacecraft. Hughes is a major subcontractor on Anik D, being built by Spar Aerospace Ltd for Telesat Canada. Hughes has also been selected by Western Union to build a new, more powerful Westar domestic communications satellite with 24 transponder channels, twice the capacity of previous Westars.

COMSTAR

Built by Hughes Aircraft, Comstar-D4, the fourth in the Comstar series of US domestic communications satellites owned by the Comsat General Corporation, was launched by an Atlas-Centaur rocket in 1981.

Now in geosynchronous orbit over the Pacific Ocean, the 1,516 kg (3,342 lb) satellite carries 12 transponders, each capable of relaying 1,500 one-way voice circuits, giving it a communications capability of 18,000 simultaneous high-quality two-way telephone transmissions. The satellite is providing communications services to major metropolitan areas of the continental US and Hawaii.

LEASAT

In September 1978, the US Navy announced a contract award to Hughes Communication Services Inc, a Hughes Aircraft Company subsidiary, to provide worldwide communications satellite service to the Department of Defense for at least five years at each of four orbital locations. The new satellite, known as LEASAT, will begin service in 1982. Users will include mobile air, surface, subsurface, and fixed Earth stations of the US Navy, Marine Corps, Air Force and Army. Hughes Space and Communications Group will build the satellites, which will occupy geostationary positions over the USA and over the Atlantic, Pacific and Indian Oceans.

The satellites will be spin-stabilised, with the spun portion containing the solar array and Sun and Earth sensors for attitude determination and Earth pointing reference, batteries for eclipse operation, and all propulsion and attitude control hardware. The de-spun platform will contain Earth-pointing communications antennae, communications repeaters, and the majority of the telemetry, tracking and command equipment.

Designed to be launched from the Space Shuttle, each satellite is approx 4·22 m (13 ft 10 in) in diameter and 6·16 m (20 ft 2½ in) long with the UHF and omnidirectional antennae deployed. Total payload weight in the Shuttle will be 7,938 kg (17,500 lb). Weight after separation from the Shuttle will be 6,849 kg (15,100 lb), and the satellite's weight on station at beginning of life will be nearly 1,315 kg (2,900 lb).

PALAPA B

Perumtel, Indonesia's state-owned telecommunications company, has chosen Hughes to build a new generation of Palapa B communications satellites, to succeed the Hughes-built system placed in orbit in 1976-77. The new satellites, based on the Hughes HS 376 spacecraft, will be more than twice as big and four times as powerful as those

Artist's impression of LEASAT

currently in use.

The Palapa B satellites will provide voice, video, telephone and high-speed data services. Each will have 24 transponders, compared with 12 for the present model, and each of the transponders will have 10W power output, compared with 5W on the earlier spacecraft. This increased capability reduces the expense of extended coverage to small rural terminals in remote locations.

The new satellites have outer cylindrical sleeves which, together with folding antennae, deploy in space. The vehicles' cylindrical bodies and their extensions are covered with solar cells. This arrangement nearly doubles the area of power-producing solar cells usable with the same-size satellite. Palapa B will have a height of 6·83 m (22 ft 5 in) when deployed in space and a diameter of 2·16 m (7 ft 1 in). In-orbit weight at the start of the satellite's eight-year lifetime will be 628 kg (1,385 lb).

SBS

In December 1977, Satellite Business Systems (a consortium of IBM, Comsat General Corporation, and Aetna Life and Casualty) selected Hughes Aircraft Company to build three SBS commercial communications satellites for service over the continental USA.

First US domestic commercial communications satellite system to utilise 12/14GHz (K-band) frequencies, SBS will provide point-to-point voice, high-speed data, facsimile and teleconferencing service to its users. With K-band communications, small, relatively inexpensive Earth stations can be used. Another advantage of SBS is that its solar panel almost doubles the power output over that obtainable with other spin-stabilised satellites of the same size.

Since the SBS satellite will operate in the 12/14GHz bands, small antennae may be used in urban areas to communicate with the satellite, without causing interference to terrestrial systems or other spacecraft.

Artist's impression of Palapa B

Designed to be launched by NASA on either the Space Shuttle or a Delta booster, the first SBS satellite was launched on 15 November 1980 by a Delta. Using a folding antenna reflector and a deployable solar panel, the compact satellite will be able to stow upright in the Shuttle bay. This configuration will reduce Shuttle charges, which are proportional to the length of payload bay used. After the satellite is separated from the Shuttle, the antenna will be erected and the outer section of the telescoping solar panel deployed.

The spacecraft is 2·16 m (7 ft 1 in) in diameter and 6·60 m (21 ft 8 in) high with the solar panel and antenna deployed. It weighs 1,057 kg (2,326 lb) following injection into elliptical transfer orbit by the perigee motor, part of the payload assist module (PAM). After the apogee engine fires, the on-station weight of the satellite will be 555 kg (1,222 lb), of which 96 kg (211 lb) will be on-orbit propellant.

TELSTAR 3

Hughes announced in September 1980 that it had been awarded a contract to construct three Telstar 3 communications satellites and related ground control facilities for the American Telephone and Telegraph Company. Compared with earlier Telstars, the new spacecraft will each have a life expectancy of ten, instead of seven years, and will have a capacity for 21,600, instead of 18,000, simultaneous telephone calls. This is made possible by increased fuel capacity, to keep them on station in orbit, and improved batteries and solid-state amplifiers.

Telstar 3 will be based on the Hughes HS 376 spacecraft, used also for the satellites supplied to Perumtel of Indonesia, Telesat Canada, Western Union and Satellite Business Systems. The first will be launched by a Delta vehicle in June 1983, the second in the Summer of 1984, and the third in 1985 or 1986. They will replace Comstar satellites leased currently by AT&T from Comsat General Corporation.

LMSC

LOCKHEED MISSILES & SPACE COMPANY INC (Subsidiary of Lockheed Corporation)

1111 Lockheed Way, Sunnyvale, California 94086
Telephone: (408) 742 6688
PRESIDENT: Robert A. Fuhrman
OTHER FACILITIES: Palo Alto and Santa Cruz, California
DIRECTOR, PUBLIC INFORMATION: George Mulhern

Lockheed Missiles & Space Company is heavily engaged in both missile work and the design, development and production of satellites and space vehicles.

AGENA D

The Agena satellite is used normally as the upper stage of a two-stage launcher, in combination with an Atlas, Thor, or Titan IIIB. The current Agena D version consists of a cylindrical body containing a Bell Aerosystems Model 8096 (YLR81-BA-11) restartable liquid-propellant rocket engine (71·2 kN; 16,000 lb st) and propellant tanks, telemetry, instrumentation, guidance and attitude control systems. It has carried most types of power supply, including a nuclear reactor electric power supply and an ion engine. The payload section (nosecone) can accommodate a wide variety of Earth-orbiting and space probes weighing up to several hundred pounds. The Agena system and its attached payload have functioned for more than six months in some missions for the US Air Force.

Agena D differs from earlier versions in being able to accept a variety of payloads, whereas its predecessors had integrated payloads. The restartable engine permits the satellite to change its orbit in space.

Since 1959, Agenas have served as satellite or booster on more missions than any other spacecraft in the world, having been used on well over 300 flights, representing approximately half of all US space missions.

Agena forms the basis of the new Seasat ocean survey satellite, described separately in this entry. A high proportion of the unidentified US satellites included in the table at the end of this section can be assumed to be Agena payloads of various kinds.

The following details refer to Agena D:

DIMENSIONS:	
Length (typical)	7·09 m (23 ft 3 in)
Diameter	1·52 m (5 ft 0 in)
WEIGHTS (typical):	
Propellant weight	6,148 kg (13,553 lb)
Vehicle weight empty	673 kg (1,484 lb)
Weight in orbit, less payload	579 kg (1,277 lb)

LOCKHEED/USAF 467 BIG BIRD

First launched on 15 June 1971, from Pt Arguello, this highly-advanced photographic reconnaissance satellite is reported to weigh about 11,340 kg (25,000 lb) and to be 15·25 m (50 ft) long. The first Big Bird was launched by a Titan IIID into an orbit of 299 × 183 km (186 × 114 miles), with an inclination of 96·41°. Its capabilities clearly included the same kind of close-look high-resolution photography as that of the recoverable type of Agena vehicle; and it is reported to have ejected a series of capsules for air-snatch recovery.

Big Bird is believed to process photographs taken by cameras and transmit information to Earth in the form of digital data by radio link. Some reports suggest that it also carries infra-red mapping and side-looking radar equipment. Its orbit takes its cameras within range of every point on the Earth twice in each period of 24 h.

Big Birds continue to be launched at four-to-six-monthly intervals and have demonstrated an endurance of up to four months in orbit.

SEASAT

Seasat, an experimental oceanographic satellite built for

Artist's impression of NASA's Space Telescope

NASA, was launched on 27 June 1978 from Vandenberg Air Force Base into a near-circular polar orbit of about 800 km (497 miles), so that it circled the Earth 14 times daily and covered 96 per cent of the world's oceans every 36 h. Radars and radiometers aboard the spacecraft were used to see through clouds and provide data on wave heights, currents, ocean temperatures, icebergs, storms, and coastal features. On 9 October, after three months of gathering extensive data, the satellite was silenced by an apparent power failure due, it is thought, to an electrical short in the slip ring assembly of the solar array. Nevertheless, the satellite proved its primary objective: that precise information on the state of the oceans could be obtained in this manner. Processing of extensive data from all sensors was completed in late 1980, and data will be made available to interested user groups. Full details of the 2,274 kg (5,000 lb) satellite and its special sensors were given in the 1978-79 *Jane's*. Future ocean survey satellites are planned for Space Shuttle launches in the mid-1980s.

SPACE TELESCOPE

Lockheed Missiles & Space Company is building for NASA a Space Telescope which will be launched in 1983 as the most ambitious space astronomy project yet undertaken. Operating well above the optically obscuring effects of Earth's atmosphere, the telescope will be able to detect objects 50 times more faint and seven times further away than those which can be seen by telescopes on Earth. The solar array and photon detector assembly are being developed and manufactured by British Aerospace Dynamics under contracts from the European Space Agency.

MARTIN MARIETTA

MARTIN MARIETTA CORPORATION

6801 Rockledge Drive, Bethesda, Maryland 20034
Telephone: (301) 897 6000
CHAIRMAN AND CHIEF EXECUTIVE OFFICER:
J. Donald Rauth
PRESIDENT AND CHIEF OPERATING OFFICER:
Thomas G. Pownall
VICE-PRESIDENT, PUBLIC RELATIONS: Roy Calvin
AEROSPACE HEADQUARTERS:
6801 Rockledge Drive, Bethesda, Maryland 20034
PRESIDENT, MARTIN MARIETTA AEROSPACE:
Laurence J. Adams
PRESIDENT, DENVER AEROSPACE: Caleb B. Hurtt
PRESIDENT, ORLANDO AEROSPACE: Robert J. Whalen
Baltimore Division
103 Chesapeake Park Plaza, Baltimore, Maryland 21220
Denver Aerospace
PO Box 179, Denver, Colorado 80201
Telephone: (303) 973 3000
Orlando Aerospace
PO Box 5837, Orlando, Florida 32805
Telephone: (305) 352 2000

Martin Marietta Aerospace produces missiles, spacecraft, launch vehicles, communications systems and electronic systems. It also fabricates precision components for the US Department of Defense and NASA, and conducts research related to defence and space systems.

Denver Aerospace produces the Titan III family of space boosters, defence systems, command and information systems, mobile transporter systems and external tanks for the Space Shuttle. Orlando Aerospace produces advanced strategic air-launched missiles, communications systems, the Pershing surface-to-surface ballistic missile, the Patriot surface-to-air missile, strategic terminal interceptors, and tactical weapons systems. Baltimore Division fabricates structural components of aircraft, space vehicles and defence systems.

TITAN III

Titan III is America's standard heavy-duty space 'workhorse' booster and is used for both military and civil missions. It provides a high frequency launch capability for a wide variety of payloads, ranging from 15,875 kg (35,000 lb) in Earth orbit to 3,175 kg (7,000 lb) for planetary missions.

Martin Marietta, in addition to its role as systems integrating contractor, builds the airframe and liquid-propellant stages, supplies the flight control system, and is integrating contractor for facilities and launch operation at Cape Canaveral. Aerojet-General produces the liquid-propellant engines. UTC's Chemical Systems Division supplies the solid-propellant boosters used in the more powerful models. Guidance systems for the Titan IIIC and D are built by General Motors Corporation's Delco Division and Western Electric respectively.

The core section of Titan III consists of elements which provide a high degree of commonality throughout all configurations. It consists of two booster stages and an upper stage, known as Transtage, that can function both in the boost phase of flight and as a restartable space tug propulsion vehicle. All stages use storable liquid propellants and have gimbal-mounted thrust chambers for vehicle control.

Titan III exists in three current configurations:

Titan IIIB. Basically the first two stages of the core section. It can accommodate a variety of specialised upper stages. First launched on 29 July 1966. Series of launches continued through 1978, all with Agena upper stages and classified US Air Force payloads.

Titan IIIC. Consists of the core section, including the Transtage upper stage, with solid-propellant rocket motors attached to each side to function as a booster stage before ignition of main engines. Payloads include US Air Force and NASA military, scientific and communications satellites, including about 80 per cent of all those placed into synchronous equatorial orbit from US launch sites.

Titan IIID. Basically similar to IIIC but has only a two-stage liquid-propellant core (without Transtage) and radio guidance instead of the standard inertial guidance. Able to accept a variety of upper stages. Reported to have been used to orbit the first Lockheed Big Bird advanced photo-reconnaissance spacecraft, weighing about 11,340 kg (25,000 lb), from Pt Arguello on 15 June 1971.

The first stage of the core section is 22·25 m (73 ft) long and 3·05 m (10 ft) in diameter. Its engines, which use a blend of hydrazine and unsymmetrical dimethylhydrazine (UDMH) for fuel, and nitrogen tetroxide as an oxidiser, have a 15 : 1 expansion ratio and are ignited at an altitude where efficiency is increased, giving a thrust of 2,339·6 kN (526,000 lb) in vacuum. The second stage is 7·10 m (23 ft 3½ in) tall and 3·05 m (10 ft) in diameter. Its engine uses the same propellants as the first stage and develops 453·7 kN (102,000 lb st).

The Transtage space propulsion vehicle is 4·57 m (15 ft) tall and 3·05 m (10 ft) in diameter and also uses UDMH/hydrazine and nitrogen tetroxide as propellants. The twin-chamber engine produces 71·17 kN (16,000 lb) of thrust and is capable of multiple restarts in space, which permits a wide variety of manoeuvres, including change of plane, change of orbit, and transfer to deep-space trajectory. Transtage also houses the control module for the entire vehicle, including the guidance system and segments of the flight control and vehicle safety systems.

Titan IIIC/D's solid-propellant booster motors are each 25·91 m (85 ft) long and 3·05 m (10 ft) in diameter. Each motor is built in five segments and develops more than 5,115 kN (1,150,000 lb st). The booster stage is steered by a thrust vector control system, which injects nitrogen tetroxide into the engine nozzle.

Titan IIIC heavy-duty launch vehicle

TITAN 34-D

Instead of Transtage, the latest Titan IIIs use the Boeing Inertial Upper Stage (which see) that has been developed for the Space Shuttle, and have a first stage lengthened by 1·75 m (5 ft 9 in) to contain more propellants. Designated Titan 34-D, these vehicles are being used for some primary launches, as well as for backup of the Space Shuttle during that vehicle's transition period.

The first Titan 34-D was completed at Denver on 28 February 1981, and was expected to launch two Defense Satellite Communications System satellites from Kennedy Space Center in July 1981.

MDAC

MCDONNELL DOUGLAS ASTRONAUTICS COMPANY

5301 Bolsa Avenue, Huntington Beach, California 92647

In March 1977 MDAC was awarded a Spacelab integration contract by NASA involving the design, development and fabrication of most of the Spacelab hardware for which NASA is responsible. This includes the crew transfer tunnel, verification flight instrumentation, mockups and ground support equipment. The contract also covers systems engineering and integration to develop Spacelab operational capability.

DSV-3 DELTA

Details of early versions of the Delta can be found in the 1971-72, 1972-73 and 1977-78 *Jane's*. Production is now centred on the DSV-3P, of which details follow:

DSV-3P Extended Long Tank Delta (also known as 'Straight-Eight' or '2000 Series Delta'). This launch vehicle has a constant 2·44 m (8 ft) diameter from the base of the boat-tail to the conical nose section of the shroud. This provides an enlarged volume, to accommodate larger payloads. The second stage, with a TRW LMDE engine, is suspended within the 2·44 m (8 ft) diameter barrel section. The first-stage length is increased by 3·05 m (10 ft) by comparison with earlier long-tank versions, providing an increase of 13,600 kg (30,000 lb) in propellant capacity. In addition, the MB-3-III main engine is replaced by a Rocketdyne RS-27 of 911·84 kN (205,000 lb st). As an alternative to the TE-364-3 motor, a higher-performing motor, the TE-364-4, is available as a third stage. The two-stage capability is increased to 1,880 kg (4,150 lb) into a 370 km (230 mile) circular orbit. The three-stage capability is increased to 700 kg (1,550 lb) into a synchronous transfer orbit.

DSV-3P (Delta 3914). First launched on 12 December 1975, this uprated version of the standard 'Straight-Eight' Delta utilises nine Castor IV solid-propellant strap-on motors in place of the nine Castor IIs used on the Delta 2914. Development costs were borne by MDAC, and will be recovered through a user charge imposed on non-government users of the vehicle. Dimensions are the same as for the standard DSV-3P Delta 2914, but firing weight is increased to 191,400 kg (422,000 lb). Delta 3914 is capable of putting a 930 kg (2,050 lb) payload into a geosynchronous transfer orbit. The payload is increased to 1,091 kg (2,400 lb) with PAM (see separate entry).

DSV-3P (Delta 3920). This version, development of which was authorised by NASA in March 1980, has an upgraded second-stage propulsion assembly produced by Aerojet. This new stage has a larger propellant tank than the present second stage, extending the burn time by 30 per cent, and uses an AJ10-118K-ITIP (Improved Transtage Injector Programme) engine supplied by Aerojet Liquid Rocket Company.

Delta 3920 with PAM (described separately) is

McDonnell Douglas Payload Assist Module

designed to place 1,250 kg (2,750 lb) into a geosynchronous transfer orbit.

Deltas, in many versions, completed 141 successful flights in 153 attempts between 13 May 1960 and 15 November 1980.

DIMENSIONS:

Length overall	35·15 m (115 ft 4 in)
Body diameter	2·44 m (8 ft 0 in)

WEIGHTS:

Firing weight of DSV-3P (2000 Series):

3 solid motors	104,330 kg (230,000 lb)
6 solid motors	117,930 kg (260,000 lb)
9 solid motors	131,540 kg (290,000 lb)

PAYLOAD ASSIST MODULE (PAM/SSUS)

PAM (Payload Assist Module) is designed to improve the load-carrying capability of the Delta and Atlas launch vehicles, and for use on the Space Shuttle. It is powered by a solid-propellant Star 48 rocket motor, developed by Thiokol Corporation, which has more thrust and a longer burning time than the normal TE-364-4 Delta third stage. PAM is spin-stabilised; small rockets on the structure that joins it to the Delta second stage set the PAM and its payload rotating before the PAM motor is fired.

The PAM that will be used for Space Shuttle launches is identical to the Delta PAM except that the rocket exhaust nozzle will be slightly shorter and the stabilising rotation will be imparted by a pair of small electric motors. NASA designates PAMs used on Shuttle missions as Spinning Solid Upper Stages (**SSUS**). As many as four satellites, with PAM stages attached, could be launched during a single Shuttle mission. When launched from the Shuttle, PAM will carry payloads of up to 1,110 kg (2,450 lb) into transfer orbit. Kick motors will then insert the payloads into 35,880 km (22,300 mile) high geosynchronous orbits.

By April 1981, MDAC had received orders for 18 PAMs. In addition, the US Air Force has ordered a version which, with two Star 48 motors, will be mounted on Atlas boosters to launch Navstar Global Positioning System satellites.

MDAC is developing a larger PAM for NASA, specifically for use from the Shuttle. Designated PAM-A, or SSUS-A, this will be able to lift payloads of up to 2,000 kg (4,410 lb) into geosynchronous transfer orbits.

First use of a PAM was on 15 November 1980 when, as the third stage of a Delta 3910, the rocket was used to place the SBS-1 communications satellite in geosynchronous orbit for Satellite Business Systems.

Delta 117 launch vehicle for Explorer 55 satellite

NASA
NATIONAL AERONAUTICS AND SPACE ADMINISTRATION

Washington, DC 20546

ADMINISTRATOR; ACTING: Dr Alan Lovelace

NASA is responsible for co-ordinating and conducting virtually all US non-military space projects. Its Office of Space Transportation Systems and its office of Space Transportation Operations are responsible for the Space Shuttle programme.

DYNAMICS EXPLORER

Dynamics Explorer is a NASA-sponsored programme which will investigate the interactive coupling between the Earth's magnetosphere, ionosphere and plasmasphere. It will utilise two coplanar polar orbiting spacecraft, constructed by RCA Astro-Electronics and described under the RCA entry in this section. Scheduled to be launched in July 1981, their mission will be to explore the boundary region between Earth and space that affects the atmosphere, auroral displays, radio transmissions and, perhaps, climate and weather. One satellite (DE-B) will be in an elliptical orbit with a low perigee of 305 km (190 miles); the second (DE-A) will be placed in a highly elliptical orbit having an apogee of 24,875 km (15,457 miles).

GALILEO

Galileo is a NASA-sponsored spacecraft designed to study the turbulent atmosphere of the planet Jupiter. It comprises two main components, a carrier and a probe. The carrier, being made by Hughes Aircraft, will transport towards Jupiter the Galileo probe, which will separate from the carrier 100 days before arrival at the planet. The probe will enter Jupiter's thick atmosphere at 160,000 km/h (100,000 mph), slowing in seconds to a few hundred kilometres per hour. At this point a parachute will open and the probe will descend slowly. In this critical manoeuvre the probe will experience deceleration forces of 400*g*. The probe will measure the atmosphere for one hour, down to a level where the pressure is ten times that on the surface of the Earth. Information from the probe will be sent to the carrier, which will transmit the data to receiving stations on Earth.

Galileo is scheduled to be launched by the Space Shuttle in March 1984, and will reach Jupiter in July 1987.

Orbiter OV-102 *Columbia* undergoing prelaunch pad validation tests

ISPM SATELLITE

In March 1979 NASA announced that it had signed an agreement with the European Space Agency for a joint International Solar Polar Mission (ISPM). Under the agreement, NASA and ESA will each provide a spacecraft, and Dornier has been named as prime contractor for the European programme. Both spacecraft will be launched simultaneously by the Space Shuttle, and will then be directed on a trajectory in the ecliptic plane (the plane containing all the planets) to Jupiter. The gravity of Jupiter will be used to redirect their paths out of the ecliptic plane, back to the Sun in trajectories—one northbound and one southbound—that are mirror images of each other, passing over the solar poles. Purpose of the mission is to extend scientific knowledge of the Sun.

In early 1981 funding for the NASA spacecraft, scheduled for launch in 1985, was under review by the Reagan Administration.

SMM

The SMM (Solar Maximum Mission) spacecraft was launched on 14 February 1980 by a Delta 3914 from Cape Canaveral Air Force Station. Carrying seven scientific instruments, SMM is the first spacecraft designed specifically for the study of solar flares. Described in the 1980-81 *Jane's*, it has returned data significantly advancing knowledge of solar flare mechanisms. The satellite has provided images of solar flares throughout their life cycle. It has also provided data on total solar energy output.

SPACE SHUTTLE TRANSPORTATION SYSTEM

The Space Shuttle is the first re-usable space vehicle, consisting basically of two stages: a booster and an Orbiter. The Orbiter has a delta wing and looks very like a conventional aeroplane, but is powered by rocket engines. The liquid propellants for these engines are carried in a large external jettisonable tank, which is attached to the Orbiter at lift-off. Two large solid-propellant jettisonable boosters are mounted on opposite sides of the propellant tank for lift-off.

Prime contractors are Rockwell International (which see) for the Orbiter, Martin Marietta for the external tank, and Thiokol for the boosters. In operation, the Shuttle is launched vertically, with all engines firing in both the boosters and Orbiter. At an altitude of about 43 km (27 miles), the booster stages separate and descend into the sea by parachute, for recovery. The Orbiter continues under its own power, and jettisons its large external propellant tank just before attaining orbit.

In space, the Orbiter manoeuvres by means of two smaller rocket engines, also mounted in the rear-fuselage propulsion cluster. For minor course corrections and adjustments of attitude, the Orbiter has a series of small thrusters.

The Orbiter's main tasks are to place satellites into orbit, retrieve satellites from orbit, and repair and service satellites in orbit. It will also be used to put propulsive stages and satellites into precise low Earth orbit, for subsequent transfer to synchronous orbit or to an 'escape' mission into space. It can also be used for short-duration scientific and applications missions, as an orbiting research laboratory or reconnaissance vehicle, for space rescue, as a tanker for space refuelling, and for support of orbiting space stations.

On some flights a pressurised Spacelab, developed by ten European countries under the leadership of the European Space Agency, will be carried in the payload bay. Spacelab will be the means by which man-associated experiments can be performed in the Orbiter payload bay. It includes a pressurised enclosure housing support equipment (to make it habitable) as well as the experimental equipment. When sensors require direct exposure to the space environment, a pallet will be used in association with the pressurised enclosure. On other types of missions, a pallet may be used alone, with control of the instruments being exercised from the Orbiter cabin or even from the ground. The Spacelabs are designed around a basic seven-day mission, which is extendable up to 30 days by trading payload weight and volume for the additional consumables necessary to accommodate the further time in orbit.

On conclusion of its mission, the Orbiter flies back into the atmosphere towards its land base, protected by a form of heat shielding which is designed to survive 100 missions, unlike ablative-type heatshields. Once through the re-entry phase, the Orbiter is able to glide up to 950 nm (1,760 km; 1,100 miles) to its base, steered by aerodynamic controls.

Special equipment developed for use on the Space Shuttle includes a new type of spacesuit and a rescue system known as a Personal Rescue Enclosure. This consists of a 0·86 m (34 in) diameter ball which contains its own short-term simplified life support and communication systems. The ball has three layers (urethane, Kevlar and an outside thermal protective layer) and a small viewing port of tough Lexan.

Orbiter OV-101 *Enterprise* completed approach and landing tests at NASA's Dryden Flight Research Center, Edwards, California, in 1977, and ground vibration tests in 1978. On 1 May 1979 the craft was moved to launch Complex 39, Pad A, at the Kennedy Space Center, Florida, where it remained until 23 July 1979. Mated with an external tank and solid rocket boosters, it was used to check out ground support equipment, procedures and launch complex facilities. The first test flight was made successfully by Orbiter OV-102 *Columbia*, which was launched from Kennedy Space Center on 12 April 1981 and landed at Edwards on 14 April after a 36-orbit, 54·5 h mission.

The second Shuttle test flight, again involving *Columbia*, was scheduled for September 1981.

The first operational orbital mission is scheduled to be flown by *Columbia* in 1982, after further test flights. It will be joined later by OV-099 *Challenger*, first flight scheduled for November 1982, OV-103 *Discovery*, scheduled to fly in December 1983, and OV-104 *Atlantis*, which will make its first flight in March 1985. Seventy-four operational flights are scheduled for the first four years, and 487 through the mid-1990s.

Basic dimensions and weights of the complete Space Transportation System are as follows:

DIMENSIONS, EXTERNAL:	
Wing span	23·79 m (78 ft 0·68 in)
Length overall	56·14 m (184 ft 2·4 in)
Length of external tank	47 m (154 ft 2·4 in)
Length of boosters	45·46 m (149 ft 2·0 in)
Height overall	23·35 m (76 ft 7·2 in)
WEIGHTS:	
Shuttle complete (OV-102)	2,022,114 kg (4,458,000 lb)
Orbiter (empty)	68,040 kg (150,000 lb)
External tank (full)	743,253 kg (1,638,565 lb)
Boosters (2), each	586,506 kg (1,293,004 lb)
THRUST:	
Total, at lift-off	30,622 kN (6,925,000 lb)
Orbiter, main engines (3), each	1,608 kN (375,000 lb)
Boosters (2), each	12,899 kN (2,900,000 lb)
PERFORMANCE:	
Payload:	
In 185 km (115 mile) orbit, due east	29,485 kg (65,000 lb)
In 500 km (310 mile) orbit, 55° inclination	11,340 kg (25,000 lb)
In 185 km (115 mile) polar orbit	14,515 kg (32,000 lb)

VOYAGER

Two Mariner-type spacecraft, designated Voyager 1 and 2, were launched on a Jupiter-Saturn mission in 1977—Voyager 1 on 5 September and Voyager 2 on 20 August. They were intended to fly by both planets, and conduct exploratory investigations of the Jupiter and Saturn planetary systems. Voyager 1, flying faster than its companion, encountered Jupiter in March 1979, passing within 357,000 km (222,000 miles) of the planet. It took the first close-up pictures of Jupiter's four largest moons. The craft arrived at Saturn in November 1980, flying past at a distance of 209,000 km (130,000 miles) and within 4,000 km (2,500 miles) of Titan, Saturn's largest moon. Voyager 2 reached Jupiter in July 1979 and Saturn in August 1981.

Jupiter-Saturn Voyager

The Voyagers discovered that the inner Galilean moon, Io, is volcanically by far the most active member of the solar system, and provided more data on Saturn than previously learned in the entire span of recorded history, including the discovery of three new moons (the 13th, 14th and 15th), and that the Saturnian ring system contains many more rings and is a much more complex system than was thought before the encounter of Voyager 2.

Voyager 2 is now targeted for the first encounter with Uranus, some 2,700 million km (1,700 million miles) from Earth, in January 1986, after which it may be targeted for an encounter with Neptune in 1989. Voyager 1 is on a trajectory that will take it outside the solar system.

The Voyagers were designed and built at NASA's Jet Propulsion Laboratory, California. Each of the 825 kg (1,820 lb) spacecraft carries wide-angle and narrow-angle television cameras and 10 scientific instruments. Major investigations include imaging, radio science, infrared and ultraviolet spectroscopy, magnetometry, charged particles, cosmic rays, photopolarimetry, planetary radio astronomy, plasma and particulate matter.

RCA

RCA CORPORATION

PRESIDENT AND CHIEF EXECUTIVE OFFICER: Edgar H. Griffiths
GROUP VICE-PRESIDENT: Dr James Vollmer
DIRECTOR, PUBLIC AFFAIRS: Nicholas F. Pensiero

Government Systems Division

Cherry Hill Offices, Camden, New Jersey 08101
Telephone: (609) 234 1234

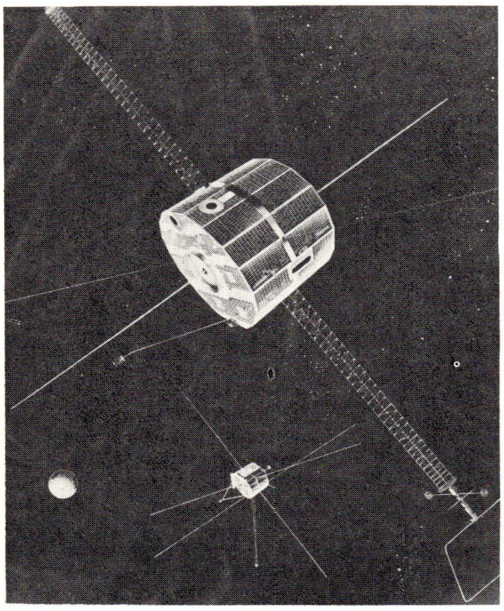

Artist's impression of Dynamics Explorer spacecraft

RCA is prime contractor for a number of major defence programmes, including the US Navy's Aegis advanced ship-to-air missile system, the US Air Force's Block 5D satellites, and Tiros meteorological satellites. It also produces many satellite subsystems and components, including the Closed Circuit Television (CCTV) camera system for the Space Shuttle.

BLOCK 5D DEFENSE METEOROLOGICAL PROGRAM

Block 5D is the name given to a series of advanced meteorological satellites forming part of the Defense Meteorological Satellite Program (DMSP) managed by the Air Force Space Division.

Twenty-one DMSP satellites had been launched successfully by early 1981, including four in the series of the advanced Block 5D Integrated Spacecraft. The first Block 5D-1 was placed into orbit on 11 September 1976, the second on 4 June 1977, the third on 30 April 1978, and the fourth on 6 June 1979. Largest and most complex in the US Air Force's weather satellite series, the Block 5D is designed to provide high-quality meteorological data to tri-service users worldwide. The spacecraft is designed to handle all manoeuvres automatically, from guiding the booster to the accurate positioning of the satellite and its sensors in orbit. This is accomplished by a high-accuracy attitude control system using a star sensor, inertial platform, and two onboard computers which can be reprogrammed from the ground if desired.

Artist's impression of Block 5D-2 Air Force meteorological satellite

The latest Block 5D-2 satellites are improved versions of the 5D-1, and are each 155 kg (342 lb) heavier. The extra weight results from added redundant systems and advanced computers that give the Block 5D-2 an in-orbit operational life of three years (twice that of 5D-1 satellites).

DYNAMICS EXPLORER

Dynamics Explorer (DE) is a NASA-sponsored programme which will investigate the interactive coupling between the Earth's magnetosphere, ionosphere and plasmasphere. It will utilise two coplanar polar orbiting spacecraft, designed by RCA Astro-Electronics; these were to be launched in 1981 by a single Delta 3913, into low and high elliptical orbits. The planned perigee altitude of both satellites is 250 km (155 miles); the apogee altitude of the low spacecraft is 1,000 km (620 miles) and that of the high spacecraft 4 to 6 Earth radii geocentric.

The Dynamics Explorer spacecraft is a derivative of the Atmosphere Explorer satellite. DE-A is estimated to weigh 258 kg (569 lb), of which 30 kg (65 lb) is science payload. DE-B will weigh approximately 305 kg (672 lb), including 70 kg (154 lb) of experiments.

Of particular interest is the variety of experiment antennae and booms carried on both spacecraft. The high mission configuration (DE-A) accommodates two wire antennae 213 m (700 ft) long tip-to-tip, two 4 m (13 ft) flexible stem antennae, and two rigid booms 6 m (20 ft) long. The low mission spacecraft (DE-B) contains six 30 mm (1·2 in) diameter flexible stem antennae 10 m (33 ft) long, and a single rigid boom 6 m (20 ft) in length.

DE-A is spin stabilised and DE-B is three-axis

stabilised with its pitch axis controlled to point continually toward the Earth's surface in a one-revolution-per-orbit 'wheel' mode.

NOVA

The Nova satellite is an advanced version of the US Navy's Transit, and is part of the Navy's Navigation Satellite System. Improvements include onboard station-seeking and station-keeping propulsion systems. The satellite also requires fewer position-updating transmissions from ground control stations than Transit. The new design was developed by the Applied Physics Laboratory of The Johns Hopkins University, with support from RCA.

In operation, Nova satellites will transmit signals that indicate their orbital positions. Using information derived from these signals, a user will be able to determine his precise longitude and latitude.

Nova will weigh approximately 136 kg (300 lb) in orbit and will utilise a 7·6 m (25 ft) long gravity-gradient boom for stabilisation. The first of three satellites was scheduled for launch in 1981 from Vandenberg Air Force Base, California, by a Scout rocket.

RCA SATCOM

RCA Satcom domestic communications satellites are designed and built by RCA Astro-Electronics, Princeton, NJ. The satellites are owned and operated by RCA American Communications, Inc.

RCA Satcom I was launched on 12 December, 1975, followed by RCA Satcom II on 26 March, 1976. The third and fourth satellites in the series were scheduled for launch in 1981, and have a design life of more than eight years, with continuous full power.

The RCA Satcoms have 24 channels, each of which is designed to carry 1,000 voice circuits, one FM/colour TV transmission, or 64 million bits per second of computer data. They are the first of a new generation of communications satellites with three-axis stabilization which gives them extra weight and power margins. The RCA Satcoms cover Alaska, Hawaii and the contiguous 48 States.

TIROS

Designed and built by RCA Astro-Electronics, Prince-

Artist's impression of Nova navigational satellite

ton, NJ, the Tiros meteorological satellite system has provided global weather observations since April 1960 for the USA and the world community. In the past two decades, the TIROS polar-orbiting satellite has progressed through four generations of design with continually increasing capabilities in terms of improved and diverse data products, and more frequent and broader coverage. The TIROS-N/NOAA configuration is the current operational system. TIROS-N was introduced in October 1978 and its complementary satellite, NOAA-6, was launched in June 1979.

The TIROS-N/NOAA series of eight satellites carry more advanced sensing devices, including the advanced very high resolution radiometer (AVHRR), the TIROS operational vertical sounder (TOVS), a data collection system (DCS), and a solar environment monitor (SEM). The last three spacecraft in this series (NOAA E, F and G) are being modified to enhance further the TIROS mission. Three new payloads are being incorporated for this advanced TIROS-N mission (the Earth radiation budget experiment, the solar backscatter ultra-violet radiometer, and an experimental payload for search and rescue of aircraft or ships in distress). Further design studies are underway to prepare TIROS-N for US Space Shuttle

Artist's impression of RCA Satcom

Artist's impression of TIROS-N weather satellite

launches.

TIROS-N has an overall diameter of 1·88 m (6 ft 2 in) and is 3·71 m (12 ft 2 in) long. Weight at lift-off is 1,421 kg (3,127 lb); in-orbit weight is 736 kg (1,620 lb). Each satellite will be launched by an Atlas E/F vehicle into a near-polar Sun-synchronous orbit at a nominal height of 833 km (517 miles) or 870 km (540 miles).

By Spring 1981, 27 TIROS satellites had been launched successfully, and all had met or exceeded their mission objectives.

ROCKWELL INTERNATIONAL

ROCKWELL INTERNATIONAL CORPORATION

Space Operations

12214 Lakewood Boulevard, Downey, California 90241

Telephone: (213) 922 2111

PRESIDENT: George W. Jeffs

Rockwell's Space Operations comprise two divisions: the Space Transportation System Development and Production Division which is responsible for design, development and test of Space Shuttle Orbiters and integration of NASA's Space Transportation System and payloads; and the Space Operations and Satellite Systems Division which has responsibility for design, development and test of the NavStar satellites of the Global Positioning System, as well as government and commercial space operations, payload/cargo integration, and manned space business areas.

NAVSTAR

The NavStar global positioning system (GBS) is intended to satisfy future precise positioning and navigation needs of all the US military services, and to have potential civil applications. NavStar is designed to provide suitably equipped users with highly accurate (to within 10 m; 30 ft) three-dimensional position and velocity information and a precise timing reference in real time. The NavStar spacecraft will be placed in subsynchronous, 12-h circular orbits of about 17,525 km (10,890 miles) in three orbital planes at 63° inclination, with eight satellites per ring. The complete system of 18 satellites should be operational by 1985.

NavStar I was launched on 22 February 1978 by an

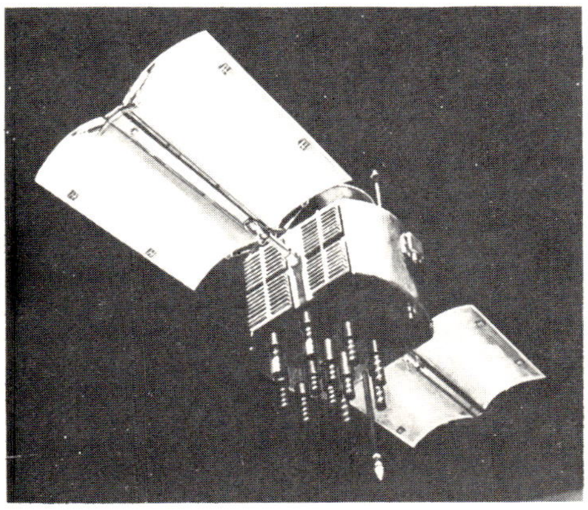

Artist's impression of NavStar satellite

Atlas F from Vandenberg AFB, and by April 1981 had been followed by NavStar II to VI. These six satellites are being used to validate the GPS concept, and its ability to provide navigational data with three-dimensional accuracy within 9 m (30 ft).

The GPS set (the user's or receiver's equipment) may be one of several models, installed or even hand-held in land, sea, air or space vehicles. The set comprises a combined radio receiver and computer, which locks on to NavStar signals from the four satellites that are most favourably located and computes the signal's time and range into navigational data. Key to the system is the use of highly accurate atomic clocks, of which three are installed on each of the 759 kg (1,673 lb) satellites; these will lose or gain only one second in 36,000 years.

In 1979, it was announced that funding had been made available for the development of electronic countermeasures technology for NavStar and other US military satellites. Work includes development of onboard sensors capable of detecting whether a spacecraft has been illuminated by laser energy or has been touched by another spacecraft.

SPACE SHUTTLE ORBITER

The National Aeronautics and Space Administration's Space Shuttle will be the world's first re-usable space transportation system, and will be the keystone of America's space programme through this century.

The Shuttle system includes the Orbiter stage, capable of carrying up to 29,484 kg (65,000 lb) of cargo into Earth orbit; an external propellant tank; and two solid-propellant rocket boosters. The Orbiter lifts off from Earth like a rocket, operates in orbit as a spacecraft, and returns to land in a manner similar to that of a conventional aeroplane.

The first Shuttle spacecraft, *Enterprise* (OV-101), was returned to Rockwell's Palmdale, California, facility in 1980, following completion of the mated vertical ground vibration tests at NASA's Marshall Space Flight Center, and fit and function checkout of the launch facilities and equipment at Kennedy Space Center. *Columbia* (OV-102), made the first successful, 36-orbit, test flight in April 1981, manned by astronauts John Young (commander) and Capt Robert Crippen (pilot). *Challenger* (STA-099) was returned to Rockwell's Palmdale facility, following successful load-limit structural testing, and is undergoing structural renovation, systems installation, and final assembly as the second Earth orbital flight spacecraft. *Discovery* (OV-103) was in the initial stage of fabrication in March 1981 and *Atlantis* (OV-104) will be in its initial stage of fabrication in 1982.

Details of the overall operation of the Space Shuttle Transportation System are given in the NASA entry in this section.

The following description applies to Orbiter vehicle OV-102:

TYPE: Re-usable space transportation vehicle.

WINGS: Cantilever low-wing monoplane, of double-delta planform. Wing section NACA 0010 (modified). Sweepback 81° on inner leading-edges, 45° on outer leading-edges. Dihedral 3° 30′ on trailing-edges. The main wing assembly, for which Grumman is responsible, is primarily a conventional aluminium alloy structure made up of a corrugated spar web, truss-type ribs, and riveted skin/stringer and honeycomb skins. Wing has a very blunt leading-edge and is more than 1·52 m (5 ft) thick at the thickest point. Two-segment hydraulically-actuated elevons on each trailing-edge, for pitch and roll control, are of aluminium honeycomb construction with a titanium rubbing strip on each of their leading-edges. Hinged panels on the wing upper surface, of titanium and Inconel sandwich, are used to seal the wing/elevon gap; these are the only areas of the wing not covered by the thermal protection system.

FUSELAGE: Conventional semi-monocoque aluminium alloy structure, built in three main portions. Forward fuselage contains the crew module, three forward electronics bays, forward reaction control system and nose-wheel unit. The mid-fuselage portion is an 18·28 m (60 ft) long section of primary load-carrying structure, built by General Dynamics (Convair), and includes the wing carry-through structure. Upper half of the mid-fuselage consists of structural payload bay doors, hinged along the side and meeting at the top centreline. These doors are of graphite epoxy bonded honeycomb sandwich construction, with a Nomex core, and are opened and closed by Curtiss-Wright actuators. The forward 9·14 m (30 ft) of each door incorporates Vought radiator panels that are hinged and latched to the inside of the door and are deployed in orbit. Fixed radiator panels are attached to the remaining inner surface on the front of each aft door. The rear of the aft doors can be fitted with fixed radiator panels if required by a specific mission. The aft fuselage interfaces with the removable orbital manoeuvring system (OMS)/reaction control system (RCS) pods, the wing rear spar, the vertical tail assembly, the underbody flap, the external tank rear supports, the main propulsion system, the launch umbilical panel, the three aft electronics bays, and other discrete system equipment. A bulkhead heatshield at the rear of the vehicle protects the main engine systems. A large body flap under the rear fuselage protects the main engine nozzles during re-entry, and is actuated hydraulically to serve also as a trimming surface.

TAIL UNIT: Vertical surfaces only, built by Fairchild Republic, of wedge-shaped section with 45° sweepback on fin leading-edge. Fin is a conventional aluminium alloy structure. The rudder/speed brake assembly has an

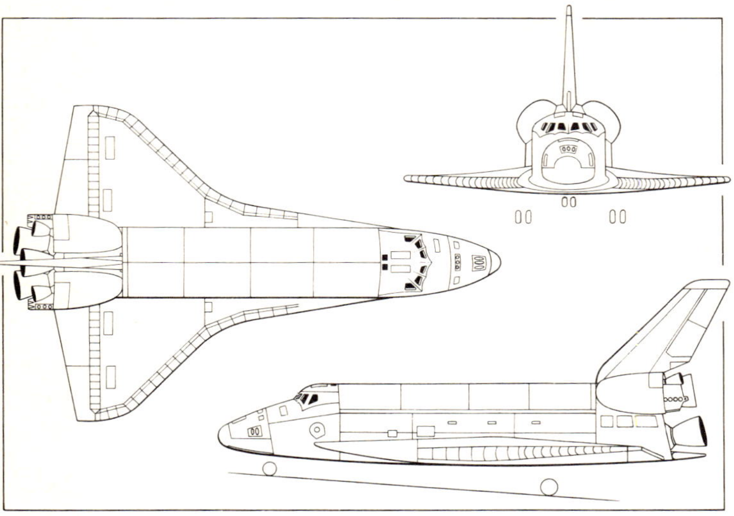

NASA/Rockwell International Space Shuttle Orbiter *(Michael A. Badrocke)*

Orbiter OV-102 *Columbia* **landing after its successful first orbital flight in April 1981**

aluminium honeycomb skin and is divided into upper and lower sections. Each of these is also split longitudinally and actuated individually to serve as both rudder and speed brake, operated by a Sundstrand hydraulic rotary actuator. The Inconel honeycomb seal over these is the only part of the vertical tail not covered by the thermal protection system. Mission requirements call for a locked rudder/speed brake during boost, orbit and re-entry. The speed brake control is provided from approx Mach 10 to Mach 5; from Mach 5 to landing the rudder and speed brake controls are combined as required. Primary system control is automatic, with manual override.

Thermal Protection System: Almost the whole of the exterior of the Orbiter, a total area of some 1,105 m² (11,895 sq ft), is covered in one of four main types of thermal insulation. Two of these, known as HRSI (high-temperature re-usable surface insulation) and LRSI (low-temperature re-usable surface insulation) are the responsibility of Lockheed Missiles and Space Co, and are in the form of silica fibre-based quartz tiles. Some 31,000 tiles cover 70 per cent of the surface (479·8 m²; 5,164 sq ft of HRSI and 254·6 m²; 2,741 sq ft of LRSI, covering most of the wings, fuselage and tail areas). The tiles are coated with reaction cured glass (RCG), the HRSI tiles giving protection from temperatures of 649-1,260°C (1,200-2,300°F) and the LRSI tiles from temperatures of 371-649°C (700-1,200°F). For temperatures below 371°C, some 332·7 m² (3,581 sq ft) of the surface, mostly on the cargo bay doors, upper wing and mid-fuselage, is covered with coated Nomex felt. The nose-cap, and most of the wing leading-edges, a total of 38·00 m² (409 sq ft), are covered in a reinforced carbon-carbon (RCC) composite for which Vought is the subcontractor. Metal insulation is used along the elevon hinge lines. Total weight of the thermal insulation is 8,074 kg (17,800 lb).

Landing Gear: Retractable tricycle type, with twin wheels and Menasco oleo-pneumatic shock-absorbers on each unit. Hydraulic release, with a pyrotechnic backup for deployment in flight. Landing gear cannot be retracted in flight, after release. Nose unit is retracted forward into fuselage and main units forward into wings before launch. Nose unit is steerable; main units are fitted with Goodrich brakes and Hydro-Aire anti-skid units. All units have Goodrich wheels and tyres. Landing gear is designed to facilitate safe landing at speeds of up to 225 knots (415 km/h; 258 mph). The main gear tyres are rated at 20,410 kg (45,000 lb) and the brakes at 240 × 10⁶ ft-lb.

Boosters: Two Thiokol solid-propellant rocket boosters (each 12,899 kN; 2,900,000 lb st for lift-off) are attached one on each side of the external propellant tank.

Main Propulsion: Three Rocketdyne SSME (Space Shuttle Main Engines) high-pressure liquid oxygen/liquid hydrogen engines, each rated at 1,668 kN (375,000 lb st) for lift-off and 2,090 kN (470,000 lb thrust) in space.

External Propellant Tank: Contains the main propellants for the Orbiter. It is of aluminium alloy semi-monocoque construction, with a 25 mm (1 in) thick foam external insulation. In the forward end of the tank is a 552 m³ (19,495 cu ft) tank holding 604,185 kg (1,332,000 lb) of liquid oxygen; in the aft end is a 1,573·1 m³ (55,552 cu ft) tank holding 101,604 kg (224,000 lb) of liquid hydrogen. Total propellant weight 743,240 kg (1,638,565 lb).

Orbit Manoeuvring Engines: Two Aerojet Liquid Rocket Company (ALRC) bipropellant liquid rocket engines, running on monomethylhydrazine (MMH) and nitrogen tetroxide (N_2O_4), are used for the Orbiter's orbit manoeuvring subsystem (OMS). These engines are housed in pods, one on each side of the Orbiter's aft fuselage. The OMS engines, for which a usable total of 10,830 kg (23,876 lb) of propellant is carried, are used to position the Orbiter in orbit; each develops 26·7 kN (6,000 lb thrust) in space.

Reaction Control Engines: The Orbiter's reaction control subsystem (RCS) utilises thirty-eight Marquardt R-40A bipropellant liquid rocket engines (each 3·87 kN; 870 lb vacuum thrust) and six Marquardt R-1E bipropellant liquid rocket vernier thrusters (each 0·11 kN; 24 lb vacuum thrust). Fourteen of the R-40A engines are on the Orbiter's nose and 24 on the aft end, 12 in each OMS/RCS pod; there are two of the R-1E verniers on the nose and two in each aft pod. Propellants are the same as for the OMS engines; 1,096 kg (2,418 lb) is carried in the RCS tanks, and there is provision for crossfeed between the OMS and RCS tanks.

Crew Compartment: Self-contained crew module has a fuselage-side hatch for access, a hatch into the airlock from the mid-section, and a hatch from the airlock into the payload bay. It is divided into three levels, the upper (flight deck) level having side-by-side seating for two flight crewmen with dual controls. Behind them are seats for one or two mission specialists. On the middle deck are seats for three more mission specialists, three bunks, galley, hygiene section, airlock, three electronics bays, and payload bay access; for rescue missions, seats for three more persons can be fitted in place of the bunks. The lower deck contains environmental control equipment and crew equipment storage.

Payload Bay: In centre of fuselage, 18·29 m (60 ft) long and 4·57 m (15 ft) in diameter. Retractable manipulator arm on left hand side (with provision for a second one on the right), for deploying and retrieving payloads. Complete closed-circuit TV system by RCA Astro-Electronics includes a colour camera in the crew compartment and several black and white cameras in the cargo bay and on the manipulator arm. These will facilitate payload handling and provide TV coverage for engineers and the general public on Earth.

Systems: Environmental control and life support system, made up of four subsystems: atmosphere revitalisation subsystem (ARS), to control atmospheric environment for occupants and thermal environment for electronics; food, water and waste subsystem (FWW), to provide hygiene, and other life support functions; active thermal control subsystem (ATCS), to maintain subsystems and components within specified temperature limits and to provide, via payload door radiator panels, active heat rejection to protect payloads; and an airlock support subsystem. Three redundant hydraulic systems, each of 207 bars (3,000 lb/sq in), supply actuators for the elevons, body flap, rudder/speed brake, and power to actuate main engine thrust vector controls, landing gear, brakes and steering. Electrical power subsystem (EPS) consists, functionally, of a fuel cell power plants (FCP) subsystem, and a power reactant storage and distribution (PRSD) subsystem. There are three FCPs, each providing power at 27·5V to 32·5V DC over a power range of 2-12kW and each connected to one of the three main DC buses; these supply the primary in-flight electrical power used by the Shuttle, generated through the chemical combination and conversion of cryogenic oxygen and hydrogen. In the PRSD subsystem, enough of these materials is stored to provide a total of 1,530kWh of electrical energy to the Orbiter, including 50kWh to a payload over a seven-day period. Westinghouse remote power control system and master timing unit. Honeywell four-channel fly-by-wire electrical flight control subsystem for operation of all control surfaces and main engine controls. APU subsystem consists of three Sundstrand independent APUs (each 100·7 kW; 135 shp), deriving their energy from the decomposition of hydrazine (N_2H_4).

Electronics and Equipment: Fully fail-operational/fail-safe guidance, navigation and control system, including three Singer-Kearfott KT-70/SKN-2600 type inertial measuring units; triplex Ku-band microwave scan beam landing system, by the AIL Division of Cutler-Hammer; three Northrop rate gyro assemblies; three Hoffman L-band Tacan; three Bendix accelerometer assemblies; two Honeywell C-band radar altimeters; four AiResearch air data transducers; three Lear Siegler attitude director indicators; two Collins horizontal situation indicators; two Sperry alpha/Mach indicators; two Bendix altitude/vertical velocity indicators; two Bendix surface position indicators; two Sperry barometric altimeters; and two Sperry ATC transponders. Communications and tracking equipment includes one (optionally two) Ku-band rendezvous radar/satellite com; two Ball star trackers; two one-way Doppler extractors; two 100W Watkins-Johnson S-band TWT amplifiers; two P-band UHF for EVA/ATC com; Conrac S-band FM for Orbiter/ground and Orbiter/payload com; and Ku-band radio for Orbiter/ground com. Central data processing is by means of five IBM Advanced System/4 Pi Model AP-101 (modified) digital computers (each with a capacity for 65,000 32-byte words) and two mass memory units (each with capacity of 134 megabits). Four of the computers are interconnected to process guidance, navigation and control inputs and to relay commands to FBW flight control systems; the fifth is provided for independent backup and systems management.

Basic dimensions and weights of the Orbiter are:

Dimensions, external:
- Wing span — 23·79 m (78 ft 0·68 in)
- Wing aspect ratio — 2·265
- Wing mean aerodynamic chord — 12·06 m (39 ft 6·81 in)
- Length — 37·19 m (122 ft 0·2 in)
- Height — 17·25 m (56 ft 7 in)

Dimensions, internal:
- Payload bay:
  - Length — 18·29 m (60 ft 0 in)
  - Diameter — 4·57 m (15 ft 0 in)
- Crew module: Volume — 74·33 m³ (2,625 cu ft)

Areas:
- Wings, gross — 249·91 m² (2,690 sq ft)
- Elevons (total) — 38·38 m² (413·14 sq ft)
- Rudder/speed brake — 9·09 m² (97·84 sq ft)
- Vertical tail surfaces (total) — 38·39 m² (413·25 sq ft)
- Body flap — 12·61 m² (135·75 sq ft)

Weights:
- Weight empty — 68,040 kg (150,000 lb)
- Design landing weight with 14,515 kg (32,000 lb) payload — 96,162 kg (212,000 lb)

Performance:
- Orbital speed — approx 15,285 knots (28,325 km/h; 17,600 mph)
- Nominal touchdown speed (unpowered) — 180 knots (334 km/h; 208 mph) EAS

TRW

TRW DEFENSE AND SPACE SYSTEMS GROUP

1 Space Park, Redondo Beach, California 90278
Telephone: (213) 535 4321
VICE-PRESIDENT AND GENERAL MANAGER: Dr George E. Solomon
PUBLIC AFFAIRS AND COMMUNICATIONS DIRECTOR: Emil Gaynor

TRW Defense and Space Systems Group has provided technical direction and systems engineering for the US Air Force Atlas/Titan/Minuteman missile programme since 1954. The company has also designed and built more than 150 spacecraft and scores of major subsystems. Its recent and current military and civilian contracts have included prime contracts for the NASA Pioneer Jupiter and Saturn, US Air Force DSCS II, international communications satellites and US Navy Fleet Communications Satellite.

DEFENSE SATELLITE COMMUNICATIONS SYSTEM, PHASE II (DSCS II)

The Phase II Defense Satellite Communications System (DSCS II) utilises synchronous-orbit, high-capacity, super high frequency communications satellites and surface terminals to provide reliable worldwide circuits for carrying military communications. The satellites are developed and produced by TRW for the US Air Force.

Protected against interference, the satellites are each equipped with steerable narrow-beam antennae that focus a portion of the satellite's energy to areas 870 nm (1,600 km; 1,000 miles) in diameter. Within these specially illuminated areas, the narrow beam antennae allow small terminals to be used in place of more costly large terminals. The narrow beams are designed to be steered in a matter of minutes to different locations on the Earth's surface, and the satellites are so designed that they can be moved in a few days to new synchronous orbital positions. In this way coverage can be tailored to fit defence contingency communications all over the world.

The Phase II satellites each weigh 590 kg (1,300 lb), are 2·75 m (9 ft) in diameter and 3·95 m (13 ft) tall with antennae extended. Electrical power is supplied by solar arrays with an output of 535W at launch, decreasing to a minimum of 358W after five years. The X-band single-frequency conversion repeater has a bandwidth of 410MHz and capacity of 1,300 voice channels or up to 100 megabits per second of data.

The Earth-coverage antennae have a transmit beamwidth of 18°, a gain of 16·8 dbi, and effective radiated power of 28 dbw. The narrow-coverage antennae have beamwidths of 2·6° and 6·5°, a gain of 33 dbi and 22 dbi, and an effective radiated power of 43 dbw and 32 dbw. They are steerable to ±10°.

First launch of two DSCS II satellites was on 3 November 1971, by Titan IIIC launch vehicle, from Kennedy Space Center. Seven more pairs had been launched by early 1979, with a further pair scheduled for launch in 1981, to maintain a total of at least 16 operational satellites.

FLTSATCOM

The US Navy-sponsored Fleet satellite communications system provides worldwide UHF communications between naval aircraft, ships, submarines, ground stations, Strategic Air Command and the Presidential command networks.

The first **FLTSATCOM** was launched by an Atlas-Centaur on 9 February 1978 and was followed by others as recorded in the Table of Satellites at the end of this section of successive editions of *Jane's*. Four of the three-axis stabilised satellites have been placed into geosynchronous equatorial orbit to provide complete Earth coverage, except for the polar regions. A fifth satellite, scheduled for

DSCS II defence communications satellite

launch in 1981, will serve as a contingency spare. The satellites have a design life of five years and provide more than 30 voice and 12 teletype channels.

The satellite consists of two major components, each with a basic 2·44 m (8 ft) hexagonal body. The payload module contains UHF and X-band communications equipment and antennae. Each of its six side panels carries related communications components. The 4·88 m (16 ft) parabolic UHF antenna is made up of ribs and mesh, and opens like an umbrella. The solar array, never shadowed, is exposed to sunlight in both folded and deployed configurations. Three nickel-cadmium batteries provide power during eclipse.

WEIGHTS:

At lift-off	1,860 kg (4,100 lb)
In orbit	987 kg (2,176 lb)

HEAO

HEAO, a High Energy Astronomy Observatory, was designed to study some of the most intriguing mysteries in the universe, including pulsars, quasars, exploding galaxies and 'black holes' in space.

Three spacecraft, HEAO-1, 2 and 3 were involved in a three-mission programme, which was described in the 1980-81 *Jane's*. HEAO-1 added more than 1,200 new sources to the catalogue of known X-ray objects. HEAO-2's X-ray telescope provided the first images of X-ray sources in galaxies outside the Milky Way, some estimated to be more than 10,000 million light years from Earth. HEAO-3 carries the only cosmic ray experiments to be flown by NASA.

IMEWS (USAF 647 SATELLITE)

The US Air Force's 647 series of satellites, for which TRW is prime contractor, is intended to provide early warning of a hostile ballistic missile launch by detecting the infra-red emission from the missiles by means of an Aerojet-General infra-red 'telescope'. The programme is believed to be known by the acronym IMEWS (Integrated Missile Early Warning Satellite).

French reports suggest that the IMEWS satellite is cylindrical, weighs 820 kg (1,800 lb) at launch, has an inertial three-axis stabilisation system and measures

Artist's impression of FLTSATCOM satellite

approximately 3 m (9 ft 10 in) in diameter and 3 m (9 ft 10 in) in height; cruciform solar panels are said to span 7 m (23 ft). The same reports stated that the satellite carries high-resolution cameras, able to transmit photographs of any missiles that are located to a ground station 500 km (300 miles) north-west of Adelaide, Australia, for onward transmission via synchronous-orbit relay satellites to NORAD headquarters, Colorado Springs, USA.

Debriefing reports of an espionage trial held in 1977 contain references to two US satellite surveillance systems known as Argus and Rhyolite. Developed by TRW and stationed in geosynchronous orbit over Asia, the Rhyolite satellites are reported to intercept telemetry signals transmitted by Soviet missiles during test launchings and when they deliver dummy payloads in Eastern Siberia or in the Pacific Ocean. The system can also monitor Chinese missile tests. Argus satellites are reported to intercept Soviet and Chinese telephonic and radio microwave communications and to have other classified capabilities. A third satellite surveillance system is reported to have the code name Keyhole.

VOUGHT

VOUGHT CORPORATION (Subsidiary of The LTV Corporation)

HEADQUARTERS: PO Box 225907, Dallas, Texas 75265
OFFICERS: See Aircraft section

Vought Corporation is prime contractor for the NASA/DoD Scout launch vehicle. It is also engaged in the fields of aircraft, advanced defence systems, ramjet propulsion systems and laser technology.

SCOUT (XRM-91)

Scout was designed to make possible space, orbital and re-entry research at comparatively low cost. On 16 February 1961, it became the first solid-propellant vehicle ever to put a satellite into orbit when it was used to launch Explorer 9. In its current form, the first-stage motor is an Algol III, manufactured by Chemical Systems Division of UTC and providing a total impulse of 3,266,000 kg/s (7,200,000 lb/s); the second stage is the 266·9 kN (60,000 lb) Castor II by Thiokol; the third stage is the 83·1 kN (18,700 lb) Antares III by Thiokol; the current Altair III fourth stage, by Thiokol, develops 26·7 kN (6,000 lb st). Honeywell provides the simplified gyro guidance system. Spin stabilisation of the fourth stage is by Vought.

In this standard four-stage form, Scout is able to put a 193 kg (425 lb) payload into a 500 km (310 mile) easterly orbit. A five-stage version is also available, which increases the Scout's hypersonic re-entry performance and makes possible highly-elliptical deep-space orbits.

In addition to its use by NASA and the Department of Defense, Scout is used for international programmes, including those of the United Kingdom, Italy, France, West Germany, the Netherlands and the European Space Agency (ESA). A total of 105 had been launched by January 1981.

DIMENSIONS:

Height overall	22·92 m (75 ft 2½ in)
Max body diameter	1·14 m (3 ft 9 in)

WEIGHT:

Launching weight	21,400 kg (47,185 lb)

ASAT

In FY 1979 The LTV Corporation was awarded a $78·2 million contract by the US Air Force to develop a new miniature anti-satellite (ASAT) weapon. The programme received a further $82·3 million in FY 1980. In January 1981 a contract valued at $268 million was awarded for continued development and flight test of the space defence weapon until 1984.

The ASAT programme involves the development of a

Scout four-stage solid-propellant launch vehicle

small satellite interceptor to be launched from an F-15 aircraft and then boosted by a two-stage rocket vehicle to orbital altitudes where it could destroy an enemy satellite.

Vought is developing the high-technology interceptor vehicle and is responsible for the second stage of the rocket booster. This upper stage of the booster employs an Altair III rocket motor, which has served for many years as the fourth stage of Vought's Scout launch vehicle. The lower stage is a modified version of the Short Range Attack Missile (SRAM), which Boeing Aerospace Company produces for the US Air Force. Boeing will develop the lower stage, provide integration services and have responsibility for the system mission control centre. McDonnell Douglas Corporation will modify its F-15 aircraft to serve as the ASAT launch platform.

It is reported that ASAT is fitted with a simple seeker and does not carry an explosive warhead, targets being destroyed by collision.

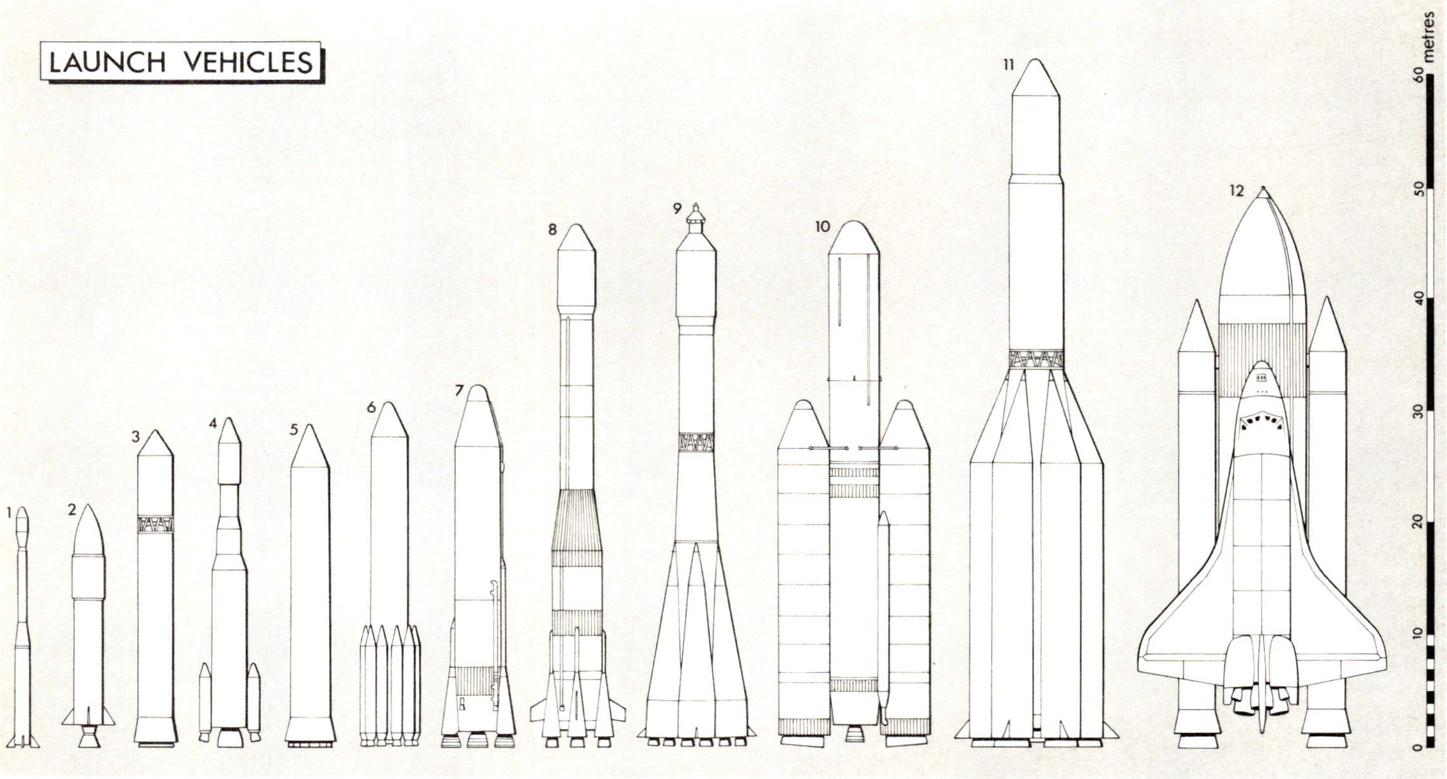

Launch vehicles: 1, Scout (USA); 2, Diamant BP-4 (France); 3, B-1 Cosmos (USSR); 4, N (Japan); 5, C-1 Cosmos (USSR); 6, Delta 3914 (USA); 7, Atlas-Centaur (USA); 8, Ariane (France); 9, A-2 Soyuz (USSR); 10, Titan IIIC (USA); 11, D-1e Zonda (USSR); 12, Space Shuttle (USA) *(Michael A. Badrocke)*

SATELLITES AND SPACECRAFT LAUNCHED DURING 1980

Note: Both the USA and USSR have withheld information on some launchings and this listing may be incomplete. Data in italics are approximate or estimated.

Date	Origin	Name	Total Weight kg	lb	Launch Vehicle	Perigee km	miles	Apogee km	miles	Inclin-ation	Lifetime	Remarks
9 Jan	USSR	Cosmos 1149	*6,000*	*13,230*	—	193	120	387	240	72·87	14 days	*Reconnaissance satellite.* Recovered.
11 Jan	USSR	Molniya-1 (46)	*1,000*	*2,205*	—	431	268	40,843	25,379	62·84	*12 years*	Communications satellite.
14 Jan	USSR	Cosmos 1150	*700*	*1,543*	—	969	602	1,015	631	82·95	*1,000 years*	*Navigation satellite.*
18 Jan	USA	FLTSATCOM 3	1,860	4,100	Atlas-Centaur	35,405	22,000	35,661	22,159	2·40	Unlimited	Military communications satellite.
23 Jan	USSR	Cosmos 1151	*700*	*1,543*	—	635	395	664	413	82·52	*50 years*	*Ocean survey satellite.*
24 Jan	USSR	Cosmos 1152	*6,000*	*13,230*	—	171	106	347	216	67·15	13 days	*Reconnaissance satellite.*
25 Jan	USSR	Cosmos 1153	*700*	*1,543*	—	965	600	1,018	633	82·93	*1,000 years*	—
30 Jan	USSR	Cosmos 1154	*2,500*	*5,510*	—	629	391	643	400	81·22	*60 years*	—
7 Feb	USSR	Cosmos 1155	*6,000*	*13,230*	—	194	121	395	245	72·87	14 days	*Reconnaissance satellite.* Recovered.
7 Feb	USA	Big Bird	13,300	29,320	Titan 3D	224	139	502	312	96·97	—	Reconnaissance satellite.
9 Feb	USA	Navstar	433	955	Atlas F	162	101	20,136	12,512	63·08	Unlimited	Navigation satellite.
11 Feb	USSR	Cosmos 1156-1163	*40*	*88*	—	*1,400*	*870*	*1,474*	*916*	74·04	*10,000 years*	*Military communications satellites.*
12 Feb	USSR	Cosmos 1164	*1,250*	*2,756*	—	435	270	40,858	25,388	62·82	*12 years*	*Failed communications satellite?*
14 Feb	USA	SMM	2,315	5,103	Delta	563	350	570	354	28·51	*10 years*	Solar Maximum Mission satellite.
17 Feb	Japan	Tansei 4 (MS-T4)	185	408	Mu-3	518	322	720	447	38·68	*10 years*	Engineering test satellite.
20 Feb	USSR	Raduga 6	—	—	—	36,535	22,702	36,720	22,817	0·55	Unlimited	Communications satellite.
21 Feb	USSR	Cosmos 1165	*6,000*	*13,230*	—	170	106	353	219	72·88	13 days	*Reconnaissance satellite.* Recovered.
22 Feb	Japan	ECS 2 (Ayame)	260	573	—	213	132	35,592	22,116	24·53	Unlimited	Experimental communications satellite (failed).
3 Mar	USA	USAF	—	—	Atlas F	1,060	658	1,147	713	63·5	*2,000 years*	Military satellite.
4 Mar	USSR	Cosmos 1166	*6,000*	*13,230*	—	197	122	380	236	72·85	14 days	*Reconnaissance satellite.* Recovered.
14 Mar	USSR	Cosmos 1167	—	—	—	142	88	595	370	65·01	6 months	*Ocean survey satellite.*
17 Mar	USSR	Cosmos 1168	*700*	*1,543*	—	962	598	1,013	629	82·95	*1,000 years*	*Navigation satellite.*
27 Mar	USSR	Cosmos 1169	—	—	—	476	296	515	320	65·84	*2 years*	*Interception test satellite.*
27 Mar	USSR	Progress 8	7,000	15,430	—	180	112	260	162	51·60	29 days	Unmanned supply craft. Carried spares and materials to Salyut 6.
1 Apr	USSR	Cosmos 1170	*6,000*	*13,230*	—	170	106	371	231	70·39	11 days	*Reconnaissance satellite.* Recovered.
3 Apr	USSR	Cosmos 1171	—	—	—	966	600	1,009	627	65·84	*1,000 years*	*Target for Cosmos 1174.*
9 Apr	USSR	Soyuz 35	*6,500*	*14,330*	—	193	120	234	145	51·62	55 days	Crew: Leonid Popov and Valery Ryumin. Docked with Salyut 6. Spacecraft returned on 3 June with Valery Kubasov and Bertalan Farkash, crew of Soyuz 36.
12 Apr	USSR	Cosmos 1172	*1,000*	*2,205*	—	609	378	40,146	24,946	62·76	*12 years*	*Missile early warning satellite.*
17 Apr	USSR	Cosmos 1173	*6,000*	*13,230*	—	176	109	355	221	70·31	11 days	*Reconnaissance satellite.* Recovered.
18 Apr	USSR	Cosmos 1174	—	—	—	362	225	1,025	637	65·84	—	*Intercepted Cosmos 1171.*
18 Apr	USSR	Cosmos 1175	*1,500*	*3,307*	—	252	157	459	285	62·84	3 months	*Failed communications satellite.*
26 Apr	USA	Navstar	450	992	Atlas F	170	106	20,288	12,606	63·02	Unlimited	Navigation satellite.
27 Apr	USSR	Progress 9	*7,000*	*15,430*	—	185	115	255	158	51·65	25 days	Unmanned supply craft. Carried supplies to Salyut 6.

Date	Origin	Name	Total weight kg	lb	Launch Vehicle	Perigee km	miles	Apogee km	miles	Inclin-ation	Lifetime	Remarks
29 Apr	USSR	Cosmos 1176	—	—	—	250	155	265	165	65	600 years	Ocean survey satellite.
29 Apr	USSR	Cosmos 1177	6,000	13,230	—	165	102	341	212	67·16	44 days	Reconnaissance satellite. Recovered.
7 May	USSR	Cosmos 1178	6,000	13,230	—	196	122	390	242	72·84	15 days	Reconnaissance satellite. Recovered.
14 May	USSR	Cosmos 1179	700	1,543	—	301	187	1,550	963	82·97	500 years	Navigation satellite?
15 May	USSR	Cosmos 1180	6,000	13,230	—	238	148	290	180	62·81	12 days	Reconnaissance satellite. Recovered.
20 May	USSR	Cosmos 1181	700	1,543	—	975	606	1,007	626	82·95	1,000 years	Navigation satellite?
23 May	USSR	Cosmos 1182	6,000	13,230	—	205	127	215	134	82·32	13 days	Reconnaissance satellite. Recovered.
26 May	USSR	Soyuz 36	6,600	14,550	—	191	119	265	165	51·60	66 days	Crew: Valery Kubasov and Bertalan Farkash. Docked with Salyut 6. Spacecraft returned on 31 July with Victor Gorbatko and Pham Tuan, crew of Soyuz 37.
28 May	USSR	Cosmos 1183	6,000	13,230	—	199	124	387	240	72·80	14 days	Reconnaissance satellite. Recovered.
29 May	USA	NOAA 7	1,405	3,097	Atlas F	265	165	1,447	899	92·23	1 year	Meteorological satellite. Failed.
4 June	USSR	Cosmos 1184	2,500	5,510	—	620	385	647	402	81·23	60 years	Electronic ferret satellite.
5 June	USSR	Soyuz T-2	7,000	15,430	—	195	121	231	144	51·63	4 days	Manned test of improved Soyuz. Crew: Yuri Malyshev and Vladimir Aksyonov. Docked with Salyut 6.
6 June	USSR	Cosmos 1185	6,000	13,230	—	212	132	282	175	82·33	14 days	Earth resources satellite.
6 June	USSR	Cosmos 1186	550	1,212	—	471	293	518	322	74·03	18 months	–
12 June	USSR	Cosmos 1187	6,000	13,230	—	198	123	305	189	72·86	14 days	Reconnaissance satellite. Recovered.
14 June	USSR	Horizont 4	5,000	11,023	—	36,382	22,606	36,637	22,765	0·96	Unlimited	Statsionar communications satellite.
14 June	USSR	Cosmos 1188	1,250	2,755	—	609	378	40,128	24,934	62·92	12 years	Missile early warning satellite?
18 June	USSR	Meteor 1 (30)	2,200	4,850	—	583	362	665	413	97·94	60 years	Earth resources and meteorological satellite.
18 June	USA	Big Bird	13,300	29,320	Titan IIID	168	104	265	165	96·45	6 months	Reconnaissance satellite.
		USAF			Titan IIID	1,330	826	1,331	827	96·63	5,000 years	Carried by Big Bird.
21 June	USSR	Molniya-1 (47)	1,000	2,205	—	630	391	40,702	25,291	62·83	12 years	Communications satellite.
26 June	USSR	Cosmos 1189	6,000	13,230	—	197	122	302	188	72·87	14 days	Reconnaissance satellite. Recovered.
29 June	USSR	Progress 10	7,000	15,430	—	183	114	264	164	51·62	20 days	Unmanned supply craft. Carried supplies to Salyut 6.
1 July	USSR	Cosmos 1190	750	1,653	—	792	492	804	500	74·05	120 years	—
2 July	USSR	Cosmos 1191	—	—	—	599	372	40,130	24,936	62·65	12 years	Missile early warning satellite.
9 July	USSR	Cosmos 1192-1199	40	88	—	1,397	868	1,474	916	74·02	10,000 years	Military communications satellites.
9 July	USSR	Cosmos 1200	6,000	13,230	—	202	125	302	188	72·81	14 days	Reconnaissance satellite. Recovered.
14 July	USSR	Ekran 5	2,000	4,410	—	In geostationary orbit					Unlimited	Domestic TV relay satellite.
15 July	USSR	Cosmos 1201	6,000	13,230	—	211	131	246	153	82·33	13 days	Reconnaissance satellite. Recovered.
18 July	India	Rohini	35	77	SLV-3	306	190	919	571	44·75	3 years	First Indian-built and launched satellite.
18 July	USSR	Molniya-3 (13)	1,500	3,307	—	457	284	40,817	25,362	62·79	12 years	—
23 July	USSR	Soyuz 37	6,500	14,330	—	190	118	273	170	51·58	80 days	Crew: Victor Gorbatko and Pham Tuan. Docked with Salyut 6. Spacecraft returned on 11 October with Leonid Popov and Valery Ryumin, crew of Soyuz 35.
24 July	USSR	Cosmos 1202	6,000	13,230	—	197	122	306	190	72·85	14 days	Reconnaissance satellite. Recovered.
31 July	USSR	Cosmos 1203	6,000	13,230	—	212	132	268	167	82·32	14 days	Reconnaissance satellite. Recovered.
31 July	USSR	Cosmos 1204	550	1,212	—	345	214	538	334	50·66	1 year	—
12 Aug	USSR	Cosmos 1205	6,000	13,230	—	198	123	305	189	72·82	14 days	Reconnaissance satellite. Recovered.
15 Aug	USSR	Cosmos 1206	2,500	5,512	—	629	391	631	392	81·20	60 years	Reconnaissance satellite.
22 Aug	USSR	Cosmos 1207	6,000	13,230	—	209	130	255	158	82·32	13 days	Reconnaissance satellite. Recovered.
26 Aug	USSR	Cosmos 1208	6,000	13,230	—	171	106	336	209	67·14	—	Reconnaissance satellite.
3 Sep	USSR	Cosmos 1209	6,000	13,230	—	259	161	272	169	82·34	14 days	Reconnaissance satellite. Recovered.
9 Sep	USSR	Meteor 2 (6)	2,500	5,512	—	846	526	892	554	81·25	500 years	Meteorological satellite.
9 Sep	USA	GOES 4	627	1,382	Delta	34,257	21,286	39,965	24,833	0·33	Unlimited	Meteorological satellite.
18 Sep	USSR	Soyuz 38	6,800	14,991	—	195	121	296	184	51·60	8 days	Crew: Yuri Romanenko and Arnoldo Tamayo Mendez. Docked with Salyut 6.
19 Sep	USSR	Cosmos 1210	6,000	13,230	—	179	111	243	151	82·29	14 days	Earth resources satellite.
23 Sep	USSR	Cosmos 1211	5,500	12,120	—	215	134	240	149	82·35	11 days	Earth resources satellite.
26 Sep	USSR	Cosmos 1212	6,000	13,230	—	208	129	247	153	82·34	13 days	Earth resources satellite.
28 Sep	USSR	Progress 11	7,000	15,432	—	189	117	247	153	51·63	74 days	Unmanned supply craft. Carried supplies to Salyut 6.
3 Oct	USSR	Cosmos 1213	6,000	13,230	—	195	121	316	196	72·87	—	—
5 Oct	USSR	Raduga 7	5,000	11,023	—	35,737	22,206	35,883	22,297	0·59	Unlimited	Statsionar communications satellite.
10 Oct	USSR	Cosmos 1214	—	—	—	172	107	345	214	67·15	13 days	Reconnaissance satellite. Recovered.
14 Oct	USSR	Cosmos 1215	750	1,653	—	496	308	549	341	74·04	10 years	—
16 Oct	USSR	Cosmos 1216	6,000	13,230	—	193	120	387	240	72·86	14 days	Reconnaissance satellite. Recovered.
24 Oct	USSR	Cosmos 1217	1,000	2,205	—	594	369	40,133	24,937	62·92	12 years	Missile early warning satellite.
30 Oct	USSR	Cosmos 1218	6,000	13,230	—	170	106	353	219	64·89	—	—
31 Oct	USA	FLTSATCOM 4	1,860	4,100	Atlas-Centaur	35,032	21,768	36,236	22,516	2·46	—	Military communications satellite.
31 Oct	USSR	Cosmos 1219	6,000	13,230	—	193	120	326	203	72·86	13 days	Reconnaissance satellite. Recovered.
4 Nov	USSR	Cosmos 1220	—	—	—	428	266	444	276	65·04	—	Manoeuvrable reconnaissance satellite.
12 Nov	USSR	Cosmos 1221	6,000	13,230	—	195	121	398	247	72·90	14 days	Reconnaissance satellite. Recovered.
15 Nov	USA	SBS 1	550	1,212	Delta	35,781	22,233	35,793	22,240	0·42	Unlimited	Commercial communications satellite.
16 Nov	USSR	Molniya-1 (48)	1,000	2,205	—	628	390	40,636	25,250	62·82	12 years	Communications satellite.
21 Nov	USSR	Cosmos 1222	2,500	5,512	—	629	391	631	392	81·22	60 years	Electronic ferret satellite.
27 Nov	USSR	Soyuz T-3	6,600	14,550	—	196	122	236	147	51·63	13 days	Crew: Leonid Kizim, Oleg Makarov, Gennady Strekalov.
27 Nov	USSR	Cosmos 1223	1,000	2,205	—	610	379	40,183	24,968	62·87	12 years	Missile early warning satellite.
1 Dec	USSR	Cosmos 1224	6,000	13,230	—	197	122	377	234	72·87	—	—
5 Dec	USSR	Cosmos 1225	700	1,543	—	948	589	1,029	145	82·94	1,000 years	Navigation satellite.
6 Dec	USA	Intelsat 5A (F2)	1,950	4,300	Atlas-Centaur	35,638	22,144	35,734	22,204	0·86	Unlimited	Communications satellite.
10 Dec	USSR	Cosmos 1226	700	1,543	—	964	599	1,012	629	82·94	1,000 years	Navigation satellite.
13 Dec	USA	USAF	—	—	—	—	—	—	—	—	—	Military satellite.
16 Dec	USSR	Cosmos 1227	6,000	13,230	—	228	142	299	186	72·84	12 days	Reconnaissance satellite.
23 Dec	USSR	Cosmos 1228-35	40	88	—	1,400	870	1,464	910	74·00	9,000 years	Military communications satellites.
25 Dec	USSR	Prognoz 8	950	2,094	—	979	608	197,369	122,639	65·83	12 years	Magnetospheric observatory.
26 Dec	USSR	Ekran 6	2,000	4,410	—	35,737	22,206	35,832	22,265	0·06	Unlimited	Domestic TV relay satellite.
26 Dec	USSR	Cosmos 1236	6,000	13,230	—	170	106	364	226	67·13	—	Reconnaissance satellite.

AERO-ENGINES

AUSTRALIA

CAC
COMMONWEALTH AIRCRAFT CORPORATION LIMITED

Head Office and Works: GPO Box 779H, Melbourne 3001, Victoria
Telephone: (03) 647 6111
Officers: See Aircraft section

This company built the Rolls-Royce Avon and Viper and SNECMA Atar 9C turbojets, and supports these engines in service with the RAAF.

BELGIUM

FN
FABRIQUE NATIONALE HERSTAL SA

Head Office and Works: B-4400 Herstal
Telephone: (041) 64 08 00

FN has been making and overhauling jet engines since 1949. Its major current programme involves the production of components for the Pratt & Whitney F100-PW-100 fan module and core engine module for Belgium, Denmark, the Netherlands and Norway, and also for the US Air Force and for third-party countries; assembly and test of these modules; and final building and testing of the complete engines for the European consortium and for third-party countries. The first engine ran at FN in July 1978, and deliveries to F-16 production centres began two months later.

Participation in the Larzac jet engine programme involved final building and testing of the engines for Alpha Jet trainers supplied to the Belgian Air Force.

FN is also a member of the consortium which has put the Rolls-Royce Tyne turboprop back into production to power the Transall C-160 (second series) and Dassault-Breguet Atlantic ANG.

FN's share in the CFM56 engine amounts to 5 per cent of the 50 per cent work-split due to SNECMA in this SNECMA-General Electric joint programme.

FN is still making parts for major engine manufacturers including Rolls-Royce (marine Tyne, Dart), SNECMA (Atar 9C, M53) and Pratt & Whitney (JT8D). FN also makes spare parts for the General Electric J79-11A and SNECMA Atar 9C, and for the FN-Boeing 553 gas turbine powering Swedish S tanks.

CANADA

ORENDA
HAWKER SIDDELEY CANADA INC, Orenda Division

PO Box 6001, Toronto AMF, Ontario L5P 1B3
Telephone: (416) 677 3250
Telex: 06-968620 and 06-968727
Vice-President and General Manager: D. J. Caple
Director of Engineering: B. A. Avery
Director of Operations: R. J. Munro
Director of Marketing: C. Johnson
Director of Finance: K. R. Church

Orenda has a 67,262 m² (724,000 sq ft) facility close to Toronto International Airport. It meets the manufacturing, repair, overhaul and technical support needs of the Canadian Forces for J79 and J85 engines, and for future engines, and performs subcontract manufacture of parts for aero engines currently in production, and for future developments. Orenda also supplies parts to Belgium, Canada, West Germany, Italy, Netherlands, Norway, Pakistan, the USA and Venezuela.

P&WC
PRATT & WHITNEY AIRCRAFT OF CANADA LTD

Head Office and Works: PO Box 10, Longueuil, Quebec J4K 4X9
Telephone: (514) 677 9411
President and Chief Executive Officer: E. L. Smith
Executive Vice-President: L. D. Caplan
Vice-Presidents:
- K. H. Sullivan (Marketing)
- G. P. Ouimet (Finance)
- J. P. Beauregard (Materials)
- A. L. Tontini (Personnel)
- R. H. Guthrie
- E. H. Schweitzer (Commuter Operations)
- C. B. Wrong (Engineering)
- R. J. Losch (Product Support)
- J. B. Haworth (Industry & Marine)
- C. P. Abraham (Production)
- C. J. C. Pascoe (Counsel)

Pratt & Whitney Aircraft of Canada is a major subsidiary of United Technologies Corporation, Connecticut, USA, and is the UTC company responsible for engines for general aviation.

Original turbine work by the company was initiated in 1957 by the concept and preliminary design of the JT12 (J60) turbojet, development and manufacture of which were completed subsequently by Pratt & Whitney Aircraft. Design, development and manufacture of the PT6, ST6, PT6T, JT15D and PW100 series of small turbine aero-engines represents more than 70 per cent of the company's activities.

P&WC is owned 97 per cent by the United Technologies Corporation. Approximately 7 per cent of its sales are linked to defence requirements. It occupies more than 153,300 m² (1·65 million sq ft) of space in three plants and employs more than 7,700 persons.

In 1980 a total of 2,680 new engines were delivered. By 1 January 1981 P&WC had delivered more than 18,000 PT6A and 6B engines, and over 1,960 JT15D turbofans.

P&WC JT15D-1 two-spool turbofan engine

P&WC JT15D

Following a comprehensive performance study of small turbofan engines carried out by P&WC during 1965, detail design of a definitive engine, the 8·9-11·12 kN (2,000 lb to 2,500 lb st) JT15D was initiated in June 1966. First run of the new turbofan was on 23 September 1967.

Designed to power business aircraft, small transports and training aircraft in the 8,000 lb to 12,500 lb AUW category, the JT15D is an advanced technology two-spool front-fan engine having a minimum number of aerodynamic components. Major design features include a significant improvement in sfc, and simplicity of construction to ensure low first cost and maintenance costs. Other advantages are low noise levels, ease of handling, and the attainment of airline standards of reliability.

Initial application for the JT15D was the twin-engined Cessna Citation. Flightworthy prototype engines were delivered in August 1969 for the Citation's first flight in mid-September. Up to 1976 Cessna produced the Citation powered by the **JT15D-1**. Late that year it announced the Citation I powered by the improved-performance **JT15D-1A** and the Citation II powered by the **JT15D-4**. More than 1,254 JT15D-1/D-1A and 592 JT15D-4 engines had been delivered for these aircraft by January 1981.

Other twin-engined business jets powered by the JT15D-4 are the Aérospatiale Corvette and Mitsubishi Diamond I. More than 116 engines had been delivered for these aircraft by January 1981. TBO is 3,000 h for the JT15D-1/D-1A, and 2,000 h for the JT15D-4. By January 1981 the total operating time was 3,115,080 h.

The **JT15D-4C** is derived from the basic D-4 model. Major differences include the incorporation of an aerobatic oil system for sustained inverted flight, and a full throttle electronic supervision fuel control unit. Engine certification is scheduled for 1982.

The **JT15D-5** is a growth version of the JT15D-4. A new fan with higher pressure ratio and flow, plus an improved boost stage and HP compressor, are combined to produce 25 per cent more altitude cruise thrust, with a 3 per cent improvement in specific fuel consumption. HP turbine blades and electronic fuel control are also improved. JT15D-5 development began in 1977, and first flight was in April 1978. Certification is planned for early 1983.

The following description relates to the JT15D-1 (JT15D-4 features in brackets):

Type: Two-shaft turbofan.

Air Intake: Direct pitot intake without inlet guide vanes. Hot-air anti-icing for nose bullet.

Fan: Single-stage axial fan, aerodynamically related to that of the JT9D but on a much smaller scale. Forged disc fitted with 28 solid titanium blades secured by dovetail fixings riveted to disc. Blades have part-span shrouds. Casing, which forms the engine air intake, of forged stainless steel. Circular splitter ring behind fan, held between two rows of 33 inner wrapped-sheet stators and single row of 66 outer stator blades. Total air mass flow, 34 kg (75 lb)/s; bypass ratio about 3·3:1; bypass flow typically 26 kg (57·5 lb)/s; primary core air flow 8 kg (17·5 lb)/s; fan pressure ratio 1·5:1.

COMPRESSOR: Primary airflow enters eye of single-stage titanium centrifugal compressor. Single-sided impeller, with 16 full vanes and 16 splitter vanes, secured to shaft by special bolt and key-washer. Two-piece casing with diffuser in form of pipes containing straightening vanes. Overall pressure ratio almost 10:1. (JT15D-4 compressor airflow augmented by axial boost stage between fan and compressor.)

COMBUSTION CHAMBER: Annular reverse-flow type. Outer casing of heat-resistant steel; flame tube of nickel alloy, supported on low-pressure turbine stator assembly. Spark igniters at 5 and 7 o'clock (viewed from rear).

FUEL SYSTEM: Engine-driven sandwich-mounted pump delivering through FCU, flow divider and dual manifolds at 44·8 bars (650 lb/sq in); DP-L2 pneumatic control unit on pump, with dual metering valve.

FUEL GRADES: JP-1, JP-4, JP-5 to PWA Spec. 522.

NOZZLE GUIDE VANES: High-pressure ring of 15, air-cooled, integrally cast in cobalt alloy.

TURBINE: Single-stage HP with 71 solid blades held in fir-tree roots. Two-stage LP with nickel alloy discs, first stage being cast integrally with 61 blades and second stage carrying 55 blades in fir-tree roots. LP fan shaft drives fan, with ball thrust bearing behind fan and roller gear and intershaft bearings; HP shaft drives centrifugal compressor, with front ball thrust bearing and rear roller bearing. Gas temperature 960°C before turbine, 562°C after turbine.

JET PIPE: Nickel alloy cone and sheet-metal pipe. Provision made for adjusting the area to match engines and to trim performance.

ACCESSORY DRIVES: Package under front of engine driven by power offtake from front of HP shaft.

LUBRICATION SYSTEM: Integral oil system, with gear-type pump delivering at up to 5·52 bars (80 lb/sq in). Capacity 9·0 litres (2·4 US gallons; 2·0 Imp gallons).

OIL SPECIFICATION: PWA521 Type II.

MOUNTING: Hard or soft, according to customers' choice. Four main pads on front casing, arranged two on each side at 30° above and below horizontal. One rear mount at top or on either side of centreline.

STARTING: Air-turbine starter or electric starter/generator.

DIMENSIONS:

Diameter:

JT15D-1, -1A — 691 mm (27·2 in)

JT15D-4, -4C, -5 — 686 mm (27·0 in)

Length overall:

JT15D-1, -1A — 1,506 mm (59·3 in)

JT15D-4, -4C, -5 — 1,600 mm (63·0 in)

Frontal area — 0·37 m² (4 sq ft)

WEIGHT, EQUIPPED:

JT15D-1, -1A — 232·5 kg (514 lb)

JT15D-4 — 253 kg (557 lb)

JT15D-4C — 261 kg (575 lb)

JT15D-5 — 291·5 kg (642 lb)

PERFORMANCE RATINGS:

T-O:

JT15D-1, -1A — 9·8 kN (2,200 lb st)

JT15D-4, -4C — 11·12 kN (2,500 lb st)

JT15D-5 — 12·89 kN (2,900 lb st)

Max continuous:

JT15D-1, -1A — 9·3 kN (2,090 lb st)

JT15D-4 — 10·56 kN (2,375 lb st)

JT15D-4C — 9·45 kN (2,125 lb st)

JT15D-5 — 12·89 kN (2,900 lb st)

SPECIFIC FUEL CONSUMPTION (T-O):

JT15D-1, -1A — 15·30 mg/Ns (0·540 lb/h/lb st)

JT15D-4, -4C, -5 — 15·92 mg/Ns (0·562 lb/h/lb st)

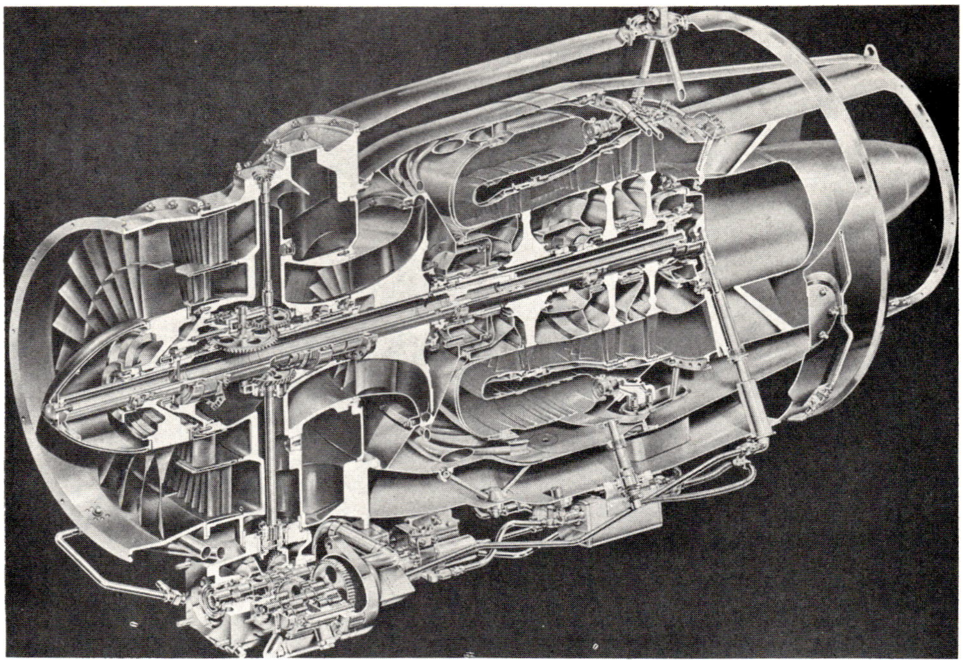

Cutaway drawing of the P&WC JT15D-1 turbofan (9·8 kN; 2,200 lb st)

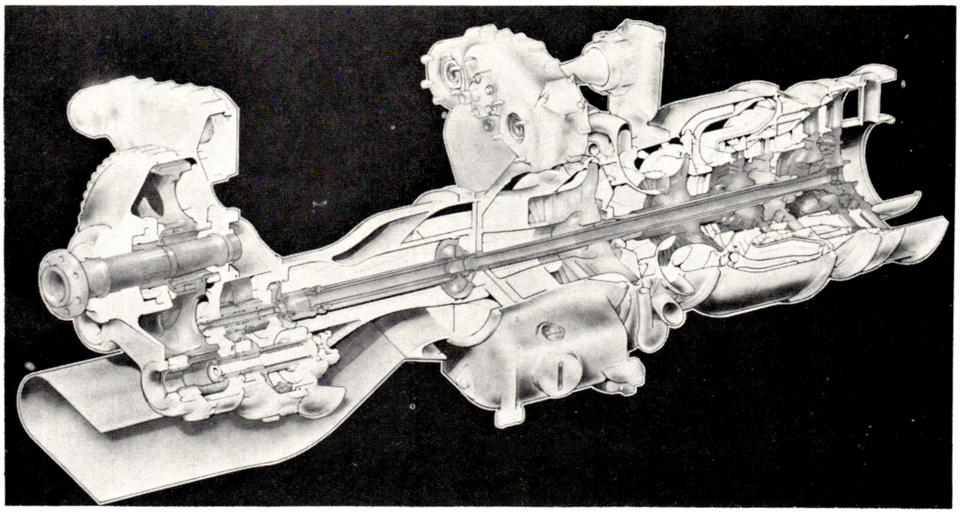

Simplified cutaway drawing of the P&WC PW100 three-shaft turboprop

P&WC PW100

The PW100 (originally PT7A) is an advanced-technology turboprop designed for the 30/40-passenger short-haul transports and executive aircraft being developed for introduction in the 1980s. An all-new free-turbine engine, the PW100 is claimed to offer 20 per cent improvement in specific fuel consumption over current engines. Features include electric starting, electronic fuel management, compatibility with new lightweight propellers, and a unique ground APU capability. The internal configuration of LP and HP centrifugal compressors each driven by a single-stage turbine, and a two-stage free power turbine, is seen in an accompanying cutaway drawing.

Principal versions of the PW100 are as follows:

PW102A. Flat rated at 1,178 ekW; 1,119 kW (1,580 ehp; 1,500 shp) at 1,300 propeller rpm to 37·2°C, this version is scheduled for certification in 1983. Selected for EMBRAER EMB-120 Brasilia.

PW108A. Reserve power rating at 1,566 ekW; 1,491 kW (2,100 ehp; 2,000 shp) at 1,200 propeller rpm to 27·7°C. Alternative T-O at 1,411 ekW; 1,342 kW (1,892 ehp; 1,800 shp). Certification scheduled for 1983. Selected for de Havilland Canada Dash-8.

P&WC PW100 three-shaft turboprop, in the 1,491 kW (2,000 shp) class

DIMENSIONS:

Length: PW102A — 2,057 mm (81 in)

PW108A — 2,134 mm (84 in)

Width: PW 102A — 635 mm (25 in)

PW108A — 635 mm (25 in)

Height: PW102A — 787 mm (31 in)

PW108A — 787 mm (31 in)

WEIGHT, DRY:

PW102A — 381·5 kg (841 lb)

PW108A — 413·7 kg (912 lb)

PERFORMANCE RATINGS (SL, static):

T-O: See under model listings

Max continuous:

PW102A

1,178 ekW; 1,119 kW (1,580 ehp; 1,500 shp) at 1,300 rpm to 37·2°C

PW108A

1,333 ekW; 1,268 kW (1,787 ehp; 1,700 shp) at 1,200 rpm to 32·7°C

Max cruise:

PW102A

1,178 ekW; 1,119 kW (1,580 ehp; 1,500 shp) at 1,300 rpm to 20°C

PW108A

1,271 ekW; 1,207 kW (1,704 ehp; 1,619 shp) at 1,200 rpm to 15°C

SPECIFIC FUEL CONSUMPTION:

T-O rating:

PW102A — 91·9 μg/J (0·544 lb/ehp/h)

PW108A — 84·5 μg/J (0·500 lb/ehp/h)

PW108A alternative T-O — 86·9 μg/J (0·514 lb/ehp/h)

P&WC PT6A

The PT6A is a free-turbine turboprop, built in many versions. By January 1981 more than 13,575 had logged 42,000,000 h in some 5,500 aircraft registered in 120 countries.

An experimental PT6 ran for the first time in November 1959 and flight trials in the nose of a Beech 18 began in

May 1961. Civil certification of the first production model, the 578 ehp PT6A-6, was granted in late 1963. Progressively higher rated versions have followed to power a wide variety of aircraft. In September 1977 P&WC introduced the PT6A-10 series, with ratings extending down to 354 kW (475 shp) for light and agricultural aviation.

Principal versions of the PT6A are as follows:

PT6A-6. Flat rated at 431 ekW; 410 kW (578 ehp; 550 shp) at 2,200 propeller rpm to 21°C, this version received civil certification in December 1963. A total of 350 PT6A-6s were built between then and November 1965. Among aircraft powered are de Havilland Canada Turbo-Beaver and early DHC-6 Twin Otter Series 100.

PT6A-10. Flat rated at 374 ekW; 354 kW (502 ehp; 475 shp) at 2,200 propeller rpm to 38°C. Certification expected early 1981.

PT6A-11. Flat rated at 394 ekW; 373 kW (528 ehp; 500 shp) at 2,200 propeller rpm to 42°C. Certificated December 1977. Fitted to Piper Cheyenne I.

PT6A-110. Flat rated at 374 ekW; 354 kW (502 ehp; 475 shp) at 1,900 propeller rpm to 38°C. Certificated January 1980. Fitted to Dornier 128-6.

PT6A-11AG. Flat rated at 394 ekW; 373 kW (528 ehp; 500 shp) at 2,200 propeller rpm to 42°C. Certificated in May 1979, it embodies design features for agricultural aviation, including operation on diesel fuel. Fitted to Ayres Turbo-Thrush and Weatherly 620 TP.

PT6A-112. Flat rated at 394 ekW; 373 kW (528 ehp; 500 shp) at 1,900 propeller rpm to 56°C. Certificated October 1978. Fitted to Cessna Corsair.

PT6A-15AG. Flat rated at 533 ekW; 507 kW (715 ehp; 680 shp) at 2,200 propeller rpm to 22°C. Certificated October 1977, including operation on diesel fuel. Embodies design features for agricultural aviation. Fitted to Ayres Turbo-Thrush, Frakes Turbo-Cat, Schweizer Turbo Ag-Cat D and Air Tractor AT-400.

PT6A-20. Flat rated at 432 ekW; 410 kW (579 ehp; 550 shp) at 2,200 propeller rpm to 21°C, the -20 offered improved reliability and increases in max continuous, max climb and max cruise power ratings over the PT6A-6. The PT6A-20 was certificated in October 1965. Between then and 1974 approximately 2,400 were built to power the Beechcraft King Air B90, Beechcraft Model 99, prototypes of the EMBRAER EMB-110 Bandeirante, de Havilland Canada DHC-6 Twin Otter Series 100 and 200, James Aviation (Fletcher FU-24) conversion, Marshall of Cambridge (Grumman) Goose conversion, McKinnon G-21C and G-21D Turbo-Goose (Grumman Goose) conversions, Pilatus PC-6/B1-H2 Turbo-Porter, prototype Pilatus PC-7 Turbo Trainer and the Swearingen Merlin IIA (which can be re-engined with the PT6A-27).

PT6A-20A. Similar to A-20; fitted to early Beechcraft King Air C90.

PT6A-21. Flat rated at 432·5 ekW; 410 kW (580 ehp; 550 shp) at 2,200 propeller rpm to 21°C, the A-21 offers improved fuel consumption and reliability, mainly by mating the A-27 power unit with the A-20A gearbox. Certificated on 10 December 1974. Fitted to current Beechcraft King Air C90. Total deliveries of -20, -20A and -21 exceed 2,990.

PT6A-25. Flat rated at 432·5 ekW; 410 kW (580 ehp; 550 shp) at 2,200 propeller rpm to 33°C. Special oil system for sustained inverted flight. Certificated May 1976. Fitted to Beechcraft T-34C. Selected for NDN-5 Firecracker.

PT6A-25A. Identical to -25 except for certain castings being made of magnesium alloy instead of aluminium alloy. Fitted to Pilatus PC-7.

PT6A-25C. Flat rated at 584 ekW; 559 kW (783 ehp; 750 shp) at 2,200 propeller rpm to 31°C. Basically an A-25 with standard A-34 hot end and A-27 first-stage reduction gearing. The A-25C, whilst maintaining aerobatic capabilities, offers an increased power rating over that of the A-25 and A-25A. Certification expected late 1981. Selected to power EMBRAER EMB-312.

PT6A-27. Flat rated at 553 ekW; 507 kW (715 ehp; 680 shp) at 2,200 propeller rpm to 22°C, attained by 12½ per cent increase in mass flow provided by larger-diameter compressor, at lower turbine temperatures than in PT6A-20. Production began in November 1967 and 2,083 had been delivered by January 1981. Production continues. Applications include the Hamilton Westwind II/III (Beech 18) conversions, Beechcraft Model 99, Beechcraft Model 99A, Beechcraft U-21A and U-21D, de Havilland Canada DHC-6 Twin Otter Series 300, Pilatus/Fairchild Industries PC-6/B2-H2 Porter, Frakes Aviation (Grumman) Mallard conversion, IAI Arava, Let L-410A Turbolet, Saunders Aircraft ST-27A (de Havilland Heron) conversion and EMBRAER EMB-110 Bandeirante and EMB-312 single-engined military trainer.

The 553 ekW (715 ehp) P&WC PT6A-27 free-turbine turboprop

PT6A-28. Similar to the PT6A-27 and with the same T-O and max continuous ratings, this version has an additional normal cruise rating of 562 ehp available up to 21°, corresponding to the max cruise rating conditions of the -27. In addition the max cruise rating of the -28 gives 652 ehp up to the higher ambient of 33°C. This model continues in production, with 2,463 engines delivered by the beginning of 1981. Applications are Beechcraft King Air E90 and A100, Piper Cheyenne II and EMBRAER Xingu.

PT6A-34. Flat rated at 584 ekW; 559 kW (783 ehp; 750 shp) at 2,200 propeller rpm to 31°C, this version has aircooled nozzle guide vanes to allow operation at higher turbine entry temperatures. For the IAI Arava, Saunders ST-28, Frakes Aviation (Grumman) Mallard conversion, Jetstream conversion and EMBRAER EMB-111.

PT6A-34B. Identical to -34 except for aluminium alloy replacing magnesium in major castings. Fitted to Beechcraft T-44A.

PT6A-34AG. First model intended specifically for agricultural use, and certificated on diesel fuel. Retrofitted to Frakes conversion of Ag-Cat and Ayres Turbo-Thrush. Selected for PZL-106AT Turbo-Kruk.

PT6A-135. Flat rated at 587 ekW; 559 kW (787 ehp; 750 shp) at 1,900 rpm. Changed drive ratio reduces propeller noise; hot-end modifications to permit higher cycle temperatures. Certificated July 1977. Fitted to Jetcrafters Taurus and Beechcraft King Air C100. Selected for Schafer Comanchero conversion.

PT6A-36. Flat rated at 586 ekW, 559 kW (786 ehp, 750 shp) at 2,200 rpm to 36°C. Similar to -34 but higher ratings. Certificated June 1977. Fitted to IAI Arava. Selected for Beechcraft C99.

PT6A-38. Derated A-41, flat rated at 597 ekW; 559 kW (801 ehp; 750 shp) to 39°C. Certificated May 1975. Installed in Beechcraft C-12A.

PT6A-41. Higher mass flow, aircooled stage-one turbine nozzle guide vanes and two-stage free turbine. T-O rating of 673 ekW; 634 kW (903 ehp; 850 shp) at 2,000 propeller rpm, available up to 41°C. Thermodynamic power is 812 ekW (1,089 ehp). By the beginning of 1981 more than 2,000 engines had been delivered for Beechcraft Super King Air 200, Piper Cheyenne III and Gulfstream American Hustler.

PT6A-42. Identical to A-41, but with detailed improvements to give approx 10 per cent increase in cruise performance. Certificated August 1979. Selected for Beechcraft Super King Air B200.

PT6A-45A. Similar to PT6A-41 but with redesigned gearbox to transmit higher powers at reduced propeller speeds. Rated at 916 ekW; 875 kW (1,229 ehp; 1,173 shp) at 1,700 rpm to 8°C, or to 21°C with water injection. Certificated February 1976. Powers Shorts 330 and Mohawk 298.

PT6A-45B. Identical to PT6A-45A, but with increased water injection to give improved thermodynamic performance. Rated at 916 ekW; 875 kW (1,229 ehp; 1,173 shp) at 1,700 rpm to 11°C or to 30°C with water injection. Certificated March 1979. Powers Shorts 330.

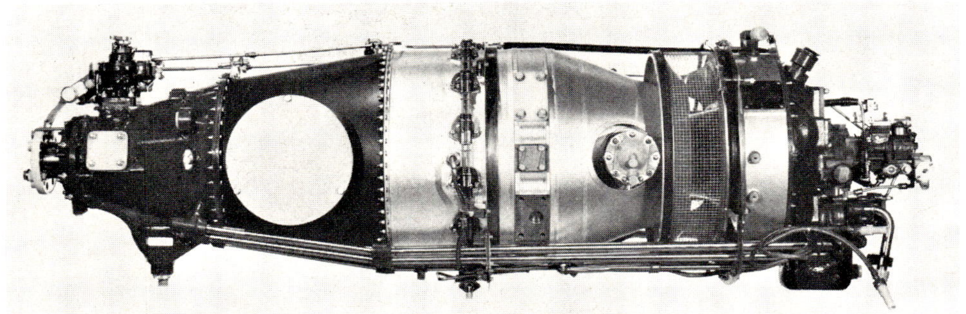

The 916 ekW (1,229 ehp) P&WC PT6A-45B free-turbine turboprop

PT6A-45R. PT6A-45B with added reserve power rating and deleted water/methanol injection system. Reserve power rated at 935 ekW; 893 kW (1,254 ehp; 1,197 shp) at 1,700 rpm to 23°C. Alternative T-O at 916 ekW; 875 kW (1,229 ehp; 1,173 shp) at 1,700 rpm to 11°C. Certificated 1980.

PT6A-50. Similar to PT6A-41 with a longer, higher-ratio reduction gear to give lower propeller tip speed for quieter operation at T-O. Rating at T-O is 875·5 ekW; 835 kW (1,174 ehp; 1,120 shp) available with water injection up to 34°C at 1,210 propeller rpm. Certificated September 1976, and delivered same month for de Havilland Canada DHC-7.

PT6A-65R: Similar to PT6A-45R, with a new four-stage compressor with jet-flap intake, fuel control unit and fuel dump. Improved hot-end hardware and exhaust duct. Reserve power rated at 1,025 ekW; 965 kW (1,375 ehp; 1,294 shp) at 1,700 rpm to 31°C. Alternative T-O at 931 ekW; 875 kW (1,249 ehp; 1,173 shp) at 1,700 rpm to 26°C. Certification scheduled for 1982.

PT6A-65B: Identical to PT6A-65R, but without reserve power rating. Flat rated at 931 ekW; 875 kW (1,249 ehp; 1,173 shp) at 1,700 rpm to 43°C. Certification scheduled for 1982. Selected to power Beechcraft 1900.

Current TBO is 3,500 h on all certificated models except 11AG, -15AG, -25, -25A, -38, -41, -42, -45A, -45B and 45R for which TBO is 3,000 h, and -50, for which TBO is 2,000 h.

The following data apply generally to the PT6A series:

Type: Free-turbine axial-plus-centrifugal turboprop engine.

Propeller Drive (all models up to and including PT6A-41): Two-stage planetary gear train. Ratio 15 : 1. Rotation clockwise when viewed from rear. Drive from free turbine. Flanged propeller shaft. Plain bearings. Higher-ratio reduced gears developed for PT6A-45R, -50 and -65.

Air Intake: Annular air intake at rear of engine, with intake screen. Aircraft-supplied alcohol anti-icing system or inertial separation anti-icing system.

Compressor: Three axial-flow stages, plus single centrifugal stage (-65 series, four axial stages). Single-sided centrifugal compressor, with 26 vanes, made from titanium forging. Axial rotor of disc-drum type, with stainless steel stator and rotor blades. The stator vanes (44 first-stage, 44 second-stage, 40 third-stage) are brazed to casing. The rotor blades (16 first-stage, 32 second-stage and 32 third-stage) are dovetailed to discs. Discs through-bolted, with centrifugal compressor, to shaft. Fabricated one-piece stainless steel casing and radial diffuser. PT6A-27: compression ratio 6·7 : 1, air mass flow 3·1 kg/s (6·8 lb/s).

Combustion Chamber: Annular reverse-flow type of stainless steel construction, with 14 simplex burners around periphery of chamber. All versions up to A-34 have two glow plug igniters with option of two spark igniters; A-38 onwards, two spark igniters. PT6A-27 has one plug at 64° on starboard side of vertical centreline and one at 90° on port side.

Fuel System: Bendix DP-F2 pneumatic automatic fuel control system. Pneumatic computing section, fuel metering and regulating section, gas generator governor and free turbine governor. Primary and secondary flow manifolds with seven nozzles per manifold. PT6A-50 has DP-F3 with starting spill valve and motive flow systems.

Fuel Grade: Commercial jet fuels JP-1, JP-4, JP-5, MIL-J-5624. Use of aviation gasolines (MIL-G-5572) grades 80/87, 91/98, 100/130 and 115/145 permitted for a period of up to 150 h during any overhaul period.

Nozzle Guide Vanes: 29 nozzle guide vanes; A-34 onward, 14 aircooled HP vanes.

Turbines: Models up to A-34 have two single-stage axial; HP turbine (with 58 blades) drives compressor, and LP turbine (with 41 shrouded blades) drives output shaft. PT6A-38 onward have two-stage LP turbine. All blades have fir-tree root fixings.

Bearings: Each main rotor (gas generator and free turbine) supported by one ball and one roller anti-friction bearing.

Jet Pipe: Collector duct surrounding free-turbine shaft, exhaust through two ports on horizontal centreline.

Accessories: Mounting pads on accessory case (rear of engine) for starter/generator, hydraulic pump, aircraft accessory drive, vacuum pump and tachometer-generator. Mounting pad on the shaft-turbine reduction gear case for propeller overspeed governor, propeller constant-speed control unit and tachometer generator.

Lubrication System: One pressure and four scavenge elements in the pump stacks. All are gear type and are driven by the gas generator rotor. Engine has an integral oil tank with a capacity of 8·75 litres (2·3 US gallons). Oil supply pressure is 5·5 bars (80 lb/sq in) on PT6A-11 to -28, 5·85 bars (85 lb/sq in) on -34 to -36, and 7·25 bars (105 lb/sq in) on -38 to -65.

Oil Specification: CPW202, PWA522 Type II (7·5 cs vis) (MIL-L-23699, MIL-L-7808 for military engines).

Mounting: Up to A-34, three-point ring suspension. A-38 onward, four-point mounting, except -50 has base mounting.

Starting: Electric starter/generator on accessory case.

Dimensions:

Max diameter	483 mm (19 in)
Length, less accessories:	
PT6A-6 to -34	1,575 mm (62 in)
PT6A-38, -41	1,701 mm (67 in)
PT6A-45	1,829 mm (72 in)
PT6A-50	2,133 mm (84 in)
PT6A-65	1,880 mm (74 in)
Frontal area	0·18 m² (1·95 sq ft)

Weight, Dry:

PT6A-10, -11, -21, -27, -34, -36	137·4 kg (303 lb)
PT6A-11AG, -34, -36	141·1 kg (311 lb)
PT6A-110, -112	144·7 kg (319 lb)
PT6A-15AG	140·2 kg (309 lb)
PT6A-20	130 kg (286 lb)
PT6A-25	150·1 kg (331 lb)
PT6A-25A, -25C, -135	146·1 kg (322 lb)
PT6A-28	136 kg (300 lb)
PT6A-38, -41, -42	172 kg (380 lb)
PT6A-45A	192 kg (423 lb)
PT6A-45B, -45R	193 kg (425 lb)
PT6A-50	263 kg (580 lb)
PT6A-65B	208 kg (458 lb)
PT6A-65R	210 kg (463 lb)

Performance Ratings (S/L, static):

T-O rating:

See under model listings

Max continuous rating:

PT6A-10 374 ekW; 354 kW (502 ehp; 475 shp) at 2,200 rpm (to 38°C)
PT6A-110 374 ekW; 354 kW (502 ehp; 475 shp) at 1,900 rpm (to 38°C)
PT6A-11 394 ekW; 373 kW (528 ehp; 500 shp) at 2,200 rpm (to 42°C)
PT6A-11AG 394 ekW; 373 kW (528 ehp; 500 shp) at 2,200 rpm (to 42°C)
PT6A-112 394 ekW; 373 kW (528 ehp; 500 shp) at 1,900 rpm (to 56°C)
PT6A-15AG, -27, -28 533 ekW; 507 kW (715 ehp; 680 shp) at 2,200 rpm (to 22°C)
PT6A-20 432 ekW; 410 kW (579 ehp; 550 shp) at 2,200 rpm (to 21°C)
PT6A-21 432·5 ekW; 410 kW (580 ehp; 550 shp) at 2,200 rpm (to 33°C)
PT6A-25 432·5 ekW; 410 kW (580 ehp; 550 shp) at 2,200 rpm (to 33°C)
PT6A-25C 584 ekW; 559 kW (783 ehp; 750 shp) at 2,200 rpm (to 31°C)
PT6A-34 584 ekW; 559 kW (783 ehp; 750 shp) at 2,200 rpm (to 30°C)
PT6A-135 587 ekW; 559 kW (787 ehp; 750 shp) at 1,900 rpm (to 29°C)
PT6A-36 586 ekW; 559 kW (786 ehp; 750 shp) at 2,200 rpm (to 36°C)
PT6A-38 597 ekW; 559 kW (801 ehp; 750 shp) at 2,200 rpm (to 39°C)
PT6A-41 673 ekW; 634 kW (903 ehp; 850 shp) at 2,000 rpm (to 41°C)
PT6A-42 674 ekW; 634 kW (904 ehp; 850 shp) at 2,000 rpm (to 41°C)
PT6A-45A 798 ekW; 761 kW (1,070 ehp; 1,020 shp) at 1,700 rpm (to 27°C)
PT6A-45B, -45R 798 ekW; 761 kW (1,070 ehp; 1,020 shp) at 1,700 rpm (to 29°C)
PT6A-50 762 ekW; 725·5 kW (1,022 ehp; 973 shp) at 1,210 rpm (to 32°C)
PT6A-65B 931 ekW; 875 kW (1,249 ehp; 1,173 shp) at 1,700 rpm (to 38°C)
PT6A-65R 931 ekW; 875 kW (1,249 ehp; 1,173 shp) at 1,700 rpm (to 26°C)

Max cruise rating:

PT6A-10 374 ekW; 354 kW (502 ehp; 475 shp) at 2,200 rpm (to 19°C)
PT6A-110 374 ekW; 354 kW (502 ehp; 475 shp) at 1,900 rpm (to 19°C)
PT6A-11 394 ekW; 373 kW (528 ehp; 500 shp) at 2,200 rpm (to 37°C)
PT6A-11AG 394 ekW; 373 kW (528 ehp; 500 shp) at 2,200 rpm (to 36°C)
PT6A-112 394 ekW; 373 kW (528 ehp; 500 shp) at 1,900 rpm (to 48°C)
PT6A-15AG as PT6A-27
PT6A-20 389 ekW; 369 kW (522 ehp; 495 shp) at 2,200 rpm
PT6A-21 390 ekW; 309 kW (523 ehp; 495 shp) at 2,200 rpm (to 15°C)
PT6A-25 432·5 ekW; 410 kW (580 ehp; 550 shp) at 2,200 rpm (to 33°C)
PT6A-25C 545 ekW; 522 kW (731 ehp; 700 shp) at 2,200 rpm (to 19°C)
PT6A-27 486 ekW; 462 kW (652 ehp; 620 shp) at 2,200 rpm (to 21°C)
PT6A-28 486 ekW; 462 kW (652 ehp; 620 shp) at 2,200 rpm (to 33°C)
PT6A-34 545 ekW; 522 kW (731 ehp; 700 shp) at 2,200 rpm (to 19°C)
PT6A-135 548 ekW; 522 kW (735 ehp; 700 shp) at 2,200 rpm (to 28°C)
PT6A-36 548 ekW; 522 kW (735 ehp; 700 shp) at 2,200 rpm (to 28°C)
PT6A-38 597 ekW; 559 kW (801 ehp; 750 shp) at 2,000 rpm (to 27°C)
PT6A-41 673 ekW; 634 kW (903 ehp; 850 shp) at 2,000 rpm (to 28°C)
PT6A-42 674 ekW; 634 kW (904 ehp; 850 shp) at 2,000 rpm (to 33°C)
PT6A-45A 749 ekW; 713 kW (1,004 ehp; 956 shp) at 1,425 rpm (to 15°C)
PT6A-45B, -45R 749 ekW; 713 kW (1,004 ehp; 956 shp) at 1,425 rpm (to 15°C)
PT6A-50 706 ekW; 671 kW (947 ehp; 900 shp) at 1,020-1,160 rpm (to 23°C)
PT6A-65B, -65R 762 ekW; 713 kW (1,022 ehp; 956 shp) at 1,425 rpm (to 27°C)

Specific Fuel Consumption:

At T-O rating:

PT6A-10, -110	111·0 μg/J (0·657 lb/h/ehp)
PT6A-11, -11AG	109·4 μg/J (0·647 lb/h/ehp)
PT6A-112	107·6 μg/J (0·637 lb/h/ehp)
PT6A-15AG, -27, -28	101·8 μg/J (0·602 lb/h/ehp)
PT6A-20	109·7 μg/J (0·649 lb/h/ehp)
PT6A-21, -25	106·5 μg/J (0·630 lb/h/ehp)
PT6A-25C, -34, -34B, -34AG	100·6 μg/J (0·595 lb/h/ehp)
PT6A-135	98·9 μg/J (0·585 lb/h/ehp)
PT6A-36, -41	99·9 μg/J (0·591 lb/h/ehp)
PT6A-38	106·3 μg/J (0·629 lb/h/ehp)
PT6A-42	101·5 μg/J (0·601 lb/h/ehp)
PT6A-45A, -45B	93·5 μg/J (0·554 lb/h/ehp)
PT6A-45R	93·4 μg/J (0·553 lb/h/ehp)
PT6A-50	94·6 μg/J (0·560 lb/h/ehp)
PT6A-65B, -65R	88·8 μg/J (0·527 lb/h/ehp)

Oil Consumption:

Max 0·091 kg (0·20 lb)/h

P&WC T74

T74 is the US designation for military versions of the PT6A turboprop and PT6B turboshaft.

T74-CP-700. US Army counterpart of the PT6A-20. More than 300 T74-CP-700s have been delivered to Beechcraft for 129 U-21A aircraft. Inertial separator system to protect against sand and dust ingestion.

T74-CP-702. Rated at 580 ekW (778 ehp) and retrofitted in Beechcraft U-21.

P&WC PT6B/PT6C

The PT6B is the commercial turboshaft version of the PT6A and has a lower-ratio reduction gear. Principal versions are:

PT6B-9. Rated at 410 kW (550 shp) at 6,230 rpm available to 25°C. Civil certification received in May 1965. Production complete.

PT6B-34. Similar to B-9 except based on PT6A-34. T-O rating 559 kW (750 shp) at 6,188 rpm to 35°C (2½-min contingency 671 kW; 900 shp to 15°C). Certificated July 1976. Installed in experimental Westland Lynx.

PT6B-35F. Based upon PT6A-135 but with single-stage reduction gearbox. T-O rating 485 kW (650 shp) at 6,000 rpm to 43°C. Engine certification expected 1981. Chosen by LearAvia Corporation to power Lear Fan 2100.

PT6C. This series of engines provides direct drive from the power turbine, with no reduction gearing.

Dimensions:

Max diameter: PT6B-34, -35F	572 mm (22·5 in)
Length, less accessories:	
PT6B-34	1,499 mm (59·0 in)
Frontal area	0·18 m² (1·95 sq ft)

Weight, Dry:

PT6B-9	116 kg (255 lb)
PT6B-34	135 kg (298 lb)
PT6B-35F	133 kg (294 lb)

Performance Ratings:

T-O:

See under model listings

Max cruise:

PT6B-9 362 kW (485 shp) at 6,230 rpm
PT6B-34 466 kW (625 shp) at 6,188 rpm (to 15°C)
PT6B-35F 485 kW (650 shp) at 6,000 rpm (to 43°C)

Specific Fuel Consumption:

At T-O rating:

PT6B-9	112·4 μg/J (0·665 lb/h/shp)
PT6B-34	104·6 μg/J (0·619 lb/h/shp)
PT6B-35F	103·0 μg/J (0·608 lb/h/shp)

Oil Consumption:

Max 0·091 kg (0·20 lb)/h

P&WC PT6T TWIN-PAC

First run in July 1968, the PT6T Twin-Pac comprises two PT6 turboshaft engines mounted side by side and driving into a combining gearbox to provide a single output drive. The engine was launched as a coupled power unit for a family of twin-engined helicopters based on the Bell Helicopter UH-1 series. First of these, jointly financed by Bell, P&WC and the Canadian government, was the 15-seat Bell Model 212, which first flew with the PT6T-3 in April 1969.

Installation of the 1,342 kW (1,800 shp) PT6T-3 in the Model 212, in addition to offering true engine-out capability, provides an additional 300 shp over the single-engined 205A and gives enhanced hot-day and high altitude performance. Qualified PT6T-3s became available in the third quarter of 1970 coincident with certification of the Model 212, which is also produced under licence by Agusta in Italy.

Another application of the PT6T-3 engine is for conversion from piston engine to turbine power of the Sikorsky S-58. The prototype S-58T flew in August 1970 and certification was received in April 1971.

In these two helicopter applications, total shaft-power output is limited by the helicopter transmission. In the Model 212 the 1,342 kW (1,800 shp) PT6T-3 is restricted to a T-O rating of 962 kW (1,290 shp) and 843 kW (1,130 shp) for continuous power. In the S-58T the limits are 1,122 kW (1,505 shp) at T-O and 935 kW (1,254 shp) for

The 1,342 kW (1,800 shp) P&WC PT6T-3 Twin-Pac coupled free-turbine turboshaft

continuous operation. The PT6T-3 is easily adapted to such power requirements by a simple setting of its torque control. In the event of a power-section failure, torquemeters in the combining gearbox signal the other power section to maximum power. A single-engine 30-minute rating is included for use, at pilot discretion, in such contingencies.

Certificated in late 1979, the PT6T-3B was introduced as an alternative to the PT6T-3 for installation in the Bell 212. The T-3B is basically a PT6T-3 with some T-6 hardware and improved single-engine performance, and powers the Bell 412.

An uprated Twin-Pac, the PT6T-6, was certificated in December 1974. The higher power is achieved by material and aerodynamic improvements to the compressor-turbine nozzle guide vanes and rotor blades. Installed in S-58T and Agusta-Bell 212. By the beginning of 1981 a total of 6·8 million equivalent PT6 h had been flown by PT6T engines in 1,760 helicopters in 62 countries.

The following details describe the main features differing from those of the standard PT6:

TYPE: Coupled free-turbine turboshaft.

SHAFT DRIVE: Combining gearbox comprises three separate gear trains, two input and one output, each contained within an individual sealed compartment and all interconnected by drive shafts. Overall reduction ratio 5 : 1. Input gear train comprising three spur gears provides speed reduction between power sections and output gearbox. The two drives into the output gearbox are via Formsprag fully-phased overrunning clutches with input third gear forming outer member of clutch, and interconnect shaft forming inner, overrunning member. Output gear train comprises three helical spur gears, two input pinions meshing with single output gear.

AIR INTAKES: Additional inertial particle separator fitted upstream of engine to reduce sand and dust ingestion. High frequency compressor noise suppressed.

FUEL SYSTEM: As PT6 with manual backup system, and dual manifold for cool starts. Automatic power sharing and torque limiting. Torquemeters provide signals to Bendix fuel system metering valves to maintain power at level set by pilot's selective-collective control. Fuel heaters.

FUEL GRADES: JP-1, JP-4 and JP-5.

JET PIPE: Single upward-facing exhaust port on each gas generator.

ACCESSORIES: Starter/generator and tacho-generator mounted on accessory drive case at front of each power section. Other accessory drives on combining gearbox, including individual power turbine speed governors and tacho-generators, and provision for blowers and aircraft accessories.

LUBRICATION SYSTEM: Independent lubrication system on each power section for maximum safety during single-engine operation. Integral oil tanks. Separate oil system for output section of combining gearbox.

OIL SPECIFICATION: PWA Spec 521. For military engines, MIL-L-7808 and -23699.

STARTING: Electrical, with cold weather starting down to −54°C.

DIMENSIONS:

Height	838 mm (33·0 in)
Width	1,118 mm (44·0 in)
Length	1,702 mm (67·0 in)

WEIGHT, DRY (standard equipment):

PT6T-3	292 kg (645 lb)
PT6T-3B, -6	298 kg (657 lb)

PERFORMANCE RATINGS:

T-O (5 min):	
Total output, at 6,600 rpm:	
PT6T-3, -3B	1,342 kW (1,800 shp)
PT6T-6	1,398 kW (1,875 shp) (to 21°C)
Single power section only, at 6,600 rpm:	
PT6T-3, -3B	671 kW (900 shp)
PT6T-6, -3B (2½ min)	764 kW (1,025 shp)
30 min power (single power section), at 6,600 rpm:	
PT6T-3B, -6	723 kW (970 shp)
Max continuous:	
Total output, at 6,600 rpm:	
PT6T-3, -3B	1,193 kW (1,600 shp)
PT6T-6	1,249 kW (1,675 shp) (to 19°C)
Single power section only, at 6,600 rpm:	
PT6T-3, -3B	596 kW (800 shp)
PT6T-6	615 kW (825 shp) (to 19°C)
Cruise A:	
Total output, at 6,600 rpm:	
PT6T-3, -3B	932 kW (1,250 shp)
PT6T-6	1,014 kW (1,360 shp)
Single power section only, at 6,600 rpm:	
PT6T-3, -3B	466 kW (625 shp)
PT6T-6	500 kW (670 shp)
Cruise B:	
Total output, at 6,600 rpm:	
PT6T-3, -3B	820 kW (1,100 shp)
PT6T-6	891 kW (1,195 shp)
Single power section only, at 6,600 rpm:	
PT6T-3, -3B	410 kW (550 shp)
PT6T-6	440 kW (590 shp)
Ground idle, at 2,200 rpm	44·7 kW (60 shp) max

SPECIFIC FUEL CONSUMPTION:

At 2½-min rating (single power section):	
PT6T-3B	100·7 μg/J (0·596 lb/h/shp)
PT6T-6	101·6 μg/J (0·602 lb/h/shp)
At 30-min rating (single power section):	
PT6T-3	101·9 μg/J (0·603 lb/h/shp)
PT6T-3B	101·0 μg/J (0·598 lb/h/shp)
PT6T-6	101·3 μg/J (0·600 lb/h/shp)
T-O, 5-min rating (total output):	
PT6T-3	100·6 μg/J (0·595 lb/h/shp)
PT6T-3B	101·3 μg/J (0·600 lb/h/shp)
PT6T-6	100·0 μg/J (0·592 lb/h/shp)
At max continuous rating (total output):	
PT6T-3	101·2 μg/J (0·599 lb/h/shp)
PT6T-3B	103·9 μg/J (0·615 lb/h/shp)
PT6T-6	101·9 μg/J (0·603 lb/h/shp)

OIL CONSUMPTION:

Max (for both gas generators)	0·18 kg (0·4 lb)/h

P&WC T400

Military version of the PT6T Twin-Pac, the T400-CP-400 has castings of aluminium instead of magnesium. For military roles, P&WC describes the T400 as producing a minimum infra-red signature. Military Qualification Tests (MQT) were completed in March 1970, and production deliveries started in the same month.

The T400 is used in the US Air Force and Navy Bell UH-1N (military version of the Model 212), the US Marine Corps Bell AH-1J, and the Canadian Armed Forces Bell CH-135. T400 field operations started in the middle of 1970.

TBO on the T400-CP-400 is 2,000 h on both the power section and reduction gearbox. By the beginning of 1981 deliveries totalled 799 CP-400 engines.

The T400-WV-402 is the military counterpart of the PT6T-6 and is used in the AH-1T. By the beginning of 1981 a total of 524 WV-402s had been delivered.

DIMENSIONS (CP-400 and WV-402):

Height	828 mm (32·6 in)
Width	1,115 mm (43·5 in)
Length	1,659 mm (65·3 in)

WEIGHT, DRY:

T400-CP-400	324 kg (714 lb)
T400-WV-402	338 kg (745 lb)

PERFORMANCE RATINGS:

Intermediate:	
T400-CP-400	1,342 kW (1,800 shp) at 6,600 rpm
T400-WV-402	1,469 kW (1,970 shp) at 6,600 rpm
Max continuous:	
T400-CP-400	1,141 kW (1,530 shp) at 6,600 rpm
T400-WV-402	1,248 kW (1,673 shp) at 6,600 rpm

SPECIFIC FUEL CONSUMPTION (Intermediate rating):

T400-WV-402	99·9 μg/J (0·591 lb/h/shp)
T400-CP-400	100·4 μg/J (0·594 lb/h/shp)

CHINA
(PEOPLE'S REPUBLIC)

NATIONAL AIRCRAFT ENGINE FACTORY

MAIN LOCATION: Shenyang

As related in the 1977-78 *Jane's*, the first aircraft engine made in numbers in the People's Republic of China was the Soviet M-11 radial.

In 1958 licences were obtained by the 2nd Ministry of Machine Building for two additional Soviet aircooled radial engines, the 194 kW (260 hp) Ivchenko AI-14R and 746 kW (1,000 hp) Shvetsov ASh-62IR (both described under Poland), fitted respectively to the locally-built Jinge (Chinko) No. 1 (Yak-12) and Fong Shou No. 2 (An-2). Both of these aircraft and their engines are believed to have been built in large numbers. By 1959 the Manchurian plants were licence-building the Soviet Mi-4 helicopter and the Czech Super Aero 45 light twin. It is thought that in each case the engine (respectively the 1,268 kW; 1,700 shp Shvetsov ASh-82V 14-cylinder radial and the 104·4 kW; 140 hp M 332 four-in-line, the latter last described in the 1971-72 *Jane's*) was also produced either at Shenyang or at one of the other national factories. One possibility is that the Czech engine was made in the works at Harbin, because it was there that the Chinese version of the Super Aero 45 was produced. Harbin may also have taken over the M-11FR programme, because from 1959-60 that factory manufactured the M-11-powered Heilongjiang (Hei Lun-kiang) No. 1, a locally designed liaison aircraft resembling the Yak-12.

GAS TURBINE ENGINES

During the Korean War (1950-53) large numbers of MiG-15 fighters were ferried through Manchuria. Chinese technicians became familiar with the aircraft and its Klimov RD-45 (Rolls-Royce Nene derivative) engine. In 1955 a licence for the manufacture of the MiG-15 fighter and MiG-15UTI trainer was signed in Moscow, and from 1958 several hundred of the latter were produced, powered by the RD-45 of 24·24 kN (5,450 lb st). The MiG-15 fighter was apparently not built in China, but in 1959 the first Chinese J-5 (F-5), a licence-built MiG-17F, began a production run of well over 1,000 aircraft, all probably powered by Chinese-built Klimov VK-1 turbojets rated at 26·47 kN (5,950 lb st).

In February 1959 the Chinese signed a licence agreement for the manufacture of the MiG-19 supersonic fighter, powered by RD-9 turbojets. Soon afterwards the relationship with the Soviet Union was severed; but the Chinese, working alone, managed to fly a locally-built J-6 (F-6) (MiG-19) in 1961, and have since constructed a number estimated to reach 1,500. Thus, probably more than 4,000 RD-9 engines have been made at Shenyang. A subsequent production programme has concerned the J-7 (F-7) MiG-21. As described in the Aircraft section, this fighter and its R-11 axial turbojet were put into production in China without a licence or any Soviet help. Deliveries of the R-11 from Shenyang are thought to have begun in 1965.

Chinese versions or developments of the RD-9 are likely to have been used in locally produced military designs. One of these is the Q-5 (A-5) (NATO 'Fantan') twin-engined strike fighter, which has been built in quantity. In 1981 the J-6 and Q-5 were still in production, but in quantities smaller than the J-7.

In 1975, the Chinese government signed a preliminary contract with Rolls-Royce Ltd for the licensed manufacture of a supersonic afterburning Spey turbofan generally similar to the Spey 202/203 used in F-4 Phantom fighters of the RAF. The value of British contracts on this programme, which includes a new production plant, is tentatively put at £100 million. It is assumed that the engines will power military aircraft, probably of Chinese design, and that the Chinese objective is total self-sufficiency in their production and operation (see Spey entry under Rolls-Royce).

CZECHOSLOVAKIA

OMNIPOL
OMNIPOL FOREIGN TRADE CORPORATION

Nekazanka 11, 112 21 Prague 1
Telephone: 268261-8

Omnipol is responsible for exporting products of the Czech aviation industry and for supplying information on those products which are available for export.

AVIA
AVIA NARODNÍ PODNIK

199 03 Praha 9, Letňany
Telephone: Prague 89 51 21

Originally a member of the Czechoslovak Aviation Industry Group, Avia National Corporation was transferred to the Czechoslovak Automotive Industry (CAZ) Group in 1960. The company is at present engaged in series production of piston engines, as well as propeller and spare parts manufacture.

AVIA M 137

Designed to power light aerobatic, training, and single-engined and multi-engined sports aircraft, the 134 kW (180 hp) M 137A piston engine is a modification of the M 337 with fuel and oil systems for aerobatic operation and without a supercharger. It powers the Zlin 42 M and Z 526 F. The M 137AZ is a modified version, with the air intake port at the rear so that a dust filter can be incorporated. Details are as for the M 337, with the following differences:

CRANKSHAFT: No oil holes for propeller control.
FUEL SYSTEM: Type LUN 5150 pump; system designed for sustained aerobatics.
STARTER: LUN 2131 electric.
DIMENSIONS:

Length	1,344 mm (52·9 in)
Width	443 mm (17·44 in)
Height	630 mm (24·80 in)
WEIGHT (including starter):	141·5 kg (312 lb)

PERFORMANCE RATINGS:

T-O	134 kW (180 hp) at 2,750 rpm
Max continuous	119 kW (160 hp) at 2,680 rpm
Max cruising	104·5 kW (140 hp) at 2,580 rpm

SPECIFIC FUEL CONSUMPTION:

At T-O rating	91·26 μg/J (0·540 lb/h/hp)
At max cruise rating	81·96 μg/J (0·485 lb/h/hp)

The 134 kW (180 hp) Avia M 137A six-cylinder aircooled piston engine

AVIA M 337

The M 337 six-cylinder aircooled supercharged engine powers several types of light aircraft that were built in Czechoslovakia, including the L-200D Morava, Zlin 43 and Zlin 726K. It can be supplied with hubs for fixed-pitch or controllable-pitch propellers. The **M 337 AK** is fitted to the Zlin 142.

TYPE: Six-cylinder inverted in-line aircooled, ungeared, supercharged and with direct fuel injection.
CYLINDERS: Bore 105 mm (4·13 in). Stroke 115 mm (4·53 in). Swept volume 5·97 litres (364·31 cu in). Compression ratio 6·3 : 1. Steel cylinders with cooling fins machined from solid. Cylinder bores nitrided. Detachable cylinder heads are aluminium alloy castings.
PISTONS: Aluminium alloy stampings with graphited surfaces. Two compression rings and two knife-shaped scraper rings in common groove above gudgeon-pin.
CONNECTING RODS: H-section aluminium alloy forgings. Two split big-ends bolted together by two bolts. Steel two-piece liner, lead-bronze plated.
CRANKSHAFT: Forged from special chrome-vanadium steel, machined all over. Nitrided crank-pins. Carried in seven steel-backed lead-bronze plated slide bearings and one ball thrust bearing.
CRANKCASE: Heat-treated magnesium alloy (Elektron) casting, with top and front covers. Bearing covers forged from aluminium alloy, with double cross webs.
VALVE GEAR: Camshaft on the cylinder heads actuates the valves by means of rocker arms. One inlet valve of heat-treated steel, one sodium-filled exhaust valve of austenitic steel with stellite seat. Nitrided valve stems.
IGNITION: Shielded type. Two vertical magnetos with automatic sparking advance. Two plugs per cylinder, 12 × 1·25 mm.
LUBRICATION: Dry sump pressure-feed type. The M 337 AK has a system for sustained inverted operation.
SUPERCHARGER: Centrifugal type mounted on engine rear flange. Driven through a damping rubber coupling from crankshaft. Planetary gear, ratio 7·4 : 1.
FUEL SYSTEM: Low-pressure injection system. LUN 5152 pump driven from camshaft. Fuel injection nozzles located in front of intake valves. Automatic control in relation to engine manifold pressure. Fuel supplied to injection pump by fuel pressure pump located in common body with injection pump. The M 337A has a unified fuel injection pump, type LUN 5150, and other minor changes.
FUEL GRADE: Minimum 72-78 octane, with maximum TEL 0·06 per cent (volume).
STARTING: Electric starter combined with supercharger. The starter dog is engaged by an electromagnet.
ACCESSORIES: One 600W 28V dynamo. Electric tachometer. Propeller control unit. Mechanical tachometer on oil pump, drive 1 : 2. High-pressure hydraulic pump type P 6121A (to special order).
MOUNTING: Four engine-bearer feet with rubber dampers.
PROPELLER DRIVE: Direct left-hand tractor.
DIMENSIONS:

Length, without propeller boss	1,410 mm (55·51 in)
Width	472 mm (18·58 in)
Height	628 mm (24·72 in)
WEIGHT, DRY:	148 kg (326·3 lb)

PERFORMANCE RATINGS:

T-O rating	157 kW (210 hp) at 2,750 rpm
Max cruising power at 1,200 m (3,940 ft)	112 kW (150 hp) at 2,400 rpm

SPECIFIC FUEL CONSUMPTION:

At T-O rating	100·6 μg/J (0·595 lb/h/hp)
At max cruising power at 1,200 m (3,940 ft)	72·7 μg/J (0·430 lb/h/hp)

MOTORLET
MOTORLET NC, ZÁVOD JANA SVERMY

Prague-Jinonice
Telephone: Prague 520714
GENERAL MANAGER: Zdenek Horcík
ASSISTANTS TO GENERAL MANAGER:
TECHNICAL DIRECTOR: Ing Josef Krca
ECONOMIC DIRECTOR: Alois Svoboda
PRODUCTION DIRECTOR: Bohumil Hamerník
HEAD OF DESIGN DEVELOPMENT: Ing Vladimír Pospísil

Motorlet National Corporation operates the main aero-engine establishment in Czechoslovakia, based on the former Walter factory at Jinonice, previously well known for its radial and in-line piston engines. Today, the Walter name continues in use only as a trade-mark for Motorlet piston and turbine engines.

Motorlet started turbine engine manufacture in 1952 with licensed production of the Soviet RD-45 centrifugal turbojet for MiG-15 fighters.

WALTER M 601B

Second of Czechoslovakia's small turbine engines to enter production, the M 601 was designed to power the Czech L-410 twin-engined light transport aircraft. It drives an Avia V 508 constant-speed three-blade propeller with hydraulically variable pitch.

The first version of the M 601, rated at 550 ehp, ran in October 1967. Development of the completely revised M 601B, of increased diameter, started during 1968. The Let L-410M, powered by M 601B engines in place of the Canadian PT6A-34s fitted to the L-410A, was in Aeroflot service in Siberia by early 1979. The M601B powers the current L-410UVP version which then superseded the L-410M in production.

The Walter M 601B free-turbine turboprop, rated at 515 kW (691 shp)

TYPE: Free-turbine combined axial-and-centrifugal turboprop.
PROPELLER DRIVE: Reduction gear at front of engine with drive from free-turbine. Reduction ratio 14·9 : 1.
AIR INTAKE: Annular intake at rear of engine, with debris screen, feeds air to compressor plenum chamber.
COMPRESSOR: Two axial stages of stainless steel, plus single centrifugal stage of titanium. Pressure ratio 6·4 : 1 at 36,660 rpm gas generator speed. Air mass flow 3·25 kg (7·17 lb)/s.
COMBUSTION CHAMBER: Annular combustor with rotary fuel injection and low-voltage LUN 2201.01-08 ignition.
COMPRESSOR TURBINE: Single-stage; inlet temperature 952°C.
POWER TURBINE: Single-stage.
FUEL SYSTEM: Low-pressure LUN 6590 regulator, with three-lever control providing gas-generator and power-turbine speed controls.
FUEL GRADE: PL4, PL5 kerosene.
JET PIPE: Collector duct surrounding power turbine shaft. Exhaust through two ports on horizontal centreline.
ACCESSORIES: Mounting pads on accessory case at rear of engine. Propeller controls mounted on reduction gear case at front of engine.
LUBRICATION SYSTEM: Pressure gear-pump circulation. Integral oil tank and cooler.
OIL SPECIFICATION: B3V synthetic oil.
MOUNTING: Three elastically-supported pins on compressor casing.
STARTING: LUN 2132-8 8kW electric starter/generator.
DIMENSIONS:

Length	1,658 mm (65·27 in)
Width	590 mm (23·23 in)
Height	650 mm (25·59 in)
WEIGHT, DRY:	193 kg (425·5 lb)

PERFORMANCE RATINGS:

T-O rating	515 kW (691 shp)

SPECIFIC FUEL CONSUMPTION:

At T-O rating:	109·55 μg/J (0·648 lb/h/ehp)

FINLAND

VALMET
VALMET OY

HEAD OFFICE: Jyskän Tehdas, 40420 Jyskä
Telephone: (941) 261 100

This famous aircraft company is marketing a small piston engine for light aircraft and many other applications. It powers the Colomban Cri-Cri.

VALMET SM 160

Small piston engine for ultralights, RPVs and many surface applications.

TYPE: Single-cylinder two-stroke.
CYLINDER: Bore 52 mm (2·05 in). Stroke 62 mm (2·44 in). Capacity 157 cc (9·6 cu in). Compression ratio 9·6 : 1.
FUEL: Gasoline (petrol)/oil in 25 : 1 ratio.
ACCESSORIES: Starter/generator 35/55W, 12V.
DIMENSIONS:
Height 280 mm (11·0 in)
Width (excl carburettor) 189 mm (7·4 in)
Length (incl drive shaft) 251 mm (9·9 in)
WEIGHT, DRY: 9 kg (19·8 lb)
POWER RATING:
S/L T-O 7·5 kW (10 hp) at 6,500 rpm

FRANCE

G2P
GROUPEMENT POUR LES GROS PROPULSEURS À POUDRE

HEAD OFFICE: 3 avenue du Général de Gaulle, 92800 Puteaux
Telephone: 778 15 15
MAIN ESTABLISHMENT: St Aubin de Médoc 33160
ADMINISTRATOR: Roger Guernon

SEP and SNPE (both listed in this section) formed G2P to ensure the close co-ordination of their activities and to act as prime contractor in the field of solid-propellant propulsion. Its activities are centred upon the motors of strategic missiles (MSBS, SSBS) and large tactical missiles (Pluton).

MICROTURBO
MICROTURBO SA

HEAD OFFICE AND WORKS: Chemin du Pont de Rupé, 31019 Toulouse Cédex
Telephone: (61) 70 07 77
Telex: 531442
DIRECTORATE:
J. G. Bayard (President)
M. E. Faury (R & D)
L. A. Pech (Commercial Director)
P. F. Calmels (Technical Director)

Microturbo was established in 1960 for the production of small gas turbines. The initial product was the Noelle 60290 free-turbine starter for the SNECMA Atar turbojet, and from this a wide range of units has evolved.

MICROTURBO TRS 18

This single-shaft turbojet was designed for installation in gliders, to impart a self-launch and climb capability, but has since been adapted for ultralight aeroplanes and unmanned vehicles.

MICROTURBO TRS 18-046

This is the basic TRS 18 model for manned applications. It is in production, and an American Type Certificate was issued in May 1976. Applications include the twin-jet NASA AD-1 slew-wing research aircraft, Chagnes Microstar version of the VariViggen homebuilt aircraft, Caproni Vizzola A-21SJ and twin-engined Microjet 200.

The TRS 18-046 is of modular construction. The forward module incorporates the air intake, gearbox, electronic governing and protection unit and the start sequencing and indication unit. The 28V 600W starter/generator is located in the nose bullet. The oil tank, with submerged pump, is on the underside, and includes provision for inverted flight. The HP oil filter and pressure transducer are on the top of this module. Adjacent to the compressor are the probes for engine speed and air temperature.

The turbine module comprises: the one-piece centrifugal compressor, with diffuser and straightener vanes; the axial turbine rotor and nozzle diaphragm; and the main frame, carrying the rotor assembly on two ball bearings between the compressor and turbine. The aft module comprises: the turbine casing backplate, carrying the annular folded combustion chamber liner, exhaust cone and nozzle; 10 spill-type burners; two igniter plugs, used only during starting; and the jet pipe with thermocouple.

The fuel pump is driven electrically. The lubrication system is a closed circuit, with pressure supply to the rotor and gearbox bearings. The engine can be shut down and restarted in flight, and incorporates automatic fault and protection systems.

DIMENSIONS:
Length 650 mm (25·59 in)
Width 325 mm (12·797 in)
Height 350 mm (13·78 in)
WEIGHT, DRY:
Basic 32·0 kg (70·5 lb)
With igniter and voltage regulator 33·4 kg (73·63 lb)
PERFORMANCE RATING (ISA, S/L):
T-O 1·10 kN (247 lb st) at 45,000 rpm
Max continuous 1·00 kN (225 lb st) at 44,000 rpm
SPECIFIC FUEL CONSUMPTION:
At max cont 35 mg/Ns (1·24 lb/h/lb st)

MICROTURBO TRS 18-056

This simplified model retains only the gas-generator section of earlier versions and was developed to power the France-Engins Mitsoubac and other RPVs. The lubrication system uses fuel, and the engine is cranked electrically for starting. An exhaust turbo drives the generator and pressurises the fuel.

DIMENSIONS
Length 600 mm (23·6 in)
Width 305 mm (12·0 in)
Height 345 mm (13·6 in)
WEIGHT, DRY
Basic, no jet pipe 23·0 kg (50·7 lb)
PERFORMANCE RATING: As TRS 18-046

Microturbo TRS 18-046 turbojet as prepared for use in Microjet 200

Microturbo TRS 18-075 for Flight Refuelling ASAT

MICROTURBO TRS 18-075, -076

These engines differ from the -056 only in detail. The -075 powers the Flight Refuelling ASAT target drone. It has a high-speed alternator driven by the gas-generator shaft, and fuel and oil pumps mechanically driven off a front hydraulic block. T-O rating is 1·15 kN (260 lb st) and max cont 1·10 kN (247 lb st). The -076 powers the Meteor Mirach-100 and has the same ratings as the 075.

MICROTURBO TRS 20/22

This turbojet is under development, based on the TRS 18 -046.

PERFORMANCE RATINGS (ISA, S/L):
T-O: RPV version 1·35 kN (303 lb st)
man-rated version 1·25 kN (281 lb st)
Max continuous:
RPV version 1·15 kN (260 lb st)
man-rated version 1·10 kN (247 lb st)

MICROTURBO TRI 60

Representing a significant French development in the propulsion of cruise-type unmanned vehicles, the TRI 60 was designed under a contract from the Direction des Recherches et Moyens d'Essais. It is an extremely simple single-shaft turbojet for use in subsonic missiles and RPVs. The design has been biased towards minimal cost and absence of any maintenance or overhaul, though engine design life exceeds 20 h.

The annular intake contains the accessory gearbox in the central bullet, together with an alternator or starter/generator; the struts house fuel and oil pipes. The simple axial compressor operates at a pressure ratio of about 4 : 1, with airflow of 5·6 kg (12·3 lb)/s, and is carried between front and rear bearings with labyrinth seals. The smokeless combustor is of the axial type, with multiple spray burners fed by a peripheral manifold. The axial turbine is overhung behind the rear bearing on the central diffuser housing.

An air bleed provides up to 1·5 per cent of total airflow. There is an engine-driven fuel pump, but lubrication is by either pre-lubricated bearings or a total-loss system from a pressurised reservoir. Speed control can be mechanical, electronic, fluidic or pneumatic, according to installation. Starting can be by impingement, electrical, cartridge or other means.

Versions of the TRI 60 announced by early 1981 are as follows:

TRI 60-1 Model 067. For British Aerospace Sea Eagle missile. Hydro-pneumatic fuel control unit to ensure a constant-range mission at low altitude. Ignition by two pyro-igniters for windmill start.

Microturbo TRS 18-076 for Meteor Mirach-100

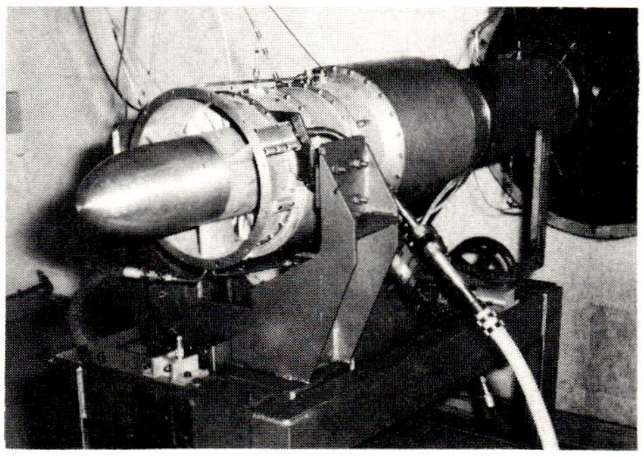
Microturbo TRI 60-2 rated at 3·7 kN (831 lb st)

TRI 60-2 Model 071. For Aérospatiale C.22 variable-speed target drone. Continuously controlled electronic system. Fully throttleable.

TRI 60-2 Model 074. For Beechcraft MQM-107A variable-speed target drone. As Model 071, but with 1·5 kVA AC generator directly driven by gas generator shaft.

TRI 60-2 Model 077. For Saab Bofors Missile Corporation RBS15 long-range anti-ship missile. As TRI 60-1 Model 067, with performance similar to that of TRI 60-2 Model 071.

TRI 60-3 series. Under development. Geometrically identical to TRI 60-2.

DIMENSIONS (gas generator, no jetpipe):
Length overall 749 mm (29·49 in)
Envelope diameter 330 mm (12·99 in)
WEIGHT, DRY: 45 kg (99·2 lb)
PERFORMANCE RATING (ISA, S/L):
T-O: Model 067 3·50 kN (787 lb st)
Model 071 3·70 kN (831 lb st)
TRI 60-3 4·00 kN (900 lb st)
SPECIFIC FUEL CONSUMPTION:
Min: Model 067, Model 071 32·7 mg/Ns (1·20 lb/h/lb st)
TRI 60-3 31·3 mg/Ns (1·18 lb/h/lb st)

MICROTURBO TRI 80

This new series of engines, under development for unmanned applications, will have a fourth stage added in front of a basic TRI 60 axial compressor, to increase performance.

DIMENSIONS (gas generator, no jetpipe):
Length 810 mm (31·89 in)
Diameter 330 mm (12·99 in)
WEIGHT, DRY: 54 kg (119·1 lb)
PERFORMANCE RATING (ISA, S/L):
T-O (min, designed) 4·50 kN (1,011 lb st)
SPECIFIC FUEL CONSUMPTION:
Min 29·94 mg/Ns (1·10 lb/h/lb st)

MICROTURBO/TURBOMÉCA SS-RPV

In collaboration with Turboméca, Microturbo is developing an engine for RPVs that are designed to cruise at Mach 1·5 to 1·8.

MUDRY

MOTEURS MUDRY-BUCHOUX (AVIONS MUDRY et CIE)

Aérodrome de Bernay, BP 47, 27300 Bernay
Telephone: (32) 43 47 34

DIRECTOR: Auguste Mudry

MUDRY MB-4-80

Described as being of highly original design, a prototype of this aircooled four-stroke flat-four was nearing completion in March 1981. It is intended to power the Mudry CAP X and Mini-Cap light aircraft, and develops 59·5 kW (80 hp) at 2,300-2,800 rpm. Fuel consumption (100LL aviation fuel or automobile 'super') is quoted as 220 g/hp/h.

RECTIMO

RECTIMO AVIATION SA

OFFICES AND WORKS:
Aérodrome de Chambéry, 73420 Savoie
Telephone: (79) 63 40 06
DIRECTOR: André Rosselot

Rectimo has manufactured over 500 Type 4 AR 1200 single-ignition derivatives of the Volkswagen four-cylinder aircooled car engine, which together with the larger 4 AR 1600 are used in the Sportavia RF4D powered glider and various ultralight aircraft. The 30 kW (40 hp) 4 AR 1200 engine has a 1,192 cc cubic capacity, 7 : 1 compression ratio and weighs 61·5 kg (136 lb). Fuel consumption under cruise conditions is 11 litres (2·4 Imp gal)/h. The 4 AR 1600 produces 45·5 kW (61 hp) at T-O and has a cubic capacity of 1,600 cc and an 8 : 1 compression ratio. Weight is 64 kg (141 lb). Both engines have a maximum speed of 3,600 rpm.

Rectimo 4 AR 1200 piston engine of 30 kW (40 hp)

SACMA

Zone Industrielle de Vinon sur Verdon (Var)
Telephone: (92) 78 80 82
PRESIDENT: M Negre

Preliminary details became available in mid-1977 of the family of light piston engines under development by this French company in a new factory inland from the Riviera. Covering a range of take-off power from 74·5 to 179 kW (100-240 hp), they utilise numerous automotive components, notably of Talbot origin, to reduce cost and ensure proven performance. All models have watercooled cylinders arranged in line horizontally on the left side of the engine, a turbocharger or rotary mixture distributor, and 2·93:1 reduction to the propeller shaft. The adjacent table lists the basic models to be offered. (The more powerful models have a turbocharger.) The 112 kW (150 hp) engine began running in a Socata Rallye in the Spring of 1977, but first flight did not take place until 11 February 1981.

CYLINDERS: Bore 91·7 mm (3·61 in). Stroke 75 mm (2·95 in). Capacity 1,981 cc (121 cu in). Compression ratio 10·3.
FUEL GRADE: 100LL, automotive grade gasoline, kerosene.
DIMENSIONS (112 kW, 150 hp):

Length	620 mm (24·4 in)
Width	510 mm (20·1 in)
Height	500 mm (19·7 in)
WEIGHT, DRY (112 kW, 150 hp):	125 kg (276 lb)

Socata Rallye testbed for the 112 kW (150 hp) SACMA engine *(Aviation Magazine International)*

T-O power	Speed (rpm)		Dry weight	Fuel consumption 75% power	Oil capacity
kW (hp)	Engine	Propeller	kg (lb)	cc/h (lb/h)	litres (Imp pints)
74·5 (100)	5,400	1,800	122 (269)	19 (30)	5 (8·8)
89·5 (120)	5,400	1,800	122 (269)	23 (37)	5 (8·8)
112 (150)	5,400	1,800	125 (276)	28 (45)	6 (10·5)
134 (180)	5,400	1,800	135 (298)	33·5 (53)	7 (12·3)
179 (240)	5,400	1,800	140 (309)	45 (72)	7 (12·3)

SEP

SOCIÉTÉ EUROPÉENNE DE PROPULSION

Tour Roussel Nobel Cédex No 3, 92080 Paris La Défense
Telephone: (1) 778 15 15
Telex: 630906 SEP Putau
ESTABLISHMENTS: Bordeaux-Le Haillan, Vernon, Melun-Villaroche, Istres and Puteaux
CHAIRMAN: P. Soufflet
INDUSTRIAL POLICY DIRECTOR: A. Garnault
MARKETING AND SALES DIRECTOR: R. Morin
TECHNICAL DIRECTOR: J. Meriguet

SEP specialises in the design and development of all categories of propulsion systems and engines for missiles, space launchers and satellites. One quarter of its 2,800 personnel are engineers and technicians specialising in research, development and testing.

SEP produces a wide range of solid- and liquid-propellant motors as sustainers and boosters for French and other European guided and unguided missiles and space launchers. Some 60 different types of motor have been designed since 1950, including the three stages of the Diamant A and B and B P 4 French space launchers, the second stage and perigee motor of the ELDO Europa II launcher, and motors for the SSBS, MSBS and Pluton nuclear-warhead missiles. The company acquired experience in cryogenic-propellant rockets through developing the HM4 and HM7 engines.

Centre National d'Etudes Spatiales entrusted SEP with the entire propulsion systems of the three stages of the Ariane launch vehicle. A major engine for manned aircraft is the SEP 844 which provides thrust boost for Dassault Mirage III fighters in service with the French and other air forces.

Other SEP developments include engines using hybrid propellants, fluorine and fluorine compounds, monopropellants and compressed gases, as well as electric 'thrusters'. The company is also applying its missile and space technology to oceanology.

SEP VIKING

The Viking series of turbopump-fed rocket engines was designed for simplicity and low cost. The thrust chamber, fed with unsymmetrical dimethyl hydrazine (UDMH) and nitrogen tetroxide (N_2O_4), has a single wall of HS 25 steel, coated with zirconium oxide, fuel-film cooled. The nozzle throat is of graphite. The light-alloy injector is of the radial type, with alternate doublets.

Mounted directly on the chamber, the turbopump has a two-stage Curtiss turbine, driven by propellant gases cooled by water. The turbine shaft carries impellers for UDMH, N_2O_4 and water. The gas-generator also provides onboard power and pressurises the tanks. Combustion pressure is regulated against a reference by varying turbine speed. Mixture ratio is maintained by controlling the flow of N_2O_4.

In 1972 SEP ran Viking I qualification tests of 150 s duration. The Ariane launch vehicle has a first stage, Lilo, with four Viking engines. Testing of the cluster began in 1976, the current engine being Viking V. The second-stage Ariane engine is Viking IV, tuned for vacuum operation, with two-axis thrust vectoring. The nozzle has a bell shape, fabricated by welding rolled and Flo-turned steel sheet.

For data see table.

SEP HM7

Developed for the Ariane third stage, the HM7 is a liquid oxygen/liquid hydrogen engine. It is also referred to in the entry for MBB of the Federal Republic of Germany.
TYPE: Liquid-propellant rocket engine.
PROPELLANTS: Liquid oxygen and liquid hydrogen.
THRUST CHAMBER ASSEMBLY: Single-chamber unit of 48 : 1 nozzle area ratio, regeneratively cooled, and of stainless steel tube construction. Operating sequence initiated by hydrogen pre-cooling and pre-opening of hydrogen injection valve. Concentric-tube propellant injection system with central oxygen flow. Pyrotechnic ignition. Combustion pressure 35 bars (507·5 lb/sq in) and temperature 2,727°C.
THRUST CHAMBER MOUNTING: Gimballed assembly, turbopump integral with chamber.
PROPELLANT PUMPS: Axial-plus-centrifugal pumps, coaxial, 60,500 rpm.

Propulsion bay of Lilo, first stage of the L 3 S Ariane launch vehicle, showing the four SEP Viking V engines. Total S/L thrust is 2,444 kN (549,420 lb)

PROPELLANT FLOWS: Liquid hydrogen flow rate 2·76 kg (6·07 lb)/s at 65 bars (942·5 lb/sq in). Liquid oxygen flow rate 14·21 kg (31·26 lb)/s at 52 bars (754 lb/sq in).
TURBINE: Two-stage axial-flow impulse unit in Inconel X 750. Gas inlet temperature 617°C.

Gas Generator: Liquid hydrogen flow rate 0·133 kg (0·29 lb)/s. Liquid oxygen flow rate 0·12 kg (0·26 lb)/s. Pyrotechnic ignition.
Lubrication System: Uses tributyl phosphate spray into gaseous hydrogen.
Starting: Solid grain primer.
Thrust Control: Thrust held constant by regulation of turbopump speed via control of gas generator propellant supply.
For data see table.

SEP HYDRAZINE THRUSTER

SEP has for a long period been engaged in the development of small monopropellant thrusters for satellite attitude and orbit control. Most of this work has been based on hydrazine, decomposed through a catalyst to serve as a monopropellant. The present SEP hydrazine propulsion system uses CNESRO 1 catalyst, developed jointly by SEP and the Faculté des Sciences of Paris.

SEP delivered to CNES (Space Centre of Toulouse) a flight model of its hydrazine micropropulsion system, comprising: surface-tension tank, engine, European CNESRO catalyst, hydrazine electrovalve, sensor and onboard electronics. This micropropulsion system was assembled on the D-5A satellite, launched by Diamant B P 4. During endurance testing a D-5A thruster operated continuously for 145,000 s, without attention to the catalyst bed.

SEP has since developed a larger thruster for mounting on the GEOS satellite. During qualification testing this thruster operated for 8,600 s (specified time 7,800 s) in 34,000 impulses, and for 3,400 s in continuous operation (specification 3,000 s). In the following data this thruster is referred to as GM, the earlier unit being referred to as D-5A.

Weight, Dry:
D-5A 0·18 kg (0·397 lb)
GM 0·35 kg (0·771 lb)

SEP hydrazine thruster system for D-5A satellite, shown with conical fairing removed

Chamber Pressure:
D-5A 15·3-30·5 bars (222-442 lb/sq in)
GM 10-30 bars (145-435 lb/sq in)
Electrical Load:
D-5A 5W
GM 6W
Thrust:
D-5A 0·0028-0·0016 kN (0·629-0·359 lb)
GM 0·014-0·006 kN (3·15-1·35 lb)

SEP HM7 high-energy upper-stage rocket engine

Specific Impulse (Vacuum):
D-5A, GM 215-230

SEP ROCKET ENGINES

	Type	Length mm (in)	Diameter mm (in)	Weight kg (lb)	Thrust (* = in vacuum)	Specific Impulse (* = in vacuum)	
Viking V	Turbopump engine for Ariane	2,873 (113·1)	990 (39)	818 (1,803)	611 kN (137,355 lb) *693 kN (155,790 lb)	*280·6 s	See text
Viking IV	Turbopump engine for Ariane	3,684 (145)	1,700 (66·9)	905 (1,995)	721 kN (162,080 lb)	*295·6 s	See text
HM 7	Oxygen/hydrogen engine	1,713 (67·4)	938 (36·9)	152 (335)	61·67 kN (13,864 lb)	*442·6 s	See text
Mage 1	Solid motor	1,287 (50·7)	766 (30·2)	348 (767)	1,890 daN (425 lb)	920,000 daNs total impulse	Developed for ESA with SNIA-Viscosa and MAN as subcontractors. Apogee motor for European geostationary satellite
Mage 2	Solid motor	1,522 (59·9)	766 (30·2)	528 (1,164)	3,250 daN (731 lb)	1,418,000 daNs total impulse	Later version under development
401	Solid motor		1,950 (76·8)	22,500 (49,600)	900 kN (202,320 lb)		Flexible bearing nozzle, Sepcarb throat, rocket-type igniter
402	Solid motor		1,950 (76·8)	8,800 (19,400)	300 kN (67,440 lb)		Flexible bearing nozzle, Sepcarb throat, rocket-type igniter
403	Solid motor		1,200 (47·2)	1,500 (3,307)	100 kN (22,480 lb)		Flexible bearing nozzle, Sepcarb throat, rocket-type igniter
902	Solid motor (SSBS 1st stage)		1,500 (59·1)	17,600 (38,800)	550 kN (123,640 lb)		Four swivelling nozzles, tungsten throat, rocket-type igniter
Rita II	Solid motor (SSBS, MSBS)		1,500 (59·1)	6,410 (14,131)	260 kN (58,450 lb)		Single nozzle, Pyrocarbon throat
904	Solid motor (MSBS)		1,500 (59·1)	11,500 (25,350)	450 kN (101,160 lb)		Four swivelling nozzles, tungsten throat, rocket-type igniter
Styx	Solid motor (Pluton)		619 (24·4)	1,470 (3,241)	boost 178 kN (40,015 lb) cruise 86 kN (19,330 lb)		Single fixed nozzle, with graphite throat, rocket-type igniter

SNECMA
SOCIÉTÉ NATIONALE D'ÉTUDE ET DE CONSTRUCTION DE MOTEURS D'AVIATION

Head Office: 2 boulevard Victor, 75724 Paris Cédex 15
Telephone: 554 92 00
Telex: 202 834 Motav
Chairman and Chief Executive: René Ravaud
Special Assistant to the Chairman:
Constantin Davidoff
Executive Vice-President, Programme Management and Marketing: Jean Péquignot
Executive Vice-President, Finance and Planning:
Jean Sollier
Vice-President, Engineering and Production:
Jean Calmon
Vice-President, Subsidiaries: Roger Abel
Vice-President, Commercial Engines:
Maurice Avramito
Production General Manager: Yves Bonnet
General Secretary: Henri Forsans
Vice-President, International Relations:
Jean Pascaud
Engineering General Manager: Pierre Lachaume
Press Advisor to the Chairman: Philippe Dreux

Villaroche Centre: 77550 Moissy-Cramayel
Design, development and ground test centre. A subsidiary establishment for flight and noise tests is located at Istres.

Evry-Corbeil:
RN 7, BP 81, 91003 Evry Cédex
Engine production, quality control, service, procurement and laboratories for research and development.

Gennevilliers:
291 avenue d'Argenteuil, BP 48, 92234 Gennevilliers
Forging and casting production, complete machining of mechanical parts.

Suresnes—ELECMA Division:
22 quai Galliéni, 92150 Suresnes
Design, development and production of electronic equipment, especially electronic control systems for the aircraft industry.

SNECMA (Société Nationale d'Etude et de Construction de Moteurs d'Aviation) was born on 28 August 1945 from the merger of several aero-engine companies: Gnome et Rhône, Société Anonyme des Moteurs Renault pour l'Aviation, Société Générale de Mécanique et d'Aviation (former Moteurs Lorraine), and Groupe d'Etudes des Moteurs à Huile Lourde.

These companies already had a long aeronautical tradition and SNECMA has always devoted its main activity to aero-engines. More than 5,000 Atar turbojets have been ordered; they have played a significant part in the world-wide success of Mirage fighters. SNECMA is developing the M53 turbojet for fighters of the 1980s, and the M88 for the 1990s.

SNECMA is also participating in the following international collaborative programmes:

The Olympus 593 for Concorde, developed and produced with Rolls-Royce; the CF6-50 for the Airbus A300B and some 747s (with MTU), and CF6-80 for the A310 under a co-production agreement with General Electric; the CFM56, which SNECMA shares equally with General Electric, but with FN associated within SNECMA's share; the Larzac, produced in co-operation with Turboméca and with production also involving the German companies MTU and KHD, and FN in Belgium; and the Tyne, produced by a consortium (SNECMA, MTU, Rolls-Royce, FN).

SNECMA ATAR

The Atar is a single-shaft military turbojet first run in 1946 and since developed and cleared for flight at Mach numbers greater than 2. Major versions are:

Atar 9C. Compared with the earlier 9B this introduced a new compressor, a self-contained starter and an improved overspeed which comes into operation automatically when the aircraft reaches Mach 1·4, giving power equivalent to a sea level thrust of 62·76 kN (14,110 lb). Equips most Mirage III and 5.

Atar 9K50. Derived from the Atar 9C. Designed to offer improved subsonic specific fuel consumption, increased thrust for supersonic acceleration and improved overhaul life. The main improvements are in an entirely redesigned turbine with blades not forged but cast and coated with refractory metal from the vapour phase. Stages 1 and 8 of the compressor have been redesigned, resulting in pressure ratio raised from 6 : 1 to 6·15 : 1, coupled with slightly augmented mass flow. The control and electronic equipment have been revised and extended to improve the security of single-engined aircraft. The 9K50 is the power plant of all production Mirage F1 versions and the Mirage 50.

Atar 8K50. This is essentially the 9K50, the latest variant in production, re-engineered to have a simple unaugmented jet-pipe and fixed nozzle, for the Super Etendard. All parts are protected against sea corrosion. The 8K50 completed its 150 h official type test in May 1975. Production deliveries began in May 1977.

DIMENSIONS:	
Diameter	1,020 mm (40·2 in)
Length overall:	
Atar 8K50	3,936 mm (155 in)
Atar 9C, 9K50	5,944 mm (234 in)
WEIGHTS:	
Dry, complete with all accessories:	
Atar 8K50	1,155 kg (2,546 lb)
Atar 9C	1,456 kg (3,209 lb)
Atar 9K50	1,582 kg (3,487 lb)

SNECMA Atar 9K50 turbojet of 70·6 kN (15,870 lb st) with afterburning

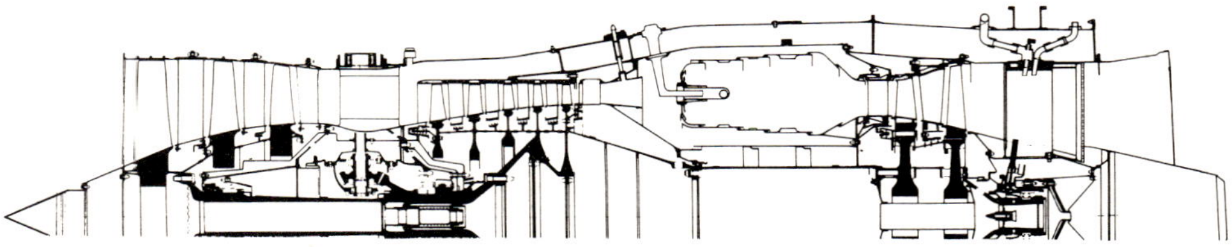

Longitudinal section through the SNECMA M53 showing LP and HP sections on single shaft (pressure ratio, 9·3 at 10,500 rpm) and bypass duct

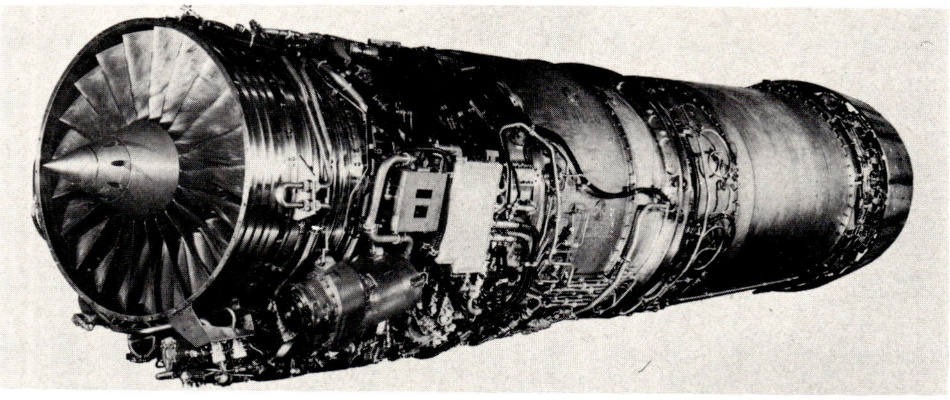

SNECMA M53 augmented bypass turbojet of 88·3 kN (19,840 lb st)

PERFORMANCE RATINGS:	
Max with afterburner:	
Atar 9C	58·9 kN (13,230 lb st) at 8,400 rpm
Atar 9K50	70·6 kN (15,870 lb st) at 8,400 rpm
Max without afterburner:	
Atar 8K50	49 kN (11,025 lb st) at 8,550 rpm
Atar 9C	42 kN (9,430 lb st) at 8,400 rpm
Atar 9K50	49·2 kN (11,055 lb st)
SPECIFIC FUEL CONSUMPTION:	
At max rating with afterburner:	
Atar 9C	57·5 mg/Ns (2·03 lb/h/lb st)
Atar 9K50	55·5 mg/Ns (1·96 lb/h/lb st)
At max rating without afterburner:	
Atar 8K50	27·5 mg/Ns (0·97 lb/h/lb st)
Atar 9C	28·6 mg/Ns (1·01 lb/h/lb st)
Atar 9K50	27·5 mg/Ns (0·97 lb/h/lb st)
OIL CONSUMPTION:	1·5 litres (2·64 Imp pints)/h

SNECMA M53

The M53 is a single-shaft turbofan—more strictly a continuous-bleed turbojet—capable of propelling fighter aircraft at high altitude at Mach 2·5. The first applications are in the Mirage 2000, being developed for the Armée de l'Air, and Super Mirage 4000. The engine is of modular construction.

The single shaft comprises a three-stage fan and five-stage compressor driven by a two-stage turbine designed for operation at high gas temperature. There are no inlet guide vanes. Max airflow is 86 kg (190 lb)/s. Between the fan and compressor is a mid-frame incorporating accessory drives and front roller bearing and ball thrust bearing. The annular combustion chamber is designed for smoke-free operation. The turbine delivery casing incorporates the third bearing. Fuel to the combustion chamber and reheat system, and the multi-flap nozzle, are controlled by a fuel system monitored by an ELECMA electronic computer.

Flight trials began on 18 July 1973 in the starboard pod of a Caravelle testbed. The supersonic flight envelope was explored with the Mirage F1-M53 flying testbed, which first flew in December 1974. The Mirage 2000 first flew in March 1978 and the Super Mirage 4000 in March 1979. By early 1981, total engine running time exceeded 12,400 h.

The initial M53-2 version is rated at 83.4 kN (18,740 lb st). Development is now centred on the M53-5, data for which are given below. This will be the engine of the production Mirage 2000. It passed its 150 h official type test in May 1979 and the first engines for the Armée de l'Air were in production in 1980. It is also the version fitted to the Super Mirage 4000.

The M53-P2 is an uprated version now under development. It is intended to power the Mirage 2000 from 1985.

DIMENSIONS:	
Overall length	4,853 mm (191 in)
Max diameter	1,055 mm (41·5 in)
WEIGHT, DRY:	1,450 kg (3,195 lb)
PERFORMANCE RATINGS:	
Max with afterburner:	
M53-5	88·3 kN (19,840 lb st)
M53-P2	95·2 kN (21,400 lb st)
Max without afterburner:	
M53-5	55 kN (12,350 lb st)
M53-P2	64·5 kN (14,500 lb st)
SPECIFIC FUEL CONSUMPTION (without afterburner):	
M53-5	24·64 mg/Ns (0·87 lb/h/lb st)
M53-P2	26·06 mg/Ns (0·92 lb/h/lb st)

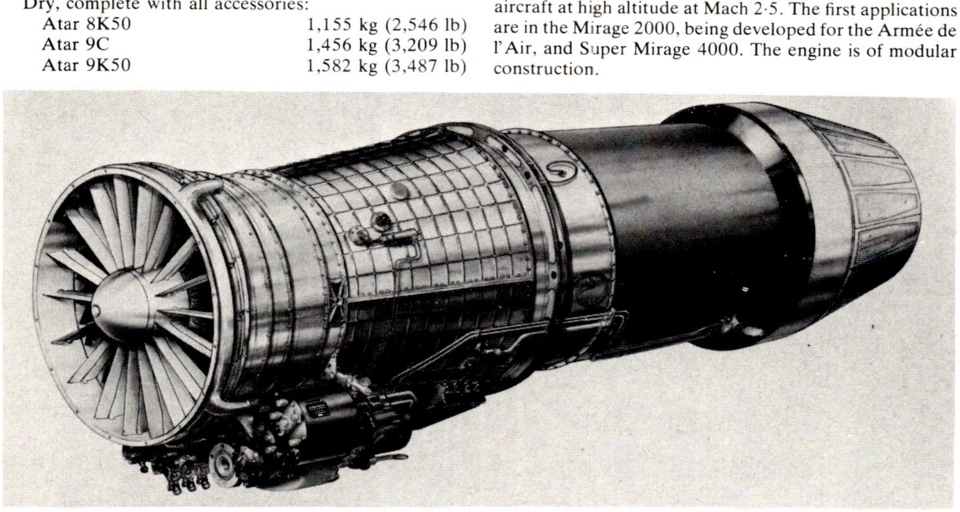

Artist's impression of the SNECMA M88 turbofan, under development for future fighter aircraft

SNECMA M88

This two-shaft turbofan will have variable inlet guide vanes, a three-stage LP compressor (fan), six-stage HP compressor, annular combustor, single-stage HP and LP turbines, and an afterburner with convergent/divergent nozzle. Intended to be the basis of a new family of engines for air-superiority and ground-attack fighters of the 1990s, it is being developed with the agreement of the French Defence Ministry; the following data relate to a demonstrator scheduled to run in 1983.

DIMENSION:

Overall length	3,800 mm (150 in)
WEIGHT, DRY:	900 kg (1,984 lb)
PERFORMANCE RATINGS:	
Max with afterburner	73·4 kN (16,500 lb st)
Max without afterburner	46 kN (10,350 lb st)
SPECIFIC FUEL CONSUMPTION:	
Max with afterburner	52·68 mg/Ns (1·86 lb/h/lb st)
Max without afterburner	22·09 mg/Ns (0·78 lb/h/lb st)

CFM INTERNATIONAL CFM56

SNECMA decided in the Autumn of 1971 to develop the CFM56, a subsonic turbofan in the ten-tonne-thrust (22,000 lb) class, in co-operation with General Electric. The project is covered under CFM in the International section.

GE/SNECMA/MTU CF6

To provide engines for the Airbus Industrie A300B and French Boeing 747s, SNECMA and MTU of Federal Germany participate in a co-production programme with General Electric to make the CF6-50 turbofan. SNECMA performs assembly and test, and manufactures parts, with a total share on a cost basis of 27 per cent. Included in the share are parts for engines for the DC-10-30. The agreement has been extended to include the CF6-80 series for the Airbus A310, A300B-600 and further Boeing 747s.

SNECMA/TURBOMÉCA LARZAC

This 13·2 kN (2,966 lb st) turbofan was designed jointly by SNECMA and Turboméca. A description of the Larzac is given under the entry for Turboméca-SNECMA GRTS, in this section.

SNECMA/MTU/RR/FN TYNE

Under Rolls-Royce licence, the Tyne 22 turboprop for the Transall C-160, was, in October 1977, put back into production by the original four-nation consortium, with the original work-split. SNECMA's share is made by Hispano-Suiza, and assembly and test take place at Sochata-SNECMA, Châtellerault. First delivery 1980. The similar Tyne 21 is also to be re-ordered, to power the ANG (Atlantic Nouvelle Génération).

SNPE
SOCIÉTÉ NATIONALE DES POUDRES ET EXPLOSIFS

HEADQUARTERS: 12 Quai Henri IV, 75181 Paris Cédex 04
Telephone: 277 15 70

SNPE continues to be the largest European producer of solid-propellant motors for missile manufacturers. Significant progress has been made in research and development, including smokeless high-impulse propellants, smokeless ramjet solid fuel, and new manufacturing processes for double-base and composite propellants. In the field of strategic missiles, the submarine-launched M4 MSBS has completed development and proved so successful that further developments are being studied. In tactical missiles, active programmes include motors for the MM 38, AM 39 and MM 40 Exocet, the new ASMP and Matra Super 530D, and production deliveries of motors for Shahine. As noted under the G2P heading, SNPE and SEP created a limited joint venture to co-ordinate their activities and act as a unified prime contractor for the propulsion of large missiles such as SSBS, MSBS and Pluton. Available details of important production motors are as follows:

SD. Double-base propellants, extruded, laminated or stamped. Theoretical specific impulse up to 235; combustion speed 5 to 44 mm/s; smokeless. Roubaix (Roland missile), Bugeat (HOT), PRPA (120 mm mortar), Lens (Crotale). Thonon (Thomson-Brandt 100 mm rocket), Amou (BAZ. 100), Cavaillon (ACL 89 mm), R4 Etoile (Thomson-Brandt 68 mm rocket), Valras (Milan).

Epictete. Cast double-base. Theoretical specific impulse, up to 230; combustion speed 3 to 30 mm/s; smokeless. Lampyre (Roland), Infra (HOT), Artus (Milan), Eole V (MM38 Exocet), Hector (Durandal).

Butalane. Composite HTPB or CTPB. Theoretical specific impulse up to 265; combustion speed 3 to 50 mm/s; smoke. Angèle (Matra Super 530), Richard (Magic II), Gerfaut (MM40 Exocet), Alain (ASMP), motor for M4.

Butalite. Composite derived from Butalane but non-aluminised. Theoretical specific impulse up to 242; combustion speed 0·5 to 45 mm/s; no primary smoke. Cosson (Patricia, Thomson-Brandt rocket), Nanesse (NRZ 96 Forges de Zeebrugge air/ground rocket), Narval (SM39 Exocet).

Nitramite. Double-base composite. Theoretical specific impulse up to 265; combustion speed 6 to 25 mm/s; smokeless. Anubis (AS 15TT), Hélios (AM39 Exocet), Aither (MM40 Exocet), Bias (Shahine).

Isolane (Isorgols). Composite polyurethanes. Theoretical specific impulse up to 258; combustion speed 3 to 20 mm/s; smoke. Madeleine (Matra R530), Venus (Malafon), Polka and Jacée (Masurca), motors for SSBS, M4 and Pluton.

Statolite. Solid propellants for ducted rockets. Theoretical specific impulse up to 1,200; combustion speed 0·5 to 5 mm/s, smokeless. Experimental systems.

Unidentified large motor, probably MSBS M4, at St Médard works

TURBOMÉCA
SOCIÉTÉ TURBOMÉCA

Bordes, 64320 Bizanos
Telephone: (15-59) 32 84 37
Telex: 560928
OTHER WORKS: Mézières S/Seine (Yvelines) and Tarnos (near Bayonne)
PARIS OFFICE: 1, Rue Beaujon, Paris 8e
PRESIDENT AND DIRECTOR-GENERAL:
J. R. Szydlowski

The Société Turboméca was formed in 1938 by Messieurs Szydlowski and Planiol to develop blowers, compressors and turbines for aeronautical use. Today, it is the leading European manufacturer of small turbine aero-engines. Since it first started development of gas turbines in 1947, the company has developed about 50 different types of power plant of which some 15 have entered production and ten types have been manufactured under licence in five countries.

By 1 January 1981 more than 18,500 Turboméca engines for fixed and rotary-wing applications and aircraft auxiliary duties had been delivered to customers in 97 countries. Approximately 14,000 more engines have been built under licence by what are today Rolls-Royce Ltd in the UK, Teledyne CAE in the USA, ENMASA in Spain, Hindustan Aeronautics Ltd in India, Bet-Shemesh in Israel and state factories in Romania and Yugoslavia.

Total covered floor area for Turboméca's three plants at Bordes, Mézières and Tarnos is 136,487 m² (1,469,134 sq ft). At 1 June 1981 the company employed a total of 4,399 people.

Turboméca has 45·6 per cent participation in CGTM, 50 per cent in RRTI, 26 per cent in CGR, 50 per cent in Turboméca SNECMA, 50 per cent in Astadyne (APU development with ABG-Semca), 50 per cent in MTU-Turboméca, 50 per cent in Rolls-Royce Turboméca, 30·22 per cent in Bet-Shemesh Engines, 22·08 per cent in Ormat Turbines and 50 per cent in RR-TM do Brasil.

ROLLS-ROYCE TURBOMÉCA ADOUR

This turbofan was developed jointly by Rolls-Royce and Turboméca for the SEPECAT Jaguar tactical strike fighter and advanced trainer. Other versions were developed subsequently for the BAe Hawk and Mitsubishi T-2 and F-1. A brief description of the Adour is given in the International section.

TURBOMÉCA ASTAFAN

The Astafan is a low-consumption lightweight turbofan of high by-pass ratio, low noise level design which made its first run during the Summer of 1969. Comprising an Astazou turboprop power section, operating at constant speed, driving a single-stage variable-pitch fan via reduction gearing, it is being developed in several versions corresponding to different development stages of the Astazou. All have the high bypass ratio of 7 : 1 and are characterised by very low specific fuel consumption. Variable-pitch blading facilitates constant-speed operation and permits off-loading of the engine during starting.

The following are the first versions under development:

Astafan IIG. Variant developed for military trainer applications.

Astafan III. Derived from the Astazou XVIII, with air-cooled turbine.

Astafan IV. Derived from the Astazou XX, with three-stage axial high-pressure compressor.

Astafan IVG. Refined engine for Aérospatiale Fouga 90.

Flight development of prototype Astafan engines began in a Hawk Commander on 8 April 1971. Two podded underwing Astafans replaced the original piston engines, conferring a substantial improvement in flight performance and a reduction in noise and vibration.

On 24 January 1976 an Aero Commander 690 flew at Pau for the first time powered by two Astafan IVF6 engines, similar to the basic Astafan IV but rated at 10·49 kN (2,359 lb) take-off thrust and with even lower noise and fuel consumption.

TYPE: Single-shaft turbofan with geared fan.

ENTRY CASING: Annular light alloy entry cowl and fan duct supported on double row of air straightener vanes downstream of fan rotor.

FAN: Single-stage fan with variable-incidence rotor blading overhung at front without entry guide vanes. Drive from gas generator section is via two-stage epicyclic gear train housed in cylindrical casing forming inner wall of fan duct. Astafan IV has fan of increased diameter (700 mm; 27·5 in).

COMPRESSOR, COMBUSTION SYSTEM AND TURBINE: Same as Astazou XVIII. Normal gas-generator operating speed, 43,000 rpm. (Astafan IV has Astazou XX compressor, running at 42,000 rpm.)

JET PIPE: Fixed type with straight frustum inner cone.

ACCESSORIES: Mounted on casing forming rear of secondary air intake.

FUEL SYSTEM: Independent control systems for starting and normal operation. Fuel regulator maintains speed

Turboméca Astafan IVG geared variable-pitch turbofan

constant with pilot operating single lever controlling fan blade pitch hydraulically to vary thrust output. Turboméca 'thermic' load limiter controls turbine entry temperature.

Fuel Grades: AIR 3404A, 3405 or 3407A.

Lubrication System: Pressure lubrication to bearings and reduction gear, with annular engine-mounted oil tank.

Starting: Automatic electrical starting with compressor blow-off valve and fan in minimum pitch.

Dimensions:

Length overall:	
Astafan IIG	1,980 mm (77·95 in)
Astafan III	2,030 mm (80·0 in)
Astafan IV	2,218 mm (87·5 in)
Astafan IVG	2,088 mm (82·2 in)
Max diameter over fan cowl:	
Astafan IIG	693 mm (27·3 in)
Astafan III	665 mm (26·2 in)
Astafan IV	780 mm (30·7 in)
Astafan IVG	700 mm (27·56 in)

Weights:

Equipped Astafan IIG	285 kg (628 lb)
Astafan III bare engine	210 kg (462 lb)
Equipped Astafan III	approx 230 kg (507 lb)
Astafan IV bare engine	220 kg (485 lb)
Equipped Astafan IVG	305 kg (672 lb)

Performance Ratings:

T-O dry:	
Astafan IIG	6·88 kN (1,543 lb st)
Astafan IVG	7·74 kN (1,740 lb st)
T-O, wet:	
Astafan III	8·34 kN (1,870 lb st)
Astafan IV	12·06 kN (2,710 lb st)

Specific Fuel Consumption:

At T-O rating:	
Astafan IIG	10·76 mg/Ns (0·38 lb/h/lb st)
Astafan III	10·34 mg/Ns (0·365 lb/h/lb st)
Astafan IV	8·78 mg/Ns (0·310 lb/h/lb st)
Astafan IVG	11·61 mg/Ns (0·41 lb/h/lb st)

TURBOMÉCA-SNECMA LARZAC

This small turbofan has been developed jointly by Turboméca and SNECMA to power military trainers and other small aircraft. A description of the Larzac is given under Turboméca-SNECMA GRTS.

TURBOMÉCA MARBORÉ

The Marboré turbojet is the most widely used of Turboméca's range of gas turbines. By 1 September 1979, a total of 4,353 Marboré II engines of 3·92 kN (880 lb st) had been delivered by Turboméca and a further 10,000 by Continental Aviation and Teledyne CAE (see US section) as the J69.

This initial version of the engine was joined in production by the 4·71 kN (1,058 lb st) Marboré VI with receipt of type approval in June 1962. By 1 September 1979, a total of 1,194 Marboré VI turbojets had been built. Production by Turboméca is now complete.

The Marboré VI was also built under licence in Spain by ENMASA as the Marboré M21. Details of the engine were given in the 1979-80 *Jane's*.

TURBOMÉCA ARBIZON

Announced in 1970, the Arbizon is a simple single-shaft turbojet, known originally as the TR281 and derived from the Turmo $IIIC_3$ turboshaft. Intended for the propulsion of missiles and RPVs, it has single-stage axial and centrifugal compressors and a single-stage turbine.

Arbizon IIIB. Original version, for Otomat cruise missile. Four-lobed bellmouth inlet ducts surround front face carrying electric starter and other accessories. Mass flow 6 kg (13·2 lb)/s; pressure ratio 5·5. Total of 164 built by 1 September 1979.

Arbizon IV. Smaller version, under development. Design life 30 h. Air-impingement starter.

Turboméca Arbizon IV short-life turbojet (underside shown, with inlet to left)

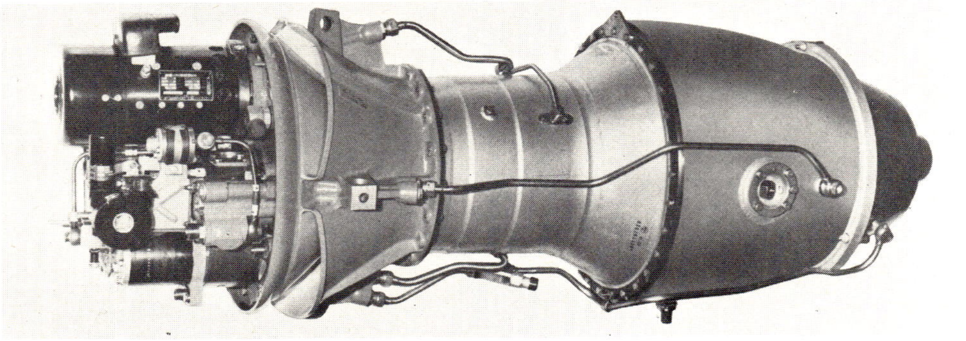

Turboméca Arbizon IIIB expendable turbojet, rated at 3·73 kN (836 lb st)

Dimensions:

Diameter of combustion chamber:	
Arbizon IIIB	405 mm (15·95 in)
Max diameter:	
Arbizon IV	330 mm (13·00 in)
Length overall, with accessories:	
Arbizon IIIB	1,361 mm (53·58 in)
Arbizon IV	1,025 mm (40·35 in)

Weight, Dry:

Arbizon IIIB	115 kg (253 lb)
Arbizon IV	60 kg (132 lb)

Performance Ratings:

T-O: Arbizon IIIB	3·73 kN (836 lb st) at 33,000 rpm
Arbizon IV	3·60 kN (809 lb st)
Max continuous:	
Arbizon IIIB	3·24 kN (727 lb st) at 32,000 rpm
Arbizon IV	3·30 kN (741 lb st)

Specific Fuel Consumption:

At max continuous rating:	
Arbizon IIIB	31·44 mg/Ns (1·11 lb/h/lb st)
Arbizon IV	29·29 mg/Ns (1·034 lb/h/lb st)

TURBOMÉCA ARRIEL

This turboshaft engine is being used initially in two Aérospatiale helicopters, the single-engined AS 350 Ecureuil and twin-engined SA 365 Dauphin. It could also power a future version of the SA 341 Gazelle.

The Arriel is intended to have low first cost, low maintenance cost and low specific weight. It is characterised by modular construction, and is expected eventually to form the basis for a single-shaft turboprop and a turbofan in the 4·90 kN (1,100 lb st) class. The gas generator ran in 1973. The first complete engine ran on the bench on 7 August 1974. Flight development began on 17 December 1974 in the SA 341-02 Gazelle, which had been converted for Arriel development by Aérospatiale and the CGTM. The twin-engined SA 365 first flew on 24 January 1975, and the AS 350 Ecureuil on 14 February 1975. By 1 September 1979 a total of 203 Arriels had been delivered and the rate of output was increasing.

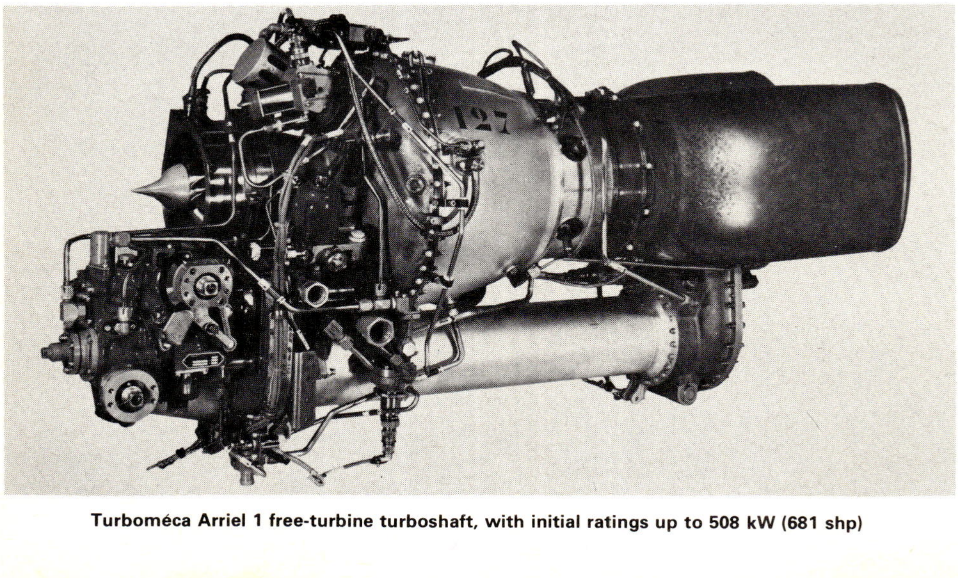

Turboméca Arriel 1 free-turbine turboshaft, with initial ratings up to 508 kW (681 shp)

Type: Single-shaft axial-plus-centrifugal free-turbine turboshaft.

Air Intake: Direct pitot entry to axial compressor.

Compressor: Single-stage axial compressor, machined from titanium forging. Supersonic centrifugal stage also machined from titanium. High rotational speed for maximum attainable pressure ratio (9:1).

Combustion Chamber: Annular chamber, with flow radially outwards and then inwards. Centrifugal fuel injection without central tube.

Gas-Generator Turbine: Two integral cast axial stages with solid blades. Turbine shield capable of disc containment.

Power Turbine: Single axial stage with inserted blades.

Jet Pipe: Exhaust diffuser fabricated by welding.

Reduction Gear: Light alloy gearbox, containing two stages of helical gears, giving drive at 6,000 rpm to output shaft extending whole length of engine, with drive connections to both front and rear. Hydraulic torquemeter.

Accessory Drives: Two bevel gears and radial quill shaft drive accessory gearbox at front end, carried between compressor case and output shaft. Main pad provides for optional 12,000 rpm alternator; other drives for oil pumps, tachometer generator, governor and starter.

Lubrication System: Independent circuit. Oil from tank passes through gear pump and metallic-cartridge filter. Return from engine via three gear scavenge pumps. Temperature probe and pressure switch to verify operation.

Oil Specification: AIR 3512 (mineral) or AIR 3513A (synthetic).

Mounting: Multi-point flanges allow easy mounting in single or twin installation.

Starting: Electric starter or starter/generator.

Dimensions:

Length, excl accessories	1,090 mm (42·91 in)
Height overall	569 mm (22·40 in)
Width	430 mm (16·93 in)

Weight, Dry:

With all engine accessories	109 kg (240 lb)

Performance Ratings:

Max contingency, initial	508 kW (681 shp)
Max contingency, later	544 kW (730 shp)
Take-off and intermediate contingency	478 kW (641 shp)
Max continuous	441 kW (592 shp)

Specific Fuel Consumption:

Max contingency	93·1 μg/J (0·551 lb/h/shp)
Intermediate contingency	96·8 μg/J (0·573 lb/h/shp)

TURBOMÉCA ASTAZOU TURBOPROP

The Astazou is the major turboprop in the Turboméca range and is in production in its 636 kW (853 ehp) Astazou XIVC and 760·6 kW (1,020 ehp) Astazou XVI versions. The Astazou XIV was certificated by the French airworthiness authorities in October 1968, followed by ARB/FAA certification of the Astazou XIVC and C1 in March 1969.

Current versions of the Astazou are:

Astazou XII. Powered Shorts Skyvan Srs 2 at 690 shp and Pilatus Turbo-Porter PC-6/A1-H2 at 700 ehp.

Astazou XIV (alias AZ14). Developed from Astazou XII. Powers early Jetstream business aircraft at 853 ehp.

Astazou XVI (alias AZ16). Higher rated version of Astazou XIV and first engine to enter production with new

Turboméca aircooled turbine. The XVID, without starter/generator, powers the former production versions of the Jetstream; the XVIZ powers the Nord 260A. The Astazou XVIG, equipped for sustained inverted flight, was certificated by the Services Officiels Français on 30 April 1971; it powers the Argentinian Pucará combat aircraft. By 1 September 1979, a total of 184 Astazou XVI engines had been built.

Astazou XVIII. Higher rated version of Astazou XVI which first ran in early 1969 with T-O rating of 860·5 kW (1,154 ehp).

Astazou XX. Under development. This engine has two transonic axial compressor stages in titanium, in addition to the centrifugal stage machined in steel. Maximum T-O rating is 1,075 ekW (1,442 ehp).

Details of the Astazou series of turboshafts are given separately.

TYPE: Single-shaft axial-plus-centrifugal turboprop.

REDUCTION GEAR: Mounted in tapered cylindrical casing at front of engine, with two-stage epicyclic reduction gear having helical primary gears and straight secondary gears. Reduction ratio 24·115 : 1 (XVIG, 21·8 : 1). (Astazou XX gearbox incorporates torquemeter.)

AIR INTAKE: Annular air intake at rear of reduction gear casing. Hot-air de-icing.

COMPRESSOR: Two-stage axial followed by single-stage centrifugal with single-sided impeller. Two rows of stator blades aft of each axial rotor. Centrifugal stage has radial and axial diffusers. (Astazou XX has a four-stage compressor, comprising three axial followed by one centrifugal.)

COMBUSTION SYSTEM: Reverse-flow annular type with centrifugal fuel injector using rotary atomiser disc. Ignition by two ventilated torch igniters.

TURBINE: Three-stage axial with blades integral with discs. (Air cooling provided for Astazou XVI, XVIII and XX.) Discs attached by curvic couplings and bolts.

JET PIPE: Fixed type with curved inner cone.

ACCESSORIES: Mounted on casing forming rear of air intake. Drive pads provided for starter/generator, oil pump, fuel pump and speed governor, tacho-generator, AC generator and hydraulic pump.

MOUNTING: Trunnion located on each side of turbine casing front flange, plus third trunnion on underside of turbine casing.

FUEL SYSTEM: Automatic constant-speed system with propeller Beta-control and Turboméca 'thermic' load limiter and speed governor.

FUEL GRADES: AIR 3404, 3405, or 3407.

LUBRICATION SYSTEM: Pressure lubrication to bearings and reduction gear, with 8 litre (14 Imp pint) oil tank mounted at front of engine.

OIL SPECIFICATION: AIR 3515 or synthetic AIR 3573.

STARTING: Electric.

DIMENSIONS:
Diameter over intake cowl 546 mm (21·5 in)
Overall length, incl propeller 2,047 mm (80·6 in)

WEIGHTS:
With accessories:
Astazou XIV approx 206 kg (454 lb)
Astazou XVID 205 kg (452 lb)
Astazou XVIG 228 kg (502 lb)
Astazou XVIZ 213 kg (468 lb)
Astazou XVIII approx 205 kg (452 lb)
Astazou XX 220 kg (484 lb)

PERFORMANCE RATINGS:
T-O:
Astazou XIV 636 ekW; 596·5 kW (853 ehp; 800 shp) at 43,000 rpm
Astazou XVID 723 ekW; 681 kW (969 ehp; 913 shp) at 43,089 rpm
Astazou XVIG, XVIZ 761 ekW; 720 kW (1,020 ehp; 965 shp) at 43,000 rpm

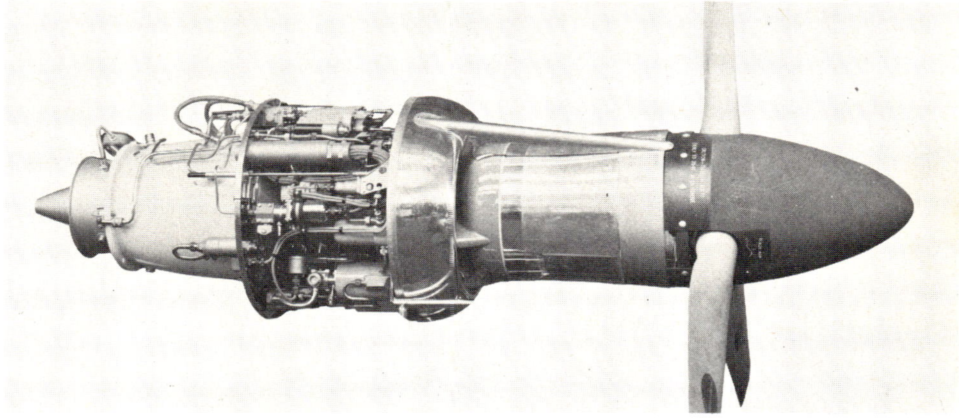

Turboméca Astazou XIVC turboprop engine, rated at 636 kW (853 ehp)

Astazou XVIII 861 ekW; 813 kW (1,154 ehp; 1,090 shp) at 43,000 rpm
Astazou XX 1,075 ekW; 1,030 kW (1,442 ehp; 1,381 shp) at 42,000 rpm

Max continuous:
Astazou XIV 574 ekW; 537 kW (770 ehp; 720 shp) at 43,000 rpm
Astazou XVID 626 ekW; 586 kW (840 ehp; 786 shp) at 43,089 rpm
Astazou XVIG, XVIZ 696 ekW; 654 kW (934 ehp; 877 shp) at 43,000 rpm
Astazou XVIII 809 ekW; 768 kW (1,085 ehp; 1,030 shp) at 43,000 rpm
Astazou XX 951 ekW; 907 kW (1,276 ehp; 1,217 shp) at 42,000 rpm

SPECIFIC FUEL CONSUMPTION:
At T-O rating:
Astazou XIV 92·4 μg/J (0·547 lb/h/shp)
Astazou XVI (all versions) 88·7 μg/J (0·525 lb/h/shp)
Astazou XVIII 85·3 μg/J (0·505 lb/h/ehp)
Astazou XX 75·9 μg/J (0·449 lb/h/ehp)

TURBOMÉCA ASTAZOU TURBOSHAFT

This turboshaft series of the Astazou family is derived from the early second-generation Astazou II turboprop fitted to the Mitsubishi MU-2 and Pilatus Turbo-Porter. Variants are as follows:

Astazou IIA. Rated at 390 kW (523 shp) and powers the Aérospatiale SA 318C Alouette II Astazou helicopter. Total of 732 built by 1977.

Astazou III. Definitive turboshaft for Anglo-French helicopter programme for production SA 341 Gazelle. Derived from Astazou IIA but with revised profile of turbine, using higher temperature alloy to match power needs of SA 341. Astazou IIIA, B, C and N for Gazelle versions all certificated in June 1972; uprated IIIA2, B2 C2 and N2 certificated May 1978. Produced jointly by Turboméca and Rolls-Royce Ltd, with 896 delivered by 1 September 1979.

Astazou XIVB and XIVF. In production for the SA 319B Alouette III; XIVB is civil and XIVF military. Flat rated to 441 kW (591 shp) (1 h) up to 4,000 m (13,125 ft) or +55°C.

Astazou XIVH. In production for SA 341 Gazelle, with much increased power. Flat-rated to transmission limit, to remove all altitude and temperature limitations. Certificated October 1974. Total of all XIV versions, 586 by 1 September 1979.

Astazou XVIIIA. Further increase in power gained by improved turbine, allowing higher gas temperature. Powers SA 360 Dauphin.

Astazou XX. Fourth axial compressor stage added. Designed for operation in hot and high countries. Intended to power SA 361 helicopter.

The following description relates to the Astazou IIIN except where indicated:

TYPE: Single-shaft axial-plus-centrifugal turboshaft.

REDUCTION GEAR: Similar to Astazou XIV turboprop. Reduction ratio 7·039 : 1 (Astazou XIVB/F, 7·345; XVIA, XVIIIA, 7·375).

AIR INTAKE: Annular air intake at rear of reduction gear casing.

COMPRESSOR: Single-stage axial (IIA, IIN, IIIN), two-stage axial (XIV, XVIII) or three-stage axial (XX) followed by single-stage centrifugal with single-sided impeller. Air mass flow 2·5 kg/s (5·5 lb/s).

COMBUSTION SYSTEM: Similar to Astazou XIV.

TURBINE: Similar to Astazou XIV.

JET PIPE: Similar to Astazou XIV.

ACCESSORIES: Five drive pads on casing forming rear of air intake.

MOUNTING: At front by flange located at power take-off section, and at rear by two lugs on accessory mounting pad section.

FUEL SYSTEM: Automatic constant speed control with speed governor.

LUBRICATION SYSTEM: Pressure type with gear type pumps. Oil tank of 8 litre (14 Imp pint) capacity mounted at front of engine.

STARTING: Electrical, automatic.

DIMENSIONS:
Height: Astazou IIA 458 mm (18 in)
Astazou III, XIVH 460 mm (18·1 in)
Astazou XVIA, XVIIIA 698 mm (27·48 in)
Astazou XX 721 mm (28·4 in)
Width: Astazou IIA 480 mm (18·8 in)
Astazou III, XIVH 460 mm (18·1 in)
Length overall: Astazou IIA 1,272 mm (50·0 in)
Astazou III, XIVB/F 1,433 mm (56·3 in)
Astazou XIVH 1,470 mm (57·9 in)
Astazou XVIA, XVIIIA 1,327 mm (52·2 in)
Astazou XX 1,529 mm (60·22 in)

WEIGHTS:
Equipped:
Astazou III 147 kg (324 lb)
Astazou III (suffix 2) 150 kg (330 lb)
Astazou XIVB/F 166 kg (366 lb)
Astazou XIVH 160 kg (353 lb)
Astazou XVIA, XVIIIA 155 kg (341 lb)
Astazou XX 195 kg (430 lb)

PERFORMANCE RATINGS:
Max power: Astazou IIA 390 kW (523 shp)
Astazou III 441 kW (592 shp)
Astazou III (suffix 2) 481 kW (645 shp)
Astazou XX 749 kW (1,005 shp)
One hour: Astazou XIVB/F 441 kW (591 shp)
Astazou XVIA 651 kW (873 shp) maintained at sea level to 30°C
Astazou XVIIIA 651 kW (873 shp) maintained at sea level to 40°C
Max continuous: Astazou IIA 353 kW (473 shp)
Astazou III 390 kW (523 shp)
Astazou III (suffix 2) 441 kW (592 shp)
Astazou XIVB/F 405 kW (543 shp)
Astazou XIVH flat-rated in SA 341 to 440·7 kW (591 shp) to 55°C or 4,000 m (13,125 ft)
Astazou XVIA, XVIIIA 600 kW (805 shp)
Astazou XX 675 kW (905 shp)

SPECIFIC FUEL CONSUMPTION:
At max power rating:
Astazou IIA 105·3 μg/J (0·623 lb/h/shp)
Astazou III 108·7 μg/J (0·643 lb/h/shp)
Astazou III (suffix 2) 109·9 μg/J (0·65 lb/h/shp)
Astazou XIVB/F 105·5 μg/J (0·624 lb/h/shp)
Astazou XVIA 93·8 μg/J (0·555 lb/h/shp)
Astazou XVIIIA 91·3 μg/J (0·540 lb/h/shp)
Astazou XX 85·9 μg/J (0·508 lb/h/shp)

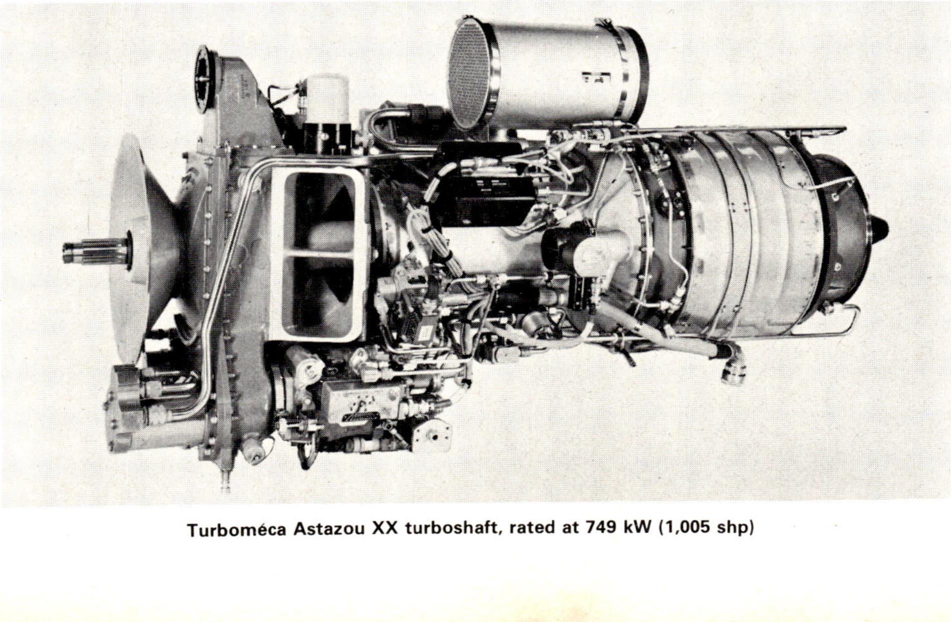

Turboméca Astazou XX turboshaft, rated at 749 kW (1,005 shp)

TURBOMÉCA ARTOUSTE III

The Artouste IIIB is a single-shaft turboshaft derived from the Artouste II. It is a member of the second generation of Turboméca engines with two-stage axial-centrifugal compressor and three-stage turbine. The Artouste IIIB has a pressure ratio of 5·2 : 1. Air mass flow is 4·3 kg/s (9·5 lb/s) at 33,300 rpm.

Type approval at the rating given below was received on 25 May 1961, following completion of a 150-h official type test. Production at Turboméca continues. In addition, Artouste IIIBs are being built under licence in India by Hindustan Aeronautics Ltd.

The Artouste IIIB, which powers the Aérospatiale SA 316B Alouette III, obtained FAA certification in March 1962 and in August 1968 similar certification of the Artouste IIC1, C2, C5 and C6, powering the SE 3130 and 313B Alouette II Artouste, was also obtained.

An uprated version, the Artouste IIID, was certificated on 30 April 1971. It differs in having a reduction gear giving 5,864 rpm at the driveshaft (instead of 5,773 rpm) and in slightly revised equipment. The IIID powers a late version of the Alouette III. A total of 2,301 Artouste III engines had been built by 1 September 1979.

DIMENSIONS:

Length	1,815 mm (71·46 in)
Height	627 mm (24·68 in)
Width	507 mm (19·96 in)

WEIGHT, DRY:

Fully equipped:	
Artouste IIIB	182 kg (400 lb)
Artouste IIID	178 kg (392 lb)

PERFORMANCE RATINGS (maintained up to 55°C at S/L or up to approximately 4,000 m; 13,150 ft):

T-O:	
Artouste IIIB	420 kW (563 shp)
Artouste IIID	440 kW (590 shp)
Max continuous (both)	405 kW (543 shp)

SPECIFIC FUEL CONSUMPTION:

T-O:	
Artouste IIIB	128·8 μg/J (0·762 lb/h/shp)
Artouste IIID	126·2 μg/J (0·747 lb/h/shp)
Max continuous (both)	130·1 μg/J (0·77 lb/h/shp)

MTU-TURBOMÉCA MTM 380

This new helicopter engine is covered in the International part of this section.

TURBOMÉCA TURMO

The Turmo free-turbine engine is in service in both turboshaft and turboprop versions.

The main variants are as follows:

Turmo IIIC$_3$. This was the original power plant of the triple-engined SA 321 Super Frelon helicopter. Maximum contingency rating is 1,104 kW (1,480 shp).

Turmo IIIC$_4$. Developed from Turmo IIIC$_3$ and with a maximum contingency rating of 1,032 kW (1,384 shp), this all-weather version is manufactured jointly by Turboméca and Rolls-Royce to power SA 330 Puma twin-engined helicopters under the Franco-British helicopter agreement of October 1967. Certificated by the Services Officiels Français on 9 October 1970. Total production of IIIC$_4$ and IV by 1 September 1979 was 1,644 engines.

Turmo IIIC$_5$, IIIC$_6$, IIIC$_7$. Similar to Turmo IIIC$_3$ but with different ratings. The SA 321F and 321J Super Frelons powered by these engines obtained French certification in June 1968. Total production of Super Frelon engines (including E series), 549.

Turmo IIID. Turboprop version, similar in basic construction to Turmo IIIC series but with output speed limited to 6,000 rpm.

Turmo IIIE$_3$. Similar to Turmo IIIC$_3$ but with different ratings. In production for SA 321 Super Frelon.

Turmo IIIE$_6$. Higher turbine temperature.

Turmo IV. The Turmo IVA is a civil engine derived from the IIIC$_4$, with a maximum contingency rating of 1,057 kW (1,417 shp). The IVB is a military version having the same ratings as the IIIC$_4$.

The following description applies generally to the Turmo IIIC$_3$, C$_4$, C$_5$ and E$_3$, except where indicated:

TYPE: Free-turbine axial-plus-centrifugal turboshaft.

The 420 kW (563 shp) Turboméca Artouste IIIB single-shaft helicopter turboshaft

The 1,032 kW (1,384 shp) Turboméca Turmo IIIC$_4$ turboshaft which powers the SA 330 Puma helicopter

REDUCTION GEAR: Turmo IIIC$_3$, C$_5$ and E$_3$ fitted with rear-mounted reduction gear mounted in bifurcated exhaust duct with rear-facing power take-off shaft. Turmo IIIC$_4$ is a direct-drive engine.

AIR INTAKE: Annular forward-facing intake, with de-icing in Turmo IIIC$_4$ and C$_5$.

COMPRESSOR: Single-stage axial followed by single-stage centrifugal with single-sided impeller. Two rows of light alloy stator blades aft of axial stage. Centrifugal stage has steel radial and axial diffusers; impeller located by lugs on turbine shaft. Axial rotor blades, titanium in Turmo IIIC$_3$, C$_5$ and E$_3$ and steel in Turmo IIIC$_4$, pin-mounted in steel disc with integral shaft. Pressure ratio 5·9 : 1 on Turmo IIIC$_3$. Air mass flow 5·9 kg (13 lb)/s.

COMBUSTION SYSTEM: Reverse-flow annular type with centrifugal fuel injector using rotary atomiser disc. Ignition by two ventilated torch igniters.

GAS GENERATOR TURBINE: Two-stage axial unit with integral rotor blades.

POWER TURBINE: Two-stage axial unit in Turmo IIIC$_3$, C$_5$ and E$_3$, and single-stage in Turmo IIIC$_4$. In all advanced production engines of IIIC$_4$ derivation the power turbine speed is 22,840 rpm under all high-power conditions.

JET PIPE: Fixed type with lateral bifurcated exhaust duct in Turmo IIIC$_3$, C$_5$ and E$_3$, and single lateral duct on Turmo IIIC$_4$.

ACCESSORIES: Mounted above and below intake casing with drive pads for oil pump, fuel control unit, electric starter, tacho-generator and, on Turmo IIIC$_4$, oil cooler fan. Control unit remote drive also provided on Turmo IIIC$_4$ from bevel gear drive on power turbine output shaft.

MOUNTING: Two lateral supports fitted to lower part of turbine casing at rear flange output shaft protection tube. On Turmo IIIC$_4$, also on reduction gear case.

FUEL SYSTEM: Fuel control unit for gas generator on Turmo IIIC$_3$, C$_5$ and E$_3$, with speed limiter for power turbine also fitted on E$_3$. Constant-speed system fitted on Turmo IIIC$_4$ power turbine, with speed limiter also fitted on gas generator.

FUEL GRADE: AIR 3405 for Turmo IIIC$_4$.

LUBRICATION SYSTEM: Pressure type with oil cooler and 13 litre (23 Imp pint) tank at front of engine on Turmo IIIC$_4$, with oil tank only around intake casing on Turmo IIIC$_3$, C$_5$ and E$_3$, and by intake accessory drive gear on Turmo IIIC$_4$.

OIL SPECIFICATION: AIR 3155A, or synthetic AIR 3513, for Turmo IIIC$_4$.

STARTING: Automatic system with electric starter motor.

DIMENSIONS:

Height:	
Turmo IIIC$_3$, C$_5$ and E$_3$	716·5 mm (28·2 in)
Turmo IIIC$_4$	719 mm (28·3 in)
Turmo IIID$_3$	926 mm (36·5 in)
Width:	
Turmo IIIC$_3$, C$_5$ and E$_3$	693 mm (27·3 in)
Turmo IIIC$_4$	637 mm (25·1 in)
Turmo IIID$_3$	934 mm (36·8 in)
Length:	
Turmo IIIC$_3$, C$_5$ and E$_3$	1,975·7 mm (78·0 in)
Turmo IIIC$_4$	2,184 mm (85·5 in)
Turmo IIID$_3$	1,868 mm (73·6 in)

WEIGHT, DRY:

Turmo IIIC$_3$ and E$_3$, fully equipped	297 kg (655 lb)
Turmo IIIC$_5$, IIIC$_6$ and IIIC$_7$	325 kg (716 lb)
Turmo IIIC$_4$, equipped engine	225 kg (496 lb)
Turmo IIID$_3$, basic engine	365 kg (805 lb)

PERFORMANCE RATINGS:

T-O: Turmo IIIC$_3$, D$_3$ and E$_3$	1,104 kW (1,480 shp)
Turmo IIIE$_6$	1,181 kW (1,584 shp)
Max contingency:	
Turmo IIIC$_4$ at 33,800 gas-generator rpm	1,032 kW (1,384 shp)
Turmo IIIC$_6$ at 33,550 gas-generator rpm	1,156 kW (1,550 shp)
Turmo IIIC$_7$ at 33,800 gas-generator rpm	1,200 kW (1,610 shp)
Turmo IVA at 33,950 gas-generator rpm	1,057 kW (1,417 shp)
Turmo IVC at 33,800 gas-generator rpm	1,163 kW (1,560 shp)
T-O and intermediate contingency:	
Turmo IIIC$_5$	1,050 kW (1,408 shp)

SPECIFIC FUEL CONSUMPTION:

At T-O rating:	
Turmo IIIC$_3$ and E$_3$	101·9 μg/J (0·603 lb/h/shp)
Turmo IIID$_3$	104·1 μg/J (0·616 lb/h/shp)
At max contingency rating:	
Turmo IIIC$_4$, C$_5$, C$_6$, C$_7$ and IV	106·8 μg/J (0·632 lb/h/shp)
Turmo IVA	106·3 μg/J (0·629 lb/h/shp)

TURBOMÉCA/ROLLS-ROYCE RTM 321

See International section.

TURBOMÉCA MAKILA

This new turboshaft engine, rated at an initial 1,368 kW (1,835 shp) for take-off and intermediate contingency, is under development to power the Aérospatiale

Turboméca Makila free-turbine turboshaft, with initial ratings up to 1,398 kW (1,875 shp)

Super Puma helicopter. Derived partly from the Turmo family, it incorporates all the latest features of the company's advanced engines, including: rapid-strip modular construction; three axial stages of compression plus one centrifugal; later fuel inlet to centrifugal atomiser; two-stage gas-generator turbine (probably with cooled blades); two-stage free power turbine; and lateral exhaust.

During 1974 this engine was confirmed as partner to the Arriel in laying the foundation for the company's marketing in the next 15 years. The world market for this size of engine is put at 10,000 units. The first engine was delivered for bench test in 1976, and gas generator testing began in November 1976. The first complete Makila ran in January and flew in June 1977. French and FAA certification followed in February 1980 and April 1981 respectively.

DIMENSIONS:
- Length, intake face to rear face: 1,395 mm (54·94 in)
- Width: 530 mm (20·9 in)
- Max diameter: 514 mm (20·25 in)

WEIGHT, DRY: Basic: 210 kg (463 lb)
- Equipped: 242 kg (533 lb)

PERFORMANCE RATINGS (ISA, S/L):
- Max contingency
  - 1,398 kW (1,875 shp) at 36,300 gas-generator rpm
- T-O and intermediate
  - 1,368 kW (1,835 shp) at 35,500 gas-generator rpm
- Max continuous
  - 1,208 kW (1,620 shp) at 34,750 gas-generator rpm

SPECIFIC FUEL CONSUMPTION:
- Max contingency: 84·0 μg/J (0·497 lb/h/shp)
- T-O and intermediate: 84·7 μg/J (0·501 lb/h/shp)
- Max continuous: 85·7 μg/J (0·507 lb/h/shp)

TURBOMÉCA-SNECMA
GROUPEMENT TURBOMÉCA-SNECMA (GRTS)

1 rue Beaujon, BP 37-08, 75362 Paris Cédex 08
Telephone: 924 18 61

ADMINISTRATORS:
R. Florentini
E. Delfour

MANAGEMENT CONTROL COMMITTEE:
R. Martin
F. Rigaud

FINANCIAL COMMISSARY: C. Hirt

Announced in March 1969, Groupement Turboméca-SNECMA is a company formed jointly by Société Turboméca and SNECMA to be responsible for the design, development, manufacture, sales and service support of the Larzac all-axial small turbofan launched in 1968 as a joint venture by the two companies. Groupement Turboméca-SNECMA has no capital at present and comprises primarily a joint management organisation to produce the engine.

Turboméca-SNECMA Larzac 04 two-shaft turbofan, rated at 13·25 kN (2,980 lb st)

TURBOMÉCA-SNECMA LARZAC

Originally this small turbofan was planned for a wide range of applications, and the first prototype was a 9·8 kN (2,200 lb st) engine aimed at the commercial market. This type of engine ran in May 1969 and began flight development in a pod carried by a Constellation in March 1971. By this time the main immediate market had shifted to military trainers, and GRTS designed the Larzac 04 for this purpose.

In February 1972 the Larzac 04 was selected for a joint Franco-German programme to provide propulsion for the Alpha Jet (see International entry in Aircraft section). In addition to the two French partners in GRTS, two West German companies, MTU and KHD, were added to the programme. Both played a part in the manufacture of prototype engines and the achievement of endurance tests. All four companies are sharing in production and post-certification development. Complete engines are assembled in both countries for the Alpha Jet programme. The first production engine was delivered in September 1977. The work split is France 56·5 per cent, West Germany 43·5.

Bench testing of the Larzac 04 began in May 1972. Flight development with the Constellation testbed began in March 1973, and with a Falcon 10 in July 1973. The first Alpha Jet flew on 26 October 1973, and qualification of the Larzac 04 was accomplished on schedule in May 1975. The rating given is at a turbine entry temperature of 1,130°C; growth thrust potential greater than 20 per cent is forecast without dimensional change.

In September 1972 the French Services Officiels approved an agreement between GRTS and Teledyne CAE covering the production, marketing and after-sales support of the Larzac in the USA and Canada.

The Larzac 04 has a two-stage fan, four-stage HP compressor, annular combustion chamber with vaporising burners, single-stage HP turbine with cooled blades and single-stage LP turbine. Maximum airflow is 28 kg (62 lb)/s, pressure ratio 10·6 and bypass ratio 1·13. A single fixed-area jet pipe is used. All accessories are driven by the HP spool and grouped under the fan case. The engine is mounted by an isostatic suspension on either side of the centre of gravity. The engine is of modular design and is intended to produce minimum noise and smoke.

DIMENSIONS:
- Overall length of basic engine: 1,179 mm (46·4 in)
- Overall diameter: 602 mm (23·7 in)

WEIGHT, DRY:
- Larzac 04: 290 kg (640 lb)

T-O THRUST (S/L, static):
- Larzac 04: 13·25 kN (2,980 lb)

SPECIFIC FUEL CONSUMPTION:
- Larzac 04: 20·1 mg/Ns (0·71 lb/h/lb st)

GERMANY
(FEDERAL REPUBLIC)

ERNO
ERNO RAUMFAHRTTECHNIK GmbH

Hünefeldstrasse 1-5, 2800 Bremen
Telephone: 0421 53 91

ERNO, a subsidiary of VFW claims to be Europe's leading producer of hydrazine-based motor systems. Most of its products in this field are small thrusters for space applications. Thrusts range from 0·05 N to 2·5 kN (0·011-557 lb) for motors operating at about 1,000°C and from 0·001 N to 4·0 N (0·0002-0·899 lb) for cold-gas thrusters. Spacecraft using ERNO thrusters include Intelsat III, Aeros A/B, ECS, OTS, TD-1A, MARECS, Telecom 1 and ISPM. ERNO also produces low-thrust electrothermal hydrazine thrusters and many other devices including complete reaction-control subsystems.

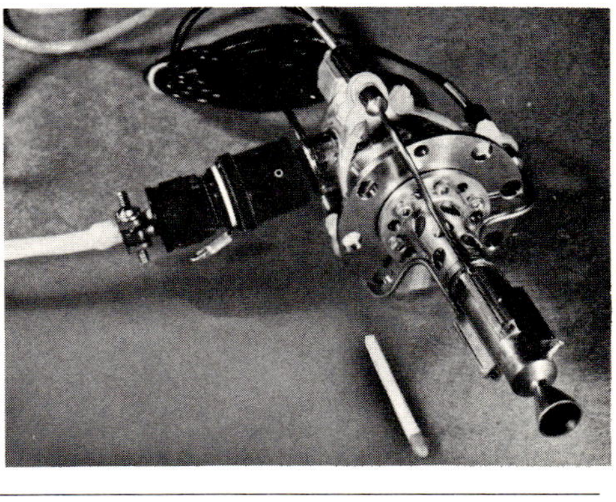

ERNO 2·0 N hydrazine thruster used on OTS, ECS, MARECS, Telecom and ISPM.

KHD
KHD LUFTFAHRTTECHNIK GmbH

6370 Oberursel, PO Box 246
Telephone: (0 61 71) 500

KHD is a leading manufacturer of diesel engines. In its subsidiary at Oberursel, given the title above in 1980, it concentrates its activities in the gas-turbine field for aircraft. Its most important gas turbine is the T312, the compact lightweight APU (auxiliary power unit) of the Panavia Tornado. KHD also supplies the complete Tornado secondary power system. It is participating in production of the Larzac 04 turbofan (see Turboméca-SNECMA in French section), and is responsible for assembly and test of these engines for the Luftwaffe. Since 1980 it has been developing the engines described below.

KHD T117

This small turbojet is being developed to power RPVs, especially for surveillance. First application is in the Canadair CL-289.

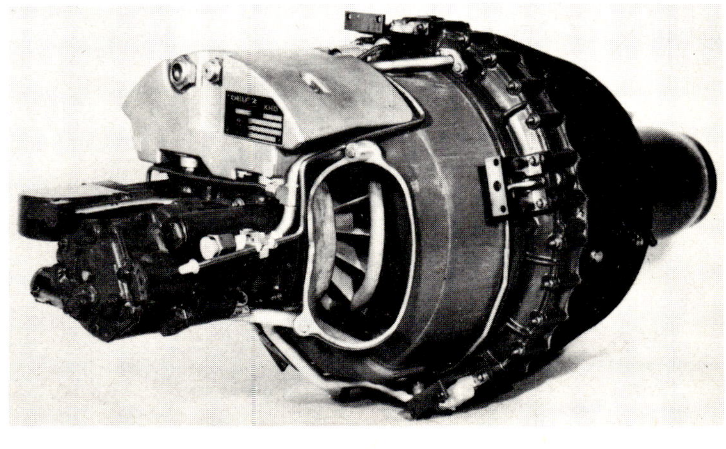

The KHD T117 turbojet for the CL-289 RPV

Compressor: Centrifugal. Mass flow 1·56 kg (3·44 lb)/s. Pressure ratio 5·5.
Turbine: Single stage. Exhaust temperature 847°C.
Fuel: JP-1, JP-4, JP-5.
Dimensions:

Length	770 mm (30·3 in)
Diameter	350 mm (13·8 in)
Weight, Dry:	21·8 kg (48 lb)

Performance Rating (S/L, static):

Max continuous	1·05 kN (236 lb st)
Specific Fuel Consumption:	
At T-O rating	33·3 mg/Ns (1·18 lb/h/lb st)

KHD T317

This small turbojet is an advanced development of the T117 intended for light manned aircraft. It has an annular intake and the same compressor performance as the T117. Fuel grade MIL-T-5624. The weight given includes 8·5 kg (18·74 lb) of electrical equipment.
Dimensions:

Length	770 mm (30·3 in)
Diameter	348 mm (13·7 in)
Weight, Dry:	32 kg (70·5 lb)

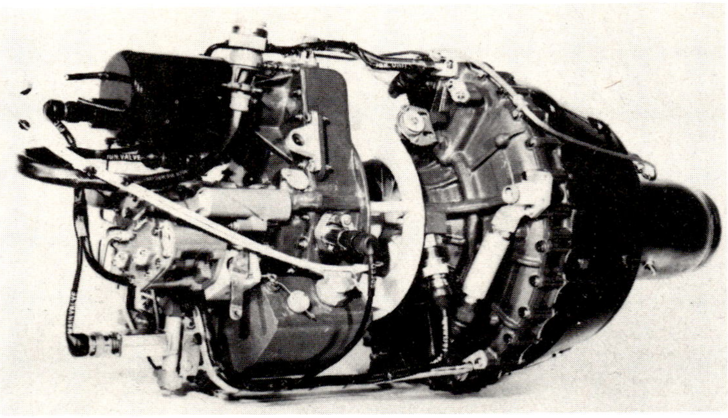

The KHD T317 turbojet for light aircraft

Performance Rating (S/L, static):

T-O	1·1 kN (247 lb st)

Specific Fuel Consumption:

At T-O rating	32·5 mg/Ns (1·15 lb/h/lb st)

LIMBACH
LIMBACH MOTORENBAU

Head Office and Works: Kotthausener Str 5, D-533 Königswinter 21, Sassenberg
Telephone: (02244) 2322 and 3031
President: P. Limbach

This company manufactures four-stroke piston engines for very light aeroplanes and powered gliders. All are of similar basic design. A two-cylinder two-stroke engine is under development.

LIMBACH SL 1700

Several variants of this engine have been certificated by the Luftfahrt-Bundesamt (Federal Office of Civil Aviation). Apart from the first subtype listed below, all are four-stroke (Otto) engines:

Limbach SL 1700D. Dual-ignition. Not certificated. Fitted to Sportavia RF7.

Sportavia-Limbach SL 1700E. Basic engine of the current range. Fitted to Sportavia RF5 and RF5B.

Limbach SL 1700EA. Differs in having front-end starter and different induction system. Fitted to Scheibe SF-25C Falke.

Limbach SL 1700EAI. Similar to EA except equipped to drive Hoffmann variable-pitch propeller. Fitted to Scheibe SF-28.

Limbach SL 1700EB. Similar to E except for having increased cylinder stroke and twin carburettors.

Limbach SL 1700EBI. Similar to EB except equipped to drive Hoffmann variable-pitch propeller. Fitted to Schleicher ASK 16.

Limbach SL 1700EC. Similar to E except for having a carburettor intake heating box.

Limbach SL 1700 ECI. Similar to EC except equipped to drive Hoffmann variable-pitch propeller.

Sportavia-Limbach SL 1700EI. Similar to E except equipped to drive Hoffmann variable-pitch propeller. Optional for Sportavia RF5B.

Unless otherwise stated, the following description refers to the SL 1700E:

Type: Four-cylinder horizontally-opposed aircooled piston engine.

Cylinders: Bore 88 mm (3·46 in). Stroke 69 mm (2·71 in) (EB, EBI, 74 mm; 2·87 in). Swept volume 1,680 cc (102·51 cu in) (EB, EBI, 1,800 cc; 108·56 cu in). Compression ratio 8 : 1.

Induction: Stromberg-Zenith 150CD carburettor (two in EB, EBI).

Fuel Grade: 90 octane.

Ignition: Single Slick 4230 magneto feeding one Bosch WB 240 ERT 1 plug in each cylinder.

Starting: One Fiat 0·37 kW (0·5 hp) starter (EA, EAI, one Bosch 0·3 kW; 0·4 hp).

Accessories: Ducellier 250W alternator; APG 17.09.001 fuel pump (EA,EAI, 17.09.001A).

Dimensions:

Length overall:	
SL 1700D	649 mm (25·6 in)
SL 1700EA, EAI	558 mm (22·0 in)
SL 1700E, EI, EC, ECI	618 mm (24·33 in)
other variants	580 mm (22·8 in)
Width overall:	
SL 1700D	800 mm (31·5 in)
SL 1700EA, EAI	770 mm (30·3 in)
other variants	764 mm (30·1 in)
Height overall:	
SL 1700D	451 mm (17·8 in)
SL 1700EA, EAI	392 mm (15·4 in)
other variants	368 mm (14·5 in)

Weight, Dry:

SL 1700E, EI	73 kg (161 lb)
SL 1700EA, EAI	70 kg (154 lb)
SL 1700EB, EBI, EC, ECI	74 kg (164 lb)

Performance Ratings:

T-O:	
SL 1700D	48·5 kW (65 hp) at 3,600 rpm
SL 1700E, EI, EC, ECI	51 kW (68 hp) at 3,600 rpm
SL 1700EA, EAI	44·7 kW (60 hp) at 3,550 rpm
SL 1700EB, EBI	53·7 kW (72 hp) at 3,600 rpm
Continuous:	
SL 1700E, EI, EC, ECI	45·5 kW (61 hp) at 3,200 rpm
SL 1700EA, EAI	41·7 kW (56 hp) at 3,300 rpm
SL 1700EB, EBI	49·2 kW (66 hp) at 3,200 rpm

LIMBACH L 2000

This family of engines is based on the SL 1700 with increased bore and stroke:

Limbach L 2000EO1. Dimensions as SL 1700EI. Installed in Fournier RF-9 and Valentin Taifun.

Limbach L 2000EA1. Dimensions as SL 1700EA1. Installed in Scheibe SF-25C Falke 2000 and SF-36.

Limbach L 2000EB1. Dimensions as SL 1700EB1. Installed in Grob G 109 and Hoffmann Dimona.

Details as for SL 1700, except for following:

Cylinders: Bore 90 mm (3·54 in). Stroke 78·4 mm (3·09 in). Swept volume 1,994 cc (120·26 cu in). Compression ratio 8·7:1 (EA1, 8·9:1).

Fuel Grade: 100L.

Weight (dry, with all accessories):

L 2000EO1	70 kg (154 lb)
L 2000EA1	69 kg (152 lb)
L 2000EB1	71·5 kg (157·5 lb)

Performance Ratings:

T-O:	
all models	59 kW (80 hp) at 3,400 rpm
Continuous:	
L 2000EO1, EA1	51 kW (70 hp) at 3,000 rpm
L 2000EB1	53 kW (72 hp) at 3,000 rpm

LIMBACH L 275E

This engine is intended for low-cost propulsion of RPVs and ultra-light aircraft.

Type: Two-cylinder horizontally-opposed two-stroke aircooled piston engine.

Cylinders: Cast aluminium alloy with Nicasil liner. Bore 66 mm (2·6 in). Stroke 40 mm (1·57 in). Swept volume 274 cc (16·72 cu in).

Induction: Two all-attitude diaphragm carburettors.

Fuel Grade: 90 octane, mixed 25:1 with two-stroke oil.

Ignition: 12V Bosch transistorised, one Bosch WK 175T6 plug per cylinder.

Accessories: Leistritz-type turbo silencer (muffler).

Dimensions:

Length overall	226 mm (8·89 in)
Width overall	390 mm (15·35 in)
Height overall	187 mm (7·36 in)
Weight (with silencer):	7·5 kg (16·5 lb)
Performance Rating:	18 kW (24 hp) at 7,300 rpm

Sportavia-Limbach SL 1700E flat-four four-stroke engine, rated at 51 kW (68 hp)

Limbach SL 1700EA flat-four four-stroke engine, rated at 44·7 kW (60 hp)

Limbach L 275E two-stroke for light aircraft and RPV applications, to be rated at 18 kW (24 hp)

MBB
MESSERSCHMITT-BÖLKOW-BLOHM GmbH

8 München 80, Postfach 801 220
Telephone: (089) 6 00 01
Development and Production Centres: Ottobrunn bei München; Lampoldhausen; Hamburg; and Bölkow-Apparatebau GmbH at Nabern and Schrobenhausen

As noted in the Aircraft section, Messerschmitt-Bölkow and Hamburger Flugzeugbau merged in 1969 to form the MBB group. The former Bölkow element of this group is engaged in the design and development of a wide variety of rocket engines. Other activities include thrust augmentation by afterburning, engine casing design, propellant insulation and mounting, and preparatory work towards series production and reliability, as well as work on advanced thermodynamic combustion engines and injection and cooling of engines of low thrust level.

MBB MONOPROPELLANT GAS-GENERATORS

MBB has produced 13 different types of gas-generator, with flow rates from 0·00018 kg (0·0004 lb) to 8·2 kg (18·2 lb)/s and using the propellants N_2H_4, N_2H_4 + H_2O, and H_2O_2 + H_2O. The bigger gas-generators are operating with catalytic and thermic decomposition. The smallest generator supplies a cold-gas satellite attitude-control system, while the bigger types are for propellant-tank pressurisation.

MBB STORABLE ROCKET ENGINES

This engine family includes motors of 0·01/0·03/0·05/0·08/0·3/0·4/0·5/97·9 kN (2·2/6·6/11/18/66/88/110/22,000 lb) thrust. Except for the 0·01 kN (2·2 lb) engine, which uses MMH/N_2O_4, all other engines run on AZ50/N_2O_4. The 0·01 kN (2·2 lb) engine is in use as the attitude-control thruster for the Franco-German Symphonie communications satellite; it is designed for steady-state and pulse-mode operation. The 0·4 kN (88 lb) engine was the vernier for the German third stage of

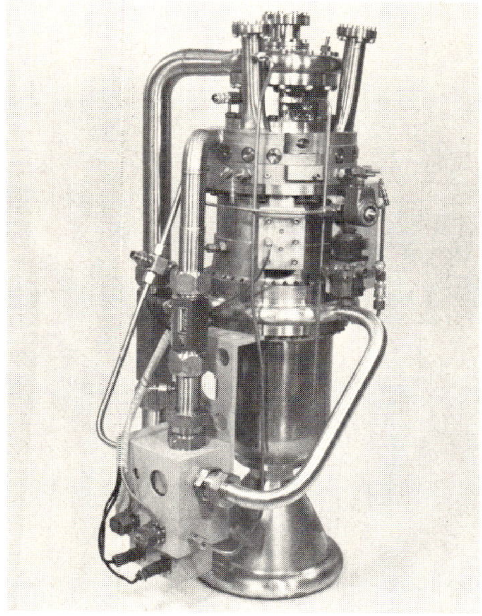

MBB 49 kN (11,000 lb) sea-level thrust lox/kerosene rocket engine

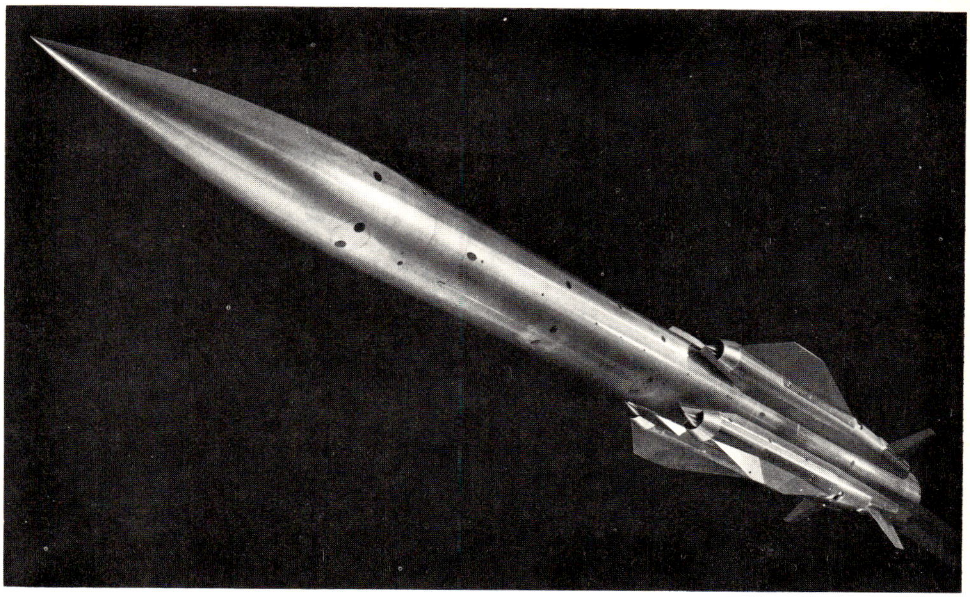

A tunnel model of a possible ASSM (Anti-Surface Ship Missile) with MBB ram rocket propulsion

the ELDO-A launcher, and in modified form it was to be used as the apogee motor for ELDO-II. This engine is highly qualified and in several ELDO-A launches operated in space without failure. A total of 135 was produced. In 1973 a new tactical missile engine was developed, operating at thrust levels of 23·5 kN (5,290 lb); 11·77 kN (2,646 lb); 7·85 kN (1,764 lb). The 98 kN (22,000 lb) engine was developed as the prototype power plant for an artillery rocket.

MBB HIGH-ENERGY ROCKET ENGINES

Since 1973, in partnership with SEP of France, MBB has developed the HM7 engine to power the third stage of the European Ariane space launch vehicle. This LH_2/LO_2 engine has a vacuum thrust of 60 kN (13,500 lb). Its regeneratively cooled thrust chamber of milled copper operates at a pressure of 30 bars (435 lb/sq in), and the specific impulse is 444. Total thrust-chamber weight is 63 kg (138·9 lb). The nozzle extension, with an area-ratio of 61, is dump-cooled with LH_2. By the start of 1977, a total of 220 firings had been completed on MBB's high-pressure test facility under sea-level and simulated altitude conditions. SEP is Ariane propulsion prime contractor (see SEP entry). MBB's role is to supply the thrust chamber and main propellant valves.

MBB HIGH-PRESSURE ENGINES

In the course of development of high-pressure liquid-rocket engines MBB developed a turbopump engine utilising staged combustion in 1963. This 49 kN (11,000 lb) lox/kerosene engine was the first to run in the western world, with turbine exhaust completely burned in the main combustor. It was designed for 96·5 bars (1,400 lb/sq in) abs and was throttleable at 14 : 1.

For the storable propellants $UDMH/N_2O_4$ a 9·8 kN (2,200 lb) pressure-fed engine was developed and tested at chamber pressures of more than 193 bars (2,800 lb/sq in) abs to demonstrate high-performance injection systems and electroformed thrust-chamber technology for corrosive propellants.

MBB developed and fabricated some 133·44 kN (30,000 lb) H_2/O_2 engines designed for 207 bars (3,000 lb/sq in) abs. These were tested at Rocketdyne's facilities in California up to chamber pressures of 276 bars (4,000 lb/sq in) abs without failure. The electroforming thrust-chamber technology demonstrated in this programme was the baseline for the Rocketdyne SSME (which see).

MBB RAM ROCKETS

Since 1965 MBB has been developing ram rocket engines with liquid and solid propellants, and liquid-fuelled ramjet engines, for the propulsion of missiles with supersonic cruise speed and long range capability.

MBB has studied high-energy boron propellants as well as medium-energy composite grains, UDMH, MMH and kerosene. For missile applications, studies were carried out with semi-integrated and fully integrated engines with many configurations of air inlet. These studies were supported by half-scale burning tests (net thrust 4·0 kN; 899 lb) and aerodynamic tests in wind-tunnels.

Solid-propellant ram rockets are inherently simple in structure and have high reliability because of hypergolic ram combustion.

By 1980 this development was increasing with a view to fields of application in anti-ship, air-to-air and surface-to-air missiles.

MTU
MOTOREN-UND TURBINEN-UNION MÜNCHEN GmbH

München-Allach, Dachauer Str 655 (postal address, 8 München 50, Postfach 50 06 40)
Telephone: (089) 1489-1

Board of Management MTU Group:
Dr Ernst Zimmermann (President)
Dr Hans Dinger (Executive Vice-President)
Hubert Dunkler
Dr Wolfgang Hansen
Günter Welsch
Dr Peter Beer (Deputy Vice-President)

MTU München is owned half by Maschinenfabrik Augsburg-Nürnberg AG (MAN) and half by Daimler-Benz AG, and manages all aircraft engine programmes formerly managed by MAN Turbo and Daimler-Benz.

MTU handles service support for the J79-MTU-J1K/17A and T64-MTU-7 engines made by MTU under General Electric licence. It also handles service support and parts manufacture of such civil engines as the JT8D and CF6, as well as Lycoming piston engines. In November 1979 MTU formed a new subsidiary, MTU Maintenance GmbH, based at Hanover-Langenhagen, to support, repair and modify the CF6, RB.211, JT8D and JT9D, LM2500 and PW2037. MTU Maintenance will also assemble CF6-80C engines for the A300B-600 and carry out testing for the CF6-80A and -80C.

GENERAL ELECTRIC CF6

Under the terms of a co-production agreement signed with General Electric, MTU has approximately a 10 per cent share in the manufacture of the CF6-50 engine for the Airbus A300, together with SNECMA. The main task of MTU is production of the complete HP turbine. MTU is now engaged in producing the same component of the CF6-80A/A1 for the A310 and B767, and -80C for the A300B-600, B747 and stretched B767, under a 7 per cent co-production agreement.

Comparison of the Turbo-Union RB.199-34R (above) and J79-MTU-17A of similar thrust

PRATT & WHITNEY PW2037

MTU is a partner, with Fiat of Italy, in this new turbofan described in the US part of this section. The company is responsible for the LP turbine, under an 11·2 per cent share.

ROLLS-ROYCE TYNE

MTU has a 28 per cent share in the production of a further batch of about 170 Tyne engines for the re-opened Transall production line. In addition MTU has assumed responsibility for service support of all Tyne 21 engines (Atlantic) and Tyne 22 (Transall), including engines used by civil operators.

TURBO-UNION RB.199

MTU München's largest programme is a 40 per cent share in this engine, described under Turbo-Union in the International part of this section.

SNECMA/TURBOMÉCA LARZAC

MTU has approximately a 25 per cent share in development, production and support of this French engine used in the Alpha Jet.

ALLISON 250-C20B

MTU is licence-producing 687 of these turboshaft engines, designated 250-MTU-C20B, for the PAH-1 and VBH helicopter programmes (see MBB in Aircraft section). MTU also supports C20 and C20B engines used by civil operators.

MTU/TURBOMÉCA MTM 385

This is a proposed new turboshaft engine for helicopters. For initial data see International part of this section.

PARODI
ROLAND PARODI

Klettgau-Erzingen

This engineer has developed a four-cylinder four-stroke aircooled piston engine for ultralight aircraft and motor gliders. With various designations in the HP 45 series, these are derived from a Honda design, with a new crankcase and lubrication system. Crankshaft speed of 8,000 rpm is reduced by belt to about 2,800 at the propeller. Maximum ratings are 33·5 or 37·3 kW (45 or 50 hp).

PIEPER
PIEPER MOTORENBAU GmbH

495 Minden/Westf, Postfach 1229
Telephone: (0571) 34088

STAMO MS 1500

Pieper manufactures the 33·5 kW (45 hp) Stamo MS 1500-1 modified Volkswagen four-cylinder aircooled piston engine, applications for which have included the Scheibe SF-25B Falke two-seat motor glider. The capacity of this is 1,500 cc, compression ratio 7·2 : 1, length 640 mm (25 in), width 745 mm (29·3 in), height 395 mm (15·5 in) and dry weight 52 kg (115 lb). The MS 1500-1 operates on either 80/86 or 90 octane fuel, and is started by a pull-cord. A variant is the MS 1500-2, with electric starter and generator. This increases overall height to 450 mm (17·7 in) and dry weight to 60 kg (132 lb). By 1981 well over 600 engines had been delivered, and the period between complete overhauls had been extended to 1,000 h.

Pieper-built Stamo MS 1500-1 four-cylinder four-stroke engine, rated at 33·5 kW (45 hp)

PULCH
OTTO PULCH

Hölderlinstrasse 21, 7513 Spöck-Stutensee
Telephone: 07249 8509

PULCH 003

As a private venture Otto Pulch (OUV, EAA-Chapter 308) has developed a four-stroke radial piston engine for light aircraft. Using BMW motorcycle cylinders and pistons, he designed a new crankcase and connecting-rod assembly, reduction gear and dual ignition system. The first prototype Pulch 003A ran in Autumn 1976. In Summer 1977 it was awarded the Wolf-Hirth-Preis by *Flug Revue* magazine as the most outstanding achievement in light aircraft technology in West Germany in 1977. The second prototype Pulch 003B, incorporating substantial improvements, ran in 1978, and has a demonstrated noise level of only 96 dBA at full power at a distance of 10 m (33 ft). The 003B is described as "an ideal engine for homebuilt aircraft," and will be available in kit form.

TYPE: Six-cylinder radial aircooled piston engine.

CYLINDERS: Pulch 003A (BMW R 90/6) Bore, 90 mm (3·55 in). Stroke, 70 mm (2·76 in). Swept volume, 2,700 cc (165 cu in). Pulch 003B (BMW R 100S) Bore, 94 mm (3·71 in). Stroke, 70 mm (2·76 in). Swept volume, 2,900 cc (177 cu in).

INDUCTION: Pulch 003A, two SUM carburettors. Pulch 003B, fuel-injection system, Bosch K-Jetronic.

IGNITION: Both versions, electronic system.

PROPELLER DRIVE: Geared, ratio (003A) 2·29:1, (003B) 2·56:1.

DIMENSIONS:
- Length overall:
  - Pulch 003A, 003B: 800 mm (31·5 in)
- Width overall:
  - Pulch 003A, 003B: 750 mm (29·55 in)

WEIGHT, DRY:
- Pulch 003A: 110 kg (242·5 lb)
- Pulch 003B: 120 kg (264·5 lb)

PERFORMANCE RATING (T-O):
- Pulch 003A: 110 kW (150 hp) at 5,000 rpm
- Pulch 003B: 132 kW (180 hp) at 5,500 rpm

PULCH THREE-CYLINDER

This new engine for motorised gliders ran in 1979. It has three four-stroke BMW cylinders driving a vertical crankshaft and with 1:3 angle box to the propeller shaft. Swept volume is 1,500 cc (91·7 cu in) and rating 55 kW (75 hp).

Pulch 003A six-cylinder radial, to be rated at 110 kW (150 hp)

RFB
RHEIN-FLUGZEUGBAU GmbH (Subsidiary of VFW GmbH)

ADDRESS: See Aircraft section

Production of the SG 85 fan pod for motor gliders (see 1980-81 *Jane's*) has been completed. RFB is planning a new pod incorporating a different engine.

INDIA

HAL
HINDUSTAN AERONAUTICS LTD

Indian Express Building, Vidhana Veedhi, PO Box 5150, Bangalore 560 001

Telephone: 76091

OFFICERS: See Aircraft section

The Bangalore and Koraput Engine Divisions of HAL comprise the main aero-engine design, development and manufacturing elements of the Indian aircraft industry.

BANGALORE COMPLEX (Engine Division)

This Division is engaged in the manufacture of gas-turbine engines. Adour 811 engines for all except the initial batch of Jaguars for the Indian Air Force will be manufactured by HAL under Rolls-Royce Turboméca licence. The Orpheus 701 to power the Ajeet and Kiran II, the Orpheus 703 to power the Marut and the Dart 531 to power the BAe (HS) 748 are made under licence from Rolls-Royce. The Artouste IIIB to power the Alouette III and SA 315 is made under licence from Turboméca. The division also overhauls Dart, Avon, Orpheus, Artouste, Gnome and J34 engines for the Indian Air Force and other customers.

KORAPUT DIVISION

This Division of HAL is located at Koraput in Orissa. It was established to manufacture under Soviet government licence the Tumansky R-11 afterburning turbojet for HAL-built MiG-21 fighters. With help from the Soviet Union, the first engine was run on the bench (which it was used to calibrate) in early 1969.

INDONESIA

LAPAN
LEMBAGA PENERBANGAN DAN ANTARIKSA NASIONAL (The National Institute of Aeronautics and Space)

HEAD OFFICE: Jalan Pemuda Persil No 1, Jakarta Timur
MAILING ADDRESS: PO Box 3048, Jakarta
Telephone: (021) 48 28 02
CHAIRMAN: Air Vice-Marshal J. Salatun
VICE-CHAIRMAN: Prof Wiranto Arismunandar

PUSAT TEKNOLOGI DIRGANTARA (Aerospace Technology Centre)

Rumpin Airfield, Bogor, West Java
HEAD, AEROSPACE TECHNOLOGY CENTRE: Dr Haryono Djojodihardjo
MANAGER, ROCKET DEVELOPMENT PROJECT: Ir Kisman Subandhi

LAPAN was established in 1963, to exploit indigenous capabilities in aeronautics and space in support of the National Five Year Plan. At the beginning of 1978 LAPAN had 527 personnel, of whom 195 were university graduates, divided among the Space Applications Centre (Jakarta), the Aerospace Technology Centre (Rumpin Airfield, near Bogor), the Atmospheric and Space Research Centre (Bandung) and the Aerospace Study Centre (Jakarta).

In 1974 LAPAN began the design of the RC-741 liquid propellant rocket engine, the first prototype of which was completed in February 1978. The RC-741 is the first liquid-propellant rocket engine made in Indonesia. Component and static tests at LAPAN's Aerospace Technology Centre were planned for 1980-81.

LAPAN RC-741

The RC-741 is designed for use in a boosted single-stage sounding rocket.

TYPE: Single-chamber liquid-propellant rocket engine.

PROPELLANTS: RFNA and xylidine, mixture ratio 3:1.

THRUST CHAMBER: Single chamber of 7:1 nozzle area ratio, using regenerative and radiative cooling, and with double-wall machined casing. Wall material nickel-chrome steel. Coaxial propellant injection, with hypergolic ignition. Combustion pressure 55·7 bars (808·28 lb/sq in) and temperature 2,700°C.

MOUNTING: Chamber bolted on four mountings concentric with engine axis.

PROPELLANT PUMPS: Centrifugal.

PROPELLANT FLOW: Fuel flow 2·45 kg (5·4 lb)/s; oxidiser flow rate 8·12 kg (17·9 lb)/s.

TURBINE: Single-stage axial impulse type.

THRUST CONTROL: Thrust held constant by regulation of turbopump speed via control of gas-generator propellant supply.

DIMENSIONS:
- Length overall: 605 mm (23·819 in)
- Diameter overall: 210 mm (8·268 in)

WEIGHT, DRY: 41 kg (90·38 lb)

PERFORMANCE (S/L):
- Max thrust: 20·6 kN (4,630 lb)
- Specific impulse: 240 s

INTERNATIONAL PROGRAMMES

CFM INTERNATIONAL

CFM INTERNATIONAL SA

160 avenue de Versailles, 75016 Paris, France
Telephone: 524 43 62
CHAIRMAN AND CHIEF EXECUTIVE: J. C. Malroux
EXECUTIVE OFFICERS:
N. Epstein (Marketing and Product Support)
P. Alesi (Project)
A. O. Kohn (Technical)
C. E. Adams (Finance)
T. G. Homburg (Logistic and Product Support Services)

CFM International, a joint company, was formed by General Electric and SNECMA in early 1974 to provide overall programme management for the CFM56 engine and a single customer interface for sales and service. Owned and managed on a 50/50 share basis, the company has been staffed with experienced people from the two parent companies.

In addition to the CFM International management team which directs the overall programme, the parent companies have their own CFM56 programme managers. For SNECMA, this position is held by J. Rossignol. His GE counterpart is R. B. Smith.

The SNECMA/GE agreement is not a profit-sharing arrangement. Responsibilities for hardware design, development and production are assigned through CFM International on an equal basis to the parent companies. Each company then assumes responsibility and funding for its assigned task throughout the life of the programme. This is a unique concept among international co-operative ventures in the aerospace field. SNECMA has from the outset been agreeable to participation of other European engine companies within its 50 per cent share. FN of Belgium has joined SNECMA for certain low-pressure system components.

GE is responsible for design integration, the core engine and the main engine control. The core engine is that of the F101 turbofan developed for the B-1 bomber. SNECMA is responsible for the low-pressure system, reverser, gearbox and accessory integration and engine installation.

Each company is responsible for its assigned hardware from design through development, production and product improvement. CFM International will provide the planning and integration of the product support programme.

CFM INTERNATIONAL CFM56

In the late 1960s General Electric and SNECMA made independent studies of the market requirement for the next generation of high bypass ratio engines. The GE studies were centered around an engine designated GE13, the core of which is now being used in the F101. SNECMA's studies were based on an engine designated M56. Each company concluded that a large market existed for a high bypass ratio engine in the ten tonne class (97·9-106·8 kN; 22,000-24,000 lb st), with low noise, low emissions and low fuel consumption, coupled with ease of maintenance and low operating costs.

In April 1971 SNECMA began a search for possible partners to undertake development of a commercial engine in this class. By December 1971 SNECMA had chosen GE as its partner and, after obtaining French government approval, detailed design activity began. Work was stopped, however, in September 1972 when the export licence restriction applied by the US State Department Office of Munitions Control made the programme untenable for the two companies. It was resumed a year later when negotiations led to a workable licensing agreement which also protected US technology. Since that time, a working agreement and management structure have been defined and the development programme led to engine certification on 8 November 1979. This marked the first time that an engine had been certificated simultaneously to both US and European standards by the French and US certification authorities. As part of its certification, the CFM56 had to comply with FAR Pt 33 for the USA and JAR-E (Joint Airworthiness Requirements-Engine) for France and most of Europe. In addition, its demonstrated noise reduction is sufficient to enable the aircraft in which it is installed to comply with the latest, more stringent noise regulations.

The first CFM56 demonstrator engine ran at the GE Evendale plant on 20 June 1974. The engine reached its full rated thrust within 10 h of running, with fuel consumption lower than specification. Since then, 10 more engines have entered the test programme. The engine first flew in a US Air Force McDonnell Douglas YC-15 transport prototype. Engine-development flight testing began in March 1977 when an engine installed with a full-length fan duct flew in a Caravelle at the SNECMA flight test centre.

CFM56-2 (106·8 kN; 24,000 lb st), due to enter service in 1981

A post-certification programme of endurance testing is being followed, to complete more than 22,000 extra-severe cycles, simulating short-haul operations, before entry into service.

The certificated engine, designated **CFM56-2** and rated at 106·8 kN (24,000 lb st) is described below. CFM International has plans for other versions of the CFM56, including a growth engine, the **CFM56-4**, with ratings up to 120 kN (27,000 lb st) within the existing fan diameter, and a reduced-thrust derivative, the **CFM56-3**, rated at 89 kN (20,000 lb st), using the same basic core engine but with a reduced-diameter fan based on that of the CF6-80; this engine is specified for the Boeing 737-300.

CFM56 production was launched in early 1979 when United, Delta and Flying Tiger announced their decision to re-engine their DC-8 Series 60 aircraft. The DC-8 re-engining programme is managed by CAMMA Corporation. Modification work on the re-engined aircraft, designated DC-8 Series 70, is carried out by Douglas Aircraft at Tulsa, Oklahoma. By the middle of 1981 six airlines had firm contracts with CFM International and CAMMA Corporation to re-engine 86 DC-8s, plus a number of options. Deliveries to airlines will begin in December 1981.

Flight evaluation of a Boeing 707 equipped with four CFM56 engines is underway at Boeing Commercial Airplane Company. First flight of the 707-CFM56 test aircraft was on 27 November 1979.

The CFM56 was selected by the US Air Force for its KC-135A tanker re-engining programme on 22 January 1980. The US Air Force has awarded an R & D contract to the Boeing Military Airplane Company. Installation of CFM56 engines on a KC-135 will begin in 1982, followed by one year of flight and operational testing.

Other potential commercial and military applications are the Airbus Industrie A320 and TA11, DC-9, MDF-100, 707 Tanker and Patrol, new Boeing commercial transport and CX programmes.

TYPE: Two-shaft turbofan for subsonic applications.

AIR INTAKE: Direct entry, without inlet guide vanes.

FAN: Single-stage axial. Forged titanium disc holding 44 inserted titanium blades, each with a tip shroud to form a continuous peripheral ring. Fan and the attached LP compressor (booster) run in front roller bearing and rear ball bearing. Pointed conical spinner rotates with fan. Alloy steel fan frame of continuous ring construction, carried by 12 radial struts of low thickness/chord ratio well downstream of fan. Max airflow 376 kg (830 lb)/sec. Bypass ratio 6:1 (CFM56-2), 5:1 (CFM56-3).

LP COMPRESSOR: Three axial stages on titanium drum bolted to fan disc serve as booster to supercharge core. Downstream flow curves inwards to match diameter of HP compressor. In this section are main fan frame and sumps and bearings for front end of both shafts. Ring of bleed doors allows core airflow to escape into fan duct at low power settings. Bleed doors are closed at all normal flight power settings.

HP COMPRESSOR: Nine-stage axial with tapering tip diameter. Rotor of high-strength corrosion-resistant alloy, with blades of titanium (to stage 3) or steel. Stators steel, with first four stators variable. Split titanium front casing with steel liners. Based upon HP compressor of F101, with minor modifications. Overall pressure ratio in 25 : 1 class.

COMBUSTION CHAMBER: Fully annular with advanced film cooling. Based upon F101 combustor but modified for reduced emissions. Level of pollution from the core is claimed to be below that of any engine at present in airline service.

HP TURBINE: Single-stage axial with aircooled stator and rotor blades. Entry gas temperature in 1,260°C class. High stage loading. HP system carried in only two bearings.

LP TURBINE: Four-stage axial.

EXHAUST UNIT (FAN): Constant-diameter duct of sound-absorbent construction. Outer cowl and engine cowl form convergent plug nozzle, with airframe-mounted reverser.

EXHAUST UNIT (CORE): Fixed-area type with convergent plug nozzle. Sound-absorbent construction.

ACCESSORY DRIVE: Gearbox in front sump transmits drive from front of HP spool, via radial shaft in fan frame, to transfer gearbox mounted on underside of fan case. Drive faces on both front and rear sides. Air starter at transfer gearbox.

FUEL SYSTEM: Hydromechanical with electronic trim.

LUBRICATION: Non-pressure-regulated system.

DIMENSIONS:

Front flange diameter	1,828 mm (72·0 in)
Length, excl spinner	2,430 mm (95·7 in)
WEIGHT, DRY (CFM56-2):	2,092 kg (4,612 lb)

PERFORMANCE RATINGS:
Max T-O 106·8 kN (24,000 lb st) flat rated to 30°C
Cruise at 9,144 m (30,000 ft) at Mach 0·80
28·26 kN (6,350 lb st)

SPECIFIC FUEL CONSUMPTION:
Cruise rating, as above 18·4 mg/Ns (0·65 lb/h/lb st)

GARRETT/VOLVO FLYGMOTOR

THE GARRETT CORPORATION

HEAD OFFICE: 9851 Sepulveda Boulevard, Los Angeles, California 90009, USA
Telephone: (213) 670 0131

VOLVO FLYGMOTOR AB

HEAD OFFICE: 461 81 Trollhättan, Sweden
Telephone: 0520 30100

Agreement on co-operative development of a low bypass ratio derivative of the TFE731 turbofan engine (described under Garrett in the US section) was signed on 2 March 1978 in Trollhättan. The agreement was approved later by the US and Swedish authorities.

GARRETT/VOLVO FLYGMOTOR TFE1042

With this new engine the two parent companies are aiming at the market for the next generation of training, ground support, lightweight strike and fighter aircraft.

The TFE1042 will be offered at two ratings, one **(1042-5)** with a thrust of 18·2 kN (4,100 lb) at S/L ISA, and another **(1042-6)** with 21·5 kN (4,840 lb). This latter version will also be offered with an afterburner, giving 37·1 kN (8,340 lb) as the **1042-7**.

The engine will be developed around the core of the commercial TFE731 turbofan and will utilise the same accessory drive system. Garrett will provide the manufacturing facilities for all parts common to the two engine models, and will be responsible for the overall design.

Volvo Flygmotor will be responsible for the detail design and manufacture of the non-common parts, such as the new two-stage fan, fan gearbox and afterburner.

The first phase of the development programme culminated in the start of testing a prototype unaugmented engine (1042-4) in August 1979. Flight prototype engines can be made available in two years from the start.

Type: Two-shaft military turbofan, with geared front fan. Optional fully modulated afterburner.

Air Intake: Direct entry, fixed, without inlet guide vanes.

Fan: Two-stage axial, with titanium inserted blades. Total airflows and bypass ratios at max rating, S/L ISA, are: (1042-5) 38·0 kg (83·7 lb)/s, 1·00:1; (1042-6, -7) 40·8 kg (89·9 lb)/s, 0·84:1.

Compressors, Combustion Chamber, Fuel System and Accessory Drives: As TFE731

Turbine: Single axial HP stage with aircooled inserted blades and cooled nozzle guide vanes (basically as TFE731-5). Turbine inlet temperature for TFE1042-5 is similar to that of TFE731-5. For 1042-6 and -7, turbine temperature is increased, requiring cooling also of guide vanes of first stage of three-stage LP turbine.

Exhaust System: Mixed fan and turbine exhaust discharged to atmosphere through common convergent nozzle. For augmented version, mixed streams pass through afterburner with outer casing of titanium and inner wall of special alloys. One high-energy spark igniter. Fuel-actuated fully-variable iris-type con-di nozzle. Aircraft-mounted digital system containing torque-motor valves for afterburner fuel metering and nozzle actuation.

Mountings: Main mount transmitting thrust, weight and moment is at front frame. Engine can be mounted from either side. Single mount on turbine casing for vertical loads only.

Shafting: HP spool consists of HP turbine and HP compressor, on shaft supported by one roller bearing and one ball bearing. This drives accessory gearbox through tower-shaft transfer box. LP spool consists of LP turbine and LP compressor, interconnected by curvic couplings and supported on one ball bearing at compressor and one roller bearing at turbine. Drives fan through reduction gear with ratio of 1·416:1. Fan shaft supported by roller bearing under first-stage fan disc and ball thrust bearing under second-stage disc.

Dimensions:

Length:	
1042-5, -6	1,625 mm (64·0 in)
1042-7	3,330 mm (131 in)
Max diameter	686 mm (27·0 in)

Weight, Dry:

1042-5, -6	395 kg (870 lb)
1042-7	560 kg (1,235 lb)

First photograph of a TFE1042 in the Volvo Flygmotor factory; this is a basic engine without afterburner

Performance Ratings (S/L, ISA):

Max T-O:	
1042-5	18·2 kN (4,100 lb st)
1042-6	21·5 kN (4,840 lb st)
1042-7 (dry)	21·4 kN (4,820 lb st)
1042-7 (augmented)	37·1 kN (8,340 lb st)

Specific Fuel Consumption (ratings as above):

1042-5	18·6 mg/Ns (0·659 lb/h/lb st)
1042-6	19·4 mg/Ns (0·685 lb/h/lb st)
1042-7 (dry)	19·7 mg/Ns (0·696 lb/h/lb st)
1042-7 (augmented)	62·6 mg/Ns (2·21 lb/h/lb st)

MTU/TURBOMÉCA SARL

MOTOREN-UND TURBINEN-UNION MÜNCHEN GmbH

München-Allach, Dachauer Str 655, Postfach 50 06 40, Federal Republic of Germany

SOCIÉTÉ TURBOMÉCA

Bordes, 64320 Bizanos, France

MTU/TURBOMÉCA MTM 385

This turboshaft engine will be developed by MTU/Turbomeca SARL, a joint company owned equally by both partners. The reason for developing the engine, with a specification almost identical to that of the Rolls-Royce Gem, is said to be to power the proposed PAH-2 Franco-German anti-tank helicopter. MTU is responsible for the combustion chamber, gas generator and power turbine, the French partner being assigned the rest, representing a 50/50 work-split by cost. The first gas generator ran before the end of 1979.

The MTM 385 is of modular design and is intended for minimum emissions. It is a simplified version of the MTM 380 described in previous editions.

Compressor: Single-shaft, with two axial stages and one centrifugal stage. Mass flow over 3 kg (6·6 lb)/s.

Combustion Chamber: Reverse-flow annular with vaporising burners.

MTU/Turboméca MTM 380 turboshaft engine, predecessor of the 385 *(Brian M. Service)*

Turbines: Single-stage aircooled gas-generator turbine. Two-stage power turbine.

Control System: Digital.

Ratings:

Thermodynamic power 700 to 800 kW (939 to 1,072 shp)

RJAE

ROLLS-ROYCE & JAPANESE AERO ENGINES LTD

Moor Lane, Derby DE2 8BJ, England

Chairman: Ashley Raeburn

This international company has been formed by Rolls-Royce Ltd (UK) and a consortium of Japanese companies to develop and produce a new turbofan engine in the 10 to 15-tonne (20,000-30,000 lb st) class. The Japanese companies are: Ishikawajima-Harima Heavy Industries (IHI), Kawasaki Heavy Industries (KHI) and Mitsubishi Heavy Industries (MHI).

RJAE is committed to the initial phase of the RJ.500 programme which includes completion of engine design, component rig testing and material provisioning for initial development testing. For this purpose a joint design team has been established at Bristol.

Full-scale development and production of the RJ.500 will be shared equally between Rolls-Royce and the Japanese consortium. While the joint company will ensure that the experience of each partner is used to the best advantage in the programme, the responsibility for component design and manufacture will be shared as in the diagram.

RJ.500

The RJ.500 is intended primarily as a power plant for short/medium-haul airliners of 120-160 seats planned for entry into service in the mid-1980s. Design is based broadly on the Rolls-Royce RB.401 and Japanese FJR.710 engines. It incorporates features from other Rolls-Royce engines such as the Spey and RB.211 series. Fuel efficiency is expected to be significantly improved by comparison with that of current turbofans of similar thrust. The RJ.500 is also designed for low acquisition and maintenance costs, particularly in highly cyclic short-haul operations, and to meet all known future requirements regarding noise and pollution.

The basic engine is designated **RJ.500-01D4**, though other versions with enhanced thrust and reduced sfc are being studied.

Full-scale display model of RJ.500-01D4

Basic features can be deduced from the diagram and photograph. It is a two-shaft engine with 26 wide-chord fan blades, without clappers, nine-stage HP compressor with three stages of variable IGVs, two-stage HP turbine and three-stage LP turbine. The following data apply to the -01D4:

Maximum airflow is 281 kg (618 lb)/s. Cruise pressure ratio and bypass ratio are respectively 20·2 and 4·94.

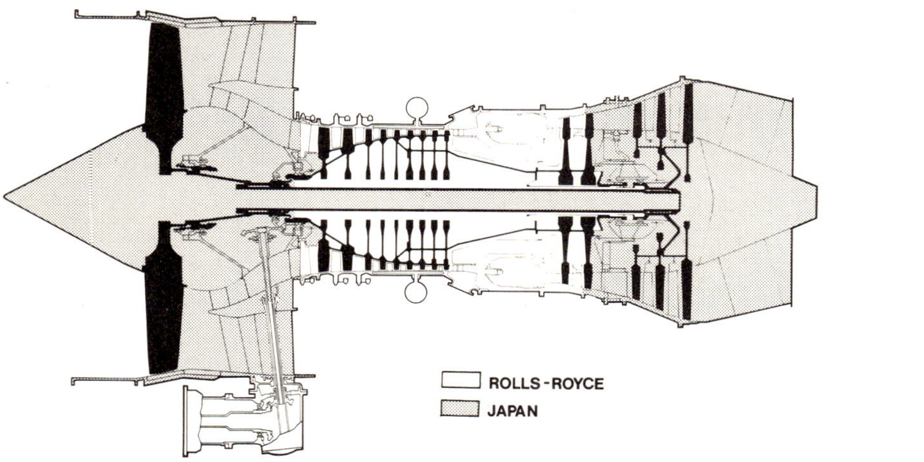

DIMENSIONS:

Fan diameter	1,499 mm (59·0 in)
Length (flange-to-flange)	2,768 mm (109·0 in)

WEIGHT, DRY:

Basic engine	1,867 kg (4,117 lb)

RATINGS:
Max T-O (S/L, static)
89·0 kN (20,000 lb st) to ISA + 15°C
Cruise, Mach 0·8 at 9,145 m (30,000 ft)
22·1 kN (4,965 lb st) to ISA + 10°C

Longitudinal section of RJ.500, showing 50:50 work split between the British and Japanese partners

ROLLS-ROYCE/ALLISON

ROLLS-ROYCE LIMITED

HEAD OFFICE: 65 Buckingham Gate, London SW1E 6AT, England

THE DETROIT DIESEL ALLISON DIVISION, GENERAL MOTORS CORPORATION

HEAD OFFICE: Detroit, Michigan, USA

Co-operation between Rolls-Royce and Allison started in November 1958, when the two companies began work on the design and development of high-performance jet engines for commercial and military applications.

Cutaway drawing of the Allison/Rolls-Royce TF41-A-1 (Spey RB.168-62) turbofan (64·5 kN; 14,500 lb st)

ALLISON/ROLLS-ROYCE TF41

Manufacturers' designations: Rolls-Royce Spey RB. 168-62 and -66, Allison Model 912-B3 and -B14

In August 1966 Allison and Rolls-Royce were awarded a joint contract by US Air Force Systems Command for the development and production of an advanced version of the RB.168-25 Spey turbofan, to power Vought A-7D Corsair II fighter-bomber aircraft for the US Air Force.

Development and production were undertaken jointly by Rolls-Royce and Allison, with Rolls-Royce supplying parts common to existing Spey variants and Allison, which is manufacturing under licence, being responsible for items peculiar to the TF41. This provided an approximately 50/50 division of manufacturing effort, but with Allison also undertaking assembly, test and delivery.

Design of the RB.168-62 started in June 1966 and, following the award of the US Air Force contract, the engine was given the US Air Force designation TF41-A-1. Major change compared with the RB.168-25 is the move forward of the bypass flow split into the LP compressor, to give a larger three-stage fan followed by a two-stage IP compressor, all five stages being driven by the two-stage LP turbine. The number of HP compressor stages is reduced from 12 to 11, the HP turbine remaining at two stages. These modifications raise the mass flow and the bypass ratio (from 0·7 : 1 to 0·76 : 1). No afterburner is fitted.

Other design changes compared with the RB.168-25 include omission of the fan inlet guide vanes, the first rotor stage being overhung on a bearing supported by the first-stage stator vanes. The fan and IP compressor are of more modern aerodynamic design, and the HP and LP turbine nozzle throat areas have been increased to pass the additional flow. The HP turbine is of modified aerodynamic design, and an annular exhaust mixer replaces the RB.168-25's chuted design.

First run of the TF41-A-1/RB.168-62 was at Rolls-Royce, Derby, in October 1967, the first Allison engine following at Indianapolis in March 1968.

A second version of the TF41 is the A-2, developed for the US Navy and ordered in 1968 to power the Vought A-7E Corsair. Differences are slight, although the thrust rating is appreciably increased by raising the engine speed. This required restressing the disc of the LP turbine and HP compressor. Mass flow is slightly increased, the bypass ratio being 0·74 : 1. The engine has additional protection against corrosion.

By 1977 more than 1,300 engines had been delivered, and production of the TF41 for the A-7 was expected to continue into the 1980s. In combat service both versions of the TF41 have shown outstanding reliability. The exceptional overhaul life (for a combat engine) of 1,500 h has been reached.

TYPE: Military turbofan.

AIR INTAKE: Direct entry, fixed, without intake guide vanes.

COMPRESSOR: Two-shaft axial. 3 fan stages, 2 intermediate stages on same shaft and 11 high-pressure stages. All rotor blades carried on separate discs. Fan and LP rotor blades of titanium, held by dovetail roots in slots broached in discs which are bolted together through curvic couplings and similarly attached to the stubshafts. HP rotor blades also of titanium, except stages 9, 10 and 11 of stainless steel, the first HP stage being pinned and the remainder being dovetailed into broached slots; discs similarly bolted together but driven through splined shaft coupling. Overall pressure ratio 20 : 1 (A-2, 21·4 : 1); mass flow 117 kg/s (258 lb/s) (A-2, 119 kg/s; 263 lb/s). HP compressor pressure ratio, 6·2 : 1. mass flow, 67 kg/s (148 lb/s).

COMBUSTION CHAMBER: Tubo-annular, with 10 interconnected Ni-Co alloy flame tubes in steel outer casing. Duple spray atomising burner at head of each chamber. High-energy 12-joule igniter plug in chambers 4 and 8.

FUEL SYSTEM: Hydromechanical HP system with automatic acceleration and speed control and emergency manual override. Variable-stroke dual fuel pump.

FUEL GRADE: JP-4 (A-2, JP-5).

NOZZLE GUIDE VANES: Two HP stages with air cooling; two LP stages uncooled.

TURBINE: Impulse-reaction axial type, two HP stages and two LP. All blades forged in Ni-Co alloy; first HP stage blades cooled internally by HP compressor air.

JET PIPE: Fixed, heat-resistant steel.

ACCESSORY DRIVES: External gearbox driven by radial shaft from HP system; provision for starter, fuel boost pump, two hydraulic pumps, HP fuel pump, fuel control, HP tachometer, CSD and alternator, permanent-magnet generator, LP fuel pump and oil pumps. Additional low-speed (LS) gearbox, driven from LP shaft, serving LP rotor governor and tachometer.

LUBRICATION SYSTEM: Self-contained, with engine-mounted tank, fuel/oil heat exchanger and gear type pump; pressure 3·45 bars (50 lb/sq in). Tank capacity: A-1, 4·5 litres (1·0 Imp gal); A-2, 10·3 litres (2·27 Imp gal).

MOUNTING: Main ball-type trunnions on compressor intermediate casing; rear tangential steady-type at rear of bypass duct.

STARTING: Integral gas turbine (air turbine).

DIMENSIONS:

Length overall	2,610 mm (102·6 in)
Intake diameter	953 mm (37·5 in)
Height overall	1,026 mm (40 in)

WEIGHT, DRY:

A-1	1,353 kg (2,980 lb)
A-2	1,370 kg (3,018 lb)

PERFORMANCE RATING (Max T-O):

A-1	64·5 kN (14,500 lb st) to ISA +8°C
A-2	66·7 kN (15,000 lb st) to ISA

SPECIFIC FUEL CONSUMPTION (Max T-O):

A-1	17·93 mg/Ns (0·633 lb/h/lb st)
A-2	18·33 mg/Ns (0·647 lb/h/lb st)

ROLLS-ROYCE TURBOMÉCA

ROLLS-ROYCE TURBOMÉCA LIMITED

4/5 Grosvenor Place, London SW1X 7HH, England
Telephone: 01 235 3641

This company was formed jointly by Rolls-Royce and Turboméca in June 1966 to control the design, development and production programmes for the Adour two-shaft turbofan. The main function of the company is to receive contracts from the British Ministry of Defence on the Adour for both the British and French governments. The company can also enter into commercial contracts for the sale of the Adour to customers other than the British and French governments, and grant licences for its manufacture.

ROLLS-ROYCE TURBOMÉCA ADOUR

The Adour was designed originally for the SEPECAT Jaguar. The whole engine is simple and robust and of modular design. The complete propulsion unit was designed for an overhaul life of 1,000 h. Temperatures and rotational speeds are moderate, and a thrust growth of the order of 40 per cent was envisaged.

Bench testing began at Derby on 9 May 1967. Engines for Jaguars are assembled at Derby (R-R) and Tarnos (Turboméca) from parts made at single sources in Britain and France. Turboméca makes the compressors, casings and external pipework (to preserve Anglo-French parity the afterburner is subcontracted to SNECMA); Rolls-Royce makes the remainder.

The Adour was next selected for the Japanese Mitsubishi T-2 trainer and F-1 fighter/support aircraft, and since 1970 Ishikawajima-Harima Heavy Industries has been producing the Adour under a licence agreement. In 1972 a non-afterburning Adour was selected to power the British Aerospace Hawk advanced trainer.

Current versions of the engine are as follows:

Mk 102. Original production engine for Jaguars in service with RAF and Armée de l'Air. Qualified in 1972.

Mk 104. Uprated RT.172-26 version similar to Mk 804; RAF Mk 102 engines were converted to this standard.

Mk 151. Non-afterburning version for Hawk. Internal components and certification temperatures identical to Mk 102 and Mk 801A. Qualified in 1975.

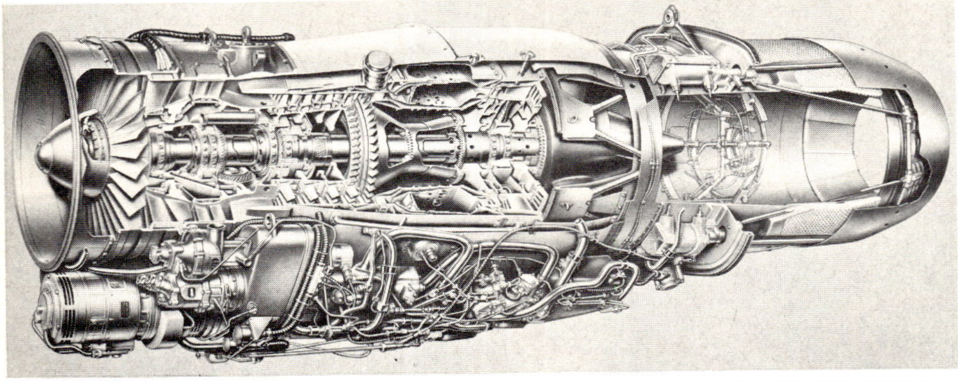

Cutaway drawing of the Rolls-Royce Turboméca Adour Mk 102 augmented turbofan (32·5 kN; 7,305 lb st)

ROLLS-ROYCE TURBOMÉCA ADOUR ENGINES

Version	Application	Basic wt dry kg (lb)	Max S/L thrust kN (lb st)
* Mk 102/801A	Jaguar, T-2, F-1	704 (1,552)	32·5 (7,305)
Mk 104	Jaguar	713 (1,571)	35·1 (7,900)
Mk 151	Hawk	553 (1,220)	23·1 (5,200)
Mk 851	Hawk	568 (1,252)	23·1 (5,200)
* Mk 804	Jaguar	715 (1,576)	35·8 (8,040)
Mk 861		577 (1,273)	25·4 (5,700)
* Mk 811		738 (1,627)	37·4 (8,400)

* afterburning versions

Rolls-Royce Turboméca Adour Mk 151 for British Aerospace Hawk T.1, rated at 23·1 kN (5,200 lb st)

Mk 801A. Japanese designation TF40-IHI-801A. For Mitsubishi T-2 and F-1. Qualified in 1972. (See Ishikawajima-Harima in Japanese section.)

Mk 804. Uprated engine for Jaguar International. Installationally interchangeable with Mk 102. General increase in thrust of some 10 per cent, with greater increase at high forward speeds (rating with full afterburner at Mach 0·9 at S/L, ISA, increased by 27 per cent). Qualified in 1976.

Mk 811. Uprated version for Jaguar International, installationally interchangeable with 804. Revised compressor aerodynamics and hot-end improvements to match higher temperatures. Qualified 1980 for Indian Jaguar; engines eventually to be assembled by Hindustan Aeronautics, with increasing Indian-manufactured content.

Mk 851. Non-afterburning version of Mk 804 for export Hawk.

Mk 861. Non-afterburning version of Mk 811, installationally interchangeable with 851; qualification due mid-1981, with deliveries same year.

Further versions of the Adour, giving substantial increases in thrust, are projected.

The following refers to afterburning versions:

TYPE: Two-shaft turbofan for subsonic and, with augmentation, supersonic aircraft.

INTAKE: Formed by forward extension of fan casing. No radial struts or inlet guide vanes.

FAN: Two-stage. Rotating spinner, anti-iced by turbine-bearing cooling air, on front of first-stage disc. Individually replaceable blades. Fixed stators and exit vanes. Unit overhung on spring-loaded ball bearing of squeeze-film type. Full-length bypass duct leading to afterburner. Bypass ratio, 0·75-0·80.

COMPRESSOR: Five-stage compressor on HP shaft. Large-diameter double-conical shaft for rigidity with bolted curvic couplings. Wide-chord blades of titanium. Steel stator blades. Overall pressure ratio 11 : 1.

COMBUSTION CHAMBER: Annular, with straight-through flow. Fitted with 18 air-spray fuel nozzles and two igniter plugs. Engine fuel system by Lucas GTE.

HP TURBINE: Single-stage, aircooled.

LP TURBINE: Single-stage. Both turbine bearings of squeeze-film type.

JET PIPE: Fully modulated afterburner of compact, short-length design incorporating four concentric but staggered spray rings and vapour gutters. Catalytic igniters between inner gutters. Variable nozzle has eight master and eight slave petals positioned by frame moved axially by four fuel-operated rams. Control system by Dowty Fuel Systems, with vapour-core pump.

DIMENSIONS:
- Inlet diameter (all): 559 mm (22 in)
- Max width (all): 762 mm (30 in)
- Max height (all): 1,041 mm (41 in)
- Length:
  - Mks 102, 801A, 804, 811: 2,970 mm (117 in)
  - Mks 151, 851, 861: 1,956 mm (77 in)

WEIGHT, DRY: See table

PERFORMANCE RATINGS: See table

SPECIFIC FUEL CONSUMPTION (Mk 102):
- S/L static, dry: 21 mg/Ns (0·74 lb/h/lb st)
- Mach 0·8, 11,890 m (39,000 ft): 27 mg/Ns (0·955 lb/h/lb st)

ROLLS-ROYCE TURBOMÉCA RTM 321

At the 1979 Paris Salon Rolls-Royce and Turboméca both exhibited full-scale display examples of this new turboshaft engine which they are to develop jointly, subject to government approval. Though Westland has ordered 11 General Electric T700 engines for the EH 101 helicopter programme, the two engine companies hope that the RTM 321 will be adopted as the standard production engine on this helicopter. It offers an alternative to the T700 in other new helicopters in the 7/15-tonne class. The RTM 321 incorporates all the experience gained with such modern engines as the Gem and Makila to achieve extreme simplicity, good sfc and low costs. It comprises four modules and can be configured for any desired installation, with front or rear drive. Features will include an inlet particle-separator and anti-IR protective measures.

COMPRESSOR: Three-stage axial, single-stage centrifugal.

COMBUSTION CHAMBER: Annular reverse-flow.

TURBINES: Two-stage (presumably aircooled, but not yet announced) gas-generator turbine. Two-stage power turbine with drive to front or rear.

DIMENSIONS:
- Length overall: 1,050 mm (41·34 in)
- Diameter: 515 mm (20·28 in)

WEIGHT: approx 180 kg (396 lb)

RATINGS:
- Max contingency: initially in 1,430 kW (1,917 shp) range, rising later to 1,865 kW (2,500 shp)
- Intermediate: initially in 783 kW (1,050 shp) range
- Long-range: initially in 582 kW (780 shp) range

SPECIFIC FUEL CONSUMPTION:
- Intermediate: 86·0 μg/J (0·509 lb/h/shp)
- Cruise, as above: 86·2 μg/J (0·51 lb/h/shp)

Rolls-Royce Turboméca RTM 321 turboshaft engine
(Brian M. Service)

TURBO-UNION

TURBO-UNION LTD

PO Box 3, Filton, Bristol BS12 7QE, England
Telephone: 0272 693871

MUNICH OFFICE: 8 München 81, Arabellastrasse 4/7, Federal Republic of Germany

CHAIRMAN: Dr G. C. Boffetta

Formed in October 1969, this international company was established to manage the entire programme for the RB.199 engine for the Panavia Tornado. Shares are held in the ratio Fiat Aviazione SpA, 20%; MTU München GmbH, 40%; Rolls-Royce Ltd, 40%.

TURBO-UNION RB.199

The RB.199 is a three-shaft augmented turbofan of extremely advanced design, offering low fuel consumption for long-range dry cruise, even at sea level, and approximately 100 per cent thrust augmentation with full afterburner for combat manoeuvre and supersonic acceleration. Further design goals included minimal weight, frontal area and volume, and moderate first cost and operating costs. Rapid strip and rebuild are facilitated by modular construction and by the use of advanced manufacturing techniques to reduce the number of separate components.

The first RB.199 ran in September 1971, less than two years after go-ahead. The early development of the engine was described in the 1979-80 *Jane's*. The 150-h FQT (Formal Qualification Test) was completed in November 1978. By the beginning of 1981, a total of 200 production engines had been delivered. Firm production orders (three batches) stand at 823, with a fourth batch of 425 engines covered by contracts for long lead-time items. Total engine production for the initial programme of 809 Tornado aircraft is just over 2,000. The overall work-share is intended to be in proportion to the number of aircraft acquired by each nation.

The basic designation RB.199-34R defined the original rating for the Tornado. The first production designation was **RB.199 Mk 101**, which applies to the production standard currently being delivered. The **RB.199 Improved Mk 101** is the second production standard, developed under the three-year Post FQT Programme, which started at the beginning of 1979; this is aimed at modest performance increases and a significant extension of life.

The Improved Mk 101 could be introduced from engine 824 onward. Development to produce higher thrusts, and to reduce afterburning fuel consumption, continues.

Considerably developed versions of the RB.199 have been considered for air-superiority projects such as the Euro-fighter. These uprated versions, if proceeded with, could be available for Tornado application. Retrofit upgrading of current production engines is facilitated by the modular construction.

Initial service experience gained by British, German and Italian aircrew, operating from the Tri-national Tornado Training Establishment (TTTE) at RAF Cottesmore, has confirmed the basic robustness of the design. The ability of the engine to cope with FOD and birdstrikes is due largely to the three-spool layout, with its consequent short, rigid rotating assemblies held between small bearing spans.

TYPE: Three-shaft turbofan with afterburner and integral thrust reverser.

INTAKE: Annular, without inlet guide vanes.

LP COMPRESSOR: Three-stage axial LP compressor of titanium alloy. Casing of three bolted sections, comprising a ring of stator vanes welded at their outer ends to form the casing and at their inner ends to form the interstage seal. Rotor of three discs welded together. Rotor blades secured by dovetail roots, all with snubbers to control vibration and provide greater resistance to foreign-object damage. Blade containment by local increase of casing thickness. Mass flow approx 70 kg (154 lb)/s. Bypass ratio about 1:1.

The first detailed longitudinal section of the RB.199 Mk 101

IP COMPRESSOR: Three-stage of titanium alloy. Rotor has welded discs in which blades are secured by dovetail roots. Stator rings bolted together, with rear pair of flanges forming attachment to intermediate casing, holding blades in dovetail grooves.

HP COMPRESSOR: Six-stage; material changes from titanium at front to heat-resisting alloy at rear, except stator blades are heat-resisting steel throughout. Casing of six rings bolted together. Each ring forms track of corresponding rotor and each pair of rings forms groove in which stators are located by dovetail roots. Rotor discs secured by ten through-bolts, carrying blades by dovetail roots. Provides bleed air for aircraft services, turbine cooling and air motors for reverser and nozzle. Bevel gear provides radial drive to gearbox. Overall pressure ratio greater than 23:1.

INTERMEDIATE CASING: Fabricated titanium inner and outer casings, with main thrust spigot and front side-mounting links. Inner casing houses IP and HP compressor bearings, supported on eight hollow vanes across bypass and core flows.

BYPASS DUCT: Fabricated in titanium, forms outer shell between intermediate casing and jetpipe.

COMBUSTION CHAMBER: Annular flame tube fabricated from nickel-based alloy, bolted at rear end between outer casing, forged and chemically milled in nickel-iron alloy, and inner casing, fabricated in nickel alloy. Carries 13 double-headed fuel vaporisers which give combustion without visible smoke. Two igniter plugs combined with primary fuel injectors. Hot-streak injector for afterburner ignition.

TURBINES: Rotor blades and stator vanes machined from precision castings in heat-resisting nickel-based alloy. All rotor blades shrouded and secured by fir-tree fixings. Stator vanes of all four stages carried in single casing machined from nickel forging.

HP TURBINE: Entry temperature over 1,327°C. Rotor blades and stator vanes aircooled. Stator cooling provided by HP air which passes through root and tip of each hollow vane into a perforated sheet-metal insert. This cooling air passes through holes in the insert before exhausting into through-holes in the vane leading and trailing edges. Remainder of cooling air is ducted to HP rotor-blade roots via pre-swirl nozzles and a turbine cover plate which increases cooling-air pressure. Air leaves blades through holes in leading edge, at tip and in trailing edge.

IP TURBINE: Aircooled stator vanes and rotor blades. Cooling air from third HP compressor stage passes into perforated central insert in stator vanes. Approximately half air cools vane and exhausts through holes in trailing edge; rest feeds via pre-swirl nozzles into rotor-blade roots and via machined holes to exit at tips.

LP TURBINE: Two-stage with hollow uncooled rotor blades. First-stage stator vanes cooled by third-stage HP compressor air, and contain support struts for IP/LP bearing.

LP SPOOL: LP compressor rotates clockwise (from rear), overhung on two widely spaced roller bearings, on bearing support welded to inner wall of third-stage stator ring. LP turbine shaft runs in ball thrust bearing in combined IP/LP bearing housing.

IP SPOOL: Front end of shaft, which also has clockwise rotation, runs in ball thrust bearing carried by intermediate casing. Rear end runs in roller bearing supported by IP/LP bearing housing.

HP SPOOL: Rotates anticlockwise in two roller bearings. Front bearing is carried in intermediate casing; rear is a contra-rotating intershaft bearing carried on IP shaft sleeve. Compensating shaft is located at front by ball thrust bearing in intermediate casing and at rear is attached to HP turbine, locating whole spool axially. This shaft of low-expansion steel matches thermal expansion of casings to maintain axial seal clearances.

AFTERBURNER: Front end of titanium fabricated jetpipe carries afterburner in which bypass air and core gas burn concurrently, without a mixing section. For core flow, two gutter flameholders fed by upstream atomisers. For bypass flow, reverse colander with radial extensions, each containing vaporising primary burner, between which multiple jets inject remainder of afterburner fuel. Provides fully-modulated augmentation. Aircooled heat shield protects jetpipe from afterburner temperature greater than 1,627°C.

NOZZLE: Variable-area, short-petal, convergent nozzle operated by shroud actuated by four screwjacks, driven by fourth-stage HP air motor via flexible shafting. Each of 14 master and 14 secondary petals is precision casting in cobalt alloy which minimises friction.

REVERSER: External two-bucket type driven via flexible shafts by motor using HP air. In stowed position outer skins form aircraft profile. Deployment takes 1 s at any thrust setting from idle to max dry.

ACCESSORY DRIVES: Accessory gearbox on underside of intermediate casing (quick attach/detach coupling) carries hydromechanical portions of main and afterburner fuel systems, oil tank and pump, and output shaft to aircraft gearbox carrying KHD gas-turbine starter/APU.

FUEL SYSTEM: The electronic Main Engine Control Unit (MECU), which is aircraft-mounted, uses signals from pilot's lever and power plant sensors to control hydromechanical engine and afterburner fuel systems, nozzle and reverser. Duplicated electronics provide integrity, with automatic switching. Flow through main gear pump determined by metering orifice commanded by MECU, with recirculation for cooling aircraft and engine systems. Afterburner fuel from engine-driven vapour-core pump controlled by main and distribution actuators, both under MECU command.

LUBRICATION SYSTEM: Main bearings, with squeeze-film damping, in three chambers; each is supplied with pressure oil and scavenged through return pipes which have filters and magnetic detector plugs. Oil samples drawn from tank for spectrographic analysis.

DIMENSIONS:

Max diameter	870 mm (34·25 in)
Overall length	3,230 mm (127 in)

WEIGHT, DRY:

Excluding reverser	approx 900 kg (1,980 lb)

PERFORMANCE RATINGS (S/L, ISA, static):

Max dry	35 kN (7,870 lb) class
Max afterburning	70 kN (15,735 lb) class

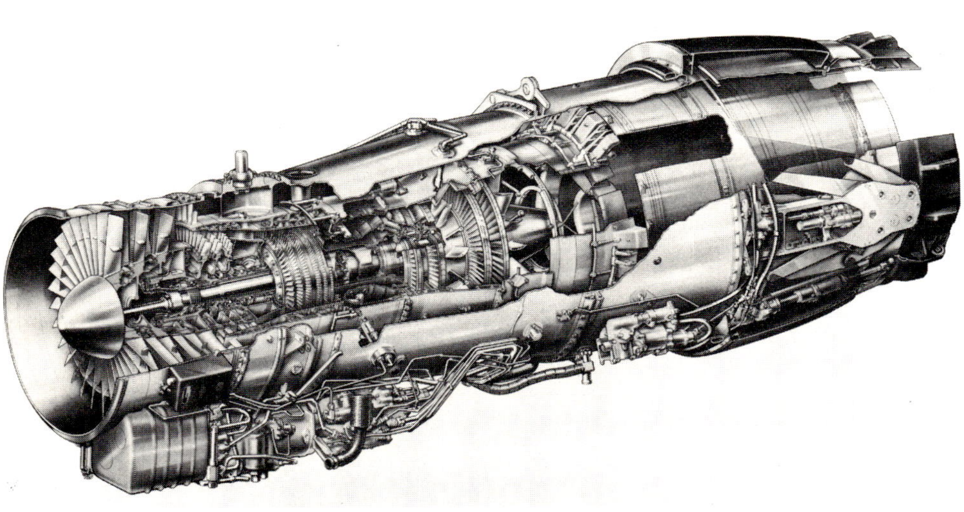

Cutaway drawing of Turbo-Union RB.199 Mk 101 three-shaft augmented turbofan

ISRAEL

BET-SHEMESH
BET-SHEMESH ENGINES LTD

Mobile Post Haela, Bet-Shemesh
Telephone: 02 911661-2
Telex: 25290

The first section of a 12,077 m² (130,000 sq ft) Israeli aero-engine factory, Bet-Shemesh Engines Ltd, was inaugurated officially on 15 January 1969. The company is owned by the Israeli government (majority holding), Turboméca SA and M J. R. Szydlowski. The manufacturing plant was based on the Turboméca factory at Tarnos. On 1 January 1977 a change in management took place. Over the subsequent year the number of employees increased from 550 to about 1,300, and plant area increased to 22,500 m² (242,000 sq ft).

At first Bet-Shemesh manufactured turboprop components on behalf of Turboméca. By 1973 complete Marboré VI turbojets for CM 170 Super Magister trainers were being produced. The company also manufactures parts of the Marboré II, Artouste II and III, Turmo II and Astazou II. Today the plant is the centralised source for all aircraft gas-turbine manufacture in Israel. It makes most portions of the General Electric J79 and will shortly produce the complete engine. It also makes parts of the Pratt & Whitney F100 and Teledyne CAE J69 and provides support for the Allison 250-C20.

Other operations include a precision foundry, and gas-turbine programmes for surface applications.

ITALY

ALFA ROMEO
SOCIETÀ PER AZIONI ALFA ROMEO

Via Gattamelata 45, 20149 Milan

Telephone: Milan 3977

Aviation Works: 80038 Pomigliano D'Arco, Naples

Telephone: 8841 344

Alfa Romeo participated in the European production programme for General Electric J79-GE-11A turbojets to power Lockheed F-104G Starfighters. It is collaborating with Rolls-Royce in the manufacture and overhaul of Gnome H.1000, H.1200 and H.1400 turboshafts. A licence agreement has also been signed for the manufacture of General Electric T58 and T700 turboshafts.

Alfa Romeo was prime contractor for the manufacture, under General Electric licence, of the J85-GE-13A turbojet to power the G91Y aircraft. It is manufacturing the hot section of the J79-GE-19 turbojet for the F-104S, as well as many parts for the J79-11B, -17 and -1K. Alfa Romeo manufactures CF6 combustors under subcontract to General Electric; assembles the kits and overhauls P&WC PT6T engines for the AB 212 helicopter; under GE licence, it is responsible for the hot section of the T64-P4D turboprop co-produced with Fiat; and participates in development and hot-section component manufacture for the Turbo-Union RB.199 for the Panavia Tornado. The company would probably be a major partner in Italian development and licence production of the Rolls-Royce Spey 807 for the AM-X aircraft. It has taken over a small turboprop that was originally a collaborative project with Rolls-Royce.

The company overhauls many engines of Rolls-Royce, General Electric and Pratt & Whitney manufacture.

Alfa Romeo is a member of the Finmeccanica-IRI group of companies.

ALFA ROMEO RB.318

In the mid-1970s Alfa Romeo's Aero Engine Division began collaboration with Rolls-Royce on a small gas turbine for general-aviation applications, known as the RB.318. This was never formally announced by either partner. By 1979 a successful study, design and development programme had led to component and engine testing, including flight testing of a turboprop in a Beechcraft King Air A90, which met or exceeded all performance targets.

By 1980 the programme was being transferred entirely to Alfa Romeo in accordance with an agreed time schedule. The definitive engine, a turboprop covering a range of powers from 298 to 596 kW (400-800 shp), will probably have a different designation. The programme is matched to the company's resources, and will draw upon Rolls-Royce under a contract for technical support. Certification is expected in 1982.

CSTM
CENTRO STUDI TRASPORTI MISSILISTICI
(Missile Transport Research Centre)

Head Office: Via Squarcialupo 19-A, 00162 Rome

Telephone: 423.833

President: Glauco A. Partel

CSTM has begun work on the P-77 self-pressurised monopropellant rocket engine. Starting and reaction is by a special heating system. This project has in turn led to a powder/powder hypergolic bipropellant motor.

DIFESA E SPAZIO
DIFESA E SPAZIO SpA

Via Sicilia 162, 00187 Rome

Telephone: (06) 4680

Telex: 610114 SNIA

Factories: Colleferro (Rome) and Ceccano (Frosinone)

Division Manager: Ing E. Svizzeretto

Chief of Rocket, Missiles and Space R & D Department: Ing P. Laurienzo

The SNIA Viscosa Defence and Space Division is engaged in the production of conventional ammunition and the research, design and production of solid propellants, solid-propellant rocket motors, complete rockets and missiles, as well as air-to-surface and surface-to-surface unguided weapons. It possesses all necessary installations for the production of double-base solid propellants. Composite solid-propellant grains and motors incorporating polyurethane or polybutadiene polymers as binder are also produced in all sizes.

Complete propulsion units of up to 1,500 mm (59 in) diameter, with combustion times ranging from a fraction of a second to about 1 minute, are manufactured by SNIA Viscosa's Defence and Space Division for use in military rockets and missiles.

SNIA's Defence and Space Division produced the motors for the Italian Sparrow air-to-air missile programme, and developed an advanced motor for the Aspide air-to-air missile, now coming into the production phase. The Division is also involved in the development of many weapon systems in co-operation with such companies as MBB, OTO Melara, Breda Meccanica Bresciana and Sistel.

In the space field, a new amagnetic high-performance apogee motor, with a titanium-alloy case and improved polybutadiene propellant, was developed and qualified for the Italian Siro experimental satellite successfully launched in August 1977. The solid-propellant apogee motor for the GEOS satellite, developed and qualified in co-operation with SEP of France, has operated twice successfully, in April 1977 and July 1978. For OTS and Meteosat, a high-performance apogee motor with carbon-fibre case was qualified by a consortium formed by SNIA, SEP and MAN. The same consortium has since begun development of a larger apogee motor for Ariane payloads. SNIA has also completed, on behalf of CNES, the development and qualification of retro and ullage motors for the separation of all three stages of the Ariane vehicle. In 1980, it signed an agreement with Aeritalia for development of the IRIS (Italian Research Interim Stage) propulsion stage for transferring payloads from low orbit (mainly ex-Shuttle) to geosynchronous or transfer orbits.

FIAT
FIAT AVIAZIONE SpA

Via Nizza 312, 10127 Torino, Italy

Telephone: (011) 63991

Telex: 221320 Fiatav

Chairman: C. E. Rossi

Managing Director: G. C. Boffetta

Production Director: S. Innocenti Torrini

Design and Engineering Director: L. La Rocca

Fiat Aviazione SpA was incorporated in 1976 as a wholly Fiat-owned company. It took over all the aero-engine activities carried out formerly by the Fiat Aviation Division, and the participation in Turbo-Union Ltd and Turbomotori Internazionale, as well as control of a plant at Brindisi where overhaul of jet and piston engines and construction of jet engine parts are carried out.

Fiat's main aircraft engine programmes now concern the Turbo-Union RB.199, Rolls-Royce Viper 600, Pratt & Whitney PW2037 and General Electric J79, J85, CF6 and T64, all of which are referred to below. In addition the company overhauls many types of engine, including the J79 and Orpheus, and the R-2800 piston engine. It is engaged in the design, development and production of the LM 2500 marine gas turbine, in collaboration with General Electric, and the design and production of main gearboxes for Aérospatiale helicopters.

TURBO-UNION RB.199

Fiat holds 20 per cent of the shares of Turbo-Union Ltd, the joint company set up to produce the RB.199-34R engine for the Panavia Tornado. Fiat's responsibility is the LP turbine and shaft, exhaust diffuser, jetpipe and nozzle. The programme is described under Turbo-Union in the International part of this section.

PRATT & WHITNEY PW2037

Since 1974 Fiat has been responsible for design and development of the accessory drive gearbox for the Pratt & Whitney PW2037 (originally JT10D) civil turbofan engine. The programme is described under Pratt & Whitney in the US part of this section.

GENERAL ELECTRIC CF6

Fiat is engaged in the manufacture of components for the CF6 civil turbofan for both General Electric and SNECMA. For GE the company supplies complete accessory gearboxes, transfer gearboxes, inlet gearboxes and shafts. SNECMA is supplied with various gearbox components and shafts for CF6-50 engines for the Airbus A300.

GENERAL ELECTRIC T64-P4D

This free-turbine turboprop powers most versions of the Aeritalia G 222 military transport aircraft. Under a licence agreement between the General Electric Company and the Italian government, the engine is being manufactured in Italy under the leadership of Fiat as prime contractor.

GENERAL ELECTRIC J79-19

This afterburning turbojet powers Lockheed F-104S fighters of the Italian Air Force. Fiat, as prime contractor for Italy, produced the engine under a licence agreement between the General Electric Company and the Italian government.

GENERAL ELECTRIC J85-13

With Alfa Romeo as prime contractor, Fiat participated in production under licence of the J85-13A, power plant of the Aeritalia G91Y aircraft.

ROLLS-ROYCE VIPER 600

Development of this turbojet for business aircraft and military trainers was undertaken in collaboration with Rolls-Royce. For most versions, components rearward of the compressor housing (except turbine discs and blades) are Fiat's responsibility. However, the Mk 632-43 engine for the Aermacchi MB.339 is licensed to Piaggio. The Viper 600 is described in the Rolls-Royce entry in this section.

PIAGGIO
INDUSTRIE AERONAUTICHE E MECCANICHE RINALDO PIAGGIO SpA

Head Office: Viale Brigata Bisagno 14, 16129 Genoa (426)

Telephone: 540 521

Works and Officers: See Aircraft section

The Aero-Engine Division of Piaggio manufactures the following engines under licence agreements: Rolls-Royce Viper 11, 526, 540 and 632-43 turbojets to power the Aermacchi M.B. 326 and 339 (a sublicence for manufacture of the Viper 11 and 540 was issued to Atlas Aircraft to power South African-built M.B. 326 aircraft); Avco Lycoming T53-L-13 turboshafts for various Bell and Agusta-Bell helicopters; Avco Lycoming T55-L-11 and derivatives for CH-47 Chinook helicopters; and Rolls-Royce Gem 2 turboshafts for Agusta A 129 anti-tank helicopters. Piaggio also participates in co-production under licence of the Rolls-Royce Spey 807 turbofan.

JAPAN

IHI
ISHIKAWAJIMA-HARIMA JUKOGYO KABUSHIKI KAISHA (Ishikawajima-Harima Heavy Industries Co Ltd)

Head Office: No 2-1, 2 chome, Ohte-Machi, Chiyoda-ku, Tokyo

Aero Engine and Space Development Group: 3-5-1, Mukodai-cho, Tanashi-shi, Tokyo 188

Telephone: (0424) 66 1225

President: Dr Hisashi Shinto

Executive Vice-President: Dr Osamu Nagano

Managing Director and General Manager, AESD Group: Dr Kaneichiro Imai

Ishikawajima-Harima J3-IHI-7C turbojet engine (13·7 kN; 3,090 lb st)

In February 1960 IHI began the licence production of General Electric J79-IHI-11A turbojet engines for Japanese-built Lockheed F-104J Starfighters.

Under further licensing agreements with General Electric, IHI is producing the J79-IHI-17 turbojet for the McDonnell Douglas F-4EJ, the T58 turboshaft for helicopters and other applications, including the propulsion of air cushion vehicles and hydrofoil boats, and the T64-IHI-10 turboprop engine to power the JMSDF's PS-1 anti-submarine flying-boat and US-1 search and rescue amphibian, and the Kawasaki P-2J maritime patrol aircraft. By March 1981 deliveries totalled 666 T58s, 381 T64s and 610 J79s.

In 1979 IHI linked with other Japanese companies in the 50/50 development with Rolls-Royce of the RJ.500, described under RJAE in the International part of this section.

The manufacture of the Rolls-Royce Turboméca Adour augmented turbofan in Japan under licence agreement received government approval in September 1970. Its production began in early 1973, under the Japanese designation TF40-IHI-801A, to power the supersonic T-2 trainer and F-1 fighter. By March 1981 IHI had delivered 330 TF40s.

IHI was nominated in 1978 as prime contractor of Pratt & Whitney F100 engines for the Japanese-built F-15 fighter and Allison T56 engines for the Japanese-built Lockheed P-3C. Kawasaki and Mitsubishi will be subcontractors. Production of 32 F100 and 12 T56 engines was ordered in Fiscal Year 1979, followed by 60 F100 and 40 T56 engines in FY 1980. Two F100s had been delivered by March 1981.

IHI undertakes overhaul and repair of Pratt & Whitney JT8D and Rolls-Royce RB.211 commercial turbofans, and General Electric J79, T58 and T64, and R-R Turboméca TF40 military engines, and Turboméca Artouste and Astazou turboshafts.

Prior to the start of licence production, in April 1959, IHI had been responsible for the J3 turbojet engine which had been under development by the Nippon Jet-Engine Company since 1956. The J3-IHI-7 version is installed in Fuji T-1B intermediate jet trainer and Kawasaki P-2J aircraft.

IHI participated in developing the XJ11 lift-jet, as well as the JR100 and JR200 built under supervision of the NAL. In addition, in collaboration with Mitsubishi and Kawasaki, IHI made the prototypes of the FJR710 turbofan. The JR100, JR200 and FJR710 are described in the NAL entry.

IHI is co-operating with JDA in studies of small turbofans in the 11·8 kN (2,645 lb) thrust class.

IHI XF3

This new military turbofan was unveiled at the International Gas Turbine Conference in England in April 1978. Development had begun just two years earlier, with funding by the JDA's Technical Research & Development Institute. The initial Phase I (XF3) form has a single-stage fan with bypass ratio of 1·9, five-stage transonic compressor, 12-burner combustion chamber and single-stage HP and LP turbines. Rating in this form is to be 11·79 kN (2,650 lb st). The more powerful Phase II engine (XF3-20) has a two-stage fan, two-stage LP turbine and higher turbine temperature. The second engine reached 16·28 kN (3,660 lb st) in 1979. Five more engines were built in 1980 and a further four were funded in FY 1981. Bench testing of the first two engines had totalled 620 h by March 1981. The eventual version is intended for the MT-X twin-engined trainer, at a rating of 15·68 kN (3,530 lb st).

IHI J3

The J3-IHI-7C is a derivative of the J3-1, of which a description appeared in the 1959-60 *Jane's*, under the entry for Nippon Jet-Engine Company. It is installed in the Kawasaki P-2J aircraft currently in service with the JMSDF. Production, which totalled 247, ended in 1980.

The J3-IHI-7C, to which the following data apply, was described fully in the 1980-81 *Jane's*.

Dimensions:

Length, less tailpipe	1,661 mm (65·4 in)
Length overall with rear cone	1,994 mm (78·5 in)
Diameter overall	627 mm (24·7 in)
Frontal area	0·28 m² (3·01 sq ft)

Weight, Dry:

Bare	380 kg (838 lb)
With accessories	430 kg (948 lb)

Performance Rating:

T-O	13·7 kN (3,090 lb st)

Specific Fuel Consumption:

At T-O rating	29·74 mg/Ns (1·05 lb/h/lb st)

Oil Consumption:

At normal rating (max) 0·60 litres (1·06 Imp pints)/h

IHI TF40-IHI-801A

This is the Rolls-Royce Turboméca Adour augmented turbofan engine, as built under licence by IHI. For details see Rolls-Royce Turboméca entry in International section.

KAWASAKI

KAWASAKI JUKOGYO KABUSHIKI KAISHA (Kawasaki Heavy Industries Ltd)

2-16-1 Nakamachi-Dori, Ikuta-ku, Kobe

Telephone: Kobe (078) 341 7731

Works: Akashi, Hyogoken

Officers: See Aircraft section

Kawasaki's factory at Akashi started repair, overhaul and component manufacturing for aircraft engines, on behalf of the US armed forces and the Japan Defence Agency, in 1953. Since then, it has overhauled more than 11,900 engines, mainly of the Allison J33, General Electric J47, Rolls-Royce Orpheus, Westinghouse J34, and Kawasaki KT5311A and KT5313B series. Since 1968 the company has also been making spare parts for the J33, and since 1973 for the Orpheus.

In 1967, under a licence agreement with Avco Lycoming, Kawasaki started manufacturing T53 turboshaft engines. Deliveries of the resulting KT5311A, KT5313B and T53-K-13B engines totalled 241 units by early 1981. In 1977, under a licence agreement with Rolls-Royce, Kawasaki began making Industrial/Marine Olympus TM-3B and Tyne RM-1C engines, with deliveries starting in September 1980.

Kawasaki shares in parts manufacturing for the Rolls-Royce Turboméca Adour, the Pratt & Whitney F100 and JT8D (for Kawasaki C-1 transport, with Kawasaki noise suppressor) and the IHI-assembled Allison T56. In 1979 Kawasaki became a member of the Japanese consortium responsible for a 50 per cent share in the development and manufacture of the RJAE RJ.500.

MHI

MITSUBISHI JUKOGYO KABUSHIKI KAISHA (Mitsubishi Heavy Industries Ltd)

Head Office: 5-1, Marunouchi 2 chome, Chiyoda-ku, Tokyo 100

Engine Works: Daiko Plant, Nagoya Aircraft Works, 1-1, Daiko-cho, Higashi-ku, Nagoya 455

Telephone: (052) 721 3111

Komaki North Plant, Nagoya Aircraft Works, 1200, Higashi-Tanaka, Komaki-Shi, Aichi 485

Telephone: (0568) 79 2111

Officers: See Aircraft section

Since 1952 Mitsubishi Heavy Industries (MHI) has repaired and overhauled engines of the Japan Defence Agency and domestic and foreign airlines. It resumed its manufacturing activity in the aviation gas-turbine field in 1967 by undertaking production of the CT63 turboshaft engine to power Hughes 369HM helicopters of the JGSDF, under a licence agreement with Allison. A total of 217 engines was delivered to the Japan Defence Agency by March 1978. In 1972, under licence agreement with Pratt & Whitney Aircraft, MHI began manufacture of the JT8D-M-9 turbofan. The first was delivered in January 1973. By June 1981 a total of 72 engines was to be delivered to the Japan Defence Agency for use in Kawasaki C-1 military transports.

In collaboration with IHI and Kawasaki, MHI participates in both the FJR710 turbofan programme (see NAL entry) and the RJ.500 programme (see RJAE in the International part of this section). It is also developing a small turbojet of its own design to power the new SSM-1 surface-to-ship missile being developed for the JGSDF.

NAL

NATIONAL AEROSPACE LABORATORY

1880 Jindaiji-machi, Chofu City, Tokyo

Telephone: 0422 47 5911

Director: Toshio Kawasaki

Head of Aero-Engine Division: Masakatsu Matsuki

The National Aerospace Laboratory (NAL) is a government establishment responsible for research and development in the field of aeronautical and space science. Since 1962 it has extended its activity to include V/STOL techniques. The decision was made in that year to initiate development of an engine to fulfil the requirement for a lightweight lift-jet power plant for VTOL aircraft.

The more advanced NAL/IHI JR200 was developed in 1966, the NAL/IHI JR220 was completed in 1971.

In 1971 the Agency of Industrial Science and Technology, Ministry of International Trade and Industry (MITI), funded a high by-pass ratio turbofan engine (FJR710) development programme. NAL has completed the basic design of this engine, and many component tests are being made at NAL. The FJR710 is being developed by an industry consortium comprising IHI, Kawasaki and Mitsubishi, in co-operation with NAL and MITI.

MITI/NAL FJR710

In the late 1960s the Japanese government and industry, seeking an engine programme that might remain competitive for many years, decided to embark on the design of a subsonic turbofan of high by-pass ratio. After a preliminary study by the NAL, funding was provided by the Ministry of International Trade and Industry in 1971 for a prototype demonstrator and test programme.

NAL has managed the design of the resulting FJR710. Manufacture of the prototype and development engines was subcontracted to IHI, Kawasaki and Mitsubishi. The first engine made its first run in May 1973. By the end of 1978 six engines (three FJR710/10 and three FJR710/20 with small changes) had run a total of 1,300 h. Tests completed included 150-h endurance and low-cycle fatigue. In November 1977 high-altitude performance testing was completed successfully at the NGTE in the UK.

Phase 2 of the FJR-710 programme began in 1976, and the first of six FJR-710/600s had been completed by December 1978. The second high altitude test (and the first for the FJR-710/600) was carried out in the UK, at the NGTE, on 13 April 1981, and further testing in the UK facility is planned in 1982. Assembly of the FJR-710/600S, scheduled to power the NAL experimental QSTOL aircraft, began in FY 1980, and flight testing of this version in a JASDF Kawasaki C-1 is due to take place in 1982. The QSTOL, with four FJR-710/600S, is due to fly in FY 1983. Total FJR-710 testing had reached 2,400 h by March 1981.

The following description applies to the prototype engine, and is provisional:

Type: Two-shaft high bypass ratio turbofan for subsonic commercial or military aircraft.

Air Intake: Direct annular entry around fan spinner.

Fan: Single-stage fan, with rotating spinner and inserted titanium blades with part-span shrouds. Metal fan duct held by eight aerofoil struts, preceded by ring of flow-straightening vanes. Bypass ratio 6·5 : 1.

Compressor: Mechanically independent HP compressor. Multi-stage axial assembly with inserted blades of titanium and, at delivery end, high-nickel alloy. Several rows of variable stator blades held in upper and lower half-casings and operated by peripheral rings scheduled by hydraulic ram.

Combustion Chamber: Smokeless annular type.

Turbine: Two-stage HP gas-generator turbine with cooled blades. Multi-stage LP fan turbine.

Jet Pipe: Fixed area.

Dimensions (approx):
- Length 3,300 mm (130 in)
- Diameter 1,520 mm (60 in)

Weight, Dry: 980 kg (2,160 lb)

Performance Ratings (ISA):
- T-O 49 kN (11,025 lb st)
- Cruise at 6,100 m (20,000 ft) at Mach 0·7 16·7 kN (3,748 lb st)

Specific Fuel Consumption:
- T-O 9·83 mg/Ns (0·347 lb/h/lb st)
- Cruise, as above 17·7 mg/Ns (0·623 lb/h/lb st)

NAL/IHI JR200 and JR220

Following work on the NAL/IHI JR100, NAL designed and developed the higher-thrust NAL/IHI JR200 of improved thrust/weight ratio, and this was manufactured by IHI. An improved version, the NAL/IHI JR220 with higher pressure ratio and higher turbine entry temperature, also began development, but both programmes have now been suspended. All known details can be found in the 1980-81 *Jane's*.

MITI/NAL FJR710 two-shaft turbofan (49 kN; 11,025 lb st)

ZENOAH

KOMATSU ZENOAH COMPANY

Head Office: 2-142-1 Sakuragaoka, Higashiyamato, Tokyo 189

President: Shiro Sakuma

This company was established in 1910 as Tokyo Gasu Denki (Tokyo Gas & Electric) and has made aircraft engines for 65 years, during the second World War as Hitachi Aircraft Co. Since 1950 it has become famous as Xenoah Co, but changed its name on 1 October 1979 to Komatsu Zenoah, a subsidiary of Komatsu Ltd.

ZENOAH G72C-C

This engine has been type-tested by the JCAB and FAA (Pacific-Asia) and was certificated by the JCAB on 24 July 1978; FAA certification followed on 28 April 1979.

Type: Three-cylinder aircooled two-stroke piston engine.

Cylinders: Bore 72·0 mm (2·835 in). Stroke 59·5 mm (2·342 in). Swept volume 726·7 cc (44·34 cu in). Compression ratio 7·0.

Induction: Three Mikuni VM34SS float-type carburettors with mixture control. Two Mikuni DF52-501 fuel pumps. One Mikuni 4M-501 oil injection pump.

Fuel Grade: 100/130 octane gasoline and Shell Super two-stroke oil injected separately into induction system.

Ignition: Dual capacitor-discharge.

Exhaust System: 3 into 1 manifold and tuned silencer.

Starting: Hitachi SI 08-62 electric starter, 12V, 1kW.

Accessories: Optional Mitsubishi 14V 35A alternator and mounting bracket for instrument vacuum pump.

Dimensions:
- Length with exhaust manifold 564 mm (22·2 in)
- Width including carburettors 443 mm (17·44 in)
- Height overall (incl plugs) 390 mm (15·35 in)

Weight, Dry:
- Bare 59·4 kg (131 lb)
- With exhaust system, silencer, starter, alternator and vacuum pump 71·2 kg (157 lb)

Performance Ratings:
- Max T-O 44·7 kW (60 hp) at 6,080 rpm
- Max continuous 40·25 kW (54 hp) at 6,080 rpm

Specific Fuel Consumption:
- Max at 75% power (33·6 kW; 45 hp) 135 μg/J (0·8 lb/h/hp)

Fuel Consumption:
- Max at 75% power 22·5 litres (4·95 Imp gallons; 5·95 US gallons)/h

The Zenoah G72C-C three-cylinder engine

POLAND

PZL

POLSKIE ZAKLADY LOTNICZE

Headquarters: ul. Miodowa 5, 00251 Warsaw

Sales Representative: Pezetel, ul. Przemyslowa 26, 00450 Warsaw

Telephone: Warsaw 285071

The entire Polish aircraft industry is subordinate to the Zjednoczenie Przemyslu Lotniczego i Silnikowego PZL (Aircraft and Engine Industry Union). Pezetel handles all export sales of Polish aeronautical material.

BORZECKI

JOZEF BORZECKI

Wroclaw, ul. Sernicka 20/4

BORZECKI JB 2 × 250

This designer's 2RB engine was described in the 1980-81 *Jane's*. Work is now concentrated on new engines comprising two single-cylinder units, each driving a propeller, coupled rigidly together through the cylinder heads by steel rods which offer extremely high stiffness axially. The dual-engine combination offers single-engine or two-engine operation without vibration problems. The engines are handed (rotate in opposite directions).

Cylinders: Capacity 250 cc (15·25 cu in) each. Compression ratio 10.

Induction: Two floatless carburettors fed by pump.

Fuel: 30:1 mix of LO 92 petrol and oil.

Dimension:
- Distance between crankshaft centrelines 820 mm (32·3 in)

Performance Ratings:
- T-O 2 × 15 = 30 kW (40·25 hp) at 5,200 rpm
- Continuous 2 × 14 = 28 kW (37·5 hp) at 4,900 rpm

The Borzecki JB 2 × 250 dual engine unit (30 kW; 40·25 hp)

Specific Fuel Consumption:
- Cont rating 123·4 μg/J (0·73 lb/h/hp)

IL
INSTYTUT LOTNICTWA (Aviation Institute)

Headquarters: Al. Krakowska 110/114, 02-256 Warsaw-Okecie
Telephone: Warsaw 460993
Managing Director: Dr Ing Andrzej Wierzba
Chief Consultant for Scientific and Technical Co-operation: Dipl Ing Jerzy Grzegorzewski
Chief of Scientific, Technical and Economic Information Division: Dr Ing Tadeusz Kostia

The Aviation Institute is an establishment concerned with aeronautical research, aerodynamic tests, strength tests, test flights of aeroplanes, helicopters and gliders, aviation equipment, materials, technical information and standardisation. The Institute has a special manufacturing plant responsible for constructing prototypes to its own design.

IL SO-1 turbojet (9·8 kN; 2,205 lb st), initial power plant of the TS-11 Iskra (Spark) trainer *(BIIL)*

IL SO-1

The Aviation Institute designed the SO-1 turbojet to power the Polish TS-11 Iskra (Spark) jet basic trainer. It was designed to permit the full range of aerobatics, including inverted flight. Guaranteed overhaul life is 200 h. Production was handled by the WSK-Rzeszów, as noted in that organisation's entry.

Type: Single-shaft axial-flow turbojet.

Air Intake: Annular intake casing manufactured as a cast shell. Fixed inlet guide vanes.

Compressor Casing: Manufactured as a cast shell in two parts, split along horizontal centreline, in aluminium alloy.

Compressor: Seven-stage axial-flow compressor. Drum-type rotor built up of disc assemblies, with constant diameter over tips of rotor blades. Carried in ball bearing at front and roller bearing at rear. Steel stator blades bonded with resinous compound into slots in carrier rings. Rotor originally of steel and duralumin, with first three blade rows of steel and remainder of aluminium alloy. Modified as a result of operating experience; entire compressor rotor and blades on all stages now made of steel. Pressure ratio 4·8.

Combustion Chamber: Annular type with 24 integral vaporisers. Outer casing made of welded steel.

Fuel System: Two independent systems supplied by one pump. Starting system consists of six injectors, with direct injection. Main system consists of twelve twin injectors with outlets towards the vaporisers.

Fuel Specification: Kerosene P-2 or TS-1.

Turbine: Single-stage axial-flow type. Blades attached to disc by fir-tree roots. Supported in roller bearing at front.

Jet Pipe: Outer tapered casing and central cone connected by streamlined struts. Nozzle area adjusted by exchangeable inserts.

Lubrication System: Open type for rear compressor and turbine bearings, supplied by separate pumps. Closed type for all other lubrication points, fed by separate pumps.

Oil Specification: Type AP-26 (synthetic).

Accessory Drives: Gearbox mounted at bottom of air intake casing and driven by bevel gear shaft from front of compressor.

Starting: 27V starter/generator and bevel gear shaft, driven by aircraft battery or ground power unit, mounted on air intake casing.

Dimensions:

Length overall	2,151 mm (84·7 in)
Width	707 mm (27·8 in)
Height	764 mm (30·1 in)
Weight, Dry:	303 kg (668 lb)

Performance Ratings:

T-O	9·8 kN (2,205 lb st) at 15,600 rpm
Max cont	8·7 kN (1,958 lb st) at 15,100 rpm

Specific Fuel Consumption:

At T-O rating	29·6 mg/Ns (1·045 lb/h/lb st)
Oil Consumption:	0·8 litres (1·4 Imp pints)/h

IL SO-3

This improved version of the SO-1 replaced the earlier type in production at the WSK-Rzeszów. The SO-3 is intended for tropical use and incorporates minor changes in compressor, combustion chamber and turbine, data remaining the same as for the SO-1. It is fitted to all current versions of the TS-11. TBO has been doubled, to 400 h. In 1978 testing began on the SO-3B, with all-steel compressor rotor, modified vaporisers and new flame tube. By 1979 this was in production for the Iskra-Bis DF version of the TS-11, alongside the SO-3, with the designation SO-3W and the following changed particulars:

Weight, Dry: 325 kg (715 lb)

Performance Ratings:

T-O	10·8 kN (2,425 lb st) at 15,600 rpm
Max cont	9·8 kN (1,958 lb st) at 15,100 rpm

JANOWSKI
JAROSLAW JANOWSKI

Lodz 11, ul. Nowomiejska 2/29

JANOWSKI SATURN 500

The Saturn 500 was designed by Mr Jaroslaw Janowski and built by Mr S. Polawski for the Janowski J-1 ultra-light amateur-built aircraft (see 1979-80 *Jane's*). The prototype Saturn 500 was built in 1969. This two-cylinder two-stroke engine may be used with tractor or pusher propeller, and is intended for ultralight aircraft built by amateurs.

A version of the Saturn 500 with new cylinder heads, improved crankshaft and dual ignition has been reported. Its rating (max T-O) is increased to 22·5 kW (30 hp); dry weight is believed to be about 25 kg (55 lb).

The following description applies to the initial 25 hp version:

Type: Two-cylinder two-stroke horizontally-opposed air-cooled piston engine.

Cylinders: Bore 70 mm (2·76 in). Stroke 65 mm (2·56 in). Swept volume 500 cc (30·5 cu in). Compression ratio 8·5 : 1. Steel barrels with aluminium alloy cylinder heads. Cylinder and head assembly attached to crankcase by four studs.

Pistons: Of aluminium alloy. Two compression rings and one oil scraper ring.

Connecting Rods: Steel forgings.

Crankshaft: Steel counterbalanced shaft, supported in two lead-bronze plain bearings and one ball-thrust bearing at the front.

Crankcase: Aluminium alloy case, split in the vertical plane, with front and aft covers.

Induction: Two BVF 28N1 carburettors.

Fuel: Petrol/oil mixture using aviation 90 octane.

Ignition: Two magnetos. One M14-250 14 mm (0·55 in) sparking plug per cylinder.

Mounting: Four rubber dampers at rear of crankcase.

Propeller Drive: Direct tractor or pusher.

Dimensions:

Length overall, with propeller boss	430 mm (16·93 in)
Width, without sparking plugs	515 mm (20·27 in)
Weight, Dry:	27 kg (59·5 lb)

Performance Rating:

T-O	18·65 kW (25 hp) at 4,000 rpm

Specific Fuel Consumption:

Max T-O rating	118·23 μg/J (0·70 lb/h/hp)
Normal cruising power	111·48 μg/J (0·66 lb/h/hp)

Saturn 500 two-cylinder two-stroke engine designed by Jaroslaw Janowski

---

WSK-PZL-KALISZ
WYTWÓRNIA SPRZETU KOMUNIKACY JNEGO-PZL-KALISZ

Head Office and Works: ul. Czestochowska 140, 62800 Kalisz
Telephone: 77351
General Manager: Dipl Ing Henryk Jaruzel

In 1952 the Soviet Union transferred responsibility for manufacture and service support of Soviet aircooled radial piston engines to the WSK (transport equipment manufacturing centre) at Kalisz.

Current production is centred on: the 1,000 hp ASh-62IR (Polish designation ASz-62IR) for all versions of the An-2; and the 260 hp AI-14RA for the PZL-104 Wilga (see Aircraft section). Production of the ASz-62IR exceeds 15,000 engines. WSK-Kalisz has developed a version of the AI-14 with electric starter, the AI-14RC.

PZL (IVCHENKO) AI-14R

The original 260 hp AI-14R version of this nine-cylinder aircooled radial engine has been produced in very large quantities, in both the Soviet Union and Poland. Since 1960 several later versions have gone into production in the Soviet Union for fixed-wing aircraft and helicopters. Both of the following were developed at Kalisz.

AI-14RA. Basic production engine, rated at 194 kW (260 hp); fitted to Yak-12, Yak-18, PZL-101A Gawron and PZL-104 Wilga 35.

AI-14RC. Version fitted with electric starter.

The following description refers to the AI-14RA:

Type: Nine-cylinder single-row aircooled radial, geared with blower.

Cylinders: Bore 105 mm (4·125 in). Stroke 130 mm (5·125 in). Displacement 10·16 litres (620 cu in). Nitrided steel barrel; cast light alloy head incorporating air starting valve. Two spark plugs and single inlet and exhaust valves. Compression ratio 5·9.

Pistons: Aluminium forgings, each with two chromium-plated compression rings and two scraper rings.

Connecting Rods: One master rod, with lead-bronze big-end bearing; eight link rods articulated by steel cemented knuckle pins fixed in the master-rod cheeks against rotation and secured laterally by retaining plates.

Crankshaft: Heat-treated steel in two parts, the front portion being gripped in the split cheek of the rear portion and held by a pinch bolt. Both portions carry a

PZL AI-14RA piston engine

counterweight, the rear counterweight being pendulous type which balances inertia forces and also serves as vibration damper. Shaft held in two main roller bearings and ball thrust bearing.

CRANKCASE: Comprises gearcase, thrust-bearing cover, mid-case, mixture chamber and rear casing.

VALVE GEAR: Each valve is opened by push/pull rod, from a cam plate geared to rotate in opposition to the crankshaft.

INDUCTION SYSTEM: Carburettor type K-14A, fed by fuel pump type 702M.

FUEL GRADE: 91 to 100 octane.

BLOWER: Aluminium forged impeller. Magnesium cast diffuser

LUBRICATION: Gear-type pressure and scavenge pumps.

IGNITION: Two M-9 magnetos; automatic timing and fully screened.

PROPELLER DRIVE: Planetary gears, ratio 0·787.

STARTING: By compressed air from airborne or ground bottle to cylinder-head valves.

DIMENSIONS:

Diameter	985 mm (38·78 in)
Length	956 mm (37·63 in)

WEIGHT, DRY: 197 kg (434 lb)

PERFORMANCE RATINGS:

T-O	194 kW (260 hp) at 2,350 rpm
Rated	164 kW (220 hp) at 2,050 rpm
Cruise	98·4 kW (132 hp) at 1,730 rpm

SPECIFIC FUEL CONSUMPTION:

T-O	95-104·3 μg/J (0·562-0·617 lb/h/shp)
Cruise	76·4-83·8 μg/J (0·452-0·496 lb/h/shp)

PZL (SHVETSOV) ASz-62R

Power plant of the An-2 transport biplane, the ASz-62R is a 1,000 hp nine-cylinder aircooled radial engine. It was developed from the Wright Cyclone R-1820 by Arkadiya Shvetsov's bureau in the Soviet Union, as the ASh-62. Several variants have been built in the Soviet Union, including the ASh-62IR/TK driving a turbo-compressor to maintain 850 hp up to a height of 9,500 m (31,000 ft).

ASz-62IR. Standard power plant of the Li-2 (Soviet DC-3) and all versions of the An-2 except for the An-2M. Transferred to WSK-PZL-Kalisz in 1959.

All versions have a cylinder bore of 155·5 mm (6⅛ in), swept volume of 29·87 litres (1,823 cu in) and compression ratio of 6·4 : 1.

The planetary reduction gear has a ratio of 0·637 : 1.

DIMENSIONS:

Length overall	1,130 mm (44·50 in)
Diameter	1,375 mm (54·13 in)

WEIGHT, DRY:

Without power take-off	579 kg (1,276 lb)

PERFORMANCE RATINGS:

T-O	746 kW (1,000 hp) at 2,200 rpm
Rated power	611 kW (820 hp) at 2,100 rpm
Cruise (50% power)	306 kW (410 hp) at 1,670 rpm

SPECIFIC FUEL CONSUMPTION:

T-O	112 μg/J (0·661 lb/h/hp)
Cruise	80 μg/J (0·474 lb/h/hp)

PZL ASz-62IR piston engine

WSK-PZL-RZESZÓW
WYTWÓRNIA SPRZETU KOMUNIKACYJNEGO-PZL-RZESZÓW

HEAD OFFICE AND WORKS: ul. Obrońców Stalingradu 120, 35078 Rzeszów, Postbox 340

Telex: 83411

GENERAL MANAGER: Dipl Ing Jozef Rokoszak

WSK-Rzeszów, founded in 1938 as PZL-Rzeszów, at first produced Bristol Pegasus and Walter Junior and Minor engines under licence. After World War 2 the works expanded considerably. The first product was the Soviet M-11D, followed by series production of the ASz-62IR and LiT-3 piston engines and HO-10 and SO-1 turbojets.

Current production is centred on: the Soviet-designed GTD-350 turboshaft, together with WR-2 reduction gear for Mi-2 helicopters; the tropicalised SO-3 turbojet for the TS-11 Iskra trainer; and the PZL-3S piston engine for agricultural aircraft. The SO-3 is described (with the SO-1) on an earlier page, under the heading IL.

PZL-10W free-turbine turboshaft (rated at 715 kW; 960 shp) designed by Glushenkov and licensed for production in Poland

PZL-10W

An accompanying photograph is the first to show the Soviet designed Glushenkov TVD-10 turboshaft engine, rated at 715 kW (960 shp), which is now made under licence in Poland. It is worth noting that in the Soviet Union the TVD-10 designation is reserved for the turboprop version, used in the An-28 STOL light transport which is soon to enter production in Poland at WSK-PZL Mielec. The turboprop is designated TVD-10B.

The Polish designation for this engine is PZL-10W, and it is the power plant of the PZL-Swidnik Sokól twin-engined helicopter. The engines are handed, with left- and right-facing jetpipes. There are probably two or three axial compressor stages followed by a centrifugal stage. The free power turbine drives to the rear.

PZL GTD-350

The GTD-350 is a free-turbine helicopter power plant. In the version used in the twin-engined Mi-2, the drive is taken from the rear, with the twin jet pipes of each engine exhausting to port (port engine) and starboard (starboard engine). The GTD-350 can be supplied with downward-facing jet pipe and with drive from the front, if required. Though developed and initially produced by the Isotov bureau in the Soviet Union, it is now in production only in Poland. WSK-Rzeszów has developed a new version rated at 331 kW (444 shp) and designated GTD-350P. TBO of the GTD-350 is 3,000 h, and work is in hand to extend this.

TYPE: Axial/centrifugal-flow free-turbine turboshaft engine.

AIR INTAKE: Annular intake casing and inlet guide vanes of stainless steel. Automatic de-icing of inlet guide vanes and central bullet by air bleed from compressor.

COMPRESSOR: Seven axial stages and one centrifugal stage, all of steel, connected together with a tie-bolt. Discs shrunk-fitted to shaft. Blades of axial stages have dovetail roots. Shaft carried in front roller bearing and rear ball bearing. Pressure ratio 6·05 : 1. Air mass flow 2·19 kg (4·83 lb)/s at 45,000 rpm.

COMPRESSOR CASING: Horizontally-split aluminium alloy casing, with stator blades brazed to semi-rings. No diffuser blades.

COMBUSTION CHAMBER: Reverse-flow type with air supply through two tubes. Centrifugal duplex single-nozzle burner. Ignition system comprises burner and semiconductor spark-plug. Eight thermocouples at gas outlet.

FUEL SYSTEM: Includes NR-40TA pump governor with shut-off cock, which feeds fuel to burner, controls gas-generator rpm and limits max output; RO-40TA power turbine rpm governor, DS-40 signal transmitter controlling bleed valves; and electromagnetic valve to provide fuel for starting. Provision against power-turbine overspeed and (GTD-350P only) helicopter transmission overload.

FUEL GRADE: TS-1, TS-2 or Jet A-1.

COMPRESSOR TURBINE: Single-stage turbine with aircooled disc. Shrouded blades with fir-tree roots. Precision-cast fixed guide vanes. Turbine casing has metal-ceramic insert in plane of blades. Shaft supported in ball bearing at rear. Temperature before turbine 940°C (GTD-350P, 985°C).

POWER TURBINE: Two-stage constant-speed type (24,000 rpm). Shrouded blades with fir-tree roots. Discs bolted together. Stator blades welded to rings. Airflow is again reversed aft of power turbine.

JET PIPES: Two fixed-area jet pipes.

REDUCTION GEARING: Two sets of gears, with ratio of 0·246 : 1, in cast magnesium alloy casing. Output shaft speed, 5,900 rpm.

LUBRICATION SYSTEM: Closed type. Gear-type pump with one pressure and four scavenge units. Nominal oil pressure 2·94±0·5 bars (43 lb/sq in). Oil cooler and oil tank, capacity 12·5 litres (2·75 Imp gallons), fitted to airframe.

OIL GRADE: B3-W (synthetic) or Aero Shell Turbine Oil-500.

ACCESSORIES: STG3 3kW starter/generator, NR-40TA governor pump, D1 tachometer generator and oil pumps mounted on reduction gear casing and driven by gas-generator. RO-40TA rotating speed governor, D1 tachometer generator and centrifugal breather, also

PZL (Isotov) GTD-350 turboshaft engine (298 kW; 400 shp)

mounted on reduction gear casing, driven by power turbine.

STARTING: STG3 starter/generator suitable for operation at up to 4,000 m (13,125 ft) altitude.

DIMENSIONS:

Length overall	1,385 mm (54·53 in)
Max width	520 mm (20·47 in)
Width (with jet pipes)	626 mm (24·65 in)
Max height	630 mm (24·80 in)
Height (with jet pipes)	760 mm (29·9 in)

WEIGHT, DRY:

Less jet pipes and accessories	135 kg (298 lb)

PERFORMANCE RATINGS:

T-O rating (6 min) at 96% max gas-generator rpm:

GTD-350	298 kW (400 shp)
GTD-350P	335·5 kW (450 shp)

Nominal rating (1 h) at 90% gas-generator rpm:

GTD-350	238·5 kW (320 shp)
GTD-350P	261 kW (350 shp)

Cruise rating (I)
212·5 kW (285 shp) at 87·5% gas-generator rpm

Cruise rating (II)
175 kW (235 shp) at 84·5% gas-generator rpm

SPECIFIC FUEL CONSUMPTION:

T-O	136 μg/J (0·805 lb/h/shp)
Nominal	146 μg/J (0·861 lb/h/shp)
Cruise (I)	154 μg/J (0·913 lb/h/shp)
Cruise (II)	165 μg/J (0·978 lb/h/shp)

OIL CONSUMPTION:

Max	0·3 litres (0·53 Imp pints)/h

PZL-F 6AS-350A with turbocharger

PZL-3S

With ancestry going back via the LiT-3 to the AI-26W, this radial was the only type in its class to be developed in the 1970s. It has FAA and Canadian DoT Type Certificates and Polish CACA certificate to BCAR. Applications include the PZL-106A Kruk, and conversions of the Grumman/Schweizer Ag-Cat A, B and C, Thrush Commander, DHC-3 Otter and IAR-827A.

TYPE: Seven-cylinder aircooled radial.

CYLINDERS: Bore 155·5 mm (6·12 in). Stroke 155 mm (6·1 in). Swept volume 20·6 litres (1,265 cu in). Compression ratio 6·4 : 1.

PISTONS: Forged aluminium.

INDUCTION SYSTEM: Float-type carburettor. Mechanically driven supercharger.

FUEL GRADE: Aviation gasoline, minimum 91 octane.

LUBRICATION: Gear-type oil pump. Oil grade Aero Shell 100 or other to MIL-L-6082.

PZL-3S seven-cylinder radial

PZL-F 2A-120C (45 kW; 60 hp)

PZL-F 4A-235B (93 kW; 125 hp)

PROPELLER DRIVE: Direct. Provision for constant-speed US-132000 or Dowty Rotol propeller.

ACCESSORIES: ANG 6423 Prestolite alternator and two output shafts, one 27 hp (35 max) for spraying pump and the other 5 hp.

STARTING: Electrical.

DIMENSIONS:

Diameter	1,267 mm (49·88 in)
Length	1,110 mm (43·72 in)

WEIGHT, DRY: 411 kg (906 lb)

PERFORMANCE RATINGS:

Max T-O	448 kW (600 hp) at 2,200 rpm
Max continuous	410 kW (550 hp) at 2,050 rpm
Cruise (75 per cent)	310 kW (415 hp) at 2,000 rpm

SPECIFIC FUEL CONSUMPTION:

T-O, max cont	105 μg/J (0·61 lb/h/hp)
Cruise	86 μg/J (0·51 lb/h/hp)

PZL-3R

This is the geared version of the PZL-3S. The following are the main differences:

PROPELLER DRIVE: Planetary gear of 0·7 ratio. Provision for constant-speed propeller, Type US-133000.

DIMENSION: Length	1,271 mm (50·06 in)
WEIGHT, DRY:	446 kg (983 lb)

PZL-F ENGINES

In 1975 Pezetel acquired rights to manufacture and market the entire range of aircooled piston engines formerly produced by the Franklin Engine Company (Aircooled Motors) of the USA. These engines, now known as PZL-F, are being produced in Poland for light aircraft of all kinds, including helicopters and motor-gliders. Current applications include the SZD-45-2 Ogar F motor glider (2A-120C), PZL-110 Koliber (4A-235B3), and PZL Mielec M-20 Mewa (6A-350C). The 2A-120C, 4A-235B3 and 6A-350C each have a Polish CACA certificate. The 6AS-350A is under test.

All except the 6V-350B are of the horizontally-opposed type, with cylinders of 117·48 mm (4·625 in) bore and 88·9 mm (3·5 in) stroke. All have direct drive and operate on 100/130 grade fuel. Accessories normally include electric starter, alternator and fuel pump. Other details are tabulated:

PZL-F ENGINES

Engine model	Cylinder arrangement	Capacity cc (cu in)	Compression ratio	Max T-O rating at S/L kW (hp) at rpm	Overall dimensions mm (in) length	width	height	Weight dry kg (lb)	Remarks
2A-120C	2 horiz	1,916 (117)	8·5	45 (60) at 3,200	581 (22·9)	795 (31·3)	515 (20·3)	75·8 (167)	
4A-235B	4 horiz	3,850 (235)	8·5	93(125) at 2,800	774 (30·5)	795 (31·3)	637 (25·1)	117·6 (259)	
6A-350C	6 horiz	5,735 (350)	10·5	164 (220) at 2,800	952 (37·5)	795 (31·3)	641 (25·25)	167 (367)	
6AS-350A	6 horiz	5,735 (350)	7·4	186 (250) at 2,800*	1,097 (43·2)	868 (34·2)	983 (38·7)	189 (417)	turbocharged
6A-350D	6 horiz	5,735 (350)	10·5	175 (235) at 3,200	825 (32·5)	795 (31·3)	642 (25·3)	145 (320)	helicopter engine
6V-350B	6 vert	5,735 (350)	10·5	175 (235) at 3,200	968 (38·0)	795 (31·3)	779 (30·6)	144 (320)	helicopter engine

*Rated at 175 kW (235 hp) at 2,800 rpm at 5,000 m (16,400 ft).

ROMANIA

TURBOMECANICA
INTREPRINDEREA TURBOMECANICA BUCURESTI

Bucharest

This factory produces under licence the Rolls-Royce Spey and Viper, and the Turboméca Turmo IVC.

SOUTH AFRICA

ATLAS
ATLAS AIRCRAFT CORPORATION OF SOUTH AFRICA (PTY) LTD

Address and Officers: See Aircraft section

Atlas is manufacturing the Rolls-Royce Viper 540 turbojet under sublicence from Piaggio of Italy, for use in Atlas Impala attack trainers.

SWEDEN

FLYGMOTOR
VOLVO FLYGMOTOR AB

Head Office and Works: S-461 81 Trollhättan
Telephone: 0520 94000

This company was founded in 1930 and began by building under licence the Bristol Pegasus I aero-engine, under the designation My VI. Since 1970 Volvo Flygmotor has been a wholly owned subsidiary of AB Volvo.

Volvo Flygmotor AB holds a licence to build Rolls-Royce Avon engines and is also engaged in research and development work on turbojet engines, ramjet engines and rocket engines. Its major current programme involves the production of two versions of the RM8 supersonic turbofan developed from the Pratt & Whitney JT8D to power different sub-types of the Saab 37 Viggen combat aircraft. Its major development programme is the TFE1042 in partnership with Garrett.

It is also engaged in the development of experimental hybrid and liquid-fuel rocket engines. The company has a technical collaboration agreement on ramjet development with Rolls-Royce Ltd.

Volvo Flygmotor is the manufacturer of the combustion (thrust) chambers for the first and second stages of the Ariane European space launcher. The engines concerned, Viking V and Viking IV, are described in the SEP entry of the French section.

Volvo Flygmotor RM8B augmented turbofan for JA 37 fighter version of Viggen

FLYGMOTOR RM8

The RM8 is a Swedish military version of the Pratt & Whitney JT8D civil subsonic turbofan which Flygmotor developed to power the Saab 37 Viggen supersonic multi-purpose combat aircraft.

Until 1970 the main effort was devoted to development and manufacture of the RM8A version for the AJ 37, SF 37, SH 37 and SK 37 versions of the Viggen. Production of this was almost complete by the beginning of 1979. The substantially modified **RM8B** was conceived to meet the propulsion requirements of the fighter Viggen, the JA 37. Research at Pratt & Whitney and Flygmotor showed that a changed design could improve the stability of operation at high altitudes and in severe manoeuvres, as well as increase thrust in all regimes. In collaboration with Pratt & Whitney the design of the RM8B was completed in late 1971. The major change to improve functional stability at high altitude involved replacing the first stage of the LP compressor by a third stage on the fan. To increase thrust the RM8B has a four-nozzle burner combustion system and a new HP turbine. Delivery of production RM8B engines began in 1978.

The following description refers to the RM8A:

Type: Axial-flow two-spool turbofan with modulated afterburner.

Air Intake: Annular, with 19 fixed inlet guide vanes.

Fan: Two-stage front fan. Titanium blades.

LP Compressor: Four-stage axial-flow, integral with fan stages, on inner of two concentric shafts. Blades of titanium. Steel casing.

HP Compressor: Seven-stage axial-flow on outer hollow shaft. Blades made of special high-temperature alloys. Overall pressure ratio 16·5 : 1. By-pass ratio approximately 1 : 1. Total air mass flow 145 kg (320 lb)/s.

Combustion Chamber: Cannular type with nine cylindrical flame tubes, each downstream of a single Duplex fuel nozzle. Two high-energy spark plugs, each with its own igniter box.

HP Turbine: Single-stage axial-flow, with cast aircooled blades.

LP Turbine: Three-stage axial-flow, with cast blades. Exit guide vanes after turbine.

Afterburner: Double-skinned to provide duct for cooling air. Outer skin of titanium. Inner skin of special alloys. One hot-streak igniter. Hydraulically-actuated fully-variable nozzle, using fuel as the operating fluid.

Bearings: Main shafts run in total of six bearings.

Control Systems: There are two systems. The main system for the gas-generator comprises a Bendix hydromechanical control of advanced design. A further Bendix unit controls the fuel flow to the afterburner and the nozzle area. A single power lever controls thrust from maximum afterburner down to idle; further movement below idle actuates the fuel cut-off valve.

Mounting: Three-point. Main mountings on each side of compressor casing; one under turbine casing.

Accessory Drive: Via gearbox, under engine, driven from HP turbine shaft.

Dimensions:

Length overall:	
RM8A	6,160 mm (242·5 in)
RM8B	6,230 mm (245·25 in)
Max diameter (both versions)	1,397 mm (55 in)
Inlet diameter (both)	1,030 mm (40·55 in)
Weight, Dry:	
RM8A	2,100 kg (4,630 lb)
RM8B	2,350 kg (5,181 lb)
Performance Ratings (ISA, S/L):	
Max T-O, augmented:	
RM8A	115·6 kN (25,990 lb st)
RM8B	125 kN (28,110 lb st)
Max T-O, dry:	
RM8A	65·6 kN (14,750 lb st)
RM8B	72 kN (16,200 lb st)
Specific Fuel Consumption:	
Max augmented:	
RM8A	70·0 mg/Ns (2·47 lb/h/lb st)
RM8B	71·4 mg/Ns (2·52 lb/h/lb st)
Max dry:	
RM8A	17·8 mg/Ns (0·63 lb/h/lb st)
RM8B	18·1 mg/Ns (0·64 lb/h/lb st)
Max continuous (both)	17·3 mg/Ns (0·61 lb/h/lb st)

GARRETT/FLYGMOTOR TFE1042

This collaborative family of turbofans is covered in the International part of this section.

FLYGMOTOR VR35

The VR35 is a prepackaged liquid-propellant rocket engine with positive expulsion of the storable inhibited red fuming nitric acid (IRFNA) and Hydyne propellants. A solid-propellant gas generator delivers the expelling gas and programmes the thrust into a boost phase followed by a sustain blow-down period.

The positive expulsion is accomplished by a gas-pressurised piston for the fuel, contained in a central tank, and an inward-collapsible aluminium bladder for the oxidiser in its concentric tank. This expulsion system enables the engine to be fired under any acceleration direction, a capability which is essential for a missile engine with a thrust programme that includes the sustain phase.

Another advantage with this type of engine is the completely smoke-free exhaust, leaving no signature at launch or during flight.

The VR35 is used in the Saab-Scania RB05 air-to-surface missile, which is one of the main weapons for the AJ 37 Viggen.

Dimensions:	
Length overall	1,770 mm (69·7 in)
Max diameter	300 mm (11·8 in)
Weight:	127 kg (280 lb)
Propellant Mass Fraction:	0·59
Operating Temperature Range:	−50°C to +65°C
Performance:	
Boost	20-25 kN (4,496-5,620 lb st) depending on temperature
Sustain	down to 5 kN (1,124 lb st)
Total impulse	160 kN-s (35,970 lb-s)

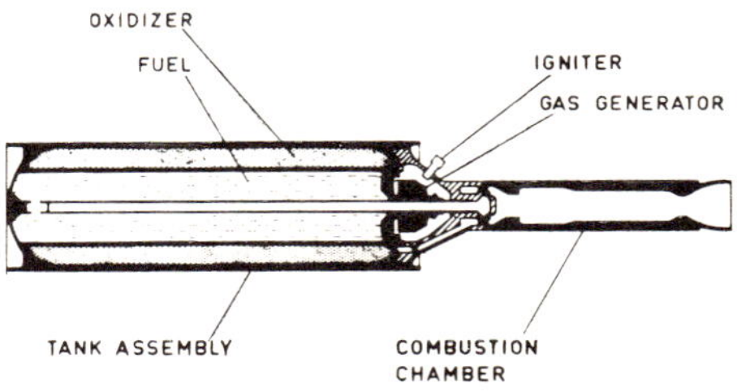

VR35 longitudinal cross-section

Volvo Flygmotor VR35 prepackaged liquid rocket

FLYGMOTOR RAMJETS

Ramjet engine research and development have been under way at Volvo Flygmotor since 1952, when the company's underground high-pressure air magazine was completed. The stored air drives the combustion test rigs and wind tunnels necessary for this kind of work.

The engine research culminated in several flight tests with the RR2, designed for a Mach number of 2·85 and intended for pod-mounting on a ground-to-air missile. The RR2 engine has a diameter of 260 mm (10·25 in) and empty weight of 42·0 kg (92·5 lb). The kerosene-type fuel is stored in the missile main body and is delivered to the ramjet combustion chamber by means of an air-driven turbopump integrated into the engine forebody.

The main research effort is now directed towards an integral rocket-ramjet. Two types have been investigated and tested. The **RRX1** has a circumferential nose air intake and is designed for a Mach number of 5·0. The body diameter is 190 mm (7·5 in). The **RRX5** has four side-mounted air intakes to free the nose for a missile warhead and guidance. It is intended for Mach numbers up to 4·0. Alternative solutions for monoplane configurations are being studied, as are solid ducted rockets using fuel-rich gases from underoxidised solid propellant.

UNION OF SOVIET SOCIALIST REPUBLICS

GDL

The Gas Dynamics Laboratory was founded in Leningrad in 1929. Since that time it has completed a very large number of rocket-engine programmes, mainly using liquid propellants. The laboratory has occupied many facilities, but is still headquartered in Leningrad. Four of its best-known engines are described below.

GDL RD-107

This four-chamber liquid-propellant rocket engine was developed during 1954 to 1957. The RD-107 and its derivatives have been in use for many years as first-stage engines for the launch vehicles for Vostok, Voskhod and derived programmes. Each complete RD-107 comprises four main chambers and two small vernier chambers.

TYPE: Four-chamber liquid-propellant rocket engine.

PROPELLANTS: Liquid oxygen and kerosene.

THRUST CHAMBERS: Four primary thrust chambers of double-wall construction, with fabricated corrugations between walls and inner walls of copper or copper-rich alloy. Conical nozzles. All-welded heads. Flat-plate injectors, with concentric rings of tubes in which propellants are pre-mixed before injection. Estimated diameters: throat 150-165 mm (6-6·5 in), nozzle 685 mm (27 in). Combustion pressure 60·8 bars (882 lb/sq in).

VERNIER CHAMBERS: Two chambers of double-wall construction, with finning between walls. Estimated diameters: throat 75 mm (3 in), nozzle 305 mm (12 in).

TURBOPUMP: One single-shaft turbopump mounted in tubular thrust frame and feeding all chambers. Assembly comprises turbine exhaust hood containing coiled heat exchanger, single-sided shrouded centrifugal kerosene pump, double-sided shrouded centrifugal liquid oxygen pump, gearbox, and two auxiliary centrifugal pumps, one of which supplies the monopropellant gas generator. Fuel lines to main chambers pass through common valve.

PERFORMANCE (in vacuum):

Rated thrust	1,000 kN (224,870 lb)
Specific impulse	314 s

GDL RD-119

This more modern single-chamber liquid-propellant engine, which has been in use since 1962, forms the second-stage engine of a launch vehicle for Cosmos research satellites. More than 1,280 of these satellites have been launched, using two-, three- and four-stage launch vehicles of various types and having lifting capacities ranging from hundreds of pounds to 7·6 tonnes (7·5 tons).

TYPE: Single-chamber liquid-propellant rocket engine.

PROPELLANTS: Liquid oxygen and dimethyl-hydrazine.

THRUST CHAMBER: Single fixed chamber, possibly of tubular-wall construction, with fuel entry above base of nozzle. Estimated diameters: throat 100 mm (4 in), nozzle 940 mm (37 in). Combustion pressure 81·1 bars (1,176 lb/sq in).

TURBOPUMP: One single-shaft turbopump, driven by monopropellant (hydrazine) gas generator. Exhaust from gas generator taken to multiple auxiliary nozzles for control in roll, pitch and yaw.

PERFORMANCE (in vacuum):

Rated thrust	108 kN (24,250 lb)
Specific impulse	352 s

GDL RD-214

This neat liquid-propellant rocket engine, developed at GDL in 1952-57, has been adopted as the standard first-stage propulsion for launching the Cosmos series of Soviet satellites. It has four thrust chambers, burning nitric acid and kerosene, each chamber being rated at 176·6 kN (39,700 lb thrust) at sea level. Vacuum rating of the engine is 725·7 kN (163,142 lb). The propellants are fed by a single large turbopump group mounted above the chamber group, the fuel being supplied straight to a bolted connection on the welded chamber heads and the nitric acid passing through part-flexible pipes to the regeneratively cooled chamber nozzle and throat. Chamber pressure is given by GDL as 3·1 bars (45 lb/sq in) and specific impulse as 246. The four chambers are rigidly fixed to the Cosmos launcher first stage, vehicle control being accomplished by four refractory deflector vanes, one per chamber, mounted on the vehicle skirt control packages and projecting into the rocket exhaust. More than 1,000 RD-214 engines have been flown from Kapustin Yar on non-recoverable Cosmos missions.

GDL RD-219

Since the mid-1950s the emphasis in Soviet development of large liquid-propellant rockets has swung from lox/kerosene to storable mixtures, particularly RFNA/UDMH. This combination offers good specific impulse, can be left in the vehicle tanks indefinitely and is hypergolic. The first large GDL engine of this type to be disclosed was the RD-219, a twin-chamber unit developed in 1958-61 and used for second-stage propulsion of large vehicles (possibly the SS-9 ICBM, or the Type G or Proton/Zond launcher). The chambers have a vacuum expansion ratio, and are fed by a single turbopump. They are fixed rigidly to a welded tubular thrust frame extending down to the vehicle skirt.

PERFORMANCE (in vacuum):

Rated thrust	883 kN (198,441 lb)
Specific impulse	293 s

The RD-219 powers the second stage of launchers; a nitric acid/UDMH engine, its twin chambers have vacuum thrust of 883 kN (198,441 lb)

Exhibition model of the RD-107 rocket engine, with two-chamber vernier engine in front *(TAM Air et Cosmos)*

The 706·4 kN (158,800 lb) thrust GDL RD-214 rocket engine used to power the first stage of the Cosmos launcher. Vacuum rating 725·7 kN (163,142 lb)

GLUSHENKOV

This design bureau became known in 1969 when it was revealed as that responsible for the TVD-10 engines fitted in the Beriev Be-30 STOL transport. In 1978 the improved TVD-10B, rated at 723 kW (970 shp) went into production for the An-28 twin-engined STOL aircraft. It is also reported that the TVD-10B powers the An-3 agricultural aircraft. In 1979 photographs were released of one of these engines mounted in a Yak-18PM fully aerobatic trainer. Further details can be found in the Polish section, under the entry for WSK-PZL-Rzeszów, which builds the TVD-10B under licence as the PZL-10.

Glushenkov is also responsible for the 671 kW (900 shp) GTD-3 turboshaft engine, two of which power the Kamov Ka-25 helicopter. This engine and the TVD-10 appear to have a generally similar layout, with a free power turbine, and it is possible that they use a common gas generator.

ISOTOV

GENERAL DESIGNER IN CHARGE OF BUREAU:
Sergei Pietrovich Isotov

This bureau was responsible for the GTD-350 and TV2-117A turboshaft engines which power the Mil Mi-2 and Mi-8 helicopters respectively. The former is in production in Poland, and is described under the WSK-PZL-Rzeszów entry in this edition. Isotov was also responsible for the TVD-850 turboprop engines which powered early Antonov An-28 transport aircraft (see Polish Aircraft section).

ISOTOV TV2-117A

The power plant of the Mi-8 comprises two TV2-117A engines coupled through a VR-8A gearbox. As is common with modern Soviet helicopters, the engines and gearbox are delivered and thereafter treated as a single unit. The complete package incorporates a control system (separate from the control system of each gas generator) which maintains desired rotor speed, synchronises the power of

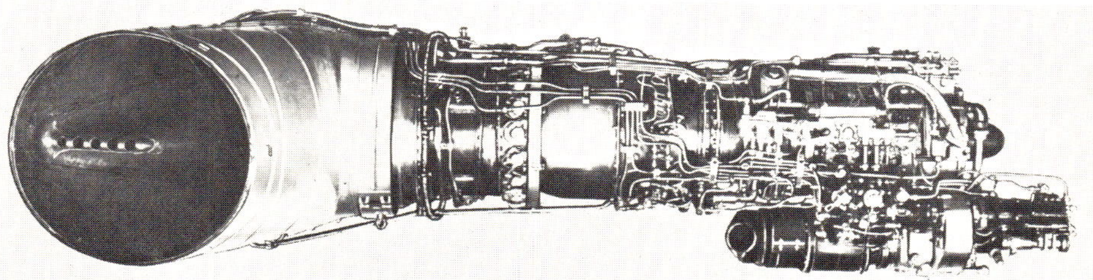
Isotov TV2-117A turboshaft engine (1,118 kW; 1,500 shp)

both engines, and increases the power of the remaining engine if the other should fail.

The TV2 engine is of conservative design, being biased in favour of long and trouble-free life rather than attempting to rival the small size and weight of some Western engines in the same power class.

Type: Free-turbine helicopter turboshaft engine.

Air Intake: Direct pitot, with main front casing providing vertical upper and lower drive-shafts to accessory packages. Main accessory group above the engine projects ahead of intake face. Casing incorporates variable-incidence inlet vanes.

Compressor: Ten-stage axial. Construction principally in titanium to reduce weight in comparison with the steel that would otherwise be used. Inlet guide vanes and stators of stages 1, 2 and 3 are of variable incidence to facilitate starting and increase compressor efficiency over a wide speed range; for the same reasons the casing incorporates automatic blow-off valves. Pressure ratio 6·6 : 1 at 21,200 rpm.

Combustion Chamber: Annular, with eight burner cones. Fabricated from inner and outer diffuser casings, flame tube, casing, burners, and anti-icing bleed air pipe.

Fuel Grade: T-1 or TS-1 to GOST 10227-62 specification (Western equivalents, DERD.2494, MIL-F-5616).

Turbine: Two-stage axial compressor turbine bolted to rear of splined shaft with front extension to drive accessories. Solid rotor blades, held by fir-tree roots in discs cooled by bleed air (first disc 10th-stage air, all other discs 8th-stage). First- and second-stage stators have 51 and 47 inserted blades respectively. Free power turbine of similar two-stage design; its rotors have 43 and 37 blades respectively.

Exhaust Unit: Large fixed-area duct which deflects the gas out at 60°. It comprises a pipe, pipe shroud and tie-band, shroud connector links and exhaust pipe attachments. The exhaust pipe and shroud together form a double-wall assembly which minimises heat transfer into the power plant nacelle, the pipe being cooled by air circulating in the double wall.

Output Shaft: The main drive-shaft is an extension of the power turbine rotor shaft. It conveys torque from the free turbine to the overrunning clutch of the helicopter main gearbox (VR-8A) and is also coupled to the speed governor of the free-turbine rotor. Max output speed 12,000 rpm; main rotor speed 192 rpm.

Accessories: Mounted on the main drive box above the intake casing, in which a train of bevel and spur gears provides drives for airframe and engine accessories. The engine automatic control system includes a fuel system, hydraulic system, anti-icing system, gas temperature restriction system, engine electric supply and starting system, and monitoring instruments. The hydraulic system positions the variable stators according to a preset programme, depending on compressor speed and air temperature at the inlet; it also sends electrical signals to control the starter/generator system, close the starting bleed air valves and restrict peak gas temperature to 600°C. Air up to 1·8 per cent of the total mass flow can be used to heat the intake and other parts liable to icing. Fire extinguishant can be released manually by the pilot, upon receipt of a fire warning, through a series of spray rings and pipes.

Lubrication: Pressure circulation type. Oil is supplied by the upper pump and scavenged from the five main bearings by the lower pump, returned through the helicopter-mounted air/oil heat exchanger and thence to the helicopter tank. The oil seals and air/oil labyrinth seals are connected to a centrifugal breathing system.

Oil Grade: Synthetic, permitting operation at oil temperatures above 200°C, combined with easy starting at minus 40°C without heating the oil. Grade B-3V to MRTU 38-1-157-65 (nearest foreign substitute Castrol 98 to DERD.2487). Consumption, not over 0·5 litre per hour per engine.

Starting: Electrical, fuel, and ignition systems are integrated. The SP3-15 system comprises DC starter/generator, six storage batteries, control panel, ground supply receptacle, and control switches and relays; of these all are airframe mounted except for the GS-18TP starter/generator which cranks the compressor during the starting cycle. The ignition unit comprises a control box, two semiconductor plugs, solenoid valve, and switch. The starting fuel system comprises an automatic starting unit on the NR-40V fuel regulating pump, constant-pressure valve, and two igniters.

Dimensions:
- Length, with accessories and exhaust pipe: 2,835 mm (111·5 in)
- Length, intake face to rotor gearbox connection: 2,391 mm (94·25 in)
- Width: 547 mm (21·5 in)
- Height: 745 mm (29·25 in)

Weight, Dry:
- Engine, without generator, transducers, etc: 330 kg (727 lb)
- VR-8A gearbox, less entrapped oil: 745 kg (1,642 lb)

Performance Ratings:
- Max: 1,267 kW (1,700 shp)
- T-O (S/L, static): 1,118 kW (1,500 shp)
- Max cont: 895 kW (1,200 shp)
- Cruise (122 knots; 225 km/h; 140 mph at 500 m; 1,640 ft): 746 kW (1,000 shp)

Specific Fuel Consumption:
- T-O, as above: 102·4 μg/J (0·606 lb/h/shp)
- Cruise, as above: 115·4 μg/J (0·683 lb/h/shp)

ISOTOV TV3-117

The A-10 (Mi-24) helicopter in which Gourguen Karapetyan set the current speed record for helicopters over a 15/25 km course, on 21 September 1978, was said to be powered by two TV3-117 turboshaft engines, each rated at 1,640 kW (2,200 shp). Similar engines are known to power the Mil Mi-14, Mi-17 and Mi-24. The designation and applications suggest that the TV3-117 is an uprated version of the TV2-117, but no other details are available.

ISOTOV TVD-850

Power plant of early Antonov An-28 STOL transports, the TVD-850 turboprop has a rating of 604 kW (810 shp). The cowled engine looks almost identical to the Turboméca Astazou, and the reduction gearbox must be housed in the centre of an annular air intake section. This engine appears to have lost in competition with the Glushenkov TVD-10B.

IVCHENKO

The collective headed by general designer Alexander G. Ivchenko until his death in June 1968 was based in a factory at Zaparozhye in the Ukraine. The bureau is now headed by General Designer Lotarev, who has his own entry on a later page.

The first engine with which Ivchenko was associated officially was the 41 kW (55 hp) AI-4G piston engine used in the Kamov Ka-10 ultra-light helicopter. He progressed ultimately, via the widely used AI-14 and AI-26 piston engines, to become one of the Soviet Union's leading producers of gas turbine engines. Since 1952 all Soviet piston engines have been assigned to Poland (see under WSK-PZL-Kalisz in Polish section).

Ivchenko AI-20M turboprop of 3,169 kW (4,250 ehp) *(courtesy of Aviation Magazine International, Paris)*

IVCHENKO AI-20

This engine was developed as the NK-4 at the Kuznetsov bureau in 1947-52, with the assistance of German engineers. Eventually preferred to the VK-2, it was ordered into production, and the Zaporozhye collective was charged with the final refinement and production.

Redesignated AI-20, it was produced from 1955 in several versions for aircraft that included the An-10, An-12, An-32, Be-12, Il-18 and Il-38.

AI-20K. Rated at 2,942 kW (3,945 ehp). Used in Il-18V, An-10A and An-12.

AI-20M. Initial T-O rating of 3,124 kW (4,190 ehp). Used in Il-18D/E, An-10A and An-12. Probably fitted to Il-38 and Beriev M-12. Capable of operation on a wide range of fuels and lubricating oils. Power increased to 3,169 kW (4,250 ehp) for later engines as in detailed description below. Uprated again to 3,862 kW (5,180 ehp) for An-32.

The AI-20 was designed to operate reliably in all temperatures from −60°C to +55°C at heights up to 10,000 m (33,000 ft). It is a constant-speed engine, the rotor speed being maintained at 12,300 rpm by automatic variation of propeller pitch. Gas temperature after turbine is 560°C in both current versions. TBO of the AI-20K was 4,000 h in the Spring of 1966; the same life was reached by the -20M in 1968.

In the Il-18 installation, the AI-20 turboprop is supplied as a complete power plant with cowling, mounting and automatically-feathering reversible-pitch four-blade propeller.

Type: Single-shaft turboprop.

Air Intake: Inner and outer cones connected by six radial struts. Outer casing carries accessories and front mountings. Centre casing carries reduction gear.

Compressor: Ten-stage assembly of discs running in roller bearing in front casing, joined to first disc by tubular extension shaft, and ball-thrust bearing in combustion chamber casing on through-bolted rear shaft. Magnesium alloy stator casing in upper and lower halves, bolted together. Pressure ratio 9·2 under altitude cruise conditions. Air mass flow 20·7 kg (45·6 lb)/s.

Combustion Chamber: Annular chamber with ten burner cones welded to front ring, and separate inner and outer shrouds. Burners anchored by flanges on chamber casing. Pilot burners and ignition plugs at top of casing.

Fuel Grade: T-1 or TS-1 to GOST-10227-62 (DERD.2492, JP-1 to MIL-F-5616).

Turbine: Three stages overhung on cantilevered shaft running in roller bearing in tapered cone of combustion chamber casing and splined to compressor drive-shaft. Rotor blades shrouded at inner and outer ends and installed in pairs in slots in aircooled discs. Stator blades secured in grooves in casing, first stage being aircooled and second stage being hollow to ensure uniform heating.

Jet Pipe: Fixed-area type with five radial struts. Nozzle area 0·225 m² (2·42 sq ft).

Reduction Gear: Planetary type, incorporating six-cylinder torquemeter and negative-thrust transmitter (type IKM), with self-checking device, for autofeathering AV-681 propeller. Ratio 0·08732 (input speed 12,300 rpm except ground-idle 10,400 rpm).

Lubrication: Pressure-feed type with full re-circulation; hourly consumption not over 0·8 litres (0·175 Imp gal).

Oil Grade: Mixture 75% transformer oil GOST 982-56 or MK-8 to GOST 6457-66 (equivalent to DERD.2490 or MIL-O-6081B) and 25% MS-20 or MK-22 to GOST 1013-49 (DERD.2472 or MIL-O-6082B).

Accessories: Engine and airframe accessories driven off compressor front extension shaft, via radial shafts at 6 and 12 o'clock. Full ice-protection and fire-extinguishing systems.

Starting: Two electric starter/generators, Type STG-12 TMO-1000, supplied from ground source or from APU Type TG-16.

Dimensions:
- Length: 3,096 mm (121·89 in)
- Width: 842 mm (33·15 in)
- Height: 1,180 mm (46·46 in)

Weight, Dry: 1,040 kg (2,292 lb)

Performance Ratings:
- T-O: 3,169 kW (4,250 ehp)
- Cruise (350 knots; 650 km/h; 404 mph at 8,000 m; 26,000 ft): 2,013 kW (2,700 ehp)

Specific Fuel Consumption:
- T-O: 104·3 μg/J (0·617 lb/h/shp)
- Cruise, as above: 73·3 μg/J (0·434 lb/h/shp)

IVCHENKO AI-24

This turboprop powers the An-24 and its derivatives. Production began in 1960 and the following data refer to engines of the second series, which were in production by the Spring of 1966.

The AI-24 of 1,875 kW (2,515 ehp) powered the An-24V Series I, and was followed by the AI-24A with provision for water injection, in the main production version of that aircraft.

The more powerful AI-24T of 2,103 kW (2,820 ehp) with water injection is used in the An-26. This engine has in-flight vibration monitoring, automatic relief of power overloads and gas temperature behind the turbine, and auto-shutdown and feathering.

The AI-24 is a constant-speed engine, maintained at 15,100 rpm by automatic variation of propeller pitch. The engine is flat-rated to maintain its nominal output to 3,500 m (11,500 ft). TBO was 3,000 h in the Spring of 1966; by 1968 the later AI-24T had reached 4,000 h.

Type: Single-shaft turboprop.

Air Intake: Large magnesium alloy casting, comprising inner and outer cones joined by four radial struts. Carries accessories, reduction gear, front mountings and compressor inlet guide vanes.

Compressor: Ten-stage axial. Stainless steel rotor, comprising rigidly-connected discs carrying dovetailed blades. Front shaft runs in roller bearing and is bolted to propeller drive-shaft of reduction gear; rear shaft runs in ball-thrust bearing and is splined to turbine shaft.

Welded steel casing in bolted left and right halves, with welded front and rear connecting flanges. Pressure ratio (max continuous, 6,000 m; 18,300 ft, 272 knots; 505 km/h; 314 mph) 7·85 : 1. Air mass flow 14·4 kg (31·7 lb)/s.

COMBUSTION CHAMBER: Annular, of spot-welded heat-resistant steel, with eight simplex burners inserted into swirl-vane heads. Contains two starting units, each comprising a body, pilot burner and igniter plug.

FUEL GRADE: T-1, TS-1 to GOST 10227-62 (DERD.2494 or MIL-F-5616).

TURBINE: Three-stage axial. Three discs carry solid blades in fir-tree roots, and are automatically centred on each other when connected by stay-bolts to the extended flange at the rear of the turbine shaft. Shaft splined to compressor rear shaft and held by tie-rod; runs in roller bearing ahead of first turbine disc. Three stator diaphragms through-bolted together and to combustion-chamber casing. First nozzle diaphragm cooled by secondary air from combustion chamber. Rotor/stator sealing effected by soft inserts mounted in grooves in nozzle assemblies. Peak exhaust temperature during starting 750°C.

JET PIPE: Fixed-area type. Inner and outer rings connected by three hollow struts carrying 12 thermocouples.

REDUCTION GEAR: Planetary type, incorporating hydraulic torquemeter and electromagnetic negative-thrust transmitter for propeller auto-feathering. Magnesium alloy casing. Front flange of propeller shaft has end splines and 12 stud holes for type AV-72 propeller (AI-24T drives AV-72T propeller). Ratio 0·08255.

LUBRICATION: Pressure circulation system; hourly consumption not over 850 g (1·87 lb).

OIL GRADE: Mixture of 75% transformer oil GOST 982-56 or MK-8 (DERD.2490 or MIL-O-6081B) and 25% MS-20 or MK-22 (DERD.2472 or MIL-O-6082B).

ACCESSORIES: Mounted on front casing are starter/generator, alternator, propeller speed governor and centrifugal breather. Below casing are oil unit, air separator and removable box containing LP and HP fuel pumps and drives to hydraulic pump and tachometer generators. Also on front casing are an aerodynamic probe, ice detector, and negative-thrust feathering valve, torque transmitter and oil filter.

STARTING: Electric STG-18TMO starter/generator supplied from ground power or from TG-16 APU.

DIMENSIONS:
- Length overall: 2,346 mm (92·36 in)
- Width: 677 mm (26·65 in)
- Height: 1,075 mm (42·32 in)

WEIGHT, DRY: 600 kg (1,323 lb)

PERFORMANCE RATINGS:
- T-O:
  - AI-24A: 1,875 kW (2,515 ehp)
  - AI-24T: 2,103 kW (2,820 ehp)
- Cruise rating at 243 knots (450 km/h; 280 mph) at 6,000 m (18,300 ft):
  - AI-24A: 1,156 kW (1,550 ehp)
  - AI-24T: 1,178 kW (1,580 ehp)

SPECIFIC FUEL CONSUMPTION:
- At cruise rating:
  - AI-24A: 91·3 μg/J (0·540 lb/h/shp)
  - AI-24T: 90·1 μg/J (0·533 lb/h/shp)

OIL CONSUMPTION: 0·85 kg (1·87 lb)/h

The 1,875 kW (2,515 ehp) Ivchenko AI-24 turboprop *(courtesy of Aviation Magazine International, Paris)*

IVCHENKO AI-25

This turbofan powers the three-engined Yakovlev Yak-40 STOL transport. The Aeroflot Yak-40 has the basic AI-25 engine, but the Yak-40B of the Soviet Air Force has an AI-25 with an aircooled HP turbine and rating of 17·13 kN (3,850 lb st). The AI-25 also powers the Czech L-39 trainer. The emergence of the Polish WSK-PZL-Mielec M-15 agricultural aircraft led to transfer of AI-25 activity to Poland, and production was expected to take place in that country to provide engines for the M-15 programme.

The Ivchenko bureau planned the engine for small transports, trainers and business jet aircraft. It is claimed to have an exceptional margin of flow stability and to be unusually robust and simple.

Ivchenko AI-25 TL turbofan (14·68 kN; 3,300 lb st), which powers the Czech Aero L-39

TYPE: Two-shaft turbofan.

AIR INTAKE: Fabricated from titanium sheet. Central bullet and intake leading-edges anti-iced by hot bleed air.

FAN: Three-stage axial. Drum/disc construction with pin-jointed blades. Casing and fan duct of magnesium alloy. Peak pressure ratio, 1·695 at 10,750 rpm. Bypass ratio 2.

COMPRESSOR: Eight-stage axial. Drum/disc construction of titanium, with aluminium and magnesium casing. Dovetailed blades. Peak pressure ratio, 4·68 at 16,640 rpm. Overall pressure ratio, 8.

COMBUSTION CHAMBER: Annular. Inner and outer casings joined upstream to 12 burner heads with stabilisers.

FUEL GRADE: T-1, TS-1 to GOST 10227-62 (DERD.2494, MIL-F-5616).

TURBINE: Single-stage HP turbine; two-stage LP turbine. Shrouded solid rotor blades held by fir-tree roots in cooled discs.

JET PIPE: Plain fixed convergent nozzles for core and bypass airflow. No mixer.

LUBRICATION: Self-contained, pressure circulating.

OIL GRADE: MK-8 to GOST 6457-66 or MK-6 to GOST 10328-63 (Western equivalents, DERD.2490 or MIL-O-6081B). Consumption 0·3 litres (0·53 Imp pints)/h.

ACCESSORIES: All shaft-driven accessories mounted on gearbox on underside of engine and driven off HP spool. Equipment includes automatic fire extinguishing (agent can be supplied into oil-contacted labyrinth cavities), ice protection, automatic starting and control system, oil-system chip detector and casing vibration monitor.

STARTING: Pneumatic. Air starter type SV-25 is supplied from ground hose coupling or from APU type AI-9 or an operating engine bleed. System claimed to develop high torque for rapid start in any climatic condition, and to be cleared for exceptional number of starts in given overhaul life.

DIMENSIONS:
- Length overall: 1,993 mm (78·46 in)
- Width overall: 820 mm (32·28 in)
- Height overall: 895 mm (35·24 in)

WEIGHT, DRY:
- Without accessories: 290 kg (639 lb)

PERFORMANCE RATINGS:
- T-O: 14·68 kN (3,300 lb st)
- Long-range cruise rating, 6,000 m (20,000 ft) and 296 knots (550 km/h; 342 mph): 3·49 kN (785 lb st)

SPECIFIC FUEL CONSUMPTION:
- T-O: 15·86 mg/Ns (0·56 lb/h/lb st)
- Cruise, as above: 23·71 mg/Ns (0·837 lb/h/lb st)

KOLIESOV

During an official tour of the Soviet aircraft industry in mid-1973, a representative of *Air Force Magazine* was told of the existence of the hitherto-unreported Koliesov engine design bureau. It is believed that this is the successor to the Dobrynin bureau. It was then developing an alternative engine to the Kuznetsov NK-144, the power plant used in the Tu-144 and, probably, in the Tupolev Tu-22M/Tu-26 bomber, known to NATO as 'Backfire'.

The Koliesov engine was described as a variable-geometry, variable by-pass ratio engine which functions as a turbojet in supersonic flight and as a turbofan in the subsonic regime. No such advanced design exists in the West. This engine is almost certainly fitted to the Tu-144D, the power plant of which has been described as 50 per cent more economical in operation than the NK-144.

The Koliesov designers are also believed to be one of the teams to have produced lift-jet engines for V/STOL aircraft.

Soviet engine designers are reported to be experimenting with hypersonic vehicles powered by scramjet propulsion systems and capable of operating in the Mach 5 to Mach 7 range.

KUZNETSOV

GENERAL DESIGNER IN CHARGE OF BUREAU:
Nikolai Dmitrievich Kuznetsov

Kuznetsov was deputy to General V. Ya. Klimov during the second World War. In the late 1940s his own bureau at Kuibyshev developed a series of large turboprop and turbofan engines. One of the first, the NK-4, was transferred to Ivchenko, and is described under that collective's heading, as the AI-20. The much larger NK-12 turboprop was used in the Tu-95 and Tu-126 military aircraft and in the Tu-114 civil transport, and was produced subsequently in small numbers for the An-22. The NK-8 and NK-144 may no longer be produced in their original forms.

KUZNETSOV NK-8

One of the first Soviet civil turbofans, the NK-8 has been developed through a number of variants, the most powerful of which is the NK-144 supersonic augmented engine for the Tupolev Tu-144 supersonic transport. Basic versions are the 99·1 kN (22,273 lb st) **NK-8-4** which originally powered the Ilyushin Il-62 four-engined transport, and the further-developed 93·2 kN (20,950 lb) **NK-8-2** which was the original engine of the Tupolev Tu-154. At one time it was planned to replace the NK-8-2 by the Soloviev D-30K, which replaced the NK-8-4 in the Il-62M. It is believed that all current Tu-154A/B aircraft are powered by the **NK-8-2U** of greater thrust (details unknown in the West). The NK-8-4 remains in service with several Il-62 (not Il-62M) operators, including LOT. It led to the NK-86 described later.

Type: Two-shaft turbofan.

Air Intake: Fabricated from outer ring, inner splitter and welded stator blades (15 in core airflow, 30 ahead of fan). Hot-air ice-protection.

Fan: Two-stage axial, with anti-flutter sweptback blades on first rotor stage. Pressure ratio 2·15 at 5,350 rpm. Bypass ratio 1·02 (NK-8-2, 1·00).

Compressor: Two IP stages on fan shaft. Six-stage HP compressor. Construction of rotors and stators, including blading, almost wholly of titanium alloy. Core pressure ratio, 10·8 at 6,950 HP rpm (NK-8-2, 10 at 6,835 rpm).

Combustion Chamber: Annular, with 139 burners. Claimed to produce no visible smoke.

Fuel Grade: T-1 and TS-1 to GOST 10227-62 or T-7 to GOST 12308-66 (equivalent to Avtur 50 to DERD.2494 or MIL-F-5616).

Turbine: Single-stage HP turbine, two-stage LP turbine, all with shrouded rotor blades, aircooled discs and hollow nozzle blades (stators). All shafting carried between shock-absorbing bearings at each end, with labyrinth and contact (rubbing) graphite seals to prevent gas leakage. Gas temperature, not over 870°C (1,143°K) ahead of turbine, not over 670°C (NK-8-2, 650°C) downstream, both values sea level, static.

Jet Pipe: Mixer leads bypass flow into common jet pipe which may be fitted with blocker/cascade-type reverser giving up to 48% (NK-8-2, 45%) reverse thrust, and noise suppressor.

Lubrication: Continuous pressure feed and recirculation. Oil consumption not over 1·3 kg (2·87 lb)/h. Pressure not less than 2·28 bars (33 lb/sq in).

Oil Grade: Mineral oil MK-8 or MK-8P to GOST 6457-66 (DERD.2490 or MIL-O-6081B). External tank on left side of front casing.

Accessories: These include automatic flight-deck warning of vibration exceeding permissible limit, ice and fire. All accessories grouped beneath fan duct casing. Engine claimed to need no attention for long periods, other than inspection of fuel and oil filters. RTA-26-9-1 turbine gas temperature controller by Smiths Industries.

Starting: HP spool driven by constant-speed drive type PPO-62M, or started pneumatically by air from APU type TA-6, from ground hose or by air from another engine (NK-8-2, pneumatic starter only). Time to idling speed not over 80 s. Engine can be windmill-started in the air under all conditions, up to altitudes of 11,000 m (36,000 ft).

Kuznetsov NK-8-2 turbofan with thrust reverser (93·2 kN; 20,950 lb st)

Dimensions:
NK-8-4:
Length, no reverser — 5,100 mm (201 in)
NK-8-2:
Length, with reverser — 5,288 mm (208·19 in)
Length, without reverser — 4,762 mm (187·48 in)
Diameter — 1,442 mm (56·8 in)

Weight, Dry:
NK-8-4:
No reverser — 2,100 kg (4,629 lb)
With reverser — 2,400 kg (5,291 lb)
NK-8-2:
No reverser — 2,100 kg (4,629 lb) max
With reverser — 2,350 kg (5,180 lb) max

Performance Ratings:
NK-8-4:
T-O rating — 99·1 kN (22,273 lb st)
Cruise rating at 11,000 m (36,000 ft) and 458 knots (850 km/h; 530 mph) — 27 kN (6,063 lb st)
NK-8-2:
T-O rating — 93·2 kN (20,950 lb st)
Cruise (as above) — 17·65 kN (3,968 lb st)

Specific Fuel Consumption:
At cruise rating at 11,000 m (36,000 ft) and 458 knots (850 km/h; 530 mph):
NK-8-4 — 22·1 mg/Ns (0·78 lb/h/lb st)
NK-8-2 — 21·53 mg/Ns (0·76 lb/h/lb st)

KUZNETSOV NK-86

Though described by the Ilyushin aircraft bureau as a new engine, this turbofan of 127·5 kN (28,660 lb st) appears to be closely related to the lower-thrust NK-8 series used in the Il-62 and Tu-154. The thrust is the same as the unaugmented rating of the NK-144 (see below) and the NK-86 may bear an even closer resemblance to the supersonic engine, which was itself derived from the NK-8. Four NK-86s power the Il-86 wide-body transport.

KUZNETSOV NK-144

This is the two-spool augmented turbofan developed for the Soviet Union's first supersonic transport aircraft, the Tu-144. It is a development of the NK-8 and the first five pre-production NK-144s completed some 1,500 h of bench-testing by October 1965. The engine flew in at least one testbed aircraft before the start of the Tu-144 flight programme in 1968.

Since 1972 the afterburner augmentation has been increased, raising maximum rating from 171·6 kN (38,580 lb) to the figure given below. A version of the NK-144 is believed to be the engine of at least the first sub-type of the Tupolev Tu-22M/Tu-26 supersonic bomber known to NATO as 'Backfire'.

The NK-144 is reported to have a two-stage titanium fan, three-stage IP compressor, eleven-stage HP compressor, annular combustion chamber, single-stage HP turbine and two-stage LP turbine. Aircooled blades are used in the HP turbine, and titanium is used extensively in construction of the engine. Bypass ratio is reported to be 1 : 1, maximum mass flow 250 kg (551 lb)/s, and pressure ratio 15 : 1. The jet pipe incorporates an afterburner, with hydraulically-actuated variable-area nozzle. Gas temperature at turbine entry is 1,050°C.

Dimensions:
Length overall — 5,200 mm (204·7 in)
Diameter — 1,500 mm (59 in)

Weight:
Without jet pipe, but with afterburner — 2,850 kg (6,283 lb)

Performance Ratings:
Max, without afterburning — 127·5 kN (28,660 lb st)
Max, with afterburning — 196·1 kN (44,090 lb st)

KUZNETSOV NK-12M

Designed at Kuibyshev under the leadership of N. D. Kuznetsov and former German engineers, the NK-12M is the most powerful turboprop engine in the world. In its original form as the NK-12M it developed 8,948 kW (12,000 ehp). The later NK-12MV is rated at 11,033 kW (14,795 ehp) and powers the Tupolev Tu-114 transport, driving four-blade contra-rotating propellers of 5·6 m (18 ft 4 in) diameter. As the NK-12MA, rated at 11,185 kW (15,000 shp), it powers the Antonov An-22 military transport, with propellers of 6·2 m (20 ft 4 in) diameter. A third application is in the Tupolev Tu-95 bomber and its derivatives, and Tu-126 'AWACS'.

The NK-12M has a single 14-stage axial-flow compressor. Compression ratio varies from 9 : 1 to 13 : 1 according to altitude, and variable inlet guide vanes and blow-off valves are necessary. A cannular-type combustion system is used: each flame tube is mounted centrally on a downstream injector, but all tubes merge at their maximum diameter to form an annular secondary region. The single turbine is a five-stage axial. Mass flow is 65 kg (143 lb)/s.

The casing is made in four portions, from sheet steel, precision welded. An electric control for variation of propeller pitch is incorporated, to maintain constant engine speed.

Dimensions:
Length — 6,000 mm (236·2 in)
Diameter — 1,150 mm (45·3 in)

Weight, Dry: — 2,350 kg (5,181 lb)

Performance Ratings:
T-O — 11,033 kW (14,795 ehp)
Nominal power — 8,826 kW (11,836 ehp) at 8,300 rpm
Idling speed — 6,600 rpm

Kuznetsov NK-12MV single-shaft turboprop of 11,033 kW (14,795 ehp)
(courtesy of Aviation Magazine International, Paris)

LOTAREV

General Designer in Charge of Bureau:
Vladimir Lotarev

LOTAREV D-36

As successor to Alexander Ivchenko at Zaparozhye, Vladimir Lotarev has developed the three-shaft turbofan engine that powers the An-72 and Yak-42. With a bypass ratio of 5·6 : 1 the D-36 is the first avowed turbofan—as distinct from a bypass turbojet—to emerge in the Soviet

Union. Bench testing was in progress in September 1973, and flights in a pod carried beneath a Tu-16 testbed aircraft preceded the Yak-42 first flight by several years.

Western engineers who have visited the Soviet Union report that the D-36 is a more advanced and modern engine than other Soviet commercial gas turbines in current production. (See Addenda for further details).

WEIGHT, DRY: 1,100 kg (2,425 lb)

PERFORMANCE RATINGS:
- T-O rating: 63·74 kN (14,330 lb st)
- Max cont: 49·0 kN (11,025 lb st)
- Max cruise rating at 8,000 m (26,200 ft) at Mach 0·75: 15·7 kN (3,525 lb st)

SPECIFIC FUEL CONSUMPTION:
- At T-O rating: 10·62 mg/Ns (0·375 lb/h/lb st)
- At max cont rating: 9·83 mg/Ns (0·347 lb/h/lb st)
- At max cruise rating as above: 18·4 mg/Ns (0·65 lb/h/lb st)

LYULKA

GENERAL DESIGNER IN CHARGE OF BUREAU:
Arkhip Mikhailovich Lyulka

During the late 1930s Arkhip Lyulka worked on the design of an axial turbojet that became an early war casualty. In 1942 he planned a more advanced engine that finally materialised as the TR-1, of 12·75 kN (2,866 lb st), run on the bench in 1944 and used in the Ilyushin Il-22 four-jet bomber and Sukhoi Su-11 twin-jet fighter prototypes, both of 1947. Ultimately, in 1948, this pioneer Soviet-designed turbojet was developed to give 14·71 kN (3,307 lb st).

In 1946 Lyulka began the design of a very ambitious axial engine to give a thrust of 44·13 kN (9,920 lb), and in 1950 this began bench trials under the designation AL-5. Although of basically simple, single-shaft configuration, with a seven-stage compressor and single-stage turbine, the AL-5 was more powerful than any Western engine apart from the prototype Olympus and J57. By 1951 it was rated at 45·1 kN (10,140 lb st) and flew in the prototype Ilyushin Il-30 twin-jet bomber; later in 1951-52 uprated AL-5 engines, giving a static thrust of 49·1 kN (11,032 lb), powered the Il-46 twin-jet bomber and the transonic Lavochkin La-190 and Yakovlev Yak-1000 fighters. An advanced civil version of the same engine, the AL-5 rated at 53·93 kN (12,125 lb st) powered the Tu-110 four-engined derivative of the Tu-104 airliner that did not go into production (at the time, in 1959, this engine was reported in the West as the 'Lu-4').

By the time the AL-5 was running, Lyulka had conducted extensive research with axial compressors having supersonic airflow through some or all of the stages. It was clear that, if problems of flow breakdown and inefficiency could be resolved, such a compressor would enable turbojets to be made much smaller and lighter for a given thrust and with greater thrust per unit frontal area, and thus much better suited to the propulsion of supersonic fighters. By 1952 a supersonic-compressor engine had been designed and built. This, the AL-7, became Lyulka's first major success.

LYULKA AL-7

The first AL-7 ran on the bench in late 1952 and the first production version was cleared for use in 1954 at a design rating of 63·74 kN (14,330 lb st). Its initial application was on the Il-54, yet another Ilyushin twin-jet bomber that failed to see production, despite the fact that its speed at low altitude of 620 knots (1,150 km/h; 714 mph) was probably unrivalled by any other bomber in 1955. In the same year the Sukhoi Su-7 single-seat ground-attack fighter was designed around the AL-7F afterburning version of this engine, with thrust increased by about 40 per cent (see data below). By 1956 the Su-7 was flying, and the AL-7F had also been chosen for the basically similar Su-9 all-weather fighter. Subsequently the -7F was also produced for the Su-11 and -11U.

By 1958 a further developed version of the basic unaugmented engine, the AL-7RV, had been chosen by Beriev for the Be-10 reconnaissance flying-boat, which—apart from being the only pure-jet flying-boat ever to go into service anywhere—set up a number of world records for speed, load-carrying and altitude. Other versions of the AL-7, in both cases of the -7F afterburning family, powered the unsuccessful Tu-98 bomber and La-250 interceptor of 1956.

The AL-7F, or a development of it, has been persistently reported to be the power plant of the Tu-28P twin-engined interceptor. This is not confirmed. The Tu-28P is considered to need greater thrust, greater even than provided by the AL-21 which was not available when the Tu-28 entered service in the early 1960s.

TYPE: Single-shaft axial-flow turbojet, available with or without afterburner.

AIR INTAKE: Central bullet fairing and 14 fixed aerofoil struts anti-iced by compressor bleed air.

COMPRESSOR: Nine-stage axial (probably eight stages in original AL-7 design). First two stages widely separated axially, with variable stators ahead of second stage. Each stage has blades inserted in centreless disc held by peripheral spacers at correct distance from adjacent discs, the whole being coupled together finally by the central drive-shaft in tension. Pressure ratio probably about 8 : 1.

COMBUSTION CHAMBER: Annular type with perforated inner flame tube. Multiple downstream fuel injectors inserted through cups in forward face of liner. Liner outer casing provided with multiple inward secondary-air injection ducts.

TURBINE: Two-stage axial-flow type. Both wheels overhung behind rear bearing; front disc bolted to flange on hollow tubular driveshaft which, in turn, is splined to rear of compressor shaft running in main centre bearing which locates compressor axially against end loads.

AFTERBURNER (AL-7F): Comprises upstream diffuser and downstream combustion section. Pilot combustor on turbine exit cone includes single nozzle ring and flame-holder; main spray ring and gutter flame-holder assembly located further downstream at greater radius. Refractory liner in combustion section. Variable-area nozzle, with multiple hinged flaps which govern nozzle size and profile according to signals from reheat control system based on turbine exit temperature and throttle lever position.

ACCESSORIES: Fuel pump and control unit, oil pumps, hydraulic pump, electric generator, tachometer and other items grouped into quickly replaceable packages beneath compressor casing.

PERFORMANCE RATINGS:
Max rating:
- AL-7F, unaugmented: 68·64 kN (15,432 lb st)
- AL-7F, afterburning: 98·1 kN (22,046 lb st)
- AL-7RV: 63·74 kN (14,330 lb st)

LYULKA AL-21

When continued development of the Su-7 family by the Sukhoi bureau was matched with increased thrust it was logical to suppose that the engine would be a derivative of the AL-7. This is reported to have happened, and the designation AL-21 has been given. The AL-21 resembles the AL-7 closely, and may be installationally interchangeable, but has significant improvements to the compressor and other components. The first production version was the AL-21F-3, for the Su-17 variable-geometry tactical aircraft.

PERFORMANCE RATINGS:
- Max S/L, unaugmented: 80·1 kN (18,000 lb st)
- Max S/L, afterburning: 109 kN (24,500 lb st)

MIKULIN

GENERAL DESIGNER IN CHARGE OF BUREAU:
Alexander Alexandrovich Mikulin

Eighty years old on 16 February 1975, Mikulin has been engaged in aircraft engine design since 1916. Notable Vee-12 water-cooled engines from his designs were the AM-13, AM-34, AM-38 (used in the Il-2 and Il-10 Shturmovik) and AM-42.

Very large numbers of M-11 five-cylinder radial engines have been built in the Soviet Union and (from 1949) Poland to power a variety of light aircraft and helicopters. Best-known variants are the 93 kW (125 hp) M-11D and the 119 kW (160 hp) M-11FR which powers the Yak-18 primary trainer. Brief descriptions appeared in the 1975-76 *Jane's*.

The large turbojet described below was designed immediately after the second World War. Though a version was fitted to M-4 four-engined prototypes, it is not known if Mikulin's bureau was involved with the Soloviev D-15 fitted to later M-4 aircraft.

MIKULIN RD-3M-500

The basic RD-3M (or AM-3M) single-spool axial-flow turbojet was developed under the design leadership of P. F. Zubets from the original Mikulin M-209 (civil RD-3 or AM-3) engine which powers the Tu-16 and early M-4 bombers and was adapted for the USSR's first jet transport, the Tu-104.

The RD-3M-500 evolved, in turn, from the RD-3M and powers the Tu-104A and Tu-104B commercial transports. It has a simple basic configuration, with an eight-stage axial-flow compressor, annular-type combustion system with 14 flame tubes, and a two-stage turbine. The compressor casing is made in front, centre and rear portions, the front casing housing a row of inlet guide vanes. A bullet fairing mounted centrally in the annular ram-air intake houses a type S-300M gas-turbine starter, developing 90-100 hp at 31,000-35,000 rpm. The jet pipe consists of a central cone and fixed nozzle with an orifice diameter of approximately 840 mm (33 in). Pressure ratio is 6·4 : 1; temperature after turbine 720°C.

DIMENSIONS:
- Length overall: 5,340 mm (210·23 in)
- Diameter: 1,400 mm (55·12 in)

PERFORMANCE RATING:
- T-O: 93·15 kN (20,940 lb st)

SHVETSOV

FOUNDER OF BUREAU: Arkadiya Dmitrievich Shvetsov

Arkadiya Shvetsov had meteoric rise to fame with his aircooled radial engines based on original US designs, which transformed Soviet fighters, bombers and transport aircraft during and after the second World War. The most important, and most powerful, of these engines was the ASh-82 series, derived from the Pratt & Whitney Twin Wasp, which at ratings up to 1,491 kW (2,000 hp) powered Lavochkin fighters, the Tu-2 and Tu-4 bombers, Il-12 and -14 transports and Mi-4 and Yak-24 helicopters (see 1975-76 *Jane's*). The ASh-62, originally based on the Wright Cyclone, was made in large numbers at ratings in the 750 kW (1,000 hp) class for the Li-2 and An-2. The ASh-62M agricultural version, developed by Vedeneev, and all An-2 engines after 1952, were transferred with all other Soviet aircooled radials to Poland (see WSK-PZL-Kalisz and WSK-PZL-Rzeszów in Polish part of this section). Another important Shvetsov piston engine was the 545 kW (730 hp) ASh-21, fitted to the Yak-11. All these engines have been described in earlier editions.

SOLOVIEV

GENERAL DESIGNER IN CHARGE OF BUREAU: P. A. Soloviev

Engines for which Soloviev's design team is responsible include the turbofans fitted in the Ilyushin Il-62M and Il-76 and Tupolev Tu-124 and Tu-134 transport aircraft, and the turboshafts which power the Mi-6, Mi-10 and Mi-26 helicopters.

SOLOVIEV D-15

This engine was first reported, in 1959, as that fitted to the four-engined Type 201-M aircraft which gained a number of world records for speed and altitude. Over the years it has become apparent that the aircraft was a special Myasishchev M-4, and that the engine was standard in the later service versions of this aircraft. Details of the D-15 are still unknown in the West, but it is probably safe to deduce that it is a two-shaft bypass turbojet (low bypass ratio turbofan). It is likely that it laid the foundation upon which Soloviev's bureau produced the civil D-20 and D-30, both of which are considerably smaller engines.

PERFORMANCE RATING:
- T-O: 128·6 kN (28,660 lb st)

SOLOVIEV D-20P

The D-20P is a two-spool turbofan fitted to the Tupolev Tu-124 twin-engined passenger transport. Of conservative design, it underwent prolonged testing before entering service. It was designed for maximum economy and reliability over the range of ambient temperatures between −40°C and 40°C.

TYPE: Two-shaft turbofan (bypass turbojet).

AIR INTAKE: Eight radial struts and central bullet fairing, de-iced by hot bleed air from fourth HP stage (from final stage at low rpm).

FAN: Three-stage axial, with supersonic blading in first stage. Mass flow 113 kg (249 lb)/s at 8,550 rpm. Pressure ratio (S/L, static at max cont 7,900 rpm) 2·4 : 1. Bypass ratio 1 : 1.

COMPRESSOR: Eight-stage axial. Automatically-controlled flap valves downstream of the third and fourth stages bleed air into the fan duct to stabilise behaviour. Pressure ratio (at max continuous, 11,170 rpm) 5 : 1; overall pressure ratio 13 : 1.

COMBUSTION CHAMBER: Can-annular, with 12 flame tubes each fitted with duplex burner.

FUEL GRADE: T-1, TS-1 to GOST 10227-62 (Avtur-50 to DERD.2494, MIL-F-5616).

TURBINE: Single-stage HP turbine with cast blades; stator blades and both sides of disc cooled by bleed air. Two-stage LP turbine with forged blades. Max gas temperature downstream of turbine 650°C.

JET PIPE: Concentric pipes for fan airflow and core gas, terminating in supersonic nozzles of fixed-area type.

LUBRICATION: Open type, with oil returned to tank. Consumption in flight, not over 1 kg (2·2 lb)/h. Typical pressure 3·4-4·5 kg/cm² (50-64 lb/sq in).

OIL GRADE: Mineral oil MK-8 or MK-8P to GOST 6457-66 (DERD.2490 or MIL-O-6081B).

ACCESSORIES: Two gearboxes provide drives for starter/generator, tachometer, air compressors, hydraulic pump, oil pump and other controls and instruments. For restarting in flight, an altitude sensing device meters fuel flow appropriate to height. An automatic fire extinguishing system is fitted. De-icing of the air intake and inlet guide vanes is controlled automatically. The engine also has oil chip detectors, vibration monitors and turbine gas temperature limiters.

STARTING: Electric (DC) system, incorporating STG-18TM starter/generator.

DIMENSIONS:
- Length overall: 3,304 mm (130 in)
- Diameter, bare: 976 mm (38·3 in)

WEIGHT, DRY: 1,468 kg (3,236 lb)

PERFORMANCE RATINGS:
- Max T-O rating: 52·96 kN (11,905 lb st)
- Long-range cruise, Mach 0·75, 11,000 m (36,000 ft): 10·79 kN (2,425 lb st)

SPECIFIC FUEL CONSUMPTION:
- Max T-O: 20·4 mg/Ns (0·72 lb/h/lb st)
- Long-range cruise, as above: 25·5 mg/Ns (0·90 lb/h/lb st)

Soloviev D-20P turbofan (52·96 kN; 11,905 lb st)

The 4,101 kW (5,500 shp) Soloviev D-25V turboshaft

SOLOVIEV D-25V

D-25V is the Soloviev bureau designation for the free-turbine turboshaft which powers the Mil Mi-6 and Mi-10 helicopters and was also fitted to the Kamov Ka-22 experimental convertiplane and V-12 helicopter. It is usually referred to by its official designation of **TV-2BM.**

The complete helicopter power plant comprises two D-25V engines, identical except for handed jet pipes, and an R-7 gearbox. The latter has four stages of large gearwheels providing an overall ratio of 69·2 : 1. The R-7 is 2,795 mm (110·04 in) high, 1,551 mm (61·06 in) wide and 1,852 mm (72·91 in) long. Its dry weight is 3,200 kg (7,054 lb), more than that of the pair of engines.

The D-25V is flat rated to maintain rated power to 3,000 m (10,000 ft) or to temperatures up to 40°C at sea level.

The D-25VF turboshafts fitted to the Mil V-12 helicopter were uprated to 6,500 shp. These engines were believed to incorporate a zero stage on the compressor and to operate at higher turbine gas temperatures. The following details apply to the basic D-25V:

TYPE: Single-shaft turboshaft with free power turbine.

AIR INTAKE: Six hollow radial struts, the two vertical struts housing splined shafts driving upper and lower accessory drive boxes. Vertical struts de-iced by oil drained from upper drive box; four inclined struts and bullet fairing de-iced by hot oil returned from engine to tank.

COMPRESSOR: Nine-stage axial. Comprises fixed inlet guide vane assembly, first-stage stator ring, upper and lower casings with dovetailed stator blades, ninth-stage stator ring and exit vanes, rotor, and air blow-off valves. Pressure ratio 5·6 at T-O power, 10,530 rpm.

COMBUSTION CHAMBER: Can-annular. Assembled from diffuser (the structural basis of the engine), inner shroud, 12 flame tubes with transition liners, diaphragm and compressor-shaft shroud.

FUEL GRADE: T-1, TS-1 to GOST 10227-62 (DERD.2494, MIL-F-5616).

TURBINE: Single-stage compressor turbine, overhung behind rear roller bearing. Two-stage power turbine, overhung on end of rear output shaft. Both turbines rotate counter-clockwise, seen from the rear. Normal power turbine rpm, 7,800-8,300; maximum 9,000. Transmission shaft in three universally-jointed sections, allowing for 10 mm (4 in) misalignment between engine and gearbox.

JET PIPE: Large fabricated assembly in heat-resistant steel, curved out to side to allow rotor transmission to pass through duct wall in aircooled protecting pipe.

LUBRICATION: Pressure circulation at 3·45-4·41 bars (50-64 lb/sq in). Separate systems for gas-generator and for power turbine, transmission and gearbox.

OIL GRADE: Gas-generator, MK-8 to GOST 6457-66 or transformer oil to GOST 982-56. Power turbine and gearbox, mixture (75-25 Summer, 50-50 Winter) of MK-22 or MS-20 to GOST 1013-49 and MK-8 or transformer oil. Hourly oil consumption, gas-generator not over 1 kg (2·2 lb), power turbine and transmission not over 2 kg (4·4 lb).

ACCESSORIES: SP3-12TV electric supply and starting system; fuel supply to separate LP and HP systems; airframe accessories driven off upper and lower gearboxes on inlet casing.

STARTING: The SP3-12TV system starts both engines and also generates electric current. It comprises an STG-12TM starter/generator on each engine, igniter unit, two spark plugs with cooling shrouds, two switch-over contactors, solenoid air valve, pressure warning, PSG-12V control panel and electro-hydraulic cutout switch of the TsP-23A centrifugal governor. In the starter mode the system draws current from a ground supply receptacle or from batteries.

DIMENSIONS:
- Length overall, bare: 2,737 mm (107·75 in)
- Length overall with transmission shaft: 5,537 mm (218·0 in)
- Width: 1,086 mm (42·76 in)
- Height: 1,158 mm (45·59 in)

WEIGHT, DRY:
- With engine-mounted accessories: 1,325 kg (2,921 lb)

PERFORMANCE RATINGS:
- T-O: 4,101 kW (5,500 shp)
- Rated power: 3,504 kW (4,700 shp)
- Cruise (1,000 m; 3,280 ft, 135 knots; 250 km/h; 155 mph): 2,983 kW (4,000 shp)

SPECIFIC FUEL CONSUMPTION:
- T-O, as above: 108 μg/J (0·639 lb/h/shp)
- Cruise, as above: 118·1 μg/J (0·699 lb/h/shp)

SOLOVIEV D-30

This two-spool turbofan powers the Tu-134 twin-engined airliner and is derived from the D-20. Major portions of the core and carcase are similar, but the complete power plant is larger than the D-20, and more powerful and efficient.

In turn the D-30 has been developed into the considerably larger D-30K, described separately. They show a continuing allegiance to the form of power plant pioneered by Rolls-Royce as the 'bypass turbojet', a term still used in the Soviet Union to describe these engines, in which the LP system comprises several stages and the bypass ratio is not greater than unity (in the West most turbofans have single-stage fans with a bypass ratio of from 3 to 8).

Soloviev D-30 turbofan (66·68 kN; 14,990 lb st) with thrust reverser

TYPE: Two-shaft turbofan (bypass turbojet).

AIR INTAKE: Titanium alloy assembly, incorporating air bleed anti-icing of centre bullet and radial struts.

FAN: Four-stage axial (LP compressor). First stage has shrouded titanium blades held in disc by pinned joints. Pressure ratio (T-O rating, 7,700 rpm, S/L, static), 2·65 : 1. Mass flow 125 kg (265 lb)/s. Bypass ratio 1 : 1.

COMPRESSOR: Ten-stage axial (HP compressor). Drum and disc construction, largely of titanium. Pressure ratio (T-O rating, 11,600 rpm, S/L, static), 7·1 : 1. Overall pressure ratio, 17·4 : 1.

COMBUSTION CHAMBER: Can-annular, with 12 flame tubes fitted with duplex burners.

FUEL GRADE: T-1 and TS-1 to GOST 10227-62 (equivalent to DERD.2494 or MIL-F-5616).

TURBINE: Two-stage HP turbine. First stage has cooled blades in both stator and rotor. LP turbine also has two stages. All discs aircooled on both sides, and all blades shrouded to improve efficiency and reduce vibration. All shaft bearings shock-mounted.

JET PIPE: Subsonic fixed-area type, incorporating main and bypass flow mixer with curvilinear ducts of optimum shape. D-30-2 engine of Tu-134A fitted with twin-clamshell reverser.

LUBRICATION: Open type, with oil returned to tank.

OIL GRADE: Mineral oil MK-8 or MK-8P to GOST 6457-66 (equivalent to DERD.2490 or MIL-O-6081B). Consumption in flight not over 1·0 kg (2·2 lb)/h.

ACCESSORIES: Automatic ice-protection system, fire extinguishing for core and bypass flows, vibration detectors on casings, oil chip detectors and automatic limitation of exhaust gas temperature to 620°C at take-off or when starting and to 630°C in flight (5 min limit). Shaft-driven accessories driven via radial bevel-gear shafts in centre casing, mainly off HP spool, accessory gearboxes being provided above and below centre casing and fan duct. D-30-2 carries constant-speed drives for alternators.

STARTING: Electric DC starting system incorporating STG-12TVMO starter/generators.

DIMENSIONS:
- Overall length: 3,983 mm (156·8 in)
- Base diameter of inlet casing: 1,050 mm (41·3 in)

WEIGHT, DRY: 1,550 kg (3,417 lb)

PERFORMANCE RATINGS:
- T-O: 66·68 kN (14,990 lb st)
- Long-range cruise rating, 11,000 m (36,000 ft) and Mach 0·75: 12·75 kN (2,866 lb st)

SPECIFIC FUEL CONSUMPTION:
- T-O: 17·56 mg/Ns (0·62 lb/h/lb st)
- Cruise, as above: 21·81 mg/Ns (0·77 lb/h/lb st)

SOLOVIEV D-30K

Despite its designation, this turbofan is very different from the D-30 described previously. It is larger, and has a much higher rating, a bypass ratio considerably higher, and very few parts (in the core) common to the D-30.

The basic **D-30KU** version, to which the specification details below apply, replaced the Kuznetsov NK-8 as power plant of the Ilyushin Il-62M long-range transport. The more powerful **D-30KP,** rated at 117·7 kN (26,455 lb st), powers the Ilyushin Il-76 freight transport. Clamshell-type thrust reversers are fitted to all four engines of this aircraft, and to the outer engines of the Il-62M. These reversers are not an integral part of the engine but are airframe assemblies incorporated in the nacelle.

Type: Two-shaft turbofan, with integral flow mixer and reverser.

Air Intake: Fabricated from titanium alloy. Fixed spinner and 26 cambered inlet guide vanes anti-iced by air bled from sixth or eleventh stage of HP compressor (depending on rpm). Integral front roller bearing for LP shaft.

Fan (LP Compressor): Three stages, mainly of titanium alloy. First-stage rotor blades held in dovetail slots, with part-span anti-vibration snubbers. Other two stages have pinned rotor blades. Spool rotates between front roller bearing and rear ball bearing, with additional roller bearing behind LP turbine. Drum/disc construction, coupled with tie-bolt and driven by splined shaft connection. Mass flow at take-off, 269 kg (593 lb)/s at 4,730 rpm (87·9 per cent), with bypass ratio of 2·42.

Division Casing: Linking the LP and HP compressors, this is the main structural attachment band to the aircraft. Magnesium-alloy casting, held by front and rear rows of peripheral bolts. Carries LP mid bearing and HP front bearing and incorporates vertical radial drive to front drive box for accessories on underside.

HP Compressor: Eleven stages. Drum/disc rotor, with discs centred on shaft by rectangular splines. Rotor blades held in dovetail slots, first two stages having part-span snubbers. Construction of titanium alloys, except for shafts, rear casing, and rotor blades and discs of stages 9-11, and stator vanes of stages 10-11, which are steel. To reduce blade vibration inlet guide vanes are turned through up to 30° according to preset programme over speed range of 7,900-9,600 rpm, while air is bled from fifth and sixth stages under transient conditions; in addition a closed peripheral chamber with perforated walls surrounds the first-stage rotor blades. HP shaft supported in front roller bearing in division casing, ball thrust bearing at rear of compressor spool and roller bearing ahead of turbine. Casing split horizontally. Overall pressure ratio (S/L, static) 20 at HP speed of 10,460 rpm (96 per cent).

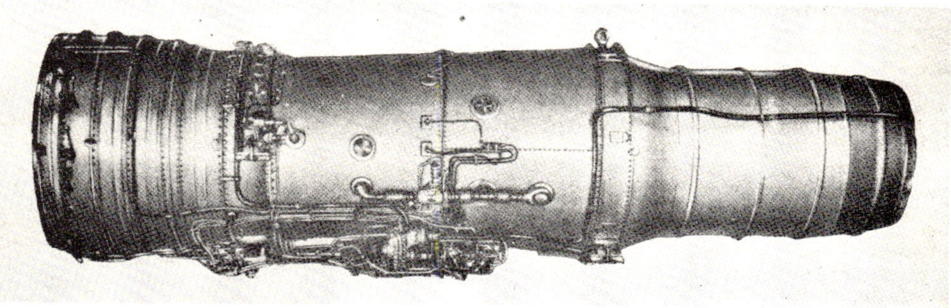

Soloviev D-30KU turbofan (108 kN; 24,250 lb st); see Addenda for equipped engine

Combustion Chamber: Cannular type with 12 flame tubes in annular chamber. Each tube comprises hemispherical head and eight short sections welded with gaps for dilution air. Single swirl-type main/pilot burner centred in each tube. Igniter plugs in two tubes. Outer casing and duct shroud provided with longitudinal joints for access to flame tubes and HP turbine nozzle ring.

Fuel Grade: T-1, TS-1, GOST-10227-62, A-1 (D1655/63t), DERD.2494 or 2498, Air 3405/B or 3-GP-23e.

Turbines: Two-stage HP turbine with first-stage nozzles, part of second-stage nozzles and both sets of discs and rotor blades cooled by HP bleed air. Second-stage rotor blades tip-shrouded. Both discs interchangeable. Take-off inlet gas temperature 1,122°C. Four-stage LP turbine with uncooled shrouded rotor blades carried in four identical discs cooled by bypass air.

Jet Pipe: Downstream of LP turbine a rear support frame serves as the rear structural band attaching the engine to the aircraft. This frame incorporates the rear LP shaft roller bearing and 12 thermocouples, and also includes the 16-chute mixer for the core and bypass flows.

Lubrication: Closed type, with oil returned to tank. Incorporates fuel/oil heat exchanger and centrifugal air separator with metal-particle warning unit.

Oil Grade: MK-8 or MK-8P to GOST 6467-66 (mineral) or BNII NP-50-1-4F to GOST 13076-67 (synthetic) or Western equivalents.

Accessories: Front and rear drive boxes underneath engine carry all shaft-driven accessories. Differential constant-speed drive to alternator and air-turbine starter.

Starting: Pneumatic air-turbine starter fed by ground supply, APU or cross-bleed from running engine. Start cycle time to idling rpm, 40-80 s depending on ambient temperature (limits, −60° to +50°C). In-flight starting up to 9,000 m (27,430 ft) by windmilling.

Dimensions:

Length with reverser	5,700 mm (224 in)
Inlet diameter	1,464 mm (57·6 in)
Maximum diameter of casing	1,560 mm (61·4 in)

Weight, Dry:

With reverser	2,650 kg (5,842 lb)
Without reverser	2,300 kg (5,071 lb)

Performance Ratings (ISA):

T-O 108 kN (24,250 lb st) to 21°C

Cruise at 11,000 m (36,000 ft) and Mach 0·8 27 kN (6,063 lb st)

Specific Fuel Consumption:

At T-O rating	13·88 mg/Ns (0·49 lb/h/lb st)
Cruise, as above	19·83 mg/Ns (0·70 lb/h/lb st)

TUMANSKY

Academician Sergei K. Tumansky, who died in 1973, left a legacy of turbojet and bypass jet engines which were the propulsion basis on which the MiG bureau created the MiG-21, in world-wide service, and the extremely high-performance MiG-25. Tumansky is believed also to have designed the engines of the MiG-23/27 and Su-15.

TUMANSKY R-11 and R-13

Designated TRD Mk R37F (turbojet R37F) by the Soviet armed forces, the R-11/R-13 family have been built in very large numbers, and two versions have been licence-built by HAL in India. At least one version is produced, without a licence, in China.

A single-shaft turbojet with afterburner, the R-11 entered production in 1956 with dry and afterburning ratings of 38·25 kN (8,600 lb) and 50 kN (11,240 lb) respectively. In 1959 a world speed record was set by the E-66, powered by an R-11-F2-300, with the same dry rating but a new afterburner giving a maximum thrust of 58·4 kN (13,120 lb). This engine also powered the production variant of the E-66, the MiG-21F, as well as the MiG-21PF, FM, FL and possibly other versions.

A very similar engine is the R-11-F2S-300, which powers the Indian-built MiG-21M. It is believed that versions of the R-11 power all known versions of the twin-engined Yak-28.

The R-13-300 incorporates major changes to handle an increased airflow, and has dry and afterburning ratings of 50 kN (11,240 lb) and 64·73 kN (14,550 lb) respectively. It powers later MiG-21 versions, including the MF, and possibly the Sukhoi Su-15. It is believed not to be installationally interchangeable with the R-11 series.

TUMANSKY R-266

In about 1961 the Tumansky bureau began the design of a jet engine with afterburner considerably more powerful than the R-11 and intended for flight at speeds higher than Mach 3. This engine was adopted by the MiG bureau for the MiG-25 twin-engined fighter. Existence of this Mach 3·2 aircraft was disclosed in April 1965 when its first world record was announced. At that time the aircraft was referred to as the E-266, and its engine as the TRD Mk 31. Thrust rating was given as 98·1 kN (22,046 lb).

Most production versions of the MiG-25 have a pair of R-266 engines. These are extremely simple afterburning turbojets similar in concept to the earlier de Havilland Gyron. The single-shaft compressor has five stages and works at a peak pressure-ratio of about 7 : 1. The single-stage turbine is reported to be uncooled, an extraordinary feature for a Mach 3 engine. Construction is mainly of steel, with a little titanium. Turbine entry temperature is estimated at only 847°C (again a scarcely credible low figure). At supersonic speeds almost all the thrust is provided by the variable inlets, three-ring afterburner and variable nozzle. Special T-6 fuel is used with freezing point of −62·2°C and flash point of 54·4°C. Dry and augmented ratings are (S/L, static) 9,300 kg (20,500 lb) and 12,250 kg (27,010 lb).

Tumansky R-11 built under licence by HAL in India

A more powerful (R-266F ?) engine powers the latest 'Foxbat' interceptor (said to be the MiG-25MP), and possibly other variants. This engine is rated at 14,000 kg (30,865 lb st), and is almost certainly the same as the RD-F engines of the same rating which power the E-266M aircraft which set time-to-height records in 1975.

TUMANSKY R-27 and R-29B

The variable-geometry MiG-27 tactical attack fighter has an engine said to be designated R-29B, designed for STOL and low-level operation, with large LP compressor (fan) and simple afterburner. Dry and augmented ratings are estimated at 66·72 kN (15,000 lb) and 88·96 kN (20,000 lb) respectively. The engine is installed with simple fixed inlets and has a short two-position nozzle. The

MiG-23MF all-weather fighter has an engine sharing a common core but having a smaller LP compressor and large afterburner. This engine is designed for supersonic operation at Mach 1·8-2·3, and is fed by fully-variable supersonic inlets and discharges through a large fully modulated primary and secondary nozzle of variable profile. Dry and augmented ratings are estimated at 78·45 kN (17,635 lb st) and 112·8 kN (25,350 lb st) respectively. As far as can be judged, an earlier and lower-rated engine, reported as the R-27, is fitted to the MiG-23U two-seat trainer and most MiG-23 aircraft (of all versions) for export.

VEDENEEV

General Designer in Charge of Bureau:
Ivan M. Vedeneev

This designer was responsible for the improvement and development of certain models of the AI-14 piston engine designed by the Ivchenko bureau. He also developed the ASh-62M, produced in Poland as the ASz-62M, from Shvetsov's ASh-621R nine-cylinder radial. He now heads his own bureau.

M-14V-26

Derived from the Ivchenko AI-14 family of engines for fixed-wing aircraft, the M-14V-26 powers the Kamov Ka-26 helicopter. In this installation the stub-wing carries an engine on each tip. Beneath the rotor an R-26 gearbox combines the power of the engines and distributes it equally between the two coaxial main rotors turning in opposite directions.

It is said to incorporate all the experience gained in many years of developing engines in this class, and shows numerous areas of refinement compared with the AI-14 series. The engine has forced cooling by an axial fan driven via a friction clutch and extension shaft ahead of the main output bevel box at 1·452 times crankshaft speed. The engine planetary gearbox has a ratio of 0·309 and incorporates friction and ratchet clutches. The central R-26 gearbox has a ratio of 0·34; it also drives the generator, hydraulic pump, oil pump and tachometer generator.

Dimensions:
- Diameter: 985 mm (38·78 in)
- Length: 1,145 mm (45·08 in)

Weight, Dry: 245 kg (540 lb)

Performance Ratings:
- T-O: 242 kW; 325 hp at 2,800 rpm
- Max continuous I: 205 kW; 275 hp at 2,450 rpm
- Max continuous II: 142 kW; 190 hp at 2,350 rpm
- Cruise I: 142 kW; 190 hp at 2,350 rpm
- Cruise II: 108 kW; 145 hp at 2,350 rpm

Specific Fuel Consumption:
- At cruise ratings: 77·7 μg/J (0·46 lb/h/hp)

VEDENEEV M-14P

With this engine Vedeneev reverted to the original fixed-wing application, apparently independently of the Ivchenko bureau. The M-14P is used with direct drive to a fixed-pitch two-blade propeller in the Moscow Aviation Institute OSKB-1-3PM and the Yak-18T.

In the former aircraft the T-O rating is given as 242 kW (325 hp), and in the Yak-18T as 269 kW (360 hp).

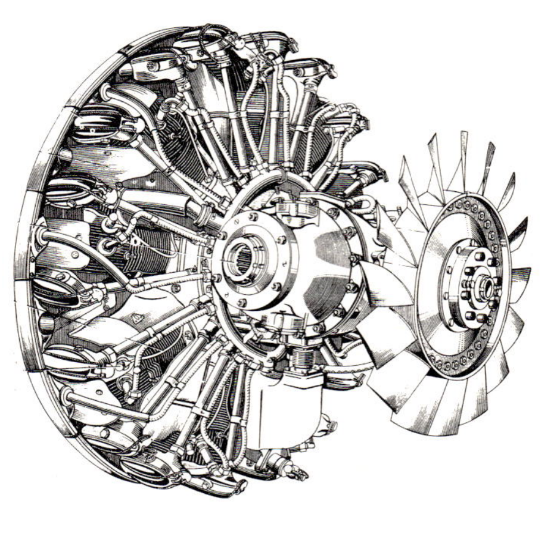

Vedeneev M-14V-26 radial piston engine, with cooling fan, for Kamov Ka-26 helicopter

ENGINES OF UNKNOWN DESIGN

1. RU 19-300. Auxiliary turbojet of 8·83 kN (1,985 lb st) for which the initial application was in the Antonov An-24/26/30 transports. Mounted in the rear of the starboard nacelle in place of the TG-16 APU, the RU 19-300 provides additional take-off thrust and also drives an integrally-mounted generator to relieve the aircraft's AI-24T turboprops of supplying electrical power during take-off. This arrangement increases the An-24RV's take-off performance under hot and high conditions, and improves single-engine handling and stability. After take-off, the auxiliary turbojet is shut down and the AI-24Ts are coupled mechanically to the engine-mounted generators. In this dual role the RU 19-300 provides 2·16 kN (485 lb st) for take-off. During flight the auxiliary turbojet is available for use as an APU. The version installed in the An-30 is designated RU 19A-300.

In 1976 it became known in the West that the RU 19 is used to power the La-17 target (see RPVs and Targets section in 1977-78 *Jane's*).

2. **Lift-Jet.** Soviet design bureaux conducted extensive research into jet lift from the late 1950s, and purpose-designed lift-jets were probably flying by 1962 (possibly much earlier). At the 1967 air show at Domodedovo three lift-jet V/STOL research aircraft were displayed. A MiG aircraft based on the MiG-21 had two lift-jets in the mid-fuselage, with a single large inlet door above. A larger MiG, with fuselage and tail closely similar to the eventual MiG-23 but with a small delta wing, had a similar arrangement, differing in detail design of the dorsal door. A Sukhoi aircraft, similar to the twin-engined Su-15, had three lift engines fed by two upper doors. All these installations had large open slots in the dorsal doors (which were hinged at the rear) and transverse louvres filling the ventral jet aperture (probably hinged and under pilot control to give variable forward thrust). Photographs of these aircraft last appeared in the 1971-72 *Jane's*. The two lift-jets used in the Yak-36MP VTOL aircraft carried on board the carriers *Kiev* and *Minsk* are probably developments of the same engines, though details are not yet known. They could be turbojets or turbofans. The inlet door has open louvres, and it is safe to assume pilot-controlled exit cascades. Thrust estimates for these engines are agreed at around 25 kN (5,600 lb) each.

3. **Lift/Cruise Engine.** The Yakovlev V/STOL research aircraft (NATO 'Freehand') demonstrated at the 1967 Domodedovo air show was powered by a twin-nozzle vectored-thrust propulsion system. The use of a large bifurcated nose intake and only two exhaust nozzles suggests that the installation may have used two standard turbojets or turbofans, each fitted with a single swivelling nozzle. On balance, however, it is likely that the 'Freehand' research aircraft used the same large single engine as the Yak-36MP combat aircraft carried on board the *Kiev* and *Minsk*. This is probably a turbofan, similar in essentials to the British Pegasus and Anglo-German RB.193 (used in the VAK 191B), but with the entire efflux discharged through left and right rear nozzles. The HP spool must be bled to provide air for the reaction control jets at the tail and wingtips. There is no afterburning in this engine, and estimates put its T-O thrust in the 78 kN (17,500 lb st) class.

UNITED KINGDOM

BAJ VICKERS
BAJ VICKERS LIMITED

Head Office and Works: Banwell, Weston super Mare, Avon BS24 8PD
Telephone: Banwell (0934) 82 2251

Directors:
- R. M. Howarth (Chairman)
- J. M. Harper (Managing Director)
- C. N. Davies
- M. A. Bowen
- R. Enticott
- A. C. Johnson
- Dr F. Llewellyn Smith, CBE

In 1952 a team was formed to meet the rocket-motor requirements of early British guided missile projects. This team constituted the nucleus of a company named Bristol Aerojet, formed in 1958 by the Bristol Aeroplane Company and the Aerojet-General Corporation of California. In March 1979 Bristol Aerojet became a wholly-owned subsidiary of Vickers Ltd, and was renamed as above.

BAJ Vickers maintains on a single site a complete range of technologies, processes and facilities for the design, development and production of conventional and advanced rocket motors. Much of the work is undertaken in collaboration with the Ministry of Defence Propellants, Explosives and Rocket Motor Establishment, for which the company undertakes applied research and provides hardware, design, development and manufacturing services.

In addition to solid-propellant motors, BAJ Vickers has, in collaboration with the MoD, developed a number of pre-filled sealed liquid-propellant motors. These are smoke-free and offer instant readiness after long storage, with a capability of complex thrust/time programmes on command, and with shutdown and re-start capability. The concept has been developed to production status.

BAJ Vickers has supplied production motors for the Bloodhound, Seaslug, Seacat, Tigercat, Rapier, Seawolf, Sea Skua, USD-501, Skylark, Skua, Petrel, Fulmar and INTA Flamenco, and for other flight and ground applications.

The company's rocket-motor manufacturing processes have been licenced to Breda Meccanica Bresciana in Italy and to Instituto Nacional de Tecnica Aerospacial (INTA) in Spain, and are used in Canada and Australia. Under an agreement with the British government the company is supplying British 'plastic' propellant manufacturing and filling technology and equipment to INTA.

BAJ Vickers produces rocket motors to customer requirements, undertakes launcher design, and provides a technical service for assembly and firing of experimental rockets at British and overseas ranges. The company also produces research rockets and ballistic targets.

Motor Name	Diam mm (in)	Length mm (in)	Burn Time (s)	Total impulse (kN-s)	Application
Bantam	125 (4·92)	1,473 (58)	33·6	52·0	Skua
Chick	68 (2·68)	554 (21·8)	0·18	4·45	Skua and Petrel boost
Cuckoo	431·8 (17)	1,312 (51·65)	4·1	360·0	Skylark boost
Goldfinch	431·8 (17)	2,224 (87·56)	3·6	701·0	Skylark boost
Lapwing	176·8 (6·96)	1,826 (71·9)	25·0	153·0	Petrel
Raven	431·8 (17)	5,232 (206)	30·0	1,510·0	Skylark
Siskin	141·2 (5·56)	600·0 (23·62)	3·58	16·5	
Waxwing	712 (28)	(spherical)	55·0	845·5	
Heron	257·3 (10·13)	3,364 (132·45)	3·1	418	Flamenco/Fulmar
Snipe	259·8 (10·23)	2,212 (87·1)	15·5	237	Flamenco/Fulmar

BAJ Vickers solid-propellant rocket motors for non-military purposes

BONNER
AERO BONNER LTD

Shoreham Airport, East Sussex BN4 5FJ
Telephone: 079 17 5764

This company has developed a watercooled V-6 piston engine for general aviation. In 1976 the first flight prototype was installed in a Chipmunk, flight testing of which began on 2 July 1979.

BONNER SUPER SAPPHIRE

This engine is expected to set a new level of lightweight, compact, quiet and smooth power for aircraft requiring engines in the 149 kW (200 hp) class. After exhaustive development, production began in 1979, about four-fifths of the engine being made at Shoreham.

Type: Six-cylinder four-stroke piston engine, turbocharged and with geared drive.

Cylinders: V-6 configuration. Watercooled. Swept volume 3,000 cc (183 cu in).

Induction: Garrett turbocharger. Pressure carburettor.

Ignition: Dual, with one magneto and electronic system.

Lubrication: Dry sump.

Propeller Drive: Reduction gear with quill shaft, 0·5 ratio.

Dimensions:

Length overall	991 mm (39·0 in)
Width overall	559 mm (22·0 in)
Height overall	609 mm (24·0 in)

Weight, Dry: 149 kg (328 lb)

Performance Rating:
Max T-O 149 kW (200 hp) at 5,500 rpm

Fuel Consumption:
At cruise (74·5 kW; 100 hp) 29·5 litres (6·5 Imp gal)/h

Looking down on a Bonner Super Sapphire, with turbocharger on the right

DOWTY ROTOL
DOWTY ROTOL LTD

Cheltenham Road East, Gloucester GL2 9QH
Telephone: 0452 712424

Known primarily as a producer of propellers and related rotary devices, Dowty Rotol has developed the propulsion system described below. Though *Jane's* does not include propellers as such, this new development—in the form of an integrated propulsion unit—is exceptional.

DOWTY DUCTED PROPULSOR

Starting with the premise that propeller-driven aircraft are too noisy to meet future noise legislation, and in many cases even existing limits (80 dBA flyover), Dowty began in 1972 to seek an answer. One is the propeller of increased diameter, turning at lower rpm and driven through a large gearbox. This was soon discarded as too costly, heavy and limited in noise reduction, and as generally inapplicable to existing general-aviation aircraft because of propeller-blade clearance and landing gear height. A much better answer was found to be the multi-blade ducted fan, looking superficially like a high bypass ratio turbofan, which can produce not only dramatic reduction in noise but also markedly increased aircraft performance.

In the Spring of 1976 a static test rig was run, driven by a 224 kW (300 hp) Continental IO-520. The fan had seven blades and a diameter of 1,219 mm (48·0 in). The target static thrust was 5·38 kN (1,200 lb st) and the achieved figure at first build was 5·56 kN (1,250 lb st). The target noise at 305 m (1,000 ft) was 65 dB; the achieved figure was 62 dB. Further development showed how to obtain substantial net thrust from the engine cooling air and exhaust, ducted to a propulsive nozzle at the rear of the nacelle. Very good engine cooling was achieved, despite the extra-tight cowling, and vibration was reduced well below that with a propeller.

In the Autumn of 1976 the Islander was selected as a suitable research and demonstrator aircraft, and Miles-Dufon at Shoreham handled the conversion of both engines (224 kW; 300 hp Avco Lycoming IO-540), each being lowered on a pylon below the wing and coupled directly to a fan with large spinner and seven aluminium variable-pitch blades running inside a specially profiled duct carried on six downstream flow-straightener vanes. At full power the 1,219 mm (48 in) fan has a tip speed of only 172 m (565 ft)/s, compared with 287 m (942 ft)/s for the original 2,032 mm (80·0 in) propeller. The reduction in noise is exceptional, and there is a gain in thrust. The Ducted Propulsor Islander flew for the first time on 10 June 1977, and has given remarkable demonstrations of quietness and improved performance in the hands of the test pilot, Neville Duke. Noise in the cabin and on the ground has been reduced from about 85 to 65 dB, far below any future environmentalist goal. There are many side benefits, such as reduced pollution from leaner mixture.

Close-up of Dowty Ducted Propulsor on the Islander

LEONIDES
HARKER & ASSOCIATES

40 Pont Street, London SW1X 0AD
Telephone: 01 584 6720, 01 584 4725

Formerly of Rolls-Royce, Mr Harker is trying to get the Alvis Leonides nine-cylinder geared aircooled radial into agricultural aviation. Over 3,000 h have been flown with the Leonides 125/7 of 418 kW (560 hp) in an Ag-Cat and a Rockwell Thrush Commander R in California, and two Ag-Cat B aircraft are certificated and in use. Approximately 300 used engines are available for use. The intention is to start new production, and negotiations to this end are in progress with companies in the USA and the UK.

Harker & Associates can offer overhauls, and hopes to offer new engines of the Leonides 531 (long-stroke 485 kW, 650 hp) type in 1983. Alvis is supplying technical assistance and spares. Plans are in hand to convert a Stearman and Air Tractor to Leonides power.

Installation of Leonides 125/7 in Ag-Cat (popularly called Leo-Cat)

LUCAS
LUCAS AEROSPACE LTD

Head Office: Shirley, Solihull, West Midlands B90 2JJ
Telephone: 021 744 8522
Director and General Manager: J. Blyth

Lucas Aerospace is supporting the propulsion needs of the MBLE Epervier RPV, in service with the Belgian Army. A description of its CT 3201 turbojet appeared in the 1980-81 *Jane's*.

NPT
NOEL PENNY TURBINES LTD

Siskin Drive, Toll Bar End, Coventry CV3 4FE
Telephone: 0203 301528
Telex: 312285 Penny G

Directors:
E. R. Ponsford (Chairman)
R. N. Penny (Managing Director)
The Earl of Minto
S. Penny (Secretary)

Noel Penny has over 30 years of small gas turbine experience, and derives a major part of its income from packaged research and development programmes for worldwide clients, across the entire spectrum of small gas turbines. Current projects include a turboprop, a single-shaft gas turbine for power generation, an advanced turbocharging system and a heat pump.

The company's range of products has been updated by the introduction of new designs, and no further manufacture of earlier products is envisaged except for special applications. An example is the NPT 401, which is still produced as an instructional turbojet. The current aerospace programme is concentrated on small, lightweight, low-cost, expendable turbojets for unmanned vehicles, notably the NPT 151 and NPT 301 described below.

NPT 151

Now in service, this low-cost turbojet has been developed for RPV propulsion. It can be adapted for other applications requiring extended life.

Type: Single-shaft turbojet.

Compressor: Centrifugal, aluminium alloy. Maximum airflow 1·2 kg (2·65 lb)/s. Pressure ratio 4·3.

Combustion Chamber: Straight-through annular with vaporising burners. Ignition by electrically-fired pyrotechnic cartridge.

Fuel System: LP single-speed pump mounted on compressor shaft. Control by pump mechanical governor and acceleration unit.

Fuel Grades: Jet A-1 or JP-5 preferred.

Turbine: Inward-radial flow, cast nickel alloy. Max inlet temperature 987°C.

Lubrication: Total-loss oil mist.

Mounting: Three brackets on front casing.

Starting: Windmilling, or by external air source on compressor.

Dimensions:

Diameter	275 mm (10·83 in)
Length	310 mm (12·20 in)

Noel Penny NPT 301 turbojet

WEIGHT, DRY: 20 kg (44 lb)
PERFORMANCE RATING: Max S/L 0·672 kN (151 lb st)
SPECIFIC FUEL CONSUMPTION: 35·4 mg/Ns (1·25 lb/h/lb st)

NPT 301

This simple expendable turbojet can be supplied at a range of outputs up to 2 kN (450 lb st). The following description refers to the definitive NPT 301.

TYPE: Single-shaft turbojet.
COMPRESSOR: Single-stage centrifugal. 1·85 kg (4·07 lb)/s. Pressure ratio 5·02.
COMBUSTION CHAMBER: Reverse-flow annular, with vaporising burners.
FUEL SYSTEM: LP pressurised-inlet electric pump. Electronic control.
FUEL GRADES: Kerosene, Jet A-1, JP-5 preferred.
TURBINE: Single-stage axial with integral blades. Cast nickel alloy.
ACCESSORY DRIVE: Alternator to customer's specification.
LUBRICATION: Total-loss oil mist.
STARTING: External air source on compressor.
DIMENSIONS:
Diameter 343 mm (13·5 in)
Length 577 mm (22·72 in)
WEIGHT, DRY: 34·02 kg (75 lb)
PERFORMANCE RATING (S/L): 1·334 kN (300 lb st)
SPECIFIC FUEL CONSUMPTION: 33·4 mg/Ns (1·18 lb/h/lb st)

ROLLASON
ROLLASON AIRCRAFT AND ENGINES LTD

Brighton, Hove and Worthing Joint Municipal Airport, Shoreham-by-Sea, Sussex BN4 5FJ
Telephone: 079 17 62680

In support of its manufacture (now ended) of the Druine Turbulent light aeroplane, Rollason Aircraft and Engines Ltd undertook the conversion of Ardem 4CO2 power plants for this aircraft, from motor car engines.

Rollason has developed versions of the Ardem engine with capacities of 1,500 or 1,600 cc. Any of these engines can be installed in Nipper aircraft. The Mk X is also used in the Australian Corby Starlet and other homebuilts. A version with single ignition and electric starter powers the Slingsby-built Motor-Falke.

ARDEM Mk X (1,500 cc)

The standard model has a compression ratio of 7·8 : 1 and gives 33·6 kW (45 hp) at 3,300 rpm for take-off, with a fuel consumption of 17 litres (3·75 Imp gallons)/h at max rating. In addition, a high-compression version is available, as follows:

CYLINDERS: Bore 83 mm (3·27 in). Stroke 69 mm (2·72 in). Cast steel barrels, light alloy heads. Compression ratio 8·5 : 1.
PISTONS: Aluminium alloy high-compression pistons, each with two compression rings and one scraper ring. Floating gudgeon pins.
CONNECTING RODS: White metal bearings in big-end. Bronze bearings in little-end.
CRANKSHAFT: Runs in four white metal bearings.
CRANKCASE: Magnesium case.
VALVE GEAR: Two valves per cylinder. Camshaft geared to crankshaft.
INDUCTION: Zenith 32 KL P10 carburettor.
FUEL GRADE: 80 octane.
IGNITION: Lucas SR4 magneto mounted below engine, with chain drive. Two Lodge LH spark plugs per cylinder.
LUBRICATION: Wet sump type, with single gear-type pump.
OIL SPECIFICATION: Shell W80.
PROPELLER DRIVE: Direct drive.
ACCESSORIES: SEV 46C fuel pump.
DIMENSIONS:
Length 426 mm (16·75 in)
Width 750 mm (29·50 in)
Height 559 mm (22·00 in)
PERFORMANCE RATING:
Max 39·5 kW (53 hp) at 3,600 rpm
FUEL CONSUMPTION:
At max rating 18·2 litres (4·0 Imp gallons)/h

ARDEM Mk XI (1,600 cc)

After extensive testing this engine was approved and is produced as the Ardem Mk XI. It differs from the Mk X in having cylinders of 85·5 mm (3·365 in) bore and dual ignition.

WEIGHT:
With accessories 71·6 kg (158 lb)
PERFORMANCE RATINGS:
Max 41 kW (55 hp) at 3,300 rpm
Cruise 26·5 kW (35·5 hp) at 2,500 rpm
FUEL CONSUMPTION:
At cruise rating 12·5 litres (2·75 Imp gallons)/h

ROLLS-ROYCE
ROLLS-ROYCE LIMITED

HEAD OFFICE: 65 Buckingham Gate, London SW1E 6AT
Telephone: 01 222 9020
MAIN LOCATIONS:
PO Box 31, Moor Lane, Derby DE2 8BJ
Telephone: 0332 42424
PO Box 3, Filton, Bristol BS12 7QE
Telephone: 0272 693871
Ansty, Coventry CV7 9JR
Telephone: 0203 613211
Leavesden, Watford WD2 7BZ
Telephone: 09273 74000
Scottish group of factories to the south of Glasgow
CHAIRMAN AND CHIEF EXECUTIVE: Lord McFadzean
MANAGING DIRECTORS:
D. A. Head (Operations)
D. J. Pepper (Commercial)
A. R. G. Raeburn (Planning and Administration)
BOARD MEMBERS:
Sir George Burton
Sir St John Elstub
S. L. Higginbottom
P. J. Molony
Sir Peter Thornton
Sir Neil Wheeler
R. T. Whitfield
COMPANY SECRETARY: A. Warrington

Rolls-Royce RB.211-524B three-shaft turbofan (222·4 kN; 50,000 lb st)

Rolls-Royce Ltd, which produces a range of gas turbines and ramjets, retains the experience in aircraft engines built up over more than 65 years by the predecessor companies that it incorporated. The company also represents British experience in lightweight high-power gas turbines for industrial and marine purposes, since such engines were first derived from aircraft gas turbines.

More than 238 million hours of operating experience have been accumulated with Rolls-Royce civil and military gas turbines, which are used by 258 airlines and 103 armed forces.

In addition to the products designed, developed and manufactured solely in Britain, the company works with partners abroad on a number of joint civil and military aircraft engine programmes. Licences for the manufacture of Rolls-Royce engines or components are also held by many countries throughout the world.

The main activities are at Derby, Glasgow, Bristol, Coventry and Leavesden, where aircraft gas turbines are produced, and at Ansty, where aircraft gas turbine techniques are applied to industrial and marine uses. Employees total 60,700.

Rolls-Royce engines in commercial service include the RB.211, Spey and Conway turbofans, Avon, Viper and RB.162 turbojets, and Dart and Tyne turboprops. New civil engines include the RB.211-535 and RB.401. Military engines include the Pegasus vectored-thrust turbofan and Odin ramjet, as well as versions of the Spey, Conway, Avon, Viper, Dart and Tyne. Helicopter engines include the Gem, Gnome and (with Turboméca) Turmo and Astazou. Under the International heading in this section are described the RJ.500 (RJAE), RB.199 (Turbo-Union), Adour and RTM.321 (with Turboméca) and TF41 (with Detroit Diesel Allison).

Versions of several Rolls-Royce aircraft gas turbines are used as power units for large and small ships, hydrofoils, air cushion vehicles, electricity-generating sets, gas and oil pumping equipment and for other industrial uses.

ROLLS-ROYCE TURBOMÉCA ADOUR

This turbofan was designed by Rolls-Royce and Turboméca. It is described in the International part of this section.

ROLLS-ROYCE RB.199

This advanced augmented turbofan is the power unit for the Panavia Tornado. The RB.199 programme is managed by Turbo-Union (see International part of this section).

ROLLS-ROYCE RB.211

The RB.211 is a family of advanced technology three-shaft turbofans of high bypass ratio and high pressure ratio, with thrusts so far announced from 166·4 kN (37,400 lb) to 249 kN (56,000 lb). The engine was selected by Lockheed in March 1968 to power the L-1011 TriStar, and later by Boeing as an alternative option on the 747. In August 1978 the lowest-thrust version was selected as launch engine for the Boeing 757. In June 1979 Rolls-Royce and Airbus Industrie agreed to initiate a programme to develop versions of the A300 and A310 powered by RB.211 engines.

Rolls-Royce initiated design studies of three-shaft turbofans in 1961 and a twin-spool engine, the RB.178, was tested in 1967 to provide relevant component and gas generator experience. Among the advantages afforded by a three-shaft layout are its ready use of a high pressure ratio with fewer compressor and turbine stages while maintaining excellent handling. The need for compressor variable stator mechanisms can also be minimised and the rotating assemblies can be made relatively short and rigid while preserving light construction. As a result the RB.211 has demonstrated outstandingly low seal and aerofoil wear, thus maintaining a high level of performance throughout engine life.

For all announced applications Rolls-Royce retains responsibility for the complete propulsion system, comprising the engine, fan airflow reverser, pod cowlings and related systems, and noise attenuation for the intake, fan cowl and turbine exhaust duct.

The engine is divided into seven modules. This permits rapid change of engine parts, and enables service life to be set up individually for each module. It also facilitates rapid repair, as a damaged or time-expired module can be replaced with the engine installed in the aircraft. Maximum provision is made for in-service monitoring of engine condition and visual inspection of all engine sections.

The RB.211 combustion chamber is of annular design, giving significant advantages over tubo-annular systems in terms of reduced cost, weight and length, and improved efficiency. The reduced length makes a two-bearing HP system possible, with both bearings located away from the high temperatures of the combustion area. Detailed design has been aimed at reducing noise and exhaust contaminants to a minimum.

The **RB.211-22B,** the standard engine of the L-1011-1 and -100 TriStar, is flat rated at 187 kN (42,000 lb st) to 28·9°C. This engine was certificated in February 1973 by the CAA and in April 1973 by the FAA. By the beginning of 1981 more than 585 engines had been delivered, and engine flight hours in service were in excess of 7 million.

The **RB.211-524** series of engines was developed from the original RB.211-22B and now covers a range of thrust from 222·4 kN (50,000 lb) to 236 kN (53,000 lb), with future development up to and beyond 267 kN (60,000 lb). The -524 entered airline service in 1977 with the L-1011 and 747. By the beginning of 1981 more than 580 engines had been ordered, of which over 325 had been delivered and service experience of nearly 1 million hours achieved.

The **RB.211-524B**, currently fitted to the L-1011-200, L-1011-500 and 747, is certificated at 222·4 kN (50,000 lb) to 28·9°C for all aircraft. Further developments at the same thrust but offering better fuel consumption are available; the first of these was the **RB.211-524B3**, giving a 3·5 per cent improvement over the -524B, and which entered service in May 1980. Next was the **RB.211-524B4**, which offers up to 4·8 per cent better sfc than the -524B and entered service in February 1981.

The **RB.211-524C2** is a 229 kN (51,500 lb) engine for the 747 which entered service in April 1980. This engine offers increased thrust ratings with no change in sfc.

The **RB.211-524D4** is an improved engine rated at 236 kN (53,000 lb) for the 747 and other aircraft, and offering a 4·8 per cent better fuel consumption relative to the current 747 service engine. Certification of the -524D4 in March 1981 was to be followed by entry into service in November 1981.

The **RB.211-524G** is an improved engine scheduled for certification in 1984-85 and suitable for application to the A300, 767 or heavier-weight 747. It will be rated at 249 kN (56,000 lb), with potential for development up to 267 kN (60,000 lb), and offers a further sfc improvement of up to 4 per cent over the -524D4.

The **RB.211-535C** is rated at 166 kN (37,400 lb) for the 757. In this application, a policy of flexible rating means that the -535C will normally operate with significant derate. It has the HP module of the -22B, a six-stage IP compressor and scaled-down version of the 33-blade -524B fan. Fan airflow is 25 per cent lower than that of the -22B and core airflow 16 per cent lower. The engine runs at lower temperatures, pressures and velocities than the -22B, resulting in low noise. Orders for the -535C totalled 173 at the beginning of 1981.

The latest of the -535 series, the **RB.211-535E4**, is derived from the -535C, incorporating the latest RB.211 advanced technology. The E4 will have a fuel burn up to 8 per cent better than the -535C, together with an increase in take-off thrust to 176 kN (39,600 lb). These improvements have been achieved by the introduction of a new wide-chord hollow fan blade without snubbers (22 per engine), together with improved core engine aerodynamics and a common nozzle exhaust assembly.

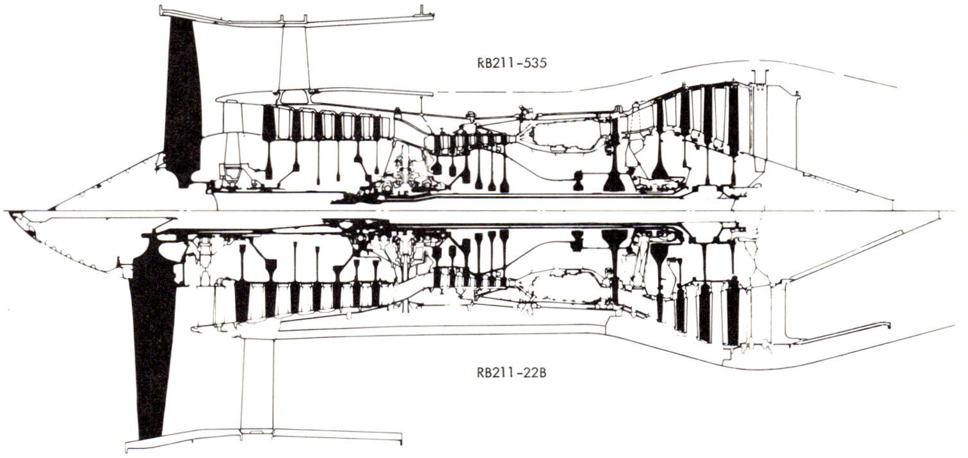

Comparative cross-sections of Rolls-Royce RB.211-22B and RB.211-535C turbofans

Scheduled for certification in late 1983 for 757 (initial orders Eastern and Monarch) and other aircraft.

The following description relates to the RB.211-22B:

TYPE: Three-shaft axial turbofan.

AIR INTAKE: Forward-facing pitot.

LP FAN: Single-stage overhung fan driven by LP turbine, the whole rotor assembly being supported on three bearings. Front bearing is large roller, squeeze-film supported behind fan. Axial location of rotor is by intershaft ball bearing in rear end of IP compressor drum. LP turbine supported on roller bearing, squeeze-film mounted in exhaust cone panel. Rotating spinner supported from fan rotor disc and hot-air anti-iced via central feed-tube within shaft. Titanium alloy used for 33 fan rotor blades, and steel for 70 fan outlet guide vanes. Titanium fan disc bolted with curvic coupling to LP shaft. Aluminium fan casing. Total fan airflow (T-O rating), 626 kg (1,380 lb)/s (-524, 657·7 kg; 1,450 lb). Bypass ratio 5 : 1 (-524, 4·4 : 1.)

IP COMPRESSOR: Seven-stage compressor rotor driven by IP turbine and supported on three bearings located directly in support panels. Front squeeze-film bearing is roller. Mid bearing at rear of IP compressor is ball bearing providing axial location for IP rotor. Rear bearing is roller, squeeze-film supported in panel between HP and IP turbines. Two drums, one of titanium discs welded together and the other of welded steel discs, are bolted to form one rotor, carrying titanium rotor blades. Aluminium and steel casings carry aluminium and steel stator blades. Single-stage titanium variable inlet guide vanes.

HP COMPRESSOR: Six-stage compressor rotor driven by HP turbine connected by large-diameter shaft and carried on ball location bearing at front and roller bearing mounted at the rear in panel behind HP turbine disc. Welded titanium discs, a single steel disc and welded nickel alloy discs are bolted together to form the rotor, carrying titanium, steel and nickel alloy blades. Steel casing carries steel and Nimonic stator blades. Overall pressure ratio 25 : 1.

COMBUSTION CHAMBER: Fully annular, with steel outer casings and Nimonic combustor. Downstream fuel injection by 18 airspray burners with annular atomisers. Ignition by high-energy igniter plugs in Nos 8 and 12 burners.

HP TURBINE: Single-stage axial unit with nozzle guide vanes and directionally solidified cast rotor blades, both convection- and film-cooled. Blades mounted in Nimonic disc by fir-tree roots.

IP TURBINE: Single-stage axial unit with Nimonic nozzle guide vanes and Nimonic rotor blades. NGVs aircooled. Rotor blades fir-tree mounted in Nimonic disc.

LP TURBINE: Three-stage axial unit with Nimonic rotor blades fir-tree mounted in steel discs.

JET PIPE: Steel jet pipe without spoiler.

ACCESSORY DRIVES: Radial drive from HP shaft to gearbox on fan casing. Accessories driven include integrated-drive generator and aircraft hydraulic pumps.

LUBRICATION SYSTEM: Continuous circulation 'dry sump' system with single gear-type pressure pump and multiple gear-type scavenge pumps. Oil tank 21 litres (37 Imp pints) capacity integral with gearbox.

MOUNTING: Two-point mounting system. Front mount on fan casing takes thrust, vertical and side-loads. Rear link mount on exhaust casing takes torsional, side and vertical loads. Both mounts are fail-safe and allow for carcase expansion.

DIMENSIONS:

Length overall	3,033 mm (119·4 in)
Intake diameter	2,182 mm (85·9 in)

WEIGHT, DRY:

RB.211-22B	4,171 kg (9,195 lb)
RB.211-524B4	4,452 kg (9,814 lb)

PERFORMANCE RATINGS:

T-O, see model listings

Cruise at 10,670 m (35,000 ft) and Mach 0·85:

RB.211-22B	43·9 kN (9,875 lb st)
RB.211-524B4	48·9 kN (11,000 lb st)

SPECIFIC FUEL CONSUMPTION:

At cruise rating, as above:

RB.211-22B	17·7 mg/Ns (0·625 lb/h/lb st)
RB.211-524B4	17·6 mg/Ns (0·620 lb/h/lb st)

ROLLS-ROYCE RJ.500

This turbofan in the 'ten-tonne' class is described under RJAE in the International section.

ROLLS-ROYCE RB.163 CIVIL SPEY

Design of the Spey RB.163 began in September 1959, and the first engine ran at the end of December 1960. Civil Speys are in service in the Trident, BAe One-Eleven, Grumman Gulfstream II and F28 Fellowship, and in the C-8A augmentor-wing research aircraft of NASA. In 1978 the Spey was selected for the Gulfstream III which, like current One-Elevens, has noise-suppressing features; it is also subject of a major collaborative production programme between Rolls-Royce and the government of Romania.

The following versions of the civil Spey are in service:

Mk 505-5F. T-O rating of 44·6 kN (10,050 lb st) at 12,150 rpm, for Trident 1C.

Mk 506-14. T-O rating of 46·3 kN (10,410 lb st) at 12,530 rpm, for One-Eleven.

Mk 506-14A. As 506-14 with smaller capacity turbine.

Mk 511-8 and 511-14. T-O rating of 50·7 kN (11,400 lb st) at 12,390 rpm. Mk 511-5 for Trident, Mk 511-8 for Gulfstream II and III, and Mk 511-14 for One-Eleven.

Mk 511-5W and 511-14W. As 511-5 and 511-14 but with water injection to maintain rating to 35°C.

Mk 512-5 and 512-14. T-O rating 53·0 kN (11,930 lb st) in Mk 512-5/50 for Trident 3B, Mk 512-5W and 5W/50 at same rating for Trident 2E, Mk 512-14E at 55·26 kN (12,420 lb st) for One-Eleven, and Mk 512DW at 55·8 kN (12,550 lb st), all at 12,640 rpm.

Mk 555-15. Lightened and simplified version for F28 Fellowship. Has changed company project number of **RB.183.** Current production version is the **555-15H**, flat rated at 44·0 kN (9,900 lb) to 29·7°C; on a 40°C day it gives 6 per cent higher T-O thrust than the 555-15.

Mk 555-15N and **-15P** are the -15 and -15H with improved internal silencing and ten-lobe mixer.

Rolls-Royce Spey 555 simplified turbofan rated at 44·0 kN (9,900 lb st)

The military versions are described separately.

The following details refer specifically to the Spey Mk 512-14DW, as fitted to the BAe One-Eleven Series 500, except where indicated:

TYPE: Two-spool axial-flow turbofan engine.

AIR INTAKE: Annular, with bleed air thermal anti-icing.

COMPRESSOR: Two spools. Five-stage (four-stage on Mks 505, 506 and 555) low-pressure (LP) and 12-stage high-pressure (HP). First-stage HP stator vanes variable-incidence. LP compressor steel drum type, pinned to shaft. HP compressor is of the steel disc type, first stage bolted to shaft, remaining stages splined. HP stator blades steel; LP stators aluminium. LP rotor blades aluminium (Mk 512 titanium 1st stage); HP blades titanium. Stators slotted into casing; rotor blades pinned or dovetailed. LP casing two-piece magnesium. HP casing steel. Pressure ratio 21·2 : 1 (15·0 on Mk 505 and 555, 17·2 on Mk 506, 18·9 on Mk 510 and 511). Air mass flow 94·4 kg (208 lb)/s (90·27 kg; 200 lb on Mk 505 and 555, 92 kg; 203 lb on Mk 506, 92·5 kg; 204 lb on Mk 510 and 511). Bypass ratio 0·64 : 1 (1·0 on Mk 505, 555 and 506).

COMBUSTION CHAMBER: Tubo-annular with 10 Nimonic sheet liners. Duplex downstream burners, one per chamber. High energy igniters in chambers 4 and 8.

FUEL SYSTEM: Plessey LP pump feeding through fuel-cooled oil cooler and Marston Excelsior fuel heater to LP filter at inlet to Lucas GD pump. HP metered by Lucas regulator, embodying combined speed and acceleration control and fed through Lucas LP governor and shut-off valve to Duple spray nozzles. Maximum pressure 124 bars (1,800 lb/sq in).

FUEL GRADE: DERD.2482 or 2486.

WATER INJECTION SYSTEM: (engines bearing 'W' suffix): Water supplied by Lucas air turbopump through engine-mounted automatic shut-off valve to injector passages in fuel spray nozzles (water sprays into primary airflow through flame tube swirlers).

NOZZLE GUIDE VANES: Hollow cast in nickel-based alloy. HP aircooled.

TURBINES: Two two-stage. First HP aircooled. HP discs nickel-based alloy, bolted to shaft (HP discs steel on Mks 505, 506 and 555). LP discs creep-resisting ferritic steel. Nickel-based alloy blades attached by fir-tree roots.

BEARINGS: LP compressor supported in roller bearings, plus ball thrust bearing. HP compressor has front roller bearing and ball thrust bearing. Turbine bearings all roller type flexibly mounted.

JET PIPE: Fixed-area stainless steel sheet.

REVERSER AND SUPPRESSOR: Normally internal clamshell (Gulfstream II target type, not Rolls-Royce supplied, and F28 has no reverser). Five- or six-chute silencing nozzles available.

ACCESSORY DRIVES: Port gearbox, driven from LP rotor, carries LP governor and LP tacho. Starboard gearbox, driven from HP rotor, carries LP and HP fuel pumps, fuel regulator, main oil pumps, airflow control rpm signal transmitter, starter and HP tacho. Provision in starboard gearbox for aircraft ancillaries.

LUBRICATION SYSTEM: Self-contained continuous circulation. Single pressure pump feeds oil from tank through fuel-cooled cooler and HP filter to gearboxes and shaft bearings. Five main scavenge pumps. Tank capacity 6·8 litres (12 Imp pints). Usable oil 5·1 litres (9 Imp pints). Normal pressure 2·41-3·45 bars (35-50 lb/sq in).

OIL SPECIFICATION: DERD.2487.

MOUNTING: Two trunnions, two saddle mountings and one rear mounting.

STARTING: Plessey 220 air-turbine starter. Rotax alternative on Mks 505, 511 and 512; Garrett on Mk 555.

DIMENSIONS:
Length, less tailpipe:
- Mk 505, 506, 555 — 2,795 mm (110·0 in)
- Mk 510, 511 — 2,911 mm (114·6 in)

Diameter:
- Mk 505, 506, 555 — 940 mm (37·0 in)
- Mk 510, 511 — 942 mm (37·1 in)

WEIGHT, DRY:

Mk 505-5	998 kg (2,200 lb)
Mk 506-14	1,024 kg (2,257 lb)
Mk 506-14AW	1,038 kg (2,288 lb)
Mk 510-5, 511-5	1,049 kg (2,312 lb)
Mk 510-14, 511-14	1,058 kg (2,332 lb)
Mk 510-14W, 511-14W	1,188 kg (2,621 lb)
Mk 511-5W	1,050 kg (2,317 lb)
Mk 512	1,168 kg (2,574 lb)
Mk 555-15	1,008 kg (2,222 lb)

PERFORMANCE RATINGS:
Max T-O: See under series descriptions
Max continuous:
- Mk 505 — 42 kN (9,450 lb st) at 12,260 rpm
- Mk 506 — 44·4 kN (9,990 lb st) at 12,385 rpm
- Mk 511 — 48·7 kN (10,940 lb st) at 12,240 rpm
- Mk 512, Mk 512DW — 51·5 kN (11,580 lb st) at 12,450 rpm
- Mk 555-15 — 42·1 kN (9,470 lb st) at 11,900 rpm

Typical cruise rating at 450 knots (834 km/h; 518 mph) at 9,750 m (32,000 ft):
- All versions — 13·7 kN (3,070 lb st)

SPECIFIC FUEL CONSUMPTION:
At T-O rating:
- Mk 505 — 15·9 mg/Ns (0·560 lb/h/lb st)
- Mk 506 — 15·95 mg/Ns (0·563 lb/h/lb st)
- Mk 511 — 17·3 mg/Ns (0·612 lb/h/lb st)
- Mk 555-15 — 15·9 mg/Ns (0·560 lb/h/lb st)

At typical cruise rating:
- Mks 505, 506 — 21·5 mg/Ns (0·760 lb/h/lb st)
- Mk 510, 511 — 22·3 mg/Ns (0·790 lb/h/lb st)
- Mks 512, 555 — 22·7 mg/Ns (0·800 lb/h/lb st)

OIL CONSUMPTION:
- Max (all Marks) — 0·42 litres (0·75 Imp pints)/h

ROLLS-ROYCE RB.168 MILITARY SPEY

The military Spey RB.168 incorporates modifications to meet higher-duty conditions.

Design of the RB.168-1, **Mk 101**, started in November 1960; this mark powers the Hawker Siddeley Buccaneer strike aircraft.

The RB.168-25R (**Mks 202/3**) supersonic engine with afterburner has a dry rating of 54·5 kN (12,250 lb) plus a 67 per cent static augmentation and powers the McDonnell Douglas Phantom FG.1 and FGR.2. Major change is the introduction of a robust shaft-and-disc LP compressor. As with Mk 101, use is made of HP compressor bleed air for aircraft BLC purposes. A Plessey gas-turbine starter is fitted. Augmentation is thrust-modulating from an initial boost, at sea level static, of ten per cent.

Supersonic augmented Spey engines were supplied to China under the terms of a contract signed in 1975, which also included licence agreement for production of engines to the same standard. Commensurate with this, pre-production engines have been manufactured and tested in China.

In 1978 the Spey was selected by Aeritalia for the AM-X aircraft. The engine chosen is the unaugmented **Mk 807**, rated at 49·0 kN (11,030 lb), derived from the Mk 101 and civil Mk 555. Certification was due in 1981. It is anticipated that production engines will be produced under licence (see Alfa Romeo and Piaggio).

The RB.168-20 **Mk 250**, based closely on the commercial Spey, powers the British Aerospace Nimrod. Embodying extensive anti-corrosion treatment, it provides a higher thrust than its civil counterpart, through operation at higher rpm and higher turbine entry temperature. Provision is made for driving a large alternator; for the Nimrod AEW.3 a modified gearbox drives further-enlarged generators.

DIMENSIONS:
- Diameter — 825 mm (32·5 in)

Length:
- Mks 101, 250 — 2,985 mm (117·5 in)
- Mks 202, 203 — 5,204 mm (204·9 in)

WEIGHT, DRY:
- Mk 101 — 1,181 kg (2,603 lb)
- Mk 250 — 1,227 kg (2,704 lb)
- Mks 202, 203 — 1,857 kg (4,093 lb)

PERFORMANCE RATING:
Max T-O:
- Mks 101, 807 — 49·0 kN (11,030 lb st)
- Mks 202, 203 — 91·25 kN (20,515 lb st)
- Mk 250 — 53·4 kN (11,995 lb st)

ROLLS-ROYCE/ALLISON SPEY TF41

Versions of the Spey produced jointly by Rolls-Royce and the Detroit Diesel Allison Division of General Motors are described in the International section.

ROLLS-ROYCE TYNE

The 4,570 ekW (6,100 ehp) Tyne 22 two-shaft turboprop has been put back into production for the additional Transall C-160 aircraft to be built by Aérospatiale. Like engines for previous Transall aircraft they are made by a consortium comprising Rolls-Royce, SNECMA, MTU and FN. Brief details are given under SNECMA.

Limited production is also being undertaken of the Tyne 20 Mk 801, derated to 4,090 ekW (5,480 ehp) for an export version of the Aeritalia G222.

ROLLS-ROYCE DART

Beginning life in 1945 at 738 kW (990 hp), this classic turboprop was developed to give 2,420 ekW (3,245 ehp) in military and 2,256 kW (3,025 ehp) in civil versions. Current production of new engines is centred on the less powerful **Mk 536-2** for the BAe 748 and **Mk 536-7R** for the Fokker F27. About 7,000 Dart engines have been delivered, and demand remains strong.

TYPE: Single-shaft centrifugal-flow turboprop engine.

REDUCTION GEAR: Double helical high-speed train and final helical drive. Ratio in current engines: 0·093.

AIR INTAKE: Circular intake with annular duct leading to impeller eye of first-stage compressor. Oil tank cast integral with casing. Secondary air intake supplies air to oil cooler.

COMPRESSOR: Two-stage centrifugal. Each impeller has nineteen vanes and steel rotating guide vanes. Mass air flow at maximum rpm typically 10·66 kg (23·5 lb)/s at 5·62 : 1 pressure ratio.

COMBUSTION CHAMBERS: Seven inclined flame tubes with atomisers for downstream injection. High-energy igniter plugs in Nos. 3 and 7 chambers.

FUEL SYSTEM: Variable-stroke pump delivers fuel to burners through flow control unit, which incorporates a filter, throttle valve, shut-off cock and barometric pressure control. Automatically-progressive injection of water/methanol to maintain take-off power under high ambient temperature. System linked with throttle lever to prohibit use except at take-off rpm. Fuel filter de-icing by hot air from compressor.

TURBINE: Three-stage axial.

EXHAUST UNIT: Inclined to suit installation.

ACCESSORY DRIVES: Gearbox drive from main-shaft centre-coupling immediately behind compressor.

LUBRICATION: Integral oil tank (14 litres; 25 Imp pints) feeds via standpipe and feathering pump through tank base, to ensure feathering possible even after prolonged system oil leak. Gear pump supplies oil to all bearings and reduction-gear jets.

CONTROLS: Throttle interconnected with propeller controller and high-pressure cock linked with feathering controls. Propeller feathered by moving shut-off cock past closed position.

MOUNTING: Four feet at 90° on compressor casing, although only three need be used.

DIMENSIONS:
- Length (no jetpipe) — 2,496 mm (98·27 in)
- Diameter — 963 mm (37·9 in)

WEIGHT, DRY: 639 kg (1,409 lb)

PERFORMANCE RATINGS:
- T-O — 1,551 kW (2,080 shp); 1,700 ekW (2,280 ehp) at 15,000 rpm
- Recommended cruise (250 knots; 463 km/h; 288 mph at 6,100 m; 20,000 ft) — 910 kW (1,220 shp)

ROLLS-ROYCE/ALFA ROMEO RB.318

This engine in the 448 kW (600 shp) class for general aviation is described briefly under the Alfa Romeo heading in the Italian part of this section.

ROLLS-ROYCE VIPER

This turbojet remains in production for civil and military customers. More than 5,500 Vipers have been ordered by 41 countries, 29 of which have chosen the engine for trainer and light attack aircraft.

The 1,700 ekW (2,280 ehp) Rolls-Royce Dart 535 single-shaft turboprop, typical of later versions

Current versions are as follows:

Viper 11 (Mk 200 Series). Single-shaft seven-stage axial-flow compressor driven by single-stage turbine. Air mass flow 20 kg/s (44 lb/s). Type-tested at 11·12 kN (2,500 lb st) and powers Jindivik Mk 3 drone, BAC Jet Provost T.4 and 5, Yugoslav Soko Galeb and Hindustan HJT-16 Kiran trainers.

A Viper 11 version, the 22-1, was built under licence in Italy by Piaggio for the Aermacchi M.B.326 trainer and by Atlas of South Africa and Commonwealth Aircraft of Australia for local versions of that aircraft.

Viper 500 Series. Development with increased airflow, achieved by zero stage on compressor. Major applications early HS 125 (Mks 521, 522) and PD-808 executive aircraft (Mk 526) and BAC Strikemaster (Mk 535), Aermacchi M.B.326GB (Mk 540) and Soko Jastreb (Mk 531) training and light combat aircraft. Mk 540 built under licence by Piaggio and Atlas.

Viper 600 Series. Eight-stage axial-flow compressor driven by two-stage turbine; annular vaporising combustion chamber. Take-off rating 16·7 kN (3,750 lb st) civil and 17·8 kN (4,000 lb st) military. Agreement signed with Fiat (Italy) in July 1969 for technical collaboration in design, development and production (see Fiat entry).

The civil Viper 601 powers the HS 125-600, and the military Viper 632 is fitted to the Aermacchi M.B.326K and M.B.339. An engine similar to the 632 powers the Soko/CNIAR Orao/IAR-93. The Viper 632 is now built under licence in Italy, Romania and Yugoslavia. The Viper 633 has an afterburner of the three-gutter type, with hot-streak ignition and a two-position nozzle. Thrust with maximum augmentation is 22·3 kN (5,000 lb st).

The following details apply to the Viper 600 series:

TYPE: Single-shaft axial turbojet.

AIR INTAKE: Direct pitot. Anti-icing by hot compressor air. No inlet guide vanes.

COMPRESSOR: Eight-stage. Steel drum-type rotor with disc assemblies. Magnesium alloy casing with blow-off valve. Stator blades mounted in carrier rings slotted into casing. All stator blades and 1st, 2nd and 8th stage rotor blades of steel; remainder aluminium alloy. Zero-stage and first-stage rotor blades attached by fir-tree roots; stages 3-8 riveted. Pressure ratio 5·8 : 1. Air mass flow 26·5 kg (58·4 lb)/s.

COMBUSTION CHAMBER: Short annular type with 24 vaporising burners and six starting atomisers. Electric ignition.

FUEL SYSTEM: Hydromechanical, consisting primarily of fuel pump, barometric fuel control and air/fuel ratio control.

FUEL GRADE: JP-1 or JP-4.

TURBINE: Two-stage axial. Shrouded blades attached to discs by fir-tree roots and locking strips. Discs attached by Hirth couplings.

BEARINGS: Ball-thrust type at forward end of compressor, roller bearings at centre section and at rear end of combustion chamber inner casing.

JET PIPE: Cone of heat-resisting steel rings butt-welded together. (Viper 601: eight-lobe convoluted nozzle to meet FAR Pt 36 noise requirements).

ACCESSORY DRIVES: Gearbox driven from front of compressor by bevel gear.

LUBRICATION SYSTEM: Self-contained. Recirculatory system supplying front bearing and gearbox, metered feed supplied to centre and rear bearings by micro-pumps. Military version fully aerobatic.

OIL SPECIFICATION: Mobil Jet 2, Shell ASTO 500 and Castrol 580.

MOUNTING: Civil: cantilevered, side mounted, single spherical bearing in centre-section casing, with top and bottom links and attachment at intake casing. Military: trunnion mounted at centre-section with additional support at intake casing.

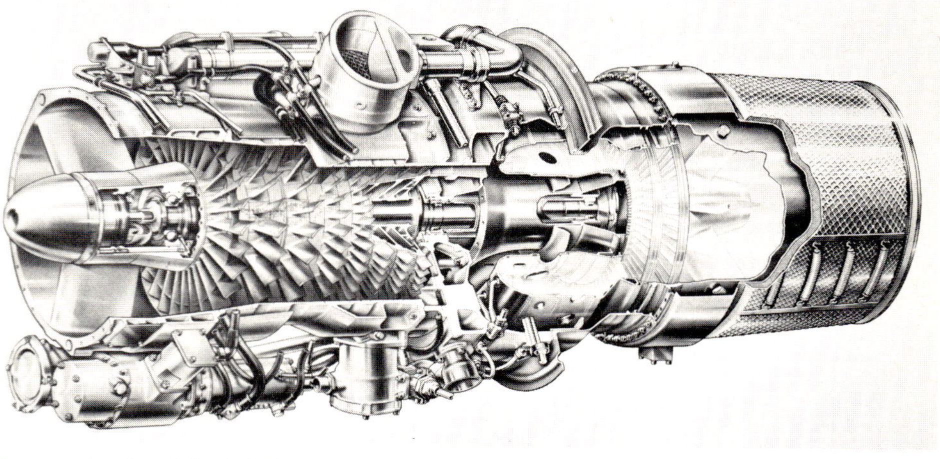

Cutaway drawing of the Rolls-Royce Viper 632 single-shaft turbojet (take-off rating 17·8 kN; 4,000 lb st)

STARTING: 24V starter/generator.

DIMENSIONS:	
Max casing diameter:	
All versions	624 mm (24·55 in)
Length (flange to flange):	
Viper 11	1,626 mm (64·0 in)
Viper 531, 535, 540, 632	1,806 mm (71·1 in)
Viper 601 (plus jet pipe)	2,270 mm (89·4 in)
WEIGHT, DRY:	
Viper 11	284 kg (625 lb)
Viper 531	345 kg (760 lb)
Viper 535	331 kg (730 lb)
Viper 540	342 kg (755 lb)
Viper 601, 632	358 kg (790 lb)
PERFORMANCE RATINGS:	
T-O:	
Viper 11	11·12 kN (2,500 lb st)
Viper 531	13·9 kN (3,120 lb st)
Viper 535, 540	14·9 kN (3,360 lb st)
Viper 601	16·7 kN (3,750 lb st)
Viper 632	17·8 kN (4,000 lb st)
SPECIFIC FUEL CONSUMPTION:	
Viper 11	30·3 mg/Ns (1·07 lb/h/lb st)
Viper 500 series	28·3 mg/Ns (1·00 lb/h/lb st)
Viper 601	26·6 mg/Ns (0·94 lb/h/lb st)
Viper 632	27·5 mg/Ns (0·97 lb/h/lb st)
OIL CONSUMPTION:	
All versions	0·57 litres (1 Imp pint)/h

ROLLS-ROYCE RB.401

The RB.401 is a two-shaft, medium bypass ratio turbofan. The 24·5 kN (5,500 lb) thrust class was chosen to meet the requirements of the next generation of business jets and military trainer/light attack aircraft. Increased pressure ratio, reduced bypass ratio and turboshaft versions have also been studied for specific applications.

A demonstrator engine, the RB.401-06, ran on the testbed on 21 December 1975. The production-standard engine for the business-aviation market is the RB.401-07. This engine first ran in November 1977, exceeding its T-O rating after six hours of testing, with fuel flow within 2·0 per cent of brochure specification.

The RB.401 incorporates components already proven in company research programmes, with technology which reads across from the RB.211 and other advanced commercial engines. It has been designed to meet all foreseeable requirements regarding noise and exhaust pollution, in addition to having an extremely low specific fuel consumption.

The following data apply to the RB.401-07:

TYPE: Two-shaft turbofan.

FAN: Single-stage axial, based on latest RB.211 technology. Low aspect ratio titanium blades without snubbers. Maximum airflow 82·5 kg (182 lb)/s. Bypass ratio 4·2 : 1.

BY-PASS DUCT: Full-length, giving installational simplicity with optimum overall nacelle performance.

INTERMEDIATE CASING: One-piece magnesium casting incorporating accessory-gearbox drive.

HP COMPRESSOR: Eight-stage axial. Titanium rotor blades and inlet guide vanes and nickel-alloy stator blades. Inlet guide vanes and first-stage stators have variable stagger. Pressure ratio of HP spool alone, 11·5 : 1.

COMBUSTION SYSTEM: Annular, with vaporising burners.

FUEL SYSTEM: Hydromechanical, with provision for electronic supervision of engine parameters.

TURBINES: Single-stage HP turbine, with aircooled rotor and stator blades. Two-stage LP turbine, with uncooled fully-shrouded blades.

LUBRICATION SYSTEM: Self-contained, including tank, pumps, filter, scavenge strainer, tubes, fittings, and cooler.

DIMENSIONS:	
Length, flange to flange	1,545 mm (60·8 in)
Diameter, fan casing	823 mm (32·4 in)
WEIGHT, DRY:	447 kg (985 lb)
PERFORMANCE RATINGS:	
T-O (flat-rated to 15°C):	
S/L, ISA	24·7 kN (5,540 lb st)
Cruise:	
Mach 0·7 and 12,000 m (40,000 ft) ISA	5·0 kN (1,130 lb st)
SPECIFIC FUEL CONSUMPTION:	
T-O	12·72 mg/Ns (0·449 lb/h/lb st)
Cruise (as above)	20·0 mg/Ns (0·707 lb/h/lb st)

ROLLS-ROYCE ORPHEUS

This single-spool turbojet was initiated in December 1953 as a private venture. The Orpheus 701 was type-tested in November 1956 at 20·1 kN (4,520 lb st) and is used in the Gnat fighter; a modified version powers the Indian HAL Ajeet. The Mk 703, rated at 21·6 kN (4,850 lb st), powers the HF-24 Marut. For all Indian applications the engine is built under licence at Bangalore by Hindustan Aeronautics Limited (HAL). A description appeared in the 1967-68 *Jane's*.

ROLLS-ROYCE PEGASUS

The Pegasus is a turbofan for V/STOL applications. It has two main rotating systems which are mechanically independent and rotate in opposite directions, thus minimising gyroscopic effects. It has a three-stage axial-flow fan of transonic design and an eight-stage high-pressure compressor, each driven by a two-stage turbine. Thrust vectoring is achieved by four rotatable nozzles simultaneously operated and symmetrically positioned on each side of the engine. The front nozzles discharge by-pass air, whilst the rear nozzles discharge the turbine efflux. The total thrust is divided between the four nozzles, and the resultant thrust passes through a fixed point irrespective of nozzle angle, thus minimising aircraft control problems. HP bleed air is used for aircraft stabilisation. The Pegasus was the first engine designed with an overhung fan, without inlet guide vanes, requiring no anti-icing.

By varying the angle of the nozzles, an aircraft powered by the Pegasus can take off vertically or with a short run or with a conventional long run. The reserve of power in the normal runway take-off allows for an appreciable increase in payload. The engine is stressed for operation up to Mach 2, ISA tropopause, at maximum thrust.

The Pegasus ran in August 1959, and flight trials in the Hawker Siddeley P.1127 prototypes began in October 1960. The **Pegasus 3,** which powered prototype P.1127 aircraft, was rated at 60 kN (13,500 lb st). The engine entered service in the Harrier with the RAF in 1969 and

Cutaway drawing of the Rolls-Royce RB.401-07 turbofan (24·7 kN; 5,540 lb st)

later with the AV-8A of the USMC and Matador of the Spanish Navy. These aircraft have the Pegasus 11 Mk 103, rated at 95·6 kN (21,500 lb st).

A maritime version of the Pegasus 11 is designated **Mk 104**. Developed for the Sea Harrier, it embodies material changes to the LP and intermediate casings, sacrificial protective coatings on ferrous components, and an increased-capacity gearbox. The Sea Harrier entered service with the Royal Navy in September 1979 and has been ordered by the Indian Navy for service from 1982.

A version with Rolls-Royce designation **11-21E** is now being developed for the McDonnell Douglas AV-8B. This engine has the US designation **F402-RR-406** and will include a new shrouded LP turbine, revised swan-neck intermediate casing and revisions to suit the AV-8B airframe. The full-scale development (FSD) version of the AV-8B will be powered by the **F402-RR-404A**, a version of the existing AV-8A powerplant with improved maintainability and reliability, and new zero-scarf cold nozzles. The first of eight FSD engines was delivered in February 1981; first flight of the AV-8B was due in October 1981.

A further version, the Pegasus **11F-35**, is being developed for an advanced version of the AV-8B, known as the AV-8B+. This engine will have a re-bladed LP compressor with a 4½ per cent increase in mass flow and a 6 per cent increase in rpm. The 11F-35 will offer a further 907 kg (2,000 lb st) over the AV-8B engine.

Work has restarted on a thrust augmentation system known as PCB (plenum-chamber burning). Proposed Pegasus 11 and Pegasus 11-35 engines with PCB are designated **11-03B** and **11-33B** and have thrusts of 120 kN (27,000 lb) and 151·25 kN (34,000 lb) respectively. These proposals are being studied by BAe and McDonnell Douglas for a supersonic V/STOL demonstrator aircraft. Advanced vectored-thrust engines with thrust:weight ratios up to 8:1 are currently being studied.

The following data apply specifically to the Pegasus 11:

Type: Two-shaft vectored-thrust turbofan.

Air Intake Casing: One-piece casting in ZRE magnesium-zirconium alloy (Mk 104, aluminium).

Fan: Three-stage axial, overhung ahead of front bearing. Titanium blades with part-span snubbers. Maximum airflow 196 kg (432 lb)/s. Pressure ratio 2·3 : 1. Bypass ratio 1·4 : 1.

Intermediate Casing: Houses front fan bearing, accessory drives and HP compressor front bearing. All engine-driven accessories mounted above this casing.

HP Compressor: Eight-stage with titanium rotor blades. Overall pressure ratio 14 : 1.

Combustion System: Annular, with low-pressure vaporising burner system.

Fuel System: Hydromechanical, comprising centrifugal backing pump, gear-type pressure pump, HP shut-off cock and overspeed governors and emergency manual control.

Turbines: Two-stage HP turbine and two-stage LP turbine. First-stage HP blades precision cast. Remaining three rotor stages have forged blades. Both HP stages aircooled.

Thrust Nozzles: Two steel cold front-thrust nozzles and two Nimonic hot thrust nozzles, actuated by duplicated air motors through shafts and chains. Vectored-thrust control by pilot command lever.

Lubrication System: Self-contained, comprising pressure pump and three scavenge pumps, with fuel-cooled oil cooler.

Mounting: Four-point suspension, with main trunnions on each side of delivery casing and tie link at rear of turbines.

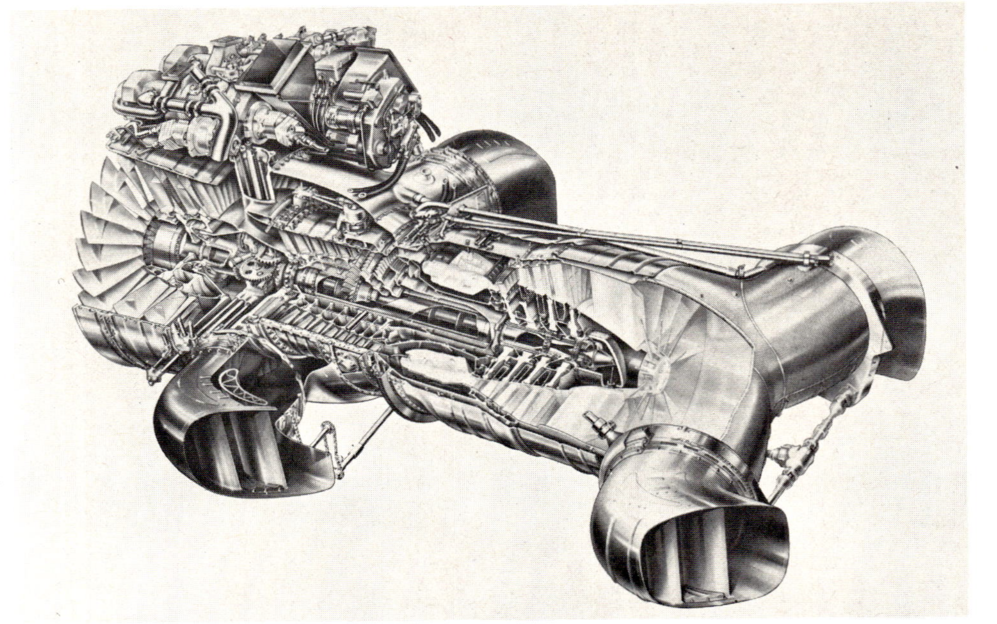

Cutaway drawing of the Rolls-Royce Pegasus 103 turbofan (95·6 kN; 21,500 lb st)

Starting: Gas-turbine starter/APU on intermediate casing.

Dimensions:

Length, without nozzles	2,510 mm (98·84 in)
Diameter, fan casing	1,220 mm (48·05 in)
Length, with nozzles	3,480 mm (137 in)

Weight, Dry (without nozzles):

Mk 103	1,404 kg (3,096 lb)
Mk 104	1,429 kg (3,148 lb)

Performance Ratings:
See under sub-type descriptions

ROLLS-ROYCE ODIN

The Odin ramjet powers the British Aerospace Sea Dart medium-range surface-to-air and surface-to-surface missile. Sea Dart is in service on HMS *Bristol* and Type 42 destroyers of the Royal Navy, and with the Armada Republica Argentina. It is also standard equipment on the anti-submarine carriers being supplied to the Royal Navy.

A lightweight Sea Dart system is being developed by BAe for ships down to 300 tonnes.

The Odin ramjet is mounted integrally with the Sea Dart body and gives longer range and more flexible performance than a rocket-powered missile. Details of the Odin are classified.

ROLLS-ROYCE TURBOMÉCA RTM.321

This turboshaft engine is described in the International section.

ROLLS-ROYCE GEM

The Gem was designed and developed at Leavesden, for the Westland Lynx helicopter. Subsequent applications are the Westland 30 and Agusta A 129 helicopters, and the Gem 2 and 41 series have been civil certificated.

Choice of a two-spool gas generator gives fast response to power demand without the need for a complex control system. Conservative stressing and thermo-dynamic loading, and use of proven design and manufacturing techniques, are features which experience has shown to contribute to engine reliability.

The design concept of the engine is based upon seven major modules, each capable of being assembled, tested and released as an interchangeable unit for service use in the interest of reducing the operator's product support commitments.

The nine main bearings each have labyrinth seals pressurised by LP compressor air which also cools the bearings and minimises heat transfer to the oil, and oil cooler and fan requirements.

Provision is made for in-flight and on-ground condition monitoring systems. Features include access ports for intrascope inspection of each LP compressor stage, HP compressor, combustor, LP turbine and power turbine, and mountings for vibration pickups.

The following versions have been announced:

Gem Mk 10001. Engine-change unit for Lynx of British services and French Navy. Rated at 671 kW (900 shp). Entered service in 1976; over 450 delivered by early 1981.

Gem 2. Export military engine, rated at 671 kW (900 shp). More than 150 delivered by early 1981, to Netherlands, Brazil, Argentina, Denmark and Norway.

Gem 2-3. Version of Gem 2 being developed for Agusta A 129. Direct-drive module in place of reduction gearbox module, and electronic fuel control instead of hydromechanical system.

Gem 41 series. Based on Gem 2, with modified compressor to increase mass flow by about 10 per cent plus small increase in TET to give maximum contingency rating of 835 kW (1,120 shp). In production for Lynx and Westland 30. In service with Royal Netherlands Navy.

Gem Mk 20101. Engine-change unit of Gem 41 for uprated Lynx of British and French navies. Deliveries from late 1981.

Gem Mk 510. Civil version of Gem 41 for commercial Westland 30. Ordered by British Airways Helicopters.

Gem 60 series. Further uprating of Gem to 1,004 kW (1,346 shp), by further increase in mass flow of approximately 26 per cent over Gem 41, derived from new-technology LP compressor incorporating new blades to achieve high work per stage. TET also increased. Planned for production in 1984.

The following description relates to the Gem Mk 10001, except where otherwise indicated:

Type: Free-turbine turboshaft, with two-spool gas generator.

Air Intake: Annular forward-facing.

Shaft Drive: Compact single-stage double-helical reduction gear with rotating planet cage carried by ball bearing at front and roller bearing at rear. Reduction gear mounted within intake casing and driven by power turbine shaft. Gearbox comprises No. 1 module, and power turbine shaft No. 2 module. Gearbox provides governed output speed of 6,000 rpm. No. 2 module provides for signal to phase-displacement torquemeter. Gem 2-3 has direct-drive No. 1 module with output speed of 27,000 rpm.

LP Compressor: Four-stage axial. LP compressor and intake case comprise No. 3 module.

HP Compressor: Single-stage centrifugal impeller having alternate inducer and radial vanes. Combined radial-and-axial diffuser feeds compressor delivery air to annular combustor. Overall pressure ratio 12·0 : 1.

Combustion Chamber: Fully annular reverse-flow with air-atomiser fuel sprays supplied by external fuel manifold. High-energy ignition box mounted on power turbine/jet pipe case.

Cutaway drawing of the Rolls-Royce Gem free-turbine turboshaft, with initial service rating of 671 kW (900 shp)

Gem Engine Ratings, kW (shp) ISA S/L Static

Engine Designation	*One Engine Inoperative* *Emergency (20 s)*	*Max Contingency (2½ min)*	*Intermediate Contingency (60 min)*	*Normal Twin Operation* *Max (T-O) (30 min)*	*Max (T-O) (5 min)*	*Max Continuous*
Gem 2, Mk 10001	N/A	671 (900)	619 (830)	N/A	619 (830)	559 (750)
Gem 2-3	772 (1,035)	716 (960)	667·5 (895)	667·5 (895)	N/A	604 (810)
Gem 41-1, Mk 20101, Mk 510	N/A	835 (1,120)	790·5 (1,060)	N/A	746 (1,000)	664 (890)
Gem 60-3	N/A	1,004 (1,346)	953 (1,278)	N/A	911 (1,222)	832 (1,115)

N/A: Not applicable

HP Turbine: Single-stage axial close-coupled to HP impeller. Rotor blades and aircooled nozzle guide vanes based on R-R Dart technology. HP spool with compressor intermediate casing and combustor comprise No. 4 module.

LP Turbine: Single-stage axial with shrouded rotor blades, drives LP compressor. LP turbine and main shaft comprise No. 6 module.

Power Turbine: Two-stage axial with shrouded rotor blades. Rear of power drive shaft drives output speed governor and overspeed fuel cut-off trip mechanism via spur and bevel gear train in exhaust cone. Power turbine and jet pipe form No. 7 module.

Jet Pipe: Short-length duct with casing extending forward to combustor rear casing. Four cruciform struts integral with exhaust cone.

Accessory Drives: Bevel gear on front of HP compressor shaft drives accessory shaft extending through compressor intermediate casing to spiral bevel gear drive to accessory wheelcase mounted atop intermediate casing. Drives provided for starter/generator, fuel pump, oil cooler fan and other accessories. Wheelcase forms No. 5 module.

Fuel System: Plessey fuel system with fluidics circuit providing fully automatic control, and power matching for multi-engine installation. Also automatic restoration of power from 'good' engine in event of single engine failure. Incorporates fuel filter. Alternative Hamilton Standard digital electronic control fitted to Gem 2-3 and later versions of Gem 41 and Gem 60.

Lubrication System: Engine-mounted oil tank and cooler to provide self-contained system. Magnetic chip detectors fitted in each scavenge line. Oil filter incorporated in accessory wheelcase.

Dimensions:
- Height overall: 596 mm (23·5 in)
- Width overall: 575 mm (22·6 in)
- Length overall: 1,099 mm (43·2 in)

Weight, Dry:
- Gem 10001, 2: 150 kg (330 lb)
- Gem 41 (Mk 20101, 510): 156 kg (343 lb)
- Gem 2-3: 140 kg (309 lb)
- Gem 60: 155 kg (342 lb)

Performance Ratings: see table

Specific Fuel Consumption:
- 50 per cent max T-O:
  - All versions: 110 μg/J (0·65 lb/h/shp)

ROLLS-ROYCE GNOME

Gnome is the name given to the versions of the General Electric T58 turboshaft which Rolls-Royce manufactures in the UK. The first ran on 5 June 1959.

More than 1,900 Gnome engines have been delivered. Four versions have been announced:

H.1000. Initial version, rated at 783 kW (1,050 shp). Power plant for military Whirlwind HAR.Mk 9, HAR.Mk 10 and HCC.Mk 12, civil S-55 Series 3 and Agusta-Bell 204B.

H.1200. Rated at 932 kW (1,250 shp). Used in Agusta-Bell 204B, Boeing Vertol 107 and some Kawasaki KV-107/II-5s. Coupled version for Wessex comprises two H.1200s driving through a coupling gearbox.

H.1400. Rated at 1,044 kW (1,400 shp). Based on the H.1200, with modified compressor to increase airflow.

H.1400-1. Rated at 1,145 kW (1,535 shp). Uprated from H.1400, without change in size or weight, by increasing gas-generator speed and using improved gas-generator turbine-blade material. In production for Westland Sea King and Commando.

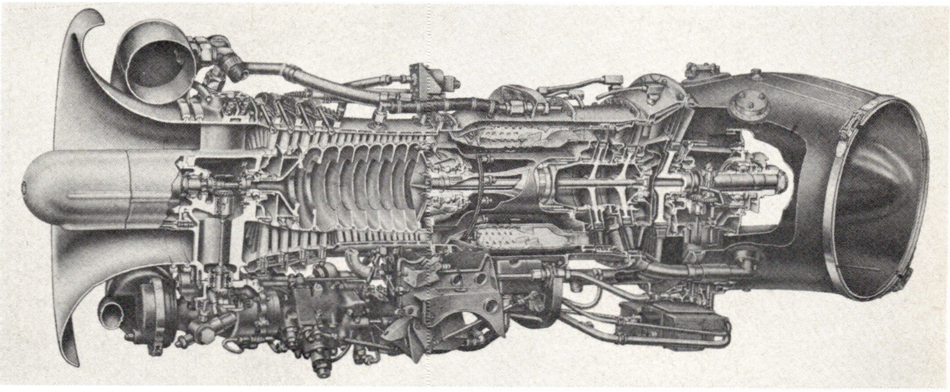

Cutaway drawing of the Rolls-Royce Gnome H.1400-1 free-turbine turboshaft

The following description refers to the H.1400-1:

Type: Axial-flow free-turbine turboshaft engine.

Air Intake: Annular forward-facing. Centre housing carrying front main bearing supported by four radial struts. Struts and inlet guide vanes anti-iced with hot compressor bleed air and oil drainage.

Compressor: Ten-stage axial. Controlled variable incidence for inlet guide vanes and first three rows of stator blades. Air mass flow 6·26 kg (13·8 lb)/s.

Combustion System: Straight-through annular chamber with outer casing split along horizontal centreline. Sixteen Simplex-type fuel injectors, eight on each of two sets of manifolds. One Lodge capacitor-discharge high-energy igniter plug.

Fuel System: Lucas hydromechanical units, comprising variable-stroke multi-plunger pump, flow control unit and throttle controlled by BAeD computer.

Fuel Grade: DERD.2452, 2453, 2454, 2486, 2494 and 2498 (NATO F44, F34, F40, F35 and F43).

Gas-Producer Turbine: Two-stage, coupled to compressor shaft by conical shaft. Extended-root blading with fir-tree attachments.

Power Turbine: Single-stage free turbine. Extended-root blading with fir-tree attachments.

Exhaust System: Curved exhaust ducting arranged to suit individual applications.

Reduction Gear: Optional double-helical gear providing reduction from nominal 19,500 rpm power turbine speed to 6,600 rpm at left or right output shaft.

Accessory Drives: Quill shaft drive through lower intake strut. Fuel and lubrication systems mounted beneath compressor casing. Power take-off shaft up to 100 shp.

Lubrication: Scavenged gear pumps. Serck oil cooler.

Mounting: Three mounting faces on intake casing. Rear mounts on gearbox or output housing.

Starting: Rotax electric starter in nose bullet.

Dimensions:
- Length:
  - H.1000, H.1200, H.1400-1: 1,392 mm (54·8 in)
  - Coupled H.1200 (Wessex): 1,747 mm (68·8 in)
- Max height:
  - H.1000, H.1200, H.1400-1: 549 mm (21·6 in)
  - Coupled H.1200 (Wessex): 1,031 mm (40·6 in)
- Max width:
  - H.1000, H.1200 (ungeared): 462 mm (18·2 in)
  - H.1400-1 (ungeared): 577 mm (22·7 in)
  - Coupled H.1200 (Wessex): 1,059 mm (41·7 in)

Weight, Dry:
- H.1000 (ungeared): 134 kg (296 lb)
- H.1200 (ungeared): 142 kg (314 lb)
- H.1400-1 (ungeared): 148 kg (326 lb)
- Reduction gearbox: 52·6 kg (116 lb)
- Coupled H.1200: 422 kg (930 lb)

Performance Ratings (at power-turbine shaft):
- Max contingency (2½ min; multi-engine aircraft only):
  - H.1200: 1,007 kW (1,350 shp)
  - H.1400-1: 1,238 kW (1,660 shp)
- Max one-hour (single engine):
  - H.1000: 783 kW (1,050 shp)
  - H.1200: 932 kW (1,250 shp)
  - H.1400-1: 1,145 kW (1,535 shp)
- Max continuous:
  - H.1000: 671 kW (900 shp)
  - H.1200: 783 kW (1,050 shp)
  - H.1400-1: 932 kW (1,250 shp)

Specific Fuel Consumption:
- At max contingency rating:
  - H.1200: 104·4 μg/J (0·618 lb/h/shp)
  - H.1400-1: 102·75 μg/J (0·608 lb/h/shp)

ROLLS-ROYCE MOTORS
ROLLS-ROYCE MOTORS LTD

Head Office: Crewe, Cheshire CW1 3PL
Telephone: 0270 55155

This company and Teledyne Continental Motors (see US section) terminated their licence agreement by mutual consent on 31 December 1980. Since then the British company's Teledyne Continental engine production, sales, product support and related activities have been transferred to Teledyne Continental's Mobile, Alabama, plant.

WESLAKE
WESLAKE AEROMARINE ENGINES LTD

Yeovil, Somerset BA20 2YB
Telephone: Yeovil (0935) 5181

Officers:
B. D. Blackwell, MA, BSc(Eng)
W. T. C. Miller, OBE, MA, CEng, MIMechE, MInstM

Secretary: K. Stansfield

Divisional Manager: D. P. Short, MA, CEng, MIMechE

On 1 May 1979 Weslake Aeromarine Engines Ltd was acquired by Westland Aircraft Ltd, whose subsidiary company, Normalair-Garrett Ltd, from that date undertook management control. In 1980 the company was relocated at Yeovil, to take advantage of the high-quality manufacturing facility available there for quantity production.

WESLAKE TYPE 116

Two-cylinder two-stroke with bore and stroke of 46 mm (1·81 in) and 35 mm (1·38 m) respectively. Weight including ignition, carburettor and stub exhausts about 4·5 kg (10 lb). Rated output 6 kW (8 hp) at 6,500 rpm.

WESLAKE TYPE 200

Two-cylinder, simultaneous-firing two-stroke for RPV applications. Bore and stroke 55 mm (2·17 in) and 42 mm (1·65 in) respectively. Weight with ignition, carburettor, stub exhausts and alternator drive, about 8·2 kg (18 lb). Rating 9 kW (12 hp) at 6,500 rpm.

WESLAKE TYPE 274

The Type 274 is the first production engine available. It is being produced in two basic versions for RPVs, ACVs (Hovercraft) and powered hang gliders. The 274-6 is used in the Westland Wideye, Shorts Skeet, Cranfield A2 and other aircraft.

Type: Two-cylinder aircooled two-stroke, simultaneous-firing (274-6 loop scavenged).

Cylinders: Die-cast aluminium with plated bore (274-6, treated bore). Bore 66 mm (2·60 in). Stroke 40 mm (1·575 in). Capacity 274 cc (42·50 cu in).

Connecting Rods: Forged steel, caged needle bearing at both ends (274-3, single-piece; 274-4, -6, split type).

CRANKSHAFT: Forged steel. 274-3, three-piece, vertical, running in caged needle and thrust bearings at both ends, with drive at both ends. 274-4 and -6, single-piece with large-diameter propeller mounting at one end.
CRANKCASE: 274-3, two-piece cast in LM25 aluminium with mounts for vertical operation. 274-4, similar but rear radial mounts. 274-6, single-piece with radial anti-vibration mounts.
INDUCTION: 274-3 and -4, twin Tillotson carburettors incorporating fuel pump. 274-6, single carburettor at rear feeding through rotary disc valve.
FUEL GRADE: 274-3, regular road-vehicle gasoline plus branded two-cycle oil in ratio 50:1. 274-4, same in ratio 25:1. 274-6, three-star gasoline plus oil in ratio 25:1 reducing to 50:1.
IGNITION: 274-3 and -4, specially developed SEM electronic flywheel magneto, type AM3/90; fully shielded, no external power. 274-6, modified AM3/90.

DIMENSIONS:	
Length: 274-3, -4	228·5 mm (9·0 in)
274-6	355·5 mm (14·0 in)
Width	355·5 mm (14·0 in)
Height	228·5 mm (9·0 in)
WEIGHT, DRY (complete to run):	
274-3, -4	7·7 kg (17·0 lb)
274-6	7·26 kg (16·0 lb)
PERFORMANCE RATING:	
274-3, -4, -6	13·43 kW (18 hp) at 6,500 rpm
SPECIFIC FUEL CONSUMPTION:	
274-3, -4, -6	below 186 μg/J (1·1 lb/h/hp)

WESLAKE TYPE 342

Two-cylinder simultaneous-firing two-stroke for RPV applications. Bore and stroke 66 mm (2·60 in) and 50 mm (1·97 in) respectively. Weight with ignition, carburettor, stub exhausts and alternator control system, about 12·7 kg (28 lb). Rating 19·4/22·4 kW (26/30 hp) at 6,000/8,000 rpm.

WESLAKE TYPE 430

This is a larger two-cylinder two-stroke engine, generally following the principles of the 274. Single-piece con-

Weslake Type 274-6 fitted in Shorts Skeet target

necting rods drive a three-piece crankshaft driving at both ends. A twin-plug version is available.

DIMENSIONS:	
Length	305 mm (12·0 in)
Width	432 mm (17·0 in)
Height	254 mm (10·0 in)
WEIGHT, DRY (complete):	11·3 kg (25·0 lb)
PERFORMANCE RATING:	up to 30 kW (40 hp)
SPECIFIC FUEL CONSUMPTION:	149 μg/J (0·8 lb/h/hp)

WESLAKE TYPE 548

This is basically a four-cylinder version of the 274-6, with an aluminium or magnesium alloy crankcase incorporating ports for reed-valve induction.

WEIGHT, DRY:	16·3 kg (36·0 lb)
PERFORMANCE RATING:	26·9 kW (36 hp)

WESLAKE TYPE 860

A four-cylinder two-stroke; details restricted.

WESLAKE TYPE 1527

Based on well-proven production components, this is a flat-four four-stroke aircooled RPV engine.

Weslake Type 200 RPV engine rated at 9 kW (12 hp)

CYLINDERS: Die-cast aluminium barrels and heads. Aluminium liner with internal coating. Bore 90·0 mm (3·54 in). Stroke 60 mm (2·36 in). Capacity 1,527 cc (93·2 cu in).
CRANKSHAFT: Single-piece, four-throw, three-bearing.
INDUCTION: Twin Stromberg or Weber carburettors.
FUEL GRADE: Pump four-star or Avgas 100R.
IGNITION: Lucas Rita breakerless system.

DIMENSIONS:	
Length	355·5 mm (14·0 in)
Width	647·4 mm (25·5 in)
Height (dry-pump version)	228·5 mm (9·0 in)
WEIGHT, DRY (incl starter):	under 45·4 kg (100·0 lb)
PERFORMANCE RATING (ISA):	74·6 kW (100 hp) at 5,500 rpm
SPECIFIC FUEL CONSUMPTION (at rating):	under 84·5 μg/J (0·5 lb/h/hp)

WESLAKE
WESLAKE & COMPANY

Harbour Road, Rye Harbour, Sussex TN31 7TH
Telephone: 07973-2755

Last appearing in the 1977-78 *Jane's,* this company is no longer connected with its former associate, Weslake Aeromarine Engines (see preceding entry). Weslake & Co has spent many years refining the design of a cylinder head for high-performance piston engines with central spark plug and four valves. The work has led to the design of a flat-twin tailored for outstanding fuel economy, adequate performance for a light two-seat aircraft (fixed- or rotary-wing) and acceptably low noise and vibration. Pent-roof heads were running in 1980 with both central and top plug locations, with rear-facing inlets and forward-facing exhausts. A conventional impulse magneto is used for starting and standby, and an electronic system with speed and manifold-pressure controlled ignition advance to assist leanest operation at all times.

The engine measures 711 mm (28·0 in) across the rocker-box covers; target weight with ignition system is 57 kg (125 lb). Rating is 56 kW (75 hp) at 3,500 rpm, with normal running at 48·5 kW (65 hp) at 2,800 rpm.

UNITED STATES OF AMERICA

AEROJET
AEROJET-GENERAL CORPORATION (Subsidiary of The General Tire & Rubber Company)

10300 N Torrey Pines Road, La Jolla, California 92037
Telephone: (714) 455 8500
CHAIRMAN OF THE BOARD: M. G. O'Neil
PRESIDENT: J. H. Vollbrecht
GROUP VICE-PRESIDENT: J. L. Heckel
Aerojet Liquid Rocket Company
PRESIDENT: Roger I. Ramseier
Aerojet Services Company
PRESIDENT: R. W. Warden
Aerojet Strategic Propulsion Company
PRESIDENT: K. R. Collins
Aerojet Solid Propulsion Company
TACTICAL SYSTEMS GROUP
VICE-PRESIDENT AND GENERAL MANAGER:
W. M. Necoechea

Aerojet-General Corporation has activities in four major areas of business: chemicals, electronics, engineering and construction, and mechanical systems and metal products.

In the chemical area, two organisations produce solid-propellant rocket motors for the Department of Defense. The Aerojet Strategic Propulsion Company designs, develops and produces solid-propulsion strategic ballistic-missile rocket motors; Aerojet Tactical Systems produces a variety of tactical missile motors.

In the mechanical systems area, the Aerojet Liquid Rocket Company is active in research, development, testing and production of liquid-propellant rocket engines, waterjet propulsors and lift fan systems for surface-effect ships. The Aerojet Services Company, formed in 1977, provides technical services and products for defence and space programmes, including rocket launch support, engine repair and overhaul, systems engineering and related management activities.

Aerojet Liquid Rocket Company (ALRC), Sacramento, California. Development and manufacture of all liquid-fuel engines for the US Air Force's Titan family of vehicles, including the improved Transtage; second-stage propulsion for the uprated Delta 3920 and Japanese Space Agency (NASDA) N-II launch vehicles; Orbital Manoeuvring System (OMS) engines for the Space Shuttle; and application of Titan engines to the Liquid Boost Module (LBM), to provide Space Shuttle thrust augmentation for significant increases in payload capability.

AEROJET TITAN III ENGINES

The production of Titan III first, second and Transtage engines for use as booster propulsion on the Titan family of vehicles has been under way continuously since 1962 by Aerojet Liquid Rocket Company and its predecessor Aerojet organisations. These engines, utilising storable propellants, develop 2,353 kN (529,000 lb), 445 kN (100,000 lb) and 71·2 kN (16,000 lb) thrust respectively. Their flight reliability is in the 90-99 per cent class. The nominal weights of these engines are 1,977, 564 and 196 kg (4,360, 1,245 and 432 lb) respectively.

In January 1978 ALRC was awarded a $38·5 million contract by the US Air Force Space and Missile Organisation for nine additional Titan propulsion systems. This extended production until June 1981.

AEROJET SPACE SHUTTLE OMS ENGINE

The Space Shuttle Orbiter has two Orbiter Manoeuvring System (OMS) engines, in pods on each side of the vertical stabiliser (fin). The OMS engine, produced by ALRC, provides thrust for orbit insertion, circularisation and plane change, rendezvous and de-orbit manoeuvres. The propellants are monomethylhydrazine (MMH) and nitrogen tetroxide. At launch the basic system carries 4,087 kg (9,010 lb) of usable MMH and 6,743 kg (14,866 lb) of usable oxidiser. The following are chief characteristics:

THRUST (in vacuo)	26·7 kN (6,000 lb)
CHAMBER PRESSURE (in vacuo)	8·62 bars (125 lb/sq in)
SPECIFIC IMPULSE	313·2 s
MIXTURE RATIO	1·65 : 1
NOZZLE AREA RATIO (flight)	55 : 1
DRY WEIGHT	118 kg (260 lb)
THRUST VECTOR LIMITS	±7°
STORAGE LIFE	10 years
FIRING LIFE (100 missions)	15 h
NUMBER OF STARTS	1,000
LONGEST SINGLE FIRING	1,250 s

ALRC Space Shuttle OMS engine

AEROJET MINUTEMAN MOTORS

Aerojet Strategic Propulsion Company is currently remanufacturing the second stage of the LGM-30G Minuteman III ICBM. This motor has polybutadiene/ammonium perchlorate propellant packaged in a titanium case, and the single submerged nozzle has liquid injection thrust-vector control. Loaded weight is 7,076 kg (15,600 lb) and average thrust 269·5 kN (60,000 lb).

Aerojet Solid Propulsion Company (ASPC), Tactical Systems Group, Sacramento, California. Development and manufacture of the motors for Hawk, Sparrow and Standard ARM, and the booster for Harpoon. Production starting on advanced Sidewinder and Maverick motors.

AEROJET MAVERICK MOTOR

The Aerojet Maverick rocket motor, Type SR115-AJ-1, utilises a reduced-smoke HTPB propellant system and is completely interchangeable in all Maverick missile systems. The primary use will be in improved systems following the AGM-65B configuration. Aerojet has completed qualification of this boost/sustain rocket motor, and was anticipating a production programme beginning in 1980.

AEROJET SIDEWINDER MOTOR

Aerojet qualified a rocket motor in 1979 for use on the AIM-9B/E/J/N/P air-to-air missile systems. The motor utilises a reduced-smoke HTPB propellant. Production was scheduled to begin in 1980.

AEROJET HAWK MOTOR

The single-chamber solid-propellant rocket motor of the Hawk surface-to-air missile was the first dual-thrust dual-grain motor to be mass-produced.

Within its single propellant mass, the motor has an inner core of propellant constituting a short-duration booster grain which launches and accelerates the missile to supersonic speed. When this inner core is consumed, a slower-burning outer core, forming the sustainer portion of the propellant, takes over and keeps the missile at the required velocity.

ASPC Minuteman III second-stage motor

AEROJET SOLID PROPULSION COMPANY MOTORS

Name	Designation	Fuel	Oxidiser	Average thrust kN (lb)	Max length m (in)	Max dia m (in)	Total weight kg (lb)	Remarks/Primary Application
Strategic Motors								
Minuteman 2nd stage	SR19-AJ-1	Polybutadiene	NH_4ClO_4	269·5 (60,000)	4·01 (162)	1·32 (52)	7,076 (15,600)	Minuteman LGM-30G second stage, titanium case, single submerged nozzle, liquid injection TVC.
Polaris 1st stage	A3P	Polyurethane	NH_4ClO_4	—	4·62 (182)	1·37 (54)	10,886 (24,000)	Polaris A3, glass case, four rotatable nozzles; nitroplasticiser additive.
Tactical Motors								
Phoenix	Mk 60 Mod 0	Polybutadiene	NH_4ClO_4	—	1·78 (70)	0·38 (15)	199 (439)	Propulsion for Navy's fleet-defence air-to-air missile.
Sparrow III, AIM-7E	Mk 52 Mod 2	Polybutadiene	NH_4ClO_4	—	1·32 (52)	0·2 (8)	68·5 (151)	Propulsion for Navy's Sparrow air-to-air missile and UK BAe Sky Flash.
Shrike	Mk 53 Mod 2	Polybutadiene	NH_4ClO_4	—	1·32 (52)	0·2 (8)	71 (157)	Propulsion for Navy's AGM-45 anti-radiation air-to-surface missile.
Shrike, Improved	Mk 78 Mod 0	Polyurethane	NH_4ClO_4	—	1·30 (51)	0·2 (8)	78 (172)	Dual-thrust propulsion for anti-radiation air-to-surface missile.
Tartar	Mk 1 Mod 0	Polyurethane	NH_4ClO_4	—	2·62 (103)	0·34 (13·5)	345 (760)	Dual-thrust propulsion for Navy ship-to-air missile.
Tartar, Improved	Mk 27 Mod 2, 3	Polyurethane	NH_4ClO_4	—	2·62 (103)	0·34 (13·5)	354 (780)	Dual-thrust propulsion for Navy ship-to-air missile.
Standard Missile	Mk 56 Mod 2	Polybutadiene Polyurethane	NH_4ClO_4	—	2·62 (103)	0·34 (13·5)	411·5 (907)	Dual-thrust propulsion for Navy ship-to-air Missile Type 1, MR.
Harpoon	MX-(TBD)B446-2	Polyurethane	NH_4ClO_4	—	0·61 (24)	0·34 (13·5)	119 (262)	Booster for Navy Harpoon anti-ship missile.
2·75 in FFAR, Improved	SR105-AJ-1	Polybutadiene	NH_4ClO_4	—	0·34 (33)	0·07 (2·75)	5·9 (13)	Air-launched forward-firing.
Hawk	XM22E8	Polyurethane	NH_4ClO_4	—	2·77 (109)	0·36 (14)	388 (856)	Dual-thrust motor for Army's surface-to-air Hawk missile.
Hawk, Improved	XM112	Polyurethane	NH_4ClO_4	—	2·77 (109)	0·36 (14)	395 (870)	Dual-thrust motor for Army's surface-to-air Hawk missile.
Sidewinder	Mk 17	HTPB	NH_4ClO_4	Classified	1·91 (75)	0·127 (5)	40·4 (89)	Propulsion for AIM-9J/9P Sidewinder air-to-air missiles.
Maverick		HTPB	NH_4ClO_4	*	1·00 (39·5)	0·28 (11)	48·4 (107)	Propulsion for AGM-65 Maverick missile.
Launch Vehicle Boosters and Space Motors								
Alcor 1B		Polybutadiene	NH_4ClO_4	44·5 (10,000)	1·43 (76)	0·51 (20)	455·5 (1,004)	Third stage of Athena test vehicle and 2nd stage for Astrobee 1500 launch vehicle.
Astrobee F		HTPB	NH_4ClO_4	169/37 (38,000/ 8,300)	7·11 (280)	0·43 (15)	1,255 (2,768)	Dual-thrust sounding rocket (200 lb to 235 miles altitude).
SVM-4A	SVM-4A	Polybutadiene	NH_4ClO_4	—	1·52 (60)	0·94 (37)	706 (1,557)	Apogee-boost motor for Intelsat IV synchronous communications satellite.
SVM-5	SVM-5	Polybutadiene	NH_4ClO_4	21·8 (4,900)	0·91 (36)	0·76 (30)	318·5 (702)	Apogee-boost motor for NASA Synchronous Meteorological Satellite.
SVM-7	SVM-7	Polybutadiene	NH_4ClO_4	43·2 (9,520)	1·45 (57)	0·76 (30)	440 (970)	Apogee-boost motor for RCA Comsat.
Gas Generators and JATOs								
Turbine Start Cartridge		Butyl Rubber	NH_4NO_3	n.a.	0·41 (16)	0·18 (7)	11 (24)	For Titan II first stage.
Turbine Start Cartridge		Butyl Rubber	NH_4NO_3	n.a.	0·48 (19)	0·10 (4)	5·5 (12)	For Titan II second stage.
Gas Gen Mk 46		Butyl Rubber	NH_4NO_3	n.a.	1·09 (43)	0·305 (12)	—	Prime power source for the Navy's Mk 46 Mod 0 torpedo.
Controllable Motors								
SCCSRM		Polyurethane	NH_4ClO_4	—	0·38 (15)	0·51 (20)	415 (915)	Single-chamber controllable solid-rocket motor.
Air Launched (VTM)		Polybutadiene	NH_4ClO_4	—	1·37 (54)	0·43 (17)	185 (407)	Variable-thrust motor; has both thrust-magnitude and vector control.
Air-to-Air Controllable (ATAC)		Polybutadiene	NH_4ClO_4	—	1·70 (67)	0·20 (8)	95 (210)	Throttling over wide temperature range.
Stop-Start Motor (SSM)		Polyurethane	NH_4ClO_4	—	2·21 (87)	0·51 (20)	396·5 (874)	Stop/start operation on command.

*Boost 44·5 kN (10,000 lb), sustainer 8·0 kN (1,800 lb)

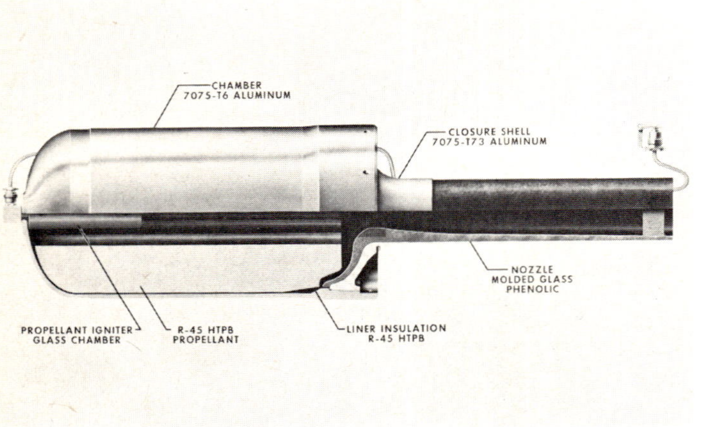

Cross-section of Aerojet Maverick motor

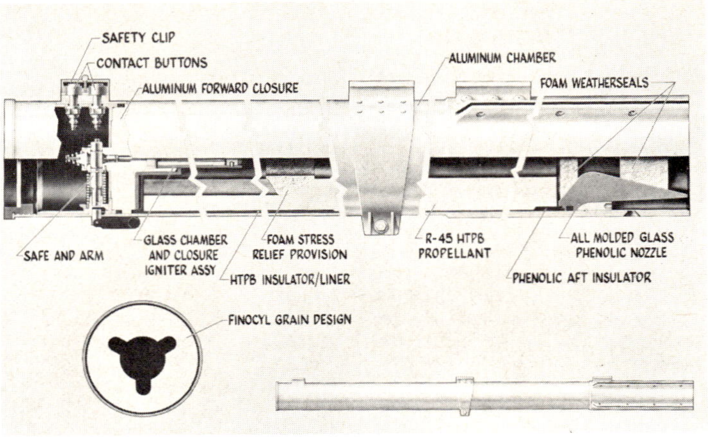

Cross-section of Aerojet motor used in advanced models of Sidewinder

An Improved Hawk propulsion system, using an upgraded and higher-impulse polyurethane propellant, passed its qualification testing at the US Army Test and Evaluation Command, White Sands Missile Range. Several hundred Improved Hawk motors have been test fired and flight tested without failure, demonstrating the same high degree of reliability and long life as the Basic Hawk motor. The Improved Hawk motor is now in production at Sacramento. Aerojet Solid Propulsion Company has signed agreements with several foreign firms for its production.

AEROJET STANDARD MISSILE MOTOR

This motor is a dual-thrust design for the boost and sustain phases of the Standard I medium-range missile trajectory. This motor fits the same envelope as the test motors that were used in an earlier version of Standard Missile. The chamber has been lengthened, and the extension or blast tube eliminated, to accommodate more propellant and increase performance. These motors have been in production for the US Navy since 1970. Improvement is expected in the future to enhance missile performance.

AEROJET SPARROW / SHRIKE MOTORS

The Mark 52 and Mark 53 motors are the sustainer or 'all boost' thrust units for the Sparrow AIM-7E and Shrike AGM-45A missiles. They are similar in construction and dimensions. The principal difference lies in the performance characteristics, which are tailored to match the air-to-air and air-to-ground missions. These motors have been in production since the mid-1960s and several thousand have been produced to meet US Navy and Air Force requirements. In 1969 an improvement programme was initiated to develop the Mark 78 motor. This is a dual-thrust (boost and sustain) motor which greatly enhances the performance of the AGM-45B Improved Shrike. Production of this motor started in 1972, with several thousand produced to date.

AEROJET HARPOON BOOSTER MOTOR

The Harpoon booster motor is used to launch the RGM-84 surface-launched version of the US Navy's Harpoon missile. It is mounted in tandem and provides thrust during the boost phase before the cruise engine starts. The motor has completed all research and development and is now in production. The design incorporates many advanced features such as a submerged nozzle, developed over many years by Aerojet Solid Propulsion Company. Use of this missile/motor combination is planned for many fleet units of the US Navy and other navies.

AEROJET ALCOR 1B

The Alcor 1B is a high-performance Aerojet Solid Propulsion Company motor featuring a high specific-impulse polybutadiene propellant and extremely lightweight inert components. The nozzle is a unique combination of laminated reinforced-plastics materials, and the chamber is a very thin, welded, high-strength titanium 6Al-4V alloy structure. This highly efficient chamber has demonstrated excellent resistance to external flight loads in structural tests.

The Alcor 1B is used as third stage on the Athena test vehicle and is the second-stage propulsion system on the Astrobee 1500 launch vehicle. It can be applied as an upper-stage sounding rocket motor, a small component test vehicle, a synchronous orbit injection motor, or an upper-stage booster for low and medium orbits. The motor can be spin-stabilised and is fabricated to a very small thrust-misalignment tolerance (0·0004 radians angular and 0·51 mm; 0·020 in linear). This motor has been used on more than 100 flights without a failure.

AEROJET ASTROBEE F

The Astrobee F is a dual-thrust sounding-rocket motor incorporating many of the design features, including HTPB propellant, of the successful Astrobee D. The boost thrust averages 169 kN (38,000 lb) for 3 s followed by a sustain thrust of 36·9 kN (8,300 lb) for 53 s. The 381 mm (15 in) diameter motor is designed to be compatible with the Aerobee payload systems and vehicle facilities. The initial Astrobee F test design is capable of delivering 90·7 kg (200 lb) payloads up to altitudes of 378 km (235 miles). Over 20 successful flights have been made.

ASPC motor for Standard I MR surface-to-air missile

ASPC Mk 78 motor for Improved Shrike

ASPC boost motor for RGM-84 Harpoon missile

ALLISON
DETROIT DIESEL ALLISON DIVISION, GENERAL MOTORS CORPORATION

Detroit, Michigan
Telephone: (313) 531 7100
Indianapolis Operations: PO Box 894, Indianapolis, Indiana 46206
Telephone: (317) 244 1511
General Manager: Donald J. Atwood
General Sales Manager: Chester B. Clum
Manager, Indianapolis Operations: Robert M. Clark

Detroit Diesel Allison's Gas Turbine Operations continue to produce T56 turboprop engines for the Lockheed C-130 Hercules, Lockheed P-3 Orion and Grumman E-2 Hawkeye. The commercial counterpart of the T56, the Model 501, powers the Lockheed Electra, commercial Hercules and Convair 580 airliners.

The Allison T63 small gas turbine, and its commercial counterpart, the Model 250, have been developed through many versions with numerous applications, as listed under each model heading. A turboprop version of the Model 250 was certificated in March 1969 for light fixed-wing aircraft applications.

It was announced in January 1976 that Allison and Rolls-Royce of England would develop and produce jointly a version of the Rolls-Royce Spey turbofan engine, under the designation TF41, to power advanced versions of the LTV A-7 Corsair II aircraft. The TF41 remains in production for the US Navy's A-7E carrier-based attack aircraft. Further details can be found in the International section.

ALLISON MODEL 250
US military designation: T63

The Model 250 is a small turboshaft engine in which power is derived from a free power turbine and is delivered through an offset gearbox which includes all accessory drive pads.

A development contract for the T63 military version was received by Allison in June 1958 and the engine was first run in the Spring of 1959. Details of early versions last appeared in the 1978-9 *Jane's*; the following are current models:

T63-A-720. Hot-end improvements, increasing T-O rating to 313 kW (420 shp). Corresponds to commercial C20B. In production for the Bell OH-58C.

250-B17. Announced in 1972, the B17 is an uprated version of the earlier B15 turboprop, corresponding to the C20 turboshaft. Current model is the **B17B,** operating at 17°C higher turbine gas temperature and with hot-end improvements similar to those of the C20B, which maintain full power at high ambient temperatures. The B17B entered production in September 1974. In production for Turbostar 402 conversions, Turbostar 414, GAF Nomad N22 and 24, Ahrens AR-404, SIAI-Marchetti SM.1019E and a growing variety of agricultural aircraft modifications.

250-C20B. Introduced in 1974, and rated at 313 kW (420 shp). In production for Bell 206B JetRanger III and 206L LongRanger, MBB BO 105D, Agusta A 109A, Hughes 500D, and Soloy UH-12E and Bell 47G conversions.

250-C28. Representing Series III of the Model 250 evolutionary process, the 373 kW (500 shp) C28 is a near-total redesign. The axial multi-stage compressor, one of the primary features of the original design, has been eliminated. Instead a single-stage front-entry centrifugal impeller is used, handling a considerably increased airflow. The philosophy behind Series III was to reduce noise and emissions, and despite the increase in power the sound pressure level of the bare engine has been reduced. The compressor-acceleration bleed is eliminated, and the exhaust leaves through a single low-velocity stack which also has a minimal infra-red signature. The main gearbox has new gears with increased helix and decreased pressure angles. Flight-cleared prototype engines were first available in September 1975 in two forms: C28B, with particle separator; C28C with plain inlet. Certificated December 1977.

250-C28B. Improved model with 2½-min rating of 410 kW (550 shp).

250-C28C. Improved model with 2½-min rating of 410 kW (550 shp).

250-C30. Representing Series IV of the Model 250 growth programme, the C30 has a more advanced single-stage compressor, handling a higher mass flow at an

The 313 kW (420 shp) Allison Model 250-C20B turboshaft engine

The 485 kW (650 shp) Allison Model 250-C30 turboshaft engine

increased pressure ratio. The engine has numerous new features, one of which is dual ignition to comply with FAR Pt 29. The C30 has an initial rating of 485 kW (650 shp). Prototype engines were delivered in 1976. Certification was completed in March 1978.

The following description applies to Model 250 engines currently in production or development:

Type: Light turboshaft or turboprop.

Compressor: The C20, C20B and B17B incorporate an axial/centrifugal compressor with six axial stages and one centrifugal. The C28B and C and the C30 have a single-stage centrifugal compressor only. Pressure ratio: C20, 7·0:1; C20B, B17B, 7·2:1; C28B/C, C30, 8·5:1. Air mass flow: C20, 1·5 kg (3·4 lb)/s; C20B, B17B, 1·56 kg (3·45 lb)/s; C28B/C, 2·02 kg (4·45 lb)/s; C30, 2·54 kg (5·6 lb)/s.

Combustion Chamber: Single can-type chamber at rear of engine. Single duplex fuel nozzle in rear face of chamber. One igniter on C20, C20B, B17B, C28B/C. Dual igniters on C30, optional on C28B/C.

Turbines: Two-stage gas-producer turbine and two-stage free power turbine. Integrally-cast rotor blades and wheels. Gas-producer turbine outlet temperature: C20B, B17B, 810°C; C28B/C, 730°C; C30, 740°C.

Gearcase: Magnesium casting which forms primary structure of engine and contains all power and accessory gear trains, torque sensor and oil pumps. Compressor and turbine assemblies bolted to front and rear faces respectively. Rated shp available at either front or rear output-shaft spline or any combination totalling both. Turboshaft output speed, 6,016 rpm. Turboprop output speed, 2,030 rpm.

Control System: Pneumatic-mechanical system consisting essentially of fuel pump and filter assembly, gas producer fuel control and power turbine governor (B17B, hydromechanical; C20B, C28, C30, pneumatic-mechanical).

Fuel: Primary fuels are ASTM-A or A-1 (Model 250-C20, ASTM D-1655) and MIL-T-5624, JP-4, JP-5 and diesel fuel.

Lubrication: Dry sump.

Oil Specification: MIL-L-7808 and MIL-L-23699.

Dimensions:	
Length:	
B17B	1,143 mm (45·0 in)
C20B	1,046 mm (40·8 in)
C28B	1,239 mm (48·78 in)
C28C	1,032 mm (40·63 in)
C30	1,097 mm (43·2 in)
Width:	
B17B, C20B	483 mm (19·0 in)
C28, C30	557 mm (21·94 in)
Height:	
B17B	572 mm (22·5 in)
C20B	589 mm (23·2 in)
C28, C30	638 mm (25·13 in)
Weight, Dry:	
B17B	88·4 kg (195 lb)
C20B	71·5 kg (158 lb)
C28	99·3 kg (219 lb)
C28B	104 kg (230 lb)
C28C	102·5 kg (226 lb)
C30	106·6 kg (235 lb)
Performance Ratings (S/L, ISA):	
T-O:	
C20B (5 min)	313 kW (420 shp)
C28, 28B, 28C (30 min)	373 kW (500 shp)
C30 (2½ min)	522 kW (700 shp) to 32·2°C
Max continuous:	
B17B	287 kW (385 shp)
C20B	298 kW (400 shp)
C28	373 kW (500 shp)
C30	485 kW (650 shp)
Cruise B (75 per cent):	
B17B	205 kW (275 shp)
C20B	207 kW (278 shp)
C28	274 kW (367 shp)
C30	312 kW (418 shp)
Specific Fuel Consumption:	
At T-O rating:	
B17B	110·7 μg/J (0·655 lb/h/shp)
C20B	110 μg/J (0·650 lb/h/shp)
C28	102·5 μg/J (0·606 lb/h/shp)
C30	100 μg/J (0·592 lb/h/shp)
At cruise B rating:	
B17B	121 μg/J (0·716 lb/h/shp)
C20B	120 μg/J (0·709 lb/h/shp)
C28	112 μg/J (0·664 lb/h/shp)
C30	111 μg/J (0·657 lb/h/shp)

ALLISON GMA 300

This is the latest core engine developed under the USAF/USN Ategg (Advanced Turbine Engine Gas Generator) programme. Projected applications include a wide range of turbofans, some with variable-pitch blades, and turboshaft engines for air and surface applications.

ALLISON GMA 500

Advanced 600 kW (800 shp) class demonstrator engine under development for US Army.

ALLISON T56

Current versions of the T56 are as follows:

T56-A-14. Rated at 3,661 ekW (4,910 ehp). Generally similar to T56-A-15, but seven-point suspension like T56-A-10W and detail changes. Powers the P-3B and C Orion.

T56-A-15. Rated at 3,661 ekW (4,910 ehp). Introduced aircooled turbine blades. Powers C-130H (all versions), C-130K, HC-130N, HC-130P and some AC-130s.

T56-A-422. Rated at 3,661 ekW (4,910 ehp). Powers Grumman E-2C Hawkeye and C-2A Greyhound.

T56-A-423. Rated at 3,661 ekW (4,910 ehp). Powers US Navy versions of the C-130.

Including the Model 501 commercial engines, production of these engines reached 11,000 by February 1976. Production is continuing.

Model 501-D22D. This civil engine was proposed in 1977 to power the Lockheed L-400 Twin Hercules. It incorporates water/alcohol augmentation and auto-feathering.

Model 501-M71. The latest version of T56/501 offered for C-130 and P-3 aircraft would have 24 per cent greater power and 10 per cent lower sfc.

The following details apply to the T56-A-15:

Type: Axial-flow turboprop engine.

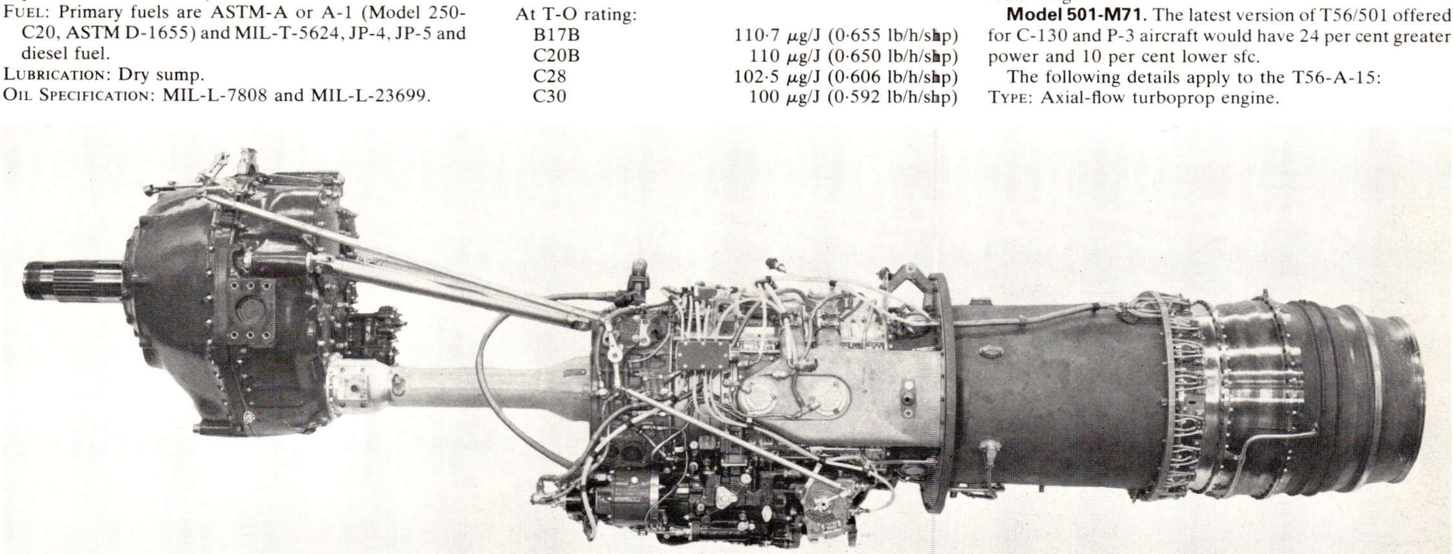

The 3,661 ekW (4,910 ehp) Allison T56-A-15 turboprop engine which powers late versions of the Lockheed C-130 Hercules transport

PROPELLER DRIVE: Combination spur/planetary gear type, primary step-down by spur, secondary by planetary. Overall gear ratio 13·54 : 1. Power section rpm 13,820. Cast magnesium reduction-gear housing. Gearbox assembly supported from power section by main drive shaft casing 711 mm (28 in) long and two inclined struts. Weight of gearbox assembly approximately 249 kg (550 lb) with pads on rear face for accessory mounting.

AIR INTAKE: Circular duct on engine face. Thermal de-icing.

COMPRESSOR: Fourteen-stage axial-flow. Series of fourteen discs with rotor blades dovetailed in peripheries and locked by adjacent discs. Rotor assembly tie-bolted to shaft which runs on one ball and one roller type bearing. Fifteen rows of stator blades, welded in rings. Disc, rotor and stator blades and four-piece cast casing of stainless steel. Compressor inlet area 1,004 cm² (155·65 sq in). Pressure ratio 9·5 : 1. Air mass flow 14·70 kg (32·4 lb)/s.

COMBUSTION CHAMBER: Six stainless steel cannular-type perforated combustion liners within one-piece stainless steel outer casing. Fuel nozzles in forward end of each combustor liner. Primary ignition by two igniters in diametrically-opposite combustors.

FUEL SYSTEM: High-pressure type. Bendix control system. Water/alcohol augmentation system available.

FUEL GRADE: MIL-J-5624, JP-4 or JP-5.

NOZZLE GUIDE VANES: Hollow aircooled blades of special high-temperature alloy.

TURBINE: Four-stage. Rotor assembly consists of four stainless steel discs, with first stage having hollow air-cooled blades of special high-temperature alloy, secured in peripheries of discs by fir-tree roots. Discs splined to rotor shaft which runs on front and rear roller bearings. Steel outer turbine casing. Gas temperature before turbine 1,076°C.

JET PIPE: Fixed. Stainless steel.

ACCESSORY DRIVES: Accessory pads on rear face of reduction-gear housing at front end of engine.

LUBRICATION SYSTEM: Low-pressure. Dry sump. Pesco dual-element oil pump. Normal oil supply pressure 3·8 bars (55 lb/sq in).

OIL SPECIFICATION: MIL-L-7808.

MOUNTING: Three-point suspension.

STARTING: Air turbine, gearbox-mounted.

DIMENSIONS:

Length (all current versions)	3,708 mm (146 in)
Width (all current versions)	686 mm (27 in)
Height:	
A-15, A-422, A-423	991 mm (39 in)
A-14	1,118 mm (44 in)

WEIGHT, DRY:

A-14	855 kg (1,885 lb)
A-15	828 kg (1,825 lb)
A-422	859 kg (1,984 lb)
A-423	836 kg (1,844 lb)
D22D	840 kg (1,851 lb)

PERFORMANCE RATINGS (S/L, ISA, static):

T-O:

D22D augmented
3,728 kW (5,000 shp) at 13,820 rpm

A-14, A-15, A-422, A-423, D22D
3,661 ekW; 3,424 kW (4,910 ehp; 4,591 shp) at 13,820 rpm

Normal:

A-14, A-15, A-422, A-423, D22D
3,255 ekW; 3,028 kW (4,365 ehp; 4,061 shp) at 13,820 rpm

SPECIFIC FUEL CONSUMPTION:

At max rating:

A-14, A-15 84·67 μg/J (0·501 lb/h/ehp)

At normal rating:

A-14, A-15, A-422, A-423
87·4 μg/J (0·517 lb/h/ehp)

OIL CONSUMPTION:

A-14, A-15 1·3 litres (0·35 US gallons)/h

ALTURDYNE

ALTURDYNE

8050 Armour Street, San Diego, California 92111
Telephone: (714) 565 2131
Telex: (910) 335-2000
PRESIDENT: Frank Verbeke

This company, "dedicated to bringing the advantages of the small gas turbine to new areas of industry and commerce", has developed small engines for RPVs and very light aircraft. It was formed in 1970 and employs 90 people.

At present it purchases the basic gas generator and incorporates this into a complete engine for various purposes. The two main products are of 112 kW (150 hp) and 157 kW (210 hp). These are based upon the Solar T62 Titan single-shaft gas turbine (last described in the 1975-76 *Jane's*). Alturdyne is marketing these units in APU, generator and air compressor applications. A company-funded aircraft turbine is under study.

ALTURDYNE SP-440

This RC (Wankel-type) engine is intended chiefly for surface applications, but its aviation appeal is self-evident. By 1981 an SP-440 was well advanced in flight development.

TYPE: Rotary-combustion engine, with oil-cooled rotor and overall air cooling.

DISPLACEMENT: 440 cc (26·9 cu in).

COMPRESSION RATIO: 7·6.

DIMENSIONS:

Length (typical mainshaft)	428 mm (16·85 in)
Width	471 mm (18·56 in)
Height	368 mm (14·49 in)
WEIGHT, DRY:	27·2 kg (60·0 lb)
MAX RATING:	29·8 kW (40 hp) at 6,000 rpm

ATLANTIC RESEARCH

ATLANTIC RESEARCH CORPORATION

5390 Cherokee Avenue, Alexandria, Virginia 22314
Telephone: (703) 642 4000

This is a privately held company of some 1,500 employees, organised into four operating divisions: Propulsion, Research and Technology, Teleproducts, and Advanced Programs. The Propulsion Division is a leading designer and manufacturer of small solid-propellant rockets and gas generators; it also produces industrial mixers and air-purification systems (for example, for all US Navy nuclear submarines). R & T includes in its work many areas relevant to aerospace propulsion. Teleproducts is concerned mainly with communications. APD's work is very diverse.

ARC SOLID ROCKET MOTORS

Atlantic Research claims to make solid propellants that "propel missiles, spin satellites, provide thrust-vector control, inflate air cushions, produce non-toxic exhaust, burn fast, burn slow, separate nose shrouds, gobble up electrons, produce smoke, produce no smoke, radiate IR or lase". The company supplies sustainer motors for Terrier and Standard ship/air missiles, the launch and sustainer motors for Redeye and Stinger infantry/air missiles, the complete Arcas, Super Arcas and Boosted Arcas meteorological and atmosphere-research rockets, motors for later Minuteman ICBMs that eject shrouds and perform other functions, the gas generator for UpSTAGE (unaffected by extreme launch acceleration), the gas generators and auxiliary rockets for the Polaris and Poseidon programmes, and the very-long-burning dual-pressure gas generator for C-4 Trident. ARC Propulsion Division is under contract to General Dynamics Convair Division to design and manufacture the boost motor for the BGM-109 sea-launched Tomahawk cruise missile, and to Vought to supply the solid propulsion system for MLRS (Multiple Launch Rocket System). R & T Division is researching alternative energy sources, including a substitute for No. 6 fuel oil. In February 1981 an Air Force contract called for research into advanced composites in solid-propellant rocket motor components.

AVCO LYCOMING

AVCO LYCOMING STRATFORD DIVISION OF AVCO CORPORATION

HEAD OFFICE: 550 South Main Street, Stratford, Connecticut 06497
Telephone: (203) 378 8211
PRESIDENT OF AVCO CORPORATION: Ross M. Hett
VICE PRESIDENTS (STRATFORD):
J. R. Myers (General Manager)
A. J. Burrows (Washington Operation)
M. J. Leff (Marketing/Product Support)
M. S. Saboe (T55 & ALF 502 Programme Management)
P. J. Foley (LT 101 & T53 Programme Management)
J. Laborde (AGT 1500/XM1 Programme Management)
R. Ainsworth (Engineering)
C. Kuintzle (Assistant General Manager)
D. Painter (Procurement & Material Control)
H. J. Frederick (Finance)

Avco Lycoming Stratford is the gas-turbine manufacturing division of Avco Corporation. Together with the Williamsport division, it forms the Avco Lycoming Engine Group.

Stratford division is producing several families of engines, including the T53, T55, LTS 101 and LTP 101, with turboshaft, turboprop, turbofan, vehicular and marine variants. It also produces the AGT 1500 engine of the XM1 battle tank. More than 23,000 turbine engines had been delivered by 1981.

Avco Lycoming is working upon an advanced-technology turboshaft under US Army contract. The engine is a high-performance low-consumption unit for use in a demonstrator programme.

AVCO LYCOMING ALF 502

The ALF 502 turbofan was developed primarily as a power plant for commercial and executive aircraft. The

Avco Lycoming ALF 502L geared turbofan rated at 33·4 kN (7,500 lb st)

ALF 502L and L-2 versions received FAA type certification in February 1980. The lower-rated ALF 502R-3 was certificated in January 1981. The core gas-generator of the ALF 502 is the T55.

The engine was launched as a commercially directed, company-sponsored effort in 1969. Its first application was in the Northrop A-9A, with the military engine designation F102. Nonetheless, the design of the ALF 502 remained directed at the commercial requirement for low cost of operation and high reliability. Construction was made totally modular, and the basic design was refined to satisfy commercial needs of long life and low cost of operation, with on-condition maintenance instead of a fixed TBO. Thus, a TBO of 4,000 h has been granted, with on-condition authorisation. The ALF 502 has flown in several aircraft, demonstrating low noise signature, and at altitudes up to 14,950 m (49,000 ft).

Two ALF 502L-2 turbofans power the Canadair CL-600 Challenger. The British Aerospace BAe 146 has four ALF 502R-3s. NASA is flying the Quiet Short-haul Research Aircraft (QSRA) with four YF102 turbofans that were used previously to power the Northrop A-9A.

Engine deliveries off production tooling were made in December 1977, with quantity production starting in 1979.

Type: High bypass ratio, two-shaft geared turbofan.

Fan Module: Cast fan frame includes four engine mounts 90° apart, and may be used for mounting thrust reverser. Fan rotor blades are base and part-span shrouded and individually replaceable. Containment ring surrounds rotor. Mounted directly behind rotor (6,700 lb st engines) is single or (7,500 lb st engines) two stages of compression. Individually replaceable fan exit stators are located to minimise noise. Inlet spinner bolted to fan rotor is anti-iced by oil flow. Anti-icing of compressor stators by bleed air. Accessory gearbox on fan frame takes HP shaft power. Additional services such as fuel, oil, pneumatic and electrical lines, are routed through other hollow struts. Reduction gear within centre of fan frame couples LP turbine to fan. Bypass ratio: 502R-3, 5·71; 502L, 5·0.

Compressor: Core supercharged by one or two LP stages, as described above. HP compressor has seven axial stages and single radial stage, driven by two-stage HP turbine. Supported by roller bearing and aft ball bearing. Acceleration bleed control between stages 6 and 7 operated by main fuel control. Upper and lower casings contain stator half-rings, removable for maintenance, and borescope ports. Overall engine pressure ratio: R-3, 11·6; L-2, 13·6.

Turbine: HP has two air-cooled stages. LP has two stages, splined to fan reduction gear. All rotor blades base shrouded: LP additionally tip shrouded.

Combustion Chamber: One-piece annular combustor wraps around turbine. Atomising nozzles inserted through outer chamber at rear. Four customer bleed ports around exterior. Combustion liner and housing bolted to compressor diffuser. Disconnecting permits removal of cumbustor/turbine module, providing access to HP turbine. Borescope ports permit inspection of combustor and turbine without removal.

Accessory Drives: Accessory gearbox carries main fuel control, oil pump and filter, tachometer (if required) and provisions for customer accessories.

Exhaust Unit: Fan and core exhaust include flanges for bolted ducts and nozzles.

Data: See table.

AVCO LYCOMING ALF 101

Supported by NASA through the QCGAT (Quiet Clean General-Aviation Turbofan) programme, the ALF 101 is a high-bypass ratio turbofan based on the same gas generator as the LTP 101 turboprop and LTS 101 turboshaft. The total engine airflow passing through the front fan is split into two paths. The larger portion bypasses the core engine and exhausts directly to atmosphere through the mixed flow or annular exhaust nozzles. The smaller portion is ducted into the core engine through the core inlet stator. The core air is further compressed by a two-stage axial compressor in tandem with a single-stage centrifugal compressor prior to entering the reverse-flow annular combustor.

A single-stage axial gas-producer turbine, coupled to the compressor shaft, drives the high-pressure compressor. A single-stage low-pressure power turbine, with a shaft extending forward concentrically through the hollow compressor shaft to a fixed-planetary reduction gear, provides the drive for the fan. The front frame includes main engine mounting provisions and ducting for the fan and core airflows. Hollow struts provide fairings for services across the fan and core streams. Accessory power from the compressor rotor is transmitted through a bevel drive in the front frame to the accessory gearbox, which is mounted externally to optimise accessibility.

The NASA QCGAT commitments were completed during 1979, with the delivery of an engine to NASA's Lewis Research Center for further testing. Available data appear in the table.

AVCO LYCOMING LTS 101

US military designation: YT702-LD-700

This series of small turboshaft engines includes seven models with maximum power ratings between 459 and 548 kW (615 and 735 shp) and output speeds of 6,000 or 9,265 rpm. All commercial models are certificated by the FAA, and the military YT702 has completed a 60-h PFRT. This simple modular engine was designed to provide low life cycle costs. It has a single axial compressor stage followed by a single centrifugal stage, a reverse-flow annular combustor, and single-stage axial gas-generator and power turbines. The front-mounted gearbox provides forward or aft drives, or a combination. The radial-flow inlet provides for an optional particle separator. Mass flow is 2·03 kg (4·8 lb)/s, and pressure ratio 8·5 : 1.

The LTS 101 first flew in a Bell JetRanger testbed in 1973. Since then it has been selected to power the Bell 222, Aérospatiale AS 350D, Aérospatiale SA 366, MBB/Kawasaki BK 117 and Aérospatiale HH-65A Dolphin for the US Coast Guard. Lycoming has also received US Government contracts which will provide a turbofan and an airborne APU both based on the LTS 101 core.

Data: see table.

AVCO LYCOMING LTP 101

The LTP 101 is a modular turboprop derived from the LTS 101. It incorporates a free power turbine, provisions for tractor or pusher installation, hydraulic propeller governor, radial screened inlet and anti-icing protection. The output shaft speed lies in the range 1,700-1,950 rpm, consistent with available three-blade reversible-pitch propellers.

FAA certification was received in 1976. The LTP 101 has since been selected for the Piaggio P.166-DL3, Air

Avco Lycoming turbine engines in the 600 shp class (from left to right): LTS 101-600B/650A (459 kW; 615 shp), LTS 101-650C (504 kW; 675 shp) and LTP 101-600 (462 ekW; 620 ehp)

AVCO LYCOMING GAS TURBINE ENGINES

Manufacturer's and civil designation	Military designation	Type*	T-O Rating kN (lb st) or max kW (hp)	SFC µg/J; ‡ mg/Ns (lb/h/hp; ‡lb/h/lb st)	Weight dry less tailpipe kg (lb)	Max dia mm (in)	Length overall mm (in)	Remarks
T5313B	—	ACFS	1,044 kW (1,400 shp)	98 (0·58)	245 (540)	584 (23)	1,209 (47·6)	Powers Bell 205A
T5317A	—	ACFS	1,119 kW (1,500 shp)	99·7 (0·59)	256 (564)	584 (23)	1,209 (47·6)	Based on T5319A
T5319A	—	ACFS	1,342 kW (1,800 shp)	96·3 (0·57)	256 (564)	584 (23)	1,209 (47·6)	Awaiting FAA certification
T5311A	—	ACFS	820 kW (1,100 shp)	115 (0·68)	225 (496)	584 (23)	1,209 (47·6)	Bell 204B
—	T53-L-13B	ACFS	1,044 kW (1,400 shp)	98 (0·58)	245 (540)	584 (23)	1,209 (47·6)	Advanced UH-1H, AH-1G
—	T53-L-702	ACFS	1,417 kW (1,900 shp)	94·6 (0·56)	254 (561)	584 (23)	1,209 (47·6)	Military T5319A
—	T53-L-703	ACFS	1,106 kW (1,485 shp)	101·4 (0·60)	247 (545)	584 (23)	1,209 (47·6)	Bell AH-1Q, AH-1S TOW/Cobra
LTC1K-4C	—	ACFS	1,119 kW (1,500 shp)	98 (0·58)	247 (545)	584 (23)	1,209 (47·6)	Canadair CL-84
T5321A	—	ACFP	1,393 ekW (1,868 ehp)	96·3 (0·57)	306 (675)	584 (23)	1,656 (65·2)	Turboprop T5319A
—	T53-L-701	ACFP	1,082 ekW (1,451 ehp)	101·4 (0·60)	312 (688)	584 (23)	1,483 (58·4	Grumman OV-1D
LTC4R-1	—	ACFP	2,837 ekW (3,804 ehp†)	87·9 (0·52)	422 (930)	615 (24·2)	1,580 (62·2)	Turboprop T55-L-11A
—	T55-L-7C	ACFS	2,125 kW (2,850 shp)	101·4 (0·60)	267 (590)	615 (24·2)	1,118 (44)	Boeing CH-47B, Bell 214A
T5508D (LTC4B-8D)	—	ACFS	1,750 kW (2,347 shp) (flat-rated)	106·5 (0·63)	274 (605)	610 (24)	1,118 (44)	Bell 214A, 214B
—	T55-L-11A§	ACFS	2,796 kW (3,750 shp)	89·6 (0·53)	322 (710)	615 (24·2)	1,181 (46·5)	Boeing CH-47
LTC4B-12	—	ACFS	3,430 kW (4,600 shp)	86·2 (0·51)	329 (725)	615 (24·2)	1,118 (44)	Improved T55-L-11A
LTS 101-600	YT702-LD-700	ACFS	459 kW (615 shp)	95·8 (0·567)	110 (241)	584 (23)	785 (30·9)	Bell 222, AS 350D, BK 117, HH-65A
LTS 101-650C-2	—	ACFS	507 kW (680 shp)	95·6 (0·566)	105 (232)	584 (23)	785 (30·9)	
LTP 101-600	—	ACFP	462 ekW (620 ehp)	93 (0·55)	148 (325)	533 (21)	914 (36)	Piaggio P.166-DL3, Air Tractor AT-302
LTP 101-700	—	ACFP	522 ekW (700 ehp)	93 (0·55)	148 (325)	533 (21)	914 (36)	
ALF 101	—	ACFF	7·2 kN (1,620 lb)	‡10·19 (‡0·36)	156 (343)	584 (23)	890 (35)	NASA QCGAT
ALF 502R-3	—	ACFF	29·8 kN (6,700 lb)	‡11·64 (‡0·411)	565 (1,245)	1,059 (41·7)	1,443 (56·8)	BAe 146
—	F102-LD-100	ACFF	35 kN (7,860 lb)	‡11·89 (‡0·42)	500 (1,100)	1,041 (41)	1,422 (56)	NASA QSRA
ALF 502L/L-2	—	ACFF	33·4 kN (7,500 lb)	‡12·1 (‡0·428)	590 (1,298)	1,059 (41·7)	1,487 (58·56)	Canadair CL-600 Challenger

*ACFS = axial plus centrifugal, free-turbine shaft; ACFP = axial plus centrifugal, free-turbine propeller; ACFF = axial plus centrifugal, free-turbine fan

†2,752 kW (3,690 shp); also has military rating of 2,574 ekW/2,494 kW (3,452 ehp/3,344 shp).

§Applies to T55-L-11A, C**, D, E** and 712**, those designated ** having 2½ min contingency rating of 3,357 kW (4,500 shp).

Tractor Turbo Tractor and New Zealand Aerospace Industries Fletcher agricultural aircraft, Riley conversion of Cessna 421, Page Aircraft Turbo Thrush and Ag-Cat conversions and OMAC I. It is also flying in Turbo-Islander and Turbo-Skyservant prototypes.
DATA: See table.

AVCO LYCOMING LTC1

US military designation: T53

The T53 is a turboshaft with a free power turbine, which was developed under a joint US Air Force/US Army contract. More than 18,000 units had logged over 29 million hours of operation, with every US armed service and in 29 other countries, by January 1981.

Licences for manufacture of the T53 are held by Klöckner-Humboldt-Deutz in Federal Republic of Germany, Piaggio in Italy, Kawasaki in Japan, and in Taiwan.

Versions currently in production or under development are as follows:

T53-L-13. Uprated version of L-11, which it superseded in production in August 1966.Redesigned 'hot end' and initial stages of compressor section to provide substantially increased power for hot-day and high-altitude performance. Four turbine stages, compared with two in earlier models, and variable-incidence inlet guide vanes combined with redesigned first two compressor stages, permit greater airflow and lower turbine temperatures. This version has atomising combustor to facilitate operation on a wider range of fuels. Powers Bell UH-1C and UH-1D and AH-1G HueyCobra. The **T5313A** commercial version of the T53-L-13 received FAA type certification in 1968 for the Bell 205A and has been superseded by the T5313B.

T53-L-701. Turboprop version of the L-13 incorporating the Lycoming 'split-power' propeller reduction gear. Produced for Grumman OV-1D previously powered by T53-L-15, and in production for T-CH-1 (Taiwan).

T53-L-703. Improved-durability L-13. Flat rated for AH-1Q and AH-1S HueyCobra.

LTC1K-4C. Generally similar to the T53-L-13, but incorporating special seals to allow operation in the attitude range from 105° nose up to 90° nose down. Has 10-minute rating of 1,500 shp. Powered Canadair CL-84-1 VTOL aircraft.

LTC1K-4K. Direct-drive version of LTC1K-4C rated at 1,800 shp; powers Bell XV-15 VTOL aircraft.

T5319A. Latest growth version of T53 turboshaft family. Improvements over L-13 include new gearing, improved cooling of first gas producer turbine nozzle plus aircooled blades in first turbine rotor. Also incorporates new materials in other turbine stages. Rated at 1,800 shp at take-off.

T5317A. Flat rated version of -19A with take-off rating limited to 1,500 shp by use of standard L-13 reduction gear.

T5321A. Turboprop version of -19A with 'split-power' gear. Uses standard SBAC No. 4 propeller shaft with through-the-shaft oil provisions.

The following details apply to the T53-L-13 and L-701:

TYPE: Free-turbine turboshaft engine.

AIR INTAKE: Annular casing of magnesium alloy, with 6 struts supporting reduction gearbox and front main bearings. Anti-icing by hot air tapped from engine.

COMPRESSOR: Five axial stages followed by a single centrifugal stage. Four-piece magnesium alloy casing with one row of variable-incidence inlet guide vanes and five rows of steel stator blades, bolted to one-piece steel alloy diffuser casing with tangential outlet to combustion chamber. Rotor comprises one stainless steel disc, and welded titanium drum construction for the four remaining axial stages, with stainless steel blades and one titanium impeller. Rotor shaft supported by front ball thrust bearing and rear roller bearing. Pressure ratio is 7·2, and mass flow 5·53 kg (12·2 lb)/s at 25,150 gas producer rpm. Four aluminium alloy discs with stainless steel blades and one titanium impeller mounted on shaft supported in forward ball thrust and rear roller bearings. Compression ratio 7·4 : 1. Air mass flow 4·85 kg/s (10·7 lb/s) at 25,240 gas producer rpm.

COMBUSTION CHAMBER: Annular reverse-flow type, with one-piece sheet steel outer shell and annular liner. Twenty-two atomising fuel injectors.

FUEL CONTROL SYSTEM: Hydromechanical controls for gas generator and for power sections. Chandler Evans TA-2S system with one dual fuel pump. Pump pressure 41·4 bars (600 lb/sq in). Main and emergency flow controls. Separate interstage air bleed control.

FUEL GRADE: ASTM A-1, MIL-J-5624, MIL-F-26005A, JP-1, JP-4, JP-5, CITE.

TURBINE: Four axial-flow turbine stages. Casing fabricated from sheet steel. First two stages, driving compressor, use hollow aircooled stator vanes and cored-out cast steel rotor blades and are mounted on outer coaxial shaft to gas producer. Second two stages, driving reduction gearing, have solid steel blades, and are spline-mounted to shaft.

EXHAUST UNIT: Fixed-area nozzle. Steel outer casing and inner cone, supported by four radial struts.

ACCESSORIES: Electric starter or starter/generator (not furnished). Bendix-Scintilla TGLN high-energy ignition unit. Two igniter plugs.

LUBRICATION: Recirculating system, with gear pump with one pressure and one scavenge unit. Filter. Pump pressure 4·83 bars (70 lb/sq in).

OIL GRADE: MIL-L-7808, MIL-L-23699.

DATA: See table.

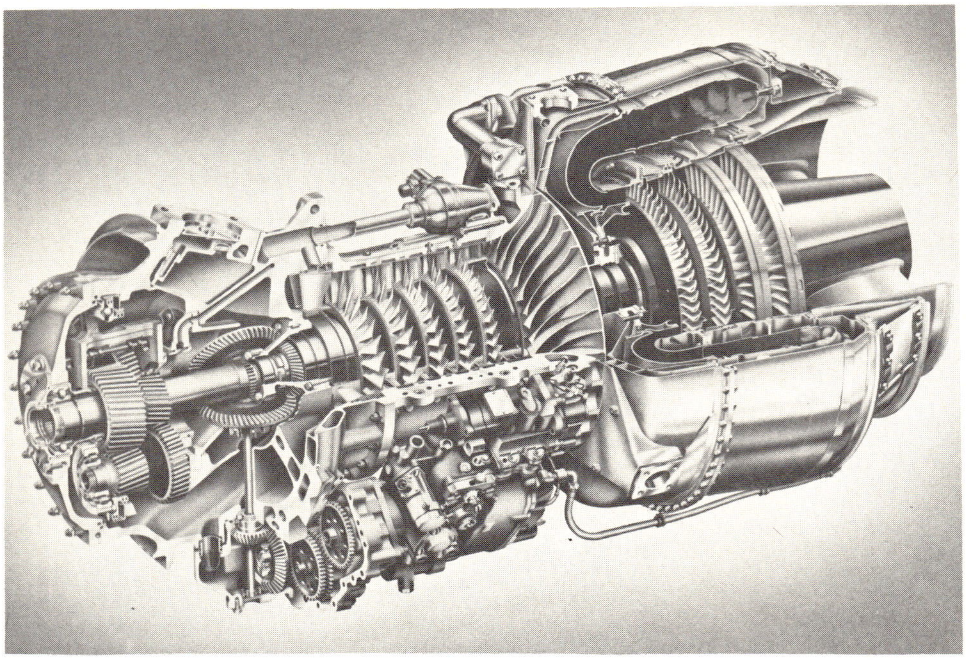

Cutaway drawing of the 1,044 kW (1,400 shp) Avco Lycoming T53-L-13 turboshaft engine

AVCO LYCOMING LTC4

US military designation: T55

This engine is based on the T53 design concept but with higher mass flow. It was developed under a joint US Air Force/US Army contract. Total operating time by early 1981 was 3·7 million hours. Most of this time was logged by L-7 versions fitted to the CH-47A and CH-47B Chinook.

Current production and development versions are as follows:

LTC4B-8D. Modified version of the T55-L-7C. Powers Bell 214A and 214C utility helicopters for Iran, flat rated to transmission limit of 1,678 kW (2,250 shp).

T5508D. Commercial version of LTC4B-8D. Powers Bell 214B.

T55-L-11 (LTC4B-11B) series. Uprated and redesigned version of L-7, with a second stage added to the compressor turbine, and variable-incidence inlet guide vanes ahead of the compressor. First two compressor stages transonic. Atomising fuel nozzles. Powers CH-47C Chinook, first deliveries having been made in August 1968.

T55-L-712. Improved T55-L-11D fitted to the US Army's modernised Boeing Vertol YCH-47D. Wide-chord compressor blades and one-piece rotor. Emergency (engine-out) rating 3,357 kW (4,500 shp). Improvements in design and materials are aimed at achieving a time between overhauls of 2,500 h.

AL5512. Commercial version of L-712, with engine-out contingency rating of 3,250 kW (4,355 shp). Certificated in 1980 for Boeing Vertol 234.

LTC4B-12. Proposed growth version with 3,430 kW (4,600 shp) maximum power rating, 3,258 kW (4,370 shp) on hot day. Higher turbine entry temperature and increased turbine cooling.

LTC4R-1. Proposed turboprop version of L-11 with Lycoming split-power reduction gear.

The following description applies to the T55-L-11 and LTC4R-1:

TYPE: Free-turbine turboshaft engine.

AIR INTAKE: Annular type casing of magnesium alloy with four struts supporting reduction gearbox and front main bearings. Anti-icing by hot air tapped from engine. Provision for intake screens.

COMPRESSOR: Seven axial stages followed by a single centrifugal stage. Two-piece magnesium alloy stator casing with one row of variable inlet guide vanes and seven rows of steel stator blades, bolted to steel diffuser casing to which combustion chamber casing is attached. Late models have wide-chord blades in first two stages and no inlet guide vanes. Rotor comprises seven steel discs and one titanium impeller mounted on shaft supported in forward ball-thrust bearing and rear roller bearing. Pressure ratio 8·2 : 1. Air mass flow 12·25 kg (27 lb)/s.

COMBUSTION CHAMBER: Annular reverse-flow type. Steel outer shell and inner liner. Twenty-eight fuel burners with downstream injection.

FUEL SYSTEM: Hamilton Standard JFC 31 fuel control system. Gear-type fuel pump, with gas producer and power shaft governors, flow control with altitude compensation and shut-off valve.

FUEL GRADE: MIL-J-5624 grade JP-4, JP-5, MIL-F-46005A or CITE.

Avco Lycoming T55-L-712 (military) or AL5512 (commercial) turboshaft

TURBINE: Two mechanically independent axial turbines. Gas-generator turbine has single stage (two stages on T55-L-11 series) with cored-out cast steel blades having inner cooling airflow. Disc flange-bolted to drive shaft. Hollow stator vanes. Two-stage power turbine has solid steel blades.

EXHAUST UNIT: Fixed-area nozzle, with inner cone, supported by six radial struts.

ACCESSORIES: Electric starter or starter/generator, or air or hydraulic starter. Bendix-Scintilla TGLN high-energy ignition unit. Four igniter plugs.

LUBRICATION: Recirculating type. Integral oil tank and cooler on L-11, external tank for 4R-1.

OIL GRADE: MIL-L-7808, MIL-L-23699.

DATA: See table.

AVCO LYCOMING WILLIAMSPORT DIVISION

Williamsport, Pennsylvania 17701
Telephone: (717) 323 6181
VICE-PRESIDENTS:
Peter J. Goodwin (General Manager)
Howard M. Knutson (Assistant General Manager)
A. E. Light (Engineering)
S. T. Jedrziewski (Product Planning)
W. R. Bower (Operations)
H. A. Schuck (Controller)

Williamsport Division is engaged primarily in the production of well-known Lycoming series of horizontally-opposed aircooled reciprocating engines ranging from 110 to 450 hp. Turbocharging is being offered on additional models and the horsepower, in some cases, has been increased. Development efforts are being directed to improvements resulting in lower cost of manufacturing and longer time between overhauls, to help offset increasing labour and material costs. During recent years FAA approval has been received for several new engine models. Turbocharged versions of four- and six-cylinder engines have been certificated, along with several versions of new high-compression, simplified-design engines.

The 231 kW (310 hp) Avco Lycoming TIO-540 six-cylinder engine fitted to the Piper Navajo C

AVCO LYCOMING O-235 SERIES

Smallest and basic engine of the Lycoming range. Four cylinders of 111 mm (4⅜ in) bore and 98·4 mm (3⅞ in) stroke. The high-compression O-235-L2C is the most recent production version of the O-235, used in several primary trainers. It requires 100-octane (minimum) aviation fuel.

AVCO LYCOMING O-320 and IO-320 SERIES

Cylinder bore increased to 130 mm (5⅛ in). The O-320 is an engine in the 112-119 kW (150-160 hp) class. Both carburetted and fuel-injected versions are produced in low- and high-compression models for use with 80/87 or 100-octane minimum grade fuels, respectively. Fully aerobatic models are available.

AVCO LYCOMING O-360 and IO-360 SERIES

The O-360 series is basically the same as the O-320 except for an increase in stroke to 111 mm (4⅜ in). Like the O-320, this engine is manufactured with low or high compression, with carburettor or fuel injection. The various models include aerobatic capability, a specific design for helicopters and a turbocharged version. The IO-360 is built in several versions; the IO-360-A series has fuel injection, tuned induction and high-output cylinders, while the IO-360-B series has continuous-flow port injection and standard cylinders.

Latest additions to this family are the TO-360-E series rated at 134 kW (180 hp). These engines have a simple turbocharger with a fixed-orifice exhaust bleed. Engine and turbo speed are both controlled by the pilot's throttle.

AVCO LYCOMING O-540 and IO-540 SERIES

The O-540 is basically a direct-drive, six-cylinder version of the four-cylinder O-360. It is available in low- and high-compression versions, and the VO-540 is in production as a helicopter powerplant with crankshaft vertical. Fuel-injected IO-540 models are manufactured in a variety of configurations, with rating of 186-224 kW (250-300 hp). An aerobatic version is available. Many of these engines, particularly in the higher horsepower range, have a tuned induction system. Continuous monitoring and computer analysis of engine service records has made it possible to increase the recommended time between overhaul (TBO) for these engines; most models have a TBO of 2,000 h. The most recent members of this family are the -J series of direct-drive engines.

AVCO LYCOMING TIO-540 SERIES

This is a turbocharged version of the fuel-injected IO-540, with tuned induction. It is manufactured in a wide range of horsepowers and is used in the Piper Turbocharged Aztec and all unpressurised models of the Piper Navajo. The latest addition to the TIO-540 series is the -S model with simple turbocharger interconnected with the pilot's throttle.

Avco Lycoming O-235-L rated at 86 kW (115 hp)

The 175 kW (235 hp) Avco Lycoming O-540-J six-cylinder engine

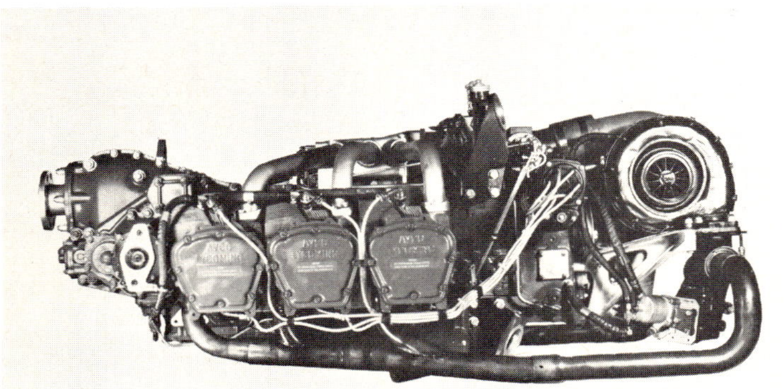
The 317 kW (425 hp) Avco Lycoming TIGO-541-E six-cylinder engine

The 298 kW (400 hp) eight-cylinder Avco Lycoming IO-720-A1A engine

AVCO LYCOMING TIO-541 SERIES

Although the displacement of this turbocharged, six-cylinder series is the same as the TIO-540, the TIO-541 and geared TIGO-541 are totally redesigned. The accessory housing is an integral part of the crankcase, and the new design features allow these engines to operate at the high end of the Lycoming power spectrum. The TIO-541-E is rated at 283 kW (380 hp) and the geared TIGO-541-E at 317 kW (425 hp). Times between overhauls are recommended at 1,600 and 1,200 h, respectively. A double-scroll blower is available to provide cabin pressurisation.

AVCO LYCOMING IO-720 SERIES

This eight-cylinder version of the IO-540 is used in a variety of aircraft at ratings from 280 to 298 kW (375 to 400 hp). Time between overhauls is 1,800 h (1,500 in agricultural applications).

AVCO LYCOMING NEW RANGE

Avco Lycoming, Williamsport, is continuing the development of a new series of high-performance, normally-aspirated and turbocharged flat opposed piston engines. In addition, extensive study and testing are being devoted to finding ways of reducing noise and improving overall efficiency.

AVCO LYCOMING HORIZONTALLY-OPPOSED PISTON ENGINES

Engine Model*	No. of Cylinders	Rated output at Sea Level kW (hp) at rpm	Capacity litres (cu in)	Compression Ratio	Fuel grade Minimum	Weight Dry kg (lb)	Length Overall mm (in)	Width Overall mm (in)	Height Overall mm (in)	Gear Ratio†
O-235-C	4	86 (115) at 2,800	3·85 (233)	6·75	80/87	97·5 (215)	751 (29·56)	812 (32·00)	569 (22·53)	D
O-235-H	4	81 (115) at 2,600	3·85 (233)	6·75	80/87	96 (213)	738 (29·05)	812 (32·00)	569 (22·40)	D
O-235-L	4	86 (115) at 2,700 78 (105) at 2,400	3·85 (233)	8·5	100	98 (218)	738 (29·05)	812 (32·00)	569 (22·40)	D
O-320-A	4	112 (150) at 2,700	5·2 (319·8)	7·0	80/87	110 (243)	751 (29·56)	819 (32·24)	584 (22·99)	D
O-320-D	4	119 (160) at 2,700	5·2 (319·8)	8·5	91/96	114 (253)	808 (31·82)	819 (32·24)	488 (19·22)	D
O-320-E	4	112 (150) at 2,700	5·2 (319·8)	7·0	80/87	113 (249)	738 (29·05)	819 (32·24)	584 (22·99)	D
O-320-H	4	119 (160) at 2,700	5·2 (319·8)	9·0	100	115 (253)	819 (32·26)	830 (32·68)	621 (24·46)	D
O-360-A	4	134 (180) at 2,700	5·92 (361)	8·5	91/96	118 (260)	808 (31·82)	848 (33·37)	488 (19·22)	D
LO-360-A	4	134 (180) at 2,700	5·92 (361)	8·5	91/96	120 (266)	808 (31·82)	848 (33·37)	488 (19·22)	D
O-360-E	4	134 (180) at 2,700	5·92 (361)	9·0	100	122 (269)	819 (32·26)	859 (33·81)	535 (21·08)	D
LO-360-E	4	134 (180) at 2,700	5·92 (361)	9·0	100	122 (269)	819 (32·26)	859 (33·81)	535 (21·08)	D
O-360-F	4	134 (180) at 2,700	5·92 (361)	8·5	100	122 (269)	808 (31·81)	859 (33·38)	507 (19·96)	D
IVO-360-A	4	134 (180) at 2,900	5·92 (361)	8·5	100	124 (274)	762 (30·00)	848 (33·37)	583 (22·95)	DV
TO-360-C	4	157 (210) at 2,575 to 3,050 m (10,000 ft)	5·92 (361)	7·3	100	154 (343)	876 (34·50)	921 (36·25)	534 (21·02)	D
TO-360-E	4	134 (180) at 2,575	5·92 (361)	8·0	100	143 (315)	972 (38·28)	859 (33·81)	539 (21·21)	D
LTO-360-E	4	134 (180) at 2,575	5·92 (361)	8·0	100	143 (315)	972 (38·28)	859 (33·81)	539 (21·21)	D
IO-360-A	4	149 (200) at 2,700	5·92 (361)	8·7	100	133 (293)	757 (29·81)	870 (34·25)	491 (19·35)	D
IO-360-B	4	134 (180) at 2,700	5·92 (361)	8·5	100	122 (268)	757 (29·81)	848 (33·37)	631 (24·84)	D
IO-360-C	4	149 (200) at 2,700	5·92 (361)	8·7	100	134 (298)	855 (33·65)	870 (34·25)	495 (19·48)	D
HIO-360-C	4	153 (205) at 2,900	5·92 (361)	8·7	100	132 (291)	791 (31·14)	870 (34·25)	495 (19·48)	D
HIO-360-D	4	142 (190) at 3,200 to 1,280 m (4,200 ft)	5·92 (361)	10·0	100	132 (290)	894 (35·23)	904 (35·62)	495 (19·48)	D
HIO-360-E	4	142 (190) at 2,900	5·92 (361)	8·0	100	132 (290)	797 (31·36)	870 (34·25)	507 (19·97)	D
HIO-360-F	4	142 (190) at 3,050	5·92 (361)	8·0	100	133 (293)	797 (31·36)	870 (34·25)	507 (19·97)	D
AEIO-360-A	4	149 (200) at 2,700	5·92 (361)	8·7	100	139 (307)	780 (30·70)	870 (34·25)	492 (19·35)	D
AEIO-360-B	4	134 (180) at 2,700	5·92 (361)	8·5	91/96	125 (277)	738 (29·05)	848 (33·37)	631 (24·84)	D
O-540-B	6	175 (235) at 2,575	8·86 (541·5)	7·2	80/87	166 (366)	945 (37·22)	848 (33·37)	624 (24·56)	D
O-540-E	6	194 (260) at 2,700	8·86 (541·5)	8·5	91/96	167 (368)	976 (38·42)	848 (33·37)	624 (24·56)	D
O-540-G	6	194 (260) at 2,700	8·86 (541·5)	8·5	91/96	174 (386)	999 (39·34)	848 (33·37)	624 (24·56)	D
O-540-J	6	175 (235) at 2,400	8·86 (541·5)	8·5	100	162 (357)	989 (38·93)	848 (33·37)	519 (20·43)	D
VO-540-B	6	227 (305) at 3,200	8·86 (541·5)	7·3	80/87	202 (446)	882 (34·73)	880 (34·70)	617 (24·29)	D V
VO-540-C	6	227 (305) at 3,200 to 915 m (3,000 ft)	8·86 (541·5)	8·7	100	200 (441)	882 (34·73)	880 (34·70)	649 (25·57)	D V
IO-540-C	6	186 (250) at 2,575	8·86 (541·5)	8·5	91/96	170 (375)	976 (38·42)	848 (33·37)	622 (24·46)	D
IO-540-K	6	224 (300) at 2,700	8·86 (541·5)	8·7	100	201 (443)	999 (39·34)	870 (34·25)	498 (19·60)	D
IO-540-S	6	224 (300) at 2,700	8·86 (541·5)	8·7	100	201 (444)	997 (39·24)	870 (34·25)	498 (19·60)	D
IO-540-T	6	194 (260) at 2,700	8·86 (541·5)	8·5	91/96	171 (381)	989 (38·93)	848 (33·37)	546 (21·50)	D
AEIO-540-D	6	194 (260) at 2,700	8·86 (541·5)	8·5	91/96	174 (386)	999 (39·34)	848 (33·37)	621 (24·46)	D
TIO-540-A	6	231 (310) at 2,575 to 4,575 m (15,000 ft)	8·86 (541·5)	7·3	100	232 (511)	1,304 (51·34)	870 (34·25)	577 (22·71)	D
TIO-540-C	6	186 (250) at 2,575 to 4,575 m (15,000 ft)	8·86 (541·5)	7·2	100	205 (456)	1,026 (40·38)	848 (33·37)	770 (30·33)	D
TIO-540-F	6	242 (325) at 2,575 to 4,575 m (15,000 ft)	8·86 (541·5)	7·3	100	233 (514)	1,304 (51·34)	870 (34·25)	570 (22·42)	D
LTIO-540-F	6	242 (325) at 2,575 to 4,575 m (15,000 ft)	8·86 (541·5)	7·3	100	233 (514)	1,304 (51·34)	870 (34·25)	570 (22·42)	D
TIO-540-J	6	261 (350) at 2,575 to 4,575 m (15,000 ft)	8·86 (541·5)	7·3	100	235 (518)	1,308 (51·50)	870 (34·25)	573 (22·56)	D
LTIO-540-J	6	261 (350) at 2,575 to 4,575 m (15,000 ft)	8·86 (541·5)	7·3	100	235 (518)	1,308 (51·50)	870 (34·25)	573 (22·56)	D
TIO-540-R	6	261 (350) at 2,500 254 (340) to 4,575 m (15,000 ft)	8·86 (541·5)	7·3	100	238 (524)	1,309 (51·52)	870 (34·25)	574 (22·60)	D
TIO-540-S	6	224 (300) at 2,700 to 3,660 m (12,000 ft)	8·86 (541·5)	7·3	100	228 (502)	1,004 (39·56)	915 (36·02)	667 (26·28)	D
TIO-541-E	6	283 (380) at 2,900 to 4,575 m (15,000 ft)	8·86 (541·5)	7·3	100	270 (596)	1,282 (50·70)	905 (35·66)	640 (25·17)	D
TIGO-541-E	6	317 (425) at 3,200 to 4,575 m (15,000 ft)	8·86 (541·5)	7·3	100	319 (704)	1,462 (57·57)	885 (34·86)	575 (22·65)	0·667
IO-720-A	8	298 (400) at 2,650	11·84 (722)	8·7	100	257 (567)	1,179 (46·41)	870 (34·25)	573 (22·53)	D
IO-720-B	8	298 (400) at 2,650	11·84 (722)	8·7	100	252 (556)	1,218 (47·97)	870 (34·25)	530 (20·88)	D
IO-720-D	8	298 (400) at 2,650	11·84 (722)	8·7	100	259 (570)	1,189 (46·80)	870 (34·25)	562 (22·11)	D

*Model designation code: A, Aerobatic; AE, Aerobatic engine; G, Geared; H, Helicopter; I, Fuel injected; L, Left-hand rotation crankshaft; O, Opposed-cylinders; S, Supercharged; T, Turbocharged; V, Vertical crankshaft; †D, Direct drive; V, Vertical mounting

BELL

BELL AEROSPACE TEXTRON

HEAD OFFICE AND WORKS: PO Box 1, Buffalo, New York 14240

Telephone: (716) 297 1000

OFFICERS: See Aircraft section

Bell has been engaged in the design, development and production of liquid-propellant rocket engines since 1946.

On 20 February 1979, Bell completed 20 years of space use of the Agena series of engines. In that time the Models 8096, described hereunder, and 8247 had been used in almost 340 missions by the US Air Force and NASA.

BELL MODEL 8096 AGENA ENGINE

This engine was first developed as the power plant for one of the weapon pods that was to be carried by the B-58 Hustler supersonic bomber. It is used in modified form, with gimballed chamber, as the power unit of the Lockheed-built Agena vehicle, forming the second stage of the Thor-Agena, Atlas-Agena and other space vehicles.

During its development the engine has undergone five major modifications, each resulting in an improvement in specific impulse. The present version, designated Model 8096, has a specific impulse of nearly 300 s, which is more than 10 per cent better than the original version of 1959, an increase equivalent to 225 kg (500 lb) in payload for Earth orbital missions.

The Model 8096 engine is a single-chamber pump-fed engine, running on red fuming nitric acid and unsymmetrical dimethyl-hydrazine (UDMH) hypergolic propellants.

It gives 71 kN (16,000 lb) S/L thrust, and has re-start capability in space. This feature can be used, for example, to change from a circular to an elliptical orbit.

The Model 8096 engine has the ability to be re-started twice in space. It powers the Agena vehicles used in many US Air Force and NASA programmes.

Bell has qualified the Agena engine to run on high-density acid (HDA), which burns at increased temperature to give increased thrust and efficiency; it requires a silicone additive to protect the thrust chamber. A higher-performance baffled injector has been developed and qualified for the current production engine.

Dimensions:	
Length overall	approx 2,134 mm (84 in)
Nozzle diameter	825·5 mm (32·5 in)
Weight:	approx 132 kg (290 lb)
Performance:	
Thrust	75·62 kN (17,000 lb)
Chamber pressure approx	34·5 bars (500 lb/sq in)
Specific impulse	approx 300 s

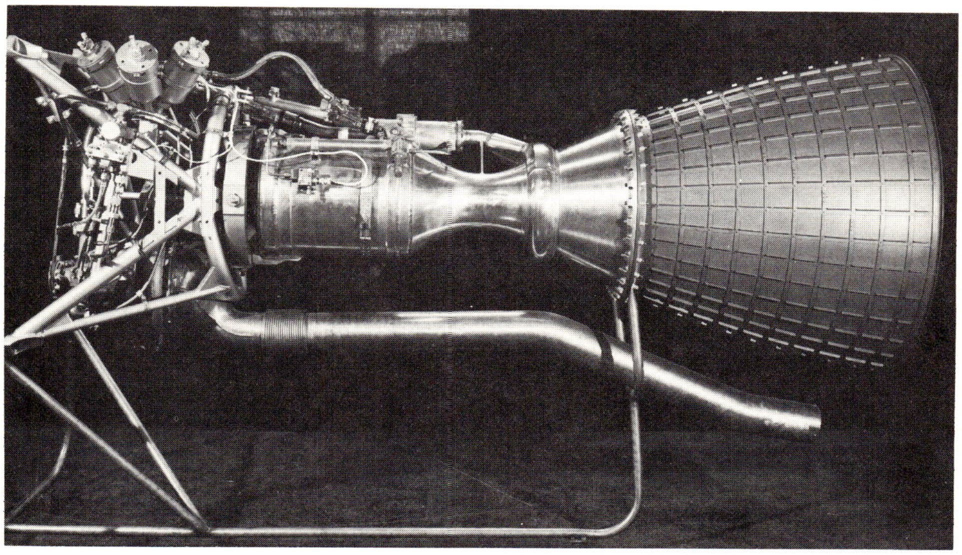

Bell Agena single-chamber liquid-propellant rocket engine

CHOTIA
WEEDHOPPER OF UTAH INC

1148 Century Drive, PO Box 2253, Ogden, Utah 84404
Telephone: (801) 621 3941

This company has produced a 'clean sheet of paper' two-stroke specifically for microlight aircraft, in the belief that a large market exists. Unlike most such engines, originally designed for other applications, the Chotia 460 gives large torque at low rpm. It is designed for long life, reliability and ease of tuning.

CHOTIA 460

Type: Single-cylinder, aircooled, direct-drive, two-stroke.
Cylinder: Bore 88·0 mm (3·47 in). Stroke 75·0 mm (2·95 in). Swept volume 456 cc (27·85 cu in). Schnurle-port design. Four-bolt exhaust flange.
Connecting Rod: Machined bar stock. Needle bearings at both ends.
Crankshaft: Three ball bearings. Reversible rotation. Direct drive.
Induction: Impulse fuel pump. Bowl-type carburettor. Fuel/oil ratio 50:1.
Ignition: Four D-size flashlight batteries giving 25 h operation before full-power misfire (then still 4 h cruise available): standard 14 mm plug.

Dimensions:	
Length	305 mm (12·0 in)
Width	178 mm (7·0 in)
Height	406 mm (16·0 in)
Weight, Dry:	14·3 kg (31·5 lb)
Performance Ratings (S/L):	
T-O	18·6 kW (25 hp) at 4,700 rpm
Continuous	14·9 kW (20 hp) at 3,500 rpm

CSD
CHEMICAL SYSTEMS DIVISION
(A division of United Technologies Corporation)

Headquarters: Sunnyvale, California 94086
Telephone: (408) 739 4880
Division Executive Vice-President and General Manager: A. J. Medica
Division Senior Vice-Presidents:
Dr David Altman
Eugene Roberts
Daniel Nuzzo

In 1980 CSD joined UTC Electronics Group as a division of Norden Systems. It was re-formed into four sections: Space and Strategic Boosters, Space Maneuvering Systems, Tactical Systems, and Advanced Technology.

CSD is currently engaged in research, development and production of rockets, rocket propellants and advanced propulsion systems, as well as a range of 'spin-off' by-products. Since its founding in the late 1950s it has conducted a continuous programme of rocket technology and is now producing a variety of advanced space propulsion systems for both the Department of Defense and the National Aeronautics and Space Administration.

CSD's largest programme is the production of 3,048 mm (120 in) diameter, segmented solid-propellant booster rockets. Other programmes described hereafter include the Algol III and FW-5 rocket motors, which are fully operational, and development of IUS.

In 1976 a specially formed subsidiary, United Space Boosters Inc (USBI), was selected by NASA as boost assembly contractor for the Space Shuttle. USBI is now responsible, under a $122 million contract, for assembly, checkout, launch, post-launch disassembly and refurbishment of the motors for re-use. The present contract covers six development flights, with an option for 21 operational flights.

A potentially important non-propulsion award is for the freon coolant tank of the MX missile.

CSD 3,048 mm (120 in) SEGMENTED MOTOR

Mass-produced at CSD's Coyote, California, rocket production facility, this 3,048 mm (120 in) diameter solid-propellant rocket motor centre segment is interchangeable with any other centre segment. Its cylindrical metal case is manufactured from high-strength steel (D6AC) and heat treated to an ultimate strength of 13,790 bars (200,000 lb/sq in). Less than 12·7 mm (0·5 in) thick, each case is equipped with clevis-type end joints with holes for cylindrical fastening pins. Loaded with propellant, a synthetic rubber (polybutadiene acrylonitrile) with aluminium additives as fuel and ammonium perchlorate as oxidiser, a segment has an overall length of 3,276 mm (10 ft 9 in) and weighs 36,400 kg (80,300 lb). Designed as the basic building block for large solid-propellant booster rockets, the segmented motor can be assembled in from one- to seven-segment configurations with thrusts ranging from 1,112-6,673 kN (250,000 to 1·5 million lb). A fully configured motor contains a destruct system, a forward-end ignition system a liquid-injection thrust-vector control system for steering and has its own staging sequence capability.

Full-scale mockup of Boeing IUS, showing the forward and aft CSD motors

CSD has made progress in the development of more efficient and economical means of fabricating the steel motor cases used in the 120 in programme, through production processes known as internal roll extrusion and shear forming. Both techniques reduce production time and result in a more reliable and economical product.

A five-segment configuration of this motor is used in pairs as the zero (launch) stage of the US Air Force Titan IIIC and D space launch vehicles. Operating together, the 26·12 m (86 ft) tall motors produce a thrust of 10,675 kN (2,400,000 lb). Launched for the first time on 18 June 1965, the boosters have flown 59 successful missions in as many launches, using 118 motors.

In 1979 CSD completed production of five-segment motors and started production of its six-segment boosters for the Titan 34D, the advanced version of the USAF's Titan standard space launch vehicle. Two of these boosters will launch the Titan 34D on its maiden flight from Cape Canaveral in 1982. The ten-tonne additional segment (actually six-tenths the length of a full segment) is located just below the rocket's forward closure and provides increased thrust (11,565 kN; 2,600,000 lb for the pair) for payloads of the 1980s. In 1978 CSD received contract awards of $89 million for eight pairs of the six-segment systems, followed by orders for three additional pairs, the last to be delivered in Spring 1983. The first pair was delivered to Cape Canaveral AFS in April 1980.

Essentially a lengthened version of the five-segment motor, a seven-segment motor has been statically test-fired four times. Producing a record 6,227 kN (1·4 million lb) of thrust, the motors operated perfectly while testing an advanced steering system, called an ullage blow-down, which is now used on all flights of Titan IIIC and IIID vehicles.

CSD IUS

In October 1976 the US Air Force awarded the Boeing/CSD team a contract to develop the Inertial Upper Stage (IUS) for the Space Shuttle and the Titan 34D space launch vehicle. CSD's all-solid IUS proposal won the award over five competitors' liquid-propellant submissions. The Air Force cited highest reliability and safety at lowest cost in its choice. The Boeing/CSD IUS completed its validation phase on 19 December 1977, after a successful test firing of the 2,310 mm (91 in) diameter rocket motor at the Arnold Engineering Development Center (AEDC). During the firing, the 9,979 kg (22,000 lb) motor encountered no difficulties during an unusually long (154 s) burn. Key components, including the advanced flight-weight nozzle, proved their survivability. Average thrust was 173·6 kN (39,000 lb) and maximum 249·2 kN (56,000 lb) at a simulated height of 33,800 m (111,000 ft). Full-scale development of the IUS propulsion system began on 1 March 1978 under a $32 million contract award from Boeing Aerospace. The propulsion system for the two-stage IUS consists of two solid rocket motors designed by CSD: a 2·31 m (91 in) aft motor containing 9,707 kg (21,400 lb) of propellant and a 1·6 m (63 in) forward stage motor with 2,722 kg (6,000 lb) of propellant. In May 1978 CSD received an award from Boeing to develop the thrust vector control (TVC) servo-actuator system to control the movable nozzle. The basic TVC system consists of two controllers and four actuators powered by 28V batteries. In December 1978 CSD received a contract to produce a servo-actuator system to raise the IUS to deployment position inside the Shuttle Orbiter's cargo bay.

The first full-scale development large motor for the IUS was test-fired on 19 December 1979, at AEDC, for approx 150 s in near-vacuum conditions. The motor's nozzle was vectored ±4° during the firing.

The first firing in the full-scale development phase of the small motor was conducted successfully on 26 June 1979. The motor fired for approx 95 s at a simulated altitude of 30,480 m (100,000 ft) and generated over 111·2 kN (25,000 lb) thrust. Its carbon/carbon nozzle was vectored to full deflection of 7° during the firing.

The first firing of the small solid spin motor was conducted at AEDC on 5 December 1979, when the motor was spun at 70 rpm for 100 s. Maximum thrust was 106·8 kN (24,000 lb). The system underwent several successful test firings during 1980. On 8 March the TVC servo-actuator system was employed for the first time during a test firing, the motor nozzle being deflected ±7°. Two more firings of the small motor were conducted in 1980,

Three of CSD's advanced rocket nozzle systems: left, cutaway model showing operation of Techroll; centre, a test EEC during firing; right, a hot ball and socket

the first on 14 May and the second, with an EEC (extensible exit cone, described separately), on 14 October. The IUS will be the first spacecraft to use an EEC operationally. In late 1980 CSD completed testing of its improved deployment system, which involves three steps: (1) the nested cones are unlatched for deployment; (2) three stainless steel booms unroll to extend the cones and latch them; (3) the booms disconnect and remain in a locked position out of the way for motor firing and steering.

On 17 April 1980 a large solid-propellant rocket motor for the IUS was test fired for 150 s at 30,480 m (100,000 ft) simulated altitude. It generated more than 222·4 kN (50,000 lb) thrust.

On 13 August 1980, the first offloaded IUS motor was test fired, with the most extreme propellant offload (50 per cent) of any solid rocket motor ever tested. The motor was fired for approximately 64 s at a simulated altitude of 30,480 m (100,000 ft). It burned 4,808 kg (1,600 lb) of HTPB propellant and generated 275·78 kN (62,000 lb) max thrust.

CSD ALGOL III

CSD is in full production of Algol III, a more powerful solid-propellant first stage for NASA's Scout space launch vehicle. 9·14 m (30 ft) tall and 1·14 m (45 in) in diameter, the improved solid booster permits an increase of 40-54 per cent in the Scout's payload/weight capability. When loaded with its propellant (PBAN with aluminium additives and ammonium perchlorate oxidiser) Algol III weighs 13,605 kg (30,000 lb). Its nozzle is fabricated from steel with a graphite cloth-phenolic and silica cloth-phenolic cone liner. Ignition is by a small nose-mounted solid rocket. Producing 623 kN (140,000 lb) of lift-off thrust, Algol III burns for 75 s to boost the Scout to an altitude of 48 km (30 miles) before burnout. The motor flew for the first time in August 1972 and by the end of 1979 had completed 19 missions. Ten more motors were ordered under a $2 million contract in 1980.

CSD FW-5

The FW-5 is designed to be interchangeable with existing upper-stage motors. While its primary application is that of an apogee motor, it can be used for orbital, probe or re-entry missions, as well as retro-rocket propulsion for space vehicles.

The FW-5 is 691 mm (27·2 in) in diameter with an overall length of 1,123 mm (44·2 in). Its case is fabricated from glassfibre and epoxy resin, with a cylindrical wall thickness of 3·3 mm (0·13 in) and end domes with solar isotensoid contours with a nominal wall thickness of 15·2 mm (0·60 in). The igniter and nozzle attachment fittings are fabricated from aluminium, and the aluminium inter-stage flange is bonded and riveted to the integrally-wound glassfibre skirt of the motor case.

The standard FW-5 has an inert weight of 29·7 kg (65·6 lb) and a propellant weight of 260 kg (575 lb), resulting in a motor mass fraction of 0·89. The average motor thrust is 25·04 kN (5,630 lb) with a burn time of 42·1 s.

FW-5 operated perfectly on its first mission in November 1972 when it placed Telesat Canada's Anik communications satellite into final orbit. The second Anik was put on station by an FW-5 in April 1973, followed by a third in May 1975.

By 1980 the FW-5 had completed 12 additional consecutive successful flights on the Delta 2914 launch vehicle. The FW-5 was used to put on station communications satellites such as the Anik, Westar, Marisat, and Palapa. CSD has also developed a version of the FW-5 with a 178 mm (7·0 in) extension to the nozzle exit cone. The flight unit was used to launch the Martin Marietta/Denver Scatha satellite in early 1979.

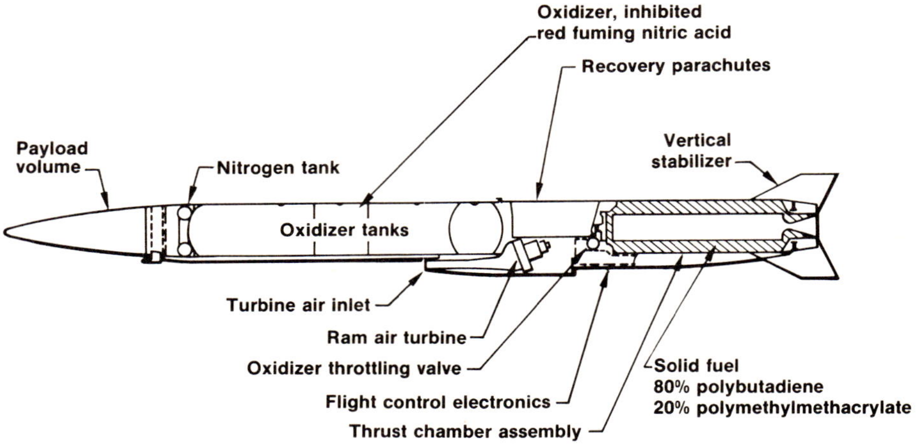

Simplified section of Firebolt showing CSD hybrid propulsion

CSD BOOSTER SEPARATION MOTOR

In late 1975 NASA selected CSD to develop and produce solid-propellant booster separation motors (BSM) to be used on the first six development flights of the Space Shuttle programme.

The BSMs are used to separate the two reusable solid rocket boosters (SRB) approximately 110 s after launch of the Space Shuttle. Each SRB requires eight BSMs, four forward and four aft, which fire for less than 1 s to achieve safe separation. In September 1978 the motor passed acceptance by NASA. The last of the 104 motors ordered was shipped in early 1980.

Each BSM has a length of 838 mm (33·0 in) and a diameter of 327 mm (12·88 in), with a weight of 73·0 kg (161 lb). It uses CSD UTP-19048 (HTPB) propellant, and will have 62·5 kN-s (14,000 lb-s) minimum total impulse WEB action time, and 67·0 kN-s (15,000 lb-s) minimum action time. Burn-time (WEB) performance time limit is 0·8 s. Maximum thrust called for is 129·5 kN (29,000 lb), with a WEB action time average thrust of 82·6 kN (18,500 lb) minimum.

Each motor consists of solid propellant, motor case with attachments, nozzle, igniter and pyrotechnic initiators. Aft motors are provided with cork insulation for protection from the Shuttle's main engine plumes and solid rocket booster exhaust. Forward separation motors are equipped with an aerodynamic heat-shield which opens instantly on motor ignition but does not discharge any debris that could damage the Orbiter.

CSD HYBRID PROPULSION

CSD's hybrid rocket for the Department of Defense High Altitude High-Speed Target (HAHST) drone is 330 mm (13 in) in diameter, has an overall length of 3,556 mm (140 in) and weighs 390 kg (860 lb). It uses a polybutadiene and methyl methacrylate fuel grain, and nitric acid as its liquid oxidiser. Ablative-cooled, its thrust ranges from 0·88-5·34 kN (200 to 1,200 lb) at altitudes between 10,670 and 30,480 m (35,000 and 100,000 ft). The HAHST successfully completed all its initial flight-readiness test-firings in 1973. In 1978 the HAHST was flown successfully three times by the US Navy at the Pacific Missile Test Center.

In 1980 CSD received a contract for full-scale development of the propulsion system of the Teledyne Ryan Firebolt, the advanced successor to HAHST. CSD will supply 14 systems, those for flight test being designed for Mach numbers up to 4.

CSD is continuing work on integral hybrid/solid propulsion for the US Air Force air-to-ground 'Mini-Missile'. In late 1980 it also received a contract to design hybrid systems optimised for medium- and short-range air-to-air missiles.

CSD LIQUID ROCKET ENGINES

The company's liquid rockets range in size from 1,321 mm (52 in) long and 660 mm (26 in) in diameter and weighing 32·6 kg (72 lb) to 1,918 mm (75·5 in) long, 1,219 mm (48 in) in diameter and weighing 83·9 kg (185 lb). Their propellant is 50/50 hydrazine and unsymmetrical dimethyl hydrazine and nitrogen tetroxide. The engine nozzles are a composite structure with a glassfibre shell and silica-phenolic liner. Ignition is hypergolic.

CSD AIR-BREATHING PROPULSION

In early 1973, CSD was selected to spearhead the efforts of its parent, United Technologies Corporation, in the research and development of ramjet propulsion systems, with the support of UTC's Hamilton Standard Division and United Technologies Research Center. In mid-1973 CSD was awarded a contract by the US Navy to research, design and develop a Modern Ramjet Engine (MRE). Called an integral rocket/ramjet, the propulsion device involves both solid rocket and liquid fuel ramjet technology. It operates as a solid rocket booster until it reaches supersonic speeds. At that point, through a series of mechanical changes that take place in flight, it becomes a ramjet. These changes involve the opening of air inlets, an increase in the nozzle diameter and a switch to the burning of liquid fuel and air in the combustion chamber within a common system. The propulsion technology acquired in carrying out this project will be applied to the development of an advanced air-to-air missile system.

In mid-1974 CSD was awarded another ramjet programme by the US Air Force, to design and ground test an advanced integral rocket/ramjet propulsion system which would satisfy the requirements of an advanced strategic air-launched missile (ASALM).

In the ASALM programme, CSD had by 1978 embarked on a project to demonstrate the feasibility of two design approaches, one by McDonnell Douglas, the other by Martin Marietta. CSD was carrying out extensive ground testing, including numerous hot firings at its ramjet test facility near San Jose. This testing also involved simulated flight-test missions, in preparation for the next phase of development. In late 1979 the ASALM combustor underwent a successful test firing at Coyote which simulated a supersonic mission lasting nearly ten minutes. Post-fire analysis revealed that the combustor, its quartz phenolic thermal-protection system and its moulded silica phenolic nozzle were in excellent condition.

In 1981 two seven-minute firings of liquid-propellant ramjets demonstrated new technology for securing insulation inside the combustor. A 'spin casting' method is used, and is being applied to other ramjet programmes.

In mid-1977, the division received a subcontract from Vought, funded by the Navy, to refine the earlier Advanced Low Volume Ramjet (ALVRJ) for the Supersonic Tactical Missile (STM) (see Vought entry in this section).

On 21 April 1979 the STM, powered by a liquid rocket/ramjet system produced by CSD, successfully completed a test flight at the Naval Pacific Missile Test Center. The missile was launched from an A-7 and achieved speeds as high as 1,735 knots (3,220 km/h; 2,000 mph) during its 71 nm (132 km; 82 mile) flight over the Pacific before executing a programmed terminal dive.

In 1979 CSD was chosen by Hughes Aircraft to supply the propulsion subsystem for the ducted rocket missile that Hughes is developing for the US Air Force. CSD is responsible for design, development, integration and ground testing of the gas generator and air induction system. Flight testing of 12 of these missiles is scheduled to begin early in 1982. The division completed a ducted rocket engine development programme in 1979 which was the precursor of its current work for Hughes.

In the ducted rocket a solid rocket fuel produces a fuel-rich gas. This is injected into a combustion (ramburner) chamber, mixed with ram air and further burned to provide sustained thrust at supersonic speeds. The ram air is introduced through inlet ducts alongside the vehicle. Compared to a conventional rocket, the ducted rocket is claimed to offer many times the thrust duration, permitting powered flight all the way to a target and a greater range with less weight. In 1981 CSD announced development of a technique for throttling ducted rockets which promises to double vehicle range and increase flight performance.

In early 1979 CSD announced that it had demonstrated the feasibility of a side-dump combustor for solid-fuel ramjet systems that may allow an increase in missile range without increasing combustor size. By late 1979 CSD had completed a series of semi-free-jet tests of its solid-fuel ramjet. The three-year programme of development of a solid-fuel ramjet for future air-to-air missiles culminated in a series of free-jet tests involving cold-flow tests and firings at speeds up to Mach 3, at NASA Lewis Laboratory.

In 1980 CSD received a contract to develop advanced high-energy solid fuels for ramjets. The study, for the Naval Weapons Center and Air Force Aeropropulsion Laboratory, calls for fuels offering a greater range of performance for tactical missiles.

In 1980 CSD began two programmes involving development of solid-fuel ramjet propulsion for gun-launched projectiles. The first was for the US Army's Advanced Indirect Fire System (AIFS), a new artillery system that will deliver cannon-launched ramjet projectiles against staging areas for tanks and other enemy positions beyond the range of conventional artillery. AIFS projectiles will be fired from conventional artillery. CSD's internal ramjet engines will then extend the weapon's range at supersonic speed, and carry warheads accurately to their targets using a novel guidance approach. The second programme, from the Naval Weapons Center, involves a feasibility study of a solid-fuel ramjet projectile for the Navy's 5-in/54-calibre gun.

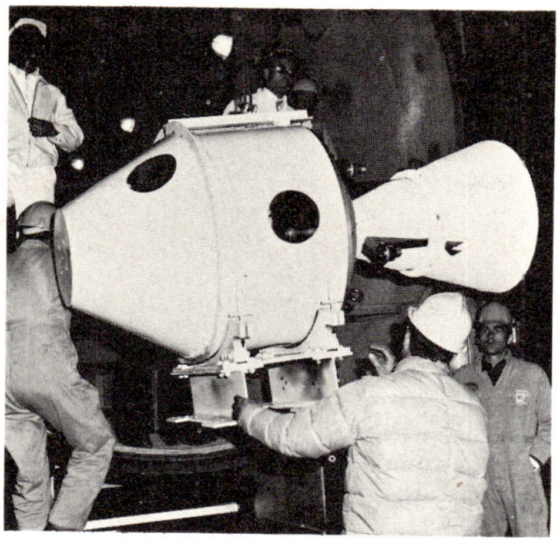

CSD/SEP Advanced Apogee Motor and Nozzle

CSD is also completing a two-year contract to develop and test advanced integral rocket/ramjet combustors which utilise the technology of swirl combustion. CSD engineers expect to realise a significant improvement in combustor performance while reducing combustor length by as much as 50 per cent.

Another programme aims at developing a ramjet capable of cruising at Mach 6.

CSD/SEP ADVANCED APOGEE MOTOR

In November 1979, the Advanced Apogee Motor and Nozzle, produced jointly by CSD and the Société Européenne de Propulsion (SEP) of France, was test-fired successfully at AFRPL.

The 762 mm (30 in) diameter motor was test-fired for 60 s at a simulated altitude of 30,500 m (100,000 ft). Maximum thrust was 22·25 kN (5,000 lb) and maximum pressure 41·5 bars (600 lb/sq in). Specific impulse of 302·9 s was the highest ever recorded for composite aluminised propellant.

The experimental motor embodies important technological advances, including 'head end' web solid-propellant grain design, free-standing carbon/carbon ITE (integral throat and entrance), and an extensible exit cone which employs reinforced rubber tube actuators.

Following State Department approval in 1980, CSD and SEP are simplifying the ITE by redesigning the filament-wound case to increase propellant load, and improve the igniter and EEC. A test firing is planned for December 1981.

ADVANCED PROPULSION TECHNOLOGY

Testing of CSD's patented Techroll joint continues. This is a fluid-filled, constant-volume bearing which allows easy pivoting of the nozzle by the steering system. It is considered to be the first major improvement in rocket steering in almost a decade. Both IUS motors use the Techroll seal. Other uses being researched include a main rotor bearing for helicopters and a joint in high-pressure and transmission lines.

The extensible exit cone (EEC) is a device which enlarges mechanically as a rocket motor fires in space, improving performance by over five per cent. The first test at operational pressures and temperatures of an EEC system for ICBM nozzle applications was successfully carried out in June 1977. This CSD system uses nested cones of an advanced composite material, which lock into place by a latching mechanism. An EEC is fitted to the IUS small motor, as described earlier.

CSD solid-fuel ramjet test vehicle mounted in supersonic tunnel at NASA Lewis Laboratory

CSD is continuing developmental work on its Bolt Extrusion Thrust Termination (BETT) system, considered the only new thrust-termination system for solid rocket motors developed in the last 20 years. BETT replaces nozzle flange bolts with a combination of explosive-released and extrudable bolts. Upon command, explosive bolts are actuated and the nozzle moves aft, dropping the motor's internal pressure drastically. This halts propellant burning and neutralises motor thrust. A 1978 test of the system used a 2·34 m (92 in) Kevlar case. Upon command all explosive bolts operated within 0·001 s and the nozzle moved 146 mm (5·76 in) rearward in 0·016 s. Two tests of BETT on third-stage Minuteman motors took place in 1979. In both, thrust-termination was initiated 5 s into the firing, resulting in motor depressurisation in 180 ms.

CSD's 'hot ball and socket' rocket steering system employs an integral carbon/carbon nozzle and thrust vector control system without the use of any seals, insulators or metallic structures. The materials are self-lubricating, and loads are transmitted between the surfaces even while white hot. This allows ball diameter to be reduced to a minimum, sharply reducing vectoring torque. The hot ball and socket completed its first test firing in 1978 under contract to the US Naval Sea Systems Command.

Under a 1980 contract, firings were planned in 1981 at Coyote Center and at the Air Force Rocket Propulsion laboratory.

In early 1979 CSD tested a nozzle employing a one-piece freestanding integral throat and entrance which needs no insulation or structural support. The FSITE replaces the entire submerged area of a solid-propellant motor nozzle, consisting typically of numerous phenolic and steel parts.

In 1979 CSD was awarded a contract by the US Army Missile Command to develop solid motor cases made of chemically treated layers of papers. The motors will be fabricated at CSD and delivered to Redstone Arsenal for test firing.

CURTISS-WRIGHT

CURTISS-WRIGHT CORPORATION, WOOD-RIDGE FACILITY

HEAD OFFICE AND WORKS: One Passaic St, Wood-Ridge, New Jersey 07075

Telephone: (201) 777 2900

CHAIRMAN OF THE BOARD AND PRESIDENT: T. Roland Berner

EXECUTIVE VICE-PRESIDENTS:
Charles E. Ehringer
Richard P. Sprigle

VICE-PRESIDENT, ENGINEERING: A. F. Kossar

VICE-PRESIDENTS:
D. Lasky
W. Figart

The Wood-Ridge facility of Curtiss-Wright Corporation is engaged in the research, development and licensing of Wankel-type rotary engines, manufacture of engine parts, aircraft and industrial engine overhaul, electric power generation systems and advanced marine and turbine technology.

Curtiss-Wright RC2-75, rated at 254 kW (340 hp), on bench with three-blade propeller

CURTISS-WRIGHT SETE

In competition with Garrett and Pratt & Whitney, Curtiss-Wright is participating in the SETE (Supersonic Expendable Turbine Engine) programme of the US Navy. The objective is the cheapest possible jet engine capable of unfailingly-reliable instant starting and flight propulsion, under severe conditions of manoeuvre, over a wide band of speeds and heights, on a single flight of a missile. Details are restricted, but several sources give the thrust class as 10·23 kN (2,300 lb st).

CURTISS-WRIGHT RC ENGINES

In 1958 Curtiss-Wright Corporation obtained a licence for the NSU-Wankel type of rotating-combustion (RC) engine and embarked on a major programme of independent development of a range of such engines aimed at a wide spectrum of applications. At first the company concentrated on large engines in the power range around 500 hp for aircraft use, but since the early 1960s smaller engines have dominated the hardware test and development programme, some of which has been funded by US military agencies, including the Naval Air Systems Command, and NASA.

Most research has been carried out on versions of the **RC2-60** (twin rotors each of about 0·983 litre; 60 cu in capacity), rated at up to 149 kW (200 hp) at 5,500 rpm.

In 1965 the 310 hp **RC2-90** was run, with helicopter applications in mind, and the 1·47 litre (90 cu in) rotor has since been used in extensive development of stratified charge engines capable of operating on a range of fuels including JP-4 and JP-5 gas turbine kerosenes.

The **RC2-75-Y3** is one of a very important family of engines, regarded as optimally sized for a wide range of general aviation aircraft. As the designation indicates, it is based on two rotors each of nominal 1·23 litres (75 cu in) capacity, and has liquid cooling and a geared drive. Current work is centred on reducing fuel consumption and pollution.

DREHER

DREHER ENGINEERING COMPANY

708 Lincoln Boulevard, Santa Monica, California 90402

Telephone: (213) 395 6510

Mr Max Dreher, an aeronautical engineer, has built a series of small turbojet engines over a period of 25 years.

DREHER TJD-76C BABY MAMBA

Type: Single-shaft turbojet.

Air Intake: At front. Air flow 0·50 kg (1·1 lb)/s.

Compressor: Single-stage mixed-flow. Stainless steel impeller with sixteen vanes. Pressure ratio 2·8 : 1.

Compressor Casing: Of 2024 alloy and 347 stainless steel.

Combustion Chamber: Annular type vaporising system with fuel/air pre-mix. One spark plug in flame tube.

Fuel System: Manual with pressurised fuel supply, or electrically-driven fuel pump. Fuel pressure 5·52 bars (80 lb/sq in). Automatic system for drone applications.

Fuel Grade: Kerosene or petrol.

Nozzle Guide Vanes: Single axial stage, with sixteen investment-cast vanes in Stellite 31.

Turbine: Single-stage axial-flow, with nineteen integrally-cast blades, of Inconel 713 LC. Gas temperature 770°C before turbine, 675°C after turbine, at continuous cruising power.

Lubrication: Air/oil mist system with total loss, using bleed air equivalent to 2·5 per cent of total mass flow.

Oil Grade: MIL-L-7808E (Turbo 15).

Two views of the Dreher Baby Mamba single-shaft turbojet: *(left)* **TJD-76C on mount;** *(right)* **TJD-76C Jet Pack installed on sailplane. Thrust rating is 0·245 kN (55 lb st)**

Starting: Compressed air 10·34 bars (150 lb/sq in), via three nozzles driving turbine wheel.

Dimensions:

Length overall	416 mm (16·38 in)
Diameter	151 mm (5·94 in)

Weights:

Dry	6·4 kg (14·1 lb)
Complete with fuel tank	10·0 kg (22 lb)

Performance Ratings:

Max	0·245 kN (55 lb st)
Continuous	0·20 kN (45 lb st)

Specific Fuel Consumption:

at max rating	42·5 mg/Ns (1·5 lb/h/lb st)

DREHER TJD-76D and E

These versions were derived from the TJD-76C in 1972 to meet a need for a very low-cost short-life unit for the propulsion of mini-RPVs and other expendable vehicles. Both have similar performance to the TJD-76C but weigh approximately 6·5 kg (14·33 lb), complete with 2 kW alternator. Envelope diameter is 165 mm (6·5 in) and length 400 mm (15·75 in).

GARRETT

GARRETT TURBINE ENGINE COMPANY (a division of The Garrett Corporation)

Sky Harbor Airport, 111 S 34th St, PO Box 5217, Phoenix, Arizona 85010

Telephone: (602) 267 3011

President: Jack Marinick

Vice-President, Sales: Larry D. Huppert

The Garrett Corporation has been called the world's largest producer of small gas turbines. Development of the first Garrett (formerly Garrett-AiResearch) small turbines began in 1946 and the division claims to have produced over 70 per cent of the total of gas-turbine units with power ratings from 60 to 2,500 hp built in the USA and Europe. It is also developing a mainly military series of derivative low-bypass-ratio engines in partnership with Volvo Flygmotor of Sweden, and in 1980 signed agreements with Volvo for development and production of uprated models of turbofan and turboprop engines.

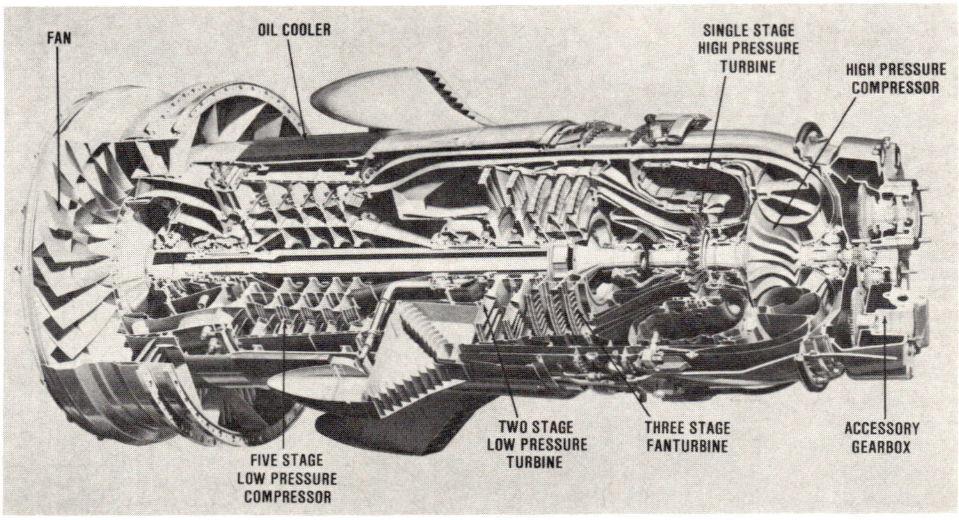

Cutaway of the Garrett ATF3-6 three-shaft turbofan with double flow-reversal (22·46 kN; 5,050 lb st)

GARRETT ATF3

US military designation: F104-GA-100

Considered to be the first three-spool engine to run in the USA, the Garrett ATF3 is the first engine in the world to combine the three-spool features with a reverse-flow combustion system and turbines, and mixed-flow exhaust.

The arrangement of components allows the fan design to be determined largely independently of the gas-generator compressor requirements, and permits operation at optimum fan speed. Omission of fan inlet guide vanes, mixing of the gas-generator exhaust with the fan airflow, and double reversal of the internal airflow enable the ATF3 to offer significant reductions in overall noise generation.

Other design considerations include reliability, maintainability and elimination of visible smoke. The accessories are revealed by removing the tailcone fairing, and their positioning at the rear of the engine is claimed to reduce installed drag.

The basic design was completed in early 1966, and testing of demonstrator engines was initiated in May 1968. Under US Air Force contract, the ATF3 successfully completed preliminary flight rating tests in 1972 at 18 kN (4,050 lb) thrust. Both the aerodynamic and mechanical design criteria were established around a sea-level, ISA +15°C take-off rating of 22·46 kN (5,050 lb).

In May 1976 it was announced that the **ATF3-6** had been selected by Dassault-Breguet to power the Falcon 20 business jet, with US coast-to-coast range. Commercial certification was achieved on 19 May 1981 after 20,000 h of testing, and production shipments began in that year. The ATF3-6 is offered under a retrofit programme for existing Falcon 20 aircraft. It powers the HU-25A Guardian and Falcon 200.

Type: Three-shaft axial-flow turbofan.

Intake: Direct pitot, fixed type. No inlet vanes or struts. Total airflow 73·5 kg/s (162 lb/s).

Low-Pressure (Fan) System: Single-stage titanium fan, driven by three-stage IP turbine. Bypass ratio 2·8 at take-off.

Intermediate-Pressure System: Five-stage titanium axial IP compressor, each stage having a separate disc, driven by two-stage LP turbine. Airflow is then delivered to rearward-facing HP compressor via eight tubes feeding into an annular duct concentric with the bypass duct. Core airflow 18·15 kg (40 lb)/s.

High-Pressure System: Single-stage titanium centrifugal compressor, driven by single-stage HP turbine. IP airflow enters the single-sided impeller from the rear. Overall pressure ratio (T-O) 21, (high-altitude cruise) 25.

Combustion System: Reverse-flow annular type.

Turbines: Single-stage HP, three-stage IP and two-stage LP turbines drive, respectively, the HP, fan (LP) and IP compressors. IP and LP turbines have fully shrouded blades. Aircooled first-stage nozzle vanes and HP rotor blades. Exhaust gases turned 180° through eight sets of cascades to mix with fan bypass flow.

Fuel System: Electromechanical, incorporating solid-state computer. Manual emergency backup system.

Accessory Drives: Three drive pads on rear-mounted gearbox driven by HP shaft, providing for hydraulic pump, starter/generator and one spare. Accessory cooling by fan discharge air which is exhausted through a nozzle at the tip of the fairing.

Exhaust System: Mixed fan and turbine exhaust discharged to atmosphere through annular nozzle surrounding combustion section.

Lubrication System: Self-contained hot-tank type; tank integral with gearbox.

Mounting: Two-plane pickup system.

Starting: Electrical or pneumatic.

Dimensions:

ATF3-6:

Length	2,591 mm (102·0 in)
Max diameter	853 mm (33·6 in)

Weight, Dry:

ATF3-6	499 kg (1,100 lb)

Performance Ratings (uninstalled):

T-O (S/L, static):

ATF3-6	22·46 kN (5,050 lb), ISA+15°C

Cruise (12,200 m; 40,000 ft at Mach 0·8):

ATF3-6	4·5 kN (1,012 lb)

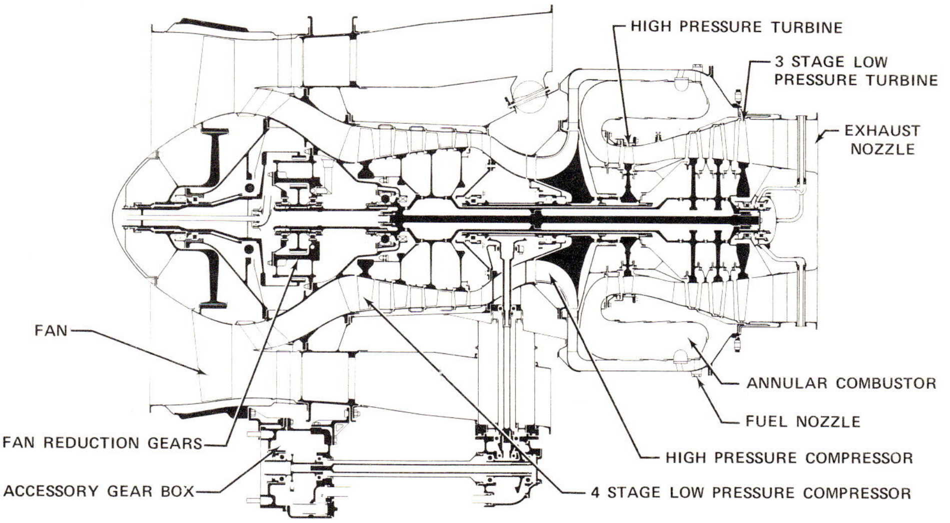

Cross-section (right) and external view (above) of the 15·57 kN (3,500 lb st) Garrett TFE731-2 geared front-fan engine

SPECIFIC FUEL CONSUMPTION:

At T-O rating (S/L, ISA static)	13·6 mg/Ns (0·48 lb/h/lb st)
At cruise (as above)	22·38 mg/Ns (0·79 lb/h/lb st)

GARRETT TFE731

Announced in April 1969, the TFE731 is a two-spool geared turbofan designed to confer US coast-to-coast range upon business jet aircraft. Use of a geared fan confers flexibility in operation and yields optimum performance both at low altitudes and at up to 15,250 m (50,000 ft).

Component testing began in March 1969. The first engine ran in September 1970, and was tested at Phoenix in a Learjet 25. FAA certification and first production deliveries to Dassault for the Falcon 10 took place in August 1972.

In October 1972 it was stated that the Lockheed JetStar would be re-engined with the **TFE731-3**, flat rated at 16·46 kN (3,700 lb st) by a modest increase in turbine inlet temperature. The TFE731-3 was certificated in 1974. The modified aircraft is designated JetStar II, and first flew in July 1974. AiResearch Aviation converted a number of JetStar I aircraft to have TFE731-3 power.

Since 1973 the TFE731, in various sub-types, has been selected for the Gates Learjet 35/36, Cessna Citation III, Dassault-Breguet Falcon 50, HS 125-700, IAI Westwind 1 and 2 and Astra, Rockwell International's Sabreliner 65A, CASA C-101 trainer and light attack aircraft and Learjet 54/56 Longhorn, and proposals have been made for RPV and other short-life programmes.

By May 1981 deliveries of TFE731-2 and -3 engines had reached 3,000. Output had risen to 60 per month.

In 1980 the **731-5** was announced, to offer higher thrust and reduced noise, while maintaining low smoke characteristics. The -5 is installationally similar to earlier models but has a higher-bypass-ratio fan driven by a new LP turbine, the design of which was demonstrated in a NASA programme. Development and production are in partnership with Volvo Flygmotor of Sweden which, on a risk-sharing basis, will produce the outer structure. Certification is due in 1983.

TYPE: Turbofan with two shafts and geared front fan.

AIR INTAKE: Direct pitot, fixed, without guide vanes.

FAN: Single-stage axial titanium fan, with inserted blades. The fan shaft is connected directly to the planetary gearbox ring gear. Max fan airflow, sea level static, 51·25 kg (113 lb)/s (-3, 53·7 kg: 118·3 lb/s). Bypass ratio 2·66 (-3, 2·80).

COMPRESSOR: Low-pressure compressor has four stages (TFE731-4, five stages), each with a separate disc. Rotors and stators have inserted blades and vanes. High-pressure compressor, carried on a separate shaft running at higher speeds, is centrifugal. Overall pressure ratio (S/L, static): -2, 14·0; -3, 14·6.

COMBUSTION CHAMBER: Annular combustion chamber of reverse-flow type, with 12 fuel nozzles inserted radially and injecting fuel tangentially.

FUEL SYSTEM: Hydro-electronic, with single-lever control to mechanical and electronic elements.

TURBINES: High-pressure turbine has a single axial stage with inserted blades. Low-pressure turbine has three axial stages, all with inserted blades. Average inlet gas temperature to HP turbine, S/L, max T-O thrust, 1,010°C (-3 and -5, higher).

JET PIPE: Short fan duct facilitating installation of fan reverser. Fixed core nozzle.

ACCESSORY DRIVES: Accessories driven from HP spool are grouped around underside of the forward section of the fan duct. Pads are provided on the front side of the accessory gearbox for the airframe-type accessories: hydraulic pump, starter/generator or starter motor and alternators. Pads on the back side of the gearbox drive the engine accessories: fuel control unit and oil pump.

DIMENSIONS:

Intake diameter	716 mm (28·2 in)
Length overall (-2, -3)	1,263 mm (49·73 in)
Width	869 mm (34·20 in)
Height overall	992 mm (39·07 in)

WEIGHT, DRY:

TFE731-2, -3	329 kg (725 lb)
TFE731-5	374 kg (825 lb)

PERFORMANCE RATINGS:

Max T-O (S/L, 24·4°C):	
TFE731-2	15·57 kN (3,500 lb st)
TFE731-3	16·46 kN (3,700 lb st)
TFE731-5	19·14 kN (4,304 lb st)
Cruise (12,200 m; 40,000 ft at Mach 0·8):	
TFE731-2	3·36 kN (755 lb)
TFE731-3	3·64 kN (817 lb)

SPECIFIC FUEL CONSUMPTION:

Max T-O (as above):	
TFE731-2	13·88 mg/Ns (0·49 lb/h/lb st)
TFE731-3	14·33 mg/Ns (0·506 lb/h/lb st)
Cruise (as above):	
TFE731-2	23·08 mg/Ns (0·815 lb/h/lb)
TFE731-3	23·65 mg/Ns (0·835 lb/h/lb)

GARRETT/FLYGMOTOR TFE1042

This low-bypass derivative of the TFE731 is a joint project with Volvo Flygmotor of Sweden and is described in the International part of this section.

GARRETT TPE331

US military designation: T76

Originally based upon extensive experience with APUs, this was the first Garrett engine for aircraft propulsion; it has since been the main product which has helped to provide funding for later engines. The military T76 has achieved only modest sales, but the civil TPE331 has been most successful and developed in a series of versions. Deliveries in mid-1980 were at the rate of more than 55 engines per month, and total deliveries then exceeded 7,000. Flight time exceeded 17,000,000 h.

The following are major versions:

TPE331 series I, II. Initial production version, FAA certificated in February 1965. Rated at 451 ekW; 429 kW plus 0·33 kN (605 ehp; 575 shp plus 75 lb st). Redesignated **TPE331-25/61** and **-25/71** and produced until 1970. Powers Mitsubishi MU-2 (A to E models), Fairchild Industries/Pilatus Porter, Carstedt Jet Liner, Volpar Super Turbo 18, Aerospace FU-24, Rockwell International Hawk Commander and 680 and DHC-2 Turbo Beaver.

TPE331-1 series. Certificated December 1967 at 526 ekW; 496 kW plus 0·44 kN (705 ehp; 665 shp plus 100 lb st). Powers Mitsubishi MU-2 (F and G), Pilatus Turbo-Porter and Fairchild Industries AU-23A Peacemaker, Mid West CJ600, Volpar Turboliner, Interceptor 400, Rockwell International Turbo Commander and (customer option) Thrush Commander, Swearingen Merlin IIB and Aerospace Fletcher 1284, Marsh Turbo Thrush and Marsh Turbo Ag-Cat.

TPE331-2 series. The -201 was certificated in December 1967 at 563 ekW; 533 kW plus 0·45 kN (755 ehp; 715 shp plus 102 lb st). Powers Shorts Skyvan, CASA 212 pre-series and Volpar Turbo Goose and Turbo Beaver.

TPE331-3 series. Certificated in March 1970 at 674 ekW; 626 kW plus 0·71 kN (904 ehp; 840 shp plus 159 lb st). Uprated gas generator with increased airflow and pressure ratio, but same turbine temperature as in original TPE 331. Powers Fairchild Swearingen Merlin III, IV and Metro, and Century Jetstream III.

TPE331-5/6 series. The -251 was certificated in March 1970: this matches the gas generator of the -3 series with the 715 shp gearbox, and is flat-rated to 2,134 m (7,000 ft). Powers Shorts Skyvan 3 (-201), Beechcraft King Air B100 (-252), CASA 212 Aviocar (-251) and Rockwell International Turbo Commander 690A (-251). The -5 designation indicates output speed of 1,591 rpm; the -6 has an output speed of 2,000 rpm.

TPE331-8. Matches compressor and gearbox of -251 with new turbine section. Thermodynamic power of 676 ekW; 645 kW (865 shp plus 47·7 kg, 105 lb st), but flat rated at 533 kW (715 shp) to 36°C. Certification was received in September 1976. Powers Cessna Conquest.

TPE331-9. Thermodynamic rating 645 kW (865 shp).

TPE331-10. Rated at 746 kW (1,000 shp). Certificated February 1978 and for use in new MU-2 models, Merlin IIIB, Jetstream and CASA 212.

TPE331-11. Certificated 1979 at thermodynamic power of 746 kW (1,000 shp). Higher gearbox limit than Dash-10 engine; wet rating 820 kW (1,100 shp). Powers Swearingen Metro III and FMA Pucará.

TPE331-14/15. These are uprated TPE331s to meet future needs. They are essentially scaled-up models, with thermodynamic power in the 932/1,230 kW (1,250/1,650

Garrett TPE331 series commercial turboprop engine. This has the air intake below the spinner, unlike the T76 military engines

shp) class respectively, and will be available either with an existing gearbox as a flat-rated engine or with a new gearbox matched to the uprated gas-generator. Development and production are in partnership with Volvo Flygmotor of Sweden. Certification and initial production are scheduled for mid-1982.

T76. Military engine, with gas generator similar to TPE331-1 series but with front end inverted, to give inlet above instead of below spinner. Two versions, originally designated T76-G-10 and G-12 and restyled G-410 and G-411, respectively giving clockwise and anticlockwise propeller rotation (seen from rear). Development near completion of revised T76-G-420/421 rated at 776 kW (1,040 shp), with certification in September 1976. All models power Rockwell International OV-10 Bronco.

Except for the TPE331-14 all versions of the TPE331 and T76 are of similar frame size, and the following data apply generally to both models:

Type: Single-shaft turboprop.

Propeller Drive: Two-stage reduction gear, one helical spur and one planetary, with overall ratio of 20·865 : 1 or 26·3 : 1. Rotation clockwise or anticlockwise.

Air Intake: Single scoop intake duct at top (T76) or bottom of engine, at front. Provision for bleed air de-icing.

Compressor: Tandem two-stage centrifugal type. Each impeller is single-sided, and is made from titanium. First-stage casing of magnesium, with aluminium diffuser. Second-stage casing and diffuser of stainless steel. Mass flow, 2·61 kg (5·78 lb)/s for 25/61, 25/71, 2·81 kg (6·2 lb)/s for -1, 2·80 kg (6·17 lb)/s for -2 and T76, 3·52 kg (7·75 lb)/s for -251 and 3·54 kg (7·8 lb)/s for -3. Pressure ratio 8·0 for 25/61, 25/71, 8·34 for -1, 8·54 for 2 and T76, 10·37 for -251 and -3.

Combustion Chamber: Annular type of high-temperature alloy. High-energy capacitor discharge ignition. Igniter plug on turbine plenum.

Fuel System: Woodward or Bendix control system for use with Beta propeller governing control system. Five radial primary nozzles in continuous operation. Ten axial simplex nozzles. Max fuel pressure 41·4 bars (600 lb/sq in).

Fuel Grade: (TPE331): Aviation turbine fuels ASTM designation D1655-64T types Jet A, Jet B and Jet A-1; MIL-F-5616-1, Grade JP-1.

Fuel Grade: (T76): MIL-L-5624F(2), Grades JP-4 and JP-5; MIL-G-5572, Grade 115/145.

Nozzle Guide Vanes: Axial vanes made from Inco 713C castings.

Turbine: Three-stage axial-flow type. Discs of first two stages of Inco 100, third stage of Inco 713C, attached to shaft by curvic couplings. Blades cast integrally with disc. Turbine inlet gas temperature, 987°C for 25/61, 25/71, 993°C for T76, 1,005°C for all other models.

Jet Pipe: Fixed type, stainless steel.

Accessories: AND 20005 Type XV-B tachometer generator, AND 20002 Type XII-D starter/generator, AND 20010 Type XX-A propeller governor and AND 20001 Type XI-B hydraulic pump, all mounted on aft face of accessories case.

Lubrication System: Medium-pressure dry sump system. Gerotor internal gear-type pressure and scavenge pumps. Normal oil supply pressure 6·90 bars (100 lb/sq in). Provision for automatic fuel filter anti-icing.

Oil Specification: MIL-L-23699-(1) or MIL-L-7808.

Mounting: Five-point suspension. Three pads on aft face of accessory case, two pads at aft end of turbine plenum.

Starting: Pad for 399A starter/generator on aft face of accessory case.

Dimensions (approx):
Length overall:
TPE331 — 1,092 to 1,333 mm (43-52·5 in)
T76 — 1,118 mm (44 in)
Width:
TPE331 — 533 mm (21 in)
T76 — 483 mm (19 in)
Height:
TPE331 — 660 mm (26 in)
T76 — 686 mm (27 in)

Weight, Dry:
TPE331-25/61, 71 — 152 kg (335 lb)
TPE331-1, -2 — 152·5 kg (336 lb)
T76 — 155 kg (341 lb)
TPE331-251 — 163 kg (360 lb)
TPE331-3 — 161 kg (355 lb)
TPE331-8 — 168 kg (370 lb)
TPE331-10 — 170 kg (375 lb)
TPE331-11 — 177 kg (390 lb)
TPE331-14 — 242·5 kg (535 lb)
TPE331-15 — 250 kg (551 lb)

Performance Ratings:
T-O — see under model listings
Military (30 min):
T76-G-410/411 — 533 kW; 563 ekW (715 shp; 755 ehp)
Normal:
T76-G-410/411 — 485 kW; 514·5 ekW (650 shp; 690 ehp)
Max cruise (ISA, 3,050 m; 10,000 ft and 250 kt; 463 km/h; 288 mph):
TPE331-25/61, 71 — 332 kW (445 shp)
TPE331-1 — 404 kW (542 shp)
TPE331-2, T76 — 430 kW (577 shp)
TPE331-251, -3 — 530 kW (710 shp)

Specific Fuel Consumption:
At T-O rating:
TPE331-25/61, 71 — 111·5 μg/J (0·66 lb/h/shp)
TPE331-1 — 102·2 μg/J (0·605 lb/h/shp)
TPE331-2 — 99·4 μg/J (0·588 lb/h/shp)
TPE331-251 — 105·8 μg/J (0·626 lb/h/shp)
TPE331-3 — 99·7 μg/J (0·59 lb/h/shp)
TPE331-14/-15 — 84·8 μg/J (0·502 lb/h/shp)
T76-G-410/411 — 101·4 μg/J (0·60 lb/h/shp)

Oil Consumption:
Max — 0·009 kg (0·02 lb)/h

GENERAL ELECTRIC

GENERAL ELECTRIC COMPANY AIRCRAFT ENGINE GROUP

Cincinnati, Ohio 45215

Telephone: (513) 243 2000

Group Locations: Lynn and Everett, Massachusetts; Cincinnati, Ohio; Rutland and Ludlow, Vermont; Hooksett, New Hampshire; Albuquerque, New Mexico and Madisonville, Kentucky. Also test facilities at Edwards Air Force Base, California, and Peebles, Ohio. Further facilities at Seattle, Washington; Arkansas City, Kansas; Ontario, California.

Senior Vice-President and Group Executive: B. H. Rowe

Military Engine Operations:
Vice-President and General Manager: J. N. Krebs

Commercial Engine Operations:
Vice-President and General Manager: J. E. Worsham

Engineering Division:
Vice-President and General Manager: F. E. Pickering

Manufacturing Division:
General Manager: W. G. Krall

Marine and Industrial Division:
Vice-President and General Manager: O. R. Bonner

Strategic Planning & Development Operation:
Manager: J. J. Kramer

Current products of the GE Aircraft Engine Group include the F101, F103, F404, J79, J85, T58, T64, T700 and TF34 for military use, and the CF6, CF34, CF700, CJ610, CT58, CT7 and CT64 for the commercial and general aviation market. In partnership with SNECMA of France a company was formed to develop and market the CFM56 turbofan, as described in the International part of this section under CFM International.

General Electric J79-GE-17 turbojet (79·24 kN; 17,820 lb st with afterburning)

GENERAL ELECTRIC J79

Development of the J79, America's first high-compression variable-stator turbojet, began in 1952. In addition to US manufacture, it was produced by Orenda of Canada to power the Canadair CF-104/F-104G (MAP), by Ishikawajima-Harima in Japan for the licence-built F-104DJ, and by MTU of West Germany, Fiat of Italy and FN of Belgium for the European-built F-104G. The Italian production team, including Alfa Romeo, also produced the J79-GE-19, an improved engine similar to the J79-GE-17 but configured for the F-104S Starfighter. More than 17,000 J79s were built by GE and licensees by January 1981. The latest model is the GE-119.

Derivatives of the J79 have been the CJ805-3 turbojet and CJ805-23 turbofan, powering the Convair 880 and 990 Coronado airliners, respectively, as well as the LM1500 industrial and marine gas turbine.

Versions of the J79 in service are as follows:

J79-GE-7A. Built under licence by Orenda (as J79-OEL-7) for Canadair CF-104.

J79-GE-8. McDonnell Douglas F-4B and RF-4B. Air mass flow 76·5 kg (169 lb)/s. Pressure ratio 12·9 : 1.

J79-GE-10. RA-5C and F-4J. Pressure ratio 13·5 : 1. Improved -10A has smoke-free combustor for F-4S.

J79-GE-11A. Lockheed F-104G. Built under licence in Japan (as J79-IHI-11A), West Germany, Italy, Belgium and Canada.

J79-GE-15. Powers F-4C, F-4D and RF-4C for USAF. Similar to J79-GE-8 except for self-contained starting.

J79-GE-17. Similar to J79-GE-10, but for F-4E, F-4G, RF-4E, F-4EJ and F-4F. The J1E, a slightly modified -17, powers all versions of IAI Kfir so far announced.

J79-GE-19. F-104S and F-104A. Differs from J79-GE-10/17 only in external characteristics. Guided expansion jet nozzle. Afterburner provides continuous fuel-flow modulation from 1,225 kg (2,700 lb)/h to 15,420 kg (34,000 lb)/h.

J79-GE-119. Developed to power F-16/79 intermediate (FX International) fighter. Derived from GE-17C with new transfer gearbox and installation changes.

Full details of the J79 can be found in the 1977-78 and 1978-79 editions of *Jane's*.

Type: Variable-stator single-shaft axial-flow turbojet.

Air Intake: Annular type.

Compressor: Seventeen-stage axial-flow.

Combustion Chamber: Cannular type consisting of 10 combustion cans.

Fuel System: Hydromechanical.

Fuel Grade: JP-4 or JP-5.

Turbine: Three-stage.

Dimensions:
Length overall:
J79-GE-7A, 11A — 5,283 mm (207·96 in)
J79-GE-8 — 5,295 mm (208·45 in)
J79-GE-10, 17, 19, 119 — 5,301 mm (208·69 in)
Diameter at compressor:
J79-GE 7A, 8, 11A, 15 — 973 mm (38·3 in)
J79-GE-10, 17, 19, 119 — 992 mm (39·06 in)

Weight, Dry:
J79-GE-7A — 1,644 kg (3,625 lb)
J79-GE-8 — 1,676 kg (3,695 lb)
J79-GE-10 — 1,749 kg (3,855 lb)
J79-GE-11A — 1,658 kg (3,655 lb)
J79-GE-15 — 1,678 kg (3,699 lb)
J79-GE-17, 19, 119 — 1,745 kg (3,847 lb)

Performance Ratings:
T-O, with afterburning:
J79-GE-7A, 11A — 70·3 kN (15,800 lb st)
J79-GE-8, 15 — 75·6 kN (17,000 lb st)
J79-GE-10, 17, 19 — 79·24 kN (17,820 lb st)
J79-GE-119 — 83·31 kN (18,730 lb st)
Military:
J79-GE-7A, 11A — 44·5 kN (10,000 lb st)
J79-GE-8, 15 — 48·5 kN (10,900 lb st)
J79-GE-10, 17, 19, 119 — 52·5 kN (11,810 lb st)
Cruise:
J79-GE-7A, 11A — 11·8 kN (2,650 lb st)
J79-GE-8, 10, 15, 17, 19, 119 — 11·6 kN (2,600 lb st)

Specific Fuel Consumption:
At T-O rating:
J79-GE-7A, 11A 55·8 mg/Ns (1·97 lb/h/lb st)
J79-GE-8 54·67 mg/Ns (1·93 lb/h/lb st)
J79-GE-15 55·1 mg/Ns (1·945 lb/h/lb st)
J79-GE-10, 17, 19 55·66 mg/Ns (1·965 lb/h/lb st)
J79-GE-119 56·08 mg/Ns (1·98 lb/h/lb st)
At military rating:
J79-GE-8, 15 24·36 mg/Ns (0·86 lb/h/lb st)
J79-GE-7A, 10, 11A, 17, 19, 119
23·79 mg/Ns (0·84 lb/h/lb st)
At cruise rating:
J79-GE-7A, 8, 11A, 15
29·74 mg/Ns (1·05 lb/h/lb st)
J79-GE-10, 17, 19 26·91 mg/Ns (0·95 lb/h/lb st)
J79-GE-119 27·76 mg/Ns (0·98 lb/h/lb st)

GENERAL ELECTRIC J85

The following are major versions of the J85 small military turbojet, the -21 being the main production version. By 1979 more than 12,000 J85 engines had been delivered to air forces in 26 nations.

J85-4A. Powers the Rockwell International T-2C Buckeye trainer.

J85-5. Afterburning version with 6·6 : 1 thrust-to-weight ratio; powers Northrop T-38 Talon supersonic trainer.

J85-13. Developed from J85-5, with increased turbine inlet temperature for Northrop F-5A/B supersonic fighter. As the J85-13A, licence-built by Alfa Romeo, also powers Aeritalia G91Y.

J85-15. Version of J85-13 with improved turbine and hydraulically actuated exhaust nozzle to power CF-5 and NF-5. Manufactured under licence in Canada by Orenda.

J85-17A/B. Powers Saab 105G attack/reconnaissance aircraft and Cessna A-37B attack aircraft. Also used as take-off and climb booster for Fairchild C-123K and AC-119K.

J85-21. Higher airflow version with zero stage to give total of nine compressor stages. Equipped with afterburner for supersonic aircraft. Powers Northrop F-5E/F Tiger II.

J85/J1. Non-afterburning derivative with nine-stage compressor.

Civil version of the J85 is the CJ610 turbojet, to which the aft-fan CF700 turbofan is closely related. Both are described separately.

The following data refer specifically to the J85-21:

Type: Single-shaft turbojet.

Air Intake: Annular type, surrounding central bullet fairing. Variable-incidence inlet guide vanes, with hot-air anti-icing.

Compressor: Nine-stage axial-flow type, with variable inlet guide vanes and first three stator stages. Titanium rotor blades, first two stages having part-span shrouds. Discs joined at periphery. Casing in upper and lower halves. Pressure ratio approximately 8·3 : 1. Air mass flow 24·0 kg (53·0 lb)/s.

Combustion Chamber: Annular type with perforated liner. Twelve duplex fuel injectors. Ports in outer casing facilitate inspection of liner.

Turbine: Two-stage axial-flow type. Casing is in halves, split horizontally. Turbine inlet temperature 977°C.

Afterburner: Consists of a diffuser and a combustor. A pilot burner with four spraybars and a main burner of 12 spraybars are located in the diffuser section. Combustion is initiated by a single igniter plug and is then self-sustained. Nozzle position governs exit area and is regulated automatically by the afterburner control system as a function of turbine exit temperature and throttle lever position.

Lubrication: Positive displacement, pressurised recirculating type.

Starting: Air impingement starter on afterburning engines. Provision for starter/generator on non-afterburning engines.

Dimensions:
Length overall 2,858 mm (112·5 in)
Max diameter 533 mm (21·0 in)

General Electric J85-21 turbojet (22·2 kN; 5,000 lb st with afterburning)

Weight, Dry:
J85-21 310 kg (684 lb)
Performance Ratings:
Max rating, with afterburner 22·2 kN (5,000 lb st)
Military rating, without afterburner:
15·6 kN (3,500 lb st)
Specific Fuel Consumption:
At max rating, with afterburner:
60·3 mg/Ns (2·13 lb/h/lb st)
At military rating, without afterburner:
28·3 mg/Ns (1·00 lb/h/lb st)

GENERAL ELECTRIC CJ610

Announced in May 1960, the CJ610 is a power plant tailored for commercial and executive aircraft of 5,700-7,500 kg (12,500-16,500 lb) gross weight. It is essentially similar to the basic J85 turbojet, without afterburner.

By January 1979, a total of 1,850 CJ610s had accumulated more than 5,000,000 h. TBO reached 5,000 h in 1980.

There are seven versions:

CJ610-1, CJ610-4. Initial production versions, differing only in accessory gearbox location.

CJ610-5, CJ610-6. Developed versions of -1 and -4 respectively, providing increased T-O thrust. Power Gates Learjet 24D, 25B and 25C, Hansa and IAI Westwind 1121.

CJ610-8, CJ610-9. Developed for production deliveries beginning in 1969. Power Hansa, IAI Westwind 1123 and NAL (Japan) experimental VTOL.

CJ610-8A. FAA certificated in April 1977 for operation at up to 15,500 m (51,000 ft), to give better economy and over-weather capability to Gates Learjet Century III 24E, 24F, 25D and 25F. Main differences are longer-life turbine and turbine-nozzle area change.

Dimensions:
Length overall:
CJ610-1, -5, -9 1,298 mm (51·1 in)
CJ610-4, -6, -8, -8A 1,153 mm (45·4 in)
Max flange diameter 449 mm (17·7 in)
Weight, Dry:
CJ610-1 181 kg (399 lb)
CJ610-4 176 kg (389 lb)
CJ610-5 183 kg (402 lb)
CJ610-6 180 kg (396 lb)
CJ610-8, -8A 185 kg (407 lb)
CJ610-9 191 kg (421 lb)
Performance Ratings (guaranteed):
T-O:
CJ610-1, -4 12·7 kN (2,850 lb st)
CJ610-5, -6, -8A 13·1 kN (2,950 lb st)
CJ610-8, -9 13·8 kN (3,100 lb st)
Max continuous:
CJ610-1, -4 12 kN (2,700 lb st)
CJ610-5, -6 12·4 kN (2,780 lb st)
CJ610-8, -9 13 kN (2,925 lb st)
CJ610-8A 12·7 kN (2,850 lb st)
Specific Fuel Consumption:
At T-O rating:
CJ610-1, -4 28·05 mg/Ns (0·99 lb/h/lb st)
CJ610-5, -6, -8, -9 27·75 mg/Ns (0·98 lb/h/lb st)
CJ610-8A 27·5 mg/Ns (0·97 lb/h/lb st)
At max continuous rating:
CJ610-1, -4, -8A 27·5 mg/Ns (0·97 lb/h/lb st)
CJ610-5, -6 27·2 mg/Ns (0·96 lb/h/lb st)
CJ610-8, -9 27·2 mg/Ns (0·96 lb/h/lb st)

GENERAL ELECTRIC CF700

Like the CJ610 turbojet, the CF700 is also derived from the J85 engine. Utilising the same gas generator, it is an aft-fan turbofan suitable for military and commercial aircraft. Since it can be tilted while in steady-state operation and operate vertically, it affords lift/cruise capability in VTOL aircraft.

FAA certification of the original version was received on 1 July 1964. The uprated CF700-2D was certificated in early 1968. The CF700-2D has an improved compressor turbine with higher thermodynamic efficiency. The CF700-2D2 incorporates a new design of tailpipe.

CF700 engines power the Dassault-Breguet Falcon 20 and Rockwell Sabre 75A executive transports. By 1974 the TBO had reached 3,000 h. By January 1979 more than 1,100 CF700s had flown 3,000,000 h.

The general description of the J85 turbojet applies also to the CF700, with the following additional assembly:

Aft Fan: Single-stage free-floating fan. Bypass ratio 1·6 : 1. Mass air flow through fan 39·9 kg (88·0 lb)/s.

Dimensions:
Overall length, compressor nose to tailcone tip
1,912 mm (75·57 in)
Length, flange to flange 1,361 mm (53·6 in)
Max diameter 840·4 mm (33·1 in)
Max diameter less fan 447 mm (17·6 in)
Weight, Dry:
CF700-2C 330 kg (725 lb)
CF700-2D, -2D2 334 kg (737 lb)
Performance Ratings:
Max T-O (flat-rated to 30°C):
CF700-2C 18·68 kN (4,200 lb st)
CF700-2D 19·24 kN (4,325 lb st)
CF700-2D2 20·02 kN (4,500 lb st)
Max continuous:
CF700-2C 17·8 kN (4,000 lb st)
CF700-2D, 2D2 18·3 kN (4,120 lb st)

General Electric CJ610-4 (12·7 kN; 2,850 lb st) and (right) CJ610-5 (13·1 kN; 2,950 lb st) turbojet engines

General Electric CF700-2D turbofan (19·24 kN; 4,325 lb st)

Specific Fuel Consumption:
Max T-O:
CF700-2C, -2D, -2D2 18·4 mg/Ns (0·65 lb/h/lb st)
Max continuous:
CF700-2C, -2D 18·4 mg/Ns (0·65 lb/h/lb st)
CF700-2D2 18·1 mg/Ns (0·64 lb/h/lb st)

GENERAL ELECTRIC F404

The F404 is an advanced technology augmented turbofan described as being "in the 16,000 lb thrust class". It is the US Navy derivative of the YJ101 engine flown in the US Air Force's YF-17 aircraft (see 1975-76 *Jane's*). The changed designation from J to F (turbofan) reflects a higher bypass ratio, and the number in the 400-series indicates funding by the US Navy.

In contrast, the YJ101 was funded by the US Air Force, in April 1972, to power the twin-engined Northrop YF-17 Air Combat Fighter. Seven engines in two YF-17 prototypes logged 719 h in 302 flights. These test flights explored a large part of the flight envelope, and a maximum of Mach 2·05 was reached at 12,500 m (41,000 ft). All engine commitments were achieved during the seven-month YF-17 flight programme.

In May 1975 the US Navy selected the McDonnell Douglas/Northrop team to develop its Navy Air Combat Fighter (NACF), designated F-18 Hornet. The F-18 is a derivative of the YF-17, powered by two F404-GE-400 engines. The F404-GE-F1G1 has been selected for the single-engined F-5G export fighter, currently under development by Northrop, and the FSW (Forward Swept Wing) demonstrators, also single-engined.

Compared with that of the J101, the F404 fan diameter is increased less than 25·4 mm (1·0 in) while increasing the bypass ratio from 0·20 to 0·34. The fan is driven by a slightly larger LP turbine. The technology, and the core, comprising the HP compressor, combustion chamber and HP turbine, remain the same.

First F404 engine test took place one month ahead of schedule in January 1977 and soon demonstrated sea-level production performance. Preliminary flight rating test took place in May 1978 and first F-18 flight took place in November 1978. Nine engines were delivered in 1978 and 24 in 1979, and MQT (Model Qualification Test) was completed in July 1979. The first production delivery took place in December 1979.

Type: Two-shaft augmented low-ratio turbofan (turbojet with continuous bypass bleed).
Air Intake: Plain annular. Fixed central bullet, fixed and variable inlet vanes.
Fan: Three-stage axial. Outer flow diverted to bypass duct. Bypass ratio 0·34. Airflow 63·5 kg (140 lb)/s.
HP Compressor: Seven-stage axial. Overall pressure ratio, 25 : 1 class.
Combustion Chamber: Single-piece annular.
HP Turbine: Single-stage axial. Highly loaded aircooled blades.
LP Turbine: Single-stage axial.
Exhaust System: Close-coupled high-augmentation afterburner. Convergent-divergent exhaust nozzle with hydraulic actuation.
Control System: Electrical-hydromechanical.
Dimensions:
Length overall 4,030 mm (158·8 in)
Max diameter 880 mm (34·8 in)
Weight, Dry: approx 908 kg (2,000 lb)
Performance Rating:
Max T-O 71·2 kN (16,000 lb st) class

GENERAL ELECTRIC TF34

It was announced in April 1968 that the US Naval Air Systems Command had awarded General Electric a contract for development of the TF34. This high bypass ratio turbofan had won a 1965 US Navy competition aimed at providing a tailor-made engine in the 40 kN (9,000 lb st) category for the VS(X) aircraft by 1972 within a budget of $96 million. In August 1972 the **TF34-GE-2,** the initial variant for the VS(X) application (the Lockheed S-3A Viking), completed its Model Qualification Test (MQT) and subsequently entered production. The S-3A entered fleet service in February 1974, and GE and the US Navy have defined a 4,000 h TBO extension programme.

In January 1975 GE began shipment of the **TF34-GE-400A,** which replaced the GE-2 as S-3A engine. The later model incorporates various improvements, with changed external piping, an adaptive control system for optimising accessory power extraction, and a simplified rocket gas ingestion system.

In 1970 the TF34 was selected to power the twin-engined Fairchild Republic A-10A Thunderbolt II attack aircraft to compete in the AX competition. The A-10A application led in July 1972 to an Air Force contract for development of the **TF34-GE-100.** This was re-engineered to minimise unit price. It has a long fan duct and side mountings. The GE-100 flew in the first A-10A in May 1972. The A-10A won the AX competition, and the TF34-GE-100 was formally qualified for production in October 1974.

In 1974 a third version of the TF34, most nearly resembling the GE-2, was selected to provide auxiliary (thrust) power for the Sikorsky S-72 RSRA (Rotor Systems Research Aircraft) for NASA and the US Army.

Type: Two-shaft high bypass ratio turbofan.
Air Intake: Plain annular intake. No fixed inlet struts or guide vanes. Small spinner rotates with fan.
Fan: Single-stage fan has blades forged in titanium, without part-span shrouds. Blades replaceable with engine installed. Performance at max S/L rating, mass flow 153 kg (338 lb)/s at 7,365 rpm with pressure ratio 1·5. Bypass ratio 6·2.
Compressor: 14-stage axial on HP shaft. Inlet guide vanes and first five stators variable. First nine rotor stages titanium, remainder high-nickel alloy. Performance at max S/L rating, core airflow 21·3 kg (47 lb)/s at 17,900 rpm with pressure ratio 14 : 1, overall engine pressure ratio 21.
Combustion Chamber: Annular Hastelloy chamber liner and front dome, providing ports for primer nozzles, igniters and 18 carburetting burners.
Turbine: Two-stage HP gas generator turbine with convection-cooled rotor blades and stator vanes, the first-stage nozzle vanes having film and impingement cooling. Four-stage LP fan turbine with tip-shrouded blades. Turbine entry gas temperature 1,225°C maximum.
Fuel System: Contamination-resistant, carburetting type. Integrated hydromechanical control unit with electronic amplifier. Fuel grade JP-4 or JP-5.
Accessory Drives: Engine and customer accessories mounted around horseshoe-shaped gearbox, fitting closely around lower half of compressor casing. Radial shaft drive from front of HP shaft.
Lubrication: Enclosed, pressurised, dual system with vent along centre shaft.

General Electric F404 augmented turbofan (71·2 kN; 16,000 lb st class)

General Electric TF34-GE-400A turbofan of 41·3 kN (9,275 lb st)

DIMENSIONS:	
Max diameter:	
TF34-GE-400A	1,326 mm (52·2 in)
TF34-GE-100	1,259 mm (48·6 in)
Basic length (both)	2,540 mm (100·0 in)
WEIGHT, DRY:	
TF34-GE-400A	670 kg (1,478 lb)
TF34-GE-100	653 kg (1,440 lb)
PERFORMANCE RATINGS:	
Max T-O (S/L, static):	
TF34-GE-400A	41·3 kN (9,275 lb st)
TF34-GE-100	40·3 kN (9,065 lb st)
SPECIFIC FUEL CONSUMPTION:	
Max T-O, S/L static:	
TF34-GE-400A	10·3 mg/Ns (0·363 lb/h/lb st)
TF34-GE-100	10·5 mg/Ns (0·370 lb/h/lb st)

GENERAL ELECTRIC CF34

In April 1976 General Electric's General Aviation Engine Department, at Lynn, announced the CF34 as a new turbofan in the 31-40 kN (7,000-9,000 lb st) class for business and commercial aircraft. A natural derivative of the military TF34 at Cincinnati, the CF34 is closely similar to the TF34-100, but with external configuration tailored to FAA and customer requirements.

Total fan airflow is 139 kg (307 lb)/s, with pressure ratio 1·4. Core pressure ratio is 12·5, giving overall pressure ratio of 17·5. Bypass ratio is 6·3.

In January 1980 the CF34 was selected by Canadair to power the Challenger E. It is also available for other Challengers.

DIMENSIONS: As TF34-100	
WEIGHT, DRY:	692 kg (1,525 lb)
PERFORMANCE RATINGS:	
T-O (S/L static)	35·5 kN (7,990 lb st) flat rated to 22·8°C
Cruise (11,000 m: 36,000 ft at Mach 0·8)	7·83 kN (1,760 lb)
SPECIFIC FUEL CONSUMPTION:	
T-O, as above	10·18 mg/Ns (0·359 lb/h/lb st)
Cruise, as above	19·48 mg/Ns (0·687 lb/h/lb)

GENERAL ELECTRIC CF6

US military designation (CF6-50E): F103-GE-100

On 11 September 1967 General Electric announced the commitment of corporate funding for development of the CF6 turbofan for the then-forthcoming generation of wide-body transports. From the initial family of 142 to 160 kN (32,000 lb to 36,000 lb st) CF6 two-shaft engines announced in September 1967 to cover the anticipated thrust requirements of the Lockheed and McDonnell Douglas airbus projects, the CF6 evolved through a series of variants to the CF6-6D, flat rated at 178 kN (40,000 lb) to 31°C and tailored to the McDonnell Douglas DC-10 Series 10 intermediate-range transport. Announcement that this engine had been selected by United Air Lines and American Airlines was made on 25 April 1968. Further orders have since been placed by many airlines for the CF6-6, 45, 50 and 80 series.

Basic configuration of the CF6-6 comprises a '1¼-stage' fan driven by a five-stage LP turbine energised by a slightly modified TF39 core engine, consisting of a 16-stage HP compressor, annular combustor and two-stage turbine. Modifications have been introduced to enable the accessory systems to suit airline installation requirements, while other changes are aimed at enhancing reliability, durability and maintainability.

CF6-6D. Initial 178 kN (40,000 lb st) version of engine in production for intermediate-range DC-10 Series 10. First ran on 21 October 1968 and 18 days later attained 203·5 kN (45,750 lb st). Following a series of successful factory and outdoor tests, engine was released for production in February 1969. The second CF6-6D, built to the production configuration, first ran in May 1969. By December 1970 a total of 30 engines had been shipped and flight testing with a single engine hung on the starboard inner pylon of a B-52 had extended to 15,250 m (50,000 ft), Mach 0·896 and 420 knots (779 km/h; 484 mph) indicated airspeed. Delivery of flight test engines to McDonnell Douglas started in late 1969, with aircraft first flight following in September 1970. Certification of the CF6-6D for commercial service was granted by the FAA in September 1970, and the engine entered airline service in the DC-10 Series 10 in August 1971.

CF6-6D1. In August 1971 this growth version was FAA certificated and offered to take advantage of the demonstrated margin of the -6D. The D1 rating is increased by 1,000 lb to 182·4 kN (41,000 lb st) at 28·9°C. By 1981 more than 400-6D and -6D1 engines had been delivered.

CF6-6K. This new version of the -6D1, rated at 184·6 kN (41,500 lb st) at 31·1°C, was to be certificated in September 1981. It features improved sfc, improved performance retention and further improved reliability, and is planned for applications currently using CF6-6D1 engines. Mechanical changes include improved-performance fan, -50 type HP turbine and external-case LP turbine case cooling.

CF6-32. This simplified, reduced-thrust engine, rated at 162 kN (36,500 lb st), was planned as a joint programme with SNECMA (France), Volvo (Sweden) and Alfa Romeo (Italy), and was intended primarily to power the Boeing 757. It was terminated in early 1981.

A typical General Electric CF6-80 series turbofan, rated at up to 262·5 kN (59,000 lb st)

CF6-45A and -45B. Economical derated CF6-50E (described later) giving flat rating of 206·8 kN (46,500 lb st) to 36·1°C (-45A for 747SR) or 43·3°C (-45B for 747SP).

CF6-50A. Announced by GE in January 1969, the 218 kN (49,000 lb st) CF6-50A is a growth version of the CF6-6. The increased thrust is achieved by increased flow through the core engine (reducing the bypass ratio from 5·9 to 4·4) at slightly decreased turbine entry temperature. A major change is the introduction of two additional stages behind the single-stage LP compressor of the CF6-6, with no change in the turbofan's external dimensions. To provide for flow matching between the two rotors, variable bypass doors are incorporated between the LP and HP compressors. FAA certification testing was completed in March 1972. The CF6-50A entered airline service in December 1972 in the DC-10 Series 30. The CF6-50 series also powers the Airbus Industrie A300 and some versions of the Boeing 747.

CF6-50C. The CF6-50C is rated at 226·8 kN (51,000 lb st) up to 30°C. Higher thrust is provided by an increase in turbine temperature, with improved cooling of hot-section components. Certificated November 1973.

CF6-50C1/E1. Rated 233·5 kN (52,500 lb st) to 30°C.

CF6-50C2/E2. Similar to C1 and E1 but improved sfc and EGT margins and new fan case and blades with improved bird-strike resistance. Certification 1978.

CF6-50E. (Military designation **F103-GE-100**). This engine is rated to give 233·5 kN (52,500 lb st) up to 26°C. Certificated November 1973. Powers some versions of Boeing 747-200 and E-4A.

CF6-80. New family of engines, with significant sfc improvement, derived from the CF6-50. Thrust ratings from 213·5 kN (48,000 lb st) (-80AR) to 262·5 kN (59,000 lb st) (-80C) are possible. The basic -80A is shorter because of elimination of the turbine mid-frame, a new combustor and compressor rear frame, and a totally new LP turbine with only four stages of increasing diameter. Fan blades are improved, the all-steel compressor has long-life variable-stator bushings and the booster is improved. The -80C has a larger (2,362 mm; 93·0 in) fan and four-stage booster driven by a five-stage LP turbine. Launched November 1977 as choice of American and Delta for 767-200 at 213 kN (48,000 lb st) (-80A) and adopted for Airbus A310. Engines for the 767 have accessories mounted on the core, while A310 engines have accessories on the fan case. First engine was run 21 October 1979, at 246 kN (55,500 lb st). Certification was due in 1981.

By January 1981 CF6-6 engines had accumulated over 7,300,000 h. Engine-attributable three-month in-flight shutdown rate was 0·02 per 1,000 h. During the same time period CF6-50 engines had accumulated over 11 million h with the same engine-attributable in-flight shutdown rate.

The following data relate to the CF6-6D, with the differing features of the CF6-50 series also detailed.

TYPE: Two-shaft high bypass ratio commercial turbofan.

AIR INTAKE: Single forward-facing annular configuration.

FAN: Single-stage fan with integrally-mounted single-stage LP compressor (described together as a 1¼-stage fan), both driven by LP turbine. Fan has rotating spinner and omits inlet guide vanes. Blade-containment shroud provided against possible blade failure. The 38 fan rotor blades are individually removable from the thick-section disc bolted to forward conical extension of LP shaft system. Blade aerofoil has anti-vibration shrouds at two-thirds span. Fan exit airflow split between LP compressor and fan slipstream. Fan frame has 12 radial struts across fan slipstream exit. Fan frame provides support for LP and HP rotor front bearings, fan being overhung ahead of large-diameter ball-thrust bearing with rear roller bearing ahead of core engine. Blades, discs, spool of titanium; exit guide vanes of aluminium; fan frame and shaft of steel; spinner and fan

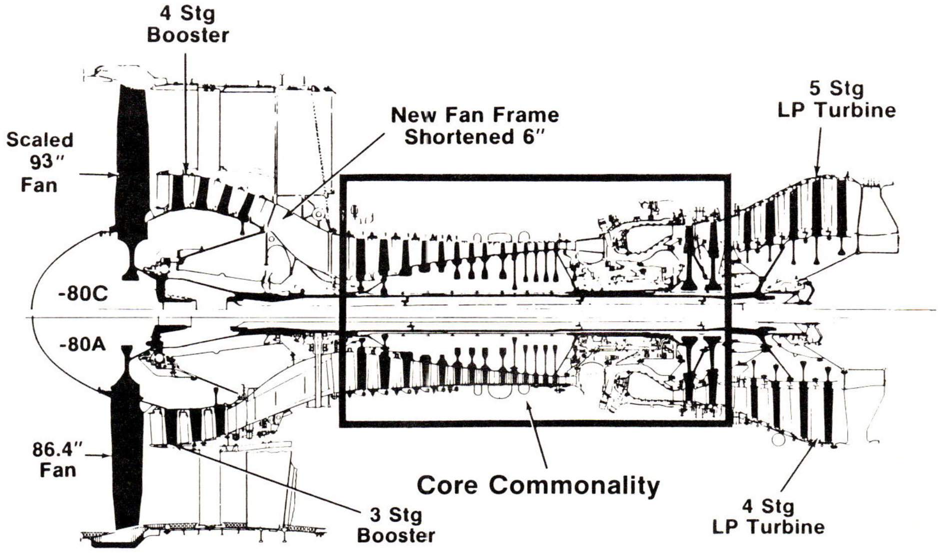

Features of the General Electric CF6-80C (upper half of drawing) compared with the CF6-80A (lower half)

case of aluminium alloy. Total airflow 591 kg (1,303 lb)/s, bypass ratio 5·7. Configuration of CF6-50 is similar but with two added LP stages and bypass doors (described below). Total airflow 658 kg (1,450 lb)/s; bypass ratio 4·3. CF6-80A, 650 kg (1,433 lb)/s; -80C, 792 kg (1,745 lb)/s. Bypass ratio, -80A, 4·7; -80C, 4·9.

LP Compressor: Single-stage compressor acting as booster to airflow into core engine. Rotor blades carried on rear rim of tapered drum bolted to rear of fan disc. Stators cantilevered off short-chord shroud ring, supported by radial outer struts and radial/tangential inner struts located on fan front frame. Compressor exit flow free to balance between core engine and fan slipstream exit. Configuration of CF6-50 modified to three compressor booster stages carried on flanged rotor drum. Continuous shroud extends to fan front frame with 12 integral bypass doors located between canted radial struts in fan exit inner casing. These doors maintain proper flow matching between the fan/LP system and core by opening at low power settings to permit LP supercharged flow to bleed into the fan airstream. The doors are closed during take-off and cruise.

HP Compressor: Sixteen-stage compressor of near-constant tip diameter, with inlet guide vanes and first six stator rows having variable incidence. Provision for interstage air bleed for airframe use and engine cooling. Rotor is of combined drum-and-disc construction with front stage and rear three stages overhung on conical shaft providing location on HP front bearing and HP main shaft. All rotor blades held in rabbeted discs and individually replaceable without rotor disassembly. Stages 1-7 blades forged titanium, 8-16 steel. Stages 1-10 disc titanium, 11-16 and aft casing Inconel 718. Casing split on horizontal centreline: stator vanes held in dovetail slots and replaceable individually. All blading now steel. Double-skin inner casing shrouds the LP main shaft. Outlet frame contains compressor diffuser and incorporates support structure for HP rotor mid-bearings. Overall pressure ratio (T-O), 24·3 (-6D), 24·9 (-6D1). Core airflow (-6D) 88 kg (194 lb)/s. CF6-50A has 15th and 16th stages removed to pass greater core airflow of 125 kg (276 lb)/s and reduce pressure and temperature of air entering combustion chamber. Titanium blades in stages 1-5, steel 6-14. Overall pressure ratio (T-O), 27·1 (-45B), 28·4 (-50A), 29·13 (-50C), 30·1 (-50E).

Combustor: Fully annular with comprehensive film-cooling. Separate snout, dome and inner/outer skirts, with nozzles, igniter, leads and manifold externally removable. Dome contains ports for two igniters and axial swirler cups for 30 fuel nozzles. Igniters of high-voltage surface-gap type with energy level of 2·0 joules, each igniter operated independently. Forged steel nozzles with liner of Hastelloy X. Nozzle and dome designed to minimise smoke, and entrance diffuser has gradual profile to assure low temperature gradient to turbine under all flight conditions. CF6-50 combustor is shorter, of improved material (HS 18-8), and can be removed with fuel nozzles in place.

HP Turbine: Two-stage aircooled turbine with 1,330°C entry temperature. Rotor blades are film and convection cooled. Rotor blades cast from René 80; discs and forward and rear shafts of Inconel 718. First-stage nozzle guide vanes supported at inner and outer ends, second-stage cantilevered from outer ends, with inner ends carrying interstage labyrinth seals. First-stage vanes cast from X40 and film cooled by compressor discharge pressure. Second-stage vanes are cast from René 80 material and are convection cooled. Vanes are welded into pairs to decrease number of gas leakage paths. Thin-section discs with heavy-section centreless hubs are bolted to front and rear conical shafts, including conical and arched inter-disc diaphragms. Configuration for CF6-50 is similar but introduces improved materials and cooling, and blades are not Siamesed but individual. CF6-80 has no turbine mid-frame and eliminates one main bearing.

LP Turbine: Five-stage constant tip-diameter turbine with nominal 871°C inlet temperature. Rotor blades tip-shrouded and cast in René 77, not aircooled. Forward and rear shafts, case and discs of Inconel 718. First-stage nozzle guide vanes supported at inner and outer ends, remaining stages are cantilevered from outer ends, with inner ends carrying interstage labyrinth seals. Stages 1-3 guide vanes cast in six-vane segments in René 77, stages 4 and 5 cast in pairs in René 41. Vanes held in slots machined in the two half-stator casing. Drum and centreless disc construction, located on LP rotor by front and rear conical shafts attached to third- and fourth-stage discs. Drive to rotor by means of long fan midshaft. On CF6-50 a four-stage LP turbine is used, all stages being modified in geometry and cooled by 7th HP-stage compressor air instead of 9th. CF6-80, new four-stage turbine with active clearance control; -80C has five-stage turbine and modified bearing.

Exhaust Unit (Fan): Fixed-area annular duct with outer cowl and engine cowl forming convergent plug nozzle for fan slipstream.

Exhaust Unit (Turbine): Short-length fixed-area exhaust duct with convergent plug nozzle. Provision for exhaust thrust reverser.

Thrust Reverser (Fan): Annular cascade reverser with blocker doors across fan duct. For reverse thrust, rear portion of fan outer cowl translates aft on rotating ballscrews to uncover cascade vanes. Blocker doors (16) flush-mounted in cowl on link arms hinged in inner cowl, rotate inwards to expose cascade vanes and block fan duct. Reverser hinged at top to open in L/R halves for access to HP casing and combustor.

Thrust Reverser (Turbine): Post nozzle exit, cascade type. Two cascade screens are mounted in vertical plane on fixed pivot aft of turbine exhaust and are enclosed in fairing forming aerofoil-shaped plug. Aft translation of fairing uncovers cascades which open across nozzle exit and divert turbine exhaust radially outward and slightly forward in horizontal plane. Configuration for CF6-50 similar to fan thrust reverser with nine blocker doors, but not split. CF6-50 available with long fixed core nozzle; short nozzle also designed, for performance improvement. Acoustic treatment is provided in the nozzle flow path. CF6-80, not fitted.

Rotor Support System: Eight bearings (four for each rotor) at seven locations. Fan and LP compressor carried on ball thrust bearing (1) behind fan disc and roller bearing (2) at front of LP main shaft; both bearings mounted in fan front frame structure, which also supports HP compressor front roller bearing (3). LP turbine carried on roller bearings at front and rear of turbine rotor assembly—rear bearing (7) being mounted in spider structure across turbine exit, and front bearing (6) on major spider structure between HP and LP turbines. HP compressor carried at rear on adjacent roller bearing (4R) and ball-thrust bearing (4B) at interconnection with HP turbine front conical shaft, both bearings being mounted on support structure integral with compressor outlet diffuser. A roller bearing (5), mounted in the inter-turbine structure, carries the aft HP conical shaft.

Accessory Drive: This consists of the inlet gearbox, radial gearbox, radial driveshaft, transfer gearbox, horizontal driveshaft and accessory gearbox. The inlet gearbox is located in the forward sump of the engine. The gearbox transfers energy from the core-engine (HP) rotor to the radial driveshaft located in a housing aft of the bottom vertical strut of the fan frame. The transfer gearbox is mounted on the bottom of the fan frame. Accessory mounting pads are provided on both the forward and aft faces of the gearbox. The engine accessories mounted on the gearbox are starter, fuel pump, main engine control, lubrication pump and tachometer. Pads are also provided for mounting the aircraft hydraulic pumps, constant-speed drive and alternator. CF6-80A gearbox in environmental enclosure on core; -80A1 on fan case.

Fuel System: Hydromechanical fuel control system regulates steady-state fuel flow and schedules acceleration and deceleration fuel flow. It also schedules and powers variable-stator vane position. A governor in the Woodward control provides core-engine speed stability during steady-state operation. During transient operation, core-engine fuel flow is scheduled on the basis of throttle position, compressor inlet temperature, compressor discharge pressure and core-engine speed. The fuel control and fuel pump are mounted in the accessory package as an integrated unit which avoids interconnecting high-pressure fuel lines and potential leakage points (they are separable for change or maintenance). This configuration provides a single drive mounting flange. The filter, fuel/oil heat exchanger and control pressurising valve may be removed individually without removing the entire assembly. The fuel manifold is double-wall constructed for safety and mounted on the exterior of the engine. For CF6-50, fuel control is modified to provide scheduling function for LP compressor variable bypass doors. CF6-80, electronic.

Fuel Grades: Fuels conforming to ASTM-1655-65T, Jet A, Jet A1 and Jet B, and MIL-T-5624G2 grades JP-4 or JP-5 are authorised, but Jet A is primary specification.

Lubrication System: Dry-sump centre-vented system in which oil is pressure-fed to each engine component requiring lubrication. Oil is removed from the sump areas by scavenge pumps, passed through a fuel/oil heat exchanger and filter to the engine tank. Nominal lubrication system pressure is 2·07-6·21 bars (30-90 lb/sq in) above sump reference pressure. All pressure and scavenge pumps and filters are located in the lubrication centre on the forward side of the gearbox.

Oil Specification: Conforming to General Electric specification D50TF1 classes A & B, equivalent to MIL-L-7808 or MIL-L-23699A.

Mounting: Main thrust mount located on the inner fan frame; aft mount located on the turbine mid-frame.

Starting: Air-turbine starter mounted on the front of the accessory gearbox at the through shaft.

Noise Suppression Equipment: Acoustic panels integrated with fan casing, fan front frame and thrust reverser.

Dimensions:

Fan tip diameter (except -80C)	2,195 mm (86·4 in)
Max width (cold)	2,390 mm (94·1 in)
Max height (over gearbox)	2,675 mm (105·3 in)
Length overall (cold):	
CF6-6D	4,775 mm (188·0 in)
CF6-50 series	4,394 mm (173·0 in)
CF6-80A	3,998 mm (157·4 in)
CF6-80C	4,061 mm (159·9 in)

Weight, Dry (basic engine):

CF6-6D, -6D1	3,582 kg (7,896 lb)
CF6-45A/B, -50E, -50E1	3,851 kg (8,490 lb)
CF6-50A, -50C, -50C1	3,956 kg (8,721 lb)
CF6-80A	3,826 kg (8,435 lb)
CF6-80A1, -80B	3,769 kg (8,310 lb)
CF6-80C	4,241 kg (9,350 lb)
Fan and turbine reverser:	
CF6-6D, -6D1	932 kg (2,054 lb)
CF6-50A, -50C	968 kg (2,135 lb)
CF6-50E	962 kg (2,121 lb)

Performance Ratings:

Max T-O, uninstalled, ideal nozzle:

See under model listings

Max altitude and Mach No:

CF6-6 and -50: 13,700 m (45,000 ft) at Mach 1·0

Max cruise thrust at 10,670 m (35,000 ft), Mach 0·85, flat rated to ISA + 10°C, uninstalled, real nozzle:

CF6-6D	40·6 kN (9,120 lb)
CF6-6D1	41·1 kN (9,250 lb)
CF6-45A	48·9 kN (11,000 lb)
CF6-50A, -50C	48·0 kN (10,800 lb)
CF6-50E, -45B	50·3 kN (11,300 lb)
CF6-80A, -80A1	45·9 kN (10,320 lb)
CF6-80B	52·7 kN (11,850 lb)
CF6-80C	52·9 kN (11,900 lb)

Specific Fuel Consumption:

At T-O thrust, as above:

CF6-6D	9·86 mg/Ns (0·348 lb/h/lb st)
CF6-6D1	9·91 mg/Ns (0·350 lb/h/lb st)
CF6-45A, -45B	10·28 mg/Ns (0·363 lb/h/lb st)
CF6-50A	10·90 mg/Ns (0·385 lb/h/lb st)
CF6-50C	11·05 mg/Ns (0·390 lb/h/lb st)
CF6-50E	10·65 mg/Ns (0·376 lb/h/lb st)

Oil Consumption: 0·9 kg (2·0 lb)/h

GENERAL ELECTRIC F101-DFE

The F101-GE-100 was designed by General Electric for the Rockwell International B-1 strategic bomber. It was qualified for production in November 1976. By early 1980 the F101 had logged over 25,000 h, including over 6,000 h in flight. It was described in the 1979-80 *Jane's*.

From it has been derived the F101-DFE advanced augmented turbofan for supersonic fighters, in particular as a candidate engine for re-engining the F-14 Tomcat and F-16. Based largely on components from the F101, YJ101 and F404, the F101-DFE was announced in June 1979. At that time development was being accelerated under a $79·7 million US Air Force contract. The first of four prototypes ran at Evendale in late 1979, and achieved full speed and maximum power on the first run.

General Electric F101-DFE engine, built at Evendale for flight-test programmes in the F-14 and F-16

First flight took place ahead of schedule in an F-16 on 19 December 1980. A crack was soon manifest in a fuel pipe, but the cause was evident and speedily rectified. After extremely successful testing ended in June 1981, F101-DFE engines were transferred to the US Navy to begin similar testing in an F-14 in July 1981.

Fan: Inlet guide vanes with variable trailing flaps. Three axial stages scaled up from F404, with solid titanium blades. Split case. Blades and vanes individually replaceable. Pressure ratio greater than 3. Mass flow approx 122 kg (270 lb)/s. Bypass ratio 0·87.

Compressor: Similar to F101, with nine stages on one-piece inertia-welded rotor. Inlet guide vanes and first three stators variable. Split casing is titanium at front and steel at rear. Blades and vanes individually replaceable. Pressure ratio over 10.

Combustion Chamber: Similar to F101, with very short annular design fed by dual cone nozzles injecting into 20 small scroll cups.

HP Turbine: Identical to F101. Advanced single stage with hollow blades and vanes both impingement and film-cooled. Blades individually replaceable without rotor disassembly. Segmented shroud providing tip clearance control.

LP Turbine: Based on F101, with two stages, tip-shrouded but uncooled. Blades individually replaceable: second-stage vanes replaceable in groups.

Afterburner: Scaled from F101. Mixed-flow with convoluted mixer of fan and core streams in plane of flameholder, where ignition begins on inner ring. Core oxygen is 90 per cent consumed upstream of fuel injection into bypass air.

Nozzle: Scaled from F404 with primary, divergent and outer flaps and seals. Translating hydraulically powered ring drives triple flaps with hinged connections running on cams and rollers.

Dimensions:

Diameter	1,270 mm (50·0 in)
Length (F-16)	5,156 mm (203·0 in)

Weight, Dry: not disclosed

Rating:

T-O 116-124·5 kN (26,000-28,000 lb st) class

GENERAL ELECTRIC T58

The T58 is a small free-turbine turboshaft engine which was developed for helicopter propulsion for the US Navy Bureau of Weapons. A civil version, the CT58, was awarded a Type Certificate by the FAA on 1 July 1959 and is described separately.

Hydromechanical constant-speed control system featured in the T58 maintains essentially constant rotor speed by regulating the engine power automatically, so eliminating the need for speed adjustment by the pilot during normal operation.

Rolls-Royce Ltd produces modified versions of the T58 under licence in the United Kingdom as the Gnome. The T58 is also licensed for manufacture in Italy and Japan. Industrial and marine version of the T58 is the LM100. By January 1981 more than 7,700 T58 engines had been manufactured.

Versions currently in service or in production are as follows:

T58-GE-3. Five-minute rating of 988 kW (1,325 shp). Powers Bell UH-1F.

T58-GE-5. Five-minute rating of 1,119 kW (1,500 shp). Powers Sikorsky CH-3E, HH-3E/F and NASA RSRA (Sikorsky S-72).

T58-GE-8E, F. Rated at 1,007 kW (1,350 shp). Powers Boeing Vertol CH-46A, Kaman SH-2, Sikorsky SH-3A/G and HH-52A.

T58-GE-10. Rated at 1,044 kW (1,400 shp). Powers Sikorsky SH-3D/H, and Boeing Vertol CH-46D/F.

T58-GE-16. Rated at 1,394 kW (1,870 shp). US military qualified. Aircooled gas-generator turbine and two-stage power turbine. Powers Boeing Vertol CH-46E.

T58-GE-100. Uprated T58-GE-5. Ten-minute rating 1,119 kW (1,500 shp) to 15°C or 1,100 kW (1,475 shp) to 26°C. Powers selected CH/HH-3E. Qualified 1976.

Type: Free-turbine turboshaft.

Air Intake: Annular intake casing with four hollow radial struts supporting central housing for starter drive clutch and front main roller bearing. Casing and struts anti-iced by air bled from compressor.

Compressor: Ten-stage axial-flow. Variable-incidence inlet guide vanes. First three of the eleven rows of stator blades also have variable incidence. One-piece steel construction for last eight stages of rotor hub. Casing divided into upper and lower halves. Pressure ratio 8·4 : 1. Air mass flow 5·62 kg (12·4 lb)/s in T58-GE-3 and -8E, 6·21 kg (13·7 lb)/s in T58-GE-5, -10 and -100, 6·30 kg (13·9 lb)s/in T58-GE-16.

Combustion Chamber: Annular type. Sixteen fuel nozzles (eight on each of two manifolds) mounted on front of inner liner. Dual capacitor discharge ignition unit. Outer casing in two halves to facilitate inspection.

Gas-Generator Turbine: Two-stage short-chord axial-flow type, coupled directly to compressor by hollow conical shaft. Centre ball thrust bearing, rear roller bearing. Cooling by air bled from compressor. T58-GE-16 has aircooled first-stage turbine nozzle and blades and second-stage nozzle.

The 1,394 kW (1,870 shp) General Electric T58-GE-16 turboshaft

Power Turbine: Single-stage (two-stage in T58-GE-16) axial-flow type, mechanically independent of gas generator turbine. Operated nominally at 19,500 rpm. Engines with single-stage power turbine can have reduction gear giving output at 6,000 rpm. Power turbine accessory drive unit and flexible feed-back cable provide a speed signal to the control.

Torque Sensor Speed Decreaser Gearbox (optional except on -16): Integral lubrication system. Reduces power speed to 6,000 rpm. Assembly includes an integral torque sensing system.

Jet Exhaust: Two positions (90° left or right) on all versions. T58-GE-16 can also be supplied with downward-ejecting or multiple-position exhaust.

Controls (except T58-GE-10 and -16): Free turbine constant-speed control. Hydromechanical controls.

Controls (T58-GE-10, -16): Integrated hydro-mechanical/electrical power control system for isochronous speed governing and twin-engine load sharing.

Accessory Drives: Engine accessories driven from compressor shaft. Airframe accessories mounted on free-turbine reduction gearbox or rotor hub.

Dimensions:

Max width:	
except T58-GE-16	526 mm (20·7 in)
T58-GE-16	607 mm (23·9 in)
Length overall:	
except T58-GE-16	1,499 mm (59·0 in)
T58-GE-16	1,626 mm (64·0 in)

Weight, Dry:

T58-GE-3	140 kg (309 lb)
T58-GE-5, -100	152 kg (335 lb)
T58-GE-8E, F	138 kg (305 lb)
T58-GE-10	159 kg (350 lb)
T58-GE-16	201 kg (443 lb)

Performance Ratings:

Five-minute:

See under model listings

Military:	
T58-GE-3	988 kW (1,325 shp) at 20,960 rpm
T58-GE-5, 10	1,044 kW (1,400 shp) at 19,500 rpm
T58-GE-8E, F	1,007 kW (1,350 shp) at 19,500 rpm
T58-GE-16	1,394 kW (1,870 shp) at 19,500 rpm
T58-GE-100	1,119 kW (1,500 shp) at 19,500 rpm
Cruise:	
T58-GE-3	798 kW (1,070 shp)
T58-GE-5, 10	932 kW (1,250 shp)
T58-GE-8E, F	857·5 kW (1,150 shp)
T58-GE-16	1,320 kW (1,770 shp)
T58-GE-100	1,015 kW (1,360 shp) at 19,500 rpm

Specific Fuel Consumption:

At military rating:	
T58-GE-3	103 μg/J (0·61 lb/h/shp)
T58-GE-5, 8E/F, 10, 100	101 μg/J (0·60 lb/h/shp)
T58-GE-16	89·5 μg/J (0·53 lb/h/shp)
At cruise rating:	
T58-GE-3	106·5 μg/J (0·63 lb/h/shp)
T58-GE-5, 100	103 μg/J (0·61 lb/h/shp)
T58-GE-8E, F	105 μg/J (0·62 lb/h/shp)
T58-GE-10	105 μg/J (0·62 lb/h/shp)
T58-GE-16	91 μg/J (0·54 lb/h/shp)

GENERAL ELECTRIC CT58

The commercial version of the T58 is designated CT58 and was the first US helicopter turbine to receive FAA certification.

Current versions are as follows:

CT58-110. Rated at 932 kW; 1,250 shp (1,007 kW; 1,350 shp for 2½ min) at 19,500 rpm. Air mass flow 5·67 kg (12·7 lb)/s. Pressure ratio 8·2 : 1.

CT58-140. Rated at 1,044 kW; 1,400 shp (1,119 kW; 1,500 shp for 2½ min) at 19,500 rpm. Air mass flow 6·21 kg (13·7 lb)/s. Pressure ratio 8·4 : 1.

The CT58 powers the Sikorsky S-61 and S-62 and Boeing Vertol 107 Model II.

Dimensions:

Max width	406 mm (16·0 in)
Length overall	1,500 mm (59·0 in)

Weight, Dry:

CT58-110	143 kg (315 lb)
CT58-140	154 kg (340 lb)

Performance Ratings:

2½ min and normal T-O:

See under model listings

Cruise:	
CT58-110	783 kW (1,050 shp)
CT58-140	932 kW (1,250 shp)

Specific Fuel Consumption:

At normal T-O rating	103 μg/J (0·61 lb/h/shp)
At cruise rating:	
CT58-110	108 μg/J (0·64 lb/h/shp)
CT58-140	105 μg/J (0·62 lb/h/shp)

GENERAL ELECTRIC T64

The T64 is a versatile aircraft gas turbine engine which was developed initially for the US Navy. The basic T64 turboshaft engine becomes a turboprop with the addition of a two-part speed-reduction gearbox.

Current versions include:

T64-GE-7. Direct-drive turboshaft rated at 2,928 kW (3,925 shp). Powers US Air Force CH-53B, CH-53C and HH-53C. Also powers VFW-built CH-53D/G. Produced under licence by MTU (see entry under Germany, Federal Republic).

T64-GE-7A. Direct-drive turboshaft flat rated at 2,935 kW (3,936 shp) to 28°C. Powers Sikorsky S-65.

T64-GE-10. Turboprop engine with propeller gearbox above centreline. Rated at 2,215 kW (2,970 shp). Produced under licence by Ishikawajima-Harima Heavy Industries in Japan for Shin Meiwa PS-1 flying-boat (four engines) and Kawasaki P-2J patrol aircraft (two engines).

T64-GE-413A. Direct-drive turboshaft rated at 2,935 kW (3,936 shp). Two engines power the US Navy CH-53D and RH-53D.

T64-GE-415. Growth version with improved combustion liner and turbine cooling. Max rating 3,266 kW (4,380 shp). Powers Sikorsky RH-53D.

T64-GE-416. As -415; powers CH-53E.

CT64-GE-820. CT64-820-1, -2 and -3 turboprops based on early turboshaft versions power DHC-5C Buffalo and prototype Aeritalia G222. FAA certificated CT64-820-4 has improved components of T64-415 and is flat-rated at 2,336 kW (3,133 shp) to 38°C, in production for DHC-5D Buffalo.

T64-P4D. Turboprop version flat rated at 2,535 kW (3,400 shp) to 45°C. Two P4D engines power most Aeritalia G222 transports. Production by Fiat, supported by Alfa Romeo, from 1975.

All T64s are qualified to operate from 100° nose-up to 45° nose-down. The T64 was designed for extensive growth: current production engines rated at 3,266 kW (4,380 shp) are a result of growth made possibly largely by aircooling of the first-stage gas-generator turbine rotor and stator. The addition of aircooling to the second turbine stage provides further horsepower growth beyond

The 2,935 kW (3,936 shp) General Electric T64-GE-413A turboshaft

3,729 kW (5,000 shp) without significant change in external dimensions.

By January 1981 a total of over 2,100 T64 engines of all kinds had been delivered by GE to customers in 27 countries. In addition licences to produce the T64 are held by MTU in West Germany, Fiat in Italy, Rolls-Royce in the UK and IHI in Japan.

TYPE: Free-turbine turboshaft/turboprop engine.

COMPRESSOR: Fourteen-stage axial-flow. Single-spool steel rotor for -10 and -820-1/2/3. Titanium and steel compressor for -7A, -413A, -415, -P4D and CT64-820-4. Inlet guide vanes and first four stages of stator blades variable. Compressor blades can be removed individually without rotor disassembly. Casing flanged along centreline. Stator blades removable. Air mass flow per second: -10, 11·6 kg (25·5 lb): -7A, -413A, 12·8 kg (28·3 lb); -415, 13·3 kg (29·4 lb): -820-4, 11·9 kg (26·2 lb): P4D, 12·2 kg (27·0 lb). Pressure ratio: -10, 820-4, 12·5; -7A, -413A, 14·1: -415, 14·8; P4D, 13·0.

COMBUSTION CHAMBER: Annular type. Double fuel manifold feeds twelve duplex-type fuel nozzles with external flow divider. Nozzles mounted on outer diffuser wall of compressor rear frame.

GAS GENERATOR TURBINE: Two-stage axial-flow type, coupled directly to compressor rotor by spline connection.

POWER TURBINE: Two-stage axial-flow type, mechanically independent of gas generator turbine.

REDUCTION GEAR: Remotely-mounted basic reduction gear for turboprop versions is offset and accessible for inspection and replacement. Gear-driven by power turbine, using co-axial shafting through the compressor. Propeller gear ratio 13·44 : 1.

STARTING: Mechanical, airframe supplied.

DIMENSIONS:

Length:	
T64-GE-7A, -413A, -415	2,006 mm (79 in)
T64-GE-10, -P4D	2,793 mm (110 in)
CT64-820	2,870 mm (113 in)
Width:	
T64-GE-7A, -413A, -415	60 mm (26·0 in)
T64-GE-10, -P4D, -820	683 mm (26·9 in)
Height:	
T64-GE-7A, -413A, -415	825 mm (32·5 in)
T64-GE-10, -P4D	1,168 mm (46 in)
CT64-820	1,026 mm (40·4 in)

WEIGHT, DRY:

T64-GE-7A, -415	327 kg (720 lb)
T64-GE-10	529 kg (1,167 lb)
T64-GE-413A	325 kg (716 lb)
CT64-820-1	513 kg (1,130 lb)
CT64-820-2, -4	520 kg (1,145 lb)
CT64-820-3	518 kg (1,140 lb)
T64-P4D	538 kg (1,188 lb)

PERFORMANCE RATINGS:

Max rating (sea level):

T64-GE-7A, -413A	2,935 kW (3,936 shp)
T64-GE-10	2,215 kW (2,970 shp) at 1,160 output rpm
T64-GE415	3,266 kW (4,380 shp)
CT64-820-4	2,336 kW (3,133 shp)
T64-P4D	2,535 kW (3,400 shp)

SPECIFIC FUEL CONSUMPTION (S/L):

At max rating:

T64-GE-7A, -415	79·4 μg/J (0·47 lb/h/shp)
T64-GE-10	84·5 μg/J (0·50 lb/h/shp)
T64-GE-413A, -P4D	81 μg/J (0·48 lb/h/shp)
CT64-820-4	83 μg/J (0·49 lb/h/shp)

GENERAL ELECTRIC T700

A competition was conducted in 1971 to provide the power plant for the US Army's projected utility tactical transport system (UTTAS) proposed as a replacement for the present Bell UH-1 family of Army helicopters. In late 1971 it was announced that the winner was the GE T700.

The first T700 engine was tested in February 1973. Shipment of ground test engines was accomplished on schedule in February 1974, and flight qualification testing was completed ahead of schedule in August 1974. Shipments for UTTAS flight aircraft began immediately, with first flights of the Boeing YUH-61A and Sikorsky YUH-60A following during the autumn of 1974. Identical T700 engines powered the two advanced attack helicopter (AAH) contenders, the Bell YAH-63A and Hughes YAH-64A, of which the latter was selected.

In December 1976 the first contract for production engines was received to power the Sikorsky UH-60A Black Hawk. Production **T700-700** engines for this helicopter have been delivered from early 1978. By 1981 more than 125,000 h had been logged, including 80,000 flying in the four competing helicopters. The Navy **T700-401**, with 95-97 per cent commonality, has been selected for the Sikorsky SH-60B (LAMPS Mk III) helicopter (Light Airborne Multi-Purpose System). It has greater corrosion protection and emergency power provisions.

Other applications for the T700 engine include the Bell 214ST, Bell AH-1T+ (AH-1T Plus) and the prototypes of the EHI EH-101.

The T700 has been designed to be compatible with the Army's special operating and environmental conditions, and embodies high reliability, simplicity of maintenance, low vulnerability to combat damage, and high performance combined with compact dimensions. Use is made of higher pressure ratios and turbine entry temperatures than with existing small turboshafts to assist in reducing size and weight.

To reduce vulnerability, all external lines and leads are short in length and are grouped compactly for minimum exposure. Self-contained electrical and lubrication systems are fitted. Multiple mounting points allow for ease of installation and the necessary airframe connections have been minimised and are located close to the engine centreline. The whole engine is of modular construction for swift field maintenance or section replacement without special tools.

The T700-401 completed severe testing in SH-60B helicopters at sea in 1981. The programme for the SH-60B for the US Navy requires some 500 engines to be delivered in 1983-87.

GE has announced studies for derivatives, including a family of turbojets for RPVs and other fixed-wing aircraft. Other possibilities include turboprops and high bypass ratio turbofans, while front and rear drive modifications make surface applications possible.

A programme of growth has been started to meet the widening spectrum of applications. As envisaged, 10 per cent greater power will be available from the existing basic engine. Addition of an LP booster (zero stage) will then provide 20-30 per cent growth, depending on the increase in turbine temperature chosen. These growth versions are planned to be interchangeable with the present T700-700 and -401.

TYPE: Ungeared free-turbine turboshaft engine.

INTAKE: Annular type, with anti-iced integral inlet particle separator containing no moving parts yet designed to remove 95 per cent of sand, dust and foreign-object ingestion. Extracted matter discharged by separator blower driven from accessory gearbox.

COMPRESSOR: Combined axial/centrifugal. Five axial stages and single centrifugal stage mounted on same shaft. Each axial stage is one-piece 'blisk' (blades plus disc) in AM355 steel highly resistant to erosion. Inlet guide vanes and first two stator stages are variable. Pressure ratio, about 15 : 1. Airflow about 4·5 kg (10 lb)/s at 44,720 rpm.

COMBUSTION CHAMBER: Fully annular. Compact short-length configuration, designed for maximum reliability and long life. Central fuel injection to maximise acceptance of contaminated fuel and give minimal smoke generation and uniform temperature profile into the turbine. Flame tube is machined ring in Hastelloy X. Ignition system obtains power from separate winding on engine-mounted alternator and serves dual plugs.

TURBINE: Two-stage gas-generator (HP) turbine operates at gas temperatures exceeding 1,100°C. First-stage nozzle investment-cast in X40. Second-stage nozzle investment-cast in two-vane segments in R80. Discs, cooling plates, and blades of both stages clamped by five short tiebolts; five larger bolts then tighten turbine to shaft, driving via curvic joint. Rated shaft speed (S/L, ISA, max T-O power), 44,720 rpm. Two-stage free power turbine, designed for high efficiency at part-power levels (especially 30 and 60% of military power), with tip-shrouded blades and segmented nozzles. Power turbine inlet temperature at intermediate power, 827°C. Nozzle guide vanes René 77, rotor discs Inco 718, rotor blades René 80 uncooled. Output speed,

General Electric T700-401 turboshaft engine for SH-60B

17,000-21,000 rpm. Power output shaft for front drive.

CONTROLS: Hydromechanical control can be replaced in less than 12 minutes and requires no adjustment or lockwire. Electrical control, coupled with hydromechanical control, provides twin-engine speed and torque matching.

ACCESSORIES: Grouped at top of engine, together with engine control system, for maximum simplicity, accessibility and combat survivability. Integral lubrication supply tank, plus an emergency supply of mist lubrication following total loss of main supply. Torque sensor provides signal to electrical control system.

DIMENSIONS:
Length overall 1,181 mm (46·5 in)
Width 635 mm (25 in)
Height overall 584 mm (23 in)

WEIGHT, DRY (with particle separator):
T700-700 192 kg (423 lb)
T700-401 194 kg (427 lb)

PERFORMANCE RATINGS (ISA, SL, static):
T700-700:
Intermediate 1,163 kW (1,560 shp)
Continuous 938 kW (1,258 shp)
T700-401:
Contingency 1,285 kW (1,723 shp)
Intermediate 1,260 kW (1,690 shp)
Continuous 1,071 kW (1,437 shp)

SPECIFIC FUEL CONSUMPTION (ISA, SL, static):
T700-700:
Intermediate 81·78 μg/J (0·484 lb/h/shp)
Continuous 80·1 μg/J (0·474 lb/h/shp)
T700-401:
Contingency 78·6 μg/J (0·465 lb/h/shp)
Intermediate 78·4 μg/J (0·464 lb/h/shp)

GENERAL ELECTRIC CT7

In September 1976 General Electric announced a new commercial helicopter engine known as the CT7, based on the T700 military engine developed for the US Army. Certification testing to Federal Aviation Administration standards was completed in April 1977.

Another GE engine, the **CT7-5**, adds a reduction gearbox for a propeller. It was selected in June 1980 to power the Saab-Fairchild 340, in this application rated at 1,600 shp and driving a Dowty-Rotol carbonfibre propeller. The **CT7-7** will power the CASA-Nurtanio CN-235.

Data below are for the **CT7-2A** turboshaft version.

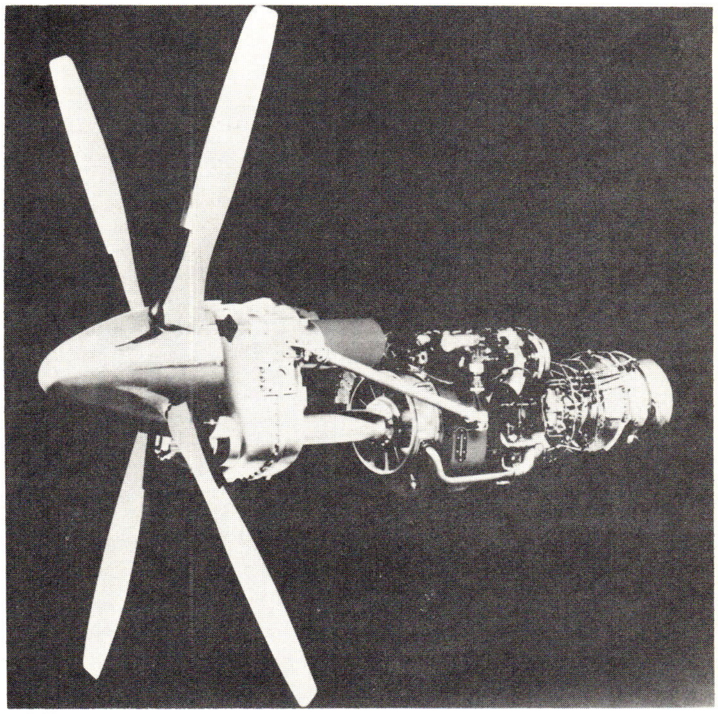

General Electric CT7-5 turboprop, power plant of the Saab-Fairchild 340 commuter transport

DIMENSIONS:
Length 1,194 mm (47·0 in)
Width 635 mm (25·0 in)
Height 584 mm (23·0 in)

WEIGHT, DRY: 195 kg (430 lb)

PERFORMANCE RATINGS (S/L, static, 15°C):
Contingency (2½ min) 1,209 kW (1,620 shp)
T-O (5 min) and en route contingency (30 min) 1,164 kW (1,560 shp)
Max cruise 940 kW (1,260 shp)

Data below are for the turboshaft versions.

PERFORMANCE RATINGS (S/L, static):
T-O: CT7-5 1,233 ekW (1,654 ehp) to 30°C
CT7-7 1,309 ekW (1,756 ehp) to 30°C
Max cruise, below 4,570 m (15,000 ft):
CT7-5 1,092 ekW (1,465 ehp)
CT7-7 1,148 ekW (1,540 ehp)
Max cruise, above 4,570 m (15,000 ft):
CT7-5 1,212·5 ekW (1,626 ehp)
CT7-7 1,273·5 ekW (1,708 ehp)

GESCHWENDER

GESCHWENDER AEROMOTIVE INC

4131 NW 36 Street, Lincoln, Nebraska 68524
Telephone: (402) 470 3333

Geschwender Aeromotive was developing liquid-cooled V-8 engines for aircraft. It ceased operating on 31 January 1981.

GLUHAREFF

EMG ENGINEERING COMPANY

18518J South Broadway Avenue, Gardena, California 90248
Telephone: (213) 321 8699

Eugene M. Gluhareff, a pioneer of ultralight rotorcraft, has been developing a unique type of air-breathing jet engine which he considers to offer notable advantages over all other systems for rotor tip-drive. The first model is the G8-2, which was designed in 1955 and fully developed over the next ten years for tip-drive or sailplane auxiliary propulsion. It has also been used for surface application in a go-kart, and in numerous static rigs sold to universities and other organisations. Production has been hard pressed to keep up with demand, a fast-growing market being radio-controlled flight vehicles. Manned platforms and small helicopters are also proving to be a large market for various sizes of the family of engines described below.

GLUHAREFF G8-2

Although extremely simple, the G8 series corresponds to no prior jet system. The design is based on propane, a readily-available volatile fuel. The pressure of the liquid propane in the tank delivers the fuel, via a needle valve serving as the throttle, to the burner unit. The fuel enters the burner duct and is immediately vaporised in a hot heat-exchanger. Vapour then passes back down an insulated pipe to the injector where its residual pressure is converted to kinetic energy. The high-velocity gas jet induces air through three 'supercharger' intakes, each synchronised to the internal flow, which gives the correct final fuel/air ratio to the mixture entering the combustion chamber. Here the mixture is initially ignited by a spark plug and thereafter burns continuously. The intake ducts are tuned to each other to create one-way flow. Resonance in the tailpipe is undesirable and is prevented by making the propelling nozzle of fishtail shape.

By 1979 three more models were available, the G8-2-20, -80 and -130, respectively rated at 0·09, 0·36 and 0·58 kN (20, 80 and 130 lb st), and a G8-2-250 (1·1 kN, 250 lb st) was on bench test. The best specific fuel consumption had been dramatically reduced from 170 mg/Ns (6 lb/h/lb st) to only 21·8 mg/Ns (0·77 lb/h/lb st).

The units are manufactured by Gluhareff's subsidiary EMG Engineering Co. Customers have the option of buying plans only, a construction package, partly prefabricated or an assembly kit or finished engine.

JACOBS

PAGE INDUSTRIES INC

HEAD OFFICE: C.E. Page Airport, PO Box 191, Yukon, Oklahoma 73099
Telephone: (405) 354 5385

PRESIDENT: O. J. Butts
VICE-PRESIDENT AND TREASURER: Currie Smith
VICE PRESIDENT, ENGINEERING: Merrill H. Bumbaugh

Page Industries is the successor to the Jacobs Aircraft Engine Co, which had manufactured aircooled radial piston engines since 1929. In the 1970s production centred upon various models of R-755, but the Type Certificate has been sold to Bill Goldman, Jacobs Service Co, 4305 Saturn Way, Chandler, Arizona 85224. A description of the R-755 appeared in *Jane's* 1980-81 edition.

MARQUARDT

THE MARQUARDT COMPANY (a division of CCI Corporation)

HEAD OFFICE: 16555 Saticoy Street, Van Nuys, California 91409
Telephone: (213) 781 2121
PRESIDENT: G. H. Hanauer
VICE-PRESIDENTS:
K. E. Woodgrift (Engineering)
A. L. Sorensen (Operations)
B. E. Huston (Finance)
C. Long (Advanced Programmes)

The Marquardt company was formed in November 1944 to undertake research and development of ramjet engines, and it produced the first American subsonic ramjet in 1945. Its main engineering business continues to be advanced aerospace propulsion and the supply of ram-air turbine power systems. A major portion of sales is currently associated with manufacture of sophisticated structures and components for the aerospace industry, and the production of clustered munitions.

Marquardt is currently developing security-classified types of composite rocket/air-breathing propulsion systems for the US Air Force and Navy. These are regarded by the company as likely to lead to a new generation of power plants for supersonic strategic missiles and expendable tactical vehicles.

Marquardt also has been developing precision rockets for spacecraft since 1959. The company developed and qualified attitude-control rockets for the service module and lunar module of the Apollo and for the Orbiter vehicle of the Space Transportation System. Marquardt has been prominent in the development of monopropellant rocket engines and systems, and an engine for a fluorine/hydrazine space-propulsion system. Advanced development activities include resistojet thrusters, a water-electrolysis system and oxy/hydrogen thrusters for long-life spacecraft reaction-control systems. Marquardt is currently producing propulsion subsystems and bipropellant rocket engines for the India satellite and the System Applications Laboratories satellite, and monopropellant engines for the US Air Force's Defense Meteorological System satellite.

Marquardt maintains extensive test facilities at Van Nuys for research and testing of air-breathing rocket engines, and controls and accessories. Total land area occupied exceeds 56 acres, and covered buildings exceed 46,400 m² (500,000 sq ft); employment exceeds 900.

MARQUARDT DUCTED ROCKET

In October 1977 Marquardt announced a new $2 million contract from the US Air Force Aero Propulsion Laboratory for further work on the ducted-rocket propulsion system. According to the company president: "It is designed to combine the best features of the ramjet engine and the solid rocket into a simple low-cost high-efficiency propulsion system. It employs a fuel-rich solid-rocket gas-generator and, like the basic ramjet, utilises the atmosphere as its primary source for oxidiser. The specific impulse is significantly higher than that of conventional solid rockets, yet it maintains the inherent simplicity desired for tactical applications".

The programme's objective was to demonstrate system performance in flightweight hardware, culminating in an integrated engine demonstration in September 1979. Supporting Marquardt are Hughes (integration of aircraft, missile and engine), and Rocketdyne McGregor Division (solid-rocket gas-generator). Applications centre on both tactical and strategic missiles.

MARQUARDT FIREBRAND RAMJET

In May 1977 Marquardt began work on a Teledyne Ryan contract, expected to be valued at about $6 million, for design and development of ramjet engines for the Teledyne Ryan Firebrand (see RPVs and Targets section). Each Firebrand will be powered by two of the new engines. At completion of the present programme in February 1982 six R & D targets and two special test vehicles will have flown; a production quantity is expected to follow. Firebrand will be highly supersonic, simulating advanced anti-ship missiles, and is designed to be reusable.

MARQUARDT R-40A

This precision control rocket has been developed and qualified for the Space Shuttle Orbiter vehicle.

Type: Liquid-propellant reaction control rocket.

Propellants: Nitrogen tetroxide and monomethyl hydrazine.

Thrust Chamber Assembly: Single chamber. Area ratio 20. Made of welded coated columbium, with welded-on orthogonal and scarfed nozzle extension in same material. Internal film cooling. Exterior insulated for buried installation. Started by electrical signal to on/off solenoid valve. Multiple doublet injector with hypergolic ignition.

Thrust Chamber Mounting: Flange bolt circle on injector head.

Propellant Feed System: Pressurised tanks, with feed of 0·526 kg (1·16 lb)/s fuel and 0·838 kg (1·85 lb)/s oxidant at 16·4 bars (238 lb/sq in abs).

Dimensions:
- Length overall: 472 mm (18·6 in)
- Nozzle exit diameter: 267 mm (10·5 in)

Weight, Dry: 9·5 kg (21·0 lb)

Performance Ratings:
- Max thrust (vacuum): 3·87 kN (870 lb)
- Chamber pressure: 10·5 bars (152 lb/sq in)
- Specific impulse: 281

MARQUARDT R-30

This precision control rocket has been qualified for use on the kick stage of the US Air Force's Defense Meteorological Satellite and NASA Tiros-No.

Type: Liquid monopropellant precision control rocket.

Propellant: Hydrazine.

Thrust Chamber: Single chamber. Area ratio 22. Fabricated in L605 stainless steel. Shell 405 catalyst bed in combustion chamber. Multiple element injection.

Thrust Chamber Mounting: Fixed, by flange bolt circle on injector head.

Propellant Feed System: Pressurised tank. Flow rate 0·297 kg (0·657 lb)/s at 31 bars (450 lb/sq in abs).

Dimensions:
- Length overall: 273 mm (10·75 in)
- Diameter: 103 mm (4·0 in)

Weight, Dry: 1·36 kg (3·0 lb)

Performance Ratings:
- Max thrust (vacuum): 0·69 kN (155 lb)
- Chamber pressure: 15·5 bars (225 lb/sq in)
- Specific impulse: 236

MARQUARDT R-1E

This small high-performance rocket was qualified for use on the reaction control subsystem of the subsequently-abandoned US Air Force Manned Orbital Laboratory, and is now qualified as the vernier engine for the Space Shuttle Orbiter.

Type: Liquid bipropellant rocket for use in space.

Propellants: Nitrogen tetroxide and monomethyl hydrazine.

Thrust Chamber: Single chamber. Minimum area ratio 26 with orthogonal and scarfed nozzles. Made of metal-coated columbium. Insulated for buried installation. Started by electrical signal to on/off solenoid valve. Single doublet injector with hypergolic ignition.

Thrust Chamber Mounting: Fixed, by flange bolt circle on injector head.

Marquardt MA212-XAA conical-inlet supersonic ramjet (net thrust 6·05 kN; 1,360 lb at Mach 2·5 at 12,200 m; 40,000 ft)

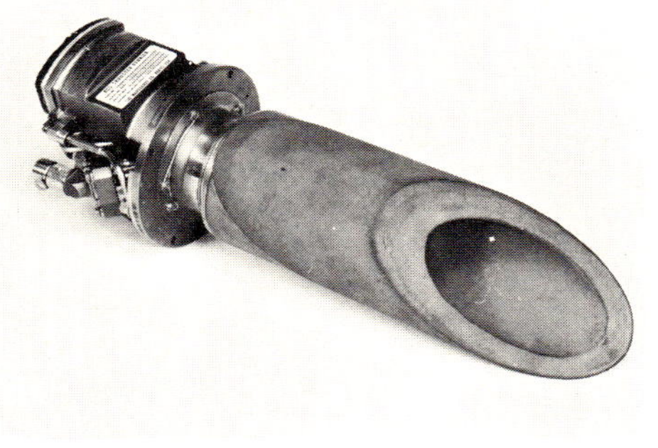

Marquardt R-1E vernier long scarf thruster

Propellant Feed System: Pressurised tank. Flow rate 0·016 kg (0·0354 lb)/s fuel and 0·0256 kg (0·565 lb)/s oxidant.

Dimensions:
- Length overall: 279 mm (11·0 in)
- Width: 147 mm (5·8 in)
- Height (depth): 145 mm (5·7 in)

Weight, Dry: 3·7 kg (8·2 lb)

Performance Ratings:
- Max thrust (vacuum): 0·11 kN (25·0 lb)
- Chamber pressure: 7·45 bars (108 lb/sq in)
- Specific impulse (area ratio 40): 272

MARQUARDT R-4D-11 AND R-4D-12

These small attitude-control and manoeuvre rockets are the latest derivatives of the R-4D engine, first developed and qualified for the Apollo programme. Both were qualified in early 1980 and the R-4D-11 is to be used on the INSAT-1 communications satellite being developed by Ford Aerospace & Communications. The R-4D-12 is intended for application on the Hughes Aircraft spin-stabilised System Applications Laboratories satellite.

Type: Liquid-propellant reaction control rocket.

Propellants: Nitrogen tetroxide and monomethyl hydrazine.

Thrust Chamber Assembly: Single chamber. Area ratio of 164. Made of welded coated columbium. Started by electrical signal to on/off solenoid valve. Multiple doublet injector with hypergolic ignition.

Thrust Chamber Mounting: Flanged bolt circle on injector head.

Propellant Feed System: Pressurised tanks with feed of 0·056 kg (0·122 lb)/s fuel and 0·091 kg (0·201 lb)/s oxidant at 15·17 bars (220 lb/sq in).

Dimensions:
- Overall length: 554·1 mm (21·8 in)
- Nozzle exit diameter: 279·4 mm (11·00 in)

Weight, Dry: 3·36 kg (7·4 lb)

Performance Ratings:
- Max thrust (vacuum): 0·445 kN (100 lb)
- Chamber pressure: 6·84 bars (100 lb/sq in)
- Specific impulse: 310

MARQUARDT R-6C

This engine is designed to provide attitude-control and station-keeping forces for satellites. It is a derivative of the engine designed by Marquardt for the Advent satellite in the early 1960s. Present applications are on the INSAT-1 and Hughes Aircraft spin-stabilised System Applications Laboratory satellites. The engine was to be qualified for these applications in early 1980. It has been tested extensively at both NASA laboratories and by Marquardt.

Type: Liquid-propellant reaction control rocket.

Propellants: Nitrogen tetroxide and monomethyl hydrazine.

Thrust Chamber Assembly: Single chamber. Area ratio of 100. Made of welded coated columbium. Started by electrical signal to on/off solenoid valve; multiple doublet injector with hypergolic ignition.

Marquardt R-40A reaction control rocket

Marquardt R-6C (0·022 kN; 5·0 lb st)

Thrust Chamber Mounting: Flanged bolt circle on injector head.

Propellant Feed System: Pressurised tanks with feed of 0·003 kg (0·006 lb)/s fuel and 0·005 kg (0·011 lb)/s oxidant at 15·17 bars (220 lb/sq in).

Dimensions:
- Overall length: 251·6 mm (9·88 in)
- Nozzle exit diameter: 55·9 mm (2·20 in)

Weight, Dry: 0·544 kg (1·2 lb)

Performance Ratings:
- Max thrust (vacuum): 0·022 kN (5 lb)
- Chamber pressure: 6·84 bars (100 lb/sq in)
- Specific impulse: 290

MARQUARDT MA210/212

Two versions of this low-cost ramjet engine have been developed. The pitot-inlet MA210-XAA is designed primarily for the high-subsonic speed regime of Mach 0·7 to 0·9, although it has been tested at Mach 2·0. The conical-inlet version, MA212-XAA, provides higher thrust and better cruise fuel consumption in supersonic flight conditions. The engines use the same combustor system with an interchangeable inlet-diffuser assembly selected for either subsonic or supersonic flight operation. Both versions are of single-wall all-steel construction with two-point mounting to the vehicle pylon. Ignition is by pyrotechnic flare.

Dimensions:
- Length overall:
  - MA210-XAA: 2,070 mm (79 in)
  - MA212-XAA: 2,134 mm (84 in)
- Diameter: 381 mm (15 in)

Weight, Dry:
- MA210-XAA: 31·75 kg (70 lb)
- MA212-XAA: 35·2 kg (77·5 lb)

Performance:
- Operating envelope:
  - MA210-XAA: S/L to 9,145 m (30,000 ft), Mach 0·7-1·5
  - MA212-XAA: S/L to 18,300 m (60,000 ft), Mach 1·0-2·5
- Design thrust (net):
  - MA210-XAA: 2·94 kN (660 lb), S/L, Mach 0·9
  - MA212-XAA: 6·05 kN (1,360 lb), 12,200 m (40,000 ft), Mach 2·5

NELSON
NELSON AIRCRAFT CORPORATION

HEAD OFFICE: PO Box 454, Irwin, Pennsylvania 15642
Telephone: (412) 863 5900
PRESIDENT: Charles R. Rhoades
VICE-PRESIDENT: Lawrence J. Rhoades

Nelson Aircraft Corporation, among its many industrial activities, produces to order the Nelson H-63 four-cylinder two-cycle aircooled engine, which has been certificated by the FAA as a power unit for single-seat helicopters, and is now available also as a power plant for propeller-driven aircraft. All these engines are capable of sustained inverted flight. Recommended overhaul period is 800 h.

Nelson four-stroke prototype engine in 89·5 kW (120 hp) size

The 32 kW (43 hp) Nelson H-63C four-cylinder two-stroke engine

NELSON FOUR-STROKE RANGE

Nelson Aircraft is developing a new range of lightweight four-stroke engines. They will be compact, and will have liquid cooling. The three sizes have outputs of 44·7 kW (60 hp), 89·5 kW (120 hp) and 112 kW (150 hp). Dry weights are, respectively, 39·95 kg (88 lb), 57·2 kg (126 lb) and 70·37 kg (155 lb).

By 1978 the 112 kW (150 hp) engine and a range of 89·5 kW (120 hp) engines had been run. Emphasis is on the latter. Future engines will include a Roots-type blower running at the 6,000 rpm of the engine, propeller speed being 2,000 rpm.

NELSON H-63
US military designation: YO-65

Developed originally as a power unit for single-seat helicopters, the H-63 is now available in five versions, as follows:

H-63C. Basic helicopter power unit for vertical installation. Battery ignition and direct drive. Certificated by FAA. Supplied as complete power package, including clutch, cooling fan and shroud.

H-63CP. Basically as H-63C, but without clutch, fan and shroud. Intended primarily for installation in horizontal position, with direct drive to propeller. FAA certificated.

H-63CPM. Magneto ignition (see 'Ignition' in detail description).

H-63CPMR, CPR. Versions with speed-reducing propeller drive (see 'Power Take-off' in detail description).

Nelson has developed a 1·07 m (42 in) wooden propeller with glassfibre covering for use with the H-63. It is suitable for either tractor or pusher installation.

TYPE: Four-cylinder horizontally-opposed aircooled, two-stroke.

CYLINDERS: Bore 68·3 mm (2[11]/16 in). Stroke 70 mm (2¾ in). Total capacity 1·03 litres (63 cu in). Compression ratio 8 : 1. Each complete cylinder is machined from an aluminium alloy casting, the bore being porous-chrome plated for wear resistance. Cylinders bolted to and detachable from crankcase.

PISTONS: Aluminium alloy casting. Two piston rings. Two needle roller bearings pressed in boss. Piston (gudgeon) pin pressed into small end of connecting rod.

CONNECTING RODS: Alloy steel forging. Caged roller bearing at big-end.

CRANKSHAFT: Four-throw. Nitralloy shaft on ball and roller bearings.

CRANKCASE: Two-piece case divided on horizontal centre-line. Each half is a magnesium alloy casting.

INDUCTION: Nelson diaphragm-type all-angle fuel control carburettor. Hot-air anti-icing. Fuel/oil mixture valves from crankcase through specially-designed rotary valve

First photograph of the Nelson H-63CPR with speed-reducing chain drive. Maximum rating will be 35·4 kW (47·5 hp)

driven by crankshaft. Intake to and exhaust from cylinders through ports. Exhaust stacks are of aluminium alloy.

FUEL: 80/87 octane gasoline and SAE 30 ash-free-base oil in 16 : 1 mixture for fuel and lubrication.

IGNITION: (except M models): Battery-type dual-ignition with automatic retard for starting. (M models): Two Slick magnetos. Two Champion D-9 or 5 COM spark plugs per cylinder.

LUBRICATION: See under 'Fuel'.

POWER TAKE-OFF: (H-63C): Hollow shaft extension from Salisbury centrifugal clutch output drive. (R models): Hy-Vo high-speed chain drive with ratio of 8 : 5 (0·625).

STARTING: 12V DC Autolite electric motor and Bendix drive.

COOLING: (H-63C): Centrifugal aluminium fan and two-piece glassfibre shrouding designed to maintain all temperatures within acceptable limits on an FAA hot day of 37·8°C, S/L. (Other versions): by propeller slipstream.

MOUNTING: Four Lord-type mounts, two on each half of crankcase.

DIMENSIONS (H-63C):

Length	508 mm (20·0 in)
Height	376 mm (14·8 in)
Width	605 mm (23·8 in)

WEIGHT, DRY:

H-63C, with accessories	34·5 kg (76 lb)
H-63CP, with accessories	30·8 kg (68 lb)
H-63CPM	30·8 kg (68 lb)
H-63CPR, H-63CPMR	38·6 kg (85 lb)

The 35·8 kW (48 hp) Nelson H-63CP for fixed-wing aircraft

POWER RATINGS:
T-O:
- H-63C — 32 kW (43 hp) at 4,000 rpm
- H-63CP, H-63CPM — 35·8 kW (48 hp) at 4,400 rpm
- H-63CPR, H-63CPMR — 35·4 kW (47·5 hp) at 2,750 output rpm

Max continuous:
- H-63C — 32 kW (43 hp) at 4,000 rpm
- H-63CP, H-63CPM 33·6 kW (45 hp) at 4,000 rpm
- H-63CPR, H-63CPMR — 33·25 kW (44·6 hp) at 2,500 output rpm

FUEL CONSUMPTION:
R models
16 litres (4·2 US gal; 3·47 Imp gal)/h at 2,500 output rpm

NORTHROP
NORTHROP CORPORATION, VENTURA DIVISION

1515 Rancho Conejo Boulevard, Newbury Park, California 91320
Telephone: (805) 498 3131

In 1972 Northrop Corporation acquired the rights to this engine from McCulloch Corporation. The 4318 series continues in production at Ventura Division, to power the KD2R-5 target (see RPVs and Targets section).

NORTHROP MODEL 4318F
US military designation: O-100-3

TYPE: Four-cylinder opposed aircooled two-stroke.

CYLINDERS: Bore 80·8 mm (3[3]/16 in). Stroke 79·4 mm (3⅛ in). Displacement 1·6 litres (100 cu in). Compression ratio 7·8 : 1. Die-cast aluminium with integral head and hard chrome plated walls.

PISTONS: Cast aluminium. Two rings above pins. Piston (gudgeon) pins of case-hardened steel.

CONNECTING RODS: Forged steel. 'Free-roll' silver-plated bearings at big-end. Small-end carries needle bearing.

CRANKSHAFT: Four-throw one-piece steel forging on four anti-friction bearings, two ball and two needle, one with split race for centre main bearing.

CRANKCASE: One-piece heat-treated permanent-mould aluminium casting, closed at rear end with cast aluminium cover which provides mounting for magneto.

VALVE GEAR: Fuel mixture for scavenging and power stroke introduced to cylinders through crankshaft-driven rotary valves and ported cylinders.

INDUCTION: Crankcase pumping type. Diaphragm-type carburettor with adjustable jet.

FUEL SPECIFICATION: Grade 100/130 aviation fuel (20 parts), two-cycle oil (1 part).

IGNITION: Single magneto and distributor. Directly connected to crankshaft through impulse coupling for easy starting. Radio noise suppressor included. BG type RB 916S, AC type 83P or Champion REM-38R spark plugs. Complete radio shielding.

PROPELLER DRIVE: RH tractor. Keyed taper shaft.

STARTING: By separate portable hydraulic starter.

MOUNTING: Three mounting lugs with rubber bushings.

DIMENSIONS:

Length	686 mm (27·0 in)
Width	711 mm (28·0 in)
Height	381 mm (15·0 in)

WEIGHT, DRY (less propeller hub): 34·9 kg (77 lb)

POWER RATING: 62·6-71·6 kW (84-96 hp) at 4,100 rpm

SPECIFIC CONSUMPTION (fuel/oil mixture at S/L at rated power): 127μg/J (0·75 lb/h/hp)

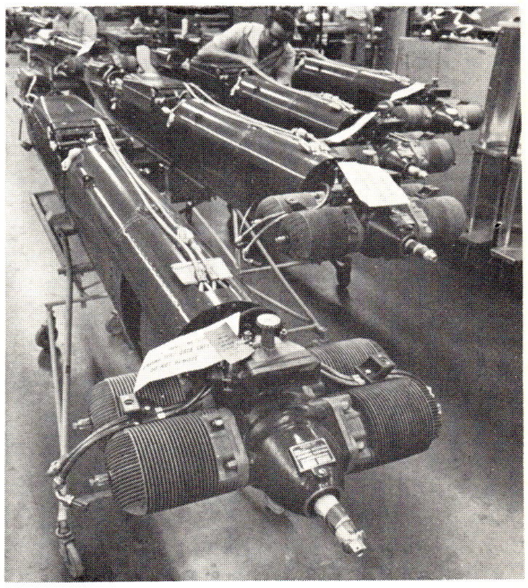

Northrop Ventura targets powered by O-100-3 piston engines rated at 62·6-71·6 kW (84-96 hp)

PRATT & WHITNEY

THE PRATT & WHITNEY AIRCRAFT GROUP OF UNITED TECHNOLOGIES CORPORATION

Group Headquarters: East Hartford, Connecticut 06108
Telephone: (203) 565 4321
Group President: Robert J. Carlson
Group Executive Vice-President: Richard J. Coar
Group Public Relations Director: James H. Lynch

Commercial Products Division
East Hartford, Connecticut
Division President: Donald C. Lowe

Government Products Division
West Palm Beach, Florida
Division President: Frank W. McAbee Jr

Manufacturing Division
East Hartford, Connecticut
Division President: Donald Nigro

Pratt & Whitney Aircraft of Canada
See separate entry under Canada

Pratt & Whitney Aircraft was formed in 1925 and rapidly became a world leader in aircraft piston engines. It is now the world's largest producer of gas turbine engines.

On 23 April 1976 Pratt & Whitney was restructured as the Pratt & Whitney Aircraft Group, composed of four divisions as listed above.

Commercial Products Division is responsible for commercial aircraft engines. Government Products Division is responsible for military engines. Manufacturing Division provides plant and facilities for making the products of the CPD and GPD. Pratt & Whitney Aircraft of Canada (P&WC), which has its own entry under that country on an earlier page, is responsible for engines for general aviation.

The Pratt & Whitney Aircraft Group has approximately 36,000 employees at locations in Connecticut, and about another 14,000 in Florida and Canada.

Excluding P&WC, the divisions of the Group had by 1981 manufactured more than 64,000 gas turbine engines, most of them for aircraft. These engines had accumulated over 583 million flight hours in military and commercial service. Most of this time has been logged by the JT3D, JT8D and JT9D turbofans on which a major part of the world's air transport is based.

Pratt & Whitney J52-P-408 two-shaft turbojet rated at 50 kN (11,200 lb st)

PRATT & WHITNEY JT8

US military designation: J52

The J52 is a medium-sized turbojet which was designed under the auspices of the US Navy Bureau of Weapons. It powers all versions of the Grumman A-6 Intruder/Prowler attack and ECM aircraft and current versions of the McDonnell Douglas A-4 Skyhawk.

J52-P-6A, 6B, 8A, 8B. Rated at 37·8 kN (8,500 lb) (6A, 6B) or 41·4 kN (9,300 lb) (8A, 8B). Powers most versions of A-4 and A-6.

J52-P-408. Rated at 50 kN (11,200 lb). Powers A-4F, A-4M, some export A-4 versions, EA-6B.

The J52 is a two-spool turbojet, with total of 12 compressor stages, a 'cannular' type combustion system fed by 36 dual-orifice injectors and independent high-pressure and low-pressure single-stage turbines. Pressure ratio ranges from 12·4 to 14·6 : 1. Burner cans include features for reduced smoke. In the P-408 advanced design features are incorporated to achieve the rating increases with a minimum change in engine envelope and weight compared to other J52 models. These include two-position inlet guide vanes and aircooled first-stage turbine vanes and blades. Data are for P-408:

Dimensions:	
Diameter	814·3 mm (32·06 in)
Length	3,020 mm (118·9 in)
Weight, Dry:	1,052 kg (2,318 lb)

PRATT & WHITNEY JT4

US military designation: J75

Last described in the 1970-71 *Jane's*, about 50 of these large two-shaft turbojets, of J75-P-13 type, are being rebuilt by Pratt & Whitney Aircraft to power Lockheed TR-1 reconnaissance aircraft. Afterburners will be removed and other changes made.

PRATT & WHITNEY JT3D

US military designation: TF33

The JT3D is a turbofan version of the J57 turbojet, handling almost 2·5 times more air than the J57 and with pressure ratio ranging from 13 : 1 on the JT3D-1 to 14·3 : 1 on the JT3D-8A (TF33-P-7).

Flight trials in a B-52 Stratofortress bomber and Boeing 707 and DC-8 transports began in 1960. The JT3D powers all late versions of these aircraft. The Lockheed C-141A StarLifter military transport uses the TF33-P-7 version, with an additional stage of compression. In January 1973 the -7 engine, modified to incorporate additional accessory drives, was selected to power the Boeing E-3A (AWACS) aircraft. Designation of the E-3A engine is TF33-PW-100A (JT3D-8B).

More than 8,480 JT3D turbofans, including converted JT3C engines, had been delivered by the end of 1980. Additional engines remain to be delivered.

Dimensions:	
Diameter:	
JT3D-3B	1,350 mm (53·14 in)
TF33-PW-100A	1,373 mm (54·06 in)
Length:	
JT3D-3B	3,479 mm (137 in)
TF33-PW-100A	3,607 mm (142 in)
Weight, Dry:	
JT3D-3B	1,969 kg (4,340 lb)
TF33-PW-100A	2,173 kg (4,790 lb)
Performance Ratings (T-O, S/L, static):	
JT3D-3B	80 kN (18,000 lb st)
TF33-PW-100A	93·4 kN (21,000 lb st)
Specific Fuel Consumption (T-O rating):	
JT3D-3B	15·5 mg/Ns (0·535 lb/h/lb st)
TF33-PW-100A	15·86 mg/Ns (0·560 lb/h/lb st)

PRATT & WHITNEY JT8D

This turbofan engine was developed as a company-sponsored project to power the Boeing 727. It was later selected for other types of aircraft, and supersonic military versions of the JT8D have been developed in Sweden by Volvo Flygmotor (see RM8 in that company's entry).

Construction of the JT8D is largely of steel and titanium. An annular bypass duct runs the full length of the engine, with balanced mixing of the hot and cold air streams in the tailpipe.

The JT8D entered commercial service on 1 February 1964. It has since become the most widely used commercial jet engine, over 10,400 having logged more than 180 million flight hours.

The following are the basic versions:

JT8D-1, -1A, -1B. Initial version rated at 62·3 kN (14,000 lb st). Powers Boeing 727-100 and -100C, McDonnell Douglas DC-9-10 and -10F, and Aérospatiale Caravelle 10B and 10R.

JT8D-7, -7A, -7B. Develops 62·3 kN (14,000 lb st) to 28·9 °C at S/L. Specified for Boeing 727-100, -100C and -200, Boeing 737-100 and -200, McDonnell Douglas DC-9-10, -30 and -30F, Aérospatiale Caravelle 10R and 11R.

JT8D-9, -9A. Develops 64·5 kN (14,500 lb st) to 28·9°C at S/L. Specified for Boeing 727-100, -100C and -200, 737-200, -200C and T-43A, McDonnell Douglas DC-9-20, -30, -40, C-9A, C-9B and VC-9C, Aérospatiale Caravelle 12 and Kawasaki C-1. Deliveries began in July 1967. Produced under licence in Japan (see entry under Mitsubishi).

JT8D-11. Develops 66·7 kN (15,000 lb st) to 28·9°C at S/L. Specified for McDonnell Douglas DC-9-20, -30 and -40 series aircraft. Deliveries began in November 1968.

JT8D-15. Develops 69 kN (15,500 lb st) to 28·9°C. FAA certification was received and deliveries began in April 1971. Powers Dassault Mercure, Boeing Advanced 727 and 737, and DC-9. Entered service 1972.

JT8D-17. Develops 71·2 kN (16,000 lb st) to 28·9°C. Certificated on 1 February 1974. Entered service July 1974. Powers Advanced versions of Boeing 727 and 737, and DC-9.

JT8D-17R. Alternative T-O rating 72·95 kN (16,400 lb) but has capability of providing 4·448 kN (1,000 lb) additional thrust in the event of significant thrust loss on any other engine. Certificated at 77·40 kN (17,400 lb st) T-O rating in April 1976. Delivery of Advanced 727-200 August 1976 and certification with reserve-thrust feature November 1976.

JT8D-200 Series. Described separately.

Since February 1970 all new JT8D engines have incorporated smoke-reduction hardware, and conversion kits are available for in-service engines. Low-emission burners using aerating nozzles are being developed for the JT8D. Two noise-reduction options are also available for all JT8D models. Maximum TBO for the JT8D is 16,800 h.

Type: Axial-flow two-spool turbofan.

Air Intake: Annular with 19 fixed inlet guide vanes.

Fan: Two-stage front fan. First stage has 27 titanium blades (30 in -1 and -7) dovetailed into discs. First-stage blades have integral shroud at about 61 per cent span. Airflow: -1, -1A, -7, -7A, 143 kg (315 lb)/s; -9, -9A, 145 kg (319 lb)/s; -11, -15, 146 kg (322 lb)/s; -17, 147 kg (324 lb)/s; -17R, 148 kg (326 lb)/s. Bypass ratio: -1, -1A, -7, -7A, 1·10; -9, -9A, 1·04; -11, 1·05; -15, 1·03; -17, 1·02; -17R, 1·00.

LP Compressor: Six-stage axial, integral with fan stages, on inner of two concentric shafts. Blades made of titanium. Shaft carried in double ball bearings, either half of each bearing being able to handle the complete loading.

HP Compressor: Seven-stage axial-flow on outer hollow shaft which, like the inner shaft, is carried in double ball bearings. One-piece casing. Blades made of steel or titanium. Overall pressure ratio: -1, -1A, -7, -7A, 15·8; -9, -9A, 15·9; -11, 16·2; -15, 16·5; -17, 16·9; -17R, 17·3.

Pratt & Whitney TF33-P-7 turbofan rated at 93·4 kN (21,000 lb st)

COMBUSTION CHAMBER: Cannular type with nine cylindrical flame-tubes, each downstream of a single Duplex burner and discharging into a single annular nozzle.

HP TURBINE: Single-stage axial-flow. Solid blades in -1 to -9, aircooled in -11 and later; guide vanes hollow and aircooled in all models.

LP TURBINE: Three-stage axial-flow. Solid blades and guide vanes.

DIMENSIONS (-1 to -17R):

Diameter	1,080 mm (42·5 in)
Length	3,137 mm (123·5 in)

WEIGHT, DRY:

JT8D-1, -1A, -1B	1,431 kg (3,155 lb)
JT8D-7, -7A, -7B	1,454 kg (3,205 lb)
JT8D-9	1,475 kg (3,252 lb)
JT8D-9A	1,520 kg (3,352 lb)
JT8D-11, -15	1,537 kg (3,389 lb)
JT8D-17	1,546 kg (3,410 lb)
JT8D-17R	1,585 kg (3,495 lb)

PERFORMANCE RATINGS:

T-O thrust (S/L, static): see model descriptions

Max cruise thrust (10,665 m; 35,000 ft at Mach 0·8):

JT8D-1, -1A	15·7 kN (3,520 lb)
JT8D-7, -7A	16·1 kN (3,630 lb)
JT8D-9, -9A	18·2 kN (4,100 lb)
JT8D-11	17·6 kN (3,950 lb)
JT8D-15	18·2 kN (4,100 lb)
JT8D-17, -17R	18·9 kN (4,240 lb)

SPECIFIC FUEL CONSUMPTION:

T-O rating:

JT8D-1, -1A, -7, -7A	16·57 mg/Ns (0·585 lb/h/lb st)
JT8D-9, 9A	16·85 mg/Ns (0·595 lb/h/lb st)
JT8D-11	17·56 mg/Ns (0·620 lb/h/lb st)
JT8D-15	17·84 mg/Ns (0·630 lb/h/lb st)
JT8D-17	18·27 mg/Ns (0·645 lb/h/lb st)
JT8D-17R (alt T-O)	18·55 mg/Ns (0·655 lb/h/lb st)

Max cruise rating, as above:

JT8D-1, -1A	22·24 mg/Ns (0·785 lb/h/lb st)
JT8D-7, -7A	22·38 mg/Ns (0·790 lb/h/lb st)
JT8D-9, -9A	22·86 mg/Ns (0·807 lb/h/lb st)
JT8D-11	23·14 mg/Ns (0·817 lb/h/lb st)
JT8D-15	22·97 mg/Ns (0·811 lb/h/lb st)
JT8D-17, -17R	23·37 mg/Ns (0·825 lb/h/lb st)

PRATT & WHITNEY JT8D-200 SERIES

The first member of this series is the JT8D-209, rated at 82·2 kN (18,500 lb st) to 25°C, for normal use and 85·6 kN (19,250 lb st) to 28·9°C following loss of thrust on any other engine. This reduced-noise derivative of the JT8D family is substantially redesigned. The JT8D-209 combines the HP compressor, HP turbine spool and combustion section of the JT8D-9 with advanced LP technology derived from the NASA JT8D Refan Programme and other recently developed P&WA engines. The 200 Series offers substantially increased thrust with reduced noise and specific fuel consumption, together with the established reliability and low maintenance cost of the JT8D HP spool. The new single-stage fan has increased diameter. The new six-stage LP compressor, integral with the fan, offers increased pressure ratio. The LP turbine has 20 per cent greater annular area and achieves a higher efficiency. Surrounding the engine is a new bypass duct. The exhaust system includes a 12-lobe internal mixer to provide forced mixing of the fan and primary streams. A JT8D-209 prototype engine began flight development in a McDonnell Douglas YC-15 AMST prototype transport aircraft on 4 March 1977. Later that month Pratt & Whitney launched the -209 as a commercial product, the first application being the McDonnell Douglas DC-9 Super 80, first flown in June 1979. FAA certification of the JT8D-209 was awarded in the same month, with first production engine delivery in July 1979.

Studies of growth beyond the JT8D-209, by further application of the same derivative engine concept, or by proven turbine section changes, are in progress. The first growth engine, the JT8D-217, is rated at 88·96 kN (20,000 lb st) for normal use and 92·75 kN (20,850 lb st) following loss of thrust on any other engine. Engine certificated October 1980; production deliveries began one month later.

Pratt & Whitney JT8D-17R two-shaft turbofan rated at 77·40 kN (17,400 lb st) to 25°C

Pratt & Whitney JT8D-209 two-shaft turbofan rated at 85·6 kN (19,250 lb st) to 28·9°C

TYPE: Axial-flow two-spool turbofan.

AIR INTAKE: Annular, with 23 fixed inlet guide vanes.

FAN: Single-stage front fan has 34 titanium blades, with part-span shrouds, dovetailed into discs. Airflow: -209, 213 kg (469 lb)/s; -217, 217 kg (483 lb)/s. Bypass ratio: -209, 1·78; -217, 1·70.

LP COMPRESSOR: Six-stage axial, integral with fan; blades of titanium.

HP COMPRESSOR: Seven-stage axial on outer hollow shaft. One-piece casing. Blades of steel or titanium. Overall pressure ratio: -209, 17·1; -217, 18·6.

COMBUSTION CHAMBER: Nine cannular low-emissions burners with aerating fuel nozzles.

HP TURBINE: Single-stage axial. Aircooled guide vanes; solid rotor blades in -209, aircooled in -217.

LP TURBINE: Three-stage axial. Solid blades and guide vanes.

DIMENSIONS:

Diameter	1,250 mm (49.2 in)
Length	3,911 mm (154 in)

WEIGHT, DRY:

JT8D-209	2,001 kg (4,410 lb)
JT8D-217	2,010 kg (4,430 lb)

PERFORMANCE RATINGS:

T-O (S/L static): see model descriptions

Max cruise thrust (10,665 m; 35,000 ft at Mach 0·8):

JT8D-209	21·9 kN (4,950 lb)
JT8D-217	23·75 kN (5,350 lb)

SPECIFIC FUEL CONSUMPTION:

Max cruise rating, as above:

JT8D-209	20.50 mg/Ns (0.724 lb/h/lb st)
JT8D-217	21·32 mg/Ns (0.753 lb/h/lb st)

PRATT & WHITNEY JT9D

US military designation (JT9D-7): F105-PW-100

Based on technology stemming from the US Air Force heavy freighter propulsion of 1961-63, the JT9D was the first of the new era of very large, high bypass ratio turbofans on which the design of the present generation of wide-body commercial transports rests.

In its basic design the JT9D is compact, being shorter than the JT3D, and has two shafts, each supported in two bearings. In cruising flight the installed sfc is 22-23 per cent lower than for the JT3D or JT8D. Careful attention has been paid to maintenance.

First run of the JT9D was in December 1966, and first engine flight test, with the engine mounted on the starboard inboard pylon of a Boeing B-52E, was in June 1968. The first flight of the Boeing 747 occurred on 9 February 1969. The DC-10-40 flew on 28 February 1972, and a JT9D-powered A300B4 in 1978.

Versions of the JT9D include:

JT9D-3. The initial production model, rated at 193·5 kN (43,500 lb) to 26·7°C. Fitted to first production Boeing

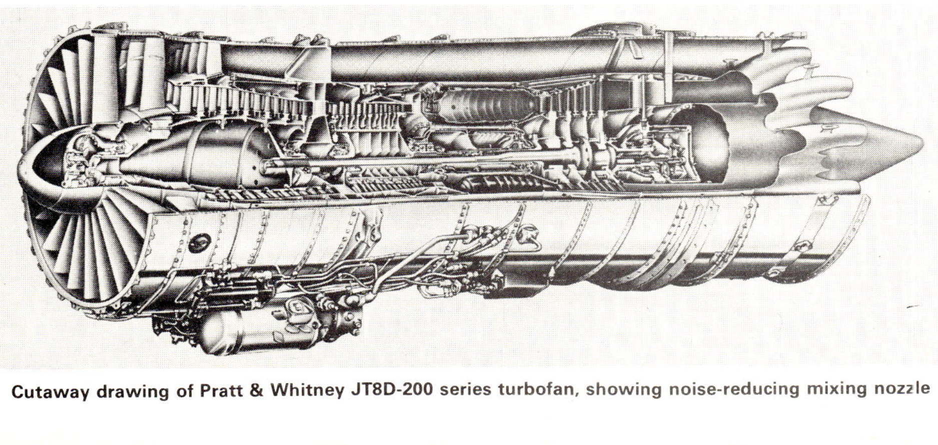

Cutaway drawing of Pratt & Whitney JT8D-200 series turbofan, showing noise-reducing mixing nozzle

747. Engines delivered from April 1969 and certificated the following month.

JT9D-3A. Incorporates water injection for wet rating of 200 kN (45,000 lb) to 26·7°C. Powers Boeing 747-100 and -200B. Engines delivered from December 1969 and certificated on 9 January 1970.

JT9D-7. This engine incorporates improvements resulting from -3A service experience. The LP compressor has blades and vanes sloped back perpendicular to the inclined core airflow for increased stability and life; pylon-matched fan exit vanes reduce sfc; HP compressor discs have a longer life, and the stators are driven through a low-friction mechanism; a short-cone hooded burner increases durability and reduces smoke emission far below the visible level; changes to HP and LP turbines increase life, and improved HP disc sealing improves performance. The -7 was certificated in June 1971. It powers the 747-200B, C, F and SR, raising the certificated take-off weight from 322,050 to 351,530 kg (710,000 to 775,000 lb). On 30 November 1971 the 747-200 was certificated at full weight and thrust, and with a fixed-inlet cowl, quieter than the original type with blow-in doors (104 EPNdB 'traded' compared with 112).

JT9D-7A. This incorporates a number of aerodynamic improvements which provide higher component efficiencies. The result is an increased thrust capability at the same turbine temperature. This has been reflected in rating increases over the JT9D-7. A version of the -7A, with improved component efficiencies to reduce specific fuel consumption, powers the 747SP.

JT9D-7F. Aerodynamically identical to the JT9D-7A, the -7F has first- and second-stage turbine rotor blades and second-stage stator vanes of directionally solidified material, allowing a rise in turbine gas temperature. This is reflected in further increase in thrust. The -7F permits operation at the Boeing 747 basic structural limit of 362,870 kg (800,000 lb). The -7F was certificated in September 1974; first deliveries were made in March 1975.

JT9D-7J. Aerodynamically identical to -7F, this has improved cooling and HP turbine to achieve -7F wet T-O rating without water injection.

JT9D-7Q, -7R. Described later.

JT9D-20. In this engine the T-O rating with water injection has been increased to 220 kN; 49,400 lb to 30°C. With this engine, the DC-10-20 was redesignated DC-10-40 and certificated at a gross weight of 530,000 lb for dry operation. At the wet rating **(JT9D-20W)** it is certificated at 251,745 kg (555,000 lb). The D-20 is similar to the D-7, except for external configuration changes such as accessory-gearbox location, thrust-transmitting points and plumbing hardware locations. The gearbox is under the fan exit casing, and the new mounting has enabled the thrust-frame yoke (added to earlier engines to prevent ovalising of the casing) to be eliminated. The D-20 was certificated in October 1972.

JT9D-59A. This engine is the first member of the family of growth versions selected to power the DC-10 and A300B. It evolved from an intensive component development programme begun in 1970, which led to the running of a complete experimental engine at 276 kN (62,000 lb st). The Dash 59A differs from earlier JT9D engines mainly in the following respects: the fan has a diameter approximately 25·4 mm (one inch) larger and reprofiled blades of higher efficiency; the low-pressure compressor has a zero (fourth) stage and is completely redesigned; and the whole hot end is entirely redesigned. The burners are recontoured, an HP turbine carbon seal is added, the HP turbine rotor blades are of directionally solidified PWA 1422 superalloy, the HP turbine annulus is of greater area, and the LP turbine is mechanically and aerodynamically redesigned. The carcase of the engine is stressed for 249 kN (56,000 lb st). With a dry rating of 236 kN (53,000 lb) the JT9D-59A was certificated on 12 December 1974; production deliveries began in January 1975. These engines are configured for installation in a common nacelle, developed jointly by P&WA and Rohr Industries, for the 747, DC-10 and A300B. The growth potential of this size of engine is predicted at 267 kN (60,000 lb).

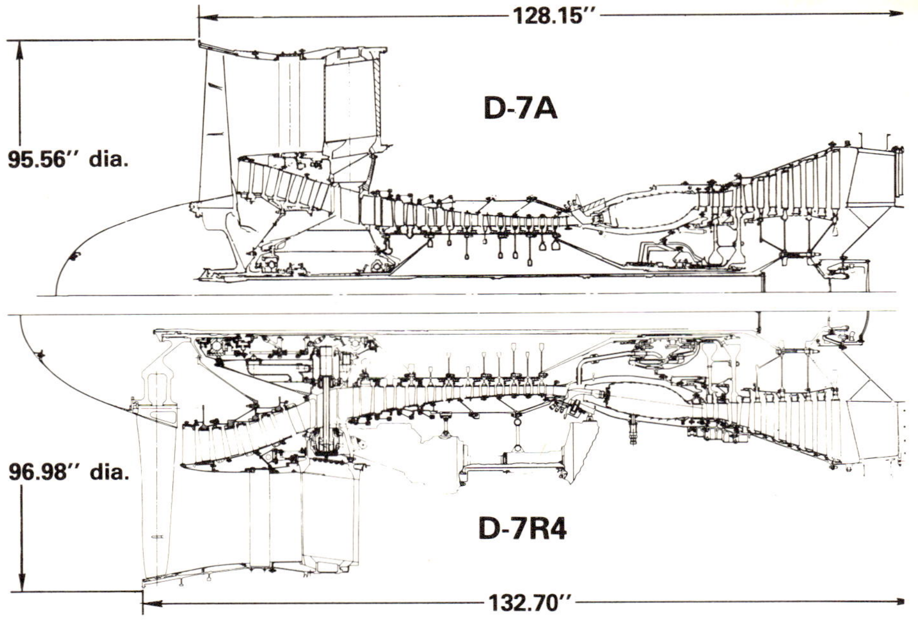

Comparative cross-sections of the Pratt & Whitney JT9D-7A *(upper half)* **and JT9D-7R4 turbofans**

JT9D-70A. This is the corresponding growth version of the JT9D for the Boeing 747. The engine was certificated on 12 December 1974, and first deliveries were made in January 1975 at a rating of 236 kN (53,000 lb). First application is the 747F certificated at 371,946 kg (820,000 lb).

JT9D-7Q Series. These have the same gas path as the -59A and -70A but an exterior configured like the -7 for installation in the Boeing 747-200 nacelle. The combination of improved performance, reduced drag and reduced propulsion-system weight significantly improved 747-200 performance. The -7Q was certificated in October 1978. This series spans a thrust range of 236-249·1 kN (53,000-56,000 lb).

JT9D-7R4 Series. This family comprises eight models (7R4A to 7R4H), with common fan, LP and HP compressor and gearbox modules, incorporating the latest technology and materials. Compared with the 1977 D-7A the 7R4 series has a larger fan with wide-chord blades, a zero-stage on the LP compressor, improved combustor, single-crystal HP turbine blades, a supervisory electronic fuel control and many smaller changes. The new engines offer a TSFC reduction of 5·7-8·0 per cent, and weight reduction of 25 kg (55 lb). Compared with the D-59A and D-70A the 7R4G to H offer weight reductions up to 184 kg (405 lb). In 1979 the 7R4D was released to manufacturing, and the engine was certificated in November 1980. Engines have been selected for Airbus A310 (7R4D1), and Boeing 767 (7R4D), 747 with stretched upper deck (7R4G2) and A300B-600 (7R4H1).

Since entry to service on 21 January 1970 the JT9D has gained experience in the 747 more rapidly than any previous engine. Within one year 653 engines had been delivered, and early in 1973 the total exceeded 1,132. Rate of delivery has since slowed but the total now exceeds 2,300 and flight time in early 1981 was in excess of 37 million h.

The following description applies to early versions of the JT9D, with data for later models given in parentheses:

Type: Two-shaft turbofan of high bypass ratio.

Intake: Direct pitot, annular fixed geometry (except that airframe inlet on early 747 aircraft has blow-in side doors around periphery). No inlet guide vanes ahead of fan. Airflow improved by rotating spinner.

Fan: Single stage, with 46 titanium blades of 4·6 aspect ratio (7R4, 40 blades of 4·0 a.r.) and two part-span shrouds (7R4, one shroud) held by dovetails in steel LP rotor. Downstream are 108 aluminium alloy exit guide vanes (96 on the -59A, -70A and -7Q, 84 on 7R4), followed by nine discharge-case radial struts. Fan case of stainless steel (7R4, titanium) and aluminium alloy, designed to contain fan blades. Discharge case lined with perforated acoustic material. Nominal airflow 684 kg (1,509 lb)/s at 3,650 rpm (-7, 698 kg; 1,540 lb/s at 3,750 rpm; -59A, -70A, 734 kg; 1,619 lb/s). Pressure ratio; typically 1·6 : 1. Bypass ratio: -3A, 5·17; -7, 5·15; -59A, -70A, 4·9; 7R4D, E, 5·0; 7R4G2, H1, 4·8.

LP Compressor: Three stages (JT9D-59A, -70A, -7Q/Q1, 7R4, four different stages), rotating with fan. Rotor made up of rings, spacers and conical disc splined to steel LP shaft and held by lock-nut ahead of fan and overhung ahead of main LP ball thrust bearing. Hydraulically opened bleed ring at LP exit to increase flight-idle stall margin and excess air during deceleration. Rotor stages have 104, 132 and 130 (-7Q, -59A, -70, 7R4, 108, 120, 112, 100) dovetailed blades of titanium alloy. First stator stage anti-iced by 9th stage bleed air. Stator stages have 88, 128, 126 and 120 (-7Q, -59, -70, 7R4, 96, 114, 116, 104, 88) titanium vanes and 120 (4th stage) nickel alloy vanes, all riveted to outer rings. Casing of aluminium alloy. Core airflow typically 118 kg (260 lb)/s (all versions).

HP Compressor: Eleven stages. All stages have rings or centreless discs with integral spacers carried on conical discs at 3rd and 11th stages on HP shaft of titanium alloy (front) and high-nickel alloy (rear), bolted at rear hub. Rotor stages have 60, 84, 102, 100, 110, 108, 104, 94 and 100 dovetailed titanium blades and 102 and 90 nickel alloy blades. Stator has 76, 70, 80, 106, 100 and 112 titanium alloy vanes and 126, 146, 154, 158 and 92 vanes of nickel alloy, all brazed to inner and outer rings. First four stator stages are variable, positioned by hydraulic actuator to provide adequate stall margin for starting, acceleration and part-power operation. Casing of titanium alloys (last two stages, nickel alloy) has bleed ports supplying 8th-stage air for airframe requirements. Max HP speed: -3A, 7,580 rpm; -7, 8,000 rpm. Overall engine pressure ratio: -3A, 21·5; -7,

Cutaway drawing of Pratt & Whitney JT9D-7R4 turbofan

22·3; -7Q, -59A, -70, 24·5; 7R4D, D1, 23·4; 7R4E, E1, 24·2; 7R4G2, 26·3; 7R4H1, 26·7.

Combustion Chamber: The diffuser case, which extends from the HP compressor to the midpoint of the combustion section, incorporates two sets of bleed ports for 15th-stage air for airframe requirements. The forward set takes air from the outside case via an integral manifold and the rear set bleeds air from the inner diameter via four of the ten radial struts. The combustor itself is fabricated in nickel alloy and is annular, with the forward end of the liner extended in 20 conical primary zones held in 20 burners fed from external fuel manifolds. In early models (-3A, -7 and -20), the outer casing can be slid forward over the diffuser for access to the HP turbine. Ignition by dual AC 4-joule capacitor system serving two plugs just above chamber centreline on each side.

Fuel System: Pressure type with hydraulic control system operating at up to 76 bars (1,100 lb/sq in). Main components are fuel control, pump, fuel/air heater and fuel/oil heat exchanger. (7R4 has digital supervisory electronic system to control hydromechanical control; engine is operational with or without electronic system functioning.) Provision for water injection, as customer option, with regulator, piping and spray nozzles, adds 18·1 kg (40 lb) to engine weight (not fitted to 7R4).

Fuel Grade: P&W specification PWA 522.

HP Turbine: Two stages. Both have high-nickel discs splined to HP shaft, secured by lock-nut, carrying high-nickel blades in fir-tree roots; first stage has 116 aircooled blades and second has 138 solid blades (aircooled in -7 and all subsequent models). Stators have 66 and 90 high-nickel alloy vanes, both rows aircooled. (7R4, single-crystal alloy in first-stage blades to 222·5 kN, 50,000 lb st, and in first and second blades and second stator (vane) for higher-thrust models.) Turbine inlet temperature (-3A, max T-O), typically 1,243°C (-59A, -70A, 1,350-1,370°C, 7R4 1,200-1,300°C).

LP Turbine: Four stages. Stages have 108, 126, 122 and 116 solid nickel alloy blades held in fir-tree roots in discs of nickel alloy (last disc, iron alloy). Stators have 122, 120, 110 and 102 solid nickel alloy vanes. An improved-efficiency LP turbine with the same number of stages will be introduced into 7R4 models in 1982. Exhaust gas temperature after turbine, typically 452°C (-3A), 482°C (-7, -20), 580°C (-59A, -70, -7Q), 625°C (7R4D), 635°C (7R4E), 660°C (7R4G, H).

Jet Pipe: Fixed Inconel assembly, with large central plug cone.

Reverser: Fan duct reverser comprises a translating sleeve (the rearmost portion of fan duct) which moves aft, causing long links to close the blocker doors and simultaneously pulling aft the cascade vanes. Primary (core) reverser, largely of Inconel 625, uses fixed cascades which are uncovered by aft movement of translating sleeves to which are hinged blocker doors pulled by links against the central nozzle plug. No primary reverser is used on -59A, -70A, -7Q or 7R4.

Accessory Drives: Main accessory gearbox driven by tower bevel shaft from front of HP spool and mounted under central diffuser case (-20, -59A, -70A, under fan discharge case). Main driven accessories include CSD (IDG on 7R4) fuel pump and control, starter, hydraulic pump, alternator and N_2 tachometer; Boeing 747 includes primary reverser motor and the DC-10-40 a second hydraulic pump and a fuel boost pump. The box also includes numerous lubrication system items, and provides for hand-turning the HP spool during borescope inspection.

Lubrication System: Pressure feed through fuel/oil cooler to four main bearings and return through scavenge pumps (-20 also centrifugal scavenge) to 18·8-37·6 litre (5-10 US gal; 4·16-8·32 Imp gal) tank.

Oil Grade: PWA 521 (blend of synthetic and/or mineral oils).

Mounting: From above, in two planes. Front mount (-3A, -7) is double flange at top of fan discharge case, absorbing vertical and side loads. On -20, -59A, -70A the mount is rectangular block above intermediate case, taking vertical and side loads, and thrust brackets at 40° each side of vertical on intermediate-case outer flange.

Starting: Pneumatic, by HamStan PS 700 or AiResearch ATS100-384 (DC-10, PS 700 only). Supplied at 2·76-3·10 bars (40-45 lb/sq in) from APU, ground cart or cross-bleed.

Dimensions:

JT9D-3A, -7, -7A, -7F, -7J, -20:	
Diameter	2,428 mm (95·6 in)
Length (flange to flange)	3,256 mm (128·2 in)
JT9D-59A, -70A, -7Q:	
Diameter	2,482 mm (97·7 in)
Length	3,358 mm (132·2 in)
JT9D-7R4D to H:	
Diameter	2,464 mm (97·0 in)
Length	3,371 mm (132·7 in)

Weight, Dry:

Guaranteed, including standard equipment:	
JT9D-3A	3,905 kg (8,608 lb)
JT9D-7, -7A	3,982 kg (8,780 lb)
JT9D-7F, -7J	4,014 kg (8,850 lb)
JT9D-20	3,833 kg (8,450 lb)
JT9D-59A	4,146 kg (9,140 lb)
JT9D-70A	4,153 kg (9,155 lb)
JT9D-7Q, -7Q1	4,262 kg (9,395 lb)
JT9D-7R4D to E	4,029 kg (8,885 lb)
JT9D-7R4G2	4,127 kg (9,100 lb)
JT9D-7R4H1	4,023 kg (8,870 lb)

Performance Ratings:*

T-O thrust, dry:	
JT9D-3A	196·9 kN (44,250 lb) to 26·7°C
JT9D-7	206·0 kN (46,300 lb) to 26·7°C
JT9D-7A	208·3 kN (46,950 lb) to 26·7°C
JT9D-7F	213·5 kN (48,000 lb) to 26·7°C
JT9D-7J	225·5 kN (50,000 lb) to 30°C
JT9D-20	206·0 kN (46,300 lb) to 28·9°C
JT9D-59A, -70A, -7Q	236·0 kN (53,000 lb) to 30°C
JT9D-7Q1	242·0 kN (54,500 lb) to 30°C
JT9D-7R4D, D1	213·5 kN (48,000 lb) to 33°C
JT9D-7R4E, E1	222·4 kN (50,000 lb) to 33°C
JT9D-7R4G2	243·4 kN (54,750 lb) to 30°C
JT9D-7R4H1	249·0 kN (56,000 lb) to 30°C
T-O thrust, wet:	
JT9D-3A	203·8 kN (45,800 lb) to 26·7°C
JT9D-7	213·1 kN (47,900 lb) to 30°C
JT9D-7A	215·8 kN (48,500 lb) to 30°C
JT9D-7F*	222·5 kN (50,000 lb) to 30°C
JT9D-20*	220·0 kN (49,400 lb) to 30°C
Max cruise thrust, 10,665 m (35,000 ft) at Mach 0·85:	
JT9D-3A, -7	45·4 kN (10,200 lb)
JT9D-7A	47·5 kN (10,680 lb)
JT9D-7F, -7J	49·2 kN (11,050 lb)
JT9D-20	47·5 kN (10,680 lb)
JT9D-59A*, -70A*, -7Q*	53·2 kN (11,950 lb)
JT9D-7Q1	54·3 kN (12,200 lb)
JT9D-7R4D, D1	50·0 kN (11,250 lb)
JT9D-7R4E, E1, G2	52·0 kN (11,670 lb)
JT9D-7R4H1	54·5 kN (12,250 lb)

Specific Fuel Consumption:

Cruise, Mach 0·85 at 10,665 m (35,000 ft):	
JT9D-3A	17·84 mg/Ns (0·630 lb/h/lb st)
JT9D-7	18·01 mg/Ns (0·636 lb/h/lb st)
JT9D-7A	18·16 mg/Ns (0·641 lb/h/lb st)
JT9D-7F*	18·55 mg/Ns (0·655 lb/h/lb st)
JT9D-20*	17·67 mg/Ns (0·624 lb/h/lb st)
JT9D-59A*, -70A*	17·67 mg/Ns (0·624 lb/h/lb st)
JT9D-7Q*	18·66 mg/Ns (0·659 lb/h/lb st)
JT9D-7Q1	18·75 mg/Ns (0·662 lb/h/lb st)
JT9D-7R4D, D1, E, E1	17·99 mg/Ns (0·635 lb/h/lb st)
JT9D-7R4G2	18·67 mg/Ns (0·659 lb/h/lb st)
JT9D-7R4H1	18·19 mg/Ns (0·642 lb/h/lb st)

**Ideal nozzles*

PRATT & WHITNEY PW2037

The PW2037, formerly designated JT10D, is a second-generation high bypass ratio turbofan, intended for medium-sized transport aircraft. Work on the engine began in 1972, and the first JT10D flightweight demonstrator engine ran in August 1974, at a thrust level of 102·3 kN (23,000 lb). Major changes to the engine have been made since that time, to keep abreast of increased thrust requirements and to incorporate advanced-technology features.

The development programme on the JT10D-232 began in January 1980. This engine was rated at 142·3 kN (32,000 lb st). In mid-1980 the engine was again scaled up to be compatible with the Boeing 757-200, at 167·2 kN (37,600 lb st). Following company policy it was given a new designation in the PW2000 series, the last two digits denoting thrust (a 25,000 lb st version would be the PW2025).

The PW2037 incorporates technology advancements such as single-crystal turbine blades, higher-strength disc material, aerodynamically superior aerofoils and an electronic control system. It therefore represents a substantial improvement over the JT10D engines which were proposed earlier.

Specific fuel consumption is reduced 30 per cent compared with first-generation turbofans. Configuration changes have lightened the engine, resulting in a thrust: weight ratio higher than 5·5. The PW2037 incorporates the low-noise features of the JT9D, including the use of a single-stage fan without inlet guide vanes, wide axial separation between the blade and vane rows, and a moderate fan tip speed.

The engine is designed to be compatible with acoustically treated nacelles, to achieve noise levels below FAR Pt 36 requirements. The engine is also configured with a low-emissions burner. The first engine test run is scheduled for late 1981. FAA certification is scheduled for December 1983.

Companies participating in development of the PW2037 are Motoren-und Turbinen Union GmbH (MTU) of Federal Germany and Fiat SpA of Italy. P&WA is expected to bear 84·8 per cent of the programme, MTU 11·2 per cent and Fiat 4 per cent. A collaboration agreement between these companies was signed in July 1977.

P&WA is also proceeding with design studies for smaller engines in the PW2000 series, to be rated at 115·6-133·4 kN (26,000-30,000 lb st), for application on both new and derivative aircraft. The PW2000 engine family is expected to consist ultimately of a series of models which will span a take-off thrust range of 111·2 to 177·9 kN (25,000 to 40,000 lb), filling the gap between the JT8D and JT9D.

Type: Two-shaft turbofan of high bypass ratio.

Air Intake: Direct front entry. No inlet guide vanes or anti-icing.

Fan: Single-stage. Titanium forged hub, with 36 inserted titanium alloy blades with part-span shrouds. Downstream are a single row of exit guide vanes and radial struts supporting the fan case. Rotating fan spinner. Bypass ratio 5·8.

LP Compressor: Four stages.

HP Compressor: Twelve stages. Variable vanes on first five stages. Overall engine pressure ratio 30·0.

Combustion Chamber: Annular, with flame-tube fabricated in nickel alloy.

HP Turbine: Two-stage axial with aircooled blades.

LP Turbine: Five-stage axial.

Weight, Dry: 3,028 kg (6,675 lb)

Performance Rating (T-O, S/L, static): See text

Pratt & Whitney PW2037 display model

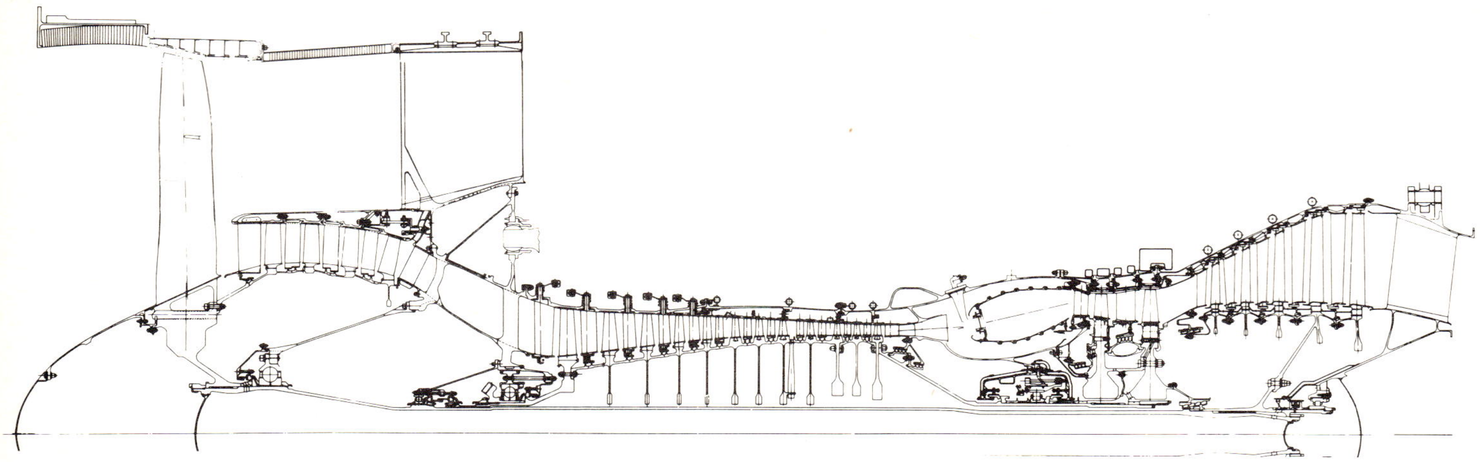

Longitudinal cross-section through Pratt & Whitney PW2037, rated at 167·2 kN (37,600 lb st)

PRATT & WHITNEY JTF10A

US military designation: TF30

Development of this high-compression two-spool turbofan was begun in 1958 as a private venture, and resulted in testing of the first turbofan with afterburning. It was chosen subsequently as the power plant for the General Dynamics F-111.

The version used initially in the F-111 was designated TF30-P-1 (JTF10A-20) which provides 82·3 kN (18,500 lb st) with afterburning. It was superseded in the F-111A by the TF30-P-3 (JTF10A-21) which provides the same thrust with reduced sea level supersonic specific fuel consumption. The F-111D is powered by the TF30-P-9 (JTF10A-36) engine with afterburning, rated at 92·7 kN (20,840 lb st). The FB-111 bomber is equipped with the TF30-P-7 (JTF10A-27D) engine, which is in the 89 kN (20,000 lb) thrust class with afterburning. The F-111F is equipped with the TF30-P-100 (JTF10A-32C) engine, an advanced version with higher thrust. The Vought A-7A and A-7B Corsair II tactical attack aircraft are powered by the TF30-P-6 (JTF10A-8) and TF30-P-408 (JTF10A-16A), these being simplified versions without afterburning and rated at 50·5 kN (11,350 lb) and 59·6 kN (13,400 lb st) respectively.

Most recent application of the TF30 is the US Navy's Grumman F-14A Tomcat fighter, powered by the **TF30-P-412A**, a modified TF30-P-12, with a revised form of afterburning nozzle. The P-412A has an afterburning rating of 93 kN (20,900 lb st). In 1977 the P-412A was modified to **TF30-P-414** standard with new first-stage compressor rotor blades and strengthened fan case to ensure containment. The US Navy has converted all F-14A engines to P-414 standard.

The most advanced TF30 production version was the USAF **TF30-P-100**, qualified in January 1971, in which weight is held below 1,815 kg (4,000 lb) while increasing thrust to the 111 kN (25,000 lb) class, with reduced fuel consumption.

Time between overhauls of the various TF30 models varies from 850 h (P-100) to 1,500 h.

Type: Two-shaft axial-flow turbofan.

Intake: Direct pitot annular type with 23 fixed inlet guide vanes (19 on P-8 and P-408). Hollow vanes pass anti-icing air.

Fan: Three stages (two on P-8 and P-408). Rotor and stator and casings all of titanium, except for steel containment case on P-412A. Three rotor stages have 28 (with part-span shrouds), 36 and 36 blades, all dovetailed; stator stages have 44, 44 and 48 vanes, all rivet-retained. Pressure ratio 2·14 to 2·4. Mass flow typically 112 kg (247 lb)/s (P-100 118 kg; 260 lb/s).

Pratt & Whitney TF30-P-414 afterburning turbofan, rated at 93 kN (20,900 lb st)

LP Compressor: Six stages (seven on P-8 and P-408), constructed integrally with fan to form nine-stage spool. Wholly of titanium construction, except stator blades of steel.

HP Compressor: Seven stages, constructed mainly of nickel-based alloy.

Combustion Chamber: Can-annular, with steel casing and eight Hastelloy X flame cans each held at the front by four dual-orifice burners. Spark igniters in chambers 4 and 5.

Fuel System: HP system (above 69 bars: 1,000 lb/sq in), with conventional hydromechanical control. Main elements comprise fuel pump, filter, fuel control, P & D valve and nozzles. Separate afterburner system for A/B engines. No water injection.

Fuel Grade: JP-4, JP-5, JP-8.

HP Turbine: Single stage, with aircooled nozzle guide vanes (stators) of cobalt alloy and aircooled rotor blades of nickel-based alloy (P-100 vanes and blades of directionally solidified alloy). Max gas temperature, early models 1,137°C, P-100 1,240°C.

LP Turbine: Three stages of nickel-based alloys. Rotor stages have 94, 96 and 80 (P-100, 72) fir-tree root blades. Gas temperature after turbine, typically 587°C.

Jet Pipe (non-A/B engine): Simple steel pipe where fan airflow and core gas mix before passing through fixed nozzle.

Afterburner: Diffuser leads to combustion section comprising double-wall outer duct and inner liner carrying five-zone combustion system. Ignition by auxiliary squirt in A/B diffuser, coupled with main squirt in No. 4 burner can which produces hot-streak of fuel through the turbine (P-100 engine, fully modulated light-up by 4-joule electrical ignition system). Max gas temperature 1,677°C.

Nozzle (A/B engines): Primary nozzle has variable area, with six hinged segments actuated by engine-fuel rams (P-100, 18 iris segments translated along curved profile by six long-stroke rams). Ejector nozzle has six blow-in doors with free tail-feathers (P-414, 18 iris segments and no blow-in doors).

Accessory Drives: Main gearbox under compressor, driven by bevel shaft from HP spool. Contains major elements of lubrication and breather systems. Drive pads at front and rear for main and A/B fuel pumps, main oil pump, N_2 tachometer, starter, fluid power pumps and power take-off.

Lubrication System: Self-contained dry-sump hot-tank system. Accessory gearbox housing forms 15 litre (4 US gal; 3·3 Imp gal) tank. Oil circulated at 3·10 bars (45 lb/sq in) through pump, filter, coolers (air/oil on airframe, fuel/oil on engine and A/B fuel/oil cooler) and three main bearing components; returned by scavenge pumps and de-aerator.

Oil Grade: MIL-L-7808, MIL-L-23699.

Mounting: Two-planar. Front peripheral pair of flanges absorbs vertical, side and thrust loads; rear pair of peripheral flanges (in line with No. 6 bearing behind LP turbine) absorbs vertical and side loads.

Starting: Air-turbine starter on left forward drive pad of accessory gearbox.

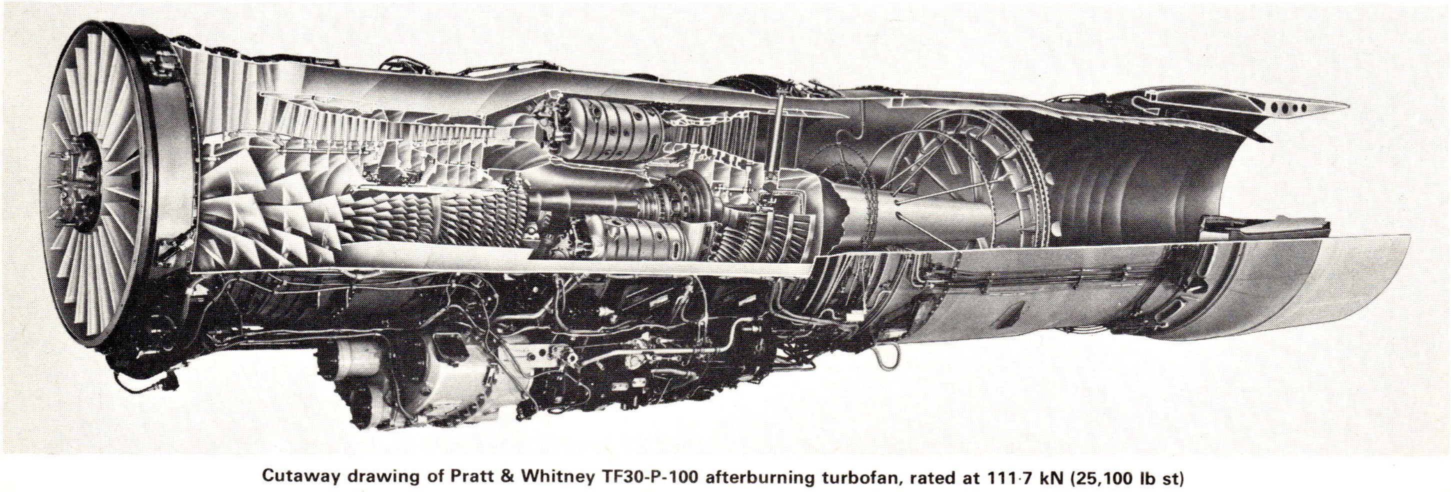

Cutaway drawing of Pratt & Whitney TF30-P-100 afterburning turbofan, rated at 111·7 kN (25,100 lb st)

22·3; -7Q, -59A, -70, 24·5; 7R4D, D1, 23·4; 7R4E, E1, 24·2; 7R4G2, 26·3; 7R4H1, 26·7.

COMBUSTION CHAMBER: The diffuser case, which extends from the HP compressor to the midpoint of the combustion section, incorporates two sets of bleed ports for 15th-stage air for airframe requirements. The forward set takes air from the outside case via an integral manifold and the rear set bleeds air from the inner diameter via four of the ten radial struts. The combustor itself is fabricated in nickel alloy and is annular, with the forward end of the liner extended in 20 conical primary zones held in 20 burners fed from external fuel manifolds. In early models (-3A, -7 and -20), the outer casing can be slid forward over the diffuser for access to the HP turbine. Ignition by dual AC 4-joule capacitor system serving two plugs just above chamber centreline on each side.

FUEL SYSTEM: Pressure type with hydraulic control system operating at up to 76 bars (1,100 lb/sq in). Main components are fuel control, pump, fuel/air heater and fuel/oil heat exchanger. (7R4 has digital supervisory electronic system to control hydromechanical control; engine is operational with or without electronic system functioning.) Provision for water injection, as customer option, with regulator, piping and spray nozzles, adds 18·1 kg (40 lb) to engine weight (not fitted to 7R4).

FUEL GRADE: P&W specification PWA 522.

HP TURBINE: Two stages. Both have high-nickel discs splined to HP shaft, secured by lock-nut, carrying high-nickel blades in fir-tree roots; first stage has 116 aircooled blades and second has 138 solid blades (air-cooled in -7 and all subsequent models). Stators have 66 and 90 high-nickel alloy vanes, both rows aircooled. (7R4, single-crystal alloy in first-stage blades to 222·5 kN, 50,000 lb st, and in first and second blades and second stator (vane) for higher-thrust models.) Turbine inlet temperature (-3A, max T-O), typically 1,243°C (-59A, -70A, 1,350-1,370°C, 7R4 1,200-1,300°C).

LP TURBINE: Four stages. Stages have 108, 126, 122 and 116 solid nickel alloy blades held in fir-tree roots in discs of nickel alloy (last disc, iron alloy). Stators have 122, 120, 110 and 102 solid nickel alloy vanes. An improved-efficiency LP turbine with the same number of stages will be introduced into 7R4 models in 1982. Exhaust gas temperature after turbine, typically 452°C (-3A), 482°C (-7, -20), 580°C (-59A, -70, -7Q), 625°C (7R4D), 635°C (7R4E), 660°C (7R4G, H).

JET PIPE: Fixed Inconel assembly, with large central plug cone.

REVERSER: Fan duct reverser comprises a translating sleeve (the rearmost portion of fan duct) which moves aft, causing long links to close the blocker doors and simultaneously pulling aft the cascade vanes. Primary (core) reverser, largely of Inconel 625, uses fixed cascades which are uncovered by aft movement of translating sleeves to which are hinged blocker doors pulled by links against the central nozzle plug. No primary reverser is used on -59A, -70A, -7Q or 7R4.

ACCESSORY DRIVES: Main accessory gearbox driven by tower bevel shaft from front of HP spool and mounted under central diffuser case (-20, -59A, -70A, under fan discharge case). Main driven accessories include CSD (IDG on 7R4) fuel pump and control, starter, hydraulic pump, alternator and N_2 tachometer; Boeing 747 includes primary reverser motor and the DC-10-40 a second hydraulic pump and a fuel boost pump. The box also includes numerous lubrication system items, and provides for hand-turning the HP spool during borescope inspection.

LUBRICATION SYSTEM: Pressure feed through fuel/oil cooler to four main bearings and return through scavenge pumps (-20 also centrifugal scavenge) to 18·8-37·6 litre (5-10 US gal; 4·16-8·32 Imp gal) tank.

OIL GRADE: PWA 521 (blend of synthetic and/or mineral oils).

MOUNTING: From above, in two planes. Front mount (-3A, -7) is double flange at top of fan discharge case, absorbing vertical and side loads. On -20, -59A, -70A the mount is rectangular block above intermediate case, taking vertical and side loads, and thrust brackets at 40° each side of vertical on intermediate-case outer flange.

STARTING: Pneumatic, by HamStan PS 700 or AiResearch ATS100-384 (DC-10, PS 700 only). Supplied at 2·76-3·10 bars (40-45 lb/sq in) from APU, ground cart or cross-bleed.

DIMENSIONS:

JT9D-3A, -7, -7A, -7F, -7J, -20:	
Diameter	2,428 mm (95·6 in)
Length (flange to flange)	3,256 mm (128·2 in)
JT9D-59A, -70A, -7Q:	
Diameter	2,482 mm (97·7 in)
Length	3,358 mm (132·2 in)
JT9D-7R4D to H:	
Diameter	2,464 mm (97·0 in)
Length	3,371 mm (132·7 in)

WEIGHT, DRY:

Guaranteed, including standard equipment:	
JT9D-3A	3,905 kg (8,608 lb)
JT9D-7, -7A	3,982 kg (8,780 lb)
JT9D-7F, -7J	4,014 kg (8,850 lb)
JT9D-20	3,833 kg (8,450 lb)
JT9D-59A	4,146 kg (9,140 lb)
JT9D-70A	4,153 kg (9,155 lb)
JT9D-7Q, -7Q1	4,262 kg (9,395 lb)
JT9D-7R4D to E	4,029 kg (8,885 lb)
JT9D-7R4G2	4,127 kg (9,100 lb)
JT9D-7R4H1	4,023 kg (8,870 lb)

PERFORMANCE RATINGS:*

T-O thrust, dry:	
JT9D-3A	196·9 kN (44,250 lb) to 26·7°C
JT9D-7	206·0 kN (46,300 lb) to 26·7°C
JT9D-7A	208·3 kN (46,950 lb) to 26·7°C
JT9D-7F	213·5 kN (48,000 lb) to 26·7°C
JT9D-7J	225·5 kN (50,000 lb) to 30°C
JT9D-20	206·0 kN (46,300 lb) to 28·9°C
JT9D-59A, -70A, -7Q	236·0 kN (53,000 lb) to 30°C
JT9D-7Q1	242·0 kN (54,500 lb) to 30°C
JT9D-7R4D, D1	213·5 kN (48,000 lb) to 33°C
JT9D-7R4E, E1	222·4 kN (50,000 lb) to 33°C
JT9D-7R4G2	243·4 kN (54,750 lb) to 30°C
JT9D-7R4H1	249·0 kN (56,000 lb) to 30°C
T-O thrust, wet:	
JT9D-3A	203·8 kN (45,800 lb) to 26·7°C
JT9D-7	213·1 kN (47,900 lb) to 30°C
JT9D-7A	215·8 kN (48,500 lb) to 30°C
JT9D-7F*	222·5 kN (50,000 lb) to 30°C
JT9D-20*	220·0 kN (49,400 lb) to 30°C
Max cruise thrust, 10,665 m (35,000 ft) at Mach 0·85:	
JT9D-3A, -7	45·4 kN (10,200 lb)
JT9D-7A	47·5 kN (10,680 lb)
JT9D-7F, -7J	49·2 kN (11,050 lb)
JT9D-20	47·5 kN (10,680 lb)
JT9D-59A*, -70A*, -7Q*	53·2 kN (11,950 lb)
JT9D-7Q1	54·3 kN (12,200 lb)
JT9D-7R4D, D1	50·0 kN (11,250 lb)
JT9D-7R4E, E1, G2	52·0 kN (11,670 lb)
JT9D-7R4H1	54·5 kN (12,250 lb)

SPECIFIC FUEL CONSUMPTION:

Cruise, Mach 0·85 at 10,665 m (35,000 ft):	
JT9D-3A	17·84 mg/Ns (0·630 lb/h/lb st)
JT9D-7	18·01 mg/Ns (0·636 lb/h/lb st)
JT9D-7A	18·16 mg/Ns (0·641 lb/h/lb st)
JT9D-7F*	18·55 mg/Ns (0·655 lb/h/lb st)
JT9D-20*	17·67 mg/Ns (0·624 lb/h/lb st)
JT9D-59A*, -70A*	17·67 mg/Ns (0·624 lb/h/lb st)
JT9D-7Q*	18·66 mg/Ns (0·659 lb/h/lb st)
JT9D-7Q1	18·75 mg/Ns (0·662 lb/h/lb st)
JT9D-7R4D, D1, E, E1	17·99 mg/Ns (0·635 lb/h/lb st)
JT9D-7R4G2	18·67 mg/Ns (0·659 lb/h/lb st)
JT9D-7R4H1	18·19 mg/Ns (0·642 lb/h/lb st)

**Ideal nozzles*

PRATT & WHITNEY PW2037

The PW2037, formerly designated JT10D, is a second-generation high bypass ratio turbofan, intended for medium-sized transport aircraft. Work on the engine began in 1972, and the first JT10D flightweight demonstrator engine ran in August 1974, at a thrust level of 102·3 kN (23,000 lb). Major changes to the engine have been made since that time, to keep abreast of increased thrust requirements and to incorporate advanced-technology features.

The development programme on the JT10D-232 began in January 1980. This engine was rated at 142·3 kN (32,000 lb st). In mid-1980 the engine was again scaled up to be compatible with the Boeing 757-200, at 167·2 kN (37,600 lb st). Following company policy it was given a new designation in the PW2000 series, the last two digits denoting thrust (a 25,000 lb st version would be the PW2025).

The PW2037 incorporates technology advancements such as single-crystal turbine blades, higher-strength disc material, aerodynamically superior aerofoils and an electronic control system. It therefore represents a substantial improvement over the JT10D engines which were proposed earlier.

Specific fuel consumption is reduced 30 per cent compared with first-generation turbofans. Configuration changes have lightened the engine, resulting in a thrust: weight ratio higher than 5·5. The PW2037 incorporates the low-noise features of the JT9D, including the use of a single-stage fan without inlet guide vanes, wide axial separation between the blade and vane rows, and a moderate fan tip speed.

The engine is designed to be compatible with acoustically treated nacelles, to achieve noise levels below FAR Pt 36 requirements. The engine is also configured with a low-emissions burner. The first engine test run is scheduled for late 1981. FAA certification is scheduled for December 1983.

Companies participating in development of the PW2037 are Motoren-und Turbinen Union GmbH (MTU) of Federal Germany and Fiat SpA of Italy. P&WA is expected to bear 84·8 per cent of the programme, MTU 11·2 per cent and Fiat 4 per cent. A collaboration agreement between these companies was signed in July 1977.

P&WA is also proceeding with design studies for smaller engines in the PW2000 series, to be rated at 115·6-133·4 kN (26,000-30,000 lb st), for application on both new and derivative aircraft. The PW2000 engine family is expected to consist ultimately of a series of models which will span a take-off thrust range of 111·2 to 177·9 kN (25,000 to 40,000 lb), filling the gap between the JT8D and JT9D.

TYPE: Two-shaft turbofan of high bypass ratio.

AIR INTAKE: Direct front entry. No inlet guide vanes or anti-icing.

FAN: Single-stage. Titanium forged hub, with 36 inserted titanium alloy blades with part-span shrouds. Downstream are a single row of exit guide vanes and radial struts supporting the fan case. Rotating fan spinner. Bypass ratio 5·8.

LP COMPRESSOR: Four stages.

HP COMPRESSOR: Twelve stages. Variable vanes on first five stages. Overall engine pressure ratio 30·0.

COMBUSTION CHAMBER: Annular, with flame-tube fabricated in nickel alloy.

HP TURBINE: Two-stage axial with aircooled blades.

LP TURBINE: Five-stage axial.

WEIGHT, DRY: 3,028 kg (6,675 lb)

PERFORMANCE RATING (T-O, S/L, static): See text

Pratt & Whitney PW2037 display model

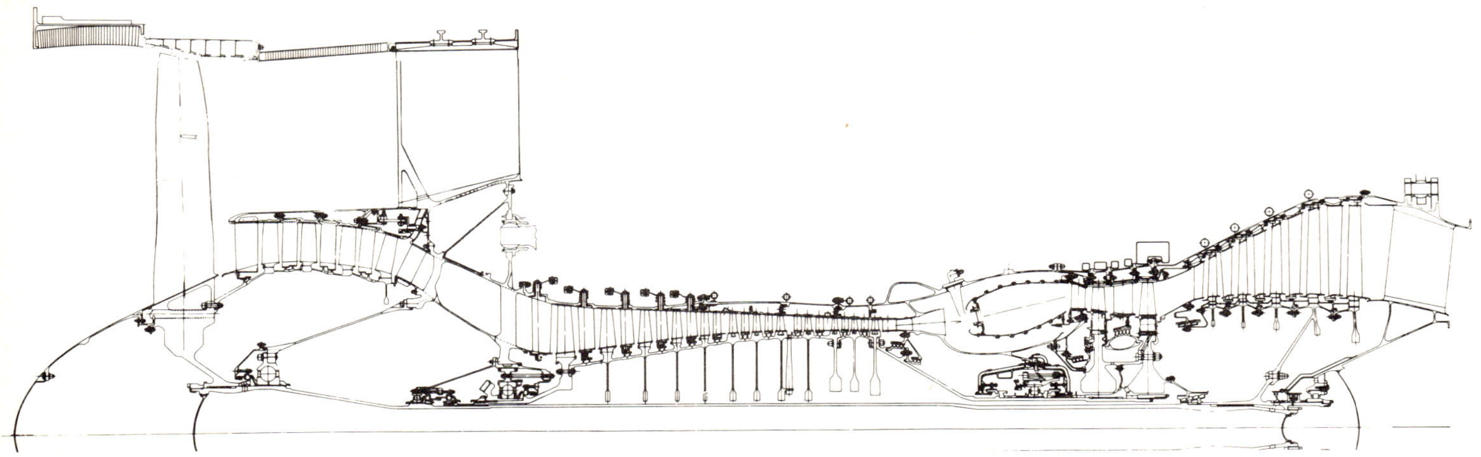

Longitudinal cross-section through Pratt & Whitney PW2037, rated at 167·2 kN (37,600 lb st)

PRATT & WHITNEY JTF10A

US military designation: TF30

Development of this high-compression two-spool turbofan was begun in 1958 as a private venture, and resulted in testing of the first turbofan with afterburning. It was chosen subsequently as the power plant for the General Dynamics F-111.

The version used initially in the F-111 was designated TF30-P-1 (JTF10A-20) which provides 82·3 kN (18,500 lb st) with afterburning. It was superseded in the F-111A by the TF30-P-3 (JTF10A-21) which provides the same thrust with reduced sea level supersonic specific fuel consumption. The F-111D is powered by the TF30-P-9 (JTF10A-36) engine with afterburning, rated at 92·7 kN (20,840 lb st). The FB-111 bomber is equipped with the TF30-P-7 (JTF10A-27D) engine, which is in the 89 kN (20,000 lb) thrust class with afterburning. The F-111F is equipped with the TF30-P-100 (JTF10A-32C) engine, an advanced version with higher thrust. The Vought A-7A and A-7B Corsair II tactical attack aircraft are powered by the TF30-P-6 (JTF10A-8) and TF30-P-408 (JTF10A-16A), these being simplified versions without afterburning and rated at 50·5 kN (11,350 lb) and 59·6 kN (13,400 lb st) respectively.

Most recent application of the TF30 is the US Navy's Grumman F-14A Tomcat fighter, powered by the **TF30-P-412A**, a modified TF30-P-12, with a revised form of afterburning nozzle. The P-412A has an afterburning rating of 93 kN (20,900 lb st). In 1977 the P-412A was modified to **TF30-P-414** standard with new first-stage compressor rotor blades and strengthened fan case to ensure containment. The US Navy has converted all F-14A engines to P-414 standard.

The most advanced TF30 production version was the USAF **TF30-P-100**, qualified in January 1971, in which weight is held below 1,815 kg (4,000 lb) while increasing thrust to the 111 kN (25,000 lb) class, with reduced fuel consumption.

Time between overhauls of the various TF30 models varies from 850 h (P-100) to 1,500 h.

Type: Two-shaft axial-flow turbofan.

Intake: Direct pitot annular type with 23 fixed inlet guide vanes (19 on P-8 and P-408). Hollow vanes pass anti-icing air.

Fan: Three stages (two on P-8 and P-408). Rotor and stator and casings all of titanium, except for steel containment case on P-412A. Three rotor stages have 28 (with part-span shrouds), 36 and 36 blades, all dovetailed; stator stages have 44, 44 and 48 vanes, all rivet-retained. Pressure ratio 2·14 to 2·4. Mass flow typically 112 kg (247 lb)/s (P-100 118 kg; 260 lb/s).

Pratt & Whitney TF30-P-414 afterburning turbofan, rated at 93 kN (20,900 lb st)

LP Compressor: Six stages (seven on P-8 and P-408), constructed integrally with fan to form nine-stage spool. Wholly of titanium construction, except stator blades of steel.

HP Compressor: Seven stages, constructed mainly of nickel-based alloy.

Combustion Chamber: Can-annular, with steel casing and eight Hastelloy X flame cans each held at the front by four dual-orifice burners. Spark igniters in chambers 4 and 5.

Fuel System: HP system (above 69 bars; 1,000 lb/sq in), with conventional hydromechanical control. Main elements comprise fuel pump, filter, fuel control, P & D valve and nozzles. Separate afterburner system for A/B engines. No water injection.

Fuel Grade: JP-4, JP-5, JP-8.

HP Turbine: Single stage, with aircooled nozzle guide vanes (stators) of cobalt alloy and aircooled rotor blades of nickel-based alloy (P-100 vanes and blades of directionally solidified alloy). Max gas temperature, early models 1,137°C, P-100 1,240°C.

LP Turbine: Three stages of nickel-based alloys. Rotor stages have 94, 96 and 80 (P-100, 72) fir-tree root blades. Gas temperature after turbine, typically 587°C.

Jet Pipe (non-A/B engine): Simple steel pipe where fan airflow and core gas mix before passing through fixed nozzle.

Afterburner: Diffuser leads to combustion section comprising double-wall outer duct and inner liner carrying five-zone combustion system. Ignition by auxiliary squirt in A/B diffuser, coupled with main squirt in No. 4 burner can which produces hot-streak of fuel through the turbine (P-100 engine, fully modulated light-up by 4-joule electrical ignition system). Max gas temperature 1,677°C.

Nozzle (A/B engines): Primary nozzle has variable area, with six hinged segments actuated by engine-fuel rams (P-100, 18 iris segments translated along curved profile by six long-stroke rams). Ejector nozzle has six blow-in doors with free tail-feathers (P-414, 18 iris segments and no blow-in doors).

Accessory Drives: Main gearbox under compressor, driven by bevel shaft from HP spool. Contains major elements of lubrication and breather systems. Drive pads at front and rear for main and A/B fuel pumps, main oil pump, N_2 tachometer, starter, fluid power pumps and power take-off.

Lubrication System: Self-contained dry-sump hot-tank system. Accessory gearbox housing forms 15 litre (4 US gal; 3·3 Imp gal) tank. Oil circulated at 3·10 bars (45 lb/sq in) through pump, filter, coolers (air/oil on airframe, fuel/oil on engine and A/B fuel/oil cooler) and three main bearing components; returned by scavenge pumps and de-aerator.

Oil Grade: MIL-L-7808, MIL-L-23699.

Mounting: Two-planar. Front peripheral pair of flanges absorbs vertical, side and thrust loads; rear pair of peripheral flanges (in line with No. 6 bearing behind LP turbine) absorbs vertical and side loads.

Starting: Air-turbine starter on left forward drive pad of accessory gearbox.

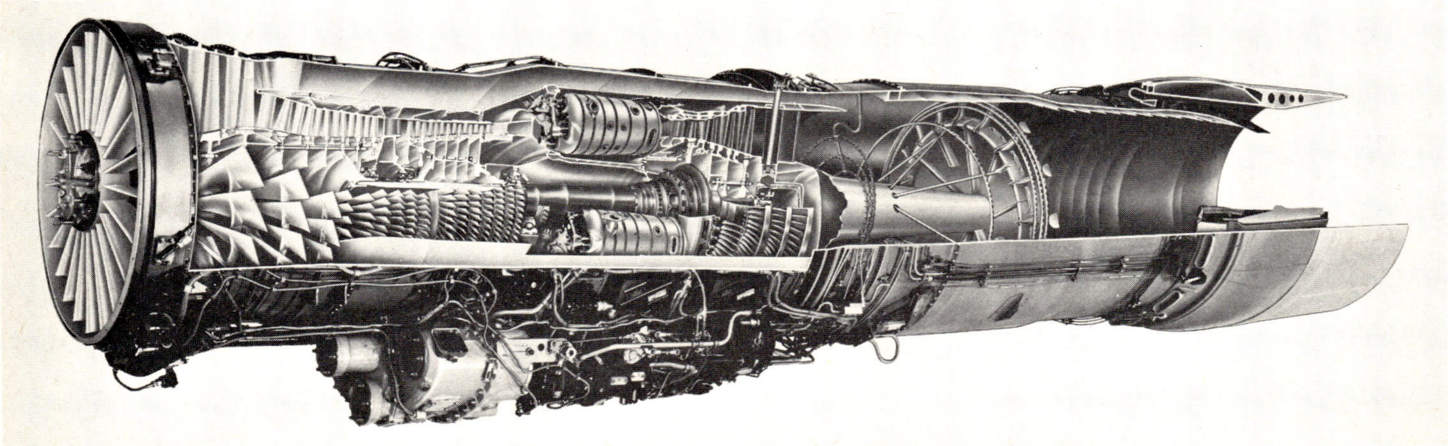

Cutaway drawing of Pratt & Whitney TF30-P-100 afterburning turbofan, rated at 111·7 kN (25,100 lb st)

Pratt & Whitney F100-PW-100 augmented turbofan; shown also below in longitudinal cross-section

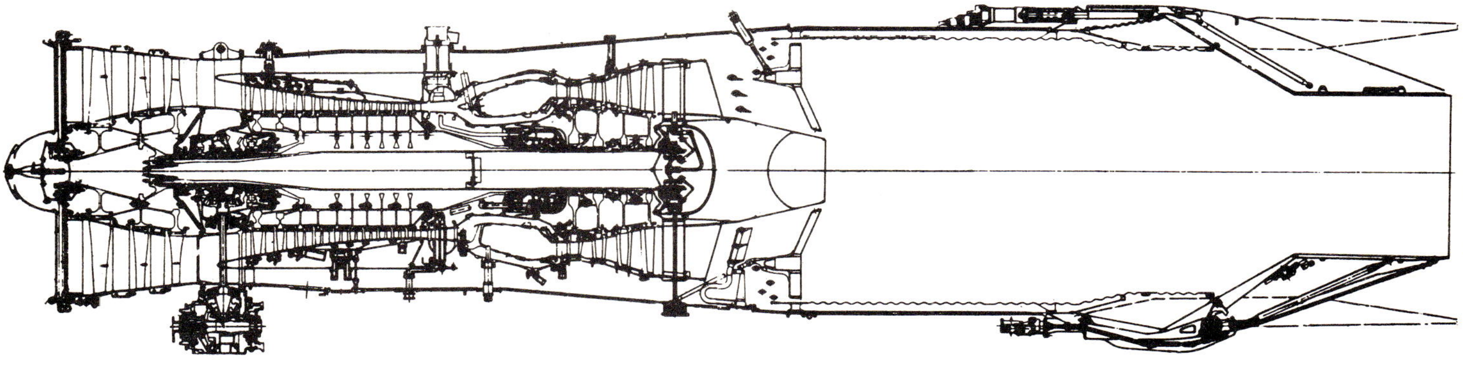

Dimensions:
- Max diameter:
  - TF30-P-414 1,293 mm (50·9 in)
  - TF30-P-100 1,242 mm (48·88 in)
- Length overall:
  - TF30-P-414 5,987 mm (235·7 in)
  - TF30-P-100 6,139 mm (241·7 in)

Weight, Dry:
- TF30-P-414 1,905 kg (4,201 lb)
- TF30-P-100 1,813 kg (3,999 lb)

Performance Ratings (T-O, S/L):
- TF30-P-414 93 kN (20,900 lb st)
- TF30-P-100 111·7 kN (25,100 lb st)

Specific Fuel Consumption (T-O):
- TF30-P-414 78·75 mg/Ns (2·78 lb/h/lb st)
- TF30-P-100 69·40 mg/Ns (2·450 lb/h/lb st)

PRATT & WHITNEY JTF22

US military designation: F100

Stemming partly from the JTF16 demonstrator engine designed in 1965-66, the JTF22 is an advanced-technology military turbofan with afterburner for supersonic applications. Basic development was funded as a demonstrator programme for the US Air Force. In February 1970 the decision was taken to use the JTF22 core engine as the basis for the **F100-PW-100** (JTF22A-25A) to power the twin-engined McDonnell Douglas F-15 Eagle fighter for the US Air Force. Subsequently, the F100 was adopted for the single-engined General Dynamics F-16, production versions of which are powered by a slightly modified engine designated **F100-PW-200**, with a backup fuel control system, revised control-system cooling and a forward extension of the compressor inlet.

Some 3,000 h of development testing were accomplished between 1968 and the 60 h PFRT (preliminary flight rating test) in February 1972. The 150 h QT (qualification test) was completed in October 1973. In January 1981 running time exceeded 1,200,000 h, of which more than 660,000 h had been in the F-15 and 40,000 in the F-16. More than 1,900 engines had been delivered, and shipments were at the rate of 40 per month. The F100 has a current maximum operating time interval of 1,350 h.

Type: Two-shaft turbofan with high-augmentation afterburner.

Intake: Direct pitot type. Fabricated titanium, with fixed nose bullet. Single row of 21 inlet guide vanes, with hot-air anti-iced leading-edges and variable-camber trailing-edge flaps.

Fan: Three stages. Fan blades have part-span shrouds. Discs of titanium 6-2-4-6, blades titanium 8-1-1. Entry diameter 928 mm (36·5 in). Bypass ratio 0·7.

Compressor: Ten-stage axial, on HP shaft. First three stages have variable stators. Discs 1-2, forged Ti 6-2-4-6; 3, forged Ti 8-1-1; 4, forged PWA 1016; 5, 7 and 9, PWA 1027; 6, 8 and 10, Gatorised (isothermal squeeze forging) IN-100. Blades 1-3, Ti 8-1-1; 4, Ti 6-2-4-6; 5-9 Incoloy 901; 10, PWA 1005. Pressure ratio 8 : 1. Overall engine pressure ratio 25 : 1.

Combustion Chamber: Annular. Fabricated in nickel alloy with film cooling throughout. Large-diameter air-blast fuel nozzles. Capacitor-discharge ignition.

HP Turbine: Two stages. Discs forged IN-100. Blades and vanes PWA 1422 directionally solidified alloy with PWA 73 coating; first rotor impingement cooled, second with convective (HP bleed air) only. Maximum gas temperature 1,399°C. Maximum speed 13,450 rpm.

LP Turbine: Two stages. Discs forged IN-100. Blades, uncooled, cast in IN-100 with PWA 73 coating. Maximum speed 10,400 rpm.

Afterburner: Five concentric spray rings in flow from core engine; two slightly farther downstream in bypass airflow. Flameholder assembly downstream of spray nozzles, with high-energy electrical ignition to give modulated light-up. Outer bypass duct and other major portions fabricated in sheet and stringer titanium. Interior liner of coated refractory material.

Nozzle: Multi-flap balanced-beam articulated nozzle giving very wide range in area and profile.

Control System: Unified hydromechanical fuel and nozzle-area control, with electronic supervisory control. The F100-PW-200 also has a hydromechanical backup control.

Dimensions:
- Overall diameter 1,180 mm (46·5 in)
- Length, excl bullet 4,851 mm (191·0 in)

Weight, Dry:
- F100-PW-100 1,391 kg (3,068 lb)
- F100-PW-200 1,398 kg (3,085 lb)

Performance Ratings (S/L, ISA):
- Max T-O, dry 66·7 kN (15,000 lb st) class
- Max T-O, augmented 111·2 kN (25,000 lb st) class

PRATT & WHITNEY PW1130

This augmented turbofan for supersonic applications comprises advanced components integrated within the existing F100 design. Using the F100 core, the PW1130 includes an increased-airflow three-stage fan, driven by a two-stage uncooled turbine to provide a 30,000 lb st-class engine (hence the designation). In addition, the engine incorporates a digital electronic engine control (DEEC) system, an advanced augmentor, light-off detector and dual ignition system. Increased turbine durability is achieved by using single-crystal blades. Sea level testing began in 1980, with altitude testing planned for 1981.

Dimensions:
- Max diameter 1,194 mm (47·0 in)
- Length 5,309 mm (209·0 in)

Weight, Dry: 1,463 kg (3,225 lb)

Performance Rating:
- Max T-O, augmented 133·4 kN (30,000 lb st)

PRATT & WHITNEY PW1120

A smaller engine in the planned PW1100 family, the PW1120 uses the F100 core in a 20,000 lb st-class turbojet cycle. Engine features include a reduced-airflow three-stage low-pressure compressor driven by a single-stage uncooled turbine. Combustor exit temperature is reduced below the current F100 operating level, and cases are cooled by low-pressure compressor air, providing a self-cooled engine. Flight clearance tests are scheduled for 1983, with a target production date of 1985.

Dimensions:
- Max diameter 876 mm (34·4 in)
- Length 4,140 mm (163·0 in)

Weight, Dry: 1,284 kg (2,830 lb)

Performance Rating:
- Max T-O, augmented 89 kN (20,000 lb st)

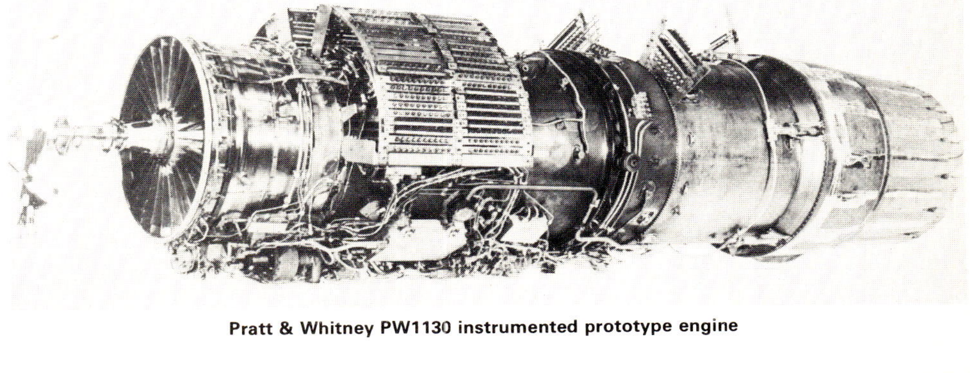

Pratt & Whitney PW1130 instrumented prototype engine

Pratt & Whitney PW1120 mock-up

Pratt & Whitney RL10A-3-3 rocket engine

PRATT & WHITNEY RL10

The RL10 rocket engine, for the propulsion of space vehicle upper stages, is a regeneratively cooled, turbopump-fed engine with a single chamber. The current RL10A-3-3 production version is rated at 66·7 kN (15,000 lb st) at an altitude of 61,000 m (200,000 ft), and has a nominal specific impulse of 444 s. Propellants are liquid oxygen and liquid hydrogen, injected at a nominal oxidiser-to-fuel mixture ratio of 5·0 : 1. Rated engine thrust is achieved at a nominal design chamber pressure of 27·6 bars (400 lb/sq in) absolute, with a nominal nozzle area ratio of 57 : 1. The engine can be used for multi-engine installation on an interchangeable basis and is capable of multiple starts after extended coast periods.

First deliveries were made in August 1960 for use in NASA's Centaur stage of the Atlas-Centaur rocket, which is powered by two RL10 engines. A six-engine cluster of RL10A-3 engines powered the S-IV stage of the Saturn I, achieving a perfect performance record for the entire launch programme. Over 9,700 RL10 firings have been accomplished, and 152 engines have flown on operational Saturn and Centaur vehicles, accomplishing 238 successful in-flight starts. In the Titan-Centaur 5 mission two RL10 engines accomplished a record seven in-flight starts, five of them after spacecraft separation. Between two starts there was a 5·25 h coast at zero-*g*.

Experimental versions of the RL10 have been tested at the Pratt & Whitney Aircraft Group's Government Products Division. These tests include variable-thrust operation, low idle operation, pumped idle operation, operation with a 205 : 1 ratio nozzle extension, operation on fluorine/hydrogen, lox/propane and flox/methane propellants. NASA has ordered additional RL10 engines to power Centaur missions during the late 1980s.

ROCKETDYNE

ROCKETDYNE DIVISION OF ROCKWELL INTERNATIONAL

HEADQUARTERS: 6633 Canoga Ave, Canoga Park, California 91304

Telephone: (213) 884 4000

OTHER FACILITIES: Santa Susana, California

PRESIDENT: N. J. Ryker

VICE-PRESIDENTS:
D. J. Sanchini (SSME Programme Manager)
R. D. Paster (Propulsion Programmes)
S. J. Domokos (Laser Programmes)
C. G. Fargo (Engineering and Test)
D. M. Kruszenski (Business Development)
A. T. Fiore (Finance and Administration)

Rocketdyne is a division of Rockwell International, devoted primarily to the design and manufacture of rocket engines for the US Air Force and the National Aeronautics and Space Administration. It was established as a separate division on 8 November 1955.

Rocketdyne liquid-propellant engines power more than three-quarters of all large US space vehicle stages.

ROCKETDYNE SSME

On 13 July 1971, the Rocketdyne Division of Rockwell International was selected by NASA to design and develop the main engine for the Orbiter stage of the US Space Shuttle. Three of these engines provide a total of 6,833 kN (1,536,000 lb) vacuum thrust.

Two large solid-propellant boosters are strapped on the sides of the Orbiter's expendable propellant tank which carries the liquid oxygen and liquid hydrogen for the three main engines in the Orbiter. The Orbiter rides piggyback on the propellant tank in a parallel configuration. The solid motors and the three Space Shuttle Main Engines (SSME) produce 29,037 kN (6,527,000 lb st) to lift the vehicle from the pad in a conventional vertical flight path. The solid motors burn out at about 40 km (25 miles) altitude, separate from the Orbiter stage, and are lowered by parachutes into the ocean for recovery. The three main engines continue to power the vehicle to near orbit; the external tank then separates and is de-orbited and disposed in a safe area of the ocean. After mission completion the Orbiter re-enters the Earth's atmosphere and manoeuvres to a landing site for an unpowered horizontal landing similar to that of a conventional jet aircraft.

In overall configuration, the SSME is slightly smaller in size than the F-1 engine used in the Saturn V vehicle first stage. It burns liquid oxygen and liquid hydrogen propellants and has been designed for high reliability, reusability, multiple re-start capability and low cost. It is capable of 7½ h of burn time, accrued during 55 flights. Modified airline maintenance procedures are used to service the engine between flights without removing it from the vehicle.

The design combines the merits of high-chamber-pressure operation, an optimum-performance contoured bell-shaped nozzle, and a regeneratively-cooled thrust chamber, capable of 11° gimballing, for maximum performance and long life. The chamber wall is cooled so efficiently that it is at 567°C, although the combustion temperature is about 3,300°C. No propellants are wasted in the cooling process. The combustion chamber wall is made of slotted metal, rather than tubes, using Rocketdyne-developed NARloy-Z, a copper alloy that is easily machined, has higher strength than pure copper, and has very high thermal conductivity. Tubes are incorporated in the lower nozzle section.

The SSME is controlled by a unique system incorporating dual-redundant digital computers. This system monitors engine parameters such as pressure and temperature and the engine is automatically adjusted to operate at the required thrust and mixture ratio. The system also develops a record of engine operating history for maintenance purposes to improve serviceability and extend total engine life.

Flight certification of the SSME was achieved in December 1980, after two certification cycles had been completed on each of two engines. Each cycle required a minimum of 13 tests and 5,000 s, including simulation of nominal and abort mission profiles. Forty tests, totalling 12,750 s, were made on one engine, and 34 tests, totalling 10,650 s, on the other.

By mid-February 1981, more than 650 single-engine tests had been conducted, bringing total time beyond 95,000 s, far exceeding NASA's earlier goal of 65,000 s prior to the first manned orbital flight (FMOF).

In addition, main propulsion test article (MPTA) testing, demonstrating three-engine cluster operation, completed FMOF certification in early January 1981. In excess of 10,000 s were accumulated during 18 tests of the MPTA. Component design verification, a form of component qualification, was also completed in January 1981, to support FMOF requirements.

The first three flight engines were acceptance-tested in early 1979 and installed in Orbiter 102 (FMOF). These engines demonstrated a specific impulse 2 s greater than the specified minimum. A spare flight engine was acceptance-tested in February 1981.

Current SSME programme emphasis is directed toward full power level (FPL) certification, production-engine delivery for the three additional Orbiters, and support of flight operations, including requirements for replacement and overhaul of engines. The FPL requirements are essentially the same as those required for FMOF. Future development and uprating are planned to extend the demonstrated engine life to its full 55-mission goal, and to increase the thrust and payload capabilities.

COMBUSTION CHAMBER: Channel-wall construction with regenerative cooling by the hydrogen fuel. Concentric-element injector.

TURBOPUMPS: Two low-pressure pumps boost the inlet pressures for two high-pressure pumps. Dual pre-burners provide turbine-drive gases to power the high-pressure pumps. Hydrogen-pump discharge pressure is 478·3 bars (6,937 lb/sq in) at 37,250 rpm; it develops 56,554 kW (75,840 hp).

CONTROLLER: Honeywell digital computer controller provides closed-loop engine control, in addition to data processing and signal conditioning for control, check-out, monitoring engine status, and maintenance data acquisition.

CONTROLS: A hydraulic-actuation control system is used. The dual-redundant self-monitoring servo-actuators respond to signals from the controller to position the ball valves. A pneumatic system provides backup for the hydraulic system for engine cut-off. Main subcontractors: Honeywell Inc (controller); Hydraulic Research and Manufacturing Co (actuators).

Production SSME on single-engine test

MAINTENANCE: Engine to be maintained using airline-type maintenance procedure for on-the-vehicle servicing. Time between overhauls is 55 flights or 7·5 h of cumulative operation.

DIMENSIONS:	
Length	4,242 mm (13 ft 11 in)
Diameter at nozzle exit	2,388 mm (7 ft 10 in)
PERFORMANCE:	
S/L thrust (one engine)	1,819 kN (409,000 lb)
Vacuum thrust	2,277 kN (512,000 lb)
Specific impulse	455 s
Chamber pressure	207 bars (3,000 lb/sq in)
Throttling ratio	2 : 1
Expansion ratio:	
Test	35 : 1
Flight	77·5 : 1

ROCKETDYNE RS-34

The Rocketdyne RS-34 is the fourth stage for the USAF's Missile X (MX). It is integral to the ground-launched ICBM, and provides roll-control torques during operation of the third stage, staging functions during stage three/stage four separation, and post-boost manoeuvres for deployment of the multiple independent re-entry vehicles.

The RS-34 comprises a shell structure (the missile skin) with internal structures to house the propulsion systems, guidance and control set (MGCS), and all related equipment and hardware. The aluminium Iso-grid shell includes

Engineering mock-up of MX Stage-IV, the Rocketdyne RS-34

The Rocketdyne MA-5 propulsion system comprises a central sustainer and two booster engines, and (not shown) two small vernier engines for fine control of vehicle velocity and trajectory

a stage three/four separation joint, with aluminium internal structures attached to it. The shell is enclosed within an external protection material, and forward and aft stage interfaces by thermal-protection tents.

Stage four complete has a diameter of 2,337 mm (92 in) and length of 1,067 mm (42 in). Its weight before staging is 1,250 kg (2,756 lb). Performance is classified.

The bipropellant liquid propulsion system uses helium pressurisation to feed the nitrogen tetroxide/monomethyl hydrazine hypergolic propellants. These are fed from separate positive-expulsion storage tanks by pressurised metal diaphragms. The propellants serve eight attitude-control engines, with beryllium chambers, and a single vectored axial thrust engine with unlike-double injector plate and ablative chamber/nozzle. Firing of ordnance-actuated isolation valves activates the propulsion system, and independent engine valves control the burn of each engine.

The RS-34 provides space and mounting provisions for the MGCS and all its related equipment, ordnance initiation sets, flight-termination ordnance system, instrumentation and cables. The engine started full-scale development in September 1979, with completion due in June 1984. The first MX test flight is scheduled for January 1983.

ROCKETDYNE RS-27

The RS-27 power plant consists of an RS2701A main engine and two LR101-NA-11 vernier engines. The verniers provide vehicle control during flight and vehicle stabilisation prior to stage separation. The RS-27 is used as the booster propulsion system for the Delta launch vehicle, replacing the Rocketdyne MB-3 (US Air Force designation LR79) propulsion system.

The RS2701A is a single-chamber bipropellant fixed-thrust gimballed engine. It utilises liquid oxygen and RP-1 propellants at a nominal mixture ratio of 2·245 : 1. Its rated thrust is 921 kN (207,000 lb) at sea level, with a maximum duration of 242 s. The thrust and mixture ratio are controlled by fixed orifices. The engine is a hybrid design which utilises the turbopump, turbine, gas generator, valves and thrust chamber of the H-1 engine, and the control system, start system and component-packaging arrangement of the MB-3 engine.

This power plant remains in production, with deliveries scheduled into 1984 to support launches into 1986.

Dimensions:	
Overall length	3,607 mm (11 ft 10 in)
Envelope max diameter	1,900 mm (6 ft 4 in)
Weight, Dry (approx):	1,028 kg (2,267 lb)

ROCKETDYNE MA-5

USAF designations: YLR89-NA-7 booster, YLR105-NA-7 sustainer and YLR101-NA-15 vernier

The MA-5 propulsion system consists of a dual-chamber liquid-propellant booster engine, a single-chamber liquid-propellant sustainer engine, and two vernier engines to control vehicle roll and to trim final velocity and directional control after burnout of the sustainer. This propulsion system powers the Atlas-Agena and Atlas-Centaur launch vehicles.

The design consists of two gimballed tubular-wall booster chambers, with twin-turbopump feed for the liquid oxygen and RP-1 propellants, and a single gimballed tubular-wall sustainer chamber, with similar feed. Ignition of both boosters and the sustainer engine takes place shortly before the vehicle is launched. Each YLR89-NA-7 booster is rated at 1,646 kN (370,000 lb) but can be derated to 1,495 kN (336,000 lb). The YLR105-NA-7 is rated at 267 kN (60,000 lb) but can be derated to 254 kN (57,000 lb). The YLR101 verniers are rated at 3 kN (669 lb) each.

Production is continuing, and the MA-5 will be used to boost launches until at least 1984.

Dimensions:	
Length	2,490 mm (98·0 in)
Diameter, nozzle exit	1,242 mm (48·88 in)
Weight, Dry:	
Booster	1,424 kg (3,140 lb)
Sustainer	471 kg (1,037 lb)

ROCKETDYNE MA-3

USAF designations: LR89-NA-5 booster, LR105-NA-5 sustainer, and LR101-NA-7 vernier

The MA-3 propulsion system consists of two independent liquid-propellant booster engines, a single-chamber liquid-propellant sustainer engine, and two vernier engines to control vehicle roll and to trim final velocity and directional control after burnout of the sustainer. This propulsion system is used to power Atlas E/F launch vehicles.

The booster and sustainer engines consist of gimballed tubular-wall thrust chambers, and independent bipropellant turbopumps for liquid oxygen and RP-1. The vernier has a gimballed solid-wall regeneratively cooled thrust chamber which receives propellants from the sustainer and propellant tanks after sustainer shutdown. Both the booster engines and sustainer engines are ignited prior to vehicle lift-off, whereas the vernier engines are ignited approximately 3·5 s after the sustainer is started. Each LR89-NA-5 booster engine is rated at 734 kN (165,000 lb) and the LR105-NA-5 sustainer is rated at 254 kN (57,000 lb). The LR101-NA-7 vernier engines are rated at 4·45 kN (1,000 lb) each.

Rated at 921 kN (207,000 lb) at sea level, the RS2701A is the main engine of the Rocketdyne RS-27 propulsion system

Although not currently in production, the MA-3 will be used to boost launch vehicles until 1987.

Dimensions:	
Length	3,371 mm (132·7 in)
Diameter, nozzle exit	1,242 mm (48·88 in)
Weight, Dry:	
Booster (each)	626 kg (1,380 lb)
Sustainer	412 kg (909 lb)

ROTORWAY

ROTORWAY INC

14805 S. Interstate 10, Tempe, Arizona 85284
Telephone: (602) 961 1001

RotorWay Inc is a builder of small helicopters (see entry in Aircraft section). For a considerable period it has been developing its own power plant, for these and for other light aircraft. The following is preliminary information:

ROTORWAY RW-100

This is the fixed-wing version of the RotorWay horizontally-opposed four-cylinder engine. Like the growing number of builders or convertors of water-cooled general-aviation engines, the company expects the RW-100 to be competitive with established aircooled units. It is demonstrating the ability to cruise at 90-95 per cent of maximum power, and thus offers slightly greater cruise power than larger aircooled engines, for fractionally less installed weight, smaller bulk, similar consumption,

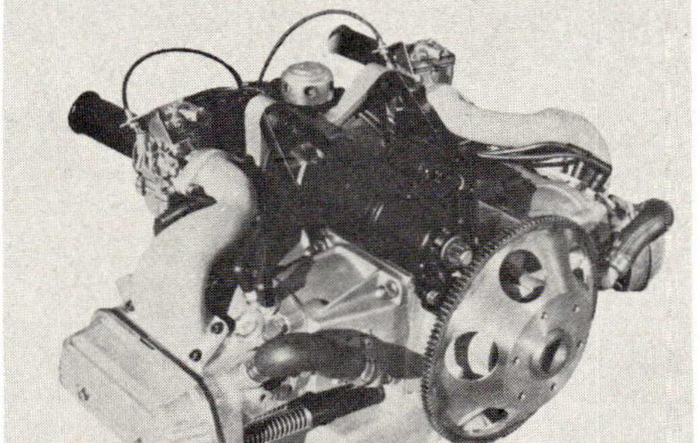

RotorWay RW-100 horizontally-opposed fixed-wing aircraft engine (74·5 kW; 100 hp)

RotorWay RW-133 vertical-crankshaft turbocharged helicopter engine (flat rated at 119 kW; 160 hp)

but using 92-octane automotive fuel, with extended aircraft range and much-reduced noise. The RW-100 is offered normally aspirated as a tractor or pusher engine.

Dimensions:
- Length, excl exhaust: 610 mm (24·0 in)
- Width overall: 781 mm (30·75 in)
- Height overall: 457 mm (18·0 in)

Weight, Dry: 74·8 kg (165 lb)

Performance Rating: 74·5 kW (100 hp) at 3,000 rpm

Specific Fuel Consumption: 64·2 μg/J (0·38 lb/h/shp)

ROTORWAY RW-133

This engine was designed to power the Scorpion Too light helicopter, but is expected to find a wide market. It was designed with the following objectives in mind: high power/weight ratio; improved fuel economy; reduced noise and emissions; smooth operation; and long, reliable life. It has been developed with and without turbocharger.

Type: Horizontally-opposed, vertical-crankshaft, water-cooled four-stroke piston-engine.

Cylinders: Offset left and right for plain connecting rods side by side. Swept volume 2·19 litres (133 cu in). Compression ratio 9·6 : 1.

Induction: Through circular air cleaner to single downdraught carburettor with fixed main jet and adjustable idle jet.

Ignition: Aircraft-type magnetos, dual ignition.

Lubrication: Oil temperature 82°-99°C. Oil pressure 2·72-4·1 bars (40-60 lb/sq in).

Cooling: Closed water system, operating temperature 82°C.

Weight, Dry (with starter):
- Without turbocharger: 77·1 kg (170 lb)
- With turbocharger: 83·9 kg (185 lb)

Performance Ratings:
- Without turbocharger 99·1 kW (133 hp) at 4,500 rpm
- With turbocharger (flat rated) 119 kW (160 hp) at 4,500 rpm

TELEDYNE CAE
TELEDYNE CAE DIVISION OF TELEDYNE INC

Head Office: 1330 Laskey Road, Toledo, Ohio 43697

Telephone: (419) 470 3000

President: James L. Murray

Executive Vice-President: Raymond Ortiz

Vice-Presidents:
- Dennis E. Barbeau (Engineering)
- Eugene R. Sullivan (Finance)
- Robert P. Schiller (Marketing, Plans, Contracts)
- Frank X. Marsh (Programmes)

The headquarters for management, marketing, finance, engineering and production is the Toledo, Ohio, facility of over 32,500 m² (350,000 sq ft).

From 1951 until 1960 almost all development was based on Turboméca designs. By far the most important of these was the Marboré, from which stemmed the J69 series of turbojets on which the manufacturing programme has depended. Since 1960 Teledyne CAE has embarked on an in-house development programme on a large scale with engine designs for RPVs, targets, cruise missiles and trainer aircraft.

In September 1978 Teledyne CAE was selected by Williams International Corporation as the latter's licensee to be second-source manufacturer of the F107-WR-101 and -400 cruise-missile engines. Full production is planned for 1982.

Teledyne CAE J69-T-29 turbojet of 7·56 kN (1,700 lb st)

TELEDYNE CAE 352 and 356
US military designation: J69

The J69 was originally the Turboméca Marboré, which has been developed to meet American requirements. Four versions are currently available as follows:

J69-T-25 (Teledyne CAE Model 352-5A). Long-life version, which powers the Cessna T-37B trainer and is FAA certificated as the Model CJ69-1025. Its air mass flow is 9 kg (19·8 lb)/s. Operational ceiling is 13,720 m (45,000 ft). In production.

J69-T-29 (Teledyne CAE Model 356-7A). Powers the Teledyne Ryan BQM-34A subsonic target drone. Operational ceiling is 18,300 m (60,000 ft). This is a Marboré II with a single-stage transonic axial compressor supercharging the centrifugal stage. In production.

J69-T-41A (Teledyne CAE Model 356-29A). Transonic axial compressor and revised centrifugal stage handling airflow of 13·5 kg (29·8 lb)/s with pressure ratio of 5·45 : 1. Operational ceiling in excess of 21,030 m (69,000 ft). Production is now spares only.

YJ69-T-406 (Teledyne CAE Model 356-34A). Produced for the US Navy's BQM-34E and the US Air Force's BQM-34F supersonic target drones. Initial qualification testing was completed during 1967 and deliveries of production engines began in 1970. The T-406 engine can propel the BQM-34E to Mach 1·5 at 18,300 m (60,000 ft) altitude. Production is now spares only.

The J69-T-29, YJ69-T-406 and J69-T-41A have a single-stage axial compressor ahead of the standard centrifugal compressor. Combustion system and turbine arrangements are basically the same as on the J69-T-25.

Teledyne CAE YJ69-T-406 turbojet of 8·54 kN (1,920 lb st)

Dimensions (nominal):

Length overall:
- J69-T-25: 899 mm (35·39 in)
- YJ69-T-406, J69-T-41A and J69-T-29: 1,138 mm (44·8 in)

Width:
- J69-T-25: 566 mm (22·30 in)
- J69-T-41A, J69-T-29: 568 mm (22·36 in)
- YJ69-T-406: 572 mm (22·52 in)

Weight, Dry:
- J69-T-25: 165 kg (364 lb)
- J69-T-29: 154 kg (341 lb)
- J69-T-41A: 159 kg (350 lb)
- YJ69-T-406: 163 kg (360 lb)

Performance Ratings:

Max rating:
- J69-T-25: 4·56 kN (1,025 lb) at 21,730 rpm
- J69-T-29: 7·56 kN (1,700 lb) at 22,000 rpm
- J69-T-41A: 8·54 kN (1,920 lb) at 22,000 rpm
- YJ69-T-406: 8·54 kN (1,920 lb) at 22,150 rpm

Normal rating:
- J69-T-25: 3·91 kN (880 lb) at 20,700 rpm
- J69-T-29: 6·12 kN (1,375 lb) at 20,790 rpm
- J69-T-41A: 7·34 kN (1,650 lb) at 20,900 rpm
- YJ69-T-406: 7·65 kN (1,719 lb) at 21,450 rpm

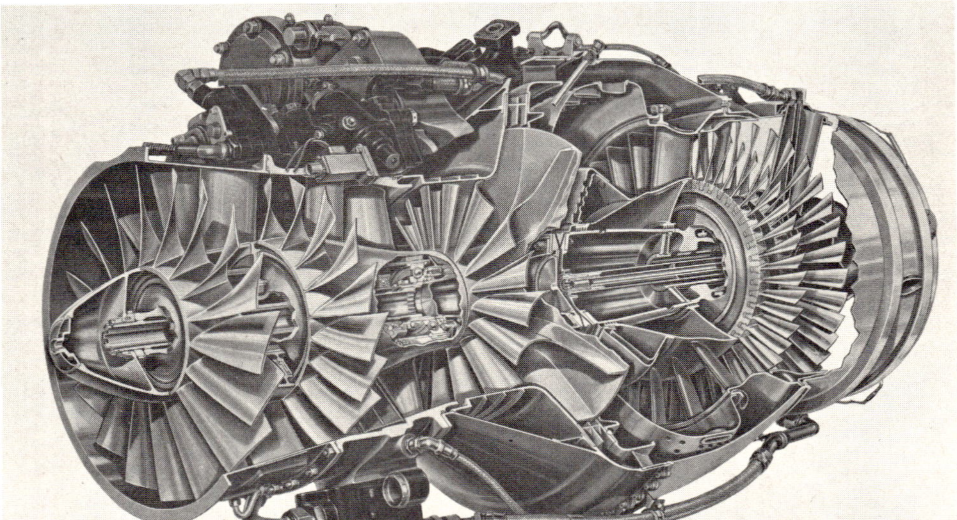

Teledyne CAE J100-CA-100 turbojet of 12·01 kN (2,700 lb st)

Specific Fuel Consumption:

At max rating:
- J69-T-25: 32·30 mg/Ns (1·14 lb/h/lb st)
- J69-T-41A, J69-T-29: 31·16 mg/Ns (1·10 lb/h/lb st)
- YJ69-T-406: 31·44 mg/Ns (1·11 lb/h/lb st)

Teledyne CAE J402-CA-400 expendable low-cost turbojet of 2·94 kN (660 lb st)

This slightly retouched photograph remains the only illustration of the Teledyne CAE 455H-2 turbofan demonstrator cleared for publication

TELEDYNE CAE 356-28A

US military designation: J100-CA-100

The Model 356-28A was developed by Teledyne CAE as a power plant for RPVs and other unmanned aircraft. The engine is derived from the J69 family but has no parts in common with that family. It has a two-stage transonic axial compressor ahead of the centrifugal stage, handling a mass flow of 20·4 kg (44·9 lb)/s with a pressure ratio of 6·3 : 1. The combustion chamber is annular with centrifugal fuel injection. The turbine has two axial stages, each fitted with replaceable blades. Fixed geometry is used throughout, although the engine is at present operating at altitudes in excess of 22,860 m (75,000 ft).

The J100-CA-100 completed a 108 h qualification test in June 1969. Applications included the Teledyne Ryan 147TE and 147TF medium-altitude intelligence-collection RPVs.

Dimensions:

Length, intake flange to jet pipe flange	1,225 mm (48·21 in)
Max width	629 mm (24·75 in)
Max height	663 mm (26·10 in)

Weight, Dry: 195 kg (430 lb)

Performance Ratings:

Max	12·01 kN (2,700 lb st) at 20,700 rpm
Normal	10·81 kN (2,430 lb st) at 20,120 rpm

Specific Fuel Consumption:

At max rating 31·16 mg/Ns (1·10 lb/h/lb st)

TELEDYNE CAE 356-28F

US military designation: J100-CA-101

This version of the J100 has been optimised for low-altitude performance with minimal cost. Changes include a slight increase in shaft speed, revised radial-diffuser vane angle and reduced turbine inlet nozzle area. Application has not been disclosed.

Dimensions:

Length overall	1,234 mm (48·6 in)
Max width	627 mm (24·7 in)
Max height	643 mm (25·3 in)

Weight, Dry: 195 kg (430 lb)

Performance Ratings:

Military S/L static 13·6 kN (3,050 lb st)

Specific Fuel Consumption:

Military S/L static 31·16 mg/Ns (1·10 lb/h/lb st)

TELEDYNE CAE 365

US military designation: LJ95

This family of engines had its inception in a lift-jet, the Model 365-7, developed for the US Air Force as the XLJ95-T-1. Details remain classified, except that the engine is in the 22·24 kN (5,000 lb st) class, has an above-average turbine gas temperature and offers a ratio of thrust to weight exceeding 20 : 1, yet is intended for propulsion of manned aircraft.

TELEDYNE CAE 370

US military designation: J402-CA-400

This low-cost expendable engine was designed for the propulsion of cruise-type missiles and is in production for the US Navy AGM-84A and RGM-84A Harpoon missiles. The J402 is noteworthy for its compact component and accessory disposition, giving minimum frontal area. Though the entire design minimises production time and cost, high reliability was a prime requirement. Flight limits are 12,200 m (40,000 ft) and Mach 0·9 continuous or Mach 1·1 for limited periods. Engine life is reported unofficially to be 1 h.

In 1974 the J402-CA-400 was selected as cruise power plant for the tactical versions of the US Navy Tomahawk Sea Launched Cruise Missile. In partnership with the competing SLCM contractors, Vought Systems Division and GD Convair, Teledyne CAE supported extensive testing, leading to selection of the GD tactical SLCM in mid-1976. Subsequently there were changes in the Tomahawk

Teledyne CAE J402-CA-700 turbojet of 2·85 kN (640 lb st) for MQM-107 variable-speed training target

programme, and though the J402 has been flown in Tomahawk test flights the missile now has a turbofan engine of different manufacture.

Type: Single-shaft turbojet.

Intake: Direct pitot inlet with four struts.

Compressor: Single transonic axial compressor with precision cast construction. Single centrifugal compressor with precision cast construction. Max airflow 4·35 kg (9·6 lb)/s. Pressure ratio 5·8.

Combustion Chamber: Annular type.

Fuel System: Low-pressure supply to centrifugal injection nozzles in compressor shaft. Electronic control system with automatic sequencing and regulation to meet demands of missile flight profile.

Turbine: Single-stage axial.

Jet Pipe: Fixed-area.

Accessories: Pyrotechnic starting and ignition systems. Optional integral alternator and alternator regulator to give 6 kW of DC power.

Mounting: Four main mountings disposed radially around main (compressor diffuser) frame.

Dimensions:

Length (excl bullet)	748 mm (29·44 in)
Overall diameter	318 mm (12·52 in)

Weight, Dry: 45·36 kg (100 lb)

Performance Rating:

Max S/L static 2·94 kN (660 lb st) at 41,200 rpm

Specific Fuel Consumption (S/L, static):

34·0 mg/Ns (1·20 lb/h/lb st)

TELEDYNE CAE 372-2

US military designation: J402-CA-700

This turbojet is in production for the Beech MQM-107 variable-speed training target. It is based on the Model 370 (J402) but differs in detail engineering and equipment, reflecting the need for repeated missions of extended duration. The electronic fuel control governs engine operation throughout the starting cycle and over the whole operating range. A shaft-mounted high-speed alternator provides 1·2 kW of DC power. Engine life is unofficially reported to be 15 h.

Dimensions:

Length (excl bullet)	753 mm (29·65 in)
Overall diameter	317 mm (12·50 in)

Weight, Dry: 52 kg (115 lb)

Performance Rating:

Max S/L static 2·85 kN (640 lb) at 40,400 rpm

Specific Fuel Consumption (S/L, static):

33·71 mg/Ns (1·19 lb/h/lb st)

TELEDYNE CAE 373

A growth version of the basic J402 described above, the CAE 373 is a new turbojet in the 4·00-4·45 kN (900-1,000 lb st) class. Addition of a second axial compressor stage increases pressure ratio to 8·7 : 1 and airflow to 6·2 kg (13·7) lb/s. The engine retains low-cost features but is designed for both expendable and long-life applications. It will be available with various starting options and either high-speed alternator or a centreline reduction gear and starter/generator. The CAE 373 is a company-funded development, and a demonstrator engine ran in 1976. It is intended for missile and RPV applications.

TELEDYNE CAE 440/555

A possible basis for a wide family of advanced small engines for the period after 1975, the 440 and 555 core engines developed as a result of the company's participation in the US Air Force Advanced Turbine Engine Gas Generator programme. Like ATEGG studies by other companies, the Models 440 and 555 have design parameters (pressure ratio, turbine entry temperature and specific fuel consumption) similar to those of the most advanced large engines. Most likely applications of these engine cores would be in turbofans in the 13·3-22·24 kN (3,000-5,000 lb) thrust class for piloted aircraft or high-performance RPVs. A photograph appeared in the 1978-79 *Jane's*.

TELEDYNE CAE 455

In 1979 preliminary details were disclosed of the Model 455H-2, first of the three rival JTDE (Joint Technology Demonstrator Engine) types to complete initial objectives for the US Air Force and Navy APSI (Aircraft Propulsion Subsystem Integration) programme. The 455H-2 uses a USAF-sponsored LP (fan) turbine, Navy-sponsored single-stage fan and HP spool based on the Model 555 described previously. It is intended to lead to a new family of durable, low-cost engines in the 31 kN (7,000 lb st) class for cruise missiles, trainers and liaison aircraft until the end of the century.

TELEDYNE CAE 490

This is the French-designed Turboméca-SNECMA Larzac (see Turboméca-SNECMA GRTS in French section), an 'exclusive agreement' for which was announced by Teledyne CAE in January 1973. The American company will "market, manufacture and service" the European turbofan for the USA and Canada.

The president of Teledyne CAE said that the Larzac, the initial US version of which is designated Model 490-4, "provides a valuable new source of flight-ready jet engines for strike and trainer aircraft, missiles and remotely piloted vehicles". A Model 490-4 demonstrator engine began running at Toledo in March 1973, and it is planned eventually to market a commercial Model 490-6. The CAE 490 is also being evaluated as a candidate engine for a future US Navy and/or US Air Force undergraduate pilot trainer.

TELEDYNE CONTINENTAL

TELEDYNE CONTINENTAL MOTORS
Aircraft Products Division
PO Box 90, Mobile, Alabama 36601
Telephone: (205) 438 3411

PRESIDENT: D. G. Bigler
EXECUTIVE VICE-PRESIDENT: J. H. Engler
SENIOR VICE-PRESIDENT: Ralph Hillard
VICE-PRESIDENTS:
S. A. Riggs (Marketing)
N. DeLaunay (Operations)
E. Wallace Jr (Quality)
L. Waters (Engineering)
J. Ishee (Finance)
K. Oehler (Industrial Relations)
A. Barkley (Procurement)
DIRECTOR OF MARKETING SUPPORT: Larry Jensen
DIRECTOR OF SERVICE: Larry Anderson
DIRECTOR OF FIELD INVESTIGATION: Robert Moore
DIRECTOR OF COMMUNICATIONS: Don Fairchilds

In 1927, the former Continental Motors Corporation, one of the largest automobile engine manufacturers in the world, produced its first aero-engine, a sleeve-valve air-cooled radial incorporating the Argyll (Burt-McCollum) patents, which had been purchased by the Corporation from the British Argyll Company in 1925.

In 1931 the 38 hp A40 flat-four was put on the market. This was followed by the A50, A65, A75, A80 and C90 engines.

On 1 January 1981 a 20-year licensing agreement with Rolls-Royce Motors was terminated. Teledyne Continental Motors is now responsible for manufacture and distribution of the engines previously covered by that agreement.

CONTINENTAL O-200 SERIES

The O-200-A engine is a four-cylinder horizontally-opposed aircooled engine. It is fitted with a single updraught carburettor, dual magnetos and starter and generator.

The 149 kW (200 hp) Teledyne Continental TSIO-360-E

The O-200-B is similar to the O-200-A, but is designed for pusher installation.

For other details see table.

CONTINENTAL IO-360 SERIES

The IO-360 is a six-cylinder horizontally-opposed air-cooled engine with fuel injection. Design and materials are generally similar to the IO-346-A (1970-71 *Jane's*), except for number and size of cylinders. Accessories include oil cooler, two magnetos, propeller governor drive, vacuum pump and 24V alternator. The IO-360 has a sandcast crankcase, with the accessory case mounted at the rear. The cylinders are shell-moulded.

The IO-360-C has dual accessory drive. The TSIO-360-A, B and C have a turbocharger pressurised induction system, revised fuel system, starter and accessory drive, scavenge pump and full-flow oil cooler. These engines power the Cessna T337 Skymaster. The TSIO-360-E, F and G are equipped with a complete exhaust system, and the turbocharger is engine mounted; the LTSIO-360-E is identical except that the crankshaft rotates in the opposite direction. They power the Piper Seneca II, Arrow III and Mooney 231.

For further details see table.

CONTINENTAL IO-368 SERIES

This large-displacement four-cylinder engine was originally a joint development of Teledyne Continental Motors and Rolls-Royce Motors. Its design incorporates many features previously available only on larger engines.

Its cylinder is based on that of the 520 series, with increased stroke, on a compact four-cylinder block. Lighter and less costly to produce, the 368 uses the same basic structural arrangement and many components of its six-cylinder predecessors. Its integrity of design and construction and use of proven parts result in an initial TBO of

REPRESENTATIVE TELEDYNE CONTINENTAL HORIZONTALLY-OPPOSED ENGINES

Engine Model	No. of Cylinders	Bore and Stroke mm (in)	Capacity litres (cu in)	Power Ratings kW (hp) at rpm: Take-off	Power Ratings kW (hp) at rpm: M.E.T.O.	Comp. Ratio	Dry Weight* kg (lb)	Dimensions: Length mm (in)	Dimensions: Width mm (in)	Dimensions: Height mm (in)	Octane Rating
O-200-A	4	103·2×98·4 ($4\frac{1}{16}\times3\frac{7}{8}$)	3·28 (201)	74·5 (100) at 2,750	74·5 (100) at 2,750	7·0	99·8 (220)	725 (28·53)	802 (31·56)	589 (23·18)	80/87
IO-360-D	6	112·5×98·4 ($4\frac{7}{16}\times3\frac{7}{8}$)	5·9 (360)	157 (210) at 2,800	157 (210) at 2,800	8·5	148·3 (327)	877 (34·53)	798 (31·40)	618 (24·33)	100/130
TSIO-360-C, D	6	112·5×98·4 ($4\frac{7}{16}\times3\frac{7}{8}$)	5·9 (360)	168 (225) at 2,800	168 (225) at 2,800	7·5	136 (300)	910† (35·84)	838 (33·03)	603 (23·75)	100/130
TSIO-360-E, F	6	112·5×98·4 ($4\frac{7}{16}\times3\frac{7}{8}$)	5·9 (360)	149 (200) at 2,575	149 (200) at 2,575	7·5	175 (385)	1,437§ (56·58)	795 (31·30)	671 (26·44)	100/130
IO-368	4	133×107·95 ($5\frac{1}{4}\times4\frac{1}{4}$)	6·02 (368)	134 (180) at 2,700	134 (180) at 2,700	8·2	136 (300)	752·4 (29·62)	850·9 (33·5)	590·6 (23·25)	100LL
IO-470-H	6	127×101·6 (5×4)	7·7 (471)	194 (260) at 2,625	194 (260) at 2,625	8·6	202·5 (446·5)	1,100 (43·31)	852 (33·56)	502 (19·75)	100/130
O-470-R, S	6	127×101·6 (5×4)	7·7 (471)	172 (230) at 2,600	172 (230) at 2,600	7·0	193·2 (426)	915 (36·03)	852 (33·56)	723 (28·42)	80/87
TSIO-470-D	6	127×101·6 (5×4)	7·7 (471)	194 (260) at 2,600	194 (260) at 2,600	7·5	231·8 (511)	1,465 (58·07)	852 (33·56)	514 (20·25)	100/130
IO-520-A	6	133×101·6 ($5\frac{1}{4}\times4$)	8·5 (520)	213 (285) at 2,700	213 (285) at 2,700	8·5	215·9 (476)	1,053 (41·41)	852 (33·56)	502 (19·75)	100/130
IO-520-BA	6	133×101·6 ($5\frac{1}{4}\times4$)	8·5 (520)	213 (285) at 2,700	213 (285) at 2,700	8·5	207·3 (457)	1,009 (39·71)	853 (33·58)	678 (26·71)	100/130
IO-520-D	6	133×101·6 ($5\frac{1}{4}\times4$)	8·5 (520)	224 (300) at 2,850	213 (285) at 2,700	8·5	208·2 (459)	949 (37·36)	901 (35·46)	604 (23·79)	100/130
IO-520-M	6	133×101·6 ($5\frac{1}{4}\times4$)	8·5 (520)	213 (285) at 2,700	213 (285) at 2,700	8·5	188 (415)	1,189 (46·80)	852 (33·56)	518 (20·41)	100/130
TSIO-520-B	6	133×101·6 ($5\frac{1}{4}\times4$)	8·5 (520)	213 (285) at 2,700	213 (285) at 2,700	8·5	219 (483)	1,490 (58·67)	852 (33·56)	516 (20·32)	100/130
TSIO-520-C	6	133×101·6 ($5\frac{1}{4}\times4$)	8·5 (520)	213 (285) at 2,700	213 (285) at 2,700	7·5	208 (458)	1,040† (40·91)	852 (33·56)	509 (20·04)	100/130
TSIO-520-E	6	133×101·6 ($5\frac{1}{4}\times4$)	8·5 (520)	224 (300) at 2,700	224 (300) at 2,700	7·5	219 (483)	1,010† (39·75)	852 (33·56)	527 (20·74)	100/130
TSIO-520-J, N	6	133×101·6 ($5\frac{1}{4}\times4$)	8·5 (520)	231 (310) at 2,700	231 (310) at 2,700	7·5	221·3 (487·8)	997 (39·25)	852 (33·56)	516 (20·32)	100/130
TSIO-520-L	6	133×101·6 ($5\frac{1}{4}\times4$)	8·5 (520)	231 (310) at 2,700	231 (310) at 2,700	7·5	244·5 (539)	1,286 (50·62)	852 (33·56)	508 (20·02)	100/130
TSIO-520-M, R	6	133×101·6 ($5\frac{1}{4}\times4$)	8·5 (520)	231 (310) at 2,700	213 (285) at 2,600	7·5	198 (436)	1,040† (40·91)	852 (33·56)	598 (23·54)	100/130
GTSIO-520-C	6	133×101·6 ($5\frac{1}{4}\times4$)	8·5 (520)	254 (340) at 3,200	254 (340) at 3,200	7·5	252·7 (557)	1,081 (42·56)	880 (34·04)	587 (23·1)	100/130
GTSIO-520-D, H	6	133×101·6 ($5\frac{1}{4}\times4$)	8·5 (520)	280 (375) at 3,400	280 (375) at 3,400	7·5	250 (550·37)	1,081 (42·56)	880 (34·04)	680 (26·78)	100/130
GTSIO-520-F, K	6	133×101·6 ($5\frac{1}{4}\times4$)	8·5 (520)	324 (435) at 3,400	324 (435) at 3,400	7·5	272·0 (600)	1,426 (56·12)	880 (34·04)	664 (26·15)	100/130
GTSIO-520-L, M, N	6	133×101·6 ($5\frac{1}{4}\times4$)	8·5 (520)	280 (375) at 3,350	280 (375) at 3,350	7·5	228 (502)‡	1,114 (43·87)	880 (34·04)	671 (26·41)	100/130

*With accessories; †Not including turbocharger;
‡N weight 220 kg (486 lb)

The 134 kW (180 hp) Teledyne Continental IO-368

The 280 kW (375 hp) Teledyne Continental GTSIO-520-L

2,000 h (1,800 h for the turbocharged TSIO-368). Available turbocharging control options cover a broad range from simplified, through single-point, to a complex variable system giving fully automatic control over the entire engine operating range.

A Teledyne Continental continuous-flow fuel-injection system, standard on all 368 series engines, improves performance and reduces fuel consumption by supplying metered fuel to the individual cylinders, for optimum performance at any altitude. The fuel-injection system can be located either above or to the rear of the engine. The Prestolite electric starter and belt-driven alternator are also located at the rear, so moving the centre of gravity aft and reducing the load on the engine mounting, which can be of the bed or rear type. Additional features include a spin-on oil filter, two independently driven Slick magnetos, provision for mounting an oil cooler, and a mounting pad for a belt-driven freon compressor.

For further details see table.

CONTINENTAL O-470 SERIES

Engines in the O-470 series (including the E-185 and E-225) are all basically similar. Engines prefixed 'IO' have direct fuel-injection.

The 168 kW and 172 kW (225 hp and 230 hp) models have a compression ratio of 7 : 1, the 186 kW (250 hp) models a ratio of 8 : 1, and the 194 kW (260 hp) models a ratio of 8·6 : 1. The exception is the O-470-U, which has a ratio of 8·6 : 1 and runs on 100LL grade fuel.

CONTINENTAL IO-520 SERIES

These engines are basically similar to the IO-470, but with cylinders of larger bore. They are fitted with an alternator driven either by a belt or by a face gear on the crankshaft. All IO-520 series engines are rated at 213 kW (285 hp) except for the IO-520-D, -E and -F which have a take-off rating of 224 kW (300 hp). IO-520 engines power the Beechcraft Baron and Bonanza, Navion and Cessna 210. New in 1970 were the generally similar IO-520-J, -K and -L, also rated at 213 kW (285 hp) (-K and -L are cleared to 224 kW (300 hp) at 2,850 rpm at take-off). The IO-520-M was developed in 1975 for use in the Cessna 310, replacing the IO-470-V.

The TSIO-520 series are turbocharged. Take-off rating is 213 kW (285 hp) except for the -E and -G, rated at 224 kW (300 hp), and the TSIO-520-J rated at 231 kW (310 hp) and equipped with an intercooler and provision for an overboost valve. These engines power the Cessna 414, 320D, T210 and 210F, and turbocharged Bonanza. The TSIO-520-L was developed for use in the Beech Pressurised Baron. It develops 231 kW (310 hp) at 2,700 rpm, has a complete exhaust system and an engine mounted turbocharger. The TSIO-520-N is used in the Cessna 340A and 414.

For other details see table

CONTINENTAL GTSIO-520

This is similar to the TSIO-520 range but is geared and uprated. The -C model, rated at 254 kW (340 hp) at 3,200 rpm, powers the Cessna 411. The GTSIO-520-D, rated at 280 kW (375 hp) at 3,400 rpm, powers the Cessna 421. The -K has an integral turbocharger and complete exhaust system: the most powerful Continental engine in production, used in the Rockwell Commander 685. The -G is used in a military application, the -H powers the Cessna 421A Golden Eagle. The -L is used in the Cessna 421C and the -M in the Cessna Titan.

THERMO-JET

THERMO-JET STANDARD INC

PO Box 55976, Houston, Texas 77055
Telephone: (713) 465 5735
MANAGER: John A. Melenric

This company specialises in the design and manufacture of valveless pulse-jet units for remotely piloted vehicles and the homebuilt aircraft market. These engines are devoid of moving parts and are characterised by multiple reverse-flow air inlets to a combustion chamber in which is burned propane, butane or compressed natural gas, obviating the need for a fuel pump. Intermittent combustion and expulsion takes place at a cycle frequency determined by the chamber size and geometry and combustion pressure.

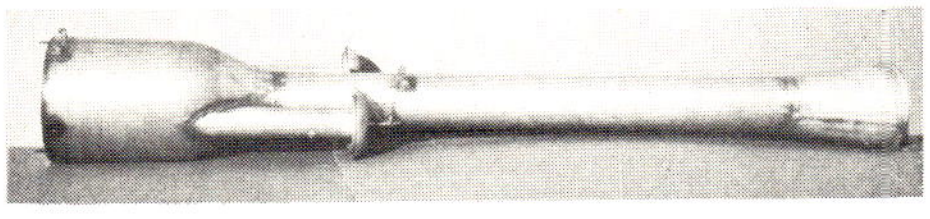

Thermo-Jet J13-202 valveless pulse-jet (0·40 kN; 90 lb st)

At present Thermo-Jet is offering four sizes of unit, the J7-300 (0·09 kN, 21 lb st at S/L), J8-200 (0·044 kN, 10 lb st at S/L), J10-200 (0·244 kN, 55 lb st at S/L), and J13-202 (0·4 kN, 90 lb st at S/L). Full details were given in the 1979-80 *Jane's*.

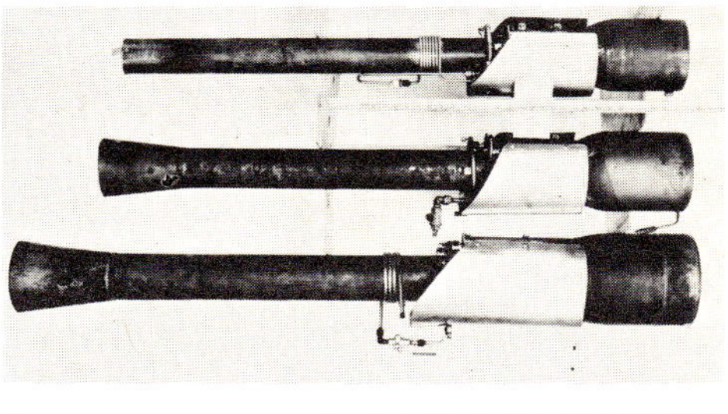

From the top: Thermo-Jet J7-300, J8-200 and J10-200

THIOKOL

THIOKOL CORPORATION

CORPORATE OFFICE: PO Box 1000, Newtown, Pennsylvania 18940
Telephone: (215) 968 5911
GOVERNMENT SYSTEMS GROUP: PO Box 9258, Ogden, Utah 84409
Telephone: (801) 626 1000
SOLID PROPELLANT ROCKET MOTOR PLANTS: Elkton, Maryland; Huntsville, Alabama; Marshall, Texas; Brigham City and Ogden, Utah
PYROTECHNIC AND ORDNANCE PLANTS: Shreveport, Louisiana; Marshall, Texas
CHAIRMAN OF THE BOARD, PRESIDENT AND CHIEF EXECUTIVE: R. E. Davis
VICE-PRESIDENT AND TREASURER: A. P. Roeper
GROUP VICE-PRESIDENTS:
R. N. Thompson (Chemical)
James M. Stone (Government Systems)
E. R. Kearney (General Products)

Organised in 1929, Thiokol Chemical Corporation produced and marketed the first synthetic rubber manufactured in the USA. In 1943, the discovery by Thiokol of liquid polymer, a new type of synthetic rubber, paved the way for the practical development of the 'case-bonded' principle of rocket power plant design. The company's polysulphide liquid polymer proved to be the catalyst for the first mass production of efficient solid-propellant rocket motors, as well as for the development of large solid-propellant motors. The firm's operations have now been organised into separate groups to serve widening areas of related products. Reflecting this diversity, the name was changed to Thiokol Corporation.

Details of some of the more important solid-propellant rocket motors used in missiles, sounding rockets, spacecraft and space launch vehicles are given below. Important current rocket motor activity, of which details cannot be reported, includes: production of gas generators for the Poseidon missile; recent completion of production of the TX-481 Maverick motor, and engineering development of the TX-486 motor for the Patriot air-defence missile.

In 1974 Thiokol Corporation was awarded a contract to develop the solid rocket motors for the NASA Space Shuttle.

THIOKOL SPACE SHUTTLE SRM

The Space Shuttle Solid Rocket Motor is a segmented design which is 3·708 m (146 in) in diameter, 38·1 m (125 ft) long, weighs 569,268 kg (1,255,000 lb), and contains 503,000 kg (1,109,000 lb) of propellant. The SRM is cast in four segments, which are rail-shipped to launch sites at Kennedy Space Center and Vandenberg Air Force Base. At the sites the motors are stacked vertically in pairs as part of the Solid Rocket Booster for the Space Shuttle.

At lift-off, the SRMs are burning in parallel with the three Space Shuttle liquid main engines mounted in the Orbiter. The two SRMs develop 10,535 kN (2,368,000 lb)

Thiokol Space Shuttle SRM being prepared for static firing

average thrust each (vacuum conditions) and burn for about two minutes. Thrust-vector control for each SRM is provided by a flexible-bearing movable nozzle, driven by two hydraulic actuators.

After burnout, the two boosters are separated from the vehicle, which continues under the power of the three main engines. The boosters fall into the Earth's atmosphere, where parachutes are deployed to control impact velocity into the ocean. The boosters are then towed back to the launch site, disassembled, preserved, and returned to the factory for propellant reloading (see under CSD in this section).

The motor case consists of 11 weld-free ring-rolled forgings fabricated from Ladish D6AC steel. The ends of the forged case sections are machined to form tang-and-clevis joints by means of which the sections are assembled into 'casting segments'. Four casting segments (consisting of a forward segment, two centre segments and one aft segment) make up the basic manufacturing and shipping configuration of a single SRM.

The propellant is a PBAN terpolymer composite, bonded to the asbestos-silica filled nitrile-butadiene rubber internal insulation by means of a compatible liner. The propellant grain in the forward end is configured to an 11-point star which transitions into a tapered centre-perforated (CP) cross-section for the centre and aft sections of the motor. The nozzle is a 20·4 per cent submerged, omni-directionally movable, flexible-bearing nozzle with a throat diameter of 1·38 m (54·5 in) and an exit-plane diameter of 3·76 m (148 in). The nozzle consists of insulated aluminium and steel components. The flexible bearing which provides thrust vector control (TVC) consists of a series of 10 semi-spherical steel shims laminated between flexible elastomer pads.

The ignition system is mounted internally at the forward end of the forward segment. The system consists of an electro-mechanical safe/arm device, an initiator which is a small, multi-nozzled rocket motor containing about 0·7 kg (1·5 lb) of fast-burning propellant, and the main igniter which is an insulated steel-case rocket motor containing about 59 kg (130 lb) of fast-burning propellant. A moulded-silica throat insert in the main igniter directs the flame pattern to the SRM's main propellant grain. Ignition time for the SRM is approximately 0·2 s.

THIOKOL SSUS-A MOTOR

The Thiokol TU-844 motor provides propulsion for the McDonnell Douglas Spinning Solid Upper Stage (SSUS-A) which launches Atlas-Centaur-class payloads from the Space Shuttle. The motor is a slightly modified third-stage Minuteman III with TVC and roll-control systems removed. Lengths of the nozzle and case are optimised to minimise Space Shuttle user changes. The motor uses an S-901 glassfibre case and is loaded with CTPB propellant. It is 2·12 m (83·4 in) long and 1·32 m (52 in) in diameter, and weighs 3,662 kg (8,074 lb).

TRIDENT MOTORS

The three propulsion stages for the Trident C-4 fleet ballistic missile are being developed and manufactured jointly by Thiokol and Hercules Inc. The motors are loaded with high-energy cross-linked double-base propellant. High-performance subsystems, including Kevlar cases, flex-seal TVC and low-erosion nozzles, have been developed and are being manufactured by the joint venture. Technical data are classified.

THIOKOL LONG-LIFE SRAM MOTOR

Thiokol is under contract to Boeing to improve the ageing and life of the motor for the US Air Force's AGM-69 SRAM missile. A motor with a projected service life of over 10 years has been developed and is currently in design-verification. The SRAM motor is 2·54 m (100 in) long, 0·44 m (17·5 in) in diameter, and is loaded with 449 kg (990 lb) of HTPB propellant. The grain configuration consists of an aft-end boost grain and a forward sustain grain. They are separated by a barrier and have separate igniters. The motor has a fixed nozzle and a steel case.

THIOKOL TX-690 (FIREBRAND BOOSTER)

The Thiokol TX-690 is being used to provide initial launch impulse for the US Navy's Firebrand aerial target. The target simulates a family of supersonic anti-ship missiles and provides threat simulation for test and evaluation of anti-ship missile defence systems of the 1980s. The booster motor, designed and built by Thiokol's Huntsville Division, accelerates the drone to supersonic speeds, at which the main ramjet engines take over. The motor is a modification of the Patriot propulsion system. The primary modifications are a canted nozzle and attachment fittings. Thirty motors were ordered for test and evaluation flights.

THIOKOL TE-M-707 (HARPOON)

The TE-M-707 is Thiokol's version of the US Navy's 642AS7000 solid-propellant rocket motor. It provides boost propulsion for the ship- and submarine-launched versions of the Harpoon anti-ship missile. The booster has an outside diameter of 0·34 m (13·5 in) and a length of 0·74 m (29 in); it contains approximately 65 kg (143 lb) of C-1 polyurethane composite propellant. The motor case and nozzle shell are of heat-treated 4130 steel. Glass phenolic insulation material is employed as nozzle insulation, and high-density graphite acts as a throat. The ignition system consists of a basket igniter in line with an electromechanical arm/fire device. The booster provides a nominal total impulse of 160 kN-s (36,000 lb-s). Qualification was completed in June 1980.

THIOKOL TX-174

The TX-174 is the first stage of the Pershing tactical weapon system. The motor case is a thin-wall flight design fabricated from modified AISI H-11 or type D6AC steel. The contoured nozzle has an expansion ratio for sea-level operation.

The TX-174 has an overall length of 2·61 m (102·61 in) and an outside diameter of 1·02 m (40 in). The nominal propellant weight is 2,019 kg (4,451 lb). The nominal total weight of the TX-174 is 2,270 kg (5,004 lb). It provides an average thrust of 117 kN (26,290 lb) and a total impulse of 1,017,200 lb-s. The web burning time is 38·30 s. The TX-174 utilises a cylindrical-core propellant configuration.

THIOKOL TX-175

The TX-175 is the sustainer stage for the Pershing. The motor case is a thin-wall flight design fabricated from modified AISI H-11 or type D6AC steel. The forward dome of the case has three ports for impulse control. The contoured nozzle is sized for altitude operation.

The TX-175 has an overall length of 2·46 m (96·72 in) and an outside diameter of 1·02 m (40·0 in). The nominal propellant weight is 1,263 kg (2,785 lb). The nominal total weight of the TX-175 is 1,471 kg (3,244 lb). At vacuum conditions, it provides an average thrust of 85·5 kN (19,220 lb) and a total impulse of 757,200 lb-s. The web burning time is 39·0 s. The TX-175 utilises a cylindrical-core propellant configuration.

Thiokol Long-life SRAM motor

THIOKOL TU-289

US military designation: SR49-TC-1

The TU-289 solid-propellant rocket motor powers the AIR-2A Genie unguided air-to-air missile. It replaced an earlier motor, and has an improved propellant which increases the storage life and permits the missile to be deployed in a wide range of environmental temperatures.

THIOKOL TX-354 CASTOR II

TX-354 motors are used in a variety of applications such as first and second stages, and as strap-on boosters for launch vehicles. The TX-354-3 has a high-altitude nozzle and is used as the second stage of the Scout vehicle. TX-354-4 has a sea-level straight nozzle for first-stage applications. TX-354-5 is the strap-on booster for the Delta vehicle. It has an 11° canted sea-level nozzle and is used in groups of three, six, or nine. TX-354 weighs 4,320 to 4,410 kg (9,525 to 9,743 lb). The case without nozzle is 5,130 mm (202 in) long and has a diameter of 787 mm (31 in). Sea-level thrust is 232 kN (52,150 lb) over a total burning time of 39 s for a total impulse of 886,043 kg-s (1,953,400 lb-s).

THIOKOL TX-481 (MAVERICK)

TX-481 motors are used in the AGM-65A and AGM-65B Maverick missiles. The motor has an overall length of 1,020 mm (40·16 in), is 272 mm (10·7 in) in diameter, and has a total weight of 47·2 kg (104 lb). The case and blast-tube are aluminium. The composite propellant employs a Thiokol polysulphide polymer binder with ammonium perchlorate oxidiser. The motor operates in a boost/sustain mode with thrust levels of 44·5 kN (10,017 lb) in boost for 0·575 s and 9·68 kN (2,175 lb) in sustain phase for 3·495 s. The overall impulse is 6,169 kg-s (13,600 lb-s). The operational temperature range for the motor is −54°C to 71·1°C. Excellent shelf life has been demonstrated.

Production of motors for the A and B models of Maverick began in 1972 and was continuous until 1978 at delivery rates varying from approximately 250 to 1,000 motors per month. The total quantity of motors delivered is 26,213; batch acceptance static firings totalled 1,299.

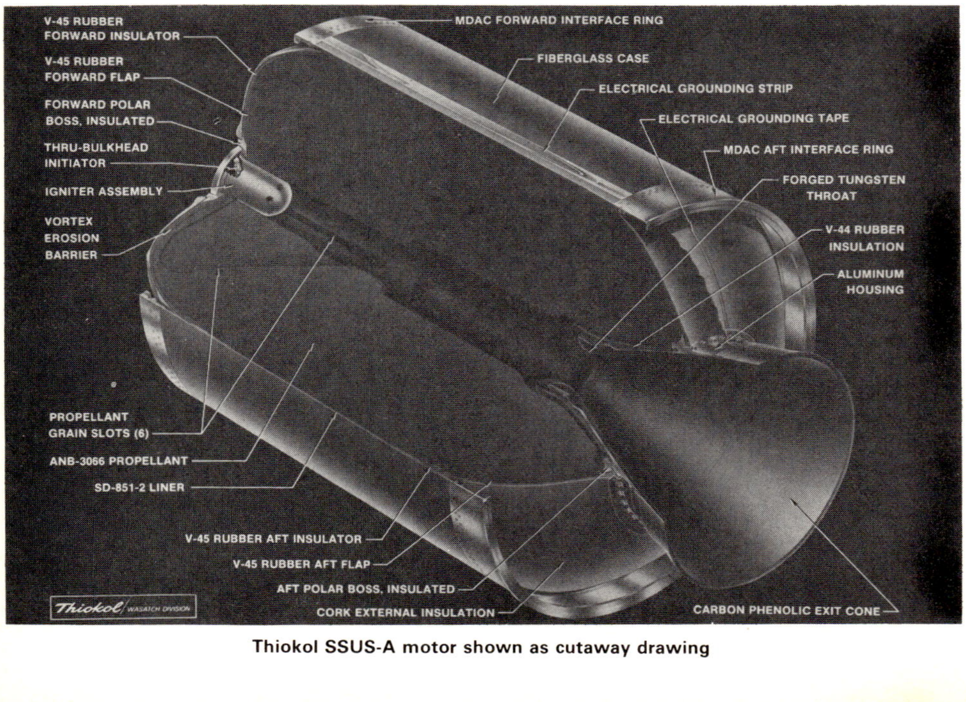

Thiokol SSUS-A motor shown as cutaway drawing

THIOKOL TX-486 (PATRIOT)

The TX-486 motor is the propulsion system for the Patriot missile which provides defence against high-performance aircraft. The overall length of the motor is 3,302 mm (130 in) and its outside diameter is 406 mm (16·0 in). It weighs approximately 590 kg (1,300 lb). Ballistic parameters for the motor are classified. The motor case is D6AC steel, and the nozzle adapter is 4130 steel. Silica phenolic materials are employed in the nozzle throat and exit cone. The propellant and liner use a hydroxyl-terminated polybutadiene polymer binder; ammonium perchlorate is the propellant oxidiser. The Engineering Development programme for the motor was successfully completed in 1978. Initial production was scheduled for early 1981.

THIOKOL TX-581

The TX-581 (Castor II-X) is an extended-length version of the TX-354 (Castor II) motor and was developed as a strap-on booster with an 11° canted nozzle. It can also be used in first- or second-stage applications with a straight nozzle. The motor is 787·4 mm (31 in) in diameter, 8,000 mm (315 in) long and weighs about 5,806 kg (12,800 lb). Sea-level performance provides an average thrust of approximately 339·4 kN (76,300 lb) over a total burning time of 35·8 s, giving a total impulse of 11,720 kN-s (2,634,900 lb-s), a 35 per cent increase over Castor II.

THIOKOL TX-526 CASTOR IV

TX-526 motors are used in a variety of booster/strap-on applications. The TX-526-0 motor, combined with four Recruit strap-on motors, is the booster for the Athena H vehicle. The TX-526-1 with 7° canted nozzle is qualified for strap-on booster applications. The TX-526-2 with 11° canted nozzle is used as a strap-on booster for the Delta 3914 vehicle, with each vehicle having nine motors. The 9,093 mm (358 in) long, 1,016 mm (40 in) diameter solid-propellant motor weighs 10,550 kg (23,250 lb). It provides an average thrust of 379·3 kN (85,270 lb), with a total burning time of approximately 58 s for a total impulse of 2,159,544 kg-s (4,760,900 lb-s) at sea-level.

THIOKOL TX-633 (MAVERICK)

US military designation: SR-114-TC-1

The TX-633 was developed as a reduced-smoke version of the Maverick rocket motor (see TX-481). Qualification testing was completed in 1977. Performance unchanged within the same envelope, varying only in the propellant and liner system, ignition system and protective closure. Motor case, blast-tube, nozzle, grain configuration and insulation are identical to the original model. The propellant and liner for this motor employ a hydroxyl-terminated polybutadiene polymer binder; the oxidiser is ammonium perchlorate. The reduced-smoke version of the motor could phase in with the next element of production.

THIOKOL (HARM)

US military designation: AGM-88A

The HARM motor is a reduced-smoke dual-thrust rocket that powers the USAF/US Navy High-speed Anti-Radiation Missile. The motor is approximately 254 mm (10 in) in diameter and 2,120 mm (83·5 in) long. It utilises a D6AC steel case loaded with 127 kg (280 lb) of non-aluminised HTPB propellant and has a single fixed nozzle. A single propellant grain provides the boost/sustain thrust profile.

THIOKOL TX657 (HELLFIRE)

The TX657 motor is used as propulsion for the Hellfire modular missile system. The motor has an overall length of 594 mm (23·38 in) and a diameter of 178 mm (7 in). The case is made of aluminium. The nozzle blast tube is of all-moulded composite construction. The reduced-smoke composite propellant employs an HTPB binder with AP oxidiser. The motor operates in an all-boost mode. The operating temperature range is −42·7 to 62·8°C, with storage range −45·6 to 65·6°C. Excellent shelf life is being demonstrated. Qualification was completed in early 1980 with production to be funded in late 1981.

THIOKOL TE-M-416 TOMAHAWK

The Tomahawk is a high-performance motor designed specifically for use in sounding rocket systems. It is used in the Tomahawk vehicle as a single stage and in several other vehicles, such as Nike-Tomahawk and Terrier-Tomahawk, as the second stage. The motor is 3,607 mm (142 in) long and has a diameter of 229 mm (9 in). It weighs 220 kg (486 lb) and produces 48·9 kN (11,000 lb) thrust at sea level. This motor has been used by NASA in scale studies of the Shuttle SRM acoustic effects on an operational launch.

THIOKOL TX-683 (AIM-9L)

This is a reduced-smoke replacement for the Mk 36 Mod 5, used in the AIM-9L Sidewinder. It uses the same case and has the same performance. Propellant is HTPB/AP; diameter is 127 mm (5 in), length 1,803 mm (71 in) and weight about 45·4 kg (100 lb). Qualification 1980; production 1981.

THIOKOL TX-732 (AMRAAM)

This motor is tailored to the Advanced Medium-Range AAM. It has an operating temperature from −54° to 62·8°C and nozzle/blast tube with composite insulation in a metal shell. The Validation Phase was completed in 1980, with engineering development to start in 1981.

THIOKOL TE-M-29-8

Super Recruit, TE-M-29-8, is an improved version of previous Recruit motor configurations. The burn-time motor total impulse is increased from 55,000 to 62,000 lb-s, while keeping burn time unchanged. Weight is increased by about 6 per cent.

THIOKOL TE-M-307 (APACHE)

The TE-M-307-3 rocket motor was designed for second-stage applications, and therefore includes a 20 second delay igniter. It is 2,741 mm (107·91 in) long, 174·2 mm (6·86 in) in diameter and is used both as a sounding rocket and as a target missile. The TE-M-307-4 version was designed for single-stage applications. It uses the same loaded case and headcap assembly as the TE-M-307-3, with a 3·32 : 1 expansion ratio nozzle and an instantaneous TE-P-415 pyrogen. It is also used as a sounding rocket and as a target missile.

THIOKOL TE-M-364 (STAR-37)

The TE-M-364-2 (STAR-37B) is a 939·8 mm (37 in) diameter spherical main retro-rocket designed for the Surveyor and modified for use on the Burner II stage. Modifications consisted of increasing propellant loading to 653 kg (1,440 lb) and strengthening the attachment structure to accommodate higher inertial loads.

The TE-M-364-3 (STAR-37D) is a Surveyor main retro-rocket modified for use as third stage propulsion on the Improved Delta vehicle. Modifications consisted of again increasing propellant load, to 1,440 lb, redesigning the attachment structure to mate with the Delta launch vehicle and changing the diameter to 952·3 mm (37·49 in).

The TE-M-364-4 (STAR-37E) is an elongated version of the Delta motor, the AP/hydrocarbon/Al propellant grain being increased in mass from 653 kg (1,440 lb) to 1,040 kg (2,290 lb) by adding a 355 mm (14 in) cylinder to the case. Average thrust is 68·8 kN (15,472 lb) for a burn time of 41·96 s. This motor provides third-stage propulsion on Improved Delta.

The TE-M-364-19 (STAR-37F) is a shorter (1,518 mm; 59·8 in) version of the Delta motor, accommodating 865·5 kg (1,908 lb) of propellant. The nozzle is of composite asbestos, glass and graphite phenolic structure. The titanium case has a 178 mm (7 in) cylindrical section. Average thrust is 59·8 kN (13,440 lb) for a burn time of 40 s. It provides the impulse to circularise the orbit of the Fleetsatcom satellite at the apogee of the launch orbit. This motor has been selected as the AKM (apogee-kick motor) for Intelsat V.

The TE-M-364-11 (STAR-37G) is a very similar extended Delta motor, likewise used for Improved Delta third-stage propulsion. Average thrust is 62·9 kN (14,145 lb) for a burn time of 45·48 s.

The TE-M-364-14 (STAR-37N) is a version having a propellant loading of 557·9 kg (1,230 lb). Average thrust is 38·4 kN (8,634 lb) for a burn time of 37·7 s. This motor provides third-stage propulsion on the Japanese N vehicle.

The TE-M-364-15 (STAR-37S) is a titanium (6Al-4V) spherical-cased version with modified attachment and propellant loading of 657·7 kg (1,450 lb). Average thrust is 43·5 kN (9,790 lb) for a burn time of 42·2 s. It provides the third-stage propulsion for weather satellites flown on the US Air Force's Thor vehicle.

THIOKOL TE-M-442-1 (STAR-26B)

This motor is spherical, 663 mm (26·1 in) in diameter and 839 mm (33·05 in) long; propellant weight is 238 kg (525 lb) and total motor weight is 261 kg (576 lb). The TE-M-442-1 was developed from the TE-M-442 of 1965 and features a case of titanium instead of steel. It flies as an additional stage to the standard Burner II launch vehicle, atop the TE-M-364-2 second stage.

THIOKOL TE-M-479 (STAR-17)

The TE-M-479 is a 442 mm (17·4 in) spherical rocket motor developed for NASA's Radio Astronomy Explorer satellite programme. The motor is 687 mm (27·06 in) long and serves as the apogee kick stage which makes the orbit of the spacecraft truly circular. Total motor weight is 78·8 kg (173·8 lb); propellant weight is 69·4 kg (153 lb). High mass-fraction and excellent performance reproducibility characterise this motor for space systems application. The TE-M-479 was first flown in July 1968.

THIOKOL TE-M-521 (STAR-17A)

This 444 mm (17·5 in) diameter and 980 mm (38·6 in) long motor was developed by adding a 175 mm (6·9 in) straight section to the spherical TE-M-479 (RAE) motor. The TE-M-521 has a propellant weight of 112 kg (247 lb) and a total weight of 123·9 kg (273·2 lb). It served to 'circularise' the orbit of the Skynet I, NATO I and IMP-H and -J satellites. The motor has a titanium case and flight-proven propellant.

THIOKOL TE-M-541/542 (STAR-6)

This small glassfibre motor measures 157 mm (6·2 in) in diameter and 356 mm (14 in) long and serves in a classified space application. Using the same hardware, with minor insulation changes, the motor is loaded to either of two configurations: 4·85 kg (10·7 lb), 1,395 kg-s (3,075 lb-s) total impulse, 5·99 kg (13·2 lb) total weight (TE-M-541); and 3·27 kg (7·2 lb), 930 kg-s (2,050 lb-s) total impulse, 4·8 kg (10·6 lb) total weight (TE-M-542). These motors have an extensive flight history.

THIOKOL TE-M-473 (SANDHAWK)

The Sandhawk TE-M-473 is a high-performance 330 mm (13 in) diameter, 5,105 mm (201 in) long rocket motor designed for sounding rocket use. It features a regressive thrust-time trace, which results in near-constant vehicle acceleration during its 15 s burn time and provides an extremely smooth flight environment. This motor is suited for use in single-stage, two-stage and three-stage vehicle configurations.

THIOKOL TE-M-236 (SARV RETRO)

This is a retrograde motor for an unmanned satellite. It uses an internal-burning case-bonded grain weighing 18·3 kg (40·34 lb) in a case of 4130 steel, with a re-entrant conical rear closure to keep overall length to only 324 mm (12·76 in). Burn-time (7·5 s) average thrust is 5·6 kN (1,250 lb).

THIOKOL TE-M-640 ALTAIR III (STAR-20)

This 0·5 m (19·67 in) diameter, 1·49 m (58·45 in) long motor is the fourth stage of the Scout launch vehicle. The 273·3 kg (602·5 lb) AP/CTPB/Al propellant grain is cast in a filament-wound glassfibre case. The external nozzle is a composite of graphite, plastics and steel. Motor thrust is 26·9 kN (6,050 lb) for a burn time of 27·8 s.

THIOKOL TE-M-458 (STAR-13)

This is a deceleration motor used in the Anchored Interplanetary Monitoring Platform (AIMP) programme. The 31 kg (68·3 lb) charge of AP/Al urethane is contained in a spherical case of 6Al-4V titanium, with graphite/vitreous silica phenolic nozzle. Loaded weight is 35·65 kg (78·6 lb) and average thrust 3·8 kN (850 lb) for a burn time of 21·8 s.

THIOKOL TE-M-516 (STAR-13A)

This apogee-boost motor is made by mating the propellant and nozzle of the TE-M-444 with the case and igniter of the TE-M-458. Average thrust is 5·8 kN (1,309 lb) for a burn time of 15·3 s. The motor was used as an injection stage of the Thor Burner II carrying two satellites: Secor and Aurora.

THIOKOL TE-M-604 (STAR-24)

This apogee-boost motor has a 199·5 kg (439·8 lb) charge of AP/hydrocarbon/Al propellant contained in a spherical case of 6Al-4V titanium with graphite/carbon phenolic nozzle. Average thrust is about 20 kN (4,500 lb) for a burn time of 30·21 s. The motor was used as the apogee kick stage on the UK Skynet II and US Air Force Space Test Platform, and successfully placed the Lageos spacecraft in orbit. It is being modified for use on the NASA IUE, and was the orbit-insertion motor for the Pioneer-Venus spacecraft which reached Venus in 1978.

THIOKOL TE-M-616 (STAR-27)

This 694 mm (27·3 in) diameter, 1,303 mm (51·3 in) long apogee boost motor has a 334 kg (736 lb) charge of AP/CTPB/Al propellant in a case of 6 Al-4V titanium and a graphite/carbon-phenolic nozzle. It has a Model 2130 remotely located electromechanical safe/arm device. Motor weight is 363 kg (800 lb); mass fraction is 0·920. Propellant weight can be reduced up to 25 per cent. Average thrust in vacuum is 27 kN (6,080 lb) for a burn time of 34·2 s. This motor was used to place in orbit the Canadian Communications Technology, Japanese Geometeorological, Japanese Broadcast and several US Air Force Navstar satellites. It was also used for the US Air Force P78-1 and NASA GOES satellites.

THIOKOL TE-M-696/TE-M-697

These qualified 35 mm (1·38 in) diameter motors were developed to provide total impulse reproducibility of 0·77 per cent. The propellant provides exhaust with low radar signature through a canted nozzle. The titanium alloy case minimises weight, and the TAM molybdenum nozzle controls thrust vector. The spin motor is 142 mm (5·6 in) long, weighs 0·28 kg (0·62 lb), and delivers 377·5 N (84·9 lb) thrust in vacuum for 0·30 s. The de-spin motor is 92 mm (3·6 in) long, weighs 0·23 kg (0·51 lb), and delivers 219·2 N (49·3 lb) thrust in vacuum for 0·29 s.

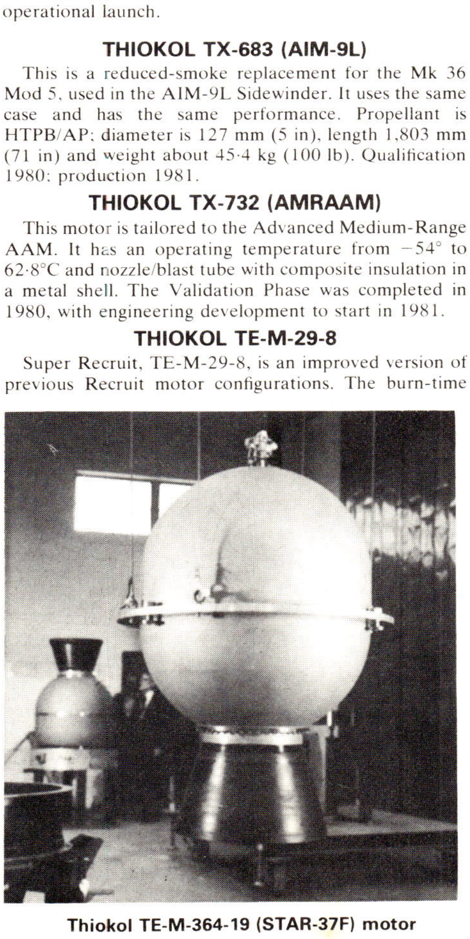

Thiokol TE-M-364-19 (STAR-37F) motor

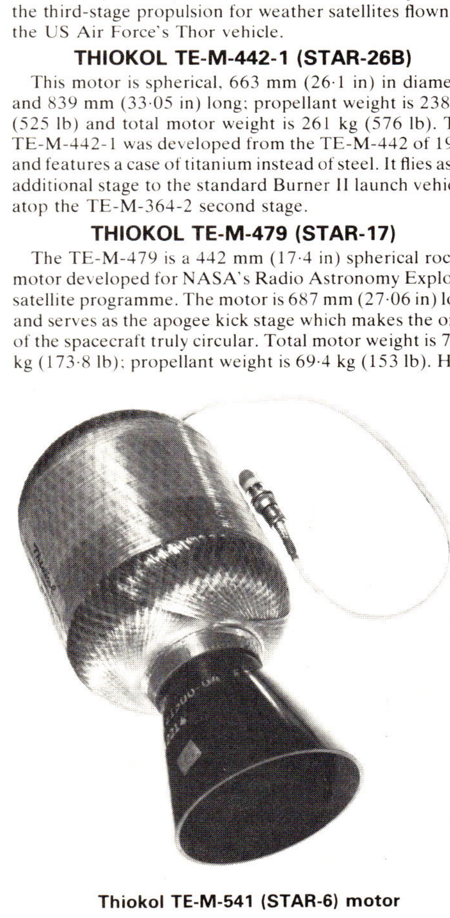

Thiokol TE-M-541 (STAR-6) motor

THIOKOL TE-M-700 (STAR-30)

This 762 mm (30 in) diameter, 1,508 mm (59·37 in) long apogee-boost motor has a 472·7 kg (1,042 lb) charge of AP/HTPB/Al propellant in a case of 6Al-4V titanium and with a graphite/carbon-carbon nozzle. The aft-end igniter is initiated by a Model 2130 remotely located safe/arm device. Motor weight is 505·3 kg (1,114 lb); propellant mass fraction is 0·935. The propellant weight can be increased 6 per cent and decreased 10 per cent. Average thrust in vacuum is 26·6 kN (5,980 lb) over a 51·5 s burn time. Qualification was completed, and first flight made, in 1980.

THIOKOL TE-M-711 (STAR-48)

This 1,247 mm (49·1 in) diameter, 1,829 mm (72 in) long perigee kick motor has 1,996 kg (4,400 lb) of AP/HTPB/Al propellant in a case of 6Al-4V titanium and has a deeply submerged nozzle with a graphite throat insert and carbon/carbon exit cone. The aft-end igniter is initiated from a remotely located Model 2130 safe/arm device. Motor weight is 2,109 kg (4,650 lb); propellant mass fraction is 0·946. Average thrust in vacuum is 66 kN (14,900 lb) over an 85 s burn time. The motor is used in the Payload Assist Module for spacecraft launches from the Delta launch vehicle and Space Shuttle. Qualification was completed, and first flight made, in 1980.

THIOKOL TE-M-762 ANTARES III (STAR-31)

This 762 mm (30 in) diameter, 2,873 mm (113 in) long motor has 1,228 kg (2,840 lb) of AP/HTPB/Al propellant in a filament-wound Kevlar case and a carbon/phenolic nozzle with a carbon/carbon throat insert. Motor weight is 1,393 kg (3,070 lb), including external insulation. Average thrust is 80 kN (18,000 lb) over a 46 s burn time. The motor was qualified as an improved third stage for the Scout launch vehicle and first flew in late 1979.

THIOKOL TE-M-714 (STAR-37X)

This is a high-performance, high-technology motor series complementing the basic TE-M-364 (STAR-37) series.

The TE-M-714-4 (STAR-37XE) is a 939·8 mm (37 in) diameter, 1,890 mm (74·4 in) elongated spherical rocket motor. Propellant weight is 1,324 kg (2,918 lb), propellant mass fraction is 0·945. The average thrust is 62·3 kN (14,000 lb) for a burn time of 60 s. It provides upper-stage propulsion for weather satellites flown on the US Air Force Thor vehicle.

The TE-M-714-6 (STAR-37XF) is a shorter, 1,514 mm (59·6 in) long version of the STAR-37XE. The propellant weight is 875 kg (1,929 lb). Average thrust is 40·0 kN (9,000 lb) for a burn time of 62 s. It provides apogee boost propulsion for Intelsat V spacecraft flown on Atlas-Centaur, Shuttle and Ariane vehicles.

TRW

TRW DEFENSE AND SPACE SYSTEMS GROUP

HEAD OFFICE: One Space Park, Redondo Beach, California 90278

Telephone: (213) 535 4321

TRW developed, built and launched the first monopropellant hydrazine propulsion system to enter and be started in space.

TRW produces a wide variety of chemical propulsion engines. One of these, the man-rated Lunar Module Descent Engine, landed astronauts on the Moon. Another engine was built to provide midcourse trajectory corrections for the Mariner '69 missions to Mars and the Mariner 10 (Venus-Mercury) missions. TRW provided the monopropellant hydrazine orbit-adjust propulsion system for three NASA Atmospheric Explorer Satellites launched in 1973, 1974 and 1975.

TRW's propulsion research programmes include low-thrust monopropellant, bipropellant, colloid, ion, radioisotope and electro-thermal engines. In addition, an active research programme in low-cost propulsion technology is being continued. Tests of TRW rocket engines are conducted at the company's test site at San Juan Capistrano, California.

TRW TR-201 (DELTA)

A bipropellant engine designed for vacuum operation, the TR-201 serves as propulsion of the second stage of the NASA/McDonnell Douglas Delta launch vehicle. This is one of NASA's standard production launch vehicles.

TYPE: Liquid bipropellant rocket engine.

PROPELLANTS: Nitrogen tetroxide and 50/50 mix of hydrazine and UDMH.

THRUST CHAMBER: Single chamber. Area ratio 43. Chamber of quartz phenolic construction with ablative cooling. Nozzle of columbium, with radiation cooling. Coaxial injector with hypergolic ignition. Starting by 28V electrical signal to on/off solenoid valves.

THRUST CHAMBER MOUNTING: Gimbal attachment above injector.

PROPELLANT FEED SYSTEM: Pressure feed system by McDonnell Douglas Astronautics. Gas pressure 15·51 bars (225 lb/sq in). Flow rate 5·62 kg (12·4 lb)/s fuel and 8·92 kg (19·7 lb)/s oxidant.

DIMENSIONS:

Length overall	2,156 mm (84·9 in)
Nozzle diameter	922 mm (36·3 in)

WEIGHT, DRY: 113 kg (250 lb)

PERFORMANCE RATING:

Max thrust (vacuum)	43·6 kN (9,800 lb)
Combustion pressure	7·03 bars (102 lb/sq in)
Combustion temperature	2,700°C
Specific impulse	302

TRW TR-201 Delta engine, with vacuum thrust of 43·6 kN (9,800 lb)

VOUGHT

VOUGHT CORPORATION

HEAD OFFICE: PO Box 225907, Dallas, Texas 75265

Telephone: (214) 266 2011

OFFICERS: See Aircraft section

VOUGHT STM

This integral rocket/ramjet vehicle, the first produced under the US Navy's Supersonic Tactical Missile programme, made its initial flight on 21 April 1979, continuing an unbroken string of successes accomplished under the previous Low-Volume Ram Jet (LVRJ) programme. In the STM programme the ramjet vehicles incorporate new low-cost propulsion features and guidance elements in the continuing development of an advanced high-performance long-range missile powered all the way to its target. In the predecessor LVRJ programme, five successful flights were made with range exceeding 87 nm (161 km; 100 miles) and sustained speeds of approx 1,735 knots (3,220 km/h; 2,000 mph). The programme is referred to in the entry for CSD in this section, under the heading 'air-breathing propulsion'.

Vought STM carried by A-7E Corsair

WILLIAMS

WILLIAMS INTERNATIONAL CORPORATION

2280 W Maple Road, Walled Lake, Michigan 48088

Telephone: (313) 624 5200

PRESIDENT: Sam Williams

VICE-PRESIDENTS:
- E. L. Klein (Executive V-P and Chief Operating Officer)
- John Jones (Technical)
- Robert Haas (Engineering)
- Robert Katz (Finance)
- Myron Goers (Operations)
- David C. Jolivette (Public Relations)

Sam Williams believed in 1955 that gas-turbine technology could be extended down to very small sizes, and that if a small turbojet were made available it would find a market. The WR2 first ran at a thrust of 0·31 kN (70 lb) in 1962 and has since been developed into the WR2-6 and WR24-6. The more advanced WR19 uses an aerodynamically similar core and Williams is also building a range of shaft-drive engines.

Versions of the US Air Force and Navy cruise missiles, ALCM and Tomahawk, are propelled by the F107 turbofan. New engine production facilities for the F107 have been established in Ogden, Utah, and in September 1978 Teledyne CAE was named as second-source producer, with mass production due in early 1982. In early 1981 the Ogden plant was expanded.

WILLIAMS WR2 and WR24

US military designation (WR24): J400

Air enters at the eye of a single-sided light alloy centrifugal compressor which handles an air mass flow of 1 kg (2·2 lb)/s at a pressure ratio of 4·1 : 1. After passing through the diffuser which provides the structural basis for the engine the air divides, part of it flowing radially inwards as primary combustion airflow and the main bulk entering the short outward-radial annular combustor, through dilution apertures around the outer and rear face of the flame tube.

Fuel is sprayed centrifugally through a group of fine holes in the main compressor drive shaft. Surrounding the fuel pipe along the centreline of the main drive shaft is a cool airflow bled from the diffuser, which escapes through holes in the drive shaft to cool the combustion flames and reduce metal shaft and bearing temperatures, the main bearing being behind the compressor. A single igniter is mounted in the chamber at 12 o'clock. The hot gas, at about 955°C, then turns inwards and exits rearwards through the single-stage axial turbine and simple jet pipe.

The first production versions of the WR2 are the WR2-6, fitted to the Canadair AN/USD-501 high-performance battlefield reconnaissance vehicle; and the WR24-6 and -7 (YJ400-WR-400 and J400-WR-401) which power, respectively, the Northrop Chukar I and II target drones. The WR2-6 has a variable-area exhaust nozzle with translating central bullet, and drives a DC generator. The WR24 family have a minimal fixed-area jet pipe and drive a 4,000 Hz alternator. The WR24-7 runs at higher temperature than the WR24-6 and incorporates detail modifications which increase mass flow to 1·36 kg (3 lb)/s and pressure ratio to 5·3. The WR24-17, not yet in production, is further uprated though similar externally. More than 5,000 small jets had been delivered by 1981.

DIMENSIONS:

Overall length:	
WR2-6	566 mm (22·3 in)
WR24-6	490 mm (19·3 in)
WR24-7, WR24-17	about 635 mm (25 in)
Max diameter:	
WR2-6, WR24-6	274 mm (10·8 in)
WR24-7, WR24-17	about 305 mm (12 in)

Left: Williams WR2-6 turbojet, for the Canadair CL-89 (AN/USD-501) reconnaissance drone (0·56 kN; 125 lb st). Right: Williams WR24-7 turbojet (0·76 kN; 170 lb st); the 0·89 kN (200 lb st) WR24-17 is visually identical

Weight, Dry:

WR2-6, WR24-6	about 13·6 kg (30 lb)
WR24-7	20 kg (44·0 lb)

Maximum Ratings (S/L, static):

WR2-6	0·56 kN (125 lb) at 60,000 rpm
WR24-6	0·54 kN (121 lb) at 60,000 rpm
WR24-7	0·76 kN (170 lb)
WR24-17	0·89 kN (200 lb)

Specific Fuel Consumption:

WR2-6, WR24-6	35·41 mg/Ns (1·25 lb/h/lb st)

WILLIAMS WR19

To produce this two-shaft turbofan Williams used the WR2 as core and added an additional fan, axial compressor and drive turbine on a separate shaft, together with a by-pass duct. The LP turbine is related to those developed for the company's shaft-drive engines.

The WR19 was the power plant used in the Bell Aerosystems Flying Belt. It has also been used in the Williams Aerial Systems Platform (WASP) and Kaman Stowable Aircrew VEhicle Rotoseat (SAVER). From it has been derived the US Air Force/Navy F107, and the WR44 for general aviation, described separately.

In early 1970 the company received a $1,400,000 contract from the US Air Force for further development of a turbofan for future decoys. The company is making great efforts to increase the maximum gas temperature, particularly in the WR19 and derived engines. At present the temperature actually used is about 955°C, with potential of the present materials (Haynes 31 cobalt-base alloy for inlet guide vanes, Inco 100 for first-stage turbine blades and Inco 713 for other hot parts) limited to about 1,010°C.

Despite the mechanical difficulty of working on such small components, with turbine rotor discs and blades cast as single units, Williams is experimenting with aircooled turbine rotor blades and expects soon to be able to operate at gas temperatures higher than 1,100°C. The WR19 would be the first engine offered with cooled blades, and it also continues the company philosophy of using specially developed alternators, governors and other accessories capable of running at the full 60,000 rpm of the main shaft.

Air Intake: Direct pitot type with four struts but no fixed inlet guide vanes. Unlike most WR2 engines the WR19 has a plain annular entry instead of a side intake downstream of an alternator or generator on the nose of the main shaft.

Compressor: Two-stage metal fan and two-stage axial IP compressor on common shaft leading to HP centrifugal compressor, handed to rotate in opposite direction to minimise gyroscopic couple. Total air mass flow, about 2 kg (4·4 lb)/s; overall pressure ratio, 8·1; by-pass ratio, approximately 1 : 1.

Combustion Chamber: Folded annular type, with fuel sprayed from revolving slinger on HP shaft. Dilution airflow admitted through perforated liner; cooling air injected through two sets of holes in HP shaft. Single igniter mounted diagonally on engine upper centreline.

Fuel System: Fuel fed at low pressure through transfer seal into pipe in HP shaft and ejected at high centrifugally-induced pressure, through calibrated fine orifices drilled radially through HP shaft in line with combustion chamber.

Turbine: Single-stage axial-flow HP turbine, with Haynes 31 nozzle guide vanes and rotor wheel cast as single unit in Inco IN 100. Two-stage LP turbine, again with both wheels cast as single units, in Inco 713. Provision to be made for aircooling to raise entry gas temperature from 955°C to above 1,100°C.

Jet Pipe: Mixer unit immediately downstream of LP turbine allows bypass flow to merge with core gas flow to pass through plain propelling nozzle.

Accessories: Fuel and control system, filters, oil pump, tacho-generator and optional other accessories grouped into flat packages around upper part of fan/IP compressor casing. Starting system, depending on application, drives HP spool.

Mounting: Depending on application, main mounting above centrifugal diffuser casing with two double-lug pickups on horizontal centreline at LP turbine casing.

Dimensions:

Length overall	787 mm (31·0 in)
Width	338 mm (13·31 in)
Height	407 mm (16·20 in)

Weight, Dry:

With starter/generator	64 kg (141 lb)

Performance Rating (S/L, static):

Standby/T-O (wet)	3·19 kN (718 lb st)
Maximum continuous	2·22 kN (500 lb st)

WILLIAMS WR19-A7

US military designation: F107

The F107 two-shaft turbofan is an advanced and uprated WR19 originally designed to propel the US Air Force/Boeing ALCM (Air-Launched Cruise Missile). Rated in the 2·67 kN (600 lb st) class, the F107 also powers the General Dynamics Convair Tomahawk Sea-Launched Cruise Missile, and General Dynamics ALCM. Williams has emerged clear winner in the competition to propel US cruise missiles, simplifying the attempt to maximise commonality exerted since 1977 by the JCMPO (Joint Cruise Missiles Project Office). All F107 versions are basically similar, but differ in mission and equipment, as follows:

F107-WR-100. Also designated WR19-A7, this was the engine selected for the Boeing AGM-86A ALCM, making the first ALCM flight at White Sands on 5 March 1976. Prototype engines only.

F107-WR-101. Selected to power the Boeing AGM-86B (ALCM-B). Improved configuration and performance. Qualification testing began in October 1978 and production deliveries in Spring 1981.

F107-WR-102. This engine powered the General Dynamics AGM-109 Tomahawk ALCM.

F107-WR-400. Selected to power the General Dynamics BGM-109 SLCM (Sub/Ship-Launched Cruise Missile) and the GLCM (Ground-Launched Cruise Missile) for the US Air Force, all versions of Tomahawk. An early -400 engine powered the first Tomahawk to fly (air-launched by A-6A) on 5 June 1976.

Type: Two-shaft turbofan.

Air Intake: Direct pitot type.

LP Compressor: Two-stage fan coupled to two-stage IP compressor.

HP Compressor: Single-stage centrifugal.

Combustion Chamber: Folded annular with rotary fuel injection.

Turbine: Single-stage HP, two-stage LP.

Accessories: Grouping varies with sub-type. Self-contained lubrication system. Solid-propellant gas-impingement starter.

Dimensions:

Length overall:

F107-WR-100	800 mm (31·5 in)
F107-WR-101	1,232 mm (48·5 in)
F107-WR-102, -400	937 mm (36·9 in)
Envelope diameter	305 mm (12 in)

Weight, Dry:

F107-WR-100	58·7 kg (130 lb)
F107-WR-101	64·0 kg (141 lb)
F107-WR-102	65·8 kg (145 lb)
F107-WR-400	64·4 kg (142 lb)
Performance Rating:	2·67 kN (600 lb st) class

WILLIAMS WR44

Development of the WR44 began in 1971, using the WR19 as a basis. By-pass ratio was increased to 3, and overall pressure ratio increased. Claimed to be outstandingly quiet and suitable for general-aviation applications, the WR44 is no longer under contract for the Foxjet but is planned for one version of Gulfstream American Hustler and the Rockwell VTX (US Navy trainer) proposal.

Williams F107-WR-400 two-shaft turbofan for BGM-109 Tomahawk SLCM and GLCM

Williams F107-WR-101 two-shaft turbofan for AGM-86B Air-Launched Cruise Missile

Williams WR34 turboshaft

Dimensions:

Length (with tailpipe)	1,143 mm (45·0 in)
Max diameter	419 mm (16·5 in)

Weight, Dry: 87·5 kg (193 lb)

Performance Rating (S/L):

T-O	3·78 kN (850 lb st)

WILLIAMS WR34

This family of simple turboshaft engines is aimed at a wide spectrum of applications, including aviation. Based on a centrifugal compressor and radial inflow turbine, it is robust and reliable, and has unchanged configuration over power outputs from 11-26 kW (15-35 hp) at output speeds of 3,600, 6,000, or 12,000 rpm. Equipped weight is 29·5 kg (65 lb). One version, the WR34-15, will power the Canadair CL-227 (see RPVs and Targets section). This model weighs only 17·25 kg (38 lb), is rated at 24 kW (32 hp), and is 239 mm (9·4 in) in diameter by 450 mm (17·7 in) long.

ADDENDA

AIRCRAFT

AUSTRALIA

AUSTRALIAN AIRCRAFT CONSORTIUM AUSTRALIAN BASIC TRAINER

Australian government approval was expected in late 1981 for development of a new turboprop military basic training aircraft, to replace the New Zealand Aerospace Industries CT4A Airtrainer in RAAF service in the late 1980s, and for export. Project definition of this aircraft, to meet AFST 5044 (Air Force Staff Target 5044) has already been completed by AAC, a consortium formed by Commonwealth Aircraft Corporation, Government Aircraft Factories, and Hawker de Havilland Australia. The initial contract is likely to cover three prototypes and two structure test airframes, followed by 69 production trainers.

As the accompanying illustration shows, the aircraft is a cantilever low-wing monoplane, with a retractable tricycle landing gear and side-by-side seating for the two occupants (instructor and pupil). The cockpit will be equipped with IFR instrumentation, and will be air-conditioned and soundproofed; the airframe makes extensive use of corrosion-resistant materials, corrosion-proofing and weatherproofing. Power plant will be a single turboprop engine (chief candidates are the Pratt & Whitney Aircraft of Canada PT6A-25 and Garrett TPE331-1 or TPE331-6; possible alternatives include the Alfa Romeo AR.318 and Avco Lycoming LTP 101), flat rated to about 410 kW (550 shp). The four underwing stores points are stressed for loads of up to 250 kg (551 lb) each inboard and 150 kg (330 lb) each outboard.

Detail design is expected to be completed by March 1983, with a first flight in December that year and entry into RAAF service in the Spring of 1987. Export versions, possibly available before then, could be configured for weapons training, light close air support, reconnaissance/surveillance, and forward air control.

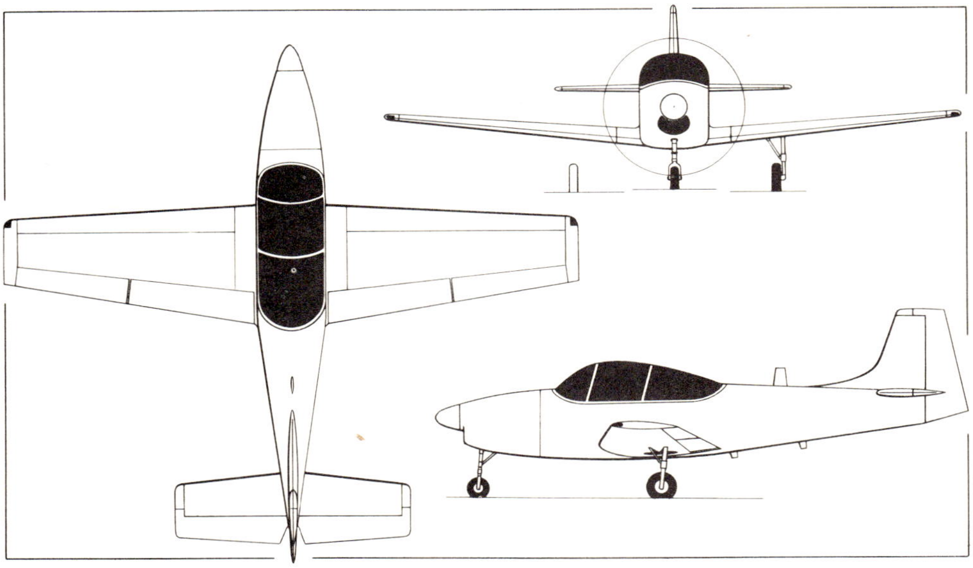

General configuration of the new Australian basic trainer designed by CAC, GAF and HDH *(Pilot Press)*

DIMENSIONS, EXTERNAL:	
Wing span	11·00 m (36 ft 1 in)
Wing chord: at c/l	2·42 m (7 ft 11¼ in)
at tip	2·21 m (7 ft 3 in)
Wing aspect ratio	6·05
Length overall	10·10 m (33 ft 1¾ in)
Height overall	3·70 m (12 ft 1¾ in)
Elevator span	4·50 m (14 ft 9¼ in)
Wheel track	4·30 m (14 ft 1¼ in)
Wheelbase	3·30 m (10 ft 10 in)
Propeller diameter	2·30 m (7 ft 6½ in)
AREAS:	
Wings, gross	20·00 m² (215·3 sq ft)
Vertical tail surfaces (total)	2·90 m² (31·2 sq ft)
Horizontal tail surfaces (total)	5·40 m² (58·1 sq ft)
WEIGHTS:	
Weight empty, equipped	1,390 kg (3,065 lb)
Max training T-O weight	2,000 kg (4,409 lb)
Max T-O weight, alternative roles	2,600 kg (5,732 lb)
PERFORMANCE (estimated, at max training T-O weight, RAAF Mean Tropical Atmosphere):	
Never-exceed speed	280 knots (519 km/h; 322 mph) EAS
Max cruising speed at S/L	195 knots (361 km/h; 224 mph)
Max speed in gusty conditions	180 knots (333 km/h; 207 mph) EAS
Approach speed	80 knots (148 km/h; 92 mph)
Stalling speed: flaps up	62 knots (115 km/h; 72 mph)
flaps down	57 knots (106 km/h; 66 mph)
Max rate of climb at S/L	579 m (1,900 ft)/min
Time to 3,050 m (10,000 ft)	less than 6 min
Service ceiling	9,145 m (30,000 ft)
T-O to, and landing from, 15 m (50 ft)	less than 500 m (1,640 ft)
Design endurance, 50 min reserves	3 h
Design *g* limits	+6·0/−2·5

BRAZIL

SÃO CARLOS (page 18)

IPAI-27 Jipe Voador, modified from the IPT SP-18 by the São Carlos Engineering School's Aeronautical Division *(Roberto Pereira de Andrade)*

FRANCE

DASSAULT-BREGUET (page 55)

Dassault-Breguet is among the major French companies designated for nationalisation by the nation's new Socialist administration. Others include the military sectors of Matra and Thomson-Brandt. Transfer to state ownership is not intended to change the identity, autonomy, executive appointments or programmes of these companies.

DASSAULT-BREGUET MIRAGE F1

The Qatar Emiri Air Force has ordered Mirage F1s, including two-seaters.

HELICOP-JET (page 67)

News was received in mid-August that R and D on the Helicop-Jet 'cold-jet' tip-driven light helicopter are completed. The final prototype was nearing completion at that time, and efforts were being made to find industrialists willing to finance the establishment of a production line.

MUDRY (page 68)

MUDRY CAP X

The prototype of this two-seat light aircraft is reported to be powered initially by a Volkswagen conversion, pending certification of the MB-4-80 engine, and to be of all-wood construction to facilitate the embodiment of any changes found desirable during flight testing. Production CAP Xs will make extensive use of composite materials, including carbonfibre for wing spars and fuselage longerons, and honeycomb sandwich for control surfaces. Fuel capacity is 70 litres (15·4 Imp gallons). Additional specification data follow:

DIMENSIONS, EXTERNAL:	
Wing area	9·00 m² (96·9 sq ft)
Wing aspect ratio	7
Height overall	2·05 m (6 ft 8¾ in)
Propeller diameter	1·55 m (5 ft 1 in)
WEIGHTS:	
Weight empty	290 kg (640 lb)
PERFORMANCE (estimated):	
Max cruising speed at 1,500 m (4,920 ft)	102 knots (190 km/h; 118 mph)
Stalling speed	45 knots (82 km/h; 51 mph)

GERMANY (FEDERAL REPUBLIC)

DORNIER (page 76)

DORNIER/BREGUET Br.1150 ATLANTIC KWS

The Federal German Navy has operated 20 Br 1150 Atlantics since 1968, with 15 of them dedicated to the maritime reconnaissance role. Early deployment revealed serious corrosion problems, which were overcome by the adoption of new honeycomb cores and the use of more effective bonding materials. This increased the life expectancy of the aircraft to around 10,000 hours which, at their normal rate of utilisation, meant that they could remain in operational service until the early 1990s. To maintain their capability throughout such a period, Dornier was awarded a contract valued at approximately DM 170 million by the Federal Defence Technology and Procurement Agency to install more advanced operational equipment in the aircraft, as mentioned on page 77 of this edition.

Under the programme, Dornier is responsible for integrating into the Atlantics equipment from a number of US and European subcontractors. Loral Corporation of Yonkers, New York, is supplying advanced ESM which is carried in wingtip pods; the navigational equipment was developed by Litton Industries and Decca; long-range high-definition unjammable radar by Texas Instruments of Dallas, Texas; sonar systems by Emerson Electric, St Louis, Missouri; an IRIG-standard tape recorder system by Bell & Howell GmbH of Friedberg; and sonobuoy launch and storage equipment by Dornier.

The programme, which began in 1978, is providing the German Navy's Atlantics with much-enhanced capability. This comes not only from the improved radar, but from ESM systems that incorporate automatic analysis, a greater frequency range, and high angular accuracy; plus a sonar system with increased capacity and a wider frequency response that is far more effective for the passive location of underwater targets. To speed the updating of the aircraft, much 'off the shelf' equipment is being used, and it is anticipated that all 15 aircraft will be fully operational by 1983.

A description of the basic Breguet Br.1150 Atlantic can be found in the 1973-74 *Jane's*.

Breguet Atlantic after modification by Dornier under the KWS programme

MBB (page 82)

MBB BO 105

Under a programme sponsored by the Federal German Ministry of Research and Technology, MBB is developing as a joint venture with various equipment manufacturers a version of the BO 105 helicopter with a stabilised mast-mounted visual aid that allows day and night observation in flight.

Intended originally for use on the projected Franco-German PAH-2 anti-tank helicopter, the system is now being developed for military scout missions and civil rescue operations by night. It was first flown on a BO 105 (D-HABV) on 21 May 1981. Because of the high viewpoint of the new sensor package, 0·89 m (2 ft 11 in) above the rotor plane, this helicopter is known sometimes as 'Giraffe'. Weight of the complete package above the rotor hub is 115 kg (254 lb).

The sensors are contained in a spherical housing, carried on a mast that passes through the rotor head and is rigidly attached to the airframe. Thus, the sight does not turn with the rotor, but its weight is supported through a thrust bearing by the rotor head, which means that the installation is not weight limited. The housing contains a package developed by SFIM as Ophelia (Optique Platforme HÉLIcoptère Allemand), comprising a two-axis stabilised platform carrying FLIR and TV cameras, and a laser rangefinder, with provisions for TV and infra-red tracking. The sphere can be traversed through ±120° in azimuth and −30°/+20° in elevation. Associated equipment in the helicopter includes a sensor steering stick on the centre console, laser rangefinder control stick on the port door pillar, FLIR and TV electronics, sensor control unit, and a 20 cm (8 in) monitor.

During the development programme the electro-optical sensor images are being tested in conjunction with various display systems. These include VDO head-up and head-down displays that are each able to provide IR and TV images using TV raster techniques, with superimposed symbology generated by stroke-writing techniques. A Thomson/CSF head-down display, able to present coloured areas and symbology in 15 different colours, is also being evaluated, as is a Ferranti helmet-mounted sight and display, used in conjunction with a second stabilised platform beneath the nose of the helicopter, carrying a wide-angle FLIR sensor for low-level navigation.

MBB BO 105 scout helicopter with SFIM Ophelia mast-mounted sight *(Air Portraits)*

INTERNATIONAL PROGRAMMES

AERITALIA/AERMACCHI/EMBRAER (page 91)

AM-X INTEGRATION MANAGER: Ing Giorgio Danieli (Aeritalia)

AM-X

This single-seat combat aircraft is under development to meet the requirements of the Italian and Brazilian Air Forces. Work on the project, at that time for the Italian Air Force only, was started by Aeritalia in mid-1977. Aermacchi became an associate in the programme in mid-1978, the AM-X designation signifying Aeritalia/Macchi—Experimental. A non-afterburning version of the Rolls-Royce Spey Mk 807 turbofan engine was selected by the Italian Air Force as the AM-X power plant in October 1978.

In March 1980, following completion of an 18-month project definition phase ordered by the Italian Air Force, the Brazilian government announced its intention of taking part in the AM-X programme, and in July of that year EMBRAER became an industrial partner of the two Italian companies. The development phase, initiated in January 1981, was followed two months later by an initial memorandum of understanding between the two air forces concerned.

Primary roles of the AM-X are short/medium-range interdiction, close air support and reconnaissance, with secondary capability for anti-shipping attack and counter-air duties. Its design, coupled with sophisticated avionics and other airborne systems, is expected to meet effectively all tactical and operational requirements of its operators during the late 1980s and 1990s. It will be capable of carrying out these missions under exacting operational conditions, at high subsonic speed and very low altitude, by day and by night, in poor visibility, if necessary from bases with poorly-equipped or damaged runways. Basic design requirements included good take-off and landing performance; high subsonic speed with good penetration capability; an internally-mounted gun; wingtip positions for Sidewinder-type air-to-air missiles; underwing pylons for external stores; comprehensive avionics systems; and a proven, in-production power plant requiring a minimum of modification.

In the Italian Air Force the AM-X is intended to take over duties performed currently by four types of aircraft: the G91, which will be phased out of its close air support role by 1985/86; the G91Y interdictor, also due for phase-out by 1985/86; and the F-104G and F-104S versions of the Starfighter, scheduled for replacement in the strike role by 1986/87 and 1990 respectively. The close air support and interdiction tasks will be undertaken fully by the AM-X; counter-air duties will be shared with the longer-range Tornado. A total of 187 aircraft, enough for eight squadrons, is needed to meet these re-equipment plans, with deliveries beginning ideally in late 1986, to ensure initial operational capability by the second half of 1987.

The Brazilian Air Force requirement is for 144 aircraft. These will differ primarily in avionics and weapon delivery systems, and will have two internally-mounted 30 mm DEFA cannon instead of the single multi-barrel 20 mm weapon of the Italian version. Procurement is planned over the six-year period 1984-89, at the rate of 24 per year.

Six prototypes are being built, including two airframes for static and fatigue testing. First flight is scheduled for late 1983/early 1984, and two flying prototypes will be assembled and flown in each country. All three participating companies are involved equally in both programme management and aircraft development. EMBRAER will build the wings and landing gear; fuselages and tail units will be manufactured by the two Italian partners. Engines will be licence-built in Italy by Fiat Aviazione and Alfa Romeo. There will be final assembly lines in both Italy and Brazil.

TYPE: Single-seat interdiction, close air support and reconnaissance aircraft, with secondary capability for anti-shipping attack and counter-air.

WINGS: Cantilever shoulder-wing monoplane, with 31° sweepback on leading-edges and thickness/chord ratio of 12%. Three-spar torsion-box structure, machined from solid aluminium alloy with integrally-stiffened skins. Three-point attachment to fuselage. Leading-edge slats (two segments each side) over most of span. Hydraulically actuated two-segment Fowler-type double-slotted flaps, of carbonfibre, over approx two-thirds of each trailing-edge. Forward of each pair of flaps is a pair of electrically controlled, hydraulically actuated spoilers, which are operated separately in inboard and outboard pairs. Hydraulically actuated ailerons, with manual reversion. No tabs.

FUSELAGE: Conventional semi-monocoque oval-section structure, built chiefly of aluminium alloy. Forward section incorporates main avionics and equipment bays, airborne systems and cockpit; central section includes main landing gear bays; rear section houses power plant and some items of equipment. Extreme rear fuselage, complete with tailplane, is detachable for access to engine. Small ventral strake under each side of rear fuselage.

TAIL UNIT: Sweptback fin and carbonfibre rudder. Variable-incidence tailplane, mid-mounted on fuselage, with carbonfibre elevators. No tabs. Fly-by-wire control of rudder. Hydraulically actuated elevators, with manual reversion.

LANDING GEAR: Retractable tricycle type, of levered-suspension design, with single wheel on each unit.

Nosewheel has hydraulic steering. Hydraulic extension and retraction, nose unit retracting forward, main units forward and inward into underside of engine air intake trunks. Main-wheel tyre pressure 9·65 bars (140 lb/sq in). Hydraulic brakes and anti-skid system.

Power Plant: One 49·1 kN (11,030 lb st) Rolls-Royce Spey Mk 807 non-afterburning turbofan engine, with lateral intakes. Fuel in compartmented fuselage tank and two integral wing tanks. Single-point pressure refuelling, via standard NATO connector. Provision for auxiliary underwing fuel tanks of up to 1,000 litres (220 Imp gallons) capacity each on inboard underwing pylons, 455 litres (100 Imp gallons) capacity on outboard pylons.

Accommodation: Pilot only, on Martin-Baker Mk 10L zero-zero ejection seat. One-piece wraparound windscreen and one-piece canopy. Cockpit pressurised and air-conditioned. Two-seat training version under study.

Systems: Environmental control system (ECS) provides air-conditioning of cockpit and reconnaissance pallets, cockpit pressurisation, air intake de-icing, windscreen demisting, and anti-*g* systems. Duplicated redundant hydraulic systems, driven by engine gearboxes, operate at pressure of 207 bars (3,000 lb/sq in) for actuation of primary flight control system, flaps, spoilers, landing gear, wheel brakes, anti-skid system, nosewheel steering and gun purging. Primary electrical system AC power (115/200V at fixed frequency of 400Hz) supplied by two 30kVA IDG generators, with two transformer-rectifier units for conversion to 28V DC; 36Ah nickel-cadmium battery for emergency use, to provide power for essential systems in the event of primary and secondary electrical system failure.

Avionics and Equipment: In terms of redundancy and monitoring, avionics system is designed to permit successful completion of mission, even in the event of initial failure. Modular design and space provisions within the aircraft will permit retrofitting of alternative systems as and when required. All avionics/equipment packages are pallet-mounted to facilitate removal and replacement, and are positioned to allow rapid access for routine maintenance and change of configuration. Avionics are divided into six main subsystems: UHF and VHF com/IFF; autonomous (INS) and assisted (VOR) navigation; computer-based weapon aiming and delivery, incorporating a range-only radar and stores management subsystem; data display (head-up, multi-function head-down, and weapons/nav); data processing, with air data computer; and active and passive ECM.

Armament and Operational Equipment: One M61A1 multi-barrel 20 mm cannon in lower forward fuselage (two 30 mm DEFA 553 cannon in aircraft for Brazilian Air Force). Single twin-pylon stores attachment point under fuselage, on centreline, plus two attachments under each wing, and wingtip rails for two AIM-9L Sidewinder or similar infra-red air-to-air missiles. Fuselage and inboard underwing hardpoints each stressed for loads of up to 907 kg (2,000 lb), outboard underwing points for 454 kg (1,000 lb) each. Total external stores load 3,800 kg (8,377 lb). Attack weapons can include bombs, air-to-surface missiles (including anti-shipping weapons) and rocket launchers. Three alternative pallet-mounted photographic systems can be carried, installed internally in forward fuselage or in an external pod. Each of these systems is fully compatible with the aircraft, and will not affect operational capability; the aircraft will therefore be able to carry out reconnaissance missions without effect upon its normal navigation, attack and self-defence capabilities. Camera bay is in lower starboard side of fuselage, forward of main-wheel bay. Provision is made to replace camera pallet with a laser ranger and marked target seeker module if required.

Dimensions, external:

Wing span: over missiles	10·00 m (32 ft 9¾ in)
excl wingtip missiles and rails	8·88 m (29 ft 1½ in)
Wing area, gross	21·00 m² (226·04 sq ft)
Wing aspect ratio	3·75
Wing taper ratio	0·5
Length overall	13·57 m (44 ft 6¼ in)
Length of fuselage	12·55 m (41 ft 2 in)
Height overall	4·58 m (15 ft 0¼ in)
Tailplane span	approx 5·20 m (17 ft 0¾ in)
Wheel track	2·15 m (7 ft 0¾ in)
Wheelbase	approx 4·75 m (15 ft 7 in)

Weights:

Operating weight empty	6,500 kg (14,330 lb)
Max external stores load	3,800 kg (8,377 lb)
Normal mission T-O weight	10,500 kg (23,148 lb)
Max T-O weight	12,000 kg (26,455 lb)

Performance (estimated):

Required runway length at normal mission T-O weight 915 m (3,000 ft)

Design radius for interdiction, lo-lo-lo with 1,360 kg (3,000 lb) of external weapons:

with two 455 litre (100 Imp gallon) drop-tanks 180 nm (333 km; 205 miles)

with two 1,000 litre (220 Imp gallon) drop-tanks 520 nm (963 km; 598 miles)

AIRBUS (page 93)

Senior Vice-President: Pierre G. Pailleret (Marketing)

M Pailleret replaces Mr George Warde who has left Airbus Industrie.

AIRBUS A300

Revisions to lists of orders and options, reflecting conversions of options into firm orders.

	Orders	*Options*
Egyptair (B4-203)	7	—
Indian Airlines (B2-100)	8	1
(B4-200)	2	—

SAAB-FAIRCHILD (page 106)

SAAB-FAIRCHILD 340

The following amended data were received just as pages 106-107 were closing for press:

Areas:

Wings, gross	41·81 m² (450·0 sq ft)
Ailerons (total)	2·03 m² (21·88 sq ft)
Trailing-edge flaps (total)	8·25 m² (88·80 sq ft)
Fin, incl dorsal fin	10·405 m² (112·00 sq ft)
Rudder, incl tab	2·72 m² (29·27 sq ft)
Tailplane	13·30 m² (143·16 sq ft)
Elevators (total, incl tabs)	3·46 m² (37·24 sq ft)

Weights and Loadings:

Max fuel	2,676 kg (5,900 lb)
Max T-O and landing weight	11,793 kg (26,000 lb)
Max zero-fuel weight	10,432 kg (23,000 lb)
Max wing loading	282·1 kg/m² (57·8 lb/sq ft)
Max power loading	4·7 kg/kW (7·76 lb/shp)

Performance (estimated at max T-O weight, ISA):

As given on page 107 except:

Max cruising speed at 4,570 m (15,000 ft) 270 knots (500 km/h; 311 mph)

Stalling speed:

0° flap	102 knots (189 km/h; 118 mph)
T-O flap setting	90 knots (167 km/h; 104 mph)
approach flap setting	84 knots (156 km/h; 97 mph)
landing flap setting	79 knots (147 km/h; 91 mph)

Max rate of climb at S/L	549 m (1,800 ft)/min
Rate of climb at S/L, one engine out	183 m (600 ft)/min
Service ceiling, one engine out	3,505 m (11,500 ft)

FAR Pt 25 required T-O field length:

at S/L, ISA	1,220 m (4,000 ft)
at S/L, ISA + 15°C	1,340 m (4,400 ft)
at 1,525 m (5,000 ft), ISA	1,585 m (5,200 ft)
at 1,525 m (5,000 ft), ISA + 15°C	1,830 m (6,000 ft)
Min ground turning radius	15·17 m (49 ft 9·2 in)

Range, allowances for 130 nm (240 km; 150 mile) diversion and 45 min hold:

with max payload 900 nm (1,665 km; 1,035 miles)

with max fuel 1,640 nm (3,035 km; 1,885 miles)

MCDONNELL DOUGLAS/FOKKER (page 103)

MCDONNELL DOUGLAS/FOKKER MDF-100

Fokker announced a number of important revisions to the MDF-100 specification on 22 September 1981, made at the request of potential airline customers. The initial design now provides for a twin-aisle cabin layout; replacement of the original twin-wheel main landing gear units by four-wheel bogies; and a T tail instead of the low tail envisaged previously. In-depth studies have shown that the T tail offers significant fuel saving. Requests for proposals for engines to power the MDF-100 have been sent to CFM International, Pratt & Whitney and Rolls-Royce/Japan Aero Engines. Requirements include a T-O rating of 104 kN (23,400 lb st), lower sfc than current production engines, and certification by late 1985.

ITALY

Following meetings at the Ministero Partecipazioni Statali (Ministry of State Participation) in the Summer of 1981, a number of major changes in the structure of the Italian aerospace industry have received government approval. The main changes involved — for which no time-scale had been announced at the time of closing for press — are reported as follows:

Aeritalia (page 118). Will take over Partenavia and the Officine Aeronavali di Venezia; may acquire minority shareholdings in Aermacchi and Meteor. Will sell to Agusta its Nerviano factory, and its controlling interest in the optical instruments and avionics company OMI. May market Partenavia aircraft via Agusta sales organisation.

Aermacchi (page 122). May become part-owned by Aeritalia.

Agusta (page 124). Will acquire BredaNardi, Caproni Vizzola, the Aeritalia factory at Nerviano, and the optronics company OMI. May market Partenavia aircraft on behalf of Aeritalia.

BredaNardi (page 134). Will be taken over by Agusta.

Caproni Vizzola (page 135). Will be taken over by Agusta, through the latter's controlling interest in SIAI-Marchetti.

Meteor (see RPVs and Targets section, page 652). May become part-owned by Aeritalia.

Partenavia (page 137). Will be taken over by Aeritalia.

JAPAN

KAWASAKI (page 143)

KAWASAKI KA-850

It was announced shortly before this edition closed for press that Kawasaki's KA-850 design had been chosen to fulfil the JASDF MT-X trainer requirement. Kawasaki, as prime contractor, will be assisted in the manufacturing programme by Fuji and Mitsubishi, which will each be allocated 30 per cent of the work. Shin Meiwa and Nippi will participate in the five-company joint design group, headed by Mr Hiroki Isozaki of Kawasaki.

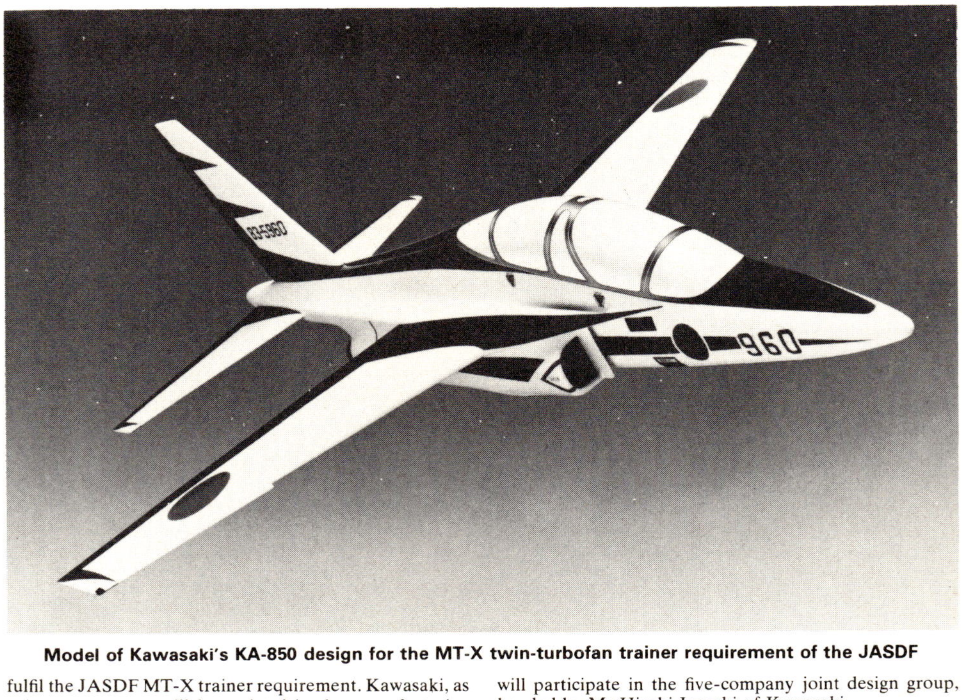

Model of Kawasaki's KA-850 design for the MT-X twin-turbofan trainer requirement of the JASDF

Airframe: Cantilever mid-wing monoplane, with 30° sweptback wings and sweptback tail surfaces. Wings and nosecone built by Fuji, centre-fuselage and tail unit by Mitsubishi, forward fuselage by Kawasaki. Final assembly and flight testing by Kawasaki. Retractable tricycle-type landing gear. Tandem two-seat cockpit, with raised rear seat.

POWER PLANT: Two 15·68 kN (3,530 lb st) Ishikawajima-Harima F3 turbofan engines, mounted side by side in fuselage.

AVIONICS: Will include head-up display.

ARMAMENT: Hardpoints for, typically, four 500 lb bombs or two 454 or 568 litre (120 or 150 US gallon) drop-tanks underwing, and one 757 litre (200 US gallon) drop-tank or a gun pod under the fuselage. Provision for two wingtip-mounted infra-red air-to-air missiles.

DIMENSIONS, EXTERNAL:
Wing span 10·00 m (32 ft 9¾ in)
Wing area, gross 20·40 m² (219·6 sq ft)
Length overall 12·00 m (39 ft 4½ in)

WEIGHTS (approx):
Max T-O weight:
trainer, 'clean' 5,500 kg (12,125 lb)
with external stores 7,300 kg (16,094 lb)

PERFORMANCE (estimated):
Max level speed Mach 0·9
Max cruising speed 530 knots (982 km/h; 610 mph)
Econ cruising speed at 7,620 m (25,000 ft)
Mach 0·75 (450 knots; 834 km/h; 518 mph)
Max rate of climb at S/L 3,050 m (10,000 ft)/min
Normal T-O run 800 m (2,625 ft)
Max range:
internal fuel only 700 nm (1,295 km; 805 miles)
with two 454 litre (120 US gallon) auxiliary tanks 900 nm (1,670 km; 1,035 miles)

NETHERLANDS

FOKKER (page 150)

FOKKER F27 FRIENDSHIP

Garuda Indonesian Airlines has ordered six F27 Mk 500 Friendship twin-turboprop transports for delivery in the first half of 1982.

FOKKER F28 FELLOWSHIP

Empire Airlines of Utica-Rome, New York, has ordered its seventh F28 Mk 4000 twin-turbofan airliner, by converting one of its four options into a firm contract.

Despite periodical rumours of an F27 in service with the US Navy, this is the first time that a photograph of the UC-27A has become available *(George Cockle)*

PERU

FUERZA AÉREA PERUANA (FAP)

Ministerio Aeronáutica, Dir Material, Campo de Marte, Lima

With the assistance of Aeronautica Macchi of Italy, the Peruvian Air Force is reported to plan the establishment of an embryo national aircraft industry in Peru. Fourteen Aermacchi M.B. 339 trainers were ordered for the FAP in 1980, and a total of 66 may be acquired eventually to replace current T-33A, T-37B and T-37C aircraft. It is expected that all but the first few of these will be assembled in Peru, with a progressively increased proportion of locally-manufactured components.

POLAND

CNPSL-PZL WARSZAWA (page 158)

PZL-106AS KRUK

To increase the performance of more than 60 PZL-106A Kruk agricultural aircraft operated by Pezetel in Egypt and the Sudan, CNPSL-PZL Warszawa has adapted the design to take a 746 kW (1,000 hp) PZL (Shvetsov) ASz-621R nine-cylinder radial engine instead of the standard 441 kW (592 hp) PZL-3S. The prototype of this uprated version (SP-PBD), designated PZL-106AS, flew for the first time on 19 August 1981.

PZL-106B KRUK

This new version of the Kruk agricultural monoplane has redesigned wings with shortened V bracing struts. The prototype (SP-PKW) flew for the first time on 15 May 1981.

Prototype of the PZL-106B Kruk with redesigned wings *(A. Pryslopski, IL)*

UNION OF SOVIET SOCIALIST REPUBLICS

ANTONOV (page 186)

First photograph of the turboprop-powered Antonov An-3 agricultural biplane

MYASISHCHEV (page 214)

Simulated decontamination of a Myasishchev M-4 (NATO 'Bison-B') after exposure to nuclear fallout

UNITED KINGDOM

WALLIS (page 264)

VINTEN WALLIS WA-116W

Until 1981, Wallis autogyros were custom-built by Wg Cdr Wallis for special purposes, and were not available for public sale. However, on 1 March 1981 Wg Cdr Wallis and W. Vinten Ltd, the well-known manufacturer of airborne reconnaissance systems, reached a collaborative agreement for the purpose of exploiting the basic single-seat Wallis design for various civil and military applications such as aerial dispatch, air observation post, photography, crop-spraying and other agricultural duties, reconnaissance, fishery survey and protection, geophysical survey, traffic control, powerline/pipeline inspection, border patrol, and sport or business flying.

The production version, designated WA-116W, is powered by a new Weslake flat-twin engine, and will be manufactured and marketed by Vinten, with Mr David Skellon of Vinten as Project and Marketing Manager. Three prototypes were under construction in mid-1981, of which the first was due to fly in the late Autumn or early Winter of that year. (All of these will be piloted aircraft: trials of a radio controlled version, referred to briefly in the RPVs and Targets section, will follow later.)

Full certification of the WA-116W was anticipated in early 1982, with deliveries to begin shortly afterwards. The following description applies to this version:

AIRFRAME: Generally as described and illustrated in main aircraft section, with streamlined nacelle and windscreen as standard. Rotor blades, designed for unlimited fatigue life, are of laminated bonded steel and Hidulignum, surfaced with plastics-filled Madapolan.

Rotor head is designed specifically to give a light and simple spin-up drive. Rotor vibration is minimal, and aircraft is an extremely stable platform for precision photography and remote sensing. It can be towed on a road trailer behind an average motorcar, or taxied under its own power with the rotor blades secured fore and aft. Main-wheel and nosewheel brakes are fitted, to allow optimum use of rotor spin-up capability. Nosewheel is steerable.

Power Plant: One 56 kW (75 hp) Weslake 65/75 flat-twin engine, driving a two-blade pusher propeller. Standard fuel tank capacity of 50 litres (11 Imp gallons). Increased tankage available optionally.

Avionics and Equipment: According to customer/mission requirements. Photographic equipment can include multispectral camera pack, with horizontal/oblique mounting bracket; Vinten 751 panoramic cross-track/oblique mounting bracket; ciné camera kit; BAe Dynamics infra-red Linescan and Steadyscope binoculars; TV and video camera and transmitter systems. Military equipment can include photographic and reconnaissance systems, ground attack weapon systems, and cable-laying gear; agricultural equipment can include low-volume crop-spraying gear and disease detection equipment. Optional avionics and equipment include 720-channel VHF com radio, night flying conversion kit, electric start facility, uprated alternator power output, engine special silencing kit, detachable Perspex hardtop, long-range fuel tank, baggage pannier, and road trailer.

Dimensions, external:
Rotor diameter 6·15 m (20 ft 2 in)
Length overall, excl rotor 3·38 m (11 ft 1 in)
Height to top of rotor head 1·85 m (6 ft 1 in)
Width over main wheels 1·52 m (5 ft 0 in)
Wheelbase 1·07 m (3 ft 6 in)

Weight:
Weight empty 140·5 kg (310 lb)

Performance (standard aircraft: may vary according to role):
Max cruising speed 100 knots (185 km/h; 115 mph)
Loiter speed 35 knots (65 km/h; 40 mph)
Min speed in level flight 15 knots (28 km/h; 17·5 mph)
Min landing speed 7 knots (13 km/h; 8 mph)
Max rate of climb at S/L 365 m (1,200 ft)/min
Service ceiling 4,570 m (15,000 ft)
T-O run (still air, average weight) 27·5 m (90 ft)
Still-air range, standard fuel, with reserves 260 nm (483 km; 300 miles)
Endurance 4 h

UNITED STATES OF AMERICA

ADAMS INDUSTRIES INC

Aircraft Division: 10001 Erwin Avenue, Detroit, Michigan 48234
Telephone: (313) 921 8211
President: John O. Adams
Marketing Director: Laurin Darrell

THORP T211

This two-seat light aircraft, designed by Mr John Thorp, originated as the Sky Scooter, which flew for the first time in 1946 with a 48·5 kW (65 hp) engine and was last described in the 1950-51 *Jane's*. Several versions were completed subsequently, but plans to market them in kit form did not materialise. Adams Industries now intends to put the type into production as the Thorp T211, with a more powerful engine, for business, recreational and training purposes. The first of 100 T211s planned for sale in 1982 will be assembled from components already in existence.

Type: Two-seat cabin monoplane.

Wings: Cantilever low-wing monoplane. All-metal structure with ribbed skins. Wide-span trailing-edge flaps, and plain ailerons, both with ribbed skins. No trim tabs.

Fuselage: All-metal semi-monocoque structure of light alloy.

Tail Unit: Cantilever structure of light alloy; elevators, tailplane and rudder with ribbed skins. Ground-adjustable tab on rudder.

Landing Gear: Non-retractable tricycle type, with single wheel on each unit.

Power Plant: One 74·5 kW (100 hp) Continental O-200-A flat-four engine, driving a two-blade metal propeller with spinner. Fuel capacity 91 litres (24 US gallons). Oil capacity 5·7 litres (1·5 US gallons).

Thorp T211, marketed by Adams Industries *(J. M. G. Gradidge)*

Accommodation: Two seats side by side beneath rearward-sliding transparent canopy. Baggage capacity 36 kg (80 lb).

Avionics: IFR avionics optional.

Dimensions, external:
Wing span 7·62 m (25 ft 0 in)
Length overall 5·49 m (18 ft 0 in)
Height overall 2·44 m (8 ft 0 in)

Area:
Wings, gross 9·75 m² (105·0 sq ft)

Weights and Loadings:
Weight empty 332 kg (733 lb)
Max T-O weight 576 kg (1,270 lb)
Max wing loading 59·1 kg/m² (12·1 lb/sq ft)
Max power loading 7·73 kg/kW (12·7 lb/hp)

Performance (at max T-O weight):
Max level speed at S/L 137 knots (254 km/h; 158 mph)
Cruising speed (75% power) 104 knots (193 km/h; 120 mph)
Stalling speed, flaps down 34 knots (63 km/h; 39 mph)
Max rate of climb at S/L 290 m (950 ft)/min
Service ceiling 5,790 m (19,000 ft)
T-O run 91 m (300 ft)
Landing run 61 m (200 ft)
Range with max fuel 417 nm (772 km; 480 miles)

BELL (page 293)

BELL MODEL 406 (AHIP)

The US Army announced on 21 September 1981 that Bell Helicopter Textron had been selected as winner of its Army Helicopter Improvement Program (AHIP) competition to develop a near-term scout helicopter. The resulting $148 million multi-year contract calls for Bell to design, modify and test five prototype aircraft. Successful development of these will lead to modification of 720 existing OH-58A Kiowa helicopters to the new Army Scout configuration at an estimated cost of $1,000 million. Brief details of the conversion are given on page 295.

BELL RING FIN ANTI-TORQUE DEVICE

Bell Helicopter Textron announced on 11 September 1981 a new helicopter anti-torque device that has been developed by Mr H. E. Lemont, the company's senior design engineer. It permits an increase of operational speed to over 43 knots (80 km/h; 50 mph) sideways, reduces noise and provides additional safety for ground personnel.

Known as the Ring Fin concept, it consists of a flat ring located near the plane of a less-than-normal-diameter tail rotor. The ring compels educed air to mix with the rotor blade tip vortices to create an expanding wake. This circulation of educed air about the ring, which results from energy normally wasted in the rotor tip vortices, creates ring lift that augments the tail rotor's anti-torque thrust. In addition, the lower tip speed associated with the smaller-diameter tail rotor reduces rotor noise emission. Flight testing on Bell helicopters has shown that this device gives a marked sound reduction in steady-state hover. Comparative FAA flyover test manoeuvres with a standard Bell helicopter indicate a sound reduction of up to 6 dbA for the Ring Fin.

The concept eliminates the normal fin/rotor aerodynamic flow interferences that occur in side and quartering flight with conventional tail rotors. It permits more accurate control, and the possibility of reducing pilot pedal effort. Test flights at more than 43 knots (80 km/h; 50 mph) sideways, and up to 41 knots (76 km/h; 47 mph) backward, were achieved with remaining control margins greater than ten per cent. Favourable ring lift levels of 30 to 50 per cent of the total anti-torque thrust required in hover were measured.

Annotated photograph of model of Bell Model 406 (AHIP)

Bell's Ring Fin anti-torque device under test on a Model 206B helicopter

BELL/SIKORSKY ACAP PROGRAMME

Under the US Army's Advanced Composite Airframe Program (ACAP), Bell Helicopter Textron and Sikorsky Aircraft have each been awarded contracts, valued at $14·9 million and $11·6 million respectively, to initiate Phase 1 of a planned programme to design and develop the prototype of an advanced composite-airframe helicopter.

The purpose of this programme is to achieve the Army's goal of reducing weight and cost, and improving military helicopter characteristics, by demonstrating the application of advanced composite materials. Phase 1 covers design and design support testing, and is scheduled to continue until September 1982. Phase 2, which is still optional, calls for the construction, flight testing, and delivery by September 1984 of prototype aircraft for evaluation by the Army's Applied Technology Laboratory. If this option is exercised, one or both companies will build three airframes, comprising a flight test vehicle, a static test airframe that will serve also for the evaluation of crash survivability, and one for tool-proofing, which will confirm ballistic survivability and airframe repairability. Each competing manufacturer will also design a duplicate aircraft in current conventional metal construction for baseline comparison against the composite-construction aircraft.

Bell's design, which has the designation **D292**, will embody the power plant, transmission and rotor system of the commercial Model 222, and seat a crew of two and two passengers in its initial form. It will have non-retractable tailwheel-type landing gear; estimated gross weight is 3,402 kg (7,500 lb). Selection of a particular composite material for each component of the helicopter will be made on the basis of required strength, toughness, environment, ballistic tolerance, crashworthiness, cost, and manufacturing methods. Thus, graphite will be incorporated in such areas as the fuselage beams and frames, compartment bulkheads, and the forward roof, where high strength and stiffness are needed. Kevlar, which is impact damage resistant, will be used for the fuselage outer skin, cockpit and compartment doors, and the nose. Glassfibre will be used for secondary structures, such as the engine compartment firewalls, rear roof, pylon and engine cowling, and compartment floors. Other materials, such as boron, will be used in special applications.

Sikorsky's design, embodying the power plant, transmission and rotor system of the S-76, has the designation **ACAP**, and will be somewhat larger than Bell's D292, with accommodation for a crew of two and six combat-equipped troops. Its gross weight is estimated at 3,842 kg (8,470 lb), and planned maximum speed is 140 knots (259 km/h; 161 mph). Like the Bell entry, it will have non-retractable landing gear, but of tricycle configuration. Sikorsky is likely to use composite materials similar to those detailed for the D292. (See also page 458.)

Bell has chosen Grumman Aerospace as its major subcontractor for the design and manufacture of various composite components. Sikorsky's team includes Hercules Corporation and Vought Corporation.

FAIRCHILD INDUSTRIES INC (page 351)

FAIRCHILD NEXT GENERATION TRAINER (NGT)

Fairchild Industries has built a full-scale mockup of its NGT design, and will take this on a tour of US Air Force bases as part of the company's campaign to win an estimated $2,300 million contract for a Cessna T-37 replacement. Pilot and engineering reactions gained during this tour will assist the company in finalising its submission, which will compete against proposals from Cessna, General Dynamics, Rockwell International, and Vought teamed with MBB of West Germany.

Fairchild's NGT mockup has a wing span of 9·80 m (32 ft 2 in), and overall length of 9·17 m (30 ft 1 in); the estimated maximum take-off weight of a production example is 2,540 kg (5,600 lb). Although no decision has yet been made regarding the twin-engine power plant of this aircraft, it is believed that the company favours the Garrett TFE76-1075-4 turbofan. The NGT's McDonnell Douglas ACES II ejection seats are similar to those fitted in the F-15, F-16 and A-10.

FAIRCHILD/AMES SCALE NGT

Completion of the detailed mockup represents only one aspect of the considerable efforts being made by Fairchild to secure this contract. More significant, perhaps, is the 62 per cent scale version of the NGT which was rolled out on 28 August 1981, prior to instrumentation and flight testing at Mojave Airport, California. The data recorded in these tests was to be incorporated in Fairchild's submission for the NGT contract.

The scaled NGT was constructed with the assistance of Mr Burt Rutan, who completed the detailed scaling from Fairchild's NGT lofting drawings, provided technical details for the aircraft's composite construction, and was also to carry out the initial flight tests. The scaled NGT was constructed by Ames Industrial Corporation of Bohemia, New York, and all available details follow:

Fairchild/Ames scaled New Generation Trainer *(Howard Levy)*

TYPE: Scale version of projected two-seat military training aircraft.

WINGS: Cantilever shoulder-wing monoplane of all-composite construction. Anhedral 4°. One-piece structure comprises a foam core, with top and bottom channels for inset spars of graphite fibres at 67 per cent chord, and skinned with two plies of unidirectional glassfibre cloth and two plies of unidirectional graphite cloth. In addition, the centre-section has five plies of bidirectional glassfibre cloth, for strength as well as to provide a means of contouring. Trailing-edge flaps and ailerons. No tabs.

FUSELAGE: A load-carrying plywood bulkhead, glassfibre-covered on each side, carries the wing, main landing gear units, and engines. Eight additional frames of urethane foam, each glassfibre-covered on both sides, are retained in place by foam planking, also with glassfibre covering on both sides. This forms a longitudinal series of glassfibre-covered foam boxes, epoxied together and strengthened by two glassfibre-covered foam longerons at cockpit sill height. Urethane foam surrounds this box-like fuselage, shaped and sanded to provide the desired contours.

TAIL UNIT: Cantilever tailplane with elevators, and end-plate fins and rudders, of similar construction to wings. Small anti-balance tab in starboard elevator.

LANDING GEAR: Retractable tricycle type. Nose unit retracts rearward manually; main units retract inward electrically, with manual backup.

POWER PLANT: Two 0·98 kN (220 lb st) Microturbo TRS 18-046 turbojet engines, mounted within ducts beneath the wing roots.

ACCOMMODATION: Pilot only, beneath transparent canopy, which is hinged to open upward and rearward. Pilot's feet protrude through cutout in forward bulkhead to reach rudder pedals in nose.

DIMENSIONS, EXTERNAL:	
Wing span	6·66 m (21 ft 10¼ in)
Wing chord at root	1·08 m (3 ft 6½ in)
Wing chord at tip	0·60 m (1 ft 11¾ in)
Length overall	5·42 m (17 ft 9½ in)
Height overall	1·73 m (5 ft 8¼ in)
Tailplane span	2·49 m (8 ft 2 in)
Wheel track	1·54 m (5 ft 0¾ in)
Wheelbase	2·18 m (7 ft 2 in)
DIMENSIONS, INTERNAL:	
Cockpit:	
Length (to fwd bulkhead)	0·83 m (2 ft 8½ in)
Max width	0·83 m (2 ft 8½ in)
Max height	0·99 m (3 ft 3 in)
WEIGHTS:	
Weight empty, approx	408 kg (900 lb)
Max T-O weight	680 kg (1,500 lb)
PERFORMANCE (estimated):	
Never-exceed speed	250 knots (463 km/h; 288 mph)
Max cruising speed	217 knots (402 km/h; 250 mph)

HUGHES HELICOPTERS (page 379)

HUGHES MODEL 77

It is understood, from a statement made in August 1981 by Maj Gen Edward M. Browne, US Army AAH Program Manager, that the AH-64A production version of this helicopter is likely to receive the name **Apache**. The 1,262 kW (1,693 shp) T700-GE-701 engines intended for the production version were due to begin a six-month flight test programme in the fifth YAH-64 in early 1982.

LOCKHEED AIRCRAFT SERVICE COMPANY (page 387)

LOCKHEED EC-130 ARE

The Airborne Radar Extension (ARE) was conceived by Lockheed Aircraft Service Company as a combination of an existing airframe and already-operational radar that would provide early detection of approaching airborne threats at lower cost than current highly-specialised AWACS aircraft. After studying various candidate aircraft and avionics system, LAS decided that a Lockheed C-130 Hercules platform for an updated version of the General Electric AN/APS-125 UHF wavelength Doppler radar used on the E-2C Hawkeye would offer the most cost-effective solution.

Choice of a Lockheed airframe was not merely partisan. The Hercules offers a large-volume fuselage for installation of the avionic equipment, providing easy access for maintenance. Mounting the 7·32 m (24 ft) diameter rotodome on a shortened fin offers an almost unhindered field of view (360° in azimuth, 21° in elevation). The fact that 51 countries already operate C-130s means that they already have experience of the airframe and engines, and trained air and ground crews, and would need only to provide for operation and maintenance of the new avionics systems. The AN/APS-125 radar offers not only proven reliability and capacity for overland detection, but is less complex than other systems, requiring a minimum of maintenance, logistics support, and personnel training.

The complete Lockheed EC-130 ARE system is already licensed by the US State Department, Office of Munitions Control, for sale to overseas customers. It will expand greatly the limited geographic coverage of ground radar sites by providing airborne detection, tracking, and identification functions in the surveillance role, and command control communications assistance for air defence intercept missions. The radar can detect and track intruding aircraft to a nominal range of 200 nm (370 km; 230 miles) over land and water. Additionally, it can be used to detect surface shipping encroaching upon coastal waters.

As well as the radar, the ARE system includes IFF and other passive subsystems. Information gathered by these onboard sensors is fed through a data processing system to complete the track and identification information. This is used by the mission operators for situation monitoring, and to integrate the national defence functions for additional area coverage. The ARE system provides a radar/IFF capacity for 300 tracks; the passive subsystem capacity is 256 threats. A complete communications system, independent of that used by the flight crew, provides data link and secure communications for ARE co-ordination, and a voice relay for ground controlled intercept functions. This allows for ARE deployment either beyond line-of-sight, or within line-of-sight, of an air defence centre (ADC) on the ground.

Artist's impression of LAS EC-130 ARE proposal for a low-cost AWACS

To meet training and logistics requirements, Lockheed can offer a two-phase programme. The first represents an interim stage, with the aircraft used to relay the tactical plot, as generated by the radar/IFF/passive system sensors, to a ground ADC site for display on a three-man console station. A two-man station on board the EC-130 monitors the radar and passive system to ensure that adequate data is being relayed to the ADC, and that the equipment is performing within its specified limits. The second phase integrates three-man consoles (similar to those at the ADC site) within the aircraft, so that the functions of mission director, passive system co-ordinator, and interceptor controller can be performed as part of the aircraft mission. Tactical data plots would also be relayed to the ADC for use by ground station controllers, to complement and extend the capability of the ARE platform.

Lockheed estimates that, with preliminary engineering substantially completed, ARE aircraft could be delivered within 18 to 24 months of contract finalisation.

LOCKHEED CORPORATION (page 388)

LOCKHEED S-3B VIKING

Lockheed announced on 18 August 1981 the receipt of a full-scale engineering development contract, from the US Naval Air Systems Command, for an improved avionics system for S-3A Vikings currently in service with the US Navy. With initial funding of $14·5 million, it follows upon a contract awarded by the Navy in 1980, under which Lockheed-California developed the specifications for an S-3A weapons system improvement programme (WSIP).

Improvements covered by the WSIP programme include increased acoustic processing capacity, expanded electronic support measure coverage, increased radar processing, a new sonobuoy receiver system, and provisions for the Harpoon missile. It is anticipated that a total of 160 S-3As could be retrofitted under the programme, after which they will be redesignated S-3B.

MARS
MISSION AVIATION RELATED SERVICES

2380 Camino Vida Roble, Suite E, Carlsbad, California 92008
Telephone: (714) 438 1005
PRESIDENT: David Whitney

MARS DHC-6-X 41

This company, established in 1980, is marketing a series of conversions of the de Havilland Canada DHC-6 Twin Otter aimed at the commuter airline market. The '41' in the designation of converted aircraft refers to US SFAR 41 regulations, under which MARS has been able to offer an increased range/payload by raising the aircraft's maximum take-off weight, at some cost in terms of field performance which is seldom critical for commuter operators. Nine versions are available:

DHC-6-1 41A. Basic conversion of DHC-6 Series 100, for increased range/payload only.

DHC-6-2 41A. Similar, but for Series 200.

DHC-6-3 41A. Similar, but for Series 300.

DHC-6-1 41B. Direct operating costs lowered by adding a retractable landing gear, with smaller tyres, to DHC-6-1 41A modifications. Port main gear retracts rearward and starboard leg forward, into underfloor wells. New in-wing fuel tanks optional to offset resultant loss of underfloor tank capacity.

DHC-6-2 41B. Similar, but for Series 200.

DHC-6-3 41B. Similar, but for Series 300.

DHC-6-1 41C. As DHC-6-1 41B, but fuselage 'stretched' by 0·76 m (2 ft 6 in) by insertion of equal-length plugs fore and aft of the wings. Seating capacity increased to 23 in addition to pilot, but one seat must be occupied by a non-revenue cabin attendant.

DHC-6-2 41C. Similar, but for Series 200.

DHC-6-3 41C. Similar, but for Series 300.

Maximum T-O weight is raised to 6,395 kg (14,100 lb) by the MARS conversion. Max zero-fuel weight remains at 5,670 kg (12,500 lb).

MCDONNELL DOUGLAS (page 402)

MCDONNELL DOUGLAS F/A-18 HORNET

Australia is to acquire 75 F-18 Hornets to replace RAAF Mirage tactical fighters. The aircraft will be assembled in Australia by Government Aircraft Factories for delivery in 1984-90.

MCDONNELL DOUGLAS RF/A-18 HORNET

McDonnell Douglas has under development a reconnaissance variant of the Hornet strike fighter designated RF/A-18. This is required in only limited numbers, to provide a substantial increase in US fleet reconnaissance capability, and an important factor in its development has been the design of a system that will allow rapid conversion from reconnaissance to the fighter/attack role, while retaining the performance features that led to selection of the F/A-18 for service with the USN/USMC.

Planned initial capability, which is identified as Baseline, will substitute a sensor pallet for the M61 20 mm gun system that is mounted in the nose of the standard F/A-18. All other weapon system interfaces and stores capability will be retained for easy conversion back to a fighter/attack role; the Sidewinder air-to-air missiles which are carried at the two wingtip stations will also be retained for self protection. The internal sensor pallet will carry a variety of cameras, to provide panoramic photo coverage from all altitudes, and an infra-red linescanner for night operations. These sensors will augment the Hughes AN/APG-65 multi-mode digital radar carried by the standard F/A-18, as well as the Ford Aerospace/Texas Instruments AAS-38 FLIR pod which it can carry.

During the development programme a variety of sensors in current USN/USMC use, as well as development sensors, have been installed in the nose pallet to obtain frame, panoramic, and linescan imagery on a day or night, overfly or standoff, clear-weather mission basis. The RF/A-18 Baseline onboard displays for radar and FLIR systems are the same as those used for the F/A-18, and provide for navigation, terrain avoidance, and some target detection/recognition capability in adverse weather conditions. The Baseline capability is designed to accept, in later development stages, high-resolution synthetic aperture radar technology for all-weather enhancement, long-range high-resolution infra-red sensor technology for standoff enhancement, and a data link interface with surface stations for real-time data transfer.

Artist's impression of McDonnell Douglas RF/A-18 Hornet reconnaissance aircraft

MCDONNELL DOUGLAS C-17

The US Air Force's C-X programme is for a long-range, heavy-lift air-refuellable cargo transport, intended primarily to provide inter-theatre airlift of outsize loads, including tanks and infantry fighting vehicles, directly into airfields in potential conflict areas. Design requirements therefore include outstanding STOL performance. The first production aircraft, if the programme receives a full go-ahead, are intended for delivery to Military Airlift Command in the late 1980s, with full operational capability planned for the early 1990s. The C-X is but one ingredient of an Air Force airlift improvement plan that also includes enhancement of current aircraft capabilities, and expanded cargo-carrying capability for the Civil Reserve Air Fleet.

The USAF request for proposals (RFP) for the C-X, issued in October 1980, stressed the need for an aircraft that could be integrated with other USAF airlift equipment, to enhance the responsiveness and flexibility of the nation's conventional military forces. On 2 September 1981 McDonnell Douglas announced that, against competing designs from Boeing and Lockheed, the Douglas Aircraft Company had been selected as prime contractor for full-scale engineering development of this aircraft, which has the USAF designation C-17. Selection did not, at that stage, represent an Air Force commitment to build, since the USAF was still evaluating alternative ways of overcoming the current shortfall in airlift capability. Award of a contract to build is therefore dependent upon Defense Department approval of the Air Force's overall plans for satisfying its airlift requirements.

The McDonnell Douglas design, making use of medium STOL transport prototypes (see 1979-80 *Jane's*), will be able to airlift outsize combat equipment which at present can be carried only by the Lockheed C-5A Galaxy, and offer a short-field performance currently provided only by the C-130 Hercules. It will be able to operate from runways only 915 m (3,000 ft) long and 18·3 m (60 ft) wide; on the ground, it will be able to execute a 180° turn in only 24·5 m (80 ft), and a fully-loaded aircraft, using thrust reversal, will be able to reverse up a 1 in 40 gradient.

Type: Long-range heavy-lift cargo transport.

Wings: Cantilever high-wing monoplane, of supercritical section with 25° sweepback. NASA-type winglet at each tip. Externally-blown flap system, developed from that used on McDonnell Douglas YC-15, to reduce final approach and landing speeds by directing engine efflux over double-slotted trailing-edge flaps to provide additional lift. Flaps constructed of titanium, using superplastic-forming/diffusion-bonding techniques.

Fuselage: Conventional semi-monocoque structure, upswept at rear. Rear-loading ramp/door in underside of rear fuselage. Twin strakes under extreme rear of fuselage.

Tail Unit: Cantilever T tailplane; two-segment rudder.

Landing Gear: Retractable tricycle type, with twin-wheel nose unit and two six-wheel main units. Main-wheel units, each consisting of two legs in tandem on each side of fuselage, with three wheels on each leg, retract into bottom of large fairings on lower fuselage sides; nose unit is forward-retracting.

Power Plant: Four 167·2 kN (37,600 lb st) Pratt & Whitney PW2037 turbofan engines, with in-flight and on-ground deployable thrust-reversers; pylon-mounted in individual underwing pods. Provision for in-flight refuelling.

Accommodation: Flight crew of two (pilot and co-pilot), plus loadmaster. Main hold can accommodate Army wheeled vehicles in two side-by-side rows, or Jeeps in triple rows. The C-17 will be the only aircraft able to airdrop outsize firepower such as the US Army's new infantry fighting vehicle (three of which comprise one deployment load); it can also carry the new M1 main battle tank in combination with other vehicles. Up to 18,145 kg (40,000 lb) of cargo can be carried on rear-loading ramp.

Avionics: Advanced digital avionics system, with six CRT displays.

Dimensions, external:
- Wing span: 50·29 m (165 ft 0 in)
- Wing area, gross: 353 m² (3,800 sq ft)
- Length overall: 52·02 m (170 ft 8 in)
- Height overall: 16·31 m (53 ft 6 in)

Dimensions, internal:
- Cargo compartment:
  - Length, incl 5·79 m (19 ft 0 in) rear-loading ramp: 26·52 m (87 ft 0 in)
  - Max width: 5·49 m (18 ft 0 in)
  - Height under wing: 3·61 m (11 ft 10 in)
  - Max height: 4·11 m (13 ft 6 in)

Weights:
- Weight empty: 117,480 kg (259,000 lb)
- Max payload: 78,110 kg (172,200 lb)
- Max T-O weight: 259,455 kg (572,000 lb)

Performance (estimated):
- T-O field length with max payload 2,320 m (7,600 ft)
- Landing field length with max payload and thrust reversal: 915 m (3,000 ft)
- Range with max payload: 2,400 nm (4,445 km; 2,765 miles)
- Ferry range, without in-flight refuelling: 5,000 nm (9,265 km; 5,755 miles)

Artist's impression of the McDonnell Douglas C-17 long-range heavy cargo transport

PAT
PIPER ADVANCED TECHNOLOGY INC

7736 Dublin Street, Wichita, Kansas 67206
Telephone: (316) 683 1570

PIPER ADVANCED TECHNOLOGY PAT-1

Piper Advanced Technology was founded during 1980 by the late Mr Howard Piper, specifically to design and produce a new-generation lightplane that would be safe to fly, comparatively cheap to acquire and maintain, and yet would provide superior performance to contemporary aircraft of similar power.

A development prototype of the resulting design, designated PAT-1, flew for the first time in July 1981. At this early stage of a three-year flight development programme, the company is not prepared to make specific forecasts of performance, but claims that low-speed handling qualities and stall characteristics appear to be superior to those of current aircraft in this category.

The PAT-1 is of canard configuration, makes extensive use of sandwich composite construction, has non-retractable tricycle landing gear, and is powered by a 119 kW (160 hp) Avco Lycoming O-320 engine. An enclosed cabin, with the door on the port side of the fuselage, accommodates four persons.

Dimensions, external:
- Wing span: 7·70 m (25 ft 3¼ in)
- Foreplane span: 5·84 m (19 ft 2 in)
- Length overall: 7·13 m (23 ft 4¾ in)
- Height overall: 2·79 m (9 ft 2 in)

Weights:
- Weight empty: 500 kg (1,102 lb)
- Max T-O weight: 910 kg (2,005 lb)

Piper Advanced Technology PAT-1 four-seat experimental light aircraft *(Howard Levy)*

PIPER AIRCRAFT CORPORATION (page 425)

PIPER ENFORCER

Piper Aircraft Corporation announced on 4 September 1981 the receipt of a contract from the US Air Force covering the design, development, and testing of two prototypes of a lightweight turboprop-powered close-support aircraft known as the Enforcer. This is based on the North American P-51 Mustang of the second World War and, as reported in the 1980-81 *Jane's*, Piper acquired the original programme from Cavalier Aircraft Corporation. Total contract value of the current two-year programme is expected to be about $12 million.

Piper has made significant changes to the original Enforcer design flight tested in 1971, including aerodynamic improvements to the tailplane, modifications to the aileron control system, and provision for modern weapons. The first flight of an updated Enforcer is scheduled for December 1982, with an operational demonstration in the late Summer of 1983. The aircraft is being evaluated primarily for use as an internal defence weapons system, but meets the basic requirements for Pave Coin missions. A tandem-seat dual-control trainer version has been developed and could be produced if the demand existed. This would retain the combat capability of the standard Enforcer, making it suitable also for armed reconnaissance. Its training capability would include bombing and gunnery, formation and instrument flight, tactical procedures, and transition. All available details of the single-seater follow:

Type: Single-seat turboprop attack aircraft.

Wings: Cantilever low-wing monoplane. NAA-NACA high-speed wing section. Dihedral 5°. All-metal stressed-skin structure. Sealed metal ailerons. Hydraulically actuated trailing-edge flaps. Non-jettisonable wingtip fuel tanks.

Second of two Enforcer prototypes tested at Piper's Vero Beach facility in 1971

FUSELAGE: All-metal semi-monocoque structure, incorporating extensive armour protection for power plant and pilot.

TAIL UNIT: Cantilever all-metal structure. Trim tabs in elevators and rudder.

LANDING GEAR: Hydraulically-retractable tailwheel type, with single wheel on each unit. Main units retract inward, tailwheel forward. Oleo-pneumatic shock-absorbers. Main wheels 68·58 cm (27 in) diameter, with high-flotation tyres. Multi-disc brakes.

POWER PLANT: One 1,823 ekW (2,445 ehp) Avco Lycoming T55-L-9 turboprop engine. Standard fuel capacity, in self-sealing and foam-protected tanks, totals 1,605 litres (424 US gallons). Auxiliary fuel can be carried optionally in four 416 litre (110 US gallon) and two 265 litre (70 US gallon) drop-tanks, to provide a maximum capacity of 3,800 litres (1,004 US gallons).

ACCOMMODATION: Pilot only, on zero-zero ejection seat, in enclosed cockpit with stretched acrylic windscreen and rearward-sliding canopy. Optional heating or cooling of cockpit.

SYSTEMS: Electrical system. Hydraulic system for actuation of landing gear, wheel brakes, and trailing-edge flaps.

AVIONICS AND EQUIPMENT: Solid-state avionics. Autopilot optional.

ARMAMENT: Standard armament comprises six wing-mounted 0·50 in calibre M3 machine-guns, with a total of 2,000 rds. Six underwing pylons have a maximum load capacity of 2,486 kg (5,480 lb), with 454 kg (1,000 lb) on each inboard pylon, and 394·5 kg (870 lb) on each of the two outer pylons. Optionally, two additional outboard pylons on each wing can each carry 158·8 kg (350 lb), with a reduced load on the inner pylons to give same total weight. Weapons can include AN-M47A4 smoke bombs; B-37K1 practice bomb containers; BLU-1C/B, -23B, -27B, -32B, -52B incendiary bombs; CBU-14A + 14A/A, -22A + 22A/A, -24A/B, -24B/B, -25A + 25A/A, -29A/B, -29B/B, -49A/B, -49B/B, -53B, -54B cluster bomb units; CBU-19A canister clusters (riot control); LAU-3A, -32A/A, -32B/A, -59A rocket launchers; M117/M117A-1, Mk 81, Mk 82, Mk 82 Snakeye GP bombs; SUU-11A/A Minigun pod; SUU-20A bomb/rocket dispenser; SUU-25A + 25A/A flare dispenser; and XM75 40 mm gun pods.

DIMENSIONS, EXTERNAL:

Wing span (over tip-tanks)	12·15 m (39 ft 10¼ in)
Wing chord at root	2·64 m (8 ft 8 in)
Wing chord at tip	1·27 m (4 ft 2 in)
Mean aerodynamic chord	2·02 m (6 ft 7½ in)
Length overall	9·94 m (32 ft 7¼ in)
Height overall	2·97 m (9 ft 9 in)
Tailplane span	4·02 m (13 ft 2¼ in)
Wheel track	3·61 m (11 ft 10 in)
Propeller diameter	3·40 m (11 ft 2 in)
Propeller ground clearance	0·20 m (7¾ in)

AREAS:

Wings, gross	23·41 m² (252 sq ft)
Ailerons (total, incl tabs)	1·18 m² (12·73 sq ft)
Trailing-edge flaps (total)	2·99 m² (32·22 sq ft)
Fin	1·44 m² (15·50 sq ft)
Rudder (incl tab)	0·87 m² (9·40 sq ft)
Tailplane	2·60 m² (27·98 sq ft)
Elevators (total, incl tabs)	1·21 m² (13·05 sq ft)

WEIGHTS (estimated):

Weight empty	3,066 kg (6,759 lb)
Operating weight, incl pilot, guns and armour plate	3,499 kg (7,714 lb)
Max T-O weight at 8*g* load factor	5,216 kg (11,500 lb)
Max T-O weight at 6*g* load factor	6,350 kg (14,000 lb)
Normal landing weight	3,674 kg (8,100 lb)

PERFORMANCE (estimated, A: 8*g* T-O weight; B: 6*g* T-O weight; C: normal landing weight):

Max level speed: A at 4,570 m (15,000 ft)	403 knots (747 km/h; 464 mph)
Cruising speed: B at 4,570 m (15,000 ft)	284 knots (526 km/h; 327 mph)
Stalling speed: C	78 knots (145 km/h; 90 mph)
Max rate of climb at S/L: A	1,460 m (4,790 ft)/min
Service ceiling: A	11,460 m (37,600 ft)
T-O run at S/L: B	335 m (1,100 ft)
T-O to 15 m (50 ft) at S/L: B	512 m (1,680 ft)
Landing from 15 m (50 ft): C	524 m (1,720 ft)
Landing run: C	268 m (880 ft)

SCHAFER (page 455)

Schafer Comanchero 500 turboprop conversion of a Piper Chieftain *(Howard Levy)*

The USAF/Boeing NKC-135 airborne laser laboratory

HOMEBUILTS

FRANCE

PIEL (page 489)

Prototype Piel C.P.1320 Saphir, built by M J. C. Lascoutounas *(Geoffrey P. Jones)*

USAF
UNITED STATES AIR FORCE

Air Force Systems Command, Andrews Air Force Base, Maryland 20334

USAF/BOEING NKC-135 ALL (AIRBORNE LASER LABORATORY)

A Boeing KC-135 tanker has been modified as shown in an accompanying illustration to serve as a research tool for the US Air Force's Weapons Laboratory, which is established at Kirtland AFB, New Mexico. This highly instrumented aircraft has the designation NKC-135 ALL.

The modification has been carried out as a part of the Defense Advanced Research Projects Agency (DARPA) programme, this organisation being responsible for certain areas of research on behalf of the US Department of Defense. One current DARPA project is investigating the capability of high-energy laser (HEL) beams to provide a new weapon system that could revolutionise tactical and strategic attack, as well as defence against airborne targets, both in the atmosphere and in space. The military capability of a HEL was based on the assumption that the beam would be able to burn through the surface of a target and destroy a vital component, ignite propulsive fuel, or activate its warhead. This was demonstrated to be practical as early as 1973, when the US Air Force destroyed a winged drone over the Sandia Optical Range at Kirtland AFB, using a gas-dynamic HEL application for this purpose. Both the US Army and Navy have since achieved similar success, the latter using a chemical laser developed jointly with DARPA to engage and destroy a TOW missile in flight.

Such experiments have proved beyond doubt the feasibility of using HEL beams to destroy moving targets, and the NKC-135 ALL will help DARPA gain a deeper insight into problems related to the operational deployment of such a weapon. HEL beams can now be directed over long ranges, and continued research and development should result in their becoming even more effective. It should be understood, however, that for a HEL weapon to be practical, it requires a guidance system that will be able to direct this high-intensity light beam over a range that could extend to several hundred kilometres (miles), focus it as an intense spot of energy, and track a small rapidly moving target with such accuracy that the laser beam is held steady. If the focused spot is blurred, or moves, then its energy is dispersed, requiring much longer contact with the target to achieve its destruction.

The task of DARPA is to develop a control system that, in a high-density threat environment, can direct the HEL methodically from target to target, focus the beam on each selected aim point, and hold it steady irrespective of speed and manoeuvre until the target is destroyed. The HEL must then be redirected automatically to the next most threatening target. The NKC-135 ALL, working together with a new high energy laser system test facility that should become operational at White Sands Missile Range in late 1982 will, it is hoped, speed the development of a fire-control system that will have just such a capability. It will, at the same time, investigate the propagation of laser beams from an airborne vehicle against an airborne target.

The first air-to-air firing test, on 1 June 1981, failed either because the laser beam missed its target, a Sidewinder missile, or because the beam had no effect. The second test, a few weeks later, was rated as 75 per cent successful, as the beam remained on target for longer. Further tests have been deferred until the HEL test facility is operational.

INDONESIA

WIWEKO SOEPONO

Akasamitra Homebuilt Aircraft Club, PO Box 167, Jakarta

While serving as head of the Indonesian Air Force Design and Construction Branch in 1948, Mr Wiweko Soepono, who is currently the President Director of Garuda Indonesian Airways, designed and built the WEL-1 (Wiweko Experimental Lightplane). Powered by a 15 kW (20 hp) Harley Davidson motorcycle engine, it had the distinction of being the first powered aircraft designed and built in the Republic of Indonesia. After making its first flight and being displayed in an aviation exhibition, the original WEL-1 was destroyed by a grenade explosion while being transported by rail.

To meet requests from both the Air Force and Armed Forces museums, an exact replica of the WEL-1 has been built, under the direction and guidance of Mr Wiweko, by Akasamitra, a homebuilt aircraft club. Another version, with a Revmaster engine, made its first flight on 13 August 1981.

WIWEKO WEL-1

Data on the WEL-1 with a Revmaster engine are as follows:

TYPE: Single-seat light aircraft.

WINGS: Strut-braced parasol monoplane. Wing section NACA 2415. Dihedral 2°. Incidence 3° 48′. Wooden two-spar structure, built in two halves, attached to fuselage by tubular centre-section pylons and braced by parallel lift struts of streamline-section steel tubing. Wings and ailerons are fabric covered. No flaps.

FUSELAGE: Welded steel tube structure, fabric covered. Curved decking aft of cockpit, aluminium covered.

TAIL UNIT: Wire-braced welded steel tube structure, fabric covered.

LANDING GEAR: Non-retractable tailwheel type, with divided main legs of tubular steel construction and rubber-block shock-absorption. Main-wheel tyres size 5·00 × 5. Fully castoring tailwheel. Cleveland Model 30-9 brakes.

POWER PLANT: One 52 kW (70 hp) Revmaster 2100 D converted Volkswagen engine, driving a two-blade fixed-pitch wooden propeller. Aluminium fuel tank in fuselage, aft of firewall, capacity 40 litres (8·8 Imp gallons). Refuelling point on upper surface of fuselage, forward of windscreen. Oil capacity 4 litres (0·8 Imp gallons).

ACCOMMODATION: Single seat in open cockpit.

DIMENSIONS, EXTERNAL:

Wing span	9·00 m (29 ft 6 in)
Wing chord	1·45 m (4 ft 9 in)
Wing area, gross	13·65 m² (147 sq ft)
Length overall	5·05 m (16 ft 7 in)
Height overall	2·40 m (7 ft 10 in)

WEIGHTS:

Weight empty	300 kg (660 lb)
Disposable load	130 kg (286 lb)
T-O weight	430 kg (946 lb)

PERFORMANCE:

Maximum level speed	57 knots (105 km/h; 65 mph)
Cruising speed	55 knots (101 km/h; 63 mph)
Stalling speed	33 knots (61 km/h; 38 mph)
Ceiling	3,660 m (12,000 ft)

Latest version of the Wiweko WEL-1 with Revmaster engine

Artomov Simurg two-seat homebuilt lightplane

UNION OF SOVIET SOCIALIST REPUBLICS

ARTOMOV
MIKHAIL ARTOMOV

Dneprodzerzinsk, Ukraine

ARTOMOV SIMURG (BIRD OF HAPPINESS)

Working in conjunction with fellow aviation enthusiast Viktor Timofeyev, Mikhail Artomov has designed and built a two-seat lightplane which he has named Simurg. An earlier aircraft of the same name crashed during flight testing, when a reversible-pitch propeller reversed inadvertently during flight. The new Simurg incorporates unusual control features, in that the wings have no control surfaces. Instead, the incidence of each wing can be varied differentially up to 3° for roll control, and from 5° to 8° collectively for slow-speed take-off and landing.

TYPE: Two-seat lightplane.

WINGS: Braced high-wing monoplane with A shape bracing struts each side. Wing section Wortmann 126. Single-spar structure, with light alloy covering forward of the spar, the remainder fabric-covered. No control surfaces. Incidence of each wing variable differentially and collectively.

FUSELAGE: Includes a forward pod to accommodate pilot and passenger, a mounting aft for the power plant, and a tailboom to carry the tail unit. Basic structure of metal tube with fabric covering, except for the use of glassfibre skins in the area of the engine installation.

TAIL UNIT: Braced T tail, with tailplane strut-mounted at tip of fin. Large fin and rudder. Structure of metal tube with fabric covering.

LANDING GEAR: Non-retractable tricycle type with single wheel on each unit. Castoring nosewheel. Main wheels, from a disabled person's motor vehicle, are mounted at the extremities of a motorcar suspension spring. Small tailskid to protect rudder in a tail-down landing.

POWER PLANT: One own-construction 26 kW (35 hp) two-cylinder inline inverted engine, strut-mounted from the basic fuselage structure aft of the cabin, and driving a two-blade fixed-pitch wooden pusher propeller.

ACCOMMODATION: Pilot and passenger, with dual controls. Windscreen and window on starboard side of cabin, of Plexiglas. Opening on port side of cabin provides access and is not enclosed in flight.

DIMENSIONS, EXTERNAL:

Wing span	9·00 m (29 ft 6½ in)
Wing area, gross	11·00 m² (118·4 sq ft)
Length overall	5·60 m (18 ft 4½ in)
Propeller diameter	1·60 m (5 ft 3 in)

WEIGHT:

Weight empty	175 kg (386 lb)

PERFORMANCE:

T-O speed: pilot only	30 knots (55 km/h; 34 mph)
pilot and passenger	32·5 knots (60 km/h; 37·5 mph)

TIMOFEYEV
VIKTOR TIMOFEYEV

Dneprodzerzinsk, Ukraine

TIMOFEYEV MUSTANG 2

Aviation enthusiast Viktor Timofeyev is currently flying an improved version of an earlier Mustang (see 1977-78 *Jane's*), designed and built in collaboration with Mikhail Artomov. (Artomov and Timofeyev have previously built and flown four other aircraft of their own design.) In overall configuration the Mustang 2 is similar to Artomov's Simurg. All available details follow:

TYPE: Single-seat lightplane.

WINGS: Braced high-wing monoplane with two bracing struts each side. Wing section TsAGI R-2-12. Ailerons for roll control.

FUSELAGE: Of similar pod and boom type to Simurg.

TAIL UNIT: Braced T tail, with tailplane strut-mounted at tip of fin.

LANDING GEAR: Non-retractable two-wheel type, with tailskid. Main wheels carried at extremities of motor car suspension spring.

POWER PLANT: One 22 kW (30 hp) two-cylinder modified motorcycle engine, from an Iz-Planeta 3 motorcycle, mounted above rear of wing centre-section and driving a two-blade wooden pusher propeller.

Timofeyev Mustang 2 single-seat homebuilt aircraft

ACCOMMODATION: Pilot only, in semi-enclosed cabin.

DIMENSIONS, EXTERNAL:

Wing span	8·40 m (27 ft 6¾ in)
Length overall	5·00 m (16 ft 5 in)
Propeller diameter	1·20 m (3 ft 11¼ in)

WEIGHT:

Weight empty	150 kg (330 lb)

PERFORMANCE:

Cruising speed	65 knots (120 km/h; 75 mph)
Landing speed	32·5 knots (60 km/h; 37·5 mph)

UNITED KINGDOM

KENDALL
DR RIDLEY KENDALL

KENDALL MAYFLY

Winner of the award for the best original design at the PFA Rally held at Leicester in July 1981, the Mayfly (G-PFAK) had flown for the first time only a few days earlier.

TYPE: Single-seat sporting lightplane.

WINGS: Strut-braced high-wing monoplane, with single streamline-section bracing strut each side. Wortmann FX-150-17 wing section. Constant chord. No dihedral. Tips chamfered at 45°. Wings removable for transport by trailer behind a motorcar. Full-span trailing-edge control surfaces. No tabs.

FUSELAGE: Pod and boom type of basically circular section, with broad pylon to carry wing and power plant.
TAIL UNIT: T tail. All-moving tailplane, with servo-tab, mounted at tip of sweptback parallel-chord fin and rudder.
LANDING GEAR: Non-retractable tailwheel type. Curved cantilever main-wheel legs. Go-kart brakes.
POWER PLANT: One 1,600 cc converted Volkswagen motorcar engine, mounted above wing trailing-edge and driving a two-blade fixed-pitch wooden pusher propeller. Fuel capacity 36 litres (8 Imp gallons).
ACCOMMODATION: Pilot only, in shallow sailplane-type cockpit under large sideways-hinged (to starboard) blister canopy.
WEIGHTS:
Weight empty 240 kg (530 lb)
Max T-O weight 340 kg (750 lb)
PERFORMANCE (provisional):
Cruising speed 80 knots (148 km/h; 92 mph)
Stalling speed 42 knots (78 km/h; 48 mph)
Design range with max fuel 260 nm (482 km; 300 miles)

Kendall Mayfly single-seat light aircraft *(Geoffrey P. Jones)*

UNITED STATES OF AMERICA

BARRETT
BARRETT AIRCRAFT CORPORATION

PO Box 695, Anoka, Minnesota 55303
Telephone: (612) 755 8100
PRESIDENT: Robert S. Barrett

Mr Barrett has designed and built a lightweight autogyro, named the Gyracar, of which plans and kits are available. Considerable effort has been devoted to making this aircraft easy to fly for complete amateurs. The control column incorporates handgrips which are turned like a car steering wheel to operate the rudder. Where rudder pedals are normally positioned, in the nose, two foot pedals duplicate the accelerator and brake pedals of a motorcar. Autorotation of the two-blade rotor is initiated by the pilot spinning it by hand before starting the engine. Construction is also extremely simple, the kit comprising a number of modules that only require holes to be drilled so that the modules can be bolted together. Four hours' dual instruction on a company two-seat Gyracar is included in the package.

BARRETT GYRACAR

TYPE: Single-seat lightweight autogyro.
ROTOR SYSTEM: Two-blade rotor, with blades and gimbal hub of metal construction.
FUSELAGE: Keel and rotor mast fabricated from light alloy. Landing gear, power plant and tail unit supported by steel tube.
TAIL UNIT: Large fin, mounted between keel structure and upper tubular support of rotor mast, with rudder hinged aft. Ground-adjustable tab on rudder.
LANDING GEAR: Non-retractable tricycle type on earlier versions; non-retractable tailwheel type on current model. Brakes on main wheels. Parking brake. Ski landing gear optional.
POWER PLANT: One 64 kW (86 hp) Volkswagen converted motorcar engine, mounted aft of the rotor mast, and driving a two-blade fixed-pitch wooden pusher propeller. Fuel tank, of moulded translucent plastics material for ease of contents gauging, forms pilot's seat. Capacity 45·4 litres (12 US gallons).
ACCOMMODATION: Pilot only, in enclosed cabin. Door each side. Accommodation heated and ventilated.
DIMENSIONS, EXTERNAL:
Rotor diameter 7·32 m (24 ft 0 in)
Rotor disc area 42·03 m² (452·4 sq ft)
WEIGHTS:
Weight empty 147 kg (324 lb)
Max payload 113·5 kg (250 lb)
PERFORMANCE (at max T-O weight):
Max level speed 87 knots (161 km/h; 100 mph)
Cruising speed 69 knots (129 km/h; 80 mph)
Service ceiling 4,875 m (16,000 ft)
T-O run 46 m (150 ft)
Landing run 2·4 m (8 ft)
Range with max fuel 130 nm (241 km; 150 miles)

EICH (page 524)

EICH JE-2

Detailed drawings and instructions for building this dual-control two-seat lightweight autogyro are now available from:
Tech Man Company, 8525 E Duarte Road, San Gabriel, California 91775

MUSSO (page 538)

Musso Real Sporty, photographed with pilot 'Chuck' Andrews (left) and designer Paul Musso at the 1980 Cleveland Air Races

PALOMINO
PALOMINO AIRCRAFT ASSOCIATES

Stinson Municipal Airport, Hangar 10, 8619 Mission Road, San Antonio, Texas 78214
Telephone: (512) 922 1203
DESIGNER: Bert Wilcut

This company is marketing plans and kits of parts for an all-metal high-performance tandem two-seat light aircraft named Palomino. This inherits features of the Midget Mustang and Mooney M-19 aircraft, rights in which were acquired by the San Antonio Aviation School, Palomino Aircraft's parent company, prior to detail design of the Palomino. It is planned eventually to produce and market completed aircraft, following FAA certification.

PALOMINO

TYPE: Two-seat all-metal light aircraft.
WINGS: Cantilever low-wing monoplane. Wing section NACA 64A-212 at root, NACA 64A-215 at tip. Dihedral 6° from roots. Incidence 3° at root, 0° at tip. Conventional two-spar structures of light alloy, bolted together at aircraft centreline. Skin flush riveted back to approx 60% chord. All-metal statically-balanced ailerons and three-position manually-actuated flaps. No tabs.
FUSELAGE: Conventional construction in three sections. Centre-fuselage, including cabin, has welded steel tube truss structure to which is attached the cabin aluminium framing covered with glassfibre. The rear fuselage is a light alloy semi-monocoque made up of pressed bulkheads, extruded longerons and stringers, and tapered rolled skins. Engine cowlings are of epoxy glassfibre.
TAIL UNIT: Cantilever all-metal structure, with sweptback vertical surfaces. Statically-balanced all-moving horizontal surface, with anti-servo trim tab. No tab on rudder.
LANDING GEAR: Manually-retractable tricycle type, with single wheel on each unit. Main wheels retract inward into wings, nosewheel rearward. Shock-absorption by neoprene 'biscuits'. All three wheels size 5·00-5, by Goodyear or Cleveland. Steerable nosewheel. Brakes operable from both seats.
POWER PLANT: One flat-four engine in the 93-149 kW (125-200 hp) range, driving a two-blade constant-speed propeller. Approved types include a Franklin 4A-235-B with McCauley propeller and Avco Lycoming engines driving Hartzell propellers. Integral fuel tanks in wing leading-edges, with total capacity of 95 litres (25 US gallons), and header tank, capacity 38 litres (10 US gallons), aft of engine firewall.
ACCOMMODATION: Two seats in tandem under large rearward-sliding one-piece canopy, with tinted glassfibre roof for protection from Sun. Front seat adjustable. Dual controls standard. Cabin heated and ventilated. Baggage compartment behind rear seat, capacity 23 kg (50 lb).
AVIONICS: King or Narco radios and full IFR instrumentation optional.
DIMENSIONS, EXTERNAL:
Wing span 8·58 m (28 ft 2 in)
Wing chord at root 1·73 m (5 ft 8 in)
Wing chord at tip 0·57 m (1 ft 10½ in)
Wing area, gross 9·92 m² (106·8 sq ft)
Wing aspect ratio 7
Length overall 6·27 m (20 ft 7 in)
Wheel track 2·13 m (7 ft 0 in)
Wheelbase 1·58 m (5 ft 2¼ in)
Propeller diameter 1·83 m (6 ft 0 in)
WEIGHTS (A: 93 kW; 125 hp Franklin 4A-235-B engine. B: 112 kW; 150 hp Avco Lycoming O-320. C: 134 kW; 180 hp Avco Lycoming O-360):
Weight empty: A 453 kg (998 lb)
B 466 kg (1,026 lb)
C 469 kg (1,033 lb)
Max T-O weight: all versions 787 kg (1,736 lb)
PERFORMANCE (A, B, C as listed for weights):
Max cruising speed (75% power) at S/L:
A 134 knots (249 km/h; 155 mph)
B 145 knots (268 km/h; 167 mph)
C 154 knots (286 km/h; 178 mph)
Stalling speed, flaps down:
A 60 knots (111 km/h; 69 mph)
B 61 knots (113 km/h; 70 mph)
C 63 knots (116 km/h; 72 mph)
Max rate of climb at S/L:
A 365 m (1,200 ft)/min
B 426 m (1,400 ft)/min
C 487 m (1,600 ft)/min
Range with max fuel:
A 521 nm (965 km; 600 miles)
B 499 nm (925 km; 575 miles)
C 478 nm (885 km; 550 miles)

Palomino tandem two-seat light aircraft with Franklin engine

SMITH
A. J. SMITH

SMITH AJ-2

Built by a former gliding champion, Mr A. J. Smith, this tandem two-seat light aircraft embodies considerable sailplane technology in its slender design and composite construction. Powered by a 149 kW (200 hp) Avco Lycoming IO-360-A1B6 flat-four engine, it has already demonstrated outstanding performance. Its designer hopes to achieve even better results in a later version of the AJ-2 with retractable landing gear.

Dimensions, external:
Wing span	7·31 m (24 ft 0 in)
Wing aspect ratio	8·5
Length overall	6·70 m (22 ft 0 in)

Weights:
Weight empty	500 kg (1,102 lb)
Max T-O weight	771 kg (1,700 lb)

Smith AJ-2, embodying considerable sailplane technology *(J. M. G. Gradidge)*

Performance:
Max level speed	259 knots (480 km/h; 298 mph)
Max cruising speed (75% power)	248 knots (460 km/h; 285 mph)
Landing speed	81 knots (150 km/h; 93 mph)

THUNDER WINGS
THUNDER WINGS (a division of Thunder Development Inc)

7326 East Evans Road, Scottsdale, Arizona 85260

Vice-President, Marketing: David A. Bratset

This company has been established by Thunder Development Inc, which was itself formed in 1975 by a group of men who wished to build and fly replicas of second World War combat aircraft. Their first product was a half-scale representation of the Focke-Wulf Fw 190A, powered by a 74·5 kW (100 hp) Continental O-200 flat-four engine, and using the carved foam and outside glassfibre layup constructional technique.

After careful consideration of the structure, flight characteristics and scale proportions of this small fighter replica, it was decided to adopt an entirely different approach to the programme before marketing kits and plans to other amateur constructors. A basic scale of approximately 80 per cent was selected to permit the installation of more powerful engines without prejudicing the scale proportions and general appearance of the aircraft. It was realised also that this larger scale would offer room for full-size military pilot seats, with provision for seat or back parachutes, instrument panels spacious enough for full gyros and IFR avionics, and systems such as oxygen and remote fire extinguishing.

To ensure availability of a suitable power plant for the representations of inline-engined aircraft, Thunder developed from the Jaguar V-12 motorcar engine what it calls the Lightning Merlin. Drawings are available for this conversion, which involves the addition of a reduction gear; future developments may add dual magneto ignition, a fuel injection system and turbocharging. For constructors of radial-engined replicas, such as the 80 per cent Fw 190A, the standard Continental W670 seven-cylinder aircraft engine provides an easily-obtainable authentic appearance.

Thunder Wings was established as a division of Thunder Development Inc to market plans and kits of the 80 per cent replicas. Initially, these aircraft were designed with laminated spruce wing spars, wooden spars for the fin, tailplane and all control surfaces, and a welded steel tube fuselage. The airframe was then covered with a set of pre-moulded skins that reproduced accurately the shape of the original full-scale fighter. Subsequently, Thunder Wings developed all-composite internal structures to supersede the wood and steel tube, resulting in considerable reduction of structure weight and simplifying construction for the homebuilder. The composite kits became available in 1981, although a limited quantity of the original wood and steel kits are still available optionally for anyone who might prefer them. All three of the current kits are designed to produce aircraft stressed to ±6*g*.

The company intends to sell only 100 kits of each replica, and each is offered in three individual component groups (fuselage, wing, and controls and accessories) to spread the purchase cost over a period of time. The kits do not include a power plant, but it was hoped to market kits of the Lightning Merlin engine by Autumn 1981.

The description which follows applies specifically to the Curtiss P-40C replica, but is applicable generally to the all-composite versions of all replica kits available from the company.

THUNDER WINGS CURTISS P-40C

Type: Eight-tenths scale combat aircraft replica.

Wings: Cantilever low-wing monoplane built in three sections. Wing section NACA 23015. Two-spar structure, with spars and ribs of moulded sandwich composite material. Preformed skins attached by structural bonding. Ailerons, of similar construction, have single spar. No trailing-edge flaps. No tabs.

Fuselage: Semi-monocoque structure, incorporating bulkheads, centre-section spars and interior bracing of composite material, covered by preformed half-shells.

Tail Unit: Conventional cantilever structure of similar construction to wings. The tailplane incorporates two spars; elevators, fin and rudder each have a single spar. No tabs.

Landing Gear: Hydraulically-retractable type with a single wheel on each unit. Main units retract aft and turn through 90° as they are raised to lie flush in the under-surface of the wing. Non-retractable tailwheel.

Power Plant: One 224 kW (300 hp) Lightning Merlin V-12 piston engine, driving a three-blade constant-speed propeller with spinner. Radiators in scale position under engine. Fuel contained in integral wing tanks with a standard capacity of 170 litres (45 US gallons). Long-range tanks of 265 litre (70 US gallon) capacity optional.

Accommodation: Pilot and optional passenger, seated in tandem. Removable rear window cover inserts can be installed when no passenger is carried, to maintain authentic appearance of replica. Rearward-sliding cockpit canopy.

Systems: Electrical system includes engine-driven alternator. Hydraulic system for landing gear actuation. Vacuum system for flight instruments. Fire extinguishing and oxygen systems optional.

Avionics and Equipment: Optional avionics can include 720-channel com transceiver, VOR/ILS, GS and marker beacon receivers, transponder with altitude encoder, digital approach timer, annunciator panel, and associated antennae. Full blind-flying instrumentation optional.

Dimensions, external:
Wing span	9·14 m (30 ft 0 in)
Wing area, gross	14·03 m² (151 sq ft)
Length overall	7·62 m (25 ft 0 in)
Height overall	3·05 m (10 ft 0 in)

Weights:
*Weight empty	1,193 kg (2,630 lb)
Max T-O weight	1,453 kg (3,204 lb)

**includes all listed optional avionics and equipment*

Performance (at max T-O weight, flown solo):
Cruising speed	174 knots (322 km/h; 200 mph)
Range with standard fuel	434 nm (805 km; 500 miles)
Range with optional long-range fuel	651 nm (1,207 km; 750 miles)

THUNDER WINGS FOCKE-WULF Fw 190A

Details of the Fw 190A replica are generally similar to those of the P-40C, except as detailed:

Tail Unit: As for P-40C, except trim tab in rudder and starboard elevator.

Landing Gear: As for P-40C, except that main units retract inward.

Power Plant: One 179 kW (240 hp) Continental W670 seven-cylinder aircooled radial piston engine, driving a three-blade fixed-pitch wooden propeller with spinner. Fuel capacity 189·25 litres (50 US gallons).

Accommodation: Pilot only, beneath transparent canopy. Baggage space with capacity of 18 kg (40 lb).

Dimensions, external:
Wing span	8·53 m (28 ft 0 in)
Wing area, gross	12·26 m² (132 sq ft)
Length overall	7·01 m (23 ft 0 in)
Height overall	2·44 m (8 ft 0 in)

Weights:
Weight empty, equipped (see P-40C entry)	897 kg (1,978 lb)
Max T-O weight	1,168 kg (2,575 lb)

Performance (at max T-O weight):
Cruising speed	161 knots (298 km/h; 185 mph)
Range	544 nm (1,009 km; 627 miles)

THUNDER WINGS SUPERMARINE SPITFIRE Mk IX

Details of the Spitfire Mk IX are generally similar to those of the P-40C, except as detailed:

Tail Unit: As for P-40C, except trim tab in rudder and port elevator.

Landing Gear: As for P-40C, except that main units retract outward.

Power Plant: As for P-40C, except that engine drives a four-blade propeller with spinner. Fuel capacity 246 litres (65 US gallons).

Accommodation: Pilot only, beneath rearward-sliding transparent canopy. Hinged access panel on port side of cockpit. Baggage space with capacity of 13·6 kg (30 lb).

Dimensions, external:
Wing span	8·23 m (27 ft 0 in)
Wing area, gross	11·89 m² (128 sq ft)
Length overall	7·01 m (23 ft 0 in)
Height overall	2·74 m (9 ft 0 in)

Weights:
*Weight empty, equipped	864 kg (1,905 lb)
Max T-O weight	1,136 kg (2,505 lb)

Performance (at max T-O weight):
Cruising speed	195 knots (362 km/h; 225 mph)
Range	705 nm (1,306 km; 812 miles)

**Empty weight quoted includes optional avionics, equipment and systems as detailed for P-40C*

Thunder Wings eight-tenths scale replica of the Spitfire Mk IX

SAILPLANES

GERMANY (Federal Republic)

MILOMEI

MILOMEI M2

The Milomei M2 takes its name from that of its designer/builder, Michel-Lorenz Meier, an engineer with MBB at Hamburg, who was assisted in its completion by Herbert Löhner and Klaus Tesch. The prototype (D-2502) flew for the first time on 7 April 1981.

Type: Single-seat Open Class sailplane.

Airframe: Cantilever shoulder-wing monoplane. Wortmann wing sections: FX-67-VC-170 at root, FX-67-VC-136 at tip. Dihedral 1° 18′. Two-spar all-metal wings, with 0·6 mm duralumin skin and foam plastics infill, fitted with Wortmann variable-geometry trailing-edge flaps, plain flaps, ailerons and upper-surface spoilers. Light alloy forward fuselage, with 0·8 mm skin, and lightweight steel tailboom. Cantilever all-moving T tailplane; light alloy rudder, with 0·4 mm duralumin skin and foam plastics infill. Retractable unsprung monowheel, size 380 × 150, with brake; small, semi-recessed tailwheel. Semi-reclining seat for pilot; canopy opens sideways to starboard. Provision for 50 kg (110 lb) of water ballast.

DIMENSIONS, EXTERNAL:
Wing span 22·00 m (72 ft 2¼ in)
Wing area, gross:
flaps in 12·20 m² (131·3 sq ft)
flaps out 16·60 m² (178·7 sq ft)
Wing aspect ratio: flaps in 39·67
flaps out 29·16
Length overall 7·85 m (25 ft 9 in)

WEIGHTS AND LOADINGS:
Weight empty 365 kg (804 lb)
Max T-O weight 510 kg (1,124 lb)
Max wing loading:
flaps in 41·8 kg/m² (8·56 lb/sq ft)
flaps out 30·7 kg/m² (6·29 lb/sq ft)

PERFORMANCE:
Best glide ratio at 51 knots (95 km/h; 59 mph) 49
Min sinking speed at 35 knots (65 km/h; 40·5 mph) without water ballast 0·50 m (1·64 ft)/s
Min sinking speed at 81 knots (150 km/h; 93 mph) with water ballast 1·23 m (4·03 ft)/s
Stalling speed 30·5 knots (56 km/h; 35 mph)
Max speed (smooth air) 156 knots (290 km/h; 180 mph)
Max speed (rough air) 105 knots (195 km/h; 121 mph)
Max aero-tow speed 92 knots (170 km/h; 105 mph)
Max winch-launching speed 65 knots (120 km/h; 74·5 mph)

SWITZERLAND

EFF (page 588)

EFF Prometheus twin-jet motor glider in its current Prometheus II form, with wingtip tanks *(Martin Fricke)*

Prototype Monnett Moni ARV single-seat motor glider for amateur construction *(Howard Levy)*

UNITED STATES OF AMERICA

MONNETT (page 592)

Latest product of Monnett Experimental Aircraft Inc is the Moni single-seat motor glider, described by the manufacturer as an ARV (air recreation vehicle).

MONNETT MONI

The general appearance of this single-seat all-metal motor glider, for amateur construction, is shown in an accompanying illustration. The prototype (N107MX) flew for the first time at the end of July 1981 and was demonstrated at the subsequent EAA Fly-in at Oshkosh. It is of aluminium alloy construction, bonded and riveted, and is powered by the new Italian 16·5 kW (22 hp) KFM Mk 107 two-cylinder engine. This has an electric starter to permit restarting of the engine in flight. The wings fold for transport and storage. Other features of the Moni include a V tail and a landing gear comprising a non-retractable monowheel inside a large fairing, tailwheel inside a smaller fairing, and two underwing balancer skids. Fuel capacity is 5·5 litres (1·5 US gallons).

TYPE: Single-seat motor glider, stressed to +6/−4*g*.

DIMENSIONS, EXTERNAL:
Wing span 8·38 m (27 ft 6 in)
Wing area, gross 6·97 m² (75 sq ft)
Length overall 4·46 m (14 ft 7½ in)

WEIGHTS:
Weight empty 118 kg (260 lb)
Max T-O weight 227 kg (500 lb)

PERFORMANCE:
Max level speed 104 knots (193 km/h; 120 mph)
Max rate of climb at S/L 122 m (400 ft)/min
Endurance in powered flight 45 min

MICROLIGHT AIRCRAFT

AUSTRALIA

WINTON (page 606)

The 38th production Winton Grasshopper, powered by a 21 kW (28 hp) Fuji Robin 440 cc engine *(Howard Levy)*

FRANCE

MOTO-DELTA (page 609)

MOTO-DELTA G11

Design of the Moto-Delta G11 was undertaken in the late 1970s by M Jean-Marc Geiser, with assistance from the Ecole Nationale Supérieure des Arts et Métiers. Development is currently being continued by Société Cofrédic, formed by M Frédéric Sotteau, and a pre-production prototype of the G11 (F-WYXS) was displayed at the Paris Air Show in June 1981. A pre-series batch of ten was then being built under subcontract by Avions Pierre Robin. The following description applies to this version:

AIRFRAME: Delta-shaped Rogallo-type wing, with leading-edges, keel, cross-member and ribs of AU4GT4 aluminium tube, is supported on inverted-V members above the fuselage and braced by a single streamline strut each side. Wings can be dismantled for transportation and storage. Moulded glassfibre/epoxy minimum fuselage, incorporating open seat, windscreen, engine mounting and fuel tank. Slender tailboom, with angular fin and rudder. Tricycle landing gear, with cantilever main legs.

POWER PLANT: One 18 kW (24 hp) 424·3 cc JPX Vibraye 72 two-cylinder two-stroke engine, driving a two-blade pusher propeller. Fuel tank capacity 17 litres (3·7 Imp gallons).

DIMENSIONS:
Wing span 8·00 m (26 ft 3 in)
Wing area 12·00 m² (129·2 sq ft)
Wing aspect ratio 5·33
Length of fuselage 3·30 m (10 ft 10 in)

WEIGHTS AND LOADING:
Weight empty 55 kg (121 lb)
Max T-O weight 105 kg (231 lb)
Max wing loading 8·75 kg/m² (1·79 lb/sq ft)

PERFORMANCE:
Typical cruising speed 32 knots (60 km/h; 37 mph)
Ceiling 3,000 m (9,850 ft)
T-O run 30 m (98·5 ft)

UNITED STATES OF AMERICA

EIPPER-FORMANCE (page 618)

QUICKSILVER QM-1 and QM-2

Eipper-Formance revealed details of these two proposed military versions of the Quicksilver at the 1981 Paris Air Show. The QM-1 is the basic foot-launchable Quicksilver, with weight-shift control; the QM-2 has standard three-axis flight controls similar to those of the Quicksilver MX. Eipper claims that the small size, manoeuvrability, and minimum radar and infra-red signatures of microlight aircraft give them certain unique advantages for combat and covert operations; and that, with the wide range of miniaturised weapons now available, a realistic mission load is feasible. Typical weapon/equipment loads for an armed reconnaissance mission, each weighing 13·6 kg (30 lb) or less, might include: a Maremont M60 7·62 mm lightweight machine-gun with 500-1,000 rounds; a 5·56 mm assault rifle with ammunition clips and rifle grenades; an M203 40 mm grenade launcher on an M16A1 rifle, with ammunition clips, rifle grenades and 40 mm grenades; two Armbrust lightweight ballistic weapons; two firing pods each with four LAWs (light anti-tank weapons); ten anti-personnel mines; twelve smoke or tear gas canisters, or white phosphorus or incendiary grenades; fifteen M67 fragmentation hand grenades; ten parachute flares; an aerial camera; plus one walkie-talkie, headset with microphone, radio, pair of microchannel night flying goggles, and two quartz halogen flying lights.

TYPICAL PERFORMANCE (QM-2 with 22·4 kW; 30 hp Cuyuna 430D engine, at max T-O weight of 227 kg; 500 lb with 72·5 kg; 160 lb pilot):
Max level speed at 1,525 m (5,000 ft) 47 knots (87 km/h; 54 mph)
Max cruising speed (75% power) at 2,440 m (8,000 ft) 41 knots (76 km/h; 47 mph)
Cruising speed for max endurance 24 knots (45 km/h; 28 mph)
Stalling speed 18 knots (34 km/h; 21 mph)
Max rate of climb at S/L 158 m (520 ft)/min
Service ceiling 3,660 m (12,000 ft)
Range at 2,440 m (8,000 ft) at max cruising speed, max fuel, no reserves 227 nm (420 km; 261 miles)
Max endurance with max fuel, no reserves 6 h 42 min
Min sinking speed, power off 107 m (350 ft)/min

JIM JAEGER (page 621)
Campbellsport, Wisconsin

JAEGER J-BIRD

Basis of the J-Bird is a UFM Easy Riser wing, to which Mr Jaeger has added a chassis of his own design, a high-mounted horizontal tail surface, and an unusual power plant. A combination of skateboard seat and fuel tank is used in controlling pitch by weight shift, with foot pedals for rudder control. The power plant comprises a 7·5 kW (10 hp) McCulloch engine, with 3 : 1 reduction gear to a 1·07 m (3 ft 6 in) diameter/0·69 m (2 ft 3 in) pitch propeller, plus an offset 5·2 kW (7 hp) Chrysler 820 engine with direct drive to a 0·69 m (2 ft 3 in) diameter/0·305 m (1 ft 0 in) pitch propeller. Both propellers are two-blade, and are in counter-rotating 'pusher' configuration.

DIMENSIONS:

Wing span	9·14 m (30 ft 0 in)
Length	2·44 m (8 ft 0 in)
Height	2·13 m (7 ft 0 in)

WEIGHTS:

Weight empty	72·5 kg (160 lb)
Max T-O weight	145 kg (320 lb)

PERFORMANCE:

Max level speed	35 knots (64 km/h; 40 mph)
Cruising speed	26 knots (48 km/h; 30 mph)
Landing speed	19 knots (35 km/h; 22 mph)
Max rate of climb at S/L	122 m (400 ft)/min
Ceiling	1,525 m (5,000 ft)
T-O run	61 m (200 ft)
Landing run	30·5 m (100 ft)
Range with max fuel	17 nm (32 km; 20 miles)

KOLB (page 621)

New 1981 version of the Kolb Flyer, with two 5·6 kW (7·5 hp) Chrysler 820 direct-drive engines and other changes
(Howard Levy)

LIGHTER-THAN-AIR: AIRSHIPS

UNITED KINGDOM

AIRSHIP INDUSTRIES (page 635)

AIRSHIP INDUSTRIES SKYSHIP 500

The first of this company's new Skyship 500 non-rigid airships (G-BIHN) made a successful 2 hour maiden flight from Cardington, Bedfordshire, on 28 September 1981.

Dornier MTC II remotely piloted mini-helicopter

RPVs and TARGETS

GERMANY (Federal Republic)

DORNIER (page 648)

DORNIER MTC II

First flown on 21 March 1981, the MTC II is being tested initially within a limited flight envelope in hovering flight, and at forward speeds of up to 13·5 knots (25 km/h; 15·5 mph) and flight altitudes of 25 m (82 ft). For these tests it is controlled from a console via a control cable approx 60 m (200 ft) in length. Conversion to remote radio control is planned for later testing.

AIRFRAME: Two coaxial, contra-rotating three-blade rotors, driven by a single-stage coaxial bevel gear governed by a centrifugal clutch. Constant-chord blades, of composite construction, are attached to hub by blade holders and flapping hinges. Clutch disengages immediately in the event of engine failure, to permit autorotation. Main airframe consists of a torsionally rigid tube on which the power plant, fuel tanks, gearbox, rotors and mission equipment are mounted, and rests on a four-leg 'spider' type sprung landing gear.

POWER PLANT: One 30 kW (40 hp) Hirth O28-276-RO-3E two-cylinder two-stroke engine, with electrical starting. Two laterally-mounted fuel tanks, with combined capacity of 30 litres (6·6 Imp gallons).

LAUNCH AND RECOVERY: Conventional vertical take-off and landing.

GUIDANCE AND CONTROL: See introductory text. Aircraft is fully stabilised by an attitude controller and equipped also with a yaw-damping circuit. Four control interventions are possible, each independent of the others. Control commands are linear in character, and there is very little cross-influencing even with multiple overriding. Heading is controlled by changing the torque balance of the rotors.

EQUIPMENT: Electrical power supplied by one 300W three-phase generator. Other equipment according to mission.

DIMENSIONS, EXTERNAL:

Rotor diameter (each)	3·20 m (10 ft 6 in)
Rotor separation	0·305 m (1 ft 0 in)
Length overall (excl rotors)	1·505 m (4 ft 11¼ in)
Width of fuselage	0·48 m (1 ft 7 in)
Height overall	1·15 m (3 ft 9¼ in)
Landing gear track (c/l of feet)	0·80 m (2 ft 7½ in)
Landing gear base (c/l of feet)	1·20 m (3 ft 11¼ in)

WEIGHTS:

Max equipment load	60 kg (132 lb)
Max T-O weight	190 kg (419 lb)

PERFORMANCE:

Max level speed	75·5 knots (140 km/h; 87 mph)
Max endurance	approx 2 h

ITALY

METEOR (page 652)

METEOR ANDROMEDA/MIRACH SYSTEM

The **Mirach 100**, already in service with the Italian armed services, is reported to have a one-way penetration range of 485 nm (900 km; 560 miles). A programme of flight tests with a Mirach 100 mounted on an Agusta A 109 helicopter was nearing completion in mid-1981, and at the Paris Air Show in June Meteor exhibited an example of the Mirach 100 equipped with a Pacific Aerosystems Sirah automatic navigation system. With this system installed, the RPV can be programmed to loiter over a battlefield for surveillance purposes, and has increased range and resistance to jamming. It also has potential applications as a tactical cruise missile, possibly launched from an Aeritalia G222.

Meteor is also understood to be developing for the Italian Army a piston-engined battlefield surveillance mini-RPV known as the **Mirach 20**, similar in general appearance to the LMSC Aquila.

UNITED STATES OF AMERICA

LMSC (page 664)

LMSC Aquila target acquisition and surveillance RPV in its latest configuration

AIR-LAUNCHED MISSILES

UNITED STATES OF AMERICA

VOUGHT

ASAT

Vought Corporation is currently under contract to the US Air Force to develop and flight test an air-launched anti-satellite (ASAT) weapon. Under this programme, Vought is developing a small, high-technology interception vehicle which would be launched from an F-15 Eagle fighter and boosted by a two-stage rocket motor to orbital altitudes, where it would destroy an enemy satellite.

ENGINES

CANADA

PRATT & WHITNEY AIRCRAFT OF CANADA (page 713)

P&WC PW100

In June 1981 it was announced that, contrary to earlier reports, the engine selected for the Aérospatiale/Aeritalia ATR 42 is a version of the PW100 free-turbine turboprop. It is expected to be flat rated to about 1,343 kW (1,800 shp) in this application.

INTERNATIONAL

CFM (page 731)

CFM International announced in June 1981 the selection of the **CFM56-2000** as the launch engine for the Airbus Industrie A320. Described as embodying experience gained by earlier models of CFM56 in service, and incorporating the most advanced technology, the Dash-2000 is the subject of discussions on funding with the French government.

ROLLS-ROYCE/JAPAN (page 732)

This company announced in June 1981 that detail design of the RJ.500 (now being written RJ500) was then almost completed, and that manufacture of the RJ500-01, the first of two bench demonstrator engines, had begun. The HP core was to be tested in October 1981. RJ500-01 is scheduled to begin test running at Rolls-Royce Bristol in late January 1982. RJ500-02 is to run at IHI's Mizuho factory in mid-March. It was revealed that the three Japanese partners (IHI, Kawasaki and Mitsubishi) are responsible for the LP spool, accessories and dressings; Rolls-Royce is responsible for the HP spool, combustion section and accessory gearbox.

ITALY

KFM

KFM Mk 104

This new four-cylinder engine, designed to run on premium grade automotive fuel or avgas, was exhibited for the first time at the 1981 EAA Fly-in at Oshkosh, where it powered the Monnett Monex light aircraft (see Homebuilts section). The light alloy cylinders are separate and have iron liners. Cylinder heads are also of light alloy; the crankshaft is a high-strength alloy forging.

General appearance of the Mk 104 is shown in an accompanying illustration. It can be supplied with a propeller governor and hydraulic feed for pitch control, and has an electric starter. It will be available in 1982 with provision for inverted fuel and oil supply, for use in aerobatic aircraft. Normal rated power is 57 kW (78 hp) at 3,600 rpm.

KFM Mk 107

This two-cylinder version of the Mk 104 powers the Monnett Moni motor glider, described in this Addenda. It is rated at 16·5 kW (22 hp) and can be used in tractor or pusher installations.

KFM Mk 104 four-cylinder four-stroke engine in Monnett Monex *(Howard Levy)*

JAPAN

IHI (page 736)

IHI revealed in June 1981 that its XF3 turbofan programme has involved complete engine testing since 1975. First application for this series of two-shaft turbofans, developed under contract to the Technology Research & Development Institute (TRDI), will be in the JASDF's Kawasaki KA-850 twin-engined trainer. IHI also announced research and experimental manufacture of a smaller turbofan, designated IHI-15, intended for future civil and military light aircraft.

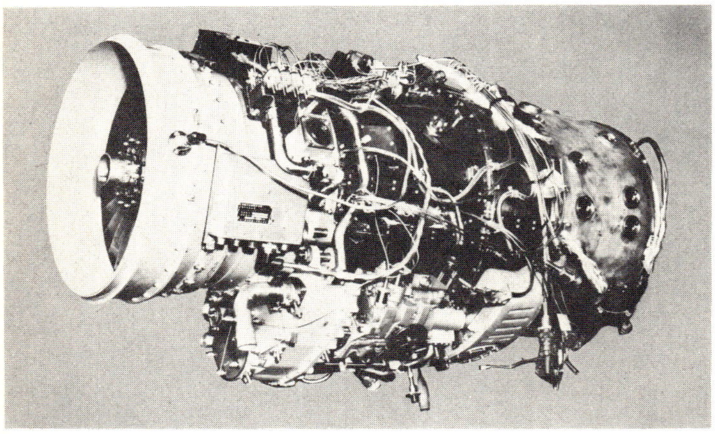

The 16·28 kN (3,660 lb st) IHI XF3-20 turbofan

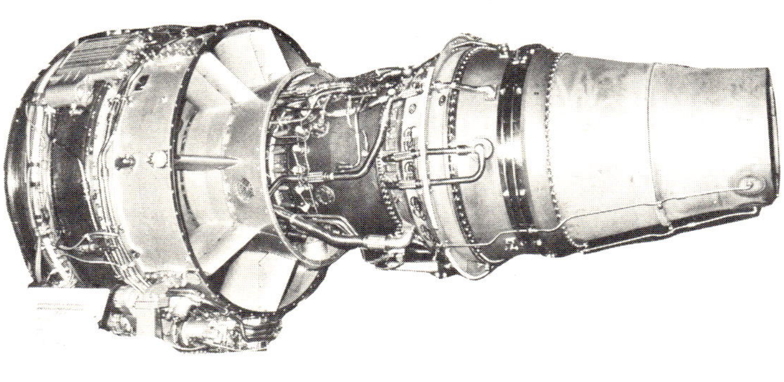

Lotarev D-36 turbofan, rated at 63·74 kN (14,330 lb st) *(Brian M. Service)*

USSR

LOTAREV (page 746)

LOTAREV D-36

It is now possible to provide a fuller description of this high bypass ratio turbofan, which powers the Yak-42 airliner and the An-72 STOL transport. Bench testing was in progress in September 1973, and flight testing in a pod carried under the converted bomb bay of a Tu-16 preceded the first flight of the Yak-42 by several years.

TYPE: Three-shaft turbofan for subsonic transport applications.

AIR INTAKE: Direct entry to fan, without inlet guide vanes.

FAN: Single stage; 29 inserted titanium blades with part-span shrouds (snubbers). Bypass ratio 5·6 at S/L ISA static.

LP COMPRESSOR: Multi-stage axial with variable inlet guide vanes.

HP COMPRESSOR: Multi-stage axial housed in intermediate case. Overall engine pressure ratio at T-O, S/L ISA static, 20.

COMBUSTION CHAMBER: Annular, with 28 burners and with integral inlet guide vanes to HP turbine. Described as a new concept for low pollution.

HP TURBINE: Single-stage; aircooled blades. Max inlet temperature 1,177°C (1,450 °K).

LP TURBINE: Single-stage. (Called IP in Western engines).

FAN TURBINE: Two stages. (Called LP in Western engines).

JET PIPE: Called rear support case. No provision for reverser.

BYPASS DUCT: Short-length, comprising forward module (called the fan contravane) and rear module (called intermediate case). Contravane case contains 49 inserted blades, divided into inner and outer sections by ring downstream of fan snubbers to remove residual twist from fan delivery. Provision for sound-reducing lining and for reverser downstream of intermediate case.

ACCESSORIES: Shaft-driven units mounted on gearbox of banana type mounted around underside of intermediate case.

DIMENSIONS: Not disclosed

WEIGHT, DRY: 1,100 kg (2,425 lb)

PERFORMANCE RATINGS(ISA):
- T-O static 63·74 kN (14,330 lb st)
- Max cruise at 8,000 m (26,250 ft) at Mach 0·75 15·7 kN (3,527 lb st)

SPECIFIC FUEL CONSUMPTION:
- T-O 10·195 mg/Ns (0·360 lb/h/lb st)
- Max cruise as above 18·4 mg/Ns (0·65 lb/h/lb st)

LOTAREV D-136

Preliminary details are available of this powerful turboshaft engine which powers the Mi-26 helicopter. Like all modern gas-turbine engines it is of modular construction, and the Soviet Aviaexport organisation stresses its simplicity, low fuel consumption, environmental acceptance and background of over 20,000 h running prior to certification to standards which included FAR and BCAR.

TYPE: Two-spool free-turbine turboshaft.

AIR INTAKE: Fabricated from sheet, with large central bullet and eight long-chord radial struts.

LP COMPRESSOR: Six stages, with one row of variable inlet guide vanes. Faces for very large bleed-air ducts, of which three normally used, with pipes facing forward (probably for anti-icing the main inlets, part of the Mi-26 airframe).

HP COMPRESSOR: Six stages, in casing provided with large bleed-air pipes and carrying main gearbox and mounting points. Overall pressure ratio 18·3.

COMBUSTION CHAMBER: Annular low-emissions type, with 28 burners and incorporating integral inlet guide vanes for HP turbine.

HP TURBINE: Single-stage with aircooled blades. Max inlet temperature 1,205°C (1,478°K).

LP TURBINE: Single-stage, separated from HP rotor by large intermediate case with 26 guide vanes (stators).

POWER TURBINE: Two stages forming separate module; casing with 12 long-chord radial struts. Large-diameter tube housing rear drive shaft.

JET PIPE: The only handed part of the engine. Large welded steel assembly with oval-section exhaust stack curved to left or right.

DRIVE SHAFT: Flexibly mounted shaft at rear; max speed 8,300 rpm.

ACCESSORIES: Shaft-driven units mounted on gearbox above intermediate casing around HP compressor.

DIMENSIONS: Not disclosed

WEIGHT, DRY: 1,050 kg (2,315 lb)

PERFORMANCE RATINGS (ISA, S/L):
- Max T-O 8,500 kW (11,400 shp)
- Max cont at 7,500 output rpm 8,280 kW (11,100 shp)

SPECIFIC FUEL CONSUMPTION:
- Max T-O 73·8 μg/J (0·4365 lb/h/shp)

Lotarev D-136 free-turbine turboshaft, rated at 8,500 kW (11,400 shp)
(Brian M. Service)

SOLOVIEV (page 747)

Soloviev D-30KU turbofan (108 kN; 24,250 lb st) complete with external dressing, tankage and reverser *(Brian M. Service)*

UNITED STATES OF AMERICA

GENERAL ELECTRIC (page 772)

GENERAL ELECTRIC CT7

In June 1981 it was announced by Airtec (CASA/P.T. Nurtanio) that the launch engine selected for the CN-235 commuter airliner (see International Aircraft section) is the CT7-7. This is a new version of the free-turbine turboprop already chosen for the Saab-Fairchild 340.

PRATT & WHITNEY (page 782)

PRATT & WHITNEY PW3000

In June 1981 Pratt & Whitney Aircraft announced this family of shaft-drive engines in the power range 2,984-5,968 kW (4,000-8,000 shp). The 3000-series engines will be planned in both turboshaft and turboprop forms for commuter-type aeroplanes, heavy-lift helicopters and advanced V/STOL vehicles. The same core is also to be mated with a fan of about 6 bypass ratio, to yield a range of turbofans in the thrust range from 17·8 to more than 28·9 kN (4,000-6,500 lb st).

The first member of the family, on which work is proceeding, is the PW3005, rated at 3,730 kW (5,000 shp). P&WA's Government Products Division is managing the programme, in partnership with Pratt & Whitney Aircraft of Canada. An initial market is foreseen for more than 10,000 PW3005-type engines, about 60 per cent in re-engining aircraft such as the CH-47 helicopter and P-3 and C-130 aeroplanes. Company funds are being used, and the first core is scheduled to run in 1983. The following data were given as estimates:

Dimensions:
- Length (turboshaft) 1,549 mm (61·0 in)
- Length (turboprop) 1,626 mm (64·0 in)
- Length (turbofan) 1,753 mm (69·0 in)
- Diameter (turboshaft/turboprop) 635 mm (25·0 in)
- Diameter (turbofan) 787 mm (31·0 in)

Weight, Dry:
- Turboshaft 358 kg (790 lb)
- Turboprop 390 kg (860 lb)
- Turbofan 431 kg (950 lb)

Performance Ratings (ISA, S/L static, max power):
- Turboshaft 4,178 kW (5,600 shp) to 15°C
- Turboprop 3,506 kW (4,700 shp) to 32°C
- Turbofan 22·25 kN (5,000 lb st) to 32°C

Specific Fuel Consumptions (as above):
- Turboshaft 70·96 μg/J (0·42 lb/h/shp)
- Turboprop 54·06 μg/J (0·32 lb/h/shp)
- Turbofan 18·4 mg/Ns (0·65 lb/h/lb st)

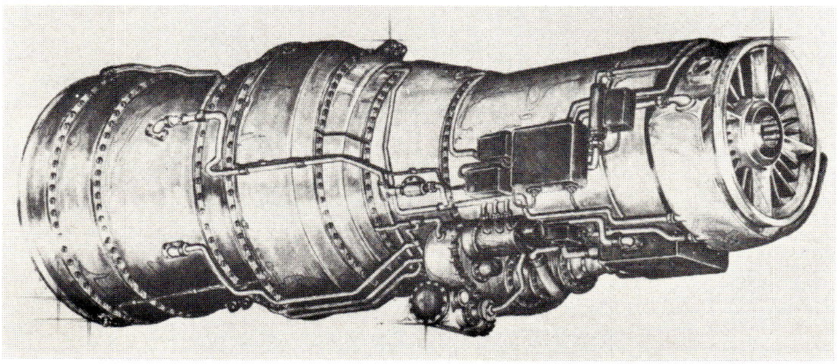

Artist's impression of 4,178 kW (5,600 shp) PW3005 turboshaft

Production Thunder TE495 V-8 piston engine with Hartzell propeller hub

THUNDER ENGINES INC

7120 Hayvenhurst Avenue, Suite 321, Van Nuys, California 91406
Telephone: (213) 997 0117

This company has used the Can-Am McLaren-Chevrolet racing car engine as the basis for an extremely competitive V-8 piston aircraft engine.

THUNDER TE495-TC700

Following long and successful development, this advanced watercooled piston engine is now in production, with a first batch of 12 assigned to certification. Present Thunder V-8s run on 100/130 grade fuel, an increasing disadvantage, and future models will have multi-fuel capability achieved mainly by electronic engine control. The cylinder head sensors will adapt fuel injection, timing and boost to any gasoline (petrol) grade or to such mixtures as JP-4/alcohol. Later a spark-assisted diesel may be developed. Initial market for the present 522 kW (700 hp) engine is put at 2,000 units, mainly for agricultural aircraft. Flight development began in January 1981 in the left position on a Rockwell Commander, the engine matching well with the TPE331 turboprop except in having much faster throttle response. The aircraft was to fly with two V-8s in Winter 1981-82, leading to certification by about April 1982. Numerous new-build installations are being negotiated for production engines.

Type: V-8 watercooled turbocharged four-stroke piston engine.

Cylinders: Blocks and pistons of Reynolds 390 Si/Al alloy. Bore 112·78 mm (4·44 in). Stroke 101·60 mm (4·00 in). Displacement 8·12 litres (495 cu in).

Induction: Direct fuel injection, twin exhaust-driven turbochargers with liquid-cooled intercooler.

Fuel: Avgas 100/130.

Ignition: Two rear-mounted magnetos, two plugs per cylinder.

Lubrication: Dry-sump system, two pressure segments and eight scavenge segments in pumps.

Propeller Drive: Reduction gear ratio 2·14.

Cooling: Twin water pumps circulating through radiator at best place on airframe (in Commander, in fuselage behind pressure bulkhead).

Accessories: Nine drive pads available.

Mounting: Four anti-vibration mounts (see photograph).

Dimensions:
- Length (with turbos) 1,625·6 mm (64·0 in)
- Length (without turbos) 1,174·75 mm (46·25 in)
- Width 704·85 mm (27·75 in)
- Height 685·80 mm (27·00 in)

Weight, Dry:
- Basic 244·27 kg (537·39 lb)
- With all accessories 323·61 kg (711·95 lb)

Power Ratings:
- Max cont 522 kW (700 hp) at 4,400 rpm
- Max recommended cruise 392 kW (525 hp) at 4,000 rpm
- Rated torque at 4,000 rpm 1,134 N-m (836 lb-ft)

Specific Fuel Consumption:
- 65% power 72·66 μg/J (0·43 lb/h/hp)

INDEXES

(Items in italics refer to the ten previous editions)

AIRCRAFT (including Homebuilt Aircraft)

A

A1 Flamingo-Trainer (FUS) 81
A1 Mk 2 (Cranfield) (1980-81) 239
A2F Intruder (Grumman) 367
A3J Vigilante (North American) (1972-73) 377
A-4 Skyhawk (Douglas) 413
A-4S Skyhawk Conversion (LAS/McDonnell Douglas) (1978-79) 358
A-5 Vigilante (North American) (1972-73) 377
A-5/F-6bis (China) (1980-81) 34
A-6 Intruder (Grumman) 367
A-6E Intruder (Grumman) 367
A-6E/TRAM (Grumman) 367
A-7 Corsair II (Vought) 469
A-9A (Northrop) (1972-73) 388
A-9B-M Quail (AAMSA) 150
A-10 (Mil) 211
A-10A (Fairchild Republic) 351
A-11 (Lockheed) (1974-75) 367
A-18 Hornet (McDonnell) 406
A 20 (Saab) (1979-80) 169
A-28F (Volpar) (1973-74) 457
A36 Bonanza (Beechcraft) 281
A36TC Bonanza (Beechcraft) 282
A-37 (Cessna) (1978-79) 311
A56 TC Turbo Baron (Beechcraft) (1972-73) 248
A65-Queen Air (Beechcraft) (1972-73) 250
A100 King Air (Beechcraft) 289
A101G (Agusta) (1971-72) 119
A 106 (Agusta) (1974-75) 126
A 109A (Agusta) 124
A 109A (Military, Naval & Police versions) (Agusta) 125
A 109A Mk II (Agusta) 126
A 119 (Agusta) (1977-78) 112
A-122A Uirapuru (Aerotec) (1980-81) 10
A-122B Uirapuru (Aerotec) (1980-81) 10
A 129 Mangusta (Agusta) 126
A-132 Tangará (Aerotec) 9
A150 Aerobat (Cessna) (1978-79) 299
A300 (Airbus) 93
A300B2-600 (Airbus) 95
A300B4-600 (Airbus) 95
A300C4/F4 (Airbus) 95
A310 (Airbus) 96
A320 (Airbus) 98
AA (see Alliance Aviation) (1974-75) 42
AA-1 Yankee (AAC) (1972-73) 236
AA-1A Trainer (AAC) (1972-73) 237
AA-1B Trainer (Grumman American) (1976-77) 297
AA-1C (Gulfstream American) (1979-80) 352
AA-2 Patriot (AAC) (1971-72) 222
AA-5A (Gulfstream American) (1980-81) 355
AA-5B (Gulfstream American) (1989-81) 355
AAC (see American Aviation Corporation) (1972-73) 236
AAMSA (See Aeronautica Agricola Mexicana SA) 150
AB 205A-1 (Agusta-Bell) (1974-75) 128
AB 206A-1 (Agusta-Bell) (1974-75) 128
ABHC (see Arab British Helicopter Company) (1980-81) 45
AC-7 Eagle 1 (Windecker) (1975-76) 480
AC-10 (Chatelain) (1971-72) 51
AC-119 Gunship (Fairchild Industries) (1974-75) 333
AC-119G (Fairchild Industries) (1974-75) 333
AC-119K (Fairchild Industries) (1974-75) 333
AC-130 Hercules (Lockheed) 395
ACAP programme (Sikorsky/US Army) 458
ACLG (Bell) (1979-80) 277
AD-1 (Ames) (1980-81) 261
AD-1 Oblique-wing aircraft (NASA) 417
ADA (see Aeronautical Development Associates Ltd) 235
AERCAB (Fairchild Industries) (1975-76) 331
AERO (see Aero Vodochody Narodni Podnik) 36
AESL (see Aero Engine Services Ltd) (1972-73) 157
AFTI-16 Programme (General Dynamics) 364
Ag-5 (Bell) (1971-72) 242
AG O2 Poussin (Gatard) 483
AG O2Sp (Gatard) 483
AG O4 Pigeon (Gatard) (1980-81) 476
AG O5 Mésange (Gatard) (1978-79) 475
AH. Mk 1 and 2 Sioux (Bell) (1971-72) 242
AH-1G HueyCobra (Bell) 296
AH-1J SeaCobra (Bell) 296
AH-1Q HueyCobra (Bell) 296
AH-1R HueyCobra (Bell) 296
AH-1S HueyCobra (Bell) 298
AH-1T SeaCobra (Bell) 297
AH-56A (Lockheed) (1973-74) 359
AH-64 (Hughes) 382
AIC (see Ames Industrial Corporation) 274
AIDC/CAF (see Aero Industry Development Center, Chinese Air Force) 183
AIRTEC (see Aircraft Technology Industries) 98
AISA (see Aeronautica Industrial SA) 173
AJ 37 Viggen (Saab) 177
AJEP (see AJEP Developments) (1978-79) 213
AJI (see American Jet Industries Inc) (1979-80) 259
AL-2 Tijerete (Al-Aire) (1974-75) 3
AL.60C5/AL.60F5 (Aermacchi/Lockheed) (1974-75) 126
AL-AIRE (see Talleres Al-Aire SCA) (1975-76) 2
ALR (see Arbeitsgruppe für Luft- und Raumfahrt) 178

AM (see École Nationale Supérieure d'Arts et Métiers) (1978-79) 470
AM.3C (Aeritalia) (1976-77) 103
AM-69 Georges Payre (AM) (1977-78) 484
AMC (see Area de Materiel Córdoba, Fuerza Aérea Argentina) 1
AM-C 111 (Air-Metal) (1978-79) 65
AMF-S14 (SAL) (1974-75) 29
AMR (see Aero-Marine Research) (1975-76) 243
AMST (Boeing) (1979-80) 301
AM-X (International) 91
An-2 (Antonov) 186
An-2 (WSK-PZL-Mielec) 161
An-3 (Antonov) 186
An-6 (Antonov) (1971-72) 427
An-10 Ukraina (Antonov) (1973-74) 463
An-12 (Antonov) 186
An-14 Pchelka (Antonov) (1980-81) 179
An-14M (Antonov) (1973-74) 465
An-22 Antheus (Antonov) 187
An-24 (Antonov) 187
An-26 (Antonov) 187
An-28 (Antonov) 188
An-28 (WSK-PZL-Mielec) 161
An-30 (Antonov) 188
An-32 (Antonov) 189
An-32 (HAL/Antonov) 90
An-40 (Antonov) (1979-80) 181
An-72 (Antonov) 190
AN-100 (Neukom) (1980-81) 492
AO-1 Mohawk (Grumman) (1973-74) 338
AP-63-31 Thunderchief (Republic) (1973-74) 326
AP 68B (Aeritalia) 121
AP.68B (Partenavia) 138
AR 404 (Ahrens) 273
ARDC (see Aeronautical Research and Development Corporation) (1972-73) 239
ARDC OF NZ (see Aeronautical Research and Development Company of New Zealand) 498
AS-1 Albee Sport (Brodhead) (1973-74) 287
AS-27 (Starck) (1978-79) 483
AS 32T Turbo Trainer (FWA) 180
AS-37 (Starck) 494
AS-61A-4 (Agusta-Sikorsky) (1980-81) 124
AS-90 Super New Look (Starck) (1979-80) 484
AS 202 Bravo (FWA) 179
AS 332, Super Puma (Aérospatiale) 48
AS 332 Super Puma (Nurtanio/Aérospatiale) 91
AS 350 Ecureuil/Astar (Aérospatiale) 51
AS 355 Ecureuil 2/Twinstar (Aérospatiale) 51
AS 365F Dauphin 2 (Aérospatiale) 54
AS 365M Dauphin 2 (Nurtanio-Aérospatiale) 54
AS 365N, Dauphin 2 (Aérospatiale) 53
AS 366G Dauphin 2 (Aérospatiale) 55
ASTEC (see Advanced Systems Technology Inc) 275
AT-19 (Greenapples) 526
AT-26 (Embraer) 15
AT-301 Air Tractor (Air Tractor) 274
AT-302 Air Tractor (Air Tractor) 274
AT-400 (Air Tractor) 275
ATCA (Douglas) (1979-80) 391
ATE-3 (Tervamäki) (1977-78) 483
ATEL (see Aviation Traders (Engineering) Ltd) (U.K.) (1974-75) 195
ATI-2 Fantrainer (RFB) 83
ATR-42 (International) 92
A.U.M. 903 (Umbra) (1971-72) 135
AU-23A (Fairchild Industries) (1977-78) 293
AU-24A (Helio) (1976-77) 301
AV-8A (BAe) 247
AV-8B (McDonnell Douglas) 405
AV-8C Harrier (BAe/McDonnell) 405
AV-8S (BAe) 247
AV-16A (McDonnell Douglas/Hawker Siddeley) (1974-75) 113
AV.44 (Fauvel) (1979-80) 474
AV.50(61) Lutin (Fauvel) (1979-80) 474
AVE (see Advanced Vehicle Engineers) (1973-74) 241
AVEX (see Asociacion Argentina de Constructores de Aviones Experimentales) (1979-80) 465
AWACS (Boeing) (1979-80) 300
AWACS aircraft (Tupolev) (1975-76) 514
AWI-2, Fantrainer (RFB) 83
AX-01 (FMA) (1972-73) 4
AX-02 (FMA) (1972-73) 4
ABE, KEUCHI (Japan) 496, 612
ABELS, L. GALE (USA) (1980-81) 500
Abe Mizet II (JEAA) (1976-77) 474
Acapella (Option Air) 538
ACE AIRCRAFT MANUFACTURING CO (U.S.A.) 506
Acey Deucy (Powell) 542
Acro Z-CH 150 (Zenair) (1980-81) 470
Acrobatique (Robin) (1977-78) 63
Acrobin (Robin) (1979-80) 68
Acroduster 1 (Stolp) 556
Acroduster Too (Stolp) 556
Acro-Sport I (EAA) 522
Acro-Sport II (EAA) 523
Acrostar Mk III (Hirth) (1977-78) 71
Acro-Zenith CH150 (Zenair) 478
Advanced Airborne Command Post (Boeing) 316

Advanced Harrier (McDonnell) (1980-81) 383
Advanced Hercules (Lockheed) (1979-80) 376
ADVANCED SYSTEMS TECHNOLOGY INC (USA) 275
Advanced Tanker/Cargo Aircraft (Douglas) 412
Advanced TriStar (Lockheed) 394
ADVANCED VEHICLE ENGINEERS (U.S.A.) (1973-74) 241
Advanced Wild Weasel F-4G (McDonnell) (1980-81) 380
Adventure (Mead) 534
Aereon 26 (Aereon) (1976-77) 205
AEREON CORPORATION (U.S.A.) (1976-77) 205
AERFER-INDUSTRIE AEROSPAZIALI MERIDIONALI SpA (Italy) (1971-72) 113
AERIAL AGRICULTURE PTY LTD (Australia) (1972-73) 5
AERIAL DISTRIBUTORS, INC (U.S.A.) (1973-74) 316
AERITALIA/AERMACCHI/EMBRAER (International) 91
AERITALIA — SOCIETÀ AEROSPAZIALE ITALIANA pa (Italy) 118
AERMACCHI (see Aeronautica Macchi SpA) 122
Aerobat (Reims) 69
Aerobat, Model 152 (Cessna) 325
AERO BOERO SRL (Argentina) 1
AEROCAR, INC (U.S.A.) 506
Aerocar Model III (Aerocar) (1971-72) 216
AEROCOM INC (U.S.A.) (1978-79) 249
AERO COMMANDER (see North American Rockwell Corporation Industrial Products Group) (1972-73) 379
AERO DESIGN ASSOCIATES (U.S.A.) (1980-81) 502
AERO-DIFUSIÓN SL (Spain) (1971-72) 164
AERO ENGINE SERVICES LTD (New Zealand) (1972-73) 157
AERO FLIGHT INC (U.S.A.) (1972-73) 232
Aerofoil Boat (RFB) (1974-75) 98
Aerofoilcraft, X-114 (RFB) (1978-79) 70
AERO INDUSTRY DEVELOPMENT CENTER — CHINESE AIR FORCE (Taiwan) 183
AERO-MARINE RESEARCH (U.S.A.) (1975-76) 243
Aero Model A (Meyer) 535
AEROMOD, INC (U.S.A.) (1971-72) 217
AERONAUTICA AGRICOLA MEXICANA SA (Mexico) 150
AERONAUTICA INDUSTRIAL SA (Spain) 173
AERONAUTICAL DEVELOPMENT ASSOCIATES LTD (U.K.) 235
AERONAUTICA MACCHI SpA (Italy) 122
AERONAUTICA UMBRA SpA (Italy) (1972-73) 137
AERONAUTICAL RESEARCH AND DEVELOPMENT COMPANY OF NEW ZEALAND (New Zealand) 498
AERONAUTICAL RESEARCH AND DEVELOPMENT CORPORATION (U.S.A.) (1972-73) 239
AERONCA INC (U.S.A.) (1979-80) 257
AERONEERING INC (U.S.A.) (1979-80) 503
Aero Knight Twister (Payne) 540
AERONICS (PTY) LTD (South Africa) 171
AERO RESOURCES INC (U.S.A.) (1975-76) 241
AEROSPACE (see New Zealand Aerospace Industries Ltd) 155
AEROSPACE GENERAL COMPANY (U.S.A.) (1980-81) 259
AERO SPACELINES, INC (U.S.A.) (1974-75) 243
AÉROSPATIALE (see Société Nationale Industrielle Aérospatiale) 43
AÉROSPATIALE/AERITALIA (International) 92
Aero Sport (D'Apuzzo) (1977-78) 533, 534
Aerosport (Olszewski-Obarewicz) (1977-78) 502
AEROSPORT INC (U.S.A.) 508
Aerostar 600A/601P, PA-60 (Piper) 441
Aerostar 602P (Piper) 442
Aerostar 700/700P (Smith) (1974-75) 464
AEROSTAR AIRCRAFT CORPORATION (U.S.A.) (1972-73) 235
AEROSTRUCTURES LTD (Australia) (1971-72) 6
Aero Subaru (Fuji) 141
AERO-TECH (U.S.A.) 509
AEROTEC S/A INDUSTRIA AERONAUTICA (Brazil) 9
AERO TECHNIK CANARY (West Germany) (1972-73) 86
AEROTECHNIK ENTWICKLUNG UND APPARATEBAU GmbH (West Germany) (1975-76) 88
AERO VODOCHODY NARODNI PODNIK (Czechoslovakia) 36
AEW Nimrod (Hawker Siddeley) (1977-78) 191
AFIC (PTY) LTD. (South Africa) (1974-75) 176
AGcarryall (Cessna) (1980-81) 313
Ag-Cat (Grumman/Schweizer) (1973-74) 430
Ag-Cat G-164/R-1820 (Serv-Aero) (1979-80) 440
Ag-Cat/Leonides (Serv-Aero) (1979-80) 439
Ag Husky (Cessna) 330
Agile (Wallis) (1971-72) 210
AgMaster (Frontier) 355

AGpickup (Cessna) .... *(1975-76)* 305
Agricopter (Bensen) .... *(1976-77)* 500
AGRICOPTEROS LTDA (Colombia) 36
Agricorallye (SOCATA) .... *(1971-72)* 79
Agricultural aircraft (NASA) 418
AGRICULTURAL AVIATION ENGINEERING COMPANY (U.S.A.) .... *(1974-75)* 245
Agricultural Islander (Britten-Norman) ... *(1977-78)* 196
AGRINAUTICS (see Agricultural Aviation Engineering Company) .... *(1974-75)* 245
Agronemair MA-1 Paymaster (Emair) ... *(1974-75)* 328
AGRUPACION AVIONES-DEPARTAMENTO INGENIERIA, GUARNICION AEREA CORDOBA (Argentina) .... *(1979-80)* 3
AgTrainer (Aviones de Colombia/Cessna) 36
Ag Truck (Cessna) 330
AGUSTA GROUP, THE (Italy) 124
AGUSTA SpA, COSTRUZIONI AERONAUTICHE GIOVANNI (Italy) 124
Ag Wagon (Cessna) 330
AHRENS AIRCRAFT CORPORATION (USA) .. 273
AHRENS AIRCRAFT INC (U.S.A.) 273
Aiglon, R 1180 (Robin) 72
Airborne Scientific Observatories (NASA) (1975-76) 405
Airbuggy (WHE) 272
Airbuggy, VoWi 10 (Airconcept) .... *(1980-81)* 77
AIRBUS INDUSTRIE (International) 93
Aircamper, B4 (Pietenpol) 542
Aircamper, GN-1 (Grega) 526
Air Car Model S-12-E (Spencer) 551
AIRCONCEPT FLUGZEUG UND GERÄTEBAU GmbH und Co KG (West Germany) 76
AIRCRAFT DESIGNS (BEMBRIDGE) LTD (U.K.) 235
AIRCRAFT GROUP, ROCKWELL INTERNATIONAL (U.S.A.) 452
AIRCRAFT HYDRO-FORMING, INC (U.S.A.) .... *(1971-72)* 269
Aircraft noise monitoring programme (NASA) .... *(1980-81)* 396
AIRCRAFT TECHNOLOGY INDUSTRIES (International) 98
Aircruiser CT/2 (AESL) .... *(1972-73)* 158
Aircruiser CT/2 (Aerospace) .... *(1974-75)* 162
Air Cushion Landing Gear (Bell) .... *(1979-80)* 277
Airkart, MW3 (Whittaker) .... *(1979-80)* 500
Airlifter CV-880 (GACC) 355
Airliner, Model 99 (Beechcraft) .... *(1979-80)* 273
AIRMARK LTD (U.K.) .... *(1972-73)* 186
AIRMASTER HELICOPTERS LTD (U.K.) *(1975-76)* 193
AIR-METAL FLUGZEUGBAU UND ENTWICKLUNGS GmbH & CO BETRIEBS-KG (West Germany) ...... *(1978-79)* 65
Airmite (Millicer) .... *(1977-78)* 478
AIR-MOD ENGINEERING COMPANY (U.S.A.) .... *(1972-73)* 235
AIR NOVA (PTY) LTD (South Africa) (1975-76) 175
AIR PARTS (N.Z.) LTD (New Zealand) (1972-73) 159
AIRPLANE ALLEY (U.S.A.) 509
Air Skimmer, X-28A (Osprey) .... *(1974-75)* 410
Airtourer (AESL) .... *(1972-73)* 157
Airtourer (Aerospace) .... *(1975-76)* 162
Air Tractor (Air Tractor) 274
AIR TRACTOR INC (U.S.A.) 274
Airtrainer CT/4 (AESL) .... *(1972-73)* 158
Airtrainer CT4A (Aerospace) 155
Airtruk PL-12 (Transavia) 7
Airtruk T-320 (Transavia) .... *(1979-80)* 8
Ajeet (HAL) 88
Ajeet Trainer (HAL) 88
AJEP DEVELOPMENTS (U.K.) .... *(1978-79)* 213
AKADEMISCHE FLIEGERGRUPPE STUTTGART EV (West Germany) .... *(1975-76)* 89
AKAFLIEG STUTTGART (see Akademische Fliegergruppe Stuttgart EV) .... *(1975-76)* 89
Akro (Stephens) 554
Akrobat (Moravan) .... *(1975-76)* 39
Akromaster (Spinks) .... *(1971-72)* 408
Aladin II (Bertin) .... *(1973-74)* 51
Albatros L-39 (Aero) 36
Albatross, G-111 (Grumman/Resorts International) 365
Albee Sport, AS-1 (Brodhead) .... *(1973-74)* 287
ALCOR AVIATION INC (U.S.A.) .... *(1972-73)* 235
Alizé (Breguet) 61
ALLIANCE AVIATION (France) .... *(1974-75)* 42
Alouette II Astazou (Aérospatiale) ...... *(1976-77)* 38
Alouette III, ICA/Aérospatiale (ICA-Brasov) 171
Alouette III, IAR-361B (ICA-Aérospatiale) 171
Alouette III, SA 316B (Aérospatiale) 45
Alouette III, SA 316B (HAL/Aérospatiale) 89
Alouette III Astazou, SA 319B (Aérospatiale) 45
Alpha (Partenavia) .... *(1976-77)* 114
Alpha 11D (Alpha) .... *(1972-73)* 236
ALPHA AVIATION COMPANY (U.S.A.) *(1972-73)* 236
Alpha Jet (Dassault-Breguet/Dornier) 98
ALPHA JET (International) 98
Alpha series (Robin) 72
Alpha Sport (Robin) 72
Alpine Commander (Rockwell International) .... *(1980-81)* 431
Altinger/Yakstas Monoplane (AVEX) .... *(1971-72)* 3
ALVAREZ, JOSEPH P. (U.S.A.) .... *(1978-79)* 504
Amazonas (EMBRAER) .... *(1972-73)* 11
AMBROSINI & C, SOCIETA AERONAUTICA ITALIANA ING A (Italy) .... *(1972-73)* 124
AMERICAN AIR RACING INC (U.S.A.) *(1978-79)* 504
AMERICAN AVIATION CORPORATION (U.S.A.) .... *(1972-73)* 236
AMERICAN JET INDUSTRIES INC (U.S.A.) *(1979-80)* 259
American Tr-2 (AAC) .... *(1972-73)* 237
American Traveler (AAC) .... *(1972-73)* 237
AMES INDUSTRIAL CORPORATION (U.S.A.) . 274
Ami Pou, GL02 (Landray) 487
Amphibian Air Car (Spencer) 551
ANAHUAC, FABRICA DE AVIONES, SA (Mexico) 150
Analogue (Mikoyan) .... *(1979-80)* 192
ANDERSON AIRCRAFT CORPORATION (U.S.A.) 510
Andover (Hawker Siddeley) .... *(1973-74)* 203
ANDREASSON, BJORN (Sweden) 498
Antares (Brandt) .... *(1972-73)* 47
Antares (Kokubun) .... *(1972-73)* 143
Antheus (Antonov) 187
ANTONOV (U.S.S.R.) 186
Apollo Jetstream (ADA) 235
Aquilottero (Silvestri) .... *(1976-77)* 474
ARAB BRITISH HELICOPTER COMPANY (Egypt) .... *(1980-81)* 45
Araguaia (EMBRAER) .... *(1978-79)* 14
Arapaho (Piper) .... *(1980-81)* 416
Arava (IAI) 114
Arbalète II (Payen) .... *(1973-74)* 75
Arbalète III (Payen) .... *(1973-74)* 76
ARBEITSGEMEINSCHAFT TRANSALL (International) 11
ARBEITSGRUPPE FÜR LUFT- UND RAUMFAHRT (Switzerland) 178
Archer II (Piper) 428
ARCTIC AIRCRAFT COMPANY (U.S.A.) 275
Arctic Tern (Arctic) 275
AREA DE MATERIAL CÓRDOBA, FUERZA AÉREA ARGENTINA (Argentina) ....
Argosy (Hawker Siddeley) .... *(1974-75)* 215
Aries T-250 (Bellanca) .... *(1980-81)* 287
Aristocraft II (O'Neill) .... *(1973-74)* 392
Armar I Gorrion (AVEX) .... *(1977-78)* 478
Armed Porter (Fairchild Hiller/Pilatus) ... *(1971-72)* 302
Arrambide/Marino Armar I Gorrion (AVEX) *(1977-78)* 478
Arrow IV (Piper) 429
ARTIOMOV, MIKHAIL (U.S.S.R.) ..... *(1977-78)* 574
ASOCIACION ARGENTINA DE CONSTRUCTORES DE AVIONES EXPERIMENTALES (Argentina) ..... *(1979-80)* 465
Astar AS 350 (Aérospatiale) 51
Astra, IAI-1125 (IAI) 117
Astro-Rocket (Maule) .... *(1976-77)* 324
Atlantic (Wassmer) .... *(1977-78)* 66
Atlantic M4 (Dassault-Breguet) .... *(1977-78)* 54
Atlantic ANG (Dassault-Breguet) 61
ATLAS AIRCRAFT CORPORATION OF SOUTH AFRICA (PTY) LIMITED (South Africa) 172
Attack aircraft system 85 (Saab) .... *(1977-78)* 166
Augmentor wing jet STOL research aircraft (NASA/DITC) .... *(1980-81)* 25
Aurora (Lockheed) 392
Autogyro GN (AISA) 173
Auto-Gyrocopter, W.S.9 (Seremet) .... *(1980-81)* 471
Autoplane (Lebouder) .... *(1975-76)* 74
Avenger (Marchetti) 534
AVIAN AIRCRAFT LTD (Canada) ..... *(1972-73)* 14
AVIATION SPECIALTIES INTERNATIONAL (U.S.A.) .... *(1976-77)* 208
AVIATION TRADERS (ENGINEERING) LTD. (U.K.) .... *(1974-75)* 195
Aviette (Hurel) .... *(1977-78)* 490
Aviocar (CASA) 173, 174, 175
Aviocar, NC-212 (Nurtanio/CASA) 91
Aviojet, C-101 (CASA) 175
AVIONES DE COLOMBIA SA (Colombia) 36
Avion-Planeur RF7 (Sportavia) .... *(1972-73)* 96
AVIONS FOURNIER (France) .... *(1977-78)* 58
AVIONS A. GATARD (France) 483
AVIONS CLAUDE PIEL (France) 489
AVIONS JODEL SA (France) 484
AVIONS MARCEL DASSAULT/BREGUET AVIATION (France) 55
AVIONS MUDRY ET COMPAGNIE (France) ... 68
AVIONS PIERRE ROBIN (France) 70
AVIONS ROBIN (Canada) 27
AVIONS ROGER DRUINE (France) .... *(1975-76)* 66
Avispa (Akaflieg Stuttgart) .... *(1974-75)* 90
Avro 748 (Hawker Siddeley) .... *(1977-78)* 181
AYRES CORPORATION (U.S.A.) 276
Ayres Thrush modification (Serv-Aero) 458
Aztec (Piper) 426

B

B-1 (Rockwell) 453
B-1, Sylkie One (Wayne) .... *(1979-80)* 506
B-1-A (Drews) .... *(1979-80)* 516
B-2 (Barker) .... *(1972-73)* 241
B-2B (Brantly-Hynes) 322
B3LA (Saab) .... *(1978-79)* 157
B4 Aircamper (Pietenpol) 542
B-5 (China) 32
B-6 (China) 33
B-8 Gyro-Glider (Bensen) 510
B-8H (Bensen) .... *(1976-77)* 500
B-8HD (Bensen) 512
B-8M Gyro-Copter (Bensen) 511
B-8MA Agricopter (Bensen) .... *(1976-77)* 500
B-8MH Hover-Gyro (Bensen) .... *(1980-81)* 506
B-8MJ (Bensen) 512
B-8MW Hydro-Copter (Bensen) 511
B-8V Gyro-Copter (Bensen) 511
B-8W Hydro-Glider (Bensen) 510
B-10 (Mitchell) .... *(1980-81)* 531
B-12 Sky-Mat (Bensen) .... *(1971-72)* 252
B-16S Gyro-Copter (Bensen) .... *(1976-77)* 500
B-18 Hover-Gyro (Bensen) 512
B19 Sport (Beechcraft) .... *(1973-74)* 246
B-52 Stratofortress (Boeing) 316
B55 Baron (Beechcraft) .... *(1977-78)* 225
B-57G (Westinghouse/Martin Marietta) ... *(1972-73)* 442
B60 Duke (Beechcraft) 285
B-80 Gyro-Copter (Bensen) 512
B-80D Gyro-Copter (Bensen) 512
B80 Queen Air (Beechcraft) .... *(1979-80)* 271
B80 Queen Airliner (Beechcraft) .... *(1979-80)* 271
B90 King Air (Beechcraft) .... *(1971-72)* 236
B99 Airliner (Beechcraft) .... *(1978-79)* 267
B100 King Air (Beechcraft) 290
B.103 Buccaneer (Hawker Siddeley) ..... *(1977-78)* 189
B200 Super King Air (Beechcraft) 290
B-377MG Mini Guppy (Aero Spacelines) .... *(1972-73)* 233
B-377PG Pregnant Guppy (Aero Spacelines) .... *(1972-73)* 233
B-377SG Super Guppy (Aero Spacelines) .... *(1972-73)* 233
BA-4B (Andreasson) 498
BA-4B (Crosby/Andreasson) .... *(1980-81)* 494
BA-11 (Andreasson) 499
BAC (see British Aircraft Corporation Ltd) *(1977-78)* 176
BAC 145 Jet Provost (BAC) .... *(1972-73)* 190
BAC 167 Strikemaster (BAe) 245
BAe (see British Aerospace Public Limited Company) 235
BD-1 Durl-E-Aire (Durley) .... *(1971-72)* 297
BD-2 Love One (WFI) .... *(1972-73)* 445
BD-4 (Bede) .... *(1977-78)* 524
BD-5 Micro (Bede) .... *(1977-78)* 524
BD-5D (Bede) .... *(1977-78)* 219
BD-5J (Bede) .... *(1977-78)* 525
BD-5JP (Bede) .... *(1977-78)* 219
BD-6 (Bede) .... *(1977-78)* 525
BD-7 (Bede) .... *(1977-78)* 526
BD-8 (Bede) .... *(1977-78)* 526
Be-12 Tchaika (Beriev) 191
Be-30 (Beriev) .... *(1974-75)* 469
Be-32 (Beriev) .... *(1977-78)* 431
B-EC-9 Paras-Cargo (Croses) 481
BGA (see British Gliding Association) 252
BGP1 Biplane (Plumb) 504
BH 125 (Hawker Siddeley) .... *(1977-78)* 180
Bi 20 (Binder) .... *(1980-81)* 486
BJ-520 (Brokaw) .... *(1977-78)* 530
BK 117 (MBB/Kawasaki) 102
BN-2A Mk III Trislander (Pilatus Britten-Norman) 258
BN-2B Islander (Pilatus Britten-Norman) 256
BN-2T Turbine Islander (Britten-Norman) 258
BN-3 Nymph (Britten-Norman) .... *(1974-75)* 202
BO 105 (MBB) 84
BO 105 (PADC/MBB) 157
BO 105 Executaire (Boeing Vertol) ...... *(1979-80)* 306
BO 105 LS (MBB) 85
BO 106 (MBB) .... *(1976-77)* 73
BO 107 (MBB) .... *(1976-77)* 73
BO 108 (MBB) .... *(1977-78)* 74
BO 115 (MBB) .... *(1975-76)* 94
BO 208 C Junior (MBB) .... *(1971-72)* 88
BO 209 Monsun (MBB) .... *(1972-73)* 92
Bo 210 Tornado (MBB) .... *(1971-72)* 89
Bo 211 Taifun (MBB) .... *(1971-72)* 89
Br 941 (Breguet) .... *(1972-73)* 56
Br 1150 Atlantic (Breguet) .... *(1972-73)* 55
BT-5 (China) 34
BT-6 (China) 34
Bü 131A Jungmann (Morand) .... *(1975-76)* 189
Bü 133 BM Antares (Brandt) .... *(1972-73)* 47
Bü 133D-1 Jungmeister (Bücker) .... *(1971-72)* 83
Bü 133F Jungmeister (Bücker) .... *(1971-72)* 83
BURD II (MIT) .... *(1977-78)* 584
BX-50A (Berger-Helikopter) 500
BX-110 (Berger-Helikopter) 500
Baby Ace Model D (Ace Aircraft) 506
'Baby' Lakes (Barney Oldfield) 510
Backfire (Tupolev) 224
BACKSTROM, AL (U.S.A.) .... *(1977-78)* 521
Badger (China) 33
Badger (Tupolev) 220
BAGALINI, WALTER (Italy) .... (1976-77) 473
Bagheera (SITAR) .... *(1971-72)* 77
BAKENG AIRCRAFT (U.S.A.) .... *(1979-80)* 504
Baldadou (Wassmer) .... *(1971-72)* 81
BALL-BARTOE AIRCRAFT CORPORATION (U.S.A.) .... *(1979-80)* 261
Bamboo Aircraft (PATS/NSDB) .... *(1980-81)* 153
Bandeirante (EMBRAER) 9
Bandido (Brueggemann) .... *(1971-72)* 267
Bandit (Practavia) .... *(1979-80)* 498
BANGALORE COMPLEX, HINDUSTAN AERONAUTICS LTD (India) 87
BANNICK COPTERS, INC (U.S.A.) .... *(1972-73)* 240
Bantam (Warwick) .... *(1973-74)* 457
Barak (IAI) .... *(1974-75)* 119
BARKER, ARTHUR (U.S.A.) .... *(1972-73)* 241
BARNETT ROTORCRAFT COMPANY (U.S.A.) *(1977-78)* 522
BARNEY OLDFIELD AIRCRAFT CO (U.S.A.) . 510
Barney Oldfield Lakes (Oldfield) .... *(1974-75)* 249
Baron (Beechcraft) 283, 284
Baron 58 (Beechcraft) 284
Baron 58P (Beechcraft) 284
Baron 58TC (Beechcraft) 285
Baron Model 95-B55 (Beechcraft) 283
BARR BROTHERS LTD (New Zealand) (1974-75) 163
Barracuda (Jeffair) 530
BARTON, WAYNE (U.S.A.) .... *(1980-81)* 504
Basant, HA-31 Mk II (Hal) 90

Baserga H.B.1 (AVEX) .... *(1971-72)* 2
Bazant (WSK-PZL Swidnik/Mil) 165
Beagle (China) 32
Bear (Tupolev) 221
BEAUJON, HERBERT (U.S.A.) .... *(1977-78)* 523
Bébé (Jodel) 484
BECKER, HORST (West Germany) ..... *(1979-80)* 484
BEDE AIRCRAFT, INC (U.S.A.) .... *(1979-80)* 506
Bede-Huffman BD-8 (Airplane Alley) 509
Bee (Scheutzow) .... *(1977-78)* 398
BEECH AIRCRAFT CORPORATION (U.S.A.) .. 277
BEECHCRAFT (see Beech Aircraft Corporation) . 277
Beechcraft Hawker BH 125 (Hawker Siddeley) .... *(1974-75)* 206
Beechcraft Model 18 conversion (Volpar) *(1974-75)* 477
BEETS, GLENN (U.S.A.) 526
Belfast conversion (Marshall) 254
BELL AEROSPACE TEXTRON (U.S.A.) .... *(1980-81)* 278
BELL HELICOPTER TEXTRON (U.S.A.) 293
BELLANCA AIRCRAFT CORPORATION (U.S.A.) 303
BELLANCA AIRCRAFT ENGINEERING INC (U.S.A.) 303
Bellanca 260C Model 14-19-3C (Bellanca) *(1971-72)* 248
Belphegor, M-15 (WSK-PZL Mielec) 163
Bengali, HN 500 (Nicollier) 488
BENSEN AIRCRAFT CORPORATION (U.S.A.) . 510
Berg-Aero 1A (Bergeson) .... *(1971-72)* 253
BERGER, FRANZ (West Germany) ..... *(1976-77)* 473
BERGER, HANS (Switzerland) .... *(1974-75)* 188
BERGER-HELIKOPTER (Switzerland) 500
BERGESON, JOHN B. (U.S.A.) .... *(1971-72)* 253
Bergwind (Van Aswegen) .... *(1979-80)* 490
BERIEV (U.S.S.R.) 191
BERLIN DOMAN HELICOPTERS, INC (U.S.A.) .... *(1971-72)* 253
BERTIN & CIE, SOCIÉTÉ (France) .... *(1973-74)* 51
Beryl (Piel) 490
BESNEUX, ALAIN (France) .... *(1976-77)* 460
Beta (Rollason/Luton Group) .... *(1974-75)* 225
Beta Bird (Hovey) 529
BEZZOLA, GION (Switzerland) 500
'Big Bird' (Quickie) 544
BigLifter (Bell) 300
Big Wing Harrier (BAe) .... *(1980-81)* 235
BINDER, HELLMUTH (West Germany) .... *(1980-81)* 486
Biplane (EAA) 522
Biplane CB-1 (Hatz) 527
Bird (Taylor) 556
BIRD CORPORATION (U.S.A.) .... *(1971-72)* 253
BIRDMAN INC (U.S.A.) .... *(1980-81)* 506
BISHOP, RAY (U.S.A.) .... *(1973-74)* 271
Bison (Myasishchev) 214
BITTNER, WILFRED (Canada) .... *(1977-78)* 480
Bi-Universal (Neiva) .... *(1973-74)* 17
Blackhawk (Sikorsky) .... *(1975-76)* 459
Black Hawk (Sikorsky) 461
Blinder (Tupolev) 223
Bock II (Bock) .... *(1971-72)* 253
BOCK, JOHN (U.S.A.) .... *(1971-72)* 253
BOEING AEROSPACE COMPANY (U.S.A.) 314
BOEING COMMERCIAL AIRPLANE COMPANY (U.S.A) 304
BOEING COMPANY, THE (U.S.A.) 303
Boeing F4B/P-12 Replica (Aero-Tech) 509
BOEING MILITARY AIRPLANE COMPANY (U.S.A.) 316
BOEING OF CANADA LTD (VERTOL DIVISION) (Canada) 18
BOEING VERTOL COMPANY (U.S.A.) 317
BOEING WICHITA COMPANY (U.S.A.) *(1979-80)* 301
BOLEN INC, RALPH (U.S.A.) 321
BÖLKOW (see Messerschmitt-Bölkow-Blohm GmbH) 82
Bonanza A36 (Beechcraft) 281
Bonanza A36TC (Beechcraft) 282
Bonanza F33A/C (Beechcraft) 281
Bonanza V35B (Beechcraft) 280
BOND, JOHN (USA) 512
Boniev Helicopter (Bulgaria) .... *(1977-78)* 480
Boomerang (Rotor Master) .... *(1979-80)* 538
Boon Jr (Compcop/Weilage) .... *(1974-75)* 320
BORCHERS, LOWELL J (U.S.A.) 512
BORG, RICHARD R (U.S.A.) 513
Bosbok, AM.3C (Aeritalia) .... *(1976-77)* 103
BOUGIE, YVAN C. (Canada) .... *(1976-77)* 456
Bouvreuil, P-50 (Pottier) 492
BOWKER AIR SERVICES LTD (U.K.) .*(1972-73)* 192
BOWERS, PETER, M. (U.S.A.) 513
Boxmoth (Thompson) 557
BOYD, GARY (U.S.A.) 514
BRANDT, MICHEL (France) .... *(1977-78)* 484
BRANTLY-HYNES (see Brantly-Hynes Helicopter Inc) 322
BRANTLY-HYNES HELICOPTER INC (U.S.A.) 322
Brasilia, EMB-120 (EMBRAER) 12
Brave (Piper) 438
Bravo (SIAI-Marchetti/FFA) .... *(1972-73)* 135
Bravo, AS 202 (FWA) 179
BREDANARDI COSTRUZIONI AERONAUTICHE SpA (Italy) 134
Breezy Model RLU-1 (RLU) 545
BREGUET AVIATION (France) .... *(1971-72)* 47
Brewer (Yakovlev) 230
Bristol F.2B Fighter replica (Ohmert) .... *(1980-81)* 534
BRITISH AEROSPACE AIRCRAFT GROUP (U.K.) 236
BRITISH AEROSPACE PUBLIC LIMITED COMPANY (U.K.) 235
BRITISH AIRCRAFT CORPORATION, LTD. (U.K.) .... *(1977-78)* 176
BRITISH AIRCRAFT CORPORATION (COMMERCIAL AIRCRAFT) LTD (U.K.) *(1977-78)* 176
BRITISH AIRCRAFT CORPORATION (FILTON) LTD. (U.K.) .... *(1971-72)* 178
BRITISH AIRCRAFT CORPORATION (HOLDINGS) LTD. (U.K.) .... *(1971-72)* 177
BRITISH AIRCRAFT CORPORATION LTD, MILITARY AIRCRAFT DIVISION (U.K.) *(1976-77)* 172
BRITISH AIRCRAFT CORPORATION (PRESTON) LTD. (U.K.) .... *(1977-78)* 177
BRITISH AIRCRAFT CORPORATION (WEYBRIDGE) LTD. (U.K.) .... *(1971-72)* 178
BRITISH GLIDING ASSOCIATION (U.K.) 252
BRITTEN-NORMAN (BEMBRIDGE) LTD (U.K.) .... *(1978-79)* 231
BRODHEAD, ARTHUR L. (U.S.A.) .... *(1973-74)* 287
BROKAW AVIATION INC (U.S.A.) 514
Bronco (Rockwell) 452
Broucek (Verner) .... *(1972-73)* 32
BRUEGGEMANN, CLARENCE (U.S.A.) *(1971-72)* 267
BRÜGGER, MAX (Switzerland) 500
BRYAN, LELAND D. (U.S.A.) .... *(1973-74)* 287
Buccaneer (BAe) .... *(1979-80)* 237
Buccaneer Conversion (Marshall) .... *(1977-78)* 201
Buccaneer, LA-4-200 (Lake) 386
BUCCIERO, RENATO (Italy) .... *(1975-76)* 131
BÜCKER (see Aero Technik Canary) .... *(1972-73)* 86
Buckeye (Rockwell) 452
'Buddy Baby' Lakes (Oldfield) 510
Buffalo (DH Canada) 22
Bulldog Series 120 (BAe) 245
Bulldog Series 200 (BAe) .... *(1978-79)* 223
Bullet (Brokaw) 514
Bullfinch (Scottish Aviation) .... *(1977-78)* 193
Busara (Lefebvre) .... *(1974-75)* 76
Busard, MP.205 (Lefebvre) 488
Bushbird (Fletcher) 474
BUSHBY AIRCRAFT, INC (U.S.A) 515
BUSHMASTER AIRCRAFT CORPORATION (U.S.A.) .... *(1975-76)* 297
Businessliner (Cessna) 337, 338
BUTLER AIRCRAFT COMPANY (U.S.A.) *(1972-73)* 282
BUTTERWORTH, G. N. (U.S.A.) 516

C

C-1 (Kawasaki) 143
C-2A Greyhound (Grumman) .... *(1971-72)* 312
C4M Kudu (Atlas) 172
C-5 (China) 34
C-5A Galaxy (Lockheed) 398
C-7 Caribou (D.H. Canada) .... *(1974-75)* 22
C-7A Caribou (D.H. Canada) .... *(1973-74)* 21
C-8A Buffalo (D.H. Canada) 22
C-9 (Northrop) .... *(1977-78)* 362
C-9A Nightingale (Douglas) 408
C-9B Skytrain II (Douglas) 408
C-10 (China) 34
C-10 (Hispano) .... *(1973-74)* 178
C-11 (China) 35
C-12 (Beechcraft) 290
C22J (Caproni Vizzola) 135
C23 Sundowner (Beechcraft) .... *(1976-77)* 211
C-42 Regente (Neiva) .... *(1974-75)* 18
C90 King Air (Beechcraft) 286
C-95 (EMBRAER) 10
C-95A (EMBRAER) 10
C-96 (EMBRAER) .... *(1971-72)* 12
C99 Commuter (Beechcraft) 288
C100 King Air (Beechcraft) .... *(1978-79)* 269
C-101 Aviojet (CASA) 175
C-102 (CASA) 176
C-112 (CASA) .... *(1978-79)* 155
C-119K (Fairchild Industries) .... *(1974-75)* 333
C-119/R4Q (Steward-Davis/Fairchild) .... *(1974-75)* 468
C-123T (Mancro/Fairchild) 400
C-130 Hercules (Lockheed) 395
C-130 Conversions (LAS) 387
C-140 Jet Star (Lockheed) .... *(1974-75)* 376
C-141 StarLifter (Lockheed) 398
C-160 (Maritime versions) (Transall) 111
C-160 (Second series) (Transall) 111
C-212-5 Aviocar (CASA) 173
C-212 Series 200 Aviocar (CASA) 174, 175
C-3605 (Federal Aircraft Factory) .... *(1974-75)* 192
CA-4 Skyhawk (Douglas) .... *(1972-73)* 368
CA-61/-61R Mini Ace (Cvjetcovic) 518
CA-65 (Cvjetkovic) 518
CA-65A (Cvjetkovic) 519
C.A.A.R.P. (see Coopérative des Ateliers Aéronautiques de la Région Parisienne) .*(1977-78)* 49
CAC (see Commonwealth Aircraft Corporation Ltd.) 4
CAC (see Commuter Aircraft Corporation) 323
CAC-100 (CAC) 323
CANARY (see Aero Technik Canary) .... *(1972-73)* 86
CAP 10 B (Mudry) 68
CAP 20 (C.A.A.R.P./Mudry) .... *(1978-79)* 49
CAP 20L (Mudry) 68
CAP 21 (Mudry) 69
CAP X (Mudry) 69
CASA (see Construcciones Aeronáuticas SA) 173
CASI (see Canadian Aeronautics and Space Institute) .... *(1978-79)* 467
CATA (see Compania Argentina de Trabajos Aéreos) .... *(1975-76)* 3
CB-1 Biplane (Hatz) 527
CC-108 Caribou (D.H. Canada) .... *(1974-75)* 22
CC-115 Buffalo (D.H. Canada) 22
CC-130 Hercules (Lockheed) 395
CC-132 (D.H. Canada) 25
CC-138 Twin Otter (D.H. Canada) 24
CCV YF-16 (General Dynamics) .... *(1978-79)* 336
CCW A-6A (Grumman) .... *(1979-80)* 346
CE-9 (Northrop) .... *(1977-78)* 362
CE.43 Guépard (CERVA) .... *(1977-78)* 49
CE.44 Cougar (CERVA) .... *(1977-78)* 50
CE.45 Léopard (CERVA) .... *(1977-78)* 50
CERVA (G.I.E.) (see Consortium Européen de Réalisation et de Ventes d'Avions (Groupement d'Intérêts Économiques) ..*(1978-79)* 45
CF-5 (Canadair) .... *(1972-73)* 17
CF-5 (Northrop) .... *(1978-79)* 392
CF-18A Hornet (McDonnell) 406
CF-104 Starfighter (Lockheed) .... *(1972-73)* 344
CH-II (Cicaré) .... *(1974-75)* 4
CH-III Colibri (Cicaré) .... *(1976-77)* 2
CH-3 (Sikorsky) 449
CH-3B (Sikorsky) .... *(1977-78)* 400
CH-46 Sea Knight (Boeing Vertol) 317
CH-47 Chinook (Boeing Vertol) 318
CH-47C (EM/Boeing Vertol) 129
CH 50 Mini Z (Zenair) 478
CH-53 Sea Stallion (Sikorsky) 460
CH-53E (Sikorsky) 460
CH-53G (VFW-Fokker/Sikorsky) .... *(1975-76)* 99
CH-54 (Sikorsky) .... *(1978-79)* 433
CH-113 Labrador (Boeing Canada) 18
CH-113 Labrador (Boeing Vertol) .... *(1973-74)* 283
CH-113A Voyageur (Boeing Canada) 18
CH-113A Voyageur (Boeing Vertol) ..... *(1973-74)* 283
CH-118 Iroquois (Bell) 293
CH-124 (Sikorsky) 458
CH-135 (Bell) 299
CH-136 (Bell) 294
CH-147 (Boeing Vertol) 318
CHSS-2 Sea King (Sikorsky) .... *(1974-75)* 453
CIAR (see Centrala Industriala Aeronautica Romania) .... *(1979-80)* 159
CJ-1 Starlet (Corby) 473
CJ600 (Texas Airplane) .... *(1976-77)* 397
C.K.1 (Cicaré) .... *(1979-80)* 3
CL-44-0 (Conroy) .... *(1972-73)* 303
CL-84-1 (Canadair) .... *(1975-76)* 18
CL-84-1D (Canadair) .... *(1972-73)* 15
CL-215 (Canadair) 20
CL-219 (Canadair) .... *(1972-73)* 17
CL-226 (Canadair) .... *(1972-73)* 17
CL-600 Challenger (Canadair) 19
CL-601 Challenger (Canadair) 19
CL-610 Challenger E (Canadair) 20
CM 170 Magister (Aérospatiale) .... *(1971-72)* 41
CM 170 Super Magister (Aérospatiale) ...*(1971-72)* 41
CN-235 (AIRTEC) 98
CNA (see Centrul National Aeronautic) 169
CNIAR (see Centrul National al Industrie Aeronautice Romåne) 168
CNPSL-PZL-WARSZAWA (see Centrum Naukowo-Produkcyjne Samolotow Lekkich PZL-Warszawa) 158
CP-16 (Turner/Carroll) .... *(1977-78)* 569
C.P. 60/601/602 Diamant (Piel) .... *(1975-76)* 76
C.P. 70/750/751 Beryl (Piel) 490
C.P.80 (Piel) 491
C.P. 90 Pinocchio (Piel) 492
CP-140 Aurora (Lockheed) 392
C.P. 301/304/305/308/320/320A/321/323A/323AB Emeraude (Piel) 489
CP 301, Mistral (Becker) .... *(1979-80)* 484
C.P.500 (Piel) 492
C.P. 604/605/605B Super Diamant (Piel) 490
C.P. 1320 (Piel) 490
CR420 (Campbell) .... *(1974-75)* 204
CR640 (Campbell) .... *(1974-75)* 204
CR.LTH-1 Grasshopper (Cierva) .... *(1973-74)* 199
CR Twin (Campbell) .... *(1974-75)* 203
CS2F Tracker (Grumman) .... *(1972-73)* 324
CSIR (see Council for Scientific and Industrial Research) 172
CT/2 Airtrainer (AESL) .... *(1972-73)* 158
CT/2 Aircruiser (Aerospace) .... *(1974-75)* 162
CT/4 Airtrainer (AESL) .... *(1972-73)* 158
CT4 Airtrainer (Aerospace) .... *(1977-78)* 145
CT4A, Airtrainer (Aerospace) 155
CT-39 Sabreliner (Rockwell) .... *(1978-79)* 424
CTA (see Centro Técnico Aerospacial) 9
CTDC (see Zaidan Hozin Minkan Yusooki Kaihatsu Kyokai) .... *(1979-80)* 127
CTN Twin Navion (Camair) .... *(1980-81)* 306
CUH-1H (Bell) .... *(1978-79)* 272
CUH-1N (Bell) .... *(1978-79)* 276
CV-2A Caribou (DH Canada) .... *(1973-74)* 21
CV-7A Buffalo (DH Canada) .... *(1973-74)* 22
CV-880 Airlifter (GACC) 355
CW 205 (CW Helicopter) .... *(1976-77)* 456
CX (McDonnell) 413
CX-84 (Canadair) .... *(1974-75)* 20
Cacique T-11 (Al-Aire) .... *(1974-75)* 2
Cadet (Robin/CEA) .... *(1972-73)* 78
Cadet M-10 (Univair) .... *(1975-76)* 474
CALIFORNIA POLYTECHNIC STATE UNIVERSITY (U.S.A.) .... *(1979-80)* 307
Cali, T-610 (PAF/AJI) 157
CAL POLY (see California Polytechnic State University) .... *(1979-80)* 307
CALSPAN ADVANCED TECHNOLOGY CENTER (USA) 323
CAMAIR AIRCRAFT CORPORATION (U.S.A.) .... *(1980-81)* 306
Camber (Ilyushin) 197
CAMPBELL AIRCRAFT LTD. (U.K.) ..*(1975-76)* 201
CANADAIR, LTD (Canada) 19
CANADIAN AERONAUTICS AND SPACE INSTITUTE (Canada) .... *(1978-79)* 467
Canard (Utva) .... *(1972-73)* 490

Canberra (BAC) *(1973-74)* 194
Candid (Ilyushin) 195
Canguro, F.600 (General Avia) *(1980-81)* 133
Canguro, SF.600 (SIAI-Marchetti) 132
CAPRONI VIZZOLA COSTRUZIONI AERONAUTICHE SpA (Italy) 135
Carajá (Neiva) *(1974-75)* 19
Caravelle (Aérospatiale) *(1972-73)* 39
Caravelle 11R (Aérospatiale) *(1971-72)* 38
Caravelle 12 (Aérospatiale) *(1974-75)* 43
Caravelle Super B (Aérospatiale) *(1971-72)* 38
Cardinal (Cessna) *(1977-78)* 267
Cardinal II (Cessna) *(1977-78)* 267
Cardinal RG (Cessna) *(1979-80)* 310
Cardinal RG II (Cessna) *(1979-80)* 310
Cardinal Classic (Cessna) *(1979-80)* 310
Careless (Tupolev) 228, 229
Caribou (DH Canada) *(1974-75)* 22
Carioca (EMBRAER/Piper) 17
Carioquinha, EMB-712 (EMBRAER) *(1978-79)* 16
CARMAM, SOCIÉTÉ (France) *(1975-76)* 70
CARMICHAEL, BOX (U.S.A.) 516
CARSTEDT, INC (U.S.A.) *(1972-73)* 283
Cash (Antonov) 188
Cash (WSK-PZL-Mielec) 161
CASSUTT (U.S.A.) 516
Cassutt Special I (Cassutt) 516
Cassutt Special II (Cassutt) 517
Cat (Antonov) *(1973-74)* 463
CAUCHY, ALAIN (France) 482
CAVALIER AIRCRAFT CORPORATION (U.S.A.) *(1972-73)* 283
Cavalier, SA 102 Point 5 (K & S) *(1975-76)* 26
Cavalier (Macfam) 476
Cayuse (Hughes) *(1977-78)* 318
CB ENTERPRISES (U.S.A.) *(1977-78)* 532
Centennial (Volpar) *(1977-78)* 420
CENTRALA INDUSTRIALA AERO-NAUTICA ROMANA (Romania) *(1979-80)* 159
CENTRE DE RECHERCHES JEAN ST-GERMAIN INC (Canada) 474
CENTRO TÉCNICO AEROSPACIAL (Brazil) 9
CENTRUL NATIONAL AERONAUTIC (Romania) 169
CENTRUL NATIONAL AL INDUSTRIE AERONAUTICE ROMANE (Romania) 168
CENTRUM NAUKOWO-PRODUKCYJNE SAMOLOTOW LEKKICH PZL-WARSZAWA (Poland) 158
Centurion I/II (Cessna) 333
Century 600 (Colemill) 346
CENTURY AIRCRAFT CORPORATION (USA) *(1978-79)* 299
CESSNA AIRCRAFT COMPANY (U.S.A.) 324
CHAGNES, LÉO (France) *(1980-81)* 472
Chaika-1 (Riga) *(1971-72)* 453
Challenger (Cook) *(1974-75)* 321
Challenger, CL-600 (Canadair) 19
Challenger E, CL-610 (Canadair) 20
Champ (Bellanca/Champion) *(1973-74)* 267
CHAMPION, KENNETH R. (U.S.A.) *(1972-73)* 302
Chancellor/II/III (Cessna) 339
Chaparral (Aerostar) *(1971-72)* 220
Chaparral (Mooney) *(1976-77)* 339
Charger (Tupolev) 227, 228
Charlie, P.66C-160 (Partenavia) 137
CHASLE, YVES (France) *(1979-80)* 470
CHATELAIN, ANDRÉ (France) *(1971-72)* 51
Cheetah (Aérospatiale) 45
Cheetah (Gulfstream American) *(1980-81)* 355
Cheetah (HAL/Aérospatiale) 89
Cherokee 140 (Piper) *(1972-73)* 400
Cherokee 180 (Piper) *(1972-73)* 401
Cherokee 235 (Piper) *(1972-73)* 402
Cherokee Archer II (Piper) *(1977-78)* 369
Cherokee Arrow III (Piper) *(1977-78)* 369
Cherokee Arrow Trainer (Chincul/Piper) 4
Cherokee Challenger (Piper) *(1973-74)* 403
Cherokee Charger (Piper) *(1973-74)* 404
Cherokee Cruiser (Piper) *(1977-78)* 367
Cherokee Flite Liner (Piper) *(1974-75)* 421
Cherokee Lance (Piper) *(1977-78)* 371
Cherokee PA-28 conversion (Isham/Piper) 384
Cherokee Pathfinder (Piper) *(1979-80)* 408
Cherokee SIX (Piper) *(1978-79)* 400
Cherokee Warrior II (Piper) *(1977-78)* 368
Chetak (Aérospatiale) 45
Chetak (HAL/Aérospatiale) 89
Chevalier, DR 400/160 (Robin) *(1979-80)* 67
Cheyenne (Lockheed) *(1973-74)* 359
Cheyenne I (Piper) 434
Cheyenne II (Piper) 435
Cheyenne IIXL, PA-31T2 (Piper) 436
Cheyenne III (Piper) 436
Chieftain (Piper) 434
Chienshou (AIDC/Pazmany) *(1976-77)* 169
CHINCUL S.A.C.A.I.F.I. (Argentina) 3
Chinook (Boeing-Vertol) 318
Chiricahua UV-20 (Pilatus) 180
CHOLOT, JACQUES (France) *(1972-73)* 49
CHRIS TENA AIRCRAFT ASSOCIATION (U.S.A.) *(1976-77)* 504
CHRISTEN INDUSTRIES INC (USA) 517
Chrysalis (MIT) 413
Chung Cheng (AIDC/CAF) *(1980-81)* 176
Chung Hsing, T-CH-1 (AIDC/CAF) 183
Cicaré II (Cicaré) *(1973-74)* 5
Cicaré III (Cicaré) *(1973-74)* 5
CICARÉ AERONÁUTICA SC (Argentina) *(1980-81)* 4
CIERVA (see Cierva Rotorcraft Ltd) *(1975-76)* 201
CIERVA ROTORCRAFT LTD. (U.K.) *(1975-76)* 201
Citabria (Bellanca/Champion) *(1980-81)* 287
Citation I (Cessna) 344
Citation I/SP Model 501 (Cessna) 344
Citation II (Cessna) 344
Citation III Series (Cessna) 345
Citation Eagle (ASTEC) *(1978-79)* 253
CITRÖEN-MARCHETTI (France) *(1978-79)* 45
CIVIL AVIATION DEPARTMENT, TECHNICAL CENTRE (India) *(1978-79)* 72
Civil Lynx (Westland) *(1976-77)* 201
CIVIL TRANSPORT DEVELOPMENT CORPORATION (Japan) *(1973-74)* 138
Clank (Antonov) 188
Classic (Ilyushin) 194
Cleat (Tupolev) *(1973-74)* 492
Cline (Antonov) 90, 189
Cline (HAL) 90
Clobber (Yak) 233
Clod (Antonov) *(1980-81)* 179
Club (Jodel) 484
Club (Reims) *(1971-72)* 73
Club (Robin/CEA) *(1973-74)* 83
Club, DM-125/DM-160 (Dalotel) *(1974-75)* 56
Club, RF-6B (Fournier) *(1977-78)* 58
CLUTTON, ERIC (U.K.) 500
CLUTTON-TABENOR (see Clutton, Eric) 500
Cmelák (Let) (1976-77) 30
Coaler (Antonov) 190
Coastguarder (BAe) 240
COATES, J. R. (U.K.) 502
Cobra, F-3 (Forbes) *(1978-79)* 520
Cochise (Beechcraft) 283
Cock (Antonov) 187
Codling (Yakovlev) 232
Coke (Antonov) 187
COLE, W. (U.K.) *(1971-72)* 188
COLEMILL ENTERPRISES INC (USA) 346
Colibri (Cicaré) *(1976-77)* 2
Colibri 1 (Brügger) *(1971-72)* 172
Colibri 2 (Brügger) 500
COLOMBAN, MICHEL (France) 480
Colt (Antonov) 186
Colt (China) 34
Colt (WSK-PZL-Mielec) 161
Comanche C (Piper) *(1974-75)* 419
Comanchero (Schafer) 455
Comanchero 500 (Schafer) 455
Comanchero 750 (Schafer) 455
Commander III (Rockwell) *(1973-74)* 423
Commander 112 (Rockwell) *(1978-79)* 426
Commander 114 (Rockwell) *(1978-79)* 427
Commander 685 (Rockwell) *(1976-77)* 377
Commander 690 (Rockwell) *(1977-78)* 395
Commander 700 (Rockwell) *(1980-81)* 432
Commander 710 (Rockwell) *(1980-81)* 432
Commando (Westland) 268
Commercial Chinook (Boeing Vertol) 320
Commercial Hercules (Lockheed) 397
Commodore Jet Eleven 23 (IAI) *(1972-73)* 113
COMMONWEALTH AIRCRAFT CORPORATION LTD. (Australia) 4
COMMONWEALTH OF AUSTRALIA DEPARTMENT OF SUPPLY (Australia) *(1974-75)* 8
COMMONWEALTH OF AUSTRALIA GOVERNMENT AIRCRAFT FACTORIES (Australia) *(1974-75)* 8
Commuter (Reims) *(1978-79)* 57
Commuter (Richard) *(1974-75)* 432
Commuter IIA (International Helicopters) 530
Commuter 1900 (Beechcraft) 289
Commuter C99 (Beechcraft) 288
COMMUTER AIRCRAFT CORPORATION (U.S.A.) 323
Commuter, Model 340 (SAAB/Fairchild) 106
Commuter Jr Model H-1A (Helicom) *(1972-73)* 330
COMPANIA ARGENTINA DE TRABAJOS AÉREOS (Argentina) *(1975-76)* 3
COMPCOP INC (U.S.A.) *(1974-75)* 320
Competitor, VM-7 (Murphy) 537
COMPOSITE AIRCRAFT CORPORATION (U.S.A.) 347
Concorde (Concorde) 101
CONCORDE SUPERSONIC TRANSPORT (International) 101
Condor (Aero Boreo) *(1974-75)* 2
Condor (Druine) *(1975-76)* 66
Condor (Rollason) *(1974-75)* 225
CONDOR AERO INC (U.S.A.) 518
Condor F.20 (General Avia) 136
Conestoga (Aermacchi/Lockheed) *(1974-75)* 126
Conquest (Cessna) 343
CONROY AIRCRAFT CORPORATION (U.S.A.) *(1973-74)* 311
CONSORTIUM EUROPÉEN DE RÉALISATION ET DE VENTES D'AVIONS (GROUPEMENT D'INTÉRETS ÉCONOMIQUES) (France) *(1978-79)* 45
CONSTRUCCIONES AERONÁUTICAS S.A. (Spain) 173
CONTINENTAL COPTERS, INC. (U.S.A.) 347
Continental Rosette (Sizer) *(1975-76)* 230
CONVAIR (see General Dynamics Corporation) 360
Convair 580 (PAC) *(1971-72)* 374
Convair 600 (General Dynamics) *(1971-72)* 310
Convair 640 (General Dynamics) *(1971-72)* 310
COOK AIRCRAFT CORPORATION (U.S.A.) *(1977-78)* 532
"Cookpot" (Tupolev) *(1974-75)* 517
COOPÉRATIVE DES ATELIERS AÉRONAUTIQUES DE LA RÉGION PARISIENNE (France) *(1977-78)* 49
Coot (Ilyushin) 192
Coot Model B (Aerocar) *(1974-75)* 242
Copter (Bannick) *(1972-73)* 240
CORBY, JOHN C. (Australia) 473
Corisco (EMBRAER/Piper) 17
CORNELL AERONAUTICAL LABORATORY INC. (U.S.A.) *(1972-73)* 305
Corsair (Cessna) 341
Corsair, F4U (WAR) 562
Corsair II (Vought) 469
Corsair II₂ (Vought) *(1980-81)* 453
Corvette (Aérospatiale) *(1978-79)* 37
Corvette 200 (Aérospatiale) *(1975-76)* 45
COSMIC AIRCRAFT CORPORATION (U.S.A.) *(1974-75)* 322
COSTRUZIONI AERONAUTICHE GENERAL AVIA SpA (Italy) 136
COSTRUZIONI AERONAUTICHE GIOVANNI AGUSTA (Italy) 124
Cougar (Campbell) *(1974-75)* 203
Cougar (EAA) *(1979-80)* 519
Cougar (Gulfstream American) *(1980-81)* 355
Cougar, CE.44 (CERVA) *(1977-78)* 50
COUNCIL FOR SCIENTIFIC AND INDUSTRIAL RESEARCH (South Africa) 172
Coupé (Ord-Hume) *(1973-74)* 216
COUPÉ-AVIATION (France) 481
Coupe RSH-1A (Rotor Sport) 545
Courier (Helio) *(1972-73)* 331
COX AIR RESOURCES LTD (Canada) *(1980-81)* 23
CRANFIELD INSTITUTE OF TECHNOLOGY (U.K.) 252
CRANWELL MPAG (see RAF Cranwell Man-Powered Aircraft Group) *(1977-78)* 508
Cresco 600 (Aerospace) 156
Cricket (Campbell) *(1975-76)* 201
Cricri, MC 10 (Colomban) 480
Cricri, MC 12 (Colomban) 480
Criquet (Croses) 481
Criquet, LC-10 (Croses) *(1977-78)* 488
Crivat (Jurca) *(1971-72)* 67
CRONK, DAVID (U.S.A.) *(1976-77)* 505
CROSBY AVIATION LTD. (U.K.) *(1980-81)* 494
CROSES, EMILIEN (France) 481
Crusader (LTV) *(1971-72)* 342
Crusader, Model T303 (Cessna) 335
Crusty (Tupolev) 226
CRUTCHLEY, S (South Africa) *(1979-80)* 489
Cub (Antonov) 186
CUBy (Wag-Aero) 561
"Cuff" (Beriev) *(1973-74)* 469
CUNNING AIRCRAFT (U.S.A.) *(1974-75)* 322
Curl (Antonov) 187
Currie Wot (PFA) 504
Currie Wot (Phoenix) *(1972-73)* 215
CUSTER CHANNEL WING CORPORATION (U.S.A.) *(1972-73)* 306
Cutlass (Cessna) *(1979-80)* 310
Cutlass RG (Cessna) 326
CVJETKOVIC, ANTON (U.S.A.) 518
CW HELICOPTER RESEARCH (Belgium) *(1976-77)* 456
Cygnet, SF-2A (Sisler) 549
Cygnus (Nihon University) *(1972-73)* 148
CZERNIEJEWSKI, CZESLAW (Poland) *(1974-75)* 170

D

D (Kelly) *(1980-81)* 526
D.1 (Jones) 475
D-1 Derringer (Wing) 471
D-1 Skyservant (Dornier) *(1975-76)* 90
D2 (Rabouyt) *(1974-75)* 81
D.5 Turbi (Druine) *(1975-76)* 66
D.9 Bébé (Jodel) 484
D.11 (Jodel) 484
D.31 Turbulent (Druine) 504
D31 Turbulent (Rollason) *(1974-75)* 224
D31A Turbulent (Rollason) *(1974-75)* 224
D.61/D.62 Condor (Druine) *(1975-76)* 66
D62 Condor (Rollason) *(1974-75)* 225
D.92 Bébé (Jodel) 484
D.112 Club (Jodel) 484
D.119 (Jodel) 484
D-200 Freshman (D'Apuzzo) 519
D-200 Junior Aero Sport (D'Apuzzo) *(1977-78)* 534
D-201 Sportwing (D'Apuzzo) 520
D-260/D-295 Senior Aerosport (D'Apuzzo) 519
D-295 (Parsons-Jocelyn) *(1974-75)* 412
D2127 (Bell) *(1971-72)* 240
DA-2A (Davis) 520
DA-3 (Davis) *(1980-81)* 514
DA-5A (Davis) 521
DAL-1 Tuholer (Spezio) 552
DC-01 Rapace (Cholot) *(1972-73)* 49
DC-1 (Delmontez-Cauchy) 482
DC-8 (Douglas) *(1973-74)* 375
DC-8 freighter conversion (McDonnell Douglas) 407
DC-8F Jet Trader (Douglas) *(1973-74)* 375
DC-8 Srs 71/72/73 (McDonnell Douglas) 407
DC-9 (Douglas) 408
DC-9 Super 80 (Douglas) 408
DC-10 (Douglas) 410
DC-10 Model 30CF (Douglas) 411
DC-130 Hercules (Lockheed) 395
DC-XX (McDonnell) 413
DF-4B Phantom II (McDonnell) *(1980-81)* 380
DG-1 (Aero Design) *(1980-81)* 502
DGCA (see Directia Generala Constructii Aeronautice) *(1972-73)* 167
D.H.2 Replica (Mason) *(1975-76)* 382
DH-2 Whirlymite (Del Mar) *(1972-73)* 308
D.H. 121 Trident (Hawker Siddeley) *(1977-78)* 184
D.H. 125 (Hawker Siddeley) *(1977-78)* 180
DHC-2/A1 Wallaroo 605 (Aerial Agriculture) *(1972-73)* 5
DHC-3-T Turbo Otter (Cox) *(1980-81)* 23
DHC-4A Caribou (DH Canada) *(1974-75)* 22
DHC-5 Buffalo (DH Canada) 22

DHC-5E Transporter (DH Canada) 23
DHC-6 Twin Otter (DH Canada) 24
DHC-7 Dash 7 (DH Canada) 25
DHC-7R Ranger (DH Canada) ....(1980-81) 28
DHC-8 Dash 8 (DH Canada) 26
DHT-2 Whirlymite Helicopter Pilot Trainer (Del Mar) ....(1972-73) 309
DK-1, Der Kricket (FLSZ) 525
D.L.5 (Livesey) ....(1979-80) 496
DM-125 Club (Dalotel) ....(1974-75) 56
DM-160 Club (Dalotel) ....(1974-75) 56
DM-160 Professional (Dalotel) ....(1974-75) 56
DM-165 (Paulet) ....(1972-73) 75
Do 24A (Dornier) ....(1978-79) 67
Do 28 D Skyservant (Dornier) ....(1975-76) 90
Do 28 D-1 Skyservant (Dornier) ....(1975-76) 90
Do 28 D-2 Skyservant (Dornier) 77
Do 31E (Dornier) ....(1971-72) 85
Do 128 Skyservant (Dornier) 77
Do 128-6 Turbo-Skyservant (Dornier) 78
Do 132 (Dornier) ....(1971-72) 85
Do 231 (Dornier) ....(1973-74) 89
DR 220 "2+2" (Robin/CEA) ....(1972-73) 77
DR 221 Dauphin (Robin/CEA) ....(1972-73) 78
Do 228 (Dornier) 78
DR 253 Regent (Robin/CEA) ....(1972-73) 78
DR 300/108 "2+2 Tricycle" (Robin/CEA)(1972-73) 79
DR 300/125 Petit Prince (Robin/CEA) ...(1972-73) 79
DR 300/140 Petit Prince (Robin/CEA) ...(1972-73) 79
DR 300/180R Remorqueur (Robin/CEA) .(1972-73) 79
DR 315 Cadet (Robin/CEA) ....(1972-73) 78
DR 340 Major (Robin/CEA) ....(1972-73) 80
DR 360 Major 160 (Robin/CEA) ....(1972-73) 80
DR 380 Prince (Robin/CEA) ....(1972-73) 80
DR 400/100 2 + 2 (Robin) ....(1980-81) 71
DR 400/120 Dauphin 80 (Robin) 70
DR 400/140B Major (Robin) ....(1979-80) 66
DR 400/160 Major 80 (Robin) 71
DR 400/180 Régent (Robin) 71
DR 400/180R Remorqueur (Robin) 71
DSA-1 Miniplane (Smith) 550
DSK-1 Hawk (DSK Airmotive) ....(1980-81) 515
DSK-2 Golden Hawk (DSK Airmotive) ..(1980-81) 515
DW-1 (Distributor Wing) ....(1971-72) 296
Dakota (Piper) 430
DALOTEL, SOCIÉTÉ (France) ....(1975-76) 56
Dalotel DM-165 (Poulet) ....(1972-73) 75
D'APUZZO, NICHOLAS E. (U.S.A.) 519
Dart, M.J.3H (Jurca) 485
Dash 7, DHC-7 (DH Canada) 25
Dash 8, DHC-8 (DH Canada) 26
Dash X (DH Canada) ....(1979-80) 26
DASSAULT-BREGUET AVIATION, AVIONS MARCEL (France) 55
DÄTWYLER & CO., MAX (Switzerland) (1976-77) 164
Dauntless, SBD-3 (Roberts) ....(1979-80) 538
Dauphin (Robin/CEA) ....(1972-73) 78
Dauphin, SA 360 (Aérospatiale) 52
Dauphin, SA 361H/HCL (Aérospatiale) ..(1979-80) 49
Dauphin 2, AS 365F (Aérospatiale) 54
Dauphin 2, AS 365M (Aérospatiale) 54
Dauphin 2, AS 365N (Aérospatiale) 53
Dauphin 2, SA 365C (Aérospatiale) 53
Dauphin 2, AS 366 (Aérospatiale) 55
Dauphin 80, DR 400/120 (Robin) 70
Dauphin, SA 366 (Aérospatiale) ....(1976-77) 45
DAVIS, LEEON D (U.S.A.) 520
Decathlon (Bellanca/Champion) ....(1980-81) 287
Dedal-2 (Polniak) ....(1977-78) 503
DEE HOWARD COMPANY, THE (USA) 347
Defender (Pilatus Britten-Norman) 257
Defender (Hughes) 381
Defiant (Rutan) 548
DE HAVILLAND AIRCRAFT OF CANADA LTD., THE (Canada) 21
DELEMONTEZ-CAUCHY (see Cauchy, Alain) .. 482
Delfin (Aero) ....(1975-76) 33
Delfin Akrobat (Aero) ....(1974-75) 33
DEL MAR ENGINEERING LABORATORIES (U.S.A.) ....(1973-74) 316
Delta, JD-2 (Dyke) 521
Delta Bird (Hovey) 529
Delta Dart (Convair) ....(1971-72) 310
Delta-Ente (Richter) 494
Delta Mirage 2000 (Dassault) 58
Delta Stinger (Rinker) ....(1971-72) 393
Delta Stingray (Borchers) 512
Delta Super Mirage (Dassault) ....(1977-78) 53
DEMCHENKO, YURI, KHITRY, DMITRY AND GUSEV, VSEVOLOD (U.S.S.R.) .....(1971-72) 435
DEPARTAMENTO DE AERONAVES (Brazil) (1972-73) 10
DEPARTEMEN ANGKATAN UDARA REPUBLIK INDONESIA, LEMBAGA INDUSTRI PENERBANGAN NURTANIO (Indonesia) ....(1977-78) 85
Der Donnerschlag (Harmon) ....(1978-79) 521
DeRic DE 13P Delta-Ente (Richter) 494
Der Jäger D.IX (White) 564
Der Kricket DK-1 (FLSZ) 525
Derringer, D-1 (Wing) 471
DEUTSCHE AIRBUS GmbH (West Germany) ... 76
Diablo 1200, MA-1B (Emair) ....(1980-81) 333
Diamant (Piel) ....(1977-78) 495
Diamond I (Mitsubishi) 146
Diamond I (Mitsubishi International) 415
DIETRICK SALES & SERVICE (U.S.A.) 347
Digital Fly-By-Wire system (NASA) .....(1978-79) 389
Diplomate (Socata) ....(1976-77) 66
DIRECTIA GENERALA CONSTRUCTII AERONAUTICE (Romania) ....(1972-73) 167
DISTRIBUTOR WING (see Aerial Distributors Inc) ....(1973-74) 316
Distributor Wing DW-1 (Aerial Distributors) (1971-72) 296
DMITRIEV, V (U.S.S.R.) ....(1976-77) 546
DOBRACZYNSKI, TADEUSZ (Poland) (1974-75) 170
Dolphinair (Omnionics) ....(1980-81) 401
Dominie (Hawker Siddeley) ....(1977-78) 180
DOMINION AIRCRAFT CORPORATION LTD (U.S.A.) ....(1979-80) 330
Don Kichot, J-1B (Janowski) ....(1979-80) 488
DORAND (see Giravions Dorand SARL) (1971-72) 60
Dormoy Bathtub Mk 1 (Kimbrel) 532
DORNIER GmbH (West Germany) 76
Double Duce (Bakeng) ....(1979-80) 505
Double Teenie (Parker) ....(1979-80) 532
DOUGLAS AIRCRAFT COMPANY (U.S.A.) 407
DOWNTOWN AIRPARK INC (U.S.A.) .(1972-73) 309
DOWTY BOULTON PAUL LTD. (U.K.)(1973-74) 200
Drake (Frizzle) ....(1979-80) 466
Draken (Saab) ....(1977-78) 168
Draggin' Fly, RDW-2 (Wier) ....(1974-75) 480
Dragonfly (Prestwick) ....(1976-77) 487
Dragonfly (Viking) 560
Dragonfly, Model 318E (Cessna) 336
DREWS, JOHN (USA) ....(1979-80) 516
Dromader, M-18 (WSK-PZL-Mielec) 164
DRUINE, AVIONS ROGER (France) ...(1975-76) 66
DSK AIRMOTIVE (U.S.A.) ....(1980-81) 515
Duce (Bakeng) ....(1979-80) 504
Duchess 76 (Beechcraft) 282
Dudus Kudlacz (Studenckie Biuro Konstrukcyine) (1974-75) 172
Duet (Knowles) ....(1980-81) 495
Duke (Beechcraft) 285
DUMOD CORPORATION, THE (U.S.A.) (1971-72) 297
DURAND, ASSOCIATES INC (U.S.A.) 521
Durl-E-Aire (Durley) ....(1971-72) 297
DURLEY, WILLIAM (U.S.A.) ....(1971-72) 297
DURUBLE, ROLAND (France) 482
DYKE AIRCRAFT (U.S.A.) 521

E

E-2 Hawkeye (Grumman) 366
E-3 Sentry (Boeing) 314
E-3A (Boeing) 314
E-4 (Boeing) 308, 316
E-14 (Hispano) ....(1971-72) 167
E-18B (Piper) ....(1975-76) 424
E.25 (CASA) 175
E-33 (Mikoyan) 201
E55 Baron (Beechcraft) 284
E-66 (Mikoyan) 201
E-76 (Mikoyan) 202
E90 King Air (Beechcraft) 287
E-133 (Mikoyan) 205
E-266 (Mikoyan) 204
E-266M (Mikoyan) 206
EA-1 Kingfisher (Anderson) 510
EA-6 Intruder (Grumman) 367
EA-6B Prowler (Grumman) 368
EA7 Optica (Edgley) 252
EAA (see Experimental Aircraft Association, Inc.) 522
EAA Biplane (EAA) 522
EAC-3 Pouplume (Croses) 481
E & C (see Elston & Pruitt) ....(1974-75) 327
E.C.2 Easy Too (Clutton-Tabenor) 502
EC-6 Criquet (Croses) 481
EC-8 Tourisme (Croses) 482
EC-95 (EMBRAER) 10
EC119J (SIAI-Marchetti/Fairchild) ......(1973-74) 138
EC-130 Hercules (Lockheed) 395
EC-137D (Boeing) 314
EF-4 Phantom II (McDonnell) ....(1980-81) 380
EF-111A (General Dynamics) 361
EF-111A (Grumman/General Dynamics) 370
EH-1H Iroquois (Bell) 293
EH-60A/B Black Hawk (Sikorsky) 461
EH 101 (EHI) 101
EHI (see EHI Industries) 101
EM (see Elicotteri Meridionali SpA) 129
EM-5A (Studenckie Biuro Konstrukcyjne) (1974-75) 172
EMA 124 (Meridionali/Agusta) ....(1974-75) 132
EMB-110 Bandeirante (EMBRAER) 9
EMB-110K1 (EMBRAER) ....(1980-81) 12
EMB-110P (EMBRAER) ....(1980-81) 12
EMB-110P1 (EMBRAER) 10
EMB-110P1/41 (EMBRAER) 10
EMB-110P1K (EMBRAER) 10
EMB-110P2 (EMBRAER) 10
EMB-110P2/41 (EMBRAER) 10
EMB-110P3 (EMBRAER) 10
EMB-111 (EMBRAER) 12
EMB-120 Brasilia (EMBRAER) 12
EMB-121 Xingu (EMBRAER) 13
EMB-121B Xingu II (EMBRAER) 14
EMB-123 Tapajos (EMBRAER) ....(1978-79) 14
EMB-130 (EMBRAER) ....(1974-75) 16
EMB-201A Ipanema (EMBRAER) 16
EMB-201R (EMBRAER) ....(1980-81) 17
EMB-312 (EMBRAER) 15
EMB-326GB Xavante (EMBRAER/Aermacchi) ... 15
EMB-500 Amazonas (EMBRAER) ......(1972-73) 11
EMB-710 Carioca (EMBRAER/Piper) 17
EMB-711 Corisco (EMBRAER/Piper) 17
EMB-712 Tupi (EMBRAER/Piper) 17
EMB-720 Minuano (EMBRAER/Piper) 17
EMB-721 Sertanejo (EMBRAER/Piper) 17
EMB-810 Senaca II (EMBRAER/Piper) 17
EMB-820 Navajo (EMBRAER/Piper) 17
EMBRAER (see Empresa Brasileira de Aeronáutica SA) 9
EOS/SFA (Frederick-Ames) 525
EP-3 Orion (Lockheed) 390
ER-2 (Lockheed) 389
ER-2 (NASA/Lockheed) 419
ESS 641 (Flugwag-Bremen) ....(1974-75) 93
ESTOL, S-220 (Vertak) ....(1973-74) 454
EWR (see Entwicklungsring Süd GmbH) .(1972-73) 88
Eagle (ASTEC/Cessna) 276
Eagle (Composite) ....(1980-81) 330
Eagle (Enstrom) ....(1978-79) 325
Eagle (Eagle) ....(1979-80) 331
Eagle, F-15 (McDonnell) 403
Eagle I (Christen) 517
Eagle I/II/III (Dietrick/Windecker) 347
Eagle II (ASTEC/Grumman) ....(1980-81) 262
Eagle II (Christen) 517
Eagle II (Composite) ....(1980-81) 331
Eagle II (Eagle Helicopter) 524
Eagle 220 (Eagle) 348
Eagle 300 (Eagle) 348
EAGLE AIRCRAFT COMPANY (U.S.A.) 348
Eagle DW-1 (Eagle) ....(1980-81) 332
EAGLE HELICOPTER CORPORATION (U.S.A.) 524
Eagle, Model 480 (Enstrom) ....(1980-81) 334
Eagle SP (ASTEC/Cessna) 276
Eaglet (Guttman) ....(1972-73) 329
Eaglet (Ward) ....(1972-73) 226
Easy Too, E.C.2 (Clutton-Tabenor) 502
ÉCOLE NATIONALE SUPÉRIEURE D'ARTS ET MÉTIERS (France) .....(1978-79) 470
ECTOR AIRCRAFT COMPANY (U.S.A.) 348
Ecureuil/Astar (Aérospatiale) 51
Ecureuil 2/Twinstar (Aérospatiale) 51
Edelweiss RD-02 (Duruble) ....(1972-73) 61
Edelweiss RD-03 (Duruble) 482
EDGLEY AIRCRAFT LTD (U.K.) 252
EDO CORPORATION (U.S.A.) ....(1971-72) 298
EDWARDS, WILLIAM (U.S.A.) 552
Egret I, NM-72 (Nihon University) ......(1973-74) 149
Egret III, NM-74 (Nihon) ....(1975-76) 152
EH INDUSTRIES (International) 101
EICH, JAMES P. (USA) 524
EIDGENÖSSISCHES FLUGZEUGWERK (Switzerland) 182
EKIN (ENGINEERING) CO. LTD., W. H. (U.K.) 272
EKSTRÖM, STAFFAN W. (Sweden) 499
Electrically-powered light aircraft (NASA) ....(1980-81) 396
El Gringo (CB Enterprises) ....(1977-78) 532
ELICOTTERI MERIDIONALI SpA (Italy) 129
ELMENDORF, LEONARD C. (U.S.A.) .(1973-74) 318
ELSTON & PRUITT (U.S.A.) ....(1974-75) 327
El Tomcat (Continental) ....(1980-81) 331, 332
EMAIR (U.S.A.) 349
Emeraude (Piel) ....(1980-81) 482
EMPRESA BRASILEIRA DE AERONÁUTICA SA (Brazil) 9
Enforcer (Piper) 443
ENSTROM HELICOPTER CORPORATION, THE (U.S.A.) 349
ENTWICKLUNGSGEMEINSCHAFT LEICHTFLUGZEUG DIPL ING HERMANN MYLIUS (West Germany) (1977-78) 74
ENTWICKLUNGSRING SÜD GmbH (West Germany) ....(1972-73) 88
EPPS, BEN T. JNR. (U.S.A.) ....(1972-73) 311
Epsilon (Aérospatiale) 44
Epsilon, TB 30 (SOCATA) ....(1980-81) 76
Equator (Pöschel) ....(1972-73) 94
Equator, P-300/-350/-400/-450 (Equator) 80
EQUATOR AIRCRAFT GESELLSCHAFT FÜR FLUGZEUGBAU mbH ULM (West Germany) 79
ERICSON, HUGO (Sweden) ....(1971-72) 168
ERNI, HANS (Switzerland) ....(1973-74) 185
Esqualo (Moura) 473
Esquilo (Helibras) 17
Eurofighter, Saab-37E (Saab) ....(1975-76) 183
Europa (Wassmer) ....(1977-78) 66
EUROPEAN AIRLINER (International) (1976-77) 91
EUROPLANE LTD (International) .....(1974-75) 113
EVANGEL AIRCRAFT CORPORATION (U.S.A.) ....(1975-76) 329
EVANS AIRCRAFT (U.S.A.) 524
EXCALIBUR AVIATION COMPANY (U.S.A.) .. 350
Excalibur (Excalibur) ....(1979-80) 334
Excalibur 800 (Excalibur) ....(1979-80) 334
Excalibur, MW2 (Trago Mills) ....(1977-78) 516
Exec (RotorWay) 546
Executaire (Boeing Vertol) ....(1979-80) 305
Executive (Aerostar) ....(1971-72) 221
Executive (Mooney) ....(1978-79) 388
Executive 600 (Colemill) 346
Executive Commuter (Cessna) ....(1975-76) 317
EXPERIMENTAL AIRCRAFT ASSOCIATION, INC. (U.S.A.) 522
Explorer/Observer (National Dynamics/Patchen) (1977-78) 161
Extender (Douglas) 412

F

F-1 (Mitsubishi) 147
F1 Mirage (Dassault) ....(1977-78) 52
F1-E Mirage (Dassault) ....(1975-76) 60
F-2 (China) ....(1978-79) 27
F.2B Fighter replica (Ohmert) ....(1980-81) 534
F-3 Cobra (Forbes) ....(1978-79) 520
F-4 Cobra (China) ....(1978-79) 27
F-4CCV (McDonnell) ....(1979-80) 382
F-4 Phantom II (McDonnell) 403
F-4G Advanced Wild Weasel (McDonnell) ....(1980-81) 380
F4U Corsair (WAR) 562
F-5 (Frebel) ....(1979-80) 484
F-5 (Northrop) ....(1978-79) 392
F-5E Tiger II (Northrop) 421
F-5F (Northrop) 421

F-5G Tigershark (Northrop) 422
F-6 (China) 29
F-6bis (China) (1980-81) 34
F-7 (China) 32
F-8 (China) 32
F-8 Crusader (LTV) (1971-72) 342
F.8L, Falco (Sequoia) 548
F8U Crusader (LTV) (1971-72) 342
F-9 (China) (1980-81) 35
F-11-2 Husky (Harrison) (1972-73) 22
F-12 (China) 32
F-14 Tomcat (Grumman) 368
F-14/TARPS (Grumman) (1978-79) 341
F-15 Eagle (McDonnell Douglas) 403
F-15 Strike Eagle (McDonnell) 404
F15E Picchio (Procaer) 140
F15F (Procaer/General Avia) 141
F-16 Fighting Falcon (General Dynamics) 361
F-16/79 (General Dynamics) 363
F-16/101 (General Dynamics) 364
F-16E (General Dynamics) 364
F-18A Hornet (McDonnell) 406
F-19 Sportsman 100 (Taylorcraft) (1979-80) 451
F.20 Pegaso (General Avia) 136
F.20 Pegaso (Italair) (1977-78) 117
F-21 (Taylorcraft) 466
F-23 (Cosmic) (1974-75) 322
F-27 (Fairchild Industries) (1974-75) 332
F.27 Friendship (Fokker) 151
F27 Maritime (Fokker) 153
F-28 (Enstrom) 349
F.28 Fellowship (Fokker) 153
F28-2 (Fokker-VFW) (1976-77) 138
F29 (Fokker) (1980-81) 149
F33A/C Bonanza (Beechcraft) 281
F-51D Mustang (Cavalier) (1971-72) 271
F90 Super King Air (Beechcraft) 288
F-104 Starfighter (Lockheed) (1977-78) 326
F-104CCV (MBB) 83
F-104RB Starfighter (Red Baron/Greenamyer) (1978-79) 535
F-104S (Aeritalia/Lockheed) 118
F-105 Thunderchief (Republic) (1973-74) 326
F-106 (Convair) (1971-72) 310
F-110 Phantom II (McDonnell) (1979-80) 381
F-111 (General Dynamics) 361
F 150 Aerobat (Reims) (1978-79) 57
F 150 Commuter (Reims) (1978-79) 57
F 152 Aerobat (Reims) 69
F 172 Reims Rocket (Reims) (1976-77) 59
F 172 Skyhawk/100 (Reims) 70
F 177RG (Reims) (1978-79) 58
F 182 (Reims) 70
F-337 (Reims) (1980-81) 70
F.600 (Foxjet) (1980-81) 338
F.600 Canguro (General Avia) (1980-81) 133
F/A-18 Hornet (McDonnell) 406
F/A-18L (Northrop) 423
FA 150 Aerobat (Reims) (1976-77) 59
FA 152 Aerobat (Reims) 69
FA-200 Aero Subaru (Fuji) 141
FA-300/Model 700 (Fuji) (1980-81) 138
FA-300/Model 710 (Fuji) 142
FA 337 (Reims) (1977-78) 60
FAC (see Fuerza Aérea Chilena) 28
FB2 (Berger) (1976-77) 473
FB-111 (General Dynamics) (1980-81) 345
FFA (see Flug- & Fahrzeugwerke AG) (1980-81) 173
FH-227 (Fairchild Industries) (1974-75) 332
FH-1100 (Fairchild Industries) (1974-75) 333
FH-1100 (Hiller) 377
FJ-Skytrac (HTN) (1974-75) 93
FLAC (Striplin) (1980-81) 552
FLSZ (see Flight Level Six Zero) 525
FMA (see Area de Material Córdoba) 1
FR 172 Hawk XP (Reims) 70
FR 172 Reims Rocket (Reims) (1976-77) 59
FR 182 RG (Reims) 70
FRA 150 Aerobat (Reims) (1978-79) 57
FRB 172 (Reims) (1976-77) 59
FRS. Mk 1 (BAe) 248
FRS. Mk 51 (BAe) 248
FS-28 Avispa (Akaflieg Stuttgart) (1974-75) 90
FS-T2-KAI (Mitsubishi) (1976-77) 130
FT 337P (Reims) (1980-81) 70
FTB 337 (Reims/Cessna) 70
FTMA-Milirole (Reims) (1973-74) 80
FU-24, Fletcher (Aerospace) (1974-75) 162
FU-24-954 (Aerospace/Fletcher) 155
FUS (see Flugzeug-Union-Süd GmbH) 81
FVA (see Flugwissenschaftliche Vereinigung Aachen (1920) (E.V.)) (1972-73) 89
FVA-18 Primitivkrähe (FVA) (1972-73) 89
F & W (see Federal Aircraft Factory) (1975-76) 191
FW-190A3 (Richard) (1971-72) 392
F + W C-3605 (Federal Aircraft Factory) (1974-75) 192
FWA (see Flugzeugwerke Altenrhein AG) 179
FABRICA DE AVIONES ANAHUAC SA (Mexico) 150
FABRICA FEDERALE D'AEROPLANI (Switzerland) 182
FÁBRICA MILITAR DE AVIONES (Argentina) 1
FABRIQUE FEDERALE D'AVIONS (Switzerland) 182
FAIRCHILD INDUSTRIES INC (U.S.A.) 351
FAIRCHILD REPUBLIC COMPANY (U.S.A.) 351
FAIRCHILD SWEARINGEN CORPORATION (U.S.A.) 353
FAIREY CANADA LTD (Canada) (1971-72) 23
FAIREY SA (Belgium) (1977-78) 10
"Faithless" (Mikoyan) (1971-72) 444
Falco F.8L (Sequoia) 548
Falcon (AFIC) (1972-73) 170
Falcon (Dassault) (1977-78) 54
Falcon (Enstrom) (1978-79) 325
Falcon (Magni) (1974-75) 133
Falcon (National Dynamics/Reed) (1979-80) 163
Falcon (Williams) (1976-77) 543
Falcon 10 (Dassault) (1979-80) 59
Falcon 20/Mystère (Dassault) (1979-80) 59
Falcon 30 (Dassault) (1974-75) 65
Falcon 50 (Dassault-Breguet) (1977-78) 57
Falcon, F-28F (Enstrom) 349
Falcon ST (Dassault-Breguet) 63
FALCONAR AIRCRAFT LTD (Canada) (1971-72) 23
Falcon Cargo Jet (Dassault) 65
Falcon Guardian (Dassault) (1979-80) 59
Falcon Series E (Dassault) (1975-76) 62
Falcon Series F (Dassault) (1977-78) 54
Falcon Series G (Dassault) (1977-78) 54
Fang, OR-70 (Stevenson/Owl) (1977-78) 550
Fan Jet 400 (Beechcraft) (1976-77) 224
Fanjet, Series 600 (Cessna) (1973-74) 311
Fanliner (RFB/Grumman American) (1978-79) 70
Fantan-A (China) 31
Fantrainer 400/600 (MBB) 83
Fantrainer AWI-2 (RFB) (1979-80) 79
FARM AVIATION LTD (U.K.) (1971-72) 188
Farmer-C (China) 29
Farmer-D (China) 29
FARRINGTON AIRCRAFT CORPORATION (U.S.A.) (1975-76) 333
Faucon, PMB-78 (Milon) (1980-81) 481
FAUVEL, CHARLES (France) (1980-81) 475
FEDERAL AIRCRAFT FACTORY (Switzerland) (1975-76) 191
Fellowship, F28 (Fokker) 153
Fencer (Sukhoi)) 219
FIAT SOCIETA PER AZIONI (Italy) (1971-72) 114
Fiddler (Tupolev) 225
Fieldmaster, Type 6 (NDN) 255
Fighting Falcon, F-16 (General Dynamics) 361
FIKE, WILLIAM J. (U.S.A.) (1980-81) 519
Firebar (Yakovlev) 230
Firecracker (NDN) 254
Firefighter (Britten-Norman) (1978-79) 232
Firefly (Statler) 553
Fishbed (China) 32
Fishbed (HAL) 90
Fishbed (Mikoyan) 200
FISHER, DICK AND VAN NORMAN, CHARLES (USA) (1980-81) 520
FISHER-VAN NORMAN (see Fisher, Dick and Van Norman, Charles) (1980-81) 520
Fishpot (Sukhoi) 215
Fitter-A (Sukhoi) 214
Fitter-B (Sukhoi) (1975-76) 510
Fitter-C, D, E, F, G, H, J (Sukhoi) 216
Fixed-wing aircraft prototype (PADC/NSDB) 157
FLAGLOR AIRCRAFT (U.S.A.) (1980-81) 520
Flaglor Scooter (Headberg) 528
Flagon (Sukhoi) 216
Flamingo (CASA) (1976-77) 159
Flamingo (Hispano) (1972-73) 174
Flamingo (MBB) (1973-74) 93
Flamingo-Trainer (FUS) 81
"Flashlight" (Yakovlev) (1971-72) 467
FLÉCHAIR SA (France) (1974-75) 77
Fleet 150 (CATA) (1975-76) 3
Fletcher 1160 (Air Parts/Fletcher) (1971-72) 155
Fletcher 1284 (Aerospace) (1974-75) 163
Fletcher FU-24 (Aerospace) (1974-75) 162
FLETCHER, DANIEL & RICHARD (Canada) 474
Fletcher Cresco (Aerospace) (1978-79) 136
FLEURY, ROBERT (France) (1975-76) 68
FLIGHT DYNAMICS INC (U.S.A.) (1980-81) 520
FLIGHT ENGINEERS LTD (New Zealand) (1980-81) 152
FLIGHT INVERT LTD-CRANFIELD (U.K.) (1978-79) 235
FLIGHT LEVEL SIX-ZERO (U.S.A.) 525
Flightsail VII (Flight Dynamics) (1977-78) 540
Floatmaster (GAF) 5
Flogger (HAL) 90
Flogger (Mikoyan) 202
Flogger-D (Mikoyan) 204
FLUGWISSENSCHAFTLICHE ARBEITS-GEMEINSCHAFT (FLUWAG) BREMEN (West Germany) (1974-75) 93
FLUGWISSENSCHAFTLICHE VEREINIGUNG AACHEN (1920) E.V. (West Germany) (1972-73) 89
FLUGZEUG-UNION-SÜD GmbH (West Germany) 81
FLUGZEUGWERKE ALTENRHEIN AG (Switzerland) 179
Flut-R-Bug (AVEX) (1971-72) 3
FLUWAG BREMEN (see Flugwissenschaftliche Arbeitsgemeinschaft (Fluwag) Bremen) (1974-75) 93
Fly Baby 1-A (Bowers) 513
Fly Baby 1-B (Bowers) 514
Fly Baby 1B (Cronk/Bowers) (1976-77) 505
Flybike (Beaujon) (1977-78) 523
Flycycle (Lane) (1977-78) 546
Flying Bathtub (MaHugh) (1980-81) 529
Flyray (Flyray) (1980-81) 520
FLYRAY (USA) (1980-81) 520
Focke-Wulf 190 (WAR) 562
FOKKER BV (Netherlands) 150
FOKKER-VFW BV (Netherlands) (1979-80) 141
Foo Fighter (Stewart) 555
FORBES ENTERPRISES (U.S.A.) (1978-79) 520
Forbes-Wolfram Special (Forbes) (1975-76) 335
Forger (Yakovlev) 231
Fouga 90 (Aérospatiale) (1980-81) 46
Four (Bittner) (1977-78) 480
FOURNIER AVIATION (France) 67
Fournier RF8 (Indraéro) (1975-76) 70
Fourtouna, M.J.14 (Jurca) 486
Foxbat (Mikoyan) 204
FOXJET INTERNATIONAL INC (U.S.A.) 355
FRAKES AVIATION INC (U.S.A.) (1978-79) 329
Francesco (Iannotta) (1971-72) 125
FRANKS, TOM (U.S.A.) (1977-78) 540
FREBEL, HERMANN (West Germany) (1979-80) 484
FREDERICK-AMES RESEARCH CORPORATION (U.S.A.) 525
Fred Series 2 (Clutton-Tabenor) 500
Fred Series 3 (Clutton-Tabenor) 500
Freehand (Yakovlev) (1976-77) 447
Frégate (Aérospatiale) (1978-79) 36
FREIBERGER, RONALD (U.S.A.) (1973-74) 329
Freshman, D-200 (D'Apuzzo) 519
Friendship, F27 (Fokker) 151
FRIZZLE, JAMES R. (Canada) (1980-81) 466
FRONTIER-AEROSPACE INC (U.S.A.) 355
FUERZA AÉREA (Chile) 28
FUJI HEAVY INDUSTRIES LTD (Japan) 141
FUJI JUKOGYO KABUSHIKI KAISHA (Japan) 141
FUKUOKA HIGH SCHOOL (Japan) (1976-77) 474
Fury II (Isaacs) 502

G

G-1 Papillon (Georges) (1973-74) 90
G2-A Galeb (Soko) 471
G-3 Galeb-3 (Soko) (1975-76) 526
G6 Antares (Kokubun) (1972-73) 143
G8 Mirage (Dassault) (1974-75) 60
G8A/ACF (Dassault) (1974-75) 61
G.10 Moto Delta (Geiser) (1977-78) 490
G11, IA 50 (FMA) (1975-76) 4
G 11 Moto-Delta (Geiser) (1980-81) 68
G 20 Moto-Delta (Geiser) (1980-81) 69
G-21C (McKinnon) (1976-77) 337
G-21D (McKinnon) (1976-77) 337
G-21G Turbo-Goose (McKinnon-Viking) (1980-81) 29
G33 Bonanza (Beechcraft) (1973-74) 248
G-89 Tracker (Grumman) (1972-73) 324
G91 (Dornier/Fiat) (1972-73) 88
G91Y (Aeritalia) (1977-78) 104
G-100-A (Galaxie) (1971-72) 306
G-111 Albatross (Grumman/Resorts International) 365
G-134 Mohawk (Grumman) (1973-74) 338
G-159 Gulfstream (Grumman) (1972-73) 328
G159C Gulfstream I-C (Gulfstream American) 371
G-164 Ag-Cat (Grumman) (1972-73) 329
G-164A Super Ag-Cat (Grumman/Schweizer) (1976-77) 381
G-164 C-T Turbo Cat (Marsh/Gulfstream American) 401
G-164D Turbo Ag-Cat D (Gulfstream American) (1980-81) 356
G222 (Aeritalia) 119, 120
G222T (Aeritalia) 121
GA-7 (Gulfstream American) (1980-81) 355
GAB (see Grupul Aeronautic Bucuresti) (1979-80) 159
GAC 159-C Gulfstream I Commuter (Gulfstream American) (1980-81) 353
GACC (see Gulfstream American Corporation of California) 355
GAF (see Government Aircraft Factories) 5
G.B.1 (Boyd) 514
G/B Special (Beets) 526
GB-1 Luftibus (Bezzola) (1978-79) 490
GB-2 Retro (Bezzola) 500
GC-130 Hercules (Lockheed) (1972-73) 350
GLO1 (Landray) 487
GLO2 Ami Pou (Landray) 487
GLO3 (Landray) 487
GL-1 (Laudray) (1980-81) 480
GN, Autogyro (AISA) 173
GN-1 Aircamper (Grega) 526
GP2 Osprey I (Osprey) (1974-75) 410
GP3 Osprey II (Osprey) 538
GV-1 Hercules (Lockheed) 395
GV 10-20 Gazelle (Gazuit-Valladeau) (1975-76) 70
GV 10-31 Gazelle (Gazuit-Valladeau) (1975-76) 70
GV 103 (Gazuit-Valladeau) (1973-74) 68
GW-1 Windwagon (Watson) 563
GY-20 Minicab (Ord-Hume) (1973-74) 216
GY 90 Mowgli (SITAR) (1971-72) 78
GY 100 Bagheera (SITAR) (1971-72) 77
GY 110 Sher Khan (SITAR) (1971-72) 78
GY-201 Minicab (Ord-Hume) (1979-80) 497
Gabier (SOCATA) 74
GafHawk 125 (Hawk) 376
Gaillard (SOCATA) 73
GALAXIE CORPORATION (U.S.A.) (1971-72) 306
Galaxy (Lockheed) 398
Galeb G2-A (Soko) 471
Galeb-3 (Soko) (1974-75) 526
Galérien (SOCATA) 73
Galopin (SOCATA) 73
Ganagobie 05 (ARDC of NZ) 498
Gardian (Dassault/Breguet) 64
Garnement (SOCATA) 73
GARRISON, PETER (U.S.A.) (1977-78) 540
GATARD, AVIONS A. (France) 483
GATES LEARJET CORPORATION (U.S.A.) 357
Gaucho (SOCATA) 74, 75
Gavião (Helibras) 17
Gawron (WSK-Okecie) (1974-75) 167
Gazelle (Aérospatiale/Westland) 49, 272
Gazelle (Gazuit-Valladeau) (1975-76) 70
GAZUIT-VALLADEAU, ETABLISSEMENTS (France) (1975-76) 70
G/B AIRCRAFT (USA) 526
Gee Bee Model Z Super Sportster (Turner) 559

GEIDE, RICHARD F (U.S.A.) 526
GEISER, JEAN MARC (France) ....(1980-81) 68
Gelatik 32 (LIPNUR) ....(1976-77) 84
GENERAL AIRCRAFT CORPORATION (U.S.A.) ....(1975-76) 339
GENERAL AVIA COSTRUZIONI AERONAUTICHE Srl (Italy) 136
GENERAL AVIATION DIVISION, ROCKWELL CORPORATION (U.S.A.) ....(1980-81) 431
GENERAL DYNAMICS CORPORATION (U.S.A.) 360
General purpose airborne simulator (NASA) (1979-80) 397
Geodetic LW 108 (Harris) ....(1976-77) 514
GEORGES, GERARD (West Germany) .(1973-74) 90
Georges Payre, AM-69 (AM) ....(1977-78) 484
GEORGIAS (U.S.A.) ....(1976-77) 513
Ghinassi helicopter (AVEX) ....(1979-80) 465
Gibkolet (KAI) ....(1976-77) 545
Gipsy Hawk, RJJ-1 (Jameson) ....(1975-76) 362
GIRAVIONS DORAND, SARL (France) (1971-72) 60
Glasair (Stoddard-Hamilton) 555
GLUHAREFF HELICOPTERS (U.S.A.) (1974-75) 334
Gnat (HAL/Hawker Siddeley) ....(1973-74) 101
Gnat Mk II (HAL) ....(1974-75) 104
Gnatsum (Jurca) 486
Goldcrest (Sizer) ....(1972-73) 224
Golden Eagle/II/III (Cessna) 340
Golden Hawk (DSK Airmotive) ....(1980-81) 515
GONSERKEVIS, ALPHONSE J. (U.S.A.) (1975-76) 343
Go-Plane (Levi) ....(1975-76) 217
Gorrion, Armar I (AVEX) ....(1977-78) 478
Gossamer Albatross (MacCready) 398
Gossamer Albatross (NASA/MacCready) 419
Gossamer Condor (MacCready) ....(1979-80) 528
GOVERNMENT AIRCRAFT FACTORIES (Australia) 5
GOVERNMENT OF AUSTRALIA, DEPARTMENT OF INDUSTRY AND COMMERCE (Australia) ....(1976-77) 5
GRADY, MELVIN (U.S.A.) 526
Gran Turismo Commander (Rockwell International) 431
Grasshopper (Cierva) ....(1972-73) 196
Grasshopper (R and B) ....(1977-78) 555
GREAT LAKES AIRCRAFT COMPANY (U.S.A.) 364
GREENAPPLES AIRCRAFT (U.S.A.) 526
GREGA, JOHN W. (U.S.A.) 526
Greyhound (Grumman) ....(1971-72) 312
GRINVALDS, JEAN (France) 484
GRUMMAN AEROSPACE CORPORATION (U.S.A.) 365
GRUMMAN AMERICAN AVIATION CORPORATION (U.S.A.) ....(1977-78) 309
GRUMMAN CORPORATION (U.S.A.) 365
Grumman/Rand Monoplane (Grady) 526
GRUPUL AERONAUTIC BUCURESTI (Romania) ....(1979-80) 159
Guarani II (FMA) ....(1974-75) 5
Guardian (Dassault-Breguet) ....(1978-79) 52
Guardian, HU-25A (Dassault-Breguet) 64
Guepard (CERVA) .... (1977-78) 49
Guerrier (SOCATA) 74
GUEX, JEAN-CLAUDE (Switzerland) ..(1979-80) 492
GULFSTREAM AMERICAN CORPORATION (U.S.A.) 371
GULFSTREAM AMERICAN CORPORATION, COMMANDER DIVISION (U.S.A.) 394
GULFSTREAM AMERICAN CORPORATION OF CALIFORNIA (U.S.A.) 355
Gulfstream I-C, G159-C (Gulfstream American) 371
Gulfstream II-B (Gulfstream American) 372
Gulfstream III (Gulfstream American) 372
Gulfstream III Maritime version (Gulfstream American) 373
Gunship (Fairchild Industries) ....(1974-75) 333
Guppy-101 (Aero Spacelines) ....(1973-74) 238
Guppy-201 (Aero Spacelines) ....(1974-75) 244
GUTTMAN, BOB (U.S.A.) ....(1972-73) 329
GUZMAN, FRANK (U.S.A.) ....(1972-73) 330
Gypsy Hawk (Jameson) ....(1975-76) 362
Gyro-Copter (Bensen) 511, 512
Gyro-Falcon 201 (Wotring) ....(1974-75) 483
GYROFLIGHT LTD (U.K.) ....(1973-74) 200
GYROFLUG INGENIEURSGESELLSCHAFT mbH (West Germany) 81
Gyro-Glider (Bensen) 510
Gyroplane (Avian) ....(1971-72) 15
Gyroplane (McCandless) ....(1972-73) 211
Gyroplane, JE-2 (Eich) 524

H

H-1A (Commuter Jr. Helicom) ....(1972-73) 330
H2-B1 (Airmaster) ....(1974-75) 194
H3-H4 (VFW-Fokker) ....(1972-73) 98
H-5 (China) 32
H5 Delta Stinger (Rinker) ....(1971-72) 393
H-6 (China) 32
H-13KH (Kawasaki) ....(1977-78) 130
H-13 Sioux (Bell) ....(1971-72) 242
H-250 Courier (Helio) ....(1972-73) 331
H-295 Super Courier (Helio) ....(1977-78) 315
H-550A Stallion (Helio) ....(1976-77) 301
H-580 Twin Courier (Helio) ....(1971-72) 320
HA-2M Sportster (Hollmann) 528
Ha-11 (Haak) ....(1972-73) 89
HA-31 Mk II Basant (HAL) 90
HA-200 Saeta (Hispano) ....(1971-72) 167
HA-220 Super Saeta (Hispano) ....(1973-74) 178
HAC/PAH-2 (MBB/Aérospatiale) 102
HAI (see Hellenic Aerospace Industry SA) 86
HAL (see Hindustan Aeronautics Ltd.) 86
H and H (see Halpin, Richard and Huf, Tom) (1975-76) 353
HASA (see La Hispano-Aviación S.A.) ..(1972-73) 174
H.B.1, Baserga (AVEX) ....(1971-72) 2
HC. Mk 1 Chinook (Boeing Vertol) 318
HC-4 (Aero) ....(1972-73) 30
HC-130 Hercules (Lockheed) 395
HEA-23B Ressu (Hietanen) ....(1972-73) 37
HELIBRAS (see Helicopteros do Brasil SA) 17
HF-24 Marut (HAL) ....(1979-80) 81
HF-73 (HAL) ....(1977-78) 80
HFB (see Messerschmitt-Bölkow-Blohm GmbH) ... 82
HFB 320 Hansa (MBB) ....(1974-75) 96
HFB 320 Hansa Jet (MBB) ....(1980-81) 83
HFB 330 Hansa Fan Jet (MBB) ....(1971-72) 92
HH-1 Hawker Hurricane (Sindlinger) 549
HH-1 Zipper (Hammer-Hunt) ....(1980-81) 358
HH-1H Iroquois (Bell) 293
HH-1K Iroquois (Bell) ....(1972-73) 261
HH-2 Seasprite (Kaman) 386
HH-3 (Sikorsky) 459
HH-3A (Sikorsky) 458
HH-3F (Agusta/Sikorsky) 129
HH-52A (Sikorsky) ....(1971-72) 402
HH-53 (Sikorsky) 460
HH-53 Pave Low (Sikorsky) ....(1976-77) 387
HJT-16 Mk I Kiran (HAL) 87
HKP 4 (Boeing-Vertol) ....(1973-74) 283
HKP 4C (Kawasaki-Boeing) 144
HKP 6 (Agusta-Bell) ....(1976-77) 109
HKP 7 (Kawasaki/Boeing) ....(1973-74) 143
HL-10 (Northrop/NASA) ....(1973-74) 391
HLH (Boeing Vertol) ....(1974-75) 294
HM 293 Pou-du-Ciel (Cole) ....(1971-72) 188
HN 433 Menestrel (Nicollier) 488
HN 500 Bengali (Nicollier) 488
HN 600 Week-End (Nicollier) 489
HO-6 (Hughes) ....(1977-78) 318
HPA (see Hertfordshire Pedal Aeronauts) (1979-80) 494
HPA 1 (Kiceniuk) ....(1978-79) 524
HPT-32 (HAL) 88
HR 100/4 + 2 (Robin) ....(1979-80) 68
HR 100/200 Royale (Robin/CEA) ....(1972-73) 80
HR 100/180 (Robin) ....(1977-78) 62
HR 100/200 Safari (Robin) ....(1975-76) 83
HR 100/210 Safari II (Robin) ....(1977-78) 62
HR 100/210R Royal (Robin/CEA) ......(1974-75) 84
HR 100/230TR (Robin) ....(1976-77) 62
HR 100/23 (Robin/CEA) ....(1974-75) 84
HR 100/250TR President (Robin) ....(1980-81) 72
HR 100/285 Tiara (Robin) ....(1978-79) 60
HR 100/285C Tiara (Robin) ....(1977-78) 63
HR 200 (Robin) ....(1977-78) 63
HR 200 Acrobin (Robin/CEA) ....(1973-74) 83
HR 200 Club (Robin/CEA) ....(1973-74) 83
HRB-1 Sea Knight (Boeing Vertol) ......(1972-73) 277
HS 9 Biplane (Schoenenberg) 494
HS 125 (Hawker Siddeley) ....(1976-77) 178
HS 125 Series 700 (BAe) 238
HS 146 (BAe) ....(1978-79) 219
HS 747 (BAe) ....(1978-79) 217
HS 748 (Hawker Siddeley) ....(1977-78) 181
HS 748 (HAL/BAe) 90
HS 748 Coastguarder (BAe) 240
HS 748 Military Transport (BAe) 239
HS 748 Series 2B (BAe) 239
HS 801 Nimrod (Hawker Siddeley) ......(1977-78) 189
HS 1127 (RAF) Harrier (Hawker Siddeley) (1977-78) 186
HS 1182 Hawk (Hawker Siddeley) ......(1977-78) 186
HSS-2 Sea King (Sikorsky) 458
HSS-2Z (Sikorsky) ....(1974-75) 453
HSX-3 (Schilling) ....(1979-80) 485
H.T./Hollsmidt 222 (Thyregod) ....(1971-72) 36
HT-295 Trigear Courier (Helio) ....(1977-78) 315
HTL (Bell) ....(1971-72) 242
HTM (see Helicopter Technik München) .(1975-76) 92
HU-1 Iroquois (Bell) ....(1972-73) 261
HU2K-1 Seasprite (Kaman) 386
HU-25A Guardian (Dassault-Breguet) 64
HW 03 (Hellbach) ....(1979-80) 485
HAAK, REINHARD (West Germany) ...(1972-73) 89
Halo (Mil) 212
HALPIN, RICHARD AND HUF, TOM (U.S.A.) (1975-76) 353
HALSMER, JOSEPH L. (U.S.A.) ......(1973-74) 342
HAMAO, Y (Japan) ....(1978-79) 485
Hamao Shiokaratombo (JAA) ....(1980-81) 488
HAMILTON AVIATION (U.S.A.) 375
HAMMER, ROBERT AND HUNT, RICHARD (U.S.A.) ....(1980-81) 358
HAMMER-HUNT (see Hammer, Robert and Hunt, Richard) ....(1980-81) 358
HAN JIN GROUP, THE (South Korea) .(1978-79) 128
Hansa (MBB) ....(1974-75) 96
Hansa Fan Jet (MBB) ....(1971-72) 92
Hansa Jet, HFB 320 (MBB) ....(1980-81) 83
HANSEN, WILLIAM (U.S.A.) ....(1976-77) 513
Harbin Y-11 (C-11) (China) 35
Harbin (Antonov) Y-5 (C-5) (China) 34
Harbin (Bell) Model 212 (China) 34
Harbin (Bell) Model 412 (China) 34
Harbin (Ilyushin) H-5 (B-5) (China) 32
Harbin (Mil) Z-5 (H-5) (China) 34
Harke (Mil) 209
HARMON ENGINEERING CO (U.S.A.) 526
Harp (Kamov) ....(1976-77) 417
Harrier (BAe) 247
Harrier, AV-8C (BAe/McDonnell) 405
HARRIS, J. WARREN (U.S.A.) ....(1976-77) 514
HARRISON AIRWAYS LTD (Canada) ..(1972-73) 22
HAST, GEORGE (U.S.A.) ....(1975-76) 354
HATZ AIRPLANE SHOP (U.S.A.) 527
Hauscat (Bougie) ....(1976-77) 456
HAVERTZ, HERMANN (West Germany) (1972-73) 98
Hawk 140 (Scheutzow) ....(1975-76) 449
Hawk, DSK-1 (DSK Airmotive) ....(1980-81) 515
Hawk (BAe) 245
Hawk (Enstrom) 350
Hawk XP (Cessna) 326
Hawk XP/II (Cessna) 326
Hawk XP, FR 172K (Reims) 70
HAWK INDUSTRIES INC (USA) 376
HAWKER DE HAVILLAND AUSTRALIA PTY., LTD. (Australia) 7
Hawker Hurricane, HH-1 (Sindlinger) 549
HAWKER SIDDELEY AVIATION LTD (U.K.) (1977-78) 179
HAWKER SIDDELEY CANADA INC (Canada) . 27
HAWKER SIDDELEY GROUP, LIMITED (U.K.) ....(1977-78) 200
Hawkeye (Grumman) 366
Haze (Mil) 210
HEADBERG AVIATION INC (U.S.A.) 528
Headwind (Stewart) 554
HEINTZ, CHRISTOPHE (Canada) .....(1974-75) 27
HEINTZ, CHRISTOPHE (France) ......(1972-73) 64
Heintz Mono Zenith-CH 100 (Zenair) ..(1977-78) 482
Heintz Tri-Z-CH 300 (Zenair) ....(1977-78) 482
Heintz Zenith-CH 200 (Zenair) ....(1980-81) 468
Heintz Zenith-CH 400 (Zenair) 28
HELIBRAS (Brazil) ....(1980-81) 18
HELICOM INC. (U.S.A.) ....(1972-73) 330
HELICOP-JET (France) 67
HELICOPTEROS DO BRASIL SA (Brazil) 17
HELICOPTER TECHNIK MUNCHEN GmbH & Co ANLAGEN KG (West Germany) (1975-76) 92
HELI-EUROPE INDUSTRIES LTD (International) ....(1978-79) 84
HELIGYRO CORPORATION (U.S.A.) .(1978-79) 522
HELIO AIRCRAFT COMPANY (U.S.A.) (1976-77) 300
HELIO COURIER LTD (USA) ....(1979-80) 358
HELIPARTS INC (U.S.A.) ....(1975-76) 357
Heli-Star (Meger) ....(1972-73) 370
Heli-Stat (Piasecki) 424
HELITEC CORPORATION (U.S.A.) ...(1979-80) 358
HELLBACH, WALTER (West Germany) (1979-80) 485
HELLENIC AEROSPACE INDUSTRY SA (Greece) 86
HELTON AIRCRAFT CORPORATION (U.S.A.) (1971-72) 320
Hercules (Lockheed) 395
Hercules conversion (Marshall) 253
Hercules Regional Airfreighter (Lockheed) ....(1980-81) 377
HERTFORDSHIRE PEDAL AERONAUTS (U.K.) (1979-80) 494
HESLEY, R. (U.S.A.) ....(1972-73) 333
HIBBARD AVIATION (U.S.A.) 377
HIETANEN, ARI AND ESKO (Finland) (1972-73) 37
HILL, JOHN S. (U.S.A.) ....(1974-75) 355
HILL, KLAUS (U.S.A.) ....(1979-80) 523
HILLER AVIATION INC (U.S.A.) 377
HILLMAN HELICOPTER ASSOCIATES (U.S.A.) 379
Hind (Mil) 211
HINDUSTAN AERONAUTICS LTD. (India) 86
Hinu (PIK) ....(1973-74) 40
Hip (Mil) 208
Hiperbipe (Sorrell) 550
Hirondelle (Dassault) ....(1971-72) 59
Hirondelle, PGK-1 (Western) 476
HIRTH GmbH, WOLF (West Germany) .(1977-78) 71
Hirundo (Agusta) ....(1974-75) 126
HISPANO-AVIACION S.A., LA (Spain) (1972-73) 174
Holiday Knight Twister SKT-1250 (Payne) 540
HOLLMANN AIRCRAFT (U.S.A.) 528
Homer (Mil) ....(1978-79) 190
Honcho 100 (Nagler) ....(1974-75) 402
Honcho 200 (Nagler) ....(1974-75) 402
Honeybee (Sinfield) ....(1979-80) 542
Hoodlum (Kamov) 199
Hook (Mil) 207
Hoplite (WSK-PZL Swidnik/Mil) 165
Hormone (Kamov) 198
Hornet, F/A-18 (McDonnell) 406
Hornisse (Wallerkowski) 495
Hornet (Gyroflight) ....(1972-73) 197
Hound (China) 34
Hound (Mil) 207
Hover-Gyro, B-18 (Bensen) 512
HOVEY, R.W. (U.S.A.) 528
HueyCobra (Bell) 296, 298
Huey Plus (Bell) ....(1974-75) 273
HueyTug (Bell) ....(1971-72) 245
HUGHES HELICOPTERS (U.S.A.) 379
HUGO, ADOLPH B., Jr. (U.S.A.) .....(1972-73) 336
Hu-go Craft (Hugo) ....(1971-72) 323
Humbug (Sky Sports) ....(1980-81) 441
Humlan 2 (Elkström) 499
Hummer (Hummer) ....(1980-81) 524
HUMMER (U.S.A.) ....(1980-81) 524
Hummingbird (Jonas) ....(1980-81) 526
HUNEAULT, JEAN PAUL (Canada) ...(1971-72) 24
Hunter Mk 8M (BAe) ....(1979-80) 237
HUREL, MAURICE (France) ....(1977-78) 490
HUREL-DUBOIS S.A., SOCIÉTÉ DE CONSTRUCTION DES AVIONS (France) .(1974-75) 69
Huron (Beechcraft) ....(1975-76) 263
Husky F-11-2 (Harrison) ....(1972-73) 22
Hustler Model 400 (AJI) ....(1978-79) 252
Hustler 500 (GACC) 355
Hydro-Copter (Bensen) 511
Hydro-Glider (Bensen) 510

I

I-11B (AISA) (1974-75) 178
I-66 San Francesco (Iannotta) (1973-74) 126
I-122 (AISA) 173
I-124 (AISA) 173
IA 50 GII (FMA) (1975-76) 4
IA 53 (FMA) (1971-72) 4
IA 58A Pucará (AMC) 2
IA 58B Pucará (AMC) 3
IA 60 (FMA) (1976-77) 4
IA 63 (FMA) 3
IA 66 (FMA) 3
IA 80 Piranha (Issoire) (1980-81) 67
IAE (see Instituto de Atividades Espaciais) 9
IAI (see Israel Aircraft Industries) 112
IAI-101 Arava (IAI) 114
IAI-102 Arava (IAI) 114
IAI-201 Military Arava (IAI) 114
IAI-202 Arava (IAI) 115
IAI 1123 Westwind (IAI) (1978-79) 103
IAI 1124 Sea Scan (IAI) 116
IAI 1124 Westwind (IAI) 115
IAI 1124 Westwind 1 (IAI) 115
IAI 1124A Westwind 2 (IAI) 116
IAI 1125 Astra (IAI) 117
IAR-93/Orao (SOKO/CNIAR) 110
IAR-316 Alouette III (ICA/Aérospatiale) 171
IAR-330 Puma (ICA/Aérospatiale) 171
IAR-361B Alouette III (ICA/Aérospatiale) 171
IAR-821 (IRMA-Bucuresti) (1974-75) 173
IAR-821B (IRMA-Bucuresti) (1974-75) 173
IAR-822 (ICA-Brasov) (1979-80) 160
IAR-822B (ICA-Brasov) (1979-80) 160
IAR-823 (ICA) 170
IAR-824 (ICA) (1980-81) 164
IAR-826 (ICA-Brasov) (1979-80) 160
IAR-827A (ICA) 171
IAv BACAU (see Intreprinderea de Avioane Bacau) 169
IAv BUCURESTI (see Intreprinderea de Avioane Bucuresti) 170
IAv CRAIOVA (see Intreprinderes de Avioane Craiova) 170
ICA (see Intreprinderea de Constructii Aeronautice) 170
IEP (see Instituto de Ensaios e Padroes) (1978-79) 10
IFI (see Instituto de Fomento e Coordenacão Industrial) 9
IHI (see Iranian Helicopter Industry) (1980-81) 101
IL (see Instytut Lotnictwa) 158
Il-18 Moskva (Ilyushin) 192
Il-28 (China) 32
Il-38 (Ilyushin) 194
Il-62 (Ilyushin) 194
Il-62M (Ilyushin) 194
Il-62MK (Ilyushin) 194
Il-76M (Ilyushin) 195
Il-76T (Ilyushin) 195
Il-86 (Ilyushin) 197
IPAI-26 Tuca (São Carlos) 18
IPAI-27 Jipe Voador (São Carlos) 18
IPAI-28 Super Surubim (São Carlos) 18
IPAI-29 Tira Prosa (São Carlos) 18
IPAI-30 (São Carlos) 18
IPD (see Instituto de Pesquisas e Desenvolvimento) 9
IPD-2601 Universal (Nieva) (1972-73) 13
IPD/PAR 6504 (IPD) (1972-73) 10
IPD-6504 Bandeirante (IPD) (1972-73) 10
IPD-6901 Ipanema (IPD) (1972-73) 12
IPT-16 Surubim Modificado (Rio Claro/IPT) (1979-80) 17
IS-23A (ICA-Brasov) (1974-75) 174
ISa-24 (ICA-Brasov) (1974-75) 174
ITA (see Instituto Tecnológico de Aeronáutica) 9
ITOH (see Shin Nihon) (1971-72) 148
IANNOTTA, DR ING ORLANDO (Italy) (1974-75) 131
Ibis A, NM-77 (Nihon) (1979-80) 486
ICX AVIATION INC (U.S.A.) (1980-81) 364
Ilse, RS-1 (Siegrist) (1976-77) 532
ILYUSHIN (U.S.S.R.) 192
Imp (Aerocar) 507
Imp (Wallis) (1973-74) 228
IMP AEROSPACE LTD (Canada) (1974-75) 28
Impala (Atlas) (1975-76) 176
Impala (Jacquet-Pottier) (1975-76) 70
Impala Mk 2 (Atlas) 172
Imperial Knight Twister M6 (Payne) 540
Improved SeaCobra (Bell) 297
INDRAÉRO SA (France) (1976-77) 58
INDUSTRIAL PRODUCTS GROUP ROCKWELL (U.S.A.) (1974-75) 441
INDUSTRIE AERONAUTICHE E MECCANICHE RINALDO PIAGGIO (Italy) 139
INSTITUTO DE ATIVIDADES ESPACIAIS (Brazil) 9
INSTITUTO DE ENSAIOS E PADROES (Brazil) (1978-79) 10
INSTITUTO DE FOMENTO E COORDENACÃO INDUSTRIAL (Brazil) 9
INSTITUTO DE PESQUISAS E DESENVOLVIMENTO (Brazil) 9
INSTITUTO TECNOLÓGICO DE AERONÁUTICA (Brazil) 9
INSTYTUT LOTNICTWA (Poland) 158
INTERCEPTOR COMPANY (U.S.A.) (1979-80) 363
INTERNATIONAL HELICOPTERS INC (U.S.A.) 529
INTREPRINDEREA DE AVIOANE BACU (Romania) 169
INTREPRINDEREA DE AVIOANE BUCURESTI (Romania) 170
INTREPRINDEREA DE AVIOANE CRAIOVA (Romania) 170
INTREPRINDEREA DE CONSTRUCTII AERONAUTICE (Romania) 170
Intruder (Grumman) 367
Ipanema (EMBRAER) 16
Ipanema (IPD) (1972-73) 12
IRANIAN HELICOPTER INDUSTRY (Iran) (1979-80) 101
Iroquois (Bell) 293
ISAACS, JOHN O. (U.K.) 502
Isfahan, Model 214A (Bell) (1978-79) 277
ISHAM AIRCRAFT (U.S.A.) 384
Iskra, TS-11 (WSK-PZL Mielec) 162
ISLAND AIRCRAFT CORPORATION (U.S.A.) (1972-73) 337
Islander (IAvB Bucuresti/Pilatus Britten-Norman) 170
Islander (PADC/Pilatus Britten-Norman) 157
Islander BN-2B (Pilatus Britten-Norman) 256
ISON, WAYNE (U.S.A.) (1974-75) 360
ISRAEL AIRCRAFT INDUSTRIES LTD (Israel) 112
ISSOIRE-AVIATION, SOCIÉTÉ (France) (1980-81) 67
ITALAIR SpA (Italy) (1977-78) 117

J

J-1/RJ-1 Jastreb (SOKO) (1980-81) 455
J-1 Jupiter (Champion) (1972-73) 302
J-1 Martin Fierro (RRA) 4
J-1 Przasniczka (Janowski) (1974-75) 171
J-1B Don Kichot (Janowski) (1979-80) 488
J-2 (Aero Resources) (1974-75) 243
J-2 (McCulloch) (1972-73) 358
J-2 Polonez (Janowski) (1979-80) 589
J-3M (Barnett) (1977-78) 522
J-4B (Barnett) (1977-78) 522
J-5 (China) 29
J-12 (F-12) (China) 32
J12, You-Too (Franks) (1977-78) 540
J 35 Draken (Saab) (1977-78) 168
JA 37 Viggen (Saab) 177
JC-01 (Coupé-Aviation) 481
JC-1 Challenger (Cook) (1974-75) 321
JC-1 Jet Cat (Continental Copters) (1980-81) 331
JC-24 Weedhopper (Weedhopper) (1980-81) 558
JC-130 Hercules (Lockheed) 395
JD$_1$HW 1.7 Headwind (Stewart) 554
JD-2 Delta (Dyke) 521
JD$_2$FF Foo Fighter (Stewart) 555
JDA (see Japan Defence Agency) (1973-74) 140
JE-2 Gyroplane (Eich) 524
JEAA (see Japan Experimental Aircraft Association) (1977-78) 500
JET (see Joint European Transport) (1979-80) 93
JF-105 Thunderchief (Republic) (1973-74) 326
JHC-130 Hercules (Lockheed) 395
JM-2 (Miller) (1977-78) 548
JP001 Wild Turkey (American Air Racing) (1978-79) 504
JP-20-90 Impala (Jacquet-Pottier) (1975-76) 70
JT.1 Monoplane (Taylor) 505
J.T.2 Titch (Taylor) 506
JT-5 (Tervamäki) 480
Ju 87B-2 Replica (Langhurst) 533
JUP-1 (AVEX) (1972-73) 3
JW-1 Jetwing (Ball-Bartoe) (1978-79) 254
JACOBS JETS (U.S.A.) 384
JACQUET-POTTIER (see Société CARMAM) (1975-76) 70
Jaguar (SEPECAT) 107
Jaguar Development (SEPECAT) 109
Jaguar International (HAL/SEPECAT) 87
JAMES-DENNIS ENTERPRISES INC (U.S.A.) (1980-81) 524
JAMESON, RICHARD J. (U.S.A.) (1975-76) 362
JANOWSKI, JAROSLAW (Poland) (1979-80) 488
JANOX CORPORATION (U.S.A.) (1972-73) 337
JAPAN AMATEUR BUILT AIRCRAFT LEAGUE (Japan) 496
JAPAN DEFENCE AGENCY (Japan) (1973-74) 140
JAPAN EXPERIMENTAL AIRCRAFT ASSOCIATION (Japan) (1977-78) 500
JARZAB, KAZIMIERZ (Poland) 498
JAS (JAS) 176
JAS, INDUSTRI GRUPPEN (Sweden) 176
Jastreb (Soko) (1980-81) 455
Jastreb Trainer (Soko) (1980-81) 455
JAVELIN AIRCRAFT COMPANY INC. (U.S.A.) 530
JEAN ST-GERMAIN INC, CENTRE DE RECHERCHES (Canada) 474
JEFFAIR CORPORATION (U.S.A.) 530
JELM AIRCRAFT COMPANY (U.K.) (1972-73) 209
Jetasen (Maule) (1976-77) 324
Jet Cat JC-1 (Continental Copters) (1980-81) 331
JETCRAFTERS INCORPORATED (U.S.A.) 384
Jet Flying Belt (Bell) (1974-75) 266
Jet Liner 600 (Carstedt) (1972-73) 283
Jet-Pak C-119/R4Q (Steward-Davis/Fairchild) (1974-75) 468
Jet Prop 340 (Riley) (1979-80) 423
Jet Prop 421 (Riley) (1978-79) 415
Jetprop 840 (Gulfstream Commander) 394
Jetprop 980 (Gulfstream Commander) 394
Jetprop 1000 (Gulfstream Commander) 395
Jet Provost (B.A.C.) (1978-79) 177
JetRanger (Agusta-Bell) (1974-75) 128
JetRanger (Bell) (1974-75) 269
JetRanger II (Agusta-Bell) (1978-79) 104
JetRanger II (Bell) (1977-78) 237
JetRanger III (Agusta-Bell) 127
JetRanger III (Bell) 294
JetStar (Lockheed) (1974-75) 376
JetStar II (Lockheed) (1980-81) 377
Jetstream (Jetstream) (1971-72) 198
Jetstream (Riley) (1977-78) 381
Jetstream III (Volpar/Century) (1980-81) 451
Jetstream 31 (BAe) 237
JETSTREAM AIRCRAFT LTD (U.K.) (1971-72) 198
Jetstream Series 200 (BAe) (1979-80) 225
Jet Trader (Douglas) (1972-73) 364
Jetwing, JW-1 (Ball-Bartoe) (1978-79) 254
Jipe Voador, IPAI-27 (São Carlos) 18
JOB 15 (Oberlerchner) (1971-72) 9
JODEL (see Avions Jodel SA) 484
JOHANSEN, DIPL-ING CARL (Denmark) (1971-72) 35
JOINT EUROPEAN TRANSPORT (International) (1979-80) 93
Jolly Green Giant (Sikorsky) 459
JONAS, GERALD (U.S.A.) (1980-81) 526
JONES, DANIEL G. (Canada) 475
Jungmann (Morand) (1975-76) 189
Jungmeister (Bücker) (1971-72) 83
Jungster I (K & S) (1975-76) 25
Jungster II (K & S) (1975-76) 25
Jungster III (Kaminskas) (1974-75) 364
Junior (MBB) (1971-72) 88
Junior Ace Model E (Ace Aircraft) 506
Junior Aero Sport (D'Apuzzo) (1977-78) 534
Junkers Ju 87B-2 Replica (Langhurst) 533
JUPITER (see Project Jupiter) (1975-76) 216
Jupiter (Champion) (1972-73) 302
Jupiter (Cranwell) (1977-78) 508
JURCA, MARCEL (France) 484
JUROM (International) (1975-76) 112

K

K-1 Flamingo-Trainer (FUS) 81
K-1 Super Fli (Kraft) 532
KA-6 (Grumman) 367
KA-7 Corsair II (Vought) 469
Ka-25 (Kamov) 198
Ka-25K (Kamov) 199
Ka-26 (Kamov) 199
Ka-32 (Kamov) 200
Ka-126 (Kamov) 200
KA-840 (Kawasaki) (1980-81) 138
KA-850 (Kawasaki) 144
KAI (see Kharkov Aviation Institute) (1976-77) 545
KAI-20 (KAI) (1972-73) 459
KAI-21 Gibkolet (KAI) (1976-77) 545
KAI-22A (KAI) (1972-73) 460
KAL (see Korean Air Lines) 149
KC-3 Skylark (Larkin) (1980-81) 528
KC-10A (Douglas) 412
KC-130 Hercules (Lockheed) 395
KC-135 winglet tests (NASA/USAF) 419
KC-135A Stratotanker (Boeing) 317
KH-4 (Kawasaki) (1977-78) 130
KH-7 (Kawasaki) (1976-77) 125
KL-35 D-160 (Hirth) (1973-74) 90
KM-2 (Fuji) 142
KM-2B (Fuji) 142
KR-1 (Rand Robinson) 544
KR-2 (Rand Robinson) 544
KR-3 (Rand Robinson) (1979-80) 534
K & S (see McLeod, Kaye and Stan) (1976-77) 456
KSA-100 SAVER (Kaman) (1975-76) 365
KT-75 Knight Twister Junior (Payne) (1980-81) 536
KT-85 Knight Twister (Payne) (1980-81) 536
KTT-90 Knight Twister (Payne) (1971-72) 376
KV-107/II (Kawasaki/Boeing) 144
KV-107/IIA (Kawasaki/Boeing) 144
KY-03 Libellula (Yager) (1975-76) 11
Kachina (Hibbard/Varga) 377
Kachina, Model 2150A (Varga) 467
Kachina, Model 2180 (Varga) 467
KAHN, DAVID (U.S.A.) (1976-77) 517
KAKO, TERUO (Japan) (1979-80) 486
KAMAN AEROSPACE CORPORATION (U.S.A.) 385
KAMINSKAS, RIM (U.S.A.) (1975-76) 365
KAMOV (U.S.S.R.) 198
Kania/Kittyhawk (WSK-PZL-Swidnik) 166
KANPUR DIVISION, HINDUSTAN AERONAUTICS LTD (India) 90
KAWASAKI JUKOGYO KABUSHIKI KAISHA (Japan) 143
KELEHER, JAMES J. (U.S.A.) 531
KELLY, DUDLEY R (USA) 531
Kelly-D (Kelly) 531
Kfir (IAI) 113
Kfir-C2 (IAI) 113
Kfir-TC2 (IAI) 113
KHARKOV AVIATION INSTITUTE (U.S.S.R.) (1976-77) 545
KICENIUK JR, TARAS (U.S.A.) (1978-79) 524
KILLINGSWORTH, RICHARD (U.S.A.) (1974-75) 353
KIMBREL, MICHAEL (U.S.A.) 532
King Air A100 (Beechcraft) 289
King Air B100 (Beechcraft) 290
King Air C90 (Beechcraft) 286
King Air C100 (Beechcraft) (1978-79) 269
King Air E90 (Beechcraft) 287
King Cat (Mid-Continent) 413
KingCobra (Bell) (1974-75) 274
Kingfisher (Anderson) 510
KINMAN, DUANE (U.S.A.) 532
Kiowa (Bell) 294
Kiran (HAL) 87
Kiran Mk II (HAL) 88
Kittiwake I (Mitchell-Procter) (1978-79) 498
Kittiwake II (Mitchell) (1973-74) 215
KIYUSKI INTERNATIONAL (Pakistan) (1977-78) 146
Klos (Jarzab) 498
K-Meyer Aero Model A (Meyer) 535
Knight Twister Imperial (Payne) (1976-77) 525
Knight Twister Junior KT-85 (Payne) (1980-81) 536
Knight Twister KT-85 (Payne) (1980-81) 536
Knight Twister KTT-90 (Payne) (1972-73) 376
KNOWLES, GP CAPT A.S. (U.K.) (1980-81) 495
KOBUBUN, MASANORI (Japan) (1972-73) 143
Kochab (Brandt) (1976-77) 461
Kodiak, Model 302 (Sequoia) 548
KOENIG, M (France) (1980-81) 479

KOKKOLA, DIPL INGS KALEVI AND SEPPO (Finland) (1971-72) 36
Koliber (CNPSL-PZL Warszawa) 160
KOREAN AIR LINES (Korea) 149
KOVACH, KIM (U.S.A.) 532
KRAFT, PHIL (U.S.A.) (1977-78) 545
KRAFT SYSTEMS INC (U.S.A.) 532
Kruk, (CNPSL-PZL Warszawa) 159
KUYBYSHEV POLYTECHNIC (U.S.S.R.) 172
Kudu (Atlas) 172
KUIBYSHEV AVIATION INSTITUTE (U.S.S.R.) (1973-74) 477
Kukulka (Pieniazek) (1973-74) 168
Kwant (MAI) (1975-76) 496

L

L5 (Lucas) 488
L-5 (SSVV) (1980-81) 135
L.8C (AISA) (1974-75) 178
L-19 (Ector) 348
L-28 Super Courier (Helio) (1977-78) 315
L-29 Delfin (Aero) (1975-76) 33
L-29A Delfin Akrobat (Aero) (1974-75) 33
L-39 (Aero) 36
L-42 Regente (Neiva) (1974-75) 18
L-70 Miltrainer (Valmet) 42
L-100 (Lockheed) 397
L 100-50 Hercules (Lockheed) (1980-81) 377
L 400 Twin Hercules (Lockheed) (1980-81) 377
L-410UVP Turbolet (Let) 39
L450F (LTV) (see also under "RPVs") (1971-72) 344
L 500-114ZC (Lockheed) (1971-72) 341
L-1011 conversions (LAS) 387
L-1011 TriStar (Lockheed) 393
LA-4-200 Buccaneer (Lake) 386
L.A.4a Luton Minor (Phoenix) (1974-75) 222
L.A.5a Luton Major (Phoenix) (1974-75) 222
L.A.4a Minor (PFA) 504
LACO-125 (LACO) 532
LAPAN (see Lembaga Penerbangan dan Antariska Nasional) 90
LAS (see Lockheed Aircraft Service Company) 387
LC-3, X-Avia (Lewis) 388
LC-10 Criquet (Croses) 481
LC-130 Hercules (Lockheed) 395
LD 20b (Lemberger) (1973-74) 92
LD 261 (Midgy) (1980-81) 481
LDA-01 (Lockspeiser) 253
LFU (see Leichtflugtechnik Union GmbH) (1974-75) 94
LIPNUR (see Nurtanio P.T.) (1978-79) 77
LLRV (Bell) (1972-73) 257
LLTV (Bell) (1972-73) 257
LMC-1 Sprintair (Chasle) (1977-78) 485
LN-11 (Larsen) (1971-72) 155
LP Dedal-2 (Polniak) (1977-78) 503
LR-1 (Mitsubishi) (1976-77) 127
LRPA (Boeing) (1976-77) 247
LT-200 (LIPNUR) (1978-79) 77
LTA (Dornier) (1979-80) 74
LTV (see LTV Aerospace Corporation) (1975-76) 379
Labrador (Boeing Vertol) (1973-74) 283
Labrador, CH-113 (Boeing Canada) 18
LACEY, JOSEPH L. (U.S.A.) (1971-72) 328
LACO (USA) 532
L'Acrobat (Robin/CEA) (1971-72) 76
LAKE AMPHIBIAN INC (U.S.A.) 386
Lala-1 (IL) (1974-75) 164
Lama (Aérospatiale) 45
Lama, SA 315B (HAL/Aérospatiale) 90
Laminar (Pazmany) (1980-81) 536
Laminar flow aircraft (NASA) 418
Lance II (Piper) (1980-81) 408
Lanceiro (Neiva) (1976-77) 16
LANDRAY, GILBERT (France) 487
LANE, DICK (U.S.A.) (1977-78) 546
LANGHURST, LOUIS (USA) 533
Lark 1B (Keleher) 531
Lark III (Fukuoka) (1976-77) 474
Lark Commander (Aero Commander) (1971-72) 366
LARKIN AIRCRAFT COMPANY (U.S.A.) (1980-81) 528
LARSEN, CAPTAIN CARL L. (Norway) (1971-72) 155
Laser 200 (Carmichael) 516
LASHER, C. W. (U.S.A.) 533
Lavi (IAI) 113
LEAR FAN CORPORATION (U.S.A.) 387
LearFan 2100 (LearFan) 253
LEARFAN LTD (UK) 253
Learjet (Calspan) 324
Learjet 24E/F (Gates Learjet) 357
Learjet 25D (Gates Learjet) 357, 358
Learjet 25F (Gates Learjet) (1979-80) 332
Learjet 28/29 Longhorn (Gates Learjet) 358
Learjet 35A/36A (Gates Learjet) 358
Learjet 55/56 (Gates Learjet) 359
Learjet Sea Patrol (Gates Learjet) 359
LearStar 600 (Canadair) (1977-78) 20
LEBOUDER, ROBERT (France) (1975-76) 74
LEDERLIN, FRANCOIS (France) 487
LEFEBVRE, ROBERT (France) 488
LEICHTFLUGTECHNIK-UNION GmbH (West Germany) (1974-75) 94
LEICHTFLUGZEUGE-ENTWICKLUNGEN DIPL ING HERMANN MYLIUS (West Germany) 85
LEISURE SPORT LTD (U.K.) (1980-81) 496
Leko-70 (Valmet) (1978-79) 35
LEMBAGA PENERBANGAN DAN ANTARISKA NASIONAL (Indonesia) 90
LEMBERGER, KARL (West Germany) (1973-74) 92
LENIN KOMSOMOL INSTITUTE, RIGA (U.S.S.R.) (1977-78) 574
Leonardino (Bagalini) (1976-77) 473
Léopard, CE.45 (CERVA) (1977-78) 50
LET NARODNI PODNIK (Czechoslovakia) 39
LEVI, R. (U.K.) (1975-76) 217
LEWIS AIRCRAFT CORPORATION (U.S.A.) 388
Libellula (Yager) (1971-72) 8
Libellula, KY-03 (Yager) (1975-76) 11
Light aircraft programme (EMBRAER/Piper) 17
Light Helicopter (HAL) 89
Lightning (BAC) (1974-75) 198
Lightning Model 400 (Smith) 464
Lightning Trainer (BAC) (1971-72) 183
Lil Quickie, OR-71 (Vin-Del/Owl) 539
Lil Rascal (Aeroneering) (1979-80) 503
LING-TEMCO-VOUGHT, INC. (U.S.A.) (1971-72) 342
Linnet IV (Nihon) (1971-72) 146
Little Toot (Meyer) (1974-75) 397
LIVERPOOL UNIVERSITY (see University of Liverpool, Department of Mechanical Engineering) (1974-75) 218
Liverpuffin (Liverpool University) (1974-75) 218
LIVESEY, DAVID M. (U.K.) (1979-80) 496
LOBET-DE-ROUVRAY AVIATION PTY. LTD. (New Zealand) (1978-79) 487
LOCKHEED AIRCRAFT SERVICE COMPANY (U.S.A.) 387
LOCKHEED-CALIFORNIA COMPANY (U.S.A.) 389
LOCKHEED CORPORATION (U.S.A.) 388
LOCKHEED-GEORGIA COMPANY (U.S.A.) 395
LOCKHEED MISSILES AND SPACE COMPANY (U.S.A.) (1973-74) 365
LOCKSPEISER AIRCRAFT LTD (U.K.) 253
Logger (Piasecki) 424
Long-EZ (Rutan) 547
Longhorn (Gates Learjet) 358, 359
LongRanger (Agusta-Bell) (1979-80) 114
LongRanger II (Agusta-Bell) 127
LongRanger II, III & IV (Bell) 295
Love One (WFI) (1972-73) 445
LTV AEROSPACE CORPORATION (U.S.A.) (1975-76) 379
LTV ELECTROSYSTEMS INC (U.S.A.) (1971-72) 344
LUCAS, EMILE (France) 488
Ludion (Aérospatiale) (1971-72) 47
LUDWIG, HARRY (U.S.A.) 533
Luftibus, GB-1 (Bezzola) (1978-79) 490
Lunar Flying Vehicle (Bell) (1971-72) 241
Lunar Landing Research Vehicle (Bell) (1972-73) 257
Lunar Landing Training Vehicle (Bell) (1972-73) 257
Lunar Rocket (Maule) 401
Lutin (Fauvel) (1979-80) 474
Luton L.A.4a Minor (PFA) 504
Luton L.A.4a Minor (Phoenix) (1974-75) 222
Luton L.A.5a Major (Phoenix) (1974-75) 222
Lynx (ABHC/Westland) (1980-81) 45
Lynx (Gulfstream American) (1979-80) 352
Lynx (Westland/Aérospatiale) 269

M

M2-F2/F3 (Northrop/NASA) (1973-74) 391
M-II Mustang II (Bushby) 515
M-4 Astro-Rocket (Maule) (1976-77) 324
M4 Gyroplane (McCandless) (1972-73) 211
M-4 Jetasen (Maule) (1976-77) 324
M-4 Rocket (Maule) (1976-77) 324
M-4 Strata-Rocket (Maule) (1976-77) 324
M-4 (Myasischev) 214
M-5 Lunar Rocket (Maule) 401
M6, Imperial Knight Twister (Payne) 540
M-10 Cadet (Univair) (1975-76) 474
M-10 Miracle of 10's (Lacey) (1971-72) 328
M-10C VW Twin (Lacey) (1971-72) 329
M-12 Tchaika (Beriev) 191
M-15 Belphegor (WSK-PZL Mielec) 163
M-17 (WSK-Mielec) (1979-80) 156
M-18 Dromader (WSK-PZL Mielec) 164
M-20 Mewa (WSK-PZL Mielec) 165
M20J (Mooney) 416
M20K (Mooney) 417
M-30 (Mooney) 417
M-74 Wasp (Texas Helicopter) (1980-81) 446
M79S Wasp II (Texas Helicopter) (1980-81) 446
M102 Scorchy (Mace) 534
M484 (Fairchild Industries) (1974-75) 333
MA-1B Diablo 1200 (Emair) (1980-81) 333
Mac 6 Mercury (Macdonald) (1976-77) 483
MAI (see Moscow Aviation Institute) (1975-76) 496
MARS (see Mission Aviation Related Services) 27
MB-2 Colibri 2 (Brügger) 500
MB-3 (Brügger) (1980-81) 492
M.B.326 (Aermacchi) 122
M.B.326H (CAC) (1973-74) 8
M.B.339A (Aermacchi) 122
M.B.339K Veltro 2 (Aermacchi) 123
M.B.340 (Aermacchi) (1977-78) 111
MBB (see Messerschmitt-Bölkow-Blohm GmbH) 82
MBB 223 Flamingo (CASA) (1974-75) 180
MC-4E (McCulloch) (1971-72) 347
MC-8 Mini-Copter (Aerospace General) (1978-79) 250
MC 10 Cricri (Colomban) 480
MC 12 Cricri (Colomban) 480
MC-130 Hercules (Lockheed) 395
MD 320 Hirondelle (Dassault) (1971-72) 69
MDF-100 (McDonnell/Fokker) 103, 413
MDV-1 (Aerospace General) (1977-78) 214
Me 108F Taifun (Taifun) (1975-76) 97
MEG-2X (Gluhareff) (1973-74) 334
MFI (see A.B. Malmö Flygindustri) (1971-72) 168
MH-53E (Sikorsky) 461
MHO.235 (JEAA) (1975-76) 144
Mi-2 (Mil) 207
Mi-2 (WSK-PZL Swidnik/Mil) 165
Mi-2M (WSK-Swidnik) (1977-78) 155
Mi-4 (Mil) 207
Mi-6 (Mil) 207
Mi-8 (Mil) 208
Mi-10 (Mil) 209
Mi-10K (Mil) 210
Mi-12 (Mil) (1978-79) 190
Mi-14 (Mil) 210
Mi-17 (Mil) 210
Mi-18 (Mil) 211
Mi-24 (Mil) 211
Mi-26 (Mil) 212
MiG (formerly Mikoyan) (U.S.S.R.) 200
MiG-19PF (China) 29
MiG-19SF (China) 29
MiG-21 (Mikoyan) 200
MiG-21*bis* (HAL) 90
MiG-21F (HAL) (1976-77) 83
MiG-21FL (HAL) (1977-78) 85
MiG-21M (HAL) 90
MiG-21MF (HAL) 90
MiG-21PF (HAL) (1976-77) 83
MiG-21U (HAL) (1977-78) 85
MiG-21 (STOL version) (Mikoyan) (1971-72) 444
MiG-23 (HAL) 90
MiG-23 (Mikoyan) 202
MiG-25 (Mikoyan) 204
MiG-27 (Mikoyan) 204
MIT (see Massachusetts Institute of Technology) 413
MIT (see MIT Manpowered Aircraft Association) (1977-78) 548
M.J.2 Tempête (Jurca) 484
M.J.3 (Jurca) (1972-73) 68
M.J.3H Dart (Jurca) 485
M.J.4 Shadow (Jurca) (1972-73) 68
M.J.5 Sirocco (Jurca) 485
M.J.5 Sirocco (Sport Wing) (Jurca) 486
M.J.6 Crivat (Jurca) (1971-72) 67
M.J.7 Gnatsum (Jurca) 486
M.J.7S Solo (Jurca) 486
M.J.8 1-Nine-O (Jurca) 486
M.J.9 One-oh-Nine (Jurca) 486
M.J.10 Spit (Jurca) 486
M.J.12 Pee-40 (Jurca) 486
M.J.14 Fourtouna (Jurca) 486
M.J.20 Tempête (Jurca) 484
M.J.50 Windy (Jurca) (1977-78) 492
M.J.51 Sperocco (Jurca) 486
M.J.66 Crivat (Jurca) (1971-72) 67
M.J.77 Gnatsum (Jurca) 486
M.J.80 1-Nine-O (Jurca) 486
Mk 1 Phillicopter (Phillips) (1974-75) 10
Mk V (Durand) 521
Mk 55 Jet Provost (BAC) (1973-74) 193
MM-1 Midget Mustang (Bushby) 515
MP (See Moto-Plane Aviation Inc) (1975-76) 26
MP.205 Busard (Lefebvre) 488
MPA Mk 1 (Wright) (1975-76) 238
MRCA (Panavia) 103
MS 880B Rallye-Club (Socata) (1972-73) 82
MS 892 Rallye Commodore 150 (Socata) (1972-73) 82
MS 893 Agricorallye (Socata) (1971-72) 78
MS 893 Rallye Commodore 180 (Socata) (1972-73) 82
MS 894A Rallye Minerva 220 (Socata) (1972-73) 82
MT5 (Magni) (1974-75) 132
MT5 (VPM) 496
MT7 (VPM) 496
MU-2 (Mitsubishi) 146
MU-2 (Mitsubishi International) 414
MU-2S (Mitsubishi) (1972-73) 144
MU-300 (Mitsubishi) 146
MW2 Excalibur (Trago Mills) (1977-78) 516
MW3 Airkart (Whittaker) (1979-80) 500
MY 102 Tornado (Mylius) (1980-81) 84
MY 103 Mistral (Mylius) (1980-81) 84
MY 104 Passatt (Mylius) (1980-81) 84
Mya-4 (Myasishchev) (1975-76) 507
MACE AIRCRAFT (U.S.A.) 534
MACE, HARVEY F. (U.S.A.) (1971-72) 345
MACCHI S.p.A., AERONAUTICA (Italy) 122
MACDONALD, D. J. (UK) (1976-77) 483
MacDONALD AIRCRAFT CO (U.S.A.) (1979-80) 528
MacCREADY, DR PAUL B. (U.S.A.) 398
MACFAM WORLD TRADERS (Canada) 475
MACHEN INC (U.S.A.) 400
Maestro (Yakovlev) 230
Magister (Aérospatiale) (1971-72) 41
MAGNI, VITTORIO (Italy) (1976-77) 112
MAHONEY, LEE (U.S.A.) (1971-72) 345
MAHUGH, G. IRVIN (U.S.A.) (1980-81) 529
Maiden (Sukhoi) 215
Mail (Beriev) 191
Major (Phoenix) (1974-75) 222
Major, DR 400/140B (Robin) (1979-80 66
Major 80, DR 400/160 (Robin) 71
Major 160 (Robin/CEA) (1972-73) 80
Majorette (Phoenix) (1971-72) 203
Makino MHO.235 (JEAA) (1975-76) 144
MALLIGA, JOSEF (Austria) (1974-75) 12
MALMÖ FLYGINDUSTRI A.B. (Sweden) (1971-72) 168
MANCRO AIRCRAFT COMPANY (U.S.A.) 400
"Mandrake" (Yakovlev) (1973-74) 498
"Mangrove" (Yakovlev) (1971-72) 467
Mangusta, A 129 (Agusta) 126
Man-powered aircraft (CASI) (1977-78) 480
Man-powered aircraft (Jupiter) (1975-76) 216
Man-powered aircraft Waybridge (MPAG) (1974-75) 240
MARCHETTI, FRANK (U.S.A.) 534
Marimo, TK-1 (Kako) (1978-79) 486
Maritime, F27 (Fokker) 153
Maritime Defender (Pilatus Britten-Norman) 258
Mark 1 (Spitfire) 465
Mark II (Merkel) (1978-79) 527
Mark II (Spitfire) 466
Mark II System (Howard/Raisbeck) (1980-81) 420
Mark IV (Spitfire) 466
Mark IV System (Raisbeck) (1980-81) 421

Mark V System (Rockwell/Raisbeck) (1980-81) 421
Mark Five System (Rockwell/Raisbeck) (1980-81) 421
Mark Four Learjet Century III Improvements (Raisbeck) (1980-81) 421
Marlin, TSC-1A3 (TAC) (1979-80) 27
Marquise (Mitsubishi International) 414
MARSH AVIATION COMPANY (U.S.A.) 400
MARSHALL OF CAMBRIDGE (ENGINEERING) LTD. (U.K.) 253
Martin Fierro, J-1 (RRA) 4
MARTIN MARIETTA CORPORATION (U.S.A.) (1976-77) 324
Marut (HAL) (1979-80) 81
Mascot (China) 32
Mashshaq (PAC) 157
MASON, JOE J. (U.S.A.) (1975-76) 382
MASSACHUSETTS INSTITUTE OF TECHNOLOGY (U.S.A.) 413
MASSCHELEIN, PAUL & STEPHAAN AND VERSTRAETE, ERIC (Belgium) (1977-78) 480
MASSCHELEIN/VERSTRAETE (see Masschelein, Paul & Stephaan and Verstraete, Eric) (1977-78) 480
MASTER-MEHRZWECKFLUGZEUG-ENTWICKLUNGS GmbH & CO KG (West Germany) (1978-79) 67
Master Porter (Poligrat) (1975-76) 95
Matador (BAe) 247
MAULE AIRCRAFT CORPORATION (U.S.A.) 401
Mav Isik (THvK/KIBM) 185
Max (China) (1980-81) 37
"Max" (Yakovlev) (1972-73) 483
May (Ilyushin) 194
"Maya" (Aero) (1975-76) 33
MBB/AÉROSPATIALE (International) 102
MBB HELICOPTER CORPORATION (USA) 402
MBB/KAWASAKI (International) 102
McCANDLESS (AVIATION) LTD, R. & W. A. C. (U.K.) (1973-74) 214
McCARLEY, MRS CHARLES E. (U.S.A.) 534
McCULLOCH AIRCRAFT CORPORATION (U.S.A.) (1972-73) 358
McDONNELL (see McDonnell Douglas Corporation) 402
McDONNELL AIRCRAFT COMPANY (U.S.A.) 403
McDONNELL DOUGLAS CORPORATION (U.S.A.) 402
MCDONNELL DOUGLAS/FOKKER (International) 103
McDONNELL DOUGLAS/HAWKER SIDDELEY (International) (1975-76) 112
McKINNON-VIKING ENTERPRISES (Canada) 27
McLAUGHLIN, M. L. (U.S.A.) (1974-75) 396
McLEOD, KAYE AND STAN (Canada) (1976-77) 456
M COMPANY (U.S.A.) (1979-80) 529
MEAD, ENGINEERING COMPANY (U.S.A.) 534
MERGER, MICHAEL F. (U.S.A.) (1972-73) 370
MELLEN, KAI (Finland) (1975-76) 40
Melmoth, OM-1 (Garrison) (1976-77) 512
Menestrel, HN 433 (Nicollier) 488
Mercure (Dassault) (1976-77) 56
Mercure 200 (Dassault-Breguet) (1977-78) 57
Mercurius (Hawker Siddeley) (1976-77) 178
Mercury (Cranwell) (1977-78) 508
Mercury (Macdonald) (1976-77) 483
Merganser (Van Dine) 559
MERIDIONALI SpA ELICOTTERI (Italy) 129
MERKEL AIRPLANE COMPANY (U.S.A.) (1978-79) 527
Merlin IIB (Swearingen) (1973-74) 449
Merlin IIIC (Fairchild Swearingen) 354
Merlin IVC (Fairchild Swearingen) 355
Merlin maritime surveillance aircraft (Swearingen) (1980-81) 445
Merlin SM (Swearingen) (1980-81) 445
Mésange (Gatard) (1978-79) 475
Mescalero (Cessna) (1980-81) 308
MESSERSCHMITT-BÖLKOW-BLOHM GmbH (West Germany) 82
Messersnipe (Messier) (1972-73) 371
MESSIER, RAOUL (U.S.A.) (1972-73) 371
Metal Hurlant (Thompson) (1979-80) 552
Meteorite (Wallis) 265
Meteorite 2 (Wallis) 266
Metro II (Swearingen) (1980-81) 443
Metro IIA (Swearingen) (1980-81) 444
Metro III (Fairchild Swearingen) 353
Metro IIIA (Fairchild Swearingen) 353
Mewa, M-20 (WSK-PZL Mielec) 165
MEYER, LES K. (U.S.A.) 535
MEYERS AIRCRAFT MANUFACTURING COMPANY (U.S.A.) (1980-81) 391
Michel 1 (Michel) (1972-73) 12
MICHEL, LUDWIG (Brazil) (1972-73) 12
Micro (Bede) (1977-78) 524
Micro-Imp (Aerocar) 508
Microjet 200 (Microturbo) 67
Microstar (Chagnes) (1980-81) 472
MICROTURBO SA (France) 67
MID-CONTINENT AIRCRAFT CORPORATION (U.S.A.) 413
Midge (Gyroflight) (1972-73) 198
Midget Mustang (Bushby/Long) 515
MIDGY, M (France) (1980-81) 481
MiG (U.S.S.R.) 200
MiG COMPLEX, HINDUSTAN AERONAUTICS LTD (India) 90
MIKOYAN, ARTEM, and MIKHAIL GUREVICH (U.S.S.R.) (1973-74) 477
Mikrosamolet (Dmitriev) (1976-77) 546
MIL (U.S.S.R.) 207
Milan (Dassault) (1973-74) 57
Milan (Sportavia) (1971-72) 95
Milanov XL-13T (Bulgaria) (1977-78) 480
Milanov/Vasileva Tangra (Bulgaria) (1977-78) 480
Military Arava (IAI) (1975-76) 117
MILLER AVIATION INC., J. W. (U.S.A.) (1977-78) 547
MILLER, MERLE B. (U.S.A.) (1975-76) 399
MILLER, WILLIAM Y. (U.S.A.) 535
MILLICER, HENRY K. (Australia) (1978-79) 466
MILON, PIERRE (France) (1980-81) 481
Miltrainer, L-70 (Valmet) 42
Mini Ace (Cvjetkovic) 518
Minicab (Ord-Hume) (1979-80) 497
Mini-Cap (Mudry) 69
Mini-Copter (Aerospace General) (1978-79) 249
Mini-Copter, W.S.4A (Seremet) (1973-74) 39
Mini Coupe (Sport Air Craft) 552
Mini Coupe TD (Sport Air Craft) (1980-81) 548
Mini Guppy (Aero Spacelines) (1972-73) 233
Mini-Gyroglider, W.S.7 (Seremet) (1973-74) 40
MiniHawk (Hawk) (1980-81) 358
Mini-Hawk 1 (Mini-Hawk) (1973-74) 382
MINI-HAWK INTERNATIONAL INC. (U.S.A.) 536
Mini-Imp (Aerocar) 507
Mini-Mac (McCarley) 534
Miniplane (Smith) 550
Miniplane + 1 (Smith) (1974-75) 463
Mini-Swat S-325 (Schapel) 455
Mini Tanker (Volpar/Beechcraft) (1974-75) 479
Minor (Ord-Hume) (1979-80) 497
Minor (Phoenix) (1974-75) 222
Minor, L.A.4A (PFA) 504
Minor III (Phoenix) (1972-73) 214
MINTY, E. R. (Australia) 7
Minuano (EMBRAER/Piper) 17
Miracle of 10's (Lacey) (1971-72) 328
Mirage III (Dassault) 55
Mirage III-D (GAF) (1972-73) 7
Mirage 5 (Dassault) 56
Mirage 5 "Moustaches" (Dassault) (1974-75) 59
Mirage 50 (Dassault) 57
Mirage 2000 (Dassault) 58
Mirage F1 (Dassault) 57
Mirage F1-E (Dassault) (1977-78) 52
Mirage G8 (Dassault) (1974-75) 60
Mirage G8A/ACF (Dassault) (1974-75) 61
Mission adaptive wing (NASA) 420
MISSION AVIATION RELATED SERVICES (Canada) 27
Missionmaster (GAF) 5
MISSISSIPPI STATE UNIVERSITY (U.S.A.) (1971-72) 359
Mister America (Harmon) 526
Mistral CP 301 (Becker) (1979-80) 484
Mistral, MY 103 (Mylius) (1980-81) 84
MITCHELL, C.G.B. (U.K.) (1974-75) 221
MITCHELL AIRCRAFT CORPORATION (U.S.A.) (1980-81) 531
MITCHELL-PROCTER (U.K.) (1974-75) 221
Mitchell Wing (M Company) (1979-80) 529
Mite (Mooney Mite) 537
MIT MANPOWERED AIRCRAFT ASSOCIATION (U.S.A.) (1977-78) 548
MITSUBISHI AIRCRAFT INTERNATIONAL INC. (U.S.A.) 414
MITSUBISHI JUKOGYO KABUSHIKI KAISHA (Japan) 146
Miyahara (JEAA) (1972-73) 139
MIYAUCHI MANUFACTURING CO LTD (Japan) (1974-75) 151
Mizar (AVE) (1973-74) 241
Mizzet No III (Abe) 496
Mod AH-1S (Bell) 298
Model 05, Ganagobie (Lobet-de Rouvray) (1975-76) 9
Model 1 (Malliga) (1974-75) 12
Model 1-A (Bowers) 513
Model 1-B (Bowers) 514
Model 1-2 Mister America (Harmon) 526
Model 1-Nine-O (Jurca) (1979-80) 477
Model II (Cicaré) (1973-74) 5
Model 2 Super Sportster (Turner) (1980-81) 554
Model "2 + 2" (Robin/CEA) (1973-74) 81
Model "2 + 2 Tricycle" (Robin/CEA) (1972-73) 79
Model 2T-1A-1 (Great Lakes) (1977-78) 302
Model 2T-1A-2 Sport Trainer (Great Lakes) 364
Model 2/3 Mustang (SAL) (1974-75) 30
Model 2/180 Gyroplane (Avian) (1971-72) 15
Model III (Aerocar) (1971-72) 216
Model III (Bryan) (1973-74) 287
Model III (Cicaré) (1973-74) 5
Model 4E (McCulloch) (1971-72) 347
Model 7ACA Champ (Bellanca/Champion) (1973-74) 267
Model 7ECA/7GCAA/7GCBCJ Citabria (Bellanca) (1979-80) 287
Model 7GCBC Scout (Bellanca/Champion) (1975-76) 275
Model 8GCBC Scout (Bellanca) (1980-81) 287
Model 8KCAB Decathlon (Bellanca/Champion) (1980-81) 287
Model 11 Akromaster (Spinks) (1971-72) 408
Model 11D (Alpha) (1972-73) 236
Model 14-19-3C Bellanca 260C (Bellanca) (1971-72) 248
Model 16H-1HT (PiAC) (1972-73) 395
Model 16H-1T (PiAC) (1974-75) 416
Model 16H-3K Pathfinder III (PiAC) (1974-75) 416
Model 17-30A (Bellanca) (1980-81) 287
Model 17-31A (Bellanca) (1979-80) 286
Model 17-31ATC (Bellanca) (1979-80) 286
Model 18 (Volpar/Beechcraft) 468
Model 18-A (Farrington) (1975-76) 333
Model 19-25 Skyrocket II (Bellanca) (1980-81) 288
Model 20-A (Farrington) (1975-76) 334
Model 24 (Gates Learjet) 357
Model 25 (Gates Learjet) 357, 358
Model 26 (Aereon) (1976-77) 205
Model 28/29 Longhorn (Gates Learjet) 358
Model 35 Bonanza (Beechcraft) (1975-76) 253, 254
Model 35/36 (Gates Learjet) 358
Model 36 Bonanza (Beechcraft) (1977-78) 224
Model 40 (Rutan) (1978-79) 538
Model 47 (Agusta-Bell) (1977-78) 112
Model 47G Trooper (Bell) (1971-72) 242
Model 47G-3B-2A (Bell) (1974-75) 267
Model 47G-4A Trooper (Bell) (1971-72) 242
Model 47G-5A (Bell) (1974-75) 267
Model 55/56 Longhorn (Gates Learjet) 359
Model 55 Baron (Beechcraft) (1977-78) 225, 226
Model 58 Baron (Beechcraft) 284
Model 58P Baron (Beechcraft) 284
Model 58TC (Beechcraft) 285
Model 60 Duke (Beechcraft) 285
Model 65-A90 (Beechcraft) (1972-73) 251
Model 75 Gyro-Copter (Bensen) (1976-77) 500
Model 76 (Beechcraft) (1977-78) 224
Model 77 Skipper (Beechcraft) 279
Model 77 (Hughes) 382
Model 82 Hercules (Lockheed) 395
Model 85 Orion (Lockheed) 390
Model 90 King Air (Beechcraft) 286
Model 95/115 (Aero Boero) (1972-73) 1
Model 95-B55 (Beechcraft) 283
Model 97-34J Logger (Piasecki) 424
Model 99 Airliner (Beechcraft) (1979-80) 273
Model 100 Honcho (Nagler) (1973-74) 385
Model 100 King Air (Beechcraft) 289
Model 105 (Spratt) (1980-81) 546
Model 106 (Agusta) (1973-74) 122
Model 107 (Boeing Vertol) 317
Model 107 (Spratt) 553
Model 109 (Agusta) (1974-75) 126
Model 111 (North American) (1972-73) 381
Model 111 (Rockwell) (1973-74) 423
Model 111m (Airmark/Cassutt) (1972-73) 186
Model 112 (Rockwell) (1974-75) 444
Model 112 (North American) (1972-73) 381
Model 114 (Boeing Vertol) 318
Model 115 BS (Aero Boero) (1977-78) 2
Model 121 Teal (Falconar) (1971-72) 23
Model 122 Uirapuru (Aerotec) (1980-81) 10
Model 125 Commuter (Richard) (1971-72) 391
Model 150 (Cessna) (1978-79) 299
Model 150 Commuter (Richard) (1974-75) 432
Model 152 (Cessna) 324
Model 152 Aerobat (Cessna) 325
Model 165 (Whitcraft) (1973-74) 459
Model 172 (Cessna) 325
Model 177 (Cessna) (1977-78) 267
Model 179 (Boeing Vertol) (1976-77) 251
Model 180 (Cessna) 327
Model 180 Condor (Aero Boreo) (1974-75) 2
Model 182 (Cessna) (1976-77) 258
Model 185 Orion (Lockheed) 390
Model 185 Skywagon (Cessna) 329
Model 186 (Lockheed) (1971-72) 334
Model 190 Sportplane (Richard) (1971-72) 392
Model 200 (Beechcraft) (1977-78) 234
Model 200D (Meyers) (1979-80) 392
Model 200T (Beechcraft) 291
Model 201 (Mooney) 416
Model 201 Gyro-Falcon (Wotring) (1974-75) 483
Model 201A (Weatherley) (1972-73) 441
Model 201C (Weatherley) (1979-80) 459
Model 202 (Nagler) (1975-76) 402
Model 204 (Bell) (1972-73) 261
Model 204AS (Agusta-Bell) (1976-77) 108
Model 204B (Agusta-Bell) (1976-77) 108
Model 204B/204B-2 (Fuji/Bell) 142
Model 205 (Agusta-Bell) 127
Model 205 (Bell) 293
Model 205A-1 (Agusta-Bell) (1976-77) 108
Model 205A-1 (Bell) 293
Model 205A-1/UH-1H (Fuji/Bell) 142
Model 206A JetRanger (Agusta-Bell) (1974-75) 128
Model 206A JetRanger (Bell) (1974-75) 269
Model 206A-1 (Agusta-Bell) (1976-77) 109
Model 206B JetRanger III (Agusta-Bell) 127
Model 206B JetRanger II (Bell) (1977-78) 237
Model 206B JetRanger III (Bell) 294
Model 206B-1 (Agusta-Bell) (1976-77) 109
Model 206L TexasRanger (Bell) 296
Model 206L-1 LongRanger II (Agusta-Bell) 127
Model 206L-1 LongRanger II, III & IV (Bell) 295
Model 207 (Cessna) (1977-78) 273
Model 209 HueyCobra (Bell) 296, 298
Model 209 Improved SeaCobra (Bell) 297
Model 209 SeaCobra (Bell) 296
Model 210 (Aero Boero) (1975-76) 1
Model 210 Centurion (Cessna) (1977-78) 274
Model 212 (Agusta-Bell) 127
Model 212 (China) 34
Model 212ASW (Agusta-Bell) 127
Model 212 Twin Two-Twelve (Bell) 299
Model 214 Huey Plus (Bell) (1974-75) 273
Model 214A (Agusta-Bell) (1978-79) 105
Model 214A/C (Bell) (1980-81) 283
Model 214B BigLifter (Bell) 300
Model 214ST (Bell) 300
Model 222 (Bell) 301
Model 223 Flamingo (CASA) (1975-76) 180
Model 223 Flamingo (Hispano) (1972-73) 174
Model 223 Flamingo (MBB) (1973-74) 93
Model 231 (Mooney) 417
Model 234 (Boeing Vertol) 320
Model 234LR (Boeing Vertol) 320
Model 235CA, Rallye (SOCATA) (1978-79) 63
Model 235G, Rallye (SOCATA) (1978-79) 63
Model 237 (Boeing Vertol) (1977-78) 262
Model 249 HueyCobra (Bell) 298
Model 260 (Aero Boero) (1975-76) 1
Model 280 (Enstrom) 349, 350
Model 285 Orion (Lockheed) 390

Model 298 (Mohawk) (1978-79) 329
Model 300 Sequoia (Sequoia) 548
Models 300 and 301 (Aeronics/Sequoia) (1980-81) 165
Model 300 (Hughes) (1977-78) 317
Model 300 Tauro (Anahuac) (1980-81) 140
Model 300 Viking (Bellanca) (1979-80) 286
Model 300C (Hughes) 379
Model 300CQ (Hughes) 380
Model 301 (Bell) 302
Model 301 (Sequoia) (1980-81) 544
Model 302 Kodiak (Sequoia) 548
Model 303 (Cessna) 335
Model 305 (Brantly-Hynes) 322
Model 309 KingCobra (Bell) (1974-75) 274
Model 310 (Cessna) 335
Model 310 II (Cessna) 335
Model 318 (Cessna) (1978-79) 311
Model 318E Dragonfly (Cessna) 336
Model 330 (Shorts) 260
Model 335/335 II (Cessna) 336
Model 337 Skymaster/Skymaster II (Cessna) 336
Model 337 Skymaster (Reims/Cessna) (1980-81) 70
Model 340 Commuter (SAAB/Fairchild) 106
Model 340A/340A II/340A III (Cessna) 336
Model 360 (Hillman) 379
Model 360 (Shorts) 261
Model 360C Regente (Neiva) (1974-75) 18
Model 369 (Kawasaki/Hughes) 145
Model 369D (Kawasaki/Hughes) 145
Model 374 (Boeing Vertol) (1974-75) 294
Model 377SGT Guppy-201 (Aero Spacelines) (1974-75) 244
Model 380-L (Lederlin) 487
Model 382 Hercules (Lockheed) 395
Model 385 TriStar (Lockheed) 393
Model 400 (Interceptor) (1978-79) 354
Model 400 Hustler (AJI) (1978-79) 252
Model 400 Hustler (GACC) (1980-81) 339
Model 400, Lightning (Smith) 464
Model 401 (Cessna) (1974-75) 315
Model 401 (General Dynamics) (1974-75) 343
Model 402/402 II/402 III (Cessna) 337, 338
Model 406 (AHIP) (Bell) 294
Model 409 (Bell) (1977-78) 245
Model 412 (Agusta-Bell) 128
Model 412 (Bell) 299
Model 412 (China) 34
Model 414 (Boeing Vertol) 318
Model 414A Chancellor (Cessna) 339
Model 420L Regente (Neiva) (1974-75) 18
Model 421 I/II Executive Commuter (Cessna) (1977-78) 282
Model 421 Golden Eagle (Cessna) 340
Model 425 Corsair (Cessna) 341
Model 441 (Cessna) (1977-78) 284
Model 460 (Wren) (1971-72) 425
Model 480 Eagle (Enstrom) (1980-81) 334
Model 500 Hustler (Gulfstream American) (1980-81) 339
Model 500 (Hughes) 380
Model 500 (Kawasaki/Hughes) 145
Model 500 (Lockheed) (1971-72) 339
Model 500C (Hughes) 380
Model 500C (Kawasaki/Hughes) 145
Model 500 Citation (Cessna) 344
Model 500D (Hughes) 380
Model 500D (Kawasaki/Hughes) 145
Model 500M (Hughes) 381
Model 500MD Defender (Hughes) 381
Model 501 (Alpha Jet) (1971-72) 104
Model 533 (Bell) (1971-72) 247
Model 550A Stallion (Helio) (1974-75) 354
Model 600 Jet Liner (Carstedt) (1972-73) 283
Model 600X (Hughes) 382
Model 606 (Westland) (1976-77) 204
Model 616 AERCAB (Fairchild Industries) (1975-76) 331
Model 620 (Weatherley) 470
Model 620 Super Pinto (Aeronca) (1973-74) 236
Model 630 Super Pinto (Aeronca) (1973-74) 236
Model 640QC/T Super Pinto (Aeronca) (1973-74) 236
Model 681B (Rockwell) (1974-75) 445
Model 685 (Rockwell) (1974-75) 445
Model 690 (Rockwell) (1974-75) 445
Model 700 (Cessna) (1976-77) 275
Model 700/FA-300 (Fuji) (1980-81) 136
Model 707 (Boeing) 304
Model 707-LRPA (Boeing) (1976-77) 247
Model 710/FA-300 (Fuji) 142
Model 727 (Boeing) 304
Model 737 (Boeing) 305
Model 737-300 (Boeing) 307
Model 747 (Boeing) 308
Model 747SP (Boeing) 310
Model 747SUD (Boeing) 311
Model 747-123 Space Shuttle Orbiter Carrier (Boeing/NASA) (1979-80) 298
Model 747-200F Freighter (Boeing) 311
Model 757 (Boeing) 312
Model 767 (Boeing) 313
Model 777 (Boeing) (1979-80) 300
Model 800/8800 (Excalibur) 350
Model 800 Skytrader (Dominion) (1979-80) 330
Model 941S (Breguet) (1973-74) 61
Model 1000C Nifty (Morrisey) (1974-75) 401
Model 1123 (IAI) (1971-72) 112
Model 1150 Atlantic (Breguet) (1976-77) 52
Model 1160 (Air Parts/Fletcher) (1971-72) 155
Model 1284 (Aerospace) (1974-75) 163
Model 1300 (Beechcraft) (1979-80) 273
Model 1329-25 JetStar II (Lockheed) (1980-81) 377
Model 1900 (Beechcraft) 289
Model 2000 (Bushmaster) (1974-75) 298
Model 2000 (Cavalier) (1971-72) 271
Model 2100 (Lear Fan) 387
Model 2150A Kachina (Varga) 467
Model 2180 Kachina (Varga) 467
Model 2707-300 (Boeing) (1971-72) 262
Model 4000 Turbofan Aerostar (Smith) (1975-76) 472
Model 4000 Turboprop Beaver (Volpar/DHC) (1974-75) 479
Model 4500-300 (Evangel) (1975-76) 329
Model A (Aerocar) 506
Model A Skybuggy (McLaughlin) (1974-75) 396
Model A Welsh Rabbit (Welsh) (1973-74) 458
Model A36 (Beechcraft) 281
Model A36TC (Beechcraft) 282
Model AT-301 (Air Tractor) 274
Model AT-302 (Air Tractor) 274
Model AT-400 (Air Tractor) 275
Model B (Aerocar) (1974-75) 242
Model B (Wickham) (1972-73) 442
Model B, Scamp (Agricopteros/Aerosport) (1979-80) 33
Model B-1, Sylkie One (Wayne) (1980-81) 504
Model B-2B (Brantly-Hynes) 322
Model B-10 (Mitchell) (1980-81) 531
Model C90 King Air (Beechcraft) 286
Model C Copter (Bannick) (1972-73) 240
Model D (Ace Aircraft) 506
Model D (Fike) (1979-80) 520
Model D2127 (Bell) (1971-72) 240
Model E (Ace Aircraft) 506
Model "E" (Fike) (1979-80) 520
Model E55, Baron (Beechcraft) 284
Model F-21 (Taylorcraft) 466
Model 33A/C (Beechcraft) 281
Model F90 Super King Air (Beechcraft) 288
Model G (Navion) (1975-76) 406
Model G-164D Turbo Ag-Cat D (Schweizer) 457
Model H (Janox/Navion) (1972-73) 337
Model H (Navion) (1978-79) 391
Model H-295 (Helio Courier) (1978-79) 348
Model HT-295 (Helio Courier) (1978-79) 348
Model I-11B (AISA) (1974-75) 178
Model P, EAA Biplane (EAA) (1975-76) 325
Model P337 Pressurised Skymaster (Cessna) (1980-81) 320
Model PD 249 (Beechcraft) (1973-74) 250
Model R22 (Robinson) 451
Model R172 Hawk XP (Cessna) 326
Model "S" Sidewinder (Smyth) 550
Model S-12-E (Spencer) 551
Model S-325 Mini-Swat (Schapel) 455
Model S-1080 Thunderbolt (Schapel) 456
Model SA227-AC Metro III (Fairchild Swearingen) 353
Model T-34C (Beechcraft) 278
Model T210 Centurion (Cessna) (1977-78) 274
Model T210 Turbo Systems Centurion (Cessna) (1971-72) 281
Model T303 Crusader (Cessna) 335
Model T310/T310 II (Cessna) 335
Model VW (Bannick) (1972-73) 240
Model V35B Bonanza (Beechcraft) 280
Model V35B TC (Beechcraft) (1971-72) 228
Model W-8 Tailwind (Wittman) 565
Model W Page-powered (O'Neill) (1973-74) 391
Model W Winner (O'Neill) (1973-74) 391
Modernised AH-1S (Bell) 298
Module Porter (Pilatus) (1975-76) 190
Mohawk (Grumman) 368
Mohawk 298 (Aérospatiale) (1977-78) 40
Mohawk 298 (Frakes) (1978-79) 329
MOHAWK AIR SERVICES (U.S.A.) (1978-79) 355
MONDAIR AVIATION LTD (Canada) (1972-73) 25
Monex (Monnett) 536
MONG, RALPH E. JNR (U.S.A.) (1974-75) 398
Mongol (Mikoyan) 200
Monika (Dobraczynski) (1974-75) 171
MONNETT (see Monnett Experimental Aircraft Inc) 536
MONNETT EXPERIMENTAL AIRCRAFT INC. (U.S.A.) 536
Monoplane (Epps) (1972-73) 311
Monoplane (Taylor) 505
Mono Z-CH 100 (Zenair) 478
Monsoon RL-3 (Western) (1977-78) 481
Monsun (MBB) (1973-74) 93
MOONEY AIRCRAFT CORPORATION (U.S.A.) 416
Mooney M-10 Cadet (Univair) (1975-76) 474
Mooney Mite (Mooney Mite) (1979-80) 531
MOONEY MITE AIRCRAFT CORPORATION (U.S.A.) 537
MORAND, ATELIERS DE PRÉCISION (Switzerland) (1975-76) 189
MORAVAN NÁRODNÍ PODNIK (Czechoslovakia) 40
MORRISEY, WILLIAM J. (U.S.A.) (1975-76) 402
MOSCOW AVIATION INSTITUTE (U.S.S.R.) (1975-76) 496
Moss (Tupolev) 222
MOTO-DELTA (see Geiser, Jean Marc) (1980-81) 68
Moto-Delta G 11 (Geiser) (1980-81) 68
Moto-Plane (Monoplane) (1975-76) 26
MOTOPLANE AVIATION INC. (Canada) (1975-76) 26
Moujik (Sukhoi) 214
Mountaineer (Ector) 348
MOURA, MAURICIO IMPELIZIERI P (Brazil) 473
"Moustaches", Mirage 5 (Dassault) (1974-75) 59
Mowgli (SITAR) (1971-72) 78
Mr Smoothie (Sawano) (1980-81) 489
MUDRY ET COMPAGNIE, AVIONS (France) 68
Muhinu, PIK-19 (PIK) (1975-76) 40
MUKAI, ISAO (Japan) 496
MURPHY, R. L. (U.S.A.) 537
MURRAYAIR LTD (U.S.A.) (1972-73) 374
Musketeer (Beechcraft) (1971-72) 227
Musketeer Super R (Beechcraft) (1971-72) 227
MUSSO, PAUL (U.S.A.) 538
Mustang (Cavalier) (1971-72) 271
Mustang (Timofeyev) (1977-78) 574
Mustang II (Bushby) 515
Mustang II (Cavalier) (1971-72) 271
Mustang 2/3 (SAL) (1974-75) 30
MYASISHCHEV (U.S.S.R.) 214
MYLIUS (see Leichtflugzeuge-Entwicklungen Dip Ing Hermann Mylius) 85
Mystère-Falcon 10 (Dassault) 65
Mystère-Falcon 10MER (Dassault) 65
Mystère-Falcon 20 Series F (Dassault-Breguet) 63
Mystère-Falcon 20 Series G (Dassault-Breguet) 64
Mystère 30/Falcon 30 (Dassault) (1973-74) 64
Mystère 40/100 (Dassault) (1974-75) 65
Mystère-Falcon 50 (Dassault) 66
Mystère-Falcon 200 (Dassault-Breguet) 64
Mystery S replica (Younkin) (1980-81) 560

N

N2 (GAF) (1971-72) 6
N22 Nomad (GAF) 5
N24 Nomad (GAF) 5
N262 (Aérospatiale) (1978-79) 36
N 500 (Aérospatiale) (1971-72) 47
N-621 Universal (Neiva) 17
N-622 Universal (Neiva) 17
NA300 (Rockwell) (1973-74) 418
NAL (see National Aerospace Laboratory) 147
NAL (see National Aircraft Leasing Ltd) (1978-79) 388
NAM (see National Aero Manufacturing Corporation) 157
NAMC (see National Aero Manufacturing Corporation) (1979-80) 147
NAMC (see Nihon Kokuki Seizo Kabushiki Kaisha) (1979-80) 137
NASA (see National Aeronautics and Space Administration) 417
NBO 105 (Nurtanio/MBB) 91
NC-130 Hercules (Lockheed) (1972-73) 350
NC-131H (Calspan) (1975-76) 298
NC-212 Aviocar (Nurtanio/CASA) 91
NF-5 (Canadair) (1972-73) 17
NF-5 (Northrop) (1978-79) 392
NF 15 (Ambrosini) (1971-72) 123
NF-104 Starfighter (Lockheed) (1972-73) 344
NF 409 (Ambrosini) (1971-72) 124
NH-300C (BredaNardi) (1977-78) 115
NH-500C (BredaNardi) (1977-78) 115
NH-500D (BredaNardi) (1977-78) 115
NH-500M (BredaNardi) (1977-78) 115
NH-500MC (BredaNardi) (1977-78) 115
NHH-2D (Kaman) 386
NIPPI (see Nihon Hikoki Kabushiki Kaisha) 148
NM-70 Linnet IV (Nihon University) (1971-72) 146
NM-72 Egret I (Nihon University) (1973-74) 149
NM-73 New Egret (Nihon University) (1974-75) 152
NM-74 Egret III (Nihon) (1975-76) 152
NM-75 Stork (Nihon University) (1977-78) 501
NM-77 Ibis A (Nihon)) (1978-79) 486
NOTAR helicopter (Hughes) 382
NT-33A (Calspan) 323
NTF-8 Crusader (LTV) (1971-72) 342
NUH-2C (Kaman) (1975-76) 364
NWI (see Northwest Industries) 27
NAGLER HELICOPTERS INC (U.S.A.) (1975-76) 402
Namu II (Bowers) (1978-79) 509
Nanzhang Q-5 (A-5) (China) 31
NARDI SA PER COSTRUZIONI AERONAUTICHE (Italy) (1974-75) 133
NASH AIRCRAFT LTD (U.K.) 254, 503
NATIONAL AERO MANUFACTURING CORPORATION (Philippines) 157
NATIONAL AERONAUTICS AND SPACE ADMINISTRATION (U.S.A.) 417
NATIONAL AEROSPACE LABORATORY (Japan) 147
NATIONAL AIRCRAFT LEASING LTD (U.S.A.) (1978-79) 388
NATIONAL DYNAMICS (PTY) LTD (South Africa) (1979-80) 163
Natural laminar flow aircraft (NASA) (1979-80) 397
Navajo (EMBRAER/Piper) 17
Navajo (Piper) 432
Navajo C/R (Piper) 433
Navajo Chieftain (Piper) (1977-78) 374
Naval helicopter (Mil) 214
NAVION RANGEMASTER AIRCRAFT CORPORATION (U.S.A.) (1979-80) 397
NDN AIRCRAFT LTD (U.K.) 254
NEIVA SA, INDUSTRIA AERONÁUTICA (Brazil) 17
Nesher (IAI) (1980-81) 110
NESMITH (U.S.A.) (1975-76) 407
Nesmith Cougar (EAA) (1979-80) 519
NEUMANN, EVERETT (USA) (1980-81) 534
New Egret (Nihon University) (1974-75) 152
New fighter (Mikoyan) 206
New heavy bomber (Tupolev) (1980-81) 213
New Technology Wing (TNT) testbed aircraft (Dornier) 78
NEW ZEALAND AEROSPACE INDUSTRIES LTD (New Zealand) 155
N-Flyer (Cal Poly) (1979-80) 307
NICOLLIER, HENRI (France) 488
NIEMI, LEONARD (USA) (1980-81) 534
Nifty (Morrisey) (1974-75) 401
Night/Adverse Weather A-10 (Fairchild Republic) 352

Nightingale (Douglas) 408
NIHON HIKOKI KABUSHIKI KAISHA (Japan) 148
NIHON KOKUKI SEIZO KABUSHIKI KAISHA (Japan) (1979-80) 137
NIHON UNIVERSITY, COLLEGE OF SCIENCE AND TECHNOLOGY (Japan) 497
Nimrod (BAe) 249
Nimrod AEW Mk 3 (BAe) 251
NIPPER (see Nipper Kits and Components Ltd) 503
NIPPER KITS AND COMPONENTS LTD (U.K.) 503
Nipper Mk IIIb (Nipper) 503
Nipper Mk IV (Nipper) (1971-72) 201
Nite Star (Olsen) 538
Nomad N22/24 (GAF) 5
NORTH AMERICAN (see Rockwell International Corporation) (1979-80) 429
NORTH AMERICAN AEROSPACE OPERATIONS (U.S.A.) 452
NORTH AMERICAN AIRCRAFT GROUP (USA) (1979-80) 429
NORTH AMERICAN ROCKWELL CORPORATION (U.S.A.) (1972-73) 376
NORTH AMERICAN SPACE OPERATIONS (U.S.A.) (1977-78) 386
NORTHROP CORPORATION (U.S.A.) 420
NORTHROP CORPORATION, AIRCRAFT GROUP (U.S.A.) 420
NORTHWEST INDUSTRIES LTD (Canada) 27
Nuri (Sikorsky) (1980-81) 435
Nymph (Britten-Norman) (1974-75) 202

O

O-2 (Cessna) 336
O2-337 Sentry (Summit) 466
OA-4M Skyhawk (Douglas) (1980-81) 390
OG-2 (Ohnishi) (1972-73) 149
OGMA (see Oficinas Gerais de Material Aeronáutico) 168
O-H4B Minor (Ord-Hume) (1979-80) 497
OH-6 (Hughes) (1977-78) 318
OH-6D (Kawasaki/Hughes) 145
OH-6J (Kawasaki/Hughes) 145
O-H-7 Coupé (Ord-Hume) (1973-74) 216
OH-13 Sioux (Bell) (1971-72) 242
OH-58 Kiowa (Bell) 294
OM-1 Melmoth (Garrison) (1976-77) 512
OR-70 Fang (Stevenson/Owl) (1977-78) 550
OR-71 Lil Quickie (Vin-Del/Owl) 539
OSKB-1-3PM Kwant (MAI) (1975-76) 496
OV-1 Mohawk (Grumman) (1973-74) 338
OV-10 Bronco (Rockwell) 452
OBERLERCHNER HOLZINDUSTRIE JOSEF (Austria) (1971-72) 9
Oblique-wing aircraft (NASA) 417
Observer (Partenavia) 138
Observer/Explorer (National Dynamics/Patchen) (1977-78) 161
OFICINAS GERAIS DE MATERIAL AERONÁUTICO (Portugal) 168
OHMERT, VERNON R (U.S.A.) (1980-81) 534
OHNISHI, YUICHI (Japan) (1972-73) 149
Ol' Ironsides (Scott) (1971-72) 397
Olive No. 2 (Mukai) 496
Olive SMG III (Mukai) 496
OLSEN, GORDON L (U.S.A.) 538
OLSZEWSKI, KAZIMIERZ & OBAREWICZ J. (Poland) (1977-78) 502
OLSZEWSKI-OBAREWICZ (see Olszewski, Kazimierz & Obarewicz J.) (1977-78) 502
Olympian (Zinno) (1977-78) 573
Omac 1 (Omac) 424
OMAC INC (U.S.A.) 424
Omega (Artiomov) (1977-78) 574
OMNIONICS (U.S.A.) 424
OMNIPOL FOREIGN TRADE CORPORATION (Czechoslovakia) 36
One-Eleven (BAe) 243
One-Eleven (IAv Bucuresti) 170
One-Eleven (NAL/BAC) (1977-78) 357
O'NEILL AIRPLANE COMPANY, INC (U.S.A.) (1975-76) 411
One-Oh-Nine (Jurca) 486
Onishi OS63 (JEAA) (1977-78) 500
ONISHI, MR (Japan) (1978-79) 486
Optica, EA7 (Edgley) 252
OPTION AIR RENO (U.S.A.) 538
Orao/IAR-93 (SOKO/CNIAR) 110
ORD-HUME, ARTHUR W. J. G. (U.K.) (1979-80) 497
Orion (Grinvalds) 484
Orion (Lockheed) 390
OSADCHY, PAUL (U.S.A.) (1976-77) 523
Osage (Hughes) (1977-78) 317
Oscar-100 (Partenavia) (1976-77) 112
Oscar-150 (Partenavia) (1976-77) 112
Oscar-180 (Partenavia) (1976-77) 112
Oscar-200 (Partenavia) (1976-77) 112
Osprey 1, GP2 (Osprey) (1974-75) 410
Osprey II, GP3 (Osprey) 538
OSPREY AIRCRAFT (U.S.A.) 538
OWL, GEORGE JR (U.S.A.) 539

P

P1 Bandeirante (EMBRAER) 10
P2 Bandeirante (EMBRAER) 10
P-2J (Kawasaki) (1980-81) 137
P2V-7 (P-2H) VSA (Kawasaki) 143
P-3 Orion (Lockheed) 390
P3V Orion (Lockheed) 390
P-9 Pober Pixie (EAA) 523
P.50 Bouvreuil (Pottier) 492
P-51D Mustang Replica (Fisher-Van Norman) (1980-81) 520
P.51R (Pottier) (1977-78) 498
P.64 Oscar Series (Partenavia) (1976-77) 112
P.66 Oscar Series (Partenavia) (1976-77) 112
P.66C-160 Charlie (Partenavia) 137
P.68 Floatplane/amphibian (Partenavia) 137
P.68 Observer (Partenavia) 138
P.68B Victor (Partenavia) 137
P.68C (Partenavia) 137
P.68C-R (Partenavia) 137
P.68C-TC (Partenavia) 137
P.68R (Partenavia) 138
P.68RTC (Partenavia) (1980-81) 127
P.68WB (Partenavia) (1980-81) 127
P-70 Acey Deucy (Powell) 542
P.70 Alpha (Partenavia) (1976-77) 114
P.70B (Besneux) (1976-77) 460
P.70S (Pottier) 492
P.78 (Aeritalia) 121
P.78 (Partenavia) (1980-81) 128
P.80S (Pottier) 492
P.84 Jet Provost (BAC) (1972-73) 190
P-95 (EMBRAER) 12
P.100TS (Pottier) 492
P.105TS (Pottier) 493
P.110TS (Pottier) 492
P.166B Portofino (Piaggio) (1972-73) 129
P.166-DAI (Piaggio) (1972-73) 130
P.166-DL2 (Piaggio) (1978-79) 110
P.166-DL3 (Piaggio) 139
P.166-DL3-MAR (Piaggio) 139
P.170S (Pottier) 493
P.180S (Pottier) 493
P-300 Equator (Equator) 80
P-300 Equator (Pöschel) (1972-73) 94
P337 Pressurised Skymaster (Cessna) (1980-81) 320
P-350 Equator (Equator) 80
P-400 Equator (Equator) 80
P-400 Turbo-Equator (Equator) (1976-77) 70
P-418/71 (Utva) (1975-76) 528
P-420 Twin Equator (Equator) 80
P-450 Equator (Equator) 80
P.531 Wasp/Scout (Westland) (1972-73) 230
P-550 Turbo Equator (Equator) 80
P-600 (Northrop) (1972-73) 388
PA-18-150 Super Cub (Piper) 425
PA-23-250 Aztec (Piper) 426
PA-23T-250 Turbo Aztec (Piper) 426
PA-24-260 Comanche C (Piper) (1974-75) 419
PA-25-235 Pawnee D (Piper) 427
PA-28 Cherokee conversion (Isham/Piper) 384
PA-28 Cherokee Series (Piper) (1977-78) 367
PA-28-140 Cherokee 140 (Piper) (1972-73) 400
PA-28-161 Warrior II (Piper) 427
PA-28-180 Cherokee 180 (Piper) (1972-73) 401
PA-28-181 Archer II (Piper) 428
PA-28-200 Cherokee Arrow II (Piper) (1976-77) 352
PA-28-201T Turbo Dakota (Piper) 430
PA-28-235 Cherokee 235 (Piper) (1972-73) 402
PA-28-235 Cherokee Pathfinder (Piper) (1977-78) 370
PA-28-236 Dakota (Piper) 430
PA-28R-201 Arrow III (Piper) (1978-79) 399
PA-28R-300XBT (Piper) 430
PA-28RT-201 Arrow IV (Piper) 429
PA-28RT-201T Turbo Arrow IV (Piper) 429
PA-31-300 Navajo (Piper) (1972-73) 403
PA-31 Navajo (Piper) 432
PA-31-310 Turbo Navajo (Piper) (1976-77) 354
PA-31-325 Navajo C/R (Piper) 433
PA-31-325 Turbo Navajo C/R (Piper) (1976-77) 355
PA-31-350 Chieftain (Piper) 434
PA-31P Pressurised Navajo (Piper) (1977-78) 373
PA-31T Cheyenne II (Piper) 435
PA-31T-1 Cheyenne I (Piper) 434
PA-31T2 Cheyenne IIXL (Piper) 436
PA-32-300 SIX 300 (Piper) (1979-80) 408
PA-32RT-300 Lance II (Piper) (1980-81) 408
PA-32RT-300T Turbo Lance II (Piper) (1980-81) 408
PA-32-301 Saratoga (Piper) 431
PA-32-301T Turbo Saratoga (Piper) 431
PA-32R-301 Saratoga SP (Piper) 432
PA-32R-301T Turbo Saratoga SP (Piper) 432
PA-34-220T Seneca III (Piper) 437
PA-36 Brave (Piper) 438
PA-38-112 Tomahawk II (Piper) 438
PA-39 Turbo Twin Comanche C/R (Piper) (1974-75) 429
PA-39 Twin Comanche C/R (Piper) (1974-75) 429
PA-40 Arapaho (Piper) (1980-81) 416
PA-42 Cheyenne III (Piper) 436
PA-44-180 Seminole (Piper) 440
PA-44-180T Turbo Seminole (Piper) 441
PA-60 Aerostar 600A/601P (Piper) 441
Pa.61F/H Arbalète II (Payen) (1973-74) 75
Pa.71 (Payen) (1973-74) 76
Pa.149 (Payen) (1973-74) 76
Pa.610 Arbalète III (Payen) (1973-74) 76
PAC (see Pacific Airmotive Corporation) (1972-73) 391
PAC (see Pakistan Aeronautical Complex) 157
PADC (see Philippine Aerospace Development Corporation) 157
PAF (see Philippine Air Force) 157
PAH-2/HAC (MBB/Aérospatiale) 102
PAR (see Departamento de Aeronaves) (1972-73) 10
PATS/NSDB (see Philippine Air Transport & Training Services Inc/National Science Development Board) (1980-81) 153
PBY-5A Catalina/Canso Conversion (Trans-Caribbean) (1971-72) 9
PC-6 Porter (Pilatus) (1974-75) 190
PC-6 Turbo-Porter (Pilatus) 180
PC-6 Turbo-Porter Agricultural (Pilatus) 181
PC-7 Turbo-Trainer (Pilatus) 181
PC-7/CH Turbo-Trainer (Pilatus) 181
PC-10 Twin Porter (Poligrat) (1975-76) 95
PD-01 Master Porter (Poligrat) (1975-76) 95
PD-02 Porter (Master) (1977-78) 71
PD 249 (Beechcraft) (1973-74) 250
PD 285 (Beechcraft) (1977-78) 221
PD 289 (Beechcraft) (1977-78) 224
PD-808 (Piaggio) (1978-79) 111
PD-808 526 (Piaggio) (1977-78) 121
PD-808 TF (Piaggio) (1976-77) 115
PDQ (see PDQ Aircraft Products) 541
PDQ-2 (PDQ) 541
PFA (see Popular Flying Association) 504
PGK-1 Hirondelle (Western) 476
PiAC (see Piasecki Aircraft Corporation) 424
PIACL (see Piasecki Aircraft of Canada Ltd.) (1971-72) 26
PIK (see Polyteknikkojen Ilmailukerho) 479
PIK-15 Hinu (PIK) (1973-74) 40
PIK-19 Muhinu (PIK) (1975-76) 40
PIK-21 Super Sytky (PIK) 479
PIK-23 Towmaster (Valmet) 43
PJ-260 (Fasons-Jocelyn) (1974-75) 412
PL-1 Laminar (Pazmany) (1979-80) 534
PL-1A (AIDC/CAF) (1971-72) 27
PL-1B (Pazmany) (1979-80) 534
PL-1B Chienshou (AIDC/Pazmany) (1976-77) 169
PL-2 (LIPNUR) (1974-75) 107
PL-2 (Pazmany) 540
PL-4A (Pazmany) 540
PL-12 Airtruk (Transavia) 7
PL-12-U (Transavia) 7
PM-3 Duet (Phoenix) (1974-75) 222
PMB-78 Faucon (Milon) (1980-81) 481
PRC (Aérospatiale) SA 365N Dauphin 2 (China) 34
PROCAER (see Progetti Costruzioni Aeronautiche) 140
PS-1 (Shin Meiwa) 148
PS-1 Mod (Shin Meiwa) (1976-77) 131
PT NURTANIO (see PT Industri Pesawat Terbang Nurtanio) 91
PX-S (Shin Meiwa) (1972-73) 150
PZL (see Zjednoczenie Przemyslu Lotniczego i Silnikowego PZL) 157
PZL-101A Gawron (WSK-Okecie) (1974-75) 167
PZL-101AF (WSK-Okecie) (1973-74) 165
PZL-104 Gelatik 32 (LIPNUR) (1976-77) 84
PZL-104 Wilga (CNPSL-PZL Warszawa) 158
PZL-104 Wilga 35R (CNPSL-PZL Warszawa) 159
PZL-104 Wilga 80 (CNPSL-PZL Warszawa) 158
PZL-106A Kruk (CNPSL-PZL Warszawa) 159
PZL-106AR Kruk (CNPSL-PZL Warszawa) 159
PZL-106AT Turbo-Kruk (CNPSL-PZL Warszawa) 159
PZL-110 Koliber (CNPSL-PZL Warszawa) 160
PZL-112 M-20 Mewa (Pezetel) (1978-79) 138
Pacific (Nicollier) (1972-73) 71
Pacific (Wassmer) (1975-76) 89
PACIFIC AIRMOTIVE CORPORATION (U.S.A.) (1972-73) 391
PAKISTAN AERONAUTICAL COMPLEX (Pakistan) 157
PAKISTAN ARMY AVIATION (No. 503 WORKSHOP) (Pakistan) (1976-77) 140
Pallich JUP-1 (AVEX) (1973-74) 3
PANAVIA AIRCRAFT GmbH (International) 103
Panther Navajo (Colemill) 346
Panther Navajo/winglets (Colemill) 346
Papillon (Georges) (1973-74) 90
Paras-Cargo, B-EC-9 (Croses) 482
Parawing (Seremet) (1972-73) 36
Paris Jet IV (Jacobs Jets) 384
PARKER, C. Y. (U.S.A.) 539
PARSONS-JOCELYN (U.S.A.) (1975-76) 413
PARTENAVIA COSTRUZIONI AERONAUTICHE S.p.A. (Italy) 137
Passatt, MY 104 (Mylius) (1980-81) 84
PATCHEN INC, MARVIN (U.S.A.) (1975-76) 413
Pathfinder III (PiAC) (1974-75) 416
Patriot (AAC) (1971-72) 222
Patroller (Maule) 401
PATTISON, ROSS (South Africa) (1972-73) 171
Pawnee D (Piper) 427
Pawnee Brave (Piper) (1977-78) 377
PAYEN (see Flèchair SA) (1974-75) 77
Payloader, S-61 (Sikorsky) (1979-80) 441
Paymaster, MA-1 (Emair) (1974-75) 328
PAYNE, VERNON W. (U.S.A.) 539
PAZMANY AIRCRAFT CORPORATION (U.S.A.) 540
Pchelka (Antonov) (1980-81) 179
PDQ AIRCRAFT PRODUCTS (U.S.A.) 541
Peacemaker (Fairchild/Pilatus) (1977-78) 293
Pee-40 (Jurca) 486
Pegaso (General Avia) (1972-73) 125
Pegaso, F.20 (General Avia) 136
Pegasus I (Frontier) 355
Pegasus II (Frontier) 355
Pegasus programme (Aérospatiale) (1979-80) 41
Pegasus, S.W.1. (Schretzmann) (1974-75) 191
Pelican (Sikorsky) 459
Peregrine/Peregrine II (GACC) 356
Perereca (Weber) 474
Petit Prince, DR 400/120 (Robin) (1979-80) 66
Petrel (Nash) 254
PEZETEL FOREIGN TRADE ENTERPRISE OF AVIATION INDUSTRY (Poland) 158
PFLUMM, MANFRED (Germany) 494
Phantom II (McDonnell) 403
PHILIPPINE AERONAUTICS TRAINING SCHOOL/NATIONAL SCIENCE DEVELOPMENT BOARD (Philippines) (1978-79) 137
PHILIPPINE AEROSPACE DEVELOPMENT CORPORATION (Philippines) 157
PHILIPPINE AIR FORCE (Philippines) 157

PHILIPPINE AIR TRANSPORT & TRAINING SERVICES INC/NATIONAL SCIENCE DEVELOPMENT BOARD (Philippines) .... *(1980-81)* 153
Phillips Phillicopter Mk 1 (VTOL) .... *(1979-80)* 9
PHILLIPS, D. A. (Australia) .... *(1974-75)* 10
Phoenix (Heligyro) .... *(1977-78)* 542
PHOENIX AIRCRAFT LTD (U.K.) .... *(1975-76)* 221
Phoenix man-powered aircraft (To) 264
PIAGGIO S.p.A., INDUSTRIE AERONAUTICHE E MECCANICHE, RINALDO (Italy) 139
PIASECKI AIRCRAFT CORPORATION (U.S.A.) 424
PIASECKI AIRCRAFT OF CANADA LTD (Canada) .... *(1971-72)* 26
Piast (Czerniejewski) .... *(1974-75)* 170
Picchio, F15E (PROCAER) 140
PIEL, AVIONS CLAUDE (France) 489
PIENIAZEK, EUGENIUSZ (Poland) .... *(1973-74)* 168
PIETENPOL, BERNARD H. (U.S.A.) .... *(1978-79)* 532
PIETENPOL, DON (U.S.A.) 541
Pigeon (Gatard) .... *(1980-81)* 476
PILATUS BRITTEN-NORMAN LTD (U.K.) 256
PILATUS FLUGZEUGWERKE A.G. (Switzerland) 180
Pilatus Porter (Fairchild Industries) .... *(1972-73)* 315
Pillan (Piper) 430
Pilot (Martin Marietta) .... *(1972-73)* 356
Pilot self-rescue system (Bell) .... *(1973-74)* 259
Pinocchio (Piel) 490
PIPER AIRCRAFT CORPORATION (U.S.A.) 425
Pipit (Sisler) .... *(1972-73)* 424
Piranha (ALR) 178
Piranha (Kahn) .... *(1976-77)* 517
Piranha, IA 80 (Issoire) .... *(1980-81)* 67
PITTS AEROBATICS (U.S.A.) 444, 542
Pixie, Pober P-9 (EAA) 523
Plank (Backstrom) .... *(1977-78)* 521
PLUMB, BARRY G (U.K.) 504
Pober P-9 Pixie (EAA) 523
"Pogo" Vehicles (Bell) .... *(1971-72)* 241
POLEN AIRCRAFT (U.S.A.) .... *(1974-75)* 430
POLIGRAT DEVELOPMENT GmbH & CO. KG (West Germany) .... *(1976-77)* 73
Polliwagen (Polliwagen) 542
POLLIWAGEN INC (U.S.A.) 542
POLNIAK, LEON (Poland) .... *(1977-78)* 503
Polonez, J-2 (Janowski) .... *(1979-80)* 489
POLSKIE ZAKLADY LOTNICZE (Poland) .... *(1980-81)* 153
POLYT (see Polyteknisk Flyvegruppe) .... *(1973-74)* 39
Polyt 5 (Polyt) .... *(1973-74)* 39
POLYTEKNIKKOJEN ILMAILUKERHO (Finland) 479
POLYTEKNISK FLYVEGRUPPE (Denmark) .... *(1973-74)* 39
Poor Man's Cherokee (Gonserkevis) .... *(1975-76)* 343
POPULAR FLYING ASSOCIATION (U.K.) 504
Porter (Fairchild Industries/Pilatus) .... *(1977-78)* 293
Porter (Pilatus) .... *(1974-75)* 190
Porter PD-02 (Master) .... *(1977-78)* 71
Portofino (Piaggio) .... *(1972-73)* 129
POSCHEL AIRCRAFT GmbH (West Germany) *(1972-73)* 94
POTTIER, JEAN (France) .... *(1980-81)* 484
Pou-du-Ciel (AVEX) .... *(1971-72)* 3
Pou-du-Ciel (Cole) .... *(1971-72)* 188
POULET (see Dalotel) .... *(1974-75)* 56
Pouplume (Croses) 481
Poussin (Gatard) 483
POWELL, JOHN C. (U.S.A.) 542
PRACTAVIA LTD (U.K.) 259, 505
Pregnant Guppy (Aero Spacelines) .... *(1972-73)* 233
President 600 (Colemill) 346
President, HR 100/250TR (Robin) .... *(1980-81)* 72
Pressurised Centurion/Centurion II (Cessna) 334
Pressurised Navajo (Piper) .... *(1977-78)* 373
Pressurised Skymaster (Reims) .... *(1978-79)* 58
Pressurised Skymaster I/II (Cessna) .... *(1980-81)* 320
PRESTWICK (see Prestwick Man-Powered Aircraft Group) .... *(1976-77)* 487
PRESTWICK MAN-POWERED AIRCRAFT GROUP (U.K.) .... *(1976-77)* 487
Primitivkrähe (FVA) .... *(1972-73)* 89
Prince (Robin/CEA) .... *(1972-73)* 80
Privrednik (Utva) .... *(1972-73)* 489
PROCAER (see Progetti Costruzioni Aeronautiche S.p.A.) 140
PROCTER (see Procter Aircraft Associates Ltd.) .... *(1980-81)* 498
PROCTER AIRCRAFT ASSOCIATES LTD. (U.K.) .... *(1980-81)* 498
Professional, DM-160 (Dalotel) .... *(1974-75)* 56
PROGETTI COSTRUZIONI AERONAUTICHE S.p.A. (Italy) 140
PROJECT JUPITER (U.K.) .... *(1975-76)* 216
Project V-519 (LTV) .... *(1972-73)* 356
Project 12X (EMBRAER) .... *(1977-78)* 15
Project 7307En (PATS/NSBDB) .... *(1978-79)* 137
Protector (BAe) 238
'Project Stealth' (Lockheed) 389
PROP-JETS INC (U.S.A.) .... *(1979-80)* 421
Prowler (Grumman) 368
Przasniczka (Janowski) .... *(1974-75)* 171
PT INDUSTRI PESAWAT TERBANG NURTANIO (Indonesia) 91
Pucará, IA 58A (AMC) 2
Pucará Bravo, IA 58B (AMC) 3
Puma (Aerospatiale/Westland) 46, 272
Puma (Nurtanio/Aérospatiale) 91
Puma, IAR-330 (ICA/Aérospatiale) 171
Pure Air Machine (Ludwig) 533
Pussy Cat (Hesley) .... *(1972-73)* 333
PÜTZER (see Sportavia-Pützer) .... *(1980-81)* 86

Q

Q-Star (Lockheed) .... *(1972-73)* 354
QF-4B Phantom II (McDonnell) .... *(1980-81)* 380
QF-104 Starfighter (Lockheed) .... *(1972-73)* 344
QSRA (DH Canada) .... *(1980-81)* 25
QSRA (NASA/Boeing) 417
QSTOL research aircraft (NAL) 147
QU-22B (Beechcraft) .... *(1973-74)* 249
Quail (Aerosport) 508
Quail A9B-M (AAMSA) 150
Queen Air 65 (Beechcraft) .... *(1972-73)* 250
Queen Air 70 (Beechcraft) .... *(1972-73)* 250
Queen Air B80 (Beechcraft) .... *(1979-80)* 271
Queenaire 800/8800 (Excalibur) 350
Queen Airliner 70 (Beechcraft) .... *(1972-73)* 250
Queen Airliner A65 (Beechcraft) .... *(1972-73)* 250
Queen Airliner B80 (Beechcraft) .... *(1979-80)* 271
Quickie (Quickie) 543
QUICKIE AIRCRAFT CORPORATION (U.S.A.) 542
Quickie Q2 (Quickie) 543
Quiet Short-Haul Research Aircraft (DH Canada) .... *(1980-81)* 25

R

R-2 Shark (Mace) .... *(1971-72)* 345
R22 (Robinson) 451
R-95 (EMBRAER) 10
R172E (Cessna) 325
R172 Hawk XP (Cessna) 326
R172K Skyhawk conversions (Isham/Cessna) 384
R 235 Guerrier (SOCATA) 74
R 1180 Aiglon (Robin) 72
R 2000 Alpha series (Robin) 72
R 2112 Alpha (Robin) 72
R 2160 Alpha Sport (Robin) 72
R 3000 series (Robin) 72
RA-5 Vigilante (North American) .... *(1972-73)* 377
RA-7E Corsair II (Vought) .... *(1979-80)* 457
RACA (see Representaciones Aero Commerciales Argentinas) SA 4
RB-1 Ray's Rebel (Bishop) .... *(1973-74)* 271
RB-57F (General Dynamics/Martin) .... *(1971-72)* 309
RBT 1 Sport Plane (Pattison) .... *(1972-73)* 171
RC-95 (EMBRAER) .... *(1977-78)* 12
RC-130 Hercules (Lockheed) 395
RD-02 Edelweiss (Duruble) .... *(1972-73)* 61
RD-03 Edelweiss (Duruble) 482
RD-2 (Rigault-Deproux) .... *(1972-73)* 77
RD-3 (Rigault-Deproux) .... *(1972-73)* 77
RDW-2 Draggin' Fly (Weir) .... *(1974-75)* 480
REP-2 (Replogle) .... *(1979-80)* 538
RF-1B Jastreb (Soko) .... *(1977-78)* 475
RF-4 Phantom II (McDonnell) 403
RF-5E (Northrop) 421
RF6 Sportsman (Sportavia) .... *(1979-80)* 79
RF6-180 Sportsman (Sportavia) .... *(1979-80)* 79
RF-6B (Fournier) 67
RF7 (Sportavia) .... *(1972-73)* 96
RF8 (Indraéro) .... *(1975-76)* 70
RF-8 Crusader (LTV) .... *(1972-73)* 342
RF-21 Trimard (Fleury) .... *(1975-76)* 68
RF-104 Starfighter (Lockheed) .... *(1972-73)* 344
RF-110 Phantom II (McDonnell) .... *(1980-81)* 380
RF-111C (General Dynamics) .... *(1980-81)* 345
RFB (see Rhein-Flugzeugbau GmbH) 86
RH-3 Sea King (Sikorsky) 458
RH-53 (Sikorsky) .... *(1979-80)* 443
RH-53D (Sikorsky) .... *(1979-80)* 443
RJ-1 (Jastreb) .... *(1979-80)* 460
RJJ-1 Gipsy Hawk (Jameson) .... *(1975-76)* 362
RL-3 Monsoon (Western) .... *(1977-78)* 481
RL-6 Go-Plane (Levi) .... *(1975-76)* 217
RLU (see Charles Roloff, Robert Liposky and Carl Unger) 545
RLU-1 Breezy (RLU) 545
ROKAF (see Republic of Korea Air Force) *(1975-76)* 155
RP-3D Orion (Lockheed) 390
RRA (see Ronchetti Razzetti Aviación SA) 4
RS 180 Sportsman (RFB) 86
RS1 Ilse (Siegrist) .... *(1976-77)* 532
RSA 200 Falcon (AFIC) .... *(1972-73)* 170
RSH-1A, Coupe (Rotor Sport) 545
RTF-104G1 Starfighter (Lockheed) .... *(1972-73)* 344
RU-21J (Beechcraft) 290
RV-3 (Van's) 560
RV-4 (Van's) 560
RV-5 Swinger (Van's) .... *(1979-80)* 550
RV-6 (Van's) 560
RABOUYT (France) .... *(1975-76)* 80
RACA SA (Argentina) .... *(1973-74)* 7
Racer (AVEX) .... *(1974-75)* 4
Racer (Jurca) .... *(1976-77)* 468
Racer (Piel) .... *(1973-74)* 78
Rafale, R 2160 A (Robin) .... *(1979-80)* 68
RAF CRANWELL MAN-POWERED AIRCRAFT GROUP (U.K.) .... *(1977-78)* 508
Rail Mk II (Aerosport) 508
RAISBECK ENGINEERING (U.S.A.) 445
RAISBECK GROUP, THE (U.S.A.) .... *(1980-81)* 420
Rallye 7-300 (Socata) .... *(1971-72)* 79
Rallye series (Socata) 73
R AND B AIRCRAFT COMPANY (U.S.A.) *(1977-78)* 555
RAND ROBINSON ENGINEERING INC (U.S.A.) 544
Rangemaster Model H (Navion) .... *(1978-79)* 391
Ranger (Aerostar) .... *(1971-72)* 220
Ranger (Mooney) .... *(1980-81)* 393
Ranger (Northwest Industries) .... *(1972-73)* 25
Ranger, DHC-7R (DH Canada) 25
Rapace (Cholot) .... *(1972-73)* 49
Rapid Transit (Taylor) .... *(1972-73)* 434
Ray's Rebel, RB-1 (Bishop) .... *(1973-74)* 271
Raz de Marée (Chasle) .... *(1972-73)* 49
Raz-Mut (St-Germain) 474
Real Sporty (Musso) 538
Red Bare-Un (Miller) .... *(1974-75)* 398
RED BARON FLYING SERVICE (U.S.A.) *(1978-79)* 535
Red Wing Black Bird (Wallis) 562
Reed Falcon (Air Nova) .... *(1975-76)* 175
REED/UNIVERSITY OF NATAL (South Africa) .... *(1974-75)* 178
Régent, DR 400/180 (Robin) 71
Regente (Neiva) .... *(1974-75)* 18
REIMS AVIATION SA (France) 69
Reims Club (Reims) .... *(1971-72)* 73
Reims Rocket (Reims) .... *(1976-77)* 59
Rekcub (Guex) .... *(1979-80)* 492
Remorqueur, DR 400/180R (Robin) 71
Renegade 1 (Lasher) 533
REPAIR AG ALTENRHEIN (Switzerland) .... *(1980-81)* 174
Replica Ike (Kovach) 532
Replica Junkers Ju 87B-2 Stuka (Langhurst) 533
Replica P-51D Mustang (Fisher-Van Norman) .... *(1980-81)* 520
REPLICA PLANS (Canada) 476
REPLOGLE, E. H. (U.S.A.) .... *(1979-80)* 538
REPRESENTACIONES AERO COMMERCIALES ARGENTINAS SA (Argentina) 4
REPUBLIC AVIATION DIVISION OF FAIRCHILD HILLER CORPORATION (U.S.A.) .... *(1971-72)* 304
REPUBLIC OF KOREA AIR FORCE (Korea) .... *(1975-76)* 155
Research Projects (PATA/NSDB) .... *(1979-80)* 149
Ressu (Hietanen) .... *(1972-73)* 37
Retro, GB-2 (Bezzola) 500
Revathi (Civil Aviation Dept., India) .... *(1973-74)* 98
Revathi Mk II (India) .... *(1978-79)* 72
RHEIN-FLUGZEUGBAU GmbH (West Germany) 86
RICHARD COMPANY, THE C.H. (U.S.A.) *(1974-75)* 432
RICHTER, KLAUS J (Germany) 494
RIGA (U.S.S.R.) .... *(1971-72)* 453
Riga-50 (Riga) .... *(1971-72)* 453
Riga-50M (Riga) .... *(1971-72)* 453
RIGAULT-DEPROUX, AUTOGYROS (France) .... *(1972-73)* 77
RILEY AIRCRAFT CORPORATION (U.S.A.) 445
RINKER, BUD (U.S.A.) .... *(1971-72)* 393
RIO CLARO, AEROCLUBE DE (Brazil) *(1979-80)* 17
Rkiiga-74 (Lenin Komsomol) .... *(1977-78)* 574
ROBERTS, DON (USA) .... *(1979-80)* 538
ROBERTSON AIRCRAFT CORPORATION (U.S.A.) 447
ROBIN, AVIONS (Canada) 27
ROBIN, AVIONS PIERRE (France) 70
ROBINSON HELICOPTER COMPANY INC (U.S.A.) 451
Rocket (Maule) .... *(1976-77)* 324
Rocket (Reims) .... *(1976-77)* 59
Rocket 340 (Riley) 446
Rocket Power 414 (Riley) 446
Rocket Powered 421 (Riley) .... *(1978-79)* 415
ROCKWELL (see Rockwell International Corporation) 452
ROCKWELL INTERNATIONAL CORPORATION (U.S.A.) 452
ROHR INDUSTRIES INC (U.S.A.) .... *(1975-76)* 447
ROLLASON AIRCRAFT & ENGINES LTD. (U.K.) .... *(1975-76)* 224
ROLOFF, CHARLES, ROBERT LIPOSKY & CARL UMGER (U.S.A.) 545
Rombac 1-11 (IAv Bucuresti) 170
RONCHETTI, RAZZETTI AVIACION SA (Argentina) 4
Ron's 1 (Freiberger) .... *(1973-74)* 329
Rooivalk (Reed) .... *(1972-73)* 171
Rose Parakeet Replica (Borg) 513
Rosette (Sizer) .... *(1975-76)* 230
Rossiya (Tupolev) .... *(1973-74)* 492
ROTOR MASTER AIRCRAFT (U.S.A.) *(1979-80)* 538
Rotorplane (AMR) .... *(1975-76)* 243
ROTOR SPORT HELICOPTERS INC (U.S.A.) 545
ROTORWAY AIRCRAFT INC (U.S.A.) 546
Royal (Robin/CEA) .... *(1974-75)* 84
ROYAL THAI AIR FORCE (Thailand) *(1975-76)* 193
RUTAN AIRCRAFT FACTORY (U.S.A.) 546
RYSON AVIATION CORPORATION (U.S.A.) .... *(1975-76)* 448

S

S1 (JEAA) .... *(1973-74)* 140
S-1 Special (Pitts) 444
S1B2 Arctic Tern (Arctic) 275
S-1D/-1E Special (Pitts) 542
S-2 (Sheppard) .... *(1980-81)* 544
S-2 Tracker (Grumman) .... *(1972-73)* 324
S-2A Special (Pitts) 444
S-2G Tracker (Martin Marietta/Grumman) *(1973-74)* 369
S-2R Turbo-Thrush (Ayres) 277
S2R-T Turbo-Thrush (Marsh/Rockwell) 400
S-2S (Pitts) 445
S-3A Viking (Lockheed) 393
S.5 replica (Leisure Sport) .... *(1979-80)* 496
S-12-E (Spencer) 551
S-20 (MacDonald) .... *(1979-80)* 528
S-21 (MacDonald) .... *(1979-80)* 528
S25X Mr Smoothie (Sawano) .... *(1980-81)* 489
S25X II Tandem Smoothie (Sawano) .... *(1980-81)* 489
S34 Majorette (Phoenix) .... *(1971-72)* 203

S 35 Draken (Saab) (1977-78) 168
S-55-T (Aviation Specialties/Sikorsky) (1975-76) 244
S-55T (Helitec/Sikorsky) (1978-79) 349
S-58T (Sikorsky) (1978-79) 431
S-61A-4 (Agusta-Sikorsky) (1979-80) 115
S-61A-4 Nuri (Sikorsky) (1980-81) 435
S-61A/B (Sikorsky) 458
S-61D-4 (Sikorsky) (1980-81) 436
S-61L/N (Sikorsky) (1980-81) 435
S-61R (Sikorsky) 459
S-61R (Agusta-Sikorsky) 129
S-62 (Sikorsky) (1971-72) 402
S-64 Skycrane (Sikorsky) (1978-79) 433
S-65A (Sikorsky) 460
S-65 (MCM) (Sikorsky) (1978-79) 435
S-65-40 (Sikorsky) (1975-76) 458
S-65-Oe (Sikorsky) (1978-79) 434
S-67 Blackhawk (Sikorsky) (1975-76) 459
S-69 (Sikorsky) 461
S-70 (UTTAS) (Sikorsky) 461
S-70L (Sikorsky) 462
S-72 (RSRA) (Sikorsky) (1980-81) 440
S-76 Spirit (Sikorsky) 463
S-78 (Sikorsky) (1977-78) 410
S.205AC (SIAI-Marchetti) (1980-81) 132
S.205/20R (SIAI-Marchetti) 132
S.208 (SIAI-Marchetti) (1976-77) 118
S.208A (SIAI-Marchetti) 132
S.208AG (SIAI-Marchetti) (1976-77) 118
S.208M (SIAI-Marchetti) (1976-77) 118
S.210 (SIAI-Marchetti) (1975-76) 140
S.210M (SIAI-Marchetti) (1973-74) 135
S.211 (SIAI-Marchetti) 133
S-220 ESTOL (Vartek) (1973-74) 454
S-325 Mini-Swat (Schapel) 455
S-525 Super Swat (Schapel) 455
S-1080 Thunderbolt (Schapel) 456
SA (Airbus) (1980-81) 97
S.A. II Swalesong (Coates) (1977-78) 508
S.A. III Swalesong (Coates) 502
SA-3-120 Bulldog (BAe) 245
SA-3-200 Bulldog (BAe) (1978-79) 223
SA-6B Flut-R-Bug (AVEX) (1971-72) 3
SA-28T (Swearingen) (1971-72) 414
SA-100 Starduster (Stolp) (1974-75) 470
SA 102.5 Cavalier (Macfam) 475
SA 103 Cavalier (Macfam) 476
SA 104 Cavalier (Macfam) 476
SA 105 Super Cavalier (Macfam) 476
SA.202 Bravo (SIAI-Marchetti/FFA) (1972-73) 135
SA-226TC Metro II (Swearingen) (1980-81) 443
SA227-AC Metro III (Fairchild Swearingen) 353
SA227-AT Merlin IVC (Fairchild Swearingen) 355
SA227-TT Merlin IIIC (Fairchild Swearingen) 354
SA-300 Starduster Too (Stolp) 556
SA 315B Lama (Aérospatiale) 45
SA 315B Lama (HAL/Aérospatiale) 89
SA 316 Alouette III (F + W) (1974-75) 192
SA 316B Alouette III (Aérospatiale) 45
SA 316B Alouette III (HAL/Aérospatiale) 89
SA 318C Alouette II Astazou (Aérospatiale) (1976-77) 38
SA 319B Alouette III Astazou (Aérospatiale) 45
SA 321 Super Frelon (Aérospatiale) 46
SA 330 Puma (Aérospatiale/Westland) 46, 272
SA 330 Puma (Nurtanio/Aérospatiale) 91
SA 331 Super Puma (Aérospatiale) (1977-78) 45
SA 332 Super Puma (Aérospatiale) (1978-79) 41
SA 341 Gazelle (Aérospatiale/Westland) 49, 272
SA 342 Gazelle (Aérospatiale/Westland) 49
SA 349-2 (Aérospatiale) (1978-79) 42
SA 360 Dauphin (Aérospatiale) 52
SA 361H/HCL Dauphin (Aérospatiale) (1979-80) 49
SA 365C Dauphin 2 (Aérospatiale) 53
SA-500 Starlet (Stolp) 556
SA-700 Acroduster 1 (Stolp) 556
SA-750 Acroduster Too (Stolp) 556
SA-900 V-Star (Stolp) 556
SAAB-SCANIA (see SAAB-Scania Aktiebolag) 176
SABCA (see Société Anonyme Belge de Constructions Aéronautiques) 8
SAH-1 (Trago Mills) 264
SAL (see Sturgeon Air Ltd) (1974-75) 29
SARA II (CSIR) (1978-79) 151
SARA III (CSIR) (1980-81) 166
Sarcup Programme (Boeing Vertol/CAF) 18
SAVER (Kaman) (1975-76) 365
SBD-3 Dauntless (Roberts) (1979-80) 538
SC.7 Skyvan (Shorts) 262
SC-130 Hercules (Lockheed) (1972-73) 350
SCS CL-84 (Canadair) (1975-76) 18
SD3-30 (Short) (1977-78) 204
SD3-M (Shorts) (1980-81) 247
SD3-MR Seeker (Shorts) (1980-81) 247
SE-5A (Replica Plans) 476
SE 210 Caravelle (Aérospatiale) (1974-75) 43
SE 3160 Alouette III (Aérospatiale) 45
SEPECAT (see Société Européenne de Production de l'Avion École de Combat et Appui Tactique) 107
SF-1 Pipit (Sisler) (1972-73) 424
SF-2A Cygnet (Sisler) 549
SF-5A (CASA) (1972-73) 173
SF-5A/B (Northrop) (1978-79) 392
SF-5B (CASA) (1972-73) 173
SF 37 Viggen (Saab) 177
SF.260A (SIAI-Marchetti) (1980-81) 132
SF.260C (SIAI-Marchetti) 131
SF-260M (SIAI-Marchetti) 130
SF.260SW Sea Warrior (SIAI-Marchetti) 130
SF.260TP (SIAI-Marchetti) 131
SF.260W Warrior (SIAI-Marchetti) 130
SF.600 Canguro (SIAI-Marchetti) 132
SFS 31 Milan (Sportavia) (1971-72) 95
SH-2 (Kaman) 386
SH-2D LAMPS (Kaman) 386
SH-2F (Kaman) 386
SH-3 Sea King (Sikorsky) 458
SH-3D (Agusta-Sikorsky) 128
SH-3D Sea King (Westland/Sikorsky) 267
SH-3D/TS (Agusta-Sikorsky) 128
SH-3H (Agusta-Sikorsky) 128
SH-4 (Silvercraft) (1977-78) 124
SH 37 Viggen (Saab) 177
SH-60B Seahawk (Sikorsky) 462
SH-200 (Silvercraft) (1978-79) 115
SISA Ha-11 (Haak) (1972-73) 89
SITAR (see Société Industrielle de Tolerie pour l'Aéronautique et Matériel Roulant) (1972-73) 8
SK-1 Trempik (Simunek/Kamaryt) 478
SK 35 Draken (Saab) (1977-78) 168
SK 37 Viggen (Saab) 177
SK 60 (Saab) (1973-74) 182
SKT-125 Sunday Knight Twister (Payne) (1979-80) 533
SKT-150, Holiday Knight Twister (Payne) 540
SL-1 (Wood) (1976-77) 544
SM-1, Winged (IL) (1974-75) 164
SM.208 (SIAI-Marchetti) (1973-74) 136
SM.1019E (SIAI-Marchetti) 133
SMIC (see Shin Meiwa Industry Co., Ltd) 148
SN 600 Corvette (Aérospatiale) (1977-78) 40
SN 601 Corvette (Aérospatiale) (1978-79) 37
SN 602 Corvette (Aérospatiale) (1975-76) 44
SNS-7 Hiperbipe (Sorrell) 550
SOCATA (see Société de Construction d'Avions de Tourisme et d'Affaires) 73
SONACA (see Société Nationale de Construction Aérospatiale) 8
SR-71 (Lockheed) (1977-78) 326
SS-1 (Scholz-Schmitz) (1979-80) 485
SS-2 (Shin Meiwa) 148
SS-2A (Shin Meiwa) 148
SSVV (see Sezione Sperimentale Volo a Vela) (1980-81) 135
ST 10 Diplomate (Socata) (1976-77) 66
ST-27 (Saunders) (1975-76) 27
ST-28 (Saunders) (1976-77) 23
ST 60 Rallye 7-300 (Socata) (1971-72) 79
ST-600-S/8 (Foxjet) (1979-80) 337
STAMP (Williams) (1975-76) 479
STOL conversions (Robertson/Cessna) 447
STOL conversions (Robertson/Piper) 447
STOL improvements (Seaplane Flying)) (1979-80) 439
STP-1 Swallow (Ryson) (1973-74) 428
Su-7B (Sukhoi) 214
Su-9 (Sukhoi) 215
Su-11 (Sukhoi) 215
Su-15 (Sukhoi) 216
Su-17 (Sukhoi) 216
Su-19 (Sukhoi) (1979-80) 207
Su-20 (Sukhoi) 216
Su-22 (Sukhoi) 216
Su-24 (Sukhoi) 219
SV-5P Pilot (Martin Marietta) (1972-73) 356
SV-20A (SIAI-Marchetti) (1973-74) 137
SV-20C (SIAI-Marchetti) (1973-74) 137
SVIT 70 (Bucciero) (1975-76) 130
S.W.1 Pegasus (Schretzmann) (1974-75) 19
SWL (Rockwell) (1978-79) 422
SY Motor Glider (Yamato) (1980-81) 490
SAAB/FAIRCHILD (International) 106
SAAB-SCANIA AKTIEBOLAG (Sweden) 176
Saab-35 Draken (Saab) (1977-78) 168
Saab 35S (Saab) (1976-77) 161
Saab 35X (Saab) (1976-77) 161
Saab 35XD (Saab) (1976-77) 161
Saab-37 Viggen (Saab) 177
Saab-37E Eurofighter (Saab) (1975-76) 183
Saab 37X (Saab) 177
Saab-105 (Saab) (1976-77) 162
Saab 105G (Saab) (1977-78) 163
Saab 105Ö (Saab) (1975-76) 186
Saab 105XH (Saab) (1975-76) 186
Saab-105XT (Saab) (1975-76) 186
Saab-114 (Saab) 173
Saab 2105 (JAS) 175
Saab-MFI 15 (Saab) (1974-75) 185
Saab-MFI 17 (Saab) (1974-75) 187
Sabre 40A (Rockwell) (1974-75) 443
Sabre 75 (North American) (1972-73) 380
Sabre 75 (Rockwell) (1974-75) 442
Sabre 75A (Rockwell) (1975-76) 443
Sabre Commander 40A (North American) (1972-73) 381
Sabreliner (North American) (1972-73) 380
Sabreliner 40R (Rockwell) 454
Sabreliner 60 (Rockwell) (1979-80) 432
Sabreliner 60A (Rockwell) (1978-79) 425
Sabreliner 65 (Rockwell) 454
Sabreliner 75A (Rockwell) (1979-80) 432
Sabreliner 80A (Rockwell) (1978-79) 425
Sabreliner 85 (Rockwell) (1979-80) 434
SABRELINER DIVISION, ROCKWELL INTERNATIONAL (U.S.A.) 454
Sabre Series 40 (Rockwell) (1974-75) 441
Sabre Series 60 (Rockwell) (1975-76) 442
Saeta (Hispano) (1971-72) 167
Safari (Saab) 178
Safari, HR 100/210 (Robin) (1975-76) 83
Safari TS (Saab) 178
Safari II, HR 100, 210 (Robin) (1977-78) 62
Safety and STOL conversions (Robertson) 447
Safety Twin (Halsmer) (1973-74) 342
Sailcraft/Hagiwara (JEAA) (1972-73) 139
SALVAY-STARK (see Skyhopper Airplanes, Inc.) (U.S.A.) (1976-77) 531
San Francesco, I-66 (Iannotta) (1973-74) 126
SÃO CARLOS ENGINEERING SCHOOL, UNIVERSITY OF SÃO PAULO (Brazil) 18
Sapphire (Sizer) (1975-76) 229
Saratoga (Piper) 431
Saratoga SP (Piper) 432
SAUNDERS AIRCRAFT CORPORATION LTD (Canada) (1977-78) 28
SAWANO, SHIRO (Japan) (1980-81) 489
SAWYER, RALPH (U.S.A.) (1978-79) 538
Scamp (Aerosport) 508
Scamp Model B (Agricopteros/Aerosport) 36
SCENIC AVIATION SERVICES (U.S.A.) (1980-81) 433
SCHAFER AIRCRAFT MODIFICATION INC (U.S.A.) 455
SCHAPEL AIRCRAFT COMPANY (U.S.A.) 455
SCHEUTZOW HELICOPTER CORPORATION (U.S.A.) (1977-78) 398
SCHILLING, ING HORST (West Germany) (1979-80) 485
SCHOENENBERG, HEINRICH (Germany) 494
SCHOLZ, MANFRED (West Germany) (1979-80) 485
SCHOLZ-SCHMITZ (see Scholz, Manfred) (1979-80) 485
SCHRETZMANN, WALTER (Switzerland) (1974-75) 191
SCHWEIZER AIRCRAFT CORPORATION (U.S.A.) 456
Scooter (Flaglor) (1979-80) 521
Scorchy, M-102 (Mace) 534
Scorpion (RotorWay) (1972-73) 413
Scorpion 133 (RotorWay) 546
SCOTT, RONALD G. (U.S.A.) (1971-72) 397
SCOTTISH AVIATION, LTD. (U.K.) (1977-78) 192
Scout (Bellanca/Champion) (1980-81) 287
Scout (Skycraft) (1980-81) 8
SCOVILLE, JOHN R. (U.S.A.) (1971-72) 397
SeaCobra (Bell) 296
Sea Harrier (BAe) 248
Seahawk SH-60B (Sikorsky) 462
Sea King (Sikorsky) 458
Sea King (Westland) 267
Sea Knight (Boeing Vertol) 317
Sea Lynx (Westland) (1977-78) 210
Sealander (Britten-Norman) (1971-72) 183
Sea Patrol (Gates Learjet) 359
SEAPLANES INC (U.S.A.) 457
SeaRanger, TH-57A (Bell) 294
Searchmaster (GAF) 5
Sea Scan, IAI 1124 (IAI) 116
Seasprite (Flight Dynamics) (1980-81) 520
Seasprite (Kaman) 386
Sea Stallion (Sikorsky) 460
Sea Warrior (SIAI-Marchetti) 130
Seeker, SD3-MR (Shorts) (1979-80) 248
SÉGUIN, M (France) (1980-81) 486
Séguin/Adams-Wilson Helicopter (Séguin) (1980-81) 486
SEGUIN AVIATION (U.S.A.) (1972-73) 416
Seminole (Beechcraft) (1972-73) 250
Seminole (Piper) 440
Seneca II (EMBRAER/Piper) 17
Seneca III, PA-34-220T (Piper) 437
Senior Aero Sport (D'Apuzzo) 519
Sentry, E-3 (Boeing) 314
Sentry 02-337 (Summit) 466
Sequoia, Model 300 (Sequoia) 548
SEQUOIA AIRCRAFT CORPORATION (U.S.A.) 548
SEREMET, W. VINCENT (Denmark) (1980-81) 471
Series 12X (EMBRAER) (1978-79) 14
Series 600 Fanjet (Cessna) (1973-74) 311
Sertanejo (EMBRAER/Piper) 17
SERV-AERO ENGINEERING INC (U.S.A.) 458
SEZIONE SPERIMENTALE VOL A VELA (Italy) (1980-81) 135
Shackleton AEW.Mk 2 (Hawker Siddeley) (1973-74) 209
Shadow (Jurca) (1972-73) 68
Shanghai (?) Y-7 (C-7) (China) 34
Shanghai Y-10 (C-10) (China) 34
Shark (Enstrom) 349
Shark (Mace) (1971-72) 345
SHEFFIELD AIRCRAFT LTD (U.S.A.) (1974-75) 450
Shenyang CJ-6 (BT-6) (China) 34
Shenyang J-5 Trainer (FT-5) (China) 29
Shenyang J-8 (F-8) (China) 32
Shenyang (Mikoyan) J-6 (F-6) (China) 29
Shenyang Yakovlev CJ-5 (BT-5) (China) 34
SHEPPARD, H. GORDON (USA) (1980-81) 544
Sheriff (Aircraft Designs/Britten) 235
Sher Khan (SITAR) (1971-72) 78
SHIELD, G. W. (U.K.) (1975-76) 226
SHIN MEIWA INDUSTRY CO., LTD. (Japan) 148
SHIN NIHON KOKU SEIBI KABUSHIKI KAISHA (Japan) (1971-72) 148
Shiokaratombo (Hamao) (1979-80) 486
SHOBER AIRCRAFT ENTERPRISES (U.S.A.) (1977-78) 560
Shoestring (Condor) 518
SHORT BROTHERS LTD. (U.K.) 259
SHOWA HIKOKI KOGYO KABUSHIKI KAISHA (Japan) 149
Shrike Commander (Rockwell) (1974-75) 444
Shrike Commander 500S (Rockwell International)) (1980-81) 431
Shrike Commander Esquire (Rockwell) (1973-74) 424
Shuttle Carrier Aircraft (NASA/Boeing) (1979-80) 395
Shuttle tile test (NASA) 419
Shuttle Training Aircraft (NASA) (1980-81) 395
SIAI-MARCHETTI SOCIETA PER AZIONI (Italy) 129
Sian B-6 (China) (1979-80) 32
Sidewinder (Smyth) 550
SIEGRIST, RUDOLF (U.S.A.) (1977-78) 560
Sierra 200 (Beechcraft) 279
SIEVERS (see Sievers, Clyde and Miller, Phillip) (1971-72) 398
SIEVERS, CLYDE AND MILLER, PHILLIP (U.S.A.) (1971-72) 398

SIKORSKY AIRCRAFT DIVISION OF UNITED TECHNOLOGIES CORPORATION (U.S.A.) 458
SILVERCRAFT SpA (Italy) (1979-80) 126
SILVESTRI, OLINDO (Italy) (1976-77) 474
SIMUNEK, JAN & KAMARÝT, JAROSLAV (Czechoslovakia) 478
SINDLINGER AIRCRAFT (U.S.A.) 548
SINFIELD, ROLAND A (USA) (1979-80) 539
Sioux (Bell) (1971-72) 242
Sirocco (Jurca) 485
Sirocco (Sport Wing) (Jurca) 486
SISLER AIRCRAFT COMPANY (U.S.A.) 549
SIZER, JOHN A. (U.K.) (1975-76) 229
Skeeter X-1 (Sheffield) (1974-75) 450
Skipper, Model 77 (Beechcraft) 279
Skybolt (Steen) 554
Skybuggy Model A (McLaughlin) (1974-75) 396
SKYCRAFT PTY LTD (Australia) (1980-81) 8
Skycrane (Sikorsky) (1978-79) 433
Skycycle (Todhunter) (1980-81) 465
Sky Dancer (Bond) 512
Skyfarmer, T-300 (Transavia) 8
Skyhawk (Cessna) 325
Skyhawk (Douglas) (1980-81) 390
Skyhawk/100 (Cessna) (1977-78) 265
Skyhawk/100 (Reims) (1978-79) 57
Skyhawk/100, F 172 (Reims) (1977-78) 59
Skyhawk II (Cessna) (1979-80) 309
Skyhawk II (Douglas) (1977-78) 353
Skyhawk II/100 (Cessna) (1977-78) 265
Skyhawk/100 (Reims/Cessna) 70
Skyhawk R172K conversions (Isham/Cessna) 384
Skyhawk Conversion (LAS/McDonnell Douglas) (1978-79) 358
Skyhook (Minty) 7
Skyhopper (Salvay-Stark) (1976-77) 531
SKYHOPPER AIRPLANES, INC. (U.S.A.) (1976-77) 531
Skyjacker II (Sawyer) (1978-79) 538
Sky Knight (Hughes) 379
Skylane (Cessna) 328
Skylane RG (Cessna) 328
Skylane II (Cessna) (1977-78) 269
Skylark (Larkin) (1980-81) 528
Skyliner (Aerocom) (1977-78) 214
Skyliner (Riley) (1975-76) 433
Skyliner (Shorts) (1980-81) 248
Skymaster (Cessna) 336
Skymaster (Reims/Cessna) (1980-81) 70
Skymaster II (Cessna) 336
Sky-Mat (Bensen) (1971-72) 252
Skyote (Skyote) 549
SKYOTE AEROMARINE LTD (U.S.A.) 549
Skyrider (HTM) (1974-75) 94
Sky-Rider (Kimberley) (1980-81) 465
Skyrocket (Bellanca) (1974-75) 278
Skyrocket II (Bellanca) (1980-81) 288
Skyservant (Dornier) 77
SKY SPORTS INTERNATIONAL (U.S.A.) (1980-81) 441
Sky-trac (Wagner) (1972-73) 98
Sky-Trac (Helicopter Technik) (1974-75) 93
Skytrader 800 (Dominion) (1979-80) 330
Skytrain II (Douglas) 408
Skyvan, SC.7 (Shorts) 262
Skywagon/Skywagon II (Cessna) 327
Skywagon 185 (Cessna) 329
Skywagon 207 (Cessna) (1977-78) 273
Slaviej (Bulgaria) (1976-77) 456
SLINGSBY ENGINEERING LTD (U.K.) 263
SMITH, MRS. FRANK W. (DOROTHY) (U.S.A.) 550
SMITH AIRCRAFT INC, MIKE (USA) 464
SMITH AND ASSOCIATES INC, TED R. (U.S.A.) (1974-75) 463
SMYTH, JERRY (U.S.A.) 550
SOCIEDADE AEROTEC LTDA (Brazil) (1977-78) 11
SOCIEDADE CONSTRUCTORA AERONAUTICA NEIVA, LTDA. (Brazil) (1979-80) 16
SOCIETA PER AZIONI SIAI-MARCHETTI (Italy) 129
SOCIÉTÉ ANONYME BELGE DE CONSTRUCTIONS AÉRONAUTIQUES (Belgium) 8
SOCIÉTÉ DE CONSTRUCTION D'AVIONS DE TOURISME ET D'AFFAIRES (France) 73
SOCIÉTÉ DALOTEL (France) (1975-76) 56
SOCIÉTÉ EUROPÉENNE DE PRODUCTION DE L'AVION ÉCOLE DE COMBAT ET APPUI TACTIQUE (International) 107
SOCIÉTÉ INDUSTRIELLE DE TOLERIE POUR L'AÉRONAUTIQUE ET MATÉRIEL ROULANT (France) (1972-73) 81
SOCIÉTÉ ISSOIRE-AVIATION (France) (1980-81) 67
SOCIÉTÉ NATIONALE DE CONSTRUCTION AÉROSPATIALE (Belgium) 8
SOCIÉTÉ NATIONALE INDUSTRIELLE AÉROSPATIALE (France) 43
SOCIÉTÉ NOUVELLE WASSMER AVIATION (France) (1978-79) 64
SOKO/CNIAR (International) 110
Sokol W-3 (WSK-PZL Swidnik) 167
SOKO VAZDUHOPLOVNA INDUSTRIJA, RO VAZDUHOPLOVSTVO (Yugoslavia) 471
Solar Challenger (MacCready) 399
Solar One (To) (1980-81) 249
SOLAR-POWERED AIRCRAFT DEVELOPMENTS (U.K.) 264
Solar Riser (UFM) (1980-81) 555
Solitaire (Mitsubishi Internatonal) 414
Solo (Jurca) 486
SOLOY CONVERSIONS LTD (U.S.A.) 465
Sonerai (Monnett) 536
Sonerai II and IIL (Monnett) 536
Sooper-Coot Model A (Aerocar) 506
Sorceress (Mahoney) (1971-72) 345
SORRELL AVIATION (U.S.A.) 550
Space Shuttle Orbiter Carrier (Boeing) (1979-80) 298
Sparrow Commander (AAMSA) (1975-76) 155
Special I (Cassutt) 516
Special II (Cassutt) 517
Special S-1 (Pitts) 444
Special, S-1D/-1E (Pitts) 542
Special S-2A (Pitts) 444
SPECIALISED AIRCRAFT CO (USA) (1978-79) 439
Special-Missions/Sea Patrol Version (Gates Learjet) 359
Spectra IV (Island Aircraft) (1972-73) 337
Speed-Canard (Gyroflug) 81
SPENCER, P. H. (U.S.A.) 551
Sperocco (Jurca) (1980-81) 478
SPEZIO (U.S.A.) 552
SPINKS INDUSTRIES INC (U.S.A.) (1971-72) 408
Spirit S-76 (Sikorsky) 463
Spit (Jurca) 486
Spitfire (Isaacs) 502
SPITFIRE HELICOPTER COMPANY LTD (U.S.A.) 465
Sport 150 (Beechcraft) (1978-79) 257
Sport (Geide) 526
Sport (Mong) (1974-75) 398
SPORT AIR CRAFT INTERNATIONAL (U.S.A.) 552
SPORTAVIA (see Sportavia-Pützer) (1980-81) 86
SPORTAVIA-PÜTZER GmbH ü CO KG (West Germany) (1980-81) 86
Sport Plane (Pattison) (1972-73) 171
Sportplane (Richard) (1971-72) 392
Sportsman, RS 180 (RFB) 86
Sportsman (Volmer) 561
Sportsman 100 (Taylorcraft) (1980-81) 451
Sportster (Hollmann) 528
Sport Trainer (Great Lakes) 364
Sportwing, D-201 (D'Apuzzo) 519
SPRATT & COMPANY INC (U.S.A.) 552
Spraymaster/Chipmunk conversion (Aero-structures) (1971-72) 6
Sprintair (Chasle) (1977-78) 485
Sprinter (Niemi) (1980-81) 534
Sprinter (VFW-Fokker) (1971-72) 98
Sprite (Gyroflight) (1972-73) 198
Sprite (Practavia) 259, 505
Stallion (Helio) (1976-77) 301
STARCK, ANDRE (France) 494
Stardust II (Scoville) (1971-72) 397
Starduster (Stolp) (1974-75) 470
Starduster Too (Stolp) 556
Starduster Too (Stolp/Neubert) (1975-76) 467
Starfighter (Aeritalia/Lockheed) 118
Starfighter (Lockheed) (1977-78) 326
Starfighter, F-104RB (Red Baron/Greenamyer) (1978-79) 535
Starlet (Corby) 473
Starlet (Stolp) 556
StarLifter, C-141 (Lockheed) 398
STATE AIRCRAFT FACTORIES (China) 28
Stationair (Cessna) (1978-79) 306
Stationair 6 (Cessna) 331
Stationair 8 (Cessna) 332
STATLER, WILLIAM H (USA) 553
Statoplan AG 02 Poussin (Gatard) 483
Statoplan AG 04 Pigeon (Gatard) (1980-81) 476
Statoplan AG 05 Mésange (Gatard) (1978-79) 475
'Stealth Fighter' (Lockheed) (1977-78) 326
STEEN, LAMAR A. (U.S.A.) (1975-76) 465
STEEN AERO LAB INC (U.S.A.) 554
STEPHENS AIRCRAFT (U.S.A.) 554
STEVENSON, BERNADINE AND JIM (U.S.A.) (1977-78) 550
STEWARD-DAVIS INC. (U.S.A.) (1975-76) 466
STEWART AIRCRAFT CORPORATION (U.S.A.) 554
ST-GERMAIN, JEAN (see Centre de Recherches Jean St-Germain Inc) 474
Stinger (Rinker) (1971-72) 393
Stinson 108-5 (Univair) (1980-81) 448
Stinson L-5 (SSVV) (1980-81) 135
STODDARD-HAMILTON (U.S.A.) 555
Stolifter (Conroy) (1972-73) 303
Stolmaster (Steward-Davis/Aircraft International) (1974-75) 469
STOL Transport (Aeritalia/Boeing) (1973-74) 119
STOLP STARDUSTER CORPORATION (U.S.A.) 556
Stork, NM-75 (Nihon University) (1977-78) 501
Strata-Rocket (Maule) (1976-77) 324
Stratofortress (Boeing) 316
Stratotanker, KC-135A (Boeing) 317
Strekoza (Kuybyshev) 500
Stretched Gazelle (Aérospatiale) 50
Strike Eagle, F-15 (McDonnell) 404
Strikemaster (BAe) 245
STRIPLIN AIRCRAFT CORPORATION (USA) 552
STUDENCKIE BURO KONSTRUKCYJNE (Poland) (1974-75) 172
STURGEON AIR LTD (Canada) (1974-75) 29
SUKHOI (U.S.S.R.) 214
SUMMIT AVIATION INC (U.S.A.) 466
Sump'n Else (Hansen) (1976-77) 513
Sunday Knight Twister SKT-125 (Payne) (1980-81) 536
Sundowner 180 (Beechcraft) 279
SUNRISE AIRCRAFT CORPORATION OF AMERICA (U.S.A.) (1971-72) 411
Super 4 Sancy (Wassmer) (1971-72) 81
Super 80 (Douglas) 408
Super 80 SF (Douglas) (1978-79) 380
Super 310 (Riley) 445
Super 340 (Riley) 445
Super 414-8 (Riley) (1976-77) 363
Super 880 (Riley) (1977-78) 381
Super BA-4B (Crosby) (1977-78) 509
Super BA-4B (Crosby/Andreasson) (1980-81) 494
Super J-2 (Aero Resources) (1974-75) 243
Super T-40A (Turner) 558
Super Acro-Sport (EAA) 523
Super Ag-Cat (Gulfstream American) 374
Super Ag-Cat B/450 (Schweizer) 456
Super Ag-Cat C/600 (Schweizer) 456
'Super Baby' Lakes (Barney Oldfield) 510
Super Bug (Bensen) 511
Super Cavalier, SA 105 (Macfam) 476
Super Courier Model H-295 (Helio Courier) (1978-79) 348
Supercritical wing (NASA) (1974-75) 402
Super Cub (Piper) 425
Super Decathlon (Bellanca) (1979-80) 288
Super Diamant (Piel) 490
Super Emeraude (Piel) 489
Super Etendard (Dassault) 60
Super F28 (VFW-Fokker) (1979-80) 146
Super Fli (Kraft) 532
Super Foxbat (MiG) 206
Super Frelon (Aérospatiale) 46
Super Guppy (Aero Spacelines) (1972-73) 233
Super Guppy (Airbus) 98
Super Interceptor (Smith) (1980-81) 441
Super King Air 200/B200 (Beechcraft) 290
Super King Air A100 (Beechcraft) (1980-81) 274
Super King Air F90 (Beechcraft) 288
Super Magister (Aérospatiale) (1971-72) 41
Super Mirage (Dassault) (1975-76) 60
Super Mirage 4000 (Dassault-Breguet) 60
Super Mirage Delta (Dassault) (1977-78) 53
Super Mountaineer (Ector) 348
Supermunk (BGA) 252
Super Mystère B-2 (IAI/Dassault) (1980-81) 102
Super New Look (Starck) 494
Super Pinto (Aeronca) (1974-75) 243
Super Pinto (American Jet Industries) (1973-74) 240
Super Pinto (Guzman) (1972-73) 330
Super Privrednik-350, Utva-65 (UTVA) (1975-76) 526
Super Puma (Nurtanio-Aérospatiale) 91
Super Puma AS 332 (Aérospatiale) 48
Super Quickie Q2 (Quickie) 544
Super Saeta (Hispano) (1973-74) 178
Super Simple 1 (Kinman) 532
Super Sixty Series (Douglas) (1972-73) 363
Super Skymaster (Cessna) (1971-72) 284
Super Sportster, Model 2 (Turner) (1980-81) 554
Super Surubim, IPAI-28 (São Carlos) 18
Super Stallion (Sikorsky) 460
Superstar (Machen) 400
Super Swat S-525 (Schapel) 455
Super Sytky (Mellen) (1975-76) 40
Super Sytky (PIK) (1980-81) 471
Super Trident (Hawker Siddeley) (1977-78) 184
Super Viking 300 (Bellanca) (1980-81) 287
Super Widgeon (McKinnon-Viking) (1980-81) 30
Supporter (Saab) (1979-80) 169
Surubim Modificado, IPT-16 (Rio Claro/IPT) (1979-80) 17
"SURVOL" CHARLES FAUVEL (France) (1979-80) 474
Sverchok-1 (Kuibyshev) (1973-74) 477
Swalesong (Coates) 502
Swallow, STP-1 (Ryson) (1973-74) 428
SWEARINGEN AVIATION CORPORATION (U.S.A.) (1980-81) 443
Swift (Globe/York) (1977-78) 504
Swinger (Van's) (1978-79) 550
SWISS FEDERAL AIRCRAFT FACTORY (Switzerland) 182
Swiss Trainer (Dätwyler) (1974-75) 188
Sylkie One Model B-1 (Wayne) (1980-81) 504
Syuan Fen (China) (1980-81) 37
Szmiel (Kuybyshev) 500

T

T1 Flamingo-Trainer (FUS) (1980-81) 81
T-1B Tucan (AVEX) (1971-72) 3
T-1 Mustang (Timofeyev) (1977-78) 574
T-2 (Mitsubishi) 146
T-2 Buckeye (Rockwell) 452
T2J Buckeye (North American) (1972-73) 377
T-3 (Fuji) 142
T3J-1 Sabreliner (North American) (1972-73) 380
T-11 Cacique (Al-Aire) (1975-76) 2
T-12B (CASA) (1979-80) 165
T-17 (Saab) (1978-79) 159
T-18 Tiger (Thorp) 557
T-19 (Trella) (1972-73) 435
T-23 (Aerotec) (1979-80) 9
T-25 (Neiva) 17
T-25B (Neiva) (1980-81) 19
T-27 (EMBRAER) 15
T-28 (Enstrom) (1971-72) 332
T-34 Turbo Mentor (Marsh/Beechcraft) 401
T-34C (Beechcraft) 278
T-34V Turbotrainer (Volpar/Beechcraft) (1980-81) 451
T-37 (Cessna) (1978-79) 311
T-38 Talon (Northrop) (1972-73) 385
T-39 Sabreliner (Rockwell) (1979-80) 432
T-40A (Turner) 558
T-40A, Super (Turner) 558
T-40B (TEDDE) (1977-78) 567
T-40C (TEDDE) (1979-80) 550
T-41A Mescalero (Cessna) (1978-79) 300
T-41B/C/D Mescalero (Cessna) (1980-81) 308
T-42A Cochise (Beechcraft) 283
T-43A (Boeing) 305
T-44A (Beechcraft) 288
T67 (Slingsby) 263
T.77 (T-40C) (Turner) 559

T207 (Cessna) (1977-78) 273
T210 (Cessna) (1977-78) 274
T-250 Aries (Bellanca) (1980-81) 287
T-300 Skyfarmer (Transavia) 8
T-300A Skyfarmer (Transavia) 8
T-320 Airtruk (Transavia) (1979-80) 8
T337 Pressurised Skymaster (Cessna) (1977-78) 278
T337 Pressurised Skymaster II (Cessna) (1977-78) 278
T-610 Cali (Super Pinto) (PAF/AJI) 157
T-610 Super Pinto (American Jet Industries) (1974-75) 240
T-1020 (Piper) 443
T-1040 (Piper) 443
TA-4 Skyhawk (Douglas) (1980-81) 390
TA-4S Skyhawk conversion (LAS) (1978-79) 358
TA-7 Corsair II (Vought) 469
TA9 (Airbus) 97
TA11 (Airbus) 97
TA12 (Airbus) 98
TAC (see Teal Aircraft Corporation) (1979-80) 27
TAV-8A (BAe) 247
T/AV-8S (BAe) 247
TB 9 Tampico (Socata) 75
TB 10 Tobago (Socata) 75
TB 20 Trinidad (Socata) 76
TB 30 Epsilon (Socata) (1980-81) 76
T-CH-1 Chung Hsing (AIDC) 183
TE-2C Hawkeye (Grumman) 366
TEDDE (see Turner Educational Development Enterprises) (1979-80) 549
TF-1 Tomcat (Tru-Flite) (1978-79) 548
TF-6 (China) (1979-80) 29
TF-8 Crusader (LTV) (1971-72) 342
TF-15 (McDonnell) 403
TF-18A Hornet (McDonnell) 406
TF-26 (EMBRAER) (1972-73) 12
TF-104 Starfighter (Lockheed) (1972-73) 344
TH-1 Iroquois (Bell) (1972-73) 261
TH-13 (Bell) (1971-72) 242
TH-13T (Bell) (1971-72) 242
TH-55A (Hughes) (1977-78) 317
TH-57A SeaRanger (Bell) 294
TH.E.01 Tiger-Hawk (Mini-Hawk) 536
THvK (KIBM) (see Türk Hava Kuvvetleri, Hava Ikmal Bakim Merkezi Komütanglici Kayseri) 185
TIFS (Calspan) 324
TJ-1 Jastreb Trainer (Soko) (1979-80) 460
TK-1 (Kako) (1979-80) 486
TL-1 (Fuji) 142
TL-1A (Birdman) (1980-81) 506
TNT (New Technology Wing) testbed aircraft (Dornier) 78
TO (see Solar-Powered Aircraft Developments) 264
TR-1 (Lockheed) 389
TR-1 Trigull (Trident) 27
Tr-2 (AAC) (1972-73) 237
Tr-2 (Grumman American) (1976-77) 297
TR.12A (CASA) (1979-80) 165
TRS-III (Gyroflug) (1972-73) 89
TS-2 Tracker (Grumman) (1972-73) 324
TS-11 Iskra (WSK-PZL Mielec) 162
TSC-1 "T-boat" (Schweizer) (1974-75) 450
TSC-1 Teal (TAC) (1979-80) 27
TSC-1A1 Teal (Schweizer) (1974-75) 449
TSC-1A2 Teal II (Schweizer) (1976-77) 380
TSC-1A3 Marlin (TAC) (1979-80) 27
TSC-2 Explorer (Thurston/Aero) (1971-72) 416
TSW (Turner) (1980-81) 499
TTI (see Tony Team Industries Inc) (1977-78) 415
Tu-16 (Tupolev) 220
Tu-22 (Tupolev) 223
Tu-22M (Tupolev) 224
Tu-26 (Tupolev) 224
Tu-28P (Tupolev) 225
Tu-95 (Tupolev) 221
Tu-104A (Meteorological) (Tupolev) 225
Tu-114 Military Version (Tupolev) (1972-73) 478
Tu-114 Rossiya (Tupolev) (1973-74) 492
Tu-124 (Tupolev) (1974-75) 517
Tu-126 (Tupolev) 222
Tu-128 (Tupolev) 225
Tu-134 (Tupolev) 226
Tu-142 (Tupolev) 221
Tu-144 (Tupolev) 227
Tu-144D (Tupolev) 228
Tu-154 (Tupolev) 228
Tu-154A/B (Tupolev) 229
TUSAS (see Türk Ucak Sanayii Anonim Ortakligi) 185
Taifun, Me 108F (Taifun) (1975-76) 97
TAIFUN FLUGZEUGBAU GmbH (West Germany) (1976-77) 76
'Taildragger conversions' (Bolen) 321
Tailwind (AJEP/Wittman) (1978-79) 213
Tailwind Model W-8 (Wittman) 565
Tailwind Model W-9 (Wittman) (1979-80) 553
Tailwind Model W-10 (Wittman) (1979-80) 556
TALLERES AL-AIRE SCA (Argentine) (1975-76) 2
Talon (Northrop) (1972-73) 385
Tampico, TB 9 (Socata) 75
Tandem Smoothie (Sawano) (1980-81) 489
Tangará, A-132 (Aerotec) 9
Tangra, Milanov/Vasileva (Bulgaria) (1977-78) 480
Tapajos (EMBRAER) (1978-79) 14
Tarhe (Sikorsky) (1978-79) 433
TaskMaster (Frontier) 355
Tauro 300 (Anahuac) (1980-81) 145
Tauro 350 (Anahuac) 150
Taurus modification (Jetcrafters) 384
Taurus (Spitfire/WSK-PZL Swidnik) 166, 465
Taurus (WSK-PZL Swidnik) 166
TAYLOR, DONALD E. (U.S.A.) (1977-78) 566
TAYLOR, C. GILBERT (U.S.A.) (1980-81) 552
TAYLOR, MRS. JOHN F. (U.K.) 565
TAYLOR AERO INDUSTRIES (U.S.A.) 556
Taylor Monoplane (Taylor) 565
TAYLOR, RALPH (U.S.A.) (1972-73) 434
TAYLORCRAFT AVIATION CORPORATION (U.S.A.) 466
"T-boat" (Schweizer) (1974-75) 450
T-cat (Gulfstream American) (1979-80) 352
Tchaika (Beriev) 191
Teal (Falconar) (1971-72) 23
Teal (Schweizer) (1976-77) 380
Teal (TAC) (1979-80) 27
TEAL AIRCRAFT CORPORATION (Canada) (1979-80) 27
TEAL AIRCRAFT CORPORATION (U.S.A.) (1979-80) 451
TED SMITH AEROSTAR CORPORATION (U.S.A.) (1977-78) 414
Teenie Two (Parker) 559
TEKNILLINEN KORKEAKOULU (Finland) (1975-76) 40
TELEDYNE RYAN AERONAUTICAL (U.S.A.) (1974-75) 474
Tempête (Jurca) 484
TENSA SA (Argentina) 4
Terminal configured vehicle (NASA) 418
Termite (Termite) (1974-75) 474
TERMITE AIRCRAFT (U.S.A.) (1974-75) 474
TERVAMÄKI, JUKKA (Finland) 480
TEXAS AIRPLANE MANUFACTURING COMPANY INC (U.S.A.) (1976-77) 397
TEXAS HELICOPTER CORPORATION (U.S.A.) 467
TexasRanger, Model 206L (Bell) 296
Thomas Monoplane (AVEX) (1972-73) 3
THOMPSON AIRCRAFT (USA) 557
THOMPSON, DR NORMAN (U.S.A.) (1979-80) 552
THOMPSON, JOHN A. (U.S.A.) (1979-80) 551
THORP ENGINEERING COMPANY (U.S.A.) 557
Three-Eleven (BAC) (1971-72) 180
Thrush (Ayres) 276
Thrush-600 (Ayres) 276
Thrush Commander (Ayres) (1978-79) 253
Thrush Commander-600 (Rockwell) (1978-79) 429
Thrush Commander-800 (Rockwell) (1978-79) 429
Thrush Commander/Leonides (Serv-Aero) (1979-80) 439
Thrush Commander/PT6 (Serv-Aero) (1978-79) 430
Thrush Commander/R-1820 (Serv-Aero) 458
Thrush Commander S-2R/Pezetal PZL-3 (Serv-Aero) (1978-79) 430
Thunderbolt, S-1080 (Schapel) 456
Thunderbolt II (Fairchild Republic) 351
Thunderchief (Republic) (1973-74) 326
THURSTON AIRCRAFT CORPORATION (U.S.A.) (1971-72) 415
THYREGOD, HARALD C. (Denmark) (1971-72) 36
Tiara, HR 100/285 (Robin) (1979-80) 60
Tiger (Gulfstream American) (1980-81) 355
Tiger (Thorp) 557
Tiger II (Northrop) 421
Tiger Hawk (Mini-Hawk) 536
Tigershark, F-5G (Northrop) 422
Tijerete AL-2 (Al-Aire) (1974-75) 3
TIMOFEYEV, VIKTOR (U.S.S.R.) (1977-78) 574
Tinker Toy (Taylor) (1977-78) 566
Tiny Hawk (Hill) (1974-75) 355
Tira Prosa, IPAI-29 (São Carlos) 18
Titan Ambassador/II/III (Cessna) 342
Titan Courier/II (Cessna) 342
Titan Freighter/II (Cessna) 342
Titch (Taylor) 506
Tobago, TB 10 (Socata) 75
TODHUNTER, R. W. (Australia) (1979-80) 465
Tomahawk (Piper) 425
Tomahawk II, PA-38-112 (Piper) 438
Tombo (JEAA) (1976-77) 474
Tomcat (Grumman) 368
Tom-Pouss (Koenig) (1980-81) 479
Tomcat, TF-1 (Tru-Flite) (1978-79) 548
TONY TEAM INDUSTRIES INC (U.S.A.) (1977-78) 415
Tornado (MBB) (1971-72) 89
Tornado (Panavia) 104
Tornado ADV (Panavia) 105
Tornado, MY 102 (Mylius) (1980-81) 84
Total In-Flight Simulator (Calspan) (1975-76) 298
Toucan Mk II (HPA) (1979-80) 494
Tourbillon (Chasle) (1979-80) 470
Tourisme, EC-8 (Croses) 482
Tourist (Demchenko/Khitry/Gusey) (1971-72) 435
Towmaster, PIK-23 (Valmet) 43
Tracker (Grumman) (1972-73) 324
Tradewind (PAC) (1971-72) 374
TRAGO MILLS LTD (U.K.) 264
Trainer (AAC) (1972-73) 237
Trainer (Bellanca) (1974-75) 275
Trainer (Grumman American) (1976-77) 297
TRANSAIR SA (Switzerland) (1972-73) 184
TRANSALL, ARBEITSGEMEINSCHAFT (International) 111
TRANSAVIA DIVISION, TRANSFIELD (NSW) PTY LTD (Australia) 7
TRANS-CARIBBEAN AIRCRAFT CO (Bahamas) (1971-72) 9
Transonic Aircraft Technology (NASA) (1979-80) 396
Transporter, DHC-5E (Canada) 23
Travel Air Mystery S replica (Younkin) (1980-81) 550
Traveler (Grumman American) (1975-76) 351
Traveler, WH-1 (Wendt) 564
TRELLA, FRED (U.S.A.) (1972-73) 435
Trempik, SK-1 (Simunek/Kamaryt) 478
Trener (Moravan) (1975-76) 38
Trident (BAe) (1979-80) 233
TRIDENT AIRCRAFT LTD (Canada) 27
Trigear Courier Model HT-295 (Helio Courier) (1978-79) 348
Trigull, TR-1 (Trident) 27
Trimard, RF-21 (Fleury) (1975-76) 68
Trinidad (SOCATA) 76
Trislander, BN-2A Mk III (Pilatus Britten-Norman) 258
TriStar (Lockheed) 393
Tri Turbo 3 (Tri-Turbo) (1979-80) 452
TRI-TURBO CORPORATION (U.S.A.) (1980-81) 447
Tri-Z-CH 300 (Zenair) 478
Trooper (Bell) (1971-72) 271
TRU-FLITE (U.S.A.) (1978-79) 548
Tuca, IPAI-26 (São Carlos) 18
Tucan T-1B (AVEX) (1971-72) 3
Tuholer (Spezio) 552
Tupi, EMB-712 (EMBRAER/Piper) 17
TUPOLEV (U.S.S.R.) 220
TURBAY, ING ALFREDO (Argentina) (1972-73) 5
Turbay T-11 Cacique (Al-Aire) (1974-75) 2
Turbi (Druine) (1975-76) 66
Turbine 404 (Riley) 446
Turbine Eagle 421 (Riley) 446
Turbine Islander (Pilatus Britten-Norman) 258
Turbine Mentor 34C (Beechcraft) 278
Turbine Rocket 421 (Riley) 446
Turbo 18 (Volpar/Beechcraft) 468
Turbo Ag-Cat D (Gulfstream American) (1980-81) 356
Turbo Ag-Cat D (Schweizer) 457
Turbo Air Tractor (Air Tractor) 275
Turbo Albatross (Conroy) (1972-73) 304
Turbo Arrow IV (Piper) 429
Turbo Aztec (Piper) 426
Turbo-Baron (Beechcraft) (1972-73) 248
Turbo Bonanza, A36TC (Beechcraft) 282
Turbo-Cat (Frakes/Grumman) (1978-79) 329
Turbo Cat, G-164 C-T (Marsh/Gulfstream American) 401
Turbo Centurion/Turbo Centurion II (Cessna) 333
Turbo Comanche C (Piper) (1973-74) 401
Turbo Commander 618B (Rockwell) (1975-76) 445
Turbo Commander 690B (Rockwell) (1979-80) 435
Turbo Dakota (Piper) 430
Turbo-Equator, P-400 (Equator) (1976-77) 70
Turbo Equator, (Equator) 80
Turbo-Exec 400 (Riley) (1975-76) 433
Turbofan Aerostar (Smith) (1975-76) 472
Turbo-Goose, G-21G (McKinnon-Viking) (1980-81) 29
Turbo-Kruk, PZL-106AT (CNPSL-PZL Warszawa) 159
Turbo Lance II (Piper) (1980-81) 408
Turbolet (Let) 39
Turboliner (Volpar/Beechcraft) 468
Turboliner II (Volpar/Beechcraft) 469
Turboliner IIA (Volpar) (1974-75) 479
Turbo Mallard (Frakes) (1975-76) 336
Turbo Mentor (Beechcraft) (1977-78) 220
Turbo Mentor (Marsh/Beechcraft) 401
Turbo Mooney 231 (Mooney) 417
Turbo Mustang III (Cavalier) (1971-72) 271
Turbo Navajo (Piper) (1976-77) 354
Turbo Navajo C/R (Piper) (1976-77) 355
Turbo Otter, DHC-3-T (Cox) (1980-81) 23
Turbo-Porter (Pilatus) 180
Turbo-Porter Agricultural (Pilatus) 181
Turboprop Beaver (Volpar/DHC) (1974-75) 479
Turboprop Goose Conversion (McKinnon-Viking) (1980-81) 30
Turbo Saratoga (Piper) 431
Turbo Saratoga SP (Piper) 432
Turbo Seminole (Piper) 441
Turbo Single Otter (Cox) (1978-79) 21
Turbo Skylane/Skylane II (Cessna) 328
Turbo Skylane RG/RG II (Cessna) 328
Turbo Skyliner (Riley) (1974-75) 433
Turbo-Skymaster/Skymaster II (Cessna) (1980-81) 319
Turbo-Skyservant, Do 128-6 (Dornier) 78
Turbo-Skywagon (Cessna) (1977-78) 273
Turbo Star 402 (Scenic/AJI) (1979-80) 438
Turbo Star Baron (AJI) (1973-74) 240
Turbo Star Pressurised 414 (Scenic/AJI) (1977-78) 398
Turbo-Stationair 6 (Cessna) 331
Turbo-Stationair 8 (Cessna) 332
Turbostream (Riley) (1975-76) 432
Turbo T310/T310 II (Cessna) 335
Turbo Three (Conroy) (1972-73) 302
Turbo-Thrush S-2R (Ayres) 277
Turbo-Thrush, S2R-T (Marsh/Rockwell) 400
Turbotrainer (Volpar/Beechcraft) (1980-81) 451
Turbo Trainer, AS 32T (FWA) 180
Turbo-Trainer, PC-7 (Pilatus) 181
Turbo Trigull (Trident) (1980-81) 31
Turbo Twin Comanche C/R (Piper) (1974-75) 429
Turbo Viking (Bellanca) (1979-80) 286
Turbulent (Druine) 504
Turbulent (Rollason) (1974-75) 224
TÜRK HAVA KUVVETLERI, HAVA IKMAL BAKIM MERKEZI KOMÜTANGLICI KAYSERI (Turkey) 185
TURKISH AIR FORCE (Turkey) 185
TURK UCAK SANAYII ANONIM ORTAKLIGI (Turkey) 185
TURNER AIRCRAFT (USA) 558
TURNER, BILL (USA) 559
TURNER, CHRISTOPHER (UK) (1980-81) 499
TURNER EDUCATIONAL DEVELOPMENT ENTERPRISES (U.S.A.) (1979-80) 549
TURNER, EUGENE L. (U.S.A.) (1975-76) 473
TURNER, ROYCE (U.S.A.) (1977-78) 569
Twin Cat (Gulfstream American/Twin Cat) (1980-81) 447
TWIN CAT CORPORATION (U.S.A.) (1980-81) 447
Twin Comanche C/R (Piper) (1974-75) 429

Twin Courier H-580 (Helio) (1971-72) 320
Twin-Engined Trainer (FMA) (1975-76) 5
Twin Equator, P-420 (Equator) 80
Twin Equator, P-420 (Equator) 80
Twin Hercules (Lockheed) (1980-81) 377
Twin Navion (Camair) (1980-81) 306
Twin Otter (DH Canada) 24
Twin Porter (Poligrat) (1975-76) 95
Twinstar (Aérospatiale) 51
Twin Two-Twelve (Bell) 299
Type 02-337 Sentry (Summit) 466
Type 05, Ganagobie (ARDC of NZ) 498
Type 1 (OMAC) 424
Type 1 Firecracker (NDN) 254
Type 1-2 Mister America (Harmon) 526
Type 2 + 2, DR 400/100 (Robin) (1980-81) 66
Type 5 (NDN) 255
Type 5 Huongzhazi (China) 32
Type 5 Qianjiji (China) 31
Type 5 Yunshuji (China) 34
Type 5 Zhishengji (China) 34
Type 6 Fieldmaster (NDN) 255
Type 6 Huongzhazi (China) 33
Type 6 Jianjiji (China) 29
Type 7 Jianjiji (China) 32
Type 8 Jianjiji (China) 32
Type 10 Yunshuji (China) 34
Type 11 Yunshuji (China) 35
Type 12X Series (EMBRAER) (1977-78) 15
Type 66-400 (HAL) 90
Type 66-600 (HAL) 90
Type 74 (HAL) 90
Type 75 (UTVA) 472
Type 76 (HAL) 90
Type 77 (HAL) 90
Type 88 (HAL) (1976-77) 83
Type 96 (HAL) 90
Type 145 Jet Provost (BAC) (1973-74) 193
Type 146 (BAe) 241
Type 146M (BAe) 243
Type 201 (Mooney) 416
Type 201-M (Myasishchev) (1972-73) 470
Type 231 (Mooney) 416
Type 330 (Shorts) 260
Type 330-200 (Shorts) 261
Type 340 (Saab-Fairchild) 106
Type 360 (Shorts) 261
Type 369/369D (Kawasaki/Hughes) 145
Type 748 (HAL/Hawker Siddeley) (1977-78) 84
Type 748 (Hawker Siddeley) (1977-78) 181
Type 941S (Breguet) (1973-74) 61
Type 1000 (Rockwell) (1980-81) 433
Type 1124 Sea Scan (IAI) 116
Type 1150 Atlantic (Breguet) (1974-75) 62
Type 1182 (Hawker Siddeley) (1977-78) 186
Type 2150A-TG Kachina (Hibbard/Varga) 377

U

U-2 (Lockheed) 389
U-2 (Mitchell) (1980-81) 531
U-2 (NASA/Lockheed) 419
U-3 (Cessna) (1978-79) 310
U-8F Seminole (Beechcraft) (1972-73) 250
U-10 (Helio) (1977-78) 315
U-11A Aztec (Piper) (1975-76) 418
U-17 Skywagon (Cessna) 329
U-19 (EMBRAER) (1980-81) 17
U-21F (Beechcraft) 289
U-42 Regente (Neiva) (1974-75) 18
U-60 (Utva) (1971-72) 472
UAC (see United Aircraft Corporation) (1974-75) 476
UC-12B Super King Air (Beechcraft) 290
UCA-1 (James-Dennis) (1979-80) 524
UFM (see Ultralight Flying Machines) (1980-81) 555
UH-1 Iroquois (Bell) 293
UH-1B (Fuji-Bell) (1974-75) 144
UH-1 D (Dornier/Bell) (1971-72) 86
UH-1D/H (Bell) 293
UH-1H (Fuji-Bell) (1977-78) 127
UH-1 HELMS (Bell) (1974-75) 268
UH-1N (Bell) 299
UH-2 Seasprite (Kaman) 386
UH-12 (Hiller) 378
UH-13 Trooper (Bell) (1971-72) 242
UH-14 Lynx (Westland/Aérospatiale) (1977-78) 210
UH-46 (Boeing Vertol) 317
UH-46A (Boeing Vertol) (1973-74) 283
UH-60A Black Hawk (Sikorsky) 461
UP-2J (Kawasaki) 143
US-1 (Shin Meiwa) 148
US-2 Tracker (Grumman) (1972-73) 324
US-3A Viking (Lockheed) (1978-79) 363
UTC (see United Technologies Corporation) (1980-81) 448
UV-18 (DH Canada) 24
UV-20 Chiricahua (Pilatus) 180
Uirapuru (Aerotec) (1980-81) 10
Uirapuru II, A-132 (Aerotec) (1980-81) 10
Ukraina (Antonov) (1973-74) 463
ULTRALIGHT FLYING MACHINES (U.S.A.) (1980-81) 555
UMBRA, AERONAUTICA (Italy) (1972-73) 137
UNITED AIRCRAFT CORPORATION (U.S.A.) (1974-75) 476
UNITED TECHNOLOGIES CORPORATION (U.S.A.) (1980-81) 448
UNIVAIR AIRCRAFT CORPORATION (U.S.A.) (1980-81) 448
Universal (Neiva) 17
Universal, Z 726 (Moravan) (1978-79) 35
Universal II, Model 622 (Nieva) 17
UNIVERSITY OF LIVERPOOL, DEPARTMENT OF MECHANICAL ENGINEERING (U.K.) (1974-75) 218
Update (Lockheed) 390
Up-gun AH-1S (Bell) 298
URDANETA Y GALVEZ LTDA (Colombia) (1978-79) 30
URMSTON, Dr J. H. B. (U.K.) (1975-76) 231
Utililiner (Cessna) 337
UTVA-SOUR METALNA INDUSTRIJA, RO FABRIKA AVIONA (Yugoslavia) 472
Utva-60 (Utva) (1971-72) 472
Utva-65 Privrednik-GO (Utva) (1972-73) 489
Utva-65 Privrednik-IO (Utva) (1972-73) 489
Utva-65 Super Privrednik-350 (UTVA) (1975-76) 527
Utva-66 (Utva) (1975-76) 527
Utva-70 (Utva) (1971-72) 473
Utva-75 (Utva) 472
Utva P-418/71 (UTVA) (1975-76) 528

V

V-2 (Mil) 207
V-8 (Mil) 208
V-10 (Mil) 209
V-12 (Mil) (1978-79) 190
V-14 (Mil) 210
V35B Bonanza (Beechcraft) 280
V35B TC Bonanza (Beechcraft) (1971-72) 228
V-519 (LTV) (1972-73) 356
V-529D (Vought) (1979-80) 458
V-530 (Vought) (1978-79) 449
V-539 Eaglet (MBB) 83
VAK 191 B (VFW-Fokker) (1974-75) 100
VC-6B King Air (Beechcraft) 286
VC-9C (Douglas) 408
VC10 K.Mk 2/Mk 3 Tankers (BAe) 244
VC-11A (Gulfstream American) (1980-81) 354
VC-137 (Boeing) 304
VC-140 JetStar (Lockheed) (1974-75) 377
VC-400 (VFW-Fokker) (1972-73) 98
VCA-1 (James-Dennis) (1980-81) 524
VFW 614 (VFW-Fokker) (1978-79) 72
VG-2C (Vertidynamics) (1972-73) 437
VG-2P (Vertidynamics) (1972-73) 437
V-G Bomber (Tupolev) (1977-78) 462
V-G interdictor (Mikoyan) (1976-77) 423
VH-3 (Sikorsky) 458
VH-53 (Sikorsky) (1976-77) 387
VJ-22 Sportsman (Volmer) 561
VM-7 Competitor (Murphy) 537
Vo Wi 8 (Wilden) (1974-75) 101
VoWi 10 Airbuggy (Airconcept) (1980-81) 77
Vo Wi 10 (Wilden) (1977-78) 77
VP-1 (Evans) 524
VP-1 (Yamazaki) (1979-80) 488
VP-2 (Evans) 524
VSA (Kawasaki) 143
VTI/CIAR (International) (1977-78) 98
VTOL Flying Testbed (NAL) (1975-76) 152
VU-9 (EMBRAER) 13
VW Twin (Lacey) (1971-72) 329
VZLU (see Vyzkumny a Zkusebni Letecky Ustav) 40
VALMET OY KUOREVESI WORKS (Finland) 42
VAN ASWEGEN, C. H. J. (South Africa) 498
VAN DINE, PETER D (U.S.A.) 559
Vanguard Merchantman Cargo Conversion (Aviation Traders) (1972-73) 187
VAN'S AIRCRAFT (U.S.A.) 560
Vansin (Van Aswegen) 498
VARGA AIRCRAFT CORPORATION (U.S.A.) 467
Variable Stability Aircraft (Kawasaki) 143
VariEze (Rutan) 546
VariViggen (Rutan) 546
Veltro 2, M.B.339K (Aermacchi) 123
Venom (F + W) (1974-75) 192
Ventura (Neiva) (1974-75) 19
VEREINIGTE FLUGTECHNISCHE WERKE-FOKKER GmbH (West Germany) 86
VERNER, VLADA (Czechoslovakia) (1972-73) 32
VERTAK CORPORATION (U.S.A.) (1974-75) 477
VERTIDYNAMICS CORPORATION (U.S.A.) (1972-73) 437
Vertigiro VG-2P (Vertidynamics) (1972-73) 437
VFW-FOKKER GmbH (West Germany) (1979-80) 80
Victor (Partenavia) 138
VIET-NAM AIR FORCE (Viet-Nam) (1974-75) 525
Viggen (Saab) 177
Vigilante (North American) (1972-73) 377
VIKING AIRCRAFT (U.S.A.) 560
Viking (Lockheed) 393
Viking 300 (Bellanca) (1972-73) 249
Viking series (Bellanca) (1979-80) 286
VIN-DEL AIRCRAFT (U.S.A.) 539
Vinka, Leko-70 (Valmet) 42
Viscount Freighter Conversion (Scottish Aviation) (1971-72) 207
Vliôm (Vliôm Development) (1979-80) 163
VLIÔM DEVELOPMENT CO (PTY) LTD (South Africa) (1979-80) 163
Volksplane (Cunning) (1974-75) 322
VOLMER AIRCRAFT (U.S.A.) 561
VOLPAR, INC (U.S.A.) 467
VOUGHT CORPORATION (U.S.A.) 469
Voyageur (Boeing Vertol) (1973-74) 283
Voyageur, CH-113A (Boeing Canada) 18
VPM SnC (Italy) 496
V-Star, SA-900 (Stolp) 556
VTOL AIRCRAFT PTY LTD (Australia) (1979-80) 9
VTOL research aircraft (NAL) (1977-78) 135
VYZKUMNY A ZKUSEBNI LETECKY USTAV (Czechoslovakia) 40

W

W-01 Broucek (Verner) (1972-73) 32
W-2 (Williams) (1973-74) 460
W2F-1 Hawkeye (Grumman) (1973-74) 336
W-3 Sokol (WSK-PZL-Swidnik) 167
W-7, Special (Westfall) (1977-78) 572
W-8 Tailwind (Wittman) 565
W-9 Tailwind (Wittman) (1979-80) 553
W-10 Tailwind (Wittman) (1979-80) 556
W-17 (Williams) (1976-77) 543
W-18 Falcon (Williams) (1976-77) 543
WA-40A Super 4 Sancy (Wassmer) (1971-72) 81
WA-41 Baladou (Wassmer) (1971-72) 81
WA 51 Pacific (Wassmer) (1976-77) 66
WA 52 Europa (Wassmer) (1977-78) 66
WA 54 Atlantic (Wassmer) (1977-78) 66
WA 70 (Wassmer) (1975-76) 88
WA 80 Piranha (Wassmer) (1977-78) 67
WA-116 (Wallis) 265
WA-116-T (Wallis) 265
WA-117/R-R (Wallis) 265
WA-118/M Meteorite (Wallis) 265
WA-119 Imp (Wallis) (1974-75) 228
WA-120/R-R (Wallis) 266
WA-121 (Wallis) 266
WA-121/M Meteorite 2 (Wallis) 266
WA-122/R-R (Wallis) 266
WAR (see War Aircraft Replicas) 562
WASP (Williams) (1976-77) 403
WC-1 (Williams/Cangie) (1976-77) 543
WC-130 Hercules (Lockheed) 395
WD-II (Hovey) 528
WE 01 (Westermayer) (1980-81) 465
WE 02 (Westermayer) (1980-81) 465
WE 03 (Westermayer) (1980-81) 465
WE 04 (Westermayer) 473
WFI (see World Flight Incorporated) (1972-73) 445
WG-30 (Westland) 271
WG-34 (Westland) (1979-80) 251
WGM 21 (Aerotechnik) (1973-74) 87
WGM 22 (Aerotechnik) (1975-76) 88
WH-1 Traveler (Wendt) 564
WHE (see W. H. Ekin (Engineering) Ltd) (1978-79) 248
WM-2 (Miller) 535
WP-3 Orion (Lockheed) 390
WPB-1 Plank (Backstrom) (1977-78) 521
W.S.4A Mini-Copter (Seremet) (1973-74) 39
W.S.7 Mini-Gyroglider (Seremet) (1973-74) 40
W.S.8 (Seremet) (1978-79) 470
W.S.9 Auto-Gyrocopter (Seremet) (1980-81) 471
W.S.12 (Seremet) (1980-81) 471
WSK-PZL MIELEC (see Wytwornia Sprzetu Komunikacyjnego-PZL Mielec) 161
WSK-PZL OKECIE (see Wytwornia Sprzetu Komunikacyjnego-PZL Okecie) (1976-77) 145
WSK-PZL SWIDNIK (see Wytwornia Sprzetu Komunikacyjnego Im. Zygmunta Pulawskiego-PZL Swidnik) 165
WU-2 (Lockheed) (1980-81) 369
WW-I Der Jäger D.IX (White) 564
WACO AIRCRAFT COMPANY, A SUBSIDIARY OF ALLIED AERO INDUSTRIES, INC (U.S.A.) (1971-72) 420
Waco Meteor (SIAI-Marchetti) (1980-81) 132
Wag-a-bond (Wag-Aero) 562
WAG-AERO INC (U.S.A.) 561
WAGNER, HELICOPTER TECHNIK (West Germany) (1971-72) 98
Wallaroo 605 (Aerial Agriculture) (1972-73) 5
Wallbro monoplane replica (Wallis) (1978-79) 242
WALLERKOWSKI, HEINZ (West Germany) 495
WALLIS, STANLEY B (U.S.A.) 562
WALLIS AUTOGYROS LTD. (U.K.) 264
WAR AIRCRAFT REPLICAS (U.S.A.) 562
WARD, M. (U.K.) (1972-73) 226
Warrior (SIAI-Marchetti) 130
Warrior II (Piper) 427
WARWICK, WILLIAM (U.S.A.) (1973-74) 457
Wasp (Texas Helicopter) 467
Wasp (Westland) (1975-76) 238
WASSMER AVIATION, SOCIÉTÉ NOUVELLE (France) (1978-79) 64
WATSON, GARY (U.S.A.) (1979-80) 554
WATSON WINDWAGON COMPANY (U.S.A.) 563
WEATHERLEY AVIATION COMPANY INC (U.S.A.) 470
WEBER, WILLIBALD (Brazil) 473
Weedhopper, JC-24 (Weedhopper) (1980-81) 558
WEEDHOPPER OF UTAH INC (U.S.A.) (1980-81) 558
Week-End, HN 600 (Nicollier) 489
WEEKS, KERMIT (U.S.A.) 563
WELSH, GEORGE T. (U.S.A.) (1973-74) 458
Welsh Rabbit (Welsh) (1973-74) 458
WENDT AIRCRAFT ENGINEERING (U.S.A.) 564
WESTAIR INTERNATIONAL INC (U.S.A.) (1972-73) 442
WESTERMAYER, OSKAR (Austria) 473
WESTERN AIRCRAFT SUPPLIES (Canada) 476
WESTFALL, MILES (U.S.A.) (1977-78) 572
WESTINGHOUSE ELECTRIC CORPORATION (U.S.A.) (1972-73) 442
Westland 30 (Westland) 271
WESTLAND AIRCRAFT, LTD. (U.K.) 266
WESTLAND HELICOPTERS LTD (U.K.) 266
Westland Whirlwind Mk II (Butterworth) 516
Westwind 1, IAI 1124 (IAI) 115
Westwind 2, IAI 1124A (IAI) 117
Westwind IISTD (Hamilton) 376
Westwind III (Hamilton) 375
Westwind IV (Hamilton) (1980-81) 358
WEYBRIDGE MAN POWERED AIRCRAFT GROUP (U.K.) (1975-76) 238

WEYBRIDGE MPAG (see Weybridge Man Powered Aircraft Group) ....(1975-76) 238
Whing Ding II (Hovey) 528
Whirlymite (Del Mar) ....(1972-73) 308
Whirlymite Helicopter Pilot Trainer (Del Mar) (1972-73) 309
Whirlymite Scout (Del Mar) ....(1972-73) 308
WHITCRAFT CORPORATION (U.S.A.) (1973-74) 459
WHITE, E. MARSHALL (U.S.A.) 564
WHITTAKER, MICHAEL W. J. (U.K.) (1979-80) 500
Wichawk (Javelin) 530
WICKHAM, J. (U.S.A.) ....(1972-73) 442
WIER, RONALD (U.S.A.) ....(1974-75) 480
WILDEN, ING HELMUT (West Germany) (1978-79) 72
Wild Turkey, JP001 (American Air Racing) (1978-79) 504
Wilga 35, PZL-104 (CNPSL-PZL Warszawa) 159
Wilga 80, PZL-104 (CNPSL-PZL Warszawa) 158
WILLIAMS, ARTHUR L. (U.S.A.) .....(1975-76) 479
WILLIAMS, BOB (U.S.A.) ....(1973-74) 460
WILLIAMS AIRCRAFT DESIGN COMPANY (U.S.A.) ....(1976-77) 543
WILLIAMS RESEARCH CORPORATION (U.S.A.) ....(1976-77) 403
Willie II (Shober) ....(1976-77) 532
WINDECKER INDUSTRIES INC (U.S.A.) (1975-76) 480
WINDSTAR (USA) ....(1978-79) 552
Windwagon, GW-1 (Watson) 563
Windy (Jurca) ....(1977-78) 492
WING AIRCRAFT COMPANY (U.S.A.) 471
Wing Charmer Model 1 (Osadchy) ....(1976-77) 523
Wingtip sails (Cranfield) ....(1980-81) 239
Winner (O'Neill) ....(1973-74) 391
WITTMAN, S. J. (U.S.A.) 565
Witt's V. (Wittman) ....(1974-75) 483
WOOD, STANLEY (U.S.A.) ....(1976-77) 544
Wooden Baby (Hast) ....(1975-76) 354
Woody Pusher (Aerosport) ....(1978-79) 502
WORLD FLIGHT INCORPORATED (U.S.A.) ....(1972-73) 445
WOTRING, MELVIN A. (U.S.A.) ......(1974-75) 483
Wren 460 (Wren) ....(1971-72) 425
WREN AIRCRAFT CORPORATION (U.S.A.) ....(1972-73) 446
WRIGHT, PETER (U.K.) ....(1975-76) 238
Wroclaw (Dobraczynski) ....(1974-75) 170
WYTWORNIA SPRZETU KOMUNIKACY-JNEGO-PZL MIELEC (Poland) 161
WYTWORNIA SPRZETU KOMUNIK-ACYJNEGO-PZL OKECIE (Poland) .(1976-77) 145
WYTWORNIA SPRZETU KOMUNIKACY-JNEGO IM. ZYGMUNTA PULAWSK-IEGO-PZL SWIDNIK (Poland) 165

X

X-1 Skeeter (Sheffield) ....(1974-75) 450
X-22A (Calspan) 323
X-24A (Martin Marietta) ....(1972-73) 356
X-24B (Martin Marietta) ....(1975-76) 382
X-25 (Bensen) ....(1977-78) 526
X-28A Air Skimmer (Osprey) ....(1974-75) 410
X-113 Am Aerofoil Boat (RFB) ....(1974-75) 98
X-114 Aerofoilcraft (RFB) ....(1978-79) 70
XAT-3 (AIDC) 184
XC-1 (NAMC) ....(1972-73) 147
XC-2 (AIDC) 184
XCH-62 (Boeing Vertol) ....(1975-76) 292
XD 110C (AMR) ....(1975-76) 243
XF4H-1 Phantom II (McDonnell) ....(1980-81) 380
XF-8 Crusader (LTV) ....(1971-72) 342
XF8U Crusader (LTV) ....(1971-72) 342
XF-104 Starfighter (Lockheed) ....(1972-73) 344
XFV-12A (Rockwell) 454
XH-51 (Lockheed) ....(1971-72) 334
XH-59A (Sikorsky) 461
XL-13T, Milanov (Bulgaria) ....(1977-78) 480
XMC Research Aircraft (Cessna) ....(1974-75) 301
XMH (Fuji) ....(1974-75) 144
XR 200 (AMR) ....(1975-76) 243
XR 207 Rotorplane (AMR) ....(1975-76) 243
XS-1 Ol' Ironsides (Scott) ....(1971-72) 397
XT-001 (PAF) ....(1979-80) 148
XT-2 (Mitsubishi) ....(1972-73) 146
XT-37 (Cessna) ....(1972-73) 282
XT-400 (LAPAN) 90
XT-CH-1A (AIDC) ....(1974-75) 193
XV-15 (Bell) 302
Xavante (EMBRAER/Aermacchi) 15
X-Avia LC-3 (Lewis) 388
Xian (Mikoyan) J-7 (F-7) (China) 32
Xian (Tupolev) H-6 (B-6) (China) 33
Xingu (EMBRAER) 13
Xingu II, EMB-121B (EMBRAER) 14
XR Learjet (Dee Howard) 347
Xuanfeng (China) 34
Xyla (Shield) ....(1974-75) 228
Xyphi (Shield) ....(1974-75) 228

Y

Y-2 (York) ....(1971-72) 477
Y-5 (China) 34
Y-10 (China) 34
Y-11 (China) 35
YA-7 Corsair II (Vought) 469
YA-7E Corsair II₂ (Vought) ....(1980-81) 453
YAC-1DH Caribou (D. H. Canada) .....(1974-75) 22
YAH-63 (Bell) ....(1977-78) 245
YAH-64 (Hughes) 382
Yak-18 (Yakovlev) ....(1972-73) 483
Yak-18T (Yakovlev) 229
Yak-25/26/27 (Yakovlev) ....(1971-72) 467
Yak-28 (Yakovlev) 230
Yak-36MP (Yakovlev) 231
Yak-40 (Yakovlev) 232
Yak-42 (Yakovlev) 233
Yak-50 (Yakovlev) 234
Yak-52 (IAv Bacau/Yakovlev) 169
Yak-52 (Yakovlev) 234
YAT-37D (Cessna) ....(1972-73) 294
YAV-8B (McDonnell) 405
YC-12 Tourbillon (Chasle) ....(1979-80) 470
YC-14 (Boeing) ....(1979-80) 301
YC-15 (Douglas) ....(1979-80) 392
YC-20 Raz de Marée (Chasle) ....(1972-73) 49
YC-119K (Fairchild Industries) ....(1973-74) 324
YC-121 (Chasle) ....(1979-80) 470
YC-122 (Chasle) ....(1979-80) 470
YC-123 (Chasle) ....(1979-80) 470
YC-130 Hercules (Lockheed) ....(1972-73) 350
YC-141B (Lockheed) ....(1977-78) 338
YCH-47D Chinook (Boeing Vertol) 318
YCH-53 (Sikorsky) 460
YCH-54A (Sikorsky) ....(1978-79) 433
YE-5 (Windecker) ....(1975-76) 480
YF-12A (Lockheed) ....(1975-76) 368
YF-12 Programme (NASA) ....(1979-80) 395
YF-16 (General Dynamics) 361
YF-17 (Northrop) ....(1976-77) 346
YF-80 (Windstar) ....(1978-79) 552
YF-105 Thunderchief (Republic) ....(1973-74) 326
YF-111A (General Dynamics) ....(1980-81) 345
YHC-1B Chinook (Boeing-Vertol) ....(1975-76) 290
YL-42 (Neiva) ....(1973-74) 16
YO-3A (Lockheed) ....(1972-73) 354
YOV-1 Mohawk (Grumman) ....(1972-73) 326
YOV-10 (Rockwell) ....(1980-81) 429
YP-3 Orion (Lockheed) 390
YP3V Orion (Lockheed) 390
YS-11 (NAMC) ....(1974-75) 152
YSH-2E (Kaman) ....(1977-78) 322
YSH-3J (Sikorsky) ....(1978-79) 431
YT-17 (Aerotec) 9
YT-25B (Neiva) 17
YT-34C (Beechcraft) ....(1974-75) 252
YUH-60 (Sikorsky) 461
YUH-61A (Boeing Vertol) ....(1976-77) 251
YX Programme (CTDC) ....(1979-80) 127
YAGER, KARL (Australia) ....(1976-77) 455
YAKOVLEV (U.S.S.R.) 229
Yakstas Racer (AVEX) ....(1977-78) 478
YAMATO, SAWAZO (Japan) ....(1980-81) 490
YAMAZAKI, TAKASHI (Japan) ....(1979-80) 488
Yankee (AAC) ....(1972-73) 236
YORK, T. R. (South Africa) ....(1977-78) 504
YOUNKIN, JIM (USA) ....(1980-81) 560
You-Too J12 (Franks) ....(1977-78) 540

Z

Z-6 (H-6) (China) 34
Z-14 (Bell) ....(1980-81) 280
Z-16 (Bell) ....(1979-80) 280
Z-37A Cmelák (Let) ....(1976-77) 30
Z-37A-2 (Let) ....(1976-77) 30
Z 50 L (Moravan) ....(1977-78) 35
Z-237 (Let) ....(1974-75) 36
Z 526 AFS Akrobat (Moravan) ....(1975-76) 39
Z 526 F Trener (Moravan) ....(1976-77) 33
Z 726 Universal (Moravan) ....(1978-79) 35
ZB-1 Olympian (Zinno) ....(1977-78) 573
ZAIDAN HOZIN MINKAN YUSOOKI KAIHATSU KYOKAI (Japan) ....(1979-80) 127
Zef, C.P.80 (Piel) ....(1979-80) 480
ZENAIR LTD (Canada) 27, 477
Zenith-CH 200 (Zenair) 477
Zénith-CH 250 (Zenair) 477
Zenith-CH 400 (Zenair) 28
ZENTRALGESELLSCHAFT VFW-FOKKER mbH (International) ....(1979-80) 101
ZINNO, JOSEPH A., & ASSOCIATES (U.S.A.) (1977-78) 573
Zipper HH-1 (Hammer-Hunt) ....(1980-81) 358
ZJEDNOCZENIE PRZEMYSLU LOTNICZEGO I SILNIKOWEGO PZL (Poland) 157
ZLIN (see Moravan Národni Podnik) 40
Zlin 42 M (Moravan) 40
Zlin 42 MU (Morovan) 40
Zlin 43 (Moravan) ....(1978-79) 34
Zlin 142 (Moravan) 40
Zlin Z 50L (Moravan) 41
Zlin Z 526 AFS Akrobat (Moravan) .....(1975-76) 39
Zlin Z 526 Trener (Moravan) ....(1976-77) 33
Zlin Z 526 L (Moravan) ....(1973-74) 39
Zlin Z 726 Universal (Moravan) ....(1978-79) 35

SAILPLANES

A

A-10 (Caproni Vizzola) (1972-73) 532
A-12 (Caproni Vizzola) (1972-73) 532
A-14 (Caproni Vizzola) (1973-74) 562
A-15 (Antonov) (1974-75) 606
A-15 (Caproni Vizzola) (1973-74) 562
A-20 (Caproni Vizzola) (1972-73) 532
A-21 (Caproni Vizzola) (1976-77) 573
A-21J (Caproni Vizzola) (1977-78) 602
A-21S (Caproni Vizzola) 584, 600
A-21SJ (Caproni Vizzola) 584, 600, 604
AA FLIGHT SYSTEMS (USA) (1979-80) 599
AFH 22 (Akaflieg Hannover) 576
AK-1 (Akaflieg Karlsruhe) (1973-74) 546
AK 2 (Akaflieg Karlsruhe) 576, 598, 604
ALPLA (see Alpla-Werke Alwin Lechner OHG) (1979-80) 554
AN-21 (Neukom) (1980-81) 580
AN-66 (Neukom) (1973-74) 574
AN-66-2 (Neukom) (1973-74) 574
AN-66C (Neukom) (1979-80) 577, 590
AR 124 (Ahrens) (1978-79) 576, 586
AS (see Aero Saladillo) (1979-80) 557
ASK 13 (Schleicher) 582, 600
ASK-14 (Schleicher) (1973-74) 556
ASK 15-B (Schleicher) (1975-76) 582
ASK 16 (Schleicher) (1977-78) 598
ASK 18 (Schleicher) (1977-78) 598
ASK 21 (Schleicher) 582, 600
ASK 23 (Schleicher) 582, 600
AS-W12 (Schleicher) (1972-73) 525
ASW 15 B (Schleicher) (1978-79) 567
ASW 15 M (Schleicher) (1977-78) 598
ASW 17 Super Orchidee (Schleicher) (1978-79) 567
ASW 19 Club (Schleicher) 582
ASW 19B (Schleicher) 582, 600
ASW 20 (Centair/Schleicher) 572, 598
ASW 20 (Schleicher) 582, 600
ASW 20F (Centrair/Schleicher) 582
ASW 20 L (Schleicher) 582, 600
ASW 22 (Schleicher) 582, 600
ATS-1 Ardhra (Civil Aviation Department) 584, 600
AV.45 (Fauvel) (1979-80) 561, 586
AV.48 (Fauvel) (1976-77) 554
AV.50 Lutin (Fauvel) (1973-74) 542
AV.221 (Fauvel) (1974-75) 567
AV.222 (Fauvel) (1979-80) 561, 586
AV.361 (Fauvel) (1979-80) 561
AV.451 (Fauvel) (1979-80) 561
AV.451 (Gross/Fauvel) 574, 598, 604
AVEX (see Asociacion Argentina de Constructores de Aviones Experimentales) (1978-79) 554
AVO 68V Samburo (Alpla) (1979-80) 557
Acroracer (Glaser-Dirks) 576
AÉRO CLUBE DE RIO CLARO (Brazil) 568
AEROCLUB VOLOVELISTICO MILANESE SEZIONE SPERIMENTALE VOLO A VELA (Italy) (1971-72) 509
AEROCLUBE DO PARANÁ (Brazil) (1972-73) 508
AEROFIBRA (see Altinger, Dipl Ing Theo) (1974-75) 558
AERONAUTICA UMBRA SpA (Italy) (1972-73) 534
AERO SALADILLO (Argentina) (1979-80) 557
AERO TEK (USA) (1979-80) 578
AHRENS AIRCRAFT CORPORATION (U.S.A.) (1979-80) 578
Aiglon J.P. 15-36A/AR (CARMAM) 572, 596
Aiglon (Jacquet) (1975-76) 568
AIRCRAFT FACULTY OF ANTONINA ZAPOTOCKEHO MILITARY ACADEMY, BRNO (Czechoslovakia) (1973-74) 539
AKADEMISCHE FLIEGERGRUPPE BERLIN EV (West Germany) (1978-79) 560
AKADEMISCHE FLIEGERGRUPPE BRAUNSCHWEIG EV (West Germany) 575
AKADEMISCHE FLIEGERGRUPPE DARMSTADT EV (West Germany) 575
AKADEMISCHE FLIEGERGRUPPE HANNOVER EV (West Germany) 576
AKADEMISCHE FLIEGERGRUPPE KARLSRUHE EV (West Germany) 576
AKADEMISCHE FLIEGERGRUPPE MÜNCHEN (West Germany) 576
AKADEMISCHE FLIEGERGRUPPE MUNSTER (West Germany) (1973-74) 547
AKADEMISCHE FLIEGERGRUPPE STUTTGART EV (West Germany) 576
AKAFLIEG BERLIN (see Akademische Fliegergruppe Berlin EV) (1978-79) 560
AKAFLIEG BRAUNSCHWEIG (see Akademische Fliegergruppe Braunschweig) 575
AKAFLIEG DARMSTADT (see Akademische Fliegergruppe Darmstadt EV) 575
AKAFLIEG HANNOVER (see Akademische Fliegergruppe Hannover EV) 576
AKAFLIEG KARLSRUHE (see Akademische Fliegergruppe Karlsruhe EV) 576
AKAFLIEG MÜNCHEN (see Akademische Fliegergruppe München) 576
AKAFLIEG MUNSTER (see Akademische Fliegergruppe Münster) (1973-74) 547
AKAFLIEG STUTTGART (see Akademische Fliegergruppe Stuttgart) 576
ALANNE, PENTTI (Finland) 570
Albatross, NP-100A (Nippi) 585
ALPLA-WERKE ALWIN LECHNER OHG (Austria) (1979-80) 557
ALTINGER, DIPL. ING THEO (Argentina) (1974-75) 558
Alto-Stratus (Borzecki) (1973-74) 570
AmEAGLE CORPORATION (U.S.A.) 590
American Eaglet (AmEagle) 590, 602
Antares, SB-11 (Akaflieg Braunschweig) (1979-80) 563, 586
ANTONOV, OLEG KONSTANTINOVICH (U.S.S.R.) (1974-75) 606
APPLEBAY SAILPLANES (USA) (1980-81) 582
Aqua Glider (Explorer) 592, 602
Araponga (Rio Claro) 568
Ardhra, ATS-1 (Civil Aviation Department) 584, 600
Argus (McBroom) (1974-75) 608
Arion (McBroom) (1974-75) 608
Arrow (Muller-Kites) (1978-79) 596
Artemis (Akaflieg Darmstadt) (1972-73) 515
ASOCIACION ARGENTINA DE CONSTRUCTORES DE AVIONES EXPERIMENTALES (Argentina) (1978-79) 554
Astir II (Grob) 578, 598
Astir IIB (Grob) 578, 598
Astir CS Jeans (Grob) (1977-78) 590
Austria-Krahe (Brditsch) (1973-74) 537
AVIAFIBER AG (Switzerland) (1980-81) 578
AVIONAUTICA RIO SpA (Italy) (1972-73) 534

B

B4-PC11AF (Nippi/Pilatus) 585, 600
B-12 (Akaflieg Berlin) (1978-79) 560, 582
BD-5S (Bede) (1977-78) 612
BF IV-BIMO (LVD) (1978-79) 564, 584
BG-6 (Briegleb) (1977-78) 612
BG-7 (Briegleb) (1977-78) 612
BG 12B (Briegleb) (1972-73) 545
BG 12BD (Briegleb) (1977-78) 612
BG 12-16 (Briegleb) (1977-78) 612
BG 100/12 (Birmingham Guild) (1972-73) 542
BG 100/13.5 (Birmingham Guild) (1972-73) 542
BG Gipsy 135 (Birmingham Guild) (1973-74) 575
BJ-1 Duster (Jannson) (1973-74) 572
BJ-1b Duster (DSK) 592, 602
BJ-3 (Beatty-Johl) (1971-72) 514
BJ-3A (Beatty-Johl) (1972-73) 539
BJ-4 (Beatty-Johl) (1972-73) 540
BK-7 Lietuva (Karvyalis) (1975-76) 603
BM 1 (Menin) (1976-77) 555
BRO-11 Zile (LAK) 588
BRO-11 Zile (Oshkinis) 588, 602
BRO-16 (Oshkinis) 588
BRO-17U Utochka (Oshkinis) 588
BS 1 (Glasflügel) (1973-74) 550
BAGALINI (Italy) (1974-75) 608
Batso (Kiceniuk) (1979-80) 599, 604
BEDE AIRCRAFT INC (U.S.A.) (1977-78) 612
BEATTY-JOHL (see W.A.T. Johl) (1973-74) 572
Bekas, SZD-35 (SZD) (1974-75) 589
Berca Nahuel (AVEX) (1974-75) 558
Bergfalke-III (Scheibe) (1973-74) 552
Bergfalke-IV (Scheibe) 580, 598
Bergfalke-IV M (Scheibe) (1977-78) 594
Bergfalke-IV conversion (LVD) (1975-76) 576
BERKSHIRE MANUFACTURING CORPORATION (U.S.A.) (1976-77) 583
BETTERIDGE, DAVID (Australia) 567
Biguá (Aero Saladillo) (1975-76) 561
Biguá, KW2 (IPE) (1975-76) 563
Bijave WA-30 (Wassmer) (1975-76) 570
BIRMINGHAM GUILD LTD (U.K.) (1975-76) 595
Bisier (Bulgaria) (1973-74) 538
Blaník 2M (Sigmund/Let) (1973-74) 558
Blaník, L-13 (Let) 570
Blaník, L-13J (Let) (1979-80) 558
BLESSING, GERHARD (West Germany) 576
Blue Wren, T-5 (Todhunter) 567, 596
Bocian 1E (SZD) (1978-79) 571
Boomerang (Schneider) (1972-73) 507
BORZECKI, JOZEF (Poland) (1973-74) 570
BRDITSCHKA OHG, H. W. (Austria) 567
BRIEGLEB (see Sailplane Corporation of America) (1978-79) 577
BRNO MILITARY ACADEMY, AIRCRAFT FACULTY (Czechoslovakia) (1974-75) 539
BRYAN AIRCRAFT INC (U.S.A.) 590
BUCHANAN, JOHN (Australia) 567
BURKHART GROB FLUGZEUGBAU GmbH & CO KG (West Germany) (1980-81) 570
Bussard (Austria) (1974-75) 560
BYTTEBIER, JOHAN F (Argentina) (1979-80) 596

C

C-5 Falke '76 (Scheibe) (1976-77) 563
C.30S Edelweiss (Siren) (1972-73) 512
C.38 (CARMAM) 572, 596
C.40 (CARMAM) 572, 596
CA-14 (WBAC) (1979-80) 601
C.34 Edelweiss IV (Siren) (1972-73) 512
CB-2/B Minuano (EEUFMG-CEA) 568, 596
CB-3 (Ultralight Flight Systems) (1979-80) 602
CE 75 Silène (CERVA) (1977-78) 581
CERVA (see Consortium Européen de Réalisation et de Ventes d'Avions, Groupement d'Intérêts Économiques) (1978-79) 558
CTA (see Centro Tecnico Aeroespacial) 567
CVT (see Centro di Volo a Vela del Politecnico di Torino) (1977-78) 603
Calif Series (Caproni Vizzola) 584
CALVEL, JACQUES (France) 571
Canard-2 FL (Aviafiber) (1980-81) 578, 592
CAPRONI VIZZOLA COSTRUZIONI AERONAUTICHE SpA (Italy) 584
CARMAM SA (France) 572
CARMICHAEL, DOUGLAS (U.S.A.) (1975-76) 608
CASES, FRANCISCO (Spain) 588
CENTRAIR, SA (France) 572
CENTRO DI VOLO A VELA DEL POLITECNICO DI TORINO (Italy) (1977-78) 603
CENTRO TECNICO AEROESPACIAL (Brazil) 567
C-Falke '80 (Scheibe) 580, 598, 604
Chanute biplane glider (1896) (Spurgeon) (1979-80) 600
Cherokee II (Hall) (1972-73) 547
CHUCK'S GLIDER SUPPLIES (USA) (1979-80) 599
Cirrus (Schempp-Hirth) (1974-75) 577
Cirrus (VTC/Schempp-Hirth) 596, 602
Cirrus 78L (SCAP) 574, 598
CIVIL AVIATION DEPARTMENT (India) 584
Cloudster, ST-100 (Ryson) 594, 602, 604
Club, ASW 19 (Schleicher) 582
Club Astir II (Grob) 578, 598
Club Elan, DG-100 (ELAN) (1979-80) 582, 592
Club Libelle (Glasflügel) (1978-79) 562
Club-Spartz (Scheibe) (1978-79) 566
Cobra 15 (SZD) (1977-78) 605
Cobra 17 (SZD) (1972-73) 538
Colibri 1 SL (Farner) 588
Compact (GEPAS) (1980-81) 566, 588
Compact (Lagarde) (1977-78) 584
Concept 70 (Berkshire) (1976-77) 583
Condor (Ferrier) 572, 598, 604
Condor (Hollman/Klepeis) 592, 602, 604
Condor (Up Inc) (1979-80) 600, 604
Condor, MG3-15L (Maggi) 584, 600
Consort, YS 55 (Yorkshire) (1974-75) 597
CONSORTIUM EUROPÉEN DE RÉALISATION ET DE VENTES D'AVIONS (GROUPEMENT D'INTÉRÊTS ÉCONOMIQUES) (France) (1978-79) 558
Cuervo, FS-25 (Akaflieg Stuttgart) (1973-74) 547
Cuervo, FS-25 S (ENSMA) 572, 598
Cuervo, FS-26 (ENSMA) (1978-79) 558
CURITIBA, GRUPO (Brazil) (1973-74) 538
Cumulus (Eipper-Formance) (1978-79) 598
Cumulus (Tohoku) (1971-72) 510
Cumulus III (Dahu Plane) (1978-79) 596
Cygnus, N-70 (Nihon University) (1974-75) 585

D

D-8 (Coward) (1972-73) 577
D-37b Artemis (Akaflieg Darmstadt) (1972-73) 515
D-38 (Akaflieg Darmstadt) (1975-76) 571
D-39 (Akaflieg Darmstadt) 575, 598, 604
D-40 (Akaflieg Darmstadt) 575
D 77 Iris (Issoire) 574, 598
D 177 Iris (Issoire) (1978-79) 560, 582
DH-HG-1 (Carmichael) (1974-75) 600
DG-100 (Glaser-Dirks) 576
DG-100 Club Elan (Elan) (1979-80) 582, 592
DG-100 Club Elan (Elan/Glaser Dirks) 595, 602
DG-100 Elan (Elan/Glaser-Dirks) 594
DG-101 (Glaser-Dirks) 576
DG-101 (G) (Elan) 595
DG-200 (Glaser-Dirks) 576, 598
DG-202 (Glaser-Dirks) 576
DG-400 (Glaser-Dirks) 576
DJ:1B (Thor) (1974-75) 605
DANIS, BERNARD (France) (1979-80) 597
Danzig (Akaflieg Braunschweig) (1976-77) 556
Delfin (VTC) (1974-75) 607
Delphin, HB-1340 (Mahrer) (1979-80) 576, 590
Delphin, SF-34 (Scheibe) 580, 600
Demoisel (MAV) (1979-80) 597, 603
DEPARTAMENTO DE AERONAVES (Brazil) (1971-72) 488
Diamant (FFA) (1972-73) 540
DIEHL-ZETTL, BAUGEMEINSCHAFT (West Germany) (1974-75) 569
Dimona, H-36 (Hoffmann) 578, 598, 604
DIVISAO DE AERONAVES (Brazil) 567
Dolphin (Projekt 8) 570, 596
Dolphin M2 (Projekt 8) 570, 596
Drag-N-Fly (Volpar/Spencer) (1980-81) 585, 592
Drifter (DSK) (1979-80) 599
DSK AIRCRAFT CORPORATION (U.S.A.) 592
DSK (DUSTER SAILPLANE KITS) (U.S.A) (1979-80) 599
Duster, BJ-1b (DSK) 592
Duster, BJ-1 (Jannson) (1973-74) 572
Duster DJ-1B (Thor) (1974-75) 605

E

E-31 (Esztergom) (1971-72) 505
E 75 Sagittaire (Siren) (1972-73) 512
E 75 Silène (Issoire) (1978-79) 559, 582
E 78 Silène (Issoire) 574, 598
EEUFMG (CEA) (see Escola de Engenharia da Universidade Federal de Minas Gerais: Centro de Estudos Aeronáuticos) 568
EFF (see Entwicklungsgemeinschaft für Flugzeugbau) 588
EIRI (see Eino Riihela KV) 571
EMBRAER (see Empresa Brasileira de Aeronáutica SA) (1972-73) 508
ENSAE (see École Nationale Supérieure de l'Aéronautique et de l'Espace) (1979-80) 560
ENSMA (see École Nationale Supérieure de Mécanique et d'Aérotechnique) 572
ES 60 Series 2 Boomerang (Schneider) (1972-73) 507
ES 60B Super Arrow (Schneider) (1975-76) 562
ES 63 (Schneider) (1977-78) 576
ESPE (see Espenlaub, Gotthold) (1973-74) 548

Eaglet (AmEagle) ....(1980-81) 581, 592
ÉCOLE NATIONALE SUPÉRIEURE DE L'AÉRONAUTIQUE ET DE L'ESPACE (France) ....(1979-80) 560
ÉCOLE NATIONALE SUPÉRIEURE DE MÉCANIQUE ET D'AÉROTECHNIQUE (France) 572
Eden (MAV) ....(1979-80) 597, 603
Espe-37T (ESPE) ....(1972-73) 517
Edelweiss (Siren) ....(1972-73) 512
Edelweiss IV (Siren) ....(1972-73) 512
EINO RIIHELA KV (Finland) ....(1980-81) 564
ELAN TOVARNA SPORTNEGA ORODJA N.SOL.O (Yugoslavia) 594
Elan, DG-100 (Elan/Glaser-Dirks) 594, 602
Elastic Wing Z-77 (Wolf) ....(1979-80) 598, 603
Elektra (Farrelly) ....(1979-80) 602
Elfe 15 (Neukom) 588, 602
Elfe 17 (Neukom) 588, 602
Elfe M 17 (Neukom) ....(1979-80) 577, 594
Elfe S-4 (Neukom) ....(1972-73) 541
Elfe Standard (Neukom) ....(1975-76) 593
EMPRESA BRASILEIRA DE AERONAUTICA SA (Brazil) ....(1972-73) 508
ENTWICKLUNGSGEMEINSCHAFT FÜR FLUGZEUGBAU (Switzerland) 588
ESCOLA DE ENGENHARIA DA UNIVERSIDADE FEDERAL DE MINAS GERAIS: CENTRO DE ESTUDOS AERONAUTICOS (Brazil) 568
Espadon, WA 28 (Wassmer) ....(1977-78) 585
ESPENLAUB, GOTTHOLD (West Germany) (1973-74) 548
ESZTERGOM (see Pestvidéki Gépgyár Esztergomi Gyáregységében) ....(1971-72) 505
Exo 7 (La Mouette) ....(1978-79) 597
EXPLORER AIRCRAFT COMPANY (U.S.A.) ... 592

F

FFA (see Flug- & Fahrzeugwerke AG) ...(1974-75) 593
FK-3 (VFW-Fokker) ....(1971-72) 504
FL-III (Flugwag Bremen) ....(1976-77) 559
FS-25 Cuervo (Akaflieg Stuttgart) ....(1973-74) 547
FS-25 S Cuervo (ENSMA) 572, 598
FS-26 Cuervo (ENSMA) ....(1978-79) 558
FS-26 Moseppl (Akaflieg Stuttgart) ....(1973-74) 548
FS-29 (Akaflieg Stuttgart) ....(1977-78) 588
FS-31 (Akaflieg Stuttgart) 576
FVA (see Flugwissenschaftliche Vereinigung Aachen (1920) EV) 576
FVA-20 (FVA) ....(1977-78) 588
FVA-20 F.B. Schmetz (FVA) 576, 598
Falcon (Chuck's Glider Supplies) ...(1979-80) 599, 604
Falcon (Waspair) ....(1979-80) 598, 603
Falcon V (Waspair) ....(1979-80) 599, 603
Falcon 15 m (Wright) 590, 602
Falke (Scheibe) 580
Falke (Slingsby/Scheibe) ....(1977-78) 610
Falke 80 (Scheibe) 580, 600, 604
Falke, T.61G (Slingsby) 590, 602, 604
Faltente (Pflumm & Richter) ....(1980-81) 570, 594
Fantail (Marshall) ....(1978-79) 597
Farfelu (ENSAE) ....(1978-79) 558, 582
FARNER, DIPL-ING ETH HANS ULRICH (Switzerland) 588
FARRELLY HANG GLIDERS (Australia) (1979-80) 596
FAUVEL, CHARLES (France) ....(1980-81) 566
FERRIER, HUBERT (France) 572
FIBERA AB, OY (Finland) ....(1973-74) 540
Firefly (UP Inc) ....(1979-80) 600, 604
Flatlander (DSK) ....(1979-80) 599
Fledgling (Manta) ....(1979-80) 599
Flexi-Flier (Eipper-Formance) ....(1978-79) 598
Flexi-Floater (Eipper-Formance) ....(1978-79) 598
Flider (Takatori) ....(1975-76) 589
FLIGHT DYNAMICS INC (U.S.A.) 592
Flying Circus (High Perspective) ....(1978-79) 596
FLUG & FAHRZEUGWERKE AG (Switzerland) ....(1974-75) 593
FLUGWISSENSCHAFTLICHE VEREINIGUNG AACHEN (1920) EV (West Germany) 576
Flyer (Seagull/Waterman) ....(1973-74) 585
Foehn (CARMAM) ....(1974-75) 564
Foka 5 (SZD) ....(1972-73) 538
FOURNIER AVIATION (France) 572
Foxbat (Solar Wings) ....(1978-79) 597
FRANKE, HANS (West Germany) ....(1975-76) 573
FREE FLIGHT HANG GLIDERS (Australia) ....(1979-80) 596
Frelon (Calvel) 571, 596, 604

G

G.10 Moto Delta (Geiser) ....(1975-76) 567
G 102 Series III (Grob) 578, 598
G 103 Twin II (Grob) 578, 598
G.104 Speed Astir (Grob) ....(1977-78) 591
G 109 (Grob) 578, 598, 604
G160, Gryphon (Waspair) ....(1979-80) 599, 603
G180, Gryphon (Waspair) ....(1979-80) 599, 603
GEP (see Groupe d'Études Georges Payre) (1979-80) 562
GEPAS (see Groupe d'Études Pour l'Aviation Sportive) ....(1980-81) 566
GHH (see Glasflugel, Holighaus & Hillenbrand GmbH & Co KG) ....(1976-77) 559
GABRIELS, LEN (U.K.) ....(1974-75) 608
GEISER, JEAN MARC (France) ....(1975-76) 567
Gemini (Marsden) 568
GÉPGYAR, PESTVIDEKI (Hungary) ...(1974-75) 582
GLASER-DIRKS FLUGZEUGBAU GmbH (West Germany) 576
GLASFASER ITALIANA SrL (Italy) 584
GLASFLÜGEL GmbH (West Germany) 578
GLASFLÜGEL ITALIANA SrL (Italy) (1977-78) 603
Globetrotter, H 121 (Start + Flug) ..(1978-79) 568, 584
Gobé, R-26 S (Gépgyar) ....(1973-74) 560
GROB FLUGZEUGBAU GmbH & Co KG (West Germany) 578
GROSS, FRANCOIS AND RIGER (France) 574
GROUPE D'ETUDES GEORGES PAYRE (France) (1979-80) 562
GROUPE D'ÉTUDES POUR L'AVIATION SPORTIVE (France) ....(1980-81) 566
Gryphon (Miles Wings) ....(1979-80) 598
Gryphon G160 (Waspair) ....(1979-80) 599, 603
Gryphon G180 (Waspair) ....(1979-80) 599, 603
Gullwing (Solar Wings) ....(1978-79) 597

H

H-36 Dimona (Hoffman) 578, 598, 604
H 101 Salto (Hänle) 578, 598
H 101 Salto (Start + Flug) ....(1978-79) 568, 584
H 121 Globetrotter (Start + Flug) ...(1978-79) 568, 584
H 301 Libelle (Glasflügel) ....(1969-70) 533
HA-18 (Pacific Gull) ....(1978-79) 598
HA-19 (Pacific Gull) ....(1978-79) 598
HA-20 (Pacific Gull) ....(1978-79) 598
HB-3B (Brditschka) ....(1977-78) 576
HB-3R (Brditschka) ....(1975-76) 562
HB-21 (Brditschka) 567, 596, 604
HB-22 (Brditschka) 567, 596, 604
HB-1340 Delphin (Mahrer) ....(1979-80) 576, 590
HBV-Diamant (FFA) ....(1972-73) 540
HF Colibri 1 SL (Farner) 588
HG-1 (Carmichael) ....(1975-76) 608
Hi 26 Mose II (Hirth) ....(1974-75) 573
HP-14 (Bryan/Schreder) ....(1973-74) 580
HP-14T (Bryan/Schreder) ....(1973-74) 580
HP-15 (Bryan/Schreder) ....(1973-74) 580
HP-16 (Bryan/Schreder) ....(1973-74) 580
HP-18 (Bryan/Schreder) 591, 602
HP-19 (Bryan/Schreder) 591, 602
HS-II Mrigasheer (India) 584
HS-3 Nimbus (Schempp-Hirth) ....(1973-74) 555
HS3 Skikarus (Spengeler) ....(1973-74) 575
HS-7 Mini-Nimbus (Schempp-Hirth) 582, 600
HALL, STANLEY (U.S.A.) ....(1976-77) 586
HANG GLIDERS PTY LTD (Australia) (1979-80) 596
Hankur 1 (Franke) ....(1975-76) 573
HÄNLE, URSULA (West Germany) 578
Happy Duck (United Hang Gliders) ....(1978-79) 597
Hawk (Manuel) ....(1974-75) 595
Hawk (Muller Kites) ....(1978-79) 596
Hawk-1, ST-1 (Takatori) ....(1975-76) 589
HAWKSWORTH SKYSPORTS LTD (UK) (1978-79) 593
Hippie (Start + Flug) ....(1979-80) 568, 584
Hirondelle Club (Véliplane) ....(1978-79) 597
HOFFMAN FLUGZEUGBAU KG, WOLF (West Germany) 578
HOLIGHAUS & HILLENBRAND GmbH & CO KG (West Germany) ....(1979-80) 564
HOLLMANN AIRCRAFT (USA) 592
HOLMES, KENNETH (UK) ....(1976-77) 581
Hornet (Glasflügel) 578, 598
Hornet C (Glasflügel) 578, 598
HUFNAGEL, BERNHARD (West Germany) 579

I

I 79 (Issoire) ....(1980-81) 566
ICA (see Intreprindere de Constructii Aeronautice) 586
IKV (see Ilmailukerho Vasama) ....(1972-73) 510
IKV-3 Kotka (IKV) ....(1971-72) 490
IPD (see Instituto de Pesquisas e Desenvolvimento) 567
IPD/PAR-6505 Urupema (IPD) ....(1971-72) 488
IPE (see Industria Paranaense de Estruturas) 568
IS-3D (ICAB) ....(1972-73) 538
IS-7/12 (ICAB) ....(1972-73) 539
IS-18 (ICAB) ....(1972-73) 539
IS-28 (ICA-Brasov) ....(1976-77) 578
IS-28B2 (ICA) 586, 600
IS-28M (ICA) 586, 600, 604
IS-29 (ICA) 587, 600
IS-29DM (ICA) 588, 600, 604
IS-29EM (ICA) 588, 600, 604
IS-29G (ICA-Brasov) ....(1974-75) 593
IS-30 (ICA) 588, 600
IS-31 (ICA) ....(1980-81) 590
IS-32 (ICA) 588, 600
IS-33 (ICA) 587, 600
ISF (see Ingenieur-Büro Dipl-Ing Strauber-Frommhold GmbH & Co KG) ....(1980-81) 570
Ibex (Hall) ....(1971-72) 521
Idel, SA-8T (SA) ....(1976-77) 591
ILMAILUKERHO VASAMA (Finland) .(1972-73) 510
INDUSTRIA PARANAENSE DE ESTRUTURAS (Brazil) 568
INGENIEUR-BÜRO DIPL-ING STRAUBER-FROMMHOLD GmbH & Co KG (West Germany) ....(1980-81) 570
INSTITUTO DE PESQUISAS E DESENVOLVIMENTO (Brazil) 567
INTREPRINDERE DE CONSTRUCTII AERONAUTICE (Romania) 586
Iris, D 77 (Issoire) 574, 598
Iris, D 177 (Issoire) ....(1978-79) 560, 582
ISSOIRE-AVIATION SA (France) 574

J

JB-1 Nahuel (AVEX) ....(1975-76) 561
JEAA (Japan) ....(1979-80) 597
JEAA (see Japan Experimental Aircraft Association) ....(1974-75) 585, 608
J.15-38 (Jacquet) ....(1975-76) 569
J.P. 15-34 (Jacquet et Pottier) ....(1974-75) 565
J.P.15-36A Aiglon (CARMAM) 572, 596
J.P.15-36AR Aiglon (CARMAM) 572
JACQUET, ROBERT (France) ....(1975-76) 568
JANSSON, BEN (Sweden) ....(1973-74) 572
Jantar-1 (SZD) ....(1977-78) 605
Jantar 2, SZD-42-1 (SZD) 586
Jantar 2B, SZD-42-2 (SZD) 586, 600
Jantar 19, SZD-37 (SZD) ....(1975-76) 589
Jantar K, SZD-49 (SZD) 586
Jantar Standard, SZD-41A (SZD) 586
Jantar Standard 2, SZD-48 (SZD) 586, 600
Janus (Schempp-Hirth) 582, 600
JAPAN EXPERIMENTAL AIRCRAFT ASSOCIATION (Japan) ....(1974-75) 585, 608
JASTREB AEROPLANE AND GLIDER FACTORY (Yugoslavia) 595
Javelin (Wings) ....(1979-80) 602
Jet (La Mouette) ....(1978-79) 597
Jian Fan, X-7 (China) 569, 596
João Grande (Sidou) 568, 596
Johan 087 (Byttebier) ....(1979-80) 602
JOHL, W.A.T. (South Africa) ....(1973-74) 572

K

K 8 (Schleicher) ....(1977-78) 597
K-14 Kirigamine (JEAA) ....(1974-75) 585
Ka 6E (Schleicher) ....(1973-74) 556
Ka 16X (Kaiser/Schleicher) ....(1975-76) 576
KAI (see Kazan Aviation Institute) ....(1973-74) 586
KAI-14 (KAI) ....(1972-73) 551
KAI-19 (KAI) ....(1972-73) 551
KBK (Kleb) ....(1975-76) 569
KH-1 (Holmes) ....(1976-77) 581
KK-1 Utu (Fibera) ....(1972-73) 509
KR-1B (Rand Robinson) 593
KS-II Kartik (India) ....(1977-78) 602
KW 1 b 2 Quero Quero II (IPE) ....(1980-81) 562
KW 2 Biguá (IPE) ....(1975-76) 563
KAISER, RUDOLF (West Germany) ....(1975-76) 576
Kartik (India) ....(1977-78) 602
KARVYALIS, BALIS (U.S.S.R.) ....(1975-76) 603
KAZAN AVIATION INSTITUTE (U.S.S.R.) (1973-74) 586
Keihikoki SS-1 (LADCO) ....(1971-72) 509
Kestrel 17 (Glasflügel) ....(1979-80) 564, 586
Kestrel 19 (Slingsby/Glasflügel) ....(1975-76) 595
Kestrel 22 Srs 2, T.58H (Vickers-Slingsby/Glasflügel) .....(1978-79) 576, 586
K-Falke '80 (Scheibe) 580, 600, 604
Kirigamine K-14 (JEAA) ....(1974-75) 585
Kit-Club 15-34 (CARMAM/Pottier) 572
Kit-Club 15-34 (Pottier) 574, 598
KLEB (France) ....(1975-76) 569
KNECHTEL, ING WILHELM J (West Germany) 579
KONDOR KITE CO (USA) ....(1978-79) 593
Kora 1 (Kortenbach & Rauh) ....(1980-81) 570
KORTENBACH & RAUH (West Germany) ....(1980-81) 570
Kotka (IKV) ....(1971-72) 490
Kranich IIIM (Schappert/Focke-Wulf) ....(1975-76) 578
KUFFNER GmbH & Co KG, LEICHTFLUGZEUGBAU (West Germany) ....(1980-81) 570

L

L-Spatz III (Scheibe) ....(1972-73) 521
L-13 Blaník (Let) 570
L-13 Wankel (Let) ....(1979-80) 558
L-13J Blaník (Let) ....(1979-80) 558
L-13SW (Let) ....(1980-81) 564
L-13-2M (Let) ....(1979-80) 558
L190, Laser (Waspair) ....(1979-80) 599, 603
L-252 Phoebus (MBB) ....(1973-74) 551
LA-11 Topaze (Loravia) ....(1979-80) 562
LAK (see Litovskaya Aviatsionnaya Konstruktsiya) 588
LAK-5 Nyamunas (LAK) 589, 602, 604
LAK-6 (Let) ....(1979-80) 558
LAK-9 Lietuva (LAK) 589, 600
LAK-10 Lietuva (LAK) 589
LAK-11 Lietuva (LAK) 589
LAK-12 Lietuva (LAK) 589, 602
LCA-11 Topaze (SCAY) ....(1975-76) 569
LCF (see Luftsportclub der Zeppelinstadt Friedrichshafen) ....(1979-80) 566
LCF II (LCF) ....(1979-80) 566
LP-15 Nugget (Laister) ....(1977-78) 615
LP-49 (Laister) ....(1977-78) 614
LS1 (Rolladen-Schneider) ....(1975-76) 577
LS1-f (Rolladen-Schneider) ....(1978-79) 564
LS2 (Rolladen-Schneider) ....(1975-76) 577
LS3 (Rolladen-Schneider) 580
LS3-a (Rolladen-Schneider) 580, 598
LS3-17 (Rolladen-Schneider) 580
LS4 (Rolladen-Schneider) 580, 598
LS5 (Rolladen-Schneider) 580
LSS-02 (Tainan) ....(1974-75) 586
LSS-02B (Tainan) ....(1974-75) 586
LVD (see Schülerfluggemeinschaft der Luftsportverein Detmold EV) ....(1979-80) 566
LADCO (see Light Aircraft Development Company, Tokyo) ....(1971-72) 509
LAGARDE, B. DE (France) ....(1977-78) 584
LAISTER SAILPLANES INC (U.S.A.) (1979-80) 578
LANAVERRE INDUSTRIE (France) ...(1978-79) 560
Lanzalone Aulanz (AVEX) ....(1978-79) 554, 588
Laser L190 (Waspair) ....(1979-80) 599, 603
Lenticular 15 S (AS) ....(1979-80) 557
Leonardino (Bagalini) ....(1974-75) 608
Leonardo, M.B.2 (Militi) ....(1973-74) 564

LET NARODNI PODNIK (Czechoslovakia) 570
Libel-lula (Cases) 588
Lietuva, Lak-9 (LAK) 589, 600
Lietuva, LAK-10 (LAK) 589
Lietuva, LAK-12 (LAK) 589, 602
LIGHT AIRCRAFT DEVELOPMENT COMPANY, TOKYO (Japan) ......(1971-72) 509
LIGHTER THAN AIR PRODUCTS (USA) (1978-79) 593
Lilienthal monoplane glider (1895) (Spurgeon) (1979-80) 600
LINDNER GmbH & Co KG, FIBERGLAS-TECHNIK RUDOLF (West Germany) 580
Liska, SL-40 (Soko) 596, 602, 604
LITOVSKAYA AVIATSIONNAYA KONSTRUKTSIYA (USSR) 588
LORAVIA (see Lorraine Aviation) ......(1979-80) 562
LORRAINE AVIATION (France) ......(1979-80) 562
LUFTSPORTCLUB DER ZEPPELINSTADT FRIEDRICHSHAFEN (West Germany) (1979-80) 566
Lutin, AV.50 (Fauvel) ......(1973-74) 542
Lynx (Wings) ......(1979-80) 602
Lynx B (McBroom) ......(1978-79) 598

M

M2, Dolphin (Projekt 8) 570, 596
M-17 Universal (Matejcek) ......(1974-75) 561
M-100 S Mésange (CARMAM) ......(1974-75) 564
M-200 Foehn (CARMAM) ......(1974-75) 564
M-300 (CVT/Morelli) ......(1973-74) 563
MAV (see Manufacture d'Ailes Volantes) (1979-80) 597
M.B.1 (Militi) ......(1973-74) 563
M.B.2 Leonardo (Militi) ......(1973-74) 564
MBB (see Messerschmitt-Bölkow-Blohm) .(1973-74) 551
MB-E1 (Militky) ......(1975-76) 577
MF (see Masinski Fakultet) 595
MFB (see Mistral Flugzeugbau) 580
MG3-15L Condor (Maggi) 585, 600
MGM-2 Working Bee (JEAA/Miyahara) .(1974-75) 585
MJ-4 (Birdman Enterprises) ......(1978-79) 596
Mü 26 (Akaflieg München) ......(1974-75) 568
Mü 27 (Akaflieg München) 576, 598
Mü 28 (Akaflieg München) 576, 598
MVTU (U.S.S.R.) ......(1977-78) 619
MAGGI, ANGELO (Italy) 585
MAHRER, FRITZ (Switzerland) ......(1979-80) 576
Mammoth (Takatori) ......(1974-75) 587
MANUEL, WILLIAM L. (U.K.) ......(1974-75) 595
MANUFACTURE D'AILES VOLANTES (France) (1979-80) 597
Marauder (Muller Kites) ......(1978-79) 596
MARSDEN, PROF DAVID J (Canada) 568
MARSHALL, FRANK (South Africa) ...(1979-80) 598
MASINSKI FAKULTET (Yugoslavia) 595
MATEJCEK, JIRI (Czechoslovakia) ....(1974-75) 561
MAUPIN, JIM (USA) 592
Maxi Strato (Sunbird) ......(1978-79) 599
McBROOM SAILWINGS LTD (UK) ...(1978-79) 593
MENIN, JEAN PAUL (France) ......(1976-77) 555
Mésange (CARMAM) ......(1974-75) 56
MESSERSCHMITT-BÖLKOW-BLOHM (West Germany) ......(1973-74) 551
Micro Stinger (Moyes) ......(1979-80) 596, 602
Midas (Chargus) ......(1978-79) 597
Midi-Stinger (Moyes) ......(1978-79) 596
Milan (Sportavia) ......(1974-75) 582
MILES WINGS (ENGINEERS) LTD (UK) (1979-80) 598
MILITI, BRUNO (Italy) ......(1975-76) 588
MILITKY, FRED (West Germany) ......(1977-78) 593
MILLER, W. TERRY (U.S.A.) ......(1977-78) 616
Mini-Nimbus C (Schempp-Hirth) 582, 600
Mini-Nimbus, HS-7 (Schempp-Hirth) 582, 600
Mini-Stinger (Moyes) ......(1979-80) 596, 602
Mini Strato (Sunbird) ......(1978-79) 599
Minuano, CB-2/B (EEUFMG/CEA) 568, 596
Mistral-C, Model 2 (ISF) ......(1980-81) 570, 588
Mistral C, Model 2 (MFB) 580
MISTRAL FLUGZEUGBAU (DIPL-ING STRAUBER GmbH & Co KG) (West Germany) 580
Mita III (Tainan) ......(1979-80) 572, 588
Moba 2C (Sunderland) 567, 596
Moba 3 Virago (Sunderland) 567, 596
Model II (LCF) ......(1978-79) 564, 584
Model 2 Mistral-C (ISF) ......(1980-81) 570, 588
Model 2 Mistral C (MFB) 580
Model 2-32 (Schweizer) ......(1978-79) 580, 586
Model 2-33 (Schweizer) ......(1977-78) 618
Model 10 (VSO) ......(1978-79) 556, 582
Model 20 (FVA) ......(1977-78) 588
Model 205 Club Libelle (GHH) ......(1976-77) 559
Model 206 Hornet/Mosquito (GHH) .....(1976-77) 560
Model 604 (Glasflügel) ......(1975-76) 575
Model 701 (Glasflügel) ......(1975-76) 575
Model 1616 (Muller Kites) ......(1978-79) 596
Model 1718 (Muller Kites) ......(1978-79) 596
Model 1820 (Muller Kites) ......(1978-79) 596
Modified FS-25 Cuervo (ENSMA) ......(1979-80) 560
Monerai P (Monnett) 592
Monerai S (Monnett) 592, 602
MONNETT EXPERIMENTAL AIRCRAFT INC (USA) 592
Montgomery monoplane glider (1883) (Spurgeon) ......(1979-80) 600
Moonraker 78 (Birdman Sports) ......(1978-79) 597
Mora (Hirth) ......(1973-74) 551
Morelli M-300 (CVT) ......(1973-74) 563
Mose II (Hirth) ......(1974-75) 573
Moseppl, FS-26 (Akaflieg Stuttgart) ......(1973-74) 548
Mosquito (Glasflügel) ......(1979-80) 565, 588
Mosquito B (Glasflügel) 578, 598
Motorblanik (Alanne) 570, 596, 604
Motorlerche (Alanne) 570, 596, 604
Motorraab IV 017 (Diehl-Zettl) ......(1974-75) 569
Mrigasheer (India) 584
MULLER KITES LTD (Canada) ......(1979-80) 596
Mustang (Muller Kites) ......(1978-79) 596

N

N-70 Cygnus (Nihon University) ......(1974-75) 585
NIPPI (see Nihon Hikoki Kabushiki Kaisha) 585
No.II My Wing (JEAA) ......(1979-80) 597
NP-100A Albatross (Nippi) 585
Nahuel (AVEX) ......(1975-76) 561
NEUKOM SEGELFLUGZEUGBAU, ALBERT (Switzerland) 588
Nhapecam, Type 02 (IPE) 568, 596
Nida (LAK) 589, 602
NIHON HIKOKI KABUSHIKI KAISHA (Japan) 585
NIHON UNIVERSITY (Japan) ......(1974-75) 585
Nimbus (Akaflieg Braunschweig) ......(1973-74) 545
Nimbus (Schempp-Hirth) ......(1973-74) 555
Nimbus 2C (Schempp-Hirth) 581, 600
Nimbus II M (Schempp-Hirth) ......(1977-78) 596
Nimbus 3 (Schempp-Hirth) 582, 600
Nugget, LP-15 (Laister) ......(1977-78) 615
Nyamunas, LAK-5 (LAK) 589, 602, 604

O

O2 (IPE) ......(1978-79) 555, 582
O-3 (Oldershaw) ......(1979-80) 581
OZ-5 (Zanner) ......(1976-77) 590
Odyssée (Favul) ......(1979-80) 597, 602
Ogar, SZD-45A (SZD) ......(1980-81) 576, 590, 594
OLDERSHAW, VERNON W (USA) ....(1979-80) 581
OMNIPOL FOREIGN TRADE CORPORATION (Czechoslovakia) 570
OPERATION SIGMA LTD (U.K.) ......(1974-75) 596
Orion, SZD-43 (SZD) ......(1974-75) 590
Orlice (VSB) ......(1972-73) 509
OSHKINIS, BRONIS (U.S.S.R.) ......(1977-78) 619
OSOKINE, A. I. (U.S.S.R.) ......(1973-74) 586
OSRODEK BADAWCZO ROZWOJOWY SZYBOWNICTWA (Poland) ......(1975-76) 589

P

P.A. 15-34 Kit Club (Pottier) 574, 598
PAR (see Divisao de Aeronaves) 567
PATS/NSDB (see Philippine Aeronautics Training School/National Science Development Board) (1978-79) 570
PC-X11 (Pilatus) ......(1974-75) 595
PC-XS (Pilatus) ......(1975-76) 595
PE-80367 Urubu (CTA) 567, 596
PG-1 Aqua Glider (Explorer) 592, 602
PIK (see Teknillinen Korkeakoulu) ......(1976-77) 552
PIK-16c Vasama (PIK) ......(1974-75) 563
PIK-17a Tumppi (PIK) ......(1972-73) 510
PIK-17b Tintti (PIK) ......(1972-73) 510
PIK-20E (EIRI) 574, 598
PIK-20E (Siren/EIRI) 574, 598, 604
PARANA (see Aeroclube da Paraná) ....(1974-75) 508
Pathfinder (Electra) ......(1978-79) 598
PAYRE, GROUPE D'ÉTUDES GEORGE (France) (1979-80) 562
Pelzner biplane glider (1920/21) (Spurgeon) (1979-80) 600
PENNER, HELMUTH (West Germany) .(1978-79) 592
PESTVIDÉKI GÉPGYAR ESZTERGOMI GYAREGYSÉGÉBEN (Hungary) ....(1971-72) 505
PETERSEN, HELGE (Denmark) ......(1974-75) 562
PFLUMM, MANFRED and RICHTER, K. J. (West Germany) ......(1980-81) 570
PHILIPPINE AERONAUTICS TRAINING SCHOOL/NATIONAL SCIENCE DEVELOPMENT BOARD (Philippines) ......(1978-79) 570
Phoebus (MBB) ......(1973-74) 551
Phoebus B 3 (Lindner) 580
Phoenix (Delta Wing) ......(1978-79) 598
PILATUS FLUGZEUGWERKE AG (Switzerland) (1978-79) 574
PINGOUIN, SOCIÉTÉ DES ATELIERS DU (France) ......(1974-75) 565
Pirat, SZD-30C (SZD) ......(1980-81) 576, 590
POLYTEKNIKKOJEN ILMAILUKERHO (Finland) ......(1972-73) 510
PONSOT, BERNARD (France) 574
POTTIER, JEAN (France) 574
Project 8 (Petersen) ......(1974-75) 562
PROJEKT 8 1/S (Denmark) 570
Prometheus (EFF) 588, 600, 604
PRZEDSIEBIORSTOW DOSWIADCZALNO-PRODUKCYJNE SZYBOWNICTWA (Poland) . 586
Pushacz, SZD-50 (SZD) 586, 600

Q

Qian Jin, X-10 (China) 570
Quero Quero II (IPE) ......(1980-81) 562

R

R-26 S Gobé (Gépgyar) ......(1973-74) 560
RFD4 (Sportavia/Avion Planeur) ......(1974-75) 581
RF5 (Sportavia/Avion Planeur) (1979-80) 570, 588, 594
RF5B Sperber (Sportavia) ......(1980-81) 573
RF5D (Sportavia) ......(1975-76) 585
RF-9 (Fournier) 572, 598, 604
RF-10 (Fournier) 572, 598
RF55 (Sportavia) ......(1974-75) 582
RFB (see Rhein-Flugzeugbau) ......(1974-75) 573
RG-1 Rohini-1 (India) ......(1975-76) 586
RS-15 (Bryan/Schreder) 590, 602
Raider (Birdman Enterprises) ......(1978-79) 596
RAND ROBINSON ENGINEERING INC (USA) 593
Ranger (Wings) ......(1979-80) 602
Rebell (Blessing) ......(1980-81) 568
REITER, HELMUT (West Germany) ....(1975-76) 577
RHEIN-FLUGZEUGBAU GmbH (West Germany) ......(1974-75) 573
Rhönlerche IIM (Knechtel) 579
Ricochet (Buchanan) 567
Ridgemaster (Electra) ......(1978-79) 598
RIO CLARO (see Aéro Clube de Rio Claro) 568
RIO SpA, AVIONAUTICA (Italy) ......(1972-73) 534
Rohini-1 (India) ......(1975-76) 586
ROLLADEN-SCHNEIDER FLUGZEUGBAU GmbH (West Germany) 580
RYSON AVIATION CORPORATION (U.S.A.) .. 593

S

S-3 Sagami (Tainan) ......(1973-74) 566
S-3 Standard Elfe (Neukom) ......(1975-76) 593
S-3K Sagami (Tainan) ......(1975-76) 588
S-4 Elfe (Neukom) ......(1972-73) 541
S-4A Elfe 15 (Neukom) 588, 602
S-4 Standard Elfe (Neukom) ......(1973-74) 574
S-15A Standard Elfe (Neukom) ......(1975-76) 593
S-17A Standard Elfe (Neukom) ......(1975-76) 593
SA (see Sportivnaya Aviatsiya) ......(1976-77) 591
SA-8T Idel (SA) ......(1976-77) 591
SAH-7 Cherokee II (Hall) ......(1972-73) 547
SAH-8 Ibex (Hall) ......(1971-72) 521
SB-5E Danzig (Akaflieg Braunschweig) ..(1976-77) 556
SB-7B Nimbus (Akaflieg Braunschweig) ..(1974-75) 545
SB-9 Stratus (Akaflieg Braunschweig) ....(1973-74) 545
SB-10 Schirokko (Akaflieg Braunschweig) (1974-75) 568
SB-11 (Akaflieg Braunschweig) .....(1979-80) 563, 586
SB-12 (Akaflieg Braunschweig) 575, 598
SCAP (see Société de Commercialisation Aéronautiques du Plessis, SaRL) 574
SCAY (see Société dé Construction Aéronautique de Yutz) ......(1975-76) 569
SD-15V (Swales) ......(1977-78) 611
SD3-15T (Swales) 590, 602
SD3-15V (Swales) ......(1978-79) 575, 586
SF-25 Falke (Scheibe) ......(1972-73) 522
SF-25B Falke (Scheibe) ......(1975-76) 579
SF-25C Falke (Scheibe) 580, 598, 604
SF-25C-S (Scheibe) 580, 598, 604
SF-25C-S Falke '76 (Scheibe) ..(1979-80) 568, 588, 594
SF-25E Super-Falke (Scheibe) 581, 600
SF-27 Zugvogel V (SLCA) ......(1973-74) 543
SF-27 Zugvogel V (Scheibe) ......(1973-74) 553
SF-27M (Scheibe) ......(1975-76) 579
SF-28A Tandem-Falke (Scheibe) 580, 600
SF-29 (Scheibe) ......(1974-75) 577
SF-30 Club-Spatz (Scheibe) ......(1978-79) 566
SF-32 (Scheibe) ......(1978-79) 566
SF-34 Delphin (Scheibe) 580, 600
SF-36 (Scheibe) 581, 600, 604
SFS 31 Milan (Sportavia) ......(1974-75) 582
SF-33 Trainer (Scheibe) ......(1978-79) 566, 584, 588
SGS 1-26E (Schweizer) 594
SGS 1-34B (Schweizer) ......(1979-80) 582, 592
SGS 1-35 (Schweizer) 594, 602
SGS 1-36 Sprite (Schweizer) 594, 602
SGS 2-32 (Schweizer) ......(1979-80) 582, 592
SGS 2-33A (Schweizer) 594, 602
SH-8 Flider (Takatori) ......(1975-76) 589
SH-10E (Takatori) ......(1975-76) 589
SH-15 (YGC) ......(1972-73) 536
SH-16 (Takatori) ......(1974-75) 587
SH-16S (Takatori) ......(1974-75) 587
SH-18 Mammoth (Takatori) ......(1974-75) 587
Sie 3 (Siebert) ......(1977-78) 599
SK 1 (Danis) ......(1979-80) 597
SK 2 (Danis) ......(1979-80) 597
SK 3 (Danis) ......(1979-80) 597
SL-2 (SCAP-Lanaverre) ......(1980-81) 567, 588
SL-40 Liska (Soko) 596, 602, 604
SLCA (see Société Lorraine de Constructions Aéronautiques) ......(1974-75) 566
SLCA-10/11 Topaze (SLCA) ......(1974-75) 566
SOLE-77 (Jastreb) 595, 602, 604
SS-01 (Tainan) ......(1973-74) 565
SST (Hawksworth) ......(1978-79) 598
SSV-17 (VTC) ......(1976-77) 592
SSVV (see Sezione Sperimentale Volo a Vela, Aeroclub Volovelistico Milanese) ......(1971-72) 509
ST-1 Hawk-1 (Takatori) ......(1975-76) 589
ST-100 Cloudster (Ryson) 594, 602, 604
SZD (see Przedsiebiorstow Doswiadczalno-Produkcyjne Szybownictwa) 586
SZD-9 bis Bocian 1E (SZD) ......(1978-79) 571
SZD-30C Pirat (SZD) ......(1980-81) 576, 590
SZD-31 Zefir 4 (SZD) ......(1972-73) 538
SZD-32A Foka 5 (SZD) ......(1972-73) 538
SZD-35 Bekas (SZD) ......(1974-75) 589
SZD-36A Cobra 15 (SZD) ......(1978-79) 571, 584
SZD-37 Jantar 19 (SZD) ......(1974-75) 589
SZD-38 Jantar-1 (SZD) ......(1977-78) 605
SZD-39 Cobra 17 (SZD) ......(1972-73) 538
SZD-40X Halny (SZD) ......(1975-76) 591
SZD-41A Jantar Standard (SZD) 586
SZD-42-1 Jantar 2 (SZD) 586
SZD-42-2 Jantar 2B (SZD) 586, 600
SZD-43 Orion (SZD) ......(1974-75) 590
SZD-45A Ogar (SZD) ......(1980-81) 576, 590, 594
SZD-48 Jantar Standard 2 (SZD) 586, 600
SZD-49 Jantar K (SZD) 586
SZD-50 Pushacz (SZD) 586, 600
Sagami (Tainan) ......(1975-76) 588
Sagittaire (Siren) ......(1972-73) 512
Sagittaire, CE.75 (CERVA) ......(1973-74) 541
SAILPLANE CORPORATION OF AMERICA (U.S.A.) ......(1978-79) 577

SAILPLANE DESIGN LTD. (U.K.) ....(1974-75) 595
SAILWINGS (SCOTLAND) LTD (UK) (1979-80) 598
Salto H 101 (Hänle) 578, 598
Salto, H 101 (Start + Flug) ....(1978-79) 568, 584
Samburo, Avo 68V (Alpla) ....(1979-80) 557
Saturn IV (Free Flight) ....(1979-80) 602
SCHAPPERT, EDUARD (West Germany) (1975-76) 578
SCHEIBE FLUGZEUGBAU GmbH (West Germany) 580
SCHEMPP-HIRTH FLUGZEUGBAU GmbH (West Germany) 581
Schirokko, SB-10 (Akaflieg Braunschweig) (1974-75) 568
SCHLEICHER SEGELFLUGZEUGBAU, ALEXANDER (West Germany) 582
SCHLEUNIGER AND CO., DR. K. (Switzerland) ....(1974-75) 595
SCHNEIDER OHG, SEGELFLUGZEUGBAU (West Germany) ....(1971-72) 504
SCHNEIDER PTY LTD, EDMUND (Australia) (1978-79) 554
Schreder HP-14 (Bryan) ....(1972-73) 546
Schreder HP-14T (Bryan) ....(1972-73) 546
Schreder HP-15 (Bryan) ....(1972-73) 546
Schreder HP-16 (Bryan) ....(1972-73) 547
SCHÜLERFLUGGEMEINSCHAFT DER LUFTSPORTVEREIN DETMOLD EV (West Germany) ....(1979-80) 566
Schülmeister (Start + Flug) ....(1977-78) 601
SCHWEIZER AIRCRAFT CORPORATION (U.S.A.) 594
SCOT-KITES (UK) ....(1979-80) 598
Seagull (Seagull) ....(1979-80) 600, 604
Seasprite (Flight Dynamics) 592, 602
SEZIONE SPERIMENTALE VOLO A VELA, AEROCLUB VOLOVELISTICO MILANESE (Italy) ....(1971-72) 509
Shadow (Birdman Enterprises) ....(1978-79) 596
SIDOU, ENG ANTONIO MENEZES (Brazil) 568
SIEBERT SPORT-UND SEGELFLUGZEUGBAU, PAUL (West Germany) ..(1977-78) 599
SIGMA (see Operation Sigma Ltd) ......(1974-75) 596
Sigma 1 (Sigma) ....(1974-75) 596
SIGMUND FLUGTECHNIK (West Germany) (1973-74) 558
Silène, E 78 (Issoire) 574, 598
SIREN, S.A. (France) 574
Sirius I (RFB) ....(1973-74) 551
Sirius II (RFB) ....(1974-75) 573
Skikarus (Spengeler) ....(1973-74) 575
SKYCRAFT PTY LTD (Australia) ......(1975-76) 605
Skydart (Ultralight Flight Systems) ....(1979-80) 602
Skyhook (Ultralight Flight Systems) ......(1979-80) 602
Skytrek (Delta Wing) ....(1978-79) 598
SLINGSBY ENGINEERING LTD (UK) 590
SOBIESZCZANSKI, WITOLD (Poland) (1978-79) 592
SOCIÉTÉ ANONYME CARMAM (France) 572
SOCIÉTÉ DE COMMERCIALISATION AÉRONAUTIQUE DU PLESSIS, SàRL (France) 572
SOCIÉTÉ DE CONSTRUCTION AÉRONAUTIQUE DE YUTZ (France) (1976-77) 569
SOCIÉTÉ ISSOIRE-AVIATION (France) 574
SOCIÉTÉ LORRAINE DE CONSTRUCTIONS AÉRONAUTIQUES (France) ....(1973-74) 543
SOCIÉTÉ NOUVELLE WASSMER AVIATION (France) ....(1977-78) 585
SOKO VAZDUHOPLOVNA INDUSTRIJA RADNA ORGANIZACIJA VAZDUHOPLOVSTVO (Yugoslavia) 596
SOLAR WINGS (South Africa) ....(1979-80) 598
Sovereign, YS 53 (Yorkshire) ....(1974-75) 597
Speed-Astir II (Grob) ....(1979-80) 566, 588
SPENGELER, HANS (Switzerland) .....(1973-74) 575
Sperber (Sportavia) ....(1980-81) 573
Spitfire (Chuck's Glider Supplies) ...(1979-80) 599, 604
SPORTAVIA-PÜTZER GmbH u Co KG (West Germany) ....(1980-81) 573
SPORTIVNAYA AVIATSIYA (U.S.S.R.) (1976-77) 591
SPORT KITES INC (USA) ....(1978-79) 594
Sport Vega, T.65C (Slingsby) 590, 602
Spring Wing Z-76 (Wolf) ....(1979-80) 598, 603
Sprint (Torva) ....(1974-75) 597
Sprite (Torva) ....(1974-75) 597
Sprite, SGS 1-36 (Schweizer) 594, 602
SPURGEON, JIM (USA) ....(1979-80) 600
Spyder (UP Inc) ....(1979-80) 600, 604
Squale (Wassmer) ....(1975-76) 570
Staff Rebel (Blessing) 576, 598, 604
Standard Astir (Grob) ....(1979-80) 566, 588
Standard Cirrus (Lanaverre) ....(1978-79) 560, 582
Standard Cirrus (Schempp-Hirth) ....(1978-79) 566
Standard Cirrus 75-VTC (VTC/Schempp-Hirth) 597, 602
Standard Elfe S-3 (Neukom) ....(1975-76) 593
Standard Elfe S-4 (Neukom) ....(1973-74) 574
Standard Elfe S-15A (Neukom) ....(1975-76) 593
Standard Elfe S-17A (Neukom) ....(1975-76) 593
Standard Libelle (Glasflügel) ....(1975-76) 574
Standard Strato (Sunbird) ....(1978-79) 599
Star (Favul) ....(1979-80) 597, 602
START + FLUG GmbH (West Germany) (1978-79) 568
STATE AIRCRAFT FACTORIES (China) 569
Sterne (Véliplane) ....(1978-79) 597
Stinger (Moyes) ....(1979-80) 596, 602
Stratus (Akaflieg Braunschweig) ....(1973-74) 545
Streaker (AA Flight) ....(1979-80) 599, 604
SUNDERLAND, GARY (Australia) 567
Sunspot (Skyhooks Sailwings) ....(1979-80) 599
Super Arrow (Schneider) ....(1975-76) 562
Super Elfe AN-66C (Neukom) ......(1979-80) 577, 590
Super-Falke (Scheibe) 580, 600, 604
Super Lynx (McBroom) ....(1978-79) 567
Super Orchidee, ASW 17 (Schleicher) ....(1978-79) 598
Super Strato (Sunbird) ....(1978-79) 599
Super Swallowtail (Sport Kites) ....(1978-79) 599
SWALES SAILPLANES (U.K.) 590

T

T-5 Blue Wren (Todhunter) 567, 596
T.53.C (Slingsby) ....(1972-73) 543
T.59H Kestrel 22 Srs 2 (Vickers-Slingsby/Glasflügel) .....(1978-79) 576, 586
T.61F Venture (Slingsby) 590, 602, 604
T.61G Falke (Slingsby) 590, 602, 604
T.65C Sport Vega (Slingsby) 590, 602
TA Series 2 Sprite (Torva) ....(1973-74) 578
TA Series 3 Sprint (Torva) ....(1973-74) 578
TCV-01 (Vaysse) ....(1972-73) 512
TCV-02 (Vaysse) ....(1972-73) 512
TCV-03 Trucavaysse (GEP) ....(1979-80) 562
TCV-04 (GEP) ....(1978-79) 558
TN-1 (Tainan) ....(1980-81) 576, 590
Taifun (Valentin) 583, 600, 604
TAINAN INDUSTRY CO LTD (Japan) 586
TAKATORI-SANKEN CO. LTD. (Japan) (1976-77) 575
Tandem-Falke (Scheibe) 580, 600, 604
Tantal Sinusöide (MAV) ....(1979-80) 597, 603
TECHNICAL CENTRE, CIVIL AVIATION DEPARTMENT (India) 584
TEKNILLINEN KORKEAKOULU (Finland) (1976-77) 552
Tern (Miller) ....(1976-77) 587
Tern II (Miller) ....(1976-77) 587
Tintti (PIK) ....(1972-73) 510
TODHUNTER, R. W. (Australia) 567
TOHOKU, UNIVERSITY OF (Japan)...(1972-73) 536
Topaze (SCAY) ....(1975-76) 569
Topaze, LA-11 (Loravia) ....(1979-80) 562
Tornado (United Hang Gliders) ....(1978-79) 597
Torva 15 Sprint (Torva) ....(1974-75) 596
Torva 15 Sprite (Torva) ....(1974-75) 596
TORVA SAILPLANES LTD (U.K.) ....(1975-76) 595
Trainer, SF-33 (Scheibe) ......(1978-79) 566, 584, 588
Trener (VTC) ....(1978-79) 580, 588
Trucavaysse TCV-01 (Vaysse) ....(1972-73) 512
Trucavaysse TCV-02 (Vaysse) ....(1972-73) 512
Trucavaysse (GEP) ....(1979-80) 562
Tumppi (PIK) ....(1972-73) 510
Tweetie (Skycraft) ....(1975-76) 605
Two-seat training sailplane (IPE) ....(1977-78) 578
Twin II, G 103 (Grob) 578, 598
Twin-Astir (Grob) ....(1980-81) 570
Twin-Astir Trainer (Grob) ....(1979-80) 566, 588
Twister V (Free Flight) ....(1979-80) 602
Type 01 Compact (GEPAS) ....(1980-81) 566, 588
Type 02 Nhapecam (IPE) 568, 596
Type 03 (IPE) ....(1980-81) 562
Type 12.80 Compact (Lagarde) ....(1977-78) 584
Type 15-34 Kit Club (Pottier) 574, 598
Type 101 (Centrair) 572, 598
Type 205 Club Libelle (Glasflügel) ......(1978-79) 562
Type 206 Hornet (Glasflügel) ....(1978-79) 562, 584
Type 303 Mosquito (Glasflügel) ....(1978-79) 562
Type 304 (Glasflügel) 578, 598
Type 402 (Glasflügel) 578, 598

U

ULTRALIGHT FLIGHT SYSTEMS (Australia) (1979-80) 596
UMBRA SpA, AERONAUTICA (Italy) ..(1972-73) 534
UNITED HANG GLIDERS (South Africa) (1979-80) 598
Universal, M-17 (Matejcek) ....(1973-74) 538
Upstart (UP Inc) ....(1978-79) 599
Urubu, PE-80367 (CTA) 567, 596
Urupema (IPD) ....(1971-72) 488
Utochka (Oshkinis) 588
Utu (Fibera) ....(1972-73) 509

V

VFW-FOKKER (see Vereinigte Flugtechnische Werke Fokker) ....(1972-73) 530
VJ-11 (Volmer) ....(1975-76) 612
VSB (see Aircraft Faculty of Antonina Zápotockého Military Academy, Brno) ......(1973-74) 539
VSB-66-S Orlice (VSB) ....(1972-73) 509
VSO (see Vyvojová Skupina Orlican) 570
VSO 10 (VSO) 570, 596
VTC (see Vazduhoplovno Tehnicki Centar-VRSAC) 596
VUK-T (MF) 595, 602
VALENTIN GmbH GERÄTE UND MASCHINENBAU (West Germany) 583
Vasama (PIK) ....(1974-75) 563
VAYSSE, DR. PIERRE (France) ....(1973-74) 544
VAZDUHOPLOVNO TEHNICKI CENTAR-VRSAC (Yugoslavia) 596
Vector 1 (Hall) ....(1976-77) 586
Vega (Slingsby) 590, 602
VEGA (U.S.S.R.) ....(1973-74) 586
Vega (Vickers-Slingsby) ....(1979-80) 578, 590
Vega II (Chargus) ....(1978-79) 597
Vega-2 (Vega) ....(1973-74) 586
VELIPLANE (France) ....(1978-79) 591
Venture, T.61F (Slingsby) 590, 602, 604
Ventus (Schempp-Hirth) 582, 600
VEREINIGTE FLUGTECHNISCHE WERKE-FOKKER GmbH (West Germany) ....(1972-73) 530
VICKERS-SLINGSBY, A DIVISION OF VICKERS LIMITED OFFSHORE ENGINEERING GROUP (U.K.) .....(1979-80) 578
Viking (Favul) ....(1979-80) 597, 602
VOLPAR INC (USA) 594
Virago, Moba 3 (Sunderland) 567, 596
VYVOJOVÁ SKUPINA ORLICAN (Czechoslovakia) 570

W

WA-26 Squale (Wassmer) ....(1975-76) 570
WA-26 CM (Wassmer) ....(1974-75) 567
WA-28 Espadon (Wassmer) ....(1977-78) 585
WA-28 Squale (Wassmer) ....(1974-75) 567
WA-30 Bijave (Wassmer) ....(1975-76) 570
WBAC (see Wright Brothers Aeroplane Co, The) (1979-80) 601
WK-1 (Kuffner) ....(1980-81) 570
WK-1 (Wala-Kralovik) ....(1972-73) 509
WM-1 (Wiegand) 584, 600
WM-2 (Miller) ....(1974-75) 602
WALA-KRALOVIK (Czechoslovakia) ...(1972-73) 509
WASPAIR LTD (UK) ....(1979-80) 599
WASSMER AVIATION, SOCIÉTÉ NOUVELLE (France) ....(1977-78) 585
Waterman & Seagull Flyer (Seagull) ......(1979-80) 600
WIEGAND, HEINZ (West Germany) 584
Windspiel (Penner) ....(1978-79) 592
WINGS (Australia) ....(1979-80) 596
Woodstock One (Maupin) 592, 602
Working Bee, MGM-2 (JEAA/Miyahara) (1974-75) 585
WRIGHT, PETER W (UK) 590
WRIGHT BROTHERS AEROPLANE CO, THE (USA) ....(1979-80) 601

X

X-7 Jian Fan (China) 569, 596
X-9 (China) 569
X 9 (Danis) ....(1979-80) 597
X-10 Qian Jin (China) 570
X-11 (China) 570
XL-13M (Let) ....(1979-80) 558
XL-13T (Let) ....(1979-80) 558
XG-001 (PATS/NSDB) ....(1979-80) 570

Y

YGC (see Yokohama Gliding Club) ......(1973-74) 566
YS 53 Sovereign (Yorkshire) ....(1974-75) 597
YS 55 Consort (Yorkshire) ....(1974-75) 597
Yarará (Aerofibra) ....(1974-75) 558
YOKOHAMA GLIDING CLUB (Japan) (1973-74) 566
YORKSHIRE SAILPLANES LTD. (U.K.) (1975-76) 595

Z

Z-75 (Wolf) ....(1978-79) 592
Z-75-7 (Wolf) ....(1978-79) 592
Z-76, Spring Wing (Wolf) ....(1979-80) 598, 603
Z-77, Elastic Wing (Wolf) ....(1979-80) 598, 603
Zamberle (Hufnagel) 579
ZANNER, OTTO (U.S.A.) ....(1976-77) 590
Zefir 4 (SZD) ....(1972-73) 538
Zile, BRO-11 (LAK) 588
Zile, BRO-11 (Oshkinis) 588
Zugvogel V (Scheibe) ....(1973-74) 553
Zuni (Applebay) ....(1980-81) 582, 592
Zuni, Model V (Aero Tek) ....(1979-80) 578, 592

MICROLIGHTS AND HANG GLIDERS

A

ACT (see Aircraft Composite Technology) 616
AFC (see Adventure Flight Centres Inc) 607
AN-20 (Neukom) 612
AN-20B (Richter) 610
AN-20M (Neukom) 612
AN-21R (Neukom) 613
AN-22 (Neukom) 613
ADVENTURE FLIGHT CENTRES INC (Canada) 607
AGRIPLANE (see La Culture de l'An 2000 SA ... 608
Agriplane 2000 (Agriplane) 608
Aiglon (Danis) 609
AIR CEFELEC AÉRONAUTIQUE (France) 608
AIRCRAFT COMPOSITE TECHNOLOGY (USA) 616
AMERICAN AEROLIGHTS (USA) 616
ANTONOV (USSR) 613
Aolus 150 (Spectra) 626, 632
Aolus 170 (Spectra) 626, 632
Aolus 190 (Spectra) 626, 632
Atlas 14 (La Mouette) 609, 630
Atlas 16 (La Mouette) 609, 630

B

BEIJING AERONAUTICS INSTITUTE (China) .. 608
BENNETT'S DELTA WING KITES & GLIDERS INC, BILL (USA) 618
BIRDMAN ENTERPRISES LTD (Canada) 607
BIRDMAN INC (USA) 616
BIRDMAN SPORTS LTD (UK) 614

C

CA-15 (Goldwing) 620
CGS (see CGS Aviation Inc) 616
CIVL (see Commission Internationale de Vol Libre) 605
CP-16 (Goldwing) 620
Canadian Skyseeker (AFC) 607
Canard-2 FL (Canard) 612
CANARD AVIATION (Switzerland) 612
CASCADE ULTRALIGHTS (USA) 616
CGS AVIATION INC (USA) 617
Challenger (Sunbird) 627
CHARGUS HANG GLIDERS (UK) 614
Cherokee (Birdman) 614, 632
Chotia Gypsy (Weedhopper) 629
CIESLAR, MARIAN AND ŚWIGOŃ, PIOTR (Poland) 612, 632
CIESLAR/ŚWIGOŃ (see Cieslar, Marian and Świgoń, Piotr) 612, 632
Cloudbuster (Kubasek) 622
Comet 135 (UP) 628, 632
Comet 165 (UP) 628, 632
Comet 185 (UP) 628, 632
COMMISSION INTERNATIONALE DE VOL LIBRE (France) 605
COOK, DAVID (UK) 614
Crechet (Leningrad) 614
Cricket (Winton) 606
Cutlass (Skyhook) 614, 632

D

DS-26 Nomad (Delta) 617
DS-27A Honcho (Delta) 617
DS-28A Super Honcho (Delta) 617
DAHU PLANE (ÉTABLISSEMENTS NOIN AÉRONAUTIQUE (France) 608
DANIS, BERNARD (France) 609
DECHOW, CURT (USA) 617
DELTA SAILPLANE CORPORATION (USA) ... 617
DELTA WING (see Bennett's Delta Wing Kites & Gliders Inc, Bill) 618
DMITRIYEV, VICTOR (USSR) 614
Doublequick (Eipper) 618

E

EEL (see Entwicklung und Erprobun von Leichtflugzeugen) 610
Eagle (American Aerolights) 616
Eaglet (Sky Sports) 625, 632
Easy Riser (UFM) 627
Easy Riser Arrow (UFM) 628
ECLIPSE (UK) 614
EIPPER-FORMANCE INC (USA) 618
EMC2 AVIONICS (Belgium) 606
ENTWICKLUNG UND ERPROBUN VON LEICHTFLUGZEUGEN (Germany) 610
EURO WING (UK) 614

F

Falcon I (Birdman) 607, 630
Falcon II (Birdman) 607, 630
Falcon V (CGS) 616, 632
Falcon 5+ (CGS) 616, 632
Feiyan (Beijing) 608
Feiyan II (Beijing) 608, 630
FISHER, MICHAEL (USA) 618
Fledge 2B (Manta) 622
Fledgling 430D (Pterodactyl) 624
Fledgling OR (Pterodactyl) 624
FLEXI-FORM SKYSAILS (UK) 614
FLIGHT DESIGNS (USA) 618
Flyer (Fisher) 618
Flyer (Kolb) 621

G

G11 (Moto Delta) 609
G20 (Moto Delta) 609
GEISER, JEAN MARC (France) 609
GEMINI INTERNATIONAL INC (USA) 619
GLA INC (USA) 619
Goldwing (Goldwing) 620
GOLDWING LTD (USA) 620
Goshawk (Highster) 620
Grach (Leningrad) 614
Grasshopper (Winton) 606
Gypsy, Chotia (Weedhopper) 629

H

H.M.81 Tomcat (Waspair) 628
HP.IV (High Perspective) 607
HALL, LARRY (USA) 620
Harrier 147 (Wills Wing) 629, 632
Harrier 177 (Wills Wing) 629, 632
HIGH PERSPECTIVE INC (Canada) 607
Highster (Highster) 620
HIGHSTER AIRCRAFT INC (USA) 620
Hi-Nuski (Lafayette) 622
HIRTH GmbH, WOLF (Germany) 610
HIWAY HANG GLIDERS LTD (UK) 614
Honcho, DS-27A (Delta) 617
Honcho, DS-28A (Delta) 617
Hummingbird (Gemini) 619
Hummingbird GT (Gemini) 619
Hummingbird Sport (Gemini) 619

I

Ibis (ACT) 616
IKARUSFLUG BODENSEE (Germany) 610

J

JC-24B Weedhopper (Weedhopper) 629
JC-24BL Weedhopper II (Weedhopper) 629
JC-24C (Weedhopper) 629
JD3 Puffin (Stewart) 625
JAEGER (USA) 621
J-Bird (Jaeger) 621
Johan SH-1 (LADAS) 630

K

Kasperwing (Cascade) 616
KAZUROV/POLUSHKIN/RUSAKOV (USSR) ... 614
KICENIUK JR, TARAS (USA) 621
KIMBERLEY, GARY J (Australia) 605
KITTY HAWK KITES INC (USA) 621
KOLB COMPANY INC (USA) 621
KUBASEK, JOHN W (USA) 622

L

LADAS (see Liga Argentina de Aerodeslizadores Superlivianos) 605
LEAF (see Leading Edge Air Foils Inc) 622
LA CULTURE DE L'AN 2000 SA (France) 608
LAFAYETTE AVIATION AND AIRCRAFT (U.S.A.) 622
LA MOUETTE (France) 609
Lancer IVL (Flight Designs) 618, 632
Lancer IVS (Flight Designs) 618, 632
Lazair (Ultraflight) 608
LEADING EDGE AIR FOILS INC (USA) 622
LIGA ARGENTINA DE AERODESLIZADORES SUPERLIVIANOS (Argentina) 605
Lightning P-38 (Mitchell) 624
Lone Ranger (Striplin) 626
Lotus 15 (La Mouette) 609, 630
Lotus 17 (La Mouette) 609, 630

M

MJ-5 (Birdman) 607, 630
MJ-6 (Birdman) 607, 630
MANTA PRODUCTS INC (USA) 622
Mars-Beryl, WT-10 (Talanchuk) 607, 630
MARSHALL, FRANK (UK) 614
MARSKE AIRCRAFT CORPORATION (USA) .. 622
Mars-Marta, WT-9 (Talanchuk) 607, 630
Maxi Stinger Mark III (Moyes) 605, 630
Maxi Stinger Mark IV (Moyes) 605
Maxi Stinger SP (Moyes) 605, 630
Mega Major (Moyes) 606, 630
Mega Mark II (Moyes) 606, 630
Mega Mark III (Moyes) 606, 630
Météor (Danis) 609
Mifeng-1 (Beijing) 608
Minibat (GLA) 619
MITCHELL AIRCRAFT CORPORATION (USA) 622
Mitchell Wing Model B-10 (Mitchell) 622
Mitchell Wing Model B-10F (Mitchell) 622
Mitchell Wing U-2 (Mitchell) 623
Model II (Monarch) 624, 632
Model V (Stratus) 626, 632
Model 140 (LEAF) 622, 632
Model 155 (LEAF) 622, 632
Model 170 (LEAF) 622, 632
Model 185 (LEAF) 622, 632
Model 200 (LEAF) 622, 632
Model 215 (LEAF) 622, 632
Model B-10, Mitchell Wing (Mitchell) 622
Model B-10F, Mitchell Wing (Mitchell) 622
Monarch C (Marske) 622
MONARCH PRODUCTS INC (USA) 624
Mono-Fly (Teman) 627
MOTO DELTA (see Geiser, Jean Marc) 609
Motor Hang IV (Abe) 612, 632
Mouette 15 (La Mouette) 609, 630
Mouette 17 (La Mouette) 609, 630
Mouette 19 (La Mouette) 609, 630
Mouette 23B (La Mouette) 609, 630
MOYES DELTA GLIDERS PTY LTD (Australia) 605

N

NFL (Pterodactyl) 624
NEUKOM SEGELFLUGZEUGBAU, ALBERT (Switzerland) 612
Nighthawk (Theis) 627
Nomad, DS-26 (Delta) 617
Nova 150 (Sunbird) 627, 632
Nova 170 (Sunbird) 627, 632
Nova 190 (Sunbird) 627, 632
Nova 210 (Sunbird) 627, 632
Nova 230 (Sunbird) 627, 632

O

ODYSSEY INC (USA) 624

P

P-38 Lightning (Mitchell) 624
PW (see Politechniki Warszawskiej) 612
PACIFIC GULL (USA) 624
Papillon (Paup) 624
PAUP, D (USA) 624
Peregrine (Sky Sports) 625
PFLUMM, MANFRED (Germany) 610
Phoenix 6D (Delta Wing) 618, 632
Phoenix Lazor 170 (Delta Wing) 618, 632
Phoenix Lazor 190 (Delta Wing) 618, 632
Phoenix Lazor II 155 (Delta Wing) 618, 632
Phoenix Lazor II 175 (Delta Wing) 618, 632
Phoenix Lazor II 195 (Delta Wing) 618, 632
Phoenix Viper (Delta Wing) 618, 632
PLETNYOV (USSR) 614
POLITECHNIKI WARSZAWSKIEJ (Poland) 612
Pro Air Series 1 (Progressive) 624
PROGRESSIVE AIRCRAFT COMPANY (USA) . 624
PTERODACTYL LTD (USA) 624
Ptraveler (Pterodactyl) 624
Puffin JD3 (Stewart) 626

Q

Quicksilver (Eipper) 618
Quicksilver MX (Eipper) 618

R

RFC (see Recreational Flight Centres Inc) 607
Rally 2B (Rotec) 624
Rally Marine (Rotec) 624
Raven 149 (Wills Wing) 629, 632
Raven 179 (Wills Wing) 629, 632
Raven 209 (Wills Wing) 629, 632
Raven 229 (Wills Wing) 629, 632
RECREATIONAL FLIGHT CENTRES INC (Canada) 607
Redtail Stinger (Moyes) 605, 630
RICHTER INGENIEURBÜRO, KLAUS J (Germany) 610
ROCHELT, GÜNTER (Germany) 611
ROTEC ENGINEERING INC (USA) 624

S

SL 180 (Flight Designs) 618, 632
SL 200 (Flight Designs) 618, 632
SL 220 (FLight Designs) 618, 632
SOFREC 'VELIPLANE' (see Société Française de Représentation et d'Échanges Commerciaux) 610
SP (see Siniaya Ptica) 614
SVAZARM (see Svazu pro spolupráci s Armádou) 608
Sabre (Danis) 609
Sabre (Skyhook) 615
Sauterelle (EMC2) 606
SCHÜRMANN, WERNER (Germany) 611
Scout (Wheeler) 606
Seaquick (Eipper) 618
SHENYANG AVIATION SCHOOL (China) 608
Shtyvnoplat (Kazurov/Polushkin/Rusakov) 614
Silhouette (Skyhook Sailwings) 615, 632
SINIAYA PTICA (USSR) 614
Sirocco III (Sky Sports) 625
SKYHOOK SAILWINGS LTD (UK) 614
Sky Ranger (Striplin) 626
Sky-Rider (Kimberley) 605
Skyrider (RFC) 607
Skytrike (Hiway) 614
SKY SPORTS INC (USA) 625
Slavutitch (Antonov) 613, 632
SOCIÉTÉ FRANÇAISE DE REPRÉSENTATION ET D'ÉXCHANGES COMMERCIAUX (France) 610
Sojourn I (Dechow) 617
Sokol (Leningrad) 614
Solair I (Rochelt) 611
Solar Riser (UFM) 628
SPECTRA AIRCRAFT CORPORATION (U.S.A.) 626, 632
Starship 170 (Ultimate Hi) 628, 632
Starship 190 (Ultimate Hi) 628, 632
Starship 220 (Ultimate Hi) 628, 632
STEWART AIRCRAFT CORPORATION (USA) 626
STRATUS (USA) 626
STRIPLIN AIRCRAFT CORPORATION (USA) . 626
SUNBIRD ULTRALIGHT GLIDERS (USA) 627
Sunfun VJ-24 (Volmer) 628
Super Eagle (Eclipse) 614
Super FLAC (Striplin) 626
SVAZU PRO SPOLUPRÁCI S ARMÁDOU (Czechoslovakia) 608
Swingwing VJ-23 (Volmer) 628

T

TL-1A (Birdman) 616
TALANCHUK, VLADIMIR (Canada) 607
TEMAN AIRCRAFT CO (USA) 627
TEMPLE, PROF DIPL-ING BERNHARD E.R.J. DE (Germany) 611
Temple Wing TE-F1 (Temple) 611
Temple Wing TE-F2 (Temple) 611
Temple Wing TE-F3 (Temple) 611
The Hummer (The Hummer) 620
THE HUMMER (USA) 620
THEIS, JAMES (USA) 627
Tomcat H.M.81 (Waspair) 628

U

UAC (see Ultralight Aviation Centre) 615
UFM (see Ultralight Flying Machines of Wisconsin) 627
ULF-1 (EEL) 610
ULM (see Ultra Léger Motorisé SARL) 610
ULM 811 (Air Cefelec) 608
UP (see UP Inc (Ultralite Products)) 628
ULTIMATE HI (USA) 628
ULTRAFLIGHT INC (Canada) 608
ULTRA LÉGER MOTORISÉ SARL (France) 610
ULTRALIGHT AVIATION CENTRE LTD (UK) 615
ULTRALIGHT FLYING MACHINES OF WISCONSIN (USA) 627
UP INC (ULTRALITE PRODUCTS) (USA) 628

V

VJ-23 Swingwing (Volmer) 628
VJ-23E (Volmer) 628
VJ-24 Sunfun (Volmer) 628
VJ-24E (Volmer) 628
Vampyr (Odyssey) 624
Vector 600 (Sky Sports) 625
Vega SST-A (Wierzbowski) 612, 632
Vega SST-B (Wierzbowski) 612, 632
Vega Star (Wierzbowski) 612, 632
VOLMER AIRCRAFT (USA) 628
Vortex (Chargus) 614

W

WT-9 Mars-Marta (Talanchuk) 607, 630
WT-10 Mars-Beryl (Talanchuk) 607, 630
WASPAIR CORPORATION (USA) 628
WHEELER AIRCRAFT (SALES) PTY LTD, RON (Australia) 606
WEEDHOPPER OF UTAH INC (USA) 629
WIERZBOWSKI, PAWEL (Poland) 612
WILLS WING INC (USA) 629
Windspiel 2 (Ikarusflug) 610, 630
WINTON AIRCRAFT (Australia) 606
Wizard (Schürmann) 611
WOLF, WILLIAM H (USA) 629

X

X-12C (Dmitriyev) 614
XC (Birdman) 607, 630

Z

ZK-1 (SVAZARM) 608
ZK-2 (SVAZARM) 608
Zhuravlic (Pletnyov) 614
ZMIEJ AERO CLUB (Bulgaria) 606
ZMIEJ-4 (Zmiej) 606, 630

LIGHTER-THAN-AIR

A

A-1 Albatross (Boland) 637
A50 (Adams') 643
A50S (Adams') 644
A55 (Adams') 644
A55S (Adams') 644
A-105 (Cameron) 644
A-140 (Cameron) 644
AD-500 (Airship Industries) 635
AG-60 (ILC Dover) 639
A.S.80 (ThunderColt) 637
AX-4 (Hare) .... (1979-80) 612
AX-5 (Hare) .... (1979-80) 612
AX-6 (Hare) .... (1979-80) 613
AX-7 (Hare) .... (1979-80) 613
AX-8 (Hare) .... (1979-80) 613
AX-9 (Hare) .... (1979-80) 613
ADAMS' BALLOON LOFT INC, MIKE (USA) .. 641
Aero-Ship Model 500 (Fuji) 635
AEROSPACE DEVELOPMENTS (LONDON) LTD (UK) .... (1979-80) 606
Air Chair (ThunderColt) 643
AIRSHIP INDUSTRIES LTD (UK) 635
Albatross, A-1 (Boland) 637
AVIAN BALLOON COMPANY (USA) 642

B

B 800/2-Ri (Ballonfabrik) 643
BALLONFABRIK SEE-UND LUFT-AUSRÜSTUNG GmbH und Co KG (West Germany) 640
BALLOON WORKS, THE (USA) 642
BALLONS CHAIZE (France) 640
Ballymurray (General) 643
BOLAND BALLOON (USA) 637, 642

C

CS.1600 (Chaize) 643
CS.1800 (Chaize) 644
CS.2000 (Chaize) 644
CS.2200 (Chaize) 644
CS.4000 (Chaize) 644
CAMERON BALLOONS LTD (UK) 636, 640
Challenger (Semco) .... (1979-80) 613
Cloudhopper Junior (ThunderColt) 643
Cloudhopper Super (ThunderColt) 643
CONRAD AIRSHIP CORPORATION (USA) .... (1979-80) 608

D

D38 (Cameron) 637
D96 (Cameron) 636
Dimples (Boland) 643

E

Eagle (Avian) .... (1979-80) 613
Enterprise (Goodyear) 638

F

Falcon II (Avian) 644
FireFly 5 (Balloon Works) 643
FireFly 6 (Balloon Works) 643
FireFly 6B (Balloon Works) 643
FireFly 7 (Balloon Works) 644
FireFly 8 (Balloon Works) 644
Fred II (Boland) 643
FUJI MANUFACTURING CO LTD (JAPAN) 635

G

GENERAL BALLOON CORPORATION (USA) . 642
GOODYEAR TIRE & RUBBER COMPANY — AIRSHIP OPERATIONS (USA) 638

H

HX-1 (Hov-Air-Ship) 638
HARE BALLOONS (USA) .... (1979-80) 611
High Fred (Boland) 643
HOV-AIR-SHIP INC (USA) 638

I

ILC DOVER (division of ILC Dover Industries Inc) (USA) 639

K

K 630/1-Ri (Ballonfabrik) 643
K 780/2-Ri (Ballonfabrik) 643
K 945/2-Ri (Ballonfabrik) 643
K 1050/3-Ri (Ballonfabrik) 643
K 1260/3-Ri (Ballonfabrik) 643
K 1680/4-Ri (Ballonfabrik) 643

L

LD (Adams') 643
LD-S (Adams') 643
Large-volume aerostat (ILC Dover) 639
Le Mans (General) 643
Lightning 33 (General) 643
Lightning 45 (General) 643

M

MA-1 (Mantainer) .... (1979-80) 606
MANTAINER PTY LTD (Australia) .... (1979-80) 606
Mayflower (Goodyear) .... (1979-80) 608
Model 500 Aero-Ship (Fuji) 635
Model 4060 (Airship Industries) 636
Model T (Semco) .... (1979-80) 613

N

N-31 (Cameron) 643
N-56 (Cameron) 643
N-65 (Cameron) 644
N-77 (Cameron) 644
N-105 (Cameron) 644
Newport (General) 644

O

O-20 (Cameron) .... (1980-81) 614
O-31 (Cameron) 643
O-42 (Cameron) 643
O-56 (Cameron) 643
O-65 (Cameron) 644
O-77 (Cameron) 644
O-84 (Cameron) 644
O-105 (Cameron) 644

P

Peaches (Boland) 643
Piccard (General) 643
Piccard A-3 (General) 643
Piccard A-4 (General) 643
Piccard A-5 (General) 643
PICCARD BALLOONS (USA) .... (1980-81) 614
Piccolo (Boland) 643
Piccup-3 (General) 643
Proposal 751 (Tucker) .... (1979-80) 609
Proposal 761 (Tucker) .... (1979-80) 609

R

RX-6 (Raven) 643
RX-7 (Raven) 644
RAVEN INDUSTRIES INC (USA) 642

S

S-40 (Raven) 643
S-50A (Raven) 643
S-55A (Raven) 644
S-60A (Raven) 644
S-66A (Raven) 644
SSP-6000 (ILC Dover) 639
SEMCO BALLOON (USA) .... (1979-80) 612
Silver Hawk, TX-1 (Tucker) 639
Sky Chariot (Maxi) (ThunderColt) 643
Sky Chariot (Midi) (ThunderColt) 643
Sky Chariot (Mini) (ThunderColt) 643
Skyhawk (Avian) 644
Skyship 500 (Airship Industries) 635
Skyship 600 (Airship Industries) 636
Sparrow (Avian) 643
Spirit of Lake Garda (Boland) 644
Stretch (General) 644

T

TC-4 (Semco) .... (1979-80) 613
TX-1 Silver Hawk (Tucker) 639
THUNDERCOLT BALLOONS LTD (UK) .. 637, 640
TUCKER AIRSHIP COMPANY (USA) 639
Type 30-AL (Semco) .... (1979-80) 612
Type 31Z (ThunderColt) 643
Type 42 (ThunderColt) 643
Type 42 Bolt (ThunderColt) 643
Type 56 (ThunderColt) 643
Type 56 Bolt (ThunderColt) 643
Type 56Z (ThunderColt) 643
Type 65 (ThunderColt) 644
Type 65 Bolt (ThunderColt) 644
Type 65Z (ThunderColt) 644
Type 77 (ThunderColt) 644
Type 77 Bolt (ThunderColt) 644
Type 77Z (ThunderColt) 644
Type 84 (ThunderColt) 644
Type 90 (ThunderColt) 644
Type 105 (ThunderColt) 644
Type 140 (ThunderColt) 644
Type 160A (ThunderColt) 644
Type 200A (ThunderColt) 644

V

Viva-20 (Cameron) 643
Viva-31 (Cameron) 643
Viva-56 (Cameron) 643
Viva-65 (Cameron) 644
Viva-77 (Cameron) 644

X

XXUS (Boland) 643

RPVs AND TARGETS

A

ACT (Lockheed-Georgia/NASA) (1976-77) 627, 628
ADE (see Aeronautical Development Establishment) (1976-77) 610
AEL (see Aero Electronics (AEL) Ltd) 654
AEL4012 Target System (AEL) (1974-75) 539
AEL4015 Camera System (AEL) (1974-75) 539
AEL4017 Composite Target/Camera System (AEL) (1974-75) 540
AEL4018 Target System (AEL) (1974-75) 539
AEL4019 Camera System (AEL) (1974-75) 539
AEL4020 Composite Target/Camera System (AEL) (1974-75) 539
AEL4024 Heron (AEL) (1974-75) 540
AEL4030 Training Drone (AEL) (1974-75) 540
AEL4041 Mossette (AEL) (1974-75) 540
AEL4111 Snipe Mk II (AEL) 654
AEL4560 Smidge (AEL) (1979-80) 624
AEL 4600 Streek (AEL) 655
AMC (see Area de Material Córdoba, Fuerza Aérea Argentina) 645
AMMR (USAF) (1975-76) 559
AMR (see Aero-Marine Research) (1977-78) 657
AN/USD-501 (Canadair) 645
AN/USD-502/CL-289 (International) 650
APL (see Applied Physics Laboratory, The Johns Hopkins University) 658
AQM-34 (Teledyne Ryan) 670
AQM-37A (Beechcraft) 659
AQM-81A (Beechcraft) (1979-80) 628
AQM-81A (Teledyne Ryan) 670
AQM-91A (Teledyne Ryan) (1975-76) 555
ARC (see Atlantic Research Corporation) (1977-78) 658
ARPV (USAF) (1977-78) 677
ASAT (FR) 656
Active Expendable Decoy (DSI/US Navy) 662
Advanced RPV (Boeing) (1977-78) 661
Advanced RPV (Northrop) (1977-78) 670
Advanced RPV (Rockwell International) (1977-78) 670
Aequare (LMSC) (1978-79) 628
Aerodyne E 1 (Dornier) (1978-79) 613
AERO ELECTRONICS (AEL) LTD (U.K.) 654
AERO-MARINE RESEARCH (U.S.A.) (1977-78) 657
AERONAUTICAL DEVELOPMENT ESTABLISHMENT (India) (1976-77) 610
AERONUTRONIC FORD CORPORATION (U.S.A.) (1976-77) 618
AÉROSPATIALE (see Société Nationale Industrielle Aérospatiale) 646
AMBROSINI & C, SOCIETA AERONAUTICA ITALIANA ING A (Italy) (1973-74) 512
Andromeda/Mirach System (Meteor) 652
AN/USD-502 (International) 650
APPLIED PHYSICS LABORATORY, THE JOHNS HOPKINS UNIVERSITY (U.S.A.) 658
Apteron (ILC Dover) (1980-81) 636
Aquila (DSI/Lockheed) (1977-78) 663
Aquila (LMSC) 664
Aquila (U.S. Army) (1979-80) 642
AREA DE MATERIAL CÓRDOBA, FUERZA AÉREA ARGENTINA (Argentina) 645
Argus (Argus) 650
ARGUS (International) 650
Asmodée (MBLE) (1979-80) 616
ATLANTIC RESEARCH CORPORATION (U.S.A.) (1977-78) 658
Axillary (USAF) (1978-79) 634

B

BAC (see British Aircraft Corporation (Guided Weapons) Ltd) (1977-78) 652
BAe (see British Aerospace Public Limited Company, Dynamics Group) 655
BATS (SPT-B) (Teledyne Brown) (1973-74) 526
BD-5J (Bede) (1975-76) 541
BQM-34A (Teledyne Ryan) 669
BQM-34E (Teledyne Ryan) (1980-81) 642
BQM-34F (Teledyne Ryan) (1980-81) 642
BQM-34S (Teledyne Ryan) 669
BQM-34T (Teledyne Ryan) (1980-81) 642
BQM-74C (Northrop) 666
BQM-90 (US Navy) (1973-74) 535
Basic sailwing RPV (Fairchild) (1975-76) 546
Basic Training Target (Northrop) 665
BEDE AIRCRAFT INC (U.S.A.) (1975-76) 541
BEECH AIRCRAFT CORPORATION (U.S.A.) 659
BEECHCRAFT (see Beech Aircraft Corporation) 659
Beemoth, HDH-11 (Hawker de Havilland) (1980-81) 617
BENDIX CORPORATION, THE (U.S.A.) (1974-75) 543
BENSEN AIRCRAFT CORPORATION (USA) 660
B-Gull (Boeing) (1978-79) 624
BIG BIRD (U.S.A.) (1979-80) 630
Big Bird (Big Bird) (1979-80) 630
Black Fly Sailwing RPV (Fairchild) (1973-74) 521
BOEING AEROSPACE COMPANY (U.S.A.) (1978-79) 624
BOEING VERTOL COMPANY (U.S.A.) (1977-78) 661
Bonanza Model A36 (Sperry/Beechcraft) (1974-75) 550
BRITISH AEROSPACE PUBLIC LIMITED COMPANY, DYNAMICS GROUP (U.K.) 655
BRITISH AIRCRAFT CORPORATION (GUIDED WEAPONS) LTD (U.K.) (1977-78) 652
BRUNSWICK DEFENSE DIVISION (U.S.A.) 660

C

C.20 (Aérospatiale) (1978-79) 611
C.20 TBA (Aérospatiale) (1977-78) 644
C.22 (Aérospatiale) 646
C.30C (Aérospatiale) (1973-74) 509
CL-89 (Canadair) 645
CL-227 (Canadair) 646
CL-289 (Canadair) (1978-79) 614
CT.20 (Aérospatiale) 646
CT.20 TBA (Aérospatiale) 646
Calere (Ford) (1977-78) 665
Calere III (Ford) (1977-78) 665
CANADAIR LIMITED (Canada) 645
Cardinal 1025 (Beechcraft) (1973-74) 516
CELESCO INDUSTRIES INC (U.S.A.) (1976-77) 622
CHINA, PEOPLE'S REPUBLIC (1977-78) 644
Chukar I, NV-105 (Northrop) (1975-76) 550
Chukar II (Northrop) 666
Chukar III (Northrop) (1979-80) 637
Chukar-R (Northrop) 666
Compass Cope B (Boeing) (1978-79) 624
Compass Cope R (Teledyne Ryan) (1978-79) 634
Condor (Ambrosini) (1973-74) 512
Constant Angel (USAF) (1976-77) 638

D

D 15 (Aérospatiale) (1977-78) 645
DATS (Dornier) (1975-76) 534
Do 32 K Experimental Kiebitz (Dornier) (1977-78) 646
Do 34 Kiebitz (Dornier) 648
DS.7 (Dorand) (1975-76) 533
DSI (see Developmental Sciences Inc) 661
Delta Dagger, F-102A (Sperry) 668
DEVELOPMENTAL SCIENCES INC (U.S.A.) 661
DORAND, GIRAVIONS (France) (1976-77) 607
DORNIER GmbH (West Germany) 648
DORNIER SYSTEMS GmbH (West Germany) (1973-74) 511

E

E 1, Aerodyne (Dornier) (1978-79) 613
E-45 (E-Systems) (1978-79) 626
E-55 (E-Systems) 662
E-55S (E-Systems) 662
E-75 (E-Systems) (1978-79) 626
E-90 (E-Systems) 662
E-100X (E-Systems) 662
E-130 (E-Systems) 662
E-150 (E-Systems) 663
E-200 (E-Systems) 663
ECA (see Société ECA) (1973-74) 509
ECA 57 (ECA) (1973-74) 509
EERM (see Etablissements d'Etudes et de Recherches Météorologiques) 647
EHI (see Eglen Hovercraft Inc) (1980-81) 635
Eagle (Eglen/USAF) (1977-78) 664
EGLEN HOVERCRAFT INC (U.S.A.) (1980-81) 635
Electronic Warfare RPV (US Army) (1980-81) 643
Enmoth (Hawker de Havilland) (1979-80) 615
Épervier System (MBLE) (1979-80) 616
E-SYSTEMS INC (U.S.A.) 662
ETABLISSEMENTS D'ETUDES ET DE RECHERCHES METEOROLOGIQUES (France) 647
Experimental Kiebitz (Dornier) (1977-78) 646

F

F-100 Super Sabre (Sperry) 668
F-102A Delta Dagger (Sperry) 668
FMA (see Area de Material Córdoba) 645
FOA (see Försvarets Forskningsanstalt) 654
FQM-117A (RS Systems) 667
FR (see Flight Refuelling Ltd) 656
FR-500 (FR) (1977-78) 653
FSI (see Flight Systems Inc) 664
FAIRCHILD AIRCRAFT SERVICE DIVISION (U.S.A.) (1978-79) 626
FAIRCHILD REPUBLIC COMPANY (U.S.A.) (1977-78) 664
FAIRCHILD SPACE & ELECTRONICS COMPANY (U.S.A.) (1975-76) 546
Falcon (RCS) (1980-81) 628
Falcon-M (EHI) (1979-80) 633
Falcon (Meridian) (1975-76) 548
Falconer (Northrop) (1972-73) 502
Firebee I (Teledyne Ryan) 669
Firebee II (Teledyne Ryan) (1980-81) 642
Firebee II RPRV (NASA/Teledyne Ryan) (1978-79) 629
Firebolt, Model 305 (Teledyne Ryan) 670
Firebrand (Teledyne Ryan) 670
Firefly (Teledyne Ryan) (1975-76) 555
FLIGHT REFUELLING LTD (U.K.) 656
FLIGHT SYSTEMS INC (U.S.A.) 664
Flybac (BAe) 655
FORD AEROSPACE & COMMUNICATIONS CORPORATION (U.S.A.) (1978-79) 627
FÖRSVARETS FORSKNINGSANSTALT (Sweden) 654
FRANCE-ENGINS, SOCIÉTÉ ANONYME (France) 648
FUJI (see Fuji Jukogyo Kabushiki Kaisha) 653
FUJI JUKOGYO KABUSHIKI KAISHA (Japan) 653

G

GAF (see Government Aircraft Factories) 645
GTD-21B (LMSC) (1979-80) 634
GTS 7901 Sky-Eye (Intergard) 656
GIRAVIONS DORAND (France) (1976-77) 607
Girun, RB06 (Beechcraft) 660
GOVERNMENT AIRCRAFT FACTORIES (Australia) 645
GOVERNMENT OF AUSTRALIA, DEPARTMENT OF MANUFACTURING INDUSTRY (Australia) (1976-77) 602
Gufetto (Meteor) (1973-74) 513
Gufo (Meteor) (1979-80) 622
Gufone (Meteor) (1978-79) 616
Gunsight (DSI) 662
GYRODYNE COMPANY OF AMERICA, INC. (U.S.A.) (1975-76) 547

H

HAHST (Beechcraft) (1980-81) 632
HASPA (US Navy) (1979-80) 642
HAST (Beechcraft) (1980-81) 632
HDH (see Hawker de Havilland Australia Pty Ltd) (1980-81) 617
HDH-10 Enmoth (Hawker de Havilland) (1979-80) 615
HDH-11 Beemoth (Hawker de Havilland) (1980-81) 617
HDH-12 Promoth (HDH) (1980-81) 617
HiMAT (NASA/Rockwell International) 666
HI-SPOT (US Navy) 670
HWS (Northrop) (1979-80) 638
HWS (US Air Force) (1980-81) 643
Harassment Weapon System (LMSC) (1978-79) 628
Harassment Weapons System (Northrop) (1979-80) 638
Harassment Weapons System ((USAF) (1980-81) 643
HAWKER DE HAVILLAND AUSTRALIA PTY LTD (Australia) (1980-81) 617
Helicopter drone kits (Kaman) 664
Heliscope (Marchetti) (1978-79) 612
Heron (RCS) (1980-81) 628
Heron HS (RCS) (1980-81) 628
Hyper III (NASA) (1973-74) 523

I

IA 59 (AMC) 645
IAI (see Israel Aircraft Industries) 650
IETS 7501 Shrike (Intergard) 656
IETS 7502 Pipit (Intergard) 657
ILC (see ILC Industries Inc) 664
ILC Dover (see ILC Industries Inc) 664
ILC INDUSTRIES INC (U.S.A.) 664
Improved BATS (SPT-B2A) (Teledyne Brown) (1973-74) 526
INTERGARD ELECTRONICS LTD (U.K.) 656
ISRAEL AIRCRAFT INDUSTRIES (Israel) 650

J

Jindivik (GAF) 645

K

KDB-1 (Beechcraft) (1973-74) 516
KD2B-1 (Beechcraft) (1973-74) 516
KD2R-5 (Northrop) 665
KDAR (Dornier) 648
KDAR (MBB) 650
KDAR (VFW) 650
KAMAN AEROSPACE CORPORATION (U.S.A.) 664
"Kamikaze" RPV (US Army) (1976-77) 639
KAWASAKI JUKOGYO KABUSHIKI KAISHA (Japan) (1971-72) 481
Kiebitz (Dornier) 648

L

L450F (E-Systems) (1977-78) 664
LAMPS (USAF) (1979-80) 642
LAPAN (see Lembaga Penerbangan dan Antariska Nasional) 650
LA-RPV (Dornier) 648
LAST (Bendix) (1973-74) 518
LCEHV (LMSC) (1978-79) 628
LCEHV (Northrop) (1978-79) 630
LCEHV (USAF) (1978-79) 634
LCT (see Laboratoire Central de Télécommunications) (1980-81) 622
LMSC (see Lockheed Missiles and Space Company Inc) 664
LOCAT (Philco-Ford) (1973-74) 525
LABORATOIRE CENTRAL DE TÉLÉCOMMUNICATIONS (France) (1980-81) 622
Large rotary-wing RPV (RCS) (1980-81) 630
LEMBAGA PENERBANGAN DAN ANTARISKA NASIONAL (Indonesia) 650
Little Scout (US Army) (1979-80) 642
LOCKHEED-CALIFORNIA COMPANY (U.S.A.) (1974-75) 547
LOCKHEED CORPORATION (USA) (1978-79) 628
LOCKHEED-GEORGIA COMPANY (U.S.A.) (1976-77) 627
LOCKHEED MISSILES AND SPACE COMPANY INC (U.S.A.) 664
Locomp (DSI) 662
Locust (Dornier) 648
Locust (MBB) 650
Locust (USAF) 670
Locust (VFW) 650

M

MAP/UV 8001 (APL) 658
MATS-A (Skyleader) (1979-80) 627
MATS-B (Shorts) (1979-80) 626
MBB (see Messerschmitt-Bölkow-Blohm GmbH) 650
MBLE (see Manufacture Belge de Lampes et de Matériel Electronique SA) (1979-80) 616
MCS (see Mid-Continent Scientific) 654
MoD (Army) (see Ministry of Defence (Army)) (1975-76) 539
MQM-33 (Northrop) (1974-75) 548
MQM-33C (Northrop) 665
MQM-34D (Teledyne Ryan) 669

MQM-34D/MOD II (Teledyne Ryan) ....(1975-76) 558
MQM-36 (Northrop) ....(1974-75) 548
MQM-39A (Beechcraft) ....(1973-74) 516
MQM-61A Cardinal (Beechcraft) ....(1973-74) 516
MQM-74A (Northrop) ....(1975-76) 550
MQM-74C (Northrop) 666
MQM-107A Streaker (Beechcraft) 660
MQM-107B (Beechcraft) 660
MTC II (Dornier) 648
Machan (Marconi) 657
MANUFACTURE BELGE DE LAMPES ET DE MATÉRIEL ÉLECTRONIQUE SA (Belgium) ....(1979-80) 616
MARCHETTI, SA CHARLES (France) (1978-79) 612
MARCONI AVIONICS LTD (U.K.) 657
Maritime Aerodyne (Dornier) ....(1977-78) 648
Mars Astroplane (DSI) ....(1980-81) 634
MARTIN MARIETTA AEROSPACE (U.S.A.) (1974-75) 547
Mastiff (Tadiran) 652
Matador (EHI) ....(1979-80) 634
Maxi-decoy (Brunswick) 660
MCDONNELL DOUGLAS ASTRONAUTICS COMPANY (U.S.A.) ....(1977-78) 667
MCDONNELL DOUGLAS CORPORATION (U.S.A.) ....(1978-79) 628
MERIDAN CORPORATION (U.S.A.) ..(1975-76) 619
Merlin (RCS) ....(1980-81) 628
MESSERSCHMITT-BÖLKOW-BLOHM GmbH (West Germany) 650
METEOR SpA AIRCRAFT AND ELECTRONICS INDUSTRY (Italy) 652
MID-CONTINENT SCIENTIFIC CO (Saudi Arabia) 654
Midge (Canadair) ....(1980-81) 617
Minipop RPV (Fairchild) ....(1975-76) 547
Mini-RPV (Boeing Vertol) ....(1977-78) 661
Mini-RPV (Dornier) ....(1979-80) 619
Mini RPV (MBB) ....(1978-79) 614
"Mini-RPV" (McDonnell Douglas) ......(1975-76) 547
Mini-RPV (VFW-Fokker) ....(1979-80) 620
Mini-RPV, 003 (ARC) ....(1977-78) 658
Mini-RPVs (Aeronutronic Ford) ....(1975-76) 540
Mini-Sniffer II (NASA) 665
MINISTRY OF DEFENCE (ARMY) (U.K.) (1975-76) 539
Mirach (Meteor) 652
Mitsoubac (France-Engins) 648
Model 124 Firebee I (Teledyne Ryan) 669
Model 147 (Teledyne Ryan) 670
Model 154 (Teledyne Ryan) ....(1975-76) 555
Model 166 Firebee II (Teledyne Ryan) ..(1980-81) 642
Model 234 (Teledyne Ryan) ....(1979-80) 641
Model 235 (Teledyne Ryan) ....(1978-79) 634
Model 239 (Teledyne Ryan) ....(1976-77) 637
Model 251 (Teledyne Ryan) ....(1975-76) 558
Model 255 (Teledyne Ryan) 670
Model 259 (Teledyne Ryan) ....(1979-80) 641
Model 262 (Teledyne Ryan) ....(1978-79) 634
Model 272 (Teledyne Ryan) ....(1975-76) 558
Model 275 (Teledyne Ryan) ....(1975-76) 558
Model 305 Firebolt (Teledyne Ryan) 670
Model 845A (Martin Marietta) ....(1974-75) 547
Model 999F/MQM-107B (Beechcraft) 660
Model 1001 (Beechcraft) ....(1973-74) 516
Model 1019 Series (Beechcraft) 659
Model 1019A (Beechcraft) 659
Model 1025 (Beechcraft) ....(1973-74) 516
Model 1070 HAST (Beechcraft) ....(1980-81) 632
Model 1072 (Beechcraft) 659
Model 1088 (Beechcraft) 660
Model 1089 (Beechcraft) 660
Model 1089E TEDS (Beechcraft) ....(1978-79) 624
Model 1089 VSTT (Beechcraft) ....(1976-77) 620
Model 1092 (Beechcraft) ....(1974-75) 542
Model 1094 (Beechcraft) 660
Model 1095 (Beechcraft) 660
Model 1098 (Beechcraft) 660
Model 1100 (Beechcraft) 660
Model 1101 (Beechcraft) 660
Model 1102 (Beechcraft) 660
Model 1104 (Beechcraft) 660
Model 1241 (Teledyne Ryan) ....(1978-79) 632
Model A36 Bonanza (Sperry/Beechcraft) (1974-75) 550
Mossette (RCS) ....(1976-77) 615

N

NASA (see National Aeronautics and Space Administration) 665
NAT (U.S. Navy) ....(1977-78) 679
ND-100 Observer (National Dynamics) ..(1979-80) 624
NEC (see Nippon Electric Company Ltd) 654
NV-105 Chukar I (Northrop) ....(1975-76) 550
NV-123 VSTT (Northrop) ....(1976-77) 620
NV-128 (Northrop) ....(1976-77) 629
NV-130 'Nutcracker' (Northrop) ....(1979-80) 638
NV-135 (Northrop) ....(1979-80) 638
NATIONAL AERONAUTICS AND SPACE ADMINISTRATION (U.S.A.) 665
NATIONAL DYNAMICS (PTY) LTD (South Africa) ....(1980-81) 626
NIPPON DENKI KABUSHIKI KAISHA (Japan) (1977-78) 652
NIPPON ELECTRIC COMPANY LTD (Japan) ... 654
Nord CT.20 (Aérospatiale) ....(1978-79) 611
Nord R.20 (Aérospatiale) ....(1977-78) 645
NORTHROP CORPORATION, VENTURA DIVISION (U.S.A.) 665
'Nutcracker' (Northrop) ....(1979-80) 638

O

Oblique wing RPRA (DSI/NASA) ....(1979-80) 628
Observer, ND-100 (National Dynamics) ..(1979-80) 624

P

P.1 (Meteor) ....(1977-78) 649
P.1/R (Meteor) ....(1975-76) 537
P.2 (Meteor) ....(1974-75) 537
PL-60 (MCS) 654
PQM-102 (Sperry) 668
P.X (Meteor) ....(1979-80) 622
PHILCO-FORD CORPORATION SUBSIDIARY OF FORD MOTOR COMPANY (U.S.A.) ....(1974-75) 550
Pipit, IETS 7502 (Intergard) 657
Praeire IIB (Ford) ....(1977-78) 665
Project DAST (NASA) 665
Project Lookout (U.S. Navy) ....(1973-74) 535
Promoth, HDH-12 (HDH) ....(1980-81) 617
Propelled-decoy (Brunswick) 660

Q

Q-2C (Teledyne Ryan) ....(1972-73) 503
QF-4B (U.S. Navy) 670
QF-86 (U.S. Navy) 670
QF-86E Sabre (FSI) 664
QF-86F Sabre (Sperry) 668
QF-100 (Sperry) 668
QF-102 (Sperry) 668
QH-50D (Gyrodyne) ....(1975-76) 547
QT-38 (US Navy) ....(1977-78) 679
QU-22A (Sperry/Beechcraft) ....(1974-75) 550
QU-22B (Sperry/Beechcraft) ....(1974-75) 550

R

R.20 (Aérospatiale) ....(1977-78) 645
RB06 Girun (Beechcraft) 660
RCMAT (RS Systems) 667
RCS (see RCS Guided Weapons System) 658
RCS4012 target system (RCS) ....(1980-81) 628
RCS4015 camera system (RCS) ....(1980-81) 628
RCS4017 composite target/camera system (RCS) ....(1980-81) 628
RCS4018 target system (RCS) ....(1980-81) 628
RCS4019 camera system (RCS) ....(1980-81) 628
RCS4020 composite target/camera system (RCS) ....(1980-81) 628
RCS4024 Heron (RCS) ....(1980-81) 628
RCS4030 Training Drone (RCS) ....(1980-81) 628
RCS4041 Mossette (RCS) ....(1976-77) 615
RED (USAF) ....(1974-75) 557
RPA-9 Scout (DSI) ....(1979-80) 630
RPA-12 Sky Eye R-4D (DSI) 661
RPAODS (U.S. Army) ....(1976-77) 639
RPD2 (APL) ....(1980-81) 630
RPMB (DSI) 662
RPV-004 (AMR) ....(1977-78) 657
RPV-007 (AMR) ....(1977-78) 657
RPV launcher and recovery system (Fairchild) (1974-75) 546
RPV testbed (Lockheed) ....(1974-75) 547
RT 910 Tucan (VFW) 650
RTV-2 (LMSC) ....(1975-76) 547
RCS GUIDED WEAPONS SYSTEM (U.K.) 658
Recovery Systems (U.S.A.F.) ....(1977-78) 678
ROCKWELL INTERNATIONAL CORPORATION, LOS ANGELES AIRCRAFT DIVISION (U.S.A.) ....(1979-80) 638
ROCKWELL INTERNATIONAL CORPORATION, MISSILE SYSTEM DIVISION (U.S.A.) (1978-79) 630
ROCKWELL INTERNATIONAL CORPORATION, NORTH AMERICAN AIRCRAFT DIVISION (U.S.A.) 666
Rotormobile (Marchetti) ....(1974-75) 534
RS SYSTEMS (U.S.A.) 667
R-Tern (Teledyne Ryan) ....(1978-79) 634
Ryan-Benson RPV-007 (AMR) ....(1977-78) 657

S

SAM-B (EERM) 647
SAM-C (EERM) 647
SAM-C MP (EERM) 647
SAM-C MAR (EERM) 647
SAM-D (EERM) 647
SATS (RCS) ....(1980-81) 628
SD.2 Stiletto (Beechcraft) 659
SPT-B (BATS) (Teledyne Brown) ....(1973-74) 526
SPT-B24 (Improved BATS) (Teledyne Brown) (1973-74) 526
STAPL (Kaman) 664
SA CHARLES MARCHETTI (France) ..(1978-79) 612
Sabre (FSI) 664
Sabre, F-86F (Sperry) 668
Scout (DSI) ....(1979-80) 630
Scout (IAI) 650
Sea Skipper (Beechcraft) ....(1979-80) 628
Shelduck (Northrop) ....(1976-77) 628
SHORT BROTHERS LTD (U.K.) 658
Shrike, IETS 7501 (Intergard) 656
Skatan (FOA) 654
Skeet (Shorts) 658
Sky-Eye, GTS 7901 (Intergard) 656
Sky Eye, RPA-12 (DSI) ....(1980-81) 632
Sky Eye R-4D (DSI) 661
SKYLEADER RADIO CONTROL LTD (U.K.) ....(1980-81) 630
Skyspy (Short) ....(1978-79) 621
Small rotary-wing drone (RCS) ....(1980-81) 629
Smidge, AEL4560 (AEL) ....(1979-80) 624
Snipe Mk II, AEL4111 (AEL) 654
SOCIÉTÉ ANONYME FRANCE-ENGINS (France) 648
SOCIÉTÉ ECA (France) ....(1973-74) 509
SOCIÉTÉ NATIONALE INDUSTRIELLE AÉROSPATIALE (France) 646
Spähplattform (Dornier) 649
SPERRY FLIGHT SYSTEMS DIVISION, SPERRY CORPORATION (U.S.A.) 668
Stabileye (BAe) 655
Stiletto (Beechcraft) 659
Streaker (Beechcraft) 660
Super Condor (CSTM/Ambrosini) ....(1973-74) 512
Super Sabre, F-100 (Sperry) 668
Swift (RCS) ....(1980-81) 628

T

Tactical Expendable Drone System (Northrop) ....(1977-78) 669, 678
Tactical High-Speed Surveillance RPV (US Army) (1976-77) 639
TADIRAN ISRAEL ELECTRONICS INDUSTRIES LTD (Israel) 652
Tailwind (Drone version) (BAC/AJEP) ..(1977-78) 652
TEDS (Beechcraft) ....(1978-79) 624
TEDS (Northrop) ....(1977-78) 669, 678
TEDS (USAF) ....(1977-78) 678
TELEAVIO SpA (Italy) ....(1973-74) 514
Teledrone II (RCS) ....(1980-81) 629
TELEDYNE BROWN (see Teledyne Brown Engineering) ....(1974-75) 550
TELEDYNE BROWN ENGINEERING (U.S.A.) ....(1974-75) 550
TELEDYNE RYAN (see Teledyne Ryan Aeronautical) 669
TELEDYNE RYAN AERONAUTICAL (U.S.A.) . 669
Tucan, RT 910 (VFW) 650
Turana (GAF) ....(1980-81) 617
Type 003 Mini-RPV (ARC) ....(1977-78) 658

U

U-2 (Lockheed) ....(1974-75) 547
UKF (Dornier) ....(1977-78) 647
UR-1 (USSR) 654
UNITED STATES AIR FORCE (USA) 670
UNITED STATES ARMY (USA) ....(1980-81) 643
UNITED STATES NAVY (USA) 670
UNIVAC DIVISION, SPERRY RAND CORPORATION (U.S.A.) ....(1973-74) 526
Universal Drone Pack (FR) ....(1979-80) 624
US DEPARTMENT OF DEFENSE (U.S.A.) 670

V

VATOL (US Navy) ....(1980-81) 643
VF-500 (VFW) ....(1980-81) 622
VF-500 (VFW-Fokker) ....(1979-80) 620
VFW (see Vereinigte Flugtechnische Werke GmbH) 650
VLCHV (USAF) ....(1976-77) 638
VSTT (Beechcraft) ....(1975-76) 542
VSTT (Northrop) ....(1976-77) 620
Vanneau (Beechcraft) 660
VENTURA DIVISION, NORTHROP CORPORATION (U.S.A.) 665
VEREINIGTE FLUGTECHNISCHE WERKE GmbH (West Germany) 650
VINTEN LTD, W (UK) 658

W

WALLIS AUTOGYROS LTD (U.K.) 658
WALLIS/VINTEN (U.K.) 658
WESTLAND HELICOPTERS LTD (U.K.) 658
Wideye/Supervisor (Westland) 658
WINDECKER INDUSTRIES INC (U.S.A.) (1975-76) 560
Wisp (Westland) ....(1978-79) 622

X

X-4 Epervier (MBLE) ....(1976-77) 604
X-5 Epervier (MBLE) ....(1976-77) 604
XBQM-8D (Bendix) ....(1973-74) 518
XBQM-8F (Bendix) ....(1973-74) 518
XBQM-106 (U.S.A.F.) 670
XBQM-108A (U.S. Navy) ....(1980-81) 643
XBQM-111A (Teledyne Ryan) ....(1979-80) 641
XD-110 (AMR) ....(1977-78) 657
XMQM-105 (LMSC) ....(1980-81) 636
XQM-93 (E-Systems) ....(1977-78) 664
XQM-103 (Teledyne Ryan) ....(1978-79) 632
XTG-01 (LAPAN) ....(1980-81) 622

Y

Yak-25RD (Yakovlev) ....(1977-78) 679
YAQM-34U (Teledyne Ryan) ....(1978-79) 632
YMQM-105 (LMSC) 664
YQM-94A (Boeing) ....(1977-78) 661
YQM-98A (Teledyne Ryan) ....(1977-78) 676

Z

ZBQM-111A (Teledyne Ryan) 670

AIR-LAUNCHED MISSILES, SPACEFLIGHT

A

AA-1 (U.S.S.R.) 676, 688
AA-2 (U.S.S.R.) 676, 688
AA-2-2 (U.S.S.R.) 677
AA-3 (U.S.S.R.) 676, 688
AA-5 (U.S.S.R.) 676, 688
AA-6 (U.S.S.R.) 678
AA-7 (U.S.S.R.) 678, 688
AA-8 (U.S.S.R.) 678, 688
AAM-1 (Mitsubishi) (1977-78) 689
AAM-2 (Mitsubishi) (1980-81) 649
AAM-N-6B Sparrow (Raytheon) (1978-79) 649, 652
AAM-N-7 (NWC) (1974-75) 632
AAM-N-11 Phoenix (Hughes) (1978-79) 646, 652
AA-X-9 (U.S.S.R.) 678
ADSM (Rockwell International) (1977-78) 702
AE-C (NASA) (1974-75) 665
AEM-B (NASA) (1980-81) 678
AGM-12 Bullpup (Martin) (1971-72) 557
AGM-22A (Aérospatiale) (1972-73) 555
AGM-45A (NWC) (1978-79) 648, 652
AGM-53 (Rockwell International) (1977-78) 702
AGM-62A (Martin Marietta) 684, 688
AGM-65 (Hughes) 683, 688
AGM-69 (Boeing) (1978-79) 644
AGM-78 (General Dynamics) (1979-80) 652
AGM-79A Blue Eye (Martin) (1971-72) 557
AGM-83A (NWC) (1974-75) 633
AGM-84A (MDAC) 684, 688
AGM-86 (Boeing) 680, 688
AGM-88A (Texas Instruments) 686, 688
AGM-109 (General Dynamics) 682
AIM-4 (Hughes) (1971-72) 551
AIM-7 Sparrow (Raytheon) 685, 688
AIM-9 Sidewinder (NWC) 684, 688
AIM-54 Phoenix (Hughes) 682, 688
AIM-82A (U.S.A.) (1971-72) 564
AIM-95 (NWC) (1976-77) 660
AIR-2A Genie (MDAC) (1974-75) 631
ALCM (Boeing) 680, 688
AM39 Exocet (Aérospatiale) 671, 687
AMRAAM (Hughes) 682
AMRAAM (Northrop) (1978-79) 648
AMRAAM (Raytheon) 686
ANS-1 (ANS) (1976-77) 680
ANS (see Industrieel Consortium Astronomische Nederlandse Satelliet) (1976-77) 680
APPLE satellite (ISRO) 692
ARMSCOR (see Armaments Corporation of South Africa) 676
AS-1 (U.S.S.R.) (1975-76) 639
AS-2 (U.S.S.R.) 678, 688
AS-3 (U.S.S.R.) 678, 688
AS-4 (U.S.S.R.) 678, 688
AS-5 (U.S.S.R.) 678, 688
AS-6 (U.S.S.R.) 678, 688
AS-7 (U.S.S.R.) 678
AS-8 (U.S.S.R.) (1980-81) 652
AS.11 (Aérospatiale) 671, 687
AS.12 (Aérospatiale) 671, 687
AS.15TT (Aérospatiale) 671, 687
AS.20 (Aérospatiale) (1975-76) 616
AS.30 (Aérospatiale) (1978-79) 636
AS.30 Laser (Aérospatiale) 671, 687
AS.36/AJ.168 Martel (Matra/BAe) (1979-80) 646
AS.37 (HSD/Matra) (1977-78) 684, 686
ASALM (MDAC) (1980-81) 657
ASAT (see Arbeitsgemeinschaft Satellitentragersystem) (1973-74) 620
ASAT (Vought) 709
ASM-1 (Mitsubishi) 676, 688
ASM-N-7 Bullpup (Martin) (1971-72) 557
ASM-N-10 (NWC) (1974-75) 633
ASMP (Aérospatiale) 672
ASRAAM programme (BAe) 680
ASTP (NASA) (1976-77) 695
ASTP/Apollo spacecraft and docking module (Rockwell International) (1975-76) 670
AS-X-9 (U.S.S.R.) 678
AS-X-10 (U.S.S.R.) (1980-81) 652
AT-2 (U.S.S.R.) 678, 688
AT-3 (U.S.S.R.) 678, 688
AT-6 (U.S.S.R.) 678
ATS-1 (Fairchild) (1977-78) 725
ATS-6 (Fairchild) (1977-78) 725
ATS F (Fairchild Industries) (1973-74) 631
ACADEMY OF SCIENCES (Czechoslovakia) 690
Acrid (U.S.S.R.) 678
Advanced Atoll (U.S.S.R.) 677
AERITALIA — SOCIETÀ AEROSPAZIALE ITALIANA pA (Italy) 694
Aerobee 150 (ALRC) (1977-78) 723
Aerobee 150A (Aerojet) (1973-74) 628
Aerobee 170 (ALRC) (1977-78) 723
Aerobee 200 (ALRC) (1977-78) 723
Aerobee 300 (ALRC) (1977-78) 723
Aerobee 300S (Aerojet) (1973-74) 628
Aerobee 350 (ALRC) (1977-78) 723
Aerobee-Hi (ALRC) (1977-78) 723
Aero High (Australia) (1975-76) 642
AEROJET-GENERAL CORPORATION (U.S.A.) (1978-79) 662
AERONUTRONIC FORD CORPORATION (U.S.A.) (1976-77) 683
Aeros satellite (Dornier) (1974-75) 644
AÉROSPATIALE (see Société Nationale Industrielle Aérospatiale) 671, 690
Agena D (Lockheed) 703
Agile (NWC) (1976-77) 660
Air-launched Minuteman (Boeing) (1976-77) 654
Airtos (Sistel) (1974-75) 621
Albatross (Matra/Oto Melara) (1972-73) 560, 565
Alfa (Aeritalia) (1977-78) 717
Alkali (U.S.S.R.) 676, 688
AMBROSINI & C, SOCIETA AERONAUTICA ITALIANA ING A (Italy) (1973-74) 623
Anab (U.S.S.R.) 677, 688
Anik (Spar) 689
Anik B (RCA) (1979-80) 676
Apex (U.S.S.R.) 678, 688
Aphid (U.S.S.R.) 678, 688
Apollo 17 (Rockwell International) (1973-74) 644
Apollo-Soyuz Test Project (NASA) (1976-77) 695
Apollo Spacecraft (Rockwell International) (1973-74) 644
Apollo Telescope Mount (U.S.A.) (1972-73) 611
Arabsat (Aérospatiale) 691
ARBEITSGEMEINSCHAFT SATELLITEN-TRAGERSYSTEM (West Germany) (1973-74) 620
Arcas (ARC) (1972-73) 639
ARC PROPULSION DIVISION (U.S.A.) (1972-73) 638
Argo B-13 Nike-Apache (Celesco) (1976-77) 685
Argo C-22 (Susquehanna) (1972-73) 638
Argo D-4 Javelin (Brunswick) (1977-78) 724
Ariane (Aérospatiale) 690
Ariel IV (UK.4) (BAC) (1972-73) 619
Ariel 5 (MSDS) (1975-76) 654
Ariel 6 (MSDS) (1979-80) 671
ARMAMENTS CORPORATION OF SOUTH AFRICA (South Africa) 676
ARMED FORCES SCIENTIFIC AND TECHNICAL RESEARCH CENTRE (Argentina) 671
Army Hornet (Rockwell International) (1973-74) 611
Aryabhata satellite (ISRO) (1977-78) 714
Ash (U.S.S.R.) 677, 688
Aspide (Selenia) 674, 688
Astrobee 200 (Space General) (1971-72) 595
Astrobee 250 (ALRC) (1977-78) 723
Astrobee 1500 (ALRC) (1977-78) 723
Astrobee D (ALRC) (1977-78) 723
Astrobee F (ALRC) (1977-78) 723
Athena (Brunswick) (1977-78) 724
Athena H (Brunswick) (1977-78) 725
Atlas E and F (General Dynamics) 702
Atlas/Agena (General Dynamics) (1972-73) 623
Atlas SLV (General Dynamics) 702
Atoll (U.S.S.R.) 676, 688
Aura (Aérospatiale) (1976-77) 672
Aura (Matra) (1976-77) 672
AVIBRAS INDUSTRIA AEROESPACIAL SA (Brazil) (1977-78) 708

B

B-13 Nike-Apache (Susquehanna) (1972-73) 638
BAC (see British Aircraft Corporation) (1977-78) 721
BAC (AT) (see British Aircraft Corporation (AT) Ltd) (1972-73) 572
BAe (see British Aerospace Public Limited Company, Dynamics Group) 680, 700
BAL (see Bristol Aerospace Limited) (1977-78) 709
BGM-71A (Hughes) 682, 688
BGM-109 (General Dynamics) (1977-78) 696
BSE satellite (NASDA) (1978-79) 658
Bantam (Bofors) (1977-78) 690
Basic Hawk (Raytheon) (1972-73) 587
BEECH AIRCRAFT CORPORATION (U.S.A.) (1973-74) 603
Beeswing (BAC) (1975-76) 626
Bélier III (Aérospatiale) (1977-78) 671
Belisama (Thomson-Brandt) (1977-78) 713
BENDIX CORPORATION, THE (U.S.A.) (1977-78) 723
BHARAT DYNAMICS LTD (India) (1977-78) 685
Bhaskara (ISRO) 693
Big Bird (Lockheed/USAF) 703
Black Arrow (Westland) (1971-72) 594
Black Brant IIIA (BAL) (1977-78) 709
Black Brant IIIB (BAL) (1977-78) 709
Black Brant IVA (BAL) (1977-78) 709
Black Brant IVB (BAL) (1977-78) 709
Black Brant VA (BAL) (1977-78) 709
Black Brant VB (BAL) (1977-78) 709
Black Brant VC (BAL) (1977-78) 709
Block 5D satellite (RCA) 706
Blue Eye (Martin) (1971-72) 557
BODENSEEWERK GERATETECHNIK GmbH (West Germany) 672
BOEING AEROSPACE COMPANY (U.S.A.) 680, 701
BOEING COMPANY, THE (U.S.A.) 701
BOFORS, AB (Sweden) (1977-78) 690
Boosted Arcas II (ARC) (1972-73) 639
Bora (Ambrosini) (1973-74) 623
Bora-Sond (CSTM) (1975-76) 650
Brazo/Pave Arm (Hughes) (1977-78) 697
BREDA MECCANICA BRESCIANA SpA (Italy) (1978-79) 640
BRISTOL AEROJET LTD. (U.K.) (1977-78) 721
BRISTOL AEROSPACE LTD. (Canada) (1977-78) 709
BRITISH AEROSPACE PUBLIC LIMITED COMPANY, DYNAMICS GROUP (U.K.) 680, 700
BRITISH AIRCRAFT CORPORATION (U.K.) (1977-78) 721
BRITISH AIRCRAFT CORPORATION (A.T.) LTD. (U.K.) (1972-73) 572
BRITISH AIRCRAFT CORPORATION (GUIDED WEAPONS) LTD. (U.K.) (1977-78) 692
BRUNSWICK CORPORATION (U.S.A.) (1977-78) 724
Bulldog (NWC) (1974-75) 633
Bullpup (Martin) (1971-72) 557
Burner II (Boeing) (1975-76) 657
Burner IIA (Boeing) (1975-76) 657

C

C.2 (CRA) (1974-75) 648
C-22 (Susquehanna) (1972-73) 638
CASA (see Construcciones Aeronauticas SA) (1977-78) 720
CAT satellite (Aeritalia) (1977-78) 717
CESAR consortium (International) (1977-78) 715
CFTH-HB (see Compagnie Française Thomson Houston—Hotchkiss-Brandt) (1972-73) 608
CIA (see Compagnia Industriale Aerospaziale) (1977-78) 717
CIFAS (see Consortium Industriel Franco-Allemand pour le satellite Symphonie) 693
CITEFA (see Instituto de Investigaciones Cientificas y Tecnicas de las Fuerzas Armadas) 671
CNA (see Compagnia Nazionale Aerospaziale) (1980-81) 669
CNS (see Compagnia Nazionale Satelliti per Telecomunicazioni SpA) 694
CONIE (see Comision Nacional de Investigacion del Espacio) (1977-78) 720
COS B satellite (CESAR) (1976-77) 675
COSMOS consortium (International) 693
CRA (see Istituto di Costruzioni Aeronautiche Centro Aerospaziali, Università Degli Studi di Roma) (1975-76) 650
CS satellite (NASDA) (1978-79) 658
CSTM (see Centro Studi Trasporti Missilistici) (1976-77) 677
CTS (Spar) (1977-78) 710
Canopus II (FMA) (1973-74) 616
Castor (FMA) (1973-74) 616
CELESCO INDUSTRIES INC. (U.S.A.) (1976-77) 685
Centaur (General Dynamics) 702
Centaure III (Aérospatiale) (1977-78) 711
CENTRO TECNICO AEROESPACIAL INSTITUTO DE ATIVIDADES ESPACIAIS (Brazil) (1977-78) 708
CENTRO STUDI TRASPORTI MISSILISTICI (Italy) (1976-77) 677
CESKOSLOVENSKÁ AKADEMIE VED, GEOFYZIKÁLNÍ ÚSTAV (Czechoslovakia) 690
CHINA (PEOPLE'S REPUBLIC) 689
CHRYSLER CORPORATION (U.S.A.) (1975-76) 658
Cockatoo (Australia) (1977-78) 708
COMISION NACIONAL DE INVEST-IGACION DEL ESPACIO (Spain) (1977-78) 720
COMISSAO CENTRAL DE MISSEIS (Brazil) (1972-73) 554
COMMONWEALTH (see Commonwealth of Australia, Department of Manufacturing Industry) (1974-75) 640
COMMONWEALTH OF AUSTRALIA, DEPART-MENT OF MANUFACTURING INDUSTRY (Australia) (1974-75) 640
Communications Technology Satellite (Spar) (1977-78) 710
COMPAGNIA INDUSTRIALE AEROSPAZIALE (Italy) (1977-78) 717
COMPAGNIE FRANÇAISE THOMSON HOUSTON—HOTCHKISS-BRANDT (France) (1976-77) 673
COMPAGNIA NAZIONALE SATELLITI PER TELECOMUNICAZIONI SpA (Italy) 694
Comstar (Hughes) 703
Condor (Rockwell International) (1977-78) 702
CONSORTIUM INDUSTRIEL FRANCO-ALLEMAND POUR LE SATELLITE SYMPHONIE (International) 693
CONSTRUCCIONES AERONAUTICAS SA (Spain) (1977-78) 720
CONTRAVES AG (Switzerland) (1971-72) 541
CONTRAVES ITALIANA S.p.A. (Italy) (1972-73) 565
CONTRAVES-OERLIKON (see Contraves AG) (1971-72) 541
CONVAIR DIVISION, GENERAL DYNAMICS CORPORATION (USA) 702
Coralie (ELDO Second stage) (Aérospatiale) (1972-73) 607
Corella (Australia) (1977-78) 708
Cosmos launchers (U.S.S.R.) 697
Cosmos satellites (U.S.S.R.) 696

D

D2A (Aérospatiale) (1972-73) 608
D-2B satellite (Aérospatiale) (1976-77) 672
D-2B satellite (Matra) (1976-77) 672
D-4 Javelin (Susquehanna) (1972-73) 638
D-5 satellite (Aérospatiale) (1976-77) 672
D-5A SRET (Aérospatiale) (1972-73) 608
D-5B Cactus (Aérospatiale) (1972-73) 608
DEPT (see Departamento de Estudos e Pesquisas Tecnologicas) (1972-73) 554
D.I.M. (FMA) (1973-74) 616
DM-19 (MDAC) (1971-72) 605
DSCS I (Philco-Ford) (1972-73) 635
DSCS II (TRW) 709
DSV-2L (MDAC) (1972-73) 605
DSV-3 (MDAC) 704
DTAT (see Direction Technique des Armements Terrestres) (1972-73) 558
DTC (see Direction Technique des Constructions (Navales), Délégation Ministérielle pour l'Armement) (1972-73) 559
Dauphin (Aérospatiale) (1977-78) 711
Defense Satellite Communications System Phase II (TRW) 709
Delta (MDAC) 704

DEPARTAMENTO DE ASSUNTOS ESPECIAIS (Brazil) ....(1972-73) 605
DEPARTAMENTO DE ESTUDOS E PESQUISAS TECNOLOGICAS (Brazil) ....(1972-73) 554
Diamant B (Aérospatiale) ....(1975-76) 644
Diamant B P-4 (Aérospatiale) ....(1977-78) 711
DIRECTION TECHNIQUE DES ARMEMENTS TERRESTRES (France) ......(1972-73) 558
DIRECTION TECHNIQUE DES CONSTRUCTIONS (NAVALES),DÉLÉGATION MINISTÉRIELLE POUR L' ARMEMENT (France) ....(1972-73) 559
Discoverer (Lockheed) ....(1974-75) 660
DORNIER GmbH (West Germany) .....(1975-76) 619
DORNIER SYSTEM GmbH (West Germany) 691
Dragon III (Aérospatiale) ....(1977-78) 711
Dynamics Explorer (NASA) 705
Dynamics Explorer (RCA) 706

E

ECS (MESH) 693
ECS satellite (NASDA) ....(1978-79) 658
ELDO (see European Space Vehicle Launcher Development Organisation) ....(1973-74) 622
ELDO First Stage (HSD) ....(1972-73) 620
ELDO Second Stage (Aérospatiale) ......(1972-73) 607
ELDO Third Stage (ASAT) ....(1972-73) 610
EMDG (see Euromissile Dynamics Group) 674
EOGB (Rockwell International) ....(1975-76) 637
ERS (TRW) ....(1977-78) 739
ERTS (General Electric) ....(1979-80) 672
ESA (see European Space Agency) ......(1973-74) 622
ESRO IV (HSD) ....(1973-74) 627
EST (see European Satellite Team) ......(1972-73) 614
ETS-2 (Ford) ....(1978-79) 663
Ekran (U.S.S.R.) 697
Engineering Test Satellite (Ford) ....(1978-79) 663
Environmental Research Satellite (TRW) ..(1973-74) 664
Eole (Aérospatiale) ....(1972-73) 607
Epona (Thomson-Brandt) ....(1977-78) 713
Eridan (Aérospatiale) ....(1977-78) 711
ERNO RAUMFAHRTTECHNIK GmbH (West Germany) 691
EUROMISSILE (International) 674
EUROMISSILE DYNAMICS GROUP (International) 674
EUROMISSILE GROUPEMENT D'INTÉRET ÉCONOMIQUE (International) ... 674
Europa I and II (ELDO) ....(1972-73) 613
EUROPEAN SATELLITE TEAM (International) ....(1972-73) 614
EUROPEAN SPACE AGENCY (International) ....(1973-74) 622
EUROPEAN SPACE VEHICLE LAUNCHER DEVELOPMENT ORGANISATION (International) ....(1973-74) 622
Exocet (Aérospatiale) 671, 687
Exosat (Cosmos) 693
Explorer 42 (SAS-A) (U.S.A.) ....(1971-72) 611
Explorer 43 (NASA) ....(1972-73) 631
Explorer 44 (NASA) ....(1972-73) 631
Explorer 45 (NASA) ....(1972-73) 631
Explorer 46 (NASA) ....(1975-76) 666
Explorer 47 (NASA) ....(1973-74) 640
Explorer 48 (NASA) ....(1974-75) 664
Explorer 49 (NASA) ....(1974-75) 664
Explorer 50 (NASA) ....(1974-75) 665
Explorer 51 (NASA) ....(1974-75) 665
Explorer 52 (NASA) ....(1975-76) 666
Explorer 53 (NASA) ....(1975-76) 666
Explorer 54 (NASA) ....(1976-77) 693
Explorer 55 (NASA) ....(1976-77) 693
Explorer 56 (NASA) ....(1976-77) 693
Explorer 57 (NASA) ....(1976-77) 693
Extended Long Tank Delta (MDAC) 704

F

FIM-43A (General Dynamics) ....(1972-73) 576
FMA (see Fábrica Militar de Aviones, Instituto de Investigaciones Aeronauticas y Espacial) (1975-76) 642
FLAME (Aerojet-General) ....(1978-79) 662
FLTSATCOM (TRW) 709
FR-2 (Aérospatiale) ....(1972-73) 607
FÁBRICA MILITAR DE AVIONES, INSTITUTO DE INVESTIGACIONES AERONAUTICAS Y ESPACIAL (Argentina) ....(1975-76) 642
FAIRCHILD SPACE AND ELECTRONICS COMPANY (USA) 701
Falcon (Hughes) ....(1971-72) 551
Fighter Launched Advanced Materials Equipment (Aerojet-General) ....(1978-79) 662
Firestreak (HSD) ....(1971-72) 544
Flamenco (Bristol Aerojet/INTA) ....(1977-78) 721
FLYGMOTOR (see Volvo Flygmotor AB) (1972-73) 681
FORD AEROSPACE & COMMUNICATIONS CORPORATION (U.S.A.) 681, 701

G

GAM-83 Bullpup (Martin) ....(1971-72) 557
GAR-1 and 2 Falcon (Hughes) ....(1971-72) 551
GAR-8 (NWC) ....(1974-75) 632
GBU-8 (Rockwell International) ....(1980-81) 658
GBU-15 (Hughes) ....(1980-81) 656
GBU-15 (Rockwell International) 686, 688
GBU-15 Pave Strike (Rockwell International) (1977-78) 703
GEOS (STAR) ....(1980-81) 668
GEOS-3 (NASA) ....(1976-77) 693
GGTS (General Electric) ....(1971-72) 601
GMS satellite (NASDA) ....(1977-78) 718
GOES (Ford) ....(1979-80) 671
Galileo (NASA) 705
Gamma Centauro I & II (FMA) ....(1973-74) 616
GENERAL DYNAMICS CORPORATION (U.S.A.) 681, 702
GENERAL ELECTRIC COMPANY SPACE DIVISION (U.S.A.) 702
Genie (MDAC) ....(1974-75) 631
Geostationary Operational Environmental Satellite (Ford) ....(1979-80) 671
GOODYEAR AEROSPACE CORPORATION (U.S.A.) ....(1974-75) 628
GOVERNMENT OF AUSTRALIA, DEPARTMENT OF DEFENCE (Australia) (1977-78) 708
Grannos (Thomson-Brandt) ....(1977-78) 713
Gravity Gradient Test Satellite (General Electric) ....(1971-72) 601
GRUMMAN AEROSPACE CORPORATION (U.S.A.) ....(1974-75) 658

H

H-1 launch vehicle (NASDA) 695
H 8 Ariane third stage (Aérospatiale) 690
HAD (Commonwealth) ....(1973-74) 616
HARM (NWC) ....(1971-72) 559
HARM (Texas Instruments) 686, 688
HAT (Commonwealth) ....(1973-74) 616
HAT Mk 2 (Commonwealth) ....(1973-74) 616
HEAD (Grumman) ....(1971-72) 602
HEAO (TRW) 709
HELIOS (MBB) ....(1977-78) 713
HEOS-A2 (MBB) ....(1972-73) 611
HM-55 Falcon (Hughes) ....(1971-72) 551
HM-58 Falcon (Hughes) ....(1971-72) 551
HOBOS (Rockwell International) ....(1980-81) 658
Hot (Euromissile) 674, 688
HSD (see Hawker Siddeley Dynamics Ltd.) (1977-78) 686, 693, 722
Harpoon (MDAC) 684, 688
HAWKER SIDDELEY DYNAMICS LTD. (U.K.) (1977-78) 686, 693, 722
Hawkeye-1 (NASA) ....(1975-76) 666
Hawkswing (BAC) ....(1977-78) 692
Hellcat (Short) ....(1971-72) 547
Hellfire (Hughes) ....(1975-76) 633
Hellfire (Rockwell International) 686, 688
High Energy Astronomy Observatory (Grumman) (1971-72) 602
HONEYWELL INC. (U.S.A.) ....(1971-72) 551
Horizont (U.S.S.R.) 697
Hornet (North American) ....(1971-72) 559
HUGHES AIRCRAFT COMPANY (U.S.A.) 682, 702

I

IAE/EDE (see Centro Técnico Aeroespacial Instituto de Atividades Espaciais) ......(1977-78) 708
IAI (see Israel Aircraft Industries Ltd) ....(1972-73) 565
I.I.A.E. (see Instituto de Investigaciones Aeronauticas y Espacial) ....(1972-73) 604
IL (see Instytut Lotnictwa) ....(1974-75) 650
IMEWS (TRW) 709
IMGW (see Instytut Meteorologii I Gospordarki Wodnej) ....(1977-78) 720
IMP-I (U.S.A.) ....(1971-72) 611
IMP-J (NASA) ....(1974-75) 665
INCOSPAR (see Indian National Committee for Space Research) ....(1973-74) 621
INPE (see Instituto de Pesquisas Espaciais) (1977-78) 709
INSAT (ISRO) 693
INTA (see Instituto Nacional de Tecnica Aeroespacial) ....(1977-78) 720
INTA-255 (CONIE) ....(1973-74) 626
INTA-300 (CONIE) ....(1977-78) 721
INTA-300 (CONIE/Bristol Aerojet) .....(1975-76) 654
IPD (see Instituto de Pesquisas e Desenvolvimento) ....(1972-73) 605
IPD/PAE (see Centro Técnico Aeroespacial Instituto de Pesquisas e Desenvolvimento) (1972-73) 605
IR-AAM (Mitsubishi) 676
IRIS (Aeritalia) 694
ISEE-2 Explorer satellite (MESH) ....(1979-80) 665
ISEE-3 (Fairchild) ....(1980-81) 674
ISEE-B (STAR) ....(1978-79) 658
ISIS-2 (Spar) ....(1971-72) 582
ISIS-B (Spar) ....(1971-72) 582
ISPM (NASA) 705
ISRO (see Indian Space Research Organisation) ... 692
ISSP (ISRO) ....(1975-76) 648
ITOS (RCA) ....(1980-81) 680
IUE satellite (NASA) ....(1980-81) 678
IUS (Boeing) 701
Improved Delta (MDAC) ....(1971-72) 605
Indian Centaure (India) ....(1973-74) 621
INDIAN NATIONAL COMMITTEE FOR SPACE RESEARCH (India) ....(1973-74) 621
INDIAN SPACE RESEARCH ORGANISATION (India) 692
Indigo (Sistel) ....(1973-74) 566
INDUSTRIEEL CONSORTIUM ASTRONOMISCHE NEDERLANDSE SATELLIET (Netherlands) ....(1976-77) 680
Inertial upper stage (Boeing) 701
INSTITUTO DE INVESTIGACIONES AERONAUTICAS Y ESPACIAL (Argentina) ....(1972-73) 604
INSTITUTO DE INVESTIGACIONES CIENTIFICAS Y TECNICAS DE LAS FUERZAS ARMADAS (Argentina) 671
INSTITUTO DE PESQUISAS E DESENVOLVIMENTO (Brazil) ......(1972-73) 605
INSTITUTO DE PESQUISAS ESPACIAIS (Brazil) ....(1977-78) 709
INSTITUTO NACIONAL DE TECNICA AEROESPACIAL (Spain) ....(1977-78) 720
INSTYTUT LOTNICTWA (Poland) .....(1974-75) 650
INSTYTUT METEOROLOGII I GOSPODARKI WODNEJ (Poland) ..(1977-78) 720
Intasat-1 (CONIE) ....(1975-76) 653
Intelsat 3 (TRW) ....(1971-72) 610
Intelsat IV (Hughes) ....(1977-78) 728
Intelsat IVA (Hughes) ....(1980-81) 676
Intelsat V (Ford) 701
Intercosmos Satellites (U.S.S.R.) 697
Interim Upper Stage (Boeing) ....(1978-79) 663
INTERNATIONAL PROGRAMMES 648, 667
ISRAEL AIRCRAFT INDUSTRIES LTD (Israel) ....(1972-73) 565
ISTITUTO DI COSTRUZIONI AERONAUTICHE CENTRO AEROSPAZIALI, UNIVERSITA DEGLI STUDI DI ROMA (Italy) .....(1975-76) 650
Italsat (Aeritalia) 694

J

JDA (see Japan Defence Agency) ....(1973-74) 599
JAPAN DEFENCE AGENCY (Japan) ..(1973-74) 599
Japanese communications satellites (Ford) (1977-78) 727
Japanese engineering test satellites (Ford) .(1977-78) 726
Javelin, Argo D-4 (Brunswick) ....(1977-78) 724
Judi Chaff Dart (Talley) ....(1972-73) 639
Jumbo (MBB) ....(1977-78) 685

K

K-9M (Nissan) ....(1977-78) 719
K-10 (Nissan) ....(1977-78) 719
K-13A (USSR) 676, 688
Kangaroo (U.S.S.R.) 678, 688
Kangaroo (UTC) ....(1972-73) 640
KAWASAKI (see Kawasaki Jukogyo Kabushiki Kaisha) ....(1972-73) 566
KAWASAKI JUKOGYO KABUSHIKI KAISHA (Japan) ....(1972-73) 566
Kelt (U.S.S.R.) 678, 688
"Kennel" (U.S.S.R.) ....(1975-76) 639
Kerry (U.S.S.R.) 678
Kingfish (U.S.S.R.) 678, 688
Kipper (U.S.S.R.) 678, 688
Kitchen (U.S.S.R.) 678, 688
KONGSBERG VAAPENFABRIK, A/S (Norway) . 676
Kookaburra (Australia) ....(1977-78) 708
Kormoran (MBB) 673, 688

L

L-3H (Nissan) ....(1977-78) 719
L3S Launch Vehicle (ESA) ....(1973-74) 622
L-17 Diamant B first stage (Aérospatiale) .(1977-78) 711
L 33 Ariane second stage (Aérospatiale) 690
L 140 Ariane first stage (Aérospatiale) 690
LEASAT (Hughes) 703
LES 8 & 9 (Lincoln) ....(1977-78) 729
LGDM (Goodyear) ....(1974-75) 629
LM (Grumman) ....(1972-73) 624
LMSC (see Lockheed Missiles & Space Company) 703
LRBA (see Laboratoire de Recherches Balistiques et Aérodynamiques) ....(1972-73) 608
LS-A (Mitsubishi) ....(1976-77) 677
L-Sat (BAe) 700
LS-C (Mitsubishi) ....(1976-77) 677
LTD (MDAC) ....(1971-72) 605
LTV (see LTV Aerospace Corporation) ..(1975-76) 663
LABORATOIRE DE RECHERCHES BALISTIQUES ET AÉRODYNAMIQUES (France) (1972-73) 608
Lageos (Bendix) ....(1977-78) 723
Landsat (General Electric) 702
Laser, AS.30 (Aérospatiale) 671
Lasso (CNS) 694
Lassoo (Aérospatiale) ....(1979-80) 643
LATÉCOERE, SOCIÉTÉ INDUSTRIELLE D'AVIATION (France) ....(1972-73) 559
LINCOLN (see Lincoln Laboratory, Massachusetts Institute of Technology) ..(1977-78) 729
LINCOLN LABORATORY, MASSACHUSETTS INSTITUTE OF TECHNOLOGY (U.S.A.) (1977-78) 729
LOCKHEED AIRCRAFT CORPORATION (U.S.A.) ....(1977-78) 730
LOCKHEED MISSILES & SPACE COMPANY (U.S.A.) 703
Lockheed/USAF 467 Big Bird (LMSC) 703
Long Tank Delta (MDAC) ....(1972-73) 629
Long Tank Thor (MDAC) ....(1977-78) 733
Lorikeet (Australia) ....(1977-78) 708
LTV AEROSPACE CORPORATION (subsidiary of THE LTV CORPORATION) (U.S.A.) (1975-76) 663
Luna Spacecraft (U.S.S.R.) ....(1977-78) 742
Lunar Module (Grumman) ....(1972-73) 624
Lunar Rover (Boeing) ....(1973-74) 629
Lunokhod 1 (U.S.S.R.) ....(1971-72) 616
Lunokhod 2 (U.S.S.R.) ....(1973-74) 647
Luz-1 (Rafael) 674

M

M-58 (Saab) ....(1975-76) 625
M-112 (Aérospatiale) ....(1971-72) 530
MARECS (MESH) 693
MAROTS-B (MESH) ....(1978-79) 657
MAROTS (MSDS) ....(1977-78) 722
MB-1 Genie (MDAC) ....(1972-73) 582
MBB (see Messerschmitt-Bölkow-Blohm) 673, 692
MDAC (see McDonnell Douglas Astronautics Company) 684, 704
MDD (CNS) 694

MESH (International) 693
MMS (Fairchild) 701
MRAM (General Dynamics) (1972-73) 576
MRASM (General Dynamics) 682
MSDS (see Marconi Space and Defence Systems Ltd.) 700
MT-135 (Nissan) (1977-78) 719
Magic (Matra) 672, 688
Magion (Czechoslovakia) 690
Magsat (NASA) (1980-81) 678
MARCONI SPACE AND DEFENCE SYSTEMS LTD. (U.K.) 700
MARINE FRANCAISE (see Direction Technique des Constructions Navales) (1971-72) 531
Mariner (NASA) (1978-79) 667
Mariner Jupiter-Saturn (NASA) (1976-77) 694
Mariner-Uranus (NASA) (1976-77) 694
Mariner 10 Venus/Mercury spacecraft (Boeing) (1975-76) 667
Marisat (Hughes) (1979-80) 673
Maritime satellite (ESA) (1973-74) 623
Mars 2 & 3 (U.S.S.R.) (1973-74) 648
Mars 4, 5, 6 and 7 (U.S.S.R.) (1974-75) 673
Marte (Sistel) 675
Martel (Matra/BAe) (1979-80) 646
Martel, P3T (BAe) (1979-80) 650
MARTIN MARIETTA CORPORATION (U.S.A.) 684, 704
Martin Pescador (CITEFA) 671, 687
MATRA/BAe (HSD) (International) (1979-80) 646
MATRA/OTO MELARA (International) (1976-77) 646
MATRA, SA (France) 672, 691
Maverick (Hughes) 683, 688
MCDONNELL DOUGLAS ASTRONAUTICS COMPANY (U.S.A.) 684, 704
MCDONNELL DOUGLAS CORPORATION (U.S.A.) (1979-80) 674
Medium-range AAM (HSD) (1973-74) 603
Menaka (India) (1973-74) 621
MESSERSCHMITT-BÖLKOW-BLOHM GmbH (West Germany) 673, 692
Meteor (U.S.S.R.) 697
Meteor-1E (IMGW) (1977-78) 720
Meteor-2H (IL) (1971-72) 591
Meteor-2K (IL) (1973-74) 626
Meteor-3E (IMGW) (1977-78) 720
Meteosat (COSMOS) 693
Military Cosmos (U.S.S.R.) 696
Miranda (HSD) (1975-76) 654
MITSUBISHI DENKI KABUSHIKI KAISHA (Japan) 676
MITSUBISHI ELECTRICAL CORPORATION (Japan) 695
MITSUBISHI JUKOGYO KABUSHIKI KAISHA (Japan) 676, 695
Molniya (U.S.S.R.) 697
Mu (Nissan) 695
Murene (Matra) (1972-73) 561

N

NAR (see North American Rockwell Corporation) (1972-73) 585, 634
NASA (see National Aeronautics and Space Administration) 705
NASDA (see National Space Development Agency of Japan) 695
NATO 2B (Philco-Ford) (1974-75) 669
NATO 3C (Ford) (1979-80) 671
N launch vehicle (NASDA) 695
NOAA (RCA) 707
"N" Rocket (NASDA) (1975-76) 651
NWC (see Naval Weapons Center) 684
NATIONAL AERONAUTICS AND SPACE ADMINISTRATION (U.S.A.) 705
NATIONAL SPACE DEVELOPMENT AGENCY OF JAPAN (Japan) 695
NAVAL WEAPONS CENTER (U.S.A.) 684
Navstar (Rockwell International) 707
Night attack missile (Hughes) (1977-78) 699
Nike-Apache (Celesco) (1976-77) 685
Nike-Black Brant V (BAL) (1977-78) 709
Nimbus (General Electric) 702
Niro (ALRC) (1977-78) 723
NISSAN JIDOSHA KABUSHIKI KAISHA (Japan) 695
Nord AS.11 (Aérospatiale) 671, 687
Nord AS.12 (Aérospatiale) 671, 687
Nord AS.20 (Aérospatiale) (1975-76) 616
Nord AS.30 (Aérospatiale) (1978-79) 636
NORTH AMERICAN ROCKWELL CORPORATION (U.S.A.) (1972-73) 585, 634
NORTHROP CORPORATION (U.S.A.) (1978-79) 648
Nova satellite (RCA) 707

O

OAO (Grumman) (1973-74) 632
OFO (U.S.A.) (1971-72) 612
ONERA (see Office National d'Études et de Recherches Aérospatiales) (1977-78) 712
OSCAR 6 (NASA) (1973-74) 641
OSCAR 7 (NASA) (1975-76) 668
OSO (Hughes) (1977-78) 729
OTRAG (see Orbital Transport-und Raketen AG) 692
OTS (MESH) 694
OV1 (General Dynamics) (1975-76) 659
OERLIKON MACHINE TOOL WORKS BUEHRLE & CO (Switzerland) (1971-72) 541
OFFICE NATIONAL D'ÉTUDES ET DE RECHERCHES AÉROSPATIALES (France) (1977-78) 712
Orbital Test Satellite (MESH) 694
ORBITAL TRANSPORT-UND RAKETEN AG (West Germany) 692
Orbital Vehicle One (General Dynamics) (1975-76) 659
Orbiting Astronomical Observatory (Grumman) (1973-74) 632
Orbiting Solar Observatory (Hughes) (1977-78) 729
Orion II (FMA) (1973-74) 616
Otomat (Matra/Oto Melara) (1976-77) 646
OTO MELARA SpA (Italy) (1976-77) 646

P

P-2 Diamant B second stage (Aérospatiale) (1973-74) 618
P3T (BAe) 680
P-4 Diamant B second stage (Aérospatiale) (1977-78) 711
P-72-2 satellite (Rockwell International) (1975-76) 671
PAE (see Departamento de Assuntos Especiais) (1972-73) 605
PAM/SSUS (MDAC) 705
PIHM (see Panstwowy Instytut hydrologiczno-Meteorologiczny) (1973-74) 626
P-O6 Diamant B third stage (Aérospatiale) (1977-78) 644
PR satellite (Lockheed/USAF) (1976-77) 690
Palapa B (Hughes) 703
PANSTWOWY INSTYTUT HYDRO-LOGICZNO-METEOROLOGICZNY (Poland) (1973-74) 626
Pave Arm/Brazo (Hughes) (1977-78) 697
Pave Strike (Rockwell International) (1976-77) 663
Paveway (Texas Instruments) 686
Paveway LGB (Texas Instruments) 686
Payload Assist Module (McDonnell Douglas) 705
Penguin (Kongsberg) 676
Penguin Mk 2 (Kongsberg) (1980-81) 649, 661
Penguin Mk 3 (Kongsberg) 676, 688
Peole (Aérospatiale) (1972-73) 607
Percheron A & B (Brunswick) (1977-78) 725
Percheron C (Brunswick) (1977-78) 725
Petrel (Bristol Aerojet) (1977-78) 722
Phebus (Aérospatiale) (1979-80) 662
PHILCO-FORD CORPORATION (U.S.A.) (1974-75) 668
Phoenix (Hughes) 682, 688
Pioneer 6 (TRW) (1976-77) 698
Pioneer 10 (TRW) (1977-78) 740
Pioneer 11 (TRW) (1977-78) 740
Pioneer-Jupiter (TRW) (1971-72) 610
Pioneer Saturn (TRW) (1977-78) 740
Pioneer Venus 1978 (NASA) (1979-80) 675
Piranha MAA 1 (Brazil) 671, 687
Prognoz (U.S.S.R.) 698
Progress (USSR) 698
Prospero (X-3) (BAC) (1972-73) 619
Python 3 (Rafael) 674

R

R.530 (Matra) (1980-81) 646, 660
R.550 Magic (Matra) 672, 688
R.A.E. (see Royal Aircraft Establishment) (1971-72) 594
RB04 (Vapenavdelningen) (1975-76) 625
RB04E (Saab) (1977-78) 690
RB05A (Saab) (1977-78) 691
RB27 (Saab) (1976-77) 649
RB28 (Saab) (1976-77) 649
RB53 (Bofors) (1977-78) 690
RB68 (BAC) (1971-72) 542
RB69 (General Dynamics) (1972-73) 576
RB71 (BAe) 680
RB72 (Saab) (1977-78) 691
RBS 15 (SBMC) 676
RBS 70 (SBMC) 676
RCA (see RCA Corporation) 706
RGM-66D (General Dynamics) (1979-80) 652
RGM-66F (General Dynamics) (1974-75) 628
RGM-84A (MDAC) 684
RIM-7H (Raytheon) 685
RM spacecraft (U.S.A.) (1971-72) 612
RMP-B Programme (Philco-Ford) (1972-73) 635
RMS (Spar) 689
Radio Astronomy Explorer B (NASA) (1975-76) 664
Raduga/Statsionar (USSR) 698
RAFAEL ARMAMENT DEVELOPMENT AUTHORITY (Israel) 674
Rasko-2R (IMGW) (1977-78) 720
RAYTHEON COMPANY (U.S.A.) 685
RCA CORPORATION (U.S.A.) 706
Red Top (HSD) (1975-76) 627
Rigel (FMA) (1973-74) 616
Robot RB04 (Vapenavdelningen) (1975-76) 625
ROCKWELL INTERNATIONAL CORPORATION (U.S.A.) 686, 707
Rohini satellite (ISRO) 693
ROYAL AIRCRAFT ESTABLISHMENT (U.K.) (1971-72) 594

S

S-1B (Saturn 1B first stage) (Chrysler) (1976-77) 655
S.IC Saturn V First Stage (Boeing) (1973-74) 629
S-II second stage for Saturn V (Rockwell International) (1973-74) 644
S-IVB Orbital Workshop (MDAC) (1974-75) 663
S-IVB (Saturn 1B second stage and Saturn V third stage) (MDAC) (1975-76) 665
S-160 JA (Nissan) (1977-78) 719
S-210 JA (Nissan) (1977-78) 719
S-310 (Nissan) (1977-78) 719
SAAB (see Saab-Scania Aktiebolag) 696
SACI (INPE) (1975-76) 642
SAGE (NASA) (1980-81) 678
SAGMI (Beech) (1973-74) 603
S-B (Mitsubishi) (1976-77) 677
SBMC (see Saab-Bofors Missile Corporation) 676
SBS (Hughes) 703
S-C (Mitsubishi) (1976-77) 677
SCAD (Boeing) (1973-74) 604
SEO satellite (ISRO) (1979-80) 664
SEP (see Société Européenne de Propulsion) (1973-74) 619
SIGME (see Societa Generale Missilistica Italiana) (1971-72) 537
SIGNE 3 (Matra) (1977-78) 712
SITE (ISRO) (1977-78) 715
SKA (Saab) (1977-78) 691
SLT (Aérospatiale) (1976-77) 672
SLV-3 (ISRO) 692
SLV-3 Atlas (General Dynamics) (1976-77) 686
SMM (NASA) 705
SMS (Ford) (1977-78) 726
SPAZ (MBB) (1971-72) 586
SPOT (Matra) 691
SR-1 (Flygmotor) (1971-72) 592
SRAAM (HSD) (1977-78) 693
SRAM (Boeing) (1978-79) 644
SRMS (Spar) (1979-80) 661
SS-4 + Cosmos stage (U.S.S.R.) 697
SS-5 + Restart stage (U.S.S.R.) 697
SS-9 + FOBS stage (U.S.S.R.) 697
SSS-A (NASA) (1972-73) 631
STAR consortium (International) 694
STEP (ISRO) (1977-78) 715
SWIK MOD A, B, C & D (Brunswick) (1977-78) 724
SYLDA (Aérospatiale) 691
SAAB-BOFORS MISSILE CORPORATION (Sweden) 676
SAAB-SCANIA AKTIEBOLAG (Sweden) 696
Saab-04E (Saab) (1977-78) 690
Saab-05A (Saab) (1977-78) 691
Saab-05B (Saab) (1977-78) 691
Saab SKA (Saab) (1977-78) 691
Saab 372 (Saab) (1977-78) 691
Sabre (BAe) (1979-80) 650
Sagger (U.S.S.R.) 678, 688
Salyut (U.S.S.R.) 698
Samos/Ferret (Lockheed) (1975-76) 662
San Marco 4 (CRA) (1974-75) 648
Satcom (RCA) 707
Saturn (NASA) (1975-76) 668
Scout (Vought) 709
Sea Eagle (P3T) (BAe) 680, 688
Sea Killer Mk 1 (Sistel) (1977-78) 689
Sea Killer Mk 2 (Sistel) 675, 688
Seasat satellite (Lockheed) 703
Sea Skua (BAe) 680, 688
Seekbat (General Dynamics) (1973-74) 605
SELENIA INDUSTRIE ELETTRONICHE ASSOCIATE SpA (Italy) 674
SENER (Spain) (1977-78) 721
Shafrir (Rafael) 674, 688
Shinsei satellite (Nissan) (1972-73) 616
Shrike (NWC) (1978-79) 648, 652
Shuttle Remote Manipulator System (Spar) 689
Sidewinder (Bodenseewerk) (1973-74) 595
Sidewinder (NWC) 684, 688
Sidewinder-HV Arcas (ARC) (1972-73) 639
Sirio (CIA) (1977-78) 717
Sirio-2 (CNS) 694
SISTEL-SISTEMI ELETTRONICI SpA (Italy) 675
Skua (Bristol Aerojet) (1977-78) 722
Sky Flash (BAe) 680, 688
Skylab (NASA) (1977-78) 734
Skylab 2 (NASA) (1974-75) 666
Skylab 3 (NASA) (1974-75) 667
Skylab 4 (NASA) (1974-75) 667
Skylark (BAC) (1977-78) 721
Skynet (Philco-Ford) (1972-73) 636
Skynet II (MSDS) (1975-76) 654
SNIA VISCOSA (Italy) (1973-74) 624
SOCIETA GENERALE MISSILISTICA ITALIANA (Italy) (1971-72) 537
SOCIÉTÉ EUROPÉENNE DE PROPULSION (France) (1973-74) 619
SOCIÉTÉ NATIONALE INDUSTRIELLE AÉROSPATIALE (France) 671, 690
Solrad 10 (NASA) (1972-73) 631
Sonate (Aérospatiale) 691
Sonda I/A (Avibras) (1976-77) 668
Sonda I/B (Avibras) (1976-77) 668
Sonda II (IAE/EDE) (1977-78) 708
Sonda III (IAE/EDE) (1977-78) 708
Sounding rockets (Aérospatiale) (1977-78) 711
Sounding rockets (Nissan) (1977-78) 719
Soyuz launcher (U.S.S.R.) 700
Soyuz spacecraft (U.S.S.R.) 699
Soyuz T spacecraft (U.S.S.R.) 700
SPACE AND ALTERNATIVE ENERGIES GROUP, AERITALIA (Italy) 694
SPACE GENERAL COMPANY (U.S.A.) (1972-73) 636
Spacelab (ERNO) 692
Space Shuttle orbiter (Rockwell) 707
Space Shuttle Transportation System (NASA) 705
Space Shuttle upper stages (MDAC) (1977-78) 733
Space Telescope (Lockheed) 704
Space Test Programme (Lockheed) (1977-78) 730
Spaerobee (ALRC) (1977-78) 723
SPAR AEROSPACE LTD (Canada) 689
Sparrow (Raytheon) 685, 688
Sparrow-HV Arcas (ARC) (1972-73) 639
Sparviero (Breda) (1978-79) 640
Spiral (U.S.S.R.) 678
Standard ARM (General Dynamics) (1979-80) 652
Standard Active (General Dynamics) (1974-75) 628
Starlette (Aérospatiale) (1976-77) 672
Statsionar 1 (U.S.S.R.) (1977-78) 743
Strategic Comsat System (Hughes) (1976-77) 689
Striker (BAC) (1975-76) 626
Subroc (Goodyear) (1972-73) 578
Super 530 (Matra) 672, 687
Super Arcas (ARC) (1972-73) 639

Super Chief (ALRC) ....(1977-78) 723
Super Six (MDAC) ....(1971-72) 605
SUSQUEHANNA CORPORATION, THE (U.S.A.) ....(1972-73) 638
Swatter (U.S.S.R.) 678, 688
Swingfire (BAC) ....(1977-78) 692
Symphonie (CIFAS) 693

T

TAD (MDAC) ....(1971-72) 605
TD-1 Satellite (MESH) ....(1972-73) 609, 614
TDF-1 (Aérospatiale) 691
TIROS-N (RCA) 707
TOW (Hughes) 682, 688
TOW 2 (Hughes) 682
TRW (see TRW Defense & Space Systems Group) 709
TALLEY INDUSTRIES OF ARIZONA (U.S.A.) ....(1972-73) 639
Tansei Satellite (Nissan) ....(1972-73) 616
Taramis (CFTH-HB) ....(1972-73) 608
Telecom 1 (Matra) 691
Telstar 3 (Hughes) 703
Terminal Homing Flight Test Vehicle (Rockwell International) ....(1973-74) 611
Tethered satellite (Aeritalia) 694
TEXAS INSTRUMENTS INC (U.S.A.) 686
THOMSON-BRANDT (see Branche Hotchkiss-Brandt Armements) ....(1977-78) 712
Thor (MDAC) ....(1977-78) 733
Tibère (ONERA) ....(1975-76) 645
Titan III (Martin Marietta) 704
Titan IIIE-Centaur (Martin Marietta) .....(1978-79) 666
Titan 34-D (Martin Marietta) 704
Tomahawk II (General Dynamics) 682
Tournesol (Aérospatiale) ....(1972-73) 608
Trailblazer II (Celesco) ....(1976-77) 685
TRW DEFENSE & SPACE SYSTEMS GROUP (U.S.A.) 709
Type 80 (Mitsubishi) 675
Type 372 (Saab) ....(1975-76) 626

U

UK.4 (BAC) ....(1972-73) 619
UK 5 (MSDS) ....(1974-75) 653
UK 6 (MSDS) ....(1979-80) 671
UOSAT (see University of Surrey) 701
UOSAT (UOSAT) 701
USAF 313 satellite (Hughes) ....(1976-77) 689
USAF 417 satellite (RCA) ....(1977-78) 736
USAF 437 anti-satellite system (MDAC) ..(1977-78) 733
USAF 647 satellite (TRW) 709
USAF 711 satellite (Hughes) ....(1976-77) 689
USAF 777 satellite (TRW) ....(1975-76) 671
USAF 922 anti-satellite system (LTV) ....(1975-76) 663
USAF 1010 programme (TRW) ....(1973-74) 646
USN 749 programme (TRW) ....(1973-74) 646
UTC (see United Technology Center) .....(1972-73) 640
UNITED TECHNOLOGY CENTER (U.S.A.) ....(1972-73) 640
UNIVAC SALT LAKE CITY (U.S.A.) ..(1971-72) 563
UNIVERSITY OF SURREY (UK) 701

V

V3 (ARMSCOR) 675
VAPENAVDELNINGEN (Sweden) .....(1976-77) 624
Venus (U.S.S.R.) ....(1980-81) 673
Venus launcher (U.S.S.R.) ....(1980-81) 673
Véronique AG1 (SEP) ....(1972-73) 609
Véronique 61 M (SEP) ....(1973-74) 619
Vesta (SEP) ....(1972-73) 609
Viking (Martin) ....(1977-78) 731
Viking (SAAB) 695
Viper (Dornier) ....(1973-74) 595
VOLVO FLYGMOTOR AB (Sweden) ...(1971-72) 592
Voskhod launcher (U.S.S.R.) ....(1974-75) 676
Vostok core + Venus stage (U.S.S.R.) 697
Vostok launcher (U.S.S.R.) ....(1974-75) 676
VOUGHT CORPORATION (U.S.A.) 709
Voyager (NASA) 706

W

WS-140A (Boeing) ....(1978-79) 644
Walleye (Martin) ....(1972-73) 583
Walleye (Martin Marietta) 684, 688
Wasp (Boeing) 681
Wasp (Hughes) 683
Westar (Hughes) ....(1975-76) 661
WESTERN ELECTRIC COMPANY INC (U.S.A.) ....(1972-73) 589
WESTLAND AIRCRAFT LTD. (U.K.) ..(1972-73) 620
Whiplash (South Africa) ....(1976-77) 649

X

X-3 satellite (BAC) ....(1972-73) 619
X-4 satellite (HSD) ....(1975-76) 654
XAIM-54A Phoenix (Hughes) 682, 688
XAIM-97A (General Dynamics) ....(1973-74) 605
XJ521 (BAe) ....(1979-80) 650
XMGM-71A TOW (Hughes) ....(1980-81) 655, 661
XRL (LTV) ....(1972-73) 581
XRM-91 (Vought) 709

Y

Yantar 1 (U.S.S.R.) ....(1972-73) 646

Z

ZAGM-64A Hornet (North American) ...(1971-72) 559
ZAGM-84A Harpoon (MDAC) ....(1972-73) 582
Zond 8 (U.S.S.R.) ....(1971-72) 618
Zonda launcher (U.S.S.R.) 700

AERO-ENGINES

A

A1 Apollo motor (LPC) ....(1975-76) 755
A2 (LPC) ....(1974-75) 755
AGM-88A (Thiokol) 795
AI-14 (IL) ....(1977-78) 778
AI-14 (Ivchenko) ....(1975-76) 777
AI-14R (PZL) 739
AI-20 (Ivchenko) 744
AI-24 (Ivchenko) 744
AI-25 (Ivchenko) 745
AI-25 (Motorlet) ....(1973-74) 662
AI-26V (Ivchenko) ....(1975-76) 778
AJ10-118 (Aerojet) ....(1971-72) 672
AL-7 (Lyulka) 747
AL-21 (Lyulka) 747
ALF 101 (Avco Lycoming) 763
ALF 301 (Avco Lycoming) ....(1974-75) 733
ALF 502 (Avco Lycoming) 762
ARC (see Atlantic Research Corporation) 762
ARDEM (see Druine, Avions Roger) ....(1975-76) 687
ASh-21 (Shvetsov) ....(1975-76) 780
ASh-62 (Shvetsov) ....(1975-76) 780
ASh-62IR (Shvetsov) ....(1975-76) 780
ASh-62M (Shvetsov) ....(1975-76) 780
ASh-82T (Shvetsov) ....(1975-76) 781
ASh-82V (Shvetsov) ....(1975-76) 781
ASz-62R (PZL) 740
ASz-621R (PZL) 740
ATF-3 (Garrett) 770
AZ 14 (Turboméca) 724
AZ 16 (Turboméca) 724
Adour (Rolls-Royce/Turboméca) 723, 733, 752
Advanced apogee motor (CSD) 769
AERO BONNER LTD. (UK) 750
AEROJET-GENERAL CORPORATION (U.S.A.) 758
AEROJET LIQUID ROCKET COMPANY (U.S.A.) 758
AEROJET NUCLEAR SYSTEMS COMPANY (U.S.A.) ....(1973-74) 702
AEROJET SOLID PROPULSION COMPANY (U.S.A.) 758
AEROJET STRATEGIC PROPULSION COMPANY (USA) 758
AÉROSPATIALE (see Société Nationale Industrielle Aérospatiale) ....(1973-74) 663
AEROSPORT INC (U.S.A.) ....(1975-76) 729
Agena engine (Bell) 766
AIRESEARCH MANUFACTURING COMPANY OF ARIZONA (U.S.A.) ....(1980-81) 747
AIRESEARCH MANUFACTURING DIVISION OF LOS ANGELES (U.S.A.) ....(1972-73) 708
Alcor 1B (Aerojet) 760
ALFA ROMEO, SOCIETA PER AZIONI (Italy) . 736
Algol (Aerojet) ....(1976-77) 752
Algol III (CSD) 768
ALLISON (see Detroit Diesel Allison Division, General Motors Corporation) 733, 760
Altair III (Thiokol) 795
ALTURDYNE (U.S.A.) 762
ALVIS LTD. (U.K.) ....(1972-73) 682
Antares III (Thiokol) 796
Apache (Thiokol) 795
Apogee Motor (MBB) ....(1971-72) 645
Apollo engine (Aerojet) ....(1972-73) 723
Apollo motor (LPC) ....(1975-76) 755
Apollo SPS (Aerojet) ....(1977-78) 792
Arbizon (Turboméca) 724
Arc (SEP) ....(1980-81) 695
Ardem Mk X (Rollason) 752
Ardem Mk XI (Rollason) 752
Arriel (Turboméca) 724
Artouste III (Turboméca) 726
Astafan (Turboméca) 723
Astazou Turboprop (Turboméca) 724
Astazou Turboshaft (Turboméca) 725
Astrobee D (Aerojet) ....(1976-77) 753
Astrobee F (Aerojet) 760
Atar (SNECMA) 722
ATLANTIC RESEARCH CORPORATION (U.S.A.) 762
ATLAS AIRCRAFT CORPORATION OF SOUTH AFRICA (PTY) LTD. (South Africa) 742
AVCO LYCOMING STRATFORD DIVISION OF AVCO CORPORATION (U.S.A.) 762
AVCO LYCOMING WILLIAMSPORT DIVISION (U.S.A.) 765
AVIA NARODNI PODNIK (Czechoslovakia) 717
AVIAZIONE FIAT SpA (Italy) 736
AVIONS ROGER DRUINE (France) ....(1975-76) 687
Avon (Flygmotor) ....(1977-78) 781
Avon (Rolls-Royce) ....(1975-76) 717

B

B-1 (Rocketdyne) ....(1974-75) 765
BS 53 Pegasus (Rolls-Royce Bristol) .....(1977-78) 788
BS.360 (Rolls-Royce) ....(1973-74) 699
BS.605 (Rolls-Royce) ....(1971-72) 669
Baby Mamba (Dreher) 770
Baby Python (Dreher) ....(1975-76) 740
BAJ VICKERS LTD (UK) 750
Ball (SEP) ....(1980-81) 695
Bastan (Turboméca) ....(1980-81) 700
BELL AEROSPACE TEXTRON (U.S.A.) 766
BET-SHEMESH ENGINES LTD. (Israel) 735
BONNER (see Aero Bonner Ltd) 750
BORZECKI, JOZEF (Poland) 738
BRISTOL AEROJET LIMITED (UK) ..(1979-80) 719
BUDWORTH LTD, DAVID (U.K.) .....(1975-76) 715
Burner II (Thiokol) ....(1972-73) 741
Burner IIA (Thiokol) ....(1972-73) 741

C

C90 (Continental) ....(1978-79) 755
C90 (Rolls-Royce/Continental) ....(1980-81) 733
CAC (see Commonwealth Aircraft Corporation Ltd). 713
CASA (see Construcciones Aeronáuticas SA) (1977-78) 780
CF6 (General Electric) 736, 775
CF6 (MTU/General Electric) 729
CF6-32 (GE/SNECMA/Alfa Romeo/Volvo) ....(1980-81) 696
CF6-50 (GE/SNECMA/MTU) 723
CF6-50E (General Electric) 775
CF34 (General Electric) 775
CF-50A (General Electric) ....(1974-75) 723
CF700 (General Electric) 773
CFJ 810 (IHI) ....(1979-80) 706
CFM56 (CFM International) 731
CFM 56 (SNECMA/GE) 723
CJ610 (General Electric) 773
CSD (see Chemical Systems Division) 767
CSTM (see Centro Studi Trasporti Missilistici) 736
CT7 (General Electric) 779
CT58 (General Electric) 777
CT64 (General Electric) 777
CT 3201 (Lucas) ....(1980-81) 726
Cajun (Thiokol) ....(1972-73) 741
Castor II (Thiokol) 294
Castor IV (Thiokol) 294
CENTRO STUDI TRASPORTI MISSILISTICI (Italy) 736
CFM INTERNATIONAL (International) 731
CHEMICAL SYSTEMS DIVISION (U.S.A.) 767
Cherokee (Thiokol) ....(1972-73) 741
CHOTIA (see Weedhopper of Utah Inc) 767
Chotia 460 (Chotia) 767
CICARÉ AERONÁUTICA SC (Argentina) ....(1980-81) 687
Ciron (SEP) ....(1980-81) 695
Civil Spey (Rolls-Royce) 753
CLUTTON, ERIC (U.K.) ....(1977-78) 782
Cobra Booster and sustainer Motor (MBB) (1971-72) 645
COMMONWEALTH AIRCRAFT CORPORATION LTD. (Australia) 713
CONSTRUCCIONES AERONÁUTICAS SA (Spain) ....(1977-78) 780
CONTINENTAL (see Teledyne Continental Motors) 792
Coralie (SEP) ....(1974-75) 689
Couguar (Microturbo) ....(1980-81) 693
CURTISS-WRIGHT CORPORATION WOOD-RIDGE FACILITY (U.S.A.) 769

D

D-15 (Soloviev) 747
D-20P (Soloviev) 747
D-25V (Soloviev) 748
D-30 (Soloviev) 748
D-30K (Soloviev) 749
D-36 (Lotarev) 746
DMU (Thiokol) ....(1972-73) 742
DSI (see Developmental Sciences Inc) ....(1979-80) 741
Dart (Rolls-Royce) 754
Delta (TRW) ....(1979-80) 767
Delta engine (Aerojet) ....(1972-73) 696
DETROIT DIESEL ALLISON DIVISION, GENERAL MOTORS CORPORATION (U.S.A.) 733, 760
DEVELOPMENTAL SCIENCES INC. (U.S.A.) (1979-80) 741
DIFESA E SPAZIO SpA (Italy) 736
Double Turmastazou (Turboméca) ....(1979-80) 697
DOWTY ROTOL LTD (UK) 751
DREHER ENGINEERING COMPANY (U.S.A.) . 770
Dropt (SEP) ....(1980-81) 695
DRUINE, AVIONS ROGER (France) ...(1975-76) 687
Ducted Propulsor (Dowty Rotol) 751
Ducted rocket (Marquardt) 780

E

ELDO-A Vernier Engine (MBB) ....(1971-72) 645
ELDO Perigee (SEP) ....(1974-75) 690
ENMASA (see Empresa Nacional de Motores de Aviacion, SA) ....(1973-74) 688
EPM 600 (Rolls-Royce/MTU/Alfa Romeo) (1974-75) 702, 706, 710, 726
ESM 600 (Rolls-Royce/MTU/Alfa Romeo) (1974-75) 702, 706, 710, 726
ETJ131 (Garrett-AiResearch) ....(1980-81) 748
ETJ331 (AiResearch) ....(1977-78) 808
Eclair (Microturbo) ....(1971-72) 633
Eject motor (LPC) ....(1976-77) 778

Elan (SEP) (1977-78) 756
Emeraude (Microturbo) (1972-73) 657
EMG ENGINEERING COMPANY (U.S.A.) 779
EMPRESA NACIONAL DE MOTORES DE AVIACION SA (Spain) (1972-73) 681
Epictete (SNPE) 723
ERNO RAUMFAHRTTECHNIK GmbH (West Germany) 727
Espadon (Microturbo) (1972-73) 657

F

F-1 (Rocketdyne) (1973-74) 739
F3 (IHI) (1979-80) 706
F 10 (Hirth) (1972-73) 669
F 20 (Hirth) (1974-75) 699
F21 (Hirth) (1974-75) 699
F100 (Pratt & Whitney) 787
F101-DFE (General Electric) 776
F102 (Avco Lycoming) (1976-77) 757
F103-GE-100 (General Electric) 775
F104-GA-100 (Garrett) 770
F105 (General Electric) (1972-73) 716
F105-PW-100 (Pratt & Whitney) 783
F106-CA-100 (Teledyne) (1975-76) 768
F107 (Williams) 797
F401 (Pratt & Whitney) (1979-80) 757
F404 (General Electric) 774
FJR710 (NAL) 737
FN (see Fabrique Nationale Herstal SA) 713
FW-4 (CSD) (1977-78) 803
FW-5 (CSD) 768
FABRIQUE NATIONALE HERSTAL SA (Belgium) 713
Fan pod SG 85 (RFB) (1980-81) 704
Faon (SEP) (1977-78) 756
FAUVEL, "SURVOL"-CHARLES (France) (1973-74) 664
FIAT AVIAZIONE SpA (Italy) 736
FICHTEL & SACHS AG (West Germany) (1974-75) 699
Firebrand booster (Thiokol) 794
Firebrand ramjet (Marquardt) 780
Flexem (Rocketdyne) (1973-74) 740
FLYGMOTOR AB, VOLVO (Sweden) 731, 742
FRANKLIN ENGINE COMPANY, INC. (U.S.A.) (1977-78) 806

G

G2P (see Groupement pour les Gros Propulseurs à Poudre 719
G8-2 (Gluhareff) 779
G72C-C (Zenoah) (1980-81) 712
GDL (U.S.S.R.) 743
GE1 (General Electric) (1973-74) 720
GE1/10 (General Electric) (1973-74) 720
GE1/J1A2 (General Electric) (1971-72) 686
GE4 (General Electric) (1971-72) 691
GE12 (General Electric) (1971-72) 692
GIO-470A (Continental) (1971-72) 705
GMA 300 (Allison) 761
GMA 500 (Allison) 761
GO-480 (Lycoming) (1977-78) 800
GPS motor (CSD) (1978-79) 731
GRD Mk. U2 (U.S.S.R.) (1977-78) 846
GTD-350 (Isotov) (1974-75) 777
GTD-350 (WSK-PZL-Rzeszów) 740
GTSIO-520 (Continental) 793
Gamma Type 2 (Rolls-Royce) (1971-72) 670
Gamma Type 8 (Rolls-Royce) (1971-72) 669
GARRETT CORPORATION, THE (USA) 731, 770
GARRETT TURBINE ENGINE COMPANY (a division of The Garrett Corporation) (USA) 770
GARRETT/VOLVO FLYGMOTOR (International) 731
Gem (Rolls-Royce) 756
GENERAL ELECTRIC COMPANY, AIRCRAFT ENGINE GROUP (U.S.A.) 772
GESCHWENDER AEROMOTIVE INC (U.S.A.) 779
GLUHAREFF (see EMG Engineering Company) 779
GLUHAREFF HELICOPTERS (U.S.A.) (1975-76) 753
GLUSHENKOV (U.S.S.R.) 743
Gnome (Rolls-Royce) 757
GROUPEMENT POUR LES GROS PROPULSEURS A POUDRE (France) 719
GROUPEMENT TURBOMÉCA-SNECMA GRTS (France) 727

H

H-1 (Rocketdyne) (1973-74) 740
H-63 (Nelson) 781
H.1000 Gnome (Rolls-Royce) 757
H.1200 Gnome (Rolls-Royce) 757
H.1400 Gnome (Rolls-Royce) 757
HAL (see Hindustan Aeronautics Ltd.) 730
HARM motor (Thiokol) 795
HIO-360 (Lycoming) (1977-78) 801
HJE-2500 (HAL) (1976-77) 728
HM4 (SEP) (1979-80) 689
HM7 (SEP) 720
HPE-2 (HAL) (1974-75) 703
HPE-4 (HAL) (1976-77) 728
HR4 (Flygmotor) (1974-75) 715
HARKER & ASSOCIATES (UK) 751
Harpoon (Thiokol) 794
Harpoon booster motor (Aerojet) 760
HAWKER SIDDELEY CANADA LTD. (Canada) 713
Hawk motor (Aerojet) 759
Hellfire (Thiokol) 795
High-energy rocket engines (MBB) 729
HINDUSTAN AERONAUTICS LTD (India) 730
HIRTH MOTOREN KG (West Germany) (1975-76) 698
Hydrazine Engine (MBB) (1971-72) 645
Hydrazine motors (LRBA) (1971-72) 632
Hydrazine Thruster (SEP) 721
Hydrogen/Fluorine Engine (MBB) (1971-72) 645

I

IDA (see Innovation et Développement Aérothermodynamiques) (1975-76) 687
IGO-540 (Avco Lycoming) (1976-77) 760
IGSO-480 (Avco Lycoming) (1977-78) 801
IGSO-540 (Avco Lycoming) (1977-78) 801
IHI (see Ishikawajima-Harima) 736
IL (see Instytut Lotnictwa) 739
IO-260 (Rolls-Royce/Continental) (1974-75) 727
IO-320 (Avco Lycoming) 765
IO-346 (Continental) (1971-72) 706
IO-360 (Continental) 792
IO-360 (Avco Lycoming) 765
IO-360 (Rolls-Royce/Continental) (1980-81) 733
IO-368 (Continental) 792
IO-368 (Rolls-Royce/Continental) (1980-81) 733
IO-470 (Continental) (1977-78) 833
IO-520 (Continental) 793
IO-540 (Lycoming) 765
IO-720 (Avco Lycoming) 766
IUS (CSD) 767
IVO-360 (Lycoming) (1977-78) 801
IVO-540 (Avco Lycoming) (1976-77) 762
INNOVATION ET DÉVELOPPEMENT AÉROTHERMODYNAMIQUES (France) (1975-76) 687
INSTYTUT LOTNICTWA (Poland) 739
INTREPRINDEREA TURBOMECANICA BUCURESTI (Romania) 741
Ion Thruster (SEP) (1979-80) 691
Iroquois (Thiokol) (1972-73) 742
ISHIKAWAJIMA-HARIMA JUKOGYO KABUSHIKI KAISHA (Japan) 736
ISOTOV (U.S.S.R.) 743
IVCHENKO (U.S.S.R.) 744

J

J-2 (Rocketdyne) (1973-74) 739
J3 (IHI) 737
J3-IHI-8 (IHI) (1977-78) 776
J7-300 (Thermo-Jet) (1980-81) 770
J8-104 (Thermo-Jet) (1972-73) 739
J8-200 (Thermo-Jet) (1980-81) 770
J10-200 (Thermo-Jet) (1980-81) 770
J13-202 (Thermo-Jet) (1980-81) 770
J13-400 (Thermo-Jet) (1972-73) 740
J52 (Pratt & Whitney) 782
J57 (Pratt & Whitney) (1973-74) 735
J58 (Pratt & Whitney) (1976-77) 781
J60 (Pratt & Whitney) (1978-79) 745
J69 (Teledyne CAE) 790
J75 (Pratt & Whitney) 782
J79 (General Electric) 772
J79 (MTU) (1978-79) 691
J79-19 (General Electric) 736
J85 (General Electric) 773
J85-13 (Fiat/General Electric) 736
J85-15 (Orenda/General Electric) (1973-74) 657
J97 (General Electric) (1976-77) 771
J100-CA-100 (Teledyne CAE) 791
J100-CA-101 (Teledyne CAE) 791
J101 (General Electric) (1975-76) 747
J400 (Williams) 796
J402-CA-400 (Teledyne CAE) 791
J402-CA-700 (Teledyne CAE) 791
JB 2 × 250 (Borzecki) 738
JFTD12 (Pratt & Whitney) (1977-78) 826
JR 100 (IHI/NAL) (1976-77) 736
JR 200 (IHI/NAL) 738
JR 220 (IHI/NAL) 738
JR 300 (IHI/NAL) (1974-75) 713
JT3C (Pratt & Whitney) (1973-74) 735
JT3D (Pratt & Whitney) 782
JT4 (Pratt & Whitney) 782
JT8 (Pratt & Whitney) 782
JT8D (Pratt & Whitney) 782
JT8D-200 (Pratt & Whitney) 783
JT9D (Pratt & Whitney) 783
JT10D (Pratt & Whitney) (1980-81) 703, 710, 762
JT10D (Pratt & Whitney/MTU/Alfa Romeo) (1976-77) 734
JT11 (Pratt & Whitney) (1976-77) 781
JT12 (Pratt & Whitney) (1978-79) 745
JT15D (P&WC) 713
JTF10A (Pratt & Whitney) 786
JTF20 (Pratt & Whitney) (1971-72) 699
JTF22 (Pratt & Whitney) 787
JACOBS (see Page Industries Inc) 779
Jaguar System (Microturbo) (1972-73) 687
JANOWSKI, JAROSLAW (Poland) 739
Jawa (Avia) (1971-72) 629

K

K-5 (WSK-Kalisz) (1978-79) 701
KHD (see Klöckner-Humboldt-Deutz AG) (1980-81) 701
KAWASAKI JUKOGYO KABUSHIKI KAISHA (Japan) 737
KHD LUFTFAHRTTECHNIK GmbH (Germany) 727
KLIMOV (U.S.S.R.) (1977-78) 842
KLÖCKNER-HUMBOLDT-DEUTZ AG (West Germany) (1980-81) 701
KOLIESOV (U.S.S.R.) 745
KOMATSU ZENOAH COMPANY (Japan) 738
KUZNETSOV (U.S.S.R.) 745

L

L 275E (Limbach) 728
L 2000 (Limbach) 728
LAPAN (see Lembaga Penerbangan dan Antariska Nasional) 730
LB600 (Aerosport-Rockwell) (1975-76) 729
LiT-35 (WSK-PZL-Rzeszów) (1975-76) 713
LJ95 (Teledyne CAE) 791
LP-117 (LPC) (1976-77) 778
LP-118 (LPC) (1976-77) 778
LP-119 (LPC) (1976-77) 778
LPC (see Lockheed Propulsion Company) (1976-77) 777
LPC-A1 (LPC) (1975-76) 755
LPC-A2 (LPC) (1975-76) 755
LPC-415 (LPC) (1976-77) 778
LR79 (Rocketdyne) (1972-73) 732
LR89-NA-5 (Rocketdyne) 789
LR101-NA-7 (Rocketdyne) 789
LR105-NA-5 (Rocketdyne) 789
LRBA (see Laboratoire de Recherches Balistiques et Aérodynamiques) (1973-74) 664
LSM 156-5 (LPC) (1976-77) 778
LSM 156-6 (LPC) (1976-77) 778
LTC1 (Avco Lycoming) 764
LCT4 (Avco Lycoming) 764
LTC4V Series (Avco Lycoming) (1975-76) 736
LTP 101 (Avco Lycoming) 763
LTS 101 (Avco Lycoming) 763
LTV (see LTV Aerospace Corporation) (1975-76) 755
LVRJ (Vought) (1979-80) 767
LABORATOIRE DE RECHERCHES BALLISTIQUES ET AÉRODYNAMIQUES (France) (1973-74) 664
Larzac (SNECMA/Turboméca) 723, 724, 727, 729
LEMBAGA PENERBANGAN DAN ANTARISKA NASIONAL (Indonesia) 730
LEONIDES (see Harker & Associates) 751
LIMBACH MOTORENBAU (West Germany) 728
LOCKHEED PROPULSION COMPANY (U.S.A.) (1976-77) 777
Long-life SRAM motor (Thiokol) 794
LOTAREV (U.S.S.R.) 746
LTV AEROSPACE CORPORATION (U.S.A.) (1975-76) 755
LUCAS AEROSPACE LTD. (U.K.) 751
Lunar Module descent engine (TRW) (1972-73) 742
LYCOMING (see Avco Lycoming) 738, 741
LYULKA (U.S.S.R.) 747

M

M-1 (LPC) (1972-73) 723
M-11 (Mikulin) (1975-76) 780
M-14P (Vedeneev) 750
M-14V (WSK-PZL-Kalisz) (1977-78) 778
M-14V-26 (Vedeneev) 750
M-14VF (Ivchenko) (1975-76) 778
M45 Series (Rolls-Royce/SNECMA) (1979-80) 723
M45H (Rolls-Royce/SNECMA) (1979-80) 723
M45S-11 (Rolls-Royce/SNECMA) (1978-79) 694
M45SD-02 (Rolls-Royce/SNECMA) (1979-80) 703, 724
M49 Larzac (SNECMA/Turboméca) 723, 724, 727, 729
M53 (SNECMA) 722
M-55 (Thiokol) (1979-80) 765
M57H (Rolls-Royce/SNECMA) (1975-76) 705
M88 (SNECMA) 722
M 137 (Avia) 718
M 150 Jawa (Avia) (1971-72) 629
M-205 (Klimov) (1973-74) 759
M 332 (Avia) (1971-72) 629
M 337 (Avia) 718
M 462RF (Avia) (1978-79) 679
M 601B (Motorlet) 718
M 701 (Motorlet) (1978-79) 680
MA-3 (Rocketdyne) 789
MA-5 (Rocketdyne) 789
MA210/212 (Marquardt) 780
MB-3 (Rocketdyne) (1972-73) 732
MBB (see Messerschmitt-Bölkow-Blohm GmbH) 728
MiG-23 engine (Tumansky) (1980-81) 723
MiG-25 engine (Tumansky) (1978-79) 710
Mk 25 JATO (Rocketdyne) (1977-78) 828
Mk 36 Mod 5 Sidewinder motor (Rocketdyne) (1977-78) 828
Mk 38 (Rocketdyne) (1977-78) 828
Mk 39 (Rocketdyne) (1977-78) 828
Mk 47 (Rocketdyne) (1977-78) 828
Mk 70 Mod 0 (Rocketdyne) (1976-77) 789
MS-1500 (Pieper) 730
MTM 380 (MTU/Turboméca) 726
MTM 385 (MTU/Turboméca) 729, 732
MTU (see Motoren-und Turbinen-Union München GmbH) 729
Mage (SEP) 721
Makila (Turboméca) 726
Marboré (Turboméca) 724
MARQUARDT COMPANY, THE (U.S.A.) 779
Maverick (Thiokol) 794, 795
Maverick Motor (Aerojet) 759
McCULLOCH CORPORATION, LOS ANGELES DIVISION (U.S.A.) (1974-75) 756
MESSERSCHMITT-BÖLKOW-BLOHM GmbH (West Germany) 728
MICROTURBO SA (France) 719
MIKULIN (U.S.S.R.) 747
Military Spey (Rolls-Royce) 754
Mini Motor (LPC) (1976-77) 778
Minor 6-III (Avia) (1971-72) 628
Minuteman first-stage motor (Thiokol) (1979-80) 765
Minuteman motor (Aerojet) 758
Minuteman third-stage motor (Thiokol) (1979-80) 765
MITSUBISHI JUKOGYO KABUSHIKI KAISHA (Japan) 737
Model 003 (Pulch) 730
Model IS/60 (Rover) (1972-73) 683
Model IS/90 (Rover) (1972-73) 683
Model 2A-120 (Franklin) (1975-76) 741

Model 2.5KS-18000 (Aerojet) (1975-76) 728
Model 2RB (Borzecki) (1980-81) 713
Model 2S/150A (748) (Rover) (1972-73) 683
Model 2S/150A (801) (Rover) (1972-73) 683
Model 4A-235 (Franklin) (1975-76) 741
Model 4AR 1200 (Rectimo) (1974-75) 688
Model 4C2T (Cicaré) (1979-80) 683
Model 4-180 (Continental) (1972-73) 737
Model 6-260 (Continental) (1974-75) 770
Model 6-285-B (Continental) (1977-78) 834
Model 6-320 (Continental) (1977-78) 834
Model 6A4-150 (Franklin) (1975-76) 741
Model 6A4-165 (Franklin) (1975-76) 741
Model 6A-335 (Franklin) (1975-76) 741
Model 6A-350 (Franklin) (1975-76) 741
Model 6AS-335 (Franklin) (1975-76) 741
Model 6AS-350 (Franklin) (1975-76) 741
Model 6V-335 (Franklin) (1975-76) 741
Model 6V4-200 (Franklin) (1975-76) 741
Model 6VS-335 (Franklin) (1975-76) 741
Model 6V-350 (Franklin) (1975-76) 741
Model 16NS-1000 (Rocketdyne) (1977-78) 828
Model 100 (NPT) (1976-77) 742
Model 101 (NPT) (1977-78) 783
Model 150 (Aerojet) (1975-76) 728
Model 156-2C1 (Thiokol) (1974-75) 772
Model 250 (Allison) 760
Model 250 (Rolls-Royce/Allison) (1972-73) 667
Model 352 (Teledyne CAE) 790
Model 356 (Teledyne CAE) 790
Model 356-28A (Teledyne CAE) 791
Model 356-28E (Teledyne) (1979-80) 761
Model 356-28F (Teledyne CAE) 791
Model 365 (Teledyne CAE) 791
Model 370 (Teledyne CAE) 791
Model 372-2 (Teledyne CAE) 791
Model 373 (Teledyne CAE) 791
Model 400 (NPT) (1976-77) 742
Model 401A (NPT) (1979-80) 720
Model 415 (LPC) (1973-74) 729
Model 440/555 (Teledyne CAE) 791
Model 455 (Teledyne CAE) 791
Model 471 (Teledyne) (1978-79) 755
Model 472 (Teledyne) (1975-76) 768
Model 490 (Teledyne CAE) 791
Model 501 (Allison) (1980-81) 737
Model 912-B3 and -B14 (Rolls-Royce/Allison) 733
Model 4318A (McCulloch) (1972-73) 724
Model 4318E (McCulloch) (1972-73) 724
Model 4318F (McCulloch) (1972-73) 724
Model 4318F (Northrop) 781
Model 6012C, E and L (MTU) (1971-72) 646
Model 6022 (MTU) (1972-73) 671
Model 6318 (McCulloch) (1971-72) 697
Model 6318K (McCulloch) (1971-72) 697
Model 8096 Agena engine (Bell) 766
Model T6-285 (Continental) (1972-73) 737
Model T6-320 (Continental) (1975-76) 771
Model T8-450 (Continental) (1972-73) 737
Model TC6150 (McCulloch) (1971-72) 697
Monopropellant engines (MBB) (1978-79) 690
Monopropellant gas-generators (MBB) 728
MOTEURS MUDRY BUCHOUX (AVIONS MUDRY ET CIE) (France) 720
MOTOREN-UND TURBINEN-UNION MÜNCHEN GmbH (West Germany) 732
MOTORLET NC, ZAVOD JANA SVERMY (Czechoslovakia) 718
MTU/TURBOMÉCA SARL (International) 732
MUDRY (see Moteurs Mudry Buchoux; Avions Mudry et Cie) 720

N

NAL (see National Aerospace Laboratory) 737
NASA (see National Aeronautics and Space Administration) (1973-74) 731
NK-8 (Kuznetsov) 745
NK-12M (Kuznetsov) 746
NK-86 (Kuznetsov) 746
NK-144 (Kuznetsov) 746
NPT (see Noel Penny Turbines Ltd) 751
NATIONAL AERONAUTICS AND SPACE ADMINISTRATION (U.S.A.) (1973-74) 731
NATIONAL AEROSPACE LABORATORY (Japan) 737
NATIONAL AIRCRAFT ENGINE FACTORY (China) 717
NATIONAL CHINESE ACADEMY OF SCIENCE (China) (1977-78) 752
NELSON AIRCRAFT CORPORATION (U.S.A.) 781
Nimbus (Rolls-Royce) (1975-76) 725
Noelle (Microturbo) (1972-73) 658
NOEL PENNY TURBINES LTD (U.K.) 751
NORTHROP CORPORATION, VENTURA DIVISION (U.S.A.) 781

O

O22 Cougar (Microturbo) (1980-81) 693
O 28 (Hirth) (1973-74) 675
O-100-1 (McCulloch) (1972-73) 724
O-100-3 (Northrop) 781
O-150-2 (McCulloch) (1971-72) 697
O-150-4A (McCulloch) (1971-72) 697
O-200 (Continental) 792
O-200-A (Rolls-Royce/Continental) (1980-81) 733
O-235 (Avco Lycoming) 765
O-240-A (Rolls-Royce/Continental) (1980-81) 733
O-290 (Avco Lycoming) (1976-77) 759
O-300 (Continental) (1972-73) 734
O-300 (Rolls-Royce/Continental) (1980-81) 733
O-320 (Avco Lycoming) 765
O-340 (Avco Lycoming) (1976-77) 760
O-360 (Avco Lycoming) 765
O-435 (Avco Lycoming) (1976-77) 760
O-470 (Continental) 793
O-480 (Avco Lycoming) (1976-77) 760
O-540 (Avco Lycoming) 765
ODA (see Ostdeutsches Automobilwerk) (1975-76) 698
Odin (Rolls-Royce) 756
Olympus (Rolls-Royce) (1979-80) 725
Olympus 593 (Rolls-Royce/SNECMA) (1980-81) 696, 707
OMNIPOL FOREIGN TRADE CORPORATION (Czechoslovakia) 717
Ordnance Turbojet (NASA/Navy) (1973-74) 731
ORENDA (see Hawker Siddeley Canada Ltd) 713
Orpheus (Rolls-Royce) 755
OSTDEUTSCHES AUTOMOBILWERK (West Germany) (1975-76) 698

P

P4 (SEP) (1980-81) 694
P8E-9 (Rocketdyne) (1980-81) 766
P16 (LPC) (1972-73) 722
P.1400 Gnome (Rolls-Royce) (1976-77) 750
P2721-77 (LPC) (1972-73) 723
P & WC (see Pratt & Whitney Aircraft of Canada Ltd) 713
PLT-27 (Avco Lycoming) (1979-80) 735
PT6A (P&WC) 714
PT6A-27 (MTU) (1971-72) 646
PT6B/PT6C (P&WC) 716
PT6T Twin Pac (P&WC) 716
PW100 (P&WC) 714
PW1120 (Pratt & Whitney) 787
PW1130 (Pratt & Whitney) 787
PW2037 (Pratt & Whitney) 729, 736, 785
PZL (see Polskie Zaklady Lotnicze) 738
PZL-3R (WSK-PZL-Rzeszów) 741
PZL-3S (WSK-PZL-Rzeszów) 741
PZL-10W (WSK-PZL-Rzeszów) 740
PZL-F engines (WSK-PZL-Rzeszów) 741
PAGE INDUSTRIES INC (U.S.A.) 779
PARODI, ROLAND (West Germany) 730
Patriot (Thiokol) 794
Pegasus (Rolls-Royce) 755
PIAGGIO SpA, INDUSTRIE AERONAUTICHE E MECCANICHE RINALDO (Italy) 736
PIEPER MOTORENBAU GmbH (West Germany) 730
PLESSEY COMPANY LTD, THE (U.K.) (1972-73) 684
POLSKIE ZAKLADY LOTNICZE (Poland) 738
Poseidon first stage motor (Thiokol) (1980-81) 771
POUDRERIE NATIONALE DE SAINT-MÉDARD (France) (1971-72) 635
PRATT & WHITNEY AIRCRAFT GROUP OF UNITED TECHNOLOGIES CORPORATION (U.S.A.) 782
PRATT & WHITNEY AIRCRAFT OF CANADA LTD (Canada) 713
PRATT & WHITNEY/MTU/ALFA ROMEO (International) (1974-75) 704
Puffin (Budworth) (1974-75) 716
PULCH, OTTO (West Germany) 730
PUSAT TEKNOLOGI DIRGANTARA (Indonesia) 730
Pygmée (De Coucy "Survol") (1972-73) 657

Q

QFT55 (Avco Lycoming) (1974-75) 735
Quiet Engine (GE/NASA) (1975-76) 749

R

R-1E (Marquardt) 780
R-4D (Marquardt) 780
R-6C (Marquardt) 780
R-11 (Tumansky) 749
R-13 (Tumansky) 749
R-27 (Tumansky) 749
R-29B (Tumansky) 749
R-30 (Marquardt) 780
R-40A (Marquardt) 780
R-266 (Tumansky) 749
R-755 (Jacobs) (1980-81) 756
RA. Avon (Rolls-Royce) (1975-76) 717
RB.146 Avon (Rolls-Royce) (1975-76) 717
RB.153 (Rolls-Royce/MAN) (1969-70) 865, 691, 702
RB.162 (Rolls-Royce) (1978-79) 713
RB.163 Civil Spey (Rolls-Royce) 753
RB.168 Military Spey (Rolls-Royce) 754
RB.168-62 and -66 Spey (Rolls-Royce/Allison) 733
RB.172/T260 Adour (Rolls-Royce Turboméca) (1977-78) 760
RB.183 Civil Spey (Rolls-Royce) (1979-80) 723
RB.199 (Turbo-Union) 729, 735, 736, 752
RB.202 (Rolls-Royce) (1972-73) 685
RB.211 (Rolls-Royce) 752
RB.235 (Rolls-Royce) (1976-77) 719
RB.318 (Rolls-Royce/Alfa Romeo) 736, 754
RB.401 (Rolls-Royce) 755
RB.410D-2 (Rolls-Royce/SNECMA) (1979-80) 703
RB.410-11 (Rolls-Royce/SNECMA) (1979-80) 703
RB.432 (Rolls-Royce) (1979-80) 722
RB eject motor (LPC) (1976-77) 778
RC engines (Wright) 770
RC2-60 (Wright) 770
RC2-75 (Wright) 770
RC2-90 (Wright) 770
RC-741 (LAPAN) 730
RD-3M-500 (Mikulin) 747
RD-5 (Mikulin) (1977-78) 844
RD-9 (Tumansky) (1977-78) 846
RD-9F (Klimov) (1973-74) 759
RD-107 (GDL) 743
RD-119 (GDL) 743
RD-214 (GDL) 743
RD-219 (GDL) 743
RDa. Dart (Rolls-Royce) 754
RFB (see Rhein-Flugzeugbau GmbH) 730
RJ.500 (RJAE) 732, 753
RJAE (see Rolls-Royce and Japanese Aero Engines Ltd) 732
RL10 (Pratt & Whitney) 788
RM6C Avon (Flygmotor) (1977-78) 781
RM8 (Flygmotor) 742
RR.605 (Rolls-Royce) (1973-74) 700
RRTI (see Rolls-Royce Turboméca International SA) (1973-74) 670, 682, 691
RRX1 (Flygmotor) 742
RRX2 (Flygmotor) (1973-74) 689
RRX5 (Flygmotor) 742
RS18 (Rocketdyne) (1972-73) 733
RS-21 (Rocketdyne) (1972-73) 733
RS-27 (Rocketdyne) 789
RS-34 (Rocketdyne) 788
RS.360 (Rolls-Royce) (1974-75) 725
RS2101C (Rocketdyne) (1976-77) 788
RSVP (LPC) (1973-74) 730
RT.172 (Rolls-Royce/Turboméca) (1979-80) 703
RTM 321 (Turboméca/Rolls-Royce) 726, 734, 756
RTy. Tyne (Rolls-Royce) (1975-76) 722
RU-19-300 (U.S.S.R.) 750
RW-100 (RotorWay) 789
RW-133 (RotorWay) 790
RZ.2 (Rolls-Royce) (1972-73) 694
RZ.12 (Rolls-Royce) (1972-73) 694
Ram rocket (MBB) 729
RECTIMO AVIATION SA (France) 720
Recruit (Thiokol) (1977-78) 836
Resonating ramjet (DSI) (1979-80) 741
RHEIN-FLUGZEUGBAU GmbH (West Germany) 730
Rita (SEP) (1974-75) 689
ROCKETDYNE DIVISION OF ROCKWELL INTERNATIONAL (U.S.A) 788
ROLLASON AIRCRAFT & ENGINES, LTD. (U.K.) 752
ROLLS-ROYCE & JAPANESE AERO ENGINES LTD (International) 732
ROLLS-ROYCE LTD (U.K.) 733, 752
ROLLS-ROYCE/ALLISON (International) 733
ROLLS-ROYCE/KHD (International) (1973-74) 680
ROLLS-ROYCE MOTORS LTD (U.K.) 757
ROLLS-ROYCE/MTU (International) (1974-75) 706
ROLLS-ROYCE/MTU/ALFA ROMEO (International) (1974-75) 706
ROLLS-ROYCE/SNECMA (International) (1980-81) 707
ROLLS-ROYCE TURBOMÉCA LTD. (International) 733
ROTORWAY INC. (U.S.A.) 789
ROVER GAS TURBINES (U.K.) (1972-73) 682
Rover TJ.125 (Lucas) (1973-74) 690

S

SARV retro (Thiokol) 795
SD (SNPE) 723
SDS 12 (SERMEL) (1973-74) 667
SE-8 (Rocketdyne) (1972-73) 733
SEP (see Société Européene de Propulsion) 720
SEP 163 (SEP) (1980-81) 695
SEP 299 (SEP) (1980-81) 695
SEP 300 (SEP) (1980-81) 695
SEP 739 (SEP) (1975-76) 690
SEP 6854 (SEP) (1980-81) 695
SEP 7342 (SEP) (1980-81) 695
SEP 7382 (SEP) (1980-81) 695
SEP 7392 (SEP) (1980-81) 695
SERMEL (see Société d'Études et de Recherches pour la Mécanique, l'Électricité et l'Électrique, SARL) (1973-74) 667
SETE (Curtiss-Wright) 770
SG 85 (RFB) 704
SL 1700 (Limbach) 728
SL 2400 (Limbach) (1975-76) 699
SM 160 (Valmet) 718
SNECMA (see Société Nationale d'Étude et de Construction de Moteurs d'Aviation) 721
SNPE (see Société Nationale des Poudres et Explosifs) 723
SO-1 (IL) 739
SO-3 (IL) 739
SP-440 (Alturdyne) 762
SPAL 30 (SEP) (1974-75) 689
SR-1 (Flygmotor) (1972-73) 682
SR13-LP-1 (LPC) (1974-75) 729
SR49-TC-1 (Thiokol) 794
SR73-AJ-1 (Thiokol) (1979-80) 765
SR75-LP-1 (LPC) (1976-77) 755
SR-114-TC-1 (Thiokol) 795
SRAM motor (Thiokol) (1977-78) 835
SSME (Rocketdyne) 788
SS-RPV (Microturbo/Turboméca) 719
SSUS-A motor (Thiokol) 794
ST6 (P&WC) (1975-76) 684
ST9 (Pratt & Whitney) (1972-73) 730
STAR-6 (Thiokol) 795
STAR-13 (Thiokol) 795
STAR-13A (Thiokol) 795
STAR-17 (Thiokol) 795
STAR-17A (Thiokol) 795
STAR-20 (Thiokol) 795
STAR-24 (Thiokol) 795
STAR-26B (Thiokol) 795
STAR-27 (Thiokol) 795
STAR-30 (Thiokol) 796
STAR-31 (Thiokol) 796
STAR-37 (Thiokol) 795
STAR-37X (Thiokol) 796

STAR-48 (Thiokol) 796
STJ442 (Pratt & Whitney) (1975-76) 758
STM (Vought) 796
SVM (Aerojet) (1979-80) 730
SACMA (France) 720
Sandhawk (Thiokol) 795
Saphir (Microturbo) (1972-73) 658
Saturn 500 (Janowski) 739
Saturn S-1C retro (Thiokol) (1977-78) 836
Segmented motor (CSD) 767
Senior (Aerojet) (1972-73) 696
SHVETSOV (U.S.S.R.) 747
Sidewinder motor (Aerojet) 759
Sidewinder motor (Rocketdyne) (1971-72) 705
SNIA VISCOSA (Italy) (1980-81) 711
SOCIETA PER AZIONI ALFA-ROMEO (Italy) 736
SOCIÉTÉ D'ÉTUDE DE LA PROPULSION PAR RÉACTION (France) (1970-71) 705
SOCIÉTÉ D'ÉTUDES ET DE RECHERCHES POUR LA MÉCANIQUE, L'ÉLECTRICITÉ ET L'ÉLECTRONIQUE SARL (France) (1973-74) 667
SOCIÉTÉ EUROPÉENNE DE PROPULSION (France) 720
SOCIÉTÉ NATIONALE INDUSTRIELLE AÉROSPATIALE (France) (1973-74) 663
SOCIÉTÉ NATIONALE DES POUDRES ET EXPLOSIFS (France) 723
SOCIÉTÉ NATIONALE D'ÉTUDE ET DE CONSTRUCTION DE MOTEURS D'AVIATION (France) 721
SOLAR DIVISION OF INTERNATIONAL HARVESTER COMPANY (U.S.A.) (1975-76) 766
Solent (Plessey) (1972-73) 684
SOLOVIEV (U.S.S.R.) 747
Space Shuttle OMS engine (Aerojet) 758
Space Shuttle SRM (Thiokol) 793
Space Vehicle Motors (SVM) (Aerojet) (1979-80) 730
Sparrow/Shrike motors (Aerojet) 760
Spey (Rolls-Royce) 753, 754
Spey 67 (Rolls-Royce) (1974-75) 720
Spey 605W (Rolls-Royce) (1974-75) 721
Spey 606 (Rolls-Royce) (1976-77) 746
Spin/despin motors (Thiokol) (1979-80) 767
Stamo MS 1500 (Pieper) 730
Standard missile motor (Aerojet) 760
Statolite (SNPE) 723
Super Cougar (IDA) (1975-76) 687
Super Sapphire (Bonner) 750
SVENSKA FLYGMOTOR (see Volvo Flygmotor AB) 731, 742

T

T6-320 (Continental) (1976-77) 794
T53 (Avco Lycoming) 764
T55 (Avco Lycoming) 764
T56 (Allison) 761
T58 (General Electric) 777
T63 (Allison) 760
T64 (General Electric) 777
T64 (MTU) (1977-78) 768
T64 (Rolls-Royce) (1971-72) 669
T64-P4D (General Electric) 736, 777
T73 (Pratt & Whitney) (1977-78) 826
T74 (P&WC) 716
T76 (Garrett) 771
T112 (Rolls-Royce/KHD) (1973-74) 680
T117 (KHD) 727
T212 (KHD) (1975-76) 698
T260 (Rolls-Royce/Turboméca) (1974-75) 693
T312 (KHD) (1975-76) 699
T317 (KHD) 727
T400 (P&WC) 717
T405 (Avco Lycoming) (1979-80) 735
T700 (General Electric) 778
T701-AD-700 (Allison) (1976-77) 756
TAA 230 (Turboméca-Agusta) (1975-76) 696
TC6150 (McCulloch) (1971-72) 697
TE-260G (Thiokol) (1980-81) 771
TE-2606 (Thiokol) (1976-77) 796
TE-M-29 Recruit (Thiokol) (1977-78) 836
TE-M-29-8 (Thiokol) 795
TE-M-82 Cajun (Thiokol) (1972-73) 741
TE-M-146 Cherokee (Thiokol) (1972-73) 741
TE-M-236 (Thiokol) 795
TE-M-307 Apache (Thiokol) 795
TE-M-364 (Thiokol) 795
TE-M-388 Iroquois (Thiokol) (1972-73) 742
TE-M-416 Tomahawk (Thiokol) 795
TE-M-424 Saturn S-1C retro (Thiokol) (1977-78) 836
TE-M-442-1 (Thiokol) 795
TE-M-458 (Thiokol) 795
TE-M-473 Sandhawk (Thiokol) 795
TE-M-479 (Thiokol) 795
TE-M-516 (Thiokol) 795
TE-M-521 (Thiokol) 795
TE-M-541/542 (Thiokol) 795
TE-M-604 (Thiokol) 795
TE-M-616 (Thiokol) 795
TE-M-640 Altair III (Thiokol) 795
TE-M-696 (Thiokol) 795
TE-M-697 (Thiokol) 795
TE-M-700 (Thiokol) 796
TE-M-707 (Thiokol) 794
TE-M-711 (Thiokol) 796
TE-M-714 (Thiokol) 796
TE-M-762 Antares III (Thiokol) 796
TF25 (Avco Lycoming) (1975-76) 734
TF30 (Pratt & Whitney) 786
TF33 (Pratt & Whitney) 782
TF34 (General Electric) 774
TF39 (General Electric) (1976-77) 772
TF40-IHI-801A (IHI) 737
TF41 Spey (Rolls-Royce/Allison) 733, 754
TFE731 (Garrett) 771
TFE1042 (Garrett/Volvo Flygmotor) 731, 742, 771
TGS 4 (Sermel) (1971-72) 637
THS 200 (Rolls-Royce/SNECMA) (1971-72) 650
THS 2000 (Rolls-Royce/SNECMA) (1974-75) 707
TIO-540 (Avco Lycoming) 765
TIO-541 (Avco Lycoming) 766
TIVO-540 (Avco Lycoming) (1976-77) 761
TJ.124 (Rover) (1972-73) 683
TJD-76A Baby Mamba (Dreher) (1977-78) 805
TJD-76C Baby Mamba (Dreher) 770
TJD-76D and E (Dreher) 770
TJD-79A Baby Python (Dreher) (1975-76) 740
TJE 341 (Garrett-AiResearch) (1980-81) 748
TM-251 (Turboméca-Agusta) (1975-76) 696
TPE331 (Garrett) 771
TR-201 (TRW) 796
TR 281 (Turboméca) (1978-79) 685
TRD Mk P.166/TRD31 (U.S.S.R.) (1977-78) 846
TRI 60 (Microturbo) 719
TRI 80 (Microturbo) 719
TRS 12 (SERMEL) (1973-74) 667
TRS 18 (Microturbo) 719
TRS 18-046 (Microturbo) 719
TRS 18-056 (Microturbo) 719
TRS 18-075 (Microturbo) 719
TRS 18-076 (Microturbo) 719
TRS 20/22 (Microturbo) 719
TRS 25 (SERMEL) (1973-74) 667
TRS 60 (IDA) (1975-76) 687
TRW (see TRW Defense and Space Systems Group) 796
TS 35 (Avco Lycoming) (1975-76) 734
TSE 36-1 (AiResearch) (1975-76) 743
TSE 231-1 (AiResearch) (1975-76) 743
TSE 331-3U (Garrett-AiResearch) (1980-81) 749
TSIO-360 (Continental) (1977-78) 833
TSIO-360 (Rolls-Royce/Continental) (1975-76) 726
TSIO-470 (Continental) (1977-78) 833
TSIO-520 (Continental) (1977-78) 833
TU-122 (Thiokol) (1978-79) 757
TU-289 (Thiokol) 794
TU-312 (Thiokol) (1975-76) 772
TU-393 (Thiokol) (1974-75) 772
TU-412 (Thiokol) (1975-76) 772
TU-562 (Thiokol) (1975-76) 772
TV-2BM (Soloviev) (1980-81) 722
TV2-117A (Isotov) 743
TV3-117 (Isotov) 744
TVD-850 (Isotov) 744
TX-30 (Thiokol) (1979-80) 765
TX-131-15 (Thiokol) (1972-73) 741
TX-174 (Thiokol) 794
TX-175 (Thiokol) 794
TX-261 (Thiokol) (1979-80) 765
TX-280 (Thiokol) (1977-78) 836
TX-306 (Thiokol) (1979-80) 765
TX-354 Castor II (Thiokol) 794
TX-481 (Thiokol) 794
TX-486 (Thiokol) 794
TX-526 Castor IV (Thiokol) 794
TX-581 (Thiokol) 794
TX-633 (Thiokol) 795
TX-657 (Thiokol) 795
TX-683 (Thiokol) 795
TX-690 (Thiokol) 794
TX-732 (Thiokol) 795
TELEDYNE CAE DIVISION OF TELEDYNE INC (U.S.A.) 790
TELEDYNE CONTINENTAL MOTORS (U.S.A.) 792
THERMO-JET STANDARD INC. (U.S.A.) 793
THIOKOL CORPORATION (U.S.A.) 793
Thor (Rolls-Royce) (1975-76) 724
Tiara Series (Continental) (1978-79) 756
Titan III engines (Aerojet) 758
Tomahawk (Thiokol) 795
Topping Cycle Engine (MBB) (1971-72) 645
Trabant engine (ODA) (1975-76) 699
Transtage (Aerojet) (1975-76) 727
Trap (SEP) (1980-81) 695
Trident motors (Thiokol) 794
TRW DEFENSE AND SPACE SYSTEMS GROUP (U.S.A.) 796
TUMANSKY (U.S.S.R.) 749
Turboméca 1800 hp (Turboméca) (1975-76) 697
TURBOMECANICA (see Intreprinderea Turbomecanica Bucuresti) 741
TURBOMECA-SNECMA GRTS, GROUPEMENT (France) 727
TURBOMÉCA, SOCIÉTÉ (France) 723, 732
TURBO-UNION LTD (International) 734
Turmastazou (Turboméca) (1979-80) 697
Turmo (Turboméca) 726
Twin Pac (P&WC) 716
Tyne (MTU) (1973-74) 678
Tyne (Rolls-Royce) 754
Tyne (SNECMA/MTU/RR/FN) 723, 729, 754
Type 003 (Pulch) 730
Type 060 (Weslake) (1978-79) 720
Type 2RB (Borzecki) (1979-80) 708
Type 4 AR 1200 (Rectimo) (1974-75) 688
Type 4C2T (Cicaré) (1979-80) 683
Type 116 (Weslake) 757
Type 151 (NPT) 751
Type 156 Booster Motors (Thiokol) (1975-76) 772
Type 200 (Weslake) 757
Type 251 (NPT) (1980-81) 726
Type 274 (Weslake) 757
Type 301 (NPT) 752
Type 342 (Weslake) 757
Type 352 (Teledyne CAE) 790
Type 356 (Teledyne CAE) 790
Type 356-28A (Teledyne CAE) 791
Type 356-28F (Teledyne CAE) 791
Type 365 (Teledyne CAE) 791
Type 370 (Teledyne CAE) 791
Type 372-2 (Teledyne CAE) 791
Type 373 (Teledyne CAE) 791
Type 401A (NPT) (1980-81) 726
Type 430 (Weslake) 757
Type 440/555 (Teledyne CAE) 791
Type 455 (Teledyne CAE) 791
Type 460 (Chotia) 767
Type 490 (Teledyne CAE) 791
Type 548 (Weslake) 757
Type 860 (Weslake) 757
Type 1527 (Weslake) 757

U

UACL (see United Aircraft of Canada Ltd.) (1974-75) 680
UTC (see United Technology Center) (1974-75) 774
UNITED AIRCRAFT OF CANADA LTD. (Canada) (1974-75) 680
UNITED TECHNOLOGY CENTER (U.S.A.) (1974-75) 774
USINE DE SAINT MÉDARD (France) (1977-78) 759

V

VK (see V. Y. Klimov) (1973-74) 759
VK-1 (Klimov) (1977-78) 842
VK-5 (Klimov) (1973-74) 759
VK-7 (Klimov) (1973-74) 759
VO-435 (Avco Lycoming) (1977-78) 801
VO-540 (Avco Lycoming) (1977-78) 801
VR-3 (Flygmotor) (1974-75) 715
VR-35 (Flygmotor) 742
VALMET OY (Finland) 718
Valois (SEP) (1974-75) 689
VEDENEEV (U.S.S.R.) 750
Vexin (SEP) (1972-73) 659
Viking (SEP) 720
Viper (Rolls-Royce) 754
Viper 600 (Fiat/Rolls-Royce) 736
VOLVO FLYGMOTOR AB (Sweden) 731, 742
VOUGHT CORPORATION (U.S.A.) 796

W

WR2 (Williams) 796
WR3 (Williams) (1976-77) 799
WR19 (Williams) 797
WR19-A7 (Williams) 797
WR24 (Williams) 796
WR34 (Williams) 797
WR44 (Williams) 797
WR79 (Williams) (1975-76) 775
WSK-PZL-KALISZ (see Wytwórnia Sprzetu Komunikacyjnego-PZL-Kalisz) 739
WSK-PZL-RZESZOW (see Wytwórnia Sprzetu Komunikacyjnego-PZL-Rzeszów) 740
Walter Titan (Motorlet) (1973-74) 662
WEEDHOPPER OF UTAH INC (USA) 767
WESLAKE AEROMARINE ENGINES LTD (U.K.) 757
WESLAKE & COMPANY (UK) 758
WILLIAMS INTERNATIONAL CORPORATION (U.S.A.) 796
WRIGHT (see Curtiss-Wright Corporation Wood-Ridge Facility) 769
WYTWÓRNIA SPRZETU KOMUNIKACYJNEGO-PZL-KALISZ (Poland) 739
WYTWÓRNIA SPRZETU KOMUNIKACYJNEGO-PZL-RZESZÓW (Poland) 740

X

XAGM-88A (Thiokol) 795
XF3 (IHI) 737
XF104-GA-100 (AiResearch) (1976-77) 736
XJ11 (IHI) (1978-79) 698
XJ99 (Rolls-Royce/Allison) (1975-76) 704, 718
XLJ95 (Teledyne) (1972-73) 739
XLR66-AJ-2 engine (Aerojet) (1971-72) 672
XM-51 (Thiokol) (1972-73) 741
XM-105 (Thiokol) (1972-73) 741
XM-106 (Thiokol) (1972-73) 741

Y

YF102 (Avco Lycoming) (1972-73) 701
YJ69 (Teledyne CAE) 790
YJ85-19 (General Electric) (1972-73) 685
YLR89-NA-7 (Rocketdyne) 789
YLR101-NA-15 (Rocketdyne) 789
YLR105-NA-7 (Rocketdyne) 789
YO-65 (Nelson) 781
YT55 (Avco Lycoming) (1975-76) 734
YT702-LD-700 (Avco Lycoming) 763

Z

ZENOAH (see Komatsu Zenoah Company) 738

Printed and made in the United Kingdom by Netherwood Dalton & Co. Ltd., Huddersfield

IL-86 WIDE-BODIED AIRLINER FOR MEDIUM-RANGE ROUTES WITH HEAVY TRAFFIC

Equipped with modern automatic flight control and navigation systems permitting flights to be performed with high regularity as well as all-weather approaches and landings at all times.

Max passenger accommodation	350 seats
Max payload	42,000 kg
Max take-off weight	206,000 kg
Cruising speed	900-950 km/h
Range with 40,000 kg payload	3,600 km
Engines: Four turbofans, thrust 13,000 kg each	

More information available from:

AVIAEXPORT, USSR, MOSCOW

Our address: 32/34 Smolenskaja-Sennaja Sq.
121200 Moscow, USSR.
Tel. 244-26-86
Telex 411257, 411335
Cable: Aviaexport Moscow

AVIAEXPORT · USSR · MOSCOW